ADX500 CAS:107-13-1 *HR: 3*

ACRYLONITRILE

DOT: UN 1093

mf: C_3H_3N mw: 53.07

PROP: Colorless, mobile liquid; mild odor. Sol in water. Mp: −82°, bp: 77.3°, fp: −83°, flash p: 30°F (TCC), lel: 3.1%, uel: 17%, d: 0.806 @ 20°/4°, autoign temp: 898°F, vap press: 100 mm @ 22.8°, vap d: 1.83, flash p: (of 5% aq sol): <50°F.

SYNS: ACRYLNITRIL (GERMAN, DUTCH) ◇ ACRYLONITRILE MONOMER ◇ AKRYLONITRYL (POLISH) ◇ CARBACRYL ◇ CIANURO di VINILE (ITALIAN) ◇ CYANOETHYLENE ◇ CYANURE de VINYLE (FRENCH) ◇ ENT 54 ◇ FUMIGRAIN ◇ NITRILE ACRILICO (ITALIAN) ◇ NITRILE ACRYLIQUE (FRENCH) ◇ PROPENENITRILE ◇ 2-PROPENENITRILE ◇ RCRA WASTE NUMBER U009 ◇ TL 314 ◇ VENTOX ◇ VINYL CYANIDE

TOXICITY DATA with REFERENCE

bfa-rat/sat 30 mg/kg TXCYAC 16,67,80
dns-rat:lvr 1 mmol/L PMRSDJ 5,371,85
slt-dmg-orl 1520 μmol/L PMRSDJ 5,325,85
skn-hmn 500 mg nse INMEAF 17,199,48
skn-rbt 10 mg/24H open JIHTAB 30,63,48
skn-rbt 500 mg MLD SCCUR* -,1,61
eye-rbt 20 mg SEV JIHTAB 30,63,48
ipr-ham TDLo:641 mg/kg (female 8D post):TER TJADAB 23,325,81
orl-rat TDLo:650 mg/kg (female 6-15D post):REP DOWCC* 03NOV76
orl-rat TDLo:18200 mg/kg/52W C:CAR FCTOD7 24,129,86
ihl-rat TCLo:5 ppm/52W-I:ETA MELAAD 68,401,77
orl-rat LD:3640 mg/kg/52W-C:NEO DOWCC* MAR77
ihl-hmn TCLo:16 ppm/20M:EYE,PUL INMEAF 17,199,48
ihl-man LCLo:1 g/m³/1H:CNS,GIT ZAARAM 16,1,66
skn-chd LDLo:2015 mg/kg:CNS,RSP,GIT DMWOAX 75,1087,50
orl-rat LD50:78 mg/kg JOHYAY 3,106,59
ihl-rat LCLo:500 ppm/4H JIDHAN 31,343,49
skn-rat LD50:148 mg/kg GISAAA 41(10)103,76
ihl-mus LCLo:315 ppm/4H NTIS** PB280-478
ipr-mus LD50:46 mg/kg TXAPA9 59,589,81
orl-mus LD50:27 mg/kg JHEMA2 3,106,59
scu-mus LD50:35 mg/kg JHEMA2 3,106,59
ihl-dog LCLo:110 ppm/4H JIHTAB 24,27,42

CONSENSUS REPORTS: NTP Fifth Annual Report on Carcinogens. IARC Cancer Review: Group 2A IMEMDT 7,79,87; Human Limited Evidence IMEMDT 19,73,79; Animal Limited Evidence IMEMDT 19,73,79. Community Right-To-Know List. EPA Extremely Hazardous Substances List. Reported in EPA TSCA Inventory.

OSHA PEL: TWA 2 ppm; CL 10 ppm/15M; Cancer Hazard.
ACGIH TLV: Suspected Human Carcinogen, TWA 2 ppm (skin).
DFG TRK: 3 ppm (7 mg/m³), Animal Carcinogen, Suspected Human Carcinogen.
NIOSH REL: TWA 1 ppm; CL 10 ppm/15M
DOT Classification: Flammable Liquid and Poison.

SAFETY PROFILE: Confirmed human carcinogen with experimental carcinogenic, neoplastigenic, and tumorigenic data. Poison by inhalation, ingestion, skin contact, and other routes. Human systemic effects by inhalation and skin contact: conjunctive irritation, somnolence, general anesthesia, cyanosis and diarrhea. An experimental teratogen. Other experimental reproductive effects. Human mutation data reported. Dangerous fire hazard when exposed to heat, flame, or oxidizers. Moderate explosion hazard when exposed to flame. Can react vigorously with oxidizing materials (see also CYANIDES).

HR: Hazard Rating indicating relative hazard for toxicity, fire, and reactivity with 3 denoting the worst hazard level.
See Introduction: paragraph 3, *p.* xiii.

CAS: – The American Chemical Society's Chemical Abstracts Service number. A complete CAS number cross-index is located in Section 1.
See Introduction: paragraph 4, *p.* xiii.

Cited Reference: – A code which represents the cited reference for the toxicity data. Complete bibliographic citations in order are listed in Section 3.
See Introduction: paragraph 16, *p.* xxiv.

Safety Profiles: – These are text summaries of the toxicity, flammability, reactivity, incompatibility and other dangerous properties of the entry.
See Introduction: paragraph 19, *p.* xxvii.

Sax's Dangerous Properties of Industrial Materials

Eighth Edition

Volume II

RICHARD J. LEWIS, SR.

DISCLAIMER

Extreme care has been taken in preparation of this work. However, neither the publisher nor the authors shall be held responsible or liable for any damages resulting in connection with or arising from the use of any of the information in this book.

Library of Congress Catalog Card Number 92-3896
ISBN VOLUME I 0-442-01276-4
ISBN VOLUME II 0-442-01277-2
ISBN VOLUME III 0-442-01278-0
ISBN SET 0-442-01132-6

Manufactured in the United States of America

Published by Van Nostrand Reinhold
115 Fifth Avenue
New York, NY 10003

Chapman and Hall
2-6 Boundary Row
London, SE 1 8HN

Thomas Nelson Australia
102 Dodds Street
South Melbourne 3205
Victoria, Australia

Nelson Canada
1120 Birchmont Road
Scarborough, Ontario M1K 5G4, Canada

16 15 14 13 12 11 10 9 8 7 6 5 4 3 2 1

Library of Congress Cataloging-in-Publication Data
Lewis, Richard J., Sr.
Sax's dangerous properties of industrial materials / Richard J. Lewis, Sr.—8th ed.
p. cm.
Rev. ed. of: Dangerous properties of industrial materials / N. Irving Sax and Richard J. Lewis, Sr. 7th ed. ©1989.
Includes bibliographical references and index.
ISBN 0-442-01132-6
1. Hazardous substance—Handbooks, manuals, etc. I. Sax, N. Irving (Newton Irving). Dangerous properties of industrial materials. II. Title. III. Title: Dangerous properties of industrial materials.
T55.3.H3S3 1992
604.7—dc20 92-3896
CIP

Contents

Preface

This eighth edition of *Dangerous Properties of Industrial Materials*, a three volume set, represents a major revision and updating of the seventh edition. The objective of the book, however, remains the same: to promote safety by providing the most up-to-date hazard information available. The growth in the availability of toxicological and hazard control reports continues unabated. This book cannot contain all the published data and continue to provide the accessibility for which it is known. To continue to provide complete hazard assessments for the maximum number of entries, data for each entry has been selectively reduced. In particular, carcinogenic and reproductive data lines above those required to establish the hazard of the entry have been excluded. Complete data for these entries are available in the books *Carcinogenically Active Chemicals* and *Reproductively Active Chemicals*, both available from the publisher.

Over 14,000 entries have been revised for this edition, and 1,500 new entries have been added. Some less useful entries have been eliminated or combined with related entries. Emphasis has been placed on including new carcinogenic and reproductive entries. A special effort has resulted in a significant increase in entries containing skin and eye irritation data.

All carcinogenic entries were reviewed and have been categorized as either confirmed, suspected, or questionable. This assessment was based on a detailed classification scheme discussed under Safety Profiles in the Introduction.

All reproductive entries have been updated and new data added.

Numerous synonyms have been added to assist in locating the many materials which are known under a variety of systematic and common names. The synonym cross-index contains the entry name as well as each synonym. This index should be consulted first to locate a material by name. Synonyms are given in English, as well as other major languages such as French, German, Dutch, Polish, Japanese, and Italian.

Additional physical and chemical properties have been added. Whenever available, physical descriptions, formulas, molecular weights, melting points, boiling points, explosion limits, flash points, densities, autoignition temperatures, and the like, have been supplied.

The following classes of data have been updated for all entries for which they apply as follows:

1. IARC Group 1-4 classes and recent assessments.
2. OSHA revised standards which were published in January 1989 and take effect on December 31, 1992.
3. ACGIH TLVs and BEIs reflect the latest recommendations and now include "intended Changes."
4. NTP Fifth Annual Report on Carcinogens entries are identified.
5. DOT classifications now have the corresponding guide number for entries having multiple guide numbers.
6. CAS numbers are provided for additional entries.

Each entry concludes with a safety profile, a textual summary of the hazards presented by the entry. The discussion of human exposures includes target organs and specific effects reported. Carcinogenic and reproductive assessments have been completely revised for this edition.

Fire and explosion hazards are briefly summarized in terms of conditions of flammable or reactive hazard. Where feasible, fire-fighting materials and methods are discussed. Materials which are known to be incompatible with an entry are listed here.

Also included in the safety profile are comments on disaster hazards which serve to alert users of materials to the dangers that may be encountered on entering storage premises during a fire or other emergency. Although the presence of water, steam, acid fumes, or powerful vibrations can cause the decomposition of many materials into dangerous compounds, of particular concerned are high temperatures (such as those resulting from a fire) since these can cause many otherwise mild chemicals to emit highly toxic gases or vapors such as NO_x, SO_x, acids, and so forth, or evolve vapors of antimony, arsenic, mercury, and the like.

The book, which consists of three volumes, is divided as follows:

The first volume contains a CAS number cross-index, a synonym cross-index, and the complete citations for bibliographic references given in the data section.

Section 1 contains the CAS Number cross-index for CAS numbers for the listed materials.

Section 2 contains the prime name and synonym cross-index for the listed materials

Section 3 contains the complete bibliographic references.

The main section of the book is contained in the Volumes II and III. It lists and describes approximately 20,000 materials in alphabetical order by entry name.

Please refer to the Introduction in Volume I for an explanation of the sources of data and codes used.

Every effort was made to include the most current and complete information. The author welcomes comments or corrections to the data presented.

Richard J. Lewis, Sr.

Key to Abbreviations

abs – absolute
ACGIH – American Conference of Governmental Industrial Hygienists
alc – alcohol
alk – alkaline
amorph – amorphous
anhyd – anhydrous
approx – approximately
aq – aqueous
atm – atmosphere
autoign – autoignition
aw – atomic weight
af – atomic formula
BEI – ACGIH Biological Exposure Indexes
bp – boiling point
b range – boiling range
CAS – Chemical Abstracts Service
cc – cubic centimeter
CC – closed cup
CL – ceiling concentration
COC – Cleveland open cup
conc – concentration, concentrated
compd(s) – compound(s)
contg – containing
cryst, crys – crystal(s), crystalline
d – density
D – day(s)
decomp, dec – decomposition
deliq – deliquescent
dil – dilute
DOT – U.S. Department of Transportation
EPA – U.S. Environmental Protection Agency
eth – ether
(F) – Fahrenheit
FCC – Food Chemical Codex
FDA – U.S. Food and Drug Administration
flam – flammable
flash p – flash point
fp – freezing point
g, gm – gram
glac – glacial
gran – granular, granules
hygr – hygroscopic
H, hr – hour(s)
HR: – hazard rating
htd – heated
htg – heating
IARC – International Agency for Research on Cancer
immisc – immiscible
incomp – incompatible
insol – insoluble
IU – International Unit
kg – kilogram (one thousand grams)
L,l – liter
lel – lower explosive limit
liq – liquid
M – minute(s)
m^3 – cubic meter
mf – molecular formula
mg – milligram
misc – miscible
μ, u – micron
mL, ml – milliliter
mm – millimeter
mod – moderately
mp – melting point
mppcf – million particles per cubic foot
mw – molecular weight
ng – nanogram
NIOSH – National Institute for Occupational Safety and Health
nonflam – nonflammable
NTP – National Toxicology Program
OBS – obsolete
OC – open cup
org – organic
OSHA – Occupational Safety and Health Administration
Pa – Pascals
PEL – permissible exposure level
petr – petroleum
pg – picogram (one trillionth of a gram)
Pk – peak concentration
pmole – picomole
powd – powder
ppb – parts per billion (v/v)
pph – parts per hundred (v/v)(percent)
ppm – parts per million (v/v)
ppt – parts per trillion (v/v)
prep – preparation
PROP – properties
refr – refractive
rhomb – rhombic

S,sec – second(s)
sl, slt, sltly – slightly
sol – soluble
soln – solution
solv(s) – solvent(s)
spont – spontaneous(ly)
STEL – short term exposure limit
subl – sublimes
TCC – Tag closed cup
tech – technical
temp – temperature
TLV – Threshold Limit Value
TOC – Tag open cup
TWA – time weighted average
U, unk – unknown, unreported
μ, u – micron
uel – upper explosive limit
μg, ug – microgram
ULC, ulc – Underwriters Laboratory Classification
USDA – U.S. Department of Agriculture
vac – vacuum
vap – vapor
vap d – vapor density
vap press – vapor pressure
vol – volume
visc – viscosity
vsol – very soluble
W – week(s)
Y – year(s)
% – percent(age)
> – greater than
< – less than
< = – equal to or less than
= > – equal to or greater than
° – degrees of temperature in Celsius (Centigrade)
(F), °F – temperature in Fahrenheit

General Chemicals
A–F

A

AAB250 ***HR: 2***
A-200 PYRINATE

PROP: Composed of pyrethrins, piperonyl butoxide, deodorized kerosene (AROPAW 68,36,62).

TOXICITY DATA with REFERENCE
eye-rbt 50 mg SEV AROPAW 68,36,62

SAFETY PROFILE: A severe eye irritant.

AAC000 CAS:1405-35-2 ***HR: 3***
ABBOTT ANTIBIOTIC M259

SYNS: A 6413 ◇ M 259

TOXICITY DATA with REFERENCE
ipr-mus LD50:1470 μg/kg 85FZAT -,739,67
ivn-mus LD50:2210 μg/kg 85FZAT -,739,67

SAFETY PROFILE: Poison by intraperitoneal and intravenous routes.

AAC250 CAS:8021-27-0 ***HR: 2***
ABIES ALBA OIL

PROP: Colorless to pale-yellow oil from the steam distillation of the crushed cones of *Abies Alba Mill* (FCTXAV 12,807,74).

SYNS: OIL of ABIES ALBA ◇ OIL of FUR ◇ OIL of SILVER FIR ◇ OIL of SILVER PINE ◇ SILVER FIR NEEDLE OIL ◇ SILVER FIR OIL ◇ SILVER PINE OIL ◇ TEMPLIN OIL

TOXICITY DATA with REFERENCE
skn-rbt 500 mg/24H MOD FCTXAV 12,807,74

CONSENSUS REPORTS: Reported in EPA TSCA Inventory.

SAFETY PROFILE: A skin irritant. When heated to decomposition it emits acrid smoke and irritating fumes.

AAC500 CAS:514-10-3 ***HR: 3***
ABIETIC ACID
mf: $C_{20}H_{30}O_2$ mw: 302.50

PROP: Yellow powder. Mp: 172-175°.

SYNS: 13-ISOPROPYLPODOCARPA-7,13-DIEN-15-OICACID ◇ SILVIC ACID

TOXICITY DATA with REFERENCE
ivn-mus LD50:180 mg/kg CSLNX* NX#02819

CONSENSUS REPORTS: Reported in EPA TSCA Inventory.

SAFETY PROFILE: Poison by intravenous route. Combustible. Slight explosion hazard as dust. When heated to decomposition it emits acrid smoke and irritating fumes.

AAC875 CAS:55077-30-0 ***HR: 3***
ABOVIS
mf: $C_{10}H_{20}NO_4 \cdot 1/2C_{10}H_6O_6S_2$ mw: 361.42

PROP: Crystals. Mp: 189-191°.

SYNS: (2-ACETYLLACTOYLOXYETHYL)TRIMETHYLAMMONIUM HEMI-1,5-NAPHTHALENEDISULFONATE◇ (2-ACETYLLACTOYLOXYETHYL)TRIMETHYLAMMONIUM1,5-NAPHTHALENEDISULFONATE ◇ 2-(2-(ACETYLOXY)-1-OXOPROPOXY)-N,N,N-TRIMETHYLETHANAMINIUM 1,5-NAPHTHALENEDISULFONATE(2:1) ◇ ACLATONIUM NAPADISILATE ◇ CHOLINE 1,5-NAPHTHALENEDISULFONATE (2:1), DILACTATE, DIACETATE ◇ TM 723

TOXICITY DATA with REFERENCE
orl-mus TDLo:600 g/kg (9W male/2W pre-6D preg):TER OYYAA2 18,923,79
orl-rat TDLo:60 g/kg (30D male):REP OYYAA2 18,749,79
orl-rat LDLo:15 g/kg IYKEDH 12,1204,81
scu-rat LD50:986 mg/kg OYYAA2 13,497,77
ivn-rat LD50:46 mg/kg OYYAA2 13,497,77
orl-mus LD50:15 g/kg USXXAM #3903137
scu-mus LD50:826 mg/kg OYYAA2 13,497,77
ivn-mus LD50:41900 μg/kg IYKEDH 12,30420,81

SAFETY PROFILE: Poison by intravenous route. Moderately toxic by subcutaneous route. An experimental teratogen. Other experimental reproductive effects. When heated to decomposition it emits toxic fumes of NO_x and SO_x. A cholinergic agent. See also SULFONATES.

AAD000 CAS:1393-62-0 ***HR: 3***
ABRIN

PROP: Yellowish-white powder. Sol in solns of sodium chloride, usually with turbidity. Incubation at 60° for 30 min fails to remove toxic effect, but at 80°, most of the toxicity is lost.

SYNS: ABRINS ◇ AGGLUTININ ◇ CRAB'S EYES ◇ INDIAN LICORICE SEED ◇ JUMBLE BEAD ◇ PRAYER BEAD ◇ TOXALBUMIN

TOXICITY DATA with REFERENCE
dni-mus-ast 50 μg/kg TOXIA6 11,379,73
orl-hmn LDLo:7 μg/kg MEIEDD 10,1,83
orl-rat LDLo:300 mg/kg AMIHAB 12,468,55
orl-mus LD50:6638 mg/kg ARZNAD 21,888,71
ipr-mus LD50:20 μg/kg 85GDA2 8(1),107,82
ivn-mus LD50:20 μg/kg MEIEDD 10,1,83
orl-rbt LDLo:21 mg/kg AMIHAB 12,468,55
orl-gpg LD50:299 mg/kg ARZNAD 21,888,71

SAFETY PROFILE: A deadly poison to humans by ingestion. Poison by ingestion, intravenous, and intraperitoneal route. Mutation data reported. When heated to decomposition it emits acrid fumes and irritating smoke. See also RICIN; LECTINS. Note: Do not confuse with abrine.

AAD100 ***HR: 3***
ABRUS PRECATORIUS L., seed kernel extract

TOXICITY DATA with REFERENCE
orl-rat TDLo:6 g/kg (male 60D pre):REP AEFTAA 18,217,87
ipr-mus LD50:550 ng/kg CTYAD8 18,196,87
scu-mus LD50:200 μg/kg TOXIA6 6,211,69
scu-gpg LDLo:430 μg/kg TOXIA6 7,211,69

SAFETY PROFILE: Poison by intraperitoneal and subcutaneous routes. Experimental reproductive effects. When heated to decomposition it emits acrid smoke and irritating fumes.

AAD125 ***HR: D***
ABRUS PRECATORIUS OIL

PROP: The oil extracted from the seeds of *Abrus precatorius* (IJPAAO 29,235,67).

TOXICITY DATA with REFERENCE
orl-mus TDLo:1250 mg/kg (10D pre/1-10D preg):REP IJPAAO 29,235,67

SAFETY PROFILE: Experimental reproductive effects. When heated to decomposition it emits acrid smoke and irritating fumes.

AAD250 CAS:93164-88-6 ***HR: 2***
ACACIA (EXTRACT)

PROP: Indian plant belonging to the family *Leguminosae* (IJEBA6 7,250,69).

SYN: BABUL STEM BARK EXTRACT

TOXICITY DATA with REFERENCE
ipr-mus LD50:500 mg/kg IJEBA6 7,250,69

SAFETY PROFILE: Moderately toxic by intraperitoneal route. When heated to decomposition it emits acrid smoke and irritating fumes.

AAD500 ***HR: 2***
ACACIA FARNESIANA (Linn.) Willd., extract excluding roots

PROP: Indian plant belonging to the family *Mimosaceae* (IJEBA6 22,487,84)

TOXICITY DATA with REFERENCE
orl-rat TDLo:150 mg/kg (female 12-14D post):REP IJEBA6 22,487,84
ipr-mus LD50:562 mg/kg IJEBA6 22,487,84

SAFETY PROFILE: Moderately toxic by intraperitoneal route. Experimental reproductive effects. When heated to decomposition it emits acrid smoke and irritating fumes.

AAD750 ***HR: 3***
ACACIA VILLOSA

PROP: Aqueous extract from the root of the plant (JNCIAM 52,1579,74).

SYN: WATAPANA SHIMARON

TOXICITY DATA with REFERENCE
scu-rat TDLo:198 mg/kg/22W-I:NEO JNCIAM 52,1579,74
imp-ham TDLo:1660 mg/kg:CAR JNCIAM 53,1259,74
scu-rat TD:300 g/kg/60W-I:NEO,REP JNCIAM 52,445,74

SAFETY PROFILE: Experimental reproductive effects. Questionable carcinogen with experimental neoplastigenic and carcinogenic data. When heated to decomposition it emits smoke and acrid fumes.

AAD875 CAS:5892-41-1 ***HR: 3***
ACAMYLOPHENINE DIHYDROCHLORIDE
mf: $C_{19}H_{32}N_2O_2 \cdot 2ClH$ mw: 393.45

SYNS: AVACAN ◇ CAMYLOFINE DIHYDROCHLORIDE ◇ CAMYLOFINE HYDROCHLORIDE ◇ CAMYLOFIN HYDROCHLORIDE

TOXICITY DATA with REFERENCE
orl-mus LD50:760 mg/kg MEIEDD 10,239,83
scu-mus LD50:1350 mg/kg MEIEDD 10,239,83
ivn-mus LD50:49200 μg/kg MEIEDD 10,239,83

SAFETY PROFILE: Poison by intravenous route. Moderately toxic by ingestion and other routes. When heated to decomposition it emits toxic fumes of NO_x and HCl. See also ESTERS.

AAE000 CAS:3697-25-4 ***HR: 3***
4,10-ACE-1,2-BENZANTHRACENE
mf: $C_{20}H_{14}$ mw: 254.34

SYN: 1,2-DIHYDROBENZ(e)ACEANTHRYLENE ◇ 5,6-DIHYDROBENZENE(e)ACEANTHRYLENE

TOXICITY DATA with REFERENCE
scu-mus TDLo:4 mg/kg:ETA AJCAA7 33,499,38

SAFETY PROFILE: Questionable carcinogen with experimental tumorigenic data. When heated to decomposition it emits acrid smoke and fumes.

AAE100 CAS:37517-30-9 ***HR: 3***
ACEBUTOLOL
mf: $C_{18}H_{28}N_2O_4$ mw: 336.48

PROP: Crystals. Mp: 119-123°.

SYNS: (±)-ACEBUTOLOL ◇ dl-ACEBUTOLOL ◇ 1-(2-ACETYL-4-n-BUTYRAMIDOPHENOXY)-2-HYDROXY-3-ISOPROPYLAMINOPROPANE ◇ 3'-ACETYL-4'-(2-HYDROXY-3-(ISOPROPYLAMINO)PROPOXY)BUTYRANILIDE ◇ (±)-N-(3-ACETYL-4-(2-HYDROXY-3-((1-METHYLETHYL)AMINO)PROPOXY)PHENYL)BUTANAMIDE ◇ 5'-BUTYRAMIDO-2'-(2-HYDROXY-3-ISOPROPYLAMINO-PROPOXY)ACETOPHENONE ◇ PRENT

TOXICITY DATA with REFERENCE
unr-wmn TDLo:1080 mg/kg (1-39W preg):REP BMJOAE 283,1077,81
orl-wmn TDLo:120 mg/kg (34-39W preg):TER DPTHDL 4(Suppl 1),109,82
orl-wmn TDLo:152 mg/kg:CVS,BPR JTCTDW 20,69,83
ivn-dog LD50:4 mg/kg MASODV 16,13,80

SAFETY PROFILE: Moderately toxic by intravenous route. Human systemic effects by ingestion: developmental abnormalities of the cardiovascular and respiratory systems; effects on newborn in biochemical and metabolic abnormalities and reduced growth statistics. A human teratogen. When heated to decomposition it emits toxic fumes of NO_x. A beta-adrenergic blocker.

AAE125 CAS:34381-68-5 ***HR: 3***
ACEBUTOLOL HYDROCHLORIDE
mf: $C_{18}H_{28}N_2O_4$•ClH mw: 372.94

SYNS: ACETOBUTOLOL HYDROCHLORIDE ◇ dl-1-(2-ACETYL-4-BUTYRAMIDOPHENOXY)-2-HYDROXY-3-ISOPROPYLAMINOPROPANE HYDROCHLORIDE ◇ 3'-ACETYL-4'-(2-HYDROXY-3-(ISOPROPYLAMINO)PROPOXY)BUTYRANILIDEHYDROCHLORIDE ◇ M&B 17,803A ◇ SECTRAL

TOXICITY DATA with REFERENCE
orl-rat TDLo:1100 mg/kg (7-17D preg):REP OYYAA2 15,885,78
orl-rat TDLo:1100 mg/kg (7-17D preg):TER OYYAA2 15,885,78
orl-rat LD50:6620 mg/kg OYYAA2 20,883,80
ipr-rat LD50:222 mg/kg OYYAA2 15,837,78
scu-rat LD50:1310 mg/kg OYYAA2 15,837,78
ivn-rat LD50:103 mg/kg OYYAA2 15,837,78
orl-mus LD50:4050 mg/kg NIIRDN 6,19,82
ipr-mus LD50:185 mg/kg OYYAA2 15,837,78
scu-mus LD50:291 mg/kg OYYAA2 15,837,78
ivn-mus LD50:53 mg/kg NIIRDN 6,19,82
orl-rbt LD50:296 mg/kg OYYAA2 15,837,78
ivn-rbt LD50:41 mg/kg OYYAA2 16,837,78

SAFETY PROFILE: Poison by ingestion, subcutaneous, intravenous, and intraperitoneal routes. An experimental teratogen. Other experimental reproductive effects. When heated to decomposition it emits toxic fumes of NO_x and HCl.

AAE250 CAS:827-61-2 ***HR: 3***
ACECLIDINE
mf: $C_9H_{15}NO_2$ mw: 169.25

SYNS: 3-ACETOXYQUINUCLIDINEGLAUCOSTAT ◇ 3-QUINUCLIDINOL ACETATE

TOXICITY DATA with REFERENCE
scu-rat LD50:225 mg/kg ARZNAD 18,320,68
ivn-rat LD50:45 mg/kg ARZNAD 18,320,68
orl-mus LD50:165 mg/kg ARZNAD 18,320,68
scu-mus LD50:102 mg/kg ARZNAD 18,320,68
ivn-mus LD50:36 mg/kg RPTOAN 35(2),55,72

SAFETY PROFILE: Poison by ingestion, subcutaneous, and intravenous routes. When heated to decomposition it emits toxic fumes of NO_x.

AAE500 CAS:3685-84-5 ***HR: 3***
ACEFEN
mf: $C_{12}H_{16}ClNO_3$•ClH mw: 294.20

SYNS: AMIPOLNE ◇ 235 ANP HYDROCHLORIDE ◇ BRENAL ◇ CELLATIVE ◇ CENTROPHENOXINE ◇ CERUTIL ◇ (p-CHLOROPHENOXY)ACETIC ACID 2-(DIMETHYLAMINO)ETHYL ESTER HYDROCHLORIDE ◇ CLOCETE ◇ DIMETHYLAMINOETHYL p-CHLOROPHENOXYACETATEHYDROCHLORIDE ◇ DIMETHYLAMINOETHYL 4-CHLOROPHENOXYACETATE HYDROCHLORIDE ◇ DIMETHYLAMINOETHYL ESTER of p-CHLOROPHENOXYACETIC ACID HYDROCHLORIDE ◇ HELFERGIN ◇ LUCIDRIL ◇ LUCIDRYL HYDROCHLORIDE ◇ MARUCOTOL ◇ MECLOFENOXATE HYDROCHLORIDE ◇ METHOXYNAL ◇ NSC 113619 ◇ PROSEROUT

TOXICITY DATA with REFERENCE
orl-rat LD50:865 mg/kg KSKZAN 16(2),59,78
ipr-mus LD50:660 mg/kg NIIRDN 6,814,82
scu-mus LD50:1560 mg/kg NIIRDN 6,814,82
orl-mus LD50:1750 mg/kg CRSBAW 153,1914,59
ipr-mus LD50:845 mg/kg CRSBAW 153,1914,59
ivn-mus LD50:350 mg/kg CRSBAW 153,1914,59
ivn-rbt LDLo:150 mg/kg CRSBAW 153,1914,59

SAFETY PROFILE: Poison by intravenous route. Moderately toxic by ingestion and intraperitoneal routes. When heated to decomposition it emits very toxic fumes of Cl^-, NO_x, and HCl.

AAE625 CAS:53164-05-9 ***HR: 3***
ACEMETACIN
mf: $C_{21}H_{18}ClNO_6$ mw: 415.83

PROP: Very fine, pale yellow crystals from petr ether. Mp: 150-153°.

SYNS: ACM ◇ 1-(p-CHLORBENZOYL)-5-METHOXY-2-METHYLINDOL-3-ACETOXY)ESSIGSAEURE(GERMAN) ◇ 1-(4-CHLOROBENZOYL)-5-METHOXY-2-METHYL-1H-INDOLE-3-ACETIC ACID CARBOXYMETHYL ESTER ◇ ((1-(4-CHLOROBENZOYL)-5-METHOXY-2-METHYLINDOLE-3-YL)ACETOXY)ACETIC ACID ◇ K-708 ◇ RANTUDIL ◇ TV 1322

TOXICITY DATA with REFERENCE
orl-rat TDLo:44 mg/kg (7-17D preg):REP OYYAA2 22,765,81
orl-rat TDLo:44 mg/kg (7-17D preg):TER OYYAA2 22,765,81
orl-rat LD50:24 mg/kg ARZNAD 30,1398,80
ipr-rat LD50:23 mg/kg ARZNAD 30,1398,80
scu-rat LD50:28 mg/kg ARZNAD 30,1398,80
ivn-rat LD50:28 mg/kg ARZNAD 30,1398,80
ims-rat LD50:19 mg/kg ARZNAD 30,1398,80
orl-mus LD50:18 mg/kg ARZNAD 30,1398,80
ipr-mus LD50:23 mg/kg ARZNAD 30,1398,80
scu-mus LD50:23 mg/kg ARZNAD 30,1398,80
ivn-mus LD50:34 mg/kg ARZNAD 30,1398,80

SAFETY PROFILE: Poison by ingestion, subcutaneous, intraperitoneal, intravenous, and intramuscular routes. An experimental teratogen. Other experimental reproductive effects. When heated to decomposition it emits toxic fumes of Cl^- and NO_x. An anti-inflammatory agent.

AAE750 ***HR: 1***
ACENAPHTHALENE
mf: $C_{10}H_6(CH_2)_2$ mw: 154.2

PROP: White, elongated crystals. Mp: 95°, bp: 277.5°, d: 1.024 @ 99°/4°, vap press.: 10 mm @ 131.2°, vap d: 5.32. Insol in water, sltly sol in hot alc, ether, and chloroform.

SYN: 1,8-ETHYLENE NAPHTHALENE.

TOXICITY DATA with REFERENCE
mma-sat 490 μmol/L/2H CNREA8 39,4152,79

SAFETY PROFILE: Mutation data reported. A skin and mucous membrane irritant. May cause acute vomiting if swallowed in large quantities. Combustible. When heated to decomposition it emits acrid smoke and irritating fumes.

AAF000 CAS:5779-79-3 ***HR: 3***
ACENAPHTHANTHRACENE
mf: $C_{20}H_{14}$ mw: 254.34

SYNS: BENZ(k)ACEPHENANTHRENE ◇ 4,5-DIHYDROBENZ(k)ACEPHENANTHRYLENE ◇ 3:4-DIMETHYLENE-1:2-BENZANTHRACENE

TOXICITY DATA with REFERENCE
skn-mus TDLo:960 mg/kg/40W-I:ETA PRLBA4 129,439,40

SAFETY PROFILE: Questionable carcinogen with experimental tumorigenic data. When heated to decomposition it emits acrid smoke and irritating fumes.

AAF250 CAS:4657-93-6 ***HR: 3***
5-ACENAPHTHENAMINE
mf: $C_{12}H_{11}N$ mw: 169.24

PROP: Colorless needles, sol in ethanol. Mp: 108°.

SYNS: 5-AMINOACENAPHTHENE ◇ 1,2-DIHYDRO-5-ACENAPHTHYLENAMINE

TOXICITY DATA with REFERENCE
ipr-mus TDLo:3744 mg/kg/78W-I:ETA NEZAAQ 24,263,69
imp-mus TDLo:160 mg/kg:CAR NEZAAQ 24,263,69
ivn-mus LD50:56 mg/kg CSLNX* NX#01911

CONSENSUS REPORTS: IARC Cancer Review: Group 3 IMEMDT 7,56,87; Animal Inadequate Evidence IMEMDT 16,243,78

SAFETY PROFILE: Poison by intravenous route. Questionable carcinogen with experimental carcinogenic and tumorigenic data. When heated to decomposition it emits toxic fumes of NO_x.

AAF500 CAS:208-96-8 ***HR: D***
ACENAPHTHYLENE
mf: $C_{12}H_8$ mw: 152.20

SYN: CYCLOPENTA(de)NAPHTHALENE

TOXICITY DATA with REFERENCE
mma-sat 1 mmol/L/2H CNREA8 39,4152,79

CONSENSUS REPORTS: Reported in EPA TSCA Inventory.

SAFETY PROFILE: Mutation data reported. When heated to decomposition it emits acrid smoke and irritating fumes.

AAF625 CAS:72064-79-0 ***HR: 3***
ACEPREVAL
mf: $C_{28}H_{38}O_7$ mw: 486.66

SYNS: 21-(ACETYLOXY)-11-β-HYDROXY-17-((1-OXOPENTYL)OXY)PREGNA-1,4-DIENE-3,20-DIONE ◇ PREDNISOLONE VALERATE ACETATE ◇ PREDNISOLONE-17-VALERATE-21-ACETATE ◇ PVA ◇ 11-β,17-α,21-TRIHYDROXY-1,4-PREGNADIENE-3,20-DIONE-21-ACETATE-17-VALERATE

TOXICITY DATA with REFERENCE
scu-rat TDLo:55 mg/kg (7-17D preg):REP OYYAA2 20,67,80
scu-rat TDLo:110 mg/kg (7-17D preg):TER OYYAA2 20,67,80
ipr-mus LD50:1360 mg/kg OYYAA2 20,195,80

scu-mus LD50:1150 mg/kg OYYAA2 20,195,80
scu-rbt LD50:100 mg/kg OYYAA2 20,195,80

SAFETY PROFILE: Poison by subcutaneous route. Moderately toxic by other routes. An experimental teratogen. Other experimental reproductive effects.

AAF750 CAS:3598-37-6 ***HR: 3***
ACEPROMAZINE MALEATE
mf: $C_{19}H_{22}N_2OS \cdot C_4H_4O_4$ mw: 442.57

SYNS: 2-ACETYL-10-(3-(DIMETHYLAMINO)PROPYL)PHENOTHIAZINE, MALEATE ◇ ACETYLPROMAZINE MALEATE (1:1) ◇ ATRAVET ◇ 10-(3-(DIMETHYLAMINO)PROPYL)PHENOTHIAZIN-2-YL METHYL KETONE MALEATE (1:1) ◇ MALEATE ACIDE de l'ACETYL-3-DIMETHYLAMINO-3-PROPYL-10-PHENOTHIAZINE (FRENCH) ◇ NOTENSIL ◇ PREGICIL ◇ SOPRONTIN

TOXICITY DATA with REFERENCE
orl-rat LD50:400 mg/kg AIPTAK 123,78,59
ivn-rat LD50:95 mg/kg MEIEDD 10,5,83
orl-mus LDLo:270 mg/kg AIPTAK 113,53,57
scu-mus LD50:175 mg/kg AIPTAK 113,53,57
ivn-mus LD50:65 mg/kg APTOA6 19,87,62

SAFETY PROFILE: Poison by ingestion, subcutaneous, and intravenous routes. When heated to decomposition it emits highly toxic fumes of NO_x and SO_x. See also KETONES.

AAG000 CAS:105-57-7 ***HR: 3***
ACETAL
DOT: UN 1088
mf: $C_6H_{14}O_2$ mw: 118.20

PROP: Colorless, volatile liquid; agreeable odor, nutty after-taste. Bp: 102.7°, flash p: −5°F (CC), lel: 1.65%, uel: 10.4%, d: 0.831, autoign temp: 446°F, vap press: 10 mm @ 8.0°, vap d: 4.08, mp: −100°. Sltly sol in water; misc in alc and ether.

SYNS: ACETAAL (DUTCH) ◇ ACETAL DIETHYLIQUE (FRENCH) ◇ ACETALE (ITALIAN) ◇ 1,1-DIAETHOXY-AETHAN (GERMAN) ◇ DIAETHYLACETAL (GERMAN) ◇ 1,1-DIETHOXY-ETHAAN (DUTCH) ◇ 1,1-DIETHOXYETHANE ◇ DIETHYL ACETAL ◇ 1,1-DIETOSSIETANO (ITALIAN) ◇ ETHYLIDENE DIETHYL ETHER ◇ USAF DO-45

TOXICITY DATA with REFERENCE
skn-rbt 10 mg/24H MLD JIHTAB 31,60,49
eye-rbt 500 mg JIHTAB 31,60,49
orl-rat LD50:4600 mg/kg MDZEAK 8,244,67
ihl-rat LCLo:4000 ppm/4H JIHTAB 31,343,49
orl-mus LD50:3500 mg/kg GISAAA (3),12,77
ipr-mus LD50:500 mg/kg NTIS** AD277-689
orl-rbt LD50:3545 mg/kg PSEBAA 29,730,32

CONSENSUS REPORTS: Reported in EPA TSCA Inventory.

DOT Classification: Label: Flammable Liquid.

SAFETY PROFILE: Moderately toxic by ingestion and intraperitoneal routes. A skin and eye irritant. A narcotic. Dangerous fire hazard when exposed to heat or flame; can react vigorously with oxidizing materials. Forms heat-sensitive explosive peroxides on contact with air. When heated to decomposition it emits acrid smoke and fumes. See also ETHERS and ALDEHYDES.

AAG250 CAS:75-07-0 ***HR: 3***
ACETALDEHYDE
DOT: UN 1089
mf: C_2H_4O mw: 44.06

PROP: Colorless, fuming liquid; pungent, fruity odor. Mp: −123.5°, bp: 20.8°, lel: 4.0%, uel: 57%, flash p: −36°F (CC), d: 0.804 @ 0°/20°, autoign temp: 347°F, vap d: 1.52. Misc in water, alc, and ether.

SYNS: ACETALDEHYD (GERMAN) ◇ ACETIC ALDEHYDE ◇ ALDEHYDE ACETIQUE (FRENCH) ◇ ALDEIDE ACETICA (ITALIAN) ◇ ETHANAL ◇ ETHYL ALDEHYDE ◇ FEMA No. 2003 ◇ NCI-C56326 ◇ OCTOWY ALDEHYD (POLISH) ◇ RCRA WASTE NUMBER U001

TOXICITY DATA with REFERENCE
eye-hmn 50 ppm/15M JIHTAB 28,262,46
skn-rbt 500 mg open MLD UCDS** 12/13/63
eye-rbt 40 mg SEV UCDS** 12/13/63
mma-sat 10 μL/plate EVHPAZ 21,79,77
dnr-esc 10 μL/plate EVHPAZ 21,79,77
sce-hmn:lym 20 ppm/48H MUREAV 58,115,78
ipr-rat TDLo:300 mg/kg (female 8-13D post):REP TJADAB 36,31A,87
ipr-rat TDLo:50 mg/kg (12D preg):TER DADEDV 9,339,82
ihl-rat TCLo:735 ppm/6H/2Y-I:CAR TXCYAC 41,213,86
ihl-ham TCLo:2040 ppm/7H/52W-I:ETA EJCAAH 18,13,82
ihl-hmn TCLo:134 ppm/30M:PUL JAMAAP 165,1908,57
orl-rat LD50:661 mg/kg AGACBH 4,125,74
ihl-rat LC50:37 g/m^3/30M APTOA6 6,299,50
ipr-rat LDLo:500 mg/kg JBCHA3 152,41,44
ihl-mus LC50:1500 ppm/4H DTLVS* 4,3,80
scu-rat LD50:640 mg/kg APTOA6 6,299,50
scu-mus LD50:560 mg/kg APTOA6 6,299,50
ivn-mus LD50:212 mg/kg JOANAY 128,65,79
ihl-ham LC50:17000 ppm/4H PEXTAR 24,162,79
itr-ham LD50:96 mg/kg PEXTAR 24,162,79

CONSENSUS REPORTS: IARC Cancer Review: Group 2B IMEMDT 7,77,87; Animal Sufficient Evidence IMEMDT 36,101,85; Human Inadequate Evidence IMEMDT 36,101,85. On Community Right-To-Know List. Reported in EPA TSCA Inventory. EPA Genetic Toxicology Program.

OSHA PEL: (Transitional: TWA 200 ppm) TWA 100 ppm; STEL 150 ppm
ACGIH TLV: TWA 100 ppm; STEL 150 ppm
DFG MAK: 50 ppm (90 mg/m^3), Suspected Carcinogen.
DOT Classification: Flammable Liquid; Label: Flammable Liquid

SAFETY PROFILE: Suspected carcinogen with experimental carcinogenic and tumorigenic data. Poison by intratracheal and intravenous routes. A human systemic irritant by inhalation. An experimental teratogen. Other experimental reproductive effects. A skin and severe eye irritant. A narcotic. Human mutation data reported. A common air contaminant. Highly flammable liquid. Mixtures of 30-60 percent of the vapor in air ignite above 100°. It can react violently with acid anhydrides, alcohols, ketones, phenols, NH_3, HCN, H_2S, halogens, P, isocyanates, strong alkalies, and amines. Reactions with cobalt chloride, mercury(II) chlorate, or mercury(II) perchlorate form sensitive, explosive products. Polymerizes violently in the presence of traces of metals or acids. Reaction with oxygen may lead to detonation. When heated to decomposition it emits acrid smoke and fumes.

AAG500 CAS:75-39-8 ***HR: 2***
ACETALDEHYDE AMMONIA
DOT: UN 1841
mf: $C_2H_4O \cdot H_3N$ mw: 61.10

PROP: White, crystalline solid. Bp: 110°, mp: 97°. Very sol in water, alc; sltly sol in ether.

SYNS: ACETALDEHYDE, AMINE SALT ◇ ALDEHYDE AMMONIA ◇ 1-AMINOETHANOL ◇ α-AMINOETHYL ALCOHOL

DOT Classification: ORM-A; Label: None.

SAFETY PROFILE: It readily decomposes into acetaldehyde and ammonia when heated, causing the hazards of these substances. Moderate fire and explosion hazard when exposed to heat or flame. Can react with oxidizing materials. When heated to decomposition it emits toxic fumes of NH_3 and NO_x.

AAG750 CAS:10143-67-6 ***HR: 2***
ACETALDEHYDE BIS(2-METHOXYETHYL ACETAL
mf: $C_8H_{18}O_4$ mw: 178.26

SYN: 1,1-DI-(2-METHOXYETHOXY)ETHANE

TOXICITY DATA with REFERENCE
eye-rbt 500 mg open AMIHBC 10,61,54
orl-rat LD50:3260 mg/kg AMIHBC 10,61,54
skn-rbt LD50:4240 mg/kg AMIHBC 10,61,54

SAFETY PROFILE: Moderately toxic by ingestion. Mildly toxic by skin contact. An eye irritant. When heated to decomposition it emits acrid smoke and fumes. See also ALDEHYDES.

AAG850 CAS:105-82-8 ***HR: 1***
ACETALDEHYDE-DI-n-PROPYL ACETAL
mf: $C_8H_{18}O_2$ mw: 146.26

SYNS: ACETALDEHYDE, DIPROPYL ACETAL ◇ 1,1-DIPROPOXYETHANE ◇ DIPROPYL ACETAL ◇ n-PROPYL ACETAL

TOXICITY DATA with REFERENCE
skn-rbt 500 mg/24H MOD FCTXAV 17,897,79

CONSENSUS REPORTS: Reported in EPA TSCA Inventory.

SAFETY PROFILE: A skin irritant. When heated to decomposition it emits acrid smoke and irritating fumes.

AAH000 CAS:16568-02-8 ***HR: 3***
ACETALDEHYDE-N-METHYL-N-FORMYLHYDRAZONE
mf: $C_4H_8N_2O$ mw: 100.14

SYNS: ACETALDEHYDE-N-FORMYL-N-METHYLHYDRAZONE ◇ ETHYLIDENE GYROMITRIN ◇ GYROMITRIN ◇ N-METHYL-N-FORMYL HYDRAZONE of ACETALDEHYDE

TOXICITY DATA with REFERENCE
scu-mus TDLo:600 mg/kg/12W-I:CAR NEOLA4 28,559,81
orl-mus TD:5200 mg/kg/52W-I:ETA FEPRA7 39(3,Pt.2),884,80
unk-chd LDLo:10 mg/kg MGLHAE 65,453,74
unk-hmn LDLo:20 mg/kg MGLHAE 65,453,74
orl-rat LD50:320 mg/kg FCTXAV 15,575,77
orl-mus LD50:344 mg/kg MUREAV 54,167,78
orl-rbt LD50:50 mg/kg NATWAY 62,395,75

CONSENSUS REPORTS: IARC Cancer Review: Group 3 IMEMDT 7,56,87; Animal Limited Evidence IMEMDT 7,391,87. EPA Genetic Toxicology Program.

SAFETY PROFILE: Poison via ingestion and possibly other routes. Questionable carcinogen with experimental carcinogenic and tumorigenic data. When heated to decomposition it emits toxic fumes of NO_x.

AAH100 CAS:17167-73-6 ***HR: 2***
ACETALDEHYDE METHYLHYDRAZONE
mf: $C_3H_8N_2$ mw: 72.13

SYNS: ACETALDEHYDE, N-METHYLHYDRAZONE ◇ AMFH

TOXICITY DATA with REFERENCE
orl-mus TDLo:208 mg/kg/1Y-I:ETA JJIND8 67,881,81
orl-mus LD50:390 mg/kg TXAPA9 45,429,78

SAFETY PROFILE: Poison by ingestion. Questionable carcinogen with experimental tumorigenic data. When heated to decomposition it emits toxic fumes of NO_x.

AAH250 CAS:107-29-9 ***HR: 3***
ACETALDEHYDE OXIME
DOT: UN 2332
mf: C_2H_5NO mw: 59.08

PROP: A water-sol, crystalline material; sol in alc, ether. Mp: (α) 46.5°, mp: (β) 12°, d: 0.966, bp: 114.5°, flash p: ≤ 72°F.

SYNS: ACETALDOXIME ◇ ALDOXIME ◇ ETHANAL OXIME ◇ ETHYLIDENEHYDROXYLAMINE ◇ USAF AM-5

TOXICITY DATA with REFERENCE
ipr-mus LD50:100 mg/kg NTIS** AD277-689
unk-mus LD50:1150 mg/kg PCJOAU 12,227,78

CONSENSUS REPORTS: Reported in EPA TSCA Inventory.

DOT Classification: Flammable Liquid; Label: Flammable Liquid

SAFETY PROFILE: Poison via intraperitoneal route. A dangerous fire hazard with a flash point at room temperature. When heated to decomposition it emits toxic fumes of NO_x. See also ALDEHYDES.

AAH500 CAS:918-04-7 ***HR: 2***
ACETALDEHYDE SODIUM SULFITE
mf: $C_2H_3O_2SO_2Na•1/2H_2O$ mw: 166.2

PROP: White crystals decomp by acid; sol in water; insol in alc.

SYNS: ACETALDEHYDE SODIUM BISULFITE ◇ AZETALDEHYDSCHWEFLIGSAUREN NATRIUMS (GERMAN) ◇ SODIUM-1-HYDROXYETHANESULFONATE

TOXICITY DATA with REFERENCE
orl-rbt LDLo:1220 mg(SO_2)/kg AHYGAJ 57,87,06

CONSENSUS REPORTS: Reported in EPA TSCA Inventory.

SAFETY PROFILE: Moderately toxic by ingestion based upon SO_2 content. When heated to decomposition it emits toxic fumes of SO_x and Na_2O. See also ALDEHYDES and SULFITES.

AAH750 CAS:107-89-1 ***HR: 3***
ACETALDOL
DOT: UN 2839
mf: $C_4H_8O_2$ mw: 88.12

PROP: Clear, white-to-yellow syrupy liquid. Bp: 83° @ 20 mm, flash p: 150°F (OC), d: 1.11, autoign temp: 482°F, vap d: 3.04.

SYNS: ALDOL ◇ 3-BUTANOLAL ◇ 3-HYDROXYBUTANA ◇ β-HYDROXYBUTYRALDEHYDE ◇ 3-HYDROXYBUTYRALDEHYDE ◇ OXYBUTANAL ◇ OXYBUTYRIC ALDEHYDE

TOXICITY DATA with REFERENCE
skn-rbt 10 mg/24H MLD JIHTAB 31,60,49
skn-rbt 10 mg/24H open MLD AIHAAP 23,95,62
eye-rbt 100 mg MLD UCDS** 4/21/67
orl-rat LD50:2180 mg/kg JIHTAB 31,60,49
skn-rbt LD50:140 mg/kg UCDS** 4/21/67

CONSENSUS REPORTS: Reported in EPA TSCA Inventory.

DOT Classification: Poison B; Label: Poison.

SAFETY PROFILE: Poison via skin contact. Moderately toxic by ingestion. A skin and eye irritant. Moderate fire hazard when exposed to heat or flame; emits crotonaldehyde and water when heated. See CROTONALDEHYDE. Can react with oxidizing materials.

AAI000 CAS:60-35-5 ***HR: 3***
ACETAMIDE
mf: C_2H_5NO mw: 59.08

PROP: Colorless crystals; mousey odor. Mp: 81°, bp: 221.2°, d: 1.159 @ 20°/4°, vap press: 1 mm @ 65°. Decomp in hot water.

SYNS: ACETIC ACID AMIDE ◇ ACETIMIDIC ACID ◇ AMID KYSELINY OCTOVE ◇ ETHANAMIDE ◇ METHANECARBOXAMIDE ◇ NCI-C02108

TOXICITY DATA with REFERENCE
oms-mus/ast 10 pph IDZAAW 51,53,76
otr-ham:emb 1 mg/L IJCNAW 19,642,77
orl-rbt TDLo:13 g/kg (6-18D post):TER ARZNAD 30,1557,80
orl-rbt TDLo:39 g/kg (6-18D post):REP ARZNAD 30,1557,80
orl-rat TDLo:431 g/kg/1Y-C:CAR JEPTDQ 3(5-6),149,80
orl-mus TDLo:517 g/kg/1Y-C:CAR JEPTDQ 3(5-6),149,80
orl-rat TD:546 g/kg/52W-C:NEO TXAPA9 14,163,69
orl-rat LD50:7000 mg/kg JRPFA4 4,219,62
ipr-rat LD50:10300 mg/kg ARZNAD 20,1242,70
scu-rat LD50:10 g/kg OYYAA2 4,451,70
ivn-rat LD50:12500 mg/kg NYKZAU 64(1),42S,68
unr-rat LD50:2300 mg/kg ARZNAD 18,645,68
orl-mus LD50:12900 mg/kg NYKZAU 64(1),42S,68
ipr-mus LD50:1000 mg/kg JJIND8 62,911,79
scu-mus LD50:8300 mg/kg OYYAA2 4,451,70
ivn-mus LD50:10 g/kg NYKZAU 64(1),42S,68
ivn-rbt LD50:7500 mg/kg NYKZAU 64(1),42S,68
ivn-ckn LDLo:33410 mg/kg ARZNAD 20,1242,70

CONSENSUS REPORTS: IARC Cancer Review: Group 2B IMEMDT 7,56,87; Animal Sufficient Evidence IMEMDT 7,389,87. On Community Right-To-Know List. Reported in EPA TSCA Inventory.

DFG MAK: Suspected Carcinogen.

SAFETY PROFILE: Suspected carcinogen with experimental carcinogenic neoplastigenic data. Moderately toxic by intraperitoneal and possibly other routes. An experimental teratogen. Other experimental reproductive effects. Mutation data reported. See also AMIDES. When heated to decomposition it emits toxic fumes of NO_x.

AAI100 CAS:103416-59-7 ***HR: 2***
ACETAMIDE, 2-(DIETHYLAMINO)-N-(1,3-DIMETHYL-4-(o-FLUOROBENZOYL)-5-PYRAZOLYL)-, MONOHYDROCHLORIDE
mf: $C_{18}H_{23}FN_4O_2 \cdot ClH$ mw: 382.91

SYN: 2-(DIETHYLAMINO)-N-(1,3-DIMETHYL-4-(o-FLUOROBENZOYL)-5-PYRAZOLYL)ACETAMI DE HYDROCHLORIDE ◇ 2-(DIETHYLAMINO)-N-(4-(2-FLUOROBENZOYL)-1,3-DIMETHYL-1H-PYRAZOL-5-YL)ACE TAMIDE HYDROCHLORIDE ◇ PD 109394

TOXICITY DATA with REFERENCE
mma-sat 1 μmol/plate CRNGDP 7,2019,86
orl-rat TD:4550 mg/kg/13W-C:ETA AJPAA4 124,392,86

SAFETY PROFILE: Questionable carcinogen with experimental tumorigenic data. Mutation data reported. When heated to decomposition it emits toxic fumes of F^-, NO_x, and HCl.

AAI125 CAS:85723-21-3 ***HR: 2***
ACETAMIDE, N-(4-(2-FLUOROBENZOYL)-1,3-DIMETHYL-1H-PYRAZOL-5-YL)-2-((3-(2-METHYL-1-PIPERIDINYL)PROPYL)AMINO)-, (Z)-2-BUTENEDIOATE (1:2)
mf: $C_{23}H_{32}FN_5O_2$ mw: 429.60

TOXICITY DATA with REFERENCE
orl-rat TD:4550 mg/kg/13W-C:ETA AJPAA4 124,392,86

SAFETY PROFILE: Questionable carcinogen with experimental tumorigenic data. When heated to decomposition it emits toxic fumes of NO_x.

AAI250 CAS:59-66-5 ***HR: 3***
5-ACETAMIDE-1,3,4-THIADIAZOLE-2-SULFONAMIDE
mf: $C_4H_6N_4O_3S_2$ mw: 222.26

SYNS: 2-ACETAMIDO-5-SULFONAMIDO-1,3,4-THIADIAZOLE ◇ ACETAMIDOTHIADIAZOLESULFONAMIDE ◇ ACETAMOX ◇ ACETAZOLAMID ◇ ACETAZOLAMIDE ◇ ACETAZOLEAMIDE ◇ ACETOZALAMIDE ◇ 2-ACETYLAMINO-1,3,4-THIADIAZOLE-5-SULFONAMIDE ◇ N-(5-(AMINOSULFONYL)-1,3,4-THIADIAZOL-2-YL)ACETAMIDE ◇ CARBONIC ANHYDRASE INHIBITOR NO. 6063 ◇ CIDAMEX ◇ DEFILTRAN ◇ DEHYDRATIN ◇ DIACARB ◇ DIAKARB ◇ DIAMOX ◇ DIDOC ◇ DILURAN ◇ DIURAMID ◇ DIURETICUM-HOLZINGER ◇ DIUTAZOL ◇ DONMOX ◇ EDEMOX ◇ EUMICTON ◇ FONURIT ◇ GLAUPAX ◇ GLUPAX ◇ MUIRAMID ◇ NATRIONEX ◇ NEPHRAMIDE ◇ PHONURIT ◇ N-(5-SULFAMOYL-1,3,4-THIADIAZOL-2-YL)ACETAMIDE ◇ VETAMOX

TOXICITY DATA with REFERENCE
orl-mus TDLo:10 g/kg (female 8-12D post):REP TCMUD8 6,361,86
orl-rat TDLo:3300 mg/kg (1-22D preg):TER TJADAB 1,51,68
orl-man TDLo:54 mg/kg/5D-I:PUL AIMDAP 143,1278,83
ipr-rat LD50:2750 mg/kg NYKZAU 56(4),134S,60
orl-mus LD50:4300 mg/kg ABMGAJ 21,193,68
ipr-mus LD50:1175 mg/kg RPTOAN 39,255,76
scu-mus LD50:3 mg/kg DRUGAY 6,15,82
ivn-mus LD50:3 mg/kg DRUGAY 6,15,82

CONSENSUS REPORTS: Reported in EPA TSCA Inventory.

SAFETY PROFILE: Poison by subcutaneous and intravenous routes. Moderately toxic by intraperitoneal route. Human systemic effects by ingestion: dyspnea. An experimental teratogen by many routes. Other experimental reproductive effects. When heated to decomposition it emits very toxic fumes of NO_x and SO_x. A carbonic anhydrase inhibitor and diuretic used to treat glaucoma.

AAI500 ***HR: 2***
ACETAMIDINE HYDROCHLORIDE
mf: $C_2H_6N_2 \cdot HCl$ mw: 94.6

PROP: Long, somewhat deliquescent prisms when crystallized from ethanol. Mp: 164°, sol in water and alcs.

SYNS: α-AMINO-o-IMINOETHANE HYDROCHLORIDE ◇ ETHANAMIDINE HYDROCHLORIDE

SAFETY PROFILE: Moderately toxic irritant. When heated to decomposition it emits toxic fumes of Cl^- and NO_x.

AAI750 CAS:440-58-4 ***HR: 1***
3-ACETAMIDO-5-(ACETAMIDOMETHYL)-2,4,6-TRIIODOBENZOIC ACID
mf: $C_{12}H_{11}I_3N_2O_4$ mw: 627.95

SYNS: 3-(ACETYLAMINO)-5-((ACETYLAMINO)METHYL)-2,4,6-TRIIODOBENZOIC ACID ◇ AMET (GERMAN) ◇ AMETRIODINIC ACID ◇ B-4130 ◇ α-5-DIACETAMIDO-2,4,6-TRIIODO-m-TOLUIC ACID ◇ IODAMIDE ◇ JODAMID (GERMAN) ◇ JODOMIRON ◇ SH 926 ◇ UROMIRO ◇ UROMIRON

TOXICITY DATA with REFERENCE
ipr-rat LD50:17900 mg/kg ARZNAD 15,222,65
ivn-rat LD50:11400 mg/kg ARZNAD 15,222,65
ivn-mus LD50:10800 mg/kg MEIEDD 10,725,83
ivn-rbt LD50:13200 mg/kg ARZNAD 15,222,65
ipr-gpg LD50:15 g/kg ARZNAD 15,222,65

SAFETY PROFILE: Mildly toxic by intraperitoneal and intravenous routes. When heated to decomposition it emits very toxic fumes of NO_x and HI.

AAJ000 CAS:53222-14-3 ***HR: 3***
N-(p-(9-(3-ACETAMIDOACRIDINYL)AMINO)PHENYL)METHANESULFONAMIDE
mf: $C_{22}H_{20}N_4O_3S$ mw: 420.52

TOXICITY DATA with REFERENCE
mmo-sat 112 μmol/L JMCMAR 23,269,80
ipr-mus LD10:19 mg/kg JMCMAR 21,430,78

SAFETY PROFILE: Poison by intraperitoneal route. Mutation data reported. See also SULFONATES. When heated to decomposition it emits very toxic fumes of NO_x and SO_x.

AAJ125 CAS:1713-07-1 ***HR: 1***
3-ACETAMIDO-5-AMINO-2,4,6-TRIIODOBENZOIC ACID
mf: $C_9H_7I_3N_2O_3$ mw: 571.88

TOXICITY DATA with REFERENCE
mma-sat 1 mg/plate PWPSA8 23,249,80
mnt-hmn:lym 40 mg/L RADLAX 129,199,78
cyt-hmn:lym 2000 ppm RADLAX 129,199,78
ivn-mus LD50:7200 mg/kg JPETAB 116,394,56

SAFETY PROFILE: Mildly toxic by intravenous route. Human mutation data reported. When heated to decomposition it emits toxic fumes of I^- and NO_x.

AAJ250 CAS:63906-75-2 ***HR: 3***
2-ACETAMIDO-4,5-BIS-(ACETOXYMERCURI)THIAZOLE
mf: $C_9H_{10}Hg_2N_2O_5S$ mw: 659.45

TOXICITY DATA with REFERENCE
ipr-mus LDLo:15 mg/kg CBCCT* 6,63,54

CONSENSUS REPORTS: Mercury and its compounds are on the Community Right-To-Know List.

OSHA PEL: (Transitional: CL 1 mg/10m^3) CL 0.1 mg(Hg)/m^3 (skin)
ACGIH TLV: TWA 0.1 mg(Hg)/m^3 (skin)
NIOSH REL: (Mercury, Inorganic) TWA 0.05 mg(Hg)/m^3

SAFETY PROFILE: Poison by intraperitoneal route. When heated to decomposition it emits very toxic fumes of Hg, NO_x, and SO_x. See also MERCURY COMPOUNDS.

AAJ350 CAS:3025-96-5 ***HR: 3***
4-ACETAMIDOBUTYRIC ACID
mf: $C_6H_{11}NO_3$ mw: 145.18

SYNS: Γ-ACETYLAMINOBUTYRIC ACID ◇ BUTYRIC ACID, 4-ACETAMIDO- ◇ DF 469

TOXICITY DATA with REFERENCE
scu-rat TDLo:105 mg/kg (male 21D pre):REP ARANDR 10,239,83
ivn-mus LD50:425 mg/kg AIPTAK 145,233,63

SAFETY PROFILE: Poison by intravenous route. Experimental reproductive effects. When heated to decomposition it emits toxic fumes of NO_x

AAJ500 CAS:55941-39-4 ***HR: D***
4-ACETAMIDO-4-CARBOXAMIDO-n-(N-NITROSO)BUTYLCYANAMIDE
mf: $C_8H_{13}N_5O_3$ mw: 227.26

SYN: ACNBC

TOXICITY DATA with REFERENCE
mmo-esc 100 nmol/plate MUREAV 49,9,78
dnr-esc 100 nmol/plate MUREAV 49,9,78

SAFETY PROFILE: Mutation data reported. When heated to decomposition it emits toxic fumes of NO_x. See also N-NITROSO COMPOUNDS.

AAJ750 CAS:35629-37-9 ***HR: 2***
2-ACETAMIDO-4,5-DIMETHYLOXAZOLE
mf: $C_7H_{10}N_2O_2$ mw: 154.19

TOXICITY DATA with REFERENCE
orl-mus LD50:1600 mg/kg JMCMAR 14,1075,71
ipr-mus LD50:800 mg/kg JMCMAR 14,1075,71

SAFETY PROFILE: Moderately toxic by ingestion and intraperitoneal routes. When heated to decomposition it emits toxic fumes of NO_x.

AAK000 CAS:35629-39-1 ***HR: 2***
2-ACETAMIDO-4,5-DIPHENYLOXAZOLE
mf: $C_{17}H_{14}N_2O_2$ mw: 278.33

TOXICITY DATA with REFERENCE
orl-mus LD50:1000 mg/kg JMCMAR 14,1075,71
ipr-mus LD50:800 mg/kg JMCMAR 14,1075,71

SAFETY PROFILE: Moderately toxic by ingestion and intraperitoneal routes. When heated to decomposition it emits toxic fumes of NO_x.

AAK250 CAS:19293-56-2 ***HR: 2***
4-(2-ACETAMIDOETHYLDITHIO)BUTANESULFINATE SODIUM
mf: $C_8H_{16}NO_3S_3 \cdot Na$ mw: 293.42

SYN: SODIUM-4-(2-ACETAMIDOETHYLDITHIO)BUTANESULFINATE

TOXICITY DATA with REFERENCE
orl-mus LD50:1050 mg/kg JMCMAR 18,798,75
ipr-mus LD50:694 mg/kg JMCMAR 15,312,72

SAFETY PROFILE: Moderate ingestion and intraperitoneal routes. When heated to decomposition it emits very toxic fumes of NO_x, Na_2O, and SO_x.

AAK400 CAS:19361-41-2 ***HR: 2***
3-ACETAMIDOFLUORANTHENE
mf: $C_{18}H_{13}NO$ mw: 259.32

SYNS: 3-ACETYLAMINO-FLUORANTHEN ◇ 3-ACETYLAMINOFLUORANTHENE ◇ N-FLUORANTHEN-3-YLACETAMIDE ◇ N-3-FLUORANTHENYLACETAMIDE

TOXICITY DATA with REFERENCE
mma-sat 1 μg/plate NTIS** PB86-213733
orl-rat TDLo:4050 mg/kg/16W-C:ETA ONCOAR 8,233,55

SAFETY PROFILE: Questionable carcinogen with experimental tumorigenic data. Mutation data reported. When heated to decomposition it emits toxic fumes of NO_x.

AAK500 CAS:73106-12-4 ***HR: D***
N-(2-ACETAMIDOFLUOREN-1-YL)-N-FLUOREN-2-YL ACETAMIDE
mf: $C_{30}H_{24}N_2O_2$ mw: 444.53

SYNS: 2AAF DIMER ◇ 1-(N-2'-FLUORENYLACETAMIDO-2-ACETYLAMINO)FLUORENE

TOXICITY DATA with REFERENCE
mma-sat 50 μg/plate DMDSAI 7,296,79

SAFETY PROFILE: Mutation data reported. When heated to decomposition it emits toxic fumes of NO_x.

AAK750 CAS:1068-90-2 ***HR: 2***
ACETAMIDOMALONIC ACID DIETHYL ESTER
mf: $C_9H_{15}NO_5$ mw: 217.25

SYN: DIETHYLESTER KYSELINY ACETYLAMINOMALONOVE (CZECH)

TOXICITY DATA with REFERENCE
eye-rbt 500 mg/24H MLD 28ZPAK -,130,72
orl-rat LD50:4220 mg/kg 28ZPAK -,130,72

CONSENSUS REPORTS: Reported in EPA TSCA Inventory.

SAFETY PROFILE: Mildly toxic by ingestion. An eye irritant. When heated to decomposition it emits toxic fumes of NO_x. See also ESTERS.

AAL000 CAS:50309-20-1 ***HR: 2***
7-ACETAMIDO-1-METHYL-4-(p-(p-((1- METHYL-PYRIDINIUM-4-YL)AMINO)BENZAMIDO) ANILINO)QUINOLINIUM DI-p-TOLUENESULFONATE
mf: $C_{31}H_{30}N_6O_2{\cdot}2C_7H_7O_3S$ mw: 861.07

TOXICITY DATA with REFERENCE
dnd-mus:lym 340 nmol/L JMCMAR 22,134,79
ipr-mus LD10:97 mg/kg JMCMAR 22,134,79

SAFETY PROFILE: Moderately toxic by intraperitoneal route. Mutation data reported. When heated to decomposition it emits very toxic fumes of NO_x and SO_x.

AAL250 CAS:65400-81-9 ***HR: 2***
2-ACETAMIDO-N-(3-METHYL-2-THIAZOLIDINYLIDENE)ACETAMIDE
mf: $C_8H_{13}N_3O_2S$ mw: 215.30

TOXICITY DATA with REFERENCE
orl-mus LD50:936 mg/kg JMCMAR 23,773,80
ivn-mus LD50:408 mg/kg JMCMAR 23,773,80

SAFETY PROFILE: Moderately toxic by ingestion and intravenous routes. When heated to decomposition it emits very toxic fumes of NO_x and SO_x.

AAL300 CAS:55123-66-5 ***HR: 2***
(S)-2-(2-ACETAMIDO-4-METHYLVALERAMIDO)-N-(1-FORMYL-4-GUANIDINOBUTYL)-4-METHYL-VALERAMIDE
mf: $C_{20}H_{38}N_6O_4$ mw: 426.64

SYN: VALERAMIDE,2-(2-ACETAMIDO-4-METHYLVALERAMIDO)-N-(1-FORMYL-4-GUANIDINOBUTYL)-4-METHYL-(S)-

TOXICITY DATA with REFERENCE
orl-mus TDLo:57600 mg/kg/69W-C:NEO GANNA2 71,913,80

SAFETY PROFILE: Questionable carcinogen with experimental neoplastigenic data. When heated to decomposition it emits toxic fumes of NO_x.

AAL500 CAS:24143-08-6 ***HR: 3***
5-ACETAMIDO-3-(5-NITRO-2-FURYL)-6H-1,2,4-OXADIAZINE
mf: $C_9H_8N_4O_5$ mw: 252.21

SYN: N-(3-(5-NITRO-2-FURYL)-6H-1,2,4-OXADIAZINYL)ACETAMIDE

TOXICITY DATA with REFERENCE
mma-sat 1 μg/plate MUREAV 40,9,76
dnr-sat 500 nmol/well CNREA8 34,2266,74
mmo-esc 300 nmol/well CNREA8 34,2266,74
mrc-esc 500 nmol/well CNREA8 34,2266,74
pic-esc 500 μg/L MUREAV 26,3,74
orl-rat TDLo:21 g/kg/28W-C:CAR CNREA8 29,2212,69

SAFETY PROFILE: Questionable carcinogen with experimental carcinogenic data. When heated to decomposition it emits toxic fumes of NO_x.

AAL750 CAS:531-82-8 ***HR: 3***
2-ACETAMIDO-4-(5-NITRO-2-FURYL)THIAZOLE
mf: $C_9H_7N_3O_4S$ mw: 253.25

SYNS: 2-ACETAMINO-4-(5-NITRO-2-FURYL)THIAZOLE ◇ 2-ACETYLAMINO-4-(5-NITRO-2- FURYL)THIAZOLE ◇ N-(4-(5-NITRO-2-FURANYL)-2-THIAZOLYL)ACETAMIDE ◇ N-(4-(5-NITRO-2-FURYL)-2-THIAZOLYL)ACETAMIDE ◇ N-(4-(5-NITRO-2-FURYL)THIAZOL-2-YL)ACETAMIDE

TOXICITY DATA with REFERENCE
mma-sat 100 ng/plate MUREAV 40,9,76
dnr-sat 500 nmol/well CNREA8 34,2266,74
mmo-esc 300 nmol/well CNREA8 34,2266,74
mrc-esc 500 nmol/well CNREA8 34,2266,74
orl-rat TDLo:43 g/kg/46W-C:CAR CNREA8 30,936,70
orl-mus TDLo:2400 mg/kg/14W-C:CAR CNREA8 30,2320,70
orl-dog TDLo:27 g/kg/2Y-C:ETA JNCIAM 45,535,70
orl-rat TD:47 g/kg/46W-C:NEO JNCIAM 54,841,75

CONSENSUS REPORTS: IARC Cancer Review: Group 2B IMEMDT 7,56,87; Animal Sufficient Evidence IMEMDT 1,181,72; IMEMDT 7,185,74

SAFETY PROFILE: Suspected carcinogen with experimental carcinogenic, tumorigenic, and neoplastigenic data. Mutation data reported. When heated to decomposition it emits very toxic fumes of SO_x and NO_x.

AAM000 CAS:52162-18-2 ***HR: D***
2-ACETAMIDO-5-(NITROSOCYANAMIDO) VALERAMIDE
mf: $C_8H_{13}N_5O_3$ mw: 227.26

SYN: ACETYL-l-ARGININE, NITROSATED

TOXICITY DATA with REFERENCE
mmo-sat 20 mmol/L GANNA2 65,45,74

SAFETY PROFILE: Mutation data reported. When heated to decomposition it emits toxic fumes of NO_x.

AAM250 CAS:4120-77-8 ***HR: 3***
2-ACETAMIDOPHENATHRENE
mf: $C_{16}H_{13}NO$ mw: 235.30

SYNS: 2-ACETAMINOPHENANTHRENE ◇ 2-ACETYLAMINO-PHENANTHRENE ◇ 2-PHENANTHRYLACETAMIDE ◇ N-2-PHENANTHRYLACETAMIDE ◇ N-(2-PHENANTHRYL)ACETAMIDE

TOXICITY DATA with REFERENCE
dnr-ham:fbr 1 μmol/L JNCIAM 54,1287,75
orl-rat TDLo:900 mg/kg/13W-C:CAR CNREA8 19,210,59
scu-rat TDLo:75 mg/kg/3W-I:ETA CNREA8 26,2239,66

SAFETY PROFILE: Questionable carcinogen with experimental carcinogenic and tumorigenic data. Mutation data reported. When heated to decomposition it emits toxic fumes of NO_x. See also AMIDES.

AAM500 CAS:35629-38-0 ***HR: 2***
2-ACETAMIDO-4-PHENYLOXAZOLE
mf: $C_{11}H_{10}N_2O_2$ mw: 202.23

TOXICITY DATA with REFERENCE
orl-mus LD50:800 mg/kg JMCMAR 14,1075,71
ipr-mus LD50:500 mg/kg JMCMAR 14,1075,71

SAFETY PROFILE: Moderately toxic by ingestion and intraperitoneal routes. When heated to decomposition it emits toxic fumes of NO_x.

AAM750 CAS:5221-42-1 ***HR: 3***
4-ACETAMIDOPYRIDINE
mf: $C_7H_8N_2O$ mw: 136.17

SYNS: 4-ACETYLAMINOPYRIDINE ◇ PHILLIPS 2038

TOXICITY DATA with REFERENCE
orl-rat LD50:446 mg/kg TXAPA9 21,315,72
orl-bwd LD50:13 mg/kg TXAPA9 21,315,72

SAFETY PROFILE: Poison by ingestion. When heated to decomposition it emits toxic fumes of NO_x. See also AMIDES.

AAM875 CAS:85-36-9 ***HR: 1***
3-ACETAMIDO-2,4,6-TRIIODOBENZOIC ACID
mf: $C_9H_6I_3NO_3$ mw: 556.86

SYNS: ACETRIZOIC ACID ◇ 3-(ACETYLAMINO)-2,4,6-TRIIODOBENZOIC ACID ◇ ACIDO 3-ACETILAMINO-2,4,6-TRIIODOBENZOICO (ITALIAN)

TOXICITY DATA with REFERENCE
unr-rat LD50:9650 mg/kg JAPMA8 42,721,53
orl-mus LD50:20 g/kg FRPSAX 18,33,63
ivn-mus LD50:8000 mg/kg FRPSAX 18,33,63

SAFETY PROFILE: Mildly toxic by ingestion and intravenous routes. When heated to decomposition it emits toxic fumes of I^- and NO_x.

AAN000 CAS:129-63-5 ***HR: 1***
3-ACETAMIDO-2,4,6-TRIIODOBENZOIC ACID SODIUM SALT
mf: $C_9H_5I_3NO_3 \cdot Na$ mw: 578.84

SYNS: ACETIODONE ◇ ACETRIZOATE SODIUM ◇ ACETRIZOIC ACID SODIUM SALT ◇ BRONCHOSELECTAN ◇ CYSTOKON ◇ DIAGINOL ◇ FORTOMBRINE-N ◇ IODOPACT ◇ IODOPAQUE ◇ JODOPAX ◇ MP 1023 ◇ PYELOKON-FR ◇ SALPIX ◇ SODIUM-3-ACETYLAMINO-2,4,6-TRIIODOBENZOATE ◇ SODIUM-3-ACETAMIDO-2,4,6-TRIIODOBENZOATE ◇ SODIUM ACETRIZOATE ◇ THIXOKON ◇ TRI-ABRODIL ◇ TRIIODRAST ◇ TRIIODYL ◇ TRIIOTRAST ◇ 2,4,6-TRIJOD-3-ACETAMINOBENZOSAEURE NATRIUM (FERMAN) ◇ TRIOPAC 200 ◇ TRIOPAS ◇ TRIUMBREN ◇ TRIUROL ◇ TRIUROPAN ◇ UROKON SODIUM ◇ VESAMIN ◇ VISOTRAST ◇ VROKON

TOXICITY DATA with REFERENCE
ivn-mus LD50:9956 mg/kg JACSAT 74,4365,52

ims-mus LD50:12156 mg/kg JPETAB 117,307,56
ivn-rat LDLo:7500 mg/kg CLDND*
ivn-dog LD50:6300 mg/kg JPETAB 116,394,56
ivn-cat LD50:5650 mg/kg CLDND*
ivn-rbt LD50:5200 mg/kg JPETAB 116,394,56

SAFETY PROFILE: Mildly toxic by intravenous route. When heated to decomposition it emits very toxic fumes of NO_x, Na_2O, and HI.

AAN250 CAS:100700-23-0 ***HR: 2***
2-((4-(3-ACETAMIDO-2,4,6-TRIIODOPHENOXY)BUTOXY)METHYL) BUTYRIC ACID SODIUM SALT
mf: $C_{17}H_{22}I_3NO_5$•Na mw: 724.09

TOXICITY DATA with REFERENCE
orl-mus LD50:2200 mg/kg FRPSAX 31,349,76
ivn-mus LD50:610 mg/kg FRPSAX 31,349,76

SAFETY PROFILE: Moderately toxic by ingestion and intravenous routes. When heated to decomposition it emits very toxic fumes of I^-, Na_2O, and NO_x.

AAN500 CAS:101651-76-7 ***HR: 2***
2-(2-(3-ACETAMIDO-2,4,6-TRIIODOPHENOXY)ETHOXY)ACETIC ACID SODIUM SALT
mf: $C_{12}H_{12}I_3NO_5$•Na mw: 653.94

TOXICITY DATA with REFERENCE
orl-mus LD50:9800 mg/kg FRPSAX 31,349,76
ivn-mus LD50:1500 mg/kg FRPSAX 31,349,76

SAFETY PROFILE: Moderately toxic by intravenous route. Mildly toxic by ingestion. When heated to decomposition it emits very toxic fumes of I^-, Na_2O, and NO_x.

AAN750 CAS:100700-24-1 ***HR: 2***
2-(2-(3-ACETAMIDO-2,4,6-TRIIODOPHENOXY)ETHOXY)BUTYRIC ACID SODIUM SALT
mf: $C_{14}H_{16}I_3NO_5$•Na mw: 682.00

TOXICITY DATA with REFERENCE
orl-mus LD50:3900 mg/kg FRPSAX 31,349,76
ivn-mus LD50:1320 mg/kg FRPSAX 31,349,76

SAFETY PROFILE: Moderately toxic by ingestion and intravenous routes. When heated to decomposition it emits very toxic fumes of I^-, Na_2O, and NO_x.

AAO000 CAS:102504-51-8 ***HR: 2***
2-((2-(3-ACETAMIDO-2,4,6-TRIIODOPHENOXY)ETHOXY)METHYL)PROPIONIC ACID SODIUM SALT
mf: $C_{14}H_{16}I_3NO_5$•Na mw: 682.00

TOXICITY DATA with REFERENCE
orl-mus LD50:5200 mg/kg FRPSAX 31,349,76
ivn-mus LD50:1150 mg/kg FRPSAX 31,349,76

SAFETY PROFILE: Moderately toxic by intravenous route. Mildly toxic by ingestion. When heated to decomposition it emits very toxic fumes of NO_x, Na_2O, and I^-.

AAO250 CAS:102504-52-9 ***HR: 2***
2-(2-(3-ACETAMIDO-2,4,6-TRIIODOPHENOXY)ETHOXY)PROPIONIC ACID SODIUM SALT
mf: $C_{13}H_{14}I_3NO_5$•Na mw: 667.97

TOXICITY DATA with REFERENCE
orl-mus LD50:9800 mg/kg FRPSAX 31,349,76
ivn-mus LD50:1480 mg/kg FRPSAX 31,349,76

SAFETY PROFILE: Moderately toxic by intravenous route. Mildly toxic by ingestion. When heated to decomposition it emits very toxic fumes of NO_x, Na_2O, and I^-.

AAO500 CAS:101651-77-8 ***HR: 2***
2-(2-(3-ACETAMIDO-2,4,6-TRIIODOPHENOXY)ETHOXY)-2-(o-TOLYL)ACETIC ACID SODIUM SALT
mf: $C_{19}H_{18}I_3NO_5$•Na mw: 744.07

TOXICITY DATA with REFERENCE
orl-mus LD50:3550 mg/kg FRPSAX 31,349,76
ivn-mus LD50:585 mg/kg FRPSAX 31,349,76

SAFETY PROFILE: Moderately toxic by ingestion and intravenous routes. When heated to decomposition it emits very toxic fumes of I^-, Na_2O, and NO_x.

AAO750 CAS:101651-78-9 ***HR: 2***
2-(2-(3-ACETAMIDO-2,4,6-TRIIODOPHENOXY)ETHOXY)-2-(p-TOLYL)ACETIC ACID SODIUM SALT
mf: $C_{19}H_{18}I_3NO_5$•Na mw: 744.07

TOXICITY DATA with REFERENCE
orl-mus LD50:3430 mg/kg FRPSAX 31,349,76
ivn-mus LD50:658 mg/kg FRPSAX 31,349,76

SAFETY PROFILE: Moderately toxic by ingestion and intravenous routes. When heated to decomposition it emits very toxic fumes of I^-, Na_2O, and NO_x.

AAP000 CAS:102584-89-4 ***HR: 2***
2-(2-(3-ACETAMIDO-2,4,6-TRIIODOPHENOXY)ETHOXY)VALERIC ACID SODIUM SALT
mf: $C_{15}H_{18}I_3NO_5$•Na mw: 696.03

TOXICITY DATA with REFERENCE
orl-mus LD50:2500 mg/kg FRPSAX 31,349,76
ivn-mus LD50:740 mg/kg FRPSAX 31,349,76

SAFETY PROFILE: Moderately toxic by ingestion and intravenous routes. See also IODIDES. When heated to decomposition it emits very toxic fumes of I^-, Na_2O, and NO_x.

AAP250 CAS:100700-25-2 ***HR: 2***
2-((3-(3-ACETAMIDO-2,4,6-TRIIODOPHENOXY)PROPOXY)METHYL)BUTYRIC ACID SODIUM SALT
mf: $C_{16}H_{20}I_3NO_5$•Na mw: 710.06

TOXICITY DATA with REFERENCE
orl-mus LD50:3700 mg/kg FRPSAX 31,349,76
ivn-mus LD50:750 mg/kg FRPSAX 31,349,76

SAFETY PROFILE: Moderately toxic by ingestion and intravenous routes. When heated to decomposition it emits very toxic fumes of I^-, Na_2O, and NO_x.

AAP500 CAS:23279-53-0 ***HR: 2***
2-(3-ACETAMIDO-2,4,6-TRIIODOPHENYL)BUTYRIC ACID
mf: $C_{12}H_{12}I_3NO_3$ mw: 598.95

TOXICITY DATA with REFERENCE
orl-mus LD50:2800 mg/kg JMCMAR 13,559,70
ivn-mus LD50:950 mg/kg JMCMAR 13,559,70

SAFETY PROFILE: Moderately toxic by ingestion and intravenous routes. When heated to decomposition it emits very toxic fumes of I^- and NO_x.

AAP750 CAS:23217-81-4 ***HR: 2***
2-(3-ACETAMIDO-2,4,6-TRIIODOPHENYL)PROPIONIC ACID
mf: $C_{11}H_{10}I_3NO_3$ mw: 584.92

TOXICITY DATA with REFERENCE
orl-mus LD50:3800 mg/kg JMCMAR 13,559,70
ivn-mus LD50:1200 mg/kg JMCMAR 13,559,70

SAFETY PROFILE: Moderately toxic by ingestion and intravenous routes. When heated to decomposition it emits very toxic fumes of NO_x and I^-.

AAQ000 CAS:23217-87-0 ***HR: 2***
2-(3-ACETAMIDO-2,4,6-TRIIODOPHENYL)VALERIC ACID
mf: $C_{13}H_{14}I_3NO_3$ mw: 612.98

TOXICITY DATA with REFERENCE
orl-mus LD50:4000 mg/kg JMCMAR 13,559,70
ivn-mus LD50:700 mg/kg JMCMAR 13,559,70

SAFETY PROFILE: Moderately toxic by ingestion and intravenous routes. See also IODIDES. When heated to decomposition it emits very toxic fumes of I^- and NO_x.

AAQ250 CAS:2832-40-8 ***HR: 3***
ACETAMINE YELLOW CG
mf: $C_{15}H_{15}N_3O_2$ mw: 269.33

SYNS: ACTIOQUINONE LIGHT YELLOW ◇ AMACEL YELLOW G ◇ CALCOSYN YELLOW GC ◇ CELLITON FAST YELLOW G ◇ C.I. 11855 ◇ CIBACET YELLOW GBA ◇ C.I. DISPERSE YELLOW 3 ◇ HISPERSE YELLOW G ◇ N-(4-((2-HYDROXY-5-METHYLPHENYL)AZO)PHENYL)ACETAMIDE ◇ 4'-((6-HYDROXY-m-TOLYL)AZO)ACETANILIDE ◇ INTRASPERSE YELLOW GBA EXTRA ◇ MICROSETILE YELLOW GR ◇ NACELAN FAST YELLOW CG ◇ NCI-C53781 ◇ YELLOW Z

TOXICITY DATA with REFERENCE
cyt-frg-par 2800 µL/7D CYTBAI 25,175,79
orl-rat TDLo:180 g/kg/2Y-C:CAR NTPTR* NTP-TR-222,82
orl-mus TDLo:433 g/kg/2Y-C:CAR NTPTR* NTP-TR-222,82
orl-mus TD:216 g/kg/2Y-C:ETA NTPTR* NTP-TR-222,82

CONSENSUS REPORTS: Community Right-To-Know List. Reported in EPA TSCA Inventory. IARC Cancer Review: Animal Inadequate Evidence IMEMDT 8,97,75; NTP Carcinogenesis Bioassay (feed); Clear Evidence: mouse, rat NTPTR* NTP-TR-222,82

SAFETY PROFILE: Suspected carcinogen with experimental tumorigenic and carcinogenic data. An allergen. Mutation data reported. When heated to decomposition it emits toxic fumes of NO_x.

AAQ500 CAS:103-84-4 ***HR: 3***
ACETANILIDE
mf: C_8H_9NO mw: 135.18

PROP: White, shining, crystalline scales. Mp: 113.5°, bp: 305°, flash p: 345°F (OC), d: 1.2105 @ 4°/4°, autoign temp: 1004°F, vap press: 1 mm @ 114.0°, vap d: 4.65. Somewhat sol in water, alc, and ether.

SYNS: ACETAMIDOBENZENE ◇ ACETANIL ◇ ACETIC ACID ANILIDE ◇ ACETOANILIDE ◇ ACETYLAMINOBENZENE ◇ ACETYLANILINE ◇ N-ACETYLANILINE ◇ AN ◇ ANTIFEBRIN ◇ PHENALGENE ◇ N-PHENYLACETAMIDE ◇ USAF EK-3

TOXICITY DATA with REFERENCE
orl-hmn TDLo:14 mg/kg/D:PUL,KID,BLD 34ZIAG -,62,69
orl-man LDLo:56 mg/kg/H-I:CNS,GIT,MET AJMSA9 122,770,01
orl-man TDLo:405 mg/kg:CNS,PUL JAMAAP 12,103,1889
ipr-rat LD50:540 mg/kg JAPMA8 48,204,59
orl-mus LD50:1210 mg/kg TXAPA9 19,20,71
ipr-mus LD50:500 mg/kg NTIS** AD277-689
orl-cat LDLo:250 mg/kg JPHAA3 28,70,39
ivn-cat LDLo:8500 µg/kg JAPMA8 30,91,41
orl-rbt LDLo:1500 mg/kg JPETAB 29,466,63
unk-man LDLo:59 mg/kg 85DCAI 2,73,70
orl-rat LD50:800 mg/kg JPETAB 54,159,35

orl-mus LD50:1210 mg/kg TXAPA9 19,20,71
ipr-mus LDLo:1000 mg/kg JACSAT 63,1437,41
orl-dog LDLo:500 mg/kg HBAMAK 4,1290,35
ivn-dog LDLo:300 mg/kg HBAMAK 4,1290,35
orl-gpg LDLo:200 mg/kg HBAMAK 4,1290,35

CONSENSUS REPORTS: Reported in EPA TSCA Inventory. EPA Genetic Toxicology Program.

SAFETY PROFILE: A human poison by an unspecified route. Poison by ingestion and intravenous routes. Moderately toxic by intraperitoneal route. Human systemic effects by ingestion: hallucinations and distorted perceptions, sleepiness, constipation, cyanosis, respiratory stimulation, kidney damage, methemoglobinemia-carboxhemoglobinemia and decreased body temperature. When heated to decomposition it emits toxic fumes of NO_x. Combustible when exposed to heat or flame. See also ANILINE.

AAQ750 CAS:588-16-9 ***HR: 2***
m-ACETANISIDIDE
mf: $C_9H_{11}NO_2$ mw: 165.21

PROP: Mp: 80-81°.

SYNS: ACETO-m-ANISIDIDE ◇ m-METHOXYACETANILIDE ◇ 3-METHOXYACETANILIDE ◇ 3'-METHOXYACETANILIDE

TOXICITY DATA with REFERENCE
orl-mus LD50:1100 mg/kg TXAPA9 19,20,71

CONSENSUS REPORTS: Reported in EPA TSCA Inventory.

SAFETY PROFILE: Moderately toxic by ingestion. When heated to decomposition emits toxic fumes of NO_x.

AAR000 CAS:93-26-5 ***HR: 2***
o-ACETANISIDIDE
mf: $C_9H_{11}NO_2$ mw: 165.21

PROP: Mp: 87-88°; bp: 303-305°. Very sol in hot water, alc, acetone, and ether.

SYNS: o-METHOXYACETANILIDE ◇ 2-METHOXYACETANILIDE ◇ 2'-METHOXYACETANILIDE

TOXICITY DATA with REFERENCE
orl-mus LD50:940 mg/kg TXAPA9 19,20,71

CONSENSUS REPORTS: Reported in EPA TSCA Inventory.

SAFETY PROFILE: Moderately toxic by ingestion. When heated to decomposition emits toxic fumes of NO_x.

AAR250 CAS:51-66-1 ***HR: 2***
p-ACETANISIDIDE
mf: $C_9H_{11}NO_2$ mw: 165.21

PROP: Mp: 127°. Sol in alc, chloroform and ether.

SYNS: ACETO-p-ANISIDIDE ◇ ACETYL-p-ANISIDINE ◇ p-METHOXYACETANILIDE ◇ 4-METHOXYACETANILIDE ◇ 4'-METHOXYACETANILIDE

TOXICITY DATA with REFERENCE
orl-mus LD50:1190 mg/kg TXAPA9 19,20,71
unk-rbt LDLo:3000 mg/kg XPHBAO 271,71,41

CONSENSUS REPORTS: Reported in EPA TSCA Inventory.

SAFETY PROFILE: Moderately toxic by ingestion and other unspecified routes. When heated to decomposition emits toxic fumes of NO_x.

AAR500 CAS:3572-06-3 ***HR: 2***
ACETATE of 4-(HYDROXYPHENYL)-2-BUTANONE
mf: $C_{12}H_{14}O_3$ mw: 206.26

SYNS: 4-(p-ACETOXYPHENYL)-2-BUTANONE ◇ ENT 32,833 ◇ 4-(p-HYDROXYPHENYL)-2-BUTANONE ACETATE ◇ p-(3-OXOBUTYL)PHENYL ACETATE

TOXICITY DATA with REFERENCE
orl-rat LD50:3038 mg/kg TXAPA9 31,421,75

CONSENSUS REPORTS: Reported in EPA TSCA Inventory.

SAFETY PROFILE: Moderately toxic by ingestion. When heated to decomposition it emits acrid smoke and irritating fumes.

AAR750 CAS:63868-93-9 ***HR: 3***
(ACETATO)BIS(HEPTYLOXY)PHOSPHINYLMERCURY
mf: $C_{16}H_{33}HgO_5P$ mw: 537.05

SYN: (BIS-(HEPTYLOXY)PHOSPHINYL)MERCURYACETATE

TOXICITY DATA with REFERENCE
ipr-mus LDLo:63 mg/kg CBCCT* 8,103,56

CONSENSUS REPORTS: Mercury and its compounds are on the Community Right-To-Know List.

OSHA PEL: (Transitional: CL 1 mg/10m³) CL 0.1 mg(Hg)/m³ (skin)
ACGIH TLV: TWA 0.1 mg(Hg)/m³ (skin)
NIOSH REL: (Mercury, Inorganic) TWA 0.05 mg(Hg)/m³

SAFETY PROFILE: Poison by intraperitoneal route. See also MERCURY COMPOUNDS. When heated to decomposition it emits very toxic fumes of Hg and PO_x.

AAS000 CAS:63868-94-0 ***HR: 3***
(ACETATO)BIS(HEXYLOXY)PHOSPHINYLMERCURY
mf: $C_{14}H_{29}HgO_5P$ mw: 508.99

SYN: (BIS(HEXYLOXY)PHOSPHINYL)MERCURY ACETATE

TOXICITY DATA with REFERENCE
ipr-mus LDLo:125 mg/kg CBCCT* 8,103,56

CONSENSUS REPORTS: Mercury and its compounds are on the Community Right-To-Know List.

OSHA PEL: (Transitional: CL 1 mg/10m^3) CL 0.1 mg(Hg)/m^3 (skin)
ACGIH TLV: TWA 0.1 mg(Hg)/m^3 (skin)
NIOSH REL: (Mercury, Inorganic) TWA 0.05 mg(Hg)/m^3

SAFETY PROFILE: Poison by intraperitoneal route. See also MERCURY COMPOUNDS. When heated to decomposition it emits very toxic fumes of Hg and PO_x.

AAS250 CAS:5421-48-7 ***HR: 3***
(ACETATO)(DIETHOXYPHOSPHINYL)MERCURY
mf: $C_6H_{13}HgO_5P$ mw: 396.75

SYN: (DIETHOXY-PHOSPHINYL)MERCURY ACETATE

TOXICITY DATA with REFERENCE
ipr-mus LDLo:8 mg/kg CBCCT* 8,103,56

CONSENSUS REPORTS: Mercury and its compounds are on the Community Right-To-Know List.

OSHA PEL: (Transitional: CL 1 mg/10m^3) CL 0.1 mg(Hg)/m^3 (skin)
ACGIH TLV: TWA 0.1 mg(Hg)/m^3 (skin)
NIOSH REL: (Mercury, Inorganic) TWA 0.05 mg(Hg)/m^3.

SAFETY PROFILE: Poison by intraperitoneal route. See also MERCURY COMPOUNDS. When heated to decomposition it emits very toxic fumes of Hg and PO_x.

AAS500 CAS:21450-81-7 ***HR: 3***
(ACETATO)(2,3,5,6-TETRAMETHYLPHENYL)MERCURY
mf: $C_{12}H_{16}HgO_2$ mw: 392.87

SYN: (2,3,5,6-TETRAMETHYLPHENYL)MERCURY ACETATE

TOXICITY DATA with REFERENCE
ivn-mus LD50:32 mg/kg CSLNX* NX#05139

CONSENSUS REPORTS: Mercury and its compounds are on the Community Right-To-Know List.

OSHA PEL: (Transitional: CL 1 mg/10m^3) CL 0.1 mg(Hg)/m^3 (skin)
ACGIH TLV: TWA 0.1 mg(Hg)m^3 (skin)
NIOSH REL: (Mercury, Inorganic) TWA 0.05 mg(Hg)/m^3

SAFETY PROFILE: Poison by intravenous route. See also MERCURY COMPOUNDS. When heated to decomposition it emits toxic fumes of Hg.

AAS750 CAS:1424-27-7 ***HR: 1***
ACETAZOLAMIDE SODIUM
mf: $C_4H_5N_4O_3S_2$•Na mw: 244.24

SYNS: ACETAZOLAMIDE SODIUM SALT ◇ SODIUM ACETAZOLAMIDE

TOXICITY DATA with REFERENCE
ipr-mus TDLo:2000 mg/kg (9D preg):TER TJADAB 20,289,79
ipr-mus TDLo:2000 mg/kg (9D preg):REP TJADAB 20,289,79
ivn-mus LD50:6 g/kg YAKUD5 21,775,79

SAFETY PROFILE: An experimental teratogen. Other experimental reproductive effects. When heated to decomposition it emits very toxic fumes of NO_x, Na_2O, and SO_x.

AAT000 CAS:2047-14-5 ***HR: 3***
ACETHION AMIDE
mf: $C_6H_{13}NO_3PS_2$ mw: 242.29

SYN: β-CARBAMIDOCARBOMETHYL-O,O-DIETHYLDITHIOPHOSPHATE

TOXICITY DATA with REFERENCE
orl-rat LDLo:10 mg/kg GISAAA 24,47,59
orl-mus LDLo:40 mg/kg GISAAA 24,47,59
ipr-mus LD50:200 mg/kg JEENAI 51,714,58
orl-rbt LDLo:15 mg/kg GISAAA 24,47,59

SAFETY PROFILE: Poison by ingestion and intraperitoneal routes. When heated to decomposition it emits very toxic fumes of PO_x, SO_x and NO_x.

AAT250 CAS:64-19-7 ***HR: 3***
ACETIC ACID
DOT: UN 2789/UN 2790
mf: $C_2H_4O_2$ mw: 60.06

PROP: Clear, colorless liquid; pungent odor. Mp: 16.7°, bp: 118.1°, flash p: 109°F (CC), lel: 5.4%, uel: 16.0% @ 212°F, d: 1.049 @ 20°/4°, autoign temp: 869°F, vap press: 11.4 mm @ 20°, vap d: 2.07. Misc in water, alc, and ether.

SYNS: ACETIC ACID (aqueous solution) (DOT) ◇ ACETIC ACID, GLACIAL (DOT) ◇ ACIDE ACETIQUE (FRENCH) ◇ ACIDO ACETICO (ITALIAN) ◇ AZIJNZUUR (DUTCH) ◇ ESSIGSAEURE (GERMAN) ◇ ETHANOIC ACID ◇ ETHYLIC ACID ◇ FEMA No. 2006 ◇ GLACIAL ACETIC ACID ◇ METHANECARBOXYLIC ACID ◇ OCTOWY KWAS (POLISH) ◇ VINEGAR ACID

TOXICITY DATA with REFERENCE
skn-hmn 50 mg/24H MLD TXAPA9 31,481,75
skn-rbt 525 mg open SEV UCDS** 8/7/63

skn-rbt 50 mg/24H MLD TXAPA9 31,481,75
eye-rbt 50 μg open SEV AMIHBC 4,119,51
eye-rbt 100 mg rns MLD TXCYAC 23,281,82
mmo-esc 300 ppm/3H AMNTA4 85,119,51
sln-dmg-ihl 1000 ppm/24H THAGA6 39,330,69
sln-dmg-orl 1000 ppm THAGA6 39,330,69
cyt-grl-par 40 μmol/L NULSAK 9,119,66
orl-rat TDLo:700 mg/kg (18D post):REP NTOTDY 4,105,82
orl-hmn TDLo:1470 μg/kg:GIT AIHAAP 33,624,72
ihl-hmn TCLo:816 ppm/3M:NOSE,EYE,PUL AMIHAB 21,28,60
unk-man LDLo:308 mg/kg 85DCAI 2,73,70
orl-rat LD50:3310 mg/kg JIHTAB 23,78,41
ihl-rat LCLo:16000 ppm/4H JIHTAB 23,78,41
ihl-mus LC50:5620 ppm/1H MELAAD 48,559,57
ivn-mus LD50:525 mg/kg APTOA6 18,141,61
orl-rbt LDLo:1200 mg/kg CRSBAW 83,136,20
skn-rbt LD50:1060 mg/kg UCDS** 8/7/63
scu-rbt LDLo:1200 mg/kg CRSBAW 83,136,20
rec-rbt LDLo:1200 mg/kg CRSBAW 83,136,20

CONSENSUS REPORTS: Reported in EPA TSCA Inventory.

OSHA PEL: TWA 10 ppm
ACGIH TLV: TWA 10 ppm; STEL 15 ppm
DFG MAK: 10 ppm (25 mg/m^3)
DOT Classification: Corrosive Material; Label: Corrosive (UN2789, UN2790); Corrosive Material; Label: Corrosive, Flammable Liquid (UN2789)

SAFETY PROFILE: A human poison by an unspecified route. Moderately toxic by various routes. A severe eye and skin irritant. Can cause burns, lachrymation, and conjunctivitis. Human systemic effects by ingestion: changes in the esophagus, ulceration or bleeding from the small and large intestines. Human systemic irritant effects and mucous membrane irritant. Experimental reproductive effects. Mutation data reported. A common air contaminant. A combustible liquid. Moderate fire and explosion hazard when exposed to heat or flame; can react vigorously with oxidizing materials. To fight fire, use CO_2, dry chemical, alcohol foam, foam and mist. When heated to decomposition it emits irritating fumes.

Potentially explosive reaction with 5-azidotetrazole; bromine pentafluoride; chromium trioxide; hydrogen peroxide; potassium permanganate; sodium peroxide; and phosphorus trichloride. Potentially violent reactions with acetaldehyde and acetic anhydride. Ignites on contact with potassium-tert-butoxide. Incompatible with chromic acid; nitric acid; 2-amino-ethanol; NH_4NO_3; ClF_3; chlorosulfonic acid; (O_3 + diallyl methyl carbinol); ethylenediamine; ethylene imine; (HNO_3 + acetone); oleum; $HClO_4$; permanganates; $P(OCN)_3$; KOH; NaOH; n-xylene.

AAT500 CAS:64046-61-3 ***HR: 3***
ACETIC ACID-3-ALLYLOXYALLYL ESTER
mf: $C_8H_{12}O_3$ mw: 156.20

TOXICITY DATA with REFERENCE
orl-rat LD50:3730 mg/kg TXAPA9 28,313,74
skn-rbt LD50:350 mg/kg TXAPA9 28,313,74

SAFETY PROFILE: Poison by skin contact. Moderately toxic by ingestion. When heated to decomposition, it emits acrid smoke and irritating fumes. See also ESTERS.

AAU000 CAS:150-84-5 ***HR: 1***
ACETIC ACID, CITRONELLYL ESTER
mf: $C_{12}H_{22}O_2$ mw: 198.34

PROP: Found in oils of Citronella Ceylon, Geranium, and about 20 other oils (FCTXAV 11,1011,73). Colorless liquid; fruity odor. D: 0.883-0.893, refr index: 1.440-1.450, flash p: +212°F. Sol in alc and fixed oils; insol in glycerin, propylene glycol, and water @229°.

SYNS: ACETIC ACID-3,7-DIMETHYL-6-OCTEN-1-YL ESTER ◇ CITRONELLYL ACETATE (FCC) ◇ 2,6-DIMETHYL-2-OCTEN-8-OL ACETATE ◇ 3,7-DIMETHYL-6-OCTEN-1-YL ACETATE ◇ FEMA No. 2311

TOXICITY DATA with REFERENCE
skn-hmn 20 mg/48H MLD FCTXAV 11,1011,73
skn-rbt 500 mg/24H FCTXAV 11,1011,73
orl-rat LD50:6800 mg/kg FCTXAV 11,1011,73

CONSENSUS REPORTS: Reported in EPA TSCA Inventory.

SAFETY PROFILE: Mildly toxic by ingestion. A human skin irritant. See also ESTERS. Combustible liquid. When heated to decomposition it emits acrid smoke and irritating fumes.

AAU250 CAS:18461-55-7 ***HR: 3***
ACETIC ACID-4,6-DINITRO-o-CRESYL ESTER
mf: $C_9H_8N_2O_6$ mw: 240.19

SYNS: 4,6-DINITRO-o-KRESYLESTER KYSELINY OCTOVE (CZECH) ◇ DNOK-ACETAT (CZECH)

TOXICITY DATA with REFERENCE
skn-rbt 500 mg/24H MOD 28ZPAK -,131,72
eye-rbt 100 mg/24H SEV 28ZPAK -,131,72
ipr-mus LDLo:63 mg/kg CBCCT* 6,146,54
orl-rat LD50:46 mg/kg 28ZPAK -,131,72

NIOSH REL: (Dinitro ortho-Cresyl) TWA 0.2 mg/m^3

SAFETY PROFILE: Poison by ingestion and intraperitoneal routes. A skin and severe eye irritant. When heated to decomposition it emits toxic fumes of NO_x.

AAU500 CAS:39920-56-4 ***HR: 1***
ACETIC ACID-3-HEPTANOL ESTER
mf: $C_9H_{18}O_3$ mw: 174.27

SYN: ACETIC ACID-3-HYDROXYHEPTYL ESTER

TOXICITY DATA with REFERENCE
skn-rbt 10 mg/24H MLD AMIHBC 10,61,54
eye-rbt 500 mg AMIHBC 10,61,54
orl-rat LD50:8350 mg/kg AMIHBC 10,61,54

SAFETY PROFILE: Mildly toxic by ingestion. A skin and eye irritant. See also ESTERS. When heated to decomposition it emits acrid smoke and irritating fumes.

AAU750 CAS:1516-17-2 ***HR: 2***
ACETIC ACID-2,4-HEXADIEN-1-OL ESTER
mf: $C_8H_{12}O_2$ mw: 140.20

SYNS: 2,4-HEXADIEN-1-OL ACETATE ◇ 2,4-HEXADIENYL ACETATE ◇ SORBYL ACETATE

TOXICITY DATA with REFERENCE
orl-rat LD50:4360 mg/kg TXAPA9 28,313,74
skn-rbt LD50:2520 mg/kg TXAPA9 28,313,74

CONSENSUS REPORTS: Reported in EPA TSCA Inventory

SAFETY PROFILE: Moderately toxic by skin contact. See also ESTERS. When heated to decomposition it emits acrid smoke and irritating fumes.

AAV250 CAS:40853-56-3 ***HR: 1***
ACETIC ACID-2-ISOPROPYL-5-METHYL-2-HEXEN-1-YL ESTER
mf: $C_{12}H_{22}O_2$ mw: 198.34

SYNS: ISODIHYDRO LAVANDULYL ACETATE ◇ 2-ISOPROPYL-5-METHYL-2-HEXEN-1-YL ACETATE ◇ 5-METHYL-2-(1-METHYLETHYL)-2-HEXEN-1-YLACETATE

TOXICITY DATA with REFERENCE
skn-rbt 500 mg/24H MLD FCTXAV 14,307,76

SAFETY PROFILE: A skin irritant. See also ESTERS. When heated to decomposition it emits acrid smoke and irritating fumes.

AAV500 CAS:3610-27-3 ***HR: 1***
ACETIC ACID, 2-(2-(2-METHOXYETHOXY)ETHOXY)ETHYL ESTER
mf: $C_9H_{18}O_5$ mw: 206.27

PROP: Liquid. Bp: 130°, flash p: 260°F (OC), d: 1.094, vap d: 7.11.

SYN: METHOXYTRIGLYCOL ACETATE

TOXICITY DATA with REFERENCE
eye-rbt 500 mg AMIHBC 10,61,54
orl-rat LD50:11 g/kg AMIHBC 10,61,54
skn-rbt LD50:8000 mg/kg AMIHBC 10,61,54

CONSENSUS REPORTS: Reported in EPA TSCA Inventory.

SAFETY PROFILE: Mildly toxic by ingestion and skin contact. An eye irritant. See also ESTERS. Combustible. To fight fire, use alcohol foam, CO_2, dry chemical. When heated to decomposition it emits acrid smoke and irritating fumes.

AAW000 CAS:56856-83-8 ***HR: 3***
ACETIC ACID METHYLNITROSAMINOMETHYL ESTER
mf: $C_4H_8N_2O_3$ mw: 132.14

SYNS: α-ACETOXY DIMETHYLNITROSAMINE ◇ N-α-ACETOXYMETHYL-N-METHYLNITROSAMINE ◇ ACETOXYMETHYL-METHYL-NITROSAMIN (GERMAN) ◇ ACETOXYMETHYL METHYLNITROSAMINE ◇ 1-ACETOXY-N-NITROSODIMETHYLAMINE ◇ AMMN ◇ ANN (GERMAN) ◇ DMN-OAC ◇ MAMN ◇ METHYL(ACETOXYMETHYL)NITROSAMINE ◇ N-NITROSO-N-(ACETOXY)METHYL-N-METHYLAMINE ◇ N-NITROSO-N-METHYL-N-ACETOXYMETHYLAMINE

TOXICITY DATA with REFERENCE
slt-dmg-par 100 μmol/L CNREA8 35,3780,75
cyt-dmg-par 100 μmol/L CNREA8 35,3780,75
mmo-esc 25 μmol/plate GANNA2 70,663,79
orl-rat TDLo:13 mg/kg:CAR JJIND8 63,93,79
ipr-rat TDLo:13 mg/kg:CAR JJIND8 63,93,79
ivn-rat TDLo:13 mg/kg:ETA JJIND8 63,93,79
ipr-mus TDLo:10 mg/kg (11D preg):TER ARTODN 52,45,83
rec-rat TDLo:12 mg/kg/46W-I:ETA,REP HEGAD4 30,30,83
ipr-rat LD:13 mg/kg:NEO JJIND8 58,1531,77
ipr-rat LD:13 mg/kg:NEO,REP VTPHAK 16,574,79
orl-rat LD50:130 mg/kg ONCOBS 38,18,81
ipr-rat LD50:25 mg/kg JNCIAM 58,1533,77
scu-rat LD50:25 mg/kg ZEKBAI 91,217,78
ivn-rat LD50:25 mg/kg ZEKBAI 91,217,78
rec-rat LD50:24 mg/kg ZEKBAI 91,217,78

SAFETY PROFILE: Suspected carcinogen with experimental carcinogenic, neoplastigenic, and tumorigenic data. Poison by ingestion, subcutaneous, intravenous, and intraperitoneal routes. Experimental teratogenic data. Human mutation data reported. When heated to decomposition it emits toxic fumes of NO_x. See also NITROSOAMINES, N-NITROSO COMPOUNDS and ESTERS.

AAW250 CAS:10476-95-6 ***HR: 3***
ACETIC ACID-2-METHYL-2-PROPENE-1,1-DIOL DIESTER
mf: $C_8H_{12}O_4$ mw: 172.20

TOXICITY DATA with REFERENCE
ipr-mus LDLo:250 mg/kg CBCCT* 5,61,53
orl-rat LD50:440 mg/kg AIHAAP 30,470,69
ihl-rat LCLo:62 ppm/1H AIHAAP 30,470,69
skn-rbt LD50:44 mg/kg AIHAAP 30,470,69

CONSENSUS REPORTS: Reported in EPA TSCA Inventory. EPA Extremely Hazardous Substances List.

SAFETY PROFILE: Poison by inhalation, skin contact and intraperitoneal routes. Moderately toxic by ingestion. See also ESTERS. When heated to decomposition it emits acrid smoke and irritating fumes.

AAW500 CAS:1118-39-4 *HR: 1*
ACETIC ACID MYRCENYL ESTER
mf: $C_{12}H_{20}O_2$ mw: 196.32

SYNS: ACETICACID-2-METHYL-6-METHYLENE-7-OCTEN-2-YL ESTER ◇ 3-METHYLENE-7-METHYL-1-OCTEN-7-YL ACETATE ◇ 2-METHYL-6-METHYLENE-7-OCTEN-2-OL ACETATE ◇ 2-METHYL-6-METHYLENE-7-OCTEN-2-YL ACETATE ◇ MYRCENYL ACETATE

TOXICITY DATA with REFERENCE
skn-rbt 500 mg/24H MOD FCTXAV 14,601,76
orl-rat LD50:6300 mg/kg FCTXAV 14(6),601,76

CONSENSUS REPORTS: Reported in EPA TSCA Inventory.

SAFETY PROFILE: Mildly toxic by ingestion. A skin irritant. See also ESTERS. When heated to decomposition it emits acrid smoke and irritating fumes.

AAW750 CAS:117-98-6 *HR: 1*
ACETIC ACID-VETIVEROL ESTER
mf: $C_{17}H_{27}O_2$ mw: 263.44

SYNS: VETIVER ACETATE ◇ VETIVEROL ACETATE ◇ VETIVERT ACETATE ◇ VETIVERYL ACETATE ◇ 1,2,3,3a,4,5,6,8a-OCTAHYDRO-2-ISOPROPYLIDENE-6-AZULENOL-4,8-DIMETHYLACETATE

TOXICITY DATA with REFERENCE
skn-rbt 500 mg/24H MOD FCTXAV 12,1011,74

CONSENSUS REPORTS: Reported in EPA TSCA Inventory.

SAFETY PROFILE: A skin irritant. See also ESTERS. When heated to decomposition it emits acrid smoke and irritating fumes.

AAX175 CAS:9003-22-9 *HR: 1*
ACETIC ACID, VINYL ESTER, POLYMER with CHLOROETHYLENE
mf: $(C_4H_6O_2{\cdot}C_2H_3Cl)n$

SYNS: ACETIC ACID ETHENYL ESTER POLYMER with CHLORETHENE (9CI) ◇ A 15 (polymer) ◇ BAKELITE LP 70 ◇ BAKELITE VLFV ◇ BAKELITE VMCC ◇ BAKELITE VYNS ◇ BREON 351 ◇ CHLOROETHYLENEVINYL ACETATE POLYMER ◇ CORVIC 236581 ◇ DENKALAC 61 ◇ DIAMOND SHAMROCK 744 ◇ EXON 450 ◇ EXON 454 ◇ GEON 135 ◇ HOSTAFLEX VP 150 ◇ LEUCOVYL PA 1302 ◇ NORVINYL P 6 ◇ OPALON 400 ◇ PLIOVAC AO ◇ POLYVINYL CHLORIDE-POLYVINYL ACETATE ◇ PVC CORDO ◇ RHODOPAS 6000 ◇ SARPIFAN HP 1 ◇ SCONATEX ◇ SOLVIC 523KC ◇ SUMILIT PCX ◇ TENNUS 0565 ◇ TYGON ◇ VAGD ◇ VINNOL H 10/60 ◇ VINYL ACETATE-VINYL CHLORIDE COPOLYMER ◇ VINYL ACETATE-VINYL CHLORIDE POLYMER ◇ VINYL CHLORIDE-VINYL ACETATE POLYMER ◇ VINYLITE VYDR 21 ◇ VLVF ◇ VMCC ◇ VYNW

TOXICITY DATA with REFERENCE
imp-mus TDLo:1200 mg/kg:ETA JNCIAM 58,1443,77

CONSENSUS REPORTS: IARC Cancer Review: Animal Limited Evidence IMEMDT 19,377,79. Reported in EPA TSCA Inventory.

SAFETY PROFILE: Suspected carcinogen with experimental tumorigenic data. When heated to decomposition it emits toxic fumes of HCl.

AAX250 CAS:9003-20-7 *HR: 3*
ACETIC ACID VINYL ESTER POLYMERS
mf: $(C_4H_6O_2)_n$

PROP: Clear, water-white solid resin. Sol in benzene, acetone; insol in water.

SYNS: ACETIC ACID ETHENYL ESTER HOMOPOLYMER ◇ ASAHISOL 1527 ◇ ASB 516 ◇ AYAA ◇ AYAF ◇ BAKELITE AYAA ◇ BAKELITE LP 90 ◇ BASCOREZ ◇ BOND CH 18 ◇ BOOKSAVER ◇ BORDEN 2123 ◇ CEVIAN A 678 ◇ D 50 ◇ DANFIRM ◇ DARATAK ◇ DCA 70 ◇ DUVILAX BD 20 ◇ ELMER'S GLUE ALL ◇ EP 1463 ◇ FORMVAR 1285 ◇ GELVA CSV 16 ◇ GOHSENYL E 50 Y ◇ KURARE OM 100 ◇ LEMAC 1000 ◇ MERCKOGEN 6000 ◇ MOVINYL 114 ◇ NATIONAL 120-1207 ◇ POLYVINYL ACETATE (FCC) ◇ PROTEX (POLYMER) ◇ RHODOPAS M ◇ SOVIOL ◇ SP 60 ESTER ◇ TOABOND 40H ◇ UCAR 130 ◇ VA 0112 ◇ VINAC B 7 ◇ VINYL ACETATE HOMOPOLYMER ◇ VINYL ACETATE POLYMER ◇ VINYL ACETATE RESIN ◇ VINYL PRODUCTS R 10688 ◇ WINACET D

CONSENSUS REPORTS: IARC Cancer Review: Animal Inadequate Evidence IMEMDT 19,341,79. Reported in EPA TSCA Inventory.

SAFETY PROFILE: When heated to decomposition it emits acrid smoke and irritating fumes. See also ESTERS.

AAX500 CAS:108-24-7 *HR: 2*
ACETIC ANHYDRIDE
DOT: UN 1715
mf: $C_4H_6O_3$ mw: 102.10

PROP: Colorless, very mobile, strongly refractive liquid; very strong acetic odor. Mp: −73.1°, bp: 140°, flash p: 129°F (CC), d: 1.082 @ 20°/4°, lel: 2.9%, uel: 10.3%, autoign temp: 734°F, vap press: 10 mm @ 36.0°, vap d: 3.52. Somewhat sol in cold water; decomp in hot water and hot alc; misc in alc and ether.

SYNS: ACETIC ACID, ANHYDRIDE ◇ ACETIC OXIDE ◇ ACETYL ANHYDRIDE ◇ ACETYL ETHER ◇ ACETYL OXIDE ◇ ANHYDRIDE ACETIQUE (FRENCH) ◇ ANIDRIDE ACETICA (ITALIAN)

◇ AZIJNZUURANHYDRIDE (DUTCH) ◇ ESSIGSAEUREANHYDRID (GERMAN) ◇ ETHANOIC ANHYDRATE ◇ OCTOWY BEZWODNIK (POLISH)

TOXICITY DATA with REFERENCE
skn-rbt 10 mg/24H open MLD AMIHBC 4,119,51
skn-rbt 540 mg open MLD UCDS** 8/7/63
eye-rbt 250 μg open SEV AMIHBC 4,119,51
orl-rat LD50:1780 mg/kg AMIHBC 4,119,51
ihl-rat LC50:1000 ppm/4H 34ZIAG -,607,69
skn-rbt LD50:4000 mg/kg UCDS** 8/7/63

CONSENSUS REPORTS: Reported in EPA TSCA Inventory.

OSHA PEL: CL 5 ppm
ACGIH TLV: CL 5 ppm
DFG MAK: 5 ppm (20 mg/m^3)
DOT Classification: IMO: Corrosive Material; Label: Corrosive, Flammable Liquid.

SAFETY PROFILE: Moderately toxic by inhalation, ingestion, and skin contact. A skin and severe eye irritant. Moderate fire and explosion hazard when exposed to heat or flame. Potentially explosive reactions with barium peroxide, boric acid, chromium trioxide, 1,3-diphenyltriazene, hydrochloric acid + water, hypochlorous acid, nitric acid, perchloric acid + water, peroxyacetic acid, potassium permanganate, tetrafluoroboric acid, 4-toluenesulfonic acid + water, and acetic acid + water. Reactions with ethanol + sodium hydrogen sulfate, and hydrogen peroxide form explosive products. Reactions with ammonium nitrate + hexamethylenetetraminium acetate + nitric acid form as products the military explosives RDX and HMX. Reacts violently with N-tert-butylphthalimic acid + tetrafluoroboric acid, chromic acid, glycerol + phosphoryl chloride, and metal nitrates (e.g., copper or sodium nitrates). Incompatible with 2-aminoethanol, aniline, chlorosulfonic acid, (CrO_3 + acetic acid), ethylenediamine, ethyleneimine, glycerol, oleum, HF, permanganates, NaOH, Na_2O_2, H_2SO_4, water, N_2O_2, (glycerol + phosphoryl chloride). When heated to decomposition it emits toxic fumes; can react vigorously with oxidizing materials, will react violently on contact with water or steam. Used in production of drugs of abuse. To fight fire, use CO_2, dry chemical, water mist, alcohol foam. See also ANHYDRIDES.

AAX750 CAS:93-29-8 ***HR: 2***
ACETISOEUGENOL
mf: $C_{12}H_{14}O_3$ mw: 206.26

PROP: White crystals; clove odor. Flash p: 153°F. Sol in alc, chloroform, ether; insol in water.

SYNS: 4-ACETOXY-3-METHOXY-1-PROPENYLBENZENE ◇ ACETYLISOEUGENOL ◇ FEMA No. 2470 ◇ ISOEUGENOL ACETATE ◇ ISOEUGENYL ACETATE (FCC) ◇ 2-METHOXY-4-PROPENYLPHENYL ACETATE

TOXICITY DATA with REFERENCE
orl-rat LD50:3450 mg/kg FCTXAV 13,681,75

CONSENSUS REPORTS: Reported in EPA TSCA Inventory.

SAFETY PROFILE: Moderately toxic by ingestion. Combustible liquid. When heated to decomposition it emits acrid smoke and irritating fumes.

AAY000 CAS:102-01-2 ***HR: 3***
ACETOACETANILIDE
mf: $C_{10}H_{11}NO_2$ mw: 177.22

PROP: White, crystalline solid. Mp: 85°, bp: decomp, flash p: 365°F (COC), d: 1.260 @ 20°, vap press: 0.01 mm @ 20°.

SYNS: ACETOACETAMIDOBENZENE ◇ ACETOACETIC ACID ANILIDE ◇ ACETOACETIC ANILIDE ◇ ((ACETOACETYL)AMINO)BENZENE ◇ ACETOACETYLANILINE ◇ ACETYLACETANILIDE ◇ α-ACETYLACETANILIDE ◇ N-(ACETYLACETYL)ANILINE ◇ β-KETOBUTYRANILIDE ◇ N-PHENYLACETOACETAMIDE ◇ USAF EK-1239

TOXICITY DATA with REFERENCE
orl-rat LD50:5400 mg/kg LONZA# 08FEB79
ipr-mus LD50:300 mg/kg NTIS** AD277-689

CONSENSUS REPORTS: Reported in EPA TSCA Inventory.

SAFETY PROFILE: Poison by intraperitoneal route. A weak allergen. See also ACETANILIDE. Combustible when exposed to heat or flame. See ANILINE and CYANIDES for disaster hazard. When heated to decomposition it emits toxic NO_x fumes. To fight fire, use alcohol foam, water mist, CO_2, dry chemical.

AAY250 CAS:101-92-8 ***HR: 2***
ACETOACET-p-CHLORANILIDE
mf: $NC_{10}H_{10}O_2Cl$ mw: 211.65

PROP: Crystals. Mp: 107°, bp: decomp, flash p: 350°F (COC), d: 1.438 @ 20°, vap press: 0.01 mm @ 20°, vap d: 7.31.

SYNS: ACETOACETYL-4-CHLOROANILIDE ◇ p-CHLOROACETO ACETANILIDE ◇ 4′-CHLOROACETO ACETANILIDE

TOXICITY DATA with REFERENCE
ipr-mus LDLo 500 mg/kg CBCCT* 4,225,52

CONSENSUS REPORTS: Reported in EPA TSCA Inventory.

SAFETY PROFILE: Moderately toxic by intraperitoneal route. See also ACETANILIDE. Combustible when exposed to heat or flame. Dangerous; see ANILINE and CYANIDE. Can react vigorously with oxidizing materials. To fight fire, use water, foam, CO_2, water mist, dry chemical. When heated to decomposition it emits toxic fumes of Cl^- and NO_x.

AAY500 CAS:10032-00-5 ***HR: 1***
ACETOACETIC ACID-3,7-DIMETHYL-2,6-OCTADIENYL ESTER
mf: $C_{13}H_{20}O_3$ mw: 224.33

SYN: GERANYL ACETOACETATE

TOXICITY DATA with REFERENCE
skn-rbt 500 mg/24H MOD FCTXAV 16,637,78

SAFETY PROFILE: A skin irritant. See also ESTERS. When heated to decomposition it emits acrid smoke and irritating fumes.

AAY750 CAS:21282-96-2 ***HR: 3***
2-ACETOACETOXYETHYL ACRYLATE
mf: $C_9H_{12}O_5$ mw: 200.21

SYNS: 3-OXO-BUTANOIC ACID 2-((1-OXO-2-PROPENYL)OXY)ETHYL ESTER (9CI) ◇ ACETOACETIC ACID-2-HYDROXYETHYL ESTER ACRYLATE

TOXICITY DATA with REFERENCE
orl-rat LD50:1300 mg/kg TXAPA9 28,313,74
skn-rbt LD50:280 mg/kg TXAPA9 28,313,74

SAFETY PROFILE: Poison by skin contact. Moderately toxic by ingestion. See also ESTERS. When heated to decomposition it emits acrid smoke and irritating fumes.

AAZ000 CAS:122-82-7 ***HR: 3***
ACETOACET-p-PHENETIDIDE
mf: $C_{12}H_{15}NO_3$ mw: 221.28

PROP: Crystals. Mp: 108.5°, bp: decomp, flash p: 325°F (OC), d: 1.220 @ 20°, vap press: 0.02 mm @ 20°, vap d: 7.63.

SYNS: p-ACETOACETOPHENETIDIDE ◇ 4-ETHOXYACETOACETANILIDE ◇ 4'-ETHOXYACETOACETANILIDE

TOXICITY DATA with REFERENCE
orl-rat LD50:176 mg/kg FRPSAX 19,822,64

CONSENSUS REPORTS: Reported in EPA TSCA Inventory.

SAFETY PROFILE: Poison by ingestion. See also ACETANILIDE. Combustible. To fight fire, use water, foam, CO_2, water spray, mist, dry chemical. When heated to decomposition it emits toxic fumes of NO_x.

ABA000 CAS:93-68-5 ***HR: 2***
ACETOACET-o-TOLUIDIDE
mf: $C_{11}H_{13}NO_2$ mw: 191.25

PROP: Crystals. Mp: 106°, bp: decomp, d: 1.300 @ 20°, vap press: 0.01 mm @ 20°, flash p: 320°F (COC).

SYNS: 2-ACETOACETYLAMINOTOLUENE ◇ ACETOACETYL-2-METHYLANILIDE ◇ 2'-METHYLACETOACETANILIDE

TOXICITY DATA with REFERENCE
orl-rat LD50:1600 mg/kg KODAK* -,N-229,76
orl-mus LD50:1600 mg/kg KODAK* -,N-229,76

CONSENSUS REPORTS: Reported in EPA TSCA Inventory.

SAFETY PROFILE: Moderately toxic by ingestion. When heated to decomposition it emits toxic fumes of NO_x.

ABA250 ***HR: 2***
ACETOACET-m-XYLIDIDE
mf: $C_{12}H_{15}O_2N$ mw: 205.25

PROP: White to light yellow crystalline solid, sol in water to 0.5% @ 25°. Mp: 89-90°, d: 1.238, flash p: 340°F (OC).

SYN: AAMX

SAFETY PROFILE: Combustible. To fight fire, use alcohol foam, water spray or mist, dry chemical (multipurpose). When heated to decomposition it emits toxic fumes of NO_x.

ABA500 CAS:92-15-9 ***HR: 2***
ACETOACETYL-o-ANISIDINE
mf: $C_{11}H_{13}NO_3$ mw: 207.25

PROP: Crystals. Mp: 86.6°, flash p: 325°F (OC), d: 1.132 @ 86.6°/20°, vap d: 7.0.

SYNS: o-ACETOACETANISIDE ◇ ACETOACET-o-ANISIDIN (CZECH) ◇ ACETOACETIC ACID-o-ANISIDIDE ◇ 2-ACETOACETYLAMINOANISOLE ◇ ACETOACETYL-o-ANISIDE ◇ ACETOACETYL-o-ANISINE ◇ o-METHOXYACETOACETANILIDE ◇ 2-METHOXYACETOACETANILIDE ◇ 2'-METHOXYACETOACETANILIDE

TOXICITY DATA with REFERENCE
skn-rbt 500 mg/24H MLD 28ZPAK -,116,72
eye-rbt 500 mg/24H MOD 28ZPAK -,116,72
orl-rat LD50:2290 mg/kg 28ZPAK -,116,72

CONSENSUS REPORTS: Reported in EPA TSCA Inventory.

SAFETY PROFILE: Moderately toxic by ingestion. A skin and eye irritant. When heated to decomposition it emits toxic fumes of NO_x. Combustible when exposed to heat or flame or oxidizing materials. To fight fire, use CO_2, mist, dry chemicals.

ABA750 CAS:1271-55-2 ***HR: 3***
ACETOFERROCENE
mf: $C_{12}H_{12}FeO$ mw: 228.09

SYNS: ACETYLFERROCENE ◇ MONACETYLFERROCENE

TOXICITY DATA with REFERENCE
ivn-mus LD50:75 mg/kg CSLNX* NX#08812

CONSENSUS REPORTS: Reported in EPA TSCA Inventory.

SAFETY PROFILE: Poison via intravenous route.

ABB000 CAS:968-81-0 ***HR: 3***
ACETOHEXAMIDE
mf: $C_{15}H_{20}N_2O_4S$ mw: 324.43

SYNS: 1-(p-ACETYLBENZENESULFONYL)-3-CYCLOHEXYLUREA ◇ 4-ACETYL-N-((CYCLOHEXYLAMINO)CARBONYL)-BENZENESULFONAMIDE ◇ CYCLAMIDE ◇ DIMELIN ◇ DIMELOR ◇ DYMELOR ◇ ORDIMEL ◇ TSIKLAMID ◇ NCI-CO03247

TOXICITY DATA with REFERENCE
unr-wmn TDLo:900 mg/kg (26-39W preg):REP BMJOAE 2,187,64
orl-rat LD50:5000 mg/kg TXAPA9 18,185,71

CONSENSUS REPORTS: NCI Carcinogenesis Bioassay (feed); No Evidence: mouse, rat NCITR* NCI-CG-TR-50,78. Reported in EPA TSCA Inventory.

SAFETY PROFILE: Human reproductive effects by an unspecified route: stillbirth. Mildly toxic by ingestion. When heated to decomposition it emits very toxic fumes of SO_x and NO_x.

ABB250 CAS:546-88-3 ***HR: D***
ACETOHYDROXAMIC ACID
mf: $C_2H_5NO_2$ mw: 75.08

SYNS: ACETHYDROXAMSAURE (GERMAN) ◇ AHA ◇ ACETIC ACID, OXIME ◇ ACETOHYDROXIMIC ACID ◇ ACETYLHYDROXAMIC ACID ◇ METHYLHYDROXAMIC ACID

TOXICITY DATA with REFERENCE
mmo-sat 160 μmol/plate JOPHDQ 3,557,80
dns-rat:lvr 5 mmol/L MUREAV 145,201,85
mma-ham:lng 20 mmol/L MUREAV 152,225,85
ipr-rat TDLo:750 mg/kg (12D preg):REP JMXSAE 2,230,74
ipr-rat TDLo:750 mg/kg (12D preg):TER JMXSAE 2,230,74

SAFETY PROFILE: An experimental teratogen. Other experimental reproductive effects. Mutation data reported. When heated to decomposition it emits toxic fumes of NO_x.

ABB500 CAS:513-86-0 ***HR: 3***
ACETOIN
DOT: UN 2621
mf: $C_4H_8O_2$ mw: 88.12

PROP: Sltly yellow liquid or crystalline solid; buttery odor. D: 1.016, bp: 147-148°, refr index: 1.417, mp: 15°, flash p: 106°F. Misc with water, alc, propylene glycol; insol in vegetable oil.

SYNS: ACETYL METHYL CARBINOL ◇ 2-BUTANOL-3-ONE ◇ DIMETHYLKETOL ◇ FEMA No. 2008 ◇ 3-HYDROXY-2-BUTANONE ◇ 1-HYDROXYETHYL METHYL KETONE ◇ γ-HYDROXY-β-OXOBUTANE

TOXICITY DATA with REFERENCE
skn-rbt 500 mg/24H MOD CNREA8 33,3069,73
orl-rat TDLo:12600 mg/kg (42D male):REP FCTXAV 10,131,72
scu-rat LDLo:14 g/kg FCTXAV 17,509,79

CONSENSUS REPORTS: Reported in EPA TSCA Inventory.

DOT Classification: Flammable Liquid; Label: Flammable Liquid.

SAFETY PROFILE: Experimental reproductive effects. Mildly toxic by subcutaneous route. A moderate skin irritant. Flammable liquid. When heated to decomposition it emits acrid smoke and fumes. See also KETONES.

ABC000 CAS:116-09-6 ***HR: 2***
ACETOL (1)
mf: $C_3H_6O_2$ mw: 74.09

$$HOCH_2CO{\cdot}CH_3$$

PROP: Colorless liquid. D: 1.084 @ 20°/4°, mp: −7°, bp: 145°-146° decomp; misc in water, alc and ether.

SYNS: HYDROXYACETONE ◇ 1-HYDROXY-2-PROPANONE

TOXICITY DATA with REFERENCE
mmo-sat 500 μg/plate ABCHA6 47,2461,83
orl-rat LD50:2200 mg/kg JIHTAB 30,63,48

CONSENSUS REPORTS: Reported in EPA TSCA Inventory.

SAFETY PROFILE: Moderately toxic by ingestion. Mutation data reported. An allergen. Implicated in aplastic anemia. A 10 gram dose may be fatal to an adult. Skin contact, inhalation, or ingestion can cause asthma, sneezing, irritation of eyes and nose, hives and eczema. Combustible when exposed to heat or flame. When heated to decomposition it emits acrid smoke and fumes.

ABC250 CAS:828-00-2 ***HR: 1***
ACETOMETHOXANE
mf: $C_8H_{14}O_4$ mw: 174.22

PROP: Yellow to amber, clear liquid. Sol in water and org solvents. D: 1.068-1.075 @ 25/25; bp: 66-68° @ 3 mm; fp: < −25°.

SYNS: ACETIC ACID-2,6-DIMETHYL-m-DIOXAN-4-YL ESTER

◇ ACETOMETHOXAN ◇ 6-ACETOXY-2,4-DIMETHYL-m-DIOXANE ◇ DDOA ◇ DIMETHOXANE ◇ 2,6-DIMETHYL-m-DIOXAN-4-OL ACETATE ◇ 2,6-DIMETHYL-m-DIOXAN-4-YL ACETATE ◇ DIOXIN (bactericide) (OBS.) ◇ GIV GARD DXN ◇ NCI-C56213

TOXICITY DATA with REFERENCE
mma-sat 5500 μg/plate ENMUDM 8(Suppl 7),1,86
sln-dmg-par 1 pph ENMUDM 7,677,85
orl-rat TDLo:948 g/kg/88W-I:CAR JNCIAM 53,791,74
orl-rat LD50:1930 mg/kg GCTB** 3/25/77

CONSENSUS REPORTS: IARC Cancer Review: Group 3 IMEMDT 7,56,87; Animal Limited Evidence IMEMDT 15,177,77.

SAFETY PROFILE: Questionable carcinogen with experimental carcinogenic data. Moderately toxic by ingestion. See also ESTERS. When heated to decomposition it emits acrid smoke and fumes.

ABC475 CAS:941-98-0 ***HR: 2***
1'-ACETONAPHTHONE
mf: $C_{12}H_{10}O$ mw: 170.22

PROP: Mp: 10.5°, bp: 302°, flash p: >230° F, d: 1.120.

SYNS: 1-ACETONAPHTHALENE ◇ α-ACETONAPHTHONE ◇ 1-ACETONAPHTHONE ◇ 1-ACETYLNAPHTHALENE ◇ ETHANONE, 1-(1-NAPHTHALENYL)-(9CI) ◇ α-METHYL NAPHTHYL KETONE ◇ METHYL α-NAPHTHYL KETONE ◇ METHYL 1-NAPHTHYL KETONE ◇ 1-(1-NAPHTHALENYL)ETHANONE ◇ α-NAPHTHYL METHYL KETONE ◇ 1-NAPHTHYL METHYL KETONE

TOXICITY DATA with REFERENCE
skn-rbt 500 mg/24H MLD FCTOD7 20,755,82
orl-rat LD50:1560 mg/kg FCTOD7 20,755,82

CONSENSUS REPORTS: Reported in EPA TSCA Inventory.

SAFETY PROFILE: Moderately toxic by ingestion. A skin irritant. A combustible liquid. When heated to decomposition it emits acrid smoke and irritating fumes.

ABC500 CAS:93-08-3 ***HR: 2***
2'-ACETONAPHTHONE
mf: $C_{12}H_{10}O$ mw: 170.22

PROP: White crystalline solid; orange blossom odor. Flash p: 264°F. Sol in fixed oils; sltly sol in propylene glycol; insol in glycerin.

SYNS: β-ACETONAPHTHALENE ◇ ACETONAPHTHONE ◇ β-ACETONAPHTHONE ◇ 2-ACETONAPHTHONE ◇ β-ACETYLNAPHTHALENE ◇ 2-ACETYLNAPHTHALENE ◇ FEMA No. 2723 ◇ METHYL-β-NAPHTHYL KETONE (FCC) ◇ METHYL-2-NAPHTHYL KETONE ◇ β-METHYL NAPHTHYL KETONE ◇ 1-(2-NAPHTHALENYL)ETHANONE ◇ β-NAPHTHYL METHYL KETONE ◇ 2-NAPHTHYL METHYL KETONE ◇ ORANGE CRYSTALS

TOXICITY DATA with REFERENCE
skn-hmn 500 mg/24H FCTXAV 13,681,75
orl-mus LD50:599 mg/kg MDZEAK 8,244,67

CONSENSUS REPORTS: Reported in EPA TSCA Inventory.

SAFETY PROFILE: Moderately toxic by ingestion. A human skin irritant. Combustible liquid. When heated to decomposition it emits acrid smoke and fumes.

ABC750 CAS:67-64-1 ***HR: 3***
ACETONE
DOT: UN 1090/UN 1091
mf: C_3H_6O mw: 58.09

PROP: Colorless liquid; fragrant mint-like odor. Mp: −94.6°, bp: 56.48°, refr index: 1.356, flash p: 0°F (CC), lel: 2.6%, uel: 12.8%, d: 0.7972 @ 15°, autoign temp: (color) 869°F, vap press: 400 mm @ 39.5°, vap d: 2.00. Misc in water, alc, and ether.

SYNS: ACETON (GERMAN, DUTCH, POLISH) ◇ DIMETHYLFORMALDEHYDE ◇ DIMETHYLKETAL ◇ DIMETHYL KETONE ◇ FEMA No. 3326 ◇ KETONE PROPANE ◇ β-KETOPROPANE ◇ METHYL KETONE ◇ PROPANONE ◇ 2-PROPANONE ◇ PYROACETIC ACID ◇ PYROACETIC ETHER ◇ RCRA WASTE NUMBER U002

TOXICITY DATA with REFERENCE
eye-hmn 500 ppm JIHTAB 25,282,43
skn-rbt 395 mg open MLD UCDS** 5/7/70
skn-rbt 500 mg/24H MLD 28ZPAK -,42,72
eye-rbt 3950 μg SEV AJOPAA 29,1363,46
eye-rbt 20 mg/24H MOD 85JCAE -,280,86
cyt-smc 200 mmol/tube HEREAY 33,457,47
sln-smc 47600 ppm ANYAA9 407,186,83
ihl-mam TCLo:31500 μg/m³/24H (1-13D preg):REP GTPZAB 26(6),24,82
orl-man TDLo:2857 mg/kg 34ZIAG -,64,69
orl-man TDLo:2857 mg/kg DIAEAZ 15,810,66
ihl-man TCLo:12000 ppm/4H:CNS AOHYA3 16,73,73
ihl-man TDLo:440 μg/m³/6M GISAAA 42(8)42,77
ihl-man TDLo:10 mg/m³/6H GISAAA 42(8)42,77
ihl-hmn TCLo:500 ppm:EYE JIHTAB 25,282,43
ihl-man TCLo:12000 ppm/4H:GIT AOHYA3 16,73,73
ivn-rat LD50:5500 mg/kg NPIRI* 1,1,74
orl-rat LD50:5800 mg/kg JTEHD6 15,609,85
ihl-rat LC50:50100 mg/m³/8H AIHAAP 20,364,59
ipr-rat LDLo:500 mg/kg JPPMAB 11,150,59
ivn-rat LD50:5500 mg/kg NPIRI* 1,1,74
orl-mus LD50:3000 mg/kg PCJOAU 14,162,80
ihl-mus LCLo:110 g/m³/1H AGGHAR 5,1,33
ipr-mus LD50:1297 mg/kg SCCUR* -,1,61
ivn-mus LDLo:4 g/kg FAONAU 48A,86,70
orl-dog LDLo:8 g/kg FAONAU 48A,86,70
orl-rbt LD50:5340 mg/kg FAONAU 48A,86,70
skn-rbt LD50:20 g/kg UCDS** 5/7/70

CONSENSUS REPORTS: On Community Right-To-Know List. Reported in EPA TSCA Inventory.

OSHA PEL: (Transitional: TWA 1000 ppm) TWA 750 ppm; STEL 1000 ppm
ACGIH TLV: TWA 750 ppm; STEL 1000 ppm
DFG MAK: 1000 ppm (2400 mg/m^3)
NIOSH REL: (Ketones) 10H TWA 590 mg/m^3
DOT Classification: Flammable Liquid; Label: Flammable Liquid.

SAFETY PROFILE: Moderately toxic by various routes. A skin and severe eye irritant. Human systemic effects by inhalation: changes in EEG, changes in carbohydrate metabolism, nasal effects, conjunctiva irritation, respiratory system effects, nausea and vomiting, and muscle weakness. Human systemic effects by ingestion: coma, kidney damage, and metabolic changes. Narcotic in high concentration. In industry, no injurious effects have been reported other than skin irritation resulting from its defatting action, or headache from prolonged inhalation. Experimental reproductive effects. A common air contaminant. Highly flammable liquid. Dangerous disaster hazard due to fire and explosion hazard; can react vigorously with oxidizing materials.

Potentially explosive reaction with nitric acid + sulfuric acid, bromine trifluoride, nitrosyl chloride + platinum, nitrosyl perchlorate, chromyl chloride, thiotrithiazyl perchlorate, and 2,4,6-trichloro-1,3,5-triazine + water. Reacts to form explosive peroxide products with 2-methyl-1,3-butadiene, hydrogen peroxide, and peroxomonosulfuric acid. Ignites on contact with activated carbon, chromium trioxide, dioxygen difluoride + carbon dioxide, and potassium-tert-butoxide. Reacts violently with bromoform, chloroform + alkalies, bromine, and sulfur dichloride. Incompatible with CrO, (nitric + acetic acid), NOCl, nitryl perchlorate, permonosulfuric acid, NaOBr, (sulfuric acid + potassium dichromate), (thio-diglycol + hydrogen peroxide), trichloromelamine, air, HNO_3, chloroform, and H_2SO_4. To fight fire, use CO_2, dry chemical, alcohol foam. Used in production of drugs of abuse.

ABD000 CAS:57-15-8 ***HR: 3***
ACETONE CHLOROFORM
mf: $C_4H_7Cl_3O$ mw: 177.46

PROP: Crystals, camphor odor. Mp: 97°, bp: 167°.

SYNS: ANHYDROUS CHLOROBUTANOL ◇ CHLORBUTANOL ◇ CHLORBUTOL ◇ CHLORETONE ◇ CHLOROBUTANOL ◇ CLORTRAN ◇ HCP ◇ METHAFORM ◇ SEDAFORM ◇ β,β,β-TRICHLORO-tert-BUTYL ALCOHOL ◇ TRICHLORO-tert-BUTYL ALCOHOL ◇ tert-TRICHLOROBUTYL ALCOHOL ◇ 1,1,1-TRICHLORO-2-METHYL-2-PROPANOL

TOXICITY DATA with REFERENCE
mmo-sat 20 μmol/plate MUREAV 90,91,81
cyt-smc 10 mmol/tube HEREAY 33,457,47
skn-rbt 850 μg MLD XEURAQ MDDC-1715
eye-rbt 9180 μg/30S MLD XEURAQ MDDC-1715
orl-dog LDLo:238 mg/kg AIPTAK 8,77,01
orl-rbt LDLo:213 mg/kg AIPTAK 8,77,01
par-frg LDLo:800 mg/kg AIPTAK 8,77,01

CONSENSUS REPORTS: Reported in EPA TSCA Inventory.

SAFETY PROFILE: Poison by ingestion. Moderately toxic by parenteral route. A narcotic. A skin and eye irritant. Mutation data reported. See also CHLORAL HYDRATE, which acts similarly. Dangerous; can react with oxidizing materials. Combustible when exposed to heat or flame. When heated to decomposition it emits toxic fumes of Cl^-. See also PHOSGENE.

ABD250 CAS:126-84-1 ***HR: 3***
ACETONE DIETHYL KETAL
mf: $C_7H_{16}O_2$ mw: 132.23

SYNS: 2,2-DIETHOXYPROPANE ◇ USAF DO-44

TOXICITY DATA with REFERENCE
ipr-mus LD50:125 mg/kg NTIS** AD277-689

CONSENSUS REPORTS: Reported in EPA TSCA Inventory.

SAFETY PROFILE: Poison by intraperitoneal route. When heated to decomposition it emits acrid smoke.

ABD500 CAS:115-24-2 ***HR: 3***
ACETONE DIETHYLSULFONE
mf: $C_7H_{16}O_4S_2$ mw: 228.35

PROP: D: 1.183, mp: 127°-128°, bp: 300° (sl decomp), Sol in water, alc, and ether.

SYNS: ACETONE BIS(ETHYL SULFONE) ◇ 2,2-BIS(ETHYLSULFONYL)PROPANE ◇ DIETHYLSULFONDIMETHYLMETHANE ◇ PROPANE DIETHYL SULFONE ◇ SULFONAL ◇ SULFONMETHANE

TOXICITY DATA with REFERENCE
unk-man LDLo:147 mg/kg 85DCAI 2,73,70
orl-dog LDLo:900 mg/kg HBAMAK 4,1404,35
orl-rbt LDLo:3000 mg/kg HBAMAK 4,1404,35
orl-gpg LDLo:8500 mg/kg HBAMAK 4,1404,35

SAFETY PROFILE: A human poison by unspecified route. Moderately toxic by ingestion. Mutation data reported. When heated to decomposition it emits toxic fumes of SO_x.

ABE000 ***HR: 3***
ACETONE PEROXIDE

PROP: Liquid or absorbed on cornstarch. The trimeric form is crystalline. Mp: 97°.

SAFETY PROFILE: Severe skin and eye irritant. Flammable by spontaneous chemical reaction; can react vigorously with reducing materials. The trimeric form is shock-sensitive and static-electricity-sensitive and may detonate.

ABE250 CAS:110-20-3 ***HR: 3***
ACETONE SEMICARBAZONE
mf: $C_4H_9N_3O$ mw: 115.16

PROP: Mp: 190-199° (decomp). Sol in cold water; sltly sol in cold alc; insol in ether.

TOXICITY DATA with REFERENCE
ivn-mus LD50:90 mg/kg JPETAB 122,110,58

CONSENSUS REPORTS: Reported in EPA TSCA Inventory.

SAFETY PROFILE: Poison by intravenous route. When heated to decomposition it emits toxic fumes of NO_x.

ABE500 CAS:75-05-8 ***HR: 3***
ACETONITRILE
DOT: UN 1648
mf: C_2H_3N mw: 41.06

PROP: Colorless liquid, aromatic odor. Mp: −45°, bp: 81.1°, flash p: 42°F (COC), d: 0.7868 @ 20°/20°, vap d: 1.42, vap press: 100 mm @ 27°, lel: 4.4%, uel: 16%, autoign temp: 975°F. Misc in water, alc, and ether.

SYNS: ACETONITRIL (GERMAN, DUTCH) ◇ CYANOMETHANE ◇ CYANURE de METHYL (FRENCH) ◇ ETHANENITRILE ◇ ETHYL NITRILE ◇ METHANECARBONITRILE ◇ METHYL CYANIDE ◇ NCI-C60822 ◇ RCRA WASTE NUMBER U003 ◇ USAF EK-488

TOXICITY DATA with REFERENCE
sln-smc 47600 ppm MUREAV 149,339,85
skn-rbt 10 mg/24H JIHTAB 30,63,48
skn-rbt 500 mg open MLD UCDS** 3/18/65
eye-rbt 20 mg SEV JIHTAB 30,63,48
orl-ham TDLo:300 mg/kg (8D preg):TER TJADAB 27,313,83
orl-ham TDLo:400 mg/kg (8D preg):REP TJADAB 27,313,83
orl-hmn TDLo:570 mg/kg:CNS APTOA6 41,340,77
ihl-hmn TCLo:160 ppm/4H 34ZIAG -,65,69
orl-rat LD50:2730 mg/kg TXAPA9 19,699,71
ihl-rat LC50:7551 ppm/8H JOCMA7 1,634,59
ipr-rat LD50:850 mg/kg JOCMA7 1,634,59
scu-rat LD50:3500 mg/kg 85GMAT -,16,82
ivn-rat LD50:1680 mg/kg JOCMA7 1,634,59
par-rat LD50:1100 mg/kg 85GMAT -,16,82
orl-mus LD50:269 mg/kg ARTODN 55,47,84
ihl-mus LC50:2693 ppm/1H CTOXAO 18,991,81
ipr-mus LD50:175 mg/kg TXAPA9 59,589,81
scu-mus LD50:4480 mg/kg 85GMAT -,16,82
ihl-dog LCLo:16000 ppm/4H JOCMA7 1,634,59

CONSENSUS REPORTS: On Community Right-To-Know List. Reported in EPA TSCA Inventory.

OSHA PEL: TWA 40 ppm; STEL 60 ppm
ACGIH TLV: TWA 40 ppm; STEL 60 ppm (skin)
DFG MAK: 40 ppm (70 mg/m^3)
NIOSH REL: (Nitriles) TWA 34 mg/m^3
DOT Classification: Flammable Liquid; Label: Flammable Liquid and Poison.

SAFETY PROFILE: Poison by ingestion and intraperitoneal routes. Moderately toxic by several routes. An experimental teratogen. Other experimental reproductive effects. A skin and severe eye irritant. Human systemic effects by ingestion: convulsions, nausea or vomiting, and metabolic acidosis. Human respiratory system effects by inhalation. Mutation data reported. Dangerous fire hazard when exposed to heat, flame, or oxidizers. Explosion Hazard: See also CYANIDE and NITRILES. When heated to decomposition it emits highly toxic fumes of CN^- and NO_x. Potentially explosive reaction with lanthanide perchlorates and nitrogen-fluorine compounds. Exothermic reaction with sulfuric acid at 53°C. Will react with water, steam, acids to produce toxic and flammable vapors. Incompatible with oleum, chlorosulfonic acid, perchlorates, nitrating agents, indium, dinitrogen tetraoxide, N-fluoro compounds (i.e., perfluorourea + acetonitrile), HNO_3, SO_3. To fight fire, use foam, CO_2, dry chemical.

ABE750 ***HR: 3***
ACETONITRILE IMIDAZOLE-5,7,7,12,14,14-HEXAMETHYL-1,4,8,11-TETRAAZA-4,11-CYCLOTETRADECA DIENE IRON (11) PERCHLORATE
mf: $C_{21}H_{39}Cl_2FeN_7O_8$ mw: 644.10

SAFETY PROFILE: An unstable and explosive compound. See also IRON COMPOUNDS. When heated to decomposition it emits toxic fumes of NO_x, Cl^-, and CN^-.

ABF000 CAS:127-06-0 ***HR: 2***
ACETONOXIME
mf: C_3H_7NO mw: 73.11

PROP: D: 0.97; mp: 60-61°; bp: 136.3°. Very sol in water, alc, and ether. Sol in ligroin ether.

SYNS: ACETOXIME ◇ β-ISONITROSOPROPANE ◇ 2-PROPANONE OXIME

TOXICITY DATA with REFERENCE
ipr-mus LD50:4000 mg/kg JPETAB 119,522,57

CONSENSUS REPORTS: Reported in EPA TSCA Inventory.

SAFETY PROFILE: Moderately toxic by intraperi-

toneal route. When heated to decomposition it emits toxic fumes of NO_x.

ABF500 CAS:117-52-2 ***HR: 3***
3-(α-ACETONYLFURFURYL)-4-HYDROXYCOUMARIN
mf: $C_{17}H_{14}O_5$ mw: 298.31

PROP: White powder; practically insol in water, sol in alcohols. Mp: 124°.

SYNS: COUMAFURYL ◇ CUMAFURYL (GERMAN) ◇ FOUMARIN ◇ 3-(α-FURYL-β-ACETYLAETHYL)-4-HYDROXYCUMARIN (GERMAN) ◇ 3-(1-FURYL-3-ACETYLETHYL)-4-HYDROXYCOUMARIN ◇ KRUMKIL ◇ RATAFIN ◇ RAT-A-WAY

TOXICITY DATA with REFERENCE
orl-rat LDLo:400 mg/kg 85GYAZ -,115,71
orl-rat LD50:25 mg/kg FMCHA2 -,D146,80
orl-mus LD50:14700 μg/kg FMCHA2 -,D146,80

SAFETY PROFILE: Poison by ingestion and possibly other routes. See also COUMADIN (WARFARIN).

ABF750 CAS:152-72-7 ***HR: 3***
3-(α-ACETONYL-p-NITROBENZYL)-4-HYDROXY-COUMARIN
mf: $C_{19}H_{15}NO_6$ mw: 353.35

SYNS: ACENOCOUMARIN ◇ ACENOCOUMAROL ◇ ASCUMAR ◇ G-23350 ◇ 4-HYDROXY-3-(1-(4-NITROPHENYL)-3-OXOBUTYL)-2H-1-BENZOPYRAN-2-ONE ◇ NICOUMALONE ◇ 3-(α-(p-NITROPHENOL)-β-ACETYLETHYL)-4-HYDROXYCOUMARIN ◇ 3-(α-p-NITROPHENYL-β-ACETYLETHYL)-4-HYDROXYCOUMARIN ◇ 3-(α-(4'-NITROPHENYL)-β-ACETYLETHYL)-4-HYDROXYCOUMARIN ◇ NITROWARFARIN ◇ SINKUMAR ◇ SINTHROME ◇ ZOTIL

TOXICITY DATA with REFERENCE
unr-wmn TDLo:24 mg/kg (1-28W preg):TER AFPEAM 36,63,79
orl-rat LD50:513 mg/kg 29ZVAB -,3,69
orl-mus LD50:1470 mg/kg THERAP 11,85,56
ipr-mus LD50:115 mg/kg MEIEDD 11,6,89

SAFETY PROFILE: Poison by intraperitoneal route. Moderately toxic by ingestion. A human teratogen by an unspecified route. When heated to decomposition it emits toxic fumes such as NO_x. See also COUMADIN (WARFARIN).

ABG000 CAS:5714-00-1 ***HR: 3***
ACETOPHENAZINE
mf: $C_{23}H_{29}N_3O_2S \cdot 2C_4H_4O_4$ mw: 643.77

SYNS: ACETOPHENAZINE MALEATE ◇ 2-ACETYL-10-(3-(4-(β-HYDROXYETHYL)PIPERAZINYL)PROPYL)PHENOTHIAZINE ◇ 1-(2-HYDROXYETHYL)-4-(3-(2-ACETYL-10-PHENOTHIAZYL)PROPYL)PIPERAZINE ◇ 1-(10-(3-(4-(2-HYDROXYETHYL)-1-PIPERAZINYL)PROPYL)-10H-PHENOTHIAZIN-2-YL)ETHANONE ◇ 10-(3-(4-(2-HYDROXYETHYL)-1-PIPERAZINYL)PROPYL)PHENOTHIAZIN-2-YL METHYL KETONE ◇ SCH 6673 ◇ TINDAL

TOXICITY DATA with REFERENCE
eye-rbt 112 mg SEV AMIHAB 14,250,56
orl-rat LD50:415 mg/kg 27ZQAG -,11,72
ipr-rat LD50:60 mg/kg 27ZQAG -,11,72
ivn-rat LD50:39 mg/kg 27ZQAG -,11,72
ivn-mus LD50:71 mg/kg CSLNX* NX#01100
orl-bwd LD50:75 mg/kg TXAPA9 21,315,72

SAFETY PROFILE: Poison by ingestion, intraperitoneal, and intravenous routes. Severe eye irritant. See also KETONES.

ABG250 CAS:591-33-3 ***HR: 2***
m-ACETOPHENETIDIDE
mf: $C_{10}H_{13}NO_2$ mw: 179.24

SYNS: m-ETHOXYACETANILIDE ◇ 3-ETHOXYACETANILIDE ◇ 3'-ETHOXYACETANILIDE ◇ N-(3-ETHOXYPHENYL)ACETAMIDE (9CI)

TOXICITY DATA with REFERENCE
orl-mus LD50:1250 mg/kg TXAPA9 19,20,71

CONSENSUS REPORTS: Reported in EPA TSCA Inventory.

SAFETY PROFILE: Moderately toxic by ingestion. See also p-ACETOPHENETIDIDE. When heated to decomposition it emits toxic fumes of NO_x.

ABG750 CAS:62-44-2 ***HR: 3***
p-ACETOPHENETIDIDE
mf: $C_{10}H_{13}NO_2$ mw: 179.24

SYNS: 1-ACETAMIDO-4-ETHOXYBENZENE ◇ ACETO-p-PHENALIDE ◇ p-ACETOPHENETIDE ◇ ACETO-p-PHENETIDIDE ◇ ACETO-4-PHENETIDINE ◇ ACETOPHENETIDIN ◇ ACETOPHENETIDINE ◇ ACETOPHENETIN ◇ ACET-p-PHENALIDE ◇ ACET-p-PHENETIDIN ◇ p-ACETPHENETIDIN ◇ ACETPHENETIDIN ◇ ACETYLPHENETIDIN ◇ N-ACETYL-p-PHENETIDINE ◇ ACHROCIDIN ◇ ANAPAC ◇ APC ◇ ASA COMPOUND ◇ BROMO SELTZER ◇ BUFF-A-COMP ◇ CITRA-FORT ◇ CODEMPIRAL ◇ COMMOTIONAL ◇ CONTRADOL ◇ CORICIDIN ◇ CORIFORTE ◇ CORYBAN-D ◇ DAPRISAL ◇ DARVON COMPOUND ◇ DASIKON ◇ EMPIRIN COMPOUND ◇ 4-ETHOXYACETANILIDE ◇ p-ETHOXYACETANILIDE ◇ N-(4-ETHOXYPHENYL)ACETAMIDE ◇ N-p-ETHOXYPHENYLACETAMIDE ◇ FENACETINA ◇ FIORINAL ◇ MELABON ◇ PARACETOPHENETIDIN ◇ PERCOBARB ◇ PERCODAN ◇ p-PHENACETIN ◇ RCRA WASTE NUMBER U187 ◇ SINUTAB ◇ TETRACYDIN ◇ XARIL ◇ ZACTIRIN COMPOUND

TOXICITY DATA with REFERENCE
mma-sat:333 μg/plate IARCCD 27,283,80
sce-mus-ipr 165 mg/kg JTEHD6 16,355,85
orl-rat TDLo:50336 mg/kg (17W male):REP JCNDBK 11,96,71
orl-rat TDLo:24 g/kg (female 1-20D post):TER KLWOAZ 43,364,65
orl-wmn TDLo:80 g/kg/63Y-I:CAR JOURAA 113,653,75
orl-man TDLo:57 g/kg/47Y I:CAR JOURAA 113,653,75

scu-mus TDLo:19200 mg/kg/24W-I:ETA VOONAW 32(5),63,86
orl-mus LD:484 g/kg/96W-C:NEO IJCNAW 29,439,82
unr-man LDLo:74 mg/kg 85DCAI 2,73,70
orl-rat LD50:3600 mg/kg ARZNAD 24,600,74
ipr-rat LD50:630 mg/kg NYKZAU 62,11,66
orl-mus LD50:866 mg/kg ARZNAD 28,1644,78
ipr-mus LD50:540 mg/kg YKKZAJ 81,659,61
scu-mus LD50:1625 mg/kg ARZNAD 8,25,58
ivn-dog LDLo:260 mg/kg NTIS** PB282-666
orl-rbt LD50:2500 mg/kg GTPZAB 21(9),53,77
scu-rbt LD50:1 g/kg ARZNAD 21,719,71
orl-gpg LD50:1870 mg/kg TXAPA9 2,23,60
orl-ham LD50:1690 mg/kg PHARAT 8,572,53

CONSENSUS REPORTS: NTP Fifth Annual Report on Carcinogens. IARC Cancer Review: Group 2A IMEMDT 7,310,87; Animal Inadequate Evidence IMEMDT 13,141,77; Human Limited Evidence IMEMDT 13,141,77; IMEMDT 24,135,80, Animal Limited Evidence IMEMDT 24,135,80; IMEMDT 24,135,80. Reported in EPA TSCA Inventory.

SAFETY PROFILE: Confirmed carcinogen producing tumors of the kidney and bladder. A human poison by an unspecified route. Poison by intravenous and possibly other routes. Moderately toxic by several routes. Human systemic effects by ingestion: cyanosis, liver damage, and methemoglobinemia-carboxhemoglobinemia. Experimental teratogenic data. Other experimental reproductive effects. Mutation data reported. Experimental reproductive effects. Chronic effects consist of weight loss, insomnia, shortness of breath, weakness and often aplastic anemia. When heated to decomposition it emits toxic fumes of NO_x.

ABH000 CAS:98-86-2 ***HR: 3***
ACETOPHENONE
mf: C_8H_8O mw: 120.16

PROP: Colorless liquid or plates; sweet, pungent odor. Mp: 19.7°, bp: 202.3°, flash p: 180°F (OC), d: 1.026 @ 20°/4°, vap d: 4.14, vap press: 1 mm @ 15°, autoign temp: 1060°F. Very sol in propylene glycol and fixed oils; sol in alc, chloroform, and ether; sltly sol in water; insol in glycerin.

SYNS: ACETYLBENZENE ◇ BENZOYL METHIDE ◇ DYMEX ◇ FEMA No. 2009 ◇ HYPNONE ◇ KETONE METHYL PHENYL ◇ METHYL PHENYL KETONE ◇ 1-PHENYLETHANONE ◇ PHENYL METHYL KETONE ◇ USAF EK-496

TOXICITY DATA with REFERENCE
skn-rbt 10 mg/24H open JIHTAB 26,269,44
skn-rbt 515 mg open MLD UCDS** 12/27/71
eye-rbt 771 μg SEV AJOPAA 29,1363,46
cyt-smc 10 mmol/tube HEREAY 33,457,47
orl-rat LD50:815 mg/kg GTPZAB 26(8),53,82
orl-mus LD50:740 mg/kg GTPZAB 26(8),53,82
scu-mus LDLo:330 mg/kg HDTU** -,-,33
ipr-mus LD50:200 mg/kg NTIS** AD277-689

CONSENSUS REPORTS: Reported in EPA TSCA Inventory.

SAFETY PROFILE: Poison by intraperitoneal and subcutaneous routes. Moderately toxic by ingestion. A skin and severe eye irritant. Mutation data reported. Narcotic in high concentration. A hypnotic. See also KETONES. Combustible liquid. To fight fire use foam, CO_2, dry chemical. When heated to decomposition it emits acrid smoke and fumes.

ABH250 CAS:2302-93-4 ***HR: 3***
ACETOPHENONE THIOSEMICARBAZONE
mf: $C_9H_{11}N_3S$ mw: 193.29

SYN: 1-(α-METHYLBENZYLIDENE)THIOSEMICARBIZIDE

TOXICITY DATA with REFERENCE
orl-rat LDLo:15 mg/kg NCNSA6 5,43,53
ivn-mus LD50:320 mg/kg CSLNX* NX#01020

SAFETY PROFILE: Poison by ingestion and intravenous routes. When heated to decomposition it emits very toxic fumes of NO_x and SO_x.

ABH500 CAS:61-00-7 ***HR: 3***
ACETOPROMAZINE
mf: $C_{19}H_{22}N_2OS$ mw: 326.49

SYNS: ACEPROMAZINA ◇ ACEPROMAZINE ◇ ACEPROMIZINA ◇ ACETAZINE ◇ ACETHYLPROMAZIN ◇ 3-ACETYL-10-(3-DIMETHYLAMINOPROPYL)PHENOTHIAZINE ◇ ACETYLPROMAZINE ◇ ANATRAN ◇ ANERGAN ◇ ATRAVET ◇ ATSETOZIN ◇ AY-57,062 ◇ AZEPROMAZINE ◇ 1522 CB ◇ 10-(3-DIMETHYLAMINOPROPYL)PHENOTHIAZINE-3-ETHYLONE ◇ 1-(10-(3-(DIMETHYLAMINO)PROPYL)-10H-PHENOTHIAZIN-2-YL)ETHANONE ◇ 10-(3-DIMETHYLAMINOPROPYL)PHENOTHIAZIN-3-YLMETHYL KETONE ◇ LISERGAN ◇ NOTENQUIL ◇ NOTENSIL ◇ NOTESIL ◇ PLEGECYL ◇ PLEGICIN ◇ PLIVAPHEN ◇ SOPRINTIN ◇ SOPRONTIN ◇ SOPROTIN ◇ SV-1522 ◇ VETRANQUIL ◇ WY-1172

TOXICITY DATA with REFERENCE
ipr-mus LD50:350 mg/kg RMNIBN 81,105,77
orl-mus LDLo:200 mg/kg AIPTAK 113,53,57
scu-mus LD50:130 mg/kg AIPTAK 113,53,57
ivn-mus LD50:59 mg/kg AIPTAK 115,1,58

SAFETY PROFILE: Poison by ingestion, intravenous, and subcutaneous routes. When heated to decomposition it emits toxic fumes of SO_x and NO_x. See also KETONES. An animal tranquilizer.

ABH750 CAS:1071-73-4 ***HR: 2***
ACETOPROPYL ALCOHOL
mf: $C_5H_{10}O_2$ mw: 102.15

TOXICITY DATA with REFERENCE
orl-rat LDLo:4180 mg/kg GISAAA 43(8),103,78
ihl-rat LCLo:2000 mg/m^3/4H TNICS* 13,119,73
orl-mus LD50:1960 mg/kg GISAAA 43(8),103,78
ihl-mus LCLo:2000 mg/m^3/4H TNICS* 13,119,73
orl-rbt LDLo:3500 mg/kg GISAAA 43(8),103,78
orl-gpg LD50:2260 mg/kg GISAAA 43(8),103,78

SAFETY PROFILE: Moderately toxic by ingestion and inhalation. When heated to decomposition it emits acrid smoke.

ABI000 CAS:350-03-8 ***HR: 2***
3-ACETOPYRIDINE
mf: C_7H_7NO mw: 121.15

SYNS: β-ACETYLPYRIDINE ◇ 3-ACETYLPYRIDINE ◇ METHYL PYRIDYL KETONE ◇ METHYL-β-PYRIDYL KETONE ◇ METHYL-3-PYRIDYL KETONE

TOXICITY DATA with REFERENCE
orl-qal LD50:422 mg/kg AECTCV 12,355,83
orl-bwd LD50:178 mg/kg AECTCV 12,355,83
ipr-mus LDLo:512 mg/kg CBCCT* 3,53,51

CONSENSUS REPORTS: Reported in EPA TSCA Inventory.

SAFETY PROFILE: Poison by ingestion. Moderately toxic by intraperitoneal route. When heated to decomposition emits toxic fumes of NO_x. See also KETONES.

ABI250 CAS:87-11-6 ***HR: 3***
ACETOPYRROTHINE
mf: $C_8H_8N_2O_2S_2$ mw: 228.30

PROP: Produced by *Streptomyces albus* (ANTCAO 2,357,52).

SYNS: 6-ACETAMIDO-4-METHYL-1,2-DITHIOLO(4,3-B)PYRROL-5(4H)-ONE ◇ 3-ACETAMIDO-5-METHYLPYRROLIN-4-ONE(4,3-D)-1,2-DITHIOLE ◇ 6-(ACETYLAMINO)-4-METHYL-1,2-DITHIOLO(4,3-B)PYRROL-5(4H)-ONE ◇ N-(4,5-DIHYDRO-4-METHYL-5-OXO-1,2-DITHIOLO(4,3-B)PYRROL-6-YL)ACETAMIDE ◇ THIOLUTIN

TOXICITY DATA with REFERENCE
orl-mus LD50:25 mg/kg MEIEDD 10,1338,83
scu-mus LD50:25 mg/kg ANTCAO 2,357,52

SAFETY PROFILE: Poison by ingestion and subcutaneous routes. When heated to decomposition it emits very toxic fumes of NO_x and SO_x.

ABI500 CAS:88-15-3 ***HR: 3***
2-ACETOTHIENONE
mf: C_6H_6OS mw: 126.18

SYNS: 2-ACETOTHIOPHENE ◇ 2-ACETYLTHIOPHENE

TOXICITY DATA with REFERENCE
ipr-mus LD50:40 mg/kg NTIS** AD691-490

CONSENSUS REPORTS: Reported in EPA TSCA Inventory.

SAFETY PROFILE: Poison by intraperitoneal route. When heated to decomposition emits toxic fumes of SO_x.

ABI750 CAS:537-92-8 ***HR: 2***
m-ACETOTOLUIDIDE
mf: $C_9H_{11}NO$ mw: 149.21

SYNS: N-ACETYL-m-TOLUIDINE ◇ 3-ACETAMIDOTOLUENE ◇ ACETO-m-AMINOTOLUENE ◇ ACETOTOLUIDE ◇ m-METHYLACETANILIDE ◇ 3-METHYLACETANILIDE ◇ 3′-METHYLACETANILIDE ◇ m-TOLYLACETAMIDE ◇ N-m-TOLYLACETAMIDE

TOXICITY DATA with REFERENCE
orl-mus LD50:1450 mg/kg TXAPA9 19,20,71

CONSENSUS REPORTS: Reported in EPA TSCA Inventory.

SAFETY PROFILE: Moderately toxic by ingestion. See also p-ACETOTOLUIDIDE. When heated to decomposition it emits toxic fumes of NO_x.

ABJ000 CAS:120-66-1 ***HR: 2***
o-ACETOTOLUIDIDE
mf: $C_9H_{11}NO$ mw: 149.21

SYNS: ACETYL-o-TOLUIDINE ◇ o-ACETOTOLUIDE ◇ o-METHYLACETANILIDE ◇ 2-METHYLACETANILIDE ◇ 2′-METHYLACETANILIDE

TOXICITY DATA with REFERENCE
mma-sat 1 mg/plate NTPTB* JAN 82
mma-sat 47 nmol/plate MUREAV 137,39,84
orl-mus LD50:1450 mg/kg TXAPA9 19,20,71

CONSENSUS REPORTS: Reported in EPA TSCA Inventory.

SAFETY PROFILE: Moderately toxic by ingestion. Mutation data reported. See also p-ACETOTOLUIDIDE. When heated to decomposition it emits toxic fumes of NO_x.

ABJ250 CAS:103-89-9 ***HR: 2***
p-ACETOTOLUIDIDE
mf: $C_9H_{11}NO$ mw: 149.21

PROP: Crystals. Bp: 307°, flash p: 335°F (CC), d: 1.212, vap d: 5.14, mp: 153°.

SYNS: p-ACETAMIDOTOLUENE ◇ p-ACETOTOLUIDE ◇ 4-ACETOTOLUIDE ◇ 4-(ACETYLAMINO)TOLUENE ◇ ACETYL-p-TOLUIDINE ◇ N-ACETYL-p-TOLUIDIDE ◇ p-METHYLACETANILIDE ◇ 4-METHYLACETANILIDE ◇ 4′-METHYLACETANILIDE

TOXICITY DATA with REFERENCE
orl-rat LD50:2640 mg/kg MarJV# 29MAR77
orl-mus LD50:980 mg/kg TXAPA9 19,20,71

CONSENSUS REPORTS: Reported in EPA TSCA Inventory.

SAFETY PROFILE: Moderately toxic by ingestion. See also ACETANILIDE. Combustible. When heated to decomposition it emits toxic fumes of NO_x. To fight fire use water, foam, CO_2, dry chemical.

ABJ750 CAS:26541-56-0 ***HR: 2***
N-ACETOXY-4-ACETAMIDOBIPHENYL
mf: $C_{16}H_{15}NO_3$ mw: 269.32

SYNS: ACETIC ACID (N-ACETYL-N-(4-BIPHENYL)AMINO) ESTER ◇ ACETIC ACID ESTER with N-4-BIPHENYLYLACETOHYDROXAMIC ACID ◇ N-ACETOXY-4-BIPHENYLACETAMIDE ◇ N-(4-BIPHENYLYL)ACETOHYDROXAMIC ACETATE

TOXICITY DATA with REFERENCE
mmo-sat 25 μg/plate CBINA8 26,11,79
mma-sat 1 μg/plate CBINA8 26,11,79
skn-mus TDLo:4309 μg/kg:NEO JNCIAM 54,491,75

SAFETY PROFILE: Questionable carcinogen with experimental neoplastigenic data. Mutation data reported. See also ESTERS. When heated to decomposition it emits toxic fumes of NO_x.

ABK000 CAS:64058-72-6 ***HR: 2***
ACETOXY(2-ACETAMIDO-5-NITROPHENYL)MERCURY
mf: $C_{10}H_{10}HgN_2O_5$ mw: 438.81

SYN: 2′-(ACETOXYMERCURI)-4′-NITROACETANILIDE

TOXICITY DATA with REFERENCE
ipr-rat LDLo:500 mg/kg NCNSA6 5,8,53

OSHA PEL: (Transitional: CL 1 mg/10m^3) CL 0.1 mg(Hg)/m^3 (skin)
ACGIH TLV: TWA 0.1 mg(Hg)m^3 (skin)
NIOSH REL: (Mercury, Inorganic) TWA 0.05 mg(Hg)/m^3

SAFETY PROFILE: Moderately toxic by intraperitoneal route. See also MERCURY COMPOUNDS. When heated to decomposition it emits very toxic fumes of Hg and NO_x.

ABK250 CAS:26541-57-1 ***HR: 3***
N-ACETOXY-2-ACETAMIDOPHENANTHRENE
mf: $C_{18}H_{15}NO_3$ mw: 293.34

SYNS: ACETIC ACID (N-ACETYL-N-(2-PHENANTHRYL)AMINO)ESTER ◇ ACETIC ACID ESTER with N-(2-PHENANTHRYL)ACETOHYDROXAMIC ACID ◇ N-ACETOXY-2-ACETYLAMINOPHENANTHRENE ◇ N-ACETOXY-4-PHENANTHRYLACETAMIDE ◇ N-(2-PHENANTHRYL)ACETOHYDROXAMIC ACETATE

TOXICITY DATA with REFERENCE
mmo-bcs 14 mol CNREA8 30,1473,70
oms-bcs 10 g/L CNREA8 30,1473,70
dns-hmn:fbr 10 mmol/L/5H IJCNAW 16,284,75
mmo-sat 5 μg/plate CBINA8 26,11,79
mma-sat 50 ng/plate CBINA8 26,11,79
dnd-mam:lym 625 mg/L CNREA8 35,1416,75
skn-mus TDLo:584 μg/kg:NEO JNCIAM 54,491,75

SAFETY PROFILE: Questionable carcinogen with experimental neoplastigenic data. Human mutation data reported. See also ESTERS. When heated to decomposition it emits toxic fumes of NO_x.

ABL000 CAS:6098-44-8 ***HR: 3***
N-ACETOXY-N-ACETYL-2-AMINOFLUORENE
mf: $C_{17}H_{15}NO_3$ mw: 281.33

SYNS: ACETIC ACID (N-ACETYL-N-(2-FLUORENYL)AMINO) ESTER ◇ N-ACETOXY-2-ACETAMIDOFLUORENE ◇ N-ACETOXY-2-ACETYLAMINOFLUORENE ◇ N-ACETOXY-2-FLUORENYLACETAMIDE ◇ N-(FLUOREN-2-YL)ACETOHYDROXAMIC ACETAMIDE

TOXICITY DATA with REFERENCE
mma-sat 1500 ng/plate CBINA8 54,71,85
sce-ham:oth 1800 nmol/L CRNGDP 6,1627,85
dns-hmn:lym 10 μmol/L CALEDQ 2,311,77
dns-hmn:leu 10 μmol/L CRNGDP 1,547,80
dni-hmn:hla 100 μmol/L/30M-C JEPTDQ 2(1),65,78
scu-rat TDLo:21 mg/kg:NEO CNREA8 37,1461,77
imp-rat TDLo:28 mg/kg:ETA CNREA8 37,111,77

CONSENSUS REPORTS: EPA Genetic Toxicology Program.

SAFETY PROFILE: Questionable carcinogen with experimental tumorigenic and neoplastigenic data. Human mutation data reported. When heated to decomposition it emits toxic fumes of NO_x.

ABL250 CAS:26488-34-6 ***HR: 3***
trans-N-ACETOXY-4-ACETYL-AMINOSTILBENE
mf: $C_{18}H_{17}NO_3$ mw: 295.36

SYN: trans-N,o-DIACETYL-N-(p-STYRYLPHENYL)HYDROXYLAMINE

TOXICITY DATA with REFERENCE
orl-rat TDLo:180 mg/kg/20W-I:CAR ZEKBAI 74,200,70
mrc-smc 10 ppm ZEKBAI 74,412,70

CONSENSUS REPORTS: EPA Genetic Toxicology Program.

SAFETY PROFILE: Questionable carcinogen with experimental carcinogenic data. Mutation data reported. When heated to decomposition it emits toxic fumes of NO_x.

ABL500 CAS:3061-65-2 ***HR: 3***
2-ACETOXYACRYLONITRILE
mf: $C_5H_5NO_2$ mw: 111.11

SYNS: α-ACETOXYACRYLONITRILE ◇ α-CYANOVINYL ACETATE

TOXICITY DATA with REFERENCE
skn-rbt 10 mg/24H open MLD AIHAAP 23,95,62
orl-rat LD50:100 mg/kg AIHAAP 23,95,62
ihl-rat LCLo:125 ppm/4H AIHAAP 23,95,62
skn-rbt LD50:140 mg/kg AIHAAP 23,95,62

CONSENSUS REPORTS: Cyanide and its compounds are on the Community Right-To-Know List.

SAFETY PROFILE: Poison by inhalation, ingestion, and skin contact. A skin irritant. See also NITRILES. When heated to decomposition it emits toxic fumes of NO_x.

ABL625 CAS:95282-98-7 ***HR: D***
19-ACEOTXY-Δ$^{(1,4)}$-ANDROSTADIENE-3,17-DIONE
mf: $C_{21}H_{26}O_4$ mw: 342.47

SYNS: 3,17-DIONE-19-ACETOXY-Δ$^{(1,3)}$-ANDROSTADIENE ◇ 19-HYDROXYANDROSTA-1,4-DIENE-3,17-DIONE ACETATE

TOXICITY DATA with REFERENCE
scu-rat TDLo:5 mg/kg (1D preg):REP PSEBAA 111,595,62

SAFETY PROFILE: Experimental reproductive effects. When heated to decomposition it emits acrid smoke and irritating fumes.

ABL750 CAS:53555-67-2 ***HR: D***
6-ACETOXY-BENZO(a)PYRENE

SYN: BENZO(a)PYRENE-6-YL ACETATE

TOXICITY DATA with REFERENCE
otr-ham:emb 2500 μg/L CBINA8 13,105,76
dnd-ham:emb 1250 μg/L/18H CBINA8 13,105,76
dns-ham:emb 10 μg/L/18H CBINA8 13,105,76

SAFETY PROFILE: Mutation data reported. When heated to decomposition it emits acrid smoke and irritating fumes.

ABL875 CAS:70490-99-2 ***HR: D***
N-α-ACETOXYBENZYL-N-BENZYLNITROSAMINE
mf: $C_{16}H_{16}N_2O_3$ mw: 284.34

SYNS: α-ACETOXY-N-NITROSODIBENZYLAMINE ◇ N-BENZYL-N-(α-ACETOXYBENZYL)NITROSAMINE ◇ α-(BENZYLNITROSAMINO) BENZYL ALCOHOL ACETATE (ester) ◇ α-(NITROSO(PHENYLMETHYL)AMINO)-BENZENEMETHANOL ACETATE (ester)

TOXICITY DATA with REFERENCE
mmo-sat 100 nmol/plate CALEDQ 6,83,79
mma-sat 1 μmol/plate CALEDQ 6,83,79
msc-ham:lng 10 μmol/L GANNA2 73,517,82

SAFETY PROFILE: Mutation data reported. When heated to decomposition it emits toxic fumes of NO_x. See also NITROSAMINES and ESTERS.

ABM000 ***HR: 3***
3-β-ACETOXY-BIS NOR-Δ^5-CHOLENIC ACID
mf: $C_{24}H_{36}O_4$ mw: 388.60

SYN: 3-β-ACETOXYPREGN-6-ENE-20-CARBOXYLICACID

TOXICITY DATA with REFERENCE
scu-mus TDLo:200 mg/kg/90D-I:ETA NATUAS 209,1026,66

SAFETY PROFILE: Questionable carcinogen with experimental tumorigenic data. When heated to decomposition it emits acrid smoke and irritating fumes.

ABM250 CAS:1515-76-0 ***HR: 3***
1-ACETOXY-1,3-BUTADIENE
mf: $C_6H_8O_2$ mw: 112.14

SYN: ACETIC ACID-1,3-BUTADIENYL ESTER

TOXICITY DATA with REFERENCE
skn-rbt 100 μg/24H open AIHAAP 23,95,62
orl-rat LDLo:710 mg/kg AIHAAP 23,95,62
ihl-rat LCLo:63 ppm/4H AIHAAP 23,95,62
skn-rbt LD50:420 mg/kg AIHAAP 23,95,62

SAFETY PROFILE: Poison by inhalation. Moderately toxic by other routes. A skin irritant. Mutation data reported. When heated to decomposition it emits acrid smoke.

ABM500 CAS:70103-79-6 ***HR: D***
N-(4-ACETOXYBUTYL)-N-(ACETOXYMETHYL)NITROSAMINE
mf: $C_9H_{16}N_2O_5$ mw: 232.24

SYN: ACETICACID-4-((ACETOXYMETHYL)NITROSAMINO)BUTYL ESTER

TOXICITY DATA with REFERENCE
mmo-sat 1 μmol/plate GANNA2 71,124,80
mmo-esc 1 μmol/plate GANNA2 71,124,80
mrc-bcs 200 nmol/plate GANNA2 71,124,80

SAFETY PROFILE: Mutation data reported. When heated to decomposition it emits toxic fumes of NO_x. See also NITROSAMINES and ESTERS.

ABN000 CAS:2885-39-4 ***HR: 3***
ACETOXYCYCLOHEXIMIDE
mf: $C_{17}H_{25}NO_6$ mw: 339.43

SYNS: ACETYLOXYCYCLOHEXIMIDE ◇ 3-(2-(5-ACETOXY-3,5-DIMETHYL-2-OXOCYCLOHEXYL)-2-HYDROXYETHYL)GLUTARIMIDE ◇ AXM ◇ E-73 ACETATE ◇ NSC 32743 ◇ STREPTOVITACIN E 73

TOXICITY DATA with REFERENCE
oms-hmn:hla 29 μmol/L BCPCA6 14,205,65
dni-hmn:hla 1 μmol/L BCPCA6 14,205,65
orl-rat LD50:158 μg/kg JPETAB 136,400,62
ipr-rat LD50:170 μg/kg JPETAB 136,400,62

scu-rat LD50:190 μg/kg JPETAB 136,400,62
ipr-mus LD50:19 mg/kg JPETAB 136,400,62
ivn-dog LDLo:920 μg/kg JPETAB 136,400,62

SAFETY PROFILE: Deadly poison by ingestion, intravenous, intraperitoneal, and subcutaneous routes. Human mutation data reported. When heated to decomposition it emits toxic fumes, such as NO_x.

ABN250 CAS:24684-58-0 ***HR: 3***
11-ACETOXY-15-DIHYDROCYCLOPENTA(a)PHENANTHRACEN-17-ONE
mf: $C_{19}H_{14}O_3$ mw: 290.33

SYN: 11-HYDROXY-15,16-DIHYDROCYCLOPENTA(a)PHENANTHRACEN-17-ONEACETATE (ESTER)

TOXICITY DATA with REFERENCE
skn-mus TDLo:108 mg/kg/1Y-I:ETA PEXTAR 11,69,69

SAFETY PROFILE: Questionable carcinogen with experimental tumorigenic data. See also ESTERS. When heated to decomposition it emits acrid smoke and irritating fumes.

ABN500 CAS:38539-23-0 ***HR: 3***
1-ACETOXY-1,4-DIHYDRO-4-(HYDROXYAMINO)QUINOLINE ACETATE (ESTER)
mf: $C_{13}H_{13}N_2O_4$ mw: 261.28

SYN: O,O'-DIACETYL4-HYDROXYAMINOQUINOLINE-1-OXIDE

TOXICITY DATA with REFERENCE
mmo-smc 50 mg/L IGSBAL 85,127,72
scu-rat TDLo:5 mg/kg/2W-I:ETA PSEBAA 136,1206,71
scu-mus TDLo:30 mg/kg/4W-I:NEO PSEBAA 136,1206,71

SAFETY PROFILE: Questionable carcinogen with experimental tumorigenic and neoplastigenic data. Mutation data reported. See also ESTERS. When heated to decomposition it emits toxic fumes of NO_x.

ABN625 ***HR: 3***
1-ACETOXYDIMERCURIO-1-PERCHLORATODIMERCURIOPROPEN-2-ONE
mf: $C_4H_3ClHg_4O_7$ mw: 1000.84

$(CH_2COOH)OHgHgC(:C:O)HgHgOClO_3$

SAFETY PROFILE: Dangerously explosive. When heated to decomposition it emits toxic fumes of Hg and Cl^-. See also MERCURY COMPOUNDS; PERCHLORATES; and EXPLOSIVES.

ABN700 CAS:66827-45-0 ***HR: 3***
β-ACETOXY-N,N-DIMETHYLPHENETHYLAMINE
mf: $C_{12}H_{17}NO_2$ mw: 207.30

SYNS: ACETIC ACID-α-(DIMETHYLAMINOMETHYL)BENZYL ESTER ◇ β-ACETYLOXY-β-PHENYLETHYL DIMETHYLAMINE ◇ N,N-DIMETHYL-β-ACETOXY β-PHENYLETHYLAMINE

TOXICITY DATA with REFERENCE
ipr-mus LD50:310 mg/kg EJMCA5 13,277,78
scu-mus LDLo:722 mg/kg AIPTAK 47,96,34
ivn-rbt LDLo:72 mg/kg AIPTAK 47,96,34

SAFETY PROFILE: Poison by intravenous and intraperitoneal routes. Moderately toxic by subcutaneous route. When heated to decomposition it emits toxic fumes of NO_x. See also AMINES and ESTERS.

ABN725 CAS:61691-82-5 ***HR: 2***
1'-ACETOXYESTRAGOLE
mf: $C_{12}H_{14}O_3$ mw: 206.26

SYN: p-METHOXY-α-VINYLBENZYL ALCOHOL ACETATE (ESTER)

TOXICITY DATA with REFERENCE
dnd-hmn:fbr 500 μmol/L CRNGDP 3,935,82
ipr-mus TDLo:20626 μg/kg:CAR CNREA8 47,2275,87

CONSENSUS REPORTS: EPA Genetic Toxicology Program.

SAFETY PROFILE: Questionable carcinogen with experimental carcinogenic data. Human mutation data reported. When heated to decomposition it emits acrid smoke and fumes. See also ESTERS.

ABN750 CAS:70103-77-4 ***HR: D***
N-(2-ACETOXYETHYL)-N-(ACETOXYMETHYL) NITROSAMINE
mf: $C_7H_{12}N_2O_5$ mw: 204.20

SYNS: ACETIC ACID-2-((ACETOXYMETHYL)NITROSAMINO)ETHYL ESTER ◇ AEAMN

TOXICITY DATA with REFERENCE
mmo-sat 1 μmol/plate GANNA2 71,124,80
mrc-bcs 100 nmol/plate GANNA2 71,124,80

SAFETY PROFILE: Mutation data reported. When heated to decomposition it emits toxic fumes of NO_x. See also NITROSAMINES and ESTERS.

ABO000 CAS:60-31-1 ***HR: 3***
2-ACETOXYETHYLTRIMETHYLAMMONIUM CHLORIDE
mf: $C_7H_{16}NO_2•Cl$ mw: 181.69

SYNS: ACECOLINE ◇ ACETYLCHOLINE CHLORIDE ◇ ACETYLCHOLINE HYDROCHLORIDE ◇ ACETYLCHOLINIUM CHLORIDE ◇ 2-(ACETYLOXY)-N,N,N-TRIMETHYLETHANAMINIUM CHLORIDE ◇ ACH CHLORIDE ◇ ARTEROCOLINE ◇ CHOLINE CHLORIDE ACE-

TATE ◇ (2-HYDROXYETHYL)TRIMETHYLAMMONIUM CHLORIDE ACETATE ◇ OVISOT ◇ TL 1505

TOXICITY DATA with REFERENCE
orl-rat LD50:2500 mg/kg JPETAB 58,337,36
scu-rat LD50:250 mg/kg JPETAB 58,337,36
ivn-rat LD50:22 mg/kg JPETAB 58,337,36
orl-mus LD50:3000 mg/kg JPETAB 58,337,36
ipr-mus LD50:162 mg/kg ATXKA8 29,39,72
scu-mus LD50:170 mg/kg JPETAB 58,337,36
ivn-mus LD50:12500 μg/kg MPHEAE 13,117,65
par-frg LDLo:200 mg/kg AEPPAE 166,437,32

CONSENSUS REPORTS: Reported in EPA TSCA Inventory.

SAFETY PROFILE: Poison by subcutaneous, intravenous, intraperitoneal, and parenteral routes. Moderately toxic by ingestion. When heated to decomposition it emits very toxic fumes of NO_x and Cl^-. A cholinergic agent. See also CHOLINE ACETATE (ESTER).

ABO250 CAS:38105-27-0 ***HR: 3***
N-ACETOXYFLUORENYLACETAMIDE
mf: $C_{17}H_{15}NO_3$ mw: 281.33

SYNS: ACETIC ACID ESTER with N-(FLUOREN-3-YL)ACETOHYDROXAMIC ACID ◇ N-ACETOXY-3-FLUORENYLACETAMIDE ◇ N-(FLUOREN-3-YL)ACETOHYDROXAMIC ACETATE

TOXICITY DATA with REFERENCE
ipr-rat TDLo:350 mg/kg/4W-I:NEO CNREA8 35,447,75
ims-rat TDLo:210 mg/kg/8W-I:ETA CNREA8 35,447,75

SAFETY PROFILE: Questionable carcinogen with experimental tumorigenic and neoplastigenic data. See also ESTERS. When heated to decomposition it emits toxic fumes of NO_x.

ABO500 CAS:55080-20-1 ***HR: 3***
N-ACETOXY-4-FLUORENYLACETAMIDE
mf: $C_{17}H_{15}NO_3$ mw: 281.33

SYNS: ACETICACID(N-ACETYL-N-(4-FLUORENYL)AMINO)ESTER ◇ ACETIC ACID, ESTER with N-(FLUOREN-4-YL)ACETOXYHYDROXAMIC ACID ◇ N-(FLUOREN-4-YL)ACETOHYDROXAMIC ACETATE

TOXICITY DATA with REFERENCE
ipr-rat TDLo:350 mg/kg/4W-I:ETA CNREA8 35,447,75

SAFETY PROFILE: Questionable carcinogen with experimental tumorigenic data. See also ESTERS. When heated to decomposition it emits toxic fumes of NO_x.

ABO750 CAS:29968-75-0 ***HR: 3***
N-ACETOXY-2-FLUORENYLBENZAMIDE
mf: $C_{22}H_{17}NO_3$ mw: 343.40

SYN: N-FLUOREN-2-YL BENZOHYDROXAMIC ACID ACETATE

TOXICITY DATA with REFERENCE
orl-rat TDLo:5200 mg/kg/21W-I:ETA CNREA8 30,1485,70

SAFETY PROFILE: Questionable carcinogen with experimental tumorigenic data. When heated to decomposition it emits toxic fumes of NO_x.

ABP250 CAS:2198-53-0 ***HR: 2***
2′,6′-ACETOXYLIDIDE
mf: $C_{10}H_{13}NO$ mw: 163.24

SYN: 2,6-DIMETHYLACETANILIDE

TOXICITY DATA with REFERENCE
orl-mus LD50:620 mg/kg TXAPA9 19,20,71

CONSENSUS REPORTS: Reported in EPA TSCA Inventory.

SAFETY PROFILE: Moderately toxic by ingestion. When heated to decomposition it emits toxic fumes of NO_x.

ABP500 CAS:2198-54-1 ***HR: 3***
3′,4′-ACETOXYLIDIDE
mf: $C_{10}H_{13}NO$ mw: 163.24

SYNS: 3,4-DIMETHYLACETANILIDE ◇ 3′,4′-DIMETHYLACETANILIDE

TOXICITY DATA with REFERENCE
orl-rat TDLo:9900 mg/kg/56W-C:ETA CNREA8 16,525,56
orl-mus LD50:1030 mg/kg TXAPA9 19,20,71

SAFETY PROFILE: Moderately toxic by ingestion. Questionable carcinogen with experimental tumorigenic data. When heated to decomposition it emits toxic fumes of NO_x.

ABP750 CAS:2050-43-3 ***HR: 2***
2′,4′-ACETOXYLIDIDE
mf: $C_{10}H_{13}NO$ mw: 163.24

SYNS: 2,4-DIMETHYLACETANILIDE ◇ 2′,4′-DIMETHYLACETANILIDE

TOXICITY DATA with REFERENCE
orl-mus LD50:1300 mg/kg TXAPA9 19,20,71

CONSENSUS REPORTS: Reported in EPA TSCA Inventory.

SAFETY PROFILE: Moderately toxic by ingestion. When heated to decomposition it emits toxic fumes of NO_x.

ABQ000 CAS:6283-24-5 ***HR: 3***
p-(ACETOXYMERCURI)ANILINE
mf: $C_8H_9HgNO_2$ mw: 351.77

PROP: Colorless crystals, insol in water. Mp: 167°.

SYNS: (ACETATO)(p-AMINOPHENYL)MERCURY ◇ p-AMINOPHENYLMERCURIC ACETATE

TOXICITY DATA with REFERENCE
ivn-mus LD50:18 mg/kg CSLNX* NX#04750

CONSENSUS REPORTS: Reported in EPA TSCA Inventory.

OSHA PEL: (Transitional: CL 1 mg/10m^3) CL 0.1 mg(Hg)/m^3 (skin)
ACGIH TLV: TWA 0.1 mg(Hg)/m^3 (skin)
NIOSH REL: (Mercury, Inorganic) TWA 0.05 mg(Hg)/m^3

SAFETY PROFILE: Poison by intravenous routes. See also MERCURY COMPOUNDS, ANILINE. When heated to decomposition it emits very toxic fumes of NO_x and Hg.

ABQ250 CAS:54481-45-7 ***HR: 3***
2-(ACETOXYMERCURI)-4-NITROANILINE
mf: $C_8H_8HgN_2O_4$ mw: 396.77

SYN: ACETATO(2-AMINO-5-NITROPHENYL)MERCURY

TOXICITY DATA with REFERENCE
ipr-rat LDLo:250 mg/kg NCNSA6 5,12,53

OSHA PEL: (Transitional: CL 1 mg/10m^3) CL 0.1 mg(Hg)/m^3 (skin)
ACGIH TLV: TWA 0.1 mg(Hg)/m^3 (skin)
NIOSH REL: (Mercury, Inorganic) TWA 0.05 mg(Hg)/m^3

SAFETY PROFILE: Poison by intraperitoneal route. See also MERCURY COMPOUNDS, NITRO COMPOUNDS of AROMATIC HYDROCARBONS. When heated to decomposition it emits very toxic fumes of Hg and NO_x.

ABQ375 ***HR: 3***
1-ACETOXYMERCURIO-1-PERCHLORATOMERCURIOPROPEN-2-ONE
mf: $C_4H_3ClHg_2O_7$ mw: 599.66

$CH_2COOH)OHgC(:C:O)HgOClO_3$

SAFETY PROFILE: Dangerously explosive. When heated to decomposition it emits toxic fumes of Cl^- and Hg. See also MERCURY COMPOUNDS, EXPLOSIVES, and PERCHLORATES.

ABQ500 CAS:55936-77-1 ***HR: D***
N-ACETOXY-N-METHYL-4-AMINOAZOBENZENE
mf: $C_{15}H_{15}N_3O_2$ mw: 269.33

SYN: N-(ACETYLOXY)-N-METHYL-4-(PHENYLAZO)BENZENAMINE (9CI)

TOXICITY DATA with REFERENCE
otr-rat:lvr 50 μmol/L JJIND8 76,95,86
cyt-rat:lvr 13400 μg/L/2H-C ITCSAF 18,501,82
mmo-sat 100 nmol/plate CALEDQ 1,91,75

CONSENSUS REPORTS: EPA Genetic Toxicology Program.

SAFETY PROFILE: Mutation data reported. When heated to decomposition it emits toxic fumes of NO_x.

ABQ600 CAS:83876-62-4 ***HR: 2***
4-ACETOXY-7-METHYLBENZ(c)ACRIDINE
mf: $C_{20}H_{15}NO_2$ mw: 301.36

SYN: BENZ(c)ACRIDIN-4-OL, 7-METHYL-, ACETATE (ESTER)

TOXICITY DATA with REFERENCE
scu-mus TDLo:72 mg/kg/12W-I:ETA JMCMAR 26,303,83

SAFETY PROFILE: Questionable carcinogen with experimental tumorigenic data. When heated to decomposition it emits toxic fumes of NO_x.

ABR125 CAS:70715-92-3 ***HR: 3***
N-(ACETOXYMETHYL)-N-ISOBUTYLNITROSAMINE
mf: $C_7H_{14}N_2O_3$ mw: 174.23

SYNS: N-ISOBUTYL-N-(ACETOXYMETHYL)NITROSAMINE ◇ N-NITROSO-N-(ACETOXYMETHYL)-N-ISOBUTYLAMINE

TOXICITY DATA with REFERENCE
mmo-sat 5 μmol/plate GANNA2 70,663,79
mmo-esc 25 μmol/plate GANNA2 70,663,79
dnr-bcs 1 μmol/plate GANNA2 70,663,79
scu-rat TDLo:50 mg/kg/10W-I:ETA JCROD7 104,13,82
scu-rat TD:66 mg/kg/10W-I:CAR IAPUDO 41,619,82

SAFETY PROFILE: Questionable carcinogen with experimental carcinogenic and tumorigenic data. Mutation data reported. When heated to decomposition it emits toxic fumes of NO_x. See also NITROSAMINES.

ABR250 CAS:2517-98-8 ***HR: 3***
7-ACETOXYMETHYL-12-METHYLBENZ(a)ANTHRACENE
mf: $C_{22}H_{18}O_2$ mw: 314.40

SYN: 12-METHYLBENZ(a)ANTHRACENE-7-METHANOLACETATE (ESTER)

TOXICITY DATA with REFERENCE
orl-rat TDLo:100 mg/kg:ETA JMCMAR 10,932,67
scu-rat TDLo:150 mg/kg/39D-I:NEO CNREA8 31,1951,71

SAFETY PROFILE: Questionable carcinogen with experimental tumorigenic and neoplastigenic data. See also ESTERS. When heated to decomposition it emits acrid smoke and fumes.

ABR500 CAS:65986-79-0 ***HR: D***
1-ACETOXY-N-METHYL-N-NITROSOETHYLAMINE
mf: $C_5H_{10}N_2O_3$ mw: 146.17

SYNS: N-(α-ACETOXY)ETHYL-N-METHYLNITROSAMINE ◇ 1-(METHYLNITROSAMINO)ETHYL ACETATE

TOXICITY DATA with REFERENCE
mmo-sat 50 nmol/plate MUREAV 49,187,78

SAFETY PROFILE: Mutation data reported. When heated to decomposition it emits toxic fumes of NO_x. See also NITROSAMINES and N-NITROSO COMPOUNDS.

ABR625 ***HR: 3***
ACETOXYMETHYLPHENYLNITROSAMINE
mf: $C_9H_{11}N_2O_3$ mw: 195.22

TOXICITY DATA with REFERENCE
mmo-sat 64400 pmol/plate CALEDQ 15,289,82
scu-ham TDLo:42 mg/kg/27W-I:ETA CALEDQ 15,289,82
scu-ham LD50:117 mg/kg CALEDQ 15,289,82

SAFETY PROFILE: Poison by subcutaneous route. Questionable carcinogen with experimental tumorigenic data. Mutation data reported. When heated to decomposition it emits toxic fumes of NO_x. See also NITROSAMINES.

ABS250 CAS:38105-25-8 ***HR: D***
N-ACETOXY-N-(1-NAPTHYL)-ACETAMIDE
mf: $C_{14}H_{13}NO_3$ mw: 243.28

SYN: N,O-DIACETYL-N-(1-NAPHTHYL)HYDROXYLAMINE

TOXICITY DATA with REFERENCE
mmo-sat 5 μg/plate CBINA8 26,11,79
mma-sat 1 μg/plate CBINA8 26,11,79

SAFETY PROFILE: Mutation data reported. When heated to decomposition it emits highly toxic fumes of NO_x.

ABS750 CAS:830-03-5 ***HR: 3***
p-ACETOXYNITROBENZENE
mf: $C_8H_7NO_4$ mw: 181.16

SYNS: p-NITROPHENOL ACETATE ◇ p-NITROPHENYL ACETATE ◇ 4-NITROPHENYL ACETATE

TOXICITY DATA with REFERENCE
ivn-mus LD50:180 mg/kg CSLNX* NX#00217

CONSENSUS REPORTS: Reported in EPA TSCA Inventory.

SAFETY PROFILE: Poison by intravenous route. When heated to decomposition it emits toxic fumes of NO_x.

ABT250 ***HR: D***
1-ACETOXY-N-NITROSODIBUTYLAMINE
mf: $C_{10}H_{20}N_2O_3$ mw: 216.32

SYNS: N-(α-ACETOXY)BUTYL-N-BUTYLNITROSAMINE ◇ 1-(BUTYLNITROSAMINO)BUTYL ACETATE

TOXICITY DATA with REFERENCE
mmo-sat 100 nmol/plate MUREAV 49,187,78
cyt-ham:fbr 125 mg/L/48H MUREAV 48,337,77

SAFETY PROFILE: Mutation data reported. See also AMINES, NITROSAMINES and N-NITROSO COMPOUNDS many of which are carcinogenic. When heated to decomposition it emits highly toxic fumes of NO_x.

ABT500 CAS:58431-24-6 ***HR: D***
1-ACETOXY-N-NITROSODIETHYLAMINE
mf: $C_6H_{12}N_2O_3$ mw: 160.20

SYNS: ACETIC ACID-1-(ETHYLNITROSAMINO)ETHYL ESTER ◇ N-(α-ACETOXY)ETHYL-N-ETHYLNITROSAMINE ◇ 1-(ETHYLNITROSAMINO)ETHYL ACETATE

TOXICITY DATA with REFERENCE
slt-dmg-par 500 μmol/L CNREA8 36,4504,76
mmo-sat 100 nmol/plate MUREAV 49,187,78
sln-dmg-par 2 mmol/L BCPCA6 24,2009,75

SAFETY PROFILE: Mutation data reported. See also AMINES, NITROSAMINES, and N-NITROSO COMPOUNDS (many of which are carcinogenic). When heated to decomposition it emits highly toxic fumes of NO_x.

ABT750 CAS:53198-41-7 ***HR: 3***
1-ACETOXY-N-NITROSODIPROPYLAMINE
mf: $C_8H_{16}N_2O_3$ mw: 188.26

SYNS: ACETIC ACID-1-(PROPYLNITROSAMINO)PROPYL ESTER ◇ N-(α-ACETOXY)PROPYL-N-N-PROPYLNITROSAMINE ◇ 1-(PROPYLNITROSAMINO)PROPYL ACETATE

TOXICITY DATA with REFERENCE
mmo-sat 100 nmol/plate MUREAV 49,187,78
scu-ham TDLo:410 mg/kg/33W-I:CAR ZKKOBW 90,127,77
scu-ham LD50:500 mg/kg ZKKOBW 90,127,77

SAFETY PROFILE: Moderately toxic by subcutaneous route. Mutation data reported. Questionable carcinogen with experimental carcinogenic data. See also N-NITROSO COMPOUNDS, NITROSAMINES, ESTERS, and AMINES. When heated to decomposition it emits toxic fumes of NO_x.

ABU000 CAS:51-98-9 ***HR: 3***
17-ACETOXY-19-NOR-17-α-PREGN-4-EN-20-YN-3-ONE
mf: $C_{22}H_{28}O_3$ mw: 340.50

SYNS: 17-β-ACETOXY-19-NOR-17-α-PREGN-4-EN-20-YN-3-ONE ◇ (17-α)-17-(ACETYLOXY)-19-NORPREGN-4-EN-20-YN-3-ONE ◇ 17-ACETYLOXY(17-α)-19-NORPREGN-4-ESTREN-17-β-OL-ACETATE-3-ONE ◇ 17-ENT ◇ 17-α-ETHINYL-19-NORTESTOSTERONE ACETATE ◇ 17-α-ETHINYL-19-NORTESTOSTERONE-17-β-ACETATE ◇ 17-α-ETHYNYL-17-β-ACETOXY-19-NORANDROST-4-EN-3-ONE ◇ 17-α-ETHYNYL-17-HYDROXYESTR-4-EN-3-ONE ACETATE ◇ 17-α-ETHYNYL-19-NORTESTOSTERONE ACETATE ◇ 17-HYDROXY-19-NOR-17-α-PREGN-4-EN-20-YN-3-ONE ACETATE ◇ 17-β-HYDROXY-19-NOR-17-α-PREGN-4-EN-20-YN-3-ONE ACETATE ◇ NORETHINDRONE-17-ACETATE ◇ 19-NORETHISTERONE ACETATE ◇ 19-NORETHYNYLTESTOSTERONE ACETATE ◇ NORETHYSTERONE ACETATE ◇ NORLUTATE ◇ NORLUTINE ACETATE ◇ ORLUTATE

TOXICITY DATA with REFERENCE
dlt-mus-orl 1120 mg/kg/4W MUREAV 26,535,74
spm-mus-orl 11200 mg/kg/4W MUREAV 26,535,74
orl-wmn TDLo:2190 μg/kg (52W pre):REP BMJOAE 2,730,69
unr-wmn TDLo:15 mg/kg (female 13-30W post):TER OBGNAS 22,210,63
orl-rat TDLo:303 μg/kg/2Y-C:ETA JTEHD6 6,895,80

CONSENSUS REPORTS: IARC Cancer Review: Animal Limited Evidence IMEMDT 21,441,79; Animal Sufficient Evidence IMEMDT 6,179,74. EPA Genetic Toxicology Program.

SAFETY PROFILE: Suspected carcinogen with experimental tumorigenic data. Human reproductive effects by ingestion and implant routes: menstrual cycle changes, postpartum effects and changes in fertility. A human teratogen by an unspecified route with developmental abnormalities of the urogenital system. Mutation data reported. When heated to decomposition it emits acrid smoke and irritating fumes. Used in the treatment of menstrual disorders and uterine bleeding.

ABU500 CAS:62-38-4 ***HR: 3***
ACETOXYPHENYLMERCURY
DOT: UN 1674
mf: $C_8H_8HgO_2$ mw: 336.75

PROP: Lustrous crystals, sltly sol in water. Mp: 149°.

SYNS: ACETATE PHENYLMERCURIQUE (FRENCH) ◇ (ACETATO)PHENYLMERCURY ◇ ACETIC ACID, PHENYLMERCURY DERIV. ◇ (ACETOXYMERCURI)BENZENE ◇ AGROSAN ◇ ALGIMYCIN ◇ ANTIMUCIN WDR ◇ BUFEN ◇ CEKUSIL ◇ CELMER ◇ CERESAN ◇ CONTRA CREME ◇ DYANACIDE ◇ FEMMA ◇ FENYLMERCURIACETAT (CZECH) ◇ FMA ◇ FUNGITOX OR ◇ GALLOTOX ◇ HL-331 ◇ HONG KIEN ◇ HOSTAQUICK ◇ KWIKSAN ◇ LEYTOSAN ◇ LIQUIPHENE ◇ MERCURIPHENYL ACETATE ◇ NORFORMS ◇ NYLMERATE ◇ OCTAN FENYLRTUTNATY (CZECH) ◇ PAMISAN ◇ PHENMAD ◇ PHENOMERCURIC ACETATE ◇ PHENYLMERCURIACETATE ◇ PHENYL MERCURIC ACETATE ◇ PHENYLMERCURY ACETATE ◇ PHENYLQUECKSILBERACETAT (GERMAN) ◇ PHIX ◇ PMA ◇ PMAC ◇ PMACETATE ◇ PMAL ◇ PMAS ◇ PURASAN-SC-10 ◇ PURATURF 10 ◇ QUICKSAN ◇ RCRA WASTE NUMBER P092 ◇ SANITIZED SPG ◇ SC-110 ◇ SEEDTOX ◇ SPOR-KIL ◇ TAG ◇ TAG FUNGICIDE ◇ ZIARNIK

TOXICITY DATA with REFERENCE
dnr-esc 2 mmol/L MJDHDW 28,F39,80
sce-ham:lym 30 mg/L DBABEF 8,105,84
scu-mus TDLo:110 μg/kg (8D preg):TER ARINAU 3,88,56
ipr-uns TDLo:125 μg/kg (female 8D post):REP TXCYAC 6,281,76
orl-rat LD50:22 mg/kg FMCHA2 -,C232,89
ivn-rat LDLo:20 mg/kg PEMNDP 8,658,87
orl-mus LD50:13250 μg/kg YAKUD5 22,291,80
ipr-mus LD50:13 mg/kg AMSVAZ 143,365,52
scu-mus LD50:12 mg/kg TOIZAG 9,101,62
ivn-mus LD50:18 mg/kg CSLNX* NX#00921

CONSENSUS REPORTS: EPA Extremely Hazardous Substances List. Reported in EPA TSCA Inventory. EPA Genetic Toxicology Program. Mercury and its compounds are on the Community Right-To-Know List.

OSHA PEL: (Transitional: CL 1 mg/10m^3) CL 0.1 mg(Hg)/m^3 (skin)
ACGIH TLV: TWA 0.1 mg(Hg)/m^3 (skin)
NIOSH REL: (Mercury, Inorganic) TWA 0.05 mg(Hg)/m^3
DOT Classification: IMO: Poison B; Label: Poison.

SAFETY PROFILE: Poison by ingestion, intravenous, intraperitoneal, subcutaneous, and possibly other routes. An experimental teratogen. Other experimental reproductive effects. Mutation data reported. See also MERCURY COMPOUNDS. When heated to decomposition it emits toxic fumes of Hg.

ABU800 CAS:2114-33-2 ***HR: 1***
2-ACETOXY-1-PHENYLPROPANE
mf: $C_{11}H_{14}O_2$ mw: 178.25

SYNS: ACETIC ACID, α-METHYL-PHENETHYL ESTER ◇ BENZYLMETHYLCARBINYL ACETATE ◇ METHYLBENZYLCARBINYL ACETATE ◇ α-METHYL-β-PHENYLETHYL ACETATE ◇ 1-PHENYL-2-PROPANOL ACETATE

TOXICITY DATA with REFERENCE
skn-rbt 500 mg/24H MLD FCTOD7 20,737,82

CONSENSUS REPORTS: Reported in EPA TSCA Inventory.

SAFETY PROFILE: A skin irritant. When heated to decomposition it emits acrid smoke and irritating fumes.

ABV250 CAS:17427-00-8 ***HR: 3***
3-ACETOXYPHENYLTRIMETHYLAMMONIUM IODIDE
mf: $C_{11}H_{16}NO_2 \cdot I$ mw: 321.18

SYN: NU 2017

TOXICITY DATA with REFERENCE
orl-mus LD50:800 mg/kg JPETAB 99,16,50

scu-mus LD50:125 mg/kg JPETAB 99,16,50
ivn-mus LD50:3700 μg/kg JPETAB 99,16,50

SAFETY PROFILE: A poison via subcutaneous and intravenous routes. When heated to decomposition it emits very toxic fumes of NO_x and I^-.

ABV500 CAS:64051-12-3 ***HR: 3***
3-ACETOXYPHENYL TRIMETHYLAMMONIUM METHYLSULFATE
mf: $C_{11}H_{16}NO_2 \cdot CH_3O_4S$ mw: 305.38

SYN: ACETIC ACID (m-TRIMETHYLAMMONIO)PHENYL ESTER METHYLSULFATE

TOXICITY DATA with REFERENCE
orl-mus LDLo:1000 mg/kg JPETAB 43,413,31
ivn-mus LDLo:8 mg/kg JPETAB 43,413,31

SAFETY PROFILE: Poison by intravenous route. Moderately toxic by ingestion. See also ESTERS and SULFATES. When heated to decomposition it emits very toxic fumes of NO_x and SO_x.

ABV600 ***HR: D***
17-α-ACETOXYPREGN-4-ENE-3-β-OL-20-ONE and MESTRANOL (20:1)
mf: $C_{23}H_{34}O_4 \cdot C_{21}H_{26}O_2$ mw: 685.04

SYN: PREGN-4-EN-20-ONE, 3-β,17-DIHYDROXY-, 17-ACETATE, and 3-METHOXY-19-NOR-17-α-PREGNA-1,3,5(10)-TRIEN-20-YN-17-OL(20:1)

TOXICITY DATA with REFERENCE
orl-rat TDLo:11760 μg/kg (female 14D pre):REP RPTOAN 48,174,85

SAFETY PROFILE: Experimental reproductive effects. When heated to decomposition it emits acrid smoke and irritating fumes.

ABV750 CAS:70103-78-5 ***HR: D***
N-(3-ACETOXYPROPYL)-N-(ACETOXYMETHYL)NITROSAMINE
mf: $C_8H_{14}N_2O_5$ mw: 218.21

SYNS: ACETICACID-4-((ACETOXYMETHYL)NITROSAMINO)PROPYL ESTER ◇ APAMN

TOXICITY DATA with REFERENCE
mmo-esc 1 μmol/plate GANNA2 71,124,80
mrc-bcs 200 nmol/plate GANNA2 71,124,80

SAFETY PROFILE: Mutation data reported. See also NITROSAMINES and ESTERS. When heated to decomposition it emits toxic fumes of NO_x.

ABW250 CAS:59901-90-5 ***HR: D***
1'-ACETOXYSAFROLE-2',3'-OXIDE
mf: $C_{12}H_{12}O_5$ mw: 236.24

SYNS: 1,3-BENZODIOXOLE-5-METHANOL, α-(OXIRANYL)-, ACETATE (ester) ◇ α-EPOXYETHYL-1,2-(METHYLENEDIOXY)BENZYL ALCOHOL ACETATE

TOXICITY DATA with REFERENCE
mmo-sat 800 nmol/plate MUREAV 60,143,79
mma-sat 250 μg/plate CNREA8 37,1883,77

CONSENSUS REPORTS: EPA Genetic Toxicology Program.

SAFETY PROFILE: Mutation data reported. See also ESTERS. When heated to decomposition it emits acrid smoke and irritating fumes.

ABW500 CAS:26594-44-5 ***HR: 3***
N-ACETOXY-N-(4-STILBENYL) ACETAMIDE
mf: $C_{18}H_{17}NO_3$ mw: 295.36

SYNS: ACETICACID-(N-ACETYL-N-(p-STYRYLPHENYL)AMINO) ESTER ◇ ACETIC ACID-ESTER with N-(p-STYRYLPHENYL)ACETOHYDROXAMIC ACID ◇ N-ACETOXY-4-ACETAMIDOSTILBENE ◇ N,O-DIACETYL-N-(p-STYRYLPHENYL)HYDROXYLAMINE ◇ N-(p-STYRYLPHENYL)ACETOHYDROXAMIC ACETATE ◇ N-(p-STYRYLPHENYL)ACETOHYDROXAMIC ACID ACETATE

TOXICITY DATA with REFERENCE
oms-omi 3 g/L CBINA8 26,47,79
dns-hmn:hla 100 nmol/L CNREA8 38,2621,78
mmo-sat 5 μg/plate CBINA8 26,11,79
skn-mus TDLo:2360 μg/kg:NEO JNCIAM 54,491,75

CONSENSUS REPORTS: EPA Genetic Toxicology Program.

SAFETY PROFILE: Questionable carcinogen with experimental neoplastigenic data by skin contact. Human mutation data reported. When heated to decomposition it emits highly toxic fumes of NO_x.

ABW600 CAS:13121-71-6 ***HR: 3***
ACETOXYTRICYCLOHEXYLSTANNANE
mf: $C_{20}H_{36}O_2Sn$ mw: 427.25

SYN: STANNANE, ACETOXYTRICYCLOHEXYL-

TOXICITY DATA with REFERENCE
orl-rat LD50:178 mg/kg PHARAT 37,801,82

OSHA PEL: TWA 0.1 mg(Sn)/m^3
ACGIH TLV: TWA 0.1 mg(Sn)/m^3 (skin)

SAFETY PROFILE: Poison by ingestion. When heated to decomposition it emits toxic fumes of Sn.

ABW750 CAS:1907-13-7 ***HR: 3***
ACETOXYTRIETHYLSTANNANE
mf: $C_8H_{18}O_2Sn$ mw: 264.95

SYNS: ACETOXYTRIETHYLTIN ◇ TIN TRIAETHYLZINNACETAT (GERMAN) ◇ TRIETHYLTIN ACETATE

TOXICITY DATA with REFERENCE
orl-rat LD50:4 mg/kg BJIMAG 15,15,58

ivn-rat LD50:4200 μg/kg BJIMAG 15,15,58
ivn-mus LD50:8 mg/kg CSLNX* NX#02839

OSHA PEL: TWA 0.1 mg(Sn)/m^3 (skin)
ACGIH TLV: TWA 0.1 mg(Sn)/m^3 (skin) (Proposed: TWA 0.1 mg(Sn)/m^3; STEL 0.2 mg(Sn)/m^3 (skin))
NIOSH REL: (Organotin Compounds) TWA 0.1 mg(Sn)/m^3

SAFETY PROFILE: Poison by ingestion and intravenous routes. See also TIN COMPOUNDS. When heated to decomposition it emits acrid smoke and irritating fumes.

ABX000 CAS:2897-46-3 ***HR: 3***
ACETOXYTRIHEXYLSTANNANE
mf: $C_{20}H_{42}O_2Sn$ mw: 433.31

SYNS: ACETOXYTRIHEXYLTIN ◇ TRIHEXYLTIN ACETATE ◇ TRI-N-HEXYLZINNACETAT (GERMAN)

TOXICITY DATA with REFERENCE
orl-rat LD50:1000 mg/kg BJIMAG 15,15,58
ivn-rat LDLo:6 mg/kg BJIMAG 15,15,58

OSHA PEL: TWA 0.1 mg(Sn)/m^3 (skin)
ACGIH TLV: TWA 0.1 mg(Sn)/m^3 (skin) (Proposed: TWA 0.1 mg(Sn)/m^3; STEL 0.2 mg(Sn)/m^3 (skin))
NIOSH REL: (Organotin Compounds) TWA 0.1 mg(Sn)/m^3

SAFETY PROFILE: Poison by intravenous route. Moderately toxic by ingestion. See also TIN COMPOUNDS. When heated to decomposition it emits acrid smoke and fumes.

ABX125 CAS:5711-19-3 ***HR: 3***
ACETOXYTRIMETHYLPLUMBANE
mf: $C_5H_{12}O_2Pb$ mw: 311.36

SYN: ACETATE de TRIMETHYLPLOMB (FRENCH)

TOXICITY DATA with REFERENCE
ipr-rat LD50:66 mg/kg APFRAD 24,17,66
orl-mus LD50:82 mg/kg APFRAD 24,17,66
ipr-mus LD50:34 mg/kg APFRAD 24,17,66

SAFETY PROFILE: Poison by ingestion and intraperitoneal routes. When heated to decomposition it emits toxic fumes of Pb.

ABX150 CAS:919-28-8 ***HR: 2***
ACETOXYTRIOCTYLSTANNANE
mf: $C_{26}H_{54}O_2Sn$ mw: 517.49

SYNS: STANNANE, ACETOXYTRIOCTYL- ◇ (ACETYLOXY)TRIOCTYLSTANNANE

TOXICITY DATA with REFERENCE
orl-rat LD50:30 g/kg PHARAT 37,801,82

OSHA PEL: TWA 0.1 mg(Sn)/m^3
ACGIH TLV: TWA 0.1 mg(Sn)/m^3 (skin)

SAFETY PROFILE: Slightly toxic by ingestion. When heated to decomposition it emits toxic fumes of Sn.

ABX175 CAS:2587-75-9 ***HR: 2***
ACETOXYTRIPENTYLSTANNANE
mf: $C_{17}H_{36}O_2Sn$ mw: 391.22

SYN: STANNANE, ACETOXYTRIPENTYL-

TOXICITY DATA with REFERENCE
orl-rat LD50:447 mg/kg PHARAT 37,801,82

OSHA PEL FINAL: TWA 0.1 mg(Sn)/m^3 (skin)
ACGIH TLV: TWA 0.1 mg(Sn)/m^3 (skin)

SAFETY PROFILE: Moderately toxic by ingestion. When heated to decomposition it emits toxic fumes of Sn.

ABX250 CAS:900-95-8 ***HR: 3***
ACETOXYTRIPHENYLSTANNANE
mf: $C_{20}H_{18}O_2Sn$ mw: 409.07

PROP: Practically insol, crystalline solid. Mp: 120°.

SYNS: ACETATE de TRIPHENYL-ETAIN (FRENCH) ◇ ACETATO di STAGNO TRIFENILE (ITALIAN) ◇ ACETATOTRIPHENYLSTANNANE ◇ ACETOXY-TRIPHENYL-STANNAN (GERMAN) ◇ ACETOXY-TRIPHENYLSTANNANE ◇ ACETOXYTRIPHENYLTIN ◇ (ACETYLOXY)TRIPHENYL-STANNANE (9CI) ◇ BATASAN ◇ BRESTAN ◇ ENT 25,208 ◇ FENOLOVO ACETATE ◇ FENTIN ACETAAT (DUTCH) ◇ FENTIN ACETAT (GERMAN) ◇ FENTIN ACETATE ◇ FENTINE ACETATE (FRENCH) ◇ FINTIN ACETATO (ITALIAN) ◇ GC 6936 ◇ HOE-2824 ◇ LIROMATIN ◇ LIROSTANOL ◇ PHENTIN ACETATE ◇ PHENTINOACETATE ◇ SUZU ◇ TINESTAN ◇ TINESTAN 60 WP ◇ TIN TRIPHENYL ACETATE ◇ TPTA ◇ TPZA ◇ TRIFENYLTINACETAAT (DUTCH) ◇ TRIPHENYLACETO STANNANE ◇ TRIPHENYLTIN ACETATE ◇ TRIPHENYL-ZINNACETAT (GERMAN) ◇ TUBOTIN ◇ VP 1940

TOXICITY DATA with REFERENCE
orl-rat TDLo:54 mg/kg (7-15D preg):REP NTOTDY 4,247,82
orl-rat TDLo:50 mg/kg:TER BECTA6 24,936,80
orl-mus TDLo:132 g/kg/78W-I:NEO NTIS** PB223-159
orl-rat LD50:125 mg/kg TIUSAD 43,9,58
skn-rat LD50:450 mg/kg ARZNAD 19,934,69
ipr-rat LD50:8500 μg/kg BJIMAG 23,222,66
ivn-rat LD50:18 mg/kg GUCHAZ 6,281,73
orl-mus LD50:81 mg/kg BJIMAG 23,222,66
ipr-mus LD50:719 mg/kg BJIMAG 23,222,66
scu-mus LD50:44 mg/kg GUCHAZ 6,281,73
ivn-mus LD50:18 mg/kg CSLNX* NX#00648
orl-rbt LD50:30 mg/kg 85DPAN -,-,71/76
ipr-rbt LD50:10 mg/kg ARZNAD 13,432,63
orl-gpg LD50:21 mg/kg 85GYAZ -,127,71

CONSENSUS REPORTS: EPA Extremely Hazardous Substances List. Reported in EPA TSCA Inventory.

OSHA PEL: TWA 0.1 mg(Sn)/m^3 (skin)
ACGIH TLV: TWA 0.1 mg(Sn)/m^3 (skin) (Proposed: TWA 0.1 mg(Sn)/m^3; STEL 0.2 mg(Sn)/m^3 (skin))
NIOSH REL: (Organotin Compounds) TWA 0.1 mg(Sn)/m^3

SAFETY PROFILE: Poison by ingestion, intraperitoneal, intravenous, and subcutaneous routes. Moderately toxic by skin contact. Questionable carcinogen with experimental neoplastigenic data. An experimental teratogen. Other experimental reproductive effects. A fungicide and algicide used as a wood preservative. When heated to decomposition it emits acrid smoke and fumes. See also TIN COMPOUNDS.

ABX325 CAS:13266-07-4 ***HR: 3***
ACETOXYTRIPROPYLPLUMBANE
mf: $C_{11}H_{24}O_2Pb$ mw: 395.54

SYN: ACETATE de TRIPROPYLPLOMB (FRENCH)

TOXICITY DATA with REFERENCE
orl-rat LD50:214 mg/kg APFRAD 24,17,66
ipr-rat LD50:17 mg/kg APFRAD 24,17,66
orl-mus LD50:236 mg/kg APFRAD 24,17,66
ipr-mus LD50:24 mg/kg APFRAD 24,17,66

SAFETY PROFILE: Poison by ingestion and intraperitoneal routes. When heated to decomposition it emits toxic fumes of Pb.

ABX500 CAS:97-44-9 ***HR: 3***
ACETPHENARSINE
mf: $C_8H_{10}AsNO_5$ mw: 275.11

PROP: Crystalline material, sltly water sol. Decomp @ 240-250°.

SYNS: 3-ACETAMIDO-4-HYDROXY-PHENYLARSONIC ACID ◇ ACETARSOL ◇ ACETARSONE ◇ 3-ACETYLAMINO-4-HYDROXYPHENYLARSONIC ACID ◇ (3-(ACETYLAMINO)-4-HYDROXYPHENYL)ARSONINE (9CI) ◇ N-ACETYL-4-HYDROXY-m-ARSANILIC ACID ◇ AMARSAN ◇ AMOEBAL ◇ ARSONIC ACID ◇ ARSPHEN ◇ DEVEGAN ◇ DISPARICIDA ◇ DYNARSAN ◇ EHRLICH 594 ◇ 190 F ◇ F 190 ◇ FOURNEAU 190 ◇ GINARSOL ◇ GOYL ◇ GYNOPLIX ◇ KHAROPHEN ◇ KUBARSOL ◇ LIMARSOL MALAGRIDE ◇ MEXYL ◇ MONARGAN ◇ NILACID ◇ ORALCID ◇ ORARSAN ◇ OSARSAL ◇ OSARSOLE ◇ OSVARSAN ◇ PALLICID ◇ PAROXYL ◇ SPIROCID ◇ SPIROZID ◇ STOVARSAL ◇ STOVARSOL ◇ STOVARSOLAN ◇ SVC ◇ VAGISEPT ◇ VAGOFLOR

TOXICITY DATA with REFERENCE
dnd-esc 20 μmol/L MUREAV 89,95,81
orl-wmn TDLo:86 mg/kg/8D:RSP,SKN,MET AJMSA9 174,819,27
orl-man TDLo:89 mg/kg/9D:RSP,END,SKN AJMSA9 174,819,27
ivg-wmn LDLo:155 mg/kg/2D-I:CNS,GIT,MET BMJOAE 1,1282,61
ivg-wmn LDLo:1576 mg/kg/2D-I:CNS,GIT,KID BMJOAE 2,242,60
ivn-rat LDLo:300 mg/kg ADSYAF 25,799,32
ivn-rbt LDLo:120 mg/kg ADSYAF 25,799,32
orl-mus LD50:4 mg/kg CLDND* NX#03309
ivn-mus LD50:180 mg/kg CSLNX* NX#03309
orl-cat LDLo:150 mg/kg PSEBAA 27,267,30
orl-rbt LDLo:125 mg/kg PSEBAA 27,267,30

CONSENSUS REPORTS: Arsenic and its compounds are on the Community Right-To-Know List.

OSHA PEL: TWA 500 μg(As)/m^3
ACGIH TLV: TWA 0.2 mg(As)/m^3

SAFETY PROFILE: Poison by ingestion and intravenous routes. Human systemic effects by ingestion: respiratory system, endocrine system, dermatitis and fever. Human systemic effects by intravaginal route: hallucinations, distorted perceptions, convulsions, nausea or vomiting, decreased urine volume, and fever. Mutation data reported. See also ARSENIC COMPOUNDS. When heated to decomposition it emits very toxic fumes of NO_x and As.

ABX750 CAS:123-54-6 ***HR: 3***
ACETYL ACETONE
DOT: UN 2310
mf: $C_5H_8O_2$ mw: 100.13

PROP: Colorless to sltly yellow liquid; pleasant odor. Mp: −23.2°, bp: 139° @ 746 mm, flash p: 105°F (OC), d: 0.952-0.962, refr index: 1.402, vap d: 3.45, autoign temp: 644°F. Misc in alc, ether, chloroform, acetone, glacial acetic acid, and propylene glycol; insol in glycerin and water.

SYNS: ACETOACETONE ◇ DIACETYLMETHANE ◇ FEMA No. 2841 ◇ PENTANEDIONE ◇ 2,4-PENTANEDIONE (FCC)

TOXICITY DATA with REFERENCE
skn-rbt 10 mg/24H JIHTAB 26,269,44
skn-rbt 488 mg open MLD UCDS** 7/8/71
eye-rbt 4760 μg SEV AJOPAA 29,1363,46
orl-rat LD50:1000 mg/kg JIHTAB 26,269,44
ihl-rat LCLo:1000 ppm/4H JIHTAB 31,343,49
ipr-mus LD50:750 mg/kg NTIS** AD691-490
skn-rbt LD50:5000 mg/kg UCDS** 7/8/71
ipr-rat LDLo:400 mg/kg BCPCA6 13,285,64

CONSENSUS REPORTS: Reported in EPA TSCA Inventory.

DOT Classification: Flammable or Combustible Liquid; Label: Flammable Liquid.

SAFETY PROFILE: Moderately toxic via ingestion, intraperitoneal and inhalation routes. A skin and severe eye irritant. Flammable liquid when exposed to heat or flame. Incompatible with oxidizing materials. To fight fire, use alcohol foam, CO_2, dry chemical.

ABX800 CAS:78600-25-6 ***HR: 3***
3-ACETYLACONITINE HYDROBROMIDE
mf: $C_{36}H_{49}NO_{12}•BrH$ mw: 768.78

TOXICITY DATA with REFERENCE
orl-mus LD50:2500 μg/kg CYLPDN 2(2),82,81
ipr-mus LD50:700 μg/kg CYLPDN 2(2),82,81
scu-mus LD50:1400 μg/kg CYLPDN 2(2),82,81

SAFETY PROFILE: Poison by ingestion, subcutaneous and intraperitoneal routes. When heated to decomposition it emits toxic fumes of NO_x and HBr.

ABY000 CAS:28322-02-3 ***HR: 3***
4-ACETYLAMINOFLUORENE
mf: $C_{15}H_{13}NO$ mw: 223.29

SYNS: 4-ACETYLAMINOFLUOREN (GERMAN) ◇ N-FLUOREN-4-YLACETAMIDE ◇ N-4-FLUORENYLACETAMIDE

TOXICITY DATA with REFERENCE
mma-sat 50 μg/plate PMRSDJ 1,285,81
otr-ham:kdy 25 mg/L PMRSDJ 1,638,81
orl-rat TDLo:4175 mg/kg/17W-C:ETA ONCOAR 8,233,55
orl-rat TD:5240 mg/kg/57W-C:ETA,REP JNCIAM 24,149,60
ipr-mus LD50:364 mg/kg PMRSDJ 1,682,81

CONSENSUS REPORTS: EPA Genetic Toxicology Program.

SAFETY PROFILE: Poison by intraperitoneal route. Questionable carcinogen with experimental tumorigenic data. Experimental reproductive effects. Mutation data reported. When heated to decomposition it emits toxic fumes of NO_x.

ABY150 CAS:57229-41-1 ***HR: 2***
2-ACETYLAMINO-9-FLUORENOL
mf: $C_{15}H_{13}NO_2$ mw: 239.29

SYNS: N-(9-HYDROXYFLUOREN-2-YL)ACETAMIDE ◇ 9-HYDROXY-2-FLUORENYLACETAMIDE

TOXICITY DATA with REFERENCE
orl-rat TDLo:4644 mg/kg/32W-C:ETA CNREA8 15,188,55

SAFETY PROFILE: Questionable carcinogen with experimental tumorigenic data. When heated to decomposition it emits toxic fumes of NO_x.

ABY250 CAS:3096-50-2 ***HR: 3***
2-ACETYLAMINOFLUORENONE
mf: $C_{15}H_{11}NO_2$ mw: 237.27

SYNS: 2-ACETYLAMINO-9-FLUORENONE ◇ 9-OXO-2-FLUORENYLACETAMIDE ◇ N-(9-OXO-2-FLUORENYL)ACETAMIDE

TOXICITY DATA with REFERENCE
orl-rat TDLo:4740 mg/kg/32W-C:CAR CNREA8 15,188,55
orl-rat TD:6075 mg/kg/65W-C:CAR JNCIAM 24,149,60

SAFETY PROFILE: Questionable carcinogen with experimental carcinogenic data. When heated to decomposition it emits toxic fumes of NO_x.

ABY900 CAS:140-40-9 ***HR: 3***
2-ACETYLAMINO-5-NITROTHIAZOLE
mf: $C_5H_5N_3O_3S$ mw: 187.19

PROP: Needles from alc, elongated plates from acetic acid. Mp: 264-265°. The commercial product may be yellow. Sol in aq solns of NaOH and NH_3 with deep orange color.

SYNS: ACETAMIDO-5-NITROTHIAZOLE ◇ ACINITRAZOLE ◇ AMINITROZOLE ◇ ENHEPTIN-A ◇ GYNOFON ◇ N-(5-NITRO-2-THIAZOLYL)ACETAMIDE ◇ PLEOCIDE ◇ TRICHORAD ◇ TRICHORAL ◇ TRITHEOM

TOXICITY DATA with REFERENCE
mmo-sat 500 nmol/L MUREAV 118,153,83
mmo-esc 20 μmol/L MUREAV 118,153,83
mmo-klp 20 μmol/L MUREAV 118,153,83
mrc-smc 200 ppm MUREAV 118,153,83
orl-mus LD50:1000 mg/kg FRPSAX 19,301,64
orl-dog LD50:125 mg/kg ANTCAO 5,540,55
orl-ckn LD50:800 mg/kg ANTCAO 5,540,55
orl-trk LD50:600 mg/kg ANTCAO 5,540,55

SAFETY PROFILE: Poison by ingestion. Mutation data reported. When heated to decomposition it emits toxic fumes of SO_x and NO_x.

ACA125 ***HR: 3***
5-ACETYLAMINO-2,4,6-TRIIODO ISOPHTHALIC ACID DI-(N-METHYL-2,3-DIHYDROXYPROPYLAMIDE)
mf: $C_{18}H_{24}I_3N_3O_7$ mw: 775.15

TOXICITY DATA with REFERENCE
ivn-mus LD50:31965 mg/kg USXXAM #4001323
ice-mus LD50:1670 mg/kg USXXAM #4001323
par-rbt LD50:165 mg/kg USXXAM #4001323

SAFETY PROFILE: Poison by parenteral route. Moderately toxic by intracerebral route. When heated to decomposition it emits toxic fumes of I^- and NO_x.

ACA750 CAS:73637-16-8 ***HR: 3***
9-ACETYL-1,7,8-ANTHRACENETRIOL
mf: $C_{16}H_{11}O_4$ mw: 267.27

SYNS: 10-ACETYL-1,8,9-ANTHRACENETRIOL ◇ 10-ACETYLANTHRALIN ◇ 1,8-DIHYDROXY-10-ACETYL-9-ANTHRONE

TOXICITY DATA with REFERENCE
skn-mus TDLo:32 mg/kg/53W-I:NEO JMCMAR 21,26,78

SAFETY PROFILE: Questionable carcinogen with experimental neoplastigenic data. When heated to decomposition it emits acrid smoke and fumes.

ACB000 ***HR: 3***
ACETYL AZIDE
mf: $C_2H_3N_3O$ mw: 85.07

SAFETY PROFILE: A sensitive explosive. See also AZIDES and EXPLOSIVES.

ACB250 CAS:460-07-1 ***HR: 3***
1-ACETYLAZIRIDINE
mf: C_4H_7NO mw: 85.12

SYN: ACETYLETHYLENEIMINE

TOXICITY DATA with REFERENCE
scu-rat TDLo:78 mg/kg/26W-I:NEO BJPCAL 9,306,54
scu-rat TD:80 mg/kg/16W-I:ETA BJPCAL 9,306,54
ipr-mus LD50:13 mg/kg NCISA* PH-43-63-1132

CONSENSUS REPORTS: Reported in EPA TSCA Inventory. EPA Genetic Toxicology Program.

SAFETY PROFILE: Poison by intraperitoneal route. Questionable carcinogen with experimental tumorigenic and neoplastigenic data. When heated to decomposition it emits toxic fumes of NO_x.

ACB750 ***HR: 3***
4-(p-(p-ACETYLBENZAMIDO)ANILINO)-6-AMINO-1-METHYLQUINOLINIUM)-p-AMIDINOHYDRAZONE-p-TOLUENESULFONATE-MONO-p-TOLUENSULFONATE
mf: $C_{26}H_{27}N_8O{\bullet}C_7H_7O_3S{\bullet}C_7H_8O_3S$ mw: 811.02

TOXICITY DATA with REFERENCE
dnd-mus:lym 400 nmol/L JMCMAR 22,134,79
ipr-mus LD10:40 mg/kg JMCMAR 22,134,79

SAFETY PROFILE: Poison by intraperitoneal route. See also SULFONATES. Mutation data reported. When heated to decomposition it emits very toxic fumes of NO_x and SO_x.

ACC000 CAS:3366-61-8 ***HR: 2***
N-ACETYLBENZIDINE
mf: $C_{14}H_{14}N_2O$ mw: 226.30

SYN: 4′-ACETAMIDOBENZIDINE ◇ N-(4′-AMINO(1,1′-BIPHENYL)-4-YL)-ACETAMIDE ◇ 4′-(p-AMINOPHENYL)ACETANILIDE

TOXICITY DATA with REFERENCE
mma-sat 5 μg/plate ENMUDM 6,145,84
dnd-rat-ipr 25 mg/kg CNREA8 42,2678,82
dnd-rat:lvr 100 mg/L CRNGDP 5,407,84
orl-rat LD50:1630 mg/kg 28ZPAK -,131,72

SAFETY PROFILE: Moderately toxic by ingestion. Mutation data reported. When heated to decomposition it emits toxic fumes of NO_x.

ACC250 CAS:644-31-5 ***HR: 3***
ACETYL BENZOYL PEROXIDE (solid)
DOT: UN 2081
mf: $C_9H_8O_4$ mw: 180.17

PROP: White crystals. Sol in oils, alc, ether, and chloroform. Mp: 36-37°, bp: 130° @ 19 mm.

DOT Classification: Forbidden (solid or > 40% in solution); Organic Peroxide; Label: Organic Peroxide (UN2081)

SAFETY PROFILE: Poison by inhalation and ingestion. Severe irritant. A powerful oxidizing agent which is corrosive to the skin and mucous membranes. See also PEROXIDES, ORGANIC. Dangerous; shock or heat will cause detonation with evolution of toxic fumes; will react with water or steam to produce heat; can react vigorously with reducing materials. Flammable by spontaneous chemical reaction. To fight fire, use CO_2 or dry chemical. When heated to decomposition it emits acrid smoke and fumes.

ACC500 CAS:644-31-5 ***HR: 3***
ACETYL BENZOYL PEROXIDE (solution)
DOT: UN 2081

PROP: Solution contains not over 40% acetyl benzoyl peroxide (FEREAC 41,15972,76).

DOT Classification: Organic Peroxide; Label: Organic Peroxide.

SAFETY PROFILE: Highly irritating to skin, eyes, and mucous membranes. When heated to decomposition it emits acrid smoke and fumes. See also ACETYL BENZOYL PEROXIDE (solid).

ACC750 CAS:63018-98-4 ***HR: 3***
2-ACETYL-3:4-BENZPHENANTHRENE
mf: $C_{20}H_{14}O$ mw: 270.34

SYN: 5-ACETYLBENZO(C)PHENANTHRENE

TOXICITY DATA with REFERENCE
skn-mus TDLo:720 mg/kg/30W-I:ETA PRLBA4 131,170,42

SAFETY PROFILE: Questionable carcinogen with experimental tumorigenic data. When heated to decomposition it emits acrid smoke and irritating fumes.

ACD000 CAS:4463-22-3 ***HR: 3***
N-ACETYL-4-BIPHENYLHYDROXYLAMINE
mf: $C_{14}H_{13}NO_2$ mw: 227.28

SYNS: 4-BIPHENYLACETHYDROXAMIC ACID ◇ N-HYDROXY-AABP ◇ N-HYDROXY-4-ACETAMIDOBIPHENYL ◇ N-4-(N-HYDROXYACETAMIDO)BIPHENYL ◇ N-HYDROXY-4-ACETAMIDO-DIPHENYL ◇ N-HYDROXY-4-ACETYLAMINOBIPHENYL ◇ N-HYDROXY-N-4-BIPHENYLACETAMIDE

TOXICITY DATA with REFERENCE
mnt-ham:ovr 290 μmol/L MUREAV 88,397,81
dns-hmn:oth 1 μmol/L JJIND8 72,847,84
dnd-rat-ipr 25 mg/kg COINAV 256,115,77
dns-rat:oth 10 μmol/L CNREA8 43,3974,82
dns-mus:oth 10 μmol/L CNREA8 43,3974,82
dns-rbt:oth 10 μmol/L CNREA8 45,221,85
dns-dog:oth 1 μmol/L CNREA8 42,3974,82
sce-ham:ovr 1440 μmol/L MUREAV 88,397,81
ipr-rat TDLo:91 mg/kg:CAR CNREA8 41,2450,81
orl-rat TDLo:2400 mg/kg/16W-C:ETA CNREA8 21,1465,61

SAFETY PROFILE: Questionable carcinogen with experimental carcinogenic and tumorigenic data. Human mutation data reported. When heated to decomposition it emits toxic fumes of NO_x.

ACD250 CAS:3733-45-7 ***HR: 3***
N-(N-ACETYL-3-(p-(BIS(2-CHLOROETHYL) AMINO)PHENYL)ALANYL-3-PHENYLALANINE ETHYL ESTER

SYN: ETHYL ESTER of N-ACETYL-dl-SARCOLYSYL-l-PHENYLALANINE

TOXICITY DATA with REFERENCE
orl-rat LD50:115 mg/kg FATOAO 33,472,70
ims-rat LD50:33 mg/kg FATOAO 33,472,70
rec-rat LD50:64 mg/kg FATOAO 33,472,70

SAFETY PROFILE: Poison by ingestion and intramuscular routes. When heated to decomposition it emits very toxic fumes of Cl^- and NO_x. See also ESTERS.

ACD500 CAS:18869-73-3 ***HR: 3***
1-ACETYL-3,3-BIS(p-HYDROXYPHENYL) OXINDOLE DIACETATE
mf: $C_{26}H_{21}NO_6$ mw: 443.48

SYNS: 1-ACETYL-3,3-BIS(4-(ACETYLOXY)PHENYL)-1,3-DIHYDRO-2H-INDOL-2-ONE ◇ ISATEX ◇ LAXAGEN ◇ LAXAGETTEN ◇ PHENISATIN ◇ TRIACETYLDIPHENOLISATIN ◇ TRISATIN ◇ UNILAX

TOXICITY DATA with REFERENCE
orl-rat LD50:500 mg/kg JAPMA8 42,468,53
ipr-rat LD50:350 mg/kg JAPMA8 42,468,53

CONSENSUS REPORTS: Reported in EPA TSCA Inventory.

SAFETY PROFILE: Poison by intraperitoneal route. Moderately toxic by ingestion. When heated to decomposition it emits toxic fumes of NO_x. A cathartic.

ACD750 CAS:506-96-7 ***HR: 3***
ACETYL BROMIDE
DOT: UN 1716
mf: C_2H_3BrO mw: 122.96

PROP: Colorless, fuming liquid; turns yellow in air. Mp: −96.5°, bp: 76.7°, d: 1.52 @ 9.5°/4°. Decomp in water and alc; misc in benzene, ether, and chloroform.

TOXICITY DATA with REFERENCE
ihl-mam LC50:48 g/m^3 GTPZAB 18(4)55,74
ipr-mam LD50:250 mg/kg GTPZAB 18(4)55,74

CONSENSUS REPORTS: Reported in EPA TSCA Inventory.

DOT Classification: Corrosive Material; Label: Corrosive.

SAFETY PROFILE: Poison by inhalation, ingestion, skin contact, and intraperitoneal routes. See also HYDROBROMIC ACID and ACETIC ACID. Violent reaction on contact with water, steam, methanol, or ethanol produces toxic and reactive HBr. When heated to decomposition it emits highly corrosive and toxic fumes of carbonyl bromide and bromine. To fight fire, use dry chemical, CO_2.

ACE000 CAS:77-66-7 ***HR: 2***
1-ACETYL-3-(2-BROMO-2-ETHYLBUTYRYL)UREA
mf: $C_9H_{15}BrN_2O_3$ mw: 279.17

SYNS: ABASIN ◇ ABSIN ◇ ACECARBROMAL ◇ ACETCARBROMAL ◇ ACETKARBROMAL ◇ ACETYL ADALIN ◇ N-((ACETYLAMINO) CARBONYL)-2-BROMO-2-ETHYLBUTANAMIDE ◇ ACETYLBROMODIETHYLACETYLCARBAMIDE ◇ N-ACETYL-N-BROMODIETHYL-ACETYLCARBAMIDE ◇ N-ACETYL-N-BROMODIETHYL-ACETYLUREA ◇ N-ACETYL-N'-α-BROMO-α-ETHYLBUTYRYL CARBAMIDE ◇ 1-ACETYL-3-(α-BROMO-α-ETHYLBUTYRYL)UREA ◇ ACETYLCARBROMAL ◇ ADITYL ◇ CARBASED ◇ DAROLON ◇ IBATRAN ◇ PAXAREL ◇ SEDAMYL ◇ SEDMYNOL ◇ SEDTRAN

TOXICITY DATA with REFERENCE
orl-hmn TDLo:7 mg/kg:PSY 27ZQAG -,423,72
orl-mus LD50:1600 mg/kg CLDND*

CONSENSUS REPORTS: Reported in EPA TSCA Inventory.

SAFETY PROFILE: Moderately toxic by injection. Human systemic effects by ingestion: toxic psychosis. When heated to decomposition it emits very toxic fumes of Br^- and NO_x. A sedative.

ACE250 CAS:77372-67-9 ***HR: D***
o-ACETYL-N-(p-BUTOXYPHENYLACETYL)-HYDROXYLAMINE
mf: $C_{14}H_{19}NO_4$ mw: 265.16

SYN: 4-N-BUTOXYPHENYLACETOHYDROXAMICACID-o-ACETATE ESTER

TOXICITY DATA with REFERENCE
mmo-sat nmol/plate PAACA3 21,126,80
sce-ham:ovr 20 μmol/L PAACA3 21,126,80

SAFETY PROFILE: Mutation data reported. See also ESTERS. When heated to decomposition it emits very toxic fumes of NO_x.

ACE500 CAS:2813-95-8 ***HR: 3***
o-ACETYL-2-sec-BUTYL-4,6-DINITROPHENOL
mf: $C_{12}H_{14}N_2O_6$ mw: 282.28

SYNS: ACETIC ACID-(2,4-DINITRO-6-sec-BUTYLPHENYL) ESTER ◇ ACETIC ACID-(4,6-DINITRO-2-sec-BUTYLPHENYL) ESTER ◇ ARETIT ◇ 2-sec-BUTYL-4,6-DINITROPHENYLACETATE ◇ 6-sec-BUTYL-2,4-DINITROPHENYLACETATE ◇ 2,4-DINITRO-6-sec-BUTYLFENYLESTER KYSELINY OCTOVE (CZECH) ◇ 2,4-DINITRO-6-sek.BUTYL-PHENYLACETAT (GERMAN) ◇ 4,6-DINITRO-2-sec-BUTYLPHENYL ACETATE ◇ DINOSEB-ACETATE ◇ HOE 2904 ◇ β-(2-HYDROXY-3,5-DINITROPHENYL)BUTANE ACETATE ◇ IVOSIT ◇ 2-(1-METHYLPROPYL)-4,6-DINITROPHENYL ACETATE ◇ PHENOTAN

TOXICITY DATA with REFERENCE
skn-rbt 500 mg/24H MOD 28ZPAK -,131,72
eye-rbt 500 mg/24H MOD 28ZPAK -,131,72
orl-rat LD50:48500 μg/kg 28ZPAK -,131,72
orl-rat LD50:60 mg/kg FMCHA2 -,D22,80
orl-ckn LD50:40 mg/kg GUCHAZ 6,229,73

SAFETY PROFILE: Poison by ingestion. A skin and eye irritant. See also ESTERS and NITRO COMPOUNDS of AROMATIC HYDROCARBONS. When heated to decomposition it emits toxic fumes of NO_x. A herbicide.

ACF000 CAS:36573-63-4 ***HR: 3***
3'-o-ACETYLCALOTROPIN
mf: $C_{31}H_{42}O_{10}$ mw: 574.73

PROP: A glycoside isolated *Asclepius cunssuica* (ARZNAD 28,1095,78).

SYN: ASCLEPIN

TOXICITY DATA with REFERENCE
ipr-mus LDLo:15 mg/kg ARZNAD 28,1095,78
ivn-pgn LDLo:400 μg/kg ARZNAD 28,1095,78
scu-frg LDLo:5 mg/kg ARZNAD 28,1095,78

SAFETY PROFILE: Poison by intravenous, subcutaneous, and intraperitoneal routes. When heated to decomposition it emits acrid smoke and irritating fumes.

ACF250 CAS:80449-58-7 ***HR: 1***
ACETYL CEDRENE

PROP: Prepared by acetylation of the hydrocarbon portion of cedarwood oil in the presence of an acid catalyst.

SYN: VERTOFIX COEUR

TOXICITY DATA with REFERENCE
skn-rbt 500 mg/24H MOD FCTXAV 16,637,78
orl-rat LD50:5200 mg/kg FCTXAV 16,637,78

SAFETY PROFILE: Mildly toxic by ingestion. A skin irritant. When heated to decomposition it emits acrid smoke and fumes.

ACF750 CAS:75-36-5 ***HR: 3***
ACETYL CHLORIDE
DOT: UN 1717
mf: C_2H_3ClO mw: 78.50

PROP: Colorless, fuming liquid. Mp: −112°, bp: 51-52°, flash p: 40°F (CC), autoign temp: 734°F, d: 1.1051 @ 20°/4°, vap d: 2.70. lel: 5%. Decomp in water and alc; misc in benzene, ether, and chloroform.

SYNS: ACETIC ACID CHLORIDE ◇ ACETIC CHLORIDE ◇ ETHANOYL CHLORIDE ◇ RCRA WASTE NUMBER U006

TOXICITY DATA with REFERENCE
ihl-hmn TCLo:2 ppm/1M:IRR TGNCDL 2,28,61

CONSENSUS REPORTS: Reported in EPA TSCA Inventory.

DOT Classification: Flammable Liquid; Label: Flammable Liquid, Corrosive.

SAFETY PROFILE: Poison by inhalation and ingestion. A human systemic irritant by inhalation. Violent hydrolysis reaction with water or steam produces heat, acetic acid, HCl, and other corrosive chlorides. May decompose during preparation. Dangerous fire hazard when exposed to heat or flame. Explosion hazard by spontaneous chemical reaction with dimethyl sulfoxide or ethanol. Also incompatible with PCl_3. When heated to decomposition it emits highly toxic fumes of phosgene and Cl^-. To fight fire, use CO_2 or dry chemical. See also CHLORIDES.

ACG125 CAS:39426-77-2 ***HR: 3***
1-(3-(3-ACETYL-4-(p-(CHLOROPHENYL)PIPERIDINO)PROPYL)-4-METHYLPIPERAZINE TRIHYDROCHLORIDE
mf: $C_{21}H_{32}ClN_3O \cdot 3ClH$ mw: 487.39

SYNS: 4-ACETYL-4-(3-CHLOROPHENYL)-1-(3-(4-METHYLPIPERAZINO)-PROPYL)PIPERIDINETRIHYDROCHLORIDE ◇ TROJCHLOROWODOREK 4-ACETYLO-4-(3-CHLOROFENYLO)-1-(3-(4-METYLOPIPERAZYNO)-PROPYLO)-PIPERYDYNY

TOXICITY DATA with REFERENCE
orl-mus LD50:2290 mg/kg APPHAX 37,579,80
scu-mus LD50:2694 mg/kg APPHAX 37,579,80
ivn-mus LD50:148 mg/kg APPHAX 37,579,80

SAFETY PROFILE: Poison by intravenous route. Moderately toxic by ingestion and other routes. When heated to decomposition it emits toxic fumes of NO_x and HCl.

ACG250 CAS:38838-26-5 ***HR: 3***
N-ACETYL COLCHINOL
mf: $C_{20}H_{23}NO_5$ mw: 357.44

SYNS: N-AZETYL-COLCHINOL (GERMAN) ◇ (S)-N-(3-HYDROXY-9,10,11-TRIMETHOXY-5H-DIBENZO(a,c)CYCLOHEPTEN-5-YL)-ACETAMIDE (9CI)

TOXICITY DATA with REFERENCE
oms-mus-ipr 28 mg/kg CANCAR 3,130,50
oms-mus-par 56 mg/kg CANCAR 3,130,50
spm-mus-par 56 mg/kg CANCAR 3,130,50
unk-rat LDLo:200 mg/kg CANCAR 3,124,50
ipr-mus LD50:56 mg/kg CANCAR 3,124,50
unk-cat LDLo:10 mg/kg CANCAR 3,124,50

SAFETY PROFILE: Poison by intraperitoneal and other unspecified routes. Mutation data reported. When heated to decomposition it emits toxic fumes of NO_x.

ACH000 CAS:616-91-1 ***HR: 3***
N-ACETYL-l-CYSTEINE
mf: $C_5H_9NO_3S$ mw: 163.21

SYNS: l-α-ACETAMIDO-β-MERCAPTOPROPIONIC ACID ◇ ACETEIN ◇ N-ACETYL-l-CYSTEINE (9CI) ◇ ACETYLCYSTEINE ◇ N-ACETYLCYSTEINE ◇ N-ACETYL-N-CYSTEINE ◇ N-ACETYL-3-MERCAPTOALANINE ◇ AIRBRON ◇ BRONCHOLYSIN ◇ FLUIMUCETIN ◇ FLUIMUCIL ◇ FLUMICIL ◇ INSPIR ◇ MERCAPTURIC ACID ◇ (R)-MERCAPTURIC ACID ◇ MUCOLYTICUM ◇ MUCOLYTICUM LAPPE ◇ MUCOMYST ◇ MUCOSOLVIN ◇ NAC ◇ NAC-TB ◇ NSC 111180 ◇ PARVOLEX ◇ RESPAIRE

TOXICITY DATA with REFERENCE
mma-sat 8 μg/plate CRNGDP 7,431,86
orl-rat LD50:5050 mg/kg TXAPA9 18,185,71
ivn-rat LD50:1140 mg/kg EJRDD2 61(Suppl 111),45,80
orl-mus LD50:7888 mg/kg THEWA6 30,1926,80
ivn-mus LD50:3800 mg/kg JMCMAR 10,1172,67
ipr-mus LD50:400 mg/kg NTIS** AD691-490

CONSENSUS REPORTS: Reported in EPA TSCA Inventory.

SAFETY PROFILE: Poison by intraperitoneal route. Moderately toxic by other routes. Mutation data reported. When heated to decomposition it emits very toxic fumes of NO_x and SO_x.

ACH075 CAS:50722-38-8 ***HR: 3***
3-ACETYLDEOXYNIVALENOL
mf: $C_{17}H_{22}O_7$ mw: 338.39

SYNS: DEHYDRONIVALENOL MONOACETATE ◇ DEOXYNIVALENOL MONOACETATE

TOXICITY DATA with REFERENCE
skn-gpg 3384 ng MLD FAATDF 4(2, Pt 2),S124,84
orl-mus LD50:34 mg/kg FAATDF 4(2, Pt 2),S124,84
ipr-mus LD50:47 mg/kg 41KEAL -,108,78

SAFETY PROFILE: Poison by ingestion and intraperitoneal routes.

ACH125 CAS:59183-18-5 ***HR: 3***
ACETYL-1,1-DICHLOROETHYL PEROXIDE
mf: $C_4H_6Cl_2O_3$ mw: 173.00

$CH_3CO{\bullet}OOCCl_2CH_3$

SAFETY PROFILE: A viscous liquid explosive, sensitive to friction and heat. When heated to decomposition it emits toxic fumes of Cl^-. See also PEROXIDES.

ACH250 CAS:69225-98-5 ***HR: 3***
1-ACETYL-3-(2,2-DICHLOROETHYL)UREA
mf: $C_5H_8Cl_2N_2O_2$ mw: 199.05

TOXICITY DATA with REFERENCE
ipr-mus LDLo:300 mg/kg JPETAB 42,1,31
scu-rbt LDLo:800 mg/kg JPETAB 42,1,31

SAFETY PROFILE: Poison by intraperitoneal route. Moderately toxic by subcutaneous route. When heated to decomposition it emits very toxic fumes of Cl^- and NO_x.

ACH375 CAS:73987-00-5 ***HR: 3***
16-ACETYLDIGITALINUM VERUM

SYN: 16-ACETATE DIGITOXIN

TOXICITY DATA with REFERENCE
ivn-cat LDLo:255 μg/kg JMPCAS 5,988,62
orl-frg LD50:36800 μg/kg JJPAAZ 9,91,60
scu-frg LD50:490 μg/kg JJPAAZ 9,91,60

SAFETY PROFILE: Poison by ingestion, subcutaneous and intravenous routes.

ACH500 CAS:1111-39-3 ***HR: 3***
ACETYLDIGITOXIN-α
mf: $C_{43}H_{66}O_{14}$ mw: 807.09

SYNS: α-ACETYLDIGITOXIN ◇ ACYLANID

TOXICITY DATA with REFERENCE
orl-cat LD50:250 μg/kg AIPTAK 159,1,66
ivn-cat LD50:514 μg/kg JPETAB 111,365,54

SAFETY PROFILE: Poison by ingestion and intrave-

nous routes. When heated to decomposition it emits acrid smoke and fumes. See also DIGITOXIN.

ACH750 CAS:1264-51-3 ***HR: 3***
ACETYLDIGITOXIN-β
mf: $C_{43}H_{66}O_{14}$ mw: 807.09

SYNS: β-ACETYLDIGOXIN ◇ DIGITOXIGENIN + 2-DIGITOXOSE + 1-ACETYL-(4)-DIGITOSE (GERMAN) ◇ DIGITOXIGENIN + 2-DIGITOXOSE + ACETYL-(3)-DIGITOXOSE (GERMAN)

TOXICITY DATA with REFERENCE
ivn-cat LD50:476 μg/kg JPETAB 111,365,54
orl-gpg LD50:50 mg/kg AIPTAK 159,1,66
ivn-gpg LDLo:1750 μg/kg ARZNAD 15,481,65

SAFETY PROFILE: Poison by ingestion and intravenous routes. When heated to decomposition it emits acrid smoke and fumes. See also DIGITOXIN.

ACI000 CAS:5511-98-8 ***HR: 3***
ACETYLDIGOXIN-α
mf: $C_{43}H_{66}O_{15}$ mw: 823.09

SYNS: α-ACETYLDIGOXIN ◇ DIGORID A ◇ DIGOXIGENIN + ZUCKERKETTE WIE BIE ACETYL-DIGITOXIN A (GERMAN)

TOXICITY DATA with REFERENCE
orl-cat LD50:200 μg/kg AIPTAK 159,1,66
ivn-cat LD50:466 μg/kg JPETAB 111,365,54
idu-cat LDLo:494 μg/kg ARZNAD 20,1765,70
orl-gpg LD50:3300 μg/kg AIPTAK 159,1,66
ivn-gpg LDLo:1380 μg/kg ARZNAD 15,483,65

SAFETY PROFILE: Deadly poison by ingestion, intravenous, and intraduodenal routes. When heated to decomposition it emits acrid smoke and fumes. See also DIGITOXIN.

ACI250 CAS:5355-48-6 ***HR: 3***
ACETYLDIGOXIN-β
mf: $C_{43}H_{66}O_{15}$ mw: 823.09

SYNS: β-ACETYLDIGOXIN ◇ DIGORID B ◇ DIGOXIGENIN + ZUCKERKETTE WIE BEI ACETYL-DIGITOXIN-α (GERMAN) ◇ HEXAMETHYLENEIMINE-3,5-DINITROBENZOATE

TOXICITY DATA with REFERENCE
orl-dog LD50:422 μg/kg ARZNAD 24,1914,74
ivn-cat LD50:430 μg/kg JPETAB 111,365,54
idu-cat LDLo:413 μg/kg ARZNAD 19,687,69
orl-gpg LD50:2400 μg/kg ARZNAD 15,481,65
ivn-gpg LDLo:1500 μg/kg ARZNAD 15,481,65

SAFETY PROFILE: Deadly poison by ingestion, intravenous, and intraduodenal routes. When heated to decomposition it emits acrid smoke and fumes. See also DIGITOXIN.

ACI375 CAS:21380-82-5 ***HR: 3***
ACETYLDIMETHYLARSINE
mf: C_4H_9AsO mw: 148.04

$(CH_2COOH)As(CH_3)_2$

CONSENSUS REPORTS: Arsenic and its compounds are on the Community Right-To-Know List.

SAFETY PROFILE: A poison. Ignites on contact with air. When heated to decomposition it emits toxic fumes of As. See also ARSENIC COMPOUNDS.

ACI500 CAS:2386-25-6 ***HR: 3***
3-ACETYL-2,4-DIMETHYLPYRROLE
mf: $C_8H_{11}NO$ mw: 137.20

SYN: 2,4-DIMETHYLPYRROL-3-YL METHYL KETONE

TOXICITY DATA with REFERENCE
ipr-rat LD50:250 mg/kg JMCMAR 11,1251,68
ipr-mus LD50:400 mg/kg JMCMAR 11,1251,68
ivn-mus LD50:71 mg/kg CSLNX* NX#04669

SAFETY PROFILE: Poison by intravenous and intraperitoneal routes. When heated to decomposition it emits toxic fumes of NO_x. See also KETONES.

ACI629 CAS:35629-40-4 ***HR: 2***
N-ACETYL-N-(4,5-DIPHENYL-2-OXAZOLYL) ACETAMIDE
mf: $C_{19}H_{16}N_2O_3$ mw: 320.37

SYN: N-(4,5-DIPHENYL-2-OXAZOLYL)DIACETAMIDE

TOXICITY DATA with REFERENCE
orl-mus LD50:1600 mg/kg JMCMAR 14,1075,71
ipr-mus LD50:400 mg/kg JMCMAR 14,1075,71

SAFETY PROFILE: Moderately toxic by ingestion and intraperitoneal routes. When heated to decomposition it emits toxic fumes of NO_x.

ACI750 CAS:74-86-2 ***HR: 3***
ACETYLENE
DOT: UN 1001
mf: C_2H_2 mw: 26.04

PROP: Colorless gas, garlic-like odor. Flammable. Bp: −84.0° (sublimes), lel: 2.5%, uel: 82%, mp: −81.8°, flash p: 0°F (CC), d: 1.173 g/L @ 0°, autoign temp: 581°F, vap press: 40 atm @ 16.8°, vap d: 0.91; d: (liquid) 0.613 @ −80°. D: (solid) 0.730 @ −85°. Quite sol in water; very sol in alc; almost misc in ether.

SYNS: ACETYLEN ◇ ACETYLENE, dissolved (DOT) ◇ ETHINE ◇ ETHYNE ◇ NARCYLEN

TOXICITY DATA with REFERENCE
ihl-hmn TCLo:20 pph:CNS,RSP 34ZIAG -,67,69

ihl-hmn LCLo:50 pph/5M TABIA2 3,231,33
ihl-mam LCLo:500000 ppm/5M AEPPAE 138,65,28

CONSENSUS REPORTS: Reported in EPA TSCA Inventory.

OSHA PEL: CL 2500 ppm
ACGIH TLV: Simple asphyxiant.
NIOSH REL: (Acetylene) 10H TWA no exposure > 2500 ppm
DOT Classification: Forbidden; Flammable Gas; Label: Flammable Gas.

SAFETY PROFILE: Mildly toxic by inhalation. Human systemic effects by inhalation: headache and dyspnea. Narcotic in high concentration. In general industrial practice, acetylene does not constitute a serious toxic hazard. It is a very dangerous fire hazard when exposed to heat, flame, or oxidizers. Moderate explosion hazard when exposed to heat or flame or by spontaneous chemical reaction. At high pressures and moderate temperatures, and in the absence of air, acetylene has been known to decompose explosively. Reacts with copper to form the explosive copper acetylide. Incompatible with brass, copper salts, copper carbide, powdered Co, Hg, Hg salts, K, Ag and Ag salts, RbH, CsH, halogens, HNO_3, NaH, oxidants. Acetylene + halide + UV can explode. Molten K ignites in C_2H_2 and then explodes. C_2H_2 reacts vigorously with trifluoromethyl hypofluorite. With O_2, C_2H_2 can detonate very powerfully. See ACETYLIDES. When ignited, it burns with an intensely hot flame; can react vigorously with oxidizing materials.

When mixed with O_2 in proportions of 40% or more, acetylene acts as a narcotic and has been used in anesthesia. Acetylene acts as a simple asphyxiant by diluting the O_2 in the air to a level which will not support life. However, the presence of impurities in commercial acetylene may result in the production of symptoms before an asphyxiant concentration is reached. Thus: 10% in air produces a slight intoxication, 20% produces a staggering gait, 30% produces general incoordination, 33% leads to unconsciousness in 7 minutes, up to 80% produces complete anesthesia, increased blood pressure, narcosis, and stimulated respiration.

Dizziness, headache, mild gastric symptoms, and (in high concentration) semi-asphyxia and brief loss of consciousness have all been reported. See ARGON for a discussion of simple asphyxiants. To fight fire, use CO_2, water spray, or dry chemical. Stop flow of gas.

ACJ000 ***HR: 3***
ACETYLENE CHLORIDE
mf: CHCCl mw: 60.48

PROP: A gas. Bp: −31°, vap d: 2.0, mp: −126°.

SYN: CHLOROETHYNE

SAFETY PROFILE: Dangerous fire hazard by spontaneous chemical reaction. Spontaneously flammable in air. Shock will explode it. When heated to decomposition it emits highly toxic fumes of phosgene; can react vigorously with oxidizing materials. See also ACETYLENE COMPOUNDS and CHLORINATED HYDROCARBONS, ALIPHATIC.

ACJ125 ***HR: 3***
ACETYLENE COMPOUNDS

SAFETY PROFILE: The carbon-carbon triple bond is explosively unstable in many acetylenic compounds. Both the lower acetylenes (i.e., propyne, butadyene, etc.) and higher compounds may undergo explosive decomposition. The presence of halogens and heavy metal derivatives may increase these explosive tendencies. See also ACETYLENE, ACETYLIDES, and specific compounds.

ACJ250 CAS:543-21-5 ***HR: 3***
ACETYLENEDICARBOXAMIDE
mf: $C_4H_4N_2O_2$ mw: 112.10

PROP: Produced by *Str. reticuli var. Aquamyceticus* and is identical to Cellocidin.

SYNS: ACETYLENEDICARBOXYLIC ACID DIAMIDE ◇ AQUAMYCIN ◇ 2-BUTYNEDIAMIDE ◇ CELLOCIDIN ◇ LENAMYCIN ◇ RENAMYCIN

TOXICITY DATA with REFERENCE
ivn-mus LD50:11 mg/kg JAJAAA 11,81,58

CONSENSUS REPORTS: Reported in EPA TSCA Inventory.

SAFETY PROFILE: Poison by intravenous route. When heated to decomposition it emits toxic fumes of NO_x. See also ACETYLENE COMPOUNDS.

ACJ500 CAS:928-04-1 ***HR: 3***
ACETYLENEDICARBOXYLIC ACID MONOPOTASSIUM SALT
mf: C_4HO_4•K mw: 152.15

SYNS: MONOPOTASSIUM SALT of ACETYLENEDICARBOXYLIC ACID ◇ U-4783

TOXICITY DATA with REFERENCE
orl-mus LD50:63 mg/kg TXAPA9 17,733,70
ipr-mus LD50:32 mg/kg TXAPA9 17,733,70
ivn-mus LD50:89 mg/kg TXAPA9 17,733,70

CONSENSUS REPORTS: Reported in EPA TSCA Inventory.

SAFETY PROFILE: Poison by ingestion, intravenous,

and intraperitoneal routes. When heated to decomposition it emits acrid smoke and fumes of KO_x. See also ACETYLENE COMPOUNDS.

ACK000 CAS:156-60-5 ***HR: 2***
trans-ACETYLENE DICHLORIDE
mf: $C_2H_2Cl_2$ mw: 96.94

PROP: Colorless liquid, pleasant odor. Mp: −50°, bp: 48°, flash p: 36°F, autoign temp: 860°F, lel: 9.7%, uel: 12.8%, d: 1.2743 @ 25°/4°, vap press: 400 mm @ 30.8°, vap d: 3.34.

SYNS: trans-DICHLOROETHYLENE ◇ trans-1,2-DICHLOROETHYLENE (MAK) ◇ RCRA WASTE NUMBER U079

TOXICITY DATA with REFERENCE
mma-smc 80 mmol/L TCMUD8 4,365,84
orl-mus LD50:2122 mg/kg DCTODJ 8,373,85
ihl-hmn TCLo:4800 mg/m^3/10M:CNS AHBAAM 116,131,36
ipr-rat LD50:7536 mg/kg TXCYAC 7(2),141,77
ihl-mus LCLo:75000 mg/m^3/2H AHBAAM 116,131,36
ipr-mus LD50:4019 mg/kg TXCYAC 7,141,77
ihl-cat LCLo:43000 mg/m^3/6H AHBAAM 116,131,36

CONSENSUS REPORTS: Reported in EPA TSCA Inventory.

DFG MAK: 200 ppm (790 mg/m^3)

SAFETY PROFILE: Mildly toxic by inhalation and other routes. Human systemic effects by inhalation: sleep, hallucinations and distorted perceptions. Mutation data reported. Exposure to high vapor concentration can cause nausea, vomiting, weakness, tremor and cramps. Recovery is usually prompt following removal from exposure. Dermatitis may result from de-fatting action on skin. Dangerous fire hazard when exposed to heat, flame or oxidizers. Moderate explosion hazard in the form of vapor when exposed to flame. Violent reaction with difluoromethylene dihypofluorite. Forms shock-sensitive explosive mixtures with dinitrogen tetraoxide. Reaction with solid caustic alkalies or their concentrated solutions produces chloracetylene gas which ignites spontaneously in air. Reacts violently with N_2O_4, KOH, Na, NaOH. Moderate explosion hazard in the form of vapor when exposed to flame. Can react vigorously with oxidizing materials. To fight fire, use water spray, foam, CO_2, dry chemical. When heated to decomposition it emits toxic fumes of Cl^-. See also CHLORIDES, CHLORINATED HYDROCARBONS, ALIPHATIC; and ACETYLENE COMPOUNDS.

ACK250 CAS:79-27-6 ***HR: 3***
ACETYLENE TETRABROMIDE
DOT: UN 2504
mf: $C_2H_2Br_4$ mw: 345.68

PROP: Colorless to yellow liquid. Bp: 151° @ 54 mm, fp: −1°, d: 2.9638 @ 20°/4°, autoign temp: 635°F.

SYNS: MUTHMANN'S LIQUID ◇ TBE ◇ 1,1,2,2-TETRABROMAETHAN (GERMAN) ◇ TETRABROMOACETYLENE ◇ 1,1,2,2-TETRABROMOETANO (ITALIAN) ◇ S-TETRABROMOETHANE ◇ 1,1,2,2-TETRABROMOETHANE ◇ 1,1,2,2-TETRABROOMETHAAN (DUTCH)

TOXICITY DATA with REFERENCE
skn-rbt 500 mg/24H MOD AIHAAP 24,28,63
eye-rbt 100 mg MLD AIHAAP 24,28,63
dnr-esc 10 μL/disc MUREAV 41,61,76
orl-rat LD50:1100 mg/kg 85GMAT -,107,82
ihl-rat LC50:549 mg/m^3/4H 85GMAT -,107,82
orl-mus LD50:269 mg/kg 85GMAT -,107,82
skn-rat LD50:5250 mg/kg 85GMAT -,107,82
orl-gpg LD50:400 mg/kg AIHAAP 30,251,69
skn-mus TDLo:130 g/kg/74W-I:NEO JJIND8 63,1433,79
orl-rbt LD50:400 mg/kg AMIHBC 2,407,50

CONSENSUS REPORTS: Reported in EPA TSCA Inventory. EPA Genetic Toxicology Program.

OSHA PEL: TWA 1 ppm
ACGIH TLV: TWA 1 ppm
DFG MAK: 1 ppm (14 mg/m^3)
DOT Classification: ORM-A; Label: None.

SAFETY PROFILE: Poison by inhalation and ingestion. An eye and skin irritant and a narcotic. Questionable carcinogen with experimental neoplastigenic data. Mutation data reported. When heated it emits highly toxic fumes of carbonyl bromide and Br^-. See also ACETYLENE COMPOUNDS and BROMIDES.

ACL000 CAS:2597-54-8 ***HR: 3***
N-ACETYL ETHYL CARBAMATE
mf: $C_6H_9NO_3$ mw: 131.15

SYN: ACETYLURETHANE

TOXICITY DATA with REFERENCE
ipr-mus TD:7300 mg/kg/10W-I:NEO IJCNAW 4,318,69
ipr-mus TD:3650 mg/kg/5W-I:ETA IJCNAW 4,318,69

SAFETY PROFILE: Questionable carcinogen with experimental neoplastigenic and tumorigenic data. See also CARBAMATES. When heated to decomposition it emits toxic fumes of NO_x.

ACL250 CAS:25614-78-2 ***HR: 3***
N-ACETYLETHYL-2-cis-CROTONYLCARBAMIDE
mf: $C_9H_{14}N_2O_3$ mw: 198.25

SYN: HOMEOSTAN

TOXICITY DATA with REFERENCE
orl-mus LD50:3500 mg/kg 27ZQAG -,423,72
ipr-mus LD50:1500 mg/kg 27ZQAG -,423,72
ivn-mus LD50:300 mg/kg 27ZQAG -,423,72

SAFETY PROFILE: Poison by intravenous route. Moderately toxic by ingestion and intraperitoneal routes. When heated to decomposition it emits toxic fumes of NO_x.

ACL500 CAS:52217-47-7 ***HR: 3***
N'-ACETYL ETHYLNITROSOUREA
mf: $C_5H_9N_3O_3$ mw: 159.17

TOXICITY DATA with REFERENCE
orl-rat TDLo:520 mg/kg/52W-I:ETA PPTCBY 2,73,72
orl-rat LD50:550 mg/kg PPTCBY 2,85,72

SAFETY PROFILE: Moderately toxic by ingestion. Questionable carcinogen with experimental tumorigenic data. When heated to decomposition it emits toxic fumes of NO_x.

ACL750 CAS:88-29-9 ***HR: 3***
ACETYL ETHYL TETRAMETHYL TETRALIN
mf: $C_{18}H_{26}O$ mw: 258.44

PROP: White crystals.

SYNS: ACETYLETHYL TETRAMETHYLTETRALIN ◇ 6-ACETYL-1,1,4,4-TETRAMETHYL-7-ETHYL-1,2,3,4,-TETRALIN◇ 7-ACETYL-1,1,4,4-TETRAMETHYL-1,2,3,4-TETRAHYDRONAPHTHALENE ◇ AETT ◇ ETHANONE-1-(3-ETHYL-5,6,7,8-TETRAHYDRO-5,5,8,8-TETRAMETHYL-2-NAPHTHALENYL)(9CI) ◇ 3'-ETHYL-5',6',7',8'-TETRAHYDRO-5',5',8',8'-TETRAMETHYL-2'-ACETONAPHTHONE ◇ 1-(3-ETHYL-5,6,7,8-TETRAHYDRO-5,5,8,8-TETRAMETHYL-2-NAPHTHALENYL)-ETHANONE ◇ MUSK 36A ◇ POLYCYCLIC MUSK ◇ VERSALIDE

TOXICITY DATA with REFERENCE
skn-rbt 500 mg/24H MLD FCTXAV 17,357,79
orl-rat LD50:260 mg/kg FCTXAV 19,753,81
orl-mus LDLo:470 mg/kg AECTCV 14,111,85
ipr-rat LD50:126 mg/kg FCTXAV 17,357,79
scu-rat LD50:584 mg/kg FCTXAV 17,357,79

CONSENSUS REPORTS: Reported in EPA TSCA Inventory.

SAFETY PROFILE: Poison by ingestion and intraperitoneal routes. Moderately toxic by other routes, especially by skin contact and subcutaneous routes. A skin irritant. Exposure causes blue coloration of internal organs and central nervous system effects, i.e., hyperexcitability, tremors, lack of coordination, hunched back and loss of weight. It is slowly metabolized and excreted via feces. Symptoms persist for 90 days after exposure. Severity of symptoms seems proportional to length of exposure. It is freely absorbed via human skin. When heated to decomposition it emits acrid smoke and fumes.

ACM000 CAS:557-99-3 ***HR: 3***
ACETYL FLUORIDE
mf: C_2H_3FO mw: 62.05

PROP: D: 1.002 @ 15°/4°; mp: −60°, bp: 20.8°. Sltly sol in alc, ether, acetone, and benzene.

SYN: METHYLCARBONYL FLUORIDE

TOXICITY DATA with REFERENCE
ihl-dog LCLo:2000 mg/m³/30M 11FYAN 3,74,63

CONSENSUS REPORTS: Reported in EPA TSCA Inventory.

OSHA PEL: TWA 2.5 mg(F)/m³
ACGIH TLV: TWA 2.5 mg(F)/m³

SAFETY PROFILE: Poison by inhalation. See also FLUORIDES. When heated to decomposition it emits toxic fumes of F^-.

ACM250 CAS:7242-07-1 ***HR: 3***
16-ACETYLGITOXIN
mf: $C_{43}H_{66}O_{15}$ mw: 823.09

TOXICITY DATA with REFERENCE
ivn-rat LD50:16500 μg/kg AIPTAK 155,165,65
ipr-mus LD50:6800 μg/kg AIPTAK 155,165,65
ipr-cat LD50:148 μg/kg AIPTAK 155,165,65
ivn-cat LDLo:110 μg/kg AIPTAK 155,165,65
orl-cat LD50:120 μg/kg AIPTAK 159,1,66
orl-gpg LD50:2500 μg/kg AIPTAK 159,1,66

SAFETY PROFILE: Deadly poison by ingestion, intraperitoneal and intravenous routes. When heated to decomposition it emits acrid smoke and fumes. See also GITOXIN.

ACM500 CAS:29362-48-9 ***HR: 3***
ACETYLGITOXIN-α
mf: $C_{43}H_{66}O_{15}$ mw: 823.09

TOXICITY DATA with REFERENCE
ivn-cat LD50:520 μg/kg 85ELDJ -,187,63
orl-gpg LD50:40 mg/kg AIPTAK 159,1,66

SAFETY PROFILE: Poison by ingestion and intravenous routes. When heated to decomposition it emits acrid smoke and fumes. See also GITOXIN.

ACM750 CAS:1068-57-1 ***HR: 3***
ACETYL HYDRAZIDE
mf: $C_2H_6N_2O$ mw: 74.10

SYNS: ACETHYDRAZIDE ◇ ACETOHYDRAZIDE ◇ N-ACETYLHYDRAZINE ◇ ENT 61,241 ◇ ETHANEHYDRAZONIC ACID ◇ MONOACETYLHYDRAZINE

TOXICITY DATA with REFERENCE
mmo-sat 500 μg/plate IJEBA6 19,939,81
mmo-omi 70 mg/L MUREAV 173,233,86
mnt-mus-ipr 120 mg/kg CALEDQ 23,235,84
dni-mus-ipr 150 mg/kg IJEBA6 19,939,81

orl-bwd LD50:42200 μg/kg AECTCV 12,355,83
ipr-mus LD50:153 mg/kg JPETAB 122,110,58

CONSENSUS REPORTS: Reported in EPA TSCA Inventory.

SAFETY PROFILE: Poison by ingestion and intraperitoneal routes. Mutation data reported. Exposure can cause hemolysis and liver damage. See also PHENYLHYDRAZINE. When heated to decomposition it emits toxic fumes of NO_x.

ACN250 CAS:534-33-8 ***HR: 3***
N-ACETYL-4-HYDROXYARSANILIC ACID compound with DIETHYLAMINE (1581)
mf: $C_8H_{10}AsNO_5 \cdot C_4H_{11}N$ mw: 348.27

SYNS: N-ACETYL-4-HYDROXY-m-ARSANILIC ACID DIETHYLAMINE SALT ◇ 2-AMINOPHENOL-4-ARSONIC ACID DIETHYLAMINE SALT ◇ DIETHYLAMINE-3-ACETYLAMINO-4-HYDROXYPHENYLARSONATE ◇ GOLARSYL ◇ SYNTHARSOL ◇ ACETARSIN ◇ ACETARSONE DIETHYLAMINE SALT ◇ ACETILARSANO ◇ ACETYLARSAN ◇ ARSAPHENAN ◇ DIETHYLAMINE ACETARSONE

TOXICITY DATA with REFERENCE
cyt-hmn:leu 1 nmol/L AEMBAP 91,117,78
cyt-hmn:fbr 1 nmol/L AEMBAP 91,117,78

CONSENSUS REPORTS: Arsenic and its compounds are on the Community Right-To-Know List.

OSHA PEL: TWA 0.5 $mg(As)/m^3$
ACGIH TLV: TWA 0.2 $mg(As)/m^3$

SAFETY PROFILE: A poison. Human mutation data reported. See also ARSENIC COMPOUNDS. When heated to decomposition it emits very toxic fumes of NO_x and As.

ACN500 CAS:65734-38-5 ***HR: 3***
N-ACETYL-N'-(p-HYDROXYMETHYL)PHENYLHYDRAZINE
mf: $C_9H_{12}N_2O_2$ mw: 180.23

TOXICITY DATA with REFERENCE
orl-mus TDLo:74 g/kg/85W-C:NEO CNREA8 38,177,78
scu-mus TDLo:13 g/kg/26W-I:ETA JTEHD6 8,1,81

SAFETY PROFILE: Questionable carcinogen with experimental neoplastigenic and tumorigenic data. When heated to decomposition it emits toxic fumes such as NO_x.

ACN875 CAS:4254-22-2 ***HR: 3***
ACETYL HYPOBROMITE
mf: $C_2H_3BrO_2$ mw: 138.95

SAFETY PROFILE: A dangerously unstable explosive. When heated to decomposition it emits toxic fumes of Br^-.

ACO000 ***HR: 3***
ACETYLIDES

SAFETY PROFILE: Severe explosion hazard when shocked or exposed to heat. Acetylides are very sensitive to shock, friction, and heat. They explode readily and are one of the few commercial explosives which contain no O_2 or N_2 and therefore produce no gas. The explosion simply results from the large amount of heat instantaneously produced. Acetylides are used for detonating compositions, or in combination with lead azide in detonating rivets where the acetylides reduce the flash point of the more insensitive azides. They are in a class with the fulminates and the azides as primary detonants. Because these materials are so sensitive to shock and temperature, they must be handled with extreme care. They must be kept cool, and should be kept wet if they are to be stored. (See FULMINATES for suggested precautions in storage and handling of acetylides.) Metal powders, such as finely divided Cu or Ag, should not be stored or kept with acetylene or acetylides since it is possible for them to react with these metal powders to form very sensitive acetylides which, while they are not dangerous in themselves, can cause enough of a flash to ignite a possibly explosive mixture of gases and thus cause an explosion in a warehouse or storage area. Examples of commercially used acetylides are silver acetylide and copper acetylide. See also ACETYLENE. See also individual compounds.

ACO250 CAS:2466-76-4 ***HR: 3***
N-ACETYLIMIDAZOLE
mf: $C_5H_6N_2O$ mw: 110.13

SYN: 1-ACETYLIMIDAZOLE

TOXICITY DATA with REFERENCE
ipr-mus TDLo:50 mg/kg/I:NEO JNCIAM 54,495,75
ipr-mus LDLo:250 mg/kg StoGD# 27May75

CONSENSUS REPORTS: Reported in EPA TSCA Inventory.

SAFETY PROFILE: Poison by intraperitoneal route. Questionable carcinogen with experimental neoplastigenic data. When heated to decomposition it emits toxic fumes of NO_x.

ACO500 CAS:507-02-8 ***HR: 3***
ACETYL IODIDE
DOT: UN 1898
mf: C_2H_3IO mw: 169.95

PROP: Brown, transparent, fuming liquid. Bp: 108°, d: 2.067 @ 20°/4°, decomp in water and alc; sol in ether.

CONSENSUS REPORTS: Reported in EPA TSCA Inventory.

DOT Classification: Corrosive Material; Label: Corrosive.

SAFETY PROFILE: A toxic, corrosive material. Reacts with water or steam to produce toxic and corrosive fumes. Dangerous to use. When heated to decomposition it emits toxic fumes of I^-. See also IODIDES.

ACO750 CAS:1078-38-2 **HR: 3**
1-ACETYL-2-ISONICOTINOYLHYDRAZINE
mf: $C_8H_9N_3O_2$ mw: 179.20

SYNS: ACETYL ISONIAZID ◇ N-ACETYLISONIAZID ◇ N-ACETYLISONICOTINYLHYDRAZIDE ◇ 4-PYRIDINECARBOXYLIC ACID-2-ACETYLHYDRAZIDE

TOXICITY DATA with REFERENCE
dni-mus-ipr 1 g/kg IJEBA6 19,939,81
orl-mus TDLo:380 g/kg/68W-C:NEO EJCAAH 9,285,73

CONSENSUS REPORTS: EPA Genetic Toxicology Program.

SAFETY PROFILE: Questionable carcinogen with experimental neoplastigenic data. Mutation data reported. When heated to decomposition it emits toxic fumes of NO_x.

ACP000 CAS:39293-24-8 **HR: 3**
ACETYLKIDAMYCIN
mf: $C_{46}H_{58}N_2O_{13}$ mw: 847.06

TOXICITY DATA with REFERENCE
oms-hmn:hla 1 mg/L JANTAJ 29,1334,76
ipr-rat LD50:35 mg/kg 85ERAY 2,1452,78
ivn-rat LD50:140 mg/kg 85ERAY 2,1452,78
orl-mus LD50:600 mg/kg 85ERAY 2,1452,78
ipr-mus LD50:50 mg/kg 85ERAY 2,1452,78
ivn-mus LD50:200 mg/kg 85ERAY 2,1452,78
ivn-rbt LD50:25 mg/kg 85ERAY 2,1452,78

SAFETY PROFILE: Poison by intravenous and intraperitoneal routes. Moderately toxic by ingestion. Human mutation data reported. When heated to decomposition it emits toxic fumes of NO_x.

ACP250 **HR: 3**
ACETYLLANATOSIDE A

SYN: ACETYL-LANATOSID A (GERMAN)

TOXICITY DATA with REFERENCE
orl-gpg LD50:25 mg/kg ARZNAD 15,481,65
ivn-gpg LDLo:1800 mg/kg ARZNAD 15,481,65

SAFETY PROFILE: Poison by ingestion. Moderately toxic by intravenous route.

ACP500 CAS:63938-24-9 **HR: 3**
1-ACETYLLYSERGIC ACID DIETHYLAMIDE BITARTRATE
mf: $C_{22}H_{27}N_3O_2 \cdot 2C_4H_4O_6$ mw: 661.68

SYN: 1-ACETYL-9,10-DIDEHYDRO-N,N-DIETHYL-6-METHYLERGOLINE-8-β-CARBOXAMIDE BITARTRATE

TOXICITY DATA with REFERENCE
orl-hmn TDLo:1500 ng/kg:PSY PSDTAP 8,59,67
ivn-rbt LD50:1600 μg/kg 27ZQAG -,93,72

SAFETY PROFILE: Deadly poison by intravenous route. Human systemic effects by ingestion of very small amounts: EEG changes, hallucinations, distorted perceptions and changes in psychophysiological test scores. When heated to decomposition it emits toxic fumes of NO_x. See also other lysergic acid derivatives.

ACP750 CAS:50485-03-5 **HR: 3**
d-1-ACETYL LYSERGIC ACID MONOETHYLAMIDE
mf: $C_{20}H_{23}N_3O_2$ mw: 337.46

SYNS: 1-ACETYL-9,10-DIDEHYDRO-N-ETHYL-6-METHYLERGOLINE-8-β-CARBOXAMIDE ◇ 1-ACETYLLYSERGIC ACID ETHYLAMIDE

TOXICITY DATA with REFERENCE
orl-hmn TDLo:75 μg/kg:PSY PSDTAP 8,59,67
ivn-rbt LD50:5 mg/kg 27ZQAG -,94,72

SAFETY PROFILE: Poison by ingestion and intravenous routes. Ingesting very small amounts produce psychotropic effects in humans. When heated to decomposition it emits toxic fumes of NO_x. See also various lysergic acid entries.

ACQ000 CAS:73118-22-6 **HR: 3**
β-ACETYLMANDELOYLOXY-β-PHENYLETHYL DIMETHYLAMINE
mf: $C_{20}H_{23}NO_4$ mw: 341.44

SYN: ACETYLMANDELICACID-(2-(DIMETHYLAMINO)-1-PHENYL)ETHYL ESTER

TOXICITY DATA with REFERENCE
scu-mus LDLo:683 mg/kg AIPTAK 47,96,34
ivn-rbt LDLo:34 mg/kg AIPTAK 47,96,34

SAFETY PROFILE: Poison by intravenous route. Moderately toxic by subcutaneous route. See also ESTERS. When heated to decomposition it emits toxic fumes of NO_x.

ACQ250 CAS:1190-93-8 **HR: 3**
ACETYLMERCAPTOACETIC ACID
mf: $C_4H_6O_3S$ mw: 134.16

SYNS: S-ACETYLTHIOGLYCOLIC ACID ◇ USAF EK-P-5430

TOXICITY DATA with REFERENCE
ipr-mus LD50:150 mg/kg NTIS** AD277-689

CONSENSUS REPORTS: Reported in EPA TSCA Inventory.

SAFETY PROFILE: Poison by intraperitoneal route. When heated to decomposition it emits toxic fumes of SO_x.

ACQ666 CAS:1477-40-3 ***HR: 3***
α-1-ACETYLMETHADOL
mf: $C_{23}H_{31}NO_2$ mw: 353.55

SYN: LAAM

TOXICITY DATA with REFERENCE
mmo-nsc 200 mg/L DCTODJ 4,19,81
cyt-hmn:lym 70 mg/L ENMUDM 1,180,79
mma-mus:lym 2 5 mg/L DCTODJ 4,19,81
trn-mus-unr 7 mg/kg DCTODJ 4,19,81
orl-man TDLo:8570 μg/kg (4W male):REP CLPTAT 19,371,76
orl-mus LD50:173 mg/kg JPETAB 110,135,54
ipr-mus LD50:56 mg/kg PBBHAU 9,195,78
scu-mus LD50:111 mg/kg ANYAA9 281,321,76

SAFETY PROFILE: Poison by ingestion, subcutaneous and intraperitoneal routes. Human reproductive effects by ingestion on male. Experimental reproductive effects. Human mutation data reported. When heated to decomposition it emits acrid smoke and fumes.

ACQ690 CAS:43033-72-3 ***HR: 3***
l-α-ACETYLMETHADOL HYDROCHLORIDE
mf: $C_{23}H_{31}NO_2{\cdot}ClH$ mw: 390.01

SYN: 3-HEPTANOL, 6-(DIMETHYLAMINO)-4,4-DIPHENYL-, ACETATE (ester), HYDROCHLORIDE, (3S,6S)-(-)- ◇ (3S,6S)-(-)-6-(DIMETHYLAMINO)-4,4-DIPHENYL-3-HEPTANOL ACETATE (ester) HYDROCHLORIDE ◇ LAAM HYDROCHLORIDE

TOXICITY DATA with REFERENCE
orl-rat TDLo:7400 μg/kg (female 15D pre):REP NETOD7 5,479,83
orl-rat TDLo:7400 μg/kg (female 15D pre):TER NETOD7 5,479,83
orl-rat TDLo:7061 mg/kg/2Y-C:CAR FAATDF 11,626,88
orl-mus LD50:71 mg/kg FAATDF 11,626,88
par-mus LD50:11 mg/kg JPETAB 145,11,64

SAFETY PROFILE: Poison by ingestion and parenteral routes. Questionable carcinogen with experimental carcinogenic data. An experimental teratogen. Other experimental reproductive effects. When heated to decomposition it emits toxic fumes of NO_x and HCl.

ACR000 CAS:62-51-1 ***HR: 3***
o-ACETYL-β-METHYLCHOLINE CHLORIDE
mf: $C_8H_{18}NO_2{\cdot}Cl$ mw: 195.72

PROP: Mp: 172°-173°. Very sol in water and alc, decomp in alkalies and ether.

SYNS: (2-HYDROXYPROPYL)TRIMETHYLAMMONIUMCHLORIDE ACETATE ◇ METHACHOLINE CHLORIDE ◇ METHACHOLINIUM CHLORIDE ◇ METHYLACETYL CHOLINE ◇ β-METHYLACETYLCHOLINE CHLORIDE ◇ TRIMETHYL-β-ACETOXYPROPYLAMMONIUM CHLORIDE ◇ AMECHOL ◇ MECHOLYL

TOXICITY DATA with REFERENCE
orl-rat LD50:750 mg/kg JPETAB 58,337,36
scu-rat LD50:75 mg/kg JPETAB 58,337,36
ivn-rat LD50:20 mg/kg JPETAB 58,337,36
orl-mus LD50:1100 mg/kg JPETAB 58,337,36
ipr-mus LD50:160 mg/kg TXAPA9 28,227,74
scu-mus LD50:90 mg/kg JPETAB 58,337,36
ivn-mus LD50:15 mg/kg JPETAB 58,337,36
ivn-gpg LDLo:3750 μg/kg AIPTAK 106,245,56

CONSENSUS REPORTS: Reported in EPA TSCA Inventory.

SAFETY PROFILE: Poison by subcutaneous, intravenous, and intraperitoneal routes. Moderately toxic by other routes. When heated to decomposition it emits very toxic fumes of Cl^- and NO_x.

ACR100 ***HR: 3***
α-ACETYL-6-METHYLERGOLINE-8-β-PROPIONAMIDE
mf: $C_{20}H_{25}N_3O_2$ mw: 339.48

SYN: ERGOLINE-8-β-PROPIONAMIDE, α-ACETYL-6-METHYL-

TOXICITY DATA with REFERENCE
orl-rat TDLo:4 mg/kg (female 5D post):REP ARZNAD 33,1094,83
orl-mus LD50:400 mg/kg ARZNAD 33,1094,83

SAFETY PROFILE: Poison by ingestion. Experimental reproductive effects. When heated to decomposition it emits toxic fumes of NO_x.

ACR300 CAS:83-63-6 ***HR: 3***
N-ACETYL-N-(2-METHYL-4-((2-METHYLPHENYL)AZO)PHENYL)ACETAMIDE
mf: $C_{18}H_{19}N_3O_2$ mw: 309.40

SYNS: DERMAGAN ◇ DERMAGEN ◇ DIACETAZOTOL ◇ DIACETOTOLUIDE ◇ o-DIACETOTOLUIDIDE, 4'?-(o-TOLYLAZO)-(8CI) ◇ DIACETYLAMINOAZOTOLUENE ◇ N,N-DIACETYL-o-TOLYLAZO-o-TOLUIDINE ◇ DIAMAZO ◇ DIMAZON ◇ EPIDERMOL ◇ EPITHELONE ◇ GRANULIN ◇ PELLIDOL ◇ PELLIDOLE ◇ PERIPHERMIN ◇ 4-o-TOLYLAZO-o-DIACETOTOLUIDE ◇ 4'-(o-TOLYLAZO)-o-DIACETOTOLUIDIDE

CONSENSUS REPORTS: IARC Cancer Review:

Group 3 IMEMDT 7,56,87; Animal Inadequate Evidence IMEMDT 8,113,75

SAFETY PROFILE: Questionable carcinogen. When heated to decomposition it emits dangerous and toxic fumes of NO_x.

ACR400 CAS:28895-91-2 ***HR: 3***
ACETYLMETHYLNITROSOUREA
mf: $C_4H_7N_3O_3$ mw: 145.14

SYNS: ACETYL-METHYL-NITROSO-HARNSTOFF(GERMAN) ◇ N'-ACETYL-METHYLNITROSOUREA ◇ N-METHYL-N-NITROSO-N'-ACETYLUREA ◇ 1-METHYL-1-NITROSOACETYLUREA

TOXICITY DATA with REFERENCE
cyt-ham:fbr 500 mg/L/20H MUREAV 48,337,77
orl-rat TDLo:468 mg/kg/47W-I:ETA ZEKBAI 74,23,70
orl-rat LD50:200 mg/kg XENOBH 3,271,73

SAFETY PROFILE: Poison by ingestion. Questionable carcinogen with experimental tumorigenic data. Mutation data reported. When heated to decomposition it emits toxic fumes of NO_x.

ACR750 CAS:1696-20-4 ***HR: 2***
4-ACETYLMORPHOLINE
mf: $C_6H_{11}NO_2$ mw: 129.18

PROP: Liquid. Mp: 14°, bp: decomp, flash p: 235°F (OC), d: 1.1164, vap press: 0.02 mm @ 20°, vap d: 4.46.

SYN: N-ACETYLMORPHOLINE

TOXICITY DATA with REFERENCE
eye-rbt 500 mg open AMIHBC 10,61,54
orl-rat LD50:6130 mg/kg AMIHBC 10,61,54
skn-rbt LD50:7500 mg/kg AMIHBC 10,61,54
par-mus LDLo:2400 mg/kg CBCCT* 7,691,55

CONSENSUS REPORTS: Reported in EPA TSCA Inventory.

SAFETY PROFILE: Moderately toxic by parenteral route. An eye irritant. See also MORPHOLINE. Combustible when exposed to heat or flame; can react vigorously with oxidizing materials. To fight fire use alcohol foam. When heated to decomposition it emits toxic fumes of NO_x.

ACS000 CAS:63224-44-2 ***HR: 3***
N-ACETYL-N-MYRISTOYLOXY-2-AMINOFLUORENE
mf: $C_{29}H_{39}NO_3$ mw: 449.69

SYNS: N-ACETYL-N-TETRADECANOYLOXY-2-AMINOFLUORENE ◇ N-(FLUOREN-2-YL)-o-TETRADECANOYLACETOHYDROXAMIC ACID ◇ N-MYRISTOYLOXY-AAF ◇ N-MYRISTOYLOXY-N-ACETYL-2-AMINOFLUORENE

TOXICITY DATA with REFERENCE
dns-hmn:fbr 10 μmol/L/5H IJCNAW 16,284,75
msc-ham:lng 50 μmol/L/3H CALEDQ 6,67,79
scu-rat TD:115 mg/kg/6W-I:CAR CRNGDP 2,655,81
scu-rat TDLo:114 mg/kg/5W-I:NEO CNREA8 37,1461,77

SAFETY PROFILE: Questionable carcinogen with experimental carcinogenic and neoplastigenic data. Human mutation data reported. When heated to decomposition it emits toxic fumes of NO_x.

ACS250 CAS:76749-37-6 ***HR: D***
o-ACETYL-N-(2-NAPHPTHOYL)HYDROXYLAMINE
mf: $C_{13}H_{11}NO_3$ mw: 229.23

SYN: 2-NAPHTHOHYDROXAMIC ACID-o-ACETATE ESTER

TOXICITY DATA with REFERENCE
mmo-sat 1 μmol/plate PAACA3 21,126,80
sce-ham:ovr 40 μmol/L/3H-C MUREAV 88,81,81
mma-sat 1 μmol/plate CBINA8 34,267,81

SAFETY PROFILE: Mutation data reported. See also ESTERS and AMINES. When heated to decomposition it emits very toxic fumes of NO_x.

ACS375 ***HR: 2***
N-ACETYLNEOMYCIN

TOXICITY DATA with REFERENCE
ipr-mus LD50:3250 mg/kg AACHAX -,227,65
scu-mus LD50:9250 mg/kg AACHAX -,227,65
ivn-mus LD50:625 mg/kg AACHAX -,227,65

SAFETY PROFILE: Moderately toxic by several routes.

ACS500 CAS:65041-92-1 ***HR: 3***
8-ACETYL NEOSOLANIOL
mf: $C_{20}H_{28}O_9$ mw: 412.48

SYNS: 4-β,8-α,15-TRIACETOXY-3-α-HYDROXY-12,13-EPOXY-TRICHOTHEC-9-ENE ◇ 12,13-EPOXY-4-β,8-α,15-TRIACETOXY-3-α-HYDROXY TRICHOTHEC-9-ENE

TOXICITY DATA with REFERENCE
orl-ckn LD50:3220 μg/kg AEMIDF 35,636,78

SAFETY PROFILE: Poison by ingestion. When heated to decomposition it emits acrid smoke and fumes.

ACS750 CAS:591-09-3 ***HR: 3***
ACETYL NITRATE
mf: $C_2H_3NO_4$ mw: 105.06

PROP: Colorless, fuming, mobile liquid. Bp: 22° @ 70 mm; d: 1.24 @ 15°/4°.

SYN: ACETIC ACID, ANHYDRIDE with NITRIC ACID (1:1)

TOXICITY DATA with REFERENCE
eye-hmn 4 ppm/12M IAPWAR 4,79,61

SAFETY PROFILE: Corrosive to the eye. Violently unstable. Reacts explosively with ethyl-3,4-dihydroxybenzenesulfonate + oleum, HgO, and other active oxides. Solutions may explode violently above 60°C and the pure material explodes above 100°C. When heated to decomposition it emits toxic fumes of NO_x and/or explodes. See also NITRATES.

ACT000 ***HR: 2***
ACETYL NITRITE
mf: $C_2H_3NO_3$ mw: 89.10

SAFETY PROFILE: Unstable liquid; decomposed by light. Vapor is violently explosive on heating. See also NITRITES. When heated to decomposition it emits toxic fumes of NO_x.

ACT250 CAS:5275-69-4 ***HR: 3***
2-ACETYL-5-NITROFURAN
mf: $C_6H_5NO_4$ mw: 155.12

SYN: (5-NITRO-2-FURYL) METHYL KETONE

TOXICITY DATA with REFERENCE
mmo-omi 1000 ppm APMBAY 6,45,58
scu-rat LD50:200 mg/kg SGOBA9 83,73,46
orl-mus LD50:400 mg/kg SGOBA9 83,73,46
mmo-sat 8 μg/plate CNREA8 35,3611,75

CONSENSUS REPORTS: EPA Genetic Toxicology Program.

SAFETY PROFILE: Poison by subcutaneous route. Mutation data reported. See also KETONES. When heated to decomposition it emits toxic fumes of NO_x.

ACU125 ***HR: 2***
7-ACETYL-5-OXO-5H-(1)BENZOPYRANO (2,3-b)PYRIDINE
mf: $C_{14}H_9NO_3$ mw: 239.24

SYNS: KETONE-METHYL-5-OXO-5H-(1)BENZOPYRANO (2,3-b)PYRIDYL ◇ Y-9000

TOXICITY DATA with REFERENCE
orl-rat LD50:1679 mg/kg NYKZAU 74,179,78
ipr-rat LD50:409 mg/kg NYKZAU 74,179,78
orl-mus LD50:2326 mg/kg NYKZAU 74,179,78
ipr-mus LD50:473 mg/kg NYKZAU 74,179,78

SAFETY PROFILE: Moderately toxic by ingestion and intraperitoneal routes. When heated to decomposition it emits toxic fumes of NO_x. See also KETONES.

ACU500 CAS:42978-43-8 ***HR: 3***
6-ACETYLOXYMETHYLBENZO(a)PYRENE
mf: $C_{23}H_{16}O_2$ mw: 324.39

SYN: 6-ACETOXY METHYL BENZO(a)PYRENE

TOXICITY DATA with REFERENCE
mmo-sat 1 nmol/plate PAACA3 24,93,83
dnd-rat:lym 500 mg/L CBINA8 25,35,79
dnd-mam:lym 500 mg/L CBINA8 25,35,79
scu-rat TDLo:584 μg/kg/60D-I:NEO CBINA8 29,159,80
scu-rat TD:100 mg/kg/40D-I:ETA JMCMAR 16,714,73

SAFETY PROFILE: Questionable carcinogen with experimental neoplastigenic and tumorigenic data. Mutation data reported. When heated to decomposition it emits acrid smoke and fumes.

ACV000 CAS:34627-78-6 ***HR: 3***
5-(1-ACETYLOXY-2-PROPENYL)-1,3-BENZODIOXOLE
mf: $C_{12}H_{12}O_4$ mw: 220.24

SYN: 1′-ACETOXYSAFROLE

TOXICITY DATA with REFERENCE
mmo-sat 25 μg/plate JJIND8 62,893,79
dnr-esc 25 mg/L JJIND8 62,873,79
dnd-hmn:oth 500 μmol/L CRNGDP 3,935,82
orl-rat TDLo:62 g/kg/36W-C:NEO CNREA8 33,590,73
scu-rat TDLo:529 mg/kg/10W-I:CAR CNREA8 43,1124,83
ipr-mus TDLo:22 mg/kg:CAR CNREA8 47,2275,87

CONSENSUS REPORTS: EPA Genetic Toxicology Program.

SAFETY PROFILE: Questionable carcinogen with experimental carcinogenic and neoplastigenic data. Human mutation data reported. When heated to decomposition it emits acrid smoke and fumes.

ACV500 CAS:110-22-5 ***HR: 3***
ACETYL PEROXIDE
DOT: UN 2084
mf: $C_4H_6O_4$ mw: 118.04

$$CH_3CO{\bullet}OOCO{\bullet}CH_3$$

PROP: Solid or colorless crystals or liquid. Sltly sol in cold water, decomp. D: 1.18, mp: 30°, bp: 63° @ 21 mm.

SYN: DIACETYL PEROXIDE (MAK)

TOXICITY DATA with REFERENCE
eye-rbt 60 mg/1M rns SEV ZAARAM 8,25,58
unk-mus TDLo:283 mg/kg:ETA RARSAM 3,193,63

CONSENSUS REPORTS: Reported in EPA TSCA Inventory.

DFG MAK: Strong Skin Effects.
DOT Classification: Forbidden (solid or > 25% in solution); Organic Peroxide; Label: Organic Peroxide (UN2084); Forbidden.

SAFETY PROFILE: Severe skin and eye irritant. Questionable carcinogen with experimental tumorigenic data. Dangerous fire hazard by spontaneous chemical reaction. A powerful oxidizing agent; can cause ignition of organic materials on contact. Severe explosion hazard when shocked or exposed to heat. It may explode spontaneously in storage and should be used as soon as prepared. It will react with water or steam to produce heat; can react vigorously with reducing materials; emits toxic fumes on contact with acid or acid fumes. To fight fire use CO_2, dry chemical.

Storage and Handling: Must be kept below 27° and not warmed over 30°. Do not add to hot materials. Do not add accelerator to this material. Store in original container with vented cap. Avoid bodily contact. This material is nearly always stored and handled as a 25% solution in an inert solvent. See also ACETYL PEROXIDE 25% solution (in dimethyl phthalate); and PEROXIDES, ORGANIC.

ACV750 CAS:110-22-5 ***HR: 3***
ACETYL PEROXIDE (solution)
DOT: UN 2084

PROP: Crystal clear liquid. Mp: −7°, flash p: 113°F (OC), d: 1.18 @ 20°. Solution contains not over 25% acetyl peroxide (FEREAC 41,15972,76).

SYN: ACETYL PEROXIDE SOLUTION, not over 25% peroxide (DOT) ◇ DIACETYL PEROXIDE (solution)

CONSENSUS REPORTS: Reported in EPA TSCA Inventory.

DOT Classification: Organic Peroxide; Label: Organic Peroxide.

SAFETY PROFILE: Strong irritant to skin, eyes, and mucous membranes. See also ACETYL PEROXIDE (solid), and PEROXIDES, ORGANIC. Flammable when exposed to heat or flame, or by spontaneous chemical reaction. An oxidizing agent. When heated to decomposition it emits toxic fumes; can react vigorously with oxidizing materials. To fight fire use foam, CO_2.

ACX500 CAS:13402-08-9 ***HR: 2***
1-ACETYL-3-PHENYLETHYLACETYLUREA
mf: $C_{13}H_{16}N_2O_3$ mw: 248.31

SYNS: N-((ACETYLAMINO)CARBONYL)-α-ETHYLBENZENEACETAMIDE ◇ ACETYLPHENETURIDE ◇ CRAMPOL ◇ CRAMPOLE ◇ N-α-ETHYLPHENYLACETYL-N'-ACETYL UREA ◇ P-398

TOXICITY DATA with REFERENCE
ipr-rat LD50:543 mg/kg NIIRDN 6,17,82
ipr-mus LD50:560 mg/kg NIIRDN 6,17,82
orl-rat LD50:1174 mg/kg ARZNAD 18,524,68
orl-mus LD50:1165 mg/kg ARZNAD 18,524,68

SAFETY PROFILE: Moderately toxic by ingestion and intraperitoneal routes. May have human reproductive effects. When heated to decomposition it emits toxic fumes of NO_x. An anticonvulsant.

ACX750 CAS:114-83-0 ***HR: 3***
ACETYLPHENYLHYDRAZINE
mf: $C_8H_{10}N_2O$ mw: 150.20

PROP: Mp: 130-132°. Sol in hot water and alc; sltly sol in ether.

SYNS: ACETIC ACID PHENYLHYDRAZONE ◇ β-ACETYLPHENYLHYDRAZINE ◇ 1-ACETYL-2-PHENYLHYDRAZINE ◇ APH ◇ HYDRACETIN ◇ N'-PHENYLACETHYDRAZIDE ◇ PYRODINE

TOXICITY DATA with REFERENCE
orl-mus LD50:270 mg/kg PCJOAU 14,162,80
orl-mus TDLo:31 g/kg/79W-I:NEO BJCAAI 39,584,79
ipr-mus LDLo:150 mg/kg NTIS** AD691-490

CONSENSUS REPORTS: Reported in EPA TSCA Inventory.

SAFETY PROFILE: Poison by ingestion and intraperitoneal routes. Questionable carcinogen with experimental neoplastigenic data. See also HYDRAZINE. When heated to decomposition it emits toxic fumes of NO_x.

ACY700 CAS:76298-68-5 ***HR: 2***
cis-2-ACETYL-3-PHENYL-5-TOSYL-3,3a,4,5-TETRAHYDROPYRAZOLO(4,3-c)QUINOLINE
mf: $C_{25}H_{23}N_3O_3S$ mw: 445.57

SYN: 2H-PYRAZOLO(4,3-c)QUINOLINE,3,3a,4,5-TETRAHYDRO-2-ACETYL-5-((4-METHYLPHENYL)SULFONYL)-3-PHENYL-,cis-

TOXICITY DATA with REFERENCE
orl-rat TDLo:100 mg/kg (female 1-5D post):REP IJOCAP 19,297,80
ipr-mus LD50:800 mg/kg IJOCAP 19,297,80

SAFETY PROFILE: Moderately toxic by intraperitoneal route. Experimental reproductive effects. When heated to decomposition it emits toxic fumes of NO_x.

ACY750 ***HR: 3***
12-O-ACETYL-PHORBOL-13-DECA-(Δ-2)-ENOATE
mf: $C_{32}H_{45}O_8$ mw: 557.77

TOXICITY DATA with REFERENCE
skn-mus 50 μg MLD PLMEAA 22,241,72
skn-mus TDLo:107 g/kg/12W-I:NEO PLMEAA 22,241,72

SAFETY PROFILE: Questionable carcinogen with experimental neoplastigenic data. A skin irritant. When heated to decomposition it emits acrid smoke and fumes.

ACZ000 CAS:20839-15-0 ***HR: 3***
12-O-ACETYL-PHORBOL-13-DECANOATE
mf: $C_{32}H_{48}O_8$ mw: 560.80

SYN: PHORBOL ACETATE, CAPRATE

TOXICITY DATA with REFERENCE
skn-mus 49 μg MLD PLMEAA 22,241,72
skn-mus TDLo:26 mg/kg/32W-I:NEO NATWAY 54,282,67

SAFETY PROFILE: A skin irritant. Questionable carcinogen with experimental neoplastigenic data. When heated to decomposition it emits acrid smoke and irritating fumes.

ADA000 CAS:17433-31-7 ***HR: 3***
1-ACETYL-2-PICOLINOLHYDRAZINE
mf: $C_8H_{11}N_3O_2$ mw: 179.20

SYNS: N-ACETYL-N'-ISONICOTINYL HYDRAZIDE ◇ 1-ACETYL-2-PICOLINOYLHYDRAZINE ◇ AZAPICYL ◇ NCI-C04739 ◇ NSC-68626 ◇ P-2292 ◇ 2-PYRIDINECARBOXYLIC ACID-2-ACETYLHYDRAZIDE (9CI)

TOXICITY DATA with REFERENCE
ipr-rat TDLo:9750 mg/kg/26W-I:NEO RRCRBU 52,1,75
ipr-mus TDLo:9750 mg/kg/26W-I:NEO,REP RRCRBU 52,1,75
orl-rat LD50:673 mg/kg NCIAL* -,169,65
ivn-rat LD50:470 mg/kg NCIAL* -,169,65
orl-mus LD50:410 mg/kg NCIAL* -,169,65
ivn-mus LD50:255 mg/kg NCIAL* -,169,65

CONSENSUS REPORTS: NCI Carcinogenesis Studies (ipr): Clear Evidence: mouse,rat RRCRBU 52,1,75

SAFETY PROFILE: Poison by ingestion and intravenous routes. Questionable carcinogen with experimental neoplastigenic data. Experimental reproductive effects. When heated to decomposition it emits toxic fumes such as NO_x.

ADA250 CAS:618-42-8 ***HR: 3***
1-ACETYLPIPERIDINE
mf: $C_7H_{13}NO$ mw: 127.21

PROP: Misc in water, sol in alc. D: 1.011, bp: 226°.

SYN: N-ACETYLPIPERIDIN (GERMAN)

TOXICITY DATA with REFERENCE
scu-rbt LDLo:300 mg/kg BDCGAS 34,2408,01

CONSENSUS REPORTS: Reported in EPA TSCA Inventory.

SAFETY PROFILE: Poison by subcutaneous route. When heated to decomposition it emits toxic fumes of NO_x.

ADA350 CAS:22047-25-2 ***HR: 2***
2-ACETYL PYRAZINE
mf: $C_6H_6N_2O$ mw: 122.13

PROP: Colorless to pale yellow crystals or liquid; sweet popcornlike odor. Mp: 75-78°, d: 1.100-1.115 @ 20°, refr index: 1.530-1.540 @ 25°. Sol in acids, alc, ether, and water @ 230°.

SYN: FEMA No. 3126

SAFETY PROFILE: A skin and eye irritant. When heated to decomposition emits toxic fumes of NO_x.

ADA725 CAS:50-78-2 ***HR: 3***
ACETYLSALICYLIC ACID
mf: $C_9H_8O_4$ mw: 180.17

PROP: Colorless needles. Mp: 135°. Very sltly sol in alc, sol in benzene. Solubility in water = 1% @ 37°, in ether = 5% @ 20°.

SYNS: AC 5230 ◇ ACENTERINE ◇ ACESAL ◇ ACETAL ◇ ACETICYL ◇ ACETILSALICILICO ◇ ACETILUM ACIDULATUM ◇ ACETISAL ◇ ACETOL ◇ ACETONYL ◇ ACETOPHEN ◇ ACETOSAL ◇ ACETOSALIC ACID ◇ ACETOSALIN ◇ 2-ACETOXYBENZOIC ACID ◇ o-ACETOXYBENZOIC ACID ◇ ACETYLIN ◇ 2-(ACETYLOXY)BENZOIC ACID ◇ ACETYLSAL ◇ ACETYLSALICYLSAURE (GERMAN) ◇ ACIDE ACETYLSALICYLIQUE (FRENCH) ◇ ACIDO o-ACETILBENZOICO (ITALIAN) ◇ ACIDO ACETILSALICILICO (ITALIAN) ◇ ACIDUM ACETYLSALICYLICUM ◇ ACIMETTEN ◇ ACISAL ◇ ACYLPYRIN ◇ ASA ◇ A.S.A. ◇ A.S.A. EMPIRIN ◇ ASAGRAN ◇ ASATARD ◇ ASPALON ◇ ASPERGUM ◇ ASPIRDROPS ◇ ASPIRIN ◇ ASPIRINE ◇ ASPRO ◇ ASTERIC ◇ BENASPIR ◇ BIALPIRINIA ◇ CAPRIN ◇ o-CARBOXYPHENYL ACETATE ◇ COLFARIT ◇ CONTRHEUMA RETARD ◇ CRYSTAR ◇ DELGESIC ◇ DOLEAN pH 8 ◇ DURAMAX ◇ ECM ◇ ECOTRIN ◇ EMPIRIN ◇ ENDYDOL ◇ ENTERICIN ◇ ENTEROPHEN ◇ ENTEROSARINE ◇ ENTROPHEN ◇ EXTREN ◇ GLOBOID ◇ HELICON ◇ IDRAGIN ◇ MEASURIN ◇ NEURONIKA ◇ NOVID ◇ POLOPIRYNA ◇ RHEUMIN TABLETTEN ◇ RHODINE ◇ SALACETIN ◇ SALCETOGEN ◇ SALETIN ◇ SOLPYRON ◇ XAXA

TOXICITY DATA with REFERENCE
oms-hmn:lym 75 mg/L NEZAAQ 37,673,82
cyt-hmn:lym 10 mg/L BEXBAN 78,1194,74
orl-wmn TDLo:17280 mg/kg (female 1-39W post):REP JOPDAB 92,478,78
orl-wmn TDLo:17280 mg/kg (female 1-39W post):TER JOPDAB 92,478,78
orl-wmn TDLo:525 mg/kg/5D-I:SYS AIMEAS 80,74,74
orl-wmn TDLo:480 mg/kg/5D-I:SYS NEJMAG 296,418,77
orl-man TDLo:1625 mg/kg CPEDAM 24,678,85
orl-inf TDLo:120 mg/kg:PUL BMJOAE 1,1081,79
orl-cld LDLo:104 mg/kg LANCAO 2,809,52
orl-cld TDLo:39 mg/kg/13D-I:SYS AJDCAI 139,453,85
orl-hmn TDLo:669 mg/kg/11D:STS AJHPA9 35,330,78
orl-hmn TDLo:2880 mg/kg/8W:GIT,EAR ARZNAD 33,631,83
orl-hmn TDLo:480 mg/kg/7D-I:GIT,EAR ARZNAD 25,281,75

unr-man LDLo:294 mg/kg 85DCAI 2,73,70
orl-rat LD50:200 mg/kg 34ZIAG -,67,69
ipr-rat LD50:340 mg/kg NYKZAU 62,11,66
orl-mus LD50:250 mg/kg ARZNAD 5,572,55
ipr-mus LD50:280 mg/kg JPPMAB 4,872,52
scu-mus LD50:1020 mg/kg DRFUD4 9,91,84
orl-dog LD50:700 mg/kg ARZNAD 21,719,71

CONSENSUS REPORTS: EPA Genetic Toxicology Program. Reported in EPA TSCA Inventory.

OSHA PEL: TWA 5 mg/m^3
ACGIH TLV: TWA 5 mg/m^3

SAFETY PROFILE: Poison by ingestion, intraperitoneal, and possibly other routes. Moderately toxic by several routes. Human systemic effects by ingestion: general anesthetic, nausea or vomiting, liver damage, hematuria and kidney damage, dehydration, metabolic effects, changes in the blood, respiratory stimulation, sputum, effects on the joints, tinnitus, bleeding from large intestine, constipation, and other gastrointestinal effects. Implicated in aplastic anemia. A 10 grams dose to an adult may be fatal. A human teratogen. Human reproductive effects by ingestion and possibly other routes: menstrual cycle changes, parturition, various effects on newborn including apgar score, developmental abnormalities of the cardiovascular and respiratory systems. Experimental animal reproductive effects. Human mutation data reported. An allergen; skin contact, inhalation, or ingestion can cause asthma, sneezing, irritation of eyes and nose, hives and eczema. Combustible when exposed to heat or flame. When heated to decomposition it emits acrid smoke and fumes.

ADA750 CAS:493-53-8 ***HR: 2***
o-ACETYLSALICYLIC ACID, SODIUM SALT
mf: $C_9H_7O_4$•Na mw: 202.15

PROP: Crystals from acetone and ether. Mp: 218° (slt decomp). Very sol in water and alc; sltly sol in acetone.

SYNS: ACETYLSALICYLIC ACID SODIUM SALT ◇ ACETYLSALICYLSAEURE NATRIUMSALZ (GERMAN) ◇ ASPIRIN-NATRIUM (GERMAN) ◇ SODIUM ASPIRIN

TOXICITY DATA with REFERENCE
ivn-hmn TDLo:306 μg/kg:BLD GWXXBX #2810425
ipr-rat LD50:1450 mg/kg NYKZAU 79,357,82
scu-mus LDLo:700 mg/kg HDTU** -,-,33
ipr-mus LDLo:500 mg/kg JACSAT 63,1437,41
scu-frg LDLo:909 mg/kg HBAMAK 4,1290,35

SAFETY PROFILE: Moderately toxic by intraperitoneal and subcutaneous routes. Human systemic effects by intravenous route: unspecified changes in the blood. When heated to decomposition it emits acrid smoke and fumes. See also ACETOL.

ADB250 CAS:58086-32-1 ***HR: 3***
o-ACETYLSTERIGMATOCYSTIN
mf: $C_{20}H_{14}O_7$ mw: 366.34

TOXICITY DATA with REFERENCE
dns-rat:lvr 1 μmol/L MUREAV 173,217,86
mmo-sat 100 μg/plate CNREA8 38,536,78
mma-sat 1 μg/plate CNREA8 38,536,78
mrc-bcs 1 μg/disc CNREA8 36,445,76
ipr-rat LD50:11300 μg/kg 41KEAL -,108,78
ipr-rat TDLo:245 mg/kg/23W-I:CAR GANNA2 69,237,78

SAFETY PROFILE: Poison by intraperitoneal route. Questionable carcinogen with experimental carcinogenic data. Mutation data reported. When heated to decomposition it emits acrid smoke and fumes.

ADC250 CAS:1013-59-8 ***HR: 3***
3-ACETYLTETRAMIC ACID SODIUM SALT
mf: $C_{10}H_{14}NO_3$•Na mw: 219.24

SYNS: 3-ACETYL-5-sec-BUTYL-4-HYDROXY-3-PYROLIN-2-ONE,MONOSODIUM SALT ◇ 3-ACETYL-5-sec-BUTYL-4-HYDROXY-3-PYRROLIN-2-ONE SODIUM SALT ◇ 3-ACETYL-1,5-DIHYDRO-4-HYDROXY-5-(1-METHYLPROPYL)-2H-PYRROL-2-ONE SODIUM SALT ◇ NSC-525816

TOXICITY DATA with REFERENCE
orl-rat LD50:168 mg/kg CNCRA6 52,579,68
ivn-rat LD50:146 mg/kg CNCRA6 52,579,68
orl-mus LD50:81 mg/kg CNCRA6 52,579,68
ivn-mus LD50:115 mg/kg CNCRA6 52,579,68

SAFETY PROFILE: Poison by ingestion and intravenous routes. When heated to decomposition it emits toxic fumes of NO_x and Na_2O.

ADC750 CAS:584-26-9 ***HR: 3***
1-ACETYL-2-THIOHYDANTOIN
mf: $C_5H_6N_2O_2S$ mw: 158.19

PROP: Insol in water and ether; sltly sol in alc. Mp: 175-176°.

SYNS: USAF B-7 ◇ USAF BE-0405

TOXICITY DATA with REFERENCE
ipr-mus LD50:200 mg/kg NTIS** AD277-689
ivn-mus LD50:320 mg/kg CSLNX* NX#00834

CONSENSUS REPORTS: Reported in EPA TSCA Inventory.

SAFETY PROFILE: Poison by intravenous and intraperitoneal routes. When heated to decomposition it emits very toxic fumes such as SO_x and NO_x.

ADD000 CAS:17433-39-5 ***HR: 3***
5-(ACETYLTHIOMETHYL)-4-AMINOMETHYL-2-METHYL-3-PYRIDINOL HYDROBROMIDE
mf: $C_{10}H_{14}N_2O_2S$•BrH mw: 307.24

SYNS: 4-AMINOMETHYL-5-MERCAPTOMETHYL-2-METHYL-3-PYRIDINOL THIO ACETATE HYDROBROMIDE ◇ PYRIDOXAMIN-5-THIOACETAT HYDROBROMID (GERMAN)

TOXICITY DATA with REFERENCE
orl-rat LD50:2000 mg/kg ARZNAD 11,922,61
scu-rat LD50:500 mg/kg ARZNAD 11,922,61
ivn-rat LD50:405 mg/kg ARZNAD 11,922,61
orl-mus LD50:1340 mg/kg ARZNAD 11,922,61
scu-mus LD50:880 mg/kg ARZNAD 11,922,61
ivn-mus LD50:232 mg/kg ARZNAD 11,922,61

SAFETY PROFILE: Poison by intravenous route. Moderately toxic by ingestion and subcutaneous routes. When heated to decomposition it emits very toxic fumes of NO_x, SO_x and HBr.

ADD250 CAS:591-08-2 ***HR: 3***
ACETYL THIOUREA
mf: $C_3H_6N_2OS$ mw: 118.17

PROP: Sol in hot water and alc; sltly sol in ether. Mp: 166-167°.

SYNS: 1-ACETYL-2-THIOUREA ◇ RCRA WASTE NUMBER P002 ◇ USAF EK-4890

TOXICITY DATA with REFERENCE
orl-mus LDLo:94 mg/kg AECTCV 14,111,85
orl-rat LD50:50 mg/kg JPETAB 90,260,47
ipr-mus LD50:100 mg/kg NTIS** AD277-689

CONSENSUS REPORTS: Reported in EPA TSCA Inventory.

SAFETY PROFILE: Poison by ingestion and intraperitoneal routes. When heated to decomposition it emits very toxic fumes of NO_x and SO_x. See also SULFIDES.

ADD750 CAS:77-89-4 ***HR: 2***
ACETYL TRIETHYL CITRATE
mf: $C_{14}H_{22}O_8$ mw: 318.36

SYNS: CITRIC ACID, ACETYL TRIETHYL ESTER ◇ TRICARBALLYLIC ACID-β-ACETOXYTRIBUTYL ESTER ◇ TRIETHYL ACETYLCITRATE

TOXICITY DATA with REFERENCE
orl-rat LD50:8000 mg/kg NPIRI* 2,2,75
ipr-mus LD50:1150 mg/kg JPMSAE 53,774,64
orl-cat LDLo:7500 mg/kg TXAPA9 1,283,59

CONSENSUS REPORTS: Reported in EPA TSCA Inventory.

SAFETY PROFILE: Moderately toxic by intraperitoneal route. Mildly toxic by ingestion. See also ESTERS. When heated to decomposition it emits acrid smoke and fumes.

ADD875 CAS:2260-08-4 ***HR: 2***
ACETYLTRIIODOTHYRONINE FORMIC ACID
mf: $C_{15}H_9I_3O_5$ mw: 649.95

SYNS: ACETIROMATE ◇ 4-(4-(ACETYLOXY)-3-IODOPHENOXY)-3,5-DIIODO-BENZOIC ACID

TOXICITY DATA with REFERENCE
orl-rat LD50:4600 mg/kg IYKEDH 5,383,74
ipr-rat LD50:500 mg/kg IYKEDH 5,383,74
scu-rat LD50:520 mg/kg IYKEDH 5,383,74
orl-mus LD50:3700 mg/kg IYKEDH 5,383,74
ipr-mus LD50:1 g/kg IYKEDH 5,383,74
scu-mus LD50:2500 mg/kg IYKEDH 5,383,74

SAFETY PROFILE: Moderately toxic by ingestion and other routes. When heated to decomposition it emits toxic fumes of I^-.

ADE000 CAS:477-27-0 ***HR: 3***
N-ACETYL TRIMETHYLCOLCHICINIC ACID
mf: $C_{21}H_{23}NO_6$ mw: 385.45

SYNS: 7-ACETAMIDO-6,7-DIHYDRO-10-HYDROXY-1,2,3-TRIMETHOXY-BENZO(a)HEPTALEN-9(5H)-ONE ◇ 7-ACETAMIDO-10-HYDROXY-1,2,3-TRIMETHOXY-6,7-DIHYDROBENZO(a)HEPTALEN-9(5H)-ONE ◇ O^{10}-DEMETHYLCOLCHICINE

TOXICITY DATA with REFERENCE
oms-mus-ipr 42 mg/kg CANCAR 3,130,50
oms-mus-par 84 mg/kg CANCAR 3,130,50
spm-mus-par 84 mg/kg CANCAR 3,130,50
orl-hmn LDLo:43 μg/kg PCOC** -,250,66
unk-rat LDLo:30 mg/kg CANCAR 3,125,50
ipr-mus LD50:84 mg/kg CANCAR 3,124,50
ivn-mus LD50:1 mg/kg COREAF 241,1889,55

SAFETY PROFILE: A deadly human poison by ingestion. An experimental poison by intravenous, intraperitoneal and possibly other routes. Mutation data reported. When heated to decomposition it emits toxic fumes of NO_x. See also COLCHICINE.

ADE075 CAS:1218-34-4 ***HR: 2***
ACETYLTRYPTOPHAN
mf: $C_{13}H_{14}N_2O_3$ mw: 246.29

SYNS: ACETYL-l-TRP ◇ ACETYL-l-TRYPTOPHAN ◇ N-ACETYLTRYPTOPHAN ◇ N-ACETYL-l-TRYPTOPHAN ◇ (S)-N-ACETYLTRYPTOPHAN ◇ AC-TRY

TOXICITY DATA with REFERENCE
ipr-rat TDLo:1650 mg/kg (female 7-17D post):REP IYKEDH 11,690,80
orl-rat TDLo:55 g/kg (7-17D preg):TER IYKEDH 11,690,80
orl-rat LD50:15000 mg/kg IYKEDH 11,635,80
ipr-rat LD50:3900 mg/kg IYKEDH 11,635,80
orl-mus LD50:10800 mg/kg IYKEDH 11,635,80
ipr-mus LD50:3580 mg/kg IYKEDH 11,635,80

CONSENSUS REPORTS: Reported in EPA TSCA Inventory.

SAFETY PROFILE: Moderately toxic by some routes. An experimental teratogen. Other experimental reproductive effects. When heated to decomposition it emits toxic fumes of NO_x.

ADE125 ***HR: D***
ACHANIA, flower extract

PROP: Belongs to the family *Malvaceae* (IJEBA6 18,561,80).

SYN: MALVAVISCUS CONZATTI Greenm., flower extract

TOXICITY DATA with REFERENCE
orl-rat TDLo:24 g/kg (male 30D pre):REP CCPTAY 31,101,85

SAFETY PROFILE: Experimental reproductive effects. When heated to decomposition it emits acrid smoke and irritating fumes.

ADE500 CAS:129-17-9 ***HR: 3***
ACID BLUE 1
mf: $C_{27}H_{31}N_2O_6S_2$•Na mw: 566.71

SYNS: ANHYDRO-4,4′-BIS(DIETHYLAMINO)TRIPHENYLMETHANOL-2′,4″-DISULPHONIC ACID, MONOSODIUM SALT ◇ BLUE 1084 ◇ C.I. ACID BLUE 1, SODIUM SALT ◇ COSMETIC GREEN BLUE R25396 ◇ 4,4′-DI(DIETHYLAMINO)-4′,6′-DISULPHOTRIPHENYLMETHANOL ANHYDRIDE, SODIUM SALT ◇ FOOD BLUE 3 ◇ LEATHER BLUE G ◇ PATENTBLAU V (GERMAN) ◇ SCHULTZ Nr. 826 (GERMAN) ◇ SODIUM PATENT BLUE V ◇ XYLENE BLUE VS

TOXICITY DATA with REFERENCE
mma-sat 1 mg/plate ENMUDM 8(Suppl 7),1,86
scu-rat TDLo:3000 mg/kg/33W-I:NEO BJCAAI 27,230,73
ims-rat TDLo:2070 mg/kg/50W-I:CAR JNCIAM 37,845,66
scu-rat TD:4050 mg/kg/45W-I:ETA FCTXAV 9,463,71
ipr-mus LD50:3000 mg/kg FCTXAV 5,165,67
ivn-man LDLo:33 μg/kg:IMM 34ZIAG -,611,69
ivn-mus LD50:1200 mg/kg SCPHA4 47,39,79

CONSENSUS REPORTS: IARC Cancer Review: Group 3 IMEMDT 7,56,87; Animal Sufficient Evidence IMEMDT 16,163,78. Reported in EPA TSCA Inventory.

SAFETY PROFILE: Deadly human poison by intravenous route: anaphylaxis. Moderately toxic by several routes. Questionable carcinogen with experimental carcinogenic, tumorigenic, and neoplastigenic data. Mutation data reported. When heated to decomposition it emits very toxic fumes of NO_x, NH_3, Na_2O and SO_x. See also SULFONATES.

ADE675 ***HR: D***
ACID BLUE 7

SYN: C.I. 42080

TOXICITY DATA with REFERENCE
mma-sat 100 μg/plate MUREAV 147,285,85
mnt-mus-ipr 38 mg/kg MUREAV 147,285,85
dlt-mus-ipr 220 mg/kg MUREAV 147,285,85

SAFETY PROFILE: Mutation data reported.

ADE750 CAS:3861-73-2 ***HR: 2***
ACID BLUE 92
mf: $C_{26}H_{16}N_3O_{10}S_3$•3Na mw: 695.60

SYNS: ACID BLUE A ◇ ACID LEATHER BLUE R ◇ ACID WOOL BLUE RL ◇ ACILAN FAST NAVY BLUE R ◇ AIREDALE BLUE RL ◇ AMACID FAST BLUE R ◇ ANAZOLENE, SODIUM ◇ 4-((4-ANILINO-5-SULFO-1-NAPHTHYL)AZO)-5-HYDROXY-2,7-NAPHTHALENEDIFULFONIC ACID TRISODIUM ◇ BENZYL BLUE R ◇ BENZYL FAST BLUE R ◇ BUCACID FAST WOOL BLUE R ◇ CALCOCID FAST BLUE SR ◇ C.I. 13390 ◇ C.I. ACID BLUE 92 ◇ C.I. ACID BLUE 92, TRISODIUM SALT ◇ CIRENE BRILLIANT BLUE R ◇ COLACID BLUE A ◇ COOMASSIE BLUE ◇ COOMASSIE BLUE MEDICINAL ◇ COOMASSIE BLUE RL ◇ CYANINE ACID BLUE R ◇ CYANINE ACID BLUE R NEW ◇ FAST ACID BLUE RL ◇ FAST WOOL BLUE R ◇ FENAZO BLUE SR ◇ HISPACID FAST BLUE R ◇ MEDIUM BLUE EMBL ◇ PONTACYL FAST BLUE R ◇ SODIUM AMAZOLENE ◇ SODIUM ANAZOLENE ◇ SULFONINE ACID BLUE R ◇ SULPHON ACID BLUE R ◇ SULPHON ACID BLUE RA ◇ TERTRACID FAST BLUE SR ◇ TRISODIUM-4′-ANILINO-8-HYDROXY-1,1′-AZONAPHTHALENE-3,6,5′-TRISULFONATE ◇ VONDAMOL FAST BLUE R ◇ WOOL BLUE RL ◇ WOOL FAST BLUE R

TOXICITY DATA with REFERENCE
dnd-esc 10 μmol/L MUREAV 89,95,81
ivn-mus LDLo:450 mg/kg BHJUAV 21,492,59

CONSENSUS REPORTS: Reported in EPA TSCA Inventory.

SAFETY PROFILE: Moderately toxic by intravenous route. Mutation data reported. When heated to decomposition it emits very toxic fumes of SO_x, NO_x, and Na_2O.

ADF000 CAS:3087-16-9 ***HR: 2***
ACID BRILLIANT GREEN BS
mf: $C_{27}H_{26}N_2O_7S_2$•Na mw: 577.66

SYNS: ACIDAL WOOL GREEN BS ◇ ACID LEATHER GREEN S ◇ BRILLIANTSAEURE GRUEN BS (GERMAN) ◇ C.I. 44090 ◇ C.I. ACID GREEN 50, MONOSODIUM SALT ◇ C.I. FOOD GREEN 4 ◇ GREEN 5 ◇ SCHULTZ Nr. 836 (GERMAN) ◇ WOOL GREEN S (BIOLOGICAL STAIN)

TOXICITY DATA with REFERENCE
mrc-smc 2840 μmol/L FCTXAV 19,419,81
mma-sat 1 mg/plate MUREAV 89,21,81
par-rat TDLo:470 mg/kg/2Y-I:ETA FAONAU 46A,57,69
orl-rat LD50:2 g/kg JPPMAB 16,65,64

CONSENSUS REPORTS: Reported in EPA TSCA Inventory. EPA Genetic Toxicology Program.

SAFETY PROFILE: Moderately toxic by ingestion. Questionable carcinogen with experimental tumorigenic data. Mutation data reported. When heated to decomposition it emits very toxic fumes of SO_x and NO_x.

ADF250 CAS:12788-93-1 ***HR: 3***
ACID BUTYL PHOSPHATE
DOT: UN 1718
mf: $C_4H_{10}O_4P$ mw: 153.1

PROP: Water-white liquid; sol in alc, acetone and toluene; insol in water, petroleum, and naphtha. D: 1.120-1.125 @ 25°/40°, flash p: 230°F (COC).

SYNS: n-BUTYL ACID PHOSPHATE ◇ BUTYL PHOSPHORIC ACID

DOT Classification: Corrosive Material; Label: Corrosive.

SAFETY PROFILE: Toxic and corrosive. Combustible when exposed to heat or flame. When heated to decomposition it emits highly toxic fumes of PO_x. See also ESTERS and PHOSPHORIC ACID.

ADF500 ***HR: 3***
ACID CARBOYS, EMPTY

SAFETY PROFILE: *Warning:* These containers may contain concentrated vapors or even some liquid acid remaining from their original contents. Therefore, they can give rise to all the hazards of their original contents.

ADG000 CAS:2429-80-3 ***HR: D***
ACID LEATHER ORANGE BZR
mf: $C_{35}H_{27}N_5O_9S_3 \cdot 2Na$ mw: 803.83

SYNS: BENZYL FAST ORANGE 2RN ◇ BUCACID ORANGE R ◇ C.I. 22195 ◇ C.I. ACID ORANGE 45, DISODIUM SALT

TOXICITY DATA with REFERENCE
mmo-sat 100 µg/plate MUREAV 68,307,79
mma-sat 100 µg/plate MUREAV 68,307,79

SAFETY PROFILE: Mutation data reported. When heated to decomposition it emits very toxic fumes of NO_x, SO_x and Na_2O.

ADG125 CAS:11119-62-3 ***HR: D***
ACID RED

SYN: XYLENE RED

TOXICITY DATA with REFERENCE
mma-sat 1 mg/plate AMONDS 3,253,80
cyt-ham:lng 10 g/L AMONDS 3,253,80
cyt-ham:fbr 12 g/L ESKHA5 96,55,78

CONSENSUS REPORTS: EPA Genetic Toxicology Program.

SAFETY PROFILE: Mutation data reported.

ADG250 CAS:18472-87-2 ***HR: 3***
ACID RED 92
mf: $C_{20}H_2Br_4Cl_4O_5 \cdot 2Na$ mw: 829.64

SYNS: C.I. ACID RED 92 ◇ D&C RED No. 28 ◇ EOSIN BLUE ◇ EOSINE BLUE ◇ EOSINE BLUISH ◇ FOOD DYE RED No. 104 ◇ FOOD RED No. 104 ◇ PHLOXINE B ◇ 11969 RED ◇ 3427 VERI PUR PINK

TOXICITY DATA with REFERENCE
mmo-omi 200 mg/L MUREAV 34,187,76
orl-rat TDLo:63 g/kg (1-22D preg):REP SKEZAP 16,34,75
orl-rat TDLo:63 g/kg (1-22D preg):TER SKEZAP 16,34,75
ivn-mus LD50:310 mg/kg TXAPA9 44,225,78

CONSENSUS REPORTS: Reported in EPA TSCA Inventory. EPA Genetic Toxicology Program.

SAFETY PROFILE: Poison by intravenous route. An experimental teratogen. Other experimental reproductive effects. When heated to decomposition it emits very toxic fumes of Br^-, Cl^- and Na_2O.

ADG400 ***HR: 3***
ACKEE

PROP: A 30- to 40-foot tall tree with 5-part compound leaves and small green-white flowers. A bright red pod contains 3 shiny black seeds in a white, waxy matrix. It grows in Florida, Hawaii and the West Indies.

SYNS: AKEE ◇ AKI ◇ ARBRE FRICASSE (HAITI) ◇ BLIGHIA SAPIDA ◇ SESO VEGETAL (CUBA, PUERTO RICO)

SAFETY PROFILE: The white matrix of the immature fruit and its attachment to the seeds contains the toxic hypoglycin A. In the ripe fruit these parts are edible. Systemic effects by ingestion may include: vomiting, convulsions, coma, hypoglycemia, and death. Symptoms may begin immediately or may appear after a delay of 6 to 10 hours. In Jamaica poisoning is common in the winter and is called "vomiting sickness". See also 2-METHYLENECYCLOPROPANYLALANINE.

ADG425 CAS:66789-14-8 ***HR: 3***
ACLACINOMYCIN Y
mf: $C_{42}H_{51}NO_{15}$ mw: 809.94

SYNS: ACLACINOMYCIN Y1 ◇ MA 144 Y

TOXICITY DATA with REFERENCE
dni-mus:leu 190 nmol/L JANTAJ 34,1596,81
oms-mus:leu 12 nmol/L JANTAJ 34,1596,81
ipr-mus LD50:40 mg/kg JANTAJ 33,80-64,80

SAFETY PROFILE: Poison by intraperitoneal route.

Mutation data reported. When heated to decomposition it emits toxic fumes of NO_x.

ADG500 CAS:509-20-6 ***HR: 3***
ACONINE

TOXICITY DATA with REFERENCE
ivn-mus LD50:117 mg/kg YHHPAL 19,641,84
ivn-cat LD50:400 mg/kg ARZNAD 5,324,55
ivn-gpg LD50:275 mg/kg ARZNAD 5,324,55

SAFETY PROFILE: Poison by intravenous route. When heated to decomposition it emits toxic fumes of NO_x. An antipyretic agent.

ADH000 CAS:499-12-7 ***HR: 3***
ACONITIC ACID
mf: $C_6H_6O_6$ mw: 174.12

PROP: White, crystalline powder. Sol in water, alc; very sltly sol in ether. Mp: 192° (decomp).

SYNS: ACHILLEIC ACID ◇ CITRIDIC ACID ◇ EQUISETIC ACID ◇ 1-PROPENE-1,2,3-TRICARBOXYLIC ACID

TOXICITY DATA with REFERENCE
ivn-mus LD50:180 mg/kg CSLNX* NX#00189

CONSENSUS REPORTS: Reported in EPA TSCA Inventory.

SAFETY PROFILE: Poison by intravenous route. A synthetic flavoring substance and adjuvant. When heated to decomposition it emits acrid smoke and fumes.

ADH500 CAS:8006-38-0 ***HR: 3***
ACONITINE, AMORPHOUS
mf: $C_{34}H_{47}NO_{11}$ mw: 645.82

SYNS: MILD ACONITATE ◇ MILD ACONITINE

TOXICITY DATA with REFERENCE
unk-man LDLo:147 μg/kg 85DCAI 2,73,70
ipr-rat LDLo:75 μg/kg JPHAA3 18,17,29
scu-rat LDLo:100 μg/kg JPHAA3 18,17,29
orl-mus LD50:1 mg/kg APTOA6 7,337,51
ipr-mus LD50:328 μg/kg APTOA6 7,337,51
ivn-mus LD50:166 μg/kg APTOA6 7,337,51

SAFETY PROFILE: Human poison by an unspecified route. Poison experimentally by ingestion, intraperitoneal and subcutaneous routes. See also ACONITINE. When heated to decomposition it emits toxic fumes of NO_x.

ADH750 CAS:302-27-2 ***HR: 3***
ACONITINE (CRYSTALLINE)
mf: $C_{34}H_{49}NO_{11}$ mw: 647.76

PROP: White, crystalline alkaloid; feeble bitter taste. Mp: 204°.

SYNS: ACETYL BENZOYL ACONINE ◇ ACONITANE ◇ ACONITIN CRISTALLISAT (GERMAN)

TOXICITY DATA with REFERENCE
ipr-rat LDLo:125 μg/kg PSEBAA 26,221,28
ivn-rat LD50:80 μg/kg ARZNAD 5,324,55
orl-hmn LDLo:28 mg/kg:CNS,GIT 34ZIAG -,72,69
orl-mus LD50:1 mg/kg 85GDA2 8(1),159,82
scu-mus LDLo:100 mg/kg HDTU** -,-,33
ivn-mus LD50:166 μg/kg 85GDA2 8(1),159,82
ipr-mus LD50:2708 μg/kg CYLPDN 2,170,81
ivn-dog LDLo:350 μg/kg HBAMAK 4,1291,35
scu-cat LDLo:400 μg/kg HBAMAK 4,1291,35
ivn-cat LD50:70 μg/kg ARZNAD 5,324,55
scu-rbt LDLo:131 μg/kg HBAMAK 4,1291,35
scu-gpg LDLo:50 μg/kg JPHAA3 12,957,23
ivn-gpg LD50:60 μg/kg ARZNAD 5,324,55
scu-pgn LDLo:66 μg/kg HBAMAK 4,1291,35
scu-frg LDLo:586 μg/kg HBAMAK 4,1291,35

SAFETY PROFILE: Poison by all routes including absorption through the skin. Human systemic effects by ingestion: excitement, diarrhea and other gastrointestinal effects. Used to produce heart arrhythmia in experimental animals and as an antipyretic agent. When heated to decomposition it emits highly toxic fumes of NO_x.

ADH875 CAS:6055-69-2 ***HR: 3***
ACONITINE HYDROCHLORIDE
mf: $C_{34}H_{47}NO_{11}\cdot ClH$ mw: 682.28

TOXICITY DATA with REFERENCE
scu-cat LDLo:134 μg/kg FDWU** -,-,31
scu-gpg LDLo:112 μg/kg FDWU** -,-,31
scu-pgn LDLo:45500 ng/kg FDWU** -,-,31
scu-frg LDLo:586 μg/kg FDWU** -,-,31

SAFETY PROFILE: Poison by subcutaneous route. When heated to decomposition it emits toxic fumes of NO_x and HCl.

ADI250 ***HR: 3***
ACONITUM CARMICHAELI

PROP: Raw tubers which are the source of processed aconite roots used as an oriental medicine in Japan (YKKZAJ 97,359,77).

TOXICITY DATA with REFERENCE
orl-mus LD50:5490 mg/kg YKKZAJ 97,359,77
ipr-mus LD50:190 mg/kg YKKZAJ 97,359,77
scu-mus LD50:200 mg/kg YKKZAJ 97,359,77
ivn-mus LD50:490 mg/kg YKKZAJ 97,359,77

SAFETY PROFILE: Poison by intraperitoneal and subcutaneous routes. Moderately toxic by intravenous route. Mildly toxic by ingestion. When heated to decomposition it emits acrid smoke and fumes.

ADI500 ***HR: 3***
ACONITUM JAPONICUM

PROP: Raw tubers which are the source of processed aconite roots used as an oriental medicine in Japan (YKKZAJ 97,359,77).

TOXICITY DATA with REFERENCE
orl-mus LD50:540 mg/kg YKKZAJ 97,359,77
ipr-mus LD50:110 mg/kg YKKZAJ 97,359,77
scu-mus LD50:120 mg/kg YKKZAJ 97,359,77
ivn-mus LD50:60 mg/kg YKKZAJ 97,359,77

SAFETY PROFILE: Poison by intraperitoneal, subcutaneous, and intravenous routes. Moderately toxic by ingestion. When heated to decomposition it emits acrid smoke and fumes.

ADI625 ***HR: 3***
ACORN TANNIN

SYN: TANNIN from ACORN

TOXICITY DATA with REFERENCE
ipr-mus LD50:100 mg/kg JPPMAB 9,98,57
scu-mus LD50:100 mg/kg JPPMAB 9,98,57
ivn-mus LD50:150 mg/kg JPPMAB 9,98,57
ims-mus LD50:75 mg/kg JPPMAB 9,98,57

SAFETY PROFILE: Poison by subcutaneous, intramuscular, intravenous and intraperitoneal routes.

ADI750 CAS:1684-42-0 ***HR: D***
ACRANIL HYDROCHLORIDE
mf: $C_{21}H_{26}ClN_3O_2•2ClH$ mw: 460.83

SYNS: ACRANIL ◇ ACRANIL DIHYDROCHLORIDE ◇ 1-((6-CHLORO-2-METHOXY-9-ACRIDYL)-AMINO)-3-(DIETHYLAMINO)-2-PROPANOL DIHYDROCHLORIDE ◇ 5-((γ-DIETHYLAMINO-β-HYDROXYPROPYL)AMINO)-3-METHOXY-8-CHLOROACRIDINE DIHYDROCHLORIDE ◇ SKF 16214-A2 ◇ SN 186

TOXICITY DATA with REFERENCE
mnt-mus-ipr 140 μmol/kg MUREAV 26,553,74
mnt-mus-ipr 140 μmol/kg MUREAV 26,553,74

CONSENSUS REPORTS: EPA Genetic Toxicology Program.

SAFETY PROFILE: Mutation data reported. When heated to decomposition it emits very toxic fumes of HCl and Cl^- and NO_x.

ADJ375 CAS:581-29-3 ***HR: 3***
3-ACRIDINAMINE (9CI)
mf: $C_{13}H_{10}N_2$ mw: 194.25

SYNS: 2-AMINOACRIDINE (EUROPEAN) ◇ 3-AMINOACRIDINE

TOXICITY DATA with REFERENCE
mmo-sat 20 μg/plate JOUOD4 6,257,84
mmo-omi 80 μg/L JMOBAK 3,762,61
scu-mus LD50:170 mg/kg BJEPA5 28,1,47

CONSENSUS REPORTS: EPA Genetic Toxicology Program.

SAFETY PROFILE: Poison by subcutaneous route. Mutation data reported. When heated to decomposition it emits toxic fumes of NO_x.

ADJ500 CAS:260-94-6 ***HR: 3***
ACRIDINE
DOT: UN 2713
mf: $C_{13}H_9N$ mw: 179.23

PROP: Small, colorless needles. Mp: 110.5°, bp: 346°, d: 1.005 @ 19.7°/4°, vap press: 1 mm @ 129.4°. Sltly sol in hot water; sol in alc, ether, and CS_2.

SYNS: 9-AZAANTHRACENE ◇ 10-AZAANTHRACENE ◇ BENZO(b)QUINOLINE ◇ 2,3-BENZOQUINOLINE ◇ DIBENZO(b,e)PYRIDINE

TOXICITY DATA with REFERENCE
mmo-sat 230 nmol/L ENMUDM 3,11,81
dnd-mam:lym 100 μmol/L JMOBAK 3,18,61
dnd-ckn:leu 100 μmol/L JMOBAK 3,18,61
scu-mus LD50:400 mg/kg BJEPA5 28,1,47
ivn-rbt LD50:100 mg/kg BJEPA5 28,1,47

OSHA PEL: TWA 0.2 mg/m^3

CONSENSUS REPORTS: Reported in EPA TSCA Inventory.

DOT Classification: IMO: Flammable Solid; Label: Flammable Solid.

SAFETY PROFILE: Poison by subcutaneous and intravenous routes. Mutation data reported. A skin, eye, and mucous membrane irritant. When heated to decomposition it emits toxic fumes of NO_x.

ADJ625 CAS:951-80-4 ***HR: 3***
3,9-ACRIDINEDIAMINE (9CI)
mf: $C_{13}H_{11}N_3$ mw: 209.27

SYNS: 2,5-DIAMINOACRIDINE (EUROPEAN) ◇ 3,9-DIAMINOACRIDINE

TOXICITY DATA with REFERENCE
mmo-sat 20 μg/plate JOUOD4 6,257,84
mmo-omi 8 mg/L JMOBAK 3,762,61
mmo-omi 19 μmol/L GENTAE 90,1,78
scu-mus LD50:140 mg/kg BJEPA5 28,1,78

SAFETY PROFILE: Poison by subcutaneous route. Mutation data reported. When heated to decomposition it emits toxic fumes of NO_x.

ADJ750 CAS:17784-47-3 ***HR: 3***
ACRIDINE HYDROCHLORIDE
mf: $C_{13}H_9N \cdot ClH$ mw: 215.69

SYNS: ACRIDINE MONOHYDROCHLORIDE ◇ ACRIDINIUM CHLORIDE

TOXICITY DATA with REFERENCE
mmo-omi 80 mg/L JMOBAK 3,762,61
dnd-mam:lym 10 pph BIPMAA 11,2537,72
scu-mus LD50:300 mg/kg QJPPAL 10,649,37

SAFETY PROFILE: Poison by subcutaneous route. Mutation data reported. When heated to decomposition it emits very toxic fumes of HCl and NO_x.

ADJ875 CAS:146-59-8 ***HR: 3***
ACRIDINE MUSTARD
mf: $C_{21}H_{25}Cl_2N_3O \cdot 2ClH$ mw: 479.31

SYNS: 6-CHLORO-9-(3-ETHYL-2-CHLOROETHYL)AMINOPROPYLAMINO)-2-METHOXYACRIDINE DIHYDROCHLORIDE ◇ 9-(3-(ETHYL (2-CHLOROETHYL)AMINO)PROPYLAMINO)-6-CHLORO-2-METHOXYACRIDINE DIHYDROCHLORIDE ◇ ICR 170 ◇ 2-METHOXY-6-CHLORO-9-(3-(ETHYL-2-CHLOROETHYL)AMINOPROPYLAMINO) ACRIDINE DIHYDROCHLORIDE

TOXICITY DATA with REFERENCE
mmo-sat 500 ng/plate MUREAV 136,185,84
slt-dmg-orl 20860 μmol/L ENMUDM 6,153,84
ipr-mus TDLo:4 mg/kg (1D pre):REP MUREAV 13,171,71
ivn-mus TDLo:4800 μg/kg/28D-I:NEO CNREA8 36,2423,76
ipr-mus LDLo:6 mg/kg MUREAV 5,417,68
ivn-mus LDLo:5 mg/kg CNREA8 36,2423,76

CONSENSUS REPORTS: EPA Genetic Toxicology Program.

SAFETY PROFILE: Poison by intravenous and intraperitoneal routes. Questionable carcinogen with experimental neoplastigenic data. Experimental reproductive effects. Human mutation data reported. When heated to decomposition it emits toxic fumes of NO_x and HCl.

ADK000 CAS:2465-29-4 ***HR: 3***
ACRIDINE RED
mf: $C_{15}H_{14}N_2O \cdot ClH$ mw: 274.77

PROP: Sltly sol in water; sol in alc; insol in ether.

SYNS: ACRIDINE RED 3B ◇ ACRIDINE RED, HYDROCHLORIDE ◇ DIMETHYLDIAMINOXANTHENYL CHLORIDE

TOXICITY DATA with REFERENCE
sln-dmg-orl 1000 ppm AMNTA4 87,295,53
scu-rat TDLo:1215 mg/kg/59W-I:ETA GANNA2 47,153,56

CONSENSUS REPORTS: Reported in EPA TSCA Inventory.

SAFETY PROFILE: Questionable carcinogen with experimental tumorigenic data. Mutation data reported. When heated to decomposition it emits very toxic fumes of HCl and NO_x.

ADK250 CAS:191-27-5 ***HR: 3***
ACRIDINO(2,1,9,8-klmna)ACRIDINE
mf: $C_{20}H_{10}N_2$ mw: 278.32

SYN: 6,12-DIAZAANTHANTHRENE

TOXICITY DATA with REFERENCE
imp-rat TDLo:600 mg/kg:ETA NEOLA4 18,591,71

SAFETY PROFILE: Questionable carcinogen with experimental tumorigenic data. When heated to decomposition it emits toxic fumes of NO_x.

ADK750 CAS:72739-00-5 ***HR: 3***
4'-(9-ACRIDINYLAMINO)-2'-AMINOMETHANESULFONANILIDE
mf: $C_{20}H_{18}N_4O_2S$ mw: 378.48

TOXICITY DATA with REFERENCE
mmo-sat 42 μmol/L JMCMAR 23,269,80
ipr-mus LD10:70 mg/kg JMCMAR 23,269,80

SAFETY PROFILE: Poison by intraperitoneal route. Mutation data reported. When heated to decomposition it emits very toxic fumes of NO_x and SO_x.

ADL000 CAS:61417-10-5 ***HR: 3***
4'-(9-ACRIDINYLAMINO)-3'-AMINOMETHANESULFONANILIDE
mf: $C_{20}H_{18}N_4O_2S$ mw: 378.48

TOXICITY DATA with REFERENCE
mmo-sat 282 μmol/L JMCMAR 23,269,80
ipr-mus LD10:50 mg/kg JMCMAR 23,269,80

SAFETY PROFILE: Poison by intraperitoneal route. Mutation data reported. When heated to decomposition it emits very toxic fumes of NO_x and SO_x.

ADL250 CAS:72738-89-7 ***HR: 2***
4'(9-ACRIDINYLAMINO)HEXANESULFONANILIDE
mf: $C_{25}H_{27}N_3O_2S$ mw: 433.61

TOXICITY DATA with REFERENCE
mmo-sat 4467 nmol/L JMCMAR 23,269,80
ipr-mus LD10:120 mg/kg JMCMAR 23,269,80

SAFETY PROFILE: Poison by intraperitoneal route. Mutation data reported. When heated to decomposition it emits very toxic fumes of NO_x and SO_x.

ADL500 CAS:54301-15-4 ***HR: 3***
4'-(9-ACRIDINYLAMINO)METHANESULFON-m-ANISIDE MONOHYDROCHLORIDE
mf: $C_{21}H_{19}N_3O_3S{\cdot}ClH$ mw: 429.95

SYNS: m-AMSA HYDROCHLORIDE ◇ NCI-C03190 ◇ NSC 141549

TOXICITY DATA with REFERENCE
dnd-mus:leu 2500 μg/L CNREA8 38,1329,78
dns-mus-ipr 5 mg/kg CNREA8 38,1329,78
oms-ham:ovr 2 mg/L JNCIAM 60,1147,78
cyt-ham:ovr 2 mg/L JNCIAM 60,1147,78
sce-ham:ovr 50 μg/L JNCIAM 60,1155,78
orl-mus LD50:181 mg/kg NCISP* JAN86
ipr-mus LD50:20560 μg/kg NCISP* JAN86
scu-mus LD50:110 mg/kg NCISP* JAN86

SAFETY PROFILE: Poison by ingestion, subcutaneous and intraperitoneal routes. Mutation data reported. See also SULFONATES. When heated to decomposition it emits very toxic fumes of NO_x, SO_x, and HCl.

ADL750 CAS:51264-14-3 ***HR: 3***
4'-(9-ACRIDINYLAMINO)METHANESULPHON-m-ANISIDIDE
mf: $C_{21}H_{19}N_3O_3S$ mw: 393.49

SYNS: 4'-(9-ACRIDINYLAMINO)-3'-METHOXYMETHANESULFONANILIDE ◇ 4'-(9-ACRIDINYLAMINO)METHYLSULFONYL-m-ANISIDINE ◇ AMSA ◇ m-AMSA ◇ AMSACRINE ◇ m-AMSA METHANESULFONATE ◇ AMSIDINE ◇ AMSINE ◇ NSC 141549 ◇ NSC 249992

TOXICITY DATA with REFERENCE
msc-mus:lym 1 μg/L ENMUDM 8(Suppl 6),23,86
dnd-ham:lng 500 nmol/L CNREA8 45,3143,85
ipr-rat TDLo:8 mg/kg (female 6-9D post):TER FAATDF 7,214,86
ipr-rat TDLo:8 mg/kg (female 6-9D post):REP FAATDF 7,214,86
ivn-man LDLo:5405 μg/kg/3H-C:BLD AIMDAP 143,165,83
ivn-hmn TDLo:12 mg/kg:GIT CNREA8 38,3712,78
orl-mus LD50:53420 μg/kg NCISP* JAN86
ipr-mus LD50:15470 μg/kg NCISP* JAN86
scu-mus LD50:110 mg/kg NCISP* JAN86
orl-dog LD50:50 mg/kg CTRRDO 66,1939,82
ivn-mus LD50:33.7 mg/kg CTRRDO 64,855,80
ivn-dog LD50:6.25 mg/kg CTRRDO 64,855,80

CONSENSUS REPORTS: EPA Genetic Toxicology Program.

SAFETY PROFILE: Poison by ingestion, intravenous, subcutaneous, and intraperitoneal routes. Human systemic effects by intravenous route: nausea or vomiting, thrombosis distant from injection site, and bone marrow changes. An experimental teratogen. Other experimental reproductive effects. Mutation data reported. When heated to decomposition it emits very toxic fumes of NO_x and SO_x.

ADM000 ***HR: 3***
4'-(9-ACRIDINYLAMINO)-2'-METHOXYMETHANESULFONANILIDE
mf: $C_{21}H_{19}N_3O_3S$ mw: 393.49

SYNS: N-(4-(9-ACRIDINYLAMINO)-3-METHOXYPHENYL)METHANESULFONAMIDE ◇ m-AMSA

TOXICITY DATA with REFERENCE
dnd-mus:leu 10 μmol/L BICHAW 20,6553,81
dnd-mus:oth 40 μmol/L ANBCA2 125,91,82
cyt-mus:lym 1 mg/L ENMUDM 8(Suppl 6),23,86
msc-mus:lym 100 μg/L ENMUDM 8(Suppl 6),23,86
dnd-mam:lym 100 mmol/L CBINA8 44,53,83
mmo-sat 162 μmol/L JMCMAR 23,269,80
sce-hmn:lym 50 μg/L MUREAV 68,295,79
ivn-hmn TDLo:34 mg/kg:CVS,BLD CTRRDO 62,1421,78
ipr-mus LD10:110 mg/kg JMCMAR 23,269,80

SAFETY PROFILE: Poison by intraperitoneal route. Human mutation data reported. Human systemic effects by intravenous route: thrombosis distant from injection site, leukopenia, and thrombocytopenia. When heated to decomposition it emits very toxic fumes of NO_x and SO_x.

ADM250 ***HR: 3***
4'-(9-ACRIDINYLAMINO)-3'-METHOXYMETHANESULFONANILIDE
mf: $C_{21}H_{19}N_3O_3S$ mw: 393.49

TOXICITY DATA with REFERENCE
dnd-mam:lym 10 800 μmol/L JMCMAR 21,658,78
mmo-sat 20 μmol/L JMCMAR 23,269,80
ipr-mus LD10:9 mg/kg JMCMAR 23,269,80

SAFETY PROFILE: Poison by intraperitoneal route. Mutation data reported. When heated to decomposition it emits very toxic fumes of NO_x and SO_x.

ADM500 CAS:59988-01-1 ***HR: 3***
N-(4-(ACRIDINYL-9-AMINO)-3-METHOXYPHENYL)ETHANESULFONAMIDE METHANESULFONATE
mf: $C_{22}H_{21}N_3O_3S{\cdot}CH_5O_3S$ mw: 504.64

TOXICITY DATA with REFERENCE
mma-sat 93200 nmol/L JMCMAR 22,251,79
ipr-mus LD10:10500 μg/kg JMCMAR 22,251,79

SAFETY PROFILE: Poison by intraperitoneal route. Mutation data reported. See also SULFONATES. When heated to decomposition it emits very toxic fumes of SO_x and NO_x.

ADN000 CAS:57164-87-1 ***HR: 2***
4'-(9-ACRIDINYLAMINO)-2-METHYLMETHANESULFONANILIDE
mf: $C_{21}H_{19}N_3O_2S$ mw: 377.49

TOXICITY DATA with REFERENCE
mmo-sat 36 μmol/L JMCMAR 23,269,80
ipr-mus LD10:200 mg/kg JMCMAR 23,269,80

SAFETY PROFILE: Mutation data reported. Moderately toxic by intraperitoneal route. See also SULFONATES. When heated to decomposition it emits very toxic NO_x and SO_x fumes.

ADN250 CAS:57164-89-3 ***HR: 3***
4'-(9 ACRIDINYLAMINO)-3'-METHYLMETHANESULFONANILIDE
mf: $C_{21}H_{19}N_3O_2S$ mw: 377.49

TOXICITY DATA with REFERENCE
mmo-sat 132 μmol/L JMCMAR 23,269,80
ipr-mus LD10:110 mg/kg JMCMAR 23,269,80

SAFETY PROFILE: Poison by intraperitoneal route. Mutation data reported. When heated to decomposition it emits very toxic fumes of NO_x and SO_x.

ADN500 CAS:72738-98-8 ***HR: D***
4'-(9-ACRIDINYLAMINO)-2'-NITROMETHANESULFONANILIDE
mf: $C_{20}H_{16}N_4O_4S$ mw: 408.46

TOXICITY DATA with REFERENCE
mmo-sat 28800 nmol/L JMCMAR 23,269,80

SAFETY PROFILE: Mutation data reported. See also SULFONATES. When heated to decomposition it emits very toxic fumes of NO_x and SO_x.

ADO250 CAS:53221-85-5 ***HR: 3***
N-(p-(ACRIDIN-9-YLAMINO)PHENYL)BUTANESULFONAMIDE, HYDROCHLORIDE
mf: $C_{23}H_{23}N_3O_2S•ClH$ mw: 442.01

TOXICITY DATA with REFERENCE
mma-sat 49300 nmol/L JMCMAR 22,251,79
ipr-mus LD10:350 mg/kg JMCMAR 22,251,79

SAFETY PROFILE: Poison by intraperitoneal route. Mutation data reported. When heated to decomposition it emits very toxic fumes of HCl, SO_x, and NO_x.

ADO500 CAS:53221-86-6 ***HR: 2***
N-(p-(9-ACRIDINYLAMINO)PHENYL)-1-ETHANESULFONAMIDE
mf: $C_{21}H_{19}N_3O_2S$ mw: 377.49

TOXICITY DATA with REFERENCE
mmo-sat 24 μmol/L JMCMAR 23,269,80
ipr-mus LD10:330 mg/kg JMCMAR 21,430,78

SAFETY PROFILE: Moderately toxic by intraperitoneal route. Mutation data reported. See also SULFONATES. When heated to decomposition it emits very toxic fumes of SO_x and NO_x.

ADO750 CAS:53221-83 ***HR: 3***
N-(p-(ACRIDIN-9-YLAMINO)PHENYL)-ETHANESULFONAMIDE, HYDROCHLORIDE-3
mf: $C_{21}H_{19}N_3O_2S•ClH$ mw: 413.95

TOXICITY DATA with REFERENCE
ipr-mus LD10:330 mg/kg JMCMAR 22,251,79
mma-sat 81600 nmol/L JMCMAR 22,251,79

SAFETY PROFILE: Poison by intraperitoneal route. See also SULFONATES. Mutation data reported. When heated to decomposition it emits very toxic fumes of Cl^-, SO_x, and NO_x.

ADP000 CAS:66147-69-1 ***HR: 2***
N-(p-(ACRIDIN-9-YLAMINO) PHENYLHEXANESULFONAMIDE HYDROCHLORIDE
mf: $C_{25}H_{27}N_3O_2S•ClH$ mw: 470.07

TOXICITY DATA with REFERENCE
mma-sat 9400 nmol/L JMCMAR 22,251,79
ipr-mus LD10:120 mg/kg JMCMAR 22,251,79

SAFETY PROFILE: Moderately toxic by intraperitoneal route. Mutation data reported. When heated to decomposition it emits very toxic fumes such as Cl^-, SO_x, and NO_x.

ADP500 CAS:75775-83-6 ***HR: 3***
N-(p-(ACRIDIN-9-YLAMINO)PHENYL) METHANESULFONAMIDE HYDROCHLORIDE
mf: $C_{20}H_{17}N_3O_2S•ClH$ mw: 399.92

TOXICITY DATA with REFERENCE
mma-sat 110 μmol/L JMCMAR 22,251,79
ipr-mus LD10:66 mg/kg JMCMAR 22,251,79

SAFETY PROFILE: Poison by intraperitoneal route. Mutation data reported. See also SULFONATES. When heated to decomposition it emits very toxic fumes of NO_x, SO_x, and HCl.

ADP750 CAS:66147-68-0 ***HR: 3***
N-(p-(ACRIDIN-9-YLAMINO)PHENYL)PENTANESULFONAMIDE HYDROCHLORIDE
mf: $C_{24}H_{25}N_3O_2S•ClH$ mw: 456.04

TOXICITY DATA with REFERENCE
mma-sat 15600 nmol/L JMCMAR 22,251,79
ipr-mus LD10:70 mg/kg JMCMAR 22,251,79

SAFETY PROFILE: Poison by intraperitoneal route. Mutation data reported. See also SULFONATES. When heated to decomposition it emits very toxic fumes of NO_x, SO_x, and HCl.

ADQ000 CAS:53221-88-8 ***HR: 2***
N-(p-(9-ACRIDINYLAMINO)PHENYL)-1-PROPANESULFONAMIDE
mf: $C_{22}H_{21}N_3O_2S$ mw: 391.52

TOXICITY DATA with REFERENCE
mmo-sat 24 μmol/L JMCMAR 23,269,80
ipr-mus LD10:350 mg/kg JMCMAR 21,430,78

SAFETY PROFILE: Poison by intraperitoneal route. See also SULFONATES. Mutation data reported. When heated to decomposition it emits very toxic fumes of SO_x and NO_x.

ADQ250 ***HR: 2***
N-(p-(ACRIDIN-9-YLAMINO)PHENYL)PROPANESULFONAMIDE HYDROCHLORIDE
mf: $C_{22}H_{21}N_3O_2S \cdot ClH$ mw: 427.98

TOXICITY DATA with REFERENCE
mma-sat 50100 nmol/L JMCMAR 22,251,79
ipr-mus LD10:350 mg/kg JMCMAR 22,251,79

SAFETY PROFILE: Poison by intraperitoneal route. See also SULFONATES. Mutation data reported. When heated to decomposition it emits very toxic fumes of HCl, SO_x, and NO_x.

ADQ500 CAS:72667-36-8 ***HR: D***
N-(9-ACRIDINYL)-N'-(2-CHLOROETHYL)-1,3-PROPANEDIAMINE
mf: $C_{18}H_{20}ClN_2$ mw: 299.65

SYN: ICR 449

TOXICITY DATA with REFERENCE
mmo-sat 500 ng/plate MUREAV 136,185,84
msc-ham:ovr 1 μmol/L CNREA8 39,4875,79

SAFETY PROFILE: Mutation data reported. When heated to decomposition it emits very toxic fumes of Cl and NO_x.

ADR000 CAS:107-02-8 ***HR: 3***
ACROLEIN
DOT: UN 1092
mf: C_3H_4O mw: 56.07

$$H_2C=CHCOH$$

PROP: Colorless or yellowish liquid; disagreeable, choking odor. Sol in water, alc, and ether. Mp: −87.7°, bp: 52.5°, flash p: <0°F, d: 0.841 @ 20°/4°, autoign temp: unstable (455°F), lel: 2.8%, uel: 31%, vap d: 1.94.

SYNS: ACQUINITE ◇ ACRALDEHYDE ◇ ACROLEINA (ITALIAN) ◇ ACROLEINE (DUTCH, FRENCH) ◇ ACRYLALDEHYD (GERMAN) ◇ ACRYLALDEHYDE ◇ ACRYLIC ALDEHYDE ◇ AKROLEIN (CZECH) ◇ AKROLEINA (POLISH) ◇ ALDEIDE ACRILICA (ITALIAN) ◇ ALDEHYDE ACRYLIQUE (FRENCH) ◇ ALLYL ALDEHYDE ◇ AQUALINE ◇ BIOCIDE ◇ CROLEAN ◇ ETHYLENE ALDEHYDE ◇ MAGNACIDE H ◇ NSC 8819 ◇ PROPENAL (CZECH) ◇ 2-PROPENAL ◇ PROP-2-EN-1-AL ◇ 2-PROPEN-1-ONE ◇ PROPYLENE ALDEHYDE ◇ RCRA WASTE NUMBER P003 ◇ SLIMICIDE

TOXICITY DATA with REFERENCE
eye-hmn 500 ppb/12M IAPWAR 4,79,61
skn-rbt 5 mg open SEV UCDS** 6/18/71
skn-rbt 500 mg/24H SEV 28ZPAK -,41,72
eye-rbt 1 mg SEV UCDS** 6/18/71
eye-rbt 50 μg/24H SEV 28ZPAK -,41,72
sce-ham:ovr 10 μmol/L CGCGBR 26,108,80
mma-sat 50 μg/plate NTPTB* JAN 82
ivn-rbt TDLo:6 mg/kg (9D preg):REP ARZNAD 30,2080,80
ihl-man TCLo:1 ppm:IRR,IMM BMJOAE 2,913,56
ihl-hmn LCLo:5500 ppb 34ZIAG -,73,69
ihl-hmn LCLo:153 ppm/10M NTIS** PB214-270
ihl-chd TCLo:300 ppb/2H:PUL NPMDAD 8,2469,79
idr-man LDLo:250 mg/kg AEXPBL 43,351,1900
orl-rat LD50:46 mg/kg FMCHA2 -,C24,89
ihl-rat LC50:300 mg/m^3/30M APTOA6 6,299,50
ipr-rat LD50:4 mg/kg TXAPA9 71,84,83
scu-rat LD50:50 mg/kg APTOA6 6,299,50
orl-mus LD50:40 mg/kg BIJOAK 34,1196,40
ihl-mus LC50:66 ppm/6H IAANBS 26,281,70
ipr-mus LD50:9008 μg/kg NCISP* JAN86
scu-mus LD50:30 mg/kg APTOA6 6,299,50
ihl-cat LCLo:1570 mg/m^3/8H APTOA6 6,299,50

CONSENSUS REPORTS: IARC Cancer Review: Group 3 IMEMDT 7,78,87; Animal Inadequate Evidence IMEMDT 36,133,85; IMEMDT 19,479,79; Human Inadequate Evidence IMEMDT 36,133,85. Community Right-To-Know List. EPA Extremely Hazardous Substances List. Reported in EPA TSCA Inventory.

OSHA PEL: (Transitional: TWA 0.1 ppm) TWA 0.1 ppm; STEL 0.3 ppm
ACGIH TLV: TWA 0.1 ppm; STEL 0.3 ppm
DFG MAK: 0.1 ppm (0.25 mg/m^3)
DOT Classification: Flammable Liquid; Label: Flammable Liquid and Poison.

SAFETY PROFILE: Human poison by inhalation and intradermal route. Poison experimentally by most routes. Human systemic irritant and pulmonary system effects by inhalation include: lacrimation, delayed hypersensitivity with multiple organ involvement, and respiratory system damage. Severe eye and skin irritant. Experimental reproductive effects. Human mutation data reported. Questionable carcinogen. Dangerous fire hazard when exposed to heat, flame, or oxidizers. An explosion hazard. Incompatible with amines, SO_2, metal salts, oxidants, (light + heat). Violent polymerization reaction on contact with strong acid, strong base, weak acid conditions (e.g., nitrous fumes, sulfur dioxide, carbon dioxide), thiourea, or dimethylamine. When heated to decomposition it emits highly toxic fumes; can react vigorously with oxidizing materials. To fight fire, use CO_2, dry chemical or alcohol foam.

ADR250 CAS:869-29-4 ***HR: 3***
ACROLEIN DIACETATE
mf: $C_7H_{10}O_4$ mw: 158.17

PROP: Liquid. Mp: −36.6°, bp: 107° @ 50 mm, flash p: 180°F (OC), d: 1.0749 @ 20°/20°, vap d: 5.46.

SYNS: ALLYLIDENE DIACETATE ◇ DIACETOXYPROPENE ◇ 1,1-DIACETOXYPROPENE-2 ◇ 3,3-DIACETOXYPROPENE ◇ SD-345 ◇ SHELL 345 ◇ SHELL SD 345

TOXICITY DATA with REFERENCE
skn-rbt 10 mg/24H JIHTAB 30,63,48
skn-rbt 500 mg open SEV UCDS** 12/27/71
eye-rbt 10 mg SEV UCDS** 12/27/71
orl-mus LD50:35 mg/kg SCCUR* -,1,61
skn-gpg LDLo:500 mg/kg SCCUR* -,1,61
ihl-mus LCLo:853 ppm/15M SCCUR* -,1,61

SAFETY PROFILE: Poison by ingestion. Moderately toxic by inhalation and skin contact. A severe skin and eye irritant. Flammable when exposed to heat or flame; can react with oxidizing materials. When heated to decomposition it emits acrid smoke and fumes. To fight fire, water may be used to blanket the fire; also, foam, CO_2, dry chemical.

ADR500 CAS:100-73-2 ***HR: 2***
ACROLEIN DIMER
DOT: UN 2607
mf: $C_6H_8O_2$ mw: 112.14

PROP: Liquid, sol in water. D: 1.0775 (20°), bp: 151.3°, fp: −100°, flash p: 118°F (OC).

SYNS: 3,4-DIHYDRO-2H-PYRAN-2-CARBOXALDEHYDE ◇ 2-FORMYL-3,4-DIHYDRO-2H-PYRAN ◇ PYRAN ALDEHYDE

TOXICITY DATA with REFERENCE
skn-rbt 500 mg open MLD UCDS** 7/27/65
eye-rbt 750 ug open SEV AMIHBC 10,61,54
skn-rbt 500 mg open MLD UCDS** 7/27/65
eye-rbt 750 μg SEV AMIHBC 10,61,54
orl-rat LD50:4920 mg/kg AMIHBC 10,61,54

DOT Classification: Flammable or Combustible Liquid; Label: Flammable Liquid.

SAFETY PROFILE: Mildly toxic by ingestion. A skin and severe eye irritant. Flammable when exposed to heat, flame or powerful oxidizing agents. To fight fire, use alcohol foam and multipurpose dry chemical. When heated to decomposition it emits acrid smoke and fumes.

ADR750 CAS:7008-42-6 ***HR: 3***
ACRONYCINE
mf: $C_{20}H_{19}NO_3$ mw: 321.40

SYNS: ACROMYCINE ◇ ACRONINE ◇ COMPOUND 42339 ◇ 3,12-DIHYDRO-6-METHOXY-3,3,12-TRIMETHYL-7H-PYRANO(2,3-C)ACRIDIN-7-ONE ◇ NCI-C01536 ◇ NSC 403169

TOXICITY DATA with REFERENCE
dni-mus:leu 1 μmol/L CNREA8 33,2310,73
orl-mus LD50:522 mg/kg NCISP* JAN 86
ipr-mus LD50:613 mg/kg NCISP* JAN 86
cyt-mus:fbr 10 mg/L/24H ARZNAD 27,1549,77
ipr-rat TDLo:1170 mg/kg/1Y-I:CAR NCITR* NCI-CG-TR-49,78
ipr-rat TD:585 mg/kg/1Y-I:NEO NCITR* NCI-CG-TR-49,78
ipr-rat TD:1800 mg/kg/1Y-I:CAR NCITR* NCI-CG-TR-49,78
ipr-mus TD:530 mg/kg/8W-I:ETA CNREA8 33,3069,73

CONSENSUS REPORTS: NCI Carcinogenesis Bioassay (ipr); Inadequate Studies: mouse NCITR* NCI-CG-TR-49,78; Clear Evidence: rat NCITR* NCI-CG-TR-49,78

SAFETY PROFILE: Moderately toxic by ingestion and intraperitoneal routes. Questionable carcinogen with experimental carcinogenic, neoplastigenic, and tumorigenic data. Mutation data reported. When heated to decomposition it emits toxic fumes of NO_x.

ADS150 ***HR: 2***
ACROSTICHUM AUREUM Linn., extract

PROP: Indian plant belonging to the family *Pteridiaceae* (IJEBA6 15,208,77)

TOXICITY DATA with REFERENCE
orl-rat TDLo:150 mg/kg (female 12-14D post):REP IJEBA6 15,208,77
ipr-mus LD50:750 mg/kg IJEBA6 15,208,77

SAFETY PROFILE: Moderately toxic by intraperitoneal route. Experimental reproductive effects. When heated to decomposition it emits acrid smoke and irritating fumes.

ADS250 CAS:79-06-1 ***HR: 3***
ACRYLAMIDE
DOT: UN 2074
mf: C_3H_5NO mw: 71.09

PROP: White, crystalline solid. Very sol in water, alc,

and ether. Mp: 84.5 ± 0.3°, bp: 125° @ 25 mm, d: 1.122 @ 30°, vap press: 1.6 mm @ 84.5°, vap d: 2.45.

SYNS: ACRYLIC AMIDE ◇ AKRYLAMID (CZECH) ◇ ETHYLENECARBOXAMIDE ◇ PROPENAMIDE ◇ 2-PROPENAMIDE ◇ RCRA WASTE NUMBER U007

TOXICITY DATA with REFERENCE
sce-rat-orl 600 mg/kg/10D-C ENMUDM 7(Suppl 3),79,85
dlt-mus-ipr 125 mg/kg MUREAV 173,35,86
orl-rat TDLo:200 mg/kg (7-16D preg):REP TOLED5 7,233,81
orl-rat TDLo:1456 mg/kg/2Y-C:CAR TXAPA9 85,154,86
ipr-mus TDLo:24 mg/kg/8W-I:NEO CNREA8 44,107,84
orl-mus TDLo:300 mg/kg/2W-I:CAR CALEDQ 24,209,84
orl-rat LD:1456 mg/kg/2Y-C:CAR,REP TXAPA9 85,154,86
orl-rat LD50:124 mg/kg AMPMAR 36,58,75
skn-rat LD50:400 mg/kg GISAAA 44(10),73,79
ipr-rat LD50:90 mg/kg AMPMAR 36,58,75
orl-mus LD50:107 mg/kg ARTODN 47,179,81
ipr-mus LD50:170 mg/kg TXAPA9 33,142,75
orl-rbt LDLo:126 mg/kg TXAPA9 6,172,64
skn-rbt LDLo:1000 mg/kg TXAPA9 6,172,64
orl-gpg LDLo:252 mg/kg TXAPA9 6,172,64
scu-gpg LD50:170 mg/kg MELAAD 47,192,56

CONSENSUS REPORTS: IARC Cancer Review: Group 2B IMEMDT 7,56,87; Animal Sufficient Evidence IMEMDT 39,41,86. EPA Extremely Hazardous Substances List. Community Right-To-Know List. Reported in EPA TSCA Inventory.

OSHA PEL: (Transitional: TWA 0.3 mg/m^3 (skin)) TWA 0.03 mg/m^3 (skin)
ACGIH TLV: Suspected Human Carcinogen, TWA 0.03 mg/m^3 (skin)
DFG MAK: Animal Carcinogen, Suspected Human Carcinogen.
NIOSH REL: TWA 0.3 mg/m^3
DOT Classification: IMO: Poison B; Label: St. Andrews Cross.

SAFETY PROFILE: Confirmed carcinogen with experimental carcinogenic and neoplastigenic data. Poison by ingestion, skin contact, and intraperitoneal routes. Experimental reproductive effects. Mutation data reported. A skin and eye irritant. Intoxication from it has caused a peripheral neuropathy, erythema and peeling of the palms. In industry, intoxication is mainly via dermal route, next via inhalation, and last via ingestion. Time of onset varied from 1-24 months to 8 years. Symptoms were, via dermal route, a numbness, tingling and touch tenderness. In a couple of weeks, coldness of extremities; later, excessive sweating, bluish-red and peeling of palms, marked fatigue and limb-weakness. It is dangerous because it can be absorbed through the unbroken skin. From animal experiments it seems to be a central nervous system toxin. Adult rats fed an average of 30 mg/kg for 14 days were all partially paralyzed and had reduced their food consumption by 50 percent. Polymerizes violently at its melting point. When heated to decomposition it emits acrid fumes and NO_x.

ADS750 CAS:79-10-7 ***HR: 3***
ACRYLIC ACID
DOT: UN 2218
mf: $C_3H_4O_2$ mw: 72.07

$$H_2C=CHCO{\cdot}OH$$

PROP: Liquid, acrid odor. Misc in water, benzene, alc, chloroform, ether, and acetone. Mp: 13°, bp: 141°, d: 1.062, vap press: 10 mm @ 39.9°, flash p: 130°F (OC), vap d: 2.45.

SYNS: ACROLEIC ACID ◇ ACRYLIC ACID (ACGIH,DOT,OSHA) ◇ ACRYLIC ACID, inhibited (DOT) ◇ ACRYLIC ACID, GLACIAL ◇ ETHYLENECARBOXYLIC ACID ◇ GLACIAL ACRYLIC ACID ◇ KYSELINA AKRYLOVA ◇ PROPENE ACID ◇ PROPENOIC ACID ◇ 2-PROPENOIC ACID (9CI) ◇ RCRA WASTE NUMBER U008 ◇ VINYLFORMIC ACID

TOXICITY DATA with REFERENCE
skn-rbt 500 mg open SEV UCDS** 2/2/65
eye-rbt 1 mg SEV UCDS** 2/2/65
eye-rbt 250 μg/24H SEV 85JCAE-,309,86
orl-rat TDLo:100 g/kg (13W pre-3W post):REP DCTODJ 6,1,83
ipr-rat TDLo:7329 μg/kg (5-15D preg):TER JDREAF 51,1632,72
skn-mus TDLo:37440 mg/kg/78W-I:CAR EPASR* 8EHQ-0386-0592
scu-mus TDLo:2912 mg/kg/52W-I:ETA CBINA8 61,189,87
orl-rat LD50:33500 μg/kg 85GMAT -,16,82
ihl-rat LCLo:4000 ppm/4H TXAPA9 28,313,74
ipr-rat LD50:22 mg/kg JDREAF 51,1632,72
orl-mus LD50:2400 mg/kg BIJOAK 34,1196,40
ihl-mus LCLo:5300 mg/m^3/2H 85GMAT -,16,82
scu-mus LD50:1590 mg/kg JPPMAB 21,85,69
skn-rbt LD50:280 mg/kg TXAPA9 28,313,74

CONSENSUS REPORTS: IARC Cancer Review: Group 3 IMEMDT 7,56,87; Human Inadequate Evidence IMEMDT 19,47,79. Community Right-To-Know List. Reported in EPA TSCA Inventory.

OSHA PEL: TWA 10 ppm (skin)
ACGIH TLV: 2 ppm (skin)
DOT Classification: IMO: Corrosive Material; Label: Corrosive, Flammable Liquid.

SAFETY PROFILE: Poison by ingestion, skin contact, and intraperitoneal routes. An experimental teratogen. Other experimental reproductive effects. A severe skin and eye irritant. Questionable carcinogen with experimental carcinogenic and tumorigenic data. Corrosive. Exothermic polymerization at room temperature may become explosive if confined. A fire hazard when exposed to heat or flame.

ADT000 CAS:2206-89-5 ***HR: 3***
ACRYLIC ACID-β-CHLOROETHYL ESTER
mf: $C_5H_7ClO_2$ mw: 134.57

SYNS: 2-CHLOROETHANOL ACRYLATE ◇ CHLOROETHYL ACRYLATE ◇ β-CHLOROETHYL ACRYLATE ◇ 2-CHLOROETHYL ACRYLATE ◇ 2-PROPENOIC ACID-2-CHLOROETHYL ESTER

TOXICITY DATA with REFERENCE
skn-rbt 10 mg/24H open SEV AMIHBC 4,119,51
eye-rbt 50 μg open SEV AMIHBC 4,119,51
orl-rat LD50:180 mg/kg AMIHBC 4,119,51
ihl-rat LCLo:250 ppm/4H AMIHBC 4,119,51
ipr-mus LDLo:64 mg/kg CBCCT* 3,360,51

SAFETY PROFILE: Poison by inhalation, ingestion and intraperitoneal routes. A severe skin and eye irritant. See also ESTERS. When heated to decomposition it emits toxic fumes of Cl^-.

ADT050 CAS:17831-71-9 ***HR: 2***
ACRYLIC ACID, DIESTER with TETRAETHYLENE GLYCOL
mf: $C_{14}H_{22}O_7$ mw: 302.36

SYNS: ACRYLIC ACID, OXYBIS(ETHYLENEOXYETHYLENE) ESTER ◇ 2-PROPENOIC ACID, OXYBIS(2,1-ETHANEDIYLOXY-2,1-ETHANEDIYL)ESTER ◇ TETRAETHYLENE GLYCOL DIACRYLATE

TOXICITY DATA with REFERENCE
skn-rbt 500 mg/24H MOD JTEHD6 19,149,86
eye-rbt 100 mg SEV JTEHD6 19,149,86
skn-mus TDLo:16 g/kg/80W-I:ETA JTEHD6 19,149,86

CONSENSUS REPORTS: Reported in EPA TSCA Inventory.

SAFETY PROFILE: Moderate skin and severe eye irritant. Questionable carcinogen with experimental tumorigenic data. When heated to decomposition it emits acrid smoke and irritating fumes.

ADT111 CAS:106-71-8 ***HR: 3***
ACRYLIC ACID ESTER with HYDRACRYLONITRILE
mf: $C_6H_7NO_2$ mw: 125.14

PROP: Liquid, sol in water, d: 1.069, bp: polymerizes, fp: −16.9°, flash p: 255°F (COC), vap d: 4.3.

SYNS: ACRYLIC ACID-2-CYANOETHYL ESTER ◇ CYANOETHYL ACRYLATE ◇ 2-CYANOETHYL ACRYLATE ◇ 2-CYANOETHYL PROPENOATE ◇ HYDRACRYLONITRILE ACRYLATE ◇ 2-PROPENOIC ACID-2-CYANOETHYL ESTER

TOXICITY DATA with REFERENCE
skn-rbt 500 mg open MOD UCDS** 9/27/60
eye-rbt 5 mg MLD UCDS** 9/27/60
orl-rat LD50:180 mg/kg UCDS** 9/27/60

CONSENSUS REPORTS: Reported in EPA TSCA Inventory. Cyanide and its compounds are on the Community Right-To-Know List.

SAFETY PROFILE: Poison by ingestion. A skin and eye irritant. See also ESTERS and NITRILES. A fire hazard when exposed to heat or flame. When heated to decomposition it emits toxic fumes of NO_x and CN^-.

ADT250 CAS:4074-88-8 ***HR: 3***
ACRYLIC ACID-2-ETHOXYETHANOL DIESTER
mf: $C_{10}H_{14}O_5$ mw: 214.24

SYNS: ACRYLIC ACID OXYDIETHYLENE ESTER ◇ DIETHYLENE GLYCOL DIACRYLATE ◇ OXYDIETHYLENE ACRYLATE ◇ OXYDIETHYLENE DIACRYLATE

TOXICITY DATA with REFERENCE
orl-rat LD50:770 mg/kg TXAPA9 28,313,74
skn-rbt LD50:180 mg/kg TXAPA9 28,313,74

CONSENSUS REPORTS: Reported in EPA TSCA Inventory.

SAFETY PROFILE: Poison by skin contact. Moderately toxic by ingestion. See also ESTERS. When heated to decomposition it emits acrid smoke and fumes.

ADT500 CAS:106-74-1 ***HR: 2***
ACRYLIC ACID-2-ETHOXYETHYL ESTER
mf: $C_7H_{12}O_3$ mw: 144.19

SYNS: ACRYLIC ACID-2-ETHOXYETHANOL ESTER ◇ CELLOSOLVE ACRYLATE ◇ ETHOXYETHYL ACRYLATE ◇ 2-ETHOXYETHYL ACRYLATE ◇ 2-ETHOXYETHYL-2-PROPENOATE ◇ ETHYLENE GLYCOL MONOETHYL ETHER ACRYLATE ◇ ETHYLENE GLYCOL MONOETHYL ETHER PROPENOATE ◇ 2-PROPENOIC ACID-2-ETHOXYETHYL ESTER

TOXICITY DATA with REFERENCE
skn-rbt 10 mg/24H open MLD AMIHBC 10,61,54
skn-rbt 500 mg open MLD UCDS** 6/6/69
eye-rbt 20 mg open SEV AMIHBC 10,61,54
orl-rat LD50:1070 mg/kg UCDS** 9/15/64
ihl-rat LCLo:500 ppm/4H AMIHBC 10,61,54
skn-rbt LD50:1010 mg/kg AMIHBC 10,61,54

CONSENSUS REPORTS: Reported in EPA TSCA Inventory.

SAFETY PROFILE: Moderately toxic by various routes. A skin and severe eye irritant. See also ESTERS. When heated to decomposition it emits acrid smoke and fumes.

ADU250 CAS:103-11-7 ***HR: 3***
ACRYLIC ACID-2-ETHYLHEXYL ESTER
mf: $C_{11}H_{20}O_2$ mw: 184.31

PROP: Bp: 130° @ 50 mm, fp: −90°, flash p: 180°F (OC), d: 0.8869 @ 20°/20°, vap press: 1 mm @ 50°, vap d: 6.35.

SYNS: 2-ETHYLHEXYL ACRYLATE ◇ 2-ETHYLHEXYL-2-PROPENOATE ◇ OCTYL ACRYLATE ◇ 2-PROPENOIC ACID-2-ETHYLHEXYL ESTER

TOXICITY DATA with REFERENCE
skn-rbt 20 mg/24H MOD 85JCAE -,372,86
skn-rbt 500 mg open MLD UCDS** 11/3/71
skn-rbt 10 mg/24H open SEV AMIHBC 4,119,51
eye-rbt 5 mg SEV AJOPAA 29,1363,46
eye-rbt 500 mg/24H MLD 85JCAE -,372,86
skn-mus TDLo:187 g/kg/78W-I:CAR JTEHD6 16,55,85
skn-mus TD:240 g/kg/2Y-C:NEO EPASR* 8EHQ-1079-0262
orl-rat LD50:5660 mg/kg UCDS** 9/15/64
ipr-rat LD50:1670 mg/kg AMPMAR 36,58,75
orl-mus LD50:4400 mg/kg GTPZAB 26(9),52,82
ihl-mus LCLo:600 mg/m^3 GTPZAB 26(9),52,82
ipr-mus LD50:1326 mg/kg JDREAF 51,526,72
skn-rbt LD50:8480 mg/kg AMIHBC 4,119,51

CONSENSUS REPORTS: Reported in EPA TSCA Inventory.

SAFETY PROFILE: Moderately toxic by inhalation and various other routes. A severe skin and eye irritant. Questionable carcinogen with experimental carcinogenic and neoplastigenic data. Flammable. A fire hazard when exposed to heat or flame. To fight fire use alcohol foam, CO_2, dry chemical. When heated to decomposition it emits acrid smoke and irritating fumes. See also ESTERS.

ADU500 ***HR: 3***
ACRYLIC ACID ETHYLHEXYL ESTER mixed with HYDROXYETHYL ESTER (505850)

SYNS: ETHYLHEXYL ACRYLATE 50:50 MIXTURE ◇ HYDROXYETHYL ACRYLATE

TOXICITY DATA with REFERENCE
orl-rat LD50:1540 mg/kg UCDS** 3/23/73
skn-rbt LD50:170 mg/kg UCDS** 3/23/73

SAFETY PROFILE: Poison by skin contact. Moderately toxic by ingestion. See also ESTERS. When heated to decomposition it emits acrid smoke and fumes.

ADU750 CAS:122-93-0 ***HR: 3***
ACRYLIC ACID-2-(5'-ETHYL-2-PYRIDYL)ETHYL ESTER
mf: $C_{12}H_{15}NO_2$ mw: 205.28

PROP: Liquid, very sltly water-sol. D: 1.0458 @ 20°, bp: 181° @ 50 mm, fp: −75°.

SYNS: 2-(5-ETHYL-2-PYRIDYL)ETHYL ACRYLATE ◇ 2-(5-ETHYL-2-PYRIDYL)ETHYL PROPENOATE

TOXICITY DATA with REFERENCE
skn-rbt 10 mg/24H open SEV AIHAAP 23,95,62
orl-rat LD50:4920 mg/kg AIHAAP 23,95,62
skn-rbt LD50:2230 mg/kg AIHAAP 23,95,62

SAFETY PROFILE: Moderately toxic by skin contact. Mildly toxic by ingestion. A severe skin irritant. See also ESTERS. Flammable. Store away from heat, sparks, or powerful oxidizers. To fight fire, use foam, CO_2, dry chemicals. When heated to decomposition it emits toxic fumes of NO_x.

ADV000 CAS:2499-95-8 ***HR: 1***
ACRYLIC ACID HEXYL ESTER
mf: $C_9H_{16}O_2$ mw: 156.25

SYNS: AGEFLEX n-HA ◇ HEXYL ACRYLATE ◇ N-HEXYL ACRYLATE ◇ HEXYL-2-PROPENOATE ◇ 2-PROPENOIC ACID, HEXEL ESTER

TOXICITY DATA with REFERENCE
orl-rat LD50:26 g/kg AIHAAP 30,470,69
skn-rbt LD50:5660 mg/kg AIHAAP 30,470,69

CONSENSUS REPORTS: Reported in EPA TSCA Inventory.

SAFETY PROFILE: Mildly toxic by skin contact. See also ESTERS. When heated to decomposition it emits acrid smoke and fumes.

ADV250 CAS:818-61-1 ***HR: 2***
ACRYLIC ACID-2-HYDROXYETHYL ESTER
mf: $C_5H_8O_3$ mw: 116.13

SYNS: 2-(ACRYLOYLOXY)ETHANOL ◇ BISOMER 2HEA ◇ ETHYLENE GLYCOL ACRYLATE ◇ ETHYLENE GLYCOL MONOACRYLATE ◇ HYDROXYETHYL ACRYLATE ◇ β-HYDROXYETHYL ACRYLATE ◇ 2-HYDROXYETHYL ACRYLATE ◇ 2-PROPENOIC ACID-2-HYDROXYETHYL ESTER (9CI)

TOXICITY DATA with REFERENCE
skn-rbt 500 mg open MOD UCDS** 3/23/73
skn-rbt 10 mg/24H open MLD AMIHBC 4,119,51
eye-rbt 1 mg SEV UCDS** 3/23/73
orl-mus LD50:601 mg/kg TOLED5 11,125,82
orl-rat LD50:650 mg/kg UCDS** 3/23/73
ihl-rat LCLo:500 ppm/4H AMIHBC 4,119,51
skn-rbt LD50:1010 mg/kg AMIHBC 4,119,51

CONSENSUS REPORTS: Reported in EPA TSCA Inventory.

SAFETY PROFILE: Moderately toxic by ingestion, inhalation, and skin contact. A moderate skin and severe eye irritant. When heated to decomposition it emits acrid smoke and fumes. See also ESTERS.

ADW000 CAS:9007-16-3 ***HR: 2***
ACRYLIC ACID, POLYMER with SUCROSEPOLYALLYL ETHER

SYNS: CARBOMER 934 ◇ CARBOPOL 934 ◇ SUCROSE, POLYALLYL ETHER, POLYMER with ACRYLIC ACID

TOXICITY DATA with REFERENCE
orl-rat LD50:4100 mg/kg JACTDZ 1(2),109,82
orl-mus LD50:4550 mg/kg JACTDZ 1(2),109,82
orl-gpg LD50:2500 mg/kg GRCSB* GC-36,54,60

CONSENSUS REPORTS: Allyl and its compounds are on the Community Right-To-Know List.

SAFETY PROFILE: Moderately toxic by ingestion. See also ETHERS and POLYMERS. When heated to decomposition it emits acrid smoke and fumes.

ADW200 CAS:9003-01-4 ***HR: D***
ACRYLIC ACID, POLYMERS
mf: $(C_3H_4O_2)_4$ mw: 168.06

SYNS: ACRYLIC ACID HOMOPOLYMER ◇ ACRYLIC ACID RESIN ◇ ACRYLIC POLYMER ◇ ACRYLIC RESIN ◇ ACRYSOL A 3 ◇ ALCOGUM ◇ ANTIPREX 461 ◇ ATACTIC POLY(ACRYLIC ACID) ◇ CARBOPOL 934 ◇ DISPEX C40 ◇ G-CURE ◇ HALOFLEX 208 ◇ JURIMER AC 10P ◇ NEOCRYL A-1038 ◇ OLD 01 ◇ PAA-25 ◇ POLYACRYLATE ◇ POLY(ACRYLIC ACID) ◇ POLYTEX 973 ◇ 2-PROPENOIC ACID HOMOPOLYMER

CONSENSUS REPORTS: IARC Cancer Review: Human Inadequate Evidence IMEMDT 19,47,79. Reported in EPA TSCA Inventory.

SAFETY PROFILE: Data on human carcinogenicity is indefinite. When heated to decomposition it emits acrid smoke and fumes.

ADW250 CAS:25916-47-6 ***HR: 3***
ACRYLIC ACID POLYMER, ZINC SALT
mf: $(C_3H_4O_2)_x$•xZn

SYNS: 2-PROPENOIC ACID, HOMOPOLYMER, ZINC SALT ◇ ZINC POLYACRYLATE ◇ ZINC POLYCARBOXYLATE

TOXICITY DATA with REFERENCE
imp-mus TDLo:1600 mg/kg/1Y-C:ETA JBMRBG 9,69,75

CONSENSUS REPORTS: Reported in EPA TSCA Inventory. Zinc and its compounds are on the Community Right-To-Know List.

SAFETY PROFILE: Questionable carcinogen with experimental tumorigenic data. See also ZINC COMPOUNDS. When heated to decomposition it emits toxic fumes of ZnO, acrid fumes, and CO.

ADW750 CAS:71073-91-1 ***HR: 3***
ACRYLIC ACID, TELOMER with TRICHLOROACETIC ACID
mf: $C_2HCl_3O_2$•3/2$C_3H_4O_2$ mw: 265.49

TOXICITY DATA with REFERENCE
orl-mus LD50:2750 mg/kg EJMCA5 14,119,79
ipr-mus LD50:300 mg/kg EJMCA5 14,119,79

SAFETY PROFILE: Poison by intraperitoneal route. Moderately toxic by ingestion. When heated to decomposition it emits toxic fumes of Cl^-.

ADX000 CAS:3076-04-8 ***HR: 2***
ACRYLIC ACID TRIDECYL ESTER
mf: $C_{16}H_{30}O_2$ mw: 254.46

SYNS: 2-PROPENOIC ACID TRIDECYL ESTER ◇ TRIDECYL ACRYLATE

TOXICITY DATA with REFERENCE
skn-rbt 10 mg/24H open SEV AIHAAP 23,95,62
orl-rat LD50:45 g/kg AIHAAP 23,95,62
skn-rbt LD50:6300 mg/kg AIHAAP 23,95,62

CONSENSUS REPORTS: Reported in EPA TSCA Inventory.

SAFETY PROFILE: Mildly toxic by skin contact and ingestion. A severe skin irritant. See also ESTERS. When heated to decomposition it emits acrid smoke and fumes.

ADX250 ***HR: 3***
ACRYLOAMIDE
mf: $C_3H_7ClN_2$ mw: 106.57

PROP: An antibiotic produced by the strain *Streptomyces sp.* No. D274-2.

TOXICITY DATA with REFERENCE
ipr-mus LD50:38 mg/kg 85ERAY 2,1158,78
scu-mus LD50:38 mg/kg 85ERAY 2,1158,78
ivn-mus LD50:44 mg/kg 85ERAY 2,1158,78

SAFETY PROFILE: Poison by intraperitoneal, subcutaneous, and intravenous routes. When heated to decomposition it emits very toxic fumes of Cl^- and NO_x.

ADX500 CAS:107-13-1 ***HR: 3***
ACRYLONITRILE
DOT: UN 1093
mf: C_3H_3N mw: 53.07

PROP: Colorless, mobile liquid; mild odor. Sol in water. Mp: −82°, bp: 77.3°, fp: −83°, flash p: 30°F (TCC), lel: 3.1%, uel: 17%, d: 0.806 @ 20°/4°, autoign temp: 898°F, vap press: 100 mm @ 22.8°, vap d: 1.83, flash p: (of 5% aq sol): <50°F.

SYNS: ACRYLNITRIL (GERMAN, DUTCH) ◇ ACRYLONITRILE MONOMER ◇ AKRYLONITRYL (POLISH) ◇ CARBACRYL ◇ CIANURO di VINILE (ITALIAN) ◇ CYANOETHYLENE ◇ CYANURE de VINYLE (FRENCH) ◇ ENT 54 ◇ FUMIGRAIN ◇ NITRILE ACRILICO (ITALIAN) ◇ NITRILE ACRYLIQUE (FRENCH) ◇ PROPENENITRILE

◇ 2-PROPENENITRILE ◇ RCRA WASTE NUMBER U009 ◇ TL 314 ◇ VENTOX ◇ VINYL CYANIDE

TOXICITY DATA with REFERENCE
bfa-rat/sat 30 mg/kg TXCYAC 16,67,80
dns-rat:lvr 1 mmol/L PMRSDJ 5,371,85
slt-dmg-orl 1520 μmol/L PMRSDJ 5,325,85
skn-hmn 500 mg nse INMEAF 17,199,48
skn-rbt 10 mg/24H open JIHTAB 30,63,48
skn-rbt 500 mg MLD SCCUR* -,1,61
eye-rbt 20 mg SEV JIHTAB 30,63,48
ipr-ham TDLo:641 mg/kg (female 8D post):TER TJADAB 23,325,81
orl-rat TDLo:650 mg/kg (female 6-15D post):REP DOWCC* 03NOV76
orl-rat TDLo:18200 mg/kg/52W C:CAR FCTOD7 24,129,86
ihl-rat TCLo:5 ppm/52W-I:ETA MELAAD 68,401,77
orl-rat LD:3640 mg/kg/52W-C:NEO DOWCC* MAR77
ihl-hmn TCLo:16 ppm/20M:EYE,PUL INMEAF 17,199,48
ihl-man LCLo:1 g/m^3/1H:CNS,GIT ZAARAM 16,1,66
skn-chd LDLo:2015 mg/kg:CNS,RSP,GIT DMWOAX 75,1087,50
orl-rat LD50:78 mg/kg JOHYAY 3,106,59
ihl-rat LCLo:500 ppm/4H JIDHAN 31,343,49
skn-rat LD50:148 mg/kg GISAAA 41(10)103,76
ihl-mus LCLo:315 ppm/4H NTIS** PB280-478
ipr-mus LD50:46 mg/kg TXAPA9 59,589,81
orl-mus LD50:27 mg/kg JHEMA2 3,106,59
scu-mus LD50:35 mg/kg JHEMA2 3,106,59
ihl-dog LCLo:110 ppm/4H JIHTAB 24,27,42

CONSENSUS REPORTS: NTP Fifth Annual Report on Carcinogens. IARC Cancer Review: Group 2A IMEMDT 7,79,87; Human Limited Evidence IMEMDT 19,73,79; Animal Limited Evidence IMEMDT 19,73,79. Community Right-To-Know List. EPA Extremely Hazardous Substances List. Reported in EPA TSCA Inventory.

OSHA PEL: TWA 2 ppm; CL 10 ppm/15M; Cancer Hazard.
ACGIH TLV: Suspected Human Carcinogen, TWA 2 ppm (skin).
DFG TRK: 3 ppm (7 mg/m^3), Animal Carcinogen, Suspected Human Carcinogen.
NIOSH REL: TWA 1 ppm; CL 10 ppm/15M
DOT Classification: Flammable Liquid and Poison.

SAFETY PROFILE: Confirmed human carcinogen with experimental carcinogenic, neoplastigenic, and tumorigenic data. Poison by inhalation, ingestion, skin contact, and other routes. Human systemic effects by inhalation and skin contact: conjunctive irritation, somnolence, general anesthesia, cyanosis and diarrhea. An experimental teratogen. Other experimental reproductive effects. Human mutation data reported. Dangerous fire hazard when exposed to heat, flame, or oxidizers. Moderate explosion hazard when exposed to flame. Can react vigorously with oxidizing materials (see also CYANIDES).

Acrylonitrile closely resembles hydrocyanic acid in its toxic action. By inhibiting the respiratory enzymes of tissue, it renders the tissue cells incapable of oxygen absorption. Poisoning is acute; there is little evidence of cumulative action on repeated exposure. Exposure to low concentration is followed by flushing of the face and increased salivation; further exposure results in irritation of the eyes and nose, photophobia, deepened respiration, and, if exposure continues, shallow respiration, nausea, vomiting, weakness, an oppressive feeling in the chest, and occasionally headache and diarrhea are other complaints. Several cases of mild jaundice accompanied by mild anemia and leucocytosis have been reported. Urinalysis is generally negative, except for an increase in bile pigment. Serum and bile thiocyanates are raised. See also HYDROCYANIC ACID. Unstable and easily oxidized. Explosive polymerization may occur on storage with silver nitrate. Potentially explosive reactions with benzyltrimethylammonium hydroxide + pyrrole, tetrahydrocarbazole + benzyltrimethylammonium hydroxide. Violent reactions with strong acids (e.g., nitric or sulfuric), strong bases, azoisobutyronitrile, dibenzoyl peroxide, di-tert-butylperoxide, or bromine. Incompatible with $AgNO_3$ and amines. To fight fire use CO_2, dry chemical or alcohol foam. When heated to decomposition it emits toxic fumes of NO_x and CN^-. See also NITRILES and CYANIDE.

ADX750 ***HR: 2***
ACRYLONITRILE POLYMER with 1,3-BUTADIENE, and STYRENE, COMBUSTION PRODUCTS
mf: $(C_3H_3N)_x$

SYNS: ABS (pyrolysis products) ◇ ACELAN, combustion products ◇ ACRIBEL, combustion products ◇ ACRYL, combustion products ◇ ACRYLONITRILE-BUTADIENE-STYRENE (pyrolysis products) ◇ AKSA, combustion products ◇ ANILANA, combustion products ◇ BI-LOFT, combustion products ◇ BULANA, combustion products ◇ CASHMILON, combustion products ◇ CRUMERON, combustion products ◇ DOLAN, combustion products ◇ EXLAN, combustion products ◇ FINA, combustion products ◇ MALON, combustion products ◇ ORLON, combustion products ◇ POLYACRYLONITRILE ◇ 2-PROPENENITRILE HOMOPOLYMER (9CI) ◇ ZEFRAN, combustion products

TOXICITY DATA with REFERENCE
ihl-mus LC50:10 g/m^3/30M PWPSA8 21,167,78

SAFETY PROFILE: Moderately to highly toxic by inhalation. Upon decomposition it emits toxic fumes of NO_x and CN^-.

ADY250 CAS:9003-00-3 ***HR: 3***
ACRYLONITRILE POLYMER with CHLOROETHYLENE
mf: $(C_3H_3N{\cdot}C_2H_3Cl)_n$

SYNS: ACROPOR ◇ DYNEL ◇ KANEKALON ◇ 2-PROPENENITRILE, POLYMER with CHLOROETHENE ◇ VINYON N

TOXICITY DATA with REFERENCE
imp-mus TDLo:18 mg/kg:ETA CNREA8 15,333,55

CONSENSUS REPORTS: Reported in EPA TSCA Inventory. Cyanide and its compounds are on the Community Right-To-Know List.

SAFETY PROFILE: Questionable carcinogen with experimental tumorigenic data. See also NITRILES. When heated to decomposition it emits very toxic fumes of Cl^-, CN^-, and NO_x.

ADY500 CAS:9003-54-7 ***HR: 3***
ACRYLONITRILE POLYMER with STYRENE
mf: $(C_8H_8{\cdot}C_3H_3N)_x$

SYNS: ACRILAFIL ◇ ACRYLONITRILE-STYRENE COPOLYMER ◇ ACRYLONITRILE-STYRENE POLYMER ◇ ACRYLONITRILE-STYRENE RESIN ◇ ACS ◇ AS 61CL ◇ BAKELITE RMD 4511 ◇ CEVIAN HL ◇ DIALUX ◇ ESTYRENE AS ◇ KOSTIL ◇ LITAC ◇ LURAN ◇ LUSTRAN ◇ POLYSTYRENE-ACRYLONITRILE ◇ 2-PROPENENITRILE POLYMER with ETHENYLBENZENE ◇ REXENE 106 ◇ SANREX ◇ SN 20 ◇ STYREN-ACRYLONITRILEPOLYMER ◇ STYRENE-ACRYLONITRILE COPOLYMER ◇ TERULAN KP 2540 ◇ TYRIL

TOXICITY DATA with REFERENCE
orl-rat LD50:1800 mg/kg CEHYAN 25,22,80
orl-mus LD50:1000 mg/kg CEHYAN 25,22,80

CONSENSUS REPORTS: Reported in EPA TSCA Inventory. Cyanide and its compounds are on the Community Right-To-Know List.

SAFETY PROFILE: Moderately to highly toxic by ingestion. See also NITRILES. When heated to decomposition it emits toxic fumes of NO_x and CN^-.

ADY750 CAS:4836-08-2 ***HR: 2***
2-ACRYLOXYETHYLDIMETHYLSULFONIUM METHYL SULFATE

TOXICITY DATA with REFERENCE
orl-rat LD50:1870 mg/kg AIHAAP 23,95,62
skn-rbt LD50:2000 mg/kg AIHAAP 23,95,62

SAFETY PROFILE: Moderately toxic by ingestion and skin contact. See also SULFATES. When heated to decomposition it emits toxic fumes of SO_x.

ADZ000 CAS:814-68-6 ***HR: 3***
ACRYLOYL CHLORIDE
mf: C_3H_3ClO mw: 90.51

SYNS: ACRYLIC ACID CHLORIDE ◇ ACRYLYL CHLORIDE ◇ 2-PROPENOYL CHLORIDE

TOXICITY DATA with REFERENCE
ihl-rat LCLo:25 ppm/4H BJIMAG 27,1,70
ihl-mus LC50:92 mg/m^3/2H 85GMAT -,17,82
ivn-mus LD50:180 mg/kg CSLNX* NX#03367

CONSENSUS REPORTS: EPA Extremely Hazardous Substances List. Reported in EPA TSCA Inventory.

SAFETY PROFILE: Poison by inhalation and intravenous routes. When heated to decomposition it emits toxic fumes of Cl^-.

ADZ125 CAS:37239-28-4 ***HR: D***
ACTIHAEMYL

SYN: SOLCOSERYL

TOXICITY DATA with REFERENCE
orl-mus TDLo:2400 mg/kg (7-12D preg):REP ARZNAD 22,1510,72
orl-rat TDLo:2400 mg/kg (9-14D preg):TER ARZNAD 22,1510,72
ipr-mus LD50:197 g/kg JOPHBO 5,209,76

CONSENSUS REPORTS: EPA Genetic Toxicology Program.

SAFETY PROFILE: An experimental teratogen. Other experimental reproductive effects. When heated to decomposition it emits acrid smoke and irritating fumes.

AEA000 ***HR: 3***
ACTINIC RADIATION

SAFETY PROFILE: Outdoor workers, such as fishermen, sailors, soldiers, and farmers, show a high incidence of skin cancer. The commonest acute manifestation of actinic radiation effects on skin is sunburn.

AEA109 CAS:24397-89-5 ***HR: 3***
ACTINOBOLIN
mf: $C_{13}H_{20}N_2O_6$ mw: 300.31

PROP: Amorphous, fluffy, very hygroscopic powder. Amphoteric. Freely sol in water; mod sol in methanol and ethanol. Unstable in basic solutions.

SYNS: 4-(2-AMINOPROPIONAMIDO)-3,4,4a,5,6,7-HEXAHYDRO-5,6,8-TRIHYDROXY-3-METHYLISOCOUMARIN ◇ NSC 31083

TOXICITY DATA with REFERENCE
pic-esc 240 mg/L ZAPOAK 8,139,68
ipr-mus LD50:2844 mg/kg NCISP* JAN86
scu-mus LD50:1828 mg/kg NCISP* JAN86
ivn-mus LD50:6250 μg/kg JANTAJ 32,1069,79
unr-mus LDLo:2000 mg/kg 85ERAY 2,1368,78

SAFETY PROFILE: Poison by intravenous route.

Moderately toxic by other routes. Mutation data reported. When heated to decomposition it emits toxic fumes of NO_x. An antimicrobial agent and experimental cariostat.

AEA250 CAS:1338-58-5 ***HR: 3***
ACTINOGAN

SYNS: NSC 53396 ◇ NSC A15920

TOXICITY DATA with REFERENCE
ipr-mus LD50:200 mg/kg ARZNAD 17,693,67
ivn-mus LDLo:17 mg/kg ARZNAD 17,693,67

SAFETY PROFILE: Poison by intraperitoneal and intravenous routes.

AEA500 CAS:1402-38-6 ***HR: 3***
ACTINOMYCIN

SYNS: AURANTIN ◇ ONCOSTATIN

TOXICITY DATA with REFERENCE
oms-omi 1250 μg/L SOGEBZ 4,100,68
cyt-hmn:lym 200 μg/L/2H CCPHDZ 3,143,79
ivn-rat TDLo:150 μg/kg (9D preg):TER TJADAB 30(1),32A,84
orl-rat LDLo:1 mg/kg JPETAB 74,25,42
ipr-rat LDLo:1 mg/kg JPETAB 74,25,42
scu-rat LDLo:1 mg/kg JPETAB 74,25,42
ivn-rat LDLo:1 mg/kg JPETAB 74,25,42
orl-mus LDLo:10 mg/kg JPETAB 74,25,42
ipr-mus LDLo:1 mg/kg TDKNAF 14,60,55
scu-mus LDLo:250 μg/kg JPETAB 74,25,42
ivn-mus LDLo:250 μg/kg JPETAB 74,25,42
orl-rbt LDLo:1 mg/kg JPETAB 74,25,42
ipr-rbt LDLo:1 mg/kg JPETAB 74,25,42
scu-rbt LDLo:1 mg/kg JPETAB 74,25,42
ivn-rbt LDLo:1 mg/kg JPETAB 74,25,42

SAFETY PROFILE: Poison by ingestion, intraperitoneal, subcutaneous and intravenous routes. An experimental teratogen. Other experimental reproductive effects. Human mutation data reported.

AEA625 CAS:85086-83-5 ***HR: 3***
ACTINOMYCIN 23-21

SYN: SINOACTINOMYCIN

TOXICITY DATA with REFERENCE
orl-mus LD50:9400 μg/kg YHTPAD 19,283,84
ipr-mus LD50:515 μg/kg YHTPAD 19,283,84
ivn-mus LD50:1070 μg/kg YHTPAD 19,283,84
ivn-dog LD50:500 μg/kg YHTPAD 19,283,84

SAFETY PROFILE: Poison by ingestion, intravenous, and intraperitoneal routes.

AEA750 CAS:8052-16-2 ***HR: 3***
ACTINOMYCIN C
mf: $C_{62}H_{89}N_{11}O_{17}$ mw: 1260.62

SYNS: ACTINOCHRYSIN ◇ CACTINOMYCIN ◇ DACTINOMYCIN (10%), ACTINOMYCIN C2 (45%), and ACTINOMYCIN C3 (45%) mixture ◇ HBF 386 ◇ NSC-18268 ◇ SANDAMYCIN

TOXICITY DATA with REFERENCE
pic-esc 120 mg/L ZAPOAK 8,139,68
slt-dmg-orl 5 ppm MUREAV 173,197,86
dni-hmn:oth 475 μg/L 26QZAP 2,395,72
oms-hmn:oth 47 μg/L 26QZAP 2,395,72
dni-hmn:oth 475 μg/L 26QZAP 2,395,72
oms-hmn:47 μg/L 26QZAP 2,395,72
scu-rat TDLo:350 μg/kg (6-10D preg):TER OSDIAF 14,107,65
scu-rat TDLo:350 μg/kg (6-10D preg):REP OSDIAF 14,107,65
ivn-rat LD50:100 mg/kg ARZNAD 20,1461,70
par-rat LD50:100 mg/kg RRCRBU 52,76,75
ipr-mus LD50:1110 μg/kg AEPPAE 230,559,57
ivn-mus LD50:1 mg/kg 85GDA2 4(2),53,80

SAFETY PROFILE: Deadly poison by intravenous, parenteral, intraperitoneal, and possibly other routes. An experimental teratogen. Other experimental reproductive effects. Human mutation data reported. When heated to decomposition it emits toxic fumes of NO_x. An antibiotic.

AEB000 CAS:50-76-0 ***HR: 3***
ACTINOMYCIN D
mf: $C_{62}H_{86}N_{12}O_{16}$ mw: 1255.60

SYNS: ACT ◇ ACTINOMYCINDIOIC D ACID, DILACTONE ◇ ACTINOMYCIN I ◇ AD ◇ COSMEGEN ◇ DACTINOMYCIN ◇ DILACTONE ACTINOMYCINDIOIC D ACID ◇ HBF 386 ◇ LYOVAC COSMEGEN ◇ MERACTINOMYCIN ◇ NCI-C04682 ◇ NSC 3053 ◇ ONCOSTATIN K

TOXICITY DATA with REFERENCE
dnd-hmn:hla 400 μg/L/15M ECREAL 103,175,76
cyt-hmn:lym 200 μg/L/2H CCPHDZ 3,143,79
ipr-rat TDLo:100 μg/kg (female 18-19D post):REP TJADAB 17,44A,78
ipr-mus TDLo:100 μg/kg (female 8D post):TER SEIJBO 11,5,71
ipr-rat TDLo:2600 μg/kg/17W-I:CAR CNREA8 30,2271,70
scu-mus TDLo:280 μg/kg/18W-I:ETA APJAAG 17,495,67
ipr-rat TD:1700 μg/kg/26W-I:NEO RRCRBU 52,1,75
ivn-hmn TDLo:40 μg/kg/4D-I:SKN NEJMAG 281,1094,69
orl-rat LD50:7200 μg/kg ANYAA9 89,348,60
ipr-rat LD50:100 μg/kg AOGLAR 23,219,76
ivn-rat LD50:460 μg/kg ANYAA9 89,348,60
scu-rat LD50:800 μg/kg ANYAA9 89,348,60
orl-mus LD50:13 mg/kg ANYAA9 89,348,60
ipr-mus LD50:750 μg/kg CTRRDO 61,103,77

CONSENSUS REPORTS: IARC Cancer Review:

Group 3 IMEMDT 7,80,87; NCI Carcinogenesis Studies (ipr); Clear Evidence: rat RRCRBU 52,1,75; No Evidence: mouse RRCRBU 52,1,75

SAFETY PROFILE: Poison by ingestion, intravenous, subcutaneous, and intraperitoneal routes. An experimental teratogen. Other experimental reproductive effects. Human systemic effects by intravenous and possibly other routes: dermatitis, bone marrow damage, and gastrointestinal effects. A human systemic skin irritant by intravenous route. Human mutation data reported. Questionable carcinogen with experimental carcinogenic, neoplastigenic, and tumorigenic data. When heated to decomposition it emits toxic fumes of NO_x.

AEB500 CAS:6980-13-8 ***HR: 3***
ACTINOMYCIN K
mf: $C_{11}H_{12}ClN_3O_4$ mw: 285.71

SYNS: 7-CHLORO-3-β-d-RIBOFURANOSYL-3H-IMIDAZO(4,5-b)PYRIDINE ◇ KASUGAMYCIN ◇ KENGSHENGMYCIN

TOXICITY DATA with REFERENCE
ivn-hmn TDLo:80 μg/kg/25D-I:BLD XPHPAW 441,116,74
orl-rat LD50:22 g/kg 28ZEAL 5,136,76
orl-mus LD50:21 g/kg 28ZEAL 5,136,76
ipr-mus LD50:745 μg/kg 85ERAY 2,1268,78

SAFETY PROFILE: Poison by intraperitoneal route. Mildly toxic by ingestion. Human blood effects by intravenous route. When heated to decomposition it emits very toxic fumes of Cl^- and NO_x.

AEB750 CAS:102488-99-3 ***HR: 3***
ACTINOMYCIN L

SYN: ACTINOMYCIN 2104L

TOXICITY DATA with REFERENCE
scu-mus TDLo:4725 μg/kg/35W-I:NEO BKNJA5 2,105,59

CONSENSUS REPORTS: IARC Cancer Review: Animal Sufficient Evidence IMEMDT 10,29,76

SAFETY PROFILE: Confirmed carcinogen with experimental neoplastigenic data.

AEC000 CAS:12623-78-8 ***HR: 3***
ACTINOMYCIN S

SYN: ACTINOMYCIN 1048A

TOXICITY DATA with REFERENCE
scu-mus TDLo:240 μg/kg/16W-I:NEO BKNJA5 2,105,59

CONSENSUS REPORTS: IARC Cancer Review: Animal Sufficient Evidence IMEMDT 10,29,76

SAFETY PROFILE: Confirmed carcinogen with experimental neoplastigenic data.

AEC175 CAS:11097-67-9 ***HR: D***
ACTINOMYCIN S3

TOXICITY DATA with REFERENCE
oms-rat-par 250 μg/kg JJPHAM 33,129,83
dni-mus:ast 20 μmol/L CPBTAL 17,105,69
dnd-mam:lym 100 μmol/L CPBTAL 17,105,69

SAFETY PROFILE: Mutation data reported.

AEC200 CAS:1402-61-5 ***HR: 3***
ACTINOMYCIN X2
mf: $C_{61}H_{89}N_{12}O_{17}$ mw: 1262.62

SYNS: ACTINOMYCIN BV ◇ ACTINOMYCIN DV ◇ ACTINOMYCIN J1 ◇ ACTINOMYCIN S3 ◇ ACTINOMYCIN-V

TOXICITY DATA with REFERENCE
ipr-mus LD50:300 μg/kg JANTAJ 38,1625,85
scu-mus LD50:300 μg/kg 38KLAC -,427,77
ivn-mus LD50:1 mg/kg 85GDA2 4(2),48,80

SAFETY PROFILE: Poison by subcutaneous, intravenous, and intraperitoneal routes. Mutation data reported. When heated to decomposition it emits toxic fumes of NO_x.

AEC250 CAS:59680-34-1 ***HR: 3***
ACTINOXANTHIN
mf: $C_{437}H_{667}N_{121}O_{155}S_4$ mw: 10219.664

PROP: An antibiotic produced by the strain *Actinomyces globisporus*.

SYN: ACTINOXANTHINE

TOXICITY DATA with REFERENCE
ipr-mus LD50:240 μg/kg 85ERAY 2,1414,78
scu-mus LD50:1800 μg/kg 85ERAY 2,1414,78

SAFETY PROFILE: Poison by intraperitoneal and subcutaneous routes. When heated to decomposition it emits very toxic fumes of SO_x and NO_x.

AEC625 CAS:58814-86-1 ***HR: 3***
ACULEACIN A
mf: $C_{50}H_{81}N_7O_{16}$ mw: 1036.38

TOXICITY DATA with REFERENCE
dni-omi 20 mg/L JANTAJ 35,210,82
ipr-mus LD50:600 mg/kg 85GDA2 4(1),361,80
ivn-mus LD50:350 mg/kg JANTAJ 30,297,77

SAFETY PROFILE: Poison by intravenous route. Moderately toxic by other routes. Mutation data reported. When heated to decomposition it emits toxic fumes of NO_x.

AEC700 CAS:59277-89-3 ***HR: 2***
ACYCLOVIR
mf: $C_8H_{11}N_5O_3$ mw: 225.24

SYNS: ACICLOVIR ◇ ACYCLOGUANOSINE ◇ 2-AMINO-1,9-DIHYDRO-9-((2-HYDROXYETHOXY)METHYL)-6H-PURIN-6-ONE ◇ BW 248U ◇ 9-(2-HYDROXYTHEOXYMETHYL)GUANINE ◇ WELLCOME-248U ◇ ZOVIRAX

TOXICITY DATA with REFERENCE
cyt-hmn:lym 250 mg/L/48H FAATDF 3,587,83
msc-mus:lym 400 mg/L/4H FAATDF 3,587,83
scu-rat TDLo:100 mg/kg (female 10D post):REP TJADAB 36,31A,87
scu-rat TDLo:400 mg/kg (female 9-11D post):TER ARTODN 61,468,88
orl-wmn TDLo:100 mg/kg/5D-I:SKN BMJOAE 289,1424,84
ivn-man TDLo:134 μg/kg/1D-I:CNS LANCAO 2,385,85
ipr-rat LD50:860 mg/kg IYKEDH 16,866,85
scu-rat LD50:620 mg/kg IYKEDH 16,866,85
ivn-rat LD50:910 mg/kg IYKEDH 16,866,85

SAFETY PROFILE: Moderately toxic. Human systemic effects by ingestion or intravenous routes: allergic dermatitis and hallucinations. An experimental teratogen. Other experimental reproductive effects. Human mutation data reported. When heated to decomposition it emits acrid smoke and irritating fumes.

AEC725 CAS:69657-51-8 ***HR: 2***
ACYCLOVIR SODIUM SALT
mf: $C_8H_{11}N_5O_3 \cdot Na$ mw: 248.23

SYNS: ACYCLOGUANOSINE SODIUM (OBS.) ◇ 2-AMINO-1,9-DIHYDRO-9-((2-HYDROXYETHOXY)METHYL)-6H-PURIN-6-ONE MONOSODIUM SALT ◇ 1,9-DIHYDRO-2-AMINO-9-((2-HYDROXYETHOXY)METHYL)-6H-PURIN-6-ONE SODIUM SALT ◇ SODIUM ACYCLOVIR ◇ ZOVIRAX SODIUM

TOXICITY DATA with REFERENCE
scu-rat TDLo:300 mg/kg (female 10D post):REP ARTODN 62,8,88
scu-rat TDLo:300 mg/kg (female 10D post):TER ARTODN 62,8,88
ivn-man TDLo:107 mg/kg/5D-I DICPBB 22,306,88
ivn-cld TDLo:248 mg/kg/80H-I:GIT DICPBB 20,371,86
ipr-rat LD50:1210 mg/kg FAATDF 3,573,83
scu-rat LD50:650 mg/kg FAATDF 3,573,83
ipr-mus LD50:999 mg/kg FAATDF 3,573,83
ivn-mus LD50:405 mg/kg FAATDF 3,573,83

SAFETY PROFILE: Moderately toxic by intraperitoneal and several other routes. Human systemic effects by intravenous route: nausea or vomiting. An experimental teratogen. Other experimental reproductive effects. When heated to decomposition it emits toxic fumes of NO_x and Na_2O.

AEC750 CAS:21829-25-4 ***HR: 3***
ADALAT
mf: $C_{17}H_{18}N_2O_6$ mw: 346.37

SYNS: BAY 1040 ◇ BAY A 1040 ◇ CITILAT ◇ CORDIPIN ◇ 1,4-DIHYDRO-2,6-DIMETHYL-4-(2-NITROPHENYL)-3,5-PYRIDINEDICARBOXYLIC ACID DIMETHYL ESTER ◇ NIFEDIN ◇ NIFEDIPINE ◇ NIFELAT ◇ 4-(2′-NITROPHENYL)-2,6-DIMETHYL-3,5-DICARBOMETHOXY-1,4-DIHYDROPYRIDINE ◇ OXCORD ◇ PROCARDIA

TOXICITY DATA with REFERENCE
orl-wmn TDLo:200 μg/kg (female 27W post):TER GOBIDS 24,151,87
orl-wmn TDLo:200 μg/kg (female 27W post):REP GOBIDS 24,151,87
orl-wmn TDLo:31 mg/kg/11W-I AHJOA2 108,611,84
orl-wmn TDLo:800 μg/kg/1D-I:BLD,GIT PGMJAO 62,1029,86
orl-hmn TDLo:143 μg/kg ARZNAD 35,518,85
orl-wmn TDLo:600 μg/kg/45M-I:BPR,GIT AIMDAP 147,556,87
orl-rat LD50:1022 mg/kg ARZNAD 22,1,72
ipr-rat LD50:230 mg/kg YKYUA6 28,1451,77
orl-mus LD50:310 mg/kg JJPAAZ 40,399,86
ipr-mus LD50:185 mg/kg PCJOAU 16,817,82
orl-rbt LD50:504 mg/kg ARZNAD 35,915,85

SAFETY PROFILE: Poison by ingestion, intravenous, and intraperitoneal routes. Human systemic effects by ingestion: nausea or vomiting, changes in regional blood flow, and decreased blood pressure. An experimental teratogen. Experimental reproductive effects. See also ESTERS. When heated to decomposition it emits toxic fumes of NO_x.

AED250 CAS:665-66-7 ***HR: 3***
1-ADAMANTANAMINE HYDROCHLORIDE
mf: $C_{10}H_{17}N \cdot ClH$ mw: 187.74

SYNS: ADAMANTANAMINE HYDROCHLORIDE ◇ ADAMANTINE HYDROCHLORIDE ◇ ADAMANTYLAMINE HYDROCHLORIDE ◇ 1-ADAMANTYLAMINE HYDROCHLORIDE ◇ AMANTADINE HYDROCHLORIDE ◇ AMAZOLON ◇ AMINOADAMANTANE HYDROCHLORIDE ◇ 1-AMINOADAMANTENE HYDROCHLORIDE ◇ EXP 105-1 ◇ MANTADAN ◇ NSC 83653 ◇ SYMMETREL ◇ TRICYCLO(3.3.1.1.$^{(3,7)}$) DECAN-1-AMINE, HYDROCHLORIDE (9CI) ◇ VIROFRAL

TOXICITY DATA with REFERENCE
orl-wmn TDLo:182 mg/kg (female 1-91D post):TER LANCAO 2,607,75
orl-rat TDLo:240 mg/kg (female 9-14D post):REP GNRIDX 4,44,70
orl-man LDLo:43 mg/kg CJPSDT 31,757,86
orl-man TDLo:13 mg/kg/5D-I AJPSAO 143,1170,85
orl-rat LD50:800 mg/kg IYKEDH 19,164,88
ipr-rat LD50:150 mg/kg TXAPA9 15,642,69
ivn-rat LD50:90 mg/kg IYKEDH 19,164,88
orl-mus LD50:700 mg/kg TXAPA9 15,642,69
ipr-mus LD50:198 mg/kg IYKEDH 19,164,88
scu-mus LD50:290 mg/kg IYKEDH 19,164,88
orl-gpg LD50:360 mg/kg TXAPA9 15,642,69

CONSENSUS REPORTS: Reported in EPA TSCA Inventory.

SAFETY PROFILE: Human poison by ingestion. Poison by ingestion, intraperitoneal, and intravenous routes. A human teratogen with developmental abnormalities of the circulatory system. Experimental reproductive effects. When heated to decomposition it emits very toxic fumes of NO_x and HCl.

AED750 CAS:54099 ***HR: 3***
1-ADAMANTANEACETIC ACID-2-(DIETHYLAMINO)ETHYL ESTER, ETHYL IODIDE-11-5
mf: $C_{20}H_{36}NO_2 \cdot I$ mw: 449.47

TOXICITY DATA with REFERENCE
orl-mus LD50:600 mg/kg FRPSAX 32,129,77
ipr-mus LD50:61 mg/kg FRPSAX 32,129,77

SAFETY PROFILE: Poison by intraperitoneal route. Moderately toxic by ingestion. See also ESTERS and IODIDES. When heated to decomposition it emits very toxic fumes of I^- and NO_x.

AEE000 CAS:54099 ***HR: 3***
1-ADAMANTANEACETIC ACID-3-(DIMETHYLAMINO)PROPYL ESTER, ETHYL IODIDE-12-6
mf: $C_{19}H_{36}NO_2 \cdot I$ mw: 437.46

TOXICITY DATA with REFERENCE
orl-mus LD50:450 mg/kg FRPSAX 32,129,77
ipr-mus LD50:30 mg/kg FRPSAX 32,129,77

SAFETY PROFILE: Poison by intraperitoneal route. Moderately toxic by ingestion. See also ESTERS and IODIDES. When heated to decomposition it emits very toxic fumes of I^- and NO_x.

AEE250 CAS:63869 ***HR: 3***
S-((N-1-ADAMANTYLAMIDINO)METHYL) HYDROGEN THIOSULFATE, HYDRATE (4581)-14-7
mf: $C_{12}H_{20}N_2O_3S_2 \cdot 1/4H_2O$ mw: 288.76

TOXICITY DATA with REFERENCE
orl-mus LD50:280 mg/kg JMCMAR 15,1313,72
ipr-mus LD50:38 mg/kg JMCMAR 15,1313,72

SAFETY PROFILE: Poison by ingestion and intraperitoneal routes. See also THIOSULFATES. When heated to decomposition it emits very toxic fumes of NO_x and SO_x.

AEE500 CAS:63765-69-5 ***HR: 3***
1-(1-ADAMANTYLAMINO)-2,2,2-TRIFLUORO-1-(TRIFLUOROMETHYL) ETHANOLSESQUIHYDRATE
mf: $C_{13}H_{17}F_6NO \cdot 3/2H_2O$ mw: 344.34

TOXICITY DATA with REFERENCE
orl-mus LD50:300 mg/kg JMCMAR 13,1215,70
ipr-mus LD50:300 mg/kg JMCMAR 13,1215,70

SAFETY PROFILE: Poison by ingestion and intraperitoneal routes. When heated to decomposition it emits very toxic fumes of F^- and NO_x.

AEF000 CAS:37033-23-1 ***HR: 3***
5-(1-ADAMANTYL)-2,4-DIAMINO-6-ETHYLPYRIMIDINE ETHYLSULFONATE
mf: $C_{16}H_{24}N_4 \cdot C_2H_6O_3S$ mw: 382.58

SYN: DAEP-ES

TOXICITY DATA with REFERENCE
ipr-rat LD50:15 mg/kg JNCIAM 60,1029,78
ipr-mus LD50:30 mg/kg JNCIAM 60,1029,78

SAFETY PROFILE: Poison by intraperitoneal route. See also SULFONATES and AMINES. When heated to decomposition it emits very toxic fumes of NO_x and SO_x.

AEF250 CAS:35507-78-9 ***HR: 3***
5-(1-ADAMANTYL)-2,4-DIAMINO-6-METHYLPYRIMIDINE ETHYLSULFONATE
mf: $C_{15}H_{22}N_4 \cdot C_2H_6O_3S$ mw: 368.55

SYN: DAMP-ES

TOXICITY DATA with REFERENCE
ipr-rat LD50:23 mg/kg JNCIAM 60,1029,78
ipr-mus LD50:40 mg/kg JNCIAM 60,1029,78

SAFETY PROFILE: Poison by intraperitoneal route. See also SULFONATES and AMINES. When heated to decomposition it emits very toxic fumes of NO_x and SO_x.

AEF500 CAS:31635-40 ***HR: 3***
N-1-ADAMANTYL-N-(2-(DIMETHYLAMINO)ETHOXY)ACETAMIDE HYDROCHLORIDE-2
mf: $C_{16}H_{28}N_2O_2 \cdot ClH$ mw: 316.92

SYN: 1-(DIMETHYLAMINOETHOXYACETAMIDO)ADAMANTANE HYDROCHLORIDE

TOXICITY DATA with REFERENCE
orl-dog LD50:170 mg/kg ARZNAD 23,577,73
orl-rat LD50:630 mg/kg ARZNAD 23,577,73
ivn-mus LD50:71 mg/kg ARZNAD 23,577,73

SAFETY PROFILE: Poison by ingestion and intravenous route. See also AMINES. When heated to decomposition it emits very toxic fumes of NO_x and HCl. An antiviral agent.

AEG000 CAS:40284-08-0 ***HR: 3***
N-(2-ADAMANTYL)-2-MERCAPTOACETAMIDINE HYDROCHLORIDE
mf: $C_{12}H_{20}N_2S \cdot ClH$ mw: 260.86

TOXICITY DATA with REFERENCE
orl-mus LD50:35 mg/kg JMCMAR 15,1313,72
ipr-mus LD50:17 mg/kg JMCMAR 15,1313,72

SAFETY PROFILE: Poison by ingestion and intraperitoneal routes. When heated to decomposition it emits very toxic fumes of NO_x, SO_x, and HCl.

AEG129 CAS:69804-02-0 ***HR: 3***
S-(N-(1-ADAMANTYLMETHYLAMIDINO)METHYL)PHOSPHOROTHIOATE MONOSODIUM SALT
mf: $C_{13}H_{22}N_2O_3PS \cdot Na$ mw: 340.22

TOXICITY DATA with REFERENCE
ipr-mus LD50:408 mg/kg PCJOAU 13,22,79
par-mus LD50:98 mg/kg PCJOAU 13,22,79

SAFETY PROFILE: Poison by parenteral and intraperitoneal routes. When heated to decomposition it emits very toxic fumes of NO_x, PO_x, Na_2O, and SO_x.

AEG250 CAS:22545-60-4 ***HR: 3***
N-(1-ADAMANTYLMETHYL)-2-MERCAPTOACETAMIDINE HYDROCHLORIDE
mf: $C_{13}H_{22}N_2S \cdot ClH$ mw: 274.89

TOXICITY DATA with REFERENCE
orl-mus LD50:65 mg/kg JMCMAR 15,1313,72
ipr-mus LD50:22 mg/kg JMCMAR 15,1313,72

SAFETY PROFILE: Poison by ingestion and intraperitoneal routes. When heated to decomposition it emits very toxic fumes of HCl, SO_x, and NO_x.

AEG500 CAS:40284-10-4 ***HR: 3***
N-(3-(1-ADAMANTYL)PROPYL)-2-MERCAPTOACETAMIDINE HYDROCHLORIDE HYDRATE (10:10:3)
mf: $C_{15}H_{26}N_2S \cdot ClH \cdot 3/10H_2O$ mw: 308.35

TOXICITY DATA with REFERENCE
orl-mus LD50:350 mg/kg JMCMAR 15,1313,72
ipr-mus LD50:25 mg/kg JMCMAR 15,1313,72

SAFETY PROFILE: Poison by ingestion and intraperitoneal routes. When heated to decomposition it emits very toxic fumes of NO_x, SO_x, and HCl.

AEG625 CAS:1225-60-1 ***HR: 3***
ADANTON HYDROCHLORIDE
mf: $C_{16}H_{19}N_3S \cdot ClH$ mw: 321.90

SYNS: ANDANTOL ◇ D 201 HYDROCHLORIDE ◇ N-DIMETHYLAMINOISOPROPYLTHIOPHENYLPYRIDYLAMINEHYDROCHLORIDE ◇ 10-(2-DIMETHYLAMINO-2-METHYLETHYL)-10H-PYRIDO(3,2-b)(1,4)BENZOTHIAZINE HYDROCHLORIDE ◇ 10-(2-DIMETHYLAMINOPROPYL)-1-AZAPHENOTHIAZINE HYDROCHLORIDE ◇ 10-(2-DIMETHYLAMINOPROPYL-(1)-4-AZAPHENTHIAZINHYDROCHLORID (GERMAN) ◇ 10-(2-DIMETHYLAMINOPROPYL)-9-THIA-1,10-DIAZAANTHRACENE HYDROCHLORIDE ◇ ISOTHIPENDYL HYDROCHLORIDE ◇ NILERGEX HYDROCHLORIDE ◇ THERUHISTIN HYDROCHLORIDE ◇ UDANTOL HYDROCHLORIDE

TOXICITY DATA with REFERENCE
orl-rat LD50:1220 mg/kg NIIRDN 6,72,82
orl-mus LD50:222 mg/kg ARZNAD 8,489,58
ipr-mus LD50:65 mg/kg ARZNAD 18,435,68

SAFETY PROFILE: Poison by ingestion and intraperitoneal routes. When heated to decomposition it emits toxic fumes of NO_x, SO_x, and HCl.

AEG750 CAS:1229-29-4 ***HR: 3***
ADAPIN
mf: $C_{19}H_{21}NO \cdot ClH$ mw: 315.87

SYNS: CIDOXEPIN HYDROCHLORIDE ◇ CURATIN ◇ 11-DIMETHYLAMINO PROPYLIDENE-6H-DIBENZ(b,e)OXEPIN ◇ 11-(3-(DIMETHYLAMINO)PROPYLIDENE)-6,11-DIHYDRODIBENZ(b,e) OXEPIN HYDROCHLORIDE ◇ N,N-DIMETHYLDIBENZ(b,e) OXEPIN-Δ^{11}(6H,Γ-PROPYLAMINE HYDROCHLORIDE ◇ DOXEPIN HYDROCHLORIDE ◇ NSC-108160 ◇ 1-PROPANAMINE, 3-DIBENZ(b,e)OXEPIN-11(6H)-YLIDENE-N,N-DIMETHYL-,HYDROCHLORIDE ◇ SINEQUAN

TOXICITY DATA with REFERENCE
orl-rat TDLo:1620 mg/kg (female 9-14D post):REP OYYAA2 5,913,71
orl-rat TDLo:540 mg/kg (female 9-14D post):TER OYYAA2 5,913,71
orl-wmn TDLo:112 mg/kg/4W-I JCLPDE 44,106,83
orl-hmn LDLo:90 mg/kg JATOD3 2,18,78
orl-hmn TDLo:9300 μg/kg:CNS JAMAAP 237,2632,77
orl-wmn TDLo:141 mg/kg/12W-I:EAR SMJOAV 76,1204,83
orl-rat LD50:147 mg/kg 27ZQAG -,72,72
ipr-rat LD50:84 mg/kg OYYAA2 6,889,72
scu-rat LD50:155 mg/kg OYYAA2 6,889,72
ivn-rat LD50:13 mg/kg 27ZQAG -,72,72
orl-mus LD50:180 mg/kg OYYAA2 6,889,72
ipr-mus LD50:79 mg/kg 27ZQAG -,72,72
scu-mus LD50:160 mg/kg 27ZQAG -,72,72

SAFETY PROFILE: A human poison by ingestion. An experimental poison by ingestion, subcutaneous, intraperitoneal, and intravenous routes. Human systemic effects by ingestion: hallucinations, distorted perceptions, muscle spasms and change in heart rate, and tinnitus. An experimental teratogen. Experimental reproductive effects. When heated to decomposition it emits toxic fumes of NO_x and HCl. A psychotherapeutic agent.

AEG875 CAS:5118-29-6 ***HR: 3***
ADAPTOL
mf: $C_{21}H_{25}N$ mw: 291.47

SYNS: 3-(10,10-DIMETHYL(10H)-ANTHRACENYLIDENE)-N,N-DIMETHYL-1-PROPANAMINE (9CI) ◇ 9-(3-DIMETILAMINOPROPYLIDEN)-10,10-DIMETIL-9,10-DIIDROANTHRACENE (ITALIAN) ◇ DIXERAN ◇ MELITRACEN ◇ MELITRACENE ◇ N 7001 ◇ N,N,10,10-TETRAMETHYL-$\Delta^{9(10),\Gamma}$)-ANTHRACENEPROPYLAMINE ◇ THYMEOL ◇ TRAUSABUM ◇ TRAUSABUN ◇ U-24973

TOXICITY DATA with REFERENCE
orl-rat LD50:170 mg/kg FRPPAO 25,519,70
ipr-rat LD50:96 mg/kg FRPPAO 25,519,70
orl-mus LD50:315 mg/kg FRPPAO 25,519,70
ipr-mus LD50:131 mg/kg FRPPAO 25,519,70
ivn-mus LD50:52 mg/kg FRPPAO 25,519,70

SAFETY PROFILE: Poison by ingestion, intravenous, and intraperitoneal routes. When heated to decomposition it emits toxic fumes of NO_x.

AEH000 CAS:73-24-5 ***HR: 3***
ADENINE
mf: $C_5H_5N_5$ mw: 135.15

SYNS: ADENINIMINE ◇ 6-AMINOPURINE ◇ 6-AMINO-1H-PURINE ◇ 6-AMINO-3H-PURINE ◇ 6-AMINO-9H-PURINE ◇ 1,6-DIHYDRO-6-IMINOPURINE ◇ 3,6-DIHYDRO-6-IMINOPURINE ◇ LEUCO-4 ◇ 1H-PURIN-6-AMINE ◇ USAF CB-18 ◇ VITAMIN B4

TOXICITY DATA with REFERENCE
pic-esc 1 g/L ZAPOAK 12,583,72
cyt-mus-ipr 10 mmol/L NULSAK 17,199,74
ipr-mus TDLo:200 mg/kg (female 10D post):TER OFAJAE 49,75,72
ipr-mus TDLo:200 mg/kg (female 10D post):REP OFAJAE 49,75,72
orl-rat LD50:745 mg/kg JPETAB 104,20,52
ipr-rat LD50:198 mg/kg JPETAB 104,20,52
orl-mus LD50:783 mg/kg DRUGAY 6,19,82
ipr-mus LD50:100 mg/kg NTIS** AD277-689
scu-mus LDLo:1 g/kg ANYAA9 60,251,54

CONSENSUS REPORTS: Reported in EPA TSCA Inventory.

SAFETY PROFILE: Poison by intraperitoneal route. Moderately toxic by ingestion. An experimental teratogen. Experimental reproductive effects. Mutation data reported. When heated to decomposition it emits toxic fumes of NO_x.

AEH100 CAS:24356-66-9 ***HR: 1***
ADENINE ARABINOSIDE
mf: $C_{10}H_{13}N_5O_4 \cdot H_2O$ mw: 285.30

SYNS: ARA-A ◇ 9-β-d-ARABINOFURANOSYLADENINEMONOHYDRATE ◇ 9-β-d-ARABINOFURANOSYL-9H-PURINE-6-AMINEMONOHYDRATE ◇ SPONGOADENOSINE ◇ VIDARABINE ◇ VIRA-A

TOXICITY DATA with REFERENCE
orl-mus LD50:>7950 mg/kg AACHAX-,180,68
ipr-mus LD50:4677 mg/kg AACHAX-,180,68

SAFETY PROFILE: Mildly toxic by ingestion and intraperitoneal routes. When heated to decomposition it emits toxic fumes of NO_x.

AEH250 CAS:700-02-7 ***HR: 3***
ADENINE-1-N-OXIDE
mf: $C_5H_5N_5O$ mw: 151.15

TOXICITY DATA with REFERENCE
scu-rat TDLo:1300 mg/kg/26W-I:NEO CNREA8 30,184,70

SAFETY PROFILE: Questionable carcinogen with experimental neoplastigenic data. When heated to decomposition it emits toxic fumes of NO_x.

AEH750 CAS:58-61-7 ***HR: 2***
ADENOSINE
mf: $C_{10}H_{13}N_5O_4$ mw: 267.28

SYNS: ADENINE RIBOSIDE ◇ ADENOSIN (GERMAN) ◇ β-ADENOSINE ◇ β-d-ADENOSINE ◇ 6-AMINO-9-β-d-RIBOFURANOSYL-9H-PURINE ◇ BONITON ◇ MYOCOL ◇ NUCLEOCARDYL ◇ 9-β-d-RIBOFURANOSIDOADENINE ◇ SANDESIN ◇ USAF CB-10

TOXICITY DATA with REFERENCE
pic-esc 1 g/L ZAPOAK 12,583,72
oms-hmn:oth 100 μmol/L JIDEAE 65,52,75
cyt-mus-ipr 20 mmol/L NULSAK 17,199,74
dnd-mam:lym 60 mmol/L PNASA6 48,686,62
ipr-mus LD50:500 mg/kg NTIS** AD277-689

CONSENSUS REPORTS: Reported in EPA TSCA Inventory.

SAFETY PROFILE: Moderately toxic by intraperitoneal route. Human mutation data reported. When heated to decomposition it emits toxic fumes of NO_x.

AEI000 CAS:53 ***HR: 3***
ADENOSINE-3'-(α-AMINO-p-METHOXYHYDROCINNAMAMIDO)-3'-DEOXY-N,N-DIMETHYL-79-2
mf: $C_{22}H_{29}N_7O_5$ mw: 471.58

SYNS: ACHROMYCIN (PURINE DERIVATIVE) ◇ 3'-(l-α-AMINO-p-METHOXYHYDROCINNAMAMIDO)-3'-DEOXY-N,N-DIMETHYLADENOSINE ◇ (S)-3'-((2-AMINO-3-(4-METHOXYPHENYL)-1-OXOPROPYL)AMINO)-3'-DEOXY-N,N-DIMETHYLADENOSINE ◇ CL 13,900 ◇ 6-DIMETHYLAMINO-9-(3'-(p-METHOXY-l-PHENYLALANYLAMINO)-β-d-RIBOFURANOSYL)-PURINE ◇ NSC-3055 ◇ PUROMYCIN ◇ STYLOMYCIN

TOXICITY DATA with REFERENCE
dnr-esc 100 μg/disc DNREA8 34,1658,74
cyt-dmg:oth 100 mg/L CLDFAT 2,97,73
dni-oin:oth 100 mg/L IJEBA6 16,1027,78
dni-mus:lym 100 μmol/L PLMEAA 34,231,78
dni-mus:fbr 420 μmol/L JCLBA3 58,410,73
unr-rat TDLo:100 mg/kg (female 7D post):TER 85DJA5 -,95,71
par-rbt TDLo:5 μg/kg (1D preg):REP ENDOAO 79,858,66
orl-mus LD50:20 mg/kg 85GDA2 5,302,81
ipr-mus LD50:25 mg/kg 85GDA2 5,302,81
ivn-mus LD50:15 mg/kg 85GDA2 5,302,81

SAFETY PROFILE: Poison by ingestion, intravenous, and intraperitoneal routes. An experimental teratogen. Experimental reproductive effects. Human mutation data reported. When heated to decomposition it emits toxic fumes of NO_x. An antibiotic.

AEI250 CAS:35788-21-7 ***HR: 3***
ADENOSINE-5'-CARBOXAMIDE
mf: $C_{10}H_{12}N_6O_4$ mw: 280.28

SYN: β-d-1-(6-AMINO-9H-PURIN-9-YL)-1-DEOXYRIBOFURANURONAMIDE

TOXICITY DATA with REFERENCE
orl-mus LD50:50 mg/kg JMCMAR 23,313,80
ipr-mus LD50:5 mg/kg JMCMAR 23,313,80

SAFETY PROFILE: Poison by ingestion and intraperitoneal routes. When heated to decomposition it emits toxic fumes of NO_x.

AEI500 ***HR: 3***
ADENOSINE-5'-(N-CYCLOBUTYL) CARBOXAMIDE
mf: $C_{14}H_{18}N_6O_4$ mw: 334.38

SYN: 1-(6-AMINO-9H-PURIN-9-YL)-N-CYCLOBUTYL-1-DEOXYRIBOFURANURONAMIDE

TOXICITY DATA with REFERENCE
orl-mus LD50:5 mg/kg JMCMAR 23,313,80
ipr-mus LD50:2 mg/kg JMCMAR 23,313,80

SAFETY PROFILE: Poison by ingestion and intraperitoneal routes. When heated to decomposition it emits toxic fumes of NO_x.

AEI750 CAS:35920-40-2 ***HR: 3***
ADENOSINE-5'-(N-CYCLOPENTYL) CARBOXAMIDE
mf: $C_{15}H_{20}N_6O_4$ mw: 348.41

TOXICITY DATA with REFERENCE
orl-mus LD50:200 mg/kg JMCMAR 23,313,80
ipr-mus LD50:200 mg/kg JMCMAR 23,313,80

SAFETY PROFILE: Poison by ingestion and intraperitoneal routes. When heated to decomposition it emits toxic fumes of NO_x.

AEJ000 CAS:50908-62-8 ***HR: 3***
ADENOSINE-5'-(N-CYCLOPROPYL) CARBOXAMIDE
mf: $C_{13}H_{16}N_6O_4$ mw: 320.35

SYN: 1-(6-AMINO-9H-PURIN-9-YL)-N-CYCLOPROPYL-1-DEOXYRIBOFURANURONAMIDE

TOXICITY DATA with REFERENCE
orl-mus LD50:5 mg/kg JMCMAR 23,313,80
ipr-mus LD50:2 mg/kg JMCMAR 23,313,80

SAFETY PROFILE: Poison by ingestion and intraperitoneal routes. When heated to decomposition it emits toxic fumes of NO_x.

AEJ250 CAS:72209-26-8 ***HR: 3***
ADENOSINE-5'-(N-CYCLOPROPYL)CARBOXAMIDE-N'-OXIDE
mf: $C_{13}H_{16}N_6O_5$ mw: 336.35

SYN: 1-(6-AMINO-9H-PURIN-9-YL)-N-CYCLOPROPYL-1-DEOXYRIBOFURANURONAMIDE-N-OXIDE

TOXICITY DATA with REFERENCE
orl-mus LD50:5 mg/kg JMCMAR 23,313,80
ipr-mus LD50:5 mg/kg JMCMAR 23,313,80

SAFETY PROFILE: Poison by ingestion and intraperitoneal routes. When heated to decomposition it emits toxic fumes of NO_x.

AEJ500 CAS:58048-25-2 ***HR: 3***
ADENOSINE-5'-(N-CYCLOPROPYLMETHYL) CARBOXAMIDE
mf: $C_{14}H_{18}N_6O_4$ mw: 334.38

SYN: 1-(6-AMINO-9H-PURIN-9-YL)-N-CYCLOPROPYLMETHYL-1-DEOXYRIBOFURANURONAMIDE

TOXICITY DATA with REFERENCE
orl-mus LD50:200 mg/kg JMCMAR 23,313,80
ipr-mus LD50:20 mg/kg JMCMAR 23,313,80

SAFETY PROFILE: Poison by ingestion and intraperitoneal routes. When heated to decomposition it emits toxic fumes of NO_x.

AEJ750 CAS:35788-31-9 ***HR: 3***
ADENOSINE-5'-(N-(2-(DIMETHYLAMINO) ETHYL))CARBOXAMIDE
mf: $C_{14}H_{21}N_7O_4$ mw: 351.42

SYN: 1-(6-AMINO-9H-PURIN-9-YL)-N-(2-(DIMETHYLAMINO)-ETHYL-1-DEOXYRIBOFURANURONAMIDE

TOXICITY DATA with REFERENCE
orl-mus LD50:500 mg/kg JMCMAR 23,313,80
ipr-mus LD50:20 mg/kg JMCMAR 23,313,80

SAFETY PROFILE: Poison by intraperitoneal route. Moderately toxic by ingestion. When heated to decomposition it emits toxic fumes of NO_x.

AEK000 CAS:39491-47-9 ***HR: 3***
ADENOSINE-5'-(N,N-DIMETHYL)CARBOXAMIDE HYDRATE
mf: $C_{12}H_{16}N_6O_4 \cdot H_2O$ mw: 326.36

SYN: 1-(6-AMINO-9H-PURIN-9-YL)-1-DEOXY-N,N-DIMETHYLRIBOFURANURONAMIDE HYDRATE

TOXICITY DATA with REFERENCE
orl-mus LD50:1000 mg/kg JMCMAR 23,313,80
ipr-mus LD50:50 mg/kg JMCMAR 23,313,80

SAFETY PROFILE: Poison by intraperitoneal route. Moderately toxic by ingestion. When heated to decomposition it emits toxic fumes of NO_x.

AEK250 CAS:35920-39-9 ***HR: 3***
ADENOSINE-5'-(N-ETHYL)CARBOXAMIDE HEMIHYDRATE
mf: $C_{12}H_{16}N_6O_4 \cdot 1/2H_2O$ mw: 317.35

SYN: 1-(6-AMINO-9H-PURIN-9-YL)-1-DEOXY-N-ETHYLRIBOFURANURONAMIDE HEMIHYDRATE

TOXICITY DATA with REFERENCE
orl-mus LD50:5 mg/kg JMCMAR 23,313,80
ipr-mus LD50:500 μg/kg JMCMAR 23,313,80

SAFETY PROFILE: Poison by ingestion and intraperitoneal routes. When heated to decomposition it emits toxic fumes of NO_x.

AEK500 CAS:72209-27-9 ***HR: 3***
ADENOSINE-5'-(N-ETHYL)CARBOXAMIDE-N'-OXIDE
mf: $C_{12}H_{16}N_6O_5$ mw: 324.34

SYN: 1-(6-AMINO-9H-PURIN-9-YL)-1-DEOXY-N-ETHYLRIBOFURANURONAMIDE-N-OXIDE

TOXICITY DATA with REFERENCE
orl-mus LD50:20 mg/kg JMCMAR 23,313,80
ipr-mus LD50:2 mg/kg JMCMAR 23,313,80

SAFETY PROFILE: Poison by ingestion and intraperitoneal routes. When heated to decomposition it emits toxic fumes of NO_x.

AEK750 CAS:57872-78-3 ***HR: 3***
ADENOSINE-5'-(N-HEXYL)CARBOXAMIDE HEMIHYDRATE
mf: $C_{16}H_{24}N_6O_4 \cdot 1/2H_2O$ mw: 373.47

SYN: 1-(6-AMINO-9H-PURIN-9-YL)-1-DEOXY-N-HEXYLRIBOFURANURONAMIDEHEMIHYDRATE

TOXICITY DATA with REFERENCE
orl-mus LD50:500 mg/kg JMCMAR 23,313,80
ipr-mus LD50:200 mg/kg JMCMAR 23,313,80

SAFETY PROFILE: Poison by intraperitoneal route. Moderately toxic by ingestion. When heated to decomposition it emits toxic fumes of NO_x.

AEL000 CAS:35788-28-4 ***HR: 3***
ADENOSINE-5'-(N-(2-HYDROXYETHYL))CARBOXAMIDE
mf: $C_{12}H_{16}N_6O_5$ mw: 324.34

SYN: 1-(6-AMINO-9H-PURIN-9-YL)-1-DEOXY-N-(2-HYDROXYETHYL)RIBOFURANURONAMIDE

TOXICITY DATA with REFERENCE
orl-mus LD50:5 mg/kg JMCMAR 23,313,80
ipr-mus LD50:2 mg/kg JMCMAR 23,313,80

SAFETY PROFILE: Poison by ingestion and intraperitoneal routes. When heated to decomposition it emits toxic fumes of NO_x.

AEL250 CAS:35788-29-5 ***HR: 3***
ADENOSINE-5'-(N-ISOPROPYL)CARBOXAMIDE
mf: $C_{13}H_{18}N_6O_4$ mw: 322.37

SYN: 1-(6-AMINO-9H-PURIN-9-YL)-1-DEOXY-N-ISOPROPYLRIBOFURANURONAMIDE

TOXICITY DATA with REFERENCE
orl-mus LD50:5 mg/kg JMCMAR 23,313,80
ipr-mus LD50:5 mg/kg JMCMAR 23,313,80

SAFETY PROFILE: Poison by ingestion and intraperitoneal routes. When heated to decomposition it emits toxic fumes of NO_x.

AEL500 CAS:54925-45-0 ***HR: 3***
ADENOSINE-5'-(N-METHOXY)CARBOXAMIDE HYDRATE
mf: $C_{11}H_{14}N_6O_5 \cdot H_2O$ mw: 328.33

SYN: 1-(6-AMINO-9H-PURIN-9-YL)-1-DEOXY-N-METHOXYRIBOFURANURONAMIDE HYDRATE

TOXICITY DATA with REFERENCE
orl-mus LD50:50 mg/kg JMCMAR 23,313,80
ipr-mus LD50:20 mg/kg JMCMAR 23,313,80

SAFETY PROFILE: Poison by ingestion and intraperitoneal routes. When heated to decomposition it emits toxic fumes of NO_x.

AEL750 CAS:35788-27-3 ***HR: 3***
ADENOSINE-5'-(N-METHYL)CARBOXAMIDE HEMIHYDRATE
mf: $C_{11}H_{14}N_6O_4 \cdot 1/2H_2O$ mw: 303.32

SYN: 1-(6-AMINO-9H-PURIN-9-YL)-1-DEOXY-N-METHYLRIBOFURANURONAMIDE HEMIHYDRATE

TOXICITY DATA with REFERENCE
orl-mus LD50:20 mg/kg JMCMAR 23,313,80
ipr-mus LD50:5 mg/kg JMCMAR 23,313,80

SAFETY PROFILE: Poison by ingestion and intraperitoneal routes. When heated to decomposition it emits toxic fumes of NO_x.

AEM000 CAS:57872-80-7 ***HR: 3***
ADENOSINE-5'-(N-PROPYL)CARBOXAMIDE
mf: $C_{13}H_{18}N_6O_4$ mw: 322.37

SYN: 1-(6-AMINO-9H-PURIN-9-YL)-1-DEOXY-N-PROPYLRIBOFURANURONAMIDE

TOXICITY DATA with REFERENCE
orl-mus LD50:200 mg/kg JMCMAR 23,313,80
ipr-mus LD50:5 mg/kg JMCMAR 23,313,80

SAFETY PROFILE: Poison by ingestion and intraperitoneal routes. When heated to decomposition it emits toxic fumes of NO_x.

AEM250 CAS:15237-44-2 ***HR: 2***
ADENOSINE-5'-(TETRAHYDROGENTRIPHOSPHATE) SODIUM SALT
mf: $C_{10}H_{16}N_5O_{13}P_3 \cdot 7Na$ mw: 668.15

SYNS: ATP Na SALT ◇ NaATP ◇ SODIUM ATP ◇ SODIUM ADENOSINE TRIPHOSPHATE ◇ SODIUM ADENOSINE-5'-TRIPHOSPHATE

TOXICITY DATA with REFERENCE
ipr-rat LD50:1379 mg/kg OYYAA2 4,689,70
ipr-mus LD50:1000 mg/kg ARZNAD 7,24,57

SAFETY PROFILE: Moderately toxic by intraperitoneal route. See also PHOSPHATES. When heated to decomposition it emits very toxic fumes of PO_x, Na_2O, and NO_x.

AEM500 CAS:35170-28-6 ***HR: 2***
5'-ADENYLIC ACID POTASSIUM SALT
mf: $C_{10}H_{14}N_5O_7P \cdot K$ mw: 386.36

SYNS: ADENOSINE-5'-MONOPHOSPHATE POTASSIUM SALT ◇ ADENOSINE-5'-MONOPHOSPHORIC ACID POTASSIUM SALT ◇ ADENOSINE-5'-PHOSPHATE POTASSIUM SALT ◇ ADENOSINE-5'-PHOSPHORIC ACID POTASSIUM SALT ◇ 5'-AMP POTASSIUM SALT

TOXICITY DATA with REFERENCE
orl-rat LD50:11250 mg/kg OYYAA2 4,689,70
ipr-rat LD50:1310 mg/kg OYYAA2 4,689,70
scu-rat LD50:1493 mg/kg OYYAA2 4,689,70
orl-mus LD50:13791 mg/kg OYYAA2 4,689,70
ipr-mus LD50:1955 mg/kg OYYAA2 4,689,70
scu-mus LD50:1937 mg/kg OYYAA2 4,689,70
ivn-mus LD50:536 mg/kg OYYAA2 4,689,70

SAFETY PROFILE: Moderately toxic by intraperitoneal, subcutaneous, and intravenous routes. Mildly toxic by ingestion. See also PHOSPHATES. When heated to decomposition it emits very toxic fumes of NO_x, K_2O, and PO_x.

AEM750 CAS:13474-03-8 ***HR: 2***
5'-ADENYLIC ACID SODIUM SALT
mf: $C_{10}H_{14}N_5O_7P \cdot 7Na$ mw: 508.19

SYNS: ADENOSINE 5'-MONOPHOSPHATE SODIUM SALT ◇ SODIUM ADENOSINE-5'-MONOPHOSPHATE

TOXICITY DATA with REFERENCE
ipr-rat LD50:2049 mg/kg OYYAA2 4,689,70
ipr-mus LD50:2000 mg/kg ARZNAD 7,24,57

SAFETY PROFILE: Moderately toxic by intraperitoneal route. When heated to decomposition it emits very toxic fumes of NO_x, Na_2O and PO_x.

AEN000 CAS:628-94-4 ***HR: 3***
ADIPAMIDE
mf: $C_6H_{12}N_2O_2$ mw: 144.20

PROP: Crystals. Mp: 220°. Sol in alc.

SYNS: ADIPIC ACID DIAMIDE ◇ ADIPIC DIAMIDE ◇ 1,4-BUTANEDICARBOXAMIDE ◇ HEXANEDIAMIDE (9CI) ◇ NCI-C02095

TOXICITY DATA with REFERENCE
orl-rat TDLo:1270 mg/kg:CAR JEPTDQ 3(5-6),149,80
orl-rat LDLo:500 mg/kg JPETAB 90,260,47
orl-mus LD50:6000 mg/kg BIJOAK 34,1196,40

CONSENSUS REPORTS: Reported in EPA TSCA Inventory.

SAFETY PROFILE: Moderately toxic by ingestion. Questionable carcinogen with experimental carcinogenic data. When heated to decomposition it emits toxic fumes of NO_x.

AEN250 CAS:124-04-9 ***HR: 3***
ADIPIC ACID
DOT: NA 9077
mf: $C_6H_{10}O_4$ mw: 146.16

PROP: White monoclinic prisms. Mp: 152°, flash p: 385°F (CC), d: 1.360 @ 25°/4°, vap press: 1 mm @ 159.5°, vap d: 5.04, autoign temp: 788°F, bp: 337.5°. Very sol in alc. Sol in acetone, water = 1.4% @15°; 0.6% @ 15° in ether.

SYNS: ACIFLOCTIN ◇ ACINETTEN ◇ ADILACTETTEN ◇ ADIPINIC ACID ◇ 1,4-BUTANEDICARBOXYLIC ACID ◇ FEMA No. 2011

◇ 1,6-HEXANEDIOIC ACID ◇ KYSELINA ADIPOVA (CZECH) ◇ MOLTEN ADIPIC ACID

TOXICITY DATA with REFERENCE
eye-rbt 20 mg/24H SEV 28ZPAK -,51,72
orl-rat LDLo:3600 mg/kg 28ZPAK -,51,72
ipr-rat LD50:275 mg/kg JAFCAU 5,759,57
orl-mus LD50:1900 mg/kg JAFCAU 5,759,57
ipr-mus LD50:275 mg/kg TXAPA9 32,566,75
ivn-mus LD50:680 mg/kg JAFCAU 5,759,57

CONSENSUS REPORTS: Reported in EPA TSCA Inventory.

SAFETY PROFILE: Poison by intraperitoneal route. Moderately toxic by other routes. A severe eye irritant. Combustible when exposed to heat or flame; can react with oxidizing materials. When heated to decomposition it emits acrid smoke and fumes.

AEN750 CAS:1985-84-8 ***HR: 1***
ADIPIC ACID BIS(3,4-EPOXY-6-METHYLCYCLOHEXYLMETHYL) ESTER
mf: $C_{22}H_{34}O_6$ mw: 394.56

SYNS: BIS(3,4-EPOXY-6-METHYLCYCLOHEXYLMETHYL)ADIPATE ◇ DI(3,4-EPOXY-6-METHYLCYCLOHEXYLMETHYL)ADIPATE ◇ HEXANEDIOIC ACID BIS(4-METHYL-7-OXABICYCLO(4.1.0)HEPT-3-YL)METHYL ESTER

TOXICITY DATA with REFERENCE
skn-rbt 500 mg open MLD UCDS** 9/23/70
orl-rat LD50:4290 mg/kg AIHAAP 24,305,63

CONSENSUS REPORTS: Reported in EPA TSCA Inventory.

SAFETY PROFILE: Mildly toxic by ingestion. A skin irritant. See also ESTERS. When heated to decomposition it emits acrid smoke and irritating fumes.

AEO000 CAS:103-23-1 ***HR: 3***
ADIPIC ACID BIS(2-ETHYLHEXYL) ESTER
mf: $C_{22}H_{42}O_4$ mw: 370.64

SYNS: ADIPOL 2EH ◇ BEHA ◇ BIS(2-ETHYLHEXYL) ADIPATE ◇ BISOFLEX DOA ◇ DEHA ◇ DI-2-ETHYLHEXYL ADIPATE ◇ DIOCTYL ADIPATE ◇ DOA ◇ EFFEMOLL DOA ◇ ERGOPLAST AdDO ◇ FLEXOL A 26 ◇ HEXANEDIOIC ACID, BIS(2-ETHYLHEXYL) ESTER ◇ HEXANEDIOIC ACID, DIOCTYL ESTER ◇ KODAFLEX DOA ◇ MONOPLEX DOA ◇ NCI-C54386 ◇ OCTYL ADIPATE ◇ PLASTOMOLL DOA ◇ PX-238 ◇ REOMOL DOA ◇ RUCOFLEX PLASTICIZER DOA ◇ SICOL 250 ◇ TRUFLEX DOA ◇ VESTINOL OA ◇ WICKENOL 158 ◇ WITAMOL 320

TOXICITY DATA with REFERENCE
eye-rbt 500 mg open AMIHBC 4,119,51
skn-rbt 500 mg open MLD UCDS** 1/12/72
dlt-mus-ipr 1000 mg/kg TXAPA9 32,566,75
ipr-rat TDLo:15 g/kg (5-15D preg):TER JPMSAE 62,1596,73
orl-mus TDLo:1038 g/kg/2Y-C:CAR NTPTR* NTP-TR-212,82
orl-mus TD:2163 g/kg/2Y-C:CAR NTPTR* NTP-TR-212,82
orl-rat LD50:9110 mg/kg AMIHBC 4,119,51
ivn-rat LD50:900 mg/kg MRLR** No.256,54
orl-mus LD50:15 g/kg JACTDZ 3(3),101,84
skn-rbt LD50:8410 mg/kg UCDS** 1/12/72
ivn-rbt LD50:540 mg/kg MRLR** #256,54

CONSENSUS REPORTS: IARC Cancer Review: Group 3 IMEMDT 7,56,87; Animal Limited Evidence IMEMDT 29,257,82. NTP Carcinogenesis Bioassay (feed); Clear Evidence: mouse NTPTR* NTP-TR-212,82; No Evidence: rat NTPTR* NTP-TR-212,82. Community Right-To-Know List. Reported in EPA TSCA Inventory.

SAFETY PROFILE: Moderately toxic by intravenous route. Mildly toxic by ingestion and skin contact. Experimental reproductive effects. Mutation data reported. An eye and skin irritant. Questionable carcinogen with experimental carcinogenic data. See also ESTERS. When heated to decomposition it emits acrid smoke and irritating fumes.

AEO250 CAS:63905-29-3 ***HR: 2***
ADIPIC ACID-3-CYCLOHEXENYLMETHANOL DIESTER
mf: $C_{20}H_{30}O_4$ mw: 334.50

TOXICITY DATA with REFERENCE
skn-rbt LD50:7070 mg/kg TXAPA9 28,313,74
orl-rat LD50:3730 mg/kg TXAPA9 28,313,74

CONSENSUS REPORTS: Reported in EPA TSCA Inventory.

SAFETY PROFILE: Moderately toxic by ingestion. When heated to decomposition it emits acrid smoke and irritating fumes.

AEO500 CAS:2998-04-1 ***HR: 3***
ADIPIC ACID DIALLYL ESTER
mf: $C_{12}H_{18}O_4$ mw: 226.30

SYNS: ALLYL ADIPATE ◇ HEXANEDIOIC ACID-DI-2-PROPENYL ESTER

TOXICITY DATA with REFERENCE
skn-rbt LDLo:1 g/kg SCCUR* -,3,61
orl-rat LDLo:420 mg/kg SCCUR* -,3,61
orl-mus LD50:180 mg/kg SCCUR* -,3,61

CONSENSUS REPORTS: Reported in EPA TSCA Inventory. Allyl compounds are on the Community Right-To-Know List.

SAFETY PROFILE: Poison by ingestion. Moderately toxic by skin contact. See also ALLYL COMPOUNDS and ESTERS. When heated to decomposition it emits acrid smoke and irritating fumes.

AEO750 CAS:105-99-7 ***HR: 2***
ADIPIC ACID DIBUTYL ESTER
mf: $C_{14}H_{26}O_4$ mw: 258.40

SYNS: BUTYL ADIPATE ◇ DIBUTYL ADIPATE ◇ DI-N-BUTYL ADIPATE ◇ DIBUTYL ADIPINATE ◇ DIBUTYL HEXANEDIOATE ◇ EXPERIMENTAL TICK REPELLENT 3 ◇ HEXANEDIOIC ACID-DIBUTYL ESTER

TOXICITY DATA with REFERENCE
skn-rbt 10 mg/24H MLD AMIHBC 4,119,51
eye-rbt 500 mg AMIHBC 4,119,51
ipr-rat TDLo:1049 mg/kg (5-15D preg):TER JPMSAE 62,1596,73
orl-rat LD50:12900 mg/kg 28ZEAL 5,72,76
ipr-rat LD50:5244 mg/kg JPMSAE 62,1596,73
skn-rbt LD50:20000 mg/kg AMIHBC 4,119,51

CONSENSUS REPORTS: Reported in EPA TSCA Inventory.

SAFETY PROFILE: Mildly toxic by several routes. An experimental teratogen. Skin and eye irritant. See also ESTERS. When heated to decomposition it emits acrid smoke and irritating fumes.

AEP000 ***HR: 1***
ADIPIC ACID DIDECYL ESTER (mixed isomers)
mf: $C_{26}H_{50}O_4$ mw: 426.76

TOXICITY DATA with REFERENCE
skn-rbt 10 mg/24H open MLD AIHAAP 23,95,62
orl-rat LD50:21 g/kg AIHAAP 23,95,62
skn-rbt LD50:8410 mg/kg AIHAAP 23,95,62

SAFETY PROFILE: Mildly toxic by ingestion and skin contact. A skin irritant. See also ESTERS. When heated to decomposition it emits acrid smoke and irritating fumes.

AEP250 CAS:7790-07-0 ***HR: 2***
ADIPIC ACID-DI-2-(2-ETHYLBUTOXY)ETHYL) ESTER
mf: $C_{22}H_{42}O_6$ mw: 402.64

SYN: DI-2-(2-ETHYLBUTOXY)ETHYL ADIPATE

TOXICITY DATA with REFERENCE
skn-rbt 10 mg/24H open MLD AMIHBC 10,61,54
eye-rbt 500 mg open AMIHBC 10,61,54
orl-rat LD50:3250 mg/kg AMIHBC 10,61,54
skn-rbt LD50:4240 mg/kg AMIHBC 10,61,54

SAFETY PROFILE: Moderately toxic by ingestion. Mildly toxic by skin contact. Skin and eye irritant. See also ESTERS. When heated to decomposition it emits acrid smoke and irritating fumes.

AEP500 CAS:10022-60-3 ***HR: 2***
ADIPIC ACID DI(2-ETHYLBUTYL) ESTER
mf: $C_{18}H_{34}O_4$ mw: 314.52

SYN: DI(2-ETHYLBUTYL) ADIPATE

TOXICITY DATA with REFERENCE
eye-rbt 500 mg open AMIHBC 10,61,54
orl-rat LD50:5620 mg/kg AMIHBC 10,61,54
skn-rbt LD50:17 g/kg AMIHBC 10,61,54

SAFETY PROFILE: Moderately toxic by skin contact. Mildly toxic by ingestion. An eye irritant. See also ESTERS. When heated to decomposition it emits acrid smoke and irritating fumes.

AEP750 CAS:141-28-6 ***HR: 2***
ADIPIC ACID DIETHYL ESTER
mf: $C_{10}H_{18}O_4$ mw: 202.28

SYNS: DIETHYL ADIPATE ◇ DIETHYL HEXANEDIOATE ◇ ETHYL ADIPATE ◇ ETHYL-Δ-CARBOETHOXYVALERATE

TOXICITY DATA with REFERENCE
dlt-mus-ipr 1100 mg/kg TXAPA9 32,566,75
ipr-mus LD50:2190 mg/kg TXAPA9 32,566,75

CONSENSUS REPORTS: Reported in EPA TSCA Inventory.

SAFETY PROFILE: Moderately toxic by intraperitoneal route. Mutation data reported. See also ESTERS. When heated to decomposition it emits acrid smoke and irritating fumes.

AEQ000 CAS:110-32-7 ***HR: 1***
ADIPIC ACID DI(2-HEXYLOXYETHYL) ESTER
mf: $C_{22}H_{42}O_6$ mw: 402.64

SYNS: BIS(2-HEXYLOXY)ETHYL)ADIPATE ◇ DIHEXYLOXYETHYL ADIPATE ◇ HEXANEDIOIC ACID, BIS(2-(HEXYLOXY)ETHYL)ESTER

TOXICITY DATA with REFERENCE
dlt-mus:ipr 1100 mg/kg TXAPA9 32,566,75
ipr-rat TDLo:837 mg/kg (female 5-15D post):TER JPMSAE 62,1596,73
ipr-mus LD50:2190 mg/kg TXAPA9 32,566,75

CONSENSUS REPORTS: Reported in EPA TSCA Inventory.

SAFETY PROFILE: Mildly toxic by intraperitoneal route. Experimental teratogenic data. Mutation data reported. When heated to decomposition it emits acrid smoke and irritating fumes.

AEQ250 CAS:1071-93-8 ***HR: 2***
ADIPIC ACID DIHYDRAZIDE
mf: $C_6H_{14}N_4O_2$ mw: 174.24

SYNS: ADIPIC DIHYDRAZIDE ◇ HEXANEDIOIC ACID DIHYDRAZIDE

TOXICITY DATA with REFERENCE
par-mus LDLo:4000 mg/kg CBCCT* 7,685,55

CONSENSUS REPORTS: Reported in EPA TSCA Inventory.

SAFETY PROFILE: Moderately toxic by parenteral route. When heated to decomposition it emits toxic fumes of NO_x.

AEQ500 CAS:6624-70-0 ***HR: 2***
ADIPIC ACID DIISOPENTYL ESTER
mf: $C_{16}H_{30}O_4$ mw: 286.46

SYNS: BIS(3-METHYLBUTYL) ADIPATE ◇ DIISOAMYL ADIPATE ◇ DI(3-METHYLBUTYL)ADIPATE ◇ HEXANEDIOIC ACID, BIS(3-METHYLBUTYL) ESTER

TOXICITY DATA with REFERENCE
orl-gpg LD50:25 g/kg GWXXBX #2703360
ivn-rat LD50:640 mg/kg MRLR** No.256,54
ivn-rbt LD50:640 mg/kg MRLR** No.256,54

SAFETY PROFILE: Moderately toxic by intravenous route. See also ESTERS. When heated to decomposition it emits acrid smoke and irritating fumes.

AEQ750 CAS:6900-06-7 ***HR: 3***
ADIPIC ACID DI-2-PROPYNYL ESTER

TOXICITY DATA with REFERENCE
orl-rat LD50:200 mg/kg AIHAAP 23,95,62
skn-rbt LD50:440 mg/kg AIHAAP 23,95,62

SAFETY PROFILE: Poison by ingestion. Moderately toxic by skin contact. See also ESTERS. When heated to decomposition it emits acrid smoke and irritating fumes.

AER000 ***HR: 2***
ADIPIC ACID, UREA mixed with CARBOXYMETHYLCELLULOSE ACIDS

PROP: Consists of 97.3% urea, 0.6% adipic acid, and 2.1% carboxymethylcellulose acids (ANYAA9 75,543, 59).

TOXICITY DATA with REFERENCE
ivg-mus TDLo:91 g/kg/76W-I:ETA ANYAA9 75,543,59

SAFETY PROFILE: Questionable carcinogen with experimental tumorigenic data.

AER250 CAS:111-69-3 ***HR: 3***
ADIPONITRILE
DOT: UN 2205
mf: $C_6H_8N_2$ mw: 108.16

PROP: Water-white liquid, practically odorless. Mp: 2.3°, bp: 295°, flash p: 199.4°F (OC), d: 0.965 @ 20°/4°, vap d: 3.73.

SYNS: ADIPIC ACID DINITRILE ◇ ADIPIC ACID NITRILE ◇ ADIPODINITRILE ◇ 1,4-DICYANOBUTANE ◇ HEXANEDINITRILE ◇ HEXANEDIOIC ACID DINITRILE ◇ NITRILE ADIPICO (ITALIAN) ◇ TETRAMETHYLENE CYANIDE

TOXICITY DATA with REFERENCE
orl-rat LD50:155 mg/kg GISAAA 49(12),40,84
ihl-rat LC50:1710 mg/m³/4H TOXID9 1,76,81
orl-mus LD50:172 mg/kg ARTODN 57,88,85
ipr-mus LD50:40 mg/kg NTIS** AD691-490
orl-rbt LD50:22 mg/kg GISAAA 49(12),40,84
scu-gpg LD50:50 mg/kg MELAAD 46,221,55

CONSENSUS REPORTS: EPA Extremely Hazardous Substances List. Reported in EPA TSCA Inventory. Cyanide and its compounds are on the Community Right-To-Know List.

NIOSH REL: TWA 18 mg/m³
DOT Classification: IMO: Poison B; Label: St. Andrews Cross.

SAFETY PROFILE: Poison by inhalation, ingestion, subcutaneous, and intraperitoneal routes. See also HYDROCYANIC ACID and NITRILES. The nitrile group will behave as a cyanide when ingested or absorbed in the body. It produces disturbances of the respiration and circulation, irritation of the stomach and intestines, and loss of weight. Its low vapor pressure at room temperature makes exposure to harmful concentrations of its vapors unlikely if handled with reasonable care in well ventilated areas. Flammable when exposed to heat or flame. When heated to decomposition it emits toxic fumes of CN^-. Can react with oxidizing materials. To fight fire, use foam, CO_2, dry chemical.

AER500 CAS:35108-88-4 ***HR: 3***
ADOBIOL
mf: $C_{18}H_{29}NO_4 \cdot ClH$ mw: 359.94

SYNS: BUFETOLOL HYDROCHLORIDE ◇ 1-(tert-BUTYLAMINO)-3-(o-((TETRAHYDROFURFURYL)OXY)PHENOXY)-2-PROPANOLHYDROCHLORIDE

TOXICITY DATA with REFERENCE
orl-rat TDLo:300 mg/kg (9-14D preg):REP OYYAA2 6,1267,72
orl-mus TDLo:60 mg/kg (female 7-12D post):TER OYYAA2 6,1267,72
orl-rat LD50:1088 mg/kg DRUGAY 6,681,82
scu-rat LD50:1814 mg/kg DRUGAY 6,681,82
ivn-rat LD50:59400 μg/kg DRUGAY 6,681,82
orl-mus LD50:402 mg/kg DRUGAY 6,681,82
scu-mus LD50:501 mg/kg DRUGAY 6,681,82

SAFETY PROFILE: Poison by intravenous route. Moderately toxic by ingestion and subcutaneous routes. An experimental teratogen. Other experimental reproductive effects. When heated to decomposition it emits very toxic fumes of HCl and NO_x. An antiarrhythmic drug.

AER666 CAS:51460-26-5 ***HR: D***
ADONA TRIHYDRATE
mf: $C_{10}H_{11}N_4O_5S{\cdot}Na{\cdot}3H_2O$ mw: 376.36

PROP: Yellow-orange needles from aq methanol. Decomp 227-228°. Soluble in water.

SYNS: AC-17 TRIHYDRATE ◇ ADENARON ◇ ADONA ◇ ADRECHROS ◇ ADRENOCHROME SULFONATE AC 17 TRIHYDRATE ◇ CARBAZOCHROME SODIUM SULFONATE ◇ CARBAZOCHROME SODIUM SULFONATE TRIHYDRATE ◇ CARBAZON ◇ DONASEVEN ◇ EMEX ◇ ODANON ◇ SODIUM-1-METHYL-5-SEMICARBAZONO-6-OXO-2,3,5,6-TETRAHYDROINDOLE-3-SULFONATETRIHYDRATE ◇ TAZIN

TOXICITY DATA with REFERENCE
ipr-rat TDLo:1280 mg/kg (female 7-14D post):TER OYYAA2 4,39,70
ipr-rat TDLo:1280 mg/kg (female 7-14D post):REP OYYAA2 4,39,70

SAFETY PROFILE: An experimental teratogen. Other experimental reproductive effects. When heated to decomposition it emits toxic fumes of SO_x, NO_x, and Na_2O.

AER750 CAS:8002-01-5 ***HR: 3***
ADONIDIN

TOXICITY DATA with REFERENCE
ivn-pgn LDLo:2829 μg/kg YHHPAL 10,561,63
ivn-cat LDLo:3 mg/kg 27ZWAY E.1,78
ivn-rbt LDLo:5 mg/kg 27ZWAY E.1,78
scu-frg LDLo:4 mg/kg 27ZWAY E.1,78

SAFETY PROFILE: Poison by intravenous and subcutaneous routes.

AES000 CAS:51-42-3 ***HR: 3***
ADRENALIN BITARTRATE
mf: $C_9H_{13}NO_3{\cdot}C_4H_6O_6$ mw: 333.33

SYNS: ADRENALINE ACID TARTRATE ◇ (−)-ADRENALINE ACID TARTRATE ◇ ADRENALINE BITARTRATE ◇ (−)-ADRENALINE BITARTRATE ◇ 1-ADRENALINE BITARTRATE ◇ 1-ADRENALINE-d-BITARTRATE ◇ ADRENALINE HYDROGEN TARTRATE ◇ 1-ADRENALINE HYDROGEN TARTRATE ◇ (−)-ADRENALINE HYDROGEN TARTRATE ◇ ADRENALINE TARTRATE ◇ (−)-ADRENALINE TARTRATE ◇ l-ADRENALINE TARTRATE ◇ ASMATANE MIST ◇ (−)-3,4-DIHYDROXY-α-((METHYLAMINO)METHYL)BENZYL) ALCOHOL (+)-TARTRATE (1:1) SALT ◇ EPINEPHRINE BITARTRATE ◇ (−)-EPINEPHRINE BITARTRATE ◇ l-EPINEPHRINE BITARTRATE ◇ EPINEPHRINE-d-BITARTRATE ◇ 1-EPINEPHRINE-d-BITARTRATE ◇ EPINEPHRINE HYDROGEN TARTRATE ◇ 1-EPINEPHRINE TARTRATE ◇ IOP ◇ LYOPHRIN ◇ MEDIHALER-EPI ◇ SUPRARENIN

TOXICITY DATA with REFERENCE
dni-mus:oth 1 μmol/L CNREA8 43,3514,83
scu-mus TDLo:2400 μg/kg (2D male):REP JRPFA4 22,375,70
scu-rat LD50:8300 μg/kg NIIRDN 6,122,82
ivn-rat LD50:82 μg/kg AIPTAK 137,155,62
orl-mus LD50:4 mg/kg APTOA6 31,49,72
ipr-mus LD50:7800 μg/kg APTOA6 31,43,72
scu-mus LD50:11100 μg/kg APTOA6 55,73,84
ivn-mus LD50:1780 μg/kg JPETAB 81,269,44

SAFETY PROFILE: Poison by ingestion, subcutaneous, intraperitoneal, and intravenous routes. Experimental reproductive effects. Mutation data reported. When heated to decomposition it emits toxic fumes of NO_x. See also VASOTONIN and other adrenalin compounds.

AES250 CAS:150-05-0 ***HR: 3***
d-ADRENALINE
mf: $C_9H_{13}NO_3$ mw: 183.23

PROP: Light brown or nearly white crystals. Mp: 211-212°. Very sltly sol in water, alc, 1:1 chloroform, and ether.

SYNS: 1-(+)-ADRENALINE ◇ d-EPINEPHRINE

TOXICITY DATA with REFERENCE
scu-rat LDLo:80 mg/kg JPHYA7 38,259,09
ivn-rat LD50:800 μg/kg JPETAB 95,502,49
scu-mus LDLo:4 mg/kg HBAMAK 4,1294,35
ivn-mus LD50:38 mg/kg JPETAB 95,502,49
scu-dog LDLo:5 mg/kg HBAMAK 4,1294,35
ivn-dog LDLo:1 mg/kg HBAMAK 4,1294,35
ivn-cat LDLo:500 μg/kg HBAMAK 4,1294,35
scu-rbt LDLo:10 mg/kg HBAMAK 4,1294,35
ivn-rbt LDLo:50 μg/kg HBAMAK 4,1294,35
scu-gpg LDLo:1 mg/kg HBAMAK 4,1294,35
scu-frg LDLo:5000 mg/kg HBAMAK 4,1294,35

SAFETY PROFILE: Poison by subcutaneous and intravenous routes. Can cause contact dermatitis. Usually the symptoms are of short duration and clear up spontaneously. Combustible when heated. Upon decomposition it emits toxic fumes of NO_x.

AES500 CAS:55-31-2 ***HR: 3***
1-ADRENALINE CHLORIDE
mf: $C_9H_{13}NO_3{\cdot}ClH$ mw: 219.69

SYNS: ADRENALIN CHLORIDE ◇ ADRENALIN HYDROCHLORIDE ◇ (−)-ADRENALINE HYDROCHLORIDE ◇ 1-ADRENALINE HYDROCHLORIDE ◇ 1,2-BENZENEDIOL, 4-(1-HYDROXY-2-(METHYLAMINO)ETHYL)-, HYDROCHLORIDE, (R)- (9CI) ◇ 1-1-(3,4-DIHYDROXYPHENYL)-2-METHYLAMINO-1-ETHANOL HYDROCHLORIDE ◇ EPINEPHRINE CHLORIDE ◇ 1-EPINEPHRINE CHLORIDE ◇ (−)-EPINEPHRINE HYDROCHLORIDE ◇ 1-EPINEPHRINE HYDROCHLORIDE ◇ GELATIN-EPINEPHRINE ◇ 1-METHYLAMINOETHANOLCATHECHOL HYDROCHLORIDE ◇ NCI-C55663 ◇ SUPRANEPHRIN SOLUTION ◇ SUPRARENIN HYDROCHLORIDE

TOXICITY DATA with REFERENCE
scu-rat TDLo:3200 μg/kg (13-20D preg):REP CJPPA3 43,473,65
scu-rat TDLo:3200 μg/kg (13-20D preg):TER CJPPA3 43,473,65
orl-rat LD50:24 mg/kg AIPTAK 180,155,69
scu-rat LD50:5 mg/kg AIPTAK 180,155,69
ivn-rat LDLo:50 μg/kg JPETAB 24,101,24
orl-mus LDLo:50 mg/kg ARZNAD 13,51,63
ipr-mus LD50:4664 μg/kg JPETAB 90,110,47
scu-mus LD50:1980 μg/kg JPETAB 87,214,46
ivn-mus LD50:140 μg/kg EJPHAZ 9,289,70

SAFETY PROFILE: Poison by ingestion, intravenous, subcutaneous, and intraperitoneal routes. An experimental teratogen. Other experimental reproductive effects. When heated to decomposition it emits very toxic fumes of Cl^- and NO_x. See also VASOTONIN and other adrenalin compounds.

AES625 CAS:329-63-5 ***HR: 3***
dl-ADRENALINE HYDROCHLORIDE
mf: $C_9H_{13}NO_3 \cdot ClH$ mw: 219.69

SYNS: (±)-ADRENALINE HYDROCHLORIDE ◇ (±)-3,4-DIHYDROXY-α-((METHYLAMINO)METHYL)BENZYL ALCOHOL HYDROCHLORIDE ◇ (±)-EPINEPHRINE HYDROCHLORIDE ◇ dl-EPINEPHRINE HYDROCHLORIDE

TOXICITY DATA with REFERENCE
ipr-rat LDLo:1800 μg/kg JAPMA8 30,183,41
orl-mus LDLo:96 mg/kg JAPMA8 30,183,41
ipr-mus LD50:7800 μg/kg JPETAB 92,369,48
ivn-mus LDLo:5 mg/kg JPETAB 92,108,48

SAFETY PROFILE: Poison by ingestion, intravenous, and intraperitoneal routes. When heated to decomposition it emits toxic fumes of NO_x and HCl.

AES639 CAS:54-06-8 ***HR: 3***
ADRENOCHROME
mf: $C_9H_9NO_3$ mw: 179.19

SYNS: ADRAXONE ◇ 2,3-DIHYDRO-3-HYDROXY-1-METHYL-1H-INDOLE-5,6-DIONE (9CI) ◇ 3-HYDROXY-1-METHYL-5,6-INDOLINEDIONE ◇ USAF UCTL-7

TOXICITY DATA with REFERENCE
ipr-rat LD50:150 mg/kg AIPTAK 106,90,56
ipr-mus LD50:100 mg/kg NTIS** AD277-689
ivn-mus LD50:128 mg/kg AIPTAK 106,90,56

SAFETY PROFILE: Poison by intravenous and intraperitoneal routes. When heated to decomposition it emits toxic fumes of NO_x.

AES650 CAS:9002-60-2 ***HR: 1***
ADRENOCORTICOTROPHIC HORMONE

PROP: White powder. Freely sol in water. Appreciably soluble in 60 to 70% alc or acetone. One U.S.P. unit, one international unit, one Armour unit, or one potency unit denotes the same activity.

SYNS: ACETHROPAN ◇ ACORTAN ◇ ACORTO ◇ ACTH ◇ ACTHAR ◇ ACTON ◇ ACTONAR ◇ ADRENAL CORTEX HORMONE ◇ ADRENOCORTICOTROPHIN ◇ ADRENOCORTICOTROPIC HORMONE ◇ ADRENOCORTICOTROPIN ◇ ADRENOMONE ◇ ADRENOTROPHIN ◇ ALFATROFIN ◇ CIBACTHEN ◇ CORSTILINE ◇ CORTICOTROPHIN ◇ CORTICOTROPIN ◇ CORTICOTROPIN-LIKE SUBSTANCES ◇ CORTIPHYSON ◇ CORTROPHIN ◇ CORTROPHYSON ◇ DYNAMONE ◇ EXACTHIN ◇ ISLACTID ◇ PITUITARY GLAND ADRENO CORTICO-TROPIC HORMONE ◇ REACTHIN ◇ SOLACTHYL ◇ TUBEX

TOXICITY DATA with REFERENCE
scu-mus TDLo:3600 μg/kg (female 12-17D post):REP TJADAB 35,229,87
par-rbt TDLo:10500 μg/kg (female 19-26D post):TER JOENAK 14,284,56
unr-inf TDLo:240 mg/kg/16W-I:SYS LANCAO 1,901,84

SAFETY PROFILE: Human systemic effects: kidney changes. An experimental teratogen. Other experimental reproductive effects.

AES750 CAS:23214-92-8 ***HR: 3***
ADRIAMYCIN
mf: $C_{27}H_{29}NO_{11} \cdot ClH$ mw: 543.57

PROP: Isolated from cultures of *Streptomyces peucetius var. Caesius*.

SYNS: ADM ◇ ADRIAMYCIN-HCl ◇ ADRIAMYCIN SEMIQUINONE ◇ ADRIBLASTINA ◇ DOXORUBICIN ◇ DX ◇ F.I 106 ◇ 14-HYDROXYDAUNOMYCIN ◇ 14'-HYDROXYDAUNOMYCIN ◇ 14-HYDROXYDAUNORUBICINE ◇ KW-125 ◇ NCI-C01514 ◇ NSC-123127

TOXICITY DATA with REFERENCE
mmo-smc 184 μmol/L MGGEAE 174,39,79
cyt-hmn:leu 20 μg/L CNREA8 31,32,71
unr-rat TDLo:4500 μg/kg (female 10-12D post):REP TJADAB 33,39C,86
ivn-mus TDLo:7 mg/kg (female 7-13D post):TER YKRYAH 6,1152,73
ivn-rat TDLo:5 mg/kg:CAR CNREA8 43,5248,83
ivn-rat TD:8 mg/kg:NEO EXPEAM 27,1209,71
ivn-rat TD:10 mg/kg:CAR JJIND8 66,81,81
ivn-rat TD:5 mg/kg:ETA PAACA3 21,309,80
ivn-hmn TDLo:15 mg/kg/D:CVS,GIT,SKN CANCAR 34,518,74
ivn-hmn TDLo:380 mg/kg/31W:CVS,GIT,SKN CANCAR 34,518,74
ipr-rat LD50:16 mg/kg OYYAA2 6,1075,72
ivn-rat LD50:10510 μg/kg TXAPA9 79,412,85
orl-mus LD50:570 mg/kg ANTBAL 28,298,83
ipr-mus LD50:11960 μg/kg KSRNAM 7,1052,73
scu-mus LD50:15980 μg/kg KSRNAM 7,1052,73
ivn-mus LD50:8950 μg/kg KSRNAM 7,1052,73
ivn-dog LD50:2400 μg/kg DCTODJ 6,21,83

CONSENSUS REPORTS: NTP Fifth Annual Report on Carcinogens. IARC Cancer Review: Group 2A IMEMDT 7,82,87; Animal Inadequate Evidence IMEMDT 10,43,76.

SAFETY PROFILE: Confirmed carcinogen with experimental carcinogenic, neoplastigenic, and tumorigenic data. Poison by intraperitoneal, subcutaneous, parenteral, and intravenous routes. Human systemic effects by intravenous route: cardiac myopathy including infarction, nausea or vomiting, and effects on the hair. An experimental teratogen. Other experimental reproductive effects. Human mutation data reported. When heated to decomposition it emits very toxic fumes of NO_x and HCl.

AET250 CAS:51898-39-6 ***HR: 3***
ADRIAMYCIN-14-OCTANOATEHYDROCHLORIDE
mf: $C_{35}H_{43}NO_{12}$ mw: 669.79

TOXICITY DATA with REFERENCE
dnd-mam:lym 3490 nmol/L CBINA8 20,97,78
ivn-mus LD50:19 mg/kg 31TFAO 3,987,74

SAFETY PROFILE: Poison by intravenous route. Mutation data reported. When heated to decomposition it emits toxic fumes of NO_x.

AET750 CAS:1402-68-2 ***HR: 3***
AFLATOXIN

TOXICITY DATA with REFERENCE
dlt-mus-ipr 68 mg/kg NATUAS 219,385,68
par-ham TDLo:4 mg/kg (8D preg):TER DABBBA 34,5251,73
par-ham TDLo:4 mg/kg (8D preg):REP DABBBA 34,5251,73
orl-rat TDLo:7788 μg/kg/13W-C:ETA NATUAS 202,1016,64
orl-rat TDLo:2250 μg/kg (10-21D preg):ETA,REP CNREA8 33,262,73
orl-hmn LDLo:229 μg/kg/8W LANCAO 1,1061,75
orl-mky LD50:1750 μg/kg FCTXAV 14,227,76
ims-mky LD50:2020 mg/kg FCTXAV 14,227,76
orl-qal LDLo:4 mg/kg BPOSA4 21,29,80

CONSENSUS REPORTS: NTP Fifth Annual Report on Carcinogens. IARC Cancer Review: Group 1 IMEMDT 7,83,87; Human Limited Evidence IMEMDT 10,51,76.

SAFETY PROFILE: Confirmed human carcinogen with experimental tumorigenic data. Human poison by ingestion. An experimental teratogen. Other experimental reproductive effects. Mutation data reported. See also various aflatoxins.

AEU250 CAS:1162-65-8 ***HR: 3***
AFLATOXIN B1
mf: $C_{17}H_{12}O_6$ mw: 312.29

PROP: A crystalline material. Mp: 268°.

SYNS: AFBI ◇ AFLATOXIN B

TOXICITY DATA with REFERENCE
pic-sat 1 μg/L ENMUDM 1,121,79
cyt-hmn:lym 19200 nmol/L TOLED5 7,245,81
sce-hmn:lym 19200 nmol/L TOLED5 7,245,81
mma-sat 10 ng/plate FCTOD7 22,355,84
scu-rat TDLo:1200 μg/kg (female 11-14D post):REP TOLED5 18(Suppl 1),113,83
ipr-ham TDLo:4 mg/kg (female 8D post):TER NATUAS 215,638,67
orl-rat TDLo:380 μg/kg/68W-C:CAR CNREA8 27,2370,67
ipr-rat TDLo:2 mg/kg (female 18-21D post):NEO,TER JJIND8 64,1349,80
ipr-rat TDLo:6 mg/kg/8W-I:CAR CNREA8 31,1936,71
ivn-rat TDLo:8 mg/kg (female 15D post):NEO,TER IARCCD 4,100,73
orl-mus TDLo:150 mg/kg/6W-I:NEO TXAPA9 82,19,86
scu-mus TDLo:30 mg/kg (female 15-22D post):NEO,TER BEXBAN 82,1687,76
orl-mky TDLo:168 mg/kg/6Y-C:ETA JJIND8 57,67,76
orl-rat LD50:5 mg/kg CNREA8 27,2370,67
ipr-rat LD50:6 mg/kg TXAPA9 25,458,73
orl-mus LD50:9 mg/kg APPYAG 12,303,74
ipr-mus LD50:9500 μg/kg LSPPAT 13,1143,73
ipr-dog LDLo:1 mg/kg PAVEAC 3,331,66
orl-mky LD50:2200 μg/kg TXAPA9 19,169,71
orl-cat LD50:550 μg/kg CNREA8 29,236,69
orl-pig LD50:620 μg/kg APPYAG 12,303,74
orl-gpg LD50:2 mg/kg TXAPA9 19,169,71
ipr-gpg LD50:1400 μg/kg JPBAA7 91,277,66
orl-ham LD50:10 mg/kg CNREA8 29,236,69
ipr-ham LD50:6 mg/kg ARPAAQ 83,53,67
orl-dck LD50:335 μg/kg PSEBAA 123,151,66
orl-dom LDLo:2 mg/kg NATUAS 225,1062,70

CONSENSUS REPORTS: IARC Cancer Review: Group 1 IMEMDT 7,83,87; Animal Sufficient Evidence IMEMDT 10,51,76; 1,145,72. EPA Genetic Toxicology Program.

SAFETY PROFILE: Confirmed human carcinogen with experimental tumorigenic, neoplastigenic, and carcinogenic data. Acute poison by ingestion, intraperitoneal, and possibly other routes. Experimental teratogenic and reproductive effects. Mutation data reported. When heated to decomposition it emits acrid smoke. See also various aflatoxins.

AEU500 CAS:58209-98-6 **HR: 2**
AFLATOXIN B1-2,3-DICHLORIDE
mf: $C_{17}H_{12}Cl_2O_6$ mw: 383.19

TOXICITY DATA with REFERENCE
dnd-hmn:fbr 4 μmol/L CBINA8 50,59,84
msc-hmn:fbr 4 nmol/L CBINA8 50,59,84
sln-dmg-par 200 nmol/L CNREA8 38,2608,78
scu-rat TDLo:197 μg/kg/10W-I:NEO CNREA8 35,3811,75

SAFETY PROFILE: Questionable carcinogen with experimental neoplastigenic data. Human mutation data reported. When heated to decomposition it emits toxic fumes of Cl^-. See also various aflatoxins.

AEU750 CAS:7220-81-7 **HR: 3**
AFLATOXIN B2
mf: $C_{17}H_{14}O_6$ mw: 314.31
SYN: DIHYDROAFLATOXIN B1

TOXICITY DATA with REFERENCE
mma-sat 370 ng/plate MUREAV 130,79,84
dnd-rat-par 40 μg/kg/2D-C BBRCA9 83,1354,78
sce-ham:lng 3100 μg/L CRNGDP 1,759,80
dnd-mam:lym 50 μmol/L CRNGDP 3,423,82
dns-rat:lvr 10 μmol/L/1H CNREA8 37,1845,77
mma-ham:lng 83 μmol/L MUREAV 46,27,77
ipr-rat TDLo:600 mg/kg/8W-I:ETA CNREA8 31,1936,71
orl-dck LD50:1700 μg/kg NATUAS 200,1101,63

CONSENSUS REPORTS: IARC Cancer Review: Animal Sufficient Evidence IMEMDT 10,51,76; Animal Limited Evidence IMEMDT 1,145,72.

SAFETY PROFILE: Confirmed human carcinogen with experimental tumorigenic data. Poison by ingestion. Mutation data reported. When heated to decomposition it emits acrid smoke and fumes. See also various aflatoxins.

AEV000 CAS:1165-39-5 **HR: 3**
AFLATOXIN G1
mf: $C_{17}H_{12}O_7$ mw: 328.29

PROP: A metabolite of *Aspergillus flavus link ex fries*.

TOXICITY DATA with REFERENCE
mma-sat 31 ng/plate MUREAV 130,79,84
cyt-mky:kdy 2 mg/L/2H-C JNCIAM 48,1647,72
orl-rat TDLo:4100 μg/kg/22W-I:CAR CNREA8 35,2469,75
orl-rat TDLo:5600 μg/kg/2W-I:CAR CNREA8 31,1936,71
scu-rat TDLo:12 mg/kg/30W-I:NEO BJCAAI 19,392,65
orl-dck LD50:785 μg/kg PSEBAA 123,151,66
ipr-rat LD50:14900 μg/kg JPTLAS 102,209,70

CONSENSUS REPORTS: IARC Cancer Review: Animal Sufficient Evidence IMEMDT 10,51,76.

SAFETY PROFILE: Confirmed human carcinogen with experimental carcinogenic and neoplastigenic data. Poison by ingestion and intraperitoneal routes. A suspected human carcinogenic. Mutation data reported. When heated to decomposition it emits acrid smoke and irritating fumes. See also various aflatoxins.

AEV250 **HR: 3**
AFLATOXIN G1 mixed with AFLATOXIN B1

PROP: Metabolites of *Aspergillus flavus link ex fries* Aflatoxin G1, 56.4%; Alfatoxin B1, 37.7%.

TOXICITY DATA with REFERENCE
orl-rat TDLo:66 mg/kg/66W-C:ETA BJCAAI 20,134,66
scu-rat TDLo:1760 μg/kg/44W-I:NEO BJCAAI 19,392,65
itr-rat TDLo:72 mg/kg/30W-I:CAR BJCAAI 20,134,66

CONSENSUS REPORTS: NTP Fifth Annual Report on Carcinogens.

SAFETY PROFILE: Confirmed human carcinogen with experimental carcinogenic, neoplastigenic, and tumorigenic data. See also various aflatoxins.

AEV500 CAS:7241-98-7 **HR: 3**
AFLATOXIN G2
mf: $C_{17}H_{14}O_7$ mw: 330.31

TOXICITY DATA with REFERENCE
dns-rat:lvr 10 μmol/L/1H CNREA8 37,1845,77
sce-ham:lng 3300 μg/L CRNGDP 1,759,80
orl-dck LD50:2450 μg/kg NATUAS 200,1101,63

CONSENSUS REPORTS: IARC Cancer Review: Animal Inadequate Evidence IMEMDT 1,145,72. EPA Genetic Toxicology Program.

SAFETY PROFILE: Suspected carcinogen. Acute poison by ingestion. Mutation data reported. When heated to decomposition it emits acrid smoke and irritating fumes. See also various aflatoxins.

AEW000 CAS:6795-23-9 **HR: 3**
AFLATOXIN M1
mf: $C_{17}H_{12}O_7$ mw: 328.29

SYN: 4-HYDROXYAFLATOXIN B1

TOXICITY DATA with REFERENCE
dnd-rat-orl 3600 ng/kg CBINA8 32,249,80
cyt-rat-orl 1 mg/kg JNCIAM 47,585,71
mma-sat 200 ng/plate JEPTDQ 2,1099,79
dns-rat:lvr 600 ng/plate TOXID9 1,42,81
orl-rat TDLo:8 mg/kg/8W-I:ETA FCTXAV 12,381,74
orl-rat LDLo:1500 μg/kg JNCIAM 47,585,71

CONSENSUS REPORTS: IARC Cancer Review: Animal Sufficient Evidence IMEMDT 10,51,76

CONSENSUS REPORTS: EPA Genetic Toxicology Program.

SAFETY PROFILE: Confirmed carcinogen with experimental tumorigenic data. Poison by ingestion. Mutation data reported. When heated to decomposition it emits acrid smoke and irritating fumes. See also various aflatoxins.

AEW500 CAS:29611-03-8 ***HR: 3***
AFLATOXIN Ro
mf: $C_{17}H_{14}O_6$ mw: 314.31

SYN: AFLATOXICOL

TOXICITY DATA with REFERENCE
mma-sat 25 ng/plate PNASA6 73,2241,76
orl-rat TDLo:1092 g/kg/1Y-C:CAR JJIND8 66,1159,81

SAFETY PROFILE: Suspected carcinogen with experimental carcinogenic data. Mutation data reported. When heated to decomposition it emits acrid smoke and irritating fumes. See also various aflatoxins.

AEW625 CAS:56287-74-2 ***HR: 3***
AFLOQUALONE
mf: $C_{16}H_{14}FN_3O$ mw: 283.33

PROP: Pale yellow prisms from 2-propanol. Mp: 195-196°.

SYNS: 6-AMINO-2-(FLUOROMETHYL)-3-(2-METHYLPHENYL)-4(3H)-QUINAZOLINONE (9CI) ◇ 6-AMINO-2-FLUOROMETHYL-3-(o-TOLYL)-4(3H)-QUINAZOLINONE ◇ AROFT ◇ AROFUTO

TOXICITY DATA with REFERENCE
orl-rat LD50:249 mg/kg IYKEDH 14,297,83
ipr-rat LD50:385 mg/kg KSRNAM 17,991,83
scu-rat LD50:823 mg/kg KSRNAM 17,991,83
orl-mus LD50:397 mg/kg IYKEDH 14,297,83
ipr-mus LD50:272 mg/kg IYKEDH 14,297,83
scu-mus LD50:591 mg/kg IYKEDH 14,297,83

SAFETY PROFILE: Poison by ingestion and intraperitoneal routes. Moderately toxic by other routes. When heated to decomposition it emits toxic fumes of F^- and NO_x.

AEW750 CAS:47897-65-4 ***HR: 3***
AFRIDOL BLUE
mf: $C_{32}H_{18}Cl_2N_6O_{14}S_4 \cdot 4Na$ mw: 1001.66

TOXICITY DATA with REFERENCE
scu-rat TDLo:50 mg/kg (8D preg):REP BIJOAK 91,14P,64
scu-rat TDLo:50 mg/kg (8D preg):TER BIJOAK 91,14P,64

SAFETY PROFILE: An experimental teratogen. Other experimental reproductive effects. When heated to decomposition it emits very toxic fumes of SO_x, NO_x, Na_2O, and Cl^-.

AEX000 CAS:2315-02-8 ***HR: 3***
AFRIN HYDROCHLORIDE
mf: $C_{16}H_{24}N_2O \cdot ClH$ mw: 296.88

SYNS: AFRAZINE ◇ AFRIN ◇ 2-(4-tert-BUTYL-2,6-DIMETHYL-3-HYDROXYBENZYL)-2-IMIDAZOLINIUM CHLORIDE ◇ 6-tert-BUTYL-3-(2-IMIDAZOLIN-2-YLMETHYL)-2,4-DIMETHYLPHENOLHYDROCHLORIDE ◇ 2,6-DIMETHYL-4-TERTIARYBUTYL-3-HYDROXYPHENYL)METHYLIMIDAZOLINEHYDROCHLORIDE ◇ DURATION ◇ H 990 ◇ ILIADIN ◇ NAFRINE ◇ OXYMETAZOLINE CHLORIDE ◇ OXYMETAZOLINE HYDROCHLORIDE ◇ SCH 9384

TOXICITY DATA with REFERENCE
ipr-rat TDLo:1800 μg/kg (1-3D preg):REP IJMRAQ 67,478,78
orl-rat LD50:680 μg/kg ARZNAD 30,1760,80
scu-rat LD50:1630 μg/kg ARZNAD 11,1016,61
ivn-rat LD50:1070 μg/kg ARZNAD 11,1016,61
orl-mus LD50:4700 μg/kg OYYAA2 1,74,67
ipr-mus LD50:48 mg/kg FRPPAO 21,204,66
scu-mus LD50:34 mg/kg ARZNAD 11,1016,61
ivn-mus LD50:2700 μg/kg OYYAA2 1,74,67

SAFETY PROFILE: Poison by ingestion, subcutaneous, intraperitoneal, and intravenous routes. Experimental reproductive effects. When heated to decomposition it emits very toxic fumes of NO_x and HCl. An adrenergic agent.

AEX250 CAS:9002-18-0 ***HR: 1***
AGAR

PROP: Extracted from the red algae *Rhodopyceae*. Unground: in thin, translucent, membranous pieces; ground: pale buff powder. Sol in boiling water; insol in cold water and organic solvents.

SYNS: AGAR-AGAR ◇ AGAR AGAR FLAKE ◇ AGAR-AGAR GUM ◇ BENGAL GELATIN ◇ BENGAL ISINGLASS ◇ CEYLON ISINGLASS ◇ CHINESE ISINGLASS ◇ DIGENEA SIMPLEX MUCILAGE ◇ GELOSE ◇ JAPAN AGAR ◇ JAPAN ISINGLASS ◇ LAYOR CARANG ◇ NCI-C50475

TOXICITY DATA with REFERENCE
orl-rat LD50:11 g/kg FDRLI* 124,-,76
orl-mus LD50:16 g/kg FDRLI* 124,-,76
orl-rbt LD50:5800 mg/kg FDRLI* 124,-,76
orl-ham LD50:6100 mg/kg FDRLI* 124,-,76

CONSENSUS REPORTS: NTP Carcinogenesis Bioassay (feed); No Evidence: mouse, rat NTPTR* NTP-TR-230,82. Reported in EPA TSCA Inventory.

SAFETY PROFILE: Mildly toxic by ingestion. When heated to decomposition it emits acrid smoke and fumes.

AEX750 CAS:39277-47-9 ***HR: 3***
AGENT ORANGE
mf: $C_{12}H_{14}Cl_2O_3 \cdot C_{12}H_{13}Cl_3O_3$ mw: 588.76

SYNS: 2,4-d,n-BUTYL ESTER mixed with 2,4,5-T,n-BUTYL ESTER (1:1)

◇ 2,4,5-T,n-BUTYL ESTER mixed with 2,4-d,n-BUTYL ESTER ◇ 2,4-DICHLOROPHENOXYACETIC ACID BUTYL ESTER and 2,4,5-TRICHLOROPHENOXYACETIC ACID (45.5%:48.2%)

TOXICITY DATA with REFERENCE
orl-mus TDLo:1180 mg/kg (12-15D preg):TER AECTCV 6,33,77

SAFETY PROFILE: Contains toxic impurities. An experimental teratogen. See also ESTERS. When heated to decomposition it emits toxic fumes of Cl^-.

AEX850 CAS:644-06-4 ***HR: 2***
AGERATOCHROMENE
mf: $C_{13}H_{16}O_3$ mw: 220.29

SYNS: 2H-1-BENZOPYRAN, 6,7-DIMETHOXY-2,2-DIMETHYL- ◇ 6,7-DIMETHOXY-2,2-DIMETHYL-2H-BENZO(b)PYRAN ◇ PRECOCENE 2 ◇ PRECOCENE II

TOXICITY DATA with REFERENCE
dnd-rat:lvr 25 μmol/L CALEDQ 26,311,85
dns-rat:lvr 1 μmol/L CALEDQ 26,311,85
ipr-mus TDLo:27536 μg/kg:CAR CNREA8 47,2275,87

CONSENSUS REPORTS: Reported in EPA TSCA Inventory.

SAFETY PROFILE: Questionable carcinogen with experimental carcinogenic data. Mutation data reported. When heated to decomposition it emits acrid smoke and irritating fumes.

AEY000 CAS:103-16-2 ***HR: 3***
AGERITE
mf: $C_{13}H_{12}O_2$ mw: 200.25

SYNS: AGERITE ALBA ◇ ALBA-DOME ◇ BENOQUIN ◇ BENZOQUIN ◇ BENZYL HYDROQUINONE ◇ p-BENZYLOXYPHENOL ◇ DEPIGMAN ◇ HYDROQUINONE BENZYL ETHER ◇ HYDROQUINONE MONOBENZYL ETHER ◇ p-HYDROXYPHENYL BENZYL ETHER ◇ MONOBENZONE ◇ MONOBENZYL ETHER HYDROQUINONE ◇ MONOBENZYL HYDROQUINONE ◇ 4-(PHENYLMETHOXY)PHENOL ◇ PIGMEX

TOXICITY DATA with REFERENCE
skn-gpg 5%/48H MLD JSCCA5 28,357,77
orl-mus TDLo:163 g/kg/78W-I:ETA NTIS** PB223-159
scu-mus TDLo:1000 mg/kg:NEO NTIS** PB223-159
ipr-rat LD50:4500 mg/kg MEIEDD 11,983,89

CONSENSUS REPORTS: Reported in EPA TSCA Inventory.

SAFETY PROFILE: Mild acute toxicity by intraperitoneal route. A skin irritant. Questionable carcinogen with experimental neoplastigenic and tumorigenic data. See also ETHERS. When heated to decomposition it emits acrid smoke and irritating fumes.

AEY125 ***HR: 3***
AGKISTRODON CONTORTRIX VENOM

SYN: VENOM, SNAKE, AGKISTRODON CONTORTRIX

TOXICITY DATA with REFERENCE
ipr-mus LD50:10500 μg/kg 14FHAR -,409,63
ivn-mus LD50:10920 μg/kg 14FHAR -,409,63
ipr-mam LD50:10500 μg/kg CLPTAT 8,849,67
ivn-mam LD50:10920 μg/kg CLPTAT 8,849,67

SAFETY PROFILE: Poison by intravenous and intraperitoneal routes.

AEY130 ***HR: 3***
AGKISTRODON PISCIVORUS VENOM

SYN: VENOM, SNAKE, AGKISTRODON PISCIVORUS

TOXICITY DATA with REFERENCE
ipr-mus LD50:5110 μg/kg 14FHAR -,409,63
scu-mus LD50:15 mg/kg JOIMA3 67,299,51
ivn-mus LDLo:2250 μg/kg TOXIA6 3,187,66
ivn-dog LDLo:750 μg/kg 19DDA6 1,269,67
ivn-rbt LD50:5 mg/kg PSEBAA 116,696,64
ipr-mam LD50:5110 μg/kg CLPTAT 8,849,67
ivn-mam LD50:4 mg/kg CLPTAT 8,849,67

SAFETY PROFILE: Poison by subcutaneous, intravenous, and intraperitoneal routes.

AEY135 ***HR: 3***
AGKISTRODON RHODOSTOMA VENOM

SYN: VENOM, SNAKE, AGKISTRODON RHODOSTOMA

TOXICITY DATA with REFERENCE
ivn-rat LDLo:300 μg/kg JCINAO 45,1202,66
ipr-mus LD50:4977 μg/kg TOXIA6 9,131,71
scu-mus LD50:16100 μg/kg 19DDA6 1,323,67
ivn-mus LD50:2820 μg/kg TOXIA6 7,239,69
ivn-dog LD50:900 μg/kg 19DDA6 1,323,67
ims-dog LD50:1900 μg/kg 19DDA6 1,323,67
ivn-rbt LDLo:13 μg/kg JCINAO 45,1202,66
ivn-mam LD50:6200 μg/kg CLPTAT 8,849,67

SAFETY PROFILE: Poison by subcutaneous, intramuscular, intravenous, and intraperitoneal routes.

AEY375 CAS:548-42-5 ***HR: 3***
AGROCLAVINE
mf: $C_{16}H_{18}N_2$ mw: 238.36

PROP: Rods from ether; decomp 198-203°. Needles from acetone; decomp @ 205-206°. Freely sol in alc, chloroform, pyridine; sol in benzene, ether; very sltly sol in water.

SYN: 8,9-DIDEHYDRO-6,3-DIMETHYLERGOLINE

TOXICITY DATA with REFERENCE
orl-rat TDLo:50 mg/kg (1-5D preg):REP BJPCAL 33,215P,68
ipr-mus LD50:25 mg/kg ARZNAD 35,1760,85
ivn-mus LD50:25500 μg/kg NYKZAU 58,386,62

SAFETY PROFILE: Poison by intraperitoneal route. Experimental reproductive effects. When heated to decomposition it emits toxic fumes of NO_x.

AEY400 CAS:13118-10-0 ***HR: 3***
AHR 376
mf: $C_{18}H_{25N}O_3 \cdot ClH$ mw: 339.90

SYNS: α-CYCLOPENTYLMANDELIC ACID-1-METHYL-3-PYRROLIDINYL ESTER HYDROCHLORIDE ◇ 1-METHYL-3-PYRROLIDYL-α-PHENYLCYCLOPENTANEGLYCOLATE HYDROCHLORIDE

TOXICITY DATA with REFERENCE
orl-hmn TDLo:14 μg/kg/D CPAJAK 11,5141,66
orl-mus LD50:500 mg/kg JMPCAS 2,523,60
ipr-mus LD50:250 mg/kg JMPCAS 2,523,60

SAFETY PROFILE: Poison by intraperitoneal route. Moderately toxic by ingestion. Human systemic effects by ingestion: hallucinations, distorted perceptions, and toxic psychosis. When heated to decomposition it emits toxic fumes of NO_x and HCl.

AFG000 ***HR: 3***
AIPYSURUS LAEVIS VENOM (AUSTRALIA)

SYN: VENOM, SEA SNAKE, AIPYSURUS LAEVIS (AUSTRALIA)

TOXICITY DATA with REFERENCE
ivn-mus LD50:250 μg/kg 85EGD4 5,357,78
ims-mus LD50:130 μg/kg 85EGD4 5,357,78

SAFETY PROFILE: Poison by intravenous and intramuscular routes.

AFG250 ***HR: 2***
AIR, compressed
DOT: UN 1003

PROP: Bluish, mobile liquid. $O_2 + N_2$. Bp: $-189°$ (liq); flash p: none; autoign temp: none.

DOT Classification: Nonflammable Gas; Label: Nonflammable Gas.

SAFETY PROFILE: Liquid air can cause tissue damage due to low temperature. Personnel exposed to compressed air may develop caisson disease (the bends, the chokes) if decompression is too rapid. Moderate explosion hazard when containers under pressure are shocked or exposed to heat, flame, or flammable materials, i.e., ethyl ether, hydrocarbons, or charcoal, which have been in contact with liquid air may explode very easily. Ordinary oxidation is greatly accelerated in compressed air. Moderately dangerous disaster hazard; can react vigorously with reducing materials.

AFG500 CAS:569-64-2 ***HR: 3***
AIZEN MALACHITE GREEN
mf: $C_{23}H_{25}N_2 \cdot Cl$ mw: 364.95

SYNS: ACRYL BRILLIANT GREEN B ◇ ADC MALACHITE GREEN CRYSTALS ◇ ANILINE GREEN ◇ BASIC GREEN 4 ◇ BENZALDEHYDE GREEN ◇ BRONZE GREEN TONER A-8002 ◇ BURMA GREEN B ◇ CHINA GREEN (BIOLOGICAL STAIN) ◇ C.I. 42000 ◇ C.I. BASIC GREEN 4 ◇ DIABASIC MALACHITE GREEN ◇ DIAMOND GREEN B ◇ FAST GREEN ◇ HIDACO MALACHITAE GREEN BASE ◇ LIGHT GREEN N ◇ NEW VICTORIA GREEN EXTRA I ◇ SOLID GREEN CRYSTALS O ◇ TETROPHENE GREEN M ◇ TETRAMETHYL DIAPARA-AMIDO-TRIPHENYL CARBINOL ◇ VICTORIA GREEN

TOXICITY DATA with REFERENCE
dnd-mam:lym 10 pph BIPMAA 11,2537,72
orl-mus LD50:80 mg/kg ARZNAD 1,5,51
ipr-mus LD50:4200 μg/kg ARZNAD 1,5,51

CONSENSUS REPORTS: Reported in EPA TSCA Inventory. Community Right-To-Know List.

SAFETY PROFILE: Poison by ingestion and intraperitoneal routes. Mutation data reported. When heated to decomposition it emits very toxic fumes of NO_x and Cl^-.

AFG625 ***HR: 1***
AJAX, LEMON (scouring powder)

SYN: LEMON AJAX

TOXICITY DATA with REFERENCE
skn-rbt 500 mg MLD FCTOD7 20,563,82
eye-rbt 100 mg MOD FCTOD7 20,573,82
eye-rbt 100 mg/45 rns MLD FCTOD7 20,573,82

SAFETY PROFILE: A skin and eye irritant.

AFG750 CAS:483-04-5 ***HR: 3***
AJMALICINE
mf: $C_{21}H_{24}N_2O_3$ mw: 352.47

SYNS: ALKALOID C ◇ ALKALOID II ◇ 16,17-DIDEHYDRO-19-METHYLOXAYOHIMBAN-16-CARBOXYLIC ACID METHYL ESTER ◇ HYDROSARPAN ◇ LAMURAN ◇ PY-TETRAHYDROSERPENTINE ◇ RANITOL ◇ RAUBASINE ◇ RAUMALINA ◇ SARPAN ◇ SUBSTANCE II ◇ TENSYL ◇ TETRAHYDROSERPENTINE ◇ VINCAIN ◇ VINCEINE ◇ Δ-YOHIMBINE

TOXICITY DATA with REFERENCE
orl-rat LDLo:750 mg/kg AEPPAE 233,72,58
ivn-gpg LDLo:20 mg/kg ARZNAD 23,600,73
orl-chd TDLo:12500 μg/kg:CNS,PUL CHETBF 76,97,79
ipr-rat LD50:200 mg/kg 27ZQAG -,117,72
ivn-rat LD50:24 mg/kg 27ZQAG -,117,72
orl-mus LD50:400 mg/kg 27ZQAG -,117,72
ipr-mus LD50:165 mg/kg 27ZQAG -,117-72

ivn-mus LD50:20 mg/kg AEPPAE 233,72,58
orl-rbt LD50:500 mg/kg 27ZQAG -,117,72
ivn-rbt LD50:20 mg/kg 27ZQAG -,117,72

SAFETY PROFILE: Poison by ingestion, intraperitoneal, and intravenous routes. Human systemic effects by ingestion: general anesthesia, convulsions and lung effects. When heated to decomposition it emits highly toxic NO_x. An antihypertensive agent and tranquilizer.

AFH000 CAS:4373-34-6 ***HR: 3***
AJMALICINE HYDROCHLORIDE
mf: $C_{21}H_{24}N_2O_3$•ClH mw: 388.93

SYNS: AJMALICINE MONOHYDROCHLORIDE ◇ RAUBASNE HYDROCHLORIDE ◇ Γ-YOHIMBINE HYDROCHLORIDE

TOXICITY DATA with REFERENCE
orl-mus LDLo:50 mg/kg LOBU** -,3,32
ivn-mus LD50:56 mg/kg CSLNX* NX#00444

SAFETY PROFILE: Poison by ingestion and intravenous routes. When heated to decomposition it emits very toxic fumes of NO_x and HCl. See also AJMALICINE.

AFH250 CAS:4360-12-7 ***HR: 3***
AJMALINE
mf: $C_{20}H_{26}N_2O_2$ mw: 326.48

SYNS: CARDIORYTHMINE ◇ GILURYTMAL ◇ IGNAZIN ◇ MERABITOL ◇ RAUGALLINE ◇ RAUWOLFIN ◇ RAUWOLFINE ◇ RHYTMATON ◇ RITMOS ◇ SIDDIQUI ◇ TACHMALIN ◇ TAJMALIN ◇ TAKYCOR

TOXICITY DATA with REFERENCE
scu-rat LD50:216 mg/kg FRPSAX 19,865,64
ivn-rat LD50:26 mg/kg PHARAT 31,36,76
orl-mus LD50:255 mg/kg FRPSAX 19,865,64
scu-mus LD50:180 mg/kg FRPSAX 19,865,64
ivn-gpg LDLo:28 mg/kg FRPSAX 19,865,64
orl-qal LDLo:316 mg/kg EESADV 6,149,82
orl-bwd LD50:178 mg/kg AECTCV 12,355,83
ipr-mus LD50:75 mg/kg NTIS** AD691-490
ivn-mus LD50:21 mg/kg JETOAS 8(3),188,75

SAFETY PROFILE: Poison by ingestion, subcutaneous, intraperitoneal, and intravenous routes. When heated to decomposition it emits toxic fumes of NO_x. An antihypertensive agent and tranquilizer.

AFH275 CAS:2552-89-8 ***HR: 3***
AJMALINE BIS(CHLOROACETATE) (ester) HYDROCHLORIDE
mf: $C_{24}H_{28}Cl_2N_2O_4$•ClH mw: 515.90

SYNS: DCAA ◇ DIMONOCLOROACETILAJMALINA CLORIDRATO (ITALIAN)

TOXICITY DATA with REFERENCE
scu-rat LD50:389 mg/kg FRPSAX 19,865,64
orl-mus LD50:570 mg/kg FRPSAX 19,865,64
scu-mus LD50:355 mg/kg FRPSAX 19,865,64
ivn-mus LD50:111 mg/kg FRPSAX 19,865,64
ivn-gpg LDLo:149 mg/kg FRPSAX 19,865,64

SAFETY PROFILE: Poison by subcutaneous and intravenous routes. Moderately toxic by ingestion. When heated to decomposition it emits toxic fumes of NO_x and HCl. See also ESTERS, other Ajmalines.

AFH280 CAS:4410-48-4 ***HR: 3***
AJMALINE HYDROCHLORIDE
mf: $C_{20}H_{26}N_2C_2$•7ClH mw: 581.70

SYN: CHLORHYDRATE de RAUGALLINE (FRENCH)

TOXICITY DATA with REFERENCE
orl-mus LD50:440 mg/kg AIPTAK 127,163,60
ipr-mus LD50:105 mg/kg AIPTAK 127,163,60
ivn-mus LD50:26 mg/kg AIPTAK 127,163,60
ivn-cat LDLo:2 mg/kg AIPTAK 216,63,75

SAFETY PROFILE: Poison by intravenous and intraperitoneal routes. Moderately toxic by ingestion. When heated to decomposition it emits toxic fumes of NO_x and HCl. See also other Ajmalines.

AFH500 ***HR: 3***
AK PS
mf: $C_7H_6N_2O_2S_2$ mw: 214.27

PROP: Produced by *Streptomyces flavochromogenes Iwayaensis*.

SYN: ANTIBIOTIC AK PS

TOXICITY DATA with REFERENCE
ipr-mus LD50:5 mg/kg JANTAJ 33,80-35,80
ivn-mus LD50:5 mg/kg JANTAJ 33,80-35,80

SAFETY PROFILE: Poison by intraperitoneal and intravenous routes. When heated to decomposition it emits very toxic fumes of NO_x and SO_x.

AFH550 CAS:25331-92-4 ***HR: 3***
AL-1612
mf: $C_{19}H_{28}N_2O_3$ mw: 332.49

SYNS: 5-(1,4-DIOXA-8-AZASPIRO(4.5)DEC-8-YLMETHYL)-3-ETHYL-6,7-DIHYDRO-2-METHYL-INDOL-4(5H)-ONE◇ 3-ETHYL-5-(4,4-ETHYLENEDIOXYPIPERIDINO-1-METHYL)-6,7-DIHYDRO-2-METHYLINDOL-4(5H)-ONE

TOXICITY DATA with REFERENCE
orl-rat LD50:240 mg/kg ARZNAD 23,1314,73
orl-mus LD50:200 mg/kg ARZNAD 23,1314,73
orl-dog LD50:50 mg/kg ARZNAD 23,1314,73
orl-rbt LD50:100 mg/kg ARZNAD 23,1314,73

SAFETY PROFILE: Poison by ingestion. When heated to decomposition it emits toxic fumes of NO_x.

AFH750 CAS:5854-93-3 ***HR: 3***
l-ALANOSINE
mf: $C_3H_7N_3O_4$ mw: 149.13

SYNS: ALANOSINE ◇ l-2-AMINO-3-(HYDROXYNITROSAMINO)PROPIONIC ACID ◇ l-2-AMINO-3-((N-NITROSO)HYDROXYLAMINO)PROPIONIC ACID ◇ 3-(HYDROXYNITROSOAMINO)-l-ALANINE (9CI)

TOXICITY DATA with REFERENCE
ipr-mus LD50:600 mg/kg USXXAM #3676490
scu-mus LD50:845 mg/kg NCISP* JAN86
ivn-mus LD50:300 mg/kg USXXAM #3676490

SAFETY PROFILE: Poison by intravenous route. Moderately toxic by intraperitoneal and subcutaneous routes. When heated to decomposition it emits toxic fumes of NO_x. An experimental insect reproduction inhibitor.

AFI500 CAS:38819-28-2 ***HR: 3***
4-N-d-ALANYL-2,4-DIAMINO-2,4-DIDEOXY-l-ARABINOSE
mf: $C_8H_{17}N_3O_4$ mw: 219.28

SYN: PRUMYCIN

TOXICITY DATA with REFERENCE
dni-hmn:hla 10 mg/L JANTAJ 33,226,80
ipr-rat LD50:70 mg/kg JANTAJ 32,347,79
orl-mus LDLo:750 mg/kg 85ERAY 2,1167,78
ipr-mus LD50:155 mg/kg JANTAJ 33,226,80
ivn-mus LDLo:160 mg/kg 85ERAY 2,1167,78

SAFETY PROFILE: Poison by intraperitoneal and intravenous routes. Moderately toxic by ingestion. Human mutation data reported. When heated to decomposition it emits toxic fumes of NO_x.

AFI625 CAS:1397-84-8 ***HR: 3***
ALAZOPEPTIN
mf: $C_{15}H_{20}N_6O_5$ mw: 364.41

SYNS: AA223 LEDERLE ◇ l-ALLYL-(6-DIAZO-5-OXO)-l-NORLEUCYL-(6-DIAZO-5-OXO)-l-NORLEUCINE◇ AMBOMYCIN ◇ LEDERLE AA223

TOXICITY DATA with REFERENCE
mmo-omi 1 mg/plate JGAMA9 11,129,65
ipr-rat TDLo:600 μg/kg (7-8D preg):REP PSEBAA 97,888,58
ipr-rat LD50:150 mg/kg PSEBAA 97,888,58

SAFETY PROFILE: Poison by intraperitoneal route. Experimental reproductive effects. Mutation data reported. When heated to decomposition it emits toxic fumes of NO_x.

AFI750 CAS:39301-00-3 ***HR: 3***
ALBITOCIN

TOXICITY DATA with REFERENCE
iut-gpg TDLo:250 μg/kg (1D pre):REP JPPMAB 16,369,64
ipr-rat LD50:800 μg/kg JPPMAB 19,760,67
ipr-mus LD50:5900 μg/kg JPPMAB 19,792,67
ivn-mus LD50:6 mg/kg JPPMAB 19,792,67
ivn-mky LD50:2500 μg/kg JPPMAB 19,792,67
ivn-rbt LD50:1800 μg/kg JPPMAB 19,792,67

SAFETY PROFILE: Poison by intravenous and intraperitoneal routes. Experimental reproductive effects. When heated to decomposition it emits acrid smoke and irritating fumes.

AFI850 CAS:70536-17-3 ***HR: 3***
ALBUMIN MACRO AGGREGATES

SYNS: ALBUMIN ◇ MAA

TOXICITY DATA with REFERENCE
ivn-rat LD50:17 mg/kg IJNMCI 5,51,78
ivn-mus LD50:18 mg/kg IJNMCI 5,51,78
inv-gpg LD50:19 mg/kg IJNMCI 5,51,78

SAFETY PROFILE: Poison by intravenous route. When heated to decomposition it emits acrid smoke and irritating fumes.

AFI900 CAS:12040-44-7 ***HR: D***
ALCIAN BLUE

TOXICITY DATA with REFERENCE
ipr-mus TDLo:250 mg/kg (female 8D post):REP TCMUD8 4,403,84

SAFETY PROFILE: Experimental reproductive effects. When heated to decomposition it emits acrid smoke and irritating fumes.

AFI950 CAS:70319-10-7 ***HR: D***
ALCLOFENAC EPOXIDE
mf: $C_{11}H_{11}ClO_4$ mw: 242.67

SYN: (3-CHLORO-4-(OXIRANYLMETHOXY)PHENYL)ACETICACID

TOXICITY DATA with REFERENCE
mmo-sat 100 μg/plate JJATDK 3,230,83
otr-ham:emb 125 mg/L CBINA8 34,95,81

SAFETY PROFILE: Mutation data reported. When heated to decomposition it emits toxic fumes of Cl^-.

AFI980 CAS:66734-13-2 ***HR: 2***
ALCLOMETASONE DIPROPIONATE
mf: $C_{28}H_{37}ClO_7$ mw: 521.10

SYNS: PREGNA-1,4-DIENE-3,20-DIONE,7-CHLORO-11-HYDROXY-16-METHYL-17,21-BIS(1-OXOPROPOXY)-,(7-α-11-β,16-α)- ◇ SCH 22219

TOXICITY DATA with REFERENCE
scu-rat TDLo:11 mg/kg (female 7-17D post):REP OYYAA2 33,301,87
scu-rat TDLo:1100 μg/kg (female 7-17D post):TER OYYAA2 33,301,87
scu-rat LD50:3593 mg/kg KSRNAM 21,1253,87
scu-mus LD50:2506 mg/kg KSRNAM 21,1253,87

SAFETY PROFILE: Moderately toxic by subcutaneous route. An experimental teratogen. Other experimental reproductive effects. When heated to decomposition it emits toxic fumes of HCl.

AFJ000 ***HR: 3***
ALCOHOL, DENATURED

PROP: Liquid. Composed of alcohol and denaturants.

SYN: DENATURED SPIRITS

SAFETY PROFILE: Potentially poisonous by ingestion. Toxicity depends upon alcohols in question, generally ethanol with methanol as a denaturant. Dangerous fire hazard; can react vigorously with oxidizing materials. Moderate explosion hazard. See ETHANOL, METHANOL, and PROPANOL.

AFJ250 ***HR: 3***
ALCOHOLS, N.O.S.

CONSENSUS REPORTS: A generic term applied to a series of compounds, the simplest of which has the general formula $C_nH_{2n+1}OH$. (See also specific compound.)

SAFETY PROFILE: No general statement can be made due to wide variations in toxic effects. Dangerous fire hazard when exposed to heat or flame. Can react violently in contact with (H_2O + H_2SO_4), HOCl, Cl_2, isocyanates, $LiAl_4$, N_2O_4, $HClO_4$, H_2SO_5 (Caro's acid), $Ba(ClO_4)_2$, $(CH_2)_2O$, acetaldehyde, diethyl aluminum bromide, hexamethylene diisocyanate, triisobutyl aluminum.

AFJ375 CAS:61711-25-9 ***HR: D***
ALCOHOL SULPHATE

SYNS: ALCOHOL SULFATE ◇ AS ◇ AS (surfactant)

TOXICITY DATA with REFERENCE
orl-mus TDLo:2 mg/kg (female 6-15D post):TER TXCYAC 3,91,75
skn-mus TDLo:1600 mg/kg (female 1-4D post):REP LIFSAK 26,49,80

SAFETY PROFILE: An experimental teratogen. Experimental reproductive effects. When heated to decomposition it emits toxic fumes of SO_x.

AFJ400 CAS:357-56-2 ***HR: 3***
ALCOID
mf: $C_{26}H_{32}N_2O_2$ mw: 392.59

PROP: Crystals. Mp: 180-184°. Practically insol in water. Sol in 0.1N HCl, ethanol, methanol, acetone, ethyl acetate, benzene, chloroform, and ether.

SYNS: DAURAN ◇ DEXTROMORAMIDE ◇ DIMORLIN ◇ (+)-2,2-DIPHENYL-3-METHYL-4-MORPHOLINOBUTYRYLPYRROLIDINE ◇ JETRIUM ◇ JETRIUM R ◇ LINFADOL ◇ MCP 875 ◇ MORAMIDE ◇ NARCOLO ◇ PALFADONNA ◇ PALFIUM ◇ PYRROLAMIDOL ◇ PYRROLAMIDOLUM ◇ R 875 ◇ SKF 5137 ◇ TROXILAN ◇ YETRIUM

TOXICITY DATA with REFERENCE
scu-wmn TDLo:150 μg/kg:CNS,PUL BMJOAE 1,211,59
ims-hmn TDLo:57 μg/kg:CNS,PUL NYSJAM 61,83,61
ims-wmn TDLo:200 μg/kg:PUL,GIT PRACAK 197,348,66
ivn-rat LD50:13 mg/kg JPPMAB 9,730,57
scu-mus LD50:140 mg/kg BJPCAL 17,433,61
ivn-mus LD50:21 mg/kg JPPMAB 9,730,57

SAFETY PROFILE: Poison by subcutaneous and intravenous routes. Human systemic effects by subcutaneous and intramuscular routes: coma, cyanosis, respiratory depression, and nausea or vomiting. Caution: May be habit forming. This is a controlled substance (opiate) listed in the U.S. Code of Federal Regulations, Title 21 Part 1308.11 (1985). When heated to decomposition it emits toxic fumes of NO_x.

AFJ500 CAS:52-01-7 ***HR: 3***
ALDACTAZIDE
mf: $C_{24}H_{32}O_4S$ mw: 416.62

SYNS: 7-α-ACETYLTHIO-3-OXO-17-α-PREGN-4-ENE-21,17-β-CARBOLACTONE ◇ 7-α-ACETYLTHIO-3-OXO-17-β-PREGN-4-ENE-21,17-β-CARBOLACTONE ◇ ALDACTIDE ◇ ALDACTONE ◇ ALDACTONE A ◇ 3-(3-KETO-7-α-ACETYLTHIO-17-β-HYDROXY-4-ANDROSTEN-17-α-YL)PROPIONIC ACID LACTONE ◇ OSIREN ◇ OSYROL ◇ 3'-(3-OXO-7-α-ACETYLTHIO-17-β-HYDROXYANDROST-4-EN-17-β-YL)PROPIONIC ACID LACTONE ◇ 17-α-PREGN-4-ENE-21-CARBOXYLIC ACID, 1-HYDROXY-7-α-MERCAPTO-3-OXO-α-LACTONE ◇ SC 9420 ◇ SC 15983 ◇ SPIRESIS ◇ SPIRIDON ◇ SPIROCTANIE ◇ SPIRO(17H-CYCLOPENTA(a)PHENAUTHRENE-17,2'-(3'H)-FURAN) ◇ SPIRO(17H-CYCLOPENTA(a)PHENANTHRENE-17,2'(5'H)FURAN), PREGN-4-ENE-21-CARBOXYLIC ACID DERIV. ◇ SPIROLACTONE ◇ SPIROLAKTON ◇ SPIROLANG ◇ SPIRONE ◇ SPIRONOLACTONE ◇ SPIRONOLACTONE A ◇ URACTONE ◇ VEROSPIRON ◇ VEROSPIRONE

TOXICITY DATA with REFERENCE
unr-wmn TDLo:280 mg/kg (female 35D pre):REP AIMEAS 69,685,68
orl-rat TDLo:360 mg/kg (female 13-21D post):TER ACENA7 95,540,80
orl-wmn TDLo:70 mg/kg/5W-I:BLD BMJOAE 289,731,84
orl-wmn TDLo:122 mg/kg/61D-I NZMJAX 83,147,76
orl-hmn TDLo:5600 mg/kg:SYS JLCMAK 80,224,72
orl-man TDLo:5714 μg/kg/4D-I:PUL,BLD DICPBB 21,974,87
ipr-rat LD50:277 mg/kg DRUGAY 6,381,82

ipr-mus LD50:260 mg/kg DRUGAY 6,381,82
ipr-rbt LD50:866 mg/kg JEPTDQ 1(5),641,78

CONSENSUS REPORTS: IARC Cancer Review: Group 3 IMEMDT 7,344,87; Animal Limited Evidence IMEMDT 24,259,80; Human Inadequate Evidence IMEMDT 24,259,80. Reported in EPA TSCA Inventory.

SAFETY PROFILE: Poison by intraperitoneal route. Human reproductive effects by ingestion and possibly other routes: men, impotence and breast development; women, menstrual cycle changes or disorders, changes in the breasts and lactation. An experimental teratogen. Other experimental reproductive effects. Other human systemic effects by ingestion: agranulocytosis, kidney tubule damage, increased urine volume, and changes in blood sodium and calcium levels. Questionable carcinogen. When heated to decomposition it emits toxic fumes of SO_x. Used to treat hypertension, edema of congestive heart failure, cirrhosis and kidney failure.

AFJ625 CAS:5534-09-8 ***HR: 2***
ALDECIN
mf: $C_{28}H_{37}ClO_7$ mw: 521.10

SYNS: BECLACIN ◇ BECLOFORTE ◇ BECLOMETASONE DIPROPIONATE ◇ BECLOMETASONE-17,21-DIPROPIONATE ◇ BECLOMETHASONE DIPROPIONATE ◇ BECLOVAL ◇ BECLOVENT ◇ BECOTIDE ◇ BENCONASE ◇ BP2 ◇ CLENIL-A ◇ DIPROPIONATE BECLOMETHASONE ◇ ENTYDERMA ◇ INALONE O ◇ INALONE R ◇ KORBUTONE ◇ PROPADERM ◇ RINO-CLENIL ◇ SANASTHYMYL ◇ SCH 18020W ◇ VANCENASE ◇ VANCERIL ◇ VIAROX

TOXICITY DATA with REFERENCE
ihl-rat TCLo:168 μg/kg/3M (female 8-17D post):REP OYYAA2 13,185,77
ihl-mus TCLo:406 μg/kg/3M (female 7-15D post):TER OYYAA2 13,195,77
orl-rat LD50:3750 mg/kg OYYAA2 19,323,80
ipr-rat LD50:1500 mg/kg OYYAA2 19,323,80
scu-rat LD50:3 g/kg OYYAA2 19,323,80

SAFETY PROFILE: Moderately toxic by ingestion and other routes. An experimental teratogen. Other experimental reproductive effects. When heated to decomposition it emits toxic fumes of Cl^-.

AFJ700 CAS:7779-41-1 ***HR: 1***
ALDEHYDE C-10 DIMETHYLACETAL
mf: $C_{12}H_{26}O_2$ mw: 202.38

SYNS: DECANAL, DIMETHYLACETAL ◇ DECYLALDEHYDE DMA ◇ 1,1-DIMETHOXYDECANE ◇ 10,10-DIMETHOXYDECANE

TOXICITY DATA with REFERENCE
skn-rbt 500 mg/24H MOD FCTXAV 17,759,79

CONSENSUS REPORTS: Reported in EPA TSCA Inventory.

SAFETY PROFILE: A skin irritant. When heated to decomposition it emits acrid smoke and irritating fumes.

AFJ800 ***HR: 2***
ALDEHYDES

PROP: A class of chemicals with the general formula R•CHO, and characterized by an unsaturated carbonyl group (C=O).

SAFETY PROFILE: Aldehydes are widely used in many industrial processes. The US production of acetaldehyde in 1982 was 281,000 tons. The world production of acrolein in 1975 was 59,000 tons. They occur in nature and are gaseous byproducts of incomplete combustion of wood and coal, in exhaust from gasoline and diesel engines, industrial waste gases and fumes, tobacco smoke, and wood fires. Formaldehyde and acetaldehyde are carcinogens. Many of the aldehydes are mutagens. They are reactive compounds participating in oxidation, reduction, addition, and polymerization reactions. All the aldehydes possess anesthetic properties, but this is obscured by their highly irritating action on the eyes and mucous membranes of the respiratory tract. The lower aldehydes, very soluble in water, act chiefly on the eyes and tissues of the upper respiratory tract. The higher aldehydes, less soluble in water, tend to penetrate more deeply into the respiratory system and may affect the lungs. Some higher aldehydes and also the aromatic aldehydes may exhibit much lower toxicity. See also specific compounds.

AFJ850 CAS:91315-15-0 ***HR: 2***
ALDIMORPH

TOXICITY DATA with REFERENCE
orl-rat TDLo:1400 mg/kg (female 4-18D post):TER WZERDH 32(1-2),19,83
orl-rat LD50:3500 mg/kg WZERDH 32(1-2),19,83

SAFETY PROFILE: Moderately toxic by ingestion. An experimental teratogen. When heated to decomposition it emits acrid smoke and irritating fumes.

AFJ875 CAS:52-39-1 ***HR: D***
ALDOCORTENE
mf: $C_{21}H_{28}O_5$ mw: 360.49

PROP: Hydrated crystals from dilute acetone. Mp: 108-112° (when anhydr mp: 164°).

SYNS: ALDOCORTEN ◇ ALDOCORTIN ◇ ALDOSTERONE (8CI) ◇ (+)-ALDOSTERONE ◇ d-ALDOSTERONE ◇ ELECTROCORTIN ◇ ELEKTROCORTIN ◇ 18-OXOCORTICOSTERONE ◇ REICHSTEIN X

TOXICITY DATA with REFERENCE
oms-nml:oth 70 nmol/L RCOCB8 1,363,70
dnd-rat-ivn 23 mg/kg PSEBAA 141,14,72

CONSENSUS REPORTS: Reported in EPA TSCA Inventory.

SAFETY PROFILE: Mutation data reported. When heated to decomposition it yields acrid smoke and fumes.

AFK000 CAS:1646-88-4 ***HR: 3***
ALDOXYCARB
mf: $C_7H_{14}N_2O_4S$ mw: 222.29

SYNS: ENT A13-29261 ◇ 2-METHYL-2-(METHYLSULFONYL)PROPANAL-o-((METHYLAMINO)CARBONYL)OXIME ◇ 2-METHYL-2-(METHYLSULFONYL)PROPIONALDEHYDE-o-(METHYLCARBAMOYL)OXIME ◇ STANDAK ◇ UC-21865

TOXICITY DATA with REFERENCE
orl-rat LD50:26800 μg/kg FMCHA2 -,D287,80
skn-rat LD50:1000 mg/kg SPEADM 78-1,61,78
skn-rbt LD50:1000 mg/kg SPEADM 78-1,61,78

SAFETY PROFILE: Poison by ingestion. Moderately toxic by skin contact. When heated to decomposition it emits very toxic fumes of NO_x and SO_x. An insecticide.

AFK250 CAS:309-00-2 ***HR: 3***
ALDRIN
DOT: UN 2761/NA 2762
mf: $C_{12}H_8Cl_6$ mw: 364.90

PROP: Crystals. Mp: 104-105°. Insol in water; sol in aromatics, esters, ketones, paraffins, and halogenated solvents.

SYNS: ALDREX ◇ ALDREX 30 ◇ ALDRIN, cast solid (DOT) ◇ ALDRINE (FRENCH) ◇ ALDRITE ◇ ALDROSOL ◇ ALTOX ◇ COMPOUND 118 ◇ DRINOX ◇ ENT 15,949 ◇ HEXACHLOROHEXAHYDRO-endo-exo-DIMETHANONAPHTHALENE ◇ 1,2,3,4,10,10-HEXACHLORO-1,4,4a,5,8,8a-HEXAHYDRO-1,4,5,8-DIMETHANONAPHTHALENE ◇ 1,2,3,4,10,10-HEXACHLORO-1,4,4a,5,8,8a-HEXAHYDRO-exo-1,4,-endo-5,8-DIMETHANONAPHTHALENE ◇ 1,2,3,4,10,10-HEXACHLORO-1,4,4a,5,8,8a-HEXAHYDRO-1,4-endo-exo-5,8-DIMETHANONAPHTHALENE ◇ HHDN ◇ NCI-C00044 ◇ OCTALENE ◇ RCRA WASTE NUMBER P004 ◇ SEEDRIN

TOXICITY DATA with REFERENCE
cyt-hmn:lym 1900 mg/L MUREAV 31,103,75
cyt-hmn:leu 19125 μg/L PHTHDT 6,147,79
orl-dog TDLo:73 mg/kg (female 44W pre):REP JAVMA4 123,28,53
orl-ham TDLo:50 mg/kg (female 7D post):TER TJADAB 9,11,74
orl-rat TDLo:200 mg/kg/2Y-C:NEO FCTXAV 2,551,64
orl-mus TDLo:270 mg/kg/80W-I:CAR NCITR* NCI-CG-TR-21,78
orl-mus TD:540 mg/kg/80W-I:CAR NCITR* NCI-CG-TR-21,78
orl-rat TD:188 mg/kg/2Y-C:ETA TXAPA9 11,88,67
orl-hmn TDLo:14 mg/kg:CNS 34ZIAG -,83,69
orl-chd LDLo:1250 μg/kg 34ZIAG -,83,69
orl-rat LD50:39 mg/kg SPEADM 74-1,-,74
ihl-rat LCLo:5800 μg/m³/4H 85GMAT -,73,82
skn-rat LD50:98 mg/kg TXAPA9 14,515,69
ipr-rat LD50:150 mg/kg TXAPA9 11,302,67
orl-mus LD50:44 mg/kg SPEADM 74-1,-,74
ipr-mus LDLo:50 mg/kg SOGEBZ 2,80,66
ivn-mus LD50:21 mg/kg 32ZDAL -,52,70
skn-rbt LDLo:15 mg/kg JEENAI 46,702,53

CONSENSUS REPORTS: IARC Cancer Review: Group 3 IMEMDT 7,88,87; Human Inadequate Evidence IMEMDT 5,25,74; Animal Inadequate Evidence IMEMDT 5,25,74; NCI Carcinogenesis Bioassay (feed); Clear Evidence: mouse NCITR* NCI-CG-TR-21,78; Inadequate Studies: rat NCITR* NCI-CG-TR-21,78. EPA Genetic Toxicology Program. EPA Extremely Hazardous Substances List. Community Right-To-Know List.

OSHA PEL: TWA 0.25 mg/m³ (skin)
ACGIH TLV: TWA 0.25 mg/m³/
DFG MAK: 0.25 mg/m³
NIOSH REL: (Aldrin) Reduce to lowest detectable level.
DOT Classification: Poison B; ORM-A.

SAFETY PROFILE: Poison by ingestion, skin contact, intravenous, and intraperitoneal. Human systemic effects by ingestion: excitement, tremors, and nausea or vomiting. An experimental teratogen. Other experimental reproductive effects. Continued acute exposure causes liver damage. Human mutation data reported. Questionable carcinogen with experimental carcinogenic, neoplastigenic, and tumorigenic data. See also CHLORINATED HYDROCARBONS. When heated to decomposition it emits toxic fumes of Cl^-.

AFK500 CAS:8067-82-1 ***HR: 3***
ALFADIONE
mf: $C_{23}H_{34}O_5 \cdot C_{21}H_{32}O_3$ mw: 723.10

SYNS: ALFATESINE (FRENCH) ◇ ALPHADIONE ◇ ALTHESIN ◇ CT-1341 ◇ 3-α-HYDROXY-5-α-PREGNANE-11,20-DIONE mixed with 3-α,21-DIHYDROXY-5-α-PREGNANE-11,20-DIONE 21-ACETATE (3:1) ◇ SAFFAN

TOXICITY DATA with REFERENCE
ivn-wmn LDLo:125 mg/kg PRACAK 222,249,79
ipr-rat LD50:79 mg/kg THERAP 32,375,77
ivn-rat LD50:36 mg/kg JZKEDZ 1,119,75
ipr-mus LD50:140 mg/kg JEMAAJ 62,191,79
ivn-mus LD50:47 mg/kg JEMAAJ 62,191,79
ivn-rbt LD50:12700 μg/kg PESTD5 16,208,75

SAFETY PROFILE: Poison by intravenous route in humans. Also poison by intraperitoneal route experimentally. When heated to decomposition it emits acrid smoke and fumes.

AFK750 ***HR: 1***
ALFALFA MEAL

SAFETY PROFILE: An allergen. Skin contact may cause dermatitis. Flammable when exposed to heat or flame; by spontaneous chemical reaction. Avoid moisture content extremes. Fires may smolder for 72 hours before becoming noticeable.

AFK875 CAS:23930-19-0 ***HR: 3***
ALFAXALONE
mf: $C_{21}H_{32}O_3$ mw: 332.53

PROP: Colorless prisms from ether. Mp: 172-174°.

SYNS: ALPHAXALONE ◇ GR 2/234 ◇ 3-α-HYDROXY-5-α-PREGNANE-11,20-DIONE ◇ 3-HYDROXYPREGNANE-11,20-DIONE

TOXICITY DATA with REFERENCE
orl-rat LD50:297 mg/kg IYKEDH 8,680,77
ipr-rat LD50:116 mg/kg IYKEDH 8,680,77
ivn-rat LD50:19400 μg/kg YKYUA6 28,1337,77
orl-mus LD50:880 mg/kg YKYUA6 28,1337,77
ipr-mus LD50:430 mg/kg IYKEDH 8,680,77
scu-mus LD50:5220 mg/kg IYKEDH 8,680,77
ivn-mus LD50:36900 μg/kg JZKEDZ 1,119,75
ivn-rbt LD50:9360 μg/kg IYKEDH 8,680,77

SAFETY PROFILE: Poison by ingestion, intravenous, and intraperitoneal routes. When heated to decomposition it emits acrid smoke and fumes.

AFK900 CAS:53988-42-4 ***HR: 1***
ALFONAL K

TOXICITY DATA with REFERENCE
skn-rbt 500 mg/24H MLD 28ZPAK -,265,72
eye-rbt 100 mg/24H MOD 28ZPAK -,265,72
orl-rat LD50:15800 mg/kg 28ZPAK -,265,72

SAFETY PROFILE: An eye irritant.

AFK950 ***HR: 1***
ALGERIAN IVY

PROP: Commonly cultivated climbing vines that produce black berries. They are used as outdoor wallcover and as house plants. They grow wild in some areas.

SYNS: CANARY IVY ◇ ENGLISH IVY ◇ HEDERA CANARIENSIS ◇ HEDERA HELIX ◇ MADEIRA IVY ◇ IVY ◇ YEDRA (CUBA)

SAFETY PROFILE: The leaves and berries contain hederin a poisonous saponin. Ingestion of these plant parts may cause a burning pain in the throat, vomiting, and diarrhea. See also SAPONIN.

AFL000 CAS:9005-32-7 ***HR: 2***
ALGINIC ACID

PROP: Extracted from brown seaweeds. White to yellow white fibrous powder; odorless and tasteless. Sol in alkaline solutions; insol in organic solvents.

SYNS: KELACID ◇ LANDALGINE ◇ NORGINE ◇ PLOYMANNURONIC ACID ◇ SAZZIO

TOXICITY DATA with REFERENCE
ipr-rat LD50:1600 mg/kg AIPTAK 111,167,57
ipr-mus LDLo:1000 mg/kg TXAPA9 23,288,72

CONSENSUS REPORTS: Reported in EPA TSCA Inventory.

SAFETY PROFILE: Moderately toxic by intraperitoneal route. When heated to decomposition it emits acrid smoke and irritating fumes.

AFL500 CAS:84-96-8 ***HR: 3***
ALIMEMAZINE
mf: $C_{18}H_{22}N_2S$ mw: 298.48

SYNS: ALIMEZINE ◇ BAYER 1219 ◇ 10-(3-(DIMETHYLAMINO)-2-METHYLPROPYL)PHENOTHIAZINE ◇ METHYLPROMAZINE ◇ REPELTIN ◇ TERALEN ◇ N,N,β-TRIMETHYL-10H-PHENOTHIAZINE-10-PROPANAMINE ◇ TRIMEPRAZINE

TOXICITY DATA with REFERENCE
orl-mus LD50:300 mg/kg DNEUDS 7,45,80
orl-rat LD50:210 mg/kg ANPBAZ 61,669,61
ivn-rat LD50:35 mg/kg ANPBAZ 61,669,61

SAFETY PROFILE: Poison by ingestion and intravenous routes. When heated to decomposition it emits very toxic fumes of NO_x and SO_x.

AFL750 CAS:3689-50-7 ***HR: 3***
ALIMEMAZINE-S,S-DIOXIDE
mf: $C_{18}H_{22}N_2O_2S$ mw: 330.48

SYNS: 10-(3-(DIMETHYLAMINO)-2-METHYLPROPYL)PHENOTHIAZINE-5,5-DIOXIDE ◇ DIOXO-9,9-(DIMETHYLAMINO-3-METHYL-2-PROPYL)-10-PHENOTHIAZINE (FRENCH) ◇ DOSEGRAN ◇ DOXERGAN ◇ DYSEDON ◇ IMAKOL ◇ OXOMEMAZINE ◇ OXYMEMAZINE ◇ 6847 R.P.

TOXICITY DATA with REFERENCE
eye-rbt 100 mg MLD FCTOD7 20,573,82
eye-rbt 100 mg/4S rns MLD FCTOD7 20,573,82
orl-mus LD50:140 mg/kg THERAP 26,1203,71
ipr-mus TDLo:185 mg/kg OCMJAJ 7,33,61
orl-mus LD50:220 mg/kg AIPTAK 135,364,62
scu-mus LD50:260 mg/kg AIPTAK 135,364,62
ivn-mus LD50:35 mg/kg AIPTAK 135,364,62

SAFETY PROFILE: Poison by ingestion, subcutaneous, and intravenous routes. An eye irritant. When heated to decomposition it emits very toxic fumes of NO_x and SO_x.

AFM000 CAS:62851-48-3 ***HR: 3***
ALIOMYCIN

PROP: An antibiotic produced by *Streptomyces Acidomyceticus.*

TOXICITY DATA with REFERENCE
orl-mus LD50:2650 mg/kg 85ERAY 2,1001,78
ipr-mus LD50:45 mg/kg 85ERAY 2,1001,78

SAFETY PROFILE: Poison by intraperitoneal route. Moderately toxic by ingestion.

AFM250 ***HR: 3***
ALIPHATIC and AROMATIC EPOXIDES

SAFETY PROFILE: Suspected carcinogen with experimental tumors of the skin, lung, and blood-forming tissues.

AFM375 CAS:8015-55-2 ***HR: D***
ALIPUR
mf: $C_{11}H_{22}N_2O \cdot C_{11}H_{10}ClNO_2$ mw: 422.02

SYNS: BUTYNYL-3N-3-CHLOROPHENYLCARBAMATE mixed with 3-CYCLOOCTYL-1,1-DIMETHYL UREA ◇ CHLORBUFAN mixed with CYCEURON ◇ CYCEURON plus CHLORBUFAN ◇ 3-CYCLOOCTYL-1,1-DIMETHYL UREA mixed with BUTYNYL-3N-3-CHLOROPHENYLCARBAMATE

TOXICITY DATA with REFERENCE
cyt-mus-unr 500 mg/kg TGANAK 14(6),41,80
cyt-mus-orl 500 mg/kg CYGEDX 14(6),38,80
orl-rat LD50:4600 mg/kg FMCHA2 -,C10,83

CONSENSUS REPORTS: EPA Genetic Toxicology Program.

SAFETY PROFILE: Mutation data reported. When heated to decomposition it emits toxic fumes of Cl^- and NO_x.

AFM500 ***HR: 1***
ALKALIES

PROP: A term loosely applied to the hydroxides and carbonates of the alkali metals and alkaline earth metals, as well as the bicarbonate and hydroxide of ammonium. They can neutralize acids, change the color of indicators, and impart a soapy taste and feel to aq solns.

SAFETY PROFILE: Variable toxicity. As a group, they constitute the commonest causes of contact dermatitis. Systemically ammonia is most troublesome. See also AMMONIA. See also specific compound.

AFM750 ***HR: 3***
ALKALOID SALTS

SYN: ALKALOIDS

SAFETY PROFILE: Nearly all alkaloid salts are poisonous. Some are also allergens. See specific alkaloid salt. Dangerous; when heated to decomposition they emits highly toxic fumes.

AFN250 ***HR: 2***
ALKANES

PROP: All colorless neutral liquids with light aromatic odors. See also individual alkanes as listed. (n-pentane, n-hexane, n-heptane, n-octane)

SAFETY PROFILE: Hexane can cause neuropathy with chronic exposure. Other alkanes or mixtures may have the same effect. Many are dangerous fire hazards when exposed to flame, heat, or oxidizers.

AFN500 CAS:72674-05-6 ***HR: 2***
α-ALKENESULFONIC ACID

SYNS: AOS ◇ α-OLEFIN SULFONATE ◇ α-OLEFIN SULPHONATE

TOXICITY DATA with REFERENCE
eye-rbt 1% SEV YKGKAM 21,334,72
orl-mus TDLo:3 g/kg (female 6-15D post):TER TXCYAC 3,107,75
orl-mus TDLo:3 g/kg (female 6-15D post):REP TXCYAC 3,107,75

SAFETY PROFILE: An experimental teratogen. Other experimental reproductive effects. A severe eye irritant. When heated to decomposition it emits toxic fumes of SO_x. See also SULFONATES.

AFN750 ***HR: 3***
ALKENYL DIMETHYLETHYL AMMONIUM BROMIDE

PROP: Alkenyl indicates a mixture of aliphatic hydrocarbon radicals with approximately 18 unsaturated carbons.

SYN: ONYXIDE (ONYX OIL & CHEM CO)

TOXICITY DATA with REFERENCE
orl-rat LD50:500 mg/kg SSCHAH 25,125,49
orl-gpg LD50:158 mg/kg SSCHAH 25,125,49

SAFETY PROFILE: Poison by ingestion. When heated to decomposition it emits very toxic fumes of NH_3, NO_x and Br^-. See also BROMIDES.

AFO200 ***HR: 2***
3-(ALKYLAMINO)PROPIONITRILE

PROP: A straight chain with 17 to 20 carbons. (GISAAA 48(1),82,83)

SYN: PROPIONITRILE, 3-(ALKYLAMINO)-

TOXICITY DATA with REFERENCE
orl-rat TDLo:1364 mg/kg (female 1-22D post):TER GISAAA 48(1),82,83
orl-rat LD50:6200 mg/kg GISAAA 48(1),82,83
orl-mus LD50:2800 mg/kg GISAAA 48(1),82,83

SAFETY PROFILE: Moderately toxic by ingestion. An experimental teratogen. When heated to decomposition it emits toxic fumes of NO_x.

AFO250 ***HR: 2***
ALKYL ARYL SULFONATE

PROP: A synthetic anionic detergent containing a minimum of 40% sodium alkyl aryl sulfonate, approximately 2% moisture, 1% unsulfonated oil, and the balance sodium sulfate.

SYNS: D-40 ◇ WITCONATE

TOXICITY DATA with REFERENCE
orl-rat LD50:2320 mg/kg JAPMA8 42,489,53
orl-mus LD50:2010 mg/kg JAPMA8 42,489,53
orl-rbt LD50:1730 mg/kg JAPMA8 42,489,53
orl-ham LD50:1131 mg/kg JAPMA8 42,489,53

SAFETY PROFILE: Moderately toxic by ingestion. When heated to decomposition it emits toxic fumes of SO_x.

AFO500 CAS:42615-29-2 ***HR: 3***
ALKYLBENZENESULFONATE

SYNS: ABS ◇ BENZENESULFONIC ACID, ALKYL DERIVATIVES ◇ LAS ◇ LINEAR ALKYLBENZENE SULFONATE ◇ LINEAR ALKYLBENZENE SULPHONATE

TOXICITY DATA with REFERENCE
skn-rat 5% MLD FCTXAV 18,55,80
orl-mus TDLo:100 mg/kg (female 6-15D post):TER SKEZAP 17,295,76
skn-mus TDLo:3200 mg/kg (female 1-4D post):REP LIFSAK 26,49,80
orl-rat LD50:437 mg/kg 34ZIAG -,690,69
orl-mus LD50:1407 mg/kg SKEZAP 4,15,63
scu-mus LD50:1989 mg/kg SKEZAP 4,15,63
ivn-mus LD50:157 mg/kg SKEZAP 4,15,63

SAFETY PROFILE: Poison by intravenous route. Moderately toxic by ingestion and subcutaneous routes. An experimental teratogen. Other experimental reproductive effects. A skin irritant. Very reactive with F_2. See also SULFONATES. When heated to decomposition it emits toxic fumes of SO_x.

AFO750 ***HR: 2***
p-n-ALKYLBENZENESULFONIC ACID DERIVATIVE, SODIUM SALT

PROP: Alkyl deriv. contains from C_{10} to C_{13} (28ZPAK -,195,72).

SYN: p-N-ALKYLBENZENSULFONAN SODNY (CZECH)

TOXICITY DATA with REFERENCE
skn-hmn 2500 μg/24H MLD AKEDAX 235,180,69
skn-rat 10 mg/16H MOD JSCCA5 22,411,71
skn-rbt 500 mg/24H MOD 28ZPAK -,195,72
skn-rbt 10 mg MLD JSCCA5 22,411,71
eye-rbt 250 μg/24H SEV 28ZPAK -,195,72
orl-rat LD50:1870 mg/kg 28ZPAK -,195,72

SAFETY PROFILE: Moderately toxic by ingestion. A skin and severe eye irritant. See also SULFONATES. When heated to decomposition it emits toxic fumes of SO_x.

AFP000 ***HR: 2***
ALKYLBENZENESULFONIC ACID SODIUM SALT

PROP: Alkyl is C_{12} derived from propylene tetramer and is highly branched (JSCCA5 22,411,71).

TOXICITY DATA with REFERENCE
skn-hmn 2500 μg/24H MLD AKEDAX 235,180,69
skn-rat 10 mg/16H MLD JSCCA5 22,411,71
skn-rbt 10 mg MLD JSCCA5 22,411,71

SAFETY PROFILE: A human skin irritant. See also SULFONATES. When heated to decomposition it emits toxic fumes of SO_x.

AFP075 ***HR: 2***
ALKYL DIMETHYL BENZALKONIUM CHLORIDE

PROP: Alkyl represents a mixture of fatty acid radicals (85DIA2 2,277,77).

SYN: BDM-CHLORIDE (RUSSIAN)

TOXICITY DATA with REFERENCE
unr-rat LD50:2020 mg/kg GISAAA 45(11),73,80
ipr-mus LD50:445 mg/kg JSCCA5 28,667,77
unr-mus LD50:1450 mg/kg GISAAA 45(11),73,80
unr-rbt LD50:750 mg/kg GISAAA 45(11),73,80
unr-gpg LD50:725 mg/kg GISAAA 45(11),73,80

SAFETY PROFILE: Moderately toxic by some routes. When heated to decomposition it emits toxic fumes of Cl^-. See also CHLORIDES.

AFP250 CAS:8001-54-5 ***HR: 3***
ALKYL DIMETHYLBENZYL AMMONIUM CHLORIDE

PROP: Alkyl group contains from C_8-C_{18}. D: (50%) 0.9884 @ 20°. Clear, mobile liquid.

SYNS: ALKYLDIMETHYL(PHENYLMETHYL)QUATERNARY AMMONIUM CHLORIDES ◇ AMMONYX ◇ ARQUAD DMMCB-75 ◇ BARQUAT MB-50 ◇ BAYCLEAN ◇ BENZALKONIUM CHLORIDE ◇ BIO-QUAT 50-24 ◇ BTC ◇ CATAMINE AB ◇ DRAPOLENE ◇ GARDIQUAT 1450 ◇ HYAMINE 3500 ◇ INTEXAN LB-50 ◇ KATAMINE AB ◇ NEO GERM-I-TOL ◇ ONYX BTC (ONYX OIL & CHEM CO) ◇ PHENEENE GERMICIDAL SOLUTION and TINCTURE ◇ QUATERNARY AMMONIUM COMPOUNDS, ALKYLBENZYLDIMETHYL, CHLORIDES ◇ RODALON ◇ TRITON K-60 ◇ VIKROL RQ ◇ ZEPHIRAN CHLORIDE

TOXICITY DATA with REFERENCE
skn-hmn 150 μg/3D-I MLD 85DKA8 -,127,77
eye-hmn 50 μg SEV AJOPAA 27,1118,44
eye-mky 2 mg/24H SEV TXAPA9 6,701,64
skn-rbt 50 mg/24H MOD 33NFA8 -,2,75
eye-rbt 100 ug AROPAW 34,99,45
eye-rbt 1 mg/24H SEV TXAPA9 6,701,64
dnr-bcs 50 μg/L MUREAV 193,21,88
sce-ham-emb 1 mg/L SHIGAZ 74,1365,87
ivg-rat TDLo:100 mg/kg (female 1D post):TER JJATDK 5,398,85
ivg-rat TDLo:50 mg/kg (female 1D post):REP JJATDK 5,398,85
orl-wmn TDLo:266 mg/kg HUTODJ 7,191,88
par-wmn LDLo:12 mg/kg DZGGAK 41,297,52
orl-rat LD50:240 mg/kg KSRNAM 4,219,70
skn-rat LD50:1560 mg/kg PCJOAU 12,1593,78
ipr-rat LD50:14500 μg/kg KSRNAM 4,219,70
ivn-rat LD50:13900 μg/kg KSRNAM 4,219,70
orl-mus LD50:175 mg/kg PCJOAU 12,1593,78
ipr-mus LDLo:10 mg/kg DZGGAK 41,297,52
scu-mus LD50:64 mg/kg KSRNAM 4,219,70
ivn-mus LD50:10 mg/kg DZGGAK 41,297,52

SAFETY PROFILE: A human poison by ingestion and parenteral route. An experimental poison by ingestion, subcutaneous, intraperitoneal, intravenous, and parenteral routes. An experimental teratogen. Other experimental reproductive effects. A human skin and severe eye irritant. Mutation data reported. When heated to decomposition it emits very toxic fumes of NO_x, NH_3, and Cl^-. See also CHLORIDES. An antimicrobial agent.

AFP750 CAS:8023 ***HR: 3***
ALKYL(C_8C_{18}NM)DIMETHYL-3,4-DICHLOROBENZYLAMMONIUM CHLORIDE-53-8

SYNS: ALKYL(C_8H_{17} to $C_{18}H_{37}$)DIMETHYL-3,4-DICHLOROBENZYL AMMONIUM CHLORIDE ◇ DICHLOROBENZALKONIUM CHLORIDE ◇ TETROSAN

TOXICITY DATA with REFERENCE
eye-rbt 1% SEV JAPMA8 38,428,49
orl-rat LD50:730 mg/kg SSCHAH 25,125,49
orl-mus LD50:2000 mg/kg JAPMA8 38,428,49
ivn-mus LD50:50 mg/kg JAPMA8 38,428,49
orl-gpg LD50:316 μg/kg SSCHAH 25,125,49

SAFETY PROFILE: A deadly poison by ingestion. Poison by intravenous route. Moderately toxic ingestion. A severe eye irritant. Can cause liver and kidney damage. A moderate allergen. Mutation data reported. See also ESTERS and CHLORIDES. When heated to decomposition it emits very toxic fumes of NO_x, NH_3, and Cl^-.

AFQ000 ***HR: 3***
ALKYLNITRILE
mf: $C_9H_{19}CN$ to $C_{17}H_{35}CN$ mw: 151.2 to 265.3

SYN: NITRIL MASTNE KYSELINY S (CZECH)

TOXICITY DATA with REFERENCE
eye-rbt 500 mg/24H MLD 28ZPAK -,160,72

CONSENSUS REPORTS: Cyanide and its compounds are on the Community Right-To-Know List.

SAFETY PROFILE: An eye irritant. See also NITRILES. When heated to decomposition it emits toxic fumes of NO_x and CN^-.

AFQ250 ***HR: 2***
ALKYL PHENYL POLYETHYLENE GLYCOL ETHER

PROP: Liquid. Mp: −5°; d: 1.0643 @ 20°/20°; autoign temp: 590°F.

CONSENSUS REPORTS: Glycol ethers are on the Community Right-To-Know List.

SAFETY PROFILE: Moderately toxic by ingestion. Mildly toxic by dermal contact and inhalation. Water solutions of less than 1% have irritating properties comparable to soap. See also GLYCOLS. Combustible when exposed to heat or flame. Incompatible with oxidizing materials. To fight fire, use water, foam, CO_2, dry chemical.

AFQ500 ***HR: 3***
ALKYL PYRIDINES R

TOXICITY DATA with REFERENCE
skn-rbt 500 mg open MLD UCDS** 8/5/71
eye-rbt 5 mg SEV UCDS** 8/5/71
orl-rat LD50:2240 mg/kg UCDS** 9/5/71
skn-rbt LD50:356 mg/kg UCDS** 8/5/71

SAFETY PROFILE: Poison by skin contact. Moderately toxic by ingestion. A skin and severe eye irritant.

AFQ575 CAS:5977-35-5 ***HR: 3***
ALKYROM
mf: $C_{12}H_{15}Cl_2NO_2$ mw: 276.18

SYNS: 3-BIS(2-CHLOROETHYL)AMINO-4-METHYLBENZOICACID ◇ 3-BIS(2-CHLOROETHYL)AMINO-p-TOLUIC ACID ◇ IOB 82 ◇ NSC-1461711

TOXICITY DATA with REFERENCE
cyt-hmn:leu 300 μg/L CCROBU 57,29,73
ipr-rat LD50:17 mg/kg CCROBU 54,319,70
unr-rat LD50:17 mg/kg NEOLA4 27,271,80

CONSENSUS REPORTS: EPA Genetic Toxicology Program.

SAFETY PROFILE: Poison by intraperitoneal and possibly other routes. Human mutation data reported. When heated to decomposition it emits toxic fumes of Cl^- and NO_x. See also NITRO COMPOUNDS of AROMATIC HYDROCARBONS and CHLORINATED HYDROCARBONS, AROMATIC.

AFQ625 ***HR: 1***
ALLAMANDA

PROP: An ornamental crawling or climbing shrub with large yellow flowers and 4- to 6-inch, lance-shaped leaves. It is commonly cultivated in Florida, Hawaii, and the West Indies.

SYNS: ALLAMANDA CATHARTICA ◇ CANARIO (PUERTO RICO) ◇ CAUTIVA (PUERTO RICO) ◇ FLOR de BARBERO (CUBA) ◇ LANI-ALI'I (HAWAII) ◇ NANI-ALI'I (HAWAII) ◇ YELLOW ALLAMANDA

SAFETY PROFILE: Most parts of the plant contain an unidentified cathartic toxin. Ingestion of any part of the plant results in mild catharsis.

AFQ750 CAS:2207-75-2 ***HR: 1***
ALLANTOXANIC ACID, POTASSIUM SALT
mf: $C_4H_3N_3O_4 \cdot K$ mw: 196.2

SYNS: POTASSIUM AZAOROTATE ◇ POTASSIUM OXONATE ◇ POTASSIUM-s-TRIAZINE-2,4-DIONE-6-CARBOXYLATE ◇ 1,4,5,6-TETRAHYDRO-4,6-DIOXO-s-TRIAZINE-2-CARBOXYLICACID, POTASSIUM SALT

TOXICITY DATA with REFERENCE
orl-mus TDLo:10800 μg/kg (female 9-11D post):REP TXCYAC 6,289,76
orl-rat TDLo:3 g/kg (female 8-9D post):TER TXCYAC 6,299,76
par-hmn TDLo:20 mg/kg:GIT CLPTAT 6,436,65

SAFETY PROFILE: Human gastrointestinal tract effects by parenteral route. An experimental teratogen. Other experimental reproductive effects. When heated to decomposition it emits toxic fumes of NO_x and K_2O.

AFR000 ***HR: 3***
ALLENE
mf: $H_2C{:}C{:}CH_2$ mw: 40.06

PROP: Colorless, unstable, flammable gas; sweet odor. D: 1.787, mp: −146°, bp: −32°, lel: 2.1%.

SAFETY PROFILE: Unknown toxicity. Probably anesthetic. Dangerous fire hazard when exposed to heat, flame or powerful oxidizers. Moderate explosion hazard when exposed to flame or compressed to >2 atm. To fight fire, stop flow of gas.

AFR250 CAS:584-79-2 ***HR: 3***
ALLETHRIN
DOT: UN 2902
mf: $C_{19}H_{26}O_3$ mw: 302.45

PROP: A viscous liquid.

SYNS: (+)-ALLELRETHONYL(+)-cis,trans-CHRYSANTHEMATE ◇ d-ALLETHRIN ◇ ALLETHRIN I ◇ ALLYL CINERIN ◇ ALLYL HOMOLOG of CINERIN I ◇ d,l-2-ALLYL-4-HYDROXY-3-METHYL-2-CYCLOPENTEN-1-ONE-d,l-CHRYSANTHEMUMMONOCARBOXYLATE ◇ 3-ALLYL-4-KETO-2-METHYLCYCLOPENTENYL CHRYSANTHEMUMMONOCARBOXYLATE ◇ 3-ALLYL-2-METHYL-4-OXO-2-CYCLOPENTEN-1-YL CHRYSANTHEMATE ◇ dl-3-ALLYL-2-METHYL-4-OXOCYCLOPENT-2-ENYL-dl-cis trans CHRYSANTHEMATE ◇ ALLYLRETHRONYL dl-cis-trans-CHRYSANTHEMATE ◇ BIOALLETHRIN ◇ CINERIN I ALLYL HOMOLOG ◇ ENT 17,510 ◇ EXTHRIN ◇ FDA 1446 ◇ FMC 249 ◇ NECARBOXYLIC ACID ◇ NIA 249 ◇ PALLETHRINE ◇ PYNAMIN ◇ PYNAMIN-FORTE ◇ PYRESIN ◇ PYRESYN ◇ SYNTHETIC PYRETHRINS

TOXICITY DATA with REFERENCE
mma-sat 500 μg/plate MUREAV 116,185,83
cyt-ham:lng 1900 ng/L/27H MUREAV 66,277,79
orl-rat LD50:310 mg/kg FMCHA2 -,C200,83
ivn-rat LDLo:4 mg/kg PCBPBS 2,308,72
unr-rat LD50:680 mg/kg 30ZDA9 -,131,71
orl-mus LD50:370 mg/kg ABCHA6 28,914,65
ipr-mus LD50:38 mg/kg JAFCAU 31,250,83
ice-mus LDLo:4 mg/kg TXAPA9 66,290,82
orl-rbt LD50:4290 mg/kg SPEADM 78-1,7,78

CONSENSUS REPORTS: Reported in EPA TSCA Inventory. EPA Genetic Toxicology Program.

DOT Classification: ORM-A; Label: None.

SAFETY PROFILE: Poison by ingestion, intravenous, intracerebral, and intraperitoneal routes. Moderately toxic by ingestion. An allergen. An insecticide. It can cause liver and kidney damage by all routes of entry into the body. Lung congestion may occur due to exposure. Local contact may cause contact dermatitis. Inhalation may cause asthma, coughing, wheezing, running nose and eyes. Mutation data reported. See also ALLYL COMPOUNDS and ESTERS. Slight fire hazard. When heated to decomposition it emits acrid fumes.

AFR500 CAS:34624-48-1 ***HR: 3***
(+)-cis-ALLETHRIN
mf: $C_{19}H_{26}O_3$ mw: 302.45

SYN: (+)-(Z)-2,2-DIMETHYL-3-(2-METHYLPROPENYL)-CYCLOPROPANECARBOXYLIC ACID ESTER with 2-ALLYL-4-HYDROXY-3-METHYL-2-CYCLOPENTEN-ONE

TOXICITY DATA with REFERENCE
ihl-rat LCLo:260 mg/m^3/2H EVHPAZ 14,15,76
orl-mus LD50:210 mg/kg EVHPAZ 14,15,76
ihl-mus LDLo:260 mg/m^3/2H EVHPAZ 14,15,76

SAFETY PROFILE: Poison by ingestion and inhalation. See also ALLYL COMPOUNDS and ESTERS. When heated to decomposition it emits acrid smoke and irritating fumes.

AFR750 CAS:28434-00-6 ***HR: 3***
trans-(+)-ALLETHRIN
mf: $C_{19}H_{26}O_3$ mw: 302.45

SYNS: AI 3-29024 ◇ d-ALLETHROLONE CHRYSANTHEMUMATE ◇ (+)-ALLETHRONYL (+)-trans-CHRYSANTHEMUMATE ◇ 2-ALLYL-4-HYDROXY-3-METHYL-2-CYCLOPENTEN-1-ONE◇ S-BIOALLETHRIN ◇ S-trans-BIOALLETHRIN ◇ (+)-2,2-DIMETHYL-3-(2-METHYLPROPENYL)-CYCLOPROPANECARBOXYLIC ACID-(E)-,ESTER with (+)- ◇ ESBIOL ◇ ESBIOL CONCENTRATE 90%

TOXICITY DATA with REFERENCE
orl-rat LD50:430 mg/kg EVHPAZ 14,15,76
ihl-rat LC50:1600 mg/m^3/3H EVHPAZ 14,15,76
orl-mus LD50:250 mg/kg EVHPAZ 14,15,76
ihl-mus LC50:2720 mg/m^3/3H EVHPAZ 14,15,76
scu-frg LD50:1700 μg/kg PCBPBS 20,217,83

SAFETY PROFILE: Deadly poison by subcutaneous route. Poison by ingestion. Moderately toxic by inhalation. See also ALLETHRIN, ALLYL COMPOUNDS, and ESTERS. When heated to decomposition it emits acrid and irritating fumes. An insecticide.

AFS000 ***HR: 3***
ALLETHRIN RACEMIC MIXTURE
mf: $C_{19}H_{26}O_3$•$4Cl_9H_{26}O_3$ mw: 1512.25

SYN: 4-HYDROXY-3-METHYL-2-CYCLOPENTEN-1-ONE, cis- mixed withtrans-2,2-DIMETHYL-3-(2-METHYL-PROPENYL)CYCLOPROPANECARBOXYLIC ACID ESTER with 2-ALLYL-4-HYDROXY-3-METHYL-2-CYCLOPENTEN-1-ONE(1:4)

TOXICITY DATA with REFERENCE
orl-rat LD50:720 mg/kg EVHPAZ 14,15,76
ihl-rat LCLo:260 mg/m^3/2H EVHPAZ 14,15,76
orl-mus LD50:500 mg/kg EVHPAZ 14,15,76
ihl-mus LCLo:260 mg/m^3/2H EVHPAZ 14,15,76

SAFETY PROFILE: Poison by inhalation. Moderately toxic by ingestion. When heated to decomposition it emits toxic fumes of Cl^-. See also ALLYL COMPOUNDS.

AFS250 ***HR: 2***
ALLICIN
mf: $C_6H_{10}OS_2$ mw: 162.3

PROP: A colorless, oily liquid; sharp garlic odor, d: 1.112 @ 20°/4°.

SAFETY PROFILE: Moderately toxic irritant by ingestion and inhalation. When heated to decomposition it emits toxic fumes of SO_x.

AFS500 CAS:52-43-7 ***HR: 3***
ALLOBARBITAL
mf: $C_{10}H_{12}N_2O_3$ mw: 208.24

SYNS: ALLOBARBITONE ◇ ALLYLBARBITURAL ◇ ALNOX ◇ ALOBARBITAL ◇ BARBALLYL ◇ BARBIDAL ◇ CURRAL ◇ DIADOL ◇ DIAL ◇ DIALLYLBARBITAL ◇ DIALLYLBARBITURIC ACID ◇ 5,5-DIALLYLBARBITURIC ACID ◇ DIALLYLMAL ◇ 5,5-DI-2-PROPENYL-2,4,6(1H,3H,5H)-PYRIMIDINETRIONE (9CI) ◇ DORM ◇ DORMALLYL ◇ MALIL ◇ MALILUM ◇ NOVALLYL ◇ NSC-9324

TOXICITY DATA with REFERENCE
ipr-rat TDLo:350 mg/kg (35D male):REP KSRNAM 16,2161,82
orl-hmn LDLo:36 mg/kg 28ZAA9 -,-,32
ipr-rat LD50:127 mg/kg APSCAX 24,7,51
scu-rat LD50:110 mg/kg AEPPAE 152,341,30
ipr-mus LD50:85 mg/kg AITDAQ 7,95,59
scu-mus LDLo:200 mg/kg HDTU** -,-,33
ivn-mus LD50:218 mg/kg KSRNAM 16,2161,82
orl-rbt LDLo:50 mg/kg HBAMAK 4,1289,35
ipr-rbt LDLo:100 mg/kg JAPMA8 25,597,36
ivn-rbt LD50:147 mg/kg KSRNAM 13,791,79

SAFETY PROFILE: A human poison by ingestion. An experimental poison by ingestion, intraperitoneal, subcutaneous, and intravenous routes. Experimental reproductive effects. When heated to decomposition it emits toxic fumes of NO_x. A sedative and hypnotic agent. See also BARBITURATES and ALLYL COMPOUNDS.

AFS625 CAS:5486-77-1 ***HR: 3***
ALLOCLAMIDE
mf: $C_{16}H_{23}ClN_2O_2$ mw: 310.86

PROP: Crystals from abs alc + ether. Mp: 125-127°. Sol in abs alc.

SYNS: 2-(ALLYLOXY)-4-CHLORO-N-(2-DIETHYLAMINO)ETHYL)BENZAMIDE ◇ 4-CHLORO-N-(2-(DIETHYLAMINO)ETHYL)-2-(2-PROPENYLOXY)-BENZAMIDE(9CI) ◇ 4-CHLORO-N-(2-DIETHYLAMINO)ETHYL)-2-(2-PROPENYLOXY)BENZAMIDE ◇ 264CE

TOXICITY DATA with REFERENCE
scu-mus LD50:155 mg/kg JJPAAZ 20,1,70
ivn-mus LD50:65 mg/kg NIIRDN 6,53,82
ivn-dog LDLo:50 mg/kg JJPAAZ 20,1,70

SAFETY PROFILE: Poison by subcutaneous and intra-

venous routes. When heated to decomposition it emits toxic fumes of Cl^- and NO_x.

AFS640 CAS:5107-01-7 ***HR: 3***
ALLOCLAMIDE HYDROCHLORIDE
mf: $C_{16}H_{23}ClN_2O_2 \cdot ClH$ mw: 347.32

SYN: 2-ALLYLOXY-4-CHLORO-N-(2-(DIETHYLAMINO)ETHYL)BENZAMIDEHYDROCHLORIDE

TOXICITY DATA with REFERENCE
orl-mus LD50:746 mg/kg IYKEDH 4,90,73
scu-mus LD50:155 mg/kg IYKEDH 4,90,73
ivn-mus LD50:61 mg/kg IYKEDH 4,90,73
ivn-dog LDLo:50 mg/kg JJPAAZ 16,342,66

SAFETY PROFILE: Poison by subcutaneous and intravenous routes. Moderately toxic by ingestion. When heated to decomposition it emits toxic fumes of NO_x and HCl.

AFS750 ***HR: 2***
ALLODAN
mf: $C_8H_6Cl_8$ mw: 385.74

TOXICITY DATA with REFERENCE
orl-rat LD50:940 mg/kg GTPZAB 8,30,64
skn-rat LD50:1000 mg/kg GTPZAB 8(4),30,64
orl-mus LD50:750 mg/kg GTPZAB 8(4),30,64

SAFETY PROFILE: Moderately toxic by ingestion and skin contact. When heated to decomposition it emits toxic fumes of Cl^-.

AFT000 CAS:77-02-1 ***HR: 3***
ALLONAL
mf: $C_{10}H_{14}N_2O_3$ mw: 210.26

SYNS: ALLIONAL ◇ 5-ALLYL-5-ISOPROPYLBARBITURATE ◇ ALLYLISOPROPYLBARBITURIC ACID ◇ 5-ALLYL-5-ISOPROPYLBARBITURIC ACID ◇ ALLYLISOPROPYLMALONYLUREA ◇ ALLYLPROPYMAL ◇ ALURATE ◇ APROBARBITAL ◇ APROBARBITONE ◇ APROZAL ◇ ISONAL ◇ ISOPROPYLALLYLBARBITURIC ACID ◇ NUMAL

TOXICITY DATA with REFERENCE
ipr-rat LDLo:100 mg/kg JPETAB 44,325,32
ipr-mus LD50:200 mg/kg ARZNAD 12,389,62
scu-mus LD50:350 mg/kg ARZNAD 8,25,58
orl-rbt LDLo:160 mg/kg JPETAB 44,325,32
ipr-rbt LDLo:90 mg/kg JPETAB 44,325,32

SAFETY PROFILE: Poison by ingestion and intraperitoneal routes. When heated to decomposition it emits toxic fumes of NO_x. See also BARBITURATES and ALLYL COMPOUNDS. A sedative and hypnotic agent.

AFT125 ***HR: D***
ALLOPREGNAN-3-β-OL-20-ISONICOTINYLHYDRAZONE
mf: $C_{27}H_{38}N_3O_2$ mw: 436.68

SYN: 3-β-HYDROXY-5-α-PREGNAN-20-ONE-20-ISONICOTINYLHYDRAZONE

TOXICITY DATA with REFERENCE
orl-rbt TDLo:10 mg/kg (1D pre):REP ACEDAB 73,17,63

SAFETY PROFILE: Experimental reproductive effects. When heated to decomposition it emits toxic fumes of NO_x.

AFT250 CAS:63732-62-7 ***HR: 3***
ALLOPSEUDOCODEINE HYDROCHLORIDE
mf: $C_{18}H_{21}NO_3 \cdot ClH$ mw: 335.86

TOXICITY DATA with REFERENCE
scu-mus LDLo:300 mg/kg JPETAB 51,35,34
scu-rbt LDLo:200 mg/kg JPETAB 51,35,34

SAFETY PROFILE: Poison by subcutaneous route. When heated to decomposition it emits very toxic fumes of NO_x and HCl.

AFT500 CAS:62-67-9 ***HR: 3***
ALLORPHINE
mf: $C_{19}H_{21}NO_3$ mw: 311.41

SYNS: N-ALLYL-7,8-DEHYDRO-4,5-EPOXY-3,6-DIHYDROXYMORPHINAN ◇ N-ALLYL-N-DESMETHYLMORPHINE ◇ N-ALLYLNORMORPHINE ◇ ANARCON ◇ ANTOFIN ◇ ANTORPHINE ◇ LETHIDROME ◇ LETIDRONE ◇ LITHIDRONE ◇ NALLINE ◇ NALORFINA ◇ NALORPHINE ◇ NALORPHINIUM ◇ NANM

TOXICITY DATA with REFERENCE
par-hmn TDLo:200 μg/kg:CNS PAREAQ 8,175,56
scu-hmn TDLo:71 μg/kg:PSY FEPRA7 15,442,56
scu-rat LD50:474 mg/kg AIPTAK 165,112,67
ivn-rat LD50:226 mg/kg AIPTAK 165,112,67
orl-mus LD50:1140 mg/kg 27ZQAG -,268,72
ipr-mus LD50:492 mg/kg PAREAQ 8,175,56
scu-mus LD50:500 mg/kg AIPTAK 165,112,67
ivn-mus LD50:127 mg/kg AIPTAK 165,112,67
par-mus LD50:670 mg/kg PAREAQ 8,175,56
scu-mky LDLo:400 mg/kg FEPRA7 13,369,54
ivn-mky LDLo:100 mg/kg FEPRA7 13,369,54
ivn-rbt LDLo:50 mg/kg PAREAQ 8,175,56

SAFETY PROFILE: Poison by intravenous and subcutaneous routes. Moderately toxic by ingestion and other routes. Human systemic effects by parenteral and subcutaneous routes: central nervous system changes, excitement, tremors, hallucinations and distorted perceptions, and antianxiety effects. When heated to decomposition it emits very toxic fumes of NO_x. See also ALLYL COMPOUNDS.

AFT750 CAS:50-71-5 **HR: 3**
ALLOXAN
mf: $C_4H_2N_2O_4$ mw: 142.08

HNCO•NHCO•CO•CO

PROP: D: 1.70; mp: 256° (decomp); sol in water, alc, benzene, and acetone.

SYNS: MESOXALYLCARBAMIDE ◇ MESOXALYLUREA ◇ 2,4,5,6(1H,3H)-PYRIMIDINETETRONE ◇ 2,4,5,6-PYRIMIDINTETRON (CZECH) ◇ 2,4,5,6-TETRAOXOHEXAHYDROPYRIMIDINE

TOXICITY DATA with REFERENCE
cyt-mus-ipr 50 mg/kg JOHEA8 65,345,74
ipr-rat TDLo:150 mg/kg (MGN):REP METAAJ 20,401,71
ipr-mus TDLo:200 mg/kg (female 8D post):TER JEEMAF 14,63,65
orl-rat LD50:5210 mg/kg 28ZOAJ -,150,72
ivn-rat LD50:300 mg/kg CRSBAW 142,1335,48
ipr-mus LDLo:300 mg/kg PSEBAA 67,154,48
scu-mus LDLo:400 mg/kg PSEBAA 67,154,48
ivn-mus LDLo:200 mg/kg PSEBAA 67,154,48
ivn-rbt LDLo:300 mg/kg AJCPAI 16,257,46
rec-rbt LDLo:180 mg/kg AJCPAI 16,257,46
ivn-pgn LDLo:150 mg/kg PSEBAA 58,31,45
ivn-dck LDLo:250 mg/kg HBTXAC 5,6,59
ivn-dom LDLo:200 mg/kg HBTXAC 5,6,59

CONSENSUS REPORTS: Reported in EPA TSCA Inventory.

SAFETY PROFILE: Poison by intraperitoneal, intravenous, subcutaneous, and rectal routes. Moderately toxic by ingestion. An experimental teratogen. Other experimental reproductive effects. Mutation data reported. Produces diabetes in experimental animals. Decomposes in storage to release CO_2. Do not store in sealed container. Explodes when heated above 170°C. When heated to decomposition it emits toxic fumes of NO_x.

AFU000 CAS:87-39-8 **HR: 3**
ALLOXAN-5-OXIME
mf: $C_4H_3N_3O_4$ mw: 157.10

SYNS: 5-HYDROXYIMINOBARBITURIC ACID ◇ 5-ISONITROSOBARBITURIC ACID ◇ 2,4,5,6(1H,3H)-PYRIMIDINETETRONE 5-OXIME ◇ VIOLURIC ACID

TOXICITY DATA with REFERENCE
ivn-mus LD50:100 mg/kg CSLNX* NX#05202

CONSENSUS REPORTS: Reported in EPA TSCA Inventory.

SAFETY PROFILE: Poison by intravenous route. When heated to decomposition it emits toxic fumes of NO_x.

AFU250 **HR: 2**
ALLOXANTIN
mf: $C_8H_6N_4O_8$•$2H_2O$ mw: 322.19

PROP: Crystalline powder; on exposure to air turns red; yellow @ 225°.

SYN: UROXIN

SAFETY PROFILE: Moderately toxic by ingestion. On a chronic basis caused disturbed carbohydrate metabolism leading to diabetes. Moderately dangerous; when heated to decomposition it emits toxic fumes of NO_x.

AFU750 CAS:591-87-7 **HR: 3**
ALLYL ACETATE
DOT: UN 2333
mf: $C_5H_8O_2$ mw: 100.13

PROP: Liquid, vap d: 3.45, bp: 104°, d: 0.928. Flash p: 72°F. Insol in water.

SYNS: ACETIC ACID ALLYL ESTER ◇ ACETIC ACID-2-PROPENYL ESTER ◇ 3-ACETOXYPROPENE

TOXICITY DATA with REFERENCE
skn-rbt 10 mg/24H MLD JIHTAB 31,60,49
eye-rbt 100 mg JIHTAB 31,60,49
orl-rat LD50:130 mg/kg JIHTAB 31,60,49
ihl-rat LC50:1000 ppm/1H AMIHAB 21,28,60
orl-mus LD50:170 mg/kg FCTXAV 2,327,64
skn-rbt LD50:1021 mg/kg JIHTAB 31,60,49

CONSENSUS REPORTS: Reported in EPA TSCA Inventory.

DOT Classification: Label: Flammable Liquid and Poison.

SAFETY PROFILE: Poison by ingestion. Moderately toxic by inhalation and skin contact. A skin and eye irritant. When heated to decomposition it emits acrid smoke and irritating fumes. Dangerous fire hazard. See also ALLYL COMPOUNDS.

AFV500 CAS:107-18-6 **HR: 3**
ALLYL ALCOHOL
DOT: UN 1098
mf: C_3H_6O mw: 58.09

H_2C=$CHCH_2OH$

PROP: Limpid liquid; pungent odor. Mp: −129°, bp: 96-97°, lel: 2.5%, uel: 18%, flash p: 70°F (CC), d: 0.854 @ 20°/4°, autoign temp: 713°F, vap press: 10 mm @ 10.5°, vap d: 2.00. Misc in water, alc, and ether.

SYNS: ALCOOL ALLILCO (ITALIAN) ◇ ALCOOL ALLYLIQUE (FRENCH) ◇ ALLILOWY ALKOHOL (POLISH) ◇ ALLYL AL ◇ ALLYLALKOHOL (GERMAN) ◇ ALLYLIC ALCOHOL ◇ 3-HYDROXYPROPENE ◇ ORVINYLCARBINOL ◇ PROPENOL ◇ PROPEN-1-OL-3 ◇ 1-PROPEN-3-OL ◇ 2-PROPEN-1-OL ◇ PROPENYL ALCO-

HOL ◇ 2-PROPENYL ALCOHOL ◇ RCRA WASTE NUMBER P005 ◇ SHELL UNDRAUTTED A ◇ VINYLCARBINOL ◇ WEED DRENCH

TOXICITY DATA with REFERENCE
eye-hmn 25 ppm SEV AMIHAB 18,303,58
skn-rbt 10 mg/24H open JIHTAB 30,63,48
eye-rbt 4270 μg SEV AJOPAA 29,1363,46
mmo-sat 100 μmol/L MUREAV 93,305,82
mma-sat 50 μg/plate TCMUD8 1,259,80
ihl-man LCLo:1000 ppm/1H 34ZIAG -,86,69
orl-rat LD50:64 mg/kg JIHTAB 30,63,48
ihl-rat LC50:165 ppm/4H AMIHAB 18,303,58
ipr-rat LD50:37 mg/kg TXAPA9 83,108,86
orl-mus LD50:96 mg/kg AMIHAB 18,303,58
ihl-mus LC50:500 mg/m³/2H 85GMAT -,17,82
ipr-mus LC50:60 mg/kg AMIHAB 18,303,58
ivn-mus LD50:78 mg/kg AIPTAK 135,330,62
ihl-mky LCLo:1000 ppm/4H CRTXB2 5,189,77
skn-rbt LD50:45 mg/kg JIHTAB 30,63,48

CONSENSUS REPORTS: EPA Extremely Hazardous Substances List. Reported in EPA TSCA Inventory.

OSHA PEL: (Transitional: TWA 2 ppm (skin)) TWA 2 ppm; STEL 4 ppm (skin)
ACGIH TLV: TWA 2 ppm; STEL 4 ppm (skin)
DFG MAK 2 ppm (5 mg/m³)
DOT Classification: Flammable Liquid; Label: Poison and Flammable Liquid.

SAFETY PROFILE: Poison by inhalation, ingestion, skin contact, subcutaneous, intraperitoneal, and possibly other routes. A skin, severe eye (human), and systemic irritant. Mutation data reported. Dangerous fire and explosion hazard when exposed to heat, flame, or oxidizers. Explosive or violent reaction with sulfuric acid, alkali + 2,4,6-trichloro-1,3,5-triazine, or 2,4,6-tris(bromoamino)-1,3,5-triazine. Reaction with carbon tetrachloride produces explosively unstable halogenated C_4 epoxides. Incompatible with chlorosulfonic acid, HNO_3, H_2SO_4, oleum, NaOH, diallyl phosphite, PCl_3, and tri-n-bromomelamine. When heated to decomposition it emits acrid smoke and fumes. To fight fire, use CO_2, alcohol foam, dry chemical. See also ALLYL COMPOUNDS.

AFV750 CAS:66941-48-8 ***HR: 3***
5-ALLYL-5-(1-(ALLYTHIO)ETHYL)BARBITURIC ACID SODIUM SALT
mf: $C_{12}H_{15}N_2O_3S{\cdot}Na$ mw: 290.34

TOXICITY DATA with REFERENCE
orl-rat LD50:346 mg/kg JAPMA8 35,231,46
ivn-rat LD50:95 mg/kg JPETAB 88,343,46

SAFETY PROFILE: Poison by ingestion and intravenous routes. When heated to decomposition it emits very toxic fumes of SO_x, Na_2O, and NO_x. See also BARBITURATES and ALLYL COMPOUNDS.

AFW000 CAS:107-11-9 ***HR: 3***
ALLYLAMINE
DOT: UN 2334
mf: C_3H_7N mw: 57.11

PROP: Colorless liquid, burning taste, sharp odor. Bp: 56.5°, d: 0.761 @ 20°/4°, flash p: −20°F, autoign temp: 705°F, vap d: 2.00, lel: 2.2%, uel: 22%. Misc in water, alc, and ether.

SYNS: 3-AMINOPROPENE ◇ 3-AMINOPROPYLENE ◇ MONOALLYLAMINE ◇ 2-PROPENAMINE ◇ 2-PROPEN-1-AMINE

TOXICITY DATA with REFERENCE
skn-rbt 500 mg/24H SEV AEHLAU 1,343,60
eye-rbt 50 mg/20S rns SEV AEHLAU 1,343,60
cyt-rat-orl 2500 ng/kg GISAAA 48(1),80,83
ihl-man TCLo:2500 ppb/5M:EYE,PUL AEHLAU 1,343,60
ihl-mam LC50:320 mg/M³ TPKVAL 14,80,75
orl-rat LD50:106 mg/kg AEHLAU 1,343,60
ihl-rat LC50:286 ppm/4H AEHLAU 1,343,60
orl-mus LD50:57 mg/kg AEHLAU 1,343,60
ipr-mus LD50:49 mg/kg AEHLAU 1,343,60
skn-rbt LD50:35 mg/kg AEHLAU 1,343,60
unr-mam LD50:783 mg/kg TPKVAL 14,80,75

CONSENSUS REPORTS: EPA Extremely Hazardous Substances List. Reported in EPA TSCA Inventory.

DOT Classification: Flammable Liquid; Label: Flammable Liquid and Poison.

SAFETY PROFILE: Poison by inhalation, ingestion, intraperitoneal, and skin contact. Human systemic effects by inhalation: lacrimation and lung effects. A systemic irritant. Mutation data reported. A severe eye and skin irritant. Extraordinary precautions against fumes are advised. Dangerous fire and explosion hazard when exposed to heat, flame, or oxidizers. Highly reactive. When heated to decomposition it emits toxic fumes of NO_x. To fight fire, use alcohol foam, CO_2, dry chemical. See also ALLYL COMPOUNDS and AMINES.

AFW250 CAS:77966-30-4 ***HR: 3***
2-(ALLYLAMINO)-6'-CHLORO-o-ACETOTOLUIDIDE HYDROCHLORIDE
mf: $C_{12}H_{15}ClN_2O{\cdot}ClH$ mw: 275.20

SYNS: 2-(ALLYLAMINO)-2'-CHLORO-6'-METHYLACETANILIDE HYDROCHLORIDE ◇ C 3124

TOXICITY DATA with REFERENCE
ipr-rat LD50:460 mg/kg ARZNAD 8,407,58
ipr-mus LD50:375 mg/kg ARZNAD 8,407,58
scu-mus LD50:1070 mg/kg ARZNAD 8,407,58

SAFETY PROFILE: Poison by intraperitoneal route. Moderately toxic by subcutaneous route. When heated to decomposition it emits very toxic fumes of Cl^-, NO_x, and HCl. See also ALLYL COMPOUNDS and AMINES.

AFW500 CAS:642-44-4 ***HR: 2***
1-ALLYL-6-AMINO-3-ETHYLURACIL
mf: $C_9H_{13}N_3O_2$ mw: 195.25

SYNS: ALACIL ◇ ALLACYL ◇ 1-ALLYL-6-AMINO-3-ETHYL-2,4(1H,3H)-PYRIMIDINEDIONE ◇ 1-ALLYL-3-ETHYL-6-AMINOTETRAHYDROPYRIMIDINEDIONE ◇ 6-AMINO-3-ETHYL-1-(2-PROPENYL)-2,4(1H,3H-)-PYRIMIDINEDIONE ◇ AMINOMETRADINE ◇ AMINOMETRAMIDE ◇ CATAPYRIN ◇ 1-ETHYL-3-ALLYL-6-AMINOURACIL ◇ KATAPYRIN ◇ MICTINE ◇ MINCARD ◇ S.C. 3497

TOXICITY DATA with REFERENCE
orl-rat LD50:2300 mg/kg FEPRA7 14,392,55
ipr-rat LD50:500 mg/kg CLDND* 13,125,58
ipr-mus LD50:560 mg/kg CLDND*

SAFETY PROFILE: Moderately toxic by ingestion and intraperitoneal routes. When heated to decomposition it emits toxic fumes of NO_x. See also ALLYL COMPOUNDS. A diuretic agent.

AFW750 CAS:140-67-0 ***HR: 3***
p-ALLYLANISOLE
mf: $C_{10}H_{12}O$ mw: 148.22

PROP: Isolated from rind of *Persea Gratissima Garth,* and from Oil of Estragon; found in oils of Russian Anise, Basil, Fennel, Turpentine, and others (FCTXAV 14,601,76). Colorless to sltly yellow liquid; anise odor. D: 0.960-0.968, refr index: 1.519-1.524, flash p: 178°F. Sol in alc; insol in water.

SYNS: 4-ALLYL-1-METHOXYBENZENE ◇ CHAVICOL METHYL ETHER ◇ ESDRAGOL ◇ ISOANETHOLE ◇ p-METHOXYALLYLBENZENE ◇ 1-METHOXY-4-(2-PROPENYL)BENZENE ◇ METHYL CHAVICOL ◇ NCI-C60946 ◇ TARRAGON

TOXICITY DATA with REFERENCE
skn-rbt 500 mg/24H MOD FCTXAV 14,601,76
mmo-sat 1 μmol/plate FCTXAV 14,603,76
mma-sat 1 μmol/plate MUREAV 60,143,79
bfa-rat/sat 2500 mg/kg NUCADQ 1,10,79
dnd-mus-ipr 80 mg/kg CRNGDP 5,1613,84
orl-mus TDLo:97 g/kg/1Y-C:NEO CNREA8 43,1124,83
scu-mus TDLo:140 mg/kg/22D-I:CAR JNCIAM 57,1323,76
ipr-mus TDLo:111 mg/kg:CAR CNREA8 47,2275,87
orl-rat LDLo:1230 mg/kg FCTXAV 14(6),601,76
ipr-rat LD50:1030 mg/kg COREAF 246,1465,58
orl-mus LD50:1250 mg/kg FCTXAV 2,327,64
ipr-mus LD50:1260 mg/kg COREAF 246,1465,58

CONSENSUS REPORTS: Reported in EPA TSCA Inventory.

SAFETY PROFILE: Moderate acute toxicity by many routes. A skin irritant. Questionable carcinogen with experimental carcinogenic and neoplastigenic data. Mutation data reported. Combustible liquid. When heated to decomposition it emits acrid smoke and irritating fumes. See also ALLYL COMPOUNDS. A spice used in foods, liqueurs, and perfumes.

AFX000 CAS:300-57-2 ***HR: 2***
ALLYLBENZENE
mf: C_9H_{10} mw: 118.19

TOXICITY DATA with REFERENCE
orl-rat LDLo:4620 mg/kg FCTXAV 2,327,64
orl-mus LD50:2900 mg/kg TXAPA9 7,18,65

CONSENSUS REPORTS: Reported in EPA TSCA Inventory.

SAFETY PROFILE: Moderately toxic by ingestion. When heated to decomposition it emits acrid smoke and irritating fumes. See also ALLYL COMPOUNDS.

AFX250 ***HR: 3***
ALLYL BENZENE SULFONATE
mf: $C_9H_{10}O_3S$ mw: 198.2

SAFETY PROFILE: A highly reactive and flammable compound. Residue from vacuum distillation @ 92-135° @ 2.6 mbar exploded after removal of heat source. When heated to decomposition it emits toxic fumes of SO_x. See also ALLYL COMPOUNDS and SULFONATES.

AFX500 CAS:64058-13-5 ***HR: 3***
5-ALLYL-5-BENZYL-2-THIOBARBITURIC ACID SODIUM SALT
mf: $C_{14}H_{14}N_2O_2S•Na$ mw: 297.35

SYN: SODIUM ALLYLBENZYL THIOBARBITURATE

TOXICITY DATA with REFERENCE
orl-rat LDLo:40 mg/kg JPETAB 60,125,37
ipr-rat LDLo:20 mg/kg JPETAB 60,125,37

SAFETY PROFILE: Poison by ingestion and intraperitoneal routes. When heated to decomposition it emits very toxic fumes of NO_x, Na_2O, and SO_x. See also BARBITURATES and ALLYL COMPOUNDS.

AFX750 CAS:63905-38-4 ***HR: 3***
ALLYL-BIS(β-CHLOROETHYL)AMINE HYDROCHLORIDE
mf: $C_7H_{13}Cl_2N•ClH$ mw: 218.57

TOXICITY DATA with REFERENCE
ipr-mus LD50:3 mg/kg CANCAR 2,1055,49
scu-mus LDLo:4 mg/kg JPETAB 91,224,47

SAFETY PROFILE: Poison by intraperitoneal and subcutaneous routes. When heated to decomposition it emits very toxic fumes of Cl^-, NO_x, and HCl. See also ALLYL COMPOUNDS and AMINES.

AFY000 CAS:106-95-6 ***HR: 3***
ALLYL BROMIDE
DOT: UN 1099
mf: C_3H_5Br mw: 120.99

PROP: Colorless liquid, pungent odor. Mp: −119°, bp: 71.3°, flash p: 30°F, d: 1.3980 @ 20°/4°, autoign temp: 563°F, vap d: 4.17, lel: 4.4%, uel: 7.3%. Insol in water.

SYNS: BROMALLYLENE ◇ 3-BROMOPROPENE ◇ 3-BROMOPROPYLENE

TOXICITY DATA with REFERENCE
dns-hmn:hla 500 μmol/L CALEDQ 20,263,83
ihl-mam LC50:4110 mg/m³ GTPZAB 18(4),55,74
ipr-mam LD50:88 mg/kg GTPZAB 18(4),55,74
mmo-sat 1 μmol/plate BCPCA6 29,993,80
ihl-rat LC50:10000 mg/m³/30M FAVUAI 7,35,75
ipr-mus LD50:108 mg/kg JPCEAO 320(1),133,78
orl-gpg LD50:30 mg/kg WQCHM* 4,-,74

CONSENSUS REPORTS: Reported in EPA TSCA Inventory.

DOT Classification: Flammable Liquid; Label: Flammable Liquid and Poison .

SAFETY PROFILE: Poison by ingestion and intraperitoneal routes. Mildly toxic by inhalation. Human mutation data reported. See also ALLYL CHLORIDE and ALLYL COMPOUNDS. Dangerous fire and explosion hazard when exposed to heat, flame, or oxidizers. When heated to decomposition it emits toxic fumes of Br^-. To fight fire, use alcohol foam, water spray or mist, CO_2, dry chemical.

AFY250 CAS:2051-78-7 ***HR: 3***
ALLYL BUTANOATE
mf: $C_7H_{12}O_2$ mw: 128.19

SYNS: ALLYL BUTYRATE ◇ VINYL CARBINYL BUTYRATE

TOXICITY DATA with REFERENCE
skn-hmn 20 mg/48H MLD FCTXAV 15,611,77
skn-rbt 500 mg/24H MOD FCTXAV 15,611,77
orl-rat LD50:250 mg/kg TXAPA9 6,378,64
skn-rbt LD50:530 mg/kg FCTXAV 15,611,77

CONSENSUS REPORTS: Reported in EPA TSCA Inventory.

SAFETY PROFILE: Poison by ingestion. Moderately toxic by skin contact. A human skin irritant. See also ESTERS and ALLYL COMPOUNDS. When heated to decomposition it emits acrid smoke and irritating fumes.

AFY300 CAS:66941-49-9 ***HR: 3***
5-ALLYL-5-(2-BUTENYL)-2-THIOBARBITURIC ACID SODIUM SALT
mf: $C_{11}H_{13}N_2O_2S \cdot Na$ mw: 260.31

TOXICITY DATA with REFERENCE
ipr-rat LD50:172 mg/kg JAPMA8 34,183,45
ivn-rbt LD50:73 mg/kg JAPMA8 34,183,45

SAFETY PROFILE: Poison by intraperitoneal and intravenous routes. When heated to decomposition it emits very toxic fumes of NO_x, SO_x, and Na_2O. See also BARBITURATES and ALLYL COMPOUNDS.

AFY500 CAS:115-44-6 ***HR: 3***
5-ALLYL-5-sec-BUTYLBARBITURIC ACID
mf: $C_{11}H_{16}N_2O_3$ mw: 224.29

SYNS: 5-ALLYL-5-(1-METHYLPROPYL) BARBITURIC ACID ◇ BUTABITAL ◇ sec-BUTYL ALLYL BARBITURIC ACID ◇ LATUSATE ◇ LOTUSATE ◇ 5-(1-METHYLPROPYL)-5-(2-PROPENYL)-2,4,6(1H,3H,5H)-PYRIMIDINETRIONE (9CI) ◇ PROFUNDOL ◇ TALBUTAL ◇ WIN 5095

TOXICITY DATA with REFERENCE
orl-rat LD50:57500 μg/kg TXAPA9 21 315,72
ipr-rat LDLo:75 mg/kg JPETAB 44,325,32
ivn-rat LDLo:68 mg/kg CLDND* 44,325,32
ipr-rbt LDLo:55 mg/kg JPETAB 44,325,32
ivn-rbt LDLo:50 mg/kg JACSAT 57,1961,35
orl-pgn LD50:56 mg/kg TXAPA9 21,315,72

SAFETY PROFILE: Poison by ingestion, intraperitoneal, and intravenous routes. Human psychotropic effects by ingestion. When heated to decomposition it emits toxic NO_x. See also BARBITURATES and ALLYL COMPOUNDS. A sedative and hypnotic agent.

AFY750 CAS:2095-58-1 ***HR: 3***
ALLYL-sec-BUTYL THIOBARBITURIC ACID
mf: $C_{11}H_{16}N_2O_2S$ mw: 240.35

SYN: 5-ALLYL-5-sec-BUTYL-2-THIOBARBITURICACID

TOXICITY DATA with REFERENCE
ivn-mus LD50:150 mg/kg ARZNAD 4,441,54
orl-rbt LDLo:500 mg/kg JPETAB 60,189,37
ivn-rbt LDLo:60 mg/kg JPETAB 60,189,37
rec-rbt LDLo:110 mg/kg JPETAB 60,189,37

SAFETY PROFILE: Poison by intravenous and rectal routes. When heated to decomposition it emits very toxic fumes of SO_x and NO_x. See also BARBITURATES and ALLYL COMPOUNDS.

AGA000 CAS:64058-14-6 ***HR: 3***
5-ALLYL-5-sec-BUTYL-2-THIOBARBITURIC ACID SODIUM SALT
mf: $C_{11}H_{15}N_2O_2S \cdot Na$ mw: 262.33

TOXICITY DATA with REFERENCE
orl-rat LDLo:125 mg/kg JPETAB 60,125,37
ipr-rat LDLo:100 mg/kg JPETAB 60,125,37
ivn-rat LDLo:120 mg/kg JPETAB 60,125,37
orl-dog LDLo:130 mg/kg JPETAB 60,125,37
ivn-dog LDLo:60 mg/kg JPETAB 60,125,37
ivn-rbt LDLo:40 mg/kg JPETAB 60,125,37

SAFETY PROFILE: Poison by ingestion, intraperitoneal, and intravenous routes. When heated to decomposition it emits very toxic fumes of NO_x, SO_x, and Na_2O. See also BARBITURATES and ALLYL COMPOUNDS.

AGA250 CAS:66941-53-5 ***HR: 3***
5-ALLYL-5-(1-BUTYLTHIO)ETHYL)BARBITURIC ACID SODIUM SALT
mf: $C_{13}H_{19}N_2O_3S{\cdot}Na$ mw: 306.39

SYN: SODIUM-5-ALLYL-5-(1-(BUTYLTHIO)ETHYL)BARBITURATE

TOXICITY DATA with REFERENCE
orl-rat LD50:639 mg/kg JAPMA8 35,231,46
ivn-rat LD50:90 mg/kg JPETAB 88,343,46
ivn-rbt LD50:35 mg/kg JAPMA8 35,244,46

SAFETY PROFILE: Poison by intravenous route. Moderately toxic by ingestion. When heated to decomposition it emits very toxic fumes of NO_x, SO_x, and Na_2O. See also BARBITURATES and ALLYL COMPOUNDS.

AGA500 CAS:123-68-2 ***HR: 3***
ALLYL CAPROATE
mf: $C_9H_{16}O_2$ mw: 156.25

PROP: Bp: 186-188°. Insol in water; sol in alc and ether.

SYNS: ALLYL HEXANOATE (FCC) ◇ FEMA No. 2032 ◇ 2-PROPENYL-N-HEXANOATE

TOXICITY DATA with REFERENCE
skn-hmn 20 mg/48H MLD FCTXAV 11,1079,73
mrc-bcs 18 μg/disc OEKSDJ 9,177,78
orl-rat LD50:218 mg/kg FCTXAV 2,327,64
skn-rbt LD50:300 mg/kg FCTXAV 11,477,73
orl-gpg LD50:280 mg/kg FCTXAV 2,327,64

CONSENSUS REPORTS: Reported in EPA TSCA Inventory.

SAFETY PROFILE: Poison by ingestion and skin contact. Mutation data reported. An irritant to human skin. When heated to decomposition it emits acrid smoke and irritating fumes. See also ALLYL COMPOUNDS and ESTERS.

AGA750 CAS:2114-11-6 ***HR: 3***
ALLYL CARBAMATE
mf: $C_4H_7NO_2$ mw: 101.12

SYN:
CARBAMIC ACID, ALLYL ESTER

TOXICITY DATA with REFERENCE
sce-mus-ipr 100 μmol/kg CNREA8 42,2165,82
ipr-mus TDLo:279 mg/kg/4W-I:NEO CNREA8 29,2184,69

CONSENSUS REPORTS: Reported in EPA TSCA Inventory.

SAFETY PROFILE: Questionable carcinogen with experimental neoplastigenic data. Mutation data reported. See also ALLYL COMPOUNDS and CARBAMATES. When heated to decomposition it emits toxic NO_x.

AGB000 CAS:63884-80-0 ***HR: 3***
ALLYLCARBAMIC ESTER of m-OXYPHENYLDIMETHYLAMINE HYDROCHLORIDE
mf: $C_{12}H_{16}N_2O_2{\cdot}ClH$ mw: 256.76

SYNS: AR-19 ◇ N-ALLYL CARBAMIC ACID-3-DIMETHYLAMINOPHENYL ESTER HYDROCHLORIDE

TOXICITY DATA with REFERENCE
orl-mus LDLo:500 mg/kg JPETAB 43,413,31
ivn-mus LD50:150 mg/kg NTIS** PB158-508

SAFETY PROFILE: Poison by intravenous route. Moderately toxic by ingestion. See also ALLYL COMPOUNDS, ESTERS, and CARBAMATES. When heated to decomposition it emits very toxic fumes of NO_x and HCl.

AGB250 CAS:107-05-1 ***HR: 3***
ALLYL CHLORIDE
DOT: UN 1100
mf: C_3H_5Cl mw: 76.53

$$H_2C{=}CHCH_2Cl$$

PROP: Colorless liquid. Mp: −136.4°, bp: 44.6°, d: 0.938 @ 20°/4°, flash p: −25°F, lel: 2.9%, uel: 11.2%, autoign temp: 905°F, vap d: 2.64. Solubility = <0.1 in water.

SYNS: ALLILE (CLORURO DI) (ITALIAN) ◇ ALLYLCHLORID (GERMAN) ◇ ALLYLE (CHLORURE D') (FRENCH) ◇ CHLORALLYLENE ◇ CHLOROALLYLENE ◇ 3-CHLOROPRENE ◇ 3-CHLORO-1-PROPENE ◇ 3-CHLOROPROPENE ◇ 1-CHLORO PROPENE-2 ◇ 1-CHLORO-2-PROPENE ◇ α-CHLOROPROPYLENE ◇ 3-CHLORO-1-PROPYLENE ◇ 3-CHLOROPROPYLENE ◇ 3-CHLORPROPEN (GERMAN) ◇ NCI-C04615 ◇ 2-PROPENYL CHLORIDE

TOXICITY DATA with REFERENCE
mmo-esc 20 μL/plate MUREAV 153,57,85
mmo-omi 10 μL/plate CBINA8 30,9,80
skn-rbt 10 mg/24H JIHTAB 30,63,48
eye-rbt 469 mg AJOPAA 29,1363,46
eye-gpg 290 ppm/6H JIHTAB 22,79,40

orl-mus TDLo:4 g/kg (female 7-14D post):TER NTIS** PB86-197605
ipr-rat TDLo:1200 mg/kg:REP EPASR* 8EHQ-0381-0386
orl-mus TDLo:50 g/kg/78W-I:ETA NCITR* NCI-CG-TR-73,78
orl-rat LD50:700 mg/kg JIDHAN 30,63,48
ihl-rat LC50:11 g/m^3/2H EESADV 6,19,82
orl-mus LD50:425 mg/kg EESADV 6,19,82
ihl-mus LC50:11500 mg/m^3/2H EESADV 6,19,82
ipr-mus LD50:155 mg/kg SCCUR* -,1,61
ivn-dog LD50:7150 μg/kg JPETAB 89,109,47
skn-rbt LD50:2066 mg/kg JIDHAN 30,63,48

CONSENSUS REPORTS: IARC Cancer Review: Group 3 IMEMDT 7,56,87; Animal Inadequate Evidence IMEMDT 36,39,85; NCI Carcinogenesis Bioassay (gavage); No Evidence: rat NCITR* NCI-CG-TR-73,78; Clear Evidence: mouse NCITR* NCI-CG-TR-73,78. Reported in EPA TSCA Inventory. EPA Genetic Toxicology Program. Community Right-To-Know List.

OSHA PEL: (Transitional: TWA 1 ppm) TWA 1 ppm; STEL 2 ppm
ACGIH TLV: TWA 1 ppm; STEL 2 ppm
DFG MAK: 1 ppm (3 mg/m^3), Suspected Carcinogen.
NIOSH REL: TWA 1 ppm; CL 3 ppm/15M
DOT Classification: Flammable Liquid; Label: Flammable Liquid.

SAFETY PROFILE: Suspected carcinogen with experimental tumorigenic data. Poison by intraperitoneal and intravenous routes. Moderately toxic by ingestion, inhalation and skin contact. Experimental teratogenic and reproductive effects. A skin and eye irritant. Human mutation data reported. Chronic exposure may cause liver and kidney damage. The vapors of allyl chloride are quite irritating to the eyes, nose, and throat. Contact of the liquid with the skin, in addition to local vasoconstriction and numbness, may lead to rapid absorption and distribution through the body. If remedial measures are not taken promptly, such contact may result in burns and internal injuries. Inhalation may cause headache, dizziness, and in high concentration, loss of consciousness; however, even in low concentration, its odor in most cases is irritating enough to give warning of its presence. Concentration of the vapors high enough to cause serious effects, including damage to the lungs, especially on repeated exposure, may not be intolerable. Consequently, the warning characteristics should never be disregarded. In general, precautions should be taken AT ALL TIMES to avoid spillage and accumulation of noticeable concentration of the vapors in the atmosphere. Acute exposure in experimental animals has resulted in marked inflammation of lungs, irritation of skin, and swelling of the kidneys. Chronically exposed animals have shown degenerative changes in the liver and kidneys. Reported human exposures have been principally cases of irritation of the eyes, skin, and respiratory tract, sometimes accompanied by aches and pains in the bones. Liver and kidney injury is possible.

Dangerous fire and explosion hazard when exposed to heat, flame, or oxidizers. Vigorous or explosive reaction above −70°C with alkyl aluminum chlorides (e.g., trichlorotriethyl dialuminum, ethyl aluminum dichloride, or diethyl aluminum chloride) + aromatic hydrocarbons (e.g., benzene or toluene). Violently exothermic polymerization reaction with Lewis acids (e.g., aluminum chloride, boron trifluoride, or sulfuric acid) and metals (e.g., aluminum, magnesium, zinc, or galvanized metals). Incompatible with HNO_3, ethylene imine, ethylenediamine, chlorosulfonic acid, oleum, NaOH. To fight fire, use CO_2, alcohol foam, dry chemical. See also CHLORINATED HYDROCARBONS, ALIPHATIC, ALLYL COMPOUNDS, and CHLORIDES.

Storage and Handling: Keep cool, away from heat sources. Maintain good ventilation. Work in a fume hood or with closed system if possible; otherwise, use adequate ventilation so that the odor of allyl chloride does not persist. If it should be necessary to enter an area in which the odor of allyl chloride is at all noticeable, use a gas mask equipped with an "organic vapor" canister. Do not disregard the warning odor or eye irritation of allyl chloride.

AGB500 CAS:2937-50-0 ***HR: 3***
ALLYL CHLOROCARBONATE
DOT: UN 1722
mf: $C_4H_5ClO_2$ mw: 120.54

PROP: Liquid. Bp: 106-114°, flash p: 88°F (CC), d: 1.14, vap d: 4.2.

SYNS: ALLYL CHLOROFORMATE (DOT)

TOXICITY DATA with REFERENCE
orl-rat LD50:244 mg/kg GTPZAB 28(5),51,84
ihl-rat LC50:32400 μg/m^3 GTPZAB 28(5),51,84
orl-mus LD50:210 mg/kg GTPZAB 28(5),51,84
ihl-mus LD50:23100 μg/m^3 GTPZAB 28(5),51,84

CONSENSUS REPORTS: Reported in EPA TSCA Inventory.

DOT Classification: Flammable Liquid; Label: Flammable Liquid; IMO: Corrosive Material; Label: Corrosive, Flammable Liquid.

SAFETY PROFILE: Poison by inhalation and ingestion. Corrosive. Dangerous when exposed to heat, open flame (or sparks), or powerful oxidizers. Can react with oxidizing materials. To fight fire, use alcohol foam, spray or mist, dry chemical. When heated to decomposition it emits toxic fumes of Cl^-. See also ALLYL COMPOUNDS and ESTERS.

AGB750 CAS:4638-03-3 ***HR: 3***
ALLYLCHLOROHYDRIN ETHER
mf: $C_6H_{11}ClO_2$ mw: 150.62

SYNS: ALLYL (3-CHLORO-2-HYDROXYPROPYL) ETHER ◇ 1-ALLYLOXY-3-CHLORO-2-PROPANOL

TOXICITY DATA with REFERENCE
skn-rbt 500 mg/24H MOD 28ZPAK -,79,72
eye-rbt 750 μg/24H SEV 28ZPAK -,79,72
orl-rat LD50:1710 mg/kg 28ZPAK -,79,72
ihl-rat LCLo:162 ppm/4H 28ZPAK -,79,72
orl-mus LD50:240 mg/kg SCCUR* -,1,61

SAFETY PROFILE: Poison by ingestion and inhalation. A moderate skin and severe eye irritant. See also ETHERS and ALLYL COMPOUNDS. When heated to decomposition it emits toxic fumes of Cl^-.

AGC000 CAS:1866-31-5 ***HR: 2***
ALLYL CINNAMATE
mf: $C_{12}H_{12}O_2$ mw: 188.24

PROP: Colorless to light yellow liquid; cherry odor. D: 1.052 @ 25°/25°; bp: 150-152° @ 15 mm. Insol in water; sol in alc; very sol in ether.

SYNS: ALLYL-3-PHENYLACRYLATE ◇ PROPENYL CINNAMATE ◇ VINYL CARBINYL CINNAMATE

TOXICITY DATA with REFERENCE
skn-hmn 20 mg/48H FCTXAV 15,611,77
orl-rat LD50:1520 mg/kg FCTXAV 2,327,64

CONSENSUS REPORTS: Reported in EPA TSCA Inventory.

SAFETY PROFILE: Moderately toxic by ingestion. Human skin irritant. When heated to decomposition it emits acrid smoke and irritating fumes. See also ALLYL COMPOUNDS and ESTERS.

AGC125 ***HR: 3***
ALLYL COMPOUNDS

PROP: Compounds containing the chemical group H_2C $CHCH_2$•R

SAFETY PROFILE: Allyl isovalerate is a poison and causes liver damage and may be experimental carcinogens. Eugenol is a skin sensitizer in humans and may be an experimental carcinogenic. Allyl isothiocyante is a poison, a severe skin and mucous irritant and may be experimental carcinogens. Many allyl compounds are dangerous. They are common in the workplace and the environment. There are numerous uses as chemical intermediates in industry (e.g., allyl chloride, 180 million kg in 1977). Chronic allyl chloride exposure causes reversible liver kidney and peripheral nerve damage in humans. Some are naturally occurring (e.g., allyl isothiocyanate and eugenol). Some are used as food additives and flavoring agents. Alkenylbenzenes, including methyl eugenol, have been found in the essential oil and juice of oranges treated with harvesting agents. Some alkenylbenzenes have carcinogenic activity. In general, allyl compounds are reactive and some have the ability to alkylate macromolecules either directly or after metabolic activation. Most are probably metabolized to allyl alcohol which is metabolized to acrolein.

Several allyl compounds are highly flammable and reactive. Triflates (trifluoromethanesulfonate esters) of allyl alcohol and its derivatives are very reactive and are storage hazards. They react violently with aprotic solvents (DMF or DMSO). See also individual entries and ACROLEIN.

AGC200 ***HR: 3***
6-ALLYL-α-CYANOERGOLINE-8-PROPIONAMIDE
mf: $C_{21}H_{24}N_4O$ mw: 348.49

SYN: ERGOLINE-8-PROPIONAMIDE, 6-ALLYL-α-CYANO-

TOXICITY DATA with REFERENCE
orl-rat TDLo:300 μg/kg (female 5D post):REP ARZNAD 33,1094,83
orl-mus LD50:400 mg/kg ARZNAD 33,1094,83

SAFETY PROFILE: Poison by ingestion. Experimental reproductive effects. When heated to decomposition it emits toxic fumes of NO_x.

AGC250 CAS:4728-82-9 ***HR: 2***
ALLYL CYCLOHEXANEACETATE
mf: $C_{11}H_{18}O_2$ mw: 182.29

SYNS: ALLYL CYCLOHEXYLACETATE ◇ CYCLOHEXYLACETIC ACID ALLYL ESTER

TOXICITY DATA with REFERENCE
skn-hmn 20 mg/48H MLD FCTXAV 15,611,77
skn-rbt 500 mg/24H MOD FCTXAV 15,611,77
orl-rat LD50:900 mg/kg FCTXAV 15,611,77
skn-rbt LD50:1250 mg/kg FCTXAV 15,611,77

CONSENSUS REPORTS: Reported in EPA TSCA Inventory.

SAFETY PROFILE: Moderately toxic by ingestion and skin contact. Irritating to human skin. See also ALLYL COMPOUNDS and ESTERS. When heated to decomposition it emits acrid smoke and irritating fumes.

AGC500 CAS:2705-87-5 ***HR: 3***
ALLYL CYCLOHEXANEPROPIONATE
mf: $C_{12}H_{20}O_2$ mw: 196.32

PROP: Colorless liquid; pineapple odor. D: 0.945-

0.950, refr index: 1.457-1.463, flash p: +212°F. Misc in alc, chloroform, ether; insol in glycerin and water.

SYNS: 3-ALLYLCYCLOHEXYL PROPIONATE ◇ ALLYL HEXAHYDROPHENYLPROPIONATE ◇ FEMA No. 2026

TOXICITY DATA with REFERENCE
orl-rat LD50:585 mg/kg FCTXAV 2,327,64
orl-gpg LD50:380 mg/kg FCTXAV 2,327,64

CONSENSUS REPORTS: Reported in EPA TSCA Inventory.

SAFETY PROFILE: Poison by ingestion. When heated to decomposition it emits acrid smoke and irritating fumes. Combustible liquid. See ALLYL COMPOUNDS and ESTERS.

AGC750 CAS:66827-50-7 ***HR: 3***
2-ALLYL-2-CYCLOHEXYLACETIC ACID-3-(DIETHYLAMINO)-2,2-DIMETHYLPROPYLESTER HYDROCHLORIDE
mf: $C_{20}H_{37}NO_2{\cdot}ClH$ mw: 360.04

SYN: CYCLOHEXYLALLYL-ESSIGSAEUREESTER DES 3-DIAETHYLAMINO-2,2-DIMETHYL-1-PROPANOL(GERMAN)

TOXICITY DATA with REFERENCE
ivn-mus LDLo:100 mg/kg AEPPAE 173,86,33
par-frg LDLo:500 mg/kg AEPPAE 173,86,33

SAFETY PROFILE: Poison by intravenous route. Moderately toxic by parenteral route. See also ALLYL COMPOUNDS and ESTERS. When heated to decomposition it emits very toxic fumes of HCl and NO_x.

AGD000 CAS:66941-60-4 ***HR: 3***
5-ALLYL-5-(2-CYCLOPENTENYL)-2-THIOBARBITURIC ACID
mf: $C_{12}H_{14}N_2O_2S$ mw: 250.34

TOXICITY DATA with REFERENCE
ipr-rat LD50:100 mg/kg JACSAT 65,2091,43
ivn-rat LD50:100 mg/kg JACSAT 65,2091,43

SAFETY PROFILE: Poison by intraperitoneal and intravenous routes. When heated to decomposition it emits very toxic fumes of NO_x and SO_x. See also BARBITURATES and ALLYL COMPOUNDS.

AGD250 CAS:142-22-3 ***HR: 3***
ALLYL DIGLYCOL CARBONATE
mf: $C_{12}H_{18}O_7$ mw: 274.30

PROP: Liquid. Bp: 162°, flash p: 378°F (OC), d: 1.14.

SYNS: BIS(ALLYL CARBONATE)DIETHYLENE GLYCOL ◇ CARBONIC ACID, ALLYL ESTER, DIESTER with DIETHYLENE GLYCOL ◇ CR 39 ◇ DAGC ◇ DIALLYL DIGLYCOL CARBONATE

TOXICITY DATA with REFERENCE
skn-hmn 2%/48H CODEDG 2,183,76
ipr-mus LD50:270 mg/kg JPETAB 90,338,47

CONSENSUS REPORTS: Reported in EPA TSCA Inventory.

SAFETY PROFILE: Poison by intraperitoneal route. A human skin irritant. The allyl compounds are generally toxic. Combustible when exposed to heat or flame; can react with oxidizing material. To fight fire, use water mist or spray, foam, CO_2, dry chemical. When heated to decomposition it emits acrid smoke and irritating fumes. See also ALLYL COMPOUNDS and ESTERS.

AGD500 CAS:130-83-6 ***HR: 3***
6-ALLYL-6,7-DIHYDRO-5H-DIBENZ(c,e)AZEPINE PHOSPHATE
mf: $C_{17}H_{17}N{\cdot}H_3O_4P$ mw: 333.35

SYNS: AZAPETINE PHOSPHATE ◇ AZEPINE PHOSPHATE ◇ 6,7-DIHYDRO-6-(2-PROPENYL)-5H-DIBENZ(c,e)AZEPINEPHOSPHATE ◇ ILIDAR ◇ ILIDAR PHOSPHATE ◇ RO 2-3248

TOXICITY DATA with REFERENCE
skn-hmn 2%/48H CODEDG 2,183,76
skn-rbt TDLo:14859 mg/kg (female 6-18D post):TER EPASR* 8EHQ-0787-0666 FLWP
ipr-mus LD50:270 mg/kg JPETAB 90,338,47

CONSENSUS REPORTS: Reported in EPA TSCA Inventory.

SAFETY PROFILE: Poison by intraperitoneal route. See also ALLYL COMPOUNDS and PHOSPHATES. When heated to decomposition it yields highly toxic fumes of PO_x and NO_x. An anti-adrenergic agent.

AGD750 CAS:63918-66-1 ***HR: 3***
6-ALLYL-6,7-DIHYDRO-3,9-DICHLORO-5H-DIBENZ(c,e)AZEPINE
mf: $C_{17}H_{15}Cl_2N$ mw: 304.23

TOXICITY DATA with REFERENCE
ipr-mus LD50:316 mg/kg JPETAB 103,10,51
ivn-mus LD50:47 mg/kg JPETAB 103,10,51

SAFETY PROFILE: Poison by intraperitoneal and intravenous routes. When heated to decomposition it yields highly toxic fumes of chlorides and NO_x. See also ALLYL COMPOUNDS.

AGE000 CAS:63918-56-9 ***HR: 3***
6-ALLYL-6,7-DIHYDRO-6-METHYL-5H-DIBENZ(c,e)AZEPINIUM IODIDE
mf: $C_{18}H_{20}N{\cdot}I$ mw: 377.29

SYN: RO 2-3742

TOXICITY DATA with REFERENCE
ipr-mus LD50:83 mg/kg JPETAB 103,10,51
ivn-mus LD50:6 mg/kg JPETAB 103,10,51

SAFETY PROFILE: Poison by intraperitoneal and intravenous routes. See also ALLYL COMPOUNDS and IODIDES. When heated to decomposition it yields highly toxic fumes of iodides and NO_x.

AGE250 CAS:93-15-2 ***HR: 3***
4-ALLYL-1,2-DIMETHOXYBENZENE
mf: $C_{11}H_{14}O_2$ mw: 178.25

PROP: Colorless to pale yellow liquid; clove, carnation odor. D: 1.032-1.036, refr index: 1.532, flash p: 212°F. Sol in fixed oils; insol in glycerin and propylene glycol.

SYNS: 1-ALLYL-3,4-DIMETHOXYBENZENE ◇ 4-ALLYLVERATROLE ◇ 1,2-DIMETHOXY-4-ALLYLBENZENE ◇ 1-(3,4-DIMETHOXYPHENYL)-2-PROPENE ◇ ENT 21,040 ◇ 1,3,4-EUGENOL METHYL ETHER ◇ EUGENYL METHYL ETHER ◇ FEMA No. 2475 ◇ METHYL EUGENOL (FCC) ◇ VERATROLE METHYL ETHER

TOXICITY DATA with REFERENCE
skn-rbt 500 mg/24H FCTXAV 13,681,75
dnd-mus-ipr 80 mg/kg FCTXAV 13,857,75
orl-rat LD50:1179 mg/kg TXAPA9 31,421,75
ipr-mus LD50:540 mg/kg AIPTAK 199,226,72
ivn-mus LD50:112 mg/kg AIPTAK 199,226,72

CONSENSUS REPORTS: Reported in EPA TSCA Inventory.

SAFETY PROFILE: Poison by intravenous route. Moderately toxic by ingestion and intraperitoneal routes. A skin irritant. Mutation data reported. Combustible liquid. When heated to decomposition it emits acrid smoke and irritating fumes. Some other alkenylbenzenes have carcinogenic activity. See also EUGENOL, ALLYL COMPOUNDS and ETHERS.

AGE500 CAS:523-80-8 ***HR: 2***
1-ALLYL-2,5-DIMETHOXY-3,4-METHYLENEDIOXYBENZENE
mf: $C_{12}H_{14}O_4$ mw: 222.26

PROP: Apiol is the essential oil from *Petroliselium sativum* seeds (BSIBAC 14,291,39).

SYNS: APIOL ◇ PARSLEY APIOL ◇ PARSLEY CAMPHOR

TOXICITY DATA with REFERENCE
dnd-mus-ipr 400 mg/kg CRNGDP 5,1613,84
scu-mus LDLo:1000 mg/kg BSIBAC 14,291,39
scu-frg LDLo:1515 mg/kg AEXPBL 35,342,1895

CONSENSUS REPORTS: Reported in EPA TSCA Inventory.

SAFETY PROFILE: Moderately toxic by subcutaneous route. Mutation data reported. When heated to decomposition it emits acrid smoke and irritating fumes. See also ALLYL COMPOUNDS.

AGE625 CAS:691-35-0 ***HR: 3***
ALLYLDIMETHYLARSINE
mf: $C_5H_{11}As$ mw: 146.06

$$H_2C{=}CHCH_2As(CH_3)_2$$

CONSENSUS REPORTS: Arsenic and its compounds are on the Community Right-To-Know List.

SAFETY PROFILE: Ignites in air if a large surface area is exposed (e.g., small particles on filter paper). When heated to decomposition it emits toxic fumes of As. See also ARSENIC COMPOUNDS.

AGE750 CAS:56717-11-4 ***HR: 3***
1-ALLYL-1-(3,7-DIMETHYLOCTYL)PIPERIDINIUM BROMIDE
mf: $C_{18}H_{36}N{\bullet}Br$ mw: 346.46

SYNS: 1-ALLYL-1-(3,7-DIMETHYLOCTYL)-PIPERIDIUMBROMID (GERMAN) ◇ ALLYL-TETRA-HYDROGERANYL-PIPERIDINIUMBROMID (GERMAN) ◇ PIPROCTANYLIUMBROMID (GERMAN)

TOXICITY DATA with REFERENCE
orl-rat LD50:360 mg/kg 85DPAN -,-,71/76
skn-rat LD50:115 mg/kg 85DPAN -,-,71/76

SAFETY PROFILE: Poison by ingestion and skin contact. See also ALLYL COMPOUNDS and BROMIDES. When heated to decomposition it emits very toxic fumes of HBr and NO_x.

AGF000 CAS:33132-87-5 ***HR: 3***
β-ALLYL-N,N-DIMETHYLPHENETHYLAMINE
mf: $C_{13}H_{19}N$ mw: 189.33

TOXICITY DATA with REFERENCE
orl-mus LD50:210 mg/kg CHTPBA 6,453,71
ivn-mus LD50:37 mg/kg CHTPBA 6,453,71

SAFETY PROFILE: Poison by ingestion and intravenous routes. When heated to decomposition it emits toxic fumes of NO_x. See also ALLYL COMPOUNDS.

AGF250 CAS:743-45-3 ***HR: 2***
5-ALLYL-1,3-DIPHENYLBARBITURIC ACID
mf: $C_{19}H_{16}N_2O_3$ mw: 320.37

SYN: 5-ALLYL-1,3-DIPHENYL-2,4,6(1H,3H,5H)-PYRIMIDINETERIONE

TOXICITY DATA with REFERENCE
ipr-rat LD50:1000 mg/kg JMCMAR 7,342,64
ipr-mus LD50:533 mg/kg ARZNAD 17,1519,67

SAFETY PROFILE: Moderately toxic by intraperitoneal route. When heated to decomposition it emits toxic fumes of NO_x. See also BARBITURATES and ALLYL COMPOUNDS.

AGF500 CAS:10138-39-3 ***HR: 2***
ALLYL-3,4-EPOXY-6-METHYLCYCLOHEXANECARBOXYLATE
mf: $C_{11}H_{16}O_3$ mw: 196.27

SYNS: 3,4-EPOXY-6-METHYLCYCLOHEXANECARBOXYLIC ACID,ALLYL ESTER ◇ 4-METHYL-7-OXABICYCLO(4.1.0)HEPTANE-3-CARBOXYLIC ACID, ALLYL ESTER

TOXICITY DATA with REFERENCE
skn-rbt 10 mg/24H open MLD AIHAAP 23,95,62
orl-rat LD50:500 mg/kg AIHAAP 23,95,62
skn-rbt LD50:2830 mg/kg AIHAAP 23,95,62

SAFETY PROFILE: Moderately toxic by skin contact and ingestion. A skin irritant. See also ALLYL COMPOUNDS and ESTERS. When heated to decomposition it emits acrid smoke and irritating fumes.

AGF750 CAS:432-60-0 ***HR: D***
ALLYLESTRENOL
mf: $C_{21}H_{32}O$ mw: 300.53

SYNS: 17-α-ALLYL-3-DEOXY-19-NORTESTOSTERONE ◇ 17-α-ALLYL-4-ESTREN-17-β-OL ◇ 17-α-ALLYLESTR-4-EN-17-β-OL ◇ 17-α-ALLYLHYDROXY-19-NOR-4-ANTROSTENE ◇ 17-α-ALLYL-17-β-HYDROXY-Δ^4-ESTREN ◇ 17-α-ALLYL-17-β-HYDROXY-4-ESTRENE ◇ ALLYLOESTRENOL ◇ 17-α-ALLYL-4-OESTRENE-17-β-OL ◇ GESTANIN ◇ GESTANOL ◇ GESTANON ◇ GESTANYN ◇ 17-HYDROXY-17-α-ALLYL-4-ESTRENE ◇ ORGANON ◇ ORAGESTON ◇ 17-(2-PROPENYL)ESTR-4-EN-17-OL ◇ TURINAL

TOXICITY DATA with REFERENCE
orl-wmn TDLo:8 mg/kg (20D pre):REP INJFA3 9,57,64
orl-rat TDLo:900 mg/kg (13-21D preg):TER ACENA7 65,29,70

SAFETY PROFILE: Human reproductive effects. An experimental teratogen. Other experimental reproductive effects. When heated to decomposition it emits acrid smoke and irritating fumes. See also ALLYL COMPOUNDS. Used as a progestin.

AGG000 ***HR: 3***
ALLYL ETHYL ETHER
mf: $C_5H_{10}O$ mw: 86.1

PROP: Liquid. Flash p: <75°F.

SAFETY PROFILE: Highly flammable and reactive. Forms explosive peroxides in storage. Can explode during distillation probably due to peroxide formation. When heated to decomposition it emits acrid, irritating fumes. See also ALLYL COMPOUNDS and ETHERS.

AGG250 CAS:6654-31-5 ***HR: 3***
2-ALLYL-5-ETHYL-2'-HYDROXY-9-METHYL-6,7-BENZOMORPHAN
mf: $C_{18}H_{25}NO$ mw: 271.44

SYN: 2,6-METHANO-3-BENZAZOCIN-8-OL-3-ALLYL-6-ETHYL-1,2,3,4,5,6-HEXAHYDRO-11-METHYL

TOXICITY DATA with REFERENCE
unk-hmn TDLo:70 μg/kg:PSY SCIEAS 137,541,62
scu-rat LD50:174 mg/kg JPETAB 143,141,64
ivn-rat LD50:32 mg/kg JPETAB 143,141,64
scu-mus LD50:182 mg/kg JPETAB 143,141,64
ivn-mus LD50:32 mg/kg 31ZPAG 2,175,66

SAFETY PROFILE: Poison by subcutaneous and intravenous routes. Human psychotropic effects by an unspecified route. When heated to decomposition it emits toxic fumes of NO_x. See also ALLYL COMPOUNDS.

AGG500 ***HR: 3***
ALLYL FLUORIDE
mf: C_3H_5F mw: 60.07

PROP: Colorless gas. Bp: −10°.

SYN: 3-FLUOROPROPENE

SAFETY PROFILE: Poison by inhalation and ingestion. A strong irritant. See also ALLYL COMPOUNDS and FLUORIDES. When heated to decomposition it emits highly toxic fumes of F^-. Incompatible with water or steam to produce toxic and corrosive fumes.

AGG750 CAS:406-23-5 ***HR: 3***
ALLYL FLUOROACETATE
mf: $C_5H_7FO_2$ mw: 118.12

TOXICITY DATA with REFERENCE
ihl-mus LCLo:500 mg/m^3 NDRC** -,9,44
scu-mus LD50:6 mg/kg JCSOA9 -,916,49

SAFETY PROFILE: Poison by inhalation and subcutaneous routes. See also ALLYL COMPOUNDS and FLUORIDES. When heated to decomposition it emits toxic fumes of F^-.

AGH000 CAS:1838-59-1 ***HR: 3***
ALLYL FORMATE
DOT: UN 2336
mf: $C_4H_6O_2$ mw: 86.10

$H_2C{=}CHCH_2OCO{\cdot}H$

PROP: Liquid, sltly water-sol, sol in organic solvents. D: 0.948 @ 18°/4°, bp: 83°, flash p: < −50°F.

SYNS: FORMIC ACID, ALLYL ESTER ◇ 3-PROPENYL METHANOATE

TOXICITY DATA with REFERENCE
orl-rat LD50:124 mg/kg FCTXAV 2,327,64
orl-mus LD50:136 mg/kg FCTXAV 2,327,64
ihl-mus LC50:14 g/m^3/3H 85GMAT -,18,82

CONSENSUS REPORTS: Reported in EPA TSCA Inventory.

DOT Classification: Flammable Liquid; Label: Flammable Liquid and Poison.

SAFETY PROFILE: Poison by ingestion. Mildly toxic by inhalation. Very flammable and reactive. Dangerous fire hazard. See also ALLYL COMPOUNDS and ESTERS. When heated to decomposition it yields irritating smoke and fumes.

AGH125 CAS:3952-98-5 ***HR: D***
ALLYL GLUCOSINOLATE
mf: $C_{10}H_{16}NO_9S_2$•K mw: 397.49

PROP: A β-glucopyranoside isolated from black mustard seeds and from the horseradish root *Alliaria officinalis Andrz.* Monohydrate: Crystals. Mp: 127-129°. Anhydrous: mp: 179°. Freely sol in water, hot alc; insol in benzene, chloroform, and ether.

SYNS: GLUCOSIDE of ALLYL ISOTHIOCYANATE ◇ SINIGRIN ◇ 1-THIO-β-d-GLUCOPYRANOSE1-(N-(SULFOOXY-3-BUTENIMIDATE), MONOPOTASSIUM

TOXICITY DATA with REFERENCE
mmo-sat 100 μg/plate ABCHA6 44,3017,80
orl-rat TDLo:644 mg/kg (8-9D preg):TER FCTXAV 18,159,80

SAFETY PROFILE: An experimental teratogen. Mutation data reported. When heated to decomposition it emits toxic fumes of SO_x, NO_x, and K_2O.

AGH150 CAS:106-92-3 ***HR: 3***
ALLYL GLYCIDYL ETHER
DOT: UN 2219
mf: $C_6H_{10}O_2$ mw: 114.16

PROP: Bp: 153.9°, fp: −100° (forms glass), flash p: 135°F (OC), d: 0.9698 @ 20°/4°, vap press: 21.59 mm @ 60°, vap d: 3.94.

SYNS: AGE ◇ ALLIL-GLICIDIL-ETERE (ITALIAN) ◇ 1-ALLILOSSI-2,3 EPOSSIPROPANO (ITALIAN) ◇ ALLYL-2,3-EPOXYPROPYL ETHER ◇ ALLYLGLYCIDAETHER (GERMAN) ◇ 1-ALLYLOXY-2,3-EPOXY-PROPAAN (DUTCH) ◇ 1-ALLYLOXY-2,3-EPOXYPROPAN (GERMAN) ◇ 1-(ALLYLOXY)-2,3-EPOXYPROPANE ◇ NCI-C56666 ◇ OXYDE d'ALLYLE et de GLYCIDYLE (FRENCH) ◇ ((2-PROPENYLOXY)METHYL)OXIRANE

TOXICITY DATA with REFERENCE
skn-rbt 500 mg/24H SEV 28ZPAK -,135,72
eye-rbt 97 mg SEV AMIHAB 14,250,56
eye-rbt 250 μg/24H SEV 28ZPAK -,135,72
mmo-esc 20 μmol/L ARTODN 46,277,80
dnd-esc 1 μmol/L ARTODN 46,277,80
mmo-klp 500 μmol/L MUREAV 89,269,81
sln-dmg-orl 5500 ppm ENMUDM 7,349,85
orl-rat LD50:922 mg/kg 28ZPAK -,135,72
ihl-rat LCLo:860 ppm/4H 28ZPAK -,135,72
orl-mus LD50:390 mg/kg AMIHAB 14,250,56
ihl-mus LC50:270 ppm/4H AMIHAB 14,250,56
skn-rbt LD50:2550 mg/kg AMIHAB 14,250,56

CONSENSUS REPORTS: Reported in EPA TSCA Inventory.

OSHA PEL: (Transitional: CL 10 ppm) TWA 5 ppm; STEL 10 ppm
ACGIH TLV: TWA 5 ppm; STEL 10 ppm (skin)
NIOSH REL: (Glycidyl Ethers) CL 45 mg/m^3/15M
DFG MAK: 10 ppm (45 mg/m^3)
DOT Classification: Flammable or Combustible Liquid; Label: Flammable Liquid.

SAFETY PROFILE: Poison by ingestion. Moderately toxic by inhalation and skin contact. Mutation data reported. A severe skin and eye irritant. Can cause central nervous system depression and pulmonary edema. Flammable when exposed to heat or flame; can react with oxidizing materials. To fight fire, use foam, CO_2, dry chemical. When heated to decomposition it emits acrid smoke and irritating fumes. See also ALLYL COMPOUNDS.

AGH250 CAS:142-19-8 ***HR: 2***
ALLYL HEPTANOATE
mf: $C_{10}H_{18}O_2$ mw: 170.28

PROP: Colorless to pale yellow liquid; fruity, sweet, pineapple odor. D: 0.880, refr index: 1.426, flash p: 154°F.

SYNS: ALLYL ENANTHATE ◇ ALLYL HEPTOATE ◇ ALLYL HEPTYLATE ◇ FEMA No. 2031 ◇ 2-PROPENYL HEPTANOATE

TOXICITY DATA with REFERENCE
skn-hmn 20 mg/48H MLD FCTXAV 15,611,77
skn-rbt 500 mg/24H MOD FCTXAV 15,611,77
orl-rat LD50:500 mg/kg TXAPA9 6,378,64
orl-mus LD50:630 mg/kg TXAPA9 7,18,65
skn-rbt LD50:810 mg/kg FCTXAV 15,611,77
orl-gpg LD50:444 mg/kg FCTXAV 2,327,64

CONSENSUS REPORTS: Reported in EPA TSCA Inventory.

SAFETY PROFILE: Moderately toxic by ingestion and skin contact. A human skin irritant. See also ALLYL COMPOUNDS and ESTERS. Combustible liquid. When heated to decomposition it emits acrid smoke and irritating fumes.

AGH500 CAS:52207-83-7 ***HR: 3***
ALLYLHYDRAZINE HYDROCHLORIDE
mf: $C_3H_8N_2$•ClH mw: 108.59

TOXICITY DATA with REFERENCE
orl-mus TDLo:9800 mg/kg/35W-C:CAR BJCAAI 34,90,76
orl-mus TDLo:25 mg/kg:ETA PAACA3 16,61,75

SAFETY PROFILE: Questionable carcinogen with experimental carcinogenic and tumorigenic data. When heated to decomposition it emits very toxic fumes of HCl and NO_x. See also ALLYL COMPOUNDS.

AGH750 ***HR: 3***
ALLYL HYDROPEROXIDE
mf: $C_3H_6O_2$ mw: 74.1

SAFETY PROFILE: Highly toxic. A potentially explosive liquid. Unstable to heat, light, and solid alkalies. Mixtures with sand are impact sensitive. Upon decomposition it emits acrid smoke and fumes. See also PEROXIDES, ORGANIC; and ALLYL COMPOUNDS.

AGI000 CAS:152-02-3 ***HR: 3***
N-ALLYL-3-HYDROXYMORPHINAN
mf: $C_{19}H_{25}NO$ mw: 283.45

SYNS: (−)-3-HYDROXY-N-ALLYLMORPHINAN ◇ l-3-HYDROXY-N-ALLYL MORPHINAN ◇ (−)-LEVALLORPHAN ◇ LORFAN ◇ NALOXIPHAN ◇ RO-1-7700

TOXICITY DATA with REFERENCE
ihl-hmn LCLo:100 $\mu g/m^3$ 27ZXA3 -,-,63
orl-rat LD50:949 mg/kg 27ZIAQ -,-,65
ipr-rat LD50:185 mg/kg 27ZIAQ -,140,73
ipr-mus LD50:184 mg/kg 27ZIAQ -,-,65
scu-mus LD50:200 mg/kg ANYAA9 281,321,76

SAFETY PROFILE: Human poison by inhalation. Poison experimentally by inhalation, subcutaneous, and intraperitoneal routes. Moderately toxic by ingestion. When heated to decomposition it emits toxic fumes of NO_x. See also ALLYL COMPOUNDS.

AGI250 ***HR: 3***
ALLYL IODIDE
mf: C_3H_5I mw: 168.0

PROP: Yellow liquid, pungent odor. Mp: −99°, bp: 103.1°, d: 1.825 @ 20°/4°, vap d: 5.8.

SAFETY PROFILE: Poison by inhalation and ingestion. A powerful irritant. When heated to decomposition it emits highly toxic fumes of I^-. Moderately flammable. Incompatible with oxidizing materials. To fight fire, use water, foam, CO_2, dry chemical. See also ALLYL COMPOUNDS and IODIDES.

AGI500 CAS:79-78-7 ***HR: 2***
ALLYL-α-IONONE
mf: $C_{16}H_{24}O$ mw: 232.40

PROP: Colorless to yellow liquid; fruity, woody odor. D: 0.928-0.935, refr index: 1.503-1.507, flash p: +212°F. Sol in alc; insol in water @ 265°.

SYNS: CETONE V ◇ FEMA No. 2033 ◇ 1-(2,6,6-TRIMETHYL-2-CYCLOHEXEN-1-YL)-1,6-HEPTADIEN-3-ONE

TOXICITY DATA with REFERENCE
skn-rbt 500 mg MLD FCTXAV 11,1079,73

CONSENSUS REPORTS: Reported in EPA TSCA Inventory.

SAFETY PROFILE: A skin irritant. Combustible liquid. When heated to decomposition it emits acrid smoke and irritating fumes. See also ALLYL COMPOUNDS.

AGI750 CAS:77-26-9 ***HR: 3***
ALLYLISOBUTYLBARBITURATE
mf: $C_{11}H_{16}N_2O_3$ mw: 224.29

SYNS: ALISOBUMAL ◇ ALLYLBARBITAL ◇ ALLYLBARBITONE ◇ ALLYLBARBITURIC ACID ◇ ALLYLISOBUTYLBARBITAL ◇ 5-ALLYL-5-ISOBUTYLBARBITURIC ACID ◇ 5-ALLYL-5-(2′-METHYL-N-PROPYL) BARBITURIC ACID ◇ BUTALBARBITAL ◇ BUTALBITAL ◇ ISO-BUTYLALLYLBARBITURIC ACID ◇ ISOBUTYLALLYLBARTURIC ACID ◇ ITOBARBITAL ◇ 5-(2-METHYLPROPYL)-5-(2-PROPENYL)-2,4,6(1H,3H,5H)-PYRIMIDINETRIONE (9CI) ◇ OPTALIDON ◇ SANDOPTAL ◇ TETRALLOBARBITAL

TOXICITY DATA with REFERENCE
orl-wmn TDLo:400 mg/kg:CVS,PUL,GIT SAVEAB 10,181A,39
orl-chd TDLo:10 mg/kg:CNS,KID SAVEAB 10,209A,39
scu-rat LD50:160 mg/kg AEPPAE 152,341,30
ipr-mus LDLo:64 mg/kg CBCCT* 2,190,50
orl-pgn LD50:75 mg/kg TXAPA9 21,315,72
orl-dck LD50:237 mg/kg TXAPA9 21,315,72
orl-bsd LD50:75 mg/kg TXAPA9 21,315,72

SAFETY PROFILE: Poison by ingestion, subcutaneous, and intraperitoneal routes. Human systemic effects by ingestion: toxic psychosis, coma, reduced blood pressure, respiratory depression, nausea and vomiting, and kidney effects. When heated to decomposition it emits toxic NO_x. See also ALLYL COMPOUNDS, BARBITURATES, and various barbital compounds. Used as a sleep aid.

AGJ000 CAS:1476-23-9 ***HR: 3***
ALLYL ISOCYANATE
mf: C_4H_5NO mw: 83.10

SYN: ISOCYANIC ACID ALLYL ESTER

TOXICITY DATA with REFERENCE
ivn-mus LD50:18 mg/kg CSLNX* NX#03769

CONSENSUS REPORTS: Reported in EPA TSCA Inventory.

SAFETY PROFILE: Poison by intravenous route. See also ALLYL COMPOUNDS and ESTERS. When heated to decomposition it emits toxic fumes of NO_x.

AGJ250 CAS:57-06-7 ***HR: 3***
ALLYL ISOTHIOCYANATE
DOT: UN 1545
mf: C_4H_5NS mw: 99.16

$H_2C{=}CHCH_2N{:}C{:}S$

PROP: Colorless to pale yellow liquid; irritating odor with mustard taste. Mp: −80°, bp: 150.7°, flash p: 115°F, d: 1.013-1.016 @ 25°/25°, vap press: 10 mm @ 38.3°, vap d: 3.41, refr index: 1.527-1.531. Misc with alc, carbon disulfide, and ether.

SYNS: AITC ◇ ALLYL ISORHODANIDE ◇ ALLYL ISOSULFOCYANATE ◇ ALLYL ISOTHIOCYANATE, stabilized (DOT) ◇ ALLYL MUSTARD OIL ◇ ALLYLSENFOEL (GERMAN) ◇ ALLYL SEVENOLUM ◇ ALLYL THIOCARBONIMIDE ◇ ARTIFICIAL MUSTARD OIL ◇ CARBOSPOL ◇ FEMA No. 2034 ◇ ISOTHIOCYANATE d'ALLYLE (FRENCH) ◇ 3-ISOTHIOCYANATO-1-PROPENE ◇ MUSTARD OIL ◇ NCI-C50464 ◇ OIL of MUSTARD, artificial ◇ OLEUM SINAPIS VOLATILE ◇ 2-PROPENYL ISOTHIOCYANATE ◇ REDSKIN ◇ SENF OEL (GERMAN) ◇ SYNTHETIC MUSTARD OIL ◇ VOLATILE OIL of MUSTARD

TOXICITY DATA with REFERENCE
eye-rbt 2 mg AEPPAE 219,119,53
mmo-sat 100 μg/plate ABCHA6 44,4017,80
mma-sat 1 μmol/plate CBINA8 38,303,82
scu-rat TDLo:100 mg/kg (8-9D preg):TER FCTXAV 18,159,80
scu-rat TDLo:200 mg/kg (8-9D preg):REP FCTXAV 18,159,80
orl-rat TDLo:12875 mg/kg/2Y-I:NEO NTPTR* NTP-TR-234,82
skn-mus TDLo:12 g/kg/12W-I:ETA AICCA6 11,699,55
orl-rat LD50:112 mg/kg AEPPAE 219,119,53
ipr-rat LDLo:80 mg/kg ARZNAD 16,870,66
scu-rat LD50:92 mg/kg FCTXAV 18,159,80
orl-mus LD50:308 mg/kg AMRL** TR-73-83,73
ipr-mus LDLo:4 mg/kg TXAPA9 23,288,72
scu-mus LD50:80 mg/kg ARZNAD 5,505,55
skn-rbt LD50:88 mg/kg TXAPA9 42,417,77
ivn-rbt LDLo:12 mg/kg BIJOAK 4,107,09

CONSENSUS REPORTS: IARC Cancer Review: Group 3 IMEMDT 7,56,87; Animal Limited Evidence IMEMDT 36,55,85; NTP Carcinogenesis Bioassay (gavage); No Evidence: mouse NTPTR* NTP-TR-234,82; Clear Evidence: rat NTPTR* NTP-TR-234,82. Reported in EPA TSCA Inventory.

DOT Classification: Poison B; Label:Flammable Liquid and Poison

SAFETY PROFILE: Suspected carcinogen with experimental neoplastigenic, and tumorigenic data. Poison by ingestion, skin contact, intravenous, subcutaneous, and intraperitoneal routes. Experimental teratogenic and reproductive effects. An eye irritant. An allergen. May cause contact dermatitis. Mutation data reported. Combustible liquid. Highly reactive. When heated to decomposition (above 250°) or on contact with acid or acid fumes it emits highly toxic fumes of CN^-, SO_x, and NO_x. To fight fire, use foam, CO_2, dry chemical. See also ALLYL COMPOUNDS and ESTERS.

AGJ375 CAS:3052-45-7 ***HR: 3***
ALLYLLITHIUM
mf: C_3H_5Li mw: 48.01

$H_2C{=}CHCH_2Li$

SAFETY PROFILE: Ignites on contact with air. See also LITHIUM COMPOUNDS.

AGJ500 ***HR: 3***
ALLYL MERCAPTAN
mf: C_3H_6S mw: 74.15

$H_2C{=}CHCH_2SH$

PROP: Water-white liquid with a strong garlic odor, darkens on standing. D: 0.925 @ 23°/4°, bp: 68°, flash p: 14°F.

SYN: 2-PROPENE-1-THIOL

SAFETY PROFILE: Poison by inhalation and ingestion. Strong irritant to skin and mucous membranes. When heated to decomposition it emits highly toxic fumes of SO_x. Very dangerous fire hazard. To fight fire, use water mist or spray, alcohol foam, CO_2, or dry chemical. See also ALLYL COMPOUNDS and MERCAPTANS.

AGJ750 CAS:64037-65-6 ***HR: 3***
2-ALLYLMERCAPTO-2-ETHYLBUTYRAMIDE
mf: $C_9H_{17}NOS$ mw: 187.33

SYNS: α-ALLYLMERCAPTO-α, α-DIETHYLACETAMIDE ◇ 2-ALLYLTHIO-2-ETHYLBUTYRAMIDE

TOXICITY DATA with REFERENCE
orl-mus LD50:891 mg/kg JMCMAR 6,351,63
ipr-mus LD50:400 mg/kg NTIS** AD691-490

SAFETY PROFILE: Poison by intraperitoneal route. Moderately toxic by ingestion. When heated to decomposition it emits very toxic fumes of SO_x and NO_x. See also ALLYL COMPOUNDS.

AGK000 CAS:63915-89-9 ***HR: 3***
2-ALLYLMERCAPTOISOBUTYRAMIDE
mf: $C_7H_{13}NOS$ mw: 159.27

SYN: α-ALLYLMERCAPTOISOBUTYRAMIDE

TOXICITY DATA with REFERENCE
orl-mus LD50:2000 mg/kg JMCMAR 6,351,63
ipr-mus LD50:250 mg/kg NTIS** AD691-490

SAFETY PROFILE: Poison by intraperitoneal route. Moderately toxic by ingestion. When heated to decomposition it emits toxic fumes of NO_x and SO_x. See also ALLYL COMPOUNDS.

AGK250 CAS:87-09-2 ***HR: 3***
ALLYLMERCAPTOMETHYLPENICILLIN
mf: $C_{13}H_{18}N_2O_4S_2$ mw: 330.45

SYNS: ALLYLMERCAPTOMETHYLPENICILLINIC ACID ◇ ALLYLTHIOMETHYLPENICILLIN ◇ ALMECILLIN ◇ AT ◇ CER-o-CILLIN ◇ PENICILLIN AT ◇ PENICILLIN O

TOXICITY DATA with REFERENCE
mul-wmn LDLo:35827 μg/kg/6D-I:CVS JOALAS 31,455,60
ice-mus LD50:45 mg/kg JLCMAK 34,126,49
ice-dog LD50:6490 μg/kg JLCMAK 34,126,49
isp-dog LD50:38 mg/kg JLCMAK 34,126,49
ice-rbt LD50:5610 μg/kg JLCMAK 34,126,49

SAFETY PROFILE: Poison by intracerebral and intraspinal routes. Human systemic effects by multiple routes: changes in blood vessels. When heated to decomposition it emits very toxic fumes of NO_x and SO_x. See also PENICILLIN and ALLYL COMPOUNDS.

AGK500 CAS:96-05-9 ***HR: 2***
ALLYL METHACRYLATE
mf: $C_7H_{10}O_2$ mw: 126.17

SYN: METHACRYLIC ACID ALLYL ESTER

TOXICITY DATA with REFERENCE
orl-rat LD50:430 mg/kg AIHAAP 30,470,69
ihl-rat LCLo:500 ppm AIHAAP 30,470,69
skn-rbt LD50:500 mg/kg AIHAAP 30,470,69

CONSENSUS REPORTS: EPA Genetic Toxicology Program. Reported in EPA TSCA Inventory.

SAFETY PROFILE: Moderately toxic by ingestion, inhalation, and skin contact. See also ALLYL COMPOUNDS and ESTERS. When heated to decomposition it emits acrid smoke and irritating fumes.

AGK750 CAS:6728-21-8 ***HR: 3***
ALLYL METHANESULFONATE
mf: $C_4H_8O_3S$ mw: 136.18

SYNS: ALLYL MESYLATE ◇ METHANESULFONIC ACID, 2-PROPENYL ESTER (9CI)

TOXICITY DATA with REFERENCE
mmo-sat 1 μmol/plate BCPCA6 29,993,80
dns-hmn:hla 50 μmol/L CALEDQ 20,263,83
skn-mus TDLo:540 mg/kg/10W-I:NEO CNREA8 17,64,57

SAFETY PROFILE: Questionable carcinogen with experimental neoplastigenic data. Human mutation data reported. See also ALLYL COMPOUNDS and SULFONATES. When heated to decomposition it emits toxic fumes of SO_x.

AGL000 CAS:10402-33-2 ***HR: 2***
4-ALLYL-2-METHOXYPHENYLPHENYLACETATE
mf: $C_{18}H_{18}O_3$ mw: 282.36

SYNS: BENZENEACETICACID-2-METHOXY-4-(2-PROPENYL)PHENYL ESTER ◇ EUGENOL PHENYLACETATE ◇ EUGENYL PHENYLACETATE

TOXICITY DATA with REFERENCE
skn-rbt 500 mg/24H MOD FCTXAV 16,637,78

CONSENSUS REPORTS: Reported in EPA TSCA Inventory.

SAFETY PROFILE: A skin irritant. See also ALLYL COMPOUNDS and ESTERS. When heated to decomposition it emits acrid smoke and irritating fumes.

AGL250 CAS:66941-77-3 ***HR: 3***
5-ALLYL-5-(1-METHYLALLYL)-2-THIOBARBITURIC ACID SODIUM SALT
mf: $C_{11}H_{13}N_2O_2S \cdot Na$ mw: 260.31

TOXICITY DATA with REFERENCE
ipr-rat LD50:126 mg/kg JAPMA8 34,183,45
ivn-rbt LD50:58 mg/kg JAPMA8 34,183,45

SAFETY PROFILE: Poison by intraperitoneal and intravenous routes. When heated to decomposition it emits very toxic fumes of SO_x, Na_2O, and NO_x. See also BARBITURATES and ALLYL COMPOUNDS.

AGL375 CAS:77-27-0 ***HR: 3***
5-ALLYL-5-(1-METHYLBUTYL)-2-THIOBARBITURIC ACID
mf: $C_{12}H_{18}N_2O_2S$ mw: 254.38

PROP: Crystals from dil ethanol. Mp: 132-133°. Often used as the sodium salt, $C_{12}H_{17}N_2NaO_2S$.

SYNS: DIHYDRO-5-(1-METHYLBUTYL)-5-(2-PROPENYL)-2-THIOXO-4,6(1H,5H)-PYRIMIDINEDIONE (9CI) ◇ SURITAL ◇ THIAMYLAL ◇ THIOSECONAL

TOXICITY DATA with REFERENCE
ivn-rat LD50:66 mg/kg PSEBAA 89,292,55
ivn-dog LD50:36300 μg/kg PSEBAA 89,292,55
ivn-rbt LD50:26 mg/kg PSEBAA 89,292,55

SAFETY PROFILE: Poison by intravenous route. Caution: Abuse may lead to habituation or addiction. When heated to decomposition it emits toxic fumes of SO_x and NO_x.

AGL500 CAS:63937-27 ***HR: 3***
2-ALLYL-3-METHYL-4-HYDROXY-2-CYCLOPENTEN-1-ONE DIMETHYLCARBAMATE-9
mf: $C_{12}H_{17}NO_3$ mw: 223.30

SYNS: ADK (CZECH) ◇ 2-ALLYL-3-METHYL-2-CYCLOPENTEN-1-ON-4-YL-N,N'-DIMETHYL-KARBAMAT (CZECH)

TOXICITY DATA with REFERENCE
skn-rbt 500 mg/24H MLD 28ZPAK -,164,72
orl-rat LD50:53200 μg/kg 28ZPAK -,164,72

SAFETY PROFILE: Poison by ingestion. A skin irritant. See also CARBAMATES and ALLYL COMPOUNDS. When heated to decomposition it emits toxic fumes of NO_x.

AGL750 CAS:55902-04-0 ***HR: 2***
N-ALLYL-3-METHYL-N-α-METHYLPHENETHYL-6-OXO-1(6H)-PYRIDAZINE ACETAMIDE
mf: $C_{19}H_{23}N_3O_2$ mw: 325.45

TOXICITY DATA with REFERENCE
orl-rat LD50:1710 mg/kg EJMCA5 9,644,74
ipr-rat LD50:628 mg/kg EJMCA5 9,644,74

SAFETY PROFILE: Moderately toxic by ingestion and intraperitoneal routes. When heated to decomposition it emits toxic fumes of NO_x. See also ALLYL COMPOUNDS.

AGL875 CAS:7651-40-3 ***HR: 3***
(±)-5-ALLYL-5-(1-METHYL-2-PENTYNYL)-2-THIOBARBITURIC ACID
mf: $C_{13}H_{16}N_2O_2S$ mw: 264.37

SYNS: (±)-DIHYDRO-5-(1-METHYL-2-PENTYNYL)-5-(2-PROPENYL)-2-THIOXO-4,6(1H,5H)-PYRIMIDINEDIONE ◇ LILLY 22113 ◇ THIOHEXITAL

TOXICITY DATA with REFERENCE
ivn-rat LD50:32 mg/kg JMCMAR 24,1241,81
ipr-mus LD50:64 mg/kg JMCMAR 24,1241,81
ivn-mus LD50:54 mg/kg JMCMAR 24,1241,81

SAFETY PROFILE: Poison by intravenous and intraperitoneal routes. When heated to decomposition it emits toxic fumes of SO_x and NO_x.

AGM000 CAS:66941-81-9 ***HR: 3***
5-ALLYL-5-(1-METHYLPROPENYL)BARBITURIC ACID
mf: $C_{11}H_{14}N_2O_3$ mw: 222.27

TOXICITY DATA with REFERENCE
orl-mus LD50:525 mg/kg JACSAT 61,353,39
ipr-mus LD50:380 mg/kg JACSAT 61,353,39

SAFETY PROFILE: Poison by intraperitoneal route. Moderately toxic by ingestion. When heated to decomposition it emits toxic fumes of NO_x. See also BARBITURATES and ALLYL COMPOUNDS.

AGM060 CAS:10045-34-8 ***HR: 3***
1-ALLYL-2-NITROIMIDAZOLE
mf: $C_6H_7N_3O_2$ mw: 153.16

TOXICITY DATA with REFERENCE
orl-mus LD50:158 mg/kg AACHAX -,478,65
ipr-mus LD50:144 mg/kg AACHAX -,478,65
scu-mus LD50:144 mg/kg AACHAX -,478,65

SAFETY PROFILE: Poison by ingestion, subcutaneous, and intraperitoneal routes. When heated to decomposition it emits toxic fumes of NO_x. See also ALLYL COMPOUNDS.

AGM125 ***HR: 3***
1-(ALLYLNITROSAMINO)-2-PROPANONE
mf: $C_6H_{10}N_2O_2$ mw: 142.18

SYNS: NAOP ◇ N-NITROSOALLYL-2-OXOPROPYLAMINE

TOXICITY DATA with REFERENCE
mma-sat 10 μg/plate TCMUE9 1,13,84
orl-rat TDLo:995 mg/kg/74W-I:ETA IAPUDO 57,617,84

SAFETY PROFILE: Questionable carcinogen with experimental tumorigenic data. Mutation data reported. When heated to decomposition it emits toxic fumes of NO_x. See also NITROSAMINES.

AGM200 ***HR: D***
17-α-ALLYL-19-NORTESTOSTERONE
mf: $C_{21}H_{30}O_2$ mw: 314.51

SYNS: 17-α-ALLYL-17-β-HYDROXYESTR-4-EN-3-ONE ◇ 17-ALLYL-19-NORTESTOSTERONE

TOXICITY DATA with REFERENCE
orl-rbt TDLo:3 mg/kg (female 1D pre):REP 85GRAA -,57,65

SAFETY PROFILE: Experimental reproductive effects. When heated to decomposition it emits acrid smoke and fumes. See also TESTOSTERONE and ALLYL COMPOUNDS.

AGM250 CAS:22235-85-4 ***HR: 3***
8-ALLYL-(±)-1-α-H,5-α-H-NORTHROPAN-3-α-OL
mf: $C_{10}H_{17}NO$ mw: 167.28

SYNS: N-ALLYNORATROPINE ◇ NALTROPINE

TOXICITY DATA with REFERENCE
ivn-rat LD50:60 mg/kg ARZNAD 13,567,63
ipr-mus LD50:165 mg/kg AIPTAK 154,210,65

SAFETY PROFILE: Poison by intravenous and intraperitoneal routes. When heated to decomposition it emits toxic fumes of NO_x. See also ALLYL COMPOUNDS.

AGM500 CAS:4230-97-1 ***HR: 2***
ALLYL OCTANOATE
mf: $C_{11}H_{20}O_2$ mw: 184.31

PROP: Colorless liquid; fruity odor. D: 0.8550.861, refr index: 1.425, flash p: +151°F. Sol in alc, fixed oils; sltly sol in propylene glycol; insol in glycerin and water @ 260°.

SYNS: ALLYL CAPRYLATE ◇ FEMA No. 2037 ◇ OCTANOIC ACID ALLYL ESTER ◇ OCTANOIC ACID-2-PROPENYL ESTER

TOXICITY DATA with REFERENCE
skn-rbt 310 mg/24H MOD FCTXAV 16,637,78
orl-rat LD50:570 mg/kg FCTXAV 16,637,78

CONSENSUS REPORTS: Reported in EPA TSCA Inventory.

SAFETY PROFILE: Moderately toxic by ingestion. A skin irritant. See also ALLYL COMPOUNDS and ESTERS. When heated to decomposition it emits acrid smoke and irritating fumes.

AGM750 CAS:14520-53-7 ***HR: 3***
2-ALLYLOXYBENZAMIDE
mf: $C_{10}H_{11}NO_2$ mw: 177.22

SYN: o-(ALLYLOXY)BENZAMIDE

TOXICITY DATA with REFERENCE
cpr-mus LD50:320 mg/kg PMDCAY 5,59,67
orl-rat LD50:780 mg/kg JPETAB 108,450,53
ipr-rat LD50:250 mg/kg JPETAB 108,450,53

SAFETY PROFILE: Poison by intraperitoneal route. When heated to decomposition it emits toxic fumes of NO_x. See also ALLYL COMPOUNDS.

AGN000 CAS:22131-79-9 ***HR: 2***
(4-ALLYLOXY-3-CHLOROPHENYL)ACETIC ACID
mf: $C_{11}H_{11}ClO_3$ mw: 226.67

SYNS: ALCLOPHENAC ◇ ALLOPYDIN ◇ ARGUN ◇ 2-CHLORO-4-(2-PROPENYLOXY)BENZENEACETIC ACID ◇ EPINAL ◇ MEDIFENAC ◇ MERVAN ◇ NEOSTEN ◇ NEOSTON ◇ PRINALGIN ◇ REUFENAC ◇ W 7320 ◇ ZUMARIL

TOXICITY DATA with REFERENCE
orl-rat TDLo:240 mg/kg (8-15D preg):TER IYKEDH 5,72,74
orl-rat TDLo:1 g/kg (6-15D preg):REP ARZNAD 20,618,70
orl-rat LD50:1050 mg/kg ARZNAD 20,618,70
ipr-rat LD50:465 mg/kg IYKEDH 5,38,74
orl-mus LD50:1100 mg/kg ARZNAD 20,618,70
ipr-mus LD50:508 mg/kg IYKEDH 5,38,74

SAFETY PROFILE: Moderately toxic by ingestion and intraperitoneal routes. An experimental teratogen. Other experimental reproductive effects. When heated to decomposition it emits toxic fumes of Cl^-. See also ALLYL COMPOUNDS. An analgesic, antipyretic, and anti-inflammatory agent.

AGN250 CAS:24049-18-1 ***HR: 2***
(4-(ALLYLOXY)-3-CHLOROPHENYL)ACETIC ACID SODIUM SALT
mf: $C_{11}H_{10}ClO_3 \cdot Na$ mw: 248.65

SYN: ALCLOFENAC SODIUM SALT

TOXICITY DATA with REFERENCE
orl-rat LD50:1050 mg/kg FRPSAX 32,286,77
ipr-rat LD50:530 mg/kg FRPSAX 32,286,77

SAFETY PROFILE: Moderately toxic by ingestion and intraperitoneal routes. When heated to decomposition it emits toxic fumes of Cl^- and Na_2O. See also ALLYL COMPOUNDS.

AGN500 CAS:63887-51-4 ***HR: 3***
o-ALLYLOXY-N,N-DIETHYLBENZAMIDE
mf: $C_{14}H_{19}NO_2$ mw: 233.34

TOXICITY DATA with REFERENCE
orl-rat LD50:325 mg/kg JPETAB 108,450,53
ipr-rat LD50:140 mg/kg JPETAB 108,450,53

SAFETY PROFILE: Poison by ingestion and intraperitoneal routes. When heated to decomposition it emits toxic fumes of NO_x. See also ALLYL COMPOUNDS.

AGN750 CAS:63887-52-5 ***HR: 3***
o-ALLYLOXY-N,N-DIMETHYLBENZAMIDE
mf: $C_{12}H_{15}NO_2$ mw: 205.28

TOXICITY DATA with REFERENCE
orl-rat LD50:740 mg/kg JPETAB 108,450,53
ipr-rat LD50:350 mg/kg JPETAB 108,450,53

SAFETY PROFILE: Poison by intraperitoneal route. Moderately toxic by ingestion. When heated to decomposition it emits toxic fumes of NO_x. See also ALLYL COMPOUNDS.

AGO000 CAS:111-45-5 ***HR: 3***
2-ALLYLOXYETHANOL
mf: $C_5H_{10}O_2$ mw: 102.15

SYNS: 2-ALLOXYETHANOL (CZECH) ◇ USAF DO-47

TOXICITY DATA with REFERENCE
skn-rbt 500 mg/24H MOD 28ZPAK -,99,72
eye-rbt 750 μg/24H SEV 28ZPAK -,99,72
orl-rat LD50:2360 mg/kg 28ZPAK -,99,72
ipr-mus LD50:250 mg/kg NTIS** AD277-689
par-mus LDLo:2000 mg/kg CBCCT* 7,687,55

CONSENSUS REPORTS: Reported in EPA TSCA Inventory.

SAFETY PROFILE: Poison by intraperitoneal route. Moderately toxic by ingestion and parenteral routes. Severe eye irritant. Moderate skin irritant. When heated to decomposition it emits acrid smoke and irritating fumes. See also ALLYL COMPOUNDS.

AGO250 CAS:63887-17-2 ***HR: 3***
o-ALLYLOXY-N-(β-HYDROXYETHYL)BENZAMIDE
mf: $C_{12}H_{15}NO_3$ mw: 221.28

TOXICITY DATA with REFERENCE
orl-rat LD50:1150 mg/kg JPETAB 108,450,53
ipr-rat LD50:350 mg/kg JPETAB 108,450,53

SAFETY PROFILE: Poison by intraperitoneal route. Moderately toxic by ingestion. When heated to decomposition it emits toxic fumes of NO_x. See also ALLYL COMPOUNDS.

AGO500 CAS:63731-92-0 ***HR: 3***
6-ALLYLOXY-2-METHYLAMINO-4-(N-METHYLPIPERAZINO)-5-METHYLTHIOPYRIMIDINE
mf: $C_{14}H_{23}N_5OS$ mw: 309.48

TOXICITY DATA with REFERENCE
orl-mus LD50:400 mg/kg JMCMAR 18,553,75
ivn-mus LD50:56 mg/kg JMCMAR 18,553,75

SAFETY PROFILE: Poison by ingestion and intravenous routes. When heated to decomposition it emits very toxic fumes of NO_x and SO_x. See also ALLYL COMPOUNDS.

AGO750 CAS:29181 ***HR: 3***
(+)-1-(o-ALLYLOXYPHENOXY)-3-ISOPROPYLAMINO-2-PROPANOLHYDROCHLORIDE-23-5
mf: $C_{15}H_{23}NO_3$•HCl mw: 301.46

SYN: CIBA 42155-BA

TOXICITY DATA with REFERENCE
orl-rat LD50:760 mg/kg ARZNAD 20,1890,70
ivn-dog LD50:20 mg/kg ARZNAD 20,1890,70
ivn-rbt LDLo:20 mg/kg ARZNAD 20,1890,70

SAFETY PROFILE: Poison by intravenous route. Moderately toxic by ingestion. When heated to decomposition it emits very toxic fumes of HCl and NO_x. See also ALLYL COMPOUNDS.

AGP000 CAS:6452 ***HR: 3***
(−)-1-(o-ALLYLOXYPHENOXY)-3-ISOPROPYLAMINO-2-PROPANOLHYDROCHLORIDE-73-9
mf: $C_{15}H_{23}NO_3$•HCl mw: 301.46

SYN: CIBA 42244-BA

TOXICITY DATA with REFERENCE
orl-rat LD50:900 mg/kg ARZNAD 20,1890,70
ivn-dog LD50:20 mg/kg ARZNAD 20,1890,70
ivn-rbt LDLo:20 mg/kg ARZNAD 20,1890,70

SAFETY PROFILE: Poison by intravenous route. Moderately toxic by ingestion. When heated to decomposition it emits very toxic fumes of HCl and NO_x. See also ALLYL COMPOUNDS.

AGP250 CAS:6452-54-6 ***HR: 2***
3-(ALLYLOXYPHENOXY)-1,2-PROPANEDIOL
mf: $C_{12}H_{16}O_4$ mw: 224.28

TOXICITY DATA with REFERENCE
ipr-mus LD50:650 mg/kg JPETAB 93,470,48
scu-mus LD50:582 mg/kg JPETAB 93,470,48

SAFETY PROFILE: Moderately toxic by intraperitoneal and subcutaneous routes. When heated to decomposition it emits acrid smoke and irritating fumes. See also ALLYL COMPOUNDS.

AGP400 CAS:69095-81-4 ***HR: D***
5-(o-(ALLYLOXY)PHENYL)-3-(o-TOLYL)-s-TRIAZOLE
mf: $C_{18}H_{17}N_3O$ mw: 291.38

SYN: s-TRIAZOLE,5-(m-(ALLYLOXY)PHENYL)-3-(o-TOLYL)-

TOXICITY DATA with REFERENCE
orl-ham TDLo:25 mg/kg (female 4-8D post):REP
JMCMAR 26,1187,83

SAFETY PROFILE: Experimental reproductive effects. When heated to decomposition it emits toxic fumes of NO_x.

AGP500 CAS:123-34-2 ***HR: 2***
3-ALLYLOXY-1,2-PROPANEDIOL
mf: $C_6H_{12}O_3$ mw: 132.18

SYNS: α-ALLYL GLYCEROL ETHER ◇ GLYCEROL-α-ALLYL ETHER

TOXICITY DATA with REFERENCE
skn-rbt 535 mg/24H MLD AMIHBC 2,574,50
eye-rbt 107 mg AMIHBC 2,574,50

orl-mus LD50:4485 mg/kg AMIHBC 2,574,50
scu-mus LD50:1135 mg/kg JPETAB 93,470,48

CONSENSUS REPORTS: Reported in EPA TSCA Inventory.

SAFETY PROFILE: Moderately toxic by subcutaneous route. Mildly toxic by ingestion. A skin and eye irritant. See also ETHERS and ALLYL COMPOUNDS. When heated to decomposition it emits acrid smoke and irritating fumes.

AGP750 CAS:3088-44-6 ***HR: 2***
3-ALLYLOXYPROPIONITRILE
mf: C_6H_9NO mw: 111.16

SYN: β-ALLYLOXY-PROPIONITRILE

TOXICITY DATA with REFERENCE
orl-rat LD50:1300 mg/kg TXAPA9 28,313,74
ipr-mus LDLo:500 mg/kg CBCCT* 4,380,52

CONSENSUS REPORTS: Cyanide and its compounds are on the Community Right-To-Know List.

SAFETY PROFILE: Moderately toxic by ingestion and intraperitoneal routes. See also NITRILES and ALLYL COMPOUNDS. When heated to decomposition it emits toxic fumes of NO_x and CN^-.

AGQ000 CAS:12012-95-2 ***HR: 2***
ALLYL PALLADIUM CHLORIDE DIMER

TOXICITY DATA with REFERENCE
skn-rbt 100 mg/24H SEV AEHLAU 30,168,75

SAFETY PROFILE: A severe skin irritant. See also PALLADIUM COMPOUNDS and ALLYL COMPOUNDS. When heated to decomposition it emits toxic fumes of Cl^-.

AGQ250 CAS:4255-24-7 ***HR: 3***
α-ALLYL PHENETHYLAMINEHYDROCHLORIDE
mf: $C_{11}H_{15}N \cdot HCl$ mw: 174.29

SYNS: ALETAMINE HYDROCHLORIDE ◇ 1-ALLYL-2-PHENYLETHYLAMINE HYDROCHLORIDE

TOXICITY DATA with REFERENCE
scu-rat LD50:280 mg/kg AIPTAK 159,442,66
orl-mus LD50:380 mg/kg TXAPA9 21,302,72
ivn-mus LD50:30 mg/kg 27ZQAG -,332,72

SAFETY PROFILE: Poison by ingestion, subcutaneous, and intravenous routes. When heated to decomposition it emits toxic fumes of HCl. See also ALLYL COMPOUNDS.

AGQ500 CAS:1745-81-9 ***HR: 3***
o-ALLYL PHENOL
mf: $C_9H_{10}O$ mw: 134.19

PROP: Mp: 10°, bp: 230°, d: 1.033 @ 18°/4°. Sol in water, alc, chloroform, and ether.

SYN: 2-ALLYL PHENOL

TOXICITY DATA with REFERENCE
skn-mus TDLo:8400 mg/kg/30W-I:CAR CNREA8 19,413,59
skn-mus TD:3360 mg/kg/12W-I:NEO CNREA8 19,413,59
ipr-mus LDLo:256 mg/kg CBCCT* 1,127,51

CONSENSUS REPORTS: Reported in EPA TSCA Inventory.

SAFETY PROFILE: Poison by intraperitoneal route. Questionable carcinogen with experimental carcinogenic and neoplastigenic data. When heated to decomposition it emits acrid smoke and fumes. See also ALLYL COMPOUNDS and PHENOLS.

AGQ750 CAS:7493-74-5 ***HR: 2***
ALLYL PHENOXYACETATE
mf: $C_{11}H_{12}O_3$ mw: 192.23

PROP: Colorless to light yellow liquid; heavy fruit odor.

SYN: ACETATE P.A.

TOXICITY DATA with REFERENCE
orl-rat LD50:475 mg/kg FCTXAV 13,681,75
skn-rbt LD50:820 mg/kg FCTXAV 13,681,75

CONSENSUS REPORTS: Reported in EPA TSCA Inventory.

SAFETY PROFILE: Moderately toxic by ingestion and skin contact. When heated to decomposition it emits acrid smoke and irritating fumes. See also ALLYL COMPOUNDS.

AGQ775 CAS:21905-27-1 ***HR: 3***
ALLYL PHENYL ARSINIC ACID
mf: $C_9H_{11}AsO_2$ mw: 226.12

SYNS: ALLYLHYDROXYPHENYLARSINE OXIDE ◇ ARSINE OXIDE, ALLYLHYDROXYPHENYL-

TOXICITY DATA with REFERENCE
ivn-mus LD50:100 mg/kg CSLNX* NX#06910

OSHA PEL: TWA 0.5 mg(As)/m^3

SAFETY PROFILE: Poison by intravenous route. When heated to decomposition it emits toxic fumes of As.

AGQ875 CAS:115-43-5 ***HR: 3***
5-ALLYL-5-PHENYLBARBITURIC ACID
mf: $C_{13}H_{12}N_2O_3$ mw: 244.27

PROP: Crystals, bitter taste. Mp: 156-157.5°. Readily sol in alc, chloroform. One gram dissolves in 580 mL water, in 10 mL ether, in 500 mL benzene, in 4000 mL carbon tetrachloride, in 17,500 mL petr ether.

SYNS: ACIDO-5-FENIL-5-ALLILBARBITURICO (ITALIAN) ◇ ALLOFENYL ◇ ALLOPHENYLUM ◇ ALLPHASEM ◇ ALPHEBA ◇ ALPHENAL ◇ ALPHENATE ◇ FENALLYMAL ◇ LUBERGAL ◇ LUXOMNIN ◇ PHENALLYMAL ◇ PHENALLYMALUM ◇ 5-PHENYL-5-ALLYLBARBITURIC ACID ◇ 5-PHENYL-5-(2-PROPENYL)-2,4,6(1H,3H,5H)-PYRIMIDINETRIONE (9CI) ◇ PHENYRAL ◇ PROPHENAL ◇ TUBERGAL

TOXICITY DATA with REFERENCE
ipr-rat LD50:233 mg/kg SFTIAE 56,31,52
orl-mus LD50:280 mg/kg FRPSAX 17,390,62
ipr-mus LD50:265 mg/kg FRPSAX 17,390,62

SAFETY PROFILE: Poison by ingestion and intraperitoneal routes. Caution: Abuse may lead to habituation or addiction. When heated to decomposition it emits toxic fumes of NO_x.

AGR000 CAS:1746-13-0 ***HR: 3***
ALLYL PHENYL ETHER
mf: $C_9H_{10}O$ mw: 134.19

PROP: D: 0.986, bp: 191.7. Insol in water.

SYNS: (2-PROPENYLOXY)BENZENE ◇ USAF DO-23

TOXICITY DATA with REFERENCE
ipr-mus LD50:100 mg/kg NTIS** AD277-689
ivn-mus LD50:63 mg/kg CSLNX* NX#01855

CONSENSUS REPORTS: Reported in EPA TSCA Inventory.

SAFETY PROFILE: Poison by intravenous and intraperitoneal routes. See also ETHERS. When heated to decomposition it emits acrid smoke and irritating fumes. See also ALLYL COMPOUNDS and ETHERS.

AGR125 CAS:2597-09-3 ***HR: 3***
2-ALLYL-2-PHENYL-4-PENTENOIC ACID 2-(DIETHYLAMINO)ETHYL ESTER HYDROCHLORIDE
mf: $C_{20}H_{29}NO_2 \cdot ClH$ mw: 351.96

SYN: CFT 1201

TOXICITY DATA with REFERENCE
orl-rat LD50:1500 mg/kg AEPPAE 225,453,55
ipr-rat LD50:170 mg/kg AEPPAE 225,453,55
scu-rat LD50:1500 mg/kg AEPPAE 225,453,55

SAFETY PROFILE: Poison by intraperitoneal route. Moderately toxic by ingestion and subcutaneous routes. When heated to decomposition it emits toxic fumes of NO_x and HCl.

AGR250 CAS:7341-63-1 ***HR: 3***
ALLYL PHENYL THIOUREA
mf: $C_{10}H_{12}N_2S$ mw: 192.30

SYN: 1-ALLYL-3-PHENYL-2-THIOUREA

TOXICITY DATA with REFERENCE
orl-rat LD50:750 mg/kg JPETAB 90,260,47
ivn-mus LD50:56 mg/kg CSLNX* NX#01120

SAFETY PROFILE: Poison by intravenous route. Moderately toxic by ingestion. When heated to decomposition it emits very toxic fumes of NO_x and SO_x. See also ALLYL COMPOUNDS.

AGR500 CAS:2179-59-1 ***HR: 1***
ALLYL PROPYL DISULFIDE
mf: $C_6H_{12}S_2$ mw: 148.30

PROP: Liquid, pungent odor.

OSHA PEL: (Transitional: TWA 2 PPM) TWA 2 ppm; STEL 3 ppm
ACGIH TLV: TWA 2 ppm; STEL 3 ppm
DFG MAK: 2 ppm (12 mg/m^3)

SAFETY PROFILE: A powerful irritant. Moderately flammable by exposure to heat, flame, or oxidizers. When heated to decomposition it emits highly toxic SO_x. To fight fire, use foam, CO_2, dry chemical. See also ALLYL COMPOUNDS.

AGR750 CAS:15151-00-5 ***HR: 3***
1-ALLYLQUINALDINUM BROMIDE
mf: $C_{13}H_{14}N \cdot Br$ mw: 264.19

TOXICITY DATA with REFERENCE
ipr-mus LDLo:64 mg/kg CBCCT* 2,190,50

CONSENSUS REPORTS: Reported in EPA TSCA Inventory.

SAFETY PROFILE: Poison by intraperitoneal route. See also ALLYL COMPOUNDS and BROMIDES. When heated to decomposition it emits very toxic fumes of NO_x and Br^-.

AGS000 CAS:7539-12-0 ***HR: 3***
ALLYLSUCCINIC ANHYDRIDE
mf: $C_7H_8O_3$ mw: 140.15

TOXICITY DATA with REFERENCE
skn-rbt 10 mg/24H open SEV AIHAAP 23,95,62
orl-rat LD50:1070 mg/kg AIHAAP 23,95,62
skn-rbt LD50:320 mg/kg AIHAAP 23,95,62

SAFETY PROFILE: Poison by skin contact. Moderately toxic by ingestion. A severe skin irritant. See also ANHYDRIDES and ALLYL COMPOUNDS. When heated to decomposition it emits acrid smoke and irritating fumes.

AGS250 CAS:592-88-1 ***HR: 3***
ALLYL SULFIDE
mf: $C_6H_{10}S$ mw: 114.22

$(H_2C=CHCH_2)_2S$

PROP: Colorless liquid, garlic odor. Mp: −83°, bp: 139°, d: 0.8881, vap d: 3.90.

SYNS: ALLYL MONOSULFIDE ◇ DIALLYL MONOSULFIDE ◇ DIALLYL SULFIDE ◇ DIALLYL THIOETHER ◇ OIL GARLIC ◇ THIOALLYL ETHER ◇ 3,3-THIOBIS(1-PROPENE)

TOXICITY DATA with REFERENCE
ipr-mus LDLo:512 mg/kg CBCCT* 1,43,49
ivn-rbt LDLo:330 mg/kg BIJOAK 4,107,09

CONSENSUS REPORTS: Reported in EPA TSCA Inventory.

SAFETY PROFILE: Poison by intravenous route. Moderately toxic by intraperitoneal route. An irritant to skin, eyes, and mucous membranes. When heated to decomposition it emits toxic SO_x. Explosive reaction with N-bromosuccinimide. See also SULFIDES and ALLYL COMPOUNDS.

AGS375 ***HR: 3***
3-(ALLYL-(TETRAHYDRONAPHTHYL)AMINO)-N,N-DIETHYL-PROPIONAMIDE
mf: $C_{20}H_{30}N_2O{\cdot}ClH$ mw: 350.98

SYNS: 3-(ALLYL-(TETRAHYDRONAPHTHYL)AMINO)-N,N-DIETHYLPROPIONAMIDE HYDROCHLORIDE ◇ N-d,l-ac-TETRAHYDRO-β-NAPHTHYL-N-ALLYL-β-ALANINDIAETHYLAMID-HYDROCHLORID(GERMAN)

TOXICITY DATA with REFERENCE
orl-rat LD50:127 mg/kg AIPTAK 128,500,60
scu-rat LD50:44500 μg/kg AIPTAK 128,500,60
ivn-rat LD50:18900 μg/kg AIPTAK 128,500,60
orl-mus LD50:210 mg/kg AIPTAK 128,500,60
ipr-mus LD50:118 mg/kg AIPTAK 128,500,60
ivn-mus LD50:23500 μg/kg AIPTAK 128,500,60
ivn-rbt LD50:1900 μg/kg AIPTAK 128,500,60

SAFETY PROFILE: Poison by ingestion, intravenous, and intraperitoneal routes. When heated to decomposition it emits toxic fumes of NO_x and HCl.

AGS500 CAS:2530-99-6 ***HR: 3***
1-ALLYLTHEOBROMINE
mf: $C_{10}H_{12}N_4O_2$ mw: 220.26

SYN: ALLYLTHEOBROMINE

TOXICITY DATA with REFERENCE
orl-hmn TDLo:26 mg/kg:CNS,GIT JPETAB 86,113,46
scu-mus LDLo:125 mg/kg AIPTAK 25,361,21
orl-mus LD50:191 mg/kg JPETAB 116,343,56
irp-mus LD50:102 mg/kg JPETAB 116,343,56
ivn-mus LD50:40 mg/kg JPETAB 86,113,46
scu-rbt LDLo:100 mg/kg AIPTAK 25,361,21
ivn-rbt LDLo:50 mg/kg AIPTAK 25,361,21

SAFETY PROFILE: Poison by ingestion, subcutaneous, intraperitoneal and intravenous routes. Human systemic effects by ingestion: changes in motor activity and nausea or vomiting. When heated to decomposition it emits highly toxic NO_x. See also ALLYL COMPOUNDS.

AGS750 CAS:764-49-8 ***HR: 3***
ALLYL THIOCYANATE
mf: C_4H_5NS mw: 99.16

PROP: Colorless oil. D: 1.024, mp: −102.5° bp: 152°, sol in water, alc, and ether.

SYNS: ALLYLRHODANID (GERMAN) ◇ ALLYL SULFOCYANIDE ◇ THIOCYANIC ACID, ALLYL ESTER

TOXICITY DATA with REFERENCE
ipr-rat LDLo:100 mg/kg ARZNAD 16,870,66
scu-rbt LDLo:12 mg/kg AEPPAE 150,257,30

SAFETY PROFILE: Poison by intraperitoneal and subcutaneous routes. See also THIOCYANATES and ALLYL COMPOUNDS. When heated to decomposition it emits very toxic NO_x, SO_x, and CN^-.

AGT000 CAS:3766-55-0 ***HR: 3***
4-ALLYLTHIOSEMICARBAZIDE
mf: $C_4H_9N_3S$ mw: 131.22

TOXICITY DATA with REFERENCE
par-rat LD50:500 mg/kg ARZNAD 12,260,62
ivn-mus LD50:56 mg/kg CSLNX* NX#00434

SAFETY PROFILE: Poison by intravenous route. Moderately toxic by parenteral route. When heated to decomposition it emits very toxic fumes of NO_x and SO_x. See also ALLYL COMPOUNDS.

AGT250 CAS:3571-74-2 ***HR: 3***
2-(ALLYLTHIO)-2-THIAZOLINE
mf: $C_6H_9NS_2$ mw: 159.28

SYNS: 4,5-DIHYDRO-2-(2-PROPENYLTHIO)THIAZOLE(9CI) ◇ EN-28,450

TOXICITY DATA with REFERENCE
orl-rat LD50:110 mg/kg 28ZEAL 4,40,69
skn-rbt LD50:340 mg/kg 28ZEAL 4,40,69

SAFETY PROFILE: Poison by ingestion and skin contact. When heated to decomposition it emits very toxic fumes of NO_x and SO_x. See also ALLYL COMPOUNDS.

AGT500 CAS:109-57-9 ***HR: 3***
1-ALLYL-2-THIOUREA
mf: $C_4H_8N_2S$ mw: 116.20

PROP: Colorless prisms. D: 1.219, mp: 77-78°. Sol in hot water; insol in benzene, alc; very sltly sol in ether.

SYNS: ALLYLTHIOCARBAMIDE ◇ 1-ALLYLTHIOUREA ◇ N-ALLYLTHIOUREA ◇ AMINOSIN ◇ (2-PROPENYL)THIOUREA ◇ RHODALLIN ◇ RHODALLINE ◇ THIOSINAMIN ◇ THIOSINAMINE ◇ THIOCYNAMINE ◇ U 19571

TOXICITY DATA with REFERENCE
mmo-sat 150 μg/plate ABCHA6 44,3017,80
ipr-mus LD50:1858 mg/kg YKKZAJ 94,1419,74
orl-rat LD50:200 mg/kg JPETAB 90,260,47
ipr-rat LD50:500 mg/kg JPETAB 89,186,47
scu-rat LDLo:50 mg/kg HBAMAK 4,1289,35
scu-mus LDLo:700 mg/kg HBAMAK 4,1289,35
ivn-dog LDLo:110 mg/kg HBAMAK 4,1289,35

CONSENSUS REPORTS: Reported in EPA TSCA Inventory.

SAFETY PROFILE: Poison by ingestion, subcutaneous, and intravenous routes. Moderately toxic by intraperitoneal route. When heated to decomposition it emits very toxic NO_x and SO_x. See also ALLYL COMPOUNDS.

AGT750 CAS:1530-48-9 ***HR: 3***
ALLYL TRI-N-BUTYLPHOSPHONIUMCHLORIDE
mf: $C_{15}H_{32}P \cdot Cl$ mw: 278.89

TOXICITY DATA with REFERENCE
ivn-mus LD50:18 mg/kg CSLNX* NX#03135

CONSENSUS REPORTS: Reported in EPA TSCA Inventory.

SAFETY PROFILE: Poison by intravenous route. When heated to decomposition it emits very toxic fumes of PO_x and Cl^-. See also ALLYL COMPOUNDS.

AGU250 CAS:107-37-9 ***HR: 3***
ALLYL TRICHLOROSILANE
DOT: UN 1724
mf: $C_3H_5Cl_3Si$ mw: 175.52

PROP: Colorless liquid, pungent, irritating odor. Bp: 117.5°, d: 1.217 @ 27°, flash p: 95°F (COC).

SYN: TRICHLOROALLYLSILANE

TOXICITY DATA with REFERENCE
ivn-mus LD50:56 mg/kg CSLNX* NX#04219

CONSENSUS REPORTS: Reported in EPA TSCA Inventory.

DOT Classification: Corrosive Material; Label: Corrosive, Flammable Liquid.

SAFETY PROFILE: Poison by intravenous route. Corrosive. See also SILANES and ALLYL COMPOUNDS. When heated to decomposition it emits toxic Cl^-. A dangerous fire hazard. To fight fire, use foam, mist, spray, dry chemical.

AGU400 CAS:71500-21-5 ***HR: 2***
ALLYL 3,5,5-TRIMETHYLHEXANOATE
mf: $C_{12}H_{22}O_2$ mw: 198.34

SYNS: ALLYL TRIMETHYLHEXANOATE ◇ HEXANOIC ACID, 3,5,5-TRIMETHYL-, ALLYL ESTER ◇ 2-PROPENYL 3,5,5-TRIMETHYLHEXANOATE ◇ 3,5,5-TRIMETHYLHEXANOIC ACID ALLYL ESTER

TOXICITY DATA with REFERENCE
skn-rbt 500 mg/24H SEV FCTOD7 20,639,82
orl-rat LD50:1400 mg/kg FCTOD7 20,639,82

SAFETY PROFILE: Moderately toxic by ingestion. A severe skin irritant. When heated to decomposition it emits acrid smoke and irritating fumes.

AGU500 CAS:76-63-1 ***HR: 3***
ALLYLTRIPHENYLTIN
mf: $C_{21}H_{20}Sn$ mw: 391.10

SYNS: ALLYLTRIPHENYL STANNANE ◇ DOWCO 187 ◇ ENT 50,909 ◇ TRIPHENYL-2-PROPENYL-STANNANE (9CI)

TOXICITY DATA with REFERENCE
ivn-mus LD50:100 mg/kg CSLNX* NX#02200

OSHA PEL: TWA 0.1 mg(Sn)/m^3 (skin)
ACGIH TLV: TWA 0.1 mg(Sn)/m^3 (skin) (Proposed: TWA 0.1 mg(Sn)/m^3; STEL 0.2 mg(Sn)/m^3 (skin))
NIOSH REL: (Organotin Compounds) TWA 0.1 mg(Sn)/m^3

SAFETY PROFILE: Poison by intravenous route. See also TIN COMPOUNDS and ALLYL COMPOUNDS. When heated to decomposition it emits toxic smoke and irritating fumes.

AGU750 CAS:37425-13-1 ***HR: 2***
9-ALLYL-2-(4-(2-TRITYLETHYL)-1-PIPERAZINYL)-9H-PURINEDIMETHANESULFONATE
mf: $C_{33}H_{34}N_6 \cdot 2CH_4O_3S$ mw: 706.95

TOXICITY DATA with REFERENCE
mmo-nsc 1 g/L GENTAE 48,597,63
orl-mus LDLo:2000 mg/kg CHTPBA 7,192,72

SAFETY PROFILE: Mutation data reported. Moderately toxic by ingestion. See also SULFONATES and ALLYL COMPOUNDS. When heated to decomposition it emits very toxic fumes of SO_x and NO_x.

AGV000 CAS:557-11-9 ***HR: 3***
ALLYLUREA
mf: $C_4H_8N_2O$ mw: 100.14

PROP: Mp: 85°. Very sol in water and alc; very sltly sol in ether.

SYNS: ALLYLCARBAMIDE ◇ MONOALLYLUREA ◇ N-2-PROPENYLUREA

TOXICITY DATA with REFERENCE
orl-mus LDLo:1070 mg/kg AECTCV 14,111,85
orl-rat LDLo:250 mg/kg NCNSA6 5,47,53

CONSENSUS REPORTS: Reported in EPA TSCA Inventory.

SAFETY PROFILE: Poison by ingestion. When heated to decomposition it emits toxic fumes of NO_x. See also ALLYL COMPOUNDS.

AGV250 CAS:3917-15-5 ***HR: 3***
ALLYL VINYL ETHER
mf: C_5H_8O mw: 84.13

PROP: Very sltly sol in water. D: 0.8, bp: 67°, flash p: <68°F (OC).

TOXICITY DATA with REFERENCE
orl-rat LD50:550 mg/kg AIHAAP 23,95,62
ihl-rat LCLo:8000 ppm/4H AIHAAP 23,95,62

SAFETY PROFILE: Moderately toxic by inhalation and ingestion. See also ALLYL COMPOUNDS and ETHERS. Dangerous fire and explosion hazard from heat, sparks,or powerful oxidizers. To fight fire, use alcohol foam, dry chemical, or mist. Water may be ineffective. When heated to decomposition it yields acrid, irritating fumes. Becomes shock and heat sensitive on storage.

AGV875 ***HR: 2***
ALOE

PROP: A perennial succulent cultivated outdoors in the tropics and a popular houseplant elsewhere. The leaves are thick and hard with spines on the edges. The sap of some aloes is used commercially to produce carthartic glycosides. The gel in the leaves is used as a skin moisturizer and palliative.

SYNS: LALOI (HAITI) ◇ PANINI-'AWA'AWA (HAWAII) ◇ SABILA (CUBA, PUERTO RICO) ◇ SEMPERVIVUM (JAMAICA) ◇ SINKLE BIBLE (JAMAICA) ◇ STAR CACTUS (HAWAII) ◇ ZABILA (MEXICO, DOMINICAN REPUBLIC) ◇ ZAVILA (PUERTO RICO)

SAFETY PROFILE: The sap contains the toxic barbaloin, an anthraquinone glycoside. Ingestion causes a strong purgative action within 12 hours and may color the urine red. Repeated doses may cause kidney damage.

AGV890 CAS:25384-17-2 ***HR: 3***
ALPERIDINE HYDROCHLORIDE
mf: $C_{18}H_{25}NO_2{\bullet}ClH$ mw: 323.90

SYNS: dl-1-ALLYL-1-METHYL-4-PHENYL-4-PIPERIDINOLPROPIONATE HYDROCHLORIDE ◇ ALLYLPRODINE HYDROCHLORIDE ◇ dl-3-METHYL-3-ALLYL-4-PROPIONOXYPIPERIDINEHYDROCHLORIDE ◇ NIH 7440 HYDROCHLORIDE ◇ Ro 2-7113

TOXICITY DATA with REFERENCE
orl-rat LD50:125 mg/kg AIPTAK 109,171,57
scu-rat LD50:80 mg/kg AIPTAK 109,171,57
ivn-rat LD50:13400 μg/kg AIPTAK 109,171,57
orl-mus LD50:536 mg/kg AIPTAK 109,171,57
scu-mus LD50:333 mg/kg AIPTAK 109,171,57
ivn-mus LD50:45800 μg/kg AIPTAK 109,171,57
ivn-dog LD50:25600 μg/kg AIPTAK 109,171,57
ivn-rbt LD50:7800 μg/kg AIPTAK 109,171,57

SAFETY PROFILE: Poison by ingestion, subcutaneous, and intravenous routes. When heated to decomposition it emits toxic fumes of NO_x and HCl. Caution: May be habit forming. This is a controlled substance (opiate) listed in the U.S. Code of Federal Regulations, Title 21 Part 1308.11 (1985).

AGW000 CAS:13707-88-5 ***HR: 3***
ALPRENOL HYDROCHLORIDE
mf: $C_{15}H_{23}NO_2{\bullet}ClH$ mw: 285.85

SYNS: 1-(o-ALLYLPHENOXY)-3-(ISOPROPYLAMINO)-2-PROPANOL HYDROCHLORIDE ◇ APTIN ◇ BETAPTIN ◇ GUBERNAL ◇ OXPRENOLOL

TOXICITY DATA with REFERENCE
orl-man TDLo:571 μg/kg:CVS KIZSB8 6(4),209,75
orl-rat LD50:590 mg/kg KSRNAM 11,1321,77
scu-rat LD50:290 mg/kg KSRNAM 11,1321,77
scu-mus LD50:215 mg/kg KSRNAM 11,1321,77
orl-mus LD50:184 mg/kg AIPTAK 202,79,73
orl-dog LD50:383 mg/kg GNRIDX 3,614,69
scu-dog LD50:92 mg/kg GNRIDX 3,614,69
ivn-dog LD50:18 mg/kg GNRIDX 3,614,69

SAFETY PROFILE: Poison by ingestion, subcutaneous, intravenous and intraperitoneal routes. Human systemic effects by ingestion: change in heart rate. When heated to decomposition it emits very toxic fumes of HCl and NO_x. See also ALLYL COMPOUNDS.

AGW250 CAS:13655-52-2 ***HR: 3***
ALPRENOLOL
mf: $C_{15}H_{23}NO_2$ mw: 249.39

SYNS: ALFEPROL (RUSSIAN) ◇ 1-(o-ALLYLPHENOXY)-3-(ISOPROPYLAMINO)-2-PROPANOL ◇ H 56/28 ◇ 1-((1-METHYLETHYL)AMINO)-3-(2-(2-PROPENYL)PHENOXY)-2-PROPANOL

TOXICITY DATA with REFERENCE
orl-wmn LDLo:210 mg/kg UGLAAD 139,2817,77

ipr-mus LD50:90 mg/kg EJMCA5 18,151,83
ivn-mus LD50:20 mg/kg ARZNAD 27,1022,77
orl-mam LD50:184 mg/kg PCJOAU 8,137,74
ipr-mam LD50:102 mg/kg PCJOAU 8,137,74

SAFETY PROFILE: A human poison by ingestion. Poison experimentally by ingestion, intravenous, and intraperitoneal routes. When heated to decomposition it emits toxic fumes of NO_x. See also ALLYL COMPOUNDS. A beta-adrenergic blocker.

AGW275 ***HR: 2***
ALPROSTADIL-α-CYCLODEXTRIN CLATHRATE
mf: $C_{20}H_{34}O_5 \cdot 7C_{36}H_{60}O_{30}$ mw: 7165.26

SYN: 11,15-DIHYDROXY-9-OXO-PROST-13-EN-1-OIC ACID, (11-α,13E,15S)-, and α-CYCLODEXTRIN

TOXICITY DATA with REFERENCE
orl-rat LD50:7600 mg/kg IYKEDH 11,181,80
ipr-rat LD50:830 mg/kg IYKEDH 11,181,80
scu-rat LD50:660 mg/kg IYKEDH 11,181,80
ivn-rat LD50:640 mg/kg IYKEDH 11,181,80
iat-rat LD50:720 mg/kg IYKEDH 11,181,80
orl-mus LD50:6200 mg/kg IYKEDH 11,181,80
ipr-mus LD50:660 mg/kg IYKEDH 11,181,80

SAFETY PROFILE: Moderately toxic by several routes. When heated to decomposition it emits acrid smoke and irritating fumes.

AGW300 CAS:8001-95-4 ***HR: 3***
ALSEROXYLON

TOXICITY DATA with REFERENCE
ipr-rat LD50:260 mg/kg NIIRDN 6,43,82
orl-mus LD50:532 mg/kg NIIRDN 6,43,82
ipr-mus LD50:172 mg/kg NIIRDN 6,43,82
ipr-dog LD50:50 mg/kg NIIRDN 6,43,82
ipr-rbt LD50:84 mg/kg NIIRDN 6,43,82

SAFETY PROFILE: Poison by intraperitoneal route. Moderately toxic by ingestion. When heated to decomposition it emits acrid smoke and irritating fumes.

AGW375 ***HR: 3***
ALSTONINE HYDROCHLORIDE
mf: $C_{21}H_{20}N_2O_3 \cdot ClH$ mw: 384.86

TOXICITY DATA with REFERENCE
ivn-rat LD50:14400 μg/kg JPETAB 90,57,47
ivn-mus LD50:8800 μg/kg JPETAB 90,57,47
ivn-dog LDLo:10 mg/kg JPETAB 90,57,47

SAFETY PROFILE: Poison by intravenous route. When heated to decomposition it emits toxic fumes of NO_x and HCl.

AGW476 CAS:641-38-3 ***HR: 3***
ALTERNARIOL
mf: $C_{14}H_{10}O_5$ mw: 258.24

SYNS: AOH ◇ 1-METHYL-3,7,9-TRIHYDROXY-6H-DIBENZO(b,d)PYRAN-6-ONE

TOXICITY DATA with REFERENCE
scu-mus TDLo:400 mg/kg (9-12D preg):TER EVHPAZ 4,87,73
scu-mus TDLo:400 mg/kg (9-12D preg):REP EVHPAZ 4,87,73
ipr-mus LDLo:100 mg/kg EVHPAZ 4,87,73

SAFETY PROFILE: Poison by intraperitoneal route. An experimental teratogen. Other experimental reproductive effects. When heated to decomposition it emits acrid smoke and fumes.

AGW500 CAS:23452-05-3 ***HR: D***
ALTERNARIOL-9-METHYL ETHER
mf: $C_{15}H_{12}O_5$ mw: 272.27

SYNS: ALTERNARIOL MONOMETHYL ETHER ◇ AME ◇ 3,7-DIHYDROXY-9-METHOXY-1-METHYL-6H-DIBENZO(b,d)PYRAN-6-ONE

TOXICITY DATA with REFERENCE
mmo-sat 500 μg/plate MUREAV 78,33,80
ipr-ham TDLo:200 mg/kg (8D preg):TER TOXID9 1,35,81
ipr-ham TDLo:200 mg/kg (8D preg):REP FCTOD7 20,899,82

SAFETY PROFILE: An experimental teratogen. Other experimental reproductive effects. Mutation data reported. When heated to decomposition it emits smoke and acrid fumes. See also ETHERS.

AGW550 ***HR: 3***
ALTERNARIOL MONOMETHYL ETHER and ALTERNARIOL (1:1)

SYNS: ALTERNARIOL and ALTERNARIOL MONOMETHYL ETHER (1:1) ◇ AME and AOH (1:1) ◇ AOH and AME (1:1) ◇ 6H-DIBENZO(b,d)PYRAN-6-ONE, 1-METHYL-3,7,9-TRIHYDROXY-and 3,9-DIHYDROXY-7-METHOXY-1-METHYL-DIBENZO(b,d)PYRAN-6-ONE (1:1)

TOXICITY DATA with REFERENCE
scu-mus TDLo:100 mg/kg (female 9-12D post):TER EVHPAZ 4,87,73
scu-mus TDLo:100 mg/kg (female 9-12D post):REP EVHPAZ 4,87,73
ipr-mus LDLo:200 mg/kg EVHPAZ 4,87,73

SAFETY PROFILE: Poison by intraperitoneal route. An experimental teratogen. Other experimental reproductive effects. When heated to decomposition it emits acrid smoke and irritating fumes.

AGW625 CAS:73309-75-8 ***HR: 3***
ALTOSIDE
mf: $C_{30}H_{40}O_{10}$ mw: 560.70

SYNS: ALTOSID (GERMAN) ◇ 3-β-((α-d-GLUCOPYRANOSYL)OXY)-14-HYDROXY-19-OXO-BUFA-4,20,22-TRIENOLIDE

TOXICITY DATA with REFERENCE
dlt-dmg-skn 10 pph CYGEDX 13(6),37,79
ivn-cat LD50:80 μg/kg 85ELDJ -,188,63
unr-cat LDLo:77 μg/kg 85ELDJ -,70,63

SAFETY PROFILE: Deadly poison by intravenous and possibly other routes. When heated to decomposition it emits acrid smoke and fumes.

AGW675 CAS:850-52-2 ***HR: D***
ALTRENOGEST
mf: $C_{21}H_{26}O_2$ mw: 310.47

SYNS: 17-α-ALLYL-ESTRATRIENE-4,9,11,17-β-OL-3-ONE ◇ ALLYL TRENBOLONE ◇ AT ◇ DRC 6246 ◇ ESTRA-4,9,11-TRIEN-3-ONE, 17-α-ALLYL-17-HYDROXY- ◇ ESTRA-4,9,11-TRIEN-3-ONE, 17-HYDROXY-17-(2-PROPENYL)-, (17-β)-(9CI) ◇ R 2267

TOXICITY DATA with REFERENCE
orl-pig TDLo:3 mg/kg (female 18D pre):REP THGNBO 16,195,81

SAFETY PROFILE: Experimental reproductive effects. When heated to decomposition it emits acrid smoke and irritating fumes.

AGW750 CAS:569-58-4 ***HR: 1***
ALUMINON
mf: $C_{22}H_{23}N_3O_9$ mw: 473.48

SYNS: AMMONIUM AURINTRICARBOXYLATE ◇ AURINE-TRICARBOXYLATE d'AMMONIUM (French) ◇ AURINTRICARBOXYLIC ACID AMMONIUM SALT ◇ C.I. MORDANT VIOLET 39, TRIAMMONIUM SALT (8CI) ◇ LYSOFON

TOXICITY DATA with REFERENCE
dni-mus-ipr 211 μmol/kg VAAZA2 23,137,77
orl-man TDLo:857 μg/kg (60D male):REP CRSBAW 156,1701,62
orl-rat LD50:9 g/kg GTPZAB 32(3),48,88

CONSENSUS REPORTS: Reported in EPA TSCA Inventory.

SAFETY PROFILE: Human reproductive effects by ingestion: changes in male fertility. Experimental reproductive effects. Mutation data reported. When heated to decomposition it emits toxic fumes of NH_3 and NO_x.

AGX000 CAS:7429-90-5 ***HR: 3***
ALUMINUM
DOT: UN 1309/UN 1383/UN 1396
af: Al aw: 26.98

PROP: A silvery ductile metal. Mp: 660°, bp: 2450°, d: 2.702, vap press 1 mm @ 1284°. Sol in HCl, H_2SO_4, and alkalies.

SYNS: A OO ◇ ADIM ◇ ALAUN (GERMAN) ◇ ALUMINA FIBRE ◇ ALUMINUM DEHYDRATED ◇ ALUMINUM FLAKE ◇ ALUMINUM, METALLIC, POWDER (DOT) ◇ ALUMINUM POWDER ◇ ALUMINUM POWDER, UNCOATED, NON-PYROPHORIC (DOT) ◇ C.I. 77000 ◇ EMANAY ATOMIZED ALUMINUM POWDER ◇ JISC 3108 ◇ JISC 3110 ◇ METANA ALUMINUM PASTE ◇ NORAL INK GRADE ALUMINUM ◇ PAP-1

CONSENSUS REPORTS: Community Right-To-Know List (fume or dust). Reported in EPA TSCA Inventory.

OSHA PEL: Total Dust: TWA 15 mg/m³; Respirable Fraction: TWA 5 mg/m³; Pyro Powders and Welding Fumes: 5 mg/m³; Soluble Salts and Alkyls: 2 mg/m³.
ACGIH TLV: Metal and Oxide: TWA 10 mg/m³ (dust); Pyro Powders and Welding Fumes: TWA 5 mg/m³; Soluble Salts and Alkyls) TWA 2 mg/m³
DFG MAK: 6 mg/m³; BAT: 170 μg/L in urine at end of shift.
DOT Classification: Flammable Solid; Label: Flammable Solid (UN1309, UN1396); Flammable Solid; Label: Spontaneously Combustible (UN1383); Flammable Solid; Label: Dangerous When Wet (UN1396).

SAFETY PROFILE: Although aluminum is not generally regarded as an industrial poison, inhalation of finely divided powder has been reported to cause pulmonary fibrosis. It is a reactive metal and the greatest industrial hazards are with chemical reactions. As with other metals the powder and dust are the most dangerous forms. Dust is moderately flammable/explosive by heat, flame, or chemical reaction with powerful oxidizers. To fight fire, use special mixtures of dry chemical.

Powdered aluminum undergoes the following dangerous interactions: explosive reaction after a delay period with $KClO_4 + Ba(NO_3)_2 + KNO_3 + H_2O$, also with $Ba(NO_3)_2 + KNO_3$ + sulfur + vegetable adhesives + H_2O. Mixtures with powdered AgCl, NH_4NO_3 or $NH_4NO_3 + Ca(NO_3)_2$ + formamide + H_2O are powerful explosives. Mixture with ammonium peroxodisulfate + water is explosive. Violent or explosive "thermite" reaction when heated with metal oxides, oxosalts (nitrates, sulfates), or sulfides, and with hot copper oxide worked with an iron or steel tool. Potentially explosive reaction with CCl_4 during ball milling operations. Many violent or explosive reactions with the following halocarbons have occurred in industry: bromomethane, bromotrifluoromethane, CCl_4, chlorodifluoromethane, chloroform, chloromethane, chloromethane + 2-methylpropane, dichlorodifluoromethane, 1,2-dichloroethane, dichloromethane, 1,2-dichloropropane, 1,2-difluorotetrafluoroethane, fluorotrichloroethane, hexachloroethane + alcohol, polytrifluoroethylene oils and greases, tetrachloroethylene, tetrafluoromethane, 1,1,1-trichloroethane, trichloroethylene, 1,1,2-trichlorotrifluoroethane, and trichlorotrifluoroethane-dichlorobenzene. Potentially explosive reaction with chloroform amidinium nitrate. Ignites on contact with vapors of $AsCl_3$, SCl_2, Se_2Cl_2, and PCl_5. Reacts violently on heating with Sb or As. Ignites on heating in $SbCl_3$ vapor. Ignites on contact with barium peroxide. Potentially violent reaction with sodium acetylide. Mixture with sodium peroxide may ignite or react violently. Spontaneously ignites in CS_2 vapor. Halogens: ignites in chlorine gas, foil reacts vigorously with liquid Br_2, violent reaction with $H_2O + I_2$. Violent reaction with hydrochloric acid, hydrofluoric acid, and hydrogen chloride gas. Violent reaction with disulfur dibromide. Violent reaction with the nonmetals phosphorus, sulfur, and selenium. Violent reaction or ignition with the interhalogens: bromine pentafluoride, chlorine fluoride, iodine chloride, iodine pentafluoride, and iodine heptafluoride. Burns when heated in CO_2. Ignites on contact with O_2 and mixtures with $O_2 + H_2O$ ignite and react violently. Mixture with picric acid + water ignites after a delay period. Explosive reaction above 800°C with sodium sulfate. Violent reaction with sulfur when heated. Exothermic reaction with iron powder + water releases explosive hydrogen gas.

Aluminum powder also forms sensitive explosive mixtures with oxidants such as: liquid Cl_2 and other halogens, N_2O_4, tetranitromethane, bromates, iodates, $NaClO_3$, $KClO_3$, and other chlorates, $NaNO_3$, aqueous nitrates, $KClO_4$ and other perchlorate salts, nitryl fluoride, ammonium peroxodisulfate, sodium peroxide, zinc peroxide, and other peroxides, red phosphorus, and powdered polytetrafluoroethylene (PTFE).

Bulk aluminum may undergo the following dangerous interactions: exothermic reaction with butanol, methanol, 2-propanol, or other alcohols, sodium hydroxide to release explosive hydrogen gas. Reaction with diborane forms pyrophoric product. Ignition on contact with niobium oxide + sulfur. Explosive reaction with molten metal oxides, oxosalts (nitrates, sulfates), sulfides, and sodium carbonate. Reaction with arsenic trioxide + sodium arsenate + sodium hydroxide produces the toxic arsine gas. Violent reaction with chlorine trifluoride. Incandescent reaction with formic acid. Potentially violent alloy formation with palladium, platinum at mp of Al, 600°C. Vigorous dissolution reaction in methanol + carbon tetrachloride. Vigorous amalgamation reaction with mercury (II) salts + moisture. Violent reaction with molten silicon steels. Violent exothermic reaction above 600°C with sodium diuranate.

AGX125 CAS:12607-92-0 ***HR: 3***
ALUMINUM ACEGLUTAMIDE
mf: $C_{35}H_{59}Al_3N_{10}O_{24}$ mw: 1084.98

SYNS: ACEGLUTAMIDE ALUMINUM ◇ N-ACETYL-l-GLUTAMINE ALUMINUM SALT ◇ GLUMAL ◇ KW 110 ◇ PENTAKIS(N(^{2})-ACETYL-l-GLUTAMINATO)TETRAHYDROXYTRIALUMINUM

TOXICITY DATA with REFERENCE
orl-rbt TDLo:20 g/kg (female 7-16D post):REP KSRNAM 8,959,74
orl-mus TDLo:28 g/kg (female 7-13D post):TER KSRNAM 8,959,74
ipr-rat LD50:4200 mg/kg USXXAM #3787466
ivn-rat LD50:400 mg/kg USXXAM #3787466
orl-mus LD50:13100 mg/kg NYKZAU 68,602,72
ipr-mus LD50:5 g/kg USXXAM #3787466
ivn-mus LD50:460 mg/kg USXXAM #3787466

SAFETY PROFILE: Poison by intravenous route. Moderately toxic by some other routes. An experimental teratogen. Other experimental reproductive effects. When heated to decomposition it emits toxic fumes of NO_x. See also ALUMINUM COMPOUNDS.

AGX250 ***HR: 1***
ALUMINUM AMMONIUM SULFATE
mf: $Al_2(SO_4)_3(NH_4)_2SO_4 \cdot 24H_2O$ mw: 906

PROP: Colorless crystals; odorless with sweet taste. D: 1.645, mp: 94.5°, bp: loses 20 waters @ 120°. Sol in water, glycerin; insol in alc.

SAFETY PROFILE: Irritating if inhaled or ingested. Upon decomposition it emits toxic fumes of NO_x and SO_x.

AGX300 CAS:39108-14-0 ***HR: 3***
ALUMINUM AZIDE
mf: AlN_9 mw: 153.04

SAFETY PROFILE: Shock sensitive explosive. When heated to decomposition it emits toxic fumes of NO_x. See also AZIDES and ALUMINUM COMPOUNDS.

AGX500 ***HR: 3***
ALUMINUM BOROHYDRIDE
mf: AlB_3H_{12} mw: 71.53

PROP: Liquid. Bp: 44.5°, mp: −64.5°, vap press: 400 mm @ 28.1°.

SYN: ALUMINUM TETRAHYDROBORATE

SAFETY PROFILE: Dangerous by spontaneous chemical reaction; ignites spontaneously in air, particularly in moist air. Explodes in O_2 at temperatures as low as 20°. An explosive range of 5 to 90%. Incompatible with water; steam; oxidizing materials; acid; acid fumes; will react with water or steam to produce heat, H_2, or toxic fumes. To fight fire, use CO_2, dry chemical. See HYDRIDES and BORON COMPOUNDS.

AGX750 CAS:7727-15-3 ***HR: 2***
ALUMINUM BROMIDE
DOT: UN 1725/UN 2580
mf: $AlBr_3$ mw: 266.71

PROP: White to yellow-red lumps. Mp: 97.5°, bp: 263.3° @ 748 mm, d: 3.2, vap press: 1 mm @ 81.3°.

SYNS: ALUMINUM BROMIDE, anhydrous ◇ ALUMINUM BROMIDE, solution (DOT) ◇ ALUMINUM TRIBROMIDE ◇ TRIBROMOALUMINUM

CONSENSUS REPORTS: Reported in EPA TSCA Inventory.

ACGIH TLV: TWA 2 mg(Al)/m^3
DOT Classification: Label: Corrosive.

SAFETY PROFILE: A toxic, corrosive material. See also BROMIDES and ALUMINUM COMPOUNDS. Mixtures with sodium or potassium explode violently upon impact. When heated to decomposition it emits toxic fumes of Br^-. Do not add H_2O to anhydrous material. Hydrolysis can be violent.

AGY000 CAS:12794-92-2 ***HR: 2***
ALUMINUM BROMIDE HYDROXIDE

SYNS: AL BROMOHYDRATE ◇ ALUMINUM BROMHYDROXIDE ◇ ALUMINUM BROMOHYDROL ◇ ALUMINUM HYDROXYBROMIDE

TOXICITY DATA with REFERENCE
skn-hmn 90 mg/3D-I MLD 85DKA8 -,127,77

ACGIH TLV: TWA 2 mg(Al)/m^3

SAFETY PROFILE: A human skin irritant. See also BROMIDES. When heated to decomposition it emits toxic fumes of Br-.

AGY100 ***HR: 1***
ALUMINUM CALCIUM SILICATE

SYN: CALCIUM ALUMINUM SILICATE

SAFETY PROFILE: A nuisance dust.

AGY250 ***HR: 2***
ALUMINUM CARBIDE
mf: Al_4C_3 mw: 143.91

PROP: Yellow crystals or powder, hygroscopic. Mp: 2100°, bp: decomp @ 2200°, d: 2.36.

SAFETY PROFILE: Decomposed by water. Mixture with lead dioxide or potassium permanganate reacts incandescently when warmed. Dust can cause pulmonary irritation. See also ALUMINUM COMPOUNDS.

AGY500 ***HR: 3***
ALUMINUM CHLORATE
mf: $Al(ClO_3)_3$ mw: 277.4

PROP: Colorless, deliquescent crystals. Mp: decomp.

SAFETY PROFILE: Flammable by spontaneous chemical reaction; a powerful oxidizer; may ignite upon contact with combustibles. Moderate explosion hazard when shocked, exposed to heat or by spontaneous chemical reaction with reducing agents. When contaminated, may become sensitized. Dangerous; shock or heat will explode it. Evaporation emits ClO_2. See also CHLORIDES and CHLORATES.

AGY750 CAS:7446-70-0 ***HR: 3***
ALUMINUM CHLORIDE
DOT: UN 1726/UN 2581
mf: $AlCl_3$ mw: 133.33

PROP: White hexagonal deliquescent crystals. D: 2.44, mp: 194° @ 5.2 atm, bp: subl @ 181°, vap press: 1 mm @ 100.0°. Violently sol in water; sol in alc and ether.

SYNS: ALLUMINIO(CLORURO DI) (ITALIAN) ◇ ALUMINUMCHLORID (GERMAN) ◇ ALUMINUM CHLORIDE (1:3) ◇ ALUMINUM CHLORIDE, anhydrous (DOT) ◇ ALUMINUM CHLORIDE, solution (DOT) ◇ ALUMINUM TRICHLORIDE ◇ CHLORURE d'ALUMINUM (FRENCH) ◇ PEARSALL ◇ TRICHLOROALUMINUM

TOXICITY DATA with REFERENCE
dnd-rat:ast 500 umol/L JBCHA3 261,3370,86
cyt-mus-ipr 444 mg/kg NULSAK 15,180,72
orl-rat TDLo:11512 mg/kg (female 8-22D post):REP NETOD7 8,115,86
ipr-rat TDLo:375 mg/kg (female 9-13D post):TER ANANAU 138,365,75
orl-rat LD50:3730 mg/kg EQSSDX 1,1,75
orl-mus LD50:3805 mg/kg BJIMAG 23,305,66

CONSENSUS REPORTS: Reported in EPA TSCA Inventory.

ACGIH TLV: TWA 2 mg(Al)/m^3
DOT Classification: IMO: Corrosive Material; Label: Corrosive.

SAFETY PROFILE: Moderately toxic by ingestion. Experimental teratogenic and reproductive effects. Mutation data reported. The dust is an irritant by ingestion, inhalation, and skin contact. Highly exothermic polymerization reactions with alkenes. Incompatible with nitrobenzenes or nitrobenzene + phenol. Highly exothermic reaction with water or steam produces toxic fumes of HCl. See also ALUMINUM COMPOUNDS, CHLORIDES, and HYDROCHLORIC ACID.

AGZ000 CAS:7784-13-6 ***HR: 2***
ALUMINUM CHLORIDE HEXAHYDRATE
mf: $AlCl_3 \cdot 6H_2O$ mw: 241.45

SYNS: ALUMINUM(III) CHLORIDE, HEXAHYDRATE ◇ ALUMINUM TRICHLORIDEHEXAHYDRATE

TOXICITY DATA with REFERENCE
skn-hmn 7500 μg/3D-I MLD 85DKA8 -,127,77

dnd-mam:lym 40 μmol/L JCHODP 7,411,76
ivn-mus TDLo:483 g/kg (8D preg):TER ENVRAL 33,22,86
orl-rat LD50:3311 mg/kg PHTXA6 60,280,87
ipr-rat LD50:728 mg/kg PHTXA6 60,280,87
orl-mus LD50:1990 mg/kg PHTXA6 60,280,87
ipr-mus LD50:940 mg/kg PHTXA6 60,280,87
ACGIH TLV: TWA 2 mg(Al)/m^3

SAFETY PROFILE: Moderately toxic by ingestion and intraperitoneal routes. An experimental teratogen. Corrosive and irritating to tissue. Mutation data reported. When heated to decomposition it emits toxic fumes of Cl^-. See also ALUMINUM COMPOUNDS and CLORIDES.

AHA000 CAS:12042-91-0 ***HR: 2***
ALUMINUM CHLORIDE HYDROXIDE
mf: $Al_2ClH_5O_5$ mw: 174.46

SYNS: ALUMINUM CHLORHYDRATE ◇ ALUMINUM CHLORHYDROL ◇ ALUMINUM CHLORHYDROXIDE ◇ ALUMINUM CHLOROHYDROXIDE ◇ ALUMINUM HYDROXIDE CHLORIDE ◇ ALUMINUM HYDROXYCHLORIDE ◇ ASTRINGEN ◇ BASIC ALUMINUM CHLORATE ◇ CHLORHYDROL ◇ CHLOROPENTAHYDROXYDIALUMINUM ◇ LOCRON EXTRA ◇ MICRO DRY ◇ WICKENOL 324

TOXICITY DATA with REFERENCE
skn-hmn 150 mg/3D-I MLD 85DKA8 -,127,77

CONSENSUS REPORTS: Reported in EPA TSCA Inventory.

ACGIH TLV: TWA 2 mg(Al)/m^3

SAFETY PROFILE: A mild human skin irritant. See also ALUMINUM COMPOUNDS, CHLORIDES, and SODIUM HYDROXIDE. When heated to decomposition it emits toxic fumes of Cl^-.

AHA125 CAS:3495-54-3 ***HR: 2***
ALUMINUM CHLORIDE NITROMETHANE
mf: $AlCl_3CH_3NO_2$ mw: 194.38

SAFETY PROFILE: Mixture with alkenes reacts explosively.

AHA150 CAS:24818-79-9 ***HR: 2***
ALUMINUM CLOFIBRATE
mf: $C_{20}H_{21}AlCl_2O_7$ mw: 471.29

SYNS: ALFIBRATE ◇ ALUFIBRATE ◇ ATHEROLIP ◇ BIS(2-(p-CHLOROPHENOXY)-2-METHYLPROPIONATO)HYDROXY-ALUMINUM ◇ HYDROXYBIS(2-(p-CHLOROPHENOXY)ISOBUTYRIC ACID) ALUMINUM

TOXICITY DATA with REFERENCE
ipr-rat LD50:3300 mg/kg JJPAAZ 23,281,73
scu-rat LD50:5100 mg/kg NIIRDN 6,237,82
ipr-mus LD50:3300 mg/kg NIIRDN 6,237,82
scu-mus LD50:5850 mg/kg NIIRDN 6,237,82

SAFETY PROFILE: Moderately toxic by some routes. When heated to decomposition it emits toxic fumes of Cl^-. See also ALUMINUM COMPOUNDS.

AHA175 ***HR: 2***
ALUMINUM COMPOUNDS

SAFETY PROFILE: Aluminum compounds have many commercial uses and are commonly found in industry. Many of these materials are active chemically and thus exhibit dangerous toxic and reactive properties. Inhalation of fine aluminum oxide particles is associated with Shaver's disease. The halides are generally irritants. See also ALUMINUM and individual compounds.

AHA250 CAS:7047-84-9 ***HR: 3***
ALUMINUM DEXTRAN
mf: $C_{18}H_{37}AlO_4$ mw: 344.48

PROP: Powder. A complex containing aluminum and dextran, a chain of molecular weight 2,500, corresponding to a chain of 15 anhydroglucose units.

SYN: ALUMINUM MONOSTEARATE ◇ ALUMINUM STEARATE (ACGIH) ◇ STEARIC ACID, ALUMINIUM SALT

TOXICITY DATA with REFERENCE
scu-mus TDLo:112 g/kg/14W-I:ETA JNCIAM 24,109,60

CONSENSUS REPORTS: EPA TSCA Chemical Inventory.

ACGIH TLV: TWA 10 mg/m^3

SAFETY PROFILE: Questionable carcinogen with experimental tumorigenic data. When heated to decomposition it emits acrid smoke and fumes. See also ALUMINUM COMPOUNDS.

AHA750 ***HR: 3***
ALUMINUM ETHYLATE
mf: $Al(OC_2H_5)_3$ mw: 162.15

PROP: Liquid. Decomp by H_2O. Bp: 200° @ 6-8 mm; mp: 140°.

SAFETY PROFILE: Strong irritant to skin, eyes, and mucous membranes by inhalation. See also ORGANO METALS.

AHA875 CAS:16449-54-0 ***HR: 2***
ALUMINUM FLUFENAMATE
mf: $C_{42}H_{27}AlF_9N_3O_6$ mw: 867.70

SYNS: ALFENAMIN ◇ ALUFENAMINE ◇ ALUMINUM TRIS(2-((3-TRIFLUOROMETHYL)PHENYL)AMINO)BENZOATO-N,o)-o-PYRIN ◇ TRIS(N-(α,α,α-TRIFLUORO-m-TOLYL)ANTHRANILATO)ALUMINUM ◇ TS 1801

TOXICITY DATA with REFERENCE
orl-mus TDLo:3 g/kg (female 7-12D post):REP
TOIZAG 17,159,70

orl-rat LD50:550 mg/kg TOIZAG 17,153,70
ipr-rat LD50:420 mg/kg NIIRDN 4,1289,35
scu-rat LD50:725 mg/kg NIIRDN 6,704,82
orl-mus LD50:1460 mg/kg NIIRDN 6,704,82
ipr-mus LD50:1560 mg/kg NIIRDN 6,704,82

SAFETY PROFILE: Moderately toxic by ingestion and other routes. Experimental reproductive effects. When heated to decomposition it emits toxic fumes of F^- and NO_x. See also ALUMINUM COMPOUNDS.

AHB000 CAS:7784-18-1 ***HR: 3***
ALUMINUM FLUORIDE
mf: AlF_3 mw: 83.98

PROP: Solid, mp: 1291°, subl. @ 1260°, d: 2.88, vap press: 1 mm @ 1238°, bp: 1537°.

SYNS: ALUMINUM FLUORURE (FRENCH) ◇ ALUMINUM TRIFLUORIDE ◇ FLUORID HLINITY (CZECH)

TOXICITY DATA with REFERENCE
eye-rbt 500 mg/24H SEV 28ZPAK -,20,72
scu-frg LDLo:1680 mg/kg CRSBAW 124,133,37

CONSENSUS REPORTS: Reported in EPA TSCA Inventory.

OSHA PEL: TWA 2.5 mg(F)/m^3
ACGIH TLV: TWA 2 mg(Al)/m^3; 2.5 mg(F)/m^3
NIOSH REL: (Fluorides, Inorganic) TWA 2.5 mg(F)/m^3

SAFETY PROFILE: A poison. Moderately toxic by subcutaneous routes. A severe eye irritant. Violently impact-sensitive when in contact with Na and K. When heated to decomposition it emits toxic fumes of F^-. See also FLUORIDES and ALUMINUM COMPOUNDS.

AHB250 CAS:73680-58-7 ***HR: 3***
ALUMINUM FLUOROSULFATE, HYDRATE

TOXICITY DATA with REFERENCE
ivn-mus LD50:56 mg/kg CSLNX* NX#00137

OSHA PEL: TWA 2.5 mg(F)/m^3
ACGIH TLV: TWA 2 mg(Al)/m^3; 2.5 mg(F)/m^3
NIOSH REL: (Fluorides, Inorganic) TWA 2.5 mg(F)/m^3

SAFETY PROFILE: Poison by intravenous route. See also ALUMINUM COMPOUNDS, SULFATES, and FLUORIDES. When heated to decomposition it emits very toxic fumes of F^- and SO_x.

AHB375 CAS:7360-53-4 ***HR: 2***
ALUMINUM FORMATE
mf: $C_3H_3AlO_6$ mw: 162.03

SAFETY PROFILE: Aqueous solution explodes when heated in air.

AHB500 CAS:7784-21-6 ***HR: 3***
ALUMINUM HYDRIDE
DOT: UN 2463
mf: AlH_3 mw: 30.01

PROP: Colorless powder.

SYNS: ALANE ◇ ALUMINUM TRIHYDRIDE ◇ α-ALUMINUM TRIHYDRIDE

CONSENSUS REPORTS: Reported in EPA TSCA Inventory.

ACGIH TLV: TWA 2 mg(Al)/m^3
DOT Classification: Label: Flammable Solid and Dangerous When Wet; IMO: Flammable Solid; Label: Spontaneously Combustible.

SAFETY PROFILE: Hydrides of some metals (such as AsH_3) are extremely toxic. Dangerous fire hazard. An unstable material which is spontaneously flammable in air or O_2. Evolves explosive H_2 upon contact with moisture. Severe explosion hazard by chemical reaction wherein H_2 gas is produced, also in contact with methyl ethers contaminated by CO_2. Mixtures with tetrazole derivatives are explosive. Reacts with oxidizing materials. On contact with acid or acid fumes, it can emit toxic fumes. See also HYDRIDES and ALUMINUM COMPOUNDS.

AHB625 CAS:26351-01-9 ***HR: 3***
ALUMINUM HYDRIDE-DIETHYL ETHER
mf: AlH_3-$C_4H_{10}O$ mw: 104.13

SAFETY PROFILE: Reacts violently with moist air or water. When heated to decomposition it emits acrid smoke and fumes. See also ALUMINUM COMPOUNDS and ETHERS.

AHB750 CAS:17013-07-9 ***HR: 3***
ALUMINUM HYDRIDE-TRIMETHYL AMINE
mf: $AlH_3 \cdot C_3H_9N$ mw: 89.1

SAFETY PROFILE: An unstable, dangerous compound. Explodes on contact with water. Ignites in moist air. When heated to decomposition it emits toxic fumes of NO_x. See also AMINES, HYDRIDES, and ALUMINUM COMPOUNDS.

AHC000 CAS:21645-51-2 ***HR: 3***
ALUMINUM HYDROXIDE
mf: AlH_3O_3 mw: 78.01

PROP: White, crystalline powder, balls or granules; insol in water; sol in mineral acids and caustic soda. D: 2.42; mp: loses H_2O @ 300°.

SYNS: AF 260 ◇ ALCOA 331 ◇ ALUMIGEL ◇ ALUMINA HYDRATE ◇ ALUMINA HYDRATED ◇ ALUMINA TRIHYDRATE ◇ α-ALUMINA TRIHYDRATE ◇ ALUMINIC ACID ◇ ALUMINUM HYDRATE ◇ ALU-

MINUM(III) HYDROXIDE ◇ ALUMINUM HYDROXIDE GEL ◇ ALUMINUM OXIDE HYDRATE ◇ ALUMINUM OXIDE TRIHYDRATE ◇ ALUMINUM TRIHYDRAT ◇ ALUMINUM TRIHYDROXIDE ◇ ALUSAL ◇ AMBEROL ST 140F ◇ AMPHOJEL ◇ BACO AF 260 ◇ BRITISH ALUMINUM AF 260 ◇ C.I. 77002 ◇ GHA 331 ◇ H 46 ◇ HIGILITE ◇ HYDRAL 705 ◇ LIQUIGEL ◇ PGA ◇ TRIHYDRATED ALUMINA

TOXICITY DATA with REFERENCE
cyt-rat-ipr 20 mg/kg MUREAV 30,407,75
orl-chd TDLo:122 g/kg/4D:GIT,MET JOPDAB 92,592,78
ipr-rat LDLo:150 mg/kg LANCAO 1,564,72

CONSENSUS REPORTS: Reported in EPA TSCA Inventory.

ACGIH TLV: TWA 2 mg(Al)/m^3

SAFETY PROFILE: Poison by intraperitoneal route. Human systemic effects by ingestion: fever and gastrointestinal effects. Mutation data reported. When coprecipitated with bismuth hydroxide and reduced by H_2, it is violently flammable in air. Incompatible with chlorinated rubber.

AHC250 CAS:24623-77-6 ***HR: 3***
ALUMINUM HYDROXIDE OXIDE
mf: $AlHO_2$ mw: 59.99

SYNS: ALUMINUM METAHYDROXIDE ◇ HYDRATED ALUMINA

TOXICITY DATA with REFERENCE
itr-rat LDLo:90 mg/kg JPBAA7 69,81,55

CONSENSUS REPORTS: Reported in EPA TSCA Inventory.

ACGIH TLV: TWA 2 mg(Al)/m^3

SAFETY PROFILE: Poison by intratracheal route. See also ALUMINUM COMPOUNDS.

AHC500 ***HR: 3***
ALUMINUM IODIDE
mf: AlI_3 mw: 407.7

PROP: White leaflets. Mp: 191°, bp: 360°, d: 3.98 @ 25°, vap press: 1 mm @ 178.0° (sublimes).

SAFETY PROFILE: Incompatible with water. See ALUMINUM COMPOUNDS and IODIDES.

AHC750 CAS:18917-91-4 ***HR: D***
ALUMINUM LACTATE
mf: $C_9H_{15}AlO_9$ mw: 294.22

SYNS: ALUMINUM, TRIS(2-HYDROXYPROPANOATO-O^1),O^2))- (9CI) ◇ ALUMINUM, TRIS(LACTATO)-

TOXICITY DATA with REFERENCE
orl-mus TDLo:41 g/kg (female 1-21D post):REP TJADAB 31,64A,85
scu-mus TDLo:1524 mg/kg (female 3-15D post):TER FAATDF 8,346,87

CONSENSUS REPORTS: Reported in EPA TSCA Inventory.

SAFETY PROFILE: An experimental teratogen. Other experimental reproductive effects. When heated to decomposition it emits acrid smoke and irritating fumes.

AHD250 ***HR: 3***
ALUMINUM MAGNESIUM PHOSPHIDE
DOT: UN 1419
mf: Mg_3AlP_3 mw: 192.8

SYN: MAGNESIUM ALUMINUM PHOSPHIDE (DOT)

ACGIH TLV: TWA 2 mg(Al)/m^3
DOT Classification: Label: Flammable Solid and Dangerous When Wet.

SAFETY PROFILE: A poison. Dangerous fire hazard. Evolves spontaneously flammable PH_3 in contact with water. See also PHOSPHIDES, PHOSPHINE, ALUMINUM COMPOUNDS, and MAGNESIUM COMPOUNDS.

AHD500 ***HR: 3***
ALUMINUM METHYL
mf: $Al(CH_3)_3$ mw: 72.07

PROP: Colorless liquid. Bp: 130°; mp: 0°.

SAFETY PROFILE: Related alkyl aluminum compounds are poisonous and strong irritants. Very flammable by spontaneous chemical reaction with air. Incompatible with water, halogenated hydrocarbons, and oxidizing materials. When heated to decomposition it emits toxic fumes. To fight fire, do not use water, foam, or halogenated extinguishing agents. Use dry chemical.

AHD750 CAS:13473-90-0 ***HR: 3***
ALUMINUM(III) NITRATE (1583)
DOT: UN 1438
mf: $N_3O_9 \cdot Al$ mw: 213.01

PROP: White crystals.

SYNS: ALUMINUM NITRATE (DOT) ◇ ALUMINUM TRINITRATE ◇ NITRIC ACID, ALUMIUM SALT ◇ NITRIC ACID, ALUMINUM(3+) SALT

TOXICITY DATA with REFERENCE
skn-rbt 500 mg MLD FCTOD7 20,563,82
eye-rbt 100 mg SEV FCTOD7 20 573,82
eye-rbt 100 mg/4S rns MLD FCTOD7 20,573,82
orl-rat LD50:4280 mg/kg 85INA8 5,22,86

CONSENSUS REPORTS: Reported in EPA TSCA Inventory.

ACGIH TLV: TWA 2 mg(Al)/m^3
DOT Classification: Oxidizer; Label: Oxidizer

SAFETY PROFILE: A poison. A severe eye and mild skin irritant. A powerful oxidizer. When heated to decomposition it emits toxic NO_x. See NITRATES and ALUMINUM COMPOUNDS. A nitrating agent.

AHD900 CAS:7784-27-2 ***HR: 3***
ALUMINUM(III) NITRATE, NONAHYDRATE (1:3:9)
mf: N_3O_9•Al•$9H_2O$ mw: 375.19

SYNS: ALUMINUM NITRATE NONAHYDRATE ◇ ALUMINUM TRINITRATE NONAHYDRATE ◇ NITRIC ACID, ALUMINUM SALT, NONAHYDRATE (8CI,9CI)

TOXICITY DATA with REFERENCE
orl-rat TDLo:7740 mg/kg (female 1-22D post):REP RCOCB8 57,129,87
orl-rat TDLo:1620 mg/kg (female 6-14D post):TER TJADAB 38,253,88
orl-rat LD50:264 mg/kg HYSAAV 31(7-9),204,66
ipr-rat LD50:901 mg/kg PHTXA6 60,280,87
orl-mus LD50:3980 mg/kg PHTXA6 60,280,87
ipr-mus LD50:1587 mg/kg PHTXA6 60,280,87

ACGIH TLV: TWA 2 mg(Al)/m^3

SAFETY PROFILE: Poison by ingestion. Moderately toxic by intraperitoneal route. An experimental teratogen. Other experimental reproductive effects.

AHE000 ***HR: 3***
ALUMINUM NITRIDE
mf: AlN mw: 41

PROP: White or colorless crystals. Mp: 2200°; bp: sublimes @ 2000°; d: 3.26.

SAFETY PROFILE: A poison. Will react with water or steam to produce toxic or corrosive fumes. Incompatible with water or steam. See also NITRIDES, AMMONIA, and ALUMINUM COMPOUNDS.

AHE250 CAS:1344-28-1 ***HR: 2***
ALUMINUM OXIDE (2583)
mf: Al_2O_3 mw: 101.96

PROP: White powder. Mp: 2050°, bp: 2977°, d: 3.5-4.0, vap press: 1 mm @ 2158°.

SYNS: A 1 (sorbent) ◇ A1-0109 P ◇ ABRAREX ◇ ACTIVATED ALUMINUM OXIDE ◇ ALCOA F 1 ◇ ALMITE ◇ ALON ◇ ALUMINA ◇ α-ALUMINA (OSHA) ◇ β-ALUMINA ◇ Γ-ALUMINA ◇ ALUMINUM OXIDE ◇ α-ALUMINUM OXIDE ◇ β-ALUMINUM OXIDE ◇ Γ-ALUMINUM OXIDE ◇ ALUMINUM SESQUIOXIDE ◇ ALUMITE ◇ ALUNDUM ◇ BROCKMANN, ALUMINUM OXIDE ◇ CAB-O-GRIP ◇ COMPALOX ◇ DIALUMINUM TRIOXIDE ◇ DISPAL ◇ DOTMENT 324 ◇ FASERTON ◇ G 2 (OXIDE) ◇ KHP 2 ◇ LUCALOX ◇ MICROGRIT WCA ◇ PS 1 ◇ RC 172DBM

TOXICITY DATA with REFERENCE
ipl-rat TDLo:90 mg/kg:ETA BJCAAI 28,173,73
imp-rat TDLo:200 mg/kg:NEO JJIND8 67,965,81

CONSENSUS REPORTS: Community Right-To-Know List. Reported in EPA TSCA Inventory.

OSHA PEL: Total Dust: (Transitional: TWA 5 mg/m^3) TWA 10 mg/m^3; Respirable Fraction: TWA 5 mg/m^3
ACGIH TLV: TWA (nuisance particulate) 10 mg/m^3 of total dust (when toxic impurities are not present, e.g., quartz $< 1\%$).
DFG MAK: 6 mg/m^3 (fume)

SAFETY PROFILE: Inhalation of finely divided particles may cause lung damage (Shaver's disease). Questionable carcinogen with experimental neoplastigenic and tumorigenic data by implantation. Exothermic reaction above 200°C with halocarbon vapors produces toxic HCl and phosgene. See also ALUMINUM COMPOUNDS.

AHE750 CAS:20859-73-8 ***HR: 3***
ALUMINUM PHOSPHIDE
DOT: UN 1397
mf: AlP mw: 57.95

PROP: Dark gray or dark yellow crystals. D: 2.85 @ 25°/4°. Mp: >1000°.

SYNS: AIP ◇ AL-PHOS ◇ ALUMINUM FOSFIDE (DUTCH) ◇ ALUMINUM MONOPHOSPHIDE ◇ CELPHIDE ◇ CELPHOS ◇ DELICIA ◇ DETIA GAS EX-B ◇ FOSFURI di ALLUMINIO (ITALIAN) ◇ FUMITOXIN ◇ PHOSPHURES d'ALUMIUM (FRENCH) ◇ RCRA WASTE NUMBER P006

TOXICITY DATA with REFERENCE
orl-hmn LD50:20 mg/kg 85ARAE 3,38,76
ihl-mam LCLo:1 ppm PCOC** -,25,66

CONSENSUS REPORTS: EPA Extremely Hazardous Substances List. Reported in EPA TSCA Inventory.

ACGIH TLV: TWA 2 mg(Al)/m^3
DOT Classification: Label: Flammable Solid and Dangerous When Wet; IMO: Flammable Solid; Label: Dangerous When Wet and Poison.

SAFETY PROFILE: A human poison by inhalation and ingestion. Dangerous; in contact with water, steam, or alkali it slowly yields PH_3, which is spontaneously flammable in air. Explosive reaction on contact with mineral acids produces phosphine. When heated to decomposition it yields toxic PO_x. See also ALUMINUM COMPOUNDS, PHOSPHIDES, and PHOSPHINE.

AHE875 CAS:24704-64-1 ***HR: 3***
ALUMINUM PHOSPHINATE
mf: $AlH_6O_6P_3$ mw: 221.95

$$Al[P(H)(O)OH]_3$$

SAFETY PROFILE: A poison. When heated to decomposition it emits toxic and spontaneously flammable fumes of phosphine. See also ALUMINUM COMPOUNDS and PHOSPHINE.

AHF000 ***HR: 3***
ALUMINUM PICRATE
mf: $Al(C_6H_2O(NO_2)_3)_3$ mw: 711.3

PROP: A solid.

SAFETY PROFILE: A poison. A powerful irritant. Very flammable by reaction with reducing materials. Severe explosion hazard when shocked or exposed to heat. See also EXPLOSIVES (HIGH). When heated to decomposition it emits highly toxic fumes of NO_x and explodes.

AHF100 CAS:10043-67-1 ***HR: 1***
ALUMINUM POTASSIUM SULFATE
mf: $AlK(SO_4)_2 \cdot 12H_2O$ mw: 474.38

PROP: Transparent crystals or white crystalline powder; odorless with sweet taste. Sol in water, glycerin; insol in alc.

SYN: POTASSIUM ALUM

SAFETY PROFILE: A nuisance dust.

AHF200 CAS:7784-24-9 ***HR: D***
ALUMINUM POTASSIUM SULFATE, DODECAHYDRATE
mf: $O_8S_2 \cdot Al \cdot K \cdot 12H_2O$ mw: 474.39

SYNS: ALUM ◇ KALINITE ◇ POTASSIUM ALUM ◇ POTASSIUM ALUM DODECAHYDRATE ◇ SULFURIC ACID, ALUMINUM POTASSIUM SALT (2:1:1), DODECAHYDRATE

TOXICITY DATA with REFERENCE
orl-rat TDLo:1120 mg/kg (female 7-14D post):TER OYYAA2 24,65,82

ACGIH TLV: TWA 2 mg(Al)/m^3

SAFETY PROFILE: An experimental teratogen.

AHF500 CAS:1302-76-7 ***HR: 3***
ALUMINUM(III) SILICATE (2581)
mf: $O_5Si \cdot 2Al$ mw: 162.05

SYNS: ALUMINUM OXIDE SILICATE ◇ CERAMIC FIBRE ◇ CYANITE ◇ DISTHENE ◇ KYANITE ◇ OIL-DRI ◇ SAFE-N-DRI ◇ SILICIC ACID ALUMINUM SALT ◇ SNOW TEX ◇ VALFOR

TOXICITY DATA with REFERENCE
ipl-rat TDLo:90 mg/kg:ETA BJCAAI 28,173,73

ACGIH TLV: TWA 2 mg(Al)/m^3

SAFETY PROFILE: Questionable carcinogen with experimental tumorigenic data by implantation. See also ALUMINUM COMPOUNDS.

AHF750 ***HR: 3***
ALUMINUM SODIUM LACTATE

SYN: ALUMINUMNATRIUMLACTAT(GERMAN)

TOXICITY DATA with REFERENCE
scu-mus LDLo:200 mg/kg HBAMAK 4,1289,35
scu-dog LDLo:15 mg/kg HBAMAK 4,1289,35
scu-cat LDLo:150 mg/kg HBAMAK 4,1289,35
scu-rbt LDLo:750 mg/kg HBAMAK 4,1289,35
scu-frg LDLo:600 mg/kg HBAMAK 4,1289,35
ACGIH TLV: TWA 2 mg(Al)/m^3

SAFETY PROFILE: Poison by subcutaneous route. See also ALUMINUM COMPOUNDS. When heated to decomposition it emits acrid smoke and irritating fumes.

AHG000 CAS:11138-49-1 ***HR: 2***
ALUMINUM SODIUM OXIDE
DOT: UN 1819/UN 2812
mf: $NaAlO_2$ mw: 82.0

PROP: White, hygroscopic powder. Mp: 1650°.

SYNS: β-ALUMINA ◇ β''-ALUMINA ◇ NALCO 680 ◇ SODIUM ALUMINATE, solid (DOT) ◇ SODIUM ALUMINUM OXIDE ◇ SODIUM POLYALUMINATE

ACGIH TLV: TWA 2 mg(Al)/m^3
DOT Classification: ORM-B; Label: None, solid; Corrosive Material; Label: Corrosive, solution.

SAFETY PROFILE: Moderate irritant to skin, eyes, and mucous membranes. A corrosive substance. When heated to decomposition it emits toxic fumes of Na_2O.

AHG500 ***HR: 2***
ALUMINUM SODIUM SULFATE
mf: $NaAl(SO_4)_2 \cdot 12H_2O$ mw: 458.29

PROP: Colorless crystals. Mp: 61°; d: 1.675. Anhydrous: sol in alc; sltly sol in water. Dodecahydrate: sol in water and alc.

SYNS: SODA ALUM ◇ SODIUM ALUMINUM SULFATE

SAFETY PROFILE: A weak sensitizer. A general-purpose food additive. Local contact may cause contact dermatitis. An irritant. See also SULFATES and ALUMINUM COMPOUNDS. When heated to decomposition it emits toxic fumes of SO_x and Na_2O.

AHG750 CAS:10043-01-3 ***HR: 2***
ALUMINUM SULFATE (2583)
DOT: UN 1760
mf: $O_{12}S_3 \cdot 2Al$ mw: 342.14

PROP: White powder; sweet taste. Mp: decomp @ 770°, d: 2.71. Solubility in water = 36.4% @ 20°.

SYNS: ALUM ◇ ALUMINUM TRISULFATE ◇ CAKE ALUM

◇ DIALUMINUM SULPHATE ◇ DIALUMINUM TRISULFATE ◇ SULFURIC ACID, ALUMINUM SALT (3:2)

TOXICITY DATA with REFERENCE
scu-mus TDLo:27371 μg/kg (30D male):REP JRPFA4 7,21,64
orl-mus LD50:6207 mg/kg BJIMAG 23,305,66
ipr-mus LD50:1735 mg/kg COREAF 256,1043,63

CONSENSUS REPORTS: Reported in EPA TSCA Inventory.

ACGIH TLV: TWA 2 mg(Al)/m^3
DOT Classification: ORM-E; Label: None, solid; ORM-B; Label: None, solution.

SAFETY PROFILE: Moderately toxic by ingestion and intraperitoneal routes. Experimental reproductive effects. Hydrolyzes to form sulfuric acid which irritates tissue, especially lungs. When heated to decomposition it emits toxic fumes of SO_x.

AHG875 CAS:16962-07-5 ***HR: 3***
ALUMINUM TETRAHYDROBORATE
DOT: UN 2870
mf: AlB_3H_{12} mw: 71.51

ACGIH TLV: TWA 2 mg(Al)/m^3
DOT Classification: Flammable Solid; Label: Spontaneously Combustible, Danger When Wet

SAFETY PROFILE: A poison. Spontaneously flammable in air. Explodes in oxygen with traces of water. Incompatible with alkenes and water. See also ALUMINUM COMPOUNDS and BORON COMPOUNDS.

AHH000 ***HR: 3***
ALUMINUM THALLIUM SULFATE
mf: $AlTl(SO_4)_2 \cdot 12H_2O$ mw: 639.6

PROP: Cubic, octagonal, colorless crystals. Mp: 91°; d: 2.32 @ 20°/4°.

CONSENSUS REPORTS: Thallium and its compounds are on the Community Right-To-Know List.

SAFETY PROFILE: A poison. See also THALLIUM COMPOUNDS and ALUMINUM COMPOUNDS.

AHH125 CAS:12003-96-2 ***HR: 2***
ALUMINUM-TITANIUM ALLOY (1581)
mf: AlTi mw: 74.8814

SAFETY PROFILE: Incompatible with chlorine; bromine; iodine; and hydrogen chloride vapors. See also ALUMINUM COMPOUNDS and TITANIUM COMPOUNDS.

AHH750 ***HR: 3***
ALUMINUM TRIPROPYL
mf: $Al(C_3H_7)_3$ mw: 156.24

PROP: Liquid.

SYN: TRIPROPYL ALUMINUM

SAFETY PROFILE: Related alkyl aluminum compounds are poisons. Very flammable by spontaneous reaction with air. See also DIISOBUTYL ALUMINUM CHLORIDE. Incompatible with halogenated hydrocarbons. Hydrolyzes to evolve flammable vapor. To fight fire, do not use water, foam or halogenated extinguishing agents. Use dry chemical or a special powder extinguisher.

AHI250 CAS:963-07-5 ***HR: 3***
ALYPIN
mf: $C_{16}H_{26}N_2O_2$ mw: 278.44

PROP: White, crystalline powder.

SYNS: ALYPINE ◇ AMYDRICAINE ◇ BENZOPROPYL ◇ 1-(DIMETHYLAMINO)-2-((DIMETHYLAMINO)METHYL)-2-BUTANOL BENZOATE,(ESTER) ◇ DIMETHYLAMINOSTOVAINE ◇ 2-ETHYL-1,3-BIS(DISMETHYLAMINO)-2-PROPANOLBENZOATE

TOXICITY DATA with REFERENCE
scu-rat LDLo:200 mg/kg PHREA7 12,262,32
ivn-rat LDLo:10 mg/kg PHREA7 12,262,32
scu-mus LDLo:260 mg/kg PHREA7 12,262,32
scu-dog LDLo:70 mg/kg PHREA7 12,262,32
scu-cat LDLo:60 mg/kg PHREA7 12,262,32
ivn-cat LDLo:10 mg/kg PHREA7 12,262,32
scu-rbt LDLo:96 mg/kg PHREA7 12,262,32
ivn-rbt LDLo:10 mg/kg PHREA7 12,262,32
ipr-gpg LDLo:100 mg/kg PHREA7 12,262,32
scu-gpg LDLo:72 mg/kg PHREA7 12,262,32
ivn-gpg LDLo:15 mg/kg PHREA7 12,262,32
scu-frg LDLo:200 mg/kg PHREA7 12,262,32

SAFETY PROFILE: Poison by subcutaneous, intravenous, and intraperitoneal routes. An allergen. See also ESTERS. When heated to decomposition it emits toxic fumes of NO_x.

AHI500 CAS:73990-29-1 ***HR: 3***
AMANITA RUBESCENS TOXIN

SYNS: MUSHROOM AMNITA RUBESCENS TOXIN ◇ RUBESCENSLYSIN

TOXICITY DATA with REFERENCE
ivn-rat LD50:1400 μg/kg TOXIA6 17(Suppl. 1),165,79
ivn-mus LD50:300 μg/kg TOXIA6 17(Suppl. 1),165,79

SAFETY PROFILE: Deadly poison by intravenous route.

AHI625 CAS:23109-05-9 ***HR: 3***
α-AMANITINE
mf: $C_{39}H_{54}N_{10}O_{14}S$ mw: 919.09

SYN: α-AMANITIN (8CI, 9CI)

TOXICITY DATA with REFERENCE
oms-omi 700 mg/L AMICCW 122,161,79
dni-dmg:oth 20 mg/L IJEBA6 15,973,77
oms-mus:ast 2 mg/L JANTAJ 36,155,83
ipr-mus LD50:100 μg/kg NEJMAG 269,223,63

CONSENSUS REPORTS: EPA Genetic Toxicology Program.

SAFETY PROFILE: Poison by intraperitoneal route. Mutation data reported. When heated to decomposition it emits toxic fumes of SO_x and NO_x.

AHI630 ***HR: 2***
AMARBEL EXTRACT

PROP: Indian plant belonging to the family *Convolvulaceae* (IJEBA6 7,250,69)

SYN: ALGUSI, extract ◇ CUSCUTA REFLEXA Roxb., extract excluding roots

TOXICITY DATA with REFERENCE
orl-rat TDLo:5600 mg/kg (female 1-7D post):REP IJMRAQ 70,517,79
ipr-mus LD50:750 mg/kg IJEBA6 7,250,69

SAFETY PROFILE: Moderately toxic by intraperitoneal route. Experimental reproductive effects. When heated to decomposition it emits acrid smoke and irritating fumes.

AHI635 ***HR: 2***
AMARYLLIS

PROP: Bulb-producing, flowering plants which are cultivated as ornamentals in gardens and indoors.

SYNS: A. BELLADONNA ◇ AZUCENA de MEJICO (MEXICO) ◇ BARBADOS LILY ◇ BELLADONNA LILY ◇ CAPE BELLADONNA ◇ HIPPEASTRUM (VARIOUS SPECIES) ◇ LIRIO ◇ NAKED LADY LILY ◇ TARARACO ◇ TARARACO DOBLE (CUBA)

SAFETY PROFILE: The bulbs contain the emetic narcissine (lycorine). Ingestion of a large amount of the bulbs may cause nausea, vomiting, and diarrhea.

AHI750 ***HR: 3***
AMATOL

PROP: A high explosive. Composition: NH_4NO_3, 80%; and TNT, 20%; d: 1.47.

SAFETY PROFILE: Moderately toxic by inhalation and ingestion routes. An allergen. May cause contact dermatitis. See also NITRATES. Dangerous fire hazard. An explosive by shock, spontaneous chemical reaction, or exposure to flame. Decomposition emits highly toxic fumes.

AHI875 CAS:539-21-9 ***HR: 3***
AMBAZONE
mf: $C_8H_{11}N_7S$ mw: 237.32

SYNS: AMBAZON ◇ 1-AMIDINOHYDRAZONO-4-THIOSEMICARBAZONO-2,5-CYCLOHEXADIENE ◇ ANGINON ◇ p-BENZOQUINONE AMIDINOHYDRAZONE THIOSEMICARBAZONE ◇ BENZOQUINONE GUANYLHYDRAZONE THIOSEMICARBAZONE ◇ DC 0572 ◇ FARINGOSEPT ◇ GUANOTHIAZON ◇ INVERSAL ◇ IVERSAL ◇ IVERTOL ◇ ((4-OXO-2,5-CYCLOHEXADIEN-1-YLIDENE)AMINO)GUANIDINE THIOSEMICARBAZONE ◇ PRIMAL ◇ PROMASSOL

TOXICITY DATA with REFERENCE
orl-rat LD50:750 mg/kg NIIRDN 6,57,82
ipr-rat LD50:200 mg/kg NIIRDN 6,57,82
orl-mus LD50:1000 mg/kg NIIRDN 6,57,82
orl-gpg LD50:80 mg/kg NIIRDN 6,57,82

SAFETY PROFILE: Poison by ingestion and intraperitoneal routes. When heated to decomposition it emits toxic fumes of SO_x and NO_x.

AHJ000 CAS:9000-02-6 ***HR: 1***
AMBERGRIS TINCTURE

PROP: Concretion from intestine of sperm whale, composed mostly of cholesterol.

SYNS: AMBER ◇ AMBRA ◇ GRAY AMBER

TOXICITY DATA with REFERENCE
skn-rbt 500 mg/24H MLD FCTXAV 14,659,76

SAFETY PROFILE: A mild skin irritant. When heated to decomposition it emits acrid smoke and irritating fumes.

AHJ250 CAS:18683-91-5 ***HR: 3***
AMBROXOL
mf: $C_{13}H_{18}Br_2N_2O$ mw: 378.15

SYNS: trans-4-((2-AMINO-3,5-DIBROMOBENCIL)AMINO CICLOHEXANOL (SPANISH) ◇ N-(2-AMINO-3,4-DIBROMOCICLOHEXIL)-trans-4-AMINOCICLOHEXANOL(SPANISH) ◇ N-(2-AMINO-3,4-DIBROMOCYCLOHEXYL)-trans-4-AMINOCYCLOHEXANOL ◇ N-(trans-4-HIDROXICICLOHEXIL)-(2-AMINO-3,5-DIBROMOBENCIL)AMINA (SPANISH) ◇ N-(trans-4-HYDROXYCYCLOHEXYL)-(2-AMINO-3,5-DIBROMOBENZYL)-AMINE ◇ NA-872

TOXICITY DATA with REFERENCE
orl-rat LD50:13400 mg/kg MDACAP 15,523,79
ipr-rat LD50:380 mg/kg MDACAP 15,523,79
orl-mus LD50:2720 mg/kg MDACAP 15,523,79
ipr-mus LD50:268 mg/kg MDACAP 15,523,79
ivn-mus LD50:138 mg/kg MDACAP 15,523,79

orl-gpg LD50:1180 mg/kg MDACAP 15,523,79
ipr-gpg LD50:280 mg/kg MDACAP 15,523,79

SAFETY PROFILE: Poison by intraperitoneal and intravenous routes. Moderately toxic by ingestion. When heated to decomposition it emits very toxic fumes of Br^- and NO_x.

AHJ500 CAS:23828-92-4 ***HR: 3***
AMBROXOL HYDROCHLORIDE
mf: $C_{13}H_{18}Br_2N_2O{\bullet}ClH$ mw: 414.61

SYNS: trans-4-((2-AMINO-3,5-DIBROMOBENZYL)AMINO)-CYCLOHEXANOL HYDROCHLORIDE ◇ MUCOSOLVAN ◇ NA 872

TOXICITY DATA with REFERENCE
orl-rat TDLo:88 mg/kg (female 7-17D post):REP OYYAA2 21,271,81
orl-rat TDLo:33 g/kg (7-17D preg):TER OYYAA2 21,271,81
orl-rat LD50:4203 mg/kg IYKEDH 12,263,81
ipr-mus LD50:268 mg/kg ARZNAD 28,889,78
scu-rat LD50:1489 mg/kg IYKEDH 12,263,81
ivn-rat LD50:100 mg/kg OYYAA2 21,281,81
orl-mus LD50:2380 mg/kg OYYAA2 21,281,81
ipr-mus LD50:268 mg/kg ARZNAD 28,889,78
scu-mus LD50:1060 mg/kg IYKEDH 12,263,81
ivn-mus LD50:138 mg/kg ARZNAD 28,889,78
orl-dog LD50:500 mg/kg ARZNAD 28,889,78

SAFETY PROFILE: Poison by intraperitoneal and intravenous routes. Moderately toxic by ingestion and subcutaneous routes. An experimental teratogen. Other experimental reproductive effects. When heated to decomposition it emits very toxic fumes of Br^-, NO_x, and HCl.

AHJ750 CAS:52645-53-1 ***HR: 3***
AMBUSH
mf: $C_{21}H_{20}Cl_2O_3$ mw: 391.31

SYNS: AI3-29158 ◇ BW-21-Z ◇ ECTIBAN ◇ EXMIN ◇ FMC 33297 ◇ FMC 41655 ◇ ICI-PP 557 ◇ KESTREL (Pesticide) ◇ NDRC-143 ◇ NIA 33297 ◇ OUTFLANK ◇ OUTFLANK-STOCKADE ◇ PERMETHRIN (USDA) ◇ PERMETRIN (HUNGARIAN) ◇ PERMETRINA (PORTUGUESE) ◇ 3-PHENOXYBENZYL (±)-3-(2,2-DICHLOROVINYL)-2,2-DIMETHYLCYCLOPROPANECARBOXYLATE ◇ (3-PHENOXYPHENYL)METHYL-3-(2,2-DICHLORETHENYL) -2,2-DIMETHYLCYCLOPROPANECARBOXYLATE ◇ POUNCE ◇ PP 557 ◇ S-3151 ◇ SBP-1513 ◇ TALCORD ◇ WL 43479

TOXICITY DATA with REFERENCE
skn-rbt 500 mg/24H MLD NTIS** AD-A047 284
cyt-mus-orl 150 mg/kg PHABDI 21,227,81
orl-rat TDLo:250 mg/kg (6-15D preg):REP BECTA6 29,84,82
orl-rat LD50:410 mg/kg NTIS** AD-A047-284
ihl-rat LC50:685 mg/m^3 YKYUA6 30,1635,79
skn-rat LD50:2500 mg/kg YKYUA6 35,1315,84
scu-rat LD50:6600 mg/kg BOCKAE 41,143,76
orl-mus LD50:540 mg/kg BOCKAE 41,143,76
ihl-mus LC50:685 mg/m^3 YKYUA6 30 1635,79
ipr-mus LD50:514 mg/kg TXAPA9 66,153,82
ivn-mus LD50:31 mg/kg TXAPA9 66,153,82
ice-mus LDLo:600 μg/kg TXAPA9 66,290,82

SAFETY PROFILE: Poison by inhalation, intravenous, and intracerebral routes. Moderately toxic by ingestion. Experimental reproductive effects. Mutation data reported. A skin irritant. When heated to decomposition it emits toxic fumes of Cl^-. See also ESTERS.

AHJ875 ***HR: 2***
AMERICAN BITTERSWEET

PROP: A climbing, vine-like shrub with oval, toothed leaves. The fruit is an orange-yellow capsule about 0.5 inch long with seeds imbedded in a red pulp. It grows in the region bounded by North Carolina, New Mexico, Saskatchewan and Quebec. It is commonly used in dried floral arrangements and is supplied by florists.

SYNS: BITTERSWEET ◇ BOURREAU DES ARBRES (CANADA) ◇ CELASTRUS SCANDENS ◇ CLIMBING BITTERSWEET ◇ CLIMBING ORANGE ROOT ◇ FALSE BITTERSWEET ◇ FEVER TWIG ◇ RED ROOT ◇ ROXBURY WAXWORK ◇ SHRUBBY BITTERSWEET ◇ STAFF TREE

SAFETY PROFILE: All parts of the plant and particulary the fruit contains unknown toxins which may cause vomiting and diarrhea upon ingestion.

AHK000 ***HR: 3***
AMERICIUM
af: Am aw: 243

PROP: A silvery, somewhat malleable radioactive metal. Mp: 994°; bp: 2607°; d: 13.67 @ 20°.

SAFETY PROFILE: A poison. Bone-seeking, long-lived radioactive element. Flammable, see POWDERED METALS. In a disaster, this highly toxic radioactive material can be disseminated over a wide area, causing a long-lived inhalation hazard which is difficult to remove from surfaces or from the body once it enters.

AHK250 ***HR: 3***
AMERICIUM TRICHLORIDE
mf: $AmCl_3$ mw: 349.4

SAFETY PROFILE: See AMERICIUM. Due to its alpha particle radioactivity, it can cause radiolysis and build pressure in sealed containers and eventually explode.

AHK500 CAS:50-07-7 ***HR: 3***
AMETYCIN
mf: $C_{15}H_{18}N_4O_5$ mw: 334.37

SYNS: 7-AMINO-9-α-METHOXYMITOSANE ◇ MIT-C ◇ MITO-C ◇ MITOCIN-C ◇ MITOMYCIN ◇ MITOMYCIN-C ◇ MITOMYCINUM ◇ MMC ◇ MUTAMYCIN ◇ MUTAMYCIN (MITOMYCIN for INJECTION) ◇ MYTOMYCIN ◇ NCI-C04706 ◇ NSC 26980 ◇ RCRA WASTE NUMBER U010

TOXICITY DATA with REFERENCE
mma-sat 5 μg/plate CNREA8 38,2148,78
cyt-hmn:hla 10 μmol/L TXCYAC 51,181,65
cyt-hmn:fbr 100 μg/L TRBMAV 27,409,69
ivn-mus TDLo:1 mg/kg (female 13D post):TER NATUAS 279,531,79
ipr-mus TDLo:1750 μg/kg (male 1D pre):REP MUREAV 13,433,71
ipr-rat TDLo:3000 μg/kg/26W-I:NEO RRCRBU 52,1,75
ivn-rat TDLo:2600 μg/kg/8W-I:CAR ARZNAD 20,1461,70
scu-mus TDLo:280 μg/kg/18W-I:CAR APHGBP 17,495,67
ivn-wmn TDLo:1800 μg/kg:PUL LANCAO 2,1037,80
unr-wmn TDLo:2100 μg/kg/40W-I:BLD AIMDAP 143,1617,83
unr-man TDLo:1350 μg/kg/21W-I:SYS AIMDAP 143,803,83
orl-rat LD50:30 mg/kg CNREA8 20,1354,60
ipr-rat LD50:2 mg/kg ADTEAS 3,181,68
scu-rat LD50:3250 μg/kg NIIRDN 6,798,82
ivn-rat LD50:3 mg/kg ARZNAD 20,1467,70
orl-mus LD50:23 mg/kg CNREA8 20,1354,60
ipr-mus LD50:4 mg/kg JAJAAA 13,27,60
scu-mus LD50:7800 μg/kg NIIRDN 6,798,82
ivn-mus LD50:4 mg/kg JAJAAA 13,27,60
ivn-dog LD50:1 mg/kg CNREA8 20,1354,60

CONSENSUS REPORTS: IARC Cancer Review: Group 2B IMEMDT 7,56,87; Animal Sufficient Evidence IMEMDT 10,171,76; NCI Carcinogenesis Studies (ipr); Clear Evidence: rat RRCRBU 52,1,75; No Evidence: mouse RRCRBU 52,1,75. EPA Extremely Hazardous Substances List. EPA Genetic Toxicology Program. Reported in EPA TSCA Inventory.

SAFETY PROFILE: Suspected carcinogen with experimental carcinogenic and neoplastigenic data. Poison by ingestion, subcutaneous, intravenous, and intraperitoneal routes. Human systemic effects by intravenous route: dyspnea and lung fibrosis, hemolysis with or without anemia, changes in tubules (including acute renal failure, acute tubular necrosis), normocytic anemia. Experimental teratogenic and reproductive effects. Human mutation data reported. When heated to decomposition it emits toxic fumes of NO_x. See also CARBAMATES and ESTERS.

AHK625 CAS:61941-56-8 ***HR: 3***
AMFENAC SODIUM MONOHYDRATE
mf: $C_{15}H_{13}NO_3 \cdot Na \cdot H_2O$ mw: 296.30

SYNS: AHR 5850D MONOHYDRATE ◇ 2-AMINO-3-BENZOYLBENZENEACETIC ACID SODIUM SALT HYDRATE ◇ SODIUM (2-AMINO-3-BENZOYLPHENYL)ACETATE MONOHYDRATE

TOXICITY DATA with REFERENCE
orl-rat TDLo:5 mg/kg (17-21D preg):REP OYYAA2 20,185,85
orl-rat TDLo:1344 mg/kg (9W male/2W pre-7D preg):TER OYYAA2 20,117,85
orl-rat LD50:311 mg/kg AGACBH 7,133,77
ipr-rat LD50:240 mg/kg JTSCDR 9,87,84
scu-rat LD50:240 mg/kg IYKEDH 16,1461,85
ivn-rat LD50:277 mg/kg JTSCDR 9,87,84
ims-rat LD50:277 mg/kg IYKEDH 16,1461,85
orl-mus LD50:615 mg/kg AGACBH 7,133,77
ipr-mus LD50:540 mg/kg IYKEDH 16,1461,85
scu-mus LD50:580 mg/kg JTSCDR 9,87,84
ivn-mus LD50:550 mg/kg JTSCDR 9,87,84
ims-mus LD50:540 mg/kg JTSCDR 9,87,84

SAFETY PROFILE: Poison by subcutaneous, intramuscular, intravenous, and intraperitoneal routes. Moderately toxic by ingestion. An experimental teratogen. Other experimental reproductive effects. When heated to decomposition it emits toxic fumes of NO_x and Na_2O.

AHK750 CAS:82-02-0 ***HR: 3***
AMICARDINE
mf: $C_{14}H_{12}O_5$ mw: 260.26

SYNS: AMIPTAN ◇ AMMICARDINE ◇ AMMI-KHELLIN ◇ AMMIPURAN ◇ AMMISPASMIN ◇ AMMIVIN ◇ AMMIVISNAGEN ◇ BENECARDIN ◇ BI-KELLINA ◇ CARDIO-KHELLIN ◇ CHELLIN ◇ CHELLINA (ITALIAN) ◇ CHORAFURONE ◇ CORONIN ◇ DELTOSIDE ◇ 5,8-DIMETHOXY-2-METHYL-4′,5′-FURANO-6,7-CHROMONE ◇ 5,8-DIMETHOXY-2-METHYL-6,7-FURANOCHROMONE ◇ 4,9-DIMETHOXY-7-METHYL-5H-FURO(3,2-G)(1)BENZOPYRAN-5-ONE ◇ 5,8-DIMETHOXY-2-METHYL-4′,5′-FURO-6,7-CHROMONE ◇ 4,9-DIMETHOXY-7-METHYL-5-OXO-1,8-DIOXABENZ-(F)INDENE◇ 4,9-DIMETHOXY-7-METHYL-5-OXOFURO(3,2-G)(1)BENZOPYRAN◇ 4,9-DIMETHOXY-7-METHYL-5-OXOFURO(3,2-G)-1,2-CHROMENE ◇ ESKEL ◇ EUPHORIN ◇ GYNOKHELLAN ◇ INTERKELLIN ◇ KELICORIN ◇ KELINCOR ◇ LYNAMINE ◇ MEFURINA ◇ METHAFRONE ◇ NORKEL ◇ SIMESKELLINA ◇ VASOKELLINA ◇ VISNAGALIN

TOXICITY DATA with REFERENCE
scu-rat TDLo:18 g/kg (female 9-14D post):REP TOIZAG 18,81,71
scu-mus TDLo:60 g/kg (female 7-12D post):TER TOIZAG 18,81,71
orl-rat LD50:68800 μg/kg ARZNAD 11,915,61
ipr-rat LD50:70 mg/kg JDGRAX 9(1-2),35,77
ivn-rat LD50:34 mg/kg ARZNAD 11,848,61
orl-mus LD50:50800 μg/kg ARZNAD 11,915,61
ipr-mus LD50:155 mg/kg FRPSAX 13,561,58
ivn-mus LD50:30600 μg/kg ARZNAD 11,915,61
ims-mus LD50:83 mg/kg JDGRAX 7(2),1,75

CONSENSUS REPORTS: EPA Genetic Toxicology Program.

SAFETY PROFILE: Poison by ingestion, intraperitoneal, intramuscular, and intravenous routes. An experimental teratogen. Other experimental reproductive effects. When heated to decomposition it emits acrid smoke and irritating fumes. A vasodilator.

AHL000 CAS:17650-86-1 ***HR: 3***
AMICETIN
mf: $C_{29}H_{42}N_6O_9$ mw: 618.77

SYNS: ALLOMYCIN ◇ D-13 ◇ NSC 5340 ◇ SACROMYCIN ◇ U-4761

TOXICITY DATA with REFERENCE
orl-rat LD50:3600 mg/kg UPJOH* 2(6),-,71
scu-rat LD50:600 mg/kg ANTCAO 3,718,53
ivn-rat LD50:200 mg/kg ANTCAO 3,718,53
orl-mus LD50:2000 mg/kg 85GDA2 5,214,81
ipr-mus LD50:90 mg/kg ANTCAO 3,718,53
scu-mus LD50:57 mg/kg 85FZAT -,120,67
scu-gpg LD50:14 mg/kg UPJOH* 2(6),-,71
mmo-eug 1500 mg/L NEOLA4 19,579,72
dni-eug 1500 mg/L NEOLA4 19,579,72

SAFETY PROFILE: Poison by subcutaneous, intravenous, and intraperitoneal routes. Moderately toxic by ingestion. Mutation data reported. When heated to decomposition it emits toxic fumes of NO_x.

AHL250 ***HR: 3***
AMICETIN CITRATE

TOXICITY DATA with REFERENCE
scu-rat LD50:600 mg/kg 85ERAY 1,155,78
ivn-rat LD50:200 mg/kg 85ERAY 1,155,78
scu-mus LD50:600 mg/kg 85ERAY 1,155,78
ivn-mus LD50:90 mg/kg 85ERAY 1,155,78

SAFETY PROFILE: Poison by intravenous route. Moderately toxic by subcutaneous route. When heated to decomposition it emits acrid smoke and irritating fumes.

AHL500 CAS:1421-68-7 ***HR: 3***
AMIDEFRINE MESYLATE
mf: $C_{10}H_{16}N_2O_3S \cdot CH_4O_3S$ mw: 340.45

SYNS: AMIDEPHRINE MESYLATE ◇ AMIDEPHRINE MONOMETHANESULFONATE ◇ DRICOL ◇ FENTRINOL ◇ 3'-(1-HYDROXY-2-(METHYLAMINO)ETHYL)METHANESULFONANILIDE METHANESULFONATE ◇ 3'-(1-HYDROXY-2-(METHYLAMINO)ETHYL)METHANESULFONANILIDE MONOMETHANESULFONATE SALT ◇ N-(3-(1-HYDROXY-2-(METHYLAMINO)ETHYL)PHENYL)-METHANESULFONAMIDE MONOMETHANESULFONATE SALT ◇ (2-METHYLAMINE-1-HYDROXYETHYL)METHANESULFONANILIDEMETHANESULFONATE ◇ 3'-(2-(METHYLAMINO)-1-HYDROXYETHYL)METHANESULFONANILIDE METHANESULFONATE ◇ MJ 5190 ◇ NALDE ◇ PRODUCT 5190

TOXICITY DATA with REFERENCE
eye-rbt 10 mg MLD TXAPA9 23,589,72
orl-rat LD50:13 mg/kg TXAPA9 23,589,72
ipr-rat LD50:5 mg/kg TXAPA9 23,589,72
orl-mus LD50:2284 mg/kg TXAPA9 23,589,72
ipr-mus LD50:780 mg/kg TXAPA9 23,589,72
scu-mus LD50:1990 mg/kg IJNEAQ 4,219,65
ivn-mus LD50:190 mg/kg IJNEAQ 4,219,65
ipr-dog LD50:4800 μg/kg IJNEAQ 4,219,65
ivn-dog LD50:1400 μg/kg TXAPA9 23,589,72
orl-cat LDLo:2 mg/kg TXAPA9 23,589,72
orl-rbt LD50:12 mg/kg TXAPA9 23,589,72
idr-rbt LD50:7500 μg/kg TXAPA9 23,589,72

SAFETY PROFILE: Poison by ingestion, intraperitoneal, intravenous, and intradermal routes. Moderately toxic by subcutaneous route. An eye irritant. When heated to decomposition it emits very toxic fumes of NO_x and SO_x. A vasoconstrictor and nasal decongestant. See also SULFONATES.

AHL750 ***HR: B***
AMIDES

PROP: Organic compounds containing the structural group $-CONH_2$, and closely related to the organic acids with the grouping $-COOH$. Common examples are: acetamide (CH_3CONH_2) and urea ($CO(NH_2)_2$).

SAFETY PROFILE: Most of the saturated amides have low toxicity, but the unsaturated and N-substituted amides are irritants and may be absorbed via skin contact. Can cause injury to the liver, kidney, and brain.

AHL875 ***HR: 3***
p-AMIDINOBENZOIC ACID BUTYL ESTER
mf: $C_{12}H_{16}N_2O_2$ mw: 220.30

TOXICITY DATA with REFERENCE
skn-rbt 1% SEV JAPMA8 41,202,52
skn-rbt 1000 ppm MOD JAPMA8 41,202,52
scu-mus LD50:350 mg/kg JAPMA8 41,202,52
ivn-mus LD50:70 mg/kg JAPMA8 41,202,52

SAFETY PROFILE: Poison by subcutaneous and intravenous routes. A severe skin irritant. When heated to decomposition it emits toxic fumes of NO_x. See also ESTERS.

AHL880 ***HR: 3***
p-AMIDINOBENZOIC ACID HEXYL ESTER
mf: $C_{14}H_{20}N_2O_2$ mw: 248.36

TOXICITY DATA with REFERENCE
skn-rbt 1% SEV JAPMA8 41,202,52
skn-rbt 1000 ppm MOD JAPMA8 41,202,52
scu-mus LD50:550 mg/kg JAPMA8 41,202,52
ivn-mus LD50:150 mg/kg JAPMA8 41,202,52

SAFETY PROFILE: Poison by intravenous route. Moderately toxic by subcutaneous route. A severe skin irritant. When heated to decomposition it emits toxic fumes of NO_x. See also ESTERS.

AHL885 ***HR: 3***
p-AMIDINOBENZOIC ACID PENTYL ESTER
mf: $C_{13}H_{18}N_2O_2$ mw: 234.33

TOXICITY DATA with REFERENCE
skn-rbt 1% SEV JAPMA8 41,202,52
scu-mus LD50:450 mg/kg JAPMA8 41,202,52
ivn-mus LD50:150 mg/kg JAPMA8 41,202,52

SAFETY PROFILE: Poison by intravenous route. Moderately toxic by subcutaneous route. A severe skin irritant. When heated to decomposition it emits toxic fumes of NO_x. See also ESTERS.

AHL890 ***HR: 3***
p-AMIDINOBENZOIC ACID PROPYL ESTER
mf: $C_{11}H_{14}N_2O_2$ mw: 206.27

TOXICITY DATA with REFERENCE
skn-rbt 1% SEV JAPMA8 41,202,52
scu-mus LD50:550 mg/kg JAPMA8 41,202,52
ivn-mus LD50:65 mg/kg JAPMA8 41,202,52

SAFETY PROFILE: Poison by intravenous route. Moderately toxic by subcutaneous route. A severe skin irritant. See also ESTERS.

AHN000 CAS:10319-70-7 ***HR: 3***
S-(AMIDINOMETHYL) HYDROGEN THIOSULFATE
mf: $C_2H_6N_2O_3S_2$ mw: 170.22

TOXICITY DATA with REFERENCE
orl-mus LD50:300 mg/kg JMCMAR 15,1313,72
ipr-mus LD50:87 mg/kg JMCMAR 15,1313,72

SAFETY PROFILE: Poison by ingestion and intraperitoneal routes. See also THIOSULFATES. When heated to decomposition it emits very toxic fumes of NO_x and SO_x.

AHN625 CAS:53142-01-1 ***HR: 3***
AMIDINOMYCIN
mf: $C_9H_{18}N_4O$ mw: 198.31

SYNS: N-(2-AMIDINOETHYL)-3-AMINOCYCLOPENTANOCARBOXAMIDE ◇ 3-AMINO-N-(3-AMINO-3-IMINOPROPYL)CYCLOPENTANECARBOXAMIDE ◇ MYXOVIROMYCIN

TOXICITY DATA with REFERENCE
orl-mus LD50:140 mg/kg 85FZAT -,127,67
scu-mus LD50:20400 μg/kg 85FZAT -,444,67
ivn-mus LD50:18 mg/kg 85ERAY 2,1214,78

SAFETY PROFILE: Poison by ingestion, subcutaneous, and intravenous routes. When heated to decomposition it emits toxic fumes of NO_x.

AHO250 CAS:57-67-0 ***HR: 2***
N^1-AMIDINOSULFANILAMIDE
mf: $C_7H_{10}N_4O_2S$ mw: 214.27

PROP: Colorless, monoclinic crystals from aqueous solns. Mp: 189-190°. Sltly sol in alc; insol in ether; solubility in water = 0.19% @ 37°.

SYNS: p-AMINOBENZENESULFONYLGUANIDINE ◇ N-p-AMINOBENZENESULPHONYLGUANIDINEMONOHYDRATE ◇ ATERIAN ◇ GUANICIL ◇ N^1-GUANYLSULFANILAMIDE ◇ RESULFON ◇ SHIGATOX ◇ SULFAGUANIDINE ◇ SULFANILGUANIDINE ◇ SULFANILYLGUANIDINE ◇ SULFOGUENIL ◇ SULFOGUANIDINE ◇ SULGIN

TOXICITY DATA with REFERENCE
orl-rat TDLo:22500 mg/kg (female 1-15D post):TER AFPEAM 7,180,50
orl-rat TDLo:22500 mg/kg (1-15D preg):REP AFPEAM 7,180,50
ipr-mus LDLo:500 mg/kg JHHBAI 67,163,40

CONSENSUS REPORTS: Reported in EPA TSCA Inventory.

SAFETY PROFILE: Moderately toxic by intraperitoneal route. An experimental teratogen. Other experimental reproductive effects. See also SULFONATES. When heated to decomposition it emits very toxic fumes of SO_x and NO_x.

AHO750 CAS:919-76-6 ***HR: 3***
AMIDITHION
mf: $C_7H_{16}NO_4PS_2$ mw: 273.33

SYNS: 2-MERCAPTO-N-(2-METHYOXYETHYL)-ACETAMIDES-ESTER with O,O-DIMETHYL PHOSPHORODITHIOATE ◇ C 2446 ◇ CIBA 2446 ◇ CIBA THIOCRON ◇ O,O-DIMETHYL-S-(2-METHOXYETHYLCARBAMOYLMETHYL)DITHIOPHOSPHATE ◇ O,O-DIMETHYL-S-(2-METHOXYETHYLCARBAMOYL METHYL)-PHOSPHORODITHIOATE ◇ ENT 27,160 ◇ S-(2-((2-METHOXYETHYL)-AMINO-2-OXOETHYL)O,O-DIMETHYL) PHOSPHORODITHIOATE ◇ S-(N-2-METHOXYETHYLCARBAMOYLMETHYL)DIMETHYL PHOPHOROTHIOLOTHIONATE ◇ THIOCRON

TOXICITY DATA with REFERENCE
orl-rat LD50:600 mg/kg WRPCA2 9,119,70
skn-rat LD50:1600 mg/kg 28ZEAL 5,9,76
orl-ckn LD50:94 mg/kg TXAPA9 11,49,67

SAFETY PROFILE: Poison by ingestion. Moderately toxic by skin contact. When heated to decomposition it emits very toxic fumes of SO_x, PO_x, and NO_x. See also ESTERS.

AHP000 CAS:137-09-7 ***HR: 3***
AMIDOL
mf: $C_6H_8N_2O \cdot 2ClH$ mw: 197.08

PROP: Grayish-white crystals.

SYNS: 2,4-DIAMINOPHENOL HYDROCHLORIDE ◇ NCI-C60026

TOXICITY DATA with REFERENCE
mma-sat 330 ng/plate ENMUDM 5(Suppl 1),3,83
ipr-mus LDLo:50 mg/kg RBPMAZ 22,1,52

CONSENSUS REPORTS: Reported in EPA TSCA Inventory. EPA Genetic Toxicology Program.

SAFETY PROFILE: Poison by intraperitoneal route. Mutation data reported. A mild irritant and allergen. When heated to decomposition it emits toxic fumes of NO_x and HCl.

AHP125 CAS:21590-92-1 ***HR: 3***
AMIDOLINE
mf: $C_{23}H_{29}N_3O_2$ mw: 379.55

PROP: Crystals from ligroin. Mp: 106-107°.

SYNS: 2-ETHYL-2,3-DIHYDRO-3-((4-(2-(1-PIPERIDINYL)ETHOXY)-PHENYL)AMINO)-1H-ISOINDOL-1-ONE ◇ 2-ETHYL-3-(β-PIPERIDINO-p-PHENETIDINO)PHTHALIMIDINE ◇ ETOMIDOLINE ◇ K 2680 ◇ SMEDOLIN

TOXICITY DATA with REFERENCE
ivn-mus TDLo:7 mg/kg (female 7-13D post):TER KSRNAM 8,1812,74
orl-rat TDLo:280 mg/kg (8-14D preg):REP KSRNAM 8,1802,74
orl-rat LD50:695 mg/kg KSRNAM 8,1730,74
scu-rat LD50:94 mg/kg KSRNAM 8,1730,74
ivn-rat LD50:42 mg/kg KSRNAM 8,1730,74
orl-mus LD50:168 mg/kg KSRNAM 8,1730,74
scu-mus LD50:109 mg/kg KSRNAM 8,1730,74
ivn-mus LD50:36 mg/kg KSRNAM 8,1730,74

SAFETY PROFILE: Poison by ingestion, subcutaneous, and intravenous routes. An experimental teratogen. Other experimental reproductive effects. When heated to decomposition it emits toxic fumes of NO_x.

AHP250 ***HR: 3***
AMIDO SULFURYL AZIDE
mf: $H_2N_4O_2S$ mw: 122.2

SAFETY PROFILE: A dangerously unstable, explosive compound. See also AZIDES and SULFATES. When heated to decomposition it emits highly toxic fumes of NO_x and SO_x. Shock sensitive.

AHP375 CAS:40709-23-7 ***HR: 3***
AMIKHELLIN HYDROCHLORIDE
mf: $C_{18}H_{21}NO_5 \cdot ClH$ mw: 367.86

SYNS: CHLORHYDRATE d'AMIKHELLINE (FRENCH) ◇ 2-METHYL-5-HYDROXY-8-(β-DIAETHYLAMINO-AETHOXY)FURANO-6,7:2'3'-CHROMON HYDROCHLORIDE ◇ NOKHEL

TOXICITY DATA with REFERENCE
dnd-omi 110 μmol/L BICMBE 55,1415,73
dnd-omi 110 μmol/L BICMBE 55,1415,73
dnd-omi 10 mmol/L BICMBE 55,1415,73
cyt-nml:oth 270 μmol/L/6H-C AAMMAU 61,210,72
cyt-nml:oth 270 μmol/L/6H-C AAMMAU 61,201,72
dnd-mam:lym 100 μmol/L BICMBE 55,1415,73
scu-mus LD50:14380 μg/kg ARZNAD 13,140,63

SAFETY PROFILE: Poison by subcutaneous route. Mutation data reported. When heated to decomposition it emits toxic fumes of NO_x and HCl.

AHP500 CAS:95-38-5 ***HR: 2***
AMINE 220
mf: $C_{22}H_{42}N_2O$ mw: 350.66

PROP: Liquid. Bp: 235° @ 1 mm, flash p: 465°F (OC), d: 0.9300 @ 20°/20°, vap d: 12.1.

SYNS: 2-(8-HEPTADECENYL)-2-IMIDAZOLINE-1-ETHANOL ◇ 1-HYDROXYETHYL-2-HEPTADECENYLGLYOXALIDINE ◇ 1-(2-HYDROXYETHYL)-2-HEPTADECENYLGLYOXALIDINE ◇ 1-(2-HYDROXYETHYL)-2-N-HEPTADECENYL-2-IMIDAZOLINE ◇ 1-(2-HYDROXYETHYL)-2-HEPTADECENYL-2-IMIDAZOLINE ◇ NALCAMINE G-13

TOXICITY DATA with REFERENCE
orl-rat LD50:3130 mg/kg UCDS**

CONSENSUS REPORTS: Reported in EPA TSCA Inventory.

SAFETY PROFILE: Moderately toxic by ingestion. Combustible; can react with oxidizing materials. To fight fire, use foam, CO_2, dry chemical. When heated to decomposition it emits toxic fumes of NO_x.

AHP750 ***HR: D***
AMINES

PROP: A large group of organic compounds containing nitrogen and considered as derived from ammonia (NH_3) by replacement of one or more hydrogen atoms by an organic radical. See also specific compounds.

SAFETY PROFILE: Variable toxicity; some are poisons, some are only slightly toxic. Many are skin irritants and some are sensitizers. See also AROMATIC AMINES and FATTY AMINES.

AHP760 ***HR: 1***
AMINES, FATTY

PROP: A normal aliphatic amine derived from fats and oils. May be saturated or unsaturated, primary, secondary or tertiary, but the alkyl groups are straight-chain and have an even nmumber of carbons in each. The length varies from 8 to 22 carbon atoms.

SAFETY PROFILE: Generally of mild toxicity. Used as organic bases, soaps, plasticizers, tire cords, fabric softeners, water-resistant asphalt, hair conditioners, cosmetics and medicinals.

AHP875 ***HR: D***
5-((2-AMINOACETAMIDO)METHYL)-1-(4-CHLORO-2-(o-CHLOROBENZOYL)PHENYL)-N,N-DIMETHYL-1H-s-TRIAZOLE-3-CARBOXAMIDE, HYDROCHLORIDE, DIHYDRATE
mf: $C_{21}H_{20}Cl_2N_6O_3$•ClH•$2H_2O$ mw: 529.81

SYN: 450191-S

TOXICITY DATA with REFERENCE
orl-rat TDLo:4400 mg/kg (7-17D preg):REP OYYAA2 30,765,85
orl-rat TDLo:4400 mg/kg (7-17D preg):TER OYYAA2 30,765,85

SAFETY PROFILE: An experimental teratogen. Other experimental reproductive effects. When heated to decomposition it emits toxic fumes of NO_x and HCl.

AHQ000 CAS:102-28-3 ***HR: 3***
3'-AMINOACETANILIDE
mf: $C_8H_{10}N_2O$ mw: 150.20

PROP: Mp: softens: bp: decomp @ 787°; sol in water, acetone, alc, and ether; sltly sol in benzene; insol in ligroin.

SYNS: m-ACETAMINOANILINE ◇ m-(ACETYLAMINO)ANILINE ◇ 3-ACETYLAMINOANILINE ◇ N-ACETYL-m-FENYLENEDIAMIN (CZECH) ◇ N-ACETYL-m-PHENYLENEDIAMINE ◇ 3-AMINOACETANILID (CZECH) ◇ m-AMINOACETANILIDE

TOXICITY DATA with REFERENCE
eye-rbt 500 mg/24H SEV 28ZPAK -,129,72
orl-rat LD50:1830 mg/kg 28ZPAK -,129,72
ivn-mus LD50:320 mg/kg CSLNX* NX#02899

CONSENSUS REPORTS: Reported in EPA TSCA Inventory.

SAFETY PROFILE: Poison by intravenous route. Moderately toxic by ingestion. A severe eye irritant. When heated to decomposition it emits toxic fumes of NO_x.

AHQ250 CAS:122-80-5 ***HR: 2***
4'-AMINOACETANILIDE
mf: $C_8H_{10}N_2O$ mw: 150.20

PROP: Mp: 164°.

SYNS: p-ACETAMIDOANILINE ◇ 4-ACETAMIDOANILINE ◇ p-ACETOAMINOANILINE ◇ p-(ACETYLAMINO)ANILINE ◇ 4-(ACETYLAMINO)ANILINE ◇ N-ACETYL-p-FENYLENDIAMIN (CZECH) ◇ ACETYL-p-PHENYLENEDIAMINE ◇ 4'-AMINOACETANILID (CZECH) ◇ p-AMINOACETANILIDE ◇ 4-AMINOACETANILIDE ◇ N-(p-AMINOPHENYL)ACETAMIDE ◇ C.I. 76005 ◇ C.I. OXIDATION BASE 19 ◇ FOURRINE 88 ◇ FOURRINE A

TOXICITY DATA with REFERENCE
eye-rbt 100 mg/24H SEV 28ZPAK -,130,72
orl-rat LD50:3350 mg/kg 28ZPAK -,130,72

CONSENSUS REPORTS: Reported in EPA TSCA Inventory.

SAFETY PROFILE: Moderately toxic by ingestion. A severe eye irritant. When heated to decomposition it emits toxic fumes of NO_x. See also AMINES and AMIDES.

AHQ750 CAS:151-63-3 ***HR: D***
AMINOACETONITRILE BISULFATE
mf: $C_2H_4N_2$•H_2O_4S mw: 154.16

SYNS: AMINOACETONITRILE HYDROGEN SULFATE ◇ AMINOACETONITRILE HYDROSULFATE

TOXICITY DATA with REFERENCE
orl-ham TDLo:400 mg/kg (7D preg):TER ACATA5 100,347,78
orl-ham TDLo:400 mg/kg (7D preg):REP ACATA5 100,347,78

CONSENSUS REPORTS: Cyanide and its compounds are on the Community Right-To-Know List.

SAFETY PROFILE: An experimental teratogen. Other experimental reproductive effects. See also NITRILES and SULFATES. When heated to decomposition it emits very toxic fumes of SO_x, NO_x, and CN^-.

AHR000 CAS:5466-22-8 ***HR: 3***
AMINOACETONITRILE SULFATE
mf: $C_4H_8N_4$•H_2O_4S mw: 210.24

TOXICITY DATA with REFERENCE
scu-rbt TDLo:500 mg/kg (female 12-31D post):TER AROPAW 69,602,63
orl-rat LDLo:100 mg/kg TJADAB 5,33,72

CONSENSUS REPORTS: Cyanide and its compounds are on the Community Right-To-Know List.

SAFETY PROFILE: Poison by ingestion. An experimental teratogen. Other experimental reproductive effects. See also NITRILES and SULFATES. When heated to decomposition it emits very toxic fumes of NO_x, SO_x, and CN^-.

AHR250 CAS:613-89-8 ***HR: 2***
2-AMINOACETOPHENONE
mf: C_8H_9NO mw: 135.18

PROP: Yellow, oily liquid. Bp: 251° (slt decomp); insol in water; sol in alc and ether.

SYNS: ω-AMINOACETOPHENONE ◇ PHENACYLAMINE

TOXICITY DATA with REFERENCE
scu-mus TDLo:2000 mg/kg/7W-I:CAR PGPKA8 14,12,69
scu-gpg TDLo:1480 mg/kg/13W-I:NEO,REP BEXBAN 84,1156,77
scu-ham TDLo:1600 mg/kg/10W-I:ETA VOONAW 25(6),81,79

SAFETY PROFILE: Experimental reproductive effects. Questionable carcinogen with experimental carcinogenic, neoplastigenic, and tumorigenic data. When heated to decomposition it emits toxic fumes of NO_x. See also AROMATIC AMINES.

AHR500 CAS:99-03-6 ***HR: 2***
3'-AMINOACETOPHENONE
mf: C_8H_9NO mw: 135.18

PROP: Yellow, oily liquid. Bp: 251°; (slt decomp); insol in water; sol in alc and ether.

SYNS: m-ACETYLANILINE ◇ 3-ACETYLANILINE ◇ β-AMINOACETOPHENONE ◇ m-AMINOACETOPHENONE ◇ m-AMINOACETYLBENZENE

TOXICITY DATA with REFERENCE
eye-rbt 500 mg open AMIHBC 10,61,54
mma-sat 10 mg/L ENMUDM 5,803,83
cyt-mus-scu 400 mg/kg TGANAK 2,538,68
orl-rat LD50:1870 mg/kg AMIHBC 10,61,54
skn-rbt LD50:4340 mg/kg AMIHBC 10,61,54

CONSENSUS REPORTS: Reported in EPA TSCA Inventory.

SAFETY PROFILE: Moderately toxic by ingestion. Mildly toxic by skin contact. An eye irritant. Mutation data reported. When heated to decomposition it emits toxic fumes of NO_x. See also AROMATIC AMINES.

AHR750 CAS:99-92-3 ***HR: 3***
p-AMINO ACETOPHENONE
mf: C_8H_9NO mw: 135.18

PROP: Crystalline. Mp: 106°, bp: 293-295°. Sol in hot water, alc, and ether.

SYNS: 4-ACETYLANILINE ◇ 4'-AMINOACETOPHENONE ◇ p-AMINOACETYLBENZENE ◇ USAF EK-631

TOXICITY DATA with REFERENCE
ipr-mus LD50:381 mg/kg GEPHDP 14,465,83 NTIS** AD277-689
ipr-rat LD50:260 mg/kg JPETAB 80,31,44
orl-mus LD50:596 mg/kg GEPHDP 14,465,83
ipr-mus LD50:300 mg/kg NTIS** AD277-689
orl-bwd LD50:133 mg/kg AECTCV 12,355,83

CONSENSUS REPORTS: Reported in EPA TSCA Inventory.

SAFETY PROFILE: Poison by ingestion and intraperitoneal routes. When heated to decomposition it emits toxic fumes of NO_x. See also AROMATIC AMINES.

AHS000 CAS:581-28-2 ***HR: 3***
2-AMINOACRIDINE
mf: $C_{13}H_{10}N_2$ mw: 194.25

SYNS: 2-ACRIDINAMINE ◇ 3-AMINOACRIDINE (EUROPEAN)

TOXICITY DATA with REFERENCE
mma-sat 600 ng/plate TXCYAC 34,247,85
mmo-omi 16 mg/L JMOBAK 3,762,61
dnd-mam:lym 15 μmol/L JMOBAK 13,138,65
scu-mus LD50:330 mg/kg BJEPA5 28,1,47

SAFETY PROFILE: Poison by subcutaneous route. Mutation data reported. See also AMINES. When heated to decomposition it emits toxic fumes of NO_x.

AHS500 CAS:90-45-9 ***HR: 3***
9-AMINOACRIDINE
mf: $C_{13}H_{10}N_2$ mw: 194.25

SYNS: 9AA ◇ 9-ACRIDINAMINE ◇ AMINACRINE ◇ 5-AMINOACRIDINE ◇ IZOACRIDINA ◇ MONACRIN

TOXICITY DATA with REFERENCE
mmo-sat 3 ug/plate KSRNAM 19,3212,85
slt-dmg-unr 5000 ppm/6H MUREAV 120,233,83
dns-ham:lvr 1 μmol/L ENMUDM 6,1,84

CONSENSUS REPORTS: EPA Genetic Toxicology Program.

SAFETY PROFILE: Poison by intraperitoneal and subcutaneous routes. Mutation data reported. When heated to decomposition it emits toxic fumes of NO_x. See also AMINES.

AHS750 CAS:134-50-9 ***HR: 3***
AMINOACRIDINE HYDROCHLORIDE
mf: $C_{13}H_{10}N_2 \cdot ClH$ mw: 230.71

SYNS: ACRAMINE YELLOW ◇ 9-ACRIDINAMINE MONOHYDROCHLORIDE ◇ AMINACRINE HYDROCHLORIDE ◇ 5-AMINOACRIDINE HYDROCHLORIDE ◇ 9-AMINOACRIDINE MONOHYDROCHLORIDE ◇ MONACRIN ◇ MONACRIN HYDROCHLORIDE ◇ NSC-7571

TOXICITY DATA with REFERENCE
mmo-sat 10 ug/plate KSRNAM 16,6240,82

mma-sat 50 ug/plate KSRNAM 17,70,83
dnr-esc 20 uL/disc MUREAV 97,1,82
dnr-bcs 20 uL/disc MUREAV 97,1,82
cyt-ham:lng 600 ug/L GMCRDC 27,95,81
dnd-man:lym 10 pph BIPMAA 11,2537,72
orl-mus LD50:78 mg/kg 29ZVAB -,7,69
ipr-mus LD50:70 mg/kg JAPMA8 38,498,49
scu-mus LD50:95 mg/kg JAPMA8 38,498,49

CONSENSUS REPORTS: Reported in EPA TSCA Inventory. EPA Genetic Toxicology Program.

SAFETY PROFILE: Poison by ingestion, intraperitoneal, and subcutaneous routes. Mutation data reported. See also AMINES. When heated to decomposition it emits very toxic fumes of HCl and NO_x.

AHT000 CAS:60566-40-7 ***HR: 3***
9-AMINOACRIDINE PENICILLIN

SYNS: 9AAP ◇ PENICILLIN, compounded with 9-AMINOACRIDINE

TOXICITY DATA with REFERENCE
orl-mus LD50:227 mg/kg JAPMA8 38,498,49
scu-mus LD50:562 mg/kg JAPMA8 38,498,49

SAFETY PROFILE: Poison by ingestion. Moderately toxic by subcutaneous route. See also PENICILLIN. When heated to decomposition it emits toxic fumes of NO_x.

AHT250 CAS:58658-27-8 ***HR: 3***
4'-((3-AMINO-9-ACRIDINYL)AMINO) METHANESULFONANILIDE
mf: $C_{20}H_{18}N_4O_2S$ mw: 378.48

TOXICITY DATA with REFERENCE
mmo-sat 209 μmol/L JMCMAR 23,269,80
ipr-mus LD10:2500 μg/kg JMCMAR 23,269,80

SAFETY PROFILE: Poison by intraperitoneal route. Mutation data reported. See also SULFONATES and AMINES. When heated to decomposition it emits very toxic fumes of NO_x and SO_x.

AHT825 ***HR: 3***
4'-(2-AMINO-9-ACRIDINYLAMINO) METHANESULFONANILIDE
mf: $C_{20}H_{18}N_4O_2S$ mw: 378.48

TOXICITY DATA with REFERENCE
mmo-sat 166 μmol/L JMCMAR 23,269,80
ipr-mus LD10:35 mg/kg JMCMAR 23,269,80

SAFETY PROFILE: Poison by intraperitoneal route. Mutation data reported. See also SULFONATES. When heated to decomposition it emits very toxic fumes of NO_x and SO_x.

AHT850 CAS:53222-25-6 ***HR: 3***
6-AMINO-4-((3-AMINO-4-(((4-((1-METHYLPYRIDINIUM-4-YL)AMINO)PHENYL)AMINO)CARBONYL)PHENYL)AMINO)-1-METHYLQUINOLINIUM),DIIODIDE
mf: $C_{29}H_{29}N_6O{\bullet}2I$ mw: 731.44

TOXICITY DATA with REFERENCE
dnd-mus:lym 490 nmol/L JMCMAR 22,134,79
ipr-mus LD10:20 mg/kg JMCMAR 22,134,79

SAFETY PROFILE: Poison by intraperitoneal route. Mutagenic data reported. When heated to decomposition it emits very toxic fumes of NO_x and I^-.

AHT900 CAS:63991-48-0 ***HR: 2***
6-AMINO-2-(3'-AMINOPHENYL) BENZIMIDAZOLE, DIHYDRO CHLORIDE
mf: $C_{13}H_{12}N_4{\bullet}2ClH$ mw: 297.21

SYN: 6-AMINO-2-(3'-AMINOFENYL)BENZIMIDAZOL HYDROCHLORID (CZECH)

TOXICITY DATA with REFERENCE
skn-rbt 500 mg/24H MOD 28ZPAK -,145,72
eye-rbt 20 mg/24H SEV 28ZPAK -,145,72
orl-rat LD50:2330 mg/kg 28ZPAK -,145,72

SAFETY PROFILE: Moderately toxic by ingestion. A skin and severe eye irritant. When heated to decomposition it emits toxic fumes of HCl and NO_x. See also AROMATIC AMINES.

AHT950 CAS:63991-49-1 ***HR: 2***
6-AMINO-2-(4'-AMINOPHENYL) BENZIMIDAZOLE DIHYDROCHLORIDE
mf: $C_{13}H_{12}N_4{\bullet}2ClH$ mw: 297.21

SYN: 6-AMINO-2-(4'-AMINOFENYL)BENZIMIDAZOL HYDROCHLORID (CZECH)

TOXICITY DATA with REFERENCE
skn-rbt 500 mg/24H MOD 28ZPAK -,145,72
eye-rbt 20 mg/24H SEV 28ZPAK -,145,72
orl-rat LD50:2060 mg/kg 28ZPAK -,145,72

SAFETY PROFILE: Moderately toxic by ingestion. Moderate skin and severe eye irritant. When heated to decomposition it emits very toxic fumes of HCl and NO_x. See also AROMATIC AMINES.

AIA250 CAS:2840-26-8 ***HR: 2***
3-AMINO-p-ANISIC ACID
mf: $C_8H_9NO_3$ mw: 167.18

SYNS: 3-AMINO-4-METHOXYBENZOIC ACID ◇ KYSELINA-3-AMINO-4-METHOXYBENZOOVA (CZECH)

TOXICITY DATA with REFERENCE
eye-rbt 500 mg/24H SEV 28ZPAK -,115,72
orl-rat LD50:8290 mg/kg 28ZPAK -,115,72

SAFETY PROFILE: Mildly toxic by ingestion. Severe eye irritant. When heated to decomposition it emits toxic fumes of NO_x.

AIA500 CAS:13244-33-2 ***HR: 2***
4-AMINOANISOLE-3-SULFONIC ACID
mf: $C_7H_9NO_4S$ mw: 203.23

SYNS: 2-AMINO-5-METHOXY BENZENESULFONIC ACID ◇ KYSELINA-4-AMINOANISOL-3-SULFONOVA ◇ 4-METHOXY-2-SULFOANILINE

TOXICITY DATA with REFERENCE
skn-rbt 500 mg/24H MLD 28ZPAK -,184,72
eye-rbt 500 mg/24H SEV 28ZPAK -,184,72
orl-rat LD50:10 g/kg 28ZPAK -,184,72

CONSENSUS REPORTS: Reported in EPA TSCA Inventory.

SAFETY PROFILE: Mildly toxic by ingestion. A mild skin and severe eye irritant. See also SULFONATES and AMINES. When heated to decomposition it emits very toxic fumes of SO_x and NO_x.

AIA750 CAS:82-45-1 ***HR: 3***
1-AMINOANTHRAQUINONE
mf: $C_{14}H_9NO_2$ mw: 223.24

PROP: Red needles. Mp: 256°, bp: subl. Insol in water; sol in HCl, alc, benzene, ether, and chloroform.

SYNS: 1-AMINO-9,10-ANTHRACENEDIONE ◇ 1-AMINOANTHRACHINON (CZECH) ◇ α-AMINOANTHRAQUINONE ◇ 1-AMINO-9,10-ANTHRAQUINONE ◇ α-ANTHRAQUINONYLAMINE ◇ C.I. 37275 ◇ DIAZO FAST RED AL

TOXICITY DATA with REFERENCE
eye-rbt 100 mg/24H MOD 28ZPAK -,121,72
dnd-mus-ipr 250 mg/kg ATSUDG (5),355,82
orl-rat TDLo:2400 mg/kg/60W-I:ETA TXAPA9 8,346,66
ipr-rat LD50:1500 mg/kg GTPZAB 21(12),27,77

CONSENSUS REPORTS: Reported in EPA TSCA Inventory.

SAFETY PROFILE: Moderately toxic by intraperitoneal route. An eye irritant. Questionable carcinogen with experimental tumorigenic data. Mutation data reported. When heated to decomposition it emits toxic NO_x. See also AMINES.

AIB000 CAS:117-79-3 ***HR: 3***
2-AMINOANTHRAQUINONE
mf: $C_{14}H_9NO_2$ mw: 223.24

PROP: Red needles from alc. Mp: 302°, bp: subl. Insol in water and ether; sol in alc and benzene.

SYNS: 2-AMINO-9,10-ANTHRACENEDIONE ◇ 2-AMINO-9,10-ANTRAQUINONE ◇ β-AMINOANTHRAQUINONE ◇ β-ANTHRAQUINONYLAMINE ◇ NCI-C01876

TOXICITY DATA with REFERENCE
mmo-sat 1 mg/plate ENMUDM 7(Suppl 5),1,85
mma-sat 1 μg/plate ENMUDM 7(Suppl 5),1,85
mma-esc 1 mg/plate ENMUDM 7(Suppl 5),1,85
orl-rat TDLo:115 g/kg/78W-C:CAR NCITR* NCI-CG-TR-144,78
orl-mus TDLo:655 g/kg/78W-C:CAR NCITR* NCI-CG-TR-144,78
orl-rat TD:225 g/kg/78W-C:NEO NCITR* NCI-CG-TR-144,78
orl-mus TD:330 g/kg/78W-C:ETA NCITR* NCI-CG-TR-144,78
ipr-rat LD50:1500 mg/kg GTPZAB 21(12),27,77

CONSENSUS REPORTS: NTP Fifth Annual Report on Carcinogens. IARC Cancer Review: Group 3 IMEMDT 7,56,87; Animal Limited Evidence IMEMDT 27,191,82. NCI Carcinogenesis Bioassay (feed); Clear Evidence: mouse, rat NCITR* NCI-CG-TR-144,78. Community Right-To-Know List. Reported in EPA TSCA Inventory.

SAFETY PROFILE: Confirmed carcinogen with experimental carcinogenic, neoplastigenic, and tumorigenic data. Moderately toxic via intraperitoneal route. Mutation data reported. When heated to decomposition it emits toxic NO_x. See also AMINES.

AIB250 CAS:81-46-9 ***HR: 1***
N-(4-AMINOANTHRAQUINONYL)BENZAMIDE
mf: $C_{21}H_{15}N_2O_3$ mw: 342.37

SYNS: 1-AMINO-4-BENZAMIDOANTHRAQUINONE ◇ 1-AMINO-4-BENZOYLAMINOANTHRACHINON (CZECH) ◇ 4-AMINO-1-BENZOYLAMINOANTHRAQUINONE ◇ 1-AMINO-4-(BENZOYLAMINO)ANTHRAQUINONE ◇ N-(4-AMINO-9,10-DIHYDRO-9,10-DIOXO-1-ANTHRACENTY)-BENZAMIDE ◇ CORINTH FLOUR

TOXICITY DATA with REFERENCE
eye-rbt 500 mg/24H MLD 28ZPAK -,124,72

CONSENSUS REPORTS: Reported in EPA TSCA Inventory.

SAFETY PROFILE: An eye irritant. When heated to decomposition it emits toxic fumes of NO_x. See also AMIDES.

AIC000 CAS:3398-09-2 ***HR: 3***
p-AMINO-2′:3-AZOTOLUENE
mf: $C_{14}H_{15}N_3$ mw: 225.32

SYN: 4′-AMINO-3,2′-AZOTOLUENE

TOXICITY DATA with REFERENCE
orl-mus TDLo:30 g/kg/57W-C:ETA BJCAAI 3,387,49

SAFETY PROFILE: Questionable carcinogen with ex-

perimental tumorigenic data. When heated to decomposition it emits toxic fumes of NO_x.

AIC250 CAS:97-56-3 ***HR: 3***
2-AMINO-5-AZOTOLUENE
mf: $C_{14}H_{15}N_3$ mw: 225.32

SYNS: AAT ◇ o-AAT ◇ o-AMIDOAZOTOLUOL (GERMAN) ◇ AMINOAZOTOLUENE (indicator) ◇ o-AMINOAZOTOLUENE (MAK) ◇ 4′-AMINO-2,3′-AZOTOLUENE ◇ o-AMINOAZOTOLUENO (SPANISH) ◇ 4′-AMINO-2:3′-AZOTOLUENE ◇ o-AMINOAZOTOLUOL ◇ 4-AMINO-2′,3-DIMETHYLAZOBENZENE ◇ 4′-AMINO-2,3′-DIMETHYLAZOBENZENE ◇ o-AT ◇ BRASILAZINA OIL YELLOW R ◇ BUTTER YELLOW ◇ C.I. 11160 ◇ C.I. 11160B ◇ C.I. SOLVENT YELLOW 3 ◇ 2′,3-DIMETHYL-4-AMINOAZOBENZENE ◇ FAST GARNET GBC BASE ◇ FAST OIL YELLOW ◇ FAST YELLOW AT ◇ FAST YELLOW B ◇ HIDACO OIL YELLOW ◇ 2-METHYL-4-((2-METHYLPHENYL)AZO)BENZENAMINE ◇ OAAT ◇ OIL YELLOW ◇ OIL YELLOW 21 ◇ OIL YELLOW 2681 ◇ OIL YELLOW AT ◇ OIL YELLOW A ◇ OIL YELLOW C ◇ OIL YELLOW I ◇ OIL YELLOW 2R ◇ OIL YELLOW T ◇ ORGANOL YELLOW 25 ◇ SOMALIA YELLOW R ◇ SUDAN YELLOW RRA ◇ o-TOLUENEAZO-o-TOLUIDINE ◇ o-TOLUOL-AZO-o-TOLUIDIN (GERMAN) ◇ 5-(o-TOLYLAZO)-2-AMINOTOLUENE ◇ 4-(o-TOLYLAZO)-o-TOLUIDINE ◇ TULABASE FAST GARNET GB ◇ TULABASE FAST GARNET GBC ◇ WAXAKOL YELLOW NL

TOXICITY DATA with REFERENCE
mma-sat 25 ng/plate CNREA8 45,6155,85
dnr-esc 25 mg/L JNCIAM 62,873,79
orl-rat TDLo:15 g/kg/57W-C:NEO BJCAAI 3,387,49
orl-mus TDLo:480 mg/kg (16-21D post):NEP,TER BEXBAN 85,201,78
scu-mus TDLo:4000 mg/kg (15-21D post):NEO,TER BEXBAN 78,1402,74
imp-mus TDLo:80 mg/kg:CAR BJCAAI 22,825,68
mul-mus TDLo:400 mg/kg/I:ETA CNREA8 1,397,41
orl-rat LDLo:1500 mg/kg CNREA8 26,619,66
orl-mus LDLo:800 mg/kg JNCIAM 10,927,50
scu-mus LDLo:1200 mg/kg JNCIAM 10,927,50
orl-dog LD50:300 mg/kg 85JCAE-,1315,86

CONSENSUS REPORTS: NTP Fifth Annual Report on Carcinogens. IARC Cancer Review: Group 2B IMEMDT 7,56,87; Animal Sufficient Evidence IMEMDT 8,61,75. Community Right-To-Know List. Reported in EPA TSCA Inventory. EPA Genetic Toxicology Program.

DFG MAK: Animal Carcinogen; Suspected Human Carcinogen.

SAFETY PROFILE: Confirmed carcinogen with experimental carcinogenic, neoplastigenic, and tumorigenic data. Poison by ingestion. Moderately toxic by subcutaneous route. An experimental teratogen. Human mutation data reported. When heated to decomposition it emits toxic fumes of NO_x. See also AROMATIC AMINES.

AIC500 CAS:3963-79-9 ***HR: 3***
4′-AMINO-4,2′-AZOTOLUENE
mf: $C_{14}H_{15}N_3$ mw: 225.32

SYN: 4-(p-TOLYLAZO)-m-TOLUIDINE

TOXICITY DATA with REFERENCE
orl-mus TDLo:30 g/kg/57W-C:CAR BJCAAI 3,387,49

SAFETY PROFILE: Questionable carcinogen with experimental carcinogenic data. When heated to decomposition it emits toxic fumes of NO_x.

AIC750 CAS:18936-75-9 ***HR: 3***
10-AMINOBENZ(a)ACRIDINE
mf: $C_{17}H_{12}N_2$ mw: 244.31

SYN: BENZ(a)ACRIDIN-10-AMINE

TOXICITY DATA with REFERENCE
scu-mus TDLo:72 mg/kg/9W-I:ETA CHDDAT 267,981,68

SAFETY PROFILE: Questionable carcinogen with experimental tumorigenic data. See also AROMATIC AMINES. When heated to decomposition it emits toxic fumes of NO_x.

AID250 CAS:6957-91-1 ***HR: 3***
p-AMINOBENZALDEHYDETHIOSEMICARBAZONE
mf: $C_8H_{10}N_4S$ mw: 194.28

SYN: 4-AMINOBENZALDEHYDETHIOSEMICARBAZONE

TOXICITY DATA with REFERENCE
orl-mus LD50:500 mg/kg JPPMAB 2,764,50
ivn-mus LD50:180 mg/kg CSLNX* NX#00784

SAFETY PROFILE: Poison by intravenous route. Moderately toxic by ingestion. See also ALDEHYDES. When heated to decomposition it emits very toxic fumes of NO_x and SO_x.

AID500 CAS:98-16-8 ***HR: 3***
m-AMINOBENZAL FLUORIDE
DOT: UN 2948
mf: $C_7H_6F_3N$ mw: 161.14

PROP: Colorless liquid with aniline-like odor. Mp: 3°, bp: 189°, d: 1.303 @ 15.5°/15.5°, vap d: 5.56.

SYNS: m-AMINOBENZOTRIFLUORIDE ◇ 3-AMINOBENZOTRIFLUORIDE ◇ m-(TRIFLUOROMETHYL)-ANILINE ◇ 3-(TRIFLUOROMETHYL)ANILINE ◇ 3-(TRIFLUOROMETHYL)BENZENAMINE ◇ USAF MA-4

TOXICITY DATA with REFERENCE
unr-rat LD50:480 g/kg TPKVAL 14,118,75
orl-rbt LD50:615 mg/kg 85GMAT -,20,82
unk-rat LD50:480 mg/kg TPKVAL 14,118,75
ihl-rat LC50:440 mg/m^3/4H TPKVAL 10,131,68
orl-mus LD50:220 mg/kg TPKVAL 10,131,68

ihl-mus LC50:690 mg/m^3/2H TPKVAL 10,131,68
ipr-mus LD50:50 mg/kg NTIS** AD277-689

CONSENSUS REPORTS: EPA Extremely Hazardous Substances List. Reported in EPA TSCA Inventory.

DOT Classification: IMO: Poison B; Label: Poison.

SAFETY PROFILE: Poison by inhalation, ingestion, and intraperitoneal routes. May be moderately toxic by other routes. See also AMINES and FLUORIDES. When heated to decomposition it emits very toxic fumes of F^- and NO_x.

AID625 CAS:3544-24-9 ***HR: 2***
3-AMINOBENZAMIDE
mf: $C_7H_8N_2O$ mw: 136.17

SYNS: m-AMINOBENZAMIDE ◇ 3-AMINO-BENZAMIDE (9CI)

TOXICITY DATA with REFERENCE
sce-hmn:lym 1 mmol/L MUREAV 122,223,83
dns-ham:ovr 10 mmol/L ECREAL 143,377,83
sce-ham:ovr 1 mmol/L ENMUDM 6,203,84
orl-bwd LD50:1 g/kg AECTCV 12,355,83

SAFETY PROFILE: Moderately toxic by ingestion. Human mutation data reported. When heated to decomposition it emits toxic fumes of NO_x.

AID650 CAS:3858-83-1 ***HR: D***
4-AMINOBENZAMIDINE
mf: $C_7H_9N_3$ mw: 135.19

SYNS: ABD ◇ p-AMINOBENZAMIDINE ◇ BENZENECARBOXIMIDAMIDE, 4-AMINO- (9CI)

TOXICITY DATA with REFERENCE
imp-mus TDLo:194 mg/kg (female 2D pre):REP CCPTAY 26,137,82

SAFETY PROFILE: Experimental reproductive effects. When heated to decomposition it emits toxic fumes of NO_x.

AID750 ***HR: 3***
p-AMINO BENZENE DIAZONIUMPERCHLORATE
mf: $C_6H_6ClN_3O_4$ mw: 219.6

SAFETY PROFILE: Extremely shock sensitive explosive. Very dangerous. When heated to decomposition it explodes and emits very toxic fumes of Cl^- and NO_x. See PERCHLORATES.

AIE000 CAS:98-44-2 ***HR: 1***
1-AMINO-2,5-BENZENEDISULFONIC ACID
mf: $C_6H_7NO_6S_2$ mw: 253.26

SYNS: 2-AMINO-p-BENZENEDISULFONIC ACID ◇ 2-AMINO-1,4-BENZENEDISULFONIC ACID ◇ 2-AMINO-BENZENE-1,4-DISULFONIC ACID ◇ 2-AMINO-1,4-DISULFOBENZENE ◇ ANILINE-2,5-DISULFONIC ACID ◇ 1-ANILINO-2,5-DISULFONIC ACID ◇ 2,5-DISULFO-1-AMINOBENZENE ◇ 2,5-DISULFOANILINE ◇ KYSELINA ANILIN-2,5-DISULFONOVA (CZECH) ◇ 4-SULFOMETANILIC ACID

TOXICITY DATA with REFERENCE
skn-rbt 500 mg/24H MLD 28ZPAK -,182,72
eye-rbt 100 mg/24H MOD 28ZPAK -,182,72
orl-rat LD50:7760 mg/kg 28ZPAK -,182,72

CONSENSUS REPORTS: Reported in EPA TSCA Inventory.

SAFETY PROFILE: Mildly toxic by ingestion. A skin and eye irritant. See also SULFONATES. When heated to decomposition it emits very toxic fumes of NO_x and SO_x.

AIE125 CAS:64058-57-7 ***HR: 3***
p-AMINOBENZENESTIBONIC ACID COMPOUND with UREA (3581)
mf: $C_{18}H_{24}N_3O_9Sb_3 \cdot CH_4N_2O$ mw: 851.77

TOXICITY DATA with REFERENCE
ipr-mus LDLo:256 mg/kg CBCCT* 2,243,50

CONSENSUS REPORTS: Antimony compounds are on the Community Right-To-Know List.

OSHA PEL: TWA 0.5 mg(Sb)/m^3
ACGIH TLV: TWA 0.5 mg(Sb)/m^3
NIOSH REL: (Antimony) TWA 0.5 mg(Sb)/m^3

SAFETY PROFILE: Poison by intraperitoneal route. See also ANTIMONY COMPOUNDS. When heated to decomposition it emits very toxic fumes of NO_x and Sb.

AIE500 CAS:2447-57-6 ***HR: 2***
6-(4-AMINOBENZENESULFONAMIDO)-4,5-DIMETHOXYPYRIMIDINE
mf: $C_{12}H_{14}N_4O_4S$ mw: 310.36

SYNS: 4-AMINO-N-(5,6-DIMETHOXY-4-PYRIMIDINYL)BENZENESULFONAMIDE ◇ N'-(5,6-DIMETHOXY-4-PYRIMIDYL)SULFANILAMIDE ◇ FANASIL ◇ FANZIL ◇ RO 4-4393 ◇ SULFADOXINE ◇ 4-SULFANILAMIDO-5,6-DIMETHOXYPYRIMIDINE ◇ SULFORTHOMIDINE ◇ SULPHORMETHOXINE

TOXICITY DATA with REFERENCE
orl-mus LD50:5200 mg/kg MEIEDD 10,1276,83
ipr-mus LD50:2900 mg/kg MEIEDD 10,1276,83
scu-mus LD50:2900 mg/kg MEIEDD 10,1276,83

SAFETY PROFILE: Moderately toxic by intraperitoneal and subcutaneous routes. Mildly toxic by ingestion. See also SULFONATES. When heated to decomposition it emits very toxic fumes of SO_x and NO_x.

AIE750 CAS:729-99-7 ***HR: 2***
2-(p-AMINOBENZENESULFONAMIDO)-4,5-DIMETHYLOXAZOLE
mf: $C_{11}H_{13}N_3O_3S$ mw: 267.33

SYNS: p-AMINOBENZENESULFONYL-2-AMINO-4,5-DIMETHYLOXAZOLE ◇ 2-(p-AMINOBENZOLSULFONAMIDO)-4,5-DIMETHYLOXAZOL (GERMAN) ◇ 4-AMINO-N-(4,5-DIMETHYL-2-OXAZOLYL)BENZENESULFONAMIDE ◇ 2-(p-AMINOPHENYLSULFONYLAMINO)-4,5-DIMETHYL-OXAZOLE ◇ DEPOMIDE ◇ N^1-(4,5-DIMETHYL-2-OXAZOLYL)-SULFANILAMIDE ◇ 4,5-DIMETHYL-2-SULFANILAMIDOOXAZOLE ◇ JUSTAMIL ◇ NUPRIN ◇ OXASULFA ◇ SDMO ◇ SULFABUTIN ◇ SULFADIMETHYLOXAZOLE ◇ SULFAMOXOLE ◇ SULFAMOXOLUM ◇ SULFANO ◇ SULFAVIGOR ◇ SULFMIDIL ◇ SULFUNE ◇ SULFUNO ◇ TARDAMID ◇ TARDAMIDE

TOXICITY DATA with REFERENCE
ipr-rat LD50:2500 mg/kg ARZNAD 26,634,76
ipr-mus LD50:1800 mg/kg ARZNAD 26,634,76
orl-mus LD50:15200 mg/kg ARZNAD 10,612,60
scu-mus LD50:1285 mg/kg ARZNAD 10,612,60
ivn-mus LD50:1000 mg/kg ARZNAD 10,612,60

SAFETY PROFILE: Moderately toxic by subcutaneous, intraperitoneal, and intravenous routes. Mildly toxic by ingestion. When heated to decomposition it emits very toxic fumes of SO_x and NO_x.

AIF000 CAS:526-08-9 ***HR: 2***
3-(p-AMINOBENZENESULFONAMIDO)-2-PHENYLPYRAZOLE
mf: $C_{15}H_{14}N_4O_2S$ mw: 314.39

SYNS: 4-AMINO-N-(1-PHENYL-1H-PYRAZOL-5-YL)BENZENESULFONAMIDE ◇ DEPOCID ◇ DEPTOSULFONAMIDE ◇ EFTOLON ◇ FIRMAZOLO ◇ INAMIL ◇ ISAROL ◇ MERIAN ◇ MICROTAN PIRAZOLO ◇ ORISUL ◇ ORISULF ◇ PAIDAZOLO ◇ N'-(1-PHENYLPYRAZOL-5-YL)SULFANILAMIDE ◇ $N(^1)$-(1-PHENYLPYRAZOL-5-YL)-SULFANILAMIDE (8CI) ◇ 1-PHENYL-5-SULFANILAMIDOPYRAZOLE ◇ PLISULFAN ◇ RAZIOSULFA ◇ SP ◇ SPP ◇ SULFAAFENAZOLO (ITALIAN) ◇ SULFABID ◇ 5-SULFANILAMIDO-1-PHENYLPYRAZOLE ◇ SULFAPHENAZOLE ◇ SULFAPHENAZON ◇ SULFAPHENYLPIPAZOL ◇ SULFAPHENYLPYRAZOLE ◇ SULPHENAZOLE

TOXICITY DATA with REFERENCE
scu-rat LD50:900 mg/kg NIIRDN 6 387,82
ivn-rat LD50:525 mg/kg NIIRDN 6,387,82
orl-mus LD50:4507 mg/kg NIIRDN 6,387,82
scu-mus LD50:660 mg/kg NIIRDN 6,387,82
ivn-mus LD50:470 mg/kg NIIRDN 6,387,82
scu-rbt LD50:950 mg/kg NIIRDN 6,387,82
ivn-rbt LD50:440 mg/kg NIIRDN 6,387,72

SAFETY PROFILE: Moderately toxic by subcutaneous and intravenous routes. Mildly toxic by ingestion. See also SULFONATES. When heated to decomposition it emits very toxic fumes of SO_x and NO_x.

AIF250 CAS:1126-34-7 ***HR: 1***
m-AMINOBENZENESULFONIC ACID SODIUM SALT
mf: $C_6H_6NO_3S{\cdot}Na$ mw: 195.18

SYN: METANILAN SODNY (CZECH)

TOXICITY DATA with REFERENCE
eye-rbt 750 μg/24H SEV 28ZPAK -,179,72
orl-rat LD50:7230 mg/kg 28ZPAK -,179,72

CONSENSUS REPORTS: Reported in EPA TSCA Inventory.

SAFETY PROFILE: Mildly toxic by ingestion. A severe eye irritant. See also SULFONATES. When heated to decomposition it emits very toxic fumes of SO_x, NO_x, and Na_2O.

AIF500 CAS:137-07-5 ***HR: 3***
2-AMINOBENZENETHIOL
mf: C_6H_7NS mw: 125.20

PROP: Liquid. Mp: 23°, bp: 227.2°, flash p: 175°F, d: 1.168, vap d: 4.3.

SYNS: o-AMINOTHIOPHENOL ◇ 2-AMINOTHIOPHENOL ◇ o-MERCAPTOANILINE ◇ USAF EK-4376

TOXICITY DATA with REFERENCE
orl-rat LDLo:500 mg/kg 34ZIAG -,90,69
ipr-mus LD50:25 mg/kg NTIS** AD277-689
ivn-mus LD50:100 mg/kg CSLNX* NX#02532

CONSENSUS REPORTS: Reported in EPA TSCA Inventory.

SAFETY PROFILE: Poison by intraperitoneal route. Moderately toxic by ingestion. Moderately flammable. Can react with oxidizing materials. To fight fire, use water, foam, CO_2, mist or spray, dry chemical.

AIF750 CAS:1193-02-8 ***HR: 3***
p-AMINOBENZENETHIOL
mf: C_6H_7NS mw: 125.20

SYNS: 4-AMINOBENZENETHIOL ◇ p-AMINOPHENYLMERCAPTAN ◇ p-AMINOTHIOPHENOL ◇ 4-AMINOTHIOPHENOL ◇ p-MERCAPTOANILINE ◇ 4-MERCAPTOANILINE

TOXICITY DATA with REFERENCE
mma-sat 50 μg/plate MUREAV 67,123,79
orl-mus LDLo:320 mg/kg AECTCV 14,111,85
orl-qal LD50:42200 μg/kg AECTCV 12,355,83
orl-bwd LD50:42200 μg/kg AECTCV 12,355,83

CONSENSUS REPORTS: EPA Genetic Toxicology Program.

SAFETY PROFILE: Poison by ingestion. Mutation data reported. When heated to decomposition it emits very toxic fumes of NO_x and SO_x.

AIG000 CAS:934-32-7 ***HR: 3***
2-AMINOBENZIMIDAZOLE
mf: $C_7H_7N_3$ mw: 133.17

PROP: Aqueous leaflets. Mp: 222-224°; sol in water, alkalies, alc, acetone; very sltly sol in ether.

SYN: USAF EK-4037

TOXICITY DATA with REFERENCE
mma-sat 710 μmol/L ENMUDM 3,11,81
mmo-sat 100 μg/plate MUREAV 15,273,72
orl-rat TDLo:426 mg/kg (8-15D preg):TER THERAP 31,505,76
orl-rat LDLo:500 mg/kg NCNSA6 5,22,53
orl-mus LD40:600 mg/kg JACSAT 67,905,45
ipr-mus LD50:100 mg/kg NTIS** AD277-689
ivn-mus LD50:126 mg/kg 29QHAQ -,246,74

CONSENSUS REPORTS: Reported in EPA TSCA Inventory.

SAFETY PROFILE: Poison by intravenous and intraperitoneal routes. Moderately toxic by ingestion. An experimental teratogen. Mutation data reported. When heated to decomposition it emits toxic NO_x. See also AROMATIC AMINES.

AIH000 CAS:52329-60-9 ***HR: 3***
2-AMINO-6-BENZIMIDAZOLYL PHENYLKETONE
mf: $C_{14}H_{11}N_3O$ mw: 237.28

SYN: 2-AMINO-5-BENZOYLBENZIMIDAZOLE

TOXICITY DATA with REFERENCE
orl-rat TDLo:755 mg/kg (8-15D preg):TER THERAP 31,505,76
orl-mus LD50:450 mg/kg MPPBAB 48,29,79
ipr-mus LD50:257 mg/kg MPPBAB 48,29,79

SAFETY PROFILE: Acute poison by intraperitoneal route. Moderately toxic by ingestion. An experimental teratogen. See also KETONES. When heated to decomposition it emits very toxic fumes of NO_x.

AIH500 CAS:99-05-8 ***HR: 2***
m-AMINOBENZOIC ACID
mf: $C_7H_7NO_2$ mw: 137.15

PROP: Needles. D: 1.511; mp: 173°-174°. Sol in water, alc, and ether.

TOXICITY DATA with REFERENCE
ipr-mus LDLo:500 mg/kg CBCCT* 6,53,54

CONSENSUS REPORTS: Reported in EPA TSCA Inventory.

SAFETY PROFILE: Moderately toxic by intraperitoneal route. When heated to decomposition it emits toxic fumes of NO_x.

AIH600 CAS:150-13-0 ***HR: 2***
p-AMINOBENZOIC ACID
mf: $C_7H_7NO_2$ mw: 137.15

PROP: Yellowish to red crystals. Mp: 187°. Sol in water, alc, and ether.

SYNS: Γ-AMINOBENZOIC ACID ◇ 4-AMINOBENZOIC ACID ◇ 1-AMINO-4-CARBOXYBENZENE ◇ ANTI-CHROMOTRICHIA FACTOR ◇ BACTERIAL VITAMIN H1 ◇ 4-CARBOXYANILINE ◇ p-CARBOXYPHENYLAMINE ◇ CHROMOTRICHIA FACTOR ◇ PABA ◇ TRICHOCHROMOGENIC FACTOR ◇ VITAMIN H

TOXICITY DATA with REFERENCE
dnd-mus-ipr 1 g/kg ARTODN 5,355,82
orl-rat TDLo:2500 mg/kg (female 1-22D post):REP AJANA2 110,29,62
orl-mus LD50:2850 mg/kg PSEBAA 49,184,42
orl-dog LD50:1000 mg/kg PSEBAA 49,184,42
orl-rbt LD50:1830 mg/kg FEPRA7 10,289,51
ivn-rbt LD50:2000 mg/kg FEPRA7 1,71,42

CONSENSUS REPORTS: IARC Cancer Review: Group 3 IMEMDT 7,56,87; Animal Inadequate Evidence IMEMDT 16,249,78. Reported in EPA TSCA Inventory.

SAFETY PROFILE: Moderately toxic by ingestion and intravenous routes. Ingesting large doses can cause nausea, vomiting, skin rash, methemoglobinemia and possibly toxic hepatitis. Experimental reproductive effects. Mutation data reported. Combustible. When heated to decomposition it emits toxic fumes of NO_x. A topical sunscreen.

AIK500 CAS:73698-75-6 ***HR: 3***
p-AMINOBENZOIC ACID-2-(2-(2-(2-(2-(DIETHYLAMINO)ETHOXY)ETHOXY)ETHOXY)ETHOXY)ETHYL ESTER, HYDROCHLORIDE
mf: $C_{21}H_{36}N_2O_6 \cdot ClH$ mw: 449.05

TOXICITY DATA with REFERENCE
eye-rbt 1% SEV JPETAB 48,371,33
unk-rat LDLo:60 mg/kg JPETAB 48,371,33

SAFETY PROFILE: A poison by unspecified route. Severe eye irritant. See also ESTERS. When heated to decomposition it emits very toxic fumes of HCl and NO_x.

AIK750 CAS:73698-76-7 ***HR: 3***
p-AMINOBENZOIC ACID-2-(2-(2-(2-(DIETHYLAMINO)ETHOXY)ETHOXY)ETHOXY)ETHYL ESTER, HYDROCHLORIDE
mf: $C_{19}H_{32}N_2O_5 \cdot ClH$ mw: 404.99

TOXICITY DATA with REFERENCE
eye-rbt 1% SEV JPETAB 48,371,33
unk-rat LDLo:30 mg/kg JPETAB 48,371,33

SAFETY PROFILE: A poison by unspecified route. Severe eye irritant. See also ESTERS. When heated to decomposition it emits very toxic fumes of HCl and NO_x.

AIL000 CAS:73698-77-8 ***HR: 3***
p-AMINOBENZOIC ACID-2-(2-(2-(DIETHYLAMINO)ETHOXY)ETHOXY)ETHYL ESTER, HYDROCHLORIDE
mf: $C_{17}H_{28}N_2O_4 \cdot ClH$ mw: 360.93

TOXICITY DATA with REFERENCE
eye-rbt 1% SEV JPETAB 48,371,33
unk-rat LDLo:30 mg/kg JPETAB 48,371,33

SAFETY PROFILE: A poison by unspecified route. Severe eye irritant. See also ESTERS. When heated to decomposition it emits very toxic fumes of HCl and NO_x.

AIL250 CAS:73698-78-9 ***HR: 3***
p-AMINOBENZOIC ACID-2-(2-(DIETHYLAMINO)ETHOXY)ETHYL ESTER, HYDROCHLORIDE
mf: $C_{15}H_{24}N_2O_3 \cdot ClH$ mw: 316.87

TOXICITY DATA with REFERENCE
eye-rbt 1% SEV JPETAB 48,371,33
unk-rat LDLo:60 mg/kg JPETAB 48,371,33

SAFETY PROFILE: A poison by an unspecified route. Severe eye irritant. See also ESTERS. When heated to decomposition it emits very toxic fumes of HCl and NO_x.

AIL500 CAS:63917-76-0 ***HR: 3***
p-AMINOBENZOIC ACID-3-(β-DIETHYLAMINO)ETHOXY)PROPYL ESTER
mf: $C_{16}H_{26}N_2O_3$ mw: 294.44

TOXICITY DATA with REFERENCE
ivn-mus LD50:55 mg/kg RCPRAN 15,143,54
scu-gpg LD50:31 mg/kg RCPRAN 15,143,54

SAFETY PROFILE: Poison by intravenous and subcutaneous routes. See also ESTERS. When heated to decomposition it emits toxic fumes of NO_x.

AIL750 CAS:59-46-1 ***HR: 3***
p-AMINOBENZOIC ACID-2-DIETHYLAMINOETHYL ESTER
mf: $C_{13}H_{20}N_2O_2$ mw: 236.35

SYNS: ALLOCAINE ◇ 4-AMINOBENZOIC ACID DIETHYLAMINOETHYL ESTER ◇ p-AMINOBENZOYLDIETHYLAMINOETHANOL ◇ DIETHYLAMINOETHYL-p-AMINOBENZOATE ◇ β-DIETHYLAMINOETHYL-4-AMINOBENZOATE ◇ 2-DIETHYLAMINOETHYL-p-AMINOBENZOATE ◇ GEROVITAL ◇ JENACAINE ◇ NEOCAINE ◇ NISSOCAINE ◇ NOROCAINE ◇ NOVOCAINE ◇ PROCAINE ◇ PROCAINE, BASE ◇ SCUROCAINE ◇ SPINOCAINE

TOXICITY DATA with REFERENCE
orl-man LDLo:147 mg/kg 85DCAI 2,73,70
ims-wmn TDLo:1600 μg/kg:CNS JNNPAU 34,20,71
ipr-rat LDLo:280 mg/kg TXAPA9 1,156,59
scu-rat LD50:600 mg/kg AIPTAK 104,388,56
ivn-rat LDLo:45 mg/kg PHREA7 12,190,32
orl-mus LD50:500 mg/kg 27ZIAQ -,212,73
ipr-mus LD50:124 mg/kg RPOBAR 2,213,70
scu-mus LD50:300 mg/kg ARZNAD 5,376,55
ivn-mus LD50:45 mg/kg JAPMA8 45,382,56

SAFETY PROFILE: Poison by ingestion, intraperitoneal, intravenous, and subcutaneous routes. Moderately toxic by parenteral route. Human systemic effects by intramuscular route: lack of muscular control, rigidity, and possibly catalepsy. See also ESTERS. When heated to decomposition it emits toxic fumes of NO_x. Used as a local anesthetic.

AIM000 CAS:17599-08-5 ***HR: 3***
p-AMINOBENZOIC ACID-3-(DIETHYLAMINO)PROPYL ESTER HYDROCHLORIDE
mf: $C_{14}H_{22}N_2O_2 \cdot ClH$ mw: 286.84

SYN: p-AMINO BENZOYL DIETHYL AMINO PROPANOL HYDROCHLORIDE

TOXICITY DATA with REFERENCE
unk-rat LD50:125 mg/kg ARZNAD 17,1012,67
scu-mus LD50:300 mg/kg JPETAB 24,160,24

SAFETY PROFILE: Poison by subcutaneous and other unspecified routes. See also ESTERS. When heated to decomposition it emits very toxic fumes of HCl and NO_x.

AIM250 CAS:15154 ***HR: 3***
p-AMINOBENZOIC ACID-3-(DIISOPROPYLAMINO)PROPYL ESTER HYDROCHLORIDE-37-7
mf: $C_{16}H_{26}N_2O_2 \cdot ClH$ mw: 314.90

SYN: p-AMINO BENZOYL DIISOPROPYL AMINO PROPANOL HYDROCHLORIDE

TOXICITY DATA with REFERENCE
unk-rat LD50:75 mg/kg ARZNAD 17,1012,67
scu-mus LD50:150 mg/kg JPETAB 24,160,24

SAFETY PROFILE: Poison by subcutaneous and other unspecified routes. See also ESTERS. When heated to decomposition it emits very toxic fumes of HCl and NO_x.

AIN000 CAS:15154-36-6 ***HR: 3***
p-AMINOBENZOIC ACID-2-(DIPROPYLAMINO)ETHYL ESTER HYDROCHLORIDE
mf: $C_{15}H_{24}N_2O_2 \cdot ClH$ mw: 300.87

SYN: p-AMINO BENZOYL DI-N-PROPYL AMINO ETHANOL HYDROCHLORIDE

TOXICITY DATA with REFERENCE
unk-rat LD50:100 mg/kg ARZNAD 17,1012,67
scu-mus LD50:550 mg/kg JPETAB 24,160,24

SAFETY PROFILE: A poison by unspecified route. Moderately toxic by subcutaneous route. See also ESTERS and AMINES. When heated to decomposition it emits very toxic fumes of HCl and NO_x.

AIQ875 CAS:72977-18-5 ***HR: 2***
p-AMINOBENZOIC ACID PHOSPHATE
mf: $C_7H_7NO_2 \cdot H_3O_4P$ mw: 235.15

TOXICITY DATA with REFERENCE
orl-rat LD50:8 g/kg GISAAA 49(10),82,84
orl-mus LD50:3100 mg/kg GISAAA 49(10),82,84
orl-rbt LD50:2500 mg/kg GISAAA 49(10),82,84
orl-gpg LD50:2580 mg/kg GISAAA 49(10),82,84

SAFETY PROFILE: Moderately toxic by ingestion. When heated to decomposition it emits toxic fumes of NO_x and PO_x.

AIQ880 CAS:69780-82-1 ***HR: 3***
m-AMINOBENZOIC ACID-2-(2-PIPERIDYL)ETHYL ESTER HYDROCHLORIDE
mf: $C_{14}H_{20}N_2O_2 \cdot ClH$ mw: 284.82

SYNS: 2-PIPERIDINEETHANOL-m-AMINOBENZOATE (ester) HYDROCHLORIDE ◇ β-2-PIPERIDYLETHYL-m-AMINOBENZOATE HYDROCHLORIDE

TOXICITY DATA with REFERENCE
scu-mus LDLo:87 mg/kg JACSAT 61,1713,39
scu-rbt LDLo:500 mg/kg ANESAV 1,305,40
ivn-rbt LDLo:20 mg/kg ANESAV 1,305,40
isp-rbt LDLo:24620 μg/kg ANESAV 1,305,40
scu-gpg LDLo:315 mg/kg ANESAV 1,305,40

SAFETY PROFILE: Poison by intravenous and intraspinal routes. Moderately toxic by subcutaneous route. When heated to decomposition it emits toxic fumes of NO_x and HCl. See also ESTERS.

AIQ885 CAS:69780-83-2 ***HR: 3***
o-AMINOBENZOIC ACID 2-(2-PIPERIDYL)ETHYL ESTER HYDROCHLORIDE
mf: $C_{14}H_{20}N_2O_2 \cdot ClH$ mw: 284.82

SYNS: 2-PIPERIDINEETHANOL-o-AMINOBENZOATE (ester) HYDROCHLORIDE ◇ β-2-PIPERIDYLETHYL-o-AMINOBENZOATE HYDROCHLORIDE

TOXICITY DATA with REFERENCE
scu-mus LDLo:23 mg/kg JACSAT 61,1713,39
scu-rbt LDLo:150 mg/kg ANESAV 1,305,40
ivn-rbt LDLo:15 mg/kg ANESAV 1,305,40
isp-rbt LDLo:29970 μg/kg ANESAV 1,305,40
scu-gpg LDLo:153 mg/kg ANESAV 1,305,40

SAFETY PROFILE: Poison by subcutaneous, intravenous, and intraspinal routes. When heated to decomposition it emits toxic fumes of NO_x and HCl. See also ESTERS.

AIQ890 CAS:69780-84-3 ***HR: 3***
p-AMINOBENZOIC ACID-2-(2-PIPERIDYL)ETHYL ESTER HYDROCHLORIDE
mf: $C_{14}H_{20}N_2O_2 \cdot ClH$ mw: 284.82

SYNS: 2-PIPERIDINEETHANOL-p-AMINOBENZOATE (ester) HYDROCHLORIDE ◇ β-2-PIPERIDYLETHYL-p-AMINOBENZOATE HYDROCHLORIDE

TOXICITY DATA with REFERENCE
scu-mus LDLo:20 mg/kg JACSAT 61,1713,39
scu-rbt LDLo:250 mg/kg ANESAV 1,305,40
ivn-rbt LDLo:17 mg/kg ANESAV 1,305,40
isp-rbt LDLo:10920 μg/kg ANESAV 1,305,40
scu-gpg LDLo:196 mg/kg ANESAV 1,305,40

SAFETY PROFILE: Poison by subcutaneous, intravenous, and intraspinal routes. When heated to decomposition it emits toxic fumes of NO_x and HCl. See also ESTERS.

AIR125 CAS:2237-30-1 ***HR: 2***
3-AMINOBENZONITRILE
mf: $C_7H_6N_2$ mw: 118.15

SYNS: m-AMINOBENZONITRILE ◇ m-ANTHRANILONITRILE ◇ m-CYANOANILINE ◇ 3-CYANOANILINE

TOXICITY DATA with REFERENCE
mma-sat 100 mg/L ENMUDM 5,803,83
orl-qal LD50:562 mg/kg AECTCV 12,355,83
orl-bwd LD50:562 mg/kg AECTCV 12,355,83

CONSENSUS REPORTS: Cyanide and its compounds are on the Community Right-To-Know List.

SAFETY PROFILE: Moderately toxic by ingestion. Mutation data reported. When heated to decomposition it emits toxic fumes of NO_x and CN^-. See also NITRILES.

AIR250 CAS:1137-41-3 ***HR: 3***
p-AMINOBENZOPHENONE
mf: $C_{13}H_{11}NO$ mw: 197.25

PROP: Leaflets from alc. Mp: 124°. Very sltly sol in cold water, very sol in alc.

SYN: USAF A-233.

TOXICITY DATA with REFERENCE
ipr-mus LD50:300 mg/kg NTIS**AD277-689
orl-bwd LD50:562 mg/kg AECTCV 12,355,83

CONSENSUS REPORTS: Reported in EPA TSCA Inventory.

SAFETY PROFILE: Poison by intraperitoneal route. See also KETONES and AMINES. When heated to decomposition it emits toxic fumes of NO_x.

AIS250 CAS:65793-50-2 ***HR: 3***
3-AMINOBENZO-6,7-QUINAZOLINE-4-ONE
mf: $C_{12}H_9N_3O$ mw: 211.24

SYN: 3-AMINOBENZO(g)QUINAZOLIN-4(3H)-ONE

TOXICITY DATA with REFERENCE
orl-man TDLo:1143 μg/kg:CNS IJEBA6 15,1131,77
orl-rat LD50:540 mg/kg IJEBA6 15,1125,77
ipr-rat LD50:266 mg/kg IJEBA6 15,1125,77
orl-mus LD50:525 mg/kg IJEBA6 15,1125,77
ipr-mus LD50:160 mg/kg IJEBA6 15,1125,77
ivn-mus LD50:60 mg/kg IJEBA6 15,1125,77
ipr-cat LD50:54 mg/kg IJEBA6 15,1125,77

SAFETY PROFILE: Poison by intraperitoneal and intravenous route. Moderately toxic by ingestion. Human central nervous system effects by ingestion. See also KETONES. When heated to decomposition it emits toxic fumes of NO_x.

AIS500 CAS:136-95-8 ***HR: 3***
2-AMINOBENZOTHIAZOLE
mf: $C_7H_6N_2S$ mw: 150.21

PROP: Crystals. Mp: 132°.

SYNS: 2-AMINOBENZTHIAZOLE ◇ USAF EK-3941 ◇ USAF XR-27

TOXICITY DATA with REFERENCE
mma-sat 2500 μg/plate FCTOD7 23,695,85
ipr-mus LD50:200 mg/kg NTIS** AD277-689
ivn-mus LD50:126 mg/kg JPETAB 105,486,52

CONSENSUS REPORTS: Reported in EPA TSCA Inventory.

SAFETY PROFILE: Poison by via intraperitoneal and intravenous routes. Mutation data reported. Dangerous; when heated to decomposition it emits highly toxic fumes of NO_x and SO_x.

AIS550 CAS:7442-07-1 ***HR: 3***
6-AMINO-2-BENZOTHIAZOLETHIOL
mf: $C_7H_6N_2S_2$ mw: 182.27

SYN: USAF XR-30

TOXICITY DATA with REFERENCE
ipr-mus LD50:150 mg/kg NTIS** AD277-689

CONSENSUS REPORTS: Reported in EPA TSCA Inventory.

SAFETY PROFILE: Poison by intraperitoneal route. When heated to decomposition it emits very toxic fumes of SO_x and NO_x.

AIS600 CAS:4570-41-6 ***HR: 3***
2-AMINOBENZOXAZOLE
mf: $C_7H_6N_2O$ mw: 134.15

TOXICITY DATA with REFERENCE
orl-rat LD50:600 mg/kg MDCHAG 4(1),336,64
ipr-rat LD50:275 mg/kg MDCHAG 4(1),336,64
orl-mus LD50:678 mg/kg MDCHAG 4(1),336,64
ipr-mus LD50:392 mg/kg MDCHAG 4(1),336,64
ivn-mus LD50:238 mg/kg JPETAB 105,486,52

SAFETY PROFILE: Poison by intravenous and intraperitoneal routes. Moderately toxic by ingestion. When heated to decomposition it emits toxic fumes of NO_x.

AIS625 ***HR: 3***
p-AMINOBENZOYLAMINOMETHYLHYDROCOTARNINE
mf: $C_{20}H_{23}N_3O_4$ mw: 369.46

TOXICITY DATA with REFERENCE
scu-mus LDLo:100 mg/kg IJMRAQ 25,713,38
scu-dog LDLo:25 mg/kg IJMRAQ 25,713,38
scu-cat LDLo:25 mg/kg IJMRAQ 25,713,38
scu-gpg LDLo:80 mg/kg IJMRAQ 25,713,38

SAFETY PROFILE: Poison by subcutaneous route. When heated to decomposition it emits toxic fumes of NO_x.

AIT000 CAS:5892-15-9 ***HR: 3***
AMINOBENZOYLDIBUTYLAMINOPROPANOL HYDROCHLORIDE
mf: $C_{18}H_{30}N_2O_2 \cdot ClH$ mw: 342.96

SYN: p-AMINOBENZOIC ACID-3-(DIBUTYLAMINO)PROPYL ESTER, HYDROCHLORIDE

TOXICITY DATA with REFERENCE
ipr-rat LDLo:200 mg/kg JPETAB 18,467,22
unk-rat LD50:75 mg/kg ARZNAD 17,1012,67
scu-mus LD50:100 mg/kg JPETAB 24,160,24
ipr-cat LDLo:200 mg/kg JPETAB 18,467,22
ivn-cat LDLo:15 mg/kg JPETAB 18,467,22

SAFETY PROFILE: Poison by subcutaneous, intraperitoneal, and possibly other routes. See also ESTERS and

AMINES. When heated to decomposition it emits very toxic fumes of HCl and NO_x.

AIT250 CAS:51-05-8 ***HR: 3***
p-AMINOBENZOYLDIETHYLAMINOETHANOL HYDROCHLORIDE
mf: $C_{13}H_{20}N_2O_2 \cdot ClH$ mw: 272.81

SYNS: ALLOCAINE ◇ 4-AMINOBENZOIC ACID 2-(DIETHYLAMINO)ETHYL ESTER, HYDROCHLORIDE ◇ p-AMINOBENZOIC ACID-2-DIETHYLAMINOETHYL ESTER, HYDROCHLORIDE ◇ AMINOCAINE ◇ ANADOLOR ◇ ANESTIL ◇ ANESTHESOL ◇ ATOXICOCAINE ◇ BERNOCAINE ◇ CETAIN ◇ CHLOROCAINE ◇ DIETHYLAMINOETHANOL-4-AMINOBENZOATE HYDROCHLORIDE ◇ 2-DIETHYLAMINOETHYL-p-AMINOBENZOATE HYDROCHLORIDE ◇ DUGERASE ◇ ETHOCAINE ◇ IROCAINE ◇ ISOCAINE-ASID ◇ ISOCAINE-HEISLER ◇ JUVOCAINE ◇ KEROCAINE ◇ LACTOCAINE ◇ NAUCAINE ◇ NEOCAINE ◇ NOVOCAIN HYDROCHLORID (GERMAN) ◇ NOVOCAIN-CHLORHYDRAT (GERMAN) ◇ NOVOCAINE HYDROCHLORIDE ◇ PARACAIN ◇ PLANOCAINE ◇ PROCAINE HYDROCHLORIDE ◇ SCUROCAINE ◇ SEVICAINE ◇ SYNCAINE ◇ TOPOKAIN ◇ WESTOCAINE

TOXICITY DATA with REFERENCE
orl-rat LD50:200 mg/kg ZENBAX 6B,183,51
ipr-rat LD50:160 mg/kg KSRNAM 13,791,79
ivn-rat LD50:38 mg/kg ANESAV 3,398,42
orl-mus LD50:175 mg/kg ZENBAX 6B,183,51
ipr-mus LD50:50 mg/kg ARZNAD 26,793,76
scu-mus LD50:339 mg/kg BCFAAI 107,310,68
ivn-mus LD50:46200 μg/kg TXAPA9 1,454,56
ims-mus LD50:500 mg/kg THERAP 9,332,54
ivn-dog LD50:63 mg/kg JDREAF 20,425,41

CONSENSUS REPORTS: Reported in EPA TSCA Inventory. EPA Genetic Toxicology Program.

SAFETY PROFILE: Poison by ingestion, subcutaneous, intravenous, and intraperitoneal routes. May have human reproductive effects. See also ESTERS. When heated to decomposition it emits very toxic fumes of HCl and NO_x. Used as a local anesthetic.

AIT500 CAS:5988-31-8 ***HR: 3***
p-AMINO BENZOYL DI-ISO-PROPYL AMINO ETHANOL HYDROCHLORIDE
mf: $C_{15}H_{24}N_2O_2 \cdot ClH$ mw: 300.87

SYN: p-AMINOBENZOICACID-2-(DIISOPROPYLAMINO)ETHYL ESTER, HYDROCHLORIDE

TOXICITY DATA with REFERENCE
unr-rat LD50:100 mg/kg ARZNAD 17,1012,67
scu-mus LD50:400 mg/kg JPETAB 24,160,24

SAFETY PROFILE: Poison by subcutaneous and other unspecified routes. See also ESTERS. When heated to decomposition it emits very toxic fumes of HCl and NO_x.

AIT750 CAS:532-62-7 ***HR: 3***
p-AMINOBENZOYLDIMETHYLAMINO-1,2-DIMETHYLPROPANOL HYDROCHLORIDE
mf: $C_{14}H_{22}N_2O_2 \cdot ClH$ mw: 286.84

SYNS: p-AMINOBENZOIC ACID 3-(DIMETHYLAMINO)-1,2-DIMETHYLPROPYL ESTER, HYDROCHLORIDE ◇ BUTAMIN ◇ 3-DIMETHYLAMINO-1,2-DIMETHYLPROPYL p-AMINOBENZOATE HYDROCHLORIDE ◇ 4-(DIMETHYLAMINO)-3-METHYL-2-BUTANOL 4-AMINOBENZOATE (ester) HYDROCHLORIDE ◇ 3-DIMETHYL-1,2-DIMETHYLPROPYL p-AMINOBENZOATE HYDROCHLORIDE ◇ TOTOCAINE HYDROCHLORIDE ◇ TUTOCAINE HYDROCHLORIDE

TOXICITY DATA with REFERENCE
scu-mus LDLo:350 mg/kg PHREA7 12,190,32
ivn-mus LDLo:50 mg/kg PHREA7 12,190,32
ipr-dog LDLo:82 mg/kg HBTXAC 1,308,55
ivn-dog LDLo:15 mg/kg PHREA7 12,190,32
scu-rbt LDLo:200 mg/kg PHREA7 12,190,32
ivn-rbt LDLo:15 mg/kg PHREA7 12,190,32
isp-rbt LDLo:16 mg/kg JPETAB 57,221,36
ipr-gpg LDLo:250 mg/kg PHREA7 12,190,32
scu-gpg LDLo:193 mg/kg PHREA7 12,190,32
ivn-gpg LDLo:30 mg/kg PHREA7 12,190,32

SAFETY PROFILE: Poison by subcutaneous, intravenous, intraperitoneal, and intraspinal routes. When heated to decomposition it emits very toxic fumes of HCl and NO_x. Used as a surface and infiltration anesthetic.

AIU000 CAS:17599-09-6 ***HR: 3***
p-AMINO BENZOYL DI-N-PROPYL AMINOPROPANOL HYDROCHLORIDE
mf: $C_{16}H_{26}N_2O_2 \cdot ClH$ mw: 314.90

SYN: p-AMINOBENZOIC ACID 3-(DIPROPYLAMINO)PROPYL ESTER, HYDROCHLORIDE

TOXICITY DATA with REFERENCE
unr-rat LD50:100 mg/kg ARZNAD 17,1012,67
scu-mus LD50:200 mg/kg JPETAB 24,160,24

SAFETY PROFILE: Poison by subcutaneous and possibly other unspecified routes. See also ESTERS and AMINES. When heated to decomposition it emits very toxic fumes of HCl and NO_x.

AIU250 CAS:67031-48-5 ***HR: 3***
β-4-AMINOBENZOYLOXY-β-PHENYLETHYL DIMETHYLAMINE
mf: $C_{17}H_{20}N_2O_2$ mw: 284.39

SYN: p-AMINOBENZOICACID-(2-(DIMETHYLAMINO)-1-PHENYL)ETHYL ESTER

TOXICITY DATA with REFERENCE
scu-mus LDLo:57 mg/kg AIPTAK 47,96,34
ivn-rbt LDLo:11 mg/kg AIPTAK 47,96,34

SAFETY PROFILE: Poison by subcutaneous and intra-

venous routes. See also ESTERS. When heated to decomposition it emits toxic fumes of NO_x.

AIU500 CAS:51579-82-9 ***HR: 3***
2-AMINO-3-BENZOYLPHENYLACETIC ACID
mf: $C_{15}H_{13}NO_3$ mw: 255.29

SYNS: AMFENACO (SPANISH) ◇ 2-AMINO-3-BENZOYL BENZENEACETIC ACID

TOXICITY DATA with REFERENCE
orl-rat LD50:615 mg/kg DRFUD4 3,340,78
orl-mus LD50:311 mg/kg DRFUD4 3,340,78

SAFETY PROFILE: Poison by ingestion. When heated to decomposition it emits toxic fumes of NO_x.

AIV500 CAS:69-53-4 ***HR: 3***
AMINOBENZYLPENICILLIN
mf: $C_{16}H_{19}N_3O_4S$ mw: 349.44

SYNS: ACILLIN ◇ ADOBACILLIN ◇ ALPEN ◇ AMBLOSIN ◇ AMCILL ◇ AMFIPEN ◇ d-(−)-α-AMINOBENZYLPENICILLIN ◇ d-(−)-α-AMINOPENICILLIN ◇ 6-(d(−)-α-AMINOPHENYLACETAMIDO)PENICILLANICACID ◇ (AMINOPHENYLMETHYL)-PENICILLIN ◇ AMIPENIX S ◇ AMPERIL ◇ AMPI-BOL ◇ AMPICILLIN (USDA) ◇ d-AMPICILLIN ◇ d-(−)-AMPICILLIN ◇ AMPICILLIN A ◇ AMPICILLIN ACID ◇ AMPICILLIN ANHYDRATE ◇ AMPICIN ◇ AMPIKEL ◇ AMPIMED ◇ AMPIPENIN ◇ AMPLISOM ◇ AMPLITAL ◇ AMPY-PENYL ◇ AUSTRAPEN ◇ AY-6108 ◇ BINOTAL ◇ BONAPICILLIN ◇ BRITACIL ◇ BRL ◇ BRL 1341 ◇ COPHARCILIN ◇ CYMBI ◇ DIVERCILLIN ◇ DOKTACILLIN ◇ GRAMPENIL ◇ GUICITRINA ◇ GUICITRINE ◇ LIFEAMPIL ◇ MARISILAN ◇ NSC-528986 ◇ NUVAPEN ◇ OMNIPEN ◇ P-50 ◇ PENBRISTOL ◇ PENBRITIN ◇ PENBRITIN PAEDIATRIC ◇ PENBRITIN SYRUP ◇ PENBROCK ◇ PENICLINE ◇ PENTREX ◇ PENTREXL ◇ PFIZERPEN A ◇ POLYCILLIN ◇ PONECIL ◇ PRINCIPEN ◇ QIDAMP ◇ RO-AMPEN ◇ SEMICILLIN ◇ SK-AMPICILLIN ◇ SYNPENIN ◇ TOKIOCILLIN ◇ TOLOMOL ◇ TOTACILLIN ◇ TOTALCICLINA ◇ TOTAPEN ◇ ULTRABION ◇ ULTRABRON ◇ VICCILLIN ◇ VICCILLIN S ◇ VICILLIN ◇ WY-5103

TOXICITY DATA with REFERENCE
dnr-esc 20 μL/plate MUREAV 97,1,82
pic-esc 10 ng/plate CNREA8 43,2819,83
orl-rat TDLo:2500 mg/kg (female 4-13D post):TER BEXBAN 91,169,81
orl-man TDLo:400 mg/kg/4W-I:BLD,MET AIMEAS 69,91,68
ipr-rat LD50:4500 mg/kg TXAPA9 18,185,71
ivn-mus LD50:4990 mg/kg NIIRDN 6,57,82
ipr-mus LD50:3250 mg/kg EKFMA7 9,83,80
ice-mus LD50:380 mg/kg NKRZAZ 26,196,80

CONSENSUS REPORTS: EPA Genetic Toxicology Program.

SAFETY PROFILE: Poison by intracerebral route. Moderately toxic by intraperitoneal route. Human systemic effects by ingestion: fever, angranulocytosis, and other blood effects. An experimental teratogen. Mutation data reported. When heated to decomposition it emits very toxic fumes of NO_x and SO_x.

AIV625 ***HR: 3***
4-AMINO-N-(1-BENZYL-4-PIPERIDYL)-5-CHLORO-o-ANISAMIDE HYDROXYSUCCINATE
mf: $C_{20}H_{24}ClN_3O_2{\bullet}C_4H_6O_5$ mw: 507.97

TOXICITY DATA with REFERENCE
orl-rat LD50:2780 mg/kg IYKEDH 16,866,85
ipr-rat LD50:159 mg/kg IYKEDH 16,866,85
scu-rat LD50:4850 mg/kg IYKEDH 16,866,85
ivn-rat LD50:39 mg/kg IYKEDH 16,866,85
ims-rat LD50:2080 mg/kg IYKEDH 16,866,85
orl-mus LD50:510 mg/kg IYKEDH 16,866,85
ipr-mus LD50:145 mg/kg IYKEDH 16,866,85
scu-mus LD50:305 mg/kg IYKEDH 16,866,85
ivn-mus LD50:51 mg/kg IYKEDH 16,866,85
ims-mus LD50:290 mg/kg IYKEDH 16,866,85

SAFETY PROFILE: Poison by subcutaneous, intramuscular, intravenous and intraperitoneal routes. Moderately toxic by ingestion. When heated to decomposition it emits toxic fumes of Cl^- and NO_x.

AIV750 CAS:4363-03-5 ***HR: 3***
4-AMINO-3-BIPHENYLOL
mf: $C_{12}H_{11}NO$ mw: 185.24

SYN: 4-AMINO-3-HYDROXYBIPHENYL

TOXICITY DATA with REFERENCE
scu-mus TDLo:216 mg/kg/3D:CAR JNCIAM 41,403,68

SAFETY PROFILE: Questionable carcinogen with experimental carcinogenic data. When heated to decomposition it emits toxic fumes of NO_x.

AIW000 CAS:1204-79-1 ***HR: 3***
4'-AMINO-4-BIPHENYLOL
mf: $C_{12}H_{11}NO$ mw: 185.24

SYN: 4-AMINO-4'-HYDROXYBIPHENOL

TOXICITY DATA with REFERENCE
scu-mus TDLo:216 mg/kg/3D:CAR JNCIAM 41,403,68

SAFETY PROFILE: Questionable carcinogen with experimental carcinogenic data. When heated to decomposition it emits toxic fumes of NO_x.

AIW250 CAS:1204-59-7 ***HR: 3***
3-AMINO-4-BIPHENYLOL HYDROCHLORIDE
mf: $C_{12}H_{11}NO{\bullet}ClH$ mw: 221.70

SYNS: 3-AMINO-4-HYDROXYDIPHENYL HYDROCHLORIDE ◇ 4-HYDROXY-3-AMINODIPHENYL HYDROCHLORIDE

TOXICITY DATA with REFERENCE
imp-mus TDLo:50 mg/kg:CAR BJCAAI 12,222,58

SAFETY PROFILE: Questionable carcinogen with experimental carcinogenic data by implantation. When heated to decomposition it emits very toxic fumes of HCl and NO_x.

AIW500 CAS:65146-47-6 ***HR: 3***
2-AMINO-5-BIPHENYLYLIMIDAZOLEHYDROCHLORIDE
mf: $C_{15}H_{13}N_3$•ClH mw: 271.77

TOXICITY DATA with REFERENCE
orl-mus LD50:720 mg/kg ARZNAD 27,1889,77
ipr-mus LD50:50 mg/kg ARZNAD 27,1889,77

SAFETY PROFILE: Poison by intraperitoneal route. Moderately toxic by ingestion. When heated to decomposition it emits very toxic fumes of HCl and NO_x.

AIW750 CAS:74039-01-3 ***HR: 1***
2-(4'-AMINO-1,1'-BIPHENYL-4-YL)-2H-NAPHTHO(1,2-d)TRIAZOLE-6,8-DISULFONIC ACID, DIPOTASSIUM SALT
mf: $C_{22}H_{14}N_4O_6S_2$•2K mw: 572.72

SYN: 2-(4'-AMINOXENYL)NAFTO-α,β-TRIAZOL-6,8-DISULFONAN DRASELNY (CZECH)

TOXICITY DATA with REFERENCE
skn-rbt 500 mg/24H MLD 28ZPAK -,196,72
eye-rbt 500 mg/24H MLD 28ZPAK -,196,72
orl-rat LD50:7210 mg/kg 28ZPAK -,196,72

SAFETY PROFILE: Mildly toxic by ingestion. A skin and eye irritant. See also SULFONATES. When heated to decomposition it emits very toxic fumes of SO_x, NO_x, and K_2O.

AIX000 CAS:1031-47-6 ***HR: 3***
5-AMINO-1-BIS(DIMETHYLAMIDE)PHOSPHORYL-3-PHENYL-1,2,4-TRIAZOLE
mf: $C_{12}H_{19}N_6OP$ mw: 294.34

SYNS: 5-AMINO-1-BIS(DIMETHYLAMIDO)PHOSPHORYL-3-PHENYL-1,2,4-TRIAZOLE ◇ 5-AMINO-1-(BIS(DIMETHYLAMINO)PHOSPHINYL)-3-PHENYL-1,2,4-TRIAZOLE ◇ 5-AMINO-3-FENIL-1-BIS(-DIMETILAMINO)-FOSFORIL-1,2,4-TRIAZOLO(ITALIAN) ◇ 5-AMINO-3-FENYL-1-BIS(DIMETHYL-AMINO)-FOSFORYL-1,2,4-TRIAZOOL (DUTCH) ◇ 5-AMINO-3-PHENYL-1-BIS (DIMETHYL-AMINO)-PHOSPHORYLE-1,2,4-TRIAZOLE(FRENCH) ◇ 5-AMINO-3-PHENYL-1-BIS(DIMETHYLAMINO)-PHOSPHORYL-1H-1,2,4-TRIAZOL (GERMAN) ◇ 5-AMINO-3-PHENYL-1,2,4-TRIAZOLE-1-YL-N,N,N',N'-TETRAMETHYLPHOSPHODIAMIDE ◇ 5-AMINO-3-PHENYL-1,2,4-TRIAZOLYL-1-BIS(DIMETHYLAMIDO)PHEOSPHATE ◇ 5-AMINO-3-PHENYL-1,2,4-TRIAZOLYL-N,N,NN'-TETRAMETHYL-PHOSPHONAMIDE ◇ p-(5-AMINO-3-PHENYL-1H-1,2,4-TRIAZOL-1-YL)-N,N,N'-TETRAMETHYL PHOSPHONIC DIAMIDE ◇ BIS(DIMETHYLAMINO)-3-AMINO-5-PHENYLTRIAZOLYL PHOSPHINE OXIDE ◇ ENT 27,223 ◇ NIAGARA 5943 ◇ 3-PHENYL-5-AMINO-1,2,4-TRIAZOLYL-(1)-(N,N'-TETRAMETHYL) DIAMIDOPHOSPHONATE ◇ TRIAMIFOS (GERMAN, DUTCH, ITALIAN) ◇ TRIAMIPHOS ◇ TRIAMPHOS ◇ WEPSIN ◇ WEPSYN ◇ WEPSYN 155 ◇ WP 155

TOXICITY DATA with REFERENCE
mrc-asn 400 ppm ENMUDM 2,359,80
orl-rat TDLo:6575 μg/kg:TER TXCYAC 2,327,74
orl-rat LD50:20 mg/kg FMCHA2 -,D333,80
skn-rat LD50:48 mg/kg WRPCA2 9,119,70
unk-rat LD50:10 mg/kg 30ZDA9 -,427,71
ipr-rat LD50:15 mg/kg EJPHAZ 16,361,71
orl-mus LD50:10 mg/kg ARSIM* 20,27,66
skn-rbt LD50:1500 mg/kg GUCHAZ 6,508,73
skn-rbt LD50:1500 mg/kg

CONSENSUS REPORTS: EPA Extremely Hazardous Substances List.

SAFETY PROFILE: Poison by ingestion, skin contact, intraperitoneal, and possibly other routes. An experimental teratogen. Mutation data reported. When heated to decomposition it emits very toxic fumes of PO_x and NO_x.

AIX250 CAS:56-18-8 ***HR: 3***
AMINOBIS(PROPYLAMINE)
DOT: UN 2269
mf: $C_6H_{17}N_3$ mw: 131.26

SYNS: BIS-(3-AMINOPROPYL)AMINE ◇ 3,3-DIAMINODIPROPYLAMINE ◇ 3,3'-DIAMINODIPROPYLAMINE ◇ DIPROPYLENETRIAMINE ◇ IMINOBIS(PROPYLAMINE) ◇ 3,3'-IMINOBIS(PROPYLAMINE) ◇ INITIATING EXPLOSIVE IMINOBISPROPYLAMINE (DOT)

TOXICITY DATA with REFERENCE
skn-rbt 470 mg open MOD UCDS** 6/13/68
eye-rbt 47 mg SEV UCDS** 6/13/68
orl-rat LD50:810 mg/kg AIHAAP 23,95,62
skn-rbt LDLo:110 mg/kg AIHAAP 23,95,62

CONSENSUS REPORTS: Reported in EPA TSCA Inventory.

DOT Classification: Corrosive Material; Label: Corrosive.

SAFETY PROFILE: Poison by skin contact. Moderately toxic by ingestion. A skin and severe eye irritant. When heated to decomposition it emits toxic fumes of NO_x. An explosive.

AIX500 CAS:64037-07-6 ***HR: 3***
2-AMINO-5-BROMOBENZOXAZOLE
mf: $C_7H_5BrN_2O$ mw: 213.05

TOXICITY DATA with REFERENCE
orl-rat LD50:1000 mg/kg MDCHAG 4(1),338,64
ipr-rat LD50:160 mg/kg MDCHAG 4(1),338,64

orl-mus LD50:819 mg/kg MDCHAG 4(1),336,64
ipr-mus LD50:180 mg/kg MDCHAG 4(1),336,64

SAFETY PROFILE: Poison by intraperitoneal route. Moderately toxic by ingestion. When heated to decomposition it emits very toxic fumes of Br^- and NO_x.

AIX750 CAS:52112-66-0 ***HR: 3***
2-AMINO-6-BROMOBENZOXAZOLE
mf: $C_7H_5BrN_2O$ mw: 213.05

TOXICITY DATA with REFERENCE
orl-rat LD50:500 mg/kg MDCHAG 4(1),338,64
ipr-rat LD50:200 mg/kg MDCHAG 4(1),338,64
orl-mus LD50:560 mg/kg MDCHAG 4(1),336,64
ipr-mus LD50:294 mg/kg MDCHAG 4(1),336,64

SAFETY PROFILE: Poison by intraperitoneal route. Moderately toxic by ingestion. See also BROMIDES. When heated to decomposition it emits very toxic fumes of Br^- and NO_x.

AIY000 CAS:64037-09-8 ***HR: 2***
2-AMINO-5-BROMO-6-CHLOROBENZOXAZOLE
mf: $C_7H_4BrClN_2O$ mw: 247.49

TOXICITY DATA with REFERENCE
orl-mus LD50:2000 mg/kg MDCHAG 4(1),336,64
ipr-mus LD50:450 mg/kg MDCHAG 4(1),336,64

SAFETY PROFILE: Moderately toxic by ingestion and intraperitoneal route. When heated to decomposition it emits very toxic fumes of Br^-, Cl^-, and NO_x.

AIY250 CAS:64037-08-7 ***HR: 3***
2-AMINO-6-BROMO-5-CHLOROBENZOXAZOLE
mf: $C_7H_4BrClN_2O$ mw: 247.49

TOXICITY DATA with REFERENCE
orl-rat LD50:140 mg/kg MDCHAG 4(1),338,64
ipr-rat LD50:140 mg/kg MDCHAG 4(1),338,64
orl-mus LD50:658 mg/kg MDCHAG 4(1),336,64
ipr-mus LD50:240 mg/kg MDCHAG 4(1),336,64

SAFETY PROFILE: Poison by ingestion and intraperitoneal routes. Moderately toxic by ingestion. When heated to decomposition it emits very toxic fumes of Cl^-, Br^-, and NO_x.

AIY500 CAS:116-82-5 ***HR: 1***
1-AMINO-2-BROMO-4-HYDROXYANTHRAQUINONE
mf: $C_{14}H_8BrNO_3$ mw: 318.14

SYN: 1-AMINO-2-BROM-4-HYDROXYANTHRACHINON(CZECH)

TOXICITY DATA with REFERENCE
eye-rbt 500 mg/24H MLD 28ZPAK -,83,72

CONSENSUS REPORTS: Reported in EPA TSCA Inventory.

SAFETY PROFILE: An eye irritant. When heated to decomposition it emits very toxic fumes of Br^- and NO_x.

AIY750 CAS:73791-29-4 ***HR: 1***
1-AMINO-2-BROMO-4-(2-(2-HYDROXYETHYL)SULFONYL-4-METHYLPHENYLAMINO)ANTHRAQUINONE
mf: $C_{23}H_{20}BrN_3O_5S$ mw: 530.43

SYN: MODR ALIZARINOVA CISTA B (CZECH)

TOXICITY DATA with REFERENCE
eye-rbt 500 mg/24H MLD 28ZPAK -,241,72

SAFETY PROFILE: An eye irritant. See also SULFONATES. When heated to decomposition it emits very toxic fumes of Br^-, NO_x, and SO_x.

AIY850 CAS:56741-95-8 ***HR: D***
2-AMINO-5-BROMO-6-PHENYL-4(1H)-PYRIMIDINONE
mf: $C_{10}H_8BrN_3O$ mw: 266.12

SYNS: BROPIRAMINE ◇ BROPIRIMINE ◇ 4(1H)-PYRIMIDINONE, 2-AMINO-5-BROMO-6-PHENYL- ◇ U-54461

TOXICITY DATA with REFERENCE
orl-rat TDLo:1450 mg/kg (female 15-22D post):REP TJADAB 35,30A,87
orl-rat TDLo:2900 mg/kg (female 15-22D post):TER TJADAB 35,30A,87

SAFETY PROFILE: An experimental teratogen. Other experimental reproductive effects. When heated to decomposition it emits toxic fumes of NO_x, Br^-.

AIZ000 CAS:66064-11-7 ***HR: 3***
N-AMINO-2-(m-BROMOPHENYL)SUCCINIMIDE
mf: $C_{10}H_9BrN_2O_2$ mw: 269.12

TOXICITY DATA with REFERENCE
orl-mus LD50:3682 mg/kg ARZNAD 29,290,79
ipr-mus LD50:387 mg/kg EJMCA5 13,465,78

SAFETY PROFILE: Poison by intraperitoneal route. Moderately toxic by ingestion. When heated to decomposition it emits very toxic fumes of Br^- and NO_x.

AJA000 CAS:5003-71-4 ***HR: 3***
1-AMINO-3-BROMOPROPANE HYDROBROMIDE
mf: $C_3H_6BrN \cdot BrH$ mw: 216.93

SYNS: 3-BROMO-1-PROPANAMINE HYDROBROMIDE ◇ 3-BROMOPROPYLAMINE HYDROBROMIDE

TOXICITY DATA with REFERENCE
ipr-mus TDLo:1150 mg/kg/8W-I:ETA CNREA8 39,391,79

CONSENSUS REPORTS: Reported in EPA TSCA Inventory.

SAFETY PROFILE: Questionable carcinogen with ex-

perimental tumorigenic data. When heated to decomposition it emits very toxic fumes of HBr and NO_x.

AJA250 CAS:96-20-8 **HR: 3**
2-AMINOBUTAN-1-OL
mf: $C_4H_{11}NO$ mw: 89.16

PROP: Water-white liquid. Mp: −2°, bp: 178°, flash p: 165°F (OC), d: 0.944 @ 20°/20°, vap d: 3.06.

SYNS: 2-AMINO-1-BUTANOL ◇ 2-AMINO-n-BUTYL ALCOHOL ◇ BUTANOL-2-AMINE

TOXICITY DATA with REFERENCE
orl-mus LD50:2300 mg/kg 20PKA3 -,-,67
ipr-mus LDLo:250 mg/kg CBCCT* 5,338,53
ivn-mus LD50:316 mg/kg CSLNX* NX#00036

CONSENSUS REPORTS: Reported in EPA TSCA Inventory.

SAFETY PROFILE: Poison by intravenous and intraperitoneal routes. Moderately toxic by ingestion. Moderately flammable when exposed to heat, flame, or oxidizing materials. To fight fire, use water spray, alcohol foam, dry chemical. When heated to decomposition it yields NO_x. See also ALCOHOLS and AMINES.

AJA375 CAS:52712-76-2 **HR: 3**
4-AMINO-2-(4-BUTANOYLHEXAHYDRO-1H-1,4-DIAZEPIN-1-YL)-6,7-DIMETHOXYQUINAZOLINE HYDROCHLORIDE
mf: $C_{19}H_{27}N_5O_3 \cdot ClH$ mw: 409.97

TOXICITY DATA with REFERENCE
orl-rat TDLo:169 mg/kg (female 17-22D post):REP KSRNAM 17,930,83
orl-rat TDLo:1100 mg/kg (7-17D preg):TER KSRNAM 17,914,83
orl-rat LD50:980 mg/kg KSRNAM 17,843,83
scu-rat LD50:365 mg/kg KSRNAM 17,843,83
ivn-rat LD50:50 mg/kg KSRNAM 17,843,83
ims-rat LD50:152 mg/kg KSRNAM 17,843,83
orl-mus LD50:1201 mg/kg KSRNAM 17,843,83

SAFETY PROFILE: Poison by subcutaneous, intravenous, and intramuscular routes. Moderately toxic by ingestion and other routes. An experimental teratogen. Other experimental reproductive effects. When heated to decomposition it emits toxic fumes of NO_x and HCl.

AJA500 CAS:3624 **HR: 3**
3-AMINO-2-BUTOXYBENZOIC ACID-2-DIETHYLAMINOETHYL ESTER HYDROCHLORIDE-87-1
mf: $C_{17}H_{28}N_2O_3 \cdot ClH$ mw: 344.93

SYNS: 2-BUTOXY-3-AMINOBENZOIC ACID β-DIETHYLAMINOETHYL ESTER HYDROCHLORIDE ◇ 2′-DIETHYLAMINOETHYL-3-AMINO-2-BUTOXYBENZOATE-HYDROCHLORIDE ◇ β-DIETHYLAMINOETHYL-2-BUTOXY-3-AMINOBENZOATE HYDROCHLORIDE ◇ METHAMBUCAINE HYDROCHLORIDE ◇ METHAMBUTOXYCAINE HYDROCHLORIDE ◇ PRIMACAINE ◇ PRIMACAINE HYDROCHLORIDE

TOXICITY DATA with REFERENCE
ivn-rat LD50:11 mg/kg CLDND*
ipr-mus LD50:192 mg/kg CLDND*
scu-mus LD50:392 mg/kg CLDND*
ivn-mus LD50:23 mg/kg CLDND*
ivn-cat LD50:8 mg/kg CLDND*
ivn-rbt LD50:17 mg/kg CLDND*
ipr-gpg LD50:212 mg/kg CLDND*

SAFETY PROFILE: Poison by subcutaneous, intravenous, and intraperitoneal routes. See also ESTERS. When heated to decomposition it emits very toxic fumes of HCl and NO_x.

AJA650 CAS:30486-72-7 **HR: D**
2-AMINO-5-BUTYLBENZIMIDAZOLE
mf: $C_{11}H_{15}N_3$ mw: 189.29

TOXICITY DATA with REFERENCE
orl-rat TDLo:605 mg/kg (female 8-15D post):TER THERAP 31,505,76

SAFETY PROFILE: An experimental teratogen. When heated to decomposition it emits toxic fumes of NO_x.

AJA750 CAS:3037-72-7 **HR: 3**
(4-AMINOBUTYL)DIETHOXYMETHYLSILANE
mf: $C_9H_{23}NO_2Si$ mw: 205.42

SYN: Δ-AMINOBUTYLMETHYLDIETHOXYSILANE

TOXICITY DATA with REFERENCE
skn-rbt 10 mg/24H open MLD AIHAAP 23,95,62
orl-rat LDLo:6500 mg/kg AIHAAP 23,95,62
skn-rat LD50:45 mg/kg JPMSAE 60,1113,71
skn-mus LD50:45 mg/kg JPMSAE 60,1113,71
skn-rbt LD50:45 mg/kg AIHAAP 23,95,62

CONSENSUS REPORTS: Reported in EPA TSCA Inventory. EPA Extremely Hazardous Substances List.

SAFETY PROFILE: Poison by skin contact. Moderately toxic by ingestion. A skin irritant. See also SILANES. When heated to decomposition it emits toxic fumes of NO_x.

AJB000 CAS:33132-75-1 **HR: 3**
p-AMINO-β-sec-BUTYL-N,N-DIMETHYLPHENETHYLAMINE
mf: $C_{14}H_{22}N_2$ mw: 218.38

TOXICITY DATA with REFERENCE
orl-mus LD50:185 mg/kg CHTPBA 6,453,71
ivn-mus LD50:45 mg/kg CHTPBA 6,453,71

SAFETY PROFILE: Poison by ingestion and intrave-

nous routes. When heated to decomposition it emits toxic fumes of NO_x.

AJB250 CAS:118-68-3 ***HR: 3***
3-(2-AMINOBUTYL)INDOLE ACETATE
mf: $C_{12}H_{16}N_2 \cdot C_2H_4O_2$ mw: 248.36

SYNS: α-ETHYLTRYPTAMINE ACETATE ◇ dl-α-ETHYLTRYPTAMINE ACETATE ◇ ETRYPTAMINE ACETATE ◇ INDOLE-3-(2-AMINOBUTYL) ACETATE

TOXICITY DATA with REFERENCE
orl-rat LD50:49 mg/kg TXAPA9 4,547,62
ipr-mus LD50:72 mg/kg TXAPA9 4,547,62
ivn-mus LD50:45 mg/kg CSLNX* NX#00376

SAFETY PROFILE: Poison by ingestion, intraperitoneal, and intravenous routes. See also ESTERS. When heated to decomposition it emits toxic fumes of NO_x.

AJB500 CAS:18237-16-6 ***HR: 3***
3-(4-AMINOBUTYL)INDOLE HYDROCHLORIDE
mf: $C_{12}H_{16}N_2 \cdot ClH$ mw: 224.76

SYN: Δ-INDOLYBUTYLAMINE HYDROCHLORIDE

TOXICITY DATA with REFERENCE
ipr-mus LD50:222 mg/kg RPTOAN 33,180,70
ivn-mus LD50:83 mg/kg RPTOAN 33,180,70

SAFETY PROFILE: Poison by intraperitoneal and intravenous routes. When heated to decomposition it emits very toxic fumes of NO_x and HCl.

AJC000 CAS:30653-83-9 ***HR: 3***
5-AMINO-N-BUTYL-2-PROPARGYLOXYBENZAMIDE
mf: $C_{14}H_{18}N_2O_2$ mw: 246.34

SYNS: 5-AMINO-N-BUTYL-2-(2-PROPYNYLOXY)BENZAMIDE ◇ MY 41-6 ◇ PARSAL ◇ PARSALMIDE ◇ 2-PROPARGILOSSI-5-AMINO-N-(n-BUTIL)-BENZAMIDE (ITALIAN)

TOXICITY DATA with REFERENCE
orl-rat LD50:864 mg/kg DRFUD4 2,55,77
orl-mus LD50:428 mg/kg DRFUD4 2,55,77
ivn-mus LD50:148 mg/kg DRFUD4 2,55,77

SAFETY PROFILE: Poison by intravenous route. Moderately toxic by ingestion. When heated to decomposition it emits toxic fumes of NO_x.

AJC250 CAS:3069-30-5 ***HR: 2***
(4-AMINOBUTYL)TRIETHOXYSILANE
mf: $C_{10}H_{25}NO_3Si$ mw: 235.45

SYN: 4-(TRIETHOXYSILYL)BUTYLAMINE

TOXICITY DATA with REFERENCE
skn-rbt 10 mg/24H open MLD AIHAAP 23,95,62
orl-rat LD50:1620 mg/kg AIHAAP 23,95,62
skn-rbt LD50:2500 mg/kg AIHAAP 23,95,62

SAFETY PROFILE: Moderately toxic by ingestion and skin contact. A skin irritant. See also SILANES. When heated to decomposition it emits toxic fumes of NO_x.

AJC375 CAS:3251-08-9 ***HR: 2***
4-AMINOBUTYRAMIDE
mf: $C_4H_{10}N_2O$ mw: 102.16

SYNS: AMGABA ◇ AMIDE of GABA

TOXICITY DATA with REFERENCE
ipr-rat LD50:900 mg/kg AITEAT 13,70,65
ipr-mus LD50:1080 mg/kg AITEAT 13,70,65
ivn-rbt LDLo:350 mg/kg AITEAT 13,70,65

SAFETY PROFILE: Moderately toxic by some routes. When heated to decomposition it emits toxic fumes of NO_x. See also AMINES and AMIDES.

AJC500 CAS:34562-99-7 ***HR: 3***
Γ-AMINOBUTYRIC ACID CETYL ESTER
mf: $C_{20}H_{41}NO_2$ mw: 327.62

SYNS: CETYL-γ-AMINOBUTYRATE ◇ CETYL GABA

TOXICITY DATA with REFERENCE
ipr-mus LD50:155 mg/kg NEPHBW 19,217,80
ivn-mus LD50:22 mg/kg NEPHBW 19,217,80

SAFETY PROFILE: Poison by intravenous and intraperitoneal routes. See also ESTERS. When heated to decomposition it emits toxic fumes of NO_x.

AJC625 CAS:3251-07-8 ***HR: 2***
4-AMINOBUTYRIC ACID METHYL ESTER
mf: $C_5H_{11}NO_2$ mw: 117.17

SYNS: MEGABA ◇ METHYL ESTER of GABA

TOXICITY DATA with REFERENCE
ipr-rat LD50:950 mg/kg AITEAT 13,70,65
ipr-mus LD50:1300 mg/kg AITEAT 13,70,64
ivn-rbt LDLo:300 mg/kg AITEAT 13,70,65

SAFETY PROFILE: Moderately toxic by some routes. When heated to decomposition it emits toxic fumes of NO_x. See also AMINES and ESTERS.

AJD000 CAS:60-32-2 ***HR: 2***
6-AMINOCAPROIC ACID
mf: $C_6H_{13}NO_2$ mw: 131.20

SYNS: ACEPRAMINE ◇ ACS ◇ AFIBRIN ◇ AMICAR ◇ AMINOCAPROIC ACID ◇ omega-AMINOCAPROIC ACID ◇ epsilon-AMINOCAPROIC ACID ◇ omega-AMINOHEXANOIC ACID ◇ AMINOKAPRON ◇ CAPRAMOL ◇ CAPRALENSE ◇ CAPROCID

◇ CAPROLISIN ◇ CL 10304 ◇ CY 116 ◇ EACA ◇ EACA KABI ◇ EACS ◇ EPSAMON ◇ EPSICAPRON ◇ HEMOCAPROL ◇ HEMOPAR ◇ HEPIN ◇ IPSILON ◇ 177 J.D. ◇ epsilon-LEUCINE ◇ epsilon-NORLEUCINE ◇ NSC-26154 ◇ RESPRAMIN

TOXICITY DATA with REFERENCE
eye-rbt 500 mg/24H MLD 28ZPAK -,128,72
orl-rat TDLo:153 g/kg (40D pre/1-21D preg):REP APTOA6 22,340,65
orl-man TDLo:1778 mg/kg/8D-I:SYS AJKDDP 8,441,86
mul-man TDLo:14400 mg/kg/59D-I BJURAN 60,81,87
ipr-rat LD50:7000 mg/kg PHMCAA 3,62,61
ivn-rat LD50:3300 mg/kg PHMCAA 3,62,61
orl-mus LD50:14300 mg/kg NIIRDN 6,79,82
ivn-mus LD50:4900 mg/kg AAREAV 22,481,65
ivn-dog LDLo:2150 mg/kg PHMCAA 3,62,61
ivn-gpg LDLo:19800 mg/kg AAREAV 22,481,65

CONSENSUS REPORTS: Reported in EPA TSCA Inventory.

SAFETY PROFILE: Moderately toxic by intravenous route. Human systemic effects by ingestion: changes in tubules (including acute renal failure, acute tubular necrosis), hematuria, and increased body temperature. Experimental reproductive effects. An eye irritant. When heated to decomposition it emits toxic fumes such as NO_x.

AJD250 CAS:38237-76-2 ***HR: 3***
p-AMINO CAPROPHENONE
mf: $C_{12}H_{17}NO$ mw: 191.30

SYN: p-HEXANOYLANILINE

TOXICITY DATA with REFERENCE
orl-rat LD50:216 mg/kg GEPHDP 14,465,83
orl-mus LD50:299 mg/kg GEPHDP 14,465,83
ipr-mus LD50:35 mg/kg JMCMAR 17,900,74

SAFETY PROFILE: Poison by ingestion and intraperitoneal routes. When heated to decomposition it emits toxic fumes such as NO_x. See also AROMATIC AMINES.

AJD375 CAS:55673-54-6 ***HR: D***
4-AMINO-4′-(2-CARBAMOYLETHYL)-1,1-DIMETHYL-N,4′-BI(PYRROLE-2-CARBOXAMIDE)
mf: $C_{15}H_{20}N_6O_3$•ClH mw: 368.87

TOXICITY DATA with REFERENCE
dnd-esc 200 μmol EJBCAI 26,81,72
dnd-mam:lym 200 μmol EJBCAI 26,81,72

SAFETY PROFILE: Mutation data reported. When heated to decomposition it emits toxic fumes of NO_x and HCl.

AJD500 CAS:64686-82-4 ***HR: 3***
4-AMINO-5-CARBAMYL-3-BENZYLTHIAZOLE-2(3H)-THIONE
mf: $C_{11}H_{11}N_3OS_2$ mw: 265.37

TOXICITY DATA with REFERENCE
orl-mus LD50:250 mg/kg ARZNAD 27,1652,77
ipr-mus LD50:100 mg/kg ARZNAD 27,1652,77

SAFETY PROFILE: Poison by ingestion and intraperitoneal routes. When heated to decomposition it emits very toxic fumes of NO_x and SO_x.

AJD750 CAS:26148-68-5 ***HR: 3***
AMINO-α-CARBOLINE
mf: $C_{11}H_9N_3$ mw: 183.2

SYNS: 2-AMINO-α-CARBOLINE ◇ 2-AMINO-9H-PYRIDO(2,3-B)INDOLE

TOXICITY DATA with REFERENCE
sce-hmn:lym 4000 μg/L MUREAV 77,65,80
mmo-sat 1 μg/plate ABCHA6 43,1155,79
mma-sat 1 μg/plate CALEDQ 10,141,80
dnr-bcs 10 uL/plate ABCHA6 45,2031,81
slt-dmg-orl 400 ng/kg JJCREP 76,468,85
orl-mus TDLo:37600 mg/kg/98W-C:CAR EVHPAZ 67,129,86
orl-mus TD:50424 mg/kg/82W-C:CAR CRNGDP 5,815,84

CONSENSUS REPORTS: Cancer Review: Group 2B IMEMDT 7,56,87; Animal Sufficient Evidence IMEMDT 40,245,86

SAFETY PROFILE: Suspected carcinogen with experimental carcinogenic data. Human mutation data reported. When heated to decomposition it emits toxic fumes of NO_x.

AJE000 CAS:3688-35-5 ***HR: 3***
AMINOCHLORAMBUCIL
mf: $C_{14}H_{20}Cl_2N_2O_2$ mw: 319.26

SYNS: 2-AMINO-4-(p-(BIS(2-CHLOROETHYL)AMINO)PHENYL) BUTYRIC ACID ◇ α-AMINO-Γ-(p-DICHLOROETHYLAMINO)-PHENYLBUTYRIC ACID ◇ CB-1385

TOXICITY DATA with REFERENCE
sln-dmg-unk 10 mmol/L ANYAA9 160,228,69
sln-dmg-par 3 mmol/L GENRA8 1,173,60
sln-dmg par 3 mmol/L GENRA8 1,173,60
ipr-rat LDLo:25 mg/kg BCPCA6 5,192,60
scu-mus LD10:20 mg/kg EJCAAH 10,667,74

CONSENSUS REPORTS: EPA Genetic Toxicology Program.

SAFETY PROFILE: Poison by subcutaneous and intraperitoneal routes. Mutation data reported. When

heated to decomposition it emits very toxic fumes of Cl^- and NO_x.

AJE250 CAS:50416-18-7 ***HR: 2***
2-AMINO-6'-CHLORO-o-ACETOTOLUIDIDE, HYDROCHLORIDE
mf: $C_9H_{11}ClN_2O•ClH$ mw: 235.13

SYNS: 2-AMINO-2'-CHLORO-6'-METHYLACETANILIDE, HYDROCHLORIDE ◇ C 3104

TOXICITY DATA with REFERENCE
eye-rbt 2% MLD ARZNAD 8,407,58
ipr-rat LD50:525 mg/kg ARZNAD 8,407,58
ipr-mus LD50:565 mg/kg ARZNAD 8,407,58
scu-mus LD50:840 mg/kg ARZNAD 8,407,58

SAFETY PROFILE: Moderately toxic by intraperitoneal and subcutaneous routes. An eye irritant. When heated to decomposition it emits very toxic fumes of Cl^-, NO_x, and HCl. See also AROMATIC AMINES.

AJE325 CAS:117-11-3 ***HR: 1***
1-AMINO-5-CHLOROANTHRAQUINONE
mf: $C_{14}H_8ClNO_2$ mw: 257.68

SYNS: 1-CHLOR-5-AMINOANTHRACHINON (CZECH) ◇ 5-CHLORO-1-AMINOANTHRAQUINONE

TOXICITY DATA with REFERENCE
eye-rbt 500 mg/24H MLD 28ZPAK -,87,72

CONSENSUS REPORTS: Reported in EPA TSCA Inventory.

SAFETY PROFILE: An eye irritant. When heated to decomposition it emits very toxic fumes of Cl^- and NO_x.

AJE350 CAS:94110-08-4 ***HR: 2***
3-AMINO-4-CHLOROBENZOIC ACID 2-((DIMETHYLAMINO)ETHYL) ESTER HYDROCHLORIDE
mf: $C_{11}H_{15}ClN_2O_2•ClH$ mw: 279.19

SYNS: 3-AMINO-4-CLORO-BENZOATO di DIMETILAMINOETILE CLORIDRATO (ITALIAN) ◇ 4-CLORO-3-AMINOBENZOATO di DIMETILAMINOETILE CLORIDRATO (ITALIAN) ◇ REC 1-0060

TOXICITY DATA with REFERENCE
orl-mus LD50:2480 mg/kg BCFAAI 97,457,58
ipr-mus LD50:880 mg/kg BCFAAI 97,457,58
unr-mam LD50:880 mg/kg FRPSAX 13,574,58

SAFETY PROFILE: Moderately toxic by ingestion and other routes. When heated to decomposition it emits toxic fumes of NO_x and HCl. See also ESTERS.

AJE500 CAS:19952-47-7 ***HR: 3***
2-AMINO-4-CHLOROBENZOTHIAZOLE
mf: $C_7H_5ClN_2S$ mw: 184.65

TOXICITY DATA with REFERENCE
orl-mus LD50:2400 mg/kg JPETAB 105,486,52
ivn-mus LD50:71 mg/kg JPETAB 105,486,52

CONSENSUS REPORTS: Reported in EPA TSCA Inventory.

SAFETY PROFILE: Poison by intravenous route. Moderately toxic by ingestion. When heated to decomposition it emits very toxic fumes of Cl^-, SO_x, and NO_x.

AJE750 CAS:95-24-9 ***HR: 3***
2-AMINO-6-CHLOROBENZOTHIAZOLE
mf: $C_7H_5ClN_2S$ mw: 184.65

TOXICITY DATA with REFERENCE
orl-mus LD50:398 mg/kg JPETAB 105,486,52
ivn-mus LD50:76 mg/kg JPETAB 105,486,52

CONSENSUS REPORTS: Reported in EPA TSCA Inventory.

SAFETY PROFILE: Poison by ingestion and intravenous route. When heated to decomposition it emits very toxic fumes of SO_x, NO_x, and Cl^-.

AJF250 CAS:64037-10-1 ***HR: 3***
2-AMINO-4-CHLOROBENZOXAZOLE
mf: $C_7H_5ClN_2O$ mw: 168.59

TOXICITY DATA with REFERENCE
orl-rat LD50:500 mg/kg MDCHAG 4(1),338,64
orl-mus LD50:378 mg/kg MDCHAG 4(1),336,64
ipr-mus LD50:54 mg/kg MDCHAG 4(1),336,64

SAFETY PROFILE: Poison by ingestion and intraperitoneal route. When heated to decomposition it emits very toxic fumes of NO_x and Cl^-.

AJF500 CAS:61-80-3 ***HR: 3***
2-AMINO-5-CHLOROBENZOXAZOLE
mf: $C_7H_5ClN_2O$ mw: 168.59

SYNS: 5-CHLORO-2-BENZOXAZOLAMINE ◇ DEFLEXOL ◇ FLEXILON ◇ FLEXIN ◇ MCN-485 ◇ USAF MA-12 ◇ ZOXAMIN ◇ ZOXAZOLAMINE ◇ ZOXINE

TOXICITY DATA with REFERENCE
orl-hmn TDLo:14 mg/kg/D:CNS JAMAAP 160,745,56
orl-rat LD50:782 mg/kg FEPRA7 16,319,57
ipr-rat LD50:102 mg/kg JPETAB 129,75,60
orl-mus LD50:540 mg/kg AIPTAK 128,112,60
ipr-mus LD50:100 mg/kg NTIS** AD277-689
ivn-mus LD50:376 mg/kg 29QHAQ -,-,74
ivn-dog LD50:117 mg/kg FEPRA7 16,319,57
orl-ham LD50:670 mg/kg JPETAB 129,75,60
ipr-ham LD50:268 mg/kg JPETAB 129,75,60

SAFETY PROFILE: Poison by intraperitoneal and intravenous routes. Moderately toxic by ingestion. Human systemic effects by ingestion: muscle rigidity. When heated to decomposition it yields toxic fumes of Cl^- and NO_x.

AJF750 CAS:52112-68-2 ***HR: 3***
2-AMINO-6-CHLOROBENZOXAZOLE
mf: $C_7H_5ClN_2O$ mw: 168.59

TOXICITY DATA with REFERENCE
orl-rat LD50:226 mg/kg MDCHAG 4(1),338,64
ipr-rat LD50:123 mg/kg MDCHAG 4(1),338,64
orl-mus LD50:600 mg/kg MDCHAG 4(1),336,64
ipr-mus LD50:347 mg/kg MDCHAG 4(1),336,64

SAFETY PROFILE: Poison by ingestion and intraperitoneal route. When heated to decomposition it emits very toxic fumes of Cl^- and NO_x.

AJG750 CAS:64037-11-2 ***HR: 3***
2-AMINO-7-CHLOROBENZOXAZOLE
mf: $C_7H_5ClN_2O$ mw: 168.59

TOXICITY DATA with REFERENCE
orl-rat LD50:410 mg/kg MDCHAG 4(1),338,64
orl-mus LD50:590 mg/kg MDCHAG 4(1),336,64
ipr-mus LD50:180 mg/kg MDCHAG 4(1),336,64

SAFETY PROFILE: Poison by intraperitoneal route. Moderately toxic by ingestion. When heated to decomposition it emits very toxic fumes of NO_x and Cl^-.

AJH000 CAS:364-62-5 ***HR: 3***
4-AMINO-5-CHLORO-N-(2-(DIETHYLAMINO) ETHYL)-N-ANISAMIDE
mf: $C_{14}H_{22}ClN_3O_2$ mw: 299.84

SYNS: 4-AMINO-5-CHLORO-N-(2-(DIETHYLAMINO)ETHYL)-2-METHOXYBENZAMIDE ◇ 5-CHLORO-2-METHOXYPROCAINAMIDE ◇ DEL ◇ N-(DIETHYLAMINOETHYL)-2-METHOXY-4-AMINO-5-CHLOROBENZAMIDE ◇ MAXOLON ◇ METACLOPROMIDE ◇ METHOCHLOPRAMIDE ◇ 2-METHOXY-5-CHLOROPROCAINAMIDE ◇ METOCHLOPRAMIDE ◇ METOCLOL ◇ MORIPERAN ◇ PLASIL ◇ PRIMPERAN ◇ RELIVERAN

TOXICITY DATA with REFERENCE
orl-man TDLo:34 mg/kg (male 60D pre):REP WJMDA2 144,359,86
orl-wmn TDLo:3600 μg/kg/6D-I AIMEAS 97,621,82
orl-cld TDLo:900 μg/kg:EYE ADCHAK 55,310,80
orl-man TDLo:111 mg/kg/37W-I:PNS JJMDAT 23,152,84
ivn-cld TDLo:2 mg/kg/1D-C JOPDAB 104,138,84
ivn-wmn TDLo:2400 μg/kg:BLD NEJMAG 307,1346,82
ivn-man TDLo:14 μg/kg:BPR,CNS AIMEAS 104,125,86
orl-rat LD50:750 mg/kg NIIRDN 6,838,82
ipr-rat LD50:114 mg/kg NIIRDN 6,838,82
scu-rat LD50:340 mg/kg NIIRDN 6,838,82
ivn-rat LD50:50 mg/kg BCFAAI 115,649,76
orl-mus LD50:270 mg/kg NIIRDN 6,838,82
ipr-mus LD50:96 mg/kg NIIRDN 6,838,82
scu-mus LD50:190 mg/kg NIIRDN 6,838,82
ivn-mus LD50:33 mg/kg NIIRDN 6,838,82

SAFETY PROFILE: Poison by ingestion, subcutaneous, and intravenous routes. Human systemic effects by ingestion or intravenous routes: tremors, high blood pressure and abnormal catecholamine levels in the sympathetic nervous system, diplopia. Experimental reproductive effects. When heated to decomposition it emits very toxic fumes of NO_x and Cl^-.

AJH125 CAS:27260-19-1 ***HR: 3***
4-AMINO-5-CHLORO-N-(2-(ETHYLAMINOETHYL)-o-ANISAMIDE
mf: $C_{12}H_{18}ClN_3O_2$ mw: 271.78

SYNS: 4-AMINO-5-CHLORO-N-(2-ETHYLAMINOETHYL)-2-METHOXYBENZAMIDE ◇ 4-AMINO-5-CHLORO-N-(2-ETILAMINOETIL)-2-METOSSIBENZAMIDE (ITALIAN) ◇ DEETILATO METOCLOPRAMIDE (ITALIAN) ◇ DEETILMETOCLOPRAMIDE (ITALIAN) ◇ DEM

TOXICITY DATA with REFERENCE
ipr-rat LD50:242 mg/kg BCFAAI 115,649,76
ivn-rat LD50:120 mg/kg BCFAAI 115,649,76
ipr-mus LD50:245 mg/kg BCFAAI 115,649,76
ivn-mus LD50:41 mg/kg BCFAAI 115,649,76

SAFETY PROFILE: Poison by intravenous and intraperitoneal routes. When heated to decomposition it emits toxic fumes of Cl^- and NO_x.

AJH129 CAS:3443-15-0 ***HR: 3***
6-AMINO-2-(2-CHLOROETHYL)-2,3-DIHYDRO-4H-1,3-BENZOXAZIN-4-ONEHYDROCHLORIDE
mf: $C_{10}H_{11}ClN_2O_2 \cdot ClH$ mw: 263.14

SYNS: A 350 ◇ AMINOCHLORTHENOXAZIN HYDROCHLORIDE ◇ 2-(β-CHLOROETHYL)-2,3-DIHYDRO-4-OXO-6-AMINO-1,3-BENZOXAZINE HYDROCHLORIDE ◇ ICI 350

TOXICITY DATA with REFERENCE
orl-rat LD50:619 mg/kg ARZNAD 14,124,64
ipr-rat LD50:607 mg/kg ARZNAD 13,884,63
orl-mus LD50:2250 mg/kg ARZNAD 13,884,63
ivn-mus LD50:293 mg/kg ARZNAD 13,884,63

SAFETY PROFILE: Poison by intravenous route. Moderately toxic by ingestion and intraperitoneal routes. When heated to decomposition it emits very toxic fumes of Cl^- and NO_x. An antipyretic and analgesic agent.

AJH250 CAS:116-84-7 ***HR: 1***
1-AMINO-5-CHLORO-4-HYDROXYANTHRAQUINONE
mf: $C_{14}H_8ClNO_3$ mw: 273.68

SYN: 1-AMINO-4-HYDROXY-5-CHLORANTHRACHINON (CZECH)

TOXICITY DATA with REFERENCE
eye-rbt 500 mg/24H MLD 28ZPAK -,83,72

SAFETY PROFILE: An eye irritant. When heated to decomposition it emits very toxic fumes of Cl^- and NO_x.

AJH500 CAS:5857-94-3 ***HR: 1***
3-AMINO-5-CHLORO-4-HYDROXYBENZENESULFONIC ACID
mf: $C_6H_6ClNO_4S$ mw: 223.64

SYNS: 5-CHLORO-4-HYDROXYMETANILIC ACID ◇ KYSELINA 2-CHLOR-6-AMINOFENOL-4-SULFONOVA(CZECH)

TOXICITY DATA with REFERENCE
eye-rbt 100 mg/24H MOD 28ZPAK -,181,72
orl-rat LD50:8980 mg/kg 28ZPAK -,181,72

CONSENSUS REPORTS: Reported in EPA TSCA Inventory.

SAFETY PROFILE: Mildly toxic by ingestion. An eye irritant. See also SULFONATES. When heated to decomposition it emits very toxic fumes of Cl^-, SO_x, and NO_x.

AJH750 CAS:1750-46-5 ***HR: 3***
2-AMINO-5-CHLORO-6-HYDROXYBENZOXAZOLE
mf: $C_7H_5ClN_2O_2$ mw: 184.59

TOXICITY DATA with REFERENCE
orl-rat LD50:1070 mg/kg MDCHAG 4(1),338,64
orl-mus LD50:960 mg/kg MDCHAG 4(1),336,64
ipr-mus LD50:357 mg/kg MDCHAG 4(1),336,64

SAFETY PROFILE: Poison by intraperitoneal route. Moderately toxic by ingestion. When heated to decomposition it emits very toxic fumes of Cl^- and NO_x.

AJI000 CAS:2139-00-6 ***HR: 2***
2-AMINO-5-CHLORO-6-METHOXYBENZOXAZOLE
mf: $C_8H_7ClN_2O_2$ mw: 198.62

TOXICITY DATA with REFERENCE
orl-mus LD50:1600 mg/kg MDCHAG 4(1),336,64
ipr-mus LD50:490 mg/kg MDCHAG 4(1),336,64

SAFETY PROFILE: Moderately toxic by ingestion and intraperitoneal routes. When heated to decomposition it emits very toxic fumes of NO_x and Cl^-.

AJI250 CAS:2797-51-5 ***HR: 3***
2-AMINO-3-CHLORO-1,4-NAPHTHOQUINONE
mf: $C_{10}H_6ClNO_2$ mw: 207.62

SYNS: 2-CHLORO-3-AMINO-1,4-NAPHTHOQUINONE ◇ 06K ◇ 06K-QUINONE ◇ 06K-50W ◇ MOGETON GRANULE

TOXICITY DATA with REFERENCE
orl-rat LD50:1360 mg/kg FMCHA2 -,C161,83
orl-mus LD50:1260 mg/kg FMCHA2 -,C161,83
ipr-mus LD50:800 mg/kg JMCMAR 26,570,83
ivn-mus LD50:320 mg/kg CSLNX* NX#03360

SAFETY PROFILE: Poison by intravenous route. Moderately toxic by ingestion and intraperitoneal routes. When heated to decomposition it emits very toxic fumes of Cl^- and NO_x. See also AROMATIC AMINES.

AJI500 CAS:50510-11-7 ***HR: 3***
2-AMINO-5-((p-CHLOROPHENYL)THIOMETHYL)-2-OXAZOLINE
mf: $C_{10}H_{11}ClN_2OS$ mw: 242.74

SYN: 4,5-DIHYDRO-5-((p-CHLOROPHENYL)THIOMETHYL)OXAZOLAMINE

TOXICITY DATA with REFERENCE
orl-mus LD50:383 mg/kg JMCMAR 16,510,73
ipr-mus LD50:215 mg/kg JMCMAR 16,510,73

SAFETY PROFILE: Poison by ingestion and intraperitoneal routes. When heated to decomposition it emits very toxic fumes of Cl^-, NO_x, and SO_x.

AJI520 ***HR: D***
l-1-AMINO-3-CHLORO-2-PROPANOL
mf: C_3H_8ClNO mw: 109.57

SYNS: (-)-1-AMINO-3-CHLORO-2-PROPANOL ◇ 2-PROPANOL, 1-AMINO-3-CHLORO-, (-)-

TOXICITY DATA with REFERENCE
orl-rat TDLo:150 mg/kg (male 5D pre):REP CBINA8 13,193,76

SAFETY PROFILE: Experimental reproductive effects. When heated to decomposition it emits toxic fumes of Cl^- and NO_x.

AJI530 CAS:59348-49-1 ***HR: D***
dl-1-AMINO-3-CHLORO-2-PROPANOL
mf: C_3H_8ClNO mw: 109.57

SYNS: (±)-1-AMINO-3-CHLORO-2-PROPANOL ◇ 2-PROPANOL, 1-AMINO-3-CHLORO-, (±)-

TOXICITY DATA with REFERENCE
orl-rat TDLo:300 mg/kg (male 5D pre):REP CBINA8 13,193,76

SAFETY PROFILE: Experimental reproductive effects. When heated to decomposition it emits toxic fumes of NO_x and Cl^-.

AJI550 CAS:34839-13-9 ***HR: D***
l-AMINO-3-CHLORO-2-PROPANOL HYDROCHLORIDE
mf: $C_3H_8ClNO•ClH$ mw: 146.03

SYNS: (−)-1-AMINO-3-CHLORO-2-PROPANOLHYDROCHLORIDE ◇ l-1-AMINO-3-CHLORO-2-PROPANOL HYDROCHLORIDE ◇ CL 88236

TOXICITY DATA with REFERENCE
orl-ham TDLo:1400 mg/kg (male 14D pre):REP CCPTAY 9,459,74

CONSENSUS REPORTS: EPA Genetic Toxicology Program.

SAFETY PROFILE: Experimental reproductive effects. When heated to decomposition it emits toxic fumes of Cl^- and NO_x.

AJI600 CAS:34839-12-8 ***HR: 3***
dl-1-AMINO-3-CHLORO-2-PROPANOL HYDROCHLORIDE
mf: $C_3H_8ClNO•ClH$ mw: 146.03

SYN: (±)-1-AMINO-3-CHLORO-2-PROPANOLHYDROCHLORIDE

TOXICITY DATA with REFERENCE
orl-rat TDLo:210 mg/kg (14D male):REP CCPTAY 9,451,74
orl-rat LD50:165 mg/kg CCPTAY 9,451,74

CONSENSUS REPORTS: EPA Genetic Toxicology Program.

SAFETY PROFILE: Poison by ingestion. Experimental reproductive effects. When heated to decomposition it emits toxic fumes of Cl^- and NO_x.

AJI650 CAS:41663-73-4 ***HR: 3***
2-AMINO-5-CHLOROTHIAZOLE
mf: $C_3H_3ClN_2S$ mw: 134.59

TOXICITY DATA with REFERENCE
mmo-sat 1 mmol/L MUREAV 118,153,83
mmo-klp 500 μmol/L MUREAV 118,153,83
ivn-mus LD50:180 mg/kg CLSNX* NX#02306

SAFETY PROFILE: Poison by intravenous route. Mutation data reported. When heated to decomposition it emits toxic fumes of Cl^-, SO_x, and NO_x.

AJJ250 CAS:88-51-7 ***HR: 1***
6-AMINO-4-CHLORO-m-TOLUENESULFONIC ACID
mf: $C_7H_8ClNO_3S$ mw: 221.67

SYNS: BRILLIANT TONING RED AMINE ◇ KYSELINA 2-CHLOR-4-TOLUIDIN-5-SULFONOVA (CZECH)

TOXICITY DATA with REFERENCE
eye-rbt 500 mg/24H MOD 28ZPAK -,184,72
orl-rat LD50:12 g/kg 28ZPAK -,184,72

CONSENSUS REPORTS: Reported in EPA TSCA Inventory.

SAFETY PROFILE: Mildly toxic by ingestion. An eye irritant. See also SULFONATES. When heated to decomposition it emits very toxic fumes of Cl^-, NO_x, and SO_x.

AJJ500 CAS:2448-39-7 ***HR: 3***
6-AMINOCOUMARIN COUMARIN-3-CARBOXYLIC ACID SALT
mf: $C_{19}H_{10}NO_6$ mw: 348.30

TOXICITY DATA with REFERENCE
orl-mus LD50:103 mg/kg YKKZAJ 83,1124,63
scu-mus LD50:55 mg/kg YKKZAJ 83,1124,63

SAFETY PROFILE: Poison by ingestion and subcutaneous routes. When heated to decomposition it emits toxic fumes of NO_x.

AJJ750 CAS:63989-79-7 ***HR: 3***
6-AMINOCOUMARIN HYDROCHLORIDE
mf: $C_9H_7NO_2•ClH$ mw: 197.63

TOXICITY DATA with REFERENCE
orl-mus LD50:623 mg/kg YKKZAJ 83,1124,63
scu-mus LD50:353 mg/kg YKKZAJ 83,1124,63

SAFETY PROFILE: Poison by subcutaneous route. Moderately toxic by ingestion. When heated to decomposition it emits very toxic fumes such as HCl and NO_x.

AJJ800 CAS:6264-93-3 ***HR: 3***
2-AMINO-2,4,6-CYCLOHEPTATRIEN-1-ONE
mf: C_7H_7NO mw: 121.15

SYN: 2-AMINOTROPONE

TOXICITY DATA with REFERENCE
ipr-mus LD50:176 mg/kg CPBTAL 20,60,72
scu-mus LD50:175 mg/kg CPBTAL 20,60,72
ivn-mus LD50:333 mg/kg CPBTAL 20,60,72

SAFETY PROFILE: Poison by subcutaneous, intravenous and intraperitoneal routes. When heated to decomposition it emits toxic fumes of NO_x.

AJJ875 CAS:3485-14-1 ***HR: 3***
(1-AMINOCYCLOHEXYL)PENICILLIN
mf: $C_{15}H_{23}N_3O_4S$ mw: 341.47

PROP: Crystals. Mp: 182-183° (anhydrate), 156-158° (decomp).

SYNS: AC-PC ◇ 6-(1-AMINOCYCLOHEXANECARBOXAMIDO)PENICILLANIC ACID ◇ AMINOCYCLOHEXYLPENICILLIN ◇ CALTHOR ◇ CICLACILLIN ◇ CICLACILLUM ◇ CITOSARIN ◇ CYCLACILLIN ◇ CYCLAPEN ◇ SYNGACILLIN ◇ ULTRACILLIN ◇ VASTCILLIN ◇ VATRACIN ◇ VIPICIL ◇ WY 4508 ◇ WYVITAL

TOXICITY DATA with REFERENCE
orl-rat TDLo:84600 mg/kg (30D male):REP TAKHAA 29,117,70

orl-hmn TDLo:210 mg/kg/7D-I:GIT,SKN CHTHBK 22,154,76
orl-rat LD50:5010 mg/kg CHTHBK 22,154,76
ipr-rat LD50:5010 mg/kg CHTHBK 22,154,76
scu-rat LD50:6500 mg/kg TAKHAA 29,117,70
orl-mus LD50:5010 mg/kg CHTHBK 22,154,76
ipr-mus LD50:3776 mg/kg CHTHBK 22,154,76
scu-mus LD50:7500 mg/kg TAKHAA 29,117,70
orl-dog LD50:2500 mg/kg CHTHBK 22,154,76

SAFETY PROFILE: Moderately toxic by ingestion and other routes. Human systemic effects by ingestion: dermatitis and diarrhea. Experimental reproductive effects. When heated to decomposition it emits toxic fumes of SO_x and NO_x.

AJK000 ***HR: 3***
1-AMINO-2-(o-CYCLOHEXYLPHENOXY) PROPIONALDOXIME
mf: $C_{15}H_{22}N_2O_2$ mw: 262.39

SYN: MG 18415.

TOXICITY DATA with REFERENCE
orl-mus LD50:620 mg/kg ARZNAD 29,729,79
ipr-mus LD50:200 mg/kg ARZNAD 29,729,79

SAFETY PROFILE: Poison by intraperitoneal route. Moderately toxic by ingestion. When heated to decomposition it emits toxic fumes of NO_x.

AJK250 CAS:52-52-8 ***HR: 3***
1-AMINOCYCLOPENTANE-1-CARBOXYLIC ACID
mf: $C_6H_{11}NO_2$ mw: 129.18

SYNS: ACPC ◇ 1-AMINO-1-CYCLOPENTANECARBOXYLIC ACID ◇ CB 1639 ◇ CYCLOLEUCINE ◇ NSC 1026 ◇ WR 14,997 ◇ X 201

TOXICITY DATA with REFERENCE
orl-hmn TDLo:60 mg/kg:CNS,GIT JMPCAS 3,1,61
orl-rat LD50:290 mg/kg JMPCAS 3,1,61
ivn-rat LD50:340 mg/kg JMCMAR 3,1,61
orl-mus LD50:309 mg/kg JMPCAS 3,1,61
ipr-mus LD50:119 mg/kg NCISP* JAN86
scu-mus LD50:375 mg/kg NCISP* JAN86
orl-dog LD50:300 mg/kg JMPCAS 3,1,61
ivn-dog LD50:300 mg/kg JMCMAR 3,1,61
orl-gpg LD50:140 mg/kg JMPCAS 3,1,61

CONSENSUS REPORTS: Reported in EPA TSCA Inventory. EPA Genetic Toxicology Program.

SAFETY PROFILE: Poison by ingestion, subcutaneous, intraperitoneal, and intravenous routes. Human systemic effects by ingestion: anorexia, nausea and vomiting. When heated to decomposition it emits toxic fumes of NO_x.

AJK500 CAS:60676-83-7 ***HR: 3***
4-AMINO-N-CYCLOPROPYL-3,5-DICHLOROBENZAMIDE
mf: $C_{10}H_{10}Cl_2N_2O$ mw: 245.12

SYNS: N-CYCLOPROPYL-4-AMINO-3,5-DICHLOROBENZAMIDE ◇ N-CYCLOPROPYL-3,5-DICHLORO-4-AMINOBENZAMIDE

TOXICITY DATA with REFERENCE
orl-rat LD50:170 mg/kg 27ZQAG -,400,72
ipr-rat LD50:160 mg/kg 27ZQAG -,400,72
orl-mus LD50:195 mg/kg 27ZQAG -,400,72
ipr-mus LD50:265 mg/kg JMCMAR 6,528,63

SAFETY PROFILE: Poison by ingestion and intraperitoneal routes. When heated to decomposition it emits very toxic fumes of Cl^- and NO_x.

AJK625 CAS:57294-74-3 ***HR: D***
N^4-AMINOCYTIDINE
mf: $C_9H_{14}N_4O_5$ mw: 258.27

TOXICITY DATA with REFERENCE
mmo-sat 50 nmol/plate NARHAD 11,5223,83
mmo-omi 10 μmol/L NARHAD 11,5223,83
mmo-omi 50 μmol/L BICHAW 24,7273,85

SAFETY PROFILE: Mutation data reported. When heated to decomposition it emits toxic fumes of NO_x.

AJK750 CAS:1951-25-3 ***HR: 3***
AMINODARONE
mf: $C_{25}H_{29}I_2NO_3$ mw: 645.35

SYNS: AMIODARONE ◇ 2-BUTYL-3-BENZOFURANYL p-((2-DIETHYLAMINO)ETHOXY)-m,m-DIIODOPHENYLKETONE ◇ 2-BUTYL-3-(3,5-DIIODO-4-(2-DIETHYLAMINOETHOXY)BENZOYL) BENZOFURAN ◇ 2-N-BUTYL-3',5'-DIIODO-4'-N-DIETHYLAMINO-ETHOXY-3-BENZOYLBENZOFURAN ◇ L. 3428 ◇ LABAZ

TOXICITY DATA with REFERENCE
orl-man TDLo:133 mg/kg/23D-I:SKN LANCAO 1,51,84
ipr-mus LD50:254 mg/kg EJTXAZ 8,122,75
ivn-mus LD50:178 mg/kg EJTXAZ 8,188,75

SAFETY PROFILE: Poison by intravenous and intraperitoneal routes. Human systemic effects by ingestion: photosensitivity of the skin. When heated to decomposition it emits very toxic fumes of I^- and NO_x. A coronary vasodilator.

AJL125 CAS:32764-43-5 ***HR: D***
2-AMINO-2-DEOXY-l-ASCORBIC ACID
mf: $C_6H_9NO_5$ mw: 175.16

SYNS: AMINO REDUCTONE ◇ SCORBAMIC ACID ◇ l-SCORBAMIC ACID ◇ SCORBAMINIC ACID

TOXICITY DATA with REFERENCE
spm-slw-par 25 μg/kg EISOAU 34,367,81

oms-mus:lvr 500 μmol/L JNSVA5 24,263,78
dnd-mam:lym 500 μmol/L JNSVA5 24,263,78

SAFETY PROFILE: Mutation data reported. When heated to decomposition it emits toxic fumes of NO_x.

AJL250 CAS:63041-30-5 ***HR: 3***
9-AMINO-1,2,5,6-DIBENZANTHRACENE
mf: $C_{22}H_{15}N$ mw: 293.38

SYN: 7-AMINODIBENZ(a,h)ANTHRACENE

TOXICITY DATA with REFERENCE
skn-mus TDLo:1250 mg/kg/52W-I:ETA PRLBA4 117,318,35

SAFETY PROFILE: Questionable carcinogen with experimental tumorigenic data. When heated to decomposition it emits toxic fumes of NO_x.

AJL500 CAS:81-49-2 ***HR: D***
1-AMINO-2,4-DIBROMOANTHRAQUINONE
mf: $C_{14}H_7Br_2NO_2$ mw: 381.04

SYNS: 1-AMINO-2,4-DIBROMANTHRACHINON(CZECH) ◇ 2,4-DIBROMO-1-ANTHRAQUINONYLAMINE ◇ NCI-C55458

TOXICITY DATA with REFERENCE
eye-rbt 500 mg/24H MLD 28ZPAK -,88,72
mmo-sat 333 μg/plate ENMUDM 5(Suppl 1),3,83
mma-sat 333 μg/plate NTPTB* JAN 82

CONSENSUS REPORTS: Reported in EPA TSCA Inventory.

SAFETY PROFILE: Mutation data reported. An eye irritant. When heated to decomposition it emits very toxic fumes of Br^- and NO_x.

AJL750 CAS:52112-67-1 ***HR: 2***
2-AMINO-5,7-DIBROMOBENZOXAZOLE
mf: $C_7H_4Br_2N_2O$ mw: 291.95

TOXICITY DATA with REFERENCE
orl-mus LD50:1050 mg/kg MDCHAG 4(1),336,64
ipr-mus LD50:780 mg/kg MDCHAG 4(1),336,64

SAFETY PROFILE: Moderately toxic by ingestion and intraperitoneal route. When heated to decomposition it emits very toxic fumes of Br^- and NO_x.

AJL875 CAS:102207-73-8 ***HR: 3***
2-AMINO-4-DIBUTYLAMINOETHOXYPYRIMIDINE
mf: $C_{14}H_{26}N_4O$ mw: 266.44

SYNS: 2-AMINO-4-(2-DIBUTYLAMINOETHOXY)PYRIMIDINE ◇ OR-1550

TOXICITY DATA with REFERENCE
orl-rat LD50:1000 mg/kg AIPTAK 106,50,56
ipr-rat LD50:75 mg/kg AIPTAK 106,50,56
ipr-mus LD50:157 mg/kg AIPTAK 106,50,56
ivn-mus LD50:44 mg/kg AIPTAK 106,50,56
orl-dog LD50:450 mg/kg AIPTAK 106,50,56
ivn-dog LD50:35 mg/kg AIPTAK 106,50,56
orl-rbt LD50:1260 mg/kg AIPTAK 106,50,56
ivn-rbt LD50:46 mg/kg AIPTAK 106,50,56

SAFETY PROFILE: Poison by intravenous and intraperitoneal routes. Moderately toxic by ingestion. When heated to decomposition it emits toxic fumes of NO_x.

AJM000 CAS:133-90-4 ***HR: 3***
3-AMINO-2,5-DICHLOROBENZOIC ACID
mf: $C_7H_5Cl_2NO_2$ mw: 206.03

SYNS: ACP-M-728 ◇ AMBIBEN ◇ AMOBEN ◇ CHLORAMBEN ◇ 2,5-DICHLORO-3-AMINOBENZOIC ACID ◇ NCI-C00055 ◇ ORNAMENTAL WEED ◇ VEGABEN

TOXICITY DATA with REFERENCE
mmo-sat 10 mg/plate ENMUDM 5(Suppl 1),3,83
mma-sat 1 mg/plate NTPTB* JAN 82
cyt-mus-ipr 58500 μg/kg CARYAB 33,527,80
cyt-mus-orl 234 mg/kg CARYAB 33,527,80
orl-mus TDLo:672 g/kg/80W-C:CAR NCITR* NCI-CG-TR-25,77
orl-mus TD :1344 g/kg/80W-C:CAR NCITR* NCI-CG-TR-25,77
orl-rat LD50:3500 mg/kg RREVAH 10,97,65
orl-mus LD50:3725 mg/kg GISAAA 45(4),74,80
skn-rbt LD50:3136 mg/kg WRPCA2 7,135,68

CONSENSUS REPORTS: NCI Carcinogenesis Bioassay Completed; Results Positive: mouse NCITR* NCI-CG-TR-25,77; Results Negative: rat NCITR* NCI-CG-TR-25,77. Community Right-To-Know List. Reported in EPA TSCA Inventory.

SAFETY PROFILE: Moderately toxic by ingestion. Questionable carcinogen with experimental carcinogenic data. Mutation data reported. When heated to decomposition it emits highly toxic fumes such as Cl^- and NO_x. See also AROMATIC AMINES.

AJM500 CAS:64037-12-3 ***HR: 3***
2-AMINO-5,6-DICHLOROBENZOXAZOLE
mf: $C_7H_4Cl_2N_2O$ mw: 203.03

TOXICITY DATA with REFERENCE
orl-mus LD50:1200 mg/kg MDCHAG 4(1),336,64
ipr-mus LD50:300 mg/kg MDCHAG 4(1),336,64

SAFETY PROFILE: Poison by intraperitoneal route. Moderately toxic by ingestion. When heated to decomposition it emits very toxic fumes of Cl^- and NO_x.

AJM750 CAS:50510-12-8 ***HR: 3***
2-AMINO-5-((3,4-DICHLOROPHENYL) THIOMETHYL)-2-OXAZOLINE
mf: $C_{10}H_{10}Cl_2N_2OS$ mw: 277.18

TOXICITY DATA with REFERENCE
orl-mus LD50:562 mg/kg JMCMAR 16,510,73
ipr-mus LD50:383 mg/kg JMCMAR 16,510,73

SAFETY PROFILE: Poison by intraperitoneal route. Moderately toxic by ingestion. When heated to decomposition it emits very toxic fumes of Cl^-, NO_x, and SO_x.

AJN250 CAS:2381-85-3 ***HR: 3***
5-AMINO-9-(DIETHYLAMINO)BENZO(a) PHENOXAZIN-7-IUM SULFATE (2:1)
mf: $C_{40}H_{40}N_6O_2{\bullet}O_4S$ mw: 732.92

SYN: NILE BLUE A

TOXICITY DATA with REFERENCE
ivn-mus LDLo:65 mg/kg TXAPA9 44,225,78

CONSENSUS REPORTS: Reported in EPA TSCA Inventory.

SAFETY PROFILE: Poison by intravenous route. See also SULFATES. When heated to decomposition it emits very toxic fumes of NO_x and SO_x.

AJN375 CAS:102207-75-0 ***HR: 3***
2-AMINO-4-DIETHYLAMINOETHOXYPYRIMIDINE
mf: $C_{10}H_{18}N_4O$ mw: 210.32

SYNS: 2-AMINO-4-(2-DIETHYLAMINOETHOXY)PYRIMIDINE ◇ OR-1556

TOXICITY DATA with REFERENCE
orl-rat LD50:2000 mg/kg AIPTAK 106,50,56
ipr-rat LD50:75 mg/kg AIPTAK 106,50,56
ipr-mus LD50:252 mg/kg AIPTAK 106,50,56
ivn-mus LD50:174 mg/kg AIPTAK 106,50,56
orl-dog LD50:2750 mg/kg AIPTAK 106,50,56
ivn-dog LD50:165 mg/kg AIPTAK 106,50,56
orl-rbt LD50:875 mg/kg AIPTAK 106,50,56
ivn-rbt LD50:174 mg/kg AIPTAK 106,50,56

SAFETY PROFILE: Poison by intravenous and intraperitoneal routes. Moderately toxic by ingestion. When heated to decomposition it emits toxic fumes of NO_x.

AJN500 CAS:51-06-9 ***HR: 3***
p-AMINO-N-(2-DIETHYLAMINOETHYL) BENZAMIDE
mf: $C_{13}H_{21}N_3O$ mw: 235.37

SYNS: p-AMINOBENZOICDIETHYLAMINOETHYLAMIDE ◇ 4-AMINO-N-(2-(DIETHYLAMINO)ETHYL)-BENZAMIDE (9CI) ◇ NOVOCAINAMIDE ◇ NOVOCAINE AMIDE ◇ NOVOCAMID ◇ PROCAINAMIDE ◇ PROCAINE AMIDE ◇ PROCAMIDE ◇ PRONESTYL

TOXICITY DATA with REFERENCE
orl-man TDLo:10 g/kg/Y:PUL BHJUAV 34,284,72
orl-wmn TDLo:1826 mg/kg/13W-I:CVS AHJOA2 83,798,72
orl-hmn TDLo:2280 mg/kg/22W:MSK BHJUAV 34,284,72
ivn-rat LD50:110 mg/kg RPTOAN 33,292,70
orl-mus LD50:890 mg/kg TXAPA9 21,253,72
ipr-mus LD50:312 mg/kg EJMCA5 11,539,76
ivn-mus LD50:103 mg/kg EJPHAZ 114,253,85
orl-dog LDLo:2210 mg/kg TXAPA9 21,253,72
ivn-rbt LD50:250 mg/kg FRPSAX 12,77,57
ivn-gpg LD50:280 mg/kg FRPSAX 12,77,57

SAFETY PROFILE: Poison by intravenous and intraperitoneal routes. Moderately toxic by ingestion. Human systemic effects by ingestion: cardiac abnormalities, joint effects, cough, dyspnea, and other lung effects. When heated to decomposition it emits toxic fumes of NO_x.

AJN750 CAS:63887-34-3 ***HR: 3***
p-AMINO-N-(2-DIETHYLAMINOETHYL) BENZAMIDE SULFATE
mf: $C_{13}H_{21}N_3O{\bullet}H_2O_4S$ mw: 333.45

SYNS: PROCAINAMIDE SULFATE ◇ PROCAINE AMIDE SULFATE ◇ SUPICAINE AMIDE SULFATE

TOXICITY DATA with REFERENCE
ivn-rat LD50:165 mg/kg RPOBAR 2,318,70
ivn-mus LD50:146 mg/kg RPOBAR 2,318,70

SAFETY PROFILE: Poison by intravenous route. See also SULFATES. When heated to decomposition it emits very toxic fumes of NO_x and SO_x.

AJO000 CAS:61827 ***HR: 3***
N-(2-AMINO-5-DIETHYLAMINOPHENETHYL) METHANE SULFONAMIDEHYDROCHLORDE-74-5
mf: $C_{13}H_{23}N_3O_2S{\bullet}HCl$ mw: 321.91

TOXICITY DATA with REFERENCE
orl-rat LDLo:400 mg/kg KODAK* -,-,71
ipr-rat LD50:50 mg/kg KODAK* -,-,71

SAFETY PROFILE: Poison by ingestion and intraperitoneal routes. See also SULFONATES. When heated to decomposition it emits very toxic fumes of NO_x, SO_x, and HCl.

AJO250 CAS:2198-58-5 ***HR: 3***
p-AMINO DIETHYLANILINE HYDROCHLORIDE
mf: $C_{10}H_{16}N_2{\bullet}ClH$ mw: 200.74

PROP: Bp: 217.5°. Sltly sol in water, sol in ether.

SYN: N,N-DIETHYL-p-PHENYLENEDIAMINEHYDROCHLORIDE

TOXICITY DATA with REFERENCE
orl-rat LDLo:200 mg/kg KODAK* -,-,71
ipr-rat LDLo:25 mg/kg KODAK* -,-,71
ivn-mus LD50:24 mg/kg CSLNX* NX#07893

CONSENSUS REPORTS: Reported in EPA TSCA Inventory.

SAFETY PROFILE: Poison by ingestion, intraperitoneal, and intravenous routes. When heated to decomposition it emits very toxic fumes of HCl and NO_x.

AJO500 CAS:134-58-7 ***HR: 3***
5-AMINO-1,6-DIHYDRO-7H-v-TRIAZOLO (4,5-d)PYRIMIDIN-7-ONE
mf: $C_4H_4N_6O$ mw: 152.14

SYNS: 8 AG ◇ 5-AMINO-1,4-DIHYDRO-7H-1,2,3-TRIAZOLO(4,5-d) PYRIMIDIN-7-ONE (9CI) ◇ 5-AMINO-7-HYDROXY-1H-v-TRIAZOLO (d)PYRIMIDINE ◇ 5-AMINO-1H-v-TRIAZOLO(d)PYRIMIDIN-7-OL ◇ 5-AMINO-v-TRIAZOLO(4,5-d)PYRIMIDIN-7-OL ◇ AZAGUANINE ◇ AZAGUANINE-8 ◇ 8-AZAGUANINE ◇ AZAN ◇ AZG ◇ B-28 ◇ GUANAZOL ◇ GUANAZOLO ◇ NSC-749 ◇ PATHOCIDIN ◇ PATHOCIDINE ◇ SF-337 ◇ SK 1150 ◇ TRIAZOLOGUANINE

TOXICITY DATA with REFERENCE
dni-mus:lym 66 μmol/L CJBBDU 62,280,84
oms-mus:leu 1 μmol/L AEZRA2 20,351,82
ipr-mus TDLo:80 mg/kg (female 7D post):REP JEEMAF 6,593,58
ipr-mus TDLo:80 mg/kg (female 8D post):TER JEEMAF 6,593,58
ipr-rat LD50:1000 mg/kg ADTEAS 3,181,68
orl-mus LD50:1500 mg/kg OSDIAF 17,491,68
ipr-mus LD50:100 mg/kg 85GDA2 5,193,81

CONSENSUS REPORTS: EPA Genetic Toxicology Program.

SAFETY PROFILE: Poison by intraperitoneal route. Moderately toxic by ingestion. Mutation data reported. An experimental teratogen. Other experimental reproductive effects. When heated to decomposition it emits toxic fumes of NO_x. Inhibits protein synthesis.

AJO625 CAS:102207-76-1 ***HR: 3***
2-AMINO-4-DI-ISOBUTYLAMINOETHOXY-PYRIMIDINE
mf: $C_{14}H_{26}N_4O$ mw: 266.44

SYNS: 2-AMINO-4-(2-DIISOBUTYLAMINOETHOXY)PYRIMIDINE ◇ OR-1578

TOXICITY DATA with REFERENCE
ipr-rat LD50:800 mg/kg AIPTAK 106,50,56
ipr-mus LD50:762 mg/kg AIPTAK 106,50,56
ivn-mus LD50:72 mg/kg AIPTAK 106,50,56
orl-dog LD50:2500 mg/kg AIPTAK 106,50,56
ivn-dog LD50:75 mg/kg AIPTAK 106,50,56
orl-rbt LD50:2000 mg/kg AIPTAK 106,50,56
ivn-rbt LD50:16 mg/kg AIPTAK 106,50,56

SAFETY PROFILE: Poison by intravenous route. Moderately toxic by ingestion and other routes. When heated to decomposition it emits toxic fumes of NO_x.

AJO750 CAS:73747 ***HR: 3***
3-AMINO-2-(2-(DIISOPROPYLAMINO)ETHOXY) BUTYROPHENONEDIHYDROCHLORIDE-29-2
mf: $C_{18}H_{30}N_2O_2$•2ClH mw: 379.42

SYN: REC 7-0591

TOXICITY DATA with REFERENCE
orl-rat LD50:238 mg/kg ARZNAD 16,1275,66
scu-rat LD50:111 mg/kg ARZNAD 16,1275,66
ivn-rat LD50:3 mg/kg ARZNAD 16,1275,66
orl-mus LD50:42 mg/kg ARZNAD 16,1275,66
ipr-mus LD50:38 mg/kg ARZNAD 16,1275,66
scu-mus LD50:42 mg/kg ARZNAD 16,1275,66
ivn-mus LD50:7800 μg/kg ARZNAD 16,1275,66
ivn-cat LD50:5 mg/kg ARZNAD 16,1275,66

SAFETY PROFILE: Poison by ingestion, subcutaneous, intravenous, and intraperitoneal routes. See also KETONES. When heated to decomposition it emits very toxic fumes of NO_x and HCl.

AJP000 CAS:19216-56-9 ***HR: 3***
1-(4-AMINO-6,7-DIMETHOXY-2-QUINAZOLINYL-4-(2-FURANYLCARBONYL) PIPERAZINE
mf: $C_{19}H_{21}N_5O_4$ mw: 383.45

SYNS: FURAZOSIN ◇ 2-(4-(2-FUROYL)PIPERAZIN-1-YL)-4-AMINO-6,7-DIMETHOXYQUINAZOLINE ◇ PRAZOSIN

TOXICITY DATA with REFERENCE
orl-man TDLo:2571 μg/kg/60D-I:REP AMSVAZ 213,319,83
orl-wmn TDLo:20 μg/kg:BPR,GIT AIMEAS 97,455,82
orl-man TDLo:1143 μg/kg:BPR AMSVAZ 213,157,83
orl-hmn TDLo:280 μg/kg:CNS,CVS BMJOAE 2,508,76
orl-hmn TDLo:1260 μg/kg:CNS,KID BMJOAE 1,622,78

SAFETY PROFILE: Human systemic effects by ingestion of very small amounts: somnolence, hallucinations, distorted perceptions, changes in motor activity, decreased blood pressure, nausea or vomiting, and kidney effects. Experimental reproductive effects. When heated to decomposition it emits toxic fumes of NO_x.

AJP125 CAS:102207-77-2 ***HR: 3***
2-AMINO-4-DIMETHYLAMINOETHOXY-PYRIMIDINE
mf: $C_8H_{14}N_4O$ mw: 182.26

SYNS: 2-AMINO-4-(2-DIMETHYLAMINOETHOXY)PYRIMIDINE ◇ OR-1549

TOXICITY DATA with REFERENCE
orl-rat LD50:1500 mg/kg AIPTAK 106,50,56
ipr-rat LD50:75 mg/kg AIPTAK 106,50,56
ipr-mus LD50:478 mg/kg AIPTAK 106,50,56
ivn-mus LD50:252 mg/kg AIPTAK 106,50,56
orl-dog LD50:1500 mg/kg AIPTAK 106,50,56
ivn-dog LD50:450 mg/kg AIPTAK 106,50,56
orl-rbt LD50:3750 mg/kg AIPTAK 106,50,56
ivn-rbt LD50:440 mg/kg AIPTAK 106,50,56

SAFETY PROFILE: Poison by intravenous and intraperitoneal routes. Moderately toxic by ingestion. When heated to decomposition it emits toxic fumes of NO_x.

AJP250 CAS:92-31-9 ***HR: 3***
3-AMINO-7-DIMETHYLAMINO-2-METHYLPHENAZATHIONIUM CHLORIDE
mf: $C_{15}H_{16}N_3S{\bullet}Cl$ mw: 305.85

SYNS: BLUTENE ◇ BLUTENE CHLORIDE ◇ C.I. 925 ◇ C.I. 52040 ◇ C.I. BASIC BLUE 17 ◇ DIMETHYLTOLUTHIONINE CHLORIDE ◇ F KLOT ◇ KLOT ◇ SCHULTZ No. 1041 ◇ TOLAZUL ◇ TOLONIUM CHLORIDE ◇ TOLUIDINE BLUE ◇ TOLUIDINE BLUE O ◇ TOLUIDENE BLUE O CHLORIDE

TOXICITY DATA with REFERENCE
cyt-ham:ovr 20 μmol/L/5H-C ENMODM 1,27,79
sln-dmg-unr 500 ppm 14ZYA8 -,115,65
unr-man TDLo:43 mg/kg/6D:GIT,BLD 34ZIAG -,597,69
unr-mus LD50:28 mg/kg JMCMAR 17,902,74
ivn-mus LD50:45 mg/kg TXAPA9 1,185,59

CONSENSUS REPORTS: Reported in EPA TSCA Inventory. EPA Genetic Toxicology Program.

SAFETY PROFILE: Poison by intravenous and possibly other routes. Human systemic effects by an unspecified route: nausea or vomiting and blood effects. Mutation data reported. When heated to decomposition it emits very toxic fumes of Cl^-, SO_x, and NO_x.

AJQ000 CAS:63731-93-1 ***HR: 3***
2-AMINO-4-γ-DIETHYLAMINOPROPYLAMINO-5,6-DIMETHYLPYRIMIDINE
mf: $C_{13}H_{25}N_5$ mw: 251.43

TOXICITY DATA with REFERENCE
orl-mus LDLo:250 mg/kg JCSOA9 -,357,46

SAFETY PROFILE: Poison by ingestion. When heated to decomposition it emits toxic fumes of NO_x.

AJQ100 CAS:109-55-7 ***HR: 2***
1-AMINO-3-DIMETHYLAMINOPROPANE
mf: $C_5H_{14}N_2$ mw: 102.21

$(CH_3)_2N(CH_2)_3NH_2$

PROP: Colorless liquid. Mp: < −70°, bp: 123°, flash p: 100°F (OC), d: 0.8100 @ 30°, vap press: 10 mm @ 30°, vap d: 3.52.

SYNS: N,N-DIMETHYL-N-(3-AMINOPROPYL)AMINE ◇ 3-(DIMETHYLAMINO)PROPYLAMINE ◇ N,N-DIMETHYL-1,3-DIAMINOPROPANE ◇ N,N-DIMETHYL-1,3-PROPANEDIAMINE ◇ N,N-DIMETHYL-1,3-PROPYLENEDIAMINE

TOXICITY DATA with REFERENCE
skn-rbt 100 μg/24H open AIHAAP 23,95,62
eye-rbt 5 mg MOD UCDS** 12/15/71
orl-rat LDLo:1870 mg/kg AIHAAP 23,95,62

CONSENSUS REPORTS: Reported in EPA TSCA Inventory.

SAFETY PROFILE: Moderately toxic by ingestion. A skin and eye irritant. Very flammable when exposed to heat, flame or oxidizers. Reaction with 1,2-dichloroethane produces explosive acetylene gas. This and other amines ignite on contact with cellulose nitrate of high surface area. To fight fire, use alcohol foam, CO_2, dry chemical. When heated to decomposition it emits toxic fumes of NO_x. See also AMINES.

AJQ250 CAS:553-24-2 ***HR: 3***
AMINODIMETHYLAMINOTOLUAMINOZINE HYDROCHLORIDE
mf: $C_{15}H_{16}N_4{\bullet}ClH$ mw: 288.81

SYNS: 3-AMINO-7-DIMETHYLAMINO-2-METHYLPHENAZINEHYDROCHLORIDE ◇ 3-AMINO-7-(DIMETHYLAMINO)-2-METHYL-PHENAZINE MONOHYDROCHLORIDE ◇ C.I. 50040 ◇ C.I. BASIC RED 5 ◇ C.I. BASIC RED 5, MONOHYDROCHLORIDE ◇ KERNECHTROT ◇ MICHROME No. 226 ◇ NEUTRAL RED ◇ NEUTRAL RED CHLORIDE ◇ NEUTRAL RED W ◇ NUCLEAR FAST RED ◇ TOLUYLENE RED ◇ $N^{(8)}$,$N^{(8)}$,3-TRIMETHYL-2,8-PHENAZINEDIAMINE MONOHYDROCHLORIDE

TOXICITY DATA with REFERENCE
cyt-ckn-par 85 μg/kg 47JMAE -,137,82
mma-sat 10 μg/plate MUREAV 48,109,77
mmo-esc 2500 ppt/3H AMNTA4 85,119,51
sln-dmg-orl 1000 ppm AMNTA4 87,295,53
cyt-ham:ovr 2 μmol/L/5H MUREAV 45,233,77
ivn-rat LD50:112 mg/kg FEPRA7 10,337,51
ivn-mus LD50:142 mg/kg FEPRA7 10,337,51
ivn-rbt LD50:97 mg/kg FEPRA7 10,337,51

CONSENSUS REPORTS: Reported in EPA TSCA Inventory. EPA Genetic Toxicology Program.

SAFETY PROFILE: Poison by intravenous route. Mutation data reported. When heated to decomposition it emits very toxic fumes of HCl and NO_x.

AJQ500 CAS:21554-20-1 ***HR: 3***
4-AMINO-3′,5′-DIMETHYL-4′-HYDROXYAZOBENZENE
mf: $C_{14}H_{15}N_3O$ mw: 241.32

TOXICITY DATA with REFERENCE
orl-rat TDLo:40 g/kg/2Y-C:ETA AABIAV 52,33,63
ipr-rat LD50:350 mg/kg AABIAV 52,33,63

SAFETY PROFILE: Poison by intraperitoneal route. Questionable carcinogen with experimental tumorigenic data. When heated to decomposition it emits toxic fumes of NO_x.

AJQ600 CAS:77094-11-2 ***HR: 2***
2-AMINO-3,4-DIMETHYLIMIDAZO(4,5-f) QUINOLINE
mf: $C_{12}H_{12}N_4$ mw: 212.28

SYN: 3,4-DIMETHYL-3H-IMIDAZO(4,5-f)QUINOLIN-2-AMINE

TOXICITY DATA with REFERENCE
mma-sat 100 ng/plate CRNGDP 7,273,86
slt-dmg-orl 100 ng/kg JJCREP 76,468,85
dns-rat:lng 3 μmol/L ENMUDM 7,245,85
dnd-mus:leu 100 μmol/L MUREAV 144,57,85
dns-ham:lng 3 μmol/L ENMUDM 7,245,85
msc-ham:lng 25 mg/L MUREAV 118,91,83
dns-gpg:lng 10 μmol/L ENMUDM 7,245,85
orl-mus TDLo:7476 mg/kg/89W-C:CAR CRNGDP 7,1889,86
orl-mus TD:22120 mg/kg/81W-C:CAR CRNGDP 7,1889,86

CONSENSUS REPORTS: IARC Cancer Review: Animal Inadequate Evidence IMEMDT 40,275,86

SAFETY PROFILE: Questionable carcinogen with experimental carcinogenic data. Mutation data reported. When heated to decomposition it emits toxic fumes of NO_x.

AJQ675 CAS:77500-04-0 ***HR: 3***
2-AMINO-3,8-DIMETHYLIMIDAZO(4,5-f) QUINOXALINE
mf: $C_{11}H_{11}N_5$ mw: 213.27

SYNS: 2-AMINO-3,8-DIMETHYL-3H-IMIDAZO(4,5-f)QUINOXALINE ◇ 3,8-DIMETHYL-3H-IMIDAZO(4,5-f)QUINOXALIN-2-AMINE

TOXICITY DATA with REFERENCE
mma-sat 5 ng/plate MUREAV 144,131,85
slt-dmg-orl 100 ng/kg JJCREP 76,468,85
msc-ham:ovr 300 mg/L MUTAEX 2,483,87
orl-rat TDLo:8580 mg/kg/61W-C:CAR CRNGDP 9,71,88
orl-mus TDLo:42336 mg/kg/84W-C:CAR CRNGDP 8,665,87

CONSENSUS REPORTS: IARC Cancer Review: Group 3 IMEMDT 7,56,87; Animal Inadequate Evidence IMEMDT 40,283,86

SAFETY PROFILE: Suspected carcinogen with experimental carcinogenic data. Mutation data reported. When heated to decomposition it emits toxic fumes of NO_x.

AJQ750 ***HR: 3***
2-AMINO-6-DIMETHYL-4-(p-(p-((p-((1-METHYLPYRIDINIUM-3-YL)CARBAMOYL)PHENYL)CARBABENZAMIDO)ANILINO)PYRIMIDIMIUM,DIIODIDE
mf: $C_{33}H_{32}N_8O_3$•2I mw: 842.53

TOXICITY DATA with REFERENCE
dnd-mus:lym 840 nmol/L JMCMAR 22,134,79
ipr-mus LD10:20 mg/kg JMCMAR 22,134,79

SAFETY PROFILE: Poison by intraperitoneal route. Mutation data reported. When heated to decomposition it emits very toxic fumes of I^- and NO_x.

AJR000 CAS:4302-87-8 ***HR: 3***
p-AMINO-N,α-DIMETHYLPHENETHYLAMINE
mf: $C_{10}H_{16}N_2$ mw: 164.28

SYNS: 1-(p-AMINOPHENYL)-2-METHYLAMINOPROPAN(GERMAN) ◇ α-(4-AMINOPHENYL)-β-METHYLAMINO-PROPANE ◇ 1-(p-AMINOPHENYL)-2-METHYLAMINOPROPANE

TOXICITY DATA with REFERENCE
orl-rat LD50:300 mg/kg AEPPAE 195,647,40
ipr-rat LDLo:85 mg/kg AEPPAE 195,647,40
scu-rat LD50:280 mg/kg AIPTAK 159,442,66

SAFETY PROFILE: Poison by ingestion, intraperitoneal, and subcutaneous routes. See also AMINES. When heated to decomposition it emits toxic fumes of NO_x.

AJR400 CAS:31272-21-6 ***HR: 2***
5-AMINO-1,3-DIMETHYL-4-PYRAZOLYL o-FLUOROPHENYL KETONE
mf: $C_{12}H_{12}FN_3O$ mw: 233.27

SYNS: (5-AMINO-1,3-DIMETHYL-1H-PYRAZOL-4-YL)(2-FLUOROPHENYL)METHANONE ◇ KETONE, 5-AMINO-1,3-DIMETHYLPYRAZOL-4-YL o-FLUOROPHENYL ◇ METHANONE, (5-AMINO-1,3-DIMETHYL-1H-PYRAZOL-4-YL)(2-FLUOROPHENYL)- ◇ PD 71627

TOXICITY DATA with REFERENCE
mma-sat 10 nmol/plate CRNGDP 7,2019,86
orl-rat TD:2366 mg/kg/13W-C:ETA AJPAA4 124,392,86

SAFETY PROFILE: Questionable carcinogen with experimental tumorigenic data. Mutation data reported. When heated to decomposition it emits toxic fumes of F^- and NO_x.

AJR500 CAS:68808-54-8 ***HR: 3***
3-AMINO-1,4-DIMETHYL-5H-PYRIDO(4,3-b)INDOLE ACETATE
mf: $C_{13}H_{13}N_3 \cdot C_2H_4O_2$ mw: 271.35

SYN: 1,4-DIMETHYL-5H-PYRIDO(4,3-b)INDOL-3-AMINEACETATE ◇ 1,4-DIMETHYL-5H-PYRIDO(4,3-b)INDOL-3-AMINE MONOACETATE ◇ TRP-P-1 (ACETATE)

TOXICITY DATA with REFERENCE
slt-dmg-orl 200 ppm MUREAV 122,315,83
mma-sat 1 μg/plate CPBTAL 26,611,78
orl-rat TDLo:1539 mg/kg/29W-C:CAR JJCREP 76,815,85
orl-mus TDLo:11 g/kg/89W-C:CAR SCIEAS 213,346,81

SAFETY PROFILE: Suspected carcinogen with experimental carcinogenic data. Mutation data reported. When heated to decomposition it emits toxic fumes of NO_x.

AJR750 CAS:35572-78-2 ***HR: 2***
2-AMINO-4,6-DINITROTOLUENE
mf: $C_7H_7N_3O_4$ mw: 197.17

SYNS: 3,5-DINITRO-o-TOLUIDINE ◇ 2-METHYL-3,5-DINITROBENZENAMINE

TOXICITY DATA with REFERENCE
orl-rat LD50:1394 mg/kg NTIS** AD-A080-146
orl-mus LD50:1522 mg/kg NTIS** AD-A080-146

SAFETY PROFILE: Moderately toxic by ingestion. See also AMINES and NITRO COMPOUNDS of AROMATIC HYDROCARBONS. When heated to decomposition it emits toxic fumes of NO_x.

AJS000 CAS:82-24-6 ***HR: D***
1-AMINO-9,10-DIOXO-9,10-DIHYDRO-2-ANTHRACENECARBOXYLIC ACID
mf: $C_{15}H_9NO_4$ mw: 267.25

SYN: 1-AMINO-2-CARBOXYLATE-4-NITRO-ANTHRAQUINONE

TOXICITY DATA with REFERENCE
mmo-sat 50 μg/plate MUREAV 40,203,76
mma-sat 50 μg/plate MUREAV 40,203,76

SAFETY PROFILE: Mutation data reported. When heated to decomposition it emits toxic fumes of NO_x.

AJS100 CAS:92-67-1 ***HR: 3***
4-AMINODIPHENYL
mf: $C_{12}H_{11}N$ mw: 169.24

PROP: Colorless crystals. Mp. 53°, bp: 302°, d: 1.160 @ 20°/20°, autoign temp: 842°F.

SYNS: p-AMINOBIPHENYL ◇ 4-AMINOBIPHENYL ◇ 4-AMINODIFENIL (SPANISH) ◇ p-AMINODIPHENYL ◇ BIPHENYLAMINE ◇ 4-BIPHENYLAMINE ◇ (1,1'-BIPHENYL)-4-AMINE ◇ p-BIPHENYLAMINE ◇ PARAAMINODIPHENYL ◇ p-PHENYLANILINE ◇ XENYLAMIN (CZECH) ◇ XENYLAMINE

TOXICITY DATA with REFERENCE
mma-sat 2 μg/plate ENMUDM 5(Suppl 1),3,83
dnd-esc 30 μmol/L MUREAV 89,95,81
msc-hmn:fbr 60 mg/L MUREAV 121,71,83
dnd-rat:lvr 30 μmol/L SinJF# 26OCT82
dns-mus-orl 200 mg/kg MUREAV 125,291,84
orl-rat TDLo:4524 mg/kg/48W-C:ETA ARZNAD 12,270,62
scu-mus TDLo:216 mg/kg/3D-I:CAR JNCIAM 41,403,68
orl-mus TD:5460 μg/kg:CAR EJCAAH 21,865,85
orl-mus TD:5460 μg/kg EJCODS 21,865,85
orl-rat LD50:500 mg/kg JIHTAB 29,1,47
orl-mus LD50:205 mg/kg EJCODS 21,865,85
ipr-mus LDLo:250 mg/kg CBCCT* 6,54,54
orl-dog LDLo:25 mg/kg SCIEAS 167,992,70
orl-rbt LD50:690 mg/kg JIHTAB 29,1,47

CONSENSUS REPORTS: NTP Fifth Annual Report on Carcinogens. IARC Cancer Review: Group 1 IMEMDT 7,91,87; Human Limited Evidence IMEMDT 1,74,72; Animal Sufficient Evidence IMEMDT 1,74,72; Human Sufficient Evidence IMEMDT 28,151,82. Reported in EPA TSCA Inventory. EPA Genetic Toxicology Program. Community Right-To-Know List.

OSHA PEL: Cancer Suspect Agent
ACGIH TLV: Confirmed Human Carcinogen.
DFG MAK: Human Carcinogen.

SAFETY PROFILE: Confirmed human carcinogen with experimental carcinogenic and tumorigenic data. Poison by ingestion and intraperitoneal routes. Human mutation data reported. An irritant. Effects resemble those of benzidine. See also BENZIDINE. Slight to moderate fire hazard when exposed to heat, flames (sparks), or powerful oxidizers. To fight fire, use water spray, mist, dry chemical. When heated to decomposition it emits toxic fumes of NO_x. See also AROMATIC AMINES.

AJS225 ***HR: 2***
2-AMINODIPYRIDO(1,2-a:3',2'-d)IMIDAZOLE HYDROCHLORIDE
mf: $C_{10}H_8N_4 \cdot ClH$ mw: 220.68

SYN: DIPYRIDO(1,2-a:3',2'-d)IMIDAZOLE, 2-AMINO-, HYDROCHLORIDE

TOXICITY DATA with REFERENCE
slt-dmg-orl 100 ng/kg JJCREP 76,468,85
orl-rat TDLo:4116 mg/kg/24W-C:CAR GANNA2 75,207,84

SAFETY PROFILE: Questionable carcinogen with experimental carcinogenic data. Mutation data reported. When heated to decomposition it emits toxic fumes of NO_x and HCl.

AJS250 CAS:16268-87-4 ***HR: 3***
2-AMINO-4,6-DIPYRROLIDINOTRIAZINE
mf: $C_{11}H_{18}N_6$ mw: 234.35

TOXICITY DATA with REFERENCE
orl-mus LD50:600 mg/kg JMCMAR 13,1081,70
ivn-mus LD50:56 mg/kg CSLNX* NX#03988

SAFETY PROFILE: Poison by intravenous route. Moderately toxic by ingestion. See also AMINES. When heated to decomposition it emits toxic fumes of NO_x.

AJS500 CAS:101-50-8 ***HR: 1***
4-AMINO-3,4'-DISULFOAZOBENZENE
mf: $C_{12}H_{11}N_3O_6S_2$ mw: 357.38

SYNS: 4-AMINOAZOBENZENE-3,4'-DISULFONIC ACID ◇ 6-AMINO-3,4'-AZODI-BENZENESULFONIC ACID ◇ 2-AMINO-5-((4-SULFOPHENYL)AZO)-BENZENESULFONIC ACID ◇ 4-(4-AMINO-3-SULFOPHENYLAZO)BENZENESULFONIC ACID ◇ KYSELINA 4-AMINOAZOBENZEN-3,4'-DISULFONOVA (CZECH)

TOXICITY DATA with REFERENCE
eye-rbt 500 mg/24H SEV 28ZPAK -,192,72
orl-rat LD50:14800 mg/kg 28ZPAK -,192,7

CONSENSUS REPORTS: Reported in EPA TSCA Inventory.

SAFETY PROFILE: Mildly toxic by ingestion. A severe eye irritant. When heated to decomposition it emits very toxic fumes of SO_x and NO_x. See also SULFONATES.

AJS750 CAS:146-37-2 ***HR: 3***
4-AMINO-1-DODECYLQUINALDINIUM ACETATE
mf: $C_{22}H_{35}N_2 \cdot C_2H_3O_2$ mw: 386.64

SYNS: 4-AMINO-1-DODECYLQUINALDINIUMACETATE ◇ 1-DODECYL-4-AMINOQUINALDINIUM ACETATE ◇ N-DODECYL-4-AMINOQUINALDINIUM ACETATE ◇ LAURODIN ◇ LAUROLINIUM ACETATE

TOXICITY DATA with REFERENCE
orl-mus LD50:132 mg/kg JPPMAB 15,129,63
ipr-mus LD50:2 mg/kg JPPMAB 15,129,63
scu-mus LD50:30 mg/kg JPPMAB 15,129,63
ivn-mus LD50:6 mg/kg JPPMAB 15,129,63

SAFETY PROFILE: Poison by ingestion, intraperitoneal, subcutaneous, and intravenous routes. When heated to decomposition it emits toxic fumes of NO_x.

AJS875 CAS:54779-53-2 ***HR: D***
9-AMINOELLIPTICINE
mf: $C_{17}H_{15}N_3$ mw: 261.35

SYN: 5,11-DIMETHYL-6H-PYRIDO(4,3-b)CARBAZOL-9-AMINE

TOXICITY DATA with REFERENCE
mmo-sat 500 ng/plate CNREA8 43,3544,83
mma-sat 500 ng/plate CNREA8 43,3544,83
dnd-mus:leu 500 µg/L BCPCA6 28,345,79
msc-ham:ovr 100 µg/L CNREA8 43,3544,83

SAFETY PROFILE: Mutation data reported. When heated to decomposition it emits toxic fumes of NO_x.

AJS900 CAS:2697-65-6 ***HR: 3***
2-AMINO-ETHANESELENOL HYDROCHLORIDE
mf: $C_2H_7NSe \cdot ClH$ mw: 160.52

SYN: ETHANESELENOL, 2-AMINO-, HYDROCHLORIDE

TOXICITY DATA with REFERENCE
ipr-mus LD50:10 mg/kg JMCMAR 12,510,69

OSHA PEL: TWA 0.2 mg(Se)/m^3
ACGIH TLV: TWA 0.2 mg(Se)/m^3

SAFETY PROFILE: Poison by ingestion. When heated to decomposition it emits toxic fumes of NO_x, Se, and HCl.

AJS950 CAS:2697-60-1 ***HR: 3***
2-AMINOETHANESELENOSULFURIC ACID
mf: $C_2H_7NO_3SSe$ mw: 204.12

SYN: SELENOSULFURIC ACID, 2-AMINOETHYL ESTER

TOXICITY DATA with REFERENCE
ipr-mus LD50:18 mg/kg JMCMAR 12,510,69

OSHA PEL: TWA 0.2 mg(Se)/m^3
ACGIH TLV: TWA 0.2 mg(Se)/m^3

SAFETY PROFILE: Poison by intraperitoneal route. When heated to decomposition it emits toxic fumes of NO_x, SO_x, and Se.

AJT250 CAS:60-23-1 ***HR: 3***
2-AMINOETHANETHIOL
mf: C_2H_7NS mw: 77.16

SYNS: 2-AMINOETHYL MERCAPTAN ◇ BECAPTAN ◇ CISTEAMINA (ITALIAN) ◇ CYCTEINAMINE ◇ CYSTEAMIDE ◇ CYSTEAMINE ◇ DECARBOXYCYSTEINE ◇ LAMBRATEN ◇ MEA ◇ MECRAMINE ◇ MERCAMINE ◇ MERCAPTAMINE ◇ β-MERCAPTOETHYLAMINE ◇ (2-MERCAPTOETHYL)AMINE ◇ THIOETHANOLAMINE

TOXICITY DATA with REFERENCE
pic-esc 50 mg/L APMBAY 12,234,64
cyt-ham:ovr 1 mmol/L CALEDQ 5,199,78
dns-ham:fbr 1 mmol/L CALEDQ 5,199,78
sce-ham:ovr 100 µmol/L MUREAV 68,351,79
orl-rat TDLo:42 g/kg (70D pre-21D post):REP TXAPA9 11,523,67
ipr-rat LD50:232 mg/kg ARZNAD 5,421,55
scu-rat LD50:84 mg/kg OSDIAF 5,128,56
orl-mus LD50:625 mg/kg JMCMAR 18,798,75
ipr-mus LD50:250 mg/kg JMCMAR 12,510,69
scu-mus LD50:84 mg/kg OSDIAF 5,128,56
ivn-mus LD50:190 mg/kg CHDDAT 262,206,66
ivn-rbt LD50:150 mg/kg ARZNAD 5,421,55

CONSENSUS REPORTS: EPA Genetic Toxicology Program.

SAFETY PROFILE: Poison by intravenous, subcutaneous, and intraperitoneal routes. Moderately toxic by ingestion. Experimental reproductive effects. Mutation data reported. When heated to decomposition it emits very toxic fumes of SO_x and NO_x.

AJT500 CAS:2937-53-3 ***HR: 3***
2-AMINOETHANETHIOSULFURIC ACID
mf: $C_2H_6NO_4S_2$ mw: 172.21

SYN: USAF EK-8413

TOXICITY DATA with REFERENCE
ipr-mus LD50:400 mg/kg NTIS** AD691-490

CONSENSUS REPORTS: Reported in EPA TSCA Inventory.

SAFETY PROFILE: Poison by intraperitoneal route. See also AMINES and SULFATES. When heated to decomposition it emits very toxic fumes of SO_x and NO_x.

AJT750 CAS:17026-81-2 ***HR: 3***
3-AMINO-4-ETHOXYACETANILIDE
mf: $C_{10}H_{14}N_2O_2$ mw: 194.26

SYNS: 2-AMINO-4-ACETAMINIFENETOL (CZECH) ◇ NCI-C01887

TOXICITY DATA with REFERENCE
eye-rbt 500 mg/24H MLD 28ZPAK -,115,72
mmo-sat 1 mg/plate ENMUDM 7(Suppl 5),1,85
mma-sat 33300 ng/plate ENMUDM 7(Suppl 5),1,85
mma-esc 333 µg/plate ENMUDM 7(Suppl 5),1,85
orl-rat TDLo:130 g/kg/78W-C:ETA NCITR* NCI-CG-TR-112,78
orl-mus TDLo:524 mg/kg/78W-C:CAR NCITR* NCI-CG-TR-112,78
orl-rat LD50:631 mg/kg NCIMR* NIH-71-E-2144

CONSENSUS REPORTS: NTP Carcinogenesis Bioassay (feed): Clear Evidence: mouse NCITR* NCI-TR-112,78; Inadequate Studies: rat NCITR* NCI-TR-112,78. Reported in EPA TSCA Inventory.

SAFETY PROFILE: Moderately toxic by ingestion. An eye irritant. Questionable carcinogen with experimental carcinogenic and tumorigenic data. Mutation data reported. When heated to decomposition it emits toxic fumes of NO_x.

AJU000 ***HR: 2***
2-AMINO-3-ETHOXYCARBONYL-5-BENZYL-4,5,6,7-TETRAHYDROTHIENO (2,3-c)PYRIDINE HYDROCHLORIDE
mf: $C_{17}H_{20}NO_2S \cdot ClH$ mw: 338.90

SYNS: ETHYL-2-AMINO-6-BENZYL-3-THIENO(2,3-c)PYRIDINECARBOXYLATE HYDROCHLORIDE ◇ Y-3642-HCl

TOXICITY DATA with REFERENCE
orl-rat LD50:4750 mg/kg YKKZAJ 90(11),1439,70
ipr-rat LD50:1520 mg/kg YKKZAJ 90(11),1439,70
orl-mus LD50:2050 mg/kg YKKZAJ 90(11),1439,70
ipr-mus LD50:620 mg/kg YKKZAJ 90(11),1439,70

SAFETY PROFILE: Moderately toxic by ingestion and intraperitoneal routes. When heated to decomposition it emits toxic fumes of NO_x, SO_x, and HCl.

AJU250 CAS:929-06-6 ***HR: 2***
2-AMINOETHOXYETHANOL
DOT: NA 1760
mf: $C_4H_{11}NO_2$ mw: 105.16

SYNS: 2-(2-AMINOETHOXY)ETHANOL ◇ DIGLYCOLAMINE

TOXICITY DATA with REFERENCE
skn-rbt 10 mg/24H open SEV AMIHBC 4,119,51
eye-rbt 250 µg open SEV AMIHBC 4,119,51
orl-rat LD50:5660 mg/kg AMIHBC 4,119,51
skn-rbt LD50:1190 mg/kg AMIHBC 4,119,51

CONSENSUS REPORTS: Reported in EPA TSCA Inventory.

DOT Classification: Corrosive Material; Label: Corrosive.

SAFETY PROFILE: Moderately toxic by skin contact. Mildly toxic by ingestion. Severe eye and skin irritant. Corrosive and a powerful irritant. When heated to decomposition it emits toxic fumes of NO_x.

AJU500 CAS:118-28-5 ***HR: 1***
5-AMINO-6-ETHOXY-2-NAPHTHALENESULFONIC ACID
mf: $C_{12}H_{13}NO_4S$ mw: 267.32

SYNS: C.I. 38480 ◇ ETHOXY CLEVE'S ACID ◇ KYSELINA 1-AMINO-2-ETHOXYNAFTALEN-6-SULFONOVA (CZECH) ◇ KYSELINA ETHOXY-CLEVE-1,6 (CZECH)

TOXICITY DATA with REFERENCE
eye-rbt 100 mg/24H MOD 28ZPAK -,191,72
orl-rat LD50:12 g/kg 28ZPAK -,191,72

CONSENSUS REPORTS: Reported in EPA TSCA Inventory.

SAFETY PROFILE: Mildly toxic by ingestion. An eye irritant. When heated to decomposition it emits very toxic fumes of NO_x and SO_x. See also SULFONATES.

AJU625 CAS:1501-84-4 ***HR: 3***
1-(1-AMINOETHYL)ADAMANTANE HYDROCHLORIDE
mf: $C_{12}H_{21}N \cdot ClH$ mw: 215.80

SYNS: EXP 126 ◇ JP 61 ◇ MERADAN ◇ MERADANE ◇ α-METHYL-1-ADAMANTANEMETHYLAMINE HYDROCHLORIDE ◇ α-METHYLTRICYCLO(3.3.1.1(^{3,7}))DECANE-1-METHANAMINE HYDROCHLORIDE ◇ REMANTADIN ◇ RIMANTADINE HYDROCHLORIDE

TOXICITY DATA with REFERENCE
orl-rat LD50:640 mg/kg VOONAW 28(9),23,82
ipr-rat LD50:135 mg/kg KHFZAN 11(6),73,77
ipr-mus LD50:135 mg/kg PCJOAU 11,798,77

SAFETY PROFILE: Poison by intraperitoneal route. Moderately toxic by ingestion. When heated to decomposition it emits toxic fumes of NO_x and HCl.

AJU875 CAS:25682-07-9 ***HR: 3***
2-AMINOETHYLAMMONIUM PERCHLORATE
mf: $C_2H_9ClN_2O_4$ mw: 160.56

SAFETY PROFILE: Explodes upon heating. When heated to decomposition it emits toxic fumes of Cl^-, NH_3, and NO_x. See also PERCHLORATES.

AJV000 CAS:132-32-1 ***HR: 3***
3-AMINO-9-ETHYLCARBAZOLE
mf: $C_{14}H_{14}N_2$ mw: 210.30

PROP: In cancer bioassay both free amine and hydrochloride salt used NCITR* NCI-CG-TR-93,78.

SYN: 3-AMINO-N-ETHYLCARBAZOLE

TOXICITY DATA with REFERENCE
orl-rat TDLo:33 g/kg/78W-C:CAR NCITR* NCI-CG-TR-93,78
orl-mus TDLo:87 g/kg/78W-C:CAR NCITR* NCI-CG-TR-93,78
orl-rat LD50:144 mg/kg NCIMR* NIH-71-E-2144
ipr-mus LD50:150 mg/kg NTIS** AD691-490

CONSENSUS REPORTS: Reported in EPA TSCA Inventory.

DFG MAK: Suspected Carcinogen.

SAFETY PROFILE: Suspected carcinogen with experimental carcinogenic data. Poison by ingestion and intraperitoneal routes. When heated to decomposition it emits toxic fumes of NO_x.

AJV250 CAS:6109-97-3 ***HR: 3***
3-AMINO-9-ETHYLCARBAZOLEHYDROCHLORIDE
mf: $C_{14}H_{14}N_2 \cdot ClH$ mw: 246.76

PROP: In cancer bioassay both free amine and hydrochloride salt used NCITR* NCI-CG-TR-93,78.

SYN: NCI-C03043

TOXICITY DATA with REFERENCE
mma-sat 1 μg/plate ENMUDM 5(Suppl 1),3,83
orl-rat TDLo:33 g/kg/78W-C:CAR NCITR* NCI-CG-TR-93,78
orl-mus TDLo:87 g/kg/78W-C:CAR NCITR* NCI-CG-TR-93,78
orl-rat LD50:234 mg/kg JPETAB 99,450,50

CONSENSUS REPORTS: NCI Carcinogenesis Bioassay Completed; Results Positive: mouse, rat NCITR* NCI-CG-TR-93,78.

SAFETY PROFILE: Suspected carcinogen with experimental carcinogenic data. Poison by ingestion. Mutation data reported. When heated to decomposition it emits very toxic fumes of NO_x and HCl.

AJV500 CAS:1197-18-8 ***HR: 2***
trans-4-AMINOETHYLCYCLOHEXANE-1-CARBOXYLIC ACID
mf: $C_8H_{15}NO_2$ mw: 157.24

SYNS: AMCHA ◇ trans-AMCHA ◇ AMIKAPRON ◇ trans-p-(AMINOMETHYL)CYCLOHEXANECARBOXYLICACID ◇ trans-1-AMINOMETHYLCYCLOHEXANE-4-CARBOXYLIC ACID ◇ trans-4-AMINOMETHYL-1-CYCLOHEXANECARBOXYLIC ACID ◇ AMSTAT ◇ ANVITOFF ◇ BAY 3517 ◇ CL 65336 ◇ CYCLOCAPRON ◇ DV-79 ◇ EMORHALT ◇ EXACYL ◇ FRENOLYSE ◇ HEXAPROMIN ◇ HEXATRON ◇ RIKAVARIN ◇ RP 18,429 ◇ SPIRAMIN ◇ TAMCHA ◇ TRANEX ◇ TRANEXAMIC ACID ◇ TRANHEXAMIC ACID ◇ TRANSAMLON ◇ UGUROL

TOXICITY DATA with REFERENCE
orl-mus TDLo:9 mg/kg (female 7-12D post):REP OYYAA2 5,415,71
orl-rat TDLo:9 mg/kg (female 9-14D post):TER OYYAA2 5,415,71
orl-rat LD50:3000 mg/kg APTOA6 22,340,65
ipr-rat LD50:4200 mg/kg MEIEDD 10,1269,83
scu-rat LD50:4620 mg/kg NIIRDN 6,512,82
ivn-rat LD50:1200 mg/kg APTOA6 22,340,65
scu-mus LD50:5310 mg/kg NIIRDN 6,512,82
ivn-mus LD50:1350 mg/kg NIIRDN 6,512,82
ivn-dog LD50:1110 mg/kg NIIRDN 6,512,82

SAFETY PROFILE: Moderately toxic by ingestion and intravenous routes. An experimental teratogen. Other experimental reproductive effects. When heated to decomposition it emits toxic fumes such as NO_x. A hemostatic agent.

AJV850 CAS:63991-14-0 ***HR: 3***
α-(1-AMINOETHYL)-2,4-DIMETHOXYBENZYL ALCOHOL HYDROCHLORIDE
mf: $C_{11}H_{17}NO_3 \cdot ClH$ mw: 247.75

SYN: BENZYL ALCOHOL-α-(1-AMINOETHYL)-2,4-DIMETHOXY HYDROCHLORIDE

TOXICITY DATA with REFERENCE
scu-rat LDLo:320 mg/kg JPETAB 71,62,41
ivn-rbt LDLo:21 mg/kg JACSAT 53,4149,31

SAFETY PROFILE: Poison by subcutaneous and intravenous routes. When heated to decomposition it emits very toxic fumes of NO_x and Cl^-.

AJW000 CAS:111-41-1 ***HR: 2***
N-AMINOETHYLETHANOLAMINE
mf: $C_4H_{12}N_2O$ mw: 104.18

$HOC_2H_4NHC_2H_4NH_2$

PROP: Colorless liquid. Bp: 243.7°, flash p: 216°F, d: 1.0304 @ 20°/20°, autoign temp: 695°F, vap press: <0.01 mm @ 20°, vap d: 3.59.

SYNS: AMINOETHYL ETHANOLAMINE ◇ ETHANOLETHYLENE DIAMINE ◇ N-HYDROXYETHYL-1,2-ETHANEDIAMINE ◇ N-(β-HYDROXYETHYL)ETHYLENEDIAMINE ◇ N-(2-HYDROXYETHYL)ETHYLENEDIAMINE ◇ MONOETHANOLETHYLENEDIAMINE

TOXICITY DATA with REFERENCE
skn-rbt 10 mg/24H open JIHTAB 26,269,44
skn-rbt 445 mg open MLD UCDS** 11/29/63
eye-rbt 50 mg SEV UCDS** 7/19/65
orl-rat LD50:3000 mg/kg UCDS** 7/19/65
skn-rat LD50:2250 mg/kg 85GMAT -,64,82
ipr-rat LD50:751 mg/kg GTPZAB 26(8),53,82
ivn-rat LD50:417 mg/kg 85GMAT -,64,82
ims-rat LD50:2 g/kg 85GMAT -,64,82
orl-mus LD50:3550 mg/kg 85GMAT -,64,82
orl-rbt LD50: 2 g/kg 85GMAT -,64,82
orl-gpg LD50:1500 mg/kg 85GMAT -,64,82
skn-gpg LD50:1800 mg/kg JIHTAB 26,269,44

CONSENSUS REPORTS: Reported in EPA TSCA Inventory.

SAFETY PROFILE: Moderately toxic by ingestion, skin contact and several other routes. A severe eye irritant and moderate skin irritant. Combustible. To fight fire use alcohol foam, mist, dry chemical. As with other amines it ignites on contact with cellulose nitrate of high surface area. When heated to decomposition it emits toxic fumes of NO_x.

AJW250 ***HR: 3***
6-AMINO-1-ETHYL-4-p-((p-((1-ETHYLPYRIDINIUM-4-YL)AMINO)2-AMINOPHENYL)CARBAMOYL)ANILINO)QUINOLINIUM DIIODIDE
mf: $C_{31}H_{33}N_7O{\bullet}2I$ mw: 773.51

TOXICITY DATA with REFERENCE
dnd-mus:lym 710 nmol/L JMCMAR 22,134,79
ipr-mus LD10:6500 μg/kg JMCMAR 22,134,79

SAFETY PROFILE: Poison by intraperitoneal route. Mutation data reported. When heated to decomposition it emits very toxic fumes of NO_x and I^-.

AJW500 CAS:50309-16-5 ***HR: 3***
6-AMINO-1-ETHYL-4-(p-(p-((1-ETHYLPYRIDINIUM-4-YL)AMINO)BENZAMIDO)ANILINO)QUINOLINIUM DIIODIDE
mf: $C_{31}H_{32}N_6O{\bullet}2I$ mw: 758.49

TOXICITY DATA with REFERENCE
dnd-mus:lym 600 nmol/L JMCMAR 22,134,79
ipr-mus LD10:9 mg/kg JMCMAR 22,134,79

SAFETY PROFILE: Poison by intraperitoneal route. Mutation data reported. When heated to decomposition it emits very toxic fumes of I^- and NO_x.

AJW750 CAS:42013-69-4 ***HR: 3***
6-AMINO-1-ETHYL-4-(p-((p-((1-ETHYLPYRIDINIUM-4-YL)AMINO)PHENYL)CARBAMOYL)ANILINOQUINOLINIUM DIBROMIDE
mf: $C_{31}H_{32}N_6O{\bullet}2Br$ mw: 664.51

TOXICITY DATA with REFERENCE
dnd-mus:lym 690 nmol/L JMCMAR 22,134,79
ipr-mus LD10:10 mg/kg JMCMAR 22,134,79

SAFETY PROFILE: Poison by intraperitoneal route. Mutation data reported. When heated to decomposition it emits very toxic fumes of Br^- and NO_x.

AJX000 CAS:61-54-1 ***HR: 3***
3-(2-AMINOETHYL)INDOLE
mf: $C_{10}H_{12}N_2$ mw: 160.24

SYNS: (AMINO-2 ETHYL)-3-INDOLE (FRENCH) ◇ 1H-INDOLE-3-ETHANAMINE ◇ INDOL-3-ETHYLAMINE ◇ 2-(3-INDOLYL)ETHYLAMINE ◇ TRYPTAMINE

TOXICITY DATA with REFERENCE
ipr-rat LD50:223200 μg/kg JPMSAE 66(12),1962,77
ipr-mus LD50:100 mg/kg EJMCA5 9,453,74
scu-mus LD50:500 mg/kg DPHFAK 22,313,70

SAFETY PROFILE: Poison by intraperitoneal route. Moderately toxic by subcutaneous route. When heated to decomposition it emits toxic fumes of NO_x.

AJX250 CAS:343-94-2 ***HR: 3***
3-(2-AMINOETHYL)INDOLE HYDROCHLORIDE
mf: $C_{10}H_{12}N_2{\bullet}ClH$ mw: 196.70

SYNS: β-INDOLAETHYLAMIN-CHLORHYDRAT (GERMAN) ◇ β-3-INDOLYLETHYLAMINE HYDROCHLORIDE ◇ INDOLE-3-ETHYLAMINE HYDROCHLORIDE ◇ β-INDOLE-ETHYLAMINE HYDROCHLORIDE ◇ TRYPTAMINE HYDROCHLORIDE

TOXICITY DATA with REFERENCE
scu-rat LDLo:1300 mg/kg JPMRAB 2,77,27
ipr-mus LD50:197 mg/kg YKKZAJ 94,1620,74
scu-mus LD50:504 mg/kg RPTOAN 33,180,70
ivn-mus LD50:109 mg/kg BJPCAL 23,43,64
scu-rbt LDLo:1000 mg/kg JPMRAB 2,77,27

CONSENSUS REPORTS: Reported in EPA TSCA Inventory.

SAFETY PROFILE: Poison by intravenous and intraperitoneal routes. Moderately toxic by subcutaneous route. When heated to decomposition it emits very toxic NO_x and HCl.

AJX500 CAS:50-67-9 ***HR: 3***
3-(2-AMINOETHYL)INDOL-5-OL
mf: $C_{10}H_{12}N_2O$ mw: 176.24

SYNS: 3-(β-AMINOETHYL)-5-HYDROXYINDOLE ◇ ANTEMOQUA ◇ ANTEMOVIS ◇ DS SUBSTANDE ◇ ENTERAMINE ◇ HIPPOPHAIN ◇ 5-HT ◇ 5-HTA ◇ 5-HYDROXY-3-(β-AMINOETHYL)INDOLE ◇ 5-HYDROXYTRYPTAMINE ◇ SEROTONIN ◇ SUBSTANCE DS ◇ SUBSTANZ DS ◇ THROMBOCYTIN ◇ THROMBOTONIN

TOXICITY DATA with REFERENCE
scu-rat TDLo:110 mg/kg (female 1-22D post):REP JOPDAB 63,394,63
ipr-rat TDLo:10 mg/kg (18D preg):TER ARPAAQ 81,257,66
orl-mus LD50:60 mg/kg MZUZA8 (3),61,85
ipr-mus LD50:160 mg/kg IJPPAZ 17,31,73
scu-mus LD50:601 mg/kg FEPRA7 23,T125,64
ivn-mus LD50:81 mg/kg FATOAO 26,10,63
ims-mus LD50:750 mg/kg AIPTAK 112,319,57
ivn-mus LD50:81 mg/kg FATOAO 26,10,63

SAFETY PROFILE: Poison by ingestion, intravenous, and intraperitoneal routes. An experimental teratogen. Other experimental reproductive effects. When heated to decomposition it emits toxic fumes of NO_x. A neurotransmitter.

AJX750 CAS:971-74-4 ***HR: 3***
3-(2-AMINOETHYL)INDOL-5-OL CREATININE SULFATE
mf: $C_{10}H_{12}N_2O \cdot C_4H_7N_3O \cdot H_2O_4S \cdot H_2O$ mw: 405.48

SYNS: CREATININE SULFATE compounded with 3-(2-AMINOETHYL)INDOLE-5-OL (1:1:1), MONOHYDRATE ◇ CREATININE SULFATE compounded with 3-(2-AMINOETHYL)INDOL-5-OL (1:1:1) ◇ 5-HYDROXYTRYPTAMINE CREATININE SULFATE ◇ 5-HYDROXYTRYPTAMINE CREATININE SULFATE MONOHYDRATE ◇ SEROTIN CREATININE SULFATE ◇ SEROTONIN CREATININE SULFATE MONOHYDRATE

TOXICITY DATA with REFERENCE
ivn-rat TDLo:10 mg/kg (female 20D post):REP AJPHAP 169,537,52
scu-mus TDLo:10 mg/kg (female 15D post):TER AJOGAH 99,250,67
scu-rat LD50:257 mg/kg JPETAB 105,80,52
ivn-rat LD50:66 mg/kg JPETAB 105,80,52
ipr-mus LD50:405 mg/kg YKKZAJ 94,1620,74
ivn-mus LD50:352 mg/kg JPETAB 105,80,52
ivn-gpg LD50:28 mg/kg JPMSAE 57,1543,68

CONSENSUS REPORTS: EPA Genetic Toxicology Program.

SAFETY PROFILE: Poison by subcutaneous and intravenous routes. Moderately toxic by intraperitoneal route. An experimental teratogen. Other experimental reproductive effects. See also SULFATES. When heated to decomposition it emits very toxic fumes of SO_x and NO_x.

AJY000 CAS:1704-04-7 ***HR: 3***
AMINOETHYLISOSELENOURONIUM BROMIDE HYDROCHLORIDE

SYNS: 2-AMINOAETHYLISOSELENOURONIUMBROMID-HYDROBROMID (GERMAN) ◇ MONOETHYLISOSELENOURONIUMBROMIDE-HYDROBROMIDE

TOXICITY DATA with REFERENCE
ipr-mus LD50:51 mg/kg STRAAA 151,78,76
scu-mus LD50:50 mg/kg STRAAA 151,78,76
ivn-mus LD50:44 mg/kg STRAAA 151,78,76

CONSENSUS REPORTS: On Community Right-To-Know List.

SAFETY PROFILE: Poison by intraperitoneal, subcutaneous, and intravenous routes. When heated to decomposition it emits very toxic fumes of HCl, Br^- and NO_x.

AJY250 CAS:56-10-0 ***HR: 3***
2-β-AMINOETHYLISOTHIOUREA
mf: $C_3H_9N_3S \cdot 2BrH$ mw: 281.05

SYNS: AET ◇ AET BROMIDE ◇ AET DIHYDROBROMIDE ◇ AET-2HBR ◇ β-AMINOAETHYL-ISOTHIURONIUM DIHYDROBROMID(GERMAN) ◇ 2-AMINOETHYL ESTER CARBAMIMIDOTHIOIC ACID DIHYDROBROMIDE ◇ 2-(β-AMINOETHYL)ISOTHIOURONIUM BROMIDE HYDROBROMIDE ◇ S-(β-AMINOETHYL)ISOTHIURONIUM BROMIDE HYDROBROMIDE ◇ S-(2-AMINOETHYL)ISOTHIURONIUM BROMIDE HYDROBROMIDE ◇ β-AMINOETHYLISOTHIURONIUM BROMIDE HYDROBROMIDE ◇ 2-AMINOETHYLISOTHIURONIUM BROMIDE HYDROBROMIDE ◇ 2-AMINOETHYLISOTHIOURONIUM DIBROMIDE ◇ 2-AMINOETHYLISOTHIURONIUM DIHYDROBROMIDE ◇ 2-(2-AMINOETHYL)-2-THIOPSEUDOUREA HYDROBROMIDE ◇ ANTIRAD ◇ ANTIRADON ◇ SURRECTAN ◇ USAF XR-31

TOXICITY DATA with REFERENCE
ipr-rat LD50:288 mg/kg AIPTAK 142,198,63
ivn-rat LD50:85 mg/kg CLCEAL 105,1165,66
ipr-mus LD50:400 mg/kg NTIS** AD277-289
ivn-mus LD50:96 mg/kg CLCEAL 105,1165,66
orl-dog LD50:177 mg/kg AIPTAK 142,510,63
ipr-dog LD50:113 mg/kg AIPTAK 142,510,63
ipr-rbt LD50:236 mg/kg AIPTAK 142,510,63

CONSENSUS REPORTS: Reported in EPA TSCA Inventory.

SAFETY PROFILE: Poison by ingestion, intraperitoneal, subcutaneous, and intravenous routes. When heated to decomposition it emits very toxic NO_x, SO_x, and HBr.

AJY500 CAS:871-25-0 ***HR: 3***
2-AMINOETHYLISOTHIOURONIUMDICHLORIDE
mf: $C_3H_9N_3S \cdot 2ClH$ mw: 192.13

SYNS: AET DICHLORIDE ◇ β-AMINOAETHYLISOTHIURONIUM-CHLORID-HYDROCHLORID (GERMAN) ◇ S-β-AMINOETHYLISOTHIOURONICDIHYDROCHLORIDE ◇ 2-AMINOETHYL-2-THIOPSEUDOUREA DICHLORIDE ◇ 2-(2-AMINOETHYL)-2-THIOPSEUDOUREA DIHYDROCHLORIDE ◇ USAF XR-32

TOXICITY DATA with REFERENCE
par-rat LD50:325 mg/kg TXAPA9 1,8,59
ipr-mus LD50:250 mg/kg NTIS** AD277-689
par-mus LD50:400 mg/kg TXAPA9 1,8,59
par-dog LD50:110 mg/kg TXAPA9 1,8,59
scu-mus LD50:266 mg/kg ARZNAD 8,72,58

SAFETY PROFILE: Poison by subcutaneous, parenteral, and intraperitoneal routes. When heated to decomposition it emits very toxic fumes of HCl, SO_x, and NO_x.

AJY750 CAS:92-09-1 ***HR: 3***
4-AMINO-N-ETHYL-m-(β-METHANESULFONAMIDOETHYL)-m-TOLUIDINE

TOXICITY DATA with REFERENCE
orl-rat LDLo:400 mg/kg KODAK* -,-,71
ipr-rat LDLo:10 mg/kg KODAK* -,-,71

CONSENSUS REPORTS: Reported in EPA TSCA Inventory.

SAFETY PROFILE: Poison by ingestion and intraperitoneal routes. When heated to decomposition it emits very toxic fumes of SO_x and NO_x.

AJZ000 CAS:22137-01-5 ***HR: 3***
3-(2-AMINOETHYL)-5-METHOXYBENZOFURAN HYDROCHLORIDE
mf: $C_{11}H_{13}NO_2 \cdot ClH$ mw: 227.71

TOXICITY DATA with REFERENCE
ivn-rat LDLo:6 mg/kg RPTOAN 33,246,70
ivn-mus LD50:55 mg/kg RPTOAN 33,246,70

SAFETY PROFILE: Poison by intravenous route. When heated to decomposition it emits very toxic fumes of HCl and NO_x.

AKA000 CAS:28089-06-7 ***HR: 3***
6-(β-AMINOETHYL)-5-METHOXYBENZOFURANHYDROCHLORIDE
mf: $C_{11}H_{12}NO_2 \cdot ClH$ mw: 226.70

SYN: 6-(2-AMINOETHYL)-5-METHOXYBENZOFURANHYDROCHLORIDE

TOXICITY DATA with REFERENCE
ivn-rat LDLo:48 mg/kg RPTOAN 33,246,70
ivn-mus LD50:60 mg/kg RPTOAN 33,246,70

SAFETY PROFILE: Poison by intravenous route. When heated to decomposition it emits very toxic fumes of HCl and NO_x.

AKA250 CAS:63991-23-1 ***HR: 3***
α-(1-AMINOETHYL)-4-METHOXYBENZYLALCOHOL HYDROCHLORIDE
mf: $C_{10}H_{15}NO_2 \cdot ClH$ mw: 217.72

SYN: α-(1-AMINOETHYL)-4-METHOXYBENZYL ALCOHOL HYDROCHLORIDE

TOXICITY DATA with REFERENCE
scu-rat LDLo:160 mg/kg JPETAB 71,62,41
ivn-rbt LDLo:35 mg/kg JACSAT 53,4149,31

SAFETY PROFILE: Poison by subcutaneous and intravenous routes. When heated to decomposition it emits very toxic fumes of Cl^- and NO_x.

AKA500 CAS:52479-18-2 ***HR: 3***
2-(AMINOETHYL)-2-METHYL-1,3-BENZODIOXOLE HYDROCHLORIDE
mf: $C_{10}H_{13}NO_2 \cdot ClH$ mw: 215.70

SYN: 2-AMINOETHYL-2-METHYL-1,3-BENZODIOXOLEHYDROCHLORIDE

TOXICITY DATA with REFERENCE
ivn-rat LD50:33 mg/kg EJMCA5 12,413,77
ipr-mus LD50:100 mg/kg EJMCA5 12,413,77

SAFETY PROFILE: Poison by intravenous and intraperitoneal routes. When heated to decomposition it emits very toxic fumes of HCl and NO_x.

AKA750 CAS:2038-03-1 ***HR: 3***
N-AMINOETHYLMORPHOLINE
mf: $C_6H_{14}N_2O$ mw: 130.22

PROP: Liquid. Mp: 25.6°, bp: 204.2°, flash p: 347°F (OC), d: 0.9915 @ 20°/20°, vap d: 4.49.

SYNS: β-AMINOAETHYL-MORPHOLIN(GERMAN) ◇ 4-MORPHOLINEETHANAMINE

TOXICITY DATA with REFERENCE
skn-rbt 10 mg/24H open JIHTAB 26,269,44
orl-rat LD50:3000 mg/kg JIHTAB 26,269,44
scu-mus LD50:2145 mg/kg JIHTAB 26,269,44
skn-gpg LD50:300 mg/kg JIHTAB 26,269,44

CONSENSUS REPORTS: Reported in EPA TSCA Inventory.

SAFETY PROFILE: Poison by skin contact. Moderately toxic by ingestion and subcutaneous routes. A skin irritant. Moderately flammable when exposed to heat, flame, or oxidizing materials. To fight fire, use alcohol foam, dry chemical. When heated to decomposition it emits toxic fumes of NO_x.

AKB000 CAS:140-31-8 ***HR: 3***
N-AMINOETHYLPIPERAZINE
DOT: UN 2815
mf: $C_6H_{15}N_3$ mw: 129.24

PROP: Light-colored liquid. D: 0.9852 @ 20°/20°, mp: −19°, bp: 220.4°, flash p: 200°F (OC), vap d: 4.4.

SYNS: AMINOETHYLPIPERAZINE ◇ N-(β-AMINOETHYL)PIPERAZINE ◇ N-(2-AMINOETHYL)PIPERAZINE ◇ 1-(2-AMINOETHYL)PIPERAZINE ◇ USAF DO-46

TOXICITY DATA with REFERENCE
skn-rbt 100 μg/24H open AIHAAP 23,95,62
skn-rbt 5 mg/24H SEV 85JCAE -,864,86
eye-rbt 20 mg/24H MOD 85JCAE -,864,86
otr-mus:lym 1 uL/L ENMUDM 4,390,82
orl-rat TDLo:1680 mg/kg (male 28D pre):REP GISAAA 51(10),66,86
orl-rat LD50:2140 mg/kg AIHAAP 23,95,62
ipr-mus LD50:250 mg/kg NTIS** AD277-689
skn-rbt LD50:880 mg/kg UCDS** 6/13/69

CONSENSUS REPORTS: Reported in EPA TSCA Inventory.

DOT Classification: Corrosive Material; Label: Corrosive.

SAFETY PROFILE: Poison by intraperitoneal routes. Moderately toxic by ingestion and skin contact. Experimental reproductive effects. A skin and eye irritant. Mutation data reported. See also AMINES. Moderately flammable when exposed to heat, flame, sparks, or powerful oxidizers. To fight fire, use alcohol foam. When heated to decomposition it emits toxic fumes of NO_x.

AKB125 ***HR: 2***
5-AMINO-2-ETHYLTETRAZOL
mf: $C_3H_7N_5$ mw: 113.12

SAFETY PROFILE: Forms an explosive complex with aluminum hydride. When heated to decomposition it emits toxic fumes of NO_x.

AKB250 CAS:13073-35-3 ***HR: 2***
2-AMINO-4-(ETHYLTHIO)BUTYRIC ACID
mf: $C_6H_{13}NO_2S$ mw: 163.26

SYNS: l-2-AMINO-4-(ETHYLTHIO)BUTYRIC ACID ◇ ETHIONINE ◇ l-ETHIONINE ◇ S-ETHYL-l-HOMOCYSTEINE

TOXICITY DATA with REFERENCE
dni-esc 2 g/L CYTOAN 50,387,85
dni-hmn:lym 2 mmol/L BBACAQ 520,139,79
otr-ham:emb 5 mg/L CRNGDP 4,291,83
orl-mus TDLo:44100 mg/kg/2Y-C:CAR CRNGDP 7,1143,86

CONSENSUS REPORTS: Reported in EPA TSCA Inventory. EPA Genetic Toxicology Program.

SAFETY PROFILE: Questionable carcinogen with experimental carcinogenic data. Human mutation data reported. When heated to decomposition it emits very toxic fumes of NO_x and SO_x.

AKB500 CAS:3724-89-8 ***HR: 3***
S-(2-AMINOETHYL)THIOPHOSPHATEMONO SODIUM SALT
mf: $C_2H_7NO_3PS{\cdot}Na$ mw: 179.12

SYNS: 2-AMINO-ETHANETHIOL DIHYDROGEN PHOSPHATE(ester), MONOSODIUM SALT ◇ CISTAPHOS ◇ CYSTAPHOS ◇ CYSTAPHOS SODIUM SALT ◇ MONOSODIUM-β-AMINOETHYL THIOPHOSPHATE ◇ SODIUM HYDROGEN-S-(2-AMINOETHYL)PHOSPHOROTHIOATE ◇ SODIUM HYDROGEN-S-(2-AMINOETHYL)PHOSPHOROTHIOIC ACID ◇ WR 638

TOXICITY DATA with REFERENCE
ipr-rat LD50:555 mg/kg RADOA8 16,249,76
ims-rat LD50:505 mg/kg RADOA8 16,249,76
orl-mus LD50:1433 mg/kg RADOA8 16,249,76
ipr-mus LD50:806 mg/kg RADOA8 16,249,76
ims-mus LD50:1003 mg/kg RADOA8 16,249,76
ipr-gpg LD50:358 mg/kg RADOA8 16,249,76

SAFETY PROFILE: Poison by intraperitoneal route. Moderately toxic by ingestion and intramuscular routes. See also PHOSPHATES and ESTERS. When heated to decomposition it emits very toxic fumes of PO_x, SO_x, NO_x, and Na_2O.

AKB875 CAS:3096-57-9 ***HR: D***
2-AMINOFLUORENONE
mf: $C_{13}H_9NO$ mw: 195.23

SYN: 2-AMINO-9-FLUORENONE

TOXICITY DATA with REFERENCE
mma-sat 500 nmol/L ENMUDM 3,11,81
mma-esc 500 nmol/L ENMUDM 3,11,81
dns-rat:lvr 5 μmol/L ENMUDM 3,11,81

SAFETY PROFILE: Mutation data reported. When heated to decomposition it emits toxic fumes of NO_x.

AKB900 CAS:4269-15-2 ***HR: D***
4-AMINOFLUORENONE
mf: $C_{13}H_9NO$ mw: 195.23

SYN: 4-AMINO-9-FLUORENONE

TOXICITY DATA with REFERENCE
mma-sat 500 nmol/L ENMUDM 3,11,81
mma-esc 500 nmol/L ENMUDM 3,11,81
dns-rat:lvr 5 µmol/L ENMUDM 3,11,81

SAFETY PROFILE: Mutation data reported. When heated to decomposition it emits toxic fumes of NO_x.

AKC000 CAS:63019-67-0 ***HR: 3***
2-AMINO-N-FLUOREN-2-YLACETAMIDE
mf: $C_{15}H_{14}N_2O$ mw: 238.31

SYN: 2-GLYCYLAMINOFLUORENE

TOXICITY DATA with REFERENCE
orl-rat TDLo:1200 mg/kg/20W-I:ETA NATUAS 184,2018,59

SAFETY PROFILE: Questionable carcinogen with experimental tumorigenic data. When heated to decomposition it emits toxic fumes of NO_x.

AKC250 CAS:1682-39-9 ***HR: 2***
2-AMINO-5-FLUOROBENZOXAZOLE
mf: $C_7H_5FN_2O$ mw: 152.14

TOXICITY DATA with REFERENCE
orl-rat LD50:1000 mg/kg MDCHAG 4(1),338,64
orl-mus LD50:700 mg/kg MDCHAG 4(1),336,64
ipr-mus LD50:450 mg/kg MDCHAG 4(1),336,64

SAFETY PROFILE: Moderately toxic by ingestion and intraperitoneal routes. When heated to decomposition it emits very toxic fumes of F^- and NO_x.

AKC500 CAS:324-93-6 ***HR: 3***
4-AMINO-4′-FLUORODIPHENYL
mf: $C_{12}H_{10}FN$ mw: 187.23

SYNS: 4′-FLUORO-4-AMINODIPHENYL ◇ 4′-FLUORO-4-BIPHENYLAMINE

TOXICITY DATA with REFERENCE
orl-rat TDLo:300 mg/kg:CAR CNREA8 26,619,66
orl-mus TDLo:520 mg/kg/26W-I:NEO BJCAAI 19,297,65
scu-rat TD:2000 mg/kg/W-I:ETA BMBUAQ 14,141,58
orl-rat LDLo:300 mg/kg CNREA8 26,619,66

CONSENSUS REPORTS: EPA Genetic Toxicology Program.

SAFETY PROFILE: Poison by ingestion. Questionable carcinogen with experimental carcinogenic, neoplastigenic, and tumorigenic data. When heated to decomposition it emits very toxic fumes of F^- and NO_x.

AKC600 CAS:125-84-8 ***HR: 2***
AMINOGLUTETHIMIDE
mf: $C_{13}H_{16}N_2O_2$ mw: 232.31

SYNS: p-AMINOGLUTETHIMIDE ◇ 2-(p-AMINOPHENYL)-2-ETHYLGLUTARIMIDE ◇ Ba-16038 ◇ CYTADREN ◇ ELIPTEN ◇ 3-ETHYL-3-(p-AMINOPHENYL)-2,6-DIOXOPIPERIDINE ◇ GLUTARIMIDE, 2-(p-AMINOPHENYL)-2-ETHYL- ◇ ORIMETEN ◇ 2,6-PIPERIDINEDIONE, 3-(4-AMINOPHENYL)-3-ETHYL-

TOXICITY DATA with REFERENCE
orl-wmn TDLo:3630 mg/kg (female 1-35W post):TER AJDCAI 124,421,72
orl-wmn TDLo:20500 mg/kg/94W-I:BLD BMJOAE 291,970,85
orl-man LDLo:21 mg/kg/3D-I:PUL AIMEAS 105,633,86
ipr-mus LD50:625 mg/kg JMCMAR 18,736,75

SAFETY PROFILE: Moderately toxic by intraperitoneal route. Human systemic effects by ingestion: agranulocytosis, dyspnea. Human teratogenic effects by ingestion: urogenital developmental abnormalities. When heated to decomposition it emits toxic fumes of NO_x.

AKC625 CAS:23734-88-5 ***HR: 2***
AMINOGLUTETHIMIDE PHOSPHATE
mf: $C_{13}H_{16}N_2O_2 \cdot H_3O_4P$ mw: 330.31

SYNS: AGP ◇ α-(p-AMINOPHENYL)-α-ETHYLGLUTARIMIDE PHOSPHATE ◇ 2-(p-AMINOPHENYL)-2-ETHYLGLUTARIMIDE PHOSPHATE

TOXICITY DATA with REFERENCE
scu-rbt TDLo:360 mg/kg (female 25-26D post):REP JANSAG 42,131,76
orl-mus LD50:1800 mg/kg PSEBAA 139,100,72

SAFETY PROFILE: Moderately toxic by ingestion. Experimental reproductive effects. When heated to decomposition it emits toxic fumes of PO_x and NO_x. An antisteroidogenic drug. See also PHOSPHATES.

AKC750 CAS:79-17-4 ***HR: 2***
AMINOGUANIDINE
mf: CH_6N_4 mw: 74.11

PROP: Crystalline. Mp: decomp. Sol in water and alc; insol in ether.

SYNS: AMINATE BASE ◇ GUANYL HYDRAZINE ◇ HYDRAZINECARBOXIMIDAMIDE

TOXICITY DATA with REFERENCE
scu-rat LD50:1258 mg/kg JPETAB 119,444,57
scu-mus LD50:963 mg/kg JPETAB 119,444,57

SAFETY PROFILE: Moderately toxic by subcutaneous route. See also AMINES. All of the oxoacid salts are potentially explosive. When heated to decomposition it emits toxic fumes of NO_x.

AKD250 CAS:2834-84-6 ***HR: 2***
AMINOGUANIDINE SULFATE
mf: $CH_6N_4 \cdot H_2O_4S$ mw: 172.19

SYN: AMINOGUANIDINE SULPHATE

TOXICITY DATA with REFERENCE
scu-mus TDLo:14400 mg/kg (1-6D preg):REP JRPFA4 6,179,63
orl-rat LD50:500 mg/kg JPETAB 90,260,47

SAFETY PROFILE: Moderately toxic by ingestion. Experimental reproductive effects. When heated to decomposition it emits toxic fumes of SO_x and NO_x.

AKD375 CAS:10308-82-4 ***HR: 3***
AMINO GUANIDINIUM NITRATE
mf: $CH_7N_5O_3$ mw: 137.1

SAFETY PROFILE: An unstable compound and powerful oxidizer. Aqueous solutions may explode violently when heated to evaporation. When heated to decomposition it emits toxic fumes of NO_x. See also NITRATES.

AKD500 CAS:543-38-4 ***HR: 3***
l,2-AMINO-4-(GUANIDINOOXY)BUTYRIC ACID
mf: $C_5H_{12}N_4O_3$ mw: 176.21

SYNS: 2-AMINO-4-(GUANIDINOOXY)-l-BUTYRICACID ◇ o-((AMINOIMINOMETHYL)AMINO)-l-HOMOSERINE ◇ CANAVANIN ◇ l-CANAVANINE

TOXICITY DATA with REFERENCE
mmo-omi 10 mg/L MUREAV 12,349,71
dnd-hmn:hla 200 µmol/L ECREAL 107,191,77
dni-ham:oth 2200 µmol/L JCLLAX 75,129,70
unr-mus LD50:200 mg/kg 85GDA2 8,100,82

SAFETY PROFILE: Poison by an unspecified route. Human mutation data reported. When heated to decomposition it emits toxic fumes of NO_x.

AKD600 CAS:6411-75-2 ***HR: 3***
2-AMINOHEPTANE SULFATE
mf: $C_{14}H_{34}N_2 \cdot H_2O_4S$ mw: 328.58

SYNS: 2-HEPTANAMINE SULFATE (2:1) ◇ 2-HEPTYLAMINE SULFATE ◇ 1-METHYLHEXYLAMINE SULFATE ◇ TUAMINE SULFATE ◇ TUAMINOHEPTANE SULFATE

TOXICITY DATA with REFERENCE
ipr-rat LD50:60 mg/kg JPETAB 81,235,44
ivn-rat LD50:47300 µg/kg JAPMA8 42,107,53
ipr-mus LD50:163 mg/kg JPETAB 98,300,50
scu-mus LD50:100 mg/kg JAPMA8 39,12,50
ivn-mus LD50:16300 µg/kg JAPMA8 42,107,53

SAFETY PROFILE: Poison by subcutaneous, intravenous, and intraperitoneal routes. When heated to decomposition it emits toxic fumes of SO_x and NO_x. See also AMINES and SULFATES.

AKD625 CAS:7790-12-7 ***HR: 2***
7-AMINOHEPTANOIC ACID, ISOPROPYL ESTER
mf: $C_{10}H_{21}NO_2$ mw: 187.32

TOXICITY DATA with REFERENCE
skn-rbt 100 µg/24H open AIHAAP 23,95,62
orl-rat LD50:4000 mg/kg AIHAAP 23,95,62
skn-rbt LD50:890 mg/kg AIHAAP 23,95,62

SAFETY PROFILE: Moderately toxic by ingestion and skin contact. A skin irritant. When heated to decomposition it emits toxic fumes of NO_x. See also ESTERS.

AKD750 CAS:2009-03-2 ***HR: 3***
3-(7-AMINOHEPTYL)INDOLE ADIPATE

TOXICITY DATA with REFERENCE
ipr-mus LD50:285 mg/kg RPTOAN 33,180,70
ivn-mus LD50:79 mg/kg RPTOAN 33,180,70

SAFETY PROFILE: Poison by intraperitoneal and intravenous routes. When heated to decomposition it emits toxic fumes of NO_x.

AKD775 CAS:90043-86-0 ***HR: 3***
9-AMINO-2,3,5,6,7,8-HEXAHYDRO-1H-CYCLO PENTA(b)QUINOLINE HYDROCHLORIDE HYDRATE
mf: $C_{12}H_{16}N_2 \cdot ClH \cdot H_20$ mw: 242.78

SYN: 2,3,5,6,7,8-HEXAHYDRO-9-AMINO-1H-CYCLOPENTA(b)QUIN OLINE HYDROCHLORIDE HYDRATE

TOXICITY DATA with REFERENCE
scu-rat LD50:60 mg/kg BAXXDU #2125696
orl-mus LD50:68 mg/kg BAXXDU #2125696
ipr-mus LD50:44 mg/kg BAXXDU #2125696
scu-mus LD50:52 mg/kg BAXXDU #2125696

SAFETY PROFILE: Poison by ingestion, subcutaneous, and intraperitoneal routes. When heated to decomposition it emits toxic fumes of NO_x and HCl. See also AMINES.

AKD875 CAS:60145-64-4 ***HR: 3***
3-AMINO-4-HOMOISOTWISTANE
mf: $C_{11}H_{19}N$ mw: 165.31

SYN: OCTAHYDRO-1,6-METHANONAPHTHALEN-1(2H)-AMINE

TOXICITY DATA with REFERENCE
orl-mus LD50:550 mg/kg JKXXAF #78-50338
ipr-mus LD50:102 mg/kg JKXXAF #78-50338
ivn-mus LD50:40 mg/kg JKXXAF #78-50338

SAFETY PROFILE: Poison by intravenous and intraperitoneal routes. Moderately toxic by ingestion. When heated to decomposition it emits toxic fumes of NO_x. See also AMINES.

AKE000 CAS:4502-10-7 ***HR: 3***
2-AMINO-3-HYDROXYACETOPHENONE
mf: $C_8H_9NO_2$ mw: 151.18

SYN: 2-AMINO-3-HYDROXYPHENYL METHYL KETONE

TOXICITY DATA with REFERENCE
imp-mus TDLo:80 mg/kg:NEO BJCAAI 11,212,57

SAFETY PROFILE: Questionable carcinogen with experimental neoplastigenic data. See also KETONES. When heated to decomposition it emits toxic fumes of NO_x.

AKE250 CAS:116-85-8 ***HR: 3***
1-AMINO-4-HYDROXYANTHRAQUINONE
mf: $C_{14}H_9NO_3$ mw: 239.24

PROP: Red-violet powder. Mp: 207-208°. Sol in water, HCl, alc, ether, and benzene.

SYNS: 1-AMINO-4-OXYANTHRAQUINONE (RUSSIAN) ◇ C.I. 60710 ◇ 1-HYDROXY-4-AMINOANTHRAQUINONE ◇ 4-HYDROXY-1-ANTHRAQUINONYLAMINE

TOXICITY DATA with REFERENCE
mmo-sat 100 μg/plate MUREAV 40,203,76
mma-sat 100 μg/plate MUREAV 40,203,76
ipr-rat LD50:2700 mg/kg GTPZAB 21(12),27,77
ipr-mus LDLo:256 mg/kg CBCCT* 2,58,50
ivn-mus LD50:56 mg/kg CSLNX* NX#00428

CONSENSUS REPORTS: Reported in EPA TSCA Inventory.

SAFETY PROFILE: Poison by intraperitoneal route. Mutation data reported. When heated to decomposition it emits toxic fumes of NO_x.

AKE500 CAS:103-18-4 ***HR: 3***
4-AMINO-4'-HYDROXYAZOBENZENE
mf: $C_{12}H_{11}N_3O$ mw: 213.26

TOXICITY DATA with REFERENCE
orl-rat TDLo:28 g/kg/2Y-C:ETA AABIAV 52,33,63
orl-rat LD50:1950 mg/kg AABIAV 52,33,63
ipr-rat LD50:300 mg/kg AABIAV 52,33,63

CONSENSUS REPORTS: Reported in EPA TSCA Inventory.

SAFETY PROFILE: Poison by intraperitoneal route. Moderately toxic by ingestion. Questionable carcinogen with experimental tumorigenic data. When heated to decomposition it emits toxic fumes of NO_x.

AKE750 CAS:548-93-6 ***HR: 3***
2-AMINO-3-HYDROXYBENZOIC ACID
mf: $C_7H_7NO_3$ mw: 153.15

SYNS: 3-HYDROXYANTHRANILIC ACID ◇ 3-HYDROXY-ANTHRANILSAEURE (GERMAN) ◇ 3-OHAA ◇ 3-OXYANTHRANILIC ACID

TOXICITY DATA with REFERENCE
cyt-hmn:emb 30 mg/L BEXBAN 67,200,69
cyt-hmn:leu 100 mg/L TSITAQ 15,1505,73
scu-mus TDLo:1600 mg/kg/8W-I:CAR VOONAW 22(6),47,76
scu-mus TDLo:185 mg/kg (13-17D post):TER JCREA8 96,163,80
scu-dog TDLo:500 mg/kg/20W-I:ETA VOONAW 26(3),93,80
imp-mus TD:80 mg/kg:NEO BJCAAI 11,212,57
scu-mus TD:2000 mg/kg/7W-I:CAR PGPKA8 14,12,69
imp-mus TDLo:160 mg/kg ANYAA9 108,924,63

SAFETY PROFILE: Questionable carcinogen with experimental carcinogenic, neoplastigenic, and tumorigenic data. An experimental teratogen. Human mutation data reported. When heated to decomposition it emits toxic fumes such as NO_x.

AKF000 CAS:536-25-4 ***HR: 3***
3-AMINO-4-HYDROXYBENZOIC ACID METHYL ESTER
mf: $C_8H_9NO_3$ mw: 167.18

SYNS: AMINOBENZ ◇ ORTHOCAINE ◇ ORTHODERM ◇ ORTHOFORM

TOXICITY DATA with REFERENCE
orl-dog LDLo:1 g/kg HBAMAK 4,1289,35
ipr-dog LDLo:250 mg/kg HBAMAK 4,1289,35

CONSENSUS REPORTS: Reported in EPA TSCA Inventory.

SAFETY PROFILE: Poison by intraperitoneal route. Moderately toxic by ingestion. See also ESTERS. When heated to decomposition it emits toxic fumes of NO_x.

AKF250 CAS:73728-82-2 ***HR: 3***
4-AMINO-3-HYDROXYBIPHENYL SULFATE
mf: $C_{12}H_{11}NO \cdot H_2O_4S$ mw: 283.32

SYNS: 4-AMINO-3-BIPHENYLOL HYDROGEN SULFATE ◇ 3-HYDROXY-4-AMINODIPHENYLSULPHATE

TOXICITY DATA with REFERENCE
imp-mus TDLo:100 mg/kg:CAR BJCAAI 10,539,56

SAFETY PROFILE: Questionable carcinogen with experimental carcinogenic data. See also SULFATES. When heated to decomposition it emits very toxic fumes of SO_x and NO_x.

AKF375 CAS:352-21-6 ***HR: 1***
4-AMINO-3-HYDROXYBUTYRIC ACID
mf: $C_4H_8NO_3$ mw: 118.13

PROP: dl-Form: Crystals from dil alc. Decomp 218°. Sol in water; very sparingly sol in methanol, alc, ether, chloroform, ethyl acetate. d(+)-Form: Crystals from water. Decomp 214°. l(−)-Form: Crystals from water or water + ethanol. Decomp 212°.

SYNS: Γ-AMINO-β-HYDROXYBUTYRIC ACID ◇ BUKSAMIN ◇ GABOB ◇ GABOMADE ◇ GAMIBETAL ◇ β-HYDROXY-α-AMINOBUTYRIC ACID ◇ β-OXY-GABA

TOXICITY DATA with REFERENCE
ipr-rat LD50:7000 mg/kg NIIRDN 6,195,82
ipr-mus LD50:7000 mg/kg NIIRDN 6,195,82
scu-mus LD50:7000 mg/kg NIIRDN 6,195,82
ivn-mus LD50:7000 mg/kg NIIRDN 6,195,82
unr-mus LD50:7080 mg/kg BTMNA7 25,297,62

SAFETY PROFILE: Mildly toxic by several routes. When heated to decomposition it emits toxic fumes of NO_x. See also AMINES.

AKF500 CAS:64058-65-7 ***HR: 2***
3-AMINO-4-(2-HYDROXY)ETHOXY BENZENARSONIC ACID
mf: $C_8H_{12}AsNO_5$ mw: 277.13

TOXICITY DATA with REFERENCE
ivn-rat LDLo:1700 mg/kg JPETAB 63,122,38
ims-rat LDLo:2000 mg/kg JPETAB 63,122,38

CONSENSUS REPORTS: Arsenic compounds are on the Community Right-To-Know List.

OSHA PEL: TWA 0.5 mg(As)/m^3

SAFETY PROFILE: Moderately toxic by intravenous and intramuscular routes. See also ARSENIC COMPOUNDS. When heated to decomposition it emits very toxic fumes of NO_x and As.

AKF750 CAS:64048-94-8 ***HR: 3***
(3-AMINO-4-(2-HYDROXYETHOXY)PHENYL) ARSINE OXIDE
mf: $C_8H_{12}AsNO_3$ mw: 245.13

TOXICITY DATA with REFERENCE
orl-rat LDLo:200 mg/kg JPETAB 63,122,38
ivn-rat LDLo:15 mg/kg JPETAB 63,122,38
ims-rat LDLo:16 mg/kg JPETAB 63,122,38

CONSENSUS REPORTS: Arsenic compounds are on the Community Right-To-Know List.

OSHA PEL: TWA 0.5 mg(As)/m^3

SAFETY PROFILE: Poison by ingestion, intravenous, and intramuscular routes. See also ARSENIC COMPOUNDS. When heated to decomposition it emits very toxic fumes of NO_x and As.

AKG000 CAS:69226-39-7 ***HR: 1***
4-AMINO-N-(2-HYDROXYETHYL)-o-TOLUENE SULFONAMIDE
mf: $C_9H_{14}N_2O_3S$ mw: 230.31

SYN: N-HYDROXYETHYLAMID KYSELINY 4-AMINOTOLUEN-2-SULFONOVE (CZECH)

TOXICITY DATA with REFERENCE
skn-rbt 500 mg/24H MLD 28ZPAK -,200,72
eye-rbt 100 mg/24H SEV 28ZPAK -,200,72
orl-rat LD50:8900 mg/kg 28ZPAK -,200,72

SAFETY PROFILE: Mildly toxic by ingestion. A skin and severe eye irritant. When heated to decomposition it emits very toxic fumes of NO_x and SO_x.

AKG250 CAS:60573-88-8 ***HR: 3***
α-AMINO-3-HYDROXY-5-ISOXAZOLEACETIC ACID HYDRATE
mf: $C_5H_6N_2O_4 \cdot H_2O$ mw: 176.15

SYNS: α-AMINO-2,3-DIHYDRO-3-OXO-5-ISOXAZOLEACETICACID ◇ α-AMINO-3-HYDROXY-5-ISOXAZOLESSIGSAURE HYDRAT (GERMAN) ◇ AMINO-(3-HYDROXY-5-ISOXAZOLYL)ACETIC ACID ◇ IBOTENIC ACID ◇ IBOTENSAURE (GERMAN) ◇ ISOTENIC ACID ◇ PRAMUSCIMOL

TOXICITY DATA with REFERENCE
orl-rat LD50:129 mg/kg ARZNAD 18,311,68
ivn-rat LD50:42 mg/kg ARZNAD 18,311,68
orl-mus LD50:38 mg/kg ARZNAD 18,311,68
ivn-mus LD50:15 mg/kg ARZNAD 18,311,68
scu-rbt LDLo:45 mg/kg AIPTAK 5,161,1899
scu-frg LDLo:4000 mg/kg AIPTAK 5,161,1899

SAFETY PROFILE: Poison by ingestion, subcutaneous, and intravenous routes. When heated to decomposition it emits toxic fumes of NO_x.

AKG500 CAS:4439-84-3 ***HR: 1***
2-AMINO-5-HYDROXYLEVULINIC ACID
mf: $C_5H_9NO_4$ mw: 147.15

SYNS: 2-AMINO-5-HYDROXY-4-OXOPENTANOIC ACID ◇ H-899 ◇ HON ◇ Δ-HYDROXY-γ-OXO-l-NORVALINE ◇ 5-HYDROXY-4-OXO-NORVALINE

TOXICITY DATA with REFERENCE
orl-mus LD50:7600 mg/kg JAJAAA 14,39,61
scu-mus LD50:8000 g/kg JAJAAA 14,39,61
ivn-mus LD50:5200 mg/kg JAJAAA 14,39,61

SAFETY PROFILE: Mildly toxic by ingestion, subcutaneous, and intravenous routes. When heated to decomposition it emits toxic fumes of NO_x.

AKH000 CAS:90-20-0 ***HR: 1***
4-AMINO-5-HYDROXY-2,7-NAPHTHALENEDISULFONIC ACID
mf: $C_{10}H_9NO_7S_2$ mw: 319.32

SYNS: C.I. 35570 ◇ H ACID ◇ KYSELINA 1-AMINO-8-NAFTOL-3,6-DISULFONOVA (CZECH) ◇ KYSELINA H (CZECH)

TOXICITY DATA with REFERENCE
eye-rbt 500 mg/24H MLD 28ZPAK -,189,72

CONSENSUS REPORTS: Reported in EPA TSCA Inventory.

SAFETY PROFILE: An eye irritant. See also SULFONATES. When heated to decomposition it emits very toxic fumes of NO_x and SO_x.

AKH250 CAS:2007-20-7 ***HR: D***
8-AMINO-7-HYDROXY-3,6-NAPHTHALENEDISULFONIC ACID, SODIUM SALT
mf: $C_{10}H_7NO_7S_2{\cdot}2Na$ mw: 363.28

SYNS: 1-AMINO-2-NAPHTHOL-3,6-DISULPHONIC ACID SODIUM SALT ◇ SODIUM SALT of 1-AMINO-2-NAPHTHOL-3,6-DISULPHONIC ACID

TOXICITY DATA with REFERENCE
orl-rat TDLo:600 mg/kg (1-20D preg):REP FCTXAV 11,355,73
orl-rat TDLo:300 mg/kg (1-20D preg):TER FCTXAV 11,355,73

SAFETY PROFILE: An experimental teratogen. Other experimental reproductive effects. When heated to decomposition it emits very toxic fumes of SO_x, NO_x, and Na_2O. See also SULFONATES.

AKH500 CAS:6837-93-0 ***HR: 1***
4-AMINO-5-HYDROXY-2,7-NAPHTHALENEDISULFONIC ACID-p-TOLUENESULFONATE (ESTER)
mf: $C_{17}H_{15}NO_9S_3$ mw: 473.51

SYNS: 1-AMINO-3,6-DISULFO-8-NAFTYLESTER KYSELINA p-TOLUENSULFONOVE (CZECH) ◇ KYSELINA o-TOSYL-H (CZECH)

TOXICITY DATA with REFERENCE
eye-rbt 100 mg/24H SEV 28ZPAK -,194,72
orl-rat LD50:11500 mg/kg 28ZPAK -,194,72

SAFETY PROFILE: Mildly toxic by ingestion. A severe eye irritant. See also ESTERS and SULFONATES. When heated to decomposition it emits very toxic fumes of NO_x and SO_x.

AKH750 CAS:83-64-7 ***HR: 1***
4-AMINO-5-HYDROXY-1-NAPHTHALENESULFONIC ACID
mf: $C_{10}H_9NO_4S$ mw: 239.26

PROP: Needles. Sol in water.

SYNS: AMINONAPHTHOL SULFONIC ACID S ◇ CHICAGO ACID S ◇ KYSELINA 1-AMINO-8-NAFTOL-4-SULFONOVA (CZECH)

TOXICITY DATA with REFERENCE
eye-rbt 500 mg/24H MLD 28ZPAK -,188,72
orl-rat LD50:6210 mg/kg 28ZPAK -,188,72

SAFETY PROFILE: Mildly toxic by ingestion. An eye irritant. See also SULFONATES. When heated to decomposition it emits very toxic fumes of NO_x and SO_x.

AKI000 CAS:87-02-5 ***HR: 1***
7-AMINO-4-HYDROXY-2-NAPHTHALENESULFONIC ACID
mf: $C_{10}H_9NO_4S$ mw: 239.26

SYNS: AMINONAPHTHOL SULFONIC ACID J ◇ I ACID ◇ ISOGAMMA ACID ◇ KYSELINA 2-AMINO-5-NAFTOL-7-SULFONOVA (CZECH)

TOXICITY DATA with REFERENCE
eye-rbt 500 mg/24H MLD 28ZPAK -,188,72
orl-rat LD50:11500 mg/kg 28ZPAK -,188,72

CONSENSUS REPORTS: Reported in EPA TSCA Inventory.

SAFETY PROFILE: Mildly toxic by ingestion. An eye irritant. See also SULFONATES. When heated to decomposition it emits very toxic fumes of NO_x and SO_x.

AKI250 CAS:96-93-5 ***HR: 1***
3-AMINO-4-HYDROXY-5-NITROBENZENESULFONIC ACID
mf: $C_6H_6N_2O_6S$ mw: 234.20

SYN: KYSELINA 6-NITRO-2-AMINOFENOL-4-SULFONOVA (CZECH)

TOXICITY DATA with REFERENCE
skn-rbt 500 mg/24H MLD 28ZPAK -,182,72
eye-rbt 20 mg/24H MOD 28ZPAK -,182,72
orl-rat LDLo:5360 mg/kg 28ZPAK -,182,72

SAFETY PROFILE: Mildly toxic by ingestion. A skin and eye irritant. See also SULFONATES. When heated to decomposition it emits very toxic fumes of SO_x and NO_x.

AKI500 CAS:63019-81-8 ***HR: 3***
4-AMINO-3-HYDROXY-4'-NITRODIPHENYLHYDROCHLORIDE
mf: $C_{12}H_{10}N_2O_3{\cdot}ClH$ mw: 266.70

SYNS: 4-AMINO-4'-NITRO-3-BIPHENYLOL HYDROCHLORIDE ◇ 4'-NITRO-4-AMINO-3-HYDROXYDIPHENYL HYDROCHLORIDE ◇ 4'-NITRO-4-AMINO-3-HYDROXYDIPHENYL HYDROGEN CHLORIDE

TOXICITY DATA with REFERENCE
imp-mus TDLo:100 mg/kg:ETA BJCAAI 10,539,56

SAFETY PROFILE: Questionable carcinogen with experimental tumorigenic data. When heated to decomposition it emits very toxic fumes of HCl and NO_x.

AKI750 CAS:17418-58-5 ***HR: 1***
1-AMINO-4-HYDROXY-2-PHENOXYANTHRAQUINONE
mf: $C_{20}H_{13}NO_4$ mw: 331.34

SYN: CERVEN BRILANTNI OSTACETOVA F-LB (CZECH)

TOXICITY DATA with REFERENCE
eye-rbt 500 mg/24H MLD 28ZPAK -,239,72

CONSENSUS REPORTS: Reported in EPA TSCA Inventory.

SAFETY PROFILE: An eye irritant. When heated to decomposition it emits toxic fumes of NO_x.

AKI900 CAS:102516-61-0 ***HR: 3***
3-(((3-AMINO-4-HYDROXYPHENYL) PHENYLARSINO)THIO)ALANINE
mf: $C_{15}H_{17}AsN_2O_3S$ mw: 380.32

SYN: ALANINE, 3-(((3-AMINO-4-HYDROXYPHENYL)PHENYLARSINO)THIO)-

TOXICITY DATA with REFERENCE
ivn-mus LDLo:40 mg/kg PHBUA9 2,19,54

OSHA PEL: TWA 0.5 mg(As)/m^3

SAFETY PROFILE: Poison by intravenous route. When heated to decomposition it emits toxic fumes of NO_x, SO_x, and AS.

AKJ000 CAS:58152-03-7 ***HR: 3***
1-N-(S-3-AMINO-2-HYDROXYPROPIONYL) BETAMYCIN
mf: $C_{22}H_{43}N_5O_{12}$ mw: 569.70

SYN: 1-N-(S-3-AMINO-2-HYDROXYPROPIONYL) GENTAMYCIN B

TOXICITY DATA with REFERENCE
ipr-mus LD50:5000 mg/kg DRFUD4 4,525,79
ivn-mus LD50:330 mg/kg DRFUD4 4,525,79

SAFETY PROFILE: Poison by intravenous route. Mildly toxic by intraperitoneal route. When heated to decomposition it emits toxic fumes of NO_x.

AKJ250 CAS:5423-12-1 ***HR: 2***
3-AMINO-4(1-(2-HYDROXY)PROPOXY)BENZENEARSONIC ACID
mf: $C_9H_{14}AsNO_5$ mw: 291.16

TOXICITY DATA with REFERENCE
ivn-rat LDLo:2500 mg/kg JPETAB 63,122,38
ims-rat LDLo:3000 mg/kg JPETAB 63,122,38

CONSENSUS REPORTS: Arsenic compounds are on the Community Right-To-Know List.

OSHA PEL: TWA 0.5 mg(As)/m^3

SAFETY PROFILE: Moderately toxic by intravenous and intramuscular routes. See also ARSENIC COMPOUNDS. When heated to decomposition it emits very toxic fumes of NO_x and As.

AKJ500 CAS:63717-25-9 ***HR: 2***
S-3-AMINO-2-HYDROXYPROPYL SODIUMHYDROGEN PHOSPHOROTHIOATETETRAHYDRATE
mf: $C_3H_9NO_4PS{\cdot}Na{\cdot}4H_2O$ mw: 281.23

TOXICITY DATA with REFERENCE
orl-mus LD50:2500 mg/kg JMCMAR 18,803,75
ipr-mus LD50:2200 mg/kg JMCMAR 18,803,75

SAFETY PROFILE: Moderately toxic by ingestion and intraperitoneal routes. When heated to decomposition it emits very toxic fumes of NO_x, PO_x, Na_2O, and SO_x.

AKJ750 CAS:2835-95-2 ***HR: 2***
4-AMINO-2-HYDROXYTOLUENE
mf: C_7H_9NO mw: 123.17

SYNS: 5-AMINO-o-CRESOL ◇ 5-AMINO-2-METHYLPHENOL

TOXICITY DATA with REFERENCE
orl-qal LD50:750 mg/kg AECTCV 12,355,83
orl-rat LD50:3600 mg/kg FCTXAV 15,607,77

CONSENSUS REPORTS: Reported in EPA TSCA Inventory. EPA Genetic Toxicology Program.

SAFETY PROFILE: Moderately toxic by ingestion. When heated to decomposition it emits toxic fumes of NO_x. See also PHENOLS.

AKK000 CAS:21644-95-1 ***HR: 3***
4-AMINO-4'-HYDROXY-2,3',5'-TRIMETHYLAZOBENZENE
mf: $C_{15}H_{17}N_3O$ mw: 255.35

TOXICITY DATA with REFERENCE
orl-rat LDLo:600 mg/kg AABIAV 52,33,63
ipr-rat LD50:142 mg/kg AABIAV 52,33,63

SAFETY PROFILE: Poison by intraperitoneal route. Moderately toxic by ingestion. When heated to decomposition it emits toxic fumes of NO_x.

AKK250 CAS:360-97-4 ***HR: 3***
5-AMINOIMIDAZOLE-4-CARBOXAMIDE
mf: $C_4H_6N_4O$ mw: 126.14

SYN: DIAZOL-C

TOXICITY DATA with REFERENCE
mma-sat 10 μg/plate JTEHD6 2,1095,77
mmo-klp 5 mmol/L/20H MUREAV 66,297,79
orl-rat TDLo:6390 mg/kg/21W-C:ETA JNCIAM 54,951,75

CONSENSUS REPORTS: Reported in EPA TSCA Inventory.

SAFETY PROFILE: Questionable carcinogen with experimental tumorigenic data. Mutation data reported.

When heated to decomposition it emits toxic fumes of NO_x.

AKK625 CAS:34879-34-0 ***HR: 2***
5-AMINO-4-IMIDAZOLECARBOXAMIDE UREIDOSUCCINATE
mf: $C_5H_8N_2O_5 \cdot CaH_6N_4O$ mw: 302.29

SYN: CARBAICA

TOXICITY DATA with REFERENCE
ivn-rat LD50:1190 mg/kg DECRDP 6,471,80
ivn-mus LD50:640 mg/kg DECRDP 6,471,80
ivn-gpg LD50:440 mg/kg DECRDP 6,471,80

SAFETY PROFILE: Moderately toxic by intravenous route. When heated to decomposition it emits toxic fumes of NO_x.

AKK750 CAS:581-64-6 ***HR: 3***
7-AMINO-3-IMINO-3H-PHENOTHIAZINEMONOHYDROCHLORIDE
mf: $C_{12}H_9N_3S \cdot ClH$ mw: 263.76

SYNS: 3,7-DIAMINOPHENOTHIAZIN-5-IUMCHLORIDE ◇ KATALYSIN ◇ LAUTHSCHES VIOLETT (GERMAN) ◇ THIONIN ◇ THIONINE

TOXICITY DATA with REFERENCE
dnr-bcs 2 mg/disc TRENAF 27,153,76
ipr-rat LDLo:100 mg/kg AEPPAE 182,118,36
ipr-mus LD50:400 mg/kg NTIS** AD691-490

CONSENSUS REPORTS: Reported in EPA TSCA Inventory. EPA Genetic Toxicology Program.

SAFETY PROFILE: Poison by intraperitoneal route. Mutation data reported. When heated to decomposition it emits very toxic fumes of NO_x, SO_x, and HCl.

AKL000 CAS:2338-18-3 ***HR: 3***
2-AMINOINDANE HYDROCHLORIDE
mf: $C_9H_{11}N \cdot ClH$ mw: 169.67

SYNS: 2-AMINOINDAN HYDROCHLORIDE ◇ 2,3-DIHYDRO-1H-INDEN-2-AMINE HYDROCHLORIDE (9CI) ◇ 2-INDANYLAMINE HYDROCHLORIDE ◇ SU 8629 HYDROCHLORIDE

TOXICITY DATA with REFERENCE
orl-mus LD50:500 mg/kg JPETAB 133,400,61
ipr-mus LD50:170 mg/kg JPETAB 133,400,61
scu-mus LD50:158 mg/kg JPETAB 133,400,61
orl-dog LDLo:50 mg/kg JPETAB 133,400,61

SAFETY PROFILE: Poison by ingestion, subcutaneous, and intraperitoneal routes. When heated to decomposition it emits toxic fumes of NO_x and HCl. See also AMINES.

AKL100 CAS:13935-78-9 ***HR: 3***
AMINOINDANOL HYDROCHLORIDE
mf: $C_9H_{11}NO \cdot ClH$ mw: 185.67

SYN: 2-AMINO-1-HYDROXYHYDROINDENE HYDROCHLORIDE

TOXICITY DATA with REFERENCE
scu-rat LD50:350 mg/kg AEPPAE 169,114,33
scu-mus LD50:800 mg/kg AEPPAE 169,114,33
scu-gpg LD50:400 mg/kg AEPPAE 169,114,33

SAFETY PROFILE: Poison by subcutaneous route. When heated to decomposition it emits very toxic fumes of NO_x and HCl.

AKL250 CAS:64309-76-8 ***HR: D***
5-AMINOINDAZOLE HYDROCHLORIDE
mf: $C_8H_7N_3 \cdot ClH$ mw: 181.64

SYN: 5-AMINO-1H-IMIDAZOLE HYDROCHLORIDE

TOXICITY DATA with REFERENCE
spm-rat-orl 1600 mg/kg/2W JRPFA4 50,371,77
spm-rat-scu 200 mg/kg JRPFA4 50,371,77
orl-rat TDLo:1600 mg/kg (8D male):REP JRPFA4 50,371,77

SAFETY PROFILE: Experimental reproductive effects. Mutation data reported. When heated to decomposition it emits very toxic fumes of HCl and NO_x.

AKL500 CAS:64037-13-4 ***HR: 3***
2-AMINO-5-IODOBENZOXAZOLE
mf: $C_7H_5IN_2O$ mw: 260.04

TOXICITY DATA with REFERENCE
orl-rat LD50:1000 mg/kg MDCHAG 4(1),338,64
orl-mus LD50:800 mg/kg MDCHAG 4(1),336,64
ipr-mus LD50:150 mg/kg MDCHAG 4(1),336,64

SAFETY PROFILE: Poison by intraperitoneal route. Moderately toxic by ingestion. See also IODIDES. When heated to decomposition it emits very toxic fumes of I^- and NO_x.

AKL625 CAS:550-28-7 ***HR: 2***
AMINOISOMETRADIN
mf: $C_9H_{13}N_3O_2$ mw: 195.25

PROP: Crystals. Mp: 175°. Solubility in water (25°): 2.0 g/100 mL. Freely sol in alc and acetone; insol in ether.

SYNS: AMINOISOMETRADINE ◇ 6-AMINO-1-METALLYL-3-METHYLPYRIMIDINE-2,4-DIONE ◇ 6-AMINO-3-METHYL-1-(2-METHYLALLYL)-2,4(1H,3H)-PYRIMIDINEDIONE ◇ 6-AMINO-3-METHYL-1-(2-METHYLALLYL)URACIL ◇ 6-AMINO-3-METHYL-1-(2-METHYL-2-PROPENYL)-2,4(1H,3H)-PYRIMIDINEDIONE ◇ AMISOMETRADIN ◇ AMISOMETRADINE ◇ 1-METHALLYL-3-METHYL-6-AMINOTETRAHYDROPYRIMIDINEDIONE ◇ ROLICTON

TOXICITY DATA with REFERENCE
orl-rat LD50:1560 mg/kg AIPTAK 126,400,60

ipr-rat LD50:515 mg/kg AIPTAK 126,400,60
orl-mus LD50:610 mg/kg MEIEDD 10,73,83
ipr-mus LD50:415 mg/kg MEIEDD 10,73,83

SAFETY PROFILE: Moderately toxic by ingestion and other routes. When heated to decomposition it emits toxic fumes of NO_x.

AKL750 CAS:55217-61-3 ***HR: 3***
trans-1-AMINO-2-MERCAPTOMETHYLCYCLO-BUTANEHYDROCHLORIDE
mf: $C_5H_{11}NS•ClH$ mw: 153.69

SYN: (trans)-2-MERCAPTOMETHYLCYCLOBUTYLAMINEHYDRO-CHLORIDE

TOXICITY DATA with REFERENCE
orl-mus LD50:500 mg/kg JMCMAR 18,323,75
ipr-mus LD50:250 mg/kg JMCMAR 18,323,75

SAFETY PROFILE: Poison by intraperitoneal route. Moderately toxic by ingestion. When heated to decomposition it emits very toxic fumes of HCl, SO_x, and NO_x.

AKM000 CAS:2349-67-9 ***HR: 3***
2-AMINO-5-MERCAPTO-1,3,4-THIADIAZOLE
mf: $C_2H_3N_3S_2$ mw: 133.20

SYNS: 5-AMINO-2-MERCAPTO-1,3,4-THIADIAZOLE ◇ 5-AMINO-1,3,4-THIADIAZOLE-2-THIOL ◇ 2-AMINO-1,3,4-THIADIAZOLE-5-THIOL ◇ 2-AMINO-$\Delta^{(2)}$-1,3,4-THIADIAZOLINE-5-THIONE ◇ 5-AMINO-1,3,4-THIADIAZOLINE-2-THIONE ◇ 2-MERCAPTO-5-AMINO-1,3,4-THIADIAZOLE ◇ NSC 21402 ◇ USAF PD-25

TOXICITY DATA with REFERENCE
ipr-mus LD50:250 mg/kg NTIS** AD691-490

CONSENSUS REPORTS: Reported in EPA TSCA Inventory.

SAFETY PROFILE: Poison by intraperitoneal route. When heated to decomposition it emits very toxic fumes of NO_x and SO_x.

AKM125 CAS:963-34-8 ***HR: 3***
6-AMINOMETHAQUALONE
mf: $C_{16}H_{15}N_3O$ mw: 265.34

SYNS: 6-AMINO-2-METHYL-3-(o-TOLYL)-1(3H)-QUINAZOLINONE ◇ HB-218 ◇ 2-METHYL-3-o-TOLYL-6-AMINO-CHINAZOLINON-4 (GERMAN)

TOXICITY DATA with REFERENCE
orl-rat LD50:60 mg/kg ARZNAD 21,362,71
ipr-rat LD50:64 mg/kg ARZNAD 21,362,71
orl-mus LD50:125 mg/kg ARZNAD 21,362,71
ipr-mus LD50:91 mg/kg ARZNAD 21,362,71

SAFETY PROFILE: Poison by ingestion and intraperitoneal routes. When heated to decomposition it emits toxic fumes of NO_x.

AKM250 CAS:10165-33-0 ***HR: 3***
1-AMINO-2-METHOXYANTHRAQUINONE
mf: $C_{15}H_{11}NO_3$ mw: 253.27

SYN: 1-A-2-MA (RUSSIAN)

TOXICITY DATA with REFERENCE
mma-mus:lym 50 mg/L/4H NTIS** AD-A064-953
msc-mus:lym 200 mg/L/4H NTIS** AD-A064-953
ipr-rat LD50:300 mg/kg GTPZAB 21(12),27,77

SAFETY PROFILE: Poison by intraperitoneal route. Mutation data reported. When heated to decomposition it emits toxic fumes of NO_x.

AKM750 CAS:5464-79-9 ***HR: 3***
2-AMINO-4-METHOXYBENZOTHIAZOLE
mf: $C_8H_8N_2OS$ mw: 180.24

SYN: 4-METHOXY-2-AMINOBENZOTHIAZOLE

TOXICITY DATA with REFERENCE
mma-sat 10 μg/L CNREA8 39,682,79
orl-mus LD50:562 mg/kg JPETAB 105,486,52
ivn-mus LD50:46 mg/kg JPETAB 105,486,52

SAFETY PROFILE: Poison by intravenous route. Moderately toxic by ingestion. Mutation data reported. When heated to decomposition it emits very toxic fumes as NO_x and SO_x.

AKN000 CAS:64037-14-5 ***HR: 3***
2-AMINO-5-METHOXYBENZOXAZOLE
mf: $C_8H_8N_2O_2$ mw: 164.18

TOXICITY DATA with REFERENCE
orl-rat LD50:1000 mg/kg MDCHAG 4(1),338,64
ipr-rat LD50:268 mg/kg MDCHAG 4(1),338,64
orl-mus LD50:1090 mg/kg MDCHAG 4(1),336,64
ipr-mus LD50:432 mg/kg MDCHAG 4(1),336,64

SAFETY PROFILE: Poison by intraperitoneal route. Moderately toxic by ingestion. When heated to decomposition it emits toxic fumes as NO_x.

AKN250 CAS:63040-25-5 ***HR: 3***
4-AMINO-4'-METHOXY-3-BIPHENYLOL-HYDROCHLORIDE
mf: $C_{13}H_{13}NO_2•ClH$ mw: 251.73

SYN: 3-HYDROXY-4'-METHOXY-4-AMINODIPHENYL HYDROCHLORIDE

TOXICITY DATA with REFERENCE
imp-mus TDLo:80 mg/kg:ETA BJCAAI 17,127,63

SAFETY PROFILE: Questionable carcinogen with experimental tumorigenic data. When heated to decomposition it emits very toxic fumes of HCl and NO_x.

AKN500 CAS:951-39-3 ***HR: 3***
2-AMINO-3-METHOXYDIPHENYLENE OXIDE
mf: $C_{13}H_{11}NO_2$ mw: 213.25

SYN: 3-METHOXY-2-AMINODIPHENYLENEOXIDE

TOXICITY DATA with REFERENCE
orl-rat TDLo:15 g/kg/70W-I:CAR ZEKBAI 61,45,56
orl-rat TD:21 g/kg/76W-C:ETA JNCIAM 39,1069,67

SAFETY PROFILE: Questionable carcinogen with experimental carcinogenic and tumorigenic data. When heated to decomposition it emits toxic fumes of NO_x.

AKN750 CAS:6504-77-4 ***HR: 3***
4-AMINO-N-(2-METHOXYETHYL)-7-((2-METHOXYETHYL)AMINO-2-PHENYL-6-PTERIDINE-CARBOXAMIDE
mf: $C_{19}H_{23}N_7O_3$ mw: 397.49

SYN: WY 5256

TOXICITY DATA with REFERENCE
orl-rat LD50:300 mg/kg TXAPA9 18,185,71
orl-mus LD50:250 mg/kg TXAPA9 18,185,71

SAFETY PROFILE: Poison by ingestion. When heated to decomposition it emits toxic fumes of NO_x.

AKO000 CAS:68772-17-8 ***HR: 2***
6-AMINO-8-METHOXY-1-METHYL-4-(p-(p-((1-METHYLPYRIDINIUM-4-YL)AMINO)BENZAMIDO)ANILINOQUINOLINIUM DI-p-TOLUENESULFONATE
mf: $C_{30}H_{30}N_6O_2 \cdot 2C_7H_7O_3S$ mw: 849.06

TOXICITY DATA with REFERENCE
dnd-mus:lym 310 nmol/L JMCMAR 22,134,79
ipr-mus LD10:56 mg/kg JMCMAR 22,134,79

SAFETY PROFILE: Moderately toxic by intraperitoneal route. Mutation data reported. See also SULFONATES. When heated to decomposition it emits very toxic fumes of NO_x and SO_x.

AKO250 CAS:68772-43-0 ***HR: 3***
7-AMINO-4-(2-METHOXY-p-(p-((1-METHYLPYRIDINIUM-4-YL)AMINO)BENZAMIDO)ANILINO)-1-METHYLQUINOLINIUM) DIBROMIDE
mf: $C_{30}H_{30}N_6O_2 \cdot 2Br$ mw: 666.48

TOXICITY DATA with REFERENCE
dnd-mus:lym 690 nmol/L JMCMAR 22,134,79
ipr-mus LD10:14 mg/kg JMCMAR 22,134,79

SAFETY PROFILE: Poison by intraperitoneal route. Mutation data reported. See also BROMIDES. When heated to decomposition it emits very toxic fumes of NO_x and Br^-.

AKO500 CAS:80-35-3 ***HR: 2***
4-AMINO-N-(6-METHOXY-3-PYRIDAZINYL)-BENZENESULFONAMIDE
mf: $C_{11}H_{12}N_4O_3S$ mw: 280.33

SYNS: ALTEZOL ◇ 3-(p-AMINOBENZENESULFAMIDO)-6-METHOXYPYRIDAZINE ◇ 3-p-AMINOBENZENESULPHONAMIDO-7-METHOXYPYRIDAZINE ◇ CL 13494 ◇ DAVOSIN ◇ DEPOVERNIL ◇ DUROX ◇ KINEKS ◇ KINEX ◇ KYNEX ◇ LEDERKYN ◇ LENTAC ◇ LISULFEN ◇ LONGIN ◇ MEDICEL ◇ N^1-(6-METHOXY-3-PYRIDAZINYL)SULFANILAMIDE ◇ 6-METHOXY-3-SULFANILAMIDOPYRIDAZINE ◇ MIDICEL ◇ MIDIKEL ◇ MYASUL ◇ MYLOSUL ◇ OPINSUL ◇ PARAMID ◇ PARAMID SUPRA ◇ PETRISUL ◇ PIRIDOLO ◇ QUINOSEPTYL ◇ RETAMID ◇ RETASULFIN ◇ RP 7522 ◇ SLOSUL ◇ SMOP ◇ SMP ◇ SPOFADAZINE ◇ SULFALEX ◇ 3-SULFA-6-METHOXYPYRIDAZINE ◇ SULFAMETOXIPIRIDAZINE ◇ 3-SULFANILAMIDE-6-METHOXYPYRIDAZINE ◇ 3-SULFANILAMIDO-6-METHOXYPYRIDAZINE ◇ 6-SULFANILAMIDO-3-METHOXYPYRIDAZINE ◇ SULFAPYRIDAZINE ◇ SULFDURAZIN ◇ SULFMETHOXIPIRIDAZINE ◇ SULFOZONA ◇ SULPHAMETHOXYPYRIDAZINE ◇ SULTIRENE ◇ SURIRENE ◇ VINCES

TOXICITY DATA with REFERENCE
orl-rat TDLo:1200 mg/kg (9-14D preg):TER SEIJBO 13,7,73
orl-rat TDLo:8400 mg/kg (male 6W pre):REP JRPFA4 81,259,87
orl-rat LD50:2739 mg/kg ARZNAD 11,459,61
orl-mus LD50:1750 mg/kg ARZNAD 15,1441,65
ipr-mus LD50:1200 mg/kg RPTOAN 37,223,74
scu-mus LD50:4500 mg/kg ARZNAD 10,440,60
scu-mus LD50:4500 mg/kg ARZNAD 10,440,60

SAFETY PROFILE: Moderately toxic by ingestion and intraperitoneal routes. An experimental teratogen. Other experimental reproductive effects. See also SULFONATES. When heated to decomposition it emits very toxic fumes of NO_x and SO_x.

AKO750 CAS:3690-12-8 ***HR: 3***
4-AMINO-2-METHOXY-5-PYRIMIDINEMETHANOL
mf: $C_6H_9N_3O_2$ mw: 155.1

SYNS: 4-AMINO-5-HYDROXYMETHYL-2-METHOXYPYRIMIDINE ◇ BACIMETHRIN ◇ 2-METHOXY-4-AMINO-5-HYDROXYMETHYLPYRIMIDINE

TOXICITY DATA with REFERENCE
ipr-mus LD50:300 mg/kg 85ERAY 3,1597,78
ivn-mus LD50:300 mg/kg 85ERAY 3,1597,78

SAFETY PROFILE: Poison by intraperitoneal and intravenous routes. When heated to decomposition it emits toxic fumes of NO_x.

AKP250 CAS:1220-94-6 ***HR: 2***
4-AMINO-1-METHYLAMINOANTHRAQUINONE
mf: $C_{15}H_{12}N_2O_2$ mw: 252.29

SYNS: C.I. 61105 ◇ C.I. DISPERSE VIOLET 4 ◇ C.I. SOLVENT VIOLET 12

TOXICITY DATA with REFERENCE
ipr-rat LD50:1000 mg/kg GTPZAB 21(12),27,77

CONSENSUS REPORTS: Reported in EPA TSCA Inventory.

SAFETY PROFILE: Moderately toxic by intraperitoneal route. When heated to decomposition it emits toxic fumes of NO_x.

AKP500 CAS:17463-44-4 ***HR: 2***
dl-α-AMINO-β-METHYLAMINOPROPIONIC ACID
mf: $C_4H_{10}N_2O_2$ mw: 118.16

PROP: First isolated from seeds of *Cycas circinalis* (FEPRA7 31,1473,72).

TOXICITY DATA with REFERENCE
ipr-rat LDLo:840 mg/kg FEPRA7 31,1473,72
ipr-mus LDLo:1680 mg/kg FEPRA7 31,1473,72
ipr-ckn LDLo:400 mg/kg FEPRA7 31,1473,72

SAFETY PROFILE: Moderately toxic by intraperitoneal route. See also AMINES. When heated to decomposition it emits toxic fumes of NO_x.

AKP750 CAS:82-28-0 ***HR: 3***
1-AMINO-2-METHYLANTHRAQUINONE
mf: $C_{15}H_{11}NO_2$ mw: 237.27

SYNS: ACETATE FAST ORANGE R ◇ ACETOQUINONE LIGHT ORANGE JL ◇ 1-AMINO-2-METHYL-9,10-ANTHRACENEDIONE ◇ ARTISIL ORANGE 3RP ◇ CELLITON ORANGE R ◇ C.I. 60700 ◇ C.I. DISPERSE ORANGE 11 ◇ CILLA ORANGE R ◇ DISPERSE ORANGE ◇ DURANOL ORANGE G ◇ 2-METHYL-1-ANTHRAQUINONYLAMINE ◇ MICROSETILE ORANGE RA ◇ NCI-C01901 ◇ NYLOQUINONE ORANGE JR ◇ PERLITON ORANGE 3R ◇ SERISOL ORANGE YL ◇ SUPRACET ORANGE R

TOXICITY DATA with REFERENCE
orl-rat TDLo:30 g/kg/78W-C:CAR NCITR* NCI-CG-TR-111,78
orl-mus TDLo:37 g/kg/73W-C:CAR NCITR* NCI-CG-TR-111,78
orl-rat TD:39 g/kg/77W-C:NEO TOLED5 4,71,79
orl-mus TD:307 g/kg/73W-C:ETA IARC** 27,199,82

CONSENSUS REPORTS: NTP Fifth Annual Report on Carcinogens. IARC Cancer Review: Group 3 IMEMDT 7,56,87; Animal Limited Evidence IMEMDT 27,199,82. NCI Carcinogenesis Bioassay (feed); Clear Evidence: mouse, rat NCITR* NCI-CG-TR-111,78. Community Right-To-Know List. Reported in EPA TSCA Inventory.

SAFETY PROFILE: Confirmed carcinogen with experimental carcinogenic, neoplastigenic, and tumorigenic data. When heated to decomposition it emits toxic fumes of NO_x.

AKQ000 CAS:88-44-8 ***HR: 1***
2-AMINO-5-METHYLBENZENESULFONIC ACID
mf: $C_7H_9NO_3S$ mw: 187.23

SYNS: 4-AMINOTOLUENE-3-SULFONIC ACID ◇ 6-AMINO-m-TOLUENESULFONIC ACID ◇ KYSELINA-4-TOLUIDIN-3-SULFONOVA (CZECH) ◇ PTMS ◇ PTMSA ◇ RED 4B ACID ◇ p-TOLUIDINE-m-SULFONIC ACID

TOXICITY DATA with REFERENCE
orl-rat LD50:11700 mg/kg 28ZPAK -,183,72
eye-rbt 500 mg/24H MOD 28ZPAK -,183,72

CONSENSUS REPORTS: Reported in EPA TSCA Inventory.

SAFETY PROFILE: Mildly toxic by ingestion. An eye irritant. See also SULFONATES. When heated to decomposition it emits very toxic fumes of NO_x and SO_x.

AKQ250 CAS:565-33-3 ***HR: D***
3-AMINO-4-METHYLBENZENESULFONYLCYCLOHEXYLUREA
mf: $C_{14}H_{21}N_3O_3S$ mw: 311.44

SYNS: 1-(3-AMINO-p-TOLYLSULFONYL)-3-CYCLOHEXYLUREA ◇ N-CYCLOHEXYL-N'-(3-AMINO-4-METHYLBENZENESULFONYL)UREA ◇ EUGLYCIN ◇ GEIGY 444E ◇ GEIGY HERBICIDE 444E ◇ GLYHEXYLAMIDE ◇ GLYHEXYLAMINE ISODIANE ◇ MELANEX ◇ METAHEXAMIDE ◇ METHAHEXAMIDE ◇ METHEXAMIDE ◇ MELONEX ◇ 1,3,5-TRIAZINE-2,4-DIAMINE, 6-CHLORO-N,N,N',N'-TETRAETHYL- (9CI)

TOXICITY DATA with REFERENCE
orl-rat TDLo:2 g/kg (female 9D post):TER PROEAS 14,89,68

SAFETY PROFILE: An experimental teratogen. When heated to decomposition it emits very toxic fumes of NO_x and SO_x.

AKQ500 CAS:1477-42-5 ***HR: 3***
2-AMINO-4-METHYLBENZOTHIAZOLE
mf: $C_8H_8N_2S$ mw: 164.24

SYN: 4-METHYL-2-AMINOBENZOTHIAZOLE

TOXICITY DATA with REFERENCE
orl-mus LD50:697 mg/kg JPETAB 105,486,52
ivn-mus LD50:54 mg/kg JPETAB 105,486,52

CONSENSUS REPORTS: Reported in EPA TSCA Inventory.

SAFETY PROFILE: Poison by intravenous route. Moderately toxic by ingestion. When heated to decomposition it emits very toxic fumes of SO_x and NO_x.

AKQ750 CAS:64037-15-6 ***HR: 3***
2-AMINO-5-METHYLBENZOXAZOLE
mf: $C_8H_8N_2O$ mw: 148.18

TOXICITY DATA with REFERENCE
orl-mus LD50:640 mg/kg MDCHAG 4(1),336,64
ipr-mus LD50:360 mg/kg MDCHAG 4(1),336,64

SAFETY PROFILE: Poison by intraperitoneal route. Moderately toxic by ingestion. When heated to decomposition it emits toxic fumes of NO_x.

AKR000 CAS:2454-37-7 ***HR: 2***
3-AMINO-α-METHYLBENZYL ALCOHOL
mf: $C_8H_{11}NO$ mw: 137.20

SYN: m-AMINO-α-METHYLBENZYL ALCOHOL

TOXICITY DATA with REFERENCE
eye-rbt 500 mg open AMIHBC 10,61,54
orl-rat LD50:3100 mg/kg AMIHBC 10,61,54

SAFETY PROFILE: Moderately toxic by ingestion. An eye irritant. When heated to decomposition it emits toxic fumes of NO_x.

AKR250 ***HR: 3***
8-((4-AMINO-1-METHYLBUTYL)AMINO)-6-METHOXYQUINOLINE DIPHOSPHATE
mf: $C_{15}H_{21}N_3O \cdot 2H_3O_4P$ mw: 455.39

SYN: PRIMACHIN (GERMAN)

TOXICITY DATA with REFERENCE
ipr-mus LD50:60 mg/kg ARZNAD 20,1775,70
ivn-brd LD50:11 mg/kg ARZNAD 20,1775,70

SAFETY PROFILE: Poison by intraperitoneal and intravenous routes. See also PHOSPHATES. When heated to decomposition it emits very toxic fumes of NO_x and PO_x.

AKR500 CAS:21452-14-2 ***HR: 3***
2-AMINO-4-METHYL-5-CARBOXANILIDOTHIAZOLE
mf: $C_{11}H_{11}N_2OS$ mw: 219.30

SYNS: ALF ◇ F 849 ◇ SEEDVAX ◇ SIDVAX ◇ UNIROYAL F849

TOXICITY DATA with REFERENCE
orl-rat LD50:1410 mg/kg FMCHA2 -,C211,83
unr-mam LD50:141 mg/kg 30ZDA9 -,419,71

SAFETY PROFILE: Poison by an unspecified route. Moderately toxic by ingestion. When heated to decomposition it emits very toxic fumes of NO_x and SO_x.

AKR750 CAS:2051-79-8 ***HR: 3***
4-AMINO-3-METHYL-N,N-DIETHYLANILINEHYDROCHLORIDE
mf: $C_{11}H_{18}N_2 \cdot ClH$ mw: 214.77

TOXICITY DATA with REFERENCE
orl-rat LDLo:200 mg/kg KODAK* -,-,71
ipr-rat LD50:25 mg/kg KODAK* -,-,71

CONSENSUS REPORTS: Reported in EPA TSCA Inventory.

SAFETY PROFILE: Poison by ingestion and intraperitoneal routes. When heated to decomposition it emits very toxic fumes of NO_x and HCl.

AKS000 CAS:4781-76-4 ***HR: 3***
2-AMINOMETHYL-2,3-DIHYDRO-4H-PYRAN
mf: $C_6H_{11}NO$ mw: 113.18

SYN: 2-AMINOMETHYL-3,4-DIHYDRO-2H-PYRAN

TOXICITY DATA with REFERENCE
orl-rat LD50:1000 mg/kg AIHAAP 30,470,69
ihl-rat LCLo:100 ppm BJIMAG 27,1,70
skn-rbt LDLo:180 mg/kg AIHAAP 30,470,69

SAFETY PROFILE: Poison by skin contact and inhalation. Moderately toxic by ingestion. When heated to decomposition it emits toxic fumes of NO_x.

AKS250 CAS:67730-11-4 ***HR: 3***
2-AMINO-6-METHYLDIPYRIDO(1,2-a:3',2'-d)IMIDAZOLE
mf: $C_{11}H_{10}N_4$ mw: 198.25

SYNS: GLU-P-1 ◇ 6-ME-GLU-P-2 ◇ 6-METHYL DIPYRIDO(1,2-a:3',2'-d)IMIDAZOL-2-AMINE

TOXICITY DATA with REFERENCE
mma-sat 250 ng/plate JJCREP 76,835,85
sce-hmn:lym 1000 μg/L MUREAV 77,65,80
dnd-mus-ipr 10 mg/kg JJCREP 76,835,85
orl-rat TDLo:9100 mg/kg/68W-C:CAR EVHPAZ 67,129,86
orl-mus TDLo:15200 mg/kg/58W-C:CAR EVHPAZ 67,129,86

CONSENSUS REPORTS: IARC Cancer Review: Group 2B IMEMDT 7,56,87; Animal Sufficient Evidence IMEMDT 40,223,86

SAFETY PROFILE: Suspected carcinogen with experimental carcinogenic data. Human mutation data reported. When heated to decomposition it emits toxic fumes of NO_x.

AKS275 ***HR: 2***
2-AMINO-6-METHYLDIPYRIDO(1,2-a:3',2'-d)IMIDAZOLE HYDROCHLORIDE
mf: $C_{11}H_{10}N_4 \cdot ClH$ mw: 234.71

SYN: DIPYRIDO(1,2-a:3',2'-d)IMIDAZOLE, 2-AMINO-6-METHYL-, HYDROCHLORIDE

TOXICITY DATA with REFERENCE
slt-dmg-orl 100 ng/kg JJCREP 76,468,85
orl-rat TDLo:3612 mg/kg/24W-C:CAR GANNA2 75,207,84

SAFETY PROFILE: Questionable carcinogen with experimental carcinogenic data. Mutation data reported. When heated to decomposition it emits toxic fumes of NO_x and HCl.

AKS500 CAS:31416-87-2 ***HR: 3***
α-AMINOMETHYL-3-FLUOROBENZYL-ALCOHOL HYDROBROMIDE
mf: $C_8H_{10}FNO•BrH$ mw: 236.11

SYN: 2-AMINO-1-(3-FLUOROPHENYL)ETHANOLHYDROBROMIDE

TOXICITY DATA with REFERENCE
ipr-mus LD50:600 mg/kg JPETAB 106,440,52
ivn-mus LD50:180 mg/kg JPETAB 106,440,52

SAFETY PROFILE: Poison by intravenous route. Moderately toxic by intraperitoneal route. When heated to decomposition it emits very toxic fumes of F^-, Br^-, and NO_x.

AKS750 CAS:63765-80-0 ***HR: 3***
4-AMINO-2-METHYL-3-HEXANOL
mf: $C_7H_{17}NO$ mw: 131.25

SYN: USAF CS-4

TOXICITY DATA with REFERENCE
ipr-hmn LDLo:25 mg/kg CURL** -,5,62
ipr-mus LD50:25 mg/kg NTIS** AD277-689

SAFETY PROFILE: A human poison by intraperitoneal route. When heated to decomposition it emits toxic fumes such as NO_x.

AKT000 CAS:536-21-0 ***HR: 3***
α-(AMINOMETHYL)-m-HYDROXYBENZYL ALCOHOL
mf: $C_8H_{11}NO_2$ mw: 153.20

SYNS: 1-(m-HYDROXYPHENYL)-2-AMINOETHANOL ◇ 1-(3'-HYDROXYPHENYL)-2-AMINOETHANOL ◇ m-HYDROXYPHENYLETHANOLAMINE ◇ 1-(3-HYDROXYPHENYL)-1-HYDROXY-2-AMINOETHANE ◇ METACARDIOL ◇ NORENOL ◇ NORMETOL ◇ NORPHENYLEPHRINE ◇ NORSYNEPHRINE ◇ NOVADRAL ◇ m-OCTOPAMINE

TOXICITY DATA with REFERENCE
orl-rat LD50:390 mg/kg OYYAA2 2,217,68
ipr-rat LD50:32 mg/kg OYYAA2 2,217,68
scu-rat LD50:28100 μg/kg OYYAA2 2,60,68
ivn-rat LD50:17400 μg/kg OYYAA2 2,217,68
orl-mus LD50:263 mg/kg RPOBAR 2,295,70
ipr-mus LD50:198 mg/kg RPOBAR 2,295,70
scu-mus LD50:459 mg/kg RPOBAR 2,295,70
ivn-mus LD50:4900 μg/kg RPOBAR 2,295,70

SAFETY PROFILE: Poison by ingestion, subcutaneous, intravenous, and intraperitoneal routes. When heated to decomposition it emits toxic fumes of NO_x.

AKT250 CAS:104-14-3 ***HR: 3***
α-(AMINOMETHYL)-p-HYDROXYBENZYL ALCOHOL
mf: $C_8H_{11}NO_2$ mw: 153.20

SYNS: 1-(p-HYDROXYPHENYL)-2-AMINOETHANOL ◇ p-HYDROXYPHENYLETHANOLAMINE ◇ NORDEN ◇ NORPHEN ◇ NORSYMPATHOL ◇ NORSYNEPHRINE ◇ OCTOPAMINE ◇ WIN 5512

TOXICITY DATA with REFERENCE
ipr-mus LD50:600 mg/kg JPETAB 106,341,52
scu-mus LDLo:1050 mg/kg AIPTAK 101,81,55
ivn-mus LD50:75 mg/kg JPETAB 106,341,52
ivn-gpg LDLo:200 mg/kg AIPTAK 101,81,55

SAFETY PROFILE: Poison by intravenous route. Moderately toxic by intraperitoneal and subcutaneous route. When heated to decomposition it emits toxic fumes of NO_x.

AKT500 CAS:4779-94-6 ***HR: 3***
α-AMINOMETHYL-3-HYDROXYBENZYL-ALCOHOL HYDROCHLORIDE
mf: $C_8H_{11}NO_2•ClH$ mw: 189.66

SYN: WIN 5501

TOXICITY DATA with REFERENCE
orl-rat LD50:390 mg/kg OYYAA2 4,561,70
ipr-rat LD50:32 mg/kg OYYAA2 4,561,70
scu-rat LD50:28 mg/kg OYYAA2 4,561,70
orl-mus LD50:3300 mg/kg OYYAA2 4,561,70
ipr-mus LD50:370 mg/kg JPETAB 106,440,52
ivn-mus LD50:113 mg/kg JPETAB 106,440,52

SAFETY PROFILE: Poison by ingestion, intraperitoneal, subcutaneous, and intravenous routes. When heated to decomposition it emits very toxic fumes of HCl and NO_x.

AKT600 CAS:76180-96-6 ***HR: 3***
2-AMINO-3-METHYLIMIDAZO(4,5-f)QUINOLINE
mf: $C_{11}H_{10}N_4$ mw: 198.25

TOXICITY DATA with REFERENCE
sln-dmg-orl 1 mmol/L MUREAV 156,93,85
dnd-mus:lvr 100 μmol/L JJCREP 76,835,85
orl-rat TDLo:4300 mg/kg/56W-C:CAR EVHPAZ 67,129,86

orl-mus TDLo:20800 mg/kg/97W-C:CAR EVHPAZ 67,129,86
orl-rat TD:3600 mg/kg/43W-C:ETA GANNA2 75,467,84

CONSENSUS REPORTS: IARC Cancer Review: Group 2B IMEMDT 7,56,87; Animal Sufficient Evidence IMEMDT 40,261,86

SAFETY PROFILE: Suspected carcinogen with experimental carcinogenic and tumorigenic data. Mutation data reported. When heated to decomposition it emits toxic fumes of NO_x.

AKT620 ***HR: 2***
2-AMINO-3-METHYLIMIDAZO(4,5-f)QUINOLINE DIHYDROCHLORIDE
mf: $C_{11}H_{10}N_4 \cdot 2ClH$ mw: 271.17 ◇ IQ DIHYDROCHLORIDE

TOXICITY DATA with REFERENCE
orl-rat TDLo:4081 mg/kg/31W-I:CAR JJCREP 76,570,85

SAFETY PROFILE: Questionable carcinogen with experimental carcinogenic data. When heated to decomposition it emits toxic fumes of NO_x.

AKT750 CAS:2763-96-4 ***HR: 3***
5-AMINOMETHYL-3-ISOXYZOLE
mf: $C_4H_6N_2O_2$ mw: 114.12

SYNS: AGARIN ◇ 5-AMINOMETHYL-3-HYDROXYISOXAZOLE ◇ 5-(AMINOMETHYL)-3-ISOXAZOLOL ◇ 5-(AMINOMETHYL)-3(2H)-ISOXAZOLONE ◇ 3-HYDROXY-5-AMINOMETHYLISOXAZOLE ◇ 3-HYDROXY-5-AMINOMETHYLISOXAZOLE-AGARIN ◇ MUSCIMOL ◇ RCRA WASTE NUMBER P007

TOXICITY DATA with REFERENCE
unk-hmn TDLo:109 μg/kg:CNS,GIT ARZNAD 18,311,68
orl-rat LD50:45 mg/kg ARZNAD 18,311,68
ivn-rat LD50:4500 μg/kg ARZNAD 18,311,68
orl-mus LD50:17 mg/kg JMCMAR 23,702,80
ipr-mus LD50:2500 μg/kg ARZNAD 18,311,68
scu-mus LD50:3800 μg/kg ARZNAD 18,311,68
ivn-mus LD50:5620 μg/kg CSLNX* NX#11824
ivn-rbt LDLo:10 mg/kg ARZNAD 18,311,68

CONSENSUS REPORTS: Reported in EPA TSCA Inventory. EPA Extremely Hazardous Substances List.

SAFETY PROFILE: Poison by ingestion, subcutaneous, intravenous, and intraperitoneal routes. Human systemic effects by an unspecified route: sleep, nausea or vomiting, hallucinations and distorted perceptions. When heated to decomposition it emits toxic fumes of NO_x.

AKX500 CAS:83-70-5 ***HR: 3***
4-AMINO-2-METHYL-1-NAPHTHOL
mf: $C_{11}H_{11}NO$ mw: 173.23

SYNS: 4-AMINO-2-METHYL-1-NAPHTHALENOL ◇ 1-HYDROXY-2-METHYL-4-AMINONAPHTHALENE ◇ KAYVISYN ◇ 2-METHYL-4-AMINO-1-HYDROXYNAPHTHALENE ◇ 2-METHYL-4-AMINO-1-NAPHTHOL ◇ 3-METHYL-4-HYDROXY-1-NAPHTHYLAMINE ◇ SYNKAMIN ◇ SYNKAMIN BASE ◇ VITAMIN K5

TOXICITY DATA with REFERENCE
ipr-mus LD50:250 mg/kg ARZNAD 17,1339,67

CONSENSUS REPORTS: Reported in EPA TSCA Inventory.

SAFETY PROFILE: Poison by intraperitoneal route. When heated to decomposition it emits toxic NO_x.

AKY000 CAS:10187-86-7 ***HR: 2***
3-AMINO-4-METHYL-5-(5-NITRO-2-FURYL)-s-TRIAZOLE

SYN: 4-METHYL-5-(5-NITRO-2-4H-1,2,4-TRIAZOL-3-AMINE

TOXICITY DATA with REFERENCE
orl-man TDLo:126 mg/kg:GIT JMCMAR 16,312,73
orl-mus LD50:1460 mg/kg JMCMAR 16,312,73
ipr-mus LD50:730 mg/kg JMCMAR 16,312,73

SAFETY PROFILE: Moderately toxic by ingestion and intraperitoneal routes. Human gastrointestinal tract effects by ingestion. When heated to decomposition it emits toxic fumes of NO_x.

AKY250 CAS:5581-52-2 ***HR: 3***
2-AMINO-6-(1′-METHYL-4′-NITRO-5′-IMIDAZOLYL)MERCAPTOPURINE
mf: $C_9H_8N_8O_2S$ mw: 292.31

SYNS: 2-AMINO-6-(1-METHYL-4-NITRO-5-IMIDAZOLYL)MERCAPTOPURINE ◇ 6-BENZYLAMINOPURINE ◇ BW 57-323 ◇ BW 57-323H ◇ GUANERAN ◇ IRG ◇ 1-METHYL-4-NITRO-5-(2′-AMINO-6′-PURINYL)MERCAPTOIMIDAZIDE ◇ NSC-38887 ◇ 1H-PURINE-2-AMINE,6-((1-METHYL-4-NITRO-1H-IMIDAZOL-5-YL)-THIO)- ◇ THIAMIPRINE ◇ TIAMIPRINE

TOXICITY DATA with REFERENCE
ipr-rbt TDLo:7500 μg/kg (female 6-8D post):TER TNEOAO 18,57,64
ipr-rat TDLo:5 mg/kg (female 7D post):REP JRPFA4 4,291,62
orl-mus LD50:450 mg/kg RPTOAN 34,284,71
ipr-mus LD50:136 mg/kg RPTOAN 34(6),284,71

SAFETY PROFILE: Poison by intraperitoneal route. Moderately toxic by ingestion. An experimental teratogen. Other experimental reproductive effects. When heated to decomposition it emits very toxic SO_x and NO_x.

AKY750 CAS:14370-50-4 ***HR: 2***
2-(AMINOMETHYL)NORBORNANE
mf: $C_8H_{15}N$ mw: 125.24

SYN: (2,5-ENDOMETHYLENECYCLOHEXYLMETHYL)AMINE

TOXICITY DATA with REFERENCE
skn-rbt 100 μg/24H open AIHAAP 23,95,62
orl-rat LD50:1410 mg/kg AIHAAP 23,95,62
skn-rbt LD50:520 mg/kg AIHAAP 23,95,62

SAFETY PROFILE: Moderately toxic by ingestion and skin contact. A skin irritant. When heated to decomposition it emits toxic fumes of NO_x.

AKY875 CAS:35629-70-0 ***HR: 2***
2-AMINO-4-METHYLOXAZOLE
mf: $C_4H_6N_2O$ mw: 98.10

SAFETY PROFILE: Potentially explosive reaction with hydrogen peroxide and iron(II) catalysts. When heated to decomposition it emits toxic fumes of NO_x.

AKZ000 CAS:2835-99-6 ***HR: 3***
4-AMINO-3-METHYLPHENOL
mf: C_7H_9NO mw: 123.17

TOXICITY DATA with REFERENCE
ipr-mus LD50:200 mg/kg NTIS** AD691-490

CONSENSUS REPORTS: Reported in EPA TSCA Inventory.

SAFETY PROFILE: Poison by intraperitoneal route. When heated to decomposition it emits toxic fumes of NO_x.

ALA000 CAS:50901-84-3 ***HR: 3***
cis-2-AMINO-5-METHYL-4-PHENYL-1-PYRROLINE
mf: $C_{11}H_{14}N_2$ mw: 174.27

TOXICITY DATA with REFERENCE
orl-rat LD50:420 mg/kg EJMCA5 13,161,78
orl-mus LD50:276 mg/kg EJMCA5 13,161,78
ivn-mus LD50:14 mg/kg EJMCA5 13,161,78

SAFETY PROFILE: Poison by ingestion and intravenous routes. When heated to decomposition it emits toxic fumes of NO_x.

ALA250 CAS:50901-87-6 ***HR: 3***
trans-2-AMINO-5-METHYL-4-PHENYL-1-PYRROLINE
mf: $C_{11}H_{14}N_2$ mw: 174.27

TOXICITY DATA with REFERENCE
orl-mus LD50:80 mg/kg EJMCA5 13,161,78
ivn-mus LD50:26 mg/kg EJMCA5 13,161,78

SAFETY PROFILE: Poison by ingestion and intravenous routes. When heated to decomposition it emits toxic fumes of NO_x.

ALA500 CAS:41394-05-2 ***HR: 2***
4-AMINO-3-METHYL-6-PHENYL-1,2,4-TRIAZIN-5(4H)-ONE
mf: $C_{10}H_{10}N_4O$ mw: 202.24

SYNS: BAY-DRW 1139 ◇ DRW 1139 ◇ GOLTIX ◇ METAMITON ◇ METAMITRON (GERMAN) ◇ 3-METHYL-4-AMINO-6-PHENYL-1,2,4-TRIAZIN(4H)-ON (GERMAN)

TOXICITY DATA with REFERENCE
mmo-nsc 1 mg/L ENMUDM 7(Suppl 3),11,85
orl-rat LD50:1447 mg/kg 85ARAE 2,133,77
orl-mus LD50:1450 mg/kg 85DPAN -,-,71/76

SAFETY PROFILE: Moderately toxic by ingestion. Mutation data reported. Moderately toxic by ingestion. When heated to decomposition it emits toxic fumes of NO_x.

ALA750 CAS:55921-66-9 ***HR: 3***
2-AMINO-4-(N-METHYLPIPERAZINO)-5-METHYLTHIO-6-CHLOROPYRIMIDINE
mf: $C_{10}H_{16}ClN_5S$ mw: 273.82

TOXICITY DATA with REFERENCE
orl-mus LD50:225 mg/kg JMCMAR 18,553,75
ivn-mus LD50:33 mg/kg JMCMAR 18,553,75

SAFETY PROFILE: Poison by ingestion and intravenous routes. When heated to decomposition it emits very toxic fumes of SO_x, NO_x, and Cl^-.

ALB000 CAS:115-69-5 ***HR: 3***
2-AMINO-2-METHYL-1,3-PROPANEDIOL
mf: $C_4H_{11}NO_2$ mw: 105.16

PROP: A clear liquid. Mp: 110°, bp: 151° @ 10 mm, vap d: 3.63.

SYNS: AMINOGLYCOL ◇ AMPD ◇ GENTIMON ◇ ISOBUTANDIOL-2-AMINE ◇ PENTAERYTHRITOL DICHLOROHYDRIN

TOXICITY DATA with REFERENCE
orl-mus LDLo:140 mg/kg AECTCV 14,111,85
orl-rbt LDLo:1500 mg/kg JIHTAB 22,315,40

CONSENSUS REPORTS: Reported in EPA TSCA Inventory.

SAFETY PROFILE: Poison by ingestion. Combustible. Can react with oxidizing materials. When heated to decomposition it emits toxic fumes of NO_x.

ALB250 CAS:2854-16-2 ***HR: 2***
1-AMINO-2-METHYL-2-PROPANOL
mf: $C_4H_{11}NO$ mw: 89.16

TOXICITY DATA with REFERENCE
orl-rat LDLo:3000 mg/kg SCCUR* -,1,61
orl-mus LD50:2450 mg/kg SCCUR* -,1,61

ihl-mus LCLo:1095 ppm/18H SCCUR* -,1,61
skn-rbt LDLo:1960 mg/kg SCCUR* -,1,61

SAFETY PROFILE: Moderately toxic by ingestion, inhalation, and skin contact. When heated to decomposition it emits toxic fumes of NO_x.

ALB500 CAS:7447-44-1 ***HR: 2***
S-2-AMINO-2-METHYLPROPYL DIHYDROGEN PHOSPHOROTHIOATE
mf: $C_4H_{12}NO_3PS$ mw: 185.20

SYN: S-(2-AMINO-2-METHYLPROPYL)PHOSPHOROTHIOATE

TOXICITY DATA with REFERENCE
orl-mus LD50:2800 mg/kg JMCMAR 18,803,75
ipr-mus LD50:750 mg/kg JMCMAR 18,803,75
unr-mus LD50:750 mg/kg JMCMAR 9,911,66

SAFETY PROFILE: Moderately toxic by ingestion, intraperitoneal, and other unspecified routes. See also PHOSPHATES. When heated to decomposition it emits very toxic fumes of SO_x, PO_x, and NO_x.

ALB625 CAS:18591-81-6 ***HR: 2***
3-AMINO-6-METHYL-4-PYRIDAZINETHIOL
mf: $C_5H_7N_3S$ mw: 141.21

SYNS: 3-AMINO-4-MERCAPTO-6-METHYLPYRIDAZIN(GERMAN) ◇ 3-AMINO-4-MERCAPTO-6-METHYLPYRIDAZINE

TOXICITY DATA with REFERENCE
orl-rat LD50:1975 mg/kg PHARAT 37,285,82
orl-mus LD50:1637 mg/kg PHARAT 37,136,82
scu-mus LD50:1637 mg/kg PHARAT 36,698,81

SAFETY PROFILE: Moderately toxic by ingestion and other routes. When heated to decomposition it emits toxic fumes of NO_x and SO_x.

ALB750 CAS:3731-51-9 ***HR: 3***
2-AMINOMETHYLPYRIDINE
mf: $C_6H_8N_2$ mw: 108.16

SYNS: 2-PICOLINAMINE ◇ 2-PICOLYLAMINE ◇ 2-PYRIDINEMETHYLAMINE ◇ (2-PYRIDYLMETHYL)AMINE

TOXICITY DATA with REFERENCE
orl-qal LD50:750 mg/kg AECTCV 12,355,83
orl-bwd LD50:562 mg/kg AECTCV 12,355,83
ivn-mus LD50:340 mg/kg APFRAD 26,345,68

CONSENSUS REPORTS: Reported in EPA TSCA Inventory.

SAFETY PROFILE: Poison by intravenous route. Moderately toxic by ingestion. When heated to decomposition it emits toxic fumes of NO_x.

ALC000 CAS:1603-40-3 ***HR: 3***
2-AMINO-3-METHYLPYRIDINE
mf: $C_6H_8N_2$ mw: 108.16

PROP: Bp: 221.2°; vap d: 3.73.

TOXICITY DATA with REFERENCE
ivn-mus LD50:10 mg/kg CSLNX* NX#01585

CONSENSUS REPORTS: Reported in EPA TSCA Inventory.

SAFETY PROFILE: Poison by intravenous route. Combustible. When heated to decomposition it emits toxic fumes of NO_x.

ALC250 CAS:695-34-1 ***HR: 3***
2-AMINO-4-METHYLPYRIDINE
mf: $C_6H_8N_2$ mw: 108.16

PROP: Crystals. Mp: 99°, bp: 230.9°, vap d: 3.73.

SYNS: α-AMINO-γ-PICOLINE ◇ 2-AMINO-4-PICOLINE ◇ ASCENSIL ◇ 4M2AP ◇ 4-METHYL-2-AMINOPYRIDINE ◇ METHYL-4-AMINO-2-PYRIDINE ◇ 4-PICOLYLAMINE ◇ RA 1226 ◇ W 45 ◇ W 45 RASCHIG

TOXICITY DATA with REFERENCE
ivn-mus LD50:39 mg/kg APFRAD 26,345,68

CONSENSUS REPORTS: Reported in EPA TSCA Inventory.

SAFETY PROFILE: Poison by intravenous route. Combustible. When heated to decomposition it emits toxic fumes of NO_x. An analgesic and cardiac stimulant.

ALC500 CAS:1824-81-3 ***HR: 3***
2-AMINO-6-METHYLPYRIDINE
mf: $C_6H_8N_2$ mw: 108.16

PROP: Mp: 43.7°, bp: 214.4, vap d: 3.73.

TOXICITY DATA with REFERENCE
ivn-mus LD50:18 mg/kg CSLNX* NX#00148

CONSENSUS REPORTS: Reported in EPA TSCA Inventory.

SAFETY PROFILE: Poison by intravenous route. When heated to decomposition it emits toxic fumes of NO_x.

ALC750 CAS:53222 ***HR: 3***
4-((3-AMINO-4-((4-((1-METHYLPYRIDINIUM-4-YL)AMINO)BENZOYL)AMINO)PHENYL)AMINO)-1-METHYLQUINOLINIUM)DIBROMIDE-52-9
mf: $C_{29}H_{28}N_6O•2Br$ mw: 636.45

TOXICITY DATA with REFERENCE
dnd-mus:lym 1870 nmol/L JMCMAR 22,134,79
ipr-mus LD10:7 mg/kg JMCMAR 22,134,79

SAFETY PROFILE: Poison by intraperitoneal route. Mutation data reported. See also BROMIDES. When heated to decomposition it emits very toxic fumes of Br^- and NO_x.

ALD500 CAS:62450-07-1 ***HR: 3***
3-AMINO-1-METHYL-5H-PYRIDO(4,3-b)INDOLE
mf: $C_{12}H_{11}N_3$ mw: 197.26

SYNS: 3-AMINO-1-METHYL-γ-CARBOLINE ◇ 1-METHYL-3-AMINO-5H-PYRIDO(4,3-b)INDOLE ◇ TRP-P-2 ◇ TRYPTOPHAN P2

TOXICITY DATA with REFERENCE
mmo-sat 50 ng/plate CRNGDP 5,505,84
mma-sat 50 ng/plate CRNGDP 5,505,84
dnd-sat 1 μg/plate ENMUDM 6,437,84
dnd-mus:lvr 50 μmol/L JJCREP 76,835,85
dnd-mus-ipr 10 mg/kg JJCREP 76,835,85
orl-rat TDLo:4350 mg/kg/2Y-C:NEO CALEDQ 13,23,81
orl-mus TDLo:9648 mg/kg/57W-C:CAR EVHPAZ 67,129,86
scu-mus TDLo:12500 μg/kg:CAR PPTCBY 37,193,85

CONSENSUS REPORTS: IARC Cancer Review: Group 2B IMEMDT 7,56,87; Animal Sufficient Evidence IMEMDT 31,255,83. EPA Genetic Toxicology Program.

SAFETY PROFILE: Suspected carcinogen with experimental carcinogenic and neoplastigenic data. Mutation data reported. When heated to decomposition it emits toxic fumes of NO_x.

ALD750 CAS:68006-83-7 ***HR: 3***
2-AMINO-3-METHYL-9H-PYRIDO(2,3-b)INDOLE
mf: $C_{12}H_{11}N_3$ mw: 197.2

SYN: 2-AMINO-3-METHYL-α-CARBOLINE

TOXICITY DATA with REFERENCE
mmo-sat 1 μg/plate ABCHA6 43,1155,79
dnr-bcs 10 μL/plate ABCHA6 45,2031,81
slt-dmg-orl 400 ng/kg JJCREP 76,468,85
mma-sat 10 ng/plate CALEDQ 10,141,80
orl-mus TDLo:35424 mg/kg/70W-C:CAR CRNGDP 5,815,84
orl-mus TD:37380 mg/kg/64W-C:CAR CRNGDP 5,815,84

CONSENSUS REPORTS: IARC Cancer Review: Group 2B IMEMDT 7,56,87; Animal Sufficient Evidence IMEMDT 40,253,86

SAFETY PROFILE: Suspected carcinogen with experimental carcinogenic data. Mutation data reported. When heated to decomposition it emits toxic fumes of NO_x.

ALE750 CAS:72254-58-1 ***HR: 3***
3-AMINO-1-METHYL-5H-PYRIDO(4,3-b)INDOLE ACETATE
mf: $C_{12}H_{11}N_3 \cdot C_2H_4O_2$ mw: 257.32

SYNS: 5H-PYRIDO(4,3-b)INDOL-3-AMINE, 1-METHYL-1, MONOACETATE ◇ TRP-P-2(ACETATE)

TOXICITY DATA with REFERENCE
slt-dmg-orl 400 ppm MUREAV 122,315,83
mma-sat 1 μg/plate CPBTAL 26,611,78
orl-mus TDLo:14 g/kg/89W-C:CAR SCIEAS 213,346,81

SAFETY PROFILE: Questionable carcinogen with experimental carcinogenic data. Mutation data reported. When heated to decomposition it emits toxic fumes of NO_x.

ALF250 CAS:127-79-7 ***HR: 3***
4-AMINO-N-(4-METHYL-2-PYRIMIDINYL)-BENZENESULFONAMIDE
mf: $C_{11}H_{12}N_4O_2S$ mw: 264.33

SYNS: A-310 ◇ (p-AMINOBENZOLSULFONYL)-2-AMINO-4-METHYLPYRIMIDIN (GERMAN) ◇ CREMOMERAZINE ◇ DEBENAL-M ◇ KELAMERAZINE ◇ MEBACID ◇ MESULFA ◇ METHYLPYRIMAL ◇ N^1-(4-METHYL-2-PYRIMIDINYL)SULFANILAMIDE ◇ METHYLSULFAZIN ◇ PERCOCCIDE ◇ PIRIMAL-M ◇ PYRALCID ◇ PYRIMAL M ◇ ROMEZIN ◇ RP 2632 ◇ 2643-RP ◇ SEPTACIL ◇ SULFAMERADINE ◇ SULFAMERAZIN ◇ SULFAMETHYLDIAZINE ◇ SULPHAMERAZINE ◇ SUMEDINE ◇ VETA-MERAZINE

TOXICITY DATA with REFERENCE
orl-mus TDLo:6 g/kg (female 6-10D post):REP JPETAB 101,362,51
orl-mus LD50:25 g/kg AIPTAK 94,338,53
ivn-mus LD50:31 mg/kg ARZNAD 3,66,53

CONSENSUS REPORTS: Reported in EPA TSCA Inventory.

SAFETY PROFILE: Poison by intravenous route. Experimental reproductive effects. When heated to decomposition it emits very toxic fumes of NO_x and SO_x.

ALF500 CAS:55661-38-6 ***HR: 3***
1-(4-AMINO-2-METHYLPYRIMIDIN-5-YL) METHYL-3-(2-CHLOROETHYL)-3-NITROSOUREA
mf: $C_9H_{13}ClN_6O_2 \cdot ClH$ mw: 309.19

SYNS: ACNU ◇ N'-((4-AMINO-2-METHYL-5-PYRIMIDINYL)METHYL)-N-(2-CHLOROETHYL)-N-NITROSOUREA HCl ◇ 3-((4-AMINO-2-METHYL-5-PYRIMIDINYL)METHYL)-1-(2-CHLOROETHYL)-1-NITROSOUREA HYDROCHLORIDE ◇ CS-439 ◇ NIDRAN ◇ NIMUSTINE HYDROCHLORIDE ◇ NSC-245382

TOXICITY DATA with REFERENCE
mmo-sat 50 μg/plate CNREA8 38,2148,78
mma-sat 50 μg/plate CNREA8 38,2148,78
ipr-rat TDLo:10 mg/kg (female 8D post):TER TJADAB 38,553,88

ipr-rat TDLo:10 mg/kg (female 8D post):REP TJADAB 38,553,88
orl-rat LD50:113 mg/kg 37XLA2 2,1233,78
ivn-rat LD50:44300 μg/kg IYKEDH 10,884,79
orl-mus LD50:83 mg/kg 37XLA2 2,1233,78
ipr-mus LD50:49300 μg/kg IYKEDH 10,884,79

SAFETY PROFILE: Poison by ingestion, intraperitoneal, and intravenous routes. An experimental teratogen. Other experimental reproductive effects. Mutation data reported. When heated to decomposition it emits very toxic fumes of Cl^- and NO_x.

ALF600 CAS:454-41-1 ***HR: 2***
2-AMINO-4-(METHYLSULFINYL)BUTYRIC ACID
mf: $C_5H_{11}NO_3S$ mw: 165.23

SYNS: BUTANOIC ACID, 2-AMINO-4-(METHYLSULFINYL)-(9CI) ◇ BUTYRIC ACID, 2-AMINO-4-(METHYLSULFINYL)- ◇ METHIONINE SULFOXIDE ◇ dl-METHIONINE SULFOXIDE

TOXICITY DATA with REFERENCE
ipr-mus TDLo:3500 mg/kg (female 9-15D post):TER ASMUAA 38,193,62
ipr-mus LD50:4000 mg/kg IJRBA3 3,41,61

SAFETY PROFILE: Moderately toxic by intraperitoneal route. An experimental teratogen. When heated to decomposition it emits toxic fumes of NO_x and SO_x

ALF750 CAS:6628-83-7 ***HR: 2***
2-AMINOMETHYLTETRAHYDROPYRAN
mf: $C_6H_{11}NO$ mw: 113.18

TOXICITY DATA with REFERENCE
orl-rat LD50:710 mg/kg AIHAAP 30,470,69
skn-rbt LD50:710 mg/kg AIHAAP 30,470,69

SAFETY PROFILE: Moderately toxic by ingestion and skin contact. When heated to decomposition it emits toxic fumes of NO_x.

ALG250 CAS:73696-62-5 ***HR: 2***
2-AMINO-N-(3-METHYL-2-THIAZOLIDINYLIDENE)ACETAMIDE
mf: $C_6H_{11}N_3OS$ mw: 173.26

TOXICITY DATA with REFERENCE
orl-mus LD50:4666 mg/kg JMCMAR 23,773,80
ivn-mus LD50:549 mg/kg JMCMAR 23,773,80

SAFETY PROFILE: Moderately toxic by intravenous route. Mildly toxic by ingestion. When heated to decomposition it emits very toxic fumes of NO_x and SO_x.

ALG375 CAS:55864-39-6 ***HR: 2***
5-AMINO-3-METHYLTHIO-1,2,4-OXADIAZOLE
mf: $C_3H_5N_3OS$ mw: 131.15

SAFETY PROFILE: Decomposes violently at its mp of 97-99°C. Upon decomposition it emits toxic fumes of SO_x and NO_x.

ALG500 CAS:21172-28-1 ***HR: 3***
α-AMINOMETHYL-m-TRIFLUOROMETHYLBENZYL ALCOHOL
mf: $C_9H_{10}F_3NO$ mw: 205.20

TOXICITY DATA with REFERENCE
orl-mus LD50:700 mg/kg ARZNAD 27,116,77
ipr-mus LD50:226 mg/kg ISYAM* -,21,70

SAFETY PROFILE: Poison by intraperitoneal route. Moderately toxic by ingestion. When heated to decomposition it emits very toxic fumes of F^- and NO_x.

ALH000 CAS:117-62-4 ***HR: 1***
2-AMINO-1,5-NAPHTHALENEDISULFONIC ACID
mf: $C_{10}H_9NO_6S_2$ mw: 303.32

SYNS: KYSELINA 2-NAFTYLAMIN-1,5-DISULFONOVA (CZECH) ◇ KYSELINA SULFO-TOBIAOVA (CZECH) ◇ 5-SULFO-TOBIAS ACID

TOXICITY DATA with REFERENCE
skn-rbt 500 mg/24H MLD 28ZPAK -,188,72
eye-rbt 20 mg/24H MOD 28ZPAK -,188,72
orl-rat LD50:5430 mg/kg 28ZPAK -,188,72

CONSENSUS REPORTS: Reported in EPA TSCA Inventory. EPA Genetic Toxicology Program.

SAFETY PROFILE: Mildly toxic by ingestion. A skin and eye irritant. See also SULFONATES. When heated to decomposition it emits very toxic fumes of NO_x and SO_x.

ALH250 CAS:131-27-1 ***HR: 1***
3-AMINO-1,5-NAPHTHALENEDISULFONIC ACID
mf: $C_{10}H_9NO_6S_2$ mw: 303.32

SYNS: ACID IV ◇ 2-AMINO-4,8-NAPHTHALENEDISULFONIC ACID ◇ 7-AMINO-1,5-NAPHTHALENEDISULFONIC ACID ◇ C ACID ◇ 4,8-DISULFO-2-NAPHTHALAMINE ◇ KYSELINA C (CZECH) ◇ KYSELINA-2-NAFTYLAMIN-4,8-DISULFONOVA (CZECH) ◇ β-NAPHTHYLAMINEDISULFONIC ACID ◇ β-NAPHTHYLAMINE-4,8-DISULFONIC ACID ◇ 2-NAPHTHYLAMINE-4,8-DISULFONIC ACID

TOXICITY DATA with REFERENCE
eye-rbt 100 mg/24H MOD 28ZPAK -,189,72
orl-rat LD50:11400 mg/kg 28ZPAK -,189,72

CONSENSUS REPORTS: Reported in EPA TSCA Inventory.

SAFETY PROFILE: An eye irritant. See also SULFONATES. When heated to decomposition it emits very toxic fumes of NO_x and SO_x.

ALH500 CAS:118-33-2 ***HR: 2***
6-AMINO-NAPHTHALENE-1,3-DISULFONIC ACID
mf: $C_{10}H_9NO_6S_2 \cdot Na$ mw: 326.31

SYN: 2-NAFTYLAMIN-5,7-DISULFONAN SODNY (CZECH)

TOXICITY DATA with REFERENCE
skn-rbt 500 mg/24H SEV 28ZPAK -,189,72
eye-rbt 750 μg/24H SEV 28ZPAK -,189,72
orl-rat LD50:2000 mg/kg 28ZPAK -,189,72

CONSENSUS REPORTS: Reported in EPA TSCA Inventory.

SAFETY PROFILE: Moderately toxic by ingestion. A severe skin and eye irritant. When heated to decomposition it emits very toxic fumes of NO_x and SO_x.

ALH750 CAS:81-16-3 ***HR: 1***
2-AMINO-1-NAPHTHALENESULFONIC ACID
mf: $C_{10}H_9NO_3S$ mw: 223.26

SYNS: KYSELINA-2-NAFTYLAMIN-1-SULFONOVA(CZECH) ◇ KYSELINA TOBIASOVA (CZECH) ◇ 2-NAPHTHYLAMINE-1-SULFONIC ACID ◇ TOBIAS ACID

TOXICITY DATA with REFERENCE
eye-rbt 500 mg/24H MLD 28ZPAK -,187,72
orl-rat LD50:19400 mg/kg 28ZPAK -,187,72

CONSENSUS REPORTS: Reported in EPA TSCA Inventory.

SAFETY PROFILE: Mildly toxic by ingestion. An eye irritant. When heated to decomposition it emits very toxic fumes of NO_x and SO_x. See also SULFONATES.

ALI000 CAS:84-86-6 ***HR: 3***
4-AMINO-1-NAPHTHALENESULFONIC ACID
mf: $C_{10}H_9NO_3S$ mw: 223.26

SYNS: 1-AMINONAPHTHALENE-4-SULFONIC ACID ◇ 1-AMINO-4-SULFONAPHTHALENE ◇ NAPHTHIONIC ACID ◇ 1,4-NAPHTHIONIC ACID ◇ α-NAPHTHYLAMINE-p-SULFONIC ACID ◇ 1-NAPHTHYLAMINE-4-SULFONIC ACID ◇ PIRIA'S ACID ◇ USAF M-5

TOXICITY DATA with REFERENCE
ipr-mus LD50:300 mg/kg NTIS** AD277-689

CONSENSUS REPORTS: Reported in EPA TSCA Inventory. EPA Genetic Toxicology Program.

SAFETY PROFILE: Poison by intraperitoneal route. See also SULFONATES. When heated to decomposition it emits very toxic fumes of NO_x and SO_x.

ALI250 CAS:119-79-9 ***HR: 1***
5-AMINO-2-NAPHTHALENESULFONIC ACID
mf: $C_{10}H_9NO_3S$ mw: 223.26

SYNS: 1-AMINO-6-NAPHTHALENESULFONIC ACID ◇ 1-AMINO-6-SULFONAPHTHALENE ◇ CLEVE'S ACID-1,6 ◇ CLEVE'S BETA-ACID ◇ KYSELINA CLEVE (CZECH) ◇ KYSELINA-1-NAFTYLAMIN-6-SULFONOVA (CZECH) ◇ 1-NAPHTHYLAMINE-6-SULFONIC ACID ◇ 5-NAPHTHYLAMINE-2-SULFONIC ACID

TOXICITY DATA with REFERENCE
eye-rbt 500 mg/24H MLD 28ZPAK -,187,72
orl-rat LD50:14200 mg/kg 28ZPAK -,187,72

CONSENSUS REPORTS: Reported in EPA TSCA Inventory.

SAFETY PROFILE: Mildly toxic by ingestion. An eye irritant. See also SULFONATES. When heated to decomposition it emits very toxic fumes of NO_x and SO_x.

ALI300 CAS:86-60-2 ***HR: 2***
7-AMINO-1-NAPHTHALENESULFONIC ACID
mf: $C_{10}H_9NO_3S$ mw: 223.26

SYNS: BADEN ACID ◇ BADISCHE ACID ◇ 2-NAPHTHYLAMINE-8-SULFONIC ACID

TOXICITY DATA with REFERENCE
ipr-mus TDLo:18 g/kg/8W-I:NEO JJIND8 67,1299,81

SAFETY PROFILE: Questionable carcinogen with experimental neoplastigenic data. When heated to decomposition it emits toxic fumes of NO_x and SO_x.

ALI500 CAS:28907-84-8 ***HR: D***
5-AMINO-2-NAPHTHALENESULFONIC ACID SODIUM SALT
mf: $C_{10}H_8NO_3S \cdot Na$ mw: 245.24

SYNS: SODIUM NAPHTHIONATE ◇ SODIUM-α-NAPHTHYLAMINE-6-SULPHONATE

TOXICITY DATA with REFERENCE
orl-rat LDLo:4 g/kg (1-20D preg):REP FCTXAV 11,355,73
orl-rat TDLo:600 mg/kg (1-20D preg):TER FCTXAV 11,355,73

SAFETY PROFILE: An experimental teratogen. Other experimental reproductive effects. See also SULFONATES. When heated to decomposition it emits very toxic fumes of NO_x, SO_x, and Na_2O.

ALI750 CAS:118-03-6 ***HR: 1***
7-AMINO-1,3,6-NAPHTHALENETRISULFONIC ACID
mf: $C_{10}H_9NO_9S_3$ mw: 383.38

SYNS: KYSELINA KOCHOVA (CZECH) ◇ KYSELINA 2-NAFTYLAMIN-3,6,8-TRISULFONOVA (CZECH)

TOXICITY DATA with REFERENCE
skn-rbt 500 mg/24H MLD 28ZPAK -,190,72
eye-rbt 500 mg/24H MLD 28ZPAK -,190,72
orl-rat LD50:13 g/kg 28ZPAK -,190,72

CONSENSUS REPORTS: Reported in EPA TSCA Inventory.

SAFETY PROFILE: Mildly toxic by ingestion. A skin and severe eye irritant. See also SULFONATES. When heated to decomposition it emits very toxic fumes of NO_x and SO_x.

ALJ000 CAS:5959-52-4 ***HR: 2***
3-AMINO-2-NAPHTHOIC ACID
mf: $C_{11}H_9NO_2$ mw: 187.21

PROP: Yellow scales from dilute alcohol, ether. Mp: 214°.

SYNS: 3-AMINOISONAPHTHOIC ACID ◇ 3-AMINO-2-NAPHTHALENECARBOXYLIC ACID

TOXICITY DATA with REFERENCE
orl-mus LD50:1600 mg/kg 14CYAT 2,1840,63

CONSENSUS REPORTS: Reported in EPA TSCA Inventory.

SAFETY PROFILE: Moderately toxic by ingestion. When heated to decomposition it emits toxic fumes of NO_x. See also AROMATIC AMINES.

ALJ250 CAS:42884-33-3 ***HR: 3***
2-AMINO-1-NAPHTHOL
mf: $C_{10}H_9NO$ mw: 159.20

PROP: Mp: 255° (decomp). Sol in alc.

SYN: AMINONAPHTHALENOL

TOXICITY DATA with REFERENCE
ipr-rat TDLo:39 mg/kg/13W-I:ETA CNREA8 28,535,68
imp-mus TDLo:56 mg/kg:CAR SAIGBL 24,186,82

SAFETY PROFILE: Questionable carcinogen with experimental carcinogenic and tumorigenic data. When heated to decomposition it emits toxic fumes of NO_x.

ALJ750 CAS:118-46-7 ***HR: 3***
8-AMINO-2-NAPHTHOL
mf: $C_{10}H_9NO$ mw: 159.20

PROP: Crystals from benzene or ligroin. Mp: 95-97° (decomp); sol in hot water, alkali, and HCl.

TOXICITY DATA with REFERENCE
ivn-mus LD50:180 mg/kg CSLNX* NX#04024

CONSENSUS REPORTS: Reported in EPA TSCA Inventory.

SAFETY PROFILE: Poison by intravenous route. When heated to decomposition it emits toxic NO_x.

ALK000 CAS:1198-27-2 ***HR: 3***
1-AMINO-2-NAPHTHOL HYDROCHLORIDE
mf: $C_{10}H_9NO{\cdot}ClH$ mw: 195.66

PROP: Needles from alc. Mp: 201°; sltly sol in water; sol in alc and ether.

SYN: 2-HYDROXY-1-NAPHTHYLAMINE HYDROCHLORIDE

TOXICITY DATA with REFERENCE
dnr-esc 500 μg/well/16H CBINA8 15,219,76
imp-mus TDLo:100 mg/kg:CAR BJCAAI 10,539,56

CONSENSUS REPORTS: Reported in EPA TSCA Inventory. EPA Genetic Toxicology Program.

SAFETY PROFILE: Questionable carcinogen with experimental carcinogenic data. Mutation data reported. When heated to decomposition it emits very toxic fumes of HCl and NO_x.

ALK250 CAS:41772-23-0 ***HR: 3***
2-AMINO-1-NAPHTHOL HYDROCHLORIDE
mf: $C_{10}H_9NO{\cdot}ClH$ mw: 195.66

PROP: Needles. Mp: 255° (decomp); sol in alc.

SYN: 1-HYDROXY-2-NAPHTHYLAMINE HYDROCHLORIDE

TOXICITY DATA with REFERENCE
scu-rat TDLo:1100 mg/kg/45W-I:ETA BJCAAI 6,412,52
imp-mus TDLo:80 mg/kg:CAR BJCAAI 17,127,63
imp-mus TD:100 mg/kg:CAR BJCAAI 10,539,56

CONSENSUS REPORTS: Reported in EPA TSCA Inventory.

SAFETY PROFILE: Questionable carcinogen with experimental carcinogenic and tumorigenic data. When heated to decomposition it emits very toxic fumes of NO_x and HCl.

ALK500 CAS:5959-56-8 ***HR: 3***
4-AMINO-1-NAPHTHOL HYDROCHLORIDE
mf: $C_{10}H_9NO{\cdot}ClH$ mw: 195.66

SYN: 1-AMINO-4-NAPHTHOL HYDROCHLORIDE

TOXICITY DATA with REFERENCE
imp-mus TDLo:50 mg/kg:NEO BJCAAI 12,222,58

CONSENSUS REPORTS: Reported in EPA TSCA Inventory.

SAFETY PROFILE: Questionable carcinogen with experimental neoplastigenic data. When heated to decomposition it emits very toxic fumes of NO_x and HCl.

ALK625 CAS:5438-85-7 ***HR: 3***
2-AMINO-1,4-NAPHTHOQUINONE IMINE HYDROCHLORIDE
mf: $C_{10}H_8N_2O{\cdot}ClH$ mw: 208.66

SYNS: 2-AMINO-4-IMINO-1(4H)-NAPHTHALENONE HYDROCHLORIDE ◇ ANQI

TOXICITY DATA with REFERENCE
dni-mus:ast 20 μmol/L CPBTAL 17,105,69
oms-mus:ast 20 μmol/L CPBTAL 17,105,69
dnd-mam:lym 100 μmol/L CPBTAL 17,113,69
ipr-mus LD50:5450 μg/kg CPBTAL 17,1432,69

SAFETY PROFILE: Poison by intraperitoneal route. Mutation data reported. When heated to decomposition it emits toxic fumes of NO_x and HCl.

ALK750 CAS:605-92-5 ***HR: 3***
2-AMINO-1-NAPHTHYL ESTER SULFURIC ACID
mf: $C_{10}H_9NO_4S$ mw: 239.26

PROP: Sltly sol in water.

SYNS: 2-AMINO-1-NAPHTHYL HYDROGEN SULFATE ◇ 2-AMINO-1-NAPHTHYL HYDROGEN SULPHATE

TOXICITY DATA with REFERENCE
imp-mus TDLo:50 mg/kg:NEO BJCAAI 12,222,58

SAFETY PROFILE: Questionable carcinogen with experimental neoplastigenic data. See also ESTERS and SULFATES. When heated to decomposition it emits very toxic fumes of NO_x and SO_x.

ALL000 CAS:63976-07-8 ***HR: 3***
2-AMINO-1-NAPHTHYLGLUCOSIDURONIC ACID
mf: $C_{16}H_{17}NO_7$ mw: 335.34

SYN: 2-NAPHTHYLAMINE-1-d-GLUCOSIDURONICACID

TOXICITY DATA with REFERENCE
imp-mus TDLo:80 mg/kg:NEO BJCAAI 11,212,57

SAFETY PROFILE: Questionable carcinogen with experimental neoplastigenic data. When heated to decomposition it emits toxic fumes of NO_x.

ALL250 CAS:329-89-5 ***HR: 3***
6-AMINONICOTINAMIDE
mf: $C_6H_7N_3O$ mw: 137.16

SYNS: AMINONICOTINAMIDE ◇ 6-AMINONIKOTINSAEUREAMID (GERMAN) ◇ 6-AMINONICOTINIC ACID AMIDE ◇ 6-AMINO-NICOTINSAEUREAMID (GERMAN) ◇ 6-AN ◇ 6-ANA ◇ FDA 0121 ◇ NSC 21206 ◇ U-8774

TOXICITY DATA with REFERENCE
dlt-rat-ipr 1 mg/kg TXAPA9 19,371,71
orl-mus TDLo:20 mg/kg (female 8-12D post):REP TCMUD8 6,361,86
ipr-rbt TDLo:5 mg/kg (female 12D post):TER TJADAB 16,297,77
ipr-rat LD50:11 mg/kg CAXXA4 #1089763
orl-mus LDLo:320 mg/kg AECTCV 14,111,85
ipr-gpg LD50:10 mg/kg TXAPA9 33,320,75

SAFETY PROFILE: Poison by ingestion and intraperitoneal routes. An experimental teratogen. Other experimental reproductive effects. Mutation data reported. When heated to decomposition it emits toxic fumes of NO_x. A central nervous system depressant.

ALL500 CAS:99-56-9 ***HR: 2***
2-AMINO-4-NITROANILINE
mf: $C_6H_7N_3O_2$ mw: 153.16

SYNS: C.I. 76020 ◇ 1,2-DIAMINO-4-NITROBENZENE ◇ NCI-C03941 ◇ 4NDB ◇ 4-NITRO-1,2-BENZENEDIAMINE ◇ 4-NITRO-1,2-DIAMINOBENZENE ◇ 4-NITRO-o-PHENYLENE-DIAMINE ◇ 4-NITRO-1,2-PHENYLENEDIAMINE ◇ p-NITRO-o-PHENYLENEDIAMINE ◇ 4-NOPD

TOXICITY DATA with REFERENCE
mmo-sat 1 μg/plate ENMUDM 8(Suppl 7),1,86
dnr-esc 10 mg/L CRNGDP 2,189,81
mmo-asn 200 mg/L MUREAV 97,293,82
ipr-rat TDLo:3 g/kg (30D male):TER SheCW# 25MAR77
scu-mus TDLo:7680 mg/kg (6-15D preg):REP TJADAB 24,253,81
orl-rat LD50:681 mg/kg NCILB* NIH-NCI-E-C-72-3252
orl-mus LD50:681 mg/kg NCILB* NIH-NCI-E-C-72-3252

CONSENSUS REPORTS: IARC Cancer Review: Group 3 IMEMDT 7,56,87; Animal Inadequate Evidence IMEMDT 16,63,78. NCI Carcinogenesis Bioassay (feed); No Evidence: mouse, rat NCITR* NCI-CG-TR-180,79. Reported in EPA TSCA Inventory. EPA Genetic Toxicology Program.

SAFETY PROFILE: Moderately toxic by ingestion. An experimental teratogen. Other experimental reproductive effects. Mutation data reported. When heated to decomposition it emits toxic fumes of NO_x.

ALL750 CAS:5307-14-2 ***HR: 3***
4-AMINO-2-NITROANILINE
mf: $C_6H_7N_3O_2$ mw: 153.16

SYNS: C.I. 76070 ◇ C.I. OXIDATION BASE 22 ◇ 1,4-DIAMINO-2-NITROBENZENE ◇ DURAFUR BROWN ◇ DURAFUR BROWN 2R ◇ DYE GS ◇ FOURAMIEN 2R ◇ FOURRINE 36 ◇ FOURRINE BROWN 2R ◇ NCI-C02222 ◇ 2NDB ◇ 2-NITRO-1,4-BENZENEDIAMINE ◇ 2-NITRO-1,4-DIAMINOBENZENE ◇ NITRO-p-PHENYLENEDIAMINE ◇ o-NITRO-p-PHENYLENEDIAMINE (MAK) ◇ 2-NITRO-1,4-PHENYLENEDIAMINE ◇ 2-NITRO-p-PHENYLENEDIAMINE ◇ 2-NP ◇ 2-NPPD ◇ 2-N-p-PDA ◇ OXIDATION BASE 22 ◇ URSOL BROWN RR ◇ ZOBA BROWN RR

TOXICITY DATA with REFERENCE
mmo-sat 5 μg/plate NATUAS 255,506,75
dns-rat:lvr 100 mg/L MUREAV 97,359,82
otr-ham:emb 500 μg/L NCIMAV 58,243,81
sce-ham-orl 125 mg/kg BLFSBY 29B,613,83
cyt-hmn:lym 50 mg/L/24H NATUAS 255,506,75
scu-mus TDLo:1600 mg/kg (female 6-15D post):TER TJADAB 19,37A,79
scu-mus TDLo:2240 mg/kg (female 6-15D post):REP TJADAB 24,253,81

orl-mus TDLo:288 g/kg/78W-C:NEO NCITR* NCI-CG-TR-169,79
orl-mus TD:144 g/kg/78W-C:ETA NCITR* NCI-CG-TR-169,79
orl-rat LD50:3080 mg/kg JTEHD6 2,657,77
ipr-rat LD50:348 mg/kg JTEHD6 2,657,77

CONSENSUS REPORTS: IARC Cancer Review: Group 3 IMEMDT 7,56,87; Animal Inadequate Evidence IMEMDT 16,73,78. NCI Carcinogenesis Bioassay (feed); No Evidence: rat NCITR* NCI-CG-TR-169,79; Clear Evidence: mouse NCITR* NCI-CG-TR-169,79. Reported in EPA TSCA Inventory. EPA Genetic Toxicology Program.

DFG MAK: Suspected Carcinogen.

SAFETY PROFILE: Suspected carcinogen with experimental carcinogenic and neoplastigenic data. Poison by intraperitoneal route. Moderately toxic by ingestion. An experimental teratogen. Other experimental reproductive effects. Mutation data reported. When heated to decomposition it emits toxic fumes of NO_x.

ALM000 CAS:1211-40-1 ***HR: 3***
4-AMINO-4'-NITROBIPHENYL
mf: $C_{12}H_{10}N_2O_2$ mw: 214.24

SYN: 4'-NITRO-4-BIPHENYLAMINE

TOXICITY DATA with REFERENCE
orl-rat TDLo:1440 mg/kg/73W-I:ETA TXAPA9 14,661,69

SAFETY PROFILE: Questionable carcinogen with experimental tumorigenic data. When heated to decomposition it emits toxic fumes of NO_x.

ALM250 CAS:3775-55-1 ***HR: 3***
2-AMINO-5-(5-NITRO-2-FURYL)-1,3,4-OXADIAZOLE
mf: $C_8H_7N_4O_4$ mw: 223.19

TOXICITY DATA with REFERENCE
orl-rat TDLo:20 g/kg/46W-C:CAR JNCIAM 54,841,75

SAFETY PROFILE: Questionable carcinogen with experimental carcinogenic data. When heated to decomposition it emits toxic fumes of NO_x.

ALM500 CAS:38514-71-5 ***HR: 3***
2-AMINO-4-(5-NITRO-2-FURYL)THIAZOLE
mf: $C_7H_5N_3O_3S$ mw: 211.21

SYN: ANFT

TOXICITY DATA with REFERENCE
mma-sat 100 ng/plate MUREAV 40,9,76
mmo-esc 300 nmol/well CNREA8 34,2266,74
dnd-esc 10 μmol/L CBINA8 31,133,80
dnd-mam:lym 50 μmol/L CRNGDP 3,1339,82
orl-rat TDLo:30212 mg/kg/Y-C:CAR CRNGDP 3,275,82
orl-mus TDLo:68 g/kg/46W-C:NEO CNREA8 33,1593,73
orl-rat TD:40894 mg/kg/46W-C:ETA PAACA3 21,75,80

CONSENSUS REPORTS: EPA Genetic Toxicology Program.

SAFETY PROFILE: Questionable carcinogen with experimental carcinogenic, neoplastigenic, and tumorigenic data. Mutation data reported. When heated to decomposition it emits very toxic fumes of NO_x and SO_x.

ALM750 CAS:7532-52-7 ***HR: 2***
5-AMINO-3-(5-NITRO-2-FURYL)-s-TRIAZOLE
mf: $C_6H_5N_5O_3$ mw: 195.16

SYN: 3-(5-NITRO-2-FURYL)-1H-1,2,4-TRIAZOL-5-AMINE

TOXICITY DATA with REFERENCE
orl-mus LD50:4800 mg/kg JMCMAR 16,312,73
ipr-mus LD50:1460 mg/kg JMCMAR 16,312,73

SAFETY PROFILE: Moderately toxic by intraperitoneal route. Mildly toxic by ingestion. When heated to decomposition it emits toxic fumes of NO_x.

ALN250 CAS:50832-74-1 ***HR: D***
3-AMINO-6-(2-(5-NITRO-2-FURYL)VINYL) PYRIDAZINE HYDROCHLORIDE
mf: $C_{10}H_8N_4O_3 \cdot ClH$ mw: 268.68

SYNS: NIFURPRAZINE HYDROCHLORIDE ◇ (1-(5-NITRO-2-FURYL)-2-(6-AMINO-3-PYRIDAZYL)-ETHYLENEHYDROCHLORIDE

TOXICITY DATA with REFERENCE
mrc-smc 1000 ppm MGGEAE 139,255,75
cyt-hmn:leu 20 ppm MUREAV 42,109,77
cyt-ham:lng 20 ppm/2H MUREAV 42,109,77
sce-ham:lng 5 ppm/24H MUREAV 42,109,77

SAFETY PROFILE: Human mutation data reported. When heated to decomposition it emits very toxic fumes of HCl and NO_x.

ALN500 CAS:16239-84-2 ***HR: D***
2-AMINO-4-(2-(5-NITRO-2-FURYL)VINYL)THIAZOLE
mf: $C_9H_7N_3O_3S$ mw: 237.25

SYN: 1-(2-AMINOTHIAZOLYL)-2-(5-NITRO-2-FURYL)ETHYLENE

TOXICITY DATA with REFERENCE
mmo-sat 1 nmol/plate FEPRA7 41,330,82
mma-sat 100 ng/plate MUREAV 40,9,76
mmo-esc 300 nmol/well CNREA8 34,2266,74

CONSENSUS REPORTS: EPA Genetic Toxicology Program.

SAFETY PROFILE: Mutation data reported. When

heated to decomposition it emits very toxic fumes of NO_x and SO_x.

ALN750 CAS:18264-75-0 ***HR: 2***
1-AMINO-3-NITRO GUANIDINE
mf: $CH_5N_5O_2$ mw: 119.2

SAFETY PROFILE: Very unstable, explosive compound. See also NITRO COMPOUNDS of AROMATIC HYDROCARBONS. When heated to decomposition it emits toxic fumes of NO_x. Detonates @ 190° (mp).

ALO000 CAS:121-88-0 ***HR: 2***
2-AMINO-5-NITROPHENOL
mf: $C_6H_6N_2O_3$ mw: 154.14

$H_2N(NO_2)C_6H_3OH$

SYN: C.I. 76535 ◇ NCI C55970 ◇ URSOL YELLOW BROWN A

TOXICITY DATA with REFERENCE
mmo-sat 20 μg/plate PNASA6 72,2423,75
mma-sat 1 μmol/plate MUREAV 58,11,78
cyt-ham:lng 1 mg/L ATSUDG (4),41,80
orl-rat TDLo:51500 mg/kg/2Y-C:ETA NTPTR* NTP-TR-334,88

CONSENSUS REPORTS: NTP Carcinogenesis Studies (gavage): Some Evidence: rat NTPTR* NTP-TR-334,88. Reported in EPA TSCA Inventory. EPA Genetic Toxicology Program.

SAFETY PROFILE: Questionable carcinogen with experimental tumorigenic data. Mutation data reported. Potentially explosive reaction with nitrous acid. When heated to decomposition it emits toxic fumes of NO_x.

ALO500 CAS:61702-43-0 ***HR: 3***
2-AMINO-4-NITROPHENOL SODIUM SALT
mf: $C_6H_5N_2O_3{\cdot}Na$ mw: 176.12

SYN: l'ORTHO, p-AMINONITROPHENOL, SEL SODIQUE (FRENCH)

TOXICITY DATA with REFERENCE
ipr-dog LDLo:500 mg/kg AIPTAK 50,20,35
ipr-pgn LDLo:95 mg/kg AIPTAK 50,20,35

CONSENSUS REPORTS: Reported in EPA TSCA Inventory.

SAFETY PROFILE: Poison by intraperitoneal route. Moderately toxic by intraperitoneal route. When heated to decomposition it emits toxic fumes of NO_x and Na_2O.

ALO750 CAS:2871-01-4 ***HR: D***
2-((4-AMINO-2-NITROPHENYL)AMINO)ETHANOL
mf: $C_8H_{11}N_3O_3$ mw: 197.22

SYNS: HC RED No. 3 ◇ NCI-C54922

TOXICITY DATA with REFERENCE
mmo-sat 100 μg/plate NTPTR* NTP-TR-281,86
mma-sat 3300 ng/plate NTPTR* NTP-TR-281,86
orl-mus TDLo:182 g/kg/2Y-C:CAR NTPTR* NTP-TR-281,86

CONSENSUS REPORTS: NTP Carcinogenesis Studies (gavage); Equivocal Evidence: mouse NTPTR* NTP-TR-281,86; No Evidence: rat NTPTR* NTP-TR-281,86. Reported in EPA TSCA Inventory.

SAFETY PROFILE: Questionable carcinogen with experimental carcinogenic data. Mutation data. When heated to decomposition it emits toxic fumes of NO_x.

ALP000 CAS:2104-09-8 ***HR: 3***
2-AMINO-4-(p-NITROPHENYL)THIAZOLE
mf: $C_9H_7N_3O_2S$ mw: 221.25

TOXICITY DATA with REFERENCE
orl-rat TDLo:2150 mg/kg/13W-C:CAR JNCIAM 54,841,75
orl-mus TDLo:9600 mg/kg/46W-C:ETA CNREA8 33,1593,73

SAFETY PROFILE: Questionable carcinogen with experimental carcinogenic data. When heated to decomposition it emits toxic NO_x and SO_x.

ALP750 CAS:119-72-2 ***HR: 1***
4-AMINO-4'-NITRO-2,2'-STILBENEDISULFONIC ACID
mf: $C_{14}H_{12}N_2O_8S_2$ mw: 400.40

SYN: KYSELINA 4-NITRO-4'-AMINOSTILBEN-2,2'-DISULFONOVA (CZECH)

TOXICITY DATA with REFERENCE
eye-rbt 500 mg/24H MLD 28ZPAK -,194,72
orl-rat LD50:14200 mg/kg 28ZPAK -,194,72

CONSENSUS REPORTS: Reported in EPA TSCA Inventory.

SAFETY PROFILE: Mildly toxic by ingestion. An eye irritant. When heated to decomposition it emits very toxic fumes of NO_x and SO_x. See also SULFONATES.

ALQ000 CAS:121-66-4 ***HR: 3***
2-AMINO-5-NITROTHIAZOLE
mf: $C_3H_3N_3O_2S$ mw: 145.15

$SC(NH_2){=}NCH{=}CNO_2$ (ring: S–C bond)

SYNS: AMINONITROTHIAZOLE ◇ AMINONITROTHIAZOLUM ◇ AMINZOL SOLUBLE ◇ ENHEPTIN ◇ ENTRAMIN ◇ NCI-C03065 ◇ NITRAMIN ◇ NITRAMINE ◇ 5-NITRO-2-AMINOTHIAZOLE ◇ NITROMIN IDO ◇ 5-NITRO-2-THIAZOLYLAMINE ◇ USAF EK-6561

TOXICITY DATA with REFERENCE
mmo-sat 500 μg/plate WTMOA3 69,19,82
mma-sat 666 μg/plate ENMUDM 7(Suppl 5),1,85
mmo-esc 50 μmol/L MUREAV 118,153,83
mma-esc 800 μg/plate ENMUDM 7(Suppl 5),1,85
mmo-klp 200 μmol/L MUREAV 118,153,83
orl-rat TDLo:700 mg/kg (14D male):REP TXAPA9 2,418,60
orl-rat TDLo:28 g/kg/2Y-C:CAR NCITR* NCI-CG-TR-53,78
orl-rat TD:23 g/kg/46W-C:NEO JNCIAM 54,841,75
orl-rat TD:12 g/kg/2Y-C:ETA NCITR* NCI-CG-TR-53,78
ipr-mus LD50:200 mg/kg NTIS** AD277-689

CONSENSUS REPORTS: IARC Cancer Review: Group 3 IMEMDT 7,56,87; Animal Limited Evidence IMEMDT 31,71,83. NCI Carcinogenesis Bioassay (feed); No Evidence: mouse NCITR* NCI-CG-TR-53,78; Clear Evidence: rat NCITR* NCI-CG-TR-53,78. Reported in EPA TSCA Inventory.

SAFETY PROFILE: Poison by intraperitoneal route. Experimental reproductive effects. Questionable carcinogen with experimental carcinogenic, tumorigenic, and neoplastigenic data. Mutation data reported. When heated to decomposition it emits very toxic fumes of NO_x and SO_x. Incompatible with HNO_3 and H_2SO_4. An antiprotozoal agent.

ALQ625 CAS:58-60-6 ***HR: D***
AMINONUCLEOSIDE PUROMYCIN
mf: $C_{12}H_{18}N_6O_3$ mw: 294.36

SYNS: AMINONUCLEOSIDE ◇ SAN ◇ SYTLOMYCIN AMINONUCLEOSIDE

TOXICITY DATA with REFERENCE
dni-hmn:lng 6 mg/L JCLLAX 81,71,73
oms-hmn:lng 6 mg/L JCLLAX 81,71,73
scu-rat TDLo:105 μg/kg (female 5-11D post):REP NATUAS 195,1209,62
scu-rat TDLo:45 mg/kg (female 8-10D post):TER NPRNAY 3,344,66

CONSENSUS REPORTS: Reported in EPA TSCA Inventory.

SAFETY PROFILE: An experimental teratogen. Other experimental reproductive effects. Human mutation data reported. When heated to decomposition it emits toxic fumes of NO_x.

ALQ650 CAS:645-88-5 ***HR: 3***
AMINOOXYACETIC ACID
mf: $C_2H_5NO_3$ mw: 91.08

SYNS: AOAA ◇ (CARBOXYMETHOXY)AMINE ◇ (o-CARBOXYMETHYL)HYDROXYLAMINE ◇ U 7524

TOXICITY DATA with REFERENCE
mmo-bcs 1 mol/L MUREAV 4,517,67
ipr-mus LD50:40 mg/kg BCPCA6 28,1397,79
scu-mus LD50:40 mg/kg BCPCA6 27,103,78

SAFETY PROFILE: Poison by subcutaneous and intraperitoneal routes. Mutation data reported. When heated to decomposition it emits toxic fumes of NO_x.

ALQ750 CAS:64046-62-4 ***HR: 3***
2-AMINOOXYACETIC ACID BUTYL ESTER, HYDROCHLORIDE
mf: $C_6H_{13}NO_3{\bullet}ClH$ mw: 183.66

TOXICITY DATA with REFERENCE
ivn-rat LDLo:50 mg/kg JMPCAS 5,464,62
ipr-mus LD50:69 mg/kg JMPCAS 5,464,62

SAFETY PROFILE: Poison by intravenous and intraperitoneal routes. See also ESTERS. When heated to decomposition it emits very toxic fumes of NO_x and HCl.

ALR250 CAS:6191-22-6 ***HR: 3***
dl-AMINOPENTAMIDE HYDROCHLORIDE
mf: $C_{19}H_{24}N_2O{\bullet}ClH$ mw: 332.91

SYN: α,α-DIPHENYL-γ-DIMETHYLAMINOVALERAMIDE HYDROCHLORIDE

TOXICITY DATA with REFERENCE
orl-mus LD50:396 mg/kg JPETAB 100,325,50
ivn-mus LD50:35 mg/kg JPETAB 100,325,50

SAFETY PROFILE: Poison by ingestion and intravenous route. When heated to decomposition it emits very toxic fumes of NO_x and HCl.

ALR500 CAS:31699-72-6 ***HR: 3***
3-(5-AMINOPENTYL)INDOLE ADIPATE
mf: $C_{13}H_{18}N_2{\bullet}C_6H_{10}O_4$ mw: 348.49

SYN: ω-3-INDOLYLAMYLAMINE ADIPINATE.

TOXICITY DATA with REFERENCE
ipr-mus LD50:215 mg/kg RPTOAN 33,180,70
ivn-mus LD50:79 mg/kg RPTOAN 33,180,70

SAFETY PROFILE: Poison by intraperitoneal and intravenous routes. When heated to decomposition it emits toxic fumes of NO_x.

ALR750 CAS:28832-64-6 ***HR: 3***
AMINOPERIMIDINE
mf: $C_{11}H_9N_3$ mw: 183.23

TOXICITY DATA with REFERENCE
unk-mus LDLo:50 mg/kg ATMPA2 32,177,38

CONSENSUS REPORTS: Reported in EPA TSCA Inventory.

SAFETY PROFILE: Poison by unspecified route. When heated to decomposition it emits toxic fumes of NO_x.

ALS000 CAS:4176-53-8 ***HR: 3***
1-AMINOPHENANTHRENE
mf: $C_{14}H_{11}N$ mw: 193.26

TOXICITY DATA with REFERENCE
mma-sat ng/plate ENMUDM 6,497,84
orl-rat TDLo:250 mg/kg:ETA ZEKBAI 72,321,69

SAFETY PROFILE: Questionable carcinogen with experimental tumorigenic data. Mutation data reported. When heated to decomposition it emits toxic fumes of NO_x.

ALS250 CAS:63307-29-9 ***HR: 3***
17-(p-AMINOPHENETHYL)-MORPHINAN-3-OL (−)-
mf: $C_{24}H_{30}N_2O$ mw: 362.56

TOXICITY DATA with REFERENCE
orl-mus LD50:70 mg/kg 31ZPAG 2,85,66
ivn-mus LD50:12 mg/kg 31ZPAG 2,85,66

SAFETY PROFILE: Poison by ingestion and intravenous routes. When heated to decomposition it emits toxic fumes of NO_x.

ALS500 CAS:63732-42-3 ***HR: 3***
(−)-17-(m-AMINOPHENETHYL)-MORPHINAN-3-OL, HYDROCHLORIDE
mf: $C_{24}H_{30}N_2O•ClH$ mw: 399.02

TOXICITY DATA with REFERENCE
scu-mus LD50:160 mg/kg 31ZPAG 2,85,66
ivn-mus LD50:8 mg/kg 31ZPAG 2,85,66

SAFETY PROFILE: Poison by subcutaneous and intravenous routes. When heated to decomposition it emits very toxic fumes of NO_x and HCl.

ALS750 CAS:63732-43-4 ***HR: 3***
(±)-17-(p-AMINOPHENETHYL)MORPHINAN-3-OL, HYDROCHLORIDE
mf: $C_{24}H_{30}N_2O•ClH$ mw: 399.02

TOXICITY DATA with REFERENCE
scu-mus LD50:207 mg/kg 31ZPAG 2,85,66
ivn-mus LD50:41 mg/kg 31ZPAG 2,85,66

SAFETY PROFILE: Poison by subcutaneous and intravenous routes. When heated to decomposition it emits very toxic fumes of NO_x and HCl.

ALT000 CAS:95-55-6 ***HR: 3***
2-AMINOPHENOL
DOT: UN 2512
mf: C_7H_7NO mw: 109.14

PROP: Colorless needles. Mp: 173°; bp: subl. Sol in water and alc; very sol in ether.

SYNS: 2-AMINO-1-HYDROXYBENZENE ◇ o-AMINOPHENOL ◇ BASF URSOL 3GA ◇ BENZOFUR GG ◇ C.I. 76520 ◇ C.I. OXIDATION BASE 17 ◇ FOURAMINE OP ◇ o-HYDROXYANILINE ◇ 2-HYDROXYANILINE ◇ NAKO YELLOW EGA ◇ PARADONE OLIVE GREEN B ◇ PELAGOL 3GA ◇ PELAGOL GREY GG ◇ ZOBA 3GA

TOXICITY DATA with REFERENCE
eye-rbt 100 mg MLD FCTOD7 20,573,82
mma-sat 100 μg/plate ENMUDM 5(Suppl 1),3,83
ipr-ham TDLo:150 mg/kg (8D preg):REP TXAPA9 63,264,82
ipr-ham TDLo:150 mg/kg (8D preg):TER TXAPA9 63,264,82
orl-rat LD50:1300 mg/kg RPTOAN 34,307,71
ipr-rat LDLo:300 mg/kg AIPTAK 131,151,61
scu-rat LD50:37 mg/kg YKYUA6 32,1093,81
orl-mus LD50:1250 mg/kg GTPZAB 25(8),50,81
ipr-mus LD50:200 mg/kg NTIS** AD691-490
scu-cat LDLo:37 mg/kg AEXPBL 72,241,13

CONSENSUS REPORTS: Reported in EPA TSCA Inventory.

DOT Classification: Poison B; Label: St. Andrews Cross

SAFETY PROFILE: Poison by intraperitoneal and subcutaneous routes. Moderately toxic by ingestion route. An experimental teratogen. Other experimental reproductive effects. An eye irritant. Mutation data reported. When heated to decomposition it emits toxic NO_x. See also AROMATIC AMINES.

ALT250 CAS:123-30-8 ***HR: 3***
4-AMINOPHENOL
DOT: UN 2512
mf: C_6H_7NO mw: 109.14

PROP: Colorless crystals; sltly sol in water, alc, and ether; insol in chloroform. Mp: 189.6-190.2°, bp: 284° (decomp).

SYNS: ACTIVOL ◇ p-AMINOFENOL (CZECH) ◇ 4-AMINO-1-HYDROXYBENZENE ◇ p-AMINOPHENOL ◇ p-AMINOPHENOL (DOT) ◇ BASF URSOL P BASE ◇ BENZOFUR P ◇ CERTINAL ◇ C.I. OXIDATION BASE 6A ◇ CITOL ◇ DURAFUR BROWN RB ◇ FOURAMINE P ◇ FOURRINE 84 ◇ FOURRINE P BASE ◇ FURRO P BASE ◇ p-HYDROXYANILINE ◇ 4-HYDROXYANILINE ◇ NAKO BROWN R ◇ PAP ◇ PARANOL ◇ PELAGOL GREY P BASE ◇ PELAGOL P BASE ◇ RENAL AC ◇ RODINAL ◇ TERTRAL P BASE ◇ URSOL P ◇ URSOL P BASE ◇ ZOBA BROWN P BASE

TOXICITY DATA with REFERENCE
skn-rbt 12500 μg/24H MLD FCTXAV 15,607,77
eye-rbt 100 mg MLD BIOFX* 29-4/73
spm-mus-ipr 500 mg/kg/5D MUREAV 69,149,80
mmo-ome 5 mg/L MUREAV 173,233,86

unr-rat TDLo:563 mg/kg (female 1-20D post):TER GISAAA 48(9),19,83
ipr-ham TDLo:200 mg/kg (female 8D post):REP TXAPA9 63,264,82
orl-rat LD50:375 mg/kg BIOFX* 29-4/73
unr-rat LD50:675 mg/kg GISAAA 50(3),4,85
orl-mus LD50:420 mg/kg GISAAA 35,28,70
ipr-mus LDLo:100 mg/kg RBPMAZ 22,1,52
scu-mus LDLo:470 mg/kg AEPPAE 188,130,38
scu-rat LDLo:37 mg/kg AEXPBL 72,241,13
orl-bwd LD50:56200 μg/kg AECTCV 12,355,83

CONSENSUS REPORTS: Reported in EPA TSCA Inventory. EPA Genetic Toxicology Program.

DOT Classification: Poison B; Label: St. Andrews Cross

SAFETY PROFILE: Poison by ingestion, subcutaneous, and intraperitoneal routes. An experimental teratogen. Other experimental reproductive effects. An allergen and skin and eye irritant. Mutation data reported. Can cause contact dermatitis, bronchial asthma, and methemoglobinemia with cyanosis. When heated to decomposition it emits toxic fumes of NO_x.

ALT500 CAS:591-27-5 ***HR: 3***
m-AMINOPHENOL
DOT: UN 2521
mf: C_6H_7NO mw: 109.14

PROP: Prisms from toluene. Mp: 123°. Sol in water and alc; sltly sol in ether.

SYNS: m-AMINOFENOL (CZECH) ◇ 3-AMINO-1-HYDROXYBENZENE ◇ 3-AMINOPHENOL ◇ m-AMINOPHENOL (DOT) ◇ BASF URSOL EG ◇ C.I. 76545 ◇ C.I. OXIDATION BASE 7 ◇ FOURAMINE EG ◇ FOURRINE 65 ◇ FOURRINE EG ◇ FURRO EG ◇ FUTRAMINE EG ◇ 3-HYDROXYANILINE ◇ NAKO TEG ◇ PELAGOL EG ◇ RENAL EG ◇ TERTRAL EG ◇ URSOL EG ◇ ZOBA EG

TOXICITY DATA with REFERENCE
skn-rbt 12500 μg/24H MLD FCTXAV 15,607,77
eye-rbt 100 mg/24H MOD 28ZPAK -,109,72
sln-nsc 220 mg/L MUREAV 167,35,86
unr-rat TDLo:1245 mg/kg (1-20D preg):TER GISAAA 48(9),19,83
unr-rat TDLo:1245 mg/kg (1-20D preg):REP GISAAA 48(9),19,83
orl-rat LD50:924 mg/kg GTPZAB 32(1),49,88
ihl-rat LC50:1162 mg/m^3 GTPZAB 32(1),49,88
ipr-rat LDLo:1 g/kg AIPTAK 131,151,61
orl-mus LD50:401 mg/kg GTPZAB 32(1),49,88
ipr-mus LD50:150 mg/kg NTIS** AD691-490
scu-cat LDLo:70 mg/kg AEXPBL 72,241,13
orl-qal LD50:750 mg/kg AECTCV 12,355,83

CONSENSUS REPORTS: Reported in EPA TSCA Inventory. EPA Genetic Toxicology Program.

DOT Classification: IMO: Poison B; Label: St. Andrews Cross.

SAFETY PROFILE: Poison by ingestion, subcutaneous, and intraperitoneal routes. An experimental teratogen. Other experimental reproductive effects. Mutation data reported. A skin and eye irritant. When heated to decomposition it emits toxic fumes of NO_x.

ALT550 ***HR: D***
m-AMINOPHENOL, chlorinated
mf: C_6H_6ClNO mw: 143.58

TOXICITY DATA with REFERENCE
unr-rat TDLo:532 mg/kg (female 1-20D post):TER GISAAA 48(9),19,83
unr-rat TDLo:532 mg/kg (female 1-20D post):REP GISAAA 48(9),19,83

SAFETY PROFILE: An experimental teratogen. Other experimental reproductive effects. When heated to decomposition it emits toxic fumes of NO_x and Cl^-.

ALT750 CAS:63957-37-9 ***HR: 3***
m-AMINOPHENOL ANTIMONYL TARTRATE
mf: $C_6H_8NO \cdot C_4H_4O_7Sb$ mw: 395.98

TOXICITY DATA with REFERENCE
ipr-mus LD50:55 mg/kg AJTMAQ 25,263,45

CONSENSUS REPORTS: Antimony compounds are on the Community Right-To-Know List.

OSHA PEL: TWA 0.5 mg(Sb)/m^3
ACGIH TLV: TWA 0.5 mg(Sb)/m^3
NIOSH REL: (Antimony) TWA 0.5 mg(Sb)/m^3

SAFETY PROFILE: Poison by intraperitoneal route. When heated to decomposition it emits very toxic fumes of NO_x and Sb. See also ANTIMONY COMPOUNDS.

ALU000 CAS:63957-38-0 ***HR: 3***
o-AMINOPHENOL ANTIMONYL TARTRATE
mf: $C_6H_8NO \cdot C_4H_4O_7Sb$ mw: 395.98

SYN: o-AMINOPHENOL-OXO(TARTRATO)ANTIMONATE(1-)-

TOXICITY DATA with REFERENCE
ipr-mus LD50:63 mg/kg AJTMAQ 25,263,45

CONSENSUS REPORTS: Antimony compounds are on the Community Right-To-Know List.

OSHA PEL: TWA 0.5 mg(Sb)/m^3
ACGIH TLV: TWA 0.5 mg(Sb)/m^3
NIOSH REL: (Antimony) TWA 0.5 mg(Sb)/m^3

SAFETY PROFILE: Poison by intraperitoneal route. See also ANTIMONY COMPOUNDS. When heated to decomposition it emits very toxic fumes of NO_x and Sb.

ALU250 CAS:63957-39-1 ***HR: 3***
p-AMINOPHENOL ANTIMONYL TARTRATE
mf: $C_6H_8NO{\bullet}C_4H_4O_7Sb$ mw: 395.98

SYN: p-AMINOPHENOL-OXO(TARTRATO)ANTIMONATE(1-)-

TOXICITY DATA with REFERENCE
ipr-mus LD50:50 mg/kg AJTMAQ 25,263,45

CONSENSUS REPORTS: Antimony compounds are on the Community Right-To-Know List.

OSHA PEL: TWA 0.5 mg(Sb)/m^3
ACGIH TLV: TWA 0.5 mg(Sb)/m^3
NIOSH REL: (Antimony) TWA 0.5 mg(Sb)/m^3

SAFETY PROFILE: Poison by intraperitoneal route. See also ANTIMONY COMPOUNDS. When heated to decomposition it emits very toxic fumes of NO_x and Sb.

ALU500 CAS:51-78-5 ***HR: 2***
p-AMINOPHENOL HYDROCHLORIDE
mf: $C_6H_7NO{\bullet}ClH$ mw: 145.60

PROP: Colorless prisms. Mp: 306° (decomp). Sol in water and alc.

SYN: 4-AMINOPHENOL HYDROCHLORIDE.

TOXICITY DATA with REFERENCE
ipr-mus LD50:750 mg/kg NTIS** AD691-490

CONSENSUS REPORTS: Reported in EPA TSCA Inventory.

SAFETY PROFILE: Moderately toxic by intraperitoneal route. When heated to decomposition it emits very toxic fumes of HCl and NO_x.

ALU750 CAS:69782-45-2 ***HR: 2***
p-AMINOPHENOL TARTRATE
mf: $C_6H_7NO{\bullet}C_4H_6O_6$ mw: 259.24

TOXICITY DATA with REFERENCE
ivn-rbt LDLo:1 g/kg AEXPBL 33,216,1894
scu-gpg LDLo:2 g/kg AEXPBL 33,216,1894
scu-frg LDLo:1515 mg/kg AEXPBL 33,216,1894

SAFETY PROFILE: Moderately toxic by subcutaneous route. Mildly toxic by intravenous and subcutaneous routes. When heated to decomposition it emits toxic fumes of NO_x.S

ALU875 CAS:76487-32-6 ***HR: 3***
1-(3-(p-AMINOPHENOXY)PROPYL)-4-(o-METHOXYPHENYL)PIPERAZINE DIHYDROCHLORIDE
mf: $C_{20}H_{27}N_3O_2{\bullet}2ClH$ mw: 414.42

SYNS: 1-(p-AMINOPHENOXY)-3-(N$^{(1)}$-(o-METHOXYPHENYL)-n$^{(4)}$-PIPERAZINYL)PROPANE 2HCl ◇ COMPOUND 74-637

TOXICITY DATA with REFERENCE
orl-rat LD50:416 mg/kg DRFUD4 6,346,81
orl-mus LD50:319 mg/kg DRFUD4 6,346,81
ipr-mus LD50:158 mg/kg DRFUD4 6,346,81

SAFETY PROFILE: Poison by ingestion and intraperitoneal routes. When heated to decomposition it emits toxic fumes of NO_x and HCl.

ALV000 CAS:15686-71-2 ***HR: 3***
7-(d-α-AMINOPHENYLACETAMIDO) DESACETOXYCEPHALOSPORANIC ACID
mf: $C_{16}H_{17}N_3O_4S$ mw: 347.42

SYNS: 7-(d-2-AMINO-2-PHENYLACETAMIDO)-3-METHYL-Δ(SUP 3)-CEPHEM-4-CARBOXYLIC ACID ◇ CEFA-ISKIA ◇ CEFALOTO ◇ CEPHALEXIN ◇ CEPOREX ◇ CEPOREXIN ◇ CEPOREXINE ◇ CEX ◇ KEFLEX ◇ KEFORAL ◇ LARIXIN ◇ LEXIBIOTICO ◇ MEDLEXIN ◇ NEOLEXINA ◇ ORACEF ◇ OROXIN ◇ ORTISPORINA ◇ S 6437 ◇ SARTOSONA ◇ SENCEPHALIN ◇ SYNCL

TOXICITY DATA with REFERENCE
orl-rat TDLo:52 g/kg (female 17-22D post):REP NKRZAZ 27(Suppl 7),865,79
orl-rat TDLo:4 g/kg (female 9-14D post):TER OYYAA2 3,249,69
orl-hmn TDLo:14 mg/kg/D:GIT AACHAX -,361,68
ipr-rat LD50:4 g/kg KSRNAM 3,390,69
scu-rat LD50:6100 mg/kg KSRNAM 3,390,69
orl-mus LD50:1495 mg/kg NKRZAZ 27(Suppl 7),765,79
ipr-mus LD50:400 mg/kg AACHAX-,489,68
scu-mus LD50:1150 mg/kg OYYAA2 3,227,69

SAFETY PROFILE: Poison by intraperitoneal route. Moderately toxic by ingestion and other routes. An experimental teratogen. Other experimental reproductive effects. Human systemic effects by ingestion: nausea, vomiting, and diarrhea. When heated to decomposition it emits very toxic fumes of NO_x and SO_x.

ALV100 CAS:73791-39-6 ***HR: 3***
p-AMINOPHENYLARSINE OXIDE DIHYDRATE
mf: $C_6H_6AsNO{\bullet}2H_2O$ mw: 219.09

SYNS: ANILINE, p-ARSENOSO-, DIHYDRATE ◇ 4-ARSENOSOANILINE, DIHYDRATE ◇ ARSINE, (p-AMINOPHENYL)OXO-, DIHYDRATE

TOXICITY DATA with REFERENCE
ipr-mus LD50:4430 μg/kg JPETAB 70,211,40
ivn-mus LD50:100 mg/kg CSLNX* NX#06293

OSHA PEL: TWA 0.5 mg(As)/m^3
ACGIH TLV: TWA 0.2 mg(As)/m^3

SAFETY PROFILE: Poison by intraperitoneal and intravenous route. When heated to decomposition it emits toxic fumes of NO_x and As.

ALV500 CAS:43087-91-8 ***HR: 2***
5-AMINO-2-PHENYLBENZOTHIAZOLE
mf: $C_{13}H_{10}N_2S$ mw: 226.31

SYNS: FABT (CZECH) ◇ 2-FENYL-5-AMINOBENZTHIAZOL (CZECH)

TOXICITY DATA with REFERENCE
eye-rbt 100 mg/24H SEV 28ZPAK -,203,72
orl-rat LD50:2940 mg/kg 28ZPAK -,203,72

SAFETY PROFILE: Moderately toxic by ingestion. Severe eye irritant. When heated to decomposition it emits very toxic fumes of NO_x and SO_x.

ALV750 CAS:20123-68-6 ***HR: 3***
1-m-AMINOPHENYL-2-CYCLOPROPYLAMINO-ETHANOLDIHYDROCHLORIDE
mf: $C_{11}H_{16}N_2O•2ClH$ mw: 265.21

SYN: AB-15

TOXICITY DATA with REFERENCE
orl-rat LD50:3250 mg/kg BCPCA6 18,2293,69
ipr-rat LD50:710 mg/kg BCPCA6 18,2293,69
ivn-rat LD50:390 mg/kg BCPCA6 18,2293,69
orl-mus LD50:1060 mg/kg BCPCA6 18,2293,69
ipr-mus LD50:470 mg/kg BCPCA6 18,2293,69
ivn-mus LD50:260 mg/kg BCPCA6 18,2293,69

SAFETY PROFILE: Poison by intravenous route. Moderately toxic by ingestion and intraperitoneal routes. When heated to decomposition it emits very toxic fumes of Cl^- and NO_x.

ALW000 CAS:63979-26-0 ***HR: 3***
1-(4-AMINOPHENYL)-4-(DIETHYLCARBOXAMIDE)-5-METHYL-1,2,3-TRIAZOLE HYDROCHLORIDE

SYN: SKF-183A

TOXICITY DATA with REFERENCE
orl-rat LD50:494 mg/kg TXAPA9 1,150,59
ipr-mus LD50:260 mg/kg TXAPA9 1,150,59

SAFETY PROFILE: Poison by intraperitoneal route. Moderately toxic by ingestion. When heated to decomposition it emits very toxic fumes of HCl and NO_x.

ALW250 CAS:98-84-0 ***HR: 2***
1-AMINO-1-PHENYLETHANE
mf: $C_8H_{11}N$ mw: 121.20

SYNS: α-METHYLBENZYLAMINE ◇ α-PHENYLETHYLAMINE ◇ 1-PHENYLETHYLAMINE

TOXICITY DATA with REFERENCE
skn-rbt 10 mg/24H SEV AMIHBC 4,119,51
eye-rbt 250 μg SEV AMIHBC 4,119,51
orl-rat LD50:940 mg/kg AMIHBC 4,119,51
skn-rbt LD50:780 mg/kg AMIHBC 4,119,51

SAFETY PROFILE: Moderately toxic by ingestion and skin contact. A skin and severe eye irritant. See also AMINES. When heated to decomposition it emits toxic fumes of NO_x.

ALW500 CAS:64038-09-1 ***HR: 3***
5-(p-AMINOPHENYL)-5-ETHYL-1-METHYL-BARBITURIC ACID
mf: $C_{13}H_{15}N_3O_3$ mw: 261.31

SYN: PAM

TOXICITY DATA with REFERENCE
ims-mus LD50:210 mg/kg TXAPA9 47,305,79
ipr-rat LD50:780 mg/kg PHMCAA 5,237,63
ivn-mus LD50:160 mg/kg ARZNAD 11,809,61

SAFETY PROFILE: Poison by intravenous and intramuscular routes. Moderately toxic by intraperitoneal route. When heated to decomposition it emits toxic fumes of NO_x.

ALW750 CAS:144-14-9 ***HR: 3***
N-β-(p-AMINOPHENYL)ETHYLNORMEPERIDINE
mf: $C_{22}H_{28}N_2O_2$ mw: 352.52

SYNS: 1-(p-AMINOPHENETHYL)-4-PHENYLISONIPECOTICACID, ETHYL ESTER ◇ 1-(p-AMINOPHENETHYL)-4-PHENYLPIPERIDINE-4-CARBOXYLIC ACID ETHYL ESTER ◇ N-(β-(p-AMINOPHENYL)ETHYL)-4-PHENYL-4-CARBETHOXYPIPERIDINE ◇ ETHYL-1-(p-AMINOPHENETHYL)-4-PHENYLISONIPECOTATE

TOXICITY DATA with REFERENCE
orl-rat LD50:175 mg/kg 27ZIAQ -,-,65
ipr-rat LD50:45 mg/kg 27ZIAQ -,44,73
scu-rat LD50:163 mg/kg 27ZIAQ -,-,65
orl-mus LD50:128 mg/kg 27ZIAQ -,44,73
ipr-mus LD50:53 mg/kg 27ZIAQ -,-,65
scu-mus LD50:100 mg/kg 27ZIAQ -,44,73
ivn-mus LD50:25 mg/kg 27ZIAQ -,44,73

SAFETY PROFILE: Poison by ingestion, subcutaneous, intravenous, and intraperitoneal routes. When heated to decomposition it emits toxic fumes of NO_x.

ALW900 CAS:66471-17-8 ***HR: 2***
2-(m-AMINOPHENYL)-3-INDOLECARBOXALDEHYDE, 4-(m-TOLYL)-3-THIOSEMICARBAZONE
mf: $C_{23}H_{21}N_5S$ mw: 399.55

SYN: INDOLE-3-CARBOXALDEHYDE, 2-(m-AMINOPHENYL)-, 4-(m-TOLYL)-3-THIOSEMICARBAZONE

TOXICITY DATA with REFERENCE
orl-rbt TDLo:80 mg/kg (female 1D pre):REP IJMRAQ 66,983,77
orl-rbt LD50:800 mg/kg IJMRAQ 66,983,77

SAFETY PROFILE: Moderately toxic by ingestion. Experimental reproductive effects. When heated to decomposition it emits toxic fumes of NO_x and SO_x.

ALX000 CAS:130-17-6 ***HR: 3***
2-(p-AMINOPHENYL)-6-METHYLBENZOTHIAZOLYL-7-SULFONIC ACID
mf: $C_{14}H_{12}N_2O_3S_2$ mw: 320.40

SYNS: 7-BENZOTHIAZOLESULFONIC ACID ◇ 6-METHYL-2-(p-AMINO PHENYL)

TOXICITY DATA with REFERENCE
ivn-mus LD50:178 mg/kg CSLNX* NX#00718

CONSENSUS REPORTS: Reported in EPA TSCA Inventory.

SAFETY PROFILE: Poison by intravenous route. See also SULFONATES. When heated to decomposition it emits very toxic fumes of SO_x and NO_x.

ALX250 CAS:13425-22-4 ***HR: 3***
2-AMINO-5-PHENYL-OXAZOLINE FORMATE
mf: $C_9H_{10}N_2O•C_4H_4O_4$ mw: 278.29

SYNS: AMINOREXFUMARATE ◇ MENOCIL

TOXICITY DATA with REFERENCE
orl-chd TDLo:1 mg/kg:CNS ATXKA8 26,117,70
orl-hmn TDLo:3 mg/kg:CNS ATXKA8 26,117,70
orl-rat LD50:25 mg/kg ATXKA8 26,117,70

SAFETY PROFILE: Poison by ingestion. Human central nervous system effects by ingestion. When heated to decomposition it emits toxic fumes of NO_x.

ALX500 CAS:61706-44-3 ***HR: 3***
2-(p-AMINOPHENYL)-2-PHENYLPROPIONAMIDE
mf: $C_{15}H_{16}N_2O$ mw: 240.33

SYNS: 2-FENIL-2-(p-AMINOFENIL)PROPIONAMMIDE(ITALIAN) ◇ 2-PHENYL-2-(p-AMINOPHENYL)PROPIONAMIDE

TOXICITY DATA with REFERENCE
orl-rat LD50:1600 mg/kg FRPSAX 31,671,76
orl-mus LD50:260 mg/kg FRPSAX 31,671,76

SAFETY PROFILE: Poison by ingestion. When heated to decomposition it emits toxic fumes of NO_x.

ALX750 CAS:3314-35-0 ***HR: 3***
3-AMINO-1-PHENYL-2-PYRAZOLINE
mf: $C_9H_{11}N_3$ mw: 161.23

SYN: 1-FENYL-3-AMINOPYRAZOLIN(CZECH)

TOXICITY DATA with REFERENCE
skn-rbt 500 mg/24H MOD 28ZPAK -,144,72
eye-rbt 100 mg/24H SEV 28ZPAK -,144,72
orl-rat LD50:78 mg/kg 28ZPAK -,144,72

SAFETY PROFILE: Poison by ingestion. Moderately toxic skin irritant. Severe eye irritant. When heated to decomposition it emits toxic fumes of NO_x.

ALX879 ***HR: 1***
4-AMINO-5-PHENYL-3-PYRAZOLYL METHYL KETONE
mf: $C_{11}H_{11}N_3O$ mw: 201.25

SYN: 1-(4-AMINO-5-PHENYL(1H)-PYRAZOL-3-YL)ETHANONE

TOXICITY DATA with REFERENCE
orl-mus LD50:1 g/kg FRPSAX 39,618,84

SAFETY PROFILE: Moderately toxic by ingestion. When heated to decomposition it emits toxic fumes of NO_x. See also KETONES.

ALY000 CAS:33421-40-8 ***HR: D***
2-AMINO-5-PHENYLPYRIDINE
mf: $C_{11}H_{10}N_2$ mw: 170.23

SYN: d,l-PHENYLALANINE, PYROLYZATE

TOXICITY DATA with REFERENCE
mma-sat 100 μg/plate CPBTAL 26,611,78

SAFETY PROFILE: Mutation data reported. When heated to decomposition it emits toxic fumes of NO_x.

ALY250 CAS:134-37-2 ***HR: 2***
1-(3-AMINOPHENYL)-2-PYRIDONE
mf: $C_{11}H_{10}N_2O$ mw: 186.23

PROP: Crystals. Mp. 182.5-184.5°.

SYNS: AMINOPHENYLPYRIDONE ◇ 1-(3-AMINOPHENYL)-2-(1H)-PYRIDINONE ◇ 1-(m-AMINOPHENYL)-2(1H)-PYRIDONE ◇ 1-m-AMINOPHENYL-2-PYRIDONE ◇ AMPHENIDONE ◇ DORNWAL ◇ DORNWALL

TOXICITY DATA with REFERENCE
orl-rat LD50:2300 mg/kg FEPRA7 19,390,60
orl-mus LD50:1300 mg/kg FEPRA7 19,390,60
ivn-mus LD50:660 mg/kg 27ZQAG -,201,72

SAFETY PROFILE: Moderately toxic by ingestion and intravenous routes. When heated to decomposition it emits toxic fumes of NO_x.

ALY500 CAS:41136-03-2 ***HR: 3***
2-AMINO-5-PHENYLTHIOMETHYL-2-OXAZOLINE
mf: $C_{10}H_{12}N_2OS$ mw: 208.30

TOXICITY DATA with REFERENCE
orl-mus LD50:147 mg/kg JMCMAR 16,510,73
ipr-mus LD50:178 mg/kg JMCMAR 16,510,73

SAFETY PROFILE: Poison by ingestion and intraperitoneal routes. When heated to decomposition it emits very toxic fumes of SO_x and NO_x.

ALY675 CAS:4922-98-9 ***HR: 3***
3-AMINO-3-PHENYL-1,2,4-TRIAZOLE
mf: $C_8H_8N_4$ mw: 160.18

SAFETY PROFILE: Reaction with nitrous acid gives a touch sensitive explosive product. Upon decomposition it emits toxic fumes of NO_x.

ALY750 CAS:59690-88-9 ***HR: 1***
1-(m-AMINOPHENYL)UREA HYDROCHLORIDE
mf: $C_7H_9N_3O{\bullet}ClH$ mw: 187.65

SYN: m-AMINOFENYLMOCOVINA HYDROCHLORID (CZECH)

TOXICITY DATA with REFERENCE
eye-rbt 100 mg/24H SEV 28ZPAK -,165,72
orl-rat LD50:4920 mg/kg 28ZPAK -,165,72

SAFETY PROFILE: Mildly toxic by ingestion. A severe eye irritant. When heated to decomposition it emits very toxic fumes of NO_x and HCl.

ALZ000 CAS:51249-05-9 ***HR: 2***
AMINOPHON
mf: $C_{18}H_{37}NO_3P$ mw: 346.53

SYNS: 1-(BUTYLAMINO)CYCLOHEXYLPHOSPHONIC ACID DIBUTYL ESTER ◇ O,O-DIBUTYL-1-BUTYLAMINO-CYCLOHEXYLPHOSPHONATE

TOXICITY DATA with REFERENCE
orl-rat LD50:3000 mg/kg EQSFAP 3,686,75
skn-rat LD50:1200 mg/kg EQSFAP 3,686,75
ipr-rat LD50:1385 mg/kg EQSFAP 3,686,75
orl-mus LD50:3475 mg/kg EQSFAP 3,686,75
skn-rbt LD50:500 mg/kg EQSFAP 3,686,75
orl-ham LD50:10000 mg/kg EQSFAP 3,686,75

SAFETY PROFILE: Moderately toxic by several routes. When heated to decomposition it emits very toxic fumes of PO_x and NO_x.

AMA000 CAS:1990-90-5 ***HR: 2***
4-AMINO-3-PICOLINE
mf: $C_6H_8N_2$ mw: 108.16

SYN: PHILLIPS 1908

TOXICITY DATA with REFERENCE
orl-rat LD50:446 mg/kg TXAPA9 21,315,72
orl-bwd LD50:2 mg/kg TXAPA9 21,315,72

SAFETY PROFILE: Moderately toxic by ingestion. When heated to decomposition it emits toxic fumes of NO_x.

AMA250 CAS:616-30-8 ***HR: 3***
3-AMINO-1,3-PROPANEDIOL
mf: $C_3H_9NO_2$ mw: 91.13

SYNS: 1-AMINOGLYCEROL ◇ 2,3-DIHYDROXYPROPYLAMINE

TOXICITY DATA with REFERENCE
orl-ham TDLo:784 mg/kg/41W-I:CAR JCROD7 109,1,85
orl-rat LD50:7500 mg/kg SCCUR* -,1,61
orl-mus LD50:2460 mg/kg SCCUR* -,1,61
ipr-mus LD50:246 mg/kg SCCUR* -,1,61
ipr-rbt LD50:198 mg/kg SCCUR* -,1,61

CONSENSUS REPORTS: Reported in EPA TSCA Inventory.

SAFETY PROFILE: Poison by intraperitoneal route. Moderately toxic by ingestion. Questionable carcinogen with experimental carcinogenic data. When heated to decomposition it emits toxic fumes of NO_x.

AMA500 CAS:78-96-6 ***HR: 3***
1-AMINOPROPAN-2-OL
mf: C_3H_9NO mw: 75.13

$H_2NCH_2CHOHCH_3$

PROP: Liquid, slt ammonia odor, sol in water. D: 0.969, mp: 1.4°, flash p: 171°F, vap d: 2.6.

SYNS: α-AMINOISOPROPYL ALCOHOL ◇ 1-AMINO-2-PROPANOL ◇ 2-HYDROXYPROPYLAMINE ◇ ISOPROPANOLAMINE ◇ MONO-ISO-PROPANOLAMINE ◇ THREAMINE

TOXICITY DATA with REFERENCE
skn-rbt 485 mg open MOD UCDS** 5/21/71
eye-rbt 970 μg SEV UCDS** 5/21/71
orl-ham TDLo:168 mg/kg/21W-I:CAR JCROD7 109,1,85
orl-rat LD50:4260 mg/kg JIHTAB 31,60,49
ipr-mus LDLo:250 mg/kg CBCCI* 4,232,52
skn-rbt LD50:1640 mg/kg UCDS** 5/21/71

CONSENSUS REPORTS: Reported in EPA TSCA Inventory.

SAFETY PROFILE: Poison by intraperitoneal route. Moderately toxic by ingestion and skin contact. A skin and severe eye irritant. Questionable carcinogen with experimental carcinogenic data. Moderately flammable by heat, flame, sparks, powerful oxidizers. Ignites on contact with cellulose nitrate of high surface area. Catalyzes the explosive polymerization of 2,4-hexadienal. To fight fire, use alcohol foam. When heated to decomposition it emits toxic fumes of NO_x.

AMA750 ***HR: 2***
3-AMINOPROPANOL
mf: C_3H_9NO mw: 75.11

PROP: Colorless liquid, fishy odor. Bp: 168° @ 500 mm, flash p: 175°F (TOC), fp: 12.4°, d: 0.9786 @ 30°, vap press: 2.1 mm @ 60°, vap d: 2.59.

SAFETY PROFILE: Moderately toxic by skin contact and ingestion. An irritant. See also AMINES. Combustible liquid. Incompatible with oxidizing materials. To fight fire, use foam, CO_2, dry chemical.

AMB000 CAS:138-61-4 ***HR: 3***
AMINOPROPANOL PYROCATECHOLHYDRO-CHLORIDE
mf: $C_9H_{13}NO_3$•ClH mw: 219.69

SYNS: 3,4-DIHYDROXYNOREPHEDRINE HYDROCHLORIDE ◇ 3,4-DIHYDROXYPHENYLAMINOPROPANOL HYDROCHLORIDE ◇ 3,4-DIHYDROXYPHENYLPROPANOLAMINE HYDROCHLORIDE ◇ ISOADRENALINE HYDROCHLORIDE ◇ α-METHYLNORADRENALINE HYDROCHLORIDE ◇ NORHOMOEPINEPHRINE HYDROCHLORIDE

TOXICITY DATA with REFERENCE
scu-rat LDLo:3 mg/kg JPETAB 71,62,41
ivn-rbt LDLo:11 mg/kg JACSAT 53,4149,31

SAFETY PROFILE: Poison by subcutaneous and intravenous routes. When heated to decomposition it emits very toxic fumes of NO_x and Cl^-.

AMB250 ***HR: 3***
2-AMINO PROPIONITRILE
mf: $C_3H_6N_2$ mw: 70.1

CONSENSUS REPORTS: Cyanide and its compounds are on the Community Right To Know List.

SAFETY PROFILE: A poison and dangerous fire hazard. Can explode in storage. See also NITRILES. Upon decomposition it emits toxic fumes of CN^- and NO_x.

AMB500 CAS:151-18-8 ***HR: 2***
3-AMINOPROPIONITRILE
mf: $C_3H_6N_2$ mw: 70.11

$$H_2NC_2H_4C \equiv N$$

PROP: Liquid, amine odor. Bp: 185°.

SYNS: β-AMINOPROPIONITRILE ◇ BAPN ◇ β-CYANOETHYLAMINE

TOXICITY DATA with REFERENCE
sce-mus:emb 1250 mg/kg ARTODN 47,305,81
ipr-mus TDLo:1250 mg/kg (female 10D post):TER ARTODN 47,305,81
orl-rat TDLo:5 g/kg (female 14-15D post):REP TJADAB 4,227,71
skn-mus LDLo:12800 mg/kg EMPSAL 39,154,83
ipr-mus LD50:1152 mg/kg EMPSAL 39,154,83

CONSENSUS REPORTS: EPA Genetic Toxicology Program. Reported in EPA TSCA Inventory. Cyanide and its compounds are on the Community Right-To-Know List.

SAFETY PROFILE: Moderately toxic by intraperitoneal route. An experimental teratogen. Other experimental reproductive effects. Mutation data reported. Nitriles usually have cyanide-like effects. See also CYANIDE. Easily oxidized and unstable. A storage hazard; it polymerizes to an explosive yellow solid. When heated to decomposition it emits toxic fumes of CN^- and NO_x. For fire and explosion hazards see CYANIDE.

AMB750 CAS:2079-89-2 ***HR: 2***
β-AMINOPROPIONITRILE FUMARATE
mf: $C_3H_6N_2$•$2C_4H_4O_4$ mw: 302.27

SYNS: β-APN ◇ BAPN FUMARATE ◇ DI-β-AMINOPROPIONITRILE FUMARATE ◇ DI-BAPN FUMARATE

TOXICITY DATA with REFERENCE
orl-rat TDLo:1200 mg/kg (female 15-20D post):TER ARPAAQ 81,60,66
orl-rat LDLo:800 mg/kg TJADAB 5,33,72
orl-ham LD50:5000 mg/kg TJADAB 14,43,76

CONSENSUS REPORTS: Reported in EPA TSCA Inventory. Cyanide and its compounds are on the Community Right-To-Know List.

SAFETY PROFILE: Moderately toxic by ingestion. An experimental teratogen. When heated to decomposition it emits toxic fumes of NO_x and CN^-. See also NITRILES.

AMC000 CAS:70-69-9 ***HR: 3***
p-AMINOPROPIOPHENONE
mf: $C_9H_{11}NO$ mw: 149.21

PROP: Mp: 140°.

SYNS: 1-(4-AMINOPHENYL)-1-PROPANONE ◇ ETHYL-p-AMINOPHENYL KETONE ◇ PAPP ◇ PARAMINOPROPIOPHENONE ◇ USAF UCTL-1856

TOXICITY DATA with REFERENCE
orl-rat LD50:177 mg/kg BECTA6 30,122,83
ipr-rat LDLo:525 mg/kg CJBPAZ 38,667,60
orl-mus LD50:168 mg/kg GEPHDP 14,465,83
ipr-mus LD50:80 mg/kg NTIS** AD277-689
orl-cat LD50:5600 μg/kg BECTA6 30,122,83
orl-mam LD50:5600 μg/kg BECTA6 30,122,83

CONSENSUS REPORTS: Reported in EPA TSCA Inventory. EPA Extremely Hazardous Substances List.

SAFETY PROFILE: Poison by ingestion and intraperitoneal routes. Ingestion of large doses can cause cyanosis. When heated to decomposition it emits toxic fumes of NO_x.

AMC250 CAS:112-33-4 ***HR: 1***
3-AMINOPROPOXY-2-ETHOXY ETHANOL
mf: $C_7H_{17}NO_3$ mw: 163.25

SYN: POLYGLYCOLAMINE H-163

TOXICITY DATA with REFERENCE
skn-rbt 500 mg open MLD UCDS** 1/20/72
orl-rat LD50:6500 mg/kg AIHAAP 30,470,69
ihl-rat LCLo:20000 ppm/30M AIHAAP 30,470,69
skn-rbt LD50:5990 mg/kg AIHAAP 30,470,69

CONSENSUS REPORTS: Reported in EPA TSCA Inventory.

SAFETY PROFILE: Mildly toxic by ingestion, inhalation and skin contact. A skin irritant. When heated to decomposition it emits toxic fumes of NO_x.

AMC500 CAS:31897-98-0 ***HR: 3***
1-(3-AMINOPROPYL)ADAMANTANEHYDROCHLORIDE
mf: $C_{13}H_{23}N•ClH$ mw: 229.83

SYN: 3-(1-ADAMANTYL)PROPYLAMINEHYDROCHLORIDE

TOXICITY DATA with REFERENCE
orl-mus LD50:600 mg/kg JMCMAR 17,602,74
ipr-mus LD50:150 mg/kg JMCMAR 17,602,74

SAFETY PROFILE: Poison by intraperitoneal route. Moderately toxic by ingestion. When heated to decomposition it emits very toxic fumes of HCl and NO_x.

AMC750 CAS:63717-27-1 ***HR: 3***
2-((3-AMINOPROPYL)AMINO) ETHANETHIOL, DIHYDROGEN PHOSPHATE (ester-HYDRATE)
mf: $C_5H_{15}N_2O_3PS•H_2O$ mw: 232.27

SYN: PHOSPHOROTHIOIC ACID, S-ESTER with 2-((3-AMINOPROPYL)AMINO)ETHANETHIOL, HYDRATE

TOXICITY DATA with REFERENCE
ipr-mus LD50:700 mg/kg JMCMAR 12,236,69
unk-mus LD50:375 mg/kg JMCMAR 9,911,66

SAFETY PROFILE: Poison by unspecified route. Moderately toxic by intraperitoneal route. When heated to decomposition it emits very toxic fumes of NO_x, PO_x, and SO_x. See also ESTERS.

AMD000 CAS:20537-88-6 ***HR: 3***
AMINOPROPYL AMINOETHYLTHIOPHOSPHATE
mf: $C_5H_{15}N_2O_3PS$ mw: 214.25

SYNS: AMINOFOSTINE ◇ 2-((3-AMINOPROPYL)AMINO)-ETHANETHIOL, DIHYDROGEN PHOSPHATE ESTER (9CI) ◇ S-omega-(3-AMINOPROPYLAMINO)ETHYL DIHYDROGEN PHOSPHOROTHIOATE ◇ S-(2-(3-AMINOPROPYLAMINO)ETHYL) PHOSPHOROTHIOATE ◇ S,2-(3-AMINOPROPYLAMINO)ETHYL-PHOSPHOROTHIOIC ACID ◇ 2-(3-AMINOPROPYLAMINO)ETHYL THIOPHOSPHATE ◇ APAETP ◇ AU-95722 ◇ ETHIOFOS ◇ GAMMAPHOS ◇ NSC-296961 ◇ SAPEP ◇ WR 2721 ◇ YM-08310

TOXICITY DATA with REFERENCE
cyt-mus-ipr 300 mg/kg CUSCAM 54,1080,85
ivn-rat TDLo:1300 mg/kg (female 17-22D post):REP OYYAA2 27,875,84
ipr-rat TDLo:300 mg/kg (female 9D post):TER RAREAE 107,49,86
ipr-rat LD50:418 mg/kg RADOA8 16,249,76
ims-rat LD50:396 mg/kg RADOA8 16,249,76
orl-mus LD50:842 mg/kg RADOA8 20,746,80
ipr-mus LD50:321 mg/kg NCISP* JAN86
ivn-mus LD50:557 mg/kg NTIS** PB81-199580
ims-mus LD50:514 mg/kg RADOA8 16,249,76
ivn-dog LDLo:279 mg/kg TOPADD 13,58,85
ipr-gpg LD50:407 mg/kg RADOA8 16,249,76

SAFETY PROFILE: Poison by intravenous, intramuscular, and intraperitoneal routes. Moderately toxic by ingestion. An experimental teratogen. Other experimental reproductive effects. Mutation data reported. See also ESTERS and PHOSPHATES. When heated to decomposition it emits very toxic fumes of SO_x, PO_x, and NO_x.

AMD250 CAS:56643-49-3 ***HR: 2***
S-3-(ω-AMINOPROPYLAMINO)-2-HYDROXYPROPYL DIHYDROGENPHOSPHOROTHIOATE
mf: $C_6H_{17}N_2O_4PS$ mw: 244.28

TOXICITY DATA with REFERENCE
orl-mus LD50:1200 mg/kg JMCMAR 18,803,75
ipr-mus LD50:875 mg/kg JMCMAR 18,803,75

SAFETY PROFILE: Moderately toxic by ingestion and intraperitoneal routes. See also PHOSPHATES. When heated to decomposition it emits very toxic fumes of PO_x, NO_x, and SO_x.

AMD500 CAS:1945-32-0 ***HR: 3***
N-(3-AMINOPROPYL)-1,4-BUTANEDIAMINE, PHOSPHATE
mf: $C_7H_{19}N_3•7H_3O_4P$ mw: 831.29

SYNS: SPD PHOSPHATE ◇ SPERMIDINE PHOSPHATE

TOXICITY DATA with REFERENCE
ipr-mus LD50:468 mg/kg LIFSAK 23,2137,78
ivn-mus LD50:92 mg/kg LIFSAK 23,2137,78

SAFETY PROFILE: Poison by intravenous route. Moderately toxic by intraperitoneal route. See also PHOSPHATES. When heated to decomposition it emits very toxic fumes of PO_x and NO_x.

AMD750 CAS:4985-85-7 ***HR: 3***
AMINOPROPYLDIETHANOLAMINE
DOT: NA 1760
mf: $C_7H_{18}N_2O_2$ mw: 162.27

CONSENSUS REPORTS: Reported in EPA TSCA Inventory.

DOT Classification: Corrosive Material; Label: Corrosive.

SAFETY PROFILE: A corrosive. A powerful skin, eye, and mucous membrane irritant. See also AMINES. When heated to decomposition it emits toxic fumes of NO_x.

AME000 CAS:3179-76-8 ***HR: 3***
(3-AMINOPROPYL)DIETHOXYMETHYLSILANE
mf: $C_8H_{21}NO_2Si$ mw: 191.39

TOXICITY DATA with REFERENCE
skn-rbt 10 mg/24H open MLD AIHAAP 23,95,62
orl-rat LD50:4760 mg/kg AIHAAP 23,95,62
ipr-mus LD50:40 mg/kg RCRVAB 38(12),975,69
skn-rbt LD50:2520 mg/kg AIHAAP 23,95,62

CONSENSUS REPORTS: Reported in EPA TSCA Inventory.

SAFETY PROFILE: Poison by intraperitoneal route. Moderately toxic by ingestion and skin contact. A skin irritant. When heated to decomposition it emits toxic fumes of NO_x.

AME500 CAS:299-26-3 ***HR: 3***
3-(2-AMINOPROPYL)INDOLE
mf: $C_{11}H_{14}N_2$ mw: 174.27

SYNS: INDOPAN ◇ α-METHYL-β-INDOLAETHYLAMINE (GERMAN) ◇ α-METHYL-β-INDOLEETHYLAMINE ◇ α-METHYLTRYPTAMINE

TOXICITY DATA with REFERENCE
ipr-rat TDLo:10 mg/kg (female 4-5D post):REP FATOAO 29,224,66
ipr-rat TDLo:20 mg/kg (female 4D post):TER FATOAO 29,224,66
orl-hmn TDLo:384 μg/kg:PSY JNMDAN 131,428,60
orl-rat LD50:22 mg/kg TXAPA9 4,547,62
scu-rat LD50:50 mg/kg FATOAO 29,224,66
ivn-rat LD50:75 mg/kg FATOAO 29,224,66
ipr-mus LD50:20 mg/kg PSYPAG 16,385,70
scu-mus LDLo:500 mg/kg JPMRAB 3,235,29
ivn-mus LDLo:120 mg/kg JPMRAB 3,235,29
scu-rbt LDLo:500 mg/kg JPMRAB 3,235,29
ivn-rbt LDLo:90 mg/kg JPMRAB 3,235,29

SAFETY PROFILE: Poison by ingestion and intraperitoneal routes. Moderately toxic by subcutaneous route. Human psychotropic effects by ingestion. An experimental teratogen. Other experimental reproductive effects. When heated to decomposition it emits toxic fumes of NO_x.

AME750 CAS:18237-15-5 ***HR: 3***
3-(γ-AMINOPROPYL)-INDOLEHYDROCHLORIDE
mf: $C_{11}H_{14}N_2$•ClH mw: 210.73

SYNS: HOMOTRYPTAMINE HYDROCHLORIDE ◇ INDOLE-3-PROPYLAMINE HYDROCHLORIDE ◇ γ-3-INDOLYLPROPYLAMINE HYDROCHLORIDE

TOXICITY DATA with REFERENCE
ipr-mus LD50:235 mg/kg RPTOAN 33,180,70
ivn-mus LD50:98 mg/kg RPTOAN 33,180,70

SAFETY PROFILE: Poison by intravenous and intraperitoneal routes. When heated to decomposition it emits very toxic fumes of HCl and NO_x.

AMF250 CAS:123-00-2 ***HR: 3***
4-AMINOPROPYLMORPHOLINE
DOT: UN 1760
mf: $C_7H_{16}N_2O$ mw: 144.25

PROP: Liquid. Mp: −15°, bp: 224.7°, flash p: 220°F (OC), d: 0.9872 @ 20°/20°, vap press: 0.06 mm @ 20°, vap d: 4.97.

SYN: N-AMINOPROPYLMORPHOLINE (DOT)

TOXICITY DATA with REFERENCE
skn-rbt 10 mg/24H SEV AMIHBC 4,119,51
skn-rbt 500 mg open SEV UCDS** 3/25/70
eye-rbt 1 mg UCDS** 3/25/70
orl-rat LD50:3560 mg/kg UCDS** 3/25/70
skn-rbt LD50:1230 mg/kg AMIHBC 4,119,51

CONSENSUS REPORTS: Reported in EPA TSCA Inventory.

DOT Classification: Corrosive Material; Label: Corrosive.

SAFETY PROFILE: A corrosive material. Moderately toxic by several routes. A severe skin and eye irritant. Combustible. Can react with oxidizing materials. To fight fire, use alcohol foam, dry chemical. When heated to decomposition it emits toxic fumes of NO_x.

AMF375 CAS:3690-04-8 ***HR: 2***
AMINOPROPYLON
mf: $C_{16}H_{22}N_4O_2$ mw: 302.42

PROP: Prisms from benzene. Mp: 181°. Very sol in water.

SYNS: AMINOPROPYLONE ◇ AMIPYLO ◇ N-ANTIPYRINYL-2-(DIMETHYLAMINO)PROPIONAMIDE ◇ N-(2,3-DIHYDRO-1,5-DIMETHYL-3-OXO-2-PHENYL-1H-PYRAZOL-4-YL)-2-(DIMETHYLAMINO)PROPANAMIDE ◇ 4-(2-(DIMETHYLAMINO)PROPIONAMIDO)ANTIPYRINE

TOXICITY DATA with REFERENCE
orl-mus LD50:2950 mg/kg OYYAA2 13,109,77

ipr-mus LD50:820 mg/kg OYYAA2 13,109,77
ims-mus LD50:2120 mg/kg OYYAA2 13,109,77

SAFETY PROFILE: Moderately toxic by ingestion and other routes. When heated to decomposition it emits toxic fumes of NO_x.

AMF500 CAS:1075-61-2 ***HR: 3***
m-(2-AMINOPROPYL)PHENOL
mf: $C_9H_{13}NO$ mw: 151.23

SYN: α-METHYL-m-TYRAMINE

TOXICITY DATA with REFERENCE
scu-mus LD50:17 mg/kg ARZNAD 15,219,65
ivn-mus LD50:82 mg/kg ARZNAD 15,219,65

SAFETY PROFILE: Poison by subcutaneous and intravenous routes. When heated to decomposition it emits toxic fumes of NO_x.

AMF750 ***HR: 3***
6-AMINO-1-PROPYL-4-(p-((p-((1-PROPYLPYRIDINIUM-4-YL)AMINO)-2-AMINOPHENYL)CARBAMOYL)ANILINO)QUINOLINIUM) DIIODIDE
mf: $C_{33}H_{37}N_7O{\bullet}2I$ mw: 801.57

TOXICITY DATA with REFERENCE
dnd-mus:lym 710 nmol/L JMCMAR 22,134,79
ipr-mus LD10:10 mg/kg JMCMAR 22,134,79

SAFETY PROFILE: Poison by intraperitoneal route. Mutation data reported. See also IODIDES. When heated to decomposition it emits very toxic fumes of NO_x and I^-.

AMG000 CAS:68772-13-4 ***HR: 3***
6-AMINO-1-PROPYL-4-(p-((p-((1-PROPYLPYRIDINIUM-4-YL)AMINO)PHENYL)CARBAMOYL)ANILINO)QUINOLINIUM)DIBROMIDE
mf: $C_{33}H_{36}N_6O{\bullet}2Br$ mw: 692.57

TOXICITY DATA with REFERENCE
dnd-mus:lym 680 nmol/L JMCMAR 22,134,79
ipr-mus LD10:10 mg/kg JMCMAR 22,134,79

SAFETY PROFILE: Poison by intraperitoneal route. See also BROMIDES. Mutation data reported. When heated to decomposition it emits very toxic fumes of NO_x and Br.

AMG500 CAS:17869 ***HR: 2***
1-(3-AMINOPROPYL)-2,8,9-TRIOXA-5-AZA-1-SILABICYCLO(3.3.3) UNDECANE-27-1
mf: $C_9H_{20}O_3Si$ mw: 204.38

SYN: 3-AMINOPROPYLSILATRAN(CZECH)

TOXICITY DATA with REFERENCE
skn-rbt 500 mg/24H SEV 28ZPAK -,220,72
eye-rbt 20 mg/24H SEV 28ZPAK -,220,72
orl-rat LD50:5800 mg/kg 28ZPAK -,220,72

SAFETY PROFILE: Mildly toxic by ingestion. A severe skin and eye irritant. When heated to decomposition it emits smoke and acrid fumes.

AMG750 CAS:54-62-6 ***HR: 3***
AMINOPTERIDINE
mf: $C_{19}H_{20}N_8O_5$ mw: 440.47

PROP: Yellow needles, sol in sodium hydroxide soln.

SYNS: 4-AMINO-4-DEOXYPTEROYLGLUTAMATE ◇ 4-AMINO-PGA ◇ AMINOPTERIN ◇ 4-AMINOPTEROYLGLUTAMIC ACID ◇ APGA ◇ ENT 26,079 ◇ FOLIC ACID, 4-AMINO- ◇ NSC 739

TOXICITY DATA with REFERENCE
spm-mus-ipr 2 mg/kg/5D PNASA6 78,4425,78
orl-wmn TDLo:2880 μg/kg (female 56-67D post):REP AMDCA5 97,274,59
orl-wmn TDLo:580 μg/kg (female 6-8W post):TER AJOGAH 84,356,62
ims-rat TDLo:1200 μg/kg/17W-C:ETA AMUK** 38,248,62
orl-wmn TDLo:120 μg/kg:GIT AJOGAH 63,1298,52
orl-rat LDLo:2500 μg/kg JPETAB 95,303,49
ipr-rat LD50:3 mg/kg CANCAR 9,955,56
orl-mus LD50:3 mg/kg CKFRAY 28,159,79
ipr-mus LD50:1900 μg/kg JPETAB 95,303,49

CONSENSUS REPORTS: EPA Extremely Hazardous Substances List.

SAFETY PROFILE: Poison by ingestion and intraperitoneal routes. Human and experimental teratogenic data. Other experimental reproductive effects. Mutation data reported. Human systemic effects by ingestion: gastrointestinal. Questionable carcinogen with experimental tumorigenic data. When heated to decomposition it emits toxic fumes of NO_x.

AMH000 CAS:452-06-2 ***HR: D***
2-AMINOPURINE
mf: $C_5H_5N_5$ mw: 135.15

SYNS: 1H-PURIN-2-AMINE ◇ SQ 22451

TOXICITY DATA with REFERENCE
mmo-sat 100 μg/PLATE MUREAV 111,283,83
mmo-asn 7400 μmol/L MUREAV 75,1,80
cyt-mus:emb 4 mmol/L MUREAV 75,1,80
otr-ham:emb 100 mg/L PNASA6 78,5685,81
sce-ham:ovr 100 μmol/L CHROAU 85,603,82

SAFETY PROFILE: Mutation data reported. When heated to decomposition it emits toxic fumes of NO_x.

AMH250 CAS:154-42-7 ***HR: 3***
2-AMINOPURINE-6-THIOL
mf: $C_5H_5N_5S$ mw: 167.21

SYNS: 2-AMINO-6-MERCAPTOPURINE ◇ 2-AMINO-6-MP ◇ 2-AMINO-6-PURINETHIOL ◇ 2-AMINOPURINE-6(1H)-THIONE ◇ BW 5071 ◇ LANVIS ◇ 6-MERCAPTO-2-AMINOPURINE ◇ 6-MERCAPTOGUANINE ◇ NSC-752 ◇ 6H-PURINE-6-THIONE, 2-AMINO-1,7-DIHYDRO- (9CI) ◇ TABLOID ◇ TG ◇ ThG ◇ THIOGUANINE ◇ 6-THIOGUANINE ◇ TIOGUANIN ◇ TIOGUANINE ◇ WELLCOME U3B

TOXICITY DATA with REFERENCE
cyt-hmn:fbr 73 mg/L MUREAV 4,353,67
sce-rat:oth 200 nmol/L BCPCA6 34,515,85
dnd-ham:ovr 50 μg/L PAACA3 24,295,83
oms-ham:ovr 50 μg/L PAACA3 24,295,83
cyt-ham:lng 50 μg/L MUREAV 139,149,84
sce-ham:lng 15 μg/L MUREAV 139,149,84
ipr-rat TDLo:25 mg/kg (12D preg):TER PSEBAA 116,685,64
ipr-rat TDLo:10 mg/kg (7D preg):REP JRPFA4 4,291,62
ipr-rat LD50:300 mg/kg JRPFA4 291,62
orl-mus LD50:160 mg/kg EKFMA7 9,56,80
ipr-mus LD50:54 mg/kg EKFMA7 9,56,80

SAFETY PROFILE: Poison by ingestion and intraperitoneal routes. Human mutation data reported. An experimental teratogen. Other reproductive effects. When heated to decomposition it emits very toxic fumes of SO_x and NO_x.

AMH500 CAS:58048-24-1 ***HR: 3***
1-(6-AMINO-9H-PURIN-9-YL)-N-CYCLOPROPYL-1-DEOXY-2,3-DIHYDROXYRIBOFURANURONAMIDE DIACETATE
mf: $C_{17}H_{20}N_6O_6$ mw: 404.43

TOXICITY DATA with REFERENCE
orl-mus LD50:20 mg/kg JMCMAR 23,313,80
ipr-mus LD50:5 mg/kg JMCMAR 23,313,80

SAFETY PROFILE: Poison by ingestion and intraperitoneal routes. When heated to decomposition it emits toxic fumes of NO_x.

AMH750 CAS:58048-26-3 ***HR: 3***
1-(6-AMINO-9H-PURIN-9-YL)-1-DEOXY-2,3-DIHYDROXY-N-ETHYLRIBOFURANURONAMIDE DIACETATE
mf: $C_{16}H_{20}N_6O_6$ mw: 392.42

TOXICITY DATA with REFERENCE
orl-mus LD50:2 mg/kg JMCMAR 23,313,80
ipr-mus LD50:2 mg/kg JMCMAR 23,313,80

SAFETY PROFILE: Poison by ingestion and intraperitoneal routes. When heated to decomposition it emits toxic fumes of NO_x.

AMH800 CAS:54262-83-8 ***HR: D***
(S)-3-(6-AMINO-9H-PURIN-9-YL)-1,2-PROPANEDIOL
mf: $C_8H_{11}N_5O_2$ mw: 209.24

SYN: (S)-DHPA ◇ (S)-9-(2,3-DIHYDROXYPROPYL)ADENINE ◇ 1,2-PROPANEDIOL, 3-(6-AMINO-9H-PURIN-9-YL)-, (S)-

TOXICITY DATA with REFERENCE
dni-rbt:kdy 326 mg/L JMCMAR 28,282,85
orl-mus TDLo:8400 mg/kg (male 21D pre):REP TXAPA9 59,441,81

SAFETY PROFILE: Experimental reproductive effects. Mutation data reported. When heated to decomposition it emits toxic fumes of NO_x.

AMI000 CAS:504-29-0 ***HR: 3***
2-AMINOPYRIDINE
DOT: UN 2671
mf: $C_5H_6N_2$ mw: 94.13

PROP: White powder or crystals. Mp: 58.1, bp: 210.6°. Sol in water and ether; very sol in alc; sltly sol in ligroin.

SYNS: α-AMINOPYRIDINE ◇ AMINO-2-PYRIDINE ◇ o-AMINOPYRIDINE ◇ α-PYRIDINAMINE ◇ α-PYRIDYLAMINE

TOXICITY DATA with REFERENCE
ihl-hmn TCLo:5 ppm/5H:CNS IMSUAI 19,317,50
ipr-mus LD50:35 mg/kg JMCMAR 8,296,65
scu-mus LD50:70 mg/kg AEPPAE 226,163,55
ivn-mus LD50:23 mg/kg APFRAD 26,345,68
orl-qal LD50:133 mg/kg AECTCV 12,355,83
orl-bwd LD50:31600 μg/kg AECTCV 12,355,83

CONSENSUS REPORTS: Reported in EPA TSCA Inventory.

OSHA PEL: TWA 0.5 ppm
ACGIH TLV: TWA 0.5 ppm
DFG MAK: 0.5 ppm (2 mg/m^3)
DOT Classification: IMO: Poison B; Label: Poison.

SAFETY PROFILE: Poison by inhalation, subcutaneous, intravenous, and intraperitoneal routes. Toxic effects resemble strychnine poisoning. Human systemic effects by inhalation: somnolence, convulsions, and antipsychotic effects. Human central nervous system effects by inhalation. When heated to decomposition it emits highly toxic fumes of NO_x.

AMI250 CAS:462-08-8 ***HR: 3***
3-AMINOPYRIDINE
DOT: UN 2671
mf: $C_5H_6N_2$ mw: 94.13

PROP: Leaflets from benzene or ligroin. Mp: 64°; bp: 251°. Very sol in water, alc, ether; insol in ligroin.

SYNS: AMINO-3-PYRIDINE ◇ m-AMINOPYRIDINE (DOT) ◇ 3-PYRIDINAMINE ◇ 3-PYRIDYLAMINE

TOXICITY DATA with REFERENCE
ipr-mus LD50:28 mg/kg JMCMAR 8,296,65
scu-mus LD50:30 mg/kg AEPPAE 226,163,55
ivn-mus LD50:24 mg/kg APFRAD 26,345,68
orl-qal LD50:178 mg/kg AECTCV 12,355,83
orl-bwd LD50:13300 μg/kg AECTCV 12,355,83

CONSENSUS REPORTS: Reported in EPA TSCA Inventory.

DOT Classification: Poison B; Label: Poison.

SAFETY PROFILE: Poison by ingestion, intraperitoneal, subcutaneous, and intravenous routes. When heated to decomposition it emits toxic fumes of NO_x.

AMI500 CAS:504-24-5 ***HR: 3***
4-AMINOPYRIDINE
DOT: UN 2671
mf: $C_5H_6N_2$ mw: 94.13

PROP: Needles from benzene. Mp: 158°; sol in water; sltly sol in benzene and ether.

SYNS: AMINO-4-PYRIDINE ◇ γ-AMINOPYRIDINE ◇ p-AMINOPYRIDINE ◇ 4-AP ◇ AVITROL ◇ 4-PYRIDINAMINE ◇ 4-PYRIDYLAMINE ◇ RCRA WASTE NUMBER P008 ◇ VMI 10-3

TOXICITY DATA with REFERENCE
orl-man LDLo:590 μg/kg CTOXAO 16,487,80
orl-rat LD50:20 mg/kg TXAPA9 26,532,73
ipr-rat LD50:6500 μg/kg TXAPA9 26,532,73
orl-mus LDLo:42 mg/kg AECTCV 14,111,85
ipr-mus LD50:10 mg/kg JMCMAR 8,296,65
scu-mus LD50:5 mg/kg APFRAD 26,345,68
orl-pgn LD50:7500 μg/kg ASTTA8 (680),157,79
orl-qal LD50:7650 μg/kg ASTTA8 (680),157,79
orl-dck LD50:4200 mg/kg TXAPA9 21,315,72

CONSENSUS REPORTS: Reported in EPA TSCA Inventory. EPA Extremely Hazardous Substances List.

SAFETY PROFILE: Poison by ingestion, subcutaneous, intravenous and intraperitoneal routes. Human systemic effects by ingestion: hallucinations and distorted perceptions, dyspnea, nausea or vomiting. When heated to decomposition it emits toxic fumes of NO_x.

AMI750 CAS:73074-20-1 ***HR: 3***
3-AMINOPYRIDINE HYDROCHLORIDE
mf: $C_5H_6N_2 \cdot ClH$ mw: 130.59

SYN: 3-PYRIDINAMINE HYDROCHLORIDE

TOXICITY DATA with REFERENCE
orl-rat LDLo:79 mg/kg 34ZIAG -,93,69
orl-dog LDLo:23 mg/kg 34ZIAG -,93,69
skn-rbt LDLo:327 mg/kg 34ZIAG -,93,69

SAFETY PROFILE: Poison by ingestion and skin contact. When heated to decomposition it emits very toxic fumes of HCl and NO_x.

AMJ000 CAS:1003-40-3 ***HR: 3***
4-AMINOPYRIDINE HYDROCHLORIDE
mf: $C_5H_6N_2 \cdot ClH$ mw: 130.59

SYN: 4-PYRIDINAMIDE HYDROCHLORIDE

TOXICITY DATA with REFERENCE
orl-rat LDLo:28 mg/kg 34ZIAG -,93,69
scu-rat LD50:10130 μg/kg EKMMA8 18,98,79
scu-mus LD50:11900 μg/kg EKMMA8 18,98,79
orl-dog LDLo:4 mg/kg 34ZIAG -,93,69
skn-rbt LDLo:327 mg/kg 34ZIAG -,93,69
orl-pig LDLo:18 mg/kg 34ZIAG -,93,69

SAFETY PROFILE: Poison by ingestion, skin contact and subcutaneous routes. An eye irritant. When heated to decomposition it emits very toxic fumes of HCl and NO_x.

AMJ250 CAS:3535-75-9 ***HR: 3***
4-AMINOPYRIDINE-1-OXIDE
mf: $C_5H_6N_2O$ mw: 110.13

SYNS: 4-AMINO-PYRIDINEN-OXIDE ◇ PHILLIPS 1863

TOXICITY DATA with REFERENCE
orl-rat LD50:75 mg/kg TXAPA9 21,315,72
orl-bwd LD50:85 mg/kg TXAPA9 21,315,72

SAFETY PROFILE: Poison by ingestion. When heated to decomposition it emits toxic fumes of NO_x.

AMJ500 CAS:30194-63-9 ***HR: 3***
4-AMINO-N-(2-(4-(2-PYRIDINYL)-1-PIPERAZINYL)ETHYL)BENZAMIDE
mf: $C_{18}H_{23}N_5O$ mw: 325.46

SYN: S 1688

TOXICITY DATA with REFERENCE
orl-mus LD50:441 mg/kg ARZNAD 24,1970,74
ipr-mus LD50:250 mg/kg ARZNAD 24,1964,74
ivn-mus LD50:162 mg/kg ARZNAD 24,1970,74

SAFETY PROFILE: Poison by intravenous and intraperitoneal routes. Moderately toxic by ingestion. When heated to decomposition it emits toxic fumes of NO_x.

AMJ600 CAS:69901-70-8 ***HR: D***
3-AMINO-5H-PYRIDO(4,3-b)INDOLE
mf: $C_{11}H_9N_3$ mw: 183.1

TOXICITY DATA with REFERENCE
mma-sat 2500 nmol/L PNASA6 77,1427,80
dnd-mam:lym 100 μmol/L PNASA6 77,1427,80

SAFETY PROFILE: Mutation data reported. When heated to decomposition it emits toxic fumes of NO_x.

AMJ625 CAS:70145-80-1 ***HR: 3***
2-AMINO-5-(4-PYRIDYL)-1,3,4-THIADIAZOLEHYDROCHLORIDE
mf: $C_7H_6N_4S{\cdot}ClH$ mw: 214.69

TOXICITY DATA with REFERENCE
orl-mus LD50:500 mg/kg JMCMAR 8,676,65
scu-mus LD50:200 mg/kg JMCMAR 8,676,65

SAFETY PROFILE: Poison by subcutaneous route. Moderately toxic by ingestion. When heated to decomposition it emits very toxic fumes of NO_x, SO_x, and HCl.

AMJ750 CAS:59985-27-2 ***HR: 2***
2-(2-AMINO-4-PYRIMIDINYLVINYL)QUINOXALINE-N,N'-DIOXIDE
mf: $C_{14}H_{11}N_5O_2$ mw: 281.30

SYN: 2-AMINO-4-((2-QUINOXALINYL-N,N'-DIOXIDE)VINYL)PYRIMIDINES

TOXICITY DATA with REFERENCE
ipr-rat LDLo:500 mg/kg RVFTBB 7,117,76
orl-mus LDLo:4000 mg/kg RVFTBB 7,117,76
ipr-mus LDLo:1000 mg/kg RVFTBB 7,117,76
orl-rbt LDLo:2000 mg/kg RVFTBB 7,117,76
orl-gpg LDLo:1000 mg/kg RVFTBB 7,117,76

SAFETY PROFILE: Moderately toxic by ingestion and intraperitoneal routes. When heated to decomposition it emits toxic fumes of NO_x.

AMK250 CAS:8015-18-7 ***HR: 2***
AMINOPYRINE-BARBITAL
mf: $C_{13}H_{17}N_3O{\cdot}C_8H_{12}N_2O_3$ mw: 415.55

PROP: Silky needles. Mp: 113-115°. Freely sol in water.

SYNS: BARBIMON ◇ COTALMON ◇ GRELAN ◇ MATANOL ◇ PFETFFER'S SUBSTANCE ◇ PYRABITAL ◇ PYRAMON ◇ SEDALON ◇ VERAMID ◇ VERAMON

TOXICITY DATA with REFERENCE
scu-mus TDLo:900 mg/kg (9-11D preg):TER TXCYAC 29,281,84
ipr-rat TDLo:400 mg/kg (1D male):REP KSRNAM 14,723,80
ipr-rat LD50:476 mg/kg OYYAA2 16,229,78
orl-mus LD50:1466 mg/kg OYYAA2 8,453,74
ipr-mus LD50:450 mg/kg OYYAA2 16,229,78
orl-mky LDLo:1 g/kg HBAMAK 4,1289,35
orl-rbt LDLo:1200 mg/kg HBAMAK 4,1289,35

SAFETY PROFILE: Moderately toxic by ingestion and other routes. An experimental teratogen. Other experimental reproductive effects. When heated to decomposition it emits toxic fumes of NO_x.

AMK500 CAS:68-89-3 ***HR: 3***
AMINOPYRINE SODIUM SULFONATE
mf: $C_{13}H_{17}N_3O_4S{\cdot}Na$ mw: 334.38

SYNS: (ANTIPYRINYLMETHYLAMINO)METHANESULFONICACID SODIUM SALT ◇ METHYLAMINOANTIPYRINE SODIUM METHANESULFONATE ◇ 4-METHYLAMINO-1,5-DIMETHYL-2-PHENYL-3-PYRAZOLONE SODIUM METHANESULFONATE ◇ METHYLAMINOPHENYLDIMETHYLPYRAZOLONE METHANESULFONATE SODIUM ◇ 1-PHENYL-2,3-DIMETHYL-5-PYRAZOLONE-4-METHYLAMINOMETHANESULFONATESODIUM ◇ 1-PHENYL-2,3-DIMETHYLPYRAZOLONE-(5)-4-METHYLAMINOMETHANESULFONICACID SODIUM ◇ PHENYL DIMETHYL PYRAZOLON METHYL AMINOMETHANE SODIUM SULFONATE ◇ 4-SODIUM METHANESULFONATE METHYLAMINE-ANTIPYRINE ◇ SODIUM METHYLAMINOANTIPYRINE METHANESULFONATE ◇ SODIUM-4-METHYLAMINO-1,5-DIMETHYL-2-PHENYL-3-PYRAZOLONE4-METHANESULFONATE ◇ SODIUM NORAMIDOPYRINE METHANESULFONATE ◇ SODIUM-1-PHENYL-2,3-DIMETHYL-4-METHYLAMINOPYRAZOLON-N-METHANESULFONATE ◇ SODIUM-1-PHENYL-2,3-DIMETHYL-5-PYRAZOLONE-4-METHYLAMINO METHANESULFONATE ◇ SODIUM PHENYLDIMETHYLPYRAZOLONMETHYLAMINOMETHANESULFONATE

TOXICITY DATA with REFERENCE
mma-sat:1 mg/plate AMONDS 3,253,80
cyt-hmn:lym 250 mg/L SOGEBZ 11,528,75
ipr-mus TDLo:1 g/kg (female 9D post):REP JMTHBU 53,550,73
ipr-mus TDLo:750 mg/kg (female 9D post):TER JMTHBU 53,550,73
orl-mus TDLo:536 mg/kg/78W-C:NEO JJIND8 71,1295,83
orl-rat LD50:3 g/kg ARZNAD 21,719,71
orl-mus LD50:2891 mg/kg ARZNAD 24,600,74
ipr-mus LD50:250 mg/kg AIPTAK 107,322,56
scu-mus LD50:69 mg/kg RPTOAN 31,53,68
orl-rbt LD50:2150 mg/kg GTPZAB 23(12),47,79

SAFETY PROFILE: Poison by subcutaneous route. Moderately toxic by several other routes. An experimental teratogen. Other experimental reproductive effects. Human mutation data reported. Questionable carcinogen with experimental neoplastigenic data. See also SULFONATES. When heated to decomposition it emits very toxic fumes of NO_x, Na_2O, and SO_x.

AML250 CAS:578-66-5 ***HR: D***
8-AMINOQUINOLINE
mf: $C_9H_8N_2$ mw: 144.19

PROP: Yellow needles from ethyl alc. Mp: 70°. Sol in hot water.

TOXICITY DATA with REFERENCE
mmo-sat 50 μg/plate MUREAV 39,285,77
dnr-sat 350 μg/disc MUREAV 39,285,77
mmo-smc 1 mg/L MUREAV 39,285,77
cyt-hmn:leu 100 μmol/L MUREAV 39,285,77

SAFETY PROFILE: Human mutation data reported. When heated to decomposition it emits toxic fumes of NO_x.

AML500 CAS:2508-86-3 ***HR: D***
4-AMINOQUINOLINE-1-OXIDE
mf: $C_9H_8N_2O$ mw: 160.19

SYN: 1-OXIDE-4-QUINOLINAMINE

TOXICITY DATA with REFERENCE
mrc-esc 500 μg/well CNREA8 32,2369,72
mmo-bcs 5 g/L MUREAV 42,19,77
dnr-bcs 5 g/L MUREAV 42,19,77
cyt-ham:lng 100 mg/L ATSUDG (4),41,80

CONSENSUS REPORTS: EPA Genetic Toxicology Program.

SAFETY PROFILE: Mutation data reported. When heated to decomposition it emits toxic fumes of NO_x.

AMM000 CAS:3131-60-0 ***HR: 1***
5-AMINO-2-β-d-RIBOFURANOSYL-as-TRIAZIN-3(2H)-ONE
mf: $C_8H_{12}N_4O_5$ mw: 244.24

SYN: 6-AZACYTIDINE

TOXICITY DATA with REFERENCE
sln-dmg-par 21 mmol/L BCPCA6 15,299,66
ivn-mus TDLo:200 mg/kg (5D preg):REP BCPCA6 14,1549,65
ipr-mus TDLo:2200 mg/kg (MGN):TER BCPCA6 14,1549,65
ipr-rat LD50:9200 mg/kg RPTOAN 50,50,87
ipr-mus LD50:14 g/kg BCPCA6 14,1517,65

SAFETY PROFILE: Mildly toxic by intraperitoneal route. An experimental teratogen. Other experimental reproductive effects. Mutation data reported. When heated to decomposition it emits toxic fumes of NO_x.

AMM125 CAS:117-55-5 ***HR: 1***
AMINO-S ACID
mf: $C_{10}H_9NO_6S_2$ mw: 303.32

TOXICITY DATA with REFERENCE
orl-rat LD50:56 g/kg GISAAA 45(3),73,80
orl-mus LD50:56 g/kg GISAAA 45(3),73,80
orl-rbt LD50:24 g/kg GISAAA 45(3),73,80
orl-gpg LD50:24 g/kg GISAAA 45(3),73,80

SAFETY PROFILE: When heated to decomposition it emits toxic fumes of SO_x and NO_x. See also SULFONATES.

AMM250 CAS:65-49-6 ***HR: 2***
4-AMINOSALICYLIC ACID
mf: $C_7H_7NO_3$ mw: 153.14

PROP: Minute crystals from alc. Mp: 150°. Sol in dil acid or base. Very sol in water and alc; sltly sol in ether.

SYNS: 4-AMINO-2-HYDROXYBENZOIC ACID ◇ AMINOPAR ◇ AMINOSALICYLIC ACID ◇ p-AMINOSALICYLIC ACID ◇ AMINOX ◇ APACIL ◇ APAS ◇ DEAPASIL ◇ ENTEPAS ◇ GABBROPAS ◇ HELLIPIDYL ◇ 2-HYDROXY-4-AMINOBENZOIC ACID ◇ 3-HYDROXY-4-CARBOXYANILINE ◇ KYSELINA-p-AMINOSALICYLOVA (CZECH) ◇ NSC 2083 ◇ OSACYL ◇ PAMACYL ◇ PAMISYL ◇ PARAMYCIN ◇ PARA-PAS ◇ PARASAL ◇ PARASALICIL ◇ PARASALINDON ◇ PAS ◇ PASA ◇ PASALON ◇ PASARA ◇ PAS-C ◇ PASCORBIC ◇ PASEM ◇ PASK ◇ PASMED ◇ PASNODIA ◇ PASOLAC ◇ PROPASA ◇ REZIPAS ◇ SANIPRIOL-4

TOXICITY DATA with REFERENCE
cyt-mus:mmr 2 mmol/L/24H-C JTSCDR 5,141,80
eye-rbt 100 mg/24H MOD 28ZPAK -,106,72
cyt-mus-orl 50 mg/kg NULSAK 22,96,79
orl-rat LD50:8930 mg/kg 28ZPAK -,106,72
orl-mus LD50:4 g/kg JPPMAB 2,764,50
ipr-mus LD50:4250 mg/kg ZENBAX 6B,183,51
scu-mus LD50:4 g/kg JPPMAB 2,764,50
ivn-mus LD50:3898 mg/kg ANTBAL 18,249,73
orl-rbt LD50:3650 mg/kg FEPRA7 10,289,51

CONSENSUS REPORTS: Reported in EPA TSCA Inventory. EPA Genetic Toxicology Program.

SAFETY PROFILE: Moderately toxic ingestion and other routes. An eye irritant. Mutation data reported. When heated to decomposition it emits toxic fumes of NO_x.

AMM500 CAS:89-57-6 ***HR: 3***
5-AMINOSALICYLIC ACID
mf: $C_7H_7NO_3$ mw: 153.14

PROP: Needles. Mp: decomp @ 260°-280°. Sol in HCl and CS_2; sltly sol in hot water; insol in alc.

SYNS: 5-AMINO-2-HYDROXYBENZOIC ACID ◇ m-AMINOSALICYLIC ACID ◇ p-AMINOSALICYLSAEURE (GERMAN)

TOXICITY DATA with REFERENCE
orl-mus LD50:5 g/kg ZENBAX 6B,183,51
ipr-mus LDLo:313 mg/kg CBCCT* 2,58,50

CONSENSUS REPORTS: Reported in EPA TSCA Inventory.

SAFETY PROFILE: Poison by intraperitoneal route. Moderately toxic. When heated to decomposition it emits toxic fumes of NO_x.

AMM750 CAS:551-36-0 ***HR: 3***
p-AMINOSALICYLIC ACID, 2-(DIETHYLAMINO) ETHYL ESTER, HYDROCHLORIDE
mf: $C_{12}H_{18}N_2O_3$•ClH mw: 274.78

SYNS: 4-AMINO-2-HYDROXYBENZOIC ACID, 2-(DIETHYLAMINO)ETHYL ESTER, HYDROCHLORIDE (9CI) ◇ 4-AMINOSALICYLIC ACID-2-(DIETHYLAMINO)ETHYL ESTER HYDROCHLORIDE ◇ p-AMINOSALICYLSAEUREDIAETHYLAMINO-AETHYLESTER-CHLORHYDRAT (GERMAN) ◇ C 4201 ◇ HCl SALZ des p-AMINO-SALICYLSAEURE-DIAETHYLAMINOAETHYLESTER(GERMAN) ◇ SALICYLIC ACID, 4-AMINO-, 2-(DIETHYLAMINO)ETHYL ESTER, HYDROCHLORIDE ◇ VERBINDUNG S 557 HCl ◇ WIN 2022

TOXICITY DATA with REFERENCE
orl-rat LD50:130 mg/kg ZENBAX 6B,183,51
ipr-rat LDLo:190 mg/kg ARZNAD 1,154,51
ivn-rat LDLo:36 mg/kg ARZNAD 1,154,51
orl-mus LD50:135 mg/kg ZENBAX 6B,183,51
ipr-mus LD50:115 mg/kg ARZNAD 1,154,51
scu-mus LD50:282 mg/kg ARZNAD 8,708,58
ivn-mus LD50:37 mg/kg JPETAB 104,40,52

SAFETY PROFILE: Poison by ingestion, subcutaneous, intravenous, and intraperitoneal routes. When heated to decomposition it emits very toxic fumes of NO_x and HCl. See also ESTERS.

AMN000 CAS:78280 ***HR: 3***
p-AMINOSALICYLIC ACID, 2-(DIMETHYLAMINO)ETHYL ESTER HYDROCHLORIDE-31-6
mf: $C_{11}H_{16}N_2O_3$•ClH mw: 260.75

SYNS: 4-AMINO-2-HYDROXYBENZOIC ACID, 2-(DIMETHYLAMINO)ETHYL ESTER, HYDROCHLORIDE ◇ C 4200 ◇ 2-(DIMETHYLAMINO)ETHYL-p-AMINOSALICYLATE ◇ HCl SALZ des p-AMINO-SALICYLSAEURE-DIMETHYLAMINOAETHYL-ESTER (GERMAN)

TOXICITY DATA with REFERENCE
ipr-rat LDLo:290 mg/kg ARZNAD 1,154,51
ivn-rat LDLo:63 mg/kg ARZNAD 1,154,51
scu-mus LD50:560 mg/kg ARZNAD 8,708,58
ivn-mus LD50:67 mg/kg JPETAB 123,269,58

SAFETY PROFILE: Poison by intraperitoneal and intravenous routes. Moderately toxic by subcutaneous route. See also ESTERS. When heated to decomposition it emits toxic fumes of NO_x and HCl.

AMN250 CAS:6946-29-8 ***HR: 3***
p-AMINOSALICYLIC ACID HYDRAZIDE
mf: $C_7H_9N_3O_2$ mw: 167.19

TOXICITY DATA with REFERENCE
orl-rbt LDLo:250 mg/kg CLDND*
unk-dog LDLo:30 mg/kg CLDND*

SAFETY PROFILE: Poison by ingestion and other unspecified routes. When heated to decomposition it emits toxic fumes of NO_x.

AMN300 CAS:7722-06-7 ***HR: 3***
4-AMINO-1,2,5-SELENADIAZOLE-3-CARBOXAMIDE
mf: $C_3H_4N_4OSe$ mw: 191.07

SYNS: NSC 84963 ◇ 1,2,5-SELENADIAZOLE-3-CARBOXAMIDE, 4-AMINO-

TOXICITY DATA with REFERENCE
ipr-mus LDLo:4 mg/kg AACHAX-,551,66

OSHA PEL: TWA 0.2 mg(Se)/m^3
ACGIH TLV: TWA 0.2 mg(Se)/m^3

SAFETY PROFILE: Poison by intraperitoneal route. When heated to decomposition it emits toxic fumes of NO_x and Se.

AMN500 CAS:15267-04-6 ***HR: 3***
2-AMINOSELENOAZOLINE
mf: $C_3H_6N_2Se$ mw: 149.07

SYNS: 2-AMINOSELENOAZOLIN (GERMAN) ◇ 2-ASe

TOXICITY DATA with REFERENCE
ipr-mus LD50:160 mg/kg STRAAA 151,78,76
scu-mus LD50:177 mg/kg STRAAA 151,78,76
ivn-mus LD50:151 mg/kg STRAAA 151,78,76

CONSENSUS REPORTS: Selenium compounds are on the Community Right-To-Know List.

OSHA PEL: TWA 0.2 mg(Se)/m^3
ACGIH TLV: TWA 0.2 mg(Se)/m^3
DFG MAK: 0.1 mg(Se)/m^3

SAFETY PROFILE: Poison by intraperitoneal, subcutaneous, and intravenous routes. When heated to decomposition it emits very toxic fumes of NO_x and Se. See also SELENIUM COMPOUNDS.

AMO000 CAS:4309-66-4 ***HR: 3***
trans-4-AMINOSTILBENE
mf: $C_{14}H_{13}N$ mw: 195.28

SYNS: 4-(2-PHENYLETHENYL)BENZENAMINE,(E)◇ trans-4-STILBENE ◇ trans-4-N-STILBENAMINE

TOXICITY DATA with REFERENCE
mma-sat 10 μg/plate PNASA6 70,2281,73
orl-rat TDLo:200 mg/kg/13W-C:CAR CNREA8 24,128,64
scu-rat TDLo:26 mg/kg/4W-I:CAR,REP CNREA8 24,128,64
scu-rat TD:63 mg/kg/6W-I:ETA PTRMAD 241,147,48

SAFETY PROFILE: Questionable carcinogen with experimental carcinogenic and tumorigenic data. Experimental reproductive effects. Mutation data reported.

When heated to decomposition it emits toxic fumes of NO_x.

AMO250 CAS:3432-10-8 ***HR: 3***
2-(p-AMINOSTYRYL)-6-(p-ACETYLAMINO-BENZOYLAMINO)QUINOLINE METHOACETATE
mf: $C_{27}H_{25}N_4O_2{\bullet}C_2H_3O_2$ mw: 496.61

SYN: STYRYL 430

TOXICITY DATA with REFERENCE
mmo-esc 1 pph CRSBAW 142,453,48
scu-mus TDLo:268 mg/kg:ETA JPBAA7 42,155,36

SAFETY PROFILE: Questionable carcinogen with experimental tumorigenic data. Mutation data reported. When heated to decomposition it emits toxic fumes of NO_x.

AMO750 CAS:74039-02-4 ***HR: 1***
2,(4'-AMINO-3'-SULFO-1,1'-BIPHENYL-4-YL)-2H-NAPHTHO(1,2-4)TRIAZOLE-6,8-DISULFONIC ACID, TRIPOTASSIUM SALT
mf: $C_{22}H_{13}N_4O_9S_3{\bullet}3K$ mw: 690.87

SYN: 2-(4'-AMINOXENYL)NAFTO-α,β-TRIAZOL-6,8,3'TRISULFONAN DRASELNY (CZECH)

TOXICITY DATA with REFERENCE
eye-rbt 500 mg/24H SEV 28ZPAK -,196,72
orl-rat LD50:8330 mg/kg 28ZPAK -,196,72

SAFETY PROFILE: Mildly toxic by ingestion. A severe eye irritant. See also SULFONATES. When heated to decomposition it emits very toxic fumes of NO_x, SO_x, and K_2O.

AMP000 CAS:29727-70-6 ***HR: 1***
6-AMINO-5-SULFOMETHYL-2-NAPHTHALENESULFONIC ACID
mf: $C_{11}H_{11}NO_6S_2$ mw: 317.35

SYN: KYSELINA 1-SULFOMETHYL-2-NAFTYLAMIN-6-SULFONOVA (CZECH)

TOXICITY DATA with REFERENCE
eye-rbt 500 mg/24H SEV 28ZPAK -,190,72
orl-rat LD50:9200 mg/kg 28ZPAK -,190,72

SAFETY PROFILE: Mildly toxic by ingestion. A severe eye irritant. See also SULFONATES. When heated to decomposition it emits very toxic fumes of NO_x and SO_x.

AMP500 CAS:16760-18-2 ***HR: 3***
3,3'-(2-AMINOTEREPHTHALOYLBIS(IMINO(3-AMINO-p-PHENYLENE)CARBONYLIMINO)) BIS(1-ETHYLPYRIDINIUM, DI-p-TOLUENESULFONATE
mf: $C_{36}H_{36}N_9O_4{\bullet}2C_7H_7O_3S$ mw: 1001.21

TOXICITY DATA with REFERENCE
dnd-mus:lym 260 nmol/L JMCMAR 22,134,79
ipr-mus LD10:14 mg/kg JMCMAR 22,134,79

SAFETY PROFILE: Poison by intraperitoneal route. See also SULFONATES. Mutation data reported. When heated to decomposition it emits very toxic fumes of SO_x and NO_x.

AMP750 CAS:16802-49-6 ***HR: 3***
3,3'-(2-AMINOTEREPHTHALOYLBIS(IMINO(3-AMINO-p-PHENYLENE)CARBONYLIMINO)) BIS(1-PROPYLPYRIDINIUM, DI-p-TOLUENESULFONATE
mf: $C_{38}H_{41}N_9O_4{\bullet}2C_7H_7O_3S$ mw: 1030.28

TOXICITY DATA with REFERENCE
dnd-mus:lym 260 nmol/L JMCMAR 22,134,79
ipr-mus LD10:10 mg/kg JMCMAR 22,134,79

SAFETY PROFILE: Poison by intraperitoneal route. Mutation data reported. When heated to decomposition it emits very toxic fumes of NO_x and SO_x.

AMQ000 CAS:16760-14-8 ***HR: 3***
3,3'-(2-AMINOTEREPHTHALOYBIS(IMINO-p-PHENYLENECARBONYLIMINO))BIS(1-ETHYL-PYRIDINIUM, DI-p-TOLUENESULFONATE
mf: $C_{36}H_{35}N_7O_4{\bullet}2C_7H_7O_3S$ mw: 972.18

TOXICITY DATA with REFERENCE
dnd-mus:lym 530 nmol/L JMCMAR 22,134,79
ipr-mus LD10:14 mg/kg JMCMAR 22,134,79

SAFETY PROFILE: Poison by intraperitoneal route. Mutation data reported. When heated to decomposition it emits very toxic fumes of SO_x and NO_x.

AMQ250 CAS:16760-13-7 ***HR: 3***
3,3'-(2-AMINOTEREPHTHALOYLBIS(IMINO-p-PHENYLENECARBONYLIMINO))BIS(1-METHYLPYRIDINIUM, DI-p-TOLUENESULFONATE
mf: $C_{34}H_{31}N_7O_4{\bullet}2C_7H_7O_3S$ mw: 944.12

TOXICITY DATA with REFERENCE
dnd-mus:lym 530 nmol/L JMCMAR 22,134,79
ipr-mus LD10:20 mg/kg JMCMAR 22,134,79

SAFETY PROFILE: Poison by intraperitoneal route. Mutation data reported. See also SULFONATES. When heated to decomposition it emits very toxic fumes of SO_x and NO_x.

AMQ500 CAS:23757-42-8 ***HR: 3***
4-AMINO-2,2,5,5-TETRAKIS (TRIFLUOROMETHYL)-3-IMIDAZOLINE
mf: $C_7H_3F_{12}N_3$ mw: 357.13

SYNS: 5-AMINO-2,2,4,4-TETRAKIS(TRIFLUOROMETHYL)IMIDAZOLIDINE ◇ EXP 338

TOXICITY DATA with REFERENCE
ipr-rat TDLo:6500 μg/kg (female 1D pre):REP TXAPA9 18,917,71
orl-rbt TDLo:26 mg/kg (female 6-18D post):TER TXAPA9 18,917,71
orl-rat LD50:19 mg/kg 27ZQAG -,265,72
ipr-rat LD50:12 mg/kg 27ZQAG -,265,72
orl-mus LD50:262 mg/kg 27ZQAG -,265,72
ipr-mus LD50:189 mg/kg 27ZQAG -,265,72
ivn-mus LD50:231 mg/kg 27ZQAG -,265,72
orl-dog LD50:150 mg/kg 27ZQAG -,265,72
orl-gpg LD50:11 mg/kg 27ZQAG -,265,72

SAFETY PROFILE: Poison by ingestion, intraperitoneal, and intravenous routes. An experimental teratogen. Other experimental reproductive effects. When heated to decomposition it emits very toxic fumes of F^- and NO_x.

AMQ750 CAS:6130-92-3 ***HR: 3***
1-AMINO-2,2,6,6-TETRAMETHYLPIPERIDINE
mf: $C_9H_{20}N_2$ mw: 156.31

TOXICITY DATA with REFERENCE
orl-mus LD50:261 mg/kg MDCHAG 7,312,67
ivn-mus LD50:44 mg/kg NATUAS 184,1707,59

SAFETY PROFILE: Poison by ingestion and intravenous routes. When heated to decomposition it emits toxic fumes of NO_x.

AMR000 CAS:4418-61-5 ***HR: 3***
AMINOTETRAZOLE
mf: CH_3N_5 mw: 85.09

SYNS: 5-AMINOTETRAZOLE ◇ 5-AMINO-1H-TETRAZOLE ◇ 1H-TETRAZOL-5-AMINE

TOXICITY DATA with REFERENCE
ipr-mus LD50:2500 mg/kg RPTOAN 41,249,78

CONSENSUS REPORTS: Reported in EPA TSCA Inventory.

SAFETY PROFILE: Moderately toxic by intraperitoneal route. An unstable material; explodes with KOH. When heated to decomposition it emits toxic fumes of NO_x.

AMR250 CAS:4005-51-0 ***HR: 3***
2-AMINO-1,3,4-THIADIAZOLE
mf: $C_2H_3N_3S$ mw: 101.14

SYNS: AMINOTHIADIAZOLE ◇ ATDA ◇ 1,3,4-THIADIAZOL-2-AMINE

TOXICITY DATA with REFERENCE
ipr-rat TDLo:100 mg/kg (female 16D post):REP TJADAB 27,29A,83
orl-rat TDLo:20 mg/kg (female 10D post):TER CPHPA5 12,212,81
ipr-rat LD50:200 mg/kg TAKHAA 35,68,76
scu-rat LD50:200 mg/kg JHMJAX 130,95,72
ipr-mus LD50:6500 mg/kg TAKHAA 35,68,76

CONSENSUS REPORTS: EPA Genetic Toxicology Program.

SAFETY PROFILE: Poison by subcutaneous and intraperitoneal routes. An experimental teratogen. Other experimental reproductive effects. When heated to decomposition it emits very toxic fumes of NO_x and SO_x.

AMR500 CAS:26861-87-0 ***HR: 3***
2-AMINO-1,3,4-THIADIAZOLEHYDROCHLORIDE
mf: $C_2H_3N_3S{\bullet}ClH$ mw: 137.60

SYNS: 2-AMINO-1,3,4-THIADIAZOLE,MONOHYDROCHLORIDE ◇ ATDA HYDROCHLORIDE

TOXICITY DATA with REFERENCE
ipr-rat TDLo:100 mg/kg (female 11D post):TER TJADAB 21,381,80
ipr-rat TDLo:100 mg/kg (female 10D post):REP TJADAB 7,65,73
ipr-rat LDLo:100 mg/kg TJADAB 7,65,73
ipr-mus LD50:250 mg/kg NTIS** AD691-490

CONSENSUS REPORTS: Reported in EPA TSCA Inventory.

SAFETY PROFILE: Poison by intraperitoneal route. An experimental teratogen. Other experimental reproductive effects. When heated to decomposition it emits very toxic fumes of HCl, SO_x, and NO_x.

AMR750 CAS:14949-00-9 ***HR: D***
2-AMINO-1,3,4-THIADIAZOLE-5-SULFONAMIDE SODIUM SALT
mf: $C_2H_4N_4O_2S_2{\bullet}Na$ mw: 203.21

SYN: CL 5343 SODIUM SALT ◇ TIO-URASIN SODIUM ◇ ZOLAMIDE SOLUTION

TOXICITY DATA with REFERENCE
scu-rat TDLo:1200 mg/kg (10-11D preg):TER JHMJAX 130,95,72

SAFETY PROFILE: An experimental teratogen. When heated to decomposition it emits very toxic fumes of NO_x, Na_2O, and SO_x.

AMS000 CAS:6630-99-5 ***HR: 3***
5-AMINO-1,2,3,4-THIATRIAZOLE
mf: CH_2N_4S mw: 102.11

SAFETY PROFILE: Very unstable. Explodes weakly at 130°C. Upon decomposition it emits toxic fumes of SO_x and NO_x.

AMS250 CAS:96-50-4 ***HR: 3***
2-AMINOTHIAZOLE
mf: $C_3H_4N_2S$ mw: 100.15

$\overline{SCH{-}CHN{=}C}NH_2$

PROP: Light brown crystals. Mp: 90°, bp: decomp. Sltly sol in water, alc, ether; sol in hot alc.

SYNS: ABADOL ◇ ABADOLE ◇ AMINOTHIAZOLE ◇ BASEDOL ◇ 2-THIAZOLAMINE ◇ 2-THIAZOLYLAMINE ◇ 2-THIAZYLAMINE ◇ USAF EK-P-5501

TOXICITY DATA with REFERENCE
mma-sat 3333 μg/plate MUREAV 155,17,85
mmo-klp 1 mmol/L MUREAV 118,153,83
mma-mus:lym 1214 mg/L MUREAV 155,17,85
msc-mus:lym 557 mg/L MUREAV 155,17,85
orl-rat LD50:480 mg/kg JIHTAB 30,71,48
ipr-mus LD50:200 mg/kg NTIS** AD277-689
orl-cat LDLo:120 mg/kg JIHTAB 30,71,48
orl-rbt LD50:370 mg/kg JIHTAB 30,71,48
orl-gpg LDLo:120 mg/kg JIHTAB 30,71,48

CONSENSUS REPORTS: Reported in EPA TSCA Inventory.

SAFETY PROFILE: Poison by ingestion and intraperitoneal routes. Mutation data reported. Spontaneous ignition occurs at 100°. Mixtures with nitric acid or nitric acid + sulfuric acid explode on heating. Incompatible with HNO_3 and H_2SO_4. When heated to decomposition it emits very toxic SO_x and NO_x fumes.

AMS625 CAS:49850-29-5 ***HR: D***
1-AMINO-2-(4-THIAZOLYL)-5-BENZIMIDAZOLECARBAMIC ACID ISOPROPYL ESTER
mf: $C_{14}H_{15}N_5O_2S$ mw: 317.40

TOXICITY DATA with REFERENCE
oms-hmn:oth 2 mg/L THERAP 31,505,76
orl-rat TDLo:170 mg/kg (8-15D preg):TER THERAP 31,505,76

SAFETY PROFILE: An experimental teratogen. Human mutation data reported. When heated to decomposition it emits toxic fumes of SO_x and NO_x. See also CARBAMATES and ESTERS.

AMT000 CAS:88-62-0 ***HR: 2***
2-AMINO-p-TOLUENESULFONIC ACID
mf: $C_7H_9NO_3S$ mw: 187.23

PROP: Needles. Sol in water.

SYN: KYSELINA 2-TOLUIDIN-4-SULFONOVA (CZECH)

TOXICITY DATA with REFERENCE
eye-rbt 100 mg/24H SEV 28ZPAK -,183,72
orl-rat LD50:8480 mg/kg 28ZPAK -,183,72

CONSENSUS REPORTS: Reported in EPA TSCA Inventory.

SAFETY PROFILE: Mildly toxic by ingestion. A severe eye irritant. See also SULFONATES. When heated to decomposition it emits very toxic fumes of NO_x and SO_x.

AMT250 CAS:133-78-8 ***HR: 2***
4-AMINO-o-TOLUENESULFONIC ACID
mf: $C_7H_9NO_3S$ mw: 187.23

PROP: Needles. Sol in water.

SYN: KYSELINA-3-TOLUIDIN-6-SULFONOVA(CZECH)

TOXICITY DATA with REFERENCE
eye-rbt 500 mg/24H SEV 28ZPAK -,183,72

SAFETY PROFILE: A severe eye irritant. See also SULFONATES. When heated to decomposition it emits very toxic fumes of NO_x and SO_x.

AMT500 CAS:139-13-9 ***HR: 3***
AMINOTRIACETIC ACID
mf: $C_6H_9NO_6$ mw: 191.16

SYNS: N,N-BIS(CARBOXYMETHYL)GLYSINE ◇ NCI-C02766 ◇ NITRILOTRIACETIC ACID ◇ TRIGLYCINE ◇ TRIGLYCOLLAMIC ACID ◇ VERSENE NTA ACID

TOXICITY DATA with REFERENCE
orl-rat TDLo:430 g/kg/75W-C:CAR NCITR* NCI-CG-TR-6,77
orl-mus TDLo:832 g/kg/66W-C:NEO NCITR* NCI-CG-TR-6,77
orl-rat LD50:1470 mg/kg NCILB* NIH-NCI-E-C-72-3252
orl-mus LD50:3160 mg/kg NCILB* NIH-NCI-E-C-72-3252
ipr-mus LDLo:125 mg/kg TXAPA9 23,288,72

CONSENSUS REPORTS: NTP Fifth Annual Report on Carcinogens. NCI Carcinogenesis Bioassay (feed); Clear Evidence: mouse, rat NCITR* NCI-CG-TR-6,77. Reported in EPA TSCA Inventory. Community Right-To-Know List.

SAFETY PROFILE: Confirmed carcinogen with experimental carcinogenic and neoplastigenic data. Poison by intraperitoneal route. Moderately toxic by ingestion. When heated to decomposition it emits toxic fumes of NO_x.

AMU000 CAS:35695-72-8 ***HR: 3***
3-AMINO-1-TRICHLORO-2-PENTANOL
mf: $C_5H_{10}Cl_3NO$ mw: 206.51

SYN: TCA-PE

TOXICITY DATA with REFERENCE
scu-mus LD50:1364 mg/kg SKIZAB 28,231,72
ivn-mus LD50:284 mg/kg SKIZAB 28,231,72

SAFETY PROFILE: Poison by intravenous route. Moderately toxic by subcutaneous route. When heated to decomposition it emits very toxic fumes of Cl^- and NO_x.

AMU125 CAS:53516-81-7 ***HR: 3***
((3-AMINO-2,4,6-TRICHLOROPHENYL)METHYLENE) HYDRAZIDE BENZENESULFONIC ACID
mf: $C_{13}H_{10}Cl_3N_3O_2S$ mw: 378.67

SYNS: DENVER RESEARCH CENTER No. DRC-4575 ◇ DRC-4575

TOXICITY DATA with REFERENCE
orl-mus TDLo:62 mg/kg (8-13D preg):REP JTEHD6 3,407,77
orl-mus LD50:33 mg/kg JTEHD6 3,407,77

SAFETY PROFILE: Poison by ingestion. Experimental reproductive effects. When heated to decomposition it emits toxic fumes of Cl^-, SO_x, and NO_x. See also SULFONATES.

AMU500 CAS:35695-70-6 ***HR: 3***
3-AMINO-1-TRICHLORO-2-PROPANOL
mf: $C_3H_6Cl_3NO$ mw: 178.45

SYN: TCA-PR

TOXICITY DATA with REFERENCE
scu-mus LD50:1556 mg/kg SKIZAB 28,231,72
ivn-mus LD50:301 mg/kg SKIZAB 28,231,72

SAFETY PROFILE: Poison by intravenous route. Moderately toxic by subcutaneous route. When heated to decomposition it emits very toxic fumes of Cl^- and NO_x.

AMU625 CAS:3119-15-1 ***HR: 2***
3-AMINO-2,4,6-TRIIODO-BENZOIC ACID
mf: $C_7H_4I_3NO_2$ mw: 514.82

SYN: ACIDO-3-AMINO-2,4,6-TRIIODOBENZOICO (ITALIAN)

TOXICITY DATA with REFERENCE
unr-rat LD50:1450 mg/kg JAPMA8 42,721,53
orl-mus LD50:600 mg/kg QJPPAL 19,483,46
ivn-mus LD50:800 mg/kg FRSPAX 18,33,63

SAFETY PROFILE: Moderately toxic by ingestion and other routes. When heated to decomposition it emits toxic fumes of I^- and NO_x.

AMU750 CAS:3115-05-7 ***HR: 2***
N-(3-AMINO-2,4,6-TRIIODOBENZOYL)-N-(2-CARBOXYETHYL)ANILINE
mf: $C_{16}H_{13}I_3N_2O_3$ mw: 662.01

SYNS: 3-((3-AMINO-2,4,6-TRIIODOBENZOYL)PHENYLAMINO)PROPIONIC ACID ◇ N-(3-AMINO-2,4,6-TRIJODBENZOYL)-N-PHENYL-β-AMINOPROPIONSAEURE(GERMAN) ◇ ISOBENZAMIC ACID ◇ ST 5066/S (GERMAN)

TOXICITY DATA with REFERENCE
orl-rat LD50:2800 mg/kg TXAPA9 14,232,69
ivn-rat LD50:500 mg/kg TXAPA9 14,232,69
orl-mus LD50:2870 mg/kg ARZNAD 11,384,61
ivn-mus LD50:530 mg/kg JMCMAR 13,997,70

SAFETY PROFILE: Moderately toxic by ingestion and intravenous routes. When heated to decomposition it emits very toxic fumes of I^- and NO_x.

AMV375 CAS:1634-73-7 ***HR: 3***
4-((3-AMINO-2,4,6-TRIIODOPHENYL)ETHYLAMINO)-4-OXO-BUTANOIC ACID
mf: $C_{12}H_{13}I_3N_2O_3$ mw: 613.97

SYNS: N-AETHYL-N-(2,4,6-TRIJOD-3-AMINOPHENYL)-SUCCINAMIDSAEURE (GERMAN) ◇ 3'-AMINO-N-ETHYL-2',4',6'-TRIIODOSUCCINANILIC ACID ◇ RG 235 ◇ SH 771

TOXICITY DATA with REFERENCE
ivn-rat LD50:370 mg/kg PHARAT 27,411,72
orl-mus LD50:650 mg/kg PHARAT 27,391,72
ivn-mus LD50:288 mg/kg PHARAT 27,411,72

SAFETY PROFILE: Poison by intravenous route. Moderately toxic by ingestion. When heated to decomposition it emits toxic fumes of I^- and NO_x.

AMV750 CAS:23217-86-9 ***HR: 3***
2-(3-AMINO-2,4,6-TRIIODOPHENYL)VALERIC ACID
mf: $C_{11}H_{12}I_3NO_2$ mw: 570.94

TOXICITY DATA with REFERENCE
orl-mus LD50:2100 mg/kg JMCMAR 13,559,70
ivn-mus LD50:170 mg/kg JMCMAR 13,559,70

SAFETY PROFILE: Poison by intravenous route. Moderately toxic by ingestion. See also IODIDES. When heated to decomposition it emits very toxic fumes of I^- and NO_x.

AMV790 CAS:698-49-7 ***HR: 3***
4-AMINOTROPOLONE
mf: $C_7H_7NO_2$ mw: 137.15

SYN: 4-AMINO-2-HYDROXY-2,4,6-CYCLOPHEPTATRIEN-1-ONE

TOXICITY DATA with REFERENCE
ipr-mus LD50:265 mg/kg CPBTAL 20,60,72

scu-mus LD50:400 mg/kg CPBTAL 20,60,72
ivn-mus LD50:177 mg/kg YKKZAJ 92,19,72

SAFETY PROFILE: Poison by subcutaneous, intravenous, and intraperitoneal routes. When heated to decomposition it emits toxic fumes of NO_x.

AMV800 CAS:7021-46-7 ***HR: 3***
5-AMINOTROPOLONE
mf: $C_7H_7NO_2$ mw: 137.15

SYN: 5-AMINO-2-HYDROXY-2,4,6-CYCLOHEPTATRIEN-1-ONE

TOXICITY DATA with REFERENCE
ipr-mus LD50:230 mg/kg YKKZAJ 91,550,71
scu-mus LD50:521 mg/kg YKKZAJ 91,550,71
ivn-mus LD50:175 mg/kg YKKZAJ 92,19,72

SAFETY PROFILE: Poison by intravenous and intraperitoneal routes. Moderately toxic by subcutaneous route. When heated to decomposition it emits toxic fumes of NO_x.

AMV875 ***HR: 1***
AMINO-TS-ACID
mf: $C_{10}H_9NO_6S_2$ mw: 303.32

TOXICITY DATA with REFERENCE
orl-rat LD50:29 g/kg GISAAA 45(3),73,80
orl-mus LD50:29 g/kg GISAAA 45(3),73,80
orl-rbt LD50:18 g/kg GISAAA 45(3),73,80
orl-gpg LD50:42 g/kg GISAAA 45(3),73,80

SAFETY PROFILE: Low order of toxicity. When heated to decomposition it emits toxic fumes of SO_x and NO_x. See also SULFONATES.

AMW000 CAS:2432-99-7 ***HR: 3***
11-AMINOUNDECANOIC ACID
mf: $C_{11}H_{23}NO_2$ mw: 201.35

SYNS: AMINOUNDECANOIC ACID ◇ 11-AMINOUNDECYLIC ACID ◇ NCI-C50613

TOXICITY DATA with REFERENCE
otr-ham:emb 2500 mmol/L ENMUDM 8,515,86
sce-ham:ovr 500 mg/L EMMUEG 10(Suppl 10),1,87
orl-rat TDLo:655 g/kg/2Y-C:CAR NTPTR* NTP-TR-216,82
orl-rat TD:328 g/kg/2Y-C:NEO NTPTR* NTP-TR-216,82

CONSENSUS REPORTS: IARC Cancer Review: Group 3 IMEMDT 7,56,87; Animal Limited Evidence IMEMDT 39,239,86. NTP Carcinogenesis Bioassay (feed): Clear Evidence: mouse, rat NTPTR* NTP-TR-216,82. Reported in EPA TSCA Inventory.

SAFETY PROFILE: Questionable carcinogen with experimental carcinogenic and neoplastigenic data. Mutation data reported. When heated to decomposition it emits toxic fumes of NO_x.

AMW250 CAS:873-83-6 ***HR: 2***
6-AMINOURACIL
mf: $C_4H_5N_3O_2$ mw: 127.12

PROP: Needles from water. Mp: decomp. Sol in water, alkalies, NH_4OH, and acids.

TOXICITY DATA with REFERENCE
par-mus LDLo:2400 mg/kg CBCCT* 7,696,55

CONSENSUS REPORTS: Reported in EPA TSCA Inventory.

SAFETY PROFILE: Moderately toxic by parenteral route. When heated to decomposition it emits toxic fumes of NO_x.

AMW500 CAS:38237-74-0 ***HR: 3***
p-AMINO VALEROPHENONE
mf: $C_{11}H_{15}NO$ mw: 177.27

TOXICITY DATA with REFERENCE
orl-rat LD50:84 mg/kg GEPHDP 14,465,83
orl-mus LD50:94 mg/kg GEPHDP 14,465,83
ipr-mus LD50:120 mg/kg FEPRA7 6,348,47

CONSENSUS REPORTS: Reported in EPA TSCA Inventory.

SAFETY PROFILE: Poison by ingestion and intraperitoneal route. See also AMINES. When heated to decomposition it emits toxic fumes of NO_x.

AMW750 CAS:6623-41-2 ***HR: 3***
2-AMINO-4,5-XYLENOL
mf: $C_8H_{11}NO$ mw: 137.20

SYN: 2-AMINO-4,5-DIMETHYLPHENOL

TOXICITY DATA with REFERENCE
imp-mus TDLo:80 mg/kg:CAR BJCAAI 11,212,57

CONSENSUS REPORTS: Reported in EPA TSCA Inventory.

SAFETY PROFILE: Questionable carcinogen with experimental carcinogenic data. See also AMINES. When heated to decomposition it emits toxic fumes of NO_x.

AMX000 CAS:5369-84-6 ***HR: 3***
3-AMINO-4-(2-(2,6-XYLYLOXY)ETHYL)-4H-1,2,4-TRIAZOLE
mf: $C_{12}H_{16}N_4O$ mw: 232.32

SYN: 3,5-BIS(2-FURYL)-1H-1,2,4-TRIAZOLE

TOXICITY DATA with REFERENCE
ipr-rat LD50:235 mg/kg JMCMAR 9,42,66
ipr-mus LD50:2000 mg/kg JMCMAR 9,22,66

SAFETY PROFILE: Poison by intraperitoneal route.

When heated to decomposition it emits toxic fumes of NO_x.

AMX250 ***HR: 2***
AMIODOXYL BENZOATE
mf: $C_7H_8INO_4$ mw: 297.06

PROP: White, odorless, sltly bitter, crystalline powder.

SYN: ARTHRYTIN OXOATE

SAFETY PROFILE: Moderately toxic by ingestion. It is a non-selective, systemic herbicide. Dangerous; when heated to decomposition it emits toxic fumes of NO_x and I^-.

AMX500 ***HR: 3***
AMIPURIMYCIN HYDRATE
mf: $C_{30}H_{20}N_7O_8 \cdot H_2O$ mw: 624.59

PROP: Isolated from culture filtrate of *Streptomyces novoguineesis* T-36496.

TOXICITY DATA with REFERENCE
skn-rbt 200 ppm/10D SEV JANTAJ 30,1,77
orl-rat LD50:20 mg/kg JANTAJ 30,1,77
ivn-rat LD50:1 mg/kg JANTAJ 30,1,77
orl-mus LD50:10 mg/kg JANTAJ 30,1,77
ivn-mus LD50:1 mg/kg JANTAJ 30,1,77

SAFETY PROFILE: Poison by ingestion and intravenous routes. A severe skin irritant. When heated to decomposition it emits toxic fumes of NO_x.

AMX750 CAS:57-43-2 ***HR: 3***
AMITAL
mf: $C_{11}H_{18}N_2O_3$ mw: 226.31

PROP: Slightly bitter crystals.

SYNS: AMOBARBITAL ◇ AMYLBARBITONE ◇ AMYLOBARBITAL ◇ AMYLOBARBITONE ◇ AMYTAL ◇ 5-ETHYL-5-ISOAMYLBARBITURIC ACID ◇ 5-ETHYL-5-ISOAMYLMALONYL UREA ◇ ETHYLISOPENTYLBARBITURIC ACID ◇ 5-ETHYL-5-ISOPENTYLBARBITURIC ACID ◇ 5-ETHYL-5-(3-METHYLBUTYL)BARBITURIC ACID ◇ ISOAMYLETHYLBARBITURIC ACID ◇ 5-ISOAMYL-5-ETHYLBARBITURIC ACID ◇ NSC 10815

TOXICITY DATA with REFERENCE
orl-hmn LDLo:43 mg/kg 28ZAA9 -,-,32
orl-rat LD50:250 mg/kg ARZNAD 21,719,71
ipr-rat LD50:115 mg/kg ARZNAD 21,719,71
scu-rat LD50:190 mg/kg AEPPAE 152,341,30
orl-mus LD50:345 mg/kg JACSAT 61,96,39
ipr-mus LD50:175 mg/kg%% JMCMAR 10,1078,67
scu-mus LD50:212 mg/kg ARZNAD 15,688,65
orl-cat LDLo:100 mg/kg PHREA7 19,472,39
ipr-rbt LDLo:90 mg/kg JPETAB 41,465,31
scu-rbt LDLo:170 mg/kg JACSAT 45,243,23
ivn-rbt LD50:49 mg/kg JPETAB 96,209,49

SAFETY PROFILE: A human poison by ingestion. An experimental poison by ingestion, intravenous, intraperitoneal, and subcutaneous routes. See also BARBITURATES. When heated to decomposition it emits toxic fumes of NO_x.

AMX825 CAS:3734-97-2 ***HR: 3***
AMITON OXALATE
mf: $C_{10}H_{24}NO_3PS \cdot C_2H_2O_4$ mw: 359.42

PROP: Mp: 98-99°C.

SYNS: ACID OXALATE ◇ CHIPMAN 6199 ◇ CHIPMAN R-6, 199 ◇ CITRAM ◇ 2-(2-DIETHYLAMINO)ETHYL)-O,O-DIETHYL ESTER, OXALATE (1:1) ◇ S-(2-DIETHYLAMINOETHYL)-O,O-DIETHYLPHOSPHOROTHIOATE HYDROGEN OXALATE ◇ O,O-DIETHYL-S-(2-DIETHYLAMINO) ETHYLPHOSPHOROTHIOATE HYDROGEN OXALATE ◇ O,O-DIETHYL-S-(β-DIETHYLAMINO)ETHYL PHOSPHOROTHIOLATE HYDROGEN OXALATE ◇ O,O-DIETHYL-S-(2-ETHYL-N,N-DIETHYLAMINO) PHOSPHOROTHIOATE HYDROGEN OXALATE ◇ ENT 20,993 ◇ HYDROGEN OXALATE of AMITON ◇ PHOSPHOROTHIOIC ACID ◇ TETRAM ◇ TETRAM MONOOXALATE

TOXICITY DATA with REFERENCE
ipr-mus LD50:500 μg/kg PAREAQ 11,636,59
orl-rat LD50:3 mg/kg 28ZEAL 4,162,69

CONSENSUS REPORTS: EPA Extremely Hazardous Substances List.

SAFETY PROFILE: A deadly poison by ingestion and intraperitoneal routes. Human systemic effects may include: headache, giddiness, nervousness, impaired vision, weakness, nausea, cramps, diarrhea, muscular weakness and loss of control, convulsions and coma. Flammable. To extinguish fire, use dry chemical, carbon dioxide, water spray, fog or foam. When heated to decomposition it emits toxic fumes of SO_x, PO_x, and NO_x. A cholinesterase inhibitor used as an insecticide.

AMY000 CAS:4317-14-0 ***HR: 3***
AMITRIPTYLINE-N-OXIDE
mf: $C_{20}H_{23}NO$ mw: 293.44

SYNS: AMITRIPTYLINOXIDE ◇ 1-PROPANAMINE, 3-(10,11-DIHYDRO-5H-DIBENZO(A,D)CYCLOHEPTEN-5-YLIDENE)-N,N-DIMETHYL-N-OXIDE

TOXICITY DATA with REFERENCE
orl-rat TDLo:2500 mg/kg (female 6-15D post):TER ARZNAD 28,1898,78
orl-rat TDLo:2500 mg/kg (female 6-15D post):REP ARZNAD 28,1898,78
orl-rat LD50:1800 mg/kg ARZNAD 28,1898,78
ipr-rat LD50:110 mg/kg ARZNAD 28,1898,78
ivn-rat LD50:25 mg/kg ARZNAD 28,1898,78
orl-mus LD50:330 mg/kg ARZNAD 28,1898,78
ipr-mus LD50:320 mg/kg ARZNAD 28,1898,78
ivn-mus LD50:87 mg/kg ARZNAD 28,1898,78

orl-dog LD50:330 mg/kg ARZNAD 28,1898,78
orl-rbt LD50:330 mg/kg ARZNAD 28,1898,78
orl-gpg LD50:330 mg/kg ARZNAD 28,1898,78

SAFETY PROFILE: Poison by ingestion, intraperitoneal, and intravenous routes. An experimental teratogen. Other experimental reproductive effects. When heated to decomposition it emits toxic fumes of NO_x.

AMY050 CAS:61-82-5 ***HR: 3***
AMITROLE
mf: $C_2H_4N_4$ mw: 84.10

SYNS: AMEROL ◇ AMINOTRIAZOLE (plant regulator) ◇ 2-AMINOTRIAZOLE ◇ 3-AMINOTRIAZOLE ◇ 3-AMINO-s-TRIAZOLE ◇ 3-AMINO-1,2,4-TRIAZOLE ◇ 2-AMINO-1,3,4-TRIAZOLE ◇ 3-AMINO-1H-1,2,4-TRIAZOLE ◇ AMINO TRIAZOLE WEEDKILLER 90 ◇ AMINOTRIAZOL-SPRITZPULVER ◇ AMITOL ◇ AMITRIL ◇ AMITRIL T.L. ◇ AMITROL ◇ AMITROL 90 ◇ AMITROL-T ◇ AMIZOL ◇ AT ◇ ATA ◇ AT LIQUID ◇ AZAPLANT ◇ AZOLAN ◇ AZOLE ◇ CAMPAPRIM A 1544 ◇ CYTROL ◇ DIUROL ◇ DOMATOL ◇ ELMASIL ◇ EMISOL ◇ ENT 25,445 ◇ FENAMINE ◇ FENAVAR ◇ HERBICIDE TOTAL ◇ HERBIZOLE ◇ KLEER-LOT ◇ ORGA-414 ◇ RADOXONE TL ◇ RAMIZOL ◇ RCRA WASTE NUMBER U011 ◇ SIMAZOL ◇ SOLUTION CONCENTREE T271 ◇ TRIAZOLAMINE ◇ 1H-1,2,4-TRIAZOL-3-AMINE ◇ USAF XR-22 ◇ VOROX ◇ WEEDAR ADS ◇ WEEDAZIN ◇ WEEDAZOL ◇ WEEDEX GRANULAT ◇ WEEDOCLOR ◇ X-ALL LIQUID

TOXICITY DATA with REFERENCE
mma-sat 50 μg/plate PMRSDJ 1,351,81
mrc-asn 600 μg/L MUREAV 147,288,85
sln-asn 600 μg/L MUREAV 147,288,85
hma-mus/sat 12 mg/kg JNCIAM 62,911,79
msc-ham:emb 1 mg/L MUREAV 140,205,84
orl-mus TDLo:1935 mg/kg (female 6-14D post):TER NTIS** PB223-160
scu-mus TDLo:4176 mg/kg (female 6-14D post):REP NTIS** PB223-160
orl-rat TDLo:4595 mg/kg/2.5Y-C:CAR TXAPA9 69,161,83
orl-mus TDLo:113 g/kg/3W-I:CAR JNCIAM 42,1101,69
orl-rat TD:3670 mg/kg/2Y-C:NEO SCIEAS 132,296,60
orl-mus TD:366 g/kg/26W-C:ETA TOLED5 29,145,85
orl-rat LD50:1100 mg/kg RREVAH 10,97,65
orl-mus LD50:14700 mg/kg PCOC** -,33,66
ipr-mus LD50:200 mg/kg NTIS** AD277-689

CONSENSUS REPORTS: NTP Fifth Annual Report on Carcinogens. IARC Cancer Review: Group 2B IMEMDT 7,92,87; Human Inadequate Evidence IMEMDT 41,293,86; IMEMDT 7,31,74; Animal Sufficient Evidence IMEMDT 7,31,74; IMEMDT 41,293,86. Reported in EPA TSCA Inventory. EPA Genetic Toxicology Program.

OSHA PEL: TWA 0.2 mg/m^3
ACGIH TLV: TWA 0.2 mg/m^3
DFG MAK: 0.2 mg/m^3

SAFETY PROFILE: Confirmed carcinogen with experimental carcinogenic, tumorigenic, and neoplastigenic data. Poison by intraperitoneal route. Moderately toxic by ingestion. An experimental teratogen. Other experimental reproductive effects. Mutation data reported. When heated to decomposition it emits toxic fumes of NO_x. An herbicide and plant growth regulator.

AMY250 ***HR: 3***
AMMINE PENTAHYDROXO PLATINUM
mf: H_8NO_5Pt mw: 297.2

SAFETY PROFILE: A poison. An explosively unstable compound. Explodes @ > 250°. Upon decomposition it emits toxic fumes of NO_x. See also PLATINUM COMPOUNDS.

AMY500 CAS:7664-41-7 ***HR: 3***
AMMONIA
DOT: UN 1005/UN 2073/UN 2672
mf: H_3N mw: 17.04

PROP: Colorless gas, extremely pungent odor, liquefied by compression. Mp: −77.7°, bp: −33.35°, lel: 16%, uel: 25%, d: 0.771 g/liter @ 0°, 0.817 g/liter @ −79°, autoign temp: 1204°F, vap press: 10 atm @ 25.7°, vap d: 0.6. Very sol in water; moderately sol in alc.

SYNS: AMMONIAC (FRENCH) ◇ AMMONIACA (ITALIAN) ◇ AMMONIA GAS ◇ AMMONIAK (GERMAN) ◇ AMONIAK (POLISH) ◇ ANHYDROUS AMMONIA ◇ SPIRIT of HARTSHORN

TOXICITY DATA with REFERENCE
mmo-esc 1500 ppm/3H AMNTA4 85,119,51
cyt-rat-ihl 19800 $\mu g/m^3$/16W BZARAZ 27,102,74
ihl-hmn LCLo:30000 ppm/5M TJSGA8 45,458,67
ihl-hmn TCLo:20 ppm:IRR AGGHAR 13,528,55
unk-man LDLo:132 mg/kg 85DCAI 2,73,70
orl-rat LD50:350 mg/kg PHIT**
ihl-rat LCLo:2000 ppm/4H JIHTAB 31,343,49
ihl-mus LD50:4837 ppm/1H NTIS** PB214-270
ihl-cat LCLo:7000 ppm/1H JIHTAB 26,29,44
ihl-cat TCLo:1000 ppm/10M AEHLAU 35,6,80
ihl-rbt LCLo:7000 ppm/1H JIHTAB 26,29,44
ihl-mam LCLo:5000 ppm/5M AEPPAE 138,65,28

CONSENSUS REPORTS: EPA Extremely Hazardous Substances List. Community Right-To-Know List. Reported in EPA TSCA Inventory.

OSHA PEL: TWA 35 ppm
ACGIH TLV: TWA 25 ppm; STEL 35 ppm
DFG MAK: 50 ppm (35 mg/m^3)
NIOSH REL: CL 50 ppm
DOT Classification: Nonflammable Gas; Label: NONFLAM. Gas (UN1005, UN2073); Poison A; Label: Poison Gas (UN1005).

SAFETY PROFILE: A human poison by an unspecified

route. Poison experimentally by inhalation, ingestion, and possibly other routes. An eye, mucous membrane, and systemic irritant by inhalation. Mutation data reported. A common air contaminant. Difficult to ignite. Explosion hazard when exposed to flame or in a fire. NH_3 + air in a fire can detonate. Potentially violent or explosive reactions on contact with interhalogens (e.g., bromine pentafluoride, chlorine trifluoride), 1,2-dichloroethane (with liquid NH_3), boron halides, chloroformamidnium nitrate, ethylene oxide (polymerization reaction), magnesium perchlorate, nitrogen trichloride, oxygen + platinum, or strong oxidants (e.g., potassium chlorate, nitryl chloride, chromyl chloride, dichlorine oxide, chromium trioxide, trioxygen difluoride, nitric acid, hydrogen peroxide, tetramethylammonium amide, thiocarbonyl azide thiocyanate, sulfinyl chloride, thiotriazyl chloride, ammonium peroxodisulfate, fluorine, nitrogen oxide, dinitrogen tetraoxide, and liquid oxygen). Forms sensitive explosive mixtures with air + hydrocarbons, 1-chloro-2,4-dinitrobenzene, 2-,or 4-chloronitrobenzene (above 160°C/30 bar), ethanol + silver nitrate, germanium derivatives, stibine, and chlorine. Reaction with silver chloride, silver nitrate, silver azide, and silver oxide form the explosive silver nitride. Reactions with chlorine azide, bromine, iodine, iodine + potassium, heavy metals and their compounds (e.g. gold(III) chloride, mercury, and potassium thallium amide ammoniate), tellurium halides (e.g., tellurium tetrabromide, and tellurium tetrachloride)and pentaborane(9) give explosive products. Incompatible in contact with Ag, acetaldehyde, acrolein, B, BI_3, halogens, $HClO_3$, ClO, chlorites, chlorosilane, (ethylene dichloride + liquid ammonia), Au, hexachloromelamine, (hydrazine + alkali metals), HBr, HOCl, $Mg(ClO_4)_2$, N_2O_4, NCl_3, NF_3, OF_2, P_2O_5, P_2O_3, picric acid, (K + AsH_3), (K + PH_3), (K + $NaNO_2$), potassium ferricyanide, potassium mercuric cyanide, (Na + CO), Sb, S, SCl_2, tellurium hydropentachloride, trichloromelamine, NO_2Cl, SbH_3, tetramethylammonium amide, $SOCl_2$, and thiotrithiazylchloride. Incandescent reaction when heated with calcium. Emits toxic fumes of NH_3 and NO_x when exposed to heat. To fight fire stop flow of gas.

AMZ125 CAS:57530-25-3 ***HR: 2***
2-AMMONIOTHIAZOLE NITRATE
mf: $C_3H_5N_3O_3S$ mw: 163.15

SAFETY PROFILE: Explosive decomposition at 142°C. Upon decomposition it emits toxic fumes of SO_x and NO_x.

ANA000 CAS:631-61-8 ***HR: 3***
AMMONIUM ACETATE
mf: $C_2H_4O_2 \cdot H_3N$ mw: 77.10

PROP: Crystals. Mp: 114°, d: 1.07.

SYN: ACETIC ACID, AMMONIUM SALT

TOXICITY DATA with REFERENCE
ipr-rat LD50:632 mg/kg ABBIA4 64,342,56
ivn-mus LD50:386 mg/kg MEIEDD 10,74,83
ivn-mus LD50:98 mg/kg MEIEDD 10,74,83

CONSENSUS REPORTS: Reported in EPA TSCA Inventory.

SAFETY PROFILE: Poison by intravenous route. Moderately toxic by intraperitoneal routes. When heated to decomposition it emits toxic fumes of NO_x and NH_3.

ANA500 ***HR: 3***
AMMONIUM (AMINYLENIUM BIS [TRIHYDROBORATE])
mf: $B_2H_{12}N_2$ mw: 61.72

SAFETY PROFILE: A highly reactive hydride. Self ignites in air when heated. When heated to decomposition it emits toxic fumes of NO_x and NH_3. Explodes on heating in air. See also BORON COMPOUNDS and HYDRIDES.

ANA750 ***HR: 3***
AMMONIUM AZIDE
mf: NH_4N_3 mw: 60.1

PROP: Colorless plates. Mp: 160°, bp: explodes, d: 1.346, vap press: 1 mm @ 59.2° (sublimes).

SAFETY PROFILE: Poison by inhalation and ingestion. See also AZIDES. Moderately flammable. Unstable. Explosion hazard upon rapid heating.

ANB000 CAS:5251-79-6 ***HR: 2***
AMMONIUM BENZAMIDOOXYACETATE
mf: $C_9H_9NO_4 \cdot H_3N$ mw: 212.23

SYNS: AMMONIUM-2-(BENZAMIDOOXY)ACETATE ◇ BENZADOX ◇ BENZAMIDOOXY ACETIC ACID, AMMONIUM SALT ◇ TOPCIDE

TOXICITY DATA with REFERENCE
orl-rat LD50:2500 mg/kg 28ZEAL 5,23,76
skn-rbt LD50:450 mg/kg 28ZEAL 5,23,76

SAFETY PROFILE: Moderately toxic by ingestion and skin contact. When heated to decomposition it emits toxic fumes of NO_x and NH_3.

ANB250 CAS:1066-33-7 ***HR: 3***
AMMONIUM BICARBONATE (1:1)
mf: $HCO_3 \cdot H_4N$ mw: 79.1

PROP: Hard, colorless to white crystals; faint ammonia odor, stable at room temp, volatile. Decomp @ 60°, mp: 107.5° (rapid heating). D: 1.586. Sol in water; insol in alc.

SYNS: ACID AMMONIUM CARBONATE ◇ AMMONIUM CARBON-

ATE ◇ AMMONIUM HYDROGEN CARBONATE ◇ CARBONIC ACID, MONOAMMONIUM SALT ◇ MONOAMMONIUM CARBONATE

TOXICITY DATA with REFERENCE
ivn-mus LD50:245 mg/kg AJVRAH 29,897,68

CONSENSUS REPORTS: Reported in EPA TSCA Inventory.

SAFETY PROFILE: Poison by intravenous route. When heated to decomposition it emits toxic fumes of NO_x and NH_3.

ANB500 CAS:7789-09-5 ***HR: 3***
AMMONIUM BICHROMATE
DOT: UN 1439
mf: $Cr_2H_8N_2O_7$ mw: 252.10

PROP: Red crystals. Mp: decomp, d: 2.936.

SYNS: AMMONIO (DICROMATO DI) (ITALIAN) ◇ AMMONIUMBICHROMAAT (DUTCH) ◇ AMMONIUMDICHROMAAT (DUTCH) ◇ AMMONIUMDICHROMAT (GERMAN) ◇ AMMONIUM DICHROMATE ◇ AMMONIUM DICHROMATE(VI) ◇ BICHROMATE d'AMMONIUM (FRENCH)

TOXICITY DATA with REFERENCE
scu-gpg LDLo:25 mg/kg EQSSDX 1,1,75

CONSENSUS REPORTS: Reported in EPA TSCA Inventory. Chromium and its compounds are on the Community Right-To-Know List.

OSHA PEL: CL 0.1 $mg(CrO_3)/m^3$
ACGIH TLV: TWA 0.05 $mg(Cr)/m^3$
NIOSH REL: (Chromium(VI)) TWA 25 $\mu g(Cr(VI))/m^3$; CL 50 $\mu g/m^3$/15M
DOT Classification: Oxidizer; Label: Oxidizer.

SAFETY PROFILE: Poison by inhalation, ingestion, skin contact, and subcutaneous routes. See also CHROMIUM COMPOUNDS. An unstable oxidizer. Moderately flammable; reacts with reducing agents.

ANC000 ***HR: 3***
AMMONIUM BROMATE
mf: NH_4BrO_3 mw: 145.96

PROP: Colorless crystals. Mp: explodes. Very sol in water.

SAFETY PROFILE: An unstable, explosive oxidizing material. See also BROMATES. Severe explosion hazard.

ANC250 CAS:12124-97-9 ***HR: D***
AMMONIUM BROMIDE
mf: BrH_4N mw: 97.96

PROP: Colorless, cubic, sltly hygroscopic crystals. Mp: subl @ 452°, bp: 235° in vac, d: 2.429, vap press: 1 mm @ 198.3°.

SYN: HYDROBROMIC ACID MONOAMMONIATE

TOXICITY DATA with REFERENCE
dnd-mam:lym 50 mmol/L CBINA8 19,197,77

CONSENSUS REPORTS: Reported in EPA TSCA Inventory.

SAFETY PROFILE: Mutation data reported. See also BROMIDES. When heated to decomposition it emits very toxic fumes of NO_x, Br^-, and NH_3. Incompatible with BrF_3; IF_7; K.

ANC750 ***HR: 3***
AMMONIUM BROMO SELENATE
mf: $(NH_4)SeBr_6$ mw: 594.5

PROP: Red octagonal crystals. D: 3.326, decomp in water; sltly sol in ether.

CONSENSUS REPORTS: Selenium and its compounds are on the Community Right-To-Know List.

SAFETY PROFILE: Poison and dangerous hazard. See also SELENIUM COMPOUNDS and BROMIDES.

AND250 ***HR: 3***
AMMONIUM CADMIUM CHLORIDE
mf: $4NH_4Cl•CdCl_2$ mw: 397.3

PROP: Colorless, rhombic crystals. D: 2.01; sol in water.

CONSENSUS REPORTS: Cadmium and its compounds are on the Community Right-To-Know List.

OSHA PEL: TWA 0.1 $mg(Cd)/m^3$; CL 0.6 $mg(Cd)/m^3$ (fume)
ACGIH TLV: TWA 0.05 $mg(Cd)/m^3$ (Proposed: TWA 0.01 $mg(Cd)/m^3$ (dust), Suspected Human Carcinogen; 0.002 $mg(Cd)/m^3$ (respirable dust), Suspected Human Carcinogen); BEI: 10 $\mu g/g$ creatinine in urine; 10 $\mu g/L$ in blood.
DFG BAT: Blood: 1.5 $\mu g/dL$; Urine: 15 $\mu g/dL$; Suspected Carcinogen.
NIOSH REL: (Cadmium) Reduce to lowest feasible level

SAFETY PROFILE: Confirmed human carcinogen. A poison. See also CADMIUM COMPOUNDS. When heated to decomposition it emits toxic fumes of NH_3, NO_x, and Cl^-.

AND500 ***HR: 3***
AMMONIUM CALCIUM ARSENATE
mf: $NH_4CaAsO_4•6H_2O$ mw: 305.1

PROP: Colorless crystals. Mp: 140° (decomp), d: 1.905 @ 15°. Sltly sol in cold water; sol in hot water; sol in NH_4Cl and NH_4OH.

CONSENSUS REPORTS: Arsenic and its compounds are on the Community Right-To-Know List.

SAFETY PROFILE: A poison. See also ARSENIC COMPOUNDS.

AND750 CAS:1111-78-0 ***HR: 3***
AMMONIUM CARBAMATE
DOT: NA 9083
mf: $CH_3NO_2 \cdot H_3N$ mw: 78.09

PROP: White, crystalline, rhombic powder; sol in water and alc; ammonia odor. Sublimates at 60°.

SYN: AMMONIUM AMINOFORMATE

TOXICITY DATA with REFERENCE
ivn-rat LD50:39 mg/kg AJVRAH 29,897,68
ivn-mus LD50:77 mg/kg AJVRAH 29,897,68

CONSENSUS REPORTS: Reported in EPA TSCA Inventory.

DOT Classification: ORM-A; Label: None

SAFETY PROFILE: Poison by intravenous route. See also CARBAMATES.

ANE000 CAS:506-87-6 ***HR: 3***
AMMONIUM CARBONATE
DOT: NA 9084
mf: $(NH_4)_2CO_3$ mw: 96.09

PROP: Colorless crystals. Decomposes on standing to ammonium bicarbonate. Sltly sol in water.

SYNS: AMMONIUMCARBONAT (GERMAN) ◇ CARBONIC ACID, AMMONIUM SALT ◇ CARBONIC ACID, DIAMMONIUM SALT ◇ DIAMMONIUM CARBONATE

TOXICITY DATA with REFERENCE
ivn-mus LD50:96 mg/kg AJVRAH 29,897,68
ivn-dog LDLo:200 mg/kg HBAMAK 4,1289,35
scu-frg LDLo:250 mg/kg HBAMAK 4,1289,35

CONSENSUS REPORTS: Reported in EPA TSCA Inventory.

DOT Classification: ORM-A; Label: None

SAFETY PROFILE: Poison by subcutaneous and intravenous routes. When heated to decomposition it emits toxic fumes of NO_x and NH_3.

ANE250 CAS:10192-29-7 ***HR: 3***
AMMONIUM CHLORATE
mf: ClH_3NO_3 mw: 100.49

PROP: White crystals or mass.

DOT Classification: Forbidden

SAFETY PROFILE: A powerful oxidizer. Moderately flammable due to spontaneous chemical reaction. Explosion hazard due to shock, chemical reaction, or exposure to heat. A storage hazard; it may explode at room temperature. Explodes when heated to 100°C. When contaminated it is very sensitive. Solution in water may explode if heated or dried. When heated to decomposition it emits highly toxic fumes of Cl^- and NO_x. Incompatible with reducing materials; BrF_3; BrF_5.

ANE500 CAS:12125-02-9 ***HR: 3***
AMMONIUM CHLORIDE
mf: $H_4N \cdot Cl$ mw: 53.50

PROP: White crystals; salty taste. Bp: 520°, mp: 337.8°, d: 1.520, vap press: 1 mm @ 160.4° (sublimes). Sol in water, alc, and glycerin.

SYNS: AMMONIUMCHLORID (GERMAN) ◇ AMMONIUM MURIATE ◇ CHLORID AMONNY (CZECH) ◇ SAL AMMONIA ◇ SAL AMMONIAC

TOXICITY DATA with REFERENCE
eye-rbt 500 mg/24H SEV 28ZPAK -,15,72
orl-rat LD50:1650 mg/kg 28ZPAK -,15,72
ims-rat LD50:30 mg/kg EMSUA8 4,223,46
orl-mus LD50:1300 mg/kg IYKEDH 21,257,90
ipr-mus LD50:1439 mg/kg COREAF 256,1043,63
scu-mus LDLo:500 mg/kg 27ZIAQ -,39,73
orl-dog LDLo:600 mg/kg HBAMAK 4,1289,35
orl-rbt LDLo:1000 mg/kg HBAMAK 4,1289,35
ivn-rbt LDLo:78 mg/kg HBAMAK 4,1289,35
scu-gpg LDLo:72 mg/kg HBAMAK 4,1289,35
ivn-gpg LDLo:220 mg/kg 27ZWAY 1,470,23

CONSENSUS REPORTS: Reported in EPA TSCA Inventory.

OSHA PEL: (Fume) TWA 10 mg/m^3; STEL 20 mg/m^3
ACGIH TLV: TWA 10 mg/m^3; STEL 20 mg/m^3
DOT Classification: ORM-E.

SAFETY PROFILE: Poison by subcutaneous, intravenous, and intramuscular routes. Moderately toxic by other routes. A severe eye irritant. Explosive reaction with potassium chlorate or bromine trifluoride. Violent reaction (ignition) with bromine pentafluoride, NH_4, NO_3, and IF_7. Reaction with hydrogen cyanide may give the explosive nitrogen trichloride. When heated to decomposition it emits very toxic fumes of NO_x, Cl^-, and NH_3.

ANE750 CAS:13820-40-1 ***HR: 2***
AMMONIUM CHLOROPALLADATE(II)
mf: $Cl_4H_8N_2Pd$ mw: 284.30

PROP: Olive-green crystals. D: 2.17, mp: decomp.

SYNS: AMMONIUM TETRACHLOROPALLADATE ◇ DIAMMONIUM TETRACHLOROPALLADATE

TOXICITY DATA with REFERENCE
skn-rbt 100 mg/24H SEV AEHLAU 30,168,75

CONSENSUS REPORTS: Reported in EPA TSCA Inventory.

SAFETY PROFILE: A severe skin irritant. See also PALLADIUM COMPOUNDS. When heated to decomposition it emits very toxic fumes of Cl^-, NO_x, and NH_3.

ANF000 CAS:19168-23-1 ***HR: 3***
AMMONIUM CHLOROPALLADATE(IV)
mf: $Cl_6H_8N_2Pd$ mw: 355.20

PROP: Red-brown crystals. D: 2.418, mp: decomp.

SYNS: AMMONIUM HEXACHLOROPALLADATE ◇ DIAMMONIUM HEXACHLOROPALLADATE

TOXICITY DATA with REFERENCE
skn-rbt 100 mg/24H SEV AEHLAU 30,168,75

CONSENSUS REPORTS: Reported in EPA TSCA Inventory.

SAFETY PROFILE: A poison skin irritant. When heated to decomposition it emits very toxic fumes of NO_x, Cl^-, and NH_3.

ANF250 CAS:16919-58-7 ***HR: 3***
AMMONIUM CHLOROPLATINATE
mf: $Cl_6Pt{\cdot}2H_4N$ mw: 443.89

PROP: Cubic, yellow crystals. D: 3.065, mp: decomposes.

SYNS: AMMONIUM HEXACHLOROPLATINATE(IV) ◇ AMMONIUM PLATINIC CHLORIDE ◇ DIAMMONIUM HEXACHLOROPLATINATE (2-) ◇ PLATINIC AMMONIUM CHLORIDE

TOXICITY DATA with REFERENCE
orl-rat LD50:1 mg/kg GTPZAB 21(7),55,77
ihl-hmn TCLo:0.9 $\mu g/m^3$:PUL BJIMAG 2,92,45

CONSENSUS REPORTS: Reported in EPA TSCA Inventory.

OSHA PEL: TWA 0.002 $mg(Pt)/m^3$
ACGIH TLV: TWA 0.002 $mg(Pt)/m^3$

SAFETY PROFILE: Poison by inhalation and ingestion. Human pulmonary system effects by inhalation. See also PLATINUM COMPOUNDS. An explosively unstable compound. Incompatible with KOH (boiling with alkali yields a product which, after drying, will explode @ 205° or if mixed with combustibles). When heated to decomposition it emits very toxic fumes of Cl^-, NO_x, and NH_3.

ANF500 ***HR: 3***
AMMONIUM CHROMATE
mf: $(NH_4)_2CrO_4$ mw: 152.1

PROP: Yellow, crystalline material. Mp: decomp @ 180°, d: 1.91 @ 12°. Sol in cold water.

CONSENSUS REPORTS: Chromium and its compounds are on the Community Right-To-Know List.

OSHA PEL: CL 0.1 $mg(CrO_3)/m^3$
ACGIH TLV: TWA 0.05 $mg(Cr)/m^3$
NIOSH REL: (Chromium(VI)) TWA 25 $\mu g(Cr(VI))/m^3$; CL 50 $\mu g/m^3$/15M

SAFETY PROFILE: A poison. See also CHROMIUM COMPOUNDS. A powerful oxidizer. An explosion hazard when shocked or heated. When heated to decomposition it emits toxic fumes of NH_3 and NO_x. Incompatible with reducing agents.

ANF625 ***HR: 3***
AMMONIUM CHROME ALUMS
mf: $H_8N_2O_4S{\cdot}Cr_2{\cdot}O_{12}S_3{\cdot}24H_2O$ mw: 956.82

SYN: AMMONIUM SULFATE, and CHROMIC SULFATE, TETRACOSAHYDRATE ◇ CHROMIC AMMONIUM SULFATE

TOXICITY DATA with REFERENCE
orl-rat LD50:720 mg/kg 85GMAT -,38,82
skn-rat LDLo:2 g/kg 85GMAT -,38,82
ihl-mus LC50:51 mg/m^3/2H 85GMAT -,38,82
skn-mus LD50:110 mg/kg 85GMAT -,38,82
ims-mus LD50:115 mg/kg 85GMAT -,38,82
skn-rbt LDLo:1 g/kg 85GMAT -,38,82

CONSENSUS REPORTS: Chromium and its compounds are on the Community Right-To-Know List.

ACGIH TLV: TWA 0.5 $mg(Cr)m^3$

SAFETY PROFILE: Poison by inhalation, skin contact, and intramuscular routes. Moderately toxic by ingestion. When heated to decomposition it emits toxic fumes of SO_x, NO_x, and NH_3. See also AMMONIUM SULFATE and CHROMIUM COMPOUNDS.

ANF750 ***HR: 3***
AMMONIUM CHROMIC SULFATE
mf: $NH_4Cr(SO_4)_2{\cdot}12H_2O$ mw: 478.4

PROP: Green or violet crystals. Mp: 94° ($-9H_2O$ @ 94°), d: 1.720, water sol.

CONSENSUS REPORTS: Chromium and its compounds are on the Community Right-To-Know List.

OSHA PEL: CL 0.1 $mg(CrO_3)/m^3$
ACGIH TLV: TWA 0.05 $mg(Cr)/m^3$
NIOSH REL: (Chromium(VI)) TWA 25 $\mu g(Cr(VI))/m^3$; CL 50 $\mu g/m^3$/15M

SAFETY PROFILE: Poison. See also CHROMIUM COMPOUNDS and SULFATES. When heated to decomposition it emits toxic fumes of NH_3, NO_x, and SO_x.

ANF800 CAS:7632-50-0 ***HR: 2***
AMMONIUM CITRATE
mf: $C_6H_8O_7{\cdot}xH_3N$ mw: 311.42

PROP: Granules or crystals. D: 1.48. Sol in water; sltly sol in alc.

SYNS: AMMONIUM CITRATE, DIBASIC (DOT) ◇ CITRIC ACID, AMMONIUM SALT ◇ DIAMMONIUM CITRATE

TOXICITY DATA with REFERENCE
ivn-mus LD50:331 mg/kg JCINAO 37,497,58

CONSENSUS REPORTS: Reported in EPA TSCA Inventory.

SAFETY PROFILE: Experimental poison by intravenous route. A skin and eye irritant. When heated to decomposition it emits acrid smoke and irritating fumes.

ANG000 ***HR: 3***
AMMONIUM CYANIDE
mf: NH_4CN mw: 44.1

PROP: Solid, white powder or crystals. Mp: 36° (decomp), bp: subl @ 40°, d: 1.002 @ 100°, vap press: 400 ppm @ 20.5°. Very sol in water and alc; decomp in hot water.

CONSENSUS REPORTS: Cyanide and its compounds are on the Community Right-To-Know List.

SAFETY PROFILE: A poison. See also CYANIDE. When heated to decomposition it emits toxic CN^-, NH_3, NO_x.

ANG125 CAS:12008-61-6 ***HR: 3***
AMMONIUM DECAHYDRODECABORATE (2−)
mf: $B_{10}H_{18}N_2$ mw: 154.26

SAFETY PROFILE: A poison. Product of the reaction with nitrous acid explodes when dry. When heated to decomposition it emits toxic fumes of NO_x and NH_3. See also BORON COMPOUNDS.

ANG250 ***HR: 3***
AMMONIUM DIFLUORIDE mixed with HYDROCHLORIC ACID
DOT: NA 1760

SYN: WHITE ACID (DOT)

DOT Classification: Corrosive Material; Label: Corrosive.

SAFETY PROFILE: A corrosive. Poison by inhalation, ingestion, and skin contact. When heated to decomposition it emits very toxic fumes of F^-, HF, and HCl.

ANG500 CAS:3226-36-6 ***HR: 2***
AMMONIUM DIMETHYL DITHIOCARBAMATE
mf: $C_3H_7NS_2{\cdot}H_3N$ mw: 138.27

SYN: DIRAM A

TOXICITY DATA with REFERENCE
orl-rat LD50:1458 mg/kg HYSAAV 32,169,67
orl-mus LD50:592 mg/kg HYSAAV 32,169,67
orl-rbt LD50:450 mg/kg HYSAAV 32,169,67
orl-gpg LD50:1680 mg/kg HYSAAV 32,169,67

SAFETY PROFILE: Moderately toxic by ingestion. See also CARBAMATES. When heated to decomposition it emits very toxic fumes of NO_x, SO_x, and NH_3.

ANG625 CAS:76556-13-3 ***HR: 3***
AMMONIUM-3,5-DINITRO-1,2,4-TRIAZOLIDE
mf: $C_2H_4N_6O_4$ mw: 176.09

SAFETY PROFILE: An explosive. Upon decomposition it emits toxic fumes of NO_x and NH_3. See also EXPLOSIVES.

ANG750 CAS:25954-13-6 ***HR: 1***
AMMONIUM ETHYLCARBAMOYLPHOSPHONATE
mf: $H_4N{\cdot}C_3H_7NO_4P$ mw: 170.13

SYNS: AMMONIUM-AETHYL-CARBAMOYL-PHOSPHONAT(GERMAN) ◇ KRENITE

TOXICITY DATA with REFERENCE
orl-rat LD50:24000 mg/kg 85ARAE 2,209,77
orl-qal LD50:10000 mg/kg 85DPAN -,-,71/76

SAFETY PROFILE: Mildly toxic by ingestion. See also PHOSPHATES. When heated to decomposition it emits very toxic fumes of NO_x, PO_x, and NH_3.

ANG925 CAS:14221-47-7 ***HR: 3***
AMMONIUM FERRIC OXALATE
mf: $C_6FeO_{12}{\cdot}3H_4N$ mw: 374.06

SYNS: AMMONIUM FERRIOXALATE ◇ AMMONIUM TRIOXALATOFERRATE(III) ◇ FERRIC AMMONIUM OXALATE ◇ FERRIC AMMONIUM OXALATE (DOT) ◇ TRIAMMONIUM TRIS(ETHANEDIOATO(2-)-O,O')FERRATE(3-1)

TOXICITY DATA with REFERENCE
skn-rbt 500 mg MOD FCTOD7 20,563,82
eye-rbt 100 mg MOD FCTOD7 20,573,82
eye-rbt 100 mg/4S rns MLD FCTOD7 20,573,82

CONSENSUS REPORTS: Reported in EPA TSCA Inventory.

SAFETY PROFILE: An eye and skin irritant. When heated to decomposition it emits toxic fumes of NH_3. See also OXALATES.

ANH000 CAS:13826-83-0 ***HR: 3***
AMMONIUM FLUOBORATE
DOT: NA 9088
mf: NH_4BF_4 mw: 104.9

PROP: White, rhombic crystals. D: 1.871 @ 15°, mp: subl. Sol in NH_4OH and water.

SYNS: AMMONIUM BOROFLUORIDE ◇ AMMONIUM FLUOROBORATE ◇ AMMONIUM TETRAFLUOROBORATE ◇ AMMONIUM TETRAFLUOROBORATE(1-)

CONSENSUS REPORTS: Reported in EPA TSCA Inventory.

OSHA PEL: TWA 2.5 mg(F)/m^3
NIOSH REL: (Fluorides, Inorganic) TWA 2.5 mg(F)/m^3
DOT Classification: ORM-B; Label: None

SAFETY PROFILE: A poison and strong irritant. See also FLUORIDES and BORON COMPOUNDS. When heated to decomposition it emits very toxic fumes of F^-, NO_x, and NH_3.

ANH250 CAS:12125-01-8 ***HR: 3***
AMMONIUM FLUORIDE
DOT: UN 2505
mf: $H_4N•F$ mw: 37.05

PROP: Colorless crystals. Mp: subl; d: 1.009 @ 25°.

SYNS: AMMONIUM FLUORURE (FRENCH) ◇ NEUTRAL AMMONIUM FLUORIDE

TOXICITY DATA with REFERENCE
ipr-rat LD50:32 mg/kg XEURAQ UR-154,1951
scu-frg LDLo:280 mg/kg CRSBAW 124,133,37

CONSENSUS REPORTS: Reported in EPA TSCA Inventory.

OSHA PEL: TWA 2.5 mg(F)/m^3
NIOSH REL: TWA 2.5 mg(F)/m^3
DOT Classification: Poison B; Label: St. Andrews Cross; ORM-B; Label: None

SAFETY PROFILE: Poison by subcutaneous and intraperitoneal routes. See also FLUORIDES. When heated to decomposition it emits very toxic fumes of F^-, NO_x, and NH_3. Incompatible with ClF_3.

ANH500 CAS:540-69-2 ***HR: 2***
AMMONIUM FORMATE
mf: $CH_2O_2•H_3N$ mw: 63.07

PROP: White, deliques crystals. Mp: 116°, bp: decomp @ 180°, d: 1.280.

SYN: FORMIC ACID AMMONIUM SALT

TOXICITY DATA with REFERENCE
orl-mus LD50:2250 mg/kg ZERNAL 9,332,69
ivn-mus LD50:410 mg/kg ZERNAL 9,332,69

CONSENSUS REPORTS: Reported in EPA TSCA Inventory.

SAFETY PROFILE: Moderately toxic by ingestion and intravenous routes. When heated to decomposition it emits toxic fumes of NO_x and NH_3.

ANH875 CAS:14481-29-9 ***HR: 3***
AMMONIUM HEXACYANOFERRATE(II)
mf: $C_6H_{16}FeN_{10}$ mw: 284.11

CONSENSUS REPORTS: Cyanide and its compounds are on the Community Right-To-Know List.

SAFETY PROFILE: A poison. Reacts explosively with metal nitrates when heated, e.g., cobalt(II) nitrate at 220°C and copper(II) nitrate at 220°C. When heated to decomposition it emits toxic fumes of CN^- and NH_3. See also CYANIDE.

ANI000 CAS:13815-28-6 ***HR: 3***
AMMONIUM HEXAFLUOROFERRATE
mf: $F_6FeH_{12}N_3$ mw: 224.00

SYN: HEXAFLUORO FERRATE (3-) TRIAMMONIUM SALT

TOXICITY DATA with REFERENCE
ivn-mus LD50:56 mg/kg CSLNX* NX#04248

OSHA PEL: TWA 2.5 mg(F)/m^3
ACGIH TLV: TWA 1 mg/(Fe)/m^3
NIOSH REL: (Fluorides, Inorganic) TWA 2.5 mg(F)/m^3

SAFETY PROFILE: Poison by intravenous route. See also FLUORIDES and IRON COMPOUNDS. When heated to decomposition it emits very toxic fumes of F^-, NO_x, and NH_3.

ANI250 CAS:16962-40-6 ***HR: 3***
AMMONIUM HEXAFLUOROTITANATE
mf: $F_6Ti•H_4N_2$ mw: 193.96

TOXICITY DATA with REFERENCE
ivn-mus LD50:56 mg/kg CSLNX* NX#00134

CONSENSUS REPORTS: Reported in EPA TSCA Inventory.

OSHA PEL: TWA 2.5 mg(F)/m^3
NIOSH REL: (Fluorides, Inorganic) TWA 2.5 mg(F)/m^3

SAFETY PROFILE: Poison by intravenous route. See also FLUORIDES, AMMONIA, and TITANIUM COMPOUNDS. When heated to decomposition it emits very toxic fumes of F^- and NO_x.

ANI500 CAS:13815-31-1 ***HR: 3***
AMMONIUM HEXAFLUOROVANADATE
mf: $F_6H_{12}N_3V$ mw: 219.09

SYN: HEXAFLUORO VANADATE (3-) TRIAMMONIUM SALT

TOXICITY DATA with REFERENCE
ivn-mus LD50:10 mg/kg CSLNX* NX#04249

OSHA PEL: TWA 2.5 mg(F)/m^3
ACGIH TLV: TWA 0.05 mg(V_2O_5)/m^3
NIOSH REL: (Vanadium Compounds) CL 0.05 mg(V)/m^3/15M

SAFETY PROFILE: Poison by intravenous route. See also FLUORIDES and VANADIUM COMPOUNDS. When heated to decomposition it emits very toxic NH_3, NO_x, VO_x, and fluorides.

ANI750 ***HR: 3***
AMMONIUM HEXANITRO COBALTATE
mf: $CoH_{12}N_9O_{12}$ mw: 389.1

CONSENSUS REPORTS: Cobalt and its compounds are on the Community Right-To-Know List.

SAFETY PROFILE: Explodes @ 230°. Also is impact sensitive. Upon decomposition it emits toxic fumes of NO_x. See also COBALT COMPOUNDS and NITRATES.

ANJ000 CAS:1341-49-7 ***HR: 3***
AMMONIUM HYDROGEN FLUORIDE
DOT: UN 1727/UN 2817
mf: F_2H_5N mw: 57.06

PROP: White crystals. D: 1.51, mp: 124.6°. Will etch glass. Water-sol.

SYNS: AMMONIUM BIFLUORIDE ◇ AMMONIUM HYDROGEN FLUORIDE, solid

CONSENSUS REPORTS: Reported in EPA TSCA Inventory.

OSHA PEL: TWA 2.5 mg(F)/m^3
ACGIH TLV: TWA 2.5 mg(F)/m^3
NIOSH REL: (Fluorides, Inorganic) TWA 2.5 mg(F)/m^3
DOT Classification: Corrosive Material; Label: Corrosive; Corrosive Material; Label: Corrosive, Poison

SAFETY PROFILE: Caustic poison and strong irritant by all routes. See also HYDROFLUORIC ACID. When heated to decomposition it emits very toxic fumes of F^-, NO_x, and NH_3.

ANJ250 CAS:1341-49-7 ***HR: 3***
AMMONIUM HYDROGEN FLUORIDE (solution)
DOT: UN 2817

SYN: AMMONIUM HYDROGEN FLUORIDE, solution (DOT)

CONSENSUS REPORTS: Reported in EPA TSCA Inventory.

OSHA PEL: TWA 2.5 mg(F)/m^3
ACGIH TLV: TWA 2.5 mg(F)/m^3
NIOSH REL: TWA 2.5 mg(F)/m^3
DOT Classification: Corrosive Material; Label: Corrosive.

SAFETY PROFILE: Caustic poison and strong irritant by all routes. See also HYDROFLUORIC ACID. When heated to decomposition it emits very toxic fumes of HF, F^-, and NO_x.

ANJ500 ***HR: 2***
AMMONIUM HYDROGEN SULFATE
mf: NH_4HSO_4 mw: 115.11

PROP: White rhombic crystals; sol in water, insol in acetone. Mp: 146.9°; d: 1.78.

SAFETY PROFILE: Moderately toxic by ingestion. See also SULFATES. Dangerous; when heated to decomposition it emits highly toxic fumes of sulfuric acid and SO_x, NH_3, and NO_x.

ANJ750 CAS:12124-99-1 ***HR: 3***
AMMONIUM HYDROSULFIDE
DOT: NA 2683
mf: NH_4HS mw: 51.11

PROP: Powder or crystals. Mp: 118° (150 atm); d: 1.17; vap press: 400 mm @ 21.8°.

SYNS: AMMONIUM BISULFIDE ◇ AMMONIUM HYDROGEN SULFIDE ◇ AMMONIUM HYDROSULFIDE, solution (DOT) ◇ AMMONIUM MERCAPTAN ◇ AMMONIUM SULFHYDRATE ◇ MONOAMMONIUM SULFIDE ◇ SIRNIK AMONNY ◇ TRUE AMMONIUM SULFIDE

TOXICITY DATA with REFERENCE
orl-rat LD50:168 mg/kg NTIS** AD-A062-138
orl-mus LDLo:80 mg/kg JPETAB 76,179,42
skn-mus LDLo:2457 mg/kg JPETAB 76,179,42
ipr-mus LDLo:10 mg/kg JPETAB 76,179,42
scu-mus LD50:132 mg/kg 28ZPAK-,18,72
ivn-mus LDLo:2 mg/kg JPETAB 76,179,42
ivn-dog LDLo:2 mg/kg JPETAB 76,179,42
skn-rbt LDLo:119 mg/kg JPETAB 76,179,42
scu-rbt LDLo:8 mg/kg JPETAB 76,179,42
ivn-rbt LDLo:1500 μg/kg JPETAB 76,179,42
idr-rbt LDLo:30 mg/kg JPETAB 76,179,42
skn-gpg LDLo:692 mg/kg JPETAB 76,179,42
par-gpg LDLo:143 mg/kg JPETAB 76,179,42

CONSENSUS REPORTS: Reported in EPA TSCA Inventory.

DOT Classification: ORM-A; Label: None

SAFETY PROFILE: Poison by ingestion, skin contact,

subcutaneous, intravenous, intradermal, parenteral, and intraperitoneal routes. Pyroforic in air. See also SULFIDES. When heated to decomposition it emits very toxic fumes of SO_x, NO_x, and NH_3. Incompatible with zinc.

ANK250 CAS:1336-21-6 ***HR: 3***
AMMONIUM HYDROXIDE
DOT: NA 2672
mf: $H_4N \cdot HO$ mw: 35.06

PROP: Clear, colorless liquid solution of ammonia; very pungent odor. D: 0.90, mp: −77°. Sol in water. Soln contains not more than 44% ammonia.

SYNS: AMMONIA, solution (DOT) ◇ AQUA AMMONIA ◇ AQUEOUS AMMONIA

TOXICITY DATA with REFERENCE
eye-rbt 750 μ AJOPAA 29,1363,46
eye-rbt 44 μg SEV AROPAW 25,839,41
mmo-sat 10 μL/plate ANYAA9 76,475,58
mmo-esc 10 μL/disc ANYAA9 76,475,58
orl-hmn LDLo:43 mg/kg 34ZIAG -,95,69
ihl-hmn LCLo:5000 ppm 34ZIAG -,95,69
ihl-hmn TCLo:700 ppm:EYE JISMAB 61,271,71
ihl-hmn TCLo:408 ppm:IRR JISMAB 61,271,71
orl-rat LD50:350 mg/kg JIHTAB 23,259,41
orl-cat LDLo:750 mg/kg HBAMAK 4,1289,35
ivn-rbt LDLo:10 mg/kg HBAMAK 4,1289,35

CONSENSUS REPORTS: Reported in EPA TSCA Inventory.

NIOSH REL: (Ammonia) CL 50 ppm
DOT Classification: Corrosive Material; Label: Corrosive.

SAFETY PROFILE: A human poison by ingestion. An experimental poison by inhalation and ingestion. A severe eye irritant. Human systemic eye and other systemic irritant effects by inhalation. Mutation data reported. Incompatible with acrolein, nitromethane, acrylic acid, chlorosulfonic acid, dimethyl sulfate, halogens, (Au + aqua regia), HCl, HF, HNO_3, oleum, β-propiolactone, propylene oxide, $AgNO_3$, Ag_2O, (Ag_2O + C_2H_5OH), $AgMnO_4$, H_2SO_4. Dangerous; liquid can inflict burns. Use with adequate ventilation. When heated to decomposition it emits NH_3 and NO_x.

ANK500 ***HR: 3***
AMMONIUM HYPOPHOSPHITE
mf: H_6NO_2P mw: 83.03

PROP: White granules or rhombic crystals. D: 1.634, mp: 200°, bp: decomp @ 240°. Sol in water, alc, NH_3; insol in acetone.

SAFETY PROFILE: When heated it can liberate highly toxic and flammable PH_3. See also PHOSPHINE. When heated to decomposition it can emit highly toxic fumes of PH_3, PO_x, NH_3, and NO_x.

ANK750 ***HR: 3***
AMMONIUM IODATE
mf: H_4INO_3 mw: 192.94

PROP: Colorless crystals. D: 3.309 @ 21°, mp: 150° (decomp). Sltly sol in cold water; insol in hot water.

SAFETY PROFILE: A powerful, unstable oxidizer. When heated to decomposition it emits very toxic fumes of I^- and NO_x. Has detonated upon contact with a scoop, possibly due to contamination by ammonium periodate. See also IODATES.

ANL000 ***HR: 2***
AMMONIUM IODIDE
mf: NH_4I mw: 145

PROP: Colorless, hygroscopic crystals. Mp: subl @ 551°, bp: 220° (vac), d: 2.514 @ 25°, vap press: 1 mm @ 210.9°.

SAFETY PROFILE: Moderately toxic. See also IODIDES. Incompatible with BrF_3; IF_7; K. When heated to decomposition it emits toxic fumes of I^-, NH_3, and NO_x.

ANL100 CAS:57267-78-4 ***HR: 1***
AMMONIUM ISETHIONATE
mf: $C_2H_5O_4S \cdot H_3N$ mw: 142.17

SYNS: ETHANESULFONIC ACID, 2-HYDROXY-, AMMONIUM SALT ◇ 2-HYDROXYETHANESULFONIC ACID AMMONIUM SALT

TOXICITY DATA with REFERENCE
eye-rbt 100 mg MLD FCTOD7 20,573,82

CONSENSUS REPORTS: Reported in EPA TSCA Inventory.

SAFETY PROFILE: An eye irritant. When heated to decomposition it emits toxic fumes of SO_x, NH_3, and NO_x.

ANL500 CAS:10169-00-3 ***HR: 2***
AMMONIUM LANTHANUM NITRATE
mf: $H_4N \cdot La \cdot 7NO_3$ mw: 591.03

SYNS: LANTHANUM AMMONIUM NITRATE ◇ NITRIC ACID, AMMONIUM LANTHANUM SALT ◇ NITRIC ACID, LANTHANUM AMMONIUM SALT

TOXICITY DATA with REFERENCE
orl-rat LD50:3400 mg/kg AIHOAX 1,637,50
ipr-rat LD50:625 mg/kg AIHOAX 1,637,50

SAFETY PROFILE: Moderately toxic by ingestion and intraperitoneal routes. See also LANTHANIDES, LANTHANUM COMPOUNDS, and NITRATES. When heated to decomposition it emits very toxic fumes of NH_3 and NO_x.

ANL750 ***HR: 3***
AMMONIUM MAGNESIUM ARSENATE
mf: $NH_4MgAsO_4•6H_2O$ mw: 289.4

PROP: Colorless crystals. Mp: decomp: d: 1.932 @ 15°. Very sltly water-sol.

CONSENSUS REPORTS: Arsenic and its compounds are on the Community Right-To-Know List.

SAFETY PROFILE: When heated to decomposition it emits very toxic fumes of As, NH_3, and NO_x. See ARSENIC COMPOUNDS and MAGNESIUM COMPOUNDS.

ANM000 ***HR: 3***
AMMONIUM MAGNESIUM CHROMATE
mf: $(NH_4)_2CrO_4•MgCrO_4•6H_2O$ mw: 400.5

PROP: Yellow crystals. Mp: decomp, d: 1.84. Very water sol.

CONSENSUS REPORTS: Chromium and its compounds are on the Community Right-To-Know List.

OSHA PEL: CL 0.1 $mg(CrO_3)/m^3$
ACGIH TLV: TWA 0.05 $mg(Cr)/m^3$
NIOSH REL: (Chromium(VI)) TWA 25 $\mu g(Cr(VI))/m^3$; CL 50 $\mu g/m^3$/15M

SAFETY PROFILE: A poison. See also CHROMIUM COMPOUNDS and MAGNESIUM COMPOUNDS. Moderately flammable; can explode. Incompatible with reducing agents. When heated to decomposition it can emit toxic fumes of NH_3 and NO_x.

ANM250 CAS:530-31-4 ***HR: 1***
AMMONIUM MANDELATE
mf: $C_8H_7O_3•H_4N$ mw: 169.20

TOXICITY DATA with REFERENCE
orl-rat LDLo:5000 mg/kg AIPTAK 64,79,40
orl-rbt LDLo:5000 mg/kg AIPTAK 64,79,40

SAFETY PROFILE: Mildly toxic by ingestion. When heated to decomposition it emits toxic fumes of NO_x and NH_3.

ANM500 CAS:5421-46-5 ***HR: 3***
AMMONIUM MERCAPTOACETATE
mf: $C_2H_3O_2S•H_3N$ mw: 108.15

PROP: Colorless liquid; strong skunk-like odor.

SYNS: AMMONIUM THIOGLYCOLATE ◇ AMMONIUM THIOGLYCOLLATE ◇ THIOGLYCOLLIC ACID, AMMONIUM SALT ◇ USAF MO-2

TOXICITY DATA with REFERENCE
ipr-mus LD50:100 mg/kg NTIS** AD277-689

CONSENSUS REPORTS: Reported in EPA TSCA Inventory.

SAFETY PROFILE: Poison by intraperitoneal route. An allergen; can cause contact dermatitis. Emits hydrogen sulfide. See also SULFIDES. When heated to decomposition it emits very toxic NO_x, SO_x, and NH_3.

ANM625 CAS:58696-86-9 ***HR: 3***
AMMONIUM-3-METHYL-2,4,6-TRINITROPHENOXIDE
mf: $C_7H_8N_4O_7$ mw: 260.16

SAFETY PROFILE: May explode spontaneously in storage. When heated to decomposition or on explosion it emits toxic fumes of NO_x and NH_3.

ANM750 CAS:13106-76-8 ***HR: 3***
AMMONIUM MOLYBDATE
mf: $MoO_4•2H_4N$ mw: 196.04

SYNS: AMMONIUM PARAMOLYBDATE ◇ DIAMMONIUM MOLYBDATE ◇ MOLYBDIC ACID DIAMMONIUM SALT

TOXICITY DATA with REFERENCE
orl-rat LD50:333 mg/kg 28ZLA8 -,214,61
ipr-rat LDLo:203 mg/kg EQSSDX 1,1,75
orl-cat LDLo:1600 mg/kg EQSSDX 1,1,75
orl-rbt LDLo:1870 mg/kg EQSSDX 1,1,75
scu-rbt LDLo:1600 mg/kg EQSSDX 1,1,75
orl-gpg LDLo:2200 mg/kg EQSSDX 1,1,75
ipr-gpg LDLo:800 mg/kg EQSSDX 1,1,75
scu-gpg LDLo:1380 mg/kg EQSSDX 1,1,75

CONSENSUS REPORTS: Reported in EPA TSCA Inventory.

OSHA PEL: TWA 5 $mg(Mo)/m^3$
ACGIH TLV: TWA 5 $mg(Mo)/m^3$

SAFETY PROFILE: Poison by ingestion and intraperitoneal route. Moderately toxic by other routes. An irritant. See also MOLYBDENUM COMPOUNDS. When heated to decomposition it emits toxic fumes of NH_3 and NO_x.

ANN000 CAS:6484-52-2 ***HR: 3***
AMMONIUM(I) NITRATE(1581)
DOT: UN 0222/NA 1942/UN 2426
mf: $HNO_3•H_3N$ mw: 80.06

PROP: Colorless crystals. Mp: 169.6°, d: 1.725 @ 25°, bp: decomp >210°. Solubility: 192/100 @ 20°.

SYNS: AMMONIUM NITRATE ◇ AMMONIUM NITRATE (DOT) ◇ NITRIC ACID, AMMONIUM SALT

CONSENSUS REPORTS: Community Right-To-Know List. Reported in EPA TSCA Inventory.

DOT Classification: Oxidizer; Label: Oxidizer (NA 1942, UN1942, UN2426); Class A Explosive; Label: EXPLOSIVE A (UN0222)

SAFETY PROFILE: A powerful oxidizer and an allergen. See also NITRATES. A relatively stable explosive which has, however, caused many industrial explosions. Violent or explosive spontaneous reactions with acetic anhydride + nitric acid, ammonium sulfate + potassium, copper iron(II) sulfide, sawdust, urea, barium nitrate, hot water, and ammonium chloride + water + zinc. Forms heat- or shock-sensitive explosive mixtures with acetic acid, aluminum + calcium nitrate + formamide (a blasting explosive), ammonia, charcoal + metal oxides (e.g., rust, copper oxide, zinc oxide above 80°C), chloride salts (e.g., ammonium chloride, calcium chloride, iron(III) chloride, and aluminum chloride), cyanoguanidine, fertilizers (e.g., super phosphate + organic materials above 90°C), hydrocarbon oils, powdered metals (e.g., aluminum, antimony, bismuth, cadmium, chromium, cobalt, copper, iron, lead, magnesium, manganese, nickel, tin, zinc, brass, stainless steel, titanium, and potassium), nonmetals (e.g., charcoal, and phosphorus), organic fuels (e.g., wax, oils, and stearates), potassium permanganate, sugar, sulfur, and trinitroanisole. Reaction with alkali metals (e.g., sodium) forms an explosive product. Ignites on contact with ammonium dichromate, potassium dichromate, potassium chromate, barium chloride, sodium chloride, potassium nitrate, and chromium (VI) salts. Can ignite when mixed with acetic acid. Use water in large amounts to fight fire. It is important that the mass of materials be kept cool and that burning be extinguished promptly. Ventilate well. May explode under confinement and high temperatures. When heated to decomposition it emits highly toxic fumes of NO_x. Can react vigorously with reducing materials. Incompatible with, (NH_4Cl + heat), (C + heat), organic matter, P, NaOCl, $NaClO_4$. Occasional explosions in presence of oil, $(NH_4)_2SO_4$ with K or Na.

ANO250 ***HR: 3***
AMMONIUM NITRITE
mf: NH_4NO_2 mw: 64.04

PROP: White to yellow crystals. Mp: explodes @ 60-70°, bp: subl @ 30° in vacuo, d: 1.69; very sol in water, dil alk.

SAFETY PROFILE: Powerful oxidizer. See also NITRITES. Severe explosion hazard when shocked or exposed to heat (60-70°). When heated to decomposition it emits toxic fumes of NO_x and NH_3.

ANO400 ***HR: 3***
AMMONIUM aci-NITROMETHANE
mf: $CH_7O_2N_2$ mw: 79.1

SAFETY PROFILE: A friction-sensitive explosive. See also NITRATES.

ANO500 CAS:135-20-6 ***HR: 3***
AMMONIUM-N-NITROSOPHENYLHYDROXYLAMINE
mf: $C_6H_6N_2O_2 \cdot H_4N$ mw: 156.19

SYNS: CUPFERRON ◇ N-HYDROXY-N-NITROSO-BENZENAMINE, AMMONIUM SALT ◇ KUPFERRON (CZECH) ◇ NCI-C03258 ◇ N-NITROSOFENYLHYDROXYLAMIN AMONNY (CZECH) ◇ N-NITROSOPHENYLHYDROXYLAMIN AMMONIUM SALZ (GERMAN) ◇ N-NITROSOPHENYLHYDROXYLAMINE AMMONIUM SALT

TOXICITY DATA with REFERENCE
cyt-grh-orl 1 ppm JCGEDO 1,75,66
eye-rbt 20 mg/24H SEV 28ZPAK -,110,72
orl-rat TDLo:123 g/kg/78W-C:CAR NCITR* NCI-CG-TR-100,78
orl-mus TDLo:437 g/kg/78W-C:CAR NCITR* NCI-CG-TR-100,78
orl-rat TD:9040 mg/kg/65W-C:ETA ZEKBAI 69,103,67
orl-rat LDLo:250 mg/kg NCNSA6 5,27,53
ipr-rat LDLo:50 mg/kg KODAK* -,-,71
ivn-mus LD50:180 mg/kg CSLNX* NX#04968

CONSENSUS REPORTS: NTP Fifth Annual Report On Carcinogens. NCI Carcinogenesis Bioassay (feed); Clear Evidence: mouse, rat NCITR* NCI-CG-TR-100,78. Reported in EPA TSCA Inventory. Community Right-To-Know List.

SAFETY PROFILE: Confirmed carcinogen with experimental carcinogenic and tumorigenic data. Poison by intravenous route. Powerful eye irritant. Solutions with thorium salts are unstable explosives above 15°C. Solutions with titanium or zirconium salts are unstable explosives above 40°C. When heated to decomposition it emits very toxic NH_3 and NO_x. See also N-NITROSO COMPOUNDS and AMINES.

ANO750 ***HR: 3***
AMMONIUM OXALATE
mf: $(NH_4)_2C_2O_4 \cdot H_2O$ mw: 142.12

PROP: Colorless crystals. Mp: decomp; d: 1.50. Sltly sol in water.

SAFETY PROFILE: A poison. See also OXALATES. Can react violently with (NaOCl + ammonium acetate). When heated to decomposition it can emit toxic fumes of NH_3 and NO_x.

ANO875 ***HR: 3***
AMMONIUM OXOFLUOROMOLYBDATE
mf: $F_4MoO_2 \cdot 2H_4N$ mw: 240.04

TOXICITY DATA with REFERENCE
orl-rat TDLo:1400 mg/kg (28D male):REP SHKKAN 23,859,81
orl-rat LD50:242 mg/kg SHKKAN 23,237,81
ipr-rat LD50:58 mg/kg SHKKAN 23,859,81
scu-rat LD50:152 mg/kg SHKKAN 23,859,81
orl-mus LD50:250 mg/kg SHKKAN 23,859,81
ipr-mus LD50:51 mg/kg SHKKAN 23,859,81
scu-mus LD50:72 mg/kg SHKKAN 23,859,81

SAFETY PROFILE: Poison by ingestion, subcutaneous, and intraperitoneal routes. Experimental reproductive effects. When heated to decomposition it emits toxic fumes of F^-, NO_x, and NH_3. See also MOLYBDENUM COMPOUNDS.

ANO900 CAS:12208-54-7 ***HR: 1***
AMMONIUM PARATUNGSTATE HEXAHYDRATE
mf: $H_{24}N_6O_{24}W_7 \cdot 6H_2O$ mw: 1887.37

TOXICITY DATA with REFERENCE
orl-rat LD50:11300 mg/kg AIHAAP 30,470,69

ACGIH TLV: TWA 1 mg(W)/m^3; STEL 3 mg(W)/m^3
NIOSH REL: 10H TWA 1 mg(W)/m^3

SAFETY PROFILE: Mildly toxic by ingestion. When heated to decomposition it emits toxic fumes of NO_x and W.

ANP000 ***HR: 3***
AMMONIUM PENTA PEROXODICHROMATE
mf: $Cr_2H_8N_2O_{12}$ mw: 332.2

CONSENSUS REPORTS: Chromium and its compounds are on the Community Right-To-Know List.

OSHA PEL: CL 0.1 mg(CrO_3)/m^3
ACGIH TLV: TWA 0.05 mg(Cr)/m^3
NIOSH REL: (Chromium(VI)) TWA 25 μg(Cr(VI))/m^3; CL 50 μg/m^3/15M

SAFETY PROFILE: An unstable compound. Detonation can be initiated by heat, friction or impact. See also CHROMIUM COMPOUNDS. Explodes @ 50°. When heated to decomposition it emits toxic fumes of NO_x.

ANP250 ***HR: 3***
AMMONIUM PERCHLORATE
mf: NH_4ClO_4 mw: 117.50

PROP: White crystals. Mp: decomp; d: 1.95.

SAFETY PROFILE: Easily ignited by friction. Can explode when mixed with sugar, charcoal, or on contact with hot copper pipes. Can be sensitized by nitryl perchlorate, KIO_4, $KMnO_4$, metals (as cocrystallized impurities). It becomes impact sensitive when contaminated by powdered carbon, ferrocene, sulfur, organic matter, or powdered metals. When heated to decomposition it emits toxic fumes of NH_3, Cl^-, and NO_x. See PERCHLORATES.

ANP500 ***HR: 3***
AMMONIUM PERCHLORYL AMIDE
mf: $H_5N_2O_3Cl$ mw: 116.6

PROP: Mp: 80°.

SAFETY PROFILE: A shock-sensitive explosive. May detonate @ 80°. When heated to decomposition it emits very toxic fumes of NH_3, NO_x, and Cl^-.

ANP625 CAS:3825-26-1 ***HR: 3***
AMMONIUM PERFLUOROOCTANOATE
mf: $C_8F_{15}O_2 \cdot H_4N$ mw: 431.13

SYNS: AMMONIUM PENTADECAFLUOROOCTANATE ◇ AMMONIUM PERFLUOROCAPRILATE ◇ AMMONIUM PERFLUOROCAPRYLATE ◇ APFO ◇ FC-143 ◇ PERFLUOROAMMONIUM OCTANOATE

TOXICITY DATA with REFERENCE
skn-rbt 500 mg MOD TXAPA9 81,348,85
eye-rbt 500 mg/24H MOD AIHAAP 41,576,80
ihl-rat TCLo:25 mg/m^3/6H (6-15D preg):REP FAATDF 4,429,84
orl-rat LD50:430 mg/kg AIHAAP 41,576,80
ihl-rat LC50:980 mg/m^3/4H FCTOD7 24,1325,86
skn-rat LD50:7 g/kg TXAPA9 81,348,85
skn-rbt LD50:4300 mg/kg TXAPA9 81,348,85

CONSENSUS REPORTS: Reported in EPA TSCA Inventory.

ACGIH TLV: 0.1 mg/m^3

SAFETY PROFILE: Poison by inhalation. Moderately toxic by ingestion. An eye and skin irritant. Experimental reproductive effects. When heated to decomposition it emits toxic fumes of F^- and NH_3.

ANP750 ***HR: 3***
AMMONIUM-m-PERIODATE
mf: NH_4IO_4 mw: 209

PROP: Colorless crystals. Mp: explodes, d: 3.056.

SAFETY PROFILE: A contact explosive. See also IODATES and IODIDES. Heat, impact, and touch as from a scoop or an abrasive impact may cause explosion. When heated to decomposition it can emit toxic fumes of NH_3, NO_x, and I^-.

ANQ250 ***HR: 3***
AMMONIUM PEROXO BORATE
mf: $BH_4NO_3 \cdot 1/2H_2O$ mw: 85.86

PROP: White crystals. Mp: decomp; sltly sol in water.

SAFETY PROFILE: Potentially explosive by heat, friction, or impact. See also BORON COMPOUNDS. When heated to decomposition it emits toxic fumes of NO_x and NH_3.

ANQ500 ***HR: 2***
AMMONIUM PEROXO DISULFATE
mf: $H_8N_2O_8S_2$ mw: 228.2

SAFETY PROFILE: See also SULFATES. An unstable compound. Detonated via heat, friction, or impact. A powerful oxidizer. Incompatible with Al, H_2O, powdered metal, Ag salts, Fe, Na_2O_2, zinc, NH_3. When heated to decomposition it emits very toxic fumes of SO_x, NO_x, and NH_3.

ANQ750 ***HR: 3***
AMMONIUM PEROXY CHROMATE
mf: $(NH_4)_3CrO_2$ mw: 234.1

PROP: Red-brown crystals. Mp: decomp @ 40°, bp: explodes @ 50°.

CONSENSUS REPORTS: Chromium and its compounds are on the Community Right-To-Know List.

OSHA PEL: CL 0.1 $mg(CrO_3)/m^3$
ACGIH TLV: TWA 0.05 $mg(Cr)/m^3$
NIOSH REL: (Chromium(VI)) TWA 25 $\mu g(Cr(VI))/m^3$; CL 50 $\mu g/m^3$/15M

SAFETY PROFILE: A poison. See also CHROMIUM COMPOUNDS. Moderately flammable by chemical reaction with reducing agents. A powerful oxidizer. Moderately explosive when heated. When heated to decomposition it emits toxic fumes of NO_x, and NH_3.

ANR000 CAS:7727-54-0 ***HR: 3***
AMMONIUM PERSULFATE
DOT: UN 1444
mf: $O_8S_2 \cdot 2H_4N$ mw: 228.22

$$H_4NOSO_2OOSO_2ONH_4$$

PROP: White crystals. Mp: decomp @ 120°, d: 1.982.

SYNS: AMMONIUM PEROXYDISULFATE ◇ AMMONIUM PERSULFATE (DOT) ◇ PERSULFATE d'AMMONIUM (FRENCH)

TOXICITY DATA with REFERENCE
ipr-rat LD50:226 mg/kg DTLVS* 4,327,80
ivn-rbt LD50:178 mg/kg DTLVS* 4,327,80
orl-rat LD50:820 mg/kg AIHAAP 30,470,69

CONSENSUS REPORTS: Reported in EPA TSCA Inventory.

ACGIH TLV: TWA 5 $mg(S_2O_8)/m^3$
DOT Classification: Oxidizer; Label: Oxidizer.

SAFETY PROFILE: Poison by intravenous and intraperitoneal routes. Moderately toxic by ingestion. A powerful oxidizer which can react vigorously with reducing agents. Releases oxygen when heated. Mixtures with sodium peroxide are explosives sensitive to friction, heating above 75°C, or contact with CO_2 or water. Mixtures with (powdered aluminum + water) or (zinc + ammonia) are explosive. Violent reaction with iron or solutions of ammonia + silver salts. Solution with sulfuric acid is a strong oxidizing cleaning solution. When heated to decomposition it emits toxic fumes of SO_x, NH_3, and NO_x.

ANR250 CAS:1074-52-8 ***HR: 3***
AMMONIUM PHENYLDITHIOCARBAMATE
mf: $C_7H_6NS_2 \cdot H_4N$ mw: 186.31

SYN: PHENYLDITHIOCARBAMIC ACID, AMMONIUM SALT

TOXICITY DATA with REFERENCE
ipr-mus LDLo:100 mg/kg JMPCAS 5,846,62

CONSENSUS REPORTS: Reported in EPA TSCA Inventory.

SAFETY PROFILE: Poison by intraperitoneal route. See also CARBAMATES. When heated to decomposition it emits very toxic fumes of NO_x, SO_x and NH_3.

ANR500 CAS:7783-28-0 ***HR: 2***
AMMONIUM PHOSPHATE DIBASIC
mf: $H_6N_2 \cdot H_3O_4P$ mw: 132.08

PROP: White crystals or powder; salty taste. D: 1.619, mp: 155° (decomp). Sol in water; insol in alc.

SYNS: AMMONIUM PHOSPHATE ◇ DIAMMONIUM HYDROGEN PHOSPHATE ◇ DIBASIC AMMONIUM PHOSPHATE ◇ SECONDARY AMMONIUM PHOSPHATE

CONSENSUS REPORTS: Reported in EPA TSCA Inventory.

SAFETY PROFILE: Low to moderate toxicity. See also PHOSPHATES. When heated to decomposition it emits very toxic fumes of PO_x, NO_x, and NH_3.

ANR750 CAS:7772-76-1 ***HR: 2***
AMMONIUM PHOSPHATE, MONOBASIC
mf: $NH_4H_2PO_4$ mw: 115

PROP: Brilliant white crystals or powder. D: 1.803 @ 19°; mp: 190°. Sol in water.

SAFETY PROFILE: Incompatible with NaOCl.

ANS000 ***HR: 3***
AMMONIUM PHOSPHIDE
mf: $P(NH_4)_3$ mw: 85.07

SAFETY PROFILE: Poison by inhalation and ingestion. See also PHOSPHINE. When heated to decomposition it emits toxic fumes of PO_x, NO_x, and NH_3.

ANS250 CAS:51503-61-8 ***HR: 3***
AMMONIUM PHOSPHITE
mf: H_6NO_3P mw: 99.04

SYN: AMMONIUM ORTHOPHOSPHITE

TOXICITY DATA with REFERENCE
ihl-rat LCLo:580 ppm/1H ZGSHAM 25,279,33
ihl-gpg LCLo:288 ppm/2H ZGSHAM 25,279,33

SAFETY PROFILE: Poison by inhalation. See also PHOSPHITES. When heated to decomposition it emits very toxic fumes of NO_x, NH_3, and PO_x.

ANS500 CAS:131-74-8 ***HR: 3***
AMMONIUM PICRATE
DOT: UN 0004/UN 1310
mf: $C_6H_3N_3O_7{\cdot}H_3N$ mw: 246.16

PROP: Red or yellow, rhombic crystals. D: 1.719, mp: decomp; bp: expl @ 423°. Solubility: 1.1/100 @ 20°.

SYNS: AMMONIUM PICRATE, wet with 10% or more water (DOT) ◇ AMMONIUM PICRATE, wet with 10% or more water, over 16 oz in one outside packaging (DOT) ◇ AMMONIUM PICRONITRATE ◇ EXPLOSIVE D ◇ OBELINE PICRATE ◇ PICRATE of AMMONIA (DOT) ◇ PICRIC ACID, AMMONIUM SALT

CONSENSUS REPORTS: Reported in EPA TSCA Inventory.

DOT Classification: Class A Explosive; Label: EXPLOSIVE A (UN0004); Flammable Solid; Label: Flammable Solid (UN1310)

SAFETY PROFILE: An allergen. Moderately irritating to skin, eyes, and mucous membranes. Moderately flammable by spontaneous chemical reaction. A powerful oxidizer which reacts vigorously with reducing materials. Dangerous explosive when shocked or heated. The presence of trace metals increases its heat sensitivity. See PICRIC ACID, NITRATES, and EXPLOSIVES, HIGH. When heated to decomposition it emits highly toxic fumes of NO_x.

ANT000 CAS:12259-92-6 ***HR: 3***
AMMONIUM POLYSULFIDE (solution)
DOT: UN 2818

SYN: AMMONIUM SULFIDE, solution, red

CONSENSUS REPORTS: Reported in EPA TSCA Inventory.

DOT Classification: ORM-A; Label: None; Corrosive Material; Label: Corrosive and Poison

SAFETY PROFILE: Poison due to presence of sulfides and H_2S. See also AMMONIUM SULFIDE. When heated to decomposition it emits very toxic fumes of NO_x, SO_x, and H_2S.

ANT250 CAS:64046 ***HR: 2***
AMMONIUM POTASSIUM SELENIDE mixed with AMMONIUM POTASSIUM SULFIDE-00-0
mf: $H_4KNSe + NH_4KS$ mw: 136.11 + 89.21 = 225.32

SYN: AMMONIUM POTASSIUM SULFIDE mixed with AMMONIUM POTASSIUM SELENIDE

TOXICITY DATA with REFERENCE
orl-rat TDLo:450 mg/kg/2Y-C:ETA CNREA8 3,230,43

CONSENSUS REPORTS: Selenium and its compounds are on the Community Right-To-Know List.

OSHA PEL: TWA 0.2 mg(Se)/m^3
ACGIH TLV: TWA 0.2 mg(Se)/m^3
DFG MAK: 0.1 mg(Se)/m^3

SAFETY PROFILE: Questionable carcinogen with experimental tumorigenic data. See also SELENIDES and SULFIDES. When heated to decomposition it emits very toxic fumes of NO_x, NH_3, SO_x, and Se.

ANT300 CAS:19441-09-9 ***HR: 3***
AMMONIUM REINECKATE HYDRATE
mf: $C_4H_{10}N_7S_4{\cdot}Cr{\cdot}H_2O$ mw: 354.47

SYN: CHROMATE(1-),DIAMMINETETRAKIS(ISOTHIOCYANATO)-, AMMONIUM, HYDRATE

TOXICITY DATA with REFERENCE
scu-mus LD50:110 mg/kg ABMGAJ 3,28,59
ivn-mus LD50:180 mg/kg CSLNX* NX#01547

OSHA PEL: CL 0.1 mg(CrO_3)/m^3

SAFETY PROFILE: Poison by subcutaneous and intravenous routes. When heated to decomposition it emits toxic fumes of NO_x, SO_x, and CR.

ANT500 CAS:6381-61-9 ***HR: 2***
AMMONIUM SACCHARIN
mf: $C_7H_8N_2O_3S$ mw: 200.23

PROP: White crystals or crystalline powder; intense sweet taste. Sol in water.

SYN: 1,2-BENZISOTHIAZOLIN-3-ONE 1,1-DIOXIDE AMMONIUM SALT ◇ DARAMIN ◇ SACCHARIN AMMONIUM ◇ SACCHARINATE AMMONIUM

TOXICITY DATA with REFERENCE
eye-rbt 100 mg SEV LPPTAK 24,598,76

SAFETY PROFILE: A severe eye irritant. When heated to decomposition emits toxic fumes of NO_x.

ANU000 ***HR: 2***
AMMONIUM SALTS of PHOSPHATIDIC ACIDS

TOXICITY DATA with REFERENCE
orl-rat LD50:5000 mg/kg FAONAU 53A,215,74
ivn-rat LD50:2000 mg/kg FAONAU 53A,215,74
orl-dog LD50:2000 mg/kg FAONAU 53A,215,74
orl-rbt LD50:5000 mg/kg FAONAU 53A,215,74
ivn-gpg LD50:2000 mg/kg FAONAU 53A,215,74

SAFETY PROFILE: Moderately toxic by ingestion and intravenous routes. When heated to decomposition it emits very toxic fumes of NH_3, PO_x, and NO_x.

ANU650 CAS:7773-06-0 ***HR: 2***
AMMONIUM SULFAMATE
mf: $H_2NO_3S•H_4N$ mw: 114.14

PROP: Deliquescent, crystalline material (white crystalline solid). Bp: 160° (decomp), mp: 131°.

SYNS: AMCIDE ◇ AMICIDE ◇ AMMAT ◇ AMMATE ◇ AMMONIUM AMIDOSULFONATE ◇ AMMONIUM AMIDOSULPHATE ◇ AMMONIUMSALZ der AMIDOSULFONSAURE (GERMAN) ◇ AMMONIUM SULPHAMATE ◇ AMS ◇ IKURIN ◇ MONOAMMONIUM SULFAMATE ◇ SULFAMATE ◇ SULFAMIC ACID, MONOAMMONIUM SALT ◇ SULFAMINSAURE (GERMAN)

TOXICITY DATA with REFERENCE
orl-rat LD50:2 g/kg AMIHAB 14,178,56
ipr-rat LDLo:800 mg/kg JIHTAB 25,26,43
orl-mus LD50:3100 mg/kg GTPZAB 7(5),56,63
orl-qal LD50:3000 mg/kg DTLVS* 4,17,80

CONSENSUS REPORTS: Reported in EPA TSCA Inventory.

OSHA PEL: (Transitional: TWA Total Dust: 15 mg/m^3; Respirable Fraction: 5 mg/m^3) TWA 10 mg/m^3; Respirable Fraction: 5 mg/m^3
ACGIH TLV: TWA 10 mg/m^3
DFG MAK: 15 mg/m^3

SAFETY PROFILE: Moderately toxic by ingestion and intraperitoneal routes. Somewhat explosive when heated or by spontaneous chemical reaction in a hot acid solution. A powerful oxidizer. When heated to decomposition it emits very toxic fumes of NH_3, NO_x, and SO_x. See also SULFONATES and SULFAMIC ACID.

ANU750 CAS:7783-20-2 ***HR: 2***
AMMONIUM SULFATE (2 : 1)
mf: $H_8N_2O_4S$ mw: 132.16

$SO_4•(NH_4)_2$

PROP: White crystals. Mp: > 280° (decomp); d: 1.77. Sol in water; insol in alc.

SYNS: AMMONIUM SULPHATE ◇ DIAMMONIUM SULFATE ◇ SULFURIC ACID, DIAMMONIUM SALT

TOXICITY DATA with REFERENCE
orl-man TDLo:1500 mg/kg GISAAA 42(2),100,77
orl-rat LD50:3000 mg/kg CNJMAQ 12,216,48
ipr-mus LD50:610 mg/kg UCPHAQ 2,1,41
orl-dom LDLo:3500 mg/kg AJVRAH 32,1229,71

CONSENSUS REPORTS: Community Right-To-Know List. Reported in EPA TSCA Inventory.

SAFETY PROFILE: Moderately toxic by several routes. See also SULFATES. Incandescent reaction on heating with potassium chlorate. Reaction with sodium hypochlorite gives the unstable explosive nitrogen trichloride. Incompatible with ($K + NH_4NO_3$), KNO_2, ($NaK + NH_4NO_3$). When heated to decomposition it emits very toxic fumes of NO_x, NH_3, and SO_x.

ANV750 CAS:13453-06-0 ***HR: D***
AMMONIUM TELLURATE
mf: $(NH_4)_2TeO_4$ mw: 227.7

PROP: White powder. Mp: decomp, d: 3.01 @ 25°.

SYN: TELLURIC ACID, AMMONIUM SALT

TOXICITY DATA with REFERENCE
cyt-hmn:fbr 1 nmol/L AEMBAP 91,117,78
cyt-hmn:leu 1 nmol/L AEMBAP 91,117,78

CONSENSUS REPORTS: Reported in EPA TSCA Inventory.

OSHA PEL: TWA 0.1 mg(Te)/m^3
ACGIH TLV: TWA 0.1 mg(Te)/m^3

SAFETY PROFILE: Human mutation data reported. See also TELLURIUM COMPOUNDS. When heated to decomposition it emits very toxic fumes of Te, NO_x, and NH_3.

ANV800 CAS:13820-41-2 ***HR: 3***
AMMONIUM TETRACHLOROPLATINATE

PROP: Crystals. Mp: 140-150° (decomp), d: 2.936.
mf: $Cl_4Pt•H_4N$ mw: 372.99

SYNS: PLATINATE(2-), TETRACHLORO-, DIAMMONIUM ◇ TETRAMINE PLATINUM(II) CHLORIDE

TOXICITY DATA with REFERENCE
ipr-mus LD50:60 mg/kg TXAPA9 49,41,79

OSHA PEL: TWA 0.002 mg(Pt)/m^3
ACGIH TLV: TWA 0.002 mg(Pt)/m^3

CONSENSUS REPORTS: Reported in EPA TSCA Inventory.

SAFETY PROFILE: Poison by intraperitoneal route. When heated to decomposition it emits toxic fumes of NH_3, Cl^-, and Pt.

ANW250 **HR: 3**
AMMONIUM TETRANITROPLATINATE(II)
mf: $H_8N_6O_8Pt$ mw: 415.3

SAFETY PROFILE: An explosively unstable compound. Sensitive to heat. See also NITRATES and PLATINUM COMPOUNDS.

ANW500 **HR: 3**
AMMONIUM TETRAPEROXO CHROMATE
mf: $CrH_{12}N_3O_8$ mw: 234.2

CONSENSUS REPORTS: Chromium and its compounds are on the Community Right-To-Know List.

OSHA PEL: CL 0.1 mg(CrO_3)/m^3
ACGIH TLV: TWA 0.05 mg(Cr)/m^3
NIOSH REL: (Chromium(VI)) TWA 25 μg(Cr(VI))/m^3; CL 50 μg/m^3/15M

SAFETY PROFILE: A poison. Impact explodes @ 50° or in contact with H_2SO_4. See also CHROMIUM COMPOUNDS. Incompatible with H_2SO_4. When heated to decomposition it emits toxic fumes of NO_x.

ANW750 CAS:1762-95-4 **HR: 3**
AMMONIUM THIOCYANATE
mf: $CNS \cdot H_4N$ mw: 76.13

PROP: Colorless solid or deliquescent crystals. Mp: 149.6°, bp: decomp @ 170°, d: 1.305.

SYNS: AMMONIUM RHODANATE ◇ AMMONIUM RHODANIDE ◇ AMMONIUM SULFOCYANATE ◇ AMMONIUM SULFOCYANIDE ◇ AMTHIO ◇ RHODANID ◇ RHODANIDE ◇ TRANS-AID ◇ USAF EK-P-433 ◇ WEEDAZOL TL

TOXICITY DATA with REFERENCE
orl-hmn TDLo:430 mg/kg:GIT,CNS DAKMAJ 102,606,11
orl-mus LDLo:330 mg/kg AEPPAE 169,429,33
ipr-mus LDLo:500 mg/kg NTIS** AD277-689
orl-gpg LDLo:600 mg/kg JAPMA8 29,152,40

CONSENSUS REPORTS: Reported in EPA TSCA Inventory. EPA Genetic Toxicology Program.

SAFETY PROFILE: Poison by ingestion. Moderately toxic by other routes. Human systemic effects by ingestion: hallucinations and distorted perceptions, nausea or vomiting, and other gastrointestinal effects. See also THIOCYANATES. When heated to decomposition it emits toxic fumes of NH_3, NO_x, SO_x, and CN^-. Incompatible with $KClO_3$ and mixtures with $Pb(NO_3)_2$.

ANX750 **HR: 3**
AMMONIUM TRICHLOROACETATE
mf: $NH_4O_2CCCl_3$ mw: 180.6

SAFETY PROFILE: Poison by inhalation and ingestion. A powerful irritant. When heated to decomposition or on contact with acid or acid fumes it emits toxic fumes of Cl^-, NH_3, and NO_x. Incompatible with water or steam.

ANX800 CAS:15660-29-4 **HR: 3**
AMMONIUM TRIFLUOROSTANNITE
mf: $F_3Sn \cdot H_4N$ mw: 193.74

TOXICITY DATA with REFERENCE
ivn-mus LD50:18 mg/kg CSLNX* NX#00130

OSHA PEL: TWA 2 mg(Sn)/m^3; TWA 2.5 mg(F)/m^3
ACGIH TLV: TWA 2 mg(Sn)/m^3
NIOSH REL: (Fluorides, Inorganic) TWA 2.5 mg(F)/m^3

SAFETY PROFILE: Poison by intravenous route. See also FLUORIDES and TIN COMPOUNDS. When heated to decomposition it emits very toxic fumes of NH_3, NO_x, and fluorides.

ANX875 CAS:63839-60-1 **HR: 3**
AMMONIUM-2,4,5-TRINITROIMIDAZOLIDE
mf: $C_3H_4N_6O_6$ mw: 220.10

SAFETY PROFILE: An explosive comparable in power to RDX, but more thermally stable. Upon decomposition it emits toxic fumes of NO_x and NH_3. See also EXPLOSIVES.

ANY250 CAS:7803-55-6 **HR: 3**
AMMONIUM VANADATE
DOT: UN 2859
mf: $O_3V \cdot H_4N$ mw: 116.99

PROP: Colorless to yellow crystals. Mp: 200° (decomp), d: 2.326.

SYNS: AMMONIUM METAVANADATE (DOT) ◇ RCRA WASTE NUMBER P119 ◇ VANADIC ACID, AMMONIUM SALT

TOXICITY DATA with REFERENCE
mrc-bcs 300 mmol/L MUREAV 77,109,80
ipr-ham TDLo:22500 μg/kg (5-10D preg):REP ENVRAL 29,256,82
ipr-ham TDLo:11280 μg/kg (5-10D preg):TER ENVRAL 29,256,82
orl-rat LD50:160 mg/kg AIHAAP 30,470,69
ipr-rat LD50:18 mg/kg ATXKA8 16,182,56
scu-rat LD50:23 mg/kg ATXKA8 16,182,56
itr-rat LDLo:8 mg/kg ATXKA8 16,182,56
scu-mus LDLo:16 mg/kg AJSNAO 1,347,17
ivn-rbt LDLo:1 mg/kg AJSNAO 1,347,17
scu-gpg LDLo:643 μg/kg AJSNAO 1,347,17

CONSENSUS REPORTS: Reported in EPA TSCA Inventory. EPA Genetic Toxicology Program.

ACGIH TLV: TWA 0.05 mg(V_2O_5)/m^3
NIOSH REL: (Vanadium Compounds) CL 0.05 mg(V)/m^3/15M
DOT Classification: IMO: Poison B; Label: Poison.

SAFETY PROFILE: Poison by ingestion, subcutaneous, intravenous, intratracheal, and intraperitoneal routes. An experimental teratogen. Other experimental reproductive effects. Mutation data reported. See also VANADIUM COMPOUNDS. When heated to decomposition it emits toxic fumes of NH_3, VO_x, and NO_x.

ANY500 CAS:69782-62-3 ***HR: 3***
AMMONIUM VANADI-ARSENATE
mf: $H_{16}N_4O_2 \cdot As_2O_5V_2$ mw: 515.92

TOXICITY DATA with REFERENCE
scu-rat LDLo:34 mg/kg AJSNAO 1,347,17
ivn-rbt LDLo:6 mg/kg AJSNAO 1,347,17

CONSENSUS REPORTS: Arsenic and its compounds are on the Community Right-To-Know List.

NIOSH REL: (Vanadium Compounds) CL 0.05 mg(V)/m^3/15M; CL 2 μg(As)/m^3/15M

SAFETY PROFILE: Poison by subcutaneous and intravenous routes. See ARSENIC and VANADIUM COMPOUNDS. When heated to decomposition it emits very toxic fumes of NO_x and As.

ANY750 ***HR: 3***
AMMONIUM VANADO-ARSENATE
mf: $H_{40}N_{10}O_5 \cdot 3As_2O_5 \cdot 4O_4V_2$ mw: 1228.78

TOXICITY DATA with REFERENCE
scu-rat LDLo:246 mg/kg AJSNAO 1,347,17
ivn-rbt LDLo:75 mg/kg AJSNAO 1,347,17

CONSENSUS REPORTS: Arsenic and its compounds are on the Community Right-To-Know List.

ACGIH TLV: TWA 0.2 mg(As)/m^3
NIOSH REL: (Vanadium Compounds) cL 0.05 mg(V)/m^3/15M; (Arsenic, Inorganic) CL 2 μg(As)/m^3/15M

SAFETY PROFILE: Poison by subcutaneous and intravenous routes. See ARSENIC and VANADIUM COMPOUNDS. When heated to decomposition it emits very toxic NO_x, NH_3, and As.

ANZ000 CAS:3566-10-7 ***HR: 3***
AMOBAM

PROP: Used to react with zinc sulfate to form Zineb (28ZEAL 5,11,76)

SYNS: AMBAM ◇ DITHANE STAINLESS

TOXICITY DATA with REFERENCE
orl-rat LD50:395 mg/kg 28ZEAL 5,11,76

CONSENSUS REPORTS: Reported in EPA TSCA Inventory.

SAFETY PROFILE: Poison by ingestion. When heated to decomposition it emits very toxic fumes of SO_x and NO_x.

AOA050 CAS:26328-53-0 ***HR: D***
AMOSCANATE
mf: $C_{13}H_9N_3O_2S$ mw: 271.29

PROP: Crystals from acetone. Mp: 196-198°.

SYNS: C 9333 GO ◇ CGP 4540 ◇ CIBA 9333 GO ◇ 4-ISOTHIOCYANATO-4'-NITRODIPHENYLAMINE ◇ 4-ISOTHIOCYANO-4'-NITRO DIPHENYLAMINE ◇ 4-ISOTHIOCYANATO-N-(4-NITROPHENYL)-BENZENAMINE (9CI) ◇ NITHIOCYAMINE

TOXICITY DATA with REFERENCE
dnr-esc 1 g/L MUREAV 164,9,86
bfa-mus/omi 250 mg/kg TCMUD8 1,129,80
hma-mus/esc 250 mg/kg/3H MUREAV 164,9,86

CONSENSUS REPORTS: EPA Genetic Toxicology Program.

SAFETY PROFILE: Mutation data reported. An anthemintic agent.

AOA075 ***HR: 3***
AMOSULALOL HYDROCHLORIDE
mf: $C_{18}H_{24}N_2O_5S \cdot ClH$ mw: 416.96

SYN: YM-09538

TOXICITY DATA with REFERENCE
orl-rat TDLo:3300 mg/kg (7-17D preg):TER KSRNAM 19,6105,85
orl-rat TDLo:1100 mg/kg (7-17D preg):REP KSRNAM 19,6105,85
scu-rat LD50:541 mg/kg KSRNAM 19,6121,85
ivn-rat LD50:105 mg/kg KSRNAM 19,6121,85
orl-mus LD50:5740 mg/kg KSRNAM 19,6121,85
scu-mus LD50:394 mg/kg KSRNAM 19,6121,85
ivn-mus LD50:104 mg/kg KSRNAM 19,6121,85

SAFETY PROFILE: Poison by subcutaneous and intravenous routes. An experimental teratogen. Experimental reproductive effects. When heated to decomposition it emits toxic fumes of SO_x, NO_x, and HCl.

AOA095 CAS:14028-44-5 ***HR: 3***
AMOXAPINE
mf: $C_{17}H_{16}ClN_3O$ mw: 313.79

PROP: Crystals from benzene/petr ether. Mp: 175-176°.

SYNS: AMOXEPINE ◇ ASENDIN ◇ 2-CHLORO-11-(1-PIPERAZINYL)DIBENZ(b,f)(1,4)OXAZEPINE ◇ CL 67772 ◇ DEMOLOX ◇ MOXADIL

TOXICITY DATA with REFERENCE
orl-rat LD50:313 mg/kg IYKEDH 11,811,80

ipr-rat LD50:201 mg/kg AIPTAK 233,107,78
scu-rat LD50:4500 mg/kg KSRNAM 5,1852,71
orl-mus LD50:122 mg/kg AIPTAK 233,107,78

SAFETY PROFILE: Poison by ingestion and intraperitoneal routes. When heated to decomposition it emits toxic fumes of NO_x and Cl^-.

AOA100 CAS:61336-70-7 ***HR: 2***
AMOXICILLIN TRIHYDDRATE
mf: $C_{16}H_{19}N_3O_5S\cdot 3H_20$ mw: 419.50

SYNS: α-AMINO-p-HYDROXYBENZYLPENICILLINTRIHYDRATE ◇ (2S-(2-α,5-α,6-β(S*)))-6-((AMINO(4-HYDROXYPHENYL)ACETYL)AMINO)-3,3-DIMETHYL-7-OXO-4-THIA-1-AZABICYCLO(3.2.0)HEPTANE-2-CARBOXYLIC ACID TRIHYDRATE ◇ BRL 2333 TRIHYDRATE

TOXICITY DATA with REFERENCE
orl-mus TDLo:9100 mg/kg (7-13D preg):TER KSRNAM 7,3113,73
orl-rat TDLo:162 g/kg (25W male):REP KSRNAM 7,3074,73
ipr-rat LD50:2870 mg/kg KSRNAM 7,3040,73
ipr-mus LD50:3590 mg/kg KSRNAM 7,3040,73

SAFETY PROFILE: Moderately toxic. An experimental teratogen. Other experimental reproductive effects. When heated to decomposition it emits toxic fumes of SO_x and NO_x.

AOA125 CAS:61-19-8 ***HR: 1***
AMP
mf: $C_{10}H_{14}N_5O_7P$ mw: 347.26

SYNS: ADENOSINE-5'-MONOPHOSPHATE ◇ ADENOSINE-5-MONOPHOSPHORIC ACID ◇ ADENOSINE-5'-MONOPHOSPHORIC ACID ◇ ADENOSINE PHOSPHATE ◇ ADENOSINE-5'-PHOSPHATE ◇ ADENOSINE-5'-PHOSPHORIC ACID ◇ ADENOVITE ◇ ADENYL ◇ ADENYLIC ACID ◇ tert-ADENYLIC ACID ◇ A5MP ◇ 5-AMP ◇ 5'-AMP ◇ AMP (nucleotide) ◇ CARDIOMONE ◇ ERGADENYLIC ACID ◇ LYCEDAN ◇ MUSCLE ADENYLIC ACID ◇ MY-B-DEN ◇ MYOSTON ◇ NSC-20264 ◇ PHOSADEN ◇ PHOSPHADEN ◇ PHOSPHENTASIDE

TOXICITY DATA with REFERENCE
oms-hmn:oth 100 μmol/L JIDEAE 65,52,75
oms-mus:oth 50 μmol/L JIDEAE 66,313,76
ipr-rat TDLo:2800 mg/kg (7-13D preg):REP OYYAA2 4,625,70
ipr-mus LD50:4 g/kg PCJOAU 20,160,86

CONSENSUS REPORTS: Reported in EPA TSCA Inventory.

SAFETY PROFILE: Slightly toxic by intraperitoneal route. Experimental reproductive effects. Human mutation data reported. When heated to decomposition it emits toxic fumes of PO_x and NO_x.

AOA130 CAS:60-92-4 ***HR: D***
cAMP
mf: $C_{10}H_{12}N_5O_6P$ mw: 329.24

SYNS: ADENOSINE CYCLIC MONOPHOSPHATE ◇ ADENOSINE-3',5'-CYCLIC MONOPHOSPHATE ◇ ADENOSINE CYCLIC-3',5'-PHOSPHATE ◇ ADENOSINE-3',5'-CYCLOPHOSPHATE ◇ ADENOSINE-3',5'-MONOPHOSPHATE ◇ ADENOSINE-3',5'-PHOSPHATE ◇ 3',5'-AMP ◇ CYCLIC ADENOSINE-3',5'-PHOSPHATE ◇ CYCLIC AMP ◇ CYCLIC-3',5'-AMP

TOXICITY DATA with REFERENCE
oms-hmn:oth 100 μmol/L JIDEAE 65,52,75
oms-mus:fbr 1 mmol/L IJCNAW 13,404,74
dns-ham:oth 130 mmol/L CNREA8 42,3669,82

CONSENSUS REPORTS: Reported in EPA TSCA Inventory.

SAFETY PROFILE: Human mutation data reported. When heated to decomposition it emits toxic fumes of PO_x and NO_x.

AOA250 CAS:60-15-1 ***HR: 3***
AMPHETAMINE
mf: $C_9H_{13}N$ mw: 135.23

SYNS: β-AMINOPROPYLBENZENE ◇ DESOXYNOREPHEDRINE ◇ ELASTONON ◇ FENOPROMIN ◇ α-METHYLPHENETHYLAMINE ◇ MYDRIAL ◇ 1-PHENYL-2-AMINO-PROPAN (GERMAN) ◇ 1-PHENYL-2-AMINOPROPANE ◇ β-PHENYLISOPROPYLAMIN (GERMAN) ◇ (PHENYLISOPROPYL)AMINE ◇ β-PHENYLISOPROPYLAMINE ◇ PROTIOAMPHETAMINE

TOXICITY DATA with REFERENCE
orl-inf TDLo:7500 mg/kg:CNS,CVS,SKN AJDCAI 130,507,76
orl-rat LDLo:50 mg/kg AEPPAE 195,647,40
ipr-rat LD50:125 mg/kg JPETAB 132,97,61
scu-rat LD50:39 mg/kg JPETAB 86,280,46
orl-mus LD50:22 mg/kg ARZNAD 32,604,82
ipr-mus LD50:16 mg/kg PSCHDL 51,209,77
scu-mus LD50:2800 μg/kg AEPPAE 233,72,58
ivn-mus LD50:18 mg/kg APTOA6 38,474,76
scu-rbt LDLo:20 mg/kg AEPPAE 192,331,39

SAFETY PROFILE: Poison by ingestion, subcutaneous, intravenous, and intraperitoneal routes. Human systemic effects by ingestion: excitement, changes in heart rate and sweating. When heated to decomposition it emits very toxic fumes of NO_x. See other amphetamine entries.

AOA500 CAS:51-64-9 ***HR: 3***
d-AMPHETAMINE
mf: $C_9H_{13}N$ mw: 135.23

SYNS: d-2-AMINO-1-PHENYLPROPANE ◇ (+)-AMPHETAMINE ◇ AMSUSTAIN ◇ DEPHADREN ◇ DEXAMPHETAMINE ◇ DEXEDRINE ◇ α-METHYLPHENETHYLAMINE, d-FORM ◇ d-1-PHENYL-2-AMINOPROPAN (GERMAN) ◇ d-1-PHENYL-2-AMINOPROPANE

TOXICITY DATA with REFERENCE
scu-rat TDLo:25 mg/kg (female 5-9D post):REP DABBBA 31,6304,71
orl-man TDLo:42 mg/kg/25W-I BIPCBF 20,1332,85
orl-cld TDLo:3600 μg/kg/10D-I AJPSAO 143,1176,85
orl-rat LD50:38 mg/kg TXAPA9 18,185,71
ipr-rat LDLo:20 mg/kg AEPPAE 195,647,40
scu-rat LD50:200 mg/kg 27ZIAQ -,84,73
orl-mus LD50:40 mg/kg TXAPA9 41,329,77
ipr-mus LD50:4400 μg/kg AIPTAK 161,206,66
scu-mus LD50:20 mg/kg AIPTAK 146,392,63
ivn-mus LD50:25 mg/kg JMCMAR 15,410,72
orl-dog LDLo:6400 μg/kg 27ZIAQ -,84,73
orl-mky LDLo:32 mg/kg 27ZIAQ -,84,73
ipr-grb LD50:17600 μg/kg GERNDJ 23,165,77
orl-mam LD50:375 mg/kg JMCMAR 8,836,65

CONSENSUS REPORTS: Reported in EPA TSCA Inventory.

SAFETY PROFILE: Poison by ingestion, subcutaneous, intravenous, and intraperitoneal routes. Experimental reproductive effects. Chronic exposure causes central nervous system damage and blood-pressure effects. When heated to decomposition it emits toxic NO_x. See other amphetamine entries.

AOA750 CAS:2706-50-5 ***HR: 3***
AMPHETAMINE HYDROCHLORIDE
mf: $C_9H_{13}N•ClH$ mw: 171.69

SYNS: dl-α-METHYL-PHENETHYLAMINE HYDROCHLORIDE ◇ dl-β-PHENYLISOPROPYLAMINE HYDROCHLORIDE

TOXICITY DATA with REFERENCE
ipr-mus LD50:40 mg/kg JMCMAR 8,100,65
ivn-rbt LD50:15 mg/kg JPETAB 79,187,43
scu-gpg LDLo:52 mg/kg JPETAB 47,339,33

SAFETY PROFILE: Poison by subcutaneous, intravenous, and intraperitoneal routes. When heated to decomposition it emits very toxic fumes of HCl and NO_x. See other amphetamine entries.

AOB000 ***HR: 3***
dl-AMPHETAMINE SALT with FINE RESIN

PROP: Amberlite XE-69 is a sulfonic acid cation exchange resin. Mesh size exceeds 200 mesh. (TXAPA9 1,42,59)

SYN: α-METHYL-PHENETHYLAMINE compounded with AMBERLITE XE-69

TOXICITY DATA with REFERENCE
orl-rat LD50:195 mg/kg TXAPA9 1,42,59
orl-mus LD50:200 mg/kg TXAPA9 1,42,59

SAFETY PROFILE: Poison by ingestion. When heated to decomposition it emits toxic fumes of NO_x and SO_x. See other amphetamine entries.

AOB250 CAS:60-13-9 ***HR: 3***
dl-AMPHETAMINE SULFATE
mf: $C_{18}H_{26}N_2•H_2O_4S$ mw: 368.54

SYNS: ACETDRON ◇ ADIPAN ◇ ADIPARTHROL ◇ AKETDRIN ◇ AKTEDRIN ◇ ALENTOL ◇ AMFETAMINA ◇ AMFETAMINE ◇ (±)-2-AMINO-1-PHENYLPROPANE SULFATE ◇ (±)-AMPHETAMINE SULFATE ◇ ANFETAMINA ◇ BENNIE ◇ BENZAMPHETAMINE ◇ BENZEDRYNA ◇ BENZIES ◇ BETAFEN ◇ CARTWHEELS ◇ DEOXYNOREPHEDRINE ◇ DESOXYNOREPHEDRINE ◇ HEARTS ◇ IBIOZEDRINE ◇ LINAMPHETA ◇ (±)-α-METHYLPHENETHYLAMINE SULFATE ◇ NCI-C55710 ◇ NOREPHEDRANE ◇ PEACHES ◇ PHARMEDRINE ◇ PHENAMINE ◇ PHENEDRINE ◇ (±)-PHENISOPROPYLAMINE SULFATE ◇ β-PHENYL ISOPROPYLAMINE SULFATE ◇ PSYCHEDRINUM ◇ PSYCHEDRYNA ◇ RACEPHEN ◇ ROSES ◇ STIMULAN

TOXICITY DATA with REFERENCE
sln-dmg-unk 1500 g/L CTOXAO 5,395,72
scu-rat TDLo:11 mg/kg (1-22D preg):REP PSCHDL 58,171,78
orl-mus TDLo:900 mg/kg (1-18D preg):TER OFAJAE 41,227,65
orl-hmn TDLo:41 mg/kg:CNS,CVS KLWOAZ 17,1580,38
orl-rat LD50:55 mg/kg ARZNAD 13,711,63
ipr-rat LD50:125 mg/kg JPETAB 132,97,61
scu-rat LD50:160 mg/kg AIMEAS 10,1874,37
orl-mus LD50:24 mg/kg ARZNAD 13,711,63
ipr-mus LD50:13 mg/kg RPTOAN 48,26,85
ipr-mus LD50:13 mg/kg FATOAO 48(1),15,85
scu-mus LD50:7 mg/kg AIPTAK 170,428,67
ivn-mus LD50:68 mg/kg JPETAB 84,12,45
orl-dog LD50:23 mg/kg PSEBAA 118,557,65
ivn-dog LD50:6 mg/kg PSEBAA 118,557,65
ivn-rbt LDLo:22 mg/kg JOPHAN 37,597,39
scu-gpg LD50:105 mg/kg AIPTAK 137,375,62

CONSENSUS REPORTS: EPA Genetic Toxicology Program.

SAFETY PROFILE: Poison by ingestion, subcutaneous, intravenous, and intraperitoneal routes. Human systemic effects by ingestion: altered sleep time, anorexia and change in heart rate. A central nervous system stimulant. An experimental teratogen. Other experimental reproductive effects. Mutation data reported. When heated to decomposition it emits very toxic NO_x and SO_x. See other amphetamine entries.

AOB500 CAS:139-10-6 ***HR: 3***
AMPHETANE PHOSPHATE
mf: $C_9H_{13}N•H_3O_4P$ mw: 233.23

SYNS: ACETMIN ◇ ACTEMIN ◇ AKTEDRON ◇ AMPHATE ◇ AMPHETAMINE PHOSPHATE ◇ dl-AMPHETAMINE PHOSPHATE ◇ AMPHOS ◇ BAR-DEX ◇ DEPUALONE ◇ DIETAMINE ◇ DYNAPHENIL ◇ dl-α-METHYL-PHENETHYLAMINE PHOSPHATE

◇ α-METHYLPHENETHYLAMINE PHOSPHATE, dl-MIXTURE ◇ MONOBASIC racemic AMPHETAMINE PHOSPHATE ◇ MONOBASIC dl-α-METHYLPHENETHYLAMINE PHOSPHATE ◇ MONOPHOR ◇ MONOPHOS ◇ OBESITABS ◇ 1-PHENYL-2-AMINOPROPANE MONOPHOSPHATE ◇ PROFETAMINE ◇ PROFETAMINE PHOSPHATE ◇ RACEPHEN ◇ RAPHETAMINE PHOSPHATE

TOXICITY DATA with REFERENCE
orl-rat LD50:175 mg/kg TXAPA9 1,42,59
orl-mus LD50:154 mg/kg TXAPA9 1,42,59
ipr-mus LDLo:52 mg/kg JPETAB 127,55,59

SAFETY PROFILE: Poison by ingestion and intraperitoneal routes. When heated to decomposition it emits very toxic fumes of PO_x and NO_x. See other amphetamine entries.

AOB875 CAS:1402-82-0 ***HR: 3***
AMPHOMYCIN
mf: $C_{58}H_{91}N_{13}O_{20}$ mw: 1290.46

PROP: Acidic, surface-active polypeptide. Sol in water and the lower alcs; insol in nonpolar solvents.

SYNS: AMFOMYCIN ◇ GLUMAMYCIN ◇ U-6658

TOXICITY DATA with REFERENCE
orl-mus LD50:500 mg/kg 85GDA2 4(1),317,80
ipr-mus LD50:233 mg/kg CNCRA6 30,9,63
ivn-mus LD50:178 mg/kg 85FZAT -,131,67

SAFETY PROFILE: Poison by intravenous and intraperitoneal routes. Moderately toxic by ingestion. Induces hemolysis. Active against gram-positive bacteria. Suggested as a topical agent for animal and plant infections. When heated to decomposition it emits acrid smoke and irritating fumes.

AOC250 ***HR: 1***
AMPHOTERIC-2

SYNS: AMPHOTERGE K-2 ◇ 1-CARBOXYMETHYL-1-CARBOXYETHOXYETHYL-2-COCO-IMIDAZOLINIUMBETAINE ◇ MIRANOL C2M-SF CONC

TOXICITY DATA with REFERENCE
skn-hmn 60 mg/3D-I MLD 85DKA8 -,127,77
skn-rbt 10 mg/24H DCTODJ 1,305,78
eye-rbt 2 mg DCTODJ 1,305,78

SAFETY PROFILE: An eye irritant. A human skin irritant. When heated to decomposition it emits toxic fumes of NO_x.

AOC275 CAS:71463-34-8 ***HR: 1***
AMPHOTERIC-17
mf: $C_{18}H_{35}N_2O_3•HO_4S•Na$ mw: 447.61

SYNS: 1H-IMIDAZOLIUM,4,5-DIHYDRO-1-(CARBOXYMETHYL)-1-(2-HYDROXYETHYL)-2-UNDECYL-, HYDROGEN SULFATE (salt), MONOSODIUM SALT ◇ MIRANOL MHT

TOXICITY DATA with REFERENCE
skn-rbt 10 mg/24H DCTODJ 1,305,78
eye-rbt 2 mg DCTODJ 1,305,78

SAFETY PROFILE: A skin and eye irritant. When heated to decomposition it emits toxic fumes of NO_x, SO_x, and H_2S.

AOC500 CAS:1397-89-3 ***HR: 3***
AMPHOTERICIN B
mf: $C_{47}H_{73}NO_{17}$ mw: 924.21

SYNS: AMB ◇ AMPHOMORONAL ◇ AMPHOTERICIN beta ◇ AMPHOTERICINE B ◇ AMPHOZONE ◇ FUNGILIN ◇ FUNGISONE ◇ FUNGIZONE ◇ IAB ◇ IODOACETAMIDE ◇ MYSTECLIN-F ◇ NSC 527017 ◇ TEGOPEN

TOXICITY DATA with REFERENCE
spm-rbt-ivn 20 mg/kg/11D JRPFA4 7,13,64
orl-dog TDLo:600 mg/kg (male 30D pre):REP INURAQ 7,90,69
ivn-wmn LDLo:22 mg/kg/4D-I:BLD SMJOAV 76,409,83
ivn-wmn TDLo:20 μg/kg:PUL NEJMAG 315,836,86
ivn-man LDLo:164 μg/kg/5H-I:CVS DICPBB 17,547,83
ivn-rat LD50:11300 μg/kg DRUGAY 6,36,82
ipr-mus LD50:27740 μg/kg NCISP* JAN86
ivn-mus LD50:1200 μg/kg PHINDQ 6,164,85
ivn-dog LD50:6 mg/kg BIORAK 43,2043,78

SAFETY PROFILE: Poison by intravenous and intraperitoneal routes. Human systemic effects by intravenous route: leukopenia, lungs changes, and cardiac changes. Experimental reproductive effects. Mutation data reported. When heated to decomposition it emits toxic fumes of NO_x.

AOC750 CAS:35375-29-2 ***HR: 3***
AMPHOTERICIN B, METHYL ESTER HYDROCHLORIDE
mf: $C_{47}H_{75}NO_{20}•ClH$ mw: 1010.69

TOXICITY DATA with REFERENCE
ipr-mus LD50:1320 mg/kg SCIEAS 179,584,73
ivn-mus LD50:75 mg/kg 85ERAY 2,1019,78
ivn-dog LD50:48 mg/kg SCIEAS 179,584,73

SAFETY PROFILE: Poison by intravenous route. See also ESTERS. Moderately toxic by intraperitoneal route. When heated to decomposition it emits very toxic fumes of NO_x and HCl.

AOC875 CAS:8067-85-4 ***HR: 2***
AMPICILLIN-OXACILLIN MIXTURE
mf: $C_{19}H_{19}N_3O_5S•C_{16}H_{19}N_3O_4S$ mw: 750.91

SYNS: (2S-(2-α,5-α,6-β(S*)))-6-((AMINOPHENYLACETYL)AMINO)-3,3-DIMETHYL-7-OXO-4-THIA-1-AZABICYCLO(3.2.0)HEPTANE-2-CARBOXYLIC ACID mixt. with (2S-(2-α,5-α,6-β))-3,3-DIMETHYL-6-(((5-METHYL-3-PHENYL-4-ISOXAZOLYL)CARBONYL)AMINO)-7-OXO-4-TH

IA-1-AZABICYCLO(3.2.0)HEPTANE-2-CARBOXYLIC ACID ◇ OXACILLIN-AMPICILLIN MIXTURE

TOXICITY DATA with REFERENCE
ivn-rat LD50:3710 mg/kg NIIRDN 6,58,82
ipr-mus LD50:4700 mg/kg NIIRDN 6,58,82
scu-mus LD50:4940 mg/kg NIIRDN 6,58,82
ivn-mus LD50:3250 mg/kg NIIRDN 6,58,82

SAFETY PROFILE: Moderately toxic. When heated to decomposition it emits toxic fumes of Cl^-, SO_x, and NO_x.

AOD000 CAS:26309-95-5 ***HR: 2***
AMPICILLIN PIVALOYLOXYMETHYL ESTER HYDROCHLORIDE
mf: $C_{22}H_{29}N_3O_6S{\bullet}ClH$ mw: 500.06

SYNS: ALPHACILINA ◇ ALPHACILLIN ◇ 6-(d-α-AMINO PHENYL ACETAMIDO) PENICILLANIC ACID PIVALOYL OXY METHYL ESTER HYDROCHLORIDE ◇ BEROCILLIN ◇ CENTURINA ◇ DEVONIUM ◇ DIANCINA ◇ INACILIN ◇ MAXIFEN ◇ PIVALOYLOXYMETHYL d-α-AMINOBENZYLPENICILLINATE HYDROCHLORIDE ◇ PIVAMPICILLIN HYDROCHLORIDE ◇ PIVATIL ◇ PONDOCIL ◇ PONDOCILLIN ◇ SANGUICILLIN

TOXICITY DATA with REFERENCE
orl-rat LD50:5 g/kg AACHAX -,341,70
scu-rat LD50:4500 mg/kg AACHAX -,341,70
orl-mus LD50:3340 mg/kg AACHAX -,341,70
scu-mus LD50:3600 mg/kg AACHAX -,341,70

SAFETY PROFILE: Moderately toxic by ingestion and subcutaneous routes. See also ESTERS. When heated to decomposition it emits very toxic fumes of NO_x, SO_x, and HCl.

AOD125 CAS:7177-48-2 ***HR: 1***
AMPICILLIN TRIHYDRATE
mf: $C_{16}H_{19}N_3O_4S{\bullet}3H_20$ mw: 403.50

SYNS: AMCAP ◇ AMCILL ◇ AMINOBENZYLPENICILLIN TRIHYDRATE ◇ α-AMINOBENZYLPENICILLIN TRIHYDRATE ◇ AMPERIL ◇ AMPICHEL ◇ AMPIKEL ◇ AMPINOVA ◇ AMPLIN ◇ ANCILLIN ◇ CYMBI ◇ DIVERCILLIN ◇ LIFEAMPIL ◇ MOREPEN ◇ NCI-C56086 ◇ PEN A ◇ PENSYN ◇ POLYCILLIN ◇ PRINCILLIN ◇ RO-AMPEN ◇ TRAFARBIOT ◇ UKOPEN ◇ VIDOPEN

TOXICITY DATA with REFERENCE
orl-rat TDLo:1500 mg/kg (6-11D preg):TER ANTBAL 18,815,73
orl-mus TDLo:9100 mg/kg (female 7-13D post):REP KSRNAM 7,3113,73
orl-rat LD50:10 g/kg ANTBAL 20,653,75
orl-mus LD50:15200 mg/kg ANTBAL 20,653,75

SAFETY PROFILE: Mildly toxic by ingestion. An experimental teratogen. Other experimental reproductive effects. When heated to decomposition it emits toxic fumes of SO_x and NO_x.

AOD250 CAS:134-53-2 ***HR: 3***
AMPROTROPINE PHOSPHATE
mf: $C_{18}H_{29}NO_3{\bullet}H_3O_4P$ mw: 405.48

SYNS: AP 407 ◇ 3-DIETHYLAMINO-2,2-DIMETHYLPROPYL TROPATE PHOSPHATE ◇ 1-PROPANOL, 3-(DIETHYLAMINO)-2,2-DIMETHYL-, TROPATE, PHOSPHATE ◇ SYNTROPAN ◇ dl-TROPASAEUREESTER DES 3-DIAETHYLAMINO-2,2-DIMETHYL-1-PROPANOL PHOSPHAT (GERMAN)

TOXICITY DATA with REFERENCE
ivn-rat LD50:43 mg/kg JLCMAK 30,700,45
orl-mus LDLo:570 mg/kg JPETAB 60,1,37
scu-mus LDLo:1250 mg/kg JPETAB 60,1,37
ivn-mus LD50:51 mg/kg JLCMAK 30,700,45
scu-cat LDLo:200 mg/kg JPETAB 60,1,37
scu-rbt LDLo:500 mg/kg JPETAB 60,1,37
ivn-rbt LD50:25 mg/kg SMWOAS 76,1282,46
scu-frg LDLo:1500 mg/kg JPETAB 60,1,37
par-frg LDLo:1000 mg/kg AEPPAE 173,86,33

SAFETY PROFILE: Poison by ingestion, subcutaneous, and intravenous routes. Moderately toxic by parenteral route. When heated to decomposition it emits very toxic fumes of PO_x and NO_x. An anticholinergic agent.

AOD375 CAS:60719-84-8 ***HR: 3***
AMRINONE
mf: $C_{10}H_9N_3O$ mw: 187.22

PROP: Crystals from DMF. Mp: 294-297° (decomp).

SYNS: 5-AMINO(3,4'-BIPYRIDIN)-6-(1H)-ONE ◇ 5-AMINO-5-(4-PYRIDINYL)-2(1H)-PYRIDINONE ◇ INOCOR ◇ WIN 40680 ◇ WINCORAM

TOXICITY DATA with REFERENCE
orl-rat TDLo:1 g/kg (female 6-15D post):TER PHARAT 41,214,86
orl-rat TDLo:1 g/kg (female 6-15D post):REP PHARAT 41,214,86
orl-man LDLo:1429 μg/kg:CVS,SYS,BLD AIMDAP 145,825,85
orl-rat LD50:102 mg/kg NDADD8 1,259,83
ivn-rat LD50:75 mg/kg PHARAT 41,209,86
orl-mus LD50:288 mg/kg TPHSDY 1,143,80
ivn-mus LD50:150 mg/kg TPHSDY 1,143,80

SAFETY PROFILE: Poison by ingestion and intravenous routes. Human systemic effects by ingestion: cardiac arrhythmias, liver function, thrombocytopenia. An experimental teratogen. Other experimental reproductive effects. When heated to decomposition it emits toxic fumes of NO_x. A cardiotonic agent.

AOD425 CAS:80277-11-8 ***HR: D***
AMSACRINE LACTATE
mf: $C_{21}H_{19}N_3O_3S{\bullet}C_3H_6O_3$ mw: 483.58

SYNS: N-(4-(9-ACRIDINYLAMINO)-3-METHOXYPHENYL) METHANESULFONAMIDE compounded with LACTIC ACID ◇ m-AMSA LACTATE ◇ METHANESULFONAMIDE, N-(4-(9-ACRIDINYLAMINO)-3-METHOXYPHENYL)-,MONO(2-HYDROXYPROPANOATE) ◇ METHANESULFON-m-ANISIDIDE, 4'-(9-ACRIDINYLAMINO)-, compounded with LACTIC ACID

TOXICITY DATA with REFERENCE
ipr-rat TDLo:10 mg/kg (female 6-15D post):TER TJADAB 35,76A,87
ipr-rat TDLo:10 mg/kg (female 6-15D post):REP TJADAB 35,76A,87

SAFETY PROFILE: An experimental teratogen. Other experimental reproductive effects. When heated to decomposition it emits toxic fumes of NO_x and SO_x.

AOD500 CAS:29883-15-6 ***HR: 3***
AMYGDALIN
mf: $C_{20}H_{27}NO_{11}$ mw: 457.48

SYNS: d(−)-MANDELONITRILE-β-d-GENTIOBIOSIDE ◇ d-MANDELONITRILE-β-d-GLUCOSIDO-6-β-d-GLUCOSIDE ◇ NSC-15780

TOXICITY DATA with REFERENCE
hma-mus/sat 250 mg/kg SCIEAS 198,625,77
orl-ham TDLo:300 mg/kg (female 8D post):TER SCIEAS 215,1513,82
orl-inf LDLo:50 mg/kg JAMAAP 238,482,77
orl-rat LD50:522 mg/kg WJMDA2 134,97,81

SAFETY PROFILE: Human poison by ingestion (infant data). Moderately toxic experimentally by ingestion. An experimental teratogen. Mutation data reported. When heated to decomposition it emits toxic fumes of NO_x.

AOD725 CAS:628-63-7 ***HR: 3***
n-AMYL ACETATE
DOT: UN 1104
mf: $C_7H_{14}O_2$ mw: 130.21

PROP: Colorless liquid; pear or banana-like odor. Mp: −78.5°, bp: 148° @ 737 mm, ULC: 55-60, lel: 1.1%, uel: 7.5%, flash p: 77°F (CC), d: 0.879 @ 20°/20°, autoign temp: 714°F, vap d: 4.5. Very sltly sol in water; misc in alc and ether.

SYNS: ACETATE d'AMYLE (FRENCH) ◇ ACETIC ACID, AMYL ESTER ◇ AMYL ACETATE (DOT) ◇ AMYL ACETIC ESTER ◇ AMYL ACETIC ETHER ◇ AMYLAZETAT (GERMAN) ◇ BIRNENOEL ◇ OCTAN AMYLU (POLISH) ◇ PEAR OIL ◇ PENT-ACETATE ◇ 1-PENTANOL ACETATE ◇ PENTYL ACETATE ◇ n-PENTYL ACETATE ◇ 1-PENTYL ACETATE ◇ PRIMARY AMYL ACETATE

TOXICITY DATA with REFERENCE
eye-hmn 300 ppm JIHTAB 25,282,43
ihl-hmn TCLo:5000 mg/m^3/30M:CNS,EYE,PUL AHYGAJ 78,260,13
ihl-hmn TCLo:200 ppm:CNS NPIRI* 1,3,74
orl-rat LD50:6500 mg/kg NPIRI* 1,3,74
ihl-rat LCLo:5200 ppm/8H DTLVS* 3,12,71
ipr-gpg LDLo:1500 mg/kg AIHAAP 35,21,74

CONSENSUS REPORTS: Reported in EPA TSCA Inventory.

OSHA PEL: TWA 100 ppm
ACGIH TLV: TWA 100 ppm
DOT Classification: Flammable or Combustible Liquid; Label: Flammable Liquid.

SAFETY PROFILE: Moderately toxic by intraperitoneal route. Human systemic effects by inhalation: conjunctiva irritation, headache, and somnolence. A human eye irritant. Apparently more toxic than butyl acetate. Chronic toxicity is of a low order. Dangerous fire hazard when exposed to heat or flame; can react with oxidizing materials. Moderately explosive in the form of vapor when exposed to flame. To fight fire, use alcohol foam, dry chemical. When heated to decomposition it emits acrid smoke and irritating fumes. See also ESTERS, AMYL ALCOHOL, and ACETIC ACID.

AOD735 CAS:626-38-0 ***HR: 3***
sec-AMYL ACETATE
DOT: UN 1104
mf: $C_7H_{14}O_2$ mw: 130.21

PROP: Colorless liquid. Bp: 120°, flash p: 73.4°F (CC), d: 0.862-0.866 @ 20°/20°, vap d: 4.48, lel: 1.1%, uel: 7.5%. Sltly sol in water; misc in alc and ether.

SYNS: 2-ACETOXYPENTANE ◇ 1-METHYLBUTYL ACETATE ◇ 2-PENTANOL, ACETATE ◇ 2-PENTYL ACETATE

TOXICITY DATA with REFERENCE
ihl-hmn TCLo:200 ppm:EYE JIHTAB 25,282,43
ihl-gpg LCLo:10000 ppm/5H PHRPA6 51,811,36

CONSENSUS REPORTS: Reported in EPA TSCA Inventory.

OSHA PEL: TWA 125 ppm
ACGIH TLV: TWA 125 ppm
DOT Classification: Flammable Liquid; Label: Flammable Liquid.

SAFETY PROFILE: Mildly toxic by inhalation. Human systemic effects by inhalation: conjunctiva irritation. Dangerous fire hazard when exposed to heat or flame; can react with oxidizing materials. Moderately explosive in the form of vapor when exposed to heat or flame. To fight fire, use alcohol foam, dry chemical. When heated to decomposition it emits acrid smoke and irritating fumes.

AOD750 ***HR: 2***
AMYL ACETATE (mixed isomers)
mf: $C_7H_{14}O_2$ mw: 130.21

PROP: Colorless liquid, pear-like odor. Mp: −78.5°, bp: 148° @ 737 mm, ULC: 55-60, lel: 1.1%, uel: 7.5%, flash p: 77°F (CC), d: 0.879 @ 20°/20°, autoign temp: 714°F, vap d: 4.5.

SYN: ACETIC ACID, AMYL ESTER

TOXICITY DATA with REFERENCE
skn-rbt 10 mg/24H open MLD AIHAAP 23,95,62
orl-rat LD50:4950 mg/kg AIHAAP 23,95,62

DFG MAK: 100 ppm (525 mg/m^3)

SAFETY PROFILE: A skin irritant. Mildly toxic by ingestion. Dangerous fire hazard; can react with oxidizing materials. Moderately explosive in the form of vapor when exposed to flame. To fight fire, use alcohol foam, dry chemical. When heated to decomposition it emits acrid smoke and irritating fumes.

AOE000 ***HR: 3***
AMYL ALCOHOL
mf: $C_5H_{12}O$ mw: 88.1

PROP: Clear liquid. Mp: −79°, bp: 137.8°, flash p: 91°F (CC), d: 0.8168 @ 20°/20°, ULC: 40, lel: 1.2%, uel: 10% @ 212°F, vap press: 1 mm @ 13.6°, 10 mm @ 44.9°, vap d: 3.04. Sol in water; misc in alc and ether.

SYNS: ALCOOL AMYLIQUE (FRENCH) ◇ N-AMYL ALCOHOL ◇ AMYL ALCOHOL, NORMAL ◇ N-AMYLALKOHOL (CZECH) ◇ N-BUTYLCARBINOL ◇ PENTANOL-1 ◇ N-PENTANOL ◇ PENTAN-1-OL ◇ PENTASOL ◇ PENTYL ALCOHOL ◇ PRIMARY AMYL ALCOHOL

TOXICITY DATA with REFERENCE
ipr-mus TDLo:1200 mg/kg/8W-I:ETA CNREA8 33,3069,73
skn-rbt 3200 mg/kg/24H SEV AIHAAP 34,493,73
skn-rbt 500 mg/24H MOD 28ZPAK -,35,72
eye-rbt 81 mg SEV AIHAAP 34,493,73
eye-rbt 20 mg/24H SEV 28ZPAK -,35,72
orl-rat LD50:3030 mg/kg FCTXAV 2,327,64
ihl-rat LCLo:14000 mg/m^3/6H AIHAAP 34,493,73
ipr-rat LDLo:490 mg/kg AEPPAE 132,214,28
orl-mus LD50:200 mg/kg GISAAA 35(9),88,70
ihl-mus LCLo:14000 mg/m^3/6H AIHAAP 34,493,73
ivn-cat LDLo:15 mg/kg JPETAB 16,1,20
skn-rbt LD50:4490 mg/kg 31ZTAS -,76,68

CONSENSUS REPORTS: Reported in EPA TSCA Inventory.

SAFETY PROFILE: Moderately toxic by ingestion and skin contact. An eye and upper respiratory irritant by inhalation. A severe skin and eye irritant. Ingestion can cause headache, nausea, vomiting, delirium, and methemoglobin formation. Questionable carcinogen with experimental tumorigenic data. Extremely flammable if exposed to heat, flame, or powerful oxidizers. Moderately explosive when exposed to flame. Incompatible with oxidizing materials, hydrogen trisulfide. To fight fire, use alcohol foam, dry chemical.

AOE500 ***HR: 2***
AMYL AZIDE
mf: $C_5H_{11}N_3$ mw: 113

SAFETY PROFILE: Moderately toxic irritant and toxic by ingestion and inhalation. Narcotic in high concentration. Can cause a fall in blood pressure. See also AZIDES. An unstable material.

AOE750 CAS:63018-99-5 ***HR: 3***
5-n-AMYL-1582-BENZANTHRACENE
mf: $C_{23}H_{22}$ mw: 298.45

SYN: 8-PENTYLBENZ(a)ANTHRACENE

TOXICITY DATA with REFERENCE
skn-mus TDLo:790 mg/kg/33W-I:ETA PRLBA4 129,439,40

SAFETY PROFILE: Questionable carcinogen with experimental tumorigenic data. When heated to decomposition it emits acrid smoke and irritating fumes.

AOF000 CAS:2049-95-8 ***HR: 1***
tert-AMYLBENZENE
mf: $C_{11}H_{16}$ mw: 148.27

PROP: Liquid. D: 0.867, bp: 189-190°, insol in water, misc in alc and ether.

SYN: tert-PENTYLBENZENE

TOXICITY DATA with REFERENCE
orl-rat LDLo:5000 mg/kg AMIHAB 19,403,59

CONSENSUS REPORTS: Reported in EPA TSCA Inventory.

SAFETY PROFILE: Mildly toxic by ingestion. When heated to decomposition it yields irritating fumes and smoke.

AOF250 CAS:63905-98-6 ***HR: 3***
4-AMYL-N-BENZOHYDRYLPYRIDINIUM BROMIDE
mf: $C_{22}H_{26}N\cdot Br$ mw: 384.40

SYN: B-45

TOXICITY DATA with REFERENCE
ipr-rat LD50:8 mg/kg FEPRA7 9,280,50
scu-rat LD50:4 mg/kg FEPRA7 9,280,50
orl-mus LD50:35 mg/kg FEPRA7 9,280,50
ipr-mus LD50:4500 μg/kg FEPRA7 9,280,50
scu-mus LD50:1600 μg/kg FEPRA7 9,280,50
ivn-mus LD50:1300 μg/kg FEPRA7 9,280,50
ivn-dog LD50:30 mg/kg FEPRA7 9,280,50
scu-rbt LD50:35 mg/kg FEPRA7 9,280,50

ivn-rbt LD50:10 mg/kg FEPRA7 9,280,50
scu-gpg LD50:15 mg/kg FEPRA7 9,280,50

SAFETY PROFILE: Poison by ingestion, intraperitoneal, subcutaneous, and intravenous routes. See also BROMIDES. When heated to decomposition it emits very toxic fumes of NO_x and Br^-.

AOF500 CAS:63990-96-5 ***HR: 1***
AMYL BIPHENYL
mf: $C_{17}H_{20}$ mw: 224.37

PROP: Liquid. Mp: −60°, bp: 305-337°, flash p: 300°F, d: 0.958 @ 20°/20°, vap d: 7.73.

SYN: PENTYLBIPHENYL

TOXICITY DATA with REFERENCE
orl-rat LDLo:5000 mg/kg AMIHAB 19,403,59

SAFETY PROFILE: Mildly toxic by ingestion and inhalation routes. Combustible when exposed to heat or flame. Moderately dangerous; when heated to decomposition it emits irritating fumes and smoke. Incompatible with oxidizing materials. To fight fire, use foam, CO_2, dry chemical.

AOF750 ***HR: 3***
d-AMYL BROMIDE
mf: $CH_3(CH_2)_4Br$ mw: 151.1

PROP: Colorless liquid. Bp: 120°, flash p: 90°F, fp: < −30°, d: 1.211 @ 25°/25°.

SAFETY PROFILE: Poison by intraperitoneal route. It can cause liver damage, is narcotic in high concentrations, and is a local irritant. See also BROMIDES. Extremely flammable. To fight fire, use alcohol foam, water mist or spray, dry chemical. When heated to decomposition it emits very toxic bromides. Incompatible with oxidizing materials.

AOG000 CAS:540-18-1 ***HR: 2***
n-AMYL BUTYRATE
DOT: UN 2620
mf: $C_9H_{18}O_2$ mw: 158.27

PROP: Colorless liquid. D: 0.871, mp: −73.2°, bp: 186.4°. Sol in water, miscible with alc and ether.

SYN: PENTYL BUTYRATE

TOXICITY DATA with REFERENCE
orl-rat LD50:12210 mg/kg FCTXAV 2,327,64
orl-gpg LD50:11950 mg/kg FCTXAV 2,327,64

CONSENSUS REPORTS: Reported in EPA TSCA Inventory.

DOT Classification: Flammable or Combustible Liquid; Label: Flammable Liquid

SAFETY PROFILE: Mildly toxic by ingestion. When heated to decomposition it emits acrid smoke and irritating fumes.

AOG500 CAS:122-40-7 ***HR: 2***
α-AMYL CINNAMALDEHYDE
mf: $C_{14}H_{18}O$ mw: 202.32

PROP: Yellow liquid; floral jasmine odor. D: 0.963, refr index: 1.554, bp: 174-175° @ 20 mm. Sol in fixed oils; insol in glycerin and propylene glycol

SYNS: α-AMYL CINNAMIC ALDEHYDE ◇ α-AMYL-β-PHENYLACROLEIN ◇ FEMA No. 2061 ◇ JASMINALDEHYDE ◇ α-PENTYLCINNAMALDEHYDE

TOXICITY DATA with REFERENCE
skn-gpg 5%/2W MLD ADVEA4 58,121,78
orl-rat LD50:3730 mg/kg FCTXAV 2,327,64

CONSENSUS REPORTS: Reported in EPA TSCA Inventory.

SAFETY PROFILE: Moderately toxic by ingestion. A mild skin irritant. See also ALDEHYDES. When heated to decomposition it emits acrid smoke and irritating fumes.

AOG600 ***HR: 1***
AMYL CINNAMATE
mf: $C_{14}H_{18}O_2$ mw: 218.28

PROP: Colorless to pale yellow liquid; slt cocoa odor. D: 0.992-0.997, refr index: 1.535, flash p: +212°F. Sol in fixed oils; sltly sol in propylene glycol; insol in glycerin @ 310°.

SYNS: FEMA No. 2063 ◇ ISOAMYL CINNAMATE ◇ ISOAMYL 3-PENTYL PROPENATE

SAFETY PROFILE: Combustible liquid. When heated to decomposition it emits acrid smoke and irritating fumes.

AOG750 CAS:7493-78-9 ***HR: 2***
AMYL CINNAMIC ACETATE
mf: $C_{16}H_{22}O_2$ mw: 246.38

SYNS: α-N-AMYL-β-PHENYLACRYL ACETATE ◇ α-PENTYL CINNAMYL ACETATE

TOXICITY DATA with REFERENCE
skn-rbt 500 mg/24H FCTXAV 14,659,76

SAFETY PROFILE: Moderately toxic by skin irritant. When heated to decomposition it emits acrid smoke and irritating fumes.

AOH000 CAS:101-85-9 ***HR: 2***
α-AMYLCINNAMIC ALCOHOL
mf: $C_{14}H_{20}O$ mw: 204.34

SYNS: α-AMYLCINNAMYL ALCOHOL ◇ 2-AMYL-3-PHENYL-2-PROPEN-1-OL ◇ 2-BENZYLIDENE-1-HEPTANOL

TOXICITY DATA with REFERENCE
orl-rat LD50:4000 mg/kg FCTXAV 12,807,74

CONSENSUS REPORTS: Reported in EPA TSCA Inventory.

SAFETY PROFILE: Moderately toxic by ingestion. See also ALCOHOLS. When heated to decomposition it emits acrid smoke and irritating fumes.

AOH100 CAS:68527-78-6 ***HR: 1***
AMYL CINNAMYLIDENE METHYL ANTHRANILATE
mf: $C_{22}H_{25}NO_2$ mw: 335.48

SYNS: ANTHRANILIC ACID, N-(2-BENZYLIDENEHEPTYLIDENE)-, METHYL ESTER ◇ METHYL N-(β-PENTYLCINNAMYLIDENE)ANTHRANILATE

TOXICITY DATA with REFERENCE
skn-rbt 500 mg/24H MOD FCTXAV 16,645,78

CONSENSUS REPORTS: Reported in EPA TSCA Inventory.

SAFETY PROFILE: A skin irritant. When heated to decomposition it emits toxic fumes of NO_x.

AOH250 CAS:53043-14-4 ***HR: 2***
6-n-AMYL-m-CRESOL
mf: $C_5H_{11}C_6H_3OHCH_3$ mw: 178.3

PROP: Bp: 258°, flash p: 240°F, d: 0.97.

TOXICITY DATA with REFERENCE
orl-rat LD50:1500 mg/kg PSEBAA 32,592,35

SAFETY PROFILE: Moderately toxic by ingestion. Combustible liquid when exposed to heat or flame. Dangerous; when heated to decomposition it emits irritating fumes. Incompatible with oxidizing materials.

AOH750 CAS:16587-71-6 ***HR: 1***
4-tert-AMYLCYCLOHEXANONE
mf: $C_{11}H_{20}O$ mw: 168.31

SYNS: 4-(1,1-DIMETHYLPROPYL)CYCLOHEXANONE ◇ 4-tert-PENTYLCYCLOHEXANONE

TOXICITY DATA with REFERENCE
skn-rbt 500 mg/24H FCTXAV 12,807,74
orl-rat LD50:4700 mg/kg FCTXAV 12,807,74

CONSENSUS REPORTS: Reported in EPA TSCA Inventory.

SAFETY PROFILE: Mildly toxic by ingestion. A skin irritant. When heated to decomposition it emits smoke and acrid fumes.

AOI000 ***HR: 2***
AMYLCYCLOHEXYL ACETATE (mixed isomers)
mf: $C_{13}H_{23}O_2$ mw: 211.2

SYN: PENTYLCYCLOHEXANOL ACETATE

TOXICITY DATA with REFERENCE
skn-rbt 500 mg/24H MOD FCTXAV 14,659,76

SAFETY PROFILE: Moderately toxic by skin irritant. When heated to decomposition it emits acrid smoke and acrid fumes.

AOI200 CAS:692-95-5 ***HR: 3***
AMYLDICHLORARSINE
mf: $C_5H_{11}AsCl_2$ mw: 216.98

SYNS: N-AMYLDICHLORARSINE ◇ ARSINE, AMYLDICHLORO- ◇ ARSINE, DICHLOROPENTYL- ◇ DICHLOROPENTYLARSINE ◇ PENTYLDICHLOROARSINE

TOXICITY DATA with REFERENCE
ihl-mus LC50:1400 mg/m^3/10M NTIS** PB158-508
skn-mus LDLo:4 mg/kg NTIS** PB158-508

OSHA PEL: TWA 0.5 mg(As)/m^3

SAFETY PROFILE: Poison by skin contact. Moderately toxic by inhalation. When heated to decomposition it emits toxic fumes of As and Cl^-.

AOI250 CAS:14779-78-3 ***HR: 1***
AMYL-p-DIMETHYLAMINOBENZOATE
mf: $C_{14}H_{21}NO_2$ mw: 235.36

SYN: AMYL DIMETHYL PABA

TOXICITY DATA with REFERENCE
skn-hmn 15 mg/3D-I MLD 85DKA8 -,127,77

CONSENSUS REPORTS: Reported in EPA TSCA Inventory.

SAFETY PROFILE: A mild human skin irritant. When heated to decomposition it emits toxic fumes of NO_x.

AOI500 CAS:58817-05-3 ***HR: 1***
AMYLDIMETHYL-p-AMINO BENZOIC ACID
mf: $C_{17}H_{27}NO_2$ mw: 277.45

SYNS: p-DIMETHYLAMINOBENZOIC ACID, PENTYL ESTER ◇ p-DIMETHYLAMINOBENZOIC ACID, OCTYL ESTER ◇ OCTYLDIMETHYL-p-AMINOBENZOIC ACID

TOXICITY DATA with REFERENCE
skn-hmn 15 mg/3D-I MLD 85DKA8 -,127,77

SAFETY PROFILE: A mild human skin irritant. See also ESTERS. When heated to decomposition it emits toxic fumes of NO_x.

AOI750 CAS:513-35-9 ***HR: 3***
α,η-AMYLENE
DOT: UN 2371/UN 2460
mf: C_5H_{10} mw: 70.15

PROP: Liquid, disagreeable odor. Mp: −124° bp: 30.1°, lel: 1.6%, uel: 8.7%, flash p: 0°F (OC), d: 0.643, vap d: 2.42, autoign temp: 527°F.

SYNS: β-ISOAMYLENE ◇ 2-METHYL-2-BUTENE ◇ TRIMETHYLETHYLENE

CONSENSUS REPORTS: Reported in EPA TSCA Inventory.

DOT Classification: Flammable Liquid; Label: Flammable Liquid.

SAFETY PROFILE: Moderately toxic by ingestion and inhalation. Narcotic in high concentration. A simple asphyxiant. Extremely flammable. Moderately explosive when exposed to heat, flame, or powerful oxidizers. To fight fire, use alcohol foam, spray, mist, dry chemical. When heated to decomposition it emits acrid smoke and irritating fumes.

AOJ000 ***HR: 3***
AMYLENES, MIXED
DOT: UN 1108/UN 2371/UN 2460
mf: C_5H_{10} mw: 70.58

PROP: Water-white liquid. Bp: 32.2°, flash p: 0°F, d: 0.66 @ 20°.

CONSENSUS REPORTS: Reported in EPA TSCA Inventory.

DOT Classification: Flammable Liquid; Label: Flammable Liquid

SAFETY PROFILE: Moderately toxic. See also α-n-AMYLENE. Very flammable; reacts with heat, flame and oxidizing materials. To fight fire, use foam, CO_2, dry chemical.

AOJ500 CAS:638-49-3 ***HR: 3***
n-AMYL FORMATE
DOT: UN 1109
mf: $C_6H_{12}O_2$ mw: 116.18

PROP: Clear liquid. D: 0.902, 0.893 @ 15°/4°, mp: −73.5°, bp: 130.4°, flash p: 80°F. Very sltly sol in water; misc in alc and ether.

SYNS: AMYL FORMATE (DOT) ◇ PENTYL FORMATE ◇ m-PENTYL FORMATE

TOXICITY DATA with REFERENCE
skn-rbt 500 mg/24H MLD FCTXAV 18,649,80

CONSENSUS REPORTS: Reported in EPA TSCA Inventory.

DOT Classification: Flammable Liquid; Label: Flammable Liquid; Flammable or Combustible Liquid; Label: Flammable Liquid

SAFETY PROFILE: A moderate irritant by ingestion and skin contact. See also ESTERS. Dangerously flammable; reacts vigorously with heat, flame, oxidizing materials. To fight fire, use foam, CO_2, dry chemical.

AOJ750 CAS:63885-68-7 ***HR: 3***
o-n-AMYL HARMOL HYDROCHLORIDE
mf: $C_{17}H_{20}N_2O{\cdot}ClH$ mw: 304.85

SYN: AMYL HARMOL HYDROCHLORIDE

TOXICITY DATA with REFERENCE
ipr-mus LDLo:200 mg/kg QJPPAL 5,56,32
scu-gpg LDLo:400 mg/kg QJPPAL 5,56,32
scu-frg LDLo:200 mg/kg QJPPAL 5,37,32

SAFETY PROFILE: Poison by intraperitoneal and subcutaneous routes. When heated to decomposition it emits very toxic fumes of HCl and NO_x.

AOK000 CAS:10484-36-3 ***HR: 2***
AMYLISOEUGENOL
mf: $C_{15}H_{22}O_2$ mw: 234.37

SYNS: AMYLOXYISOEUGENOL ◇ ISOEUGENOL AMYL ETHER ◇ 2-METHOXY-1-(PENTYLOXY)-4-(1-PROPENYL)-BENZENE◇ 1-PENTOXY-2-METHOXY-4-PROPENYLBENZENE

TOXICITY DATA with REFERENCE
skn-rbt 500 mg/24H MOD FCTXAV 17,509,79

CONSENSUS REPORTS: Reported in EPA TSCA Inventory.

SAFETY PROFILE: A skin irritant. See also ETHERS. When heated to decomposition it emits acrid smoke and irritating fumes.

AOK250 ***HR: 1***
AMYL LACTATE
mf: $C_8H_{16}O_3$ mw: 160.2

PROP: Colorless liquid. Bp: 210°; flash p: 175°F; d: 0.960 @ 20°.

SAFETY PROFILE: An irritant by inhalation and ingestion. See also ESTERS. Moderately flammable. Incompatible with heat, flame, oxidizing materials. To fight fire, use foam, CO_2, dry chemical.

AOK500 ***HR: 2***
AMYL LAURATE
mf: $C_5H_{11}O_2C(CH_2)_{10}CH_3$ mw: 270.44

PROP: Bp: 290°, flash p: 300°F, d: 0.86.

SAFETY PROFILE: It may defat skin and cause contact

dermatitis. Combustible. Incompatible with oxidizing materials. To fight fire, use CO_2, dry chemical.

AOK750 CAS:105-30-6 ***HR: 2***
AMYL METHYL ALCOHOL
mf: $C_6H_{14}O$ mw: 102.20

PROP: Liquid. Bp: 130°, flash p: 114°F (CC), d: 0.804, vap d: 3.52.

SYNS: 1,3-DIMETHYL BUTANOL ◇ ISOHEXYL ALCOHOL ◇ ISOPROPYL DIMETHYL CARBINOL ◇ METHYLAMYL ALCOHOL ◇ METHYL ISOBUTYL CARBINOL ◇ 2-METHYLPENTANOL-1 ◇ 2-METHYL-2-PROPYLETHANOL

TOXICITY DATA with REFERENCE
skn-rbt 10 mg/24H open MLD AIHAAP 23,95,62
ihl-hmn TCLo:50 ppm:IRR JIHTAB 28,262,46
orl-rat LD50:1410 mg/kg AMIHBC 10,61,54
skn-rbt LD50:3560 mg/kg AMIHBC 10,61,54

CONSENSUS REPORTS: Reported in EPA TSCA Inventory.

SAFETY PROFILE: Moderately toxic by ingestion and skin contact. A skin irritant. Human systemic irritant by inhalation. Combustible liquid; can react with oxidizing materials. To fight fire, use CO_2, dry chemical. When heated to decomposition it emits smoke and acrid fumes.

AOL000 CAS:13256-07-0 ***HR: 3***
n-AMYL-N-METHYLNITROSAMINE
mf: $C_6H_{14}N_2O$ mw: 130.22

SYNS: AMN ◇ METHYLAMYLNITROSAMIN (GERMAN) ◇ METHYLAMYLNITROSAMINE ◇ METHYL-N-AMYLNITROSAMINE ◇ N-METHYL-N-NITROSOPENTYLAMINE ◇ METHYL-N-PENTYLNITROSAMINE ◇ N-NITROSO-N-METHYL-N-AMYLAMINE ◇ NITROSOMETHYL-N-PENTYLAMINE

TOXICITY DATA with REFERENCE
mma-sat 10 μg/plate TCMUE9 1,13,84
mma-esc 1 μmol/plate GANNA2 75,8,84
dnr-esc 25 μL/well CBINA8 15,219,76
orl-rat TDLo:168 mg/kg/8W-C:CAR NIPAA4 78,1889,81
ipr-rat TDLo:50 mg/kg:NEO CNREA8 39,3644,79
scu-rat TDLo:240 mg/kg/40W-I:CAR CCLCDY 2,263,80
orl-rat TD:330 mg/kg/31W-C:ETA ARZNAD 19,1077,69
orl-rat LD50:120 mg/kg ZEKBAI 69,103,67
ipr-rat LD50:85 mg/kg CNREA8 39,3644,79
scu-rat LD50:120 mg/kg ZEKBAI 69,103,67

CONSENSUS REPORTS: EPA Genetic Toxicology Program.

SAFETY PROFILE: Suspected carcinogen with experimental carcinogenic, neoplastigenic, and tumorigenic data. Poison by ingestion, subcutaneous, and intraperitoneal routes. Mutation data reported. When heated to decomposition it emits toxic NO_x. See also NITROSAMINES and N-NITROSO COMPOUNDS.

AOL250 CAS:1002-16-0 ***HR: 2***
AMYL NITRATE
DOT: UN 1112
mf: $C_5H_{11}NO_3$ mw: 133.17

PROP: Liquid. Bp: 145°, flash p: 125°F (OC), d: 0.99.

SYN: NITRATE d'AMYLE (FRENCH)

TOXICITY DATA with REFERENCE
ihl-rat LCLo:3593 ppm AMIHAB 11,290,55
ihl-mus LCLo:1374 ppm AMIHAB 11,290,55
ihl-rbt LCLo:1703 ppm AMIHAB 11,290,55
ihl-gpg LCLo:1703 ppm AMIHAB 11,290,55

DOT Classification: Flammable or Combustible Liquid; Label:Flammable Liquid

SAFETY PROFILE: Moderately toxic by inhalation. Combustible liquid. An oxidizing agent. When heated to decomposition it emits toxic fumes of NO_x.

AOL500 CAS:463-04-7 ***HR: 3***
n-AMYL NITRITE
DOT: UN 1113
mf: $C_5H_{11}NO_2$ mw: 117.17

PROP: Clear, yellowish liquid; peculiar, ethereal, fruity odor and pungent, aromatic taste. Bp: 96°-99°, d: 0.8528 @ 20°/4°, autoign temp: 408°F, vap d: 4.0.

SYNS: AMYL NITRITE (DOT) ◇ 1-NITROPENTANE ◇ NITROUS ACID, PENTYL ESTER ◇ PENTYL NITRITE

TOXICITY DATA with REFERENCE
mmo-sat 1 mg/plate BSIBAC 56,816,80

CONSENSUS REPORTS: Reported in EPA TSCA Inventory.

DOT Classification: Flammable Liquid; Label: Flammable Liquid.

SAFETY PROFILE: Moderately toxic by inhalation and ingestion. Causes flushing of skin, rapid pulse, headache, and fall in blood pressure. Mutation data reported. See also NITRITES and ESTERS. Flammable when exposed to heat or flame or by spontaneous chemical reaction. To fight fire, use alcohol foam. An oxidizing material. Vapors explode when heated. It will react with oxidizing or reducing materials. When heated to decomposition it emits toxic fumes of NO_x.

AOL750 CAS:64005-62-5 ***HR: 3***
n-AMYL-N-NITROSOURETHANE
mf: $C_8H_{16}N_2O_3$ mw: 188.26

SYN: N-NITROSO-N-PENTYLCARBAMIC ACID-ETHYL ESTER

TOXICITY DATA with REFERENCE
cyt-ham:fbr 63 mg/L/48H MUREAV 48,337,77

orl-rat TDLo:2625 mg/kg/50W-C:CAR GANNA2 73,48,82
orl-rat TD:7350 mg/kg/35W-C:CAR GANNA2 73,48,82
orl-rat TDLo:5880 mg/kg/24W-C:ETA GANNA2 70,653,79

SAFETY PROFILE: Questionable carcinogen with experimental carcinogenic and tumorigenic data. Mutation data reported. See also ESTERS. When heated to decomposition it emits toxic fumes of NO_x. See also N-NITROSO COMPOUNDS.

AOM000 CAS:644-26-8 ***HR: 3***
AMYLOCAINE
mf: $C_{14}H_{21}NO_2$ mw: 235.36

SYNS: AMYLEINE ◇ 1-(DIMETHYLAMINO)-2-METHYL-2-BUTANOL BENZOATE (ESTER) ◇ STOVAINE

TOXICITY DATA with REFERENCE
ivn-rat LDLo:25 mg/kg PHREA7 12,262,32
ipr-mus LDLo:170 mg/kg HBAMAK 4,1289,35
scu-mus LDLo:170 mg/kg HBAMAK 4,1289,35
ipr-dog LDLo:100 mg/kg HBAMAK 4,1289,35
scu-dog LDLo:100 mg/kg HBAMAK 4,1289,35

SAFETY PROFILE: Poison by intravenous, subcutaneous, and intraperitoneal routes. See also ESTERS. When heated to decomposition it emits toxic fumes of NO_x.

AOM150 CAS:9047-13-6 ***HR: 2***
AMYLOPECTINE SULPHATE

SYNS: AMYLOPECTIN, HYDROGEN SULFATE ◇ AMYLOPECTIN SULFATE ◇ AMYLOPECTIN SULFATE (SN-263) ◇ SULFATED AMYLOPECTIN

TOXICITY DATA with REFERENCE
orl-rat TDLo:621 g/kg/24W-C:CAR JUIZAG 32,479,86
ipr-rat LD50:30 mg/kg TOIZAG 17,111,70
scu-rat LD50:1051 mg/kg TOIZAG 17,111,70
ipr-mus LD50:133 mg/kg TOIZAG 17,111,70
scu-mus LD50:935 mg/kg TOIZAG 17,111,70

SAFETY PROFILE: Poison by subcutaneous and intraperitoneal route. Questionable carcinogen with experimental carcinogenic data. When heated to decomposition it emits toxic fumes of NO_x and SO_x.

AOM250 CAS:14938-35-3 ***HR: 3***
4-n-AMYLPHENOL
mf: $C_{11}H_{16}O$ mw: 164.27

PROP: Liquid. Bp: 342°, vap d: 5.66, flash p: 219°F (OC), d: 0.966.

SYN: p-PENTYLPHENOL

TOXICITY DATA with REFERENCE
skn-mus TDLo:4100 mg/kg/12W-I:NEO CNREA8 19,413,59

CONSENSUS REPORTS: Reported in EPA TSCA Inventory.

SAFETY PROFILE: Questionable carcinogen with experimental neoplastigenic data. Moderately flammable. To fight fire, use foam, CO_2, dry chemical. When heated to decomposition it emits acrid smoke and irritating fumes.

AOM500 ***HR: 3***
2-sec-AMYLPHENOL
mf: $C_{11}H_{16}O$ mw: 164.27

PROP: Clear, straw-colored liquid. D: 0.955-0.971 @ 30°/30°, bp: 235-250°, flash p: 200°F. Very sltly sol in water; sol in oils and organic solvents.

SYN: o-(sec-PENTYL) PHENOL

TOXICITY DATA with REFERENCE
skn-mus TDLo:4100 mg/kg/12W-I:NEO CNREA8 19,413,59
ivn-mus LD50:100 mg/kg JMCMAR 23,1350,80

SAFETY PROFILE: Poison by intravenous route. Questionable carcinogen with experimental neoplastigenic data by skin contact. Moderately flammable when exposed to heat or flame. To fight fire, use foam, fog, dry chemical, water mist or spray, multi-purpose dry chemical. When heated to decomposition it emits acrid smoke and irritating fumes.

AOM750 CAS:25735-67-5 ***HR: 3***
4-sec-AMYLPHENOL
mf: $C_{11}H_{16}O$ mw: 164.27

PROP: D: <1.0, bp: 482°-516°F, flash p: 270°F.

SYN: p-(sec-PENTYL) PHENOL

TOXICITY DATA with REFERENCE
skn-mus TDLo:4080 mg/kg/12W-I:ETA CNREA8 19,413,59

SAFETY PROFILE: Questionable carcinogen with experimental tumorigenic data. Combustible when exposed to heat or flame. To fight fire, use dry chemical, water mist, CO_2. When heated to decomposition it emits acrid smoke and fumes.

AON000 CAS:80-46-6 ***HR: 2***
4-tert-AMYLPHENOL
mf: $C_{11}H_{16}O$ mw: 164.27

PROP: Colorless needles. Bp: 250°, mp: 92°-93°, flash p: 232°F (OC).

SYNS: AMILPHENOL ◇ AMYL PHENOL 4T ◇ p-tert-AMYLPHENOL ◇ p-(α,α-DIMETHYLPROPYL)PHENOL ◇ p-(1,1-DIMETHYLPROPYL)PHENOL ◇ 2-METHYL-2-p-HYDROXYPHENYLBUTANE ◇ PENTAPHEN ◇ p-tert-PENTYLPHENOL ◇ PTAP ◇ UCAR AMYL PHENOL 4T

TOXICITY DATA with REFERENCE
skn-rbt 100 μg/24H open AIHAAP 23,95,62
eye-rbt 1% SEV UCDS** 8/13/64
eye-rbt 500 mg SEV IHFCAY 6,1,67
orl-rat LD50:1830 mg/kg IHFCAY 6,1,67
skn-rbt LD50:2000 mg/kg UCDS** 8/13/64

CONSENSUS REPORTS: Reported in EPA TSCA Inventory.

SAFETY PROFILE: Moderately toxic by ingestion and skin contact. A skin and severe eye irritant. Combustible. When heated to decomposition it emits toxic fumes. To fight fire, use dry chemical, water mist, CO_2. Incompatible with oxidizing materials.

AON250 CAS:2282-34-0 ***HR: 3***
3-sec-AMYLPHENYL-N-METHYLCARBAMATE
mf: $C_{13}H_{19}NO_2$ mw: 221.33

SYNS: ENT 27,127 ◇ m-(1-METHYLBUTYL)PHENYL METHYLCARBAMATE

TOXICITY DATA with REFERENCE
orl-rat LD50:87 mg/kg 28ZEAL 5,31,76
skn-rbt LD50:680 mg/kg 28ZEAL 5,31,76
orl-ckn LD50:44 mg/kg TXAPA9 11,49,67

SAFETY PROFILE: Poison by ingestion. Moderately toxic by skin contact. See also CARBAMATES. When heated to decomposition it emits toxic fumes of NO_x.

AON300 CAS:131-18-0 ***HR: D***
AMYL PHTHALATE
mf: $C_{18}H_{26}O_4$ mw: 306.44

SYNS: AMOIL ◇ 1,2-BENZENEDICARBOXYLIC ACID, DIPENTYL ESTER ◇ DIAMYL PHTHALATE ◇ DIPENTYL PHTHALATE ◇ DI-n-PENTYLPHTHALATE ◇ DPP ◇ PHTHALIC ACID, DIPENTYL ESTER

TOXICITY DATA with REFERENCE
orl-rat TDLo:2206 mg/kg (male 1D pre):REP TOLED5 15,265,83

CONSENSUS REPORTS: Reported in EPA TSCA Inventory.

SAFETY PROFILE: Experimental reproductive effects. When heated to decomposition it emits acrid smoke and irritating fumes.

AON350 ***HR: 2***
AMYL PROPIONATE
mf: $C_8H_{16}O_2$ mw: 144.21

PROP: Colorless liquid; fruity, apricot-pineapple odor. D: 0.866, refr index: 1.405-1.409, flash p: 106°F. Sol in alc, fixed oils; insol in glycerine, propylene glycol, water @ 160°.

SYNS: FEMA No. 2082 ◇ ISOAMYL PROPIONATE

SAFETY PROFILE: Combustible liquid. When heated to decomposition it emits acrid smoke and irritating fumes.

AON500 CAS:32446-40-5 ***HR: 3***
n-AMYL THIOCYANATE
mf: $C_6H_{11}NS$ mw: 129.24

PROP: Pale yellow oil. D: 0.905, bp: 197°. Insol in water; sol in alc and ether.

SYN: THIOCYANIC ACID, AMYL ESTER

TOXICITY DATA with REFERENCE
ipr-mus LD50:75 mg/kg JACSAT 78,3843,56
scu-mus LD50:75 mg/kg CLDND*

SAFETY PROFILE: Poison by subcutaneous and intraperitoneal routes. See also THIOCYANATES, ESTERS. When heated to decomposition it emits toxic fumes of NO_x and SO_x.

AON750 CAS:64-43-7 ***HR: 3***
AMYTAL SODIUM
mf: $C_{11}H_{17}N_2O_3$•Na mw: 248.29

SYNS: 5-ETHYL-5-ISOPENTYLBARBITURIC ACID SODIUM SALT ◇ 5-ETHYL-5-(3-METHYLBUTYL)BARBITURIC ACID SODIUM DERIVATIVE ◇ 5-ISOAMYL-5-ETHYLBARBITURIC ACID, SODIUM DERIVATIVE ◇ SODIUM AMYLOBARBITONE ◇ SODIUM ETHYLISOAMYLBARBITURATE ◇ SODIUM ISOAMYLETHYL BARBITURATE

TOXICITY DATA with REFERENCE
orl-rat LD50:275 mg/kg JPETAB 68,22,40
scu-rat LDLo:90 mg/kg JPETAB 31,1,27
ivn-rat LD50:128 mg/kg JAPMA8 44,152,55
orl-mus LD50:505 mg/kg FRPSAX 14,845,59
ipr-mus LDLo:200 mg/kg JPETAB 31,455,27
scu-mus LDLo:280 mg/kg JPHAA3 26,1248,37
ivn-mus LDLo:200 mg/kg JPHAA3 26,1248,37
orl-dog LD50:99 mg/kg JPETAB 68,22,40

SAFETY PROFILE: Poison by ingestion, subcutaneous, intravenous, and intraperitoneal routes. When heated to decomposition it emits toxic NO_x and Na_2O.

AON825 ***HR: 3***
ANABAENA FLOS-AQUAE TOXIN

SYNS: A. FLOS-AQUAE TOXIN ◇ TOXIN, ANABAENA FLOS-AQUAE NRC-44-1

TOXICITY DATA with REFERENCE
orl-rat LDLo:7500 μg/kg SCIEAS 187,542,75
ipr-mus LDLo:300 μg/kg SCIEAS 187,542,75

orl-dck LDLo:1880 μg/kg SCIEAS 187,542,75
orl-ctl LDLo:1800 μg/kg SCIEAS 187,542,75

SAFETY PROFILE: Poison by ingestion and intraperitoneal routes.

AON875 CAS:494-52-0 ***HR: 3***
ANABASINE
mf: $C_{10}H_{14}N_2$ mw: 162.26

PROP: Liquid. Bp: 270-272°, fp 9°, d: 1.0455. Sol in water and in most organic solvents.

SYNS: ANABASIN ◇ (−)-ANABASIN ◇ ANABAZIN ◇ NEONICOTINE ◇ NEONIKOTIN ◇ 1-3-(2′-PIPERIDYL)PYRIDINE ◇ 3-(2-PIPERIDINYL)PYRIDINE ◇ 3-(2-PIPERIDYL)-PYRIDINE ◇ 2-(3-PYRIDYL)-PIPERIDINE ◇ 2-(3′-PYRIDYL) PIPERIDINE ◇ (−)-2-(3′-PYRIDYL)PIPERIDINE

TOXICITY DATA with REFERENCE
orl-pig TDLo:20800 μg/kg (30-37D preg):TER TJADAB 30,61,84
orl-dog LDLo:50 mg/kg JPETAB 48,95,33
ivn-dog LDLo:3 mg/kg JPETAB 48,95,33
ivn-rbt LDLo:1 mg/kg JPETAB 48,95,33
skn-gpg LDLo:100 mg/kg JPETAB 48,95,33

SAFETY PROFILE: Poison by ingestion, subcutaneous, and intravenous routes. Moderately toxic by skin contact. An experimental teratogen. Insecticide. Acute and subacute toxicity: increased salivation, vertigo, confusion, disturbed vision and hearing, photophobia, cold extremities, nausea, vomiting, diarrhea, syncope, clonic spasms. When heated to decomposition it emits toxic fumes of NO_x.

AOO000 ***HR: 3***
ANAGESTONE ACETATE mixed with MESTRANOL (10:1)
mf: $C_{24}H_{36}O_3$ mw: 372.60

SYNS: ANATROPIN mixed with MESTRANOL (10:1) ◇ MESTRANOL mixed with ANAGESTONE ACETATE (1:10)

TOXICITY DATA with REFERENCE
orl-dog TDLo:259 mg/kg/2Y-I:ETA JJIND8 65,137,80

SAFETY PROFILE: Questionable carcinogen with experimental tumorigenic data. When heated to decomposition it emits acrid smoke and irritating fumes.

AOO120 CAS:64285-06-9 ***HR: 3***
ANATOXIN I
mf: $C_{10}H_{15}NO$ mw: 165.26

SYNS: ANATOXIN-a ◇ ANTX-a ◇ ETHANONE, 1-(9-AZABICYCLO(4.2.1)NON-2-EN-2-YL)-,(1R)-

TOXICITY DATA with REFERENCE
ipr-ham TDLo:1125 μg/kg (female 12-14D post):TER TOXIA6 18,684,80
ipr-mus LDLo:250 μg/kg TOXIA6 27,79,89

SAFETY PROFILE: Poison by intraperitoneal route. An experimental teratogen. When heated to decomposition it emits toxic fumes of NO_x.

AOO125 CAS:53-39-4 ***HR: 2***
ANAVAR
mf: $C_{19}H_{30}O_3$ mw: 306.49

PROP: Crystals. Mp: 235-238°.

SYNS: LONAVAR ◇ OXANDROLONE ◇ PROTIVAR ◇ PROVITAR ◇ VASOROME

TOXICITY DATA with REFERENCE
orl-rat TDLo:1800 mg/kg (2-19D preg):REP FESTAS 22,735,71
ipr-rat LD50:4893 mg/kg NYKZAU 65,418,69
orl-mus LD50:1832 mg/kg NYKZAU 65,418,69
ipr-mus LD50:922 mg/kg NYKZAU 65,418,69

SAFETY PROFILE: Moderately toxic by ingestion and intraperitoneal routes. Experimental reproductive effects. When heated to decomposition it emits acrid smoke and fumes.

AOO135 ***HR: 3***
ANCISTRODON PISCIVORUS VENOM

SYN: VENOM, SNAKE, ANCISTRODON PISCIVORUS

TOXICITY DATA with REFERENCE
ipr-mus LD50:6200 μg/kg ANREAK 139,305,61
ivn-mus LDLo:7500 μg/kg 14FHAR -,373,63
ipr-frg LD50:40 mg/kg ANREAK 139,305,61

SAFETY PROFILE: Poison by intravenous and intraperitoneal routes.

AOO150 CAS:56470-64-5 ***HR: D***
ANDORDRIN DIPROPIONATE
mf: $C_{28}H_{38}O_4$ mw: 438.66

SYNS: ANORDRIN ◇ 17-β-2-epsilon,17-α-DIETHYNYL, A-NOR-ANDROSTANE-2-epsilon, DIHYDROXYDIPROPINATE ◇ 2-α-17-α-DIETHYNYL-A-NOR-5-α-ANDROSTANE-2-β,17-β-DIOL DIPROPIONATE ◇ F-53

TOXICITY DATA with REFERENCE
orl-rat TDLo:2 mg/kg (1D preg):REP CCPTAY 32,301,85

SAFETY PROFILE: Experimental reproductive effects. When heated to decomposition it emits acrid smoke and fumes. An oral contraceptive.

AOO250 ***HR: 3***
ANDROCTONUS AMOREUXI VENOM

SYNS: A. AMOREUXI VENOM ◇ VENOM, SCORPION, ANDROCTONUS AMOREUXI

TOXICITY DATA with REFERENCE
ipr-rat TDLo:1 mg/kg (8-12D preg):TER TOXIA6 21,177,83
ipr-rat TDLo:1 mg/kg (8-12D preg):REP TOXIA6 21,177,83
ims-mus LD50:880 μg/kg TOXIA6 13,253,75
unr-mus LD50:600 μg/kg TOXIA6 9,1,71

SAFETY PROFILE: Deadly poison by intramuscular and unspecified routes. An experimental teratogen. Other experimental reproductive effects. When heated to decomposition it emits acrid smoke and irritating fumes.

AOO265 ***HR: 3***
ANDROCTONUS AUSTRALIS HECTOR VENOM

SYNS: A. AUSTRALIS HECTOR VENOM ◇ VENOM, SCORPION, ANDROCTONUS AUSTRALIS HECTOR

TOXICITY DATA with REFERENCE
ipr-mus LD50:97 μg/kg TOXIA6 22,308,84
scu-mus LD50:420 μg/kg EJBCAI 16,514,70
ice-mus LD50:700 ng/kg TOXIA6 22,308,84
unr-mus LD50:9 μg/kg TOXIA6 20,9,82

SAFETY PROFILE: Deadly poison by subcutaneous, intraperitoneal, intracerebral and possibly other routes.

AOO275 CAS:76-43-7 ***HR: 2***
ANDROFLUORENE
mf: $C_{20}H_{29}FO_3$ mw: 336.49

PROP: Crystals. Decomp @ 270°. Sol in pyridine; sltly sol in acetone, chloroform; sparingly sol in methanol; practically insol in water, ether, benzene, and hexanes.

SYNS: ANDROFLUORONE ◇ ANDROSTEROLO ◇ 11-β,17-β-DIHYDROXY-9-α-FLUORO-17-α-METHYL-4-ANDROSTER-3-ONE ◇ FLUORO-9-α DIHYDROXY-11-β,17-β METHYL-17-α ANDROSTENE-4 ONE-3 (FRENCH) ◇ 9-FLUORO-11-β-,17-β-DIHYDROXY-17-METHYLANDROST-4-EN-3-ONE ◇ 9-α-FLUORO-11-β,17-β-DIHYDROXY-17-α-METHYL-4-ANDROSTENE-3-ONE ◇ 9-α-FLUORO-11-β-HYDROXY-17-METHYLTESTOSTERONE ◇ 9-α-FLUORO-17-α-METHYL-11-β,17-DIHYDROXY-4-ANDROSTEN-3-ONE ◇ FLUOTESTIN ◇ FLUOXIMESTERONE ◇ FLUOXYMESTERONE ◇ FLUOXYMESTRONE ◇ FLUSTERON ◇ FLUTESTOS ◇ HALOTESTIN ◇ 17-α-METHYL-9-α-FLUORO-11-β-HYDROXYTESTERONE ◇ NEO-ORMONAL ◇ NSC-12165 ◇ ORALSTERONE ◇ ORATESTIN ◇ ORA-TESTRYL ◇ TESTORAL ◇ U 6040 ◇ ULTANDREN ◇ ULTANDRENE

TOXICITY DATA with REFERENCE
orl-man TDLo:12 mg/kg (84D male):REP JCEMAZ 44,121,77
scu-rat TDLo:6 mg/kg (female 17-20D post):TER AVBIB9 13,71,74
orl-hmn TDLo:400 μg/kg:BIO,SKN,PUL CANCAR 41,758,78
ipr-mus LD50:2350 mg/kg OYYAA2 14,623,77

CONSENSUS REPORTS: EPA Genetic Toxicology Program.

SAFETY PROFILE: Moderately toxic by some routes. Human systemic effects by ingestion: dermatitis, changes in respiratory system and transaminase activity. Human reproductive effects by ingestion on spermatogenesis. An experimental teratogen. Other experimental reproductive effects. When heated to decomposition it emits toxic fumes of F^-.

AOO300 CAS:1239-29-8 ***HR: 2***
ANDROFURAZANOL
mf: $C_{20}H_{30}N_2O_2$ mw: 330.52

PROP: Needles from methanol. Mp: 152-153°.

SYNS: DH 245 ◇ FRAZALON ◇ FURAZABOL ◇ 17-β-HDYROXY-17-α-METHYL-5-α-ANDROSTANO(2,3-c)FURAZAN ◇ 17-METHYL-5-α-ANDROSTANO(2,3-c)(1,2,5)OXADIAZOL-17-β-OL ◇ 17-α-METHYL-5-α-ANDROSTANO(2,3-c)(1,2,5)OXADIAZOL-17-β-OL ◇ MIOTOLON ◇ MYOTOLON

TOXICITY DATA with REFERENCE
scu-mus TDLo:2100 mg/kg (female 7-12D post):REP KSRNAM 4,2088,70
orl-rat TDLo:6 g/kg (13-18D preg):TER KSRNAM 4,2088,70
orl-mus LD50:1731 mg/kg OYYAA2 3,187,69
ipr-mus LD50:494 mg/kg CPBTAL 14,285,66

SAFETY PROFILE: Moderately toxic by ingestion and other routes. An experimental teratogen. Other experimental reproductive effects. When heated to decomposition it emits toxic fumes of NO_x.

AOO375 CAS:4720-09-6 ***HR: 3***
ANDROMEDOTOXIN
mf: $C_{22}H_{36}O_7$ mw: 412.58

SYNS: ACETYLLANDROMEDOL ◇ ASEBOTOXIN ◇ G-I ◇ GRAYANOTOXANE-3,5,6,10,14,16-HEXOL 14 ACETATE ◇ GRAYANOTOXIN I ◇ RHODOTOXIN

TOXICITY DATA with REFERENCE
ipr-mus LD50:1310 μg/kg TXAPA9 35,303,76
scu-mus LD50:148 μg/kg JJPAAZ 6,46,56
ivn-cat LDLo:400 μg/kg JJPAAZ 6,46,56
ivn-rbt LDLo:270 μg/kg JJPAAZ 6,46,56
ivn-gpg LD50:1300 μg/kg ARTODN 44,259,80
par-frg LD50:3899 μg/kg JJPAAZ 6,46,56

SAFETY PROFILE: Poison by subcutaneous, parenteral, intravenous and intraperitoneal routes. When heated to decomposition it emits acrid smoke and fumes.

AOO400 CAS:302-96-5 ***HR: 1***
ANDROSTANAZOL
mf: $C_{21}H_{32}N_2O$ mw: 328.55

SYNS: ANDROSTANAZOLE ◇ 17-β-HYDROXY-17-α-

METHYLANDROSTANO(3,2-c)PYRAZOLE ◇ STANOZOLOL ◇ WIN 14833 ◇ WINSTROL ◇ WINSTROL V

TOXICITY DATA with REFERENCE
orl-rat TDLo:3 g/kg (13-18D preg):TER KSRNAM 4,2088,70
scu-rat TDLo:14 mg/kg (14D pre):REP CCPTAY 5,489,72
orl-wmn TDLo:24 mg/kg/17W-I:SYS BMJOAE 294,612,87
orl-man TDLo:4285 mg/kg/30D-I:SYS BMJOAE 294,612,87

SAFETY PROFILE: Human systemic effects by ingestion: jaundice. An experimental teratogen. Other experimental reproductive effects. When heated to decomposition it emits toxic fumes of NO_x.

AOO410 CAS:2297-30-5 ***HR: 2***
ANDROSTENEDIOL DIPROPIONATE
mf: $C_{25}H_{38}O_4$ mw: 402.63

SYNS: ANDROST-5-ENE-3-β,17-β-DIOL, DIPROPIONATE ◇ ANDROST-5-ENE-3,17-DIOL, DIPROPANOATE, (3-β,17-β)- (9CI) ◇ BISEXOVIS ◇ BISEXOVISTER ◇ GINANDRIN ◇ STENANDIOL

TOXICITY DATA with REFERENCE
scu-rat TDLo:125 mg/kg (female 16D post):TER ANENAG 16,283,55
scu-rat TDLo:125 mg/kg (female 16D post):REP ANENAG 16,283,55
orl-mus LD50:1185 mg/kg PCJOAU 17,30,83

SAFETY PROFILE: Moderately toxic by ingestion. An experimental teratogen. Other experimental reproductive effects. When heated to decomposition it emits acrid smoke and irritating fumes.

AOO425 CAS:63-05-8 ***HR: 2***
ANDROSTENEDIONE
mf: $C_{19}H_{26}O_2$ mw: 286.45

PROP: Dimorphous: Needles from acetone. Mp: 142-144°. Crystals from hexane, mp: 173-174°.

SYNS: Δ-(^{4})-ANDROSTEN-3,17-DIONE ◇ Δ(^{4})-ANDROSTENE-3,17-DIONE ◇ Δ-4-ANDROSTENEDIONE ◇ 4-ANDROSTENE-3,17-DIONE ◇ ANDROTEX ◇ SKF 2170

TOXICITY DATA with REFERENCE
ims-rat TDLo:80 mg/kg (female 14-21D post):REP JCPPAV 92,13,78
ims-rat TDLo:80 mg/kg (female 14-21D post):TER JCPPAV 92,13,78
scu-mus TDLo:600 mg/kg/72W-I:ETA JNCIAM 19,977,57

SAFETY PROFILE: An experimental teratogen. Other experimental reproductive effects. Questionable carcinogen with experimental tumorigenic data. When heated to decomposition it emits acrid smoke and irritating fumes.

AOO450 CAS:53-43-0 ***HR: D***
ANDROSTENOLONE
mf: $C_{19}H_{28}O_2$ mw: 288.47

PROP: Dimorphous: Needles. Mp: 140-141°. Leaflets. mp: 152-153°. Sol in benzene, alc, and ether; sparingly sol in chloroform and petr ether.

SYNS: 17-CHETOVIS ◇ trans-DEHYDROANDROSTERONE ◇ DEHYDROEPIANDROSTERONE ◇ 5-DEHYDROEPIANDROSTERONE ◇ DEHYDROISOANDROSTERONE ◇ 5,6-DEHYDROISOANDROSTERONE ◇ DHA ◇ DIANDRON ◇ DIANDRONE ◇ 5,6-DIDEHYDROISOANDROSTERONE ◇ 17-HORMOFORIN ◇ 3-β-HYDROXY-5-ANDROSTEN-17-ONE ◇ PRASTERONE ◇ PSICOSTERONE

TOXICITY DATA with REFERENCE
scu-rat TDLo:300 mg/kg (female 15-20D post):TER ANENAG 17,118,56
orl-rat TDLo:1800 mg/kg (2-19D preg):REP FESTAS 22,735,71
orl-mus TDLo:25 g/kg/52D-C:NEO CNREA8 48,2788,88

SAFETY PROFILE: An experimental teratogen. Experimental reproductive effects. Questionable carcinogen with experimental neoplastigenic data. When heated to decomposition it emits acrid smoke and irritating fumes.

AOO475 CAS:521-10-8 ***HR: 3***
ANDROSTESTONE-M
mf: $C_{20}H_{32}O_2$ mw: 304.52

PROP: Crystals from ethyl acetate. Mp: 205.5-206.5°. Insol in water. Sltly sol in some organic solvents.

SYNS: ANDRODIOL ◇ ANDROSTESTON-M ◇ CRESTABOLIC ◇ DIOLANDRONE ◇ DIOLOSTENE ◇ ESJAYDIOL ◇ MAD ◇ MADIOL ◇ MASDIOL ◇ MEGABION (JAPANESE) ◇ MESTENEDIOL ◇ METANDIOL ◇ METANDRIOL ◇ METENDIOL ◇ METHANABOL ◇ METHANDIOL ◇ METHANDRIOL ◇ METHANDROLAN ◇ METHOSTAN ◇ METHYLANDROSTENDIOL ◇ METIDIONE ◇ METILDIOLO ◇ METOCRYST ◇ NABADIAL ◇ NEOSTENE ◇ NEOSTERON ◇ NEUTRORMONE ◇ NEUTROSTERON ◇ NOTANDRON ◇ NOTANDRON-DEPOT ◇ PROTANDREN ◇ STENEDIOL ◇ STENIBELL ◇ STENOSTERONE ◇ TESTODIOL ◇ TROFORMONE

TOXICITY DATA with REFERENCE
unr-wmn TDLo:64 mg/kg (female 26-39W post):TER GEFRA2 13,216,53

SAFETY PROFILE: Human reproductive effects by an unspecified route: developmental abnormalities of the urogenital system. A human and experimental teratogen. When heated to decomposition it emits acrid smoke and irritating fumes.

AOO490 CAS:532-11-6 ***HR: 3***
ANETHOLE TRITHIONE
mf: $C_{10}H_8OS_3$ mw: 240.36

PROP: Orange-colored prisms from butyl acetate, very bitter taste. Mp: 111°. Practically insol in water. Sol in pyridine, chloroform, benzene, dioxane, and carbon disulfide; sltly sol in ether, acetone, ethyl acetate, acetic acid, alc, cyclohexane, and petr ether.

SYNS: ANETHOLTRITHION ◇ FELVITEN ◇ HEPORAL ◇ 5-(p-METHOXYPHENYL)-1,2-DITHIOCYCLOPENTEN-3-THIONE ◇ 5-(p-METHOXYPHENYL)-3H-1,2-DITHIOLE-3-THIONE ◇ 5-(4-METHOXYPHENYL)-3H-1,2-DITHIOLE-3-THIONE(9CI) ◇ 5-(p-METHOXYPHENYL)TRITHIONE ◇ MUCINOL ◇ SKF 1717 ◇ SUFRALEM ◇ SULFARLEM ◇ SULFOGAL ◇ SULFRALEM ◇ TIOPROPEN ◇ TIOTRIFAR ◇ TRITHIO ◇ TRITHIOANETHOLE ◇ TRITHIO-(p-METHOXYPHENYL)PROPENE

TOXICITY DATA with REFERENCE
ipr-rat TDLo:20160 mg/kg (24W pre):REP OYYAA2 12,79,76
ims-rat LD50:35 mg/kg AEPPAE 222,244,54
orl-mus LD50:3850 mg/kg NIIRDN 6,25,82
ipr-mus LD50:1780 mg/kg NIIRDN 6,25,82
orl-gpg LD50:6000 mg/kg NIIRDN 6,25,82

SAFETY PROFILE: Poison by intramuscular route. Moderately toxic by ingestion and intraperitoneal routes. Experimental reproductive effects. When heated to decomposition it emits toxic fumes of SO_x.

AOO500 CAS:956-90-1 ***HR: 3***
ANGEL DUST
mf: $C_{17}H_{25}N \cdot ClH$ mw: 279.89

PROP: Crystals. Mp: 46-46.5°, bp: 135-137°.

SYNS: CI395 ◇ CN-25,253-2 ◇ DOA ◇ ELEPHANT TRANQUILIZER ◇ ELYSION ◇ GP-121 ◇ HOG ◇ NSC-40902 ◇ PCP HYDROCHLORIDE ◇ PEACE PILL ◇ PHENCYCLIDINE HYDROCHLORIDE ◇ 1-(1-PHENYLCYCLOHEXYL)PIPERIDINE HYDROCHLORIDE ◇ SERNYL ◇ SERNYLAN ◇ SERNYL HYDROCHLORIDE ◇ TRANK

TOXICITY DATA with REFERENCE
scu-mus TDLo:50 mg/kg (female 6-15D post):REP FEPRA7 42,1157,83
ipr-rat TDLo:240 mg/kg (female 15-20D post):TER PBBHAU 11(Suppl),39,79
orl-hmn TDLo:71 μg/kg:CNS JAMAAP 238,515,77
orl-hmn LDLo:14 mg/kg JAMAAP 238,515,77
ivn-hmn TDLo:10 μg/kg:CNS,PNS CPAJAK 6,150,61
orl-rat LD50:135 mg/kg NETOD7 3,11,81
orl-mus LD50:77 mg/kg JPPMAB 28,713,76
ipr-mus LD50:59558 μg/kg SAAMDZ 2,143,81
scu-mus LD50:38 mg/kg JMCMAR 24,496,81
ivn-mus LD50:16 mg/kg LIFSAK 31,803,82
ivn-dog LDLo:50 mg/kg TXCYAC 19,11,81
orl-pgn LD50:237 mg/kg TXAPA9 21,315,72
orl-dck LD50:75 mg/kg TXAPA9 21,315,72
orl-bwd LD50:5600 μg/kg TXAPA9 21,315,72

SAFETY PROFILE: Poison by ingestion, subcutaneous, intravenous, and intraperitoneal routes. Human systemic effects by ingestion and intravenous routes: euphoria, hallucinations, distorted perceptions, excitement, paresthesia, and changes in motor activity. An experimental teratogen. Other experimental reproductive effects. Often mixed with other drugs of abuse yielding totally unpredictable effects. A controlled substance. When heated to decomposition it emits very toxic fumes of HCl and NO_x.

AOO750 CAS:591-12-8 ***HR: 2***
α-ANGELICA LACTONE
mf: $C_5H_6O_2$ mw: 98.11

SYN: 4-HYDROXYPENT-3-ENOIC ACID LACTONE

TOXICITY DATA with REFERENCE
orl-mus LD50:2800 mg/kg DCTODJ 3,249,80
ipr-mus LD50:3000 mg/kg APTOA6 2,109,46

CONSENSUS REPORTS: Reported in EPA TSCA Inventory.

SAFETY PROFILE: Moderately toxic by ingestion and intraperitoneal routes. When heated to decomposition it emits acrid smoke and irritating fumes.

AOO760 CAS:8015-64-3 ***HR: 2***
ANGELICA OIL, root

SYNS: ANGELIKA OEL ◇ ANGELICA ROOT OIL ◇ OILS, ANGELICA ROOT

TOXICITY DATA with REFERENCE
orl-rat LD50:11000 mg/kg FCTXAV 13,713,75
orl-mus LD50:2200 mg/kg FCTXAV 13,713,75

PROP: Extracted from roots of *Angelica archangelica L*. A pale yellow to amber liquid; pungent odor with bitter-sweet taste. Sol in fixed oils; sltly sol in mineral oil; insol in glycerin, propylene glycol.

CONSENSUS REPORTS: Reported in EPA TSCA Inventory.

SAFETY PROFILE: Moderately toxic by ingestion. When heated to decomposition it emits acrid smoke and irritating fumes.

AOO800 CAS:554-18-7 ***HR: 3***
ANGELI'S SULFONE
mf: $C_{24}H_{34}N_2O_{18}S_3 \cdot 2Na$ mw: 780.76

PROP: White, amorphous powder. Sol in water; sltly sol in alc; insol in ether, benzene, methanol, ethyl acetate, and pyridine.

SYNS: ACEPROSOL ◇ ANGELI SULFONE ◇ p,p'-DIAMINODIPHENYLSULFONE-N,N-DI(DEXTROSE SODIUM SULFONATE) ◇ DISODIUM p,p'-DIAMINODIPHENYLSULFONE-N,N'-DIGLUCOSE SULFONATE ◇ d-GLUCITOL, 1,1'-(SULFONYLBIS(4,1-PHENYLENEIMINO))BIS(1-DEOXY-1-SULFO-, DISODIUM SALT (9CI) ◇ GLUCOSULFONE ◇ GLUCOSULFONE SODIUM ◇ 501 P ◇ PROMANIDE ◇ PROMIN ◇ PROMIN SODIUM ◇ PROMOTIN ◇ PROTOMIN ◇ 501 SIEGFRIED ◇ S. N. 166 ◇ SODIUM GLU-

COSULFONE ◇ SULFONA P ◇ SOLFONE ◇ 1,1'-(SULFONYLBIS(4,1-PHENYLENEIMINO))BIS(1-DEOXY-1-SULFO-d-GLUCITOL)DISODIUM SALT ◇ p,p'-SULFONYLDIANILINE N,N'-DIGLUCOSIDE DISODIUM DISULFONALTE ◇ p,p'-SULFONYLDIANILINE-N,N'-DI-d-GLUCOSE SODIUM BISULFITE ◇ TASMIN

TOXICITY DATA with REFERENCE
orl-mus LD50:3930 μg/kg NIIRDN 6,225,82
scu-mus LD50:6500 μg/kg NIIRDN 6,225,82
ivn-mus LD50:5250 μg/kg NIIRDN 6,225,82

SAFETY PROFILE: Poison by ingestion, subcutaneous, and intravenous routes. When heated to decomposition it emits toxic fumes of SO_x, NO_x, and Na_2O.

AOO825 ***HR: 2***
ANGEL'S TRUMPET

PROP: A small tree or large shrub which may grow to 20 feet. The large flowers are funnel shaped, grow to 10 or 12 inches long and may be white, yellow-white or pink.

SYNS: BELLADONA (HAWAII) ◇ BRUGMANSIA ARBOREA ◇ BRUGMANSIA X CANDIDA ◇ BRUGMANSIA SANGUINEA ◇ BRUGMANSIA SUAVEOLENS ◇ CAMPANA (CUBA, PUERTO RICO) ◇ CORNUCOPIA ◇ FLORIPONDIO (PUERTO RICO) ◇ NANA-HONUA (HAWAII)

SAFETY PROFILE: The whole plant contains poisonous belladonna alkaloids. The seeds and dried leaves are used as hallucinogens. Ingestion may cause fever, increased heart rate, dilated pupils, delirium, and high blood pressure. See also BELLADONNA.

AOO875 CAS:131-49-7 ***HR: 3***
ANGIGRAFIN
mf: $C_{11}H_9I_3N_2O_4 \cdot C_7H_{17}NO_5$ mw: 809.17

PROP: Rhombic needles, sltly sweet taste. Mp: 189-193° (decomp). Solubility in water at 20°: 89 g/100 mL.

SYNS: AMIDOTRIZOATE MEGLUMINE ◇ ANGIOGRAFIN ◇ BENZOIC ACID, 3,5-DIACETAMIDO-2,4,6-TRIIODO-, compd. with 1-DEOXY-1-(METHYLAMINO)-d-GLUCITOL ◇ CARDIOGRAFIN ◇ CYSTOGRAFIN ◇ DIATRIZOATE MEGLUMINE ◇ DIATRIZOATE METHYLGLUCAMINE ◇ DITRIZOATE METHYLGLUCAMINE ◇ GASTROGRAFIN ◇ d-GLUCITOL, 1-DEOXY-1-(METHYLAMINO)-, 3,5-BIS(ACETYLAMINO)-2,4,6-TRIIODOBENZOATE (SALT) ◇ HYPAQUE 13.4 ◇ HYPAQUE 60 ◇ HYPAQUE CYSTO ◇ HYPAQUE M 30 ◇ HYPAQUE MEGLUMINE ◇ MEGLUMINE AMIDOTRIZOATE ◇ MEGLUMINE DIATRIZOATE ◇ METHYLGLUCAMINE DIATRIXOATE ◇ RENOGRAFFIN M-76 ◇ RENOGRAFIN ◇ RENO M ◇ RENO M 60 ◇ RENO-M-DIP ◇ RENURIX ◇ UNIPAQUE ◇ UROVIST

TOXICITY DATA with REFERENCE
scu-man TDLo:214 mg/kg 34ZIAG -,392,69
ivn-rat LD50:15300 mg/kg YACHDS 12(Suppl 1),11,84
ivn-mus LD50:21200 mg/kg NIIRDN 6,32,82
ice-mus LD50:80 mg/kg THERAP 26,595,71

SAFETY PROFILE: Poison by intracerebral route. Human systemic effects by subcutaneous route: kidney damage and reduced urine volume. When heated to decomposition it emits toxic fumes of I^- and NO_x.

AOO900 CAS:1407-47-2 ***HR: 3***
ANGIOTONIN

PROP: Hydrolyzed by strong acids and bases and above pH 9.5. Sol in organic solvents, in aq solns pH 5-8.

SYNS: ANGIOTENSIN ◇ HYPE[illegible]ENSIN

TOXICITY DATA with REFERENCE
scu-ham TDLo:200 μg/kg (female 8D post):TER LIFSAK 8,525,69
scu-ham TDLo:20 μg/kg (female 8D post):REP LIFSAK 8,525,69
ivn-rat LDLo:8 mg/kg 27ZIAQ -,43,73

SAFETY PROFILE: Poison by intravenous route. An experimental teratogen. Other experimental reproductive effects. When heated to decomposition it emits acrid smoke and irritating fumes.

AOO925 CAS:1402-83-1 ***HR: 3***
ANGOLAMYCIN
mf: $C_{46}H_{77}NO_{17}$ mw: 916.24

TOXICITY DATA with REFERENCE
ipr-mus LDLo:280 mg/kg 85FZAT -,135,67
scu-mus LD50:500 mg/kg 85FZAT -,135,67
unr-mus LDLo:1000 mg/kg 85ERAY 1,74,78

SAFETY PROFILE: Poison by several routes. When heated to decomposition it emits toxic fumes of NO_x.

AOP250 CAS:2270-40-8 ***HR: 3***
ANGUIDIN
mf: $C_{18}H_{26}O_7$ mw: 354.44

SYNS: ANG 66 ◇ ANGUIDINE ◇ DAS ◇ 4-β,15-DIACETOXY-3-α-HYDROXY-12,13-EPOXYTRICHOTHEC-9-ENE ◇ DIACETOXYSCIRPENOL ◇ 4,15-DIACETOXYSCIRPEN-3-OL ◇ DIAZETOXYSKIRPENOL (GERMAN) ◇ 12,13-EPOXY-4-β,15-DIAZETOXY-3-α-HYDROXY-TRICHOTHEC-9-ENE ◇ (3-α,4-β)-12,13-EPOXY-4,15-DIACETATE-TRICHOTHEC-9-ENE-3,4,15-TRIOL ◇ MM 4462 ◇ NSC-141537

TOXICITY DATA with REFERENCE
skn-gpg 284 ng MLD FAATDF 4,S124,84
dns-rat-orl 3 mg/kg CALEDQ 38,199,87
dnd-mus-ivn 5600 μg/kg PAACA3 19,65,78
ipr-mus TDLo:1 mg/kg (9D post):TER TXCYAC 45,245,87
ipr-mus TDLo:1 mg/kg (9D post):REP TXCYAC 45,245,87
ipr-wmn TDLo:12 mg/kg/5D:GIT CTRRDO 63,789,79
orl-rat LD50:7 mg/kg VHTODE 25,335,83
ipr-rat LD50:750 μg/kg DFSCDX 4,135,83
ivn-rat LD50:1300 μg/kg ARZNAD 18,989,68
orl-mus LD50:7300 μg/kg BIBIAU 10,445,68
ihl-mus LD50:11300 μg/kg TOXID9 4,12,84
ipr-mus LD50:7839 μg/kg NCISP* JAN86

CONSENSUS REPORTS: EPA Genetic Toxicology Program.

SAFETY PROFILE: A deadly poison by ingestion, inhalation, intravenous, intraperitoneal, and subcutaneous routes. Human systemic effects by intraperitoneal route: muscle weakness, nausea or vomiting, and fever. An experimental teratogen. Other experimental reproductive effects. Mutation data reported. A skin irritant. When heated to decomposition it emits acrid smoke and fumes.

AOP500 ***HR: 2***
ANHYDRIDES

PROP: Chemical compounds derived from acids by elimination of a molecule of water. Thus, sulfur trioxide (SO_3) is the anhydride of sulfuric acid (H_2SO_4); carbon dioxide (CO_2) is the anhydride of carbonic acid (H_2CO_3); phthalic acid ($C_6H_4(CO_2H)_2$) minus water gives phthalic anhydride ($C_6H_4(CO_2)O$). This term should not be confused with anhydrous, meaning without water.

SAFETY PROFILE: Anhydrides are acidic and react with bases in tissue. Thus, they tend to attack and irritate tissue.

AOP750 CAS:35891-69-1 ***HR: 3***
ANHYDROMYRIOCIN
mf: $C_{21}H_{37}NO_5$ mw: 383.4

TOXICITY DATA with REFERENCE
ipr-rat LD50:37 mg/kg 85ERAY 3,2067,78
orl-mus LD50:100 mg/kg 85ERAY 3,2067,78
ipr-mus LD50:75 mg/kg 85ERAY 3,2067,78

SAFETY PROFILE: Poison by ingestion and intraperitoneal routes. When heated to decomposition it emits toxic fumes such as NO_x.

AOQ000 CAS:62-53-3 ***HR: 3***
ANILINE
DOT: UN 1547
mf: C_6H_7N mw: 93.14

PROP: Colorless, oily liquid; characteristic odor. Bp: 184.4°, lel: 1.3%, ULC: 20-25, flash p: 158°F (CC), fp: −6.2°, d: 1.02 @ 20°/4°, autoign temp: 1139°F, vap press: 1 mm @ 34.8°, vap d: 3.22.

SYNS: AMINOBENZENE ◇ AMINOPHEN ◇ ANILIN (CZECH) ◇ ANILINA (ITALIAN, POLISH) ◇ ANILINE OIL ◇ BENZENAMINE ◇ BLUE OIL ◇ C.I. 76000 ◇ HUILE d'ANILINE (FRENCH) ◇ NCI-C03736 ◇ PHENYLAMINE

TOXICITY DATA with REFERENCE
skn-rbt 500 mg/24H MOD 28ZPAK -,65,72
eye-rbt 102 mg SEV BIOFX* 1-5/69
mma-sat 100 μg/plate PJABDW 53,34,77
dnr-esc 25 μL/well/16H CBINA8 15,219,76
bfa-rat/sat 300 mg/kg MUREAV 79,173,80
orl-mus TDLo:4480 mg/kg (female 6-13D post):REP TCMUD8 7,29,87
orl-rat TDLo:11 g/kg/29W-C:NEO APMIAL 26,473,49
unk-hmn LDLo:357 mg/kg JIDHAN 13,87,31
unk-man LDLo:150 mg/kg 85DCAI 2,73,70
orl-rat LD50:250 mg/kg JPETAB 90,260,47
ihl-rat LCLo:250 ppm/4H JIHTAB 31,343,49
skn-rat LD50:1400 mg/kg AGGHAR 15,447,57
ipr-rat LD50:420 mg/kg AGGHAR 15,447,57
ihl-mus LC50:175 ppm/7H NTIS** PB214-270
ipr-mus LD50:492 mg/kg IZSBAI 3,91,65
scu-muc LD50:200 mg/kg ARZNAD 8,107,58
orl-dog LD50:195 mg/kg NTIS** PB214-270
skn-dog LDLo:1540 mg/kg NTIS** PB214-270
ihl-cat LCLo:180 ppm/8H XPHBAO 271,4,41

CONSENSUS REPORTS: IARC Cancer Review: Group 3 IMEMDT 7,99,87; Animal Inadequate Evidence IMEMDT 4,27,74; Human No Evidence IMEMDT 4,27,74. EPA Extremely Hazardous Substances List. Community Right-To-Know List. Reported in EPA TSCA Inventory.

OSHA PEL: (Transitional: TWA 5 ppm (skin)) TWA 2 ppm (skin)
ACGIH TLV: TWA 2 ppm (skin); BEI: 50 mg/L total p-aminophenol in urine at end of shift.
DFG MAK: 2 ppm (8 mg/m^3), Suspected Carcinogen; BAT: 1 mg/L in urine at end of shift.
DOT Classification: Poison B; Label: Poison

SAFETY PROFILE: Suspected carcinogen with experimental neoplastigenic data. A human poison by an unspecified route. Poison experimentally by most routes including inhalation and ingestion. Experimental reproductive effects. A skin and severe eye irritant, and a mild sensitizer. In the body, aniline causes formation of methemoglobin, resulting in prolonged anoxemia and depression of the central nervous system; less acute exposure causes hemolysis of the red blood cells, followed by stimulation of the bone marrow. The liver may be affected with resulting jaundice. Long-term exposure to aniline dye manufacture has been associated with malignant bladder growths. A common air contaminant. Moderately flammable when exposed to heat or flame. To fight fire, use alcohol foam, CO_2, dry chemical. It can react vigorously with oxidizing materials. When heated to decomposition it emits highly toxic fumes of NO_x. Spontaneously explosive reactions occur with benzenediazonium-2-carboxylate, dibenzoyl peroxide, fluorine nitrate, nitrosyl perchlorate, red fuming nitric acid, peroxodisulfuric acid, and tetranitromethane. Violent reactions with boron trichloride, peroxyformic acid, diisopropyl peroxydicarbonate, fluorine, trichloronitromethane (145°C), acetic anhydride, chlorosulfonic acid, hexachloromelamine, (HNO_3 + N_2O_4 + H_2SO_4), (nitrobenzene + glycerin), oleum, (HCHO + $HClO_4$), perchromates, K_2O_2, β-propiolactone, $AgClO_4$, Na_2O_2, H_2SO_4, trichloromelamine, acids, peroxydisulfuric acid, FO_3Cl, diisopropyl peroxy-dicarbonate, n-haloimides, and trichloronitromethane. Ignites on contact with sodium peroxide + water. Forms heat- or shock-sensitive explosive mixtures with anilinium chloride (detonates at 240°C/7.6 bar), nitromethane, hydrogen peroxide, 1-chloro-2,3-epoxypropane, and peroxomonosulfuric acid. Reactions with perchloryl fluoride, perchloric acid, and ozone form explosive products.

AOQ250 CAS:1300-14-7 ***HR: 3***
ANILINE ANTIMONYL TARTRATE
mf: $C_6H_8N \cdot C_4H_4O_7Sb$ mw: 379.98

PROP: White crystals.

SYN: ANTIMONYL ANILINE TARTRATE

TOXICITY DATA with REFERENCE
ipr-mus LD50:81 mg/kg AJTMAQ 25,263,45

CONSENSUS REPORTS: Antimony and its compounds are on the Community Right-To-Know List.

OSHA PEL: TWA 0.5 mg(Sb)/m^3
ACGIH TLV: TWA 0.5 mg(Sb)/m^3
NIOSH REL: (Antimony) TWA 0.5 mg(Sb)/m^3

SAFETY PROFILE: Poison by intraperitoneal route. See also ANTIMONY COMPOUNDS and ANILINE. When heated to decomposition it emits very toxic fumes of Sb and NO_x.

AOQ500 ***HR: 2***
ANILINE DYES

SAFETY PROFILE: The finished dyes are generally very much less toxic than many of the intermediates occurring or used in the manufacture of the dyes. Some of the aniline dyes cause local irritating effects to the eyes, mucous membranes, and skin; the basic dyes are believed to be more irritating than the acid dyes. Allergic responses to aniline dyes have been known to occur. See also specific compounds. When heated to decomposition they emit toxic fumes of NO_x and possibly SO_x.

AOQ875 CAS:553-27-5 ***HR: 3***
ANILINE MUSTARD
mf: $C_{10}H_{13}Cl_2N$ mw: 218.14

PROP: Stout prisms from methanol. Mp: 45°, bp: (14) 164°. Sol in hot methanol and ethanol; very sltly sol in ether.

SYNS: N,N-BIS(2-CHLOROETHYL)ANILINE ◇ N,N-BIS(2-CHLOROETHYL)BENZENAMINE ◇ β,β'-DICHLORODIETHYLANILINE ◇ N,N-DI(2-CHLOROETHYL)ANILINE ◇ LYMPHCHIN ◇ LYMPHOCIN ◇ LYMPHOQUIN ◇ NSC-18429 ◇ PHENYLBIS(2-CHLOROETHYLAMINE) ◇ TL 476

TOXICITY DATA with REFERENCE
dnd-mus:lym 30 μmol/L CNREA8 44,78,84
orl-rat LD50:239 mg/kg NCIMR* -,469,69
ipr-rat LD50:141 mg/kg BCPCA6 13,969,64
orl-mus LD50:123 mg/kg NCIMR* -,469,69
ihl-mus LCLo:500 mg/m^3/10M NDRC** NDCRC-132,Dec,42
ipr-mus LD50:52 mg/kg NCIMR* -,469,69

CONSENSUS REPORTS: EPA Genetic Toxicology Program.

SAFETY PROFILE: Poison by inhalation, ingestion, and intraperitoneal routes. Mutation data reported. When heated to decomposition it emits toxic fumes of Cl^- and NO_x. See also ANILINE DYES.

AOR000 ***HR: 3***
ANILINE OIL DRUMS, EMPTY

SAFETY PROFILE: Combustible if full of vapors, such drums may ignite under the proper conditions. A dangerous disaster hazard if many drums are involved. They emit highly toxic fumes of aniline. See ANILINE.

AOR250 ***HR: 3***
ANILINE VANADATE, DIHYDRATE
mf: $C_6H_7N_2O_5V_2 \cdot 2H_2O$ mw: 325.07

TOXICITY DATA with REFERENCE
scu-rat LDLo:68 mg/kg AJSNAO 1,347,17
scu-mus LDLo:128 mg/kg AJSNAO 1,347,17
scu-gpg LDLo:2 mg/kg AJSNAO 1,347,17

ACGIH TLV: TWA 0.05 mg(V_2O_5)/m^3
NIOSH REL: (Vanadium Compounds) CL 0.05 mg(V)/m^3/15M

SAFETY PROFILE: Poison by subcutaneous route. See also VANADIUM COMPOUNDS. When heated to decomposition it emits toxic fumes of NO_x and VO_x.

AOR500 CAS:548-62-9 ***HR: 3***
ANILINE VIOLET
mf: $C_{25}H_{30}N_3 \cdot Cl$ mw: 408.03

SYNS: AIZEN CRYSTAL VIOLET EXTRA PURE ◇ GENTIAN VIOLET ◇ HEXAMETHYL-p-ROSANILINE HYDROCHLORIDE ◇ HEXAMETHYL VIOLET ◇ METHYLROSANILINE CHLORIDE ◇ NCI-C55969

TOXICITY DATA with REFERENCE
skn-hmn 3 mg/3D-I MLD 85DKA8 -,127,77
skn-hmn 2 mg/2D-I MLD ADVEA4 52,55,72
skn-gpg 6 mg/3D-I ADVEA4 52,55,72
mmo-sat 100 ng/plate MUREAV 89,21,81
cyt-hmn:hla 500 μg/L MUREAV 58,269,78

cyt-hmn:lym 500 μg/L MUREAV 58,269,78
orl-rbt TDLo:7 mg/kg (6-19D post):TER NTIS** PB83-182519
orl-rat TDLo:40 mg/kg (6-15D post):REP NTIS** PB83-155754
orl-mus TDLo:25750 mg/kg/2Y-C:CAR FAATDF 5,902,85
orl-rat LD50:420 mg/kg ARZNAD 1,5,51
ipr-rat LD50:8900 μg/kg ARZNAD 1,5,51
orl-mus LD50:96 mg/kg ARZNAD 1,5,51
ipr-mus LD50:5100 μg/kg ARZNAD 1,5,51
ivn-mus LDLo:20 mg/kg PSEBAA 31,825,34

CONSENSUS REPORTS: Reported in EPA TSCA Inventory.

SAFETY PROFILE: Poison by ingestion, intravenous, and intraperitoneal routes. An experimental teratogen. Other experimental reproductive effects. A human skin irritant. Human mutation data reported. Questionable carcinogen with experimental carcinogenic data. When heated to decomposition it emits very toxic fumes of NO_x and Cl^-.

AOR625 ***HR: 3***
ANILINIUM NITRATE
mf: $C_6H_8N_2O_3$ mw: 156.14

SAFETY PROFILE: Potentially hypergolic reaction with concentrated nitric acid is promoted by: ammonium or sodium vanadates; copper(I) chloride; potassium permanganate; sodium pentacyanonitrosylferrate; and vanadium (V) oxide. When heated to decomposition it emits toxic fumes of NO_x. See also ANILINE and NITRATES.

AOR630 ***HR: 3***
ANILINIUM PERCHLORATE
mf: $C_6H_8ClNO_4$ mw: 193.59

SAFETY PROFILE: Mixtures with metal oxides (e.g., manganese dioxide, copper oxide, and nickel oxide) are heat-sensitive explosives. When heated to decomposition it emits toxic fumes of Cl^- and NO_x. See also ANILINE and PERCHLORATES.

AOR640 CAS:5410-78-6 ***HR: 3***
4-ANILINODICHLOROARSINE, HYDROCHLORIDE
mf: $C_6H_6AsCl_2N{\cdot}ClH$ mw: 274.41

SYNS: ANILINE, p-DICHLOROARSINO-, HYDROCHLORIDE ◇ ARSINE, (p-AMINOPHENYL)DICHLORO-, HYDROCHLORIDE ◇ p-DICHLOROARSINOANILINE HYDROCHLORIDE

TOXICITY DATA with REFERENCE
ivn-mus LD50:6300 μg/kg CSLNX* NX#05113

OSHA PEL: TWA 0.5 mg(As)/m^3

SAFETY PROFILE: Poison by intravenous route. When heated to decomposition it emits toxic fumes of NO_x, As, Cl^-, and HCl.

AOR750 CAS:122-98-5 ***HR: 3***
2-ANILINOETHANOL
mf: $C_8H_{11}NO$ mw: 137.20

PROP: D: 1.1, bp: 268°, flash p: 305°F (OC).

SYNS: N-(2-HYDROXYETHYL)PHENYLAMINE ◇ PHENYL ETHANOLAMINE ◇ N-PHENYLETHANOLAMINE ◇ 2-(PHENYLAMINO)ETHANOL

TOXICITY DATA with REFERENCE
skn-rbt 545 mg open MLD UCDS** 8/21/61
eye-rbt 5 mg SEV AJOPAA 29,1363,46
orl-rat LD50:2230 mg/kg UCDS** 8/21/61
ipr-mus LDLo:176 mg/kg JAMAAP 123,761,43
scu-dog LDLo:220 mg/kg JAMAAP 123,761,43
ivn-dog LDLo:165 mg/kg JAMAAP 123,761,43
skn-rbt LD50:63 mg/kg AIHAAP 23,95,62
ivn-rbt LDLo:44 mg/kg JAMAAP 123,761,43

CONSENSUS REPORTS: Reported in EPA TSCA Inventory.

SAFETY PROFILE: Poison by skin contact, intraperitoneal, and intravenous routes. Moderately toxic by ingestion. A skin and severe eye irritant. Combustible when exposed to heat or flame. To fight fire, use dry chemical, water mist. When heated to decomposition it emits toxic fumes of NO_x.

AOS000 CAS:67227-20-7 ***HR: 3***
(2-ANILINOETHYL)HYDRAZONE DIHYDROCHLORIDE
mf: $C_8H_{13}N_3{\cdot}2ClH$ mw: 224.16

TOXICITY DATA with REFERENCE
orl-mus LD50:200 mg/kg JMCMAR 6,63,63
ipr-mus LD50:175 mg/kg JMCMAR 6,63,63

SAFETY PROFILE: Poison by ingestion and intraperitoneal routes. When heated to decomposition it emits very toxic fumes of NO_x and HCl.

AOS500 CAS:88-35-7 ***HR: 2***
2-ANILINO-5-NITROBENZENESULFONIC ACID
mf: $C_{12}H_9N_2O_5S{\cdot}Na$ mw: 316.28

SYN: 4-NITRODIFENYLAMIN-2-SULFONAN SODYN (CZECH)

TOXICITY DATA with REFERENCE
skn-rbt 500 mg/24H MLD 28ZPAK -,191,72
eye-rbt 250 μg/24H SEV 28ZPAK -,191,72
orl-rat LD50:2200 mg/kg 28ZPAK -,191,72

SAFETY PROFILE: Moderately toxic by ingestion. A mild skin and severe eye irritant. When heated to decomposition it emits very toxic fumes of SO_x, Na_2O, and NO_x.

AOS750 CAS:101-59-7 ***HR: 3***
ANILINO (p-NITROPHENYL) SULFIDE
mf: $C_{12}H_{10}N_2O_2S$ mw: 246.30

SYN: 4-AMINO-4'-NITRODIPHENYL SULFIDE

TOXICITY DATA with REFERENCE
mmo-sat 25 μg/plate MUREAV 67,123,79
mma-sat 10 μg/plate MUREAV 67,123,79
ivn-mus LD50:180 mg/kg CSLNX* NX#00435

SAFETY PROFILE: Poison by intravenous route. Mutation data reported. See also SULFIDES and NITRO COMPOUNDS of AROMATIC HYDROCARBONS. When heated to decomposition it emits very toxic fumes of NO_x and SO_x.

AOT000 CAS:122-37-2 ***HR: 2***
p-ANILINOPHENOL
mf: $C_{12}H_{11}NO$ mw: 185.24

PROP: Gray, solid leaflets. Mp: 70°, bp: 330°.

SYNS: 4-ANILINOPHENOL ◇ p-HYDROXYDIFENYLAMIN (CZECH) ◇ p-HYDROXYDIPHENYLAMINE ◇ 4-HYDROXYDIPHENYLAMINE ◇ p-OXYDIPHENYLAMINE ◇ 4-(PHENYLAMINO)-PHENOL ◇ N-PHENYL-p-AMINOPHENOL ◇ PHENYL-p-AMINOPHENOL ◇ VTI 1

TOXICITY DATA with REFERENCE
eye-rbt 250 μg/24H SEV 28ZPAK -,111,72
orl-rat LD50:1220 mg/kg 28ZPAK -,111,72
orl-mus LD50:2310 mg/kg GTPZAB 13(7),28,69

CONSENSUS REPORTS: Reported in EPA TSCA Inventory.

SAFETY PROFILE: Moderately toxic by ingestion. A severe eye irritant. See also AROMATIC AMINES. When heated to decomposition it emits toxic fumes of NO_x.

AOT125 CAS:17615-73-5 ***HR: 2***
6-(p-ANILINOSULFONYL)METANILAMIDE
mf: $C_{12}H_{13}N_3O_4S_2$ mw: 327.40

SYNS: SDDS ◇ 2-SOLFAMONYL-4,4'-DIAMINOPHENYLSULFONE

TOXICITY DATA with REFERENCE
orl-rat TDLo:24 g/kg (9-14D preg):TER OYYAA2 9,695,75
orl-rat TDLo:1800 mg/kg (9-14D preg):REP OYYAA2 9,695,75
ipr-mus LD50:5150 mg/kg OYYAA2 2,184,68

SAFETY PROFILE: Moderately toxic by intraperitoneal route. An experimental teratogen. Other experimental reproductive effects. When heated to decomposition it emits toxic fumes of SO_x and NO_x.

AOT250 ***HR: 3***
ANILITE

SAFETY PROFILE: A highly explosive mixture composed of liquid NO_2 and carbon disulfide or gasoline. Extremely sensitive to shock.

AOT525 CAS:135-02-4 ***HR: 2***
o-ANISALDEHYDE

PROP: Mp: 37-39°, bp: 238°, d: 1.127, flash p: 244° F.
mf: $C_8H_8O_2$ mw: 136.16

SYNS: 2-ANISALDEHYDE ◇ BENZALDEHYDE, 2-METHOXY-(9CI) ◇ o-METHOXYBENZALDEHYDE ◇ 2-METHOXYBENZALDEHYDE ◇ 6-METHOXYBENZALDEHYDE ◇ 2-METHOXYBENZENECARBOXALDEHYDE ◇ SALICYLALDEHYDE METHYL ETHER

TOXICITY DATA with REFERENCE
skn-rbt 500 mg/24H MOD FCTXAV 17,855,79
sce-hmn:lyms 125 umol/L MUREAV 206,17,88
orl-rat LD50:2500 mg/kg FCTXAV 17,855,79

CONSENSUS REPORTS: Reported in EPA TSCA Inventory.

SAFETY PROFILE: Moderately toxic by ingestion. A skin irritant. Mutation data reported. Combustible liquid. When heated to decomposition it emits acrid smoke and irritating fumes.

AOT530 CAS:123-11-5 ***HR: 2***
p-ANISALDEHYDE
mf: $C_8H_8O_2$ mw: 136.15

PROP: Colorless oil; hawthorn odor. D: 1.123 @ 20°/4°, refr index: 1.571-1.574, mp: 2.5°, bp: 247-248°, flash p: 250°F. Misc in alc, ether, fixed oils; sol in propylene glycol; insol in glycerin and water.

SYNS: ANISIC ALDEHYDE ◇ FEMA No. 2670 ◇ 4-METHOXYBENZALDEHYDE ◇ p-METHOXYBENZALDEHYDE (FCC)

TOXICITY DATA with REFERENCE
skn-rbt 500 mg/24H MOD FCTXAV 12,807,74
mmo-sat 400 μL/plate BECTA6 24,590,80
orl-rat LD50:1510 mg/kg FCTXAV 2,327,64
orl-gpg LD50:1260 mg/kg FCTXAV 2,327,64

CONSENSUS REPORTS: Reported in EPA TSCA Inventory.

SAFETY PROFILE: Moderately toxic by ingestion. A skin irritant. Mutation data reported. Combustible liquid. When heated to decomposition it emits acrid smoke and irritating fumes.

AOT750 CAS:2439-77-2 ***HR: 2***
o-ANISAMIDE
mf: $C_8H_9NO_2$ mw: 151.18

SYNS: o-METHOXYBENZAMIDE ◇ 2-METHOXYBENZAMIDE

TOXICITY DATA with REFERENCE
ipr-rat LD50:450 mg/kg JPETAB 108,450,53
orl-mus LD50:1200 mg/kg JPPMAB 4,872,52
ipr-mus LD50:900 mg/kg JPPMAB 4,872,52

SAFETY PROFILE: Moderately toxic by ingestion and intraperitoneal routes. When heated to decomposition it emits toxic fumes of NO_x.

AOU250 CAS:8007-70-3 ***HR: 2***
ANISE OIL

PROP: Consists of (80-90%) of Anethole and small quantities of methyl chavicol, p-methoxyacetophenone, and other materials. Found in the dried ripe fruit of *Impinella anisum L.* (FCTXAV 11,855,73). D: 0.978-0.988 @ 25°/25°.

SYNS: ANISEED OIL ◇ ANIS OEL (GERMAN) ◇ OIL of ANISE ◇ STAR ANISE OIL

TOXICITY DATA with REFERENCE
orl-rat LD50:2250 mg/kg FCTXAV 11,855,73

CONSENSUS REPORTS: Reported in EPA TSCA Inventory.

SAFETY PROFILE: Moderately toxic by ingestion. A weak sensitizer. May cause contact dermatitis. Combustible liquid. When heated to decomposition it emits acrid smoke and irritating fumes.

AOU500 CAS:586-38-9 ***HR: 3***
m-ANISIC ACID
mf: $C_8H_8O_3$ mw: 152.15

PROP: Needles from aq solns. Mp: 107-109°, bp: 170-172° @ 10 mm. Sol in hot water, alc, and ether.

SYNS: m-METHOXYBENZOIC ACID ◇ 3-METHOXYBENZOIC ACID

TOXICITY DATA with REFERENCE
ipr-mus LDLo:250 mg/kg CBCCT* 6,144,54

CONSENSUS REPORTS: Reported in EPA TSCA Inventory.

SAFETY PROFILE: Poison by intraperitoneal route. When heated to decomposition it emits acrid smoke and irritating fumes.

AOV000 CAS:94-30-4 ***HR: 2***
p-ANISIC ACID, ETHYL ESTER
mf: $C_{10}H_{12}O_3$ mw: 180.21

PROP: Colorless liquid; fruity, anise odor. D: 1.103 @ 25/25, refr index: 1.522-1.526, mp: 7-8°, bp: 269-270°, flash p: +212°F. Sol in alc and ether; sltly sol in water.

SYNS: ETHYL ANISATE ◇ ETHYL-p-ANISATE (FCC) ◇ ETHYL-4-METHOXYBENZOATE ◇ ETHYL-p-METHOXYBENZOATE ◇ FEMA No. 2420

TOXICITY DATA with REFERENCE
orl-rat LD50:2040 mg/kg FCTXAV 14,659,76

CONSENSUS REPORTS: Reported in EPA TSCA Inventory.

SAFETY PROFILE: Moderately toxic by ingestion. See also ESTERS. Combustible liquid. When heated to decomposition it emits acrid smoke and irritating fumes.

AOV250 CAS:7466-54-8 ***HR: 3***
o-ANISIC ACID, HYDRAZIDE
mf: $C_8H_{10}N_2O_2$ mw: 166.20

SYNS: o-METHOXYBENZOHYDRAZIDE ◇ o-METHOXYBENZOIC ACID HYDRAZIDE ◇ 2-METHOXYBENZOIC ACID HYDRAZIDE ◇ o-METHOXYBENZOYLHYDRAZIDE ◇ 2-METHOXYBENZOYL HYDRAZIDE ◇ 2-METHOXYBENZOYLHYDRAZINE

TOXICITY DATA with REFERENCE
orl-mus TDLo:6000 mg/kg/22W-I:NEO 34ZRA9 -,869,66

SAFETY PROFILE: Questionable carcinogen with experimental neoplastigenic data. When heated to decomposition it emits toxic fumes of NO_x.

AOV500 CAS:3290-99-1 ***HR: 3***
p-ANISIC ACID, HYDRAZIDE
mf: $C_8H_{10}N_2O_2$ mw: 166.20

SYNS: ANISIC ACID HYDRAZIDE ◇ ANISIC HYDRAZIDE ◇ ANISOYLHYDRAZINE ◇ p-ANISOYLHYDRAZINE ◇ p-METHOXYBENZOIC ACID HYDRAZIDE ◇ 4-METHOXYBENZOIC ACID HYDRAZIDE ◇ p-METHOXYBENZOIC HYDRAZIDE ◇ 4-METHOXYBENZOYL HYDRAZIDE ◇ (p-METHOXYBENZOYL)HYDRAZINE ◇ 4-METHOXYBENZOYLHYDRAZINE

TOXICITY DATA with REFERENCE
orl-mus TDLo:12 g/kg/22W-I:NEO 34ZRA9 -,869,66
ivn-mus LD50:178 mg/kg CSLNX* NX#00894

SAFETY PROFILE: Poison by intravenous route. Questionable carcinogen with experimental neoplastigenic data. When heated to decomposition it emits toxic fumes of NO_x.

AOV750 CAS:121-98-2 ***HR: 2***
p-ANISIC ACID, METHYL ESTER
mf: $C_9H_{10}O_3$ mw: 166.19

PROP: Plates from alcohol. Mp: 48-49°, bp: 255-256°. Insol in water; sol in alc and ether.

SYNS: METHYL-p-ANISATE ◇ METHYL-p-METHOXYBENZOATE

TOXICITY DATA with REFERENCE
skn-rbt 500 mg/24H MLD FCTXAV 14,443,76

CONSENSUS REPORTS: Reported in EPA TSCA Inventory.

SAFETY PROFILE: A skin irritant. See also ESTERS. When heated to decomposition it emits acrid smoke and irritating fumes.

AOV875 CAS:122-84-9 ***HR: 2***
ANISIC KETONE
mf: $C_{10}H_{12}O_2$ mw: 164.22

SYNS: p-ACETONYLANISOLE ◇ ANISKETONE ◇ ANISYL METHYL KETONE ◇ p-METHOXYBENZYL METHYL KETONE ◇ 4-METHOXYBENZYL METHYL KETONE ◇ p-METHOXYPHENYLACETONE ◇ 1-(p-METHOXYPHENYL)-2-PROPANONE

TOXICITY DATA with REFERENCE
skn-rbt 500 mg/24H MLD FCTXAV 17,857,79
orl-rat LD50:3330 mg/kg FCTXAV 17,857,79
ipr-mus LD50:560 mg/kg FCTXAV 17,857,79

CONSENSUS REPORTS: Reported in EPA TSCA Inventory.

SAFETY PROFILE: Moderately toxic by ingestion and other routes. A skin irritant. See also KETONES.

AOV900 CAS:90-04-0 ***HR: 3***
o-ANISIDINE
DOT: UN 2431
mf: C_7H_9NO mw: 123.17

SYNS: o-AMINOANISOLE ◇ 2-AMINOANISOLE ◇ 1-AMINO-2-METHOXYBENZENE ◇ 2-ANISIDINE ◇ o-ANISYLAMINE ◇ 2-METHOXY-1-AMINOBENZENE ◇ o-METHOXYANILINE ◇ 2-METHOXYBENZENAMINE (9CI) ◇ o-METHOXYPHENYLAMINE

TOXICITY DATA with REFERENCE
mma-sat 333 μg/plate IMEMDT 27,63,82
dni-mus-orl 200 mg/kg MUREAV 46,305,77
orl-rat LD50:2000 mg/kg IMEMDT 27,63,82
orl-mus LD50:1400 mg/kg IMEMDT 27,63,82
orl-rbt LD50:870 mg/kg IMEMDT 27,63,82
orl-bwd LD50:422 mg/kg AECTCV 12,355,83

CONSENSUS REPORTS: NTP Fifth Annual Report on Carcinogens. IARC Cancer Review: Group 2B IMEMDT 7,56,87; Human Limited Evidence IMEMDT 27,63,82. EPA Genetic Toxicology Program. Reported in EPA TSCA Inventory. Community Right-To-Know List.

OSHA PEL: TWA 0.5 mg/m^3
ACGIH TLV: TWA 0.5 mg/m^3 (skin)
DFG MAK: 0.1 ppm (0.5 mg/m^3)
DOT Classification: DOT-IMO: Poison B; Label: St. Andrews Cross.

SAFETY PROFILE: Confirmed carcinogen. Moderately toxic by ingestion. Mutation data reported. When heated to decomposition it emits toxic fumes of NO_x.

AOW000 CAS:104-94-9 ***HR: 2***
p-ANISIDINE
mf: C_7H_9NO mw: 123.16

PROP: Plates from aq soln. D: 1.089 @ 55°/55°, mp: 57.2°, bp: 243°, vap d: 4.28. Sol in hot water, alc, and ether.

SYNS: p-AMINOANISOLE ◇ 4-AMINOANISOLE ◇ 1-AMINO-4-METHOXYBENZENE ◇ 4-ANISIDINE ◇ p-ANISYLAMINE ◇ p-METHOXYANILINE ◇ 4-METHOXYANILINE ◇ 4-METHOXYBENZENAMINE ◇ 4-METHOXYBENZENEAMINE ◇ p-METHOXYPHENYLAMINE

TOXICITY DATA with REFERENCE
orl-rat LD50:1400 mg/kg AGGHAR 15,447,57
skn-rat LD50:3200 mg/kg AGGHAR 15,447,57
ipr-rat LD50:1400 mg/kg AGGHAR 15,447,57
unk-mus LD50:806 mg/kg JMCMAR 17,900,74

CONSENSUS REPORTS: IARC Cancer Review: Human Inadequate Evidence IMEMDT 27,63,82. Community Right-To-Know List. Reported in EPA TSCA Inventory.

OSHA PEL: TWA 0.5 mg/m^3
ACGIH TLV: TWA 0.5 mg/m^3 (skin)
DFG MAK: 0.1 ppm (0.5 mg/m^3)

SAFETY PROFILE: Moderately toxic by several routes. A mild sensitizer. May cause a contact dermatitis. See also ANILINE. When heated to decomposition it evolves toxic fumes of NO_x.

AOW500 CAS:64090-82-0 ***HR: 3***
m-ANISIDINE ANTIMONYL TARTRATE

TOXICITY DATA with REFERENCE
ipr-mus LD50:28 mg(Sb)/kg AJTMAQ 25,263,45

CONSENSUS REPORTS: Antimony and its compounds are on the Community Right-To-Know List.

OSHA PEL: TWA 0.5 mg(Sb)/m^3
ACGIH TLV: TWA 0.5 mg(Sb)/m^3
NIOSH REL: (Antimony) TWA 0.5 mg(Sb)/m^3

SAFETY PROFILE: Poison by intraperitoneal route. See also ANTIMONY COMPOUNDS. When heated to decomposition it emits very toxic fumes of NO_x and Sb.

AOW750 CAS:64070-14-0 ***HR: 3***
o-ANISIDINE ANTIMONYL TARTRATE

TOXICITY DATA with REFERENCE
ipr-mus LD50:29 mg(Sb)/kg AJTMAQ 25,263,45

CONSENSUS REPORTS: Antimony and its compounds are on the Community Right-To-Know List.

OSHA PEL: TWA 0.5 mg(Sb)/m^3
ACGIH TLV: TWA 0.5 mg(Sb)/m^3
NIOSH REL: (Antimony) TWA 0.5 mg(Sb)/m^3

SAFETY PROFILE: Poison by intraperitoneal route. See also ANTIMONY COMPOUNDS. When heated to decomposition it emits very toxic fumes of Sb and NO_x.

AOX000 CAS:64070-15-1 ***HR: 3***
p-ANISIDINE ANTIMONYL TARTRATE

TOXICITY DATA with REFERENCE
ipr-mus LD50:28 mg(Sb)/kg AJTMAQ 25,263,45

CONSENSUS REPORTS: Antimony and its compounds are on the Community Right-To-Know List.

OSHA PEL: TWA 0.5 mg(Sb)/m^3
ACGIH TLV: TWA 0.5 mg(Sb)/m^3
NIOSH REL: (Antimony) TWA 0.5 mg(Sb)/m^3

SAFETY PROFILE: Poison by intraperitoneal route. See also ANTIMONY COMPOUNDS. When heated to decomposition it emits very toxic fumes of NO_x and Sb.

AOX250 CAS:134-29-2 ***HR: 3***
o-ANISIDINE HYDROCHLORIDE
mf: $C_7H_9NO{\cdot}ClH$ mw: 159.63

SYNS: C.I. 37115 ◇ 2-METHOXYANILINE HYDROCHLORIDE ◇ NCI-C03747

TOXICITY DATA with REFERENCE
mmo-sat 333 μg/plate IARCCD 27,283,80
orl-rat TDLo:180 g/kg/2Y-C:CAR NCITR* NCI-CG-TR-89,78
orl-mus TDLo:721 g/kg/2Y-C:CAR NCITR* NCI-CG-TR-89,78
orl-mus TD:216 g/kg/78W-C:ETA NCITR* NCI-CG-TR-89,78
orl-mus TD:1803 g/kg/1Y-C:NEO IMEMDT 27,63,82

CONSENSUS REPORTS: NTP Fifth Annual Report on Carcinogens. IARC Cancer Review: Animal Sufficient Evidence IMEMDT 27,63,82. NCI Carcinogenesis Bioassay (feed); Clear Evidence: mouse, rat NCITR* NCI-CG-TR-89,78. Community Right-To-Know List.

SAFETY PROFILE: Confirmed carcinogen with experimental carcinogenic, neoplastigenic, and tumorigenic data. Mutation data reported. When heated to decomposition it emits very toxic fumes of NO_x and HCl.

AOX500 CAS:20265-97-8 ***HR: 3***
p-ANISIDINE HYDROCHLORIDE
mf: $C_7H_9NO{\cdot}ClH$ mw: 159.63

SYN: NCI-C03758

TOXICITY DATA with REFERENCE
orl-rat TDLo:2163 g/kg/1Y-C:CAR IMEMDT 27,63,82
orl-rat TDLo:116 g/kg/92W-C:ETA NCITR* NCI-CG-TR-116,78

CONSENSUS REPORTS: IARC Cancer Review: Animal Inadequate Evidence IMEMDT 27,63,82; NCI Carcinogenesis Bioassay (feed); No Evidence: mouse NCITR* NCI-CG-TR-116,78; Inadequate Studies: rat NCITR* NCI-CG-TR-116,78. Reported in EPA TSCA Inventory.

SAFETY PROFILE: Questionable carcinogen with experimental carcinogenic and tumorigenic data. When heated to decomposition it emits very toxic fumes of NO_x and HCl.

AOX750 CAS:100-66-3 ***HR: 2***
ANISOLE
DOT: UN 2222
mf: C_7H_8O mw: 108.15

PROP: Mobile liquid, clear straw color; phenol, anise odor. Vapor d: 3.72, mp: −37.3°, bp: 153.8°, flash p: 125°F (COC), d: 0.983-0.988, refr index: 1.513-1.518, vap press: 10 mm @ 42.2°, autoign temp: 887°F. Insol in water; sol in alc and ether.

SYNS: FEMA No. 2097 ◇ METHOXYBENZENE ◇ METHYL PHENYL ETHER ◇ PHENYL METHYL ETHER

TOXICITY DATA with REFERENCE
skn-rbt 500 mg/24H MOD FCTXAV 17,241,79
orl-rat LD50:3700 mg/kg TXAPA9 6,378,64
orl-mus LD50:2800 mg/kg JPETAB 88,400,46

CONSENSUS REPORTS: Reported in EPA TSCA Inventory.

DOT Classification: Flammable or Combustible Liquid; Label: Flammable Liquid

SAFETY PROFILE: Moderately toxic by ingestion. A skin irritant. Combustible liquid. To fight fire, use foam, CO_2, dry chemical. When heated to decomposition it emits acrid fumes.

AOY000 CAS:22862-76-6 ***HR: 3***
ANISOMYCIN
mf: $C_{14}H_{19}NO_4$ mw: 265.34

SYNS: ANTIBIOTIC PA-106 ◇ FLAGECIDIN

TOXICITY DATA with REFERENCE
orl-rat LD50:72 mg/kg ANTCAO 5,490,55
ipr-rat LD50:345 mg/kg ANTCAO 5,490,55
scu-rat LD50:230 mg/kg ANTCAO 5,490,55
ivn-rat LD50:167 mg/kg ANTCAO 5,490,55
orl-mus LD50:148 mg/kg ANTCAO 5,490,55
ipr-mus LD50:400 mg/kg ANTCAO 5,490,55
scu-mus LD50:600 mg/kg ANTCAO 5,490,55
ivn-mus LD50:140 mg/kg ANTCAO 5,490,55
orl-gpg LDLo:300 mg/kg ANTCAO 5,490,55

SAFETY PROFILE: Poison by ingestion, intraperi-

toneal, subcutaneous, and intravenous routes. When heated to decomposition it emits toxic fumes of NO_x.

AOY250 CAS:100-07-2 ***HR: 3***
ANISOYL CHLORIDE
DOT: UN 1729
mf: $C_8H_7ClO_2$ mw: 170.60

$CH_3OC_6H_4CO \bullet Cl$

PROP: Needle-like crystals. Mp: 22°, bp: 262-263° (slt decomp). Insol in water; sol in ether and acetone.

SYNS: p-ANISYOL CHLORIDE ◇ METHOXYBENZOYL CHLORIDE

CONSENSUS REPORTS: Reported in EPA TSCA Inventory.

DOT Classification: Corrosive Material; Label: Corrosive.

SAFETY PROFILE: Corrosive to skin, eyes, mucous membranes, and other tissue. Evolves HCl by hydrolysis. A storage hazard; can explode spontaneously at room temperature. When heated to decomposition it emits toxic fumes of Cl^- and may explode.

AOY400 ***HR: 1***
ANISYL ACETATE
mf: $C_{10}H_{12}O_3$ mw: 180.20

PROP: Colorless to slt yellow liquid; fruity, balsamic odor. D: 1.104, refr index: 1.511-1.516, flash p: +210°F. Sol in alc and most oils; insol in glycerin and propylene glycol.

SYNS: FEMA No. 2098 ◇ p-METHOXYBENZYL ACETATE

SAFETY PROFILE: Combustible liquid. When heated to decomposition it emits acrid smoke and irritating fumes.

APA000 CAS:27471-67-6 ***HR: 3***
N-(o-ANISYL)-2-(p-BUTOXYPHENOXY)-N-(2(DIETHYLAMINO)ETHYL)ACETAMIDE HYDROCHLORIDE
mf: $C_{25}H_{36}N_2O_4 \bullet ClH$ mw: 465.09

TOXICITY DATA with REFERENCE
orl-mus LD50:350 mg/kg EJMCA5 10,286,75
ivn-mus LD50:22 mg/kg EJMCA5 10,286,75

SAFETY PROFILE: Poison by ingestion and intravenous routes. When heated to decomposition it emits very toxic fumes of NO_x and HCl.

APE000 CAS:102-17-0 ***HR: 2***
ANISYL PHENYLACETATE
mf: $C_{16}H_{16}O_3$ mw: 256.3

SYNS: p-METHOXYBENZYL PHENYLACETATE ◇ PHENYLACETIC ACID, p-METHOXYBENZYL ESTER

TOXICITY DATA with REFERENCE
skn-rbt 500 mg/24H MLD FCTXAV 18,649,80

SAFETY PROFILE: A skin irritant. See also ESTERS. When heated to decomposition it emits acrid smoke and irritating fumes.

APE100 ***HR: 2***
ANNATTO EXTRACT

PROP: From solvent extraction of *Bixa orellana* L. seeds (JAPMA8 49,218,60). Yellow red solutions or powder.

SYNS: ACHIOTE ◇ BIXA ORELLANA

TOXICITY DATA with REFERENCE
orl-hmn TDLo:357 mg/kg:SKN ARTODN (Suppl.1),141,78
ipr-mus LD50:700 mg/kg JAPMA8 49,218,60

SAFETY PROFILE: Moderately toxic by intraperitoneal route. Human systemic effects by skin contact. When heated to decomposition it emits acrid smoke and irritating fumes.

APE529 CAS:66547-10-2 ***HR: 3***
ANSAMITOCIN P-4
mf: $C_{33}H_{45}N_2O_9Cl$ mw: 649.25

SYNS: 3-DE(2-(ACETYLMETHYLAMINO)PROPIONYLOXY)-3-HYDROXYMAYTANSINE ISOVALERATE (ESTER) ◇ MAYTANSINOL ISOVALERATE

TOXICITY DATA with REFERENCE
oms-mus/ast 1 μg/kg CNREA8 40,1707,80
cyt-mus/ast 4 μg/kg CNREA8 40,1707,80
ipr-mus LDLo:625 μg/kg JANTAJ 31,78-143,78

SAFETY PROFILE: Poison by intraperitoneal route. Mutation data reported. When heated to decomposition it emits toxic fumes of Cl^- and NO_x. See also ESTERS.

APE625 CAS:5968-79-6 ***HR: 3***
ANTHALLAN HYDROCHLORIDE
mf: $C_{17}H_{25}NO_5 \bullet ClH$ mw: 359.89

SYN: 3-((DIBUTYLAMINO)METHYL)-4,5,6-TRIHYDROXYPHTHALIDE HYDROCHLORIDE

TOXICITY DATA with REFERENCE
ipr-mus LD50:300 mg/kg JAPMA8 38,433,49
scu-mus LD50:1080 mg/kg JAPMA8 38,433,49
ivn-mus LD50:80 mg/kg JAPMA8 38,433,49

SAFETY PROFILE: Poison by intravenous and intraperitoneal routes. Moderately toxic by other routes. When heated to decomposition it emits toxic fumes of HCl and NO_x.

APE750 CAS:191-26-4 ***HR: 3***
ANTHANTHRENE
mf: $C_{22}H_{12}$ mw: 276.34

SYNS: ANTHANTHREN (GERMAN) ◇ ANTHRANTHRENE ◇ DIBENZO-(drf,mno)CHRYSENE ◇ DIBENZO(cd,mk)PYRENE

TOXICITY DATA with REFERENCE
mma-sat 1 μg/plate MUREAV 51,311,78
imp-rat TDLo:4150 μg/kg:CAR JJIND8 71,539,83
skn-mus TDLo:263 mg/kg/30W-I:CAR ZKKOBW 89,113,77
skn-mus TD:2100 mg/kg/88W-I:ETA PRLBA4 129,439,40

CONSENSUS REPORTS: IARC Cancer Review: Group 3 IMEMDT 7,56,87; Animal Limited Evidence IMEMDT 32,95,83

SAFETY PROFILE: Questionable carcinogen with experimental carcinogenic and tumorigenic data. Mutation data reported. A polycyclic hydrocarbon found in polluted air. When heated to decomposition it emits acrid fumes.

APF000 CAS:12706-94-4 ***HR: 3***
ANTHELMYCIN
mf: $C_{21}H_{37}N_5O_{14}$ mw: 583.57

SYNS: ANTIBIOTIC 33876 ◇ HIKIZIMYCIN ◇ 2(1H)-PYRIMIDINONE-4-AMINO-1-(4-AMINO-6-0-(3-AMINO-3-DEOXY-β-d-GLUCOPYRANOSYL)-4-DEOXY-d-GLYCERO-d-GALACTO-β-d-GLUCO-UNDECAPYRANOSYL)

TOXICITY DATA with REFERENCE
orl-mus LD50:125 mg/kg 85GDA2 5,227,81
ivn-mus LD50:5 mg/kg 85ERAY 2,1107,78

SAFETY PROFILE: Poison by ingestion and intravenous routes. When heated to decomposition it emits toxic fumes of NO_x.

APF750 CAS:189-58-2 ***HR: 3***
ANTHRA(9,1,2-cde)BENZO(h)CINNOLINE
mf: $C_{22}H_{12}N_2$ mw: 304.36

SYN: 1,2-DIAZA-3,4:9,10-DIBENZPYRENE

TOXICITY DATA with REFERENCE
scu-mus TDLo:80 mg/kg:ETA IJMRAQ 53,638,65

SAFETY PROFILE: Questionable carcinogen with experimental tumorigenic data. When heated to decomposition it emits toxic fumes of NO_x.

APG050 CAS:610-49-1 ***HR: 3***
1-ANTHRACENAMINE
mf: $C_{14}H_{11}N$ mw: 193.26

PROP: Yellow needles from alc. Mp: 130°. Insol in HCl; sol in alc.

SYNS: α-AMINOANTHRACENE ◇ 1-AMINOANTHRACENE ◇ 1-ANTHRACYLAMINE ◇ 1-ANTHRAMINE

TOXICITY DATA with REFERENCE
mma-sat 20 μg/plate PNASA6 72,5135,75
dnr-esc 100 mg/L JNCIAM 62,873,79
mrc-smc 5 pph JNCIAM 62,901,79
orl-rat TDLo:7200 mg/kg/27D-I:ETA CNREA8 28,924,68

SAFETY PROFILE: Questionable carcinogen with experimental tumorigenic data. Mutation data reported. When heated to decomposition it emits toxic fumes of NO_x.

APG100 CAS:613-13-8 ***HR: 3***
2-ANTHRACENAMINE
mf: $C_{14}H_{11}N$ mw: 193.26

PROP: Yellow leaflets from alc. Mp: 238°; bp: subl @ 93° @ 9 mm. Insol in water; sltly sol in alc and ether.

SYNS: β-AMINOANTHRACENE ◇ 2-AMINOANTHRACENE ◇ 2-ANTHRACYLAMINE ◇ 2-ANTHRAMINE ◇ 2-ANTHRYLAMINE

TOXICITY DATA with REFERENCE
mmo-sat 6 nmol/plate BBRCA9 89,259,79
mma-sat 2 μg/plate PNASA6 72,5135,75
dnr-esc 100 mg/L JNCIAM 62,873,79
hma-mus/sat 125 mg/kg JNCIAM 62,911,79
orl-mus TDLo:210 mg/kg (female 1-21D post):TER TJADAB 25(2),61A,82
orl-rat TDLo:45 mg/kg/30D-I:CAR CNREA8 28,924,68
skn-rat TDLo:260 μg/kg/33W-I:ETA BJCAAI 9,631,55
skn-mus TDLo:62 mg/kg/2Y-I:CAR NTIS** CONF-801143

SAFETY PROFILE: Suspected carcinogen with experimental carcinogenic and tumorigenic data. An experimental teratogen. Mutation data reported. See also AMINES. When heated to decomposition it emits toxic fumes of NO_x.

APG500 CAS:120-12-7 ***HR: 3***
ANTHRACENE
mf: $C_{14}H_{10}$ mw: 178.24

$C_6H_4:(CH)_2:C_6H_4$

PROP: Colorless crystals, violet fluorescence. Mp: 217°, lel: 0.6%, flash p: 250°F (CC), d: 1.24 @ 27°/4°, autoign temp: 1004°F, vap press: 1 mm @ 145.0°, (sublimes), vap d: 6.15, bp: 339.9°. Insol in water. Solubility in alc @ 1.9/100 @ 20°; in ether 12.2/100 @ 20°.

SYNS: ANTHRACEN (GERMAN) ◇ ANTHRACIN ◇ GREEN OIL ◇ PARANAPHTHALENE ◇ TETRA OLIVE N2G

TOXICITY DATA with REFERENCE
mma-sat 100 μg/plate ABCHA6 43,1433,79
skn-mus 118 μg MLD CALEDQ 4,333,78
dns-hmn:fbr 10 mg/L CNREA8 38,2091,78
hma-mus/sat 125 mg/kg JNCIAM 62,911,79

dnd-mam:lym 100 μmol BIPMAA 9,689,70
orl-rat TDLo:20 g/kg/79W-I:ETA ZEKBAI 60,697,55
scu-rat TDLo:3300 mg/kg/33W-I:NEO NATWAY 42,159,55

CONSENSUS REPORTS: IARC Cancer Review: Animal No Evidence IMEMDT 32,105,83.Reported in EPA TSCA Inventory. Community Right-To-Know List.

OSHA PEL: TWA 0.2 mg/m^3

SAFETY PROFILE: A skin irritant and allergen. Questionable carcinogen with experimental neoplastigenic and tumorigenic data. Mutation data reported. Combustible when exposed to heat, flame, or oxidizing materials. Moderately explosive when exposed to flame, $Ca(OCl)_2$, chromic acid. To fight fire, use water, foam, CO_2, water spray or mist, dry chemical. Explodes on contact with fluorine.

APH250 CAS:480-22-8 ***HR: 3***
1,8,9-ANTHRACENETRIOL
mf: $C_{14}H_{10}O_3$ mw: 226.24

PROP: Yellow powder. Mp: 178-180°. Insol in water; sol in fat, hot alc, benzene, and dilute alkalies.

SYNS: ANTHRALIN ◇ 1,8,9-ANTHRATRIOL ◇ DIHYDROXYANTHRANOL ◇ 1,8-DIHYDROXYANTHRANOL ◇ 1,8-DIHYDROXY-9-ANTHRANOL ◇ 1,8-DIHYDROXY-9-ANTHRONE ◇ DIOXYANTHRANOL ◇ 1,8,9-TRIHYDROXYANTHRACENE

TOXICITY DATA with REFERENCE
mmo-sat 100 μg/plate BCSTB5 5,1489,77
mma-sat 100 μg/plate BCSTB5 5,1489,77
dnr-esc 250 μg/plate JNCIAM 62,873,79
mmo-smc 165 nmol/L ADVEA4 51,45,71
skn-mus TDLo:509 mg/kg/53W-I:NEO JMCMAR 21,26,78
skn-mus TD:73 mg/kg/11W-I:ETA GANNA2 59,187,68

CONSENSUS REPORTS: IARC Cancer Review: Group 3 IMEMDT 7,56,87; Animal Limited Evidence IMEMDT 13,75,77

SAFETY PROFILE: Questionable carcinogen with experimental neoplastigenic and tumorigenic data. Mutation data reported. Skin contact can cause folliculitis. Absorption can cause kidney damage and intestinal disturbances. Combustible when heated. When heated to decomposition it emits acrid smoke and irritating fumes.

APH500 CAS:16203-97-7 ***HR: 2***
1,8,9-ANTHRACENETRIOL TRIACETATE
mf: $C_{20}H_{16}O_6$ mw: 352.36

SYNS: EXOLAN ◇ 1,8,9-TRIACETOXYANTHRACENE

TOXICITY DATA with REFERENCE
eye-rbt 330 μg SEV BJOPAL 53,819,69

SAFETY PROFILE: A severe eye irritant. When heated to decomposition it emits acrid smoke and irritating fumes.

API000 CAS:4803-27-4 ***HR: 3***
ANTHRAMYCIN
mf: $C_{16}H_{17}N_3O_4$ mw: 315.4

SYNS: ANTRAMYCIN ◇ (E)-1H-PYRROLO(2,1-C)(1,4)BENZODIAZEPINE-2-ACRYLAMIDE,5,10,11,11a-TETRAHYDRO-9,11-DIHYDROXY-8-METHYL-5-OXO

TOXICITY DATA with REFERENCE
pic-esc 60 ng/plate CNREA8 43,2819,83
dnd-mam:lym 400 μmol/L BBACAQ 475,521,77
mmo-eug 60 mg/L NEOLA4 19,579,72
dns-hmn:fbr 200 μmol/L JBCHA3 254,605,79
dnd-hmn:fbr 10 μmol/L JBCHA3 254,605,79
ipr-mus LD50:650 μg/kg TOLED5 18,337,83

CONSENSUS REPORTS: EPA Genetic Toxicology Program.

SAFETY PROFILE: Deadly poison by intraperitoneal route. Human mutation data reported. When heated to decomposition it emits toxic fumes of NO_x.

API125 CAS:5544-25-2 ***HR: 3***
ANTHRAMYCIN METHYL ETHER
mf: $C_{17}H_{19}N_3O_4 \cdot H_2O$ mw: 347.41

SYNS: ANTHRAMYCIN-11-METHYL ETHER ◇ ANTIBIOTIC A ◇ ROCHE 5-9000 ◇ (E)-5,10,11,11a-TETRAHYDRO-9-HYDROXY-11-METHOXY-8-METHYL-5-OXO-1H-PYRROLO(2,1-c)(1,4)BENZODIAZEPINE-2-ACRYLAMIDE MONOHYDRATE

TOXICITY DATA with REFERENCE
orl-mus LD50:3400 μg/kg 15MBAH -,303,64
ipr-mus LD50:1200 μg/kg 15MBAH -,303,64
scu-mus LD50:2900 μg/kg 15MBAH -,303,64

SAFETY PROFILE: Poison by ingestion, subcutaneous, and intraperitoneal routes. When heated to decomposition it emits toxic fumes of NO_x.

API500 CAS:118-92-3 ***HR: 2***
ANTHRANILIC ACID
mf: $C_7H_7NO_2$ mw: 137.15

PROP: Needle-like crystals. Mp: 146°, bp: subl, d: 1.412 @ 20°. Solubility: in water = 0.35/100 @ 14°, in 90% alc = 10.7/100 @ 10°, in ether = 16/100 @ 70°.

SYNS: o-AMIDOBENZOIC ACID ◇ o-AMINOBENZOIC ACID ◇ 2-AMINOBENZOIC ACID ◇ 1-AMINO-2-CARBOXYBENZENE ◇ CARBOXYANILINE ◇ o-CARBOXYANILINE ◇ 2-CARBOXYANILINE ◇ NCI-C01730 ◇ VITAMIN L

TOXICITY DATA with REFERENCE
msc-hmn:lyms 1667 mg/L MUREAV 196,61,88
mma-mus:lyms 250 mg/L EMMUEG 12(Suppl 13),37,88

orl-mus TDLo:34800 mg/kg (male 8D pre):REP MPHEAE 15,7,66
orl-rat TDLo:16 g/kg/25W-C:ETA APMIAL 26,447,49
scu-mus TDLo:2040 mg/kg (13-17D preg):ETA,REP JCROD7 96,163,80
orl-mus LD50:1400 mg/kg QJPPAL 19,483,46
ipr-mus LD50:2500 mg/kg RPTOAN 37,105,74

CONSENSUS REPORTS: IARC Cancer Review: Group 3 IMEMDT 7,56,87; Animal Inadequate Evidence IMEMDT 16,265,78. NTP Carcinogenesis Bioassay (feed): No Evidence: mouse, rat NCITR* NCI-TR-36,78. Reported in EPA TSCA Inventory.

SAFETY PROFILE: Moderately toxic by ingestion and intraperitoneal route. Experimental reproductive effects. Human mutation data reported. Questionable carcinogen with experimental tumorigenic data. Combustible. When heated to decomposition it emits toxic fumes of NO_x.

API750 CAS:87-29-6 ***HR: 3***
ANTHRANILIC ACID, CINNAMYL ESTER
mf: $C_{16}H_{15}NO_2$ mw: 253.32

PROP: Reddish yellow powder; balsamic odor. Mp: 60°, flash p: +212°F. Sol in alc, chloroform, ether; insol in water.

SYNS: 2-AMINOBENZOIC ACID-3-PHENYL-2-PROPENYL ESTER ◇ CINNAMYL ALCOHOL ANTHRANILATE ◇ CINNAMYL-2-AMINOBENZOATE ◇ CINNAMYL-o-AMINOBENZOATE ◇ CINNAMYL ANTHRANILATE (FCC) ◇ FEMA No. 2295 ◇ NCI-C03510 ◇ 3-PHENYL-2-PROPENYLANTHRANILATE ◇ 3-PHENYL-2-PROPEN-1-YL ANTHRANILATE

TOXICITY DATA with REFERENCE
orl-rat TDLo:546 g/kg/2Y-C:CAR NCITR* NCI-TR-196,80
orl-mus TDLo:1310 g/kg/2Y-C:CAR NCITR* NCI-TR-196,80
ipr-mus TDLo:12 g/kg/8W-I:NEO CNREA8 33,3069,73
orl-rat LD50:5000 mg/kg FCTXAV 13,681,75
skn-rbt LD50:5000 mg/kg FCTXAV 13,681,75

CONSENSUS REPORTS: IARC Cancer Review: Group 3 IMEMDT 7,56,87; Animal Limited Evidence IMEMDT 31,133,83; Animal Inadequate Evidence IMEMDT 16,287,78; NCI Carcinogenesis Bioassay (feed); Clear Evidence: mouse, rat NCITR* NCI-CG-TR-196,80. Reported in EPA TSCA Inventory.

SAFETY PROFILE: Suspected carcinogen with experimental carcinogenic and neoplastigenic data. See also ESTERS. Combustible liquid. When heated to decomposition it emits toxic fumes of NO_x.

APJ000 CAS:7149-26-0 ***HR: 1***
ANTHRANILIC ACID, LINALYL ESTER
mf: $C_{17}H_{23}NO_2$ mw: 273.41

SYNS: 3,7-DIMETHYL-1,6-OCTADIEN-3-YL-o-AMINOBENZOATE ◇ 1,5-DIMETHYL-1-VINYL-4-HEXEN-1-YL-o-AMINOBENZOATE ◇ LINALYL-o-AMINOBENZOATE ◇ LINALYL ANTHRANILATE

TOXICITY DATA with REFERENCE
orl-rat LD50:4250 mg/kg FCTXAV 14(5),443,76

CONSENSUS REPORTS: Reported in EPA TSCA Inventory.

SAFETY PROFILE: Mildly toxic by ingestion. See also ESTERS. When heated to decomposition it emits toxic fumes of NO_x.

APJ250 CAS:134-20-3 ***HR: 3***
ANTHRANILIC ACID, METHYL ESTER
mf: $C_8H_9NO_2$ mw: 151.18

PROP: Plates from alc or colorless liquid; grape odor. D: 1.161-1.169, mp: 23.8°, bp: 225-230° @ 15 mm, flash p: 219°F. Very sol in water, propylene glycol, hot abs alc (23/100); insol in ether, chloroform, and glycerin.

SYNS: o-AMINOBENZOIC ACID METHYL ESTER ◇ 2-AMINOBENZOIC ACID METHYL ESTER ◇ 2-CARBOMETHOXYANILINE ◇ o-CARBOMETHOXYANILINE ◇ FEMA No. 2682 ◇ 2-(METHOXYCARBONYL) ANILINE ◇ METHYL 2-AMINOBENZOATE ◇ METHYL o-AMINOBENZOATE ◇ METHYL ANTHRANILATE (FCC) ◇ METHYLESTER KYSELINY ANTHRANILOVE ◇ NEROLI OIL, ARTIFICAL

TOXICITY DATA with REFERENCE
skn-rbt 500 mg/24H MOD FCTXAV 12,807,74
dnr-bcs 23 mg/disc OIGZDE 34,267,85
orl-mus TDLo:34800 mg/kg (8D male pre)(21D pre):REP MPHEAE 15,7,66
orl-rat LD50:2910 mg/kg FCTXAV 2,327,64
orl-mus LD50:3900 mg/kg FCTXAV 2,327,64
orl-gpg LD50:2780 mg/kg FCTXAV 2,327,64

CONSENSUS REPORTS: Reported in EPA TSCA Inventory.

SAFETY PROFILE: Moderately toxic by ingestion. Experimental reproductive effects. A skin irritant. See also ESTERS. Combustible liquid. When heated to decomposition it emits toxic fumes of NO_x.

APJ500 CAS:133-18-6 ***HR: 1***
ANTHRANILIC ACID, PHENETHYL ESTER
mf: $C_{15}H_{15}NO_2$ mw: 241.31

PROP: White to yellow crystals; grape odor.

SYNS: BENZYLCARBINYL ANTHRANILATE ◇ β-PHENETHYL-o-AMINOBENZOATE ◇ PHENETHYL ANTHRANILATE ◇ 2-PHENYLETHYL-o-AMINOBENZOATE ◇ 2-PHENYLETHYL ANTHRANILATE

TOXICITY DATA with REFERENCE
skn-rbt 500 mg/24H MOD FCTXAV 14,659,76

CONSENSUS REPORTS: Reported in EPA TSCA Inventory.

SAFETY PROFILE: A skin irritant. See also ESTERS. When heated to decomposition it emits toxic fumes of NO_x.

APJ750 CAS:1885-29-6 ***HR: 3***
ANTHRANILONITRILE
mf: $C_7H_6N_2$ mw: 118.15

PROP: Needles from CS_2. Mp: 51°, bp: 267-268° @ 777 mm. Very sltly sol in water; sol in alc and ether.

SYNS: o-AMINOBENZONITRILE ◇ 2-AMINOBENZONITRILE ◇ o-CYANOANILINE ◇ 2-CYANOANILINE

TOXICITY DATA with REFERENCE
ivn-mus LD50:180 mg/kg CSLNX* NX#00381

CONSENSUS REPORTS: Reported in EPA TSCA Inventory. Cyanide and its compounds are on the Community Right-To-Know List.

SAFETY PROFILE: Poison by intravenous route. See also NITRILES. When heated to decomposition it emits toxic fumes of NO_x and CN^-.

APK000 CAS:129-56-6 ***HR: 3***
ANTHRA(1,9-cd)PYRAZOL-6(2H)-ONE
mf: $C_{14}H_8N_2O$ mw: 220.24

SYNS: C.I. 70300 ◇ PYRAZOLANTHRONE ◇ PYRAZOLEANTHRONE ◇ 1,9-PYRAZOLOANTHRONE

TOXICITY DATA with REFERENCE
orl-rat LD50:8 g/kg 28ZPAK -,146,72
ivn-mus LD50:178 mg/kg CSLNX* NX#00640

CONSENSUS REPORTS: Reported in EPA TSCA Inventory.

SAFETY PROFILE: Poison by intravenous route. Mildly toxic by ingestion. When heated to decomposition it emits toxic fumes of NO_x.

APK250 CAS:84-65-1 ***HR: 2***
ANTHRAQUINONE
mf: $C_{14}H_8O_2$ mw: 208.22

PROP: Yellow crystals. Mp: 286°, bp: 376.9°, flash p: 365°F (CC), d: 1.438, vap press: 1 mm @ 190.0°, vap d: 7.16. Insol in water; very sltly sol in ether. Solubility in alc = 0.05/100 @ 18°, in hot alc = 2.25/100.

SYNS: 9,10-ANTHRACENEDIONE ◇ ANTHRADIONE ◇ 9,10-ANTHRAQUINONE ◇ 9,10-DIOXOANTHRACENE

TOXICITY DATA with REFERENCE
ipr-rat LD50:3500 mg/kg GTPZAB 21(12),27,77
unk-mus LD50:3500 mg/kg GTPZAB 16,44,72

CONSENSUS REPORTS: Reported in EPA TSCA Inventory.

SAFETY PROFILE: Moderately toxic by several routes. A mild allergen. Combustible when exposed to heat or flame. To fight fire, use water, foam, CO_2, water spray or mist, dry chemical. When heated to decomposition it emits acrid smoke and irritating fumes.

APK500 CAS:25704-81-8 ***HR: 1***
ANTHRAQUINONE BRILLIANT GREEN CONCENTRATE ZH
mf: $C_{36}H_{18}Br_2O_4$ mw: 674.36

SYNS: CALEDON JADE GREEN 2G ◇ C.I. 59830 ◇ C.I. VAT GREEN 2

TOXICITY DATA with REFERENCE
eye-rbt 500 mg/24H MOD 28ZPAK -,249,72

CONSENSUS REPORTS: Reported in EPA TSCA Inventory.

SAFETY PROFILE: An eye irritant. See also BROMIDES. When heated to decomposition it emits very toxic fumes of Br^-.

APK625 CAS:117-14-6 ***HR: 2***
1,5-ANTHRAQUINONEDISULFONIC ACID
mf: $C_{14}H_8O_8S_2$ mw: 368.34

SYN: 1,5-DISULFOANTHRAQUINONE

TOXICITY DATA with REFERENCE
orl-rat LD50:2357 mg/kg GISAAA 45(3),73,80
orl-mus LD50:2357 mg/kg GISAAA 45(3),73,80
orl-rbt LD50:5500 mg/kg GISAAA 45(3),73,80
orl-gpg LD50:2357 mg/kg GISAAA 45(3),73,80

CONSENSUS REPORTS: Reported in EPA TSCA Inventory.

SAFETY PROFILE: Moderately toxic by ingestion. When heated to decomposition it emits toxic fumes of SO_x.

APK635 CAS:82-48-4 ***HR: 2***
1,8-ANTHRAQUINONEDISULFINIC ACID
mf: $C_{14}H_8O_8S_2$ mw: 368.34

SYN: 1,8-DISULFOANTHRAQUINONE

TOXICITY DATA with REFERENCE
orl-rat LD50:1870 mg/kg GISAAA 45(3),73,80
orl-mus LD50:2800 mg/kg GISAAA 45(3),73,80
orl-rbt LD50:4200 mg/kg GISAAA 45(3),73,80
orl-gpg LD50:1870 mg/kg GISAAA 45(3),73,80

CONSENSUS REPORTS: Reported in EPA TSCA Inventory.

SAFETY PROFILE: Moderately toxic by ingestion. When heated to decomposition it emits toxic fumes of SO_x.

APK750 CAS:62399-48-8 ***HR: 2***
((N-ANTHRAQUINON-2-YL)AMINOMETHYLENE)DIMETHYLAMMONIUM CHLORIDE
mf: $C_{17}H_{15}N_2O_2 \cdot Cl$ mw: 314.79

SYN: N,N-DIMETHYL-N'-(1-ANTHRACHINONYL)FORMAMIDINIUMCHLORID(GERMAN)

TOXICITY DATA with REFERENCE
skn-rbt 500 mg/24H MLD 28ZPAK -,124,72
eye-rbt 5 mg/24H SEV 28ZPAK -,124,72
orl-rat LD50:3190 mg/kg 28ZPAK -,124,72

SAFETY PROFILE: Moderately toxic by ingestion. Skin and severe eye irritant. When heated to decomposition it emits very toxic fumes of NO_x, Cl^-, and NH_3.

APK850 CAS:131-14-6 ***HR: 1***
2,6-ANTHRAQUINONYLDIAMINE
mf: $C_{14}H_{10}N_2O_2$ mw: 238.26

SYNS: 9,10-ANTHRACENEDIONE, 2,6-DIAMINO- ◇ ANTHRAQUINONE, 2,6-DIAMINO- ◇ 2,6-DIAMINOANTHRACHINON ◇ 2,6-DIAMINOANTHRAQUINONE ◇ 2,6-DIAMINO-9,10-ANTHRAQUINONE

TOXICITY DATA with REFERENCE
eye-rbt 500 mg/24H MLD 28ZPAK-,122,72
mma-sat 100 μg/plate MUREAV 40,203,76

CONSENSUS REPORTS: Reported in EPA TSCA Inventory.

SAFETY PROFILE: An eye irritant. Mutation data reported. When heated to decomposition it emits toxic fumes of NO_x.

APL250 CAS:61907-23-1 ***HR: 2***
N,N'''-(2,6-ANTHRAQUINONYLENE)BIS(N,N-DIETHYLACETAMIDE)
mf: $C_{26}H_{32}N_4O_2$ mw: 432.62

SYNS: BISAMIDINE ◇ N',N'''-(9,10-DIHYDRO-9,10-DI-OXO-2,6-ANTHRACENEDIYL)BIS(N,N-DIETHYLETHANIMIDAMIDE)

TOXICITY DATA with REFERENCE
orl-rat LD50:416 mg/kg DRFUD4 4,705,79
orl-ham LD50:850 mg/kg DRFUD4 4,705,79

SAFETY PROFILE: Moderately toxic by ingestion. When heated to decomposition it emits toxic fumes of NO_x.

APL500 CAS:4403-90-1 ***HR: 2***
2,2'-(1,4-ANTHRAQUINONYLENEDIIMINO)BIS(5-METHYLBENZENESULFONIC ACID) DISODIUM SALT
mf: $C_{28}H_{20}N_2O_8S_2 \cdot 2Na$ mw: 622.60

SYNS: C.I. 61 570 (CZECH) ◇ ZELEN ALIZARINOVA BRILANTNI G-EXTRA (CZECH)

TOXICITY DATA with REFERENCE
skn-rbt 500 mg/24H MLD 28ZPAK -,247,72
eye-rbt 500 mg/24H SEV 28ZPAK -,247,72

CONSENSUS REPORTS: Reported in EPA TSCA Inventory.

SAFETY PROFILE: A skin and severe eye irritant. See also SULFONATES. When heated to decomposition it emits very toxic fumes of NO_x, SO_x, and Na_2O.

APL750 CAS:116-76-7 ***HR: 1***
1,1'-(ANTHRAQUINON-1,4-YLENEDIIMINO)DIANTHRAQUINONE
mf: $C_{42}H_{22}N_2O_6$ mw: 650.66

SYNS: 1,4-BIS-1'-ANTHRACHINONYLAMINO-ANTHRACHINON (CZECH) ◇ 1,4-TRIANTHRIMID (CZECH)

TOXICITY DATA with REFERENCE
eye-rbt 500 mg/24H SEV 28ZPAK -,127,72

CONSENSUS REPORTS: Reported in EPA TSCA Inventory.

SAFETY PROFILE: A severe eye irritant. When heated to decomposition it emits toxic fumes of NO_x.

APM000 CAS:117-03-3 ***HR: 2***
1,1'-(ANTHRAQUINON-1,5-YLENEDIIMINO)DIANTHRAQUINONE
mf: $C_{42}H_{22}N_2O_6$ mw: 650.66

SYNS: 1,5-BIS-1'-ANTHRACHINONYLAMINO-ANTHRACHINON (CZECH) ◇ 1,5-TRIANTHRIMID (CZECH)

TOXICITY DATA with REFERENCE
eye-rbt 500 mg/24H SEV 28ZPAK -,128,72

CONSENSUS REPORTS: Reported in EPA TSCA Inventory.

SAFETY PROFILE: A severe eye irritant. When heated to decomposition it emits toxic fumes of NO_x.

APM250 CAS:73688-63-8 ***HR: 1***
4,4'-(1,4-ANTHRAQUINONYLENEDIIMINODI-PHENYL-1,4-ENEDIOXO)BENZENESULFONIC ACID
mf: $C_{38}H_{26}N_2O_{10}S_2$ mw: 734.78

SYN: ZELEN MIDLONOVA BLS (CZECH)

TOXICITY DATA with REFERENCE
eye-rbt 20 mg/24H SEV 28ZPAK -,247,72
orl-rat LD50:8550 mg/kg 28ZPAK -,247,72

SAFETY PROFILE: Mildly toxic by ingestion. A severe eye irritant. See also SULFONATES. When heated to decomposition it emits very toxic fumes of SO_x and NO_x.

APM750 CAS:1715-81-7 ***HR: 3***
9-ANTHRONOL
mf: $C_{14}H_{10}O_2$ mw: 210.24

TOXICITY DATA with REFERENCE
skn-mus TDLo:700 mg/kg/73W-I:NEO JMCMAR 21,26,78

SAFETY PROFILE: Questionable carcinogen with experimental neoplastigenic data. When heated to decomposition it emits acrid smoke and irritating fumes.

APM875 ***HR: 2***
ANTHURIUM

PROP: A berry-producing plant with thick, heart-shaped, dark green leaves. The "flower" consists of a leaf that has turned a bright color with a spike emerging from it. Brightly colored berries eventually grow from the spike. Native to the tropical areas of America, they are grown in gardens in southern Florida and Hawaii, and as house plants elsewhere.

SYNS: ANTURIO ◇ FLAMINGO FLOWER ◇ FLAMINGO LILY ◇ FLOR de CULEBRA (PUERTO RICO) ◇ GUINDA (PUERTO RICO) ◇ HOJA GRANDE (CUBA) ◇ LENGUNA de VACA (CUBA, PUERTO RICO) ◇ LOMBRICERO (CUBA) ◇ PIGTAIL PLANT ◇ TAIL FLOWER

SAFETY PROFILE: The leaves and stems contain toxic calcium oxalate raphides. Chewing these plant parts results in burning pain in the lips, mouth and throat, possibly followed by inflammation and blistering. Systemic effects are usually not seen because of the insolubility of calcium oxalate. See also OXALATES.

APN000 CAS:23605-05-2 ***HR: 3***
α-ANTIARBIN
mf: $C_{29}H_{42}O_{11}$ mw: 566.71

TOXICITY DATA with REFERENCE
ivn-cat LD50:90 μg/kg 85ELDJ -,188,63
ivn-rbt LDLo:1 mg/kg 27ZWAY E.1,78,-

SAFETY PROFILE: Poison by intravenous route. When heated to decomposition it emits acrid smoke and irritating fumes.

APP500 CAS:1263-89-4 ***HR: 3***
ANTIBIOTIC 1600
mf: $C_{23}H_{45}N_5O_{14}•H_2O_4S$ mw: 713.81

SYNS: AMINOSIDINE SULFATE ◇ AMINOSIDINE SULPHATE ◇ AMINOSIDIN SULFATE ◇ FARMIGLUCIN ◇ FARMINOSIDIN ◇ Fi 5853 ◇ GABBROMICINA ◇ GABBROMYCIN ◇ GABBRORAL ◇ GABBROROL ◇ HUMATIN ◇ HUMYCIN SULFATE ◇ PARAMICINA ◇ PARICINA ◇ PAROMOMYCIN SULFATE

TOXICITY DATA with REFERENCE
dnd-esc 30 μmol/L MUREAV 89,95,81
orl-rat LD50:21620 mg/kg NIIRDN 6,595,82
scu-rat LD50:1740 mg/kg JJANAX 36,644,83
ivn-rat LD50:181 mg/kg ANTCAO 9,730,59
ims-rat LD50:1200 mg/kg NIIRDN 6,595,82
orl-mus LD50:23500 mg/kg JJANAX 36,644,83
scu-mus LD50:490 mg/kg ANTCAO 9,730,59
ivn-mus LD50:98 mg/kg JJANAX 36,644,83
ims-mus LD50:438 mg/kg JJANAX 36,644,83
ivn-mky LDLo:93 mg/kg ANTCAO 9,730,59

SAFETY PROFILE: Poison by intravenous and intramuscular routes. Moderately toxic by subcutaneous route. Mildly toxic by ingestion. Mutation data reported. See also SULFATES. When heated to decomposition it emits very toxic fumes of SO_x and NO_x.

APS750 CAS:37517-28-5 ***HR: 3***
ANTIBIOTIC BB-K 8
mf: $C_{22}H_{43}N_5O_{13}$ mw: 585.70

SYN: AMIKACIN

TOXICITY DATA with REFERENCE
ims-rbt TDLo:90 mg/kg (8-16D preg):TER JJANAX 28,366,75
ipr-mus LD50:750 mg/kg AMACCQ 13,41,78
scu-mus LD50:2 g/kg NKRZAZ 26(Suppl 3),36,78
ivn-mus LD50:300 mg/kg JANTAJ 27,677,74

CONSENSUS REPORTS: EPA Genetic Toxicology Program.

SAFETY PROFILE: Poison by intravenous route. Moderately toxic by intraperitoneal route. An experimental teratogen. When heated to decomposition it emits toxic fumes of NO_x.

APT000 CAS:39831-55-5 ***HR: 3***
ANTIBIOTIC BB-K8 SULFATE
mf: $C_{22}H_{43}N_5O_{13}•2H_2O_4S$ mw: 781.86

SYNS: AMIKACIN SULFATE ◇ AMIKIN ◇ AMIKLIN ◇ BB-K8 ◇ BIKLIN ◇ FABIANOL ◇ NOVAMIN

TOXICITY DATA with REFERENCE
scu-rat TDLo:175 mg/kg (female 8-14D post):REP JJANAX 28,372,75
scu-rat TDLo:175 mg/kg (female 8-14D post):TER JJANAX 28,372,75
ipr-rat LD50:3500 mg/kg JJANAX 28,415,75
scu-rat LD50:3604 mg/kg PBPSDY 1,125,77
ivn-rat LD50:234 mg/kg PBPSDY 1,125,77
ims-rat LD50:2244 mg/kg JJANAX 39,3164,86
ipr-mus LD50:2930 mg/kg JJANAX 28,415,75
scu-mus LD50:2470 mg/kg PBPSDY 1,125,77
ivn-mus LD50:181 mg/kg JJANAX 39,3164,86

SAFETY PROFILE: Poison by intravenous route. Moderately toxic by intraperitoneal and subcutaneous routes. An experimental teratogen. Other experimental reproductive effects. When heated to decomposition it emits very toxic fumes of NO_x and SO_x. See also SULFATES.

APT250 CAS:51627-14-6 ***HR: 2***
ANTIBIOTIC BL-640
mf: $C_{18}H_{18}N_6O_5S_2$ mw: 462.54

SYNS: ANTIBIOTIC BL-S 640 ◇ BLS 640 ◇ CEFATRIZINE ◇ SKF 60771 ◇ S 640P

TOXICITY DATA with REFERENCE
orl-rat TDLo:175 mg/kg (8-14D preg):REP JJANAX 29,129,76
orl-rat TDLo:5600 mg/kg (female 8-14D post):TER JJANAX 29,129,76
ipr-rat LD50:4325 mg/kg JJANAX 29,612,76
ipr-mus LD50:6410 mg/kg JJANAX 29,612,76
ipr-rbt LD50:1500 mg/kg JJANAX 29,612,76
scu-rbt LD50:3000 mg/kg JJANAX 29,612,76

SAFETY PROFILE: Moderately toxic by intraperitoneal and subcutaneous routes. An experimental teratogen. Other experimental reproductive effects. When heated to decomposition it emits very highly toxic fumes of NO_x and SO_x.

APT375 CAS:69866-21-3 ***HR: 3***
ANTIBIOTIC CC 1065
mf: $C_{37}H_{32}N_7O_8$ mw: 702.76

SYNS: CC-1065 ◇ NSC 298223

TOXICITY DATA with REFERENCE
dni-mus:leu 4 μg/L CNREA8 42,999,82
oms-mus:leu 45 μg/L CNREA8 42,999,82
dni-ham:ovr 150 ng/L CNREA8 42,3532,82
oms-ham:ovr 5 μg/L CNREA8 42,3532,82
dnd-mam:lym 7400 nmol/L CNREA8 42,999,82
ipr-mus LD50:6900 ng/kg JANTAJ 37,63,84
ivn-mus LD50:9 μg/kg JANTAJ 37,63,84
ivn-rbt LDLo:1 μg/kg JANTAJ 37,63,84

SAFETY PROFILE: Deadly poison by intravenous and intraperitoneal routes. Mutation data reported. When heated to decomposition it emits toxic fumes of NO_x.

APT750 CAS:39391-39-4 ***HR: 2***
ANTIBIOTIC FR 1923
mf: $C_{23}H_{24}N_4O_9$ mw: 500.51

SYN: NOCARDICIN A

TOXICITY DATA with REFERENCE
ipr-rat LD50:2600 mg/kg 85ERAY 1,897,78
scu-rat LD50:3100 mg/kg 85ERAY 1,897,78
ipr-mus LD50:2500 mg/kg 85ERAY 1,897,78
scu-mus LD50:2900 mg/kg 85ERAY 1,897,78
ivn-mus LD50:2100 mg/kg 85ERAY 1,897,78

SAFETY PROFILE: Moderately toxic by intraperitoneal, subcutaneous, and intravenous routes. When heated to decomposition it emits toxic fumes of NO_x.

APU000 ***HR: 3***
ANTIBIOTIC G-52 SULFATE

PROP: Aminoglycoside produced by a species of the genus *Micromonospora* (JANTAJ 29,483,76).

TOXICITY DATA with REFERENCE
ipr-mus LD50:200 mg/kg JANTAJ 29,483,76
scu-mus LD50:400 mg/kg JANTAJ 29,483,76
ivn-mus LD50:50 mg/kg JANTAJ 29,483,76

SAFETY PROFILE: Poison by intraperitoneal, subcutaneous, and intravenous routes. See also SULFATES. When heated to decomposition it emits very toxic fumes of SO_x and NO_x.

APU500 CAS:57576-44-0 ***HR: 3***
ANTIBIOTIC MA 144A1
mf: $C_{42}H_{53}NO_{15}$ mw: 811.96

SYNS: ACLACINOMYCIN A ◇ ANTIBIOTIC MA 144A

TOXICITY DATA with REFERENCE
eye-rbt 100 μg MLD JJANAX 33,453,80
scu-gpg 200 μg MLD JJANAX 33,453,80
dnd-rat:lvr 6300 nmol/L MOPMA3 14,290,78
ipr-rat TDLo:27 mg/kg (17-22D preg/21D post):REP OYYAA2 19,855,80
ipr-rat TDLo:2200 μg/kg (female 7-17D post):TER OYYAA2 19,783,80
ipr-rat LD50:17930 μg/kg JJANAX 33,138,80
scu-rat LD50:20 mg/kg JJANAX 33,138,80
ivn-rat LD50:25710 μg/kg JJANAX 33,138,80
orl-mus LD50:30100 μg/kg ANTBAL 30,918,85
ipr-mus LD50:16100 μg/kg ANTBAL 30,918,85
scu-mus LD50:22100 μg/kg ANTBAL 30,918,85
ivn-mus LD50:16500 μg/kg ANTBAL 30,918,85

SAFETY PROFILE: Poison by ingestion, intraperitoneal, subcutaneous, and intravenous routes. An experimental teratogen. Other experimental reproductive effects. Mutation data reported. An eye and subcutaneous irritant. When heated to decomposition it emits toxic fumes of NO_x.

APV000 CAS:63710-09-8 ***HR: 3***
ANTIBIOTIC MA 144S2
mf: $C_{36}H_{45}NO_{14}$ mw: 715.82

SYNS: 1-HYDROXY MA144 S1 ◇ MA144 S2 ◇ MUSETTAMYCIN

TOXICITY DATA with REFERENCE
dnd-rat:lvr 10 μmol/L MOPMA3 14,290,78
dni-mus:leu 310 nmol/L JANTAJ 34,1596,81

oms-mus:leu 59 nmol/L JANTAJ 34,1596,81
ipr-mus LD50:13 mg/kg JANTAJ 31,78-95,78

SAFETY PROFILE: Poison by intraperitoneal route. Mutation data reported. When heated to decomposition it emits toxic fumes of NO_x.

APV750 CAS:3734-60-9 ***HR: 3***
ANTIBIOTIC PA147
mf: $C_6H_6O_3$ mw: 126.12

PROP: Created from the Streptomyceten strain, A-415-Z3 with qualities similar to the strains *Str. filipinensis* and *Str. roseochromogenes W.* and *H* (ARZNAD 17,693,67).

SYNS: ANTIBIOTICUM PA147 (GERMAN) ◇ 3-CARBOXY-2,4-PENTADIENALLACTOL

TOXICITY DATA with REFERENCE
ivn-mus LDLo:250 mg/kg ARZNAD 17,693,67
unk-mus LD50:20 mg/kg ARZNAD 17,693,67

SAFETY PROFILE: Poison by intravenous and other unspecified routes. When heated to decomposition it emits acrid smoke and irritating fumes.

APY500 CAS:53179-09-2 ***HR: 3***
ANTIBIOTIC 66-40 SULFATE
mf: $C_{19}H_{37}N_5O_7 \cdot 5/2\ H_2O_4S$ mw: 692.81

SYNS: EXTRAMYCIN ◇ MENSISO ◇ PATHOMYCIN ◇ RICKAMICIN SULFATE ◇ SISOMICIN SULFATE ◇ SISOMIN

TOXICITY DATA with REFERENCE
scu-rat LD50:500 mg/kg JZKEDZ 4,107,78
ivn-rat LD50:49 mg/kg JZKEDZ 4,107,78
ims-rat LD50:404 mg/kg JZKEDZ 4,107,78
ipr-mus LD50:221 mg/kg JANTAJ 23,551,70
scu-mus LD50:272 mg/kg JZKEDZ 4,107,78
ivn-mus LD50:34 mg/kg JANTAJ 23,551,70
ims-mus LD50:280 mg/kg JZKEDZ 4,107,78

SAFETY PROFILE: Poison by intravenous, intramuscular, intraperitoneal, and subcutaneous routes. See also SULFATES. When heated to decomposition it emits very toxic fumes of SO_x and NO_x.

AQB000 CAS:31282-04-9 ***HR: 3***
ANTIHELMYCIN
mf: $C_{20}H_{37}N_3O_{13}$ mw: 527.60

SYNS: HYGROMIX-8 ◇ HYGROMYCIN B (USDA)

TOXICITY DATA with REFERENCE
ipr-rat LD50:63 mg/kg GISAAA 38,11,73
ipr-gpg LD50:13 mg/kg GISAAA 38,11,73

SAFETY PROFILE: Poison by intraperitoneal route. When heated to decomposition it emits toxic fumes of NO_x.

AQB250 CAS:11112-10-0 ***HR: D***
ANTIMONIC ACID, SODIUM SALT

SYNS: SODIUM ANTIMONATE ◇ SODIUM ANTIMONY ◇ SODIUM POLYANTIMONATE

TOXICITY DATA with REFERENCE
cyt-hmn:fbr 1 nmol/L AEMBAP 91,117,78
cyt-hmn:leu 1 nmol/L AEMBAP 91,117,78

CONSENSUS REPORTS: Antimony and its compounds are on the Community Right-To-Know List.

SAFETY PROFILE: Mutation data reported. See also ANTIMONY COMPOUNDS. When heated to decomposition it emits toxic fumes of Sb and Na_2O.

AQB750 CAS:7440-36-0 ***HR: 3***
ANTIMONY
DOT: UN 2871
af: Sb aw: 121.75

PROP: Silvery or gray, lustrous metal. Mp: 630°, bp: 1635°, d: 6.684 @ 25°, vap press: 1 mm @ 886°. Insol in water; sol in hot concentrated H_2SO_4.

SYNS: ANTIMONY BLACK ◇ ANTIMONY REGULUS ◇ ANTYMON (POLISH) ◇ C.I. 77050 ◇ STIBIUM

TOXICITY DATA with REFERENCE
ihl-rat TCLo:50 mg/m³/7H/52W-I:CAR JTEHD6 18,607,86
orl-rat LD50:7 g/kg EQSFAP 1,1,75
ipr-rat LD50:100 mg/kg 85GMAT -,22,82
ipr-mus LD50:90 mg/kg 85GMAT -,22,82
ipr-gpg LD50:150 mg/kg EQSFAP 1,1,75

CONSENSUS REPORTS: Antimony and its compounds are on the Community Right-To-Know List. Reported in EPA TSCA Inventory.

OSHA PEL: TWA 0.5 mg(Sb)/m³
ACGIH TLV: TWA 0.5 mg(Sb)/m³
DFG MAK: 0.5 mg(Sb)/m³
NIOSH REL: TWA 0.5 mg(Sb)/m³
DOT Classification: Poison B; LABEL: St. Andrews Cross.

SAFETY PROFILE: An experimental poison by intraperitoneal route. Questionable carcinogen with experimental carcinogenic data. Moderate fire and explosion hazard in the forms of dust and vapor, when exposed to heat or flame. See also POWDERED METALS. When heated or on contact with acid it emits toxic fumes of SbH_3. Electrolysis of acid sulfides and stirred Sb halide yields explosive Sb. It can react violently with NH_4NO_3, halogens, BrN_3, BrF_3, $HClO_3$, ClO, ClF_3, HNO_3, KNO_3, $KMnO_4$, K_2O_2, $NaNO_3$, oxidants.

AQC000 CAS:72017-60-8 ***HR: 3***
ANTIMONY AMMONIA TRIACETIC ACID
mf: $C_{12}H_{14}N_2O_{12}Sb•2H_2O$ mw: 536.07

SYN: ATA-Sb

TOXICITY DATA with REFERENCE
ivn-hmn TDLo:9 mg/kg/15D-I:SYS XPHPAW 411,158,74
ipr-mus LD50:62 mg/kg XPHPAW 441,131,74

CONSENSUS REPORTS: Antimony and its compounds are on the Community Right-To-Know List.

OSHA PEL: TWA 0.5 mg(Sb)/m^3
ACGIH TLV: TWA 0.5 mg(Sb)/m^3
NIOSH REL: (Antimony) TWA 0.5 mg(Sb)/m^3

SAFETY PROFILE: Poison by intraperitoneal route. Human systemic effects by intravenous route. See also ANTIMONY COMPOUNDS. When heated to decomposition it emits toxic fumes of NO_x, Sb, and NH_3.

AQC250 CAS:64046-93-1 ***HR: 3***
ANTIMONY, BIS(TRICHLORO) compounded with 1 mole of OCTAMETHYL PYROPHOSPHORAMIDE
mf: $Cl_3Sb•1/2(C_8H_{24}N_4O_3P_2)$ mw: 371.25

TOXICITY DATA with REFERENCE
ipr-mus LD50:35 mg/kg JAFCAU 14,512,66

CONSENSUS REPORTS: Antimony and its compounds are on the Community Right-To-Know List.

OSHA PEL: TWA 0.5 mg(Sb)/m^3
ACGIH TLV: TWA 0.5 mg(Sb)/m^3
NIOSH REL: (Antimony) TWA 0.5 mg(Sb)/m^3

SAFETY PROFILE: Poison by intraperitoneal route. See also ANTIMONY COMPOUNDS and PHOSPHATES. When heated to decomposition it emits very toxic fumes of Cl^-, Sb, PO_x, and NO_x.

AQC500 CAS:10025-91-9 ***HR: 3***
ANTIMONY(III) CHLORIDE
DOT: UN 1733
mf: Cl_3Sb mw: 228.10

PROP: Colorless, rhombic, deliq crystals. D: 3.06, mp: 73.4°, bp: 220°, vap press: 1 mm @ 49.2° (subl). Sol in water @ 20°; sol in alc, benzene, and chloroform.

SYNS: ANTIMOINE (TRICHLORURE d') ◇ ANTIMONIO (TRICLORURO di) ◇ ANTIMONOUS CHLORIDE ◇ ANTIMONOUS CHLORIDE (DOT) ◇ ANTIMONTRICHLORID ◇ ANTIMONY BUTTER ◇ ANTIMONY CHLORIDE ◇ ANTIMONY CHLORIDE (DOT) ◇ ANTIMONY TRICHLORIDE ◇ ANTIMONY TRICHLORIDE, liquid (DOT) ◇ ANTIMONY TRICHLORIDE, solid (DOT) ◇ ANTIMONY TRICHLORIDE, solution (DOT) ◇ ANTIMOONTRICHLRIDE ◇ BUTTER of ANTIMONY ◇ CHLORID ANTIMONITY ◇ CHLORURE ANTIMONIEUX ◇ C.I. 77056 ◇ STIBINE, TRICHLORO- ◇ TRICHLOROSTIBINE ◇ TRICHLORURE d'ANTIMOINE

TOXICITY DATA with REFERENCE
dnr-bcs 10 mmol/L MUREAV 77,109,80
orl-rat TDLo:4400 μg/kg (female 1-22D post):REP TCMUD8 7,491,87
ihl-hmn TDLo:73 mg/kg:PUL BJIMAG 23,318,66
orl-rat LD50:525 mg/kg MARJV# 29MAR77
ipr-mus LD50:13 mg/kg 85GMAT-,23,82
orl-gpg LD50:574 mg/kg HYSAAV 29(12),16,64

CONSENSUS REPORTS: Reported in EPA TSCA Inventory. Antimony and its compounds are on the Community Right-To-Know List.

OSHA PEL: TWA 500 μg(Sb)/m^3
ACGIH TLV: TWA 0.5 mg(Sb)/m^3
NIOSH REL: (Antimony) TWA 0.5 mg(Sb)/m^3
DOT Classification: Corrosive Material; Label: Corrosive.

SAFETY PROFILE: Moderately toxic by ingestion. Human pulmonary system effects by inhalation. Corrosive by vigorous reaction with moisture, generating heat and hydrogen chloride gas (a strong irritant) which can cause pulmonary edema when inhaled. Systemic effects can be caused by the antimony. See also ANTIMONY COMPOUNDS. Experimental reproductive effects. Mutation data reported. When heated to decomposition it emits very toxic fumes of chlorine and antimony. It can react violently with aluminum, potassium, sodium.

AQD000 CAS:7647-18-9 ***HR: 3***
ANTIMONY(V) CHLORIDE
DOT: UN 1730/UN 1731
mf: Cl_5Sb mw: 299.01

PROP: Red-yellow oil, liquid, offensive odor. Mp: 2.8°, bp: 140°, d: 2.336, vap press: 1 mm @ 22.7°. Decomp in water; sol in HCl, HBr, and CS_2.

SYNS: ANTIMONIC CHLORIDE ◇ ANTIMONIO (PENTACLORURO DI) (ITALIAN) ◇ ANTIMONPENTACHLORID (GERMAN) ◇ ANTIMONY PENTACHLORIDE ◇ ANTIMONY PENTACHLORIDE (DOT) ◇ ANTIMONY PERCHLORIDE ◇ ANTIMOONPENTACHLORIDE (DUTCH) ◇ BUTTER of ANTIMONY ◇ PENTACHLOROANTIMONY ◇ PENTACHLORURE d'ANTIMOINE (FRENCH) ◇ PERCHLORURE d'ANTIMOINE (FRENCH)

TOXICITY DATA with REFERENCE
mrc-bcs 30 μL/disc MUREAV 77,109,80
orl-rat LD50:1115 mg/kg HYSAAV 29(12),16,64
orl-gpg LD50:900 mg/kg HYSAAV 29(12),16,64

CONSENSUS REPORTS: Reported in EPA TSCA Inventory. Antimony and its compounds are on the Community Right-To-Know List.

OSHA PEL: TWA 500 μg(Sb)/m^3
ACGIH TLV: TWA 0.5 mg(Sb)/m^3
NIOSH REL: (Antimony) TWA 0.5 mg(Sb)/m^3
DOT Classification: Corrosive Material; Label: Corrosive.

SAFETY PROFILE: Poison by ingestion. Corrosive. Mutation data reported. See ANTIMONY COMPOUNDS and ANTIMONY(III) CHLORIDE. When heated to decomposition it emits very toxic fumes of Cl^- and Sb.

AQD250 CAS:7647-18-9 ***HR: 3***
ANTIMONY(V) CHLORIDE (solution)
DOT: UN 1733

SYNS: ANTIMONIC CHLORIDE ◇ ANTIMONY PENTACHLORIDE, solution (DOT) ◇ ANTIMONY PERCHLORIDE

CONSENSUS REPORTS: Antimony and its compounds are on the Community Right-To-Know List.

NIOSH REL: TWA 0.5 mg(Sb)/m^3
DOT Classification: Corrosive Material; Label: Corrosive.

SAFETY PROFILE: A poison, corrosive and an irritant. See also ANTIMONY(V) CHLORIDE. When heated to decomposition it emits very toxic fumes of Sb and Cl^-.

AQD500 ***HR: 3***
ANTIMONY COMPOUNDS

CONSENSUS REPORTS: On Community Right-To-Know List.

SAFETY PROFILE: Most antimony compounds are poisons by ingestion, inhalation, and intraperitoneal routes. See also ANTIMONY. Locally antimony compounds irritate the skin and mucous membranes. (Sb^{+++} and hot $HClO_3$) can form an explosive mixture.

AQD750 CAS:3064-61-7 ***HR: 2***
ANTIMONY DIMERCAPTOSUCCINATE
mf: $C_{12}H_6O_{12}S_6Sb_2$•6Na mw: 915.98

SYNS: ANTIMONY DIMERCAPTOSUCCINATE(IV) ◇ SODIUM ANTIMONY-2,3-meso-DIMERCAPTOSUCCINATE ◇ STIBOCAPTATE

TOXICITY DATA with REFERENCE
ipr-mus LD50:2500 mg/kg FAZMAE 17,108,73
scu-mus LD50:2 g/kg BWHOA6 45,371,71

CONSENSUS REPORTS: Antimony and its compounds are on the Community Right-To-Know List.

OSHA PEL: TWA 0.5 mg(Sb)/m^3
ACGIH TLV: TWA 0.5 mg(Sb)/m^3
NIOSH REL: (Antimony) TWA 0.5 mg(Sb)/m^3

SAFETY PROFILE: Moderately toxic by intraperitoneal and subcutaneous routes. See also ANTIMONY COMPOUNDS and SULFIDES. When heated to decomposition it emits toxic fumes of SO_x, Na_2O, and Sb.

AQE000 CAS:7783-56-4 ***HR: 3***
ANTIMONY(III) FLUORIDE (1:3)
DOT: NA 1549
mf: F_3Sb mw: 178.75

PROP: Colorless, rhombic, deliq crystals. Mp: 292°, bp: 376° (subl), d: 4.379 @ 20.9°. Sol in water @ 20°.

SYNS: ANTIMOINE FLUORURE (FRENCH) ◇ ANTIMONOUS FLUORIDE ◇ ANTIMONY TRIFLUORIDE ◇ TRIFLUOROANTIMONY

TOXICITY DATA with REFERENCE
scu-frg LDLo:224 mg/kg CRSBAW 124,133,37

CONSENSUS REPORTS: Reported in EPA TSCA Inventory. Antimony and its compounds are on the Community Right-To-Know List.

OSHA PEL: TWA 0.5 mg(Sb)/m^3; TWA 2.5 mg(F)/m^3
ACGIH TLV: TWA 2.5 mg(F)/m^3; TWA 0.5 mg(Sb)/m^3
NIOSH REL: TWA 0.5 mg(Sb)/m^3
DOT Classification: Corrosive Material; Label: Corrosive

SAFETY PROFILE: Poison by subcutaneous route. Corrosive to skin and eyes. See also FLUORIDES and ANTIMONY COMPOUNDS. When heated to decomposition it emits very toxic fumes of F^- and Sb.

AQE250 CAS:58164-88-8 ***HR: 3***
ANTIMONY LACTATE
DOT: UN 1550
mf: $C_9H_{15}O_9$•Sb mw: 388.99

PROP: Tan-colored mass, water-sol.

SYNS: ANTIMONY LACTATE, solid (DOT) ◇ LACTIC ACID, ANTIMONY SALT

CONSENSUS REPORTS: Reported in EPA TSCA Inventory. Antimony and its compounds are on the Community Right-To-Know List.

OSHA PEL: TWA 0.5 mg(Sb)/m^3
ACGIH TLV: TWA 0.5 mg(Sb)/m^3
NIOSH REL: (Antimony) TWA 0.5 mg(Sb)/m^3
DOT Classification: ORM-A; Label: None; Poison B; Label: St. Andrews Cross

SAFETY PROFILE: A poison. See also ANTIMONY COMPOUNDS. When heated to decomposition it emits toxic fumes of Sb.

AQE300 CAS:77824-42-1 ***HR: 2***
ANTIMONYL-2,4-DIHYDROXY-5-HYDROXYMETHYL PYRIMIDINE
mf: $C_{10}H_{10}N_4O_8Sb_2$ mw: 557.74

SYN: 2,4,10,12-TETRAOXA-6,16,17,18-TETRAAZA-3,11-DISTIBATRICYCLO(11.3.1.1^{5,9})OCTADECA-1(17),5,7,9(18),13,15-HEXAENE-8,14-DIMETHANOL, 3,11-DIHYDROXY-

TOXICITY DATA with REFERENCE
ipr-mus LD50:660 mg/kg JEMAAJ 62,1,79

OSHA PEL: TWA 0.5 mg(Sb)/m^3
ACGIH TLV: TWA 0.5 mg(Sb)/m^3
NIOSH REL: 10H TWA 0.5 mg(Sb)/m^3

SAFETY PROFILE: Moderately toxic by intraperitoneal route. When heated to decomposition it emits toxic fumes of NO_x and Sb.

AQE305 CAS:77824-44-3 ***HR: 3***
ANTIMONYL-2,4-DIHYDROXY PYRIMIDINE
mf: $C_8H_6N_4O_6Sb_2$ mw: 497.68

SYN: 2,4,10,12-TETRAOXA-6,16,17,18-TETRAAZA-3,11-DISTIBATRICYCLO(11.3.1.1 ^{5,9})OCTADECA-1(17),5,7,9(18),13,15-HEXAENE, 3,11-DIHYDROXY-

TOXICITY DATA with REFERENCE
ipr-mus LD50:300 mg/kg JEMAAJ 62,1,79

OSHA PEL: TWA 0.5 mg(Sb)/m^3
ACGIH TLV: TWA 0.5 mg(Sb)/m^3
NIOSH REL: 10H TWA 0.5 mg(Sb)/m^3

SAFETY PROFILE: Poison by intraperitoneal route. When heated to decomposition it emits toxic fumes of NO_x and Sb.

AQE320 CAS:77824-43-2 ***HR: 3***
ANTIMONYL-7-FORMYL-8-HYDROXYQUINOLINE-5-SULPHONATE
mf: $C_{20}H_{11}N_2O_{11}S_2Sb{\bullet}2Na$ mw: 687.18

SYN: 5-QUINOLINESULFONIC ACID, 8,8′-((HYDROXYSTIBYLENE)BIS(OXY))BIS(7-FORMYL-, DISODIUM SALT

TOXICITY DATA with REFERENCE
ipr-mus LD50:75 mg/kg JEMAAJ 62,1,79

OSHA PEL: TWA 0.5 mg(Sb)/m^3
ACGIH TLV: TWA 0.5 mg(Sb)/m^3
NIOSH REL: 10H TWA 0.5 mg(Sb)/m^3

SAFETY PROFILE: Poison by intraperitoneal route. When heated to decomposition it emits toxic fumes of NO_x, SO_x, and Sb.

AQE500 CAS:6169-12-6 ***HR: 2***
ANTIMONY LITHIUM THIOMALATENONAHYDRATE
mf: $C_{12}H_9O_{12}S_3{\bullet}9H_2O{\bullet}6Li{\bullet}Sb$ mw: 766.96

SYNS: ANTHIOMALINE NONAHYDRATE ◇ LITHIUM ANTIMONY THIOMALATE NONAHYDRATE ◇ THIOANTIMONIC(III) ACID, TRIESTER with MERCAPTO SUCCINIC ACID DILITHIUM SALT, NONAHYDRATE

TOXICITY DATA with REFERENCE
ims-chd TDLo:73 mg/kg:GIT JAMAAP 125,952,44
ims-hmn TDLo:11 mg/kg:GIT JAMAAP 125,952,44
ipr-mus LDLo:512 mg/kg CBCCT* 2,136,50

CONSENSUS REPORTS: Antimony and its compounds are on the Community Right-To-Know List.

OSHA PEL: TWA 0.5 mg(Sb)/m^3
ACGIH TLV: TWA 0.5 mg(Sb)/m^3
NIOSH REL: (Antimony) TWA 0.5 mg(Sb)/m^3

SAFETY PROFILE: Human gastrointestinal tract effects by intramuscular route. Moderately toxic by intraperitoneal route. See also LITHIUM; ANTIMONY COMPOUNDS; and ESTERS. When heated to decomposition it emits very toxic fumes of SO_x and Sb.

AQE750 ***HR: 3***
ANTIMONY NITRIDE
mf: NSb mw: 135.76

CONSENSUS REPORTS: Antimony and its compounds are on the Community Right-To-Know List.

SAFETY PROFILE: See ANTIMONY COMPOUNDS and NITRIDES. Explosively decomposes upon warming in a vacuum. When heated to decomposition it emits very toxic fumes of Sb, NO_x, and NH_3.

AQF000 CAS:1309-64-4 ***HR: 3***
ANTIMONY OXIDE
mf: O_3Sb_2 mw: 291.50

PROP: White cubes. D: 5.2, mp: 650°, bp: 1550° subl. Very sltly sol in water; sol in KOH and HCl.

SYNS: ANTIMONIOUS OXIDE ◇ ANTIMONY PEROXIDE ◇ ANTIMONY SESQUIOXIDE ◇ ANTIMONY TRIOXIDE (MAK) ◇ ANTIMONY WHITE ◇ C.I. PIGMENT WHITE 11 ◇ DECHLORANE-A-O ◇ DIANTIMONY TRIOXIDE ◇ FLOWERS of ANTIMONY ◇ NCI-C55152

TOXICITY DATA with REFERENCE
mrc-bcs 50 mmol/L MUREAV 77,109,80
ihl-rat TCLo:270 μg/m^3 (1-21D post):TER GISAAA 52(10),85,87
ihl-rat TCLo:270 μg/m^3 (1-21D post):REP GISAAA 52(10),85,87
ihl-rat TCLo:4200 μg/m^3/52W-I:CAR AIHAM* 20,1,80
ihl-rat TC:4 mg/m^3/1Y-I:ETA PESTC* 8,16,80
ihl-rat TD:1600 μg/m^3/52W-I:NEO AIHAM* 20,1,80
ihl-rat TC:50 mg/m^3/7H/52W-I:CAR JTEHD6 18,607,86
ipr-rat LD50:3250 mg/kg EQSSDX 1,1,75
ipr-mus LD50:172 mg/kg 85GMAT-,23,82
ivn-dog LDLo:3 mg/kg HBAMAK 4,1289,35
scu-rbt LDLo:2500 μg/kg HBAMAK 4,1289,35

CONSENSUS REPORTS: Reported in EPA TSCA Inventory. Antimony and its compounds are on the Community Right-To-Know List.

OSHA PEL: TWA 0.5 mg(Sb)/m^3
ACGIH TLV: TWA C.5 mg(Sb)/m^3; Suspected Carcinogen
DFG MAK: Animal Carcinogen, Suspected Human Carcinogen.
NIOSH REL: TWA 0.5 mg(Sb)/m^3

SAFETY PROFILE: Confirmed carcinogen with experimental carcinogenic and neoplastigenic data. Poison by intravenous and subcutaneous routes. Moderately toxic by other routes. An experimental teratogen. Other experimental reproductive effects. Mutation data reported. See also ANTIMONY COMPOUNDS. When heated to decomposition it emits toxic Sb fumes. Incompatible with chlorinated rubber and heat of 216° and with BrF_3.

AQF250 CAS:7783-70-2 ***HR: 3***
ANTIMONY(V) PENTAFLUORIDE
DOT: UN 1732
mf: F_5Sb mw: 216.75

PROP: Oily, colorless liquid. Very reactive. Mp: 7.0°, bp: 149.5°, d: (liq) 2.99 @ 23°. Sol in water and KF.

SYNS: ANTIMONY FLUORIDE ◇ ANTIMONY(V) FLUORIDE ◇ PENTAFLUOROANTIMONY

CONSENSUS REPORTS: Reported in EPA TSCA Inventory. Antimony and its compounds are on the Community Right-To-Know List. EPA Extremely Hazardous Substances List.

OSHA PEL: TWA 0.5 mg(Sb)/m^3
ACGIH TLV: TWA 0.5 mg(Sb)/m^3
NIOSH REL: (Antimony) TWA 0.5 mg(Sb)/m^3
DOT Classification: Corrosive Material; Label: Corrosive, Poison

SAFETY PROFILE: A poison. A very reactive, corrosive liquid to skin, eyes, mucous membranes. See also FLUORIDES and ANTIMONY COMPOUNDS. Violent reaction with phosphates. When heated to decomposition it emits very toxic fumes of F^- and Sb.

AQF500 CAS:1315-04-4 ***HR: 3***
ANTIMONY PENTASULFIDE
mf: S_5Sb_2 mw: 403.80

PROP: Orange-yellow powder. Mp: (decomp), d: 4.120.

SYNS: ANTIMONIAL SAFFRON ◇ ANTIMONIC SULFIDE ◇ ANTIMONY RED ◇ ANTIMONY SULFIDE ◇ C.I. 77061 ◇ GOLDEN ANTIMONY SULFIDE

TOXICITY DATA with REFERENCE
ipr-rat LD50:1500 mg/kg EQSFAP 1,1,75
ipr-mus LD50:458 mg/kg 85GMAT-,23,82

CONSENSUS REPORTS: Reported in EPA TSCA Inventory. Antimony and its compounds are on the Community Right-To-Know List.

OSHA PEL: TWA 0.5 mg(Sb)/m^3
ACGIH TLV: TWA 0.5 mg(Sb)/m^3
NIOSH REL: (Antimony) TWA 0.5 mg(Sb)/m^3

SAFETY PROFILE: Moderately toxic by intraperitoneal route. See also ANTIMONY COMPOUNDS and SULFIDES. Flammable when exposed to heat or by chemical reaction with powerful oxidizers. Use water to fight fire. Moderately explosive when shocked or by spontaneous chemical reaction in contact with powerful oxidizers. When heated to decomposition or on contact with acid or acid fumes it emits highly toxic fumes of oxides of sulfur and antimony. Incompatible with water or steam to produce toxic and flammable vapors and with oxidizers, i.e., $Ag(ClO_3)_2$, $HClO_3$, ClO_2, $Mg(ClO_3)_2$, TlO, $Zn(ClO_3)_2$.

AQF750 CAS:1314-60-9 ***HR: 1***
ANTIMONY PENTOXIDE
mf: O_5Sb_2 mw: 323.50

PROP: Yellowish-white powder. D: 3.78, mp: decomp @ 380°. Very sltly sol in water, sltly sol in KOH.

SYNS: ANTIMONIC "ACID" ◇ ANTIMONIC OXIDE ◇ ANTIMONY PENTAOXIDE ◇ DIANTIMONY PENTOXIDE ◇ STIBIC ANHYDRIDE

TOXICITY DATA with REFERENCE
ipr-rat LD50:4 g/kg EQSFAP 1,1,75

CONSENSUS REPORTS: Reported in EPA TSCA Inventory. Antimony and its compounds are on the Community Right-To-Know List.

OSHA PEL: TWA 500 μg(Sb)/m^3
ACGIH TLV: TWA 0.5 mg(Sb)/m^3
NIOSH REL: (Antimony) TWA 0.5 mg(Sb)/m^3

SAFETY PROFILE: Toxic by intraperitoneal route. See also ANTIMONY COMPOUNDS.

AQG000 CAS:35743-94-3 ***HR: 3***
ANTIMONY POTASSIUM DIMETHYLCYSTEINOTARTRATE

PROP: A dimethyl cysteine chelate of sodium antimonyl tartrate with 14.7% Sb content (PSEBAA 129,284,68).

TOXICITY DATA with REFERENCE
scu-mus LD50:350 mg/kg PSEBAA 129,284,68
ivn-mus LD50:450 mg/kg PSEBAA 129,284,68

CONSENSUS REPORTS: Antimony and its compounds are on the Community Right-To-Know List.

OSHA PEL: TWA 500 μg(Sb)/m^3
NIOSH REL: TWA 0.5 mg(Sb)/m^3

SAFETY PROFILE: Poison by subcutaneous route.

Moderately toxic by intravenous route. See also ANTIMONY COMPOUNDS. When heated to decomposition it emits very toxic fumes of NO_x and Sb.

AQG250 CAS:28300-74-5 ***HR: 3***
ANTIMONY POTASSIUM TARTRATE
DOT: UN 1551
mf: $C_8H_4O_{12}Sb_2 \cdot 3H_2O \cdot 2K$ mw: 635.88

PROP: Colorless crystals to white powder. D: 2.607, mp: loses H_2O @ 100°.

SYNS: ANTIMONYL POTASSIUM TARTRATE ◇ EMETIQUE (FRENCH) ◇ ENT 50,434 ◇ POTASSIUM ANTIMONYL TARTRATE ◇ POTASSIUM ANTIMONYL-d-TARTRATE ◇ POTASSIUM ANTIMONY TARTRATE ◇ TARTAR EMETIC ◇ TARTARIZED ANTIMONY ◇ TARTRATE ANTIMONIO-POTASSIQUE (FRENCH) ◇ TARTRATED ANTIMONY

TOXICITY DATA with REFERENCE
dni-esc 19 umol/L BCPCA6 23,1451,74
cyt-hmn:fbr 100 umol/L JDGRAX 7(3),27,75
orl-hmn LDLo:2 mg/kg PCOC** -,1097,66
ivn-hmn TDLo:1392 μg/kg:EYE,PUL LANCAO 210,227,26
ivn-man LDLo:12 mg/kg/1W-I:LIV,KID LANCAO 210,227,26
ivn-man LD50:249 mg/kg/9D-I JTMHA9 21,38,18
orl-rat LD50:115 mg/kg ARSIM* 20,24,66
ipr-rat LD50:11 mg/kg EQSSDX 1,1,75
ims-rat LDLo:33 mg/kg EQSSDX 1,1,75
orl-mus LDLo:600 mg/kg EQSFAP 1,1,75
ipr-mus LD50:33 mg/kg BWHOA6 53,379,76
scu-mus LD50:55 mg/kg PSEBAA 129,284,68
ivn-mus LD50:45 mg/kg FAZMAE 17,108,73

CONSENSUS REPORTS: Antimony and its compounds are on the Community Right-To-Know List.

OSHA PEL: TWA 0.5 mg(Sb)/m^3
ACGIH TLV: TWA 0.5 mg(Sb)/m^3
NIOSH REL: (Antimony) TWA 0.5 mg(Sb)/m^3
DOT Classification: ORM-A; Label: None; Poison B; Label: St. Andrews Cross

SAFETY PROFILE: Human poison by intravenous route producing liver and kidney changes, somnolence, dyspnea, and pupillary dilation. Poison by ingestion, subcutaneous, intravenous, intramuscular, and intraperitoneal routes. Large doses cause severe liver damage. Human mutaton data reported. Used medicinally, the therapeutic dose is close to the toxic dose. Upon decomposition it emits toxic fumes of K_2O and Sb.

AQG500 CAS:64070-11-7 ***HR: 3***
d-ANTIMONY POTASSIUM TARTRATE
mf: $C_4H_4O_7Sb \cdot K$ mw: 324.93

PROP: White crystals.

SYN: POTASSIUM ANTIMONYL-d-TARTRATE

TOXICITY DATA with REFERENCE
ipr-mus LD50:48.8 mg/kg AJTMAQ 30,591,50

CONSENSUS REPORTS: Antimony and its compounds are on the Community Right-To-Know List.

OSHA PEL: TWA 0.5 mg(Sb)/m^3
ACGIH TLV: TWA 0.5 mg(Sb)/m^3
NIOSH REL: (Antimony) TWA 0.5 mg(Sb)/m^3

SAFETY PROFILE: Poison by intraperitoneal route. See also ANTIMONY COMPOUNDS. When heated to decomposition it emits toxic fumes of Sb and K_2O.

AQG750 CAS:64070-12-8 ***HR: 3***
dl-ANTIMONY POTASSIUM TARTRATE
mf: $C_4H_4O_7Sb \cdot K$ mw: 324.93

SYNS: POTASSIUM ANTIMONYL-d,l-TARTRATE ◇ dl-TARTARIC ACID, ANTIMONY POTASSIUM SALT

TOXICITY DATA with REFERENCE
ipr-mus LD50:48.8 mg/kg AJTMAQ 30,591,50

CONSENSUS REPORTS: Antimony and its compounds are on the Community Right-To-Know List.

OSHA PEL: TWA 0.5 mg(Sb)/m^3
ACGIH TLV: TWA 0.5 mg(Sb)/m^3
NIOSH REL: (Antimony) TWA 0.5 mg(Sb)/m^3

SAFETY PROFILE: Poison by intraperitoneal route. See also ANTIMONY COMPOUNDS. When heated to decomposition it emits toxic fumes of Sb and K_2O.

AQH000 CAS:11071-15-1 ***HR: 3***
l-ANTIMONY POTASSIUM TARTRATE
mf: $C_4H_5O_7Sb \cdot K$ mw: 325.94

SYNS: POTASSIUM ANTIMONYL-l-TARTRATE ◇ l-TARTARIC ACID, ANTIMONY POTASSIUM SALT

TOXICITY DATA with REFERENCE
ipr-mus LD50:51 mg/kg AJTMAQ 30,591,50

CONSENSUS REPORTS: Reported in EPA TSCA Inventory. Antimony and its compounds are on the Community Right-To-Know List.

OSHA PEL: TWA 0.5 mg(Sb)/m^3
ACGIH TLV: TWA 0.5 mg(Sb)/m^3
NIOSH REL: (Antimony) TWA 0.5 mg(Sb)/m^3

SAFETY PROFILE: Poison by intraperitoneal route. See also ANTIMONY COMPOUNDS. When heated to decomposition it emits toxic fumes of Sb and K_2O.

AQH250 CAS:64070-10-6 ***HR: 3***
meso-ANTIMONY POTASSIUM TARTRATE
mf: $C_4H_4O_7Sb \cdot K$ mw: 324.93

SYN: POTASSIUMANTIMONYL-meso-TARTRATE

TOXICITY DATA with REFERENCE
ipr-mus LD50:51 mg/kg AJTMAQ 30,591,50

CONSENSUS REPORTS: Antimony and its compounds are on the Community Right-To-Know List.

OSHA PEL: TWA 0.5 mg(Sb)/m^3
ACGIH TLV: TWA 0.5 mg(Sb)/m^3
NIOSH REL: (Antimony) TWA 0.5 mg(Sb)/m^3

SAFETY PROFILE: Poison by intraperitoneal route. See also ANTIMONY COMPOUNDS. When heated to decomposition it emits toxic fumes of Sb and K_2O.

AQH500 CAS:15489-16-4 ***HR: 3***
ANTIMONY PYROCATECHOL SODIUM DISULFONATE
mf: $C_{12}H_4O_{16}S_4Sb•7H_2O•5Na$ mw: 799.24

SYNS: ANTIMONYLBRENZEATECHINDISULFOSAURESNATRIUM (GERMAN) ◇ ANTIMOSAN ◇ CORYSTIBIN ◇ FOUADIN ◇ NEOANTIMOSAN ◇ PYROSTIB ◇ REPRODAL ◇ SODIUM ANTIMONY BIS(PYROCATECHOL-2,4-DISULFONATE) ◇ SODIUM ANTIMONY (III) BIS-PYROCATECHOL-3,5-DISULFONATE HEPTAHYDRATE ◇ SODIUM ANTIMOSAN ◇ STIBOPHEN ◇ TRIMON

TOXICITY DATA with REFERENCE
ipr-mus LD50:260 mg/kg AJTMAQ 25,263,45
scu-mus LD50:670 mg/kg BWHOA6 45,371,71
ivn-mus LD50:1050 mg/kg FAZMAE 17,108,73
ivn-rbt LD50:90 mg/kg JPETAB 89,196,47
ims-rbt LD50:91 mg/kg JPETAB 87,119,46

CONSENSUS REPORTS: Antimony and its compounds are on the Community Right-To-Know List.

OSHA PEL: TWA 0.5 mg(Sb)/m^3
ACGIH TLV: TWA 0.5 mg(Sb)/m^3
NIOSH REL: (Antimony) TWA 0.5 mg(Sb)/m^3

SAFETY PROFILE: Poison by intraperitoneal, intramuscular, and intravenous routes. Moderately toxic by intraperitoneal route. See also ANTIMONY COMPOUNDS and SULFONATES. When heated to decomposition it emits toxic fumes of Sb and Na_2O.

AQH750 ***HR: 3***
ANTIMONY SODIUM DIMETHYL CYSTEINO TARTRATE

PROP: Made up of 5.8 parts of sodium antimony tartrate and 10 parts of dimethyl cysteine (FAZMAE 17,108,73).

SYN: SODIUM ANTIMONYL DIMETHYLCYSTEINE TARTRATE

TOXICITY DATA with REFERENCE
ipr-mus LD50:450 mg/kg PSEBAA 129,284,68
scu-mus LD50:385 mg/kg PSEBAA 129,284,68
ivn-mus LD50:435 mg/kg PSEBAA 129,284,68
ims-mus LD50:325 mg/kg PSEBAA 129,284,68

CONSENSUS REPORTS: Antimony and its compounds are on the Community Right-To-Know List.

OSHA PEL: TWA 500 μg(Sb)/m^3
NIOSH REL: (Antimony) TWA 10H 0.5 mg(Sb)/m^3

SAFETY PROFILE: Poison by subcutaneous and intramuscular routes. Moderately toxic by other routes. See also ANTIMONY COMPOUNDS. When heated to decomposition it emits very toxic Sb fumes and Na_2O.

AQH800 CAS:16037-91-5 ***HR: 3***
ANTIMONY SODIUM GLUCONATE
mf: $C_{12}H_{20}O_{17}Sb_2•3Na•9H_2O$ mw: 1048.91

SYNS: ESTIBOGLUCONATO SODICO ◇ d-GLUCONIC ACID, CYCLIC ESTER with ANTIMONIC ACID (H8Sb2O9) (2:1),TRISODIUM SALT, NONAHYDRATE ◇ d-GLUCONIC ACID, 2,4:2′,4′-O-(OXYDISTIBYLIDYNE)BIS-, Sb,Sb′-DIOXIDE, TRISODIUM SALT, NONAHYDRATE ◇ MYOSTIBIN ◇ PENTOSTAM ◇ SODIUM STIBOGLUCONATE ◇ SOLUSTIBOSAN ◇ SOLUSTIN ◇ SOLUSURMIN ◇ SOLYUSURMIN ◇ STIBANATE ◇ STIBANOSE ◇ STIBATIN ◇ STIBINOL

TOXICITY DATA with REFERENCE
ipr-mus LD50:33 mg/kg CLDND* 11,155,49

OSHA PEL: TWA 0.5 mg(Sb)/m^3
ACGIH TLV: TWA 0.5 mg(Sb)/m^3
NIOSH REL: (Antimony) 10H TWA 0.5 mg(Sb)/m^3

SAFETY PROFILE: Poison by intraperitoneal route. When heated to decomposition it emits toxic fumes of Sb.

AQI000 CAS:12550-17-3 ***HR: 3***
ANTIMONY(III) SODIUM GLUCONATE
mf: $C_6H_8O_7Sb•Na$ mw: 336.88

SYNS: SODIUM ANTIMONY GLUCONATE ◇ SODIUM ANTIMONY(III) GLUCONATE ◇ TRIVALENT SODIUM ANTIMONYL GLUCONATE

TOXICITY DATA with REFERENCE
ipr-mus LD50:3440 μg/kg CLDND* 81,224,44

CONSENSUS REPORTS: Antimony and its compounds are on the Community Right-To-Know List.

OSHA PEL: TWA 0.5 mg(Sb)/m^3
ACGIH TLV: TWA 0.5 mg(Sb)/m^3
NIOSH REL: TWA 0.5 mg(Sb)/m^3

SAFETY PROFILE: Poison by intraperitoneal route. See also ANTIMONY COMPOUNDS. When heated to decomposition it emits toxic fumes of Sb and Na_2O.

AQI250 CAS:16037-91-5 ***HR: 3***
ANTIMONY(V) SODIUM GLUCONATE
mf: $C_{12}H_{20}O_{17}Sb_2•3Na•9H_2O$ mw: 1048.91

SYN: SODIUM ANTIMONY(V) GLUCONATE ◇ ESTIBOGLUCONATO SODICO ◇ D-GLUCONIC ACID, CYCLIC ESTER with ANTIMONIC ACID (H8Sb2O9) (2:1),TRISODIUM SALT, NONAHYDRATE

◇ MYOSTIBIN ◇ PENTOSTAM ◇ SODIUM STIBOGLUCONATE ◇ SOLUSTIBOSAN ◇ SOLUSTIN ◇ SOLUSURMIN ◇ SOLYUSURMIN ◇ STIBANATE ◇ STIBANOSE ◇ STIBATIN ◇ STIBINOL

TOXICITY DATA with REFERENCE
ipr-mus LD50:33 mg/kg CLDND* 11,155,49

CONSENSUS REPORTS: Antimony and its compounds are on the Community Right-To-Know List.

OSHA PEL: TWA 0.5 mg(Sb)/m^3
ACGIH TLV: TWA 0.5 mg(Sb)/m^3
NIOSH REL: TWA 0.5 mg(Sb)/m^3

SAFETY PROFILE: Poison by intraperitoneal route. See also ANTIMONY COMPOUNDS. When heated to decomposition it emits toxic fumes of Sb.

AQI500 CAS:66922-79-0 ***HR: 3***
ANTIMONY SODIUM PROPYLENE DIAMINE TETRAACETIC ACID DIHYDRATE
mf: $C_{11}H_{14}N_2O_8Sb \cdot Na \cdot 2H_2O$ mw: 483.05

SYN: PDTA-Sb

TOXICITY DATA with REFERENCE
ivn-hmn TDLo:9 mg/kg/15D-I:SYS XPHPAW 441,186,74
ipr-mus LD50:62 mg/kg SSINAV 13,789,64

CONSENSUS REPORTS: Antimony and its compounds are on the Community Right-To-Know List.

OSHA PEL: TWA 0.5 mg(Sb)/m^3
ACGIH TLV: TWA 0.5 mg(Sb)/m^3
NIOSH REL: (Antimony) TWA 0.5 mg(Sb)/m^3

SAFETY PROFILE: Poison by intraperitoneal route. Human systemic effects by intravenous route. See also ANTIMONY COMPOUNDS. When heated to decomposition it emits toxic fumes of NO_x and Na_2O.

AQI750 CAS:34521-09-0 ***HR: 3***
ANTIMONY SODIUM TARTRATE
mf: $C_8H_4O_{12}Sb_2 \cdot 2Na$ mw: 581.60

SYNS: ANTIMONY SODIUM OXIDE-l-(+)-TARTRATE ◇ NATRIUMANTIMONYLTARTRAT (GERMAN) ◇ SODIUM ANTIMONYL TARTRATE ◇ SODIUM ANTIMONY TARTRATE ◇ STIBNAL ◇ STIBUNAL

TOXICITY DATA with REFERENCE
cyt-hmn:leu 2300 pmol/L/48H MUREAV 16,332,72
ivn-hmn TDLo:79 mg/kg/3D-I TEARAI 34(4),62,62
ipr-mus LD50:60 mg/kg PSEBAA 129,284,68
scu-mus LD50:48 mg/kg PSEBAA 129,284,68
ivn-mus LD50:25 mg/kg MEIEDD 10,103,83

CONSENSUS REPORTS: Antimony and its compounds are on the Community Right-To-Know List.

OSHA PEL: TWA 0.5 mg(Sb)/m^3
ACGIH TLV: TWA 0.5 mg(Sb)/m^3
NIOSH REL: (Antimony) TWA 0.5 mg(Sb)/m^3

SAFETY PROFILE: Poison by subcutaneous, intravenous, and intraperitoneal routes. Human toxic effects by intravenous route. Human mutation data reported. See also ANTIMONY COMPOUNDS. When heated to decomposition it emits toxic fumes of Sb.

AQJ250 CAS:7446-32-4 ***HR: 2***
ANTIMONY(III) SULFATE (2:3)
mf: $O_{12}S_3Sb_2$ mw: 531.68

PROP: White powder. Mp: decomp, d: 3.625 @ 4°.

SYNS: ANTIMONOUS SULFATE ◇ ANTIMONY TRISULFATE ◇ DIANTIMONY TRISULFATE

CONSENSUS REPORTS: Antimony and its compounds are on the Community Right-To-Know List.

OSHA PEL: TWA 500 μg(Sb)/m^3
ACGIH TLV: TWA 0.5 mg(Sb)/m^3
NIOSH REL: (Antimony) TWA 0.5 mg(Sb)/m^3

SAFETY PROFILE: See ANTIMONY COMPOUNDS and SULFATES. When heated to decomposition it emits very toxic fumes of Sb and SO_x.

AQJ500 ***HR: 3***
ANTIMONY TARTRATE
mf: $C_{12}H_{12}O_{18} \cdot 2Sb$ mw: 687.74

PROP: White crystals.

SYNS: BRECHWEINSTEIN

TOXICITY DATA with REFERENCE
orl-rbt LDLo:115 mg/kg UDHU** -,-,37
ims-rbt LD50:90 mg/kg JPETAB 87,119,46

CONSENSUS REPORTS: Antimony and its compounds are on the Community Right-To-Know List.

OSHA PEL: TWA 0.5 mg(Sb)/m^3
ACGIH TLV: TWA 0.5 mg(Sb)/m^3
NIOSH REL: (Antimony) TWA 0.5 mg(Sb)/m^3

SAFETY PROFILE: Poison by intramuscular route. See also ANTIMONY COMPOUNDS. When heated to decomposition it emits toxic fumes of Sb.

AQJ750 CAS:6923-52-0 ***HR: 3***
ANTIMONY TRIACETATE
mf: $C_6H_9O_6 \cdot Sb$ mw: 298.90

SYNS: ACETIC ACID, TRIANHYDRIDE with ANTIMONIC ACID ◇ ANTIMONY(III) ACETATE ◇ OCTAN ANTIMONITY (CZECH)

TOXICITY DATA with REFERENCE
skn-rbt 500 mg/24H SEV 28ZPAK -,17,72
eye-rbt 20 mg/24H SEV 28ZPAK -,17,72
orl-rat LD50:4480 mg/kg 28ZPAK -,17,72

CONSENSUS REPORTS: Reported in EPA TSCA In-

ventory. Antimony and its compounds are on the Community Right-To-Know List.

OSHA PEL: TWA 0.5 mg(Sb)/m^3
ACGIH TLV: TWA 0.5 mg(Sb)/m^3
NIOSH REL: (ANtimony) TWA 0.5 mg(Sb)/m^3

SAFETY PROFILE: Mildly toxic by ingestion. A severe skin and eye irritant. See also ANTIMONY COMPOUNDS. When heated to decomposition it emits acrid smoke and irritating fumes.

AQK000 CAS:7789-61-9 ***HR: 3***
ANTIMONY TRIBROMIDE
mf: $SbBr_3$ mw: 361.51

PROP: Yellow, deliquescent, crystalline mass. Decomp by water. Mp: 96.6°, bp: 280°, d: 4.145, vap press: 1 mm @ 93.9°.

SYNS: ANTIMONY BROMIDE ◇ TRIBROMO STIBINE

CONSENSUS REPORTS: Antimony and its compounds are on the Community Right-To-Know List.

SAFETY PROFILE: A poison. Corrosive to skin, eyes, and mucous membranes. Reaction with water liberates HBr and antimony trioxide. Can cause severe burns. See also ANTIMONY COMPOUNDS.

AQK250 ***HR: 3***
ANTIMONY TRICHLORIDE OXIDE
mf: Cl_3OSb mw: 244.1

CONSENSUS REPORTS: Antimony and its compounds are on the Community Right-To-Know List.

SAFETY PROFILE: Often a component of violent hazardous materials reactions. See also ANTIMONY COMPOUNDS. When heated to decomposition it emits very toxic fumes of Cl^- and Sb. Incompatible with BF_3.

AQK500 ***HR: 3***
ANTIMONY TRIETHYL
mf: $Sb(C_2H_5)_3$ mw: 209.0

PROP: Liquid, water-insol. D: 1.324 @ 16°, mp: −29°, bp: 159.5°.

CONSENSUS REPORTS: Antimony and its compounds are on the Community Right-To-Know List.

SAFETY PROFILE: Alkyl metal compounds are often highly toxic. See also ANTIMONY COMPOUNDS. Dangerous fire hazard by spontaneous chemical reaction. Explodes in air, water, carbon tetrachloride, other halogenated hydrocarbons, dimethyl formamide, and triethyl borine. When heated to decomposition it emits highly toxic fumes of Sb.

AQK750 ***HR: 3***
ANTIMONY TRIIODIDE
mf: SbI_3 mw: 502.5

PROP: Red-to-yellow crystals. Mp: 170°, bp: 401°, d: 4.768 @ 22°, vap press: 1 mm @ 163.6°.

CONSENSUS REPORTS: Antimony and its compounds are on the Community Right-To-Know List.

SAFETY PROFILE: Poison by ingestion. See also IODIDES and ANTIMONY COMPOUNDS. Incompatible with sodium, potassium. When heated to decomposition it emits highly toxic Sb fumes and I^-.

AQL000 ***HR: 3***
ANTIMONY TRIMETHYL
mf: $Sb(CH_3)_3$ mw: 166.9

PROP: Liquid, sltly sol in water. Bp: 80.6°, d: 1.523 @ 15°

SYN: TRIMETHYL STIBINE

CONSENSUS REPORTS: Antimony and its compounds are on the Community Right-To-Know List.

SAFETY PROFILE: Toxic. See also ANTIMONY TRIETHYL and ANTIMONY COMPOUNDS. Dangerous fire hazard by spontaneous reaction in air. Explodes in water. When heated to decomposition it emits highly toxic fumes of antimony. Incompatible with oxidizing materials, halogenated hydrocarbons.

AQL250 ***HR: 3***
ANTIMONY TRIPHENYL
mf: $C_{18}H_{15}Sb$ mw: 353.11

CONSENSUS REPORTS: Antimony and its compounds are on the Community Right-To-Know List.

SAFETY PROFILE: Poison by intraperitoneal and ingestion routes. Upon heating it burns in air. See also ANTIMONY COMPOUNDS. Incompatible with BrF_3.

AQL500 CAS:1345-04-6 ***HR: 3***
ANTIMONY TRISULFIDE
DOT: NA 1325
mf: S_3Sb_2 mw: 339.68

PROP: Red-to-black crystals. Mp: 546°, d: 4.64, bp: ca. 1150°. Sol in H_2SO_4, solubility in water = 0.002/100 @ 20° (decomp).

SYNS: ANTIMONOUS SULFIDE ◇ ANTIMONY GLANCE ◇ ANTIMONY ORANGE ◇ ANTIMONY SULFIDE ◇ C.I. 77060 ◇ CRIMSON ANTIMONY ◇ LYMPHOSCAN ◇ NEEDLE ANTIMONY

TOXICITY DATA with REFERENCE
ihl-hmn TCLo:580 μg/m^3/35W:BLD,GIT IMSUAI 23,521,54

ipr-rat LDLo:1390 mg/kg INMEAF 10,15,41
ipr-mus LD50:209 mg/kg 85GMAT-,23,82

CONSENSUS REPORTS: IARC Cancer Review: Group 3 IMEMDT 47,291,89; Animal Limited Evidence IMEMDT 47,291,89; Human Inadequate Evidence IMEMDT 47,291,89. Reported in EPA TSCA Inventory. Antimony and its compounds are on the Community Right-To-Know List.

OSHA PEL: TWA 500 μg(Sb)/m^3
ACGIH TLV: TWA 0.5 mg(Sb)/m^3
NIOSH REL: (Antimony) TWA 0.5 mg(Sb)/m^3
DOT Classification: ORM-A; Label: None.

SAFETY PROFILE: Poison by intraperitoneal route. Human blood and gastrointestinal system effects by inhalation. Questionable carcinogen. See also ANTIMONY COMPOUNDS and SULFIDES. Spontaneously flammable when exposed to strong oxidizers. Flammable when exposed to heat or flame. Moderately explosive by spontaneous reaction with chlorates, perchlorates, ClO, thallic oxide. When heated to decomposition or on contact with acid or acid fumes it emits highly toxic fumes of oxides of sulfur and antimony. Will react with water or steam to produce toxic and flammable vapors.

AQL750 ***HR: 3***
ANTIMONY TRITELLURIDE
mf: Sb_2Te_3 mw: 626.4

PROP: Gray powder. Mp: 629°; d: 6.50 @ 13°.

SYN: ANTIMONY TELLURIDE

CONSENSUS REPORTS: Antimony and its compounds are on the Community Right-To-Know List.

SAFETY PROFILE: Probably a poison. See also ANTIMONY COMPOUNDS and TELLURIUM COMPOUNDS. Flammable by spontaneous reaction with strong oxidizers. Moderately explosive by chemical reaction in contact with chlorates and perchlorates. When heated to decomposition or on contact with acid or acid fumes it emits highly toxic fumes of Sb and tellurium. Incompatible with water or steam and oxidizing materials.

AQM000 CAS:11118-72-2 ***HR: 3***
ANTIMYCIN

SYN: FINTROL

TOXICITY DATA with REFERENCE
orl-rat LD50:28 mg/kg TAFSAI 96,320,67
ipr-rat LD50:1600 μg/kg TAFSAI 96,320,67
orl-mus LD50:55 mg/kg TAFSAI 96,320,67
ipr-mus LD50:1700 μg/kg TAFSAI 96,320,67
orl-rbt LD50:10 mg/kg TAFSAI 96,320,67
orl-gpg LD50:1800 μg/kg TAFSAI 96,320,67
orl-qal LD50:39 mg/kg

SAFETY PROFILE: Poison by ingestion and intraperitoneal routes.

AQM250 CAS:1397-94-0 ***HR: 3***
ANTIMYCIN A
mf: $C_{28}H_4N_2O_9$ mw: 512.34

SYNS: ANTIPIRICULLIN ◇ VIROSIN

TOXICITY DATA with REFERENCE
orl-rat LDLo:30 mg/kg 85ERAY 2,1078,78
ipr-rat LD50:800 μg/kg CNREA8 13,49,53
scu-rat LD50:25 mg/kg 85ERAY 2,1078,78
ipr-mus LD50:820 μg/kg TDKNAF 14,60,55
scu-mus LD50:21 mg/kg JAJAAA 9,63,56
ivn-mus LD50:893 μg/kg JAJAAA 9,63,56

CONSENSUS REPORTS: EPA Extremely Hazardous Substances List.

SAFETY PROFILE: Poison by ingestion, intraperitoneal, subcutaneous, and intravenous routes. When heated to decomposition it emits toxic fumes of NO_x.

AQM260 CAS:27220-59-3 ***HR: 3***
ANTIMYCIN A4
mf: $C_{25}H_{34}N_2O_9$ mw: 506.61

TOXICITY DATA with REFERENCE
ipr-mus LD50:7600 μg/kg 85FZAT -,146,67
scu-mus LD50:25 mg/kg 85FZAT -,146,67
ivn-mus LD50:900 μg/kg 85FZAT -,146,67

SAFETY PROFILE: Poison by subcutaneous, intravenous, and intraperitoneal routes. When heated to decomposition it emits toxic fumes of NO_x.

AQM500 ***HR: 3***
ANTIMYCOIN

PROP: An antifungal agent produced by the strain *Streptomyces aureus* 3569 (85ERAY 2,959,78).

TOXICITY DATA with REFERENCE
ipr-mus LD50:204 mg/kg 85ERAY 2,959,78
scu-mus LD50:532 mg/kg 85ERAY 2,959,78

SAFETY PROFILE: Poison by intraperitoneal route. Moderately toxic by subcutaneous route.

AQN000 CAS:60-80-0 ***HR: 3***
ANTIPYRINE
mf: $C_{11}H_{12}N_2O$ mw: 188.23

PROP: Fine, white crystals. Mp: 113°, bp: 319° @ 174 mm, d: 1.19. Very sol in water and alc; sltly sol in ether.

SYNS: DIMETHYLOXYQUINAZINE ◇ 2,3-DIMETHYL-1-PHENYL-3-PYRAZOLIN-5-ONE ◇ 2,3-DIMETHYL-1-PHENYL-5-PYRAZOLONE ◇ OXYDIMETHYLQUINAZINE ◇ PHENAZONE (pharmaceutical) ◇ 1-PHENYL-2,3-DIMETHYLPYRAZOLE-5-ONE ◇ 1-PHENYL-2,3-DIMETHYL-5-PYRAZOLONE

TOXICITY DATA with REFERENCE
orl-rat TDLo:361 g/kg/92W-C:ETA IJCNAW 27,521,81
unk-man LDLo:74 mg/kg 85DCAI 2,73,70
orl-rat LD50:1800 mg/kg JPETAB 89,205,47
scu-rat LDLo:1570 mg/kg AEPPAE 186,195,37
orl-mus LD50:1700 mg/kg ARZNAD 18,1404,68
ipr-mus LD50:750 mg/kg AIPTAK 135,376,62
scu-mus LD50:1000 mg/kg ARZNAD 17,214,67
ivn-mus LD50:500 mg/kg ARZNAD 10,686,60
orl-dog LDLo:500 mg/kg HBAMAK 4,1304,35

CONSENSUS REPORTS: Reported in EPA TSCA Inventory.

SAFETY PROFILE: A human poison by an unspecified route. Moderately toxic via ingestion, subcutaneous, and intravenous routes. Questionable carcinogen with experimental tumorigenic data. When heated to decomposition it emits toxic fumes of NO_x.

AQN250 CAS:520-07-0 ***HR: 2***
ANTIPYRINE SALICYLATE
mf: $C_{11}H_{12}N_2O{\cdot}C_7H_6O_3$ mw: 326.38

SYNS: ANSAL ◇ SALAZOLON ◇ SALIPHENAZON ◇ SALIPYRAZOLAN ◇ SALIPYRINE

TOXICITY DATA with REFERENCE
unk-mus LDLo:1200 mg/kg HBAMAK 4,1289,35
unk-gpg LDLo:1600 mg/kg HBAMAK 4,1289,35

SAFETY PROFILE: Moderately toxic by unspecified routes. When heated to decomposition it emits toxic fumes of NO_x. An analgesic and antipyretic.

AQN500 CAS:15387-10-7 ***HR: 3***
N-((ANTIPYRINYLISOPROPYLAMINO) METHYL)NICOTINAMIDE
mf: $C_{21}H_{25}N_5O_2$ mw: 379.51

SYN: NICOTINAMIDOMETHYLAMINOPYRAZOLONE

TOXICITY DATA with REFERENCE
ipr-rat LD50:400 mg/kg ARZNAD 20,1024,70
orl-mus LD50:1460 mg/kg ARZNAD 20,1024,70
ipr-mus LD50:1060 mg/kg ARZNAD 20,1024,70
ipr-ham LD50:853 mg/kg ARZNAD 20,1024,70

SAFETY PROFILE: Poison by intraperitoneal route. Moderately toxic by ingestion and intraperitoneal routes. When heated to decomposition it emits toxic fumes of NO_x.

AQN625 CAS:3270-78-8 ***HR: 3***
ANTRYCIDE METHYL SULFATE
mf: $C_{17}H_{21}N_6{\cdot}CH_4O_4S{\cdot}CH_3O_4S$ mw: 532.65

SYNS: 4-AMINO-6-((2-AMINO-1,6-DIMETHYLPYRIMIDINIUM-4-YL)AMINO)-1-METHYL-QUINALDINIUM BIS(METHYL SULFATE) ◇ ANTRYCIDE ◇ QUINAPYRAMINE ◇ QUINAPYRAMINE METHYL SUFLATE

TOXICITY DATA with REFERENCE
dns-omi 50 μmol/L CNREA8 45,112,85
scu-rat LD50:18 mg/kg BJPCAL 5,25,50
ipr-mus LD50:15 mg/kg BJPCAL 5,25,50
scu-mus LD50:20 mg/kg BJPCAL 5,25,50
ivn-mus LD50:10 mg/kg MEIEDD 10,1162,83
scu-rbt LD50:15 mg/kg BJPCAL 5,25,50
ivn-rbt LD50:5 mg/kg BJPCAL 5,25,50

SAFETY PROFILE: Poison by subcutaneous, intravenous, and intraperitoneal routes. Mutagenic data. When heated to decomposition it emits toxic fumes of NO_x and SO_x.

AQN635 CAS:86-88-4 ***HR: 3***
ANTU
DOT: UN 1651
mf: $C_{11}H_{10}N_2S$ mw: 202.29

PROP: Crystals; bitter taste. Mp: 198°. Sltly sol in hot alc.

SYNS: ALPHANAPHTHYL THIOUREA ◇ ALPHANAPHTYL THIOUREE (FRENCH) ◇ ALRATO ◇ ANTURAT ◇ CHEMICAL 109 ◇ DIRAX ◇ KILL KANTZ ◇ KRYSID ◇ 1-NAFTIL-TIOUREA (ITALIAN) ◇ 1-NAFTYLTHIOUREUM (DUTCH) ◇ 1-NAPHTHALENYLTHIOUREA ◇ α-NAPHTHALTHIOHARNSTOFF (GERMAN) ◇ α-NAPHTHOTHIOUREA ◇ α-NAPHTHYLTHIOCARBAMIDE ◇ 1-NAPHTHYL-THIOHARNSTOFF (GERMAN) ◇ 1-NAPHTHYL THIOUREA (MAK) ◇ α-NAPHTHYLTHIOUREA ◇ 1-(1-NAPHTHYL)-2-THIOUREA ◇ N-(1-NAPHTHYL)-2-THIOUREA ◇ α-NAPHTHYLTHIOUREA (DOT) ◇ 1-NAPHTHYL-THIOUREE (FRENCH) ◇ NAPHTOX ◇ RATTRACK ◇ RCRA WASTE NUMBER P072 ◇ SMEESANA ◇ U-5227 ◇ USAF EK-P-5976

TOXICITY DATA with REFERENCE
mma-sat 500 μmol/L ENMUDM 3,11,81
otr-ham:emb 1600 μg/L NCIMAV 58,243,81
scu-mus TDLo:5 mg/kg:ETA NTIS** PB223-159
unr-man LDLo:588 mg/kg 85DCAI 2,73,70
orl-rat LD50:6 mg/kg AFDOAQ 16,47,52
ipr-rat LD50:2470 μg/kg JPETAB 97,432,49
ipr-mus LD50:10 mg/kg NTIS** AD 277-689
orl-dog LD50:380 μg/kg PCOC** -,57,66
ipr-dog LD50:16 mg/kg PSEBAA 62,22,46
orl-mky LD50:4250 mg/kg 85DPAN -,-,71/76

CONSENSUS REPORTS: IARC Cancer Review: Animal Inadequate Evidence IMEMDT 30,347,83. Reported in EPA TSCA Inventory. EPA Extremely Hazardous Substances List. EPA Genetic Toxicology Program.

OSHA PEL: TWA 0.3 mg/m^3
ACGIH TLV: TWA 0.3 mg/m^3
DFG MAK: 0.3 mg/m^3
DOT Classification: Poison B; Label: Poison.

SAFETY PROFILE: Poison by ingestion and intraperitoneal routes. Moderately toxic to humans by an unspecified route. Questionable carcinogen with experimental tumorigenic data. Mutagenic data. A rodenticide used extensively. Death is caused by pulmonary edema. Chronic toxicity has been known to cause dermatitis and a decrease in the white blood cells. When heated to decomposition it emits toxic fumes of NO_x and SO_x.

AQN650 CAS:24345-16-2 *HR: 3*
APAMINE
mf: $C_{79}H_{131}O_{24}S_4$ mw: 2027.65

PROP: Highly basic compd.

SYN: APAMIN

TOXICITY DATA with REFERENCE
ipr-mus LD50:3800 μg/kg TOXIA6 22,308,84
ivn-mus LD50:4 mg/kg NSAPCC 300,189,77
ice-mus LD50:1800 ng/kg TOXIA6 22,308,84
par-mus LD50:600 mg/kg TOXIA6 20,157,82

SAFETY PROFILE: Poison by intravenous, parenteral, intracerebral, and intraperitoneal routes. When heated to decomposition it emits toxic fumes of SO_x and NO_x.

AQN750 CAS:13539-59-8 *HR: 2*
APAZONE
mf: $C_{16}H_{20}N_4O_2$ mw: 300.40

SYNS: AZAPROPAZON (GERMAN) ◇ AZAPROPAZONE (anhydrous) ◇ CINNAMIN ◇ CINNOPROPAZONE ◇ 1,2-DIHYDRO-3-DIMETHYLAMINO-7-METHYL-1,2-(PROPYLMALONYL)-1,2,4-BENZOTRIAZINE ◇ 3-DIMETHYLAMINO-7-METHYL-1,2-(n-PROPYLMALONYL)-1,2-DIHYDRO-1,2,4-BENZOTRIAZINE ◇ 5-DIMETHYLAMINO-9-METHYL-2-PROPYL-1H-PYRAZOLO(1,2-a)(1,2,4)BENZOTRIAZINE-1,3(2H)-DIONE ◇ MI 85 ◇ MSC-102824 ◇ PROLIXAN ◇ RHEUMOX ◇ SINNAMIN

TOXICITY DATA with REFERENCE
orl-rat LD50:1800 mg/kg OYYAA2 15,41,78
ipr-rat LD50:650 mg/kg CMROCX 4,17,76
ivn-rat LD50:660 mg/kg CMROCX 4,17,76
orl-mus LD50:1080 mg/kg CMROCX 4,17,76
ipr-mus LD50:920 mg/kg CMROCX 4,17,76
ivn-mus LD50:680 mg/kg CMROCX 4,17,76
ivn-cat LD50:500 mg/kg CMROCX 4,17,76

SAFETY PROFILE: Moderately toxic by ingestion, intraperitoneal and intravenous routes. When heated to decomposition it emits toxic fumes of NO_x.

AQO000 CAS:52-46-0 *HR: 3*
APHOLATE
mf: $C_{12}H_{24}N_9P_3$ mw: 387.36

SYNS: APN ◇ AZIRIDINE-1,3,5,2,4,6-TRIAZATRIPHOSPHORINE DERIVATIVE ◇ 1-AZIRIDINYLPHOSPHONITRILE TRIMER ◇ ENT 26,316 ◇ HEXA(1-AZIRIDINYL)TRIPHOSPHOTRIAZINE ◇ 2,2,4,4,6,6-HEXAHYDRO-2,2,4,4,6,6-HEXAKIS(1-AZIRIDINYL)-1,3,5,2,4,6-TRIAZATRIPHOSPHORINE ◇ 2,2,4,4,6,6-HEXAKIS(1-AZIRIDINYL) CYCLOTRIPHOSPHAZA-1,3,5-TRIENE ◇ 2,2,4,4,6,6-HEXAKIS (1-AZIRIDINYL)-2,2,4,4,6,6-HEXAHYDRO-1,3,5,2,4,6-TRIAZATRIPHOSPHORINE ◇ HEXAKIS-(1-AZIRIDINYL)PHOSPHONITRILE ◇ HEXAKIS(AZIRIDINYL)PHOSPHOTRIAZINE ◇ NSC-26812 ◇ OLIN MO. 2174 ◇ PHOLATE ◇ PN6 ◇ SQ 8388

TOXICITY DATA with REFERENCE
cyt-mus-ipr 15 mg/kg PISCAD 59(Pt.3),417,72
cyt-hmn:leu 100 μmol/L CHROAU 24,314,68
ipr-rat TDLo:10 mg/kg (11D preg):TER AEHLAU 16,805,68
ipr-mus TDLo:5 mg/kg (male 5D pre):REP EXPEAM 24,924,68
orl-rat LD50:98 mg/kg TXAPA9 14,515,69
orl-mus LD50:110 mg/kg JAFCAU 14,301,66
ipr-mus LD50:50 mg/kg JMCMAR 29,1341,86

CONSENSUS REPORTS: IARC Cancer Review: Group 3 IMEMDT 7,56,87; Animal Inadequate Evidence IMEMDT 9,31,75. EPA Genetic Toxicology Program.

SAFETY PROFILE: Poison by ingestion and intramuscular routes. An experimental teratogen. Other experimental reproductive effects. Human mutagenic data. When heated to decomposition it emits very toxic fumes of NO_x and PO_x.

AQO250 CAS:500-55-0 *HR: 3*
APOATROPINE
mf: $C_{17}H_{21}NO_2$ mw: 271.39

SYNS: APOATROPIN ◇ ATROPAMIN ◇ ATROPAMINE ◇ ATROPYLTROPEINE ◇ endo-α-METHYLENEBENZENEACETIC ACID 8-METHYL-8-AZABICYCLO(3.2.1)OCT-3-YL ESTER ◇ 1-α-H,5-α-H-TROPAN-3-α-OL, ATROPATE (ESTER) ◇ TROPIC ACID, 3-α-TROPANYL ESTER ◇ TROPINE, ATROPATE (ESTER)

TOXICITY DATA with REFERENCE
orl-mus LD50:160 mg/kg MEIEDD 11,117,89
ipr-mus LD50:10400 μg/kg NTIS** PB85-203544

SAFETY PROFILE: Poison by ingestion and intraperitoneal routes. See also ESTERS. When heated to decomposition it emits toxic fumes of NO_x. An antispasmodic agent.

AQO500 CAS:641-81-6 *HR: 2*
APOCHOLIC ACID
mf: $C_{24}H_{38}O_4$ mw: 390.62

SYN: 3-α,12-α-DIHYDROXY-5-β-CHOL-8(14)-EN-24-OIC ACID

TOXICITY DATA with REFERENCE
scu-mus TDLo:200 mg/kg/13W-I:ETA NATUAS 190,1007,61

SAFETY PROFILE: Questionable carcinogen with experimental tumorigenic data. When heated to decomposition it emits acrid smoke and irritating fumes.

AQO750 ***HR: 3***
APOCODEINE
mf: $C_{18}H_{19}NO_2$ mw: 281.34

PROP: White, crystalline solid. Mp: 124°.

SAFETY PROFILE: Poison by inhalation and ingestion. A weak sensitizer and may cause contact dermatitis. See also CODEINE. When heated to decomposition it emits highly toxic fumes of NO_x.

AQP000 CAS:1937-37-7 ***HR: 3***
APOMINE BLACK GX
mf: $C_{34}H_{25}N_9O_7S_2•2Na$ mw: 781.78

SYNS: AHCO DIRECT BLACK GX ◇ AIREDALE BLACK ED ◇ AIZEN DIRECT DEEP BLACK GH ◇ AMANIL BLACK GL ◇ ATLANTIC BLACK BD ◇ ATUL DIRECT BLACK E ◇ AZINE DEEP BLACK EW ◇ AZOCARD BLACK EW ◇ AZOMINE BLACK EWO ◇ BELAMINE BLACK GX ◇ BENCIDAL BLACK E ◇ BENZAMIL BLACK E ◇ BENZO DEEP BLACK E ◇ BENZOFORM BLACK BCN-CF ◇ BLACK 2EMBL ◇ BRASILAMINA BLACK GN ◇ CALCOMINE BLACK ◇ CARBIDE BLACK E ◇ CERN PRIMA 38 ◇ CHLORAMINE BLACK C ◇ CHLORAZOL BLACK E (biological stain) ◇ CHLORAZOL BLACK EA ◇ CHLORAZOL BLACK EN ◇ CHROME LEATHER BLACK EM ◇ C.I. 30235 ◇ C.I. DIRECT BLACK 38 ◇ COIR DEEP BLACK C ◇ COLUMBIA BLACK EP ◇ DIACOTTON DEEP BLACK ◇ DIAMINE DEEP BLACK EC ◇ DIAPHTAMINE BLACK V ◇ DIAZINE BLACK E ◇ DIAZOL BLACK 2V ◇ DIPHENYL DEEP BLACK G ◇ DIRECT BLACK A ◇ DIRECT BLACK META ◇ ENIANIL BLACK CN ◇ ERIE BLACK B ◇ FENAMIN BLACK E ◇ FIBRE BLACK VF ◇ FIXANOL BLACK E ◇ FORMALINE BLACK C ◇ FORMIC BLACK C ◇ HISPAMIN BLACK EF ◇ INTERCHEM DIRECT BLACK Z ◇ KAYAKU DIRECT DEEP BLACK EX ◇ LURAZOL BLACK BA ◇ META BLACK ◇ MITSUI DIRECT BLACK EX ◇ NCI-C54557 ◇ NIPPON DEEP BLACK ◇ PAPER BLACK BA ◇ PARAMINE BLACK B ◇ PEERAMINE BLACK E ◇ PHENO BLACK EP ◇ PONTAMINE BLACK E ◇ SANDOPEL BLACK EX ◇ SERISTAN BLACK B ◇ TELON FAST BLACK E ◇ TETRAZO DEEP BLACK G ◇ TERTRODIRECT BLACK E ◇ TETRODIRECT BLACK EFD ◇ UNION BLACK EM ◇ VONDACEL BLACK N

TOXICITY DATA with REFERENCE
mmo-sat 10 μg/plate TOLED5 4,519,79
mma-sat 10 μg/plate TOLED5 4,519,79
mnt-rat-orl 500 mg/kg/36H-C MUREAV 187,227,87
dns-rat-orl 500 mg/kg/12H-C MUREAV 187,227,87
bfa-rat:sat 300 mg/kg IAEHDW 49,177,81
orl-rat TDLo:6825 mg/kg/13W-C:CAR NCITR* NCI-TR-108,78
orl-mus TDLo:34 mg/kg/60W-C:ETA TJIDAH 88,467,73

CONSENSUS REPORTS: NTP Fifth Annual Report On Carcinogens, 1984. IARC Cancer Review: Animal Sufficient Evidence IMEMDT 29,295,82, Human Limited Evidence IMEMDT 29,295,82. Reported in EPA TSCA Inventory. NTP Carcinogenesis Bioassay (feed): Clear Evidence: rat NCICTR* NCI-TR-108,78; No Evidence: mouse NCICTR NCI-TR-108,78. On Community-Right-To-Know List.

SAFETY PROFILE: Confirmed carcinogen with carcinogenic and tumorigenic data. Mutation data reported. When heated to decomposition it emits very toxic fumes of NO_x, Na_2O, and SO_2.

AQP250 CAS:58-00-4 ***HR: 3***
APORMORPHINE
mf: $C_{17}H_{17}NO_2$ mw: 267.35

PROP: White, crystalline alkaloid. Mp: 195° (decomp).

SYNS: APOMORFIN ◇ APOMORPHINE ◇ 6A-β-APORPHINE-10,11-DIOL

TOXICITY DATA with REFERENCE
scu-rat TDLo:7 mg/kg (female 13-19D post):REP TJADAB 33,100C,86
ivn-rat LDLo:40 mg/kg ARZNAD 10,1003,60
orl-mus LD50:300 mg/kg FRPSAX 35,951,80
ipr-mus LD50:160 mg/kg JMCMAR 15,348,72
scu-mus LDLo:13 mg/kg HBAMAK 4,1289,35
ivn-mus LD50:56 mg/kg CSLNX* NX#03170

SAFETY PROFILE: Poison by ingestion, subcutaneous, intravenous, and intraperitoneal routes. Experimental reproductive effects. Central nervous system effects. A powerful emetic. A weak sensitizer and may cause contact dermatitis. When heated to decomposition it emits highly toxic fumes of NO_x.

AQP500 CAS:314-19-2 ***HR: 3***
APORMORPHINE CHLORIDE
mf: $C_{17}H_{17}NO_2•ClH$ mw: 303.81

SYNS: 6A-β-APORMPHINE-10,11-DIOL HYDROCHLORIDE ◇ N-METHYLNORAPORMORPHINE HYDROCHLORIDE

TOXICITY DATA with REFERENCE
ivn-mus LD50:38 mg/kg TXAPA9 6,334,64
ipr-mus LD50:247 mg/kg NCIAL* -,60,69
ivn-mus LD50:71 mg/kg JMCMAR 18,1000,75

CONSENSUS REPORTS: Reported in EPA TSCA Inventory.

SAFETY PROFILE: Poison by intravenous and intraperitoneal routes. See also APORMORPHINE. When heated to decomposition it emits very toxic fumes of NO_x and HCl.

AQP750 CAS:4361-80-2 ***HR: 3***
APOTHESINE
mf: $C_{16}H_{23}NO_2$ mw: 261.40

SYN: CINNAMIC ACID-3-(DIETHYLAMINO) PROPYL ESTER

TOXICITY DATA with REFERENCE
ivn-rat LDLo:20 mg/kg AJPHAP 68,120,24
ipr-mus LDLo:700 mg/kg JLCMAK 11,1082,26
scu-mus LDLo:700 mg/kg JLCMAK 11,1082,26
ivn-cat LDLo:20 mg/kg AJPHAP 68,120,24
scu-gpg LDLo:250 mg/kg JLCMAK 11,1082,26

SAFETY PROFILE: Poison by intravenous and subcutaneous routes. Moderately toxic by intraperitoneal route. When heated to decomposition it emits toxic fumes of NO_x.

AQP800 ***HR: 3***
APPLE of SODOM (extract)

SYNS: BEC 001 ◇ SOLANUM SODOMEUM, extract

TOXICITY DATA with REFERENCE
unr-rat LD50:41 mg/kg EPXXDW #20029
orl-mus LD50:550 mg/kg EPXXDW #20029
ipr-mus LD50:30 mg/kg EPXXDW #20029

SAFETY PROFILE: Poison by intraperitoneal and possibly other routes. Moderately toxic by ingestion.

AQP875 ***HR: 3***
APPLE SEEDS

PROP: Seeds of a deciduous tree widely cultivated in many temperate regions. The fruit is commonly available.

SYNS: MALUS (VARIOUS SPECIES) ◇ MANZANA (SPANISH) ◇ POMMIER (FRENCH)

SAFETY PROFILE: The seeds contain a cyanogenetic glycoside. Ingestion of small quantities of the seeds is harmless. After a delay period systemic effects from ingestion may include abdominal pain, vomiting, lethargy and coma. An adult died from chewing and swallowing a cup of seeds. See also CYANIDE.

AQP890 ***HR: 3***
APRICOT PITS

PROP: Fruit bearing trees and shrubs widely cultivated in temperate regions. Various varieties of fruit are commonly available.

SYNS: ALBARICOQUE (SPANISH) ◇ CEREZA (SPANISH) ◇ CHERRY ◇ CHOKE CHERRY ◇ MELOCOTON (SPANISH) ◇ PEACH ◇ PLUM ◇ PRUNUS (VARIOUS SPECIES) ◇ SLOE

SAFETY PROFILE: The seed kernel contains cyanogenetic glycosides. After a delay period systemic effects from ingestion may include abdominal pain, vomiting, lethargy and coma. Most fatalities result from ingestion of apricot pits or their products. See also CYANIDE.

AQQ000 CAS:50650-74-3 ***HR: 3***
APTROL SULFATE
mf: $C_{10}H_{15}N{\cdot}1/2H_2O_4S$ mw: 198.30

SYNS: 4-METHYLPHENISOPROPYLAMINE SULFATE ◇ 1-(4-METHYLPHENYL)-2-PROPYLAMINE SULFATE ◇ β-p-TOLYL-ISOPROPYLAMINE SULFATE ◇ p-XYLYLMETHYLCARBINAMINE SULFATE

TOXICITY DATA with REFERENCE
orl-man TDLo:1500 μg/kg:GIT JPETAB 100,298,50
ipr-mus LD50:136 mg/kg JPETAB 100,298,50

SAFETY PROFILE: Poison by intraperitoneal route. Human gastrointestinal tract effects by ingestion. See also SULFATES. When heated to decomposition it emits very toxic fumes of SO_x and NO_x.

AQQ100 CAS:17168-82-0 ***HR: 3***
AQUA-1,2-DIAMINOETHANE DIPEROXO CHROMIUM(IV)
mf: $C_2H_{10}CrN_2O_5$ mw: 194.11

$$Cr(C_2H_{10}N_2)(O_2)_2{\cdot}H_2O$$

CONSENSUS REPORTS: Chromium and its compounds are on the Community Right-To-Know List.

SAFETY PROFILE: A poison. A light-sensitive explosive which may explode when heated above 96°C. When heated to decomposition it emits toxic fumes of NO_x. See also CHROMIUM COMPOUNDS and PEROXIDES.

AQQ125 CAS:17185-68-1 ***HR: 3***
AQUA-1,2-DIAMINOPROPANEDIPEROXO-CHROMIUM(IV) DIHYDRATE
mf: $C_3H_{12}CrN_2O_5{\cdot}2H_2O$ mw: 244.16

CONSENSUS REPORTS: Chromium and its compounds are on the Community Right-To-Know List.

OSHA PEL: CL 0.1 mg(CrO_3)/m^3
ACGIH TLV: TWA 0.05 mg(Cr)/m^3
NIOSH REL: (Chromium(VI)) TWA 25 μg(Cr(VI))/m^3; CL 50 μg/m^3/15M

SAFETY PROFILE: May explode spontaneously at room temperature. Upon decomposition it emits toxic fumes of NO_x. See also CHROMIUM COMPOUNDS.

AQQ250 CAS:6091-11-8 ***HR: 3***
AR-45
mf: $C_{15}H_{21}N_3O_2{\cdot}CH_3I$ mw: 417.33

TOXICITY DATA with REFERENCE
orl-mus LDLo:250 mg/kg JPETAB 43,413,31
ivn-mus LDLo:750 μg/kg JPETAB 43,413,31

SAFETY PROFILE: Poison by ingestion and intravenous routes. When heated to decomposition it emits very toxic fumes of NO_x and I^-.

AQQ500 CAS:9000-01-5 ***HR: 2***
ARABIC GUM
mw: 240,000

PROP: A gum from the stems and branches of *Acacia senegal (L.)* Willd. or of *Acacia* (Fam. *Leguminosae*). Sol in water; insol in alc.

SYNS: ACACIA ◇ ACACIA DEALBATA GUM ◇ ACACIA GUM ◇ ACACIA SENEGAL ◇ ACACIA SYRUP ◇ AUSTRALIAN GUM ◇ GUM ARABIC ◇ GUM OVALINE ◇ GUM SENEGAL ◇ INDIAN GUM ◇ NCI-C50748 ◇ SENEGAL GUM ◇ STARSOL No. 1 ◇ WATTLE GUM

TOXICITY DATA with REFERENCE
eye-rbt 36 mg/5H SEV AROPAW 78,384,67
dlt-rat-orl 54600 mg/kg/10W-C ENMUDM 8,357,86
orl-rat TDLo:350 g/kg (male 10W pre):REP ENMUDM 8,357,86
orl-rbt LD50:8000 mg/kg FDRLI* 124,-,76

CONSENSUS REPORTS: NTP Carcinogenesis Bioassay (feed); No Evidence: mouse, rat NTPTR* NTP-TR-227,82. Reported in EPA TSCA Inventory.

SAFETY PROFILE: Mildly toxic by ingestion. Inhalation or ingestion has produced hives, eczema, and angiodema. Experimental reproductive effects. A severe eye irritant. A weak allergen. Mutation data reported. Combustible. When heated to decomposition it emits acrid smoke.

AQQ750 CAS:147-94-4 ***HR: 3***
ARABINOCYTIDINE
mf: $C_9H_{13}N_3O_5$ mw: 243.25

SYNS: 4-AMINO-1-ARABINOFURANOSYL-2-OXO-1,2-DIHYDROPYRIMIDINE ◇ 4-AMINO-1-β-d-ARABINOFURANOSYL-2(1H)-PYRIMIDINONE (9CI) ◇ 1-β-d-ARABINOFURANOSYL-4-AMINO-2(1H)PYRIMIDINONE ◇ 1-ARABINOFURANOSYLCYTOSINE ◇ 1-β-ARABINOFURANOSYLCYTOSINE ◇ 1-(β-d-ARABINOFURANOSYL)CYTOSINE ◇ β-d-ARABINOSYLCYTOSINE ◇ CYTOSINE-β-ARABINOSIDE ◇ CYTOSINE β-d-ARABINOSIDE ◇ NCI-C04728 ◇ NSC 63878

TOXICITY DATA with REFERENCE
skn-hmn 45 mg/3W CTRRDO 63,619,79
eye-hmn 105 mg/7D-I AROPAW 72,535,64
cyt-hmn:leu 50 μmol/L/6H ECREAL 46,276,67
cyt-hmn:lym 3 mg/L/4H SOGEBZ 12,1552,76
mnt-ham-ipr 2 mg/kg/24H MUREAV 40,325,76
ipr-mus TDLo:50 mg/kg (female 12D post):REP TCMUD8 7,7,87
ipr-rat TDLo:100 mg/kg (female 11D post):TER DEBIAO 45,103,75
ipr-rat TDLo:2500 mg/kg/7W-I:ETA CANCAR 40,1935,77
scu-man TDLo:60 mg/kg/90W-I:EAR,BLD DICPBB 21,798,87
scu-wmn TDLo:6480 μg/kg/12D-I:BLD NEJMAG 310,1328,84
ivn-chd TDLo:33200 μg/kg/240D-I:CNS CANCAR 42,53,78
ivn-hmn TDLo:17241 mg/kg/6D-I:SKN AIMEAS 102,556,85
ivn-wmn TDLo:720 mg/kg/3D-I:CNS NEURAI 35,1475,85
ivn-man TDLo:649 mg/kg/4D-I:PNS DICPBB 21,177,87
ivn-man LDLo:1536 mg/kg/43W-I:PNS DICPBB 21,177,87
orl-mus LD50:3150 mg/kg DRUGAY 6,321,82
ipr-mus LD50:6400 mg/kg CNREA8 39,3575,79
ivn-mus LD50:7000 mg/kg DRUGAY 6,321,82

CONSENSUS REPORTS: NCI Carcinogenesis Studies (ipr); No Evidence: mouse, rat CANCAR 40,1935,77

SAFETY PROFILE: Human systemic effects: peripheral nerve fasciculations, degenerative brain changes, allergic dermatitis (after systemic exposure), blood changes, hearing acuity change, ataxia, spleen changes, central nervous system effects. An experimental teratogen. Other experimental reproductive effects. A human skin and eye irritant. Questionable carcinogen with experimental tumorigenic data. Human mutagenic data. When heated to decomposition it emits toxic fumes of NO_x.

AQQ900 CAS:5536-17-4 ***HR: 3***
9-β-d-ARABINO FURANOSYL ADENINE
mf: $C_{10}H_{13}N_5O_4$ mw: 267.28

SYNS: ADENINE ARABINOSIDE ◇ β-d-ARABINOSYLADENINE ◇ ARABINOSYLADENINE ◇ 9-ARABINOSYLADENINE ◇ VIDARABIN ◇ VIDARABINE

TOXICITY DATA with REFERENCE
ims-rat TDLo:2 g/kg (female 6-15D post):TER TJADAB 15,231,77
ivg-rat TDLo:1316 mg/kg (female 15-21D post):REP TJADAB 15,231,77
ivn-wmn LDLo:105 mg/kg/1W-I CMAJAX 132,392,85
ivn-hmn TDLo:3 mg/kg:CNS JIDIAQ 134,75,76
ivn-hmn TDLo:300 μg/kg:SYS JIDIAQ 133,A192,76
ivn-hmn TDLo:2 mg/kg:BLD JIDIAQ 134,75,76
ipr-rat LD50:1476 mg/kg IYKEDH 15,688,84
scu-rat LD50:8914 mg/kg IYKEDH 15,688,84
ivn-rat LD50:302 mg/kg IYKEDH 15,688,84
orl-mus LD50:7800 μg/kg 37ASAA 2,962,78
ipr-mus LD50:3057 mg/kg IYKEDH 15,688,84
scu-mus LD50:5086 mg/kg IYKEDH 15,688,84
ivn-mus LD50:442 mg/kg IYKEDH 15,688,84

CONSENSUS REPORTS: Reported in EPA TSCA Inventory.

SAFETY PROFILE: Poison by ingestion and intravenous routes. Moderately toxic by intraperitoneal route. An experimental teratogen. Other experimental reproductive effects. Human systemic effects by intravenous route: central nervous system, blood, and other effects. When heated to decomposition it emits toxic fumes of NO_x.

AQQ905 CAS:29984-33-6 ***HR: 2***
9-(β-d-ARABINOFURANOSYL)ADENINE-5′-(DIHYDROGEN PHOSPHATE)
mf: $C_{10}H_{14}N_5O_7P$ mw: 347.26

SYNS: ADENINE ARABINOSIDE MONOPHOSPHATE ◇ ARABINOSYLADENINE MONOPHOSPHATE ◇ ADENINE ARABINOSIDE 5′-MONOPHOSPHATE ◇ 5′-ARABINOSYLADENINE MONOPHOSPHATE ◇ 9-(5-o-PHOSPHONO-β-d-ARABINOFURANOSYL)-9H-PURIN-6-AMINE

TOXICITY DATA with REFERENCE
ipr-rat LD50:1700 mg/kg DRFUD4 4,547,79
ipr-mus LD50:1200 mg/kg DRFUD4 4,547,79

SAFETY PROFILE: Moderately toxic by intraperitoneal route. When heated to decomposition it emits very toxic fumes of PO_x and NO_x.

AQR000 CAS:69-74-9 ***HR: 3***
1-β-d-ARABINOFURANOSYLCYTOSINE HYDROCHLORIDE
mf: $C_9H_{13}N_3O_5{\cdot}ClH$ mw: 279.71

SYNS: ARABINOSYLCYTOSINE HYDROCHLORIDE ◇ CYLOCIDE ◇ CYTARABINE HYDROCHLORIDE ◇ CYTOSAR HYDROCHLORIDE ◇ CYTOSINE ARABINOSIDE HYDROCHLORIDE ◇ IRETIN ◇ NSC 63878 ◇ SPONGOCYTIDINE HYDROCHLORIDE

TOXICITY DATA with REFERENCE
eye-hmn 21 mg/7D-I MLD AJOPAA 60,1074,65
eye-mky 35 mg/15D-I MOD AJOPAA 60,1074,65
eye-rbt 42 mg/14D-I MOD AJOPAA 60,1074,65
dni-mus:lym 10 mg/L EJCAAH 6,379,70
msc-mus:leu 100 umol/L CNREA8 29,1881,69
ipr-rat TDLo:280 mg/kg (female 15D post):REP JONRA9 34,950,80
ipr-rat LD50:5500 mg/kg OYYAA2 6,1255,72
orl-mus LD50:826 mg/kg NCISP* JAN86
ipr-mus LD50:825 mg/kg OYYAA2 6,1255,72
scu-mus LD50:2262 mg/kg NCISP* JAN86
ivn-dog LD50:172 mg/kg OYYAA2 8,353,74
ivn-mky LD50:396 mg/kg OYYAA2 8,353,74

SAFETY PROFILE: Poison by intravenous route. Moderately toxic by intraperitoneal and subcutaneous routes. Experimental reproductive effects. A human eye irritant. Mutation data reported. When heated to decomposition it emits very toxic fumes of NO_x and HCl.

AQR250 CAS:4298-10-6 ***HR: D***
1-β-d-ARABINOFURANOSYL-5-FLUOROCYTOSINE
mf: $C_9H_{12}FN_3O_5$ mw: 261.24

TOXICITY DATA with REFERENCE
dns-hmn:hla 3800 nmol/L CNREA8 26,1661,66
oms-hmn:hla 3800 nmol/L CNREA8 26,1661,66
dni-mus-ipr 10 mg/kg CNREA8 29,1790,69

SAFETY PROFILE: Human mutation data reported. When heated to decomposition it emits very toxic fumes of F^- and NO_x.

AQR500 CAS:6742-07-0 ***HR: 2***
1-β-d-ARABINOFURANOSYL-2′,3′,5′-TRIACETATE
mf: $C_{15}H_{19}N_3O_8$ mw: 369.37

SYN: NSC-93150

TOXICITY DATA with REFERENCE
ipr-mus LD50:4400 mg/kg NCIHL* -,353,67
ivn-mus LD50:680 mg/kg NCIHL* -,353,67

SAFETY PROFILE: Moderately toxic by intravenous route. Mildly toxic by intraperitoneal route. When heated to decomposition it emits toxic fumes of NO_x.

AQS750 CAS:506-32-1 ***HR: 3***
ARACHIDONIC ACID
mf: $C_{20}H_{32}O_2$ mw: 304.52

SYNS: (ALL-Z)-5,8,11,14-EICOSATETRAENOIC ACID ◇ ARCHIDONATE

TOXICITY DATA with REFERENCE
dns-mus:mmr 10 mg/L CRNGDP 5,1123,84
sce-ham:ovr 320 umol/L PAACA3 27,95,86
scu-mus TDLo:80 μg/kg (female 1D pre):REP FESTAS 25,636,74
ivn-rat LDLo:100 mg/kg THBRAA 9,67,76
ivn-mus LD50:33 mg/kg JPETAB 224,369,83
ivn-rbt LDLo:1 mg/kg THBRAA 9,67,76

SAFETY PROFILE: Poison by intravenous route. Experimental reproductive effects. Mutation data reported. When heated to decomposition it emits acrid smoke and irritating fumes.

AQS875 CAS:41948-17-8 ***HR: 3***
ARA-C PALMITATE
mf: $C_{25}H_{43}N_3O_6$ mw: 481.71

SYNS: 1-β-d-ARABINOFURANOSYLCYTOSINE-5′-PALMITATE ◇ 1-β-d-ARABINOFURANOSYLCYTOSINE-5′-PALMITOYL ESTER ◇ ARABINOSYLCYTOSINE PALMITATE ◇ ARA-CP ◇ ARACYTIDINE-5′-PALMITATE ◇ CYTOSINE ARABINOSIDE PALMITATE ◇ PALMO-ARA-C

TOXICITY DATA with REFERENCE
dni-rat-ipr 200 mg/kg TJADAB 7,219,73
ipr-rat TDLo:200 mg/kg (12D preg):TER TJADAB 7,219,73
ipr-mus LD50:155 mg/kg NCISP* JAN86

SAFETY PROFILE: Poison by intraperitoneal route. An experimental teratogen. Mutation data reported. When heated to decomposition it emits toxic fumes of NO_x.

AQT250 CAS:1446-17-9 ***HR: 3***
ARATEN PHOSPHATE
mf: $C_{18}H_{26}ClN_3 \cdot H_3O_4P$ mw: 417.92

SYNS: 7-CHLORO-4((4-(DIETHYLAMINO)-1-METHYLBUTYL)AMINO)-QUINOLINE PHOSPHATE (1:1) ◇ CHLOROQUINE PHOSPHATE

TOXICITY DATA with REFERENCE
orl-wmn TDLo:2660 mg/kg (2-39W preg):REP AROTAA 80,407,64
orl-wmn TDLo:2660 mg/kg (2-39W preg):TER AROTAA 80,407,64
orl-hmn LDLo:43 mg/kg JFSCAS 5,201,65
orl-wmn TDLo:2740 mg/kg/39W:CNS ARPAAQ 93,209,72
orl-chd LDLo:38 mg/kg PEDIAU 27,95,61

SAFETY PROFILE: Human poison by ingestion. Human systemic effects by ingestion: muscle weakness. A human teratogen. Human reproductive effects: termination of pregnancy; developemental abnormalities of the eye, ear, and musculoskeletal system; and effects on newborn and postnatal development. Moderately toxic to humans by ingestion with musculo-skeletal effects. When heated to decomposition it emits very toxic fumes of Cl^-, NO_x, and PO_x.

AQT500 CAS:39300-45-3 ***HR: 3***
ARATHANE
mf: $C_{18}H_{24}N_2O_6$ mw: 364.44

PROP: Liquid.

SYNS: CAPRYLDINITROPHENYL CROTONATE ◇ 2-CAPRYL-4,6-DINITROPHENYL CROTONATE ◇ CROTONATE de 2,4-DINITRO 6-(1-METHYL-HEPTYL)-PHENYLE (FRENCH) ◇ 4,6-DINITRO-2-CAPRYLPHENYL CROTONATE ◇ 4,6-DINITRO-2-(2-CAPRYL)PHENYL CROTONATE ◇ DINITRO(1-METHYLHEPTYL)PHENYL CROTONATE ◇ 2,4-DINITRO-6-(1-METHYLHEPTYL)PHENYL CROTONATE ◇ 2,4-DINITRO-6-(2-OCTYL)PHENYL CROTONATE ◇ ENT 24,727 ◇ (6-(1-METHYL-HEPTYL)-2,4-DINITRO-FENYL)-CROTONAAT(DUTCH) ◇ (6-(1-METHYL-HEPTYL)-2,3-DINITRO-PHENYL)-CROTONAT (GERMAN) ◇ 2-(1-METHYLHEPTYL)-4,6-DINITROPHENYL CROTONATE ◇ (6-(1-METIL-EPITL)-2,4-DINITRO-FENIL)-CROTONATO (ITALIAN)

TOXICITY DATA with REFERENCE
mmo-sat 500 μg/plate MUREAV 116,185,83
mmo-smc 5 ppm RSTUDV 6,161,76
orl-mus TDLo:120 mg/kg (female 7-16D post):REP TCMUD8 6,33,86
orl-mus TDLo:50 mg/kg (female 7-16D post):TER TCMUD8 6,375,86
scu-mus TDLo:10 mg/kg:NEO NTIS** PB223-159
orl-rat LD50:980 mg/kg PCOC** -,636,66
ivn-rat LD50:23 mg/kg AIPTAK 119,31,59
orl-mus LD50:49500 μg/kg BCTKAG 8,373,75
orl-dog LD50:100 mg/kg SPEADM 78-1,25,78
orl-rbt LD50:2000 mg/kg SPEADM 78-1,25,78
skn-rbt LD50:9400 mg/kg SPEADM 74-1,-,74

SAFETY PROFILE: Poison by ingestion and intravenous routes. An experimental teratogen. Other experimental reproductive effects. Mutation data reported. Questionable carcinogen with experimental neoplastigenic data. See NITRATES. When heated to decomposition it emits toxic fumes of NO_x.

AQT575 CAS:55028-70-1 ***HR: 2***
ARBAPROSTIL
mf: $C_{21}H_{34}O_5$ mw: 366.55

SYNS: 15(R)-METHYLPROSTAGLANDIN E2 ◇ 15(R)-15-METHYLPROSTAGLANDIN E2

TOXICITY DATA with REFERENCE
ims-wmn TDLo:600 ng/kg (12W preg):REP CCPTAY 9,523,74

SAFETY PROFILE: Human reproductive effects by intramuscular route: terminates pregnancy. Other experimental reproductive effects.

AQT625 CAS:499-04-7 ***HR: D***
ARECAIDINE
mf: $C_7H_{11}NO_2$ mw: 141.19

PROP: Plates from dil alc. Decomp @ 232°. Freely sol in water and dilute alc; almost insol in abs alc, chloroform, ether, and benzene.

SYNS: ARECAINE ◇ METHYLGUVACINE ◇ N-METHYLGUVACINE ◇ 1,2,3,4-TETRAHYDRO-1-METHYL-3-PYRIDINECARBOXYLIC ACID

TOXICITY DATA with REFERENCE
sce-mus-ipr 1 g/kg/5D-I CALEDQ 23,189,84
otr-ham:kdy 2500 μg/L LANCAO 1,112,79
msc-ham:lng 10 mg/L CRNGDP 5,501,84

SAFETY PROFILE: Mutation data reported. When heated to decomposition it emits toxic fumes of NO_x.

AQT650 ***HR: 3***
ARECA NUT

PROP: From the Areca palm tree, a native to South Asia. Orange-yellow in color when ripe. The seed, the size of a small egg, is separated from the fibrous pericarp and used fresh, after sun drying, or curing. It is chewed either alone or as a component of mixtures including Betel leaf and/or tobacco. Also known as Betel nut and supari. The nut contains several alkaloids, primarily arecoline, arecaidine, arecolidine, guvacoline, and guvacine.

SYNS: BETEL NUT ◇ SUPARI (INDIA)

SAFETY PROFILE: 3-(methylnitrosamino)propionaldehyde is an experimental carcinogen. Arecoline is one of the agents responsible for betel quid addiction. It mimics the action of acetylcholine and acts as a stimulant. It is a poison by intraperitoneal route. Reactions in the mouth and during processing can produce from these alkaloids several nitrosamines: N-nitrosoguvacoline, 3-(methylnitrosamino)propionitrile, 3-(methylnitrosamino)propionaldehyde, and N-nitrosoguvacine. N-nitrosoguvacoline and N-nitrosoguvacine have been found in the mouths of betel quid users. Areca nut extracts are experimental carcinogens and mutagens. See also BETEL QUID and SMOKELESS TOBACCO.

AQT750 CAS:63-75-2 ***HR: 3***
ARECOLINE
mf: $C_8H_{13}NO_2$ mw: 155.22

PROP: Oily liquid. Bp: 209°.

SYNS: ARECAIDINE METHYL ESTER ◇ ARECOLINE BASE ◇ METHYL-1,2,5,6-TETRAHYDRO-1-METHYLNICOTINATE ◇ N-METHYL-Δ-TETRAHYDRONICOTINIC ACID METHYL ESTER ◇ N-METHYLTETRAHYDROPYRIDINE-β-CARBOXYLIC ACID METHYL ESTER ◇ 1,2,5,6-TETRAHYDRO-1-METHYLNICOTINIC ACID, METHYL ESTER

TOXICITY DATA with REFERENCE
skn-ham TDLo:2698 mg/kg/65W-I:NEO JNCIAM 53,1259,74
scu-mus LDLo:65 mg/kg JPETAB 35,75,29
ivn-mus LD50:36 mg/kg FATOAO 28,33,65
unk-mus LDLo:100 mg/kg HBAMAK 4,1289,35
scu-dog LD50:5 mg/kg FAZMAE 17,108,73
unk-dog LDLo:5 mg/kg HBAMAK 4,1289,35

CONSENSUS REPORTS: IARC Cancer Review: Animal Inadequate Evidence IMEMDT 37,141,85

SAFETY PROFILE: Poison by inhalation, ingestion, and subcutaneous routes. Questionable carcinogen with experimental neoplastigenic data. It mimics the action of acetylcholine, a neuro transmitter, and is a parasympathetic nervous system stimulant. Its action on the central nervous system can cause tremors. A mutagen. It is easily nitrosated to several nitrosamines. See also ESTERS and NITROSAMINES. It is the major alkaloid found in betel quid. Combustible, can react with oxidizing materials. When heated to decomposition it emits highly toxic fumes of NO_x.

AQU000 CAS:300-08-3 ***HR: 3***
ARECOLINE BROMIDE
mf: $C_8H_{13}NO_2$•BrH mw: 236.14

SYNS: ARECOLINE HYDROBROMIDE ◇ METHYL-1,2,5,6-TETRAHYDRO-1-METHYLNICOTINATE, HYDROBROMIDE ◇ 1,2,5,6-TETRAHYDRO-1-METHYLNICOTINIC ACID, METHYL ESTER, HYDROBROMIDE

TOXICITY DATA with REFERENCE
par-rat LD50:270 mg/kg ABMGAJ 28,681,72
scu-mus LDLo:65 mg/kg JPETAB 35,75,29
ivn-mus LD50:18 mg/kg CSLNX* NX#11778

SAFETY PROFILE: Poison by parenteral, subcutaneous, and intravenous routes. When heated to decomposition it emits very toxic fumes of HBr and NO_x.

AQU250 CAS:61-94-9 ***HR: 3***
ARECOLINE HYDROCHLORIDE
mf: $C_8H_{13}NO_2$•ClH mw: 191.68

SYNS: NICOTINICACID-1,2,5,6-TETRAHYDRO-1-METHYL-, METHYL ESTER, HYDROCHLORIDE ◇ 3-PYRIDINECARBOXYLIC ACID-1,2,5,6-TETRAHYDRO-1-METHYL ESTER, HYDROCHLORIDE

TOXICITY DATA with REFERENCE
dni-mus-ipr 60 mg/kg IJEBA6 17,1141,79
oth-mus-ipr 60 mg/kg IJEBA6 17,1141,79
orl-mus TDLo:10400 mg/kg/1Y-I:CAR JCREA8 107,169,84
ipr-mus LD50:154 mg/kg TXAPA9 28,227,74
ivn-mus LD50:32 mg/kg CSLNX* NX#12238

SAFETY PROFILE: Poison by intraperitoneal and intravenous routes. Questionable carcinogen with experimental carcinogenic data. See also ESTERS. When heated to decomposition it emits very toxic fumes of NO_x and HCl.

AQU500 CAS:30233-80-8 ***HR: 3***
ARESKAP 100

PROP: Monobutylphenyl-phenol sodium monosulfonate (JAPMA8 38,428,49)

TOXICITY DATA with REFERENCE
eye-rbt 1% MLD JAPMA8 38,428,49
orl-mus LD50:3800 mg/kg JAPMA8 38,428,49
ivn-mus LD50:180 mg/kg JAPMA8 38,428,49

SAFETY PROFILE: Poison by intravenous route. Mildly toxic by ingestion. An eye irritant. See also SULFONATES. When heated to decomposition it emits toxic fumes of SO_x.

AQU750 CAS:30233-81-9 ***HR: 3***
ARESKET 300

SYN: MONOBUTYL DIPHENYL SODIUM MONOSULFONATE

TOXICITY DATA with REFERENCE
eye-rbt 1% SEV JAPMA8 38,428,49
orl-mus LD50:3500 mg/kg JAPMA8 38,428,49
ivn-mus LD50:250 mg/kg JAPMA8 38,428,49

SAFETY PROFILE: Poison by intravenous route. Moderately toxic by ingestion. A severe eye irritant. See also SULFONATES. When heated to decomposition it emits toxic fumes of SO_x.

AQV000 ***HR: 3***
ARESKLENE 400

SYN: DIBUTYLPHENYL-PHENOL SODIUM DISULFONATE

TOXICITY DATA with REFERENCE
eye-rbt 1% SEV JAPMA8 38,428,49
orl-mus LD50:2200 mg/kg JAPMA8 38,428,49
ivn-mus LD50:200 mg/kg JAPMA8 38,428,49

SAFETY PROFILE: Poison by intravenous route. Moderately toxic by ingestion. A severe eye irritant. See also SULFONATES. When heated to decomposition it emits toxic fumes of SO_x.

AQW000 CAS:1119-34-2 ***HR: 2***
l-ARGININE MONOHYDROCHLORIDE
mf: $C_6H_{14}N_4O_2{\cdot}ClH$ mw: 210.70

PROP: White crystalline powder; odorless. Mp: 222-235° (decomp). Very sol in water; sltly sol in alc.

SYNS: ARGAMINE ◇ ARGININE HYDROCHLORIDE ◇ l-ARGININE HYDROCHLORIDE ◇ ARGININE MONOHYDROCHLORIDE ◇ ARGIVENE ◇ DETOXARGIN ◇ l-HYDROCHLORIDE ARGININE ◇ LEVARGIN ◇ MINOPHAGEN A ◇ R-GENE

TOXICITY DATA with REFERENCE
ipr-rat TDLo:90 mg/kg (1-6D preg):TER AJEBAK 51,553,73
orl-rat LD50:12 g/kg JPMSAE 62,49,73
ipr-rat LD50:3793 mg/kg ABBIA4 58,253,55

CONSENSUS REPORTS: Reported in EPA TSCA Inventory.

SAFETY PROFILE: Moderately toxic by intraperitoneal route. Mildly toxic by ingestion. An experimental teratogen. When heated to decomposition it emits very toxic fumes of NO_x and HCl.

AQW125 CAS:113-80-4 ***HR: D***
ARGIPRESTOCIN
mf: $C_{43}H_{67}N_{15}O_{12}S_2$ mw: 1050.37

SYNS: 8-l-ARGININEOXYTOCIN ◇ ARGININE-VASOTOCIN ◇ 8-ARGININE VASOTOCIN ◇ (ARG^8)OXYTOCIN ◇ ARG-VASOTOCIN ◇ 8-ARG-VASOTOCIN ◇ AVT ◇ NATRIURETIC HORMONE (bovine pineal)

TOXICITY DATA with REFERENCE
ipr-mus TDLo:80 μg/kg (female 1-21D post):REP INJFA3 21,65,76

SAFETY PROFILE: Experimental reproductive effects. When heated to decomposition it emits toxic fumes of SO_x and NO_x.

AQW250 CAS:7440-37-1 ***HR: 1***
ARGON
af: Ar aw: 39.94
DOT: UN 1006/UN 1951

PROP: Colorless, inert gas. Mp: −189.2°, bp: −185.7°, d: 1.784 g/L @ 0°, 1.40 @ −186°, 1.65 @ −233°. Solubility in water 3.36 mL/100 g @ 20°.

CONSENSUS REPORTS: Reported in EPA TSCA Inventory.

DOT Classification: Nonflammable Gas; Label: Nonflammable Gas.

SAFETY PROFILE: A simple asphyxiant gas. As an inert gas, it has no specific inherent dangerous properties. Gases of this type have no specific toxicity effect, but they act by excluding O_2 from the lungs. The effect of simple asphyxiant gases is proportional to the extent to which they diminish the amount (partial pressure) of O_2 in the air that is breathed. The oxygen may be diminished to 0.75% of its normal percentage in air before appreciable symptoms develop, and this in turn requires the presence of a simple asphyxiant in a concentration of 33% in the mixture of air and gas. When the simple asphyxiant reaches a concentration of 50%, marked symptoms can be produced. A concentration of 75% is fatal in a matter of minutes. The first symptoms produced by simple asphyxiant gases such as argon are rapid respirations and air hunger. Mental alertness is diminished and muscular coordination is impaired. Later, judgment becomes faulty and all sensations are depressed. Emotional instability often results and fatigue occurs rapidly. As the asphyxia progresses, there may be nausea and vomiting, prostration, and loss of consciousness, and finally, convulsions, deep coma and death.

AQX250 CAS:124-94-7 ***HR: 3***
ARISTOCORT
mf: $C_{21}H_{27}FO_6$ mw: 394.48

SYNS: 9-α-FLUORO-16-α-HYDROXYPREDNISOLONE ◇ 9-α-FLUORO-11-β,16-α,17,21-TETRAHYDROXYPREGNA-1,4-DIENE-3,20-DIONE ◇ 9-α-FLUORO-11-β,16-α,17,21-TETRAHYDROXY-1,4-PREGNADIENE-3,20-DIONE ◇ 9-α-FLUORO-11-β,16-α,17-α,21-TETRAHYDROXYPREGNA-1,4-DIENE-3,20-DIONE ◇ FLUOXYPREDNISOLONE ◇ KENACORT ◇ PREGNA-1,4-DIENE-3,20-DIONE,9-FLUORO-11,16,17,21-TETRAHYDROXY-,(11-β,16-α) ◇ RODINOLONE ◇ SK-TRIAMCINOLONE ◇ 11-β,16-α,17-α,21-TETRAHYDROXY-9-α-FLUORO-1,4-PREGNADIENE-3,20-DIONE ◇ TRIAMCINOLONE

TOXICITY DATA with REFERENCE
oms-hmn-orl 428 μg/kg ARDEAC 103,39,71
unr-wmn TDLo:12600 μg/kg (female 1-39W post):TER AJOGAH 96,985,66
orl-pig TDLo:10500 μg/kg (female 15-16W post):REP THGNBO 30,137,88
scu-rat LD50:99 mg/kg TXAPA9 8,250,66

SAFETY PROFILE: Poison by subcutaneous route. An experimental teratogen. Other experimental reproductive effects. Human mutation data reported. When heated to decomposition it emits toxic fumes of F^-. An anti-inflammatory and antiallergic agent.

AQX500 CAS:76-25-5 ***HR: 3***
ARISTOCORT ACETONIDE
mf: $C_{24}H_{31}O_6F$ mw: 434.55

SYNS: ACETOSPAN ◇ ARISTODERM ◇ ARISTOGEL ◇ 9-α-FLUORO-11-β,21-DIHYDROSY-16-α-ISOPROYLIDENEDIOXY-1,4-PREGNADIENE, 3,20-DIONE ◇ 9-α-FLUORO-16-HYDROXYPREDNISOLONE ACETONIDE ◇ 9-α-FLUORO-16-α-17-α-ISOPROPYLEDENE DIOXY PREDNISOLONE ◇ 9-α-FLUORO-16-α-17-α-ISOPROPYLIDENE-DIOXY-Δ-1-HYDROCORTISONE ◇ FLUTONE ◇ KENACORT-A ◇ KENALOG ◇ TRAMACIN ◇ TRIAMCINCOLONE ACETONIDE ◇ TRIAMCINOLONE ACETONIDE ◇ TRIAMCINOLONE-16,17-ACETONIDE ◇ VETALOG

TOXICITY DATA with REFERENCE
dns-hmn:oth 1 nmol/L CNREA8 43,2664,83
oms-hmn-skn 5000 ppm ARDEAC 103,39,71
ims-ham TDLo:100 μg/kg (female 11D post):REP ANREAK 199,135A,81
ims-mus TDLo:10 mg/kg (female 11D post):TER PNASA6 67,779,70
scu-rat LD50:13100 μg/kg DRFUD4 6,44,81
orl-mus LD50:5 g/kg YAKUD5 21,2117,79
ipr-mus LD50:105 mg/kg NIIRDN 6,516,82
scu-mus LD50:132 mg/kg NIIRDN 6,516,82

CONSENSUS REPORTS: Reported in EPA TSCA Inventory.

SAFETY PROFILE: Poison by subcutaneous and intraperitoneal routes. An experimental teratogen. Other experimental reproductive effects. Human mutation data reported. When heated to decomposition it emits acrid smoke and toxic fumes of F^-.

AQX750 CAS:67-78-7 ***HR: D***
ARISTOCORT DIACETATE
mf: $C_{25}H_{31}FO_8$ mw: 478.56

SYNS: ARISTOCORT FORTE PARENTERAL ◇ ARISTOCORT SYRUP ◇ 9-FLUORO-11-β,16-α,17,21-TETRAHYDROXYPREGNA-1,4-DIENE-3,20-DIONE-16,21-DIACETATE ◇ KENACORT DIACETATE SYRUP ◇ TRIAMCINOLONE DIACETATE

TOXICITY DATA with REFERENCE
scu-rat TDLo:87500 μg/kg (15-17D preg):TER BBACAQ 385,257,75

SAFETY PROFILE: An experimental teratogen. When heated to decomposition it emits toxic fumes of F^-.

AQX825 CAS:35142-05-3 ***HR: D***
ARISTOLIC ACID
mf: $C_{17}H_{12}O_5$ mw: 296.29

SYN: 8-METHOXYPHENANTHRO(3,4-d)-1,3-DIOXOLE-5-CARBOXYLIC ACID

TOXICITY DATA with REFERENCE
mmo-sat 10 μg/plate MUREAV 206,447,88
msc-ham:ovr 50 umol/L MUREAV 206,447,88
orl-rbt TDLo:60 mg/kg (9D preg):TER EXPEAM 34,1277,78
orl-mus TDLo:90 mg/kg (6D preg):REP CCPTAY 25,639,82

SAFETY PROFILE: An experimental teratogen. Other experimental reproductive effects. Mutation data reported. When heated to decomposition it emits acrid smoke and fumes.

AQY000 ***HR: 3***
ARISTOLICHIA INDICA L., ALCOHOLIC EXTRACT

PROP: Obtained from extracts of the bitter roots of the Indian shrub *Aristolochia indica L.* (IJEBA6 15,428,77).

TOXICITY DATA with REFERENCE
orl-mus TDLo:15 mg/kg (female 6D post):REP IJMRAQ 66,991,77
orl-mus LDLo:100 mg/kg IJMRAQ 66,991,77

SAFETY PROFILE: Poison by ingestion. Experimental reproductive effects. See also ARISTOLOCHINE.

AQY125 ***HR: 3***
ARISTOLOCHIC ACID SODIUM SALT
mf: $C1_7H_{10}NO_7$•Na mw: 363.27

SYN: 8-METHOXY-6-NITROPHENANTHRO(3,4-d)-1,3-DIOXOLE-5-CARBOXYLIC ACID SODIUM SALT

TOXICITY DATA with REFERENCE
orl-rat TDLo:37 mg/kg/1Y-C:CAR ARTODN 51,107,82
orl-rat TD:90 mg/kg/13W-C:CAR ARTODN 51,107,82

SAFETY PROFILE: Questionable carcinogen with experimental carcinogenic data. When heated to decomposition it emits toxic fumes of NO_x and Na_2O.

AQY250 CAS:313-67-7 ***HR: 3***
ARISTOLOCHINE
mf: $C_{17}H_{11}NO_7$ mw: 341.29

PROP: From alcoholic extract of *Aristolochia indico* (CNCRA6 42,35,64).

SYNS: ARISTOLOCHIC ACID ◇ BIRTHWORT ◇ 8-METHOXY-6-NITROPHENANTHOL-(3,4-d)-1,3-DIOXOLE-5-CARBOXYLICACID ◇ NSC-50413

TOXICITY DATA with REFERENCE
mmo-sat 200 μg/plate MUREAV 113,259,83
msc-rat-orl 45 mg/kg MUREAV 143,143,85
ivn-man LDLo:3 mg/kg/2D-I CNCRA6 42,35,64

orl-rat LD50:184 mg/kg ARTODN 59,328,87
ivn-rat LD50:74 mg/kg ARTODN 59,328,87
orl-mus LD50:55900 μg/kg ARTODN 59,328,87
ipr-mus LD50:14320 μg/kg KIHSDM (6),2,81
ivn-mus LD50:38400 μg/kg ARTODN 59,328,87

SAFETY PROFILE: Confirmed carcinogen. Poison by intravenous route. When heated to decomposition it emits toxic fumes of NO_x.

From International Register of Potentially Toxic Chemicals: April 1982. Vol 5 No. 1: The Ministry of Health of the Federal Republic of Germany has withdrawn from the national market drugs containing aristolochic acid. The decision resulted from the demonstration of a carcinogenic potential in a three-month ingestion toxicity study undertaken in rats. Aristolochic acid is claimed to promote phagocytosis and to have immunostimulant activity. A growth-inhibiting effect on experimentally induced tumors has been described, but this effect has not been shown to have any clinical relevance. Extracts of species of *Aristolochiacea* have traditionally been used as a bitter, and a broad range of therapeutic effects has been claimed.

AQY375 CAS:5611-51-8 ***HR: D***
ARISTOSPAN
mf: $C_{30}H_{41}FO_7$ mw: 532.71

PROP: Fine, white, needle-like crystals. Mp: 295-296° (decomp). Solubility in g/100 mL @ 25°: chloroform and dimethylacetamide >5; ethyl acetate 0.77, methanol 0.59, diethyl carbonate 0.50, glycerin 0.42, propylene glycol 0.13; abs alc 0.03; water 0.0004.

SYNS: CL 34433 ◇ 9-FLUORO-11-β,16-α,17,21-TETRAHYDROXYPREGNA-1,4-DIENE-3,20-DIONE, CYCLIC 16,17-ACETAL with 21-(3,3-DIMETHYLBUTYRATE)ACETONE ◇ TATBA ◇ TRIAMCINOLONE HEXACETONIDE

TOXICITY DATA with REFERENCE
ims-rat TDLo:40 mg/kg (female 10-13D post):REP JOANAY 128,747,79
ims-rat TDLo:40 mg/kg (female 10-13D post):TER JOANAY 128,747,79

SAFETY PROFILE: An experimental teratogen. Other experimental reproductive effects. When heated to decomposition it emits toxic fumes of F^-.

AQY400 CAS:81-04-9 ***HR: 1***
ARMSTRONG'S S ACID
mf: $C_{10}H_8O_6S_2$ mw: 288.30

PROP: Crystals. Sol in water and alc; practically insol in ether.

SYNS: ARMSTRONG'S ACID ◇ 1,5-NAPHTHYLENE DISULFONIC ACID

TOXICITY DATA with REFERENCE
orl-rat LD50:30 g/kg GISAAA 45(3),73,80
orl-mus LD50:47 g/kg GISAAA 45(3),73,80
orl-rbt LD50:30 g/kg GISAAA 45(3),73,80
orl-gpg LD50:47 g/kg GISAAA 45(3),73,80

CONSENSUS REPORTS: Reported in EPA TSCA Inventory.

SAFETY PROFILE: Very low oral toxicity. When heated to decomposition it emits toxic fumes of SO_x. See also SULFONATES.

AQY500 ***HR: 3***
ARNICA

PROP: An alcoholic infusion

SYNS: MOUNTAIN TOBACCO ◇ WOLFSBANE

SAFETY PROFILE: Poison by inhalation and ingestion. A moderate irritant and allergen. It can cause gastroenteritis, nervous disturbances, and collapse. May cause contact dermatitis. Combustible when exposed to heat or flame. Incompatible with oxidizing materials.

AQY750 ***HR: 3***
AROMATIC AMINES

PROP: Amines which contain one or more rings of unsaturated or cyclic HC, such as benzene. There are vast numbers of such amines. The term is largely due to the characteristic odor.

SAFETY PROFILE: Many of these aromatic amines are recognized as carcinogenic to the human bladder, ureter, and renal pelvis, intestines, lung, liver, and prostate. See also AMINES.

AQZ000 ***HR: 2***
AROMATIC SPIRITS of AMMONIA

PROP: Colorless liquid, suffocating odor of ammonia. Composition: 10% by weight of NH_3 in alcohol.

SAFETY PROFILE: See AMMONIA. A dangerous fire hazard due to its alcohol content. Moderately explosive. When heated, it emits toxic fumes of ammonia. Incompatible with oxidizing materials.

AQZ100 CAS:88845-25-4 ***HR: D***
AROMATOL

TOXICITY DATA with REFERENCE
ihl-rat TCLo:1 g/m^3/24H (female 7-15D post):TER ARTODN 8,425,85
ihl-rbt TCLo:1 g/m^3/24H (female 7-20D post):REP ARTODN 8,425,85

SAFETY PROFILE: An experimental teratogen. Other

experimental reproductive effects. When heated to decomposition it emits acrid smoke and irritating fumes.

AQZ200 CAS:71441-28-6 ***HR: D***
AROTINOIC ACID
mf: $C_{24}H_{28}O_2$ mw: 348.52

SYNS: RO 13-7410 ◇ (E)-4-(2-(5,6,7,8-TETRAHYDRO-5,5,8,8-TETRAMETHYL-2-NAPHTHALENYL)-1-PROPEN-1-YL)BENZOIC ACID

TOXICITY DATA with REFERENCE
orl-ham TDLo:290 μg/kg (female 8D post):REP JJIND8 78,533,87
orl-ham TDLo:75 μg/kg (female 8D post):TER JJIND8 78,533,87

SAFETY PROFILE: An experimental teratogen. Other experimental reproductive effects. When heated to decomposition it emits acrid smoke and irritating fumes.

AQZ300 CAS:71441-30-0 ***HR: D***
AROTINOIC METHANOL
mf: $C_{24}H_{30}O$ mw: 334.54

SYNS: BENZENEMETHANOL,4-(2-(5,6,7,8-TETRAHYDRO-5,5,8,8-TETRAMETHYL-2-NAPHTHALENYL)-1-PROPENYL)-, (E)- ◇ RO 13-8320 ◇ (E)-4-(2-(5,6,7,8-TETRAHYDRO-5,5,8,8-TETRAMETHYL-2-NAPHTHALENYL)-1-PROPENYL)BENZENEMETHANOL ◇ (E)-4-(2-(5,6,7,8-TETRAHYDRO-5,5,8,8-TETRAMETHYL-2-NAPHTHALENYL)-1-PROPEN-1-YL)PHENYLMETHANOL

TOXICITY DATA with REFERENCE
orl-ham TDLo:560 μg/kg (female 8D post):REP JJIND8 78,533,87
orl-ham TDLo:140 μg/kg (female 8D post):TER JJIND8 78,533,87

SAFETY PROFILE: An experimental teratogen. Other experimental reproductive effects. When heated to decomposition it emits acrid smoke and irritating fumes.

AQZ400 ***HR: D***
AROTINOID ETHYL ESTER
mf: $C_{26}H_{32}O_2$ mw: 376.58

SYNS: BENZOIC ACID, p-((E)-2-(5,6,7,8-TETRAHYDRO-5,5,8,8-TETRAMETHYL-2-NAPHTHYL)-1-PROPENYL)-, ETHYL ESTER ◇ ETHYL-p-((E)-2-(5,6,7,8-TETRAHYDRO-5,5,8,8-TETRAMETHYL-2-NAPHTHYL)-1-PROPENYL)BENZOATE ◇ RO 13-6298

TOXICITY DATA with REFERENCE
dns-mus-skn
Skin 400 nmol/kg CNREA8 43,5174,83
orl-ham TDLo:310 μg/kg (female 8D post):TER JJIND8 78,533,87
orl-ham TDLo:310 μg/kg (female 8D post):REP JJIND8 78,533,87

SAFETY PROFILE: An experimental teratogen. Other experimental reproductive effects. Mutation data reported. When heated to decomposition it emits acrid smoke and irritating fumes.

AQZ900 CAS:585-54-6 ***HR: 3***
ARSACETIN SODIUM SALT
mf: $C_8H_9AsNO_4{\cdot}Na$ mw: 281.09

SYNS: ARSACETIN ◇ ARSANILIC ACID, N-ACETYL-, SODIUM SALT ◇ ARSONIC ACID, (4-(ACETYLAMINO)PHENYL)-, MONOSODIUM SALT (9CI) ◇ SODIUM ACETYLARSANILATE

TOXICITY DATA with REFERENCE
scu-rat LDLo:550 mg/kg BIZEA2 184,360,27
ivn-rbt LDLo:300 mg/kg JPETAB 23,107,24

OSHA PEL: TWA 0.5 mg(As)/m^3
ACGIH TLV: TWA 0.2 mg(As)/m^3

SAFETY PROFILE: Poison by intravenous route. Moderately toxic by subcutaneous route. When heated to decomposition it emits toxic fumes of NO_x and As.

ARA000 CAS:6018-32-2 ***HR: 2***
ARSACETIN TETRAHYDRATE
mf: $C_8H_9AsNO_4{\cdot}Na{\cdot}4H_2O$ mw: 353.17

PROP: White, crystalline powder; odorless and tasteless.

SYNS: p-ACETAMIDOBENZENEARSONIC ACID, SODIUM SALT, TETRAHYDRATE ◇ N-ACETYL-p-AMINOBENZENEARSONIC ACID, SODIUM SALT, TETRAHYDRATE ◇ N-ACETYLARSANILIC ACID, SODIUM SALT, TETRAHYDRATE ◇ ARSACETIN SODIUM SALT, TETRAHYDRATE ◇ SODIUM ACETYL ARSANILATE

TOXICITY DATA with REFERENCE
ivn-rbt LD50:550 mg/kg MEIEDD 11,125,89

CONSENSUS REPORTS: Arsenic and its compounds are on the Community Right-To-Know List.

OSHA PEL: TWA 0.5 mg(As)/m^3
ACGIH TLV: TWA 0.2 mg(As)/m^3

SAFETY PROFILE: Moderately toxic by intravenous route. See also ARSENIC COMPOUNDS. When heated to decomposition it emits very toxic fumes of As and NO_x.

ARA250 CAS:98-50-0 ***HR: 3***
ARSANILIC ACID
mf: $C_6H_8AsNO_3$ mw: 217.06

PROP: Needles from aq solns. Mp: 232°, bp: decomp, $-H_2O$ @ 15°. Very sol in hot water and alc; insol in ether and benzene.

SYNS: 4-AMINOBENZENEARSONIC ACID ◇ p-AMINOBENZENEARSONIC ACID ◇ AMINOPHENYLARSINE ACID ◇ p-AMINOPHENYLARSINE ACID ◇ p-AMINOPHENYLARSINIC ACID ◇ 4-AMINOPHENYLARSONIC ACID ◇ p-AMINOPHENYLARSONIC ACID ◇ p-ANILINEARSONIC ACID ◇ ANTOXYLIC ACID ◇ 4-ARSANILIC ACID ◇ p-ARSANILIC ACID ◇ ATOXYLIC ACID

TOXICITY DATA with REFERENCE
orl-rat LD50:216 mg/kg TXAPA9 18,185,71
ipr-rat LDLo:400 mg/kg JPETAB 80,393,44
ipr-mus LD50:291 mg/kg JMCMAR 9,221,66
ivn-mus LD50:100 mg/kg CSLNX* NX#06774

CONSENSUS REPORTS: IARC Cancer Review: Animal Inadequate Evidence IMEMDT 23,39,80. Reported in EPA TSCA Inventory. Arsenic and its compounds are on the Community Right-To-Know List.

OSHA PEL: TWA 0.5 mg(As)/m^3
ACGIH TLV: TWA 0.2 mg(As)/m^3

SAFETY PROFILE: Poison by ingestion, intravenous, and intraperitoneal routes. See also ARSENIC COMPOUNDS and ANILINE. Flammable, decomposes with heat to yield flammable vapors. When heated to decomposition or on contact with acid or acid fumes it emits highly toxic fumes of As and NO_x.

ARA500 CAS:127-85-5 ***HR: 3***
ARSANILIC ACID, MONOSODIUM SALT
DOT: UN 2473
mf: $C_6H_7AsNO_3 \cdot Na$ mw: 239.05

PROP: Tetrahydrate: white, odorless, crystalline powder; faint salty taste. Sol in water; somewhat sol in alc.

SYNS: (4-AMINOPHENYL)ARSONIC ACID SODIUM SALT ◇ ARSANILIC ACID SODIUM SALT ◇ ATOXYL ◇ NCI-C61176 ◇ SODIUM AMINARSONATE ◇ SODIUM-p-AMINOBENZENEARSONATE ◇ SODIUM AMINOPHENOL ARSONATE ◇ SODIUM-p-AMINOPHENYLARSONATE ◇ SODIUM ANILARSONATE ◇ SODIUM-ANILINE ARSONATE ◇ SODIUM ARSANILATE ◇ SODIUM-p-ARSANILATE ◇ SODIUM ARSONILATE

TOXICITY DATA with REFERENCE
scu-rat LD50:75 mg/kg BIZEA2 184,360,27
scu-mus LD50:400 mg/kg MEIEDD 10,1230,83
scu-dog LDLo:5 mg/kg HBAMAK 4,1289,35
scu-rbt LDLo:200 mg/kg HBAMAK 4,1289,35

CONSENSUS REPORTS: Arsenic and its compounds are on the Community Right-To-Know List.

OSHA PEL: 8H TWA 0.5 mg(As)/m^3
ACGIH TLV: TWA 0.2 mg(As)/m^3
DOT Classification: Poison B; Label: St. Andrews Cross

SAFETY PROFILE: Poison by subcutaneous route. Can cause blindness. When heated to decomposition it emits very toxic fumes of As and NO_x.

ARA750 CAS:7440-38-2 ***HR: 3***
ARSENIC
DOT: UN 1558
af: As aw: 74.92

PROP: Silvery to black, brittle, crystalline and amorphous metalloid. Mp: 814° @ 36 atm, bp: sublimes @ 612°, d: black crystals 5.724 @ 14°; black amorphous 4.7, vap press: 1 mm @ 372° (sublimes). Insol in water; sol in HNO_3.

SYNS: ARSEN (GERMAN, POLISH) ◇ ARSENIC, metallic (DOT) ◇ ARSENICALS ◇ ARSENIC-75 ◇ ARSENIC BLACK ◇ COLLOIDAL ARSENIC ◇ GREY ARSENIC ◇ METALLIC ARSENIC

TOXICITY DATA with REFERENCE
cyt-mus-ipr 4 mg/kg/48H-I EXPEAM 37,129,81
orl-rat TDLo:605 μg/kg (35 W preg):REP GISAAA (8)30,77
orl-rat TDLo:580 μg/kg (female 30W pre):TER FATOAO 41,620,78
orl-man TDLo:76 mg/kg/12Y-I:CAR RMCHAW 99,664,71
imp-rbt TDLo:75 mg/kg:ETA ZEKBAI 52,425,42
orl-man TDLo:7857 mg/kg/55Y:SKN CMAJAX 120,168,79
orl-man TDLo:7857 mg/kg/55Y:GIT CMAJAX 120,168,79
orl-rat LD50:763 mg/kg GTPZAB 31(12),53,87
orl-mus LD50:145 mg/kg GTPZAB 31(12),53,87
ipr-mus LD50:46200 μg/kg GTPZAB 31(12),53,87
scu-rbt LDLo:300 mg/kg ASBIAL 24,442,38
ipr-gpg LDLo:10 mg/kg CRSBAW 81,164,18
scu-gpg LDLo:300 mg/kg ASBIAL 24,442,38

CONSENSUS REPORTS: NTP Fifth Annual Report on Carcinogens. IARC Cancer Review: Group 1 IMEMDT 7,100,87; Human Sufficient Evidence IMEMDT 23,39,80; Human Inadequate Evidence IMEMDT 2,48,73. Reported in EPA TSCA Inventory. Arsenic and its compounds are on the Community Right-To-Know List.

OSHA PEL: TWA 0.01 mg(As)/m^3; Cancer Hazard
ACGIH TLV: TWA 0.2 mg(As)/m^3
DFG TRK: 0.2 mg/m^3 calculated as arsenic in that portion of dust that can possibly be inhaled.
NIOSH REL: CL 2 μg(As)/m^3
DOT Classification: Poison B; Label: Poison.

SAFETY PROFILE: Confirmed human carcinogen producing liver tumors. Poison by subcutaneous, intramuscular, and intraperitoneal routes. Human systemic skin and gastrointestinal effects by ingestion. An experimental teratogen. Other experimental reproductive effects. Mutation data reported. Flammable in the form of dust when exposed to heat or flame or by chemical reaction with powerful oxidizers such as bromates, chlorates, iodates, peroxides, lithium, NCl_3, KNO_3, $KMnO_4$, Rb_2C_2, $AgNO_4$, NOCl, IF_5, CrO_3, ClF_3, ClO, BrF_3, BrF_5, BrN_3, RbC_3BCH, CsC_3BCH. Slightly explosive in the form of dust when exposed to flame. When heated or on contact with acid or acid fumes, it emits highly toxic fumes; can react vigorously on contact with oxidizing materials. Incompatible with bromine azide, dirubidium acetylide, halogens, palladium, zinc, platinum, NCl_3, $AgNO_3$, CrO_3, Na_2O_2, hexafluoro isopropylideneamino lithium.

ARB000 CAS:10102-53-1 ***HR: 3***
m-ARSENIC ACID
mf: $AsHO_3$ mw: 123.93

SYN: METAARSENIC ACID

CONSENSUS REPORTS: Reported in EPA TSCA Inventory. Arsenic and its compounds are on the Community Right-To-Know List.

OSHA PEL: TWA 0.01 mg(As)/m^3; Cancer Hazard
ACGIH TLV: TWA 0.2 mg(As)/m^3
DFG MAK: Human Carcinogen.
NIOSH REL: (Arsenic, Inorganic) CL 2 μg(As)/m^3/15M

SAFETY PROFILE: Confirmed human carcinogen. See ARSENIC COMPOUNDS. When heated to decomposition it emits toxic fumes of arsenic.

ARB250 CAS:7778-39-4 ***HR: 3***
o-ARSENIC ACID
DOT: UN 1553/UN 1554
mf: AsH_3O_4 mw: 141.95

SYNS: ACIDE ARSENIQUE LIQUIDE (FRENCH) ◇ ARSENATE ◇ ARSENIC ACID, liquid (DOT) ◇ ARSENIC ACID, solid (DOT) ◇ DESICCANT L-10 ◇ HI-YIELD DESSICANT H-10 ◇ ORTHOARSENIC ACID ◇ RCRA WASTE NUMBER P010 ◇ ZOTOX ◇ ZOTOX CRAB GRASS KILLER

TOXICITY DATA with REFERENCE
cyt-hmn:leu 7200 nmol/L MUREAV 88,73,81
cyt-hmn:fbr 100 ppb MUREAV 88,73,81
ipr-rat TDLo:30 mg/kg (9D preg):TER JTSCDR 4,405,79
orl-rat LD50:48 mg/kg FMCHA2 -,C18,83
orl-dog LDLo:10 mg/kg FDWU** -,-,31
orl-rbt LDLo:5 mg/kg FDWU** -,-,31
orl-pgn LDLo:100 mg/kg FDWU** -,-,31
orl-ckn LDLo:125 mg/kg FDWU** -,-,31

CONSENSUS REPORTS: Reported in EPA TSCA Inventory. Arsenic and its compounds are on the Community Right-To-Know List.

OSHA PEL: TWA 0.01 mg(As)/m^3; Cancer Hazard
ACGIH TLV: TWA 0.2 mg(As)/m^3
DFG MAK: Human Carcinogen.
NIOSH REL: (Arsenic, Inorganic) CL 2 μg(As)/m^3/15M

DOT Classification: Poison B; Label: Poison.

SAFETY PROFILE: Confirmed human carcinogen. Poison by ingestion. An experimental teratogen. Human mutation data reported. When heated to decomposition it emits toxic fumes of arsenic. See also ARSENIC COMPOUNDS.

ARB750 CAS:7778-44-1 ***HR: 3***
ARSENIC ACID, CALCIUM SALT (2583)
DOT: UN 1573
mf: $As_2O_8 \cdot 3Ca$ mw: 398.08

PROP: Colorless, amorphous powder. D: 3.620. Solubility in water = 0.013/100 @ 25°.

SYNS: ARSENIATE de CALCIUM (FRENCH) ◇ CALCIUMARSENAT ◇ CALCIUM ARSENATE (MAK) ◇ CALCIUM ORTHOARSENATE ◇ KALZIUMARSENIAT (GERMAN) ◇ TRICALCIUMARSENAT (GERMAN) ◇ TRICALCIUM ARSENATE

TOXICITY DATA with REFERENCE
itr-rat TDLo:1600 μg/kg:ETA IJCNAW 24,786,79
itr-ham TDLo:120 mg/kg/15W-C:NEO CALEDQ 27,99,85
orl-rat LD50:20 mg/kg AFDOAQ 15,122,51
skn-rat LD50:2400 mg/kg 28ZEAL 5,35,76
orl-mus LD50:794 mg/kg AMRL** TR-72-62,72
orl-dog LD50:38 mg/kg 85DPAN -,-,71/76
orl-rbt LDLo:50 mg/kg JPETAB 39,246,30
orl-mam LD50:35 mg/kg PCOC** -,170,66

CONSENSUS REPORTS: NTP Fifth Annual Report on Carcinogens. IARC Cancer Review: Group 1 IMEMDT 7,100,87; Human Sufficient Evidence IMEMDT 23,39,80; Animal No Evidence IMEMDT 2,48,73; Animal Inadequate Evidence IMEMDT 23,39,80. Reported in EPA TSCA Inventory. Arsenic and its compounds are on the Community Right-To-Know List. EPA Extremely Hazardous Substances List.

OSHA PEL: TWA 0.01 mg(As)/m^3; Cancer Hazard
ACGIH TLV: TWA 0.2 mg(As)/m^3
DFG MAK: Human Carcinogen.
NIOSH REL: CL 2 μg(As)/m^3/15M
DOT Classification: Poison B; Label: Poison.

SAFETY PROFILE: Confirmed human carcinogen. Poison by ingestion. Moderately toxic by skin contact. When heated to decomposition it emits toxic fumes of arsenic.

ARC000 CAS:7778-43-0 ***HR: 3***
ARSENIC ACID, DISODIUM SALT
mf: $Na_2HAsO_4 \cdot 7H_2O$ mw: 312.01

PROP: Colorless powder, effloresces. D: 1.88, mp: $-7H_2O$ @ 130°, bp: decomp @ 150°. Solubility in water = 61/100 @ 15°, sol in glycerol.

SYNS: DISODIUM ARSENATE ◇ DISODIUM ARSENIC ACID ◇ DISODIUM HYDROGEN ARSENATE ◇ DISODIUM HYDROGEN ORTHOARSENATE ◇ DISODIUM MONOHYDROGEN ARSENATE ◇ SODIUM ACID ARSENATE ◇ SODIUM ARSENATE ◇ SODIUM ARSENATE DIBASIC, anhydrous

TOXICITY DATA with REFERENCE
cyt-hmn:leu 7200 μmol/L MUREAV 88,73,81
mrc-bcs 100 mmol/L MUREAV 77,109,80
ipr-rat LDLo:30 mg/kg JPETAB 58,454,36

CONSENSUS REPORTS: Reported in EPA TSCA Inventory. Arsenic and its compounds are on the Community Right-To-Know List.

OSHA PEL: TWA 0.5 mg(As)/m^3: Cancer Hazard
ACGIH TLV: TWA 0.2 mg(As)/m^3
NIOSH REL: (Arsenic, Inorganic) CL 2 μg(As)/m^3/15M
DFG MAK: Human Carcinogen.

SAFETY PROFILE: Confirmed human carcinogen. Poison by intraperitoneal route. Human mutation data reported. See ARSENIC COMPOUNDS. When heated to decomposition it emits toxic fumes of arsenic.

ARC250 CAS:10048-95-0 ***HR: 3***
ARSENIC ACID, DISODIUM SALT, HEPTAHYDRATE
mf: $AsHO_4$•2Na•$7H_2O$ mw: 427.05

SYNS: DISODIUM ARSENATE, HEPTAHYDRATE ◇ SODIUM ACID ARSENATE, HEPTAHYDRATE ◇ SODIUM ARSENATE, DIBASIC, HEPTAHYDRATE ◇ SODIUM ARSENATE HEPTAHYDRATE

TOXICITY DATA with REFERENCE
orl-mus TDLo:120 mg/kg (female 10D post):TER EVHPAZ 19,219,77
ipr-mus TDLo:45 mg/kg (female 9D post):REP AEHLAU 24,62,72
ims-mus LD50:87360 μg/kg EXMDA4 (440),312,78
scu-gpg LDLo:50 mg/kg BMJOAE 2,217,13

CONSENSUS REPORTS: NTP Fifth Annual Report on Carcinogens. Arsenic and its compounds are on the Community Right-To-Know List.

OSHA PEL: TWA 0.01 mg(As)/m^3; Cancer Hazard
ACGIH TLV: TWA 0.2 mg(As)/m^3
NIOSH REL: (Arsenic, Inorganic) CL 2 μg(As)/m^3/15M
DFG MAK: Human Carcinogen.

SAFETY PROFILE: Confirmed human carcinogen. Poison by subcutaneous route. An experimental teratogen. Other experimental reproductive effects. See also ARSENIC COMPOUNDS. When heated to decomposition it emits toxic fumes of arsenic.

ARC500 CAS:7774-41-6 ***HR: 3***
ARSENIC ACID, HEMIHYDRATE
mf: AsH_3O_4•$1/2H_2O$ mw: 150.96

PROP: White, translucent crystals. Mp: 35.5°, bp: $-H_2O$ @ 160°, d: 2.0-2.5.

SYNS: ARSENIC ACID, solid (DOT) ◇ ORTHOARSENIC ACID HEMIHYDRATE

TOXICITY DATA with REFERENCE
ivn-rbt LD50:6 mg/kg MEIEDD 11,126,89

CONSENSUS REPORTS: Arsenic and its compounds are on the Community Right-To-Know List.

OSHA PEL: TWA 0.01 mg(As); Cancer Hazard
ACGIH TLV: TWA 0.2 mg(As)/m^3
NIOSH REL: (Arsenic, Inorganic) CL 2 μg(As)/m^3/15M
DFG MAK: Human Carcinogen.
DOT Classification: Poison B; Label: Poison.

SAFETY PROFILE: Confirmed human carcinogen. Poison by intravenous route. See also ARSENIC COMPOUNDS. When heated to decomposition it emits toxic fumes of arsenic.

ARC750 CAS:7645-25-2 ***HR: 3***
ARSENIC ACID, LEAD SALT
DOT: UN 1617
mf: AsH_3O_4•7Pb mw: 1592.28

SYNS: ARSENIATE de PLOMB (FRENCH) ◇ LEAD ARSENATE

TOXICITY DATA with REFERENCE
orl-rat LD50:100 mg/kg AFDOAQ 15,122,51
orl-mus LD50:1000 mg/kg JPETAB 93,407,48
orl-rbt LD50:125 mg/kg JAPMA8 37,122,48
orl-ckn LD50:450 mg/kg JAPMA8 37,122,48

CONSENSUS REPORTS: Arsenic compounds and Lead compounds are on the Community Right-To-Know List.

OSHA PEL: TWA 0.01 mg(As)/m^3; Cancer Hazard
ACGIH TLV: TWA 0.15 mg(Pb)/m^3
NIOSH REL: (Lead, Inorganic): 10H TWA 0.10 mg(Pb)/m^3; (Arsenic, Inorganic): CL 0.002 mg(As)/m^3/15M
DFG MAK: Human Carcinogen.
DOT Classification: Poison B; Label: Poison

SAFETY PROFILE: Confirmed human carcinogen. Poison by ingestion. See also LEAD COMPOUNDS and ARSENIC COMPOUNDS. When heated to decomposition it emits very toxic fumes of lead and arsenic.

ARD000 CAS:10103-50-1 ***HR: 3***
ARSENIC ACID, MAGNESIUM SALT
DOT: UN 1622
mf: AsH_3O_4•7Mg mw: 312.12

PROP: Monoclinic, white crystals. D: 2.60-2.61.

SYNS: ARSENIATE de MAGNESIUM (FRENCH) ◇ MAGNESIUM ARSENATE ◇ MAGNESIUM ARSENATE PHOSPHOR

TOXICITY DATA with REFERENCE
orl-rat LDLo:280 mg/kg TXAPA9 1,156,59
orl-mus LD50:315 mg/kg IRGGAJ 20,21,63
orl-rbt LDLo:80 mg/kg AIHAAP 19,504,58

CONSENSUS REPORTS: Reported in EPA TSCA Inventory. Arsenic and its compounds are on the Community Right-To-Know List.

OSHA PEL: TWA 0.01 mg(As)/m^3; Cancer Hazard
ACGIH TLV: TWA 0.2 mg(As)/m^3
DFG MAK: Human Carcinogen.
NIOSH REL: (Arsenic, Inorganic) CL 2 μg(As)/m^3/15M
DOT Classification: Poison B; Label: Poison.

SAFETY PROFILE: Confirmed human carcinogen. Poison by ingestion. See also ARSENIC COMPOUNDS. When heated to decomposition it emits toxic fumes of arsenic.

ARD250 CAS:7784-41-0 ***HR: 3***
ARSENIC ACID, MONOPOTASSIUM SALT
DOT: UN 1677
mf: AsH_2O_4•K mw: 180.04

SYNS: MACQUER'S SALT ◇ MONOPOTASSIUM ARSENATE ◇ MONOPOTASSIUM DIHYDROGEN ARSENATE ◇ POTASSIUM ACID ARSENATE ◇ POTASSIUM ARSENATE ◇ POTASSIUM DIHYDROGEN ARSENATE ◇ POTASSIUM HYDROGEN ARSENATE

TOXICITY DATA with REFERENCE
cyt-hmn:leu 1 μmol/L CNREA8 25,980,65

CONSENSUS REPORTS: NTP Fifth Annual Report on Carcinogens. IARC Cancer Review: Human Sufficient Evidence IMEMDT 23,39,80. Reported in EPA TSCA Inventory. Arsenic and its compounds are on the Community Right-To-Know List.

OSHA PEL: TWA 0.01 mg(As)/m^3, Cancer Hazard
ACGIH TLV: TWA 0.2 mg(As)/m^3
NIOSH REL: (Arsenic, Inorganic) CL 2 μg(As)/m^3/15M
DOT Classification: Poison B; Label: Poison.

SAFETY PROFILE: Confirmed human carcinogen. Mutation data reported. See also ARSENIC COMPOUNDS. When heated to decomposition it emits toxic fumes of arsenic.

ARD500 CAS:15120-17-9 ***HR: 3***
ARSENIC ACID, MONOSODIUM SALT
mf: AsO_3•Na mw: 145.91

SYNS: ARSENIC ACID, SODIUM SALT (9CI) ◇ SODIUM ARSENATE ◇ SODIUM METAARSENATE ◇ SODIUM MONOHYDROGEN ARSENATE

TOXICITY DATA with REFERENCE
sln-dmg-orl 2 μmol/L CNJGA8 11,677,69
slt-dmg-orl 100 μmol CNJGA8 17,55,75

CONSENSUS REPORTS: Arsenic and its compounds are on the Community Right-To-Know List.

OSHA PEL: TWA 0.5 mg(As)/m^3: Cancer Hazard
ACGIH TLV: TWA 0.2 mg(As)/m^3
DFG MAK: Human Carcinogen.
NIOSH REL: (Arsenic, Inorganic) CL 2 μg(As)/m^3/15M

SAFETY PROFILE: Confirmed human carcinogen. A poison. Mutation data reported. See also ARSENIC COMPOUNDS. When heated to decomposition it emits toxic fumes of arsenic.

ARD600 CAS:10103-60-3 ***HR: 3***
ARSENIC ACID, MONOSODIUM SALT
mf: AsH_2O_4•Na mw: 163.93

SYNS: MONOSODIUM ARSENATE ◇ SODIUM ARSENATE ◇ SODIUM DIHYDROGEN ARSENATE ◇ SODIUM DIHYDROGEN ORTHOARSENATE

TOXICITY DATA with REFERENCE
ivn-rbt LDLo:45 mg/kg JPETAB 23,107,24

OSHA PEL: TWA 0.5 mg(As)/m^3: Cancer Hazard
ACGIH TLV: TWA 0.2 mg(As)/m^3
NIOSH REL: CL 2 μg(As)/m^3/15M
DFG MAK: Human Carcinogen.

CONSENSUS REPORTS: Arsenic and its compounds are on the Community Right-To-Know List.

SAFETY PROFILE: Confirmed human carcinogen. Poison by intravenous route. When heated to decomposition it emits toxic fumes of arsenic.

ARD750 CAS:7631-89-2 ***HR: 3***
ARSENIC ACID, SODIUM SALT
DOT: UN 1685
mf: AsH_3O_4•7Na mw: 202.94

SYNS: FATSCO ANT POISON ◇ SODIUM ARSENATE (DOT) ◇ SODIUM ORTHOARSENATE ◇ SWEENEY'S ANT-GO

TOXICITY DATA with REFERENCE
cyt-hmn:lym 2 μmol/L ADREDL 267,91,80
orl-mus TDLo:475 mg/kg (female 8-12D post):REP TCMUD8 6,361,86
orl-mus TDLo:120 mg/kg (female 11D post):TER JEPTDQ 1(6),857,78
scu-mus TDLo:10 mg(As)/kg/20D-C:ETA VDGPAN 55,289,71
ipr-rat LDLo:49 mg/kg JPETAB 58,454,36
ivn-rat LDLo:85 mg/kg JPETAB 33,270,28
orl-rbt LDLo:51 mg/kg JPETAB 33,270,28
ivn-rbt LDLo:28 mg/kg JPETAB 33,270,28

CONSENSUS REPORTS: NTP Fifth Annual Report on Carcinogens. IARC Cancer Review: Human Sufficient Evidence IMEMDT 23,39,80; Animal Inadequate Evidence IMEMDT 2,48,73; IMEMDT 23,39,80. Reported in EPA TSCA Inventory. Arsenic and its compounds are on the Community Right-To-Know List.

OSHA PEL: TWA 0.01 mg(As)/m^3; Cancer Hazard
ACGIH TLV: TWA 0.2 mg(As)/m^3
NIOSH REL: CL 2 μg(As)/m^3/15M
DOT Classification: Poison B; Label: Poison.

SAFETY PROFILE: Confirmed human carcinogen with experimental tumorigenic data. Poison by ingestion, intravenous, and intraperitoneal routes. An experimental teratogen. Other experimental reproductive effects. Mutation data reported. See also ARSENIC COMPOUNDS. When heated to decomposition it emits toxic fumes of As and Na_2O.

ARE000 CAS:64070-83-3 ***HR: 3***
ARSENIC(V) ACID, TRISODIUM SALT, HEPTAHYDRATE (1583587)
mf: AsO_4•3Na•7H_2O mw: 334.03

SYN: TRISODIUM ARSENATE, HEPTAHYDRATE

TOXICITY DATA with REFERENCE
ipr-mus LD50:9 mg/kg COREAF 257,791,63

CONSENSUS REPORTS: Arsenic and its compounds are on the Community Right-To-Know List.

OSHA PEL: TWA 0.01 mg(As)/m^3; Cancer Hazard
ACGIH TLV: TWA 0.2 mg(As)g/m^3
NIOSH REL: (Arsenic, Inorganic) CL 2 μg(As)/m^3/15M
DFG MAK: Human Carcinogen.

SAFETY PROFILE: Confirmed human carcinogen. Poison by intraperitoneal route. See also ARSENIC COMPOUNDS. When heated to decomposition it emits toxic fumes of arsenic.

ARE250 CAS:8028-75-9 ***HR: 3***
ARSENICAL DIP
DOT: NA 1557

SYNS: ARSENICAL DIP, liquid (DOT) ◇ SHEEP DIP

CONSENSUS REPORTS: Arsenic and its compounds are on the Community Right-To-Know List.

OSHA PEL: TWA 0.5 mg(As)/m^3
DOT Classification: Poison B; Label: Poison.

SAFETY PROFILE: A poison. See also ARSENIC COMPOUNDS.

ARE500 CAS:8028-73-7 ***HR: 3***
ARSENICAL DUST
DOT: UN 1562

SYNS: ARSENICAL FLUE DUST ◇ FLUE DUST, ARSENIC containing

TOXICITY DATA with REFERENCE
itr-rat TDLo:120 mg/kg/15W-I:ETA EVHPAZ 19,191,77

CONSENSUS REPORTS: Reported in EPA TSCA Inventory. Arsenic and its compounds are on the Community Right-To-Know List.

OSHA PEL: TWA 0.5 mg(As)/m^3
ACGIH TLV: TWA 0.2 mg(As)/m^3
NIOSH REL: CL 2 μg(As)/m^3/15M
DOT Classification: Poison B; Label: Poison.

SAFETY PROFILE: A poison. Questionable carcinogen with experimental tumorigenic data. See also ARSENIC COMPOUNDS.

ARE750 CAS:8028-73-7 ***HR: 3***
ARSENICAL FLUE DUST (DOT)
DOT: UN 1562

TOXICITY DATA with REFERENCE
itr-rat TDLo:120 mg/kg/15W-I:ETA EVHPAZ 19,191,77

CONSENSUS REPORTS: Arsenic and its compounds are on the Community Right-To-Know List. Reported in EPA TSCA Inventory.

OSHA PEL: TWA 0.5 mg(As)/m^3
DFG MAK: Human Carcinogen.
DOT Classification: Poison B; Label: Poison.

SAFETY PROFILE: Confirmed human carcinogen with experimental tumorigenic data. Poison by inhalation and ingestion. See also ARSENIC COMPOUNDS.

ARF000 ***HR: 3***
ARSENIC BISULFIDE
mf: As_2S_2 mw: 214

PROP: Red-brown crystals. Bp: 565°, mp: (β) 307°, d: (α) 3.506 @ 19°, (β) 3.254 @ 19°.

SYN: REALGAR

CONSENSUS REPORTS: Arsenic and its compounds are on the Community Right-To-Know List.

SAFETY PROFILE: A poison. See also ARSENIC COMPOUNDS and SULFIDES. Flammable in the form of dust when exposed to heat or flame. Explosion hazard when intimately mixed with powerful oxidizers such as Cl_2; KNO_3; chlorates. It will react with water or steam to produce toxic and flammable vapors.

ARF250 CAS:7784-33-0 ***HR: 3***
ARSENIC(III) BROMIDE
DOT: UN 1555
mf: $AsBr_3$ mw: 314.65

PROP: Colorless, rhombic crystals. Mp: 32.8°, bp: 220.0°, vap press: 1 mm @ 41.8°, d: 3.3972 @ 25°, (liq), 3.3282.

SYNS: ARSENIC TRIBROMIDE ◇ ARSENOUS BROMIDE ◇ ARSENOUS TRIBROMIDE ◇ TRIBROMOARSINE

CONSENSUS REPORTS: Reported in EPA TSCA Inventory. Arsenic and its compounds are on the Community Right-To-Know List.

OSHA PEL: TWA 0.01 mg(As)/m^3; Cancer Hazard
ACGIH TLV: TWA 0.2 mg(As)/m^3
NIOSH REL: (Arsenic, Inorganic) CL 2 μg(As)/m^3/15M

DOT Classification: Poison B; Label: Poison.

SAFETY PROFILE: Confirmed carcinogen. A poison. See also ARSENIC COMPOUNDS and BROMIDES. When heated to decomposition it emits very toxic fumes of As and Br^-.

ARF500 CAS:7784-34-1 ***HR: 3***
ARSENIC CHLORIDE
DOT: UN 1560
mf: $AsCl_3$ mw: 181.28

PROP: Colorless, oily liquid. D: 2.15 @ 25°, mp: −16°, bp: 130°. Decomp in water and by UV light; misc in chloroform, CCl_4, ether, iodine, P, S, alkali iodides, oils and fats. Vap d: 6.25, vap press: 10 mm @ 23.5°.

SYNS: ARSENIC BUTTER ◇ ARSENIC(III) CHLORIDE ◇ ARSENIOUS CHLORIDE ◇ ARSENOUS CHLORIDE ◇ ARSENOUS TRICHLORIDE (9CI) ◇ CHLORURE d'ARSENIC (FRENCH) ◇ CHLORURE ARSENIEUX (FRENCH) ◇ FUMING LIQUID ARSENIC ◇ TRICHLOROARSINE ◇ TRICHLORURE d'ARSENIC (FRENCH)

TOXICITY DATA with REFERENCE
cyt-hmn:leu 600 nmol/L MUREAV 88,73,81
mrc-bcs 30 μL/disc MUREAV 77,109,80
otr-ham:emb 3 μmol/L CNREA8 39,193,79
ihl-mus LCLo:338 ppm/10M HBTXAC 1,324,56
ihl-cat LCLo:100 mg/m^3/1H ZGEMAZ 13,523,21

CONSENSUS REPORTS: Reported in EPA TSCA Inventory. Arsenic and its compounds are on the Community Right-To-Know List. EPA Extremely Hazardous Substances List.

OSHA: Cancer Hazard
ACGIH TLV: TWA 0.2 mg(As)/m^3
NIOSH REL: (Arsenic, Inorganic) CL 2 μg(As)/m^3/15M
DOT Classification: Poison B; Label: Poison.

SAFETY PROFILE: A poison via inhalation. See also ARSENIC COMPOUNDS and CHLORIDES. Very poisonous; fumes in air. Mutation data reported. When heated to decomposition it emits very toxic fumes of As and Cl^-. Highly reactive. Explodes with Na, K, and Al on impact.

ARF750 ***HR: 3***
ARSENIC COMPOUNDS

SYN: ARSENICALS

CONSENSUS REPORTS: Arsenic and its compounds are on the Community Right-To-Know List.

OSHA PEL: Inorganic: TWA 0.01 mg(As)/m^3; Cancer Hazard; Organic: TWA 0.5 mg(As)/m^3
ACGIH TLV: TWA 0.2 mg(As)/m^3
NIOSH REL: CL 2 μg(As)/m^3/15M

SAFETY PROFILE: Inorganic compounds are confirmed human carcinogens producing tumors of the mouth, esophagus, larynx, bladder, and para nasal sinus. A recognized carcinogen of the skin, lungs, and liver. Used as insecticides, herbicides, silvicides, defoliants, desiccants and rodenticides. Poisoning from arsenic compounds may be acute or chronic. Acute poisoning usually results from swallowing arsenic compounds; chronic poisoning from either swallowing or inhaling. Acute allergic reactions to arsenic compounds used in medical therapy have been fairly common, the type and severity of reaction depending upon the compound. Inorganic arsenicals are more toxic than organics. Trivalent is more toxic than pentavalent. Acute arsenic poisoning (from ingestion) results in marked irritation of the stomach and intestines with nausea, vomiting, and diarrhea. In severe cases, the vomitus and stools are bloody and the patient goes into collapse and shock with weak, rapid pulse, cold sweats, coma, and death. Chronic arsenic poisoning, whether through ingestion or inhalation, may manifest itself in many different ways. There may be disturbances of the digestive system such as loss of appetite, cramps, nausea, constipation, or diarrhea. Liver damage may occur, resulting in jaundice. Disturbances of the blood, kidneys, and nervous system are not infrequent. Arsenic can cause a variety of skin abnormalities including itching, pigmentation, and even cancerous changes. A characteristic of arsenic poisoning is the great variety of symptoms that can be produced. Dangerous; when heated to decomposition, or when metallic arsenic contacts acids or acid fumes, or when water solutions of arsenicals are in contact with active metals such as Fe; Al; Zn; they emits highly toxic fumes of arsenic.

In treating acute poisoning from ingestion BAL (dimercaptol) is of questionable effectiveness for acute and chronic poisoning with trivalent arsenicals, such as arsenic trioxide, arsine, and arsenites. It is of no value for pentavalent arsenicals, such as cacodylic acid, methanearsonic acid, sodium, cacodylate, MSMA, DSMA, arsanilic acid, arsenic acid, and arsenates. Vomiting and gastric lavage are the preferred emergency treatments for acute arsenical poisoning. More recent medical treatment of arsenical poisoning uses exchange transfusion and dialysis. Note: Arsenic compounds are common air contaminants.

ARG000 ***HR: 3***
ARSENIC DIETHYL
mf: $[As(C_2H_5)_2$ mw: 266.2

PROP: Liquid or oil. Bp: 185-190°, d: about 1.

CONSENSUS REPORTS: Arsenic and its compounds are on the Community Right-To-Know List.

SAFETY PROFILE: A poison. A dangerous fire hazard by spontaneous chemical reaction. Dangerous when heated. Incompatible with oxidizing materials. See also ARSENIC COMPOUNDS.

ARG250 ***HR: 3***
ARSENIC DIMETHYL
mf: $[As(CH_3)_2$ mw: 210.0

PROP: Colorless to yellow oily liquid. Mp: −6°; bp: 186°; d: 1.15.

CONSENSUS REPORTS: Arsenic and its compounds are on the Community Right-To-Know List.

SAFETY PROFILE: Poison by inhalation and ingestion. See also ARSENIC COMPOUNDS. Flammable. Evolves dangerous fumes of arsenic when heated.

ARG500 ***HR: 3***
ARSENIC HEMISELENIDE
mf: As_2Se mw: 228.78

CONSENSUS REPORTS: Arsenic compounds and its compounds as well as selenium and its compounds are on the Community Right-To-Know List.

OSHA PEL: TWA 0.01 mg(As)/m^3; Cancer Hazard; TWA 0.2 mg(Se)/m^3
ACGIH TLV: TWA 0.2 mg(As)/m^3; TWA 0.2 mg(Se)/m^3
DFG TRK: 0.2 mg/m^3 calculated as arsenic in that portion of dust that can possibly be inhaled; 0.1 mg(Se)/m^3
NIOSH REL: CL 2 μg(As)/m^3

SAFETY PROFILE: When heated to decomposition it emits fumes of As and Se. Incompatible with oxidizing materials. When heated to decomposition it emits highly toxic fumes of Se and arsenic. See ARSENIC COMPOUNDS and SELENIUM COMPOUNDS.

ARG750 CAS:7784-45-4 ***HR: 3***
ARSENIC IODIDE
DOT: NA 1557
mf: AsI_3 mw: 455.62

PROP: Red hexagonal crystals. Mp: 141.8°; bp: 403°; d: 4.38 @ 13°. Solubility: in water = 6/100 @ 25°, in CS_2 = 5.2/100.

SYNS: ARSENIC TRIIODIDE ◇ ARSENOUS IODIDE ◇ ARSENOUS TRIIODIDE (9CI) ◇ TRIIODOARSINE

CONSENSUS REPORTS: Reported in EPA TSCA Inventory. Arsenic and its compounds are on the Community Right-To-Know List.

OSHA PEL: TWA 0.01 mg(As)/m^3; Cancer Hazard
ACGIH TLV: TWA 0.2 mg(As)/m^3
NIOSH REL: (Arsenic, Inorganic) CL 2 μg(As)/m^3/15M
DOT Classification: Poison B; Label: Poison.

SAFETY PROFILE: A poison. See also ARSENIC COMPOUNDS and IODIDES. Can form a shock sensitive compound with sodium or potassium. When heated to decomposition it emits very toxic fumes of I^- and arsenic.

ARH250 ***HR: 3***
ARSENIC PENTASULFIDE
mf: As_2S_5 mw: 310.2

PROP: Brownish-yellow glassy, amorphous, highly refractive mass. Mp: 500° (subl).

CONSENSUS REPORTS: Arsenic and its compounds are on the Community Right-To-Know List.

OSHA PEL: TWA 0.01 mg(As)/m^3
ACGIH TLV: TWA 0.2 mg(As)/m^3
NIOSH REL: CL 2 μg(As)/m^3/15M

SAFETY PROFILE: See also ARSENIC COMPOUNDS and SULFIDES. Flammable in the form of dust when exposed to heat or flame. Explosive when intimately mixed with powerful oxidizers, such as Cl_2; KNO_3; chlorates. Will react with water and steam to produce toxic and flammable vapors. Incompatible with water, steam, and strong oxidizers.

ARH500 CAS:1303-28-2 ***HR: 3***
ARSENIC PENTOXIDE
DOT: UN 1559
mf: As_2O_5 mw: 229.84

PROP: White, amorphous, deliquescent solid. Mp: decomp @ 800°, d: 4.32. Sol in alc. Solubility in water = 65.8/100 @ 20°.

SYNS: ANHYDRIDE ARSENIQUE (FRENCH) ◇ ARSENIC ACID ◇ ARSENIC ACID ANHYDRIDE ◇ ARSENIC ANHYDRIDE ◇ ARSENIC OXIDE ◇ ARSENIC(V) OXIDE ◇ DIARSENIC PENTOXIDE ◇ RCRA WASTE NUMBER P011 ◇ ZOTOX

TOXICITY DATA with REFERENCE
cyt-hmn:leu 1200 nmol/L MUREAV 88,73,81
mrc-bcs 50 mmol/L MUREAV 77,109,80

itt-rat TDLo:4597 μg/kg (male 1D pre):REP JRPFA4 7,21,64
orl-rat LD50:8 mg/kg 28ZEAL 4,50,69
orl-mus LD50:55 mg/kg IRGGAJ 20,21,63
ivn-rbt LDLo:6 mg/kg NTIS** PB214-270

CONSENSUS REPORTS: NTP Fifth Annual Report on Carcinogens. IARC Cancer Review: Human Sufficient Evidence IMEMDT 23,39,80. Reported in EPA TSCA Inventory. Arsenic and its compounds are on the Community Right-To-Know List. EPA Extremely Hazardous Substances List.

OSHA PEL: Cancer Hazard
ACGIH TLV: TWA 0.2 mg(As)/m^3
DFG MAK: Human Carcinogen.
NIOSH REL: CL 2 μg(As)/m^3/15M
DOT Classification: Poison B; Label: Poison.

SAFETY PROFILE: Confirmed human carcinogen. Poison by ingestion and intravenous routes. Experimental reproductive effects. Mutation data reported. See also ARSENIC COMPOUNDS. Reacts vigorously with Rb_2C_2. When heated to decomposition it emits toxic fumes of arsenic.

ARH750 ***HR: 3***
ARSENIC PHOSPHIDE
mf: AsP mw: 105.9

PROP: Brown to red powder. Mp: sublimes with decomp.

CONSENSUS REPORTS: Arsenic and its compounds are on the Community Right-To-Know List.

SAFETY PROFILE: Flammable by spontaneous chemical reaction. Phosphine is liberated upon contact with moisture. Dangerous when heated. Incompatible with water or steam; oxidizing materials. See ARSENIC COMPOUNDS and PHOSPHINE.

ARI000 CAS:1303-33-9 ***HR: 3***
ARSENIC SULFIDE
DOT: NA 1557
mf: As_2S_3 mw: 246.04

PROP: Yellow or red crystals. Bp: 707°, d: 3.43; mp: 312°. Insol in water; sol in alkalies.

SYNS: ARSENIC SESQUISULFIDE ◇ ARSENIC SULFIDE YELLOW ◇ ARSENIC SULPHIDE ◇ ARSENIC TRISULFIDE ◇ ARSENIC YELLOW ◇ ARSENIOUS SULPHIDE ◇ ARSENOUS SULFIDE ◇ C.I. 77086 ◇ DIARSENIC TRISULFIDE ◇ KING'S YELLOW ◇ ORPIMENT

TOXICITY DATA with REFERENCE
scu-rat TDLo:125 mg/kg:ETA BJCAAI 20,190,66

CONSENSUS REPORTS: IARC Cancer Review: Human Sufficient Evidence IMEMDT 23,39,80. Reported in EPA TSCA Inventory. Arsenic and its compounds are on the Community Right-To-Know List.

OSHA PEL: Cancer Hazard
ACGIH TLV: TWA 0.2 mg(As)/m^3
NIOSH REL: (Arsenic, Inorganic) CL 2 μg(As)/m^3/15M
DOT Classification: Poison B; Label: Poison.

SAFETY PROFILE: Confirmed human carcinogen with experimental tumorigenic data. A poison. Reacts violently with H_2O_2, (KNO_3 + S). When heated to decomposition or contact with acid or acid fumes it emits highly toxic fumes of SO_2, H_2S, and As. Reacts with water or steam to emit toxic and flammable vapors.

ARI250 CAS:7784-35-2 ***HR: 3***
ARSENIC TRIFLUORIDE
mf: AsF_3 mw: 131.92

PROP: Colorless liquid. D: 3.01, mp: −5.95, bp: 51°, vap press: 100 mm @ 13.2°, 400 mm @ 41.5°. Insol in water; sol in alc, benzene, and mercury.

SYNS: ARSENIC FLUORIDE ◇ ARSENOUS FLUORIDE ◇ TRIFLUOROARSINE

TOXICITY DATA with REFERENCE
ihl-mus LCLo:2000 mg/m^3/10M NDRC** NDCrc-132,Aug,42

CONSENSUS REPORTS: Reported in EPA TSCA Inventory. Arsenic and its compounds are on the Community Right-To-Know List.

OSHA PEL: TWA 0.01 mg(As)/m^3; Cancer Hazard
ACGIH TLV: TWA 0.2 mg(As)/m^3
NIOSH REL: (Arsenic, Inorganic) CL 2 μg(As)/m^3/15M

SAFETY PROFILE: Confirmed human carcinogen. A poison by inhalation. See also FLUORIDES and ARSENIC COMPOUNDS. Strong reaction with P_2O_3. When heated to decomposition it emits very toxic fumes of As and F^-.

ARI500 CAS:8012-54-2 ***HR: 3***
ARSENIC TRIIODIDE mixed with MERCURIC IODIDE
DOT: NA 2810

SYNS: ARSENIOUS and MERCURIC IODIDE, solution (DOT) ◇ DONOVAN'S SOLUTION

CONSENSUS REPORTS: Arsenic compounds and Mercury compounds are on the Community Right-To-Know List.

ACGIH TLV: TWA 0.1 mg(Hg)/m^3 (skin)
NIOSH REL: (Arsenic, Inorganic): CL 0.002 mg(As)/m^3/15M; (Mercury, Inorganic): 8H TWA 0.05 mg(Hg)/m^3
DOT Classification: Poison B; Label: Poison.

SAFETY PROFILE: A poison. See also ARSENIC COMPOUNDS, MERCURY COMPOUNDS, and IODIDES. When heated to decomposition it emits very toxic fumes of Hg, As, and I^-.

ARI750 CAS:1327-53-3 ***HR: 3***
ARSENIC TRIOXIDE
DOT: UN 1561
mf: As_2O_3 mw: 197.84

PROP: Colorless, rhombic crystals (dimer, claudetite). D: 4.15, mp: 278°, bp: 460°. Solubility in water = 1.82/100 @ 20°; sol in alc. Cubes: Colorless. D: 3.865, mp: 309°. Solubility in water = 1.2/100 @ 20°.

SYNS: ACIDE ARSENIEUX (FRENCH) ◇ ANHYDRIDE ARSENIEUX (FRENCH) ◇ ARSENIC, white, solid (DOT) ◇ ARSENIC BLANC (FRENCH) ◇ ARSENIC OXIDE ◇ ARSENIC(III) OXIDE ◇ ARSENIC SESQUIOXIDE ◇ ARSENIGEN SAURE (GERMAN) ◇ ARSENIOUS ACID (MAK) ◇ ARSENIOUS OXIDE ◇ ARSENIOUS TRIOXIDE ◇ ARSENOUS ACID ◇ ARSENOUS ACID ANHYDRIDE ◇ ARSENOUS ANHYDRIDE ◇ ARSENOUS OXIDE ◇ ARSENOUS OXIDE ANHYDRIDE ◇ CLAUDELITE ◇ CLAUDETITE ◇ CRUDE ARSENIC ◇ DIARSENIC TRIOXIDE ◇ RCRA WASTE NUMBER P012 ◇ WHITE ARSENIC

TOXICITY DATA with REFERENCE
mrc-bcs 50 mmol/L MUREAV 77,109,80
orl-wmn TDLo:600 mg/kg (female 30W post):REP AJDCAI 117,328,69
ihl-mus TCLo:28500 $\mu g/m^3$/4H (female 9-12D post):TER JJATDK 5,61,85
itr-rat TDLo:16 mg/kg/15W-I:ETA EVHPAZ 19,191,77
itr-ham TDLo:45 mg/kg/15W-I:NEO ARTODN 7,403,84
orl-man LDLo:29 mg/kg:GIT AEMED3 16,702,87
orl-hmn LDLo:1429 μg/kg YKYUA6 31,1247,80
orl-rat LD50:14600 μg/kg GISAAA 52(1),21,87
ipr-rat LD50:871 mg/kg GTPZAB 19(3),30,75
scu-rat LDLo:8 mg/kg JPETAB 19,337,22
orl-mus LD50:31500 μg/kg CHYCDW 14,86,80
scu-mus LD50:9800 μg/kg PSEBAA 78,392,51
ivn-mus LD50:10700 μg/kg PSEBAA 78,392,51
orl-dog LDLo:10 mg/kg HBAMAK 4,1306,35
orl-rbt LDLo:4 mg/kg NTIS** PB214-270
ivn-rbt LDLo:10560 μg/kg BIZEA2 70,144,15

CONSENSUS REPORTS: NTP Fifth Annual Report on Carcinogens. IARC Cancer Review: Group 1 IMEMDT 7,100,87; Human Limited Evidence IMEMDT 2,48,73; Human Sufficient Evidence IMEMDT 23,39,80; Animal Inadequate Evidence IMEMDT 2,48,73; IMEMDT 23,39,80. Reported in EPA TSCA Inventory. Arsenic and its compounds are on the Community Right-To-Know List. EPA Extremely Hazardous Substances List.

OSHA PEL: TWA 0.01 mg(As)/m^3: Cancer Hazard
ACGIH TLV: Production: Suspected Human Carcinogen
DFG MAK: Human Carcinogen.
NIOSH REL: CL 2 μg(As)/m^3/15M
DOT Classification: Poison B; Label: Poison.

SAFETY PROFILE: Confirmed human carcinogen with experimental neoplastigenic and tumorigenic data. Poison by ingestion, subcutaneous, and intravenous routes. Human gastrointestinal effects by ingestion. An experimental teratogen. Other experimental reproductive effects. Mutation data reported. Reacts vigorously with Rb_2C_2, ClF_3, F_2, Hg, OF_2, $NaClO_3$. See also ARSENIC COMPOUNDS.

ARJ000 ***HR: 3***
ARSENIC TRIOXIDE mixed with SELENIUM DIOXIDE (1:1)
mf: $AsO_3 \cdot O_2Se$ mw: 233.88

SYN: SELENIUM DIOXIDE mixed with ARSENIC TRIOXIDE (1:1)

TOXICITY DATA with REFERENCE
orl-mus TDLo:16 mg/kg/35W-C:ETA BICHBX 9(3),245,78

CONSENSUS REPORTS: Arsenic and its compounds, as well as selenium and its compounds, are on the Community Right-To-Know List.

OSHA PEL: TWA 0.01 mg(As)/m^3; Cancer Hazard
ACGIH TLV: TWA 0.2 mg(As)/m^3; Suspected Carcinogen; 0.2 mg(Se)/m^3
DFG MAK: 0.1 mg(Se)/m^3
NIOSH REL: (Arsenic, Inorganic): CL 0.002 mg(As)/m^3/15M

SAFETY PROFILE: Confirmed human carcinogen with experimental tumorigenic data. See also ARSENIC COMPOUNDS and SELENIUM COMPOUNDS. When heated to decomposition it emits very toxic fumes of As and Se.

ARJ250 ***HR: 3***
ARSENIDES

CONSENSUS REPORTS: Arsenic and its compounds are on the Community Right-To-Know List.

SAFETY PROFILE: Compounds of arsenic and hydrogen or metals, (i.e., transitional, alkaline earth, or rare-earth). These materials are dangerous because they readily emit very toxic arsine and arsenic fumes when exposed to heat, moisture, acids, and acid fumes.

ARJ500 CAS:14060-38-9 ***HR: 3***
ARSENIOUS ACID SODIUM SALT
mf: $AsH_3O_3 \cdot 7Na$ mw: 286.88

PROP: Colorless or grayish-white powder. D: 1.87.

SYNS: ARSONIC ACID, SODIUM SALT (9CI) ◇ ARSENIOUS ACID, SODIUM SALT POLYMERS ◇ NATRIUMARSENIT (GERMAN) ◇ SODIUM ORTHOARSENITE

TOXICITY DATA with REFERENCE
ipr-rat LDLo:9 mg/kg JPETAB 58,454,36

orl-frg LDLo:600 mg/kg HBAMAK 4,1289,35
scu-frg LDLo:200 mg/kg HBAMAK 4,1289,35

CONSENSUS REPORTS: Arsenic and its compounds are on the Community Right-To-Know List.

OSHA PEL: TWA 0.01 mg(As)/m^3; Cancer Hazard
ACGIH TLV: TWA 0.2 mg(As)/m^3
NIOSH REL: (Arsenic, Inorganic) CL 2 μg(As)/m^3/15M

SAFETY PROFILE: Confirmed human carcinogen. Poison by intraperitoneal and subcutaneous routes. Moderately toxic by ingestion. When heated to decomposition it emits toxic fumes of arsenic.

ARJ750 CAS:1303-18-0 ***HR: 3***
ARSENOPYRITE
mf: AsFeS mw: 162.83

SYNS: ARSENOMARCASITE ◇ MISPICKEL

TOXICITY DATA with REFERENCE
ivn-mus LDLo:200 mg/kg JNCIAM 1,241,40

CONSENSUS REPORTS: Arsenic and its compounds are on the Community Right-To-Know List.

OSHA PEL: TWA 0.01 mg(As)/m^3
ACGIH TLV: TWA 0.2 mg(As)/m^3
NIOSH REL: (Arsenic, Inorganic) CL 2 μg(As)/m^3/15M

SAFETY PROFILE: Poison by intravenous route. When heated to decomposition it emits very toxic fumes of As and SO_x.

ARJ755 CAS:1122-90-3 ***HR: 3***
p-ARSENOSOANILINE
mf: C_6H_6AsNO mw: 183.05

SYN: ANILINE, p-ARSENOSO-

TOXICITY DATA with REFERENCE
ipr-mus LD50:4 mg/kg JMCMAR 9,221,66

OSHA PEL: TWA 0.5 mg(As)/m^3

SAFETY PROFILE: Poison by intraperitoneal route. When heated to decomposition it emits toxic fumes of NO_x and As.

ARJ760 CAS:4164-07-2 ***HR: 3***
p-ARSENOSO-N,N-BIS(2-CHLOROETHYL)ANILINE
mf: $C_{10}H_{12}AsCl_2NO$ mw: 308.05

SYN: ANILINE,p-ARSENOSO-N,N-BIS(2-CHLOROETHYL)-

TOXICITY DATA with REFERENCE
ipr-mus LD50:5545 μg/kg JMCMAR 9,221,66

OSHA PEL: TWA 0.5 mg(As)/m^3

SAFETY PROFILE: Poison by intraperitoneal route. When heated to decomposition it emits toxic fumes of NO_x, As, and Cl^-.

ARJ770 CAS:5185-80-8 ***HR: 3***
p-ARSENOSO-N,N-BIS(2-HYDROXYETHYL)ANILINE
mf: $C_{10}H_{14}AsNO_3$ mw: 271.17

SYN: ANILINE,p-ARSENOSO-N,N-BIS(2-HYDROXYETHYL)-

TOXICITY DATA with REFERENCE
ipr-mus LD50:7593 μg/kg JMCMAR 9,221,66

OSHA PEL: TWA 0.5 mg(As)/m^3

SAFETY PROFILE: Poison by intraperitoneal route. When heated to decomposition it emits toxic fumes of NO_x and As.

ARJ800 CAS:4164-06-1 ***HR: 3***
p-ARSENOSO-N,N-DIETHYLANILINE
mf: $C_{10}H_{14}AsNO$ mw: 239.17

SYN: ANILINE, p-ARSENOSO-N,N-DIETHYL-

TOXICITY DATA with REFERENCE
ipr-mus LD50:2809 μg/kg JMCMAR 9,221,66

OSHA PEL: TWA 0.5 mg(As)/m^3

SAFETY PROFILE: Poison by intraperitoneal route. When heated to decomposition it emits toxic fumes of NO_x and As.

ARJ900 CAS:63951-03-1 ***HR: 3***
ARSENOXIDE SODIUM
mf: $C_6H_5AsNO_2$•Na mw: 221.03

SYN: PHENOL, 2-AMINO-4-ARSENOSO-, SODIUM SALT

TOXICITY DATA with REFERENCE
ivn-rat LDLo:20 mg/kg MADCAJ 6,195,37

OSHA PEL: TWA 0.5 mg(As)/m^3
ACGIH TLV: TWA 0.2 mg(As)/m^3

SAFETY PROFILE: Poison by intravenous route. When heated to decomposition it emits toxic fumes of NO_x and As.

ARK250 CAS:7784-42-1 ***HR: 3***
ARSINE
DOT: UN 2188
mf: AsH_3 mw: 77.95

PROP: Colorless gas, mild garlic odor. D: 2.695 g/L; bp: −62.5°; vap d: 2.66; mp: −116°. Solubility in water = 28 mg/100 @ 20°. Sol in benzene and chloroform.

SYNS: ARSENIC HYDRIDE ◇ ARSENIC TRIHYDRIDE ◇ ARSENIURETTED HYDROGEN ◇ ARSENOUS HYDRIDE ◇ ARSENOWODOR (POLISH) ◇ ARSENWASSERSTOFF (GERMAN) ◇ HYDROGEN ARSENIDE

TOXICITY DATA with REFERENCE
ihl-hmn TCLo:3 ppm:RBC AMIHAB 21,132,60
ihl-hmn LCLo:25 ppm/30M AEHLAU 19,133,69
ihl-man TDLo:338 ppt:GIT AEHLAU 34,224,79
ihl-man TDLo:338 ppt:SYS AEHLAU 34,224,79
ihl-man TDLo:338 ppt:CNS AEHLAU 34,224,79
ihl-rat LCLo:300 mg/m^3/15M FATOAO 30(2),226,67
ihl-mus LCLo:70 mg/m^3/3H AEXPBL 80,288,17
ihl-dog LCLo:400 mg/m^3/15M FATOAO 30(2),226,67
ihl-mky LCLo:70 mg/m^3/15M HBAMAK 4,1289,35
ihl-cat LCLo:150 mg/m^3/20M HBAMAK 4,1289,35
ihl-rbt LCLo:500 mg/m^3/15M FATOAO 30(2),226,67
ihl-frg LCLo:4500 mg/m^3/3H AEXPBL 80,288,17

CONSENSUS REPORTS: IARC Cancer Review: Human Sufficient Evidence IMEMDT 23,39,80. Reported in EPA TSCA Inventory. Arsenic and its compounds are on the Community Right-To-Know List. EPA Extremely Hazardous Substances List.

OSHA PEL: TWA 0.05 ppm
ACGIH TLV: TWA 0.05 ppm
DFG MAK: 0.05 ppm (0.2 mg/m^3)
NIOSH REL: (Arsine) CL 2 μg(As)/m^3/15M
DOT Classification: Poison A; Label: Poison Gas and Flammable Gas.

SAFETY PROFILE: Confirmed human carcinogen. Poison by inhalation. Human red blood cell, gastrointestinal system, central nervous system, and other systemic effects by inhalation. Flammable when exposed to flame. Moderately explosive when exposed to Cl_2, HNO_3, (K + NH_3), open flame, or powerful shock. Dangerous, more toxic than its oxidation product. When heated to decomposition it emits highly toxic fumes of arsenic. See also ARSENIC, ARSENIC COMPOUNDS, and HYDRIDES.

ARK500 ***HR: 3***
ARSINE BORON TRIBROMIDE
mf: AsH_3•BBr_3 mw: 328.6

CONSENSUS REPORTS: Arsenic and its compounds are on the Community Right-To-Know List.

SAFETY PROFILE: A poison. See BROMIDES, ARSENIC COMPOUNDS, and BORON COMPOUNDS. A highly unstable compound. Ignites in air. When heated to decomposition it emits very toxic fumes of As and Br^-.

ARK750 CAS:67360-94-5 ***HR: 3***
ARSINE-TRI-1-PIPERIDINIUM CHLORIDE
mf: $C_{15}H_{33}AsN_3$•3Cl mw: 436.78

SYN: ARSINOTRIS PIPERIDINIUM TRICHLORIDE

TOXICITY DATA with REFERENCE
ivn-rbt LDLo:200 mg/kg JPETAB 28,233,26
ims-rbt LDLo:500 mg/kg JPETAB 28,233,26

CONSENSUS REPORTS: Arsenic and its compounds are on the Community Right-To-Know List.

OSHA PEL: TWA 0.5 mg(As)/m^3
ACGIH TLV: TWA 0.2 mg(As)/m^3
NIOSH REL: CL 2 μg(As)/m^3/15M

SAFETY PROFILE: Poison by intravenous route. Moderately toxic by intramuscular route. When heated to decomposition it emits very toxic fumes of As, NO_x and Cl^-. See also ARSENIC COMPOUNDS.

ARK780 CAS:52740-16-6 ***HR: 3***
ARSONIC ACID, CALCIUM SALT (1:1)
mf: AsH_2O_3•Ca mw: 165.02

CONSENSUS REPORTS: NTP Fifth Annual Report on Carcinogens

SAFETY PROFILE: Confirmed carcinogen. When heated to decomposition it emits toxic fumes of As.

ARK800 ***HR: 3***
(3-(p-ARSONOPHENYL)UREIDO) DITHIOBENZOIC ACID
mf: $C_{14}H_{13}AsN_2O_4S_2$ mw: 412.33

SYN: DITHIOCARBOXYPHENYL-p-CARBAMIDOPHENYLARSENOUS OXIDE

TOXICITY DATA with REFERENCE
orl-rat LD50:850 mg/kg FEPRA7 6,306,47
ipr-rat LD50:76 mg/kg FEPRA7 6,306,47
ipr-mus LD50:265 mg/kg FEPRA7 6,306,47

CONSENSUS REPORTS: Arsenic and its compounds are on the Community Right-To-Know List.

SAFETY PROFILE: Poison by intraperitoneal route. Moderately toxic by ingestion and other routes. When heated to decomposition it emits toxic fumes of SO_x, NO_x, and arsenic. See also ARSENIC COMPOUNDS.

ARL000 CAS:538-03-4 ***HR: 3***
ARSPHENOXIDE
mf: $C_6H_6AsNO_2$•ClH mw: 235.51

SYNS: 2-AMINO-4-ARSENOSOPHENOL HYDROCHLORIDE ◇ 3-AMINO-4-HYDROXY-PHENARSINE HYDROCHLORIDE ◇ 3-AMINO-4-HYDROXYPHENYLARSINE OXIDE HYDROCHLORIDE ◇ 3-AMINO-4-HYDROXYPHENYL ARSINOXIDE HYDROCHLORIDE ◇ ARSENO 39 ◇ ARSENOSAN ◇ ARSENOXIDE ◇ EHRLICH 5 ◇ FONTARSAN ◇ MAPHARSAL ◇ MAPHARSEN ◇ MAPHARSIDE ◇ OXIARSOLAN ◇ OXOPHENARSINE HYDROCHLORIDE

TOXICITY DATA with REFERENCE
ivn-hmn TDLo:400 μg/kg:CNS,BRN AJSGA3 28,218,44
orl-rat LDLo:500 mg/kg NCNSA6 5,12,53

ivn-rat LD50:18500 μg/kg JACSAT 70,1762,48
orl-mus LD50:110 μg/kg CLDND* 81,284,44
scu-mus LD50:25 mg/kg JPETAB 81,284,44
ivn-mus LD50:20 mg/kg JPETAB 81,284,44
ivn-dog LDLo:12 mg/kg JPETAB 50,198,34

CONSENSUS REPORTS: Arsenic and its compounds are on the Community Right-To-Know List.

OSHA PEL: TWA 0.5 mg(As)/m^3
ACGIH TLV: TWA 0.2 mg(As)/m^3

SAFETY PROFILE: Poison by ingestion, intravenous, and intraperitoneal routes. Human systemic effects by ingestion: stroke, convulsions, and coma. See also ARSENIC COMPOUNDS. When heated to decomposition it emits very toxic fumes of As, NO_x, and HCl. An antirickettsial and antitrypanosomal agent.

ARL250 CAS:8022-37-5 ***HR: 2***
ARTEMISIA OIL

PROP: Chief constituent is Thujone, and found in the plant *Artemisia absinthium* L. (FCTXAV 13,681,75).

SYNS: ABSINTHIUM ◇ ARTEMISIA OIL (WORMWOOD) ◇ OIL, ARTEMISIA

TOXICITY DATA with REFERENCE
orl-rat LD50:960 mg/kg FCTXAV 13,681,75

CONSENSUS REPORTS: Reported in EPA TSCA Inventory.

SAFETY PROFILE: Moderately toxic by ingestion. An allergen. Habitual users develop, "absinthism" with tremors, vertigo, vomiting, and hallucinations. May cause a contact dermatitis. When heated to decomposition it emits acrid smoke and irritating fumes.

ARL375 CAS:63968-64-9 ***HR: 2***
ARTEMISINE
mf: $C_{15}H_{22}O_5$ mw: 282.37

PROP: Needles. Mp: 156-157°. Sol in most aprotic solvents. Sltly sol in oil.

SYNS: ARTEANNUIN ◇ ARTEMISININ ◇ OCTAHYDRO-3,6,9-TRIMETHYL-3,12-EPOXY-12H-PYRANO(4,3-j)-1,2-BENZODIOXEPIN-10(3H)-ONE ◇ QINGHAOSU (CHINESE) ◇ QING HAU SAU (CHINESE)

TOXICITY DATA with REFERENCE
orl-mus LD50:5105 mg/kg CMJODS 92,811,79
ipr-mus LD50:1558 mg/kg CMJODS 92,811,79
ims-mus LD50:2800 mg/kg CMJODS 92,811,79

SAFETY PROFILE: Moderately toxic by ingestion, intramuscular, and intraperitoneal routes. When heated to decomposition it emits acrid smoke and fumes.

ARL425 CAS:71963-77-4 ***HR: 3***
ARTEMISININELACTOL METHYL ETHER
mf: $C_{16}H_{26}O_5$ mw: 298.42

SYNS: ARTEMETHER ◇ 3,12-EPOXY-12H-PYRANO(4,3-j)-1,2-BENZODIOXEPIN, DECAHYDRO-10-METHOXY-3,6,9-TRIMETHYL-, (3-α-5a-β,6-β,8a-β,9-α-12-β,12aR)-, (+)- ◇ METHYL-DIHYDROARTEMISININE

TOXICITY DATA with REFERENCE
orl-rat TDLo:200 mg/kg (female 6-15D post):REP CYLPDN 5,118,84
ims-mus LD50:263 mg/kg CYLPDN 2,138,81

SAFETY PROFILE: Poison by intramuscular route. Experimental reproductive effects. When heated to decomposition it emits acrid smoke and irritating fumes.

ARL500 CAS:149-95-1 ***HR: 3***
d-ARTERENOL
mf: $C_8H_{11}NO_3$ mw: 169.20

SYN: α-(AMINOMETHYL)-3,4-DIHYDROXYBENZYLALCOHOL

TOXICITY DATA with REFERENCE
ivn-rat LD50:1400 μg/kg JPETAB 95,502,49
ivn-mus LD50:18 mg/kg JPETAB 95,502,49

SAFETY PROFILE: Poison by intravenous route. When heated to decomposition it emits toxic fumes of NO_x.

ARL750 CAS:138-65-8 ***HR: 3***
dl-ARTERENOL
mf: $C_8H_{11}NO_3$ mw: 169.20

SYN: α-(AMINOMETHYL)-3,4-DIHYDROXYBENZYLALCOHOL

TOXICITY DATA with REFERENCE
ivn-rat LD50:130 μg/kg JPETAB 95,502,49
ivn-mus LD50:4700 μg/kg JPETAB 95,502,49

SAFETY PROFILE: Poison by intravenous route. When heated to decomposition it emits toxic fumes of NO_x.

ARL875 ***HR: 2***
ARYL ALKYL POLYETHER ALCOHOL

SYNS: ALKYL ARYL POLYETHER ALCOHOLS ◇ ALKYLARYLPOLYGLYKOLAETHER (GERMAN) ◇ ALKYL PHENOL POLYGLYCOL ETHERS ◇ ALKYL PHENOXY POLYETHOXY ETHANOLS

TOXICITY DATA with REFERENCE
orl-rat LD50:3200 mg/kg 85GYAZ -,144,71
ipr-mus LD50:3100 mg/kg PSTGAW 3,1,45
orl-rbt LD50:3 g/kg 85GYAZ -,144,71

SAFETY PROFILE: Moderately toxic by ingestion and other routes.

ARM000 CAS:13425-94-0 ***HR: 3***
ASALIN
mf: $C_{22}H_{33}Cl_2N_3O_4$ mw: 474.48

SYN: N-ACETYL-SARCOLYSIL VALINE ETHYL ETHER ◇ ASALINE ◇ AZALINE ◇ ETHYL ESTER of N-ACETYL-dl-SARCOSYLYL-dl-VALINE

TOXICITY DATA with REFERENCE
ipr-rat TDLo:50 mg/kg (female 6D post):TER VOONAW 13(11),79,67
orl-rat LD50:59 mg/kg FATOAO 33,472,70
ipr-rat LD50:187 mg/kg PCJOAU 12,25,78
ims-rat LD50:17 mg/kg FATOAO 33,472,70
rec-rat LD50:40 mg/kg FATOAO 33,472,70

SAFETY PROFILE: Poison by ingestion, intramuscular, rectal, and intraperitoneal routes. An experimental teratogen. See also ESTERS.

ARM250 CAS:1332-21-4 ***HR: 3***
ASBESTOS
DOT: UN 2212/UN 2590

SYNS: AMIANTHUS ◇ AMOSITE (OBS.) ◇ AMPHIBOLE ◇ ASBEST (GERMAN) ◇ ASBESTOS FIBER ◇ FIBROUS GRUNERITE ◇ NCI-C08991 ◇ SERPENTINE

TOXICITY DATA with REFERENCE
ipr-rat TDLo:50 mg/kg:ETA STRHAV 39,386,79
ipl-rat TD:100 mg/kg:NEO BJCAAI 45,352,82
ihl-hmn TCLo:1.2 fibers/cc/19Y-C:PUL ARDSBL 104,576,71

CONSENSUS REPORTS: IARC Cancer Review: Group 1 IMEMDT 7,106,87; Human Sufficient Evidence IMEMDT 2,17,73; IMEMDT 14,11,77; Animal Sufficient Evidence IMEMDT 2,17,73; IMEMDT 14,11,77. Reported in EPA TSCA Inventory. On Community Right-To-Know List. EPA Genetic Toxicology Program.

OSHA PEL: TWA 2 million fb/m^3; CL 10 million fb/m^3; Cancer Hazard
ACGIH TLV: TWA 2 fb/cc; Confirmed Human Carcinogen; (Proposed: TWA 0.2 fb/cc)
DFG TRK: (Fine dust particles which are able to reach the alveolar area of the lung) crocidolite: 0.05×10^6 fibers/m^3 (0.025 mg/m^3) (definition of fiber: length greater than 5 μm; diameter less than 3 μm; length/diameter greater than 3:1, equivalent to 1 fiber/cc); chrysotile, amosite, anthophyllite, tremolite, actinolite: 1×10^6 fibers/m^3 (0.05 mg/m^3) applicable when there is more than 2.5% asbestos in the dust; 2.0 mg/m^3 applicable when there is less than or equal to 2.5 wt % asbestos in fine dust.
NIOSH REL: TWA 100,000 fb/m^3 over 5 μm in length
DOT Classification: ORM-C.

SAFETY PROFILE: Confirmed human carcinogen producing lung tumors. Experimental neoplastigenic and tumorigenic data. Human pulmonary system effects by inhalation. Usually at least 4 to 7 years of exposure are required before serious lung damage (fibrosis) results. A common air contaminant.

ARM260 CAS:77536-66-4 ***HR: 3***
ASBESTOS, ACTINOLITE

SYNS: ACTINOLITE ASBESTOS ◇ ASBESTOS (ACGIH)

CONSENSUS REPORTS: IARC Cancer Review: Group 1 IMEMDT 7,106,87; Animal Sufficient Evidence IMEMDT 14,11,77.

OSHA PEL: TWA 2 million fb/m^3; CL 10 million fb/m^3; Cancer Hazard
ACGIH TLV: TWA 2 fb/cc; Confirmed Human Carcinogen; (Proposed: TWA 0.2 fb/cc)
DFG TRK: (Fine dust particles which are able to reach the alveolar area of the lung) 1×10^6 fibers/m^3 (0.05 mg/m^3) applicable when there is more than 2.5% asbestos in the dust.
NIOSH REL: TWA 100,000 fb/m^3 over 5 μm in length

SAFETY PROFILE: Confirmed human carcinogen. See also other asbestos entries.

ARM262 CAS:12172-73-5 ***HR: 3***
ASBESTOS, AMOSITE

SYNS: AMOSITE ASBESTOS ◇ ASBESTOS (ACGIH) ◇ MYSORITE ◇ NCI-C60253A

TOXICITY DATA with REFERENCE
cyt-ham:ovr 10 mg/L CSHCAL 4,941,77
sce-ham:ovr 10 mg/L JEPTDQ 4(2-3),373,80
msc-ham:lng 10 mg/L MUREAV 68,265,79
ihl-rat TCLo:11 mg/m^3/2Y-I:CAR BJCAAI 29,252,74
ipl-rat TDLo:80 mg/kg:CAR TOLED5 13,143,82
itr-rat TDLo:12 mg/kg/12W-I:ETA TOLED5 13,143,82
imp-rat TDLo:200 mg/kg:NEO JJIND8 67,965,81

CONSENSUS REPORTS: NTP Fifth Annual Report on Carcinogens. IARC Cancer Review: Group 1 IMEMDT 7,106,87; Animal Sufficient Evidence IMEMDT 2,17,73; IMEMDT 14,11,77; Human Sufficient Evidence IMEMDT 2,17,73; IMEMDT 14,11,77. NTP Carcinogenesis Studies (feed); No Evidence: hamster NTPTR* NTP-TR-249,83. EPA Genetic Toxicology Program.

OSHA PEL: TWA 2 million fb/m^3; CL 10 million fb/m^3; Cancer Hazard
ACGIH TLV: TWA 0.5 fb/cc; Confirmed Human Carcinogen; (Proposed: TWA 0.2 fb/cc)
DFG TRK: (Fine dust particles which are able to reach the alveolar area of the lung) 1×10^6 fibers/m^3 (0.05 mg/m^3) applicable when there is more than 2.5% asbestos in the dust.
NIOSH REL: TWA 100,000 fb/m^3 over 5 μm in length

SAFETY PROFILE: Confirmed human carcinogen with experimental carcinogenic, neoplastigenic, and tumorigenic data. Mutation data reported.

ARM264 CAS:77536-67-5 ***HR: 3***
ASBESTOS, ANTHOPHYLITE

SYNS: ANTHOPHYLITE ◇ ASBESTOS (ACGIH) ◇ AZBOLEN ASBESTOS ◇ FERROANTHOPHYLLITE

TOXICITY DATA with REFERENCE
cyt-ham:ovr 10 mg/L CSHCAL 4,941,77
ihl-rat TCLo:11 mg/m^3/1Y-I:CAR BJCAAI 29,252,74
ipl-rat TDLo:200 mg/kg:NEO BJCAAI 28,173,73
ipl-rat TD:2400 mg/kg/34W-I:ETA IAPUDO 30,343,80

CONSENSUS REPORTS: NTP Fifth Annual Report on Carcinogens. IARC Cancer Review: Animal Sufficient Evidence IMEMDT 2,17,73; IMEMDT 14,11,77; Human Sufficient Evidence IMEMDT 14,11,77. EPA Genetic Toxicology Program.

OSHA PEL: TWA 2 million fb/m^3; CL 10 million fb/m^3; Cancer Hazard
ACGIH TLV: TWA 2 fb/cc; Confirmed Human Carcinogen; (Proposed: TWA 0.2 fb/cc)
DFG TRK: (Fine dust particles which are able to reach the alveolar area of the lung) 1×10^6 $fibers/m^3$ (0.05 mg/m^3) applicable when there is more than 2.5% asbestos in the dust
NIOSH REL: TWA 100,000 fb/m^3 over 5 μm in length

SAFETY PROFILE: Confirmed human carcinogen with experimental carcinogenic, neoplastigenic, and tumorigenic data. Mutation data reported.

ARM266 CAS:17068-78-9 ***HR: 3***
ASBESTOS, ANTHOPHYLLITE
SYNS: AZBLLEN ASBESTOS ◇ 16 F

TOXICITY DATA with REFERENCE
ipr-rat TDLo:250 mg/kg:ETA ZHYGAM 32,89,86

CONSENSUS REPORTS: NTP Fifth Annual Report on Carcinogens.

ACGIH TLV: TWA 2 fb/cc; Confirmed Human Carcinogen; (Proposed: TWA 0.2 fb/cc)

SAFETY PROFILE: Confirmed carcinogen with experimental tumorigenic data.

ARM268 CAS:12001-29-5 ***HR: 3***
ASBESTOS, CHRYSOTILE
DOT: UN 2590

SYNS: 7-45 ASBESTOS ◇ ASBESTOS (ACGIH) ◇ ASBESTOS, WHITE (DOT) ◇ AVIBEST C ◇ CALIDRIA RG 100 ◇ CALIDRIA RG 144 ◇ CALIDRIA RG 600 ◇ CASSIAR AK ◇ CHRYSOTILE (DOT) ◇ CHRYSOTILE ASBESTOS ◇ HOOKER No. 1 CHRYSOTILE ASBESTOS ◇ METAXITE ◇ NCI-C61223A ◇ PLASTIBEST 20 ◇ SERPENTINE ◇ SERPENTINE CHRYSOTILE ◇ SYLODEX ◇ WHITE ASBESTOS

TOXICITY DATA with REFERENCE
oms-hmn:fbr 10 mg/L MUREAV 116,369,83
oms-ham:ovr 10 mg/L MUREAV 116,369,83
ihl-man TCLo:400 mppcf/1Y-C:CAR,PUL AEHLAU 28,61,74
orl-rat TDLo:7100 mg/kg/39W-C:CAR ARGEAR 46,437,76
ihl-rat TCLo:11 mg/m^3/26W-I:CAR BJCAAI 29,252,74
itr-rat TDLo:13 mg/kg:ETA ENVRAL 21,63,80
scu-mus TDLo:2400 mg/kg/13W-I:NEO FCTXAV 6,566,68
ihl-hmn TCLo:2.8 fb/cc/5Y:PUL ENVRAL 23,292,80
ipr-rat LDLo:300 mg/kg AJPAA4 70,291,73

CONSENSUS REPORTS: NTP Fifth Annual Report on Carcinogens. IARC Cancer Review: Human Sufficient Evidence IMEMDT 2,17,73; Animal Sufficient Evidence IMEMDT 2,17,73. NTP Carcinogenesis Studies (feed); Some Evidence: rat NTPTR* NTP-TR-295,85. EPA Genetic Toxicology Program.

OSHA PEL: TWA 2 million fb/m^3; CL 10 million fb/m^3:; Cancer Hazard
ACGIH TLV: TWA 2 fb/cc; Confirmed Human Carcinogen; (Proposed: TWA 0.2 fb/cc)
DFG TRK: (Fine dust particles which are able to reach the alveolar area of the lung) 1×10^6 $fibers/m^3$ (0.05 mg/m^3) applicable when there is more than 2.5% asbestos in the dust.
NIOSH REL: TWA 100,000 fb/m^3 over 5 μm in length

SAFETY PROFILE: Confirmed human carcinogen producing tumors of the lung. Human mutation data reported. Poison by intraperitoneal route. Human systemic effects by inhalation: lung fibrosis, dyspnea, and cough.

ARM275 CAS:12001-28-4 ***HR: 3***
ASBESTOS, CROCIDOLITE
DOT: UN 2212
mf: $ONa_2Fe_2O_{33}FeO_8SiO_2H_2O$ mw: 765.98

SYNS: AMORPHOUS CROCIDOLITE ASBESTOS ◇ ASBESTOS (ACGIH) ◇ BLUE ASBESTOS (DOT) ◇ CROCIDOLITE (DOT) ◇ CROCIDOLITE ASBESTOS ◇ FIBROUS CROCIDOLITE ASBESTOS ◇ KROKYDOLITH (GERMAN) ◇ NCI C09007

TOXICITY DATA with REFERENCE
oms-hmn:fbr 10 mg/L MUREAV 116,369,83
dns-ham:oth 1280 ng/cm^2 CNREA8 42,3669,82
ihl-rat TCLo:11 mg/m^3/1Y-I:CAR BJCAAI 29,252,74
ipr-rat TDLo:100 mg/kg:CAR PBPHAW 14,47,78
imp-rat TDLo:200 mg/kg:NEO JJIND8 67,965,81
ipl-mus TDLo:200 mg/kg:ETA 31BYAP -,92,74

CONSENSUS REPORTS: NTP Fifth Annual Report on Carcinogens. IARC Cancer Review: Animal Sufficient

Evidence IMEMDT 14,11,77, IMEMDT 2,17,73; Human Sufficient Evidence IMEMDT 14,11,77. EPA Genetic Toxicology Program.

OSHA PEL: TWA 2 million fb/m^3; CL 10 million fb/m^3; Cancer Hazard
ACGIH TLV: TWA 0.2 fb/cc; Confirmed Human Carcinogen
DFG TRK: (Fine dust particles which are able to reach the alveolar area of the lung) crocidolite: 0.05 × 10^6 fibers/m^3 (0.025 mg/m^3) (definition of fiber: length greater than 5 μm; diameter less than 3 μm; length/diameter greater than 3:1, equivalent to 1 fiber/cc).
NIOSH REL: TWA 100,000 fb/m^3 over 5 μm in length
DOT Classification: ORM-C; LABEL: none

SAFETY PROFILE: Confirmed human carcinogen with experimental carcinogenic, neoplastigenic, and tumorigenic data by inhalation. Human mutation data reported.

ARM280 CAS:77536-68-6 ***HR: 3***
ASBESTOS, TREMOLITE

SYNS: ASBESTOS (ACGIH) ◇ FIBROUS TREMOLITE ◇ NCI-C08991 ◇ TREMOLITE ASBESTOS

TOXICITY DATA with REFERENCE
ipl-rat TDLo:100 mg/kg:NEO BJCAAI 45,352,82

CONSENSUS REPORTS: IARC Cancer Review: Human Sufficient Evidence IMEMDT 14,11,77; Animal Sufficient Evidence IMEMDT 14,11,77.

OSHA PEL: TWA 2 million fb/m^3; CL 10 million fb/m^3; Cancer Hazard
ACGIH TLV: TWA 2 fb/cc; Confirmed Human Carcinogen; (Proposed: TWA 0.2 fb/cc)
DFG TRK: (Fine dust particles which are able to reach the alveolar area of the lung) 1 × 10^6 fibers/m^3 (0.05 mg/m^3) applicable when there is more than 2.5% asbestos in the dust.
NIOSH REL: TWA 100,000 fb/m^3 over 5 μm in length

SAFETY PROFILE: Confirmed human carcinogen with experimental tumorigenic and neoplastigenic data.

ARM500 CAS:512-85-6 ***HR: 3***
ASCARIDOLE
mf: $C_{10}H_{16}O_2$ mw: 168.26

PROP: Colorless unstable liquid. Mp: 3.3°, bp: 40° @ 2 mm; 115° @ 15 mm, d: 1.011 @ 13°/15°.

SYNS: ASCARIDOL ◇ ASCARISIN ◇ 1,4-PEROXIDO-p-MENTHENE-2

TOXICITY DATA with REFERENCE
orl-dog LDLo:250 mg/kg JPETAB 24,359,25
skn-mus TDLo:25 g/kg/42W-I:NEO JNCIAM 35,707,65
skn-mus TD:38 g/kg/63W-I:ETA 14JTAF -,275,64
orl-rat LDLo:250 mg/kg NCNSA6 5,24,53

DOT Classification: Forbidden.

SAFETY PROFILE: Poison by ingestion. Questionable carcinogen with experimental neoplastigenic and tumorigenic data. Flammable by spontaneous chemical reaction. An oxidizer. Explodes when heated >130° or when exposed to organic acids. Dangerous; heating emits toxic fumes and may explode; reacts with reducing materials. See also CHENOPODIUM OIL, and PEROXIDES, ORGANIC.

ARM750 CAS:38462-04-3 ***HR: 2***
ASCOFURANONE

SYN: (S-(e,e))-3-CHLORO-4,6-DIHYDROXY-2-METHYL-5-(3-METHYL-7-(TETRAHYDRO-5,5-DIMETHYL-4-OXO-2-FURANYL)-2,6-OG TADIENYL)-BENZALDEHYDE

TOXICITY DATA with REFERENCE
ipr-rat LD50:1350 mg/kg JANTAJ 26,681,73
ipr-mus LD50:2220 mg/kg JANTAJ 26,681,73

SAFETY PROFILE: Moderately toxic by intraperitoneal route. See also ALDEHYDES. When heated to decomposition it emits toxic fumes of Cl^-.

ARN000 CAS:50-81-7 ***HR: 2***
l-ASCORBIC ACID
mf: $C_6H_8O_6$ mw: 176.14

PROP: White crystals. Mp: 192°. Sol in water; sltly sol in alc; insol in ether, chloroform, benzene, petroleum ether, fixed oils, and fats.

SYNS: ASCORBIC ACID ◇ l(+)-ASCORBIC ACID ◇ ASCORBUTINA ◇ CEVITAMIC ACID ◇ CEVITAMIN ◇ FEMA No. 2109 ◇ 3-KETO-l-GULOFURANOLACTONE ◇ l-3-KETOTHREOHEXURONIC ACID LACTONE ◇ NATRASCORB INJECTABLE ◇ NCI-C54808 ◇ 3-OXO-l-GULOFURANOLACTONE ◇ VITACIN ◇ VITAMIN C ◇ VITAMISIN ◇ VITASCORBOL ◇ XITIX ◇ l-XYLOASCORBIC ACID

TOXICITY DATA with REFERENCE
mmg-sat:500 μg/plate ABCHA6 45,327,81
mmo-nsc 2 mmol/L MAGDA3 10,249,79
orl-gpg TDLo:19500 mg/kg (female 30-58D post):REP ANYAA9 258,401,75
ivn-mus TDLo:800 mg/kg (female 8D post):TER TOIZAG 8,175,61
ivn-man TDLo:2300 mg/kg/2D:BLD AIMEAS 82,810,75
ivn-wmn LDLo:900 mg/kg:SYS AIMDAP 145,950,85
orl-rat LD50:11900 mg/kg OYYAA2 19,323,80
orl-mus LD50:3367 mg/kg NCISP* JAN86
ivn-mus LD50:518 mg/kg RPOBAR 2,269,70

CONSENSUS REPORTS: NTP Carcinogenesis Bioassay (feed); No Evidence: mouse, rat NTPTR* NTP-TR-247,83; NTPTR* NTP-TR-214,82. Reported in EPA TSCA Inventory.

SAFETY PROFILE: Moderately toxic by ingestion and intravenous routes. Human systemic effects by intravenous route: blood, changes in tubules (including acute renal failure, acute tubular necrosis). An experimental teratogen. Other experimental reproductive effects. Mutation data reported. When heated to decomposition it emits acrid smoke and irritating fumes.

ARN125 CAS:134-03-2 ***HR: D***
ASCORBIC ACID SODIUM SALT
mf: $C_6H_8O_6{\cdot}Na$ mw: 199.13

PROP: Minute white to yellow crystals; odorless. Decomp at 218°. Freely sol in water; very sltly sol in alc; insol in chloroform, ether.

SYNS: l-ASCORBIC ACID SODIUM SALT ◇ ASCORBICIN ◇ ASCORBIN ◇ CEBITATE ◇ CENOLATE ◇ ISKIA-C ◇ MONOSODIUM ASCORBATE ◇ NATRASCORB ◇ NATRI-C ◇ SODASCORBATE ◇ SODIUM ASCORBATE (FCC) ◇ SODIUM-l-ASCORBATE ◇ VITAMIN C ◇ VITAMIN C SODIUM

TOXICITY DATA with REFERENCE
sce-hmn:lym 100 μmol/L MUREAV 60,321,79
oms-ham:ovr 1 mmol/L CNREA8 39,4145,79
cyt-ham:ovr 20 mmol/L CNREA8 39,4145,79
sce-ham:ovr 100 μmol/L MUREAV 60,321,79

CONSENSUS REPORTS: Reported in EPA TSCA Inventory.

SAFETY PROFILE: Human mutation data reported. When heated to decomposition it emits toxic fumes of Na_2O.

ARN250 CAS:1402-88-6 ***HR: 3***
ASCOSIN

PROP: Produced by *Streptomycete canescus* (ANTCAO 2,472,52)

TOXICITY DATA with REFERENCE
orl-mus LD50:500 mg/kg 85ERAY 2,1035,78
ipr-mus LD50:9 mg/kg ANTCAO 2,472,52
scu-mus LD50:72 mg/kg 85ERAY 2,1035,78
ivn-mus LD50:13 mg/kg ANTCAO 2,472,52

SAFETY PROFILE: Poison by intravenous, subcutaneous, and intraperitoneal routes. Moderately toxic by ingestion.

ARN500 CAS:16830-15-2 ***HR: 3***
ASIATICOSIDE

PROP: A glycoside terpene from the plant *Centella asiatica* (CNREA8 32,1463,72)

SYNS: BLASTOESTIMULINA ◇ CENTELASE ◇ DERMATOLOGICO ◇ MADECASSOL

TOXICITY DATA with REFERENCE
skn-mus TDLo:400 mg/kg/52W-I:ETA CNREA8 32,1463,72

SAFETY PROFILE: Questionable carcinogen with experimental tumorigenic data. When heated to decomposition it emits acrid smoke and fumes. Promotes healing of wounds.

ARN700 CAS:35844-94-1 ***HR: 3***
ASPAMINOL HYDROCHLORIDE
mf: $C_{20}H_{25}NO{\cdot}ClH$ mw: 331.92

SYNS: 1,1-DIPHENYL-3-N-PIPERIDINOBUTANOL-1HYDROCHLORIDE ◇ (±)-Γ-METHYL-α,α-DIPHENYL-1-PYRROLIDINEPROPANOL HYDROCHLORIDE

TOXICITY DATA with REFERENCE
orl-mus LD50:642 mg/kg NIIRDN 6,333,82
scu-mus LD50:354 mg/kg NIIRDN 6,333,82
ivn-mus LD50:42700 μg/kg NIIRDN 6,333,82

SAFETY PROFILE: Poison by subcutaneous and intravenous routes. Moderately toxic by ingestion. When heated to decomposition it emits toxic fumes of NO_x and HCl.

ARN800 CAS:9015-68-3 ***HR: 3***
l-ASPARAGINASE

SYNS: ASPARAGINASE ◇ l-ASPARAGINASE X ◇ l-ASPARAGINASI (ITALIAN) ◇ l-ASPARAGINE AMIDOHYDROLASE ◇ LEUCOGEN ◇ NSC-109229

TOXICITY DATA with REFERENCE
ivn-mus TDLo:31815 μg/kg (female 7-13D post):TER YIKUAO 18,271,69
ivn-rat TDLo:3000 iu/kg (female 6-15D post):REP RRCRBU 33,174,70
ipr-mus TDLo:10 iu/kg/4D:NEO BSIBAC 47,418,71
ims-chd TDLo:8145 iu/kg/1W:SYS CANCAR 34,780,74
ipr-rat LD50:8204 mg/kg YIKUAO 18,271,69
scu-rat LD50:8204 mg/kg YIKUAO 18,271,69
ivn-rat LD50:7568 mg/kg YIKUAO 18,271,69
ivn-dog LD50 50000 iu/kg RRCRBU 33,174,70
ivn-cat LD50 50000 iu/kg RRCRBU 33,174,70
ivn-rbt LDLo 500 iu/kg RRCRBU 33,174,70

SAFETY PROFILE: Human (child) systemic effects by intramuscular route. An experimental teratogen. Other experimental reproductive effects. Questionable carcinogen with experimental neoplastigenic data.

ARN825 CAS:22839-47-0 ***HR: 1***
ASPARTAME
mf: $C_{14}H_{18}N_2O_5$ mw: 294.34

PROP: White crystalline powder; odorless with a sweet taste. Sltly sol in water, alc.

SYNS: 3-AMINO-N-(α-CARBOXYPHENETHYL)SUCCINAMIC ACID N-METHYL ESTER, stereoisomer ◇ ASPARTYLPHENYLALANINE METHYL ESTER ◇ N-l-α-ASPARTYL-l-PHENYLALANINE 1-METHYL ESTER (9CI) ◇ CANDEREL ◇ DIPEPTIDE SWEETENER ◇ EQUAL

◇ METHYL ASPARTYLPHENYLALANATE ◇ 1-METHYL N-l-α-ASPARTYL-l-PHENYLALANINE ◇ NUTRASWEET ◇ SWEET DIPEPTIDE

TOXICITY DATA with REFERENCE
orl-mus TDLo:4 g/kg (15-18D preg):REP RCPBDC 9,385,85
orl-wmn TDLo:3710 μg/kg:SKN AIMEAS 104,207,86

SAFETY PROFILE: Human systemic effects by ingestion: allergic dermatitis. Experimental reproductive effects. When heated to decomposition it emits toxic fumes of NO_x.

ARN875 ***HR: 3***
ASPERASE

TOXICITY DATA with REFERENCE
ipr-rat LD50:65300 μg/kg FATOAO 45(6),78,82
ivn-rat LD50:52500 μg/kg FATOAO 45(6),78,82
ipr-mus LD50:167 mg/kg FATOAO 45(6),78,82
ivn-mus LD50:107 mg/kg FATOAO 45(6),78,82
ivn-gpg LD50:8200 μg/kg FATOAO 45(6),78,82

SAFETY PROFILE: Poison by intravenous and intraperitoneal routes.

ARO000 CAS:490-02-8 ***HR: 3***
ASPERGILLIC ACID
mf: $C_{12}H_{20}N_2O_2$ mw: 224.34

SYNS: 2-HYDROXY-3-ISOBUTYL-6-(1-METHYLPROPYL)PYRAZINE 1-OXIDE ◇ 3-ISOBUTYL-6-sec-BUTYL-2-HYDROXYPYRAZINE-1-OXIDE

TOXICITY DATA with REFERENCE
orl-mus LDLo:200 mg/kg JOBAAY 45,433,43
ipr-mus LDLo:150 mg/kg JOBAAY 45,433,43

SAFETY PROFILE: Poison by ingestion and intraperitoneal routes. When heated to decomposition it emits toxic fumes of NO_x.

ARO250 CAS:67-99-2 ***HR: 3***
ASPERGILLIN
mf: $C_{13}H_{14}N_2O_4S_2$ mw: 326.41

SYN: GLIOTOXIN

TOXICITY DATA with REFERENCE
ipr-mus LDLo:45 mg/kg 85ERAY 3,1919,78
ivn-mus LDLo:45 mg/kg 85ERAY 3,1919,78
ivn-rbt LDLo:45 mg/kg JACSAT 65,2005,43

SAFETY PROFILE: Poison by intraperitoneal and intravenous routes. When heated to decomposition it emits very toxic fumes such as SO_x and NO_x.

ARO500 CAS:8052-42-4 ***HR: 3***
ASPHALT
DOT: NA 1999

PROP: Black or dark brown mass. Bp: <470°, flash p: 400+ °F (CC), d: 0.95 – 1.1, autoign temp: 905°F.

SYNS: ASPHALTUM ◇ BITUMEN (MAK) ◇ JUDEAN PITCH ◇ MINERAL PITCH ◇ PETROLEUM PITCH ◇ ROAD ASPHALT (DOT) ◇ ROAD TAR (DOT)

TOXICITY DATA with REFERENCE
skn-mus TDLo:130 g/kg/81W-I:CAR HYSAAV 33(4-6),180,68
skn-mus TD:69 g/kg/43W-I:ETA HYSAAV 33(4-6),180,68

CONSENSUS REPORTS: IARC Cancer Review: Group 3 IMEMDT 7,133,87; Human Inadequate Evidence IMEMDT 35,39,85. Reported in EPA TSCA Inventory.

ACGIH TLV: TWA 5 mg/m^3
DFG MAK: Suspected Carcinogen.
NIOSH REL: (Asphalt Fumes) CL 5 mg/m^3/15M
DOT Classification: ORM-C; Label: None; Flammable Liquid; Label: Flammable Liquid (cut-back); Combustible Liquid; Label: None (cut-back)

SAFETY PROFILE: Suspected carcinogen with experimental carcinogenic and tumorigenic data. A moderate irritant. May contain carcinogenic components. Combustible when exposed to heat or flame. To fight fire, use foam, CO_2, or dry chemical.

ARO750 CAS:8052-42-4 ***HR: 3***
ASPHALT (CUT BACK)
DOT: UN 1999

PROP: A liquid petroleum product, solubility of residue from distillation in carbon tetrachloride = 99.5%. Flash p: < 50°F.

SYNS: ROAD ASPHALT (DOT) ◇ ROAD TAR, liquid (DOT)

DOT Classification: Flammable liquid. Label: Flammable liquid.

SAFETY PROFILE: Contains carcinogenic components. A dangerous fire hazard when exposed to heat or flame. To fight fire, use dry chemical, water mist, fog. When heated to decomposition it emits smoke and irritating acrid fumes.

ARP000 CAS:2096-42-6 ***HR: 3***
ASPICULAMYCIN
mf: $C_{16}H_{25}N_7O_8$ mw: 443.48

SYNS: ASTEROMYCIN ◇ 1-(4-DEOXY-4-(SARCOSYL-d-SERYL)AMINO-β-d-GLUCOPYRANURONAMIDE)CYTOSINE ◇ GOUGEROTIN ◇ QUINGFENGMYCIN

TOXICITY DATA with REFERENCE
ipr-mus LD50:57 mg/kg 85FZAT-,313,67
ivn-mus LD50:57 mg/kg JAJAAA 15,93,62

SAFETY PROFILE: Poison by intraperitoneal and in-

travenous routes. When heated to decomposition it emits toxic fumes of NO_x.

ARP125 **HR: 2**
ASPIRIN-dl-LYSINE
mf: $C_{15}H_{22}N_2O_6$ mw: 326.39

SYNS: dl-LYSINE ACETYLSALICYLATE ◇ dl-LYSINE ACETYLSALICYLIC ACID SALT ◇ dl-LYSINE MONO(2-(ACETYLOXY)BENZOATE

TOXICITY DATA with REFERENCE
orl-rat LD50:4350 mg/kg IYKEDH 13,1128,82
scu-rat LD50:1860 mg/kg IYKEDH 13,1128,82
ivn-rat LD50:1525 mg/kg IYKEDH 13,1128,82
orl-mus LD50:3270 mg/kg IYKEDH 13,1128,82
scu-mus LD50:2100 mg/kg IYKEDH 13,1128,82
ivn-mus LD50:950 mg/kg IYKEDH 13,1128,82

SAFETY PROFILE: Moderately toxic by ingestion and other routes. When heated to decomposition it emits toxic fumes of NO_x.

ARP250 CAS:8003-03-0 **HR: 2**
ASPIRIN, PHENACETIN and CAFFEINE
mf: $C_{10}H_{13}NO_2 \cdot C_9H_8O_4 \cdot C_8H_{10}N_4O_2$ mw: 553.63

PROP: Composed of 50% aspirin, 46% phenacetin, and 4% caffeine (NCIMR* NIH-71-E-2144)

SYNS: 2-(ACETYLOXY)BENZOIC ACID, mixed with 3,7-DIHYDRO-1,3,7-TRIMETHYL-1H-PURINE-2,6-DIONE and N-(4-ETHOXYPHENYL)ACETAMIDE ◇ APC (pharmaceutical) ◇ ASCOPHEN ◇ CITRAMON ◇ EMPIRIN COMPOUND ◇ NCI-C02697 ◇ OSCOPHEN ◇ THOMAPYRIN

TOXICITY DATA with REFERENCE
ipr-mus LD50:44 mg/kg PCJOAU 15,139,81
orl-rat TDLo:382 g/kg/78W-C:ETA NCITR* NCI-CG-TR-67,78
orl-rat LD50:1420 mg/kg NCIMR* NIH-71-E-2144

CONSENSUS REPORTS: NCI Carcinogenesis Bioassay (feed); Inadequate Studies: mouse, rat NCITR* NCI-CG-TR-67,78

SAFETY PROFILE: Moderately toxic by ingestion. Questionable carcinogen with experimental tumorigenic data. See also CAFFEINE, p-ACETOPHENETIDIDE, and 2-(ACETYLOXY)BENZOIC ACID. When heated to decomposition it emits toxic fumes of NO_x.

ARP500 **HR: 3**
ASSAM TEA

PROP: Tannin containing fraction of leaf used (JNCIAM 57,207,76).

SYN: CAMELLIA SINENSIS

TOXICITY DATA with REFERENCE
scu-rat TDLo:1850 mg/kg/58W-I:NEO JNCIAM 57,207,76

SAFETY PROFILE: Questionable carcinogen with experimental neoplastigenic data. When heated to decomposition it emits acrid smoke and irritating fumes.

ARP625 CAS:88746-71-8 **HR: 2**
ASTA Z 7557
mf: $C_9H_{18}Cl_2N_2O_5PS_2 \cdot C_6H_{13}N$ mw: 499.48

SYN: Z 7557

TOXICITY DATA with REFERENCE
dnd-hmn:lym 50 mg/L INNDDK 2,161,84
dnd-mus-ipr 250 μmol/kg INNDDK 2,181,84
dnd-mus:lym 12500 μg/kg INNDDK 2,161,84
orl-rat LD50:1 g/kg INNDDK 2,201,84
ivn-rat LD50:250 mg/kg INNDDK 2,201,84
orl-mus LD50:2310 mg/kg INNDDK 2,201,84
ipr-mus LD50:315 mg/kg INNDDK 2,253,84
ivn-mus LD50:500 mg/kg INNDDK 2,201,84

SAFETY PROFILE: Poison by intravenous and intraperitoneal routes. Moderately toxic by ingestion. Human mutation data reported. When heated to decomposition it emits toxic fumes of Cl^-, PO_x, SO_x, and NO_x. See also ESTERS and SULFONATES.

ARP675 CAS:68844-77-9 **HR: 3**
ASTEMIZOLE
mf: $C_{28}H_{31}FN_4O$ mw: 458.63

PROP: Crystals. Mp: 149.1°.

SYNS: ASTEMIZOL (GERMAN) ◇ 1-(p-FLUOROBENZYL)-2-((1-(2-(p-METHOXYPHENYL)ETHYL)PIPERID-4-YL)AMINO)BENZIMIDAZOLE ◇ HISMANAL ◇ HISTAMINOS ◇ PARALERGIN

TOXICITY DATA with REFERENCE
scu-rat LD50:355 mg/kg ARZNAD 33,381,83
ivn-rat LD50:28 mg/kg ARZNAD 33,381,83
orl-mus LD50:3560 mg/kg ARZNAD 33,381,83
ivn-dog LD50:22 mg/kg ARZNAD 33,381,83
orl-gpg LD50:933 mg/kg ARZNAD 33,381,83

SAFETY PROFILE: Poison by subcutaneous and intravenous routes. Moderately toxic by ingestion. When heated to decomposition it emits toxic fumes of F^- and NO_x.

ARP875 CAS:14698-07-8 **HR: 3**
ASVERINE CITRATE
mf: $C_{15}H_{17}NS_2 \cdot C_6H_8O_7$ mw: 467.59

SYNS: ASVERIN-C ◇ ASVERIN CITRATE ◇ AT 327 CITRATE ◇ BITHIODINE ◇ 3-(DI-2-THIENYLMETHYLENE)-1-METHYLPIPERIDINE CITRATE ◇ 1-METHYL-3-(DI-2-THIENYLMETHYLENE)PIPERIDINE CITRATE

TOXICITY DATA with REFERENCE
orl-mus LD50:867 mg/kg KSRNAM 7,3279,73
scu-mus LD50:376 mg/kg CPBTAL 7,372,59
ivn-dog LD50:75 mg/kg CPBTAL 7,372,59

SAFETY PROFILE: Poison by subcutaneous and intravenous routes. Moderately toxic by ingestion. When heated to decomposition it emits toxic fumes of NO_x and SO_x.

ARQ000 CAS:2185-98-0 ***HR: 3***
AT-581
mf: $C_{14}H_{20}Cl_2N_2O_2 \cdot 2ClH$ mw: 392.18

SYNS: 3-(o-((BIS(2-CHLOROETHYL)AMINO)METHYL)PHENYL) ALANINE DIHYDROCHLORIDE ◇ o-BIS(2-CHLOROETHYL) AMINOMETHYLPHENYLALANINE HYDROCHLORIDE ◇ 2-((BIS(2-CHLOROETHYL)AMINO)METHYL)-PHENYLALANINEDIHYDROCHLORIDE (9CI)

TOXICITY DATA with REFERENCE
ivn-hmn TDLo:2 mg/kg/10D-I:BLD XPHPAW 441,186,74
ipr-mus LD50:5900 μg/kg SSINAV 13,789,64

SAFETY PROFILE: Poison by intraperitoneal route. Human blood effects by intravenous route. When heated to decomposition it emits very toxic fumes of HCl, NO_x, and Cl^-.

ARQ250 CAS:83-89-6 ***HR: 3***
ATABRINE
mf: $C_{23}H_{30}ClN_3O$ mw: 400.01

PROP: Bright yellow crystals. Mp: decomp @ 248°.

SYNS: ACRICHINE ◇ ACRINAMINE ◇ ACRIQUINE ◇ AKRICHIN ◇ ANTIMALARINA ◇ 6-CHLORO-9-((4-(DIETHYL AMINO)-1-METHYL BUTYL)AMINO)-2-METHOXYACRIDINE ◇ 3-CHLORO-7-METHOXY-9-(1-METHYL-4-DIETHYLAMINOBUTYLAMINO)ACRIDINE ◇ ERION ◇ HAFFKININE ◇ MEPACRINE ◇ 2-METHOXY-6-CHLORO-9-DIETHYLAMINOPENTYLAMINOACRIDINE ◇ QUINACRINE

TOXICITY DATA with REFERENCE
mma-sat 500 μg/plate TXAPA9 52,237,80
mmo-omi 50 mg/L GENTAE 90,1,78
mmo-sat 200 mg/L JOBAAY 122,549,75
mnt-mus-ipr 280 μmol/kg MUREAV 26,553,74
dns-mam:lym 1 μmol/L CBINA8 8,113,74
iut-wmn TDLo:13600 μg/kg (female 1D pre):REP CCPTAY 7,333,73
orl-mus LD50:1320 mg/kg FAZMAE 17,108,73
scu-mus LD50:239 mg/kg JPETAB 119,444,57
orl-ckn LD50:714 mg/kg AEPPAE 201,402,43

SAFETY PROFILE: Poison by subcutaneous route. Moderately toxic by ingestion. Mutation data reported. Experimental reproductive effects. Has been implicated in aplastic anemia. When heated to decomposition, it emits very toxic fumes of Cl^- and NO_x.

ARQ325 CAS:5140-35-2 ***HR: 3***
ATHERILINE
mf: $C_{19}H_{15}NO_5$ mw: 337.35

SYNS: ATHEROLINE ◇ 4,5,6,6a-TETRADEHYDRO-9-HYDROXY-1,2,10-TRIMETHOXYNORAPORPHIN-7-ONE

TOXICITY DATA with REFERENCE
orl-mus LD50:450 mg/kg APFRAD 38,537,80
ipr-mus LD50:170 mg/kg APFRAD 38,537,80
ivn-mus LD50:90 mg/kg APFRAD 38,537,80

SAFETY PROFILE: Poison by intravenous and intraperitoneal routes. Moderately toxic by ingestion. When heated to decomposition it emits toxic fumes of NO_x.

ARQ500 CAS:56-65-5 ***HR: 3***
ATP
mf: $C_{10}H_{16}N_5O_{13}P_3$ mw: 507.22

PROP: Freely sol in water. The anhydrous barium salt is stable, but the hydrated salt slowly decomp forming 5'-adenylic acid and barium pyrophosphate.

SYNS: ADENOSINE TRIPHOSPHATE ◇ ADENOSINE-5'-TRIPHOSPHATE ◇ ADENOSINE-5'-TRIPHOSPHORIC ACID ◇ ADENYLPYROPHOSPHORIC ACID ◇ ADEPHOS ◇ ADETOL ◇ ADYNOL ◇ ARA-ATP ◇ 9-β-d-ARABINOFURANOSYLADENINE 5'-TRIPHOSPHATE ◇ ATIPI ◇ 5'-ATP ◇ ATP (nucleotide) ◇ ATRIPHOS ◇ GLUCOBASIN ◇ MYOTRIPHOS ◇ STRIADYNE ◇ TRIADENYL ◇ TRIPHOSADEN ◇ TRIPHOSPHADEN ◇ TRIPHOSPHORIC ACID ADENOSINE ESTER

TOXICITY DATA with REFERENCE
dni-hmn:lym 10 μmol/L CNREA8 42,2092,82
oms-hmn:oth 10 μmol/L JIDEAE 65,52,75
dni-mus:leu 2 μmol/L CHTHBK 27,61,81
ipr-rat LD50:200 mg/kg BJANAD 53,305,81

CONSENSUS REPORTS: EPA Genetic Toxicology Program. Reported in EPA TSCA Inventory.

SAFETY PROFILE: Poison by intraperitoneal route. Human mutation data reported. When heated to decomposition it emits toxic fumes of PO_x and NO_x.

ARQ700 ***HR: 2***
ATRATOL 80W

PROP: Contains 75% atrazine and 5% prometon (FMCHA2 -,D24,80)

TOXICITY DATA with REFERENCE
orl-rat LD50:2436 mg/kg FMCHA2 -,D24,80
ihl-rat LC50:3190 mg/kg/4H FMCHA2-,D24,80
ACGIH TLV: TWA 5 mg/m^3

SAFETY PROFILE: Moderately toxic by ingestion and inhalation. When heated to decomposition it emits acrid smoke and irritating fumes.

ARQ725 CAS:1912-24-9 ***HR: 3***
ATRAZINE
mf: $C_8H_{14}ClN_5$ mw: 215.72

PROP: Crystals. Mp: 171-174°. Solubility at 25°: in water, 70 ppm; ether, 12,000 ppm; chloroform, 52,000 ppm; methanol, 18,000 ppm.

SYNS: A 361 ◇ AATREX ◇ AATREX 4L ◇ AATREX NINE-O ◇ AATREX 80W ◇ 2-AETHYLAMINO-4-CHLOR-6-ISOPROPYLAMINO-1,3,5-TRIAZIN (GERMAN) ◇ 2-AETHYLAMINO-4-ISOPROPYLAMINO-6-CHLOR-1,3,5-TRIAZIN (GERMAN) ◇ AKTIKON ◇ AKTIKON PK ◇ AKTINIT A ◇ AKTINIT PK ◇ ARGEZIN ◇ ATAZINAX ◇ ATRANEX ◇ ATRASINE ◇ ATRATOL A ◇ ATRAZIN ◇ ATRED ◇ ATREX ◇ CANDEX ◇ CEKUZINA-T ◇ 2-CHLORO-4-ETHYLAMINEISOPROPYLAMINE-s-TRIAZINE ◇ 1-CHLORO-3-ETHYLAMINO-5-ISOPROPYLAMINO-s-TRIAZINE ◇ 1-CHLORO-3-ETHYLAMINO-5-ISOPROPYLAMINO-2,4,6-TRIAZINE ◇ 2-CHLORO-4-ETHYLAMINO-6-ISOPROPYLAMINO-s-TRIAZINE ◇ 2-CHLORO-4-ETHYLAMINO-6-ISOPROPYLAMINO-1,3,5-TRIAZINE ◇ 6-CHLORO-N-ETHYL-N'-(1-METHYLETHYL)-1,3,5-TRIAZINE-2,4-DIAMINE (9CI) ◇ 2-CHLORO-4-(2-PROPYLAMINO)-6-ETHYLAMINO-s-TRIAZINE ◇ CRISATRINA ◇ CRISAZINE ◇ CYAZIN ◇ FARMCO ATRAZINE ◇ FENAMIN ◇ FENAMINE ◇ FENATROL ◇ G 30027 ◇ GEIGY 30,027 ◇ GESAPRIM ◇ GESOPRIM ◇ GRIFFEX ◇ HUNGAZIN ◇ HUNGAZIN PK ◇ INAKOR ◇ OLEOGESAPRIM ◇ PRIMATOL ◇ PRIMAZE ◇ RADAZIN ◇ RADIZINE ◇ SHELL ATRAZINE HERBICIDE ◇ STRAZINE ◇ TRIAZINE A 1294 ◇ VECTAL ◇ VECTAL SC ◇ WEEDEX A ◇ WONUK ◇ ZEAZIN ◇ ZEAZINE

TOXICITY DATA with REFERENCE
skn-rbt 38 mg open MLD CIGET* -,-,77
eye-rbt 6320 μg SEV CIGET* -,-,77
skn-mam 500 mg MLD VRDEA5 (5),133,77
eye-mam 100 mg SEV VRDEA5 (5),133,77
sln-nsc 10 mg/L MUREAV 167,35,86
dns-hmn:fbr 3 mmol/L MUREAV 74,77,80
hma-rat/esc 100 mg/kg CECED9 6388,328,80
mma-ham:lng 3 mmol/L MUREAV 74,77,80
orl-rbt TDLo:975 mg/kg (female 7-19D post):TER JTEHD6 24,307,88
scu-mus TDLo:418 mg/kg (female 6-14D post):REP NTIS** PB223-160
orl-mus TDLo:9000 mg/kg/78W-I:ETA NTIS** PB223-159
orl-rat LD50:672 mg/kg FAATDF 7,299,86
ihl-rat LC50:5200 mg/m³/4H FMCHA2 -,C3,83
ipr-rat LD50:235 mg/kg PESTD5 17,351,76
orl-mus LD50:850 mg/kg 85GMAT -,36,82
ipr-mus LD50:626 mg/kg PESTD5 17,351,76
orl-rbt LD50:750 mg/kg 85DPAN -,-,71/76
skn-rbt LD50:7500 mg/kg 28ZEAL 5,15,76
orl-ham LD50:1000 mg/kg TXAPA9 48,A192,79

CONSENSUS REPORTS: EPA Genetic Toxicology Program. Reported in EPA TSCA Inventory.

OSHA PEL: TWA 5 mg/m³
ACGIH TLV: TWA 5 mg/m³
DFG MAK: 2 mg/m³
SAFETY PROFILE: Poison by intraperitoneal route. Moderately toxic by ingestion. Mildly toxic by inhalation and skin contact. An experimental teratogen. Other experimental reproductive effects. Human mutation data reported. A skin and severe eye irritant. Questionable carcinogen with experimental tumorigenic data. When heated to decomposition it emits toxic fumes of Cl^- and NO_x.

ARQ750 CAS:637-07-0 ***HR: 3***
ATROMID S
mf: $C_{12}H_{15}ClO_3$ mw: 242.72

SYNS: AMOTRIL ◇ ANGIOKAPSUL ◇ ANPARTON ◇ ANTILIPID ◇ APOLAN ◇ ARTERIOFLEXIN ◇ ARTEROSOL ◇ ARTES ◇ ARTEVIL ◇ ATECULON ◇ ATERIOSAN ◇ ATHEBRATE ◇ ATHEROMIDE ◇ ATHEROPRONT ◇ ATHRANID-WIRKSTOFF ◇ ATROLEN ◇ ATROMID ◇ ATROMIDIN ◇ ATROVIS ◇ AY 61123 ◇ AZIONYL ◇ BIOSCLERAN ◇ BRESIT ◇ CARTAGYL ◇ α-p-CHLOROPHENOXYISOBUTYRYL ETHYL ESTER ◇ 2-(4-CHLOROPHENOXY)-2-METHYLPROPANOIC ACID ETHYL ESTER ◇ 2-(p-CHLOROPHENOXY)-2-METHYLPROPIONIC ACID ETHYL ESTER ◇ CINNARIZIN ◇ CITIFLUS ◇ CLARIPEX ◇ CLOBERAT ◇ CLOBRAT ◇ CLOBREN-SF ◇ CLOFAR ◇ CLOFIBRAM ◇ CLOFIBRAT ◇ CLOFIBRATO (SPANISH) ◇ CLOFINIT ◇ CLOFIPRONT ◇ CPIB ◇ DELIVA ◇ DURA CLOFIBRAT ◇ ELPI ◇ EPIB ◇ ETHYL CHLOROPHENOXYISOBUTYRATE ◇ ETHYL-p-CHLOROPHENOXYISOBUTYRATE ◇ ETHYL-α-p-CHLOROPHENOXYISOBUTYRATE ◇ ETHYL-α-(4-CHLOROPHENOXY)ISOBUTYRATE ◇ ETHYL-2-(p-CHLOROPHENOXY)ISOBUTYRATE ◇ ETHYL-α-(p-CHLOROPHENOXY)-α-METHYLPROPIONATE ◇ ETHYL-α-(4-CHLOROPHENOXY)-α-METHYLPROPIONATE ◇ ETHYL 2-(p-CHLOROPHENOXY)-2-METHYLPROPIONATE ◇ ETHYL 2-(4-CHLOROPHENOXY)-2-METHYLPROPIONATE ◇ ETHYL CLOFIBRATE ◇ FIBRALEM ◇ GERASTOP ◇ HYCLORATE ◇ ICI 28257 ◇ KLOFIRAN ◇ LEVATROM ◇ LIPAMID ◇ LIPAVIL ◇ LIPAVLON ◇ LIPIDE 500 ◇ LIPIDSENKER ◇ LIPOFACTON ◇ LIPOMID ◇ LIPONORM ◇ LIPOREDUCT ◇ LIPORIL ◇ LIPOSID ◇ LIPRIN ◇ LIPRINAL ◇ LOBETRIN ◇ MISCLERON ◇ NEGALIP ◇ NEO-ATOMID ◇ NORMALIP ◇ NORMAT ◇ NORMOLIPOL ◇ NSC-79389 ◇ OXAN 600 ◇ PERSANTINAT ◇ RECOLIP ◇ REGARDIN ◇ REGELAN ◇ ROBIGRAM ◇ SCROBIN ◇ SEROFINEX ◇ SEROTINEX ◇ SKEROLIP ◇ SKLEROMEX ◇ SKLEROMEXE ◇ SKLERO-TABLINEN ◇ SKLERO-TABULS ◇ TICLOBRAN ◇ VINCAMIN COMPOSITUM ◇ XYDURIL ◇ YOCLO

TOXICITY DATA with REFERENCE
dns-rat:lvr 100 pmol/L CRNGDP 5,1547,84
sce-ham:ovr 100 μmol/L/1H CRNGDP 5,703,84
scu-mus TDLo:960 mg/kg (female 17D post):REP HCMYAL 63,7,79
orl-rbt TDLo:1300 mg/kg (female 6-18D post):TER ARZNAD 31,1831,81
orl-rat TDLo:100 g/kg/72W-C:ETA CNREA8 39,3419,79
orl-man TDLo:1071 μg/kg NEURAI 37,881,87
orl-wmn TDLo:80 mg/kg/2D-I:MET NEJMAG 301,1345,79
orl-man TDLo:171 mg/kg/6D-I IJMDAI 20,1082,84
orl-rat LD50:940 mg/kg AFTOD7 6,255,80
ipr-rat LD50:910 mg/kg JETOAS 5,239,72
scu-rat LD50:2000 mg/kg NIIRDN 6,367,82
ivn-rat LD50:15 mg/kg NIIRDN 6,367,82

orl-mus LD50:1220 mg/kg CPBTAL 32,1568,84
ipr-mus LD50:540 mg/kg ARZNAD 31,1816,81
scu-mus LD50:2000 mg/kg NIIRDN 6,367,82
ivn-mus LD50:30 mg/kg NIIRDN 6,367,82
ipr-dog LD50:500 mg/kg NIIRDN 6,367,82
scu-dog LD50:500 mg/kg NIIRDN 6,367,82

CONSENSUS REPORTS: IARC Cancer Review: Group 3 IMEMDT 7,171,87; Animal Limited Evidence IMEMDT 24,39,80; Human Inadequate Evidence IMEMDT 24,39,80

SAFETY PROFILE: Poison by intravenous route. Moderately toxic by ingestion and other routes. An experimental teratogen. Other experimental reproductive effects. Reduces plasma lipid levels. Human systemic effects by ingestion: muscle weakness, muscle spasms, and fever. Questionable carcinogen with experimental tumorigenic data. When heated to decomposition it emits toxic fumes of Cl^-.

ARR000 CAS:51-55-8 ***HR: 3***
ATROPINE
mf: $C_{17}H_{23}NO_3$ mw: 289.41

PROP: Colorless crystalline alkaloid.

SYNS: ATROPIN (GERMAN) ◇ EYEULES ◇ dl-HYOSCYAMINE ◇ 2-PHENYLHYDRACRYLIC ACID-3-α-TROPANYL ESTER ◇ β-PHENYL-Γ-OXYPROPIONSAEURE-TROPYL-ESTER (GERMAN) ◇ 1-α-H,5-α-H-TROPAN-3-α-OL (±)-TROPATE (ESTER) ◇ dl-TROPANYL-2-HYDROXY-1-PHENYLPROPIONATE ◇ TROPIC ACID, ESTER with TROPINE ◇ TROPIC ACID-3-α-TROPANYL ESTER ◇ TROPINE TROPATE ◇ dl-TROPYLTROPATE ◇ (±)-TROPYL TROPATE

TOXICITY DATA with REFERENCE
ivn-wmn TDLo:20 μg/kg (female 26-39W post):TER AJOGAH 82,1055,61
unr-mus TDLo:1600 μg/kg (female 1D pre):REP FESTAS 12,346,61
orl-hmn TDLo:33 μg/kg:EYE JTCTDW 22,581,84/85
ivn-man TDLo:14 μg/kg AIMEAS 101,720,84
ims-man TDLo:175 μg/kg FEPRA7 32,250,73
ims-hmn TDLo:1 μg/kg:EYE 85IVAW 1,L1,82
orl-rat LD50:500 mg/kg AIPTAK 155,393,65
ipr-rat LD50:280 mg/kg JPETAB 105,166,52
scu-rat LD50:250 mg/kg AIPTAK 68,339,42
ivn-rat LD50:73 mg/kg SMWOAS 76,1282,46
ivn-mus LD50:44 mg/kg PCJOAU 11,905,77
ims-rat LD50:920 mg/kg DCTODJ 1,355,78
orl-mus LD50:75 mg/kg AIPTAK 59,149,38
ipr-mus LD50:30 mg/kg JMCMAR 31,683,88
scu-mus LD50:510 mg/kg NYKZAU 53,84S,57
ivn-mus LD50:30 mg/kg JMCMAR 28,1760,85

CONSENSUS REPORTS: Reported in EPA TSCA Inventory.

SAFETY PROFILE: Poison by ingestion, subcutaneous, intravenous, and intraperitoneal routes. Human systemic effects by ingestion and intramuscular routes: visual field changes, mydriasis (pupillary dilation), and muscle weakness. An experimental teratogen. Other experimental reproductive effects. An alkaloid. When heated to decomposition it emits toxic fumes of NO_x.

ARR250 CAS:2472-17-5 ***HR: 3***
ATROPINE SULFATE (1:1)
mf: $C_{17}H_{23}NO_3 \cdot H_2O_4S$ mw: 387.49

TOXICITY DATA with REFERENCE
dnd-rat-ipr 200 mg/kg/25D EJPHAZ 7,73,69
ivn-mus LD50:78 mg/kg JLCMAK 30,700,45

SAFETY PROFILE: Poison by intravenous route. Mutation data reported. See also ATROPINE and SULFATES. When heated to decomposition it emits very toxic fumes of NO_x and SO_x.

ARR500 CAS:55-48-1 ***HR: 3***
ATROPINE SULFATE (2:1)
mf: $C_{34}H_{46}N_2O_6 \cdot H_2O_4S$ mw: 676.90

SYNS: ATROPIN SIRAN (CZECH) ◇ ATROPINSULFAT (GERMAN) ◇ SULFATE d'ATROPINE (FRENCH) ◇ 1-α-H,5-α-H-TROPAN-3-α-OL (±)-TROPATE (ESTER), SULFATE (2:1) SALT ◇ dl-TROPANYL-2-HYDROXY-1-PHENYLPROPIONATE SULFATE ◇ TROPINTRAN

TOXICITY DATA with REFERENCE
ipr-rat TDLo:40 mg/kg (male 1D pre):REP ANENAG 24(Suppl 3),1,63
par-mus TDLo:50 mg/kg (female 8D post):TER JPMSAE 62,1626,73
orl-cld TDLo:20 μg/kg ADCHAK 54,222,79
ivn-man TDLo:28 μg/kg/11H-I:SYS AJDCAI 137,291,83
ims-hmn TDLo:28 μg/kg JAPYAA 8,635,56
ocu-wmn TDLo:20 μg/kg/4H-I:CVS AIMDAP 146,45,86
mul-wmn TDLo:44 μg/kg/1D-I AIMDAP 146,45,86
orl-rat LD50:600 mg/kg AIPTAK 155,393,65
ipr-rat LD50:215 mg/kg TXAPA9 11,511,67
scu-rat LD50:540 mg/kg AIPTAK 155,393,65
orl-mus LD50:468 mg/kg AIPTAK 156,467,65
ipr-mus LD50:180 mg/kg JPMSAE 55,849,66
scu-mus LD50:400 mg/kg THERAP 12,412,57
ivn-mus LD50:31 mg/kg ARZNAD 16,637,66

CONSENSUS REPORTS: Reported in EPA TSCA Inventory.

SAFETY PROFILE: Poison by subcutaneous, intravenous, and intraperitoneal routes. Moderately toxic by ingestion. Human (child) pulmonary system effects by ingestion. Human systemic effects: decreased body temperature, cardiac arrhythmias. An experimental teratogen. Other experimental reproductive effects. See also ATROPINE. When heated to decomposition it emits very toxic fumes of NO_x and SO_x.

ARR750 CAS:67-92-5 ***HR: 3***
ATUMIN
mf: $C_{19}H_{35}NO_2 \cdot ClH$ mw: 346.01

SYNS: BIS(CYCLOHEXYL)CARBOXYLIC ACID DIETHYLAMINOETHYL ESTER HYDROCHLORIDE ◇ DICYCLOMINE HYDROCHLORIDE ◇ DIETHYLAMINOCARBETHOXYBICYCLOHEXYL HYDROCHLORIDE ◇ β-DIETHYLAMINOETHYL-1-CYCLOHEXYLCYCLOHEXANECARBOXYLATE HYDROCHLORIDE ◇ β-DIETHYLAMINOETHYL-1-CYCLOHEXYLHEXAHYDROBENZOATEHYDROCHLORIDE

TOXICITY DATA with REFERENCE
orl-mus LD50:625 mg/kg JAPMA8 39,305,50
ivn-rbt LD50:35 mg/kg JAPMA8 39,305,50

CONSENSUS REPORTS: Reported in EPA TSCA Inventory.

SAFETY PROFILE: Poison by intravenous route. Moderately toxic by ingestion. When heated to decomposition it emits very toxic fumes of HCl and NO_x.

ARR875 CAS:1674-96-0 ***HR: 2***
ATURBANE HYDROCHLORIDE
mf: $C_{17}H_{24}N_2O_2 \cdot ClH$ mw: 324.89

SYN: CHLORHYDRATE de α-PHENYL-α-(β'-DIETHYLAMINOETHYL)GLUTARIMIDE(FRENCH)

TOXICITY DATA with REFERENCE
scu-rat TDLo:480 mg/kg (1-12D preg):REP COREAF 258,2666,64
orl-mus LD50:1200 mg/kg ARZNAD 15,534,65

SAFETY PROFILE: Moderately toxic by ingestion. Experimental reproductive effects. When heated to decomposition it emits toxic fumes of NO_x and HCl. An anticholinergic and antispasmodic.

ARS000 CAS:60748-45-0 ***HR: 3***
ATX II

PROP: A polypeptide isolated from the sea anemone, *Anemonia sulcata* (TOXIA6 16,561,78)

SYN: SEA ANEMONE TOXIN II

TOXICITY DATA with REFERENCE
ivn-mus LD50:310 μg/kg TOXIA6 16,561,78
par-mus LD50:1 μg/kg NSAPCC 309,165,79

SAFETY PROFILE: A deadly poison by intravenous and parenteral routes.

ARS125 CAS:74469-00-4 ***HR: 2***
AUGMENTIN
mf: $C_{16}H_{19}N_3O_5S \cdot C_8H_9NO_5 \cdot K$ mw: 603.72

SYNS: AMOXICILLIN mixed with POTASSIUM CLAVULANATE (2:1) ◇ AUGMENTIN (antibiotic) ◇ BRL 25000 ◇ POTASSIUM CLAVULANATE mixed with AMOXICILLIN (1:2)

TOXICITY DATA with REFERENCE
orl-rat TDLo:1500 mg/kg (6-15D preg):REP NKRZAZ 31(Suppl 2),238,83
ipr-rat LD50:2774 mg/kg NKRZAZ 31(Suppl 2),113,83
scu-rat LD50:3487 mg/kg NKRZAZ 31(Suppl 2),113,83
ipr-mus LD50:3925 mg/kg NKRZAZ 31(Suppl 2),113,83
scu-mus LD50:6433 mg/kg NKRZAZ 31(Suppl 2),113,83

SAFETY PROFILE: Moderately toxic by some routes. Experimental reproductive effects. When heated to decomposition it emits toxic fumes of SO_x, NO_x, and K_2O.

ARS130 CAS:78173-92-9 ***HR: 3***
AURAMYCIN A
mf: $C_{41}H_{51}NO_{15}$ mw: 797.93

TOXICITY DATA with REFERENCE
dni-mus:leu 630 nmol/L JANTAJ 34,1596,81
oms-mus:leu 120 nmol/L JANTAJ 34,1596,81
ipr-mus LD50:100 mg/kg JANTAJ 35,82-59,82

SAFETY PROFILE: Poison by intraperitoneal route. Mutation data reported. When heated to decomposition it emits toxic fumes of NO_x.

ARS135 CAS:78173-91-8 ***HR: 3***
AURAMYCIN B
mf: $C_{41}H_{49}NO_{15}$ mw: 795.91

TOXICITY DATA with REFERENCE
dni-mus:leu 750 nmol/L JANTAJ 34,1596,81
oms-mus:leu 120 nmol/L JANTAJ 34,1596,81
ipr-mus LD50:100 mg/kg JANTAJ 35,82-60,82

SAFETY PROFILE: Poison by intraperitoneal route. Mutation data reported. When heated to decomposition it emits toxic fumes of NO_x.

ARS150 CAS:34031-32-8 ***HR: 3***
AURANOFIN
mf: $C_{20}H_{34}AuO_9PS$ mw: 678.54

SYNS: RIDAURA ◇ SK&F 39162 ◇ 2,3,4,6-TETRA-o-ACETYL-1-THIO-β-d-GLUCOPYRANOSATO-S-(TRIETHYLPHOSPHINE)GOLD ◇ TRIETHYLPHOSPHINE GOLD

TOXICITY DATA with REFERENCE
dni-hmn:oth 41 umol/L BCPCA6 34,3243,85
orl-rat TDLo:100 mg/kg (female 6-15D post):TER VTPHAK 15(Suppl 5),89,78
orl-rat TDLo:100 mg/kg (female 6-15D post):REP VTPHAK 15(Suppl 5),89,78
orl-man TDLo:1200 μg/kg/2W-I:GIT JRHUA9 13,228,86
orl-wmn TDLo:5400 μg/kg/10D-I ARHEAW 27,1316,84
orl-rat LD50:265 mg/kg VTPHAK 15(Suppl.5),1,78
ipr-rat LD50:25500 μg/kg IYKEDH 17,1106,86
scu-rat LD50:235 mg/kg IYKEDH 17,1106,86
ivn-rat LD50:39 mg/kg IYKEDH 17,1106,86
orl-mus LD50:310 mg/kg VTPHAK 15(Suppl.5),1,78

SAFETY PROFILE: Poison by ingestion, intraperitoneal, and intravenous routes. Human systemic effects by ingestion: ulceration or bleeding from stomach. An experimental teratogen. Other experimental reproductive effects. Human mutation data reported. When heated to decomposition it emits very toxic fumes of SO_x and PO_x. See also GOLD COMPOUNDS.

ARS250 CAS:522-16-7 ***HR: D***
AURANTINE
mf: $C_{20}H_{20}O_7$ mw: 372.40

SYNS: AURANETIN ◇ AURANTHINE ◇ AURANTIN (FLAVONE)

TOXICITY DATA with REFERENCE
dnd-omi 10 mg/L BIORAK 39,587,74
dni-mam:lym 10 mg/L BIORAK 39,587,74

SAFETY PROFILE: Mutation data reported. When heated to decomposition it emits acrid smoke and fumes.

ARS500 CAS:130-01-8 ***HR: 3***
AUREINE
mf: $C_{18}H_{25}NO_5$ mw: 335.44

SYNS: 12-HYDROXYSENECIONAN-11,16-DIONE ◇ SENECIONINE

TOXICITY DATA with REFERENCE
sln-dmg-par 20 µmol/L ZEVBA5 91,74,60
dns-rat:lvr 700 nmol/L CNREA8 45,3125,85
dnd-mus-ipr 90 mg/kg TOXID9 1,42,81
orl-rat TDLo:140 mg/kg (female 4-10D post):TER JPMSAE 77,461,88
orl-rat TDLo:140 mg/kg (female 4-10D post):REP JPMSAE 77,461,88
ipr-rat LDLo:33 mg/kg CBINA8 12,299,76
ivn-rat LD50:41200 µg/kg JPETAB 87,382,46
unr-rat LD50:40 mg/kg CNREA8 28,2237,68
ivn-mus LD50:64 mg/kg JPETAB 75,69,42
ivn-ham LD50:61 mg/kg RETOAE 5,53,49

CONSENSUS REPORTS: EPA Genetic Toxicology Program.

SAFETY PROFILE: Poison by intravenous, intraperitoneal, and possibly other routes. An experimental teratogen. Other experimental reproductive effects. Mutation data reported. When heated to decomposition it emits toxic fumes of NO_x.

ARS750 ***HR: 3***
AUREMETINE

PROP: Percentage composition: 28% emetine, 16% auramine, and 56% iodine (AJTMAQ 10,249,30)

TOXICITY DATA with REFERENCE
orl-cat LDLo:20 mg/kg AJTMAQ 10,249,30
orl-rbt LDLo:75 mg/kg AJTMAQ 10,249,30

SAFETY PROFILE: Poison by ingestion. When heated to decomposition it emits very toxic fumes of I^- and NO_x.

ART000 CAS:58194-38-0 ***HR: 3***
AUREOFUSCIN
mf: $C_{25}H_{37}NO_{10}$ mw: 511.63

TOXICITY DATA with REFERENCE
ipr-mus LD50:25 mg/kg JANTAJ 30,77-12,77
ivn-mus LD50:28 mg/kg JANTAJ 30,77-12,77

SAFETY PROFILE: Poison by intraperitoneal and intravenous routes. When heated to decomposition it emits toxic fumes of NO_x.

ART125 CAS:70213-45-5 ***HR: 3***
AUROMOMYCIN

TOXICITY DATA with REFERENCE
dnd-omi 100 mg/L CNREA8 39,2787,79
dnd-hmn:lym 500 µg/L CNREA8 44,3202,84
dnd-mus:lym 60 ng/L/10M CNREA8 39,2787,79
oms-mus:lym 60 µg/L CNREA8 39,2787,79
ivn-mus LD50:3 mg/kg JANTAJ 32,330,79

SAFETY PROFILE: Poison by intravenous route. Human mutation data reported. When heated to decomposition it emits acrid smoke and irritating fumes.

ART250 CAS:12192-57-3 ***HR: 3***
1-AUROTHIO-d-GLUCOPYRANOSE
mf: $C_6H_{11}O_5S{\cdot}Au$ mw: 392.20

SYNS: AUREOTAN ◇ AUROMYOSE ◇ AUROTAN ◇ AUROTHIOGLUCOSE ◇ AURUMINE ◇ AUTHRON ◇ BRENOL ◇ (d-GLUCOPYRANOSYLTHIO)GOLD ◇ (1-d-GLUCOSYLTHIO)GOLD ◇ GLYSANOL B ◇ GOLD THIOGLUCOSE ◇ GTG ◇ ORONOL ◇ ROMOSOL ◇ SOLGANAL ◇ SOLGANAL B ◇ (1-THIO-d-GLUCOPYRANOSATO)GOLD ◇ 1-THIO-GLUCOPYRANOSE, MONOGOLD(1+) SALT ◇ THIOGLUCOSE d'OR (FRENCH)

TOXICITY DATA with REFERENCE
ipr-mus TDLo:400 mg/kg (1D pre):TER PSEBAA 124,1190,67
par-mus TDLo:800 mg/kg (1D pre):REP ENDOAO 78,845,66
par-mus TDLo:400 mg/kg:CAR PAACA3 3,37,59
par-mus TD:750 mg/kg:NEO RFECAC 13,40,68
par-mus TD:750 mg/kg:CAR RFECAC 11,828,66
ims-wmn TDLo:2600 µg/kg/15D-I:GIT ARHEAW 27,230,84
ims-man TDLo:3357 µg/kg/4W-I:EYE JRHUA9 12,619,85
par-wmn TDLo:2700 µg/kg/4W-I:SYS JRHUA9 11,843,84
unr-man LDLo:3 mg/kg SAVEAB 10,101A,39
scu-mus LDLo:1650 mg/kg EMSUA8 3,146,45
ivn-ckn LD50:1000 mg/kg POSCAL 52,926,73
ims-ckn LDLo:300 mg/kg TXAPA9 35,223,76

CONSENSUS REPORTS: IARC Cancer Review: Group 1 IMEMDT 7,56,87; Animal Limited Evidence IMEMDT 13,39,77

SAFETY PROFILE: Confirmed carcinogen with experimental carcinogenic and neoplastigenic data. A deadly human poison by an unspecified route. An experimental poison by intramuscular route. Moderately toxic by subcutaneous and intravenous routes. Human systemic effects: nausea or vomiting, cholestatic jaundice, and eye effects. An experimental teratogen. Other experimental reproductive effects. See also GOLD COMPOUNDS. When heated to decomposition it emits very toxic fumes of SO_x. Used to treat rheumatoid arthritis.

ART500 CAS:11002-90-7 ***HR: 3***
AUROVERTIN
mf: $C_{26}H_{34}O_9$ mw: 490.3

TOXICITY DATA with REFERENCE
ivn-mus LD50:1650 μg/kg 85ERAY 3,2003,78
ivn-dog LDLo:1 mg/kg 85ERAY 3,2003,78
ivn-rbt LDLo:1 mg/kg 85ERAY 3,2003,78

SAFETY PROFILE: Poison by intravenous route. When heated to decomposition it emits acrid smoke and irritating fumes.

ARU250 CAS:55256-53-6 ***HR: 3***
AUSTOCYSTIN D
mf: $C_{22}H_{22}O_8$ mw: 414.44

TOXICITY DATA with REFERENCE
mma-sat 500 ng/plate MUREAV 58,193,78
orl-mus LD50:300 mg/kg OYYAA2 3,187,69

SAFETY PROFILE: Poison by ingestion. Mutation data reported. When heated to decomposition it emits acrid smoke and irritating fumes.

ARU500 ***HR: 3***
AUSTRALIAN BROWN SNAKE VENOM

SYNS: PSEUDONAJA TEXTILIS (AUSTRALIA) VENOM ◇ P. TEXTILIS (AUSTRALIA) VENOM

TOXICITY DATA with REFERENCE
ipr-mus LD50:310 μg/kg TOXIA6 17(Suppl. 1),121,79
scu-mus LD50:41 μg/kg TOXIA6 17,661,79

SAFETY PROFILE: Poison by intraperitoneal and subcutaneous routes.

ARU750 ***HR: 3***
AUSTRALIAN COPPERHEAD SNAKE VENOM

SYNS: A. SUPERBA (AUSTRALIA) VENOM ◇ AUSTRELAPS SUPERBA (AUSTRALIA) VENOM ◇ VENOM, AUSTRALIAN ELAPIDAE SNAKE, AUSTRELAPS SUPERBA

TOXICITY DATA with REFERENCE
ipr-mus LD50:280 μg/kg 85EGD4 5,368,78
scu-mus LD50:500 μg/kg TOXIA6 17,661,79

SAFETY PROFILE: Poison by intraperitoneal and subcutaneous routes.

ARU875 ***HR: 3***
AUSTRALIAN DEATH ADDER SNAKE VENOM

SYNS: ACANTHOPHIA ANTARCTICUS VENOM ◇ VENOM, AUSTRALIAN ELAPIDAE SNAKE, ACANTHOPHIS ANTARCTICUS

TOXICITY DATA with REFERENCE
ipr-mus LD50:160 μg/kg TOXIA6 17,609,79
scu-mus LD50:338 μg/kg TOXIA6 17,661,79
ims-mus LD50:80 μg/kg BIJOAK 199,211,81
scu-gpg LDLo:130 μg/kg MJAUAJ 2,801,71

SAFETY PROFILE: Deadly poison by subcutaneous, intramuscular, and intraperitoneal routes.

ARV000 ***HR: 3***
AUSTRALIAN KING BROWN SNAKE VENOM

SYNS: AUSTRALIS ◇ P. AUSTRALIS VENOM ◇ PSEUDECHIS AUSTRALIS VENOM ◇ VENOM, AUSTRALIAN SNAKE, PSEUDECHIS

TOXICITY DATA with REFERENCE
ipr-mus LD50:520 μg/kg TOXIA6 17(Suppl. 1),121,79
scu-mus LD50:2380 μg/kg TOXIA6 17,661,79
ivn-mus LD50:230 μg/kg TOXIA6 23,73,85

SAFETY PROFILE: Deadly poison by intravneous, intraperitoneal, and subcutaneous routes.

ARV125 ***HR: 3***
AUSTRALIAN KING COBRA SNAKE VENOM

SYNS: OPHIOPHAGUS HANNAH VENOM ◇ VENOM, AUSTRALIAN SNAKE, OPHIOPHAGUS HANNAH

TOXICITY DATA with REFERENCE
ipr-mus LD50:355 μg/kg YHHPAL 19,721,84
scu-mus LD50:1091 μg/kg TOXIA6 17,661,79
ivn-mus LD50:125 μg/kg JICMBE 61,791,79

SAFETY PROFILE: Poison by subcutaneous, intravenous, and intraperitoneal routes.

ARV250 ***HR: 3***
AUSTRALIAN RED-BELLIED BLACK SNAKE VENOM

SYNS: P. PORPHYRIACUS (AUSTRALIA) VENOM ◇ PSEUDECHIS PORPHYRIACUS (AUSTRALIA) VENOM ◇ PSEUDECHIS PORPHYRIACUS VENOM ◇ VENOM, AUSTRALIAN ELAPIDAE SNAKE, PSEUDECHIS PORPHYRIACUS

TOXICITY DATA with REFERENCE
ipr-mus LD50:700 μg/kg 85EGD4 5,368,78

scu-mus LD50:2530 μg/kg TOXIA6 17,661,79
ims-frg LDLo:500 ug/kg TOXIA6 19,749,81

SAFETY PROFILE: Deadly poison by intramuscular, subcutaneous, and intraperitoneal routes.

ARV375 ***HR: 3***
AUSTRALIAN ROUGH SCALED SNAKE VENOM

SYNS: TROPIDECHIS CARINATUS VENOM ◇ VENOM, AUSTRALIAN SNAKE, TROPIDECHIS CARINATUS

TOXICITY DATA with REFERENCE
ipr-mus LD50:125 μg/kg TOXIA6 20,1085,82
scu-mus LD50:1090 μg/kg TOXIA6 17,661,79
scu-dog LDLo:500 μg/kg MJAUAJ 2,801,71
scu-gpg LDLo:75 μg/kg MJAUAJ 2,801,71
ivn-dom LDLo:21800 μg/kg MJAUAJ 2,801,71

SAFETY PROFILE: Poison by subcutaneous, intravenous, and intraperitoneal routes.

ARV500 ***HR: 3***
AUSTRALIAN TAIPAN SNAKE VENOM

SYNS: O. SCUTELLATUS (AUSTRALIA) VENOM ◇ OXYURANUS SCUTELLATUS (AUSTRALIA) VENOM ◇ VENOM, AUSTRALIAN ELAPIDAE SNAKE, OXYURANUS SCUTELLATUS

TOXICITY DATA with REFERENCE
ipr-mus LD50:9 μg/kg 85EGD4 5,372,78
scu-mus LD50:64 μg/kg TOXIA6 17,661,79
scu-gpg LDLo:25 ug/kg MJAUAJ 2,801,71

SAFETY PROFILE: Deadly poison by intraperitoneal and subcutaneous routes.

ARV550 ***HR: 3***
AUSTRALIAN TIGER SNAKE VENOM

SYNS: NOTECHIS SCUTATUS VENOM ◇ VENOM, AUSTRALIAN SNAKE, NOTECHIS SCUTATUS

TOXICITY DATA with REFERENCE
ipr-mus LD50:50 μg/kg AJTHAB 9,284,60
scu-mus LD50:118 μg/kg TOXIA6 17(Suppl 1),121,79
ivn-rbt LDLo:1 μg/kg PSEBAA 116,696,64
scu-gpg LDLo:20 μg/kg MJAUAJ 2,801,71
ims-frg LDLo:500 μg/kg TOXIA6 19,749,81
ipr-mam LD50:40 μg/kg CLPTAT 8,849,67

SAFETY PROFILE: Deadly poison by subcutaneous, intramuscular, intravenous, and intraperitoneal routes.

ARV625 ***HR: 3***
AUSTRELAPS SUPERBA VENOM

SYNS: A. SUPERBA VENOM ◇ AUSTRELAPS SUPERBUS VENOM ◇ VENOM, AUSTRALIAN SNAKE, AUSTRELAPS SUPERBA

TOXICITY DATA with REFERENCE
ipr-mus LD50:280 μg/kg 85EGD4 -,368,78
scu-mus LD50:6200 μg/kg TOXIA6 18,443,80
ims-frg LDLo:400 μg/kg TOXIA6 19,749,81

SAFETY PROFILE: Poison by subcutaneous, intramuscular, and intraperitoneal routes.

ARV750 CAS:7437-53-8 ***HR: 3***
AVACAN HYDROCHLORIDE
mf: $C_{19}H_{32}N_2O_2 \cdot ClH$ mw: 356.99

TOXICITY DATA with REFERENCE
ivn-hmn TDLo:357 μg/kg:EYE DMWOAX 76,479,51
ims-hmn TDLo:357 μg/kg:EYE DMWOAX 76,479,51

SAFETY PROFILE: A human poison by intravenous and intramuscular routes. Very irritating to experimental animals and humans. When heated to decomposition it emits toxic fumes such as Cl^- and NO_x.

ARW000 CAS:43222-48-6 ***HR: 3***
AVENGE
mf: $C_{16}H_{17}N_2 \cdot CH_3O_4S$ mw: 348.45

SYNS: AC 84777 ◇ DIFENZOQUAT METHYL SULFATE ◇ 1,2-DIMETHYL-3,5-DIPHENYL-1-H-PYRAZOLIUM METHYL SULFATE ◇ FINAVEN ◇ MATAVEN ◇ YEH-YAN-KU

TOXICITY DATA with REFERENCE
orl-rat LD50:270 mg/kg FMCHA2 -,C20,83
skn-rbt LD50:3540 mg/kg FMCHA2 -,C20,83

SAFETY PROFILE: Poison by ingestion. Moderately toxic by skin contact. See also SULFATES. When heated to decomposition it emits very toxic fumes of NO_x and SO_x.

ARW250 CAS:75-80-9 ***HR: 3***
AVERTIN
mf: $C_2H_3Br_3O$ mw: 282.78

PROP: Crystals; ethereal odor and aromatic taste. Mp: 70-82°, bp: 92-93° @ 10 mm. Sltly water-sol; sol in alc and organic solvents.

SYNS: BROMETHOL ◇ ETHOBROM ◇ NARCOLAN ◇ NARKOLAN ◇ RENARCOL ◇ TRIBROMETHANOL ◇ 2,2,2-TRIBROMOETHANOL ◇ TRIBROMOETHYL ALCOHOL ◇ 2,2,2-TRIBROMOETHYL ALCOHOL

TOXICITY DATA with REFERENCE
orl-rat LDLo:1 g/kg JPETAB 63,183,38
scu-rat LDLo:530 mg/kg AEPPAE 182,348,36
orl-mus LD50:930 mg/kg JPMSAE 56,920,67
ipr-mus LDLo:600 mg/kg 27ZWAY E.2,130,-
scu-mus LDLo:500 mg/kg 27ZWAY 2,-,36
orl-rbt LDLo:1100 mg/kg AEPPAE 132,214,28
ipr-rbt LDLo:450 mg/kg AEPPAE 132,214,28
ivn-rbt LDLo:120 mg/kg 27ZWAY E.2,130,-

CONSENSUS REPORTS: Reported in EPA TSCA Inventory.

SAFETY PROFILE: Poison by intravenous route. Moderately toxic by ingestion and other routes. Dangerous when heated; see also BROMIDES.

ARW750 CAS:151-06-4 ***HR: 3***
AVICOL
mf: $C_{10}H_{14}ClN•ClH$ mw: 220.16

SYNS: p-CHLORO-α,α-DIMETHYLPHENETHYLAMINE HYDROCHLORIDE ◇ 4-CHLORO-α,α-DIMETHYLPHENETHYLAMINE HYDROCHLORIDE ◇ CHLOROPHENTERMINE HYDROCHLORIDE ◇ 1-(p-CHLOROPHENYL)-2-METHYL-2-AMINOPROPANE HYDROCHLORIDE ◇ CHLORPHENTERMINE HYDROCHLORIDE ◇ α,α-DIMETHYL-p-CHLOROPHENETHYLAMINE HYDROCHLORIDE ◇ LUCOFEN ◇ LUCOFENE ◇ NSC-76098 ◇ PRE-SATE ◇ PRESATE HYDROCHLORIDE ◇ S-62 ◇ S 62-2 ◇ W 2426

TOXICITY DATA with REFERENCE
scu-rat TDLo:150 mg/kg (16-20D preg):REP TOXID9 1,31,81
orl-rat LD50:150 mg/kg CHTPBA 5,247,70
ipr-rat LD50:144 mg/kg APTOA6 17,121,60
orl-mus LD50:225 mg/kg ARZNAD 13,711,63
ipr-mus LD50:88 mg/kg CHTPBA 5,247,70
ivn-mus LD50:55 mg/kg JPETAB 137,365,62

SAFETY PROFILE: Poison by ingestion, subcutaneous, intraperitoneal, and intravenous routes. Experimental reproductive effects. When heated to decomposition it emits very toxic fumes of HCl and NO_x. An anorexic agent.

ARX125 ***HR: 3***
AYUSH-47

PROP: An Indian indigenous preparation containing equal parts of the bark of *Saraca indica, Areca catechu, Coccus lacca,* gold and sugar (IJMRAQ 70,504,79)

TOXICITY DATA with REFERENCE
orl-rat TDLo:25 mg/kg (1-5D preg):TER IJMRAQ 70,504,79
orl-rat TDLo:12500 μg/kg (1-5D preg):REP IJMRAQ 70,504,79
orl-mus LDLo:200 mg/kg IJMRAQ 70,504,79

SAFETY PROFILE: Poison by ingestion. Experimental reproductive effects. When heated to decomposition it emits acrid smoke and irritating fumes.

ARX150 CAS:21650-02-2 ***HR: 3***
AZABICYCLANE CITRATE
mf: $C_{16}H_{23}NO•C_6H_8O_7$ mw: 437.54

SYNS: 4-β-METHOXY-1-METHYL-4-α-PHENYL-3-α,5-α-PROPANOPIPERIDINE HYDROGEN CITRATE ◇ 9-β-METHOXY-9-α-PHENYL-3-METHYL-3-AZABICYCLO(3.3.1)NONANECITRATE

TOXICITY DATA with REFERENCE
ipr-mus TDLo:24 mg/kg (7-12D preg):TER IYKEDH 3,195,72
scu-rat LD50:130 mg/kg TXAPA9 17,344,70
orl-mus LD50:300 mg/kg TXAPA9 17,344,70
scu-mus LD50:200 mg/kg TXAPA9 17,344,70
ivn-mus LD50:56 mg/kg TXAPA9 17,344,70

SAFETY PROFILE: Poison by ingestion, subcutaneous, and intravenous routes. An experimental teratogen. When heated to decomposition it emits toxic fumes of NO_x.

ARX500 CAS:69766-49-0 ***HR: 3***
1-AZABICYCLO(3.2.1)OCTAN-6-OL DIPHENYLACETATE HYDROCHLORIDE
mf: $C_{21}H_{23}NO_2•ClH$ mw: 357.91

SYN: 6-DIPHENYLACETOXY-1-AZABICYCLO(3.2.1)OCTANEHYDROCHLORIDE

TOXICITY DATA with REFERENCE
ipr-mus LD50:105 mg/kg JPETAB 104,284,52
ivn-mus LD50:33 mg/kg JPETAB 104,284,52
ivn-dog LD50:30 mg/kg JPETAB 104,284,52

SAFETY PROFILE: Poison by intraperitoneal and intravenous routes. When heated to decomposition it emits very toxic fumes of HCl and NO_x.

ARX750 CAS:69766 ***HR: 3***
1-AZABICYCLO(3.2.1)OCTAN-6-OL-9-FLUORENECARBOXYLATE HYDROCHLORIDE-48-9
mf: $C_{21}H_{21}NO_2•ClH$ mw: 355.89

SYN: RO 2-3245

TOXICITY DATA with REFERENCE
ipr-mus LD50:137 mg/kg JPETAB 104,284,52
ivn-mus LD50:23 mg/kg JPETAB 104,284,52

SAFETY PROFILE: Poison by intraperitoneal and intravenous routes. When heated to decomposition it emits very toxic fumes of HCl and NO_x.

ARX770 CAS:69766-47-8 ***HR: 3***
AZABICYCLOOCTANOL METHYL BROMIDE DIPHENYLACETATE
mf: $C_{22}H_{26}NO_2•Br$ mw: 416.40

SYN: RO 2-3951

TOXICITY DATA with REFERENCE
ipr-mus LD50:46 mg/kg JPETAB 104,284,52
ivn-mus LD50:4 mg/kg JPETAB 104,284,52

SAFETY PROFILE: Poison by intraperitoneal and intravenous routes. See also BROMIDES. When heated to decomposition it emits toxic fumes of Br^- and NO_x.

ARX800 CAS:1249-84-9 ***HR: 3***
AZACOSTEROL DIHYDROCHLORIDE
mf: $C_{25}H_{44}N_2O•2ClH$ mw: 461.63

SYNS: AZACOSTEROL HYDROCHLORIDE ◇ AZASTEROL ◇ DIAZACOSTEROL HYDROCHLORIDE ◇ 20,25-DIAZOCHOLESTEROL DIHYDROCHLORIDE ◇ 17-β-((3-(DIMETHYLAMINO)-PROPYL)METHYLAMINO)ANDROST-5-EN-3β-OL DIHYDROCHLORIDE ◇ IMD 760 ◇ ORNITROL ◇ SC 12937

TOXICITY DATA with REFERENCE
oms-pgn-orl 600 mg/kg/10D-I JRPFA4 15,145,68
spm-pgn-orl 600 mg/kg/10D-I JRPFA4 15,145,68
ipr-rat TDLo:100 mg/kg (male 10D pre):REP JOAND3 7,277,86
orl-rat LD50:470 mg/kg 85ARAE 3,100,76/77
ipr-rat LD50:60 mg/kg FMCHA2 -,C174,83
orl-mus LD50:380 mg/kg FMCHA2 -,C174,83
ipr-mus LD50:92 mg/kg FMCHA2 -,C174,83

SAFETY PROFILE: Poison by ingestion and intraperitoneal routes. Experimental reproductive effects. Mutation data reported. When heated to decomposition it emits toxic fumes of NO_x and HCl.

ARX875 CAS:78110-38-0 ***HR: 2***
AZACTAM
mf: $C_{13}H_{17}N_5O_8S_2$ mw: 435.47

PROP: The first totally synthetic monocyclic β-lactam (monobactam) antibiotic.

SYNS: 2-(((1-(2-AMINO-4-THIAZOLYL)-2-((2-METHYL-4-OXO-1-SULFO-3-AZETIDINYL)AMINO)-2-OXOETHYLIDENE)AMINO)OXY)-2-METHYLPROPANOIC ACID, (2S-(2-α,3β(Z)))- ◇ AZTHREONAM ◇ AZTREONAM ◇ PRIMBACTAM ◇ SQ 26,776

TOXICITY DATA with REFERENCE
ivn-rat TDLo:2970 mg/kg (7-17D preg):REP NKRZAZ 33(Suppl 1),203,85
ivn-rat TDLo:1100 mg/kg (7-17D preg):TER NKRZAZ 33(Suppl 1),203,85
ipr-rat LD50:2549 mg/kg NKRZAZ 33(Suppl 1),143,85
scu-rat LD50:3154 mg/kg NKRZAZ 33(Suppl 1),143,85
ivn-rat LD50:2001 mg/kg KSRNAM 19,468,85
ipr-mus LD50:2897 mg/kg NKRZAZ 33(Suppl 1),143,85
scu-mus LD50:3906 mg/kg NKRZAZ 33(Suppl 1),143,85
ivn-mus LD50:1963 mg/kg NKRZAZ 33(Suppl 1),143,85

SAFETY PROFILE: Moderately toxic by several routes. An experimental teratogen. Other experimental reproductive effects. When heated to decomposition it emits toxic fumes of NO_x and SO_x.

ARY000 CAS:320-67-2 ***HR: 3***
AZACYTIDINE
mf: $C_8H_{12}N_4O_5$ mw: 244.24

SYNS: 5-AC ◇ 5-ACZ ◇ 4-AMINO-1-β-d-RIBOFURANOSYL-d-TRIAZIN-2(1H)-ONE ◇ 4-AMINO-1-β-d-RIBOFURANOSYL-1,3,5-TRIAZIN-2(1H)-ONE ◇ ANTIBIOTIC U 18496 ◇ AZACITIDINE ◇ 5-AZACYTIDINE ◇ 5'-AZACYTIDINE ◇ LADAKAMYCIN ◇ MYLOSAR ◇ NCI-C01569 ◇ NSC-102816 ◇ U 18496

TOXICITY DATA with REFERENCE
dnd-hmn:fbr 1 μmol/L PNASA6 79,2352,82
dni-hmn:leu 3 μmol/L CNREA8 43,763,83
ipr-rat TDLo:4 mg/kg (female 17D post):REP FCTOD7 22,963,84
ipr-mus TDLo:1 mg/kg (female 7D post):TER TJADAB 30(1),9A,84
ipr-rat TDLo:190 mg/kg/38W-I:ETA CRNGDP 5,1583,84
ipr-mus TDLo:100 mg/kg/50W-I:CAR CALEDQ 37,51,87
unr-mus TDLo:1 mg/kg (16D post):CAR TJADAB 32,33A,85
ipr-mus TD:284 mg/kg/43W-I:NEO NCITR* NCI-TR-42,78
ipr-mus LD:1 mg/kg (female 12D post):NEO,TER CALEDQ 27,81,85
ivn-wmn TDLo:6 mg/kg/10D-I:BLD CCROBU 56,413,72
ivn-wmn TDLo:500 μg/kg:GIT CCROBU 56,413,72
orl-mus LD50:572 mg/kg TXAPA9 19,382,71
ipr-mus LD50:68 mg/kg EXPEAM 22,53,66
ivn-mus LD50:229 mg/kg NTIS** PB84-211432
ivn-dog LD50:7200 μg/kg AVPCAQ 14,285,77
orl-bwd LD50:100 mg/kg AECTCV 12,355,83

CONSENSUS REPORTS: IARC Cancer Review: Group 3 IMEMDT 7,56,87; Animal Limited Evidence IMEMDT 26,37,81; NCI Carcinogenesis Bioassay (ipr); Inadequate Studies: rat NCITR* NCI-CG-TR-42,78; Clear Evidence: mouse NCITR* NCI-CG-TR-42,78. EPA Genetic Toxicology Program.

SAFETY PROFILE: Poison by ingestion, intravenous, and intraperitoneal routes. Human systemic effects by intravenous route: nausea, vomiting and diarrhea, reduction in white cell count (luekopenia and agranulocytosis). An experimental teratogen. Other experimental reproductive effects. Questionable carcinogen with experimental carcinogenic, neoplastigenic, tumorigenic data. Human mutation data reported. A skin irritant. When heated to decomposition it emits toxic fumes of NO_x.

ARY125 CAS:2353-33-5 ***HR: 3***
5-AZADEOXYCYTIDINE
mf: $C_8H_{12}N_4O_4$ mw: 228.24

SYNS: 4-AMINO-1-(2-DEOXY-β-d-erythro-PENTOFURANOSYL)-s-TRIAZIN-2(1H)-ONE ◇ 5-AZA-2'-DEOXYCYTIDINE

TOXICITY DATA with REFERENCE
dns-hmn:leu 5 μmol/L/1H CNREA8 42,519,82
dns-mus:leu 5 μmol/L/1H CNREA8 42,519,82
dni-mus-ipr 4 mg/kg BCPCA6 29,2929,80
dni-ham:ovr 1 μmol/L BBACAQ 697,286,82
ivn-mus LD50:22 mg/kg DCTODJ 4,373,81

SAFETY PROFILE: Poison by intravenous route.

Human mutation data reported. When heated to decomposition it emits toxic fumes of NO_x.

ARY500 CAS:63907-29-9 *HR: 3*
2-AZAHYPOXANTHINE
mf: $C_4H_3N_5O$ mw: 137.12

SYN: 4-OXO-4H-IMIDAZO(4,5-D)-v-TRIAZINE

TOXICITY DATA with REFERENCE
orl-rat TDLo:3005 mg/kg/21W-C:ETA JNCIAM 54,951,75

SAFETY PROFILE: Questionable carcinogen with experimental tumorigenic data. When heated to decomposition it emits toxic fumes of NO_x.

ARY625 CAS:4746-36-5 *HR: D*
AZALEUCINE
mf: $C_5H_{12}N_2O_2$ mw: 132.19

SYNS: 4-AZALEUCINE ◇ 3-(DIMETHYLAMINO)-ALANINE (9CI)

TOXICITY DATA with REFERENCE
oms-esc 100 mmol/L RCOCB8 2,271,71
oms-rat:lvr 100 mmol/L RCOCB8 2,271,71
dni-mus:lym 100 μmol/L BCPCA6 21,1977,72
dni-ham:ovr 1 mmol/L RCOCB8 2,271,71

SAFETY PROFILE: Mutation data reported. When heated to decomposition it emits toxic fumes of NO_x.

ARY750 CAS:11003-24-0 *HR: 3*
AZALOMYCIN F

PROP: An antibiotic produced by *Streptomyces hygroscopicus var. azalomycetic* first isolated from soil sample collected from around root of azaleas (ARZNAD 18,1396,68).

TOXICITY DATA with REFERENCE
orl-mus LD50:580 mg/kg ARZNAD 18,1396,68
ipr-mus LD50:25900 μg/kg ARZNAD 18,1396,68
scu-mus LD50:162 mg/kg ARZNAD 18,1396,68
ivn-mus LD50:12500 μg/kg ARZNAD 18,1396,68

SAFETY PROFILE: Poison by intraperitoneal, subcutaneous, and intravenous routes. Moderately toxic by ingestion.

ARZ000 CAS:146-36-1 *HR: 3*
AZAPETINE
mf: $C_{17}H_{17}N$ mw: 235.35

SYN: 6-ALLYL-6,7-DIHYDRO-5H-DIBENZ(c,e)AZEPINE◇ ILIDAR ◇ ILIDAR BASE

TOXICITY DATA with REFERENCE
orl-mus LD50:460 mg/kg BCFAAI 98,702,59
ipr-mus LD50:210 mg/kg CLDND*
scu-mus LD50:725 mg/kg CLDND*
ivn-mus LD50:27 mg/kg CLDND*
ims-mus LD50:600 mg/kg CLDND*
ivn-dog LD50:50 mg/kg CLDND*
ivn-rbt LD50:26 mg/kg CLDND*

SAFETY PROFILE: Poison by intraperitoneal and intravenous routes. Moderately toxic by ingestion, subcutaneous, and intramuscular routes. When heated to decomposition it emits toxic fumes of NO_x.

ASA000 CAS:22304-30-9 *HR: 2*
AZAPROPAZONE
mf: $C_{16}H_{20}N_4O_2$•$2H_2O$ mw: 336.44

SYNS: APAZONE DIHYDRATE ◇ AZAPROPAZON DIHYDRAT (GERMAN) ◇ 1,2-DIHYDRO-3-DIMETHYLAMINO-7-METHYL-1,2-(PROPYLMALONYL)-1,2,4-BENZOTRIAZINE DIHYDRATE ◇ 3-DIMETHYLAMINO-7-METHYL-1,2-(N-PROPYLMALONYL)-1,2-DIHYDRO-1,2,4-BENZOTRIAZINE DIHYDRATE ◇ MI 85 DI ◇ PROLIXAN

TOXICITY DATA with REFERENCE
eye-rbt 100 mg MLD CMROCX 4,17,76
orl-rat LD50:1950 mg/kg ARZNAD 23,1215,73
ipr-rat LD50:1101 mg/kg TOIZAG 19,242,72
scu-rat LD50:1101 mg/kg TOIZAG 19,242,72
ivn-rat LD50:710 mg/kg ARZNAD 19,36,69
orl-mus LD50:1080 mg/kg ARZNAD 23,1215,73
ipr-mus LD50:1107 mg/kg TOIZAG 19,242,72
scu-mus LD50:1190 mg/kg TOIZAG 19,242,72
ivn-mus LD50:750 mg/kg ARZNAD 19,36,69

SAFETY PROFILE: Moderately toxic by ingestion, intraperitoneal, subcutaneous and intravenous routes. An eye irritant. When heated to decomposition it emits toxic fumes of NO_x. An anti-inflammatory and analgesic agent.

ASA250 CAS:74037-31-3 *HR: 2*
AZAPROPAZONE SODIUM
mf: $C_{15}H_{20}N_4O_2$•Na mw: 322.39

SYNS: AZAPROPAZON NATRIUMSALZ (GERMAN) ◇ 1,2-DIHYDRO-3-DIMETHYLAMINO-7-METHYL-1,2-(PROPYLMALONYL)-1,2,4-BENZOTRIAZINE SODIUM SALT ◇ 3-DIMETHYLAMINO-7-METHYL-1,2-(PROPYLMALONYL)-1,2-DIHYDRO-1,2,4-BENZOTRIAZINE SODIUM SALT ◇ SODIUM AZAPROPAZONE

TOXICITY DATA with REFERENCE
orl-rat LD50:1900 mg/kg CMROCX 4,17,76
ivn-rat LD50:810 mg/kg CMROCX 4,17,76
orl-mus LD50:1950 mg/kg CMROCX 4,17,76
ivn-mus LD50:960 mg/kg CMROCX 4,17,76

SAFETY PROFILE: Moderately toxic by ingestion and intravenous routes. When heated to decomposition it emits toxic fumes of NO_x.

ASA500 CAS:115-02-6 ***HR: 3***
AZASERINE
mf: $C_5H_7N_3O_4$ mw: 173.15

PROP: Produced by the strain *Streptomyces fragilis* (85ERAY 2,1249,78)

SYNS: AZASERIN ◇ l-AZASERINE ◇ AZS ◇ CI-337 ◇ CL 337 ◇ CN-15,757 ◇ DIAZOACETATE (ESTER)-l-SERINE ◇ l-DIAZOACETATE (ESTER) SERINE ◇ DIAZO-ACETIC ACID ESTER with SERINE ◇ o-DIAZOACETYL-l-SERINE ◇ NSC-742 ◇ P-165 ◇ RCRA WASTE NUMBER U015 ◇ l-SERINE DIAZOACETATE ◇ l-SERINE DIAZOACETATE (ester)

TOXICITY DATA with REFERENCE
mma-sat 200 ng/plate PNASA6 72,5135,75
mmo-bcs 500 µmol/L EXPEAM 39,530,83
ipr-ham TDLo:2 mg/kg (female 8D post):TER EXPEAM 39,324,83
ipr-rat TDLo:10 mg/kg (female 7-8D post):REP PSEBAA 94,27,57
orl-rat TDLo:150 mg/kg/5W-C:NEO CANCAR 47,1562,81
ipr-rat TDLo:30 mg/kg:CAR CRNGDP 10,311,89
ipr-rat TD:440 mg/kg/13W-I:ETA CNREA8 40,592,80
ipr-rat LD:260 mg/kg/26W-I:CAR CNREA8 35,2249,75
orl-rat LD50:170 mg/kg CANCAR 10,889,57
ipr-rat LD50:70 mg/kg PSEBAA 94,27,57
orl-mus LD50:150 mg/kg CANCAR 10,889,57
ipr-mus LD50:100 mg/kg CANCAR 10,889,57
scu-mus LD50:50 mg/kg 85GDA2 4(1),432,80
ivn-mus LD50:62 mg/kg 85ERAY 2,1249,78
ivn-dog LDLo:30 mg/kg TXAPA9 22,595,72

CONSENSUS REPORTS: IARC Cancer Review: Group 2B IMEMDT 7,56,87; Animal Limited Evidence IMEMDT 10,73,76. EPA Genetic Toxicology Program.

SAFETY PROFILE: Suspected carcinogen with experimental carcinogenic, neoplastigenic, and tumorigenic data. Poison by ingestion, intraperitoneal, and subcutaneous routes. An experimental teratogen. Other experimental reproductive effects. Human mutation data reported. When heated to decomposition it emits toxic fumes of NO_x.

ASA750 CAS:1497-16-1 ***HR: 2***
6-AZASPIRO(3,4)OCTANE-5,7-DIONE
mf: $C_7H_9NO_2$ mw: 139.17

SYN: AZA-6-SPIRO(3,4)OCTANE-DIONE-5,7(FRENCH)

TOXICITY DATA with REFERENCE
orl-mus LD50:1750 mg/kg BSCFAS 3,1119,66
ipr-mus LD50:1450 mg/kg BSCFAS 3,1119,66

SAFETY PROFILE: Moderately toxic by ingestion and intraperitoneal routes. When heated to decomposition it emits toxic fumes of NO_x.

ASA875 CAS:64-60 ***HR: 3***
4-(3-AZASPIRO(5.5)UNDEC-3-YL)-4'-FLUOROBUTYROPHENONE HYDROCHLORIDE-8
mf: $C_{20}H_{28}FNO•ClH$ mw: 353.95

TOXICITY DATA with REFERENCE
orl-rat LD50:200 mg/kg JMCMAR 8,62,65
ipr-rat LD50:50 mg/kg JMCMAR 8,62,65
orl-mus LD50:120 mg/kg JMCMAR 8,62,65
ipr-mus LD50:45 mg/kg JMCMAR 8,62,65

SAFETY PROFILE: Poison by ingestion and intraperitoneal routes. When heated to decomposition it emits toxic fumes of F^-, NO_x, and HCl.

ASB250 CAS:446-86-6 ***HR: 3***
AZATHIOPRINE
mf: $C_9H_7N_7O_2S$ mw: 277.29

SYNS: AZANIN ◇ AZATIOPRIN ◇ AZOTHIOPRINE ◇ BW 57-322 ◇ CCUCOL ◇ IMURAN ◇ IMUREK ◇ IMUREL ◇ METHYLNITROIMIDAZOLYLMERCAPTOPURINE ◇ 6-(1'-METHYL-4'-NITRO-5'-IMIDAZOLYL)-MERCAPTOPURINE ◇ 6-(METHYL-p-NITRO-5-IMIDAZOLYL)-THIOPURINE ◇ 6-((1-METHYL-4-NITROIMIDAZOL-5-YL)THIO)PURINE ◇ 6-(1-METHYL-p-NITRO-5-IMIDAZOLYL)-THIOPURINE ◇ 6-(1-METHYL-4-NITROIMIDAZOL-5-YLTHIO)PURINE ◇ 6-((1-METHYL-4-NITRO-1H-IMIDAZOL-5-YL)THIO)-1H-PURINE ◇ NCI-C03474 ◇ NSC-39084 ◇ RORASUL

TOXICITY DATA with REFERENCE
mma-sat 300 µmol/L EXPEAM 40,370,84
cyt-hmn-unr 1074 mg/kg/4Y MUREAV 94,501,82
orl-rat TDLo:240 mg/kg (female 10-15D post):REP OYYAA2 2,401,68
orl-mus TDLo:50 mg/kg (female 10D post):TER NSAPCC 298,93,77
orl-wmn TDLo:273 mg/kg/13W-C:CAR,BLD BMJOAE 4,235,72
orl-rat TDLo:1932 mg/kg/46W-C:ETA ESKHA5 (102),66,84
scu-mus TDLo:1200 mg/kg/30W-I:CAR ARZNAD 29,662,79
ims-mus TDLo:3500 mg/kg/26W-I:CAR BLOOAW 31,396,68
orl-wmn TD:3 g/kg/3.5Y-C:CAR,KID JAMAAP 237,152,77
orl-wmn TD:5460 mg/kg/6Y-I:CAR,BLD AMSVAZ 207,315,80
orl-man TD:728 mg/kg/43W-C:CAR,BLD AJMSA9 273,335,77
orl-man TD:1565 mg/kg/4Y-C:CAR,BLD AJMEAZ 57,885,74
orl-man TD:3266 mg/kg/3Y-C:CAR,BLD AJMEAZ 57,885,74
orl-wmn TD:2 g/kg/3Y-C:CAR,BLD PGMJAO 53,173,77
orl-man LDLo:395 mg/kg/56W-I:SYS GASTAB 90,446,86
orl-wmn TDLo:500 µg/kg:GIT,SYS JRHUA9 13,1117,86

orl-man TDLo:7500 μg/kg/1W-I:BPR,SYS ARHEAW 24,1453,81
orl-man TDLo:243 mg/kg/1Y-I:SYS GASTAB 90,446,86
unr-cld TDLo:2500 μg/kg/D-I:BLD AJDCAI 134,377,80
orl-rat LD50:535 mg/kg NIIRDN 6,3,82
ipr-rat LD50:300 mg/kg JRPFA4 4,297,62
idu-rat LD50:630 mg/kg RPTOAN 31,223,68
orl-mus LD50:1389 mg/kg NIIRDN 6,3,82
ipr-mus LD50:273 mg/kg RPTOAN 34,284,71
scu-mus LD50:350 mg/kg JMCMAR 18,320,75
idu-mus LD50:2437 mg/kg RPTOAN 31,223,68
idu-rbt LDLo:100 mg/kg RPTOAN 31,223,68

CONSENSUS REPORTS: NTP Fifth Annual Report on Carcinogens. IARC Cancer Review: Group 1 IMEMDT 7,119,87 Human Sufficient Evidence IMEMDT 26,47,81; Animal Limited Evidence IMEMDT 26,47,81. NCI Carcinogenesis Studies (ipr); No Evidence: rat CANCAR 40,1935,77; Clear Evidence: mouse CANCAR 40,1935,77. EPA Genetic Toxicology Program.

SAFETY PROFILE: Confirmed human carcinogen producing bladder tumors and leukemia. Poison by subcutaneous, intradermal, and intraperitoneal routes. Moderately toxic by ingestion. Human systemic effects: liver changes, hypermotility, diarrhea, nausea or vomiting, increased body temperature, BP lowering, decreased urine volume or anuria, normocytic anemia, bone marrow changes. An experimental teratogen. Other experimental reproductive effects. Human mutation data reported. When heated to decomposition it emits very toxic fumes of NO_x and SO_x. An immunosuppressant.

ASB750 CAS:123-99-9 ***HR: 2***
AZELAIC ACID
mf: $C_9H_{16}O_4$ mw: 188.23

PROP: Leaflets or needles. Mp: 106.5°, bp: 286.5° @ 100 mm, d: 1.029 @ 20°/4°, vap press: 1 mm @ 178.3°. Solubility: in water = 0.2/100, very sol in alc, in ether = 2.7/100 @ 15°.

SYNS: ANCHOIC ACID ◇ HEPTANEDICARBOXYLIC ACID ◇ 1,7-HEPTANEDICARBOXYLIC ACID ◇ LEPARGYLIC ACID ◇ NONANEDIOIC ACID

TOXICITY DATA with REFERENCE
skn-rbt 500 mg/24H MLD EMERY* S3B,-,64
eye-rbt 3 mg MLD EMERY* S3B,-,64

CONSENSUS REPORTS: Reported in EPA TSCA Inventory.

SAFETY PROFILE: A skin and eye irritant. Closely related to glutaric acid and adipic acid. Combustible when exposed to heat or flame; can react with oxidizing materials.

ASC000 CAS:109-31-9 ***HR: 2***
AZELAIC ACID DIHEXYL ESTER
mf: $C_{21}H_{40}O_4$ mw: 356.61

SYNS: DI-N-HEXYL AZELATE ◇ NONANEDIOTIC ACID, DIHEXYL ESTER

TOXICITY DATA with REFERENCE
orl-rat LD50:16 g/kg AIHAAP 30,470,69
orl-mus LDLo:15000 mg/kg 34ZIAG -,226,69
orl-rbt LDLo:1000 mg/kg 34ZIAG -,226,69
orl-gpg LDLo:6000 mg/kg TXAPA9 4,247,62

CONSENSUS REPORTS: Reported in EPA TSCA Inventory.

SAFETY PROFILE: Moderately toxic by ingestion. See also ESTERS. When heated to decomposition it emits acrid smoke and irritating fumes.

ASC125 CAS:58581-89-8 ***HR: 3***
AZELASTINE
mf: $C_{22}H_{24}ClN_3O$ mw: 381.94

SYN: 4-(p-CHLOROBENZYL)-2-(HEXAHYDRO-1-METHYL-1H-AZEPIN-4-YL)-1-(2H)-PHTHALAZINONE

TOXICITY DATA with REFERENCE
orl-rat LD50:310 mg/kg ARZNAD 31,1184,81
ipr-rat LD50:43200 μg/kg ARZNAD 31,1184,81
scu-rat LD50:59600 μg/kg ARZNAD 31,1184,81
ivn-rat LD50:26900 μg/kg ARZNAD 31,1184,81
orl-mus LD50:124 mg/kg ARZNAD 31,1184,81
ipr-mus LD50:42800 μg/kg ARZNAD 31,1184,81
scu-mus LD50:54200 μg/kg ARZNAD 31,1184,81
ivn-mus LD50:35500 μg/kg ARZNAD 31,1184,81
orl-dog LD50:51300 μg/kg ARZNAD 31,1184,81
ivn-dog LD50:13700 μg/kg ARZNAD 31,1184,81

SAFETY PROFILE: Poison by ingestion, subcutaneous, intravenous, and intraperitoneal routes. Experimental reproductive effects. When heated to decomposition it emits toxic fumes of NO_x and Cl^-.

ASC130 CAS:79307-93-0 ***HR: 3***
AZELASTINE HYDROCHLORIDE
mf: $C_{22}H_{24}ClN_3O \cdot ClH$ mw: 418.40

SYNS: AZELASTIN ◇ 4-(p-CHLOROBENZYL)-2-(HEXAHYDRO-1-METHYL-1H-AZEPIN-4-YL)-1-(2H)-PHTHALAZINONEHCl

TOXICITY DATA with REFERENCE
orl-rat TDLo:755 mg/kg (7-17D preg):TER ARZNAD 31,1225,81
orl-rbt TDLo:390 mg/kg (6-18D preg):REP ARZNAD 31,1225,81
orl-rat LD50:580 mg/kg KSRNAM 20,5231,86
ivn-rat LD50:24600 μg/kg KSRNAM 20,5231,86
orl-mus LD50:143 mg/kg KSRNAM 20,5231,86

ivn-mus LD50:25400 μg/kg KSRNAM 20,5231,86
orl-dog LD50:107 mg/kg KSRNAM 20,5235,86

SAFETY PROFILE: Poison by ingestion and intravenous routes. An experimental teratogen. Other experimental reproductive effects. When heated to decomposition it emits toxic fumes of NO_x and HCl.

ASC250 CAS:24853-80-3 ***HR: 3***
AZEPHEN
mf: $C_{16}H_{19}N_4O \cdot ClH$ mw: 333.86

SYNS: 5-METHYL-3-(4-METHYL-1-PIPERAZINYL)-5H-PYRIDAZINO(3,4-b)(1,4)BENZOXAZINE HYDROCHLORIDE ◇ 2-(4-METHYL-1-PIPERAZINYL)-10-METHYL-3,4-DIAZAPHENOXAZINDIHYDROCHLORID(GERMAN)

TOXICITY DATA with REFERENCE
orl-rat LD50:1 g/kg ZPPLBF 114,787,75
scu-rat LD50:490 mg/kg ZPPLBF 114,787,75
ivn-rat LD50:48 mg/kg ZPPLBF 114,787,75
orl-mus LD50:700 mg/kg PCJOAU 4,118,70
scu-mus LD50:330 mg/kg ZPPLBF 114,787,75
ivn-mus LD50:63 mg/kg RPTOAN 37,2,74
orl-cat LD50:200 mg/kg ZPPLBF 114,787,75

SAFETY PROFILE: Poison by ingestion, intravenous, and subcutaneous routes. When heated to decomposition it emits very toxic fumes of HCl and NO_x. An antidepressant.

ASC500 CAS:2133-34-8 ***HR: 2***
l-2-AZETIDINECARBOXYLIC ACID
mf: $C_4H_7NO_2$ mw: 101.12

TOXICITY DATA with REFERENCE
ipr-mus TDLo:500 mg/kg (female 11D post):TER TJADAB 16,123,77
ipr-rat TDLo:300 mg/kg (female 8D post):REP SEIJBO 16,263,76
scu-mus LD50:1000 mg/kg 85GDA2 8(1),101,82

SAFETY PROFILE: Moderately toxic by subcutaneous route. An experimental teratogen. Other experimental reproductive effects. When heated to decomposition it emits toxic fumes of NO_x.

ASC750 ***HR: 3***
AZIDES

SAFETY PROFILE: Variable toxicity. Many azides are poisonous, cause a fall in blood pressure and some inhibit enzyme action, thus resembling nitrites and cyanides. An azide is a compound of hydrogen or a metal ion and the monovalent $-N_3$ radical. All of its salts and the acid are unstable and some decompose explosively; although lead azide, which is one of the most important azides, is not very sensitive. Dangerous; shock and heat will explode it. When heated to decomposition it emits highly toxic fumes. If exposed to CS_2, it forms violently explosive salts. Organic azides are sensitized by metal salts or traces of strong acid. (See also specific compound).

ASD000 CAS:78-57-9 ***HR: 2***
AZIDITHION
mf: $C_6H_{12}N_5O_2PS_2$ mw: 281.32

SYNS: 2,4-DIAMINO-6-DIMETHOXYPHOSPHINOTHIONYLTHIOMETHYL-s-TRIAZINE ◇ 4,6-DIAMINO-s-TRIAZINE-2-METHANETHIOL S-ESTER with O,O-DIMETHYLPHOSPHORODITHIOATE ◇ S-(4,6-DIAMINO-1,3,5-TRIAZIN-2-YL)-METHYL)-O,O-DIMETHYL-DITHIOFOSFAAT (DUTCH) ◇ S-(4,6-DIAMINO-1,3,5-TRIAZIN-2-YL)-METHYL)-O,O-DIMETHYL-DITHIOPHOSPHAT(GERMAN) ◇ 4,6-DIAMINO-1,3,5-TRIAZIN-2-YLMETHYL-O,O-DIMETHYL PHOSPHORODITHIOATE ◇ S-((4,6-DIAMINO-s-TRIAZIN-2-YL) METHYL)-O,O-DIMETHYL PHOSPHORODITHIOATE ◇ S-(4,6-DIAMINO-1,3,5-TRIAZIN-2-YLMETHYL)-O,O-DIMETHYLPHOSPHORODITHIOATE ◇ S-(4,6-DIAMINO-1,3,5-TRIAZIN-2-YLMETHYL) DIMETHYL PHOSPHOROTHIOLOTHIONATE ◇ 2-DIMETHOXYPHOSPHINOTHIOYLTHIOMETHYL-4,6-DIAMINO-s-TRIAZINE ◇ O,O-DIMETHYL-S-(4,6-DIAMINO-1,3,5-TRIAZINYL-2-METHYL) DITHIOPHOSPHATE ◇ O,O-DIMETHYL- S-(4,6-DIAMINO-s-TRIAZIN-2-YLMETHYL)PHOSPHORODITHIOATE ◇ O,O-DIMETHYL-S-(4,6-DIAMINO-1,3,5-TRIAZIN-2-YL)METHYLPHOSPHORODITHIOATE ◇ O,O-DIMETHYL-S-(4,6-DIAMINO-1,3,5-TRIAZIN-2-YL)METHYL PHOSPHOROTHIOLOTHIONATE ◇ DITHIOPHOSPHATE de O,O-DIMETHYLE et de S-(4,6-DIAMINO-1,3,5-TRIAZINE-2-YL)-METHYLE) (FRENCH) ◇ ENT 25,760 ◇ MENAZON ◇ PP175 ◇ R 15,175 ◇ SAIPHOS ◇ SAPHICOL ◇ SAPHIZON ◇ SAPHIZON-DP ◇ SAPHOS ◇ SAYFOR ◇ SAYFOS ◇ SAYPHOS ◇ SYPHOS

TOXICITY DATA with REFERENCE
mmo-smc 5 ppm RSTUDV 6,161,76
orl-rat LD50:890 mg/kg ARSIM* 20,14,66
orl-mus LD50:427 mg/kg SPEADM 78-1,48,78
orl-ckn LD50:487 mg/kg TXAPA9 7,606,65
unr-mam LD50:900 mg/kg 30ZDA9 -,373,71

CONSENSUS REPORTS: EPA Genetic Toxicology Program.

SAFETY PROFILE: Moderately toxic by ingestion and possibly other routes. Mutation data reported. When heated to decomposition it emits very toxic fumes of NO_x, PO_x, and SO_x.

ASD375 CAS:18523-48-3 ***HR: 3***
AZIDOACETIC ACID
mf: $C_2H_3N_3O_2$ mw: 101.06

SAFETY PROFILE: The acid in contact with iron or iron salts undergoes rapid exothermic decomposition at 25°C and explodes at 90°C. Upon decomposition it emits toxic fumes of NO_x. See also AZIDES.

ASD500 CAS:4504-27-2 ***HR: 3***
AZIDOACETONE
mf: $C_3H_5N_3O$ mw: 99.10

SAFETY PROFILE: A dangerous fire and storage hazard. It can explode in the dark or when heated. When heated to decomposition it emits toxic fumes of NO_x. See also AZIDES.

ASE000 CAS:57707-64-9 ***HR: 3***
AZIDOACETO NITRILE
mf: $C_2H_2N_4$ mw: 82.10

CONSENSUS REPORTS: Cyanide and its compounds are on the Community Right-To-Know List.

SAFETY PROFILE: An unstable explosive sensitive to impact and heat. When heated to decomposition it emits toxic fumes of NO_x and CN^-. See also NITRILES and AZIDES.

ASE250 ***HR: 3***
N-AZIDO CARBONYL AZEPINE
mf: $C_7H_6N_4O$ mw: 162.20

SAFETY PROFILE: Highly unstable, explosive compound. See also azides. Can explode on distillation. When heated to decomposition it emits toxic fumes of NO_x.

ASE500 CAS:54567-24-7 ***HR: 3***
AZIDOCARBONYL GUANIDINE
mf: $C_2H_4N_6O$ mw: 128.10

SAFETY PROFILE: Violently explosive on rapid heating. When heated to decomposition it emits toxic fumes of NO_x. See also AZIDES.

ASE875 CAS:22958-08-3 ***HR: 3***
AZIDOCODEINE
mf: $C_{18}H_{22}N_4O_2$ mw: 326.44

SYN: 6-DEOXY-6-AZIDODIHYDROISOCODEINE

TOXICITY DATA with REFERENCE
orl-rat LD50:120 mg/kg JPPMAB 25,929,73
scu-rat LD50:125 mg/kg JPPMAB 25,929,73
ivn-rat LD50:52 mg/kg JPPMAB 25,929,73

SAFETY PROFILE: Poison by ingestion, subcutaneous, and intravenous routes. When heated to decomposition it emits toxic fumes of NO_x. See also CODEINE and AZIDES.

ASF500 ***HR: 3***
AZIDODIMETHYL BORANE
mf: $C_2H_6BN_3$ mw: 82.9

SAFETY PROFILE: Very unstable. Explodes when heated. When heated to decomposition it emits toxic fumes of NO_x. See also BORANES, BORON COMPOUNDS, and AZIDES.

ASF625 CAS:70664-49-2 ***HR: 3***
2-AZIDO-3,5-DINITROFURAN
mf: $C_4HN_5O_5$ mw: 199.08

SAFETY PROFILE: A heat- and impact-sensitive explosive. Upon decomposition it emits toxic fumes of NO_x. See also AZIDES.

ASF750 CAS:4472-06-4 ***HR: 3***
AZIDODITHIOFORMIC ACID
mf: CHN_3S_2 mw: 119.20

DOT Classification: Forbidden

SAFETY PROFILE: Very unstable. The acid and its salts are shock- and heat-sensitive explosives. Upon decomposition it emits very toxic fumes of NO_x and SO_x. See also AZIDES.

ASG250 CAS:4658-28-0 ***HR: 2***
2-AZIDO-4-ISOPROPYLAMINO-6-METHYLTHIO-s-TRIAZINE
mf: $C_7H_{11}N_7S$ mw: 225.31

SYNS: 2-AZIDO-4-ISOPROPYLAMINO-6-METHYLTHIO-1,3,5-TRIAZINE ◇ 4-AZIDO-N-(1-METHYLETHYL)-6-(METHYLTHIO)-1,3,5-TRIAZIN-2-AMINE ◇ 4-AZIDO-N-(1-METHYLETHYL)-6-(METHYLTHIO)-1,3,5-TRIAZIN-2-AMINI ◇ AZIPROTRYN ◇ AZIPROTRYNE ◇ AZIRPOTRYNE ◇ BRASORAN ◇ C 7019 ◇ CIBA C 7019 ◇ ISOPROPYLAMINO-4-AZIDO-6-METHYLTHIO-1,3,5-TRIAZIN (GERMAN) ◇ MESORANIL ◇ MEZARONIL ◇ MEZURON

TOXICITY DATA with REFERENCE
orl-rat LD50:3600 mg/kg FMCHA2 -,C150,830
orl-rbt LD50:1800 mg/kg GUCHAZ 6,26,73

SAFETY PROFILE: Moderately toxic by ingestion. When heated to decomposition it emits very toxic fumes of NO_x and SO_x.

ASG500 ***HR: 3***
N-AZIDO METHYL AMINE
mf: $C_2H_6N_4$ mw: 86.1

SAFETY PROFILE: Very unstable, explosive compound. When heated to decomposition it emits toxic fumes of NO_x. See also AZIDES, AMINES and EXPLOSIVES. See also AZIDES and AMINES.

ASG625 CAS:59327-98-9 ***HR: 3***
2-AZIDOMETHYLBENZENEDIAZONIUM TETRAFLUOROBORATE
mf: $C_7H_6BF_4N_5$ mw: 246.96

SAFETY PROFILE: Explosive reaction with trichloroacetonitrile. When heated to decomposition it emits toxic fumes of F^- and NO_x. See also AZIDES and BORON COMPOUNDS.

ASG675 CAS:22952-87-0 ***HR: 3***
AZIDOMORPHINE
mf: $C_{17}H_{20}N_4O_2$ mw: 312.41

SYN: 6-DEOXY-6-AZIDODIHYDROISOMORPHINE

TOXICITY DATA with REFERENCE
orl-rat LD50:62 mg/kg JPPMAB 25,929,73
scu-rat LD50:13 mg/kg JPPMAB 25,929,73
ivn-rat LD50:8100 μg/kg JPPMAB 25,929,73
orl-mus LD50:58 mg/kg JPPMAB 27,99,75
scu-mus LD50:16 mg/kg JPPMAB 27,99,75
ivn-mus LD50:13 mg/kg JPPMAB 27,99,75

SAFETY PROFILE: Poison by ingestion, subcutaneous and intravenous routes. When heated to decomposition it emits toxic fumes of NO_x. See also AZIDES and MORPHINE.

ASH000 CAS:35038-46-1 ***HR: 3***
5-AZIDOTETRAZOLE
mf: CHN_7 mw: 111.10

SAFETY PROFILE: An unstable explosive. Explodes on contact with acetic acid. Its sodium, potassium, and silver salts are impact- and friction-sensitive explosives. The ammonium salt is a heat sensitive explosive. When heated to decomposition it emits toxic fumes of NO_x. See also AZIDES.

ASH250 ***HR: 3***
3-AZIDO-1,2,4-TRIAZOLE
mf: $C_2H_2N_6$ mw: 110.1

SAFETY PROFILE: Very unstable, explosive compound. Very sensitive to heat. Samples have exploded during analytical combustion. When heated to decomposition it emits toxic fumes of NO_x. See AZIDES.

ASH375 ***HR: D***
2,2'-AZINOBIS(3-ETHYL-7-BENZOTHIAZOLINESULFONIC ACID, DIAMMONIUM SALT
mf: $C_{18}H_{16}N_4O_6S_4 \cdot 2H_4N$ mw: 548.72

SYNS: ABTS ◇ 2,2'-AZINO-DI(3-ETHYL-BENZTHIAZOLINE SULPHONIC ACID (6)), AMMONIUM SALT

TOXICITY DATA with REFERENCE
mmo-sat 100 mg/L JIMMBG 36,55,80
mma-sat 500 mg/L JIMMBG 36,55,80
mmo-klp 11 g/L JIMMBG 36,55,80

SAFETY PROFILE: Mutation data reported. When heated to decomposition it emits toxic fumes of SO_x and NO_x. See also SULFONATES.

ASH425 ***HR: 3***
AZINOTHRICIN

SYN: α-ETHYL-6-(3-ETHYL-1,5-DIMETHYL-4-OXO-1,5-HEPTADIENYL)-N-(1,8,14,15,18,21,27-HEPTAAZA-21-HYDROXY-7-(1-HYDROXYETHYL-2,6,9.16,19,22-HEXAOXO-4-ISOPROPYL-20-(METHOXYMETHYL)-17,18-DIMETHYL-5-OXATRICYCLO(21.4.0.0(sup 10,15)HEPTACOSAN-3-6L)TETRAHYDRO-α-2-DIHYDROXY-5-METHYL-2H-PYRAN-2-ACETAMIDE

TOXICITY DATA with REFERENCE
ipr-mus LD50:3200 μg/kg JANTAJ 39,17,86
scu-mus LD50:420 mg/kg JANTAJ 39,17,86
ivn-mus LD50:10 mg/kg JANTAJ 39,17,86

SAFETY PROFILE: Poison by intravenous and intraperitoneal routes. Moderately toxic by other routes. When heated to decomposition it emits toxic fumes of NO_x.

ASH500 CAS:86-50-0 ***HR: 3***
AZINPHOS METHYL
DOT: NA 2783
mf: $C_{10}H_{12}N_3O_3PS_2$ mw: 317.34

PROP: Crystals or brown, waxy solid. D: 1.44, mp: 74°. Sltly sol in water; sol in organic solvents.

SYNS: AZINFOS-METHYL (DUTCH) ◇ AZINPHOS-METILE (ITALIAN) ◇ AZINPHOS METHYL, liquid (DOT) ◇ BAY 9027 ◇ BAYER 17147 ◇ BENZOTRIAZINE derivative of a METHYL DITHIOPHOSPHATE ◇ BENZOTRIAZINEDITHIOPHOSPHORIC ACID DIMETHOXY ESTER ◇ CARFENE ◇ COTNION METHYL ◇ CRYSTHION 2L ◇ CRYSTHYON ◇ DBD ◇ S-(3,4-DIHYDRO-4-OXO-BENZO(α)(1,2,3)TRIAZIN-3-YLMETHYL)-O,O-DIMETHYL PHOSPHORODITHIOATE ◇ S-(3,4-DIHYDRO-4-OXO-1,2,3-BENZOTRIAZIN-3-YLMETHYL)-O,O-DIMETHYL PHOSPHORODITHIOATE ◇ O,O-DIMETHYL-S-(BENZAZIMINOMETHYL) DITHIOPHOSPHATE ◇ O,O-DIMETHYL-S-(1,2,3-BENZOTRIAZINYL-4-KETO)METHYLPHOSPHORODITHIOATE ◇ O,O-DIMETHYL-S-(3,4-DIHYDRO-4-KETO-1,2,3-BENZOTRIAZINYL-3-METHYL) DITHIOPHOSPHATE ◇ DIMETHYLDITHIOPHOSPHORIC-ACID N-METHYLBENZAZIMIDE ESTER ◇ O,O-DIMETHYL-S-(4-OXO-3H-1,2,3-BENZOTRIZIANE-3-METHYL)PHOSPHORODITHIOATE ◇ O,O-DIMETHYL-S-(4-OXOBENZOTRIAZINO-3-METHYL)PHOSPHORODITHIOATE ◇ O,O-DIMETHYL-S-(4-OXO-1,2,3-BENZOTRIAZINO(3)-METHYL) THIOTHIONOPHOSPHATE ◇ O,O-DIMETHYL-S-((4-OXO-3H-1,2,3-BENZOTRIAZIN-3-YL)-METHYL)-DITHIOFOSFAAT (DUTCH) ◇ O,O-DIMETHYL-S-((4-OXO-3H-1,2,3-BENZOTRIAZIN-3-YL)-METHYL)-DITHIOPHOSPHAT(GERMAN) ◇ O,O-DIMETHYL-S-4-OXO-1,2,3-BENZOTRIAZIN-3(4H)-YLMETHYL PHOSPHORODITHIOATE ◇ O,O-DIMETIL-S-((4-OXO-3H-1,2,3-BENZOTRIAZIN-3-IL)-METIL)-DITIOFOSFATO (ITALIAN) ◇ ENT 23,233 ◇ GOTHNION ◇ GUSATHION ◇ GUTHION (DOT) ◇ GUTHION, liquid (DOT) ◇ 3-(MERCAPTOMETHYL)-1,2,3-BENZOTRIAZIN-4(3H)-ONE-O,O-DIMETHYL PHOSPHORODITHIOATE ◇ 3-(MERCAPTOMETHYL)-1,2,3-BENZOTRIAZIN-4(3H)-ONE-O,O-DIMETHYLPHOSPHORODITHIOATE-S-ESTER ◇ METHYLAZINPHOS ◇ N-METHYLBENZAZIMIDE, DIMETHYLDITHIOPHOSPHORIC ACID ESTER ◇ METHYL GUTHION ◇ METILTRIAZOTION ◇ NCI-C00066

TOXICITY DATA with REFERENCE
mmo-ssp 25 mmol/L MUREAV 117,139,83
mma-ssp 25 mmol/L MUREAV 117,139,83
cyt-hmn:lng 120 mg/L CNJGA8 17,455,75

cyt-hmn:oth 120 mg/L CNJGA8 17,455,75
orl-rat TDLo:190 mg/kg (6-22D preg/21D post):REP ARTODN 43,177,80
orl-mus TDLo:20 mg/kg (female 8D post):TER TCMUD8 5,3,85
orl-rat TDLo:5110 mg/kg/78W-C:ETA NCITR* NCI-CG-TR-69,78
orl-rat LD50:7 mg/kg JPPMAB 13,435,61
ihl-rat LC50:69 mg/m³/1H NTIS** PB277-077
skn-rat LD50:88 mg/kg 85JCAE -,1182,86
ipr-rat LD50:4900 μg/kg PSEBAA 114,509,63
ivn-rat LD50:7500 μg/kg NTIS** PB277-077
orl-mus LD50:15 mg/kg JPFCD2 15,867,80
skn-mus LD50:65 mg/kg KUMJAX 12,313,59

CONSENSUS REPORTS: NCI Carcinogenesis Bioassay (feed); Inadequate Studies: rat NCITR* NCI-CG-TR-69,78; No Evidence: mouse NCITR* NCI-CG-TR-69,78. EPA Genetic Toxicology Program. EPA Extremely Hazardous Substances List.

OSHA PEL: TWA 0.2 mg/m³ (skin)
ACGIH TLV: TWA 0.2 mg/m³ (skin)
DFG MAK: 0.2 mg/m³
DOT Classification: Poison B; Label: Poison, liquid mixture.

SAFETY PROFILE: Poison by inhalation, ingestion, skin contact, intravenous, and intraperitoneal routes. An experimental teratogen. Other experimental reproductive effects. Human mutation data reported. Questionable carcinogen with experimental tumorigenic data. See also PARATHION and ESTERS. When heated to decomposition it emits very toxic fumes of PO_x, SO_x, and NO_x.

ASH750 CAS:671-51-2 ***HR: 3***
AZIRIDINE CARBOXYLIC ACID ETHYL ESTER
mf: $C_5H_9NO_2$ mw: 115.15

SYNS: N-CARBETHOXYETHYLENIMINE ◇ N-(ETHOXYCARBONYL)AZIRIDINE ◇ N-ETHOXYCARBONYLETHYLENEIMINE ◇ ETHOXYCARBONYL-1-ETHYLENIMINE ◇ ETHYL AZIRIDINECARBOXYLATE ◇ ETHYL-1-AZIRIDINECARBOXYLATE ◇ ETHYL AZIRIDINOCARBOXYLATE ◇ ETHYL-1-AZIRIDINYLCARBOXYLATE ◇ ETHYL AZIRIDINYLFORMATE

TOXICITY DATA with REFERENCE
cyt-rat-ipr 40 mg/kg BJPCAL 6,357,51
ipr-rat LD50:10 mg/kg BCPCA6 14,915,65
ivn-mus LD50:180 mg/kg CSLNX* NX#04621

SAFETY PROFILE: Poison by intravenous and intraperitoneal routes. Mutation data reported. See also ESTERS. When heated to decomposition it emits toxic fumes of NO_x.

ASI000 CAS:1072-52-2 ***HR: 3***
1-AZIRIDINE ETHANOL
mf: C_4H_9NO mw: 87.14

SYNS: 2-(1-AZIRIDINYL)ETHANOL ◇ β-HYDROXY-1-ETHYLAZIRIDINE ◇ 2-HYDROXY-1-ETHYLAZIRIDINE ◇ N-(β-HYDROXYETHYL)AZIRIDINE ◇ N-(2-HYDROXYETHYL)AZIRIDINE ◇ N-HYDROXYETHYL ETHYLENE IMINE ◇ N-(2-HYDROXYETHYL)ETHYLENIMINE ◇ 1-(2-HYDROXYETHYL)ETHYLENIMINE

TOXICITY DATA with REFERENCE
skn-rbt 545 mg open MOD UCDS** 11/17/64
eye-rbt 1090 mg SEV UCDS** 11/17/64
scu-mus TDLo:900 mg/kg/75W-I:NEO JNCIAM 46,143,71
orl-rat LD50:74 mg/kg JIHTAB 31,60,49
ivn-mus LD50:56 mg/kg CSLNX* NX#03613
skn-rbt LD50:280 mg/kg JIHTAB 31,60,49

CONSENSUS REPORTS: IARC Cancer Review: Group 3 IMEMDT 7,56,87; Animal Limited Evidence IMEMDT 9,47,75. Reported in EPA TSCA Inventory.

SAFETY PROFILE: Poison by ingestion, skin contact, and intravenous routes. A skin and eye irritant. Questionable carcinogen with experimental neoplastigenic data. When heated to decomposition it emits toxic fumes of NO_x.

ASI250 CAS:1072-66-8 ***HR: 3***
1-AZIRIDINE PROPIONITRILE
mf: $C_5H_8N_2$ mw: 96.15

SYNS: 1-AZIRIDINEPROPANENITRILE ◇ N-(2-CYANOETHYL) AZIRIDINE ◇ 1-(2-CYANOETHYL)AZIRIDINE ◇ N-(β-CYANOETHYL) ETHYLENIMINE ◇ 1-(2-CYANOETHYL)ETHYLENIMINE

TOXICITY DATA with REFERENCE
cyt-rat-ipr 20 mg/kg BJPCAL 6,357,51
orl-rat LDLo:100 mg/kg NCNSA6 5,22,53

CONSENSUS REPORTS: Cyanide and its compounds are on the Community Right-To-Know List.

SAFETY PROFILE: Poison by ingestion. Mutation data reported. See also NITRILES. When heated to decomposition it emits toxic fumes of NO_x.

ASK500 CAS:1195-67-1 ***HR: D***
1-AZIRIDINYL-BIS(DIMETHYLAMINO)PHOSPHINE OXIDE
mf: $C_6H_{16}N_3OP$ mw: 177.22

SYN: p-1-AZIRIDINYL-N,N,N',N'-TETRAMETHYLAMINO-PHOSPHINE OXIDE ◇ p-1-AZIRIDINYL-N,N,N',N'-TETRAMETHYLPHOSPHINIC DIAMIDE ◇ DI-N,N'-DIMETHYLAMIDE ETHYLENEIMIDO PHOSPHATE ◇ ENT 50,991

TOXICITY DATA with REFERENCE
pic-esc 230 μmol/L HEREAY 68,245,71

cyt-oin-par 24 mmol/L CNJGA8 11,648,69
cyt-ham:lng 670 μmol/L HEREAY 68,255,71

CONSENSUS REPORTS: EPA Genetic Toxicology Program.

SAFETY PROFILE: Mutation data reported. When heated to decomposition it emits very toxic fumes of PO_x and NO_x.

ASK875 CAS:57998-68-2 ***HR: 3***
AZIRIDINYLQUINONE
mf: $C_{16}H_{20}N_4O_6$ mw: 364.40

SYNS: AZQ ◇ 2,5-BIS(1-AZIRIDINYL)-3,6-DIOXO-1,4-CYCLOHEXADIENE-1,4-DICARBAMIC ACID DIETHYL ESTER ◇ DIAZIQUONE ◇ 3,6-DIAZIRIDINYL-2,5-BIS(CARBOETHOXYAMINO)-1,4-BENZOQUINONE ◇ NSC 182986 ◇ USAN-DIAZIQUONE

TOXICITY DATA with REFERENCE
mmo-sat 10 μg/plate TCMUD8 5,319,85
dnd-omi 1 μmol/L PAACA3 24,322,83
dnd-hmn:emb 50 μmol/L PAACA3 24,246,83
dnd-hmn:oth 75 μmol/L CNREA8 44,4447,84
dnd-mus:leu 57 μmol/L PAACA3 24,246,83
dnd-ham:ovr 50 μmol/L CNREA8 44,5634,84
dns-ham:ovr 300 nmol/L PAACA3 24,322,83
ipr-mus LD50:11290 μg/kg NICSP* JAN86
ivn-mus LD50:10300 μg/kg NTIS** PB80-177934

SAFETY PROFILE: Poison by intravenous and intraperitoneal routes. Human mutation data reported. When heated to decomposition it emits toxic fumes of NO_x. See also CARBAMATES and ESTERS.

ASL250 CAS:103-33-3 ***HR: 3***
AZOBENZENE
mf: $C_{12}H_{10}N_2$ mw: 182.23

PROP: Orange, monoclinic crystals. Mp: 68°, bp: 297°, d: 1.203 @ 20°/4°, vap press: 1 mm @ 103.5°. Insol in water. Solubility in alc = 4.2/100 @ 20° in ether (ligroin) = 12/100 @ 20°.

SYNS: AZOBENZEEN (DUTCH) ◇ AZOBENZIDE ◇ AZOBENZOL ◇ AZOBISBENZENE ◇ AZODIBENZENE ◇ AZODIBENZENEAZOFUME ◇ BENZENEAZOBENZENE ◇ DIAZOBENZENE ◇ DIPHENYLDIAZENE ◇ 1,2-DIPHENYLDIAZENE ◇ DIPHENYLDIIMIDE ◇ ENT 14,611 ◇ NCI-C02926 ◇ USAF EK-704

TOXICITY DATA with REFERENCE
mma-sat 40 μg/plate PNASA6 72,5135,75
orl-rat TDLo:7350 mg/kg/2Y-C:CAR NCITR* NCI-CG-TR-154,79
scu-rat TDLo:17 g/kg/2Y-I:ETA CANCAR 3,789,50
orl-mus TDLo:300 mg/kg/8W-I:NEO TXAPA9 82,19,86
orl-rat TD:15 g/kg/2Y-C:CAR NCITR* NCI-CG-TR-154,79
orl-rat LD50:1000 mg/kg ARSIM* 20,2,66
ipr-mus LD50:500 mg/kg NTIS** AD277-689
unk-mam LDLo:50 mg/kg BESAAT 12,117,66

CONSENSUS REPORTS: IARC Cancer Review: Group 3 IMEMDT 7,56,87; Animal Limited Evidence IMEMDT 8,75,75; NCI Carcinogenesis Bioassay (feed); Clear Evidence: rat NCITR* NCI-CG-TR-154,79; No Evidence: mouse NCITR* NCI-CG-TR-154,79. Reported in EPA TSCA Inventory.

SAFETY PROFILE: Moderately toxic by ingestion and possibly other routes. Questionable carcinogen with experimental carcinogenic, neoplastigenic, and tumorigenic data. When heated to decomposition it emits toxic fumes of NO_x.

ASL750 CAS:78-67-1 ***HR: 3***
AZOBISISOBUTYLONITRILE
DOT: UN 2952
mf: $C_8H_{12}N_4$ mw: 164.24

SYNS: ACETO AZIB ◇ AIBN ◇ α,α′-AZOBISISOBUTYLONITRILE ◇ AZOBISISOBUTYRONITRILE ◇ 2,2′-AZOBIS(ISOBUTYRONITRILE) ◇ 2,2′-AZOBIS(2-METHYLPROPIONITRILE) ◇ AZODIISOBUTYRONITRILE ◇ α,α′-AZODIISOBUTYRONITRILE ◇ 2,2′-AZODIISOBUTYRONITRILE ◇ AZODIISOBUTYRONITRILE (DOT) ◇ 2,2′-DICYANO-2,2′-AZOPROPANE ◇ POLY-ZOLE AZDN ◇ POROFOR 57 ◇ VAZO 64

TOXICITY DATA with REFERENCE
orl-rat LDLo:670 mg/kg 34ZIAG 0,117,69
orl-mus LD50:700 mg/kg MEIEDD 10,132,83
ipr-mus LD50:25 mg/kg NTIS** AD691-490

CONSENSUS REPORTS: Cyanide and its compounds are on the Community Right-To-Know List. Reported in EPA TSCA Inventory.

DOT Classification: Flammable Solid; Label: Flammable Solid

SAFETY PROFILE: Poison by intraperitoneal route. Moderately toxic by ingestion. Easily oxidized, unstable. Violent exothermic decomposition when heated. Solution in acetone may decompose explosively. Explodes when heated with heptane. When heated to decomposition it emits toxic fumes of NO_x and CN^-. See also NITRILES. A free radical generator.

ASM000 CAS:64037-73-6 ***HR: 3***
AZOBIS ISOBUTYRAMIDE HYDROCHLORIDE
mf: $C_{10}H_{22}N_6 \cdot ClH$ mw: 262.84

TOXICITY DATA with REFERENCE
orl-rat LDLo:400 mg/kg KODAK* -,-,71
ipr-rat LD50:200 mg/kg KODAK* -,-,71

SAFETY PROFILE: Poison by ingestion and intraperitoneal routes. When heated to decomposition it emits very toxic fumes of NO_x and HCl.

ASM100 CAS:6420-06-0 ***HR: D***
AZO BLUE
mf: $C_{34}H_{26}N_4O_8S_2 \cdot 2Na$ mw: 728.74

SYNS: C.I. 23685 ◇ C.I. DIRECT VIOLET 28 ◇ C.I. DIRECT VIOLET 28, DISODIUM SALT ◇ 1-NAPHTHALENESULFONIC ACID, 4-HYDROXY-3-((4′-((1-HYDROXY-5-SULFO-2-NAPHTHALENYL)AZO)-3,3′-DIMETHYL(1,1′-BIPHENYL)-4-YL)AZO)-, DISODIUM SALT

TOXICITY DATA with REFERENCE
scu-rat TDLo:150 mg/kg (female 7-9D post):TER ANREAK 118,369,54

SAFETY PROFILE: An experimental teratogen. When heated to decomposition it emits toxic fumes of NO_x, SO_x, and NaO_2

ASM250 ***HR: 3***
AZOCHLORAMIDE
mf: $C_2H_4Cl_2N_6$ mw: 183.0

PROP: Bright yellow crystals. Mp: explodes @ 155°.

SYN: CHLOROZODIN

SAFETY PROFILE: Mild allergen. Severe explosion hazard when shocked or exposed to heat; explodes at 155° or more, particularly in the presence of metals. Incompatible with metals. Moderately dangerous; when heated to decomposition it emits toxic fumes of NO_x, Cl^-, and may explode.

ASM270 CAS:123-77-3 ***HR: D***
AZODICARBAMIDE
mf: $C_2H_4N_4O_2$ mw: 116.10

PROP: Orange-red powder. Decomp @ 180-200°. Very sltly sol in hot water; insol in alc. Decomp in hot HCl.

SYNS: 1,1′-AZOBISCARBAMIDE ◇ AZOBISCARBONAMIDE ◇ AZOBISCARBOXAMIDE ◇ 1,1′-AZOBIS(FORMAMIDE) ◇ AZODICARBOAMIDE ◇ AZODICARBONAMIDE ◇ AZODICARBOXAMIDE ◇ AZODICARBOXYLIC ACID DIAMIDE ◇ Δ(1,1′)-BIUREA ◇ CELOSEN AZ ◇ ChKhZ 21 ◇ ChKhZ 21R ◇ DIAZENEDICARBOXAMIDE ◇ GENITRON AC ◇ GENITRON AC 2 ◇ GENITRON AC 4 ◇ KEMPORE ◇ KEMPORE 125 ◇ KEMPORE R 125 ◇ LUCEL ADA ◇ NCI-C55981 ◇ NITROPORE ◇ PINHOLE AK 2 ◇ POROFOR 505 ◇ POROFOR ADC/R ◇ POROFOR ChKhZ 21 ◇ POROFOR ChKhZ 21R ◇ UNIFOAM AZ ◇ UNIFORM AZ ◇ YUNIHOMU AZ

TOXICITY DATA with REFERENCE
mmo-sat 100 μg/plate ENMUDM 8(Suppl 7),1,86
mma-sat 333 μg/plate ENMUDM 8(Suppl 7),1,86

CONSENSUS REPORTS: Reported in EPA TSCA Inventory.

SAFETY PROFILE: Mutation data reported. When heated to decomposition it emits toxic fumes of NO_x.

ASM300 CAS:123-77-3 ***HR: 3***
AZODICARBONAMIDE
mf: $C_2H_4N_4O_2$ mw: 116.08

PROP: Yellow to orange-red crystalline powder. Mp: above 180° (decomp). Sltly sol in dimethyl sulfoxide; insol in water, organic solvents.

SAFETY PROFILE: Flammable solid. When heated to decomposition emits toxic fumes of NO_x.

ASN000 ***HR: 3***
"AZODRIN"
mf: $C_6H_{14}O_5NP$ mw: 211.2

PROP: Reddish-brown solid, mild ester odor. Bp: 125°.

SYN: MONOCROTOPHOS

SAFETY PROFILE: Poison by ingestion and skin contact. See also ESTERS. A dangerous fire hazard. When heated to decomposition it evolves highly toxic fumes of NO_x and PO_x.

ASN250 CAS:821-14-7 ***HR: 3***
AZO ETHANE
mf: $C_4H_{10}N_2$ mw: 86.16

SYN: AZOAETHAN (GERMAN)

TOXICITY DATA with REFERENCE
ihl-rat TCLo:1250 mg/kg/1H (female 15D post):TER IARCCD 4,45,73
ihl-rat TDLo:37 mg/kg (22D preg):NEO,TER XENOBH 3,271,73
ihl-rat TCLo:37 mg/kg/1H (22D post):CAR IARCCD 4,45,73
scu-rat TDLo:1250 mg/kg/26W-I:ETA ZEKBAI 67,31,65
ihl-rat TC:4000 ppm/1H (15D post):NEO,TER FCTXAV 6,584,68
ihl-rat TC:300 mg/kg/1H (15D post):NEO,TER EXPEAM 24,561,68
scu-rat LDLo:2200 mg/kg ZEKBAI 67,31,65

SAFETY PROFILE: Moderate acute toxicity. An experimental teratogen. Questionable carcinogen with experimental carcinogenic and tumorigenic data. When heated to decomposition it emits toxic fumes of NO_x. An unstable, dangerously explosive material in concentrated state.

ASN375 ***HR: 3***
AZOFORMALDOXIME
mf: $C_2H_2N_4O_2$ mw: 114.06

PROP: Mp: 140°C.

SAFETY PROFILE: Explodes at its melting point. Upon decomposition it emits toxic fumes of NO_x.

ASN400 CAS:503-28-6 ***HR: 3***
AZOMETHANE
mf: $C_2H_6N_2$ mw: 58.08

SAFETY PROFILE: A heat-sensitive explosive. When heated to decomposition it emits toxic fumes of NO_x. See AZOETHANE.

ASN500 CAS:487-10-5 ***HR: 3***
1,1'-AZONAPHTHALENE
mf: $C_{20}H_{14}N_2$ mw: 282.35

PROP: Red needles from acetic acid. Mp: 190°, bp: subl > 190°. Insol in water; sol in acetic acid; very sltly sol in alc; very sol in benzene.

TOXICITY DATA with REFERENCE
scu-mus TDLo:6300 mg/kg/63W-I:ETA AJCAA7 40,62,40

SAFETY PROFILE: Questionable carcinogen with experimental tumorigenic data. When heated to decomposition it emits toxic fumes of NO_x.

ASN750 CAS:582-08-1 ***HR: 3***
2,2'-AZONAPHTHALENE
mf: $C_{20}H_{14}N_2$ mw: 282.35

PROP: Red prisms from chloroform. Mp: 204°, bp: subl @ 210°. Insol in water; sltly sol in alc; sol in benzene.

SYN: DI-β-NAPHTHYLDIIMIDE

TOXICITY DATA with REFERENCE
orl-mus TDLo:8400 mg/kg/42W-I:ETA AJCAA7 40,62,40

SAFETY PROFILE: Questionable carcinogen with experimental tumorigenic data. When heated to decomposition it emits toxic fumes of NO_x.

ASO375 CAS:27589-33-9 ***HR: 3***
AZOSEMIDE
mf: $C_{12}H_{11}ClN_6O_2S_2$ mw: 370.86

PROP: Crystals. Mp: 218-221°.

SYNS: 2-CHLORO-5-(1H-TETRAZOL-5-YL)-N(sup4)-2-THENYLSULFANILAMIDE ◇ DIURAPID ◇ PLE 1053

TOXICITY DATA with REFERENCE
orl-mus TDLo:12500 mg/kg (6-15D preg):TER SEIJBO 24,111,84
orl-rat LD50:2545 mg/kg IYKEDH 18,666,87
ipr-rat LD50:287 mg/kg IYKEDH 18,666,87
ivn-rat LD50:252 mg/kg IYKEDH 18,666,87
orl-mus LD50:6350 mg/kg IYKEDH 18,666,87
scu-mus LD50:762 mg/kg IYKEDH 18,666,87

SAFETY PROFILE: Poison by intraperitoneal and intravenous routes. Moderately toxic by ingestion and subcutaneous routes. An experimental teratogen. When heated to decomposition it emits toxic fumes of Cl^-, SO_x, and NO_x.

ASO501 CAS:7644-67-9 ***HR: 3***
AZOTOMYCIN
mf: $C_{17}H_{23}N_7O_8$ mw: 453.47

SYNS: ANTIBIOTIC 1719 ◇ DUAZOMYCIN B

TOXICITY DATA with REFERENCE
ivn-hmn TDLo:192 mg/kg/4D:GIT CTRRDO 61,1719,77
orl-mus LDLo:105 mg/kg ANTBAL 18,332,73
ipr-mus LDLo:27 mg/kg ANTBAL 18,332,73
scu-mus LDLo:31 mg/kg ANTBAL 18,332,73
ivn-mus LDLo:99 mg/kg ANTBAL 18,332,73

SAFETY PROFILE: Poison by ingestion, intraperitoneal, subcutaneous, and intravenous routes. Human gastrointestinal tract effects by intravenous route. When heated to decomposition it emits toxic fumes of NO_x.

ASO510 ***HR: 3***
AZOTOMYCIN SODIUM
mf: $C_{17}H_{23}N_7O_8{\cdot}Na$ mw: 475.47

SYNS: ANTIBIOTIC 1719 SODIUM SALT ◇ AZOTOMYCIN ◇ AZOTOMYCIN SODIUM SALT ◇ N-(1-((1-CARBOXY-5-DIAZO-4-OXOPENTYL)CARBAMOYL)-5-DIAZO-4-OXOPENTYL)-GLUTAMINESODIUM SALT ◇ 6-DIAZO-2-(2-(4-AMINO-4-CARBOXYBUTYRAMIDO)-6-DIAZO-5-OXOHEXANAMIDO)-HEXANOIC ACID SODIUM ◇ DIAZOMYCIN B ◇ DUAZOMYCIN B ◇ NSC 56654 ◇ 1719 SODIUM ◇ SODIUM AZOTOMYCIN

TOXICITY DATA with REFERENCE
orl-mus LD50:388 mg/kg NCISP* JAN86
ipr-mus LD50:471 mg/kg NCISP* JAN86
scu-mus LD50:271 mg/kg NCISP* JAN86

SAFETY PROFILE: Poison by ingestion and subcutaneous routes. Moderately toxic by intraperitoneal route. When heated to decomposition it emits toxic fumes of NO_x and Na_2O.

ASO750 CAS:495-48-7 ***HR: 3***
AZOXYBENZENE
mf: $C_{12}H_{10}N_2O$ mw: 198.23

PROP: Yellow, rhombic crystals. D: 1.248 @ 20°/20°; mp: 36°; bp: decomp. Insol in water; solubility in alc = 11.4/100 @ 15°, solubility in ether (ligroin) = 43.5/100 @ 15°.

SYNS: AZOBENZENE OXIDE ◇ AZOSSIBENZENE (ITALIAN) ◇ AZOXYBENZEEN (DUTCH) ◇ AZOXYBENZIDE ◇ AZOXYBENZOL (GERMAN) ◇ AZOXYDIBENZENE ◇ ORDINARY AZOXYBENZENE

TOXICITY DATA with REFERENCE
skn-rbt 10 mg/24H MLD AMIHBC 10,61,54
eye-rbt 500 mg AMIHBC 10,61,54
orl-rat LD50:620 mg/kg AMIHBC 10,61,54

orl-mus LD50:515 mg/kg KYDKAJ 21,25,76
ipr-mus LDLo:500 mg/kg CBCCT* 7,389,55
skn-rbt LD50:1090 mg/kg AMIHBC 10,61,54
scu-rbt LDLo:250 mg/kg AEXPBL 35,401,1895

CONSENSUS REPORTS: Reported in EPA TSCA Inventory.

SAFETY PROFILE: Poison by subcutaneous route. Moderately toxic by ingestion, skin contact, and other routes. A skin and eye irritant. Combustible. When heated to decomposition it emits toxic fumes of NO_x.

ASP000 CAS:16301-26-1 ***HR: 3***
AZOXYETHANE
mf: $C_4H_{10}N_2O$ mw: 102.16

SYNS: AZOXYAETHAN (GERMAN) ◇ DIETHYLDIAZENE-1-OXIDE

TOXICITY DATA with REFERENCE
ivn-rat TDLo:25 mg/kg (10D preg):TER XENOBH 3,271,73
orl-rat TDLo:500 mg/kg/20W-I:ETA CNREA8 47,3968,87
scu-rat TDLo:30 mg/kg (11D post):NEO,TER NATWAY 60,555,73
ivn-rat TDLo:50 mg/kg (15D post):NEO,TER IARCCD 4,45,73
ivn-rat TD:50 mg/kg (15D post):NEO,TER XENOBH 3,271,73
scu-rat LD50:240 mg/kg XENOBH 3,271,73
ivn-rat LD50:210 mg/kg IARCCD 4,45,73

SAFETY PROFILE: Poison by subcutaneous and intravenous routes. An experimental teratogen. Questionable carcinogen with experimental carcinogenic and tumorigenic data. When heated to decomposition it emits toxic fumes of NO_x.

ASP250 CAS:25843-45-2 ***HR: 3***
AZOXYMETHANE
mf: $C_2H_6N_2O$ mw: 74.10

SYN: AOM

TOXICITY DATA with REFERENCE
mma-sat 13600 μmol/L/20M CNREA8 38,4585,78
sln-dmg-unk 1 mmol/L/3D-C DRISAA 50,138,73
dnd-rat-scu 30 mg/kg CNREA8 38,1589,78
ivn-rat TDLo:30 mg/kg (15D post):TER IARCCD 4,45,73
orl-rat TDLo:20 mg/kg/(22D:ETA XENOBH 3,271,73
scu-rat TDLo:3200 μg/kg:CAR VTPHAK 12,165,75
ivn-rat TDLo:20 mg/kg (22D post):CAR IARCCD 4,45,73
scu-rat LD50:27 mg/kg XENOBH 3,271,73

SAFETY PROFILE: Suspected carcinogen with experimental carcinogenic and tumorigenic data. Poison by subcutaneous route. An experimental teratogen. Mutation data reported. When heated to decomposition it emits toxic fumes of NO_x.

ASP500 CAS:17697-55-1 ***HR: 2***
1-AZOXYPROPANE
mf: $C_6H_{14}N_2O$ mw: 130.22

SYNS: 1,1'-AZOXYPROPANE ◇ DIPROPYLDIAZENE 1-OXIDE

TOXICITY DATA with REFERENCE
orl-rat TDLo:773 mg/kg/26W-I:CAR CRNGDP 8,1947,87

SAFETY PROFILE: Questionable carcinogen with experimental carcinogenic data. When heated to decomposition it emits toxic fumes of NO_x.

ASP510 ***HR: 2***
2-AZOXYPROPANE
mf: $C_6H_{14}N_2O$ mw: 130.22

SYNS: AZOXYISOPROPANE ◇ BIS(1-METHYLETHYL)DIAZENE 1-OXIDE

TOXICITY DATA with REFERENCE
orl-rat TDLo:773 mg/kg/26W I:CAR CRNGDP 8,1947,87

SAFETY PROFILE: Questionable carcinogen with experimental carcinogenic data. When heated to decomposition it emits toxic fumes of NO_x.

ASP750 CAS:6580-41-2 ***HR: 2***
AZULENO(5,6,7-cd)PHENALENE
mf: $C_{20}H_{12}$ mw: 252.32

TOXICITY DATA with REFERENCE
scu-mus TDLo:80 mg/kg/4W-I:ETA PAACA3 10,12,69

SAFETY PROFILE: Questionable carcinogen with experimental tumorigenic data. When heated to decomposition it emits acrid smoke and irritating fumes.

B

BAB250 CAS:37661-08-8 *HR: 3*
BACAMPICILLIN HYDROCHLORIDE
mf: $C_{21}H_{27}N_3O_7S{\cdot}ClH$ mw: 502.03

SYNS: AMBACAMP ◇ BACACIL ◇ BAPC ◇ BECAMPICILLIN ◇ CAMBAXIN ◇ PENGLOBE ◇ SPECTROBID

TOXICITY DATA with REFERENCE
ivn-rat LD50:176 mg/kg NIIRDN 6,575,82
orl-mus LD50:8529 mg/kg NKRZAZ 27(Suppl 4),17,79
ipr-mus LD50:176 mg/kg NKRZAZ 27(Suppl 4),17,79
scu-mus LD50:9475 mg/kg NKRZAZ 27(Suppl 4),17,79
ivn-mus LD50:184 mg/kg NKRZAZ 27(Suppl 4),17,79

SAFETY PROFILE: Poison by intraperitoneal and intravenous routes. Mildly toxic by ingestion and subcutaneous routes. When heated to decomposition it emits very toxic fumes of NO_x, SO_x, and HCl.

BAB625 CAS:70458-96-7 *HR: 3*
BACCIDAL
mf: $C_{16}H_{18}FN_3O_3$ mw: 319.37

PROP: Crystals from methylene chloride/methanol. Mp: 227-228°. Hygroscopic in air, forms a hemihydrate.

SYNS: AM-715 ◇ BARAZAN ◇ 1,4-DIHYDRO-1-ETHYL-6-FLUORO-4-OXO-7-(1-PIPERAZINYL)-3-QUINOLINECARBOXYLICACID ◇ 1-ETHYL-6-FLUORO-1,4-DIHYDRO-4-OXO-7-(1-PIPERAZINYL)-3-QUINOLINECARBOXYLIC ACID ◇ FLOXACIN ◇ FULGRAM ◇ MK-366 ◇ NORFLOXACIN ◇ NOROXIN ◇ SEBERCIM ◇ ZOROXIN

TOXICITY DATA with REFERENCE
dnr-bcs 62500 μg/L NKRZAZ 29(Suppl 4),938,81
orl-mus TDLo:1250 mg/kg (female 6-15D post):REP NKRZAZ 29(Suppl 4),895,81
orl-rbt TDLo:1300 mg/kg (female 6-18D post):TER FAATDF 7,272,86
orl-man TDLo:94 mg/kg/13D-I:MSK NZMJAX 96,590,83
ivn-rat LD50:245 mg/kg NKRZAZ 29(Suppl 4),766,81
orl-mus LD50:4 g/kg JMCMAR 30,2163,87
ivn-mus LD50:220 mg/kg NKRZAZ 29(Suppl 4),766,81
ims-mus LD50:470 mg/kg NKRZAZ 29(Suppl 4),766,81

SAFETY PROFILE: Poison by intravenous route. Moderately toxic by other routes. Human systemic effects by ingestion: musculoskeletal changes. An experimental teratogen. Other experimental reproductive effects. Mutation data reported. When heated to decomposition it emits toxic fumes of F^- and NO_x.

BAB650 *HR: 3*
BACILLUS CEREUS exo-ENTEROTOXIN

TOXICITY DATA with REFERENCE
ivn-mus LDLo:15 mg/kg APMBAY 29,201,75
par-mus LDLo:14 mg/kg BIORAK 38,113,73
ivn-rbt LDLo:3 mg/kg BEXBAN 73,78,72

SAFETY PROFILE: Poison by intravenous and parenteral routes.

BAB750 CAS:1395-21-7 *HR: 3*
BACILLUS SUBTILIS BPN

PROP: A commercial raw proteolytic enzyme used in laundry detergents (FCTXAV 7,581,69).

SYNS: BACILLOMYCIN (8CI, 9CI) ◇ BACILLOMYCIN R ◇ FUNGOCIN ◇ SUBTILISINS (ACGIH) ◇ SUBTILISINS BPN

TOXICITY DATA with REFERENCE
eye-rbt 3 mg SEV FCTXAV 7,581,69
ipr-mus LD50:75 mg/kg 85ERAY 3,1606,78

OSHA PEL: CL 0.00006 mg/m^3
ACGIH TLV: CL 0.00006 mg/m^3

SAFETY PROFILE: A poison via intraperitoneal route. A severe eye irritant. When heated to decomposition it emits toxic fumes of NO_x.

BAC000 CAS:9014-01-1 *HR: 3*
BACILLUS SUBTILIS CARLSBERG

PROP: A commercial raw proteolytic enzyme used in laundry detergents (FCTXAV 7,581,69).

SYNS: ALCALASE ◇ ALK-ENZYME ◇ BACILLOPEPTIDASE A ◇ BACILLOPEPTIDASE B ◇ BIOPRASE ◇ COLISTINASE ◇ E.C. 3.4.4.16 ◇ E.C. 3.4.21.14 ◇ MAXATASE ◇ NAGARSE ◇ SUBTILISIN (9CI, ACGIH) ◇ SUBTILISIN CARLSBURG ◇ SUBTILISIN NOVO ◇ SUBTILOPEPTIDASE A ◇ SUBTILOPEPTIDASE B ◇ SUBTILOPEPTIDASE BPN' ◇ SUBTILOPEPTIDASE C ◇ THERMOASE PC-10

TOXICITY DATA with REFERENCE
eye-rbt 3 mg MOD FCTXAV 7,581,69
orl-rat LD50:3700 mg/kg FCTXAV 7,581,69

CONSENSUS REPORTS: Reported in EPA TSCA Inventory.

ACGIH TLV: CL 0.00006 mg/m^3

SAFETY PROFILE: Moderately toxic by ingestion. An

eye irritant. When heated to decomposition it emits toxic fumes of NO_x.

BAC125 CAS:23526-02-5 **HR: D**
BACILLUS THURINGIENSIS EXOTOXIN

TOXICITY DATA with REFERENCE
sln-dmg-orl 1 pph HEREAY 85,113,77
cyt-hmn:leu 20 pph HEREAY 85,105,77
cyt-rat-orl 4500 g/kg/90D-C HEREAY 85,105,77

SAFETY PROFILE: Human mutation data reported.

BAC175 CAS:29393-20-2 **HR: 2**
BACILYSIN
mf: $C_{12}H_{18}N_2O_5$ mw: 270.32

SYNS: N-l-ALANYL-3-(5-OXO-7-OXABICYCLO(4.1.0)HEPT-2-YL)-l-ALANINE ◇ α-((2-AMINO-1-OXOPROPYL)AMINO)-5-OXO-7-OX-ABICYCLO(4.1.0)HEPTANE-2-PROPANOIC ACID ◇ α-(2-AMINO-PROPIONAMIDO)-5-OXO-7-OXABICYCLO(4.1.0)HEPTANE-2-PRO-PIONIC ACID ◇ ANTIBIOTIC KM 208 ◇ BACILLIN ◇ KM-208 ◇ TETAINE

TOXICITY DATA with REFERENCE
dni-hmn:hla 320 μmol/L BBACAQ 825,199,85
oms-hmn:hla 150 μmol/L BBACAQ 825,199,85
ivn-mus LD50:450 mg/kg 85GDA2 4(1),221,80

SAFETY PROFILE: Moderately toxic by intravenous route. Human mutation data reported. When heated to decomposition it emits toxic fumes of NO_x. An antibiotic.

BAC250 CAS:1405-87-4 **HR: 3**
BACITRACIN

PROP: White to pale buff, hygroscopic powder; odorless or slt odor. Freely sol in water, alc, methanol, and glacial acetic acid; insol in acetone, chloroform, and ether. When heated to decomposition it emits acrid smoke and irritating fumes.

SYNS: AYFIVIN ◇ BACIGUENT ◇ BACI-JEL ◇ BACILIQUIN ◇ BACITEK OINTMENT ◇ FORTRACIN ◇ PARENTRACIN ◇ PENITRACIN ◇ TOPITRACIN ◇ USAF CB-7 ◇ ZUTRACIN

TOXICITY DATA with REFERENCE
dnd-esc 5 μmol/L MUREAV 89,95,81
ipr-rat LD50:190 mg/kg PSEBAA 64,503,47
ipr-mus LD50:300 mg/kg NTIS** AD277-689
scu-mus LDLo:1300 mg/kg PSEBAA 64,503,47
ivn-mus LD50:360 mg/kg PSEBAA 64,503,47
orl-gpg LD50:2 g/kg ANTCAO 4,304,54

CONSENSUS REPORTS: Reported in EPA TSCA Inventory.

SAFETY PROFILE: A poison by intraperitoneal and intravenous routes. Moderately toxic by ingestion and subcutaneous routes. Mutation data reported.

BAC275 CAS:1134-47-0 **HR: 3**
BACLOFEN
mf: $C_{10}H_{12}ClNO_2$ mw: 213.68

PROP: Crystals from water. Mp: 206-208°.

SYNS: β-(AMINOMETHYL)-4-CHLOROBENZENEPROPANOICACID ◇ β-(AMINOMETHYL)-p-CHLOROHYDROCINNAMIC ACID ◇ Γ-AMINO-β-(p-CHLOROPHENYL)BUTYRIC ACID ◇ Ba 34647 ◇ BACLON ◇ C 34647Ba ◇ β-(p-CHLOROPHENYL)-Γ-AMINOBUTYRIC ACID ◇ β-(4-CHLOROPHENYL)GABA ◇ CIBA 34,647-Ba ◇ LIORESAL

TOXICITY DATA with REFERENCE
orl-wmn TDLo:18 mg/kg:CNS,PUL JTCTDW 22,11,84
orl-rat LD50:145 mg/kg NIIRDN 6,576,82
scu-rat LD50:115 mg/kg IYKEDH 11,181,80
ivn-rat LD50:78 mg/kg IYKEDH 11,181,80
orl-mus LD50:200 mg/kg NIIRDN 6,576,82
scu-mus LD50:103 mg/kg IYKEDH 11,181,80
ivn-mus LD50:31 mg/kg YKYUA6 31,871,80

SAFETY PROFILE: Poison by ingestion, subcutaneous and intravenous routes. Human systemic effects by ingestion: coma and respiratory depression. When heated to decomposition it emits toxic fumes of Cl^- and NO_x. A muscle relaxant.

BAC325 CAS:50846-45-2 **HR: D**
BACMECILLINAM
mf: $C_{20}H_{31}N_3O_6S$ mw: 441.54

SYNS: (2S-(2-α,5-α,6-β))-6-(((HEXAHYDRO-1H-AZEPIN-1-YL)METHYLENE)AMINO)-3,3-DIMETHYL-7-OXO-4-THIA-1-AZABICYCLO (3.2.0)HEPTANE-2-CARBOXYLIC ACID 1-((ETHOXYCARBONYL) OXY0ETHYL ESTER ◇ KW-1100

TOXICITY DATA with REFERENCE
orl-rat TDLo:5850 mg/kg (7-17D preg):REP NKRZAZ 33 (Suppl 3),78,85
orl-rat TDLo:5850 mg/kg (7-17D preg):TER NKRZAZ 33 (Suppl 3),78,85

SAFETY PROFILE: An experimental teratogen. Other experimental reproductive effects. When heated to decomposition it emits toxic fumes of SO_x and NO_x.

BAC750 CAS:25155-18-4 **HR: 1**
BACTINE

TOXICITY DATA with REFERENCE
orl-rat LD50 : 55 g/kg 34ZIAG -,118,69
orl-cat LD50:55 g/kg 34ZIAG -,118,69

SAFETY PROFILE: Mildly toxic by ingestion.

BAD000 CAS:64550-80-7 **HR: 3**
BA-10,11-DIOL-8,9-EPOXIDE-1
mf: $C_{18}H_{14}O_3$ mw: 278.32

SYN: 8,9,10,11-TETRAHYDRO-10,11-DIHYDROXY-8α,9-α-EPOXY-BENZ(a)ANTHRACENE

TOXICITY DATA with REFERENCE
skn-mus TDLo:22 mg/kg:ETA CNREA8 38,1699,78

SAFETY PROFILE: Questionable carcinogen with experimental tumorigenic data. When heated to decomposition it emits acrid smoke and irritating fumes.

BAD250 ***HR: 1***
BAGASSE DUST

SAFETY PROFILE: A nuisance dust from the fibrous residue of cane sugar manufacture. Inhalation can cause bronchial asthma, sneezing, rhinorrhea, pneumonitis, etc. See also COTTON DUST. Fire and explosion hazard when exposed to heat, flame, or oxidizers. See also DUST EXPLOSIONS.

BAD625 CAS:10309-37-2 ***HR: 3***
BAKUCHIOL
mf: $C_{18}H_{24}O$ mw: 256.42

TOXICITY DATA with REFERENCE
orl-mus LD50:2560 mg/kg MZHUDX 42,646,80
ipr-mus LD50:94 mg/kg MZHUDX 42,646,80
ivn-mus LD50:31 mg/kg MXHUDX 42,646,80

SAFETY PROFILE: Poison by intravenous and intraperitoneal routes. Moderately toxic by ingestion. When heated to decomposition it emits acrid smoke and fumes.

BAD750 CAS:59-52-9 ***HR: 3***
BAL
mf: $C_3H_8OS_2$ mw: 124.23

PROP: Viscous, oily liquid; pungent odor, bp: 140° @ 40 mm, vap d: 4.3, d: 1.2385 @ 25°/4°.

SYNS: BRITISH ANTILEWISITE ◇ DICAPTOL ◇ DIMERCAPROL PROPANOL ◇ DIMERCAPTOL ◇ 2,3-DIMERCAPTOL-1-PROPANOL ◇ DIMERCAPTOPROPANOL ◇ 2,3-DIMERCAPTOPROPANOL ô 2,3-DIMERCAPTOPROPAN-1-OL ◇ DITHIOGLYCEROL ◇ 1,2-DITHIOGLYCEROL ◇ 2,3-DITHIOPROPANOL ◇ SULFACTIN ◇ USAF ME-1

TOXICITY DATA with REFERENCE
scu-mus TDLo:100 mg/kg (11D preg):TER ANREAK 135,261,59
ims-hmn TDLo:3 mg/kg:BLD,SKN SCIEAS 102,601,45
ipr-rat LD50:105 mg/kg APFRAD 5,172,47
scu-rat LD50:2 g/kg APFRAD 5,172,47
ims-rat LD50:87 mg/kg TXAPA9 36,297,76
ipr-mus LD50:25 mg/kg NTIS** AD277-689
ivn-mus LD50:56 mg/kg CSLNX* NX#04985
ims-mus LD50:113 mg/kg AEPPAE 223,408,54
ivn-rbt LD50:50 mg/kg BIJOAK 41,325,47
ims-rbt LD50:50 mg/kg BIJOAK 41,325,47
par-rbt LD50:40 mg/kg APFRAD 5,172,47

CONSENSUS REPORTS: EPA Genetic Toxicology Program. Reported in EPA TSCA Inventory.

SAFETY PROFILE: Poison via intramuscular, parenteral, intraperitoneal, and intravenous routes. Experimental teratogenic effects. Human systemic effects by intramuscular route: hemorrhage and dermatitis. Human blood and systemic skin effects by intramuscular route. It causes redness and swelling when applied locally to the skin, but does not produce blisters or ulcers. Intensely irritating to eyes and mucous membranes. Systemic symptoms are caused by injection. When heated to decomposition, it emits toxic fumes of SO_x. Used as an antidote to arsenic, gold and mercury poisoning.

BAE000 ***HR: 1***
BALATA

PROP: Dried juice of the bully tree, *mimusops balata*. Resembles gutta percha.

SAFETY PROFILE: A mild irritant and allergen. Combustible when exposed to heat or flame.

BAE325 ***HR: 1***
BALSAM APPLE

PROP: A tree that may grow to 20 or 30 feet on rocks or other trees. The oval, leathery leaves are 3 to 8 inches across. It produces white flowers with pink edges and a gold center, and a golf-ball sized fruit which turns brown and opens when ripe. The trees grow wild in Hawaii, southern Florida, and the West Indies.

SYNS: CLUSIA ROSEA ◇ COPEY ◇ CUPEY ◇ FIGUIER MAUDIT MARRON (HAITI) ◇ PITCH APPLE ◇ SCOTCH ATTORNEY ◇ WILD MAMEE

SAFETY PROFILE: The fruit and sap contain an unidentified poison which causes profuse diarrhea after ingestion.

BAE750 ***HR: 1***
BALSAM of PERU

PROP: Dark brown, viscid liquid; vanilla odor. Sol in fixed oils; sltly sol in propylene glycol; insol in glycerin. Extracted from *Myroxylon pereirae Klotzsch*.

SYNS: BALSAM PERU OIL (FCC) ◇ PERUVIAN BALSAM

SAFETY PROFILE: A mild allergen. Combustible when heated. When heated to decomposition it emits acrid smoke and irritating fumes.

BAF000 CAS:9000-64-0 ***HR: 1***
BALSAM TOLU

PROP: Resin derived from *Toluifera balsamam* (FCTXAV 14,659,76).

SYNS: BALSAMS, TOLU ◇ OPOBALSAM ◇ RESIN TOLU ◇ THOMAS BALSAM ◇ TOLU ◇ TOLU BALSAM GUM ◇ TOLU BALSAM TINCTURE ◇ TOLU RESIN

TOXICITY DATA with REFERENCE
skn-rbt 500 mg/24H MLD FCTXAV 14,689,76

CONSENSUS REPORTS: Reported in EPA TSCA Inventory.

SAFETY PROFILE: A mild skin irritant. When heated to decomposition it yields toxic and irritating fumes and smoke.

BAF250 CAS:8029-29-6 ***HR: 2***
BANDANE

PROP: A mixture of isomers containing 60-62% chlorine used as preemergent herbicide (27ZTAP 3,20,69).

SYNS: HALTS ◇ POLYCHLORODICYCLOPENTADIENE ◇ POLYCHLORODICYCLOPENTADIENE ISOMERS

TOXICITY DATA with REFERENCE
orl-rat LD50:504 mg/kg WRPCA2 9,119,70
skn-rat LD50:12 g/kg 27ZTAP 3,20,69

SAFETY PROFILE: Moderately toxic by ingestion. Mildly toxic by skin contact. When heated to decomposition it emits toxic fumes of Cl^-. See also CHLORINATED HYDROCARBONS, ALIPHATIC.

BAF325 ***HR: 3***
BANEBERRY

PROP: Perennial herbs 1 to 2 feet tall with large compound leaves. It grows small white flowers in the spring and berries in the summer. The color of the berries depends on the species: *A. pachypoda*, white; *A. rubra*, red; *A. spicata*, purple-black. Various species are found in the temperate zones of North America from Canada to Georgia and New Mexico. Some are cultivated.

SYNS: ACTAEA (VARIOUS SPECIES) ◇ A. PACHYPODA ◇ A. RUBRA ◇ A. SPICATA ◇ COHOSH ◇ DOLLS EYES ◇ HERB-CHRISTOPHER ◇ NECKLACEWEED ◇ PAIN de COULEUVRE (CANADA) ◇ POISON de COULEUVRE (CANADA) ◇ SNAKEBERRY

SAFETY PROFILE: The toxin, whose identity is not known, is found only in the berries and roots. Liquid from these is a strong irritant and forms blisters on the skin and mucous membranes. Ingestion causes pain and inflammation of the lips, mouth, and throat, vomiting and diarrhea with blood, abdominal cramps, kidney damage, and central nervous system effects including dizziness, confusion, fainting, and convulsions.

BAF500 ***HR: 3***
BANOMITE
mf: $C_{13}H_8Cl_4N_2$ mw: 334

PROP: White to yellow crystals, almost insol in water, sol in organic solvents, mp: 98°.

SAFETY PROFILE: Poison by ingestion and dermal routes. A skin irritant and allergen. Dangerous; when heated to decomposition it evolves highly toxic fumes of NO_x and Cl^-. An acaricide.

BAF825 CAS:1415-73-2 ***HR: 3***
BARBALOIN
mf: $C_{21}H_{22}O_9$ mw: 418.43

SYNS: 10-(1′,5′-ANHYDROGLUCOSYL)ALOE-EMODIN-9-ANTHRONE ◇ 1,8-DIHYDROXY-3-HYDROXYMETHYL-10-(6-HYDROXYMETHYL-3,4,5-TRIHYDROXY-2-PYRANYL)ANTHRONE ◇ 10-GLUCOPYRANOSYL-1,8-DIHYDROXY-3-(HYDROXYMETHYL)-9(10H)-ANTHRACENONE

TOXICITY DATA with REFERENCE
ivn-mus LD50:200 mg/kg 85GDA2 8(2),314,82
orl-cat LDLo:500 mg/kg HBAMAK 4,1298,35
scu-rbt LDLo:200 mg/kg HBAMAK 4,1298,35
scu-pgn LDLo:200 mg/kg HBAMAK 4,1298,35

SAFETY PROFILE: Poison by subcutaneous and intravenous routes. Moderately toxic by ingestion. When heated to decomposition it emits acrid smoke and fumes. A carthartic and purgative.

BAG000 CAS:57-44-3 ***HR: 3***
BARBITAL
mf: $C_8H_{12}N_2O_3$ mw: 184.22

SYNS: BARBITONE ◇ DEBA ◇ DIEMAL ◇ DIETHYLBARBITONE ◇ DIETHYL-BARBITURIC ACID ◇ 5,5-DIETHYLBARBITURIC ACID ◇ DIETHYLMALONYLUREA ◇ 5,5-DIETHYL-2,4,6(1H,3H,5H)-PYRIMIDINETRIONE ◇ DORMONAL ◇ ETHYLBARBITAL ◇ HYPNOGENE ◇ MALONAL ◇ SEDEVAL ◇ URONAL ◇ VEROLETTIN ◇ VERONAL ◇ VESPERAL

TOXICITY DATA with REFERENCE
cyt-mus-ipr 33 g/kg IJMRAQ 61,1568,73
cyt-ham:lng 1 g/L ATSUDG (4),41,80
ipr-rat LDLo:300 mg/kg JPETAB 44,325,32
scu-rat LD50:450 mg/kg AEPPAE 152,341,30
orl-mus LD50:600 mg/kg NIIRDN 6,590,82
ipr-mus LD50:178 mg/kg FRPSAX 14,269,59
scu-mus LD50:630 mg/kg YKKZAJ 74,122,54
orl-cat LDLo:280 mg/kg PHREA7 19,472,39
scu-rbt LDLo:250 mg/kg JACSAT 45,243,23

CONSENSUS REPORTS: EPA Genetic Toxicology Program.

SAFETY PROFILE: Poison by ingestion, intravenous, intraperitoneal, and subcutaneous routes. Ingestion causes psychological effects in humans. Mutation data reported. When heated to decomposition it emits toxic fumes of NO_x. See also BARBITURATES. An hypnotic and sedative.

BAG250 CAS:144-02-5 ***HR: 3***
BARBITAL SODIUM
mf: $C_8H_{12}N_2O_3$•Na mw: 207.21

PROP: Bitter crystals or powder.

SYNS: BARBITAL Na ◇ BARBITAL SOLUBLE ◇ BARBITONE SODIUM ◇ DIETHYLBARBITURATE MONOSODIUM ◇ 5,5-DIETHYLBARBITURIC ACID SODIUM deriv. ◇ DIETHYLMALONYLUREA SODIUM ◇ EMBINAL ◇ MEDINAL ◇ NATRINAL ◇ NATRIUMBARBITALS (GERMAN) ◇ NERVOSETON ◇ 2,4,6(1H,3H,5H)-PYRIMIDINETRIONE, 5,5-DIETHYL-, MONOSODIUM SALT (9CI) ◇ SODIUM BARBITAL ◇ SODIUM BARBITONE ◇ SODIUM DIETHYLBARBITURATE ◇ SODIUM-5,5-DIETHYLBARBITURATE ◇ SODIUM ETHYLBARBITAL ◇ SODIUM MALONYLUREA ◇ SODIUM VERONAL ◇ SOLUBLE BARBITAL ◇ SOPRINAL ◇ THYALONE ◇ VERONAL SODIUM

TOXICITY DATA with REFERENCE
sce-ham:emb 100 μg/L IJCNAW 20,768,77
unr-rat TDLo:125 mg/kg (female 19D post):REP JCPPAV 45,146,52
scu-rat TDLo:2400 mg/kg (8-19D preg):TER CUSCAM 36,3,67
orl-rat LD50:600 mg/kg ARTODN 54,275,83
ipr-rat LDLo:300 mg/kg 27ZWAY 2,-,36
scu-rat LDLo:300 mg/kg JPETAB 31,1,27
ivn-rat LD50:280 mg/kg JPETAB 135,213,62
scu-mus LD50:700 mg/kg JPETAB 109,268,53
ivn-mus LD50:830 mg/kg TXAPA9 27,70,74
orl-dog LDLo:350 mg/kg 27ZWAY 2,-,36
ivn-dog LDLo:300 mg/kg JPETAB 60,125,37
orl-cat LDLo:275 mg/kg 27ZWAY 2,-,36

SAFETY PROFILE: Poison by ingestion, subcutaneous, intravenous, and intraperitoneal routes. Large doses cause marked depression (sometimes preceded by excitation), prolonged coma, and death. Experimental teratogenic and reproductive effects. Allergic skin reactions may occur on contact. Implicated in development of aplastic anemia. A truly habit-forming drug. Other experimental reproductive effects. Mutation data reported. Combustible. When heated to decomposition it emits toxic fumes of NO_x and Na_2O. See also BARBITURATES.

BAG500 ***HR: 3***
BARBITURATES

SYNS: BARBITAL ◇ BARBITAL SODIUM ◇ BARBITONE

SAFETY PROFILE: Salts or derivatives of barbituric acid are central nervous system depressants, and are used as hypnotics, sedatives, and anesthetics. Usually administered orally. They are strongly habit forming. Several compounds including amo-, seco-, and pentabarbital are restricted chemicals. Their use can cause a reaction called barbiturism which is marked by chills, headache, fever, and cutaneous eruptions. See BARBITAL SODIUM.

BAG750 ***HR: 1***
BARBITURIC ACID
mf: $C_4H_4O_3N_2$ mw: 128.1

PROP: Crystals or white to yellow-white powder, mp: 245°, bp: 260° (decomp).

SAFETY PROFILE: Mildly toxic. Irritating to skin, eyes, and mucous membranes. An allergen. Has no hypnotic properties. Combustible.

BAH250 CAS:7440-39-3 ***HR: 3***
BARIUM
DOT: UN 1399/UN 1400/UN 1854
af: Ba aw: 137.36

PROP: Silver-white, sltly lustrous, somewhat malleable metal. Mp: 725°, bp: 1640°, d: 3.5 @ 20°, vap press: 10 mm @ 1049°.

CONSENSUS REPORTS: Reported in EPA TSCA Inventory. Community Right-To-Know List.

OSHA PEL: TWA 0.5 mg(Ba)/m^3
ACGIH TLV: TWA 0.5 mg/m^3
DOT Classification: Flammable Solid; Label: Dangerous When Wet (UN1399, UN1400); Flammable Solid; Label: Spontaneously Combustible (UN1854).

SAFETY PROFILE: Water and stomach acids solubilize barium salts and can cause poisoning. Symptoms are vomiting, colic, diarrhea, slow irregular pulse, transient hypertension, and convulsive tremors and muscular paralysis. Death may occur from a few hours to a few days. Half-life of barium in bone has been estimated at 50 days. Dust is dangerous and explosive when exposed to heat, flame, or chemical reaction. Violent or explosive reaction with water, CCl_4, fluorotrichloromethane, trichloroethylene, and C_2Cl_4. Incompatible with acids, $C_2Cl_3F_3$, $C_2H_2FCl_3$, C_2HCl_3 and water, 1,1,2-trichloro trifluoro ethane, and fluorotrichloroethane. The powder may ignite or explode in air or other oxidizing gases. See also BARIUM COMPOUNDS.

BAH500 CAS:543-80-6 ***HR: 3***
BARIUM ACETATE
mf: $C_4H_6O_4$•Ba mw: 255.44

PROP: White crystals. Water-sol.

SYNS: ACETIC ACID, BARIUM SALT ◇ BARIUM DIACETATE ◇ OCTAN BARNATY (CZECH)

TOXICITY DATA with REFERENCE
orl-rat LD50:921 mg/kg MarJV# 29MAR77
ivn-mus LD50:11 mg/kg TXAPA9 22,150,72
orl-rbt LDLo:236 mg/kg EQSSDX 1,1,75
scu-rbt LDLo:96 mg/kg EQSSDX 1,1,75
ivn-rbt LDLo:12 mg/kg EQSSDX 1,1,75

CONSENSUS REPORTS: Reported in EPA TSCA Inventory. Barium and its compounds are on the Community Right-To-Know List.

OSHA PEL: TWA 0.5 mg(Ba)/m^3
ACGIH TLV: TWA 0.5 mg(Ba)/m^3

SAFETY PROFILE: Poison via ingestion, intravenous, and subcutaneous routes. When heated to decomposition it emits acrid smoke and fumes. See also BARIUM COMPOUNDS.

BAH750 CAS:12070-27-8 ***HR: 3***
BARIUM ACETYLIDE
mf: C_2Ba mw: 161.35

CONSENSUS REPORTS: Barium and its compounds are on the Community Right-To-Know List.

SAFETY PROFILE: Ignites on contact with vapors of water or ethanol in air. Incandescent reaction when heated with: hydrogen @ 150°C; chlorine @ 140°C; bromine @ 130°C; iodine @ 122°C; and selenium @ 150°C. See also BARIUM COMPOUNDS and ACETYLIDES.

BAI000 CAS:18810-58-7 ***HR: 3***
BARIUM AZIDE
DOT: UN 0224/UN 1571
mf: BaN_6 mw: 221.40

PROP: Monoclinic prisms. Mp: $-N_2$ @ about 120°, bp: explodes, d: 2.936.

SYN: BARIUM AZIDE, dry or containing less than 50% water (DOT)

CONSENSUS REPORTS: Reported in EPA TSCA Inventory. Barium and its compounds are on the Community Right-To-Know List.

OSHA PEL: TWA 0.5 mg(Ba)/m^3
ACGIH TLV: TWA 0.5 mg/(Ba)m^3
DOT Classification: Flammable Solid; Label: Flammable Solid (UN1571); Class A Explosive; Label: Explosive A and Poison (UN0224); Flammable Solid; Label: Flammable Solid and Poison (UN1571)

SAFETY PROFILE: A poison. Moderate explosion hazard when shocked or heated to 275°. Spontaneously flammable in air. Very unstable. When heated to decomposition it emits toxic fumes of NO_x. See also BARIUM COMPOUNDS (soluble) and AZIDES.

BAI250 CAS:18810-58-7 ***HR: 3***
BARIUM AZIDE (wet)
mf: BaN_6 mw:221.40
DOT: UN 1571

PROP: Compound contains 50% or more water (FEREAC 41,15972,76).

CONSENSUS REPORTS: Reported in EPA TSCA Inventory. Barium and its compounds are on the Community Right-To-Know List.

OSHA PEL: TWA 0.5 mg/(Ba)/m^3
ACGIH TLV: TWA 0.5 mg(Ba)/m^3
DOT Classification: Flammable Solid; Label: Flammable Solid and Poison.

SAFETY PROFILE: A poison. Flammable and possibly explosive. When heated to decomposition it emits toxic fumes of NO_x. See also BARIUM COMPOUNDS (soluble) and AZIDES.

BAI500 ***HR: 3***
BARIUM BENZOATE
mf: $Ba(C_7H_5O_2)_2 \cdot 2H_2O$ mw: 415.61

PROP: White, nacreous leaflets. Mp: loses $2H_2O$ @ 100°.

CONSENSUS REPORTS: Barium and its compounds are on the Community Right-To-Know List.

SAFETY PROFILE: Deadly poison. See also BARIUM COMPOUNDS (soluble).

BAI750 CAS:13967-90-3 ***HR: 3***
BARIUM BROMATE
DOT: UN 2719
mf: $Ba(BrO_3)_2 \cdot H_2O$ mw: 411.21

PROP: White crystals or crystalline powder, mp: decomp @ 260°, d: 3.99 @ 18°.

CONSENSUS REPORTS: Barium and its compounds are on the Community Right-To-Know List.

OSHA PEL: TWA 0.5 mg(Ba)/m^3
ACGIH TLV: TWA 0.5 mg(Ba)/m^3
DOT Classification: Oxidizer; Label: Oxidizer and Poison

SAFETY PROFILE: Very toxic. Fire hazard by chemical reaction with easily oxidized materials. Explodes at 300°. Mixtures with sulfur are unstable storage hazards; igniting immediately at 91°C and after a 2-11 day delay period at room temperature. Incompatible with Al, As, C, Cu, metal sulfides, organic matter, P, and reducing materials. When heated to decomposition, it emits toxic fumes of Br^-. See also BARIUM COMPOUNDS (soluble) and BROMINE.

BAI800 CAS:1191-79-3 ***HR: 3***
BARIUM CADMIUM STEARATE
mf: $C_{72}H_{140}O_8$•Ba•Cd mw: 1383.86

SYNS: CADMIUM BARIUM STEARATE ◇ OCTADECANOIC ACID, BARIUM CADMIUM SALT (4:1:1) (9CI) ◇ STEARIC ACID, BARIUM CADMIUM SALT (4:1:1)

TOXICITY DATA with REFERENCE
orl-rat LD50:3171 mg/kg GISAAA 35(2),98,70
orl-mus LD50:1381 mg/kg 41HTAH -,14,78

ACGIH TLV: TWA 0.01 mg(Cd)/m^3; Suspected Carcinogen
NIOSH REL: TWA reduce to lowest feasible level

CONSENSUS REPORTS: Reported in EPA TSCA Inventory.

SAFETY PROFILE: A suspected carcinogen. Moderately toxic by ingestion. When heated to decomposition it emits toxic fumes of Ba and Cd.

BAI825 CAS:4696-54-2 ***HR: 2***
BARIUM CAPRYLATE
mf: $C_{16}H_{32}O_4$•Ba mw: 425.82

SYNS: BARIUM OCTANOATE ◇ BARIUM OCTOATE

TOXICITY DATA with REFERENCE
orl-rat LD50:1 g/kg GISAAA 39(11),91,74
orl-mus LD50:1100 mg/kg GISAAA 39(11),91,74
orl-gpg LD50:1250 mg/kg GISAAA 39(11),91,74

CONSENSUS REPORTS: Barium and its compounds are on the Community Right-To-Know List.

SAFETY PROFILE: Moderately toxic by ingestion.

BAJ000 ***HR: 3***
BARIUM CARBIDE
mf: BaC_2 mw: 161.4

PROP: Gray crystals. D: 3.75.

CONSENSUS REPORTS: Barium and its compounds are on the Community Right-To-Know List.

SAFETY PROFILE: A poison. A fire and explosion hazard by chemical reaction with moisture to form acetylene. Incompatible with Se; S; H_2O. To fight fire, use CO_2, dry chemical. See also BARIUM COMPOUNDS (soluble)

BAJ250 CAS:513-77-9 ***HR: 3***
BARIUM CARBONATE (1:1)
mf: CO_3•Ba mw: 197.35

PROP: White powder. Mp: 1740 @ 90 atm, bp: decomp; d: 4.43.

SYNS: BARIUM CARBONATE ◇ CARBONIC ACID, BARIUM SALT (1:1) ◇ C.I. 77099 ◇ C.I. PIGMENT WHITE 10

TOXICITY DATA with REFERENCE
ihl-rat TCLo:3130 $\mu g/m^3$/24H (female 16W pre):REP GTPZAB 20(7),33,76
orl-man LDLo:800 mg/kg YKYUA6 28,329,77
orl-wmn TDLo:800 mg/kg:GIT BMJOAE 289,882,84
orl-hmn TDLo:11 mg/kg:GIT YKYUA6 31,1247,80
orl-hmn LDLo:17 mg/kg YKYUA6 28,329,77
orl-hmn TDLo:29 mg/kg:PNS IJMDAI 3,565,67
orl-rat LD50:418 mg/kg 85GMAT -,23,82
ivn-rat LDLo:20 mg/kg EQSSDX 1,1,75
orl-mus LD50:200 mg/kg 85GMAT -,23,82
ipr-mus LD50:50 mg/kg 85GMAT -,23,82
orl-dog LDLo:400 mg/kg PCOC** -,95,66

CONSENSUS REPORTS: Reported in EPA TSCA Inventory. Barium and its compounds are on the Community Right-To-Know List.

OSHA PEL: TWA 0.5 mg(Ba)/m^3
ACGIH TLV: TWA 0.5 mg(Ba)/m^3

SAFETY PROFILE: Poison by ingestion, intravenous, and intraperitoneal routes. Human systemic effects by ingestion: stomach ulcers, muscle weakness, paresthesias and paralysis, hypermotility, diarrhea, nausea or vomiting, lung changes. Experimental reproductive effects. Incompatible with BrF_3 and 2-furanpercarboxylic acid. See also BARIUM COMPOUNDS (soluble).

BAJ500 CAS:13477-00-4 ***HR: 3***
BARIUM CHLORATE
DOT: UN 1445
mf: Cl_2O_6•Ba mw: 304.24

PROP: Colorless prisms or white powder. Mp: loses H_2O @ 414°, d: 3.18.

SYN: CHLORIC ACID, BARIUM SALT

CONSENSUS REPORTS: Reported in EPA TSCA Inventory. Barium and its compounds are on the Community Right-To-Know List.

OSHA PEL: TWA 0.5 mg(Ba)/m^3
ACGIH TLV: TWA 0.5 mg(Ba)/m^3
DOT Classification: Oxidizer; Label: Oxidizer.

SAFETY PROFILE: A poison. For fire and explosion hazards, see CHLORATES. Incompatible with Al, As, C, charcoal, Cu, MnO_2, metal sulfides, S_4N_4, organic matter, P, S. See also BARIUM COMPOUNDS (soluble).

BAJ750 CAS:13477-00-4 *HR: 3*
BARIUM CHLORATE (wet)
DOT: UN 1445

SYN: CHLORIC ACID, BARIUM SALT (wet)

CONSENSUS REPORTS: Reported in EPA TSCA Inventory. Barium and its compounds are on the Community Right-To-Know List.

OSHA PEL: TWA 0.5 mg(Ba)/m^3
ACGIH TLV: TWA 0.5 mg(Ba)/m^3
DOT Classification: Oxidizer; Label: Oxidizer.

SAFETY PROFILE: A poison. A powerful oxidizer. When heated to decomposition it emits toxic fumes of Cl^-. See also BARIUM CHLORATE, BARIUM COMPOUNDS (soluble), and CHLORATES.

BAK000 CAS:10361-37-2 *HR: 3*
BARIUM CHLORIDE
mf: $BaCl_2$ mw: 208.24

PROP: Colorless, flat crystals. Mp: transition @ 925° to cubic crystals, bp: 1560°, d: 3.856 @ 24°.

SYNS: BARIUM DICHLORIDE ◇ NCI-C61074 ◇ SBa 0108E

TOXICITY DATA with REFERENCE
mrc-smc 14 mmol/L MUTAEX 1,21,86
itt-rat TDLo:16659 μg/kg (1D male):REP JRPFA4 7,21,64
orl-hmn LDLo:11400 μg/kg 28ZAA9 7,478,48
orl-rat LD50:118 mg/kg FOREAE 7,313,42
scu-rat LD50:178 mg/kg 27ZIAQ -,53,73
ivn-rat LDLo:20 mg/kg JLCMAK 15,35,29
orl-mus LDLo:70 mg/kg EQSSDX 1,1,75
scu-mus LDLo:10 mg/kg NTIS** AEC-TR-6710
orl-dog LDLo:90 mg/kg 27ZIAQ -,53,73
scu-dog LDLo:10 mg/kg EQSSDX 1,1,75

CONSENSUS REPORTS: Reported in EPA TSCA Inventory. Barium and its compounds are on the Community Right-To-Know List. EPA Genetic Toxicology Program.

OSHA PEL: TWA 0.5 mg(Ba)/m^3
ACGIH TLV: TWA 0.5 mg(Ba)/m^3

SAFETY PROFILE: Poison to humans by ingestion. Also poison by subcutaneous, intravenous, and intraperitoneal routes. Inhalation absorption of barium chloride equals 60-80%; oral absorption equals 10-30%. Experimental reproductive effects. Mutation data reported. See also BARIUM COMPOUNDS (soluble). When heated to decomposition it emits toxic fumes of Cl^-.

BAK125 CAS:14674-74-9 *HR: 3*
BARIUM CHLORITE
mf: $BaCl_2O_4$ mw: 172.23

CONSENSUS REPORTS: Barium and its compounds are on the Community Right-To-Know List.

SAFETY PROFILE: A poison. Decomposes explosively at 190°C. Ignites on contact with dimethyl sulfate. When heated to decomposition it emits toxic fumes of Cl^-. See also BARIUM COMPOUNDS and CHLORITES.

BAK250 CAS:10294-40-3 *HR: 3*
BARIUM CHROMATE(VI)
mf: Ba•CrH_2O_4 mw: 255.36

PROP: Heavy, yellow, crystalline powder. D: 4.498 @ 15°.

SYNS: BARIUM CHROMATE (1:1) ◇ BARIUM CHROMATE OXIDE ◇ BARYTA YELLOW ◇ CHROMIC ACID, BARIUM SALT (1:1) ◇ C.I. 77103 ◇ C.I. PIGMENT YELLOW 31 ◇ LEMON CHROME ◇ LEMON YELLOW ◇ PERMANENT YELLOW ◇ STEINBUHL YELLOW ◇ ULTRAMARINE YELLOW

TOXICITY DATA with REFERENCE
sce-ham:ovr 100 μg/L MUREAV 156,219,85

CONSENSUS REPORTS: NTP Fifth Annual Report on Carcinogens. IARC Cancer Review: Group 1 IMEMDT 7,165,87, Animal Inadequate Evidence IMEMDT 2,100, 73; Human Sufficient Evidence IMEMDT 23,205,80. Reported in EPA TSCA Inventory. Barium and its compounds are on the Community Right-To-Know List.

OSHA PEL: TWA 0.1 mg (C_3O_3)m^3; 0.5 mg(Ba)/m^3
ACGIH TLV: TWA 0.5 mg(Ba)/m^3; 0.05 mg(Cr)/m^3; Confirmed Human Carcinogen
NIOSH REL: TWA 0.001 mg(Cr(VI))/m^3

SAFETY PROFILE: Confirmed human carcinogen. A poison. Mutation data reported. For fire hazard, see CHROMATES. Reacts vigorously with reducing materials. See also BARIUM COMPOUNDS (soluble) and CHROMIUM COMPOUNDS. Used in pyrotechnics and as an explosive initiator.

BAK500 *HR: 3*
BARIUM COMPOUNDS (soluble)

CONSENSUS REPORTS: Barium and its compounds are on the Community Right-To-Know List.

OSHA PEL: Soluble Compounds: TWA 0.5 mg(Ba)/m^3
ACGIH TLV: Soluble Compounds:TWA 0.5 mg/m^3
DFG MAK: Soluble Compounds: 0.5 mg/m^3
DOT Classification: Some barium compounds are flammable or explosive.

SAFETY PROFILE: The chromate is a human carcinogen. The soluble barium salts, such as the chloride and sulfide, are poisonous when ingested. The insoluble sulfate used in radiography is not acutely toxic. See also BARIUM SULFATE. Few cases of industrial systemic poisoning have been reported, but one investigator describes a fatal case of poisoning attributed to barium oxide, the symptoms being severe abdominal pain with vomiting, dyspnoea, rapid pulse, paralysis of the arm and leg, and eventually cyanosis and death. The same investigator produced paralysis in animals with barium oxide and carbonate. The usual result of exposure to the sulfide, oxide, and carbonate is irritation of the eyes, nose, and throat, and of the skin, producing dermatitis. The salts mentioned are somewhat caustic.

BAK750 CAS:542-62-1 ***HR: 3***
BARIUM CYANIDE
DOT: UN 1565
mf: C_2BaN_2 mw: 189.38

PROP: White, crystalline powder.

SYNS: BARIUM CYANIDE, solid (DOT) ◇ BARIUM DICYANIDE ◇ RCRA WASTE NUMBER P013

CONSENSUS REPORTS: Reported in EPA TSCA Inventory. Cyanide and its compounds, as well as barium and its compounds, are on the Community Right-To-Know List.

OSHA PEL: TWA 0.5 mg(Ba)/m^3
ACGIH TLV: TWA 0.5 mg(Ba)/m^3
DOT Classification: Poison B; Label: Poison.

SAFETY PROFILE: A deadly poison. See also CYANIDE and BARIUM COMPOUNDS (soluble). When heated to decomposition it emits toxic fumes of CN^-.

BAL000 ***HR: 3***
BARIUM CYANOPLATINITE
mf: $BaPt(CN)_4 \cdot 4H_2O$ mw: 508.6

PROP: (a) Monoclinic, yellow crystals; (b) rhombic crystals. Mp: loses $2H_2O$ @ 100°; d: (a) 2.076, (b) 2.085.

CONSENSUS REPORTS: Cyanide and its compounds, as well as barium and its compounds, are on the Community Right-To-Know List.

OSHA PEL: TWA 5 mg(CN)/m^3
ACGIH TLV: TWA 5 mg(CN)/m^3 (skin)
DFG MAK: 5 mg/m^3
NIOSH REL: (Cyanide) CL 5 mg(CN)/m^3/10M

SAFETY PROFILE: A poison. See also BARIUM COMPOUNDS (soluble), CYANIDE, and PLATINUM COMPOUNDS. When heated to decomposition it emits highly toxic fumes of CN^- and NO_x.

BAL250 ***HR: 3***
BARIUM DIAZIDE
mf: BaN_6 mw: 221.38

CONSENSUS REPORTS: Barium and its compounds are on the Community Right-To-Know List.

SAFETY PROFILE: A poison. Impact-sensitive when dry; avoid contact with Pb, acids. See also BARIUM COMPOUNDS and AZIDES.

BAL275 CAS:6332-68-9 ***HR: 3***
BARIUM DIBENZYLPHOSPHATE
mf: $C_{28}H_{28}BaO_8P_2$ mw: 691.84

TOXICITY DATA with REFERENCE
ivn-mus LD50:56 mg/kg CSLNX* NX#04099

OSHA PEL: TWA 0.5 mg(Ba)/m^3
ACGIH TLV: TWA 0.5 mg(Ba)/m^3

SAFETY PROFILE: Poison by intravenous route. When heated to decomposition it emits toxic fumes of PO_x and Ba.

BAL500 ***HR: 3***
BARIUM DICHROMATE
mf: $BaCr_2O_7$ mw: 353.38

PROP: Brownish-red, crystalline masses.

SYN: BARIUM BICHROMATE

CONSENSUS REPORTS: Barium and its compounds, as well as chromium and its compounds, are on the Community Right-To-Know List.

SAFETY PROFILE: A poison. Some chromates are carcinogenic. A moderate fire hazard by chemical reaction with easily oxidized materials. A powerful oxidizer. Incompatible with reducing materials. See also BARIUM COMPOUNDS and CHROMIUM COMPOUNDS.

BAL625 CAS:12047-11-9 ***HR: D***
BARIUM FERRITE
mf: $Ba \cdot Fe_{12}O_{19}$ mw: 1111.54

SYNS: BARIUM FERRATE ◇ BARIUM HEXAFERRITE ◇ FERRATE (Fe_{12}-$O_{19}^{(2-)}$) BARIUM (1:1) (9CI) ◇ FERROXDURE

TOXICITY DATA with REFERENCE
ihl-rat TCLo:750 μg/m^3/24H (female 1-22D post):TER GISAAA 49(4),72,84
ihl-rat TCLo:88 μg/m^3/24H (female 1-22D post):REP GISAAA 49(4),72,84

CONSENSUS REPORTS: Barium and its compounds are on the Community Right-To-Know List.

SAFETY PROFILE: An experimental teratogen. Other experimental reproductive effects. See also BARIUM COMPOUNDS.

BAL750 CAS:13862-62-9 ***HR: 3***
BARIUM FLUOBORATE
mf: B_2F_8•Ba mw: 310.96

SYNS: BARIUM BIS(TETRAFLUOROBORATE) ◇ BARIUM TETRAFLUOROBORATE

TOXICITY DATA with REFERENCE
orl-rat LDLo:250 mg/kg NCNSA6 5,27,53

CONSENSUS REPORTS: Barium and its compounds are on the Community Right-To-Know List.

OSHA PEL: TWA 0.5 mg(Ba)/m^3; 2.5 mg(F)/m^3
ACGIH TLV: TWA 0.5 mg(Ba)/m^3
NIOSH REL: (Fluorides, Inorganic) TWA 2.5 mg(F)/m^3

SAFETY PROFILE: Poison by ingestion. See also BARIUM COMPOUNDS, BORON COMPOUNDS, and FLUORIDES. When heated to decomposition it emits toxic fumes of F^-.

BAM000 CAS:7787-32-8 ***HR: 3***
BARIUM FLUORIDE
mf: BaF_2 mw: 175.34

PROP: White powder. Mp: 1280°, bp: 2137°, d: 4.89.

SYN: BARYUM FLUORURE (FRENCH)

TOXICITY DATA with REFERENCE
ipr-mus TDLo:656 mg/kg (female 1-21D post):TER DZZEA7 34,484,79
orl-rat LD50:250 mg/kg VAMNAQ (2),28,77
ipr-mus LD50:29910 μg/kg DZZEA7 34,484,79
scu-frg LDLo:1540 mg/kg CRSBAW 124,133,37

CONSENSUS REPORTS: Reported in EPA TSCA Inventory. Barium and its compounds are on the Community Right-To-Know List.

OSHA PEL: TWA 0.5 mg(Ba)/m^3; 2.5 mg(F)/m^3
ACGIH TLV: TWA 0.5 mg(Ba)/m^3; 2.5 mg(F)/m^3
NIOSH REL: (Fluorides, Inorganic) TWA 2.5 mg(F)/m^3

SAFETY PROFILE: A poison by ingestion and intraperitoneal routes. Moderately toxic by subcutaneous route. An experimental teratogen. See also FLUORIDES and BARIUM COMPOUNDS (SOLUBLE). When heated to decomposition it emits toxic fumes of F^-.

BAM250 CAS:13477-09-3 ***HR: 3***
BARIUM HYDRIDE
mf: BaH_2 mw: 139.38

PROP: Gray crystals or lumps. Mp: decomp @ 675°, bp: 1400°, d: 4.21 @ 0°.

CONSENSUS REPORTS: Barium and its compounds are on the Community Right-To-Know List.

SAFETY PROFILE: A poison. Rapidly decomposed by water and acids. In powder form, it ignites spontaneously in air and reacts vigorously with water. Coarser material ignites when heated in oxygen. It is incompatible with water; acids; and metal halogenates. A dangerous fire hazard because moisture may cause it to ignite. To fight fire, use dry chemical, graphite, CO_2. See also BARIUM COMPOUNDS (soluble) and HYDRIDES.

BAM500 ***HR: 3***
BARIUM HYDROXIDE
mf: $Ba(OH)_2$ mw: 171.35

CONSENSUS REPORTS: Barium and its compounds are on the Community Right-To-Know List.

SAFETY PROFILE: A poison. See also BARIUM COMPOUNDS (soluble). Incompatible with chlorinated rubber.

BAM750 ***HR: 3***
BARIUM HYPOPHOSPHITE
mf: $Ba(H_2PO_2)_2$•H_2O mw: 285.38

PROP: Crystalline powder. Mp: decomp, d: 2.90 @ 17°.

CONSENSUS REPORTS: Barium and its compounds are on the Community Right-To-Know List.

SAFETY PROFILE: A poison. When heated to decomposition it emits highly toxic fumes of PO_x. Incompatible with $KClO_3$. When heated to decomposition it emits toxic fumes of PO_x. See also BARIUM COMPOUNDS (soluble) and HYPOPHOSPHITES.

BAN000 ***HR: 3***
BARIUM IODATE
mf: $Ba(IO_3)_2$ mw: 487.20

PROP: White, crystalline powder. Mp: decomp, d: 4.998.

CONSENSUS REPORTS: Barium and its compounds are on the Community Right-To-Know List.

SAFETY PROFILE: A poison. A powerful oxidizer. Incompatible with Al; As; C; Cu; metal sulfides; organic matter. When heated to decomposition it emits toxic fumes of I^-. See also BARIUM COMPOUNDS (soluble) and IODATES.

BAN250 CAS:10022-31-8 ***HR: 3***
BARIUM(II) NITRATE (1:2)
DOT: UN 1446
mf: N_2O_6•Ba mw: 261.36

PROP: Lustrous crystals. Mp: 592°, bp: decomp, d: 3.24 @ 23°.

SYNS: BARIUM DINITRATE ◇ BARIUM NITRATE (DOT) ◇ DUSICNAN BARNATY (CZECH) ◇ NITRATE de BARYUM (FRENCH) ◇ NITRIC ACID, BARIUM SALT

TOXICITY DATA with REFERENCE
skn-rbt 500 mg/24H MLD 28ZPAK -,10,72
eye-rbt 100 mg/24H MOD 28ZPAK -,10,72
orl-rat LD50:355 mg/kg 28ZPAK -,10,72
scu-mus LDLo:10 mg/kg NTIS** AEC-TR-6710
ivn-mus LD50:8500 μg/kg TXAPA9 22,150,72
orl-dog LDLo:800 mg/kg YKYUA6 31,1247,80
orl-rbt LDLo:150 mg/kg YKYUA6 31,1247,80
par-rbt LDLo:30 mg/kg MELAAD 30,44,39

CONSENSUS REPORTS: Reported in EPA TSCA Inventory. Barium and its compounds are on the Community Right-To-Know List.

OSHA PEL: TWA 0.5 mg(Ba)/m^3
ACGIH TLV: TWA: 0.5 mg(Ba)/m^3
DOT Classification: Oxidizer; Label: Oxidizer; DOT-IMO: Oxidizer; Label: Oxidizer and Poison.

SAFETY PROFILE: A poison via ingestion, subcutaneous, parenteral, and intravenous routes. An irritant to skin and eyes. When heated to decomposition it emits very toxic fumes of NO_x. An oxidizer. Mixtures with finely divided aluminum-magnesium alloys are easily ignitable and extremely sensitive to friction or impact. Such mixtures are used in chemical photoflash applications. Incompatible with (Mg + BaO_2 + Zn), Al, and Mg alloys. When heated to decomposition it emits toxic fumes of NO_x. See also BARIUM COMPOUNDS (soluble) and NITRATES.

BAN500 CAS:12047-79-9 ***HR: 3***
BARIUM NITRIDE
mf: Ba_3N_2 mw: 440.10

PROP: Colorless crystals. Bp: 1000° (vac), d: 4.783 @ 25°/4°.

CONSENSUS REPORTS: Barium and its compounds are on the Community Right-To-Know List.

SAFETY PROFILE: A poison. Flammable by spontaneous chemical reaction with water to liberate explosive ammonia gas. Dangerous; explodes upon heating and by spontaneous chemical reaction, liberating NH_3 vapor which can form explosive mixtures with air. Violent reaction with air or moisture. See also BARIUM COMPOUNDS (soluble) and AMMONIA.

BAO000 CAS:1304-28-5 ***HR: 3***
BARIUM OXIDE
DOT: UN 1884
mf: BaO mw: 153.34

PROP: White to yellowish-white powder. Mp: 1923°, bp: 2000° (approx), d: 5.72.

SYNS: BARIUM MONOXIDE ◇ BARIUM PROTOXIDE ◇ BARYTA ◇ CALCINED BARYTA ◇ OXYDE de BARYUM (FRENCH)

TOXICITY DATA with REFERENCE
scu-mus LD50:50 mg/kg ZVKOA6 19,186,74

CONSENSUS REPORTS: Reported in EPA TSCA Inventory. Barium and its compounds are on the Community Right-To-Know List.

OSHA PEL: TWA 0.5 mg(Ba)/m^3
ACGIH TLV: TWA 0.5 mg(Ba)/m^3
DOT Classification: ORM-B; Label: None; DOT-IMO: Poison B; Label: St. Andrews Cross.

SAFETY PROFILE: A poison via subcutaneous route. See also BARIUM COMPOUNDS (soluble). Combustible by spontaneous chemical reaction; produces heat on contact with water or steam. Incompatible with H_2S, hydroxylamine, N_2O_4, triuranium octaoxide, SO_3.

BAO250 CAS:1304-29-6 ***HR: 3***
BARIUM PEROXIDE
DOT: UN 1449
mf: BaO_2 mw: 169.34

PROP: Grayish-white powder. Mp: 450°, bp: −0 @ 800°, d: 4.96.

SYNS: BARIO (PEROSSIDO di) (ITALIAN) ◇ BARIUM BINOXIDE ◇ BARIUM DIOXIDE ◇ BARIUMPEROXID (GERMAN) ◇ BARIUMPEROXYDE (DUTCH) ◇ BARIUM SUPEROXIDE ◇ DIOXYDE de BARYUM (FRENCH) ◇ PEROXYDE de BARYUM (FRENCH)

TOXICITY DATA with REFERENCE
scu-mus LD50:50 mg/kg ZVKOA6 19,186,74

CONSENSUS REPORTS: Reported in EPA TSCA Inventory. Barium and its compounds are on the Community Right-To-Know List.

OSHA PEL: TWA 0.5 mg(Ba)/m^3
ACGIH TLV: TWA 0.5 mg(Ba)/m^3
DOT Classification: Oxidizer; Label: Oxidizer and Poison.

SAFETY PROFILE: A poison via subcutaneous route. A powerful oxidizer. Explodes on contact with acetic anhydride. Ignites when mixed with calcium-silicon alloys, powdered aluminum, powdered magnesium, water + organic compounds. Mixtures with propane react violently when heated. The powder ignites when heated to 265°C with selenium. Wood ignites with friction from the peroxide. Incompatible with H_2S, water, peroxy formic acid, hydroxylamine solution, mixture of (Mg + Zn + $Ba(NO_3)_2$), and organic matter. See also BARIUM COMPOUNDS (soluble) and PEROXIDES, INORGANIC.

BAO300 CAS:50864-67-0 ***HR: 3***
BARIUM POLYSULFIDE

SYNS: BARIUMPOLYSULFID ◇ BARIUM SULFIDE ◇ SOLABAR ◇ SOLBAR

TOXICITY DATA with REFERENCE
orl-man TDLo:226 mg/kg AIMDAP 132,891,73
orl-rat LD50:375 mg/kg FMCHA2-,C34,89

OSHA PEL: TWA 0.5 mg(Ba)/m^3
ACGIH TLV: TWA 0.5 mg(Ba)/m^3

SAFETY PROFILE: Poison by ingestion. Human systemic effects by ingestion: flaccid paralysis without anesthesia, muscle weakness, and dyspnea. When heated to decomposition it emits toxic fumes of SO_x and Ba.

BAO500 ***HR: 3***
BARIUM RHODANIDE
mf: $BaC_6H_4O_2N_2S_4$ mw: 401.6

CONSENSUS REPORTS: Barium and its compounds are on the Community Right-To-Know List.

SAFETY PROFILE: A poison. Explosive. When heated to decomposition it can emit highly toxic fumes of SO_x and NO_x. See also BARIUM COMPOUNDS.

BAO750 CAS:17125-80-3 ***HR: 3***
BARIUM SILICOFLUORIDE
mf: $F_6Si•Ba$ mw: 279.43

PROP: White, crystalline powder. D: 4.29 @ 21°/4°, mp: 300° (decomp).

SYNS: BARIUM FLUOROSILICATE ◇ BARIUM FLUOSILICATE ◇ BARIUM HEXAFLUOROSILICATE ◇ BARIUM HEXAFLUOROSILICATE(2-) ◇ BARIUM SILICON FLUORIDE

TOXICITY DATA with REFERENCE
orl-rat LD50:175 mg/kg AFDOAQ 15,122,51
orl-rbt LDLo:175 mg/kg JPETAB 39,246,30

CONSENSUS REPORTS: Reported in EPA TSCA Inventory. Barium and its compounds are on the Community Right-To-Know List.

OSHA PEL: 8H TWA 0.5 mg(Ba)/m^3; TWA 2.5 mg(F)/m^3
ACGIH TLV: TWA 0.5 mg(Ba)/m^3
NIOSH REL: (Fluorides, Inorganic) TWA 2.5 mg(F)/m^3

SAFETY PROFILE: A poison by ingestion. When heated to decomposition it emits toxic fumes of F^-. See also BARIUM COMPOUNDS (soluble).

BAO825 CAS:6865-35-6 ***HR: 2***
BARIUM STEARATE
mf: $C_{36}H_{72}O_4•Ba$ mw: 706.42

SYNS: BARIUM DISTEARATE ◇ OCTADECANOIC ACID, BARIUM SALT (9CI) ◇ STAVINOR 40 ◇ STEARIC ACID, BARIUM SALT

TOXICITY DATA with REFERENCE
orl-rat LD50:4 g/kg GISAAA 39(11),91,74
orl-mus LD50:3500 mg/kg GISAAA 39(11),91,74
orl-gpg LD50:3600 mg/kg GISAAA 39(11),91,74

CONSENSUS REPORTS: Barium and its compounds are on the Community Right-To-Know List.

SAFETY PROFILE: Moderately toxic by ingestion. When heated to decomposition it emits acrid smoke and fumes. See also BARIUM COMPOUNDS.

BAP000 CAS:7727-43-7 ***HR: 2***
BARIUM SULFATE
mf: $O_4S•Ba$ mw: 233.40

PROP: White, heavy, odorless powder. D: 4.50 @ 15°, mp: 1580°. Insol in water or dilute acids.

SYNS: ACTYBARYTE ◇ ARTIFICIAL BARITE ◇ ARTIFICIAL HEAVY SPAR ◇ BAKONTAL ◇ BARIDOL ◇ BARITE ◇ BARITOP ◇ BAROSPERSE ◇ BAROTRAST ◇ BARYTA WHITE ◇ BARYTES ◇ BAYRITES ◇ BLANC FIXE ◇ C.I. 77120 ◇ C.I. PIGMENT WHITE 21 ◇ CITOBARYUM ◇ COLONATRAST ◇ ENAMEL WHITE ◇ ESOPHOTRAST ◇ EWEISS ◇ E-Z-PAQUE ◇ FINEMEAL ◇ LACTOBARYT ◇ LIQUIBARINE ◇ MACROPAQUE ◇ NEOBAR ◇ ORATRAST ◇ PERMANENT WHITE ◇ PRECIPITATED BARIUM SULPHATE ◇ RAYBAR ◇ REDI-FLOW ◇ SOLBAR ◇ SULFURIC ACID, BARIUM SALT (1:1) ◇ SUPRAMIKE ◇ TRAVAD ◇ UNIBARYT

TOXICITY DATA with REFERENCE
ipl-rat TDLo:200 mg/kg:ETA BJCAAI 28,173,73

CONSENSUS REPORTS: Reported in EPA TSCA Inventory. Barium and its compounds are on the Community Right-To-Know List.

OSHA PEL: (Transitional: Total Dust: TWA 15 mg/m^3; Respirable Fraction: 5 mg/m^3) Total Dust: TWA 10 mg/m^3; Respirable Fraction: 5 mg/m^3
ACGIH TLV: TWA (nuisance particulate) 10 mg/m^3 of total dust (when toxic impurities are not present, e.g., quartz < 1%).

SAFETY PROFILE: Questionable carcinogen with experimental tumorigenic data. A relatively insoluble salt used as an opaque medium in radiography. Soluble impurities can lead to toxic reactions. Heating with aluminum can produce an explosion. Incompatible with aluminum and potassium. When heated to decomposition it emits toxic fumes of SO_x.

BAP250 CAS:21109-95-5 ***HR: 3***
BARIUM SULFIDE
mf: BaS mw: 169.4

PROP: Cubic, colorless crystals. D: 4.25 @ 15°, mp: 1200°.

CONSENSUS REPORTS: Barium and its compounds are on the Community Right-To-Know List.

SAFETY PROFILE: A poison. Flammable by spontaneous chemical reaction, air, moisture, or acid fumes may cause it to ignite. For explosion and disaster hazards, see SULFIDES. To fight fire, use CO_2, dry chemical. Reacts violently with phosphorous (V) oxide. Mixtures with lead dioxide, potassium chlorate, or potassium nitrite explode when heated. Incompatible with Cl_2O, $Ca(NO_3)_2$, $Sr(NO_3)_2$, $Ca(ClO_3)_2$, $Sr(ClO_3)_2$, $(ClO_3)_2$. See also BARIUM COMPOUNDS (soluble) and SULFIDES.

BAP500 ***HR: 3***
BARIUM THIOCYANATE
mf: $C_2BaN_2S_2$ mw: 253.52

CONSENSUS REPORTS: Barium and its compounds are on the Community Right-To-Know List.

SAFETY PROFILE: A deadly poison. Incompatible with potassium chlorate, sodium nitrate. When heated to decomposition it emits toxic fumes of SO_x and NO_x. See also BARIUM COMPOUNDS and THIOCYANATES.

BAP750 CAS:12009-21-1 ***HR: 2***
BARIUM ZIRCONIUM(IV) OXIDE
mf: O_4Zr_4•Ba mw: 566.22

PROP: Light gray-buff powder. D: 5.52, mp: 2510°. Insol in water and alkalies; sltly sol in acid.

SYNS: BARIUM ZIRCONATE ◇ BARIUM ZIRCONIUM OXIDE ◇ BARIUM ZIRCONIUM TRIOXIDE ◇ ZIRCONATE, BARIUM (1:1)

TOXICITY DATA with REFERENCE
orl-rat LD50:1980 mg/kg AIHAAP 24,131,63
ipr-rat LD50:420 mg/kg AIHAAP 24,131,63

CONSENSUS REPORTS: Reported in EPA TSCA Inventory. Barium and its compounds are on the Community Right-To-Know List.

OSHA PEL: (Transitional: TWA 5 mg(Zr)/m^3) TWA 5 mg(Zr)/m^3; STEL 10 mg(Zr)/m^3
ACGIH TLV: TWA 5 mg(Zr)/m^3; STEL 10 mg(Zr)/m^3
DFG MAK: 5 mg(Zr)/m^3

SAFETY PROFILE: Moderately toxic by ingestion and intraperitoneal routes. Inhalation produces interstitial pneumonitis. See also ZIRCONIUM COMPOUNDS and BARIUM COMPOUNDS.

BAQ250 CAS:65-61-2 ***HR: 3***
BASIC ORANGE 3RN
mf: $C_{17}H_{19}N_3$•ClHZnCl$_2$ mw: 438.12

PROP: Mp: 182°.

SYNS: ACRIDINE ORANGE ◇ ACRIDINE ORANGE NO ◇ ACRIDINE ORANGE R ◇ C.I. 46005 ◇ C.I. BASIC ORANGE 14 ◇ RHODULINE ORANGE NO ◇ N,N,N′,N′-TETRAMETHYL-3,6-ACRIDINEDIAMINE MONOHYDROCHLORIDE (9CI)

TOXICITY DATA with REFERENCE
dnd-mus:ast 20 μmol/L BBACAQ 374,96,74
dnd-mam:lym 10 pph BIPMAA 11,2537,72
dnd-sal:spr 40 μmol/L BBRCA9 40,1239,70
ipr-mus LDLo:64 mg/kg CBCCT* 2,187,50

CONSENSUS REPORTS: EPA Genetic Toxicology Program. Reported in EPA TSCA Inventory. Zinc and its compounds are on the Community Right-To-Know List.

SAFETY PROFILE: Poison by intraperitoneal route. Mutation data reported. When heated to decomposition it emits very toxic fumes of HCl, Cl^-, and NO_x. See also ZINC COMPOUNDS.

BAQ750 CAS:14097-03-1 ***HR: D***
BASIC RED 18
mf: $C_{19}H_{25}ClN_5O_2$ mw: 390.94

SYNS: AIZEN CATHILON RED GTLH ◇ AMMONIUM (2-(p-((2-CHLORO-4-NITROPHENYL)AZO)PHENETHYLAMINO)ETHYL)-TRIMETHYL

CONSENSUS REPORTS: Reported in EPA TSCA Inventory.

SAFETY PROFILE: Mutation data reported. When heated to decomposition it emits very toxic fumes such as Cl^- and NO_x.

BAR250 CAS:8015-73-4 ***HR: 2***
BASIL OIL

PROP: Contains about 55% methyl chavicol and 35% of alcohols calculated as lenatoal and other compounds found in the leaves of *Ocimum resilium* L. (FCTXAR 11,855,73). a pale yellow liquid; floral, spicy odor. Sol in fixed oils and propylene glycol; insol in glycerin.

SYNS: BASIL OIL, EUROPEAN TYPE (FCC) ◇ OCIMUM BASILICUM OIL ◇ OIL of BASIL

TOXICITY DATA with REFERENCE
skn-mus 100 % MLD FCTXAV 11,867,73
orl-rat LD50:1400 mg/kg FCTXAV 11,855,73

CONSENSUS REPORTS: On EPA Extremely Hazardous Substances List by error. Reported in EPA TSCA Inventory.

SAFETY PROFILE: Moderately toxic by ingestion. A skin and eye irritant. When heated to decomposition it emits acrid smoke and irritating fumes.

BAR325 ***HR: 2***
BASKET FLOWER

PROP: Bulb-producing ornamental plants. The long, thin leaves emerge from the ground not from a stem. The flowers are white or yellow and grow from a leafless stem. The seeds are carried in a capsule. They are native to the southeastern United States and tropical areas of the Americas, and are commonly cultivated.

SYNS: ALLIGATOR LILY ◇ CROWN BEAUTY ◇ HYMENOCALLIS (VARIOUS SPECIES) ◇ LIRIO (SPANISH) ◇ SEA DAFFODIL ◇ SPIDER LILY ◇ TARARACO BLANCO (CUBA)

SAFETY PROFILE: The bulb contains the poison lycorine and similar alkaloids. Ingestion of large amounts may cause nausea, vomiting, and diarrhea.

BAR500 ***HR: 3***
BASORA CORRA

PROP: Aqueous extract from the root of the plant (JNCIAM 52,445,74).

SYN: MELOCHIA TOMENTOSA

TOXICITY DATA with REFERENCE
scu-rat TDLo:300 g/kg/60W-I:NEO JNCIAM 52,445,74

SAFETY PROFILE: Questionable carcinogen with experimental neoplastigenic data.

BAR750 CAS:23509-16-2 ***HR: 3***
BATRACHOTOXIN
mf: $C_{31}H_{42}N_2O_6$ mw: 538.75

PROP: Active principle from the skin of the Columbian arrow poison frog.

SYNS: 20-(2,4-DIMETHYL-1H-PYRROLE-3-CARBOXYLATE) BATRACHOTOXININ A ◇ 20-α-(2,4-DIMETHYL-1H-PYRROLE-3-CARBOXYLATE) BETRACHOTOXININ A

TOXICITY DATA with REFERENCE
ipr-mus LD50:2 μg/kg TOXIA6 7,315,69
scu-mus LD50:2 μg/kg CTOXAO 4,331,71

SAFETY PROFILE: A deadly poison by intraperitoneal and subcutaneous routes. When heated to decomposition it emits toxic fumes of NO_x.

BAR800 CAS:41621-49-2 ***HR: 3***
BATRAFEN
mf: $C_{12}H_{17}NO_2 \cdot C_2H_7NO$ mw: 268.40

SYNS: 2-AMINOETHANOL componded with 6-CYCLOHEXYL-1-HYDROXY-4-METHYL-2(1H)-PYRIDINONE (1:1) ◇ CIC ◇ CICLOPIROX ETHANOLAMINE SALT (1:1) ◇ CICLOPIROXOLAMIN ◇ CICLOPIROXOLAMINE ◇ 6-CYCLOHEXYL-1-HYDROXY-4-METHYL-2(1H)-PYRIDINONE compounded with 2-AMINOETHANOL (1:1) ◇ 6-CYCLOHEXYL-1-HYDROXY-4-METHYL-2(1H)-PYRIDON,2-AMINOETHANOL-SALZ (GERMAN) ◇ 6-CYCLOHEXYL-1-HYDROXY-4-METHYL-2(1H)-PYRIDONE, 2-AMINOETHANOL-SALT ◇ 6-CYCLOHEXYL-1-HYDROXY-4-METHYL-2(1H)-PYRIDONEETHANOLAMINE SALT ◇ HOE 296 ◇ LORPOX ◇ TERIT

TOXICITY DATA with REFERENCE
orl-rat TDLo:8400 mg/kg (28D pre):REP OYYAA2 9,67,75
orl-rat LD50:2350 mg/kg IYKEDH 8,107,77
ipr-rat LD50:146 mg/kg OYYAA2 9,57,75
scu-rat LD50:9800 mg/kg YKYUA6 28,115,77
ivn-rat LD50:72 mg/kg IYKEDH 8,107,77
orl-mus LD50:1740 mg/kg IYKEDH 8,107,77
ipr-mus LD50:83 mg/kg OYYAA2 9,57,75
scu-mus LD50:1730 mg/kg IYKEDH 8,107,77
ivn-mus LD50:71 mg/kg OYYAA2 9,57,75

SAFETY PROFILE: Poison by intravenous and intraperitoneal routes. Moderately toxic by ingestion and subcutaneous routes. Experimental reproductive effects. When heated to decomposition it emits toxic fumes of NO_x.

BAR825 CAS:64314-28-9 ***HR: 3***
BAUMYCIN A1
mf: $C_{34}H_{43}NO_{13}$ mw: 673.78

TOXICITY DATA with REFERENCE
pic-esc 50 ng/plate CNREA8 43,2819,83
dni-mus:leu 1700 nmol/L JANTAJ 34,1596,81
oms-mus:leu 560 nmol/L JANTAJ 34,1596,81
ipr-mus LD50:1500 μg/kg JANTAJ 31,78-67,78

SAFETY PROFILE: Poison by intraperitoneal route. Mutation data reported. When heated to decomposition it emits toxic fumes of NO_x.

BAR830 CAS:64253-71-0 ***HR: 3***
BAUMYCIN A2
mf: $C_{34}H_{43}NO_{13}$ mw: 673.78

TOXICITY DATA with REFERENCE
pic-esc 50 ng/plate CNREA8 43,2819,83
dni-mus:leu 1900 nmol/L JANTAJ 34,1596,81
oms-mus:leu 710 nmol/L JANTAJ 34,1596,81
ipr-mus LD50:15 mg/kg JANTAJ 31,78-68,78

SAFETY PROFILE: Poison by intraperitoneal route. Mutation data reported. When heated to decomposition it emits toxic fumes of NO_x.

BAS000 CAS:7682-90-8 ***HR: 3***
BAY 75546
mf: $C_{12}H_{17}BrN_3O_3PS$ mw: 394.26

SYN: 3-BROMO-5,7-DIMETHYLPYRAZOLYL-2-PYRIMIDINEPHOSPHOROTHIOIC ACID-O,O-DIETHYL ESTER.

TOXICITY DATA with REFERENCE
orl-rat LD50:2000 mg/kg TXAPA9 21,315,72
orl-bwd LD50:2400 μg/kg TXAPA9 21,315,72

SAFETY PROFILE: Poison by ingestion. When heated to decomposition it emits very toxic fumes of Br^-, NO_x, PO_x, and SO_x. See also ESTERS.

BAS500 CAS:39457-24-4 ***HR: 3***
BAY COE 3664

TOXICITY DATA with REFERENCE
orl-bwd LD50:2400 μg/kg TXAPA9 26,154,73
skn-bwd LD50:2400 μg/kg TXAPA9 26,154,73

SAFETY PROFILE: Poison by ingestion and skin contact.

BAT000 CAS:145-63-1 ***HR: 3***
BAYER 205
mf: $C_{51}H_{40}N_6O_{23}S_6$ mw: 1297.33

SYNS: ANTRYPOL ◇ BELGANYL ◇ FARMA ◇ FARMA 939 ◇ FOURNEAU ◇ FOURNEAU 309 ◇ GERMANIN ◇ SURAMINE

TOXICITY DATA with REFERENCE
unr-mus TDLo:195 mg/kg (female 9-11D post):REP CRSBAW 167,1717,73
scu-rat TDLo:250 mg/kg (female 9D post):TER CBINA8 58,149,86
ivn-man TDLo:46 mg/kg/5W-I:EYE NEJMAG 314,1455,86
ivn-mus LD50:620 mg/kg ADVPA3 15,289,78

SAFETY PROFILE: Moderately toxic intravenous route. Human systemic effects by intravenous route: eye effects. An experimental teratogen. Other experimental reproductive effects. When heated to decomposition it emits very toxic fumes of SO_x and NO_x.

BAT500 ***HR: 2***
BAY OIL

PROP: Consists mainly of eugenol and chavicol (55-65%), major portion of balance consists of terpenes (alpha-pinene, myrcene, and dipentene) small quantities of citro, nerol, cineol, and other terpenoids have also been found (FCTXAV 11,855,73). Yellow or brown liquid; aromatic odor, pungent, spicy taste. Sol in alc and glacial acetic acid.

SYNS: BAY LEAF OIL ◇ BOIS d'INDE ◇ LAUREL LEAF OIL ◇ MYRCIA OIL ◇ MYRICIA OIL ◇ OIL of BAY ◇ OIL of MYRCIA

TOXICITY DATA with REFERENCE
orl-rat LD50:1800 mg/kg FCTXAV 11,855,73

CONSENSUS REPORTS: Reported in EPA TSCA Inventory.

SAFETY PROFILE: Moderately toxic by ingestion. When heated to decomposition it emits acrid smoke.

BAT750 CAS:14816-18-3 ***HR: 3***
BAYTHION
mf: $C_{12}H_{15}N_2O_3PS$ mw: 298.32

SYNS: B 77488 ◇ BAY 5621 ◇ BAY 77488 ◇ BAYRE 77488 ◇ BENZOYL CYANIDE-o-(DIETHOXYPHOSPHINOTHIOYL)OXIME◇ O,O-DI-AETHYL-o-(α-CYANBENZYLIDEN-AMINO)-THIONPHOSPHAT(GERMAN) ◇ O,O-DIAETHYL-o-(α-CYANO-BENZYLIDENAMINO)-MONO-THIOPHOSPHAT (GERMAN) ◇ α-(((DIETHOXYPHOSPHINOTHIOYL)-OXY)IMINO)BENZENEACETONITRILE◇ (DIETHOXY-THIOPHO-SPHORYLOXYIMINO)-PHENYL ACETONITRILE ◇ O,O-DIETHYL PHOSPHOROTHIOATE, o-ESTER with PHENYLGLYOXYLONITRILE OXIME ◇ ENT 27,488 ◇ 4-ETHOXY-7-PHENYL-3,5-DIOXA-6-AZA-4-PHOSPHAOCT-6-ENE-8-NITRILE 4 SULFIDE ◇ PHENYLGLYO-XYLONITRILE OXIME-O,O-DIETHYL PHOSPHOROTHIOATE ◇ PHOXIME ◇ PHOXIN ◇ SEBACIL ◇ VALEXONE ◇ VOLATON

TOXICITY DATA with REFERENCE
orl-rat TDLo:49 mg/kg (74D male):TER GJSAAA 45(7),77,80
orl-rat LD50:300 mg/kg FAATDF 7,299,86
skn-rat LD50:1000 mg/kg 28ZEAL 5,181,76
orl-mus LD50:1050 mg/kg 52OLAC -,230,83
orl-dog LD50:250 mg/kg 28ZEAL 5,181,76
orl-cat LD50:250 mg/kg 28ZEAL 5,181,76
orl-rbt LD50:250 mg/kg 85DPAN -,-,71/76

CONSENSUS REPORTS: Cyanide and its compounds are on the Community Right-To-Know List.

SAFETY PROFILE: Poison by ingestion. An experimental teratogen. When heated to decomposition it emits very toxic fumes of CN^-, NO_x, PO_x, and SO_x. See also NITRILES.

BAT795 CAS:7548-44-9 ***HR: D***
BDH 2700
mf: $C_{22}H_{27}ClO_2$ mw: 358.94

SYNS: 21-CHLORO-3,17-DIMETHOXY-19-NOR-17-α-PREGN-1,3,5(10)-TRIEN-20-YNE ◇ 17-α-CHLOROETHYNYL-3,17-β-DIMETHOXY-OESTRA-1,3,5(10)-TRIENE

TOXICITY DATA with REFERENCE
orl-mus TDLo:30 μg/kg (2D preg):REP ACENA7 53,443,66

SAFETY PROFILE: Experimental reproductive effects. When heated to decomposition it emits toxic fumes of Cl^-.

BAT800 CAS:7548-46-1 ***HR: D***
BDH 6140
mf: $C_{22}H_{27}BrO_2$ mw: 403.40

SYN: 21-BROMO-3,17-DIMETHOXY-19-NOR-17-α-PREGNA-1,3,5(10)-TRIEN-20-YNE

TOXICITY DATA with REFERENCE
orl-mus TDLo:180 μg/kg (1-5D preg):REP ACENA7 53,443,66

SAFETY PROFILE: Experimental reproductive effects. When heated to decomposition it emits toxic fumes of Br^-.

BAT850 CAS:8021-39-4 ***HR: D***
BEECHWOOD CRESOATE

SYN: CRESOATE, WOOD ◇ RCRA WASTE NUMBER U051

TOXICITY DATA with REFERENCE
orl-mus TDLo:106 mg/kg (52W pre):REP OYYAA2 29,909,84

CONSENSUS REPORTS: Reported in EPA TSCA Inventory.

SAFETY PROFILE: Experimental reproductive effects. When heated to decomposition it emits acrid smoke and irritating fumes.

BAU000 CAS:8012-89-3 ***HR: 1***
BEESWAX

PROP: Yellow to brownish-yellow, soft to brittle wax. Mp: 62-65°, d: 0.95-0.96. Sol in chloroform, ether, fixed oils; sltly sol in alc.

SYNS: BEESWAX, WHITE ◇ BEESWAX, YELLOW

SAFETY PROFILE: A mild allergen. Combustible when heated.

BAU250 ***HR: 2***
"BEETLE"

PROP: Urea-formaldehyde condensation product.

SAFETY PROFILE: A mild irritant by inhalation. An allergen. Inhalation of dust may cause allergic response or irritation of lungs. Combustible. See also FORMALDEHYDE.

BAU255 CAS:39543-79-8 ***HR: 3***
BEFUNOLOL HYDROCHLORIDE
mf: $C_{16}H_{21}NO_4 \cdot ClH$ mw: 327.84

SYNS: 2-ACETYL-7-((2-HYDROXY-3-ISOPROPYLAMINO)PROPOXY)BENZOFURAN HYDROCHLORIDE ◇ BENTOX ◇ BFE 60 ◇ 1-(7-(2-HYDROXY-3-((1-METHYLETHYL)AMINO)PROPOXY)-2-BENZOFURANYL)ETHANONEHYDROCHLORIDE

TOXICITY DATA with REFERENCE
orl-mus TDLo:1300 mg/kg (female 15-21D post):REP KSRNAM 13,3740,79
orl-rat TDLo:1400 mg/kg (female 9-15D post):TER KSRNAM 13,3678,79
orl-rat LD50:922 mg/kg KSRNAM 13,4138,79
ipr-rat LD50:182 mg/kg IYKEDH 14,484,83
scu-rat LD50:498 mg/kg IYKEDH 14,484,83
orl-mus LD50:950 mg/kg IYKEDH 14,484,83
ipr-mus LD50:184 mg/kg IYKEDH 14,484,83
scu-mus LD50:434 mg/kg IYKEDH 14,484,83
ivn-mus LD50:65 mg/kg IYKEDH 14,484,83

SAFETY PROFILE: Poison by intravenous and intraperitoneal routes. Moderately toxic by ingestion and subcutaneous routes. An experimental teratogen. Other experimental reproductive effects. When heated to decomposition it emits toxic fumes of NO_x and HCl. A beta-adrenergic blocker.

BAU270 CAS:4696-76-8 ***HR: 3***
BEKANAMYCIN
mf: $C_{18}H_{37}N_5O_{10}$ mw: 483.60

SYNS: AMINODEOXYKANAMYCIN ◇ 2'-AMINO-2'-DEOXYKANAMYCIN ◇ KANAMYCIN B ◇ KANENDOMYCIN ◇ KDM ◇ NEBRAMYCIN FACTOR 5 ◇ NEBRAMYCIN V ◇ NK 1006 ◇ o-3-AMINO-3-DEOXY-α-d-GLUCOPYRANOSYL-(1-4)-o-(2,6-DIAMINO-2,6-DIDEOXYα-d-GLUCOPYRANOSYL-(1-6)-2-DEOXY-d-STREPTAMINE

TOXICITY DATA with REFERENCE
ipr-mus LD50:800 mg/kg 85GDA2 1,159,80
scu-mus LD50:750 mg/kg 85GDA2 1,159,80
ivn-mus LD50:132 mg/kg JANTAJ 27,677,74

CONSENSUS REPORTS: EPA Genetic Toxicology Program.

SAFETY PROFILE: Poison by intravenous route. Moderately toxic by intraperitoneal and subcutaneous routes. When heated to decomposition it emits toxic fumes of NO_x.

BAU325 ***HR: 2***
BELGENINE

PROP: Extracted from *Mallotus japonicus merel arg* (NIIRDN 6,768,82).

TOXICITY DATA with REFERENCE
ipr-rat LD50:3040 mg/kg NIIRDN 6,768,82
ivn-rat LD50:2800 mg/kg NIIRDN 6,768,82
ipr-mus LD50:6410 mg/kg NIIRDN 6,768,82
ivn-mus LD50:5400 mg/kg NIIRDN 6,768,82

SAFETY PROFILE: Moderately toxic by intraperitoneal and intravenous routes.

BAU500 ***HR: 3***
BELLADONNA

PROP: An extract from the deadly nightshade plant. The alkaloids atropine and belladonnine are derivatives.

SYN: DEADLY NIGHTSHADE

SAFETY PROFILE: A deadly poison. See also HYOSCYAMINE and ATROPINE. Local contact may cause a contact dermatitis. A poisonous constituent of some berries and plants, and of some folk remedies.

BAU750 CAS:147-24-0 ***HR: 3***
BENADRYL HYDROCHLORIDE
mf: $C_{17}H_{21}NO•ClH$ mw: 291.85

SYNS: AMBENYL ◇ BAX ◇ BENA ◇ BENADRYL ◇ BENDYLATE ◇ BENOCTEN ◇ BENZEHIST ◇ BENZHYDRAMINE HYDROCHLORIDE ◇ 2-(BENZHYDRYLOXY)-N,N-DIMETHYLETHYLAMINEHYDROCHLORIDE ◇ DABYLEN ◇ DIFENHYDRAMINE HYDROCHLORIDE ◇ DIMETHYLAMINE BENZHYDRYL ESTER HYDROCHLORIDE ◇ β-DIMETHYLAMINOETHYL BENZHYDRYL ETHER HYDROCHLORIDE ◇ DIPHENYLHYDRAMINE HYDROCHLORIDE ◇ 2-(DIPHENYLMETHOXY)-N,N-DIMETHYL-ETHANAMINE HYDROCHLORIDE ◇ 2-DIPHENYLMETHOXY-N,N-DIMETHYLETHYLAMINE HYDROCHLORIDE ◇ DOLESTAN ◇ ELDADRYL ◇ FELBEN ◇ FENYLHIST ◇ HALBMOND ◇ α-HYDROXYDIPHENYLMETHANE-β-DIMETHYLAMINOETHYL ETHER HYDROCHLORIDE ◇ NCI-C56075 ◇ ROHYDRA ◇ SK-DIPHENHYDRAMINE ◇ VALDRENE ◇ WEHYDRYL

TOXICITY DATA with REFERENCE
orl-mus TDLo:800 mg/kg (female 8-12D post):REP TCMUD8 6,361,86
orl-rat TDLo:1 g/kg (6-15D preg):TER NTIS** PB83-180612
skn-cld TDLo:60 mg/kg/6H-I:EYE,PSY CPEDAM 25,163,86
orl-rat LD50:500 mg/kg NIIRDN 6,334,82
ipr-rat LD50:82 mg/kg JPETAB 102,250,51
scu-rat LD50:362 mg/kg TXAPA9 18,185,71
ivn-rat LD50:35 mg/kg YACHDS 12,2769,84
orl-mus LD50:114 mg/kg JPETAB 113,72,55
ipr-mus LD50:56 mg/kg JPETAB 112,318,54
scu-mus LD50:99200 μg/kg NYKZAU 54,33,58
ivn-mus LD50:20 mg/kg ARZNAD 5,72,55
ivn-dog LD50:24 mg/kg JPETAB 89,227,47

SAFETY PROFILE: Poison by ingestion, subcutaneous, intravenous, and intraperitoneal routes. Human systemic effects by skin contact: eye effects, hallucinations, distorted perceptions, and ataxia. Experimental teratogenic and reproductive effects. When heated to decomposition it emits very toxic fumes of NO_x and HCl. See also ESTERS and ETHERS.

BAV000 CAS:3813-05-6 ***HR: 2***
BENAZOLIN
mf: $C_9H_6O_3NClS$ mw: 243.6

PROP: White, crystalline solid. Mp: 193°.

SYNS: BEN-30 ◇ BENAZALOX ◇ BEN-CORNOX ◇ BENOPAN ◇ BENSECAL ◇ BENZAR ◇ 4-CHLORO-2-OXO-3(2H)-BENZOTHIAZOLEACETIC ACID ◇ 4-CHLORO-2-OXOBENZOTHIAZOLIN-3-YL ACETIC ACID ◇ CORNOX CWK ◇ CRESOPUR ◇ EUNASIN ◇ EX10781 ◇ GALIPAN ◇ GERBITOX ◇ GRASSLAND WEEDKILLER ◇ HERBAZOLIN ◇ KEROPUR ◇ LEGUMEX EXTRA ◇ LEY-CORNOX ◇ LEYMIN ◇ METIZOLIN ◇ RD7693 ◇ TRI-CORNOX SPECIAL

TOXICITY DATA with REFERENCE
orl-rat LD50:3000 mg/kg 85ARAE 2,26,77

SAFETY PROFILE: Moderately toxic by ingestion. An herbicide. When heated to decomposition it emits toxic fumes of SO_x, Cl^-, and NO_x. See also CHLORIDES.

BAV250 CAS:14286-84-1 ***HR: 3***
BENCYCLANE FUMARATE
mf: $C_{19}H_{31}NO•C_4H_2O_4$ mw: 403.57

SYNS: BENCICLANE ◇ BENCYCLANE ◇ 3-((1-BENZYLCYCLOHEPTYL)OXY)-N,N-DIMETHYLPROPYLAMINEFUMARATE ◇ N-(3-(1-BENZYL-CYCLOHEPTYLOXY)-PROPYL)-N,N-DIMETHYL-AMMONIUM-HYDROGENFUMARAT (GERMAN) ◇ EGYT 201 ◇ FLUDILAT ◇ HALIDO

TOXICITY DATA with REFERENCE
orl-rat LD50:414 mg/kg 27ZQAG -,383,72
ipr-rat LD50:86 mg/kg 27ZQAG -,383,72
scu-rat LD50:257 mg/kg 27ZQAG -,383,72
ivn-rat LD50:41 mg/kg 27ZQAG -,383,72
orl-mus LD50:446 mg/kg 27ZQAG -,383,72
ipr-mus LD50:132 mg/kg 27ZQAG -,383,72
scu-mus LD50:203 mg/kg 27ZQAG -,383,72
ivn-mus LD50:45 mg/kg AITEAT 15,415,67
ims-mus LD50:150 mg/kg AITEAT 15,415,67
orl-dog LDLo:300 mg/kg ARZNAD 20,1385,70

SAFETY PROFILE: Poison by ingestion, intramuscular, intraperitoneal, subcutaneous, and intravenous routes. When heated to decomposition it emits toxic fumes of NO_x.

BAV275 CAS:53716-43-1 ***HR: D***
BENDACORT
mf: $C_{37}H_{42}N_2O_7$ mw: 626.81

SYNS: AF 2071 ◇ 11-β,17-DIHYDROXY-21-((((1-PHENYLMETHYL)-1H-INDAZOL-3-YL)OXY)ACETYLOXY)PREGN-4-ENE-3,20-DIONE

TOXICITY DATA with REFERENCE
orl-rat TDLo:3460 mg/kg (6-15D preg):TER BCFAAI 119,391,80

SAFETY PROFILE: An experimental teratogen. When heated to decomposition it emits toxic fumes of NO_x.

BAV325 CAS:20187-55-7 ***HR: 3***
BENDAZOLIC ACID
mf: $C_{16}H_{14}N_2O_3$ mw: 282.32

PROP: Crystals from ethanol. Mp: 160°. Practically insol in water; sol in chloroform, acetone.

SYNS: AF 983 ◇ BENDAZAC ◇ ((1-BENZYL-1H-INDAZOL-3-YL)-OXY)ACETIC ACID ◇ BINDAZAC ◇ ((1-(PHENYLMETHYL)-1H-INDAZOL-3-YL)OXY)-ACETIC ACID (9CI) ◇ VERSUS ◇ ZILDASAC

TOXICITY DATA with REFERENCE
orl-rat LD50:1200 mg/kg MEIEDD 10,146,83
ipr-rat LD50:319 mg/kg IYKEDH 10,884,79
scu-rat LD50:714 mg/kg IYKEDH 10,884,79
ivn-rat LD50:304 mg/kg MEIEDD 10,146,83
orl-mus LD50:1105 mg/kg MEIEDD 10,146,83

ipr-mus LD50:339 mg/kg IYKEDH 10,884,79
scu-mus LD50:406 mg/kg IYKEDH 10,884,79
ivn-mus LD50:380 mg/kg MEIEDD 10,146,83

SAFETY PROFILE: Poison by intravenous and intraperitoneal routes. Moderately toxic by ingestion and subcutaneous routes. When heated to decomposition it emits toxic fumes of NO_x.

BAV350 CAS:8064-77-5 ***HR: 3***
BENDECTIN
mf: $C_{19}H_{35}NO_2 \cdot C_{17}H_{22}N_2O \cdot C_8H_{11}NO_3 \cdot C_4H_6O_4 \cdot 2ClH$
mw: 940.18

SYNS: DEBENDOX ◇ LENOTAN

TOXICITY DATA with REFERENCE
orl-wmn TDLo:75600 μg/kg (female 6-11W post):REP BMJOAE 1,691,78
orl-wmn TDLo:60 mg/kg (female 4-18W post):TER BMJOAE 1,691,78

SAFETY PROFILE: Human reproductive effects by ingestion: developmental abnormalities of the gastrointestinal system. Human and experimental teratogenic and reproductive effects. When heated to decomposition it emits toxic fumes of NO_x and HCl.

BAV500 ***HR: 3***
BENLATE and SODIUM NITRITE

SYNS: 1-(BUTYLCARBAMOYL)-2-BENZIMIDAZOLECARBAMIC ACID METHYL ESTER and SODIUM NITRITE (1:6) ◇ SODIUM NITRITE and BENLATE

TOXICITY DATA with REFERENCE
orl-mus TDLo:31 g/kg/26W-I:CAR NEOLA4 24,119,77

SAFETY PROFILE: Questionable carcinogen with experimental carcinogenic data. When heated to decomposition it emits toxic fumes of Na_2O and NO_x. See also CARBAMATES, ESTERS, and NITRITES.

BAV575 CAS:17804-35-2 ***HR: 3***
BENOMYL
mf: $C_{14}H_{18}N_4O_3$ mw: 290.36

SYNS: ARILATE ◇ BBC ◇ BENLATE 50 ◇ BENOMYL 50W ◇ BNM ◇ 1-(BUTYLCARBAMOYL)-2-BENZIMIDAZOLECARBAMIC ACID, METHYL ESTER ◇ 1-(BUTYLCARBAMOYL)-2-BENZIMIDAZOL-METHYLCARBAMAT (GERMAN) ◇ 1-(N-BUTYLCARBAMOYL)-2-(METHOXY-CARBOXAMIDO)-BENZIMIDAZOL (GERMAN) ◇ DU PONT 1991 ◇ FUNDASOL ◇ FUNGICIDE 1991 ◇ MBC ◇ METHYL-1-(BUTYLCARBAMOYL)-2-BENZIMIDAZOLYLCARBAMATE ◇ TERSAN 1991

TOXICITY DATA with REFERENCE
skn-man 0.1% MLD LANCAO 2,1252,80
sln-smc 123 ppm ANYAA9 407,186,83
sln-hmn:lym 10 mg/L MUREAV 121,139,83
mmo-asn 250 μg/L MUREAV 91,115,81
orl-rat TDLo:936 mg/kg (female 7-22D post):REP TXAPA9 62,44,82
orl-mus TDLo:2200 mg/kg (female 7-17D post):TER TXAPA9 62,44,82
orl-rat LD50:10 g/kg JHEMA2 24,295,80
ihl-rat LD50:9920 mg/kg EQSFAP 3,618,75
orl-mus LD50:5600 mg/kg 17QLAD 12,85,77
unr-mam LD50:10 g/kg GISAAA 41(8),82,76
orl-bwd LD50:100 mg/kg TXAPA9 21,315,72

CONSENSUS REPORTS: Reported in EPA TSCA Inventory. EPA Genetic Toxicology Program.

OSHA PEL: (Transitional: Total Dust: TWA 15 mg/m^3; Respirable Fraction: 5 mg/m^3) Total Dust: TWA 10 mg/m^3; Respirable Fraction: 5 mg/m^3
ACGIH TLV: TWA 10 mg/m^3

SAFETY PROFILE: Poison by ingestion. Mildly toxic by inhalation. Experimental teratogenic and reproductive effects. Human mutation data reported. A human skin irritant. When heated to decomposition it emits toxic fumes of NO_x. See also CARBAMATES.

BAV625 CAS:29462-18-8 ***HR: 2***
BENTAZEPAM
mf: $C_{17}H_{16}N_2OS$ mw: 296.41

SYNS: 1,3,6,7,8,9-HEXAHYDRO-5-PHENYL-2H-(1)BENZOTHIENO-(2,3-e)-1,4-DIAZEPIN-2-ONE ◇ QM-6008 ◇ 6,7-TETRAMETHYLENE-5-PHENYL-1,2-DIHYDRO-3H-THIENO(2,3-e)(1,4)DIAZEPIN-2-ONE ◇ THIADIPONE ◇ TIADIPONE

TOXICITY DATA with REFERENCE
orl-rat LD50:2 g/kg ARZNAD 26,926,75
orl-mus LD50:980 mg/kg ARZNAD 25,926,75
ipr-mus LD50:630 mg/kg ARZNAD 25,926,75

SAFETY PROFILE: Moderately toxic by ingestion and intraperitoneal routes. When heated to decomposition it emits toxic fumes of SO_x and NO_x.

BAV750 CAS:1302-78-9 ***HR: 1***
BENTONITE

PROP: A clay containing appreciable amounts of the clay mineral montmorillonite; light yellow or green, cream, pink, gray to black solid. Insol in water and common organic solvents.

SYNS: ALBAGEL PREMIUM USP 4444 ◇ BENTONITE 2073 ◇ BENTONITE MAGMA ◇ HI-JEL ◇ IMVITE I.G.B.A. ◇ MAGBOND ◇ MONTMORILLONITE ◇ PANTHER CREEK BENTONITE ◇ SOUTHERN BENTONITE ◇ TIXOTON ◇ VOLCLAY ◇ VOLCLAY BENTONITE BC ◇ WILKINITE

TOXICITY DATA with REFERENCE
orl-mus TDLo:12000 g/kg/28W-C:ETA ANYAA9 57,678,54
ivn-rat LD50:35 mg/kg BSIBAC 44,1685,68

CONSENSUS REPORTS: Reported in EPA TSCA Inventory.

SAFETY PROFILE: Poison by intravenous route causing blood clotting. Questionable carcinogen with experimental tumorigenic data.

BAW000 CAS:7093-10-9 ***HR: 3***
BENZ(1)ACEANTHRENE
mf: $C_{20}H_{14}$ mw: 254.34

SYNS: 8:9-ACE-1:2-BENZANTHRACENE ◇ 1,2-DIHYDROBENZ-(1)ACEANTHRYLENE ◇ 8:9-DIMETHYLENE-1:2-BENZANTHRACENE

TOXICITY DATA with REFERENCE
scu-mus TDLo:800 mg/kg/13W-I:ETA AJCAA7 28,334,36

SAFETY PROFILE: Questionable carcinogen with experimental tumorigenic data. When heated to decomposition it emits acrid smoke and fumes.

BAW125 CAS:211-91-6 ***HR: D***
BENZ(1)ACEANTHRYLENE
mf: $C_{20}H_{12}$ mw: 252.32

TOXICITY DATA with REFERENCE
mma-sat 30 μg/plate CNREA8 44,4993,84
cyt-mus:lym 2500 μg/L ENMUDM 8(Suppl 6),24,86
msc-ham:lng 5 mg/L CNREA8 44,4993,84

SAFETY PROFILE: Mutation data reported. When heated to decomposition it emits acrid smoke and fumes.

BAW250 CAS:205-99-2 ***HR: 3***
BENZ(e)ACEPHENANTHRYLENE
mf: $C_{20}H_{12}$ mw: 252.32

PROP: Mp: 168°.

SYNS: 3,4-BENZ(e)ACEPHENANTHRYLENE ◇ 2,3-BENZFLUORANTHENE ◇ 3,4-BENZFLUORANTHENE ◇ BENZO(b)FLUORANTHENE ◇ BENZO(e)FLUORANTHENE ◇ 2,3-BENZOFLUORANTHENE ◇ 3,4-BENZOFLUORANTHENE ◇ 2,3-BENZOFLUORANTHRENE ◇ B(b)F

TOXICITY DATA with REFERENCE
mma-sat 31 nmol/plate CRNGDP 6,1023,85
otr-ham:lng 100 μg/L TXCYAC 17,149,80
sce-ham-ipr 900 mg/kg/24H MUREAV 66,65,79
imp-rat TDLo:5 mg/kg:ETA JJIND8 71,539,83
skn-mus TDLo:88 ng/kg/120W-I:CAR ARGEAR 50,266,80
ipr-mus TDLo:5046 μg/kg/15D-I:NEO CALEDQ 34,15,87

CONSENSUS REPORTS: NTP Fifth Annual Report on Carcinogens. IARC Cancer Review: Group 2B IMEMDT 7,56,87, Animal Sufficient Evidence IMEMDT 32,147,83; IMEMDT 3,69,73. EPA Genetic Toxicology Program.

ACGIH TLV: (Proposed: Suspected Carcinogen)

SAFETY PROFILE: Confirmed carcinogen with experimental carcinogenic and tumorigenic data. Mutation data reported. When heated to decomposition it emits acrid smoke and irritating fumes.

BAW500 CAS:71-79-4 ***HR: 3***
BENZACINE HYDROCHLORIDE
mf: $C_{18}H_{21}O_3ClH$ mw: 403.28

SYNS: BENZACIN ◇ BENZACINE ◇ BENZACIN HYDROCHLORIDE ◇ DIMETHYLAMINOETHYL BENZILATE, HYDROCHLORIDE ◇ β-DIMETHYLAMINOETHYL BENZILATE HYDROCHLORIDE ◇ 2-(DIMETHYLAMINO)ETHYL BENZILATE HYDROCHLORIDE ◇ DIMETHYLAMINOETHYL BENZYLATE HYDROCHLORIDE ◇ DIMETHYLAMINOETHYL DIPHENYLHYDROXYACETATE HYDROCHLORIDE ◇ HK-141

TOXICITY DATA with REFERENCE
orl-rat LD50:1035 mg/kg JLCMAK 30,700,45
ivn-rat LD50:30 mg/kg JLCMAK 30,700,45
orl-mus LD50:281 mg/kg JLCMAK 30,700,45
ipr-mus LD50:137 mg/kg PCJOAU 2,201,68
ivn-mus LD50:40 mg/kg JLCMAK 30,700,45

SAFETY PROFILE: Poison by ingestion, intravenous, and intraperitoneal routes. When heated to decomposition it emits toxic fumes of HCl.

BAW750 CAS:225-51-4 ***HR: 3***
BENZ(c)ACRIDINE
mf: $C_{17}H_{11}N$ mw: 229.29

PROP: Mp: 108°.

SYNS: 12-AZABENZ(a)ANTHRACENE ◇ B(c)AC ◇ 3,4-BENZACRIDINE ◇ 7,8-BENZACRIDINE (FRENCH) ◇ 3,4-BENZOACRIDINE ◇ α-CHRYSIDINE ◇ α-NAPHTHACRIDINE ◇ RCRA WASTE NUMBER U016

TOXICITY DATA with REFERENCE
mma-sat 1 nmol/plate GANNA2 70,749,79
sce-ham:ovr 10 μmol/L MUREAV 118,103,83
sce-ham:lng 1 μmol/L MUREAV 118,103,83
skn-mus TDLo:2400 mg/kg/67W-I:ETA IJCAAR 5,183,68
ipr-mus TDLo:9630 mg/kg/3D-I:NEO CNREA8 44,5161,84

CONSENSUS REPORTS: IARC Cancer Review: Group 3 IMEMDT 7,56,87; Animal Sufficient Evidence IMEMDT 3,241,73; Animal Limited Evidence IMEMDT 32,129,83

SAFETY PROFILE: Questionable carcinogen with experimental neoplastigenic and tumorigenic data. Mutation data reported. When heated to decomposition it emits toxic fumes of NO_x.

BAX000 CAS:3123-27-1 ***HR: 3***
BENZ(c)ACRIDINE-7-CARBONITRILE
mf: $C_{18}H_{10}N_2$ mw: 254.30

SYNS: 7-CYANOBENZ(c)ACRIDINE ◇ 7-CYANOBENZO(c)ACRIDINE

TOXICITY DATA with REFERENCE
scu-mus TDLo:120 mg/kg/9W-I:ETA CHDDAT 267,981,68

CONSENSUS REPORTS: Cyanide and its compounds are on the Community Right-To-Know List.

SAFETY PROFILE: Questionable carcinogen with experimental tumorigenic data. See also NITRILES. When heated to decomposition it emits toxic fumes of NO_x and CN^-.

BAX250 CAS:3301-75-5 ***HR: 3***
BENZ(c)ACRIDINE-7-CARBOXALDEHYDE
mf: $C_{18}H_{11}NO$ mw: 257.30

SYNS: 3,4-BENZACRIDINE-9-ALDEHYDE ◇ 7-FORMYLBENZ(c)ACRIDINE ◇ 7-FORMYLBENZO(c)ACRIDINE

TOXICITY DATA with REFERENCE
mma-sat 10 μg/plate CRNGDP 7,23,86
scu-mus TDLo:200 mg/kg:ETA VOONAW 1,52,55

SAFETY PROFILE: Questionable carcinogen with experimental tumorigenic data. Mutation data reported. When heated to decomposition it emits toxic fumes of NO_x. See also ALDEHYDES.

BAY250 CAS:63019-50-1 ***HR: 3***
α-(BENZ(c)ACRIDIN-7-YL)-N-(p-(DIMETHYLAMINO)PHENYL)NITRONE
mf: $C_{26}H_{21}N_3O$ mw: 391.50

SYN: α-(9-(3,4-BENZACRIDYL)-N-(p-DIMETHYLAMINO-PHENYL)-NITRONE

TOXICITY DATA with REFERENCE
scu-mus TDLo:200 mg/kg:ETA VOONAW 1,52,55

SAFETY PROFILE: Questionable carcinogen with experimental tumorigenic data. When heated to decomposition it emits toxic fumes of NO_x.

BAY300 CAS:98-87-3 ***HR: 3***
BENZAL CHLORIDE
DOT: UN 1886
mf: $C_7H_6Cl_2$ mw: 161.03

PROP: Very refractive liquid. Mp: −16°, bp: 214°, d: 1.29.

SYNS: BENZYL DICHLORIDE ◇ BENZYLENE CHLORIDE ◇ BENZYLIDENE CHLORIDE (DOT) ◇ CHLORURE de BENZYLIDENE (FRENCH) ◇ α,α-DICHLOROTOLUENE ◇ RCRA WASTE NUMBER U017

TOXICITY DATA with REFERENCE
mma-sat 600 nmol/plate/20M MUREAV 54,143,78
mma-esc 600 nmol/plate/20M MUREAV 54,143,78
mrc-bcs 31 μmol/disc MUREAV 54,143,78
skn-mus TDLo:9200 mg/kg/50W-I:CAR GANNA2 72,655,81
skn-mus TD:35200 mg/kg/42W-I:NEO GANNA2 72,655,81
orl-rat LD50:3249 mg/kg NTIS** PB214-270
ihl-rat LD50:61 ppm/2H IARC** 29,65,82
orl-mus LD50:2462 mg/kg AMRL** TR-72-62/72
ihl-mus LD50:32 ppm/2H IARC** 29,65,82

CONSENSUS REPORTS: IARC Cancer Review: Human Inadequate Evidence IMEMDT 29,65,82; Animal Limited Evidence IMEMDT 29,65,82. Reported in EPA TSCA Inventory. EPA Genetic Toxicology Program. EPA Extremely Hazardous Substances List. Community Right-To-Know List.

DFG MAK: Suspected Carcinogen.
DOT Classification: Poison B; Label: Poison.

SAFETY PROFILE: Suspected carcinogen with experimental carcinogenic and neoplastigenic data. Poison by inhalation. Moderately toxic by ingestion. A suspected human carcinogenic. A strong irritant and lachrymator. Causes central nervous system depression. Mutation data reported. When heated to decomposition it emits toxic fumes of Cl^-. See also CHLORINATED HYDROCARBONS, AROMATIC.

BAY500 CAS:100-52-7 ***HR: 3***
BENZALDEHYDE
DOT: UN 1989
mf: C_7H_6O mw: 106.13

PROP: Colorless liquid; burning taste with bitter almond odor. Mp: −26°, bp: 179°, flash p: 148°F, d: 1.041, autoign temp: 377°F, vap press: 1 mm @ 26.2°, vap d: 3.65, refr index: 1.544. Sltly sol in water; misc in alc, ether, oils.

SYNS: ALMOND ARTIFICIAL ESSENTIAL OIL ◇ ARTIFICIAL ALMOND OIL ◇ BENZENECARBALDEHYDE ◇ BENZENECARBONAL ◇ BENZOIC ALDEHYDE ◇ FEMA No. 2127 ◇ NCI-C56133

TOXICITY DATA with REFERENCE
skn-rbt 500 mg/24H MOD FCTXAV 14,659,76
orl-rat LD50:1300 mg/kg FCTXAV 2,327,64
scu-rat LDLo:5000 mg/kg AIPTAK 27,163,22
orl-mus LD50:28 mg/kg EJTXAZ 9,99,76
ipr-mus LD50:9 mg/kg EJTXAZ 9,99,76
scu-rbt LD50:5000 mg/kg FCTXAV 14,693,76
orl-gpg LD50:1000 mg/kg FCTXAV 2,327,64

CONSENSUS REPORTS: NTP Carcinogenesis Studies (gavage): Some Evidence: Mouse; NTP-TR-378,90; No

Evidence: Rat NTP-TR-378,90. EPA Genetic Toxicology Program. Reported in EPA TSCA Inventory.

DOT Classification: Combustible Liquid; Label: None.

SAFETY PROFILE: Poison by ingestion and intraperitoneal routes. Moderately toxic by subcutaneous route. An allergen. Acts as a feeble local anesthetic. Local contact may cause contact dermatitis. Causes central nervous system depression in small doses and convulsions in larger doses. A skin irritant. Questionable carcinogen. Mutation data reported. Combustible liquid. To fight fire, use water (may be used as a blanket), alcohol, foam, dry chemical. A strong reducing agent. Reacts violently with peroxyformic acid and other oxidizers. See also ALDEHYDES.

BAY750 CAS:633-03-4 ***HR: 3***
BENZALDEHYDE GREEN
mf: $C_{27}H_{33}N_2 \cdot HO_4S$ mw: 482.69

PROP: Bright green crystals.

SYNS: ADC BRILLIANT GREEN CRYSTALS ◇ AIZEN DIAMOND GREEN GH ◇ ANILINE GREEN ◇ ASTRA DIAMOND GREEN GX ◇ AVON GREEN A-4379 ◇ BASIC BRIGHT GREEN ◇ BRILLIANT GREEN SULFATE ◇ CALCOZINE BRILLIANT GREEN G ◇ C.I. 42040 ◇ C.I. BASIC GREEN 1, SULFATE (1:1) ◇ DEORLENE GREEN JJO ◇ DIAMOND GREEN G ◇ EMERALD GREEN ◇ ETHYL GREEN ◇ FAST GREEN JJO ◇ HIDACO BRILLIANT GREEN ◇ MALACHITE GREEN G ◇ MITSUI BRILLIANT GREEN G ◇ TERTROPHENE BRILLIANT GREEN G ◇ TOKYO ANILINE BRILLIANT GREEN

TOXICITY DATA with REFERENCE
skn-hmn 2 mg/2D-I MLD ADVEA4 52,55,72
skn-gpg 6 mg/3D-I ADVEA4 52,55,72
ipr-rat LDLo:8 mg/kg PSEBAA 31,825,34
ipr-mus LDLo:5 mg/kg PSEBAA 31,825,34
ivn-mus LDLo:3 mg/kg PSEBAA 31,825,34
ipr-gpg LDLo:3 mg/kg PSEBAA 31,825,34

CONSENSUS REPORTS: Reported in EPA TSCA Inventory.

SAFETY PROFILE: Poison by intraperitoneal and intravenous routes. A mild human skin irritant. See also ALDEHYDES and SULFATES. When heated to decomposition it emits very toxic fumes of NO_x, NH_3, and SO_x.

BAZ000 CAS:1627-73-2 ***HR: 3***
BENZALDEHYDE THIOSEMICARBAZONE
mf: $C_8H_9N_3S$ mw: 179.26

TOXICITY DATA with REFERENCE
orl-rat LDLo:500 mg/kg NCNSA6 5,44,53
orl-mus LD50:100 mg/kg JPPMAB 2,764,50

SAFETY PROFILE: Poison by ingestion. See also ALDEHYDES. When heated to decomposition it emits very toxic fumes of NO_x and SO_x.

BBA000 CAS:1708-39-0 ***HR: 2***
BENZAL GLYCERYL ACETAL
mf: $C_{10}H_{12}O_3$ mw: 180.22

PROP: Colorless to pale yellow liquid; mild almond odor. D: 1.183-1.193, refr index: 1.535-1.541, flash p: 165°F.

SYNS: BENZALDEHYDE GLYCERYL ACETAL (FCC) ◇ BENZYLIDENE GLYCEROL ◇ BUTYL PHENYL ACETATE ◇ FEMA No. 2209 ◇ 2-PHENYL-m-DIOXAN-5-OL

TOXICITY DATA with REFERENCE
orl-rat LD50:3150 mg/kg FCTXAV 14,699,76
ipr-mus LD50:1296 mg/kg AIPTAK 85,474,51
skn-rbt LD50:5000 mg/kg FCTXAV 14,699,76

SAFETY PROFILE: Moderately toxic by ingestion and intraperitoneal routes. Mildly toxic by skin contact. Combustible liquid. When heated to decomposition it emits acrid smoke and irritating fumes.

BBA500 ***HR: 3***
BENZALKONIUM CHLORIDE

PROP: White or yellowish-white powder, aromatic odor, very bitter taste.

SYNS: ALKYLDIMETHYLETHYLBENZYL AMMONIUM CHLORIDE ◇ ALKYL((ETHYLPHENYL)METHYL)DIMETHYL QUATERNARY AMMONIUM CHLORIDES ◇ BENIROL ◇ BTC 471 ◇ CEQUARTYL ◇ DRAPOLEX ◇ ENUCLEN ◇ GERMINOL ◇ GERMITOL ◇ OCTYL-OCTADECYL DIMETHYL ETHYLBENZYL AMMONIUM CHLORIDES ◇ PARALKAN ◇ ROCCAL ◇ RODALON ◇ ZEPHIRAN CHLORIDE ◇ ZEPHIROL

TOXICITY DATA with REFERENCE
eye-rat 2 mg FCTXAV 15,131,77
eye-mus 2 mg SEV FCTXAV 15,131,77
eye-dog 2 mg FCTXAV 15,131,77
eye-rbt mg SEV FCTXAV 15,131,77
eye-rbt 8 ug SEV AJOPAA 78,98,74
eye-rbt 10 mg MLD TXAPA9 55,501,80
eye-gpg 2 mg FCTXAV 15,131,77
eye-ham 2 mg SEV FCTXAV 15,131,77
orl-rat LD50:300 mg/kg 28ZEAL 4,38,69
skn-rat LD50:1420 mg/kg PCJOAU 12,1593,78
orl-mus LD50:150 mg/kg PCJOAU 12,1593,78

SAFETY PROFILE: Poison by ingestion. Moderately toxic by skin contact. A severe eye irritant. A bactericide and fungicide. Dangerous; when heated to decomposition it emits toxic fumes of Cl^- and NO_x. See also CHLORIDES.

BBA625 CAS:39387-42-3 ***HR: 3***
BENZALKONIUM SACCHARINATE

SYNS: AKYL DIMETHYL BENZYL AMMONIUM SACCHARINATE ◇ ALKYL DIMETHYL BENZALKONIUM SACCHARINATE ◇ HOLLICHEM HQ 3300 ◇ ONYXIDE 3300

TOXICITY DATA with REFERENCE
orl-rat LD50:990 mg/kg KSRNAM 4,219,70
ipr-rat LD50:37 mg/kg KSRNAM 4,219,70
scu-rat LD50:720 mg/kg KSRNAM 4,219,70
ivn-rat LD50:14500 μg/kg KSRNAM 4,219,70
orl-mus LD50:920 mg/kg KSRNAM 4,219,70
ipr-mus LD50:33 mg/kg KSRNAM 4,219,70
scu-mus LD50:790 mg/kg KSRNAM 4,219,70
ivn-mus LD50:23 mg/kg KSRNAM 4,219,70

SAFETY PROFILE: Poison by intravenous and intraperitoneal routes. Moderately toxic by ingestion and subcutaneous routes. When heated to decomposition it emits toxic fumes of NO_x.

BBA750 ***HR: 3***
BENZALMALONONITRILE
mf: $C_6H_5CH_2CH(CN)_2$ mw: 156.2

CONSENSUS REPORTS: Cyanide and its compounds are on the Community Right-To-Know List.

SAFETY PROFILE: Poison. See also NITRILES and CYANIDES.

BBB000 CAS:55-21-0 ***HR: 2***
BENZAMIDE
mf: C_7H_7NO mw: 121.15

SYNS: BENZOIC ACID AMIDE ◇ BENZOYLAMIDE ◇ PHENYLCARBOXYAMIDE

TOXICITY DATA with REFERENCE
sce-hmn:lym 1 mmol/L MUREAV 122,223,83
sce-ham:ovr 1 mmol/L MUREAV 123,63,84
mnt-mam:kdy 1 g/L IJEBA6 18,329,80
cyt-mam:kdy 1 g/L IJEBA6 18,329,80
orl-mus LD50:1160 mg/kg TXAPA9 19,20,71.

CONSENSUS REPORTS: Reported in EPA TSCA Inventory. Community Right-To-Know List. Human mutation data reported.

SAFETY PROFILE: Moderately toxic by ingestion. When heated to decomposition it emits toxic fumes of NO_x. See also AMIDES.

BBB250 CAS:1499-54-3 ***HR: D***
(2-BENZAMIDO)ACETOHYDROXAMIC ACID
mf: $C_9H_{10}N_2O_3$ mw: 194.21

SYNS: BENZOYLAMINOACETOHYDROXAMIC ACID ◇ HIPPUROHYDROXAMIC ACID

TOXICITY DATA with REFERENCE
dnr-bcs 10 μmol/disc JOPHDQ 3,557,80
mma-sat 5 μmol/plate MUREAV 56,7,77
mmo-sat 1 mg/plate AMACCQ 11,753,77

SAFETY PROFILE: Mutation data reported. When heated to decomposition it emits toxic fumes of NO_x.

BBB500 CAS:63018-69-9 ***HR: 3***
BENZ(a)ANTHRACEN-7-ACETONITRILE
mf: $C_{20}H_{13}N$ mw: 267.34

SYN: 10-CYANOMETHYL-1,2-BENZANTHRACENE

TOXICITY DATA with REFERENCE
scu-mus TDLo:600 mg/kg:ETA JNCIAM 1,303,40

CONSENSUS REPORTS: Cyanide and its compounds are on the Community Right-To-Know List.

SAFETY PROFILE: Questionable carcinogen with experimental tumorigenic data. See also NITRILES. When heated to decomposition it emits toxic fumes of NO_x and CN^-.

BBB750 CAS:2381-18-2 ***HR: 3***
BENZ(a)ANTHRACEN-7-AMINE
mf: $C_{18}H_{13}N$ mw: 243.32

SYN: 10-AMINO-1,2-BENZANTHRACENE

TOXICITY DATA with REFERENCE
scu-mus TDLo:1500 mg/kg/23W-I:ETA PRLBA4 129,439,40

SAFETY PROFILE: Questionable carcinogen with experimental tumorigenic data. When heated to decomposition it emits toxic fumes of NO_x. See also AROMATIC AMINES.

BBC000 CAS:56961-60-5 ***HR: 3***
BENZ(a)ANTHRACEN-8-AMINE
mf: $C_{18}H_{13}N$ mw: 243.32

SYN: 5-AMINO-1:2-BENZANTHRACENE

TOXICITY DATA with REFERENCE
scu-mus TDLo:400 mg/kg/3W-I:ETA PRLBA4 131,170,42

SAFETY PROFILE: Questionable carcinogen with experimental tumorigenic data. When heated to decomposition it emits toxic fumes of NO_x. See also AROMATIC AMINES.

BBC250 CAS:56-55-3 ***HR: 3***
BENZ(a)ANTHRACENE
mf: $C_{18}H_{12}$ mw: 228.30

PROP: Colorless leaflets or plates. Bp: 400°, mp: 160°.

SYNS: BA ◇ BENZANTHRACENE ◇ 1,2-BENZANTHRACENE ◇ 1,2-BENZ(a)ANTHRACENE ◇ 1,2-BENZANTHRAZEN (GERMAN) ◇ BENZANTHRENE ◇ 1,2-BENZANTHRENE ◇ BENZOANTHRACENE ◇ 1,2-BENZOANTHRACENE ◇ BENZO(a)ANTHRACENE ◇ BENZO(a)PHENANTHRENE ◇ BENZO(b)PHENANTHRENE ◇ 2,3-BENZOPHENANTHRENE ◇ 2,3-BENZPHENANTHRENE ◇ NAPHTHANTHRACENE ◇ RCRA WASTE NUMBER U018 ◇ TETRAPHENE

TOXICITY DATA with REFERENCE
mma-sat 4 μg/plate CRNGDP 5,747,84
msc-hmn:lym 9 μmol/L DTESD7 10,277,82
dni-hmn:oth 10 μmol/L CNREA8 42,3676,82
dnd-mus-skn 192 μmol/kg CRNGDP 5,231,84
skn-mus TDLo:18 mg/kg:NEO CNREA8 38,1699,78
scu-mus TDLo:2 mg/kg:ETA CNREA8 15,632,55
imp-mus TDLo:80 mg/kg:CAR BJCAAI 22,825,68
ivn-mus LDLo:10 mg/kg JNCIAM 1,225,40

CONSENSUS REPORTS: NTP Fifth Annual Report on Carcinogens. IARC Cancer Review: Group 2A IMEMDT 7,56,87, Animal Sufficient Evidence IMEMDT 32,135,83; IMEMDT 3,45,73. EPA Genetic Toxicology Program. Reported in EPA TSCA Inventory.

SAFETY PROFILE: Confirmed carcinogen with experimental carcinogenic, neoplastigenic, tumorigenic data by skin contact and other routes. Poison by intravenous route. Human mutation data reported. It is found in oils, waxes, smoke, food, drugs. When heated to decomposition it emits acrid smoke and irritating fumes.

BBC500 CAS:63018-40-6 ***HR: 3***
1,2-BENZANTHRACENE-10-ACETIC ACID, METHYL ESTER
mf: $C_{21}H_{16}O_2$ mw: 300.37

SYN: BENZ(a)ANTHRACEN-7-ACETIC ACID, METHYL ESTER

TOXICITY DATA with REFERENCE
scu-mus TDLo:600 mg/kg:ETA JNCIAM 1,303,40

SAFETY PROFILE: Questionable carcinogen with experimental tumorigenic data. When heated to decomposition it emits acrid smoke and fumes. See also ESTERS.

BBC750 CAS:7505-62-6 ***HR: 3***
BENZ(a)ANTHRACENE-7-CARBOXALDEHYDE
mf: $C_{19}H_{12}O$ mw: 256.31

SYN: 1,2-BENZANTHRACENE-10-ALDEHYDE

TOXICITY DATA with REFERENCE
scu-mus TDLo:280 mg/kg:ETA JNCIAM 1,303,40

SAFETY PROFILE: Questionable carcinogen with experimental tumorigenic data. When heated to decomposition it emits acrid smoke and fumes. See also ALDEHYDES.

BBD000 CAS:19926-22-8 ***HR: 3***
BENZ(a)ANTHRACENE-7,12-DICARBOXALDEHYDE
mf: $C_{20}H_{12}O_2$ mw: 284.32

SYN: 7,12-DIFORMYLBENZ(a)ANTHRACENE

TOXICITY DATA with REFERENCE
dnd-omi 2 mg/L PNASA6 74,1378,77
skn-mus TDLo:8000 mg/kg:NEO JJIND8 61,135,78

SAFETY PROFILE: Questionable carcinogen with experimental neoplastigenic data by skin contact. Mutation data reported. When heated to decomposition it emits acrid smoke and irritating fumes. See also ALDEHYDES.

BBD250 CAS:60967-88-6 ***HR: 3***
BENZ(a)ANTHRACENE-1,2-DIHYDRODIOL
mf: $C_{18}H_{14}O_2$ mw: 262.32

SYNS: BA-1,2-DIHYDRODIOL ◇ trans-1,2-DIHYDROXY-1,2-DIHYDROBENZ(a)ANTHRACENE

TOXICITY DATA with REFERENCE
mma-sat 10 μmol/L CNREA8 42,1620,82
msc-ham:lng 1200 μg/L/3H BJCAAI 39,540,79
skn-mus TDLo:2100 μg/kg:ETA CNREA8 38,1699,78

CONSENSUS REPORTS: EPA Genetic Toxicology Program.

SAFETY PROFILE: Questionable carcinogen with experimental tumorigenic data by skin contact. Mutation data reported. When heated to decomposition it emits acrid smoke and irritating fumes.

BBD500 CAS:60967-89-7 ***HR: 3***
BENZ(a)ANTHRACENE-3,4-DIHYDRODIOL
mf: $C_{18}H_{14}O_2$ mw: 262.32

SYNS: BA-3,4-DIHYDRODIOL ◇ trans-3,4-DIHYDRO-3,4-DIHYDROXYBENZO(a)ANTHRACENE ◇ trans-3,4-DIHYDROXY-3,4-DIHYDROBENZ(a)ANTHRACENE

TOXICITY DATA with REFERENCE
mma-sat 25 μmol/L BBRCA9 72,680,76
msc-ham:lng 2500 μg/L/3H BJCAAI 39,540,79
skn-mus TDLo:2100 μg/kg:NEO CNREA8 38,1699,79

CONSENSUS REPORTS: EPA Genetic Toxicology Program.

SAFETY PROFILE: Questionable carcinogen with experimental neoplastigenic data by skin contact. Mutation data reported. When heated to decomposition it emits acrid smoke and irritating fumes.

BBD750 CAS:67335-43-7 ***HR: 3***
(+)-(3S,4S)trans-BENZ(a)ANTHRACENE-3,4-DIHYDRODIOL
mf: $C_{18}H_{10}O_3$ mw: 274.28

SYNS: (+)-(3S,4S)-trans-3,4-DIHYDRO-3,4-DIHYDROXYBENZ(a)ANTHRACENE ◇ (+)-(3S,4S)-trans-3,4-DIHYDRO-3,4-DIHYDROXYBENZO(a)ANTHRACENE

TOXICITY DATA with REFERENCE
skn-mus TDLo:4390 μg/kg:ETA CNREA8 38,1705,78

SAFETY PROFILE: Questionable carcinogen with experimental tumorigenic data by skin contact. When heated to decomposition it emits acrid smoke and irritating fumes.

BBE000 CAS:32373-17-4 ***HR: D***
BENZ(a)ANTHRACENE-5,6-cis-DIHYDRODIOL
mf: $C_{16}H_{14}O_2$ mw: 238.30

SYNS: BA-5,6-cis-DIHYDRODIOL ◇ cis-5,6-DIHYDRO-5,6-DIHYDROXYBENZ(a)ANTHRACENE

TOXICITY DATA with REFERENCE
otr-ham:emb 2500 μg/L CNREA8 32,1391,72
dns-ham:emb 5 mg/L CBINA8 5,69,72

CONSENSUS REPORTS: EPA Genetic Toxicology Program.

SAFETY PROFILE: Mutation data reported. When heated to decomposition it emits acrid smoke and irritating fumes.

BBE250 CAS:3719-37-7 ***HR: 3***
BENZ(a)ANTHRACENE-5,6-DIHYDRODIOL
mf: $C_{18}H_{14}O_2$ mw: 262.32

SYNS: BA-5,6-DIHYDRODIOL ◇ BA-5,6-trans-DIHYDRODIOL ◇ BENZ(a)ANTHRACENE-5,6-trans-DIHYDRODIOL ◇ trans-5,6-DIHYDROXY-5,6-DIHYDROBENZ(a)ANTHRACENE

TOXICITY DATA with REFERENCE
otr-ham:emb 4 mg/L CNREA8 32,1391,72
skn-mus TDLo:21 mg/kg:NEO PNASA6 74,3176,77
skn-mus TD:2100 μg/kg:ETA CNREA8 38,1699,78

CONSENSUS REPORTS: EPA Genetic Toxicology Program.

SAFETY PROFILE: Questionable carcinogen with experimental tumorigenic and neoplastigenic data by skin contact. Mutation data reported. When heated to decomposition it emits acrid smoke and irritating fumes.

BBE750 CAS:34501-24-1 ***HR: 3***
trans-BENZ(a)ANTHRACENE-8,9-DIHYDRODIOL
mf: $C_{18}H_{14}O_2$ mw: 262.32

SYNS: BA-8,9-DIHYDRODIOL ◇ trans-8,9-DIHYDROXY-8,9-DIHYDROBENZ(a)ANTHRACENE

TOXICITY DATA with REFERENCE
mma-sat 25 μmol/L BBRCA9 72,680,76
skn-mus TDLo:2100 μg/kg:NEO CNREA8 38,1699,78
skn-mus TD:4200 μg/kg:ETA PNASA6 74,3176,77

CONSENSUS REPORTS: EPA Genetic Toxicology Program.

SAFETY PROFILE: Questionable carcinogen with experimental tumorigenic and neoplastigenic data by skin contact. Mutation data reported. When heated to decomposition it emits acrid smoke and irritating fumes.

BBF000 CAS:60967-90-0 ***HR: 3***
BENZ(a)ANTHRACENE-10,11-DIHYDRODIOL
mf: $C_{18}H_{14}O_2$ mw: 262.32

SYNS: BA-10,11-DIHYDRODIOL ◇ trans-10,11-DIHYDROXY-10,11-DIHYDROBENZ(a)ANTHRACENE

TOXICITY DATA with REFERENCE
mma-sat 100 μmol/L CNREA8 42,1620,82
skn-mus TDLo:2100 μg/kg:ETA PNASA6 74,3176,77

CONSENSUS REPORTS: EPA Genetic Toxicology Program.

SAFETY PROFILE: Questionable carcinogen with experimental tumorigenic data by skin contact. When heated to decomposition it emits acrid smoke and irritating fumes.

BBF500 CAS:2564-65-0 ***HR: 3***
BENZ(a)ANTHRACENE-7,12-DIMETHANOL
mf: $C_{20}H_{16}O_2$ mw: 288.36

SYNS: 9:10-BISHYDROXYMETHYL-1:2-BENZANTHRACENE ◇ 7:12-DIHYDROXYMETHYLBENZ(a)ANTHRACENE

TOXICITY DATA with REFERENCE
mma-sat 20 nmol/plate 46OJAN -,675,81
mmo-esc 1 g/L/2H GENTAE 39,141,54
scu-mus TDLo:2600 mg/kg/40W-I:ETA PRLBA4 129,439,40

CONSENSUS REPORTS: EPA Genetic Toxicology Program.

SAFETY PROFILE: Questionable carcinogen with experimental tumorigenic data. Mutation data reported. When heated to decomposition it emits acrid smoke and irritating fumes.

BBF750 CAS:63018-62-2 ***HR: 3***
BENZ(a)ANTHRACENE-7,12-DIMETHANOLDIACETATE
mf: $C_{24}H_{20}O_4$ mw: 372.44

SYNS: ACETIC ACID, BENZ(a)ANTHRACENE-7,12-DIMETHANOL DIESTER ◇ 9,10-BISACETOXYMETHYL-1,2-BENZANTHRACENE

TOXICITY DATA with REFERENCE
skn-mus TDLo:1700 mg/kg/71W-I:ETA PRLBA4 129,439,40

SAFETY PROFILE: Questionable carcinogen with experimental tumorigenic data. See also ESTERS. When heated to decomposition it emits acrid smoke and irritating fumes.

BBG000 CAS:67335-42-6 ***HR: 3***
(−)(3R,4R)-trans-BENZ(a)ANTHRACENE-3,4-DIOL
mf: $C_{18}H_{10}O_3$ mw: 274.28

SYNS: (−)(3R,4R)-trans-3,4-DIHYDRO-3,4-DIHYDROXYBENZ(a)ANTHRACENE ◇ (−)(3R,4R)trans-3,4-DIHYDRO-3,4-DIHYDROXYBENZO-(a)ANTHRACENE

TOXICITY DATA with REFERENCE
skn-mus TDLo:1100 μg/kg:NEO CNREA8 38,1705,78

SAFETY PROFILE: Questionable carcinogen with experimental neoplastigenic data by skin contact. When heated to decomposition it emits acrid smoke and irritating fumes.

BBG200 CAS:56614-97-2 ***HR: D***
BENZ(a)ANTHRACENE-3,9-DIOL
mf: $C_{18}H_{12}O_2$ mw: 260.30

SYN: 3,9-DIHYDROXYBENZ(a)ANTHRACENE

TOXICITY DATA with REFERENCE
scu-rat TDLo:25 mg/kg (female 1D pre):REP JMCMAR 25,323,82

SAFETY PROFILE: Experimental reproductive effects. When heated to decomposition it emits acrid smoke and irritating fumes.

BBG500 CAS:63020-45-1 ***HR: 3***
BENZ(a)ANTHRACENE-7-ETHANOL
mf: $C_{20}H_{16}O$ mw: 272.36

SYN: 10-β-HYDROXYETHYL-1:2-BENZANTHRACENE

TOXICITY DATA with REFERENCE
skn-mus TDLo:1220 mg/kg/51W-I:ETA PRLBA4 131,170,42

SAFETY PROFILE: Questionable carcinogen with experimental tumorigenic data. When heated to decomposition it emits acrid smoke and irritating fumes.

BBG750 CAS:17012-91-8 ***HR: 3***
BENZ(a)ANTHRACENE-7-METHANEDIOLDIACETATE (ester)
mf: $C_{23}H_{18}O_4$ mw: 358.41

SYN: 7-DIACETOXYMETHYLBENZ(a)ANTHRACENE

TOXICITY DATA with REFERENCE
scu-mus TDLo:120 mg/kg/6W-I:CAR IJCNAW 2,500,67

SAFETY PROFILE: Questionable carcinogen with experimental carcinogenic data. See also ESTERS. When heated to decomposition it emits acrid smoke and irritating fumes.

BBH000 CAS:63018-59-7 ***HR: 3***
BENZ(a)ANTHRACENE-7-METHANETHIOL
mf: $C_{19}H_{14}S$ mw: 274.39

SYN: 1,2-BENZANTHRYL-10-METHYLMERCAPTAN

TOXICITY DATA with REFERENCE
scu-mus TDLo:80 mg/kg:ETA CNREA8 6,454,46

SAFETY PROFILE: Questionable carcinogen with experimental tumorigenic data. When heated to decomposition it emits toxic fumes of SO_x. See also MERCAPTANS.

BBH250 CAS:16110-13-7 ***HR: 3***
BENZ(a)ANTHRACENE-7-METHANOL
mf: $C_{19}H_{14}O$ mw: 258.33

SYNS: 7-HMBA ◇ 7-HYDROXYMETHYLBENZ(a)ANTHRACENE ◇ 10-HYDROXYMETHYL-1,2-BENZANTHRACENE

TOXICITY DATA with REFERENCE
dnd-omi 30 μmol/L CBINA8 31,51,80
otr-mus:oth 100 μg/L IJCNAW 13,304,74
dnd-mus:emb 800 μg/L CNREA8 33,2386,73
dnd-mam:lym 30 μmol/L CBINA8 31,51,80
skn-mus TDLo:56 mg/kg/60W-I:ETA CNREA8 43,2034,83

CONSENSUS REPORTS: EPA Genetic Toxicology Program.

SAFETY PROFILE: Questionable carcinogen with experimental tumorigenic data. Mutation data reported. When heated to decomposition it emits acrid smoke and irritating fumes.

BBH500 CAS:17526-24-8 ***HR: 3***
BENZ(a)ANTHRACENE-7-METHANOL ACETATE
mf: $C_{21}H_{16}O_2$ mw: 300.37

SYNS: ACETIC ACID, BENZ(a)ANTHRACENE-7-METHANOL ESTER ◇ 10-ACETOXYMETHYL-1,2-BENZANTHRACENE

TOXICITY DATA with REFERENCE
skn-mus TDLo:16 mg/kg/17W-I:ETA VOONAW 21(10),50,75

SAFETY PROFILE: Questionable carcinogen with experimental tumorigenic data. See also ESTERS. When heated to decomposition it emits acrid smoke and irritating fumes.

BBH750 CAS:63018-57-5 ***HR: 3***
BENZ(a)ANTHRACENE-7-THIOL
mf: $C_{18}H_{12}S$ mw: 260.36

SYNS: 1,2-BENZANTHRYL-10-MERCAPTAN ◇ 7-MERCAPTOBENZ-(a)ANTHRACENE

TOXICITY DATA with REFERENCE
scu-mus TDLo:80 mg/kg:ETA CNREA8 6,454,46

SAFETY PROFILE: Questionable carcinogen with experimental tumorigenic data. When heated to decomposition it emits toxic fumes of SO_x. See also MERCAPTANS.

BBI000 CAS:960-92-9 ***HR: 3***
BENZ(a)ANTHRACEN-5-OL
mf: $C_{18}H_{12}O$ mw: 244.30

SYNS: 3-HYDROXY-1,2-BENZANTHRACENE ◇ 5-HYDROXYBENZ-(a)ANTHRACENE

TOXICITY DATA with REFERENCE
dnd-ham:kdy 5 mg/L BCPCA6 20,1297,71
dnd-ham:lng 1 mg/L CBINA8 4,389,71/72
scu-mus TDLo:1240 mg/kg:ETA JNCIAM 1,303,40

CONSENSUS REPORTS: EPA Genetic Toxicology Program.

SAFETY PROFILE: Questionable carcinogen with experimental tumorigenic data. Mutation data reported. When heated to decomposition it emits acrid smoke and irritating fumes.

BBI250 CAS:82-05-3 ***HR: 3***
7H-BENZ(de)ANTHRACEN-7-ONE
mf: $C_{17}H_{10}O$ mw: 230.27

PROP: Pale yellow needles. Mp: 174°, vap press: 1 mm @ 225.0°.

SYNS: 7H-BENZ(de)ANTHRACENE-7-ONE ◇ BENZANTHRENONE ◇ BENZANTHRONE ◇ 7H-BENZO(de)ANTHRACEN-7-ONE ◇ BENZOANTHRONE ◇ MS-BENZANTHRONE ◇ NAPHTHANTHRONE ◇ 7-OXOBENZ(de)ANTHRACENE

TOXICITY DATA with REFERENCE
skn-rbt 500 mg/24H MLD 28ZPAK -,60,72
eye-rbt 100 mg/24H MOD 28ZPAK -,60,72
ipr-rat LD50:1500 mg/kg RPTOAN 40,137,77
ipr-mus LD50:290 mg/kg RPTOAN 40,137,77

CONSENSUS REPORTS: Reported in EPA TSCA Inventory.

SAFETY PROFILE: Poison by intraperitoneal route. Skin and eye irritant. Combustible when heated. Incompatible with nitrobenzene and potassium hydroxide. When heated to decomposition it emits acrid smoke and irritating fumes.

BBI750 CAS:63018-49-5 ***HR: 3***
1,2-BENZANTHRYL-3-CARBAMIDOACETIC ACID
mf: $C_{21}H_{16}N_2O_3$ mw: 344.39

SYN: N-(BENZ(a)ANTHRACEN-5-YLCARBAMOYL)GLYCINE

TOXICITY DATA with REFERENCE
scu-mus TDLo:120 mg/kg:ETA CNREA8 6,454,46

SAFETY PROFILE: Questionable carcinogen with experimental tumorigenic data. When heated to decomposition it emits toxic fumes such as NO_x.

BBJ000 CAS:63018-50-8 ***HR: 3***
1,2-BENZANTHRYL-10-CARBAMIDOACETIC ACID
mf: $C_{21}H_{16}N_2O_3$ mw: 344.39

SYN: N-(BENZ(a)ANTHRACEN-7-YLCARBAMOYL)GLYCINE

TOXICITY DATA with REFERENCE
scu-mus TDLo:160 mg/kg:ETA CNREA8 6,454,46

SAFETY PROFILE: Questionable carcinogen with experimental tumorigenic data. When heated to decomposition it emits toxic fumes of NO_x.

BBJ250 CAS:63018-56-4 ***HR: 3***
1,2-BENZANTHRYL-10-ISOCYANATE
mf: $C_{19}H_{11}NO$ mw: 269.31

SYN: ISOCYANIC ACID, BENZ(a)ANTHRACEN-7-YL ESTER

TOXICITY DATA with REFERENCE
scu-mus TDLo:40 mg/kg:ETA CNREA8 6,454,46

SAFETY PROFILE: Questionable carcinogen with experimental tumorigenic data. See also ESTERS. When heated to decomposition it emits toxic fumes of NO_x.

BBJ500 CAS:1477-19-6 ***HR: 3***
BENZARONE
mf: $C_{17}H_{14}O_3$ mw: 266.31

SYNS: 2-ETHYL-3-BENZOFURANYL-p-HYDROXYPHENYLKETONE ◇ 2-ETHYL-3-(p-HYDROXYBENZOYL)BENZOFURAN ◇ 2-ETHYL-4′-HYDROXY-3-BENZOYLBENZOFURAN ◇ FRAGIVIX

TOXICITY DATA with REFERENCE
orl-mus TDLo:600 mg/kg (7-12D preg):REP KSRNAM 3,961,69
orl-rat TDLo:6 g/kg (female 9-14D post):TER KSRNAM 3,961,69
ipr-mus LD50:200 mg/kg AIPTAK 154,94,65

CONSENSUS REPORTS: Reported in EPA TSCA Inventory.

SAFETY PROFILE: Poison by intraperitoneal route. An experimental teratogen. Other experimental reproductive effects. When heated to decomposition it emits acrid and irritating smoke and fumes. See also KETONES.

BBJ750 CAS:59-97-2 ***HR: 3***
BENZAZOLINE HYDROCHLORIDE
mf: $C_{10}H_{12}N_2 \cdot ClH$ mw: 196.70

SYNS: ARTERODY ◇ BENZYLIMIDAZOLINE HYDROCHLORIDE ◇ 2-BENZYL-2-IMIDAZOLINE MONOHYDROCHLORIDE ◇ IMIDALINE HYDROCHLORIDE ◇ PRISCOL ◇ PRISCOLINE HYDROCHLORIDE ◇ TOLAVAD ◇ TOLAZOLINE CHLORIDE ◇ TOLAZOLINE HYDROCHLORIDE ◇ TOLPAL

TOXICITY DATA with REFERENCE
ivn-hmn TDLo:150 μg/kg:CVS,SKN FOMDAK 27,729,41
orl-rat LD50:1200 mg/kg NIIRDN 6,511,82
ipr-rat LD50:100 mg/kg NIIRDN 6,511,82
ivn-rat LD50:85 mg/kg NIIRDN 6,511,82
orl-mus LD50:400 mg/kg ARZNAD 21,1992,71
ipr-mus LD50:130 mg/kg ARZNAD 21,1992,71
ivn-mus LD50:60 mg/kg CLDND*

CONSENSUS REPORTS: Reported in EPA TSCA Inventory.

SAFETY PROFILE: Poison by ingestion, intravenous, and intraperitoneal routes. Human systemic effects by intravenous route: change in heart rate, unspecified vascular effects, and sweating. When heated to decomposition it emits very toxic fumes of NO_x and HCl.

BBK000 CAS:300-62-9 ***HR: 3***
BENZEDRINE
mf: $C_9H_{13}N$ mw: 135.23

PROP: Liquid. Bp: 200°, flash p: < 212°F (OC), d: 0.931, vap d: 4.65.

SYNS: ACTEDRON ◇ ADIPAN ◇ ALLODENE ◇ dl-AMPHETAMINE ◇ ANOREXIDE ◇ (±)-BENZEDRINE ◇ dl-BENZEDRINE ◇ DEOXYNOREPHEDRINE ◇ (±)-DESOXYNOREPHEDRINE ◇ racemic-DESOXYNOR-EPHEDRINE ◇ ELASTONON ◇ ISOAMYCIN ◇ ISOMYN ◇ MECODRIN ◇ α-METHYLBENZENEETHANEAMINE ◇ dl-α-METHYLPHENETHYLAMINE ◇ (±)-α-METHYLPHENETHYLAMINE ◇ NOREPHEDRANE ◇ NOVYDRINE ◇ ORTEDRINE ◇ PHENEDRINE ◇ dl-1-PHENYL-2-AMINOPROPANE ◇ PROFAMINA ◇ PROPISAMINE ◇ PSYCHEDRINE ◇ RAPHETAMINE ◇ SIMPATEDRIN ◇ SYMPAMINE ◇ SYMPATEDRINE ◇ WECKAMINE

TOXICITY DATA with REFERENCE
dnd-esc 40 μmol/L MUREAV 89,95,81
scu-rat TDLo:11 mg/kg (1-22D preg):REP PSYPAG 40,25,74
unr-man LDLo:2206 μg/kg 85DCAI 2,73,70
orl-rat LD50:30 mg/kg ARZNAD 23,810,73
scu-rat LD50:180 mg/kg JPETAB 85,119,45
orl-mus LD50:21 mg/kg ARZNAD 23,810,73
ipr-mus LD50:5500 μg/kg AIPTAK 161,206,66
scu-mus LD50:15 mg/kg FEPRA7 4,139,45
ivn-mus LD50:15 mg/kg AIPTAK 145,392,63

CONSENSUS REPORTS: Reported in EPA TSCA Inventory. EPA Extremely Hazardous Substances List.

SAFETY PROFILE: A deadly human poison by an unspecified route. An experimental poison by ingestion, subcutaneous, intraperitoneal, and intravenous routes. Experimental reproductive effects. Mutation data reported. A central nervous system stimulant. Overdoses cause hyperactivity, restlessness, insomnia, rapid pulse, rise in blood pressure, dilated pupils, dryness of the throat. Combustible when exposed to heat, flame, or oxidizers. When heated to decomposition it emits toxic fumes of NO_x. To fight fire, use CO_2, dry chemical, alcohol foam, water mist, fog. See other benzedrine entries.

BBK250 CAS:156-31-0 ***HR: 3***
BENZEDRINE SULFATE
mf: $C_{18}H_{26}N_2{\cdot}H_2O_4S$ mw: 368.54

SYNS: AMITRENE ◇ AMPHOIDS S ◇ AMPHORDS S ◇ BAR-TIME ◇ DIAMPHETAMINE SULFATE ◇ KLINE ◇ dl-α-METHYLPHENETHYLAMINE SULFATE ◇ PHENETHYLAMINE, α-METHYL-, SULFATE (2:1) ◇ 1-PHENYL-2-AMINOPROPANE SULFATE

TOXICITY DATA with REFERENCE
ipr-rat LDLo:25 mg/kg JPETAB 100,267,50
scu-rat LDLo:10 mg/kg JPETAB 71,62,41
ipr-mus LD50:75 mg/kg JPETAB 93,114,48
scu-mus LD50:14 mg/kg JPETAB 87,214,46
orl-dog LDLo:20 mg/kg AJMSA9 198,785,39

SAFETY PROFILE: A poison via ingestion, intraperitoneal, and subcutaneous routes. When heated to decomposition it emits very toxic fumes of SO_x and NO_x. See also BENZEDRINE and SULFATES.

BBK500 CAS:51-63-8 ***HR: 3***
d-BENZEDRINE SULFATE
mf: $C_{18}H_{26}N_2{\cdot}H_2O_4S$ mw: 368.54

SYNS: ACEDRON ◇ ADJUDETS ◇ ADRIXINE ◇ AFATIN ◇ ALBEMAP ◇ AMDEX ◇ d-AMFETASUL ◇ AMITRENE ◇ AMPHAETEX ◇ AMPHEDRINE ◇ AMPHEREX ◇ (+)-AMPHETAMINE SULFATE ◇ d-AMPHETAMINE SULFATE ◇ AMSUSTAIN ◇ APETAIN ◇ ARDEX ◇ BETAFEDRINA ◇ BETAFEDRINE ◇ d-BETAPHEDRINE ◇ CARR-TIME ◇ CRADEX ◇ DADEX ◇ DADOX d-CITRAMINE ◇ DELLIPSOIDS ◇ DEPHADREN ◇ DESOXYN ◇ DEXAIME ◇ DEXALINE ◇ DEXALME ◇ DEXAMED ◇ DEXAMINE ◇ DEXAMPHAMINE ◇ DEXAMPHETAMINE ◇ DEXAMPHETAMINE SULFATE ◇ DEXAMYL ◇ DEXEDRINA ◇ DEXEDRINE SULFATE ◇ DEXIES ◇ DEXTROAMPHETAMINE SULFATE ◇ DEXTRO-α-METHYLPHENETHYLAMINE SULFATE ◇ DEXTRO-1-PHENYL-2-AMINOPROPANE SULFATE ◇ DEXTRO-β-PHENYLISOPROPYLAMINE SULFATE ◇ FASTBALLS ◇ HEARTS ◇ (S)-α-METHYL-BENZENEETHANAMINE SULFATE (2:1) ◇ d-α-METHYLPHENETHYLAMINE SULFATE ◇ ORANGES ◇ PELLCAFS ◇ PELLCAP ◇ PELLCAPS ◇ PERKE ◇ PHENOPROMIN ◇ d-1-PHENYL-2-AMINOPROPANE SULFATE ◇ d-β-PHENYLISOPROPYLAMINE SULFATE ◇ PHETADEX ◇ PSYCHODRINE ◇ REVIDEX ◇ SIMPAMINA-D ◇ SYMPAMINA-D ◇ TEMPODEX ◇ TUPHETAMINE ◇ TYDEX ◇ ZAMINE

TOXICITY DATA with REFERENCE
orl-wmn TDLo:96 mg/kg (5-39W preg):TER INTSAO 50,79,68
orl-rat TDLo:146 mg/kg (female 30D pre-21D post):REP DABBBA 39,5121,79
orl-rat LD50:32 mg/kg ARZNAD 33,1411,83
ipr-rat LD50:43200 μg/kg TXAPA9 29,397,74
ivn-rat LD50:30 mg/kg JPETAB 110,180,54
orl-mus LD50:10 mg/kg JMCMAR 18,71,75
ipr-mus LD50:9700 μg/kg JPETAB 135,240,62
scu-mus LD50:16 mg/kg AIPTAK 184,34,70
ivn-mus LD50:30 mg/kg JPETAB 137,365,62

orl-dog LD50:10 mg/kg PSEBAA 118,557,65
ivn-dog LD50:3 mg/kg PSEBAA 118,557,65
ivn-rbt LD50:10 mg/kg JPETAB 110,180,54
orl-bwd LD50:56200 μg/kg AECTCV 12M355,83

SAFETY PROFILE: Poison by ingestion, intraperitoneal, subcutaneous, and intravenous routes. A human teratogen which causes developmental abnormalities of the central nervous system. Experimental reproductive effects including other teratogenic effects. A habit-forming stimulant. When heated to decomposition it emits very toxic fumes of SO_x and NO_x. See also other benzidrine compounds and SULFATES.

BBK750 CAS:51-62-7 ***HR: 3***
l-BENZEDRINE SULFATE
mf: $C_{18}H_{26}N_2{\cdot}H_2O_4S$ mw: 368.54

SYNS: (−)-AMPHETAMINE SULFATE ◇ l-AMPHETAMINE SULFATE ◇ LEVEDRINE ◇ l-1-PHENYL-2-AMINOPROPANE SULFATE

TOXICITY DATA with REFERENCE
scu-rat LDLo:160 mg/kg JPETAB 71,62,41
ipr-mus LD50:232 mg/kg JPETAB 158,135,67

SAFETY PROFILE: A poison via subcutaneous and intraperitoneal routes. See also SULFATES. When heated to decomposition it emits very toxic fumes of SO_x and NO_x.

BBL000 CAS:142-04-1 ***HR: 3***
BENZENAMINE HYDROCHLORIDE
DOT: UN 1548
mf: $C_6H_7N{\cdot}ClH$ mw: 129.60

PROP: Crystals. Vap d: 4.46, d: 1.22, mp: 198°, bp: 245°, flash p: 380°F (OC).

SYNS: ANILINE CHLORIDE ◇ ANILINE HYDROCHLORIDE (DOT) ◇ "ANILINE SALT" ◇ ANILINIUM CHLORIDE ◇ CHLORHYDRATE d'ANILINE (FRENCH) ◇ CHLORID ANILINU (CZECH) ◇ NCI-C03736 ◇ PHENYLAMINE HYDROCHLORIDE ◇ SUL ANILINOVA (CZECH) ◇ USAF EK-442

TOXICITY DATA with REFERENCE
skn-rbt 500 mg/24H MOD 28ZPAK -,65,72
eye-rbt 20 mg/24H MOD 28ZPAK -,65,72
sce-hmn:lym 50 μmol/L BLFSBY 29b,561,84
otr-rat:emb 79500 ng/plate JJATDK 1,190,81
sce-ham:fbr 10 μmol/L JNCIAM 58,1635,77
orl-rat TDLo:1400 mg/kg (7-20D preg):TER TXAPA9 77,465,85
orl-rat TDLo:130 g/kg/2Y-C:CAR NCITR* NCI-CG-TR-130,78
orl-rat TD:238 g/kg/2Y-C:CAR NCITR* NCI-CG-TR-130,78
orl-rat TD:137 g/kg/60W-C:ETA IARC** 27,39,82
orl-rat LD50:840 mg/kg TXAPA9 42,417,77
ipr-rat LDLo:500 mg/kg NCNSA6 5,11,53
orl-mus LD50:841 mg/kg NTIS** PB214-270
ipr-mus LD50:300 mg/kg NTIS** AD277-689

CONSENSUS REPORTS: IARC Cancer Review: Animal Limited Evidence IMEMDT 27,39,82. NCI Carcinogenesis Bioassay Completed; Results Positive: rat NCITR* NCI-CG-TR-130,78; Results Negative: Mouse NCITR* NCI-CG-TR-130,78. Reported in EPA TSCA Inventory. EPA Genetic Toxicology Program.

DOT Classification: Poison B; Label: St. Andrews Cross

SAFETY PROFILE: Suspected carcinogen with experimental carcinogenic and tumorigenic data. Poison by intraperitoneal route. Moderately toxic by ingestion. Experimental teratogenic effects. Human mutation data reported. A skin and eye irritant. Combustible when exposed to heat or flame. When heated to decomposition or on contact with acid or acid fumes, it emits highly toxic fumes of aniline and chlorine compounds. Reacts explosively with aniline at 240°C/7.6 bar. Can react vigorously with oxidizing materials. To fight fire, use water, CO_2, water mist or spray, dry chemical. See also ANILINE.

BBL250 CAS:71-43-2 ***HR: 3***
BENZENE
DOT: UN 1114
mf: C_6H_6 mw: 78.12

PROP: Clear, colorless liquid. Mp: 5.51°, bp: 80.093-80.094°, flash p: 12°F (CC), d: 0.8794 @ 20°, autoign temp: 1044°F, lel: 1.4%, uel: 8.0%, vap press: 100 mm @ 26.1°, vap d: 2.77, ULC: 95-100.

SYNS: (6)ANNULENE ◇ BENZEEN (DUTCH) ◇ BENZEN (POLISH) ◇ BENZIN (OBS.) ◇ BENZINE (OBS.) ◇ BENZOL (DOT) ◇ BENZOLE ◇ BENZOLENE ◇ BENZOLO (ITALIAN) ◇ BICARBURET of HYDROGEN ◇ CARBON OIL ◇ COAL NAPHTHA ◇ CYCLOHEXATRIENE ◇ FENZEN (CZECH) ◇ MINERAL NAPHTHA ◇ MOTOR BENZOL ◇ NCI-C55276 ◇ NITRATION BENZENE ◇ PHENE ◇ PHENYL HYDRIDE ◇ PYROBENZOL ◇ PYROBENZOLE ◇ RCRA WASTE NUMBER U019

TOXICITY DATA with REFERENCE
skn-rbt 15 mg/24H open MLD AIHAAP 23,95,62
skn-rbt 20 mg/24H MOD 85JCAE-,25,86
eye-rbt 88 mg MOD AMIHAB 14,387,56
eye-rbt 2 mg/24H SEV 28ZPAK -,23,72
oms-hmn:lym 5 μmol/L CNREA8 45,2471,85
mma-mus:emb 2500 mg/L PMRSDJ 5,639,85
orl-mus TDLo:6500 mg/kg (female 8-12D post):REP TCMUD8 6,361,86
ihl-mus TCLo:5 ppm (female 6-15D post):TER TXCYAC 42,171,86
ihl-man TCLo:200 mg/m³/78W-I:CAR,BLD EJCAAH 7,83,71
ihl-hmn TCLo:10 ppm/8H/10Y-I:CAR,BLD TRBMAV 37,153,78

orl-rat TDLo:52 g/kg/52W-I:CAR MELAAD 70,352,79
ihl-rat TCLo:1200 ppm/6H/10W-I:ETA PAACA3 25,75,84
orl-mus TDLo:18250 mg/kg/2Y-C:CAR NTPTR* NTP-TR-289,86
skn-mus TDLo:1200 g/kg/49W-I:NEO BJCAAI 16,275,62
ihl-hmn TC:150 ppm/15M/8Y-I:CAR,BLD BLOOAW 52,285,78
ihl-man TC:600 mg/m^3/4Y-I:CAR,BLD NEJMAG 271,872,64
ihl-man TC:150 ppm/11Y-I:CAR,BLD BLUTA9 28,293,74
ihl-hmn TC:8 ppb/4W-I:CAR,BLD NEJMAG 316,1044,87
ihl-hmn TC:10 mg/m^3/11Y-I:CAR,BLD BJIMAG 44,124,87
ihl-hmn LCLo:2 pph/5M TABIA2 3,231,33
orl-man LDLo:50 mg/kg YAKUD5 22,883,80
ihl-hmn LCLo:20000 ppm/5M 29ZUA8 -,-,53
ihl-man TCLo:150 ppm/1Y-I:BLD BLUTA9 28,293,74
ihl-hmn TCLo:100 ppm INMEAF 17,199,48
ihl-hmn LCLo:65 mg/m^3/5Y:BLD ARGEAR 44,145,74
orl-rat LD50:3306 mg/kg TXAPA9 19,699,71
ihl-rat LC50:10000 ppm/7H 28ZRAQ -,113,60
ipr-rat LD50:2890 μg/kg 36YFAG -,302,77
orl-mus LD50:4700 mg/kg HYSAAV 32,349,67
ihl-mus LC50:9980 ppm JIHTAB 25,366,43
ipr-mus LD50:340 mg/kg ANYAA9 243,104,75
orl-dog LDLo:2000 mg/kg HBAMAK 4,1313,35
ihl-dog LCLo:146000 mg/m^3 HBTXAC 1,324,56
ihl-cat LCLo:170000 mg/m^3 HBTXAC 1,324,56
ivn-rbt LDLo:88 mg/kg JTEHD6 -(Suppl.2),45,77

CONSENSUS REPORTS: NTP Fifth Annual Report on Carcinogens. IARC Cancer Review: Group 1 IMEMDT 7,120,87, Human Limited Evidence IMEMDT 7,203,74; Animal Inadequate Evidence IMEMDT 7,203,74; IARC Cancer Review: Animal Limited Evidence IMEMDT 29,93,82; Human Sufficient Evidence IMEMDT 29,93,82. NTP Carcinogenesis Studies (gavage); Clear Evidence: mouse, rat NTPTR* NTP-TR-289,86. EPA Genetic Toxicology Program. Reported in EPA TSCA Inventory. On Community Right-To-Know List.

OSHA PEL: (Transitional: TWA 10 ppm; CL 25 ppm; Pk 50 ppm/10M) TWA 1 ppm; STEL 5 ppm; Pk 5 ppm/15M/8H; Cancer Hazard
ACGIH TLV: TWA 10 ppm; Suspected Human Carcinogen (Proposed: TWA 0.1 ppm; Confirmed Human Carcinogen); BEI: 50 mg(total phenol)/L in urine at end of shift recommended as a mean value.
DFG TRK: 5 ppm (16 mg/m^3) Human Carcinogen.
NIOSH REL: TWA 0.32 mg/m^3; CL 3.2 mg/m^3/15M
DOT Classification: Flammable Liquid; Label: Flammable Liquid.

SAFETY PROFILE: Confirmed human carcinogen producing myeloid leukemia, Hodgkin's disease, and lymphomas by inhalation. Experimental carcinogenic, neoplastigenic, and tumorigenic data. A human poison by inhalation. An experimental poison by skin contact, intraperitoneal, intravenous, and possibly other routes. Moderately toxic by ingestion and subcutaneous routes. A severe eye and moderate skin irritant. Human systemic effects by inhalation and ingestion: blood changes,increased body temperature. Experimental teratogenic and reproductive effects. Human mutation data reported. A narcotic. In industry, inhalation is the primary route of chronic benzene poisoning. Poisoning by skin contact has been reported. Recent (1987) research indicates that effects are seen at less than 1 ppm. Exposures needed to be reduced to 0.1 ppm before no toxic effects were observed. Elimination is chiefly through the lungs. A common air contaminant.

A dangerous fire hazard when exposed to heat or flame. Explodes on contact with diborane, bromine pentafluoride, permanganic acid, peroxomonosulfuric acid, and peroxodisulfuric acid. Forms sensitive, explosive mixtures with iodine pentafluoride, silver perchlorate, nitryl perchlorate, nitric acid, liquid oxygen, ozone, arsenic pentafluoride + potassium methoxide (explodes above 30°C). Ignites on contact with sodium peroxide + water, dioxygenyl tetrafluoroborate, iodine heptafluoride, and dioxygen difluoride. Vigorous or incandescent reaction with hydrogen + Raney nickel (above 210°C), uranium hexafluoride, and bromine trifluoride. Can react vigorously with oxidizing materials, such as Cl_2, CrO_3, O_2, $NClO_4$, O_3, perchlorates, ($AlCl_3$ + $FClO_4$), (H_2SO_4 + permanganates), K_2O_2, ($AgClO_4$ + acetic acid), Na_2O_2. Moderate explosion hazard when exposed to heat or flame. Use with adequate ventilation. To fight fire, use foam, CO_2, dry chemical.

Poisoning occurs most commonly via inhalation of the vapor, although benzene can penetrate the skin and cause poisoning. Locally, benzene has a comparatively strong irritating effect, producing erythema and burning, and, in more severe cases, edema and even blistering. Exposure to high concentrations of the vapor (3000 ppm or higher) may result from failure of equipment or spillage. Such exposure, while rare in industry, may cause acute poisoning, characterized by the narcotic action of benzene on the central nervous system. The anesthetic action of benzene is similar to that of other anesthetic gases, consisting of a preliminary stage of excitation followed by depression and, if exposure is continued, death through respiratory failure. The chronic, rather than the acute form, of benzene poisoning is important in industry. It is a recognized leukemogen. There is no specific blood picture occurring in cases of chronic benzol poisoning. The bone marrow may be hypoplastic, normal, or hyperplastic, the changes reflected in the peripheral blood. Anemia, leucopenia, macrocytosis, reticulocytosis, thrombocytopenia, high color index, and prolonged bleeding time may be present. Cases of myeloid leukemia have been reported. For the worker, repeated blood examinations are necessary, including hemoglobin determinations, white and red cell counts, and differential smears. Where a worker shows a progressive drop in either red or white cells, or where the white count remains low, 5,000/mm_3 or the red count <4.0 million/mm_3, on two successive monthly examinations, the worker should be immediately removed from benzene exposure. Elimination is chiefly through the lungs, when fresh air is breathed. The portion that is absorbed is oxidized, and the oxidation products are combined with sulfuric and glycuronic acids and eliminated in the urine. This may be used as a diagnostic sign. Benzene has a definite cumulative action, and exposure to a relatively high concentration is not serious from the point of view of causing damage to the blood-forming system, provided the exposure is not repeated. In acute poisoning, the worker becomes confused and dizzy, complains of tightening of the leg muscles and of pressure over the forehead, then passes into a stage of excitement. If allowed to remain exposed, he quickly becomes stupefied and lapses into coma. In non-fatal cases, recovery is usually complete with no permanent disability. In chronic poisoning the onset is slow, with the symptoms vague, fatigue, headache, dizziness, nausea and loss of appetite, loss of weight and weakness are common complaints in early cases. Later, pallor, nosebleeds, bleeding gums, menorrhagia, petechiae and purpura may develop. There is great individual variation in the signs and symptoms of chronic benzene poisoning.

BBL500 CAS:122-78-1 ***HR: 2***
BENZENEACETALDEHYDE
mf: C_8H_8O mw: 120.16

PROP: Oily, colorless liquid which polymerizes and grows more viscous on standing; odor similar to lilac and hyacinth. Has been crystallized, mp: 33-34°, d:(25/25) 1.023-1.030, refr index: 1.525-1.545, bp: (10) 78°, n (20/D) 1.524-1.528, flash p: 154°F. Sltly sol in water; sol in alc, ether, propylene glycol. One part is sol in two parts of 80% alc forming a clear solution.

SYNS: FEMA No. 2874 ◇ HYACINTHIN ◇ PAA ◇ PHENYLACETALDEHYDE (FCC) ◇ PHENYLACETIC ALDEHYDE ◇ PHENYLETHANAL ◇ α-TOLUALDEHYDE ◇ α-TOLUIC ALDEHYDE

TOXICITY DATA with REFERENCE
skn-hmn 2%/48H FCTXAV 17,377,79
orl-rat LD50:1550 mg/kg FCTXAV 17,377,79
orl-mus LD50:3890 mg/kg FCTXAV 17,377,79
orl-gpg LD50:3890 mg/kg FCTXAV 17,377,79

CONSENSUS REPORTS: Reported in EPA TSCA Inventory.

SAFETY PROFILE: Moderately toxic by ingestion. Human skin irritant. Combustible liquid. When heated to decomposition it emits acrid smoke and irritating fumes. See also ALDEHYDES.

BBL750 CAS:98-05-5 ***HR: 3***
BENZENEARSONIC ACID
mf: $C_6H_7AsO_3$ mw: 202.05

PROP: Colorless crystals, water sol. D: 1.760, mp: 160° decomp.

SYNS: PHENYL ARSENIC ACID ◇ PHENYLARSONIC ACID

TOXICITY DATA with REFERENCE
orl-rat LDLo:50 mg/kg JPETAB 93,287,48
orl-mus LD50:270 μg/kg CLDND* 80,93,44
ivn-rbt LD50:16 mg/kg JPETAB 80,93,44

CONSENSUS REPORTS: Reported in EPA TSCA Inventory. EPA Extremely Hazardous Substances List. Arsenic and its compounds are on the Community Right-To-Know List.

OSHA PEL: TWA 0.5 mg/(As)m^3
ACGIH TLV: TWA 0.2 mg(As)/m^3

SAFETY PROFILE: A deadly poison by ingestion and intravenous routes. See also ARSENIC COMPOUNDS. When heated to decomposition it emits toxic fumes of As.

BBL825 CAS:4547-69-7 ***HR: 3***
BENZENE-1,3-BIS(SULFONYL AZIDE)
mf: $C_6H_4N_6O_4S_2$ mw: 288.26

$C_6H_4(SO_2N_3)_2$

SAFETY PROFILE: An explosive. Upon decomposition it emits toxic fumes of SO_x and NO_x. See also EXPLOSIVES and AZIDES.

BBM000 CAS:98-80-6 ***HR: 3***
BENZENEBORONIC ACID
mf: $C_6H_7BO_2$ mw: 121.94

SYNS: ACIDE PHENYLBORIQUE (FRENCH) ◇ BOROPHENYLIC ACID ◇ PHENYLBORIC ACID ◇ USAF BO-2

TOXICITY DATA with REFERENCE
orl-rat LD50:740 mg/kg 14KTAK -,693,64
ipr-mus LD50:500 mg/kg NTIS** AD277-689
ivn-mus LD50:320 mg/kg CSLNX* NX#02033
ivn-dog LDLo:450 mg/kg BANMAC 135,314,51
orl-rbt LDLo:600 mg/kg 14KTAK -,693,64
skn-rbt LDLo:4500 mg/kg 14KTAK -,693,64
ipr-gpg LD50:284 mg/kg BANMAC 135,314,51

CONSENSUS REPORTS: Reported in EPA TSCA Inventory.

SAFETY PROFILE: Poison by intravenous and intra-

peritoneal routes. Moderately toxic by ingestion. Mildly toxic by skin contact. See also BORON COMPOUNDS. When heated to decomposition it emits acrid smoke and irritating fumes.

BBM250 CAS:2227-79-4 ***HR: 3***
BENZENECARBOTHIOAMIDE
mf: C_7H_7NS mw: 137.21

SYNS: BENZOTHIAMIDE ◇ BENZOTHIOAMIDE ◇ THIOBENZAMIDE ◇ TIOBENZAMIDE (ITALIAN)

TOXICITY DATA with REFERENCE
mnt-mus-orl 180 μmol/kg MUREAV 192,141,87
orl-rat TDLo:6300 mg/kg/15W-C:ETA BSIBAC 54,1027,78
orl-mus LD50:95 mg/kg THERAP 8,237,53
ipr-mus LD50:500 mg/kg PCJOAU 11,1383,77

CONSENSUS REPORTS: Reported in EPA TSCA Inventory.

SAFETY PROFILE: Poison by ingestion. Moderately toxic by intraperitoneal route. Questionable carcinogen with experimental tumorigenic data. Mutation data reported. When heated to decomposition it emits very toxic fumes of NO_x and SO_x.

BBM500 CAS:63021-32-9 ***HR: 3***
BENZENECARBOXALDEHYDE
mf: $C_{19}H_{15}N$ mw: 257.35

SYNS: BENZALDEHYDE FFC ◇ 7-ETHYLBENZ(c)ACRIDINE ◇ 9-ETHYL-3,4-BENZACRIDINE ◇ PHENYLMETHANAL

TOXICITY DATA with REFERENCE
scu-mus TDLo:200 mg/kg:ETA VOONAW 1,52,55

SAFETY PROFILE: Questionable carcinogen with experimental tumorigenic data. See also ALDEHYDES. When heated to decomposition it emits toxic fumes of NO_x.

BBN000 ***HR: 3***
n-BENZENE-n-CYCLOPENTADIENYL IRON(II)PERCHLORATE
mf: $C_{11}H_{11}ClFeO_4$ mw: 298.51

SAFETY PROFILE: A shock-sensitive explosive. The dry material detonates on touching with spatula. Upon decomposition it emits toxic fumes of Cl^-. See also PERCHLORATES.

BBN250 CAS:17333-86-7 ***HR: 3***
BENZENE DIAZONIUM-2-CARBOXYLATE
mf: $C_7H_4N_2O_2$ mw: 148.12

SAFETY PROFILE: A heat- and shock-sensitive explosive. Explosive or violent reaction with aniline, arylisocyanides, and 1-pyrrolidinylcyclohexene. When heated to decomposition it emits toxic fumes of NO_x. See also EXPLOSIVES.

BBN500 CAS:100-34-5 ***HR: 3***
BENZENE DIAZONIUM CHLORIDE
mf: $C_6H_5ClN_2$ mw: 140.57

DOT Classification: Forbidden (dry)

SAFETY PROFILE: Potentially explosive when dry. Potentially explosive reaction with potassium o-methyldithiocarbonate. When heated to decomposition it emits toxic fumes of Cl^- and NO_x.

BBN650 CAS:36211-73-1 ***HR: 3***
BENZENEDIAZONIUM HYDROGEN SULFATE
mf: $C_6H_6N_2O_4S$ mw: 202.18

SAFETY PROFILE: Explodes at 100°C. When heated to decomposition it emits toxic fumes of SO_x and NO_x.

BBN750 ***HR: 3***
BENZENE DIAZONIUM NITRATE
mf: $C_6H_5N_3O_3$ mw: 167.12

SAFETY PROFILE: An explosive sensitive to friction, impact and heating to 90°. Upon decomposition it emits toxic fumes of NO_x. See also EXPLOSIVES and NITRATES.

BBN850 CAS:6925-01-5 ***HR: 3***
BENZENEDIAZONIUM-4-OXIDE
mf: $C_6H_4N_2O$ mw: 120.11

SAFETY PROFILE: Decomposes violently at 75°C. When heated to decomposition it emits toxic fumes of NO_x.

BBO000 ***HR: 3***
BENZENE DIAZONIUM SALTS
mf: $C_6H_5N_2^+X^-$

SAFETY PROFILE: Spontaneously explosive. Incompatible with ammonium sulfide, hydrogen sulfide, and disodium sulfide. Upon decomposition it emits toxic fumes of NO_x.

BBO125 CAS:612-31-7 ***HR: 3***
BENZENEDIAZONIUM-2-SULFONATE
mf: $C_6H_4N_2O_3S$ mw: 184.17

SAFETY PROFILE: Explodes on contact with flame or on impact. Upon decomposition it emits toxic fumes of SO_x and NO_x.

BBO250 CAS:305-80-6 ***HR: 3***
BENZENE DIAZONIUM-4-SULFONATE
mf: $C_6H_4N_2O_3S$ mw: 184.17

SAFETY PROFILE: An unstable explosive which may explode when touched. Incompatible with metals. Store in small quantities under refrigeration in loosely plugged containers. Upon decomposition it emits toxic fumes of NO_x and SO_x.

BBO325 CAS:369-57-3 ***HR: 3***
BENZENEDIAZONIUM TETRAFLUOROBORATE
mf: $C_6H_5N_2{\cdot}BF_4$ mw: 191.94

SYNS: BENZENEDIAZONIUM FLUOBORATE ◇ BENZENEDIAZONIUM FLUOROBORATE ◇ PHENYLDIAZONIUM FLUOROBORATE (SALT) ◇ PHENYLDIAZONIUM TETRAFLUOROBORATE

TOXICITY DATA with REFERENCE
mmo-sat 10 μmol/L CNREA8 42,1446,82
scu-ham TDLo:85 mg/kg/71W-I:ETA CALEDQ 15,289,82
orl-ham LD50:354 mg/kg CALEDQ 15,289,82
scu-ham LD50:166 mg/kg CALEDQ 15,289,82

SAFETY PROFILE: Poison by ingestion and subcutaneous routes. Questionable carcinogen with experimental tumorigenic data. Mutation data reported. When heated to decomposition it emits toxic fumes of NO_x and F^-. See also BORON COMPOUNDS.

BBO400 CAS:19521-84-7 ***HR: 3***
BENZENEDIAZONIUM TRIBROMIDE
mf: $C_6H_5Br_3N_2$ mw: 344.83

SAFETY PROFILE: A sensitive explosive. Upon decomposition it emits toxic fumes of Br^- and NO_x. See also EXPLOSIVES.

BBP000 CAS:123-61-5 ***HR: 3***
BENZENE-1,3-DIISOCYANATE
mf: $C_8H_4N_2O_2$ mw: 160.14

SYNS: 1,3-DIISOCYANATOBENZENE ◇ NACCONATE 400 ◇ m-PHENYLENE DIISOCYANATE ◇ m-PHENYLENE ISOCYANATE

TOXICITY DATA with REFERENCE
ivn-mus LD50:5600 μg/kg CSLNX* NX#07804

CONSENSUS REPORTS: Reported in EPA TSCA Inventory. Cyanide and its compounds are on the Community Right-To-Know List.

NIOSH REL: TWA (Diisocyanates) 0.005 ppm; CL 0.02 ppm/10M

SAFETY PROFILE: Deadly poison by intravenous route. When heated to decomposition it emits toxic fumes of NO_x and CN^-. See also ESTERS.

BBP250 CAS:623-26-7 ***HR: 2***
p-BENZENEDINITRILE
mf: $C_8H_4N_2$ mw: 128.14

PROP: Crystals, vap d: 4.42.

SYNS: 4-CYANOBENZONITRILE ◇ p-DICYANOBENZENE ◇ 1,4-DICYANOBENZENE ◇ NITRIL KYSELINY TEREFTALOVE (CZECH) ◇ p-PDN ◇ p-PHTHALODINITRILE ◇ TEREFTALODINITRIL (CZECH) ◇ TEREPHTHALONITRILE

TOXICITY DATA with REFERENCE
eye-rbt 500 mg/24H MLD 28ZPAK -,159,72
orl-rat LD50:20800 mg/kg 28ZPAK -,159,72
ipr-mus LD50:699 mg/kg INHEAO 4,11,66

CONSENSUS REPORTS: Reported in EPA TSCA Inventory. Cyanide and its compounds are on the Community Right-To-Know List.

SAFETY PROFILE: Moderately toxic by intraperitoneal route. Slightly toxic by ingestion. A skin and eye irritant. When heated to decomposition it emits toxic fumes of CN^- and NO_x. See also NITRILES.

BBP750 CAS:608-73-1 ***HR: 3***
BENZENE HEXACHLORIDE
mf: $C_6H_6Cl_6$ mw: 290.82

ClCH(CHCl)$_4$CHCl (ring)

PROP: Technical grade contains 68.7% α-BHC, 6.5% β-BHC and 13.5% Γ-BHC (JPFCD2 14,305,79). White, crystalline powder. Mp: 113°, vap press: 0.0317 mm @ 20°.

SYNS: BHC (USDA) ◇ COMPOUND-666 ◇ DBH ◇ ENT 8,601 ◇ GAMMEXANE ◇ HCCH ◇ HEXA ◇ HEXACHLOR ◇ HEXACHLORAN ◇ HEXACHLOROCYCLOHEXANE ◇ 1,2,3,4,5,6-HEXACHLOROCYCLOHEXANE ◇ HEXYLAN ◇ LINDANE

TOXICITY DATA with REFERENCE
mmo-omi 100 mg/L MILEDM 5,103,77
otr-rat-orl 875 mg/kg/7W-I CRNGDP 5,479,84
orl-mus TDLo:9120 mg/kg (22W male):REP BECTA6 26,508,81
orl-mus TDLo:6720 mg/kg/80W-C:CAR JPFCD2 14(3),305,79
skn-mus TDLo:1600 mg/kg/80W-I:ETA JPFCD2 14(3),305,79
orl-mus TD:12600 mg/kg/30W-C:CAR JCROD7 99,143,81
orl-mus TD:5400 mg/kg/13W-C:NEO TUMOAB 69,383,83
ihl-man TCLo:400 μg/kg/3D:CNS,GIT,MET GISAAA 49(10),26,84
orl-rat LD50:100 mg/kg ATXKA8 22,115,66
skn-rat LD50:0.9 mg/kg 85DPAN -,-,71/76
orl-mus LD50:59 mg/kg PEMNDP 8,443,87
scu-rbt LD50:75 mg/kg XPHPAW 414,273,55
orl-gpg LDLo:1400 mg/kg MEMOAQ 4,25,50
orl-ckn LD50:597 mg/kg POSCAL 60,2599,81
orl-brd LD50:56 mg/kg TXAPA9 21,315,72

CONSENSUS REPORTS: NTP Fifth Annual Report on Carcinogens. IARC Cancer Review: Animal Sufficient Evidence IMEMDT 5,47,74

SAFETY PROFILE: Confirmed carcinogen with experimental carcinogenic, neoplastigenic, and tumorigenic data by ingestion and skin contact. Poison by ingestion, skin contact, and subcutaneous routes. Human systemic effects by inhalation: headache, nausea or vomiting, and fever. Implicated in aplastic anemia. Experimental reproductive effects. Mutation data reported. Lindane is more toxic than DDT or dieldrin. When heated to decomposition it emits highly toxic fumes of phosgene, HCl, and Cl^-. Potentially violent reaction with dimethylformamide + iron. When heated to decomposition it emits highly toxic fumes of phosgene, HCl, and Cl^-. See other benzenehexachloride entries.

A toxic organochlorine which is persistent in the environment and accumulates in mammalian tissue. For cattle, the oral LD50 < = 100 mg/kg. The various isomers have different actions; the Γ (lindane) and α isomers are central nervous system stimulants, the principal symptom being convulsions. The β and Δ isomers are central nervous system depressants. The use of thermal vaporizers with lindane has caused acute poisoning by inhalation.

The dangerous acute dose of the technical mixture has been estimated at about 30 grams and the dangerous dose of lindane at about 7 to 15 grams. However, as already mentioned, a single dose of 45 mg (or approximately 0.65 mg/kg) of lindane caused convulsions. Lindane shows a marked difference in toxicity to different species. Its toxic effect on laboratory animals compares favorably with that of DDT, but for several domestic animals, notably calves, lindane is more toxic than DDT or dieldrin. On a chronic systemic basis the α, β and Γ isomers are experimental carcinogens. Has been implicated in aplastic anemia.

Dermatitis and perhaps other manifestations based on sensitivity represent a sort of chronic, though probably not systemic intoxication, which has been observed in humans.

The signs and symptoms of confirmed acute poisoning in humans have paralleled those of experimental animals. These signs and symptoms are: excitation, hyperirritability, loss of equilibrium, clonic-tonic convulsions, and later depression.

There is some evidence that the pulmonary edema and vascular collapse may be of neurogenic origin also. The symptoms in animals systemically poisoned by the Γ-isomer alone are essentially similar to those caused by mixtures, although the onset may be earlier. Workers acutely exposed to high air concentrations of lindane and its decomposition products show headache, nausea, and irritation of eyes, nose, and throat.

In rare instances, urticaria has followed exposure to lindane vapor. Unlike the signs and symptoms already mentioned, this allergic manifestation occurs only in susceptible individuals, and usually only after a period of sensitization.

BBQ000 CAS:319-84-6 ***HR: 3***
BENZENE HEXACHLORIDE-α-isomer
mf: $C_6H_6Cl_6$ mw: 290.82

SYNS: α-BENZENEHEXACHLORIDE ◇ α-BHC ◇ ENT 9,232 ◇ α-HCH ◇ α-HEXACHLORANE ◇ HEXACHLORCYCLOHEXAN (GERMAN) ◇ α-HEXACHLOROCYCLOHEXANE ◇ α-1,2,3,4,5,6-HEXACHLOROCYCLOHEXANE (MAK) ◇ 1-α,2-α,3-β,4-α,5-β,6-β-HEXACHLOROCYCLOHEXANE ◇ α-LINDANE

TOXICITY DATA with REFERENCE
dns-rat:lvr 1 μmol/L CNREA8 42,3010,82
cyt-rat-orl 756 mg/kg/3W JNCIAM 54,1245,75
orl-rat TDLo:20 g/kg/48W-C:NEO JNCIAM 54,801,75
orl-mus TDLo:5 g/kg/24W-C:CAR JNCIAM 51,1637,73
orl-mus TD:10 g/kg/24W-C:CAR NAIZAM 25,635,74
orl-rat TD:11040 mg/kg/86W-C:ETA CNREA8 41,4140,81
orl-hmn LDLo:14 g/kg 85GYAZ -,54,71
orl-rat LD50:177 mg/kg FATOAO 39,455,76

CONSENSUS REPORTS: NTP Fifth Annual Report on Carcinogens. IARC Cancer Review: Animal Sufficient Evidence IMEMDT 20,195,79; IMEMDT 5,47,74. EPA Genetic Toxicology Program. Reported in EPA TSCA Inventory.

DFG MAK: 0.5 mg/m^3

SAFETY PROFILE: Confirmed carcinogen with experimental carcinogenic, tumorigenic, and neoplastigenic data. Poison by ingestion. Mutation data reported. When heated to decomposition it emits toxic fumes of Cl^-. See also BENZENE HEXACHLORIDE and other benzenehexachloride entries.

BBQ500 CAS:58-89-9 ***HR: 3***
BENZENE HEXACHLORIDE-Γ-isomer
DOT: NA 2761
mf: $C_6H_6Cl_6$ mw: 290.82

SYNS: AALINDAN ◇ AFICIDE ◇ AGRISOL G-20 ◇ AGROCIDE ◇ AGRONEXIT ◇ AMEISENATOD ◇ AMEISENMITTEL MERCK ◇ APARSIN ◇ APHTIRIA ◇ APLIDAL ◇ ARBITEX ◇ BBH ◇ BEN-HEX ◇ BENTOX 10 ◇ Γ-BENZENE HEXACHLORIDE ◇ BEXOL ◇ BHC ◇ Γ-BHC ◇ CELANEX ◇ CHLORESENE ◇ CODECHINE ◇ DBH ◇ DETMOL-EXTRAKT ◇ DETOX 25 ◇ DEVORAN ◇ DOL GRANULE ◇ DRILL TOX-SPEZIAL AGLUKON ◇ ENT 7,796 ◇ ENTOMOXAN ◇ EXAGAMA ◇ FORLIN ◇ GALLOGAMA ◇ GAMACID ◇ GAMAPHEX ◇ GAMENE ◇ GAMISO ◇ GAMMA-COL ◇ GAMMAHEXA ◇ GAMMAHEXANE ◇ GAMMALIN ◇ GAMMOPAZ ◇ HCCH ◇ HCH ◇ Γ-HCH ◇ HECLOTOX ◇ HEXACHLORAN ◇ Γ-HEXACHLORAN ◇ Γ-HEXACHLORANE ◇ Γ-HEXACHLOROBENZENE ◇ 1-α,2-α,3-β,4-α,5-α,6-β-HEXACHLOROCYCLOHEXANE ◇ Γ-HEXACHLOROCYCLOHEXANE (MAK) ◇ 1,2,3,4,5,6-HEXACHLOROCYCLOHEXANE, Γ-ISOMER ◇ HEXATOX ◇ HEXICIDE ◇ HGI ◇ INEXIT ◇ ISOTOX ◇ JACUTIN ◇ KOKOTINE ◇ KWELL ◇ LENDINE ◇ LENTOX ◇ LIDENAL ◇ LINDAGRAIN ◇ LINDANE (ACGIH, DOT, USDA) ◇ LINTOX ◇ MILBOL 49 ◇ MSZYCOL ◇ NCI-C00204 ◇ NEO-SCABICIDOL ◇ NEXIT ◇ NOVIGAM ◇ OVADZIAK ◇ PEDRACZAK ◇ QUELLADA ◇ RCRA WASTE NUMBER U129 ◇ SANG gamma ◇ STREUNEX ◇ TAP 85 ◇ VITON

TOXICITY DATA with REFERENCE
dns-ofs:lvr 45 μmol/L HKXUDL 4,268,84
msc-ham:lng 200 mg/L GISAAA 49(5),82,84
orl-rbt TDLo:60 mg/kg (female 9D post):REP BVIPA7 21,85,77

orl-rat TDLo:100 mg/kg (female 9D post):TER BVIPA7 21,85,77
orl-mus TDLo:14 g/kg/2Y-C:CAR CRNGDP 8,1889,87
orl-mus TD:25 g/kg/73W-C:NEO FCTXAV 11,433,73
orl-chd LDLo:180 mg/kg:CNS,PUL CMEP** -,1,56
orl-chd TDLo:111 mg/kg:CNS AEHLAU 25,374,72
skn-man TDLo:20 mg/kg/6W I:EYE,CNS AJDCAI 141,125,87
orl-rat LD50:76 mg/kg SPEADM 74-1,-,74
skn-rat LD50:500 mg/kg WRPCA2 9,119,70
ipr-rat LDLo:35 mg/kg AEPPAE 212,463,51
orl-mus LD50:44 mg/kg JEENAI 65,632,72
ipr-mus LD50:125 mg/kg SOGEBZ 2(1),80,66
orl-dog LD50:40 mg/kg SPEADM 74-1,-,74
ivn-dog LDLo:8 mg/kg TIEUA7 5,61,50
orl-rbt LD50:60 mg/kg JHEMA2 22,115,78
skn-rbt LD50:50 mg/kg AFDOAQ 16,3,52
ivn-rbt LDLo:4500 μg/kg JPETAB 92,140,48
orl-gpg LD50:127 mg/kg FEPRA7 6,386,47
orl-ham LD50:360 mg/kg JETOAS 7,159,74
ipr-ham LD50:640 mg/kg ARTODN 58,152,85
ims-bwd LDLo:26 mg/kg TIEUA7 5,61,50

CONSENSUS REPORTS: NTP Fifth Annual Report on Carcinogens. IARC Cancer Review: Animal Sufficient Evidence IMEMDT 5,47,74; IMEMDT 20,195,79. NCI Carcinogenesis Bioassay (feed); No Evidence: mouse, rat NCITR* NCI-CG-TR-14,77. EPA Extremely Hazardous Substances List. EPA Genetic Toxicology Program. Community Right-To-Know List. Reported in EPA TSCA Inventory.

OSHA PEL: TWA 0.5 mg/m^3 (skin)
ACGIH TLV: TWA 0.5 mg/m^3 (skin)
DFG MAK: 0.5 mg/m^3
DOT Classification: ORM-A; Label: None.

SAFETY PROFILE: Confirmed carcinogen with experimental carcinogenic neoplastigenic data. A human systemic poison by ingestion. Also a poison by ingestion, skin contact, intraperitoneal, intravenous, and intramuscular routes. Human systemic effects by ingestion: convulsions, dyspnea, and cyanosis. Experimental teratogenic and reproductive effects. Mutation data reported. See also BENZENE HEXACHLORIDE and other benzene hexachloride entries. When heated to decomposition it emits toxic fumes of Cl^-, HCl, and phosgene.

BBQ750 **HR: 3**
BENZENEHEXACHLORIDE (mixed isomers)
mf: $C_6H_6Cl_6$ mw: 290.82

PROP: Technical BHC contains about 64% α, 10% β, 13% Γ, 9% Δ and 1% ε isomers of 1,2,3,4,5,6-hexachlorocyclohexane (IARC** 5,47,74).

SYNS: BENZAHEX ◇ BENZEX ◇ DOL ◇ DOLMIX ◇ FBHC ◇ FHCH ◇ 1,2,3,4,5,6-HEXACHLOROCYCLOHEXANE (mixture of isomers) ◇ HEXYCLAN ◇ KOTOL ◇ SOPROCIDE ◇ TECHNICAL BHC ◇ TECHNICAL HCH

TOXICITY DATA with REFERENCE
orl-mus TD:11 g/kg/26W-C:ETA TXCYAC 19,31,81
orl-mus TDLo:13 g/kg/24W-C:NEO GANNA2 62,431,71
unr-man TDLo:643 μg/kg:CNS CMEP** -,1,56
orl-rat LD50:400 mg/kg 85GMAT -,73,82
orl-mus LD50:500 mg/kg 85GMAT -,73,82
orl-cat LDLo:300 mg/kg 85GMAT -,73,82
ihl-cat LCLo:20 mg/m^3/6H 85GMAT -,73,82

CONSENSUS REPORTS: IARC Cancer Review: Animal Sufficient Evidence IMEMDT 5,47,74; IMEMDT 20,195,79.

SAFETY PROFILE: Confirmed carcinogen with experimental tumorigenic and neoplastigenic data. Poison by inhalation and ingestion. Human systemic effects by an unspecified route: convulsions. Potentially dangerous reaction with DMF in presence of Fe, also CCl_4. When heated to decomposition it emits highly toxic fumes of Cl^-, HCl, and phosgene. See also BENZENE HEXACHLORIDE and other benzenehexachloride entries.

BBR000 CAS:319-85-7 **HR: 3**
trans-α-BENZENEHEXACHLORIDE
mf: $C_6H_6Cl_6$ mw: 290.82

SYNS: β-BENZENEHEXACHLORIDE ◇ β-BHC ◇ ENT 9,233 ◇ β-HCH ◇ β-HEXACHLOROBENZENE ◇ 1-α,2-β,3-α,4-β,5-α,6-β-HEXACHLOROCYCLOHEXANE ◇ β-HEXACHLOROCYCLOHEXANE ◇ β-1,2,3,4,5,6-HEXACHLOROCYCLOHEXANE (MAK) ◇ β-ISOMER ◇ β-LINDANE

TOXICITY DATA with REFERENCE
orl-mus TDLo:18 g/kg/2Y-C:NEO FCTXAV 11,433,73
orl-rat LD50:6000 mg/kg ALLVAR 43,-,55

CONSENSUS REPORTS: NTP Fifth Annual Report on Carcinogens. IARC Cancer Review: Animal Sufficient Evidence IMEMDT 5,47,74; Animal Limited Evidence IMEMDT 20,195,79. Reported in EPA TSCA Inventory.

DFG MAK: 0.5 mg/m^3

SAFETY PROFILE: Confirmed carcinogen with experimental neoplastigenic data. Mildly toxic by ingestion. When heated to decomposition it emits very toxic fumes of Cl^-, HCl, and phosgene. See also BENZENE HEXACHLORIDE and other benzenehexachloride entries.

BBR325 CAS:6996-92-5 **HR: 3**
BENZENESELENIC ACID
mf: $C_6H_6O_2Se$ mw: 189.07

CONSENSUS REPORTS: Selenium and its compounds are on the Community Right-To-Know List.

OSHA PEL: TWA 0.2 mg(Se)/m^3
ACGIH TLV: TWA 0.2 mg(Se)/m^3
DFG MAK: 0.1 mg(Se)/m^3

SAFETY PROFILE: Reacts violently with hydrazine derivatives (e.g., benzohydrazide). When heated to decomposition it emits toxic fumes of Se. See also SELENIUM COMPOUNDS.

BBR380 CAS:21230-20-6 ***HR: 3***
BENZENESULFINYL AZIDE
mf: $C_6H_5N_3OS$ mw: 167.18

SAFETY PROFILE: Explodes at room temperature. Upon decomposition it emits toxic fumes of SO_x and NO_x. See also AZIDES.

BBR390 CAS:4972-29-6 ***HR: 3***
BENZENE SULFINYL CHLORIDE
mf: C_6H_5ClOS mw: 160.56

SAFETY PROFILE: May explode if stored in a sealed container. When heated to decomposition it emits toxic fumes of SO_x and Cl^-.

BBR500 CAS:98-10-2 ***HR: 2***
BENZENESULFONAMIDE
mf: $C_6H_7NO_2S$ mw: 157.20

SYNS: BENZENESULPHONAMIDE ◇ BENZOSULFONAMIDE ◇ BSA

TOXICITY DATA with REFERENCE
orl-rat LD50:991 mg/kg MarJV# 29MAR77
orl-mus LD50:740 mg/kg GTPZAB 23(12),47,79
ipr-mus LD50:1000 mg/kg JMCMAR 8,548,65

CONSENSUS REPORTS: Reported in EPA TSCA Inventory.

SAFETY PROFILE: Moderately toxic by ingestion and intraperitoneal routes. When heated to decomposition it emits very toxic fumes of SO_x and NO_x.

BBR750 CAS:1678-25-7 ***HR: 3***
BENZENESULFONANILIDE
mf: $C_{12}H_{11}NO_2S$ mw: 233.30

SYN: BENZENESULFANILIDE

TOXICITY DATA with REFERENCE
ipr-rat TDLo:593 mg/kg/4W-I:ETA CNREA8 30,1485,70

SAFETY PROFILE: Questionable carcinogen with experimental tumorigenic data. When heated to decomposition it emits very toxic fumes of SO_x and NO_x.

BBS250 CAS:98-11-3 ***HR: 3***
BENZENESULFONIC ACID
mf: $C_6H_6O_3S$ mw: 158.18

PROP: Deliquescent plates or tablets. Mp: 43-44°.

SYN: PHENYLSULFONIC ACID

TOXICITY DATA with REFERENCE
skn-rbt 100 μg/24H open AIHAAP 23,95,62
orl-rat LD50:890 mg/kg AIHAAP 23,95,62
orl-bwd LD50:75 mg/kg TXAPA9 21,315,72
skn-cat LDLo:10 g/kg JPETAB 84,358,45

CONSENSUS REPORTS: Reported in EPA TSCA Inventory.

SAFETY PROFILE: Poison by ingestion, skin contact, and probably inhalation. A skin irritant. See also SULFATES and SULFONATES.

BBS500 ***HR: 3***
BENZENE SULFONYL AZIDE
mf: $C_6H_5N_3O_2S$ mw: 183.09

SAFETY PROFILE: The crude material explodes violently on heating. The pure material decomposes rapidly but smoothly at 105°. When heated to decomposition it emits toxic fumes of SO_x and NO_x. See also AZIDES.

BBS750 CAS:98-09-9 ***HR: 3***
BENZENESULFONYL CHLORIDE
DOT: UN 2225
mf: $C_6H_5ClO_2S$ mw: 176.62

SYNS: BENZENE SULFONCHLORIDE ◇ BENZENESULFONIC (ACID) CHLORIDE ◇ BENZENE SULPHONYL CHLORIDE (DOT) ◇ BENZENOSULFOCHLOREK (POLISH) ◇ BENZENOSULPHOCHLORIDE ◇ BSC-REFINE D ◇ RCRA WASTE NUMBER U020

TOXICITY DATA with REFERENCE
orl-rat LD50:1960 mg/kg MEPAAX 20,513,69
ipr-rat LD50:76 mg/kg MEPAAX 20,513,69

CONSENSUS REPORTS: Reported in EPA TSCA Inventory.

DOT Classification: IMO: Corrosive Material; Label: Corrosive.

SAFETY PROFILE: Poison by intraperitoneal route. A dangerous storage hazard. It may explode in a sealed bottle. Explosive reaction with dimethyl sulfoxide. Reacts vigorously with methyl formamide. When heated to decomposition it emits toxic fumes of Cl^- and SO_x. See also SULFONATES.

BBT000 CAS:20611-21-6 ***HR: 1***
2-(BENZENESULFONYL)ETHANOL
mf: $C_8H_{10}O_3S$ mw: 186.24

SYNS: FENYL-β-HYDROXYETHYLSULFON (CZECH) ◇ 2-(PHENYLSULFONYL)ETHANOL

TOXICITY DATA with REFERENCE
skn-rbt 500 mg/24H MLD 28ZPAK -,200,72
eye-rbt 20 mg/24H MOD 28ZPAK -,200,72
orl-rat LD50:5830 mg/kg 28ZPAK -,200,72

SAFETY PROFILE: Mildly toxic by ingestion. A skin and eye irritant. When heated to decomposition it emits toxic fumes of SO_x.

BBT250 CAS:368-43-4 ***HR: 3***
BENZENESULPHONYL FLUORIDE
mf: $C_6H_5FO_2S$ mw: 160.17

PROP: Clear liquid. Bp: 209°, fp: −5°, flash p: 196°F, d: 1.329, vap press: 8 mm @ 80°, vap d: 5.52.

TOXICITY DATA with REFERENCE
ipr-rat LD50:100 mg/kg NATUAS 173,33,54

CONSENSUS REPORTS: Reported in EPA TSCA Inventory.

SAFETY PROFILE: A poison by intraperitoneal routes. Slightly irritating to skin. Flammable when exposed to heat or flame. It can react vigorously with oxidizing materials. To fight fire, use water, foam, CO_2, water spray or mist, dry chemical. When heated to decomposition it emits toxic fumes of F^- and SO_x. See also FLUORIDES and SULFATES.

BBU125 CAS:3470-17-5 ***HR: 2***
BENZENETRIFUROXAN
mf: $C_6N_6O_6$ mw: 252.12

SYNS: BENZOTRIFUROXAN ◇ BENZOTRIS(c)FURAZAN-2-OXIDE ◇ BTF

TOXICITY DATA with REFERENCE
skn-rbt 500 mg/24H MLD NTIS**DE 83013231
eye-rbt 100 mg/24H SEV NTIS** DE83013231
eye-rbt 100 mg/30S rns SEV NTIS** DE83013231
orl-rat LD50:2884 mg/kg NTIS** DE83013231

CONSENSUS REPORTS: Reported in EPA TSCA Inventory.

SAFETY PROFILE: Moderately toxic by ingestion. A mild skin and severe eye irritant. When heated to decomposition it emits acrid smoke and fumes.

BBU250 CAS:533-73-3 ***HR: 3***
1,2,4-BENZENETRIOL
mf: $C_6H_6O_3$ mw: 126.12

SYNS: HYDROXYHYDROQUINONE ◇ HYDROXYQUINOL ◇ OXYHYDROCHINON (GERMAN) ◇ OXYHYDROQUINONE ◇ 1,2,4-TRIHYDROXYBENZENE

TOXICITY DATA with REFERENCE
oms-hmn:lym 50 μmol/L CNREA8 45,2471,85
sce-hmn:lym 5 μmol/L CNREA8 45,2471,85
scu-mus LD50:120 mg/kg INHEAO 5,143,67
ipr-mus LDLo:125 mg/kg CBCCT* 6,145,54

CONSENSUS REPORTS: EPA Genetic Toxicology Program. Reported in EPA TSCA Inventory.

SAFETY PROFILE: Poison by subcutaneous and intraperitoneal routes. Human mutation data reported. When heated to decomposition it emits acrid smoke and irritating fumes.

BBU500 ***HR: 3***
BENZENE TRIOZONIDE
mf: $C_6H_6O_9$ mw: 222.11

SAFETY PROFILE: An unstable explosive, sensitive to the slightest touch. Upon decomposition it emits acrid smoke and fumes.

BBU625 CAS:3691-78-9 ***HR: 3***
BENZETHIDIN
mf: $C_{23}H_{29}NO_3$ mw: 367.53

SYNS: BENZETHIDINE ◇ ETHYL-1-(2-BENZYLOXYETHYL)-4-PHENYLPIPERIDINE-4-CARBOXYLATE ◇ NIH 7574 ◇ 4-PHENYL-1-(2-(PHENYLMETHXY)ETHYL)-4-PIPERIDINECARBOXYLICACID ETHYL ESTER

TOXICITY DATA with REFERENCE
orl-rat LD50:284 mg/kg BJPCAL 15,254,60
scu-rat LD50:600 mg/kg BJPCAL 15,254,60
ivn-mus LD50:10900 μg/kg BJPCAL 15,254,60

SAFETY PROFILE: Poison by ingestion and intravenous routes. Moderately toxic by subcutaneous route. When heated to decomposition it emits toxic fumes of NO_x. See also ESTERS.

BBU750 CAS:5929-09-9 ***HR: 3***
BENZETHONIUM CHLORIDE MONOHYDRATE
mf: $C_{27}H_{42}NO_2•Cl•H_2O$ mw: 466.17

SYNS: p-DIISOBUTYLPHENOXYETHOXYETHYLDIMETHYLBENZYLAMMONIUM CHLORIDE MONOHYDRATE ◇ HYAMINE 1622 ◇ PHEMEROL CHLORIDE MONOHYDRATE

TOXICITY DATA with REFERENCE
orl-rat LD50:420 mg/kg PCOC** -,121,66
ipr-rat LD50:33 mg/kg PCOC** -,121,66
ivn-rat LD50:19 mg/kg PCOC** -,121,66
ivn-mus LD50:32 mg/kg CSLNX* NX#00430

SAFETY PROFILE: Poison by intraperitoneal and intravenous routes. Moderately toxic by ingestion. When heated to decomposition it emits toxic fumes of NO_x.

BBU800 CAS:5633-14-7 ***HR: 3***
BENZETIMIDE
mf: $C_{23}H_{26}N_2O_2$•ClH mw: 398.97

SYNS: dl-1-BENZYL-4-(2,6-DIOXO-3-PHENYL-3-PIPERIDYL)PIPERIDINE HYDROCHLORIDE ◇ BZ ◇ DIOXATRINE ◇ R 4929 ◇ SPASMENTRAL

TOXICITY DATA with REFERENCE
ivn-rat LD50:37600 μg/kg ARZNAD 21,1365,71
orl-mus LD50:680 mg/kg OYYAA2 3,283,69
ivn-mus LD50:46 mg/kg ARZNAD 21,1365,71

SAFETY PROFILE: Poison by intravenous route. Moderately toxic by ingestion. When heated to decomposition it emits toxic fumes of NO_x and HCl. An anticholinergic.

BBU810 ***HR: D***
trans-BENZ(a,e)FLUOROANTHENE-3,4-DIHYDRODIOL
mf: $C_{24}H_{16}O_2$ mw: 336.40

SYN: trans-3,4-DIHYDRO-3,4-DIHYDROXYDIBENZO(a,e)FLUORANTHENE

TOXICITY DATA with REFERENCE
mma-sat 1200 nmol/L CRNGDP 5,1263,84
dns-mus:emb 1 μmol/L CRNGDP 5,379,84
dnd-man:lym 208 nmol CRNGDP 4,27,83

SAFETY PROFILE: Mutation data reported. When heated to decomposition it emits acrid smoke and fumes.

BBU825 ***HR: D***
trans-BENZ(a,e)FLUORANTHENE-12,13-DIHYDRODIOL
mf: $C_{24}H_{16}O_2$ mw: 336.40

SYN: trans-12,13-DIHYDRO-12,13-DIHYDROXYDIBENZO(a,e)FLUORANTHENE

TOXICITY DATA with REFERENCE
mma-sat 100 nmol/L CRNGDP 5,1263,84
dns-mus:emb 1 μmol/L CRNGDP 5,379,84
dnd-mam:lym 297 nmol/L CRNGDP 4,27,83

SAFETY PROFILE: Mutation data reported. When heated to decomposition it emits acrid smoke and fumes.

BBV000 CAS:52-49-3 ***HR: 3***
BENZHEXOL HYDROCHLORIDE
mf: $C_{20}H_{31}NO$•ClH mw: 337.98

SYNS: APARKAN ◇ ARTANE ◇ ARTANE HYDROCHLORIDE ◇ ARTANE TRIHEXYPHENIDYL ◇ BENZHEXOL CHLORIDE ◇ CYCLODOL ◇ α-CYCLOHEXYL-α-PHENYL-1-PIPERIDINEPROPANOL HYDROCHLORIDE ◇ PACITANE ◇ PARALEST ◇ PARGITAN ◇ PARKINSAN ◇ PARKOPAN ◇ PERAGIT ◇ 1-PHENYL-1-CYCLOHEXYL-3-PIPERIDYL-1-PROPANOL HYDROCHLORIDE ◇ PIPANOL ◇ 3-(1-PIPERIDYL)-1-CYCLOHEXYL-1-PHENYL-1-PROPANOL HYDROCHLORIDE ◇ ROMPARKIN ◇ SEDRENA ◇ TREMIN ◇ TRIESIFENIDILE ◇ TRIEXIFENIDILA ◇ TRIHEXYLPHENIDYL HYDROCHLORIDE ◇ TRIPHEDINON ◇ TRIPHENIDYL ◇ TSIKLODOL

TOXICITY DATA with REFERENCE
ipr-rat LD50:195 mg/kg 27ZQAG -,311,72
ivn-rat LD50:30 mg/kg 27ZQAG -,311,72
orl-mus LD50:217 mg/kg NIIRDN 6,525,82
ipr-mus LD50:150 mg/kg PHARAT 37,483,82
scu-mus LD50:152 mg/kg 27ZQAG -,311,72
ivn-mus LD50:39 mg/kg 27ZQAG -,311,72
scu-gpg LD50:320 mg/kg AIPTAK 137,375,62

SAFETY PROFILE: Poison by ingestion, intraperitoneal, intravenous, and subcutaneous routes. An anticholinergic agent which causes human psychotropic effects. When heated to decomposition it emits very toxic fumes of NO_x and HCl.

BBV250 CAS:613-94-5 ***HR: 3***
BENZHYDRAZIDE
mf: $C_7H_8N_2O$ mw: 136.17

SYNS: BENZOHYDRAZIDE ◇ BENZOHYDRAZINE ◇ BENZOIC HYDRAZIDE ◇ BENZOYL HYDRAZIDE

TOXICITY DATA with REFERENCE
orl-mus TDLo:15 g/kg/77W-C:CAR EJCAAH 8,341,72
orl-mus TD:13 g/kg/30W-I:NEO 34ZRA9 -,869,65
scu-mus LD50:122 mg/kg JPETAB 122,110,58

CONSENSUS REPORTS: Reported in EPA TSCA Inventory.

SAFETY PROFILE: Poison by subcutaneous and intraperitoneal routes. Questionable carcinogen with experimental carcinogenic and neoplastigenic data. Violent reaction with benzeneseleninic acid. When heated to decomposition it emits toxic fumes of NO_x.

BBV500 CAS:58-73-1 ***HR: 3***
BENZHYDRYL
mf: $C_{17}H_{21}NO$ mw: 255.39

SYNS: ALERYL ◇ ALLEDRYL ◇ ALLERGAN B ◇ ALLERGEVAL ◇ ALLERGICAL ◇ ALLERGIN ◇ ALLERGINA ◇ ALLERGIVAL ◇ AMIDRYL ◇ ANTISTOMINUM ◇ ANTOMIN ◇ AUTOMIN ◇ BAGAODRYL ◇ BARAMINE ◇ BENA ◇ BENACHLOR ◇ BENADON ◇ BENADRIN ◇ BENADRYL ◇ BEN-ALLERGIN ◇ BENAPON ◇ BENODIN ◇ BENODINE ◇ BENYLAN ◇ BENZANTINE ◇ BENZHYDRAMINE ◇ BENZHYDRAMINUM ◇ BENZHYDRIL ◇ o-BENZHYDRYLDIMETHYLAMINOETHANOL ◇ 2-(BENZHYDRYLOXY)-N,N-DIMETHYLETHYLAMINE ◇ 2-(BENZOHYDRYLOXY)-N,N-DIMETHYLETHYLAMINE ◇ BETRAMIN ◇ DABYLEN ◇ DEBENDRIN ◇ DERMISTINE ◇ DERMODRIN ◇ DESENTOL ◇ DIABENYL ◇ DIABYLEN ◇ DIBONDRIN ◇ DIFEDRYL ◇ DIFENHYDRAMIN ◇ DIFENIDRAMINA (ITALIAN) ◇ DIHIDRAL ◇ DIMEDROL ◇ DIMEDRYL ◇ β-DIMETHYLAMINO-AETHYL-BENZHYDRYL-AETHER (GERMAN) ◇ β-DIMETHYLAMINOETHANOL DIPHENYLMETHYL ETHER ◇ α-(2-DIMETHYLAMINOETHOXY)-DIPHENYLMETHANE ◇ β-DIMETHYLAMINOETHYLBENZHYDRYLETHER ◇ DIPHANTINE ◇ DIPHENYLHYDRAMINE ◇ 2-(DIPHENYLMETHOXY)-N,N-DIMETHYLETHYLAMINE ◇ DRYISTAN

◇ DRYLISTAN ◇ DYLAMON ◇ ETANAUTINE ◇ HISTAXIN ◇ HYADRINE ◇ IBIODRAL ◇ MEDIDRYL ◇ MEPHADRYL ◇ NAUSEN ◇ PROBEDRYL ◇ RESTAMIN ◇ RESTAMINE ◇ RIGIDIL ◇ RIGIDYL ◇ S51 ◇ SYNTEDRIL ◇ SYNTODRIL ◇ VENA

TOXICITY DATA with REFERENCE
dnd-esc 1 mg/L KHFZAN 16(10),11,82
dni-hmn:fbr 12500 μg/L DNSYAG 29,829,68
oms-hmn:fbr 12500 μg/L DNSYAG 29,829,68
cyt-hmn:fbr 100 mg/L ACYTAN 16,41,72
scu-mus TDLo:240 mg/kg (6D pre):REP PSEBAA 89,629,55
unr-man LDLo:7353 μg/kg 85DCAI 2,73,70
orl-rat LD50:390 mg/kg RPTOAN 40,42,77
ipr-rat LD50:280 mg/kg IJMRAQ 59,614,71
orl-mus LD50:160 mg/kg CHTPBA 7,224,72
ipr-mus LD50:56 mg/kg YKYUA6 34,27,83
scu-mus LD50:50 mg/kg BCFAAI 111,293,72
ivn-mus LD50:29 mg/kg RPTOAN 40,42,77
ipr-gpg LD50:75 mg/kg THERAP 28,767,73
scu-gpg LD50:56 mg/kg ARZNAD 4,189,54

CONSENSUS REPORTS: Reported in EPA TSCA Inventory.

SAFETY PROFILE: Deadly human poison by an unspecified route. Poison by ingestion, intravenous, intraperitoneal, and subcutaneous routes. Experimental reproductive effects. Human mutation data reported. When heated to decomposition it emits toxic fumes of NO_x. See also ETHERS.

BBV750 CAS:3733-63-9 ***HR: 3***
1-BENZHYDRYL-4-(2-(2-HYDROXYETHOXY)ETHYL)PIPERAZINE
mf: $C_{21}H_{28}N_2O_2$ mw: 340.51

SYNS: DECLOXIZINE ◇ 1-(DIPHENYLMETHYL)-4-(2-(2-HYDROXYETHOXY)ETHYL)PIPERAZINE ◇ 2-(2-((4-DIPHENYLMETHYL)-1-PIPERAZINYL)ETHOXY)ETHANOL HYDROXYDIETHYLPHENAMINE ◇ UCB 1402

TOXICITY DATA with REFERENCE
orl-rat TDLo:1120 mg/kg (female 8-15D post):TER PSDTAP 9,134,68
orl-rat TDLo:840 mg/kg (female 8-15D post):REP PSDTAP 9,134,68
orl-rat LD50:840 mg/kg ARZNAD 18,1002,68
ipr-rat LD50:103 mg/kg ARZNAD 18,1002,68
ivn-rat LD50:47 mg/kg ARZNAD 18,1002,68
orl-mus LD50:470 mg/kg ARZNAD 18,1002,68
ipr-mus LD50:135 mg/kg ARZNAD 18,1002,68
ivn-mus LD50:45 mg/kg ARZNAD 18,1002,68

SAFETY PROFILE: Poison by intraperitoneal and intravenous routes. Moderately toxic by ingestion. An experimental teratogen. Other experimental reproductive effects. When heated to decomposition it emits toxic fumes of NO_x.

BBW000 CAS:19974-69-7 ***HR: 3***
2-BENZHYDRYL-3-HYDROXY-N-METHYLPIPERIDINE HYDROCHLORIDE
mf: $C_{19}H_{23}NO{\cdot}ClH$ mw: 317.89

SYN: SCH 5472

TOXICITY DATA with REFERENCE
orl-rat LD50:1 mg/kg 27ZQAG -,301,72
orl-dog LD50:4 mg/kg 27ZQAG -,301,72

SAFETY PROFILE: Poison by ingestion. When heated to decomposition it emits very toxic fumes of NO_x and HCl.

BBW250 CAS:16136-32-6 ***HR: 3***
2-(BENZHYDRYLOXYETHYL)GUANIDINE
mf: $C_{16}H_{19}N_3O$ mw: 269.38

TOXICITY DATA with REFERENCE
orl-mus LD50:375 mg/kg JMCMAR 6,705,63
scu-mus LD50:125 mg/kg JMCMAR 6,705,63

SAFETY PROFILE: Poison by ingestion and subcutaneous route. When heated to decomposition it emits toxic fumes of NO_x.

BBW500 CAS:132-69-4 ***HR: 3***
BENZIDAMINE HYDROCHLORIDE
mf: $C_{19}H_{23}N_3O{\cdot}ClH$ mw: 345.91

SYNS: AF 864 ◇ BENALGIN ◇ BENZINDAMINE HYDROCHLORIDE ◇ BENZYDAMINE HYDROCHLORIDE ◇ 1-BENZYL-3-Γ-DIMETHYLAMINOPROPOXY-1H-INDAZOLE HYDROCHLORIDE ◇ 1-BENZYL-3-(3-(DIMETHYLAMINO)PROPOXY)-1H-INDAZOLEHYDROCHLORIDE ◇ BENZYRIN ◇ DIFFLAM ◇ N,N-DIMETHYL-3((1-PHENYLMETHYL)-1H-INDAZOL-3-YL)OXY)-1-PROPANAMINEHYDROCHLORIDE ◇ DORINAMIN ◇ ENZAMIN ◇ EPIROTIN ◇ IMOTRYL ◇ INDOLIN ◇ RIRILIM ◇ RIRIPEN ◇ SALYZORON ◇ TAMAS ◇ TANTUM ◇ VERAX

TOXICITY DATA with REFERENCE
eye-rbt 200 mg rns MOD ARZNAD 22,724,72
orl-mus TDLo:240 mg/kg (7-12D preg):REP SKNEA7 22,109,72
orl-rat TDLo:1200 mg/kg (10-15D preg):TER OYYAA2 3,271,69
orl-cld TDLo:50 mg/kg ATXKA8 23,215,68
orl-rat LD50:740 mg/kg YKKZAJ 99,240,79
scu-rat LD50:720 mg/kg OYYAA2 2,70,68
ipr-rat LD50:100 mg/kg TXAPA9 10,148,67
ivn-rat LD50:43500 μg/kg OYYAA2 2,70,68
orl-mus LD50:440 mg/kg OYYAA2 16,1011,78
ipr-mus LD50:110 mg/kg TXAPA9 10,148,67
scu-mus LD50:218 mg/kg TXAPA9 10,148,67
ivn-mus LD50:33 mg/kg TXAPA9 10,148,67

SAFETY PROFILE: Poison by intraperitoneal, subcutaneous, and intravenous routes. Moderately toxic by ingestion. An experimental teratogen. Other experimental animal reproductive effects. An eye irritant. A nonsteroidal anti-inflammatory analgesic. When heated to decomposition it emits very toxic fumes of HCl and NO_x.

BBW750 CAS:59-98-3 ***HR: 3***
BENZIDAZOL
mf: $C_{10}H_{12}N_2$ mw: 160.24

SYNS: ARTONIL ◇ BENZAZOLINE ◇ 2-BENZYL-2-IMIDAZOLINE ◇ 2-BENZYL-4,5-IMIDAZOLINE ◇ 2-BENZYL-4,5-IMIDAZOLINE HYDROCHLORIDE ◇ CLORIDRATO DI-2-BENZIL-4,5-IMIDAZOLINA (ITALIAN) ◇ DIVASCOL ◇ IMIDALIN ◇ KASIMID ◇ LAMBRIL ◇ OLITENSOL ◇ PERIPHERINE ◇ PHENYLMETHYLIMIDAZOLINE ◇ PREFAXIL ◇ PRISCOL ◇ PRISCOLINE ◇ TOLAZOLINE ◇ VASIMID ◇ VASODIL ◇ VASODILATAN

TOXICITY DATA with REFERENCE
cyt-ham:lng 62500 μg/L GMCRDC 27,95,81
ivn-rat LDLo:67 mg/kg PSEBAA 76,847,51
orl-mus LD50:350 mg/kg CPBTAL 22,514,74
ipr-mus LD50:500 mg/kg 27ZIAQ -,263,73
ivn-mus LD50:40 mg/kg RPTOAN 37,198,74

SAFETY PROFILE: Poison by ingestion and intravenous routes. Moderately toxic by intraperitoneal route. Mutation data reported. When heated to decomposition it emits toxic fumes of NO_x.

BBX000 CAS:92-87-5 ***HR: 3***
BENZIDINE
DOT: UN 1885
mf: $C_{12}H_{12}N_2$ mw: 184.26

PROP: Grayish-yellow, crystalline powder; white or sltly reddish crystals, powder, or leaf. Mp: 127.5-128.7° @ 740 mm, bp: 401.7°, d: 1.250 @ 20°/4°.

SYNS: BENZIDIN (CZECH) ◇ BENZIDINA (ITALIAN) ◇ BENZYDYNA (POLISH) ◇ p,p-BIANILINE ◇ 4,4′-BIANILINE ◇ (1,1′-BIPHENYL)-4,4′-DIAMINE (9CI) ◇ 4,4′-BIPHENYLDIAMINE ◇ 4,4′-BIPHENYLENEDIAMINE ◇ C.I. 37225 ◇ C.I. AZOIC DIAZO COMPONENT 112 ◇ p,p′-DIAMINOBIPHENYL ◇ 4,4′-DIAMINOBIPHENYL ◇ 4,4′-DIAMINO-1,1′-BIPHENYL ◇ p-DIAMINODIPHENYL ◇ 4,4′-DIAMINODIPHENYL ◇ p,p′-DIANILINE ◇ 4,4′-DIPHENYLENEDIAMINE ◇ FAST CORINTH BASE B ◇ NCI-C03361 ◇ RCRA WASTE NUMBER U021

TOXICITY DATA with REFERENCE
dnd-hmn:fbr 3 mmol/L ENMUDM 7,267,85
dnd-rat-ipr 63 mg/kg CRNGDP 6,1285,85
msc-mus:lym 500 μg/L MUREAV 125,291,84
dns-ham:lvr 20 nmol/L MUREAV 136,255,84
oms-dog:oth 100 μmol/L CNREA8 44,1893,84
ihl-man TCLo:17600 μg/m^3/14Y-C:CAR,KID AEHLAU 27,1,73
orl-rat TDLo:108 mg/kg/27D-I:CAR CNREA8 28,924,68
ihl-rat TCLo:10 mg/m^3/56W-I:ETA BEXBAN 69,68,70
scu-rat TDLo:2025 mg/kg/27W-I:CAR CANCAR 3,789,50
orl-rat LD50:309 mg/kg NTIS** PB214-270
orl-mus LD50:214 mg/kg NTIS** PB214-270
ipr-mus LD50:110 mg/kg PMRSDJ 1,682,81
orl-dog LDLo:200 mg/kg AEXPBL 58,167,1907
orl-rbt LDLo:200 mg/kg AEXPBL 58,167,1907

CONSENSUS REPORTS: NTP Fifth Annual Report on Carcinogens. IARC Cancer Review: Human Limited Evidence IMEMDT 1,80,72; Human Sufficient Evidence IMEMDT 29,149,82; Animal Sufficient Evidence IMEMDT 1,80,72; IMEMDT 29,149,82. EPA Genetic Toxicology Program. Community Right-To-Know List. Reported in EPA TSCA Inventory.

OSHA: Cancer Suspect Agent
ACGIH TLV: Confirmed Human Carcinogen
DFG MAK: Human Carcinogen.
DOT Classification: Poison B; Label: Poison.

SAFETY PROFILE: Confirmed human carcinogen producing bladder tumors. Experimental carcinogenic and tumorigenic data. Poison by ingestion and intraperitoneal routes. Human mutation data reported. Can cause damage to blood, including hemolysis and bone marrow depression. On ingestion causes nausea and vomiting which may be followed by liver and kidney damage. Any exposure is considered extremely hazardous. When heated to decomposition it emits highly toxic fumes of NO_x. See also AROMATIC AMINES.

BBX250 CAS:16993-94-5 ***HR: 3***
3,3′-BENZIDINE DICARBOXYLIC ACID, DISODIUM SALT
mf: $C_{14}H_{10}N_2O_4 \cdot 2Na$ mw: 316.24

SYN: 4,4′-DIAMINO-3,3′-BIPHENYLDICARBOXYLIC ACID DISODIUM SALT

TOXICITY DATA with REFERENCE
scu-rat TDLo:7 g/kg/77W-I:ETA VOONAW 15(5),60,69

SAFETY PROFILE: Questionable carcinogen with experimental tumorigenic data. When heated to decomposition it emits toxic fumes of NO_x and Na_2O.

BBX500 CAS:117-61-3 ***HR: 2***
2,2′-BENZIDINEDISULFONIC ACID
mf: $C_{12}H_{12}N_2O_6S_2$ mw: 344.38

SYNS: 6,6′-BIMETANILIC ACID ◇ 4,4′-DIAMINOBIPHENYL-2,2′-DISULFONIC ACID ◇ 4,4′-DIAMINO-2,2′-BIPHENYLDISULFONIC ACID ◇ 4,4′-DIAMINODIPHENYL-2,2′-DISULFONIC ACID ◇ 2,2′-DISULFOBENZIDINE ◇ KYSELINA BENZIDIN-2,2′-DISULFONOVA (CZECH)

TOXICITY DATA with REFERENCE
eye-rbt 500 mg/24H SEV 28ZPAK -,191,72

CONSENSUS REPORTS: Reported in EPA TSCA Inventory.

SAFETY PROFILE: A severe eye irritant. See also SULFONATES. When heated to decomposition it emits very toxic fumes of SO_x and NO_x.

BBX750 CAS:531-85-1 ***HR: 3***
BENZIDINE HYDROCHLORIDE
mf: $C_{12}H_{12}N_2$•2ClH mw: 257.18

SYNS: (1,1'-BIPHENYL)-4,4'-DIAMINE, DIHYDROCHLORIDE ◇ DIHIDROCLORURO de BENZIDINA (SPANISH)

TOXICITY DATA with REFERENCE
mmo-sat 100 nmol/plate MUREAV 136,33,84
sce-ham-ipr 12500 μg/kg MUREAV 113,33,83
ipr-rat TDLo:62 mg/kg/4W-I:CAR CRNGDP 2,747,81
orl-mus TDLo:3360 mg/kg/80W-C:CAR TXAPA9 64,171,82
orl-mus TD:1600 mg/kg/84W-C:ETA CNREA8 35,2814,75

CONSENSUS REPORTS: Reported in EPA TSCA Inventory. EPA Genetic Toxicology Program.

SAFETY PROFILE: Suspected carcinogen with experimental carcinogenic and tumorigenic data. Human mutation data reported. When heated to decomposition it emits very toxic fumes of HCl and NO_x.

BBY000 CAS:531-86-2 ***HR: 3***
BENZIDINE SULFATE
mf: $C_{12}H_{12}N_2$•H_2O_4S mw: 282.34

SYN: (1,1'-BIPHENYL)-4,4'-DIAMINE SULFATE (1:1)

TOXICITY DATA with REFERENCE
scu-rat TDLo:2475 mg/kg/33W-I:CAR CANCAR 3,789,50
scu-rat TD:3900 mg/kg/52W-I:CAR GTPZAB 19(6),28,75

OSHA: Carcinogen

SAFETY PROFILE: Confirmed human carcinogen with experimental carcinogenic data. See also BENZIDINE and SULFATES. When heated to decomposition it emits toxic fumes of SO_x and NO_x.

BBY250 CAS:2051-89-0 ***HR: 3***
BENZIDINE-3-SULFURIC ACID
mf: $C_{12}H_{12}N_2O_3S$ mw: 264.32

SYNS: BENZIDINE-3-SULPHURIC ACID ◇ 4,4'-DIAMINO-3-BIPHENYL-3-SULFONIC ACID ◇ 4:4'-DIAMINO-3-DIPHENYLYL HYDROGEN SULFATE ◇ 3-SULFOBENZIDINE

TOXICITY DATA with REFERENCE
imp-mus TDLo:80 mg/kg:ETA BJCAAI 17,127,63

CONSENSUS REPORTS: Reported in EPA TSCA Inventory.

SAFETY PROFILE: Questionable carcinogen with experimental tumorigenic data. See also SULFATES. When heated to decomposition it emits very toxic fumes of NO_x and SO_x.

BBY300 ***HR: 2***
BENZIDINE SULPHATE and HYDRAZINE-BENZENE
mf: $C_6H_8N_2$•$C_{12}H_{12}N_2$•H_2O_4S mw: 390.50

SYN: HYDRAZINE-BENZENE and BENZIDINE SULFATE

TOXICITY DATA with REFERENCE
scu-rat TDLo:9100 mg/kg/52W-I:CAR GTPZAB 19(6),28,75

SAFETY PROFILE: Suspected carcinogen with experimental carcinogenic data. When heated to decomposition it emits toxic fumes of NO_x and SO_x.

BBY500 CAS:3365-94-4 ***HR: 3***
BENZIDIN-3-YL ESTER SULFURIC ACID
mf: $C_{12}H_{12}N_2O_4S$ mw: 280.32

SYNS: BENZIDIN-3-YL HYDROGEN SULFATE ◇ 4,4'-DIAMINO-3-DIPHENYLYL HYDROGEN SULFATE

TOXICITY DATA with REFERENCE
imp-mus TDLo:80 mg/kg:ETA BJCAAI 17,127,63

SAFETY PROFILE: Questionable carcinogen with experimental tumorigenic data. See also ESTERS and SULFURIC ACID. When heated to decomposition it emits very toxic fumes of NO_x and SO_x.

BBY750 CAS:134-81-6 ***HR: 2***
BENZIL
mf: $C_{14}H_{10}O_2$ mw: 210.24

PROP: Yellow crystals. Mp: 95°, bp: 346-348°, d: 1.23 @ 15°/4°, vap press: 1 mm @ 128.4°.

SYNS: DIBENZOYL ◇ DIPHENYL-α,β-DIKETONE ◇ 1,2-DIPHENYLETHANEDIONE ◇ DIPHENYLGLYOXAL

TOXICITY DATA with REFERENCE
eye-rbt 100 mg/24H SEV 28ZPAK -,43,72
orl-rat LD50:2710 mg/kg 28ZPAK-,43,72

CONSENSUS REPORTS: Reported in EPA TSCA Inventory.

SAFETY PROFILE: Moderately toxic by ingestion. An eye irritant. Combustible. When heated to decomposition it emits acrid smoke and irritating fumes. See also KETONES.

BCA000 CAS:57-37-4 ***HR: 3***
BENZILIC ACID-β-DIETHYLAMINOETHYL ESTER HYDROCHLORIDE
mf: $C_{20}H_{25}NO_3$•ClH mw: 363.92

SYNS: ACTOZINE ◇ AMIOYL ◇ AMISYL ◇ AMITAKON ◇ AMIZIL HYDROCHLORIDE ◇ ARCADINE ◇ AY-5406 ◇ BENACTIZINE HYDROCHLORIDE ◇ BENACTYZIN (CZECH) ◇ BENACTYZINE CHLORIDE ◇ BENACTYZINE HYDROCHLORIDE ◇ BENAKTIN ◇ BENZILATE DU DIETHYLAMINO-ETHANOL CHLORHYDRATE (FRENCH)

◇ CAFRON ◇ CEDAD ◇ CEVANOL ◇ DESTENDO ◇ β-DIETHYLAMINOETHYL BENZILATE HYDROCHLORIDE ◇ 2-DIETHYLAMINOETHYL BENZILATE HYDROCHLORIDE ◇ 2-DIETHYLAMINOETHYL DIPHENYLGLYCOLATE HYDROCHLORIDE ◇ 2-(DIFENYL-HYDROXYACETOXY)ETHYL-DIETHYLAMMONIUMCHLORID(CZECH) ◇ DIPHENYLGLYCOLLIC ACID-2-(DIETHYLAMINO)ETHYL ESTER HYDROCHLORIDE ◇ FOBEX ◇ IBIOTYZIL ◇ KATRON ◇ LEUCIDIL ◇ NERVACTON ◇ NERVATIL ◇ NEURAKTIL ◇ NEUROBENZIL ◇ NEUROLEPTONE ◇ NUTINAL ◇ PARASAN ◇ PARPON ◇ PHOBEX ◇ PROCALM ◇ STOIKON ◇ SUAVITIL ◇ TRANQUILLIN ◇ VALLADAN ◇ WIN 5606

TOXICITY DATA with REFERENCE
scu-rat TDLo:500 μg/kg (ID male):REP PSYPAG 10,44,66
orl-hmn TDLo:14 μg/kg:CNS 27ZQAG -,363,72
orl-rat LD50:166 mg/kg TXAPA9 1,42,59
ipr-rat LD50:100 mg/kg APTOA6 11,405,55
orl-mus LD50:160 mg/kg 27ZQAG -,363,72
ipr-mus LD50:76 mg/kg JPETAB 74,274,42
scu-mus LD50:250 mg/kg 27ZQAG -,363,72
ivn-mus LD50:14300 μg/kg 28ZPAK -,253,72
idr-mus LD50:350 mg/kg AIPTAK 59,149,38
ipr-rbt LD50:100 mg/kg APTOA6 11,405,55
ivn-rbt LD50:15 mg/kg 27ZQAG -,363,72
ipr-gpg LD50:100 mg/kg APTOA6 11,405,55

CONSENSUS REPORTS: Reported in EPA TSCA Inventory.

SAFETY PROFILE: Poison by ingestion, intraperitoneal, subcutaneous, intradermal, and intravenous routes. Human systemic effects by ingestion of very small amounts: toxic psychosis. Experimental reproductive effects. When heated to decomposition it emits very toxic fumes of NO_x and HCl.

BCA250 CAS:55798-64-6 ***HR: 3***
BENZILIC ACID,-3-(2,5-DIMETHYL-1-PYRROLIDINYL)PROPYL ESTER, HYDROCHLORIDE
mf: $C_{23}H_{29}NO_3{\cdot}ClH$ mw: 403.99

TOXICITY DATA with REFERENCE
ipr-mus LD50:77 mg/kg EJMCA5 9,404,74
ivn-mus LD50:8500 μg/kg EJMCA5 9,404,74

SAFETY PROFILE: Poison by intraperitoneal and intravenous routes. See also ESTERS. When heated to decomposition it emits very toxic fumes of NO_x and HCl.

BCA375 CAS:73954-17-3 ***HR: 3***
8-BENZILOYLOXY-6,10-ETHANO-5-AZONIASPIRO(4.5)DECDANE CHLORIDE
mf: $C_{25}N_{30}NO_3{\cdot}Cl$ mw: 428.01

SYNS: 6,10-ETHANO-5-AZONIASPIRO(4.5)DECAN-8-OLCHLORIDE BENZILATE ◇ 3-HYDROXY-SPIRO(8-AZONIABICYCLO(3.2.1)OCTANE-8,1′-PYRROLIDINIUM CHLORIDE BENZILATE

TOXICITY DATA with REFERENCE
orl-rat LD50:1501 mg/kg IYKEDH 4,90,73
ipr-rat LD50:103 mg/kg ARZNAD 16,1581,66
scu-rat LD50:707 mg/kg IYKEDH 4,90,73
ivn-rat LD50:15500 μg/kg IYKEDH 4,90,73
orl-mus LD50:750 mg/kg ARZNAD 16,1581,66
ipr-mus LD50:50 mg/kg IYKEDH 4,90,73
scu-mus LD50:203 mg/kg IYKEDH 4,90,73
ivn-mus LD50:11200 μg/kg IYKEDH 4,90,73
ims-mus LD50:89 mg/kg ARZNAD 16,1581,66

SAFETY PROFILE: Poison by subcutaneous, intramuscular, intravenous, and intraperitoneal routes. Moderately toxic by ingestion. When heated to decomposition it emits toxic fumes of Cl^- and NO_x.

BCB000 CAS:67360-95-6 ***HR: 3***
4-BENZILOYLOXY-1,1,2,2,6-PENTAMETHYLPIPERIDINIUM CHLORIDE (β FORM)
mf: $C_{24}H_{32}NO_3{\cdot}Cl$ mw: 418.02

SYNS: 4-BENZILYLOXY-1,2,2,6-TETRAMETHYLPIPERIDINEMETHOCHLORIDE (β FORM) ◇ 4-((HYDROXYDIPHENYLACETYL)OXY)-1,1,2,2,6-PENTAMETHYLPIPERIDINIUM CHLORIDE (β FORM)

TOXICITY DATA with REFERENCE
orl-mus LD50:1000 mg/kg JPETAB 85,85,45
ipr-mus LD50:75 mg/kg JPETAB 85,85,45
scu-mus LD50:325 mg/kg JPETAB 85,85,45

SAFETY PROFILE: Poison by intraperitoneal and subcutaneous routes. Moderately toxic by ingestion. When heated to decomposition it emits very toxic fumes of Cl^- and NO_x.

BCB250 CAS:67360-95-6 ***HR: 3***
4-BENZILOYLOXY-1,1,2,6,6-PENTAMETHYLPIPERIDINIUM CHLORIDE (α FORM)
mf: $C_{24}H_{32}NO_3{\cdot}Cl$ mw: 418.02

SYNS: 4-BENZILYLOXY-1,2,2,6-TETRAMETHYLPIPERIDINEMETHOCHLORIDE (α FORM) ◇ 4-((HYDROXYDIPHENYLACETYL)OXY)-1,1,2,2,6-PENTAMETHYLPIPERIDINIUM CHLORIDE (α FORM)

TOXICITY DATA with REFERENCE
ipr-mus LD50:800 mg/kg JPETAB 85,85,45
scu-mus LD50:375 mg/kg JPETAB 85,85,45

SAFETY PROFILE: Poison by subcutaneous route. Moderately toxic by intraperitoneal route. When heated to decomposition it emits very toxic fumes of Cl^- and NO_x.

BCB750 CAS:51-17-2 ***HR: 3***
BENZIMIDAZOLE
mf: $C_7H_6N_2$ mw: 118.15

PROP: Tabular crystals. Mp: 170.5°, bp: >360°. Sol in alc; sparingly sol in water.

SYNS: 3-AZAINDOLE ◇ AZINDOLE ◇ o-BENZIMIDAZOLE ◇ 1H-BENZIMIDAZOLE (9CI) ◇ BENZIMINAZOLE ◇ 1,3-BENZODIAZOLE ◇ BENZOIMIDAZOLE ◇ BZI ◇ 1,3-DIAZAINDENE ◇ N,N'-METHENYL-o-PHENYLENEDIAMINE ◇ NSC 759

TOXICITY DATA with REFERENCE
mmo-sat 250 μg/plate CHIMAD 27,68,73
mmo-esc 1 mg/disc APMBAY 6,23,58
dnd-esc 15 mmol/L/48H ANBCA2 75,45,76
pic-esc 1 g/L ZAPOAK 12,583,72
ipr-mus LD50:445 mg/kg PCJOAU 14,130,80
ipr-rat LD50:385 mg/kg AIPTAK 95,123,53
orl-rat LDLo:500 mg/kg NCNSA6 5,22,53
orl-mus LD50:2910 mg/kg JPETAB 105,486,52
ivn-mus LD50:280 mg/kg 29QHAQ -,246,74

CONSENSUS REPORTS: Reported in EPA TSCA Inventory.

SAFETY PROFILE: Poison by intravenous and intraperitoneal routes. Moderately toxic by ingestion. Mutation data reported. When heated to decomposition it emits highly toxic fumes of NO_x.

BCC000 CAS:4414-88-4 ***HR: 3***
2-BENZIMIDAZOLEACETONITRILE
mf: $C_9H_7N_3$ mw: 157.19

SYNS: 2-BENZIMIDAZOLYLACETONITRILE ◇ 2-KYANMETHYLBENZIMIDAZOL (CZECH)

TOXICITY DATA with REFERENCE
orl-rat LD50:795 mg/kg MarJV# 29MAR77
ivn-mus LD50:56 mg/kg CSLNX* NX#04148

CONSENSUS REPORTS: Reported in EPA TSCA Inventory. Cyanide and its compounds are on the Community Right-To-Know List.

SAFETY PROFILE: Poison by intravenous route. Moderately toxic by ingestion. See also NITRILES. When heated to decomposition it emits toxic fumes of NO_x.

BCC100 CAS:18538-45-9 ***HR: D***
2-BENZIMIDAZOLECARBAMIC ACID
mf: $C_8H_7N_3O_2$ mw: 177.18

SYNS: BENZIMIDAZOLE CARBAMATE ◇ 2-BENZIMIDAZOLYLCARBAMIC ACID ◇ CARBAMIC ACID, 1H-BENZIMIDAZOL-2-YL-

TOXICITY DATA with REFERENCE
orl-rat TDLo:600 mg/kg (female 7-8D post):TER VETNAL 59(5),59,83
orl-rat TDLo:600 mg/kg (female 7-8D post):REP VETNAL 59(5),59,83

SAFETY PROFILE: An experimental teratogen. Other experimental reproductive effects. When heated to decomposition it emits toxic fumes of NO_x.

BCC250 CAS:6898-43-7 ***HR: 3***
BENZIMIDAZOLE METHYLENE MUSTARD
mf: $C_{14}H_{19}Cl_2N_3 \cdot ClH$ mw: 336.72

SYNS: BENZIMIDAZOLE MUSTARD ◇ 2-(BIS(2-CHLOROETHYL)AMINOMETHYL)-5,5-DIMETHYLBENZIMIDAZOLE HYDROCHLORIDE ◇ 2-(DI-2-CHLOROETHYL)AMINOMETHYL-5,6-DIMETHYLBENZIMIDAZOLE ◇ NSC-23892

TOXICITY DATA with REFERENCE
ipr-mus TDLo:12 mg/kg/4W:CAR JNCIAM 36,915,66

SAFETY PROFILE: Questionable carcinogen with experimental carcinogenic data. When heated to decomposition it emits very toxic fumes HCl and NO_x.

BCC500 CAS:583-39-1 ***HR: 3***
2-BENZIMIDAZOLETHIOL
mf: $C_7H_6N_2S$ mw: 150.21

SYNS: ANTIEGENE MB ◇ ANTIOXIDANT MB (CZECH) ◇ AOMB ◇ ASM MB ◇ 2-MERCAPTOBENZIMIDAZOLE ◇ MERCAPTBENZOIMIDAZOLE ◇ 2-MERCAPTOBENZOIMIDAZOLE ◇ MERKAPTOBENZIMIDAZOL (CZECH) ◇ NCI-C60980 ◇ o-PHENYLENETHIOUREA ◇ USAF EK-6540 ◇ USAF XF-21

TOXICITY DATA with REFERENCE
skn-rbt 500 mg/24H MLD 28ZPAK -,168,72
eye-rbt 500 mg/24H MLD 28ZPAK -,168,72
orl-rat LD50:476 mg/kg 28ZPAK -,168,72
orl-mus LD50:1250 mg/kg GTPZAB 8(7),39,64
ipr-mus LD50:200 mg/kg NTIS** AD277-689
ivn-mus LD50:180 mg/kg CSLNX* NX#04376

CONSENSUS REPORTS: Reported in EPA TSCA Inventory.

SAFETY PROFILE: Poison by intraperitoneal and intravenous routes. Moderately toxic by ingestion. Skin and eye irritant. When heated to decomposition it emits toxic fumes of SO_x and NO_x. See also MERCAPTANS.

BCD125 CAS:52096-22-7 ***HR: 3***
BENZIMIDAZOLIUM-1-NITROIMIDATE
mf: $C_7H_5N_4O_2$ mw: 177.14

SAFETY PROFILE: Explodes at its mp: 169°C. Upon decomposition it emits toxic fumes of NO_x.

BCD325 CAS:21035-25-6 ***HR: 3***
1-(2-BENZIMIDAZOLYL)-3-METHYLUREA
mf: $C_9H_{10}N_4O$ mw: 190.23

SYN: BCM (NH)

TOXICITY DATA with REFERENCE
oms-hmn:leu 1 mg/L THERAP 31,505,76
oms-hmn:oth 2 mg/L THERAP 31,505,76

orl-rat TDLo:152 mg/kg (8-15D preg):TER THERAP 31,505,76
ivn-mus LD50:56 mg/kg CSLNX* NX#03246

SAFETY PROFILE: Poison by intravenous route. Human mutation data reported. An experimental teratogen. When heated to decomposition it emits toxic fumes of NO_x.

BCD750 CAS:642-72-8 ***HR: 3***
BENZINDAMINE
mf: $C_{19}H_{23}N_3O$ mw: 309.45

SYNS: BENZYDAMINE ◇ 1-BENZYL-3-(3-(DIMETHYLAMINO)PROPOXY)-1H-INDAZOLE

TOXICITY DATA with REFERENCE
orl-rat LD50:9500 mg/kg ARZNAD 22,711,72
orl-mus LD50:460 mg/kg JMCMAR 15,923,72
ipr-mus LD50:109 mg/kg JMCMAR 15,471,72
scu-mus LD50:445 ng/kg OYYAA2 6,1285,72
ivn-mus LD50:25 mg/kg OYYAA2 6,1285,72

SAFETY PROFILE: Poison by intraperitoneal and intravenous routes. Moderately toxic by ingestion and subcutaneous routes. When heated to decomposition it emits toxic fumes of NO_x.

BCE000 CAS:208-07-1 ***HR: 3***
BENZ(e)INDENO(1,2-b)INDOLE
mf: $C_{19}H_{11}N$ mw: 253.31

SYN: 4,5-BENZO-2,3-1′,2′-INDENOINDOLE (FRENCH)

TOXICITY DATA with REFERENCE
scu-mus TDLo:120 mg/kg/9W-I:ETA BAFEAG 42,3,55
scu-mus LDLo:40 mg/kg BAFEAG 42,3,55

SAFETY PROFILE: Poison by subcutaneous route. Questionable carcinogen with experimental tumorigenic data. When heated to decomposition it emits toxic fumes of NO_x.

BCE250 CAS:5585-71-7 ***HR: 3***
BENZINDOPYRINE HYDROCHLORIDE
mf: $C_{22}H_{20}N_2 \cdot ClH$ mw: 348.90

SYNS: 4-(1-BENZYL-3-INDOLETHYL)PYRIDINEHYDROCHLORIDE ◇ 1-BENZYL-3-(2-(4-PYRIDYL)ETHYL)INDOLE HYDROCHLORIDE

TOXICITY DATA with REFERENCE
orl-hmn TDLo:714 μg/kg/D:PSY AMCTAH 6,521,59
ipr-mus LD50:520 mg/kg JPETAB 125,122,59
ivn-mus LD50:98 mg/kg JPETAB 125,122,59

SAFETY PROFILE: Poison by intravenous route. Moderately toxic by intraperitoneal route. Human psychotropic effects via ingestion. When heated to decomposition it emits very toxic fumes of NO_x and HCl.

BCE500 CAS:81-07-2 ***HR: 3***
1,2-BENZISOTHIAZOL-3(2H)-ONE-1,1-DIOXIDE
mf: $C_7H_5NO_3S$ mw: 183.19

PROP: White crystals or powder; odorless with sweet taste. Mp: 228° (decomp), bp: subl. Sol in water, alc, chloroform, and ether.

SYNS: ANHYDRO-o-SULFAMINEBENZOIC ACID ◇ 3-BENZISOTHIAZOLINONE-1,1-DIOXIDE ◇ o-BENZOIC SULPHIMIDE ◇ o-BENZOSULFIMIDE ◇ BENZOSULPHIMIDE ◇ BENZO-2-SULPHIMIDE ◇ o-BENZOYL SULFIMIDE ◇ o-BENZOYL SULPHIMIDE ◇ 1,2-DIHYDRO-2-KETOBENZISOSULFONAZOLE ◇ 1,2-DIHYDRO-2-KETOBENZISOSULPHONAZOLE ◇ 2,3-DIHYDRO-3-OXOBENZISOSULFONAZOLE ◇ 2,3-DIHYDRO-3-OXOBENZISOSULPHONAZOLE ◇ GARANTOSE ◇ GLUCID ◇ GLUSIDE ◇ HERMESETAS ◇ 3-HYDROXYBENZISOTHIAZOL-S,S-DIOXIDE ◇ INSOLUBLE SACCHARINE ◇ KANDISET ◇ NATREEN ◇ RCRA WASTE NUMBER U202 ◇ SACARINA ◇ SACCAHARIMIDE ◇ SACCHARINA ◇ SACCHARIN ACID ◇ SACCHARINE ◇ SACCHARINOL ◇ SACCHARINOSE ◇ SACCHAROL ◇ SAXIN ◇ SUCRE EDULCOR ◇ SUCRETTE ◇ o-SULFOBENZIMIDE ◇ o-SULFOBENZOIC ACID IMIDE ◇ 2-SULPHOBENZOIC IMIDE ◇ SYKOSE ◇ SYNCAL ◇ ZAHARINA

TOXICITY DATA with REFERENCE
cyt-smc 200 mg/L NATUAS 294,263,81
dnd-rat:lvr 3 mmol/L SinJF# 26OCT82
dns-rat:lvr 100 pmol/L CRNGDP 5,1547,84
dnd-mus-ipr 100 mg/kg ATSUDG (5),355,82
sce-ham:lng 100 mg/L BJCAAI 45,769,82
orl-mus TDLo:101 g/kg (female 1-21D post):REP DBTEAD 17,103,69
orl-mus TDLo:155 mg/kg (female 7D post):TER IIZAAX 16,330,64
orl-rat TDLo:2008 g/kg/2Y-C:ETA JAPMA8 40,583,51
imp-mus TDLo:80 mg/kg:NEO BJCAAI 11,212,57
orl-mus LD50:17 g/kg EXPEAM 35,1364,79

CONSENSUS REPORTS: NTP Fifth Annual Report on Carcinogens. IARC Cancer Review: Group 2B IMEMDT 7,334,87, Human Inadequate Evidence IMEMDT 22,111,80; Animal Sufficient Evidence IMEMDT 22,111,80. EPA Genetic Toxicology Program. Reported in EPA TSCA Inventory. Community Right-To-Know List.

SAFETY PROFILE: Confirmed carcinogen with experimental neoplastigenic and tumorigenic data. Mild acute toxicity by ingestion. Experimental teratogenic and reproductive effects. Mutation data reported. When heated to decomposition it emits toxic NO_x and SO_x.

BCE750 CAS:68291-97-4 ***HR: 2***
1,2-BENZISOXAZOLE-3-METHANESULFONAMIDE
mf: $C_8H_8N_2O_3S$ mw: 212.24

SYNS: AD-810 ◇ 3-SULFAMOYLMETHYL-1,2-BENZISOXAZOLE

TOXICITY DATA with REFERENCE
orl-rat TDLo:220 mg/kg (female 7-17D post):REP YACHDS 15,4399,87
orl-dog TDLo:1320 mg/kg (female 14-35D post):TER YACHDS 15,4435,87
orl-rat LD50:1992 mg/kg YACHDS 15,4337,87
ipr-rat LD50:733 mg/kg ARZNAD 30,477,80
scu-rat LD50:925 mg/kg YACHDS 15,4337,87
ivn-rat LD50:672 mg/kg YACHDS 15,4337,87
orl-mus LD50:1892 mg/kg ARZNAD 30,477,80
ipr-mus LD50:699 mg/kg ARZNAD 30,477,80
scu-mus LD50:1009 mg/kg YACHDS 15,4337,87
orl-dog LD50:1 g/kg YACHDS 15,4337,87

SAFETY PROFILE: Moderately toxic by ingestion, intraperitoneal, subcutaneous, and intravenous routes. An experimental teratogen. Other experimental reproductive effects. When heated to decomposition it emits very toxic fumes of SO_x and NO_x. An anticonvulsant.

BCE825 CAS:15301-48-1 ***HR: 3***
BENZITRAMIDE
mf: $C_{31}H_{32}N_4O_2$ mw: 492.67

PROP: White, crystalline powder. Mp: 145-149°. Also reported as pale yellow amorphous powder, mp: 124.5-126°. Solubility above 1 g/100 mL in ethyl acetate, acetone, benzene, chloroform. Almost insol in water and dilute acids.

SYNS: BEZITRAMIDE ◇ BURGODIN ◇ 1-(3-CYANO-3,3-DIPHENYLPROPYL)-4-(2-OXO-3-PROPIONYL-1-BENZIMIDAZOLINYL)PIPERIDINE ◇ 1-(1-(3-CYANO-3,3-DIPHENYLPROPYL)-4-PIPERIDYL)-3-PROPIONYL-2-BENZIMIDAZOLINONE ◇ R-4845

TOXICITY DATA with REFERENCE
orl-rat LD50:141 mg/kg MEIEDD 10,170,83
orl-mus LD50:2101 mg/kg MEIEDD 10,170,83
orl-dog LD50:80 mg/kg ARZNAD 21,862,71
orl-gpg LD50:60400 μg/kg ARZNAD 21,862,71

CONSENSUS REPORTS: Cyanide and its compounds are on the Community Right-To-Know List.

SAFETY PROFILE: Poison by ingestion. Caution: May be habit forming. This is a controlled substance (opiate) listed in the U.S. Code of Federal Regulations, Title 21 Part 1308.12 (1985). When heated to decomposition it emits toxic fumes of NO_x and CN^-.

BCF500 CAS:1491-10-7 ***HR: 3***
BENZO(f)(1)BENZOTHIENO(3,2-b)QUINOLINE
mf: $C_{19}H_{11}NS$ mw: 285.37

SYN: NAPHTHO(1,2-e)THIANAPHTHENO(3,2-b)PYRIDINE

TOXICITY DATA with REFERENCE
scu-mus TDLo:72 mg/kg/9W-I:ETA EJCAAH 4,123,68

SAFETY PROFILE: Questionable carcinogen with experimental tumorigenic data. When heated to decomposition it emits very toxic fumes of SO_x and NO_x.

BCF750 CAS:1491-09-4 ***HR: 3***
BENZO(h)(1)BENZOTHIENO(3,2-b)QUINOLINE
mf: $C_{19}H_{11}NS$ mw: 285.37

SYN: NAPHTHO(2,1-e)THIANAPHTHENO(3,2-b)PYRIDINE

TOXICITY DATA with REFERENCE
scu-mus TDLo:72 mg/kg/9W-I:ETA EJCAAH 4,123,68

SAFETY PROFILE: Questionable carcinogen with experimental tumorigenic data. When heated to decomposition it emits very toxic fumes of SO_x and NO_x.

BCG000 CAS:846-35-5 ***HR: 3***
BENZO(e)(1)BENZOTHIOPYRANO(4,3-b)INDOLE
mf: $C_{19}H_{11}NS$ mw: 285.37

TOXICITY DATA with REFERENCE
scu-mus TDLo:360 mg/kg/25W-I:NEO JNCIAM 46,1257,71

SAFETY PROFILE: Questionable carcinogen with experimental neoplastigenic data. When heated to decomposition it emits very toxic fumes of SO_x and NO_x.

BCG250 CAS:239-01-0 ***HR: 3***
11H-BENZO(a)CARBAZOLE
mf: $C_{16}H_{11}N$ mw: 217.28

SYN: 1,2-BENZCARBAZOLE

TOXICITY DATA with REFERENCE
skn-mus TDLo:840 mg/kg/21W-I:ETA HSZPAZ 236,79,35

SAFETY PROFILE: Questionable carcinogen with experimental tumorigenic data. When heated to decomposition it emits toxic fumes such as NO_x.

BCG500 CAS:214-17-5 ***HR: 3***
BENZO(b)CHRYSENE
mf: $C_{22}H_{14}$ mw: 278.36

SYNS: 2,3-BENZOCHRYSENE ◇ 3,4-BENZOTETRACENE ◇ 3,4-BENZOTETRAPHENE ◇ BENZO(c)TETRAPHENE ◇ 1,2:6,7-DIBENZOPHENANTHRENE ◇ 2,3:7,8-DIBENZOPHENANTHRENE ◇ DIBENZO-2,3,7,8-PHENANTHRENE

TOXICITY DATA with REFERENCE
mma-sat 50 μg/plate MUREAV 174,247,86
skn-mus TDLo:28 mg/kg:NEO JNCIAM 50,1717,73

SAFETY PROFILE: Questionable carcinogen with experimental neoplastigenic data by skin contact. Mutation data reported. When heated to decomposition it emits acrid smoke and irritating fumes.

BCG750 CAS:194-69-4 ***HR: 3***
BENZO(c)CHRYSENE
mf: $C_{22}H_{14}$ mw: 278.36

SYN: 1,2,5,6-DIBENZPHENANTHRENE

TOXICITY DATA with REFERENCE
skn-mus TDLo:1630 mg/kg/68W-I:ETA PRLBA4 129,439,40

SAFETY PROFILE: Questionable carcinogen with experimental tumorigenic data. When heated to decomposition it emits acrid smoke and irritating fumes.

BCH000 CAS:196-78-1 ***HR: 3***
BENZO(g)CHRYSENE
mf: $C_{22}H_{14}$ mw: 278.36

SYNS: 1,2,3,4-DIBENZOPHENANTHRENE ◇ 1,2,3,4-DIBENZPHENANTHRENE

TOXICITY DATA with REFERENCE
orl-mus TDLo:15 g/kg/74W-I:ETA PRLBA4 129,439,40

SAFETY PROFILE: Questionable carcinogen with experimental tumorigenic data. When heated to decomposition it emits acrid smoke and irritating fumes.

BCH250 CAS:5096-19-5 ***HR: 3***
N-6-(3,4-BENZOCOUMARINYL)ACETAMIDE
mf: $C_{15}H_{10}NO_3$ mw: 252.26

SYN: N-(6-OXO-6H-DIBENZO(b,d)PYRAN-1-YL)ACETAMIDE

TOXICITY DATA with REFERENCE
orl-rat TDLo:5000 mg/kg:CAR CNREA8 26,619,66
orl-rat TD:9000 mg/kg/27D-I:ETA CNREA8 28,924,68

SAFETY PROFILE: Questionable carcinogen with experimental carcinogenic and tumorigenic data. When heated to decomposition it emits very toxic fumes of NO_x.

BCH750 CAS:10085-81-1 ***HR: 3***
BENZOCTAMINE HYDROCHLORIDE
mf: $C_{18}H_{19}N•ClH$ mw: 285.84

SYNS: BA 30,803 ◇ 1-METHYLAMINOMETHYLDIBENZO(b,c)-BICYCLO(2,2,2)OCTADIENE HYDROCHLORIDE ◇ N-METHYLETHANOANTHRACENE-9-(10H)-METHYLAMINEHYDROCHLORIDE ◇ TACITIN

TOXICITY DATA with REFERENCE
ims-rat TDLo:51 mg/kg (2-18D preg):TER BSIBAC 49,1309,73
orl-rat LDLo:700 mg/kg TXAPA9 18,185,71
ivn-rat LD50:26 mg/kg 27ZQAG -,336,72

SAFETY PROFILE: Poison by intravenous route. Moderately toxic by ingestion. Experimental teratogenic effects. A sedative and muscle relaxant. When heated to decomposition it emits very toxic fumes of NO_x and HCl^-.

BCI000 CAS:198-46-9 ***HR: 3***
BENZO(de)CYCLOPENT(a)ANTHRACENE
mf: $C_{20}H_{12}$ mw: 252.32

SYN: Δ^3-DEHYDRO-3,4-TRIMETHYLENE-ISOBENZANTHRENE-2

TOXICITY DATA with REFERENCE
imp-mus TDLo:600 mg/kg/40W-I:ETA JNCIAM 2,241,41

SAFETY PROFILE: Questionable carcinogen with experimental tumorigenic data. When heated to decomposition it emits acrid smoke and irritating fumes.

BCI250 CAS:240-44-8 ***HR: 3***
1H-BENZO(a)CYCLOPENT(b)ANTHRACENE
mf: $C_{21}H_{16}$ mw: 268.37

SYN: 6,7-CYCLOPENTENO-1,2-BENZANTHRACENE

TOXICITY DATA with REFERENCE
skn-mus TDLo:820 mg/kg/34W-I:ETA PRLBA4 117,318,35

SAFETY PROFILE: Questionable carcinogen with experimental tumorigenic data. When heated to decomposition it emits acrid smoke and irritating fumes.

BCI500 CAS:135-87-5 ***HR: 3***
BENZODIOXANE HYDROCHLORIDE
mf: $C_{14}H_{19}NO_2•ClH$ mw: 269.80

SYNS: BENODAINE HYDROCHLORIDE ◇ 1-(1,4-BENZODIOXAN-2-YLMETHYL)PIPERIDINEHYDROCHLORIDE ◇ F 933 ◇ FOURNEAU 933 ◇ 2-PIPERIDINOMETHYL-1,4-BENZODIOXAN HYDROCHLORIDE ◇ 2-(1-PIPERIDYLMETHYL)-1,4-BENZODIOXAN HYDROCHLORIDE ◇ PIPEROXANE HYDROCHLORIDE

TOXICITY DATA with REFERENCE
orl-rat TDLo:735 mg/kg (3-9D preg):REP PSEBAA 100,555,59
orl-mus LD50:502 mg/kg JAPMA8 48,409,59
ipr-mus LD50:175 mg/kg JAPMA8 48,409,59
scu-mus LD50:500 mg/kg THERAP 13,17,58
ivn-mus LD50:26 mg/kg AIPTAK 105,221,56

SAFETY PROFILE: Poison by intraperitoneal and intravenous routes. Moderately toxic by ingestion and subcutaneous routes. Experimental reproductive effects. When heated to decomposition it emits very toxic fumes of NO_x and HCl.

BCI750 CAS:325-23-5 ***HR: 3***
1-(1,4-BENZODIOXAN-2-YLMETHYL-1-BENZYL)-HYDRAZINE TARTRATE
mf: $C_{16}H_{18}N_2O_2•C_4H_4O_6$ mw: 418.44

TOXICITY DATA with REFERENCE
scu-rat LD50:700 mg/kg 27ZQAG -,317,72

orl-mus LD50:250 mg/kg 27ZQAG -,317,72
scu-mus LD50:225 mg/kg 27ZQAG -,317,72
scu-mky LD50:200 mg/kg 27ZQAG -,317,72
scu-rbt LD50:325 mg/kg 27ZQAG -,317,72
scu-gpg LD50:350 mg/kg 27ZQAG -,317,72

SAFETY PROFILE: Poison by ingestion and subcutaneous route. When heated to decomposition it emits toxic fumes of NO_x.

BCJ000 CAS:5208-87-7 ***HR: 3***
1,3-BENZODIOXOLE-5-(2-PROPEN-1-OL)
mf: $C_{10}H_{10}O_3$ mw: 178.20

SYNS: 1′-HYDROXYSAFROLE ◇ 1,2-METHYLENEDIOXY-4-(1-HYDROXYALLYL)BENZENE ◇ α-VINYLPIPERONYL ALCOHOL

TOXICITY DATA with REFERENCE
mma-sat 1 μmol/plate MUREAV 60,143,79
dnd-rat-ipr 100 mg/kg CNREA8 36,1686,76
oms-mus-ipr 400 μmol/kg CNREA8 41,2664,81
orl-rat TDLo:77 g/kg/73W-C:CAR CNREA8 37,1883,77
scu-rat TDLo:1426 mg/kg/10W-I:CAR CNREA8 43,1124,83
orl-mus TD:117 g/kg/1Y-C:ETA PAACA3 24,79,83
orl-mus TD:61 g/kg/52W-C:NEO CNREA8 43,5163,83

SAFETY PROFILE: Suspected carcinogen with experimental carcinogenic, neoplastigenic, and tumorigenic data. Human mutation data reported. When heated to decomposition it emits acrid smoke and irritating fumes.

BCJ125 CAS:32283-21-9 ***HR: 3***
1,3-BENZODITHIOLIUM PERCHLORATE
mf: $C_7H_5ClO_4S_2$ mw: 252.69

SAFETY PROFILE: A friction and heat-sensitive explosive. Upon decomposition it emits toxic fumes of Cl^- and SO_x. See also PERCHLORATES.

BCJ150 CAS:54531-52-1 ***HR: 3***
BENZODOL
mf: $(C_6H_7AsO_4{\cdot}CH_2O)_n$

PROP: Sol in water, alc, and NaOH.

SYNS: ARSONIC ACID, (4-HYDROXYPHENYL)-, polymer with FORMALDEHYDE ◇ (4-HYDROXYPHENYL)ARSONIC ACID polymer with FORMALDEHYDE ◇ POLYBENZARSOL

TOXICITY DATA with REFERENCE
ipr-mus LD50:235 mg/kg MEIEDD 11,1203,89

OSHA PEL: TWA 0.5 mg(As)/m^3
ACGIH TLV: TWA 0.2 mg(As)/m^3

SAFETY PROFILE: Poison by intraperitoneal route. When heated to decomposition it emits toxic fumes of As.

BCJ500 CAS:205-82-3 ***HR: 3***
BENZO(j)FLUORANTHENE
mf: $C_{20}H_{12}$ mw: 252.32

SYNS: 7,8-BENZOFLUORANTHENE ◇ 10,11-BENZFLUORANTHENE ◇ BENZ(j)FLUOROANTHRENE ◇ BENZO(1)FLUORANTHENE ◇ B(j)F ◇ DIBENZO(a,jk)FLUORENE

TOXICITY DATA with REFERENCE
mma-sat 10 μg/plate CNREA8 40,4258,80
dnd-mus-skn 3760 nmol/kg PAACA3 25,121,84
imp-rat TDLo:25 mg/kg:CAR JJIND8 71,539,83
skn-mus TDLo:312 mg/kg/26W-I:ETA CANCAR 12,1194,59
ipr-mus TDLo:11102 μg/kg/15D-I:NEO CALEDQ 34,15,87

CONSENSUS REPORTS: NTP Fifth Annual Report on Carcinogens. IARC Cancer Review: Group 2B IMEMDT 7,56,87; Animal Limited Evidence IMEMDT 3,82,73; Animal Sufficient Evidence IMEMDT 32,155,83

SAFETY PROFILE: Confirmed carcinogen with experimental carcinogenic, neoplastigenic, and tumorigenic data. Mutation data reported. When heated to decomposition it emits acrid smoke and irritating fumes.

BCJ750 CAS:207-08-9 ***HR: 3***
BENZO(k)FLUORANTHENE
mf: $C_{20}H_{12}$ mw: 252.32

SYNS: 8,9-BENZOFLUORANTHENE ◇ 11,12-BENZOFLUORANTHENE ◇ 11,12-BENZO(k)FLUORANTHENE ◇ 2,3,1′,8′-BINAPHTHYLENE ◇ DIBENZO(b,jk)FLUORENE

TOXICITY DATA with REFERENCE
mma-sat 10 μg/plate CNREA8 40,4528,80
imp-rat TDLo:5 mg/kg:ETA 50NNAZ 7,571,83

CONSENSUS REPORTS: NTP Fifth Annual Report on Carcinogens. IARC Cancer Review: Animal Sufficient Evidence IMEMDT 32,163,83

SAFETY PROFILE: Confirmed carcinogen with experimental tumorigenic data. Mutation data reported. When heated to decomposition it emits acrid smoke and irritating fumes.

BCK250 CAS:271-89-6 ***HR: 2***
BENZOFURAN
mf: C_8H_6O mw: 118.14

SYNS: BENZO(b)FURAN ◇ 2,3-BENZOFURAN ◇ BENZOFURFURAN ◇ COUMARONE ◇ NCI-C56166 ◇ 1-OXINDENE

TOXICITY DATA with REFERENCE
ipr-mus LD50:500 mg/kg EJMCA5 12,383,77

CONSENSUS REPORTS: Reported in EPA TSCA Inventory.

SAFETY PROFILE: Moderately toxic by intraperitoneal route. When heated to decomposition it emits acrid smoke and fumes.

BCK750 CAS:5149-69-9 ***HR: 3***
2-BENZO-FURANCETONITRILE

SYNS: 2-CYANOCETYLCOUMARONE ◇ USAF KF-4

TOXICITY DATA with REFERENCE
orl-rat LDLo:3200 mg/kg KODAK* -,-,71
ipr-rat LDLo:800 mg/kg KODAK* -,-,71
ipr-mus LD50:100 mg/kg NTIS** AD277-689

CONSENSUS REPORTS: Cyanide and its compounds are on the Community Right-To-Know List.

SAFETY PROFILE: Poison by intraperitoneal route. Moderately toxic by ingestion. See also NITRILES. When heated to decomposition it emits toxic fumes of CN^- and NO_x.

BCL100 CAS:42242-58-0 ***HR: 2***
p-(7-BENZOFURYLAZO)-N,N-DIMETHYLANILINE
mf: $C_{16}H_{15}N_3O$ mw: 265.34

SYNS: ANILINE, p-(7-BENZOFURYLAZO)-N,N-DIMETHYL- ◇ N,N-DIMETHYL-p-(7-BENZOFURYLAZO)ANILINE

TOXICITY DATA with REFERENCE
orl-rat TDLo:1620 mg/kg/13W-C:ETA JMCMAR 16,717,73

SAFETY PROFILE: Questionable carcinogen with experimental tumorigenic data. When heated to decomposition it emits toxic fumes of NO_x.

BCL250 CAS:23844-24-8 ***HR: 3***
BENZOGUANAMINE
mf: $C_{22}H_{32}N_2O_5$ mw: 404.56

PROP: Crystals. Mp: 227°, d: 1.4.

SYNS: 2-ACETOXY-3-DIETHYLCARBAMYL-9,10-DIMETHOXY-1,2,3,4,6,7-HEXAHYDRO-11B-BENZO(a)QUINOLIZINE ◇ BENZOCHINAMIDE ◇ BENZOQUINAMIDE ◇ BENZQUINAMIDE ◇ BENZQUINAMIDU (POLISH) ◇ BZQ ◇ P 2647 ◇ QUANTRIL ◇ QUANTRYL

TOXICITY DATA with REFERENCE
orl-rat LD50:1050 mg/kg 27ZQAG -,208,72
ivn-rat LD50:100 μg/kg 27ZQAG -,208,72
orl-mus LD50:580 mg/kg 27ZQAG -,208,72
ipr-mus LD50:321 mg/kg DIPHAH 17,145,65
ivn-mus LD50:100 mg/kg 27ZQAG -,208,72
orl-bwd LD50:100 mg/kg TXAPA9 21,315,72

SAFETY PROFILE: Poison by ingestion, intraperitoneal, and intravenous routes. When heated to decomposition it emits toxic fumes of NO_x.

BCL500 CAS:495-18-1 ***HR: D***
BENZOHYDROXAMIC ACID
mf: $C_7H_7NO_2$ mw: 137.15

SYNS: BENZOHYDROXAMATE ◇ BENZOYLHYDROXAMIC ACID ◇ N-HYDROXYBENZAMIDE ◇ PHENYLHYDROXAMIC ACID

TOXICITY DATA with REFERENCE
mmo-sat 2500 nmol/plate MUREAV 135,139,84
mma-sat 1 μmol/plate MUREAV 56,7,77

CONSENSUS REPORTS: Reported in EPA TSCA Inventory. EPA Genetic Toxicology Program.

SAFETY PROFILE: Mutation data reported. When heated to decomposition it emits toxic fumes of NO_x.

BCL750 CAS:65-85-0 ***HR: 3***
BENZOIC ACID
mf: $C_7H_6O_2$ mw: 122.13

PROP: White crystalline powder. Mp: 121.7°, bp: 249°, flash p: 250°F (CC), d: 1.316, autoign temp: 1060°F, vap press: 1 mm @ 96.0° (sublimes), vap d: 4.21. Moderately sol in water; sol in alc, ether, chloroform, and fixed oils.

SYNS: ACIDE BENZOIQUE (FRENCH) ◇ BENZENECARBOXYLIC ACID ◇ BENZENEFORMIC ACID ◇ BENZENEMETHANOIC ACID ◇ BENZOATE ◇ BENZOESAEURE (GERMAN) ◇ BENZOIC ACID (DOT) ◇ CARBOXYBENZENE ◇ DRACYLIC ACID ◇ KYSELINA BENZOOVA (CZECH) ◇ PHENYL CARBOXYLIC ACID ◇ PHENYLFORMIC ACID ◇ RETARDER BA ◇ RETARDEX ◇ SALVO LIQUID ◇ SALVO POWDER ◇ TENN-PLAS

TOXICITY DATA with REFERENCE
skn-hmn 22 mg/3D-I MOD 85DKA8 -,127,77
skn-rbt 500 mg/24H MLD BIOFX* 28-4/73
eye-rbt 100 mg SEV BIOFX* 28-4/73
mmo-esc 10 mmol/L ZBPIA9 112,226,59
dni-hmn:lym 5 mmol/L PNASA6 79,1171,82
orl-man LDLo:500 mg/kg FCTXAV 17,715,79
skn-hmn TDLo:6 mg/kg:PUL,SKN JOALAS 16,195,45
orl-rat LD50:2530 mg/kg MarJV# 29MAR77
orl-mus LD50:1940 mg/kg IYKEDH 15,359,84
ipr-mus LD50:1460 mg/kg CRSBAW 160,1097,66
orl-dog LD50:2000 mg/kg 27ZTAP 3,22,69
orl-cat LD50:2000 mg/kg 27ZTAP 3,22,69
orl-rbt LDLo:2000 mg/kg HBTXAC 5,23,59
scu-rbt LDLo:2000 mg/kg HBTXAC 5,23,59
orl-gpg LDLo:2 g/kg MMWOAU 77,13,30
ipr-gpg LDLo:1400 mg/kg HBTXAC 5,23,59
scu-frg LDLo:100 mg/kg HBTXAC 5,23,59

CONSENSUS REPORTS: Reported in EPA TSCA Inventory. EPA Genetic Toxicology Program.

DOT Classification: ORM-E; Label: None.

SAFETY PROFILE: Poison by subcutaneous route. Moderately toxic by ingestion and intraperitoneal routes. Human systemic effects by inhalation: dyspnea and allergic dermatitis. Severe eye irritant. A human skin irritant. Combustible when exposed to heat or flame; can react with oxidizing materials. The powder burns rapidly in oxygen. To fight fire, use water, CO_2, water spray or mist, dry chemical. When heated to decomposition it emits acrid smoke and irritating fumes.

BCM000 CAS:120-51-4 ***HR: 2***
BENZOIC ACID, BENZYL ESTER
mf: $C_{14}H_{12}O_2$ mw: 212.26

PROP: Found in Peru and Tolu Balsams, in Ylang-Ylang and in about 20 other essential oils (FCTXAV 11,1011,73). Colorless oily liquid; slt aromatic odor. Mp. 21°, bp: 324°, flash p: 298°F (CC), d: 1.116, refr index: 1.568, vap d: 7.3, autoign temp: 898°F. Misc with alc, chloroform, ether; insol in glycerin, water.

SYNS: ASCABIN ◇ ASCABIOL ◇ BENYLATE ◇ BENZOIC ACID, PHENYLMETHYL ESTER ◇ BENZYL ALCOHOL BENZOIC ESTER ◇ BENZYL BENZENECARBOXYLATE ◇ BENZYL BENZOATE (FCC) ◇ BENZYLETS ◇ BENZYL PHENYLFORMATE ◇ COLEBENZ ◇ FEMA No. 2138 ◇ NOVOSCABIN ◇ PERUSCABIN ◇ SCABANCA ◇ VANZOATE ◇ VENZONATE

TOXICITY DATA with REFERENCE
orl-rat LD50:500 mg/kg FMCHA2 -,C30,83
skn-rat LD50:4000 mg/kg JPETAB 93,26,48
orl-mus LD50:1400 mg/kg JPETAB 93,26,48
orl-cat LD50:2240 mg/kg JPETAB 84,358,45
orl-rbt LD50:1680 mg/kg FCTXAV 11,1015,73
skn-rbt LD50:4000 mg/kg FCTXAV 11,1015,73
orl-gpg LD50:1000 mg/kg JPETAB 93,26,48

CONSENSUS REPORTS: Reported in EPA TSCA Inventory.

SAFETY PROFILE: Moderately toxic by ingestion and skin contact. Combustible liquid. Can react with oxidizing materials. To fight fire, use CO_2, water spray or mist, dry chemical. When heated to decomposition it emits acrid and irritating fumes and smoke. See also ESTERS.

BCM250 CAS:1696-17-9 ***HR: 2***
BENZOIC ACID-N,N-DIETHYLAMIDE
mf: $C_{11}H_{15}NO$ mw: 177.27

SYNS: BENZOIC ACID DIETHYLAMIDE ◇ BENZOYLDIETHYLAMINE ◇ N,N-DIETHYLBENZAMIDE ◇ R 2 ◇ REBEMID ◇ REP

TOXICITY DATA with REFERENCE
orl-rat LD50:2000 mg/kg FMCHA2 -,D219,80
orl-mus LD50:780 mg/kg MPPBAB 47,77,78
ihl-mus LC50:142 g/m^3 MPPBAB 47,77,78
skn-mus LD50:1700 mg/kg MPPBAB 47,77,78

CONSENSUS REPORTS: Reported in EPA TSCA Inventory.

SAFETY PROFILE: Moderately toxic by ingestion and skin contact. When heated to decomposition it emits toxic fumes of NO_x.

BCP000 CAS:67011-39-6 ***HR: 1***
BENZOIC-3-CHLORO-N-ETHOXY-2,6-DIMETHOXYBENZIMIDIC ANHYDRIDE
mf: $C_{18}H_{18}ClNO_5$ mw: 363.82

SYNS: BENZOMATE ◇ BENZOXAMATE ◇ CITRAZON ◇ ETHYL-o-BENZOYL-3-CHLORO-2,6-DIMETHOXY-BENZOHYDROXIMATE ◇ NA-53

TOXICITY DATA with REFERENCE
orl-rat LD50:15000 mg/kg 85ARAE 1,92,77
ipr-rat LD50:4217 mg/kg NYKGA7 3,123,76
orl-mus LD50:12 g/kg SPEADM 78-1,23,78
ipr-mus LD50:4264 mg/kg NYKGA7 3,123,76

SAFETY PROFILE: Mildly toxic by ingestion and intraperitoneal routes. When heated to decomposition it emits very toxic fumes of Cl^- and NO_x.

BCP250 CAS:119-53-9 ***HR: D***
BENZOIN
mf: $C_{14}H_{12}O_2$ mw: 212.26

SYNS: BENZOYLPHENYLCARBINOL ◇ BITTER ALMOND OIL CAMPHOR ◇ α-HYDROXYBENZYL PHENYL KETONE ◇ α-HYDROXY-α-PHENYLACETOPHENONE ◇ 2-HYDROXY-2-PHENYLACEsSTOPHENONE ◇ NCI-C50011

TOXICITY DATA with REFERENCE
mmo-sat 750 μg/plate PMRSDJ 5,187,85
mma-smc 25 mg/L PMRSDJ 5,247,85
dns-rat:lvr 1 mmol/L PMRSDJ 5,371,85
msc-mus:lym 62500 μg/L PMRSDJ 5,587,85

CONSENSUS REPORTS: NCI Carcinogenesis Bioassay (feed); No Evidence: mouse, rat NCITR* NCI-CG-TR-204,80. Reported in EPA TSCA Inventory.

SAFETY PROFILE: Mutation data reported. When heated to decomposition it emits acrid smoke and irritating fumes. See also KETONES.

BCP500 CAS:441-38-3 ***HR: 3***
α-BENZOIN OXIME
mf: $C_{14}H_{13}NO_2$ mw: 227.28

SYNS: BENZOINOXIM (CZECH) ◇ CUPRON (CZECH) ◇ CUPRONE ◇ α-OXIME BENZOIN ◇ USAF FA-5

TOXICITY DATA with REFERENCE
eye-rbt 500 mg/24H MLD 28ZPAK -,111,72
orl-rat LD50:5960 mg/kg 28ZPAK -,111,72
ipr-mus LD50:150 mg/kg NTIS** AD691-490

CONSENSUS REPORTS: Reported in EPA TSCA Inventory.

SAFETY PROFILE: Poison by intraperitoneal route.

Mildly toxic by ingestion. An eye irritant. When heated to decomposition it emits toxic fumes of NO_x.

BCP650 CAS:53-89-4 ***HR: 3***
BENZOMETAN
mf: $C_{22}H_{25}N_3O$ mw: 347.50

PROP: Crystals from ethanol. Decomp 181-183°.

SYNS: BENZOPIPERILONE (ITALIAN) ◇ BENZPIPERILONE ◇ BENZPIPERYLON ◇ 4-BENZYL-1-(1-METHYL-4-PIPERIDYL)-3-PHENYL-3-PYRAZOLIN-5-ONE ◇ 1,2-DIHYDRO-2-(1-METHYL-4-PIPERIDINYL)-5-PHENYL-4-(PHENYLMETHYL)-3H-PYRAZOL-3-ONE (9CI) ◇ HUMEDIL ◇ KB 95 ◇ 1-(N-METIL-PIPERIDIL-4')-3-FENIL-4-BENZIL-PIRAZOLONE-5 (ITALIAN) ◇ PPBP ◇ REUBLONIL ◇ TELON

TOXICITY DATA with REFERENCE
orl-rat LD50:2700 mg/kg BCFAAI 102,602,63
ivn-rat LD50:160 mg/kg BCFAAI 102,602,63
orl-mus LD50:1880 mg/kg BCFAAI 102,602,63
scu-mus LD50:615 mg/kg BCFAAI 102,602,63
ivn-mus LD50:160 mg/kg BCFAAI 102,602,63
orl-rbt LD50:1700 mg/kg BCFAAI 102,602,63
ivn-rbt LD50:83 mg/kg BCFAAI 102,602,63

SAFETY PROFILE: Poison by intravenous route. Moderately toxic by ingestion and subcutaneous routes. When heated to decomposition it emits toxic fumes of NO_x.

BCP685 CAS:3811-10-7 ***HR: 3***
BENZOMETHAMINE BROMIDE
mf: $C_{22}H_{31}N_2O_2 \cdot Br$ mw: 435.46

SYNS: N,N-DIETHYL-2-((HYDROXYDIPHENYLACETYL)METHYLAMINO)-N-METHYL-ETHANAMINIUM BROMIDE (9CI) ◇ DIETHYLMETHYL(2-(N-METHYLBENZILAMIDO)ETHYL)AMMONIUM BROMIDE ◇ MC 3199

TOXICITY DATA with REFERENCE
orl-mus LD50:2700 mg/kg JPETAB 114,54,55
ipr-mus LD50:136 mg/kg JPETAB 114,54,55
ivn-mus LD50:31800 μg/kg JPETAB 114,54,55

SAFETY PROFILE: Poison by intravenous and intraperitoneal routes. Moderately toxic by ingestion. When heated to decomposition it emits toxic fumes of Br^- and NO_x.

BCP750 CAS:192-70-1 ***HR: 3***
BENZO(a)NAPHTHO(8,1,2-cde)NAPHTHACENE
mf: $C_{28}H_{16}$ mw: 352.44

SYN: NAPHTO(1,2-c-d-e)NAPHTACENE(FRENCH)

TOXICITY DATA with REFERENCE
scu-mus TDLo:72 mg/kg/9W-I:ETA CHDDAT 266,301,68

SAFETY PROFILE: Questionable carcinogen with experimental tumorigenic data. When heated to decomposition it emits acrid smoke and irritating fumes.

BCQ000 CAS:196-79-2 ***HR: 3***
BENZO(h)NAPHTHO(1,2-f,s-3)QUINOLINE
mf: $C_{21}H_{13}N$ mw: 279.35

SYN: PYRIDO(3',2':5,6)CHRYSENE

TOXICITY DATA with REFERENCE
scu-mus TDLo:72 mg/kg/9W-I:ETA COREAF 252,1711,61

SAFETY PROFILE: Questionable carcinogen with experimental tumorigenic data. When heated to decomposition it emits toxic fumes such as NO_x.

BCQ250 CAS:100-47-0 ***HR: 3***
BENZONITRILE
DOT: UN 2224
mf: C_7H_5N mw: 103.13

PROP: Transparent, colorless oil; almond-like odor. D: 1.246 @ 20°/4°, bp: 191°, mp: −12.8°.

SYNS: BENZENENITRILE ◇ BENZOIC ACID NITRILE ◇ BENZONITRILE (DOT) ◇ CYANOBENZENE ◇ PHENYL CYANIDE

TOXICITY DATA with REFERENCE
skn-rbt 500 mg/24H MOD FCTXAV 17(Suppl.),695,79
orl-rat LDLo:720 mg/kg AMRL** TR-74-78,74
ihl-rat LCLo:950 ppm/8H AMRL** TR-74-78,74
skn-rat LD50:1200 mg/kg AMRL** TR-74-78,74
orl-mus LD50:971 mg/kg NEZAAQ 39,423,84
ihl-mus LC50:6000 mg/m^3 AZMZA6 52(11),60,75
scu-mus LD50:180 mg/kg MEIEDD 10,156,83
scu-rbt LDLo:200 mg/kg AIPTAK 5,161,1899
scu-frg LDLo:1700 mg/kg AIPTAK 5,161,1899
ipr-mus LD50:400 mg/kg FCTXAV 17,723,79

CONSENSUS REPORTS: Reported in EPA TSCA Inventory. Cyanide and its compounds are on the Community Right-To-Know List.

DOT Classification: Combustible Liquid; Label: None; DOT-IMO: Poison B; Label: Poison.

SAFETY PROFILE: Poison by intraperitoneal and subcutaneous routes. Moderately toxic by ingestion, inhalation, and skin contact. See also NITRILES. A skin irritant. Combustible liquid. When heated to decomposition it emits toxic fumes of CN^- and NO_x.

BCQ500 CAS:189-55-9 ***HR: 3***
BENZO(rst)PENTAPHENE
mf: $C_{24}H_{14}$ mw: 302.38

PROP: Green-yellow needles. Mp: 280-282°.

SYNS: DB(a,i)P ◇ DIBENZO(a,i)PYRENE ◇ DIBENZO(b,h)PYRENE ◇ 1,2,7,8-DIBENZOPYRENE ◇ 3,4:9,10-DIBENZOPYRENE ◇ DIBENZ(a,i)PYRENE ◇ 1,2:7,8-DIBENZPYRENE ◇ 3,4:9,10-DIBENZPYRENE ◇ RCRA WASTE NUMBER U064

TOXICITY DATA with REFERENCE
mma-sat 20 μg/plate PNASA6 72,5135,75

mrc-esc 600 μg/well MUREAV 46,53,77
dnd-esc 10 μmol/L MUREAV 89,95,81
otr-ham:kdy 80 μg/L BJCAAI 37,873,78
msc-ham:lng 30 μg/L CNREA8 42,1646,82
skn-mus TDLo:47 mg/kg/39W-I:ETA CANCAR 12,1079,59
scu-ham TDLo:2 mg/kg:NEO NATUAS 203,308,64

CONSENSUS REPORTS: NTP Fifth Annual Report on Carcinogens. IARC Cancer Review: Group 2B IMEMDT 7,56,87; Animal Sufficient Evidence IMEMDT 3,215,73; IMEMDT 32,337,83. EPA Genetic Toxicology Program.

SAFETY PROFILE: Confirmed with experimental neoplastigenic and tumorigenic data. Mutation data reported. When heated to decomposition it emits acrid smoke and irritating fumes.

BCQ750 CAS:63040-53-9 ***HR: 3***
BENZO(rst)PENTAPHENE-5-CARBOXALDEHYDE
mf: $C_{25}H_{14}O$ mw: 330.39

SYN: 5-FORMYL-3,4:9,10-DIBENZOPYRENE

TOXICITY DATA with REFERENCE
scu-mus TDLo:72 mg/kg/9W-I:ETA COREAF 252,1236,61

SAFETY PROFILE: Questionable carcinogen with experimental tumorigenic data. When heated to decomposition it emits acrid smoke and irritating fumes.

BCR000 CAS:191-24-2 ***HR: D***
BENZO(ghi)PERYLENE
mf: $C_{22}H_{12}$ mw: 276.34

SYNS: 1,12-BENZPERYLENE ◇ 1,12-BENZOPERYLENE

TOXICITY DATA with REFERENCE
mma-sat 2 μg/plate/48H FCTXAV 17,141,79

CONSENSUS REPORTS: IARC Cancer Review: Group 3 IMEMDT 7,56,87, Animal Inadequate Evidence IMEMDT 32,195,83. EPA Genetic Toxicology Program.

SAFETY PROFILE: Questionable carcinogen. Mutation data reported. When heated to decomposition it emits acrid smoke and irritating fumes.

BCR250 CAS:190-07-8 ***HR: 3***
BENZO(a)PHENALENO(1,9-hi)ACRIDINE
mf: $C_{27}H_{15}N$ mw: 353.43

SYN: BENZO(c)PHENALENO(1,9-I,j)ACRIDINE

TOXICITY DATA with REFERENCE
scu-mus TDLo:72 mg/kg/9W-I:ETA BAFEAG 52,49,65

SAFETY PROFILE: Questionable carcinogen with experimental tumorigenic data. When heated to decomposition it emits toxic fumes such as NO_x.

BCR500 CAS:190-03-4 ***HR: 3***
BENZO(a)PHENALENO(1,9-i,j)ACRIDINE
mf: $C_{27}H_{15}N$ mw: 353.43

SYN: BENZO(h)PHENALENO(1,9-bc)ACRIDINE

TOXICITY DATA with REFERENCE
scu-mus TDLo:72 mg/kg/9W-I:ETA BAFEAG 52,49,65

SAFETY PROFILE: Questionable carcinogen with experimental tumorigenic data. When heated to decomposition it emits toxic fumes of NO_x.

BCR750 CAS:195-19-7 ***HR: 3***
BENZO(c)PHENANTHRENE
mf: $C_{18}H_{12}$ mw: 228.30

SYNS: 3,4-BENZOPHENANTHRENE ◇ 3,4-BENZPHENANTHRENE ◇ TETRAHELICENE

TOXICITY DATA with REFERENCE
mma-sat 25 nmol/plate CNREA8 40,2876,80
skn-mus TDLo:940 mg/kg/39W-I:ETA PRLBA4 117,318,35

CONSENSUS REPORTS: IARC Cancer Review: Group 3 IMEMDT 7,56,87, Animal Inadequate Evidence IMEMDT 32,205,83

SAFETY PROFILE: Questionable carcinogen with experimental tumorigenic data. Mutation data reported. When heated to decomposition it emits acrid and irritating fumes.

BCS000 CAS:4466-76-6 ***HR: 3***
BENZO(c)PHENATHRENE-8-CARBOXALDEHYDE
mf: $C_{19}H_{12}O$ mw: 256.31

SYN: 2-FORMYL-3:4-BENZPHENANTHRENE

TOXICITY DATA with REFERENCE
scu-mus TDLo:5200 mg/kg/52W-I:ETA PRLBA4 131,170,42

SAFETY PROFILE: Questionable carcinogen with tumorigenic data. When heated to decomposition it emits acrid smoke and irritating fumes.

BCS100 ***HR: 2***
(±)-BENZO(c)PHENANTHRENE-3,4-DIHYDRODIOL
mf: $C_{18}H_{14}O_2$ mw: 262.32

TOXICITY DATA with REFERENCE
ipr-mus TDLo:11 mg/kg/15D-I:NEO CNREA8 46,2257,86

SAFETY PROFILE: Questionable carcinogen with ex-

perimental neoplastigenic data. When heated to decomposition it emits toxic and irritating fumes.

BCS103 ***HR: 2***
(+)-BENZO(c)PHENANTHRENE-3,4-DIOL-1,2-EPOXIDE-1
mf: $C_{18}H_{14}O_3$ mw: 278.32

SYNS: BENZO(c)PHENANTHRENE-3,4-DIOL,1,2,3,4-TETRAHYDRO-1,2-EPOXY-, (Z)-(+)-(1R,2S,3R,4S)- ◇ cis-1-β,2-β-EPOXY-1,2,3,4-TETRAHYDROBENZO(c)PHENANTHRENE-3α-4-β-DIOL

TOXICITY DATA with REFERENCE
skn-mus TDLo:111 μg/kg:NEO CNREA8 46,2257,86

SAFETY PROFILE: Questionable carcinogen with experimental neoplastigenic data. When heated to decomposition it emits toxic and irritating fumes.

BCS105 ***HR: 2***
(+)-BENZO(c)PHENANTHRENE-3,4-DIOL-1,2-EPOXIDE-2
mf: $C_{18}H_{14}O_3$ mw: 278.32

SYN: BENZO(c)PHENANTHRENE-3,4-DIOL,1,2,3,4-TETRAHYDRO-1,2-EPOXY-1, (E)-(+)-(1S,2R,3R,4S)- ◇ trans-1-α-2-α-EPOXY-1,2,3,4-TETRAHYDROBENZO(c)PHENANTHRENE-3α,4-β-DIOL

TOXICITY DATA with REFERENCE
skn-mus TDLo:278 μg/kg:NEO CNREA8 46,2257,86

SAFETY PROFILE: Questionable carcinogen with experimental neoplastigenic data. When heated to decomposition it emits toxic and irritating fumes.

BCS110 ***HR: 2***
(−)-BENZO(c)PHENANTHRENE-3,4-DIOL-1,2-EPOXIDE-2
mf: $C_{18}H_{14}O_3$ mw: 278.32

SYNS: BENZO(c)PHENANTHRENE-3,4-DIOL,1,2,3,4-TETRAHYDRO-1,2-EPOXY-, (E)-(−)-(1R,2S,3S,4R)- ◇ trans-1-β,2-β-EPOXY-1,2,3,4-TETRAHYDROBENZO(c)PHENANTHRENE-3β, 4-α-DIOL

TOXICITY DATA with REFERENCE
skn-mus TDLo:111 μg/kg:NEO CNREA8 46,2257,86

SAFETY PROFILE: Questionable carcinogen with experimental neoplastigenic data. When heated to decomposition it emits toxic and irritating fumes.

BCS250 CAS:119-61-9 ***HR: 3***
BENZOPHENONE
mf: $C_{13}H_{10}O$ mw: 182.23

PROP: Rhombic, white crystals; persistent rose-like odor. mp (α): 49°, mp (β): 26°, mp (Γ): 47°, bp: 305.4°, d (α): 1.0976 @ 50°/50°, d (β): 1.108 @ 23°/40°, vap press: 1 mm @ 108.2. Sol in fixed oils; sltly sol in propylene glycol; insol in glycerol.

SYNS: BENZOYLBENZENE ◇ DIPHENYL KETONE ◇ DIPHENYLMETHANONE ◇ FEMA No. 2134 ◇ α-OXODIPHENYLMETHANE ◇ PHENYL KETONE

TOXICITY DATA with REFERENCE
orl-mus LD50:2895 mg/kg JETOAS 9,99,76
ipr-mus LD50:727 mg/kg JETOAS 9,99,76

CONSENSUS REPORTS: Reported in EPA TSCA Inventory.

SAFETY PROFILE: Moderately toxic by ingestion and intraperitoneal routes. Combustible when heated. Incompatible with oxidizers. When heated to decomposition it emits acrid and irritating fumes. See also KETONES.

BCS325 CAS:131-55-5 ***HR: 2***
BENZOPHENONE-2
mf: $C_{13}H_{10}O_5$ mw: 246.23

SYNS: 2,2′,4,4′-TETRAHYDROXY BENZOPHENONE ◇ 2,4,2′,4′-TETRAHYDROXYBENZOPHENONE ◇ THBP ◇ UVINOL D-50 ◇ UVINUL D-50

TOXICITY DATA with REFERENCE
eye-rbt 100 mg MOD JACTDZ 2(5),35,83
mma-sat 100 μg/plate FCTOD7 20,427,82
orl-rat LD50:1220 mg/kg JACTDZ 2(5),35,83

CONSENSUS REPORTS: Reported in EPA TSCA Inventory.

SAFETY PROFILE: Moderately toxic by ingestion. An eye irritant. Mutation data reported. When heated to decomposition it emits toxic fumes of NO_x.

BCS750 CAS:50-32-8 ***HR: 3***
BENZO(a)PYRENE
mf: $C_{20}H_{12}$ mw: 252.32

PROP: Yellow crystals. Mp: 179°, bp: 312° @ 10 mm. Insol in water; sol in benzene, toluene, and xylene.

SYNS: BENZO(d,e,f)CHRYSENE ◇ 3,4-BENZOPIRENE (ITALIAN) ◇ 3,4-BENZOPYRENE ◇ 6,7-BENZOPYRENE ◇ 3,4-BENZPYREN (GERMAN) ◇ BENZ(a)PYRENE ◇ 3,4-BENZ(a)PYRENE ◇ 3,4-BENZYPYRENE ◇ B(a)P ◇ RCRA WASTE NUMBER U022

TOXICITY DATA with REFERENCE
skn-mus 14 μg MLD CALEDQ 4,333,78
dnd-sal:spr 3 g/L BIPMAA 5,477,67
dnd-hmn:oth 1500 nmol/L TCMUD8 1,3,80
msc-hmn:oth 100 nmol/L CRNGDP 1,765,80
ipr-rat TDLo:60 mg/kg (female 16-18D post):REP BNEOBV 38,291,80
orl-mus TDLo:1280 mg/kg (female 16D pre-5D post):TER DOESD6 54,410,81
orl-rat TDLo:15 mg/kg:CAR EXPTAX 18,288,80
ipr-rat TDLo:16 mg/kg:ETA BJCAAI 12,65,58
scu-rat TDLo:455 μg/kg/60D-I:NEO CBINA8 29,159,80
ims-rat TDLo:2400 μg/kg:CAR NTIS** DOE/EV/03140-5

scu-rat LD50:50 mg/kg ZEKBAI 69,103,67
ipr-mus LDLo:500 mg/kg TXAPA9 23,288,72
irn-frg LDLo:11 mg/kg CNREA8 24,1969,64

CONSENSUS REPORTS: NTP Fifth Annual Report on Carcinogens. IARC Cancer Review: Group 2A IMEMDT 7,56,87; Animal Sufficient Evidence IMEMDT 32,211,83; IMEMDT 3,91,73. Reported in EPA TSCA Inventory.

OSHA PEL: TWA 0.2 mg/m^3

SAFETY PROFILE: Confirmed carcinogen with experimental carcinogenic, neoplastigenic, and tumorigenic data. A poison via subcutaneous, intraperitoneal, and intrarenal routes. Experimental teratogenic and reproductive effects. Human mutation data reported. A skin irritant. A common air contaminant of water, food, and smoke. When heated to decomposition it emits acrid smoke and fumes. See other benzo pyrenes.

BCT000 CAS:192-97-2 ***HR: 3***
BENZO(e)PYRENE
mf: $C_{20}H_{12}$ mw: 252.32

SYNS: 1,2-BENZOPYRENE ◇ 4,5-BENZOPYRENE ◇ 1,2-BENZPYRENE ◇ B(e)P

TOXICITY DATA with REFERENCE
mmo-sat 1 nmol/plate CNREA8 40,1985,80
mma-sat 1 μg/plate ENMUDM 6(Suppl 2),1,84
msc-hmn:oth 12 μmol/L MUREAV 130,127,84
dns-rat:lvr 79 nmol/L CNREA8 42,3010,82
dnd-mus-skn 192 μmol/kg CRNGDP 5,231,84
otr-ham:kdy 25 μg/L TOLED5 7,143,80
ipr-mus TDLo:150 mg/kg (8D preg):TER SEIJBO 21,97,81
ipr-mus TDLo:300 mg/kg (8D preg):REP SEIJBO 21,97,81
orl-mus TDLo:360 mg/kg/43W-I:ETA VRRAAT 20,276,38

CONSENSUS REPORTS: IARC Cancer Review: Group 3 IMEMDT 7,56,87; Animal Inadequate Evidence IMEMDT 32,225,83; Animal Limited Evidence IMEMDT 3,137,73. EPA Genetic Toxicology Program.

SAFETY PROFILE: Questionable carcinogen with experimental tumorigenic data. Experimental teratogenic and reproductive effects. Human mutation data reported. When heated to decomposition it emits acrid smoke and irritating fumes.

BCT250 CAS:13312-42-0 ***HR: 3***
BENZO(a)PYRENE-6-CARBOXYALDEHYDE
mf: $C_{21}H_{12}O$ mw: 280.33

SYNS: 3,4-BENZPYRENE-5-ALDEHYDE ◇ 6-FORMYLBENZO(a)PYRENE

TOXICITY DATA with REFERENCE
scu-rat TDLo:17 mg/kg/60D-I:NEO CBINA8 29,159,80
unk-mus TDLo:80 mg/kg/8D-I:ETA BEBMAE 88(11),592,79

SAFETY PROFILE: Questionable carcinogen with experimental tumorigenic and neoplastigenic data. When heated to decomposition it emits acrid smoke and fumes. See also ALDEHYDES.

BCT500 CAS:64048-70-0 ***HR: 3***
BENZO(a)PYRENE-6-CARBOXALDEHYDE THIOSEMICARBAZONE
mf: $C_{22}H_{15}N_3S$ mw: 353.46

SYN: 3,4-BENZPYRENE-5-ALDEHYDETHIOSEMICARBAZONE

TOXICITY DATA with REFERENCE
scu-mus TDLo:200 mg/kg:ETA COREAF 245,876,57

SAFETY PROFILE: Questionable carcinogen with experimental tumorigenic data. When heated to decomposition it emits very toxic fumes of SO_x and NO_x. See also ALDEHYDES.

BCT750 CAS:13345-25-0 ***HR: 3***
BENZO(a)PYRENE-7,8-DIHYDRODIOL
mf: $O_2C_{20}H_{14}$ mw: 286.34

SYN: BP-7,8-DIHYDRODIOL

TOXICITY DATA with REFERENCE
mma-sat 8 μmol/L CALEDQ 24,281,84
dnd-hmn:fbr 30 μmol/L CBINA8 41,155,82
skn-mus TDLo:4580 μg/kg:NEO CCSUDL 3,371,78
par-mus TDLo:4 mg/kg:ETA BJCAAI 37,657,78

CONSENSUS REPORTS: EPA Genetic Toxicology Program.

SAFETY PROFILE: Questionable carcinogen with experimental neoplastigenic and tumorigenic data. Human mutation data reported. When heated to decomposition it emits acrid smoke and irritating fumes.

BCU000 CAS:60268-85-1 ***HR: 3***
anti-BENZO(a)PYRENE-7,8-DIHYDRODIOL-9,10-OXIDE
mf: $C_{20}H_{14}O_3$ mw: 302.34

SYNS: BENZO(a)PYRENE-7,8-DIHYDRODIOL-9,10-EPOXIDE(anti) ◇ BP-7,8-DIHYDRODIOL-9,10-EPOXIDE (anti) ◇ anti-BP-7,8-DIHYDRODIOL-9,10-OXIDE

TOXICITY DATA with REFERENCE
dnd-hmn:lym 800 μg/L CRNGDP 3,1107,82
msc-ham:lng 100 μg/L IJCNAW 24,203,79
skn-mus TDLo:2400 μg/kg:ETA CALEDQ 2,115,76

SAFETY PROFILE: Questionable carcinogen with experimental tumorigenic data by skin contact. Human mutation data reported. When heated to decomposition it emits acrid smoke and irritating fumes.

BCU250 CAS:58917-67-2 ***HR: D***
BENZO(a)PYRENE DIOL EPOXIDE ANTI

SYNS: anti(±)BENZO(a)PYRENE-DIOL-EPOXIDE ◇ anti-BPDE ◇ anti-r-7,trans-8-DIHYDROXY-trans-9,10-OXY-7,8,9,10-TETRAHYDROBENZO(a)PYRENE ◇ BPDE ◇ BP DIOL EPOXIDE ANTI ◇ trans-7,8-DIHYDROXY-9,10-OXY-7,8,9,10-TETRAHYDROBENZO(a)PYRENE

TOXICITY DATA with REFERENCE
mmo-sat 200 pmol/plate MUREAV 125,95,84
dnd-hmn:fbr 1 μmol/L ENMUDM 7,267,85
dns-hmn:fbr 1500 nmol/L BBACAQ 824,146,85
msc-hmn:fbr 200 nmol/L MUREAV 125,95,84
dnd-ham:ovr 1 mg/L MUREAV 129,365,84

CONSENSUS REPORTS: EPA Genetic Toxicology Program.

SAFETY PROFILE: Human mutation data reported. When heated to decomposition it emits acrid smoke and irritating fumes.

BCU500 CAS:3067-13-8 ***HR: 3***
BENZO(a)PYRENE-1,6-DIONE
mf: $C_{20}H_{10}O_2$ mw: 282.30

SYNS: 1,6-BENZO(a)PYRENEDIONE ◇ BENZO(a)PYRENE-1,6-QUINONE ◇ PB-1,6-QUINONE

TOXICITY DATA with REFERENCE
msc-ham:lng 2 mg/L CNREA8 36,3350,76
skn-mus TDLo:4520 μg/kg:NEO CCSUDL 3,371,78

CONSENSUS REPORTS: EPA Genetic Toxicology Program.

SAFETY PROFILE: Questionable carcinogen with experimental neoplastigenic data by skin contact. Mutation data reported. When heated to decomposition it emits acrid smoke and irritating fumes.

BCU750 CAS:3067-14-9 ***HR: 3***
BENZO(a)PYRENE-3,6-DIONE
mf: $C_{20}H_{10}O_2$ mw: 282.30

SYNS: 3,6-BENZO(a)PYRENEDIONE ◇ BENZO(a)PYRENE-3,6-QUINONE ◇ BP-3,6-QUINONE

TOXICITY DATA with REFERENCE
mma-sat 5 μg/plate ENMUDM 7,839,85
dnd-hmn:fbr 1 μmol/L TOLED5 28,37,85
msc-ham:lng 2 mg/L CNREA8 36,3350,76
skn-mus TDLo:4520 μg/kg:NEO CCSUDL 3,371,78

CONSENSUS REPORTS: EPA Genetic Toxicology Program.

SAFETY PROFILE: Questionable carcinogen with experimental neoplastigenic data by skin contact. Human mutation data reported. When heated to decomposition it emits acrid smoke and irritating fumes.

BCV000 CAS:3067-12-7 ***HR: 3***
BENZO(a)PYRENE-6,12-DIONE
mf: $C_{20}H_{10}O_2$ mw: 282.30

SYNS: 6,12-BENZO(a)PYRENEDIONE ◇ BENZO(a)PYRENE-6,12-QUINONE ◇ 6,12-BENZOPYRENE QUINONE ◇ BP-6,12-QUINONE

TOXICITY DATA with REFERENCE
msc-ham:lng 4 mg/L CNREA8 36,3350,76
skn-mus TDLo:4520 μg/kg:NEO CCSUDL 3,371,78

CONSENSUS REPORTS: EPA Genetic Toxicology Program.

SAFETY PROFILE: Questionable carcinogen with experimental neoplastigenic data. Mutation data reported. When heated to decomposition it emits acrid smoke and irritating fumes.

BCV125 CAS:71382-50-8 ***HR: D***
BENZO(a)PYRENE-4,5-IMINE
mf: $C_{20}H_{13}N$ mw: 267.34

SYN: 3b,4a-DIHYDRO-4H-BENZO(1,2)PYRENO(4,5-b)AZIRINE

TOXICITY DATA with REFERENCE
mmo-sat 100 ng/plate CNREA8 45,2600,85
mmo-esc 1 μg/plate CNREA8 45,2600,85
dnr-bcs 100 ng/plate CNREA8 45,2600,85
sce-hmn:fbr 10 nmol/L CNREA8 45,2600,85
msc-ham:lng 30 nmol/L CNREA8 45,2600,85

SAFETY PROFILE: Human mutation data reported. When heated to decomposition it emits toxic fumes of NO_x.

BCV250 CAS:21247-98-3 ***HR: 3***
BENZO(a)PYRENE-6-METHANOL
mf: $C_{21}H_{14}O$ mw: 282.35

SYN: 6-HYDROXYMETHYLBENZO(a)PYRENE

TOXICITY DATA with REFERENCE
dnd-omi 30 μmol/L CBINA8 31,51,80
dnd-mam:lym 500 mg/L CBINA8 25,35,79
bfa-rat/sat 1mg/kg MUREAV 173,251,86
scu-rat TDLo:100 mg/kg/40D-I:CAR JMCMAR 16,714,73
skn-mus TDLo:180 mg/kg/20W-I:CAR CBINA8 22(1),53,78
scu-mus TDLo:20 mg/kg:NEO BJCAAI 26,506,72
skn-mus TD:180 mg/kg/20W-I:ETA PAACA3 18,59,77

CONSENSUS REPORTS: EPA Genetic Toxicology Program.

SAFETY PROFILE: Suspected carcinogen with experimental carcinogenic, neoplastigenic, and tumorigenic data. Mutation data reported. When heated to decomposition it emits acrid smoke and fumes.

BCV500 CAS:37574-47-3 ***HR: 3***
BENZO(a)PYRENE-4,5-OXIDE
mf: $C_{20}H_{12}O$ mw: 268.32

SYNS: BENZO(1,2)PYRENO(4,5-b)OXIRENE-3b,4b-DIHYDRO ◇ BENZO(a)PYRENE-4,5-EPOXIDE ◇ BENZ(a)PYRENE 4,5-OXIDE ◇ BP-4,5-EPOXIDE ◇ BP 4,5-OXIDE

TOXICITY DATA with REFERENCE
mmo-sat 250 ng/plate ENMUDM 7,839,85
mma-sat 1 μg/plate ENMUDM 7,839,85
mmo-esc 1 μg/plate TCMUD8 5,339,85
dnr-bcs 1- μg/plate CNREA8 45,2600,85
dnd-mam:lym 800 nmol CRNGDP 3,267,82
scu-rat TDLo:40 mg/kg/50D-I:ETA IJCNAW 18,351,76
skn-mus TDLo:2144 μg/kg:NEO CCSUDL 3,371,78

CONSENSUS REPORTS: EPA Genetic Toxicology Program.

SAFETY PROFILE: Questionable carcinogen with experimental tumorigenic and neoplastigenic data. Mutation data reported. When heated to decomposition it emits acrid and irritating fumes.

BCV750 CAS:36504-65-1 ***HR: 3***
BENZO(a)PYRENE-7,8-OXIDE
mf: $C_{20}H_{12}O$ mw: 268.32

SYNS: BENZO(10,11)CHRYSENO(1,2-b)OXIRENE-6β,7-α-DIHYDRO ◇ BENZO(a)PYRENE-7,8-DIHYDRO-7,8-EPOXY ◇ BENZO(a)PYRENE-7,8-EPOXIDE ◇ 6-β,7-α-DIHYDROBENZO(10,11)CHRYSENO(1,2-b)-OXIRENE ◇ BP 7,8-EPOXIDE ◇ BP 7,8-OXIDE ◇ 7,8-EPOXY-7,8-DIHYDROBENZO(a)PYRENE

TOXICITY DATA with REFERENCE
mmo-sat 250 ng/plate CNREA8 36,3350,76
mma-sat 25 μmol/L JBCHA3 251,4882,76
skn-mus TDLo:32 mg/kg/60W-I:CAR PNASA6 73,243,76
scu-mus TDLo:10 mg/kg:ETA JJIND8 64,617,80
skn-mus TD:2144 μg/kg:NEO CCSUDL 3,371,78

CONSENSUS REPORTS: EPA Genetic Toxicology Program.

SAFETY PROFILE: Questionable carcinogen with experimental carcinogenic, neoplastigenic, and tumorigenic data. Mutation data reported. When heated to decomposition it emits irritating fumes.

BCW000 CAS:36504-66-2 ***HR: 3***
BENZO(a)PYRENE-9,10-OXIDE
mf: $C_{20}H_{12}O$ mw: 268.32

SYN: BP-9,10-OXIDE

TOXICITY DATA with REFERENCE
mmo-sat 250 ng/plate CNREA8 36,3350,76
skn-mus TDLo:2140 μg/kg:ETA CCSUDL 3,371,78

CONSENSUS REPORTS: EPA Genetic Toxicology Program.

SAFETY PROFILE: Questionable carcinogen with experimental tumorigenic data by skin contact. Mutation data reported. When heated to decomposition it emits acrid smoke and irritating fumes.

BCW250 CAS:60448-19-3 ***HR: 3***
BENZO(a)PYRENE-11,12-OXIDE
mf: $C_{20}H_{12}O$ mw: 268.32

SYN: BP-11,12-OXIDE

TOXICITY DATA with REFERENCE
mmo-sat 1 μg/plate CNREA8 36,3350,76
msc-ham:lng 5 mg/L CNREA8 36,3350,76
skn-mus TDLo:2140 μg/kg:NEO CCSUDL 3,371,78
skn-mus TD:2160 μg/kg:ETA CNREA8 37,4130,77

CONSENSUS REPORTS: EPA Genetic Toxicology Program.

SAFETY PROFILE: Questionable carcinogen with experimental tumorigenic and neoplastigenic data by skin contact. Mutation data reported. When heated to decomposition it emits acrid smoke and irritating fumes.

BCW750 CAS:13345-23-8 ***HR: D***
BENZO(a)PYREN-1-OL
mf: $C_{20}H_{12}O$ mw: 268.32

SYN: 1-HYDROXYBENZO(a)PYRENE

TOXICITY DATA with REFERENCE
mmo-sat 7 μg/plate ENMUDM 7,839,85
mma-sat 2500 ng/plate BCPCA6 28,1615,79
dnd-hmn:fbr 30 μmol/L CBINA8 41,155,82
mma-ham:lng 3700 nmol/L PNASA6 73,607,76
msc-ham:lng 15 mg/L CNREA8 36,3350,76

CONSENSUS REPORTS: EPA Genetic Toxicology Program.

SAFETY PROFILE: Human mutation data reported. When heated to decomposition it emits acrid smoke and irritating fumes.

BCX000 CAS:56892-30-9 ***HR: 3***
BENZO(a)PYREN-2-OL
mf: $C_{20}H_{12}O$ mw: 268.32

SYN: 2-HYDROXYBENZO(a)PYRENE

TOXICITY DATA with REFERENCE
mmo-sat 8500 pmol/L RRBCAD 18,291,81
dnd-hmn:fbr 30 μmol/L CBINA8 41,155,82
mma-sat 2 nmol/plate CNREA8 39,2660,79
mma-ham:lng 25 nmol/plate CNREA8 39,2660,79

msc-ham:lng 10 mg/L CNREA8 36,3350,76
skn-mus TDLo:69 mg/kg/32W-I:NEO CNREA8 37,2608,77
scu-mus TDLo:10 mg/kg:ETA JJIND8 64,617,80

CONSENSUS REPORTS: EPA Genetic Toxicology Program.

SAFETY PROFILE: Questionable carcinogen with experimental tumorigenic and neoplastigenic data. Human mutation data reported. When heated to decomposition it emits acrid smoke and irritating fumes.

BCX250 CAS:13345-21-6 ***HR: 3***
BENZO(a)PYREN-3-OL
mf: $C_{20}H_{12}O$ mw: 268.32

SYNS: BP-3-HYDROXY ◇ 3-HYDROXYBENZO(a)PYRENE ◇ 8-HYDROXY-3,4-BENZPYRENE

TOXICITY DATA with REFERENCE
dnd-hmn:fbr 30 μmol/L CBINA8 41,155,82
dnr-esc 250 mg/L JNCIAM 62,873,79
msc-ham:lng 12 μmol/L PNASA6 73,607,76
skn-mus TDLo:117 μg/kg:NEO CNREA8 38,678,78
scu-mus TDLo:160 mg/kg/21W-I:ETA BJCAAI 6,400,52

CONSENSUS REPORTS: EPA Genetic Toxicology Program.

SAFETY PROFILE: Questionable carcinogen with experimental tumorigenic and neoplastigenic data by skin contact. Human mutation data reported. When heated to decomposition it emits acrid smoke and irritating fumes.

BCX500 CAS:24027-84-7 ***HR: 3***
BENZO(a)PYREN-5-OL
mf: $C_{20}H_{12}O$ mw: 268.32

SYN: 5-HYDROXYBENZO(a)PYRENE

TOXICITY DATA with REFERENCE
skn-mus TDLo:4280 μg/kg:ETA CCSUDL 3,371,78

CONSENSUS REPORTS: EPA Genetic Toxicology Program.

SAFETY PROFILE: Questionable carcinogen with experimental tumorigenic data by skin contact. When heated to decomposition it emits acrid smoke and irritating fumes.

BCX750 CAS:33953-73-0 ***HR: 3***
BENZO(a)PYREN-6-OL
mf: $C_{20}H_{12}O$ mw: 268.32

SYN: 6-HYDROXYBENZO(a)PYRENE

TOXICITY DATA with REFERENCE
mma-sat 25 μmol/L JBCHA3 251,4882,76
mma-ham:lng 3700 nmol/L PNASA6 73,607,76
mmo-sat 7 μg/plate ENMUDM 7,839,85
msc-ham:lng 5 mg/L CNREA8 36,3350,76
skn-mus TDLo:4280 μg/kg:NEO CCSUDL 3,371,78
idr-mus TDLo:2400 μg/kg:ETA GANNA2 62,419,71

CONSENSUS REPORTS: EPA Genetic Toxicology Program.

SAFETY PROFILE: Questionable carcinogen with experimental neoplastigenic and tumorigenic data. Mutation data reported. When heated to decomposition it emits acrid smoke and fumes.

BCY000 CAS:37994-82-4 ***HR: 3***
BENZO(a)PYREN-7-OL
mf: $C_{20}H_{12}O$ mw: 268.32

SYN: 7-HYDROXYBENZO(a)PYRENE

TOXICITY DATA with REFERENCE
mmo-sat 16 μg/plate MUREAV 36,379,76
mma-sat 7 μg/plate ENMUDM 7,839,85
dni-omi 200 μg/L PNASA6 74,1378,77
dnd-hmn:fbr 30 μmol/L CBINA8 41,155,82
msc-ham:lng 12 μmol/L PNASA6 73,607,76
skn-mus TDLo:4280 μg/kg:NEO CCSUDL 3,371,78
skn-mus TD:4290 μg/kg:ETA CNREA8 38,678,78

CONSENSUS REPORTS: EPA Genetic Toxicology Program.

SAFETY PROFILE: Questionable carcinogen with experimental tumorigenic and neoplastigenic data by skin contact. Human mutation data reported. When heated to decomposition it emits acrid smoke and fumes.

BCY250 CAS:17573-21-6 ***HR: 3***
BENZO(a)PYREN-9-OL
mf: $C_{20}H_{12}O$ mw: 268.32

SYN: 9-HYDROXYBENZO(a)PYRENE

TOXICITY DATA with REFERENCE
mma-sat 7 μg/plate ENMUDM 7,839,85
dnd-hmn:fbr 30 μmol/L CBINA8 41,155,82
dnd-mam:lym 447 nmol CRNGDP 3,267,82
skn-mus TDLo:4280 μg/kg:NEO CCSUDL 3,371,78

CONSENSUS REPORTS: EPA Genetic Toxicology Program.

SAFETY PROFILE: Questionable carcinogen with experimental neoplastigenic data by skin contact. Human mutation data reported. When heated to decomposition it emits acrid smoke and irritating fumes.

BCY500 CAS:56892-31-0 ***HR: 3***
BENZO(a)PYREN-10-OL
mf: $C_{20}H_{12}O$ mw: 268.32

SYN: 10-HYDROXYBENZO(a)PYRENE

TOXICITY DATA with REFERENCE
mmo-sat 18600 pmol/L RRBCAD 18,291,81
skn-mus TDLo:4280 μg/kg:ETA CCSUDL 3,371,78

CONSENSUS REPORTS: EPA Genetic Toxicology Program.

SAFETY PROFILE: Questionable carcinogen with experimental tumorigenic data by skin contact. Mutation data reported. When heated to decomposition it emits acrid smoke and irritating fumes.

BCY750 CAS:56892-32-1 ***HR: 3***
BENZO(a)PYREN-11-OL
mf: $C_{20}H_{12}O$ mw: 268.32

SYN: 11-HYDROXYBENZO(a)PYRENE

TOXICITY DATA with REFERENCE
dnd-hmn:fbr 30 μmol/L CBINA8 41,155,82
skn-mus TDLo:82 mg/kg/38W-I:ETA CNREA8 37,2608,77

CONSENSUS REPORTS: EPA Genetic Toxicology Program.

SAFETY PROFILE: Questionable carcinogen with experimental tumorigenic data by skin contact. Human mutation data reported. When heated to decomposition it emits acrid smoke and irritating fumes.

BCZ000 CAS:56892-33-2 ***HR: 3***
BENZO(a)PYREN-12-OL
mf: $C_{20}H_{12}O$ mw: 268.32

SYN: 12-HYDROXYBENZO(a)PYRENE

TOXICITY DATA with REFERENCE
mmo-sat 1 μg/plate CNREA8 36,3350,76
mma-sat 7 μg/plate ENMUDM 7,839,85
msc-ham:lng 15 mg/L CNREA8 36,3350,76
skn-mus TDLo:4280 μg/kg:NEO CCSUDL 3,371,78

CONSENSUS REPORTS: EPA Genetic Toxicology Program.

SAFETY PROFILE: Questionable carcinogen with experimental neoplastigenic data by skin contact. Mutation data reported. When heated to decomposition it emits acrid smoke and irritating fumes.

BDA000 CAS:207-89-6 ***HR: 3***
7H-BENZO(a)PYRIDO(3,2-g)CARBAZOLE
mf: $C_{19}H_{12}N_2$ mw: 268.33

SYN: 1,2-BENZOPYRIDO(3′,2′:5,6)CARBAZOLE

TOXICITY DATA with REFERENCE
scu-mus TDLo:72 mg/kg/9W-I:ETA NATUAS 191,1005,61

SAFETY PROFILE: Questionable carcinogen with experimental tumorigenic data. When heated to decomposition it emits toxic fumes such as NO_x.

BDA250 CAS:194-62-7 ***HR: 3***
7H-BENZO(c)PYRIDO(2,3-g)CARBAZOLE
mf: $C_{19}H_{12}N_2$ mw: 268.33

SYN: 5,6-BENZOPYRIDO(3′,2′:3,4)CARBAZOLE

TOXICITY DATA with REFERENCE
scu-mus TDLo:72 mg/kg/9W-I:ETA COREAF 257,818,63

SAFETY PROFILE: Questionable carcinogen with experimental tumorigenic data. When heated to decomposition it emits toxic fumes such as NO_x.

BDA500 CAS:194-60-5 ***HR: 3***
7H-BENZO(c)PYRIDO(3,2-g)CARBAZOLE
mf: $C_{19}H_{12}N_2$ mw: 268.33

SYN: 3,4-BENZOPYRIDO(3′,2′:5,6)CARBAZOLE

TOXICITY DATA with REFERENCE
scu-mus TDLo:72 mg/kg/9W-I:ETA NATUAS 191,1005,61

SAFETY PROFILE: Questionable carcinogen with experimental tumorigenic data. When heated to decomposition it emits toxic fumes such as NO_x.

BDA750 CAS:239-67-8 ***HR: 3***
13H-BENZO(a)PYRIDO(3,2-i)CARBAZOLE
mf: $C_{19}H_{12}N_2$ mw: 268.33

SYN: 7,8-BENZOPYRIDO(2′,3′:1,2)CARBAZOLE

TOXICITY DATA with REFERENCE
scu-mus TDLo:72 mg/kg/9W-I:ETA COREAF 257,818,63

SAFETY PROFILE: Questionable carcinogen with experimental tumorigenic data. When heated to decomposition it emits toxic fumes such as NO_x.

BDB000 CAS:207-88-5 ***HR: 3***
13H-BENZO(g)PYRIDO(2,3-a)CARBAZOLE
mf: $C_{19}H_{12}N_2$ mw: 268.33

SYN: 5,6-BENZOPYRIDO(2′,3′:1,2)CARBAZOLE

TOXICITY DATA with REFERENCE
orl-mus TDLo:2880 mg/kg/24W-I:ETA COREAF 257,818,63

SAFETY PROFILE: Questionable carcinogen with experimental tumorigenic data. When heated to decomposition it emits toxic fumes as NO_x.

BDB250 CAS:207-85-2 ***HR: 3***
13H-BENZO(g)PYRIDO(3,2-a)CARBAZOLE
mf: $C_{19}H_{12}N_2$ mw: 268.33

SYN: 5,6-BENZOPYRIDO(3′,2′:1,2)CARBAZOLE

TOXICITY DATA with REFERENCE
orl-mus TDLo:3720 mg/kg/31W-I:ETA COREAF 257,818,63

SAFETY PROFILE: Questionable carcinogen with experimental tumorigenic data. When heated to decomposition it emits toxic fumes such as NO_x.

BDB500 CAS:318-03-6 ***HR: 3***
11H-BENZO(g)PYRIDO(4,3-b)INDOLE
mf: $C_{15}H_{10}N_2$ mw: 218.27

SYN: 8,9-BENZO-Γ-CARBOLINE

TOXICITY DATA with REFERENCE
scu-mus TDLo:72 mg/kg/9W-I:NEO CHDDAT 271,1474,70

SAFETY PROFILE: Questionable carcinogen with experimental neoplastigenic data. When heated to decomposition it emits toxic fumes such as NO_x.

BDB750 CAS:85-02-9 ***HR: D***
BENZO(f)QUINOLINE
mf: $C_{13}H_9N$ mw: 179.23

SYNS: 1-AZAPHENANTHRENE ◇ 5,6-BENZOQUINOLINE ◇ β-NAPHTHOQUINALDINE

TOXICITY DATA with REFERENCE
mma-sat 50 μg/plate SONNAZ 7,73,83

CONSENSUS REPORTS: EPA Genetic Toxicology Program.

SAFETY PROFILE: Mutation data reported. When heated to decomposition it emits toxic fumes of NO_x.

BDC000 CAS:230-27-3 ***HR: D***
BENZO(h)QUINOLINE
mf: $C_{13}H_9N$ mw: 179.23

SYNS: 4-AZAPHENANTHRENE ◇ α-BENZOQUINOLINE ◇ 7,8-BENZOQUINOLINE ◇ α-NAPHTHOQUINOLINE

TOXICITY DATA with REFERENCE
mma-sat 50 μg/plate SONNAZ 7,73,83

SAFETY PROFILE: Mutation data reported. When heated to decomposition it emits toxic fumes of NO_x.

BDC250 CAS:583-63-1 ***HR: 3***
o-BENZOQUINONE
DOT: UN 2587
mf: $C_6H_4O_2$ mw: 108.10

SYNS: 1,2-BENZOQUINONE ◇ BENZOQUINONE (DOT) ◇ 3,5-CYCLOHEXADIENE-1,2-DIONE ◇ o-QUINONE

TOXICITY DATA with REFERENCE
mmo-sat 100 ng/plate BECTA6 24,590,80

DOT Classification: Poison B; Label: Poison.

SAFETY PROFILE: A poison. Mutation data reported. When heated to decomposition it emits acrid smoke and irritating fumes.

BDC750 CAS:800-24-8 ***HR: 3***
BENZOQUINONE AZIRIDINE
mf: $C_{16}H_{22}N_2O_6$ mw: 338.40

SYNS: A-139 ◇ AZIRIDYL BENZOQUINONE ◇ BAYER A 139 ◇ BAYER R39 SOLUBLE ◇ 2,5-BIS(1-AZIRIDINYL)-3,6-BIS(2-METHOXYETHOXY)-p-BENZOQUINONE ◇ 2,5-BIS(1-AZIRIDINYL)-3,6-BIS(2-METHOXYETHOXY)-2,5-CYCLOHEXADIENE-1,4-DIONE ◇ 2,5-BISMETHOXYETHOXY-3,6-BISETHYLENEIMINO-1,4-BENZOQUINONE ◇ 3,6-BIS(β-METHOXYETHOXY)-2,5-BIS(ETHYLENEIMINO)-p-BENZOQUINONE ◇ 3,6-BIS(β-METHOXYETHOXY)-2,5-BIS(ETHYLENIMINO)-p-BENZOQUINONE ◇ E 39 SOLUBLE ◇ NSC-17262

TOXICITY DATA with REFERENCE
dlt-dmg-orl 1 μmol/L MUREAV 14,250,72
ipr-mus TDLo:4 mg/kg/4W:CAR JNCIAM 36,915,66
ivn-dog LDLo:250 μg/kg CCSUBJ 2,203,65
ivn-mky LDLo:500 μg/kg CCSUBJ 2,203,65

CONSENSUS REPORTS: IARC Cancer Review: Group 3 IMEMDT 7,56,87; Animal Limited Evidence IMEMDT 9,51,75. EPA Genetic Toxicology Program.

SAFETY PROFILE: Deadly poison by intravenous route. Questionable carcinogen with experimental carcinogenic data. Mutation data reported. When heated to decomposition it emits toxic fumes of NO_x.

BDD000 CAS:495-73-8 ***HR: 3***
1,4-BENZOQUINONE-N'-BENZOYLHYDRAZONE OXIME
mf: $C_{13}H_{11}N_3O_2$ mw: 241.27

SYNS: BAYER 15080 ◇ BENCHINOX ◇ BENGUINOX ◇ BENQUINOX ◇ BENZOIC ACID(4-(HYDROXYIMINO)-2,5-CYCLOHEXADIEN-1-YLIDENE) HYDRAZIDE ◇ p-BENZOQUINONE OXIME BENZOYLHYDRAZONE ◇ CEREDON ◇ CERELINE ◇ CERENOX ◇ CHINONOXIM-BENZOYLHYDRAZON (GERMAN) ◇ CHINONOXIME-BENZOYLHYDRAZONE ◇ COBH ◇ GBH ◇ LERENOX ◇ QGH ◇ QUINONE OXIME BENZOYLHYDRAZONE ◇ TILLANTOX ◇ TSERENOX

TOXICITY DATA with REFERENCE
orl-rat LD50:100 mg/kg FMCHA2 -,C48,83
orl-mus LD50:100 mg/kg GUCHAZ 6,34,73
unk-mus LD50:100 mg/kg 30ZDA9 -,116,71

SAFETY PROFILE: Poison by ingestion and possibly other routes. When heated to decomposition it emits toxic NO_x.

BDD125 CAS:637-70-7 ***HR: 3***
BENZOQUINONE-1,4-BIS(CHLOROIMINE)(1,4-BIS(CHLORIMIDO)-2,5-CYCLOHEXADIENE)
mf: $C_6H_4Cl_2N_2$ mw: 175.02

CH=CHC(:NCl)CH=CHC:NCl

SAFETY PROFILE: Explodes on heating. When heated to decomposition it emits toxic fumes of Cl^- and NO_x.

BDD200 CAS:4377-73-5 ***HR: 3***
1,4-BENZOQUINONE DIIMINE
mf: $C_6H_6N_2$ mw: 106.13

HN:C_6H_4:NH

SYN: 1,4-DIIMIDO-2,5-CYCLOHEXADIENE

SAFETY PROFILE: Explosive decomposition on contact with concentrated acids (e.g., sulfuric or nitric acid). Upon decomposition it emits toxic fumes of NO_x.

BDD250 ***HR: 3***
p-BENZOQUINONE DIIMINE
mf: $C_6H_6N_2$ mw: 106.13

SAFETY PROFILE: It decomposes and explodes on contact with concentrated hydrochloric or sulfuric acids. Incompatible with acids. When heated to decomposition it emits toxic fumes of NO_x.

BDD500 CAS:3009-34-5 ***HR: 3***
p-BENZOQUINONE MONOIMINE
mf: C_6H_5NO mw: 107.11

SYNS: p-BENZOQUINONE IMINE ◇ p-BENZOQUINONIMINE ◇ 2,5-CYCLOHEXADIEN-1-ONE, 4-IMINO- ◇ 4-IMINO-2,5-CYCLOHEXADIEN-1-ONE ◇ PBQI ◇ p-QUINONIMINE

TOXICITY DATA with REFERENCE
ipr-ham TDLo:200 mg/kg (female 8D post):REP TXAPA9 63,264,82
ipr-ham TDLo:200 mg/kg (female 8D post):TER TXAPA9 63,264,82

SAFETY PROFILE: Experimental reproductive effects. The solid decomposes violently (nearly explosive). When heated to decomposition it emits toxic fumes of NO_x.

BDE000 CAS:37150-27-9 ***HR: 3***
BENZO-1,2,3-THIADIAZOLE-1,1-DIOXIDE
mf: $C_6H_4N_2O_2S$ mw: 168.17

$C_6H_4SO_2N{=}N$ (ring)

SAFETY PROFILE: The solid may explode spontaneously or on impact, friction or heating to 60°C. Upon decomposition it emits toxic fumes of NO_x and SO_x.

BDE250 CAS:91-33-8 ***HR: 3***
BENZOTHIAZIDE
mf: $C_{15}H_{14}ClN_3O_4S_3$ mw: 431.95

SYNS: AQUATAG ◇ 3-((BENZYLTHIO)METHYL)-6-CHLORO-1,2,4-BENZOTHIADIAZINE-7-SULFONAMIDE-1,1-DIOXIDE ◇ 3-BENZYLTHIOMETHYL-6-CHLORO-2H-1,2,4-BENZOTHIADIAZINE-7-SULFONAMIDE-1,1-DIOXIDE ◇ 3-BENZYLTHIOMETHYL-6-CHLORO-7-SULFAMOYL-1,2,4-BENZOTHIADIAZINE-1,1-DIOXIDE ◇ 3-BENZYLTHIOMETHYL-6-CHLORO-7-SULFAMYL-1,2,4-BENZOTHIADIAZINE-1,1-DIOXIDE ◇ 3-BENZYLTHIOMETHYL-6-CHLORO-7-SULFAMYL-2H-1,2,4-BENZOTHIADIAZINE-1,1-DIOXIDE ◇ 6-CHLORO-3-(((PHENYLMETHYL)THIO)METHYL)-2H-1,2,4-BENZOTHIADIAZINE-7-SULFONAMIDE DIOXIDE ◇ EDEMEX ◇ EXNA ◇ EXOSALT ◇ FOVANE ◇ FREEURIL ◇ NACLEX ◇ P 1393 ◇ PFIZER 1393 ◇ URESE

TOXICITY DATA with REFERENCE
ivn-rat LD50:422 mg/kg JPETAB 128,122,60
ivn-mus LD50:410 mg/kg JPETAB 128,122,60
ivn-dog LDLo:200 mg/kg JPETAB 128,122,60

SAFETY PROFILE: Poison by intravenous route. A diuretic and antihypertensive agent. When heated to decomposition it emits very toxic fumes of SO_x, NO_x, and Cl^-.

BDE500 CAS:95-16-9 ***HR: 3***
BENZOTHIAZOLE
mf: C_7H_5NS mw: 135.19

PROP: Liquid, odor of quinoline, sltly water-sol. D: 1.246 @ 20°/4°, bp: 228° @ 765 mm.

SYNS: BENZOSULFONAZOLE ◇ O-2857 ◇ 1-THIA-3-AZAINDENE ◇ USAF EK-4812

TOXICITY DATA with REFERENCE
orl-mus LD50:900 mg/kg DCTODJ 3,249,80
ipr-mus LD50:100 mg/kg NTIS** AD277-689
ivn-mus LD50:95 mg/kg JPETAB 105,486,52
unr-mus LD50:310 mg/kg KHFZAN 9(12),11,75

CONSENSUS REPORTS: Reported in EPA TSCA Inventory.

SAFETY PROFILE: Poison by intraperitoneal, intravenous, and possibly other routes. When heated to decomposition it emits very toxic fumes of SO_x, CN^-, and NO_x.

BDE750 CAS:120-78-5 ***HR: 3***
BENZOTHIAZOLE DISULFIDE
mf: $C_{14}H_8N_2S_4$ mw: 332.48

PROP: Cream to light yellow powder; mp: 175°, d: 1.5.

SYNS: ALTAX ◇ BENZOTHIAZOLYL DISULFIDE ◇ 2-BENZOTHIAZOLYL DISULFIDE ◇ BIS(BENZOTHIAZOLYL)DISULFIDE ◇ BIS(2-BENZOTHIAZYL) DISULFIDE ◇ DI-2-BENZOTHIAZOLYLDISULFIDE ◇ DIBENZOTHIAZYL DISULFIDE ◇ 2,2'-DIBENZOTHIAZYLDISULFIDE ◇ DIBENZOYLTHIAZYL DISULFIDE ◇ DIBENZTHIAZYL DISULFIDE ◇ 2,2'-DITHIOBIS(BENZOTHIAZOLE) ◇ DWUSIARCZEK DWUBENZOTIAZYLU (POLISH) ◇ MBTS ◇ MBTS RUBBER ACCELERATOR ◇ 2-MERCAPTOBENZOTHIAZOLEDISULFIDE ◇ 2-MERCAPTOBENZOTHIAZYLDISULFIDE ◇ ROYAL MBTS ◇ THIOFIDE ◇ USAF B-33 ◇ USAF CY-5 ◇ USAF EK-5432 ◇ VULKACIT DM ◇ VULKACIT DM/MGC

TOXICITY DATA with REFERENCE
mma-mus:lym 15 mg/L ENMUDM 5,193,83
par-rat TDLo:400 mg/kg (female 4-11D post):TER BEXBAN 93,107,82
par-rat TDLo:400 mg/kg (4-11D preg):REP BEXBAN 93,107,82

orl-mus TDLo:172 g/kg/78W-I:ETA NTIS** PB223-159
ipr-rat LD50:2600 mg/kg IPSTB3 3,93,76
orl-mus LD50:7 g/kg IPSTB3 3,93,76
ipr-mus LD50:100 mg/kg NTIS** AD277-689
ivn-mus LD50:180 mg/kg CSLNX* NX#02251

CONSENSUS REPORTS: Reported in EPA TSCA Inventory.

SAFETY PROFILE: Poison by intravenous and intraperitoneal routes. Slightly toxic by ingestion. Experimental teratogenic and reproductive effects. Questionable carcinogen with experimental tumorigenic data. Mutation data reported. When heated to decomposition it emits very toxic fumes of SO_x and NO_x. See also SULFIDES.

BDF000 CAS:149-30-4 ***HR: 3***
2-BENZOTHIAZOLETHIOL
mf: $C_7H_5NS_2$ mw: 167.25

PROP: Light yellow powder. Mp: 170°, d: 1.42 @ 25°.

SYNS: CAPTAX ◇ MBT ◇ MERCAPTOBENZOTHIAZOLE ◇ 2-MERCAPTOBENZOTHIAZOLE ◇ 2-MERKAPTOBENZOTIAZOL (POLISH) ◇ NCI-C56519 ◇ PENNAC MBT POWDER ◇ ROKON ◇ ROTAX ◇ SULFADENE ◇ USAF GY-3 ◇ USAF XR-29

TOXICITY DATA with REFERENCE
par-rat TDLo:400 mg/kg (female 4-11D post):TER BEXBAN 93,107,82
par-rat TDLo:800 mg/kg (2D male/2D pre):REP BEXBAN 93,107,82
orl-rat TDLo:195 g/kg/2Y-I:CAR NTPTR* NTP-TR-332,88
orl-mus TDLo:35 g/kg/78W-I:ETA NTIS** PB223-159
scu-mus TDLo:215 mg/kg:CAR NTIS** PB223-159
orl-rat LD50:100 mg/kg IPSTB3 3,93,76
ipr-rat LD50:300 mg/kg MEPAAX 16,35,65
orl-mus LD50:1851 mg/kg VCTDC* 10/12/82
ipr-mus LD50:100 mg/kg NTIS** AD277-689

CONSENSUS REPORTS: NTP Carcinogenesis Studies (gavage); Some Evidence rat NTPTR* NTP-TR-332,88: (gavage); Equivocal Evidence: mouse NTPTR* NTP-TR-332,88. Reported in EPA TSCA Inventory.

SAFETY PROFILE: Suspected carcinogen with experimental carcinogenic tumorigenic data. Poison by ingestion and intraperitoneal routes. Experimental teratogenic and reproductive effects. Incompatible with oxidizers. When heated to decomposition or on contact with acids or acid fumes it emits toxic SO_x and NO_x. See also MERCAPTANS.

BDF250 CAS:95-30-7 ***HR: 2***
2-BENZOTHIAZOLYL-N,N-DIETHYLTHIOCARBAMYL SULFIDE
mf: $C_{12}H_{14}N_2S_3$ mw: 282.46

SYNS: 2-(N,N-DIETHYLDITHIOCARBAMYL)BENZOATHIAZOLE ◇ ETHYLAC

TOXICITY DATA with REFERENCE
orl-rbt LD50:2700 mg/kg RCTEA4 44(2),513,71

CONSENSUS REPORTS: Reported in EPA TSCA Inventory.

SAFETY PROFILE: Moderately toxic by ingestion. See also CARBAMATES. When heated to decomposition it emits very toxic fumes of NO_x and SO_x.

BDF750 CAS:95-32-9 ***HR: 2***
2-BENZOTHIAZOLYL MORPHOLINODISULFIDE
mf: $C_{11}H_{12}N_2OS_3$ mw: 284.43

SYNS: MORFAX ◇ MORPHOLINO-2-BENZOTHIAZOLYL DISULFIDE ◇ 2-(MORPHOLINODITHIO)BENZOTHIAZOLE ◇ N-MORPHOLINYL-2-BENZOTHIAZOLYL DISULFIDE ◇ 4-MORPHOLINYL-2-BENZOTHIAZYL DISULFIDE ◇ N-OXYDIETHYL-2-BENZOTHIAZOLSULFENAMID (CZECH) ◇ SULFENAX MOB (CZECH) ◇ VULCUREN 2

TOXICITY DATA with REFERENCE
eye-rbt 100 mg/24H MOD 28ZPAK -,203,72
orl-rat LD50:11500 mg/kg 28ZPAK -,203,72

CONSENSUS REPORTS: Reported in EPA TSCA Inventory.

SAFETY PROFILE: Mildly toxic by ingestion. An eye irritant. When heated to decomposition it emits very toxic fumes of NO_x and SO_x. See also SULFIDES.

BDG000 CAS:102-77-2 ***HR: 3***
2-BENZOTHIAZOLYL-N-MORPHOLINOSULFIDE
mf: $C_{11}H_{12}N_2OS_2$ mw: 252.37

SYNS: AMAX ◇ 2-BENZOTHIAZOLYLSULFENYL MORPHOLINE ◇ 4-(2-BENZOTHIAZOLYLTHIO)MORPHOLINE ◇ 2-(MORPHOLINOTHIO)BENZOTHIAZOLE ◇ MORPHOLINYLMERCAPTOBENZOTHIAZOLE ◇ 2-(4-MORPHOLINYLTHIO)BENZOTHIAZOLE ◇ N-(OXYDIETHYLENE)BENZOTHIAZOLE-2-SULFENAMIDE ◇ SANTOCURE MOR ◇ SULFENAMIDE M ◇ USAF CY-7 ◇ VULCAFOR BSM

TOXICITY DATA with REFERENCE
dnr-esc 10 μg/tube ENMUDM 5,193,83
mma-mus:lym 15 mg/L ENMUDM 5,193,83
otr-mus:emb 200 μg/L ENMUDM 5,193,83
msc-mus:lym 10 mg/L ENMUDM 5,193,83
par-rat TDLo:400 mg/kg (female 4-11D post):TER BEXBAN 93,107,82
scu-mus TDLo:464 mg/kg:NEO NTIS** PB223-159
orl-rat LD50:1980 mg/kg IPSTB3 3,93,76
orl-mus LD50:1870 mg/kg 20ZJAG -,64,68 NTIS** AD277-689

CONSENSUS REPORTS: Reported in EPA TSCA Inventory.

SAFETY PROFILE: Poison by intraperitoneal route.

Moderately toxic by ingestion. Questionable carcinogen with experimental neoplastigenic data. Experimental teratogenic effects. Mutation data reported. See also MERCAPTANS and SULFIDES. When heated to decomposition it emits very toxic fumes of NO_x and SO_x.

BDG250 CAS:1079-33-0 ***HR: 3***
BENZO(b)THIEN-4-YL METHYLCARBAMATE
mf: $C_{10}H_9NO_2S$ mw: 207.26

SYNS: 4-BENZOTHIENYL METHYLCARBAMATE ◇ BENZO(b)-THIOPHENE-4-OL METHYLCARBAMATE ◇ ENT 27,041 ◇ MCA-600 ◇ MOBAM ◇ MOBAM PHENOL ◇ MOBIL MC-A-600 ◇ MOS-708 ◇ OMS-708

TOXICITY DATA with REFERENCE
orl-rat LD50:70 mg/kg TXAPA9 11,546,67
ipr-rat LD50:40800 μg/kg BWHOA6 44(1-3),241,71
ivn-rat LD50:24800 μg/kg BWHOA6 44(1-3),241,71
orl-gpg LDLo:50 mg/kg JEENAI 60,733,67
scu-gpg LDLo:25 mg/kg JEENAI 60,733,67
orl-pgn LD50:52600 μg/kg ASTTA8 (680),157,79
orl-bwd LD50:17800 μg/kg ASTTA8 (680),157,79

SAFETY PROFILE: Poison by ingestion, intraperitoneal, intravenous, subcutaneous, and possibly other routes. See also CARBAMATES. When heated to decomposition it emits very toxic fumes of NO_x and SO_x.

BDG325 CAS:724-34-5 ***HR: 3***
6-BENZOTHIOPURINE
mf: $C_{12}H_{10}N_4S$ mw: 242.32

SYNS: 6-BENZYLMERCAPTOPURINE ◇ 6-BENZYL-MP ◇ 6-(BENZYLTHIO)PURINE ◇ NSC 29421 ◇ 6-((PHENYLMETHYL)THIO)-1H-PURINE (9CI)

TOXICITY DATA with REFERENCE
orl-rat TDLo:100 mg/kg (7D preg):REP JRPFA4 4,291,62
ipr-mus LD50:501 mg/kg NCISP* JAN86
par-mus LD50:180 mg/kg JPMSAE 71,618,82

SAFETY PROFILE: Poison by parenteral route. Moderately toxic by other routes. Experimental reproductive effects. When heated to decomposition it emits toxic fumes of NO_x and SO_x.

BDH000 CAS:90-16-4 ***HR: 3***
1,2,3-BENZOTRIAZIN-4(1H)-ONE
mf: $C_7H_5N_3O$ mw: 147.15

SYNS: BENZAZIMIDE ◇ BENZAZIMIDONE ◇ BENZOKETCTRIAZINE ◇ 3H-1,2,3-BENZOTRIAZIN-4-ONE ◇ 4-KETOBENZOTRIAZINE ◇ USAF MA-2

TOXICITY DATA with REFERENCE
orl-mus TDLo:3480 mg/kg (8D male/21D pre):REP MPHEAE 15,7,66
ipr-mus LD50:50 mg/kg NTIS** AD277-689

CONSENSUS REPORTS: Reported in EPA TSCA Inventory.

SAFETY PROFILE: Poison by intraperitoneal route. Experimental reproductive effects. When heated to decomposition it emits toxic fumes of NO_x.

BDH250 CAS:95-14-7 ***HR: 3***
1H-BENZOTRIAZOLE
mf: $C_6H_5N_3$ mw: 119.14

$C_6H_4NHN{=}N$ (ring closure between C_6H_4 and terminal N)

PROP: Needle-like crystals. Mp: 100°, bp: 204° @ 15 mm.

SYNS: 1,2,-AMINOZOPHENYLENE ◇ AZIMIDOBENZENE ◇ AZIMINOBENZENE ◇ BENZENE AZIMIDE ◇ BENZISOTRIAZOLE ◇ 1,2,3-BENZOTRIAZOLE ◇ COBRATEC #99 ◇ 2,3-DIAZAINDOLE ◇ NCI-C03521 ◇ NSC-3058 ◇ 1,2,3-TRIAZAINDENE ◇ U-6233

TOXICITY DATA with REFERENCE
mma-sat 100 μg/plate IARCCD 27,283,80
mmo-esc 333 μg/plate ENMUDM 7(Suppl 5),1,85
mma-esc 33300 ng/plate ENMUDM 7(Suppl 5),1,85
otr-rat:emb 94 μg/plate JJATDK 1,190,81
orl-rat TDLo:220 g/kg/78W-I:ETA NCITR* NCI-CG-TR-88,78
orl-rat LDLo:500 mg/kg NCNSA6 5,22,53
ipr-mus LD50:1000 mg/kg CNCRA6 30,9,63
ivn-mus LD50:238 mg/kg JPETAB 105,486,52

CONSENSUS REPORTS: NCI Carcinogenesis Bioassay (feed); Inadequate Studies: mouse, rat NCITR* NCI-CG-TR-88,78. Reported in EPA TSCA Inventory.

SAFETY PROFILE: Poison by intravenous route. Moderately toxic by ingestion and intraperitoneal routes. Questionable carcinogen with experimental tumorigenic data. Mutation data reported. May detonate at 220°C or during vacuum distillation. When heated to decomposition it emits toxic fumes of NO_x.

BDH500 CAS:98-08-8 ***HR: 3***
BENZOTRIFLUORIDE
DOT: UN 2338
mf: $C_7H_5F_3$ mw: 146.12

PROP: Water-white liquid, aromatic odor. Mp: −29.1°, bp: 104°, flash p: 54°F (CC), d: 1.197 @ 15.5°/15.5°, vap d: 5.04, vap press: 11 mm @ 0°.

SYNS: BENZENYL FLUORIDE ◇ BENZYLIDYNE FLUORIDE ◇ PHENYLFLUOROFORM ◇ (TRIFLUOROMETHYL)BENZENE ◇ α,α,α-TRIFLUOROTOLUENE ◇ ω-TRIFLUOROTOLUENE ◇ USAF MA-16

TOXICITY DATA with REFERENCE
orl-rat LD50:15000 mg/kg TPKVAL 10,131,68
ihl-rat LC50:70810 mg/m³/4H 85GMAT -,25,82

orl-mus LD50:10000 mg/kg TPKVAL 10,131,68
ihl-mus LC50:92240 mg/m^3/2H 85GMAT -,25,82
ipr-mus LD50:100 mg/kg NTIS** AD277-689
scu-frg LDLo:870 mg/kg AEPPAE 130,250,28

CONSENSUS REPORTS: Reported in EPA TSCA Inventory.

DOT Classification: Flammable Liquid; Label: Flammable Liquid.

SAFETY PROFILE: Poison by intraperitoneal route. Moderately toxic by subcutaneous route. See also FLUORIDES. Dangerous fire hazard. To fight fire, use water, foam, CO_2, spray mist, dry chemical. When heated to decomposition it emits toxic fumes of F^-. Incompatible with oxidizing materials.

BDH750 CAS:215-58-7 ***HR: 3***
BENZO(b)TRIPHENYLENE
mf: $C_{22}H_{14}$ mw: 278.36

PROP: Clear plates or leaflets. Mp: 267°.

SYNS: DB(a,c)A ◇ DIBENZ(a,c)ANTHRACENE ◇ 1,2:3,4-DIBENZANTHRACENE ◇ DIBENZO(a,c)ANTHRACENE ◇ 1,2:3,4-DIBENZOANTHRACENE

TOXICITY DATA with REFERENCE
mma-sat 10 μg/plate PNASA6 72,5135,75
dnd-hmn:emb 360 nmol/L CBINA8 22,257,78
dns-hmn:hla 100 nmol/L CNREA8 38,2625,78
msc-ham:lng 1 mg/L PNASA6 73,188,76
skn-mus TDLo:440 mg/kg/65W-I:ETA JNCIAM 44,641,70

CONSENSUS REPORTS: EPA Genetic Toxicology Program. IARC Cancer Review: Group 3 IMEMDT 7,56,87; Animal Limited Evidence IMEMDT 32,289,83

SAFETY PROFILE: Questionable carcinogen with experimental tumorigenic data. Human mutation data reported. When heated to decomposition it emits acrid smoke and irritating fumes.

BDI000 CAS:86-13-5 ***HR: 3***
BENZOTROPINE
mf: $C_{21}H_{25}NO$ mw: 307.47

SYNS: BENZTROPINE ◇ 3-α-(DIPHENYLMETHOXY)-1-α-H,5-α-H-TROPANE

TOXICITY DATA with REFERENCE
scu-mus LD50:60 mg/kg JMPCAS 4,215,61
ivn-mus LD50:25 mg/kg JMPCAS 4,215,61

SAFETY PROFILE: Poison by subcutaneous and intravenous routes. When heated to decomposition it emits toxic fumes of NO_x.

BDI500 CAS:273-53-0 ***HR: 3***
BENZOXAZOLE
mf: C_7H_5NO mw: 119.13

SYNS: 1-OXA-3-AZAINDENE ◇ USAF EK-5017

TOXICITY DATA with REFERENCE
orl-mus LD50:750 mg/kg MDCHAG 4(1),336,64
ipr-mus LD50:250 mg/kg MDCHAG 4(1),336,64
ivn-mus LD50:179 mg/kg JPETAB 105,486,52

CONSENSUS REPORTS: Reported in EPA TSCA Inventory.

SAFETY PROFILE: Poison by intraperitoneal and intravenous routes. Moderately toxic by ingestion. When heated to decomposition it emits toxic fumes such as NO_x.

BDJ000 CAS:59-49-4 ***HR: 3***
2-BENZOXAZOLINONE
mf: $C_7H_5NO_2$ mw: 135.13

SYNS: 2-BENZOXAXOLOL ◇ BENZOXAZOLINONE ◇ BENZOXAZOLONE ◇ 2(3H)-BENZOXAZOLONE ◇ 2-HYDROXYBENZOXAZOLE ◇ USAF EK-5429

TOXICITY DATA with REFERENCE
orl-rat LD50:700 mg/kg MDCHAG 4(1),308,64
orl-mus LD50:830 mg/kg AIPTAK 94,211,53
ipr-mus LD50:400 mg/kg NTIS** AD277-689

CONSENSUS REPORTS: Reported in EPA TSCA Inventory.

SAFETY PROFILE: Poison by intraperitoneal route. Moderately toxic by ingestion. When heated to decomposition it emits toxic fumes of NO_x.

BDJ250 CAS:2310-17-0 ***HR: 3***
S-((3-BENZOXAZOLINYL-6-CHLORO-2-OXO)METHYL) O,O-DIETHYLPHOSPHORODITHIOATE
mf: $C_{12}H_{15}ClNO_4PS_2$ mw: 367.82

SYNS: AZOFENE ◇ BENZOPHOSPHATE ◇ BENZPHOS ◇ CHIPMAN 11974 ◇ S-(6-CHLORO-3-(MERCAPTOMETHYL)-2-BENZOXAZOLINONE)-O,O-DIETHYL PHOSPHORODITHIOATE ◇ 3-(6-CHLORO-2-OXOBENZOXAZOLIN-3-YL)METHYL-O,O-DIETHYLPHOSPHOROTHIOLOTHIONATE ◇ O,O-DIAETHYL-S-(6-CHLOR-2-OXO-BEN(b)-1,3-OXALIN-3-YL)-METHYL-DITHIOPHOSPHAT (GERMAN) ◇ O,O-DIETHYL-S-((6-CHLOOR-2-OXO-BENZOXAZOLIN-3-YL)-METHYL)-DITHIO FOSFAAT (DUTCH) ◇ O,O-DIETHYL-S-(6-CHLOROBENZOXAZOLINYL-3-METHYL)DITHIOPHOSPHATE ◇ O,O-DIETHYL-S-((6-CHLORO-2-OXOBENZOXAZOLIN-3-YL)-METHYL) PHOSPHORODITHIOATE ◇ O,O-DIETHYL-S-(6-CHLORO-2-OXO-BENZOXAZOLIN-3-YL)METHYL-PHOSPHOROTHIOLOTHIONATE ◇ 3-DIETHYLDITHIOPHOSPHORYLMETHYL-6-CHLOROBENZOXAZOLONE-2 ◇ O,O-DIETIL-S-((6-CLORO-2-OXO-BENZOSSAZOLIN-3-IL)-METIL)-DITIOFOSFATO (ITALIAN) ◇ ENT 27,163 ◇ FOZALON ◇ NIA-9241 ◇ NIAGARA 9241 ◇ NPH-1091 ◇ PHASOLON ◇ PHOSALON ◇ PHOSALONE ◇ PHOZALON ◇ RHODIA RP 11974 ◇ RUBITOX ◇ ZOLON ◇ ZOLONE ◇ ZOLONE PM ◇ ZOOLON

TOXICITY DATA with REFERENCE
orl-rat LD50:85 mg/kg KSKZAN 16(2),59,78
skn-rat LD50:390 mg/kg WRPCA2 9,119,70
unr-rat LD50:135 mg/kg 30ZDA9 -,371,71
orl-mus LD50:73 mg/kg GTPZAB 19(9),55,75
skn-rbt LD50:1000 mg/kg 85DPAN -,-,71/76
orl-gpg LD50:150 mg/kg GUCHAZ 6,408,73
orl-ckn LD50:661 mg/kg VETNAL 54(11),75,78

CONSENSUS REPORTS: EPA: Farm Worker Field Reentry FEREAC 39,16888,74.

SAFETY PROFILE: Poison by ingestion, skin contact, and possibly other routes. A cholinesterase inhibitor. See also PARATHION. When heated to decomposition it emits very toxic fumes of Cl^-, NO_x, PO_x, and SO_x.

BDJ500 CAS:790-60-3 ***HR: D***
BENZ(a)OXIRENO(c)ANTHRACENE
mf: $C_{18}H_{10}O$ mw: 242.28

SYNS: BENZ(a)ANTHRACENE-5,6-EPOXIDE ◇ BENZ(a)ANTHRACENE-5,6-OXIDE ◇ BENZ(3,4)ANTHRA(1,2-6)OXIRENE

TOXICITY DATA with REFERENCE
cyt-hmn:fbr 1500 ng/L EXMDA4 350,58,75
oms-mus:oth 1 mg/L CBINA8 4,389,71,72
dnd-ham:lng 1 mg/L CBINA8 4,389,71/72
otr-mus:oth 500 μg/L PNASA6 68,1098,71

SAFETY PROFILE: Human mutation data reported. When heated to decomposition it emits acrid smoke and irritating fumes.

BDJ750 CAS:92-16-0 ***HR: 2***
BENZOYLACET-o-ANISIDIDE
mf: $C_{16}H_{15}NO_3$ mw: 269.32

TOXICITY DATA with REFERENCE
orl-rat LDLo:1600 mg/kg KODAK* -,-,71
ipr-rat LDLo:800 mg/kg KODAK* -,-,71

CONSENSUS REPORTS: Reported in EPA TSCA Inventory.

SAFETY PROFILE: Moderately toxic by ingestion and intraperitoneal routes. When heated to decomposition it emits toxic fumes of NO_x.

BDK750 CAS:117-05-5 ***HR: 2***
5-BENZOYLAMINO-1-CHLOROANTHRAQUINONE
mf: $C_{21}H_{12}ClNO_3$ mw: 361.79

SYNS: 1-BENZAMIDO-5-CHLORO-ANTHRAQUINONE ◇ 1-CHLOR-5-BENZOYLAMINOANTHRACHINON (CZECH) ◇ 1-CHLORO-5-BENZAMIDO-ANTHRAQUINONE ◇ N-(5-CHLORO-9,10-DIHYDRO-9,10-DIOXO-1-ANTHRACENYL)-BENZAMIDE ◇ 1-X-5-BAA (RUSSIAN)

TOXICITY DATA with REFERENCE
eye-rbt 500 mg/24H MLD 28ZPAK -,89,72
mmo-sat 50 μg/plate MUREAV 40,203,76
mma-sat 50 μg/plate MUREAV 40,203,76
ipr-rat LD50:3000 mg/kg GTPZAB 21(12),27,77

CONSENSUS REPORTS: Reported in EPA TSCA Inventory.

SAFETY PROFILE: Moderately toxic by intraperitoneal route. An eye irritant. Mutation data reported. When heated to decomposition it emits very toxic fumes of NO_x and Cl^-.

BDL750 CAS:582-61-6 ***HR: 3***
BENZOYL AZIDE
mf: $C_7H_5N_3O$ mw: 147.14

SYN: BENZAZIDE, BENZOIC ACID AZIDE

DOT Classification: Forbidden.

SAFETY PROFILE: May explode when heated above 120°C. See also AZIDES.

BDM500 CAS:98-88-4 ***HR: 3***
BENZOYL CHLORIDE
DOT: UN 1736
mf: C_7H_5ClO mw: 140.57

PROP: Colorless, fuming, pungent liquid; decomposes in water. Mp: −0.5°, bp: 197°, flash p: 162°F (CC), d: 1.2187 @ 15°/15°, vap press: 1 mm @ 32.1°, vap d: 4.88.

SYNS: BENZENECARBONYL CHLORIDE ◇ BENZOIC ACID, CHLORIDE ◇ BENZOYL CHLORIDE (DOT) ◇ α-CHLOROBENZALDEHYDE

TOXICITY DATA with REFERENCE
mmo-sat 1 μmol/plate MUREAV 58,11,78
skn-mus TDLo:9200 mg/kg/50W-I:ETA GANNA2 72,655,81
ihl-hmn TCLo:2 ppm/1M:NOSE,PUL TGNCDL 2,31,61
orl-rat LDLo:1900 mg/kg 85GMAT-,25,82
ihl-rat LC50:1870 mg/m^3/2H 85GMAT-,25,82

CONSENSUS REPORTS: IARC Cancer Review: Group 3 IMEMDT 7,56,87, Human Inadequate Evidence IMEMDT 29,83,82; Animal Inadequate Evidence IMEMDT 29,83,82. Community Right-To-Know List. Reported in EPA TSCA Inventory. EPA Genetic Toxicology Program.

DOT Classification: Corrosive Material; Label: Corrosive.

SAFETY PROFILE: Questionable carcinogen with experimental tumorigenic data by skin contact. Human systemic effects by inhalation: unspecified effects on olfaction, and respiratory systems. Corrosive effects on the skin, eyes, and mucous membranes by inhalation. Flammable when exposed to heat or flame. Will react with water or steam to produce heat and toxic and corrosive fumes. Violent or explosive reaction with dimethyl sulfoxide, and aluminum chloride + naphthalene. To fight fire, use alcohol foam, CO_2, dry chemical. Incompatible with dimethyl sulfoxide, (NaN_3 + KOH), water, steam, and oxidizers. When heated to decomposition it emits toxic fumes of Cl^-. See also CHLORIDES and ALDEHYDES.

BDN125 CAS:62303-19-9 ***HR: 3***
2-BENZOYLHYDRAZONO-1,3-DITHIOLANE
mf: $C_{10}H_{10}N_2OS_2$ mw: 238.34

SYNS: BHD ◇ YU 7802

TOXICITY DATA with REFERENCE
sce-hmn:lym 200 μg/L CIYPDA 15,318,84
orl-rat LD50:72800 μg/kg CIYPDA 15,318,84
orl-mus LD50:111 mg/kg CIYPDA 15,318,84
orl-gpg LD50:383 mg/kg CIYPDA 15,318,84

SAFETY PROFILE: Poison by ingestion. Human mutation data reported. When heated to decomposition it emits toxic fumes of SO_x and NO_x.

BDN500 CAS:6786-32-9 ***HR: 3***
BENZOYL NITRATE
mf: $C_7H_5NO_4$ mw: 167.12

SAFETY PROFILE: An unstable heat- and light-sensitive explosive. Decomposes explosively on contact with trace amounts of water. Upon decomposition it emits toxic fumes of NO_x. See also NITRATES.

BDO199 CAS:55398-24-8 ***HR: 3***
N-BENZOYLOXY-N-ETHYL-4-AMINOAZOBENZENE
mf: $C_{21}H_{19}N_3O_2$ mw: 345.43

TOXICITY DATA with REFERENCE
scu-rat TDLo:262 mg/kg/8W-I:NEO CNREA8 35,880,75

SAFETY PROFILE: Questionable carcinogen with experimental neoplastigenic data. See also AZIDES. When heated to decomposition it emits toxic fumes of NO_x.

BDO500 CAS:55398-26-0 ***HR: 3***
N-BENZOYLOXY-4'-ETHYL-N-METHYL-4-AMINOAZOBENZENE
mf: $C_{22}H_{21}N_3O_2$ mw: 359.46

TOXICITY DATA with REFERENCE
scu-rat TDLo:272 mg/kg/8W-I:NEO CNREA8 35,880,75

SAFETY PROFILE: Questionable carcinogen with experimental neoplastigenic data. When heated to decomposition it emits toxic fumes such as NO_x.

BDP000 CAS:6098-46-0 ***HR: 3***
N-BENZOYLOXY-N-METHYL-4-AMINOAZOBENZENE
mf: $C_{20}H_{17}N_3O_2$ mw: 331.40

SYNS: o-BENZOYL-N-METHYL-N-(p-(PHENYLAZO)PHENYL)HYDROXYLAMINE ◇ N-(BENZOYLOXY)-N-METHYL-4-(PHENYLAZO)-BENZENAMINE

TOXICITY DATA with REFERENCE
mmo-sat 100 nmol/plate CALEDQ 1,91,75
dnd-esc 60 mmol/L CNREA8 40,2493,80
dnd-rat:lvr 50 mmol/L CBINA8 31,1,80
scu-rat TD:375 mg/kg/12W-I:NEO CNREA8 27,1600,67

CONSENSUS REPORTS: EPA Genetic Toxicology Program.

SAFETY PROFILE: Questionable carcinogen with experimental neoplastigenic data. Mutation data reported. See also AZIDES. When heated to decomposition it emits toxic fumes of NO_x.

BDP500 CAS:42978-42-7 ***HR: 3***
6-BENZOYLOXYMETHYLBENZO(a)PYRENE
mf: $C_{28}H_{18}O_2$ mw: 386.46

TOXICITY DATA with REFERENCE
scu-rat TDLo:100 mg/kg/40D-I:CAR JMCMAR 16,714,73
scu-rat TD:2898 μg/kg/60D-I:NEO CBINA8 29,159,80

SAFETY PROFILE: Questionable carcinogen with experimental carcinogenic and neoplastigenic data. When heated to decomposition it emits acrid smoke and irritating fumes.

BDP899 CAS:67371-65-7 ***HR: D***
N-BENZOYLOXY-3'-METHYL-4-METHYLAMINOAZOBENZENE
mf: $C_{21}H_{19}N_3O_2$ mw: 345.40

TOXICITY DATA with REFERENCE
otr-rat:lvr 20 μmol/L JJIND8 76,95,86
dns-rat:lvr 1 μmol/L JJIND8 76,95,86
cyt-rat:lvr 50 μmol/L JJIND8 76,95,86

SAFETY PROFILE: Mutation data reported. When heated to decomposition it emits toxic fumes of NO_x.

BDQ000 CAS:55398-25-9 ***HR: 3***
N-BENZOYLOXY-4'-METHYL-N-METHYL-4-AMINOAZOBENZENE
mf: $C_{21}H_{19}N_3O_2$ mw: 345.43

TOXICITY DATA with REFERENCE
scu-rat TDLo:262 mg/kg/8W-I:NEO CNREA8 35,880,75

SAFETY PROFILE: Questionable carcinogen with experimental neoplastigenic data. See also AZIDES. When heated to decomposition it emits toxic fumes of NO_x.

BDQ250 CAS:31012-29-0 ***HR: 3***
7-BENZOYLOXYMETHYL-12-METHYLBENZ(a)ANTHRACENE
mf: $C_{27}H_{20}O_2$ mw: 376.47

SYN: 12-METHYLBENZ(a)ANTHRACENE-7-METHANOLBENZOATE (ESTER)

TOXICITY DATA with REFERENCE
scu-rat TDLo:20 mg/kg/39D-I:NEO CNREA8 31,1951,71

SAFETY PROFILE: Questionable carcinogen with experimental neoplastigenic data. See also ESTERS. When heated to decomposition it emits very acrid smoke and irritating fumes.

BDR750 CAS:4342-36-3 ***HR: 3***
BENZOYLOXYTRIBUTYLSTANNANE
mf: $C_{19}H_{32}O_2Sn$ mw: 411.20

SYNS: TRIBUTYLTIN BENZOATE ◇ TRI-N-BUTYL-ZINN BENZOATE (GERMAN)

TOXICITY DATA with REFERENCE
orl-rat LD50:132 mg/kg ARZNAD 19,934,69
scu-rat LD50:505 mg/kg TRIPA7 -,1,73
orl-mus LD50:108 mg/kg ATXKA8 23,283,68
ivn-mus LD50:178 mg/kg CSLNX* NX#00090

CONSENSUS REPORTS: Reported in EPA TSCA Inventory.

OSHA PEL: TWA 0.1 mg(Sn)/m^3 (skin)
ACGIH TLV: TWA 0.1 mg(Sn)/m^3 (skin) (Proposed: TWA 0.1 mg(Sn)/m^3; STEL 0.2 mg(Sn)/m^3 (skin))
NIOSH REL: (Organotin Compounds) TWA 0.1 mg(Sn)/m^3

SAFETY PROFILE: Poison by ingestion and intravenous routes. Moderately toxic by subcutaneous route. See also TIN COMPOUNDS. When heated to decomposition it emits acrid smoke and irritating fumes.

BDS000 CAS:94-36-0 ***HR: 3***
BENZOYL PEROXIDE
DOT: UN 2085/UN 2086/UN 2087/UN 2088/UN 2089/UN 2090
mf: $C_{14}H_{10}O_4$ mw: 242.24

PROP: White, granular, tasteless, odorless powder. Mp: 103-106° (decomp), bp: decomposes explosively, autoign temp: 176°F. Sol in benzene, acetone, chloroform; sltly sol in alc; insol in water.

SYNS: ACETOXYL ◇ ACNEGEL ◇ AZTEC BPO ◇ BENOXYL ◇ BENZAC ◇ BENZAKNEW ◇ BENZOIC ACID, PEROXIDE ◇ BENZOPEROXIDE ◇ BENZOYL ◇ BENZOYLPEROXID (GERMAN) ◇ BENZOYLPEROXYDE (DUTCH) ◇ BENZOYL SUPEROXIDE ◇ BZF-60 ◇ CADET ◇ CADOX ◇ CLEARASIL BENZOYL PEROXIDE LOTION ◇ CLEARASIL BP ACNE TREATMENT ◇ CUTICURA ACNE CREAM ◇ DEBROXIDE ◇ DIBENZOYLPEROXID (GERMAN) ◇ DIBENZOYL PEROXIDE (MAK) ◇ DIBENZOYLPEROXYDE (DUTCH) ◇ DIPHENYLGLYOXAL PEROXIDE ◇ DRY AND CLEAR ◇ EPI-CLEAR ◇ FOSTEX ◇ GAROX ◇ INCIDOL ◇ LOROXIDE ◇ LUCIDOL ◇ LUPERCO ◇ LUPEROX FL ◇ NAYPER B and BO ◇ NOROX BZP-250 ◇ NOVADELOX ◇ OXY-5 ◇ OXY-10 ◇ OXYLITE ◇ OXY WASH ◇ PANOXYL ◇ PEROSSIDO di BENZOILE(ITALIAN) ◇ PEROXYDE de BENZOYLE (FRENCH) ◇ PERSADOX ◇ QUINOLOR COMPOUND ◇ SULFOXYL ◇ SUPEROX ◇ THERADERM ◇ TOPEX ◇ VANOXIDE ◇ XERAC

TOXICITY DATA with REFERENCE
eye-rbt 500 mg/24H MLD 28ZPAK -,52,72
dnd-hmn:oth 100 μmol/L CNREA8 45,2522,85
dni-ham:oth 56 μmol/L CNREA8 45,2522,85
dns-rat:lvr 100 pmol/L CRNGDP 5,1547,84
skn-mus TDLo:24 g/kg/30W-I:ETA SCIEAS 213,1023,81
orl-rat LD50:7710 mg/kg 28ZPAK -,52,72
ipr-mus LDLo:250 mg/kg YKYUA6 31,855,80

CONSENSUS REPORTS: IARC Cancer Review: Group 3 IMEMDT 7,56,87, Animal Inadequate Evidence IMEMDT 36,267,85; Human Inadequate Evidence IMEMDT 36,267,85. Reported in EPA TSCA Inventory. EPA Genetic Toxicology Program. Community Right-To-Know List.

OSHA PEL: TWA 5 mg/m^3
ACGIH TLV: TWA 5 mg/m^3
DFG MAK: 5 mg/m^3
NIOSH REL: (Benzoyl Peroxide) TWA 5 mg/m^3
DOT Classification: Organic Peroxide; Label: Organic Peroxide.

SAFETY PROFILE: Poison by ingestion and intraperitoneal routes. Can cause dermatitis, asthmatic effects, testicular atrophy, and vasodilation. An allergen and eye irritant. Human mutation data reported. Questionable carcinogen with experimental tumorigenic data. Moderate fire hazard by spontaneous chemical reaction in contact with reducing agents. It ignites readily and burns rapidly. A powerful oxidizer. Dangerous explosion hazard; may explode spontaneously, when heated to above melting point, or when overheated under confinement. It is moderately sensitive to heat, shock, friction, or contact with combustible materials. Explosive decomposition above the mp (103°) forms flammable products.

Explosive or violent reaction on contact with N,N-dimethylaniline, aniline, dimethyl sulfide, lithium tetrahydroaluminate, and N-bromosuccinimide + 4-toluic acid. Mixture with carbon tetrachloride + ethylene explodes at elevated temperatures and pressures. Reacts violently in contact with various organic or inorganic acids, alcohols, amines, metallic naphthenates, as well as with polymerization accelerators, i.e., dimethylaniline, and (CCl_4 + C_2H_4). Violent reaction with charcoal when heated above 50°. Decomposition produces dense white smoke of benzoic acid, phenyl benzoate, terphenyls, biphenyls, benzene and carbon dioxide. Vigorous reaction leading to ignition with methylmethacrylate, and vinyl acetate + ethyl acetate. To fight fire, use water spray, foam. All precautions must be taken to guard against fire and explosion hazards. Keep in a cool place, out of the direct rays of the sun, away from sparks, open flames, and other sources of heat, avoid shock, rough handling, friction from grinding, etc. Isolated storage is required; keep away from possible contact with acids, alcohols, ethers, or other reducing agents or polymerization catalysts such as dimethylaniline. Complete instructions on storage and handling available from manufacturer. See also PEROXIDES.

BDS250 ***HR: 2***
BENZOYL PEROXIDE, WET

PROP: A paste or wetted granular material containing at least 30% water. Autoign temp 176°F.

SAFETY PROFILE: Moderate fire hazard by chemical reaction with reducing agents; a powerful oxidizer. Mixed with a large surplus of water (i.e., 30%), this material is relatively safe. It is most dangerous when it contains very little water (1% or less). To fight fire, use water, foam or spray. Care must be taken to prevent drying out of wet material. See BENZOYL PEROXIDE.

BDS300 CAS:744-80-9 ***HR: 2***
BENZOYLPHENOBARBITAL
mf: $C_{19}H_{16}N_2O_4$ mw: 336.37

SYNS: BENZOBARBITAL ◇ BENZONAL ◇ 1-BENZOYL-5-ETHYL-5-PHENYLBARBITURIC ACID ◇ 1-BENZOYL-5-ETHYL-5-PHENYL-2,4,6-TRIOXOHEXAHYDROPYRIMIDINE ◇ BENZOYLLUMINAL ◇ BENZOYLUMINAL ◇ 2,4,6(1H,3H,5H)-PYRIMIDINETRIONE, 1-BENZOYL-5-ETHYL-5-PHENYL-(9CI)

TOXICITY DATA with REFERENCE
dlt-mus-orl 20 mg/kg CYGEDX 10(2),1,76
orl-mus TDLo:20 mg/kg (male 1D pre):REP CYGEDX 10(2),1,76
orl-mus LD50:982 mg/kg APSXAS 6,177,69

SAFETY PROFILE: Moderately toxic by ingestion. Experimental reproductive effects. Mutation data reported. When heated to decomposition it emits toxic fumes of NO_x.

BDS500 CAS:23107-96-2 ***HR: 3***
o-BENZOYL PHENYLACETIC ACID
mf: $C_{15}H_{12}O_3$ mw: 240.27

SYNS: ACIDE BENZOYL-2-PHENYLACETIQUE (FRENCH) ◇ 2-BENZOYLPHENYLACETIC ACID

TOXICITY DATA with REFERENCE
orl-mus LD50:2700 mg/kg EJMCA5 9,397,74
ipr-mus LDLo:300 mg/kg EJMCA5 11,7,76

SAFETY PROFILE: Poison by intraperitoneal route. Moderately toxic by ingestion. When heated to decomposition it emits acrid smoke and irritating fumes.

BDS750 CAS:38940-50-0 ***HR: 3***
(4-BENZOYL-4-PHENYLBUTYL)TRIETHYLAMMONIUM IODIDE
mf: $C_{23}H_{32}NO•I$ mw: 465.46

SYN: DEOXY(α-(3-TRIETHYLAMMONIO)PROPYL)BENZOIN, IODIDE

TOXICITY DATA with REFERENCE
orl-mus LD50:800 mg/kg CHTPBA 7,287,72
ivn-mus LD50:6500 μg/kg CHTPBA 7,287,72

SAFETY PROFILE: Poison by intravenous route. Moderately toxic by ingestion. See also IODIDES. When heated to decomposition it emits very toxic fumes of I^-, NH_3, and NO_x.

BDT000 ***HR: 3***
1,1-BENZOYL PHENYL DIAZOMETHANE
mf: $C_{14}H_{10}N_2O$ mw: 222.25

SAFETY PROFILE: It may explode above 40°. When heated to decomposition it emits toxic fumes of NO_x.

BDT500 CAS:24026-35-5 ***HR: 3***
2-(3-BENZOYLPHENYL)-N,N-DIMETHYLACETAMIDE
mf: $C_{17}H_{17}NO_2$ mw: 267.35

TOXICITY DATA with REFERENCE
orl-mus LDLo:300 mg/kg EJMCA5 11,7,76
ipr-mus LDLo:100 mg/kg EJMCA5 11,7,76

SAFETY PROFILE: Poison by ingestion and intraperitoneal routes. When heated to decomposition it emits toxic fumes of NO_x.

BDT750 CAS:59512-21-9 ***HR: 3***
2-(3-BENZOYLPHENYL)-N,N-DIMETHYLPROPIONAMIDE
mf: $C_{18}H_{19}NO_2$ mw: 281.38

TOXICITY DATA with REFERENCE
orl-mus LDLo:300 mg/kg EJMCA5 11,7,76
ipr-mus LDLo:100 mg/kg EJMCA5 11,7,76

SAFETY PROFILE: Poison by ingestion and intraperitoneal routes. When heated to decomposition it emits toxic fumes of NO_x.

BDU000 CAS:72596-00-0 ***HR: D***
(4-BENZOYL-o-PHENYLENEDIAMMINE)DICHLOROPLATINUM(II)
mf: $C_{13}H_{12}Cl_2N_2OPt$ mw: 478.26

TOXICITY DATA with REFERENCE
mmo-sat 30 nmol/L JMCMAR 23,459,80
mma-sat 30 nmol/L JMCMAR 23,459,80

SAFETY PROFILE: Mutation data reported. When heated to decomposition it emits very toxic fumes of Cl^- and NO_x. See also PLATINUM COMPOUNDS.

BDU250 CAS:38940-51-1 ***HR: 3***
(5-BENZOYL-5-PHENYLPENTYL)TRIETHYLAMMONIUM IODIDE
mf: $C_{24}H_{34}NO•I$ mw: 479.49

SYN: DEOXY(α-(4-TRIETHYLAMMONIO)BUTYL)BENZOINIODIDE

TOXICITY DATA with REFERENCE
orl-mus LD50:1000 mg/kg CHTPBA 7,287,72
ivn-mus LD50:1900 μg/kg CHTPBA 7,287,72

SAFETY PROFILE: Poison by intravenous route. Moderately toxic by ingestion. See also IODIDES. When heated to decomposition it emits very toxic fumes of I^-, NH_3, and NO_x.

BDU500 CAS:22071-15-4 ***HR: 3***
2-(m-BENZOYLPHENYL)PROPIONIC ACID
mf: $C_{16}H_{14}O_3$ mw: 254.30

SYNS: ALRHEUMAT ◇ ALRHEUMUM ◇ m-BENZOYLHYDRATROPIC ACID ◇ 3-BENZOYLHYDRATROPIC ACID ◇ 2-(3-BENZOYLPHENYL)PROPIONIC ACID ◇ CAPISTEN ◇ FASTUM ◇ ISO-K ◇ KEFENID ◇ KETOPROFEN ◇ KETOPRON ◇ LERTUS ◇ MEPROFEN ◇ ORUDIS ◇ ORUVAIL ◇ PROFENID ◇ 19583 RP

TOXICITY DATA with REFERENCE
orl-rat TDLo:720 mg/kg (female 30D pre):REP YACHDS 12,1015,84
orl-rat TDLo:1 mg/kg (female 21D post):TER OYYAA2 27,117,84
orl-hmn TDLo:714 μg/kg:GIT JCPCBR 24,486,84
orl-wmn TDLo:80 mg/kg/10D-I:SYS BMJOAE 292,97,86
unr-chd TDLo:300 mg/kg/15D-I:BRN,CNS,GIT NEJMAG 300,796,79
orl-rat LD50:62400 μg/kg ARZNAD 34,280,84
ipr-rat LD50:80 mg/kg NIIRDN 6,265,82
scu-rat LD50:100 mg/kg JNPHAG 2,259,71
ivn-rat LD50:350 mg/kg IYKEDH 9,222,78
rec-rat LD50:84 mg/kg JTSCDR 6,209,81
orl-mus LD50:360 mg/kg PJPPAA 38,107,86
ipr-mus LD50:300 mg/kg EJMCA5 11,7,76
scu-mus LD50:550 mg/kg JNPHAG 2,259,71
ivn-mus LD50:500 mg/kg JNPHAG 2,259,71

CONSENSUS REPORTS: Reported in EPA TSCA Inventory.

SAFETY PROFILE: Poison by ingestion, subcutaneous, intravenous, rectal, and intraperitoneal routes. Human systemic effects by an unspecified route: headache, nausea or vomiting, and degenerative changes in the brain, changes in kidney tubules. An experimental teratogen. Other experimental reproductive effects. When heated to decomposition it emits acrid smoke and irritating fumes. An anti-inflammatory and analgesic agent.

BDV250 CAS:63989-75-3 ***HR: 3***
N-BENZOYL TRIMETHYL COLCHICINIC ACID METHYL ETHER
mf: $C_{27}H_{27}NO_6$ mw: 461.55

SYNS: N-BENZOYL-N-DEACETYL COLCHICINE ◇ N-BENZOYL TMCA METHYL ETHER

TOXICITY DATA with REFERENCE
oms-mus-ipr 8 mg/kg CANCAR 3,130,50
oms-mus-par 32 mg/kg CANCAR 3,134,50
spm-mus-par 32 mg/kg CANCAR 3,134,50
ipr-mus LD50:32 mg/kg MDREP* No.204,49
ims-mus LD50:27924 mg/kg JMCMAR 26,1365,83
scu-cat LDLo:12500 μg/kg AEXPBL 72,228,13

SAFETY PROFILE: Poison by intraperitoneal and subcutaneous routes. Mutation data reported. See also COLCHICINE and ETHERS. When heated to decomposition it emits toxic fumes of NO_x.

BDV500 CAS:1027-30-1 ***HR: 3***
BENZPHETAMINE HYDROCHLORIDE
mf: $C_{17}H_{21}N•ClH$ mw: 275.85

SYN: (+)-N-BENZYL-N,α-DIMETHYLPHENETHYLAMINE HYDROCHLORIDE

TOXICITY DATA with REFERENCE
orl-rat LD50:160 mg/kg CTCEA9 2,33,60
orl-mus LD50:227 mg/kg CTCEA9 2,33,60
ipr-mus LD50:153 mg/kg CTCEA9 2,33,60

SAFETY PROFILE: Poison by ingestion and intraperitoneal routes. When heated to decomposition it emits very toxic fumes of HCl and NO_x.

BDV750 CAS:5929-01-1 ***HR: 3***
1:2-BENZPYRENE PICRATE
mf: $C_{20}H_{12}•C_6H_3N_3O_7$ mw: 481.44

SYN: BENZO(a)PYRENE MONOPICRATE

TOXICITY DATA with REFERENCE
skn-mus TDLo:1200 mg/kg/50W-I:ETA PRLBA4 117,318,35

SAFETY PROFILE: Questionable carcinogen with experimental tumorigenic data by skin contact. See also NITRATES. When heated to decomposition it emits toxic fumes of NO_x.

BDW000 CAS:113-69-9 ***HR: 3***
BENZQUINAMIDE HYDROCHLORIDE
mf: $C_{22}H_{32}N_2O_5$•ClH mw: 441.02

SYNS: EMETE-CON ◇ EMETICON ◇ NSC 64375

TOXICITY DATA with REFERENCE
orl-rat LD50:990 mg/kg TXAPA9 18,185,71
ipr-mus LD50:376 mg/kg TXAPA9 18,185,71

SAFETY PROFILE: Poison by intraperitoneal route. Moderately toxic by ingestion. When heated to decomposition it emits very toxic fumes of Cl^- and NO_x. A tranquilizer and antiemetic.

BDW650 ***HR: 3***
BENZVALENE
mf: C_6H_6 mw: 78.11

SAFETY PROFILE: This strained ring compound is a friction-sensitive explosive. It may be handled safely in an ether solution. Upon decomposition it emits acrid smoke and fumes.

BDW750 CAS:29193-35-9 ***HR: 3***
(3-(N-BENZYLACETAMIDO)-2,4,6-TRIIODOPHENYL)ACETIC ACID
mf: $C_{17}H_{14}I_3NO_3$ mw: 661.02

TOXICITY DATA with REFERENCE
orl-mus LD50:1550 mg/kg JMCMAR 13,559,70
ivn-mus LD50:235 mg/kg JMCMAR 13,559,70

SAFETY PROFILE: Poison by intravenous route. Moderately toxic by ingestion. When heated to decomposition it emits very toxic fumes of NO_x and I^-.

BDX000 CAS:140-11-4 ***HR: 3***
BENZYL ACETATE
mf: $C_9H_{10}O_2$ mw: 150.19

PROP: Colorless liquid; sweet, floral fruity odor. Mp: −51.5°, bp: 213.5°, flash p: 216°F (CC), d: 1.06, autoign temp: 862°F, vap press: 1 mm @ 45°, vap d: 5.1, refr index: 1.501. Sol in alc, most fixed oils, propylene glycol; insol in glycerin and water @ 214°.

SYNS: ACETIC ACID BENZYL ESTER ◇ ACETIC ACID PHENYLMETHYL ESTER ◇ α-ACETOXYTOLUENE ◇ BENZYL ETHANOATE ◇ FEMA No. 2135 ◇ NCI-C06508

TOXICITY DATA with REFERENCE
skn-rbt 100 mg/24H MOD CTOIDG 94(8),41,79
dnr-bcs 21 mg/disc OIGZDE 34,267,85
mma-hmn:lyms 1500 mg/L MUREAV 196,61,88
mma-mus:lyms 500 mg/L MUREAV 196,61,88
msc-mus:lyms 700 mg/L SCIEAS 236,933,87
orl-rat TDLo:258 g/kg/2Y-I:NEO NTPTR* NTP-TR-250,86
ihl-hmn TCLo:50 ppm:PSY,PUL,GLN TGNCDL 2,31,61
orl-rat LD50:2490 mg/kg FCTXAV 2,327,64
orl-mus LD50:830 mg/kg GISAAA 50(7),17,85
ihl-mus LCLo:1300 mg/m³/22H AGGHAR 5,1,33
ihl-cat LC50:245 ppm/8H AMIHAB 21,28,60
skn-cat LDLo:10 g/kg JPETAB 84,358,45
orl-rbt LD50:2200 mg/kg GISAAA 50(7),17,85
scu-rbt LDLo:3000 mg/kg AGGHAR 5,1,33
orl-gpg LD50:2200 mg/kg GISAAA 50(7),17,85
scu-gpg LDLo:3000 mg/kg AGGHAR 5,1,33

CONSENSUS REPORTS: IARC Cancer Review: Group 3 IMEMDT 7,56,87; Animal Limited Evidence IMEMDT 40,109,86. NTP Carcinogenesis Studies (gavage); Some Evidence: mouse, rat NTPTR* NTP-TR-250,86. Reported in EPA TSCA Inventory.

SAFETY PROFILE: A poison by inhalation. Moderately toxic by ingestion and subcutaneous routes. Human systemic effects by inhalation: an antipsychotic, unspecified respiratory and urinary system effects. Questionable carcinogen with experimental tumorigenic data. See also ESTERS. Combustible liquid. To fight fire, use alcohol foam, CO_2. When heated to decomposition it emits irritating fumes. See also ESTERS.

BDX100 CAS:4261-14-7 ***HR: D***
9-BENZYLADENINE
mf: $C_{12}H_{11}N_5$ mw: 225.28

SYNS: ADENINE, 9-BENZYL- ◇ 9-BAP ◇ N^9-BENZYLADENINE ◇ 9-BENZYLAMINOPURINE ◇ 9-BENZYL-6-AMINOPURINE ◇ 9H-PURIN-6-AMINE, 9-(PHENYLMETHYL)- (9CI) ◇ SQ 21611

TOXICITY DATA with REFERENCE
orl-rat TDLo:5040 mg/kg (male 4W pre):REP TOIZAG 32,36,85

SAFETY PROFILE: Experimental reproductive effects. When heated to decomposition it emits toxic fumes of NO_x.

BDX500 CAS:100-51-6 ***HR: 3***
BENZYL ALCOHOL
mf: C_7H_8O mw: 108.15

PROP: Found in jasmine, hyacinth, ylang-ylang oils and at least two dozen other essential oils (FCTXAV 11,1011,73). Water-white liquid; faint, aromatic odor, sharp burning taste. Mp: −15.3°, bp: 205.7°, flash p: 213°F (CC), d: 1.042, autoign temp: 817°F, vap press: 1 mm @ 58.0°, vap d: 3.72, refr index: 1.540. Misc with alc, chloroform, ether, and water @ 206°(decomp).

SYNS: BENZAL ALCOHOL ◇ BENZENECARBINOL ◇ BENZENEMETHANOL ◇ BENZOYL ALCOHOL ◇ FEMA No. 2137 ◇ HYDROXYTOLUENE ◇ α-HYDROXYTOLUENE ◇ NCI-C06111 ◇ PHENOLCARBINOL ◇ PHENYLCARBINOL ◇ PHENYLMETHANOL ◇ PHENYLMETHYL ALCOHOL ◇ α-TOLUENOL

TOXICITY DATA with REFERENCE
skn-man 16 mg/48H MLD CTOIDG 94(8),41,79

skn-rbt 10 mg/24H open MLD AMIHBC 4,119,51
eye-rbt 750 μg open SEV AMIHBC 4,119,51
skn-pig 100% MOD FCTXAV 11,1011,73
dnr-bcs 21 mg/disc OIGZSE 34,267,85
orl-mus TDLo:6 g/kg (female 6-13D post):REP TCMUD8 7,29,87
orl-rat LD50:1230 mg/kg FCTXAV 2,327,64
ihl-rat LCLo:2000 ppm/4H JIDHAN 31,343,49
ipr-rat LD50:400 mg/kg NPIRI* 1,6,74
scu-rat LDLo:1700 mg/kg RMSRA6 15,561,1895
ivn-rat LD50:53 mg/kg TXAPA9 18,60,71
orl-mus LD50:1580 mg/kg FCTXAV 2,327,64
ivn-mus LD50:324 mg/kg AIPTAK 135,330,62
ivn-dog LDLo:50 mg/kg TXAPA9 18,60,71
par-dog LDLo:9 mg/kg TXAPA9 25,153,73
skn-cat LDLo:10 g/kg JPETAB 84,358,45

CONSENSUS REPORTS: EPA Genetic Toxicology Program. Reported in EPA TSCA Inventory.

SAFETY PROFILE: Poison by ingestion, intraperitoneal, intravenous, parenteral routes. Moderately toxic by inhalation, skin contact, and subcutaneous routes. A moderate skin and severe eye irritant. Mutation data reported. Combustible liquid. Mixtures with sulfuric acid decompose explosively at 180°. Exothermic polymerization is catalyzed by HBr + iron when heated above 100°. To fight fire, use alcohol foam, CO_2, dry chemical. When heated to decomposition it emits acrid smoke and fumes. See also ALCOHOLS.

BDX750 ***HR: 2***
BENZYLAMINE
mf: C_7H_9N mw: 107.2

PROP: Strongly alkaline liquid; miscible with water, alcohol and ether. D: 0.983 @ 19°/4°, bp: 185°.

SAFETY PROFILE: An irritant to skin, eyes and mucous membranes. See also AMINES. When heated to decomposition it emits toxic fumes. Violent or explosive reaction with N-chlorosuccinimide. See also ALKALIES.

BDY000 CAS:3287-99-8 ***HR: 3***
BENZYLAMINE HYDROCHLORIDE
mf: $C_7H_9N{\cdot}ClH$ mw: 143.63

SYNS: BENZYENEMETHAMAMINE HYDROCHLORIDE ◇ BENZYLAMMONIUM CHLORIDE ◇ USAF EL-82

TOXICITY DATA with REFERENCE
ipr-mus LD50:500 mg/kg NTIS** AD277-689
ivn-mus LD50:220 mg/kg APFRAD 9,390,51

CONSENSUS REPORTS: Reported in EPA TSCA Inventory.

SAFETY PROFILE: Poison by intravenous route. Moderately toxic by intraperitoneal route. When heated to decomposition it emits very toxic fumes of HCl, NH_3, and NO_x. See also AROMATIC AMINES.

BDY250 CAS:77966-31-5 ***HR: 3***
2-(BENZYLAMINO)-6'-CHLORO-o-ACETOTOLUIDIDE HYDROCHLORIDE
mf: $C_{16}H_{17}ClN_2O{\cdot}ClH$ mw: 325.26

SYN: C 3117

TOXICITY DATA with REFERENCE
eye-rbt 2% MLD ARZNAD 8,407,58
ipr-rat LD50:280 mg/kg ARZNAD 8,407,58
scu-mus LD50:1175 mg/kg ARZNAD 8,407,58

SAFETY PROFILE: Poison by intraperitoneal route. Moderately toxic by subcutaneous route. An eye irritant. When heated to decomposition it emits very toxic fumes of Cl^-, NO_x, and HCl.

BDY500 CAS:52400-76-7 ***HR: 3***
2-(2-(BENZYLAMINO)ETHYL)-2-METHYL-1,3-BENZODIOXOLE HYDROCHLORIDE
mf: $C_{17}H_{19}NO_2{\cdot}ClH$ mw: 305.83

TOXICITY DATA with REFERENCE
ivn-rat LD50:15 mg/kg EJMCA5 12,413,77
ipr-mus LD50:110 mg/kg EJMCA5 12,413,77

SAFETY PROFILE: Poison by intravenous and intraperitoneal routes. When heated to decomposition it emits very toxic fumes of HCl and NO_x.

BDY669 CAS:61-33-6 ***HR: 3***
BENZYL-6-AMINOPENICILLINIC ACID
mf: $C_{16}H_{18}N_2O_4S$ mw: 334.42

SYNS: ABBOCILLIN ◇ (5R,6R)-BENXYLPENICILLIN ◇ BENZOPENICILLIN ◇ BENZYLPENICILLIN ◇ BENZYLPENICILLIN G ◇ BENZYLPENICILLINIC ACID ◇ CILLORAL ◇ CILOPEN ◇ COMPOCILLIN G ◇ COSMOPEN ◇ DROPCILLIN ◇ FREE BENZYLPENICILLIN ◇ GALOFAK ◇ GELACILLIN ◇ LIQUACILLIN ◇ PENICILLIN G ◇ PHENYLACETAMIDOPENICILLANIC ACID ◇ (PHENYLMETHYL) PENICILLINIC ACID ◇ PRADUPEN ◇ SPECILLINE G

TOXICITY DATA with REFERENCE
dnr-esc 20 μL/disc MUREAV 97,1,82
dnr-bcs 100 μL/plate MUREAV 97,1,82
mmo-omi 12 μg/L ARMKA7 81,1,72
oms-omi 20 μg/L AMACCQ 17,572,80
scu-rat TDLo:2600 mg/kg/65W-I:ETA LANCAO 1,394,86
par-chd TDLo:15000 units/kg:NOSE,CNS,PUL BJCAAI 17,100,63
unk-rat LD50:9 g/kg ANTBAL 23, 317,78
ivn-mus LD50:329 mg/kg BCPCA6 16,1365,67
ice-mus LD50:5700 μg/kg JLCMAK 34,126,49
unk-mus LD50:7800 mg/kg ANTBAL 23,317,78
ice-rbt LD50:1118 μg/kg JLCMAK 34,126,49
isp-dog LD50:4940 μg/kg JLCMAK 34,126,49

ice-rbt LD50:653 μg/kg JLCMAK 34,126,49
orl-ham LD50:24 mg/kg TXAPA9 14,510,69
scu-ham LD50:96 mg/kg TXAPA9 14,510,69

CONSENSUS REPORTS: EPA Genetic Toxicology Program.

SAFETY PROFILE: Poison by ingestion, intravenous, intracerebral, intraspinal, subcutaneous, and possibly other routes. Human (child) systemic effects by parenteral route: changes in cochlear (inner ear) structure or function, convulsions, and dyspnea. Questionable carcinogen with experimental tumorigenic data. Mutation data reported. When heated to decomposition it emits very toxic fumes of NO_x and SO_x. See other penicillin entries.

BEA000 CAS:67465-04-7 ***HR: 3***
2-BENZYLAMINOPYRIDINE HYDROCHLORIDE
mf: $C_{12}H_{12}N_2$•ClH mw: 220.70

SYNS: 2-BAP HYDROCHLORIDE ◇ N-(2-PYRIDYL)BENZYLAMINE HYDROCHLORIDE

TOXICITY DATA with REFERENCE
orl-mus LD50:1187 mg/kg TXAPA9 37,165,76
ipr-mus LD50:220 mg/kg TXAPA9 37,165,76
ivn-mus LD50:90 mg/kg JPETAB 84,16,45

SAFETY PROFILE: Poison by intravenous and intraperitoneal routes. Moderately toxic by ingestion. When heated to decomposition it emits very toxic fumes of NO_x and HCl. See also AROMATIC AMINES.

BEA250 CAS:64059-29-6 ***HR: 2***
BENZYL AMMONIUM TETRACHLOROIODATE
mf: $C_7H_{10}N$•Cl_4IO mw: 392.88

TOXICITY DATA with REFERENCE
orl-rat LD50:1230 mg/kg TXAPA9 28,313,74
skn-rbt LD50:840 mg/kg TXAPA9 28,313,74

SAFETY PROFILE: Moderately toxic by ingestion and skin contact. When heated to decomposition it emits very toxic fumes of Cl^-, I^-, NH_3 and NO_x.

BEA275 CAS:101997-51-7 ***HR: 3***
1-(2-(N-BENZYLANILINO)ETHYL)PIPERIDINE HYDROCHLORIDE
mf: $C_{20}H_{26}N_2$•ClH mw: 330.94

SYNS: N-β-(BENZILFENILAMINO)ETILPIPERIDINACLORIDRATO (ITALIAN) ◇ N-β-(BENZYL-PHENYLAMINO)ETHYLPIPERIDINE HYDROCHLORIDE

TOXICITY DATA with REFERENCE
ims-rat LDLo:380 mg/kg FRPSAX 15,562,60
orl-mus LD50:1500 mg/kg FRPSAX 13,3,58
ipr-mus LD50:180 mg/kg FRPSAX 15,562,60
ivn-mus LD50:25 mg/kg FRPSAX 13,3,58
ipr-gpg LD50:110 mg/kg BSCIA3 31,520,49

SAFETY PROFILE: Poison by intravenous, intramuscular, and intraperitoneal routes. Moderately toxic by ingestion. When heated to decomposition it emits toxic fumes of NO_x and HCl.

BEA325 CAS:622-79-7 ***HR: 3***
BENZYL AZIDE
mf: $C_7N_7N_3$ mw: 133.15

SAFETY PROFILE: A heat-sensitive explosive. Explosive reaction with bis(trifluoromethyl)nitroxide. Upon decomposition it emits toxic fumes of NO_x. See also AZIDES.

BEA500 CAS:36226-64-9 ***HR: 3***
BENZYLBARBITAL
mf: $C_{13}H_{14}N_2O_3$ mw: 246.29

SYNS: 5-BENZYL-5-ETHYLBARBITURIC ACID ◇ ETHYLBENZYLBARBITURIC ACID ◇ 5-ETHYL-5-(PHENYLMETHYL)-2,4,6(1H,3H,5H)-PYRIMIDINETRIONE (9CI)

TOXICITY DATA with REFERENCE
ipr-mus LD50:73 mg/kg JPETAB 89,356,47
orl-cat LDLo:400 mg/kg JPETAB 26,371,25
scu-rbt LDLo:60 mg/kg JACSAT 45,243,23

SAFETY PROFILE: Poison by ingestion, intraperitoneal, and subcutaneous routes. When heated to decomposition it emits toxic fumes of NO_x. An hypnotic agent. See also BARBITURATES.

BEA825 CAS:621-72-7 ***HR: 3***
2-BENZYLBENZIMIDAZOLE
mf: $C_{14}H_{12}N_2$ mw: 208.28

PROP: Needles from benzene. Mp: 187°. Practically insol in water; freely sol in glacial acetic acid; sol in alc, hot benzene, and propylene glycol.

SYNS: BENDAZOL ◇ BENDAZOLE ◇ 2-BENZYLBENZIMINAZOLE ◇ DIBASOL ◇ DIBAZOL ◇ DIBAZOLE ◇ 2-(PHENYLMETHYL)-1H-BENZIMIDAZOLE ◇ TROMASEDAN

TOXICITY DATA with REFERENCE
unr-rat TDLo:2 mg/kg (9D preg):REP AKGIAO 43(12),10,67
unr-rat TDLo:2 mg/kg (9D preg):TER AKGIAO 43(12),10,67
orl-mus LD50:100 mg/kg FRZKAP (1),44,83
ipr-mus LD50:240 mg/kg PCJOAU 19,544,85
scu-mus LDLo:504 mg/kg PCJOAU 13,829,79

SAFETY PROFILE: Poison by ingestion and intraperitoneal routes. Moderately toxic by subcutaneous route. Experimental reproductive effects. When heated to decomposition it emits toxic fumes of NO_x.

BEA850 CAS:1421-23-4 ***HR: 3***
N-BENZYLBIGUANIDE HYDROCHLORIDE
mf: $C_9H_{13}N_5 \cdot ClH$ mw: 227.73

SYNS: 1-BENZILBIGUANIDE CLORIDRATO (ITALIAN) ◇ BENZYLBIGUANIDE HYDROCHLORIDE ◇ 1-BENZYLBIGUANIDE HYDROCHLORIDE

TOXICITY DATA with REFERENCE
orl-rat LD50:481 mg/kg FRPSAX 15,521,60
ipr-rat LD50:108 mg/kg FRPSAX 15,521,60
ipr-mus LD50:195 mg/kg JAJAAA 18,196,65

SAFETY PROFILE: Poison by intraperitoneal route. Moderately toxic by ingestion. When heated to decomposition it emits toxic fumes of NO_x and HCl.

BEB000 CAS:58050-46-7 ***HR: 3***
BENZYL BIS(2-CHLOROETHYL)AMINOMETHYLCARBAMATE
mf: $C_{13}H_{18}Cl_2N_2O_2$ mw: 305.23

SYN: N-(BIS-(2-CHLORAETHYL)AMINOMETHYLBENZYLURETHAN) (GERMAN)

TOXICITY DATA with REFERENCE
ims-rat LD50:40 mg/kg ZKKOBW 84,227,75
ipr-mus LD50:50 mg/kg ZKKOBW 84,227,75

SAFETY PROFILE: Poison by intramuscular and intraperitoneal routes. See also CARBAMATES. When heated to decomposition it emits very toxic fumes of Cl^- and NO_x.

BEB500 CAS:23111-70-8 ***HR: 3***
1-BENZYL-1,4-BIS(α-METHYLPHENETHYL)PIPERAZINIUM BROMIDE
mf: $C_{29}H_{37}N_2 \cdot Br$ mw: 493.59

TOXICITY DATA with REFERENCE
scu-mus LD50:350 mg/kg JPETAB 97,25,49
ivn-mus LD50:29 mg/kg ARZNAD 18,1431,68

SAFETY PROFILE: Poison by subcutaneous and intravenous routes. See also BROMIDES. When heated to decomposition it emits very toxic fumes of NO_x and Br^-.

BEB750 CAS:101834-51-9 ***HR: 3***
5-BENZYL-2,2-BIS(TRIFLUOROMETHYL)-4-METHYLOXAZOLIDINE HYDRATE
mf: $C_{12}H_{11}F_6NO \cdot H_2O$ mw: 317.26

TOXICITY DATA with REFERENCE
orl-mus LD50:200 mg/kg JMCMAR 13,1215,70
ipr-mus LD50:300 mg/kg JMCMAR 13,1215,70

SAFETY PROFILE: Poison by ingestion and intraperitoneal routes. See also FLUORIDES. When heated to decomposition it emits very toxic fumes of F^- and NO_x.

BEC000 CAS:100-39-0 ***HR: 2***
BENZYL BROMIDE
DOT: UN 1737
mf: C_7H_7Br mw: 171.05

PROP: Clear, refractive liquid; pleasant odor, lachrymator, insol in water. Mp: −4.0°, bp: 198°, d: 1.438 @ 22°/0°, vap d: 5.8.

SYNS: (BROMOMETHYL)BENZENE ◇ p-(BROMOMETHYL)NITROBENZENE ◇ BROMOPHENYLMETHANE ◇ ω-BROMOTOLUENE ◇ α-BROMOTOLUENE (DOT)

TOXICITY DATA with REFERENCE
dns-esc 1300 μmol/L ZKKOBW 92,177,78

CONSENSUS REPORTS: Reported in EPA TSCA Inventory.

DOT Classification: Corrosive Material; Label: Corrosive.

SAFETY PROFILE: Intensely irritating and corrosive to skin, eyes, and mucous membranes. Large doses cause central nervous system depression. Mutation data reported. Reaction with molecular sieve produces toxic hydrogen bromide gas. See also BROMIDES.

BEC250 CAS:103-05-9 ***HR: 2***
BENZYL-tert-BUTANOL
mf: $C_{11}H_{16}O$ mw: 164.27

SYNS: DIMETHYLPHENYLETHYL CARBINOL ◇ 1,1-DIMETHYL-3-PHENYLPROPANOL ◇ 1,1-DIMETHYL-3-PHENYL-1-PROPANOL ◇ α,α-DIMETHYL-Δ-PHENYLPROPYL ALCOHOL ◇ 2-METHYL-4-PHENYL-2-BUTANOL ◇ PHENYLETHYL DIMETHYL CARBINOL

TOXICITY DATA with REFERENCE
orl-rat LD50:2200 mg/kg FCTXAV 12,517,74
skn-rbt LD50:3500 mg/kg FCTXAV 12,517,74

CONSENSUS REPORTS: Reported in EPA TSCA Inventory.

SAFETY PROFILE: Moderately toxic by ingestion and skin contact. When heated to decomposition it emits acrid smoke and irritating fumes. See also ALCOHOLS.

BEC500 CAS:85-68-7 ***HR: 3***
BENZYL BUTYL PHTHALATE
mf: $C_{19}H_{20}O_4$ mw: 312.39

PROP: Clear, oily liquid. Mp: < −35°, bp: 370°, flash p: 390°F, d: 1.116 @ 25°/25°, vap d: 10.8.

SYNS: BBP ◇ 1,2-BENZENEDICARBOXYLIC ACID, BUTYL PHENYLMETHYL ESTER ◇ BUTYL BENZYL PHTHALATE ◇ n-BUTYL BENZYL PHTHALATE ◇ NCI-C54375 ◇ PALATINOL BB ◇ SANTICIZER 160 ◇ SICOL 160 ◇ UNIMOLL BB

TOXICITY DATA with REFERENCE
orl-rat TDLo:21 g/kg (14D male):REP TOXID9 4,136,84
orl-rat TDLo:433 g/kg/2Y-C:CAR NTPTR* NTP-TR-213,82
orl-rat TD:437 g/kg/2Y-C:CAR EVHPAZ 65,271,86
orl-rat LD50:2330 mg/kg IARC** 29,193,82
orl-mus LD50:4170 mg/kg IARC** 29,193,82
ipr-mus LD50:3160 mg/kg EVHPAZ 4,3,73
orl-gpg LD50:13750 mg/kg GTPZAB 24(3),25,80

CONSENSUS REPORTS: IARC Cancer Review: Group 3 IMEMDT 7,56,87, Animal Inadequate Evidence IMEMDT 29,193,82; NTP Carcinogenesis Bioassay (feed); No Evidence: mouse NTPTR* NTP-TR-213,82; Clear Evidence: rat NTPTR* NTP-TR-213,82. Reported in EPA TSCA Inventory. Community Right-To-Know List.

SAFETY PROFILE: Questionable carcinogen with experimental carcinogenic data. Moderately toxic by ingestion and intraperitoneal routes. Experimental reproductive effects. See also ESTERS. Combustible when exposed to heat or flame; can react with oxidizers. To fight fire, use spray or mist, CO_2, dry chemical. When heated to decomposition it emits acrid smoke and irritating fumes.

BED000 CAS:103-37-7 ***HR: 2***
BENZYL n-BUTYRATE
mf: $C_{11}H_{14}O_2$ mw: 178.25

PROP: Colorless liquid; floral plum-like odor. D: 1.006, refr index: 1.492, flash p: +212°F. Sol in fixed oils; insol in glycerin, propylene glycol, water @ 239°.

SYNS: BENZYL n-BUTANOATE ◇ FEMA No. 2140

TOXICITY DATA with REFERENCE
orl-rat LD50:2330 mg/kg FCTXAV 2,327,64

CONSENSUS REPORTS: Reported in EPA TSCA Inventory.

SAFETY PROFILE: Moderately toxic by ingestion. See also ESTERS. Combustible liquid. When heated to decomposition it emits acrid smoke and irritating fumes.

BED250 CAS:63884-81-1 ***HR: 3***
BENZYLCARBAMIC ESTER of 3-OXY-PHENYLDIMETHYLAMINE HYDROCHLORIDE
mf: $C_{16}H_{18}N_2O_2$•ClH mw: 306.82

SYNS: AR-22 ◇ 3-(N'-BENZYLCARBAMOYLOXY)-N,N-DIMETHYLANILINE HYDROCHLORIDE

TOXICITY DATA with REFERENCE
orl-mus LDLo:500 mg/kg JPETAB 43,413,31
ivn-mus LDLo:50 mg/kg JPETAB 43,413,31

SAFETY PROFILE: Poison by intravenous route. Moderately toxic by ingestion. See also ESTERS and CARBAMATES. When heated to decomposition it emits very toxic fumes of NO_x and HCl.

BED500 CAS:64051-16-7 ***HR: 3***
BENZYLCARBAMIC ESTER of 3-OXY-PHENYLTRIMETHYLAMMONIUM METHYLSULFATE
mf: $C_{17}H_{21}N_2O_2$•CH_3O_4S mw: 396.50

SYNS: AMMONIUM(3-N-BENZYLCARBAMOYLOXY)PHENYL)TRIMETHYL METHYLSULFATE ◇ AR-23 ◇ N-BENZYL-CARBAMIC ACID-3-(TRIMETHYLAMMONIO)PHENYL ESTER, METHYLSULFATE ◇ (m-HYDROXYPHENYL)TRIMETHYLAMMONIUM METHYLSULFATE BENZYLCARBAMATE

TOXICITY DATA with REFERENCE
orl-mus LDLo:33 mg/kg JPETAB 43,413,31
ivn-mus LDLo:100 μg/kg NTIS** PB158-508

SAFETY PROFILE: Poison by ingestion and intravenous routes. See also CARBAMATES; ESTERS; and SULFATES. When heated to decomposition it emits very toxic fumes of SO_x, NH_3 and NO_x.

BED750 CAS:14504-15-5 ***HR: 3***
3-BENZYL-4-CARBAMOYLMETHYLSYDNONE
mf: $C_{11}H_{11}N_3O_3$ mw: 233.25

SYN: 3-BENZYLSYDNONE-4-ACETAMIDE

TOXICITY DATA with REFERENCE
orl-rat TDLo:23 mg/kg/13W-I:NEO GANNA2 65,273,74
orl-rat LD50:4450 mg/kg GANNA2 65,273,74

SAFETY PROFILE: Mildly toxic by ingestion. Questionable carcinogen with experimental neoplastigenic data. When heated to decomposition it emits toxic fumes of NO_x.

BEE250 CAS:103-53-7 ***HR: 1***
BENZYLCARBINYL CINNAMATE
mf: $C_{17}H_{16}O_2$ mw: 252.33

SYNS: PHENETHYL CINNAMATE ◇ β-PHENETHYL CINNAMATE ◇ PHENYLETHYL CINNAMATE ◇ β-PHENYLETHYL CINNAMATE

TOXICITY DATA with REFERENCE
skn-rbt 500 mg/24H MLD FCTXAV 16,637,78
orl-rat LD50:5 g/kg FCTXAV 16,637,78
orl-mus LD50:4500 mg/kg FCTXAV 16,637,78
orl-gpg LD50:4500 mg/kg FCTXAV 16,637,78

CONSENSUS REPORTS: Reported in EPA TSCA Inventory.

SAFETY PROFILE: Mildly toxic by ingestion. A skin

irritant. See also ESTERS. When heated to decomposition it emits acrid smoke and irritating fumes.

BEE375 CAS:100-44-7 ***HR: 3***
BENZYL CHLORIDE
DOT: UN 1738
mf: C_7H_7Cl mw: 126.59

PROP: Colorless liquid, very refractive; irritating, unpleasant odor. Mp: −43°, bp: 179°, lel: 1.1%, flash p: 153°F, d: 1.1026 @ 18/4°, autoign temp: 1085°F, vap d: 4.36.

SYNS: BENZILE (CLORURO di) (ITALIAN) ◇ BENZYLCHLORID (GERMAN) ◇ BENZYLE (CHLORURE de) (FRENCH) ◇ CHLOROMETHYLBENZENE ◇ CHLOROPHENYLMETHANE ◇ α-CHLOROTOLUENE ◇ ω-CHLOROTOLUENE ◇ α-CHLORTOLUOL (GERMAN) ◇ CHLORURE de BENZYLE (FRENCH) ◇ NCI-C06360 ◇ RCRA WASTE NUMBER P028 ◇ TOLYL CHLORIDE

TOXICITY DATA with REFERENCE
dnd-hmn:fbr 1 mmol/L MUREAV 145,209,85
dnd-hmn:oth 1 mmol/L MUREAV 145,209,85
otr-ham:emb 1600 μg/L CRNGDP 1,323,80
orl-rat TDLo:1 g/kg (female 6-15 post):TER JTEHD6 17,51,86
ipr-mus TDLo:250 mg/kg (male 5D pre):REP MUREAV 100,345,82
scu-rat TDLo:50 mg/kg:ETA FCTXAV 6,576,68
orl-mus TDLo:31 g/kg/2Y-I:CAR JJIND8 76,1231,86
orl-rat LD50:1231 mg/kg NTIS** PB214-270
ihl-rat LC50:150 ppm/2H IARC** 11,217,76
scu-rat LD50:1 g/kg ZEKBAI 74,241,70
orl-mus LD50:1500 mg/kg 85GMAT -,25,82
ihl-mus LC50:80 ppm/2H IARC** 11,217,76

CONSENSUS REPORTS: IARC Cancer Review: Animal Limited Evidence IMEMDT 29,49,82; Animal Sufficient Evidence IMEMDT 11,217,76; Human Inadequate Evidence IMEMDT 29,49,82. EPA Genetic Toxicology Program. Community Right-To-Know List. Reported in EPA TSCA Inventory. EPA Extremely Hazardous Substances List.

OSHA PEL: TWA 1 ppm
ACGIH TLV: TWA 1 ppm
DFG MAK: 1 ppm (5 mg/m^3); Suspected Carcinogen.
NIOSH REL: (Benzyl Chloride) CL 5 mg/m^3/15M
DOT Classification: Corrosive Material; Label: Corrosive.

SAFETY PROFILE: Suspected carcinogen with experimental carcinogenic and tumorigenic data. Poison by inhalation. Moderately toxic by ingestion and subcutaneous routes. Experimental reproductive effects. Human mutation data reported. A corrosive irritant to skin, eyes, and mucous membranes. Flammable and moderately explosive when exposed to heat or flame. Can react vigorously with oxidizing materials. May explode during distillation. The decomposition rate can reach explosive violence in presence of metals such as iron. Catalytic impurities (e.g., aluminum, iron, rust) or sodium acetate + pyridine + iron (at 115°C) may cause violent polymerization reactions. Will react with water or steam to produce toxic and corrosive fumes. Incompatible with dimethyl sulfoxide. Used in production of drugs of abuse. When heated to decomposition it emits toxic fumes of Cl^-. See also CHLORINATED HYDROCARBONS, AROMATIC.

BEE500 CAS:140-18-1 ***HR: 2***
BENZYL CHLOROACETATE
mf: $C_9H_9ClO_2$ mw: 184.63

SYNS: BENZYL-α-CHLOROACETATE ◇ BENZYL MONOCHLORACETATE ◇ CHLOROACETIC ACID BENZYL ESTER

TOXICITY DATA with REFERENCE
ipr-mus LDLo:500 mg/kg CBCCT* 8,99,56

CONSENSUS REPORTS: Reported in EPA TSCA Inventory.

SAFETY PROFILE: Moderately toxic by intraperitoneal route. See also ESTERS. When heated to decomposition it emits toxic fumes of Cl^-.

BEE750 CAS:77966-32-6 ***HR: 3***
N-BENZYL-6'-CHLORO-2-(DIETHYLAMINO)-o-ACETOTOLUIDIDE HYDROCHLORIDE
mf: $C_{20}H_{25}ClN_2O•ClH$ mw: 381.38

SYN: C 3136

TOXICITY DATA with REFERENCE
eye-rbt 2% MOD ARZNAD 8,609,58
ipr-rat LD50:96 mg/kg ARZNAD 8,609,58
scu-mus LD50:360 mg/kg ARZNAD 8,609,58

SAFETY PROFILE: Poison by intraperitoneal and subcutaneous routes. An eye irritant. When heated to decomposition it emits very toxic fumes of NO_x, HCl, and Cl^-.

BEF500 CAS:501-53-1 ***HR: 3***
BENZYL CHLOROFORMATE
DOT: UN 1739
mf: $C_8H_7ClO_2$ mw: 170.60

PROP: Colorless to pale yellow liquid, odor of phosgene.

SYNS: BENZYLCARBONYL CHLORIDE ◇ BENZYL CHLOROCARBONATE (DOT) ◇ BENZYL CHLOROFORMATE (DOT) ◇ BENZYLOXYCARBONYL CHLORIDE ◇ BZCF ◇ CARBOBENZOXY CHLORIDE ◇ CARBOBENZYLOXY CHLORIDE ◇ CHLOROFORMIC ACID BENZYL ESTER

CONSENSUS REPORTS: Reported in EPA TSCA Inventory.

DOT Classification: Corrosive Material; Label: Corrosive.

SAFETY PROFILE: Poison by ingestion and inhalation routes. A powerful corrosive irritant. Thermally unstable. Will react with water or steam to produce toxic and corrosive fumes and heat. Iron salts catalyze the explosive decomposition of the ester. When heated to decomposition it emits toxic fumes of Cl^- and phosgene. See also PHOSGENE, ESTERS, and CHLORIDES.

BEF750 CAS:1322-48-1 ***HR: 3***
2-BENZYL-4-CHLOROPHENOL
mf: $C_{13}H_{11}ClO$ mw: 218.69

PROP: Nearly colorless flakes. Mp: 49°; bp: 175° @ 5 mm; d: 1.2 @ 55°/25°.

SYNS: BENZYLCHLOROPHENOL ◇ 4-CHLORO-α-PHENYLCRESOL

TOXICITY DATA with REFERENCE
orl-rat LD50:1700 mg/kg JPMSAE 63,1068,74

CONSENSUS REPORTS: Chlorophenols are on the Community Right-To-Know List.

SAFETY PROFILE: Moderately toxic by ingestion. When heated to decomposition it emits toxic fumes of Cl^-. See also CHLOROPHENOLS.

BEG000 CAS:501-68-8 ***HR: 2***
N-BENZYL-β-CHLOROPROPANAMIDE
mf: $C_{10}H_{12}ClNO$ mw: 197.68

SYNS: BECLAMID ◇ BECLAMIDE ◇ BEKLAMID ◇ BENZCHLOROPROPAMIDE ◇ BENZCHLORPROPAMID ◇ BENZOCHLORPROPAMID ◇ BENZYLAMIDE ◇ N-BENZYL-β-CHLOROPROPIONAMIDE ◇ N-BENZYL-3-CHLOROPROPIONAMIDE ◇ CHLORACON ◇ CHLORAKON ◇ CHLOROETHYLPHENAMIDE ◇ 3-CHLORO-N-(PHENYLMETHYL)PROPANAMIDE ◇ N-(3-CHLOROPROPIONYL)BENZYLAMINE ◇ HIBICON ◇ KHLORAKON ◇ NEURACEN ◇ NIDRANE ◇ NYDRAN ◇ NYDRANE ◇ POSEDRAN ◇ POSEDRINE ◇ SECLAR

TOXICITY DATA with REFERENCE
orl-rat LD50:3200 mg/kg JPETAB 107,403,53
ipr-rat LD50:770 mg/kg JPETAB 107,403,53
ivn-rat LD50:770 mg/kg 27ZQAG -,384,72
orl-mus LD50:1000 mg/kg PCJOAU 14,99,80
ipr-mus LD50:650 mg/kg 27ZQAG -,384,72

SAFETY PROFILE: Moderately toxic by ingestion, intraperitoneal and intravenous routes. When heated to decomposition it emits very toxic fumes of Cl^- and NO_x. An anticonvulsant.

BEG750 CAS:103-41-3 ***HR: 2***
BENZYL CINNAMATE
mf: $C_{16}H_{14}O_2$ mw: 238.30

PROP: Found in balsams of Peru, Tolu, Styrax, Copaiba and others (FCTXAV 11,1011,73). White crystals; aromatic odor. Mp: 39°, bp: 350.0°, vap press: 1 mm @ 173.8°, flash p: +212°F. Sol in fixed oils; insol in glycerin and propylene glycol.

SYNS: BENZYL ALCOHOL CINNAMIC ESTER ◇ BENZYL Γ-PHENYLACRYLATE ◇ CINNAMEIN ◇ trans-CINNAMIC ACID BENZYL ESTER ◇ FEMA No. 2142 ◇ 3-PHENYL-2-PROPENOIC ACID PHENYLMETHYL ESTER (9CI)

TOXICITY DATA with REFERENCE
skn-rbt 500 mg MLD FCTXAV 11,1011,73
orl-rat LDLo:5530 mg/kg FCTXAV 2,327,64
orl-gpg LD50:3760 mg/kg FCTXAV 2,327,64

CONSENSUS REPORTS: Reported in EPA TSCA Inventory.

SAFETY PROFILE: Moderately toxic by ingestion. A mild allergen and skin irritant. Combustible liquid. See also ESTERS. When heated to decomposition it emits acrid smoke and irritating fumes.

BEH000 CAS:363-13-3 ***HR: 3***
1-BENZYL-2(1H)-CYCLOHEPTIMIDAZOLONE
mf: $C_{15}H_{12}N_2O$ mw: 236.29

SYNS: BENZYLCYCLOHEPTIMIDAZOL-2(1H)-ONE ◇ 1-(PHENYLMETHYL)-2(1H)-CYCLOHEPTIMIDAZOLONE

TOXICITY DATA with REFERENCE
orl-mus LD50:358 mg/kg ARZNAD 18,939,68
ipr-mus LD50:119 mg/kg ARZNAD 18,939,68

SAFETY PROFILE: Poison by ingestion and intraperitoneal routes. When heated to decomposition it emits toxic fumes of NO_x.

BEH250 CAS:40502-72-5 ***HR: 3***
β-BENZYL-α-CYCLOPENTYL-4-METHYL-α-PHENYL-1-PIPERAZINEPROPANOL DIHYDROCHLORIDE
mf: $C_{26}H_{36}N_2O•2ClH$ mw: 465.56

TOXICITY DATA with REFERENCE
orl-mus LD50:2000 mg/kg CHTPBA 7,293,72
ipr-mus LD50:207 mg/kg EJMCA5 9,408,74

SAFETY PROFILE: Poison by intraperitoneal route. Moderately toxic by ingestion. When heated to decomposition it emits very toxic fumes of NO_x and HCl.

BEI000 CAS:5372-17-8 ***HR: 3***
1-BENZYL-5-(2-(DIETHYLAMINO)ETHOXY)-3-METHYLPYRAZOLE
mf: $C_{17}H_{25}N_3O$ mw: 287.45

SYN: B-314

TOXICITY DATA with REFERENCE
orl-mus LD50:385 mg/kg ARZNAD 17,214,67
scu-mus LD50:262 mg/kg ARZNAD 17,214,67

SAFETY PROFILE: Poison by ingestion and subcutaneous routes. When heated to decomposition it emits toxic fumes of NO_x.

BEI250 CAS:101651-55-2 ***HR: 3***
2-(BENZYL(2-(DIETHYLAMINO)ETHYL) AMINO)ACETANILIDE DIHYDROCHLORIDE
mf: $C_{21}H_{29}N_3O{\cdot}2ClH$ mw: 412.45

SYN: C 5348

TOXICITY DATA with REFERENCE
eye-rbt 2% MLD ARZNAD 9,167,59
scu-mus LD50:375 mg/kg ARZNAD 9,167,59

SAFETY PROFILE: Poison by subcutaneous route. An eye irritant. When heated to decomposition it emits very toxic fumes of HCl and NO_x.

BEI500 CAS:102489-46-3 ***HR: 3***
2-(BENZYL(2-(DIETHYLAMINO)ETHYL)AMINO)-o-ACETOTOLUIDIDE DIHYDROCHLORIDE
mf: $C_{22}H_{31}N_3O{\cdot}2ClH$ mw: 426.48

SYN: C 5351

TOXICITY DATA with REFERENCE
eye-rbt 2% MLD ARZNAD 9,167,59
scu-mus LD50:160 mg/kg ARZNAD 9,167,59

SAFETY PROFILE: Poison by subcutaneous route. An eye irritant. When heated to decomposition it emits very toxic fumes of NO_x and HCl. See also AMINES.

BEI750 CAS:102489-45-2 ***HR: 3***
2-(BENZYL(2-(DIETHYLAMINO)ETHYL)AMINO)-6'-CHLORO-o-ACETOTOLUIDIDE DIHYDROCHLORIDE
mf: $C_{22}H_{30}ClN_3O{\cdot}2ClH$ mw: 460.92

SYN: C 5296

TOXICITY DATA with REFERENCE
eye-rbt 2% MLD ARZNAD 9,167,59
scu-mus LD50:32 mg/kg ARZNAD 9,167,59

SAFETY PROFILE: Poison by subcutaneous route. An eye irritant. When heated to decomposition it emits very toxic fumes of NO_x, HCl, and Cl^-.

BEJ250 CAS:77966-34-8 ***HR: 3***
2-(BENZYL(3-DIETHYLAMINO)PROPYL) AMINO)-o-ACETOTOLUIDIDE DIHYDROHLORIDE
mf: $C_{23}H_{33}N_3O{\cdot}2ClH$ mw: 440.51

SYN: C 5353

TOXICITY DATA with REFERENCE
eye-rbt 2% MLD ARZNAD 9,167,59
scu-mus LD50:260 mg/kg ARZNAD 9,167,59

SAFETY PROFILE: Poison by subcutaneous route. An eye irritant. When heated to decomposition it emits very toxic fumes of NO_x and HCl.

BEJ500 CAS:77966-75-7 ***HR: 3***
2-(BENZYL(3-(DIETHYLAMINO)PROPYL) AMINO)-2',6'-ACETOXYLIDIDE DIHYDROCHLORIDE
mf: $C_{24}H_{35}N_3O{\cdot}2ClH$ mw: 454.54

SYN: C 5354

TOXICITY DATA with REFERENCE
eye-rbt 2% MLD ARZNAD 9,167,59
scu-mus LD50:35 mg/kg ARZNAD 9,167,59

SAFETY PROFILE: Poison by subcutaneous route. An eye irritant. When heated to decomposition it emits very toxic fumes of NO_x and HCl.

BEJ825 ***HR: 3***
N-BENZYL-N',N'-DIETHYL-N-1-NAPHTHYLETHYLENEDIAMINE
mf: $C_{23}H_{28}N_2$ mw: 332.53

TOXICITY DATA with REFERENCE
ipr-rat LDLo:45 mg/kg BJPCAL 11,1,56
ipr-mus LD50:63 mg/kg BJPCAL 11,1,56
scu-mus LD50:300 mg/kg BJPCAL 11,1,56

SAFETY PROFILE: Poison by subcutaneous and intraperitoneal routes. When heated to decomposition it emits toxic fumes of NO_x. See also AMINES.

BEJ830 ***HR: 3***
N-BENZYL-N',N'-DIETHYL-N-2-NAPHTHYLETHYLENEDIAMINE
mf: $C_{23}H_{28}N_2$ mw: 332.53

TOXICITY DATA with REFERENCE
ipr-rat LDLo:90 mg/kg BJPCAL 11,1,56
ipr-mus LD50:128 mg/kg BJPCAL 11,1,56
scu-mus LD50:459 mg/kg BJPCAL 11,1,56

SAFETY PROFILE: Poison by intraperitoneal route. Moderately toxic by subcutaneous route. When heated to decomposition it emits toxic fumes of NO_x. See also AMINES.

BEK250 CAS:32871-90-2 ***HR: 3***
4-BENZYL-6,7-DIMETHOXYISOQUINOLINE HYDROBROMIDE
mf: $C_{18}H_{17}NO_2{\cdot}BrH$ mw: 360.28

TOXICITY DATA with REFERENCE
orl-mus LD50:2300 mg/kg CHTPBA 6,358,71
ipr-mus LD50:400 mg/kg EJMCA5 11,271,76

SAFETY PROFILE: Poison by intraperitoneal route. Moderately toxic by ingestion. When heated to decomposition it emits very toxic fumes of NO_x and HBr.

BEK500 CAS:15565-25-0 ***HR: 3***
4-BENZYL-l-(3,4-DIMETHOXYPHENETHYL)PIPERIDINE HYDROCHLORIDE
mf: $C_{22}H_{29}NO_2$•ClH mw: 375.98

TOXICITY DATA with REFERENCE
orl-mus LD50:200 mg/kg ARZNAD 17,1145,67
ivn-mus LD50:20 mg/kg ARZNAD 17,1145,67

SAFETY PROFILE: Poison by ingestion and intravenous routes. When heated to decomposition it emits very toxic fumes of NO_x and HCl.

BEL500 CAS:15090-13-8 ***HR: 3***
1-BENZYL-5-(3-(DIMETHYLAMINO)PROPOXY)-3-METHYLPYRAZOLE
mf: $C_{16}H_{23}N_3O$ mw: 273.42

SYN: B-329

TOXICITY DATA with REFERENCE
orl-mus LD50:339 mg/kg ARZNAD 17,214,67
scu-mus LD50:197 mg/kg ARZNAD 17,214,67

SAFETY PROFILE: Poison by ingestion and subcutaneous routes. When heated to decomposition it emits toxic fumes of NO_x.

BEL550 CAS:73747-22-5 ***HR: 3***
BENZYLDIMETHYLAMMONIUM HEXAFLUOROARSENATE
mf: $C_9H_{13}N$•AsF_6H mw: 325.16

SYNS: BENZYLAMINE, N,N-DIMETHYL-, HEXAFLUOROARSENATE (1-) ◇ N,N-DIMETHYLBENZYLAMINE HEXAFLUOROARSENATE

TOXICITY DATA with REFERENCE
ivn-mus LD50:180 mg/kg CSLNX* NX#04251

OSHA PEL: TWA 0.5 mg(As)/m^3

SAFETY PROFILE: Poison by intravenous route. When heated to decomposition it emits toxic fumes of NO_x, F^-, and As.

BEL750 CAS:151-05-3 ***HR: 2***
BENZYLDIMETHYL CARBINYL ACETATE
mf: $C_{12}H_{16}O_2$ mw: 192.28

SYNS: DIMETHYLBENZYL CARBINOLACETATE ◇ α,α-DIMETHYLPHENETHYL ACETATE ◇ α,α-DIMETHYLPHENETHYL ALCOHOL ACETATE ◇ DMBCA

TOXICITY DATA with REFERENCE
skn-rbt 500 mg/24H MOD FCTXAV 12,533,74
orl-rat LD50:3300 mg/kg FCTXAV 12,533,74

CONSENSUS REPORTS: Reported in EPA TSCA Inventory.

SAFETY PROFILE: Moderately toxic by ingestion. A skin irritant. When heated to decomposition it emits acrid smoke and irritating fumes. See also ALCOHOLS.

BEL850 CAS:10094-34-5 ***HR: 1***
BENZYL DIMETHYLCARBINYL n-BUTYRATE
mf: $C_{14}H_{20}O_2$ mw: 220.34

SYNS: BENZYL DIMETHYLCARBINYL BUTYRATE ◇ BUTYRIC ACID, α-α-DIMETHYLPHENETHYL ESTER ◇ DIMETHYLBENZYLCARBINYL BUTYRATE ◇ α-α-DIMETHYLPHENETHYL BUTYRATE

TOXICITY DATA with REFERENCE
skn-rbt 500 mg/24H MOD FCTXAV 18,667,80

CONSENSUS REPORTS: Reported in EPA TSCA Inventory.

SAFETY PROFILE: A skin irritant. When heated to decomposition it emits acrid smoke and irritating fumes.

BEM000 CAS:139-07-1 ***HR: 2***
BENZYLDIMETHYLDODECYLAMMONIUM CHLORIDE
mf: $C_{21}H_{38}N$•Cl mw: 340.05

SYN: DODECYL DIMETHYL BENZYLAMMONIUM CHLORIDE

TOXICITY DATA with REFERENCE
skn-rbt 1 mg/24H OYYAA2 6,329,72
eye-rbt 1 mg OYYAA2 6,329,72

SAFETY PROFILE: A skin and eye irritant. When heated to decomposition it emits very toxic fumes of NO_x, NH_3, and Cl^-.

BEM250 CAS:37557-89-4 ***HR: 1***
BENZYLDIMETHYLEICOSANYLAMMONIUM CHLORIDE
mf: $C_{29}H_{54}N$•Cl mw: 452.29

SYN: EICOSANYL DIMETHYL BENZYLAMMONIUM CHLORIDE

TOXICITY DATA with REFERENCE
skn-rbt 1 mg/24H OYYAA2 6(2),329,72
eye-rbt 1 mg OYYAA2 6(2),329,72

SAFETY PROFILE: A skin and eye irritant. When heated to decomposition it emits very toxic fumes of NH_3, NO_x, and Cl^-.

BEM325 ***HR: 3***
N-BENZYL-N',N'-DIMETHYL-N-1-NAPHTHYLETHYLENEDIAMINE
mf: $C_{21}H_{24}N_2$ mw: 304.47

TOXICITY DATA with REFERENCE
ipr-rat LDLo:90 mg/kg BJPCAL 11,1,56
ipr-mus LD50:135 mg/kg BJPCAL 11,1,56
scu-mus LD50:463 mg/kg BJPCAL 11,1,56

SAFETY PROFILE: Poison by intraperitoneal route. Moderately toxic by subcutaneous route. When heated to decomposition it emits toxic fumes of NO_x. See also AMINES.

BEM330 ***HR: 3***
N-BENZYL-N',N'-DIMETHYL-N-2-NAPHTHYLETHYLENEDIAMINE
mf: $C_{21}H_{24}N_2$ mw: 304.47

TOXICITY DATA with REFERENCE
ipr-rat LDLo:135 mg/kg BJPCAL 11,1,56
ipr-mus LD50:265 mg/kg BJPCAL 11,1,56
scu-mus LD50:740 mg/kg BJPCAL 11,1,56

SAFETY PROFILE: Poison by intraperitoneal route. Moderately toxic by subcutaneous route. When heated to decomposition it emits toxic fumes of NO_x. See also AMINES.

BEM500 CAS:961-71-7 ***HR: 3***
N-BENZYL-N',N'-DIMETHYL-N-PHENYLETHYLENEDIAMINE
mf: $C_{17}H_{22}N_2$ mw: 254.41

SYNS: ANTERGAN ◇ BRIDAL ◇ DIMETINA ◇ N,N-DIMETHYL-N'-PHENYL-N'-(PHENYLMETHYL)-1,2-ETHANEDIAMINE(9CI) ◇ LERGITIN ◇ NCI-C60719 ◇ PHENBENZAMINE ◇ PM245 ◇ 2339 RP

TOXICITY DATA with REFERENCE
ipr-rat LDLo:120 mg/kg BJPCAL 11,1,56
ims-rat LDLo:350 mg/kg FRPSAX 13,3,58
ipr-mus LD50:170 mg/kg FRPSAX 13,3,58
scu-mus LD50:400 mg/kg BJPCAL 11,1,56

SAFETY PROFILE: Poison by subcutaneous, intraperitoneal, and intramuscular routes. When heated to decomposition it emits toxic fumes of NO_x.

BEM750 CAS:525-02-0 ***HR: 3***
1-BENZYL-2,5-DIMETHYL SEROTONIN HYDROCHLORIDE
mf: $C_{19}H_{22}N_2O•ClH$ mw: 330.89

SYNS: 3-(2-AMINOETHYL)-1-BENZYL-5-METHOXY-2-METHYLINDOLE HYDROCHLORIDE ◇ BAS ◇ BENANSERIN HYDROCHLORIDE ◇ BENZYL ANTISEROTONIN ◇ 1-BENZYL-2-METHYL-3-(2-AMINOETHYL)-5-METHOXYINDOLEHYDROCHLORIDE ◇ 1-BENZYL-2-METHYL-5-METHOXYTRYPTAMINE HYDROCHLORIDE ◇ SEROTONIN BENZYL ANALOG ◇ WOOLLEY'S ANTI-SEROTONIN

TOXICITY DATA with REFERENCE
ipr-rat TDLo:70 mg/kg (8-14D preg):REP PEDIAU 27,318,61
ipr-mus LD50:250 mg/kg JMCMAR 9,819,66

SAFETY PROFILE: Poison by intraperitoneal route. A serotonin antagonist which causes psychotropic effects in humans. Experimental reproductive effects. When heated to decomposition it emits very toxic fumes of HCl and NO_x.

BEN000 CAS:121-54-0 ***HR: 3***
BENZYLDIMETHYL(2-(2-(p-(1,1,3,3-TETRAMETHYLBUTYL)PHENOXY)ETHOXY)ETHYL) AMMONIUM CHLORIDE
mf: $C_{27}H_{42}NO_2•Cl$ mw: 448.15

PROP: Colorless crystals. Sol in water.

SYNS: ANTI-GERM 77 ◇ ANTISEPTOL ◇ BENZETHONIUM CHLORIDE ◇ BENZETONIUM CHLORIDE ◇ BENZYLDIMETHYL-p-(1,1,3,3-TETRAMETHYLBUTYL)PHENOXYETHOXY-ETHYLAMMONIUM CHLORIDE ◇ BZT ◇ DIAPP ◇ DIISOBUTYLPHENOXYETHOXYETHYLDIMETHYL BENZYL AMMONIUM CHLORIDE ◇ DISILYN ◇ HYAMINE ◇ HYAMINE 1622 ◇ NCI-C61494 ◇ p-tert-OCTYLPHENOXYETHOXYETHYLDIMETHYLBENZYLAMMONIUM CHLORIDE ◇ PHEMERIDE ◇ PHEMEROL CHLORIDE ◇ PHEMITHYN ◇ POLYMINE D ◇ QUATRACHLOR ◇ SOLAMINE

TOXICITY DATA with REFERENCE
eye-rbt 30 μg SEV PSTGAW 20,16,53
scu-rat TDLo:104 mg/kg/1Y-I:NEO CTOXAO 4,185,71
orl-rat LD50:368 mg/kg PSEBAA 120,511,65
ipr-rat LD50:16500 μg/kg FSDZD4 9,729,83
scu-rat LD50:119 mg/kg NTIS** PB195-158
ivn-rat LD50:19 mg/kg SCHSAV 30,147,54
unr-rat LD50:420 mg/kg MEIEDD 10,152,83
orl-mus LD50:338 mg/kg PSEBAA 120,511,65
ipr-mus LD50:15500 μg/kg FSDZD4 9,729,83
ivn-mus LD50:30 mg/kg JAPMA8 40,267,51

CONSENSUS REPORTS: Reported in EPA TSCA Inventory.

SAFETY PROFILE: Poison by ingestion, subcutaneous, intraperitoneal, and intravenous routes. A severe eye irritant. Questionable carcinogen with experimental neoplastigenic data. When heated to decomposition it emits very toxic fumes of Cl^-, NH_3, and NO_x. A topical anti-infective agent.

BEN250 CAS:101-49-5 ***HR: 2***
2-BENZYLDIOXOLAN
mf: $C_{10}H_{12}O_2$ mw: 164.22

SYN: PHENYLACETALDEHYDE ETHYLENEGLYCOL ACETAL

TOXICITY DATA with REFERENCE
orl-rat LD50:2200 mg/kg FCTXAV 14,827,76
skn-rbt LD50:2600 mg/kg FCTXAV 14,827,76

CONSENSUS REPORTS: Reported in EPA TSCA Inventory.

SAFETY PROFILE: Moderately toxic by ingestion and skin contact. When heated to decomposition it emits acrid smoke and irritating fumes. See also ALDEHYDES.

BEN750 CAS:15090-16-1 ***HR: 3***
1-BENZYL-5-(3-(DIPROPYLAMINO)PROPOXY)-3-METHYLPYRAZOLE
mf: $C_{20}H_{31}N_3O$ mw: 329.54

SYN: B-331

TOXICITY DATA with REFERENCE
orl-mus LD50:376 mg/kg ARZNAD 17,214,67
scu-mus LD50:280 mg/kg ARZNAD 17,214,67

SAFETY PROFILE: Poison by ingestion and subcutaneous routes. When heated to decomposition it emits toxic fumes of NO_x.

BEO000 CAS:7281-04-1 ***HR: 3***
BENZYLDODECYLDIMETHYL AMMONIUM BROMIDE
mf: $C_{21}H_{37}N•Br$ mw: 383.50

SYNS: BROMEKDWUMETYLOLAURYLOBENZYLOAMONIOWY (POLISH) ◇ DIMETHYL LAURYLBENZENE AMMONIUM BROMIDE ◇ STERINOLU (POLISH)

TOXICITY DATA with REFERENCE
orl-rat LD50:250 mg/kg RPZHAW 17,543,66

CONSENSUS REPORTS: Reported in EPA TSCA Inventory.

SAFETY PROFILE: Poison by ingestion. See also BROMIDES. When heated to decomposition it emits very toxic fumes of NH_3, NO_x, and Br^-.

BEO250 CAS:103-50-4 ***HR: 2***
BENZYL ETHER
mf: $C_{14}H_{14}O$ mw: 198.28

PROP: Colorless to pale yellow liquid. Mp: 5°, bp: 298°, flash p: 275°F (CC), d: 1.039, vap d: 6.84, refr index: 1.557.

SYNS: BENZYL OXIDE (CZECH) ◇ DIBENZYLETHER (CZECH) ◇ FEMA No. 2371

TOXICITY DATA with REFERENCE
skn-rbt 500 mg/24H MLD 28ZPAK -,38,72
eye-rbt 500 mg/24H MLD 28ZPAK -,38,72
orl-rat LD50:2500 mg/kg FCTXAV 16,637,78

CONSENSUS REPORTS: Reported in EPA TSCA Inventory.

SAFETY PROFILE: Moderately toxic by ingestion. Vapors are probably narcotic in high concentration. A skin and eye irritant. Combustible when exposed to heat or flame; can react with oxidizing materials. Moderate explosion hazard by spontaneous chemical reaction. To fight fire, use CO_2, dry chemical. See also ETHERS.

BEO750 CAS:2016-63-9 ***HR: 3***
8′-BENZYL-7(2-(ETHYL(2-HYDROXYETHYL)AMINO)ETHYL) THEOPHYLLINE HYDROCHLORIDE
mf: $C_{20}H_{27}N_5O_3•ClH$ mw: 421.98

SYNS: BAMIFYLLINE HYDROCHLORIDE ◇ BAMIPHYLLINE HYDROCHLORIDE ◇ BAX 2793Z ◇ BENZETAMOPHYLLINE HYDROCHLORIDE ◇ 8-BENZYL-7-(N-ETHYL-N-(β-HYDROXYETHYL)-AMINOETHYL)THEOPHYLLINE HYDROCHLORIDE ◇ 8102 CB HYDROCHLORIDE ◇ 7-(N-(β-HYDROXYETHYL)-N-ETHYL)-AMINOETHYL-8-BENZYL-THEOPHYLLINE ◇ TRENTADIL HYDROCHLORIDE

TOXICITY DATA with REFERENCE
orl-rat LD50:1139 mg/kg ARZNAD 18,460,68
ipr-rat LD50:131 mg/kg ARZNAD 18,460,68
ivn-rat LD50:65 mg/kg ARZNAD 18,460,68
orl-mus LD50:246 mg/kg ARZNAD 18,460,68
ipr-mus LD50:89 mg/kg ARZNAD 18,460,68
ivn-mus LD50:67 mg/kg ARZNAD 18,460,68
orl-dog LD50:614 mg/kg ARZNAD 18,460,68
ivn-dog LD50:29 mg/kg ARZNAD 18,460,68
orl-gpg LD50:165 mg/kg ARZNAD 18,460,68

SAFETY PROFILE: Poison by ingestion, intraperitoneal, and intravenous routes. When heated to decomposition it emits toxic fumes of HCl and NO_x. A bronchodilator. See other theophylline entries.

BEP250 CAS:104-57-4 ***HR: 2***
BENZYL FORMATE
mf: $C_8H_8O_2$ mw: 136.16

SYNS: BENZYL ALCOHOL FORMATE ◇ BENZYL METHANOATE

TOXICITY DATA with REFERENCE
orl-rat LD50:1400 mg/kg FCTXAV 11,1019,73
skn-rbt LD50:2000 mg/kg FCTXAV 11,1019,73

CONSENSUS REPORTS: Reported in EPA TSCA Inventory.

SAFETY PROFILE: Moderately toxic by ingestion and skin contact. Probably narcotic in high concentrations. See also ESTERS. When heated to decomposition it emits acrid, irritating fumes.

BEP500 CAS:10453-86-8 ***HR: 3***
5-BENZYL-3-FURYL METHYL(±)-cis,trans-CHRYSANTHEMATE
mf: $C_{22}H_{26}O_3$ mw: 338.48

SYNS: BENZOFUROLINE ◇ BENZYFUROLINE ◇ (5-BENZYL-3-FURYL)METHYL-2,2-DIMETHYL-3-(2-METHYLPROPENYL)-CYCLOPROPANECARBOXYLATE ◇ CHRYSON ◇ CHRYSRON ◇ DIMETHYL-3-(2-METHYL-1-PROPENYL)CYCLOPROPANE-CARBOXYLATE ◇ ENT 27,474 ◇ FMC 17370 ◇ FOR-SYN ◇ NIA 17170 ◇ NRDC 104 ◇ NSC 195022 ◇ OMS-1206 ◇ PREMGARD ◇ PYNOSECT ◇ PYRETHERM ◇ RESMETHRIN ◇ RESMETRINA (PORTUGUESE) ◇ SBP-1382 ◇ S.B. PENICK 1382 ◇ SYNTHRIN

TOXICITY DATA with REFERENCE
orl-rat LD50:1347 mg/kg PCBPBS 2,308,72
ivn-rat LDLo:160 mg/kg BIOGAL 41(10),283,75
orl-bwd LD50:75 mg/kg AECTCV 12,355,83
orl-mus LD50:1390 mg/kg BECTA6 19,113,78
ihl-mus LD50:99 mg/kg BECTA6 19,113,78
skn-rbt LD50:2500 mg/kg SPEADM 78-1,9,78

CONSENSUS REPORTS: EPA Genetic Toxicology Program.

SAFETY PROFILE: Poison by inhalation, ingestion, and intravenous routes. Moderately toxic by skin contact. When heated to decomposition it emits acrid and irritating fumes. See also ESTERS.

BEP750 CAS:28434-01-7 ***HR: 3***
5-BENZYL-3-FURYLMETHYL(+)-trans-CHRYSANTHEMATE
mf: $C_{22}H_{26}O_3$ mw: 338.48

SYNS: BIORESMETHRIN ◇ BIORESMETHRINE ◇ BIORESMETRINA (PORTUGUESE) ◇ NIA-18739 ◇ NRDC 107 ◇ (+)-trans-RESMETHRIN ◇ d-trans-RESMETHRIN ◇ RU-11484 ◇ SBP-1390

TOXICITY DATA with REFERENCE
ivn-rat LD50:340 mg/kg BIOGAL 41,283,75
orl-mus LD50:590 mg/kg EVHPAZ 14,15,76

SAFETY PROFILE: Poison by intravenous route. Moderately toxic by ingestion. When heated to decomposition it emits acrid smoke and irritating fumes. A pesticide. See also ESTERS.

BEQ000 CAS:555-96-4 ***HR: 3***
BENZYLHYDRAZINE
mf: $C_7H_{10}N_2$ mw: 122.19

TOXICITY DATA with REFERENCE
oms-bcs 10 mmol/L MUREAV 5,343,68
mmo-omi 4 mg/L MUREAV 173,233,86
ipr-mus LD50:100 mg/kg RPTOAN 36,27,73
scu-mus LD50:68 mg/kg ANYAA9 80,568,59

SAFETY PROFILE: Poison by intraperitoneal and subcutaneous routes. Mutation data reported. When heated to decomposition it emits toxic fumes such as NO_x.

BEQ250 CAS:20570-96-1 ***HR: 3***
BENZYLHYDRAZINE DIHYDROCHLORIDE
mf: $C_7H_{10}N_2$•2ClH mw: 195.11

TOXICITY DATA with REFERENCE
orl-mus TDLo:10 g/kg/29W-C:NEO ZEKBAI 87,267,76
ipr-mus LD50:11 mg/kg JMCMAR 18,20,75

SAFETY PROFILE: Poison by intraperitoneal route. Questionable carcinogen with experimental neoplastigenic data. When heated to decomposition it emits very toxic fumes of HCl and NO_x.

BEQ500 CAS:1073-62-7 ***HR: 3***
BENZYLHYDRAZINE HYDROCHLORIDE
mf: $C_7H_{10}N_2$•ClH mw: 158.65

SYNS: P 1297 ◇ USAF EL-54 ◇ Z 102

TOXICITY DATA with REFERENCE
orl-mus LD50:90 mg/kg JMPCAS 5,221,62
ipr-mus LD50:50 mg/kg NTIS** AD277-689

SAFETY PROFILE: Poison by ingestion and intraperitoneal route. When heated to decomposition it emits very toxic fumes of NO_x and Cl^-.

BEQ625 CAS:73-48-3 ***HR: 3***
BENZYLHYDROFLUMETHIAZIDE
mf: $C_{15}H_{14}F_3N_3O_4S_2$ mw: 421.44

PROP: Crystals. Mp: 221-223°. Insol in water, chloroform, benzene, and ether; sol in acetone and alc.

SYNS: APRINOX ◇ Be 724-A ◇ BENDROFLUAZIDE ◇ BENDROFLUMETHIAZIDE ◇ BENTRIDE ◇ BENURON ◇ BENZYDROFLUMETHIAZIDE ◇ 3-BENZYL-3,4-DIHYDRO-6-(TRIFLUOROMETHYL)-2H-1,2,4-BENZOTHIADIAZINE-7-SULFONAMIDE 1,1-DIOXIDE ◇ BENZYLRODIURAN ◇ 3-BENZYL-6-TRIFLUOROMETHYL-7-SULFAMOYL-3,4-DIHYDRO-1,2,4-BENZOTHIADIAZINE-1,1-DIOXIDE ◇ BERKOZIDE ◇ BHFT ◇ BL H368 ◇ BRISTURIC ◇ BRISTURON ◇ CENTYL ◇ FLUMESIL ◇ FT 8 ◇ INTOLEX ◇ NATERETIN ◇ NATURETIN ◇ NATURINE ◇ NEO-NACLEX ◇ NEORONTYL ◇ NIAGARIL ◇ NIKION ◇ ORSILE ◇ PLURYL ◇ PLURYLE ◇ PLUSURIL ◇ POLIURON ◇ RELAN BETA ◇ REPICIN ◇ SALURAL ◇ SALURES ◇ SINESALIN ◇ SODIURETIC ◇ THIAZIPIDICO ◇ 6-TRIFLUOROMETHYL-3-BENZYL-7-SULFAMYL-3,4-DIHYDRO-1,2,4-BENZOTHIADIAZINE-1,1-DIOXIDE ◇ URLEA

TOXICITY DATA with REFERENCE
cyt-ham:lng 200 mg/L GMCRDC 27,95,81
orl-wmn TDLo:3 mg/kg:CNS LANCAO 1,564,82
ipr-mus LD50:4800 mg/kg AEPPAE 238,435,60
ivn-mus LD50:395 mg/kg JPETAB 134,273,61

SAFETY PROFILE: Poison by intravenous route. Human systemic effects by ingestion: convulsions and somnolence. Mutation data reported. When heated to decomposition it emits toxic fumes of F^-, SO_x, and NO_x.

BER500 CAS:3268-19-7 **HR: 3**
4,6-o-BENZYLIDENE-β-d-GLUCOPYRANOSIDE PODOPHYLLOTOXIN
mf: $C_{35}H_{36}O_{13}$ mw: 664.71

SYNS: NSC 42076 ◇ PODOPHYLLOTOXIN-BENZILIDEN-GLUCOSID (GERMAN) ◇ PODOPHYLLOTOXIN-o-BENXYLIDENE-β-d-GLUCOPYRANOSIDE ◇ PRORESIDOR ◇ SP G ◇ SPG 827

TOXICITY DATA with REFERENCE
orl-mus LD50:280 mg/kg ARZNAD 11,549,61
ipr-mus LD50:280 mg/kg ARZNAD 11,549,61

SAFETY PROFILE: Poison by ingestion and intraperitoneal routes. When heated to decomposition it emits acrid smoke and irritating fumes.

BES250 CAS:2782-70-9 **HR: 3**
BENZYLIDENEMETHYLPHOSPHORODITHIOATE
mf: $C_{11}H_{18}O_4P_2S_4$ mw: 404.47

SYNS: S,S'-BENZYLIDENE BIS(O,O-DIMETHYL PHOSPHORODITHIOATE) ◇ ENT 25,739 ◇ SD 7438 ◇ SHELL SD 7,438 ◇ TOLUENE-α,α-DITHIOL BIS(O,O-DIMETHYL PHOSPHORODITHIOATE)

TOXICITY DATA with REFERENCE
orl-rat LD50:280 mg/kg 28ZEAL 4,371,69
orl-mus LD50:176 mg/kg ARSIM* 20,19,66
skn-rbt LD50:2500 mg/kg BESAAT 12,161,66
orl-ckn LD50:5096 mg/kg TXAPA9 11,49,67

SAFETY PROFILE: Poison by ingestion. Moderately toxic by skin contact. When heated to decomposition it emits very toxic fumes of PO_x and SO_x. A pesticide. See also ESTERS.

BES500 CAS:122-73-6 **HR: 1**
BENZYL ISOAMYL ETHER
mf: $C_{12}H_{18}O$ mw: 178.30

PROP: Liquid. Bp: 235°, d: 0.965 @ 15.5°/15.5°.

SYNS: BENZYL ISOPENTYL ETHER ◇ ISOAMYL BENZYL ETHER

TOXICITY DATA with REFERENCE
skn-rbt 500 mg/24H MLD FCTXAV 16,647,78

CONSENSUS REPORTS: Reported in EPA TSCA Inventory.

SAFETY PROFILE: A skin irritant. See also ETHERS. Flammable when exposed to heat or flame; can react with oxidizing materials. To fight fire, use foam, CO_2, dry chemical.

BES750 CAS:120-11-6 **HR: 1**
BENZYL ISOEUGENOL ETHER
mf: $C_{17}H_{18}O_2$ mw: 254.35

SYNS: BENZYL ALCOHOL ETHER with ISOEUGENOL ◇ BENZYL ISOEUGENOL ◇ BENZYL-2-METHOXY-4-PROPENYLPHENYL ETHER

TOXICITY DATA with REFERENCE
skn-rbt 500 mg/24H MLD FCTXAV 11,1025,73
orl-rat LD50:4900 mg/kg FCTXAV 11,1025,73

CONSENSUS REPORTS: Reported in EPA TSCA Inventory.

SAFETY PROFILE: Mildly toxic by ingestion. A skin irritant. See also ETHERS. When heated to decomposition it emits acrid smoke and irritating fumes.

BET000 CAS:51-12-7 **HR: 3**
N-BENZYL-β-(ISONICOTINYLHYDRAZINO)PROPIONAMIDE
mf: $C_{16}H_{18}N_4O_2$ mw: 298.38

SYNS: N[1]-β-BENZYLCARBAMOYLETHYL-N[2]-ISONICOTINOYLHYDRAZINE ◇ 1-(2-(BENZYLCARBAMOYL)ETHYL)-2-ISONICOTINOYLHYDRAZINE ◇ 2-(2-BENZYLCARBAMYL)ETHYL)-HYDRAZIDE ISONICOTINIC ACID ◇ N-BENZYL-β-(ISONICOTINOYLHYDRAZINE)PROPIONAMIDE ◇ DELMONEURINA ◇ ESPRIL ◇ ISALIZINA ◇ N-ISONICOTINOYL-N'(β-N-BENZYLCARBOXAMIDOETHYL)HYDRAZINE ◇ MYGAL ◇ NIALAMIDE ◇ NIAMID ◇ NIAMIDAL ◇ NIAQUITIL ◇ NUREDAL ◇ NYAZIN ◇ P 1133 ◇ PSICODISTEN ◇ 4-PYRIDINECARBOXYLIC ACID 2-(3-OXO-3-((PHENYLMETHYL)AMINO)PROPYL)HYDRAZIDE ◇ SURGEX

TOXICITY DATA with REFERENCE
mma-sat 10 μg/plate PLRCAT 12,423,80
mmo-sat 10 μg/plate MUREAV 40,305,76
dnr-esc 27 μmol/plate JTEHD6 9,287,82
oms-bcs 10 mmol/L MUREAV 5,343,68
dnd-mus-ipr 2450 μmol/kg CNREA8 41,1469,81
sce-mus-ipr 435 mg/kg JTEHD6 9,287,82
scu-mus TDLo:24 mg/kg (1-6D preg):REP JOENAK 27,147,63
ipr-rat LD50:760 mg/kg TXAPA9 1,524,59
orl-mus LD50:590 mg/kg 27ZQAG -,269,72
ipr-mus LD50:200 mg/kg MPHEAE 16,267,67
ivn-mus LD50:120 mg/kg 27ZQAG -,269,72

CONSENSUS REPORTS: EPA Genetic Toxicology Program.

SAFETY PROFILE: Poison by intravenous and intraperitoneal routes. Moderately toxic by ingestion. Mutation data reported. Experimental reproductive effects. An antidepressant. When heated to decomposition it emits toxic fumes of NO_x.

BET750 CAS:55165-33-8 **HR: 3**
4-(3-BENZYLISOPROPYLAMINO-2-HYDROXYPROPOXY)-9-METHOXY-7-METHYL-FURO(3,2-g)CHROMONE, HYDROCHLORIDE
mf: $C_{26}H_{29}NO_6•ClH$ mw: 488.02

TOXICITY DATA with REFERENCE
orl-mus LD50:380 mg/kg EJMCA5 9,563,74
ipr-mus LD50:69 mg/kg EJMCA5 9,563,74

SAFETY PROFILE: Poison by ingestion and intraperitoneal routes. When heated to decomposition it emits very toxic fumes of Cl^- and NO_x.

BEU250 CAS:622-78-6 ***HR: 3***
BENZYL-ISOTHIOCYANATE
mf: C_8H_7NS mw: 149.22

PROP: Orange-red, crystalline solid. Mp: 41°, bp: 230°, d: 1.125

SYNS: BENZYL MUSTARD OIL ◇ BENZYLSENFOEL (GERMAN) ◇ ISOTHIOCYANIC ACID BENZYL ESTER

TOXICITY DATA with REFERENCE
mmo-sat 150 μg/plate ABCHA6 44,3017,80
ipr-rat LDLo:100 mg/kg ARZNAD 16,870,66
ipr-mus LDLo:100 mg/kg ARZNAD 21,121,71
scu-mus LD50:150 mg/kg ARZNAD 5,505,55

CONSENSUS REPORTS: Reported in EPA TSCA Inventory.

SAFETY PROFILE: Poison by intraperitoneal and subcutaneous routes. Intensely irritating. Mutation data reported. Moderate fire hazard via heat, flame, and oxidizers. To fight fire, use water, spray, foam, dry chemical. When heated to decomposition it emits very toxic NO_x and SO_x. See also ESTERS and THIOCYANATES.

BEU500 CAS:538-28-3 ***HR: 3***
BENZYLISOTHIOUREA HYDROCHLORIDE
mf: $C_8H_{10}N_2S{\cdot}ClH$ mw: 202.72

SYNS: BENZYLISOTHIOURONIUM CHLORIDE ◇ 2-BENZYLISOTHIOURONIUM CHLORIDE ◇ BENZYL THIOPSEUDOUREA HYDROCHLORIDE ◇ 2-BENZYL-2-THIO-PSEUDOUREA HYDROCHLORIDE ◇ BENZYLTHIURONIUM CHLORIDE ◇ S-BENZYLTHIURONIUM CHLORIDE ◇ BTKH ◇ ISOTHIOURONIUM CHLORIDE, BENZYL ◇ 2-THIO-2-BENZYL-PSEUDOUREA HYDROCHLORIDE ◇ TL 944 ◇ USAF EK-2124

TOXICITY DATA with REFERENCE
orl-rat LD50:150 mg/kg JPETAB 90,260,47
ipr-mus LD50:50 mg/kg NTIS** AD277-689
scu-mus LDLo:80 mg/kg NDRC** No.9-4-1-9,43
ivn-mus LD50:32 mg/kg CSLNX* NX#00167

CONSENSUS REPORTS: Reported in EPA TSCA Inventory.

SAFETY PROFILE: Poison by ingestion, intraperitoneal, subcutaneous, and intravenous routes. When heated to decomposition it emits very toxic fumes of HCl, SO_x, and NO_x.

BEU750 CAS:140-25-0 ***HR: 1***
BENZYL LAURATE
mf: $C_{19}H_{30}O_2$ mw: 290.49

SYNS: BENZYL DODECANOATE ◇ DODECANOIC ACID BENZYL ESTER

TOXICITY DATA with REFERENCE
skn-rbt 500 mg/kg24H MOD FCTXAV 16,649,78

CONSENSUS REPORTS: Reported in EPA TSCA Inventory.

SAFETY PROFILE: A skin irritant. See also ESTERS. When heated to decomposition it emits acrid smoke and irritating fumes.

BEU800 CAS:35133-55-2 ***HR: 3***
4-BENZYL-α-(4-METHOXYPHENYL)-β-METHYL-1-PIPERIDINEETHANOL
mf: $C_{22}H_{29}NO_2$ mw: 339.52

SYN: RC 61-96

TOXICITY DATA with REFERENCE
orl-mus LD50:120 mg/kg ARZNAD 21,1992,71
ipr-mus LD50:45 mg/kg ARZNAD 21,1992,71
ivn-mus LD50:13 mg/kg ARZNAD 21,1992,71

SAFETY PROFILE: Poison by ingestion, intravenous, and intraperitoneal routes. When heated to decomposition it emits toxic fumes of NO_x.

BEW000 CAS:1085-32-1 ***HR: 3***
1-BENZYL-2-(3-METHYLISOXAZOL-5-YL)CARBONYL HYDRAZINE
mf: $C_{12}H_{13}N_3O_2$ mw: 231.28

SYN: 3-METHYL-5-ISOXAZOLECARBOXYLIC ACID 2-BENZYLHYDRAZIDE

TOXICITY DATA with REFERENCE
orl-mus LD50:178 mg/kg SKNEA7 14,58,64
ipr-mus LD50:245 mg/kg SKNEA7 14,58,64
orl-cat LD50:100 mg/kg SKNEA7 14,58,64

SAFETY PROFILE: Poison by ingestion and intraperitoneal routes. When heated to decomposition it emits toxic fumes of NO_x.

BEW500 CAS:15090-12-7 ***HR: 3***
1-BENZYL-3-METHYL-5-(2-(4-METHYL-1-PIPERAZINYL)ETHOXY)PYRAZOLE
mf: $C_{18}H_{26}N_4O$ mw: 314.48

SYN: B-324

TOXICITY DATA with REFERENCE
orl-mus LD50:362 mg/kg ARZNAD 17,214,67
scu-mus LD50:230 mg/kg ARZNAD 17,214,67

SAFETY PROFILE: Poison by ingestion and subcutaneous routes. When heated to decomposition it emits toxic fumes of NO_x.

BEW750 CAS:15090-10-5 ***HR: 3***
1-BENZYL-3-METHYL-5-(2-(2-METHYLPIPERIDINO)ETHOXY)PYRAZOLE
mf: $C_{19}H_{27}N_3O$ mw: 313.49

SYN: B-322

TOXICITY DATA with REFERENCE
orl-mus LD50:335 mg/kg ARZNAD 17,214,67
scu-mus LD50:182 mg/kg ARZNAD 17,214,67

SAFETY PROFILE: Poison by ingestion and subcutaneous routes. When heated to decomposition it emits toxic fumes of NO_x.

BEX500 CAS:306-07-0 ***HR: 3***
BENZYLMETHYLPROPYNYLAMINE HYDROCHLORIDE
mf: $C_{11}H_{13}N{\cdot}ClH$ mw: 195.71

SYNS: A 19120 ◇ N-BENZYL-N-METHYL-2-PROPYNYLAMINE HYDROCHLORIDE ◇ EUDATINE ◇ N-METHYL-N-(2-PROPYNYL)-BENZYLAMINE HYDROCHLORIDE ◇ PARGYLINE HYDROCHLORIDE ◇ USAF A-19120

TOXICITY DATA with REFERENCE
scu-mus TDLo:12 mg/kg (1-6D preg):REP JOENAK 27,146,63
orl-man TDLo:108 mg/kg/26W-I JCLPDE 44,25,83
orl-wmn TDLo:1500 μg/kg/D:CNS,PSY AJPSAO 118,255,61
orl-rat LD50:250 mg/kg 27ZQAG -,401,72
ipr-rat LD50:142 mg/kg ANYAA9 107,1068,63
ivn-rat LD50:175 mg/kg 27ZQAG -,401,72
orl-mus LD50:680 mg/kg ANYAA9 107,1068,63
ipr-mus LD50:300 mg/kg NTIS** AD277-689
orl-dog LD50:175 mg/kg ANYAA9 107,1068,63

SAFETY PROFILE: Poison by ingestion, intraperitoneal, and intravenous routes. Human systemic effects by ingestion: effects on fluid intake, psychological effects. Experimental reproductive effects. When heated to decomposition it emits very toxic fumes of HCl and NO_x.

BEX750 CAS:7368-12-9 ***HR: 3***
(1-BENZYL-3-METHYL-5-PYRAZOLYLOXYETHYL)TRIMETHYLAMMONIUM IODIDE
mf: $C_{16}H_{24}N_3O{\cdot}I$ mw: 401.33

SYN: B-325

TOXICITY DATA with REFERENCE
orl-mus LD50:4013 mg/kg ARZNAD 17,214,67
scu-mus LD50:181 mg/kg ARZNAD 17,214,67

SAFETY PROFILE: Poison by subcutaneous route. Moderately toxic by ingestion. See also IODIDES. When heated to decomposition it emits very toxic fumes of NH_3, NO_x, and I^-.

BEY800 CAS:62064-66-8 ***HR: 3***
N-BENZYL-α-METHYL-m-TRIFLUOROMETHYLPHENETHYLAMINE
mf: $C_{17}H_{18}F_3N$ mw: 293.36

SYN: N-BENZYL-α-METHYL-3-TRIFLUOROMETHYLPHENETHYLAMINE

TOXICITY DATA with REFERENCE
orl-mus LD50:300 mg/kg ARZNAD 27,116,77
ipr-mus LD50:144 mg/kg ISYAM* -,21,70

SAFETY PROFILE: Poison by ingestion and intraperitoneal routes. When heated to decomposition it emits very toxic fumes of F^- and NO_x.

BFA000 CAS:63906-64-9 ***HR: 3***
8-BENZYL-7-(1'-MORPHOLINO-2'-AMINO)ETHYLTHEOPHYLLINE HYDROCHLORIDE
mf: $C_{20}H_{23}N_6O_3{\cdot}ClH$ mw: 431.95

SYN: AC 3092

TOXICITY DATA with REFERENCE
orl-rat LD50:1498 mg/kg ARZNAD 19,1113,69
ipr-rat LD50:140 mg/kg ARZNAD 19,1113,69
scu-rat LD50:331 mg/kg ARZNAD 19,1113,69
ivn-rat LD50:54 mg/kg ARZNAD 19,1113,69
orl-mus LD50:247 mg/kg ARZNAD 19,1113,69
ipr-mus LD50:83 mg/kg ARZNAD 19,1113,69
scu-mus LD50:126 mg/kg ARZNAD 19,1113,69

SAFETY PROFILE: Poison by ingestion, intraperitoneal, subcutaneous, and intravenous routes. When heated to decomposition it emits very toxic fumes of NO_x and HCl. See other theophylline entries.

BFA250 CAS:15285-42-4 ***HR: 3***
BENZYL NITRATE
mf: $C_7H_7NO_3$ mw: 153.14

SAFETY PROFILE: Explodes above 180°C. Violent reaction with Lewis acids (e.g., sulfuric acid, tin(IV) chloride, boron trifluoride) results in gas evolution. When heated to decomposition it emits toxic fumes of NO_x. See also NITRATES.

BFA899 ***HR: 3***
BENZYLOXY ACETYLENE
mf: C_9H_8O mw: 132.16

SAFETY PROFILE: Explodes if heated above 60° in vacuo. When heated to decomposition it emits acrid smoke and fumes.

BFC000 CAS:40283-91-8 ***HR: 3***
S-((N-(2-BENZYLOXYETHYL)AMIDINO) METHYL) HYDROGEN THIOSULFATE
mf: $C_{11}H_{16}N_2O_4S_2$ mw: 304.41

TOXICITY DATA with REFERENCE
orl-mus LD50:90 mg/kg JMCMAR 15,1313,72
ipr-mus LD50:60 mg/kg JMCMAR 15,1313,72

SAFETY PROFILE: Poison by ingestion and intraperitoneal routes. When heated to decomposition it emits very toxic fumes of SO_x and NO_x.

BFC250 CAS:14226-68-7 ***HR: 3***
5-BENZYLOXY-3-(1-METHYL-2-PYRROLIDINYL) INDOLE
mf: $C_{20}H_{22}N_2O$ mw: 306.44

TOXICITY DATA with REFERENCE
ipr-rat LD50:71 mg/kg JMCMAR 7,415,64
ipr-mus LD50:100 mg/kg JMCMAR 7,415,64

SAFETY PROFILE: Poison by intraperitoneal route. When heated to decomposition it emits toxic fumes of NO_x.

BFC400 CAS:1252-18-2 ***HR: D***
1-(p-(BENZYLOXY)PHENYL)-2-(o-FLUOROPHENYL)-1-PHENYLETHYLENE
mf: $C_{27}H_{21}FO$ mw: 380.48

SYNS: ETHER, BENZYL p-(o-FLUORO-α-PHENYLSTYRYL)PHENYL ◇ ETHYLENE, 1-(p-(BENZYLOXY)PHENYL)-2-(o-FLUOROPHENYL)-1-PHENYL- ◇ STILBENE, 4-(BENZYLOXY)-2'-FLUORO-α-PHENYL-

TOXICITY DATA with REFERENCE
scu-rat TDLo:300 μg/kg (female 3D pre):REP AIPTAK 151,475,64

SAFETY PROFILE: Experimental reproductive effects. When heated to decomposition it emits toxic fumes of F^-.

BFC450 ***HR: D***
2-(m-(BENZYLOXY)PHENYL)PYRAZOLO(1,5-a)QUINOLINE
mf: $C_{24}H_{20}N_2O$ mw: 352.46

SYN: PYRAZOLO(1,5-a)QUINOLINE,2-(m-(BENZYLOXY)PHENYL)-

TOXICITY DATA with REFERENCE
scu-rat TDLo:100 mg/kg (female 6-10D post):REP EJMCA5 19,215,84

SAFETY PROFILE: Experimental reproductive effects. When heated to decomposition it emits toxic fumes of NO_x.

BFC500 CAS:40283-92-9 ***HR: 3***
S-((N-(3-BENZYLOXYPROPYL)AMIDINO) METHYL) HYDROGEN THIOSULFATE
mf: $C_{12}H_{18}N_2O_4S_2$ mw: 318.44

TOXICITY DATA with REFERENCE
orl-mus LD50:300 mg/kg JMCMAR 15,1313,72
ipr-mus LD50:30 mg/kg JMCMAR 15,1313,72

SAFETY PROFILE: Poison by ingestion and intraperitoneal routes. When heated to decomposition it emits very toxic fumes of NO_x and SO_x.

BFC750 CAS:1538-09-6 ***HR: 2***
BENZYLPENCILLINDIBENZYLETHYLENE-DIAMINE SALT
mf: $C_{32}H_{36}N_4O_8S_2 \cdot C_{16}H_{20}N_2$ mw: 909.22

SYNS: BEACILLIN ◇ BEN-P ◇ BENZACILLIN ◇ BENZATHINE BENZYLPENICILLIN ◇ BENZATHINE PENICILLIN ◇ BENZATHINE PENICILLIN G ◇ BENZETHACIL ◇ BENZYLPENICILLIN BENZATHINE ◇ BICA-PENICILLIN ◇ BICILLIN ◇ CEPACILINA ◇ CEPACILLINA ◇ CILLENTA ◇ DBED DIPENCILLIN G ◇ DBED PENICILLIN ◇ DEBECILLIN ◇ DEBECYLINA ◇ DIAMINE DIPENICILLIN G ◇ DIAMINOCILLIAN ◇ DIBENCIL ◇ DIBENCILLIN ◇ N,N'-DIBENZYLETHYLENEDIAMINE BIS(BENZYL PENICILLIN) ◇ DIBENZYLETHYLENEDIAMINE-DI-PENICILLIN G ◇ N,N'-DIBENZYLETHYLENEDIAMINE, compounded with PENICILLIN G (1:2) ◇ DIPO-SAFT ◇ DURABIOTIC ◇ DURA-PENITA ◇ DUROPENIN ◇ EXTENCILLINE ◇ EXTENICILLINE ◇ LENTOCILLIN ◇ LENTOPENIL ◇ LEOMYPEN ◇ LONGACILIAN ◇ LONGICIL ◇ LPG ◇ MEGACILLIN SUSPENSION ◇ MOLDAMIN ◇ NCI-C56100 ◇ NEOLIN ◇ PENADUR ◇ PENADUR L-A ◇ PENDEPON ◇ PEN-DI-BEN ◇ PENDITAN ◇ PENDURAN ◇ PENICILLIN G, compounded with N,N'-DIBENZYLETHYLENEDIAMINE (2:1) ◇ PENICILLIN G SALT of N,N'-DIBENZYLETHYLENEDIAMINE ◇ PENIDURAL ◇ PENIDURE ◇ PENILENTE ◇ PERMAPEN ◇ RETARPEN ◇ TARDOCILLIN ◇ VETARCILLIN ◇ VICIN ◇ WYCILLINA

TOXICITY DATA with REFERENCE
par-rat TDLo:22905 μg/kg (4D preg):REP BEXBAN 82,1076,76
orl-mus LD50:2000 mg/kg NIIRDN 6,774,82
ipr-mus LDLo:460 mg/kg NIIRDN 6,774,82

CONSENSUS REPORTS: Reported in EPA TSCA Inventory.

SAFETY PROFILE: Moderately toxic by ingestion and intraperitoneal routes. Experimental reproductive effects. When heated to decomposition it emits very toxic fumes of NO_x and SO_x. See other penicillin entries.

BFD000 CAS:113-98-4 ***HR: 3***
BENZYLPENICILLINIC ACID POTASSIUM SALT
mf: $C_{16}H_{17}N_2O_4S \cdot K$ mw: 372.51

SYNS: BENZYLPENICILLIN POTASSIUM ◇ BENZYLPENICILLIN POTASSIUM SALT ◇ CILLORAL ◇ COSMOPEN ◇ CRISTAPEN ◇ CRYSTAPEN ◇ ESKACILLIN ◇ FALAPEN ◇ FORPEN ◇ HIPERCILINA ◇ HYASORB ◇ HYLENTA ◇ MEGACILLIN TABLETS ◇ MONOPEN ◇ NOTARAL ◇ PENALEV ◇ PENICILLIN G POTASSIUM

◇ PENICILLIN G POTASSIUM SALT ◇ PENISEM ◇ PENTID ◇ PENTIDS ◇ PFIZERPEN ◇ POTASSIUM BENZYLPENICILLIN ◇ POTASSIUM BENZYLPENICILLINATE ◇ POTASSIUM BENZYLPENICILLIN G ◇ POTASSIUM PENICILLIN G ◇ POTASSIUM SALT of BENZYLPENICILLIN ◇ QIDPEN G ◇ SCOTCIL ◇ SK-PENICILLIN G ◇ SUGRACILLIN ◇ TABILIN ◇ TU CILLIN

TOXICITY DATA with REFERENCE
spm-rat-unr 200 mg/kg/8D JOURAA 112,348,74
orl-rat LD50:6700 mg/kg AIPTAK 123,295,60
scu-rat LD50:11250 mg/kg TXAPA9 9,445,66
orl-mus LD50:6257 mg/kg AIPTAK 125,83,60
ivn-mus LD50:400 mg/kg ANTCAO 12,192,62
ice-mus LDLo:2 mg/kg PLMEAA 49,103,83
orl-rbt LD50:5848 mg/kg ANTCAO 10,376,60
orl-gpg LDLo:1 g/kg ANTCAO 5,463,55
ipr-gpg LDLo:500 mg/kg ANTCAO 5,463,55
ivn-gpg LD50:303 mg/kg RPOBAR 2,306,70

CONSENSUS REPORTS: EPA Genetic Toxicology Program. Reported in EPA TSCA Inventory.

SAFETY PROFILE: Poison by intracerebral and intravenous routes. Moderately toxic by intravenous route. Mutation data reported. See other penicillin entries. When heated to decomposition it emits toxic fumes of NO_x and SO_x.

BFD250 CAS:69-57-8 ***HR: 3***
BENZYL PENICILLINIC ACID SODIUM SALT
mf: $C_{16}H_{17}N_2O_4S•Na$ mw: 356.40

SYNS: AMERICAN PENICILLIN ◇ BENZYLPENICILLIN SODIUM ◇ CRYSTAPEN ◇ MYCOFARM ◇ NOVOCILLIN ◇ PEN-A-BRASIVE ◇ PENICILLIN-G, MONOSODIUM SALT ◇ PENICILLIN G, SODIUM ◇ PENICILLIN G, SODIUM SALT ◇ PENILARYN ◇ PENZYLPENICILLIN SODIUM SALT ◇ SODIUM BENZYLPENICILLIN ◇ SODIUM BENZYLPENICILLIN G ◇ SODIUM BENZYLPENICILLINATE ◇ SODIUM PENICILLIN ◇ SODIUM PENICILLIN G ◇ SODIUM PENICILLIN II ◇ VETICILLIN

TOXICITY DATA with REFERENCE
scu-mus TDLo:30 mg/kg (female 14D post):REP CRSBAW 158,528,64
orl-rat TDLo:1500 mg/kg (6-11D preg):TER ANTBAL 18,815,73
scu-rat TDLo:1840 mg/kg/46W-I:ETA BJCAAI 15,85,61
orl-rat LD50:6916 mg/kg AMPMAR 39,259,78
par-rat LD50:2900 μg/kg AACHAX -,863,65
scu-mus LD50:4750 mg/kg NYKZAU 55,23,59
ivn-mus LD50:1500 mg/kg ARZNAD 9,31,59
ims-mus LD50:2800 mg/kg ARZNAD 9,31,59
ice-mus LD50:3800 μg/kg NYKZAU 55,23,59
ims-gpg LDLo:60 mg/kg LBASAE 30,524,80

CONSENSUS REPORTS: EPA Genetic Toxicology Program.

SAFETY PROFILE: Poison by intracerebral, parenteral, and intramuscular routes. Moderately toxic via intravenous route. Mildly toxic by ingestion. Experimental teratogenic and reproductive effects. Questionable carcinogen with experimental tumorigenic data. When heated to decomposition it emits very toxic fumes of NO_x, Na_2O, and SO_x. An antibiotic. See other penicillin entries.

BFD400 ***HR: 1***
BENZYL PHENYLACETATE
mf: $C_{15}H_{14}O_2$ mw: 226.27

PROP: Colorless liquid; sweet, floral odor with honey undertone. D: 1.095-1.099, refr index: 1.553-1.558, flash p: +212°F. Sol in alc, chloroform, ether.

SYN: FEMA No. 2149

SAFETY PROFILE: Combustible liquid. When heated to decomposition it emits acrid smoke and irritating fumes.

BFE750 CAS:612-98-6 ***HR: D***
BENZYLPHENYL NITROSAMINE
mf: $C_{13}H_{12}N_2O$ mw: 212.27

SYNS: BENZENEMETHANAMINE-N-NITROSO-N-PHENYL ◇ N-NITROSOPHENYLBENZYLAMINE

TOXICITY DATA with REFERENCE
mma-esc 2μmol/plate GANNA2 75,8,84
mrc-esc 6 μg/well MUREAV 46,53,77

CONSENSUS REPORTS: Reported in EPA TSCA Inventory.

SAFETY PROFILE: Mutation data reported. When heated to decomposition it emits toxic fumes of NO_x. See also NITROSAMINES and N-NITROSO COMPOUNDS.

BFG500 CAS:59177-78-5 ***HR: 3***
N-BENZYL-4-PROTOADAMANTANEMETHANAMINE MALEATE
mf: $C_{18}H_{25}N•C_4H_4O_4$ mw: 371.52

TOXICITY DATA with REFERENCE
orl-mus LD50:186 mg/kg JMCMAR 19,967,76
ipr-mus LD50:74 mg/kg JMCMAR 19,967,76

SAFETY PROFILE: Poison by ingestion and intraperitoneal routes. When heated to decomposition it emits toxic fumes of NO_x.

BFG750 CAS:2116-65-6 ***HR: 3***
4-BENZYLPYRIDINE
mf: $C_{12}H_{11}N$ mw: 169.24

TOXICITY DATA with REFERENCE
orl-bwd LD50:18 mg/kg TXAPA9 21,315,72
ivn-mus LD50:25 mg/kg CSLNX* NX#12240

CONSENSUS REPORTS: Reported in EPA TSCA Inventory.

SAFETY PROFILE: Poison by ingestion and intravenous routes. A flammable material. Incompatible with oxidizers. When heated to decomposition it emits toxic fumes of NO_x.

BFH000 CAS:2876-13-3 ***HR: 3***
1-BENZYLPYRIDINIUM CHLORIDE
mf: $C_{12}H_{12}N \cdot Cl$ mw: 205.70

TOXICITY DATA with REFERENCE
scu-mus LDLo:80 mg/kg NDRC** -,21,43

CONSENSUS REPORTS: Reported in EPA TSCA Inventory.

SAFETY PROFILE: Poison by subcutaneous route. When heated to decomposition it emits very toxic fumes of Cl^- and NO_x. See also CHLORIDES.

BFI250 ***HR: 3***
2-(BENZYL(2-(PYRROLIDINYL)ETHYL)AMINO)-2'-CHLOROACETANILIDE DIHYDROCHLORIDE
mf: $C_{21}H_{26}ClN_3O \cdot 2ClH$ mw: 444.87

SYN: C 5399

TOXICITY DATA with REFERENCE
eye-rbt 2% SEV ARZNAD 9,167,59
scu-mus LD50:287 mg/kg ARZNAD 9,167,59

SAFETY PROFILE: Poison by subcutaneous route. A severe eye irritant. When heated to decomposition it emits very toxic fumes of Cl^-, NO_x, and HCl.

BFJ750 CAS:118-58-1 ***HR: 2***
BENZYL SALICYLATE
mf: $C_{14}H_{12}O_3$ mw: 228.26

PROP: Thick colorless liquid, pleasant odor. Bp: 208° @ 26 mm, d: 1.175 @ 20°, refr index: 1.579. Sol in fixed oils; insol in glycerin and propylene glycol.

SYNS: BENZYL-o-HYDROXYBENZOATE ◇ FEMA No. 2151

TOXICITY DATA with REFERENCE
orl-rat LD50:2227 mg/kg FCTXAV 11,1029,73

CONSENSUS REPORTS: Reported in EPA TSCA Inventory.

SAFETY PROFILE: Moderately toxic by ingestion. See also BENZYL ALCOHOL, SALICYLIC ACID, and ESTERS. Combustible when exposed to heat or flame. When heated to decomposition it emits acrid smoke and irritating fumes. Incompatible with oxidizing materials.

BFJ825 CAS:766-06-3 ***HR: 3***
BENZYL SILANE
mf: $C_7H_{10}Si$ mw: 122.24

SAFETY PROFILE: Ignites spontaneously in air. Upon decomposition it emits acrid smoke and fumes. See also SILANE.

BFJ850 CAS:1121-53-5 ***HR: 3***
BENZYL SODIUM
mf: C_7H_7Na mw: 114.12

SAFETY PROFILE: Ignites spontaneously in air. Upon decomposition it emits toxic fumes of Na_2O.

BFK000 CAS:35506-85-5 ***HR: 3***
BENZYL SULFITE
mf: $C_{14}H_{14}O_3S$ mw: 262.34

SYN: SULFUROUS ACID, DIBENZYL ESTER

TOXICITY DATA with REFERENCE
ivn-mus LD50:178 mg/kg CSLNX* NX#02156

SAFETY PROFILE: Poison by intravenous route. See also SULFITES. When heated to decomposition it emits toxic fumes of SO_x.

BFK325 CAS:1090-53-5 ***HR: 3***
1-BENZYL-2-(3-(4,5,6,7-TETRAHYDROBENZISOXAZOYLYL)CARBONYL)HYDRAZINE HYDROCHLORIDE
mf: $C_{15}H_{17}N_3O_4 \cdot ClH$ mw: 307.81

TOXICITY DATA with REFERENCE
orl-mus LD50:2082 mg/kg SKNEA7 14,58,64
ipr-mus LD50:723 mg/kg SKNEA7 14,58,64
orl-cat LD50:118 mg/kg SKNEA7 14,58,64

SAFETY PROFILE: Poison by ingestion. Moderately toxic by intraperitoneal route. When heated to decomposition it emits toxic fumes of NO_x and HCl.

BFK750 CAS:13402-51-2 ***HR: 3***
S-BENZYL THIOBENZOATE
mf: $C_{14}H_{12}OS$ mw: 228.32

SYN: TIBENZATE

TOXICITY DATA with REFERENCE
orl-mus LD50:1550 mg/kg YKKZAJ 89,1179,69
ivn-mus LD50:180 mg/kg CSLNX* NX#02522

SAFETY PROFILE: Poison by intravenous route. Moderately toxic by ingestion. When heated to decomposition it emits toxic fumes of SO_x. See also ESTERS.

BFL000 CAS:3012-37-1 ***HR: 3***
BENZYL THIOCYANATE
mf: C_8H_7NS mw: 149.22

PROP: Orange-red, crystalline solid. Mp: 41°; bp: 230°; d: 1.125.

SYNS: BENZYL MUSTARD OIL ◇ PHENYLMETHYL ESTER THIOCYANIC ACID (9CI) ◇ SOLVAT 14 ◇ α-THIOCYANATOTOLUENE ◇ TROPEOLIN

TOXICITY DATA with REFERENCE
ipr-rat LDLo:40 mg/kg ARZNAD 16,870,66
ipr-mus LD50:30 mg/kg JJPAAZ 3,99,54
scu-mus LD50:100 mg/kg JJPAAZ 3,99,54

CONSENSUS REPORTS: Reported in EPA TSCA Inventory.

SAFETY PROFILE: Poison by subcutaneous and intraperitoneal routes. See also THIOCYANATES. When heated to decomposition it emits very toxic fumes of NO_x, SO_x, and CN^-.

BFL125 CAS:1874-58-4 ***HR: 3***
BENZYLTHIOGUANINE
mf: $C_{12}N_{11}N_5S$ mw: 257.34

SYNS: 2-AMINO-6-BENZYLMERCAPTOPURINE ◇ 2-AMINO-6-BENZYL-MP ◇ 2-AMINO-6-(BENZYLTHIO)PURINE ◇ 6-BENZYLTHIOGUANINE ◇ NSC 15747 ◇ 6-((PHENYLMETHYL)THIO)-1H-PURIN-2-AMINE (9CI) ◇ SRI 702

TOXICITY DATA with REFERENCE
orl-rat TDLo:100 mg/kg (7D preg):REP JRPFA4 4,291,62
ipr-mus LD50:222 mg/kg NCISP* JAN86

SAFETY PROFILE: Poison by intraperitoneal route. Experimental reproductive effects. When heated to decomposition it emits toxic fumes of NO_x and SO_x.

BFL250 CAS:98-07-7 ***HR: 3***
BENZYL TRICHLORIDE
DOT: UN 2226
mf: $C_7H_5Cl_3$ mw: 195.47

PROP: Clear, colorless to yellowish liquid; penetrating odor. Mp: −5°, bp: 221°, d: 1.38 @ 15.5°/15.5°, vap d: 6.77.

SYNS: BENZENYL CHLORIDE ◇ BENZENYL TRICHLORIDE ◇ BENZOIC TRICHLORIDE ◇ BENZOTRICHLORIDE (DOT, MAK) ◇ BENZYLIDYNE CHLORIDE ◇ CHLORURE de BENZENYLE (FRENCH) ◇ PHENYL CHLOROFORM ◇ PHENYLTRICHLOROMETHANE ◇ RCRA WASTE NUMBER U023 ◇ TOLUENE TRICHLORIDE ◇ TRICHLOORMETHYLBENZEEN (DUTCH) ◇ TRICHLORMETHYLBENZOL (GERMAN) ◇ TRICHLOROMETHYLBENZENE ◇ 1-(TRICHLOROMETHYL)BENZENE ◇ TRICHLOROPHENYLMETHANE ◇ α,α,α-TRICHLOROTOLUENE ◇ ω,ω,ω-TRICHLOROTOLUENE ◇ TRICLOROMETILBENZENE (ITALIAN) ◇ TRICLOROTOLUENE (ITALIAN)

TOXICITY DATA with REFERENCE
skn-rbt 10 mg/24H SEV AMIHBC 4,119,51
eye-rbt 50 μg SEV AMIHBC 4,119,51
mma-esc 500 nmol/plate/20M MUREAV 54,143,78
mrc-bcs 2600 nmol/disc MUREAV 54,143,78
ihl-mus TCLo:1620 ppb/30M/22W-I:NEO SAIGBL 28,352,86
skn-mus TDLo:9200 mg/kg/50W-I:CAR GANNA2 130,250,28
ihl-uns LC50:60 mg/m^3 GTPZAB 30(3),6,86

CONSENSUS REPORTS: NTP Fifth Annual Report on Carcinogens. IARC Cancer Review: Human Limited Evidence IMEMDT 29,73,82; Animal Sufficient Evidence IMEMDT 29,73,82. EPA Genetic Toxicology Program. EPA Extremely Hazardous Substances List. Reported in EPA TSCA Inventory.

DFG MAK: Suspected Carcinogen.
DOT Classification: IMO: Corrosive Material; Label: Corrosive.

SAFETY PROFILE: Confirmed carcinogen with experimental carcinogenic data by skin contact and neoplastiginic data by inhalation. Experimental poison by inhalation. Corrosive to the skin, eyes, and mucous membranes. Large doses can cause central nervous system depression. Mutation data reported. When heated to decomposition it emits toxic fumes of Cl^-. See also CHLORINATED HYDROCARBONS, AROMATIC.

BFM000 CAS:2971-75-7 ***HR: 3***
1-BENZYL-2-TRIMETHYLACETYLHYDRAZINE HYDROCHLORIDE
mf: $C_{12}H_{18}N_2O \cdot ClH$ mw: 169.56

TOXICITY DATA with REFERENCE
orl-mus LD50:400 mg/kg 27ZQAG -,403,72
scu-mus LD50:290 mg/kg 27ZQAG -,403,72
ivn-mus LD50:280 mg/kg 27ZQAG -,403,72

SAFETY PROFILE: Poison by ingestion, subcutaneous, and intravenous routes. When heated to decomposition it emits very toxic fumes of NO_x and HCl.

BFM250 CAS:56-93-9 ***HR: 3***
BENZYLTRIMETHYLAMMONIUM CHLORIDE
mf: $C_{10}H_{16}N \cdot Cl$ mw: 185.72

PROP: Bp: > 135° (some decomp), fp: < −50° (for 61% sol), d: 1.07 @ 20°/20° (61% sol).

SYN: TRIMETHYLBENZYLAMMONIUM CHLORIDE

TOXICITY DATA with REFERENCE
orl-rat LDLo:250 mg/kg NCNSA6 5,39,53

CONSENSUS REPORTS: Reported in EPA TSCA Inventory.

SAFETY PROFILE: Poison by ingestion. Combustible. When heated to decomposition it emits very toxic fumes of NH_3, NO_x, and Cl^-.

BFM500 CAS:100-85-6 ***HR: 3***
BENZYLTRIMETHYLAMMONIUM HYDROXIDE
mf: $C_{10}H_{16}N\bullet HO$ mw: 167.28

PROP: Solid

SYN: TRIMETHYLBENZYLAMMONIUMHYDROXIDE

TOXICITY DATA with REFERENCE
scu-mus LDLo:35 mg/kg JPETAB 28,367,26

CONSENSUS REPORTS: Reported in EPA TSCA Inventory.

SAFETY PROFILE: Poison by subcutaneous route. A strong base. When heated to decomposition it emits toxic fumes of NH_3 and NO_x. See also ALKALIES.

BFM750 CAS:4525-46-6 ***HR: 3***
BENZYL TRIMETHYL AMMONIUM IODIDE
mf: $C_{10}H_{16}N\bullet I$ mw: 277.17

SYNS: BENZYLDIMETHYLAMINE METHIODIDE ◇ PHENMETHYL TRIMETHYLAMMONIUM IODIDE

TOXICITY DATA with REFERENCE
ipr-mus LD50:41 mg/kg UCPHAQ 2,161,44
ivn-mus LD50:5600 μg/kg CSLNX* NX#00844

CONSENSUS REPORTS: Reported in EPA TSCA Inventory.

SAFETY PROFILE: Poison by intraperitoneal and intravenous routes. See also IODIDES. When heated to decomposition it emits very toxic fumes of NO_x, NH_3, and I^-.

BFN125 CAS:538-32-9 ***HR: 2***
BENZYLUREA
mf: $C_8H_{10}N_2O$ mw: 150.20

PROP: Crystals. Mp: 147-148°, decomp at 200°. One gram dissolves in 60 mL warm water, 33 mL acetone; sltly sol in benzene, ether.

SYNS: BENZYLCARBAMIDE ◇ N-BENZYLUREA ◇ 1-BENZYLUREA ◇ PHENYLMETHYLUREA

TOXICITY DATA with REFERENCE
orl-rat LD50:4410 mg/kg GISAAA 44(3),68,79
orl-mus LD50:570 mg/kg JMCMAR 11,814,68
orl-rat LD50:2700 mg/kg GISAAA 44(3),68,79

SAFETY PROFILE: Moderately toxic by ingestion. When heated to decomposition it emits toxic fumes of NO_x.

BFN500 CAS:2086-83-1 ***HR: 3***
BERBERINE
mf: $C_{20}H_{18}NO_4$ mw: 336.39

PROP: White to yellow crystals. Mp (anhyd): 145°.

SYNS: BERBERIN ◇ 9,10-DIMETHOXY-2,3-(METHYLENEDIOXY)-7,8,13,13A-TETRAHYDROBERBINIUM

TOXICITY DATA with REFERENCE
mmo-sat 100 μmol/L AMACCQ 9,77,76
dnd-esc 10 μmol/L MUREAV 89,95,81
orl-mus LD50:329 mg/kg YKKZAJ 82,726,62
scu-mus LD50:18 mg/kg RPTOAN 31,129,68
scu-rbt LDLo:100 mg/kg HBAMAK 4,1289,35

SAFETY PROFILE: An alkaloid poison by ingestion and subcutaneous routes. In humans, toxic doses lower the body temperature, increase peristalsis, and cause death by central paralysis. Mutation data reported. Should carry a poison label. Should never be ingested without the advice of a physician. Should not be handled excessively since it may be absorbed through the skin and have a toxic effect upon the body. An antimalarial agent. When heated to decomposition it emits highly toxic fumes of NO_x.

BFN550 ***HR: 3***
BERBERINE CHLORIDE DIHYDRATE
mf: $C_{20}H_{18}NO_4\bullet Cl\bullet 2H_2O$ mw: 407.88

SYNS: BERBERINE HYDROCHLORIDE BIHYDRATE ◇ 5,6-DIHYDRO-9,10-DIMETHOXYBENZO(g)-1,3-BENZODIOXOLO-(5,6-a)QUINOLIZINIUM CHLORIDE DIHYDRATE

TOXICITY DATA with REFERENCE
ipr-rat LD50:138 mg/kg KSRNAM 8,654,74
scu-rat LD50:7970 mg/kg KSRNAM 8,654,74
ivn-rat LD50:46200 μg/kg KSRNAM 8,654,74
ipr-mus LD50:30 mg/kg KSRNAM 8,654,74
scu-mus LD50:13900 μg/kg KSRNAM 8,654,74
ivn-mus LD50:7600 μg/kg KSRNAM 8,654,74

SAFETY PROFILE: Poison by subcutaneous, intravenous, and intraperitoneal routes. When heated to decomposition it emits toxic fumes of Cl^- and NO_x. See also CHLORIDES.

BFN625 CAS:316-41-6 ***HR: 3***
BERBERINE SULFATE
mf: $C_{40}H_{36}N_2O_8\bullet O_4S$ mw: 768.84

SYNS: BERBERINE SULFATE (2:1) ◇ BERBERIN SULFATE ◇ 5,6-DIHYDRO-9,10-DIMETHOXY-BENZO(g)-1,3-BENZODIOXOLO-(5,6-a)QUINOLIZINIUM SULFATE (2:1) ◇ NEUTRAL BERBERINE SULFATE

TOXICITY DATA with REFERENCE
ipr-mus LD50:26400 μg/kg NIIRDN 6,770,82
scu-mus LD50:13200 μg/kg NIIRDN 6,770,82
ivn-mus LD50:8200 μg/kg NIIRDN 6,770,82

SAFETY PROFILE: Poison by subcutaneous, intravenous, and intraperitoneal routes. When heated to decomposition it emits toxic fumes of NO_x and SO_x. See also SULFATES.

BFN750 CAS:6190-33-6 ***HR: 3***
BERBERINE SULFATE TRIHYDRATE
mf: $C_{40}H_{36}N_2O_8 \cdot O_4S \cdot 3H_2O$ mw: 822.90

SYNS: 5,6-DIHYDRO-9,10-DIMETHOXYBENZO(g)-1,3-BENZODIOXOLO(5,6-a)QUINOLIZINIUM SULFATE TRIHYDRATE ◇ 7,8,13,13A-TETRADEHYDRO-9,10-DIMETHOXY-2,3-(METHYLENEDIOXY)BERBINIUM SULFATE TRIHYDRATE ◇ UMBELLATINE SULFATE TRIHYDRATE

TOXICITY DATA with REFERENCE
scu-frg LDLo:20 mg/kg HBAMAK 4,1289,35

SAFETY PROFILE: Poison by subcutaneous route. See also BERBERINE and SULFATES. When heated to decomposition it emits very toxic SO_x and NO_x.

BFO000 CAS:8007-75-8 ***HR: 1***
BERGAMOT OIL rectified

PROP: Yellow-green liquid; agreeable odor. *Composition:* 1-linalyl acetate, 1-linalool, d-limonene, dipentene, bergaptene. By rectification of bergamot oil expressed, under vacuum, to remove completely the furocoumarins and other related nonvolatile residues; found in the fruit of citrus *Bergamia risso et poiteau (Fam. rutaceae)* (FCTXAV 11,1011,73). D: 0.875-0.880 @ 25°/25°. Misc with alc, glacial acetic acid; sol in fixed oils; insol in glycerin, propylene glycol.

SYNS: BERGAMOTTE OEL (GERMAN) ◇ OIL of BERGAMOT coldpressed ◇ OIL of BERGAMOT, rectified

TOXICITY DATA with REFERENCE
skn-rbt 500 mg/24H MLD FCTXAV 11,1035,73
orl-rat LD50:11520 mg/kg PHARAT 14,435,59

CONSENSUS REPORTS: Reported in EPA TSCA Inventory.

SAFETY PROFILE: Mildly toxic by ingestion. A mild skin irritant and allergen. Combustible. When heated to decomposition it emits acrid smoke and irritating fumes.

BFO100 CAS:5956-63-8 ***HR: 2***
BERGENIN HYERATE
mf: $C_{14}H_{16}O_9 \cdot 7H_2 0$ mw: 454.44

SYN: 2,4,4a,10b-TETRAHYDRO-3,4,8,10-TETRAHYDROXY-2-(HYDROXYMETHYL)-9-METHOXY-PYRANO(3,2-o)(2)BENZOPYRAN-6(2H)-ONE HYDRATE

TOXICITY DATA with REFERENCE
ipr-rat TDLo:2688 mg/kg (84D pre):REP KSRNAM 9,1198,75
ipr-rat LD50:3040 mg/kg KSRNAM 9,1198,75
ivn-rat LD50:2800 mg/kg KSRNAM 9,1198,75
ipr-mus LD50:6410 mg/kg KSRNAM 9,1198,75
ivn-mus LD50:5400 mg/kg KSRNAM 9,1198,75

SAFETY PROFILE: Moderately toxic by several routes. Experimental reproductive effects. When heated to decomposition it emits acrid smoke and fumes.

BFO125 ***HR: 3***
BERSAMA ABYSSINICA Fres. ssp. ABYSSINICA, leaf extract

PROP: African plant belonging to the family *Melianthaceae* (JPPMAB 14,496,62).

TOXICITY DATA with REFERENCE
orl-mus LD50:840 μg/kg JPPMAB 14,496,62
ipr-mus LD50:510 μg/kg JPPMAB 14,496,62
ivn-mky LD50:90 μg/kg JPPMAB 14,496,62
ivn-cat LD50:119 μg/kg JPPMAB 14,496,62

SAFETY PROFILE: Deadly poison by ingestion, intravenous, and intraperitoneal routes.

BFO250 CAS:12161-82-9 ***HR: 3***
BERTRANDITE
mf: $H_{10}O_9Si_2 \cdot H_2O \cdot Be_4$ mw: 264.34

SYN: BERYLLIUM SILICATE HYDRATE

CONSENSUS REPORTS: IARC Cancer Review: Animal Sufficient Evidence IMEMDT 1,17,72; Animal Inadequate Evidence IMEMDT 23,143,80. Reported in EPA TSCA Inventory. Beryllium and its compounds are on the Community Right-To-Know List.

OSHA PEL: (Transitional: TWA 0.002 mg(Be)/m^3; CL 0.005; Pk 0.025/30M/8H) TWA 0.002 mg(Be)/m^3; STEL 0.005 mg(Be)/m^3/30M; CL 0.025 mg(Be)/m^3
ACGIH TLV: TWA 0.002 mg(Be)/m^3, Suspected Human Carcinogen.
NIOSH REL: (Beryllium) CL not to exceed 0.0005 mg(Be)/m^3

SAFETY PROFILE: Confirmed carcinogen. See also BERYLLIUM and BERYLLIUM COMPOUNDS. When heated to decomposition it emits very toxic fumes of BeO.

BFO500 CAS:1302-52-9 ***HR: 3***
BERYL
mf: $Al_2O_{18}Si_6 \cdot 3Be$ mw: 537.53

PROP: Green, blue, yellow, or white crystals. D: 2.63-2.91.

SYNS: BERYL ORE ◇ BERYLLIUM ALUMINOSILICATE ◇ BERYLLIUM ALUMINUM SILICATE

TOXICITY DATA with REFERENCE
ihl-rat TCLo:15 mg/m^3/74W-I:NEO TXAPA9 15,10,69
ihl-rat TC:15 mg/m^3/6H/73W-I:ETA TXAPA9 8,361,66

CONSENSUS REPORTS: NTP Fifth Annual Report on Carcinogens. IARC Cancer Review: Group 2A IMEMDT 7,127,87; Animal Sufficient Evidence IMEMDT

23,143,80; IMEMDT 1,17,72. Reported in EPA TSCA Inventory. Beryllium and its compounds are on the Community Right-To-Know List.

OSHA PEL: (Transitional: TWA 0.002 mg(Be)/m^3; CL 0.005; Pk 0.025/30M/8H) TWA 0.002 mg(Be)/m^3; STEL 0.005 mg(Be)/m^3/30M; CL 0.025 mg(Be)/m^3
ACGIH TLV: TWA 0.002 mg(Be)/m^3, Suspected Human Carcinogen.
NIOSH REL: (Beryllium) CL not to exceed 0.0005 mg(Be)/m^3

SAFETY PROFILE: Confirmed carcinogen with experimental carcinogenic, neoplastigenic, and tumorigenic data. See also BERYLLIUM COMPOUNDS and SILICATES. When heated to decomposition it emits toxic fumes of BeO.

BFO750 CAS:7440-41-7 ***HR: 3***
BERYLLIUM
DOT: UN 1567
af: Be aw: 9.01

PROP: A grayish-white, hard, light metal. Mp: 1278°, bp: 2970°, d: 1.85.

SYNS: BERYLLIUM-9 ◇ BERYLLIUM, metal powder (DOT) ◇ GLUCINUM ◇ RCRA WASTE NUMBER P015

TOXICITY DATA with REFERENCE
dnd-esc 30 μmol/L MUREAV 89,95,81
dni-nml-ivn 30 μmol/kg PHMCAA 12,298,70
dnd-hmn:hla 30 μmol/L MUREAV 89,95,81
dnd-mus:ast 30 μmol/L MUREAV 89,95,81
itr-rat TDLo:13 mg/kg:NEO ENVRAL 21,63,80
ivn-rbt TDLo:20 mg/kg:ETA LANCAO 1,463,50
ihl-hmn TCLo:300 mg/m^3:PUL AEHLAU 9,473,64
ivn-rat LD50:496 μg/kg LAINAW 15,176,66

CONSENSUS REPORTS: NTP Fifth Annual Report on Carcinogens. IARC Cancer Review: Group 2A IMEMDT 7,127,87; Human Limited Evidence IMEMDT 23,143,80; Animal Sufficient Evidence IMEMDT 23,143,80; IMEMDT 1,17,72. Beryllium and its compounds are on the Community Right-To-Know List. Reported in EPA TSCA Inventory.

OSHA PEL: (Transitional: TWA 0.002 mg(Be)/m^3; CL 0.005; Pk 0.025/30M/8H) TWA 0.002 mg(Be)/m^3; STEL 0.005 mg(Be)/m^3/30M; CL 0.025 mg(Be)/m^3
ACGIH TLV: TWA 0.002 mg/m^3, Suspected Human Carcinogen.
DFG TRK: Animal Carcinogen, Suspected Human Carcinogen. Grinding of beryllium metal and alloys: 0.005 mg/m^3 calculated as beryllium in that portion of dust that can possibly be inhaled; other beryllium compounds: 0.002 mg/m^3 calculated as beryllium in that portion of dust that can possibly be inhaled
NIOSH REL: CL not to exceed 0.0005 mg(Be)/m^3

DOT Classification: Poison B, Flammable Solid Powder and Poison (metal).

SAFETY PROFILE: Confirmed carcinogen with experimental carcinogenic, neoplastigenic, and tumorigenic data. A deadly poison by intravenous route. Human systemic effects by inhalation: lung fibrosis, dyspnea, and weight loss. Human mutation data reported. See also BERYLLIUM COMPOUNDS. A moderate fire hazard in the form of dust or powder, or when exposed to flame or by spontaneous chemical reaction. Slight explosion hazard in the form of powder or dust. Incompatible with halocarbons. Reacts incandescently with fluorine or chlorine. Mixtures of the powder with CCl_4 or trichloroethylene will flash or spark on impact. When heated to decomposition in air it emits very toxic fumes of BeO. Reacts with Li and P.

BFP000 CAS:543-81-7 ***HR: 3***
BERYLLIUM ACETATE
mf: $C_4H_6O_4$•Be mw: 127.11

PROP: Plates. Mp: decomp @ 300°.
SYN: BERYLLIUM ACETATE, NORMAL

TOXICITY DATA with REFERENCE
ipr-rat LD50:317 mg/kg XEURAQ UR-70,1949

CONSENSUS REPORTS: IARC Cancer Review: Animal Inadequate Evidence IMEMDT 23,143,80. Beryllium and its compounds are on the Community Right-To-Know List.

OSHA PEL: (Transitional: TWA 0.002 mg(Be)/m^3; CL 0.005; Pk 0.025/30M/8H) TWA 0.002 mg(Be)/m^3; STEL 0.005 mg(Be)/m^3/30M; CL 0.025 mg(Be)/m^3
ACGIH TLV: TWA 0.002 mg(Be)/m^3, Suspected Human Carcinogen.
DFG MAK: Animal Carcinogen, Suspected Human Carcinogen.
NIOSH REL: (Beryllium) CL not to exceed 0.0005 mg(Be)/m^3

SAFETY PROFILE: Confirmed carcinogen. Poison by intraperitoneal route. See also BERYLLIUM COMPOUNDS. When heated to decomposition it emits toxic fumes of BeO.

BFP250 CAS:12770-50-2 ***HR: 3***
BERYLLIUM ALUMINUM ALLOY

PROP: Alloy is 62% beryllium and 38% aluminum (ENVRAL 21,63,80).

SYNS: ALUMINUM ALLOY, Al,Be ◇ ALUMINUM BERYLLIUM ALLOY

TOXICITY DATA with REFERENCE
itr-rat TDLo:13 mg/kg:ETA ENVRAL 21,63,80

CONSENSUS REPORTS: NTP Fifth Annual Report on

Carcinogens. IARC Cancer Review: Group 2A IMEMDT 7,127,87; Animal Sufficient Evidence IMEMDT 23,143,80. Beryllium and its compounds are on the Community Right-To-Know List.

OSHA PEL: (Transitional: TWA 0.002 mg(Be)/m^3; CL 0.005; Pk 0.025/30M/8H) TWA 0.002 mg(Be)/m^3; STEL 0.005 mg(Be)/m^3/30M; CL 0.025 mg(Be)/m^3
ACGIH TLV: TWA 0.002 mg(Be)/m^3, Suspected Human Carcinogen.
DFG MAK: Animal Carcinogen, Suspected Human Carcinogen.
NIOSH REL: (Beryllium) CL not to exceed 0.0005 mg(Be)/m^3

SAFETY PROFILE: Confirmed carcinogen with experimental carcinogenic and tumorigenic data. See also BERYLLIUM COMPOUNDS. When heated to decomposition it emits very toxic BeO.

BFP500 CAS:66104-24-3 ***HR: 3***
BERYLLIUM CARBONATE
mf: $C_2H_2Be_3O_8$ mw: 181.07

SYNS: BERYLLIUM CARBONATE, BASIC ◇ BERYLLIUMOXIDE CARBONATE ◇ BIS(CARBONATO(2-))DIHYDROXYTRIBERYLLIUM

TOXICITY DATA with REFERENCE

CONSENSUS REPORTS: NTP Fifth Annual Report on Carcinogens. IARC Cancer Review: Group 2A IMEMDT 7,127,87; Animal Sufficient Evidence IMEMDT 23,143,80. Reported in EPA TSCA Inventory. Beryllium and its compounds are on the Community Right-To-Know List.

OSHA PEL: (Transitional: TWA 0.002 mg(Be)/m^3; CL 0.005; Pk 0.025/30M/8H) TWA 0.002 mg(Be)/m^3; STEL 0.005 mg(Be)/m^3/30M; CL 0.025 mg(Be)/m^3
ACGIH TLV: TWA 0.002 mg(Be)/m^3, Suspected Human Carcinogen.
DFG MAK: Animal Carcinogen, Suspected Human Carcinogen.
NIOSH REL: CL not to exceed 0.0005 mg(Be)/m^3

SAFETY PROFILE: Confirmed carcinogen. See also BERYLLIUM COMPOUNDS. When heated to decomposition it emits toxic BeO dust.

BFP750 CAS:13106-47-3 ***HR: 3***
BERYLLIUM CARBONATE (1:1)
mf: $CO_3 \cdot Be$ mw: 69.02

SYN: CARBONIC ACID BERYLLIUM SALT (1:1)

TOXICITY DATA with REFERENCE
ipr-gpg LDLo:300 mg/kg NIHBAZ 181,20,43

CONSENSUS REPORTS: Reported in EPA TSCA Inventory. Beryllium and its compounds are on the Community Right-To-Know List.

OSHA PEL: (Transitional: TWA 0.002 mg(Be)/m^3; CL 0.005; Pk 0.025/30M/8H) TWA 0.002 mg(Be)/m^3; STEL 0.005 mg(Be)/m^3/30M; CL 0.025 mg(Be)/m^3
ACGIH TLV: TWA 0.002 mg(Be)/m^3, Suspected Human Carcinogen.
DFG MAK: 50 ppm (90 mg/m^3)
NIOSH REL: (Beryllium) CL not to exceed 0.0005 mg(Be)/m^3

SAFETY PROFILE: Confirmed carcinogen. Poison by intraperitoneal route. See also BERYLLIUM COMPOUNDS. When heated to decomposition it emits highly toxic fumes of BeO.

BFQ000 CAS:7787-47-5 ***HR: 3***
BERYLLIUM CHLORIDE
DOT: NA 1566
mf: $BeCl_2$ mw: 79.91

PROP: Colorless, deliquescent needles. Mp: 440°, bp: 520°, d: 1.899 @ 25°, vap press: 1 mm @ 291° (subl).

SYN: BERYLLIUM DICHLORIDE

TOXICITY DATA with REFERENCE
mmo-esc 10 μmol/L MUREAV 126,9,84
msc-ham:lng 2 mmol/L MUREAV 68,259,79
itr-rat TDLo:1685 μg/kg (3D post):TER GISAAA 51(8),44,86
itr-rat TDLo:1685 μg/kg (5D post):REP GISAAA 51(8),44,86
ihl-rat TCLo:20 μg/m^3/1H/17W-I:ETA GTPZAB 19(7),34,75
orl-rat LD50:86 mg/kg HYSAAV 30,169,65
ipr-rat LD50:44 mg/kg EQSSDX 1,1,75
orl-mus LD50:92 mg/kg HYSAAV 30,169,65
ipr-mus LD50:106 mg/kg COREAF 256,1043,63
ipr-gpg LD50:50 mg/kg EQSSDX 1,1,75

CONSENSUS REPORTS: NTP Fifth Annual Report on Carcinogens. IARC Cancer Review: Group 2A IMEMDT 7,127,87; Animal Sufficient Evidence IMEMDT 23,143,80. EPA Genetic Toxicology Program. Reported in EPA TSCA Inventory. Beryllium and its compounds are on the Community Right-To-Know List.

OSHA PEL: (Transitional: TWA 0.002 mg(Be)/m^3; CL 0.005; Pk 0.025/30M/8H) TWA 0.002 mg(Be)/m^3; STEL 0.005 mg(Be)/m^3/30M; CL 0.025 mg(Be)/m^3
ACGIH TLV: TWA 0.002 mg(Be)/m^3, Suspected Human Carcinogen.
DFG MAK: Animal Carcinogen, Suspected Human Carcinogen.
NIOSH REL: (Beryllium) CL not to exceed 0.0005 mg(Be)/m^3
DOT Classification: Poison B; Label: Poison.

SAFETY PROFILE: Confirmed carcinogen with experimental tumorigenic data. Poison by ingestion and intraperitoneal routes. An experimental teratogen. Other experimental reproductive effects. Mutation data reported. When heated to decomposition it emits very toxic fumes of BeO and Cl^-. See also BERYLLIUM COMPOUNDS and CHLORIDES.

BFQ500 ***HR: 3***
BERYLLIUM COMPOUNDS

CONSENSUS REPORTS: Beryllium and its compounds are on the Community Right-To-Know List.

OSHA PEL: (Transitional: TWA 0.002 mg(Be)/m^3; CL 0.005; Pk 0.025/30M/8H) TWA 0.002 mg(Be)/m^3; STEL 0.005 mg(Be)/m^3/30M; CL 0.025 mg(Be)/m^3
ACGIH TLV: TWA 0.002 mg/m^3, Suspected Human Carcinogen.
DFG TRK: Animal Carcinogen, Suspected Human Carcinogen. Grinding of beryllium metal and alloys: 0.005 mg/m^3 calculated as beryllium in that portion of dust that can possibly be inhaled; other beryllium compounds: 0.002 mg/m^3 calculated as beryllium in that portion of dust that can possibly be inhaled

SAFETY PROFILE: Confirmed carcinogens. Beryllium compounds can enter the body through inhalation of dusts and fumes, and may act locally on the skin. Even alloys of low beryllium content have been shown to be dangerous. In industry, inhalation of the dust can cause severe lung damage with symptoms appearing within months. Effects have been reported in persons living near processing plants and in families of beryllium workers. The fluoride, ammonium fluoride, sulfate, oxide, and hydroxide occur during extraction from beryllium ore. Exposure to the oxide may occur in processing of beryllium alloys and beryllium ceramics.

The extraction of Be from its ore is attended by exposure to acid salts of the metal, particularly the fluoride (BeF_2), the ammonium fluoride and sulfate ($BeSO_4$), and also to beryllium oxide (BeO), and hydroxide [$Be(OH)_2$]. Exposure to the oxide also occurs in the casting of beryllium alloys and in operations with beryllia ceramics. In the manufacture of fluorescent powders, lamps, and sign tubes there may be exposure to beryllium carbonate and to more complex salts, such as ZnMnBe silicate. Exposure to beryllium compounds encountered in the extraction of the metal or its oxide from the ore, particularly the halide salts, has been attended, in certain individuals, by the development of dermatitis of an edematous and papulovesicular type, chronic skin ulcers, rhinitis, nasopharyngitis, epistaxis, bronchitis and in severe cases, by the development of an acute pneumonitis, with cough, scanty sputum, low-grade fever, rales, dyspnea and substernal pain. Radiographs show diffuse haziness throughout both lungs, followed by the appearance of soft, ill-defined opacities. The condition occurs while the worker is exposed, sometimes within 1 or 2 months of starting work, and recovery occurs within 2 months, as a rule, though radiographic changes sometimes persist for longer periods. Occasionally, recovery may not occur and lung fibrosis results. In severe cases of pneumonitis the patient may die. Necropsies have revealed diffuse pulmonary edema, hemorrhagic extravasation, large numbers of plasma cells, and a relative absence of polymorphonuclear infiltration. On the basis of experimental work with animals, certain investigators are of the opinion that the acute upper and lower respiratory effects are due chiefly to the acid radical present in the dust or fume, but this view has little support. A delayed form of lung disease, characterized by the occurrence of granulomatous areas in the lung tissue, has been reported in workers manufacturing fluorescent powders, lamps and sign tubes, casting beryllium master alloys, and in the production of beryllium from beryl ore. Symptoms can start during exposure, but they might be delayed up to 5 years or more after leaving work. The commonest symptoms are coughing, shortness of breath, loss of appetite, loss of weight, and fatigue. Rales are usually present in the bases and axillae, and the red cell count is frequently elevated. Cyanosis is common and the pulse and respiratory rates are often increased. Radiographically, three stages of the disease are described: (1) a diffuse, uniform granular shadowing extending throughout both lung fields; (2) a diffuse reticular pattern on the granular background; (3) the appearance of distinct nodules scattered through the lungs, with some enlargement and blurring of the hilar shadows. The intensity of the shadowing is usually greater in the middle third of the lung fields. The prognosis is poor. Clinical improvement may occur gradually over a period of several years, but there appears to be little tendency for the radiographic shadowing to clear. In certain cases, the disease has progressed gradually for some months or years, with death resulting from respiratory and cardiac failure. In several instances necropsies have shown the presence of a diffuse fibrosis with coarse strands of hyalinized collagen between the alveoli and, in some places, replacing them. The hyalinized areas contained granulomatous foci, the alveolar walls are thickened and fibrosed, the blood vessels being engorged and dilated. In some cases the hilar lymph nodes show granulomatous change and fibrosis. Granulomatous change has also been noted in the liver and hyaline fibrosis in the spleen. Two cases of delayed lung disease coming to autopsy have presented papular lesions on the dorsum of the hands; on the biopsy these showed "sarcoid-like" lesions with central necrosis.

Several cases have been reported in which localized granulomatous lesions developed following penetrating wounds caused by splinters of glass from broken fluorescent light tubes. Several weeks or months following the accident, swellings were noted in the injured areas and excision revealed granulomatous tumors, which in one case was shown to contain beryllium.

There is no specific treatment, but temporary remissions have been produced by ACTH and cortisone.

BFQ750 CAS:12010-12-7 ***HR: 3***
BERYLLIUM COMPOUND with NIOBIUM (12 : 1)
mf: $Be_{12}Nb$ mw: 201.03

TOXICITY DATA with REFERENCE
itr-rat TDLo:2500 μg/kg:ETA ENVRAL 21,63,80

CONSENSUS REPORTS: Beryllium and its compounds are on the Community Right-To-Know List.

OSHA PEL: (Transitional: TWA 0.002 mg(Be)/m^3; CL 0.005; Pk 0.025/30M/8H) TWA 0.002 mg(Be)/m^3; STEL 0.005 mg(Be)/m^3/30M; CL 0.025 mg(Be)/m^3
ACGIH TLV: TWA 0.002 mg(Be)/m^3, Suspected Human Carcinogen.
NIOSH REL: (Beryllium) CL not to exceed 0.005 mg(Be)/m^3

SAFETY PROFILE: Confirmed carcinogen with experimental tumorigenic data. When heated to decomposition in air it emits very toxic fumes of BeO. See also BERYLLIUM COMPOUNDS and NIOBIUM.

BFR000 CAS:12232-67-6 ***HR: 3***
BERYLLIUM COMPOUND with TITANIUM (12 : 1)
mf: $Be_{12}Ti$ mw: 156.02

SYN: TITANIUM compounded with BERYLLIUM (1:12)

TOXICITY DATA with REFERENCE
itr-rat TDLo:2500 μg/kg:ETA ENVRAL 21,63,80

CONSENSUS REPORTS: Beryllium and its compounds are on the Community Right-To-Know List.

OSHA PEL: (Transitional: TWA 0.002 mg(Be)/m^3; CL 0.005; Pk 0.025/30M/8H) TWA 0.002 mg(Be)/m^3; STEL 0.005 mg(Be)/m^3/30M; CL 0.025 mg(Be)/m^3
ACGIH TLV: TWA 0.002 mg(Be)/m^3, Suspected Human Carcinogen.
NIOSH REL: (Beryllium) CL not to exceed 0.0005 mg(Be)/m^3

SAFETY PROFILE: Confirmed carcinogen with experimental tumorigenic data. See also BERYLLIUM COMPOUNDS and TITANIUM COMPOUNDS. When heated to decomposition it emits very toxic fumes of BeO.

BFR250 CAS:12400-16-7 ***HR: 3***
BERYLLIUM COMPOUND with VANADIUM (12 : 1)
mf: $Be_{12}V$ mw: 159.06

SYN: TITANIUM compounded with BERYLLIUM (1:12)

TOXICITY DATA with REFERENCE
itr-rat TDLo:2500 μg/kg:NEO ENVRAL 21,63,80

CONSENSUS REPORTS: Beryllium and its compounds are on the Community Right-To-Know List.

OSHA PEL: (Transitional: TWA 0.002 mg(Be)/m^3; CL 0.005; Pk 0.025/30M/8H) TWA 0.002 mg(Be)/m^3; STEL 0.005 mg(Be)/m^3/30M; CL 0.025 mg(Be)/m^3
ACGIH TLV: TWA 0.002 mg(Be)/m^3, Suspected Human Carcinogen
NIOSH REL: (Beryllium) CL not to exceed 0.0005 mg(Be)/m^3; (REL to Vanadium) 1.0 mg(V)/m^3

SAFETY PROFILE: Confirmed carcinogen with experimental tumorigenic data. See also BERYLLIUM COMPOUNDS and VANADIUM COMPOUNDS. When heated to decomposition it emits very toxic fumes of BeO and VO_x.

BFR500 CAS:7787-49-7 ***HR: 3***
BERYLLIUM FLUORIDE
DOT: NA 1566
mf: BeF_2 mw: 47.01

PROP: Amorphous, colorless mass. Mp: 800°, d: 1.986 @ 25°.

SYN: BERYLLIUM DIFLUORIDE

TOXICITY DATA with REFERENCE
ihl-rat TCLo:20 μg/m^3/1H/17W-I:ETA GTPZAB 19(7),34,75
orl-rat LD50:98 mg/kg XEURAQ UR-154,1951
orl-mus LD50:100 mg/kg XPHPAW 2173,23,72
scu-mus LD50:20 mg/kg XPHPAW 2173,23,72
ivn-mus LD50:1800 μg/kg XPHPAW 2173,23,72
ipr-ham LD50:21 mg/kg XEURAQ UR-154,1951

CONSENSUS REPORTS: NTP Fifth Annual Report on Carcinogens. IARC Cancer Review: Group 2A IMEMDT 7,127,87; Animal Sufficient Evidence IMEMDT 23,143,80. Beryllium and its compounds are on the Community Right-To-Know List. Reported in EPA TSCA Inventory.

OSHA PEL: (Transitional: TWA 0.002 mg(Be)/m^3; CL 0.005; Pk 0.025/30M/8H) TWA 0.002 mg(Be)/m^3; STEL 0.005 mg(Be)/m^3/30M; CL 0.025 mg(Be)/m^3
ACGIH TLV: TWA 0.002 mg(Be)/m^3, Suspected Human Carcinogen; 2.5 mg(F)/m^3
NIOSH REL: (Beryllium) CL not to exceed 0.0005 mg(Be)/m^3
DOT Classification: Poison B; Label: Poison.

SAFETY PROFILE: Confirmed carcinogen with experimental carcinogenic and tumorigenic data by inhalation. Poison by ingestion, subcutaneous, intravenous, and intraperitoneal routes. See also BERYLLIUM COMPOUNDS and FLUORIDES. Incompatible with Mg. When heated to decomposition, it emits very toxic fumes of BeO and F^-.

BFR750 CAS:7787-52-2 ***HR: 3***
BERYLLIUM HYDRIDE
mf: BeH_2 mw: 11.03

PROP: White solid.

CONSENSUS REPORTS: Beryllium and its compounds are on the Community Right-To-Know List.

SAFETY PROFILE: Confirmed carcinogen. A dangerous fire hazard. When heated to 220°C it liberates explosive hydrogen gas. Reacts violently with methanol, water, and dilute acids. When heated to decomposition it emits toxic fumes of BeO. See BERYLLIUM COMPOUNDS and HYDRIDES.

BFS000 CAS:13598-15-7 ***HR: 3***
BERYLLIUM HYDROGEN PHOSPHATE (1:1)
mf: $BeHO_4P$ mw: 104.99

SYNS: BERYLLIUM PHOSPHATE ◇ PHOSPHORIC ACID, BERYLLIUM SALT (1:1) ◇ PHOSPHOROUS ACID, BERYLLIUM SALT

TOXICITY DATA with REFERENCE
ihl-rat TCLo:3571 μg/m³/17W:ETA PEXTAR 2,203,61
ivn-mus LD50:16 mg/kg TXAPA9 24,497,73

CONSENSUS REPORTS: NTP Fifth Annual Report on Carcinogens. IARC Cancer Review: Group 2A IMEMDT 7,127,87; Animal Sufficient Evidence IMEMDT 23,143,80; IMEMDT 1,17,72. Beryllium and its compounds are on the Community Right-To-Know List.

OSHA PEL: (Transitional: TWA 0.002 mg(Be)/m³; CL 0.005; Pk 0.025/30M/8H) TWA 0.002 mg(Be)/m³; STEL 0.005 mg(Be)/m³/30M; CL 0.025 mg(Be)/m³
ACGIH TLV: TWA 0.002 mg(Be)/m³, Suspected Human Carcinogen.
NIOSH REL: (Beryllium) CL not to exceed 0.0005 mg(Be)/m³

SAFETY PROFILE: Confirmed carcinogen with experimental carcinogenic and tumorigenic data. Poison by intravenous route. See also BERYLLIUM COMPOUNDS and PHOSPHATES. When heated to decomposition it emits very toxic fumes of BeO and PO_x.

BFS250 CAS:13327-32-7 ***HR: 3***
BERYLLIUM HYDROXIDE
mf: H_2O_2•Be mw: 43.03

PROP: Amorphous powder or crystals. Mp: decomp @ 138°, d(cr): 1.909.

SYNS: BERYLLIUM DIHYDROXIDE ◇ BERYLLIUM HYDRATE

TOXICITY DATA with REFERENCE
itr-rat TDLo:1125 μg/kg:ETA TXAPA9 17,299,70
ivn-rat LDLo:3821 μg/kg XEURAQ UR-70,1949

CONSENSUS REPORTS: NTP Fifth Annual Report on Carcinogens. IARC Cancer Review: Group 2A IMEMDT 7,127,87; Animal Sufficient Evidence IMEMDT 23,143,80. Beryllium and its compounds are on the Community Right-To-Know List. Reported in EPA TSCA Inventory.

OSHA PEL: (Transitional: TWA 0.002 mg(Be)/m³; CL 0.005; Pk 0.025/30M/8H) TWA 0.002 mg(Be)/m³; STEL 0.005 mg(Be)/m³/30M; CL 0.025 mg(Be)/m³
ACGIH TLV: TWA 0.002 mg(Be)/m³, Suspected Human Carcinogen.
NIOSH REL: (Beryllium) CL not to exceed 0.0005 mg(Be)/m³

SAFETY PROFILE: Confirmed carcinogen with experimental carcinogenic and tumorigenic data. Poison by intravenous route. See also BERYLLIUM COMPOUNDS. When heated to decomposition it emits very toxic fumes of BeO.

BFS750 ***HR: 3***
BERYLLIUM MANGANESE ZINC SILICATE
mf: $BeMnO_4SiZn$ mw: 221.41

SYNS: MANGANESE ZINC BERYLLIUM SILICATE ◇ ZINC MANGANESE BERYLLIUM SILICATE

TOXICITY DATA with REFERENCE
ihl-rat TCLo:20 mg/m³/4W:ETA PEXTAR 2,203,61

CONSENSUS REPORTS: Beryllium, manganese, zinc, and their compounds are on the Community Right-To-Know List.

OSHA PEL: (Transitional: TWA 0.002 mg(Be)/m³; CL 0.005; Pk 0.025/30M/8H) TWA 0.002 mg(Be)/m³; STEL 0.005 mg(Be)/m³/30M; CL 0.025 mg(Be)/m³
ACGIH TLV: TWA 0.002 mg(Be)/m³, Suspected Human Carcinogen; TWA 5 mg(Mn)/m³
NIOSH REL: (Beryllium) CL Not to exceed 0.0005 mg(Be)/m³

SAFETY PROFILE: Confirmed carcinogen with experimental tumorigenic data. When heated to decomposition it emits very toxic fumes of BeO and ZnO. See also BERYLLIUM COMPOUNDS, MANGANESE COMPOUNDS, and ZINC COMPOUNDS.

BFT000 CAS:13597-99-4 ***HR: 3***
BERYLLIUM NITRATE
mf: BeN_2O_6 mw: 133.03
DOT: UN 2464

PROP: White-yellowish crystals, deliquescent. Mp: 60°, bp: decomp @ 100-200°.

SYNS: BERYLLIUM DINITRATE ◇ NITRIC ACID, BERYLLIUM SALT

TOXICITY DATA with REFERENCE
itt-rat TDLo:10803 μg/kg (1D male):REP JRPFA 7,21,64-
ipr-mus LD50:500 μg/kg EQSSDX 1,1,75
scu-mus LDLo:50 mg/kg RDWU** -,-,30
ipr-gpg LDLo:100 mg/kg EQSSDX 1,1,75
scu-frg LDLo:1041 mg/kg REWU** -,-,30

CONSENSUS REPORTS: Beryllium and its compounds are on the Community Right-To-Know List. Reported in EPA TSCA Inventory.

OSHA PEL: (Transitional: TWA 0.002 mg(Be)/m^3; CL 0.005; Pk 0.025/30M/8H) TWA 0.002 mg(Be)/m^3; STEL 0.005 mg(Be)/m^3/30M; CL 0.025 mg(Be)/m^3
ACGIH TLV: TWA 0.002 mg(Be)m^3, Suspected Human Carcinogen.
NIOSH REL: CL not to exceed 0.0005 mg(Be)/m^3
DOT Classification: Label: Oxidizer and Poison.

SAFETY PROFILE: Confirmed carcinogen. Poison by intraperitoneal and subcutaneous routes. Experimental reproductive effects. When heated to decomposition it emits very toxic fumes of BeO and NO_x. See also BERYLLIUM COMPOUNDS and NITRATES.

BFT250 CAS:1304-56-9 ***HR: 3***
BERYLLIUM OXIDE
mf: BeO mw: 25.01

PROP: White, amorph powder. Mp: 2530° ± 30°, bp: 3900° (approx), d: 3.025.

SYNS: BERYLLIA ◇ BERYLLIUM MONOXIDE ◇ THERMALOX

TOXICITY DATA with REFERENCE
itr-rat TDLo:139 mg/kg (3D post):TER GISAAA 51(8),44,86
itr-rat TDLo:139 mg/kg (3D post):REP GISAAA 51(8),44,86
ihl-rat TCLo:28 mg/m^3/17W-C:ETA IMSUAI 40,23,71

CONSENSUS REPORTS: NTP Fifth Annual Report on Carcinogens. IARC Cancer Review: Group 2A IMEMDT 7,127,87; Animal Sufficient Evidence IMEMDT 1,17,72; IMEMDT 23,143,80. Beryllium and its compounds are on the Community Right-To-Know List. Reported in EPA TSCA Inventory.

OSHA PEL: (Transitional: TWA 0.002 mg(Be)/m^3; CL 0.005; Pk 0.025/30M/8H) TWA 0.002 mg(Be)/m^3; STEL 0.005 mg(Be)/m^3/30M; CL 0.025 mg(Be)/m^3
ACGIH TLV: TWA 0.002 mg(Be)/m^3, Suspected Human Carcinogen.
NIOSH REL: (Beryllium) CL not to exceed 0.0005 mg(Be)/m^3

SAFETY PROFILE: Confirmed carcinogen with experimental tumorigenic data. An experimental teratogen. Other experimental reproductive effects. See also BERYLLIUM COMPOUNDS. Incompatible with (Mg + heat). When heated to decomposition it emits very toxic fumes of BeO.

BFT500 CAS:19049-40-2 ***HR: 3***
BERYLLIUM OXYACETATE
mf: $C_{12}H_{18}Be_4O_{13}$ mw: 406.34

SYNS: BERYLLIUM ACETATE, BASIC ◇ BERYLLIUM OXIDE ACETATE ◇ HEXAKIS(μ-ACETATO-O:O'))-μ^4-OXOTETRABERYLLIUM ◇ HEXAKIS(μ-ACETATO)-μ^4-OXOTETRABERYLLIUM

CONSENSUS REPORTS: Beryllium and its compounds are on the Community Right-To-Know List.

OSHA PEL: (Transitional: TWA 0.002 mg(Be)/m^3; CL 0.005; Pk 0.025/30M/8H) TWA 0.002 mg(Be)/m^3; STEL 0.005 mg(Be)/m^3/30M; CL 0.025 mg(Be)/m^3
ACGIH TLV: TWA 0.002 mg(Be)/m^3, Suspected Human Carcinogen.
NIOSH REL: (Beryllium) CL not to exceed 0.0005 mg(Be)/m^3

SAFETY PROFILE: Confirmed carcinogen. See BERYLLIUM COMPOUNDS. When heated to decomposition it emits toxic fumes BeO.

BFT750 CAS:63990-88-5 ***HR: 3***
BERYLLIUM OXYFLUORIDE
mf: BeF_2O_2 mw: 79.01

TOXICITY DATA with REFERENCE
orl-rat LD50:146 mg/kg XEURAQ UR-154,1951
scu-mus LDLo:5 mg/kg BJEPA5 30,375,49
ivn-mus LDLo:3500 μg/kg BJEPA5 30,375,49
ipr-gpg LDLo:10 mg/kg NIHBAZ 181,20,43

CONSENSUS REPORTS: Beryllium and its compounds are on the Community Right-To-Know List.

OSHA PEL: (Transitional: TWA 0.002 mg(Be)/m^3; CL 0.005; Pk 0.025/30M/8H) TWA 0.002 mg(Be)/m^3; STEL 0.005 mg(Be)/m^3/30M; CL 0.025 mg(Be)/m^3
ACGIH TLV: TWA 0.002 mg(Be)/m^3, Suspected Human Carcinogen; 2.5 mg(F)/m^3
NIOSH REL: (Beryllium) CL not to exceed 0.0005 mg(Be)/m^3

SAFETY PROFILE: Confirmed carcinogen. Poison by ingestion, subcutaneous, intravenous, and intraperitoneal routes. See also BERYLLIUM COMPOUNDS and FLUORIDES. When heated to decomposition it emits very toxic fumes of BeO and F^-.

BFU000 CAS:13597-95-0 ***HR: 3***
BERYLLIUM PERCHLORATE
mf: $Be(ClO_4)_2$ mw: 207.91

PROP: Very hygroscopic crystals, sol in water: 148.6 g/100 mL.

OSHA PEL: (Transitional: TWA 0.002 mg(Be)/m^3; CL 0.005; Pk 0.025/30M/8H) TWA 0.002 mg(Be)/m^3; STEL 0.005 mg(Be)/m^3/30M; CL 0.025 mg(Be)/m^3
ACGIH TLV: TWA 0.002 mg(Be)/m^3, Suspected Human Carcinogen.
NIOSH REL: CL not to exceed 0.0005 mg(Be)/m^3

SAFETY PROFILE: Confirmed carcinogen. A powerful oxidant used in propellant and igniter systems. When heated to decomposition it emits toxic fumes of Cl^- and BeO. See also BERYLLIUM COMPOUNDS and PERCHLORATES.

BFU250 CAS:13510-49-1 ***HR: 3***
BERYLLIUM SULFATE (1:1)
mf: $O_4S{\bullet}Be$ mw: 105.07

PROP: Crystals. Mp: 550-600° (decomp), d: 2.443.

SYN: SULFURIC ACID, BERYLLIUM SALT (1:1)

TOXICITY DATA with REFERENCE
mrc-bcs 10 mmol/L MUREAV 77,109,80
otr-mus:fbr 200 μg/L JJIND8 67,1303,81
ihl-rat TCLo:432 μg/m^3/26W:ETA PEXTAR 2,203,61
orl-rat LD50:82 mg/kg HYSAAV 30,169,65
ihl-rat LCLo:10 mg/m^3 EQSSDX 1,1,75
ipr-rat LDLo:18 mg/kg XPHPAW 2173,23,72
scu-rat LD50:1500 μg/kg XPHPAW 2173,23,72
ivn-rat LD50:7200 μg/kg XPHPAW 2173,23,72
itr-rat LDLo:10 mg/kg XPHPAW 2173,23,72
orl-mus LD50:80 mg/kg HYSAAV 30,169,65
ihl-mus LCLo:47 mg/m^3 EQSSDX 1,1,75
ipr-mus LD50:1200 mg/kg JJIND8 62,911,79
scu-mus LD50:1500 μg/kg XPHPAW 2173,23,72
ivn-mus LD50:500 μg/kg XPHPAW 2173,23,72
ivn-dog LDLo:600 μg/kg XPHPAW 2173,23,72

CONSENSUS REPORTS: NTP Fifth Annual Report on Carcinogens. IARC Cancer Review: Group 2A IMEMDT 7,127,87; Animal Sufficient Evidence IMEMDT 23,143,80. Beryllium and its compounds are on the Community Right-To-Know List. Reported in EPA TSCA Inventory.

OSHA PEL: (Transitional: TWA 0.002 mg(Be)/m^3; CL 0.005; Pk 0.025/30M/8H) TWA 0.002 mg(Be)/m^3; STEL 0.005 mg(Be)/m^3/30M; CL 0.025 mg(Be)/m^3
ACGIH TLV: TWA 0.002 mg(Be)/m^3, Suspected Human Carcinogen.
NIOSH REL: (Beryllium) CL not to exceed 0.0005 mg(Be)/m^3

SAFETY PROFILE: Confirmed carcinogen with experimental tumorigenic data. Acute poison by inhalation, ingestion, intraperitoneal, subcutaneous, intravenous, and intratracheal routes. See also BERYLLIUM COMPOUNDS and SULFATES. Mutation data reported. When heated to decomposition it emits very toxic fumes of SO_x and BeO.

BFU500 CAS:7787-56-6 ***HR: 3***
BERYLLIUM SULFATE TETRAHYDRATE (1:1:4)
mf: $O_4S{\bullet}Be{\bullet}4H_2O$ mw: 177.15

SYNS: BERYLLIUM SULPHATE TETRAHYDRATE ◇ SULFURIC ACID, BERYLLIUM SALT (1:1), TETRAHYDRATE

TOXICITY DATA with REFERENCE
mma-sat 3300 ng/plate ENMUDM 6(Suppl 2),1,84
sce-hmn:lym 1 mg/L ENMUDM 3,597,81
ihl-rat TCLo:668 μg/m^3/40W-C:CAR CNREA8 27,439,67
ivn-mus LD50:4971 μg/kg TXAPA9 24,497,73

CONSENSUS REPORTS: NTP Fifth Annual Report on Carcinogens. IARC Cancer Review: Group 2A IMEMDT 7,127,87; Animal Sufficient Evidence IMEMDT 23,143,80; IMEMDT 1,17,72. Beryllium and its compounds are on the Community Right-To-Know List.

OSHA PEL: (Transitional: TWA 0.002 mg(Be)/m^3; CL 0.005; Pk 0.025/30M/8H) TWA 0.002 mg(Be)/m^3; STEL 0.005 mg(Be)/m^3/30M; CL 0.025 mg(Be)/m^3
ACGIH TLV: TWA 0.002 mg(Be)/m^3, Suspected Human Carcinogen.
NIOSH REL: CL not to exceed 0.0005 mg(Be)/m^3

SAFETY PROFILE: Confirmed carcinogen with experimental carcinogenic data by inhalation. Deadly poison by subcutaneous and intravenous routes. Human mutation data reported. See also BERYLLIUM COMPOUNDS and SULFATES. When heated to decomposition it emits very toxic fumes of BeO and SO_x.

BFU750 ***HR: 3***
BERYLLIUM TETRAHYDROBORATE
mf: B_2BeH_8 mw: 38.70

$$Be(BH_4)_2$$

CONSENSUS REPORTS: Beryllium and its compounds are on the Community Right-To-Know List.

OSHA PEL: (Transitional: TWA 0.002 mg(Be)/m^3; CL 0.005; Pk 0.025/30M/8H) TWA 0.002 mg(Be)/m^3; STEL 0.005 mg(Be)/m^3/30M; CL 0.025 mg(Be)/m^3
ACGIH TLV: TWA 0.002 mg(Be)/m^3, Suspected Human Carcinogen.
NIOSH REL: CL not to exceed 0.0005 mg(Be)/m^3

SAFETY PROFILE: Confirmed carcinogen. Ignites and then explodes in air or on contact with water. Upon decomposition it emits toxic fumes of BeO and BO_x. See also BERYLLIUM COMPOUNDS and BORON COMPOUNDS.

BFV000 ***HR: 3***
BERYLLIUM TETRAHYDROBORATETRI-METHYLAMINE
mf: $C_3H_{17}B_2BeN$ mw: 97.78

$Be(BH_4(2 \cdot N(CH_3)_3$

CONSENSUS REPORTS: Beryllium and its compounds are on the Community Right-To-Know List.

OSHA PEL: (Transitional: TWA 0.002 mg(Be)/m^3; CL 0.005; Pk 0.025/30M/8H) TWA 0.002 mg(Be)/m^3; STEL 0.005 mg(Be)/m^3/30M; CL 0.025 mg(Be)/m^3
ACGIH TLV: TWA 0.002 mg(Be)/m^3, Suspected Human Carcinogen.
NIOSH REL: CL not to exceed 0.0005 mg(Be)/m^3

SAFETY PROFILE: Confirmed carcinogen. It will ignite in contact with air or water. When heated to decomposition it emits toxic fumes of BeO, BO_x, and NO_x. See also BERYLLIUM COMMPOUNDS and BORON COMPOUNDS.

BFV250 CAS:39413-47-3 ***HR: 3***
BERYLLIUM ZINC SILICATE
mf: $O_2Si \cdot Zn \cdot Be$ mw: 134.47

SYN: ZINC BERYLLIUM SILICATE

TOXICITY DATA with REFERENCE
ivn-rbt TDLo:100 mg/kg/2W-I:ETA LANCAO 1,519,50

CONSENSUS REPORTS: NTP Fifth Annual Report on Carcinogens. IARC Cancer Review: Group 2A IMEMDT 7,127,87; Animal Sufficient Evidence IMEMDT 23,143,80; IMEMDT 1,17,72. Beryllium and its compounds, as well as zinc and its compounds, are on the Community Right-To-Know List.

OSHA PEL: (Transitional: TWA 0.002 mg(Be)/m^3; CL 0.005; Pk 0.025/30M/8H) TWA 0.002 mg(Be)/m^3; STEL 0.005 mg(Be)/m^3/30M; CL 0.025 mg(Be)/m^3
ACGIH TLV: TWA 0.002 mg(Be)/m^3, Suspected Human Carcinogen.
NIOSH REL: (Beryllium) CL not to exceed 0.0005 mg(Be)/m^3

SAFETY PROFILE: Confirmed carcinogen with experimental tumorigenic data. When heated to decomposition it emits toxic fumes of BeO and ZnO. See also BERYLLIUM COMPOUNDS, ZINC COMPOUNDS, and SILICATES.

BFV300 CAS:58970-76-6 ***HR: 3***
BESTATIN
mf: $C_{16}H_{24}N_2O_4$ mw: 308.42

SYNS: 3-(R)-AMINO-2-(s)-HYDROXY-4-PHENYLBUTANOYL-(2)-LEUCINE ◇ (S-(4*,s*))-N-(3-AMINO-2-HYDROXY-1-OXO-4-PHENYLBUTYL)-l-LEUCINE ◇ NK 421

TOXICITY DATA with REFERENCE
orl-rat TDLo:31850 mg/kg (91D pre):REP OYYAA2 27,401,84
ipr-rat LD50:780 mg/kg JJANAX 36,2971,83
scu-rat LD50:1900 mg/kg JJANAX 36,2971,83
ipr-mus LD50:190 mg/kg JJANAX 36,2971,83
scu-mus LD50:1300 mg/kg JJANAX 36,2971,83

SAFETY PROFILE: Poison by intraperitoneal route. Moderately toxic by subcutaneous route. Experimental reproductive effects. When heated to decomposition it emits toxic fumes of NO_x.

BFV325 CAS:75219-46-4 ***HR: D***
BESTRABUCIL
mf: $C_{41}H_{47}Cl_2NO_6$ mw: 720.79

SYN: KM 2210

TOXICITY DATA with REFERENCE
dni-hmn:hlas 5 mg/L KSRNAM 20,1974,86
oth-hmn:hlas 50 mg/L KSRNAM 20,1974,86
orl-rat TDLo:135 mg/kg (female 17-22D post):REP KSRNAM 19,6349,85
orl-rat TDLo:10 mg/kg (7-17D preg):TER KSRNAM 19,6306,85

SAFETY PROFILE: An experimental teratogen. Other experimental reproductive effects. Human mutation data reported. When heated to decomposition it emits toxic fumes of Cl^- and NO_x.

BFV350 CAS:54856-23-4 ***HR: 3***
BETAHISTINE MESYLATE
mf: $C_8H_{12}N_2 \cdot 2CH_4O_3S$ mw: 328.44

SYNS: BETAHISTINE MESILATE ◇ N-METHYL-2-PYRIDINEETHANAMINE DIMETHANESULFONATE

TOXICITY DATA with REFERENCE
orl-rat LD50:3030 mg/kg NIIRDN 6,750,82
scu-rat LD50:940 mg/kg NIIRDN 6,750,82
ivn-rat LD50:604 mg/kg NIIRDN 6,750,82
orl-mus LD50:500 mg/kg NIIRDN 6,750,82
scu-mus LD50:1630 mg/kg NIIRDN 6,750,82
ivn-mus LD50:505 mg/kg NIIRDN 6,750,82
orl-gpg LD50:1400 mg/kg NIIRDN 6,750,82
scu-gpg LD50:120 mg/kg NIIRDN 6,750,82
ivn-gpg LD50:22900 μg/kg NIIRDN 6,750,82

SAFETY PROFILE: Poison by subcutaneous route. Moderately toxic by ingestion and other routes. When heated to decomposition it emits toxic fumes of NO_x and SO_x.

BFV750 CAS:378-44-9 ***HR: D***
BETAMETHASONE
mf: $C_{22}H_{29}FO_5$ mw: 392.51

SYNS: BETNELAN ◇ BETSOLAN ◇ CELESTONE ◇ 9-α-FLUORO-16-β-METHYLPREDNISOLONE ◇ 9-α-FLUORO-16-β-METHYL- 1,4-PREGNADIENE-11-β,17-α,21-TRIOL-3,20-DIONE ◇ 9-FLUORO-11-β,17,21-TRIHYDROXY-16-β-METHYLPREGNA-1,4-DIENE-3,20-DIONE ◇ 9-α-FLUORO-11-β,17,21-TRIHYDROXY-16-β-METHYLPREGNA-1,4-DIENE- 3,20-DIONE ◇ 16-β-METHYL-1,4-PREGNADIENE-9-α-FLUORO-11-β,17-α,21-TRIOL- 3,20-DIONE ◇ NSC-39470 ◇ Sch 4831

TOXICITY DATA with REFERENCE
scu-uns TDLo:138 mg/kg (female 8-18D post):TER JIDOAA 35,387,86
scu-uns TDLo:69 mg/kg (female 8-18D post):REP JIDOAA 35,387,86

SAFETY PROFILE: An experimental teratogen. Other experimental reproductive effects. When heated to decomposition it emits toxic fumes of F^-.

BFV755 ***HR: D***
BETAMETHASONE ACETATE and BETAMETHASONE PHOSPHATE
mf: $C_{24}H_{31}FO_6 \cdot C_{22}H_{30}FO_8P$ mw: 907.04

SYNS: BETAMETHASONE PHOSPHATE and BETAMETHASONE ACETATE ◇ CELESTONE SOLUSPAN ◇ PREGNA-1,4-DIENE-3,20-DIONE, 9-FLUORO-11-β,17,21-TRIHYDROXY-16-β-METHYL-, 21-ACETATE and 9-FLUORO-11-β,17,21-TRIHYDROXY-16-β-METHYL-PREGNA-1,4-DIENE-3,20-DIO NE 21-(DIHYDROGEN PHOSPHATE)

TOXICITY DATA with REFERENCE
ims-mky TDLo:5200 μg/kg (female 17-19W post):TER AJOGAH 141,1053,81
ims-mky TDLo:5600 μg/kg (female 17-19W post):REP AJOGAH 159,233,88

SAFETY PROFILE: An experimental teratogen. Other experimental reproductive effects. When heated to decomposition it emits toxic fumes of F^- and PO_x.

BFV760 CAS:2135-14-0 ***HR: 3***
BETAMETHASONE BENZOATE
mf: $C_{29}H_{33}FO_6$ mw: 496.62

SYNS: BETAMETHASONE 17-BENZOATE ◇ BETHAMETHASONE 17-BENZOATE ◇ 21-DEOXYTRIAMCINOLINE ACETONIDE ◇ DESCINOLONE ACETONIDE ◇ 9-α-FLUORO-16-α,17-α-ISOPROPYLIDENEDIOXY-11-β-HYDROXYPREGNA-1,4-DIENE-3,20-DIONE ◇ MS-1112

TOXICITY DATA with REFERENCE
scu-rat TDLo:80 μg/kg (female 9-18D post):REP OYYAA2 10,661,75
skn-rbt TDLo:7500 μg/kg (female 7-18D post):TER OYYAA2 10,685,75
scu-rat LD50:194 mg/kg TXAPA9 8,250,66

SAFETY PROFILE: Poison by subcutaneous route. An experimental teratogen. Other experimental reproductive effects. When heated to decomposition it emits toxic fumes of F^-.

BFV765 CAS:5593-20-4 ***HR: 3***
BETAMETHASONE DIPROPIONATE
mf: $C_{28}H_{37}FO_7$ mw: 504.65

SYNS: BETAMETHASONE 17,21-DIPROPIONATE ◇ DIPROSONE ◇ 9-FLUORO-11-β,17,21-TRIHYDROXY-16-β-METHYLPREGNA-1,4-DIENE-3,20-DIONE, 17,21-DIPROPIONATE ◇ S-3440

TOXICITY DATA with REFERENCE
scu-rat TDLo:7500 μg/kg (female 9-14D post):REP OYYAA2 8,705,74
scu-mus TDLo:3750 μg/kg (female 7-12D post):TER OYYAA2 8,705,74
ipr-mus LD50:103 mg/kg NIIRDN 6,753,82
scu-mus LD50:78100 μg/kg NIIRDN 6,753,82

SAFETY PROFILE: Poison by subcutaneous and intraperitoneal routes. An experimental teratogen. Other experimental reproductive effects. When heated to decomposition it emits toxic fumes of F^-.

BFV770 CAS:151-73-5 ***HR: 2***
BETAMETHASONE DISODIUM PHOSPHATE
mf: $C_{22}H_{30}FO_8P \cdot 2Na$ mw: 518.47

SYNS: BETAMETHASONE-21-DISODIUMPHOSPHATE ◇ BETAMETHASONE SODIUM PHOSPHATE ◇ BETNESOL ◇ 9-FLUORO-11-β,17,21-TRIHYDROXY-16-β-METHYLPREGNA-1,4-DIENE-3,20-DIONE, 21-(DIHYDROGEN PHOSPHATE), DISODIUM SALT

TOXICITY DATA with REFERENCE
scu-rat TDLo:468 μg/kg (9-14D preg):REP OYYAA2 8,705,75
scu-mus TDLo:60 mg/kg (female 7-12D post):TER OYYAA2 8,705,74
ivn-rat LD50:1276 mg/kg YAKUD5 21,2117,79
orl-mus LD50:1607 mg/kg SKIZAB 29,153,73
ipr-mus LD50:1166 mg/kg SKIZAB 29,153,73
scu-mus LD50:1363 mg/kg SKIZAB 29,153,73

SAFETY PROFILE: Moderately toxic by ingestion and other routes. An experimental teratogen. Other experimental reproductive effects. When heated to decomposition it emits toxic fumes of F^-, PO_x and Na_2O.

BFV900 CAS:37717-82-1 ***HR: 3***
BETANIDINE SULFATE
mf: $C_{18}H_{14}N_2O_8 \cdot 7H_2O_4S$ mw: 1072.90

SYNS: BETANIDIN SULFATE ◇ 2-CARBOXY-1-((2,6-DICARBOXY-2,3-DIHYDRO-4(1H)-PYRIDINYLIDENE)ETHYLIDENE)5,6-DIHYDROXY-1H-INDOLIUM, HYDROXIDE, INNER SALT, SULFATE (SALT)

TOXICITY DATA with REFERENCE
orl-rat LD50:3142 mg/kg NIIRDN 6,750,82
ipr-rat LD50:135 mg/kg NIIRDN 6,750,82
scu-rat LD50:681 mg/kg NIIRDN 6,750,82
orl-mus LD50:1059 mg/kg NIIRDN 6,750,82
ipr-mus LD50:170 mg/kg NIIRDN 6,750,82
scu-mus LD50:398 mg/kg NIIRDN 6,750,82

SAFETY PROFILE: Poison by subcutaneous and intraperitoneal routes. Moderately toxic by ingestion. When heated to decomposition it emits toxic fumes of SO_x and NO_x. See also SULFATES.

BFV975 ***HR: 2***
BETEL LEAVES

TOXICITY DATA with REFERENCE
orl-rat TDLo : 3000 g/kg/43W-C:ETA EXPEAM 35,384,79

CONSENSUS REPORTS: IARC Cancer Review: Animal Inadequate Evidence IMEMDT 37,141,85

SAFETY PROFILE: Questionable carcinogen with experimental tumorigenic data. When heated to decomposition it emits toxic and irritating fumes.

BFW000 CAS:39323-48-3 ***HR: 3***
BETEL NUT

PROP: Mottled brown with fawn color. Extract of 50 grams sun-dried betel nut in 100 mL boiling water (IJCNAW 17,469,76).

SYNS: ARECA CATECHU ◇ ARECA CATECHU Linn., fruit extract ◇ ARECA CATECHU Linn., nut extract ◇ BN ◇ PINANG ◇ POOGIPHALAM, nut extract ◇ SUPARI, nut extract

TOXICITY DATA with REFERENCE
mnt-mus-ipr 1600 mg/kg CRNGDP 5,501,84
sce-mus-ipr 62500 μg/kg/5D-C CRNGDP 7,37,86
msc-ham:lng 5 mg/L CRNGDP 5,501,84
orl-mus TDLo:400 mg/kg (6-15D preg):TER TXCYAC 37,315,85
orl-rat TDLo:1750 mg/kg (female 1-7D post):REP PLMEAA 26,391,74
scu-rat TDLo:2016 mg/kg/42W-I:NEO JNCIAM 60,683,78
orl-mus TDLo:340 g/kg/17W-I:CAR BJCAAI 40,922,79
scu-mus TDLo:1728 mg/kg/13W-I:CAR IJEBA6 18,1159,80
ipr-mus LD50:681 mg/kg IJEBA6 18,594,80

CONSENSUS REPORTS: IARC Cancer Review: Animal Limited Evidence IMEMDT 37,141,85

SAFETY PROFILE: Suspected carcinogen with experimental carcinogenic and neoplastigenic data. Moderately toxic by intraperitoneal route. Experimental teratogenic and reproductive effects. When heated to decomposition it emits toxic fumes of NO_x. See also ARECA NUT, other betel entries, and SMOKELESS TOBACCO.

BFW010 CAS:89957-52-8 ***HR: 3***
BETEL NUT, polyphenol fraction

SYN: POLYPHENOL FRACTION OF BETEL NUT

TOXICITY DATA with REFERENCE
orl-mus TDLo:380 mg/kg/W-C:ETA BJCAAI 40,922,79
scu-mus TDLo:988 mg/kg/13W-I:CAR IJEBA6 18,1159,80

SAFETY PROFILE: Questionable carcinogen with experimental carcinogenic and tumorigenic data. When heated to decomposition it emits acrid smoke and irritating fumes.

BFW050 ***HR: 2***
BETEL NUT TANNIN

SYN: TANNIN from BETEL NUT

TOXICITY DATA with REFERENCE
sce-mus-ipr 1500 mg/kg/15D C CRNGDP 7,37,86
orl-mus TDLo:27740 mg/kg/1Y-I:ETA IJEBA6 24,229,86

SAFETY PROFILE: Questionable carcinogen with experimental tumorigenic data. Mutation data reported. When heated to decomposition it emits toxic and irritating fumes.

BFW120 ***HR: 3***
BETEL QUID

PROP: Composed of Areca nut, lime, catechu, and possibly tobacco and spices wrapped in a betel leaf. Used throughout the Orient and many Pacific islands, it is chewed in a manner similar to chewing tobacco. The major ingredients are:

ARECA NUT (betel nut, supari in India) is the fruit of the areca palm (*Areca catechu L.*). The ripe, orange-yellow nut is separated from its fibrous pericarp and may be cured in boiling water and then dried.

BETEL LEAF comes from the Betel vine (*Piper betle L.*).

LIME (chuna or chunam in India) is prepared from seashells or quarried stone and mixed with water (slaked) to release calcium hydroxide.

CATECHU (kattha in India) is a resinous extract from the heartwood of the Acacia tree (*A. catechu* or *A. suma).*

TOBACCO is the leaf from the tobacco plant (*N. rustica* or *N. Tabacum).*

SPICES and flavorings, such as cardamom, cloves, grated fresh coconut, and sugar, are sometimes added.

SYN: PAN

SAFETY PROFILE: The areca nut contains several alkaloids which are the primary cause of habituation. The most abundant alkaloid is arecoline which mimics the action of acetylcholine and acts as a stimulant.

Betel quid toxicity is due to the presence of areca nut alkaloids, nitrosamines derived from these compounds, polyphenols, and, when used, tobacco specific nitrosamines. Several nitrosamines derived from areca nut alkaloids have been found in the saliva of quid chewers: N-nitrosoguvacoline, N-nitrosoproline, N'-nitrosonornicotine, N'-nitrosoanatabine and 4-(methylnitrosamino)1-(3-pyridyl)-1-butanone. Users of betel quid with tobacco also have nitrosamines derived from tobacco alkaloids in their saliva. Catechu has a high percentage of polyphenols such as kaempferol, dihydroxykaempferol, taxifolin, isorhamnetin, (+)afzelchin, dimeric procyanidin and (−)eipcatechin. Many nitrosamines have been shown to be experimental carcinogens.

There is sufficient evidence that betel quid with tobacco is carcinogenic to humans. There is inadequate evidence that betel quid without tobacco is carcinogenic to humans. There is limited evidence that extracts of betel quid with and without tobacco are carcinogenic to experimental animals. There is limited evidence that areca nut with and without tobacco is carcinogenic to experimental animals. Extracts from the areca nut are mutatgenic. Chewing the quid may cause mouth ulcerations and periodontal disease. There is a high incidence of oral leukoplakia (a precancerous lesion) in betel quid users.

See also specific compounds; ARECA NUT, other betel entries, SMOKELESS TOBACCO, N-NITROSO COMPOUNDS, and NITROSAMINES.

BFW125 ***HR: 3***
BETEL QUID EXTRACT

TOXICITY DATA with REFERENCE
dni-hmn : lym 25000 ppm CNREA8 39,4802,79
dni-rat:mmr 25000 ppm CNREA8 39,4802,79
dni-mus:fbr 25000 ppm CNREA8 39,4802,79
skn-mus TDLo:720 g/kg/26W-C:ETA BJCAAI 14,597,60
scu-mus TDLo:3376 mg/kg/13W-I:CAR IJEBA6 18,1159,80

CONSENSUS REPORTS: IARC Cancer Review: Human Inadequate Evidence IMEMDT 37,141,85; Animal Limited Evidence IMEMDT 37,141,85

SAFETY PROFILE: Suspected carcinogen with experimental carcinogenic and tumorigenic data by skin contact. Human mutation data reported. See other betel entries.

BFW135 ***HR: 3***
BETEL TOBACCO EXTRACT

SYN: JAFFNA TOBACCO

TOXICITY DATA with REFERENCE
sce-hmn:lym 10 mg/L TOLED5 8,17,81
mnt-mus:ipr 24 mg/kg CRNGDP 5,501,84
otr-ham:emb 50 mg/L TOLED5 8,17,81
msc-ham:lng 5 mg/L CRNGDP 5,501,84

CONSENSUS REPORTS: IARC Cancer Review: Human Sufficient Evidence IMEMDT 37,141,85; Animal Limited Evidence IMEMDT 37,141,85

SAFETY PROFILE: Confirmed human carcinogen. Human mutation data reported. See also SMOKELESS TOBACCO and other betel entries.

BFW250 CAS:114-85-2 ***HR: 3***
BETHANIDINE SULFATE
mf: $C_{20}H_{30}N_6{\cdot}H_2O_4S$ mw: 452.64

SYNS: BATEL ◇ BENTANIDOL ◇ BENZAIDIN ◇ BENZANIDINE ◇ BENZOXINE ◇ 1-BENZYL-2,3-DIMETHYL-GUANIDINE SULFATE (1:1/2) ◇ N-BENZYL-N',N''-DIMETHYLGUANIDINE SULFATE ◇ BETALING ◇ BETANIDOLE ◇ BETHANID ◇ BETHANIDINE, HEMISULFATE ◇ BW 467-C-60 ◇ 467-C-60 ◇ N,N-DIMETHYL''-(PHENYLMETHYL)GUANIDINE SULPHATE (2:1) ◇ ESBATAL ◇ ESTABAL ◇ EUSMANID ◇ HYPERSIN ◇ NSC-106563 ◇ REGULIN ◇ TENATHAN

TOXICITY DATA with REFERENCE
ivn-rat LD50:20 mg/kg NYKZAU 72,837,76
orl-mus LD50:520 mg/kg BJPCAL 20,36,63
ipr-mus LD50:150 mg/kg BJPCAL 20,36,63
scu-mus LD50:260 mg/kg BJPCAL 20,36,63
ivn-mus LD50:12 mg/kg BJPCAL 20,36,63
ivn-ckn LDLo:100 mg/kg BJPCAL 20,36,63

SAFETY PROFILE: Poison by intraperitoneal, subcutaneous, and intravenous routes. Moderately toxic by ingestion. See also SULFATES. When heated to decomposition it emits toxic fumes of NO_x and SO_x.

BFW325 CAS:312-93-6 ***HR: D***
BETNELAN PHOSPHATE
mf: $C_{22}H_{30}FO_8P$ mw: 472.49

SYNS: DEXAMETHASONE-21-ORTHOPHOSPHATE ◇ DEXAMETHASONE PHOSPHATE ◇ DEXAMETHASONE-21-PHOSPHATE ◇ NEODECADRON ◇ ORADEXON PHOSPHATE ◇ WYMESONE

TOXICITY DATA with REFERENCE
dni-rat-ipr 3 mg/kg JOENAK 62,527,74
ipr-rat TDLo:200 μg/kg (19-20D preg):REP BNEOBV 36,1,79
par-rbt TDLo:1 mg/kg (female 22-25D post):TER ANOBAU 49,465,65

SAFETY PROFILE: An experimental teratogen. Other experimental reproductive effects. Mutation data reported. When heated to decomposition it emits toxic fumes of F^- and PO_x.

BFW500 CAS:319-86-8 ***HR: 2***
Δ-BHC
mf: $C_6H_6Cl_6$ mw: 290.82

SYNS: Δ-BENZENEHEXACHLORIDE ◇ ENT 9,234 ◇ 1-α,2-α,3-α,4-β,5-α,6-β-HEXACHLOROCYCLOHEXANE ◇ Δ-HEXACHLOROCYCLOHEXANE ◇ Δ-1,2,3,4,5,6-HEXACHLOROCYCLOHEXANE ◇ Δ-LINDANE

TOXICITY DATA with REFERENCE
orl-rat LD50:1000 mg/kg ARSIM* 20,5,66

CONSENSUS REPORTS: Reported in EPA TSCA Inventory.

SAFETY PROFILE: Moderately toxic by ingestion. When heated to decomposition it emits toxic fumes of Cl^-. See also CHLORINATED HYDROCARBONS, ALIPHATIC.

BFW750 CAS:128-37-0 ***HR: 2***
BHT (food grade)
mf: $C_{15}H_{24}O$ mw: 220.39

PROP: White, crystalline solid; faint characteristic odor. Bp: 265°, fp: 68°, flash p: 260°F (TOC), d: 1.048 @ 20°/4°, vap d: 7.6. Sol in alc; insol in water and propylene glycol.

SYNS: ADVASTAB 401 ◇ AGIDOL ◇ ANTIOXIDANT DBPC ◇ ANTIOXIDANT 29 ◇ AO 29 ◇ AO 4K ◇ 2,6-BIS(1,1-DIMETHYLETHYL)-4-METHYLPHENOL ◇ BUKS ◇ BUTYLATED HYDROXYTOLUENE ◇ BUTYLHYDROXYTOLUENE ◇ CAO 1 ◇ CAO 3 ◇ CATALIN CAO-3 ◇ CHEMANOX 11 ◇ DBMP ◇ DBPC (technical grade) ◇ DIBUTYLATED HYDROXYTOLUENE ◇ 2,6-DI-tert-BUTYL-p-CRESOL (OSHA, ACGIH) ◇ 2,6-DI-tert-BUTYL-1-HYDROXY-4-METHYLBENZENE ◇ 3,5-DI-tert-BUTYL-4-HYDROXYTOLUENE ◇ 2,6-DI-terc. BUTYL-p-KRESOL (CZECH) ◇ 2,6-DI-tert-BUTYL-p-METHYLPHENOL ◇ 2,6-DI-tert-BUTYL-4-METHYLPHENOL ◇ FEMA No. 2184 ◇ 4-HYDROXY-3,5-DI-tert-BUTYLTOLUENE ◇ IMPRUVOL ◇ IONOL ◇ IONOL (antioxidant) ◇ 4-METHYL-2,6-DI-terc. BUTYLFENOL (CZECH) ◇ METHYL DI-tert-BUTYLPHENOL ◇ 4-METHYL-2,6-DI-tert-BUTYLPHENOL ◇ NCI-C03598 ◇ NONOX TBC ◇ PARABAR 441 ◇ SUSTANE ◇ TENOX BHT ◇ TOPANOL ◇ VANLUBE PCX

TOXICITY DATA with REFERENCE
skn-hmn 500 mg/48H MLD AMIHBC 5,311,52
skn-rbt 500 mg/48H MOD AMIHBC 5,311,52
eye-rbt 100 mg/24H MOD 28ZPAK -,57,72
dni-hmn:lym 20 μmol/L BBRCA9 80,963,78
dns-rat:lvr 100 pmol/L CRNGDP 5,1547,84
spm-mus-ipr 350 mg/kg/5D-I CMMUAO 5,257,78
orl-mus TDLo:12600 mg/kg (female 1-21D post):REP FEPRA7 31,596,72
orl-mus TDLo:1200 mg/kg (female 9D post):TER TRENAF 28(2),45,77
orl-rat TDLo:134 g/kg/32W-C:CAR CRNGDP 4,895,83
orl-mus TDLo:435 mg/kg/69W-C:CAR FCTXAV 12,367,74
orl-rat TD:247 g/kg/3Y-C:CAR,REP FCTOD7 24,1,86
orl-rat TD:247 g/kg/3Y-C:NEO,REP FCTOD7 24,1,86
orl-mus TD:1423 mg/kg/43W-C:NEO TXCYAC 38,151,86
orl-wmn TDLo:80 mg/kg:PSY,GIT NEJMAG 314,648,86
orl-rat LD50:890 mg/kg NEOLA4 24,253,77
orl-mus LD50:1040 mg/kg JAPMA8 38,366,49
ipr-mus LD50:138 mg/kg TXAPA9 61,475,81
ivn-mus LD50:180 mg/kg JMCMAR 23,1350,80
orl-cat LDLo:940 mg/kg AMIHAB 11,93,55
orl-rbt LDLo:2100 mg/kg AMIHAB 11,93,55
orl-gpg LD50:10700 mg/kg AMIHAB 11,93,55

CONSENSUS REPORTS: IARC Cancer Review: Group 3 IMEMDT 7,56,87; Animal Limited Evidence IMEDT 40,161,86. NCI Carcinogenesis Bioassay Completed; (feed): No Evidence: mouse,rat NCITR* NCI-CG-TR-150,79. Reported in EPA TSCA Inventory. EPA Genetic Toxicology Program.

OSHA PEL: TLV 10 mg/m^3
ACGIH TLV: TLV 10 mg/m^3

SAFETY PROFILE: Poison by intraperitoneal and intravenous routes. Moderately toxic by ingestion. An experimental teratogen. Other experimental reproductive effects. A human skin irritant. A skin and eye irritant. Questionable carcinogen with experimental carcinogenic and neoplastigenic data. Combustible when exposed to heat or flame. It can react with oxidizing materials. To fight fire, use CO_2, dry chemical. When heated to decomposition it emits acrid smoke and fumes.

BFX000 CAS:613-35-4 ***HR: 3***
4′,4‴-BIACETANILIDE
mf: $C_{16}H_{16}N_2O_2$ mw: 268.34

PROP: Mp: 329°.

SYNS: N,N′-(1,1′-BIPHENYL)-4,4′-DIYLBIS-ACETAMIDE 4′,4‴-BIACETANILIDE ◇ N,N′-4,4′-BIPHENYLYLENEBISACETAMIDE ◇ 4,4′-DIACETYLAMINOBIPHENYL ◇ N,N′-DIACETYL BENZIDINE ◇ 4,4′-DIACETYLBENZIDINE

TOXICITY DATA with REFERENCE
mma-sat 5 μg/plate ENMUDM 6,145,84
dnd-rat:lvr 100 mg/L CRNGDP 5,407,84
orl-rat TDLo:6300 mg/kg/35W-C:ETA CNREA8 16,525,56
ipr-rat TDLo:64 mg/kg/4W-I:CAR CRNGDP 2,747,81
scu-rat TDLo:900 mg/kg:NEO ARPAAQ 81,146,66

CONSENSUS REPORTS: IARC Cancer Review: Group 2B IMEMDT 7,56,87, Animal Sufficient Evidence IMEMDT 16,293,78. Reported in EPA TSCA Inventory.

SAFETY PROFILE: Suspected carcinogen with experimental carcinogenic, neoplastigenic, and tumorigenic data. Mutation data reported. When heated to decomposition it emits toxic fumes of NO_x.

BFX125 CAS:3624-96-2 ***HR: 3***
BIALLYLAMICOL DIHYDROCHLORIDE
mf: $C_{28}H_{40}N_2O_2{\cdot}2ClH$ mw: 509.62

SYNS: BIALAMICOL HYDROCHLORIDE ◇ BIALLYLAMICOL HYDROCHLORIDE ◇ α,α′-BIS(DIETHYLAMINO)-5,5′-DIALLYL-m,m′-BITOLYL-4,4′-DIOL DIHYDROCHLORIDE ◇ CAMOFORM HYDROCHLORIDE ◇ PAA-701 DIHYDROCHLORIDE ◇ SN 6771 DIHYDROCHLORIDE

TOXICITY DATA with REFERENCE
orl-hmn TDLo:243 mg/kg/9D:MET,GIT 85GLAQ 1,304,46
orl-hmn TDLo:43 mg/kg:GIT 85GLAQ 1,304,46
orl-rat LD50:1649 mg/kg ANTCAO 7,113,57
orl-mus LD50:3950 mg/kg ANTCAO 7,113,57

SAFETY PROFILE: Poison to humans by ingestion with systemic effects: fever and nausea or vomiting. When heated to decomposition it emits toxic fumes of NO_x and HCl. See also ALLYL COMPOUNDS.

BFX250 CAS:2130-56-5 ***HR: 3***
5,5'-BIANTHRANILIC ACID
mf: $C_{14}H_{12}N_2O_4$ mw: 272.28

SYNS: 3,3'-BENZIDINEDICARBOXYLIC ACID ◇ 4,4'-DIAMINO-3,3'-BIPHENYLDICARBOXYLIC ACID ◇ 4,4'-DIAMINOBIPHENYL-3,3'-DICARBOXYLIC ACID ◇ 3,3'-DICARBOXYBENZIDINE ◇ KWAS BENZYDYNODWUKAROKSYLOWY (POLISH)

TOXICITY DATA with REFERENCE
pic-esc 100 mmol/L MDMIAZ 31,11,79
scu-rat TDLo:7 g/kg/77W-I:ETA VOONAW 15(5),60,69

CONSENSUS REPORTS: Reported in EPA TSCA Inventory.

SAFETY PROFILE: Questionable carcinogen with experimental tumorigenic data. Mutation data reported. When heated to decomposition it emits toxic fumes of NO_x.

BFX325 CAS:4388-03-8 ***HR: 3***
1,1'-BIAZIRIDINYL
mf: $C_4H_8N_2$ mw: 84.12

$(CH_2CH_2N-)_2$

SAFETY PROFILE: When heated it reacts violently with oxygen. When heated to decomposition it emits toxic fumes of NO_x.

BFX500 CAS:103-29-7 ***HR: 3***
BIBENZYL
mf: $C_{14}H_{14}$ mw: 182.28

PROP: Flash p: 264°F, autoign temp: 896°F, d: 1.0, vap d: 6.29, bp: 285°.

SYNS: DIBENZYL ◇ 1,2-DIPHENYLETHANE

TOXICITY DATA with REFERENCE
ipr-mus LD50:2500 mg/kg ARZNAD 19,617,69
ivn-mus LD50:78 mg/kg ARZNAD 19,617,69

CONSENSUS REPORTS: Reported in EPA TSCA Inventory.

SAFETY PROFILE: Poison by intravenous route. Moderately toxic by intraperitoneal route. Combustible. To fight fire, use water, spray, mist, alcohol foam, dry chemical. When heated to decomposition it emits acrid smoke and fumes.

BFY000 CAS:826-62-0 ***HR: 2***
BICYCLO(2.2.1)-HEPT-5-ENE-2,3-DICARBOXYLIC ANHYDRIDE
mf: $C_9H_8O_3$ mw: 164.17

SYN: ANHYDRID KYSELINY 3,6-ENDOMETHYLEN-Δ(SUP 4)-TETRAHYDROFTALOVE (CZECH)

TOXICITY DATA with REFERENCE
skn-rbt 500 mg/24H MLD 28ZPAK -,140,72
eye-rbt 5 mg/24H SEV 28ZPAK -,140,72
orl-rat LD50:3250 mg/kg 28ZPAK -,140,72

CONSENSUS REPORTS: Reported in EPA TSCA Inventory.

SAFETY PROFILE: Moderately toxic by ingestion. Mild skin and severe eye irritant. When heated to decomposition it emits acrid smoke and irritating fumes. See also ANHYDRIDES.

BFY250 CAS:95-39-6 ***HR: 2***
BICYCLO(2.2.1)HEPT-5-ENE-2-METHYLOL ACRYLATE
mf: $C_{11}H_{14}O_2$ mw: 178.25

SYNS: ACRYLIC ACID-5-NORBORNEN-2-METHYL ESTER ◇ ACRYLIC ACID-5-NORBORNEN-2-YLMETHYL ESTER ◇ CYCLOL ACRYLATE ◇ 2,5-endo-METHYLENE-Δ^3-TETRAHYDROBENZYL ACRYLATE ◇ 5-NORBORNENE-2-METHANOL ACRYLATE ◇ 5-NORBORNENE-2-METHYLOLACRYLATE ◇ 2-PROPENOIC ACID BICYCLO(2,2,1)HEPT-5-EN-2-YLMETHYLESTER

TOXICITY DATA with REFERENCE
orl-rat LD50:1410 mg/kg TXAPA9 28,313,74
skn-rbt LD50:2830 mg/kg TXAPA9 28,313,74

CONSENSUS REPORTS: Reported in EPA TSCA Inventory.

SAFETY PROFILE: Moderately toxic by ingestion and skin contact. See also ESTERS. When heated to decomposition it emits acrid smoke and irritating fumes.

BFY750 CAS:2886-89-7 ***HR: 2***
BICYCLONONADIENE DIEPOXIDE
mf: $C_9H_{12}O_2$ mw: 152.21

SYNS: 1,2:5,6-DIEPOXYHEXAHYDROINDAN ◇ 4,9-DIOXATETRACYCLO(5.4.0.$0^{3,5}$.$0^{8,10}$)UNDECANE ◇ 4,10-DIOXATETRACYCLO(5.4.$0^{3,5}$.$0^{1,7}$.$0^{9,11}$)UNDECANE ◇ OCTAHYDRO-2H-BISOXIRENO(a,f)INDENE

TOXICITY DATA with REFERENCE
orl-rat LD50:2140 mg/kg AIHAAP 30,470,69
skn-rbt LDLo:1770 mg/kg AIHAAP 30,470,69

SAFETY PROFILE: Moderately toxic by ingestion and

skin contact. When heated to decomposition it emits acrid smoke and irritating fumes.

BFZ000 ***HR: 3***
3,3'-(BICYCLO(2.2.2)OCTANE-1,4-DIYLBIS(CARBONYLIMINO-4,1-PHENYLENECARBONYL IMINO))BIS(1-ETHYLPYRIDINIUM, SALT with 4-METHYLBENZENESULFONIC ACID (1:2)
mf: $C_{38}H_{42}N_6O_4•2C_7H_7O_3S$ mw: 989.26

TOXICITY DATA with REFERENCE
dnd-mus:lym 6800 nmol/L JMCMAR 22,134,79
ipr-mus LD10:27 mg/kg JMCMAR 22,134,79

SAFETY PROFILE: Poison by intraperitoneal route. See also SULFONATES. Mutation data reported. When heated to decomposition it emits very toxic fumes of SO_x and NO_x.

BGA250 CAS:81-21-0 ***HR: 3***
BICYCLOPENTADIENE DIOXIDE
mf: $C_{10}H_{12}O_2$ mw: 164.22

SYNS: DICYCLOPENTADIENE DIEPOXIDE ◇ DICYCLOPENTADIENE DIOXIDE ◇ 1,2:5,6-DIEPOXYHEXAHYDRO-4,7-METHANOINDAN ◇ 1,2:5,6-DIEPOXY-3a,4,5,6,7,7a-HEXAHYDRO-4,7-METHANOINDAN

TOXICITY DATA with REFERENCE
skn-rbt 500 mg open MLD UCDS** 10/5/61
orl-rat LD50:210 mg/kg UCDS** 10/5/61
ivn-mus LD50:56 mg/kg CSLNX* NX#04159
skn-rbt LD50:8000 mg/kg UCDS** 10/5/61

CONSENSUS REPORTS: Reported in EPA TSCA Inventory.

SAFETY PROFILE: Poison by ingestion and intravenous routes. Mildly toxic by skin contact. A skin irritant. When heated to decomposition it emits acrid smoke and irritating fumes.

BGA500 ***HR: 2***
BICYCLOPENTADIENYLBIS(TRICARBONYLIRON)
mf: $C_{16}Fe_2H_8O_6$ mw: 407.94

TOXICITY DATA with REFERENCE
orl-rat LD50:2900 mg/kg SCCUR* -,1,61
ipr-rat LD50:1000 mg/kg SCCUR* -,1,61

SAFETY PROFILE: Moderately toxic by ingestion and intraperitoneal routes. When heated to decomposition it emits acrid smoke and irritating fumes. See also IRON COMPOUNDS and CARBONYLS.

BGA650 CAS:5164-35-2 ***HR: 3***
BICYCLO(2.1.0)PENT-2-ENE
mf: C_5H_6 mw: 66.10

SAFETY PROFILE: This strained ring compound may explode spontaneously. When heated to decomposition it emits acrid smoke and fumes.

BGA750 CAS:1464-53-5 ***HR: 3***
1,1'-BI(ETHYLENE OXIDE)
mf: $C_4H_6O_2$ mw: 86.10

PROP: Colorless liquid. Bp: 142°, mp: 19°, d: 1.113 @ 18°/4°.

SYNS: BIOXIRANE ◇ 2,2'-BIOXIRANE ◇ BUTADIENDIOXYD (GERMAN) ◇ BUTADIENE DIEPOXIDE ◇ 1,3-BUTADIENE DIEPOXIDE ◇ BUTADIENE DIOXIDE ◇ BUTANE DIEPOXIDE ◇ DEB ◇ DIEPOXYBUTANE ◇ 2,4-DIEPOXYBUTANE ◇ 1,2:3,4-DIEPOXYBUTANE ◇ DIOXYBUTADIENE ◇ ENT 26,592 ◇ ERYTHRITOL ANHYDRIDE ◇ RCRA WASTE NUMBER U085

TOXICITY DATA with REFERENCE
skn-rbt 10 mg/24H open SEV AMIHBC 10,61,54
skn-rbt 50 mg open SEV UCDS** 4/25/58
eye-rbt 250 μg open SEV AMIHBC 10,61,54
eye-rbt 3 ppm/3D JPCAAC 10,17,60
mmo-asn 20 μmol/L MUREAV 132,161,84
cyt-hmn:bmr 100 μg/L CGCYDF 9,51,83
sce-mus-ivn 193 μmol/kg MUREAV 108,251,83
sce-ham:lng 1 mg/L CNREA8 44,3270,84
ipr-rat TDLo:380 mg/kg/13W-I:ETA BJPCAL 6,235,51
orl-rat LD50:78 mg/kg AMIHBC 10,61,54
ihl-rat LC50:90 ppm/4H SCCUR* -,2,61
orl-mus LD50:72 mg/kg SCCUR* -,2,61
ipr-mus LD50:31 mg/kg AEPPAE 230,559,57
skn-rbt LD50:80 mg/kg AMIHBC 10,61,54

CONSENSUS REPORTS: NTP Fifth Annual Report on Carcinogens. EPA Extremely Hazardous Substances List. EPA Genetic Toxicology Program. Community Right-To-Know List. Reported in EPA TSCA Inventory.

SAFETY PROFILE: Confirmed carcinogen with experimental tumorigenic data. Poison by ingestion, inhalation, skin contact and intraperitoneal routes. Human mutation data reported. A severe skin and eye irritant. When heated to decomposition it emits acrid smoke and irritating fumes.

BGA825 ***HR: 3***
BIFONAZOLE
mf: $C_{22}H_{18}N_2$ mw: 310.42

SYNS: BAY H 4502 ◇ BIFONAZOL ◇ 1-(α-(4-BIPHENYLYL)BENZYL)IMIDAZOLE ◇ 1-((4-BIPHENYLYL)PHENYLMETHYL)-1H-IMIDAZOLE ◇ MYCOSPOR

TOXICITY DATA with REFERENCE
orl-rat TDLo:1160 mg/kg (female 16-22D post):REP ARZNAD 33,739,83
orl-rat LD50:1463 mg/kg OYYAA2 28,23,84
ivn-rat LD50:63 mg/kg OYYAA2 28,23,84

orl-mus LD50:2629 mg/kg
ivn-mus LD50:57 mg/kg OYYAA2 28,23,84
orl-rbt LD50:4000 mg/kg ARZNAD 33,739,83

SAFETY PROFILE: Poison by intravenous route. Moderately toxic by ingestion. Experimental reproductive effects. When heated to decomposition it emits toxic fumes of NO_x.

BGB250 CAS:37247-90-8 ***HR: 3***
BIHOROMYCIN (crystalline)
mf: $C_{41}H_{76}O_{13}$ mw: 777.17

TOXICITY DATA with REFERENCE
orl-mus LD50:7 mg/kg 85ERAY 2,1077,78
ipr-mus LD50:19 mg/kg 85ERAY 2,1077,78

SAFETY PROFILE: Poison by ingestion and intraperitoneal routes. When heated to decomposition it emits acrid smoke and irritating fumes.

BGB315 CAS:3521-84-4 ***HR: 2***
BILIGRAFIN FORTE
mf: $C_{20}H_{14}I_6N_2O_6$•$2C_7H_{17}NO_5$ mw: 1530.26

SYNS: ADIPIODONE MEGLUMINE ◇ CAVUMBREN ◇ CHOLOGRAF-N-METHYLGLUCAMINE ◇ ENDOCISTOBIL ◇ ENDOGRAFIN ◇ ENDOGRAPHIN ◇ INTRABLIX ◇ IODIPAMIDE MEGLUMINE ◇ IODIPAMIDE MEGLUMINE SALT ◇ IODIPAMIDE METHYLGLUCAMINE SALT ◇ MEGLUMINE IODIPAMIDE ◇ METHYL GLUCAMINE BILIGRAFIN ◇ METHYL GLUCAMINE IODIPAMIDE ◇ ULTRABIL

TOXICITY DATA with REFERENCE
ivn-rat LD50:5000 mg/kg NIIRDN 6,5,82
par-rat LD50:1921 mg/kg FRPSAX 28,1011,73
ivn-mus LD50:3195 mg/kg INVRAV 15(Suppl),142,80
ivn-dog LD50:1200 mg/kg FRPSAX 28,996,73
par-rbt LD50:1446 mg/kg FRPSAX 28,1011,73

SAFETY PROFILE: Moderately toxic by several routes. When heated to decomposition it emits toxic fumes of I^- and NO_x.

BGB325 ***HR: 2***
BILIGRAFIN SODIUM
mf: $C_{20}H_{12}I_6N_2O_6$•2Na mw: 1183.72

SYNS: ADIPIC ACID DI-(3-CARBOXY-2,4,6-TRIIODOANILIDE) DISODIUM ◇ ADIPINSAEURE-DI-(3-CARBOXY-2,4,6-TRIJOD-ANILID) DINATRIUM (GERMAN) ◇ BILIGRAFIN NATRIUM (GERMAN)

TOXICITY DATA with REFERENCE
ivn-rat LD50:3800 mg/kg ARZNAD 14,451,64
ivn-mus LD50:3200 mg/kg ARZNAD 14,451,64
ivn-dog LD50:1900 mg/kg ARZNAD 14,451,64

SAFETY PROFILE: Moderately toxic by intravenous route. When heated to decomposition it emits toxic fumes of I^-, NO_x, and Na_2O.

BGB350 ***HR: 1***
BILIVISTAN SODIUM
mf: $C_{18}H_8I_6N_2O_7$•2Na mw: 1171.66

SYNS: BILIVISTAN NATRIUM (GERMAN) ◇ DIGLYCOLIC ACID DI-(3-CARBOXY-2,4,6-TRIIODOANILIDE)DISODIUM ◇ DIGLYCOLSAEURE-DI-(3-CARBOXY-2,4,6-TRIJOD-ANILID) DINATRIUM (GERMAN) ◇ 3,3'-(OXYDIMETHYLENEBIS(CARBONYLIMINO)BIS(2,4,6-TRIIODOBENZOIC ACID DISODIUM SALT)

TOXICITY DATA with REFERENCE
ivn-rat LD50:6000 mg/kg ARZNAD 14,451,64
ivn-mus LD50:5300 mg/kg ARZNAD 14,451,64
ivn-dog LD50:5500 mg/kg ARZNAD 14,451,64

SAFETY PROFILE: Mildly toxic by intravenous route. When heated to decomposition it emits toxic fumes of I^-, NO_x, and Na_2O.

BGB400 CAS:55268-74-1 ***HR: 2***
BILTRICIDE
mf: $C_{19}H_{24}N_2O_2$ mw: 312.45

PROP: Crystals. Mp: 136-138°. Solubility (g/100 ml): ethanol 9.7; chloroform 56.7; water 0.04.

SYNS: CESOL ◇ 2-CYCLOHEXYLCARBONYL-1,2,3,6,7,11b-HEXAHYDRO-4H-PYRAZINO(2,1-a)ISOQUINOLIN-4-ONE◇ DRONCIT ◇ EMBAY 8440 ◇ PRAZIQUANTEL ◇ PYQUITON

TOXICITY DATA with REFERENCE
mmo-sat 150 mg/L ENMUDM 2,234,80
dni-hmn:hla 5 mmol/L MUREAV 93,447,82
bfa-mus/sat 1200 mg/kg CNREA8 38,4478,78
msc-ham:lng 10 mg/L CNREA8 42,2692,82
orl-rat LD50:2840 mg/kg ARZNAD 31,555,81
ipr-rat LD50:796 mg/kg ARZNAD 31,555,81
orl-mus LD50:2454 mg/kg ARZNAD 31,555,81
ipr-mus LD50:1062 mg/kg VETNAL 59(5),64,83
scu-mus LD50:7172 mg/kg ARZNAD 31,555,81
orl-rbt LD50:1050 mg/kg ARZNAD 31,555,81

CONSENSUS REPORTS: EPA Genetic Toxicology Program.

SAFETY PROFILE: Moderately toxic by ingestion and other routes. Human mutation data reported. When heated to decomposition it emits toxic fumes of NO_x.

BGB500 CAS:485-31-4 ***HR: 3***
BINAPACRYL
mf: $C_{15}H_{18}N_2O_6$ mw: 322.35

SYNS: ACRICID ◇ AMBOX ◇ 2-sec-BUTYL-4,6-DINITROPHENYL-3,3-DIMETHYLACRYLATE ◇ 2-sec-BUTYL-4,6-DINITROPHENYL-3-METHYL-2-BUTENOATE ◇ 2-sec-BUTYL-4,6-DINITROPHENYL-3-METHYLCROTONATE ◇ 2-sec-BUTYL-4,5-DINITROPHENYL SENECIOATE ◇ DAPACRYL ◇ 3,3-DIMETHYL-ACRYLATE de 2,4-DINITRO-6-(1-METHYLPROPYLE) PHENYLE (FRENCH) ◇ 3,3-DIMETHYLACRYLIC ACID 2-sec-BUTYL-4,5-DINITROPHENYL ESTER ◇ DINAPACRYL ◇ 4,6-DINITRO-2-sec-BUTYLPHENYL β,β-DIMETHYLACRYLATE ◇ 2,4-DINITRO-6-sec-BUTYLPHENYL-2-

METHYLCROTONATE ◇ 4,6-DINITROPHENYL-2-sec-BUTYL-3-METHYL-2-BUTENONATE ◇ DINOSEB METHACRYLATE ◇ ENDOSAN ◇ ENT 25,793 ◇ FMC 9044 ◇ HOE 2784 ◇ 3-METHYLCROTONIC ACID 2-sec-BUTYL-4,6-DINITROPHENYL ESTER ◇ (6-(1-METHYL-PROPYL)-2,4-DINITRO-FENYL)-3,3-DIMETHYL ACRYLAAT (DUTCH) ◇ (6-(1-METHYL-PROPYL)-2,4-DINITRO-PHENYL)-3,3-DIMETHYL ACRYLAT (GERMAN) ◇ 2-(1-METHYLPROPYL)-4,6-DINITROPHENYL-β,β-DIMETHACRYLATE ◇ (6-(1-METIL-PROPIL)-2,4-DINITRO-FENIL)-3,3-DIMETIL-ACRILATO(ITALIAN) ◇ MOROCIDE ◇ MORROCID ◇ NIA 9044 ◇ NIAGARA 9044

TOXICITY DATA with REFERENCE
mmo-sat 5 mg/plate MUREAV 116,185,83
orl-rat LD50:58 mg/kg TXAPA9 14,515,69
skn-rat LD50:720 mg/kg WRPCA2 9,119,70
orl-mus LD50:1600 mg/kg TXAPA9 7,353,65
orl-dog LD50:50 mg/kg GUCHAZ 6,42,73
skn-rbt LD50:750 mg/kg GUCHAZ 6,42,73
orl-gpg LD50:200 mg/kg TXAPA9 7,353,65

SAFETY PROFILE: Poison by ingestion. Moderately toxic by skin contact. Mutation data reported. A cholinesterase inhibitor. When heated to decomposition it emits fumes of NO_x. See also PARATHION and PHOSPHORUS COMPOUNDS.

BGB750 CAS:4488-22-6 ***HR: 3***
(1,1'-BINAPHTHALENE)-2,2'-DIAMINE
mf: $C_{20}H_{16}N_2$ mw: 284.38

SYN: 2,2'-DIAMINO-1,1'-DINAPHTHYL

TOXICITY DATA with REFERENCE
skn-mus TDLo:590 mg/kg/25W-I:ETA AJCAA7 40,62,40

SAFETY PROFILE: Questionable carcinogen with experimental tumorigenic data. When heated to decomposition it emits toxic fumes of NO_x.

BGC000 CAS:795-95-9 ***HR: 3***
(1,2'-BINAPHTHALENE)-1,2'-DIAMINE
mf: $C_{20}H_{16}N_2$ mw: 284.38

SYN: 1:2'-DIAMINO-1':2-DINAPHTHYL

TOXICITY DATA with REFERENCE
skn-mus TDLo:2880 mg/kg/60W-I:ETA PRLBA4 131,170,72

SAFETY PROFILE: Questionable carcinogen with experimental tumorigenic data by skin contact. When heated to decomposition it emits toxic fumes of NO_x.

BGC250 CAS:69382-20-3 ***HR: 3***
BINDON ETHYL ETHER
mf: $C_{20}H_{14}O_3$ mw: 302.34

SYNS: BINDON ATHYLATHER ◇ 2-(3-ETHOXY-1-INDANYLIDENE)-1,3-DINDANDIONE

TOXICITY DATA with REFERENCE
ipr-mus TDLo:2500 μg/kg (9D-preg):REP ARTODN 33,191,75
ipr-mus TDLo:5 mg/kg (9D preg):TER ARTODN 33,191,75
ipr-mus LDLo:40 mg/kg ARTODN 33,191,75

SAFETY PROFILE: Poison by intraperitoneal route. An experimental teratogen. Other experimental reproductive effects. When heated to decomposition it emits acrid smoke and irritating fumes. See also ETHERS.

BGC500 CAS:57647-35-5 ***HR: 3***
BINODALINE HYDROCHLORIDE
mf: $C_{19}H_{23}N_3 \cdot ClH$ mw: 329.87

SYNS: BINODALIN HYDROCHLORID (GERMAN) ◇ 1-(DIMETHYLAMINOETHYL-METHYL)AMINO-3-PHENYLINDOLEHYDROCHLORIDE ◇ 1-(omega-DIMETHYLAMINOETHYLMETHYL)AMINO-3-PHENYLINDOLE HYDROCHLORIDE ◇ SGD-SCHA 1059 ◇ N,N,N'-TRIMETHYL-N'-(3-PHENYL-1H-INDOL-1-YL)-1,2-ETHANEDIAMINEMONOHYDROCHLORIDE

TOXICITY DATA with REFERENCE
orl-rat LD50:1160 mg/kg ARZNAD 33,726,83
ivn-rat LD50:26 mg/kg ARZNAD 22,726,83
orl-mus LD50:760 mg/kg ARZNAD 33,726,83
ivn-mus LD50:54 mg/kg ARZNAD 33,726,83
ivn-cat LDLo:50800 μg/kg ARZNAD 33,726,83

SAFETY PROFILE: Poison by intravenous route. Moderately toxic by ingestion. When heated to decomposition it emits very toxic fumes of HCl and NO_x.

BGC625 CAS:6620-60-6 ***HR: 1***
BINOSIDE
mf: $C_{18}H_{26}N_2O_4$ mw: 334.46

PROP: Crystals. Mp: 142-145°.

SYNS: dl-4-BENZAMIDO-N,N-DIPROPYLGLUTARAMICACID ◇ (±)-4-(BENZOYLAMINO)-5-(DIPROPYLAMINO)-5-OXO-PENTANOIC ACID ◇ CR 242 ◇ 242 DL ◇ GASTRIDENE ◇ GASTROTOPIC ◇ MIDELID ◇ MILID ◇ MILIDE ◇ NULSA ◇ PROGLUMIDE ◇ PROMIDE (parasympatholytic) ◇ ULCUTIN ◇ W 5219 ◇ XYDE ◇ XYLAMIDE ◇ XYLAMIDE (gastroprotective agent)

TOXICITY DATA with REFERENCE
orl-rat TDLo:1350 mg/kg (9-14D preg):TER OYYAA2 5,225,71
orl-rat LD50:20 g/kg NIIRDN 6,722,82
ipr-rat LD50:1420 mg/kg NIIRDN 6,722,82
orl-mus LD50:8070 mg/kg MIMEAO 58,3653,67
ipr-mus LD50:1480 mg/kg NIIRDN 6,722,82
ivn-mus LD50:2250 mg/kg MIMEAO 58,3653,67

SAFETY PROFILE: Moderately toxic by several routes. An experimental teratogen. When heated to decomposition it emits toxic fumes of NO_x.

BGC750 ***HR: 3***
BIOALLETHRIN
mf: $C_{19}H_{26}O_3$ mw: 302.45

SYNS: d-trans ALLETHRIN ◇ ALLYL HONOLOG of CINERIN I ◇ BIOALETRINA (PORTUGUESE) ◇ (+)-trans-CHRYSANTHEMUMIC ACID ESTER of (+ −)-ALLETHROLONE ◇ ENT 16275

TOXICITY DATA with REFERENCE
orl-rat LD50:425 mg/kg SPEADM 78-1,7,78
ivn-rat LDLo:4 mg/kg BIOGAL 41(10),283,75
orl-mus LD50:330 mg/kg EVHPAZ 14,15,76

CONSENSUS REPORTS: EPA Genetic Toxicology Program. Reported in EPA TSCA Inventory.

SAFETY PROFILE: Poison by ingestion and intravenous routes. When heated to decomposition it emits acrid and irritating fumes. An insecticide. See other allethrin entries and ALLYL COMPOUNDS.

BGC825 ***HR: 2***
BIODIASTASE 1000

PROP: Extracted from *Aspergillus* (KSRNAM 8,2378,74).

TOXICITY DATA with REFERENCE
orl-rat TDLo:600 g/kg (30D pre):REP KSRNAM 8,2378,74
ipr-rat LD50:678 mg/kg KSRNAM 8,2378,74
scu-rat LD50:7320 mg/kg KSRNAM 8,2378,74
ipr-mus LD50:445 mg/kg KSRNAM 8,2378,74
scu-mus LD50:2220 mg/kg KSRNAM 8,2378,74

SAFETY PROFILE: Moderately toxic by intraperitoneal and sunscutaneous routes. Experimental reproductive effects. When heated to decomposition it emits acrid smoke and irritating fumes.

BGD000 CAS:5697-56-3 ***HR: 3***
BIOGASTRONE
mf: $C_{34}H_{50}O_7$ mw: 570.84

SYNS: BIORAL ◇ CARBENOXOLONE ◇ 3-β-(3-CARBOXYPROPIONYLOXY)-11-OXO-OLEAN-12-EN-30-OICACID ◇ 3-β-HYDROXY-11-OXO-OLEAN-12-EN-30-OIC ACID, HYDROGEN SUCCINATE

TOXICITY DATA with REFERENCE
orl-man TDLo:120 mg/kg/56D-I BMJOAE 2,150,76
orl-rat LD50:2450 mg/kg OYYAA2 19,323,80
ipr-rat LD50:128 mg/kg OYYAA2 19,323,80
scu-rat LD50:1720 mg/kg OYYAA2 19,323,80
scu-dog LD50:1060 mg/kg OYYAA2 19,323,80
ivn-dog LD50:371 mg/kg OYYAA2 19,323,80

CONSENSUS REPORTS: Reported in EPA TSCA Inventory.

SAFETY PROFILE: Poison by intravenous and intraperitoneal routes. Moderately toxic by ingestion and subcutaneous routes. Human systemic effects by ingestion: muscle weakness and flaccid paralysis. When heated to decomposition it emits acrid smoke and fumes.

BGD250 CAS:20354-26-1 ***HR: 2***
BIOXONE
mf: $C_9H_6Cl_2N_2O_3$ mw: 261.07

SYNS: 2-(3,4-DICHLOROPHENYL)-4-METHYL-1,2,4-OXADIAZOLIDINE-3,5-DIONE ◇ METHAZOLE ◇ OXYDIAZOL ◇ PAXILON ◇ PROBE ◇ TUNIC ◇ VCS 438

TOXICITY DATA with REFERENCE
orl-rat LD50:2501 mg/kg FMCHA2 -,C195,83

SAFETY PROFILE: Moderately toxic by ingestion. When heated to decomposition it emits very toxic fumes of Cl^- and NO_x.

BGD500 CAS:514-65-8 ***HR: 3***
BIPERIDEN
mf: $C_{21}H_{29}NO$ mw: 311.51

SYNS: AKINETON ◇ AKINOPHYL ◇ BEPERIDEN ◇ α-(BICYCLO-(2.2.1)HEPT-5-EN-2-YL)-α-PHENYL-1-PIPERIDINO PROPANOL ◇ 1-BICYCLOHEPTENYL-1-PHENYL-3-PIPERIDINO-PROPANOL-1 ◇ KL 373 ◇ 3-PIPERIDINO-1-PHENYL-1-BICYCLOHEPTENYL-1-PROPANOL ◇ 3-PIPERIDINO-1-PHENYL-1-BICYCLO(2.2.1)HEPTEN-(5)-YL-PROPANOL-(1)(GERMAN)

TOXICITY DATA with REFERENCE
orl-rat LD50:750 mg/kg MIIRDN 6,636,82
orl-mus LD50:530 mg/kg NIIRDN 6,636,82
ipr-mus LD50:161 mg/kg NIIRDN 6,636,82
orl-dog LD50:340 mg/kg NIIRDN 6,636,82
scu-mus LD50:195 mg/kg AIPTAK 128,204,60
ivn-mus LD50:56 mg/kg AIPTAK 128,204,60

SAFETY PROFILE: Poison by ingestion, subcutaneous, intraperitoneal, and intravenous routes. When heated to decomposition, it emits toxic fumes of NO_x.

BGD750 CAS:1235-82-1 ***HR: 3***
BIPERIDINE HYDROCHLORIDE
mf: $C_{21}H_{29}NO \cdot ClH$ mw: 347.97

SYNS: AKINETON HYDROCHLORIDE ◇ AKINOPHYL ◇ α-BICYCLO(2.2.1)HEPT-5-EN-1-YL-α-PHENYL-PIPERIDINE-PROPANOL HYDROCHLORIDE ◇ α-(BICYCLO(2.2.1)HEPT-5-EN-2-YL)-α-PHENYL-1-PIPERIDINEPROPANOL HYDROCHLORIDE ◇ 1-BICYCLOHEPTENYL-1-PHENYL-3-PIPERIDINOPROPANOL-1HYDROCHLORIDE ◇ BIPERIDEN HYDROCHLORIDE ◇ α-5-NORBORNEN-2-YL-α-PHENYL-PIPERIDINE PROPANOL HYDROCHLORIDE

TOXICITY DATA with REFERENCE
orl-rat LD50:750 mg/kg TXAPA9 2,379,60
orl-mus LD50:545 mg/kg 29ZVAB -,17,69
ivn-mus LD50:56 mg/kg MEIEDD 10,175,83
orl-dog LD50:340 mg/kg TXAPA9 2,379,60

SAFETY PROFILE: Poison by ingestion and intrave-

nous routes. When heated to decomposition it emits very toxic fumes of NO_x and HCl. See also BIPERIDIN.

BGE000 CAS:92-52-4 ***HR: 3***
BIPHENYL
mf: $C_{12}H_{10}$ mw: 154.22

PROP: White scales, pleasant odor. Mp: 70°, bp: 255°, flash p: 235°F (CC), d: 0.991 @ 75°/4°, autoign temp: 1004°F, vap d: 5.31, lel: 0.6% @ 232°, uel: 5.8% @ 331°F.

SYNS: BIBENZENE ◇ 1,1′-BIPHENYL ◇ DIPHENYL (OSHA) ◇ LEMONENE ◇ PHENADOR-X ◇ PHENYLBENZENE ◇ PHPH ◇ XENENE

TOXICITY DATA with REFERENCE
sce-ham:fbr 100 μmol/L JNCIAM 58,1635,77
orl-mus TDLo:56 g/kg:ETA NTIS** PB223-159
scu-mus TDLo:46 mg/kg:NEO NTIS** PB223-159
ihl-hmn TCLo:4400 μg/m^3:IRR AEHLAU 26,70,73
orl-rat LD50:3280 mg/kg JIHTAB 29,1,47
ivn-mus LD50:56 mg/kg CSLNX* NX#00198
orl-rbt LD50:2400 mg/kg NASDA6 28,983,77

CONSENSUS REPORTS: EPA Genetic Toxicology Program. Reported in EPA TSCA Inventory. Community Right-To-Know List.

OSHA PEL: TWA 0.2 ppm
ACGIH TLV: TWA 0.2 ppm
DFG MAK: 0.2 ppm (1 mg/m^3)

SAFETY PROFILE: Poison by intravenous route. Moderately toxic by ingestion. A powerful irritant by inhalation in humans. Human systemic effects by inhalation of very small amounts: flaccid paralysis, nausea or vomiting, and other unspecified gastrointestinal effects. Questionable carcinogen with experimental tumorigenic and neoplastigenic data. Mutation data reported. Combustible when exposed to heat or flame; can react with oxidizing materials. To fight fire, use CO_2, dry chemical, water spray, mist, fog. When heated to decomposition it emits acrid smoke and fumes.

BGE125 CAS:5728-52-9 ***HR: 3***
4-BIPHENYLACETIC ACID
mf: $C_{14}H_{12}O_2$ mw: 212.25

SYNS: (4-BIPHENYL) ACETIC ACID ◇ (1,1′-BIPHENYL)-4-ACETIC ACID ◇ p-BIPHENYLACETIC ACID ◇ 4-BIPHENYLACETIC ACID ◇ 4-CARBOXYMETHYLBIPHENYL ◇ LY 61017

TOXICITY DATA with REFERENCE
scu-rat TDLo:540 mg/kg (17-22D preg/21D post):REP KSRNAM 20,2153,86
scu-rat TDLo:360 mg/kg (female 6-17D post):TER KSRNAM 20,2142,86
orl-rat LD50:410 mg/kg KSRNAM 20,2107,86
ipr-rat LD50:495 mg/kg KSRNAM 20,2107,86
scu-rat LD50:148 mg/kg KSRNAM 20,2107,86
orl-mus LD50:675 mg/kg KSRNAM 20,2107,86
ipr-mus LD50:508 mg/kg KSRNAM 20,2107,86
scu-mus LD50:730 mg/kg KSRNAM 20,2107,86
scu-dog LD50:320 mg/kg KSRNAM 20,2107,86
scu-rbt LD50:1280 mg/kg KSRNAM 20,2107,86

SAFETY PROFILE: Poison by subcutaneous route. Moderately toxic by ingestion and intraperitoneal routes. An experimental teratogen. Other experimental reproductive effects. When heated to decomposition it emits acrid smoke and fumes.

BGE250 CAS:90-41-5 ***HR: 2***
2-BIPHENYLAMINE
mf: $C_{12}H_{11}N$ mw: 169.24

SYNS: o-AMINOBIPHENYL ◇ 2-AMINOBIPHENYL ◇ o-AMINODIPHENYL ◇ 2-AMINODIPHENYL ◇ o-BIPHENYLAMINE ◇ (1,1′-BIPHENYL)-2-AMINE (9CI) ◇ o-PHENYLANILINE ◇ 2-PHENYLANILINE

TOXICITY DATA with REFERENCE
mmo-sat 33 μg/plate ENMUDM 5(Suppl 1),3,83
pic-esc 250 mg/L CNREA8 41,532,81
orl-rat LD50:2340 mg/kg JIHTAB 29,1,47
orl-rbt LD50:1020 mg/kg JIHTAB 29,1,47

CONSENSUS REPORTS: EPA Genetic Toxicology Program. Reported in EPA TSCA Inventory.

SAFETY PROFILE: Moderately toxic by ingestion. Mutation data reported. When heated to decomposition, it emits toxic fumes of NO_x. See also AROMATIC AMINES.

BGE300 ***HR: 2***
4-BIPHENYLAMINE, DIHYDROCHLORIDE
mf: $C_{12}H_{11}N{\cdot}2ClH$ mw: 242.16

SYN: 4-AMINOBIPHENYL DIHYDROCHLORIDE

TOXICITY DATA with REFERENCE
orl-rat TDLo:2238 mg/kg/31W-I:CAR JJCREP 76,570,85

SAFETY PROFILE: Questionable carcinogen with experimental carcinogenic data. When heated to decomposition it emits toxic fumes of NO_x and HCl.

BGE325 CAS:2185-92-4 ***HR: 2***
2-BIPHENYLAMINE, HYDROCHLORIDE
mf: $C_{12}H_{11}N{\cdot}ClH$ mw: 205.70

SYN: NCI-C50282

TOXICITY DATA with REFERENCE
mma-sat 10 μg/plate SCIEAS 236,933,87
mma-mus:lyms 110 mg/L EMMUEG 12,85,88
cyt-ham:ovr 200 mg/L SCIEAS 236,933,87

orl-mus TDLo:260 g/kg/2Y-C:CAR NTPTR* NTP-TR-233,82
orl-mus TD:262 g/kg/2Y-C:CAR FAATDF 2,201,82

CONSENSUS REPORTS: NCI Carcinogenesis Studies (feed): Clear Evidence: mouse NTPTR* NTP-TR-233,82; No Evidence: rat NTPTR* NTP-TR-233,82

SAFETY PROFILE: Questionable carcinogen with experimental carcinogenic data. Mutation data reported. When heated to decomposition it emits toxic fumes of NO_x and HCl.

BGF000 CAS:20743-57-1 ***HR: 3***
N-4-BIPHENYLBENZAMIDE
mf: $C_{19}H_{15}NO$ mw: 273.35

SYNS: N-4-BIPHENYLYLBENZAMIDE ◇ 4′-PHENYLBENZANILIDE

TOXICITY DATA with REFERENCE
ipr-rat TDLo:508 mg/kg/4W-I:ETA CNREA8 30,1485,70

SAFETY PROFILE: Questionable carcinogen with experimental tumorigenic data. When heated to decomposition it emits toxic fumes of NO_x.

BGF109 CAS:492-17-1 ***HR: 3***
2,4′-BIPHENYLDIAMINE
mf: $C_{12}H_{12}N_2$ mw: 184.26

PROP: Needles, very sltly sol in alc and ether. Mp: 54.4°, bp: 363°.

SYNS: o,p′-BIANILINE ◇ (1,1′-BIPHENYL)-2,4′-DIAMINE ◇ o,p′-DIAMINOBIPHENYL ◇ 2,4′-DIAMINODIPHENYL ◇ o,p′-DIANILINE ◇ DIFENYLIN ◇ 2,4′-DIPHENYLDIAMINE ◇ DIPHENYLINE

TOXICITY DATA with REFERENCE
mma-sat 100 μg/plate MUREAV 149,9,85
orl-dog TDLo:7020 mg/kg/5Y-I:ETA NEOLA4 15,3,68
orl-rat LD50:311 mg/kg NEOLA4 15,3,68

CONSENSUS REPORTS: IARC Cancer Review: Group 3 IMEMDT 7,56,87; Animal Inadequate Evidence IMEMDT 16,313,78

SAFETY PROFILE: A poison by ingestion. Questionable carcinogen with experimental tumorigenic data. Mutation data reported. When heated to decomposition it emits toxic fumes of NO_x. See also AROMATIC AMINES.

BGF250 CAS:1591-30-6 ***HR: 3***
4,4′-BIPHENYLDICARBONITRILE
mf: $C_{14}H_8N_2$ mw: 204.24

SYN: NCR DR DCN

TOXICITY DATA with REFERENCE
ipr-mus LD50:75 mg/kg NTIS** AD691-490

CONSENSUS REPORTS: Reported in EPA TSCA Inventory. Cyanide and its compounds are on the Community Right-To-Know List.

SAFETY PROFILE: Poison by intraperitoneal route. See also NITRILES. When heated to decomposition it emits toxic fumes of NO_x and CN^-.

BGF500 ***HR: 3***
2,2-BIPHENYL DICARBONYL PEROXIDE
mf: $C_{14}H_8O_4$ mw: 200.21

SAFETY PROFILE: A explosive which detonates violently on impact or on heating to 70°. Upon decomposition it emits acrid smoke and fumes. See also PEROXIDES.

BGF899 CAS:1137-79-7 ***HR: 3***
4-BIPHENYLDIMETHYLAMINE
mf: $C_{14}H_{15}N$ mw: 197.30

SYN: 4-DIMETHYLAMINOBIPHENYL

TOXICITY DATA with REFERENCE
orl-rat TDLo:19 g/kg/43W-C:CAR CNREA8 16,525,56

SAFETY PROFILE: Questionable carcinogen with experimental carcinogenic data. When heated to decomposition it emits toxic fumes of NO_x. See also AROMATIC AMINES.

BGG000 CAS:1806-29-7 ***HR: 3***
2,2′-BIPHENYLDIOL
mf: $C_{12}H_{10}O_2$ mw: 186.22

SYNS: o,o′-BIPHENOL ◇ 2,2′-BIPHENOL ◇ 2,2′-DIHYDROXYBIPHENYL

TOXICITY DATA with REFERENCE
oms-hmn:lym 5 μmol/L CNREA8 45,2471,85
sce-hmn:lym 300 μmol/L CNREA8 45,2471,85
ipr-mus LD50:150 mg/kg NTIS** AD691-490
ivn-mus LD50:56 mg/kg CSLNX* NX#07870

CONSENSUS REPORTS: Reported in EPA TSCA Inventory.

SAFETY PROFILE: Poison by intraperitoneal and intravenous routes. Human mutation data reported. When heated to decomposition it emits acrid smoke and irritating fumes.

BGG250 CAS:1079-21-6 ***HR: 3***
2,5-BIPHENYLDIOL
mf: $C_{12}H_{10}O_2$ mw: 186.22

SYNS: 2,5-DIHYDROXYBIPHENYL ◇ PHENYLHYDROQUINONE

TOXICITY DATA with REFERENCE
ipr-mus LDLo:250 mg/kg CBCCT* 6,222,54
ivn-mus LD50:22 mg/kg BJPCAL 22,221,64

CONSENSUS REPORTS: Reported in EPA TSCA Inventory.

SAFETY PROFILE: Poison by intraperitoneal and intravenous routes. When heated to decomposition it emits acrid smoke and irritating fumes.

BGG500 CAS:92-88-6 ***HR: 3***
4,4'-BIPHENYLDIOL
mf: $C_{12}H_{10}O_2$ mw: 186.22

SYNS: p,p'-BIPHENOL ◇ USAF DO-30

TOXICITY DATA with REFERENCE
oms-hmn:lym 100 nmol/L CNREA8 45,2471,85
sce-hmn:lym 5 μmol/L CNREA8 45,2471,85
orl-rat LD50:9850 mg/kg TXAPA9 28,313,74
ipr-mus LD50:100 mg/kg NTIS** AD277-689
skn-rbt LD50:1780 mg/kg TXAPA9 28,313,74

CONSENSUS REPORTS: Reported in EPA TSCA Inventory.

SAFETY PROFILE: Poison by intraperitoneal route. Moderately toxic by skin contact. Mildly toxic by ingestion. Human mutation data reported. When heated to decomposition it emits acrid smoke and irritating fumes.

BGH000 CAS:20275-19-8 ***HR: 3***
1,1'-(p'p'-BIPHENYLENEBIS(CARBONYLMETHYL))DI-2-PICOLINIUM DIBROMIDE
mf: $C_{28}H_{26}N_2O_2$•2Br mw: 582.38

SYN: (4,4'-BIPHENYLYLENEBIS(2-OXOETHYLENE)-2-PICOLINIUM DIBROMIDE

TOXICITY DATA with REFERENCE
ipr-mus LD50:1500 μg/kg JAPMA8 43,79,54
ivn-mus LD50:2900 μg/kg TXAPA9 27,666,74

SAFETY PROFILE: Deadly poison by intraperitoneal and intravenous routes. See also BROMIDES. When heated to decomposition it emits very toxic fumes of NO_x and Br^-.

BGH250 CAS:73-51-8 ***HR: 3***
4,4'-BIPHENYLENEBIS(2-OXOETHYLENE)BIS(DIMETHYL(2-HYDROXYETHYL)AMMONIUM) DIBROMIDE
mf: $C_{24}H_{34}N_2O_4$•2Br mw: 574.42

TOXICITY DATA with REFERENCE
ipr-rat LD50:45 μg/kg JPETAB 115,127,55
ipr-mus LD50:20 μg/kg JPETAB 115,127,55
ivn-dog LDLo:75 μg/kg JPETAB 115,127,55
ivn-rbt LDLo:50 μg/kg JPETAB 115,127,55
ipr-gpg LDLo:30 μg/kg JPETAB 115,127,55

SAFETY PROFILE: Deadly poison by intraperitoneal and intravenous routes. See also BROMIDES. When heated to decomposition it emits very toxic fumes of NO_x, NH_3, and Br^-.

BGH500 CAS:77967-05-6 ***HR: 3***
4,4'-BIPHENYLENEBIS(3-OXOPROPYLENE) BIS(DIMETHYL(2-HYDROXYETHYL)AMMONIUM)DIBROMIDE
mf: $C_{26}H_{38}N_2O_4$•2Br mw: 602.48

TOXICITY DATA with REFERENCE
ipr-mus LD50:10 mg/kg JPETAB 115,127,55
ipr-rbt LD50:20 mg/kg JPETAB 115,127,55

SAFETY PROFILE: Poison by intraperitoneal route. See also BROMIDES. When heated to decomposition it emits very toxic fumes of NO_x, NH_3, and Br^-.

BGI250 CAS:6810-26-0 ***HR: 3***
4-BIPHENYLHYDROXYLAMINE
mf: $C_{12}H_{11}NO$ mw: 185.24

SYNS: N-4-BIPHENYLYLHYDROXYLAMINE ◇ 4-HYDROXYLAMINOBIPHENYL

TOXICITY DATA with REFERENCE
mmo-sat 5 μg/plate MUREAV 151,201,85
mmo-esc 2500 nmol/L MUREAV 151,201,85
dns-hmn:oth 1 μmol/L JJIND8 72,847,84
dns-rat:lvr 5 μmol/L ENMUDM 3,11,81
dns-rbt:oth 10 μmol/L CNREA8 45,221,85
scu-mus TDLo:216 mg/kg/3D:CAR JNCIAM 41,403,68

CONSENSUS REPORTS: EPA Genetic Toxicology Program.

SAFETY PROFILE: Questionable carcinogen with experimental carcinogenic data. Human mutation data reported. When heated to decomposition it emits highly toxic fumes of NO_x. See also AROMATIC AMINES.

BGJ250 CAS:90-43-7 ***HR: 3***
2-BIPHENYLOL
mf: $C_{12}H_{10}O$ mw: 170.22

SYNS: o-BIPHENYLOL ◇ (1,1'-BIPHENYL)-2-OL ◇ o-DIPHENYLOL ◇ DOWCIDE 1 ◇ DOWCIDE 1 ANTIMICROBIAL ◇ 2-HYDROXYBIFENYL (CZECH) ◇ o-HYDROXYBIPHENYL ◇ 2-HYDROXYBIPHENYL ◇ o-HYDROXYDIPHENYL ◇ 2-HYDROXYDIPHENYL ◇ KIWI LUSTR 277 ◇ NCI-C50351 ◇ OPP ◇ ORTHOHYDROXYDIPHENYL ◇ ORTHOPHENYLPHENOL ◇ ORTHOXENOL ◇ o-PHENYLPHENOL ◇ 2-PHENYLPHENOL ◇ PREVENTOL O EXTRA ◇ REMOL TRF ◇ TETROSIN OE ◇ TORSITE ◇ TUMESCAL OPE ◇ USAF EK-2219 ◇ o-XENOL

TOXICITY DATA with REFERENCE
skn-rbt 250 mg MccSB# 15JUN84
skn-rbt 500 mg/24H MOD 28ZPAK -,56,72
eye-rbt 50 μg/24H SEV 28ZPAK -,56,72
mmo-sat 60 μg/plate ENMUDM 5(Suppl 1),3,83
cyt-hmn:fbr 200 μg/L MUREAV 54,255,78

msc-hmn:emb 20 mg/L MUREAV 156,123,85
msc-hmn:oth 15 mg/L TRENAF 35,399,84
cyt-ham:ovr 100 mg/L MUREAV 141,95,84
orl-rat TDLo:6 g/kg (female 6-15D post):TER NNGADV 3,365,78
orl-rat TDLo:52168 mg/kg (male 13W pre):REP TRENAF 32-2,33,81
orl-rat TDLo:478 g/kg/91W-C:CAR FCTOD7 22,865,84
orl-rat LD:135 g/kg/26W-C:NEO FCTOD7 25,359,87
orl-rat LD50:2000 mg/kg NNGADV 3,365,78
unr-rat LD50:2700 mg/kg TRENAF 29,89,78
orl-mus LD50:1050 mg/kg NAIZAM 32,425,81
ipr-mus LD50:50 mg/kg NTIS** AD277-689

CONSENSUS REPORTS: IARC Cancer Review: Group 3 IMEMDT 7,56,87, Animal Inadequate Evidence IMEMDT 30,329,83; NTP Carcinogenesis Studies (dermal); No Evidence: mouse NTPTR* NTP-TR-301,86. Reported in EPA TSCA Inventory. On Community Right-To-Know List.

SAFETY PROFILE: A poison by intraperitoneal route. Moderately toxic by ingestion and possibly other routes. An experimental teratogen. Other experimental reproductive effects. Human mutation data reported. Severe eye and moderate skin irritant. Questionable carcinogen with experimental carcinogenic data. When heated to decomposition it emits acrid smoke and irritating fumes.

BGJ500 CAS:92-69-3 ***HR: 3***
4-BIPHENYLOL
mf: $C_{12}H_{10}O$ mw: 170.22

SYNS: p-HYDROXYBIPHENYL ◇ 4-HYDROXYBIPHENYL ◇ p-HYDROXYDIPHENYL ◇ 4-HYDROXYDIPHENYL ◇ PARAXENOL ◇ p-PHENYLPHENOL ◇ 4-PHENYLPHENOL

TOXICITY DATA with REFERENCE
orl-mus TDLo:153 g/kg/78W-I:ETA NTIS** PB223-159
scu-mus TDLo:1000 mg/kg:CAR NTIS** PB223-159
ipr-mus LD50:150 mg/kg NTIS** AD691-490

CONSENSUS REPORTS: Reported in EPA TSCA Inventory.

SAFETY PROFILE: Acute poison by intraperitoneal route. Questionable carcinogen with experimental carcinogenic and tumorigenic data. When heated to decomposition it emits acrid, irritating fumes.

BGJ750 CAS:132-27-4 ***HR: 3***
2-BIPHENYLOL, SODIUM SALT
mf: $C_{12}H_9O$•Na mw: 192.20

SYNS: BACTROL ◇ D.C.S. ◇ DORVICIDE A ◇ DOWICIDE s◇ 2-HYDROXYDIPHENYL SODIUM ◇ MIL-DU-RID ◇ MYSTOX WFA ◇ NATRIPHENE ◇ OPP-Na ◇ ORPHENOL ◇ o-PHENYLPHENOL SODIUM SALT ◇ 2-PHENYLPHENOL SODIUM SALT ◇ PREVENTOL-ON ◇ SODIUM-2-HYDROXYDIPHENYL ◇ SODIUM-o-PHENYLPHENATE ◇ SODIUM-2-PHENYLPHENATE ◇ SODIUM-o-PHENYLPHENOLATE ◇ SODIUM-o-PHENYLPHENOXIDE ◇ SOPP ◇ STOMOLD B ◇ TOPANE

TOXICITY DATA with REFERENCE
skn-hmn 1 mg MccSB# 15JUN84
skn-rbt 50 mg/24H SEV MccSB# 15JUN84
mmo-asn 16 μmol/L PHYTAJ 66,217,76
sln-asn 52 μmol/L EVHPAZ 31,81,79
orl-mus TDLo:144 g/kg (male 60D pre):TER TRENAF 29,99,78
orl-mus TDLo:72 g/kg (60D male):REP TRENAF 29,99,78
orl-rat TDLo:109 g/kg/13W-C:CAR FCTOD7 24,207,86
orl-rat TD:269 g/kg/32W-C:ETA GANNA2 74,625,83
orl-rat TD:126 g/kg/13W-C:CAR FCTXAV 19,303,81
orl-rat TD:223 g/kg/26W-C:NEO FCTOD7 25,359,87
orl-rat LD50:656 mg/kg TRENAF 30(2),57,79
orl-mus LD50:683 mg/kg TRENAF 30,54,79

CONSENSUS REPORTS: IARC Cancer Review: Group 2B IMEMDT 7,56,87, Animal Limited Evidence IMEMDT 30,329,83. Reported in EPA TSCA Inventory.

SAFETY PROFILE: Suspected carcinogen with experimental carcinogenic, neoplastigenic, and tumorigenic data. Moderately toxic by ingestion. Experimental teratogenic and reproductive effects. A human skin irritant. A severe skin irritant to experimental animals. When heated to decomposition it emits toxic fumes of Na_2O. See also 2-BIPHENYLOL.

BGK000 CAS:3644-37-9 ***HR: 3***
(2-BIPHENYLOXY)TRIBUTYLTIN
mf: $C_{24}H_{37}OSn$ mw: 460.30

SYNS: (2-BIPHENYLYLOXY)TRIBUTYL)STANNANE ◇ ((1,1′-BIPHENYL)-2-YLOXY)TRIBUTYL-(9CI)STANNANE ◇ TRIBUTYL-o-PHENYLPHENOXYTIN ◇ TRIBUTYLTIN-o-PHENYLPHENOXIDE

TOXICITY DATA with REFERENCE
ivn-mus LD50:100 mg/kg CSLNX* NX#01826

OSHA PEL: TWA 0.1 mg(Sn)/m^3 (skin)
ACGIH TLV: TWA 0.1 mg(Sn)/m^3 (skin) (Proposed: TWA 0.1 mg(Sn)/m^3; STEL 0.2 mg(Sn)/m^3 (skin))
NIOSH REL: (Organotin Compounds) TWA 0.1 mg(Sn)/m^3

SAFETY PROFILE: Poison by intravenous route. See also TIN COMPOUNDS. When heated to decomposition it emits acrid smoke and irritating fumes.

BGK250 CAS:304-43-8 ***HR: 2***
2-BIPHENYLPENICILLIN SODIUM
mf: $C_{21}H_{20}N_2O_4S$ mw: 396.49

SYNS: (2-BIPHENYLYL)PENICILLIN ◇ DIPHENICILLIN ◇ SKF 12141

TOXICITY DATA with REFERENCE
orl-mus LD50:5600 mg/kg 85ERAY 3,1663,78
ivn-mus LD50:1030 mg/kg FATOAO 31,232,68

SAFETY PROFILE: Moderately toxic by intravenous route. Mildly toxic by ingestion. When heated to decomposition it emits very toxic fumes of NO_x and SO_x. See other penicillin entries.

BGK500 CAS:91-95-2 ***HR: 3***
3,3′,4,4′-BIPHENYLTETRAMINE
mf: $C_{12}H_{14}N_4$ mw: 214.30

SYNS: 3,3′-DIAMINOBENZIDENE ◇ 3,3′,4,4′-DIPHENYLTETRAMINE ◇ 3,3′,4,4′-TETRAAMINOBIPHENYL

TOXICITY DATA with REFERENCE
mma-sat 100 μg/plate BJCAAI 37,873,78
dnd-esc 20 μmol/L MUREAV 89,95,81
mmo-smc 140 μmol/L MGGEAE 174,39,79
dns-rat:lvr 500 μmol/L ENMUDM 3,11,81
orl-rat TDLo:9000 mg/kg/27D-I:ETA CNREA8 28,924,68
orl-rat LDLo:3000 mg/kg CNREA8 26,619,66
orl-mus LD50:1834 mg/kg GISAAA 46(1),94,81

CONSENSUS REPORTS: Reported in EPA TSCA Inventory.

SAFETY PROFILE: Questionable carcinogen with experimental tumorigenic data. Moderately toxic by ingestion. Mutation data reported. When heated to decomposition it emits toxic fumes of NO_x. See also AROMATIC AMINES.

BGK750 CAS:7411-49-6 ***HR: 3***
3,3′,4,4′-BIPHENYLTETRAMINE TETRAHYDROCHLORIDE
mf: $C_{12}H_{14}N_4{\cdot}4ClH$ mw: 360.14

SYNS: 3,3′-DIAMINOBENZIDINETETRAHYDROCHLORIDE ◇ 3,3′,4,4′-TETRAAMINOBIPHENYL TETRAHYDROCHLORIDE

TOXICITY DATA with REFERENCE
orl-rat TDLo:260 g/kg/78W-C:ETA JEPTDQ 2,325,78
orl-mus TDLo:260 g/kg/78W-C:NEO JEPTDQ 2,325,78
ipr-mus LD50:330 mg/kg NCIBR* NIH-NCI-E-68-1311,10,73

CONSENSUS REPORTS: Reported in EPA TSCA Inventory.

SAFETY PROFILE: Poison by intraperitoneal route. Questionable carcinogen with experimental neoplastigenic and tumorigenic data. When heated to decomposition it emits very toxic fumes of HCl and NO_x. See also AROMATIC AMINES.

BGL000 CAS:13607-48-2 ***HR: 3***
N-4-BIPHENYLYLBENZENESULFONAMIDE
mf: $C_{18}H_{15}NO_2S$ mw: 309.40

SYN: N-4-BIPHENYLYLBENZENESULFONAMIDE

TOXICITY DATA with REFERENCE
ipr-rat TDLo:634 mg/kg/4W-I:ETA CNREA8 30,1485,70

SAFETY PROFILE: Questionable carcinogen with experimental tumorigenic data. See also SULFONATES. When heated to decomposition it emits very toxic fumes of SO_x and NO_x.

BGL250 CAS:36330-85-5 ***HR: 3***
3-(4-BIPHENYLYLCARBONYL)PROPIONIC ACID
mf: $C_{16}H_{14}O_3$ mw: 254.30

SYNS: 4-(4-BIPHENYLYL)-4-OXOBUTYRIC ACID ◇ BUFEMID ◇ CINOPAL ◇ CINOPOP ◇ CL82204 ◇ DIPHENYL-4-Γ-OXO-Γ-BUTRIC ACID ◇ FENBUFEN ◇ LEDERFEN ◇ Γ-OXO(1,1-BIPHENYL)-4-BUTANOIC ACID ◇ β,p-PHENYLBENZOYLPROPIONIC ACID

TOXICITY DATA with REFERENCE
orl-rat TDLo:1120 mg/kg (female 16-22D post):REP ARZNAD 30,725,80
orl-rat TDLo:280 mg/kg (female 16-22D post):TER ARZNAD 30,725,80
orl-wmn TDLo:12 mg/kg/1D-I BMJOAE 290,822,85
orl-man TDLo:90 mg/kg/1W-I:PUL HUTODJ 7,35,88
orl-rat LD50:200 mg/kg ARZNAD 30,725,80
ipr-rat LD50:265 mg/kg ARZNAD 30,721,80
scu-rat LD50:247 mg/kg ARZNAD 30,721,80
orl-mus LD50:795 mg/kg ARZNAD 30,721,80
ipr-mus LD50:482 mg/kg PCIPDV 15,132,83
scu-mus LD50:1189 mg/kg IYKEDH 10,884,79

SAFETY PROFILE: Poison by ingestion, intraperitoneal, and subcutaneous routes. Human systemic effects by ingestion: cough, sweating, body temperature. An experimental teratogen. Other experimental reproductive effects. An anti-inflammatory agent. When heated to decomposition it emits acrid smoke and irritating fumes.

BGL500 CAS:7203-95-4 ***HR: 3***
1-BIPHENYLYL-3,3-DIMETHYLTRIAZENE
mf: $C_{14}H_{15}N_3$ mw: 225.32

SYNS: 3,3-DIMETHYL-1-XENYL-TRIAZENE ◇ 1-XENYL-3,3-DIMETHYLTRIAZIN (CZECH)

TOXICITY DATA with REFERENCE
orl-rat LD50:347 mg/kg JMCMAR 22,473,79
orl-rat LD50:347 mg/kg 28ZPAK -,77,72
ipr-mus LD50:344 mg/kg JMCMAR 19,1299,76

SAFETY PROFILE: Poison by ingestion and intraperitoneal routes. Mutation data reported. When heated to decomposition it emits toxic fumes of NO_x.

BGM000 CAS:18355-50-5 ***HR: 3***
7,7′-(p,p′-BIPHENYLYLENEBIS(CARBONYLIMINO))BIS(2-ETHYLQUINOLINIUM) DITOSYLATE
mf: $C_{36}H_{32}N_4O_2{\cdot}2C_7H_7O_3S$ mw: 895.12

SYN: 7,7′-(4,4′-BIPHENYLYLENEBIS(CARBONYLIMINO))BIS(1-ETHYLQUINOLINIUM)DI-p-TOLUENESULFONATE

TOXICITY DATA with REFERENCE
dnd-mus:lym 170 nmol/L JMCMAR 22,134,79
ipr-mus LD10:106 mg/kg JMCMAR 22,134,79

SAFETY PROFILE: Poison by intraperitoneal route. See also SULFONATES. Mutation data reported. When heated to decomposition it emits very toxic fumes of NO_x and SO_x.

BGN000 CAS:29968-68-1 ***HR: 3***
N-4-BIPHENYLYL-N-HYDROXYBENZENESULFONAMIDE
mf: $C_{18}H_{15}NO_3S$ mw: 325.40

SYN: HYDROXY-4-BIPHENYLYLBENZENESULFONAMIDE

TOXICITY DATA with REFERENCE
ipr-rat TDLo:618 mg/kg/4W-I:NEO CNREA8 30,1485,70

SAFETY PROFILE: Questionable carcinogen with experimental neoplastigenic data. See also SULFONATES. When heated to decomposition it emits very toxic fumes of SO_x and NO_x.

BGO000 CAS:1734-91-4 ***HR: 3***
2-(2-BIPHENYLYLOXY)TRIETHYLAMINE HYDROCHLORIDE
mf: $C_{18}H_{23}NO{\cdot}ClH$ mw: 305.88

SYNS: DACORENE HYDROCHLORIDE ◇ 2-(DIETHYLAMINOETHOXY)DIPHENYL HCl ◇ 1262 F ◇ F 1262

TOXICITY DATA with REFERENCE
ipr-mus LD50:125 mg/kg BJPCAL 1,90,46
scu-mus LDLo:125 mg/kg APFRAD 5,7,47
ivn-mus LD50:27 mg/kg BJPCAL 1,90,46

SAFETY PROFILE: Poison by subcutaneous, intraperitoneal and intravenous routes. When heated to decomposition it emits very toxic fumes of NO_x and HCl.

BGO325 CAS:75318-62-6 ***HR: D***
2-(4-BIPHENYLYL)-s-TRIAZOLO(5,1-a)ISOQUINOLINE
mf: $C_{22}H_{15}N_3$ mw: 321.40

SYNS: 2-(1,1′-BIPHENYL-4-YL)-s-TRIAZOLE(5,1-a)ISOQUINOLINE ◇ 2-(1,1′-BIPHENYL)-4-YL-(1,2,4)TRIAZOLO(5,1-a)ISOQUINOLINE (9CI) ◇ L 14105

TOXICITY DATA with REFERENCE
orl-rat TDLo:150 mg/kg (female 6-10D post):REP JVPTD9 9,246,86

SAFETY PROFILE: Experimental reproductive effects. When heated to decomposition it emits toxic fumes of NO_x.

BGO500 CAS:366-18-7 ***HR: 3***
2,2′-BIPYRIDINE
mf: $C_{10}H_8N_2$ mw: 156.20

PROP: White crystals. Mp: 69.7°, bp: 272°-273°. Sol in 2200 parts water; very sol in alc, ether, benzene, chloroform, and petroleum ether.

SYNS: BIPYRIDINE ◇ α,α′-BIPYRIDINE ◇ α,α′-BIPYRIDYL ◇ 2,2′-BIPYRIDYL ◇ 2,2′-BYPYRIDIN ◇ CI-588 ◇ α,α′-DIPYRIDYL ◇ 2,2′-DIPYRIDYL

TOXICITY DATA with REFERENCE
mmo-sat 20 μg/plate ABCHA6 45,327,81
mma-sat 20 μg/plate ABCHA6 45,327,81
ipr-rat TDLo:60 mg/kg (12D preg):TER TJADAB 18,63,78
scu-mus TDLo:8000 mg/kg/40W-I:ETA JNCIAM 24,109,60
orl-rat LD50:100 mg/kg JTEHD6 10,363,82
scu-rat LD50:131 mg/kg JPETAB 135,317,62
ipr-mus LD50:200 mg/kg JPETAB 196,478,76

CONSENSUS REPORTS: Reported in EPA TSCA Inventory.

SAFETY PROFILE: Poison by ingestion, subcutaneous, and intraperitoneal route. Experimental teratogenic data. Questionable carcinogen with experimental tumorigenic data. Mutation data reported. When heated to decomposition it emits toxic fumes of NO_x.

BGO750 CAS:8001-88-5 ***HR: 2***
BIRCH TAR OIL

PROP: Brown liquid; leather-like odor. D: 0.886-0.950. Found in the tar of the bark and wood of *Betula pendula* Roth (Fam. *Betulaceae*) and prepared by steam distillation of the tar obtained by dry distillation of the bark and wood (FCTXAV 11,1011,73). Sol in fixed oils; insol in glycerin, mineral oil, and propylene glycol.

SYN: BIRCH TAR OIL, RECTIFIED (FCC)

TOXICITY DATA with REFERENCE
skn-rbt 500 mg/24H FCTXAV 11,1037,73

CONSENSUS REPORTS: Reported in EPA TSCA Inventory.

SAFETY PROFILE: A skin irritant. Moderately irritating to eyes and mucous membranes. A mild allergen. Combustible when exposed to heat or flame; can react with oxidizing materials.

BGP250 CAS:304-28-9 ***HR: 3***
2,7-BIS(ACETAMIDO)FLUORENE
mf: $C_{17}H_{16}N_2O_2$ mw: 280.35

SYNS: 2,7-DIACETAMIDOFLUORENE ◇ 2,7-DIACETYLAMINOFLUORENE ◇ 2,7-FAA ◇ 2,7-FLUORENYLBISACETAMIDE ◇ N,N′-FLUOREN-2,7-YLBISACETAMIDE ◇ N,N′-FLUOREN-2,7-

YLENEBISACETAMIDE ◇ N,N'-2,7-FLUORENYLENEBISACETAMIDE ◇ N,N'-(FLUOREN-2,7-YLENE)BIS(ACETYLAMINE) ◇ N,N'-2,7-FLUORENYLENEDIACETAMIDE

TOXICITY DATA with REFERENCE
mmo-sat 100 μg/plate PNASA6 69,3128,72
mma-sat 10 μg/plate PNASA6 72,5135,75
dns-rat:lvr 500 nmol/L ENMUDM 3,11,81
cyt-rat-orl 315 mg/kg/3W JNCIAM 54,1245,75
orl-rat TDLo:4830 mg/kg/46W-C:CAR GANNA2 60,211,69
ipr-rat TDLo:2044 mg/kg/29W-I:CAR JNCIAM 29,977,62
orl-uns TDLo:1200 mg/kg/34W-C:ETA JJIND8 74,909,85
orl-mus TD:6480 mg/kg/26W-C:NEO NAIZAM 33,545,83

CONSENSUS REPORTS: EPA Genetic Toxicology Program.

SAFETY PROFILE: Suspected carcinogen with experimental carcinogenic, neoplastigenic, and tumorigenic data. Mutation data reported. When heated to decomposition it emits toxic fumes of NO_x.

BGP500 CAS:63981-20-4 ***HR: 3***
BIS-4-ACETAMINO PHENYL SELENIUMDIHYDROXIDE
mf: $C_8H_{11}NO_3 \cdot Se$ mw: 248.16

TOXICITY DATA with REFERENCE
orl-rat TDLo:2890 mg/kg/15W-C:ETA SCIEAS 103,762,46

CONSENSUS REPORTS: Selenium and its compounds are on the Community Right-To-Know List

OSHA PEL: TWA 0.2 mg(Se)/m^3
ACGIH TLV: TWA 0.2 mg(Se)m^3
DFG MAK: 0.1 mg(Se)/m^3

SAFETY PROFILE: Questionable carcinogen with experimental tumorigenic data. When heated to decomposition it emits very toxic fumes of NO_x and Se. See also SELENIUM COMPOUNDS.

BGP750 CAS:15172-86-8 ***HR: 3***
4,4'-BISACETOPHENONE-α,α'-DI(3-METHYLPYRIDINIUM) DIBROMIDE
mf: $C_{28}H_{26}N_2O_2 \cdot 2Br$ mw: 582.38

SYN: (4,4'-BIPHENYLYLENEBIS(2-OXOETHYLENE)-3-PICOLINIUM DIBROMIDE

TOXICITY DATA with REFERENCE
ipr-mus LD50:50 μg/kg JAPMA8 43,79,54
ivn-mus LD50:69 μg/kg TXAPA9 27,666,74

SAFETY PROFILE: Deadly poison by intraperitoneal and intravenous routes. See also BROMIDES. When heated to decomposition it emits very toxic fumes of Br^- and NO_x.

BGQ000 CAS:5967-09-9 ***HR: 3***
BIS(ACETOXYDIBUTYLSTANNANE) OXIDE
mf: $C_{20}H_{42}O_5Sn_2$ mw: 600.00

SYNS: BIS(DIBUTYLACETOXYTIN)OXIDE ◇ DIACETOXYTETRABUTYLDISTANNOXANE

TOXICITY DATA with REFERENCE
ivn-mus LD50:320 mg/kg CSLNX* NX#02081

CONSENSUS REPORTS: Reported in EPA TSCA Inventory.

OSHA PEL: TWA 0.1 mg(Sn)/m^3 (skin)
ACGIH TLV: TWA 0.1 mg(Sn)/m^3 (skin) (Proposed: TWA 0.1 mg(Sn)/m^3; STEL 0.2 mg(Sn)/m^3 (skin))
NIOSH REL: (Organotin Compounds) TWA 0.1 mg(Sn)/m^3

SAFETY PROFILE: Poison by intravenous route. See also TIN COMPOUNDS. When heated to decomposition it emits acrid smoke and irritating fumes.

BGQ250 CAS:64058-74-8 ***HR: 3***
2,6-BIS(ACETOXYMERCURI)-4-NITROACETANILIDE
mf: $C_{12}H_{12}Hg_2N_2O_7$ mw: 697.44

SYN: BIS(ACETATO-O)(ω-(2-(ACETYLAMINO)-5-NITRO-1,3-PHENYLENE)DI-MERCURY

TOXICITY DATA with REFERENCE
ipr-rat LDLo:500 mg/kg NCNSA6 5,10,53

CONSENSUS REPORTS: Mercury and its compounds are on the Community Right-To-Know List.

OSHA PEL: (Transitional: CL 1 mg/10m^3) CL 0.1 mg(Hg)/m^3 (skin)
ACGIH TLV: TWA 0.1 mg(Hg)/m^3 (skin)
NIOSH REL: (Mercury, Inorganic) TWA 0.05 mg(Hg)/m^3

SAFETY PROFILE: Moderately toxic by intraperitoneal route. See also MERCURY COMPOUNDS. When heated to decomposition it emits very toxic fumes of NO_x and Hg.

BGQ325 CAS:21327-74-2 ***HR: D***
BIS(p-ACETOXYPHENYL)-2-METHYLCYCLOHEXYLIDENEMETHANE
mf: $C_{24}H_{26}O_4$ mw: 378.50

SYNS: 4-((4-(ACETYLOXY)PHENYL)(2-METHYLCYCLOHEXYLIDENE)METHYL)PHENOL ACETATE ◇ F 6103 ◇ α-(p-HYDROXYPHENYL)-α-(2-METHYLCYCLOHEXYLIDENE)-p-CRESOLDIACETATE

TOXICITY DATA with REFERENCE
orl-rbt TDLo:7200 μg/kg (female 10-18D post):TER APTOA6 26(1),1,68
orl-mus TDLo:500 μg/kg (female 3-4D post):REP APTOA6 26(1),1,68

SAFETY PROFILE: An experimental teratogen. Other experimental reproductive effects. When heated to decomposition it emits acrid smoke and fumes.

BGQ750 CAS:14024-64-7 ***HR: 3***
BIS(ACETYLACETONATO) TITANIUM OXIDE
mf: $C_{10}H_{14}O_5Ti$ mw: 262.14

SYNS: BIS(2,4-PENTANEDIONATO)TITANIUM OXIDE ◇ TITANIUM ACETONYL ACETONATE ◇ TITANIUM OXIDE BIS(ACETYLACETONATE) ◇ TITANIUM, OXOBIS(2,4-PENTANEDIONATO-O,O') ◇ TITANYL BIS(ACETYLACETONATE)

TOXICITY DATA with REFERENCE
ims-rat TDLo:360 mg/kg/69W-I:ETA NCIUS* PH 43-64-886,AUG,69
ipr-rat LD50:650 mg/kg NCIUS* PH 43-64-886,JUL,68

SAFETY PROFILE: Moderate toxic by intraperitoneal route. Questionable carcinogen with experimental tumorigenic data. When heated to decomposition it emits acrid smoke and irritating fumes. See also TITANIUM COMPOUNDS.

BGR000 CAS:13395-16-9 ***HR: 3***
BIS(ACETYL ACETONE)COPPER
mf: $C_{10}H_{14}O_4•Cu$ mw: 261.78

SYNS: BIS(2,4-PENTANEDIONATO)COPPER ◇ COPPER(II) ACETYLACETONATE ◇ COPPER BIS(ACETYLACETONATE) ◇ COPPER BIS(ACETYLACETONE) ◇ COPPER BIS(2,4-PENTANEDIONATE) ◇ COPPER DIACETYLACETONATE ◇ CUPRIC ACETYLACETONATE

TOXICITY DATA with REFERENCE
ivn-mus LD50:10 mg/kg CSLNX* NX#00604

CONSENSUS REPORTS: Reported in EPA TSCA Inventory. Copper and its compounds are on the Community Right-To-Know List.

SAFETY PROFILE: Poison by intravenous route. See also COPPER COMPOUNDS. When heated to decomposition it emits acrid smoke and fumes of Cu.

BGR250 CAS:22750-65-8 ***HR: 3***
2,5-BIS(ACETYLAMINO)FLUORENE
mf: $C_{17}H_{16}N_2O_2$ mw: 280.35

SYNS: N,N'-FLUOREN-2,5-YLENEBISACETAMIDE ◇ 2,5-FLUORENYLENEBISACETAMIDE

TOXICITY DATA with REFERENCE
orl-rat TDLo:4550 mg/kg/26W-C:CAR CNREA8 22,1002,62

SAFETY PROFILE: Questionable carcinogen with experimental carcinogenic data. When heated to decomposition it emits toxic fumes of NO_x.

BGR325 CAS:50588-13-1 ***HR: 3***
1,1'-((2-β,3-α,5-α,16-β,17-β)-3,17-BIS(ACETYLOXY) ANDROSTANE-2,16-DIYL)BIS(1-METHYLPIPERIDINIUM, DIBROMIDE
mf: $C_{35}H_{60}N_2O_4•2Br$ mw: 732.79

TOXICITY DATA with REFERENCE
orl-rat LD50:202 mg/kg IYKEDH 4,90,73
ipr-rat LD50:479 μg/kg IYKEDH 4,90,73
scu-rat LD50:436 μg/kg IYKEDH 4,90,73
ivn-rat LD50:129 μg/kg IYKEDH 4,90,73
orl-mus LD50:21200 μg/kg IYKEDH 4,90,73
ipr-mus LD50:116 μg/kg IYKEDH 4,90,73
scu-mus LD50:168 μg/kg IYKEDH 4,90,73
ivn-mus LD50:36 μg/kg IYKEDH 4,90,73

SAFETY PROFILE: Poison by ingestion, subcutaneous, intravenous, and intraperitoneal routes. When heated to decomposition it emits toxic fumes of NO_x and Br^-. See also BROMIDES.

BGR500 CAS:12266-58-9 ***HR: 3***
BIS(ACRYLONITRILE) NICKEL (O)
mf: $C_6H_6N_2Ni$ mw: 164.84

CONSENSUS REPORTS: Cyanide and its compounds, as well as nickel and its compounds, are on the Community Right-To-Know List.

SAFETY PROFILE: Ignites spontaneously in air. When heated to decomposition it emits toxic fumes of CN^- and NO_x. See also NITRILES and NICKEL COMPOUNDS.

BGR750 CAS:63906-14-9 ***HR: 3***
1,4-BIS(4-ALDOXIMINOPYRIDINIUM)BUTANEDIOL-2,3-BIBROMIDE
mf: $C_{16}H_{20}N_4O_4•2Br$ mw: 492.22

SYNS: 1,4-BIS(4-HYDROXYIMINOMETHYL-PYRIDINIUM-(1))-BUTANEDIOL-2,3 DIBROMID (GERMAN) ◇ R 21

TOXICITY DATA with REFERENCE
ipr-mus LD50:130 mg/kg ARZNAD 14,870,64
ivn-mus LD50:64 mg/kg ARZNAD 14,870,64
ims-mus LD50:148 mg/kg ARZNAD 14,870,64

SAFETY PROFILE: Poison by intraperitoneal, intravenous, and intramuscular routes. See also BROMIDES. When heated to decomposition it emits very toxic fumes of Br^- and NO_x.

BGS250 CAS:114-90-9 ***HR: 3***
1,3-BIS(4-ALDOXIMINOPYRIDINIUM) DIMETHYL ETHER BICHLORIDE
mf: $C_{14}H_{16}N_4O_3•Cl_2$ mw: 359.24

SYNS: BH 6 ◇ 1,3-BIS(4-HYDROXYIMINOMETHYL-1-PYRIDINIO)-2-OXAPROPANE DICHLORIDE ◇ BIS(4-HYDROXYIMINOMETHY-

LPYRIDINIUM-1-METHYL)ETHER DICHLORIDE ◇ BIS(ISONICOTINALDOXIME 1-METHYL) ETHER DICHLORIDE ◇ BU-6 ◇ N,N-DIMETHYLENEOXIDEBIS(PYRIDINIUM-4-ALDOXIME) DICHLORIDE ◇ N,N-DIMETHYLENOXID-BIS-(PYRIDINIUM-4-ALDOXIM)-DICHLORID (GERMAN) ◇ ETHER BIS-14-HYDROXY-IMINOMETHYLOPYRIDINE-(1)-METYLODICHLORIDE(POLISH) ◇ LUEH 6 ◇ LUH6 ◇ LUH(6) ◇ LUH(6)-Cl2 ◇ LUH6-CHLORIDE ◇ OBIDOXIME CHLORIDE ◇ OBIDOXIME DICHLORIDE ◇ OBIDOXIME HYDROCHLORIDE ◇ 1,1'-(OXYBIS(METHYLENE))BIS(4-(HYDROXYIMINO)METHYL)PYRIDINIUMDICHLORIDE ◇ 1,1'-(OXYDIMETHYLENE)BIS(4-FORMYLPYRIDINIUM)DICHLORIDE DIOXIME ◇ 1,1'-(OXYDIMETHYLENE)BIS(4-FORMYLPYRIDINIUM) DIOXIME DICHLORIDE ◇ TOKSOBIDIN ◇ TOXOBIDIN ◇ TOXOGONIN ◇ TOXOGONIN DICHLORIDE ◇ TOXOGONINE

TOXICITY DATA with REFERENCE
ipr-rat LD50:189 mg/kg RPTOAN 38,168,75
ivn-rat LD50:133 mg/kg ARZNAD 14,5,64
ims-rat LD50:205 mg/kg RPTOAN 38,168,75
orl-mus LD50:2240 mg/kg 28ZEAL 5,168,76
ipr-mus LD50:111 mg/kg FAATDF 3,533,83
scu-mus LD50:183 mg/kg RPTOAN 38,168,75
ivn-mus LD50:70 mg/kg ARZNAD 14,870,64
ims-mus LD50:172 mg/kg ARZNAD 14,870,64

SAFETY PROFILE: Poison by intraperitoneal, intravenous, intramuscular, and subcutaneous, routes. Moderately toxic by ingestion. When heated to decomposition it emits toxic fumes of Cl^- and NO_x. See also BROMIDES.

BGS500 CAS:27469-53-0 ***HR: 3***
1-(4,6-BISALLYLAMINO-s-TRIAZINYL)-4-(p,p'-DIFLUOROBENZHYDRYL)-PIPERAZINE
mf: $C_{26}H_{29}F_2N_7$ mw: 477.62

SYNS: ALMITRINA (SPANISH) ◇ 2,4-BIS(ALLYLAMINO)-6-(4-(BIS-(p-FLUOROPHENYL)METHYL)-1-PIPERAZINYL)-s-TRIAZINE

TOXICITY DATA with REFERENCE
ipr-mus LD50:390 mg/kg DRFUD4 3,717,78
ivn-mus LD50:210 mg/kg DRFUD4 3,717,78

SAFETY PROFILE: Poison by intraperitoneal and intravenous routes. When heated to decomposition it emits very toxic fumes of F^- and NO_x. See also ALLYL COMPOUNDS.

BGS750 CAS:5975-73-5 ***HR: 1***
BIS(3-ALLYLOXY-2-HYDROXYPROPYL) FUMARATE
mf: $C_{16}H_{24}O_8$ mw: 344.40

SYN: BIS-3-ALLOXY-2-HYDROXYPROPYL-1-ESTERKYSELINY FUMAROVE (CZECH)

TOXICITY DATA with REFERENCE
skn-rbt 500 mg/24H MLD 28ZPAK -,100,72
eye-rbt 100 mg/24H MOD 28ZPAK -,100,72
orl-rat LD50:9710 mg/kg 28ZPAK -,100,72

SAFETY PROFILE: Mildly toxic by ingestion. A skin and eye irritant. When heated to decomposition it emits acrid smoke and irritating fumes. See also ESTERS and ALLYL COMPOUNDS.

BGS825 CAS:90566-09-9 ***HR: 2***
4,5-BIS(ALLYLOXY)-2-IMIDAZOLINDINONE
mf: $C_9H_{14}N_2O_3$ mw: 198.25

TOXICITY DATA with REFERENCE
orl-mus LD50:1850 mg/kg CPBTAL 12,843,64
ipr-mus LD50:1600 mg/kg CPBTAL 12,843,64
scu-mus LD50:1650 mg/kg CPBTAL 12,843,64

SAFETY PROFILE: Moderately toxic by ingestion, subcutaneous, and intraperitoneal routes. When heated to decomposition it emits toxic fumes of NO_x. See also ALLYL COMPOUNDS.

BGT000 CAS:28434-86-8 ***HR: 3***
BIS(4-AMINO-3-CHLOROPHENYL) ETHER
mf: $C_{12}H_{10}Cl_2N_2O$ mw: 269.14

SYNS: 3,3'-DICHLOR-4,4'-DIAMINO-DIPHENYLAETHER(GERMAN) ◇ 3,3'-DICHLORO-4,4'-DIAMINODIPHENYL ETHER ◇ 4,4'-OXYBIS(2-CHLOROANILINE) ◇ 4,4'-OXYBIS(2-CHLORO-BENZENAMINE)

TOXICITY DATA with REFERENCE
scu-rat TDLo:11 g/kg/27W-I:CAR NATWAY 57,676,70
scu-rat TD:14 g/kg/96W-I:CAR NATWAY 64,394,77

CONSENSUS REPORTS: IARC Cancer Review: Group 2B IMEMDT 7,56,87; Animal Sufficient Evidence IMEMDT 16,309,78

SAFETY PROFILE: Suspected carcinogen with experimental carcinogenic data. When heated to decomposition it emits toxic fumes of Cl^- and NO_x. See also ETHERS.

BGT125 CAS:26493-63-0 ***HR: 3***
BIS(2-AMINOETHYL)AMINE COBALT(III) AZIDE
mf: $C_4H_{13}CoN_{12}$ mw: 288.16

$$(HN(C_2H_4NH_2)_2Co)(N_3)_3$$

CONSENSUS REPORTS: Cobalt and its compounds are on the Community Right-To-Know List.

SAFETY PROFILE: A dangerous shock-sensitive explosive. Upon decomposition it emits toxic fumes of NO_x. See COBALT COMPOUNDS and AZIDES.

BGT150 CAS:59419-71-5 ***HR: 3***
BIS(2-AMINOETHYL)AMINEDIPEROXOCHROMIUM(IV)
mf: $C_4H_{13}CrN_3O_4$ mw: 219.16

$$HN(C_2H_4NH_2)_2Cr(O_2)_2$$

CONSENSUS REPORTS: Chromium and its compounds are on the Community Right-To-Know List.

SAFETY PROFILE: Decomposes explosively when heated to 110°C. Upon decomposition it emits toxic fumes of NO_x. See also CHROMIUM COMPOUNDS and PEROXIDES.

BGT250 CAS:314-13-6 ***HR: 3***
4,4'-BIS(1-AMINO-8-HYDROXY-2,4-DISULFO-7-NAPHTHYLAZO)-3,3'-BITOLYL, TETRASODIUM SALT
mf: $C_{34}H_{24}N_6O_{14}S_4 \cdot 4Na$ mw: 960.84

SYNS: 4,4'-BIS(7-(1-AMINO-8-HYDROXY-2,4-DISULFO)NAPHTHYLAZO)-3,3'-BITOLYL, TETRASODIUM SALT ◇ 4,4'-BIS(1-AMINO-8-HYDROXY-2,4-DISULPHO-7-NAPHTHYLAZO)-3,3 -BITOLYL, TETRASODIUM SALT ◇ BLEKIT EVANSA (POLISH) ◇ CHLORAZOL SKY BLUE FF ◇ C.I. 23860 ◇ C.I. DIRECT BLUE 53 ◇ DIAMINE SKY BLUE FF ◇ DIAZOBLEU ◇ DIAZOL PURE BLUE FF ◇ DYE EVANS BLUE ◇ EB ◇ EVABLIN ◇ EVANS BLUE DYE ◇ GEIGY-BLAU 536 ◇ T 1824

TOXICITY DATA with REFERENCE
mma-sat 33 μg/plate CRNGDP 3,21,82
dns-rat:lvr 100 μmol/L MUREAV 136,255,84
dnd-mus-skn 192 μmol/kg CRNGDP 5,231,84
ipr-rat TDLo:70 mg/kg (female 8D post):TER PSEBAA 127,215,68
ipr-rat TDLo:70 mg/kg (8D preg):REP PSEBAA 127,215,68
ipr-rat TDLo:850 mg/kg/34W-I:ETA APHGBP 33,1,53
ivn-rat LDLo:5 g/kg ARSUAX 48,17,44
ipr-mus LDLo:200 mg/kg BHJUAV 21,492,59
ivn-dog LDLo:3 g/kg ARSUAX 48,17,44
ivn-cat LDLo:1 g/kg ARSUAX 48,17,44
ivn-rbt LDLo:1 g/kg ARSUAX 48,17,44

CONSENSUS REPORTS: IARC Cancer Review: Group 3 IMEMDT 7,56,87, Animal Limited Evidence IMEMDT 8,151,75. Reported in EPA TSCA Inventory. EPA Genetic Toxicology Program.

SAFETY PROFILE: Poison by intraperitoneal route. Moderately toxic by intravenous route. An experimental teratogen. Other experimental reproductive effects. Questionable carcinogen with experimental tumorigenic data. Mutation data reported. When heated to decomposition it emits very toxic fumes of SO_x, Na_2O, and NO_x.

BGT500 CAS:2579-20-6 ***HR: 3***
1,3-BIS(AMINOMETHYL)CYCLOHEXANE

SYN: KODAK SILVER HALIDE SOLVENT HS-103

TOXICITY DATA with REFERENCE
orl-rat LD50:880 mg/kg HURC** -,-,73
skn-rat LDLo:100 mg/kg KODAK* -,-,71
ipr-rat LDLo:25 mg/kg KODAK* -,-,71

CONSENSUS REPORTS: Reported in EPA TSCA Inventory.

SAFETY PROFILE: Poison by skin contact and intraperitoneal routes. Moderately toxic by ingestion. When heated to decomposition it emits toxic fumes of NO_x.

BGT750 CAS:2549-93-1 ***HR: 2***
1,4-BIS(AMINOMETHYL)CYCLOHEXANE
mf: $C_8H_{18}N_2$ mw: 142.28

SYN: SILVER HALIDE SOLVENT (HS103)

TOXICITY DATA with REFERENCE
skn-rbt 500 mg SEV SUNCO* 10/78
orl-rat LD50:530 mg/kg SUNCO* 10/78
skn-rbt LD50:420 mg/kg SUNCO* 10/78

CONSENSUS REPORTS: Reported in EPA TSCA Inventory.

SAFETY PROFILE: Moderately toxic by ingestion and skin contact. A severe skin irritant. When heated to decomposition it emits toxic fumes of NO_x.

BGU000 CAS:63077-09-8 ***HR: 3***
BIS(2-AMINO-1-NAPHTHYL)SODIUM PHOSPHATE
mf: $C_{20}H_{17}N_2O_4P \cdot Na$ mw: 403.35

SYN: 2-AMINO-1-NAPHTHOL PHOSPHATE (ESTER) SODIUM SALT

TOXICITY DATA with REFERENCE
imp-mus TDLo:80 mg/kg:CAR BJCAAI 17,127,63

SAFETY PROFILE: Questionable carcinogen with experimental carcinogenic data. When heated to decomposition it emits very toxic fumes of PO_x, NO_x, and Na_2O. See also PHOSPHATES and ESTERS.

BGU500 CAS:4485-25-0 ***HR: 3***
2,2-BIS(p-AMINOPHENYL)-1,1,1-TRICHLOROETHANE
mf: $C_{14}H_{13}Cl_3N_2$ mw: 315.64

SYNS: 2,2,-BIS(p-ANILINE)-1,1,1-TRICHLOROETHANE ◇ p,p'-DIAMINODIPHENYLTRICHLOROETHANE

TOXICITY DATA with REFERENCE
orl-rat LDLo:1000 mg/kg JAPMA8 37,461,48
orl-mus LDLo:250 mg/kg JAPMA8 37,461,48

SAFETY PROFILE: Poison by ingestion. When heated to decomposition it emits very toxic fumes such as Cl^- and NO_x.

BGU750 CAS:105-83-9 ***HR: 3***
BIS(Γ-AMINOPROPYL)METHYLAMINE
mf: $C_7H_{19}N_3$ mw: 145.29

PROP: Liquid, completely miscible in water. D: 0.9307 @ 20°/20°, bp: 240.6°, fp: −29.6°, flash p: 220°F.

SYNS: BIS(ω-AMINOPROPYL)METHYLAMINE ◇ BIS(3-AMINOPROPYL)METHYLAMINE ◇ N,N-BIS(Γ-AMINOPROPYL)METHYLAMINE ◇ N,N-BIS(3-AMINOPROPYL)METHYLAMINE ◇ 3,7′-DIAMINO-N-METHYLDIPROPYLAMINE ◇ METHYLBIS(3-AMINOPROPYL)AMINE

TOXICITY DATA with REFERENCE
skn-rbt 100 μg/24H open AIHAAP 23,95,62
eye-rbt 5 mg SEV UCDS** 2/28/67
orl-rat LD50:1540 mg/kg UCDS** 2/28/67
ihl-rat LCLo:333 ppm/1H AIHAAP 23,95,62
skn-rbt LDLo:140 mg/kg AIHAAP 23,95,62

CONSENSUS REPORTS: Reported in EPA TSCA Inventory.

SAFETY PROFILE: Poison by inhalation and skin contact. Moderately toxic by ingestion. A skin and severe eye irritant. See also AMINES. Combustible when exposed to heat or flame. To fight fire, use foam, fog, dry chemical. When heated to decomposition it emits toxic fumes of NO_x.

BGV000 CAS:7209-38-3 ***HR: 3***
1,4-BIS(AMINOPROPYL)PIPERAZINE
DOT: NA 1760
mf: $C_{10}H_{24}N_4$ mw: 200.38

SYN: BIS(AMINOPROPYL)PIPERAZINE(DOT)

TOXICITY DATA with REFERENCE
ivn-mus LD50:3500 μg/kg CPBTAL 20,2459,72

CONSENSUS REPORTS: Reported in EPA TSCA Inventory.

DOT Classification: Corrosive Material; Label: Corrosive.

SAFETY PROFILE: Poison by intravenous route. A corrosive material and a powerful irritant to skin, eyes, and mucous membranes. When heated to decomposition it emits toxic fumes of NO_x.

BGV500 CAS:14650-81-8 ***HR: 2***
BIS(2-AMINOTHIOPHENOL), ZINC SALT
mf: $C_{12}H_{12}N_2S_2Zn$ mw: 313.75

SYNS: o-AMINOTHIOFENOLAT ZINECNATY (CZECH) ◇ BIS(2-AMINOPHENYLTHIO)ZINC

TOXICITY DATA with REFERENCE
skn-rbt 500 mg/24H MOD 28ZPAK -,11,72
eye-rbt 2 mg/24H SEV 28ZPAK -,11,72

CONSENSUS REPORTS: Zinc and its compounds are on the Community Right-To-Know List.

SAFETY PROFILE: A skin and severe eye irritant. See also ZINC COMPOUNDS. When heated to decomposition it emits very toxic fumes of ZnO, NO_x and SO_x.

BGV750 CAS:3751-44-8 ***HR: 3***
1,3-BIS(5-AMINO-1,3,4-TRIAZOL-2-YL)TRIAZENE
mf: $C_4H_7N_{11}$ mw: 209.18

SAFETY PROFILE: Decomposes explosively when heated to its melting point (187°C). When heated to decomposition it emits toxic fumes of NO_x.

BGW000 CAS:4193-55-9 ***HR: 3***
4,4′-BIS((4-ANILINO-6-BIS(2-HYDROXYETHYL)AMINO-w-TRIAZIN-2-YL)AMINO)-2,2′-STILBENEDISULFONIC ACID DISODIUM SALT
mf: $C_{40}H_{42}N_{12}O_{10}S_2$•2Na mw: 961.04

TOXICITY DATA with REFERENCE
eye-rbt 35 mg MOD MVCRB3 2,193,73
orl-rat LD50:14530 mg/kg MVCRB3 2,193,73
ipr-rat LD50:350 mg/kg MVCRB3 2,193,73
scu-mus LD50:1000 mg/kg MVCRB3 2,193,73
orl-gpg LD50:250 mg/kg MVCRB3 2,193,73

SAFETY PROFILE: Poison by ingestion and intraperitoneal routes. Moderately toxic by subcutaneous route. An eye irritant. See also SULFONATES. When heated to decomposition it emits very toxic fumes of NO_x, Na_2O, and SO_x.

BGW100 CAS:3426-43-5 ***HR: 3***
4,4′-BIS((4-ANILINO-6-METHOXY-s-TRIAZIN-2-YL)AMINO)-2,2′-STILBENEDISULFONIC ACID, DISODIUM SALT
mf: $C_{34}H_{28}N_{10}O_8S_2$•2Na mw: 814.82

SYN: DISODIUM-4,4′-BIS((4-ANILINO-6-METHOXY-s-TRIAZIN-2-YL)AMINO)STILBENE-2,2′-DISULFONATE

TOXICITY DATA with REFERENCE
eye-rbt 100 mg MOD MVCRB3 2,193,73
ipr-rat LD50:330 mg/kg GISAAA 51(1),87,86

CONSENSUS REPORTS: Reported in EPA TSCA Inventory.

SAFETY PROFILE: Poison by intraperitoneal route. An eye irritant. When heated to decomposition it emits toxic fumes of SO_x and Cl^-.

BGW325 CAS:71439-68-4 ***HR: D***
BISANTRENE HYDROCHLORIDE
mf: $C_{22}H_{22}N_8$•2ClH mw: 471.44

SYNS: CL 216942 ◇ NSC 337766

TOXICITY DATA with REFERENCE
dnd-mus:leu 620 μg/L CNREA8 42,2660,82

dni-mus:leu 2300 nmol/L CNREA8 42,440,82
oms-mus:leu 1 μmol/L CNREA8 42,440,82

SAFETY PROFILE: Mutation data reported. When heated to decomposition it emits toxic fumes of NO_x and HCl.

BGW500 ***HR: 3***
BIS-o-AZIDO BENZOYL PEROXIDE
mf: $C_{14}H_8N_6O_4$ mw: 324.26

SAFETY PROFILE: It has exploded violently in contact with metals. Upon decomposition it emits toxic fumes of NO_x. See also PEROXIDES and AZIDES.

BGW650 CAS:68979-48-6 ***HR: 3***
1,2-BIS(AZIDOCARBONYL)CYCLOPROPANE
mf: $C_5H_4N_6O_2$ mw: 180.13

$N_3CO{\cdot}CHCH_2CHCO{\cdot}N_3$

SAFETY PROFILE: Spontaneously explosive. When heated to decomposition it emits toxic fumes of NO_x. See also AZIDES.

BGW700 ***HR: 3***
BIS(2-AZIDOETHOXYMETHYL)NITRAMINE
mf: $C_6H_{12}N_8O_4$ mw: 260.21

$(N_3C_2H_4OCH_2)_2NNO_2$

SAFETY PROFILE: An impact-sensitive explosive. Upon decomposition it emits toxic fumes of NO_x. See also AZIDES.

BGW710 CAS:17607-20-4 ***HR: 3***
3,3-BIS(AZIDOMETHYL)OXETANE
mf: $C_5H_8N_6O$ mw: 168.16

$CH_2OCH_2C(CH_2N_3)_2$

SAFETY PROFILE: A sensitive explosive. Upon decomposition it emits toxic fumes of NO_x. See also AZIDES.

BGW720 CAS:5284-80-0 ***HR: 1***
1,5-BIS(p-AZIDOPHENYL)-1,4-PENTADIEN-3-ONE
mf: $C_{17}H_{12}N_6O$ mw: 316.35

SYNS: 1,5-BIS-(4-AZIDOFENYL)-1,4-PENTADIEN-3-ON ◇ DIAZIDODIBENZALACETON ◇ 1,4-PENTADIEN-3-ONE, 1,5-BIS(4-AZIDOPHENYL)-(9CI)

TOXICITY DATA with REFERENCE
skn-rbt 500 mg/24H MLD 85JCAE-,733,86
eye-rbt 500 mg/24H MLD 85JCAE-,733,86

SAFETY PROFILE: A skin and eye irritant. When heated to decomposition it emits acrid smoke and irritating fumes.

BGW739 ***HR: 3***
BIS(AZIDOTHIOCARBONYL)DISULFIDE
mf: $C_2N_6S_4$ mw: 236.31

SAFETY PROFILE: An explosive sensitive to mechanical impact or heating to 40°. Slow decomposition during storage increases the sensitivity. When heated to decomposition it emits toxic fumes of SO_x and NO_x. See also AZIDES and SULFIDES.

BGW750 CAS:526-62-5 ***HR: 3***
2,5-BIS(AZIRIDINO)BENZOQUINONE
mf: $C_{10}H_{10}N_2O_2$ mw: 190.22

SYNS: BAYER G4073 ◇ 2,5-BIS-ATHYLENIMINOBENZOCHINON-1,4 (GERMAN) ◇ 2,5-BIS(1-AZIRIDYNYL)BENZOQUINONE ◇ 2,5-BIS-ETHYLENIMINOBENZOQUINONE ◇ CHINON I (GERMAN) ◇ QUINON I

TOXICITY DATA with REFERENCE
mmo-sat 10 μL/plate ANYAA9 76,475,58
mma-sat 200 μg/plate SYSWAE 12,41,79
mmo-esc 50 μg/disc APMBAY 6,23,58
sln-dmg-orl 50000 ppm MUREAV 2,29,65
cyt-hmn:leu 200 μg/L/4H CHROAU 26,475,69
ipr-mus LD50:29500 μg/kg AEPPAE 230,559,57

CONSENSUS REPORTS: EPA Genetic Toxicology Program.

SAFETY PROFILE: Poison by intraperitoneal route. Human mutation data reported. When heated to decomposition it emits toxic fumes of NO_x.

BGX500 CAS:1553-36-2 ***HR: 3***
N,N'-BIS(AZIRIDINYLACETYL)-1,8-OCTAMETHYLENE DIAMINE
mf: $C_{16}H_{30}N_4O_2$ mw: 310.50

SYN: N,N'-BIS(AZIRIDINEACETYL)-1,8-OCTAMETHYLENEDIAMINE

TOXICITY DATA with REFERENCE
mmo-sat 6410 mg/L MUREAV 31,115,75
cyt-rat-orl 200 μg/kg MUREAV 31,115,75
dlt-rat-orl 100 mg/kg MUREAV 31,115,75
ipr-mus TDLo:100 mg/kg (5D male):REP EXPEAM 24,924,68
orl-rat LD50:225 mg/kg MUREAV 31,115,75
orl-mus LD50:1070 mg/kg EXPEAM 24,924,68
ipr-mus LD50:88 mg/kg EXPEAM 24,924,68

SAFETY PROFILE: Poison by ingestion and intraperitoneal routes. Experimental reproductive effects. Mutation data reported. When heated to decomposition it emits toxic fumes of NO_x.

BGX750 CAS:24279-91-2 ***HR: 3***
2,5-BIS(1-AZIRIDINYL)-3-(2-CARBAMOYLOXY-1-METHOXYETHYL)-6-METHYL-1,4-BENZOQUINONE
mf: $C_{15}H_{19}N_3O_5$ mw: 321.37

PROP: Red to reddish-brown crystals. Mp: 202° (decomp). Sltly sol in chloroform, acetone, and abs alc. Practically insol in water.

SYNS: 2,5-BIS(1-AZIRIDINYL)-3-(2-HYDROXY-1-METHOXYETHYL)-6-METHYL-p-BENZOQUINONE CARBAMATE (ESTER) ◇ CARBAZILQUINONE ◇ CARBOQUONE ◇ ESQUINON

TOXICITY DATA with REFERENCE
mmo-sat 2500 ng/plate TAKHAA 44,96,85
mma-sat 100 μg/plate CNREA8 38,2148,78
mmo-esc 2500 ng/plate TAKHAA 44,96,85
dnr-bcs 4 μg/plate TAKHAA 44,96,85
sce-ham:lng 10 μg/L CNREA8 44,3270,84
orl-rat TDLo:35 mg/kg (female 35D pre):REP OYYAA2 8,501,74
orl-rat LD50:27300 μg/kg IYKEDH 6,119,75
ipr-rat LD50:3070 μg/kg IYKEDH 6,119,75
scu-rat LD50:3990 μg/kg OYYAA2 8,501,74
ivn-rat LD50:3620 μg/kg IYKEDH 6,119,75
orl-mus LD50:28600 μg/kg IYKEDH 6,119,75
ipr-mus LD50:3440 μg/kg IYKEDH 6,119,75
scu-mus LD50:4900 μg/kg OYYAA2 8,501,74
ivn-mus LD50:5430 μg/kg OYYAA2 6,119,75

SAFETY PROFILE: A poison via ingestion, intraperitoneal, subcutaneous, and intravenous routes. Experimental reproductive effects. Mutation data reported. When heated to decomposition it emits toxic NO_x. See also CARBAMATES.

BGX775 CAS:302-48-7 ***HR: 3***
P,P-BIS(1-AZIRIDINYL)-N-ETHYLPHOSPHINIC AMIDE
mf: $C_6H_{14}N_3OP$ mw: 175.20

SYNS: P,P-BIS(1-AZIRIDINYL)-N-ETHYLAMINOPHOSPHINE OXIDE ◇ ENT 50787 ◇ PHOSPHINIC AMIDE, P,P-BIS(1-AZIRIDINYL)-N-ETHYL-

TOXICITY DATA with REFERENCE
pic-esc 11500 umol/L HEREAY 68,245,71
mmo-ssp 70 mmol/L HEREAY 68,245,71
ipr-mus TDLo:500 μg/kg (male 1D pre):REP FOBLAN 20,1,74
ipr-mus LDLo:41 mg/kg FATOAO 28,70,65

SAFETY PROFILE: Poison by intraperitoneal route. Experimental reproductive effects. Mutation data reported. When heated to decomposition it emits toxic fumes of NO_x and PO_x.

BGX850 CAS:5774-35-6 ***HR: D***
p,p-BIS(1-AZIRIDINYL)-N-ISOPROPYLAMINOPHOSPHINE OXIDE
mf: $C_7H_{16}N_3OP$ mw: 189.23

SYNS: p,p-BIS(1-AZIRIDINYL)-N-ISOPROPYLPHOSPHINICAMIDE ◇ ENT 51256 ◇ PHOSPHINIC AMIDE, p,p-BIS(1-AZIRIDINYL)-N-ISOPROPYL- ◇ PHOSPHINIC AMIDE, p,p-BIS(1-AZIRIDINYL)-N-(1-METHYLETHYL)-(9CI)

TOXICITY DATA with REFERENCE
pic-esc 2300 umol/L HEREAY 68,245,71
cyt-ham:lng 33 umol/L HEREAY 68,255,71
ipr-mus TDLo:500 μg/kg (male 1D pre):REP FOBLAN 20,1,74

SAFETY PROFILE: Experimental reproductive effects. Mutation data reported. When heated to decomposition it emits toxic fumes of NO_x and PO_x.

BGY000 CAS:1078-79-1 ***HR: 3***
BIS(1-AZIRIDINYL)(2-METHYL-3-THIAZOLIDINYL)PHOSPHINE OXIDE
mf: $C_8H_{16}N_3OPS$ mw: 233.30

SYNS: IMIPHOS ◇ MARCOPHANE ◇ MARKOFANE

TOXICITY DATA with REFERENCE
ipr-rat LD50:50 mg/kg 21ACAB -,129,68
orl-mus LD50:225 mg/kg 21ACAB -,129,68
ipr-mus LD50:142 mg/kg 21ACAB -,129,68

SAFETY PROFILE: Poison by ingestion and intraperitoneal routes. When heated to decomposition it emits very toxic fumes of SO_x, PO_x, and NO_x.

BGY125 CAS:27807-69-8 ***HR: 3***
N-(BIS(1-AZIRIDINYL)PHOSPHINYL)-p-CHLOROBENZAMIDE
mf: $C_{11}H_{13}ClN_3O_2P$ mw: 285.69

TOXICITY DATA with REFERENCE
unr-rat LD50:26 mg/kg PCJOAU 16,626,82
scu-mus LD50:50 mg/kg 85GDA2 1,263,80
unr-rbt LDLo:20 mg/kg PCJOAU 6,475,72

SAFETY PROFILE: Poison by subcutaneous and possibly other routes. When heated to decomposition it emits toxic fumes of Cl^-, NO_x, and PO_x.

BGY140 CAS:27807-51-8 ***HR: 3***
N-(BIS(1-AZIRIDINYL)PHOSPHINYL)-p-IODOBENZAMIDE
mf: $C_{11}H_{13}IN_3O_2P$ mw: 377.14

SYNS: A-19 ◇ N-p-IODOBENZOYL-N',N',N',N'-DIETHYLENETRIAMIDE of PHOSPHORIC ACID

TOXICITY DATA with REFERENCE
unr-rat LD50:50 mg/kg PCJOAU 16,626,82

ivn-mus LD50:572 mg/kg IJEBA6 21,31,83
unr-rbt LDLo:35 mg/kg PCJOAU 6,475,72

SAFETY PROFILE: Poison by unspecified routes. When heated to decomposition it emits toxic fumes of I^-, NO_x, and PO_x.

BGY500 CAS:2275-81-2 ***HR: 3***
p,p-BIS(1-AZIRIDINYL)-N-PROPYLPHOSPHINIC AMIDE
mf: $C_7H_{16}N_3OP$ mw: 189.23

SYNS: p,p-BIS(1-AZIRIDINYL)-N-PROPYLAMINOPHOSPHINE OXIDE ◇ ENT 51253 ◇ PHOSPHINIC AMIDE, p,p-BIS(1-AZIRIDINYL)-N-PROPYL- ◇ PROPYLAMINO-BIS(1-AZIRIDINYL)PHOSPHINE OXIDE

TOXICITY DATA with REFERENCE
pic-esc 2300 umol/L HEREAY 68,245,71
mmo-ssp 70 mmol/L HEREAY 68,245,71
ipr-mus TDLo:500 μg/kg (male 1D pre):REP FOBLAN 20,1,74
ipr-mus LDLo:25 mg/kg FATOAO 28,70,65

SAFETY PROFILE: Poison by intraperitoneal route. Experimental reproductive effects. Mutation data reported. When heated to decomposition it emits toxic fumes of NO_x and PO_x.

BGY700 CAS:1271-54-1 ***HR: 3***
BIS-BENZENE CHROMIUM
mf: $C_{12}H_{12}Cr$ mw: 208.24

SYNS: CHROMIUM, BIS(BENZENE)-(8CI) ◇ CHROMIUM, BIS(eta^6)-BENZENE)-(9CI) ◇ CHROMIUM(II), DIPHENYL- ◇ DIBENZENECHROMIUM ◇ DIPHENYLCHROMIUM

TOXICITY DATA with REFERENCE
ivn-mus LD50:17800 μg/kg CSLNX* NX#02380

OSHA PEL: TWA 0.5 $mg(Cr)/m^3$
ACGIH TLV: TWA 0.5 $mg(Cr)/m^3$

SAFETY PROFILE: Poison by intravenous route. When heated to decomposition it emits toxic fumes of Cr.

BGY720 CAS:12089-29-1 ***HR: 3***
BIS(BENZENE)CHROMIUM IODIDE
mf: $C_{12}H_{12}Cr{\bullet}I$ mw: 335.14

SYNS: BIS(BENZENE)CHROMIUM(1+)IODIDE ◇ CHROMIUM(1+), BIS(BENZENE)-, IODIDE (8CI) ◇ CHROMIUM(1+), BIS(eta^6)-BENZENE)-, IODIDE (9CI) ◇ CHROMIUM, BIS(BENZENE)IODO- ◇ CHROMIUM (III), DIPHENYL-, IODIDE ◇ DIBENZENECHROMIUM IODIDE ◇ DIPHENYLCHROMIUM (III) IODIDE

TOXICITY DATA with REFERENCE
ivn-mus LD50:18 mg/kg CSLNX* NX#02011

OSHA PEL: TWA 0.5 $mg(Cr)/m^3$
ACGIH TLV: TWA 0.5 $mg(Cr)/m^3$

SAFETY PROFILE: Poison by intravenous route. When heated to decomposition it emits toxic fumes of Cr and I^-.

BGY750 ***HR: 3***
BIS BENZENE DIAZO OXIDE
mf: $C_{12}H_{10}N_4O$ mw: 226.24

SAFETY PROFILE: A very unstable explosive. Upon decomposition it emits toxic fumes of NO_x. See also AZIDES.

BGZ000 ***HR: 3***
BIS(n-BENZENE)IRON(O)
mf: $C_{12}H_{12}Fe$ mw: 212.08

SAFETY PROFILE: An explosive gas which can detonate at −40°.

BHA000 CAS:63950-89-0 ***HR: 2***
BIS(BENZOATO)DIOXOCHROMIUM TRIHYDRATE
mf: $C_{14}H_{10}CrO_6{\bullet}3H_2O$ mw: 380.30

SYN: KYSELINA CEROMSALICYLOVA (CZECH)

TOXICITY DATA with REFERENCE
eye-rbt 100 mg/24H MOD 28ZPAK -,19,72
orl-rat LD50:4810 mg/kg 28ZPAK -,19,72

CONSENSUS REPORTS: Chromium and its compounds are on The Community Right-To-Know List.

SAFETY PROFILE: Mildly toxic by ingestion. An eye irritant. See also CHROMIUM COMPOUNDS. When heated to decomposition it emits acrid smoke and irritating fumes.

BHA500 CAS:95-35-2 ***HR: 1***
N,N'-BIS(2-BENZOTHIAZOLYLTHIOMETHYLENE)UREA
mf: $C_{17}H_{14}N_4OS_4$ mw: 418.59

SYN: 1,3-BIS((2-BENZOTHIAZOLYLTHIO)METHYL)UREA

TOXICITY DATA with REFERENCE
skn-hmn 500 mg/48H MLD AMIHBC 5,311,52
skn-rbt 500 mg MOD AMIHBC 5,311,52
orl-rat LD50:6000 mg/kg AMIHBC 5,311,52

CONSENSUS REPORTS: Reported in EPA TSCA Inventory.

SAFETY PROFILE: Mildly toxic by ingestion. A human skin irritant. When heated to decomposition it emits very toxic fumes of NO_x and SO_x.

BHA750 CAS:155-04-4 ***HR: 3***
BIS(2-BENZOTHIAZOLYLTHIO)ZINC
mf: $C_{14}H_8N_2S_4 \cdot Zn$ mw: 397.85

SYNS: 2-BENZOTHIAZOLETHIOL, ZINC SALT (2:1) ◇ BIS(MERCAPTOBENZOTHIAZOLATO)ZINC ◇ HERMAT Zn-MBT ◇ 2-MERCAPTOBENZOTHIAZOLE ZINC SALT ◇ OXAF ◇ PENNAC ZT ◇ TISPERSE MB-58 ◇ USAF GY-7 ◇ VULKACIT ZM ◇ ZENITE ◇ ZENITE SPECIAL ◇ ZETAX ◇ ZINC-2-BENZOTHIAZOLETHIOLATE ◇ ZINC BENZOTHIAZOLYL MERCAPTIDE ◇ ZINC BENZOTHIAZOL-2-YLTHIOLATE ◇ ZINC BENZOTHIAZYL-2-MERCAPTIDE ◇ ZINC MERCAPTOBENZOTHIAZOLATE ◇ ZINC-2-MERCAPTOBENZOTHIAZOLE ◇ ZINC MERCAPTOBENZOTHIAZOLE SALT ◇ ZMBT ◇ ZnMB

TOXICITY DATA with REFERENCE
scu-mus TDLo:1000 mg/kg:CAR NTIS** PB223-159
orl-rat LD50:540 mg/kg VCTDC* 12/9/76
ipr-mus LD50:200 mg/kg NTIS** AD277-689

CONSENSUS REPORTS: Reported in EPA TSCA Inventory. Zinc compounds are on the Community Right-To-Know List.

SAFETY PROFILE: Poison by intraperitoneal route. Moderately toxic by ingestion and subcutaneous routes. Questionable carcinogen with experimental carcinogenic data. When heated to decomposition it emits very toxic fumes of SO_x, NO_x, and ZnO. See also ZINC COMPOUNDS and MERCAPTANS.

BHB000 CAS:64092-23-5 ***HR: 3***
BIS(2-BENZOYLBENZOATO)BIS(3-(1-METHYL-2-PYRROLIDINYL)PYRIDINE) NICKEL TRIHYDRATE
mf: $C_{48}H_{46}N_4NiO_6 \cdot 3H_2O$ mw: 887.75

SYN: NICOTINE, COMPOUND, with NICKEL(II)-o-BENZOYL BENZOATE TRIHYDRATE (2:1)

TOXICITY DATA with REFERENCE
orl-rat LDLo:150 mg/kg NCNSA6 5,22,53
ipr-rat LDLo:75 mg/kg NCNSA6 5,22,53

CONSENSUS REPORTS: Nickel and its compounds are on The Community Right-To-Know List.

OSHA PEL: (Transitional: TWA 1 mg/m^3) TWA 0.1 mg (Ni)/m^3
ACGIH TLV: TWA 0.1 mg (Ni)/m^3; (Proposed: TWA 0.05 mg(Ni)/m^3; Human Carcinogen)
NIOSH REL: (Inorganic Nickel) TWA 0.015 mg(Ni)/m^3

SAFETY PROFILE: Suspected carcinogen. Poison by ingestion and intraperitoneal routes. See also NICKEL COMPOUNDS and NICOTINE. When heated to decomposition it emits toxic fumes of NO_x.

BHB250 ***HR: 3***
1,1-BIS(BENZOYLPEROXY)CYCLOHEXANE
mf: $C_{20}H_{20}O_6$ mw: 356.37

SAFETY PROFILE: Explodes violently in contact with a flame. When heated to decomposition it emits acrid smoke and fumes. See also PEROXIDES.

BHB500 CAS:74037-60-8 ***HR: 1***
(4,6-BIS(BIS(BUTOXYMETHYL)AMINO)-s-TRIAZIN-2-YLIMINO)DIMETHANOL
mf: $C_{25}H_{50}N_6O_6$ mw: 530.81

SYN: DIMETHYLOL-TETRAKIS-BUTOXYMETHYLMELAMIN (CZECH)

TOXICITY DATA with REFERENCE
skn-rbt 500 mg/24H MLD 28ZPAK -,157,72
eye-rbt 500 mg/24H MLD 28ZPAK -,157,72

SAFETY PROFILE: A skin and eye irritant. When heated to decomposition it emits toxic fumes of NO_x.

BHB750 CAS:4420-79-5 ***HR: 3***
2,5-BIS(BIS-(2-CHLOROETHYL)AMINOMETHYL) HYDROQUINONE
mf: $C_{16}H_{24}Cl_4N_2O_2$ mw: 418.22

SYNS: HYDROQUINONE MUSTARD ◇ NSC 18321 ◇ WEATHERBEE MUSTARD

TOXICITY DATA with REFERENCE
ipr-mus TDLo:28 mg/kg/4W:CAR JNCIAM 36,915,66
ipr-rat LD10:4700 μg/kg CNCRA6 17,1,62
ivn-dog LDLo:900 μg/kg CCSUBJ 2,201,65
ivn-mky LDLo:1800 μg/kg CCSUBJ 2,201,65

SAFETY PROFILE: Deadly poison by intravenous and intraperitoneal routes. A powerful irritant. Questionable carcinogen with experimental carcinogenic data. When heated to decomposition it emits highly toxic fumes of NO_x and Cl^-.

BHB950 CAS:4028-32-4 ***HR: 1***
4,4′-BIS((4-BIS((2-HYDROXYETHYL)AMINO)-6-CHLORO-s-TRIAZIN-2-YL)AMINO)-2,2′-STILBENEDISULFONIC ACID, DISODIUM SALT
mf: $C_{28}H_{30}Cl_2N_{10}O_{10}S_2 \cdot 2Na$ mw: 847.68

TOXICITY DATA with REFERENCE
eye-rbt 100 mg MOD MVCRB3 2,193,73

CONSENSUS REPORTS: Reported in EPA TSCA Inventory.

SAFETY PROFILE: An eye irritant. When heated to decomposition it emits toxic fumes of NO_x, SO_x, and Cl^-.

BHC500 CAS:4470-72-8 ***HR: 2***
4,4'-BIS((4-BIS(2-HYDROXYETHYL)AMINO-6-METHOXY-s-TRIAZIN-2-YL)AMINO)-2,2'-STILBENEDISULFONIC ACID DISODIUM SALT
mf: $C_{30}H_{36}N_{10}O_{12}S_2$•2Na mw: 838.86

TOXICITY DATA with REFERENCE
eye-rbt 100 mg SEV MVCRB3 2,193,73

SAFETY PROFILE: A severe eye irritant. See also SULFONATES. When heated to decomposition it emits very toxic fumes of NO_x, Na_2O, and SO_x.

BHC750 CAS:12224-02-1 ***HR: 2***
4,4'-BIS((4-BIS((2-HYDROXYETHYL)AMINO)-6-(m-SULFOANILINO)-s-TRIAZIN-2-YL)AMINO)-2,2'-STILBENEDISULFONIC ACID TETRASODIUM SALT
mf: $C_{40}H_{40}N_{12}O_{16}S_4$•4Na mw: 1165.12

TOXICITY DATA with REFERENCE
skn-rbt 500 mg/24H MLD MVCRB3 2,193,73
eye-rbt 100 mg MOD MVCRB3 2,193,73
ipr-rat LD50:1750 mg/kg MVCRB3 2,193,73
scu-mus LD50:1500 mg/kg MVCRB3 2,193,73
ivn-mus LD50:900 mg/kg MVCRB3 2,193,73

CONSENSUS REPORTS: Reported in EPA TSCA Inventory.

SAFETY PROFILE: Moderately toxic by intraperitoneal, subcutaneous, and intravenous routes. A skin and eye irritant. See also SULFONATES. When heated to decomposition it emits very toxic fumes of SO_x, Na_2O, and NO_x.

BHD000 CAS:64036-79-9 ***HR: 3***
BIS(BIS(β-HYDROXYETHYL)SULFONIUMETHYL)SULFIDE DICHLORIDE
mf: $C_{12}H_{28}O_4S_3$•2Cl mw: 403.48

SYN: (THIOETHYLENE)BIS(BIS(2-HYDROXYETHYL)SULFONIUM DICHLORIDE

TOXICITY DATA with REFERENCE
orl-rat LDLo:250 mg/kg NCNSA6 5,9,53
scu-mus LD50:200 mg/kg NTIS** PB158-507

SAFETY PROFILE: Poison by ingestion and subcutaneous routes. When heated to decomposition it emits very toxic fumes of SO_x and Cl^-. See also SULFIDES.

BHD250 CAS:3785-34-0 ***HR: 3***
1,2-BIS(BROMOACETOXY)ETHANE
mf: $C_6H_8Br_2O_4$ mw: 303.96

SYNS: BROMOACETIC ACID ETHYLENE ESTER ◇ ETHYLENE BIS(BROMOACETATE) ◇ ETHYLENE BROMOACETATE ◇ ETHYLENE GLYCOL BIS(BROMOACETATE) ◇ PANDUROL ◇ S 13

TOXICITY DATA with REFERENCE
ipr-mus LD50:39 mg/kg JNCIAM 31,297,63
ivn-mus LD50:56 mg/kg CSLNX* NX#03918
ivn-dog LD50:15 mg/kg JNCIAM 31,297,63

SAFETY PROFILE: Poison by intraperitoneal and intravenous routes. When heated to decomposition it emits toxic fumes of Br^-. See also BROMIDES and ESTERS.

BHG000 CAS:22953-41-9 ***HR: 3***
p-(BIS(2-BROMOETHYL)AMINO)PHENOL-m-(α,α,α-TRIFLUOROMETHYL)BENZOATE
mf: $C_{18}H_{16}Br_2F_3NO_2$ mw: 495.17

TOXICITY DATA with REFERENCE
ipr-rat LD50:3800 μg/kg JMCMAR 12,491,69
ipr-mus LD50:3900 μg/kg JMCMAR 12,491,69

SAFETY PROFILE: Poison by intraperitoneal route. See also ESTERS and BROMIDES. When heated to decomposition it emits very toxic fumes of Br^-, F^-, and NO_x.

BHI750 CAS:3138-86-1 ***HR: 3***
2,3-BIS(BROMOMETHYL)QUINOXALINE
mf: $C_{10}H_8Br_2N_2$ mw: 316.02

TOXICITY DATA with REFERENCE
ivn-mus LD50:18 mg/kg CSLNX* NX#02400

CONSENSUS REPORTS: Reported in EPA TSCA Inventory.

SAFETY PROFILE: Poison by intravenous route. See also BROMIDES. When heated to decomposition it emits very toxic fumes of Br^- and NO_x.

BHJ000 CAS:2050-47-7 ***HR: 3***
BIS(p-BROMOPHENYL) ETHER
mf: $C_{12}H_8Br_2O$ mw: 328.02

SYN: USAF DO-61

TOXICITY DATA with REFERENCE
ipr-mus LD50:125 mg/kg NTIS** AD277-689

CONSENSUS REPORTS: Reported in EPA TSCA Inventory.

SAFETY PROFILE: Poison by intraperitoneal route. See also ETHERS and BROMIDES. When heated to decomposition it emits toxic fumes of Br^-.

BHJ250 CAS:54-91-1 ***HR: 3***
1,4-BIS(3-BROMOPROPIONYL)-PIPERAZINE
mf: $C_{10}H_{16}Br_2N_2O_2$ mw: 356.10

SYNS: A 1803 ◇ A-8103 ◇ AMEDEL ◇ NSC-25154 ◇ PIPOBROMAN ◇ VERCYTE

TOXICITY DATA with REFERENCE
mmo-sat 1 mg/plate CNREA8 38,2148,78
mmo-esc 400 µg/plate TAKHAA 44,96,85
pic-esc 500 mg/L APMBAY 12,234,64
sce-hmn:lum 1 µmol/L CTRRDO 69,505,85
cyt-hmn:leu 1 µmol/L CNREA8 25,275,65
orl-rat TDLo:20 mg/kg (female 7-10D post):TER BTDCAV 13,103,72
orl-mus TDLo:160 mg/kg (female 6-9D post):REP BTDCAV 13,103,72
orl-rat LD50:220 mg/kg IYKEDH 4,467,73
ipr-rat LD50:140 mg/kg IYKEDH 4,467,73
scu-rat LD50:139 mg/kg NIIRDN 6,638,82
orl-mus LD50:382 mg/kg IYKEDH 4,467,73
ipr-mus LD50:285 mg/kg APPHAX 37,249,80
scu-mus LD50:353 mg/kg IYKEDH 4,467,73

SAFETY PROFILE: Poison by ingestion, subcutaneous and intraperitoneal routes. An experimental teratogen. Other experimental reproductive effects. Human mutation data reported. When heated to decomposition it emits very toxic fumes of Br^- and NO_x. See also BROMIDES.

BHJ500 CAS:126-15-8 ***HR: 2***
BISBUTENYLENETETRAHYDROFURFURAL
mf: $C_{13}H_{16}O_2$ mw: 204.29

SYNS: AC-R-11 ◇ BUTADIEN-FURFURAL COPOLYMER ◇ 2,3:4,5-BIS(2-BUTYLENE)TETRAHYDRO-2-FURFURAL ◇ 2,3,4,5-BIS(Δ^2-BUTENYLENE)TETRAHYDROFURFURAL ◇ 2,3,4,5-BIS(2-BUTENYLENE)TETRAHYDROFURFURAL ◇ 2,3,4,5-BIS(2-BUTYLENE)TETRAHYDRO-2-FURALDEHYDE ◇ BIS-Δ^2-BUTYLENETETRAHYDROFURFURAL ◇ 2,3,4,5-BIS(Δ^2-BUTYLENE)TETRAHYDROFURFURAL ◇ 2,3:4,5-DI(2-BUTENYL)TETRAHYDROFURFURAL ◇ ENT 17,596 ◇ 4A-FORMYL-1,4,4A,5A,6,9,9A,9B-OCTAHYDRODIBENZOFURAN ◇ 2-FURALDEHYDE, 2,3:4,5-BIS(2-BUTENYLENE)TETRAHYDRO- ◇ 1,5A,6,9,9A,9B-HEXAHYDRO-4A(4H)-DIBENZOFURANCARBOXALDEHYDE ◇ MGK 11 ◇ MGK REPELLENT 11 ◇ PHILLIPS R-11 ◇ R-11

TOXICITY DATA with REFERENCE
orl-rat LD50:2500 mg/kg MEIEDD 10,1170,83
ivn-rat LD50:2 g/kg YKYUA6 32,605,81

CONSENSUS REPORTS: Reported in EPA TSCA Inventory.

SAFETY PROFILE: Moderately toxic by ingestion. An insect repellant. When heated to decomposition it emits acrid smoke and fumes. See also ALDEHYDES.

BHJ625 CAS:91216-69-2 ***HR: 2***
4,5-BIS(2-BUTENYLOXY)-2-IMIDAZOLIDINONE
mf: $C_{11}H_{18}N_2O_3$ mw: 226.31

SYN: SRC-15

TOXICITY DATA with REFERENCE
orl-mus LD50:1700 mg/kg CPBTAL 12,843,64
ipr-mus LD50:770 mg/kg CPBTAL 12,843,64
scu-mus LD50:1750 mg/kg CPBTAL 12,843,64

SAFETY PROFILE: Moderately toxic by ingestion, subcutaneous, and intraperitoneal routes. When heated to decomposition it emits toxic fumes of NO_x.

BHJ750 CAS:141-18-4 ***HR: 2***
BIS(2-BUTOXYETHYL) ADIPATE
mf: $C_{18}H_{34}O_6$ mw: 346.52

SYNS: ADIPIC ACID, DIBUTOXYETHYL ESTER ◇ BUTYL "CELLOSOLVE" ADIPATE ◇ DIBUTOXYETHYL ADIPATE ◇ DI(2-BUTOXYETHYL) ADIPATE ◇ DIBUTYL CELLOSOLVE ADIPATE ◇ HEXANEDIOIC ACID, BIS(2-BUTOXYETHYL) ESTER

TOXICITY DATA with REFERENCE
ipr-rat LD50:600 mg/kg 14CYAT 2,1882,63

CONSENSUS REPORTS: Reported in EPA TSCA Inventory.

SAFETY PROFILE: Moderately toxic by intraperitoneal route. See also ESTERS. When heated to decomposition it emits acrid smoke and irritating fumes.

BHK000 CAS:117-83-9 ***HR: 1***
BIS(2-BUTOXYETHYL)PHTHALATE
mf: $C_{20}H_{30}O_6$ mw: 366.50

SYNS: 2-BUTOXYETHANOL PHTHALATE (2:1) ◇ β-BUTOXYETHYL PHTHALATE ◇ BUTYL "CELLOSOLVE" PHTHALATE ◇ BUTYL GLYCOL PHTHALATE ◇ DI(BUTOXYETHYL)PHTHALATE ◇ DIBUTYL CELLOSOLVE PHTHALATE ◇ DIBUTYLGLYCOL PHTHALATE ◇ KESSCOFLEX ◇ KRONISOL

TOXICITY DATA with REFERENCE
orl-rat LD50:8380 mg/kg JIHTAB 30,63,48
orl-gpg LDLo:6000 mg/kg 29ZWAE -,336,68

CONSENSUS REPORTS: Reported in EPA TSCA Inventory.

SAFETY PROFILE: Mildly toxic by ingestion. When heated to decomposition it emits acrid smoke and irritating fumes. See also ESTERS.

BHK250 CAS:15546-16-4 ***HR: 3***
BIS(BUTOXYMALEOYLOXY)DIBUTYLSTANNANE
mf: $C_{24}H_{40}O_8Sn$ mw: 575.33

SYNS: DI-N-BUTYLTIN DI(MONOBUTYL)MALEATE ◇ DI-N-BUTYLZINN-DI(MONOBUTYL)MALEINAT(GERMAN)

TOXICITY DATA with REFERENCE
orl-rat LD50:120 mg/kg ARZNAD 19,934,69

CONSENSUS REPORTS: Reported in EPA TSCA Inventory.

OSHA PEL: TWA 0.1 mg(Sn)/m^3 (skin)
ACGIH TLV: TWA 0.1 mg(Sn)/m^3 (skin) (Proposed: TWA 0.1 mg(Sn)/m^3; STEL 0.2 mg(Sn)/m^3 (skin))
NIOSH REL: (Organotin Compounds) TWA 0.1 mg(Sn)/m^3

SAFETY PROFILE: Poison by ingestion. See also TIN COMPOUNDS. When heated to decomposition it emits acrid smoke and irritating fumes.

BHK500 CAS:29575-02-8 ***HR: 2***
BIS(BUTOXYMALEOYLOXY)DIOCTYLSTANNANE
mf: $C_{32}H_{56}O_8Sn$ mw: 687.57

SYNS: DI-N-OCTYLTIN BIS(BUTYL MALEATE) ◇ DI-N-OCTYLTIN DIMONOBUTYLMALEATE ◇ DI-N-OCTYLZINN-DIMONOBUTYLMALEINAT (GERMAN)

TOXICITY DATA with REFERENCE
orl-rat LD50:2030 mg/kg ARZNAD 19,934,69
orl-mus LD50:3750 mg/kg FCTXAV 8,655,70

CONSENSUS REPORTS: Reported in EPA TSCA Inventory.

OSHA PEL: TWA 0.1 mg(Sn)/m^3 (skin)
ACGIH TLV: TWA 0.1 mg(Sn)m^3 (skin) (Proposed: TWA 0.1 mg(Sn)/m^3; STEL 0.2 mg(Sn)/m^3 (skin))
NIOSH REL: (Organotin Compounds) TWA 0.1 mg(Sn)/m^3

SAFETY PROFILE: Moderately toxic by ingestion. See also TIN COMPOUNDS. When heated to decomposition it emits acrid smoke and irritating fumes.

BHK750 CAS:143-29-3 ***HR: 2***
BIS(BUTYLCARBITOL)FORMAL
mf: $C_{17}H_{36}O_6$ mw: 336.53

SYNS: BUTYLCARBITOL FORMAL ◇ CRYOFLEX ◇ DIBUTYLCARBITOLFORMAL ◇ 5,8,11,13,16,19-HEXAOXATRICOSANE (9CI) ◇ TP 90B

TOXICITY DATA with REFERENCE
orl-rat LD50:1746 mg/kg NPIRI* 2,238,75
orl-mus LD50:2700 mg/kg GISAAA 46(5),87,81

CONSENSUS REPORTS: Reported in EPA TSCA Inventory.

SAFETY PROFILE: Moderately toxic by ingestion. When heated to decomposition it emits acrid smoke and irritating fumes.

BHL000 CAS:63979-95-3 ***HR: 2***
BIS(4-tert-BUTYL-m-CRESYL)SULFIDE

SYN: DI-(4-tert-BUTYL-m-CRESOL)SULFIDE

TOXICITY DATA with REFERENCE
skn-hmn 125 mg/48H MOD AMIHBC 5,311,52
skn-rbt 500 mg SEV AMIHBC 5,311,52
ipr-rat LD50:5000 mg/kg AMIHBC 5,311,52

SAFETY PROFILE: Mildly toxic by intraperitoneal route. A moderate human skin irritant; a severe experimental skin irritant. See also SULFIDES. When heated to decomposition it emits toxic fumes of SO_x.

BHL100 CAS:25155-25-3 ***HR: 1***
α-α′-BIS(tert-BUTYLPEROXY)DIISOPROPYLBENZENE
mf: $C_{20}H_{34}O_4$ mw: 338.54

SYNS: PEROXIDE,(PHENYLENEBIS(1-METHYLETHYLIDENE))BIS(1,1-DIMETHYLETHYL)- ◇ PEROXIDE, (PHENYLENEDIISOPROPYLIDENE)BIS(tert-BUTYL- ◇ (PHENYLENEDIISOPROPYLIDENE)BIS(tert-BUTYLPEROXIDE) ◇ VUL-CUP ◇ VUL-CUP 40KE ◇ VUL-CUP R

TOXICITY DATA with REFERENCE
skn-rbt 100%/24H MLD HERBU* PRC-304

CONSENSUS REPORTS: Reported in EPA TSCA Inventory.

SAFETY PROFILE: A skin irritant. When heated to decomposition it emits acrid smoke and irritating fumes.

BHL500 CAS:1000-40-4 ***HR: 3***
BIS(BUTYLTHIO)DIMETHYLTIN
mf: $C_{10}H_{24}S_2Sn$ mw: 327.15

SYN: BIS(BUTYLTHIO)DIMETHYLSTANNANE

TOXICITY DATA with REFERENCE
ivn-mus LD50:320 mg/kg CSLNX* NX#01865

OSHA PEL: TWA 0.1 mg(Sn)/m^3 (skin)
ACGIH TLV: TWA 0.1 mg(Sn)/m^3 (skin) (Proposed: TWA 0.1 mg(Sn)/m^3; STEL 0.2 mg(Sn)/m^3 (skin))
NIOSH REL: (Organotin Compounds) TWA 0.1 mg(Sn)/m^3

SAFETY PROFILE: Poison by intravenous route. See also TIN COMPOUNDS. When heated to decomposition it emits toxic fumes of SO_x.

BHL750 CAS:15263-52-2 ***HR: 3***
1,3-BIS(CARBAMOYLTHIO)-2-(N,N-DIMETHYLAMINO)PROPANE HYDROCHLORIDE
mf: $C_7H_{15}N_3O_2S_2•ClH$ mw: 273.83

SYNS: CALDAN ◇ CARBAMOTHIOIC ACID-S,S′-(2-(DIMETHYLAMINO)-1,3-PROPANEDIYL) ESTER, MONOHYDROCHLORIDE (9CI) ◇ CARTAP HYDROCHLORIDE ◇ S,S′-(2-(DIMETHYLAMINO)TRIMETHYLENE)BIS(THIOCARBAMATE) HYDROCHLORIDE ◇ NTD 2 ◇ PADAN ◇ PATAP ◇ SANVEX ◇ THIOBEL ◇ THIOCARBAMIC ACID-S,S-(2-(DIMETHYLAMINO)TRIMETHYLENE)ESTERHYDROCHLORIDE ◇ TI-1258 ◇ VEGETOX

TOXICITY DATA with REFERENCE
orl-rat TDLo:700 mg/kg (9-15D preg):TER TAKHAA 30,776,71
orl-rat LD50:250 mg/kg SPEADM 78-1,61,78
orl-mus LD50:165 mg/kg SPEADM 78-1,61,78
ivn-mus LD50:59 mg/kg JJPAAZ 17,491,67

CONSENSUS REPORTS: EPA Genetic Toxicology Program.

SAFETY PROFILE: Poison by ingestion and intravenous routes. An experimental teratogen. An insecticide. When heated to decomposition it emits very toxic fumes of NO_x, SO_x, and HCl. See also CARBAMATES.

BHM000 CAS:111-17-1 ***HR: 3***
BIS(2-CARBOXYETHYL) SULFIDE
mf: $C_6H_{10}O_4S$ mw: 178.22

PROP: Very sol in alc, hot water, acetate; sltly sol in water. Mp: 134°.

SYNS: DIETHYL SULFIDE-2,2'-DICARBOXYLIC ACID ◇ KYSELINA-β,β'-THIODIPROPIONOVA (CZECH) ◇ TDPA ◇ 2-(2,3,5,6-TETRAMETHYLPHENOXY)PROPIONIC AICD ◇ 4-THIAHEPTANEDIOIC ACID ◇ THIODIPROPIONIC ACID ◇ β,β'-THIODIPROPIONIC ACID ◇ 3,3'-THIODIPROPIONIC ACID ◇ TYOX A

TOXICITY DATA with REFERENCE
skn-rbt 500 mg/24H MLD 28ZPAK -,171,72
eye-rbt 20 mg/24H MOD 28ZPAK -,171,72
orl-rat LD50:3980 mg/kg 28ZPAK -,171,72
ipr-rat LD50:500 mg/kg AFREAW 3,197,51
orl-mus LD50:2000 mg/kg AFREAW 3,197,51
ipr-mus LD50:250 mg/kg AFREAW 3,197,51
ivn-mus LD50:175 mg/kg AFREAW 3,197,51

CONSENSUS REPORTS: Reported in EPA TSCA Inventory.

SAFETY PROFILE: A poison by intraperitoneal and intravenous routes. Moderately toxic by ingestion. A skin and eye irritant. When heated to decomposition it emits toxic fumes of SO_x. See also SULFIDES.

BHM250 CAS:95-99-8 ***HR: 3***
N,N'-BIS(CARBOXYMETHYL)DITHIOOXAMIDE
mf: $C_6H_8N_2O_4S_2$ mw: 236.28

SYN: USAF MK-3

TOXICITY DATA with REFERENCE
ipr-mus LD50:200 mg/kg NTIS** AD277-689

CONSENSUS REPORTS: Reported in EPA TSCA Inventory.

SAFETY PROFILE: Poison by intraperitoneal route. When heated to decomposition it emits very toxic fumes of NO_x and SO_x.

BHM500 ***HR: 3***
BIS(3-CARBOXYPROPIONYL)PEROXIDE
DOT: UN 2113
mf: $C_8H_{10}O_8$ mw: 234.16

DOT Classification: Organic Peroxide; Label:Organic Peroxide

SAFETY PROFILE: Explodes on contact with flame. Commercial grade (dry 95%) is highly hazardous. When heated to decomposition it emits acrid smoke and fumes. See also PEROXIDES, ORGANIC.

BHM750 CAS:94-17-7 ***HR: 2***
BIS(p-CHLOROBENZOYL) PEROXIDE
DOT: UN 2113/UN 2114/UN 2115
mf: $C_{14}H_8Cl_2O_4$ mw: 311.12

PROP: A white, granular material. Insol in water, sol in organic solvents.

SYNS: CADPX PS ◇ p-CHLOROBENZOYL PEROXIDE ◇ p-CHLOROBENZOYL PEROXIDE (DOT) ◇ p,p'-DICHLOROBENZOYL PEROXIDE ◇ DI-(4-CHLOROBENZOYL) PEROXIDE

TOXICITY DATA with REFERENCE
ipr-mus LDLo:500 mg/kg CBCCT* 4,110,52

CONSENSUS REPORTS: Reported in EPA TSCA Inventory.

DOT Classification: Organic Peroxide; Label: Organic Peroxide.

SAFETY PROFILE: Moderately toxic by intraperitoneal route. Probably an irritant to skin and mucous membranes. Dangerous fire hazard; a powerful oxidizer. Store in a cool place away from fire hazards, sparks, open flames, and out of the direct rays of the sun. Dangerous explosion hazard; this material may explode by heat (over 38°) or contamination. Any contaminant which acts as an accelerator to the polymerization or decomposition of this material can cause an explosion. Heat or contact with certain fumes or mists can cause it to explode. To fight small fires, use CO_2 or foam extinguishers. Water spray or mist may also be used. Dry chemical is effective. When heated to decomposition it emits toxic fumes of Cl^-. See also PEROXIDES, ORGANIC.

BHN000 CAS:366-93-8 ***HR: 3***
trans-N,N'-BIS(2-CHLOROBENZYL)-1,4-CYCLOHEXANEBIS(METHYLAMINE) DIHYDROCHLORIDE
mf: $C_{22}H_{28}Cl_2N_2{\cdot}2ClH$ mw: 464.34

SYNS: AY 9944 ◇ trans-1,4-BIS(2-DICHLOROBENZYLAMINOETHYL)CYCLOHEXANEDICHLORHYDRATE (FRENCH) ◇ trans-N,N'-(1,4-

CYCLOHEXYLENEDIMETHYLENE)BIS(2-CHLOROBENZYLAMINE) DIHYDROCHLORIDE

TOXICITY DATA with REFERENCE
unr-rat TDLo:225 mg/kg (female 2-4D post):REP TJADAB 29(3),32A,84
orl-mus TDLo:700 mg/kg (female 1-14D post):TER CRSBAW 163,327,69
orl-mus LD50:155 mg/kg PSEBAA 139,100,72

SAFETY PROFILE: Poison by ingestion. Experimental teratogenic and reproductive effects. Inhibits cholesterol synthesis. When heated to decomposition it emits very toxic fumes of NO_x and Cl^-.

BHN250 CAS:50570-59-7 ***HR: 2***
4,4'-BIS(4-CHLORO-6-BIS(2-HYDROXYETHYLAMINO)-s-TRIAZIN-2-YL-AMINO-2,2'-STILBENEDISULFONIC ACID
mf: $C_{28}H_{32}Cl_2N_{10}O_{10}S_2$ mw: 803.72

SYN: RYLUX BSP (CZECH)

TOXICITY DATA with REFERENCE
eye-rbt 500 mg/24H SEV 28ZPAK -,251,72

SAFETY PROFILE: A severe eye irritant. See also SULFONATES. When heated to decomposition it emits very toxic fumes of Cl^-, NO_x, and SO_x.

BHN500 CAS:3374-04-7 ***HR: 3***
N,N-BIS(β-CHLOROETHYL)-dl-ALANINE HYDROCHLORIDE
mf: $C_7H_{13}Cl_2NO_2 \cdot ClH$ mw: 250.57

SYNS: ALANINE MUSTARD ◇ NSC 17663

TOXICITY DATA with REFERENCE
ice-rat LD50:225 μg/kg JPPMAB 18,760,66
unk-man TDLo:900 μg/kg:UNS CCROBU 50,219,66
ivn-dog LDLo:1 mg/kg CCSUBJ 2,201,65
ivn-mky LDLo:1 mg/kg CCSUBJ 2,201,65

SAFETY PROFILE: Deadly poison by intracerebral and intravenous routes. Human systemic effects by an unspecified route: bone marrow changes. When heated to decomposition it emits very toxic fumes of Cl^-, NO_x and HCl.

BHN750 CAS:334-22-5 ***HR: 3***
BIS-β-CHLOROETHYLAMINE
mf: $C_4H_9Cl_2N$ mw: 142.04

SYNS: N,N-BIS-(β-CHLORAETHYL)-AMIN (GERMAN) ◇ NH-LOST ◇ NOR-NITROGEN MUSTARD ◇ NSC-10873

TOXICITY DATA with REFERENCE
mmo-sat 100 μmol/L CNREA8 41,2967,81
cyt-hmn:lym 1 mg/L CRNGDP 5,1637,84
ipr-mus TDLo:50 mg/kg (female 12D post):REP TCMUD8 7,7,87
ipr-rat LD50:97 mg/kg JMCMAR 8,167,65
ivn-rat LD50:100 mg/kg ARZNAD 24,1149,74
scu-mus LD50:20 mg/kg JPETAE 91,224,47
ivn-dog LDLo:6 mg/kg CCSUBJ 2,201,65
ivn-mky LDLo:11 mg/kg CCSUBJ 2,201,65

SAFETY PROFILE: Poison by intraperitoneal, subcutaneous, and intravenous routes. Experimental reproductive effects. Human mutation data reported. When heated to decomposition it emits very toxic NO_x and Cl^-.

BHO000 ***HR: 3***
BIS(2-CHLOROETHYL)AMINE
mf: $C_2H_5Cl_2N$ mw: 113.95

SAFETY PROFILE: Can explode violently during evaporation of an ethereal solution at 260 mbar from a bath at 80-90°. When heated to decomposition it emits toxic fumes of Cl^- and NO_x. See also CHLORIDES and AMINES.

BHO250 CAS:821-48-7 ***HR: 3***
BIS(2-CHLOROETHYL)AMINE HYDROCHLORIDE
mf: $C_4H_9Cl_2N \cdot ClH$ mw: 178.50

SYNS: BIS(β-CHLOROETHYL)AMINE HYDROCHLORIDE ◇ N,N-BIS(2-CHLOROETHYL)AMINE HYDROCHLORIDE ◇ BIS(2-CHLOROETHYL)AMMONIUM CHLORIDE ◇ 2-CHLORO-N-(2-CHLOROETHYL)ETHANAMINE HYDROCHLORIDE ◇ β,β'-DICHLORODIETHYLAMINE HYDROCHLORIDE ◇ 2,2'-DICHLORO DIETHYLAMINE HYDROCHLORIDE ◇ DI-2-CHLOROETHYLAMINE HYDROCHLORIDE ◇ LEO 72a ◇ NC 26 ◇ NOR-HN2 ◇ NOR-HN2 HYDROCHLORIDE ◇ NOR-LOST HYDROCHLORID (GERMAN) ◇ NORNITROGEN MUSTARD HYDROCHLORIDE ◇ NSC 10873 ◇ SK 555 ◇ TL 161

TOXICITY DATA with REFERENCE
hmn-lym 1 mg/L CRNGDP 5,163,84
hmn-lym 250 μg/L CRNGDP 5,163,84
mmo-sat 50 μg/plate PNASA6 72,979,75
ipr-mus TDLo:2500 μg/kg (11D preg):TER TJADAB 4,141,71
ipr-rat LD50:100 mg/kg ARZNAD 11,143,61
ims-rat LD50:160 mg/kg ZKKOBW 84,227,75
ihl-mus LCLo:1000 mg/m^3/10M NDRC** NDCrc-132,July,42
scu-mus LD50:20 mg/kg JPETAB 91,224,47

CONSENSUS REPORTS: EPA Genetic Toxicology Program. Reported in EPA TSCA Inventory.

SAFETY PROFILE: A poison by inhalation, intraperitoneal, intramuscular, and subcutaneous routes. An experimental teratogen. Human mutation data reported. When heated to decomposition it emits toxic fumes of NH_3, NO_x, and Cl^-.

BHO500 CAS:1215-16-3 ***HR: 3***
4'-(BIS(2-CHLOROETHYL)AMINO)ACETANILIDE
mf: $C_{12}H_{16}Cl_2N_2O$ mw: 275.20

SYN: p-ACETYLAMINOPHENYL DERIVATIVE of NITROGEN MUSTARD ◇ LONIN 3

TOXICITY DATA with REFERENCE
scu-mus TDLo:36 mg/kg (female 7-9D post):TER JEEMAF 9,492,61
ipr-rat LD50:28 mg/kg JMCMAR 8,167,65
ipr-mus LD50:27 mg/kg JMCMAR 8,167,65

SAFETY PROFILE: Poison via intraperitoneal route. An experimental teratogen. When heated to decomposition it emits very toxic fumes of Cl^- and NO_x.

BHP125 CAS:1141-37-3 ***HR: 3***
4-(BIS(2-CHLOROETHYL)AMINO)BENZOIC ACID
mf: $C_{11}H_{13}Cl_2NO_2$ mw: 262.15

SYN: p-(BIS(2-CHLOROETHYL)AMINO)BENZOICACID

TOXICITY DATA with REFERENCE
ipr-rat LD50:96 mg/kg JMCMAR 8,167,65
unr-rat LD50:63 mg/kg NEOLA4 27,261,80
ipr-mus LD50:87 mg/kg JMCMAR 8,167,65

SAFETY PROFILE: Poison by intraperitoneal and possibly other routes. When heated to decomposition it emits toxic fumes of Cl^- and NO_x.

BHP250 CAS:4213-30-3 ***HR: 3***
p-(BIS-(β-CHLOROETHYL)AMINO)BENZYLIDENE MALONONITRILE
mf: $C_{14}H_{13}Cl_2N_3$ mw: 294.20

SYN: NSC-48841

TOXICITY DATA with REFERENCE
ivn-dog LDLo:10 mg/kg CCSUBJ 2,201,65
ivn-mky LDLo:20 mg/kg CCSUBJ 2,201,65

CONSENSUS REPORTS: Cyanide and its compounds are on the Community Right-To-Know List.

SAFETY PROFILE: Poison by intravenous route. When heated to decomposition it emits very toxic fumes of Cl^-, CN^-, and NO_x. See also NITRILES.

BHP500 CAS:63978-55-2 ***HR: 3***
2-(BIS(2-CHLOROETHYL)AMINO)ETHYL VINYL SULFONE
mf: $C_8H_{15}Cl_2NO_2S$ mw: 260.20

SYN: VINYL(β-BIS(β-CHLOROETHYL)AMINO)ETHYLSULFONE

TOXICITY DATA with REFERENCE
scu-mus LD50:9 mg/kg JPETAB 93,1,48
ivn-rbt LD50:2550 μg/kg JPETAB 93,1,48

SAFETY PROFILE: Poison by subcutaneous and intravenous routes. See also SULFONATES. When heated to decomposition it emits very toxic fumes of Cl^-, NO_x, and SO_x.

BHP750 CAS:1492-93-9 ***HR: 3***
4'-(BIS(2-CHLOROETHYL)AMINO)-2-FLUORO ACETANILIDE
mf: $C_{12}H_{15}Cl_2FN_2O$ mw: 293.19

SYN: p-FLUOROACETYLAMINOPHENYL DERIVATIVE of NITROGEN MUSTARD

TOXICITY DATA with REFERENCE
scu-mus TDLo:45 mg/kg (female 7-9D post):TER JEEMAF 9,492,61
ipr-rat LD50:7916 μg/kg JMCMAR 8,167,65
ipr-mus LD50:34 mg/kg JMCMAR 8,167,65

SAFETY PROFILE: Poison by intraperitoneal route. An experimental teratogen. When heated to decomposition it emits very toxic fumes of Cl^-, F^-, and NO_x.

BHQ000 CAS:20982-36-9 ***HR: 3***
2-(BIS(2-CHLOROETHYL)AMINO)HEXAHYDRO-1,3,2-DIAZAPHOSPHORINE- 2-OXIDE
mf: $C_7H_{16}Cl_2N_3OP$ mw: 260.13

SYNS: N,N-BIS-(β-CHLOROETHYL)-N',N''-PROPYLENEPHOSPHORICACIDTRIAMIDE ◇ CYCLIC N,N'-TRIMETHYLENE-N''-BIS(2-CHLOROETHYL)-PHOSPHORIC TRIAMIDE

TOXICITY DATA with REFERENCE
sln-dmg-orl 10 mmol/L DRISAA 41,102,66
ipr-rat LD50:75 mg/kg ARZNAD 11,143,61

SAFETY PROFILE: Poison by intraperitoneal route. Mutation data reported. When heated to decomposition it emits toxic fumes of NO_x and Cl^-.

BHQ250 CAS:66902-62-3 ***HR: 3***
3-(4-(BIS(2-CHLOROETHYL)AMINO)-3-METHOXYPHENYL)ALANINE
mf: $C_{14}H_{20}Cl_2N_2O_3$ mw: 335.26

SYN: p-BIS(2-CHLOROETHYL)AMINO-o-METHOXYPHENYLALANINE

TOXICITY DATA with REFERENCE
orl-hmn TDLo:10 mg/kg/20D-I:GIT XPHPAW 441,185,74
ipr-mus LD50:39 mg/kg SSINAV 13,789,64

SAFETY PROFILE: Poison by intraperitoneal route. Human gastrointestinal effects by ingestion. When heated to decomposition it emits very toxic fumes of Cl^- and NO_x.

BHQ750 CAS:7751-31-7 ***HR: 3***
3-(BIS(2-CHLOROETHYL)AMINOMETHYL)-2-BENZOXAZOLINONE
mf: $C_{12}H_{14}Cl_2N_2O_2$ mw: 289.18

SYN: 3-(BIS-(2-CHLORAETHYL)AMINOMETHYL)BENZOXAZOLON-(2) (GERMAN)

TOXICITY DATA with REFERENCE
ims-rat LD50:8 mg/kg ZKKOBW 84,227,75
ipr-mus LD50:42 mg/kg ZKKOBW 84,227,75

SAFETY PROFILE: Poison by intramuscular and intraperitoneal routes. When heated to decomposition it emits very toxic fumes as Cl^- and NO_x.

BHR500 CAS:10070-95-8 ***HR: 3***
4-(BIS(2-CHLOROETHYL)AMINOMETHYL)-2,3-DIMETHYL-1-PHENYL-3-PYRAZOLIN-5-ONE HYDROCHLORIDE
mf: $C_{16}H_{21}Cl_2N_3O•ClH$ mw: 378.76

SYN: 4-(BIS-(2-CHLORAETHYL)AMINOMETHYL)-1-PHENYL-2,3-DIMETHYLPYRAZOLON HYDROCHLORID (GERMAN)

TOXICITY DATA with REFERENCE
ims-rat LD50:30 mg/kg ZKKOBW 84,227,75
ipr-mus LD50:580 mg/kg ARZNAD 16,634,66

SAFETY PROFILE: Poison by intramuscular route. Moderately toxic by intraperitoneal route. When heated to decomposition it emits very toxic fumes of HCl and NO_x.

BHR750 CAS:2089-46-5 ***HR: 3***
4-(BIS(2-CHLOROETHYL)AMINO)PHENOL
mf: $C_{27}H_{38}N_2O•2BrH$ mw: 568.51

SYNS: 2,6-BIS(1-PIPERIDYLMETHYL)-4-(α,α-DIMETHYLBENZYL)PHENOL DIHYDROBROMIDE ◇ 4-α,α-DIMETHYLBENZYL-α,α′-DIPIPERIDINO-2,6-XYLENOL DIHYDROBROMIDE ◇ 4-(1-METHYL-1-PHENYLETHYL)-2,6-BIS-(1-PIPERIDINYLMETHYL)PHENOL DIHYDROBROMIDE ◇ RO 2-5803 ◇ RYTHMOL

TOXICITY DATA with REFERENCE
ipr-rat LD50:17 mg/kg JMCMAR 8,167,65
orl-mus LD50:330 mg/kg AIPTAK 132,295,61
ipr-mus LD50:15 mg/kg JMCMAR 8,167,65
ivn-mus LD50:30 mg/kg AIPTAK 132,295,61
ims-gpg LD50:48720 μg/kg ARPMAS 313,142,80

SAFETY PROFILE: Poison by ingestion, intravenous, intramuscular and intraperitoneal routes. When heated to decomposition it emits very toxic fumes of NO_x and HBr.

BHS250 CAS:66232-25-5 ***HR: 3***
2-(N,N-BIS(2-CHLOROETHYL)AMINOPHENYL) ACETIC ACID BUTYL ESTER
mf: $C_{16}H_{23}Cl_2NO_2$ mw: 332.1

TOXICITY DATA with REFERENCE
orl-rat LD50:20 mg/kg PCJOAU 12,205,78
orl-mus LD50:15 mg/kg PCJOAU 12,205,78

SAFETY PROFILE: Poison by ingestion. See also ESTERS. When heated to decomposition it emits very toxic fumes of Cl^- and NO_x.

BHS500 CAS:66276-87-7 ***HR: 3***
2-(N,N-BIS(2-CHLOROETHYL)AMINOPHENYL) ACETIC ACID DECYL ESTER
mf: $C_{22}H_{35}Cl_2NO_2$ mw: 416.2

TOXICITY DATA with REFERENCE
orl-rat LD50:150 mg/kg PCJOAU 12,205,78
orl-mus LD50:50 mg/kg PCJOAU 12,205,78

SAFETY PROFILE: Poison by ingestion. See also ESTERS. When heated to decomposition it emits very toxic fumes of Cl^- and NO_x.

BHS750 CAS:66232-30-2 ***HR: 3***
2-(N,N-BIS(2-CHLOROETHYL)AMINOPHENYL) ACETIC ACID OCTADECYL ESTER
mf: $C_{30}H_{51}Cl_2NO_2$ mw: 529.2

TOXICITY DATA with REFERENCE
orl-rat LD50:200 mg/kg PCJOAU 12,205,78
orl-mus LD50:140 mg/kg PCJOAU 12,205,78

SAFETY PROFILE: Poison by ingestion. See also ESTERS. When heated to decomposition it emits very toxic fumes of Cl^- and NO_x.

BHT000 CAS:66232-28-8 ***HR: 3***
2-(N,N-BIS(2-CHLOROETHYL)AMINOPHENYL) ACETIC ACID TETRADECYL ESTER
mf: $C_{26}H_{43}Cl_2NO_2$ mw: 472.2

TOXICITY DATA with REFERENCE
cyt-rat:oth 150 mg/L/24H-C TXAPA9 22,355,72
orl-rat LD50:46 mg/kg PCJOAU 12,205,78
orl-mus LD50:10 mg/kg PCJOAU 12,205,78

SAFETY PROFILE: Poison by ingestion. Mutation data reported. See also ESTERS. When heated to decomposition it emits very toxic fumes of Cl^- and NO_x.

BHT250 CAS:342-95-0 ***HR: 3***
3-(o-(BIS-(β-CHLOROETHYL)AMINO)PHENYL)-dl-ALANINE
mf: $C_{13}H_{18}Cl_2N_2O_2$ mw: 305.23

SYNS: CB 1729 ◇ o-DI-2-CHLOROETHYLAMINO-dl-PHENYLALANINE ◇ FDA 0109 ◇ MEROPHAN ◇ o-MEROPHAN ◇ NSC-57199 ◇ ORTHOPHENYLALANINE MUSTARD ◇ (±)-o-PHENYLALANINE MUSTARD ◇ o-PHENYLALANINE MUSTARD ◇ o-dl-SARCOLYSIN

TOXICITY DATA with REFERENCE
ipr-rat LD50:3510 μg/kg BCPCA6 13,969,64
scu-mus LD50:18480 μg/kg NICSP* JAN86
ivn-dog LDLo:190 μg/kg CCSUBJ 2,201,65
ivn-mky LDLo:380 μg/kg CCSUBJ 2,201,65

SAFETY PROFILE: Deadly poison by intraperitoneal,

subcutaneous, and intravenous routes. When heated to decomposition it emits very toxic fumes of Cl^- and NO_x.

BHT750 CAS:531-76-0 ***HR: 3***
dl-3-(p-(BIS(2-CHLOROETHYL)AMINO)PHENYL)ALANINE
mf: $C_{13}H_{18}Cl_2N_2O_2$ mw: 305.23

SYNS: 4-(BIS(2-CHLOROETHYL)AMINO)-dl-PHENYLALANINE ◇ 3-(p-(BIS(2-CHLOROETHYL)AMINO)PHENYL)ALANINE ◇ CB-3307 ◇ p-DI-(2-CHLORAETHYL)-AMINO-dl-PHENYL-ALANIN (GERMAN) ◇ p-DI(2-CHLOROETHYL)AMINO-dl-PHENYLALANINE ◇ MERFALAN ◇ MERPHALAN ◇ o-MERPHALAN ◇ NCI-C04944 ◇ NSC-14210 ◇ PHENYLALANIN-LOST (GERMAN) ◇ dl-PHENYLALANINE MUSTARD ◇ SAKOLYSIN (GERMAN) ◇ SARCOCLORIN ◇ dl-SARCOLYSIN ◇ dl-SARCOLYSINE

TOXICITY DATA with REFERENCE
mmo-omi 10 mmol/L MUREAV 23,5,74
pic-omi 5 mmol/L MUREAV 1,355,64
ipr-rat TDLo:5 mg/kg (female 6D post):TER DKBSAS 171,801,66
ipr-rat TDLo:8 mg/kg (female 4D post):REP DKBSAS 171,801,66
ipr-mus TDLo:98 mg/kg/26W-I:ETA CANCAR 40S,1935,77
orl-rat LD50:105 mg/kg AICCA6 20,144,64
ipr-rat LD50:18 mg/kg DKBSAS 171,801,66
ivn-rat LD50:25 mg/kg ARZNAD 8,340,58
ice-rat LD50:250 μg/kg JPPMAB 18,760,66
orl-mus LD50:35 mg/kg XPHPAW 441,165,74
ipr-mus LD50:26 mg/kg ARZNAD 16,634,66

CONSENSUS REPORTS: IARC Cancer Review: Group 2B IMEMDT 7,56,87; Animal Limited Evidence IMEMDT 9,167,75; NCI Carcinogenesis Studies (ipr); Clear Evidence: mouse CANCAR 40,1935,77; No Evidence: rat CANCAR 40,1935,77

SAFETY PROFILE: Suspected carcinogen with experimental tumorigenic data. A poison by ingestion, intraperitoneal, intravenous, and intracerebral routes. An experimental teratogen. Other experimental reproductive effects. Mutation data reported. An antineoplastic agent. When heated to decomposition it emits very toxic fumes of Cl^- and NO_x.

BHU500 CAS:4213-34-7 ***HR: 3***
3-(m-(BIS(β-CHLOROETHYL)AMINO)PHENYL)-dl-ALANINE HYDROCHLORIDE
mf: $C_{13}H_{18}Cl_2N_2O_2$•ClH mw: 341.69

SYNS: METAPHENYLALANINE MUSTARD ◇ NCS-27381

TOXICITY DATA with REFERENCE
ivn-dog LDLo:430 μg/kg CCSUBJ 2,201,65
ivn-mky LDLo:860 μg/kg CCSUBJ 2,201,65

SAFETY PROFILE: Poison by intravenous route. When heated to decomposition it emits very toxic fumes of Cl^-, NO_x and HCl.

BHU750 CAS:4213-32-5 ***HR: 3***
3-(p-(BIS(β-CHLOROETHYL)AMINO)PHENYL)-d-ALANINE HYDROCHLORIDE
mf: $C_{13}H_{18}Cl_2N_2O_2$•ClH mw: 341.69

SYNS: 4-(BIS(2-CHLOROETHYL)AMINO)-d-PHENYLALANINE MONOHYDROCHLORIDE ◇ NSC-35051 ◇ PHENYLALANINE MUSTARD ◇ d-PHENYLALANINE MUSTARD

TOXICITY DATA with REFERENCE
ivn-dog LDLo:2 mg/kg CCSUBJ 2,201,65
ivn-mky LDLo:2 mg/kg CCSUBJ 2,201,65

SAFETY PROFILE: An intravenous poison. When heated to decomposition it emits very toxic fumes of Cl^-, NO_x, and HCl.

BHV000 CAS:1465-26-5 ***HR: 3***
3-(p-(BIS(β-CHLOROETHYL)AMINO)PHENYL)-dl-ALANINE HYDROCHLORIDE
mf: $C_{13}H_{18}Cl_2N_2O_2$•ClH mw: 341.69

SYNS: ALKERAN (RUSSIAN) ◇ 4-(BIS(2-CHLOROETHYL)AMINO)-dl-PHENYLALANINE MONOHYDROCHLORIDE ◇ CB 3008 ◇ MELPHALAN (RUSSIAN) ◇ MERPHALAN HYDROCHLORIDE ◇ NCS-14210 ◇ dl-PHENYLALANINE MUSTARD HYDROCHLORIDE ◇ dl-SARCOLYSINE HYDROCHLORIDE ◇ SARCOLYSIN HYDROCHLORIDE ◇ SARKOKLORIN ◇ SKI 21739

TOXICITY DATA with REFERENCE
mmo-sat 100 μg/plate KHFZAN 16(10),11,82
mma-sat 100 μg/plate KHFZAN 16(10),11,82
dnd-esc 20 μmol/L MUREAV 89,95,81
pic-omi 5 mmol/L MUREAV 1,355,64
otr-mus:emb 1 μmol/L CBINA8 38,75,81
ivn-dog LDLo:430 μg/kg CCSUBJ 2,201,65
ivn-mky LDLo:860 μg/kg CCSUBJ 2,201,65

SAFETY PROFILE: Deadly poison by intravenous route. Human mutation data reported. When heated to decomposition it emits very toxic fumes of Cl^- and NO_x.

BHV250 CAS:3223-07-2 ***HR: 3***
l-3-(p-(BIS(2-CHLOROETHYL)AMINO)PHENYL)ALANINE MONOHYDROCHLORIDE
mf: $C_{13}H_{18}Cl_2N_2O_2$•ClH mw: 341.69

SYNS: ALANINE NITROGEN MUSTARD ◇ CB 3025 ◇ MELPHALAN HYDROCHLORIDE ◇ NSC-8806 ◇ l-PHENYLALANINE MUSTARD HYDROCHLORIDE ◇ l-SARCOLYSINE HYDROCHLORIDE

TOXICITY DATA with REFERENCE
ipr-mus TDLo:1300 μg/kg/4W:CAR JNCIAM 36,915,66
orl-hmn TDLo:1200 μg/kg/5D-I:GIT CCROBU 57,369,73
ivn-dog LDLo:430 μg/kg CCSUBJ 2,201,65
ivn-mky LDLo:430 μg/kg CCSUBJ 2,201,65

SAFETY PROFILE: Deadly poison by intravenous

route. Human systemic effects by ingestion: nausea and vomiting. Questionable carcinogen with experimental carcinogenic data. When heated to decomposition it emits very toxic fumes of Cl^-, NO_x, and HCl.

BHV500 CAS:1233-89-2 ***HR: 3***
p-(BIS(2-CHLOROETHYL)AMINO)PHENYL BENZOATE
mf: $C_{17}H_{17}Cl_2NO_2$ mw: 338.25

SYN: p-(BIS(2-CHLOROETHYL)AMINO)PHENOLBENZOATE

TOXICITY DATA with REFERENCE
ipr-rat LD50:20 mg/kg JMCMAR 12,491,69
ipr-mus LD50:18500 μg/kg JMCMAR 12,491,69

SAFETY PROFILE: Poison by intraperitoneal route. See also ESTERS. When heated to decomposition it emits very toxic fumes of Cl^- and NO_x.

BHV750 CAS:22953-53-3 ***HR: 3***
p-(BIS(2-CHLOROETHYL)AMINO)PHENYL-p-BROMOBENZOATE
mf: $C_{17}H_{16}BrCl_2NO_2$ mw: 417.15

SYN: p-(BIS(2-CHLOROETHYL)AMINO)PHENOL-p-BROMOBENZOATE

TOXICITY DATA with REFERENCE
ipr-rat LD50:27 mg/kg JMCMAR 12,491,69
ipr-mus LD50:19 mg/kg JMCMAR 12,491,69

SAFETY PROFILE: Poison by intraperitoneal route. When heated to decomposition it emits very toxic fumes of Br^-, Cl^-, and NO_x.

BHW000 CAS:22953-54-4 ***HR: 3***
p-(BIS(2-CHLOROETHYL)AMINO)PHENYL-m-CHLOROBENZOATE
mf: $C_{17}H_{16}Cl_3NO_2$ mw: 372.69

SYN: p-(BIS(2-CHLOROETHYL)AMINO)PHENOL-m-CHLOROBENZOATE

TOXICITY DATA with REFERENCE
ipr-rat LD50:42 mg/kg JMCMAR 12,491,69
ipr-mus LD50:21 mg/kg JMCMAR 12,491,69

SAFETY PROFILE: Poison by intraperitoneal route. When heated to decomposition it emits very toxic fumes of Cl^- and NO_x.

BHW250 CAS:21667-01-6 ***HR: 3***
p-(BIS(2-CHLOROETHYL)AMINO)PHENYL-2,6-DIMETHYLBENZOATE
mf: $C_{19}H_{21}Cl_2NO_2$ mw: 366.31

SYN: p-(BIS(2-CHLOROETHYL)AMINO)PHENOL-2,6-DIMETHYLBENZOATE

TOXICITY DATA with REFERENCE
ipr-rat LD50:290 mg/kg JMCMAR 12,491,69
ipr-mus LD50:500 mg/kg JMCMAR 12,491,69

SAFETY PROFILE: Poison by intraperitoneal route. When heated to decomposition it emits very toxic fumes of Cl^- and NO_x.

BHW300 CAS:5185-77-3 ***HR: 3***
2-(p-BIS(2-CHLOROETHYL)AMINOPHENYL)-1,3,2-DITHIARSENOLANE
mf: $C_{12}H_{16}AsCl_2NS_2$ mw: 384.23

SYN: 1,3,2-DITHIARSENOLANE,2-(p-BIS(2-CHLOROETHYL)AMINOPHENYL)-

TOXICITY DATA with REFERENCE
ipr-mus LD50:15 mg/kg JMCMAR 9,221,66

OSHA PEL: TWA 0.5 mg(As)/m^3

SAFETY PROFILE: Poison by intraperitoneal route. When heated to decomposition it emits toxic fumes of NO_x, SO_x, As, and Cl^-.

BHW500 CAS:4465-92-3 ***HR: 3***
4-(p-BIS(β-CHLOROETHYLAMINO)PHENYLETHYLAMINO)-7-CHLOROQUINOLINE MONOHYDROCHLORIDE

SYN: NSC-50982

TOXICITY DATA with REFERENCE
ivn-dog LDLo:11 mg/kg CCSUBJ 2,202,65
ivn-mky LDLo:11 mg/kg CCSUBJ 2,202,65

SAFETY PROFILE: Poison by intravenous route. When heated to decomposition it emits very toxic fumes of Cl^- and NO_x.

BHX250 CAS:35849-41-3 ***HR: 3***
l-3-(p-(BIS(2-CHLOROETHYL)AMINO)PHENYL)-N-FORMYLALANINE
mf: $C_{14}H_{18}Cl_2N_2O_3$ mw: 333.24

SYN: N-FORMYL-l-p-SARCOLYSIN

TOXICITY DATA with REFERENCE
orl-hmn TDLo:3 mg/kg:GIT XPHPAW 441,9,74
orl-rat LD50:700 mg/kg XPHPAW 441,9,74
ipr-rat LD50:80 mg/kg XPHPAW 441,9,74
orl-mus LD50:730 mg/kg XPHPAW 441,9,74
ipr-mus LD50:242 mg/kg NCISA* PH-43-63-1132

SAFETY PROFILE: Poison by intraperitoneal route. Moderately toxic by ingestion. Human gastrointestinal effects by ingestion. When heated to decomposition it emits very toxic fumes of Cl^- and NO_x.

BHY500 CAS:857-95-4 ***HR: 3***
o-(4-(BIS(2-CHLOROETHYL)AMINO)PHENYL-dl-TYROSINE
mf: $C_{19}H_{22}Cl_2N_2O_3$ mw: 397.26

SYN: PHENTYRIN

TOXICITY DATA with REFERENCE
orl-rat LD50:620 mg/kg FATOAO 43,100,80
ipr-rat LD50:115 mg/kg FATOAO 43,100,80
ivn-rat LD50:62 mg/kg FATOAO 43,100,80
orl-mus LD50:360 mg/kg FATOAO 43,100,80
ipr-mus LD50:110 mg/kg FATOAO 43,100,80
ivn-mus LD50:30 mg/kg FATOAO 43,100,80

SAFETY PROFILE: A poison by ingestion, intravenous, and intraperitoneal routes. When heated to decomposition it emits very toxic fumes of Cl^- and NO_x.

BHY625 CAS:64508-90-3 ***HR: 3***
5-(p-(BIS(2-CHLOROETHYL)AMINO)PHENYL)VALERIC ACID
mf: $C_{15}H_{21}Cl_2NO_2$ mw: 318.27

SYNS: 4-(BIS(2-CHLOROETHYL)AMINO)-BENZENEPENTANOIC ACID (9CI) ◇ 5-(4-BIS(2-CHLOROETHYL)AMINOPHENYL)PENTANOIC ACID ◇ CB 1356 ◇ p-N,N-DI-(2-CHLOROETHYL)AMINOPHENYLVALERIC ACID

TOXICITY DATA with REFERENCE
sln-dmg-par 5 mmol/L GENRA8 1,173,60
sln-dmg-unr 10 μmol/L ANYAA9 160,228,69
ipr-rat LDLo:50 mg/kg BCPCA6 5,192,60

CONSENSUS REPORTS: EPA Genetic Toxicology Program.

SAFETY PROFILE: Poison by intraperitoneal route. Mutation data reported. When heated to decomposition it emits toxic fumes of Cl^- and NO_x.

BHY750 CAS:63815-37-2 ***HR: 3***
β-(BIS(2-CHLOROETHYLAMINO))PROPIONITRILE
mf: $C_7H_{13}Cl_2N_3$ mw: 210.13

SYN: USAF UCTL-958

TOXICITY DATA with REFERENCE
ihl-mus LCLo:660 mg/m^3/10M NDRC** No.9-4-1-19,44
ipr-mus LD50:10 mg/kg NTIS** AD277-689

SAFETY PROFILE: Poison by inhalation, intraperitoneal route. When heated to decomposition it emits very toxic fumes of Cl^- and NO_x.

BHZ000 CAS:38915-00-3 ***HR: 3***
9-((3-(BIS(2-CHLOROETHYL)AMINO)PROPYL)AMINO)ACRIDINE DIHYDROCHLORIDE
mf: $C_{20}H_{23}Cl_2N_3 \cdot 2ClH$ mw: 449.28

SYN: ICR 220

TOXICITY DATA with REFERENCE
msc-ham:ovr 1 g/L CNREA8 39,4875,79
ipr-mus LD20:1 mg/kg JMCMAR 15,739,72

SAFETY PROFILE: Mutation data reported. Poison by intraperitoneal route. When heated to decomposition it emits very toxic fumes of HCl, NO_x, and Cl^-.

BIA000 CAS:4213-40-5 ***HR: 3***
o-(4-BIS(β-CHLOROETHYL)AMINO-o-TOLYLAZO)BENZOIC ACID
mf: $C_{18}H_{19}Cl_2N_3O_2$ mw: 380.30

SYN: NSC-16498

TOXICITY DATA with REFERENCE
sln-dmg-par 35 μg BCPCA6 5,206,60
ipr-rat LD50:45 mg/kg JNCIAM 50,243,73
ivn-dog LDLo:950 μg/kg CCSUBJ 2,202,65
ivn-mky LDLo:950 μg/kg CCSUBJ 2,202,65

SAFETY PROFILE: Poison by intraperitoneal and intravenous routes. Mutation data reported. When heated to decomposition it emits very toxic fumes of Cl^- and NO_x.

BIA250 CAS:66-75-1 ***HR: 3***
5-(BIS(2-CHLOROETHYL)AMINO)URACIL
mf: $C_8H_{11}Cl_2N_3O_2$ mw: 252.12

SYNS: AMINOURACIL MUSTARD ◇ 5-(BIS(2-CHLOROETHYL)-AMINO)-2,4(1H,3H)PYRIMIDINEDIONE ◇ 5-N,N-BIS(2-CHLOROETHYL)AMINOURACIL ◇ CB-4835 ◇ CHLORETHAMINACIL ◇ DEMETHYLDOPAN ◇ DESMETHYLDOPAN ◇ 5-(DI-(β-CHLOROETHYL)AMINO)URACIL ◇ 5-(DI-2-CHLOROETHYL)AMINOURACIL ◇ 2,6-DIHYDROXY-5-BIS(2-CHLOROETHYL)AMINOPYRAMIDINE ◇ ENT 50,439 ◇ NCI-C04820 ◇ NORDOPAN ◇ NSC-34462 ◇ RCRA WASTE NUMBER U237 ◇ SK-19849 ◇ U-8344 ◇ URACILLOST ◇ URACILMOSTAZA ◇ URACIL MUSTARD ◇ URAMUSTIN ◇ URAMUSTINE

TOXICITY DATA with REFERENCE
mmo-sat 125 μg/plate JNCIAM 62,893,79
msc-mus:lym 150 μg/L/2H MUREAV 59,61,79
ipr-rat TDLo:1000 μg/kg/26W-I:NEO RRCRBU 52,1,75
ipr-mus TDLo:240 μg/kg/4W:CAR JNCIAM 36,915,66
ipr-mus TD:30 mg/kg/39W-I:CAR SCIEAS 147,1443,65
orl-rat LD50:3550 μg/kg NYKZAU 60,413,64
ipr-rat LD50:1250 μg/kg ADTEAS 3,181,68
ipr-mus LDLo:3 mg/kg TXAPA9 23,288,72

CONSENSUS REPORTS: IARC Cancer Review: Group 2B IMEMDT 7,370,87; Animal Sufficient Evidence IMEMDT 9,235,75; NCI Carcinogenesis Studies (ipr); Clear Evidence: mouse, rat RRCRBU 52,1,75. EPA Genetic Toxicology Program.

SAFETY PROFILE: Suspected carcinogen with experimental carcinogenic and neoplastigenic data. A deadly poison by ingestion and intraperitoneal routes. Mutation data reported. When heated to decomposition it emits very toxic fumes of Cl^- and NO_x.

BIA300 CAS:5185-71-7 *HR: 3*
N,N-BIS(2-CHLOROETHYL)-p-ARSANILIC ACID
mf: $C_{10}H_{14}AsCl_2NO_3$ mw: 342.07

SYN: p-ARSANILIC ACID, N,N-BIS(2-CHLOROETHYL)-

TOXICITY DATA with REFERENCE
ipr-mus LD50:8789 μg/kg JMCMAR 9,221,66

OSHA PEL: TWA 0.5 mg(As)/m^3

SAFETY PROFILE: Poison by intraperitoneal route. When heated to decomposition it emits toxic fumes of NO_x, As, and Cl^-.

BIA750 CAS:55-51-6 *HR: 3*
N,N-BIS(2-CHLOROETHYL)BENZYLAMINE
mf: $C_{11}H_{15}Cl_2N$ mw: 232.17

SYNS: BENZYLBIS(β-CHLOROETHYL)AMINE ◇ BENZYL NORMECHLORETHAMINE ◇ N,N-BIS(2-CHLOROETHYL)BENZENEMETHANAMINE ◇ BIS(2-CHLOROETHYL)BENZYLAMINE ◇ DCBA ◇ DI-(2-CHLOROETHYL)BENZYLAMINE ◇ TL 965

TOXICITY DATA with REFERENCE
dni-mus-ivg 5000 ppm JIDEAE 62,378,74
mmo-asn 2500 μmol/L MUREAV 14,115,72
unr-rat LD50:10 mg/kg PHBUA9 1,297,53
scu-mus LDLo:80 mg/kg NDRC** No.9-4-1-9,43

CONSENSUS REPORTS: EPA Genetic Toxicology Program.

SAFETY PROFILE: Poison by ingestion and possibly other routes. Mutation data reported. When heated to decomposition it emits very toxic fumes of Cl^- and NO_x. See also AROMATIC AMINES.

BIB250 CAS:55112-89-5 *HR: 3*
N,N-BIS(2-CHLOROETHYL)BUTYLAMINE HYDROCHLORIDE
mf: $C_8H_{17}Cl_2N \cdot ClH$ mw: 234.62

SYNS: BUTYLBIS(β-CHLOROETHYL)AMINE HYDROCHLORIDE ◇ N-BUTYL-BIS(2-CHLOROETHYLAMINE) HYDROCHLORIDE ◇ TL 513 HYDROCHLORIDE

TOXICITY DATA with REFERENCE
ipr-mus LD50:4890 μg/kg CANCAR 2,1055,49
scu-mus LDLo:2 mg/kg NTIS** PB158-507

SAFETY PROFILE: Poison by intraperitoneal and subcutaneous routes. When heated to decomposition it emits very toxic fumes of NO_x and HCl.

BIB750 CAS:3597-20-4 *HR: D*
N,N-BIS(2-CHLOROETHYL)-p-CHLOROBENZYLAMINE HYDROCHLORIDE
mf: $C_{11}H_{14}Cl_3N \cdot ClH$ mw: 303.07

SYN: p-CHLORO-DI-(2-CHLOROETHYL)BENZYLAMINEHYDROCHLORIDE

TOXICITY DATA with REFERENCE
mmo-asn 2500 μmol/L SOGEBZ 6,220,70
ipr-mus LD50:123 mg/kg NCISA* PH-43-63-1132

SAFETY PROFILE: Mutation data reported. When heated to decomposition it emits very toxic fumes as Cl^- and NO_x.

BIC325 *HR: 3*
1,4-BIS(2-CHLOROETHYL)-1,4-DIAZONIABICYCLO(2.2.1)HEPTANE (Z)-2-BUTENEDIOATE (1:2)
mf: $C_9H_{18}Cl_2N_2 \cdot 2C_4H_4O_4$ mw: 457.35

SYN: NSC 262666

TOXICITY DATA with REFERENCE
orl-mus LD50:746 mg/kg NCISP* JAN86
ipr-mus LD50:210 mg/kg NCISP* JAN86
ivn-mus LD50:73290 μg/kg NCISP* JAN86

SAFETY PROFILE: Poison by intraperitoneal and intravenous routes. Moderately toxic by ingestion. When heated to decomposition it emits toxic fumes of Cl^- and NO_x.

BIC500 CAS:63918-36-5 *HR: 3*
N,N'-BIS(2-CHLOROETHYL)-N,N'-DIETHYLETHYLENEDIAMINE DIHYDROCHLORIDE
mf: $C_{10}H_{22}Cl_2N_2 \cdot 2ClH$ mw: 314.16

SYN: N,N'-ETHYL-N,N'-(β-CHLOROETHYL)ETHYLENEDIAMINEDIHYDROCHLORIDE

TOXICITY DATA with REFERENCE
ipr-rat LD50:2300 μg/kg JPETAB 100,398,50
ipr-mus LD50:3141 μg/kg JPETAB 94,249,48

SAFETY PROFILE: Deadly poison by intraperitoneal route. When heated to decomposition it emits very toxic fumes of Cl^- and NO_x.

BIC600 CAS:4213-41-6 *HR: 2*
N,N-BIS(2-CHLOROETHYL)-2,3-DIMETHOXYANILINE
mf: $C_{12}H_{17}Cl_2NO_2$ mw: 278.20

SYNS: ANILINE,N,N-BIS(2-CHLOROETHYL)-2,3-DIMETHOXY- ◇ 2,3-DIMETHOXYANILINE MUSTARD ◇ NSC-18439

TOXICITY DATA with REFERENCE
ipr-mus TDLo:27 mg/kg/4W:CAR JNCIAM 36,915,66

SAFETY PROFILE: Questionable carcinogen with experimental carcinogenic data. When heated to decomposition it emits toxic fumes of NO_x.

BID000 CAS:6986-48-7 ***HR: 3***
BIS(α-CHLOROETHYL) ETHER
mf: $C_4H_8Cl_2O$ mw: 143.02

SYN: 1,1'-OXYBIS(1-CHLOROETHANE)

TOXICITY DATA with REFERENCE
scu-mus TDLo:648 mg/kg/54W-I:ETA JNCIAM 48,1431,72

SAFETY PROFILE: Questionable carcinogen with experimental tumorigenic data. See also ETHERS. When heated to decomposition it emits toxic fumes of Cl^-.

BID250 CAS:538-07-8 ***HR: 3***
BIS(2-CHLOROETHYL)ETHYLAMINE
mf: $C_6H_{13}Cl_2N$ mw: 170.10

SYNS: 2,2'-DICHLOROTRIETHYLAMINE ◇ ETHYLBIS(β-CHLOROETHYL)AMINE ◇ ETHYLBIS(2-CHLOROETHYL)AMINE ◇ ETHYL-S ◇ HN1 ◇ TL 329 ◇ TL 1149

TOXICITY DATA with REFERENCE
orl-rat LD50:2500 μg/kg NTIS** PB158-507
ihl-rat LC50750 mg/m^3/10M NTIS** PB158-508
ihl-rat LC50:1600 mg/m^3 NDRC** 11,91,43
skn-rat LD50:17 mg/kg JPETAB 91,224,47
ivn-rat LD50:500 μg/kg NTIS** PB158-507
ihl-mus LC50:900 mg/m^3/10M NTIS** PB158-508
skn-mus LD50:13 mg/kg JPETAB 91,224,47
ipr-mus LDLo:1030 μg/kg NTIS** PB158-507
scu-mus LDLo:1100 μg/kg NTIS** PB158-507
ihl-dog LC50:800 mg/m^3/10M NTIS** PB158-508
ihl-mky LC50:1500 mg/m^3/10M NTIS** PB158-508
ihl-cat LC50:400 mg/m^3/10M NTIS** PB158-508
ihl-rbt LC50:900 mg/m^3/20M NTIS** PB158-508
ihl-gpg LC50:1500 mg/m^3/30M NTIS** PB158-508
ihl-dom LC50:1500 mg/m^3/10M NTIS** PB158-508
skn-dog LDLo:40 mg/kg NTIS** PB158-507
skn-rbt LD50:15 mg/kg JPETAB 91,224,47
ivn-rbt LDLo:2 mg/kg NTIS** PB158-507

CONSENSUS REPORTS: Reported in EPA TSCA Inventory. EPA Extremely Hazardous Substances List.

SAFETY PROFILE: Deadly poison by inhalation, skin contact, ingestion, intravenous, subcutaneous, and intraperitoneal routes. When heated to decomposition it emits very toxic fumes of Cl^- and NO_x.

BID750 CAS:111-91-1 ***HR: 3***
BIS(β-CHLOROETHYL)FORMAL
mf: $C_5H_{10}Cl_2O_2$ mw: 173.05

PROP: Liquid. Bp: 217.5°, flash p: 230°F (OC), d: 1.23, vap d: 5.9.

SYNS: BIS(2-CHLOROETHOXY)METHANE ◇ BIS(2-CHLOROETHYL)FORMAL ◇ DICHLOROETHYL FORMAL ◇ DI-2-CHLOROETHYL FORMAL ◇ FORMALDEHYDE BIS(β-CHLOROETHYL) ACETAL ◇ 1,1'-(METHYLENEBIS(OXY)BIS(2-CHLOROETHANE) ◇ RCRA WASTE NUMBER U024

TOXICITY DATA with REFERENCE
skn-rbt 10 mg/24H open JIHTAB 30,63,48
eye-rbt 500 mg AJOPAA 29,1363,46
orl-rat LD50:65 mg/kg JIHTAB 30,63,48
ihl-rat LCLo:62 ppm/4H JIHTAB 31,343,49
skn-gpg LD50:170 mg/kg JIHTAB 30,63,48

CONSENSUS REPORTS: Reported in EPA TSCA Inventory.

SAFETY PROFILE: Poison by ingestion, inhalation, and skin contact. A skin and eye irritant. Combustible when exposed to heat or flame. Incompatible with oxidizers. To fight fire, use alcohol foam, foam, CO_2, dry chemical. When heated to decomposition it emits toxic fumes of Cl^-. See also CHLORIDES.

BIE250 CAS:51-75-2 ***HR: 3***
BIS(β-CHLOROETHYL)METHYLAMINE
mf: $C_5H_{11}Cl_2N$ mw: 156.07

PROP: Dark liquid. Mp: 1° @ 10 mm, d: 1.09 @ 25°, vap press: 0.17 mm @ 25°, vap d: 5.9.

SYNS: BIS(2-CHLOROETHYL)METHYLAMINE ◇ N,N-BIS(2-CHLOROETHYL)METHYLAMINE ◇ CARYOLYSIN ◇ CHLORMETHINE ◇ CLORAMIN ◇ DICHLOR AMINE ◇ DICHLOREN (GERMAN) ◇ β,β'-DICHLORODIETHYL-N-METHYLAMINE ◇ DI(2-CHLOROETHYL)METHYLAMINE ◇ 2,2'-DICHLORO-N-METHYLDIETHYLAMINE ◇ EMBICHIN ◇ ENT 25,294 ◇ HN2 ◇ MBA ◇ MECHLORETHAMINE ◇ N-METHYL-BIS-CHLORAETHYLAMIN (GERMAN) ◇ METHYLBIS(β-CHLOROETHYL)AMINE ◇ N-METHYL-BIS(β-CHLOROETHYL)AMINE ◇ N-METHYL-BIS(2-CHLOROETHYL)AMINE (MAK) ◇ N-METHYL-2,2'-DICHLORODIETHYLAMINE ◇ METHYLDI(2-CHLOROETHYL)AMINE ◇ N-METHYL-LOST ◇ MUSTARGEN ◇ MUSTINE ◇ MUTAGEN ◇ NITROGEN MUSTARD ◇ N-LOST (GERMAN) ◇ NSC 762 ◇ TL 146

TOXICITY DATA with REFERENCE
eye-rbt 400 μg SEV AJOPAA 29,1553,46
eye-rbt 20 μg/30M INOPAO 15,308,76
mmo-sat 40 μg/plate CNREA8 37,2209,77
dnr-bcs 10 μg/plate TAKHAA 44,96,85
dns-hmn:fbr 160 μg/L TXCYAC 21,151,81
sce-hmn:lym 6250 ng/L CRNGDP 5,1637,84
ipr-mus TDLo:4 mg/kg (female 6-7D post):REP JPETAB 101,362,51
scu-rat TDLo:1 mg/kg (female 12D post):TER JEEMAF 18,215,67
skn-man TDLo:153 mg/kg/3Y-C:CAR ADVEA4 58,421,78

skn-wmn TDLo:5840 mg/kg/8Y-I:CAR ADVEA4 58,421,78
skn-mus TDLo:60 mg/kg/14W-I:ETA BJCAAI 9,177,55
ivn-mus TDLo:10 mg/kg/42D-I:NEO CNREA8 25,20,65
orl-rat LD50:10 mg/kg NTIS** PB158-507
ihl-rat LC50:600 mg/m^3/2M NTIS** PB158-508
skn-rat LD50:12 mg/kg FAATDF 5,S160,85
ivn-rat LD50:1100 μg/kg NTIS** PB158-507
orl-mus LD50:10 mg/kg NTIS** PB158-507
ihl-mus LC50:1500 mg/m^3/30M NTIS** PB158-508
skn-mus LD50:29 mg/kg JPETAB 91,224,47
ihl-dog LC50:2 g/m^3/10M NTIS** PB158-508
ihl-rbt LC50:1 g/m^3/5M NTIS** PB158-508
skn-rbt LD50:12 mg/kg NTIS** PB158-507

CONSENSUS REPORTS: EPA Genetic Toxicology Program. Reported in EPA TSCA Inventory. EPA Extremely Hazardous Substances List. Community Right-To-Know List.

DFG MAK: Human Carcinogen.

SAFETY PROFILE: Confirmed human carcinogen producing skin tumors by skin contact. Experimental carcinogenic, tumorigenic, and neoplastigenic data. A deadly poison by inhalation, ingestion, skin contact, and most other routes. Experimental teratogenic and reproductive effects. A powerful skin and eye irritant. Human mutation data reported. It has been used as a blistering agent in chemical warfare. When heated to decomposition it emits very toxic fumes of Cl^- and NO_x.

BIE500 CAS:55-86-7 ***HR: 3***
BIS(2-CHLOROETHYL)METHYLAMINE HYDROCHLORIDE
mf: $C_5H_{11}Cl_2N•ClH$ mw: 192.53

SYNS: ANTIMIT ◇ AZOTOYPERITE ◇ C 6866 ◇ CAROLYSINE ◇ CARYOLYSINE ◇ CARYOLYSINE HYDROCHLORIDE ◇ CHLORAMIN ◇ CHLORAMINE ◇ CHLORAMIN HYDROCHLORIDE ◇ CHLORETHAMINE ◇ CHLORETHAZINE ◇ CHLORMETHINE HYDROCHLORIDE ◇ CHLORMETHINUM ◇ 2-CHLORO-N-(2-CHLOROETHYL)-N-METHYLETHANAMINEHYDROCHLORIDE ◇ DEMA ◇ DICHLOREN ◇ DICHLOREN HYDROCHLORIDE ◇ β,β'-DICHLORODIETHYL-N-METHYLAMINE HYDROCHLORIDE ◇ DI(2-CHLOROETHYL)METHYLAMINE HYDROCHLORIDE ◇ 1,5-DICHLORO-3-METHYL-3-AZAPENTANE HYDROCHLORIDE ◇ 2,2'-DICHLORO-N-METHYLDIETHYLAMINE HYDROCHLORIDE ◇ DIMITAN ◇ EMBECHINE ◇ EMBICHIN ◇ EMBICHIN HYDROCHLORIDE ◇ EMBIKHINE ◇ ERASOL ◇ ERASOL HYDROCHLORIDE ◇ ERASOL-IDO ◇ HN2.HCl ◇ HN2 HYDROCHLORIDE ◇ KLORAMIN ◇ N-LOST ◇ MBA HYDROCHLORIDE ◇ MEBICHLORAMINE ◇ MECHLORETHAMINE HYDROCHLORIDE ◇ MERCHLORETHANAMINE ◇ METHYLBIS(β-CHLOROETHYL)AMINE HYDROCHLORIDE ◇ N-METHYL-BIS-β-CHLORETHYLAMINE HYDROCHLORIDE ◇ METHYLBIS(2-CHLOROETHYL)AMINE HYDROCHLORIDE ◇ N-METHYLBIS(2-CHLOROETHYL)AMINE HYDROCHLORIDE ◇ N-METHYL-2,2'-DICHLORODIETHYLAMINE HYDROCHLORIDE ◇ N-METHYL-DI-2-CHLOROETHYLAMINE HYDROCHLORIDE ◇ METHYLDI(β-CHLOROETHYL)AMINE HYDROCHLORIDE ◇ METHYLDI(2-CHLOROETHYL)AMINE HYDROCHLORIDE ◇ MITOXINE ◇ N-MUSTARD (GERMAN) ◇ MUSTARGEN ◇ MUSTARGEN HYDROCHLORIDE ◇ MUSTINE HYDROCHLOR ◇ MUSTINE HYDROCHLORIDE ◇ NCI-C56382 ◇ NITOL ◇ NITOL "TAKEDA" ◇ NITROGEN MUSTARD HYDROCHLORIDE ◇ NITROGRANULOGEN ◇ NITROGRANULOGEN HYDROCHLORIDE ◇ NSC 762 ◇ NSC-762 HYDROCHLORIDE ◇ PLIVA ◇ STICKSTOFFLOST ◇ ZAGREB

TOXICITY DATA with REFERENCE
sln-dmg-orl 5 mmol/L MUREAV 95,237,82
dni-hmn:hla 1 μmol/L MUREAV 92,427,82
msc-mus:lym 20 μg/L FCTOD7 23,115,85
orl-rat TDLo:1130 mg/kg (male 8W pre):REP TJADAB 35,43A,87
scu-mus TDLo:2500 μg/kg (female 10D post):TER AMUK** 36,20,59
ipr-mus TDLo:6 mg/kg/4W:CAR JNCIAM 36,915,66
scu-mus TDLo:22 mg/kg/21W-I:CAR BJCAAI 3,118,49
ivn-mus TDLo:4 mg/kg/6D-I:NEO JNCIAM 11,415,50
scu-mus TD:6 mg/kg/6W-I:ETA JNCIAM 14,131,53
ivn-hmn TDLo:400 μg/kg:CNS CLPTAT 6,50,65
orl-rat LD50:10 mg/kg JPETAB 91,224,47
scu-rat LD50:1900 μg/kg JPETAB 91,224,47
ivn-rat LD50:1100 μg/kg MEIEDD 10,822,83
par-rat LD50:1700 μg/kg RRCRBU 52,76,75
orl-mus LD50:20 mg/kg JPETAB 91,224,47
ipr-mus LD50:2900 μg/kg NYKZAU 62,96,66
scu-mus LD50:2600 μg/kg JPETAB 91,224,47
ivn-mus LD50:2 mg/kg JPETAB 91,224,47

CONSENSUS REPORTS: NTP Fifth Annual Report on Carcinogens. IARC Cancer Review: Group 2A IMEMDT 7,269,87; Animal Sufficient Evidence IMEMDT 9,193,75. EPA Genetic Toxicology Program.

SAFETY PROFILE: Confirmed carcinogen with experimental carcinogenic, neoplastigenic, and tumorigenic data. Deadly poison by ingestion, intravenous, subcutaneous, intraperitoneal, and parenteral routes. Experimental teratogenic and reproductive effects. Human systemic effects by intravenous route: nausea or vomiting, reduction in the number of white blood cells and blood platelates. Other experimental reproductive effects. Human mutation data reported.

BIE750 CAS:63905-44-2 ***HR: 3***
(N,N-BIS(2-CHLOROETHYL))-2-METHYLPROPYLAMINE HYDROCHLORIDE
mf: $C_8H_{17}Cl_2N•ClH$ mw: 234.62

SYNS: N,N-BIS(2-CHLOROETHYL)ISOBUTYLAMINEHYDROCHLORIDE ◇ N,N-BIS(2-CHLOROETHYL)-2-METHYL-1-PROPANAMINE HYDROCHLORIDE ◇ 2,2'-DICHLORO-N-ISOBUTYL-DIETHYLAMINE HYDROCHLORIDE ◇ ISOBUTYLBIS(β-CHLOROETHYL)AMINE HYDROCHLORIDE ◇ ISOBUTYLBIS(2-CHLOROETHYL)AMINE HYDROCHLORIDE ◇ TL 525

TOXICITY DATA with REFERENCE
ipr-mus LD50:4420 μg/kg CANCAR 2,1075,49
scu-mus LDLo:5 mg/kg NDRC** No.9-4-1-9,43

SAFETY PROFILE: Poison by intraperitoneal and subcutaneous routes. When heated to decomposition it emits very toxic fumes of Cl^- and NO_x.

BIF250 CAS:494-03-1 ***HR: 3***
N,N-BIS(2-CHLOROETHYL)-2-NAPHTHYLAMINE
mf: $C_{14}H_{15}Cl_2N$ mw: 268.20

SYNS: 2-BIS(2-CHLOROETHYL)AMINONAPHTHALENE ◇ BIS(2-CHLOROETHYL)-β-NAPHTHYLAMINE ◇ CHLORNAFTINA ◇ CHLORNAPHAZIN ◇ CHLORNAPHTHIN ◇ CHLORONAFTINA ◇ CHLORONAPHTHINE ◇ CLORNAPHAZINE ◇ DICHLOROETHYL-β-NAPHTHYLAMINE ◇ DI(2-CHLOROETHYL)-β-NAPHTHYLAMINE ◇ N,N-DI(2-CHLOROETHYL)-β-NAPHTHYLAMINE ◇ 2-N,N-DI(2-CHLOROETHYL)NAPHTHYLAMINE ◇ ERYSAN ◇ NAPHTHYLAMINE MUSTARD ◇ β-NAPHTHYL-BIS-(β-CHLOROETHYL)AMINE ◇ 2-NAPHTHYLBIS(2-CHLOROETHYL)AMINE ◇ β-NAPHTHYL-DI-(2-CHLOROETHYL)AMINE ◇ NSC-62209 ◇ R48 ◇ RCRA WASTE NUMBER U026

TOXICITY DATA with REFERENCE
mmo-sat 40 μg/plate CNREA8 37,2209,77
mma-sat 10 μg/plate PNASA6 72,5135,75
dnd-dmg-orl 260 μmol/L CNREA8 30,195,70
orl-man TDLo:2468 mg/kg/6Y-I:CAR AMSVAZ 176,45,64
orl-wmn TDLo:3132 mg/kg/10Y-I:CAR AMSVAZ 175,64
ipr-rat LD50:1086 mg/kg BCPCA6 13,969,64

CONSENSUS REPORTS: IARC Cancer Review: Group 1 IMEMDT 7,130,87; Animal Sufficient Evidence IMEMDT 4,119,74; Human Sufficient Evidence IMEMDT 4,119,74. EPA Genetic Toxicology Program.

SAFETY PROFILE: Confirmed human carcinogen producing bladder tumors. Human and experimental carcinogenic data. Moderately toxic by intraperitoneal route. When heated to decomposition it emits very toxic fumes of Cl^- and NO_x.

BIF500 CAS:67856-68-2 ***HR: 3***
BIS(2-CHLOROETHYL)NITROSOAMINE
mf: $C_4H_8Cl_2N_2O$ mw: 171.04

SYNS: NITROSOBIS(2-CHLOROETHYL)AMINE ◇ N-NITROSO-2,2'-DICHLORODIETHYLAMINE

TOXICITY DATA with REFERENCE
mmo-sat-10 μg/plate MUREAV 66,1,79
mma-sat 10 μg/plate MUREAV 66,1,79
orl-rat TDLo:345 mg/kg/30W-I:ETA CNREA8 38,2391,78

SAFETY PROFILE: Questionable carcinogen with experimental tumorigenic data. Mutation data reported. When heated to decomposition it emits toxic fumes of Cl^- and NO_x. See also NITROSAMINES and N-NITROSO COMPOUNDS.

BIF625 CAS:77469-44-4 ***HR: 3***
N,N'-BIS((2-CHLOROETHYL)-N-NITROSOCARBAMOYL)CYSTAMINE
mf: $C_{10}H_{18}Cl_2N_6O_4S_2$ mw: 421.36

SYNS: CNCC ◇ 13-CHLORO-N-(2-CHLOROETHYL)-N,11-DINITROSO-10-OXO-5,6-DITHIA-2,9,11-TRIAZATRIDECANAMIDE ◇ DI((CHLORO-2-ETHYL)-2-N-NITROSO-N-CARBAMOYL)-N,N-CYSTAMINE ◇ 1,1'-DITHIODIETHYLENEBIS(3-(2-(CHLOROETHYL)-3-NITROSOUREA ◇ I.C.I.G. 1325

TOXICITY DATA with REFERENCE
mmo-sat 200 μg/plate INSSDM 19,165,81
mma-sat 200 μg/plate INSSDM 19,165,81
oms-mus:oth 20 mg/L INSSDM 19,229,81
orl-mus LD50:280 mg/kg INSSDM 19,123,81
ipr-mus LD50:75 mg/kg INSSDM 19,123,81

SAFETY PROFILE: Poison by ingestion and intraperitoneal routes. Mutation data reported. When heated to decomposition it emits toxic fumes of Cl^-, SO_x and NO_x.

BIF750 CAS:154-93-8 ***HR: 3***
N,N'-BIS(2-CHLOROETHYL)-N-NITROSOUREA
mf: $C_5H_9Cl_2N_3O_2$ mw: 214.07

SYNS: BCNU ◇ BiCNU ◇ BIS(2-CHLOROETHYL)NITROSOUREA ◇ 1,3-BIS(β-CHLOROETHYL)-1-NITROSOUREA ◇ 1,3-BIS-(2-CHLOROETHYL)-1-NITROSOUREA ◇ BISCHLOROETHYL-NITROSOUREA ◇ CARMUBRIS ◇ CARMUSTIN ◇ CARMUSTINE ◇ FDA 0345 ◇ NCI-C04773 ◇ NITRUMON ◇ NSC-409962 ◇ SK 27702 ◇ SRI 1720

TOXICITY DATA with REFERENCE
mmo-sat 33 μg/plate TCMUD8 5,319,85
sce-hmn:lym 25 μmol/L CNREA8 45,4798,85
ipr-rat TDLo:4 mg/kg (female 6-9D post):TER TXAPA9 30,422,74
ipr-rat TDLo:9 mg/kg (male 9W pre):REP TXAPA9 30,422,74
ipr-rat TDLo:15 mg/kg/7W-I:ETA CANCAR 40,1935,77
ivn-cld LDLo:78 mg/kg/52W I CANCAR 42,74,78
ivn-hmn TDLo:125 mg/kg:BLD,GIT ACRSAJ 16,273,72
ivn-hmn TDLo:6 mg/kg:BLD,GIT CTRRDO 60,709,76
par-wmn LDLo:1566 mg/kg:PUL JAMAAP 244,687,80
orl-rat LD50:20 mg/kg JPETAB 166,104,69
ipr-rat LD50:17420 μg/kg NCISP* JAN86
scu-rat LD50:83200 μg/kg IYKEDH 9,766,78
ivn-rat LD50:13800 μg/kg ONCOBS 37,177,80
ims-rat LD50:79600 μg/kg IYKEDH 9,766,78
orl-mus LD50:19 mg/kg TXAPA9 21,405,72
ipr-mus LD50:21260 μg/kg NCISP* JAN86
scu-mus LD50:24 mg/kg TXAPA9 21,405,72
ivn-mus LD50:45 mg/kg PSEBAA 118,756,65

CONSENSUS REPORTS: NTP Fifth Annual Report on Carcinogens. IARC Cancer Review: Group 2A IMEMDT 7,150,87; Human Limited Evidence IMEMDT

26,79,81; Animal Sufficient Evidence IMEMDT 26,79,81. NCI Carcinogenesis Studies (ipr); Some Evidence: rat CANCAR 40,1935,77; Clear Evidence: mouse CANCAR 40,1935,77. EPA Genetic Toxicology Program.

SAFETY PROFILE: Confirmed carcinogen with experimental carcinogenic and tumorigenic data. A human poison by parenteral route. An experimental poison by ingestion, intravenous, intraperitoneal, parenteral, and subcutaneous routes. Human systemic effects by parenteral, intravenous, and possibly other routes: nausea or vomiting, reduced white blood cell and blood platelet counts, bone marrow damage and potentially fatal respiratory system effects including lung fibrosis, dyspnea, and cyanosis. Experimental teratogenic and reproductive effects. Human mutation data reported. When heated to decomposition it emits very toxic fumes of Cl^- and NO_x. See also N-NITROSO COMPOUNDS.

BIG000 CAS:68060-50-4 ***HR: D***
1,3-BIS(2-CHLOROETHYL)-1-NITROSOUREA-DIPHENYLMETHANE
mf: $C_{19}H_{20}Cl_2N_6O_4$ mw: 467.35

SYN: 1,1'-DIPHENYLMETHYLENEBIS(3-(2-CHLOROETHYL)-3-NITROSOUREA)

TOXICITY DATA with REFERENCE
sln-dmg-orl 100 μmol/L MUREAV 57,297,78

SAFETY PROFILE: Mutation data reported. When heated to decomposition it emits very toxic fumes of Cl^- and NO_x.

BIG250 CAS:2067-58-5 ***HR: 3***
N,N-BIS(2-CHLOROETHYL)-p-PHENYLENEDIAMINE
mf: $C_{10}H_{14}Cl_2N_2$ mw: 233.16

SYN: p-AMINOPHENYL DERIVATIVE of NITROGEN MUSTARD

TOXICITY DATA with REFERENCE
scu-mus TDLo:12 mg/kg (female 7-9D post):TER JEEMAF 9,492,61
ipr-rat LD50:2200 μg/kg DBANAD 33,1005,80
ipr-mus LD50:7927 μg/kg JMCMAR 8,167,65

SAFETY PROFILE: Poison by intraperitoneal route. An experimental teratogen. When heated to decomposition it emits very toxic fumes of Cl^- and NO_x.

BIG500 CAS:1070-42-4 ***HR: 3***
BIS(2-CHLOROETHYL)PHOSPHITE
mf: $C_4H_9Cl_2O_3P$ mw: 207.00

TOXICITY DATA with REFERENCE
orl-rat LD50:260 mg/kg AIHAAP 30,470,69
ipr-mus LDLo:250 mg/kg CBCCT* 7,790,55
skn-rbt LD50:141 mg/kg AIHAAP 30,470,69

SAFETY PROFILE: Poison by ingestion and skin contact. When heated to decomposition it emits very toxic fumes of PO_x and Cl^-.

BIG750 CAS:63980-44-9 ***HR: 3***
N,N'-BIS(2-CHLOROETHYL)-1,4-PIPERAZINE HYDROCHLORIDE
mf: $C_8H_{16}Cl_2N_2$• ClH mw: 247.62

TOXICITY DATA with REFERENCE
ipr-rat LD50:1100 μg/kg JPETAB 100,398,50
ipr-mus LD50:5700 μg/kg JPETAB 100,398,50

SAFETY PROFILE: Poison by intraperitoneal route. When heated to decomposition it emits very toxic fumes of HCl and NO_x.

BIH000 CAS:2045-41-2 ***HR: 3***
N^4,N^4-BIS(2-CHLOROETHYL)SULFANILAMIDE
mf: $C_{10}H_{14}Cl_2N_2O_2S$ mw: 297.22

TOXICITY DATA with REFERENCE
ipr-rat LD50:336 mg/kg JMCMAR 8,167,65
ipr-mus LD50:410 mg/kg JMCMAR 8,167,65

SAFETY PROFILE: Poison by intraperitoneal route. When heated to decomposition it emits very toxic fumes of Cl^-, NO_x, and SO_x.

BIH250 CAS:505-60-2 ***HR: 3***
BIS(2-CHLOROETHYL)SULFIDE
mf: $C_4H_8Cl_2S$ mw: 159.08

PROP: Colorless (if pure), to light yellow, oily liquid. Bp: 228°, fp: 14.4°, flash p: 221°F, d: 1.2741 @ 20°/4°, vap d: 5.4, vap press: 0.09 mm @ 30°.

SYNS: BIS(β-CHLOROETHYL)SULFIDE ◇ BIS(2-CHLOROETHYL)SULPHIDE ◇ 1-CHLORO-2-(β-CHLOROETHYLTHIO)ETHANE ◇ β,β-DICHLOR-ETHYL-SULPHIDE ◇ 2,2'-DICHLORODIETHYL SULFIDE ◇ DI-2-CHLOROETHYL SULFIDE ◇ β,β'-DICHLOROETHYL SULFIDE ◇ 2,2'-DICHLOROETHYL SULPHIDE (MAK) ◇ DISTILLED MUSTARD ◇ KAMPSTOFF "LOST" ◇ MUSTARD GAS ◇ MUSTARD HD ◇ MUSTARD VAPOR ◇ SCHWEFEL-LOST ◇ S-LOST ◇ S MUSTARD ◇ SULFUR MUSTARD ◇ SULFUR MUSTARD GAS ◇ SULPHUR MUSTARD GAS ◇ 1,1'-THIOBIS(2-CHLOROETHANE) ◇ YELLOW CROSS LIQUID ◇ YPERITE

TOXICITY DATA with REFERENCE
skn-man 2000 mg/m^3/1H SEV NTIS** AD-A011-260
eye-man 100 mg/m^3/6H MOD NTIS** AD-A011-260
eye-rbt 200 mg/m^3 NTIS** AD-A011-260
eye-rbt 200 mg/m^3/2M MLD NTIS** AD-A011-260
dnd-smc 500 μmol/L CBINA8 44,27,83
cyt-mam:lym 750 nmol/L CHRTBC 3,162,72
oms-hmn:hla 75 mg/L IUSMDJ 9,41,79
orl-rat TDLo:20 mg/kg (female 6-15D post):TER TJADAB 33,70C,86
ihl-rat TCLo:100 $μg/m^3$/1Y-I:CAR NTIS** AD-A011-260
ihl-mus TCLo:1250 mg/m^3/15M-C:NEO PSEBAA 82,457,53

scu-mus TDLo:6 mg/kg/6W-I:ETA JNCIAM 14,131,53
ihl-hmn LC50:1500 mg/m^3/M NTIS** AD-A011-260
ihl-hmn LCLo:23 ppm/10M NTIS** PB214-270
skn-hmn LDLo:64 mg/kg WHOTAC -,24,70
ihl-rat LC50:100 mg/m^3/10M NTIS** PB158-507
skn-rat LD50:5 mg/kg CNRMAW 25,141,47
scu-rat LD50:1500 μg/kg CNRMAW 25,141,47
ivn-rat LD50:700 μg/kg JPETAB 93,1,48
ihl-mus LC50:120 mg/m^3/10M NTIS** PB158-507
skn-mus LD50:92 mg/kg JPETAB 93,1,48
scu-mus LD50:20 mg/kg NTIS** PB158-507
ivn-mus LD50:8600 μg/kg JPETAB 93,1,48
ihl-dog LC50:70 mg/m^3/10M NTIS** PB158-507
skn-dog LD50:20 mg/kg NTIS** PB158-507
ivn-dog LD50:200 μg/kg NTIS** PB158-507
ihl-mky LC50:80 mg/m^3/10M NTIS** PB158-507
skn-rbt LD50:40 mg/kg NTIS** PB158-507

CONSENSUS REPORTS: NTP Fifth Annual Report on Carcinogens. IARC Cancer Review: Group 1 IMEMDT 7,259,87; Animal Sufficient Evidence IMEMDT 9,181,75; Human Limited Evidence IMEMDT 9,181,75. EPA Extremely Hazardous Substances List. Community Right-To-Know List. EPA Genetic Toxicology Program. Reported in EPA TSCA Inventory.
DFG MAK: Human Carcinogen.

SAFETY PROFILE: Confirmed human carcinogen with experimental carcinogenic, neoplastigenic, and tumorigenic data. A human poison by inhalation and subcutaneous routes. An experimental poison by inhalation, skin contact, subcutaneous, and intravenous routes. An experimental teratogen. A severe human skin and eye irritant. Human mutation data reported. A military blistering gas. Strongly effects the skin, eyes, lungs, and gastric system. Pulmonary lesions are often fatal. It penetrates the skin deeply and injures blood vessels. Minute amounts can cause inflammation. Secondary infections are common. Combustible when exposed to heat or flame; can be ignited by a large explosive charge. It will react with water or steam to produce toxic and corrosive fumes. Vigorous reaction with oxidizing materials. Incompatible with bleaching powder. To fight fire, use water, foam, CO_2, dry chemical. Dangerous; when heated to decomposition or on contact with acid or acid fumes it emits highly toxic fumes of SO_x and Cl^-. See also SULFIDES and CHLORIDES.

BIH325 ***HR: 3***
1,1-BIS(2-CHLOROETHYL)-2-SULFINYLHYDRAZINE
mf: $C_4H_8Cl_2N_2OS$ mw: 203.10

SYN: NSC 78409

TOXICITY DATA with REFERENCE
orl-mus LD50:37240 μg/kg NCISP* JAN86
ipr-mus LD50:56650 μg/kg NCISP* JAN86
scu-mus LD50:53660 μg/kg NCISP* JAN86

SAFETY PROFILE: Poison by ingestion, subcutaneous and intraperitoneal routes. When heated to decomposition it emits toxic fumes of Cl^-, SO_x, and NO_x.

BIH500 CAS:471-03-4 ***HR: 3***
BIS(2-CHLOROETHYL)SULFONE
mf: $C_4H_8Cl_2O_2S$ mw: 191.08

SYNS: BIS(β-CHLOROETHYL)SULFONE ◇ MUSTARD GAS SULFONE ◇ MUSTARD SULFONE ◇ YPERITE SULFONE

TOXICITY DATA with REFERENCE
scu-rat LD50:50 mg/kg JPETAB 93,1,48
scu-mus LD50:35 mg/kg JPETAB 93,1,48
ivn-mus LD50:50 mg/kg JPETAB 93,1,48
ihl-cat LCLo:1430 mg/m^3/10M NDRC** NDCrc-132,Jan,42
ihl-rbt LCLo:1430 mg/m^3/10M NDRC** NDCrc-132,Jan,42

SAFETY PROFILE: A poison via intravenous and subcutaneous routes. Moderately toxic via inhalation. See also SULFONATES. When heated to decomposition it emits very toxic fumes of Cl^- and SO_x.

BIH750 ***HR: 3***
BIS(1-CHLOROETHYL THALLIUM CHLORIDE) OXIDE
mf: $C_4H_8Cl_4OTl_2$ mw: 622.66

CONSENSUS REPORTS: Thallium and its compounds are on The Community Right-To-Know List.

SAFETY PROFILE: An unstable explosive. When heated to decomposition it emits toxic fumes of Cl^-. See also THALLIUM COMPOUNDS.

BII000 CAS:14742-53-1 ***HR: D***
N,N-BIS(2-CHLOROETHYL)-2-THENYLAMINE HYDROCHLORIDE
mf: $C_9H_{13}Cl_2NS•ClH$ mw: 274.6

SYN: DI-(2-CHLOROETHYL)THENYLAMINE HYDROCHLORIDE

TOXICITY DATA with REFERENCE
mmo-asn 2500 μmol/L SOGEBZ 6,220,70

SAFETY PROFILE: Mutation data reported. When heated to decomposition it emits very toxic fumes of Cl^-, NO_x, and SO_x.

BII250 CAS:108-60-1 ***HR: 3***
BIS(2-CHLOROISOPROPYL) ETHER
DOT: UN 2490
mf: $C_6H_{12}Cl_2O$ mw: 171.08

PROP: Colorless liquid. Bp: 187.8°, fp: > −20°, flash p: 185°F (OC), d: 1.11 @ 25°/25°, vap d: 6.0, vap press: 0.10 mm @ 20°.

SYNS: BIS(2-CHLORO-1-METHYLETHYL) ETHER ◇ (2-CHLORO-1-

METHYLETHYL) ETHER ◇ DICHLORODIISOPROPYL ETHER ◇ DICHLOROISOPROPYL ETHER (DOT) ◇ 2,2'-DICHLOROISOPROPYL ETHER ◇ NCI-C50044 ◇ RCRA WASTE NUMBER U027

TOXICITY DATA with REFERENCE
eye-rbt 500 mg open AMIHBC 4,119,51
mmo-sat 1 mL/plate/3H DHEFDK FDA-78-1046,78
mma-sat 333 μg/plate ENMUDM 8(Suppl 7),1,86
orl-rat LD50:240 mg/kg AMIHBC 4,119,51
ihl-rat LCLo:700 ppm/5H BJIMAG 27,1,70
skn-rbt LD50:3000 mg/kg AMIHBC 4,119,51

CONSENSUS REPORTS: IARC Cancer Review: Group 3 IMEMDT 7,56,87, Animal Limited Evidence IMEMDT 41,149,86. NCI Carcinogenesis Bioassay (gavage); No Evidence: rat NCITR* NCI-CG-TR-191,79. Community Right-To-Know List. Reported in EPA TSCA Inventory.

DOT Classification: Corrosive Material; Label: Corrosive; IMO: Poison B; Label: Poison.

SAFETY PROFILE: Poison by ingestion. Moderately toxic by skin contact and inhalation. An eye irritant. Questionable carcinogen. Mutation data reported. A corrosive material. Moderate fire hazard when exposed to heat, flame, or powerful oxidizers. Incompatible with oxidizing materials. To fight fire, use water to blanket fire; foam, CO_2, dry chemical. When heated to decomposition it emits highly toxic fumes of Cl^-. See also ETHERS.

BII500 CAS:67465-41-2 ***HR: 3***
2,5-BIS(CHLOROMERCURI)FURAN
mf: $C_4H_2Cl_2Hg_2O$ mw: 538.14

SYN: USAF UCTL-974

TOXICITY DATA with REFERENCE
ipr-mus LD50:20 mg/kg NTIS** AD277-689

CONSENSUS REPORTS: Mercury and its compounds are on The Community Right-To-Know List.

OSHA PEL: (Transitional: CL 1 mg/10m^3) CL 0.1 mg(Hg)/m^3 (skin)
ACGIH TLV: TWA 0.1 mg(Hg)/m^3 (skin)
NIOSH REL: (Mercury, Inorganic) TWA 0.05 mg(Hg)/m^3

SAFETY PROFILE: Poison by intraperitoneal route. See also MERCURY COMPOUNDS and CHLORIDES. When heated to decomposition it emits very toxic fumes of Cl^- and Hg.

BII750 ***HR: 3***
N,N-BIS(CHLOROMERCURI)HYDRAZINE
mf: $Cl_2H_2Hg_2N_2$ mw: 502.01

CONSENSUS REPORTS: Mercury and its compounds are on The Community Right-To-Know List.

SAFETY PROFILE: An explosive. When heated to decomposition it emits toxic fumes of Cl^-, NO_x, and Hg. See also MERCURY COMPOUNDS.

BIJ000 CAS:64050-46-0 ***HR: 3***
4,5-BIS(CHLOROMERCURI)-2-THIAZOLECARBAMIC ACID BENZYL ESTER
mf: $C_{11}H_8Cl_2Hg_2N_2O_2S$ mw: 704.35

TOXICITY DATA with REFERENCE
ipr-mus LDLo:125 mg/kg CBCCT* 8,752,56

CONSENSUS REPORTS: Mercury and its compounds are on The Community Right-To-Know List.

OSHA PEL: (Transitional: CL 1 mg/10m^3) CL 0.1 mg(Hg)/m^3 (skin)
ACGIH TLV: TWA 0.1 mg(Hg)/m^3 (skin)
NIOSH REL: (Mercury, Inorganic) TWA 0.05 mg(Hg)/m^3

SAFETY PROFILE: Poison by intraperitoneal route. See also CARBAMATES; MERCURY COMPOUNDS; and ESTERS. When heated to decomposition it emits very toxic fumes of Cl^-, Hg, NO_x, and SO_x.

BIJ250 CAS:13483-18-6 ***HR: 3***
BIS-1,2-(CHLOROMETHOXY)ETHANE
mf: $C_4H_8Cl_2O_2$ mw: 159.02

PROP: Viscous liquid. Bp: 99-100° @ 22 mm, d: 1.2879 @ 14°/15°.

SYN: ETHYLENE GLYCOL BIS(CHLOROMETHYL)ETHER

TOXICITY DATA with REFERENCE
skn-mus TDLo:8640 mg/kg/72W-I:NEO CNREA8 35,2553,75

CONSENSUS REPORTS: IARC Cancer Review: Group 3 IMEMDT 7,56,87; Animal Sufficient Evidence IMEMDT 15,31,77. Reported in EPA TSCA Inventory. Glycol ethers are on the Community Right-To-Know List.

SAFETY PROFILE: Questionable carcinogen with experimental neoplastigenic data. See also GLYCOL ETHERS. When heated to decomposition it emits toxic fumes of Cl^-.

BIJ500 CAS:56894-91-8 ***HR: 2***
1,4-BIS(CHLOROMETHOXYMETHYL)BENZENE
mf: $C_{10}H_{12}Cl_2O_2$ mw: 235.12

SYN: BIS-1,4-(CHLOROMETHOXY)-p-XYLENE

TOXICITY DATA with REFERENCE
skn-mus TDLo:2590 mg/kg/72W-I:NEO CNREA8 35,2553,75
ipr-mus TDLo:310 mg/kg/78W-I:ETA CNREA8 35,2553,75

CONSENSUS REPORTS: IARC Cancer Review: Group 3 IMEMDT 7,56,87; Animal Sufficient Evidence IMEMDT 15,37,77

SAFETY PROFILE: Questionable carcinogen with experimental neoplastigenic and tumorigenic data. When heated to decomposition it emits toxic fumes of Cl^-.

BIJ750 CAS:10387-13-0 ***HR: 3***
9,10-BIS(CHLOROMETHYL)ANTHRACENE
mf: $C_{16}H_{12}Cl_2$ mw: 275.18

SYNS: 9,10-DI(CHLOROMETHYL)ANTHRACENE ◇ ICR-450

TOXICITY DATA with REFERENCE
mma-sat 100 ng/plate PNASA6 72,5135,75
ivn-mus TDLo:1100 μg/kg:NEO CNREA8 36,2423,76
ivn-mus LD50:56 mg/kg CSLNX* NX#00245

CONSENSUS REPORTS: Reported in EPA TSCA Inventory. EPA Genetic Toxicology Program.

SAFETY PROFILE: Poison by intravenous route. Questionable carcinogen with experimental neoplastigenic data. Mutation data reported. When heated to decomposition it emits toxic fumes of Cl^-. See also CHLORINATED HYDROCARBONS, AROMATIC.

BIK000 CAS:542-88-1 ***HR: 3***
BIS(CHLOROMETHYL) ETHER
DOT: UN 2249
mf: $C_2H_4Cl_2O$ mw: 114.96

PROP: Volatile liquid. Bp: 105°, d: 1.315 @ 20°, vap d: 4.0. flash p: < 19°.

SYNS: BCME ◇ BIS-CME ◇ CHLORO(CHLOROMETHOXY)METHANE ◇ DICHLORDIMETHYLAETHER (GERMAN) ◇ sym-DICHLORODIMETHYL ETHER (DOT) ◇ sym-DICHLOROMETHYL ETHER ◇ DIMETHYL-1,1'-DICHLOROETHER ◇ OXYBIS(CHLOROMETHANE) ◇ RCRA WASTE NUMBER P016

TOXICITY DATA with REFERENCE
otr-ham:kdy 80 μg/L BJCAAI 37,873,78
mma-sat 20 μg/plate BJCAAI 37,873,78
dns-hmn:fbr 160 μg/L TXCYAC 21,151,81
dns-mus-skn 360 μmol/kgL CNREA8 33,769,73
dni-mus-skn 360 μmol/kg CNREA8 33,769,73
ihl-rat TCLo:100 ppb/6H/4W-I:CAR AEHLAU 30,73,75
scu-rat TDLo:375 mg/kg/43W-I:CAR JNCIAM 43,481,69
ihl-mus TCLo:100 ppb/6H/26W-I:NEO TXAPA9 58,269,81
skn-mus TDLo:5520 mg/kg/23W-I:ETA AEHLAU 16,472,68
ihl-man TCLo:3 ppm:EYE TJSGA8 51,596,73
ihl-man LCLo:100 ppm/3M:PUL TJSGA8 51,596,73
orl-rat LD50:210 mg/kg AIHAAP 30,470,69
ihl-rat LC50:7 ppm/7H AEHLAU 30,61,75
ihl-mus LC50:25 mg/m³/6H AEHLAU 22,663,71
skn-rbt LD50:280 mg/kg AIHAAP 30,470,69
ihl-ham LC50:7 ppm/7H AEHLAU 30,61,75

CONSENSUS REPORTS: NTP Fifth Annual Report on Carcinogens. IARC Cancer Review: Group 1 IMEMDT 7,131,87; Animal Sufficient Evidence IMEMDT 4,231, 74; Human Sufficient Evidence IMEMDT 4,231,74. Community Right-To-Know List. EPA Extremely Hazardous Substances List. Reported in EPA TSCA Inventory.

OSHA: Cancer Suspect Agent
ACGIH TLV: TWA 0.001 ppm; Confirmed Human Carcinogen
DFG MAK: Human Carcinogen.
DOT Classification: Poison B; Label: Flammable Liquid and Poison

SAFETY PROFILE: Confirmed human carcinogen with experimental carcinogenic, neoplastigenic, and tumorigenic data. Poison by inhalation, ingestion, and skin contact. Human systemic effects by inhalation: irritation of the conjunctiva, unspecified nasal and respiratory effects. Human mutation data reported. A dangerous fire hazard. When heated to decomposition it emits very toxic fumes of Cl^-. See also ETHERS.

BIK100 ***HR: 2***
BIS(2-CHLORO-1-METHYLETHYL)ETHER mixed with 2-CHLORO-1-METHYLETHYL-(2-CHLOROPROPYL) ETHER
mf: $C_6H_{12}Cl_2O$ mw: 171.08

SYN: ETHER, BIS(2-CHLORO-1-METHYLETHYL), mixed with 2-CHLORO-1-METHYLETHYL-(2-CHLOROPROPYL)ETHER(7:3)

TOXICITY DATA with REFERENCE
orl-mus TDLo:51500 mg/kg/2Y-I:CAR NTPTR* NTP-TR-239,82
orl-mus TD:103 g/kg/2Y-I:CAR NTPTR* NTP-TR-239,82

CONSENSUS REPORTS: NTP CARCINOGENESIS BIOASSAY (gavage); Clear Evidence: mouse NTPTR* NTP-TR-239,83

SAFETY PROFILE: Questionable carcinogen with experimental carcinogenic data. When heated to decomposition it emits toxic fumes of Cl^-.

BIK250 CAS:534-07-6 ***HR: 3***
BIS(CHLOROMETHYL)KETONE
DOT: UN 2649
mf: $C_3H_4Cl_2O$ mw: 126.97

PROP: Crystals. Mp: 45°, bp: 173°, d: 1.3826 @ 46°/4°, vap d: 4.38.

SYNS: sym-DICHLOROACETONE ◇ α,α'-DICHLOROACETONE ◇ α,Γ-DICHLOROACETONE ◇ 1,3-DICHLOROACETONE ◇ 1,3-DICHLOROACETONE (DOT) ◇ 1,3-DICHLORO-2-PROPANONE

TOXICITY DATA with REFERENCE
mmo-sat 1250 ng/plate MUREAV 157,111,85
mma-smc 5 μg/L MUREAV 155,53,85
ihl-rat LC50:29 mg/m^3/2H 85GMAT -,44,82
ihl-mus LC50:27 mg/m^3/2H 85GMAT -,44,82

CONSENSUS REPORTS: EPA Genetic Toxicology Program. Reported in EPA TSCA Inventory. EPA Extremely Hazardous Substances List.

DOT Classification: Poison B; Label: Poison.

SAFETY PROFILE: Poison by inhalation. Mutation data reported. A sytemic irritant by ingestion and inhalation routes. See also KETONES. Dangerous; when heated to decomposition it emits highly toxic fumes of Cl^-.

BIK325 CAS:78-71-7 ***HR: 3***
3,3-BIS(CHLOROMETHYL)OXETANE
mf: $C_5H_8Cl_2O$ mw: 155.03

SYN: 3,3-DICHLOROMETHYLOXYCYCLOBUTANE

TOXICITY DATA with REFERENCE
orl-mus LD50:420 mg/kg 85GMAT -,47,82
ihl-mus LC50:200 mg/m^3/2H 85GMAT -,47,82
unr-mus LD50:545 mg/kg RPTOAN 48,67,85

CONSENSUS REPORTS: EPA Extremely Hazardous Substances List.

SAFETY PROFILE: Poison by inhalation. Moderately toxic by ingestion and other unspecified routes. When heated to decomposition it emits toxic fumes of Cl^-.

BIK500 CAS:2209-86-1 ***HR: 2***
2,2-BIS(CHLOROMETHYL)-1,3-PROPANEDIOL
mf: $C_5H_{10}Cl_2O_2$ mw: 173.05

SYN: DISPRANOL

TOXICITY DATA with REFERENCE
orl-rat LD50:1285 mg/kg BCFAAI 99,67,60
ipr-rat LD50:920 mg/kg BCFAAI 99,67,60
ipr-mus LD50:812 mg/kg BCFAAI 99,67,60
unr-mus LD50:1 g/kg RPTOAN 48,67,85

SAFETY PROFILE: Moderately toxic by ingestion, intraperitoneal, and possibly other routes. When heated to decomposition it emits toxic fumes of Cl^-.

BIK750 CAS:12712-28-6 ***HR: 3***
2,2-BIS(CHLOROMETHYL)-1,3-PROPANEDIOL SULFATE
mf: $C_5H_{10}Cl_2O_2 \cdot H_2O_4S$ mw: 271.13

SYN: PHILIPS 2605

TOXICITY DATA with REFERENCE
orl-rat LD50:20 mg/kg TXAPA9 21,315,72
orl-bwd LD50:2400 μg/kg TXAPA9 21,315,72

SAFETY PROFILE: Poison by ingestion. See also SULFATES. When heated to decomposition it emits very toxic fumes of SO_x and Cl^-.

BIL000 CAS:52444-01-6 ***HR: D***
BIS(2-CHLOROMETHYL-2-PROPYL)SULFIDE
mf: $C_8H_{16}Cl_2S$ mw: 215.20

SYN: α,α,α',α'-TETRAMETHYL-β,β'-BISCHLOROETHYL SULFIDE

TOXICITY DATA with REFERENCE
dnd-ckn:leu 30 mmol/L TELEAY (29),2477,75

SAFETY PROFILE: Mutation data reported. See also SULFIDES. When heated to decomposition it emits very toxic fumes of Cl^- and SO_x.

BIL250 CAS:14579-91-0 ***HR: 3***
1,3-BIS(CHLOROMETHYL)-1,1,3,3-TETRAMETHYLDISILAZANE
mf: $C_6H_{17}Cl_2N\text{-}Si_2$ mw: 230.32

SYN: 1-(CHLOROMETHYL)-N-((CHLOROMETHYL)DIMETHYLSILYL)-1,1-DIMETHYL-SILANAMINE

TOXICITY DATA with REFERENCE
ipr-mus TDLo:40 mg/kg/I:NEO JNCIAM 54,495,75
ipr-mus LDLo:250 mg/kg StoGD# 27May75

CONSENSUS REPORTS: Reported in EPA TSCA Inventory.

SAFETY PROFILE: Poison by intraperitoneal route. Questionable carcinogen with experimental neoplastigenic data. When heated to decomposition it emits very toxic fumes of Cl^- and NO_x.

BIL500 CAS:83-05-6 ***HR: 2***
BIS(p-CHLOROPHENYL)ACETIC ACID
mf: $C_{14}H_{10}Cl_2O_2$ mw: 281.14

SYNS: BIS(4-CHLOROPHENYL)ACETIC ACID ◇ BIS(p-CHLORPHENYL)ESSIGSAEURE (GERMAN) ◇ DICHLORODIPHENYLACETIC ACID ◇ p,p'-DICHLORODIPHENYLACETIC ACID ◇ DI(p-CHLOROPHENYL)ACETIC ACID

TOXICITY DATA with REFERENCE
sln-dmg-orl 3700 μmol/L MUREAV 16,157,72
cyt-rat:oth 150 mg/L/24H C TXAPA9 22,355,72
ipr-rat TDLo:350 mg/kg (7D pre):REP ENDOAO 91,1095,72

orl-rat TDLo:250 mg/kg (15-19D post):TER BNEOBV 26,283,75
orl-mus LD50:590 mg/kg AIPTAK 73,128,46

SAFETY PROFILE: Moderately toxic by ingestion. An experimental teratogen. Other experimental reproductive effects. Mutation data reported. When heated to decomposition it emits toxic fumes of Cl^-.

BIM000 CAS:4104-14-7 ***HR: 3***
O,O-BIS(p-CHLOROPHENYL)ACETIMIDOYL-PHOSPHORAMIDOTHIOATE
mf: $C_{14}H_{13}Cl_2N_2O_2PS$ mw: 375.22

SYNS: BAY 33819 ◇ BAYER 38819 ◇ DRC-714 ◇ GOPHACIDE ◇ PHOSAZETIM

TOXICITY DATA with REFERENCE
orl-rat LD50:3700 μg/kg FMCHA2 -,C117,83
skn-rat LD50:25 mg/kg FMCHA2 -,C117,83
ipr-rat LD50:3500 μg/kg AIPTAK 169,108,67
orl-mus LD50:12 mg/kg TXAPA9 25,42,73
ipr-mus LD50:5500 μg/kg AIPTAK 169,108,67
orl-dog LD50:23 mg/kg PCOC** -,107,66
orl-gpg LD50:20 mg/kg AIPTAK 169,108,67
ipr-gpg LD50:14 mg/kg AIPTAK 169,108,67

CONSENSUS REPORTS: EPA Extremely Hazardous Substances List.

SAFETY PROFILE: Poison by ingestion, skin contact, and intraperitoneal routes. A pesticide. When heated to decomposition it emits very toxic fumes of SO_x, PO_x, Cl^-, and NO_x. See also ESTERS.

BIM250 CAS:55-56-1 ***HR: 2***
1,6-BIS(5-(p-CHLOROPHENYL)BIGUANIDINO)HEXANE
mf: $C_{22}H_{30}Cl_2N_{10}$ mw: 505.52

SYNS: 1,6-BIS(p-CHLOROPHENYLDIGUANIDO)HEXANE ◇ CHLORHEXIDIN (CZECH) ◇ CHLORHEXIDINE ◇ 1,6-DI(4'-CHLOROPHENYLDIGUANIDO)HEXANE ◇ 1,1'-HEXAMETHYLENEBIS(5-(p-CHLOROPHENYL)BIGUANIDE ◇ HIBITANE ◇ OLVASAN ◇ ROTERSEPT ◇ STERIDO

TOXICITY DATA with REFERENCE
skn-hmn 1500 μg/3D-I MLD 85DKA8 -,127,77
mma-sat 400 nmol/L CBINA8 28,249,79
dnr-esc 7 μmol/disc CBINA8 28,249,79
orl-mus TDLo:1680 mg/kg (7D pre):REP MEXPAG 10,361,64
orl-rat LD50:9200 mg/kg YACHDS 6,2599,78
orl-mus LD50:9850 mg/kg YACHDS 6,2599,78

SAFETY PROFILE: Mildly toxic by ingestion. Experimental reproductive effects. A human skin irritant. Mutation data reported. When heated to decomposition it emits very toxic fumes of Cl^- and NO_x.

BIM500 CAS:72-54-8 ***HR: 3***
1,1-BIS(4-CHLOROPHENYL)-2,2-DICHLOROETHANE
DOT: NA 2761
mf: $C_{14}H_{10}Cl_4$ mw: 320.04

PROP: Crystalline solid. Mp: 110°, vap d: 11.

SYNS: 1,1-BIS(p-CHLOROPHENYL)-2,2-DICHLOROETHANE ◇ 2,2-BIS(p-CHLOROPHENYL)-1,1-DICHLOROETHANE ◇ 2,2-BIS(4-CHLOROPHENYL)-1,1-DICHLOROETHANE ◇ DDD ◇ p,p'-DDD ◇ 1,1-DICHLOOR-2,2-BIS(4-CHLOOR FENYL)-ETHAAN (DUTCH) ◇ 1,1-DICHLOR-2,2-BIS(4-CHLOR-PHENYL)-AETHAN (GERMAN) ◇ 1,1-DICHLORO-2,2-BIS(p-CHLOROPHENYL)ETHANE ◇ 1,1-DICHLORO-2,2-BIS(4-CHLOROPHENYL)-ETHANE (FRENCH) ◇ 1,1-DICHLORO-2,2-BIS(p-CHLOROPHENYL)ETHANE (DOT) ◇ 1,1-DICHLORO-2,2-BIS(PARACHLOROPHENYL)ETHANE (DOT) ◇ 1,1-DICHLORO-2,2-DI(4-CHLOROPHENYL)ETHANE ◇ DICHLORODIPHENYL DICHLOROETHANE ◇ p,p'-DICHLORODIPHENYLDICHLOROETHANE ◇ 1,1-DICLORO-2,2-BIS(4-CLORO-FENIL)-ETANO (ITALIAN) ◇ DILENE ◇ ENT 4,225 ◇ ME-1700 ◇ NCI-C00475 ◇ RCRA WASTE NUMBER U060 ◇ RHOTHANE ◇ RHOTHANE D-3 ◇ ROTHANE ◇ p,p'-TDE ◇ TDE (DOT) ◇ TETRACHLORODIPHENYLETHANE

TOXICITY DATA with REFERENCE
cyt-rat:oth 10 μg/L 34LXAP -,555,76
otr-mus:emb 28400 nmol/L JNCIAM 54,981,75
orl-rat TDLo:54 g/kg/78W-C:ETA NCITR* NCI-CG-TR-131,78
orl-mus TDLo:39 g/kg/2Y-C:NEO JNCIAM 52,883,74
orl-rat LD50:113 mg/kg GUCHAZ 6,154,73
orl-mus LDLo:600 mg/kg JPETAB 88,400,46
skn-rbt LD50:1200 mg/kg AFDOAQ 16,3,52

CONSENSUS REPORTS: IARC Cancer Review: Animal Sufficient Evidence IMEMDT 5,83,74. NCI Carcinogenesis Bioassay (feed); Clear Evidence: rat NCITR* NCI-CG-TR-131,78; No Evidence: mouse NCITR* NCI-CG-TR-131,78. EPA Genetic Toxicology Program.

DOT Classification: ORM-A; Label: None.

SAFETY PROFILE: Confirmed carcinogen with experimental carcinogenic, neoplastigenic, and tumorigenic data. Poison by ingestion. Moderately toxic by skin contact. Mutation data reported. An insecticide. When heated to decomposition it emits toxic fumes of Cl^-. See also DDT.

BIM750 CAS:72-55-9 ***HR: 3***
2,2-BIS(p-CHLOROPHENYL)-1,1-DICHLOROETHYLENE
mf: $C_{14}H_8Cl_4$ mw: 318.02

SYNS: DDE ◇ p,p'-DDE ◇ DDT DEHYDROCHLORIDE ◇ 1,1-DICHLORO-2,2-BIS(p-CHLOROPHENYL)ETHYLENE ◇ p,p'-DICHLORODIPHENYLDICHLOROETHYLENE ◇ 1,1'-DICHLOROETHENYLIDENE)BIS(4-CHLOROBENZENE) ◇ NCI-C00555

TOXICITY DATA with REFERENCE
sln-dmg-orl 1 pph ENMUDM 7,325,85
msc-mus:lym 40 mg/L/4H MUREAV 59,61,79
ipr-rat TDLo:3500 μg/kg (7D pre):REP ENDOAO 91,1095,72
orl-mus TDLo:9700 mg/kg/78W-C:CAR NCITR* NCI-TR-131,78
orl-ham TDLo:36 g/kg/86W-C:NEO CNREA8 43,776,83
orl-mus TD:17 g/kg/78W-C:CAR NCITR* NCI-TR-131,78
orl-rat LD50:880 mg/kg TXAPA9 14,515,69
orl-mus LD50:700 mg/kg JPETAB 88,400,46

CONSENSUS REPORTS: IARC Cancer Review: Animal Limited Evidence IMEMDT 5,83,74. NCI Carcinogenesis Bioassay (feed); Clear Evidence: mouse NCITR* NCI-CG-TR-131,78; No Evidence: rat NCITR* NCI-CG-TR-131,78. EPA Genetic Toxicology Program.

SAFETY PROFILE: Suspected carcinogen with experimental carcinogenic and neoplastigenic data. Poison by ingestion. Experimental reproductive effects. Mutation data reported. An insecticide. When heated to decomposition it emits very toxic fumes of Cl^-. See also CHLORINATED HYDROCARBONS, ALIPHATIC.

BIN000 CAS:80-06-8 ***HR: 2***
1,1-BIS(p-CHLOROPHENYL)METHYLCARBINOL
mf: $C_{14}H_{12}Cl_2O$ mw: 267.16

SYNS: BCPE ◇ 1,1-BIS(p-CHLOROPHENYL)ETHANOL ◇ 1,1-BIS(4-CHLOROPHENYL)ETHANOL ◇ BIS(p-CHLOROPHENYL)METHYL CARBINOL ◇ 1,1-BIS(4-CHLORPHENYL)-AETHANOL (GERMAN) ◇ CHLORFENETHOL ◇ DCPC ◇ DCPE ◇ DICHLORODIPHENYLETHANOL ◇ p,p'-DICHLORODIPHENYLMETHYLCARBINOL ◇ 4,4'-DICHLORO(METHYL BENZHYDROL) ◇ 4,4'-DICHLORO-α-METHYLBENZHYDROL ◇ 4,4'-DICHLORO-α-METHYLBENZOHYDROL ◇ DI-(p-CHLOROPHENYL)ETHANOL ◇ DI(p-CHLOROPHENYL) METHYLCARBINOL ◇ DIMITE ◇ DMC ◇ ENT 9,624 ◇ QIKRON

TOXICITY DATA with REFERENCE
orl-rat LD50:500 mg/kg ARSIM* 20,8,66
ipr-rat LD50:725 mg/kg OYYAA2 2,148,68

SAFETY PROFILE: Moderately toxic by ingestion and intraperitoneal routes. A pesticide. When heated to decomposition it emits toxic fumes of Cl^-.

BIN500 CAS:117-27-1 ***HR: 2***
1,1-BIS(p-CHLOROPHENYL)-2-NITROPROPANE
mf: $C_{15}H_{13}Cl_2NO_2$ mw: 310.19

SYNS: C.I. AZOIC DIAZO COMPONENT 37 ◇ CS 645A ◇ DNP ◇ ENT 22,784 ◇ 2-NITRO-1,1-BIS(p-CHLOROPHENYL)PROPANE ◇ 1,1'-(2-NITROPORPYLIDENE)BIS(4-CHLOROBENZENE) ◇ PROLAN ◇ PROLAN (CSC)

TOXICITY DATA with REFERENCE
orl-rat LD50:750 mg/kg MEIEDD 10,946,83

SAFETY PROFILE: Moderately toxic by ingestion. An insecticide. When heated to decomposition it emits very toxic fumes of Cl^- and NO_x. See also AROMATIC AMINES.

BIN750 CAS:80-07-9 ***HR: 1***
BIS(p-CHLOROPHENYL)SULFONE
mf: $C_{12}H_8Cl_2O_2S$ mw: 287.16

TOXICITY DATA with REFERENCE
orl-mus LD50:24 g/kg HCACAV 29,1317,46

CONSENSUS REPORTS: Reported in EPA TSCA Inventory.

SAFETY PROFILE: Mildly toxic by ingestion. See also SULFONATES. When heated to decomposition it emits very toxic fumes of Cl^- and SO_x.

BIO500 CAS:55216-04-1 ***HR: 3***
BIS(p-CHLOROPHENYLTHIO)DIMETHYLTIN
mf: $C_{14}H_{14}Cl_2S_2Sn$ mw: 435.99

SYN: BIS(p-CHLOROPHENYLTHIO)DIMETHYLSTANNANE

TOXICITY DATA with REFERENCE
ivn-mus LD50:56 mg/kg CSLNX* NX#01645

OSHA PEL: TWA 0.1 mg(Sn)/m^3 (skin)
ACGIH TLV: TWA 0.1 mg(Sn)/m^3 (skin) (Proposed: TWA 0.1 mg(Sn)/m^3; STEL 0.2 mg(Sn)/m^3 (skin))
NIOSH REL: (Organotin Compounds) TWA 0.1 mg(Sn)/m^3

SAFETY PROFILE: Poison by intravenous route. See also TIN COMPOUNDS. When heated to decomposition it emits very toxic fumes of Cl^- and SO_x.

BIO625 CAS:789-02-6 ***HR: 2***
2,2-BIS(o,p-CHLOROPHENYL)-1,1,1-TRICHLOROETHANE
mf: $C_{14}H_9Cl_5$ mw: 354.48

SYNS: o,p'-DDT ◇ 1,1,1-TRICHLORO-2-(o-CHLOROPHENYL)-2-(p-CHLOROPHENYL)ETHANE

TOXICITY DATA with REFERENCE
dns-rat-ipr 10 mg/kg JTEHD6 16,493,85
cyt-rat:oth 10 μg/L 34LXAP -,555,76
orl-rat TDLo:2150 mg/kg (female 1-22D post):REP BECTA6 12,373,74
orl-rat TDLo:250 mg/kg (female 15-19D post):TER BNEOBV 26,283,75
orl-mus LDLo:1000 mg/kg JPETAB 88,400,46
ipr-mus LD50:1577 mg/kg BECTA6 11,359,74

CONSENSUS REPORTS: EPA Genetic Toxicology Program.

SAFETY PROFILE: Moderately toxic by ingestion and intraperitoneal routes. An experimental teratogen. Other experimental reproductive effects. Mutation data reported. When heated to decomposition it emits toxic fumes of Cl^-. See also CHLORINATED HYDROCARBONS.

BIO750 CAS:115-32-2 **HR: 3**
1,1-BIS(p-CHLOROPHENYL)-2,2,2-TRICHLOROETHANOL
mf: $C_{14}H_9Cl_5O$ mw: 370.48

PROP: Material used in cancer bioassay was 40-60% pure NCITR* NCI-CG-TR-90,78.

SYNS: ACARIN ◇ 1,1-BIS(CHLOROPHENYL)-2,2,2-TRICHLOROETHANOL ◇ 1,1-BIS(4-CHLOROPHENYL)-2,2,2-TRICHLOROETHANOL ◇ CARBAX ◇ CEKUDIFOL ◇ 4-CHLORO-α-(4-CHLOROPHENYL)-α-(TRICHLOROMETHYL)BENZENEMETHANOL ◇ CPCA ◇ DECOFOL ◇ DICHLOROKELTHANE ◇ DI-(p-CHLOROPHENYL)-TRICHLOROMETHYLCARBINOL ◇ 4,4′-DICHLORO-α-(TRICHLOROMETHYL)BENZHYDROL ◇ DICOFOL ◇ DTMC ◇ ENT 23,648 ◇ FW 293 ◇ HIFOL ◇ KELTANE ◇ p,p′-KELTHANE ◇ KELTHANE (DOT) ◇ KELTHANE DUST BASE ◇ KELTHANETHANOL ◇ MILBOL ◇ MITIGAN ◇ NCI-C00486 ◇ 2,2,2-TRICHLOOR-1,1-BIS(4-CHLOORFENYL)-ETHANOL (DUTCH) ◇ 1,1,1-TRICHLOR-2,2-BIS(4-CHLORPHENYL)-AETHANOL (GERMAN) ◇ 2,2,2-TRICHLOR-1,1-BIS(4-CHLOR-PHENYL)-AETHANOL (GERMAN) ◇ 2,2,2-TRICHLORO-1,1-BIS(4-CHLOROPHENYL)-ETHANOL (FRENCH) ◇ 2,2,2-TRICHLORO-1,1-BIS(4-CLORO-FENIL)-ETANOLO (ITALIAN) ◇ 2,2,2-TRICHLORO-1,1-DI-(4-CHLOROPHENYL)ETHANOL

TOXICITY DATA with REFERENCE
sce-hmn:lym 1 μmol/L ARTODN 52,221,83
orl-rat TDLo:430 mg/kg (female 6-15D post):TER CHYCDW 21,238,87
orl-rat TDLo:430 mg/kg (female 6-15D post):REP CHYCDW 21,238,87
orl-mus TDLo:17 g/kg/78W-C:CAR NCITR* NCI-CG-TR-90,78
orl-mus TD:35 g/kg/78W-C:CAR NCITR* NCI-CG-TR-90,78
orl-rat LD50:575 mg/kg WRPCA2 9,119,70
skn-rat LD50:100 mg/kg WRPCA2 9,119,70
ipr-rat LD50:1150 mg/kg TXAPA9 15,30,69
orl-mus LD50:420 mg/kg GTPZAB 19(9),55,75
orl-rbt LD50:1810 mg/kg TXAPA9 1,119,59
skn-rbt LD50:1870 mg/kg GUCHAZ 6,195,73
orl-gpg LD50:1810 mg/kg 85DPAN -,-,71/76
orl-ckn LD50:4365 mg/kg VETNAL 60(10),64,84

CONSENSUS REPORTS: IARC Cancer Review: Group 3 IMEMDT 7,56,87; Animal Limited Evidence IMEMDT 30,87,83. NCI Carcinogenesis Bioassay (feed); Clear Evidence: mouse NCITR* NCI-CG-TR-90,78; No Evidence: rat NCITR* NCI-CG-TR-90,78. Community Right-To-Know List.

SAFETY PROFILE: Poison by ingestion and skin contact. Moderately toxic by intraperitoneal route. Human mutation data reported. Questionable carcinogen with experimental carcinogenic data. An experimental teratogen. Other experimental reproductive effects. When heated to decomposition it emits toxic fumes of Cl^-.

BIP250 **HR: 3**
BIS-5-CHLORO TOLUENE DIAZONIUM ZINC TETRACHLORIDE
mf: $C_{14}H_{12}Cl_6N_4Zn$ mw: 514.37

CONSENSUS REPORTS: Zinc and its compounds are on The Community Right-To-Know List.

SAFETY PROFILE: A shock-sensitive explosive. When heated to decomposition it emits toxic fumes of Cl^-, NO_x, and ZnO. See also ZINC COMPOUNDS.

BIP500 CAS:78371-84-3 **HR: 3**
1,3-BIS(6-CHLORO-o-TOLYL)-1-(2-(DIETHYLAMINO)ETHYL)UREA HYDROCHLORIDE
mf: $C_{21}H_{27}Cl_2N_3O•ClH$ mw: 444.87

SYN: C 3183

TOXICITY DATA with REFERENCE
eye-rbt 2% MLD ARZNAD 8,664,58
ipr-rat LD50:65 mg/kg ARZNAD 8,664,58
scu-mus LD50:70 mg/kg ARZNAD 8,664,58

SAFETY PROFILE: Poison by intraperitoneal and subcutaneous routes. An eye irritant. When heated to decomposition it emits very toxic fumes of Cl^- and NO_x.

BIP750 CAS:78371-85-4 **HR: 3**
1,3-BIS(6-CHLORO-o-TOLYL)-1-(2-PYRROLIDINYLETHYL)UREA HYDROCHLORIDE
mf: $C_{21}H_{25}Cl_2N_3O•ClH$ mw: 442.85

SYN: C 3218

TOXICITY DATA with REFERENCE
eye-rbt 2% MLD ARZNAD 8,664,58
ipr-rat LD50:42 mg/kg ARZNAD 8,664,58
scu-mus LD50:71 mg/kg ARZNAD 8,664,58

SAFETY PROFILE: Poison by intraperitoneal and subcutaneous routes. An eye irritant. When heated to decomposition it emits very toxic fumes of Cl^-, NO_x, and HCl.

BIQ250 CAS:40334-69-8 **HR: 3**
BIS(2-CHLOROVINYL)CHLOROARSINE
mf: $C_4H_4AsCl_3$ mw: 233.35

SYN: LEWISITE II

TOXICITY DATA with REFERENCE
skn-gpg LD50:8 mg/kg JPBAA7 58,411,46
scu-gpg LD50:200 μg/kg JPBAA7 58,411,46

OSHA PEL: TWA 0.5 mg(As)/m^3

CONSENSUS REPORTS: Arsenic and its compounds are on The Community Right-To-Know List.

SAFETY PROFILE: A poison by skin contact and subcutaneous routes. When heated to decomposition it emits very toxic fumes of As and Cl^-. See also ARSENIC COMPOUNDS.

BIQ500 CAS:111-94-4 ***HR: 3***
BIS(β-CYANOETHYL)AMINE
mf: $C_6H_9N_3$ mw: 123.18

$(NCC_2H_4)_2NH$

PROP: Liquid. Mp: −5.5°, bp: 173° @ 10 mm, d: 1.0165 @ 30°, vap d: 3.3.

SYNS: BBCE ◇ BIS-(2-CYANOETHYL)AMINE ◇ N,N-BIS(2-CYANOETHYL)AMINE ◇ 2-CYANO-N-(2-CYANOETHYL)ETHANAMINE ◇ DI(2-CIANOETIL)AMMINA (ITALIAN) ◇ DI-(2-CYANOETHYL)AMINE ◇ 2,2′-DICYANODIETHYLAMINE ◇ IDPN ◇ 3,3′-IMINOBISPROPANENITRILE ◇ IMINO-β,β′-DIPROPIONITRILE ◇ β,β-IMINODIPROPIONITRILE ◇ β,β′-IMINODIPROPIONITRILE ◇ 3,3′-IMINODIPROPIONITRILE ◇ 2341 I.S. ◇ USAF A-8564

TOXICITY DATA with REFERENCE
eye-rbt 100 mg AJOPAA 29,1363,46
orl-rat TDLo:840 mg/kg (1-21D preg):REP SEIJBO 21,407,81
orl-rat TDLo:840 mg/kg (1-21D preg):TER SEIJBO 21,407,81
orl-rat LD50:2700 mg/kg JIHTAB 31,60,49
ipr-mus LD50:200 mg/kg NTIS** AD277-689
skn-rbt LD50:2520 mg/kg AMIHBC 10,61,54

CONSENSUS REPORTS: Reported in EPA TSCA Inventory. Cyanide and its compounds are on the Community Right-To-Know List.

SAFETY PROFILE: A poison by intraperitoneal route. Moderately toxic by ingestion and skin contact. Experimental teratogenic and reproductive effects. An eye irritant. A storage hazard, may explode in a sealed container. When heated to decomposition it emits toxic fumes of NO_x and CN^-. See also NITRILES and AMINES.

BIR000 ***HR: 3***
BIS(n-CYCLOOCTATETRANENE)URANIUM(O)
mf: $C_{16}H_{16}U$ mw: 446.33

SAFETY PROFILE: Ignites spontaneously in air. See also URANIUM.

BIR250 ***HR: 3***
BIS(CYCLOPENTADIENYL)BIS(PENTAFLUOROPHENYL)ZIRCONIUM
mf: $C_{22}H_{16}F_{10}Zr$ mw: 555.52

SAFETY PROFILE: Explodes in air (but not nitrogen) above its melting point (219°). When heated to decomposition it emits toxic fumes of F^-. See also ZIRCONIUM COMPOUNDS.

BIR500 CAS:12194-11-5 ***HR: 3***
BIS(CYCLOPENTADIENYLCHROMIUM TRICARBONYL)MERCURY
mf: $C_{16}H_{10}Cr_2HgO_6$ mw: 602.85

SYN: HEXACARBONYLDI-PI-CYCLOPENTADIENYL-MU-MERCURIODI CHROMIUM

TOXICITY DATA with REFERENCE
ivn-mus LD50:56 mg/kg CSLNX* NX#04754

CONSENSUS REPORTS: Mercury and its compounds, as well as chromium and its compounds, are on the Community Right-To-Know List.

NIOSH REL: (Inorganic Mercury) TWA 0.05 mg(Hg)/m^3

SAFETY PROFILE: Poison by intravenous route. See also CHROMIUM COMPOUNDS, MERCURY COMPOUNDS, and CARBONYLS. When heated to decomposition it emits toxic fumes of Hg.

BIR529 CAS:1277-43-6 ***HR: 3***
BIS(CYCLOPENTADIENYL)COBALT
mf: $C_{10}H_{10}Co$ mw: 189.13

SYNS: COBALTOCENE ◇ DICYCLOPENTADIENYLCOBALT

TOXICITY DATA with REFERENCE
ims-rat TDLo:200 mg/kg/60W-I:ETA NCIUS* PH 42-64-886,SEPT,71
ipr-rat LD50:55 mg/kg NCIUS* PH 43-64-886,JAN,65
ipr-mus LD50:80 mg/kg NCIUS* PH 43-64-886,JAN,65

CONSENSUS REPORTS: Reported in EPA TSCA Inventory. Cobalt and its compounds are on the Community Right-To-Know List.

SAFETY PROFILE: Poison by intraperitoneal route. Questionable carcinogen with experimental tumorigenic data. When heated to decomposition it emits acrid smoke and fumes. See also COBALT.

BIR750 ***HR: 3***
BIS(n-CYCLOPENTADIENYL)MAGNESIUM
mf: $C_{10}H_{10}Mg$ mw: 154.41

SAFETY PROFILE: Ignites spontaneously in air. When heated to decomposition it emits acrid smoke and fumes. See also MAGNESIUM COMPOUNDS.

BIS200 CAS:15131-55-2 ***HR: 2***
BIS(2-CYCLOPENTENYL)ETHER
mf: $C_{10}H_{14}O$ mw: 150.24

TOXICITY DATA with REFERENCE
orl-rat LD50:11300 mg/kg AIHAAP 30,470,69
skn-rbt LD50:1590 mg/kg AIHAAP 30,470,69

SAFETY PROFILE: Moderately toxic by skin contact. Mildly toxic by ingestion. See also ETHERS. When heated to decomposition it emits acrid smoke and irritating fumes.

BIS250 CAS:38780-36-8 ***HR: 2***
cis-BIS(CYCLOPENTYLAMMINE)PLATINUM(II)
mf: $C_{10}H_{22}Cl_2N_2Pt$ mw: 436.33

SYNS: cis-DICHLOROBIS(CYCLOPENTYLAMMINE)PLATINUM(II) ◇ cis-DICYCLOPENTYLAMMINEDICHLOROPLATINUM(II)

TOXICITY DATA with REFERENCE
mma-sat 10 μg/plate MUREAV 95,79,82
dni-ham:ovr 26 mg/L CBINA8 14,217,76
scu-rat TDLo:109 mg/kg/6W-I:CAR CNREA8 39,913,79
ipr-mus TDLo:189 mg/kg/10W-I:CAR CNREA8 39,913,79
ipr-mus LD50:480 mg/kg CBINA8 11,145,75

SAFETY PROFILE: Moderately toxic by intraperitoneal route. Questionable carcinogen with experimental carcinogenic data. Mutation data reported. See also PLATINUM COMPOUNDS. When heated to decomposition it emits very toxic fumes of Cl^- and NO_x.

BIS500 CAS:3465-75-6 ***HR: 3***
BIS(DECANOYLOXY)DI-n-BUTYLSTANNANE
mf: $C_{28}H_{56}O_4Sn$ mw: 575.53

SYN: BIS(DECANOYLOXY)DI-N-BUTYLTIN

TOXICITY DATA with REFERENCE
orl-rat LD50:153 mg/kg 28ZPAK -,229,72

OSHA PEL: TWA 0.1 mg(Sn)/m^3 (skin)
ACGIH TLV: TWA 0.1 mg(Sn)/m^3 (skin) (Proposed: TWA 0.1 mg(Sn)/m^3; STEL 0.2 mg(Sn)/m^3 (skin))
NIOSH REL: (Organotin Compounds) TWA 0.1 mg(Sn)/m^3

SAFETY PROFILE: Poison by ingestion. See also TIN COMPOUNDS. When heated to decomposition it emits acrid and irritant fumes.

BIS750 CAS:64024-07-3 ***HR: D***
dl-cis-BISDEHYDRODOISYNOLIC ACID METHYL ETHER
mf: $C_{19}H_{22}O_3$ mw: 298.41

SYNS: 3-METHOXY-16,17-SECOESTRA-1,3,5(10),6,8-PENTAEN-17-OIC ACID ◇ RS 2874

TOXICITY DATA with REFERENCE
orl-mky TDLo:2 mg/kg (female 1-6D post):REP AJOGAH 115,101,73

SAFETY PROFILE: Experimental reproductive effects. When heated to decomposition it emits acrid smoke and irritating fumes.

BIT000 CAS:5684-13-9 ***HR: 3***
BISDEHYDROISYNOLIC ACID METHYL ESTER
mf: $C_{19}H_{22}O_3$ mw: 298.41

SYNS: DEHYDROFOLLICULINIC ACID ◇ DOISYNOESTROL ◇ 1-ETHYL-2-METHYL-7-METHOXY-1,2,3,4-TETRAHYDRO-PHENANTHRYL-2-CARBOXYLIC ACID ◇ FENOCYCLIN ◇ FENOCYCLINE ◇ 7-METHYLBISDEHYDRODOISYNOLIC ACID ◇ METILESTER del ACIDO BISDEHIDROISYNOLICO (SPANISH) ◇ 16,17-SECO-13-α-ESTRA-1,3,5,6,7,9-PENTAEN-17-OIC ACID, METHYL ESTER ◇ SURESTRINE ◇ SURESTRYL ◇ TETRADEHYDRODOISYNOLIC ACID METHYL ETHER

TOXICITY DATA with REFERENCE
orl-rat TDLo:500 μg/kg (5D preg):REP CCPTAY 14,487,76
imp-gpg TDLo:3952 μg/kg:ETA BSBSAS 8,142,51

SAFETY PROFILE: Questionable carcinogen with experimental tumorigenic data. Experimental reproductive effects. See also ETHERS and ESTERS. When heated to decomposition it emits acrid smoke and irritating fumes.

BIT030 ***HR: 2***
BISDEHYDRODOISYNOLIC ACID 7-METHYL ETHER
mf: $C_{19}H_{22}O_3$ mw: 298.41

SYNS: 7-METILETER del ACIDO BISDEHIDRODOISYNOLICO ◇ 16,17-SECOESTRA-1,3,5(10),6,8-PENTAEN-17-OIC ACID, 3-METHOXY-

TOXICITY DATA with REFERENCE
imp-gpg TDLo:21 mg/kg:ETA,REP BSBSAS 8,142,51

SAFETY PROFILE: Questionable carcinogen with experimental tumorigenic data. Experimental reproductive effects. When heated to decomposition it emits toxic fumes of NO_x.

BIT250 CAS:37333-40-7 ***HR: 3***
BIS(DIALKYLPHOSPHINOTHIOYL)DISULFIDE
mf: $C_{11}H_{26}O_4P_2S_4 \cdot C_9H_{22}O_4P_2S_4$ mw:797.04

SYNS: BIO 1,137 ◇ ENT 23,584 ◇ NIAGARA 1,137 ◇ PHOSTEX

TOXICITY DATA with REFERENCE
orl-rat LD50:265 mg/kg TXAPA9 14,515,69
skn-rat LD50:480 mg/kg TXAPA9 14,515,69

SAFETY PROFILE: Poison by ingestion. Moderately toxic by skin contact. When heated to decomposition it emits very toxic fumes of SO_x and PO_x.

BIT350 CAS:55870-36-5 ***HR: 3***
BIS(1,2-DIAMINOETHANE)DIAQUACOBALT(III) PERCHLORATE
mf: $C_4H_{20}Cl_3CoN_4O_{14}$ mw: 514.51

$((C_2H_8N_2)_2Co(H_2O)_2)(ClO_4)_3$

CONSENSUS REPORTS: Cobalt and its compounds are on the Community Right-To-Know List.

SAFETY PROFILE: The dry perchlorate is violently explosive. Upon decomposition it emits toxic fumes of Cl^- and NO_x. See also COBALT COMPOUNDS and PERCHLORATES.

BIT500 CAS:26388-78-3 ***HR: 3***
BIS-1,2-DIAMINO ETHANE DICHLORO COBALT(III) CHLORATE
mf: $C_4H_{16}Cl_3CoN_4O_3$ mw: 333.49

$((C_2H_8N_2)_2CoCl_2)ClO_3$

CONSENSUS REPORTS: Cobalt and its compounds are on the Community Right-To-Know List.

SAFETY PROFILE: Explodes when heated to 320°C. When heated to decomposition it emits toxic fumes of Cl^- and NO_x. See also COBALT COMPOUNDS and CHLORATES.

BIT750 CAS:14932-06-0 ***HR: 3***
BIS-1,2-DIAMINO ETHANE DICHLORO COBALT(III) PERCHLORATE
mf: $C_4H_{16}Cl_3CoN_4O_4$ mw: 317.49

$((C_2H_8N_2)_2CoCl_2)ClO_4$

CONSENSUS REPORTS: Cobalt and its compounds are on the Community Right-To-Know List.

SAFETY PROFILE: The perchlorate has low impact sensitivity but explodes when heated to 300°C. When heated to decomposition it emits toxic fumes of Cl^- and NO_x. See also COBALT COMPOUNDS and PERCHLORATES.

BIU000 ***HR: 3***
cis-BIS-1,2-DIAMINO ETHANE DINITRO COBALT(III) IODATE
mf: $C_4H_{16}CoIN_6O_7$ mw: 446.04

CONSENSUS REPORTS: Cobalt and its compounds are on the Community Right-To-Know List.

SAFETY PROFILE: Explodes on heating. When heated to decomposition it emits toxic fumes of I^- and NO_x. See also COBALT COMPOUNDS.

BIU125 CAS:14781-32-9 ***HR: 3***
BIS(1,2-DIAMINOETHANE)DINITROCOBALT(III) PERCHLORATE
mf: $C_4H_{16}ClCoN_6O_8$ mw: 370.59

$((C_2H_8N_2)_2Co(NO_2)_2)ClO_4$

CONSENSUS REPORTS: Cobalt and its compounds are on the Community Right-To-Know List.

SAFETY PROFILE: A dangerous explosive. Upon decomposition it emits toxic fumes of Cl^- and NO_x. See also COBALT COMPOUNDS, PERCHLORATES, and EXPLOSIVES.

BIU250 ***HR: 3***
BIS(1,2-DIAMINO ETHANE)HYDROXOOXO RHENIUM(V) DIPERCHLORATE
mf: $C_4H_{17}Cl_2N_4O_{10}Re$ mw: 538.30

SAFETY PROFILE: Explodes violently when dried at above room temperature. Shock sensitive. When heated to decomposition it emits toxic fumes of Cl^- and NO_x. See also RHENIUM and PERCHLORATES.

BIU260 CAS:19267-68-6 ***HR: 3***
BIS(1,2-DIAMINOETHANE)HYDROXOOXORHENIUM(V) PERCHLORATE
mf: $C_4H_{17}Cl_2N_4O_{10}Re$ mw: 538.31

$((C_2H_8N_2)_2Re(OH)O)(ClO_4)_2$

SAFETY PROFILE: A heat- and shock-sensitive explosive. Upon decomposition it emits toxic fumes of Cl^- and NO_x. See also RHENIUM and PERCHLORATES.

BIU500 ***HR: 3***
BIS-1,2-DIAMINO PROPANE-cis-DICHLORO CHROMIUM(III) PERCHLORATE
mf: $C_6H_{20}Cl_3CrN_4O_4$ mw: 370.61

$((C_3H_{10}N_2)_2CrCl_2)ClO_4$

CONSENSUS REPORTS: Chromium and its compounds are on the Community Right-To-Know List.

SAFETY PROFILE: Mixture with concentrated perchloric acid explodes violently. When heated to decomposition it emits toxic fumes of Cl^- and NO_x. See CHROMIUM COMPOUNDS and PERCHLORATES.

BIU750 CAS:1448-16-4 ***HR: D***
1,8-BIS(DIAZO)-2,7-OCTANEDIONE
mf: $C_8H_{10}N_4O_2$ mw: 194.22

SYNS: BIS(DIAZOACETYL)BUTANE ◇ 1,4-BIS(DIAZOACETYL)BUTANE ◇ DAB

TOXICITY DATA with REFERENCE
mmo-esc 60 nmol/L SOGEBZ 10,81,74
slt-dmg-orl 170 mmol/L SOGEBZ 17,468,81
dlt-dmg-orl 170 mmol/L SOGEBZ 17,468,81

CONSENSUS REPORTS: EPA Genetic Toxicology Program.

SAFETY PROFILE: Mutation data reported. When heated to decomposition it emits toxic fumes of NO_x.

BIU900 ***HR: 3***
BIS(DI-n-BENZENE CHROMIUM(IV)DICHROMATE:
mf: $C_{24}H_{24}Cr_4O_7$ mw: 632.44

CONSENSUS REPORTS: Chromium and its compounds are on the Community Right-To-Know List.

SAFETY PROFILE: An explosive catalyst. Upon decomposition it emits acrid smoke and fumes. See also CHROMIUM COMPOUNDS.

BIV000 CAS:58451-87-9 ***HR: 2***
2,6-BIS-(DIBENZYLHYDROXYMETHYL)PIPERIDINE
mf: $C_{35}H_{39}NO_2$ mw: 505.75

SYNS: 2,6-BIS-(DWUBENZYLOHYDROKSYMETYLO)-PIPERYDYNA (POLISH) ◇ α,α,α′,α′-TETRAKIS(PHENYLMETHYL)-2,6-PIPERIDINEDIMETHANOL

TOXICITY DATA with REFERENCE
orl-mus LD50:5000 mg/kg PJPPAA 27,549,75
ipr-mus LD50:3125 mg/kg PJPPAA 27,549,75

SAFETY PROFILE: Moderately toxic by intraperitoneal route. Mildly toxic by ingestion. When heated to decomposition it emits toxic fumes of NO_x.

BIV500 CAS:3072-84-2 ***HR: 1***
2,2-BIS(3,5-DIBROMO-4-(2,3-EPOXYPROPOXY)PHENYL)PROPANE
mf: $C_{21}H_{24}Br_4O_4$ mw: 660.09

SYN: 2,2-BIS-(3′,5′-DIBROM-4′-GLYCIDOXYFENYL)PROPAN (CZECH)

TOXICITY DATA with REFERENCE
skn-rbt 500 mg/24H MLD 28ZPAK -,137,72
eye-rbt 500 mg/24H MOD 28ZPAK -,137,72
orl-rat LD50:7160 mg/kg 28ZPAK -,137,72

SAFETY PROFILE: Mildly toxic by ingestion. A skin and eye irritant. When heated to decomposition it emits toxic fumes of Br^-.

BIV825 CAS:5412-25-9 ***HR: D***
BIS(2,3-DIBROMOPROPYL)PHOSPHATE
mf: $C_6H_{11}Br_4O_4P$ mw: 497.78

SYNS: BIS-BP ◇ 2,3-DIBROMO-1-PROPANOL HYDROGEN PHOSPHATE

TOXICITY DATA with REFERENCE
mmo-sat 50 μmol/plate APTOA6 51,76,72
mma-sat 25 nmol/plate TXAPA9 63,105,82
msc-ham:lng 20 μmol/L MUREAV 124,213,83

CONSENSUS REPORTS: EPA Genetic Toxicology Program.

SAFETY PROFILE: Mutation data reported. When heated to decomposition it emits toxic fumes of Br^- and PO_x.

BIV900 CAS:73926-85-9 ***HR: 3***
BIS(DIBUTYLAMMONIUM)HEXACHLOROSTANNATE
mf: $C_{16}H_{40}N_2{\bullet}Cl_6Sn$ mw: 591.97

SYNS: AMMONIUMYL, DIBUTYL-, HEXACHLOROSTANNATE(2-) (2:1) ◇ DIBUTYLAMINE, HEXACHLOROSTANNANE (2:1)

TOXICITY DATA with REFERENCE
ivn-mus LD50:56 mg/kg CSLNX* NX#06251

OSHA PEL: TWA 2 mg(Sn)/m^3
ACGIH TLV: TWA 2 mg(Sn)/m^3

SAFETY PROFILE: Poison by intravenous route. When heated to decomposition it emits toxic fumes of NO_x, Sn, and Cl^-.

BIW000 ***HR: 3***
BIS(DIBUTYLBORINO)ACETYLENE
mf: $C_{18}H_{36}B_2$ mw: 468.71

SAFETY PROFILE: Ignites spontaneously in air. When heated to decomposition it emits acrid smoke and fumes. See also BORON COMPOUNDS and ACETYLENE COMPOUNDS.

BIW250 CAS:64653-03-8 ***HR: 3***
BIS(DIBUTYLDITHIOCARBAMATO)DIBENZYLSTANNANE
mf: $C_{32}H_{50}N_2S_4Sn$ mw: 709.77

SYNS: BIS((DIBUTYLDITHIOCARBAMOYL)OXY)DIBENZYLSTANNANE ◇ DIBENZYLTIN BIS(DIBUTYLDITHIOCARBAMATE)

TOXICITY DATA with REFERENCE
ivn-mus LD50:180 mg/kg CSLNX* NX#02082

OSHA PEL: TWA 0.1 mg(Sn)/m^3 (skin)
ACGIH TLV: TWA 0.1 mg(Sn)/m^3 (skin) (Proposed: TWA 0.1 mg(Sn)/m^3; STEL 0.2 mg(Sn)/m^3 (skin))
NIOSH REL: (Organotin Compounds) TWA 0.1 mg(Sn)/m^3

SAFETY PROFILE: Poison by intravenous route. See also CARBAMATES and TIN COMPOUNDS. When heated to decomposition it emits very toxic fumes of NO_x and SO_x.

BIW500 CAS:66009-08-3 ***HR: 3***
BIS(DIBUTYLDITHIOCARBAMATO)DIMETHYLSTANNANE
mf: $C_{20}H_{42}N_2S_4Sn$ mw: 557.57

SYNS: BIS((DIBUTYLDITHIOCARBAMOYL)OXY)DIMETHYLSTANNANE ◇ DIMETHYLTIN BIS(DIBUTYLDITHIOCARBAMATE)

TOXICITY DATA with REFERENCE
ivn-mus LD50:180 mg/kg CSLNX* NX#02075

OSHA PEL: TWA 0.1 mg(Sn)/m^3 (skin)
ACGIH TLV: TWA 0.1 mg(Sn)/m^3 (skin) (Proposed: TWA 0.1 mg(Sn)/m^3; STEL 0.2 mg(Sn)/m^3 (skin))
NIOSH REL: (Organotin Compounds) TWA 0.1 mg(Sn)/m^3

SAFETY PROFILE: Poison by intravenous route. See also CARBAMATES and TIN COMPOUNDS. When heated to decomposition it emits very toxic fumes of NO_x and SO_x.

BIW750 CAS:13927-77-0 ***HR: 3***
BIS(DIBUTYLDITHIOCARBAMATO)NICKEL
mf: $C_{18}H_{36}N_2S_4$•Ni mw: 467.51

SYNS: DIBUTYLDITHIOCARBAMIC ACID, NICKEL SALT ◇ NICKEL DIBUTYLDITHIOCARBAMATE ◇ UV CHEK AM 104 ◇ VANGUARD N

TOXICITY DATA with REFERENCE
orl-mus TDLo:22 mg/kg/78W-I:ETA NTIS** PB223-159

CONSENSUS REPORTS: Reported in EPA TSCA Inventory. Nickel and its compounds are on The Community Right-To-Know List.

SAFETY PROFILE: Questionable carcinogen with experimental tumorigenic data. See also NICKEL COMPOUNDS and CARBAMATES. When heated to decomposition it emits very toxic fumes of SO_x and NO_x.

BIX000 CAS:136-23-2 ***HR: 3***
BIS(DIBUTYLDITHIOCARBAMATO)ZINC
mf: $C_{18}H_{38}N_2S_4Zn$ mw: 476.19

PROP: White powder. Mp: 104-108°; d: 1.24 @ 20°/20°.

SYNS: ACETO ZDBD ◇ BUTAZATE ◇ BUTAZATE 50-D ◇ BUTYL ZIMATE ◇ BUTYL ZIRAM ◇ DIBUTYLDITHIO-CARBAMIC ACID ZINC COMPLEX ◇ DIBUTYLDITHIOCARBAMIC ACID ZINC SALT ◇ USAF GY-5 ◇ VULCACURE ◇ VULKACIT LDB/C ◇ ZINC-BIBUTYLDITHIOCARBAMATE ◇ ZINC-DIBUTYLDITHIOCARBAMATE ◇ ZINC-N,N-DIBUTYLDITHIOCARBAMATE

TOXICITY DATA with REFERENCE
orl-mus TDLo:290 g/kg/78W-I:ETA NTIS** PB223-159
ipr-mus LD50:100 mg/kg NTIS** AD277-689

CONSENSUS REPORTS: Reported in EPA TSCA Inventory. Zinc and its compounds are on the Community Right-To-Know List.

SAFETY PROFILE: Poison by intraperitoneal route. Questionable carcinogen with experimental tumorigenic data. When heated to decomposition it emits very toxic fumes of NO_x, ZnO and SO_x. See also ZINC COMPOUNDS and CARBAMATES.

BIX125 CAS:16054-41-4 ***HR: D***
BIS(DICHLOROACETYL)DIAMINE
mf: $C_4H_4Cl_4N_2O_2$ mw: 253.90

SYN: 1,2-BIS(DICHLOROACETYL)HYDRAZINE

TOXICITY DATA with REFERENCE
orl-rat TDLo:1 g/kg (female 9D post):TER SEIJBO 23,267,83
orl-rat TDLo:1 g/kg (female 11D post):REP SEIJBO 23,267,83

SAFETY PROFILE: An experimental teratogen. Other experimental reproductive effects. When heated to decomposition it emits toxic fumes of Cl^-.

BIX250 CAS:1477-57-2 ***HR: 3***
N,N′-BIS(DICHLOROACETYL)-1,8-DIAMINOOCTANE
mf: $C_{12}H_{20}Cl_4N_2O_2$ mw: 366.14

SYNS: N,N′-BIS(DICHLOROACETYL)-1,8-OCTAMETHYLENEDIAMINE ◇ FERTILYSIN ◇ N,N′-OCTAMETHYLENEBIS(2,2-DICHLOROACTAMIDE) ◇ R-010-TK ◇ WIN 18441 ◇ WIN 18,446

TOXICITY DATA with REFERENCE
spm-hmn-orl 1150 mg/kg/23W TXAPA9 3,1,61
mnt-mus:oth 500 mg/kg NKEZA4 33,165,86
orl-man TDLo:150 mg/kg (42D male):REP TXAPA9 3,1,61
orl-mus TDLo:4 g/kg (female 10D post):TER NKEZA4 33,165,86
orl-man TDLo:943 mg/kg/50D:GIT 15QWAW -,93,65
ipr-mus LDLo:150 mg/kg TXAPA9 23,288,72

CONSENSUS REPORTS: EPA Genetic Toxicology Program.

SAFETY PROFILE: Poison by intraperitoneal route. Human systemic effects by ingestion: nausea and vomiting. An experimental teratogen. Human reproductive effects by ingestion: changes in spermatogenesis. An experimental teratogen. Other experimental reproductive effects. Human mutation data reported. When heated to decomposition it emits very toxic fumes of Cl^- and NO_x.

BIX500 CAS:15442-77-0 ***HR: 3***
BIS(3,4-DICHLOROBENZOATO)NICKEL
mf: $C_{14}H_6Cl_4NiO_4$ mw: 438.71

TOXICITY DATA with REFERENCE
ivn-mus LD50:100 mg/kg CSLNX* NX#03268

CONSENSUS REPORTS: Nickel and its compounds are on The Community Right-To-Know List.

OSHA PEL: (Transitional: TWA 1 mg/m^3) TWA 0.1 mg (Ni)/m^3
ACGIH TLV: TWA 0.1 mg(Ni)/m^3; (Proposed: TWA 0.05 mg(Ni)/m^3; Human Carcinogen)
NIOSH REL: (Inorganic Nickel) TWA 0.015 mg(Ni)/m^3

SAFETY PROFILE: Suspected carcinogen. Poison by intravenous route. See also NICKEL COMPOUNDS and CHLORIDES. When heated to decomposition it emits toxic fumes of Cl^-.

BIX750 CAS:133-14-2 ***HR: 3***
BIS(2,4-DICHLORO BENZOYL)PEROXIDE
DOT: UN 2137/UN 2138/UN 2139
mf: $C_{14}H_6Cl_4O_4$ mw: 380.00

SYNS: BIS(2,4-DICHLOROBENZOYL)PEROXIDE ◇ CADOX TS ◇ CADOX TS 40,50 ◇ 2,4-DICHLOROBENZOYL PEROXIDE (DOT) ◇ DI-2,4-DICHLOROBENZOYL PEROXIDE (DOT) ◇ LUPERCO CST

TOXICITY DATA with REFERENCE
ipr-mus LD50:225 mg/kg IPSTB3 3,93,76

CONSENSUS REPORTS: Reported in EPA TSCA Inventory.

DOT Classification: Organic Peroxide; Label: Organic Peroxide; Forbidden (> 75% with water).

SAFETY PROFILE: Poison by intraperitoneal route. Explosion Hazard: Pure compound is extremely shock sensitive and decomposes rapidly @ 80°. When heated to decomposition it emits toxic fumes of Cl^-. See also PEROXIDES, ORGANIC and ESTERS.

BIY000 CAS:2589-02-8 ***HR: 2***
2,2-BIS(3,5-DICHLORO-4-(2,3-EPOXYPROPOXY) PHENYL)PROPANE
mf: $C_{21}H_{24}Cl_4O_4$ mw: 482.25

SYN: 2,2-BIS-(3',5'-DICHLOR-4'-GLYCIDOXYFENYL)PROPAN (CZECH)

TOXICITY DATA with REFERENCE
skn-rbt 500 mg/24H MOD 28ZPAK -,137,72
eye-rbt 500 mg/24H MOD 28ZPAK -,137,72

SAFETY PROFILE: A skin and eye irritant. When heated to decomposition it emits toxic fumes of Cl^-.

BIY250 CAS:19721-74-5 ***HR: 3***
BIS(1,2-DICHLOROETHYL)SULFONE
mf: $C_4H_6Cl_4O_2S$ mw: 259.96

TOXICITY DATA with REFERENCE
orl-rat LD50:250 mg/kg AIHAAP 30,470,69
skn-rbt LD50:1000 mg/kg AIHAAP 30,470,69

SAFETY PROFILE: Poison by ingestion. Moderately toxic by skin contact. See also SULFONATES. When heated to decomposition it emits very toxic fumes of Cl^- and SO_x.

BIY500 ***HR: 3***
BIS(1,3-DICHLORO-1,1,3,3-TETRAETHYL-DISTANNOXANE)
mf: $C_{16}H_{40}Cl_4O_2{\cdot}2Sn$ mw: 643.72

SYN: DI-o-(CHLORODIETHYLSTANNYLOXO)BIS(CHLORODIETHYLTIN)

TOXICITY DATA with REFERENCE
ivn-mus LD50:22 mg/kg CSLNX* NX#03157

OSHA PEL: TWA 0.1 mg(Sn)/m^3 (skin)
ACGIH TLV: TWA 0.1 mg(Sn)/m^3 (skin) (Proposed: TWA 0.1 mg(Sn)/m^3; STEL 0.2 mg(Sn)/m^3 (skin))
NIOSH REL: (Organotin Compounds) TWA 0.1 mg(Sn)/m^3

SAFETY PROFILE: Poison by intravenous route. See also TIN COMPOUNDS and CHLORIDES. When heated to decomposition it emits toxic fumes of Cl^-.

BIZ000 ***HR: 3***
1,3-BIS(DI-n-CYCLOPENTADIENYL IRON)-2-PROPEN-1-ONE
mf: $C_{23}H_{20}Fe_2O$ mw: 324.11

SAFETY PROFILE: The dry material is a powerful explosive and detonator. Incompatible with perchloric acid; acetic anhydride; ether; methanol.

BJA000 CAS:73771-52-5 ***HR: 1***
1,5-BIS(4-(2,3-DIDEHYDROTRIAZIRIDINYL) PHENYL)-1,4-PENTADIEN-3-ONE
mf: $C_{17}H_{12}N_6O$ mw: 316.35

SYN: DIAZIDODIBENZALACETON (CZECH)

TOXICITY DATA with REFERENCE
skn-rbt 500 mg/24H MLD 28ZPAK -,123,72
eye-rbt 500 mg/24H MLD 28ZPAK -,123,72

SAFETY PROFILE: A skin and eye irritant. See also KETONES. When heated to decomposition it emits toxic fumes of NO_x.

BJA200 CAS:90466-79-8 ***HR: 3***
BIS(2,2-DIETHOXYETHYL)DISELENIDE
mf: $C_{12}H_{26}O_4Se_2$ mw: 392.30

SYN: DISELENIDE,BIS(2,2-DIETHOXYETHYL)-

TOXICITY DATA with REFERENCE
ivn-mus LD50:1800 μg/kg CSLNX* NX#09262

OSHA PEL: TWA 0.2 mg(Se)/m^3
ACGIH TLV: TWA 0.2 mg(Se)/m^3

SAFETY PROFILE: Poison by intravenous route.

When heated to decomposition it emits toxic fumes of Se.

BJA500 CAS:35697-34-8 ***HR: 3***
2,6-BIS(2-(DIETHYLAMINO)ETHOXY)-9,10-ANTHRACENEDIONE DIHYDROCHLORIDE
mf: $C_{26}H_{34}N_2O_4$•2ClH mw: 511.54

SYN: RMI 10024DA

TOXICITY DATA with REFERENCE
orl-mus LD50:1560 mg/kg ALACBI 12,77,79
scu-mus LD50:110 mg/kg ALACBI 12,77,79

SAFETY PROFILE: Poison by subcutaneous route. Moderately toxic by ingestion. When heated to decomposition it emits very toxic fumes of NO_x and HCl.

BJA750 CAS:57665-49-3 ***HR: 2***
1-(BIS(2-(DIETHYLAMINO)ETHYL)AMINO)-5-CHLORO-3-(p-CHLOROPHENYL)INDOLE DIHYDROCHLORIDE HEMIHYDRATE
mf: $C_{26}H_{36}Cl_2N_4$•2ClH•1/2H_2O mw: 557.49

TOXICITY DATA with REFERENCE
orl-rat LD50:780 mg/kg ARZNAD 30,919,80
orl-mus LD50:870 mg/kg ARZNAD 30,919,80

SAFETY PROFILE: Moderately toxic by ingestion. When heated to decomposition it emits very toxic fumes of Cl^- and NO_x.

BJA809 CAS:57647-13-9 ***HR: 2***
1-(BIS(2-(DIETHYLAMINO)ETHYL)AMINO)-3-PHENYLINDOLE DIHYDROCHLORIDE
mf: $C_{26}H_{38}N_4$•2ClH mw: 479.1

TOXICITY DATA with REFERENCE
orl-rat LD50:800 mg/kg ARZNAD 30,919,80
orl-mus LD50:540 mg/kg ARZNAD 30,919,80

SAFETY PROFILE: Moderately toxic by ingestion. When heated to decomposition it emits very toxic fumes of Cl^- and NO_x.

BJA825 CAS:3572-35-8 ***HR: 3***
3,6-BIS(3-DIETHYLAMINOPROPOXY)PYRIDAZINE BISMETHIODIDE
mf: $C_{20}H_{40}N_4O_2$•2I mw: 622.44

SYNS: 3,3'-(3,6-PYRIDAZINEDIYLBIS(OXY)BIS(N,N-DIETHYL-N-METHYL-1-PROPANAMINIUM DIIODIDE (9CI) ◇ (3,6-PYRIDAZINEDIYLBIS(OXYTRIMETHYLENE))BIS(DIETHYLMETHYLAMMONIUM IODIDE) ◇ WIN 4981

TOXICITY DATA with REFERENCE
orl-mus LD50:49 mg/kg JPETAB 118,395,56
ivn-mus LD50:610 μg/kg JPETAB 125,323,59
orl-cat LD50:5 mg/kg JPETAB 118,395,56
ivn-cat LD50:500 μg/kg JPETAB 118,395,56
ivn-rbt LD50:400 μg/kg JPETAB 118,395,56

SAFETY PROFILE: Poison by ingestion and intravenous routes. When heated to decomposition it emits toxic fumes of I^-, NO_x, and NH_3.

BJB500 CAS:14239-68-0 ***HR: 3***
BIS(DIETHYLDITHIOCARBAMATO)CADMIUM
mf: $C_{10}H_{20}CdN_2S_4$ mw: 408.96

SYNS: CADMIUM DIETHYL DITHIOCARBAMATE ◇ ETHYL CADMATE ◇ ETHYL TUADS

TOXICITY DATA with REFERENCE
mmo-sat 10 μg/plate MUREAV 68,313,79
dnd-esc 1 μmol/L ARTODN 46,277,80
orl-mus TDLo:7100 mg/kg/78W-I:ETA NTIS** PB223-159

CONSENSUS REPORTS: Reported in EPA TSCA Inventory. Cadmium and its compounds are on the Community Right-To-Know List.

OSHA PEL: TWA 0.1 mg(Cd)/m^3; CL 0.6 mg(Cd)/m^3 (fume)
ACGIH TLV: TWA 0.05 mg(Cd)/m^3 (Proposed: TWA 0.01 mg(Cd)/m^3 (dust), Suspected Human Carcinogen; 0.002 mg(Cd)/m^3 (respirable dust), Suspected Human Carcinogen); BEI: 10 μg/g creatinine in urine; 10 μg/L in blood.
DFG BAT: Blood 1.5 μg/dL; Urine 15 μg/dL, Suspected Carcinogen.
NIOSH REL: (Cadmium) Reduce to lowest feasible level

SAFETY PROFILE: Confirmed human carcinogen with experimental tumorigenic data. Mutation data reported. See also CADMIUM COMPOUNDS and CARBAMATES. When heated to decomposition it emits very toxic fumes of NO_x and SO_x.

BJB750 CAS:14239-51-1 ***HR: 3***
BIS(DIETHYLDITHIOCARBAMATO)MERCURY
mf: $C_{10}H_{20}HgN_2S_4$ mw: 497.15

TOXICITY DATA with REFERENCE
ipr-rat LDLo:100 mg/kg NCNSA6 5,30,53
ivn-mus LD50:18 mg/kg CSLNX* NX#02505

CONSENSUS REPORTS: Mercury and its compounds are on the Community Right-To-Know List.

OSHA PEL: (Transitional: CL 1 mg/10m^3) CL 0.1 mg(Hg)/m^3 (skin)
ACGIH TLV: TWA 0.1 mg(Hg)/m^3
NIOSH REL: (Mercury, Inorganic) TWA 0.05 mg(Hg)/m^3

SAFETY PROFILE: Poison by intravenous and intraperitoneal routes. See also MERCURY COM-

POUNDS and CARBAMATES. When heated to decomposition it emits very toxic fumes of NO_x, SO_x, and Hg.

BJC000 CAS:14324-55-1 ***HR: 3***
BIS(DIETHYLDITHIOCARBAMATO)ZINC
mf: $C_{10}H_{22}N_2S_4 \cdot Zn$ mw: 363.95

PROP: White powder. D: 1.47 @ 20°/20°.

SYNS: DIETHYLDITHIOCARBAMIC ACID ZINC SALT ◇ ETHAZATE ◇ ETHYL CYMATE ◇ ETHYL ZIMATE ◇ ETHYL ZIRUM ◇ VULCACURE ◇ VULKACIT LDA ◇ ZINC DIETHYLDITHIOCARBAMATE ◇ ZINC-N,N-DIETHYLDITHIOCARBAMATE

TOXICITY DATA with REFERENCE
eye-rbt 100 mg/24H MOD 28ZPAK -,11,72
mmo-sat 25 μg/plate MUREAV 68,313,79
mma-sat 25 μg/plate MUREAV 68,313,79
orl-mus TDLo:28 g/kg/78W-I:ETA NTIS** PB223-159
scu-mus TDLo:464 mg/kg:CAR NTIS** PB223-159
orl-rat LD50:3340 mg/kg 28ZPAK -,11,72
ipr-mus LD50:142 mg/kg KOKABN 26,358,77
orl-rbt LD50:570 mg/kg INMEAF 16,473,47

CONSENSUS REPORTS: Reported in EPA TSCA Inventory. Zinc and its compounds are on the Community Right-To-Know List.

SAFETY PROFILE: Poison by intraperitoneal route. Moderately toxic by ingestion and subcutaneous routes. Severe irritant to eyes, nose, and throat. Questionable carcinogen with experimental carcinogenic and tumorigenic data. Mutation data reported. When heated to decomposition it emits very toxic fumes of NO_x and SO_x. See also ZINC COMPOUNDS and CARBAMATES.

BJC250 CAS:738-99-8 ***HR: 3***
1,4-BIS(N,N'-DIETHYLENE PHOSPHAMIDE)PIPERAZINE
mf: $C_{12}H_{24}N_6O_2P_2$ mw: 346.36

SYNS: 1,4-BIS(BIS(1-AZIRIDINYL)PHOSPHINYL)PIPERAZINE ◇ DIPIN ◇ DIPINE ◇ ENT 50,107 ◇ 1,4-PIPERAZINEDIYLBIS(BIS(1-AZIRIDINYL)PHOSPHINE OXIDE ◇ TETRAETHYLENEIMIDEPIPERAZINE-N,N'-DIPHOSPHORIC ACID

TOXICITY DATA with REFERENCE
dlt-oin-unr 1 pph/3H-C AESAAI 62,790,69
cyt-hmn:lym 29 μmol/L SOGEBZ 10,1580,74
sce-hmn:lym 10 mg/L TGANAK 16(2),34,82
cyt-rat-ipr 60 mg/kg SOGEBZ 11,1347,75
ipr-mus LD50:90 mg/kg PCJOAU 14,363,80
orl-mus LD50:68 mg/kg RPTOAN 36,240,73
scu-mus LD50:58 mg/kg ANTBAL 21,262,76

SAFETY PROFILE: Poison by ingestion, intraperitoneal, and subcutaneous routes. Human mutation data reported. When heated to decomposition it emits very toxic fumes of PO_x and NO_x.

BJC500 ***HR: 3***
BISDIETHYLENE TRIAMINE COBALT(III) PERCHLORATE
mf: $C_8H_{26}Cl_3CoN_6O_{12}$ mw: 562.54

CONSENSUS REPORTS: Cobalt and its compounds are on the Community Right-To-Know List.

SAFETY PROFILE: Very sensitive to impact. Explodes @ 325°. When heated to decomposition it emits toxic fumes of NO_x. See also COBALT COMPOUNDS and PERCHLORATES.

BJD000 CAS:34491-12-8 ***HR: 3***
BIS(DIETHYLTHIO)CHLORO METHYL PHOSPHONATE
mf: $C_5H_{12}ClOPS_2$ mw: 218.71

SYNS: CHEMAGRO 5461 ◇ CHEMAGRO R-5461 ◇ S,S-DIETHYL-(CHLOROMETHYL)PHOSPHONODITHIOATE ◇ ENT 27,267 ◇ R-5461

TOXICITY DATA with REFERENCE
orl-rat LD50:35 mg/kg ARSIM* 20,7,66
skn-rat LD50:79 mg/kg TXAPA9 12,286,68
ipr-rat LD50:23 mg/kg TXAPA9 12,286,68
orl-mus LDLo:210 mg/kg AECTCV 14,111,85
ipr-mus LD50:43 mg/kg TXAPA9 12,286,68
orl-gpg LD50:224 mg/kg TXAPA9 12,286,68
ipr-gpg LD50:109 mg/kg TXAPA9 12,286,68

SAFETY PROFILE: Poison by ingestion, skin contact, and intraperitoneal routes. When heated to decomposition it emits very toxic fumes of SO_x, PO_x, and Cl^-.

BJD250 CAS:4394-93-8 ***HR: 3***
BIS(DIFLUOROAMINO)DIFLUOROMETHANE
mf: CF_6N_2 mw: 154.02

SAFETY PROFILE: An unstable explosive which may be initiated by phase changes. Upon decomposition it emits toxic fumes of F^- and NO_x. For preparation, handling, and storage, use protective equipment.

BJD375 CAS:30957-47-2 ***HR: 3***
1,1-BIS(DIFLUOROAMINO)-2,2-DIFLUORO-2-NITROETHYL METHYL ETHER
mf: $C_3H_3F_6N_3O_3$ mw: 243.07

SAFETY PROFILE: A shock-sensitive explosive. When heated to decomposition it emits toxic fumes of F^- and NO_x. See also ETHERS.

BJD500 CAS:13084-47-4 ***HR: 3***
1,2-BIS(DIFLUOROAMINO)ETHANOL
mf: $C_2H_4F_4N_2O$ mw: 148.05

SAFETY PROFILE: An impact-sensitive explosive. When heated to decomposition it emits toxic fumes of F^- and NO_x. See also EXPLOSIVES.

BJD750 CAS:13084-45-2 ***HR: 3***
1,2-BIS(DIFLUOROAMINO)ETHYL VINYL ETHER
mf: $C_4H_6F_4N_2O$ mw: 174.10

$F_2NCH_2CH(NF_2)OCH{=}CH_2$

SAFETY PROFILE: An impact-sensitive explosive. When heated to decomposition it emits toxic fumes of F^- and NO_x. See also EXPLOSIVES and ETHERS.

BJE000 CAS:33364-51-1 ***HR: 3***
4,4-BIS(DIFLUOROAMINO)-3-FLUOROIMINO-1-PENTENE
mf: $C_5H_6F_5N_3$ mw: 203.06

$H_2C{=}CHC(N{:}F)C(NF_2)_2CH_3$

SAFETY PROFILE: May explode if heated. When heated to decomposition it emits toxic fumes of F^- and NO_x. See also EXPLOSIVES.

BJE250 CAS:18273-30-8 ***HR: 3***
1,2-BIS(DIFLUOROAMINO)-N-NITROETHYLAMINE
mf: $C_2H_4F_4N_4O_2$ mw: 192.07

SAFETY PROFILE: May explode when heated above 75°C. When heated to decomposition it emits toxic fumes of F^- and NO_x. See also EXPLOSIVES.

BJE325 CAS:55124-14-6 ***HR: 3***
BIS(DIFLUOROBORYL)METHANE
mf: $CH_2B_2F_4$ mw: 111.64

SAFETY PROFILE: Highly reactive. Explodes in air or on contact with water. When heated to decomposition it emits toxic fumes of F^-. See also BORON COMPOUNDS and BORANES.

BJE500 CAS:52578-56-0 ***HR: 2***
BIS(DIHYDROXYPHENYL)SULFIDE
mf: $C_{12}H_{10}O_2S$ mw: 218.28

SYN: DIRESORCYL SULFIDE

TOXICITY DATA with REFERENCE
eye-rbt 500 mg SEV IHFCAY 6,1,67
orl-rat LD50:4290 mg/kg IHFCAY 6,1,67

SAFETY PROFILE: Mildly toxic by ingestion. A severe eye irritant. See also SULFIDES. When heated to decomposition it emits toxic fumes of SO_x.

BJE600 CAS:49773-64-0 ***HR: D***
2,5-BIS(3,4-DIMETHOXYPHENYL)-1,3,4-THIADIAZOLE
mf: $C_{18}H_{18}N_2O_4S$ mw: 358.44

SYN: 1,3,4-THIADIAZOLE,2,5-BIS(3,4-DIMETHOXYPHENYL)-

TOXICITY DATA with REFERENCE
scu-ham TDLo:30 mg/kg (female 3-8D post):REP IJMRAQ 86,256,87

SAFETY PROFILE: Experimental reproductive effects. When heated to decomposition it emits toxic fumes of NO_x and SO_x.

BJE700 CAS:107572-59-8 ***HR: D***
3,5-BIS(3,4-DIMETHOXYPHENYL)-1H-1,2,4-TRIAZOLE
mf: $C_{18}H_{19}N_3O_4$ mw: 341.40

SYN: 1H-1,2,4-TRIAZOLE,3,5-BIS(3,4-DIMETHOXYPHENYL)-

TOXICITY DATA with REFERENCE
scu-ham TDLo:30 mg/kg (female 3-8D post):REP IJMRAQ 86,256,87

SAFETY PROFILE: Experimental reproductive effects. When heated to decomposition it emits toxic fumes of NO_x.

BJE750 CAS:115-26-4 ***HR: 3***
BIS(DIMETHYLAMIDO)FLUORO PHOSPHATE
mf: $C_4H_{12}FN_2OP$ mw: 154.15

SYNS: BFP ◇ BFPO ◇ BIS(DIMETHYLAMINO)FLUOROPHOSPHATE ◇ BISDIMETHYLAMINOFLUOROPHOSPHINE OXIDE ◇ BIS(DIMETHYLAMIDO)PHOSPHORYL FLUORIDE ◇ CR 409 ◇ DIFO ◇ DIMEFOX ◇ DMF ◇ ENT 19,109 ◇ FLUOPHOSPHORIC ACID DI(DIMETHYLAMIDE) ◇ FLUORURE de N,N,N',N'-TETRAMETHYLE PHOSPHORO-DIAMIDE (FRENCH) ◇ HANANE ◇ PESTOX IV ◇ PESTOX XIV ◇ PESTOX 14 ◇ T-2002 ◇ TERRA-SYSTAM ◇ TERRASYTAM ◇ TERRASYTUM ◇ N,N,N',N'-TETRAMETHYL-DIAMIDOFOSFORZUUR-FLUORIDE (DUTCH) ◇ TETRAMETHYLDIAMIDOPHOSPHORIC FLUORIDE ◇ N,N,N',N'-TETRAMETHYL-DIAMIDOPHOSPHORSAEURE-FLUORID (GERMAN) ◇ TETRAMETHYLPHOSPHORODIAMIDIC FLUORIDE ◇ N,N,N,N-TETRAMETHYLPHOSPHORODIAMIDIC FLUORIDE ◇ N,N,N',N'-TETRAMETIL-FOSFORODIAMMIDO-FLUORURO (ITALIAN) ◇ TETRA SYTAM ◇ TL 792 ◇ WACKER S 14/10

TOXICITY DATA with REFERENCE
orl-rat LD50:1 mg/kg NTIS** PB158-508
ihl-rat LC50:2 mg/m^3/10M NTIS** PB158-508
skn-rat LD50:2 mg/kg WRPCA2 9,119,70
ipr-rat LD50:5 mg/kg AMIHBC 6,9,52
scu-rat LDLo:300 μg/kg NTIS** PB158-508
orl-mus LD50:2 mg/kg BESAAT 12,161,66
ihl-mus LC50:950 mg/m^3/10M NTIS** PB158-508
ipr-mus LD50:1400 μg/kg JPETAB 112,231,54
scu-mus LD50:1 mg/kg NTIS** PB158-508
ivn-dog LD50:5 mg/kg JPETAB 112,231,54

CONSENSUS REPORTS: EPA Extremely Hazardous Substances List.

SAFETY PROFILE: Poison by ingestion, skin contact, intraperitoneal, subcutaneous, and intravenous routes. When heated to decomposition it emits very toxic fumes of F^-, NO_x, and PO_x.

BJF000 CAS:494-38-2 ***HR: 3***
3,6-BIS(DIMETHYLAMINO)ACRIDINE
mf: $C_{17}H_{19}N_3$ mw: 265.39

SYNS: ACRIDINE ORANGE ◇ ACRIDINE ORANGE FREE BASE ◇ BASIC ORANGE 3RN ◇ 2,8-BISDIMETHYLAMINOACRIDINE ◇ BRILLIANT ACRIDINE ORANGE E ◇ C.I. 46005 ◇ C.I. No. 46005:1 ◇ C.I. BASIC ORANGE 14 ◇ C.I. SOLVENT ORANGE 15 ◇ 3,6-DI-(DIMETHYLAMINO)ACRIDINE ◇ EUCHRYSINE ◇ RHODULINE ORANGE ◇ SOLVENT ORANGE 15 ◇ N,N,N'-TETRAMETHYL-3,6-ACRIDINEDIAMINE ◇ WAXOLINE ORANGE A

TOXICITY DATA with REFERENCE
mmo-omi 10 μg/L MIBLAO 49,223,80
dns-rat:lvr 1 mmol/L ENMUDM 3,11,81
otr-ham:emb 1 μg/L NCIMAV 58,243,81
skn-mus TDLo:6630 mg/kg:ETA BJCAAI 23,587,69
scu-mus LD50:250 mg/kg BJEPA5 28,1,47

CONSENSUS REPORTS: IARC Cancer Review: Group 3 IMEMDT 7,56,87; Animal Inadequate Evidence IMEMDT 16,145,78

SAFETY PROFILE: Poison by subcutaneous route. Questionable carcinogen with experimental tumorigenic and carcinogenic data. Mutation data reported. When heated to decomposition it emits toxic fumes of NO_x.

BJF500 CAS:100-22-1 ***HR: 3***
p-BIS(DIMETHYLAMINO)BENZENE
mf: $C_{10}H_{16}N_2$ mw: 164.28

PROP: Leaflets. Mp: 51°, bp: 260°. Sltly sol in cold water; more sol in hot water; freely sol in alc, chloroform, ether, and petroleum ether.

SYNS: 1,4-BIS(DIMETHYLAMINO)BENZENE ◇ TETRAMETHYL-p-PHENYLENEDIAMINE ◇ N,N,N',N'-TETRAMETHYL-p-PHENYLENEDIAMINE ◇ TL 85 ◇ TMPD ◇ WURSTER'S BLUE ◇ WURSTER'S REAGENT

TOXICITY DATA with REFERENCE
orl-rat LDLo:500 mg/kg JPETAB 90,260,47
ihl-mus LCLo:780 mg/m³/10M NDRC** NDCrc-132,Dec,42
orl-qal LD50:42 mg/kg EESADV 6,149,82
orl-bwd LD50:23700 μg/kg AECTCV 12,355,83

CONSENSUS REPORTS: Reported in EPA TSCA Inventory.

SAFETY PROFILE: Poison by ingestion. Moderately toxic by inhalation. When heated to decomposition it emits toxic fumes of NO_x.

BJG000 CAS:39047-21-7 ***HR: 3***
BIS(DIMETHYLAMINOBORANE)ALUMINUM TETRAHYDROBORATE
mf: $C_4H_{22}AlB_3N_2$ mw: 157.66

$((CH_3)_2NBH_3)_2AlBH_4$

SAFETY PROFILE: Ignites on contact with air. Violent reaction on contact with water. When heated to decomposition it emits toxic fumes of NO_x. See also ALUMINUM COMPOUNDS, BORANES, and BORON COMPOUNDS.

BJG125 CAS:993-74-8 ***HR: 3***
BIS(DIMETHYLAMINO)DIMETHYLSTANNANE
mf: $C_6H_{18}N_2Sn$ mw: 236.93

$((CH_3)_2N)_2Sn(CH_3)_2$

SAFETY PROFILE: Mixture with chloroform explodes when heated. When heated to decomposition it emits toxic fumes of NO_x. See also TIN COMPOUNDS.

BJG150 CAS:1007-22-3 ***HR: 3***
3,5-BIS-DIMETHYLAMINO-1,2,4-DITHIAZOLIUM CHLORIDE
mf: $C_6H_{12}N_3S_2 \cdot Cl$ mw: 225.78

SYN: ORF 5513

TOXICITY DATA with REFERENCE
orl-rat TDLo:4 mg/kg (female 16-19D post):TER CCPTAY 21,529,80
orl-rat TDLo:150 μg/kg (3D pre):REP CCPTAY 21,529,80
ivn-mus LD50:56 mg/kg CSLNX* NX#07197
orl-qal LD50:24 mg/kg JRPFA4 48,371,76

SAFETY PROFILE: Poison by ingestion and intravenous routes. An experimental teratogen. Other experimental reproductive effects. When heated to decomposition it emits toxic fumes of Cl^-, SO_x and NO_x. See also CHLORIDES.

BJG250 CAS:3065-46-1 ***HR: 2***
BIS(2-DIMETHYLAMINOETHOXY)ETHANE
mf: $C_{10}H_{24}N_2O_2$ mw: 204.36

TOXICITY DATA with REFERENCE
orl-rat LD50:2830 mg/kg AIHAAP 30,470,69
skn-rbt LD50:1200 mg/kg AIHAAP 30,470,69

SAFETY PROFILE: Moderately toxic by ingestion and skin contact. When heated to decomposition it emits toxic fumes such as NO_x.

BJG500 CAS:54593-27-0 ***HR: 3***
3,6-BIS(2-(DIMETHYLAMINO)ETHOXY)-9H-XANTHEN-9-ONE DIHYDROCHLORIDE
mf: $C_{21}H_{26}N_2O_4 \cdot 2ClH$ mw: 443.41

SYN: RMI 10874DA

TOXICITY DATA with REFERENCE
orl-mus LD50:1780 mg/kg ALACBI 12,77,79
scu-mus LD50:353 mg/kg ALACBI 12,77,79

SAFETY PROFILE: Poison by subcutaneous route. Moderately toxic by ingestion. When heated to decomposition it emits very toxic fumes of NO_x and HCl.

BJG750 CAS:52673-65-1 ***HR: 3***
1,3-BIS(2-DIMETHYLAMINOETHYL)ADAMANTANE DIHYDROCHLORIDE
mf: $C_{18}H_{34}N_2 \cdot 2ClH$ mw: 351.46

SYN: 2,2'-(1,3-ADAMANTYLENE)N,N,N,N'-TETRAMETHYLETHYLAMINE DIHYDROCHLORIDE

TOXICITY DATA with REFERENCE
orl-mus LD50:1600 mg/kg JMCMAR 17,602,74
ipr-mus LD50:150 mg/kg JMCMAR 17,602,74

SAFETY PROFILE: Poison by intraperitoneal route. Moderately toxic by ingestion. When heated to decomposition it emits very toxic fumes of HCl and NO_x.

BJH000 CAS:57647-53-7 ***HR: 2***
1-(BIS(2-(DIMETHYLAMINO)ETHYL)AMINO)-5-METHYL-3-PHENYLINDOLE DIHYDROCHLORIDE
mf: $C_{23}H_{32}N_4 \cdot 2ClH$ mw: 437.3

TOXICITY DATA with REFERENCE
orl-rat LD50:980 mg/kg ARZNAD 30,919,80
orl-mus LD50:650 mg/kg ARZNAD 30,919,80

SAFETY PROFILE: Moderately toxic by ingestion. When heated to decomposition it emits very toxic fumes of Cl^- and NO_x.

BJH500 CAS:74758-19-3 ***HR: 2***
1-(BIS(2-(DIMETHYLAMINO)ETHYL)AMINO)-3-PHENYLINDOLE DIHYDROCHLORIDEHYDRATE
mf: $C_{22}H_{30}N_4 \cdot 2ClH \cdot H_2O$ mw: 441.2

TOXICITY DATA with REFERENCE
orl-rat LD50:1250 mg/kg ARZNAD 30,919,80
orl-mus LD50:810 mg/kg ARZNAD 30,919,80

SAFETY PROFILE: Moderately toxic by ingestion. When heated to decomposition it emits very toxic fumes of Cl^- and NO_x.

BJH750 CAS:3033-62-3 ***HR: 3***
BIS(2-DIMETHYLAMINOETHYL) ETHER
mf: $C_8H_{20}N_2O$ mw: 160.30

SYN: NIAX CATALYST AL

TOXICITY DATA with REFERENCE
skn-rbt 500 mg open SEV UCDS** 12/27/71
eye-rbt 1 mg SEV UCDS** 12/27/71
orl-rat LD50:1230 mg/kg DCTODJ 2,223,79
skn-rbt LD50:280 mg/kg AIHAAP 30,470,69

CONSENSUS REPORTS: Reported in EPA TSCA Inventory.

SAFETY PROFILE: Poison by skin contact. Moderately toxic by ingestion. A severe skin and eye irritant. See also ETHERS. When heated to decomposition it emits toxic fumes of NO_x.

BJI000 CAS:541-19-5 ***HR: 3***
BIS(β-DIMETHYLAMINOETHYL)SUCCINATE BIS(METHYLIODIDE)
mf: $C_{14}H_{30}N_2O_4 \cdot 2I$ mw: 517.92

SYNS: ASCURON ◇ CELOCURINE ◇ CHOLINE IODIDE SUCCINATE (2:1) ◇ CURACIT ◇ DIACETYLCHOLINE DIIODIDE ◇ DITILIN IODIDE ◇ SUCCINIC ACID BIS(β-DIMETHYLAMINOETHYL) ESTER BISMETHIODIDE ◇ SUCCINIC ACID, DIESTER with CHOLINE IODIDE ◇ SUCCINYLDICHOLINE IODIDE ◇ o,o-SUCCINYLDICHOLINE IODIDE ◇ SUXAMETHONIUM IODIDE

TOXICITY DATA with REFERENCE
ipr-mus LD50:5 mg/kg AEPPAE 228,371,56
ivn-mus LD50:550 μg/kg JPETAB 99,458,50
ivn-rbt LD50:15 mg/kg AIPTAK 88,1,51

SAFETY PROFILE: Poison by intravenous and intraperitoneal routes. When heated to decomposition it emits very toxic fumes of NO_x and I^-. See also IODIDES and ESTERS.

BJI125 CAS:21476-57-3 ***HR: 3***
BIS(DIMETHYLAMINO)ISOPROPYLMETHACRYLATE
mf: $C_{11}H_{22}N_2O_2$ mw: 214.35

SYN: 2-METHYL-2-PROPENOIC ACID 2-(DIMETHYLAMINO)-1-((DIMETHYLAMINO)METHYL)ETHYL ESTER (9CI)

TOXICITY DATA with REFERENCE
orl-rat LD50:1605 mg/kg 85GMAT -,26,82
ihl-rat LC50:110 mg/m^3/4H 85GMAT -,26,82
ihl-mus LC50:220 mg/m^3/2H 85GMAT -,26,82

SAFETY PROFILE: Poison by inhalation. Moderately toxic by ingestion. When heated to decomposition it emits toxic fumes of Cl^-. See also ESTERS.

BJI250 CAS:61-73-4 ***HR: 3***
3,7-BIS(DIMETHYL AMINO)PHENAZA THIONIUM CHLORIDE
mf: $C_{16}H_{18}N_3S \cdot Cl$ mw: 319.88

SYNS: AIZEN METHYLENE BLUE BH ◇ BASIC BLUE 9 ◇ 3,7-BIS-(DIMETHYLAMINO)PHENOTHIAZIN-5-IUM CHLORIDE ◇ CALCOZINE BLUE ZF ◇ CHROMOSMON ◇ C.I. BASIC BLUE 9 ◇ C.I. 52 015 (CZECH) ◇ D&C BLUE NUMBER 1 ◇ EXTERNAL BLUE 1 ◇ HIDACO METHYLENE BLUE SALT FREE ◇ LEATHER PURE BLUE HB ◇ METHYLENE BLUE ◇ METHYLENE BLUE A ◇ METHYLENE BLUE BB ◇ METHYLENE BLUE BB ZINC FREE ◇ METHYLENE BLUE CHLORIDE ◇ METHYLENE BLUE CHLORIDE (biological stain) ◇ METHYLENE BLUE D ◇ METHYLENE BLUE (medicinal) ◇ METHYLENE BLUE I (medicinal) ◇ METHYLENE BLUE NF (medicinal) ◇ METHYLENE BLUE POLYCHROME ◇ METHYLENE BLUE USP (medicinal) ◇ METHYLENE BLUE USP XII (medicinal) ◇ METHYLENIUM CERULEUM ◇ METHYLTHIONINE CHLORIDE ◇ METHYLTHIONIUM CHLORIDE ◇ MITSUI METHYLENE BLUE ◇ MODR METHYLENOVA (CZECH) ◇ SANDOCRYL BLUE BRL ◇ SCHULTZ No. 1038 ◇ SWISS BLUE

◇ TETRAMETHYLTHIONINE CHLORIDE ◇ YAMAMOTO METHYLENE BLUE B

TOXICITY DATA with REFERENCE
mma-sat 20 μg/plate ABCHA6 45,327,81
mmo-sat 100 μmol/L AMACCQ 9,77,76
dnr-sat 10 pph AGACBH 4,286,74
mmo-esc 2 μmol/L MUREAV 137,1,84
mrc-smc 10 pph AGACBH 4,286,74
orl-rat TDLo:2500 mg/kg (1-22D preg):REP AJANA2 110,29,62
unr-inf TDLo:15 mg/kg:PUL,BLD 34ZIAG -,390,69
orl-rat LD50:1180 mg/kg MarJV# 29MAR77
ipr-rat LD50:180 mg/kg AEPPAE 204,288,47
ivn-rat LD50:1250 mg/kg ARZNAD 18,678,68
orl-mus LD50:3500 mg/kg CKFRAY 12,94,63
ipr-mus LD50:150 mg/kg NTIS** AD691-490
ivn-mus LD50:77 mg/kg CKFRAY 12,94,63
orl-dog LDLo:500 mg/kg HBAMAK 4,1366,35
ivn-dog LDLo:50 mg/kg HBAMAK 4,1366,35
ivn-mky LDLo:10 mg/kg HBAMAK 4,1366,35

CONSENSUS REPORTS: EPA Genetic Toxicology Program. Reported in EPA TSCA Inventory.

SAFETY PROFILE: Poison by ingestion, intraperitoneal, intravenous, and subcutaneous routes. Human systemic effects: cyanosis, blood changes. Experimental reproductive effects. Mutation data reported. When heated to decomposition it emits very toxic fumes of NO_x, SO_x, and Cl^-.

BJI750 CAS:52673-66-2 ***HR: 3***
1,3-BIS(2-DIMETHYLAMINOPROPYL)ADAMANTANE DIHYDROCHLORIDE
mf: $C_{20}H_{38}N_2{\cdot}2ClH$ mw: 379.52

TOXICITY DATA with REFERENCE
orl-mus LD50:1200 mg/kg JMCMAR 17,602,74
ipr-mus LD50:75 mg/kg JMCMAR 17,602,74

SAFETY PROFILE: Poison by intraperitoneal route. Moderately toxic by ingestion. When heated to decomposition it emits very toxic fumes of NO_x and HCl.

BJJ000 CAS:62778-13-6 ***HR: 3***
N,N'-BIS(3-DIMETHYLAMINOPROPYL)DITHIOOXAMIDE
mf: $C_{12}H_{26}N_4S_2$ mw: 290.54

SYN: USAF MK-43

TOXICITY DATA with REFERENCE
ipr-mus LD50:100 mg/kg NTIS** AD277-689

CONSENSUS REPORTS: Reported in EPA TSCA Inventory.

SAFETY PROFILE: Poison by intraperitoneal route. When heated to decomposition it emits very toxic fumes of NO_x and SO_x.

BJJ125 CAS:3768-60-3 ***HR: 3***
BIS(DIMETHYLAMINO)SULFOXIDE
mf: $C_4H_{12}N_2OS$ mw: 136.21

$(CH_3)_2NS(:O)N(CH_3)_2$

SAFETY PROFILE: Violent reaction with sulfinyl chloride; becomes explosive above 90°C. When heated to decomposition it emits toxic fumes of SO_x and NO_x.

BJJ200 CAS:63382-64-9 ***HR: 3***
BIS(DIMETHYLARSINYLDIAZOMETHYL)MERCURY
mf: $C_6H_{12}As_2HgN_4$ mw: 589.62

$((CH_3)_2AsCN_2)_2Hg$

CONSENSUS REPORTS: Arsenic and its compounds, as well as mercury and its compounds, are on the Community Right-To-Know List.

SAFETY PROFILE: An explosive. When heated to decomposition it emits toxic fumes of As, Hg, and NO_x. See also MERCURY COMPOUNDS and ARSENIC COMPOUNDS.

BJJ250 CAS:503-80-0 ***HR: 3***
BIS-DIMETHYL ARSINYL OXIDE
mf: $C_4H_{12}As_2O$ mw: 225.9

$((CH_3)_2As)_2O$

CONSENSUS REPORTS: Arsenic and its compounds are on the Community Right-To-Know List.

SAFETY PROFILE: Ignites spontaneously in air. When heated to decomposition it emits toxic fumes of As. See also ARSENIC COMPOUNDS.

BJJ500 CAS:591-10-6 ***HR: 3***
BIS-DIMETHYL ARSINYL SULFIDE
mf: $C_4H_{12}As_2S$ mw: 242

$((CH_3)_2As)_2S$

CONSENSUS REPORTS: Arsenic and its compounds are on the Community Right-To-Know List.

SAFETY PROFILE: Ignites spontaneously in air. When heated to decomposition it emits toxic fumes of SO_x and As. See also ARSENIC COMPOUNDS and SULFIDES.

BJJ750 CAS:69402-04-6 ***HR: 1***
1,2-BIS(3,7-DIMETHYL-5-n-BUTOXY-1-AZA-5-BORA-4,6-DIOXOCYCLOOCTYL)ETHANE
mf: $C_{26}H_{46}B_2N_2O_6$ mw: 504.36

SYN: 1,1'-ETHYLENEBIS(5-BUTOXY-3,7-DIMETHYL-1,5-AZABOROCINE-4,6-DIONE

TOXICITY DATA with REFERENCE
skn-rbt 500 mg IHFCAY 6,1,67
eye-rbt 100 mg IHFCAY 6,1,67
orl-rat LD50:5660 mg/kg IHFCAY 6,1,67

SAFETY PROFILE: Mildly toxic by ingestion. A skin and eye irritant. When heated to decomposition it emits toxic fumes of NO_x.

BJK000 CAS:60605-72-3 ***HR: D***
BIS(DIMETHYLCARBAMODITHIOATO)((1,2-ETHANEDIYLBIS(CARBAMODITHIOATO))(2-)) DIZINC
mf: $C_{10}H_{18}N_4S_8Zn_2$ mw: 581.54

SYNS: BISDITHANE ◇ DIZINC BIS(DIMETHYLDITHIOCARBAMATE)ETHYLENEBIS(DITHIOCARBAMATE) ◇ POLYCARBAMATE

TOXICITY DATA with REFERENCE
mmo-sat 50 µg/plate MUREAV 116,185,83
mrc-bcs 4 µg/disc/24H MUREAV 40,19,76

CONSENSUS REPORTS: Zinc and its compounds are on the Community Right-To-Know List.

SAFETY PROFILE: Mutation data reported. See also ZINC COMPOUNDS and CARBAMATES. When heated to decomposition it emits very toxic fumes of NO_x and SO_x.

BJK250 CAS:15521-65-0 ***HR: 1***
BIS(DIMETHYLDITHIOCARBAMATO)NICKEL
mf: $C_6H_{12}N_2S_4 \cdot Ni$ mw: 299.15

TOXICITY DATA with REFERENCE
orl-rat LD50:17 g/kg RCTEA4 45(3),627,72

CONSENSUS REPORTS: Reported in EPA TSCA Inventory. Nickel and its compounds are on the Community Right-To-Know List.

SAFETY PROFILE: Mildly toxic by ingestion. See also NICKEL COMPOUNDS and CARBAMATES. When heated to decomposition it emits very toxic fumes of NO_x and SO_x.

BJK500 CAS:137-30-4 ***HR: 3***
BIS(DIMETHYLDITHIOCARBAMATO)ZINC
mf: $C_6H_{12}N_2S_4 \cdot Zn$ mw: 305.81

PROP: White powder. Mp: 248-250°; d: 1.65 @ 20°/20°.

SYNS: AAPROTECT ◇ AAVOLEX ◇ AAZIRA ◇ ACCELERATOR L ◇ ACETO ZDED ◇ ACETO ZDMD ◇ ALCOBAM ZM ◇ AMYL ZIMATE ◇ ANTENE ◇ BIS(DIMETHYLCARBAMODITHIOATO-S,S')ZINC ◇ BIS(DIMETHYLDITHIOCARBAMATE de ZINC) (FRENCH) ◇ BIS(N,N-DIMETIL-DITIOCARBAMMATO) DI ZINCO (ITALIAN) ◇ CARBAMIC ACID, DIMETHYLDITHIO-, ZINC SALT (2:) ◇ CARBAZINC ◇ CIRAM ◇ CORONA COROZATE ◇ COROZATE ◇ CUMAN ◇ CUMAN L ◇ CYMATE ◇ DIMETHYLCARBAMODITHIOIC ACID, ZINC COMPLEX ◇ DIMETHYLCARBAMODITHIOIC ACID, ZINC SALT ◇ DIMETHYLDITHIOCARBAMATE ZINC SALT ◇ DIMETHYLDITHIOCARBAMIC ACID, ZINC SALT ◇ DRUPINA 90 ◇ ENT 988 ◇ EPTAC 1 ◇ FUCLASIN ◇ FUCLASIN ULTRA ◇ FUKLASIN ◇ FUNGOSTOP ◇ HERMAT ZDM ◇ HEXAZIR ◇ KARBAM WHITE ◇ METHASAN ◇ METHAZATE ◇ METHYL ZIMATE ◇ METHYL ZINEB ◇ METHYL ZIRAM ◇ MEXENE ◇ MEZENE ◇ MILBAM ◇ MILBAN ◇ MOLURAME ◇ MYCRONIL ◇ NCI-C50442 ◇ ORCHARD BRAND ZIRAM ◇ POMARSOL Z FORTE ◇ PRODARAM ◇ RHODIACID ◇ SOXINAL PZ ◇ SOXINOL PZ ◇ TRICARBAMIX Z ◇ TSIMAT ◇ TSIRAM (RUSSIAN) ◇ USAF P-2 ◇ VANCIDE MZ-96 ◇ ZERLATE ◇ ZIMATE ◇ ZIMATE METHYL ◇ ZINC BIS(DIMETHYLDITHIOCARBAMATE) ◇ ZINC BIS(DIMETHYLDITHIOCARBAMOYL)DISULPHIDE ◇ ZINC DIMETHYLDITHIOCARBAMATE ◇ ZINC N,N-DIMETHYLDITHIOCARBAMATE ◇ ZINCMATE ◇ ZINK-BIS(N,N-DIMETHYL-DITHIOCARBAMAAT) (DUTCH) ◇ ZINK-BIS(N,N-DIMETHYL-DITHIOCARBAMAT) (GERMAN) ◇ ZINKCARBAMATE ◇ ZINK-(N,N-DIMETHYL-DITHIOCARBAMAT) (GERMAN) ◇ ZIRAM ◇ ZIRAMVIS ◇ ZIRASAN ◇ ZIRBERK ◇ ZIREX 90 ◇ ZIRIDE ◇ ZIRTHANE ◇ ZITOX

TOXICITY DATA with REFERENCE
mmo-sat 5 µg/plate MUREAV 68,313,79
cyt-hmn:lym 10 nmol/L TXCYAC 4,331,75
orl-rat TDLo:250 mg/kg (female 6-15D post):TER EESADV 7,531,83
orl-rat TDLo:500 mg/kg (female 6-15D post):REP EESADV 7,531,83
orl-rat TDLo:12978 mg/kg/2Y-I:CAR NTPTR* NTP-TR-238,82
imp-rat TDLo:60 mg/kg:ETA VPITAR 29,71,70
orl-rat TD:25956 mg/kg/2Y-I:CAR NTPTR* NTP-TR-238,82
orl-rat LD50:1400 mg/kg FMCHA2 -,C259,80
ipr-rat LD50:23 mg/kg JAPMA8 41,662,52
ihl-rat LD50:1230 mg/kg EQSFAP 3,618,75

CONSENSUS REPORTS: IARC Cancer Review: Group 3 IMEMDT 7,56,87, Animal Inadequate Evidence IMEMDT 12,259,76; NTP Carcinogenesis Bioassay (feed); Clear Evidence: mouse, rat NTPTR* NTP-TR-238,83. EPA Genetic Toxicology Program. Reported in EPA TSCA Inventory. Zinc and its compounds are on the Community Right-To-Know List.

SAFETY PROFILE: Poison by ingestion, intraperitoneal, and intravenous routes. Moderately toxic by inhalation. Questionable carcinogen with experimental carcinogenic and tumorigenic data. An experimental teratogen. Other experimental reproductive effects. Human mutation data reported. See also ZINC COMPOUNDS and CARBAMATES. Severe irritant to eyes, nose, and throat. When heated to decomposition it emits very toxic fumes of NO_x and SO_x.

BJK600 CAS:80387-97-9 ***HR: D***

(((3,5-BIS(1,1-DIMETHYLETHYL)-4-HYDROXYPHENYL)METHYL)THIO)ACETIC ACID 2-ETHYLHEXYL ESTER

mf: $C_{25}H_{42}O_3S$ mw: 422.73

SYN: ACETIC ACID, (((3,5-BIS(1,1-DIMETHYLETHYL)-4-HYDROXYPHENYL)METHYL)THIO)-, 2-ETHYLHEXYL ESTER

TOXICITY DATA with REFERENCE
orl-rat TDLo:3 g/kg (female 6-15D post):REP EPASR* 8EHQ-0285-0544
orl-rat TDLo:3 g/kg (female 6-15D post):TER EPASR* 8EHQ-0285-0544

CONSENSUS REPORTS: Reported in EPA TSCA Inventory.

SAFETY PROFILE: An experimental teratogen. Other experimental reproductive effects. When heated to decomposition it emits toxic fumes of SO_x.

BJK750 CAS:4636-83-3 ***HR: 3***

1,1'-BIS(3,5-DIMETHYLMORPHOLINOCARBONYLMETHYL)-4,4'-BIPYRIDYNIUM DICHLORIDE

mf: $C_{26}H_{36}N_4O_4 \cdot 2Cl$ mw: 539.56

SYNS: 1,1'-BIS(3,5-DIMETHYLMORPHOLINOCARBONYLMETHYL)-4,4'-BIPYRIDINIUM-DICHLORID (GERMAN) ◇ 1,1'-BIS(2-(3,5-DIMETHYL-4-MORPHOLINYL)-2-OXOETHYL)-4,4-BIPYRIDINIUM DICHLORIDE ◇ CEROXONE ◇ MORFAMQUAT ◇ MORFOXONE ◇ MORPHANQUAT DICHLORIDE ◇ PP 745

TOXICITY DATA with REFERENCE
orl-rat LD50:345 mg/kg GUCHAZ 6,367,73
orl-mus LD50:325 mg/kg 28ZEAL 5,158,76
orl-cat LD50:160 mg/kg 28ZEAL 5,158,76
orl-ckn LD50:367 mg/kg 31ZOAD 1,311,68

SAFETY PROFILE: Poison by ingestion. When heated to decomposition it emits very toxic fumes of Cl^- and NO_x.

BJL000 CAS:3081-14-9 ***HR: 2***

N,N'-BIS(1,4-DIMETHYLPENTYL)-p-PHENYLENEDIAMINE

mf: $C_{20}H_{36}N_2$ mw: 304.58

SYNS: N,N-DI(1,4-DIMETHYLPENTYL)-p-PHENYLDIAMINE ◇ EASTOZONE ◇ EASTOZONE 33 ◇ NCI-C56337 ◇ SANTOFLEX 77 ◇ TENAMENE

TOXICITY DATA with REFERENCE
orl-rat LD50:750 mg/kg 85GMAT -,59,82
ipr-rat LDLo:800 mg/kg RCTEA4 45,627,72
orl-mus LD50:1700 mg/kg 85GMAT -,59,82
ipr-mus LDLo:400 mg/kg RCTEA4 45,627,72

CONSENSUS REPORTS: Reported in EPA TSCA Inventory.

SAFETY PROFILE: Moderately toxic by ingestion and intraperitoneal routes. When heated to decomposition it emits toxic fumes of NO_x.

BJL250 ***HR: 3***

BISDIMETHYL STIBINYL OXIDE

mf: $C_4H_{12}OSb_2$ mw: 319.6

CONSENSUS REPORTS: Antimony and its compounds are on the Community Right-To-Know List.

SAFETY PROFILE: Antimony compounds are generally highly toxic. Ignites spontaneously in air. When heated to decomposition it emits acrid smoke and fumes. See also ANTIMONY COMPOUNDS.

BJL500 ***HR: 3***

BIS(DIMETHYL THALLIUM)ACETYLIDE

mf: $C_6H_{12}Tl_2$ mw: 492.90

CONSENSUS REPORTS: Thallium and its compounds are on the Community Right-To-Know List.

SAFETY PROFILE: An extremely heat- and friction-sensitive explosive. Heat and friction sensitive. When heated to decomposition it emits acrid smoke and fumes. See also THALLIUM COMPOUNDS and ACETYLIDES.

BJL600 CAS:97-74-5 ***HR: 3***

BIS(DIMETHYLTHIOCARBAMOYL)SULFIDE

mf: $C_6H_{12}N_2S_3$ mw: 208.38

SYNS: ACETO TMTM ◇ BIS(DIMETHYLTHIOCARBAMYL) MONOSULFIDE ◇ CARBAMIC ACID, DIMETHYLDITHIO-, ANHYDROSULFIDE ◇ MONEX ◇ MONO-THIURAD ◇ MONOTHIURAM ◇ PENNAC MS ◇ TETRAMETHYLTHIURAMMONIUM SULFIDE ◇ TETRAMETHYLTHIURAM MONOSULFIDE ◇ TETRAMETHYLTHIURAMONOSULFIDE ◇ TETRAMETHYLTHIURAM SULFIDE ◇ TETRAMETHYLTRITHIO CARBAMIC ANHYDRIDE ◇ 1,1'-THIOBIS-(N,N-DIMETHYLTHIO)FORMAMIDE ◇ THIONEX ◇ THIONEX RUBBER ACCELERATOR ◇ TMTM ◇ TMTMS ◇ UNADS ◇ USAF B-32 ◇ USAF EK-P-6255 ◇ VULKACIT THIURAM MS/C

TOXICITY DATA with REFERENCE
mmo-sat 100 μg/plate MUREAV 68,313,79
sce-ham:ovr 100 nmol/L SWEHDO 9(Suppl 2),27,83
scu-mus TDLo:900 mg/kg (6-14D preg):TER NTIS** PB223-160
scu-mus TDLo:100 mg/kg:ETA NTIS** PB223-159
ipr-rat LD50:383 mg/kg JNPHAG 9,35,78
orl-mus LD50:818 mg/kg ENVRAL 28(1),199,82
ipr-mus LD50:300 mg/kg NTIS** AD277-689
orl-dog LDLo:100 mg/kg RCTEA4 44,513,71

CONSENSUS REPORTS: Reported in EPA TSCA Inventory.

SAFETY PROFILE: Poison by ingestion and intraperitoneal routes. Questionable carcinogen with experimental tumorigenic data. Mutation data reported. An experimental teratogen. When heated to decomposition it emits very toxic fumes of NO_x and SO_x. See also SULFIDES.

BJM250 CAS:58451-85-7 ***HR: 2***
2,6-BIS(DIPHENYLHYDROXYMETHYL)PIPERIDINE
mf: $C_{31}H_{31}NO_2$ mw: 449.63

TOXICITY DATA with REFERENCE
orl-mus LD50:5000 mg/kg PJPPAA 27,549,75
ipr-mus LD50:2000 mg/kg PJPPAA 27,549,75

SAFETY PROFILE: Moderately toxic by intraperitoneal route. Mildly toxic by ingestion. When heated to decomposition it emits toxic fumes of NO_x.

BJM500 CAS:58451-82-4 ***HR: 2***
2,6-BIS(DIPHENYLHYDROXYMETHYL)PYRIDINE
mf: $C_{31}H_{25}NO_2$ mw: 443.57

TOXICITY DATA with REFERENCE
orl-mus LD50:5000 mg/kg PJPPAA 27,549,75
ipr-mus LD50:3000 mg/kg PJPPAA 27,549,75

SAFETY PROFILE: Moderately toxic by intraperitoneal route. Mildly toxic by ingestion. When heated to decomposition it emits toxic fumes of NO_x.

BJM625 ***HR: 3***
3-(BIS(3,3-DIPHENYLPROPYL)AMINO)PROPANE-1-OL
mf: $C_{33}H_{37}NO$ mw: 463.71

SYN: 3-(BIS(3,3-DIPHENYLPROPYL)AMINO)-1-PROPANOL

TOXICITY DATA with REFERENCE
ipr-rat LD50:167 mg/kg ARZNAD 25,632,75
ivn-rat LD50:40300 µg/kg ARZNAD 25,632,75
ipr-mus LD50:129 mg/kg ARZNAD 25,632,75
ivn-mus LD50:30200 µg/kg ARZNAD 25,632,75

SAFETY PROFILE: Poison by intravenous and intraperitoneal routes. When heated to decomposition it emits toxic fumes of NO_x.

BJM700 CAS:38998-91-3 ***HR: 3***
BIS(1,3-DITHIOCYANATO-1,1,3,3-TETRABUTYLDISTANNOXANE)
mf: $C_{36}H_{72}N_4O_2S_4Sn_4$ mw: 1196.12

SYNS: DISTANNOXANE,BIS(1,3-DITHIOCYANATO-1,1,3,3-TETRABUTYL- ◇ DI-mu-(THIOCYANATODI-n-BUTYLSTANNYLOXO)BIS(THIOCYANATODI-n-BUTYLTIN)

TOXICITY DATA with REFERENCE
ivn-mus LD50:180 mg/kg CSLNX* NX#03006

OSHA PEL: TWA 0.1 mg(Sn)/m^3
ACGIH TLV: TWA 0.1 mg(Sn)/m^3; STEL 0.2 mg/m^3 (skin)
NIOSH REL: 10H TWA 0.1 mg(Sn)/m^3

SAFETY PROFILE: Poison by intravenous route. When heated to decomposition it emits toxic fumes of NO_x, SO_x, and Sn.

BJM750 CAS:10171-76-3 ***HR: 3***
BIS(2,5-ENDOMETHYLENECYCLOHEXYLMETHYL)AMINE
mf: $C_{16}H_{27}N$ mw: 233.44

TOXICITY DATA with REFERENCE
skn-rbt 100 µg/24H open AIHAAP 23,95,62
orl-rat LD50:1410 mg/kg AIHAAP 23,95,62
skn-rbt LD50:110 mg/kg AIHAAP 23,95,62

SAFETY PROFILE: Poison by skin contact. Moderately toxic by ingestion. A skin irritant. When heated to decomposition it emits toxic fumes of NO_x.

BJN000 CAS:10580-77-5 ***HR: 3***
BIS(3,4-EPOXYBUTYL) ETHER
mf: $C_8H_{14}O_3$ mw: 158.22

TOXICITY DATA with REFERENCE
orl-rat LD50:1070 mg/kg AIHAAP 30,470,69
skn-rbt LD50:250 mg/kg AIHAAP 30,470,69

SAFETY PROFILE: Poison by skin contact. Moderately toxic by ingestion. See also ETHERS. When heated to decomposition it emits acrid smoke and fumes.

BJN250 CAS:2386-90-5 ***HR: 3***
BIS(2,3-EPOXYCYCLOPENTYL) ETHER
mf: $C_{10}H_{14}O_3$ mw: 182.24

SYNS: EP-205 ◇ ERR 4205 ◇ 2,2'-OXYBIS-6-OXABICYCLO-(3.1.0)HEXANE

TOXICITY DATA with REFERENCE
skn-rbt 500 mg open MLD UCDS** 12/13/63
mmo-sat 5700 µg/plate CIHPDR 6,210,84
mma-sat 5700 µg/plate CIHPDR 6,210,84
sce-hmn:lym 50 mg/L CIHPDR 6,210,84
mnt-mus-orl 1 g/kg CIHPDR 6,210,84
skn-mus TDLo:156 g/kg/2Y-I:CAR NTIS** ORNL-5375
skn-mus TD:312 g/kg/2Y-I:NEO,REP CNREA8 39,1718,79
skn-mus TD:395 g/kg/132W-I:ETA NTIS** ORNL-5762
orl-rat LDLo:2140 mg/kg AIHAAP 23,95,62
skn-mus LDLo:2000 mg/kg NTIS** ORNL-5375

CONSENSUS REPORTS: EPA Genetic Toxicology Program. Reported in EPA TSCA Inventory.

SAFETY PROFILE: Moderately toxic by ingestion. A systemic irritant by skin contact and ingestion. Experimental reproductive effects. Questionable carcinogen with experimental carcinogenic and neoplastigenic data. See also ETHERS. When heated to decomposition it emits acrid smoke and irritating fumes.

BJN500 CAS:7487-28-7 ***HR: 2***
BIS(2,3-EPOXY-2-METHYLPROPYL)ETHER
mf: $C_8H_{14}O_3$ mw: 158.22

SYN: BIS(2-METHYLGLYCIDYL)ETHER

TOXICITY DATA with REFERENCE
skn-rbt 500 mg open MLD UCDS** 4/21/67
orl-rat LD50:1680 mg/kg AIHAAP 24,305,63
skn-rbt LD50:1250 mg/kg UCDS** 4/21/67

SAFETY PROFILE: Moderately toxic by ingestion and skin contact. A skin irritant. See also ETHERS. When heated to decomposition it emits acrid and irritating fumes and smoke.

BJN750 CAS:10043-09-1 ***HR: 2***
2,3-BIS(2,3-EPOXYPROPOXY)-1,4-DIOXANE
mf: $C_{10}H_{16}O_6$ mw: 232.26

SYN: 2,3-BIS(GLYCIDYLOXY)-1,4-DIOXANE

TOXICITY DATA with REFERENCE
orl-rat LD50:1070 mg/kg AIHAAP 24,305,63
skn-rbt LD50:1590 mg/kg AIHAAP 23,95,62

SAFETY PROFILE: Moderately toxic by ingestion and skin contact. When heated to decomposition it emits acrid smoke and irritating fumes.

BJN850 CAS:63951-08-6 ***HR: 2***
N,N-BIS(2-(2,3-EPOXYPROPOXY)ETHOXY)ANILINE
mf: $C_{16}H_{23}NO_6$ mw: 325.40

SYNS: ANILINE,N,N-BIS(2-(2,3-EPOXYPROPOXY)ETHOXY)- ◇ DIGLYCIDYL ETHER of N,N-BIS(2-HYDROXYETHOXYETHYL)ANILINE

TOXICITY DATA with REFERENCE
scu-mus TDLo:5600 mg/kg/60W-I:ETA FCTXAV 4,365,66

SAFETY PROFILE: Questionable carcinogen with experimental tumorigenic data. When heated to decomposition it emits toxic fumes of NO_x.

BJN875 CAS:7329-29-5 ***HR: 2***
N,N-BIS(2-(2,3-EPOXYPROPOXY)ETHYL)ANILINE
mf: $C_{16}H_{23}NO_4$ mw: 293.40

SYNS: ANILINE,N,N-BIS(2-(2,3-EPOXYPROPOXY)ETHYL)- ◇ DIGLYCIDYL ETHER of PHENYLDIETHANOLAMINE

TOXICITY DATA with REFERENCE
scu-rat TDLo:2820 mg/kg/43W-I:ETA BECCAN 42,37,64

SAFETY PROFILE: Questionable carcinogen with experimental tumorigenic data. When heated to decomposition it emits toxic fumes of NO_x.

BJO000 CAS:13561-08-5 ***HR: 2***
BIS(2,6-(2,3-EPOXYPROPYL))PHENYL GLYCIDYL ETHER
mf: $C_{15}H_{18}O_4$ mw: 262.33

SYN: 2,6-BIS(2,3-EPOXYPROPYL)PHENYL-2,3-EPOXYPROPYLETHER

TOXICITY DATA with REFERENCE
orl-rat LD50:1620 mg/kg TXAPA 9,28,313,74
skn-rbt LD50:2520 mg/kg TXAPA9 28,313,74

CONSENSUS REPORTS: Reported in EPA TSCA Inventory.

SAFETY PROFILE: Moderately toxic by ingestion and skin contact. See also ETHERS. When heated to decomposition it emits acrid smoke and irritating fumes.

BJO125 CAS:20539-85-9 ***HR: 3***
BIS(ETHOXYCARBONYLDIAZOMETHYL)MERCURY
mf: $C_8H_{10}HgN_4O_4$ mw: 426.78

$(CH_3CH_2CO \cdot CN_2)_2Hg$

CONSENSUS REPORTS: Mercury and its compounds are on the Community Right-To-Know List.

SAFETY PROFILE: An impact sensitive explosive which decomposes at its mp: 104°C. When heated to decomposition it emits toxic fumes of NO_x and Hg. See also MERCURY COMPOUNDS.

BJO225 CAS:109-44-4 ***HR: 2***
BIS(2-ETHOXYETHYL) ADIPATE
mf: $C_{14}H_{26}O_6$ mw: 290.40

SYNS: ADIPIC ACID, BIS(2-ETHOXYETHYL) ESTER ◇ DIETHOXY ETHYL ADIPATE ◇ HEXANOIC ACID, BIS(2-ETHOXYETHYL) ESTER

TOXICITY DATA with REFERENCE
skn-man 50 mg/24H MLD CTOIDG 94(8),41,79
skn-rat 100 mg/24H MLD CTOIDG 94(8),41,79
skn-rbt 100 mg/24H SEV CTOIDG 94(8),41,79
skn-gpg 100 mg/24H MLD CTOIDG 94(8),41,79

CONSENSUS REPORTS: Reported in EPA TSCA Inventory.

SAFETY PROFILE: A severe skin irritant. When heated to decomposition it emits acrid smoke and irritating fumes.

BJO250 CAS:67856-66-0 ***HR: 3***
BIS(2-ETHOXYETHYL)NITROSOAMINE
mf: $C_8H_{18}N_2O$ mw: 158.28

SYN: N-NITROSOBIS(2-ETHOXYETHYL)AMINE

TOXICITY DATA with REFERENCE
mma-sat 50 μg/plate MUREAV 66,1,79
orl-rat TDLo:3250 mg/kg/50W-I:ETA CNREA8 38,2391,78

SAFETY PROFILE: Questionable carcinogen with experimental tumorigenic data. Mutation data reported. When heated to decomposition it emits highly toxic fumes of NO_x. See also NITROSAMINES.

BJO500 CAS:101-93-9 ***HR: 3***
N,N′-BIS(p-ETHOXYPHENYL)ACETAMIDINE
mf: $C_{18}H_{22}N_2O_2$ mw: 298.42

SYNS: N′,N(sup 2)-BIS(p-ETHOXYPHENYL)ACETAMIDINE ◇ N,N′-BIS(4-ETHOXYPHENYL)ETHANIMIDAMIDE ◇ FENACAINE ◇ HOLOCAINE ◇ PHENACAINE ◇ TANICAINE

TOXICITY DATA with REFERENCE
ivn-cat LDLo:10 mg/kg PHREA7 12,190,32
unr-rbt LDLo:5 mg/kg HBAMAK 4,1289,35
ipr-gpg LDLo:50 mg/kg PHREA7 12,190,32
scu-gpg LDLo:53 mg/kg PHREA7 12,190,32
ivn-gpg LDLo:15 mg/kg PHREA7 12,190,32

SAFETY PROFILE: Poison by subcutaneous and possibly other routes. When heated to decomposition it emits toxic fumes of NO_x.

BJP000 CAS:122-34-9 ***HR: 3***
2,4-BIS(ETHYLAMINO)-6-CHLORO-s-TRIAZINE
mf: $C_7H_{12}ClN_5$ mw: 201.69

SYNS: AKTINIT S ◇ AQUAZINE ◇ BATAZINA ◇ 2,4-BIS(AETHYLAMINO)-6-CHLOR-1,3,5-TRIAZIN (GERMAN) ◇ BITEMOL ◇ BITEMOL S 50 ◇ CAT (herbicide) ◇ CDT ◇ CEKUSAN ◇ CEKUZINA-S ◇ CET ◇ 1-CHLORO-3,5-BISETHYLAMINO-2,4,6-TRIAZINE ◇ 2-CHLORO-4,6-BIS(ETHYLAMINO)-s-TRIAZINE ◇ 2-CHLORO-4,6-BIS(ETHYLAMINO)-1,3,5-TRIAZINE ◇ FRAMED ◇ GEIGY 27,692 ◇ GESARAN ◇ GESATOP ◇ HERBAZIN ◇ HERBEX ◇ HERBOXY ◇ HUNGAZIN DT ◇ PREMAZINE ◇ PRIMATOL S ◇ RADOCON ◇ RADOKOR ◇ SIMANEX ◇ SIMAZIN ◇ SIMAZINE (USDA) ◇ SIMAZINE 80W ◇ TAFAZINE ◇ TAPHAZINE ◇ ZEAPUR

TOXICITY DATA with REFERENCE
skn-rbt 500 mg open MLD CIGET* -,-,77
eye-rbt 80 mg MOD CIGET* -,-,77
sln-dmg-orl 2000 ppm JPFCD2 15,867,80
dlt-dmg-orl 6000 ppm JTEHD6 3,691,77
orl-rat TDLo:25 g/kg (female 6-15D post):TER CHYCDW 15,83,81
orl-rat TDLo:3120 mg/kg (female 6-15D post):REP CHYCDW 15,83,81
scu-rat TDLo:16 g/kg/61W-I:ETA VOONAW 16(1),82,70
orl-rat LD50:971 mg/kg FAATDF 7,299,86
ihl-rat LC50:9800 mg/m³/1H FMCHA2 -,C261,89
ivn-mus LD50:100 mg/kg CSLNX* NX#04003

CONSENSUS REPORTS: EPA Genetic Toxicology Program. Reported in EPA TSCA Inventory.

SAFETY PROFILE: Poison by intravenous route. Moderately toxic by ingestion. Questionable carcinogen with experimental tumorigenic data. An experimental teratogen. Other experimental reproductive effects. A skin and eye irritant. Mutation data reported. When heated to decomposition it emits very toxic fumes of Cl^- and NO_x.

BJP250 CAS:673-04-1 ***HR: 3***
2,4-BIS(ETHYLAMINO)-6-METHOXY-s-TRIAZINE
mf: $C_8H_{15}N_5O$ mw: 197.28

SYNS: 4,6-BIS(ETHYLAMINO)-2-METHOXY-s-TRIAZINE ◇ GEIGY 30,044 ◇ GESADURAL ◇ 2-METHOXY-4,6-BIS(ETHYLAMINO)-s-TRIAZINE ◇ METHOXY SIMAZINE ◇ PIMETON ◇ SIMETON ◇ SIMETONE

TOXICITY DATA with REFERENCE
ivn-mus LD50:180 mg/kg CSLNX* NX#03978
orl-rat LD50:535 mg/kg RREVAH 10,97,65

SAFETY PROFILE: Poison by intravenous route. Moderately toxic by ingestion. When heated to decomposition it emits toxic fumes of NO_x.

BJP300 CAS:53213-78-8 ***HR: 3***
1,2-BIS(ETHYLAMMONIO)ETHANE PERCHLORATE
mf: $C_6H_{18}Cl_2N_2O_8$ mw: 317.12

SAFETY PROFILE: An impact-sensitive mild explosive. When heated to decomposition it emits toxic fumes of NO_x, Cl^-, and NH_3. See also PERCHLORATES.

BJP325 CAS:64693-33-0 ***HR: D***
N,N′-BIS(ETHYLENE)-p-(1-ADAMANTYL)PHOSPHONIC DIAMIDE
mf: $C_{14}H_{23}N_2OP$ mw: 266.36

SYNS: P,P-BIS(1-AZIRIDINYL)-p-(1-ADAMANTYL)-PHOSPHINE OXIDE ◇ 1,1′-(TRICYCLO(3.3.2.$2^{3,7}$)DEC-1-YLPHOSPHINYLIDENE)BIS-AZIRIDINE

TOXICITY DATA with REFERENCE
orl-mus TDLo:75 mg/kg (5D male):REP JMCMAR 21,143,78

SAFETY PROFILE: Experimental reproductive effects. When heated to decomposition it emits toxic fumes of PO_x and NO_x.

BJP425 CAS:29471-80-5 ***HR: 3***
BIS(ETHYLENEDIAMINE)(MERCURICTETRATHIOCYANATO)COPPER
mf: $(C_4H_{16}CuN_4 \cdot C_4HgN_4S_4)x$

SYNS: COPPER, BIS(ETHYLENEDIAMINE)(MERCURICTETRATHIOCYANATO)- ◇ COPPER(2+), BIS(ETHYLENEDIAMINE)-, TETRAKIS(THIOCYANATO)MERCURATE(2-), POLYMERS

TOXICITY DATA with REFERENCE
ipr-mus LD50:26100 μg/kg IJEBA6 19,1187,81

ACGIH TLV: TWA 0.1 mg(Hg)/m^3 (skin)

SAFETY PROFILE: Poison by intraperitoneal route. When heated to decomposition it emits toxic fumes of NO_x, SO_x, Hg, and Cl^-.

BJP500 CAS:6708-69-6 ***HR: 3***
2,6-BIS(ETHYLEN-IMINO)-4-AMINO-s-TRIAZINE
mf: $C_7H_{10}N_6$ mw: 178.23

TOXICITY DATA with REFERENCE
ipr-rat LD50:700 μg/kg JPETAB 100,398,50
ipr-mus LD50:1800 μg/kg JPETAB 100,398,50
ivn-dog LDLo:400 μg/kg JPETAB 100,398,50

SAFETY PROFILE: A poison by intraperitoneal and intravenous routes. When heated to decomposition it emits toxic fumes of NO_x.

BJP899 CAS:19218-16-7 ***HR: 3***
1,3-BIS(ETHYLENIMINOSULFONYL)PROPANE
mf: $C_7H_{14}N_2O_4S_2$ mw: 254.35

SYNS: BEP ◇ omega,omega'-BIS-(ETHYLENEIMINOSULPHONYL)PROPANE ◇ 1,3,-DI(ETHYLENESULPHAMOYL)PROPANE

TOXICITY DATA with REFERENCE
oms-rat-ipr 4 mg/kg BJPCAL 6,357,51
cyt-rat-ipr 4 mg/kg BJPCAL 6,357,51
scu-rat TDLo:80 mg/kg/I:NEO ANYAA9 68,750,58

SAFETY PROFILE: Questionable carcinogen with experimental neoplastigenic data. When heated to decomposition it emits very toxic fumes of SO_x and NO_x.

BJQ250 CAS:2781-10-4 ***HR: 3***
BIS(2-ETHYLHEXANOYLOXY)DIBUTYL STANNANE
mf: $C_{24}H_{48}O_4Sn$ mw: 519.41

SYNS: DIBUTYLBIS((2-ETHYLHEXANOYL)OXY)-STANNANE ◇ DIBUTYLBIS((2-ETHYL-1-OXOHEXYL)OXY)-STANNANE (9CI) ◇ DIBUTYLTIN BIS(α-ETHYLHEXANOATE) ◇ DIBUTYLTIN BIS(2-ETHYLHEXANOATE) ◇ DIBUTYLTIN DI(2-ETHYLHEXANOATE) ◇ DI-n-BUTYLTIN DI-2-ETHYLHEXANOATE ◇ DIBUTYLTIN DI(2-ETHYLHEXOATE)

TOXICITY DATA with REFERENCE
orl-rat LD50:200 mg/kg JPMSAE 56,240,67
orl-mus LDLo:1075 mg/kg AECTCV 14,111,85
ivn-mus LD50:178 mg/kg CSLNX* NX#00178

CONSENSUS REPORTS: Reported in EPA TSCA Inventory.

OSHA PEL: TWA 0.1 mg(Sn)/m^3 (skin)
ACGIH TLV: TWA 0.1 mg(Sn)/m^3 (skin) (Proposed: TWA 0.1 mg(Sn)/m^3; STEL 0.2 mg(Sn)/m^3 (skin))
NIOSH REL: (Organotin Compounds) TWA 0.1 mg(Sn)/m^3

SAFETY PROFILE: Poison by ingestion and intravenous route. See also TIN COMPOUNDS. When heated to decomposition it emits acrid smoke and irritating fumes.

BJQ500 CAS:103-24-2 ***HR: 2***
BIS(2-ETHYLHEXYL)AZELATE
mf: $C_{25}H_{48}O_4$ mw: 412.73

SYNS: AZELAIC ACID DI(2-ETHYLHEXYL)ESTER ◇ DIOCTYL AZELATE ◇ PLASTOLEIN 9058 ◇ PLASTOLEIN 9058 DOZ ◇ STAFLEX DOX ◇ TRUFLEX DOX

TOXICITY DATA with REFERENCE
skn-rbt 10 mg/24H open MLD AIHAAP 23,95,62
orl-rat LD50:8720 mg/kg AIHAAP 23,95,62
ivn-rat LD50:1060 mg/kg MRLR** No.256,54
skn-rbt LD50:20 g/kg AIHAAP 23,95,62
ivn-rbt LD50:640 mg/kg MRLR** No.256,54

CONSENSUS REPORTS: Reported in EPA TSCA Inventory.

SAFETY PROFILE: Moderately toxic by intravenous route. Mildly toxic by ingestion and skin contact. A skin irritant. See also ESTERS. When heated to decomposition it emits acrid smoke and irritating fumes.

BJQ709 CAS:3658-48-8 ***HR: 3***
BIS(2-ETHYLHEXYL) HYDROGEN PHOSPHITE
mf: $C_{16}H_{35}O_3P$ mw: 306.42

TOXICITY DATA with REFERENCE
eye-rbt 25 mg MLD AMIHAB 18,464,58
orl-rat LD50:11900 mg/kg ALBRW* #OPB-3,84
ipr-rat LD50:1500 mg/kg AMIHAB 18,464,58
ipr-mus LD50:620 mg/kg AMIHAB 18,464,58
skn-rbt LD50:4500 mg/kg ALBRW* #OPB-3,84
ivn-rbt LD50:100 mg/kg AMIHAB 18,464,58
ipr-gpg LD50:700 mg/kg AMIHAB 18,464,58

CONSENSUS REPORTS: Reported in EPA TSCA Inventory.

SAFETY PROFILE: Poison by intravenous route. Moderately toxic by intraperitoneal route. An eye irritant. When heated to decomposition it emits toxic fumes of PO_x. See also ESTERS.

BJQ750 CAS:137-89-3 ***HR: 1***
BIS(2-ETHYLHEXYL) ISOPHTHALATE
mf: $C_{24}H_{38}O_4$ mw: 390.62

SYNS: DI-2-ETHYLHEXYL ISOPHTHALATE ◇ DIOCTYL ISOPHTHALATE

TOXICITY DATA with REFERENCE
skn-rbt 500 mg open MLD UCDS** 5/17/66
orl-rat LD50:17300 mg/kg AIHAAP 23,95,62
skn-rbt LD50:7940 mg/kg AIHAAP 23,95,62

CONSENSUS REPORTS: Reported in EPA TSCA Inventory.

SAFETY PROFILE: Mildly toxic by ingestion and skin contact. When heated to decomposition it emits acrid smoke and irritating fumes.

BJR000 CAS:142-16-5 ***HR: 1***
BIS(2-ETHYLHEXYL)MALEATE
mf: $C_{20}H_{36}O_4$ mw: 340.56

PROP: Liquid. Mp: −60°, bp: 164° @ 10 mm, flash p: 365°F, d: 0.9436 @ 20°/20°, vap d: 11.7.

SYNS: DI-(2-ETHYLHEXYL)MALEATE ◇ "DIOCTYL" MALEATE ◇ DOM ◇ RC COMONOMER DOM

TOXICITY DATA with REFERENCE
skn-rbt 10 mg/24H open MLD JIHTAB 31,60,49
eye-rbt 500 mg open JIHTAB 31,60,49
orl-rat LD50:14 g/kg JIHTAB 31,60,49
skn-rbt LD50:15 g/kg JIHTAB 31,60,49

CONSENSUS REPORTS: Reported in EPA TSCA Inventory.

SAFETY PROFILE: Mildly toxic by ingestion and skin contact. A skin and eye irritant. Combustible when exposed to heat or flame; can react with oxidizing materials. To fight fire, use alcohol foam, dry chemical, mist or spray. When heated to decomposition it emits acrid smoke and irritating fumes. See also ESTERS.

BJR250 CAS:15546-12-0 ***HR: 3***
BIS((2-(ETHYL)HEXYLOXY)MALEOYLOXY)DI(n-BUTYL)STANNANE
mf: $C_{32}H_{56}O_8Sn$ mw: 687.57

SYNS: BIS(HYDROGEN MALEATO)DIBUTYL-TIN BIS(2-ETHYLHEXYL) ESTER ◇ 2-ETHYLHEXYLMALEINAN DI-N-BUTYLCINICITY (CZECH)

TOXICITY DATA with REFERENCE
skn-rbt 500 mg/24H MOD 28ZPAK -,230,72
eye-rbt 100 mg/24H SEV 28ZPAK -,230,72
orl-rat LD50:284 mg/kg 28ZPAK -,230,72

OSHA PEL: TWA 0.1 mg(Sn)/m^3 (skin)
ACGIH TLV: TWA 0.1 mg(Sn)/m^3 (skin) (Proposed: TWA 0.1 mg(Sn)/m^3; STEL 0.2 mg(Sn)/m^3 (skin))
NIOSH REL: (Organotin Compounds) TWA 0.1 mg(Sn)/m^3

SAFETY PROFILE: Poison by ingestion. A skin and eye irritant. See also TIN COMPOUNDS. When heated to decomposition it emits acrid smoke and irritating fumes.

BJR625 CAS:16368-97-1 ***HR: 2***
BIS(2-ETHYLHEXYL)PHENYL PHOSPHATE
mf: $C_{22}H_{39}O_4P$ mw: 398.58

SYNS: DAFF ◇ DEPP ◇ DI(2-ETHYLHEXYL)PHENYL PHOSPHATE

TOXICITY DATA with REFERENCE
orl-mus LD50:9333 mg/kg GTPZAB 15(8),30,71
ihl-mus LC50:5 g/m^3 GTPZAB 15(8),30,71
ipr-mus LD50:473 mg/kg GTPZAB 15(8),30,71

SAFETY PROFILE: Moderately toxic by intraperitoneal route. Mildly toxic by inhalation and ingestion. When heated to decomposition it emits toxic fumes of PO_x. See also PHOSPHATES.

BJR750 CAS:298-07-7 ***HR: 3***
BIS(2-ETHYLHEXYL)PHOSPHATE
DOT: NA 1902
mf: $C_{16}H_{35}O_4P$ mw: 322.48

SYNS: BIS(2-ETHYLHEXYL)HYDROGEN PHOSPHATE ◇ BIS(2-ETHYLHEXYL)ORTHOPHOSPHORIC ACID ◇ BIS(2-ETHYLHEXYL)PHOSPHORIC ACID ◇ DEHPA EXTRACTANT ◇ DI(2-ETHYLHEXYL)PHOSPHATE ◇ DI-2(ETHYLHEXYL)PHOSPHORIC ACID ◇ DI-(2-ETHYLHEXYL)PHOSPHORIC ACID (DOT) ◇ 2-ETHYL-1-HEXANOL HYDROGEN PHOSPHATE ◇ HDEHP

TOXICITY DATA with REFERENCE
skn-rbt 500 mg open MOD UCDS** 5/18/72
eye-rbt 5 mg MOD UCDS** 5/18/72
orl-rat LD50:4940 mg/kg UCDS** 5/18/42
ipr-rat LD50:50 mg/kg HYDRDA 3,201,78
ipr-mus LDLo:63 mg/kg CBCCT* 9,132,57
skn-rbt LD50:1250 mg/kg UCDS** 5/18/72

CONSENSUS REPORTS: Reported in EPA TSCA Inventory.

DOT Classification: Corrosive Material; Label: Corrosive.

SAFETY PROFILE: Poison by intraperitoneal route. A corrosive material. A skin and eye irritant. When heated to decomposition it emits toxic fumes of PO_x.

BJS250 CAS:122-62-3 ***HR: 2***
BIS(2-ETHYLHEXYL)SEBACATE
mf: $C_{26}H_{50}O_4$ mw: 426.76

PROP: Light, clear liquid; mild odor. Bp: 248° @ 9 mm, fp: −55°, flash p: 410°F, d: 0.913 @ 25°/25°, vap d: 14.7.

SYNS: BISOFLEX DOS ◇ DECANEDIOIC ACID, BIS(2-ETHYLHEXYL) ESTER ◇ DI(2-ETHYLHEXYL)SEBACATE ◇ DIOCTYL SEBACATE ◇ DOS ◇ 2-ETHYLHEXYL SEBACATE ◇ MONOPLEX DOS ◇ OCTOIL S ◇ OCTYL SEBACATE ◇ PX 438 ◇ STALFLEX DOS ◇ UNIFLEX DOS

TOXICITY DATA with REFERENCE
orl-rat LD50:12800 mg/kg 14CYAT 2,1883,63
ivn-rat LD50:900 mg/kg MRLR** No.256,54
ivn-rbt LD50:540 mg/kg MRLR** No.256,54

CONSENSUS REPORTS: Reported in EPA TSCA Inventory.

SAFETY PROFILE: Moderately toxic by ingestion and intravenous routes. See also ESTERS. Combustible when exposed to heat or flame; can react with oxidizing materials. To fight fire, use foam, CO_2, dry chemical. When heated to decomposition it emits acrid and irritating fumes.

BJT250 CAS:2440-45-1 ***HR: 3***
BIS(ETHYLMERCURI)PHOSPHATE
mf: $C_4H_{11}Hg_2O_4P$ mw: 555.30

PROP: Solid.

SYNS: ETHYLMERCURIC PHOSPHATE ◇ ETHYLMERCURY PHOSPHATE ◇ LIGNASAN FUNGICIDE ◇ LIGNASAN-X ◇ NEW IMPROVED CERESAN ◇ NEW IMPROVED GRANOSAN

TOXICITY DATA with REFERENCE
scu-mus TDLo:40 mg/kg (10D preg):TER NISFAY 20,1479,68
orl-rat LD50:30 mg/kg PCOC** -,516,66
unk-rat LD50:30 mg/kg 30ZDA9 -,288,71
orl-mus LD50:56 mg/kg NYKZAU 58,235,62
scu-mus LD50:88 mg/kg KUMJAX 14,65,61

CONSENSUS REPORTS: Mercury and its compounds are on the Community Right-To-Know List.

OSHA PEL: (Transitional: CL 1 mg/10m^3) TWA 0.01 mg(Hg)/m^3; STEL 0.03 mg/m^3 (skin)
ACGIH TLV: TWA 0.01 mg(Hg)/m^3; STEL 0.03 mg(Hg)/m^3
NIOSH REL: TWA 0.05 mg(Hg)/m^3

SAFETY PROFILE: Poison by ingestion and subcutaneous routes. See also MERCURY COMPOUNDS, ORGANIC. An experimental teratogen. When heated to decomposition it emits very toxic fumes of Hg and PO_x.

BJT500 CAS:139-60-6 ***HR: 2***
N,N'-BIS(1-ETHYL-3-METHYLPENTYL)-p-PHENYLENEDIAMINE
mf: $C_{22}H_{40}N_2$ mw: 332.64

SYNS: N,N'-BIS(5-METHYL-3-HEPTYL)-p-PHENYLENEDIAMINE ◇ N,N'-DI(1-ETHYL-3-METHYLPENTYL)-p-PHENYLENEDIAMINE ◇ EASTOZONE 31 ◇ ELASTOZONE 31 ◇ SANTOFLEX 17 ◇ TENAMENE 31 ◇ UOP 88

TOXICITY DATA with REFERENCE
orl-rat LD50:2400 mg/kg RCTEA4 45(3),627,72
skn-rbt LD50:1800 mg/kg RCTEA4 45(3),627,72

CONSENSUS REPORTS: Reported in EPA TSCA Inventory.

SAFETY PROFILE: Moderately toxic by ingestion and skin contact. When heated to decomposition it emits toxic fumes of NO_x.

BJT750 CAS:76-20-0 ***HR: 3***
2,2-BIS(ETHYLSULFONYL)BUTANE
mf: $C_8H_{18}O_4S_2$ mw: 242.38

SYNS: DIETHYLSULFONMETHYLETHYLMETHANE ◇ ETHYLSULFONAL ◇ METHYLSULFONAL ◇ METHYLSULPHONAL ◇ SULFONETHYLMETHANE ◇ TIONAL ◇ TRIONAL

TOXICITY DATA with REFERENCE
skn-mus TDLo:1900 mg/kg/1W-I:ETA BJCAAI 9,177,55

SAFETY PROFILE: Questionable carcinogen with experimental tumorigenic data. When heated to decomposition it emits toxic fumes of SO_x.

BJU000 CAS:502-55-6 ***HR: 3***
BISETHYL XANTHOGEN DISULFIDE
mf: $C_6H_{10}O_2S_4$ mw: 242.40

SYNS: AULIGEN ◇ BEK ◇ BEXIDE ◇ BEXT ◇ BIETHYLXANTHOGENTRISULFIDE ◇ BIS(ETHYLXANTHIC)DISULFIDE ◇ DEX ◇ DIETHYLDITHIO BIS(THIONOFORMATE) ◇ DIETHYL DIXANTHOGEN ◇ DIETHYL XANTHOGENATE ◇ DIETHYLXANTHOGEN DISULFIDE ◇ DITHIOBIS(THIOFORMIC ACID)-o,o-DIETHYL ESTER ◇ DIXANTHOGEN ◇ ETHYL XANTHOGEN DISULFIDE ◇ EXD ◇ K PREPARATION ◇ THIOPEROXYDICARBONIC ACID DIETHYL ESTER

TOXICITY DATA with REFERENCE
orl-rat LD50:480 mg/kg RREVAH 10,97,65
skn-rat LDLo:2100 mg/kg PCOC** -,578,66
orl-mus LD50:1200 mg/kg FATOAO 28,230,65
orl-rbt LD50:620 mg/kg PCOC** -,578,66
ipr-rbt LD50:320 mg/kg APTOA6 8,329,52
orl-gpg LD50:400 mg/kg PCOC** -,578,66
unk-mam LD50:600 mg/kg 30ZDA9 -,180,71

CONSENSUS REPORTS: Reported in EPA TSCA Inventory.

SAFETY PROFILE: Poison by ingestion and intraperitoneal routes. Moderately toxic by skin contact and possibly other routes. See also ESTERS and SULFIDES. When heated to decomposition it emits highly toxic fumes of SO_x.

BJU250 CAS:1851-71-4 ***HR: 3***
BIS(ETHYLXANTHOGEN) TETRASULFIDE
mf: $C_6H_{10}O_2S_6$ mw: 306.52

SYN: TETRASULFIDE, BIS(ETHOXYTHIOCARBONYL)

TOXICITY DATA with REFERENCE
orl-rat LD50:275 mg/kg 28ZEAL 5,26,76
orl-mus LD50:275 mg/kg 28ZEAL 5,26,76

SAFETY PROFILE: Poison by ingestion. When heated to decomposition it emits toxic fumes such as SO_x. See also SULFIDES.

BJU350 CAS:73526-98-4 ***HR: 3***
BIS(2-FLUORO-2,2-DINITROETHOXY)DIMETHYLSILANE
mf: $C_6H_{10}F_2N_4O_{10}Si$ mw: 364.25

$(F(O_2N)_2CCH_2O)_2Si(CH_3)_2$

SAFETY PROFILE: An explosive plasticizer sensitive to shock. When heated to decomposition it emits toxic fumes of F^- and NO_x. See also SILANE.

BJU500 CAS:18139-03-2 ***HR: 3***
BIS(2-FLUORO-2,2-DINITROETHYL)AMINE
mf: $C_4H_5F_2N_5O_8$ mw: 289.11

$(FC(NO_2)_2CH_2)_2NH$

SAFETY PROFILE: An explosive. When heated to decomposition it emits toxic fumes of F^- and NO_x. See also EXPLOSIVES and AMINES.

BJV625 CAS:72985-54-7 ***HR: 3***
1,1-BIS(FLUOROOXY)HEXAFLUOROPROPANE
mf: $C_3F_8O_2$ mw: 220.02

SAFETY PROFILE: Decomposes explosively. When heated to decomposition it emits toxic fumes of F^-.

BJV630 CAS:16329-93-4 ***HR: 3***
2,2-BIS(FLUOROOXY)HEXAFLUOROPROPANE
mf: $C_3F_8O_2$ mw: 220.02

SAFETY PROFILE: An unstable explosive. When heated to decomposition it emits toxic fumes of F^-.

BJV635 CAS:16329-92-3 ***HR: 3***
1,1-BIS(FLUOROOXY)TETRAFLUOROETHANE
mf: $C_2F_6O_2$ mw: 170.01

SAFETY PROFILE: Potentially explosive at room temperature. Upon decomposition it emits toxic fumes of F^-.

BJV750 CAS:63698-38-4 ***HR: 3***
trans-4-(4,4-BIS(p-FLUOROPHENYL)BUTYL)-1-(2-(4'-PHENYLCYCLOHEXYLAMINO)ETHYL)PIPERAZINE TRIHYDROCHLORIDE
mf: $C_{34}H_{43}F_2N_3{\bullet}3ClH$ mw: 641.18

SYN: M.G. 18001-3HCl

TOXICITY DATA with REFERENCE
orl-rat LD50:389 mg/kg FRPSAX 32,461,77
ipr-rat LD50:42 mg/kg FRPSAX 32,461,77

SAFETY PROFILE: Poison by ingestion and intraperitoneal routes. When heated to decomposition it emits very toxic fumes of F^-, HCl, and NO_x.

BJW000 CAS:63698-37-3 ***HR: 3***
trans-2-(4-(4,4-BIS(p-FLUOROPHENYL)BUTYL) PIPERAZINYL)-N-(4'-PHENYLCYCLOHEXYL) ACETAMIDE DIHYDROCHLORIDE
mf: $C_{34}H_{41}F_2N_3O{\bullet}2ClH$ mw: 618.70

SYN: M.G. 8948-2HCl

TOXICITY DATA with REFERENCE
orl-rat LD50:301 mg/kg FRPSAX 32,461,77
ipr-rat LD50:108 mg/kg FRPSAX 32,461,77

SAFETY PROFILE: Poison by ingestion and intraperitoneal routes. When heated to decomposition it emits very toxic fumes of F^-, NO_x, and HCl.

BJW250 CAS:20929-99-1 ***HR: 3***
1,1-BIS(4-FLUOROPHENYL)-2-PROPYNYL-N-CYCLOHEPTYLCARBAMATE
mf: $C_{23}H_{23}F_2NO_2$ mw: 383.47

SYN: CYCLOHEPTANECARBAMIC ACID-1,1-BIS(p-FLUOROPHENYL)-2-PROPYNYL ESTER

TOXICITY DATA with REFERENCE
orl-rat TDLo:620 mg/kg/13W-C:CAR CNREA8 30,2881,70
ipr-rat LD50:215 mg/kg HarPN# 21Oct74
orl-mus LD50:405 mg/kg HarPN# 21Oct74
ipr-mus LD50:318 mg/kg HarPN# 21Oct74

SAFETY PROFILE: Poison by ingestion and intraperitoneal routes. Questionable carcinogen with experimental carcinogenic data. See also ESTERS. When heated to decomposition it emits very toxic fumes of F^- and NO_x.

BJW500 CAS:20930-00-1 ***HR: 3***
1,1-BIS(4-FLUOROPHENYL)-2-PROPYNYL-N-CYCLOOCTYL CARBAMATE
mf: $C_{24}H_{25}F_2NO_2$ mw: 397.50

SYN: CYCLOOCTANECARBAMIC ACID-1,1-BIS(p-FLUOROPHENYL)-2-PROPYNYL ESTER

TOXICITY DATA with REFERENCE
orl-rat TDLo:1900 mg/kg/66D-C:CAR CNREA8 30,2881,70
ipr-rat LD50:617 mg/kg HarPN# 21Oct74
ipr-mus LD50:456 mg/kg HarPN# 21Oct74

SAFETY PROFILE: Moderately toxic by intraperitoneal route. Questionable carcinogen with experimental carcinogenic data. See also ESTERS and CARBAMATES. When heated to decomposition it emits very toxic fumes of F^- and NO_x.

BJW600 CAS:76674-14-1 ***HR: D***
α-α-BIS(4-FLUOROPHENYL)-1H-1,2,4-TRIAZOLE-1-ETHANOL
mf: $C_{16}H_{13}F_2N_3O$ mw: 301.32

SYNS: 1,1-DI(4-FLUOROPHENYL)-2-(1,2,4-TRIAZOLE-1-YL)-ETHANOL ◇ ICI 151291 ◇ R 151885 ◇ 1H-1,2,4-TRIAZOLE-1-ETHANOL, α-α-BIS(4-FLUOROPHENYL)-

TOXICITY DATA with REFERENCE
orl-rat TDLo:5 mg/kg (female 1D pre):REP TXAPA9 90,427,87

SAFETY PROFILE: Experimental reproductive effects. When heated to decomposition it emits toxic fumes of NO_x and F^-.

BJW750 CAS:6784-25-4 ***HR: 3***
BIS(N-FORMYL-p-AMINOPHENYL)SULFONE
mf: $C_{14}H_{12}N_2O_4S$ mw: 304.34

SYN: N,N'-DIFORMYL-p,p'-DIAMINODIPHENYLSULFONE

TOXICITY DATA with REFERENCE
ivn-rat LDLo:450 mg/kg TXAPA9 18,469,71
ipr-mus LD50:760 mg/kg EXPAAA 20,88,67
ivn-dog LDLo:98 mg/kg IJLEAG 36,432,68
ivn-cat LDLo:255 mg/kg IJLEAG 36,432,68

SAFETY PROFILE: Poison by intravenous route. Moderately toxic by intravenous and intraperitoneal routes. See also SULFONATES. When heated to decomposition it emits very toxic fumes of NO_x and SO_x.

BJW800 CAS:4387-13-7 ***HR: 3***
BIS(FORMYLMETHYL) MERCURY
mf: $C_4H_6HgO_2$ mw: 286.69

SYN: MERCURIDIACETALDEHYDE

TOXICITY DATA with REFERENCE
ivn-mus LD50:18 mg/kg CSLNX* NX#05651

OSHA PEL: (Transitional: CL 1 mg/10m^3) CL 0.1 mg(Hg)/m^3 (skin)
ACGIH TLV: TWA 0.01 mg(Hg)/m^3; STEL 0.03 mg(Hg)/m^3

SAFETY PROFILE: Poison by intravenous route. When heated to decomposition it emits toxic fumes of Hg.

BJW825 CAS:5188-42-1 ***HR: 3***
BIS(GUANIDINIUM) CHROMATE
mf: $C_2H_{10}N_6 \cdot CrH_2O_4$ mw: 236.20

SYN: BIGUANIDINE, CHROMATE

TOXICITY DATA with REFERENCE
ivn-mus LD50:180 mg/kg CSLNX* NX#02828

OSHA PEL: CL 0.1 mg(CrO3)/m^3
ACGIH TLV: TWA 0.05 mg(Cr)/m^3
NIOSH REL: (Chromium(VI)) TWA 0.025 mg/m^3; CL 0.05 mg/15M

SAFETY PROFILE: Poison by intravenous route. When heated to decomposition it emits toxic fumes of NO_x and Cr.

BJX750 CAS:19704-60-0 ***HR: 3***
BIS(HEXANOYLOXY)DI-n-BUTYLSTANNANE
mf: $C_{20}H_{40}O_4Sn$ mw: 463.29

SYNS: BIS(HEXANOYLOXY)DI-n-BUTYL-TIN ◇ KAPRONAN DI-n-BUTYLCINICITY (CZECH)

TOXICITY DATA with REFERENCE
skn-rbt 500 mg/24H SEV 28ZPAK -,229,72
eye-rbt 20 mg/24H MOD 28ZPAK -,229,72
orl-rat LD50:94 mg/kg 28ZPAK -,229,72

OSHA PEL: TWA 0.1 mg(Sn)/m^3 (skin)
ACGIH TLV: TWA 0.1 mg(Sn)/m^3 (skin) (Proposed: TWA 0.1 mg(Sn)/m^3; STEL 0.2 mg(Sn)/m^3 (skin))
NIOSH REL: (Organotin Compounds) TWA 0.1 mg(Sn)/m^3

SAFETY PROFILE: Poison by ingestion. A skin and eye irritant. See also TIN COMPOUNDS. When heated to decomposition it emits acrid smoke and fumes.

BJX800 CAS:63270-67-7 ***HR: 3***
BIS(l-HISTIDINATO)MANGANESE TETRAHYDRATE
mf: $C_{12}H_{16}MnN_6O_4 \cdot 4H_2O$ mw: 435.36

SYNS: MANGANESE, BIS(l-HISTIDINATO)-, TETRAHYDRATE ◇ MANGANESE, BIS(l-HISTIDINATO-N,O)-, TETRAHYDRATE

TOXICITY DATA with REFERENCE
uns-mus LD50:160 mg/kg FRMBAZ 29,215,81

OSHA PEL: CL 5 mg(Mn)/m^3
ACGIH TLV: TWA 5 mg(Mn)/m^3

SAFETY PROFILE: Poison by an unspecified route. When heated to decomposition it emits toxic fumes of NO_x and Mn.

BJY000 CAS:14873-10-0 ***HR: 3***
BIS(l-HISTIDINE)COBALT
mf: $C_{12}H_{14}N_6O_5 \cdot Co$ mw: 365.25

SYNS: α-AMINOIMIDAZOLE-4-PROPIONIC ACID, COBALT(2+) SALT ◇ BIS(l-HISTIDINATO)COBALT ◇ COBALT-HISTIDINE ◇ KOBALT HISTIDIN (GERMAN)

TOXICITY DATA with REFERENCE
ipr-rat LD50:134 mg/kg AEPPAE 243,254,62
ivn-rat LD50:104 mg/kg AIPTAK 143,219,63
ivn-cat LD50:50 mg/kg AIPTAK 143,219,63

CONSENSUS REPORTS: Cobalt and its compounds are on the Community Right-To-Know List.

SAFETY PROFILE: Poison by intraperitoneal and intravenous routes. See also COBALT COMPOUNDS. When heated to decomposition it emits toxic fumes of NO_x.

BJY125 CAS:53532-37-9 ***HR: D***
BIS-HM-A-TDA
mf: $C_4H_7N_3O_2S$ mw: 161.20

SYN: 2-N,N-BIS(HYDROXYMETHYL)AMINO-1,3,4-THIADIAZOLE

TOXICITY DATA with REFERENCE
orl-rat TDLo:20 mg/kg (female 10D post):TER YHHPAL 16,654,81
orl-mus TDLo:150 mg/kg (female 8D post):REP SYSWAE 12,289,79

SAFETY PROFILE: An experimental teratogen. Other experimental reproductive effects. When heated to decomposition it emits toxic fumes of SO_x and NO_x.

BJY250 ***HR: 3***
BISHYDRAZINE NICKEL(II)PERCHLORATE
mf: $Cl_2H_8N_4NiO_8$ mw: 323.7

CONSENSUS REPORTS: Nickel and its compounds are on the Community Right-To-Know List.

SAFETY PROFILE: Exploded by heat and dilute aqueous suspension. Upon decomposition it emits toxic fumes of Cl^- and NO_x. See also NICKEL COMPOUNDS and PERCHLORATES.

BJY500 ***HR: 3***
BISHYDRAZINE TIN(II)CHLORIDE
mf: $Cl_2H_8N_4Sn$ mw: 253.69

SAFETY PROFILE: Explodes on heating. Upon decomposition it emits toxic fumes of Cl^-. See also TIN COMPOUNDS and CHLORIDES.

BJY750 ***HR: 3***
BIS(1-HYDROPEROXY CYCLOHEXYL)PEROXIDE
mf: $C_{12}H_{22}O_6$ mw: 252.20

SAFETY PROFILE: Fire causes violent explosion. When heated to decomposition it emits acrid smoke and fumes. See also PEROXIDES, ORGANIC.

BJY825 CAS:2614-76-8 ***HR: 3***
2,2-BIS(HYDROPEROXY)PROPANE
DOT: UN 2178
mf: $C_3H_8O_4$ mw: 108.09

$(CH_3)_2C(OOH)_2$

DOT Classification: Organic Peroxide; Label:Organic Peroxide

SAFETY PROFILE: Ignites or explodes when heated. When heated to decomposition it emits acrid smoke and fumes. See also PEROXIDES.

BJZ000 CAS:66-76-2 ***HR: 3***
BISHYDROXYCOUMARIN
mf: $C_{19}H_{12}O_6$ mw: 336.31

PROP: Very small crystals, slight pleasant odor, bitter taste, sol in alkali. Mp: 287-293°.

SYNS: ACADYL ◇ ACAVYL ◇ ANTITROMBOSIN ◇ BARACOUMIN ◇ BHC ◇ BIS(4-HYDROXYCOUMARIN-3-YL)METHANE ◇ CUMA ◇ CUMID ◇ DICOUMARIN ◇ DICOUMAROL ◇ DICUMAN ◇ DICUMARINE ◇ DI-(4-HYDROXY-3-COUMARINYL)METHANE ◇ DI-4-HYDROXY-3,3'-METHYLENEDICOUMARIN ◇ DUFALONE ◇ KUMORAN ◇ MELITOXIN ◇ 3,3'-METHYLEEN-BIS(4-HYDROXY-CUMARINE) (DUTCH) ◇ 3,3'-METHYLEN-BIS(4-HYDROXY-CUMARIN) (GERMAN) ◇ 3,3'-METHYLENEBIS(4-HYDROXY-1,2-BENZOPYRONE) ◇ 3,3'-METHYLENEBIS(4-HYDROXYCOUMARIN) ◇ 3,3'-METHYLENE-BIS(4-HYDROXYCOUMARINE) (FRENCH) ◇ 3,3'-METILEN-BIS(4-IDROSSI-CUMARINA) (ITALIAN) ◇ TEMPARIN ◇ TROMBOSAN

TOXICITY DATA with REFERENCE
orl-wmn TDLo:110 mg/kg (31-40W preg):REP AJOGAH 57,965,49
orl-wmn TDLo:110 mg/kg (31-40W preg):TER AJOGAH 57,965,49
orl-rat LD50:250 mg/kg SMWOAS 83,471,53
ivn-rat LD50:52 mg/kg PSEBAA 50,228,42
orl-mus LD50:233 mg/kg PSEBAA 50,228,42
ipr-mus LD50:91 mg/kg DIPHAH 17,163,65
scu-mus LD50:50 mg/kg 85GDA2 8(1),360,82
ivn-mus LD50:42 mg/kg AEPPAE 222,107,54

CONSENSUS REPORTS: Reported in EPA TSCA Inventory.

SAFETY PROFILE: Poison by ingestion, subcutaneous, intravenous, and intraperitoneal routes. An experimental teratogen. Human reproductive effects by ingestion and possibly other routes: fetal death, unspecified developmental abnormalities, stillbirth, and unspecified neonatal effects. An anticoagulant. Excessive doses can cause hemorrhages. When heated to decomposition it emits acrid smoke and fumes. See also COUMADIN (warfarin).

BKA000 CAS:548-00-5 ***HR: 3***
BIS(4-HYDROXY-3-COUMARIN) ACETIC ACID ETHYL ESTER
mf: $C_{22}H_{16}O_8$ mw: 408.38

SYNS: BIS-3,3'-(4-HYDROXYCOUMARINYL)ACETIC ACID ETHYL ESTER ◇ BIS-(4-HYDROXY-3-COUMARINYL)ETHYL ACETATE ◇ BIS(4-HYDROXY-2-OXO-2H-1-BENZOPYRAN-3-YL)ACETIC ACID ETHYL ESTER ◇ BOEA ◇ B.O.E.A. ◇ 3,3'-(CARBOXYMETHYLENE)BIS(4-HYDROXYCOUMARIN) ETHYL ESTER ◇ DICUMACYL ◇ ETHYL BISCOUMACETATE ◇ ETHYL BIS(4-HYDROXYCOUMARINYL)ACETATE ◇ ETHYL BIS(4-HYDROXY-3-COUMARINYL)ACETATE ◇ ETHYLDICOUMAROL ◇ ETHYLDICOUMAROL ACETATE ◇ ETHYL-4,4'-DIHYDROXYDICOUMARINYL-3,3'-ACETATE ◇ NEODICOUMARIN ◇ NEODICOUMAROL ◇ NEODICUMARINUM ◇ PELENTAN ◇ STABILENE ◇ TROMBARIN ◇ TROMBIL ◇ TROMBOLYSAN ◇ TROMEXAN ◇ TROMEXAN ETHYL ACETATE

TOXICITY DATA with REFERENCE
orl-wmn TDLo:48 mg/kg (36-37W preg):REP BMJOAE 2,719,55
orl-wmn TDLo:48 mg/kg (36-37W preg):TER BMJOAE 2,719,55
orl-rat LD50:840 mg/kg FEPRA7 10,303,51
ipr-rat LD50:260 mg/kg AIPTAK 87,402,51
orl-mus LD50:750 mg/kg AEPPAE 222,107,54
orl-mus LD50:750 mg/kg AEPPAE 222,107,54
scu-mus LD50:750 mg/kg LANCAO 2,611,51

SAFETY PROFILE: Poison by intraperitoneal route. Moderately toxic by ingestion and subcutaneous routes. An experimental teratogen. Human reproductive effects by ingestion: developmental abnormalities of the cardiovascular system, stillbirth, and unspecified neonatal effects. An anticoagulant. See also COUMADIN (warfarin) and ESTERS. When heated to decomposition it emits acrid and irritating fumes.

BKA250 ***HR: 3***
BIS(1-HYDROXYCYCLOHEXYL)PEROXIDE
mf: $C_{22}H_{22}O_4$ mw: 230.3

SAFETY PROFILE: Explodes in vacuum. When heated to decomposition it emits acrid smoke and fumes. See also PEROXIDES, ORGANIC.

BKB000 CAS:21615-29-2 ***HR: 1***
3'-BIS(2-HYDROXYETHYL)AMINO)-p-ACETOPHENETIDIDE
mf: $C_{14}H_{22}N_2O_4$ mw: 282.38

SYNS: 2,2'-((5-ACETAMIDO-2-ETHOXYPHENYL)IMINO)DIETHANOL ◇ 2-BIS-HYDROXYETHYLAMINO-4-ACETAMINOFENETOL(CZECH)

TOXICITY DATA with REFERENCE
eye-rbt 100 mg/24H MOD 28ZPAK -,100,72
orl-rat LD50:8560 mg/kg 28ZPAK -,100,72

CONSENSUS REPORTS: Reported in EPA TSCA Inventory.

SAFETY PROFILE: Mildly toxic by ingestion. An eye irritant. When heated to decomposition it emits toxic fumes of NO_x.

BKB250 CAS:63867-52-7 ***HR: 2***
2-(BIS(β-HYDROXYETHYL)AMINO)-4,5-DIPHENYLOXAZOLE MONOHYDRATE
mf: $C_{19}H_{20}N_2O_3 \cdot H_2O$ mw: 342.43

SYNS: AGEROPLAS ◇ DIETHAMPHENAZOL MONOHYDRATE ◇ 2,2'-DIHYDROXY-N-(4,5-DIPHENYLOXAZOLE-2-YL)DIETHYLAMINE MONOHYDRATE ◇ N-(4,5-DIPHENYLOXAZOL-2-YL)DIETHANOLAMINE MONOHYDRATE ◇ 2,2'-((4,5-DIPHENYL-2-OXAZOLYL) IMINO)-DIETHANOLMONOHYDRATE ◇ DITAZOL MONOHYDRATE ◇ S 222

TOXICITY DATA with REFERENCE
orl-rat LD50:11380 mg/kg ARZNAD 23,1283,73
ipr-rat LD50:7770 mg/kg ARZNAD 23,1283,73
orl-mus LD50:9621 mg/kg ARZNAD 23,1283,73
ipr-mus LD50:3390 mg/kg ARZNAD 23,1283,73

SAFETY PROFILE: Moderately toxic by intraperitoneal route. Mildly toxic by ingestion. An anti-inflammatory agent. When heated to decomposition it emits toxic fumes of NO_x.

BKB300 CAS:70711-40-9 ***HR: 3***
1,4-BIS((2-((2-HYDROXYETHYL)AMINO) ETHYL)AMINO)-9,10-ANTHRACENEDIONE DIACETATE
mf: $C_{22}H_{28}N_4O_4 \cdot 2C_2H_4O_2$ mw: 532.66

SYNS: AMETANTRONE ACETATE ◇ 1,4-BIS((2-((HYDROXYETHYL) AMINO)ETHYL)AMINO)-9,10-ANTHRACENEDIONEDIACETATE (SALT) (9CI) ◇ CI 881 ◇ HAQ ◇ NSC 287513

TOXICITY DATA with REFERENCE
oms-hmn:leu 400 μg/L CNREA8 39,2574,79
cyt-hmn:leu 50 μg/L CNREA8 39,2574,79
oms-ham:ovr 10 nmol/L CNREA8 39,2574,79
ipr-rbt TDLo:5200 μg/kg (6-18D preg):TER TJADAB 29(2),41A,84
orl-mus LD50:495 mg/kg NCISP* JAN86
ipr-mus LD50:62830 μg/kg NCISP* JAN86
scu-mus LD50:297 mg/kg NCISP* JAN86

SAFETY PROFILE: Poison by subcutaneous and intraperitoneal routes. Moderately toxic by ingestion. An experimental teratogen. Human mutation data reported.

When heated to decomposition it emits toxic fumes of NO_x.

BKB325 ***HR: D***
1,4-BIS((2-((2-HYDROXYETHYL)AMINO)ETHYL) AMINO)ANTHRAQUINONE
mf: $C_{22}H_{28}N_4O_4$ mw: 412.54

SYNS: 1,4-BIS(2-((2-HYDROXYETHYL)AMINO)ETHYLAMINO)-9,10-ANTHRACENEDIONE ◇ HAQ ◇ NSC 287513

TOXICITY DATA with REFERENCE
mma-sat 6500 μmol/L CNREA8 41,376,81
dni-mus:oth 1 μmol/L CNREA8 43,1951,83
cyt-ham:ovr 10 nmol/L CNREA8 41,376,81
sce-ham:ovr 10 nmol/L CNREA8 41,376,81

SAFETY PROFILE: Mutation data reported. When heated to decomposition it emits toxic fumes of NO_x.

BKB500 CAS:27464-23-9 ***HR: 3***
3-(BIS(2-HYDROXYETHYL)AMINO)-6-HYDRAZINOPYRIDAZINEDIHYDROCHLORIDE
mf: $C_8H_{15}N_5O_2$•2ClH mw: 286.20

SYNS: 3-HYDRAZINO-6-(N,N-BIS-(2-HYDROXYETHYL)AMINO) PYRIDAZINE DIHYDROCHLORIDE ◇ 2-IDRAZINO-6-(N,N-BIS(2-IDROSSIETIL)-AMINO)-PIRIDAZINA CLORIDRATO (ITALIAN) ◇ L 6150

TOXICITY DATA with REFERENCE
ivn-hmn TDLo:29 μg/kg:CNS DRFUD4 2,172,77
orl-rat LD50:1800 mg/kg BCFAAI 111,480,72
ipr-rat LD50:335 mg/kg BCFAAI 111,480,72
orl-mus LD50:1520 mg/kg BCFAAI 111,480,72
ipr-mus LD50:263 mg/kg DRFUD4 2,172,77
orl-dog LD50:1 g/kg ARZNAD 23,1591,73
ivn-dog LD50:75 mg/kg ARZNAD 23,1591,73
orl-rbt LD50:5 g/kg ARZNAD 23,1591,73
orl-gpg LD50:188 mg/kg BCFAAI 111,480,72

SAFETY PROFILE: Poison by ingestion and intraperitoneal routes. Human systemic effects by intravenous route: somnolence and unspecified pulmonary system effects. When heated to decomposition it emits very toxic fumes of HCl and NO_x.

BKB750 CAS:5055-20-9 ***HR: 3***
4-BIS(2-HYDROXYETHYL)AMINO-2-(5-NITRO-2-FURYL)QUINAZOLINE
mf: $C_{16}H_{16}N_4O_5$ mw: 344.36

TOXICITY DATA with REFERENCE
orl-rat TDLo:8437 mg/kg/22W-C:CAR JNCIAM 57,277,76

SAFETY PROFILE: Questionable carcinogen with experimental carcinogenic data. When heated to decomposition it emits toxic fumes of NO_x.

BKC000 CAS:52551-67-4 ***HR: D***
2-(BIS(2-HYDROXYETHYL)AMINO)-5-NITROPHENOL
mf: $C_{10}H_{14}N_2O_5$ mw: 242.26

SYNS: HC YELLOW No. 4 ◇ NCI-C56019

TOXICITY DATA with REFERENCE
mmo-sat 333 μg/plate ENMUDM 8(Suppl 7),1,86
mma-sat 33 μg/plate ENMUDM 8(Suppl 7),1,86
sln-dmg-par 1 pph ENMUDM 7,677,85

CONSENSUS REPORTS: Reported in EPA TSCA Inventory.

SAFETY PROFILE: Mutation data reported. See also NITRO COMPOUNDS of AROMATIC HYDROCARBONS. When heated to decomposition it emits toxic fumes of NO_x.

BKC250 CAS:33372-39-3 ***HR: 3***
4-BIS(2-HYDROXYETHYL)AMINO-2-(5-NITRO-2-THIENYL)QUINAZOLINE
mf: $C_{16}H_{16}N_4O_4S$ mw: 360.42

TOXICITY DATA with REFERENCE
mma-sat 1250 μg/plate CNREA8 35,3611,75
orl-rat TDLo:3800 mg/kg/15W-C:CAR JNCIAM 57,277,76

CONSENSUS REPORTS: EPA Genetic Toxicology Program.

SAFETY PROFILE: Questionable carcinogen with experimental carcinogenic data. Mutation data reported. When heated to decomposition it emits very toxic fumes of NO_x and SO_x.

BKC500 CAS:78109-79-2 ***HR: 3***
N-(3-(BIS(2-HYDROXYETHYL)AMINO)PROPYL) BENZAMIDE HYDROCHLORIDE
mf: $C_{18}H_{30}N_2O_4$•ClH mw: 374.96

SYN: D-695

TOXICITY DATA with REFERENCE
scu-mus LD50:800 mg/kg ARZNAD 10,743,60
ivn-mus LD50:80 mg/kg ARZNAD 10,743,60

SAFETY PROFILE: Poison by intravenous route. Moderately toxic by subcutaneous route. When heated to decomposition it emits very toxic fumes of HCl and NO_x.

BKC750 CAS:101651-88-1 ***HR: 3***
10-(3-(BIS(2-HYDROXYETHYL)AMINO)PROPYL)-7-CHLOROISOALLOXAZINE SULFATE
mf: $C_{17}H_{20}ClN_5O_4$•H_2O_4S mw: 491.95

TOXICITY DATA with REFERENCE
scu-mus LD50:80 mg/kg CMTRAG 2,96,61
ivn-mus LD50:67 mg/kg CMTRAG 2,96,61

SAFETY PROFILE: Poison by subcutaneous and intravenous routes. See also SULFATES. When heated to decomposition it emits very toxic fumes of SO_x, Cl^-, and NO_x.

BKD000 CAS:78128-69-5 ***HR: 3***
N-(3-(BIS(2-HYDROXYETHYL)AMINO)PROPYL)-o-PROPOXYBENZAMIDE HYDROCHLORIDE
mf: $C_{17}H_{28}N_2O_4 \cdot ClH$ mw: 360.93

SYN: D-701

TOXICITY DATA with REFERENCE
ipr-mus LD50:330 mg/kg ARZNAD 10,743,60
scu-mus LD50:785 mg/kg ARZNAD 10,743,60
ivn-mus LD50:110 mg/kg ARZNAD 10,743,60

SAFETY PROFILE: Poison by intraperitoneal and intravenous routes. Moderately toxic by subcutaneous route. When heated to decomposition it emits very toxic fumes of NO_x and HCl.

BKD250 CAS:20182-56-3 ***HR: 1***
4,4'-BIS((4-(2-HYDROXYETHYL)AMINO-6-(p-SULFOANILINO)-s-TRIAZIN-2-YL)AMINO)-2,2'-STILBENEDISULFONIC ACID TETRASODIUM SALT
mf: $C_{36}H_{32}N_{12}O_{14}S_4 \cdot 4Na$ mw: 1077.00

TOXICITY DATA with REFERENCE
skn-rbt 500 mg/24H MLD MVCRB3 2,193,73
eye-rbt 100 mg MLD MVCRB3 2,193,73

CONSENSUS REPORTS: Reported in EPA TSCA Inventory.

SAFETY PROFILE: A skin and eye irritant. See also SULFONATES. When heated to decomposition it emits toxic fumes of NO_x, Na_2O and SO_x.

BKD500 CAS:120-07-0 ***HR: 2***
N,N-BIS(2-HYDROXYETHYL)ANILINE
mf: $C_{10}H_{15}NO_2$ mw: 181.26

SYNS: DIETHANOLAMINOBENZENE ◇ DIETHANOLANILINE ◇ N,N-DIETHANOLANILINE ◇ DIHYDROXYETHYLANILINE ◇ N,N-DI(β-HYDROXYETHYL)ANILINE ◇ N,N-DI(2-HYDROXYETHYL)ANILINE ◇ N,N-DIOXYETHYLANILINE ◇ EMERY 5703 ◇ 2,2'-(PHENYLAMINO)DIETHANOL ◇ PHENYL DIETHANOLAMINE ◇ N-PHENYLDIETHANOLAMINE ◇ 2,2'-(PHENYLIMINO)DIETHANOL

TOXICITY DATA with REFERENCE
skn-rbt 500 mg open MLD UCDS** 6/13/60
eye-rbt 100 mg SEV UCDS** 6/13/60
orl-rat LD50:1430 mg/kg LONZA# 03JAN78

CONSENSUS REPORTS: Reported in EPA TSCA Inventory.

SAFETY PROFILE: Moderately toxic by ingestion. A severe eye and mild skin irritant. When heated to decomposition it emits toxic fumes of NO_x. See also AROMATIC AMINES.

BKD600 CAS:5185-70-6 ***HR: 2***
N,N-BIS(2-HYDROXYETHYL)-p-ARSANILIC ACID
mf: $C_{10}H_{16}AsNO_5$ mw: 305.19

SYN: p-ARSANILIC ACID, N,N-BIS(2-HYDROXYETHYL)-

TOXICITY DATA with REFERENCE
ipr-mus LD50:1053 mg/kg JMCMAR 9,221,66

OSHA PEL: TWA 0.5 mg(As)/m^3
ACGIH TLV: TWA 0.2 mg(As)/m^3

SAFETY PROFILE: Moderately toxic by intraperitoneal route. When heated to decomposition it emits toxic fumes of NO_x and As.

BKD750 CAS:64036-91-5 ***HR: 3***
BIS(2-HYDROXYETHYL)-2-(2-CHLORO ETHYL THIO)ETHYL SULFONIUM) CHLORIDE
mf: $C_8H_{18}ClO_2S_2 \cdot Cl$ mw: 281.28

SYNS: β-CHLOROETHYL-β-(BIS(β-HYDROXYETHYL)SULFONIUM) ETHYL SULFIDE CHLORIDE ◇ 2-(2-CHLOROETHYL)THIOETHYL-BIS(2-HYDROXYETHYL)-CHLORIDE

TOXICITY DATA with REFERENCE
skn-rat LD50:10 mg/kg JPETAB 93,1,48
skn-mus LD50:15 mg/kg JPETAB 93,1,48
scu-mus LDLo:25 mg/kg NTIS** PB158-507
ivn-dog LD50:6 mg/kg JPETAB 93,1,48
ivn-rbt LD50:4500 μg/kg JPETAB 93,1,48

SAFETY PROFILE: A poison by skin contact, subcutaneous, and intravenous routes. See also SULFONATES. When heated to decomposition it emits very toxic fumes of Cl^- and SO_x.

BKE500 CAS:120-40-1 ***HR: 2***
N,N-BIS(2-HYDROXYETHYL)DODECAN AMIDE
mf: $C_{16}H_{33}NO_3$ mw: 287.50

PROP: Solid. Mp: 36°

SYNS: BIS(2-HYDROXYETHYL)LAURAMIDE ◇ N,N-BIS(HYDROXYETHYL)LAURAMIDE ◇ N,N-BIS(β-HYDROXYETHYL)LAURAMIDE ◇ N,N-BIS(2-HYDROXYETHYL)LAURAMIDE ◇ CLINDROL 101CG ◇ CLINDROL SUPERAMIDE 100L ◇ COCO DIETHANOLAMIDE ◇ COCONUT OIL AMIDE of DIETHANOLAMINE ◇ COMPERLAN LD ◇ CONDENSATE PL ◇ CRILLON L.D.E. ◇ DIETHANOLLAURAMIDE ◇ N,N-DIETHANOLLAURAMIDE ◇ N,N-DIETHANOLLAURIC ACID AMIDE ◇ EMID 6511 ◇ EMID 6541 ◇ ETHYLAN MLD ◇ HETAMIDE ML ◇ LAURAMIDE DEA ◇ LAURIC ACID DIETHANOLAMIDE ◇ LAURIC DIETHANOLAMIDE ◇ LAUROYL DIETHANOLAMIDE

◇ LAURYL DIETHANOLAMIDE ◇ LDA ◇ LDE ◇ MONAMID 150-LW ◇ NCI-C55323 ◇ NINOL AA-62 EXTRA ◇ NINOL 4821 ◇ NINOL AA62 ◇ ONYXOL 345 ◇ REWOMID DLMS ◇ RICHAMIDE 6310 ◇ ROLAMID CD ◇ STANDAMIDD LD ◇ STEINAMID DL 203 S ◇ SUPER AMIDE L-9A ◇ SYNOTOL L-60 ◇ UNAMIDE J-56 ◇ VARAMID ML 1

TOXICITY DATA with REFERENCE
orl-rat LD50:2700 mg/kg JSCCA5 13,469,62

CONSENSUS REPORTS: Reported in EPA TSCA Inventory.

SAFETY PROFILE: Moderately toxic by ingestion. When heated to decomposition it emits toxic fumes of NO_x. See also AMIDES.

BKE750 CAS:64058-26-0 ***HR: 3***
1,1-BIS(β-HYDROXYETHYL)ETHYLENINONIUM CHLORIDE
mf: $C_6H_{14}NO_2•Cl$ mw: 167.66

SYN: 1,1-BIS(2-HYDROXYETHYL)AZIRIDINIUMCHLORIDE

TOXICITY DATA with REFERENCE
orl-hmn TDLo:342 μg/kg:CNS NTIS** PB158-507
ipr-mus LD50:5 mg/kg NTIS** PB158-507
ivn-rbt LD50:20 mg/kg NTIS** PB158-507

SAFETY PROFILE: Poison by intraperitoneal and intravenous routes. Human systemic effects by ingestion: nausea and vomiting. When heated to decomposition it emits very toxic fumes of NO_x and Cl^-.

BKF250 CAS:2784-94-3 ***HR: 3***
N',N'-BIS(2-HYDROXYETHYL)-N-METHYL-2-NITRO-p-PHENYLENEDIAMINE
mf: $C_{11}H_{17}N_3O_4$ mw: 255.31

SYNS: HC BLUE 1 ◇ NCI-C04159

TOXICITY DATA with REFERENCE
mmo-sat 333 μg/plate NTPTR* NTP-TR-271,85
mma-sat 100 μg/plate NTPTR* NTP-TR-271,85
dns-rat:lvr 50 mg/L NTPTR* NTP-TR-271,85
msc-mus:lym 30 mg/L NTPTR* NTP-TR-271,85
orl-rat TDLo:66 g/kg/2Y-C:CAR NTPTR* NTP-TR-271,85
orl-mus TDLo:98280 mg/kg/39W-C:CAR FCTOD7 25,703,87

CONSENSUS REPORTS: NTP Carcinogenesis Studies (feed); Some Evidence: rat NTPTR* NTP-TR-271,85;(feed); Clear Evidence: mouse NTPTR* NTP-TR-271,85. Reported in EPA TSCA Inventory.

SAFETY PROFILE: Suspected carcinogen with experimental carcinogenic data. Mutation data reported. See also AMINES. When heated to decomposition it emits toxic fumes of NO_x.

BKF500 CAS:56863-02-6 ***HR: 2***
N,N-BIS(2-HYDROXYETHYL)-9,12-OCTADECADIENAMIDE
mf: $C_{22}H_{41}NO_3$ mw: 367.64

SYNS: CLINDROL LT 15-73-1 ◇ CYCLOMIDE DIN 295/S ◇ LINOLEIC DIETHANOLAMIDE

TOXICITY DATA with REFERENCE
skn-rbt 500 mg/24H SEV TXAPA9 19,276,71
eye-rbt 100 mg TXAPA9 19,276,71

CONSENSUS REPORTS: Reported in EPA TSCA Inventory.

SAFETY PROFILE: A severe skin and eye irritant. When heated to decomposition it emits toxic fumes of NO_x.

BKF750 CAS:63886-75-9 ***HR: 3***
N,N-BIS(2-HYDROXYETHYL)-p-PHENYLENEDIAMINE SULFATE (1:1)
mf: $C_{10}H_{16}N_2O_2•H_2O_4S$ mw: 294.36

SYN: N,N-BIS-2-HYDROXYETHYL-p-FENYLENDIAMINSTRAN (CZECH)

TOXICITY DATA with REFERENCE
eye-rbt 500 mg/24H MOD 28ZPAK -,110,72
orl-rat LD50:131 mg/kg 28ZPAK -,110,72

SAFETY PROFILE: Poison by ingestion. An eye irritant. See also SULFATES. When heated to decomposition it emits very toxic fumes of SO_x and NO_x.

BKG250 ***HR: 3***
BISHYDROXYL AMINE ZINC(II)CHLORIDE
mf: $Cl_2H_6N_2O_2Zn$ mw: 202.33

CONSENSUS REPORTS: Zinc and its compounds are on the Community Right-To-Know List.

SAFETY PROFILE: Explodes at 170°. When heated to decomposition it emits toxic fumes of Cl^-, NO_x, and ZnO. See also ZINC COMPOUNDS and CHLORIDES.

BKG500 CAS:73118-23-7 ***HR: 3***
3,5-BIS(3-HYDROXYMERCURI-2-METHOXYPROPYL)BARBITURIC ACID SODIUM SALT
mf: $C_{12}H_{19}Hg_2N_2O_7•7Na$ mw: 865.44

SYN: (1,5-(2,4,6-TRIOXO-(1H,3H,5H)-PYRIMIDYLENE))BIS((2-METHOXYPROPYL)HYDROXYMERCURY SODIUM SALT

TOXICITY DATA with REFERENCE
ipr-rat LD50:13500 μg/kg JAPMA8 39,297,50

CONSENSUS REPORTS: Mercury and its compounds are on the Community Right-To-Know List.

OSHA PEL: (Transitional: CL 1 mg/10m^3) CL 0.1 mg(Hg)/m^3 (skin)
ACGIH TLV: TWA 0.1 mg(Hg)/m^3 (skin)
NIOSH REL: (Mercury, Inorganic) TWA 0.05 mg(Hg)/m^3

SAFETY PROFILE: Poison by intraperitoneal route. See also MERCURY COMPOUNDS. When heated to decomposition it emits very toxic fumes of Na_2O, Hg, and NO_x.

BKG750 CAS:73118-24-8 ***HR: 3***
5,5-BIS(3-HYDROXYMERCURI-2-METHOXY-PROPYL)BARBITURIC ACIDSODIUM SALT
mf: $C_{12}H_{19}Hg_2N_2O_7$•xNa mw: 865.44

TOXICITY DATA with REFERENCE
ipr-rat LD50:30500 μg/kg JAPMA8 39,297,50

CONSENSUS REPORTS: Merucury and its compounds are on the Community Right-To-Know List.

OSHA PEL: (Transitional: CL 1 mg/10m^3) CL 0.1 mg(Hg)/m^3 (skin)
ACGIH TLV: TWA 0.1 mg(Hg)/m^3 (skin)
NIOSH REL: (Mercury, Inorganic) TWA 0.05 mg(Hg)/m^3

SAFETY PROFILE: Poison by intraperitoneal route. See also MERCURY COMPOUNDS. When heated to decomposition it emits very toxic fumes of Na_2O, Hg and NO_x.

BKH000 CAS:63951-09-7 ***HR: 3***
2,6-BIS(HYDROXYMERCURI)-4-NITROANILINE
mf: $C_6H_6Hg_2N_2O_4$ mw: 571.32

TOXICITY DATA with REFERENCE
ipr-rat LDLo:250 mg/kg NCNSA6 5,12,53

CONSENSUS REPORTS: Mercury and its compounds are on the Community Right-To-Know List.

OSHA PEL: (Transitional: CL 1 mg/10m^3) CL 0.1 mg(Hg)/m^3 (skin)
ACGIH TLV: TWA 0.1 mg(Hg)/m^3 (skin)
NIOSH REL: (Mercury, Inorganic) TWA 0.05 mg(Hg)/m^3

SAFETY PROFILE: Poison by intraperitoneal route. See also MERCURY COMPOUNDS. When heated to decomposition it emits very toxic fumes of NO_x and Hg.

BKH125 CAS:67536-44-1 ***HR: 3***
1,2-BIS(HYDROXOMERCURIO)-1,1,2,2-BIS(OXY-DIMERCURIO)ETHANE
mf: $C_2H_2Hg_6O_4$ mw: 1293.58

SYN: ETHANE HEXAMERCARBIDE

CONSENSUS REPORTS: Mercury and its compounds are on the Community Right-To-Know List.

SAFETY PROFILE: Explodes violently when heated to 230°C. When heated to decomposition it emits toxic fumes of Hg. See also MERCURY COMPOUNDS.

BKH325 CAS:105-08-8 ***HR: 2***
1,4-BIS(HYDROXYMETHYL)CYCLOHEXANE
mf: $C_8H_{16}O_2$ mw: 144.24

SYNS: 1,4-CHIDM ◇ HEXAHYDRO-2-OXO-1,4-CYCLOHEXANEDIMETHANOL

TOXICITY DATA with REFERENCE
orl-rat LDLo:3200 mg/kg KODAK* 21MAY71
ipr-rat LDLo:800 mg/kg 34ZIAG -,194,69
orl-mus LDLo:1600 mg/kg KODAK* 21MAY77
ipr-mus LDLo:1600 mg/kg 34ZIAG -,194,69

CONSENSUS REPORTS: Reported in EPA TSCA Inventory.

SAFETY PROFILE: Moderately toxic by intraperitoneal route. When heated to decomposition it emits acrid smoke and fumes.

BKH500 CAS:794-93-4 ***HR: 3***
BIS(HYDROXYMETHYL)FURATRIZINE
mf: $C_{11}H_{11}N_5O_5$ mw: 293.27

SYNS: 3-BIS(HYDROXYMETHYL)AMINO-6-(5-NITRO-2-FURYLETHENYL)-1,2,4-TRIAZINE ◇ DHNT ◇ 3-DI(HYDROXYMETHYL)AMINO-6-(5-NITRO-2-FURYLETHNEYL)-1,2,4-TRIAZINE ◇ 3-DI(HYDROXYMETHYL)AMINO-6-(2-(5-NITRO-2-FURYL)VINYL)-1,2,4-TRIAZINE ◇ DIHYDROXYMETHYL FURATRIZINE ◇ FURATONE ◇ FURATONE-S ◇ N-(6-(5-NITROFURFURYLIDENEMETHYL)-1,2,4-TRIAZIN-3-YL)IMINODIMETHANOL ◇ 6-(5-NITRO-2-FURYLVINYL)-3-(DIHYDROXYDIMETHYLAMINO)-1,2,4-TRIAZENE ◇ N-(6-(2-(5-NITRO-2-FURYL)VINYL)-1,2,4-TRIAZIN-3-YL)IMINODIMETHANOL ◇ ((6-(2-(5-NITRO-2-FURYL)VINYL)-as-TRIAZIN-3-YL)IMINO)DIMETHANOL ◇ PANFURAN-S

TOXICITY DATA with REFERENCE
mmo-esc 125 μg/L MUREAV 146,243,85
pic-esc 800 μg/L MUREAV 146,243,85
orl-rat TDLo:25725 mg/kg/35W-C:CAR NAIZAM 31,31,80
orl-mus TDLo:12740 mg/kg/35W-C:CAR NAIZAM 31,31,80
orl-mus TD:20580 mg/kg/35W-C ONKOD2 2,41,79
orl-mus TD:10290 mg/kg/35W-C ONKOD2 2,41,79
orl-rat TD:63210 mg/kg/43W-C EXPADD 26,213,84
orl-mus LD50:2690 mg/kg PMDCAY 5,320,67
ipr-mus LD50:1296 mg/kg PMDCAY 5,320,67
scu-mus LD50:1602 mg/kg PMDCAY 5,320,67

CONSENSUS REPORTS: IARC Cancer Review:

Group 2B IMEMDT 7,56,87; Animal Sufficient Evidence IMEMDT 24,77,80

SAFETY PROFILE: Suspected carcinogen with experimental carcinogenic and tumorigenic data. Moderately toxic by ingestion, intraperitoneal, and subcutaneous route. Mutation data reported. An antibacterial agent. When heated to decomposition it emits toxic fumes of NO_x.

BKH625 CAS:115-84-4 ***HR: 2***
3,3-BIS(HYDROXYMETHYL)HEPTANE
mf: $C_9H_{20}O_2$ mw: 160.29

SYNS: BEP ◇ 2-BUTYL-2-ETHYL-1,3-PROPANEDIOL ◇ 2-ETHYL-2-BUTYL-1,3-PROPANEDIOL

TOXICITY DATA with REFERENCE
eye-rbt 100 mg SEV 34ZIAG -,731,69
orl-rat LD50:5040 mg/kg 34ZIAG -,731,69
skn-rbt LD50:3810 mg/kg 34ZIAG -,731,69

SAFETY PROFILE: Moderately toxic by skin contact. Severe eye irritant. When heated to decomposition it emits acrid smoke and fumes.

BKH750 ***HR: 3***
BIS HYDROXYMETHYL PEROXIDE
mf: $C_2H_6O_4$ mw: 94.06

SAFETY PROFILE: Highly explosive. Sensitive to friction. When heated to decomposition it emits acrid smoke and fumes. See also PEROXIDES, ORGANIC.

BKH800 CAS:101564-54-9 ***HR: D***
3,4-BIS(p-HYDROXYPHENYL)-2-HEXANONE
mf: $C_{18}H_{20}O_3$ mw: 284.38

TOXICITY DATA with REFERENCE
scu-rat TDLo:175 μg/kg (35D pre):REP RPHRA6 20,395,64

SAFETY PROFILE: Experimental reproductive effects. When heated to decomposition it emits acrid smoke and fumes. See also KETONES.

BKI250 CAS:620-92-8 ***HR: 1***
BIS(p-HYDROXYPHENYL)METHANE
mf: $C_{13}H_{12}O_2$ mw: 200.25

SYNS: BIS(4-HYDROXYPHENYL)METHANE ◇ p,p'-BIS(HYDROXYPHENYL)METHANE ◇ 4,4'-METHYLENEBISPHENOL ◇ 4,4'-METHYLENE DIPHENOL

TOXICITY DATA with REFERENCE
orl-rat LD50:4950 mg/kg AIHAAP 23,95,62

CONSENSUS REPORTS: Reported in EPA TSCA Inventory.

SAFETY PROFILE: Mildly toxic by ingestion. When heated to decomposition it emits acrid smoke and irritating fumes.

BKI500 CAS:2971-36-0 ***HR: 2***
2,2-BIS(p-HYDROXYPHENYL)-1,1,1-TRICHLOROETHANE
mf: $C_{14}H_{11}Cl_3O_2$ mw: 317.60

SYN: 1,1,1-TRICHLORO-2,2-BIS(p-HYDROXYPHENYL)ETHANE

TOXICITY DATA with REFERENCE
orl-mus LDLo:600 mg/kg JPETAB 88,400,46

CONSENSUS REPORTS: Reported in EPA TSCA Inventory.

SAFETY PROFILE: Moderately toxic by ingestion. When heated to decomposition it emits very toxic fumes such as Cl^-.

BKI750 CAS:65-14-5 ***HR: 3***
2,3-BIS(p-HYDROXYPHENYL)VALERONITRILE
mf: $C_{17}H_{17}NO_2$ mw: 267.35

SYN: SC-3402

TOXICITY DATA with REFERENCE
scu-rat TDLo:7 mg/kg (35D pre):REP RPHRA6 20,395,64
ipr-rat LD50:70 mg/kg JPETAB 112,176,54
orl-mus LD50:2850 mg/kg JPETAB 112,176,54
ipr-mus LD50:93 mg/kg JPETAB 112,176,54
ivn-dog LDLo:100 mg/kg JPETAB 112,176,54

CONSENSUS REPORTS: Cyanide and its compounds are on the Community Right-To-Know List.

SAFETY PROFILE: Poison by intraperitoneal and intravenous routes. Moderately toxic by ingestion. Experimental reproductive effects. See also NITRILES. When heated to decomposition it emits toxic fumes of NO_x and CN^-.

BKJ250 CAS:62374-53-2 ***HR: 3***
BIS(3-HYDROXY-1-PROPYNYL)MERCURY
mf: $C_6H_6HgO_2$ mw: 310.71

SYN: 3,3'-MERCURIDI-2-PROPYN-1-OL

TOXICITY DATA with REFERENCE
ivn-mus LD50:4500 μg/kg CSLNX* NX#05895

CONSENSUS REPORTS: Mercury and its compounds are on the Community Right-To-Know List.

OSHA PEL: (Transitional: CL 1 mg/10m^3) CL 0.1 mg(Hg)/m^3 (skin)
ACGIH TLV: TWA 0.1 mg(Hg)/m^3 (skin)
NIOSH REL: (Mercury, Inorganic) TWA 0.05 mg(Hg)/m^3

SAFETY PROFILE: Poison by intravenous route. See

also MERCURY COMPOUNDS. When heated to decomposition it emits toxic vapors of Hg.

BKJ275 CAS:15702-65-5 ***HR: 3***
BIS(8-HYDROXYQUINOLINE-5-SULFONIC ACID) MANGANESE(II)
mf: $C_{18}H_{12}N_2O_8S_2 \cdot Mn$ mw: 503.38

SYNS: BIS(5-SULFO-8-QUINOLINOLATO-N[1],O[8]) MANGANESE(II) ◇ MANGANESE, BIS(5-SULFO-8-QUINOLINOLATO)-

TOXICITY DATA with REFERENCE
ivn-mus LD50:56 mg/kg CSLNX* NX#01222

OSHA PEL: CL 5 mg(Mn)/m^3
ACGIH TLV: TWA 5 mg(Mn)/m^3

SAFETY PROFILE: Poison by intravenous route. When heated to decomposition it emits toxic fumes of NO_x, SO_x, and Mn.

BKJ325 CAS:3286-46-2 ***HR: 3***
BISIBUTIAMINE
mf: $C_{32}H_{46}N_8O_6S_2$ mw: 702.98

SYN: o,o'-DIISOBUTYRYLTHIAMINE DISULFIDE

TOXICITY DATA with REFERENCE
ipr-rat LD50:660 mg/kg NIIRDN 6,606,82
scu-rat LD50:850 mg/kg NIIRDN 6,606,82
ivn-rat LD50:110 mg/kg NIIRDN 6,606,82

SAFETY PROFILE: Poison by intravenous route. Moderately toxic by some other routes. When heated to decomposition it emits toxic fumes of SO_x and NO_x. See also ESTERS.

BKJ500 CAS:73816-43-0 ***HR: 3***
BIS(3-INDOLEMETHYLENEMORPHOLINIUM)HEXACHLOROSTANNATE
mf: $C_{26}H_{30}N_4O_2 \cdot Cl_6Sn$ mw: 761.99

SYN: MORPHOLINIUM, (3-INDOLYLMETHYLENE)-, HEXACHLOROSTANNATE(2-)(2:1)

TOXICITY DATA with REFERENCE
ivn-mus LD50:100 mg/kg CSLNX* NX#02753

OSHA PEL: TWA 2 mg(Sn)/m^3
ACGIH TLV: TWA 2 mg(Sn)/m^3

SAFETY PROFILE: Poison by intravenous route. When heated to decomposition it emits toxic fumes of NO_x, Sn, and Cl^-.

BKK250 CAS:25168-24-5 ***HR: 2***
BIS(ISOOCTYLOXYCARBONYLMETHYLTHIO)DIBUTYL STANNANE
mf: $C_{28}H_{56}O_4S_2Sn$ mw: 639.65

SYNS: BIS(2-ETHYLHEXYLOXYCARBONYLMETHYLTHIO)DIBUTYLSTANNANE ◇ DIBUTYL-TIN BIS(ISOOCTYLTHIOGLYCOLLATE) ◇ DIBUTYLZINN-S,S'-BIS(ISOOCTYLTHIOGLYCOLAT)(GERMAN)

TOXICITY DATA with REFERENCE
orl-rat LD50:500 mg/kg TRIPA7 -,1,73

CONSENSUS REPORTS: Reported in EPA TSCA Inventory.

OSHA PEL: TWA 0.1 mg(Sn)/m^3 (skin)
ACGIH TLV: TWA 0.1 mg(Sn)/m^3 (skin) (Proposed: TWA 0.1 mg(Sn)/m^3; STEL 0.2 mg(Sn)/m^3 (skin))
NIOSH REL: (Organotin Compounds) TWA 0.1 mg(Sn)/m^3

SAFETY PROFILE: Moderately toxic by ingestion. See also TIN COMPOUNDS. When heated to decomposition it emits toxic fumes of SO_x.

BKK500 CAS:26636-01-1 ***HR: 2***
BIS(ISOOCTYLOXYCARBONYLMETHYLTHIO) DIMETHYLSTANNANE
mf: $C_{22}H_{44}O_4S_2Sn$ mw: 555.47

SYNS: BIS(2-ETHYLHEXYLOXYCARBONYLMETHYLTHIO)DIMETHYLSTANNANE ◇ DIMETHYL-TIN BIS(ISOOCTYLTHIOGLYCOLLATE) ◇ DIMETHYLZINN-S,S'-BIS(ISOOCTYLTHIOGLYCOLAT) (GERMAN)

TOXICITY DATA with REFERENCE
orl-rat LD50:1380 mg/kg TRIPA7 -,1,73

CONSENSUS REPORTS: Reported in EPA TSCA Inventory.

OSHA PEL: TWA 0.1 mg(Sn)/m^3 (skin)
ACGIH TLV: TWA 0.1 mg(Sn)/m^3 (skin) (Proposed: TWA 0.1 mg(Sn)/m^3; STEL 0.2 mg(Sn)/m^3 (skin))
NIOSH REL: (Organotin Compounds) TWA 0.1 mg(Sn)/m^3

SAFETY PROFILE: Moderately toxic by ingestion. See also TIN COMPOUNDS. When heated to decomposition it emits toxic fumes of SO_x.

BKK750 CAS:26401-97-8 ***HR: 2***
BIS(ISOOCTYLOXYCARBONYLMETHYLTHIO)DIOCTYL STANNANE
mf: $C_{36}H_{72}O_4S_2Sn$ mw: 751.89

SYNS: ADVASTAB 17 MO ◇ BIS(MERCAPTOACETATE)DIOCTYLTIN BIS(ISOOCTYL) ESTER ◇ DIISOOCTYL ((DIOCTYLSTANNYLENE)DITHIO)DIACETATE ◇ DIOCTYLTIN BIS(ISOOCTYL MERCAPTOACETATE) ◇ DIOCTYLTIN-S,S'-BIS(ISOOCTYL MERCAPTOACETATE) ◇ DIOCTYLTIN BIS(ISOOCTYL THIOGLYCOLATE) ◇ DIOCTYL-TIN BIS(ISOOCTYLTHIOGLYCOLLATE) ◇ DI-n-OCTYLTIN DIISOOCTYL THIOGLYCOLATE ◇ DI-n-OCTYL-ZINN-DI-ISOOCTYLTHIOGLYKOLAT (GERMAN) ◇ DOTG ◇ THERMOLITE 831

TOXICITY DATA with REFERENCE
orl-rat TDLo:420 mg/kg (1-21D preg):TER TXAPA9 26,253,73

orl-rat LD50:1277 mg/kg ARZNAD 19,934,69
skn-rat LD50:2250 mg/kg ARZNAD 19,934,69

CONSENSUS REPORTS: Reported in EPA TSCA Inventory.

OSHA PEL: TWA 0.1 mg(Sn)/m^3 (skin)
ACGIH TLV: TWA 0.1 mg(Sn)/m^3 (skin) (Proposed: TWA 0.1 mg(Sn)/m^3; STEL 0.2 mg(Sn)/m^3 (skin))
NIOSH REL: (Organotin Compounds) TWA 0.1 mg(Sn)/m^3

SAFETY PROFILE: Moderately toxic by ingestion and skin contact. An experimental teratogen. See also TIN COMPOUNDS and MERCAPTANS. When heated to decomposition it emits toxic fumes of SO_x.

BKL000 CAS:33568-99-9 ***HR: 2***
BIS(ISOOCTYLOXYMALEOYLOXY)DIOCTYL-STANNANE
mf: $C_{40}H_{72}O_8Sn$ mw: 799.81

SYNS: (Z,Z)-BIS((3-CARBOXYACRYLOYL)OXY)DIOCTYL-STANNANE DIISOOCTYL ESTER (8CI) ◇ (Z,Z)-4,4'-(DIOCTYLSTANNYLENE)BIS(OXY))BIS(4-OXO-2-BUTANOIC ACID DIISOOCTYL ESTER ◇ DIOCTYLTINBIS(ISOOCTYL MALEATE)

TOXICITY DATA with REFERENCE
orl-rat LD50:2760 mg/kg TRIPA7 -,1,73

CONSENSUS REPORTS: Reported in EPA TSCA Inventory.

OSHA PEL: TWA 0.1 mg(Sn)/m^3 (skin)
ACGIH TLV: TWA 0.1 mg(Sn)/m^3 (skin) (Proposed: TWA 0.1 mg(Sn)/m^3; STEL 0.2 mg(Sn)/m^3 (skin))
NIOSH REL: (Organotin Compounds) TWA 0.1 mg(Sn)/m^3

SAFETY PROFILE: Moderately toxic by ingestion. See also TIN COMPOUNDS. When heated to decomposition it emits acrid smoke and irritating fumes.

BKL250 CAS:7287-19-6 ***HR: 2***
2,4-BIS(ISOPROPYLAMINO)-6-METHYLMERCAPTO-s-TRIAZINE
mf: $C_{10}H_{19}N_5S$ mw: 241.40

SYNS: 4,6-BIS(ISOPROPYLAMINO)-2-METHYLMERCAPTO-s-TRIAZINE ◇ 2,4-BIS(ISOPROPYLAMINO)-6-METHYLTHIO-s-TRIAZINE ◇ 2,4-BIS(ISOPROPYLAMINO)-6-METHYLTHIO-1,3,5-TRIAZINE ◇ N,N'-BIS(1-METHYLETHYL)-6-METHYL-THIO-1,3,5-TRIAZINE-2,4-DIAMINE ◇ CAPAROL ◇ G 34161 ◇ GESAGARD ◇ MERKAZIN ◇ 2-METHYLMERCAPTO-4,6-BIS(ISOPROPYLAMINO)-s-TRIAZINE ◇ 2-METHYLTHIO-4,6-BIS(ISOPROPYLAMINO)-s-TRIAZINE ◇ POLISIN ◇ PRIMATOL Q ◇ PROMETREX ◇ PROMETRIN ◇ PROMETRYN ◇ PROMETRYNE (USDA) ◇ SELEKTIN ◇ SESAGARD

TOXICITY DATA with REFERENCE
eye-rbt 80 mg MLD CIGET* -,-,77
MUT mrc-smc 500 μg/L CYGEDX 21(2),59,87
orl-rat TDLo:2400 mg/kg (female 18-20D post):REP ZHYGAM 34,116,88
orl-rat LD50:1800 mg/kg 85GMAT -,87,82
orl-mus LD50:2 g/kg 85GMAT -,87,82

SAFETY PROFILE: Moderately toxic by ingestion. Experimental reproductive effects. An eye irritant. Mutation data reported. An herbicide. When heated to decomposition it emits very toxic fumes of NO_x and SO_x. See also MERCAPTANS.

BKL500 ***HR: 2***
2,4-BIS(ISOPROPYLAMINO)-6-(METHYLTHIO)-s-TRIAZINE mixed with METHANEARSONIC ACID MONOSODIUM SALT (1:4)

TOXICITY DATA with REFERENCE
orl-rat LD50:2500 mg/kg CIGET* -,-,77
skn-rbt LD50:3700 mg/kg CIGET* -,-,77

SAFETY PROFILE: Moderately toxic by ingestion and skin contact. See also ARSENIC COMPOUNDS. When heated to decomposition it emits very toxic fumes of SO_x, As, and NO_x.

BKL750 CAS:3006-93-7 ***HR: 3***
1,3-BISMALEIMIDO BENZENE
mf: $C_{14}H_8N_2O_4$ mw: 268.24

SYNS: 1,3-DIMALEIMIDOBENZENE ◇ HVA 2 ◇ HVA-2 CURING AGENT ◇ M-PHDM ◇ N,N'-(m-PHENYLENE)BISMALEIMIDE ◇ 1,1'-(m-PHENYLENE)BIS-1H-PYROLE-2,5-DIONE (9CI) ◇ N,N'-(m-PHENYLENEDIMALEIMIDE)

TOXICITY DATA with REFERENCE
ipr-rat LDLo:50 mg/kg NCNSA6 5,22,53
orl-rat LD50:1370 mg/kg GISAAA 40(11),109,75
orl-mus LD50:250 mg/kg GISAAA 40(11),109,75

CONSENSUS REPORTS: Reported in EPA TSCA Inventory.

SAFETY PROFILE: Poison by ingestion and intraperitoneal routes. When heated to decomposition it emits toxic fumes of NO_x.

BKM000 CAS:10193-95-0 ***HR: 3***
BIS(MERCAPTOACETATE)-1,4-BUTANEDIOL
mf: $C_8H_{14}O_4S_2$ mw: 238.34

SYN: BUTYLENE GLYCOL BIS(MERCAPTOACETATE)

TOXICITY DATA with REFERENCE
orl-rat LD50:405 mg/kg TRIPA7 -,1,73

CONSENSUS REPORTS: Reported in EPA TSCA Inventory.

SAFETY PROFILE: Poison by ingestion. When heated to decomposition it emits toxic fumes of SO_x. See also MERCAPTANS.

BKM100 CAS:73085-26-4 ***HR: D***
BIS(MESCALINIUM)TETRACHLOROMANGANATE(II)
mf: $C_{44}H_{62}Cl_4MnN_2O_{12}$ mw: 1007.82

SYN: BIS-3,4,5-TRIMETHOXY-β-PHENETHYLAMMONIUM TETRACHLOROMANGANATE(II) ◇ MANGANATE, TETRACHLORO-, BIS(3,4,5-TRIMETHOXYPHENETHYLAMMONIUM)

TOXICITY DATA with REFERENCE
par-rat TDLo:750 mg/kg (female 8-12D post):REP JINCAO 41,1507,79

ACGIH TLV: TWA 5 mg(Mn)/m^3
OSHA PEL: CL 5 mg(Mn)/m^3
OSHA PEL: FINAL:CL 5 mg(Mn)/m^3

SAFETY PROFILE: Experimental reproductive effects. When heated to decomposition it emits toxic fumes of Mn, NO_x, and Cl^-.

BKM125 CAS:4672-49-5 ***HR: 3***
1,2-BIS(MESYLOXY)ETHANE
mf: $C_4H_{10}O_6S_2$ mw: 218.26

SYNS: 1,2-ETHANEDIOL DIMETHANESULFONATE (9CI) ◇ 1,2-ETHANEDIYL DIMETHANESULFONATE ◇ ETHYLENE BIS-(METHANESULFONATE) ◇ ETHYLENE DIMETHANESULFONATE ◇ ETHYLENE DIMETHANESULPHONATE

TOXICITY DATA with REFERENCE
orl-rat TDLo:250 mg/kg (10D male):REP 85GUAJ -,55,66
ipr-rat LD50:150 mg/kg BJPCAL 24,24,65

SAFETY PROFILE: Poison by intraperitoneal route. Experimental reproductive effects. When heated to decomposition it emits toxic fumes of SO_x. See also SULFONATES.

BKM250 CAS:97-90-5 ***HR: 2***
1,2-BIS(METHACRYLOYLOXY)ETHANE
mf: $C_{10}H_{14}O_4$ mw: 198.1

SYNS: AGEFLEX EGDM ◇ DIGLYCOL DIMETHACRYLATE ◇ ETHANEDIOL DIMETHACRYLATE ◇ 1,2-ETHANEDIOL DIMETHACRYLATE ◇ ETHYLDIOL METACRYLATE ◇ ETHYLENE GLYCOL BIS(METHACRYLATE) ◇ ETHYLENE GLYCOL DIMETHACRYLATE ◇ ETHYLENE METHACRYLATE ◇ GLYCOL DIMETHACRYLATE ◇ SARTOMER SR 206 ◇ SR 206

TOXICITY DATA with REFERENCE
orl-rat LD50:3300 mg/kg GTPZAB 24(4),58,80
ipr-rat LD50:2800 mg/kg AMPMAR 36,58,75
orl-mus LD50:2000 mg/kg GTPZAB 24(4),58,80

CONSENSUS REPORTS: Reported in EPA TSCA Inventory.

SAFETY PROFILE: Moderately toxic by ingestion and intraperitoneal routes. When heated to decomposition it emits acrid smoke and irritating fumes. See also ESTERS.

BKM500 CAS:1187-00-4 ***HR: 3***
BIS(METHANE SULFONYL)-d-MANNITOL
mf: $C_8H_{18}O_{10}S_2$ mw: 338.38

SYNS: 1,6-BIS-o-METHYLSULFONYL-d-MANNITOL ◇ CB 2511 ◇ 1,6-DIMESYL-d-MANNITOL ◇ 1,6-DIMETHANESULFONATE-d-MANNITOL ◇ 1,6-DIMETHANE-SULFONOXY-d-MANNITOL ◇ 1,6-DIMETHANESULPHONOXY-1,6-DIDEOXY-d-MANNITOL ◇ DMM ◇ d-MANNITOL BUSULFAN ◇ MANNITOL MYLERAN ◇ MANNOGRANOL ◇ MM ◇ NSC-37538

TOXICITY DATA with REFERENCE
sln-dmg-unk 160 mmol/L ANYAA9 160,228,69
ipr-mus TDLo:1000 mg/kg/4W:NEO JNCIAM 36,915,66
ipr-rat LD50:2000 mg/kg EJCAAH 4,617,68
orl-mus LD50:6000 mg/kg ARZNAD 17,145,67
ivn-dog LDLo:135 mg/kg CCSUBJ 2,203,65
ivn-mky LDLo:135 mg/kg CCSUBJ 2,203,65

CONSENSUS REPORTS: EPA Genetic Toxicology Program.

SAFETY PROFILE: Poison by intravenous route. Moderately toxic by intraperitoneal route. Mildly toxic by ingestion. Questionable carcinogen with experimental neoplastigenic data. Mutation data reported. When heated to decomposition it emits toxic fumes of SO_x.

BKM750 CAS:7306-46-9 ***HR: 3***
3,4-BIS(METHOXY)BENZYL CHLORIDE
mf: $C_9H_{11}ClO_2$ mw: 186.65

SYNS: 3,4-DIMETHOXYBENZYL CHLORIDE ◇ VERATRYL CHLORID (GERMAN) ◇ VERATRYL CHLORIDE

TOXICITY DATA with REFERENCE
scu-rat TDLo:2100 mg/kg/42W-I:ETA ZEKBAI 74,241,70
orl-rat LD50:4700 mg/kg GTPZAB 26(2),55,82
scu-rat LD50:3000 mg/kg ZEKBAI 74,241,70
orl-mus LD50:5g/kg GTPZAB 26(2),55,82

SAFETY PROFILE: Questionable carcinogen with experimental tumorigenic data. Moderately toxic by subcutaneous route. Mildly toxic by ingestion. When heated to decomposition it emits toxic fumes of Cl^-.

BKN000 CAS:3965-55-7 ***HR: 2***
3,5-BIS(METHOXYCARBONYL)BENZENESULFONIC ACID, SODIUM SALT
mf: $C_{10}H_9O_7S{\cdot}Na$ mw: 296.24

SYN: 3,5-BIS-METHYLKARBOXY-BENZENSULFONANSODNY (CZECH)

TOXICITY DATA with REFERENCE
eye-rbt 100 mg/24H SEV 28ZPAK -,185,72

CONSENSUS REPORTS: Reported in EPA TSCA Inventory.

SAFETY PROFILE: A severe eye irritant. When heated

to decomposition it emits toxic fumes of SO_x and Na_2O. See also SULFONATES.

BKN250 CAS:58306-30-2 **HR: 2**
N-(2-(2,3-BIS-(METHOXYCARBONYL)-GUANIDINO)-5-(PHENYLTHIO)-PHENYL)-2-METHOXYACETAMIDE
mf: $C_{20}H_{22}N_4O_6S$ mw: 446.52

SYN: FEBANTEL

TOXICITY DATA with REFERENCE
orl-rat LD50:10605 mg/kg ARZNAD 28,2193,78
orl-rbt LD50:1250 mg/kg ARZNAD 28,2193,78

SAFETY PROFILE: Moderately toxic by ingestion. When heated to decomposition it emits very toxic fumes of NO_x and SO_x.

BKN500 CAS:60397-73-1 **HR: 1**
4,4'-BIS((4-(2-METHOXYETHOXY)-6-(N-METHYL-N-2-SULFOETHYL)AMINO-s-TRIAZIN-2-YL)AMINO)-2,2'-STILBENEDISULFONIC ACID
mf: $C_{32}H_{42}N_{10}O_{16}S_4$ mw: 951.08

TOXICITY DATA with REFERENCE
eye-rbt 100 mg MOD MVCRB3 2,193,73

SAFETY PROFILE: An eye irritant. See also SULFONATES. When heated to decomposition it emits very toxic fumes of NO_x and SO_x.

BKN750 CAS:111-96-6 **HR: 3**
BIS(2-METHOXY ETHYL)ETHER
mf: $C_6H_{14}O_3$ mw: 134.48

PROP: Liquid. Bp: 162°, d: 0.9451, mp: −68°, flash p: 158°F (70°C) (OC), n (20/D) 1.4097. Misc with water, alc, ether, and hydrocarbon solvents.

SYNS: DIETHYLENE GLYCOL DIMETHYL ETHER ◇ DIETHYL GLYCOL DIMETHYL ETHER ◇ DIGLYME

TOXICITY DATA with REFERENCE
sln-dmg-ihl 250 ppm/165M NTIS** PB83-138198
dlt-rat-ihl 1000 ppm/5D-C NTIS** PB83-138198
spm-mus-ihl 1000 ppm/5D-C NTIS** PB83-138198
orl-mus TDLo:1250 mg/kg (female 6-15D post):TER NTIS** PB86-135233
orl-mus TDLo:5 g/kg (female 6-15D post):REP NTIS** PB86-135233

SAFETY PROFILE: An experimental teratogen. Other experimental reproductive effects. Mutation data reported. When heated to decomposition it emtis toxic fumes of NO_x. Readily forms explosive peroxides upon exposure to air, light, or heat. Solution containing carbon dioxide may react with aluminum hydride to form an explosive product. Other metal hydrides may react similarly. See also ETHERS.

BKO000 CAS:67856-65-9 **HR: 3**
BIS(2-METHOXYETHYL)NITROSOAMINE
mf: $C_6H_{14}N_2O_3$ mw: 162.22

SYN: N-NITROSOBIS(2-METHOXYETHYL)AMINE

TOXICITY DATA with REFERENCE
orl-rat TDLo:2750 mg/kg/50W-I:ETA CNREA8 38,2391,78

SAFETY PROFILE: Questionable carcinogen with experimental tumorigenic data. When heated to decomposition it emits toxic fumes of NO_x. See also NITROSAMINES.

BKO250 CAS:15546-11-9 **HR: 3**
BIS(METHOXYMALEOYLOXY)DIBUTYLSTANNANE
mf: $C_{18}H_{28}O_8Sn$ mw: 491.15

SYNS: DIBUTYLBIS((3-CARBOXYACRYLOYL)OXY)-STANNANE DIMETHYL ESTER (Z,Z) (8CI) ◇ DIBUTYLTIN BIS(METHYL MALEATE) ◇ DIBUTYLTIN BIS(MONOMETHYL MALEATE) ◇ DIBUTYLTIN METHYL MALEATE ◇ 6,6-DIBUTYL-4,8,11-TRIOXO-5,7,12-TRIOXA-6-STANNATRIDECA-2,9-DIENOIC ACID METHYL ESTER ◇ DI-n-BUTYLZINN-DIMONOMETHYLMALEINAT (GERMAN) ◇ STAN-GUARD 156

TOXICITY DATA with REFERENCE
orl-rat LD50:62 mg/kg TRIPA7 -,1,73

CONSENSUS REPORTS: Reported in EPA TSCA Inventory.

OSHA PEL: TWA 0.1 mg(Sn)/m^3 (skin)
ACGIH TLV: TWA 0.1 mg(Sn)/m^3 (skin) (Proposed: TWA 0.1 mg(Sn)/m^3; STEL 0.2 mg(Sn)/m^3 (skin))
NIOSH REL: (Organotin Compounds) TWA 0.1 mg(Sn)/m^3

SAFETY PROFILE: Poison by ingestion. See also TIN COMPOUNDS. When heated to decomposition it emits acrid smoke and irritating fumes.

BKO500 CAS:60494-19-1 **HR: 2**
BIS(METHOXYMALEOYLOXY)DIOCTYLSTANNANE
mf: $C_{26}H_{44}O_8Sn$ mw: 603.39

SYN: DI-n-OCTYLZINN-DIMONOMETHYLMALEINAT(GERMAN)

TOXICITY DATA with REFERENCE
orl-rat LD50:1673 mg/kg TRIPA7 -,1,73

OSHA PEL: TWA 0.1 mg(Sn)/m^3 (skin)
ACGIH TLV: TWA 0.1 mg(Sn)/m^3 (skin) (Proposed: TWA 0.1 mg(Sn)/m^3; STEL 0.2 mg(Sn)/m^3 (skin))
NIOSH REL: (Organotin Compounds) TWA 0.1 mg(Sn)/m^3

SAFETY PROFILE: Moderately toxic by ingestion. See also TIN COMPOUNDS. When heated to decomposition it emits acrid smoke and irritating fumes.

BKO750 CAS:7342-13-4 ***HR: 2***
4,4'-BIS(4-METHOXY-6-PHENYLAMINO-2-s-TRIAZINYLAMINO)-2,2'-STILBENEDISULFONIC ACID
mf: $C_{34}H_{30}N_{10}O_8S_2$ mw: 770.86

SYN: RYLUX PRS (CZECH)

TOXICITY DATA with REFERENCE
eye-rbt 100 mg/24H SEV 28ZPAK -,251,72

CONSENSUS REPORTS: Reported in EPA TSCA Inventory.

SAFETY PROFILE: A severe eye irritant. When heated to decomposition it emits very toxic fumes of NO_x and SO_x.

BKO800 CAS:56622-38-9 ***HR: D***
3,4-BIS(p-METHOXYPHENYL)-3-BUTEN-2-ONE
mf: $C_{18}H_{18}O_3$ mw: 282.36

SYNS: 3-BUTEN-2-ONE, 3,4-BIS(p-METHOXYPHENYL)- ◇ 3-BUTEN-2-ONE, 3,4-BIS(4-METHOXYPHENYL)-(9CI)

TOXICITY DATA with REFERENCE
unr-rat TDLo:75 mg/kg (female 30D pre):REP RPHRA6 20,395,64

SAFETY PROFILE: Experimental reproductive effects. When heated to decomposition it emits acrid smoke and irritating fumes.

BKO825 CAS:33406-36-9 ***HR: D***
2-(p-(1,2-BIS(p-METHOXYPHENYL)-1-BUTENYL)PHENOXY)TRIETHYLAMINE
mf: $C_{30}H_{37}NO_3$ mw: 459.68

SYNS: 2-(4-(1,2-BIS(4-METHOXYPHENYL)-1-BUTENYL)PHENOXY)-N,N-DIETHYL-ETHANAMINE (9CI) ◇ H 774

TOXICITY DATA with REFERENCE
orl-mus TDLo:40 mg/kg (1D pre):REP JRPFA4 34,29,73

SAFETY PROFILE: Experimental reproductive effects. When heated to decomposition it emits toxic fumes of NO_x. See also AMINES.

BKO835 CAS:42920-39-8 ***HR: D***
2-(p-(1,2-BIS(p-METHOXYPHENYL)-1-BUTENYL)PHENOXY)TRIETHYLAMINE CITRATE
mf: $C_{30}H_{37}NO_3 \cdot C_6H_8O_7$ mw: 651.82

TOXICITY DATA with REFERENCE
orl-mus TDLo:1800 μg/kg (female 4-6D post):REP JMCMAR 14,952,71

SAFETY PROFILE: Experimental reproductive effects. When heated to decomposition it emits toxic fumes of NO_x.

BKO840 CAS:35258-12-9 ***HR: D***
trans-2-(p-(1,2-BIS(p-METHOXYPHENYL)-1-BUTENYL)PHENOXY)TRIETHYLAMINE HYDROCHLORIDE
mf: $C_{30}H_{37}NO_3 \cdot ClH$ mw: 496.14

SYNS: trans-1,2-BIS(p-METHOXYPHENYL)-1-(p-(2-(N,N-DIETHYLAMINO)ETHOXY)PHENYL)BUT-1-ENE HCl ◇ H-1067

TOXICITY DATA with REFERENCE
orl-mus TDLo:240 μg/kg (female 1-3D post):REP JRPFA4 34,23,73

SAFETY PROFILE: Experimental reproductive effects. When heated to decomposition it emits toxic fumes of NO_x and HCl.

BKP000 CAS:69352-67-6 ***HR: 3***
1,5-BIS(o-METHOXYPHENYL)-3,7-DIAZAADAMANTAN-9-ONE
mf: $C_{22}H_{24}N_2O_3$ mw: 364.48

TOXICITY DATA with REFERENCE
ipr-mus LD50:20 mg/kg JMPCAS 5,1293,62
ivn-rbt LD50:66 mg/kg JMPCAS 5,1293,62

SAFETY PROFILE: Poison by intraperitoneal and intravenous routes. When heated to decomposition it emits toxic fumes of NO_x.

BKP200 CAS:24407-55-4 ***HR: 3***
BIS-(B-o-METHOXYPHENYL-ISOPROPYL)-AMINE LACTATE
mf: $C_{20}H_{27}NO_2 \cdot C_3H_6O_3$ mw: 403.57

SYNS: o,o'-DIMETHOXY-α-α'-DIMETHYL-DIPHENETHYLAMINE compounded with LACTICACID ◇ DIPHENETHYLAMINE, o,o'-DIMETHOXY-α-α'-DIMETHYL-, compounded with LACTIC ACID ◇ U-0045

TOXICITY DATA with REFERENCE
skn-rbt 2500 ppm MLD AIPTAK 137,410,62
eye-rbt 1 pph AIPTAK 137,410,62
ipr-mus LD50:51200 μg/kg AIPTAK 137,410,62

SAFETY PROFILE: Poison by intraperitoneal route. A skin and eye irritant. When heated to decomposition it emits toxic fumes of NO_x.

BKP300 CAS:53-64-5 ***HR: D***
2,3-BIS(p-METHOXYPHENYL)-2-PENTENONITRILE
mf: $C_{19}H_{19}NO_2$ mw: 293.39

SYNS: 2,3-BIS(4-METHOXYPHENYL)PENT-2-ENENITRILE ◇ SC-3296

TOXICITY DATA with REFERENCE
orl-rat TDLo:500 μg/kg (1D preg):REP STEDAM 4,657,64

CONSENSUS REPORTS: Cyanide and its compounds are on the Community Right-To-Know List.

SAFETY PROFILE: Experimental reproductive effects.

When heated to decomposition it emits toxic fumes of NO_x and CN^-. See also NITRILES.

BKP325 CAS:2041-74-9 ***HR: D***
2-(p-(1,2-BIS(p-METHOXYPHENYL)PROPENYL) PHENOXY)TRIETHYLAMINE, CITRATE, MONOHYDRATE
mf: $C_{29}H_{34}NO_3 \cdot C_6H_8O_7 \cdot H_2O$ mw: 655.81

SYN: 2-(p-(1,2-BIS(p-METHOXYPHENYL)PROPENYL)PHENOXY)TRIETHYLAMINE CITRATE HYDRATE

TOXICITY DATA with REFERENCE
orl-rat TDLo:150 μg/kg (1-3D preg):REP JMCMAR 14,952,71

SAFETY PROFILE: Experimental reproductive effects. When heated to decomposition it emits toxic fumes of NO_x.

BKQ250 CAS:65210-37-9 ***HR: 3***
N,N'-(BIS(2-(2-METHYL-1,3-BENZODIOXOL-2-YL)ETHYL))ETHYLENEDIAMINE DIHYDROCHLORIDE
mf: $C_{22}H_{28}N_2O_4 \cdot 2ClH$ mw: 457.44

TOXICITY DATA with REFERENCE
ivn-rat LD50:20 mg/kg EJMCA5 12,413,77
ipr-mus LD50:90 mg/kg EJMCA5 12,413,77

SAFETY PROFILE: Poison by intravenous and intraperitoneal routes. When heated to decomposition it emits very toxic fumes of HCl and NO_x.

BKQ500 CAS:10024-74-5 ***HR: 2***
BIS(α-METHYLBENZYL)AMINE
mf: $C_{16}H_{19}N$ mw: 225.36

PROP: Liquid. Mp: −65°, bp: 188.5°, flash p: 175°F (OC), d: 0.9535, vap press: 0.5 mm @ 20°, vap d: 4.18.

TOXICITY DATA with REFERENCE
skn-rbt 100 μg/24H open AIHAAP 23,95,62
orl-rat LD50:2930 mg/kg AIHAAP 23,95,62

SAFETY PROFILE: Moderately toxic by ingestion and skin contact. Cmbustible when exposed to heat or flame. To fight fire, use alcohol foam, CO_2, dry chemical. Incompatible with oxidizers. When heated to decomposition it emits toxic fumes of NO_x.

BKQ750 CAS:74927-02-9 ***HR: 3***
2,6-BIS(1-METHYLBUTYL)PHENOL
mf: $C_{16}H_{26}O$ mw: 234.2

TOXICITY DATA with REFERENCE
ivn-mus LD50:160 mg/kg JMCMAR 23,1350,80
ivn-rbt LDLo:30 mg/kg JMCMAR 23,1350,80

SAFETY PROFILE: Poison by intravenous route. When heated to decomposition it emits acrid smoke and irritating fumes.

BKR000 CAS:63982-52-5 ***HR: 3***
1,4-BIS(METHYLCARBAMYLOXY)-2-ISOPROPYL-5-METHYLBENZENE
mf: $C_{14}H_{20}N_2O_4$ mw: 280.36

SYN: TL-1350

TOXICITY DATA with REFERENCE
orl-rat LD50:60 mg/kg 85ALAU -,107,76
orl-mus LD50:40 mg/kg 85ALAU -,107,76
scu-mus LDLo:20 mg/kg NTIS** PB158-508

SAFETY PROFILE: Poison by ingestion and subcutaneous routes. When heated to decomposition it emits toxic fumes of NO_x.

BKR250 CAS:66903-23-9 ***HR: 2***
BIS(3-METHYLCYCLOHEXYL PEROXIDE)
mf: $C_{14}H_{22}O_4$ mw: 254.36

SYN: 3-METHYLCYKLOHEXANONPEROXID (CZECH)

TOXICITY DATA with REFERENCE
skn-rbt 500 mg/24H SEV 28ZPAK -,142,72
eye-rbt 250 μg/24H SEV 28ZPAK -,142,72
orl-rat LD50:1500 mg/kg 28ZPAK -,142,72

SAFETY PROFILE: Moderately toxic by ingestion. A severe eye and skin irritant. When heated to decomposition it emits acrid smoke and irritating fumes. See also PEROXIDES, ORGANIC.

BKR500 CAS:64246-03-3 ***HR: 3***
1,1-BIS((3,4-METHYLENEDIOXYPHENOXY) METHYL)-N,N-DIMETHYL-1-BUTANOL CITRATE
mf: $C_{22}H_{27}NO_7 \cdot C_6H_8O_7$ mw: 609.64

SYN: 1,3-BIS-(3,4-METILENDIOSSIFENOSSI)-2-(3-DIMETILAMINOPROPIL)PROPAN-2-OLO CITRATO (ITALIAN)

TOXICITY DATA with REFERENCE
orl-mus LD50:780 mg/kg FRPSAX 32,502,77
ivn-mus LD50:94 mg/kg FRPSAX 32,502,77

SAFETY PROFILE: Poison by intravenous route. Moderately toxic by ingestion. When heated to decomposition it emits toxic fumes of NO_x.

BKR750 CAS:64246-13-5 ***HR: 3***
α,α-BIS(3,4-(METHYLENEDIOXY)PHENOXY) METHYL)-1-PIPERIDINEBUTANOLACETATE CITRATE
mf: $C_{27}H_{33}NO_8 \cdot C_6H_8O_7$ mw: 691.75

TOXICITY DATA with REFERENCE
orl-mus LD50:700 mg/kg FRPSAX 32,502,77
ivn-mus LD50:32 mg/kg FRPSAX 32,502,77

SAFETY PROFILE: Poison by intravenous route. Moderately toxic by ingestion. When heated to decomposition it emits toxic fumes of NO_x.

BKS500 ***HR: 3***
1-(1,3-BIS(3,4-(METHYLENEDIOXY)PHENOXY)-2-PROPYL)PYRROLIDINE CITRATE
mf: $C_{21}H_{23}NO_6 \cdot C_6H_8O_7$ mw: 577.59

SYN: 1,3-BIS-(3,4-METILENDIOSSIFENOSSI)-2-PIRROLIDINOPROPANO CITRATO (ITALIAN)

TOXICITY DATA with REFERENCE
orl-mus LD50:195 mg/kg FRPSAX 32,502,77
ivn-mus LD50:58 mg/kg FRPSAX 32,502,77

SAFETY PROFILE: Poison by ingestion and intravenous routes. When heated to decomposition it emits toxic fumes of NO_x.

BKS750 CAS:26087-47-8 ***HR: 2***
O,O-BIS(1-METHYLETHYL)-S-(PHENYLMETHYL)PHOSPHOROTHIOATE
mf: $C_{13}H_{21}O_3PS$ mw: 288.37

SYNS: O,O-DIISOPROPYL-S-BENZYLPHOSPHOROTHIOLATE ◇ O,O-DIISOPROPYL-S-BENZYL THIOPHOSPHATE ◇ IBP ◇ KITAZIN P

TOXICITY DATA with REFERENCE
orl-rat LD50:490 mg/kg FMCHA2 -,D175,80
orl-mus LD50:1760 mg/kg FMCHA2 -,D175,80
skn-mus LD50:5000 mg/kg FMCHA2 -,D175,80

SAFETY PROFILE: Moderately toxic by ingestion. Mildly toxic by skin contact. When heated to decomposition it emits very toxic fumes of SO_x and PO_x. See also ESTERS.

BKS800 CAS:78313-59-4 ***HR: 3***
BIS(2-METHYL-3-HYDROXY-4-METHOXYMETHYL-5-METHYLPYRIDYL)DISULFIDE DIHYDROCHLORIDE
mf: $C_{18}H_{24}N_2O_4S_2 \cdot 2ClH$ mw: 469.48

TOXICITY DATA with REFERENCE
orl-mus LD50:1450 mg/kg PCJOAU 15,79,81
ipr-mus LD50:434 mg/kg PCJOAU 15,79,81
ivn-mus LD50:111 mg/kg PCJOAU 15,79,81

SAFETY PROFILE: Poison by intravenous route. Moderately toxic by ingestion and other routes. When heated to decomposition it emits toxic fumes of NO_x, SO_x, and HCl.

BKS810 CAS:3810-81-9 ***HR: 3***
BIS(METHYLMERCURIC)SULFATE
mf: $C_2H_6Hg_2O_4S$ mw: 527.32

SYNS: ARETAN-NIEUW ◇ B 4992 ◇ BIS-(METHYLMERCURY)-SULFATE ◇ BIS-(METHYLMERKURI)SULFAT ◇ CERESAN UNIVERSAL-FEUCHTBEIZE ◇ CEREWET ◇ COMPOUND-4992 ◇ MERCURY, SULFATOBIS(METHYL- ◇ METHYLMERCURIC SULFATE ◇ SULFURIC ACID, BIS(METHYLMERCURY) SALT

TOXICITY DATA with REFERENCE
orl-rat LD50:50 mg/kg FMCHA2-,C63,89

ACGIH TLV: TWA 0.01 mg(Hg)/m^3; STEL 0.03 mg(Hg)/m^3

SAFETY PROFILE: Poison by ingestion. When heated to decomposition it emits toxic fumes of SO_x and Hg.

BKS825 CAS:13018-50-3 ***HR: 3***
N,N'-BIS(1-METHYL-4-PHENYL-4-PIPERIDYLMETHYL)SEBACAMIDE
mf: $C_{36}H_{54}N_4O_2$ mw: 574.94

SYNS: N,N'-BIS((1-METHYL-4-PHENYL-4-PIPERIDINYL)METHYL)-DECANEDIAMIDE (9CI) ◇ DIAMMIDE SEBACICA della 4-FENIL-4-AMMINOMETIL-N-METILPIPERIDINA (ITALIAN) ◇ 1665 I.S.

TOXICITY DATA with REFERENCE
ipr-rat LD50:17500 mg/kg FRPSAX 17,24,62
ivn-rat LD50:12 mg/kg FRPSAX 17,24,62
ipr-mus LD50:5 mg/kg FRPSAX 17,24,62
ivn-mus LD50:2820 μg/kg CSLNX* NX#12224
ipr-gpg LD50:40 mg/kg FRPSAX 17,24,62
ivn-gpg LD50:4 mg/kg FRPSAX 17,24,62

SAFETY PROFILE: Poison by intravenous and intraperitoneal routes. When heated to decomposition it emits toxic fumes of NO_x.

BKT250 ***HR: 3***
BIS(2-METHYL PYRIDINE)SODIUM
mf: $C_{12}H_{14}N_2Na$ mw: 209.25

SAFETY PROFILE: Ignites spontaneously in air. When heated to decomposition it emits toxic fumes of NO_x and Na_2O.

BKU000 CAS:73696-64-7 ***HR: 2***
N,N'-BIS(3-METHYL-2-THIAZOLIDINYLIDENE)UREA
mf: $C_9H_{14}N_4S_2O$ mw: 258.39

TOXICITY DATA with REFERENCE
orl-mus LD50:2000 mg/kg JMCMAR 23,773,80
ipr-mus LD50:1140 mg/kg JMCMAR 23,773,80

SAFETY PROFILE: Moderately toxic by ingestion and intraperitoneal routes. When heated to decomposition it emits very toxic fumes of SO_x and NO_x.

BKU250 CAS:14024-75-0 ***HR: 3***
BIS(4-MORPHOLINECARBODITHIOATO)MERCURY
mf: $C_{10}H_{16}HgN_2O_2S_4$ mw: 525.11

TOXICITY DATA with REFERENCE
ivn-mus LD50:56 mg/kg CSLNX* NX#02530

CONSENSUS REPORTS: Mercury and its compounds are on the Community Right-To-Know List.

OSHA PEL: (Transitional: CL 1 mg/10m^3) CL 0.1 mg(Hg)/m^3 (skin)
ACGIH TLV: TWA 0.1 mg(Hg)/m^3 (skin)
NIOSH REL: (Mercury, Inorganic) TWA 0.05 mg(Hg)/m^3

SAFETY PROFILE: Poison by intravenous route. See also MERCURY COMPOUNDS. When heated to decomposition it emits very toxic fumes of SO_x and NO_x and Hg vapors.

BKU500 CAS:103-34-4 ***HR: 3***
N,N'-BISMORPHOLINE DISULFIDE
mf: $C_8H_{16}N_2O_2S_2$ mw: 236.38

PROP: Tan to gray powder. Mp: 122° min, d: 1.36 @ 25°.

SYNS: ACCEL R ◇ BISMORPHOLINO DISULFIDE ◇ DIMORPHOLINE DISULFIDE ◇ DIMORPHOLINO DISULFIDE ◇ DITHIOBISMORPHOLINE ◇ 4,4'-DITHIOBIS(MORPHOLINE) ◇ N,N-DITHIODIMORPHOLINE ◇ 4,4'-DITHIODIMORPHOLINE ◇ 4,4'-DITHIOMORPHOLINE ◇ MORPHOLINE DISULFIDE ◇ MORPHOLINODISULFIDE ◇ SULFASAN ◇ SULFASAN R POWDER ◇ USAF B-17 ◇ USAF EK-T-6645

TOXICITY DATA with REFERENCE
mma-sat 100 μg/plate PCBRD2 141,407,84
dnr-bcs 1 mg/disc SAIGBL 26,147,84
orl-mus LD50:1660 mg/kg ARZNAD 11,797,61
ipr-mus LD50:50 mg/kg NTIS** AD277-689
ivn-mus LD50:100 mg/kg CSLNX* NX#02252

CONSENSUS REPORTS: Reported in EPA TSCA Inventory.

SAFETY PROFILE: Poison by intraperitoneal and intravenous routes. Moderately toxic by ingestion. Mutation data reported. See also MORPHOLINE. When heated to decomposition it emits very toxic fumes of NO_x and SO_x.

BKU750 CAS:7440-69-9 ***HR: 3***
BISMUTH
af: Bi aw: 208.98

PROP: Hexagonal silver-white or reddish metallic crystals. Mp: 271.3°, bp: 1420-1560°, d: 9.80, vap press: 1 mm @ 1021°.

SYN: BISMUTH-209

TOXICITY DATA with REFERENCE
unr-man LDLo:221 mg/kg 85DCAI 2,73,70

CONSENSUS REPORTS: Reported in EPA TSCA Inventory.

SAFETY PROFILE: Poisonous to humans. See also BISMUTH COMPOUNDS. Flammable when exposed to flame. Reaction with [$Bi(OH)_3$ + $Al(OH)_3$], coprecipitated and H_2 reduced produces a spontaneously flammable product. Moderately dangerous, can react with acid or acid fumes to emit toxic fumes. Incompatible with Al; BrF_3; acids; NOF; NH_4NO_3; $HClO_3$; Cl_2; IF_5; HNO_3; $HClO_4$.

BKV000 ***HR: 3***
BISMUTH AMIDE OXIDE
mf: BiH_2NO mw: 241

SAFETY PROFILE: Stable in liquid NH_3. Very unstable when free of NH_3. Upon decomposition it emits toxic fumes of Bi and NO_x. See also BISMUTH COMPOUNDS.

BKV250 CAS:12001-47-7 ***HR: 3***
BISMUTH ARSPHENAMINE SULFONATE
mf: $C_{21}H_{24}As_3Bi_2N_3O_{12}S_3$•3Na mw: 1318.35

SYNS: BISMARSEN ◇ SULFARSPHENAMINE BISMUTH

TOXICITY DATA with REFERENCE
ims-rat LDLo:500 mg/kg ADSYAF 28,389,33
ipr-mus LDLo:128 mg/kg CBCCI* 2,241,50
ims-rbt LDLo:150 mg/kg ADSYAF 28,389,33

CONSENSUS REPORTS: Arsenic and its compounds are on the Community Right-To-Know List.

OSHA PEL: TWA 0.5 mg(As)/m^3

SAFETY PROFILE: A poison by intraperitoneal and intramuscular routes. See also ARSENIC COMPOUNDS and BISMUTH COMPOUNDS. When heated to decomposition it emits very toxic fumes of Na_2O, NO_x, SO_x, As, and Bi.

BKV750 ***HR: 3***
BISMUTH COMPOUNDS

SAFETY PROFILE: Bismuth and its salts can cause kidney damage, although the degree of such damage is usually mild. Large doses can be fatal. Industrially it is considered one of the less toxic of the heavy metals, although intoxication has occurred from its use in medicine. The similarity between the pharmacologic and toxic behavior of lead and bismuth has been pointed out in the literature. Like lead, bismuth may be liberated from tissue deposits during periods of acidosis. Serious and sometimes fatal poisoning may occur from the injection of large doses into closed cavities and from extensive application to burns. Death of animals from bismuth nephritis following injections of soluble salts occurs within several hours to 24 days, the time being generally inversely proportional to the dose, and it appears to be in the order of 5-10 times higher than the dose by slow intravenous injection for rabbits. It is stated that the administration of bismuth should be stopped when gingivitis appears, for otherwise serious ulcerative stomatitis is likely to result. Other toxic results may develop, such as malaise, albuminuria, diarrhea, skin reactions and sometimes serious exodermatitis. Industrial bismuth poisoning has not been reported, although bismuth absorbed in industrial cases may complicate a diagnosis of plumbism, since the dark line in the gums, which is often present in lead poisoning, is also produced by bismuth. All bismuth compounds do not have equal toxicity. See also individual entries.

Treatment and Antidotes: Personnel showing some of the symptoms noted above which might indicate that they were absorbing too much bismuth into the body should be removed from exposure as soon as possible. Get medical advice. Personnel should be cautioned against careless handling of these materials.

BKW000 CAS:21260-46-8 ***HR: 3***
BISMUTH DIMETHYL DITHIOCARBAMATE
mf: $C_9H_{18}N_3S_6 \cdot Bi$ mw: 569.64

SYNS: BISMATE ◇ TRIS(DIMETHYLDITHIOCARBAMATO)BISMUTH

TOXICITY DATA with REFERENCE
scu-mus TDLo:1000 mg/kg:ETA NTIS** PB223-159

CONSENSUS REPORTS: Reported in EPA TSCA Inventory.

SAFETY PROFILE: Questionable carcinogen with experimental tumorigenic data. See also BISMUTH COMPOUNDS and CARBAMATES. When heated to decomposition it emits very toxic fumes of SO_x and NO_x.

BKW250 CAS:10361-44-1 ***HR: 3***
BISMUTH NITRATE
mf: BiN_3O_9 mw: 395.01

PROP: Triclinic, colorless, sltly hygroscopic crystals. Bp: $-5H_2O$ @ 80°, d: 2.83, mp: 30° (decomp).

SYN: NITRIC ACID, BISMUTH(3+) SALT

TOXICITY DATA with REFERENCE
itt-rat TDLo:31601 μg/kg (male 1D pre):REP JRPFA4 7,21,64
ipr-mus LDLo:2500 mg/kg APFRAD 34,173,76
ivn-mus LDLo:21 mg/kg APFRAD 34,173,76

CONSENSUS REPORTS: Reported in EPA TSCA Inventory.

SAFETY PROFILE: Poison by intravenous route. Moderately toxic by intraperitoneal route. Experimental reproductive effects. When heated to decomposition it emits toxic fumes of Bi and NO_x. See also BISMUTH COMPOUNDS and NITRATES.

BKW500 CAS:12232-97-2 ***HR: 3***
BISMUTH NITRIDE
mf: BiN mw: 222.99

SAFETY PROFILE: Very unstable; explodes when shaken, heated, or on contact with water and dilute acids. When heated to decomposition it emits toxic fumes of Bi and NO_x. See also BISMUTH COMPOUNDS and NITRIDES.

BKW750 CAS:7787-62-4 ***HR: 3***
BISMUTH PENTAFLUORIDE
mf: BiF_5 mw: 303.98

PROP: Crystals. Sublimes @ 550°.

SAFETY PROFILE: An irritant poison via ingestion and inhalation routes. Decomposes vigorously and sometimes ignites on contact with moisture to yield O_3 and bismuth trifluoride. Very dangerous. Incompatible with water and petrolatum above 50°, and acids. Reacts violently with water and petrolatum above 50°, and acids at room temperature, liberating much heat and ozone. When heated to decomposition it emits highly toxic fumes of F^-. See also FLUORIDES and OZONE.

BKW850 CAS:32707-10-1 ***HR: 3***
BISMUTH PERCHLORATE
mf: $BrClO_4$ mw: 308.43

SAFETY PROFILE: A shock-sensitive explosive. When heated to decomposition it emits toxic fumes of Bi and Cl^-. See also BISMUTH COMPOUNDS and PERCHLORATES.

BKX000 ***HR: 3***
BISMUTH PLUTONIDE
mf: BiPu mw: 451

SAFETY PROFILE: Ignites spontaneously in air. All plutonium compounds are extremely dangerous; when heated to decomposition it emits toxic fumes of Pu and Bi. See also BISMUTH COMPOUNDS and PLUTONIUM COMPOUNDS.

BKX250 CAS:63732-98-9 ***HR: 3***
BISMUTH POTASSIUM SODIUM TARTRATE (SOLUBLE)

SYNS: SODIUM POTASSIUM BISMUTH TARTRATE (SOLUBLE) ◇ SOLUBLE TARTRO-BISMUTHATE ◇ TREPOL (FRENCH)

TOXICITY DATA with REFERENCE
ims-rbt LD50:55 mg/kg JPETAB 87,119,46
ims-mam LDLo:3000 mg/kg JPETAB 28,109,26

SAFETY PROFILE: Poison by intramuscular route. See also BISMUTH COMPOUNDS. When heated to decomposition it emits toxic fumes of oxides of Na_2O, K_2O, and Bi.

BKX500 CAS:5806-84-8 ***HR: 3***
BISMUTH SODIUM-p-AMINOPHENYLARSONATE
mf: $C_6H_6AsNO_3$•BiO•Na mw: 463.02

SYNS: p-ARSANILIC ACID, BISMUTH, SODIUM SALT ◇ ARSENOBISMULAK

TOXICITY DATA with REFERENCE
ivn-rat LD50:631 mg/kg UCREAR 48,183,44
ims-rat LDLo:875 mg/kg UCREAR 48,183,44
ivn-rbt LD50:312 mg/kg UCREAR 48,183,44
ims-rbt LDLo:750 mg/kg UCREAR 48,183,44

CONSENSUS REPORTS: Arsenic and its compounds are on The Community Right-To-Know List.

OSHA PEL: TWA 0.5 mg(As)/m^3

SAFETY PROFILE: Poison by intravenous route. Moderately toxic by intramuscular route. See also BISMUTH COMPOUNDS and ARSENIC COMPOUNDS. When heated to decomposition it emits very toxic fumes of As, Bi, Na_2O and NO_x.

BKX750 CAS:150-49-2 ***HR: 3***
BISMUTH SODIUM THIOGLYCOLLATE
mf: $C_6H_6BiNa_3O_6S_3$ mw: 548.25

SYNS: BISTRIMATE ◇ MERCAPTOACETIC ACID, SODIUM-BISMUTH SALT ◇ SODIUM BISMUTH THIOGLYCOLATE ◇ SODIUM BISMUTH THIOGLYCOLLATE ◇ THIOBISMOL

TOXICITY DATA with REFERENCE
orl-chd LD50:47222 mg/kg/22W-I JAMAAP 198,187,66
ims-chd LDLo:2500 μg/kg JOPDAB 28,498,46
ims-chd TDLo:43 mg/kg:SYS JOPDAB 31,580,47
ims-chd TDLo:650 μg/kg AMDCA5 97,384,59

SAFETY PROFILE: Systemic toxic effects in children: somnolence, nausea or vomiting, kidney damage, and decreased urine volume. Poison by intramuscular route. See also BISMUTH COMPOUNDS. When heated to decomposition it emits very toxic fumes of SO_x and Na_2O.

BKY000 CAS:1304-82-1 ***HR: 3***
BISMUTH TELLURIDE
mf: Bi_2Te_3 mw: 800.76

PROP: Gray crystals. Mp: 573°, d: 7.7.

SYN: BISMUTH SESQUITELLURIDE ◇ BISMUTH TELLURIDE, UNDOPED

CONSENSUS REPORTS: Reported in EPA TSCA Inventory.

OSHA PEL: Total Dust: TWA 0.1 mg(Te)/m^3; Respirable Fraction: TWA 5 mg/m^3; Se doped: 5 mg/^{3}
ACGIH TLV: TWA 10 mg/^{3}; (Se doped: 5 mg/^{3})

SAFETY PROFILE: Moderate fire hazard by spontaneous chemical reaction with powerful oxidizers. Reacts with moisture to evolve a toxic gas. Slight explosion hazard by chemical reaction with powerful oxidizers; reacts with moisture. When heated to decomposition it emits toxic fumes of Te. See also BISMUTH COMPOUNDS and TELLURIUM COMPOUNDS.

BKY250 CAS:12010-67-2 ***HR: 3***
BISMUTH TIN OXIDE
mf: $Bi_2O_9Sn_3$•$5H_2O$ mw: 1008.13

SYN: BISMUTH STANNATE PENTAHYDRATE

TOXICITY DATA with REFERENCE
ivn-mus LD50:178 mg/kg CSLNX* NX#02286

OSHA PEL: TWA 2 mg(Sn)/m^3
ACGIH TLV: TWA 2 mg(Sn)/m^3
NIOSH REL: (Organotin Compounds) TWA 0.1 mg(Sn)/m^3

SAFETY PROFILE: Poison by intravenous route. See also TIN COMPOUNDS and BISMUTH COMPOUNDS. When heated to decomposition it emits acrid smoke and irritating fumes.

BKY500 CAS:19025-95-7 ***HR: 3***
BISMUTH TRISODIUM THIOGLYCOLLATE
mf: $C_6H_6BiO_6S_3$•3Na mw: 548.25

SYNS: THIOBISMOL ◇ TRIS(MERCAPTOACETATO(2-1))BISMUTHATE(3-) TRISODIUM

TOXICITY DATA with REFERENCE
unr-chd TDLo:8163 μg/kg AJSGA3 21,674,37
ipr-rat LDLo:26 mg/kg ADSYAF 15,550,27
ivn-rat LDLo:23 mg/kg ADSYAF 15,550,27
ims-rat LDLo:29 mg/kg ADSYAF 15,550,27
ipr-gpg LDLo:26 mg/kg ADSYAF 15,550,27

SAFETY PROFILE: Poison by intraperitoneal, intravenous, and intramuscular routes. Human systemic effects by an unspecified route: convulsions and kidney damage. See also BISMUTH COMPOUNDS. When heated to decomposition it emits toxic fumes of SO_x, Bi, and Na_2O.

BKZ000 ***HR: 3***
2,5-BIS-(NITRATOMERCURIMETHYL)-1,4-DIOXANE
mf: $C_6H_{10}Hg_2N_2O_8$ mw: 639.36

SYN: 1,4-DIOXOLAN-2,5-DIYLDIMETHYLENEBIS(NITROMERCURY)

TOXICITY DATA with REFERENCE
ivn-mus LD50:18 mg/kg CSLNX* NX#06958

CONSENSUS REPORTS: Mercury and its compounds are on the Community Right-To-Know List.

OSHA PEL: (Transitional: CL 1 mg/10m^3) CL 0.1 mg(Hg)/m^3 (skin)
ACGIH TLV: TWA 0.1 mg(Hg)/m^3 (skin)
NIOSH REL: (Mercury, Inorganic) TWA 0.05 mg(Hg)/m^3

SAFETY PROFILE: Poison by intravenous route. See also MERCURY COMPOUNDS and NITRATES. When heated to decomposition it emits very toxic fumes of NO_x and Hg vapors.

BLA000 CAS:13826-66-9 ***HR: 2***
BIS(NITRATO-O)OXOZIRCONIUM
mf: N_2O_7Zr mw: 231.24

SYN: ZIRCONYL NITRATE

TOXICITY DATA with REFERENCE
orl-rat LD50:2500 mg/kg AIHOAX 1,637,50
ipr-rat LD50:1250 mg/kg AIHOAX 1,637,50

CONSENSUS REPORTS: Reported in EPA TSCA Inventory.

OSHA PEL: (Transitional: TWA 5 mg(Zr)/m^3) TWA 5 mg(Zr)/m^3; STEL 10 mg(Zr)/m^3
ACGIH TLV: TWA 5 mg(Zr)/m^3; STEL 10 mg(Zr)/m^3
DFG MAK: 5 mg(Zr)/m^3

SAFETY PROFILE: Moderately toxic by ingestion and intraperitoneal routes. See also ZIRCONIUM COMPOUNDS and NITRATES. When heated to decomposition it emits toxic fumes of NO_x.

BLA250 ***HR: 3***
BIS-p-NITRO BENZENE DIAZO SULFIDE
mf: $C_{12}H_8N_6O_4S$ mw: 332.3

SAFETY PROFILE: Explosion Hazard: The dry material is extremely sensitive; avoid even light friction. When heated to decomposition it emits toxic fumes of SO_x and NO_x. See also SULFIDES and NITRO COMPOUNDS of AROMATIC HYDROCARBONS.

BLA500 CAS:100-32-3 ***HR: D***
BIS(p-NITROPHENYL)DISULFIDE
mf: $C_{12}H_8N_2O_4S_2$ mw: 308.34

SYNS: BIS(4-NITROPHENYL)DISULFIDE ◇ p,p'-DINITRODIPHENYL DISULFIDE ◇ 4,4'-DINITRODIPHENYL DISULFIDE ◇ DI-4-NITROPHENYL DISULFIDE

TOXICITY DATA with REFERENCE
mmo-sat 50 μg/plate MUREAV 67,123,79
mma-sat 25 μg/plate MUREAV 67,123,79

CONSENSUS REPORTS: Reported in EPA TSCA Inventory.

SAFETY PROFILE: Mutation data reported. When heated to decomposition it emits very toxic fumes of NO_x and SO_x.

BLA750 CAS:1223-31-0 ***HR: 2***
BIS(p-NITROPHENYL)SULFIDE
mf: $C_{12}H_8N_2O_4S$ mw: 276.28

TOXICITY DATA with REFERENCE
orl-rat LD50:1490 mg/kg MarJV# 29MAR77

CONSENSUS REPORTS: Reported in EPA TSCA Inventory.

SAFETY PROFILE: Moderately toxic by ingestion. When heated to decomposition it emits very toxic fumes of NO_x and SO_x.

BLB250 CAS:4731-77-5 ***HR: 3***
BIS(OCTANOYLOXY)DI-n-BUTYL STANNANE
mf: $C_{24}H_{48}O_4Sn$ mw: 519.41

SYNS: BIS(OCTANOYLOXY)DI-n-BUTYLTIN ◇ DIBUTYLBIS(OCTANOYLOXY)STANNANE ◇ DIBUTYLBIS((1-OXOOCTYL)OXY) STANNANE ◇ DIBUTYLTIN DICAPRYLATE ◇ DIBUTYLTIN DIOCTANOATE ◇ DIBUTYLTIN DIOCTATE ◇ DIBUTYLTIN OCTANOATE ◇ KAPRYLAN DI-N-BUTYLCINICITY (CZECH)

TOXICITY DATA with REFERENCE
skn-rbt 500 mg/24H SEV 28ZPAK -,229,72
eye-rbt 20 mg/24H MOD 28ZPAK -,229,72
orl-rat LD50:130 mg/kg 28ZPAK -,229,72

OSHA PEL: TWA 0.1 mg(Sn)/m^3 (skin)
ACGIH TLV: TWA 0.1 mg(Sn)/m^3 (skin) (Proposed: TWA 0.1 mg(Sn)/m^3; STEL 0.2 mg(Sn)/m^3 (skin))
NIOSH REL: (Organotin Compounds) TWA 0.1 mg(Sn)/m^3

SAFETY PROFILE: Poison by ingestion. A severe skin and eye irritant. See also TIN COMPOUNDS. When heated to decomposition it emits acrid and irritating fumes.

BLB500 CAS:19546-20-4 ***HR: 1***
2,2-BIS(3'-tert-OCTYL-4'-HYDROXYPHENYLPROPANE
mf: $C_{31}H_{48}O_2$ mw: 452.79

SYNS: ANTIOXIDANT TOD (CZECH) ◇ 2,2-BIS-3'-TERC. OKTYL-4'-HYDROXYFENYLPROPAN (CZECH)

TOXICITY DATA with REFERENCE
skn-rbt 500 mg/24H MOD 28ZPAK -,58,72
eye-rbt 100 mg/24H SEV 28ZPAK -,58,72
orl-rat LD50:4920 mg/kg 28ZPAK -,58,72

SAFETY PROFILE: Mildly toxic by ingestion. A skin and severe eye irritant. When heated to decomposition it emits acrid smoke and irritating fumes.

BLB750 CAS:131-15-7 ***HR: 1***
BIS(2-OCTYL)PHTHALATE
mf: $C_{24}H_{38}O_4$ mw: 390.62

SYNS: CAPRYL-o-PHTHALATE ◇ DICAPRYL-1,2-BENZENEDICARBOXYLATE ◇ DICAPRYL PHTHALATE

TOXICITY DATA with REFERENCE
ipr-mus LD50:14 g/kg JPMSAE 55,158,66

CONSENSUS REPORTS: Reported in EPA TSCA Inventory.

SAFETY PROFILE: Mildly toxic by intraperitoneal route. When heated to decomposition it emits acrid smoke and irritating fumes.

BLC000 CAS:868-18-8 ***HR: 2***
BISODIUM TARTRATE
mf: $C_4H_4O_6 \cdot 2Na$ mw: 194.06

PROP: Transparent crystals; colorless and odorless. Sol in water.

SYNS: 2,3-DIHYDROXY-(R-(R*,R*))-BUTANEDIOIC ACID DISODIUM SALT (9CI) ◇ DISODIUM TARTRATE ◇ DISODIUM l-(+)-TARTRATE ◇ SODIUM TARTRATE (FCC) ◇ SODIUM l-(+)-TARTRATE

TOXICITY DATA with REFERENCE
orl-mus LDLo:3686 mg/kg JAPMA8 31,12,42
orl-rbt LDLo:5290 mg/kg FAONAU 53A,512,74

CONSENSUS REPORTS: Reported in EPA TSCA Inventory.

SAFETY PROFILE: Moderately toxic by ingestion. When heated to decomposition it emits acrid smoke and irritating fumes.

BLC250 CAS:10380-28-6 ***HR: 3***
BIS(8-OXYQUINOLINE)COPPER
mf: $C_{18}H_{12}CuN_2O_2$ mw: 351.86

PROP: Yellow-green powder.

SYNS: BIOQUIN ◇ BIOQUIN 1 ◇ BIS(8-QUINOLINATO)COPPER ◇ BIS(8-QUINOLINOLATO)COPPER ◇ BIS(8-QUINOLINOLATO-N(^{1}),O(^{8}))-COPPER ◇ CELLU-QUIN ◇ COPPER-8 ◇ COPPER HYDROXYQUINOLATE ◇ COPPER-8-HYDROXYQUINOLATE ◇ COPPER-8-HYDROXYQUINOLINATE ◇ COPPER-8-HYDROXYQUINOLINE ◇ COPPER OXINATE ◇ COPPER (2+) OXINATE ◇ COPPER OXINE ◇ COPPER OXYQUINOLATE ◇ COPPER OXYQUINOLINE ◇ COPPER QUINOLATE ◇ COPPER-8-QUINOLATE ◇ COPPER-8-QUINOLINOL ◇ COPPER QUINOLINOLATE ◇ COPPER-8-QUINOLINOLATE ◇ CUNILATE ◇ CUNILATE 2472 ◇ CUPRIC-8-HYDROXYQUINOLATE ◇ CUPRIC-8-QUINOLINOLATE ◇ DOKIRIN ◇ FRUITDO ◇ 8-HYDROXYQUINOLINE COPPER COMPLEX ◇ MILMER ◇ OXIME COPPER ◇ OXINE COPPER ◇ OXINE CUIVRE ◇ OXYQUINOLINOLEATE de CUIVRE (FRENCH) ◇ QUINONDO

TOXICITY DATA with REFERENCE
mma-sat 5 μg/plate MUREAV 116,185,83
scu-mus TDLo:156 mg/kg/39W-I:ETA JNCIAM 24,109,60
ipr-mus LD50:67 mg/kg TXAPA9 5,599,63

CONSENSUS REPORTS: IARC Cancer Review: Group 3 IMEMDT 7,56,87, Animal Inadequate Evidence IMEMDT 15,103,77. Reported in EPA TSCA Inventory. Copper and its compounds are on the Community Right-To-Know List.

SAFETY PROFILE: Poison by intraperitoneal route. Questionable carcinogen with experimental tumorigenic data. Mutation data reported. See also COPPER COMPOUNDS. When heated to decomposition it emits toxic fumes of NO_x.

BLC500 CAS:117-97-5 ***HR: 2***
BIS(PENTACHLOROPHENOL), ZINC SALT
mf: $C_{12}Cl_{10}S_2Zn$ mw: 628.11

SYN: PENTACHLOROTHIOFENOLAT ZINECNATY (CZECH)

TOXICITY DATA with REFERENCE
skn-rbt 500 mg/24H SEV 28ZPAK -,11,72
eye-rbt 250 μg/24H SEV 28ZPAK -,11,72

CONSENSUS REPORTS: Reported in EPA TSCA Inventory. Zinc and its compounds are on the Community Right-To-Know List.

SAFETY PROFILE: A severe eye and skin irritant. See also ZINC COMPOUNDS and CHLORINATED HYDROCARBONS, AROMATIC. When heated to decomposition it emits very toxic fumes of ZnO, Cl^- and SO_x.

BLC750 ***HR: 3***
BIS(PENTA FLUORO PHENYL)ALUMINUM BROMIDE
mf: $C_{12}AlBrF_{10}$ mw: 441

SAFETY PROFILE: Ignites spontaneously in air. Hydrolysis causes explosion. When heated to decomposition it emits toxic fumes of F^- and Br^-. See also ALUMINUM COMPOUNDS.

BLD000 CAS:42310-84-9 ***HR: 3***
BISPENTAFLUOROSULFUR OXIDE
mf: $F_{10}OS_2$ mw: 270.12

SYN: SULFUR FLUORIDE OXIDE

TOXICITY DATA with REFERENCE
ihl-rat LCLo:20 ppm/6H BJIMAG 27,1,70

OSHA PEL: TWA 2.5 mg(F)/m^3
NIOSH REL: TWA 2.5 mg(F)/m^3

SAFETY PROFILE: Poison by inhalation. See also FLUORIDES. When heated to decomposition it emits very toxic fumes of F^- and SO_x.

BLD250 ***HR: 3***
BIS(2,4-PENTANEDIONATO)CHROMIUM
mf: $C_{10}H_{14}CrO_4$ mw: 250.21

CONSENSUS REPORTS: Chromium and its compounds are on the Community Right-To-Know List.

SAFETY PROFILE: Ignites spontaneously in air. When heated to decomposition it emits acrid smoke and fumes. See also CHROMIUM COMPOUNDS.

BLD325 CAS:93431-23-3 ***HR: 2***
4,5-BIS(4-PENTENYLOXY)-2-IMIDAZOLIDINONE
mf: $C_{13}H_{22}N_2O_3$ mw: 254.37

SYN: SRC-16

TOXICITY DATA with REFERENCE
orl-mus LD50:740 mg/kg CPBTAL 12,843,64
ipr-mus LD50:450 mg/kg CPBTAL 12,843,64
scu-mus LD50:621 mg/kg CPBTAL 12,843,64

SAFETY PROFILE: Moderately toxic by ingestion, subcutaneous, and intraperitoneal routes. When heated to decomposition it emits toxic fumes of NO_x.

BLD500 CAS:80-05-7 ***HR: 3***
BISPHENOL A
mf: $C_{15}H_{16}O_2$ mw: 228.31

PROP: White flakes, mild phenolic odor. Insol in water; sol in alcohol and dilute alkalies; sltly sol in CCl_4.

SYNS: BISFEROL A (GERMAN) ◇ 2,2-BIS-4'-HYDROXYFENYLPROPAN (CZECH) ◇ BIS(4-HYDROXYPHENYL) DIMETHYLMETHANE ◇ BIS(4-HYDROXYPHENYL)PROPANE ◇ 2,2-BIS(p-HYDROXYPHENYL)PROPANE ◇ 2,2-BIS(4-HYDROXYPHENYL)PROPANE ◇ DIAN ◇ p,p'-DIHYDROXYDIPHENYLDIMETHYLMETHANE ◇ 4,4'-DIHYDROXYDIPHENYLDIMETHYLMETHANE ◇ p,p'-DIHYDROXYDIPHENYLPROPANE ◇ 2,2-(4,4'-DIHYDROXYDIPHENYL)PROPANE ◇ 4,4'-DIHYDROXYDIPHENYLPROPANE ◇ 4,4'-DIHYDROXYDIPHENYL-2,2-PROPANE ◇ 4,4'-DIHYDROXY-2,2-DIPHENYLPROPANE ◇ β-DI-p-HYDROXYPHENYLPROPANE ◇ 2,2-DI(4-HYDROXYPHENYL)PROPANE ◇ DIMETHYL BIS(p-HYDROXYPHENYL) METHANE ◇ DIMETHYLMETHYLENE-p,p'-DIPHENOL ◇ 2,2-DI(4-PHENYLOL)PROPANE ◇ p,p'-ISOPROPYLIDENEBISPHENOL ◇ 4,4'-ISOPROPYLIDENEBISPHENOL ◇ p,p'-ISOPROPYLIDENEDIPHENOL ◇ NCI-C50635

TOXICITY DATA with REFERENCE
skn-rbt 250 mg open MLD UCDS** 7/14/65
eye-rbt 20 mg/24H SEV 28ZPAK -,58,72
orl-mus TDLo:12500 mg/kg (female 6-15D post):TER NTIS** PB85-205102
ipr-rat TDLo:1275 mg/kg (female 1-15D post):REP SWEHDO 7(Suppl 4),66,81
orl-rat LD50:3250 mg/kg AIHAAP 28,301,67
ihl-rat LC50:200 ppm 85JCAE -,238,86
orl-mus LD50:2500 mg/kg AIHAAP 28,301,67
ipr-mus LD50:150 mg/kg NTIS** AD691-490
orl-rbt LD50:2230 mg/kg AIHAAP 28,301,67
skn-rbt LD50:3000 mg/kg AMIHBC 4,119,51

CONSENSUS REPORTS: NTP Carcinogenesis Bioassay (feed); Inadequate Studies: mouse, rat NTPTR* NTP-TR-215,82. Community Right-To-Know List. Reported in EPA TSCA Inventory.

SAFETY PROFILE: Poison by intraperitoneal route. Moderately toxic by ingestion, inhalation, and skin contact. Experimental teratogenic and reproductive effects. A skin and eye irritant. When heated to decomposition it emits acrid and irritating fumes.

BLD750 CAS:1675-54-3 ***HR: 3***
BISPHENOL A DIGLYCIDYL ETHER
mf: $C_{21}H_{24}O_4$ mw: 340.45

SYNS: 2,2-BIS(4-(2,3-EPOXYPROPYLOXY)PHENYL)PROPANE ◇ BIS(4-GLYCIDYLOXYPHENYL)DIMETHYAMETHANE ◇ 2,2-BIS(p-GLYCIDYLOXYPHENYL)PROPANE ◇ BIS(4-HYDROXYPHENYL)DIMETHYLMETHANE DIGLYCIDYL ETHER ◇ 2,2-BIS(4-HYDROXYPHENYL)PROPANE, DIGLYCIDYL ETHER ◇ 2,2-BIS(p-HYDROXYPHENYL)PROPANE, DIGLYCIDYL ETHER ◇ D.E.R. 332 ◇ DIGLYCIDYL BISPHENOL A ETHER ◇ DIGLYCIDYL ETHER of 2,2-BIS(p-HYDROXYPHENYL)PROPANE ◇ DIGLYCIDYL ETHER of 2,2-BIS(4-HYDROXYPHENYL)PROPANE ◇ DIGLYCIDYL ETHER of BISPHENOL A ◇ DIGLYCIDYL ETHER of 4,4'-ISOPROPYLIDENEDIPHENOL ◇ 4,4'-DIHYDROXYDIPHENYLDIMETHYLMETHANE DIGLYCIDYL ETHER ◇ p,p'-DIHYDROXYDIPHENYLDIMETHYLMETHANE DIGLYCIDYL ETHER ◇ EPI-REZ 508 ◇ EPI-REZ 510 ◇ EPON 828 ◇ EPOXIDE A ◇ ERL-2774 ◇ 4,4'-ISOPROPYLIDENEDIPHENOL DIGLYCIDYL ETHER ◇ 2,2'-((1-METHYLETHYLIDENE) BIS(4,1-PHENYLENEOXYMETHYLENE))BISOXIRANE

TOXICITY DATA with REFERENCE
skn-rbt 500 mg open MLD UCDS** 4/21/67
eye-rbt 2 mg/24H SEV 28ZPAK -,137,72
mmo-sat 50 μg/plate MUREAV 66,367,79
mma-sat 50 μg/plate MUREAV 66,367,79
mmo-esc 20 μmol/L ARTODN 46,277,80
dnd-esc 1 μmol/L ARTODN 46,277,80
skn-mus TDLo:312 g/kg/2Y-I:CAR NTIS** ORNL-5375
skn-mus TD:312 g/kg/2Y-I:CAR,REP CNREA8 39,1718,79
skn-mus TD:16480 mg/kg/2Y-I:ETA FCTOD7 23,1081,85

orl-rat LD50:11 g/kg UCDS** 4/21/67
ipr-rat LD50:2200 mg/kg 38MKAJ 2A,2219,81
orl-mus LD50:15600 mg/kg 38MKAJ 2A,2219,81
ipr-mus LD50:4 g/kg 38MKAJ 2A,2219,81
orl-rbt LD50:1980 mg/kg 38MKAJ 2A,2219,81
skn-rbt LD50:20 mg/kg 38MKAJ 2A,2219,81

CONSENSUS REPORTS: EPA Genetic Toxicology Program. Reported in EPA TSCA Inventory.

SAFETY PROFILE: Poison by skin contact. Mildly toxic by ingestion. Mutation data reported. A skin and severe eye irritant. Experimental reproductive effects. Questionable carcinogen with experimental carcinogenic and tumorigenic data. See also ETHERS. When heated to decomposition it emits acrid and irritating fumes.

BLE000 ***HR: 3***
BISPHENOL DIGLYCIDYL ETHER, MODIFIED

TOXICITY DATA with REFERENCE
skn-mus TDLo:470 g/kg/39W-I:ETA AIHAAP 24,305,63

SAFETY PROFILE: Questionable carcinogen with experimental tumorigenic data. See also BISPHENOL DIGLYCIDYL ETHER and ETHERS.

BLE250 CAS:17601-12-6 ***HR: 3***
BIS(p-PHENOXYPHENYL)DIPHENYLTIN
mf: $C_{36}H_{28}O_2Sn$ mw: 611.33

SYN: BIS(p-PHENOXYPHENYL)DIPHENYLSTANNANE

TOXICITY DATA with REFERENCE
ivn-mus LD50:56 mg/kg CSLNX* NX#01351

OSHA PEL: TWA 0.1 mg(Sn)/m^3 (skin)
ACGIH TLV: TWA 0.1 mg(Sn)/m^3 (skin) (Proposed: TWA 0.1 mg(Sn)/m^3; STEL 0.2 mg(Sn)/m^3 (skin))
NIOSH REL: (Organotin Compounds) TWA 0.1 mg(Sn)/m^3

SAFETY PROFILE: Poison by intravenous route. See also TIN COMPOUNDS. When heated to decomposition it emits acrid smoke and irritating fumes.

BLE500 CAS:74-31-7 ***HR: 3***
1,4-BIS(PHENYL AMINO)BENZENE
mf: $C_{18}H_{16}N_2$ mw: 260.36

PROP: A solid. D: 1.20, vap d: 9.0.

SYNS: AGERITE ◇ AGERITEDPPD ◇ N,N'-DIFENYL-p-FENYLENDIAMIN (CZECH) ◇ N,N'-DIPHENYL-p-PHENYLENEDIAMINE ◇ DIPHENYL-p-PHENYLENEDIAMINE ◇ DPPD ◇ FLEXAMINE G ◇ JZF ◇ NONOX DPPD ◇ p-PHENYLAMINODIPHENYLAMINE ◇ 4-PHENYLAMINODIPHENYLAMINE ◇ USAF GY-2

TOXICITY DATA with REFERENCE
eye-rbt 500 mg/24H SEV 28ZPAK -,73,72
mma-sat 10 µg/plate PCBRD2 141,407,84
msc-ham:lng 30 mg/L SWEHDO 9(Suppl 2),27,83
orl-rat TDLo:450 mg/kg (14D pre/1-22D preg):REP JAFCAU 4,796,56
orl-mus TDLo:4176 mg/kg (female 6-14D post):TER NTIS** PB223-160
scu-mus TDLo:1000 mg/kg:ETA NTIS** PB223-159
orl-rat LD50:2370 mg/kg 28ZPAK -,73,72
orl-mus LD50:18 g/kg GTPZAB 10(3),49,66
ipr-mus LD50:300 mg/kg NTIS** AD277-689

CONSENSUS REPORTS: Reported in EPA TSCA Inventory.

SAFETY PROFILE: Poison by intraperitoneal route. Moderately toxic by ingestion. A weak allergen. Experimental teratogenic and reproductive effects. An eye irritant. Questionable carcinogen with experimental tumorigenic data. Mutation data reported. Combustible when exposed to heat or flame; can react with oxidizing materials. When heated to decomposition it emits toxic fumes of NO_x.

BLF250 CAS:13754-23-9 ***HR: 3***
BIS-N,N'-(3-PHENYLPROPYL-2)-PIPERAZINE DIHYDROCHLORIDE
mf: $C_{22}H_{30}N_2$•2ClH mw: 395.46

SYNS: N,N'-BIS(PHENYLISOPROPYL)PIPERAZINEDIHYDROCHLORIDE ◇ DIPHENAZINE DIHYDROCHLORIDE

TOXICITY DATA with REFERENCE
orl-mus LD50:230 mg/kg ARZNAD 7,225,57
scu-mus LD50:193 mg/kg 27ZQAG -,223,72
ivn-mus LD50:25 mg/kg ARZNAD 7,225,57

SAFETY PROFILE: Poison by ingestion, intravenous, and subcutaneous routes. When heated to decomposition it emits very toxic fumes of HCl and NO_x.

BLF500 CAS:1666-13-3 ***HR: 3***
BIS(PHENYLSELENIDE)
mf: $C_{12}H_{10}Se_2$ mw: 312.14

SYN: PHENYL DISELENIDE

TOXICITY DATA with REFERENCE
ivn-mus LD50:28 mg/kg CSLNX* NX#05657

CONSENSUS REPORTS: Reported in EPA TSCA Iventory. Selenium and its compounds are on the Community Right-To-Know List.

OSHA PEL: TWA 0.2 mg(Se)/m^3
ACGIH TLV: TWA: 0.2 mg(Se)/m^3
DFG MAK: 0.1 mg(Se)/m^3

SAFETY PROFILE: Poison by intravenous route. See also SELENIUM COMPOUNDS. When heated to decomposition it emits toxic fumes of Se.

BLF750 CAS:4848-63-9 ***HR: 3***
BIS(PHENYLTHIO)DIMETHYLTIN
mf: $C_{14}H_{16}S_2Sn$ mw: 367.11

SYN: DIMETHYLBIS(PHENYLTHIO)STANNANE

TOXICITY DATA with REFERENCE
ivn-mus LD50:56 mg/kg CSLNX* NX#01670

OSHA PEL: TWA 0.1 mg(Sn)/m^3 (skin)
ACGIH TLV: TWA 0.1 mg(Sn)/m^3 (skin) (Proposed: TWA 0.1 mg(Sn)/m^3; STEL 0.2 mg(Sn)/m^3 (skin))
NIOSH REL: (Organotin Compounds) TWA 0.1 mg(Sn)/m^3

SAFETY PROFILE: Poison by intravenous route. See also TIN COMPOUNDS. When heated to decomposition it emits toxic fumes of SO_x.

BLG000 ***HR: 3***
1,3-BIS(PHENYL)TRIAZENO)BENZENE
mf: $C_{18}H_{16}N_6$ mw: 316.37

SAFETY PROFILE: Explodes on rapid heating. When heated to decomposition it emits toxic fumes of NO_x.

BLG100 CAS:52237-03-3 ***HR: 1***
4,4'-BIS(4-PHENYL-2H-1,2,3-TRIAZOL-2-YL)-2,2'-STILBENEDISULFONIC ACID DIPOTASSIUM SALT
mf: $C_{30}H_{20}N_6O_6S_2{\cdot}2K$ mw: 702.88 2

SYNS: DIPOTASSIUM4,4'-BIS(4-PHENYL-1,2,3-TRIAZOL-2-YL)STILBENE-2,2'-DISULFONATE ◇ DIPOTASSIUM 4,4'-BIS(4-PHENYL-1,2,3-TRIAZOL-2-YL)STILBENE-2,2'-SULFONATE ◇ 2,2'-STILBENEDISULFONIC ACID, 4,4'-BIS(4-PHENYL-2H-1,2,3-TRIAZOL-2-YL)-,DIPOTASSIUM SALT ◇ 2,2'-STILBENEDISULFONIC ACID, 4,4'-BIS(4-PHENYL-1,2,3-TRIAZOL-2-YL), DIPOTASSIUM SALT

TOXICITY DATA with REFERENCE
eye-rbt 100 mg MOD CTOXAO 13,171,78
orl-rat TDLo:3 g/kg (female 6-15D post):REP EQSFAP 4,223,75

SAFETY PROFILE: Experimental reproductive effects. An eye irritant. When heated to decomposition it emits toxic fumes of NO_x, and SO_x.

BLG250 CAS:2439-99-8 ***HR: 2***
N,N-BIS(PHOSPHONOMETHYL)GLYCINE
mf: $C_4H_{11}NO_8P_2$ mw: 263.10

SYNS: GLYPHOSINE ◇ POLARIS

TOXICITY DATA with REFERENCE
orl-mam LD50:3925 mg/kg FMCHA2 -,C191,83
orl-rat LD50:3925 mg/kg 85ARAE 3,60,76

CONSENSUS REPORTS: Reported in EPA TSCA Inventory.

SAFETY PROFILE: Moderately toxic by ingestion. When heated to decomposition it emits very toxic fumes of NO_x and PO_x.

BLG325 CAS:60012-89-7 ***HR: 2***
3,8-BIS(1-PIPERIDINYLMETHYL)-2,7-DIOXASPIRO(4.4)NONANE-1,6-DIONE
mf: $C_{19}H_{30}N_2O_4$ mw: 350.51

TOXICITY DATA with REFERENCE
ipr-rat LD50:1160 mg/kg PJPPAA 28,157,76
orl-mus LD50:6 g/kg PJPPAA 28,157,76
ipr-mus LD50:820 mg/kg PJPPAA 28,157,76

SAFETY PROFILE: Moderately toxic by intraperitoneal route. When heated to decomposition it emits toxic fumes of NO_x.

BLG500 CAS:1113-14-0 ***HR: 3***
trans-1,2-BIS(n-PROPYLSULFONYL)ETHYLENE
mf: $C_8H_{16}O_4S_2$ mw: 240.36

SYNS: B-1843 ◇ C-272 ◇ CHEMAGRO B-1843 ◇ VANCIDE PA ◇ VANCIDE PA DISPERSION

TOXICITY DATA with REFERENCE
orl-rat LD50:200 mg/kg FMCHA2 -,C49,83
ipr-rat LD50:11500 μg/kg 34ZIAG -,161,69
ipr-mus LDLo:11500 mg/kg 34ZIAG -,162,69
ipr-gpg LDLo:11500 mg/kg 34ZIAG -,162,69

SAFETY PROFILE: Poison by ingestion and intravenous routes. See also SULFONATES. When heated to decomposition it emits toxic fumes of SO_x.

BLH250 CAS:14167-18-1 ***HR: 3***
BIS(SALICYLALDEHYDE)ETHYLENEDIIMINE COBALT(II)
mf: $C_{16}H_{14}CoN_2O_2$ mw: 325.25

SYNS: N,N'-ETHYLENEBIS(SALICYLIDENEIMINATO)COBALT(II) ◇ SALCOMIN ◇ SALCOMINE POWDER ◇ SALICYLALDEHYDE ETHYLENEDIIMINE COBALT

TOXICITY DATA with REFERENCE
ihl-mus LCLo:390 mg/m^3/5.5H AMRL** TR-74-78,74

CONSENSUS REPORTS: Reported in EPA TSCA Inventory. Cobalt and its compounds are on the Community Right-To-Know List. EPA Extremely Hazardous Substances List.

SAFETY PROFILE: Poison by inhalation. See also COBALT COMPOUNDS and ALDEHYDES. When heated to decomposition it emits toxic fumes of NO_x.

BLH309 CAS:28660-67-5 ***HR: 3***
BIS(TETRADECANOYLOXY)DIBUTYLSTANNANE
mf: $C_{36}H_{72}O_4Sn$ mw: 687.77

SYNS: DI-n-BUTYL-TIN DI(TETRADECANOATE) ◇ MYRISTAN DI-n-BUTYLCINICITY (CZECH)

TOXICITY DATA with REFERENCE
skn-rbt 500 mg/24H SEV 28ZPAK -,230,72
eye-rbt 5 mg/24H SEV 28ZPAK -,230,72
orl-rat LD50:138 mg/kg 28ZPAK -,230,72

OSHA PEL: TWA 0.1 mg(Sn)/m^3 (skin)
ACGIH TLV: TWA 0.1 mg(Sn)/m^3 (skin) (Proposed: TWA 0.1 mg(Sn)/m^3; STEL 0.2 mg(Sn)/m^3 (skin))
NIOSH REL: (Organotin Compounds) TWA 0.1 mg(Sn)/m^3

SAFETY PROFILE: Poison by ingestion. A severe skin and eye irritant. See also TIN COMPOUNDS. When heated to decomposition it emits smoke and acrid fumes.

BLH325 CAS:62987-05-7 ***HR: 3***
1,3-BIS(TETRAHYDRO-2-FURYL)-5-FLUOROURACIL
mf: $C_{12}H_{15}FN_2O_4$ mw: 270.29

SYNS: 1,3-BIS(TETRAHYDRO-2-FURANYL)-5-FLUORO-2,4-PYRIMIDINEDIONE ◇ FD-1 ◇ 5-FLUORO-1,3-BIS(TETRAHYDRO-2-FURANYL)-2,4(1H,3H)-PYRIMIDINEDIONE

TOXICITY DATA with REFERENCE
cyt-ham:fbr 100 nmol/L MUREAV 88,241,81
orl-rat LD50:1730 mg/kg GANNA2 71,30,80
orl-mus LD50:2664 mg/kg JMCMAR 21,738,78
orl-dog LD50:88100 μg/kg OYYAA2 16,303,78

SAFETY PROFILE: Poison by ingestion. Mutation data reported. When heated to decomposition it emits toxic fumes of F^- and NO_x.

BLI000 CAS:68594-19-4 ***HR: 3***
1,6-BIS(5-TETRAZOLYL)HEXAAZ-1,5-DIENE
mf: $C_2H_4N_{14}$ mw: 224.14

SAFETY PROFILE: An explosive extremely sensitive to pressure or heating to 90°C. Upon decomposition it emits toxic fumes of NO_x. See also EXPLOSIVES and AZIDES.

BLI250 CAS:1656-16-2 ***HR: 3***
3,4-BIS(1,2,3,4-THIATRIAZOL-5-YL THIO) MALEIMIDE
mf: $C_4HN_7O_2S_4$ mw: 307.34

SAFETY PROFILE: Explodes on impact or when heated to its melting point. When heated to decomposition it emits toxic fumes of NO_x and SO_x. See also EXPLOSIVES.

BLI500 ***HR: 3***
BIS(1,2,3,4-THIATRIAZOL-5-YL THIO)METHANE
mf: $C_3H_2N_6S_4$ mw: 250.33

SAFETY PROFILE: On impact or on heating to its melting point it explodes loudly with a flash. Upon decomposition it emits toxic fumes of SO_x and NO_x. See also EXPLOSIVES.

BLJ250 CAS:142-46-1 ***HR: 3***
BIS(THIOUREA)
mf: $C_2H_6N_4S_2$ mw: 150.24

SYNS: BISTHIOCARBAMYL HYDRAZINE ◇ 2,5-DITHIOBIUREA ◇ 1,2-HYDRAZINEDICARBOTHIOAMIDE ◇ NCI-C03009 ◇ USAF B-44 ◇ USAF EK-P-6281

TOXICITY DATA with REFERENCE
orl-mus TDLo:655 g/kg/78W-C:ETA NCITR* NCI-CG-TR-132,79
ipr-mus LD50:100 mg/kg NTIS** AD277-689

CONSENSUS REPORTS: NCI Carcinogenesis Bioassay (feed); No Evidence: mouse, rat NCITR* NCI-CG-TR-132,79. Reported in EPA TSCA Inventory.

SAFETY PROFILE: Poison by intraperitoneal route. Questionable carcinogen with experimental tumorigenic data. When heated to decomposition it emits very toxic fumes of NO_x and SO_x.

BLJ500 ***HR: 3***
BISTOLUENE DIAZO OXIDE
mf: $C_{14}H_{14}N_4O$ mw: 254.29

SAFETY PROFILE: Ignites spontaneously. Explosion Hazard: Very unstable. Shock and friction sensitive. Incompatible with toluene. Upon decomposition it emits toxic fumes of NO_x. See also AZIDES.

BLK000 CAS:128-80-3 ***HR: 2***
1,4-BIS(p-TOLYLAMINO)ANTHRAQUINONE
mf: $C_{28}H_{22}N_2O_2$ mw: 418.52

SYNS: ALIZARINE CYANINE GREEN BASE ◇ AMAPLAST GREEN OZ ◇ ARLOSOL GREEN B ◇ BIS-1,4-p-TOLYLAMINOANTHRCHINON (CZECH) ◇ C-GREEN 10 ◇ C.I. 61565 ◇ C.I. SOLVENT GREEN 3 ◇ CYANINE GREEN G BASE ◇ D&C GREEN No. 6 ◇ 1,4-DI-p-TOLUIDINOANTHRAQUINONE ◇ FAT SOLUBLE GREEN ANTHRAQUINONE ◇ 11091 GREEN ◇ GREEN No. 2 ◇ MICRO-LEX GREEN 5B ◇ NITRO FAST GREEN GB ◇ ORGANOL FAST GREEN J ◇ QUINIZARINE GREEN BASE ◇ SUDAN GREEN 4B ◇ TOYO ORIENTAL OIL BLUE G ◇ WAXOLINE GREEN

TOXICITY DATA with REFERENCE
eye-rbt 20 mg/24H MOD 28ZPAK -,124,72
orl-rat LD50:3660 mg/kg 28ZPAK -,124,72

CONSENSUS REPORTS: Reported in EPA TSCA Inventory.

SAFETY PROFILE: Moderately toxic by ingestion. An eye irritant. When heated to decomposition it emits toxic fumes of NO_x.

BLK250 CAS:63869-05-6 ***HR: 3***
N-BIS(p-TOLYLSULFONYL)AMIDOMETHYL MERCURY
mf: $C_{15}H_{17}HgNO_4S_2$ mw: 540.04

SYNS: N-METHYLMERCURI-BIS-p-TOLUENSULFONAMID(CZECH) ◇ METHYL(4-METHYL-N-((4-METHYLPHENYL)SULFONYL)BENZENESULFONAMIDATO-N)-MERCURY

TOXICITY DATA with REFERENCE
skn-rbt 500 mg/24H MLD 28ZPAK -,223,72
eye-rbt 50 μg/24H SEV 28ZPAK -,223,72
orl-rat LD50:99 mg/kg 28ZPAK -,223,72

CONSENSUS REPORTS: Mercury and its compounds are on the Community Right-To-Know List.

OSHA PEL: (Transitional: CL 1 mg/10m^3) TWA 0.01 mg(Hg)/m^3; STEL 0.03 mg/m^3 (skin)
ACGIH TLV: TWA 0.01 mg(Hg)/m^3; STEL 0.03 mg(Hg)/m^3

SAFETY PROFILE: Poison by ingestion. A skin and severe eye irritant. See also MERCURY COMPOUNDS and SULFONATES. When heated to decomposition it emits very toxic fumes of NO_x, SO_x, and Hg.

BLK500 CAS:85681-49-8 ***HR: D***
3,5-BIS(o-TOLYL)-s-TRIAZOLE
mf: $C_{16}H_{15}N_3$ mw: 249.34

SYN: s-TRIAZOLE,3,5-BIS(o-TOLYL)-

TOXICITY DATA with REFERENCE
scu-ham TDLo:3500 μg/kg (female 4-8D post):REP JMCMAR 26,1187,83

SAFETY PROFILE: Experimental reproductive effects. When heated to decomposition it emits toxic fumes of NO_x.

BLK750 CAS:10347-38-3 ***HR: 3***
BIS(TRIBENZYLSTANNYL)SULFIDE
mf: $C_{42}H_{42}SSn_2$ mw: 816.28

SYNS: BIS(TRIBENZYLTIN) SULFIDE ◇ DISTANNATHIANE, HEXAKIS(PHENYLMETHYL)-(9CI) ◇ SIRNIK TRIBENZYLCINICTY (CZECH) ◇ THIOBIS(TRIBENZYL-TIN (8CI)

TOXICITY DATA with REFERENCE
skn-rbt 500 mg/24H MOD 28ZPAK -,233,72
eye-rbt 500 mg/24H SEV 28ZPAK -,233,72
orl-rat LD50:314 mg/kg 28ZPAK -,233,72

OSHA PEL: TWA 0.1 mg(Sn)/m^3 (skin)
ACGIH TLV: TWA 0.1 mg(Sn)/m^3 (skin) (Proposed: TWA 0.1 mg(Sn)/m^3; STEL 0.2 mg(Sn)/m^3 (skin))
NIOSH REL: (Organotin Compounds) TWA 0.1 mg(Sn)/m^3

SAFETY PROFILE: Poison by ingestion. A skin and eye irritant. See also TIN COMPOUNDS and SULFIDES. When heated to decomposition it emits toxic fumes of SO_x.

BLL000 CAS:30099-72-0 ***HR: 3***
BIS(TRIBUTYL(SEBACOYLDIOXY))TIN
mf: $C_{34}H_{70}O_4Sn_2$ mw: 780.42

SYN: SEBACOYLDIOXYBIS(TRIBUTYLSTANNANE)

TOXICITY DATA with REFERENCE
ivn-mus LD50:18 mg/kg CSLNX* NX#03600

OSHA PEL: TWA 0.1 mg(Sn)/m^3 (skin)
ACGIH TLV: TWA 0.1 mg(Sn)/m^3 (skin) (Proposed: TWA 0.1 mg(Sn)/m^3; STEL 0.2 mg(Sn)/m^3 (skin))
NIOSH REL: (Organotin Compounds) TWA 0.1 mg(Sn)/m^3

SAFETY PROFILE: Poison by intravenous route. See also TIN COMPOUNDS. When heated to decomposition it emits acrid smoke and irritating fumes.

BLL250 CAS:12291-11-1 ***HR: 3***
BIS((TRI-n-BUTYLSTANNYL)CYCLOPENTADIENYL)IRON
mf: $C_{34}H_{62}FeSn_2$ mw: 764.19

SYN: 1,1′-BIS(TRIBUTYLSTANNYL)FERROCENE

TOXICITY DATA with REFERENCE
ivn-mus LD50:56 mg/kg CSLNX* NX#05870

OSHA PEL: TWA 0.1 mg(Sn)/m^3 (skin)
ACGIH TLV: TWA 0.1 mg(Sn)/m^3 (skin) (Proposed: TWA 0.1 mg(Sn)/m^3; STEL 0.2 mg(Sn)/m^3 (skin))
NIOSH REL: (Organotin Compounds) TWA 0.1 mg(Sn)/m^3

SAFETY PROFILE: Poison by intravenous route. See also TIN COMPOUNDS. When heated to decomposition it emits acrid smoke and irritating fumes.

BLL500 CAS:25711-26-6 ***HR: 3***
BIS(TRIBUTYLTIN) ITACONATE
mf: $C_{29}H_{58}O_4Sn_2$ mw: 708.25

SYN: METHYLENESUCCINYLOXYBIS(TRIBUTYLSTANNANE)

TOXICITY DATA with REFERENCE
ivn-mus LD50:180 mg/kg CSLNX* NX#03635

OSHA PEL: TWA 0.1 mg(Sn)/m^3 (skin)
ACGIH TLV: TWA 0.1 mg(Sn)/m^3 (skin) (Proposed: TWA 0.1 mg(Sn)/m^3; STEL 0.2 mg(Sn)/m^3 (skin))
NIOSH REL: (Organotin Compounds) TWA 0.1 mg(Sn)/m^3

SAFETY PROFILE: Poison by intravenous route. See also TIN COMPOUNDS. When heated to decomposition it emits acrid smoke and irritating fumes.

BLL750 CAS:56-35-9 ***HR: 3***
BIS(TRIBUTYL TIN)OXIDE
mf: $C_{24}H_{54}OSn_2$ mw: 596.16

SYNS: BIOMET TBTO ◇ BIS-(TRI-N-BUTYLCIN)OXID (CZECH) ◇ BIS(TRIBUTYLOXIDE) of TIN ◇ BIS(TRIBUTYLSTANNYL)OXIDE ◇ BIS(TRI-N-BUTYLZINN)-OXYD (GERMAN) ◇ BTO ◇ BUTINOX ◇ C-Sn-9 ◇ ENT 24,979 ◇ HEXABUTYLDISTANNOXANE ◇ HEXABUTYLDITIN ◇ KYSLICNIK TRI-N-BUTYLCINICITY (CZECH) ◇ L.S. 3394 ◇ OTBE (FRENCH) ◇ OXYBIS(TRIBUTYLTIN) ◇ OXYDE de TRIBUTYLETAIN ◇ TBOT ◇ TBTO ◇ TRI-n-BUTYL-STANNANE OXIDE ◇ TRIBUTYLTIN OXIDE

TOXICITY DATA with REFERENCE
eye-rbt 50 μg/24H SEV 28ZPAK -,232,72
eye-rbt 460 ug BJIMAG 26,165,69
dni-omi 56200 ppb AEMIDF 45,48,83
dni-omi 56200 ppb AEMIDF 45,48,83
orl-rat TDLo:150 mg/kg (female 6-20D post):REP TXAPA9 97,113,89
orl-rat TDLo:150 mg/kg (female 6-20D post):TER TXAPA9 97,113,89
orl-rat LD50:87 mg/kg MarJV# 29MAR77
skn-rat LD50:11700 mg/kg PCOC** -,1150,66
ipr-rat LD50:7210 μg/kg FCTXAV 7,47,69
scu-rat LD50:11700 mg/kg TRIPA7 -,1,73
orl-mus LD50:55 mg/kg GISAAA 41(5),10,76
ipr-mus LD50:12500 μg/kg RPTOAN 42,73,79
ivn-mus LD50:6 mg/kg EJTXAZ 9,31,76
orl-rbt LDLo:50 mg/kg SAIGBL 15,3,73
skn-rbt LD50:900 mg/kg EJTXAZ 9,31,76

CONSENSUS REPORTS: Reported in EPA TSCA Inventory.

OSHA PEL: TWA 0.1 mg(Sn)/m^3 (skin)
ACGIH TLV: TWA 0.1 mg(Sn)/m^3 (skin) (Proposed: TWA 0.1 mg(Sn)/m^3; STEL 0.2 mg(Sn)/m^3 (skin))
NIOSH REL: (Organotin Compounds) TWA 0.1 mg(Sn)/m^3

SAFETY PROFILE: A poison by ingestion, intraperitoneal, and intravenous routes. Moderately toxic by skin contact. An experimental teratogen. Other experimental reproductive effects. Mutation data reported. A severe eye irritant. See also TIN COMPOUNDS. When heated to decomposition it emits acrid and irritating fumes.

BLL825 CAS:881-99-2 ***HR: 2***
m-BIS(TRICHLORMETHYL)BENZENE
mf: $C_8H_4Cl_6$ mw: 312.82

SYNS: m-BIS(TRICHLOROMETHYL)BENZENE ◇ 1,3-BIS(TRICHLOROMETHYL)BENZENE ◇ 1,3-DI(TRICHLOROMETHYL)BENZENE ◇ α,α′-HEXACHLORO-m-XYLENE ◇ α,α,α,α′,α′,α′-HEXACHLOROXYLENE ◇ α,α,α,α′,α′,α′-HEXACHLORO-m-XYLENE

TOXICITY DATA with REFERENCE
skn-rbt 500 mg MOD 34ZIAG -,308,69
eye-rbt 3 mg MLD 34ZIAG -,308,69
orl-rat TDLo:2129 mg/kg (26W male):REP GNAMAP 21,34,82
orl-rat LD50:2924 mg/kg GNAMAP 21,34,82

SAFETY PROFILE: Moderately toxic by ingestion. Experimental reproductive effects. A skin irritant. Violent reaction when heated with oxidants (e.g., potassium nitrate, selenium dioxide, and sodium chlorate). When heated to decomposition it emits toxic fumes of Cl^-. See CHLORINATED HYDROCARBONS, AROMATIC.

BLM000 CAS:2629-78-9 ***HR: 3***
BIS(TRICHLOROACETYL)PEROXIDE
mf: $C_4Cl_6O_4$ mw: 324.76

$Cl_3CCO{\bullet}OOCO{\bullet}CCl_3$

SAFETY PROFILE: A very shock-sensitive explosive which may detonate at room temperature. Upon decomposition it emits toxic fumes of Cl^-. See also PEROXIDES.

BLM250 ***HR: 3***
BIS-2,4,5-TRICHLORO BENZENE DIAZO OXIDE
mf: $C_{12}H_4Cl_6N_4O$ mw: 212.72

SAFETY PROFILE: Ignites spontaneously. Explodes on impact or on contact with benzene. Upon decomposition it emits toxic fumes of Cl^- and NO_x.

BLM500 CAS:3064-70-8 ***HR: 3***
BIS(TRICHLOROMETHYL)SULFONE
mf: $C_2Cl_6O_2S$ mw: 300.78

SYN: N-1386 BIOCIDE

TOXICITY DATA with REFERENCE
orl-rat LD50:691 mg/kg TOXID9 4,16,84
ivn-mus LD50:18 mg/kg CSLNX* NX#04617

CONSENSUS REPORTS: Reported in EPA TSCA Inventory.

SAFETY PROFILE: Poison by intravenous route. Moderately toxic by ingestion. When heated to decomposition it emits very toxic fumes of Cl^- and SO_x.

BLM750 CAS:2532-50-5 ***HR: 3***
BIS(TRICHLORO METHYL)TRISULFIDE
mf: $C_2Cl_6S_3$ mw: 332.90

SYNS: BISTRICHLOROMETHYLTRISULFID (CZECH) ◇ TRITHIOBIS(TRICHLOROMETHANE)

TOXICITY DATA with REFERENCE
skn-rbt 500 mg/24H MOD 28ZPAK -,170,72
eye-rbt 100 mg/24H SEV 28ZPAK -,170,72
orl-rat LD50:676 mg/kg 28ZPAK -,170,72
ivn-mus LD50:56 mg/kg CSLNX* NX#04597

SAFETY PROFILE: Poison by intravenous route. Moderately toxic by ingestion. A skin and eye irritant. See also SULFIDES. When heated to decomposition it emits very toxic fumes of Cl^- and SO_x.

BLN000 CAS:63885-02-9 ***HR: 2***
BIS(2,3,5-TRICHLOROPHENYLTHIO)ZINC
mf: $C_{12}H_4Cl_6S_2Zn$ mw: 490.35

SYN: 2,3,5-TRICHLOROFENOLAT ZINECNATY (CZECH)

TOXICITY DATA with REFERENCE
skn-rbt 500 mg/24H MOD 28ZPAK -,11,72
eye-rbt 50 μg/24H SEV 28ZPAK -,11,72
orl-rat LD50:4260 mg/kg 28ZPAK -,11,72

CONSENSUS REPORTS: Zinc and its compounds are on the Community Right-To-Know List.

SAFETY PROFILE: Mildly toxic by ingestion. A severe eye and skin irritant. See also ZINC COMPOUNDS and CHLORINATED HYDROCARBONS, AROMATIC. When heated to decomposition it emits very toxic fumes of ZnO, Cl^- and SO_x.

BLN100 CAS:80660-68-0 ***HR: 3***
BIS(TRIETHYLENETETRAMINE)TUNG-STATONICKEL
mf: $C_{12}H_{36}N_8O_4W•Ni$ mw: 599.12

SYNS: NICKEL(2+), BIS(N,N'-BIS(2-AMINOETHYL)-1,2-ETHANEDIAMINE-N,N',N^N)-, (T-4)-TETRAOXOTUNGSTATE(2-) (1:1) ◇ NICKEL, BIS(TRIETHYLENETETRAMINE)TUNGSTATO-

TOXICITY DATA with REFERENCE
ipr-mus LD50:82500 μg/kg IJEBA6 19,1187,81

OSHA PEL: TWA 1 mg(Ni)/m^3
ACGIH TLV: TWA 5 mg(W)/m^3; STEL 10 mg(W)/m^3
SAFETY PROFILE: Poison by intraperitoneal route. When heated to decomposition it emits toxic fumes of NO_x, Ni, and W.

BLN250 ***HR: 3***
BIS(TRIETHYL TIN)ACETYLENE
mf: $C_{14}H_{30}Sn_2$ mw: 435.77

SAFETY PROFILE: A sensitive, powerful explosive. Incompatible with stannic chloride. When heated to decomposition it emits acrid smoke and fumes. See also TIN COMPOUNDS and ACETYLENE COMPOUNDS.

BLN500 CAS:57-52-3 ***HR: 3***
BIS(TRIETHYLTIN) SULFATE
mf: $C_{12}H_{30}O_4SSn_2$ mw: 507.86

SYNS: TRIAETHYLZINNSULFAT (GERMAN) ◇ TRIETHYLHYDROXY-STANNANE SULFATE (2:1) (8CI) ◇ TRIETHYLHYDROXYTIN SULFATE ◇ TRIETHYLTIN SULPHATE

TOXICITY DATA with REFERENCE
orl-rat LDLo:10 mg/kg BJPCAL 10,16,55
ipr-rat LD50:5700 μg/kg BJPCAL 10,16,55
scu-rat LDLo:25 mg/kg BJPCAL 10,16,55
ivn-rat LD50:9050 μg/kg AEPPAE 242,370,61
par-rat LD50:6 mg/kg BIJOAK 61,406,55
orl-rbt LDLo:10 mg/kg BJPCAL 10,16,55
ivn-rbt LDLo:3 mg/kg BJPCAL 10,16,55
ipr-gpg LD50:3 mg/kg BJIMAG 23,222,66
ivn-brd LDLo:3 mg/kg BJPCAL 10,16,55

OSHA PEL: TWA 0.1 mg(Sn)/m^3 (skin)
ACGIH TLV: TWA 0.1 mg(Sn)/m^3 (skin) (Proposed: TWA 0.1 mg(Sn)/m^3; STEL 0.2 mg(Sn)/m^3 (skin))
NIOSH REL: (Organotin Compounds) TWA 0.1 mg (Sn)/m^3

SAFETY PROFILE: Poison by ingestion, intraperitoneal, subcutaneous, intravenous, and parenteral routes. See also TIN COMPOUNDS and SULFATES. When heated to decomposition it emits toxic fumes of SO_x.

BLN750 CAS:52112-09-1 ***HR: 3***
BIS(TRIFLUOROACETOXY)DIBUTYLTIN
mf: $C_{12}H_{18}F_6O_4Sn$ mw: 458.99

SYNS: DIBUTYLTIN BIS(TRIFLUOROACETATE) ◇ DIBUTYLBIS (TRIFLUOROACETOXY)STANNANE ◇ DTBT ◇ STANNOUS DIBUTYLDITRIFLUOROACETATE ◇ TIN DIBUTYLDITRI-FLUOROACETATE

TOXICITY DATA with REFERENCE
orl-rat LD50:55 mg/kg GISAAA 46(7),18,81
skn-rat LD50:1 g/kg GISAAA 46(7),18,81
orl-mus LD50:53600 μg/kg GISAAA 41(5),10,76

OSHA PEL: TWA 0.1 mg(Sn)/m^3 (skin)
ACGIH TLV: TWA 0.1 mg(Sn)/m^3 (skin) (Proposed: TWA 0.1 mg(Sn)/m^3; STEL 0.2 mg(Sn)/m^3 (skin))
NIOSH REL: (Organotin Compounds) TWA 0.1 mg(Sn)/m^3

SAFETY PROFILE: Poison by ingestion. Moderately toxic by skin contact. See also TIN COMPOUNDS and FLUORIDES. When heated to decomposition it emits toxic fumes of F^-.

BLO000 CAS:383-73-3 ***HR: 3***
BIS(TRIFLUOROACETYL)PEROXIDE
mf: $C_4F_6O_4$ mw: 226.03

SAFETY PROFILE: A poison. May explode spontaneously at room temperature. Upon decomposition it emits toxic fumes of F^-. See also PEROXIDES.

BLO250 CAS:328-74-5 ***HR: 3***
3,5-BIS(TRIFLUOROMETHYL)ANILINE
mf: $C_8H_5F_6N$ mw: 229.14

SYN: α,α,α,α,α,α-HEXAFLUORO-3,5-XYLIDINE

TOXICITY DATA with REFERENCE
ipr-mus LDLo:31 mg/kg CBCCT* 4,323,52
ivn-mus LD50:25 mg/kg CBCCT* 6,143,54

CONSENSUS REPORTS: Reported in EPA TSCA Inventory.

SAFETY PROFILE: Poison by intraperitoneal and intravenous routes. See also FLUORIDES. When heated to decomposition it emits very toxic fumes of F^- and NO_x.

BLO270 CAS:402-31-3 ***HR: 3***
1,3-BIS(TRIFLUOROMETHYL)BENZENE
mf: $C_8H_4F_6$ mw: 214.11

SAFETY PROFILE: When heated to 90°C a mixture with nitric and sulfuric acids emits spark-sensitive explosive vapors. When heated to decomposition it emits toxic fumes of F^-.

BLO280 CAS:650-52-2 ***HR: 3***
BIS(TRIFLUOROMETHYL)CHLOROPHOSPHINE
mf: C_2ClF_6P mw: 204.44

SAFETY PROFILE: Ignites spontaneously in air. When heated to decomposition it emits toxic fumes of F^-, Cl^-, and PO_x. See also PHOSPHINE.

BLO300 CAS:431-97-0 ***HR: 3***
BIS(TRIFLUOROMETHYL)CYANOPHOSPHINE
mf: C_3F_6NP mw: 195.00

CONSENSUS REPORTS: Cyanide and its compounds are on the Community Right-To-Know List.

SAFETY PROFILE: Ignites spontaneously on contact with air. When heated to decomposition it emits toxic fumes of F^-, PO_x, CN^-, and NO_x. See also CYANIDE and PHOSPHINE.

BLO325 CAS:372-64-5 ***HR: 3***
BIS(TRIFLUOROMETHYL)DISULFIDE
mf: $C_2F_6S_2$ mw: 202.13

SAFETY PROFILE: Mixtures of the solid with chlorine mono- or tri- fluorides are explosive. Dilute with halogenated solvents. When heated to decomposition it emits toxic fumes of F^- and SO_x. See also SULFIDES.

BLP250 CAS:30184-88-4 ***HR: 3***
2,2-BIS(TRIFLUOROMETHYL)-4-METHYL-5-PHENYLOXAZOLIDINE HYDRATE
mf: $C_{12}H_{11}F_6NO•H_2O$ mw: 317.26

TOXICITY DATA with REFERENCE
orl-mus LD50:200 mg/kg JMCMAR 13,1215,70
ipr-mus LD50:300 mg/kg JMCMAR 13,1215,70

SAFETY PROFILE: Poison by ingestion and intraperitoneal routes. When heated to decomposition it emits very toxic fumes of F^- and NO_x.

BLP300 CAS:2154-71-4 ***HR: 3***
BIS(TRIFLUOROMETHYL)NITROXIDE
mf: C_2F_6NO mw: 168.02

SAFETY PROFILE: Explodes violently at room temperature. Upon decomposition it emits toxic fumes of F^- and NO_x.

BLP325 CAS:24095-80-5 ***HR: 3***
2-(3,5-BIS(TRIFLUOROMETHYL)PHENYL)-N-METHYL-HYDRAZINECARBOTHIOAMIDE (9CI)
mf: $C_{10}H_9F_6N_3S$ mw: 317.28

SYNS: 1-((3,5-BIS-TRIFLUOROMETHYL)PHENYL)-4-METHYL-THIOSEMICARBAZIDE ◇ CIBA 2696GO ◇ 1-(α,α,α,α′,α′,α′-HEXA-FLUORO-3,5-XYLYL)-4-METHYL-3-THIO-SEMICARBAZIDE

TOXICITY DATA with REFERENCE
orl-rat TDLo:50 mg/kg (6-15D preg):TER ARZNAD 23,797,73
orl-rat TDLo:25 mg/kg (6-15D preg):REP ARZNAD 23,797,73
orl-rat LD50:414 mg/kg ARZNAD 23,797,73
ipr-rat LD50:212 mg/kg ARZNAD 23,797,73
orl-mus LD50:269 mg/kg ARZNAD 23,797,73
orl-dog LDLo:1500 mg/kg ARZNAD 23,797,73

SAFETY PROFILE: Poison by ingestion and intraperitoneal routes. An experimental teratogen. Other experimental reproductive effects. When heated to decomposition it emits toxic fumes of F^-, SO_x, and NO_x.

BLP500 ***HR: 3***
BIS(TRIFLUOROMETHYL)PHOSPHORUS(III) AZIDE
mf: $C_2F_6N_3P$ mw: 211.01

SAFETY PROFILE: Explosive, very unstable even at −196°. Upon decomposition it emits toxic fumes of F^-, PO_x, and NO_x. See also AZIDES.

BLQ250 CAS:30192-67-7 ***HR: 3***
α,α-BIS(TRIFLUOROMETHYL)-1-PIPERIDINEMETHANOL HYDRATE
mf: $C_8H_{11}F_6NO•H_2O$ mw: 269.22

TOXICITY DATA with REFERENCE
orl-mus LD50:300 mg/kg JMCMAR 13,1215,70
ipr-mus LD50:300 mg/kg JMCMAR 13,1215,70

SAFETY PROFILE: Poison by ingestion and intraperitoneal routes. When heated to decomposition it emits very toxic fumes of F^- and NO_x.

BLQ325 CAS:371-78-8 ***HR: 3***
BIS(TRIFLUOROMETHYL)SULFIDE
mf: C_2F_6S mw: 170.07

SAFETY PROFILE: Mixtures of the solid with chlorine mono- or tri- fluorides are explosive. Dilute with halogenated solvents. When heated to decomposition it emits toxic fumes of F^- and SO_x. See also SULFIDES.

BLQ500 CAS:28399-14-6 ***HR: 3***
2,2-BIS(TRIFLUOROMETHYL)THIAZOLIDINE HYDRATE
mf: $C_5H_5F_6NS•H_2O$ mw: 243.19

TOXICITY DATA with REFERENCE
orl-mus LD50:300 mg/kg JMCMAR 13,1215,70
ipr-mus LD50:300 mg/kg JMCMAR 13,1215,70

SAFETY PROFILE: Poison by ingestion and intraperitoneal routes. When heated to decomposition it emits very toxic fumes of F^-, NO_x, and SO_x.

BLQ600 ***HR: 2***
2,3-BISTRIMETHYLACETOXYMETHYL-1-METHYLPYRROLE
mf: $C_{17}H_{27}NO_4$ mw: 309.45

TOXICITY DATA with REFERENCE
skn-mus TDLo:291 mg/kg/47W-I:CAR CALEDQ 17,61,82

SAFETY PROFILE: Questionable carcinogen with experimental carcinogenic data. When heated to decomposition it emits toxic fumes of NO_x.

BLQ750 CAS:64011-39-8 ***HR: 3***
BIS(TRIMETHYLHEXYL)TIN DICHLORIDE
mf: $C_{18}H_{38}Cl_2Sn$ mw: 444.15

SYNS: DICHLORODIISONONYL STANNANE ◇ DIISONONYLTIN DICHLORIDE

TOXICITY DATA with REFERENCE
ivn-rat LDLo:10 mg/kg BJIMAG 15,15,58

OSHA PEL: TWA 0.1 mg(Sn)/m^3 (skin)
ACGIH TLV: TWA 0.1 mg(Sn)/m^3 (skin) (Proposed: TWA 0.1 mg(Sn)/m^3; STEL 0.2 mg(Sn)/m^3 (skin))
NIOSH REL: (Organotin Compounds) TWA 0.1 mg(Sn)/m^3

SAFETY PROFILE: Poison by intravenous route. See also TIN COMPOUNDS. When heated to decomposition it emits toxic fumes of Cl^-.

BLQ850 CAS:73452-31-0 ***HR: 3***
N,N'-BIS(TRIMETHYLSILYL)AMINOBORANE
mf: $C_6H_{20}BNSi_2$ mw: 173.21

$((CH_3)_3SI)_2NBH_2$

SAFETY PROFILE: Ignites on contact with air. When heated to decomposition it emits toxic fumes of NO_x. See also BORANES, BORON COMPOUNDS, and SILANE.

BLQ900 CAS:86045-52-5 ***HR: 3***
cis-BIS(TRIMETHYLSILYLAMINO)TELLURIUM TETRAFLUORIDE
mf: $C_6H_{20}F_4N_2Si_2Te$ mw: 380.00

$((CH_3)_3SiNH)_2TeF_4$

SAFETY PROFILE: Explodes when heated to 100°C. During storage it converts to an explosive solid. When heated to decomposition it emits toxic fumes of F^-, SO_x, NO_x, and Te. See also TELLURIUM COMPOUNDS.

BLR000 CAS:1746-09-4 ***HR: 3***
BIS(TRIMETHYLSILYL)CHROMATE
mf: $C_6H_{18}CrO_4Si_2$ mw: 262.57

$((CH_3)_3SiO)_2CrO_2$

CONSENSUS REPORTS: Chromium and its compounds are on the Community Right-To-Know List.

SAFETY PROFILE: May explode if heated above 75°C. When heated to decomposition it emits acrid smoke and fumes. See also SILANE and CHROMIUM COMPOUNDS.

BLR125 CAS:692-56-8 ***HR: 3***
1,2-BIS(TRIMETHYLSILYL)HYDRAZINE
mf: $C_6H_{20}N_2Si_2$ mw: 176.41

$(CH_3)_3SiNHNHSi(CH_3)_3$

SAFETY PROFILE: Hypergolic reaction with strong oxidants (e.g., fluorine or fuming nitric acid). When heated to decomposition it emits toxic fumes of NO_x. See also HYDRAZINE and SILANE.

BLR140 CAS:4656-04-6 ***HR: 3***
BIS(TRIMETHYLSILYL)MERCURY
mf: $C_6H_{18}HgSi_2$ mw: 346.97

$((CH_3)_3Si)_2Hg$

CONSENSUS REPORTS: and its compounds are on the Community Right-To-Know List.

SAFETY PROFILE: May ignite spontaneously in air. When heated to decomposition it emits toxic fumes of Hg. See also MERCURY COMPOUNDS and SILANE.

BLR250 ***HR: 3***
BISTRIMETHYL SILYL OXIDE
mf: $C_6H_{18}OSi_2$ mw: 162.44

PROP: Flash p: −1°C.

SAFETY PROFILE: A very dangerous fire hazard. When heated to decomposition it emits acrid smoke and fumes.

BLR500 CAS:23115-33-5 ***HR: 3***
BIS(TRIMETHYLSILYL)PEROXOMONOSULFATE
mf: $C_6H_{18}O_5SSi_2$ mw: 258.61

$(CH_3)_3SiOSO_2OOSi(CH_3)_3$

SAFETY PROFILE: May decompose violently at room temperature and evolve toxic sulfur trioxide. See also PEROXIDES, ORGANIC, SILANE, and SULFATES.

BLR625 CAS:918-99-0 ***HR: 3***
N,N′-BIS(2,2,2-TRINITROETHYL)UREA
mf: $C_5H_6N_8O_{13}$ mw: 386.15

$((O_2N)_3CCH_2NH)_2CO$

SAFETY PROFILE: Mixtures with sodium hydroxide are storage hazards due to the formation of unstable reaction products. When heated to decomposition it emits toxic fumes of NO_x.

BLR750 CAS:28930-30-5 ***HR: 2***
BIS(TRINITROPHENYL)SULFIDE
mf: $C_{12}H_4N_6O_{12}S$ mw: 456.28

SYNS: HEXANITRODIPHENYLSULFIDE ◇ PICRYL SULFIDE

TOXICITY DATA with REFERENCE
orl-rat LD50:1200 mg/kg TNICS* 13,132,73
orl-mus LD50:470 mg/kg TNICS* 13,132,73

SAFETY PROFILE: Moderately toxic by ingestion. See also SULFIDES and NITRO COMPOUNDS of AROMATIC HYDROCARBONS. See NITRATES for fire and explosion hazard. This material is a powerful explosive and has an added military advantage in that its explosive gases contain irritating and very toxic SO_x. See also EXPLOSIVES, HIGH.

BLS000 ***HR: 3***
BISTRIPERCHLORATO SILICON OXIDE
mf: $Cl_6O_{25}Si_2$ mw: 508.88

SAFETY PROFILE: Heating to decomposition may form an explosive product. When heated to decomposition it emits toxic fumes of Cl^-. See also PERCHLORATES.

BLS250 CAS:14264-16-5 ***HR: 3***
BIS(TRIPHENYLPHOSPHINE)DICHLORONICKEL
mf: $C_{24}H_{54}P_2{\bullet}Cl_2Ni$ mw: 534.33

SYNS: BIS(TRI-N-BUTYLPHOSPHINE)DICHLORONICKEL ◇ TRIBUTYL-PHOSPHINE compounded with NICKELCHLORIDE (2:1)

TOXICITY DATA with REFERENCE
ivn-mus LD50:56 mg/kg CSLNX* NX#03119

CONSENSUS REPORTS: Reported in EPA TSCA Inventory. Nickel and its compounds are on the Community Right-To-Know List.

OSHA PEL: (Transitional: TWA 1 mg/m^3) TWA 0.1 mg (Ni)/m^3
ACGIH TLV: TWA 0.1 mg(Ni)/m^3; (Proposed: TWA 0.05 mg(Ni)/m^3; Human Carcinogen)

SAFETY PROFILE: Suspected carcinogen. Poison by intravenous route. See also NICKEL COMPOUNDS and PHOSPHOROUS COMPOUNDS. When heated to decomposition it emits very toxic fumes of Cl^- and PO_x.

BLS500 CAS:15709-62-3 ***HR: 3***
BIS(TRIPHENYL PHOSPHINE)NICKEL DITHIOCYANATE
mf: $C_{38}H_{30}N_2NiP_2S_2$ mw: 699.47

SYN: NICKEL BISTRIPHENYLPHOSPHINE DITHIOCYANATE

TOXICITY DATA with REFERENCE
ivn-mus LD50:180 mg/kg CSLNX* NX#01983

CONSENSUS REPORTS: Nickel and its compounds are on the Community Right-To-Know List.

SAFETY PROFILE: Poison by intravenous route. See also NICKEL COMPOUNDS and THIOCYANATES. When heated to decomposition it emits very toxic fumes of SO_x, PO_x, NO_x, and CN^-.

BLS750 CAS:1624-02-8 ***HR: 2***
BIS(TRIPHENYL SILYL)CHROMATE
mf: $C_{36}H_{30}CrO_4Si_2$ mw: 634.84

SYN: CHROMIC ACID, BIS(TRIPHENYLSILYL) ESTER

TOXICITY DATA with REFERENCE
orl-rat LD50:3360 mg/kg TXAPA9 28,313,74
skn-rbt LD50:710 mg/kg TXAPA9 28,313,74

CONSENSUS REPORTS: Reported in EPA TSCA Inventory. Chromium and its compounds are on the Community Right-To-Know List.

OSHA PEL: CL 0.1 mg(Cr)$_3$)/m^3
ACGIH TLV: TWA 0.05 mg(Cr)/m^3
NIOSH REL: (Chromium(VI) TWA 0.025 mg(Cr(VI))/m^3; CL 0.05/15M

SAFETY PROFILE: Moderately toxic by ingestion and skin contact. See also CHROMIUM COMPOUNDS and ESTERS. When heated to decomposition it emits toxic fumes of CrO_3 particulates.

BLS900 CAS:73940-87-1 ***HR: 3***
BIS(TRIPHENYLTIN)ACETYLENEDICARBOXYLATE
mf: $C_{40}H_{30}O_4Sn_2$ mw: 812.08

SYNS: ETHYNYLENEBIS(CARBONYLOXY)BIS(TRIPHENYLSTANNANE) ◇ STANNANE, ETHYNYLENEBIS(CARBONYLOXY)BIS(TRIPHENYL-

TOXICITY DATA with REFERENCE
ivn-mus LD50:18 mg/kg CSLNX* NX#05963

OSHA PEL: TWA 0.1 mg(Sn)/m^3
ACGIH TLV: TWA 0.1 mg(Sn)/m^3; STEL 0.2 mg/m^3 (skin)
NIOSH REL: (Organotin compound): 10H TWA 0.1 mg(Sn)/m^3

SAFETY PROFILE: Poison by intravenous route. When heated to decomposition it emits toxic fumes of Sn.

BLT000 CAS:3021-41-8 ***HR: 3***
BIS(TRIPHENYLTIN)SULFATE
mf: $C_{36}H_{30}Sn_2 \cdot O_4S$ mw: 796.10

SYN: TRIPHENYLSTANNANE SULFATE (2:1)

TOXICITY DATA with REFERENCE
ivn-mus LD50:18 mg/kg CSLNX* NX#04819

OSHA PEL: TWA 0.1 mg(Sn)/m^3 (skin)
ACGIH TLV: TWA 0.1 mg(Sn)/m^3 (skin) (Proposed: TWA 0.1 mg(Sn)/m^3; STEL 0.2 mg(Sn)/m^3 (skin))
NIOSH REL: (Organotin Compounds) TWA 0.1 mg(Sn)/m^3

SAFETY PROFILE: Poison by intravenous route. See also TIN COMPOUNDS and SULFATES. When heated to decomposition it emits toxic fumes of SO_x.

BLT250 CAS:77-80-5 ***HR: 3***
BIS(TRIPHENYLTIN)SULFIDE
mf: $C_{36}H_{30}SSn_2$ mw: 732.10

SYN: 1,1,1,3,3,3-HEXAPHENYLDISTANNTHIANE

TOXICITY DATA with REFERENCE
orl-mus LD50:710 mg/kg AECTCV 14,111,85
ivn-mus LD50:180 mg/kg CSLNX* NX#05814

OSHA PEL: TWA 0.1 mg(Sn)/m^3 (skin)
ACGIH TLV: TWA 0.1 mg(Sn)/m^3 (skin) (Proposed: TWA 0.1 mg(Sn)/m^3; STEL 0.2 mg(Sn)/m^3 (skin))
NIOSH REL: (Organotin Compounds) TWA 0.1 mg(Sn)/m^3

SAFETY PROFILE: A poison via intravenous route. Moderately toxic by ingestion. See also TIN COMPOUNDS and SULFIDES. When heated to decomposition it emits toxic fumes of SO_x.

BLT300 CAS:1067-29-4 ***HR: 3***
BIS(TRIPROPYLTIN)OXIDE
mf: $C_{18}H_{42}OSn_2$ mw: 511.98

SYNS: DISTANNOXANE, 1,1,1,3,3,3-HEXAPROPYL- ◇ 1,1,1,3,3,3-HEXAPROPYLDISTANNOXANE

TOXICITY DATA with REFERENCE
ivn-mus LD50:5600 μg/kg CSLNX* NX#03791

OSHA PEL: 8H TWA 0.1 mg(Sn)/m^3
ACGIH TLV: TWA 0.1 mg(Sn)/m^3; STEL 0.2 mg/m^3 (skin)
NIOSH REL: (Organotin compound): 10H TWA 0.1 mg(Sn)/m^3

CONSENSUS REPORTS: Reported in EPA TSCA Inventory.

SAFETY PROFILE: Poison by intravenous route. When heated to decomposition it emits toxic fumes of Sn.

BLT500 CAS:74039-78-4 ***HR: 3***
BIS(TRIS(p-CHLOROPHENYL)PHOSPHINE)MERCURIC CHLORIDE COMPLEX
mf: $C_{36}H_{24}Cl_6P_2 \cdot Cl_2Hg$ mw: 1002.73

SYN: TRIS(p-CHLOROPHENYL)PHOSPHINE COMPLEX with MERCURIC CHLORIDE (2:1)

TOXICITY DATA with REFERENCE
ivn-mus LD50:56 mg/kg CSLNX* NX#02647

CONSENSUS REPORTS: Mercury and its compounds are on the Community Right-To-Know List.

OSHA PEL: (Transitional: CL 1 mg/10m^3) CL 0.1 mg(Hg)/m^3 (skin)
ACGIH TLV: TWA 0.1 mg(Hg)/m^3 (skin)
NIOSH REL: (Inorganic Mercury) TWA 0.05 mg(Hg)/m^3

SAFETY PROFILE: Poison by intravenous route. See also MERCURY COMPOUNDS and CHLORIDES. When heated to decomposition it emits very toxic fumes of Cl^-, PO_x and Hg.

BLT750 CAS:74039-79-5 ***HR: 3***
BIS(TRIS(p-DIMETHYLAMINOPHENYL)PHOSPHINE)MERCURIC CHLORIDE COMPLEX
mf: $C_{48}H_{60}N_6P_2 \cdot Cl_2Hg$ mw: 1054.57

SYN: TRIS(p-DIMETHYLAMINOPHENYL) PHOSPHINE COMPLEX with MERCURIC CHLORIDE (2:1)

TOXICITY DATA with REFERENCE
ivn-mus LD50:100 mg/kg CSLNX* NX#02644

CONSENSUS REPORTS: Mercury and its compounds are on the Community Right-To-Know List.

OSHA PEL: (Transitional: CL 1 mg/10m^3) CL 0.1 mg(Hg)/m^3 (skin)
ACGIH TLV: TWA 0.1 mg(Hg)/m^3 (skin)
NIOSH REL: (Inorganic Mercury) TWA 0.05 mg(Hg)/m^3

SAFETY PROFILE: Poison by intravenous route. See also MERCURY COMPOUNDS and CHLORIDES. When heated to decomposition it emits very toxic fumes of NO_x, PO_x, Cl^-, and Hg.

BLT775 CAS:38402-95-8 ***HR: 3***
BIS(TRIS(p-DIMETHYLAMINOPHENYL)PHOSPHINE OXIDE)STANNIC CHLORIDE COMPLEX
mf: $C_{48}H_{60}N_6O_2P_2{\bullet}Cl_4Sn$ mw: 1075.57

SYN: PHOSPHINE OXIDE, TRIS(p-DIMETHYLAMINOPHENYL)-, compd. with STANNIC CHLORIDE (2:1)

TOXICITY DATA with REFERENCE
ivn-mus LD50:180 mg/kg CSLNX* NX#02651

OSHA PEL: TWA 2 mg(Sn)/m^3
ACGIH TLV: TWA 2 mg(Sn)/m^3

SAFETY PROFILE: Poison by intravenous route. When heated to decomposition it emits toxic fumes of NO_x, PO_x, Sn, and Cl^-.

BLU000 CAS:13356-08-6 ***HR: 2***
BIS(TRIS(β,β-DIMETHYLPHENETHYL)TIN) OXIDE
mf: $C_{60}H_{78}OSn_2$ mw: 1052.76

SYNS: BENDEX ◇ BIS(TRIS(2-METHYL-2-PHENYLPROPYL) TIN)OXIDE ◇ DI(TRI-(2,2-DIMETHYL-2-PHENYLETHYL)TIN)OXIDE ◇ ENT 27,738 ◇ FENBUTATIN OXIDE ◇ HEXAKIS(β,β-DIMETHYLPHENETHYL)DISTANNOXANE ◇ HEXAKIS(2-METHYL-2-PHENYLPROPYL)DISTANNOXANE ◇ SD 14114 ◇ SHELL SD-14114 ◇ TORQUE ◇ VENDEX

TOXICITY DATA with REFERENCE
orl-rat LD50:2630 mg/kg 85ARAE 1,17,77
skn-rat LD50:1000 mg/kg TIUSAD 110,6,76

OSHA PEL: TWA 0.1 mg(Sn)/m^3 (skin)
ACGIH TLV: TWA 0.1 mg(Sn)/m^3 (skin) (Proposed: TWA 0.1 mg(Sn)/m^3; STEL 0.2 mg(Sn)/m^3 (skin))
NIOSH REL: (Organotin Compounds) TWA 0.1 mg(Sn)/m^3

SAFETY PROFILE: Moderately toxic by ingestion and skin contact. See also TIN COMPOUNDS. When heated to decomposition it emits acrid smoke and irritating fumes.

BLU250 CAS:74039-80-8 ***HR: 3***
BIS(TRIS(p-METHOXYPHENYL)PHOSPHINE) MERCURIC CHLORIDE COMPLEX
mf: $C_{42}H_{42}O_6P_2{\bullet}Cl_2Hg$ mw: 976.27

SYN: TRIS(p-METHOXYPHENYL) PHOSPHINE COMPLEX with MERCURIC CHLORIDE (2:1)

TOXICITY DATA with REFERENCE
ivn-mus LD50:180 mg/kg CSLNX* NX#02645

CONSENSUS REPORTS: Mercury and its compounds are on the Community Right-To-Know List.

OSHA PEL: (Transitional: CL 1 mg/10m^3) CL 0.1 mg(Hg)/m^3 (skin)
ACGIH TLV: TWA 0.1 mg(Hg)/m^3 (skin)
NIOSH REL: (Inorganic Mercury) TWA 0.05 mg(Hg)/m^3

SAFETY PROFILE: Poison by intravenous route. See also MERCURY COMPOUNDS and CHLORIDES. When heated to decomposition it emits very toxic vapors of PO_x, Cl^-, and Hg.

BLU500 CAS:74039-81-9 ***HR: 3***
BIS(TRIS(p-METHYLTHIOPHENYL)PHOSPHINE)MERCURIC CHLORIDE COMPLEX
mf: $C_{42}H_{42}P_2S_6{\bullet}Cl_2Hg$ mw: 1072.63

SYN: TRIS(p-METHYLTHIOPHENYL) PHOSPHINE COMPLEX with MERCURIC CHLORIDE (2:1)

TOXICITY DATA with REFERENCE
ivn-mus LD50:180 mg/kg CSLNX* NX#02646

CONSENSUS REPORTS: Mercury and its compounds are on the Community Right-To-Know List.

OSHA PEL: (Transitional: CL 1 mg/10m^3) CL 0.1 mg(Hg)/m^3 (skin)
ACGIH TLV: TWA 0.1 mg(Hg)/m^3 (skin)
NIOSH REL: (Inorganic Mercury) TWA 0.05 mg(Hg)/m^3

SAFETY PROFILE: Poison by intravenous route. See also MERCURY COMPOUNDS and CHLORIDES. When heated to decomposition it emits very toxic fumes of PO_x, SO_x, Cl^-, and Hg.

BLV000 CAS:5169-78-8 ***HR: 3***
BITIODIN
mf: $C_{15}H_{17}NS_2$ mw: 275.45

SYNS: AT 327 ◇ CR/662 ◇ 3-(DI-2-THIENYLMETHYLENE)-1-METHYLPIPERIDINE ◇ 1-METHYL-3-PIPERIDYLIDENEDI(2-THIENYL)METHANE ◇ TIPEDINE ◇ TIPEPIDINE

TOXICITY DATA with REFERENCE
orl-mus LD50:867 mg/kg MEIEDD 11,1490,89
ipr-mus LD50:294 mg/kg MEIEDD 11,1490,89
scu-mus LD50:222 mg/kg CPBTAL 7,372,59
ivn-mus LD50:55 mg/kg PCJOAU 10,1482,76
ims-mus LD50:308 mg/kg MEIEDD 11,1490,89
ivn-dog LD50:44 mg/kg CPBTAL 7,372,59

SAFETY PROFILE: A poison via subcutaneous, intraperitoneal, intravenous, and intramuscular routes. Moderately toxic by ingestion. An antitussive. When heated to decomposition it emits very toxic fumes of NO_x and SO_x.

BLV075 ***HR: 3***
BITIS ARIETANS VENOM

SYNS: B. ARIETANS VENOM ◇ SNAKE VENOM BITIS ARIETANS

TOXICITY DATA with REFERENCE
ipr-mus LD50:560 μg/kg TOXIA6 18,384,80
scu-mus LD50:600 μg/kg TOXIA6 20,509,82
ivn-mus LD50:1055 μg/kg TOXIA6 2,5,64
ims-mus LD50:2 mg/kg TOXIA6 6,175,69
ivn-rbt LDLo:660 μg/kg TOXIA6 2,5,64
ipr-mam LD50:3680 μg/kg CLPTAT 8,849,67

SAFETY PROFILE: Deadly poison by subcutaneous, intramuscular, intravenous, and intraperitoneal routes.

BLV080 ***HR: 3***
BITIS GABONICA VENOM

SYNS: B. GABONICA VENOM ◇ SNAKE VENOM BITIS GABONICA

TOXICITY DATA with REFERENCE
ipr-mus LD50:960 μg/kg TOXIA6 18,384,80
scu-mus LD50:5 mg/kg JOIMA3 67,299,51
ivn-mus LD50:550 μg/kg TOXIA6 14,146,76
ims-mus LD50:5200 μg/kg TOXIA6 6,175,69
ivn-rbt LDLo:1065 μg/kg SCIEAS 117,47,53

SAFETY PROFILE: Deadly poison by subcutaneous, intramuscular, intravenous, and intraperitoneal routes.

BLV125 CAS:30392-41-7 ***HR: 3***
BITOLTEROL MESILATE
mf: $C_{29}H_{31}NO_5 \cdot CH_4O_3S$ mw: 557.71

SYNS: BITOLTEROL MESYLATE ◇ 4-(2-(tert-BUTYLAMINO)-1-HYDROXYETHYL)-o-PHENYLENE DI-p-TOLUATE MESILATE ◇ WIN 32784

TOXICITY DATA with REFERENCE
ivn-rat LD50:44 mg/kg NIIRDN 6,620,82
orl-mus LD50:4116 mg/kg IYKEDH 10,884,79
ivn-mus LD50:31400 μg/kg IYKEDH 10,884,79

SAFETY PROFILE: Poison by intravenous route. When heated to decomposition it emits toxic fumes of SO_x and NO_x.

BLV250 CAS:13394-86-0 ***HR: 3***
(m,o'-BITOLYL)-4-AMINE
mf: $C_{14}H_{15}N$ mw: 197.30

SYNS: 2',3-DIMETHYL-4-AMINOBIPHENYL ◇ 3,2'-DIMETHYL-4-AMINOBIPHENYL ◇ 3,2'-DIMETHYL-4-AMINODIPHENYL ◇ 3,2'-DIMETHYL-4-BIPHENYLAMINE ◇ 3,2'-DMAB

TOXICITY DATA with REFERENCE
mma-sat 10 μg/plate PNASA6 72,5135,75
dns-rat:lvr 10 μmol/L CALEDQ 4,69,78
cyt-mus-orl 50 mg/kg JJIND8 71,133,83
otr-ham:emb 100 μg/L NCIMAV 58,243,81
scu-rat TDLo:680 mg/kg/I:CAR ARPAAQ 86,475,68
par-rat TDLo:280 mg/kg/14W-I:ETA 23HZAR-,280,70
scu-ham TDLo:2300 mg/kg/37W-I:CAR,REP JNCIAM 48,1733,72
scu-ham TD:1095 mg/kg/37W-I:CAR CALEDQ 20,349,83
ipr-mus LD50:1130 mg/kg JJIND8 62,911,79

CONSENSUS REPORTS: EPA Genetic Toxicology Program.

SAFETY PROFILE: Suspected carcinogen with experimental carcinogenic and tumorigenic data. Moderately toxic by intraperitoneal route. Experimental reproductive effects. Mutation data reported. When heated to decomposition it emits toxic fumes of NO_x. See also AROMATIC AMINES.

BLV500 CAS:8013-76-1 ***HR: 3***
BITTER ALMOND OIL

PROP: Volatile oil from dried ripe kernels of bitter almonds or from other kernels containing amygdalin, such as apricots, cherries, plums, and especially peaches. Colorless liquid; strong almond odor. Bp: 179°, d: 1.045-1.070 @ 15°. Sltly sol in water; sol in fixed oils and propylene glycol; insol in glycerin.

SYNS: ALMOND OIL BITTER, FFPA (FCC) ◇ OIL, BITTER ALMOND

TOXICITY DATA with REFERENCE
skn-rbt 500 mg/24H MOD FCTXAV 17,705,79
orl-hmn LDLo:107 mg/kg FCTXAV 17,705,79
orl-rat LD50:960 mg/kg FCTXAV 17,705,79
skn-rbt LD50:1220 mg/kg FCTXAV 17,705,79

CONSENSUS REPORTS: Reported in EPA TSCA Inventory.

SAFETY PROFILE: A human poison by ingestion. Moderately toxic by skin contact. A skin irritant. When heated to decomposition it emits toxic fumes of CN^-.

BLV750 CAS:68916-04-1 ***HR: 1***
BITTER ORANGE OIL

PROP: Main constituent is d-limonene (FCTXAV 12,703,74). Pale yellow liquid, bitter taste. D: 0.842-0.848 @ 25°/25°. Very sltly sol in water; misc with abs alc; sol in 4 vols alc, in 1 vol glacial acetic acid. Keep well closed, cool, and protected from light.

TOXICITY DATA with REFERENCE
skn-mus 100 % MLD FCTXAV 12,703,74
skn-rbt 500 mg/24H MOD FCTXAV 17,509,74

SAFETY PROFILE: A skin irritant. See also d-LIMONENE. When heated to decomposition it emits acrid smoke and irritating fumes.

BLW250 CAS:8006-82-4 ***HR: 1***
BLACK PEPPER OIL

PROP: From steam distillation of dried fruit of *Piper nigrum L.* (Fam. *Piperaceae*). Main constituents include α- and β-pinene, β-caryophyllene, l-limonene, d-hydrocarveol, piperidine and piperrine (FCTXAV 16,637,78). A colorless to greenish liquid; odor and taste of pepper. Sol in fixed oils, mineral oil, propylene glycol; sltly sol in glycerin.

TOXICITY DATA with REFERENCE
skn-rbt 500 mg/24H MOD FCTXAV 16,637,78
dnr-bcs 20 mg/disc TOFOD5 8,91,85

CONSENSUS REPORTS: Reported in EPA TSCA Inventory.

SAFETY PROFILE: A moderate skin irritant. Mutation data reported. When heated to decomposition it emits acrid smoke and irritating fumes.

BLW500 ***HR: 3***
BLACK WIDOW SPIDER VENOM

SYN: LATRODECTUS M. MACTANS VENOM

TOXICITY DATA with REFERENCE
scu-mus LDLo:10 mg/kg SCNEBK 110,355,76
ivn-mus LDLo:5500 μg/kg SCNEBK 110,355,76

SAFETY PROFILE: Poison by subcutaneous and intravenous routes.

BLW750 CAS:21725-46-2 ***HR: 3***
BLADEX
mf: $C_9H_{13}ClN_6$ mw: 240.73

PROP: A white, crystalline material. Mp: 167°.

SYNS: BLADEX 80WP ◇ 2-CHLORO-4-(1-CYANO-1-METHYLETHYLAMINO)-6-ETHYLAMINO-1,3,5-TRIAZINE ◇ 2-CHLORO-4-ETHYLAMINO-6-(1-CYANO-1-METHYL)ETHYLAMINO-s-TRIAZINE ◇ 2-(4-CHLORO-6-ETHYLAMINO-s-TRIAZINE-2-YLAMINO)-2-METHYL-PROPIONITRILE ◇ 2-((4-CHLORO-6-(ETHYLAMINO)-1,3,5-TRIAZIN-2-YL)AMINO)-2-METHYL-PROPANENITRILE ◇ 2-((4-CHLORO-6-(ETHYLAMINO)-s-TRIAZIN-2-YL)AMINO)-2-METHYLPROPIONITRILE ◇ 2-(4-CHLORO-6-ETHYLAMINO-1,3,5-TRIAZINE-2-YLAMINO)-2-METHYLPROPIONITRILE ◇ CYANAZINE ◇ DW3418 ◇ FORTROL ◇ PAYZE ◇ SD 15418 ◇ WL 19805

TOXICITY DATA with REFERENCE
mma-sat 5170 μmol/L MUREAV 136,233,84
dlt-dmg-par 332 μmol/L JTEHD6 3,691,77
dlt-dmg-orl 100 ppm JTEHD6 3,691,77
sln-nsc 250 mg/L EVHPAZ 31,75,79
orl-rat TDLo:250 mg/kg (6-15D preg):TER TJADAB 25(2),59A,82
orl-rat LD50:149 mg/kg 85ARAE 2,132,77
skn-rat LD50:1200 mg/kg 28ZEAL 5,62,76
orl-mus LD50:380 mg/kg 28ZEAL 5,62,76
orl-rbt LD50:141 mg/kg 85DPAN -,-,71/76
orl-qal LD50:400 mg/kg PEMNDP 8,198,87
orl-dck LD50:750 mg/kg PSSCBG 5,153,74

CONSENSUS REPORTS: EPA Genetic Toxicology Program. Cyanide and its compounds are on the Community Right-To-Know List.

SAFETY PROFILE: Poison by ingestion. Moderately toxic by skin contact. An experimental teratogen. Mutation data reported. See also NITRILES. An herbicide. When heated to decomposition it emits very toxic fumes of Cl^-, NO_x, and CN^-.

BLX000 CAS:9084-06-4 ***HR: 2***
BLANCOL
mf: $(C_{10}H_8O_3S \cdot CH_{2O})_x \cdot xNa$

SYNS: BLANCOL DISPERSANT ◇ SODIUM SALT of SULFONATED NAPHTHALENEFORMALDEHYDECONDENSATE

TOXICITY DATA with REFERENCE
orl-rat LD50:3800 mg/kg FMCHA2 -,D44,80

CONSENSUS REPORTS: Reported in EPA TSCA Inventory.

SAFETY PROFILE: Moderately toxic by ingestion. See also ALDEHYDES and SULFONATES. When heated to decomposition it emits very toxic fumes of SO_x and Na_2O.

BLX250 CAS:63732-07-0 ***HR: 3***
BLASTICIDEN-S-LAURYLSULFONATE

TOXICITY DATA with REFERENCE
orl-rat LD50:39500 μg/kg GUCHAZ 6,48,73
scu-rat LD50:220 mg/kg GUCHAZ 6,48,73
orl-rbt LD50:48500 μg/kg GUCHAZ 6,48,73
orl-mam LD50:32 mg/kg GUCHAZ 6,48,73

SAFETY PROFILE: Poison by ingestion and subcutaneous routes. See also SULFONATES. When heated to decomposition it emits toxic fumes of SO_x.

BLX500 CAS:2079-00-7 ***HR: 3***
BLASTICIDIN S
mf: $C_{17}H_{26}N_8O_5$ mw: 422.51

PROP: From *Streptomyces griseochromogenes* (JANTAJ 11,1,58).

SYNS: BABS ◇ BLA-S ◇ BLASTICIDIN ◇ CYTOVIRIN ◇ TOA BLA-S

TOXICITY DATA with REFERENCE
orl-rat LD50:16 mg/kg GUCHAZ 6,48,73
skn-rat LD50:3100 mg/kg 28ZEAL 5,27,76
orl-mus LD50:38 mg/kg JANTAJ 30,1022,77
skn-mus LD50:220 mg/kg 28ZEAL 5,27,76
ivn-mus LD50:2820 μg/kg JAJAAA 11,1,58

SAFETY PROFILE: Poison by ingestion, skin contact, and intravenous routes. When heated to decomposition it emits toxic fumes of NO_x. See other blastomycin entries.

BLX750 CAS:522-70-3 ***HR: 3***
BLASTOMYCIN
mf: $C_{26}H_{36}N_2O_9$ mw: 520.64

SYNS: ANTIMYCIN A3 ◇ BLASTMYCIN

TOXICITY DATA with REFERENCE
ipr-mus LD50:1800 μg/kg JAJAAA 10,39,57
scu-mus LD50:1600 μg/kg JAJAAA 10,39,57

SAFETY PROFILE: Poison by intraperitoneal and subcutaneous routes. When heated to decomposition it emits toxic fumes of NO_x.

BLY000 CAS:11056-06-7 ***HR: 3***
BLEOMYCIN

PROP: A group of related glycopeptide antibiotics isolated from *Streptomyces verticillus*.

SYNS: BLENOXANE ◇ BLEO ◇ BLEOCIN ◇ BLM

TOXICITY DATA with REFERENCE
eye-rbt 1 mg MLD JJANAX 31,859,78
mnt-hmn:lym 1250 ug/L MUREAV 130,395,84
dnd-hmn:fbr 10 mg/L ENMUDM 7,267,84
dns-hmn:hla 110 μmol/L CRNGDP 7,77,86
cyt-mus:oth 4 nmol/L IPPABX 20,1,84
sce-ham-ipr 7500 μg/kg CNREA8 43,577,83
ipr-mus TDLo:200 mg/kg (1D male):REP CNREA8 39,3575,79
ivn-hmn LDLo:351 mg/kg:PUL AJCPAI 58,501,72
ims-hmn LDLo:418 mg/kg:PUL AJCPAI 58,501,72
ipr-rat LD50:168 mg/kg 40WDA5 -,311,78
ipr-mus LD50:35 mg/kg JANTAJ 37,239,84
ivn-mus LD50:53 mg/kg JANTAJ 31,667,78

CONSENSUS REPORTS: IARC Cancer Review: Group 2B IMEMDT 7,134,87, Human Inadequate Evidence IMEMDT 26,97,81. EPA Genetic Toxicology Program.

SAFETY PROFILE: Suspected carcinogen. A human poison by intravenous route; moderately toxic to humans by intramuscular route. Poison experimentally by intravenous and intraperitoneal routes. Human systemic effects by ingestion and intramuscular routes: dyspnea and fibrosing alveolitis (lung). Experimental reproductive effects. An eye irritant. Human mutation data reported. When heated to decomposition it emits toxic fumes of NO_x. See other bleomycin entries.

BLY250 CAS:11116-31-7 ***HR: 3***
BLEOMYCIN A2

SYN: ZHENGGUANGMYCIN A2 (CHINESE)

TOXICITY DATA with REFERENCE
mmo-sat 5 μg/plate MUREAV 117,9,83
dnd-esc 50 μg/L CNREA8 38,3900,78
dnd-rat:ast 13 mg/L PLCHB4 7,177,75
dnd-rat:lng 100 μmol/L CBINA8 45,65,83
dnd-mam:lym 100 μmol/L JPETAB 221,152,82
ivn-man TDLo:2143 μg/kg PUL CNREA8 36,1267,76
ipr-mus LD50:130 mg/kg YHHPAL 14,83,79
ivn-mus LD50:100 mg/kg YHHPAL 14,83,79

SAFETY PROFILE: Poison by intravenous and intraperitoneal routes. Noted for adverse pulmonary effects in humans. Mutation data reported. When heated to decomposition it emits toxic fumes of NO_x. See other bleomycin entries.

BLY500 CAS:11116-32-8 ***HR: 3***
BLEOMYCIN A5
mf: $C_{57}H_{86}N_{18}O_{21}S_2$ mw: 1423.73

SYNS: N^1-3-((4-AMINOBUTYL)AMINO)PROPYL)BLEOMYCINAMIDE ◇ BLEOMYCETIN ◇ PINGYANGMYCIN (CHINESE) ◇ ZHENGGUANGMYCIN A5 (CHINESE)

TOXICITY DATA with REFERENCE
cyt-ham:ovr 400 μg/L HKXUDL 3,78,83
sce-ham:ovr 400 μg/L HKXUDL 3,78,83
unr-hmn TDLo:192 mg/kg:GIT,PUL,MET ANTBAL 28(8),632,83
ipr-rat LD50:117 mg/kg ANTBAL 24(5),363,79
ivn-rat LD50:100 mg/kg ANTBAL 24(5),363,79
ims-rat LD50:102 mg/kg ANTBAL 24(5),363,79
orl-mus LD50:840 mg/kg ANTBAL 24(5),363,79
ipr-mus LD50:88 mg/kg ANTBAL 24(5),363,79
scu-mus LD50:77 mg/kg ANTBAL 24(5),363,79
ivn-mus LD50:61500 μg/kg ANTBAL 24(5),363,79

SAFETY PROFILE: Poison by intraperitoneal, intravenous, intramuscular, and subcutaneous routes. Moderately toxic by ingestion. Human systemic effects by an unspecified route: nausea or vomiting, dyspnea, and fever. Mutation data reported. When heated to decomposition it emits very toxic fumes of SO_x and NO_x. See other bleomycin entries.

BLY750 ***HR: 3***
BLEOMYCIN A COMPLEX

PROP: Antibiotics produced by a strain of *Streptomyces verticillus* (JAJAAA 20,15,67).

TOXICITY DATA with REFERENCE
mmo-smc 10 mg/L/30M MUREAV 58,107,78
mrc-smc 30 mg/L/15M MUREAV 58,41,78
cyt-hmn-par 430 μg/kg MUREAV 56,341,78
cyt-hmn:lym 10 mg/L MUREAV 56,341,78
msc-ham:fbr 1 mg/L/24H MUREAV 40,325,76
ipr-mus LDLo:125 mg/kg JAJAAA 20,15,67
scu-mus LDLo:125 mg/kg JAJAAA 20,15,67
ivn-mus LDLo:125 mg/kg CANCAR 20,891,67
ivn-rbt LDLo:200 mg/kg JAJAAA 20,15,67

SAFETY PROFILE: Poison by intraperitoneal, subcutaneous, and intravenous routes. Human mutation data reported. When heated to decomposition it emits toxic fumes of NO_x. See other bleomycin entries.

BLY760 CAS:9060-10-0 ***HR: 3***
BLEOMYCIN B2
mf: $C_{55}H_{84}N_{20}O_{21}S_2$ mw: 1425.71

SYNS: DEHYDROPHELOMYCIN D1 ◇ PHLEOMYCIN D2

TOXICITY DATA with REFERENCE
mmo-sat 1 μg/plate MRUEAV 117,9,83
dnd-omi 100 mg/L JANTAJ 28,537,75
dnd-omi 10700 pmol/L CNREA8 40,4173,80
itr-mus LDLo:14 μg/kg TXAPA9 56,326,80

SAFETY PROFILE: Poison by intratracheal route. Mutation data reported. When heated to decomposition it emits toxic fumes of SO_x and NO_x. See other bleomycin entries.

BLY770 CAS:68247-85-8 ***HR: 3***
BLEOMYCIN PEP
mf: $C_{61}H_{88}N_{18}O_{21}S_2$ mw: 1473.79

SYNS: BLM-PEP ◇ NK 631 ◇ PEP ◇ PEPLEOMYCIN ◇ PEPLOMYCIN

TOXICITY DATA with REFERENCE
eye-rbt 1 mg MLD JJANAX 31,859,78
mmo-sat 1 μg/plate TAKHAA 44,96,85
pic-esc 56 ng/plate MUREAV 88,325,81
dnr-bcs 800 ng/plate TAKHAA 44,96,85
dnd-ham:ovr 25 mg/L JANTAJ 38,1257,85
ipr-rat LD50:155 mg/kg 40WDA5 -,311,78
scu-rat LD50:234 mg/kg NIIRDN 6,762,82
ivn-rat LD50:215 mg/kg NIIRDN 6,762,82
ipr-mus LD50:25 mg/kg JANTAJ 37,239,84
scu-mus LD50:80 mg/kg NIIRDN 6,762,82
ivn-mus LD50:45 mg/kg NIIRDN 6,762,82
itr-mus LDLo:15 μg/kg TXAPA9 56,326,80

SAFETY PROFILE: Poison by subcutaneous, intravenous, intraperitoneal, and intratracheal routes. An eye irritant. Mutation data reported. When heated to decomposition it emits toxic fumes of SO_x and NO_x. See other bleomycin entries.

BLY780 CAS:9041-93-4 ***HR: 3***
BLEOMYCIN SULFATE

SYNS: BLENOXANE ◇ BLEOMYCIN, SULFATE (salt) (9CI) ◇ BLEXANE

TOXICITY DATA with REFERENCE
dnr-esc 20 μmol/L MUREAV 164,19,86
hma-mus/esc 10 mg/kg MUREAV 164,19,86
ipr-rat TDLo:8700 μg/kg (female 15-22D post):REP NTIS** PB261-972
ipr-rat TDLo:8 mg/kg (female 6-9D post):TER NTIS** PB261-972
scu-rat TDLo:14 mg/kg/68W-I:CAR ONCOBS 41,114,84
par-rat TDLo:18 mg/kg/52W-I:CAR PAACA3 24,96,83
par-wmn TDLo:20 μg/kg:PUL,SKN ARHEAW 28,459,85
ipr-rat LD50:240 mg/kg IYKEDH 7,108,76
scu-rat LD50:86 mg/kg JJANAX 29,894,76
ipr-mus LD50:210 mg/kg YAKUD5 17,455,75
scu-mus LD50:103 mg/kg JJANAX 29,894,76

CONSENSUS REPORTS: IARC Cancer Review: Human Inadequate Evidence IMEMDT 26,97,81. EPA Genetic Toxicology Program.

SAFETY PROFILE: Poison by subcutaneous and intraperitoneal routes. Human systemic effects: cyanosis, allergic dermatitis. Questionable carcinogen with experimental carcinogenic data. An experimental teratogen. Other experimental reproductive effects. Mutation data reported. When heated to decomposition it emits toxic fumes of SO_x. See other bleomycin entries.

BMA000 CAS:2519-30-4 ***HR: 2***
BLUE BLACK BN
mf: $C_{28}H_{21}N_5O_{14}S_4{\cdot}4Na$ mw: 871.74

SYNS: 1743 BLACK ◇ BLACK PN ◇ BRILLIANT ACID BLACK BNA EXPORT ◇ BRILLIANT ACID BLACK BN EXTRA PURE A ◇ BRILLIANT BLACK ◇ BRILLIANT BLACK A ◇ BRILLIANT BLACK BN ◇ BRILLIANT BLACK NAF ◇ BRILLIANT BLACK N.FQ ◇ BRILLIANTSCHWARZ BN (GERMAN) ◇ CERTICOL BLACK PNW ◇ C.I. 28440 ◇ C.I. FOOD BLACK 1, TETRASODIUM SALT ◇ CILEFA BLACK B ◇ E 151 ◇ EDICOL SUPRA BLACK BN ◇ HEXACOL BLACK PN ◇ MELAN BLACK ◇ NOIR BRILLANT BN (FRENCH) ◇ L-SCHWARZ 1 ◇ XYLENE BLACK F

TOXICITY DATA with REFERENCE
ipr-rat LD50:900 mg/kg FCTXAV 5,171,67
ivn-rat LD50:25000 mg/kg APFRAD 15,402,57

orl-mus LD50:1100 mg/kg FCTXAV 5,171,67
ipr-mus LD50:500 mg/kg FCTXAV 5,171,67

SAFETY PROFILE: Moderately toxic by ingestion, intravenous, and intraperitoneal routes. When heated to decomposition it emits very toxic fumes of NO_x, Na_2O, and SO_x.

BMA125 CAS:7210-92-6 ***HR: 3***
BLUECAIN
mf: $C_{15}H_{22}N_2O_3 \cdot ClH$ mw: 314.85

SYNS: BAJKAIN ◇ BAYCAIN ◇ BAYCAINE ◇ BAYCALNE ◇ 2-(((DIETHYLAMINO)ACETYL)AMINO)-3-METHYL-BENZOIC ACID METHYL ESTER, MONOHYDROCHLORIDE ◇ TOLYCAINE HYDROCHLORIDE

TOXICITY DATA with REFERENCE
ivn-rat LD50:44 mg/kg NIIRDN 6,570,82
scu-mus LD50:450 mg/kg NIIRDN 6,570,82
ivn-mus LD50:60 mg/kg NIIRDN 6,570,82
ivn-rbt LD50:40 mg/kg NIIRDN 6,570,82

SAFETY PROFILE: Poison by intravenous route. Moderately toxic by other routes. When heated to decomposition it emits toxic fumes of NO_x and HCl.

BMA150 ***HR: 2***
BLUE COHOSH

PROP: An erect herb, 1 to 3 feet tall, with clusters of small, yellow-green or purple-green flowers. It produces small, blue berries. It grows wild in damp woods in the region bounded by Alabama, Missouri, Manitoba and New Brunswick.

SYNS: BLUEBERRY ROOT ◇ BLUE GINSENG ◇ CAULOPHYLLUM THALICTROIDES ◇ PAPOOSE ROOT ◇ SQUAW ROOT ◇ YELLOW GINSENG

SAFETY PROFILE: The berries and roots contain the poison N-methylcytisine (an alkaloid similar to nicotine) and saponins. The bitter taste usually limits ingestion which could cause inflammation of the stomach and intestines. See also SAPONIN.

BMA500 ***HR: 2***
BOILER COMPOUND, (liquid)

DOT Classification: Corrosive Material; Label: Corrosive.

SAFETY PROFILE: A caustic solution. See also SODIUM HYDROXIDE.

BMA600 CAS:8022-81-9 ***HR: 3***
BOLDO LEAF OIL

TOXICITY DATA with REFERENCE
skn-rbt 500 mg/24H MOD FCTOD7 20(Suppl),643,82
orl-rat LD50:130 mg/kg FCTOD7 20(Suppl),643,82
ipr-mus LD50:420 mg/kg JPBEAJ 32,13,77
skn-rbt LD50:625 mg/kg FCTOD7 20(Suppl),643,82

CONSENSUS REPORTS: Reported in EPA TSCA Inventory.

SAFETY PROFILE: Poison by ingestion. Moderately toxic by skin contact and intraperitoneal routes. A skin irritant.

BMA625 CAS:21535-47-7 ***HR: 3***
BOLVIDON
mf: $C_{18}H_{20}N_2 \cdot ClH$ mw: 300.86

SYNS: ATHYMIL ◇ GB 94 ◇ 1,2,3,4,10,14b-HEXAHYDRO-2-METHYLDIBENZO(c,f)PYRAZINO(1,2-a)AZEPINEHYDROCHLORIDE ◇ MIANSERINE HYDROCHLORIDE ◇ MIANSERIN HYDROCHLORIDE ◇ NORVAL ◇ ORG GB 94 ◇ TOLUON ◇ TOLVIN ◇ TOLVON

TOXICITY DATA with REFERENCE
orl-rat TDLo:5 mg/kg (1D pre):REP DCTODJ 7,41,84
orl-wmn TDLo:28 mg/kg/5W:CNS,CVS,PUL BMJOAE 284,1912,82
orl-rat LD50:780 mg/kg PBPSDY 3,56,81
ipr-rat LD50:262 mg/kg SKKNAJ 31,112,79
ivn-rat LD50:31850 μg/kg IYKEDH 14,484,83
orl-mus LD50:224 mg/kg SKKNAJ 31,112,79
ipr-mus LD50:117 mg/kg IYKEDH 14,484,83
scu-mus LD50:118 mg/kg SKKNAJ 31,112,79
ivn-mus LD50:31 mg/kg PBPSDY 3,56,81

SAFETY PROFILE: Poison by ingestion, subcutaneous, intravenous, and intraperitoneal routes. Experimental reproductive effects. Human systemic effects by ingestion: hallucinations and distorted perceptions, change in heart rate, and unspecified respiratory system effects. Experimental reproductive effects. A serotonin inhibitor and antihistamine. When heated to decomposition it emits toxic fumes of NO_x and HCl.

BMA650 CAS:24543-59-7 ***HR: D***
BOMT
mf: $C_{19}H_{29}BrO_3$ mw: 385.39

SYNS: 6-α-BROMO-17-β-HYDROXY-17-α-METHYL-4-OXA-5-α-ANDROSTAN-3-ONE ◇ 6-α-BROMO-17-β-METHYL-4-OXA-5-α-ANDROSTAN-3-ONE ◇ RO 7-2340

TOXICITY DATA with REFERENCE
orl-rat TDLo:700 mg/kg (14-20D preg):REP AAMMAU 63,239,74
scu-rat TDLo:1750 mg/kg (female 14-20D post):TER AAMMAU 64,27,75

SAFETY PROFILE: An experimental teratogen. Other experimental reproductive effects. When heated to decomposition it emits toxic fumes of Br^-.

BMA750 CAS:8001-85-2 ***HR: 2***
BONE OIL

PROP: Product of destructive distillation of bones in preparation of bone charcoal containing nitrogenous compounds such as pyridine, aniline, methylamine, and pyrrole (27ZTAP 3,25,69).

SYNS: ANIMAL OIL ◇ DIPPEL'S OIL ◇ OIL of HARTSHORN

TOXICITY DATA with REFERENCE
orl-rat LDLo:800 mg/kg 27ZTAP 3,25,69

CONSENSUS REPORTS: Reported in EPA TSCA Inventory.

SAFETY PROFILE: Moderately toxic by ingestion. When heated to decomposition it emits toxic fumes of NO_x.

BMB000 CAS:1098-97-1 ***HR: 3***
BONIFEN
mf: $C_{16}H_{20}N_2O_4S_2$•2ClH•H_2O mw: 459.44

SYNS: 3,3'-DITHIOBIS(METHYLENE)BIS(5-HYDROXY-6-METHYL-4-PYRIDINEMETHANOL) DIHYDROCHLORIDE ◇ 3,3'-DITHIODIMETHYLENEBIS(5-HYDROXY-6-METHYL-4-PYRIDINEMETHANOL)DIHYDROCHLORIDE HYDRATE ◇ 5,5'-DITHIODIMETHYLENEBIS-(2-METHYL-3-HYDROXY-4-HYDROXYMETHYLPYRIDINE)DIHYDROHLORIDE HYDRATE ◇ EPOCAN ◇ PYRIDOXIN-5'-DISULFID DIHYDROCHLORID HYDRAT (GERMAN) ◇ PYRITHIOXIN

TOXICITY DATA with REFERENCE
orl-rat LD50:6000 mg/kg ARZNAD 11,922,61
scu-rat LD50:3000 mg/kg ARZNAD 11,922,61
ivn-rat LD50:500 mg/kg ARZNAD 11,922,61
orl-mus LD50:5786 mg/kg TMPBAX 54,156,78
ipr-mus LD50:790 mg/kg ARZNAD 29,479,79
scu-mus LD50:3170 mg/kg ARZNAD 11,922,61
ivn-mus LD50:221 mg/kg TMPBAX 54,156,78
ivn-cat LD50:124 mg/kg ARZNAD 11,922,61

SAFETY PROFILE: Poison by intravenous route. Moderately toxic by subcutaneous and intraperitoneal routes. Mildly toxic by ingestion. When heated to decomposition it emits very toxic fumes of HCl, SO_x, and NO_x.

BMB125 ***HR: 3***
BONNECOR
mf: $C_{21}H_{25}N_3O_3$•ClH mw: 403.91

SYNS: AWD 19-166 ◇ 3-CARBETHOXYAMINO-5-DIMETHYLAMINOACETYL-10,11-DIHYDRODIBENZ(b,f)AZEPINEHYDROCHLORIDE ◇ GS 015

TOXICITY DATA with REFERENCE
orl-rat LD50:78 mg/kg PHARAT 40,871,85
ivn-rat LD50:10900 μg/kg PHARAT 40,871,85
orl-mus LD50:48 mg/kg PHARAT 40,871,85
ivn-mus LD50:5400 μg/kg PHARAT 40,871,85

SAFETY PROFILE: Poison by ingestion and intravenous routes. When heated to decomposition it emits toxic fumes of NO_x and HCl. See also CARBAMATES and ESTERS.

BMB150 CAS:17596-45-1 ***HR: 3***
BORANE-AMMONIA
mf: BH_3-NH_3 mw: 30.86

SAFETY PROFILE: Complex may explode on rapid heating. When heated to decomposition it emits toxic fumes of NH_3. See also BORANES, BORON COMPOUNDS, and AMMONIA.

BMB250 CAS:75-22-9 ***HR: 3***
BORANE, COMPOUND with TRIMETHYLAMINE (1:1)
mf: C_3H_9N•BH_3 mw: 72.97

SYNS: BORANE, COMPOUND with N,N-DIMETHYLMETHANAMINE (1:1) ◇ TMAB ◇ TRIMETHYLAMINE BORANE ◇ TRIMETHYLAMINE, COMPOUND with BORANE (1:1)

TOXICITY DATA with REFERENCE
ipr-rat LD50:175 mg/kg JOCMA7 1,46,59
ipr-mus LD50:740 mg/kg JPMSAE 19,1025,80

CONSENSUS REPORTS: Reported in EPA TSCA Inventory.

SAFETY PROFILE: Poison by intraperitoneal route. See also BORANES. When heated to decomposition it emits toxic fumes of NO_x.

BMB260 ***HR: 3***
BORANE-HYDRAZINE
mf: BH_3-N_2H_4 mw: 45.88

SAFETY PROFILE: Complex is highly flammable and a shock-sensitive explosive. Upon decomposition it emits toxic fumes of NO_x. See also BORANES, BORON COMPOUNDS, and HYDRAZINE.

BMB270 ***HR: 3***
BORANE-PHOSPHORUS TRIFLUORIDE
mf: BH_3-PF_3 mw: 101.80

SAFETY PROFILE: An unstable explosive complex which ignites spontaneously upon exposure to air. When heated to decomposition it emits toxic fumes of F^- and PO_x. See also BORANES, BORON COMPOUNDS, and PHOSPHORUS TRIFLUORIDE.

BMB280 ***HR: 3***
BORANES

PROP: A series of boron hydrides (BH_3, B_2H_6,..., $B_{20}H_{26}$).

SAFETY PROFILE: Generally poisons. Most are unstable and react with water to produce explosive hydrogen gas. Many react violently with air. Many organoboranes are used as reducing agents. Haloboranes are highly reactive. Potentially explosive reaction with carbon tetrachloride.

BMB300 CAS:14044-65-6 ***HR: 3***
BORANE-TETRAHYDROFURAN
mf: BH_3-C_4H_8O mw: 85.93

SAFETY PROFILE: The complex is an unstable explosive in tetrahydrofuran at room temperature. When heated to decomposition it emits acrid smoke and fumes. See also BORANES, BORON COMPOUNDS, and TETRAHYDROFURAN.

BMB325 ***HR: 2***
BORASSUS FLABELLIFER Linn., extract

PROP: Indian plant belonging to the family *Lalniae* (IJEBA6 16,228,78).

TOXICITY DATA with REFERENCE
mmo-sat 530 μg/plate CALEDQ 26,113,85
mma-sat 1590 μg/plate CALEDQ 26,113,85
mmo-esc 530 μg/plate CALEDQ 26,113,85
ipr-rat LD50:850 mg/kg IJEBA6 16,228,78

SAFETY PROFILE: Moderately toxic by intraperitoneal route. Mutation data reported.

BMB500 CAS:6569-51-3 ***HR: 3***
BORAZINE
mf: $B_3H_6N_3$ mw: 80.5

SYN: BORAZOLE

PROP: Colorless liquid. Mp: −58°, bp: 53°, d: 0.824 @ 0°.

SAFETY PROFILE: A powerful irritant to skin, eyes, and mucous membranes. May explode spontaneously when stored in the light. Reacts with water to form toxic and flammable boron hydrides. A dangerous fire hazard. When heated to decomposition it emits toxic fumes of NO_x. See also BORON COMPOUNDS.

BMB750 ***HR: 3***
BORDEAUX ARSENITE
DOT: NA 2759

SYN: BORDEAU ARSENITE, liquid or solid (DOT)

CONSENSUS REPORTS: Arsenic and its compounds, as well as copper and its compounds, are on the Community Right-To-Know List.

OSHA PEL: TWA 0.01 mg(As)/m^3
NIOSH REL: CL 0.002 mg(As)/m^3/15M
DOT Classification: Poison B; Label: Poison, liquid or solid.

SAFETY PROFILE: A poison. See also ARSENIC COMPOUNDS and COPPER COMPOUNDS. When heated to decomposition it emits toxic fumes of As.

BMC000 CAS:10043-35-3 ***HR: 3***
BORIC ACID
mf: BH_3O_3 mw: 61.84

PROP: White crystals or powder. Mp: 185° (decomp), $-1.5H_2O$ @ 300°, d: 1.435 @ 15°.

SYNS: BORACIC ACID ◇ BOROFAX ◇ BORSAURE (GERMAN) ◇ NCI-C56417 ◇ ORTHOBORIC ACID ◇ THREE ELEPHANT

TOXICITY DATA with REFERENCE
skn-hmn 15 mg/3D-I MLD 85DKA8 -,127,77
mmo-esc 17000 ppm/24H AMNTA4 85,119,51
spm-rat-orl 6 mg/kg EVHPAZ 13,69,76
orl-rat TDLo:45 g/kg (90D male):REP TXAPA9 23,351,72
orl-cld TDLo:500 mg/kg:GIT JTCTDW 24,269,86
orl-wmn LDLo:200 mg/kg LANCAO 2,162,17
orl-inf TDLo:800 mg/kg/4W-I ADCHAK 58,737,83
orl-inf LDLo:934 mg/kg JAMAAP 90,382,28
skn-inf LDLo:1200 mg/kg JAMAAP 129,332,45
skn-chd LDLo:4 g/kg/4D MMWOAU 52,763,05
skn-man LDLo:2430 mg/kg JAMAAP 128,266,45
skn-cld LDLo:1500 mg/kg QJPPAL 6,714,33
scu-inf LDLo:1100 mg/kg QJPPAL 6,714,33
unr-man TDLo:170 mg/kg:GIT RTPCAT 1,472,29
unr-man LDLo:147 mg/kg 85DCAI 2,73,70
orl-rat LD50:2660 mg/kg JAMAAP 128,266,45
ihl-rat LCLo:28 mg/m^3/4H 85GMAT -,27,82
scu-rat LD50:1400 mg/kg 14KTAK -,694,64
ivn-rat LD50:1330 mg/kg MDSR** No.2,50
orl-mus LD50:3450 mg/kg JAMAAP 128,266,45
ipr-mus LDLo:800 mg/kg 14KTAK -,693,64
scu-mus LD50:1740 mg/kg JAMAAP 128,266,45
ivn-mus LD50:1240 mg/kg 14KTAK -,693,64
scu-dog LDLo:1000 mg/kg JAMAAP 128,266,45
par-dog LDLo:1 g/kg RTPCAT 1,472,29

CONSENSUS REPORTS: Reported in EPA TSCA Inventory.

SAFETY PROFILE: A human poison by ingestion and possibly other routes. Moderately toxic by skin contact and subcutaneous routes in humans. Poison experimentally by inhalation and subcutaneous routes. Moderately toxic experimentally by intraperitoneal and intravenous routes. Human systemic effects by an unspecified route: wakefulness, anorexia, nausea and vomiting. Ingestion or absorption by other routes may also cause diarrhea, abdominal cramps, erythematous lesions on skin and mucous membranes, circulatory collapse, tachycardia, cyanosis, delirium, convulsions, and coma. Death has occurred from ingestion of less than 5 grams in infants, and from 5 to 20 grams in adults. Chronic exposure may result in borism (dry skin, eruptions, and gastrointestinal disturbances). Experimental reproductive effects. Mutation data reported. A human skin irritant. See also BORON COMPOUNDS. Incompatible with K; $(CH_3CO)_2O$.

BMC250 CAS:34099-73-5 ***HR: 3***
BORIC ACID, ETHYL ESTER
DOT: UN 1176
mf: $C_2H_7BO_3$ mw: 89.90

PROP: Colorless liquid, mild odor, decomp in water. Bp: 120°, flash p: 52°F (CC), d: 0.864 @ 26.5°, vap d: 5.04.

SYN: ETHYL BORATE (DOT)

TOXICITY DATA with REFERENCE
eye-rbt 5 mg SEV AJOPAA 29,1363,46

DOT Classification: Flammable Liquid; Label: Flammable Liquid.

SAFETY PROFILE: A severe eye irritant. See also BORON COMPOUNDS and ESTERS. Dangerous fire hazard when exposed to heat or flame; will react with water or steam to produce flammable vapors. Incompatible with oxidizers, heat, and open flame. To fight fire, use CO_2, dry chemical.

BMC500 CAS:5337-42-8 ***HR: 1***
BORIC ACID, TRIOLEYL ESTER
mf: $C_{54}H_{108}BO_3$ mw: 816.43

SYN: TRIOLEYL BORATE

TOXICITY DATA with REFERENCE
eye-rbt 100 mg MLD 14KTAK -,693,64
orl-mus LD50:6200 mg/kg 14KTAK -,693,64

SAFETY PROFILE: Mildly toxic by ingestion. An eye irritant. When heated to decomposition it emits acrid smoke and irritating fumes. See also ESTERS and BORON COMPOUNDS.

BMC750 CAS:5337-37-1 ***HR: 2***
BORIC ACID, TRIS(4-METHYL-2-PENTYL) ESTER
mf: $C_{18}H_{39}BO_3$ mw: 314.38

SYN: TRI(METHYLISOBUTYLCARBINYL)BORATE

TOXICITY DATA with REFERENCE
eye-rbt 100 mg SEV 14KTAK -,706,64
orl-mus LD50:1320 mg/kg USBCC*

SAFETY PROFILE: Moderately toxic by ingestion. A severe eye irritant. See also ESTERS and BORON COMPOUNDS. When heated to decomposition it emits acrid smoke and irritating fumes.

BMD000 CAS:507-70-0 ***HR: 3***
BORNEOL
DOT: UN 1312
mf: $C_{10}H_{18}O$ mw: 154.28

PROP: Hexagonal crystals, peppery odor and burning taste. Mp: 208°, bp: 212°, flash p: 150°F, d: 1.01 @ 20°/4°, vap d: 5.31.

SYNS: BAROS CAMPHOR ◇ BHIMSAIM CAMPHOR ◇ BORNEO CAMPHOR ◇ trans-BORNEOL ◇ BORNEOL (DOT) ◇ BORNYL ALCOHOL ◇ 2-CAMPHANOL ◇ DRYOBALANOPS CAMPHOR ◇ 2-HYDROXYCAMPHANE ◇ MALAYAN CAMPHOR ◇ SUMATRA CAMPHOR ◇ endo-1,7,7-TRIMETHYL-BICYCLO(2.2.1)HEPTAN-2-OL

TOXICITY DATA with REFERENCE
cyt-smc 1 mmol/tube HEREAY 33,457,47
orl-rat LD50:500 mg/kg FRXXBL #2448856
orl-mus LD50:1059 mg/kg SHGKA3 75,934,75
orl-rbt LDLo:2000 mg/kg AEXPBL 17,363,1883

CONSENSUS REPORTS: Reported in EPA TSCA Inventory.

DOT Classification: Flammable Solid; Label: None.

SAFETY PROFILE: Moderately toxic by ingestion. Mutation data reported. A mild irritant. Flammable when exposed to heat or flame; can react with oxidizing materials. To fight fire, use water, CO_2, water spray, dry chemical. When heated to decomposition it emits acrid smoke and fumes.

BMD100 CAS:76-49-3 ***HR: 1***
BORNYL ACETATE
mf: $C_{12}H_{20}O_2$ mw: 196.29

PROP: Colorless liquid or white crystalline solid; sweet, piney odor. D: 0.981-0.985, refr index: 1.462. flash p: 192°F. Sol in alc, fixed oils; sltly sol in water; insol in glycerin, propylene glycol @ 226°.

SYNS: l-BORNYL ACETATE ◇ FEMA No. 2159

SAFETY PROFILE: Combustible liquid. When heated to decomposition it emits acrid smoke and irritating fumes.

BMD250 CAS:40283-68-9 ***HR: 3***
S-((N-BORNYLAMIDIN)METHYL) HYDROGEN THIOSULFATE
mf: $C_{12}H_{22}N_2O_3S_2$ mw: 306.48

TOXICITY DATA with REFERENCE
orl-mus LD50:225 mg/kg JMCMAR 15,1313,72
ipr-mus LD50:30 mg/kg JMCMAR 15,1313,72

SAFETY PROFILE: Poison by ingestion and intraperitoneal routes. See also THIOSULFATES. When heated to decomposition it emits very toxic fumes of NO_x and SO_x.

BMD500 CAS:7440-42-8 ***HR: 3***
BORON
af: B aw: 10.81

PROP: Monoclinic crystals, yellow or brown amorphous powder. Mp: 2300°, bp: 2550°, d: 3.33 @ 20°.

TOXICITY DATA with REFERENCE
orl-mus LD50:2000 mg/kg PCOC** -,498,66
orl-mam LD50:300 mg/kg GISAAA 40(9),49,75

CONSENSUS REPORTS: Reported in EPA TSCA Inventory.

SAFETY PROFILE: A poison by ingestion. See also BORON COMPOUNDS. A relatively inert metal except in the form of powder or when exposed to highly oxidizing agents. Flammable in the form of dust when exposed to air or by chemical reaction. An explosion hazard in the form of dust which ignites on contact with air. Reacts explosively when ground with lead fluoride or silver fluoride. Ignites in contact with gaseous chlorine or fluorine at room temperature. Incompatible with NH_3, Br_2, BrF_3, Cs_2C_2, Cl_2, CuO, HIO_3, PbO_2, HNO_3, NO, NOF, N_2O, $KClO_3$, KNO_3, Rb_2C_2, S, BrF_5, IF_5, metal fluorides, inter halogens, nitryl fluoride (FNO_2), OF_2, KNO_2, NO_x, Na_2O_2, PbO, air. See also POWDERED METALS.

BMD750 ***HR: 3***
BORON AZIDE DICHLORIDE
mf: BCl_2N_3 mw: 123.74

SAFETY PROFILE: Crust of sublimed compound explodes when crushed by spatula or upon removing solvent. When heated to decomposition it emits toxic fumes of Cl^- and NO_x. See also AZIDES, BORON COMPOUNDS, and CHLORIDES.

BMD825 CAS:68533-38-0 ***HR: 3***
BORON AZIDE DIIODIDE
mf: BI_2N_3 mw: 306.64

SAFETY PROFILE: Explodes on contact with water. When heated to decomposition it emits toxic fumes of I^- and NO_x. See also BORON COMPOUNDS, AZIDES, and IODIDES.

BME250 CAS:14355-21-6 ***HR: 3***
BORON BROMIDE DIIODIDE
mf: $BBrI_2$ mw: 344.53

PROP: Colorless liquid. Bp: 180°.

SAFETY PROFILE: Dangerous. Violent reaction with water. When heated to decomposition it emits toxic fumes of Br^- and I^-. See BORON COMPOUNDS, BROMIDES, and IODIDES.

BME500 ***HR: 3***
BORON COMPOUNDS

SAFETY PROFILE: Very toxic and therefore considered an industrial poison. Used in medicine as sodium borate, boric acid, or borax, which is a common cleanser. Fatal poisoning of children has been caused by the accidental substitution of boric acid for powdered milk. The medical literature reveals instances of accidental poisoning due to boric acid; ingestion of borates or boric acid; and presumably absorption of boric acid from wounds and burns. The fatal dose of orally ingested boric acid for an adult is somewhat greater than 15 to 20 grams and for an infant, 5 to 6 grams. Boron is one of a group of elements, such as Pb, Mn, As, which affects the central nervous system. Boron poisoning causes depression of the circulation, persistent vomiting and diarrhea, followed by profound shock and coma. The temperature becomes subnormal and a scarletinaform rash may cover the entire body. Containers of boric acid should be plainly labeled and should differ radically from those which contain powdered milk, particularly in institutions such as hospitals.

BME750 ***HR: 3***
BORON DIBROMIDE IODIDE
mf: BBr_2I mw: 297.53

PROP: Colorless liquid. Bp: 125°, vap d: 10.3.

SAFETY PROFILE: Reaction with water or steam produces toxic and corrosive fumes. See BORON COMPOUNDS, BROMIDES, and IODIDES.

BMG000 CAS:1303-86-2 ***HR: 2***
BORON OXIDE
mf: B_2O_3 mw: 69.62

PROP: Vitreous, colorless crystals. Mp: 450° (approx), bp: 1860°, d: 2.46.

SYNS: BORIC ANHYDRIDE ◇ BORON SESQUIOXIDE ◇ BORON TRIOXIDE ◇ FUSED BORIC ACID

TOXICITY DATA with REFERENCE
skn-rbt 1 g AIHAAP 20,284,59
eye-rbt 50 mg AIHAAP 20,284,59
orl-mus LD50:3163 mg/kg 85GMAT -,27,82
ipr-mus LD50:1868 mg/kg 85GMAT -,27,82

CONSENSUS REPORTS: Reported in EPA TSCA Inventory.

OSHA PEL: (Transitional: Total Dust: TWA 15 mg/m^3; Respirable Fraction: TWA 5 mg/m^3) Total Dust: TWA 10 mg/m^3; Respirable Fraction: TWA 5 mg/m^3
ACGIH TLV: TWA 10 mg/m^3
DFG MAK: 15 mg/m^3

SAFETY PROFILE: Moderately toxic by ingestion and intraperitoneal routes. An eye and skin irritant. A pesticide. Mixed with CaO and put into fused $CaCl_2$, the mixture incandesces. See also BORON COMPOUNDS.

BMG250 ***HR: 3***
BORON PHOSPHIDE
mf: BP mw: 41.79

PROP: Maroon powder. Mp: 200°.

SAFETY PROFILE: A poison. Ignites @ 200°. Deflagrates with fused alkali nitrates. Incompatible with HNO_3; oxidants; i.e., nitrates. When heated to decomposition it emits toxic fumes of PO_x. See also BORON COMPOUNDS and PHOSPHIDES.

BMG325 ***HR: 3***
BORON TRIAZIDE
mf: BN_9 mw: 136.87
SYN: TRIAZIDOBORANE

SAFETY PROFILE: An explosive which detonates by heat or contact with ether or water. See also BORON COMPOUNDS, AZIDES, and EXPLOSIVES.

BMG400 CAS:10294-33-4 ***HR: 3***
BORON TRIBROMIDE
DOT: UN 2692
mf: BBr_3 mw: 250.54

PROP: Colorless, fuming liquid. Mp: −45°, bp: 91.7°, d: 2.650 @ 0°, vap press: 40 mm @ 14.0°, 100 mm @ 33.5°.

SYNS: BORON BROMIDE ◇ TRONA

CONSENSUS REPORTS: Reported in EPA TSCA Inventory.

OSHA PEL: CL 1 ppm
ACGIH TLV: CL 1 ppm
DOT Classification: Corrosive Material; Label: Corrosive.

SAFETY PROFILE: A poison. Corrosive. A skin, eye, and mucous membrane irritant. Dangerous; may explode when heated. This and other boron halides react with water or steam to produce toxic and corrosive fumes and may explode. Incompatible with K; Na. When heated to decomposition it emits toxic fumes of Br^-. See also BORON COMPOUNDS and HYDROBROMIC ACID.

BMG500 CAS:10294-34-5 ***HR: 3***
BORON TRICHLORIDE
DOT: UN 1741
mf: BCl_3 mw: 117.16

PROP: Colorless, fuming liquid. Pungent, irritating odor. Mp: −107°, bp: 12.5°, d: 1.434 @ 0°, vap press: 1 atm @ 12.7°, vap d: 4.03.

SYNS: BORON CHLORIDE ◇ CHLORURE de BORE (FRENCH)

TOXICITY DATA with REFERENCE
ihl-rat LCLo:20 ppm/7H 14KTAK -,726,64
ihl-mus LCLo:20 ppm/7H 14KTAK -,726,64

CONSENSUS REPORTS: Reported in EPA TSCA Inventory. EPA Extremely Hazardous Substances List.

DOT Classification: Corrosive Material; Label: Corrosive; IMO: Nonflammable Gas; Label: Nonflammable Gas, Corrosive.

SAFETY PROFILE: A poison by inhalation and probably other routes. An irritant to skin, eyes, and mucous membranes. Reacts with water or steam to produce heat, toxic and corrosive fumes. Violent reaction with aniline or phosphine. Incompatible with hexafluorisopropylidene amino lithium, NO_2, grease, organic matter, O_2. When heated to decomposition it emits toxic fumes of Cl^-. See also BORON COMPOUNDS and HYDROCHLORIC ACID.

BMG700 CAS:7637-07-2 ***HR: 3***
BORON TRIFLUORIDE
DOT: UN 1008
mf: BF_3 mw: 67.81

PROP: Colorless gas; pungent, irritating odor. Mp: −126.8°, bp: −99.9°, d: 2.99 g/L.

SYNS: BORON FLUORIDE ◇ FLUORURE de BORE (FRENCH)

TOXICITY DATA with REFERENCE
ihl-rat LC50:1180 mg/m^3/4H 85GMAT -,27,82
ihl-mus LC50:3460 mg/m^3/2H FATOAO 35,369,72
ihl-gpg LC50:109 mg/m^3/4H FATOAO 35,369,72

CONSENSUS REPORTS: Reported in EPA TSCA Inventory. EPA Extremely Hazardous Substances List.

OSHA PEL: CL 1 ppm
ACGIH TLV: CL 1 ppm
DFG MAK: 1 ppm (3 mg/m^3)
NIOSH REL: (Boron Trifluoride) No Exposure Limit
DOT Classification: Nonflammable Gas; Label: Nonflammable Gas and Poison; IMO: Poison A; Label: Poison Gas.

SAFETY PROFILE: A poison by inhalation. A strong irritant. See also BORON COMPOUNDS and FLUO-

RIDES. Dangerous; when heated to decomposition or upon contact with water or steam, will produce toxic and corrosive fumes of F^-. Incompatible with alkali metals, alkaline earth metals (except Mg), alkyl nitrates, and CaO.

BMG750 CAS:7578-36-1 ***HR: 3***
BORON TRIFLUORIDE-ACETIC ACID COMPLEX
DOT: UN 1742

SYN: ACETIC ACID, compd. with BORON FLUORIDE (BF3) (8CI) ◇ BORON TRIFLUORIDE-ACETIC ACID COMPLEX (DOT)

CONSENSUS REPORTS: Reported in EPA TSCA Inventory.

DOT Classification: Corrosive Material; Label: Corrosive.

SAFETY PROFILE: A very corrosive material. When heated to decomposition it emits very toxic fumes of F^-, B oxides. See BORON COMPOUNDS, ACETIC ACID, and FLUORIDES.

BMH000 CAS:353-42-4 ***HR: 3***
BORON TRIFLUORIDE-DIMETHYL ETHER
DOT: UN 2965
mf: $C_2H_6O{\bullet}BF_3$ mw: 113.89

SYN: BORON TRIFLUORIDE DIMETHYL ETHERATE (DOT)

TOXICITY DATA with REFERENCE
ihl-gpg LCLo:50 ppm/4H 14KTAK -,726,64

CONSENSUS REPORTS: Reported in EPA TSCA Inventory.

DOT Classification: IMO: Flammable Solid; Label: Dangerous When Wet, Flammable Liquid, Corrosive.

SAFETY PROFILE: Poison by inhalation. Corrosive. When heated to decomposition it emits toxic fumes of F^-. See also ETHERS and BORON COMPOUNDS.

BMH250 CAS:109-63-7 ***HR: 3***
BORON TRIFLUORIDE ETHERATE
DOT: UN 2604
mf: $C_4H_{10}BF_3O$ mw: 141.93

PROP: Comp: Boron trifluoride, ether. D: 1.1, bp: 126°. flash p: < 22°.

SYN: BORON TRIFLUORIDE DIETHYL ETHERATE

DOT Classification: Flammable Solid; Label: Danger When Wet, Flammable Liquid, Corrosive

SAFETY PROFILE: Corrosive. A dangerous fire hazard. Peroxide containing etherate reacts explosively with solid lithium tetrahydroaluminate. Incompatible with water or steam to produce toxic, corrosive and flammable vapors; oxidizing materials. To fight fire, use Dry chemical, CO_2, fog or mist. See BORON COMPOUNDS, FLUORIDES, and ETHER.

BMH500 CAS:13517-10-7 ***HR: 3***
BORON TRIIODIDE
mf: BI_3 mw: 391.52

PROP: Colorless, hygroscopic plates. Mp: 43°, bp: 210°, d: 3.35 @ 50°.

SAFETY PROFILE: A poison. Reacts violently with water. Incandescent reaction with red or white phosphorous. Exothermic reaction with ammonia. Incompatible with ethers, carbohydrates, POCl. When heated to decomposition it emits toxic fumes of I^-. See also BORON COMPOUNDS and IODIDES.

BMH659 CAS:12007-33-9 ***HR: 3***
BORON TRISULFIDE
mf: B_2S_3 mw: 117.80

SAFETY PROFILE: Reacts violently with water. When heated to decomposition it emits toxic fumes of SO_x. See also BORON COMPOUNDS and SULFIDES.

BMH750 CAS:7184-60-3 ***HR: 3***
BORRELIDIN
mf: $C_{28}H_{43}NO_6$ mw: 489.72

TOXICITY DATA with REFERENCE
scu-rat LD50:1780 μg/kg JCINAO 28,1047,49
ivn-rat LD50:2 mg/kg 85ERAY 2,1198,78
scu-mus LD50:75 mg/kg JCINAO 28,1047,49
ivn-mus LD50:39 mg/kg JCINAO 28,1047,49
ims-ckn LD50:74 mg/kg 85ERAY 2,1198,78

SAFETY PROFILE: Poison by subcutaneous, intravenous, and intramuscular routes. When heated to decomposition it emits toxic fumes of NO_x.

BMI000 ***HR: 3***
BOTHROPS ASPER VENOM

SYNS: B. ASPER VENOM ◇ VENOM, COSTA RICAN SNAKE, BOTHROPS ASPER

TOXICITY DATA with REFERENCE
ipr-mus LD50:469 μg/kg AJTHAB 21,360,72
ivn-mus LD50:1175 μg/kg AJTHAB 21,360,72

SAFETY PROFILE: Poison by intraperitoneal and intravenous routes.

BMI125 ***HR: 3***
BOTHROPS ATROX VENOM

SYNS: B. ATROX VENOM ◇ VENOM, COSTA RICAN SNAKE, BOTHROPS ATROX

TOXICITY DATA with REFERENCE
scu-mus LD50:22140 μg/kg AJTMAQ 31,489,51
ivn-mus LD50:1400 μg/kg TXAPA9 16,73,70
ivn-rbt LDLo:5 μg/kg SCIEAS 117,47,53
ipr-mam LD50:3800 μg/kg CLPTAT 8,849,67
ivn-mam LD50:4270 μg/kg CLPTAT 8,849,67

SAFETY PROFILE: Poison by subcutaneous, intravenous, and intraperitoneal routes.

BMI250 ***HR: 3***
BOTHROPS COLOMIBIENSIS VENOM

SYN: VENOM, SNAKE, BOTHROPS COLOMBIENSIS

TOXICITY DATA with REFERENCE
ipr-mus LD50:4 mg/kg TOXIA6 17(Suppl. 1),161,79
ivn-mus LD50:2 mg/kg TOXIA6 17(Suppl. 1),161,79

SAFETY PROFILE: Poison by intraperitoneal and intravenous routes.

BMI500 ***HR: 3***
BOTHROPS GODMANI VENOM

SYN: VENOM, COSTA RICAN SNAKE, BOTHROPS GODMANI

TOXICITY DATA with REFERENCE
ipr-mus LD50:375 μg/kg AJTHAB 21,360,72
ivn-mus LD50:4750 μg/kg AJTHAB 21,360,72
ipr-mus LD50:375 μg/kg AJTHAB 21,360,72
ivn-mus LD50:4750 μg/kg AJTHAB 21,360,72

SAFETY PROFILE: A deadly poison by intraperitoneal and intravenous routes.

BMI750 ***HR: 3***
BOTHROPS LATERALIS VENOM

SYN: VENOM, COSTA RICAN SNAKE, BOTHROPS LATERALIS

TOXICITY DATA with REFERENCE
ipr-mus LD50:644 μg/kg AJTHAB 21,360,72
ivn-mus LD50:5144 μg/kg AJTHAB 21,360,72

SAFETY PROFILE: A deadly poison by intraperitoneal and intravenous routes.

BMJ000 ***HR: 3***
BOTHROPS NASUTUS VENOM

SYN: VENOM, COSTA RICAN SNAKE, BOTHROPS NASUTUS

TOXICITY DATA with REFERENCE
ipr-mus LD50:438 μg/kg AJTHAB 21,360,72
ivn-mus LD50:9063 μg/kg AJTHAB 21,360,72

SAFETY PROFILE: A deadly poison by intraperitoneal and intravenous routes.

BMJ250 ***HR: 3***
BOTHROPS NIGROVIRIDIS NEGROVIRIDIS VENOM

SYN: VENOM, COSTA RICAN SNAKE, BOTHROPS NIGROVIRIDIS NIGROVIRIDIS

TOXICITY DATA with REFERENCE
ipr-mus LD50:875 μg/kg AJTHAB 21,360,72
ivn-mus LD50:4438 μg/kg AJTHAB 21,360,72

SAFETY PROFILE: A deadly poison by intraperitoneal and intravenous routes.

BMJ500 ***HR: 3***
BOTHROPS NUMMIFER MEXICANUS VENOM

SYNS: VENOM, COSTA RICAN SNAKE, BOTHROPS NUMMIFER MEXICANUS ◇ B. N. MEXICANUS VENOM

TOXICITY DATA with REFERENCE
ipr-mus LD50:1063 μg/kg AJTHAB 21,360,72
ivn-mus LD50:5656 μg/kg AJTHAB 21,360,72

SAFETY PROFILE: A deadly poison by intraperitoneal and intravenous routes.

BMJ750 ***HR: 3***
BOTHROPS OPHYOMEGA VENOM

SYN: VENOM, COSTA RICAN SNAKE, BOTHROPS OPHRYOMEGA

TOXICITY DATA with REFERENCE
ipr-mus LD50:719 μg/kg AJTHAB 21,360,72
ivn-mus LD50:6813 μg/kg AJTHAB 21,360,72

SAFETY PROFILE: A deadly poison by intraperitoneal and intravenous routes.

BMK000 ***HR: 3***
BOTHROPS PICADOI VENOM

SYN: VENOM, COSTA RICAN SNAKE, BOTHROPS PICADOI

TOXICITY DATA with REFERENCE
ipr-mus LD50:375 μg/kg AJTHAB 21,360,72
ivn-mus LD50:1419 μg/kg AJTHAB 21,360,72

SAFETY PROFILE: A deadly poison by intraperitoneal and intravenous routes.

BMK250 ***HR: 3***
BOTHROPS SCHLEGLII VENOM

SYN: VENOM, COSTA RICAN SNAKE, BOTHROPS SCHLEGLII

TOXICITY DATA with REFERENCE
ivn-mus LD50:1600 μg/kg TXAPA9 16,73,70
ipr-mus LD50:531 μg/kg AJTHAB 21,360,72
ivn-mus LD50:2125 μg/kg AJTHAB 21,360,72

SAFETY PROFILE: A deadly poison by intraperitoneal and intravenous routes.

BMK290 CAS:27098-03-9 ***HR: D***
BOTRYODIPLODIN
mf: $C_7H_{12}O_3$ mw: 144.19

SYNS: (–)-BOTRYODIPLODIN ◇ 2-HYDROXY-3-METHYL-4-ACETYLTETRAHYDROFURANE ◇ METHYL TETRAHYDRO-5-HYDROXY-4-METHYL-3-FURYL KETONE ◇ 1-(TETRAHYDRO-5-HYDROXY-4-METHYL-3-FURANYL)-ETHANONE(9CI)

TOXICITY DATA with REFERENCE
dnd-hmn:fbr 30 μmol/L CRNGDP 5,1375,84
dns-rat:lvr 10 μmol/L CRNGDP 5,907,84
sce-ham:lng 300 μg/L CRNGDP 3,587,82

SAFETY PROFILE: Human mutation data reported. When heated to decomposition it emits acrid smoke and fumes. See also KETONES.

BMK325 CAS:64755-14-2 ***HR: 3***
BOUVARDIN
mf: $C_{40}H_{48}N_6O_{10}$ mw: 772.94

SYN: NSC 259968

TOXICITY DATA with REFERENCE
dni-mus:lym 3160 nmol/L TUMOAB 71,261,85
oms-mus:lym 1 μmol/L TUMOAB 71,261,85
ipr-mus LD50:12430 μg/kg NCISP* JAN86

SAFETY PROFILE: Poison by intraperitoneal route. Mutation data reported. When heated to decomposition it emits toxic fumes of NO_x.

BMK400 ***HR: D***
BOVINE PINEAL GLAND EXTRACT

PROP: Dose is expressed as amount of extract equivalent to the starting wet weight of the pineal gland before extraction (LIFSAK 17,531,75)

TOXICITY DATA with REFERENCE
par-rat TDLo:2500 mg/kg (4D pre):REP LIFSAK 17,531,75

SAFETY PROFILE: Experimental reproductive effects.

BMK500 CAS:774-64-1 ***HR: 3***
BOVOLIDE
mf: $C_{11}H_{16}O_2$ mw: 180.27

TOXICITY DATA with REFERENCE
scu-rat TDLo:2600 mg/kg/65W-I:NEO BJCAAI 19,392,65

SAFETY PROFILE: Questionable carcinogen with experimental neoplastigenic data. It is found in butter made from cow's milk and many other places. When heated to decomposition it emits acrid smoke and irritating fumes.

BMK620 CAS:63323-31-9 ***HR: 3***
(+)-BP-7-β,8-α-DIOL-9-α,10-α-EPOXIDE 2
mf: $C_{20}H_{14}O_3$ mw: 302.34

SYNS: (+)-trans-7-β,8-α-DIHYDROXY-9-α,10-α-EPOXY-7,8,9,10-TETRAHDYROBENZO(a)PYRENE ◇ (+)-E-7,8,9,10-TETRAHYDRO-7-α,8-β-DIHYDROXY-9-β,19-β-EPOXY-BENZO(a)PYRENE

TOXICITY DATA with REFERENCE
mmo-sat 100 pmol/plate BBRCA9 77,1389,77
msc-ham:lng 300 nmol/L BBRCA9 77,1389,77
skn-mus TDLo:1200 μg/kg:NEO CNREA8 39,67,79

SAFETY PROFILE: Questionable carcinogen with experimental neoplastigenic data by skin contact. Mutation data reported. When heated to decomposition it emits acrid smoke and fumes.

BMK630 ***HR: 3***
B(a)P EPOXIDE II
mf: $C_{20}H_{14}O_3$ mw: 302.34

SYN: anti-(±)-7-β,8-α-DIHYDROXY-9-α,10-α-EPOXY-7,8,9,10-TETRAHYDROBENZO(a)PYRENE

TOXICITY DATA with REFERENCE
dnd-hmn:fbr 1500 nmol/L/15M-C CBINA8 38,261,82
dnd-hmn:lym 5 μmol/L PAACA3 24,70,83
msc-hmn:fbr 50 nmol/L MUREAV 94,435,82
dnd-mus:emb 1800 μg/L SCIEAS 209,297,80
skn-mus TDLo:4830 μg/kg:NEO CCSUDL 3,371,78
scu-mus TDLo:11 mg/kg:ETA JJIND8 64,617,80

CONSENSUS REPORTS: EPA Genetic Toxicology Program.

SAFETY PROFILE: Questionable carcinogen with experimental neoplastigenic and tumorigenic data by skin contact. Human mutation data reported. When heated to decomposition it emits acrid smoke and fumes.

BMK634 CAS:75410-89-8 ***HR: 2***
B(c)PH DIOL EPOXIDE-1
mf: $C_{18}H_{14}O_3$ mw: 278.32

SYNS: (±)-BENZO(c)PHENANTHRENE-3,4-DIOL-1,2-EPOXIDE-1 ◇ BENZO(c)PHENANTHRENE-3-α-4-β-DIOL, 1,2,3,4-TETRAHYDRO-1-β,2-β-EPOXY-, (±)- ◇ (±)-3-α-4-β-DIHYDROXY-1-β,2-β-EPOXY-1,2,3,4-TETRAHYDROBENZO(c)PHENANTHRENE

TOXICITY DATA with REFERENCE
mmo-sat 100 pmol/plate CNREA8 40,2876,80
msc-ham:lng 200 nmol/L CNREA8 40,2876,80
skn-mus TDLo:278 μg/kg:NEO CNREA8 46,2257,86

SAFETY PROFILE: Questionable carcinogen with experimental neoplastigenic data. Mutation data reported. When heated to decomposition it emits toxic and irritating fumes

BMK635 ***HR: D***
B(c)PH DIOL EPOXIDE-2
mf: $C_{18}H_{14}O_3$ mw: 278.32

SYN: (±)-3-α,4-β-DIHYDROXY-1-α,2-α-EPOXY-1,2,3,4-TETRAHYDROBENZ(c)PHENANTHRACENE

TOXICITY DATA with REFERENCE
mmo-sat 100 pmol/plate CNREA8 40,2876,80
mma-sat 300 pmol/plate CRNGDP 4,1631,83
msc-ham:lng 200 nmol/L CNREA8 40,2876,80
skn-mus TDLo:278 μg/kg:NEO CNREA8 46,2257,86

SAFETY PROFILE: Questionable carcinogen with experimental neoplastigenic data. Mutation data reported. When heated to decomposition it emits toxic fumes of NO_x.

BMK750 ***HR: 3***
BRACKEN FERN, CHLOROFORM FRACTION

PROP: Chloroform fraction of tannin isolated from Bracken Fern (*Pteridium aquilinum*).

TOXICITY DATA with REFERENCE
orl-rat TDLo:1000 g/kg/56W-C:CAR JJIND8 65,131,80

SAFETY PROFILE: Questionable carcinogen with experimental carcinogenic data. Mutation data reported. See also SHIKIMIC ACID.

BML000 ***HR: 3***
BRACKEN FERN, DRIED

SYNS: 1-CYCLOHEXENE-1-CARBOXYLIC ACID, 3,4,5 ◇ S. EGRELTRI ATUNUN (TURKISH) ◇ PTERIDIUM AQUILINUM ◇ PTERIS AQUALINA

TOXICITY DATA with REFERENCE
sln-dmg-orl 15 pph MUREAV 92,89,82
bfa-rat/sat 1000 g/kg JJIND8 65,131,80
bfa-ctl/sat 623 g/kg/2Y-C CNREA8 38,1556,78
orl-mus TDLo:1782 g/kg (4W pre/1-7D preg):TER TXAPA9 28,264,74
orl-rat TDLo:2209 g/kg/17W-C:CAR JNCIAM 45,179,70
orl-mus TDLo:7140 g/kg/17W-C:ETA GANMAX 17,205,75
orl-ctl TDLo:600 g/kg/2Y-C:CAR VTPHAK 13,110,76
orl-rat TD:1350 g/kg/81D-C:NEO GANNA2 69,383,78

CONSENSUS REPORTS: IARC Cancer Review: Human Inadequate Evidence IMEMDT 40,47,86; Animal Sufficient Evidence IMEMDT 40,47,86

SAFETY PROFILE: Confirmed carcinogen with experimental carcinogenic, neoplastigenic, and tumorigenic data. Experimental teratogenic and reproductive effects. Mutation data reported.

BML250 ***HR: 3***
BRACKEN FERN TANNIN

SYNS: PTERIDIUM AQUILINUM TANNIN ◇ TANNIN from BRACKEN FERN

TOXICITY DATA with REFERENCE
bfa-rat/sat 2000 g/kg/56W-C JJIND8 65,131,80
scu-rat TDLo:1595 mg/kg/38W-I:NEO JJIND8 65,131,80
ipr-mus LD50:160 mg/kg JNCIAM 56,33,76

SAFETY PROFILE: Poison by intraperitoneal route. Questionable carcinogen with experimental neoplastigenic data. Mutation data reported.

BML500 CAS:58-82-2 ***HR: D***
BRADYKININ
mf: $C_{50}H_{73}N_{15}O_{11}$ mw: 1060.38

SYNS: BK ◇ BRADYKININ (synthetic) ◇ BRS 640 ◇ KALLIDIN ◇ PRS 640 ◇ SYNTHETIC BRADYKININ

TOXICITY DATA with REFERENCE
cyt-mus:emb 500 μg/L DANKAS 282,173,85
ivn-mus TDLo:25 μg/kg (7-12D preg):TER JPMSAE 58,406,69
ivn-mus TDLo:25 μg/kg (7-12D preg):REP JPMSAE 58,406,69

SAFETY PROFILE: Experimental teratogenic and reproductive effects. Mutation data reported. When heated to decomposition it emits toxic fumes of NO_x.

BML750 CAS:11011-73-7 ***HR: 3***
BRAMYCIN
mf: $C_{32}H_{55}NO_{11}$ mw: 629.88

PROP: An antibiotic produced by *Streptomyces diastatochromogenes var. bracus*.

TOXICITY DATA with REFERENCE
orl-mus LD50:4800 μg/kg 85ERAY 2,1134,78
ipr-mus LD50:490 μg/kg 85ERAY 2,1134,78

SAFETY PROFILE: Poison by ingestion and intraperitoneal routes. When heated to decomposition it emits toxic fumes of NO_x.

BML825 ***HR: 3***
BRAXORONE
mf: $C_{21}H_{27}BrO_3$ mw: 407.39

SYNS: 9-α-BROMO-11-KETOPROGESTERONE ◇ 9-BROMOPREGN-4-ENE-3,11,20-TRIONE ◇ 9-α-BROMOPREGN-4-ENE-3,11,20-TRIONE

TOXICITY DATA with REFERENCE
orl-wmn TDLo:13 mg/kg (21D pre):REP AJOGAH 76,626,58

SAFETY PROFILE: Human female reproductive effects by ingestion: disorders of the menstrual cycle and changes in the uterus, cervix, or vagina. When heated to decomposition it emits toxic fumes of Br^-.

BMM000 CAS:50924-49-7 ***HR: 2***
BREDININ
mf: $C_9H_{13}N_3O_6$ mw: 259.25

SYNS: ANHYDRO-4-CARBAMOYL-5-HYDROXY-1-β-d-RIBOFURANOSYL-IMIDAZOLIUMHYDROXIDE ◇ BREDININE ◇ 4-CARBAMOYL-1-β-d-RIBOFURANOSYL-IMIDAZOLIUM-5-OLATE ◇ 5-HYDROXY-1-β-d-RIBOFURANOSYL-1H-IMIDAZOLE-4-CARBOXAMIDE ◇ MIZORIBINE

TOXICITY DATA with REFERENCE
mnt-mus-ipr 5 g/kg OYYAA2 24,703,82
cyt-mus:leu 20 μmol/L IDZAAW 51,61,76
cyt-mus:lym 20 μmol/L CNREA8 35,1643,75
sce-mus:lym 2500 nmol/L OYYAA2 24,703,82
dlt-mus-orl 160 mg/kg OYYAA2 24,711,82
cyt-ham:fbr 100 μmol/L CNREA8 35,1643,75
orl-rat TDLo:48 mg/kg (17-22D preg):REP OYYAA2 26,397,83
orl-mus TDLo:75 mg/kg (male 3D pre):TER OYYAA2 24,711,82
orl-rat LD50:3100 mg/kg IYKEDH 15,668,84
scu-rat LD50:4161 mg/kg IYKEDH 15,688,84
ivn-rat LD50:1500 mg/kg DRFUD4 3,567,78
ipr-mus LD50:5000 mg/kg 85GDA2 5,275,81
ivn-mus LD50:500 mg/kg 85GDA2 5,275,81

CONSENSUS REPORTS: EPA Genetic Toxicology Program.

SAFETY PROFILE: Moderately toxic by ingestion and intravenous route. An experimental teratogen. Other experimental reproductive effects. Mutation data reported. An immunosuppressive agent. When heated to decomposition it emits toxic fumes of NO_x.

BMM125 CAS:2179-16-0 ***HR: D***
BREMFOL
mf: $C_{20}H_{21}N_7O_6$ mw: 455.48

SYNS: 9-ME-PGA ◇ 9-METHYL-PGA ◇ 9-METHYL PTEROYLGLUTAMIC ACID ◇ NINOPTERIN

TOXICITY DATA with REFERENCE
orl-rat TDLo:405 mg/kg (11-14D preg):TER BNEOBV 41,52,82
orl-rat TDLo:20 mg/kg (female 10-12D post):REP TJADAB 1,375,68

SAFETY PROFILE: An experimental teratogen. Other experimental reproductive effects. When heated to decomposition it emits toxic fumes of NO_x.

BMM500 CAS:2580-78-1 ***HR: 3***
BRILLIANT BLUE R
mf: $C_{22}H_{16}N_2O_{11}S_3{\cdot}2Na$ mw: 626.56

SYNS: CAVALITE BRILLIANT BLUE R ◇ C.I. 61200 ◇ C.I. REACTIVE BLUE 19 ◇ C.I. REACTIVE BLUE 19, DISODIUM SALT ◇ REACTIVE BLUE 19 ◇ REMALAN BRILLIANT BLUE R ◇ REMAZOL BRILLIANT BLUE R

TOXICITY DATA with REFERENCE
orl-rat TDLo:87 g/kg/2Y-I:ETA TKORAS 3,53,67

CONSENSUS REPORTS: Reported in EPA TSCA Inventory.

SAFETY PROFILE: Questionable carcinogen with experimental tumorigenic data. When heated to decomposition it emits very toxic fumes of Na_2O, NO_x, and SO_x. See also SULFONATES.

BMM600 CAS:555-65-7 ***HR: D***
BROCRESINE
mf: $C_7H_8BrNO_2$ mw: 218.07

SYNS: α-(AMINOOXY)-6-BROMO-m-CRESOL ◇ BROCRESIN ◇ CL 54998 ◇ m-CRESOL, α-(AMINOOXY)-6-BROMO- ◇ NSD 1055 ◇ PHENOL, 5-((AMINOOXY)METHYL)-2-BROMO-(9CI)

TOXICITY DATA with REFERENCE
ipr-mus TDLo:1200 mg/kg (female 3-4D post):REP CCPTAY 23,457,81

SAFETY PROFILE: Experimental reproductive effects. When heated to decomposition it emits toxic fumes of NO_x and Br^-.

BMM625 CAS:63638-90-4 ***HR: 3***
BROFAREMINE HYDROCHLORIDE
mf: $C_{14}H_{16}BrNO_2{\cdot}ClH$ mw: 346.68

SYNS: 4-(7-BROMO-5-METHOXY-2-BENZOFURANYL)PIPERIDINE HYDROCHLORIDE ◇ CGP-11305A

TOXICITY DATA with REFERENCE
orl-rat LD50:310 mg/kg DRFUD4 10,371,85
orl-mus LD50:190 mg/kg DRFUD4 10,371,85
orl-dog LD50:100 mg/kg DRFUD4 10,371,85

SAFETY PROFILE: Poison by ingestion. When heated to decomposition it emits toxic fumes of Br^-, NO_x, and HCl.

BMM650 CAS:314-40-9 ***HR: 2***
BROMACIL
mf: $C_9H_{13}BrN_2O_2$ mw: 261.15

SYNS: BOREA ◇ BROMAZIL ◇ 5-BROMO-3-sec-BUTYL-6-METHYLURACIL ◇ 5-BROMO-6-METHYL-3-(1-METHYLPROPYL)-2,4(1H,3H)-PYRIMIDINEDIONE ◇ 5-BROMO-6-METHYL-3-(1-METHYLPROPYL) URACIL ◇ 3-sek.BUTYL-5-BROM-6-METHYLURACIL (GERMAN) ◇ CYNOGAN ◇ DU PONT HERBICIDE 976 ◇ EEREX GRANULAR WEED KILLER ◇ EEREX WATER SOLUBLE CONCENTRATE WEED KILLER ◇ HERBICIDE 976 ◇ HYVAR ◇ HYVAREX ◇ HYVAR X ◇ HYVAR X BROMACIL ◇ HYVAR X WEED KILLER ◇ KROVAR II ◇ NALKIL ◇ URAGAN ◇ URAGON ◇ UROX B WATER SOLUBLE CONCENTRATE WEED KILLER ◇ UROX HX GRANULAR WEED KILLER

TOXICITY DATA with REFERENCE
sln-dmg-orl 2000 ppm JPFCD2 15,867,80
sln-nsc 10 mg/L MUREAV 167,35,86
msc-mus:lym 750 mg/L NTIS** PB84-138973
ihl-rat TCLo:38 mg/m^3/2H (7-14D preg):TER NTIS** PB277-077
orl-rat LD50:641 mg/kg FAATDF 7,299,86
orl-mus LD50:3040 mg/kg JPFCD2 15,867,80

CONSENSUS REPORTS: EPA Genetic Toxicology Program.

OSHA PEL: TWA 1 ppm
ACGIH TLV: TWA 1 ppm

SAFETY PROFILE: Moderately toxic by ingestion. An experimental teratogen. Mutation data reported. An herbicide. When heated to decomposition it emits very toxic fumes of Br^- and NO_x.

BMN000 CAS:28772-56-7 ***HR: 3***
BROMADIALONE
mf: $C_{30}H_{23}BrO_4$ mw: 527.11

SYNS: BROMADIOLONE ◇ 3-(3-(4'-BROMO(1,1'-BIPHENYL)-4-YL)3-HYDROXY-1-PHENYLPROPYL)-4-HYDROXY-2H-1-BENZOPYRAN-2-ONE ◇ BROMONE ◇ 3-(α-(p-(p-BROMOPHENYL)-β-HYDROXYPHENETHYL)BENZYL)-4-HYDROXYCOUMARIN ◇ CANADIEN 2000 ◇ CONTRAC ◇ (HYDROXY-4 COUMARINYL 3)-3 PHENYL-3 (BROMO-4 BIPHENYLYL-4)-1 PROPANOL-1 (FRENCH) ◇ LM-637 ◇ MAKI ◇ RATIMUS ◇ SUPER-CAID ◇ SUPER-ROZOL ◇ SUP'OPERATS ◇ TEMUS

TOXICITY DATA with REFERENCE
orl-rat LD50:1125 mg/kg FMCHA2 -,D265,80
orl-mus LD50:1750 μg/kg PHPHA6 25,69,76
orl-rbt LD50:1 mg/kg PHPHA6 25,69,76

CONSENSUS REPORTS: EPA Extremely Hazardous Substances List.

SAFETY PROFILE: A deadly poison by ingestion. Used as a rodent poison. When heated to decomposition it emits toxic fumes of Br^-. See also COUMADIN (warfarin).

BMN250 CAS:13977-28-1 ***HR: 3***
BROMADRYL
mf: $C_{18}H_{22}BrNO{\cdot}ClH$ mw: 384.78

SYNS: 2-(1-(4-BROMODIPHENYL)ETHOXY)-N,N-DIMETHYLETHYLAMINE HYDROCHLORIDE ◇ p-BROMO-α-METHYLBENZHYDRYL-2-DIMETHYLAMINOETHYL ETHER HYDROCHLORIDE ◇ 2-((p-BROMO-α-METHYL-α-PHENYLBENZYL)OXY)-N,N-DIMETHYLETHYLAMINE HYDROCHLORIDE ◇ 2-(1-(4-BROMOPHENYL)-1-PHENYLETHOXY)-N,N-DIMETHYLETHANAMINEHYDROCHLORIDE ◇ 1-(p-BROMOPHENYL)-1-PHENYL-1-(2-DIMETHYLAMINOETHOXY) ETHANE HYDROCHLORIDE ◇ (2-(1-p-BROMOPHENYL-1-PHENYLETHOXY) ETHYL)DIMETHYLETHYLAMINE HYDROCHLORIDE ◇ β-DIMETHYLAMINOETHYL-p-BROMO-α-METHYLBENZHYDRYL ETHER HYDROCHLORIDE ◇ EMBRAMINE HYDROCHLORIDE ◇ MEBROPHENHYDRAMINE ◇ MEBROPHENHYDRAMINE HYDROCHLORIDE ◇ MEBRYL

TOXICITY DATA with REFERENCE
orl-mus LD50:330 mg/kg MEIEDD 10,513,83
ivn-mus LD50:80 mg/kg MEIEDD 10,513,83

SAFETY PROFILE: Poison by ingestion and intravenous routes. An antihistaminic agent. When heated to decomposition it emits very toxic fumes of HCl, Br^-, and NO_x. See also ETHERS.

BMN350 CAS:332-69-4 ***HR: 2***
BROMANYLPROMIDE
mf: $C_{11}H_{15}BrN_2O$ mw: 271.19

SYNS: BROMAMID ◇ BROMAMIDE ◇ BROMAMIDE (PHARMACEUTICAL) ◇ N,N-DIMETHYL-β-(p-BROMOANILINO)PROPIONAMIDE ◇ 3-((4-BROMOPHENYL)AMINO)-N,N-DIMETHYL-PROPANAMIDE (9CI)

TOXICITY DATA with REFERENCE
orl-rat LD50:1810 mg/kg OYYAA2 2,70,68
orl-mus LD50:2431 mg/kg OYYAA2 2,70,68
scu-mus LD50:1375 mg/kg OYYAA2 2,70,68

SAFETY PROFILE: Moderately toxic by ingestion and other routes. When heated to decomposition it emits toxic fumes of Br^- and NO_x.

BMN500 ***HR: 3***
BROMATES

SAFETY PROFILE: Generally considered to be more toxic than chlorates; cause central nervous system paralysis. They may form methemoglobin, but less actively than chlorates. See also specific compounds. Flammable in the form of gas, vapor, or dust by chemical reaction with (powdered metals + acids); Al; As; CaH_2; C; Cu; powdered metals; metal sulfides; organic matter; PH_4I; P; SrH; S; (H_2SO_4 + metals). When heated to decomposition they emit toxic fumes of Br^-; can react with reducing materials.

BMN750 CAS:1812-30-2 ***HR: 3***
BROMAZEPAM
mf: $C_{14}H_{10}BrN_3O$ mw: 316.18

SYNS: 7-BROMO-1,3-DIHYDRO-5-(2-PYRIDYL)-2H-1,4-BENZDIAZEPIN-2-ONE ◇ 7-BROMO-5-(2-PYRIDYL)-3H-1,4-BENZODIAZEPIN-2(1H)-ONE ◇ COMPENDIUM ◇ 1,3-DIHYDRO-7-BROMO-5-(2-PYRIDYL)-2H-1,4-BENZODIAZEPIN-2-ONE ◇ KL-001 ◇ LA XVII ◇ LECTOPAM ◇ LEXOMIL ◇ LEXOTAN ◇ LEXOTANIL ◇ RO 4-9253 ◇ RO 5-3350

TOXICITY DATA with REFERENCE
skn-hmn 3 mg/3D-I MOD 85DKA8 -,127,77
rec-rat TDLo:11 mg/kg (female 7-17D post):REP OYYAA2 26,111,83
rec-rbt TDLo:130 mg/kg (female 6-18D post):TER OYYAA2 26,99,83

orl-rat LD50:1950 mg/kg KSRNAM 7,2413,73
ipr-rat LD50:1660 mg/kg OYYAA2 17,115,79
scu-rat LD50:8800 mg/kg OYYAA2 17,115,79
orl-mus LD50:879 mg/kg OYYAA2 17,115,79
ipr-mus LD50:200 mg/kg JDGRAX 16,7,85
scu-mus LD50:6870 mg/kg OYYAA2 17,115,79
orl-rbt LD50:1690 mg/kg 27ZQAG -,158,72

SAFETY PROFILE: Poison by intraperitoneal route. Moderately toxic by ingestion. An experimental teratogen. Other experimental reproductive effects. Human skin irritant. A tranquilizer. When heated to decomposition it emits very toxic fumes of NO_x and HBr.

BMO000 CAS:9001-00-7 ***HR: 3***
BROMELAIN

PROP: From pineapples *Ananas comosus* and *Ananas bracteatus* L. White to tan amorphous powder. Sol in water; insol in alc, chloroform, ether.

SYNS: ANANASE ◇ BROMELAINS ◇ BROMELIN ◇ E.C. 3.4.4.24 ◇ EXTRANASE ◇ INFLAMEN ◇ PLANT PROTEASE CONCENTRATE ◇ TRAUMANASE

TOXICITY DATA with REFERENCE
ipr-rat LD50:85200 μg/kg AIPTAK 145,166,63
ipr-rat LD50:85 mg/kg AIPTAK 145,166,63
ipr-mus LD50:37 mg/kg AIPTAK 145,166,63
ivn-mus LD50:30 mg/kg AIPTAK 145,166,63

CONSENSUS REPORTS: Reported in EPA TSCA Inventory.

SAFETY PROFILE: A poison via intraperitoneal and intravenous routes. When heated to decomposition it emits acrid smoke and fumes.

BMO250 CAS:15086-94-9 ***HR: 1***
BROMEOSIN
mf: $C_{20}H_8Br_4O_5$ mw: 647.92

SYNS: BROMOEOSIN ◇ BROMOFLUORESCEIC ACID ◇ C.I. 45380:2 ◇ C.I. SOLVENT RED 43 ◇ D&C RED No. 21 ◇ EOSIN ◇ EOSINE ◇ 2,4,5,7-TETRABROMO-3,6-FLUORANDIOL ◇ TETRABROMOFLUORESCEIN ◇ 2′,4′,5′,7′-TETRABROMOFLUORESCEIN

TOXICITY DATA with REFERENCE
dnr-bcs 2 mg/disc TRENAF 27,153,76
scu-mus LDLo:450 mg/kg HBAMAK 4,1289,35
scu-frg LDLo:1 g/kg HBAMAK 4,1289,35

CONSENSUS REPORTS: IARC Cancer Review: Group 3 IMEMDT 7,56,87, Animal Inadequate Evidence IMEMDT 15,183,77. Reported in EPA TSCA Inventory.

SAFETY PROFILE: Mutation data reported. Incompatible with reducing agents. When heated to decomposition it emits very toxic fumes of Br^-. See also BROMIDES.

BMO325 CAS:611-75-6 ***HR: 3***
BROMHEXINE CHLORIDE
mf: $C_{14}H_{20}Br_2N_2•ClH$ mw: 412.64

SYNS: 2-AMINO-3,5-DIBROMO-N-CYCLOHEXYL-N-METHYLBENZENEMETHANAMINE MONOHYDROCHLORIDE (9CI) ◇ BISOLVON ◇ BISOLVON HYDROCHLORIDE ◇ BROMHEXINE HYDROCHLORIDE

TOXICITY DATA with REFERENCE
orl-rat LD50:6 g/kg GNRIDX 3,259,69
ipr-rat LD50:1680 mg/kg GNRIDX 3,259,69
orl-mus LD50:4800 mg/kg GNRIDX 3,259,69
ipr-mus LD50:2210 mg/kg GNRIDX 3,259,69
ivn-mus LD50:44 mg/kg GNRIDX 3,259,69

SAFETY PROFILE: Poison by intravenous route. Moderately toxic by other routes. When heated to decomposition it emits toxic fumes of Br^-, NO_x, and HCl.

BMO750 ***HR: 3***
BROMIDES

SAFETY PROFILE: The most common inorganic bromides are Na, K, NH_4, Ca and Mg bromides. Methyl and ethyl bromides are among the most common organic bromides. The inorganic bromides produce depression, emaciation, and, in severe cases, psychosis and mental deterioration. Bromide rashes (bromoderma), especially of the face and resembling acne and furunculosis, often occur when bromide inhalation or administration is prolonged. Organic bromides, such as methyl bromide and ethyl bromide, are volatile liquids of relatively high toxicity. See also specific compounds. When strongly heated they emit highly toxic fumes of Br^-.

BMO825 ***HR: D***
BROMINATED VEGETABLE (SOYBEAN) OIL

PROP: Pale yellow to dark brown viscous, oily liquid; bland or fruity odor and bland taste. Sol in alc, chloroform, ether, hexane, fixed oils; insol in water.

SYN: VEGETABLE (SOYBEAN) OIL, brominated

TOXICITY DATA with REFERENCE
orl-rat TDLo:9 g/kg (2W male/2W pre-14D post):REP TJADAB 28,309,83

SAFETY PROFILE: Experimental reproductive effects. When heated to decomposition it emits toxic fumes of Br^-.

BMP000 CAS:7726-95-6 ***HR: 3***
BROMINE
DOT: UN 1744
mf: Br_2 mw: 159.82

PROP: Rhombic crystals or dark red liquid. Fp: −7.3°,

bp: 58.73°, d: 2.928 @ 59°, 3.12 @ 20°, vap press: 175 mm @ 21°, 1 atm @ 58.2°, vap d: 5.5.

SYNS: BROM (GERMAN) ◇ BROME (FRENCH) ◇ BROMINE, solution (DOT) ◇ BROMO (ITALIAN) ◇ BROOM (DUTCH)

TOXICITY DATA with REFERENCE
orl-hmn LDLo:14 mg/kg 34ZIAG -,645,69
ihl-hmn LCLo:1000 ppm 34ZIAG -,645,69
ihl-mus LC50:750 ppm/9M AIHAAP 39,129,78
ihl-cat LCLo:140 ppm/7H AHYGAJ 7,233,1887
ihl-rbt LCLo:180 ppm/6.5H HBTXAC 1,324,56
ihl-gpg LCLo:140 ppm/7H AHYGAJ 7,233,1887

CONSENSUS REPORTS: Reported in EPA TSCA Inventory. EPA Genetic Toxicology Program.

OSHA PEL: (Transitional: TWA 0.1 ppm) TWA 0.1 ppm; STEL 0.3 ppm
ACGIH TLV: TWA 0.1 ppm; STEL 0.3 ppm
DFG MAK: 0.1 ppm (0.7 mg/m^3)
DOT Classification: Corrosive Material; Label: Corrosive; IMO: Corrosive Material; Label: Corrosive and Poison.

SAFETY PROFILE: A human poison by ingestion and moderately toxic by inhalation. A poison by ingestion and inhalation experimentally. Corrosive. The action of bromine is essentially the same as that of chlorine, irritating the mucous membranes of the eyes and upper respiratory tract. Severe exposure may result in pulmonary edema. Usually, however, the irritant qualities of the chemical force the worker to leave the exposure area before serious poisoning can result. Chronic exposure is similar to the therapeutic ingestion of excessive bromides. See also BROMIDES. Regular physical examinations should be made of people who work with bromine or bromides. Flammable in the form of liquid or vapor by spontaneous chemical reaction with reducing materials. A very powerful oxidizer. Highly dangerous; when heated it emits highly toxic fumes; will react with water or steam to produce toxic and corrosive fumes. Reacts explosively with diethylzinc, germane, disilane, dimethylformamide, hydrogen, isobutyrophenone, metal azides (particularly silver or sodium azide), potassium, silane and homologs, praseodymium, antimony, trimethylamine, ammonia. Mixtures with lithium or sodium are shock-sensitive explosives. Ignition on contact with germanium, mono- or di-alkali metal acetylides, trialkyl boranes, copper acetylides. Violent reaction with carbonyl compounds (aldehydes, ketones, carboxylic acids), diethyl ether, phosphine, natural rubber, aluminum, mercury, titanium. Vigorous reaction with methanol and other alcohols, tetrahydrofuran, mixtures of ethanol and phosphorous. Incompatible with acetaldehyde, C_2H_2, acrylonitrile, NH_3, Sb, B, Ca_3N_2, Cs_2O, Cs_2C_2, CsC_2H, ClF_3C_2, CuH_2, dimethyl formamide, ethyl phosphine, F_2, Fe_2C, isobutyrophenone, Li_2C_2, Li_2Si_2, Mg_3P_2, $Ni(CO)_4$, NI_3, olefins, OF_2, O_3, P, PO_x, Rb_2C_2, RbC_2H, Na_2C_2, NaC_2H, Sr_3P, Sn, UC_2, ZrC2, reducing materials.

BMP250 ***HR: 3***
BROMINE AZIDE
mf: BrN_3 mw: 121.93

PROP: Crystals or red liquid. Mp: 45°, bp: explodes.

SYN: BROMOAZIDE

SAFETY PROFILE: A poison. Can explode spontaneously. The solid, liquid and vapor are shock-sensitive explosives. Concentrated solutions in organic solvents may explode. Moderate fire hazard in the form of vapor by chemical reaction. A powerful oxidant. Moderately explosive when exposed to heat. The liquid explodes on contact with arsenic, sodium, silver foil, or phosphorous. Incompatible with Sb, ethyl ether, Ag, metals. When heated to decomposition it emits highly toxic fumes of Br^- and explodes. See also BROMINE and AZIDES.

BMP500 CAS:21255-83-4 ***HR: 3***
BROMINE DIOXIDE
mf: BrO_2 mw: 111.91

PROP: Light yellow crystals. Mp: 0° (decomp).

CONSENSUS REPORTS: EPA Extremely Hazardous Substances List.

SAFETY PROFILE: Very unstable material. Flammable in the form of vapor by chemical reaction with reducing agents. Potentially explosive if heated rapidly. A strong oxidant. Reaction with water, steam, or reducing materials produces toxic and corrosive fumes. Must be stored at low temperatures. When heated to decomposition it emits toxic fumes of Br^-. See also BROMINE.

BMP750 CAS:13863-59-7 ***HR: 3***
BROMINE FLUORIDE
mf: BrF mw: 98.91

SAFETY PROFILE: A poison and powerful irritant. Very reactive. Ignites on contact with H_2. Incompatible with organic matter and water. When heated to decomposition it emits toxic fumes of F^- and Br^-. See also BROMINE and FLUORIDE.

BMQ000 CAS:7789-30-2 ***HR: 3***
BROMINE PENTAFLUORIDE
DOT: UN 1745
mf: BrF_5 mw: 174.91

PROP: Colorless fuming liquid. Mp: −61.3, bp: 40.5, d: 2.466 @ 25°, vap d: 6.05.

OSHA PEL: TWA 0.1 ppm; TWA 2.5 mg(F)/m^3
ACGIH TLV: TWA 0.1 ppm
NIOSH REL: (Inorganic Fluorides) TWA 2.5 mg(F)/m^3
DOT Classification: Oxidizer; Label: Oxidizer; IMO: Oxidizer; Label: Oxidizer, Poison, Corrosive.

SAFETY PROFILE: A poisonous, corrosive, and extremely reactive gas. It is a powerful oxidizer. Will react with water or steam to produce toxic and corrosive fumes. The liquefied gas reacts violently with many organic compounds and some inorganic compounds. Explodes or ignites on contact with hydrogen-containing materials (i.e., acetic acid, ammonia, benzene, ethanol, hydrogen, hydrogen sulfide, methane, cork, grease, paper, wax, chloromethane). Reacts violently and may ignite on contact with acids, halogens, non-metals, metal halides, metals, oxides, concentrated nitric or sulfuric acids, aluminum powder, ammonium chloride, antimony, arsenic, arsenic pentoxide, barium, bismuth, boron powder, boron trioxide, calcium oxide, carbon monoxide, charcoal, chlorine, chromium, chromium trioxide, cobalt powder, iodine, iodine pentoxide, iridium powder, iron powder, lithium powder, manganese, magnesium oxide, molybdenum, molybdenum trioxide, nickel powder, red phosphorous, phosphorous pentoxide, potassium iodide, rhodium powder, selenium, sulfur, sulfur dioxide, tellurium, tungsten, tungsten trioxide, water, zinc. When heated to decomposition it emits very toxic fumes of F^- and Br^-. See also BROMINE and FLUORIDES.

BMQ250 **HR: 3**
BROMINE PERCHLORATE
mf: $BrClO_4$ mw: 179.36

SAFETY PROFILE: A shock-sensitive explosive. Upon decomposition it emits toxic fumes of Cl^- and Br^-. See also PERCHLORATES and BROMIDES.

BMQ325 CAS:7787-71-5 **HR: 3**
BROMINE TRIFLUORIDE
DOT: UN 1746
mf: BrF_3 mw: 136.91

PROP: Colorless, fuming liquid. Mp: 8.8°, bp: 127°, d: 2.84.

CONSENSUS REPORTS: Reported in EPA TSCA Inventory.

OSHA PEL: TWA 2.5 mg(F)/m^3
ACGIH TLV: TWA 2.5 mg(F)/m^3
NIOSH REL: (Inorganic Fluorides) TWA 2.5 mg(F)/m^3
DOT Classification: Oxidizer; Label: Oxidizer and Poison; IMO: Oxidizer; Label: Oxidizer, Poison, Corrosive.

SAFETY PROFILE: Poisonous and corrosive. Very reactive, a powerful oxidizer. Explosive or violent reaction with organic materials, water, acetone, ammonium halides, antimony, antimony trichloride oxide, arsenic, benzene, boron, bromine, carbon, carbon monoxide, carbon tetrachloride, carbon tetraiodide, chloromethane, cobalt, ether, halogens, iodine, powdered molybdenum, niobium, 2-pentanone, phosphorus, potassium hexachloroplatinate, pyridine, silicon, silicone grease, sulfur, tantalum, tin dichloride, titanium, toluene, vanadium, uranium, uranium hexafluoride. Incompatible with Sb_2O_3, $BaCl_2$, Bi_2O_5, $CdCl_2$, $CaCl_2$, CsCl, LiCl, $MnIO_3$, metals, Nb_2O_5, $PtBr_4$, $PtCl_4$, (Pt + KFO), KBr, KCl, KI, $RhBr_4$, RbCl, AgCl, NaBr, NaCl, NaI, Ta_2O_5, Sn, W, UO_x, rubber, plastics. The product of reaction with pyridine ignites when dry. When heated to decomposition it emits toxic fumes of F^- and Br^-. Very dangerous. See also BROMINE PENTAFLUORIDE, FLUORIDES, and BROMINE.

BMQ500 CAS:70142-16-4 **HR: 3**
BROMINE(I) TRIFLUOROMETHANESULFONATE
mf: $CBrF_3O_3S$ mw: 228.97

SAFETY PROFILE: A strong oxidizer which may react explosively with readily oxidizable materials. When heated to decomposition it emits toxic fumes of F^-, Br^-, and SO_x. See also SULFONATES.

BMQ750 **HR: 3**
BROMINE TRIOXIDE
mf: BrO_3 mw: 127.91

SAFETY PROFILE: The solid produced at −5° is only stable at −80° or in the presence of ozone. Decomposition can be violently explosive in the presence of trace impurities. Upon decomposition it emits toxic fumes of Br^-. See also BROMINE.

BMR000 CAS:17157-48-1 **HR: D**
BROMOACETALDEHYDE
mf: C_2H_3BrO mw: 122.96

SYNS: α-BROMOACETALDEHYDE ◇ 2-BROMOACETALDEHYDE ◇ MONOBROMOACETALDEHYDE

TOXICITY DATA with REFERENCE
dnr-esc 10 μL/plate EVHPAZ 21,79,77
dnd-mam:lym 16 nmol CNREA8 41,4391,81

CONSENSUS REPORTS: EPA Genetic Toxicology Program.

SAFETY PROFILE: Mutation data reported. When heated to decomposition it emits toxic fumes of Br^-. See also ALDEHYDES and BROMIDES.

BMR750 CAS:79-08-3 ***HR: 3***
α-BROMOACETIC ACID
DOT: UN 1938
mf: $C_2H_3O_2Br$ mw: 138.04

PROP: Hygroscopic crystals, sol in water and alc. D: 1.93, mp: 50°, bp: 208°.

SYNS: ACIDE BROMACETIQUE (FRENCH) ◇ BROMOACETIC ACID, solid (DOT) ◇ BROMOACETIC ACID, solution (DOT) ◇ BROMOETHANIOC ACID ◇ α-BROMOETHANIOC ACID ◇ MONOBROMESSIGSAEURE (GERMAN) ◇ MONOBROMOACETIC ACID ◇ TO NTU

TOXICITY DATA with REFERENCE
dnd-mus:leu 100 μmol/L BCPCA6 30,1497,81
orl-mus LD50:100 mg/kg JPETAB 86,336,46
ipr-mus LD50:66 mg/kg JNCIAM 31,297,63
ivn-rbt LDLo:45 mg/kg AEPPAE 160,551,31
ihl-rat LCLo:114 g/m^3/30M RPTOAN 41,113,78
ipr-rat LD50:50 mg/kg RPTOAN 41,113,78

CONSENSUS REPORTS: Reported in EPA TSCA Inventory.

DOT Classification: Corrosive Material; Label: Corrosive, solid; Corrosive Material; Label: Corrosive Solution.

SAFETY PROFILE: Poison by ingestion, intraperitoneal, and intravenous routes. Irritating and corrosive to skin and mucous membranes. Mutation data reported. When heated to decomposition it emits toxic fumes of Br^-. See also BROMIDES.

BMS000 CAS:62116-25-0 ***HR: 3***
BROMOACETONE OXIME
mf: C_3H_6BrNO mw: 151.98

$BrCH_2C(CH_3)=NOH$

SYN: 1-BROMO-2-OXIMINOPROPANE

SAFETY PROFILE: Decomposes explosively during distillation. When heated to decomposition it emits toxic fumes of Br^- and NO_x. See also CHLOROACETONE and BROMIDES.

BMS250 CAS:4189-47-3 ***HR: 2***
1-BROMOACETOXY-2-PROPANOL
mf: $C_5H_9BrO_3$ mw: 197.05

SYN: NALCON 240

TOXICITY DATA with REFERENCE
orl-rat LD50:664 mg/kg PCOC** -,152,66
skn-rbt LD50:813 mg/kg PCOC** -,152,66

SAFETY PROFILE: Moderately toxic by ingestion and skin contact. When heated to decomposition it emits toxic fumes of Br^-.

BMS500 ***HR: 3***
BROMOACETYLENE
mf: C_2HBr mw: 104.9

PROP: Gas. Bp: −2°, vap d: 4.684

SYNS: BROMACETYLENE ◇ BROMOETHYNE

SAFETY PROFILE: Toxicity is probably similar to dibromoacetylene. A dangerous fire hazard by spontaneous chemical reaction. A spontaneously flammable gas. Highly explosive. May explode or ignite on contact with air. Incompatible with oxidizing materials, even when solid at −196°. When heated to decomposition it burns and emits toxic fumes of Br^-. See also ACETYLENE COMPOUNDS and BROMIDES.

BMS750 CAS:58682-45-4 ***HR: 3***
4′-(3-BROMO-9-ACRIDINYLAMINO)METHANESULFONANILIDE
mf: $C_{20}H_{16}BrN_3O_2S$ mw: 442.36

TOXICITY DATA with REFERENCE
mmo-sat 25 μmol/L JMCMAR 23,269,80
ipr-mus LD10:33 mg/kg JMCMAR 23,269,80

SAFETY PROFILE: Poison by intraperitoneal route. Mutation data reported. When heated to decomposition it emits very toxic fumes of Br^-, NO_x, and SO_x. See also SULFONATES.

BMT000 CAS:14925-39-4 ***HR: D***
2-BROMOACROLEIN
mf: C_3H_3BrO mw: 134.96

SYN: 2-BROMOPROPENALDEHYDE

TOXICITY DATA with REFERENCE
mmo-sat 1 nmol/plate MUREAV 78,113,80
mma-sat 1 nmol/plate MUREAV 78,113,80
dnd-rat:oth 1 μmol/L CRNGDP 6,705,85
otr-ham:emb 500 nmol/L CRNGDP 6,705,85

SAFETY PROFILE: Mutation data reported. See also ALDEHYDES and BROMIDES. When heated to decomposition it emits toxic fumes of Br^-.

BMT250 CAS:14519-10-9 ***HR: 3***
BROMOAMINE
mf: BrH_2N mw: 95.93

SYN: BROMAMIDE

SAFETY PROFILE: Decomposes violently @ −70°. Upon decomposition it emits toxic fumes of Br^- and NO_x. See also BROMIDES and AMINES.

BMT300 ***HR: 2***
3′-BROMO-trans-ANETHOLE
mf: $C_{10}H_{11}BrO$ mw: 227.12

SYNS: ANISOLE, p-(3-BROMOPROPENYL)-, (E)- ◇ (E)-p-(3-BROMOPROPENYL)ANISOLE

TOXICITY DATA with REFERENCE
mmo-sat 2 μmol/plate CRNGDP 7,2089,86
ipr-mus TDLo:69600 μg/kg/4D-I:CAR CNREA8 47,2275,87

SAFETY PROFILE: Questionable carcinogen with experimental carcinogenic data. Mutation data reported. When heated to decomposition it emits toxic fumes of Br^-.

BMT325 CAS:106-40-1 ***HR: 3***
4-BROMOANILINE
mf: C_6H_6BrN mw: 172.04

PROP: Rhombic crystals from dil alc. Mp: 66-66.5°, d: 1.4970 (liq). Very sol in alc and ether; insol in cold water.

SYNS: 4-BROMANILINU (CZECH) ◇ p-BROMOANILINE ◇ 4-BROMOBENZENAMINE (9CI) ◇ p-BROMOPHENYLAMINE

TOXICITY DATA with REFERENCE
dns-rat:lvr 50 μmol/L ENMUDM 3,11,81
orl-rat LD50:456 mg/kg CEHYAN 23,168,78
orl-mus LD50:289 mg/kg GISAAA 44(12),19,79
ipr-mus LD50:248 mg/kg GISAAA 44(12),19,79

CONSENSUS REPORTS: EPA Genetic Toxicology Program. Reported in EPA TSCA Inventory.

SAFETY PROFILE: Poison by ingestion and intraperitoneal routes. Mutation data reported. When heated to decomposition it emits toxic fumes of Br^- and NO_x. See also ANILINE DYES.

BMT500 CAS:19816-89-8 ***HR: 3***
1-BROMOAZIRIDINE
mf: C_2H_4BrN mw: 121.96

SAFETY PROFILE: An unstable material which may spontaneously explode. When heated to decomposition it emits toxic fumes of Br^- and NO_x. See other aziridine compounds.

BMT750 CAS:32795-84-9 ***HR: 3***
10-BROMO-1,2-BENZANTHRACENE
mf: $C_{18}H_{11}Br$ mw: 307.20

SYN: 10-BROM-1,2-BENZANTHRACEN(GERMAN)

TOXICITY DATA with REFERENCE
scu-rat TDLo:25 mg/kg:ETA SCPHA4 22,224,54

SAFETY PROFILE: Questionable carcinogen with experimental tumorigenic data. When heated to decomposition it emits toxic fumes of Br^-.

BMU000 CAS:81-96-9 ***HR: 3***
3-BROMOBENZ(d,e)ANTHRONE
mf: $C_{17}H_9BrO$ mw: 309.17

SYNS: 3-BROMBENZANTHRONE ◇ 3-BROMO-7H-BENZ(DE)ANTHRACEN-7-ONE ◇ 7-BROMOMESOBENZANTHRONE

TOXICITY DATA with REFERENCE
eye-rbt 500 mg/24H MLD 28ZPAK -,89,72
ipr-rat LD50:2400 mg/kg RPTOAN 40,137,77
ipr-mus LD50:300 mg/kg RPTOAN 40,137,77

CONSENSUS REPORTS: Reported in EPA TSCA Inventory.

SAFETY PROFILE: Poison by intraperitoneal route. An eye irritant. When heated to decomposition it emits toxic fumes of Br^-.

BMU500 CAS:21248-00-0 ***HR: 3***
6-BROMOBENZO(a)PYRENE
mf: $C_{20}H_{11}Br$ mw: 331.22

TOXICITY DATA with REFERENCE
scu-mus TDLo:40 mg/kg:ETA BJCAAI 26,506,72

SAFETY PROFILE: Questionable carcinogen with experimental tumorigenic data. When heated to decomposition it emits toxic fumes of HBr. See also BROMIDES.

BMU750 CAS:14733-73-4 ***HR: 3***
5-BROMO-2-BENZOXAZOLINONE
mf: $C_7H_4BrNO_2$ mw: 214.03

TOXICITY DATA with REFERENCE
orl-rat LD50:1050 mg/kg MDCHAG 4(1),308,64
orl-mus LD50:1440 mg/kg MDCHAG 4(1),308,64
ipr-mus LD50:262 mg/kg MDCHAG 4(1),308,64

SAFETY PROFILE: Poison by intraperitoneal route. Moderately toxic by ingestion. When heated to decomposition it emits very toxic fumes of Br- and NO_x.

BMV000 CAS:19932-85-5 ***HR: 2***
6-BROMO-2-BENZOXAZOLINONE
mf: $C_7H_4BrNO_2$ mw: 214.03

TOXICITY DATA with REFERENCE
orl-rat LD50:1000 mg/kg MDCHAG 4(1),308,64
orl-mus LD50:935 mg/kg MDCHAG 4(1),308,64
ipr-mus LD50:445 mg/kg MDCHAG 4(1),308,64

SAFETY PROFILE: Moderately toxic by ingestion and intraperitoneal routes. When heated to decomposition it emits very toxic fumes of Br^- and NO_x.

BMV250 CAS:14917-59-0 ***HR: 3***
p-BROMOBENZOYL AZIDE
mf: $C_7H_4BrN_3O$ mw: 229.04

$BrC_6H_4CO{\cdot}N_3$

PROP: Mp: 46°.

SAFETY PROFILE: Explodes when heated above its melting point. When heated to decomposition it emits toxic fumes of Br^- and NO_x. See also AZIDES.

BMV750 CAS:61-75-6 ***HR: 3***
(o-BROMOBENZYL)ETHYLDIMETHYLAMMONIUM-p-TOLUENESULFONATE
mf: $C_{11}H_{17}BrN{\cdot}C_7H_7O_3S$ mw: 414.40

SYNS: ASL-603 ◇ BRETYLAN ◇ BRETYLATE ◇ BRETYLIUM-p-TOLUENESULFONATE ◇ BRETYLIUM TOSYLATE ◇ BRETYLOL ◇ 2-BROMO-N-ETHYL-N,N-DIMETHYLBENZENEMETHANAMINIUM 4-METHYLBENZENESULFONATE ◇ DARENTHIN ◇ N-ETHYL-N-o-BROMOBENZYL-N,N-DIMETHYLAMMONIUM TOSYLATE ◇ ORNID

TOXICITY DATA with REFERENCE
ipr-rat LD50:57 mg/kg PHMGBN 21,256,80
ims-rat LD50:250 mg/kg TXAPA9 18,185,71
orl-mus LD50:400 mg/kg BJPCAL 14,536,59
ipr-mus LD50:39 mg/kg AIPTAK 155,69,65
scu-mus LD50:72 mg/kg BJPCAL 14,536,59
ivn-mus LD50:20 mg/kg BJPCAL 14,536,59

SAFETY PROFILE: A poison by ingestion, intraperitoneal, subcutaneous, intravenous, and intramuscular routes. An anti-adrenergic agent and antiarrhythmic cardiac depressant. When heated to decomposition it emits very toxic fumes of SO_x, NH_3, NO_x, and Br^-. See also SULFONATES.

BMW000 CAS:33855-47-9 ***HR: 3***
N-p-BROMOBENZYL-N'-ETHYL-N'-METHYL-N-2-PYRIDYLETHYLENEDIAMINE MALEATE
mf: $C_{17}H_{22}BrN_3{\cdot}C_4H_4O_4$ mw: 464.41

SYNS: N-p-BROMBENZYL-N-α-PYRIDYL-N'-METHYL-N'-AETHYL-AETHYLENDIAMIN-MALEINAT (GERMAN) ◇ WV 761

TOXICITY DATA with REFERENCE
orl-mus LD50:620 mg/kg ARZNAD 14,940,64
scu-mus LD50:119 mg/kg ARZNAD 14,940,64
ivn-mus LD50:16700 μg/kg ARZNAD 14,940,64

SAFETY PROFILE: Poison by subcutaneous and intravenous routes. Moderately toxic by ingestion. When heated to decomposition it emits very toxic fumes of Br^- and NO_x.

BMW250 CAS:5798-79-8 ***HR: 3***
BROMOBENZYLNITRILE
DOT: UN 1694
mf: C_8H_6BrN mw: 196.06

PROP: Pure: Yellowish-white crystals. Tech: brown, oily liquid with pungent odor of sour fruit; mp: 29°, bp: 242°, fp: 25.5°, flash p: none, d: 1.5160 @ 20°, vap d: 6.8, vap press: 0.011 mm @ 20°.

SYNS: BBC ◇ BBN ◇ BROMBENZYL CYANIDE ◇ α-BROMOBENZYL CYANIDE ◇ α-BROMOBENZYLNITRILE ◇ α-BROMOPHENYLACETONITRILE ◇ α-BROMO-α-TOLUNITRILE ◇ CA ◇ CAMITE

TOXICITY DATA with REFERENCE
ihl-hmn LC50:3500 mg/m³ SCJUAD 4,33,67
orl-rat LDLo:100 mg/kg NCNSA6 5,32,53

CONSENSUS REPORTS: Cyanide and its compounds are on the Community Right-To-Know List.

DOT Classification: IMO: Poison B; Label: Poison.

SAFETY PROFILE: Poison by ingestion. Moderately toxic to humans by inhalation. When heated to decomposition it emits very toxic fumes of NO_x, Br^-, and CN^-. See also NITRILES.

BMX000 CAS:60883-74-1 ***HR: 3***
α-BROMO-β,β-BIS(p-ETHOXYPHENYL)STYRENE
mf: $C_{24}H_{23}BrO_2$ mw: 423.38

SYN: α,α-DI(p-ETHOXYPHENYL)-β-BROMO-β-PHENYLETHYLENE

TOXICITY DATA with REFERENCE
scu-mus TDLo:94 mg/kg/26W-I:CAR MMJJAI 11,95,61

SAFETY PROFILE: Questionable carcinogen with experimental carcinogenic data. When heated to decomposition it emits toxic fumes of Br^-.

BMX250 CAS:34346-98-0 ***HR: 3***
4-BROMO-7-BROMOMETHYLBENZ(a)ANTHRACENE
mf: $C_{19}H_{12}Br_2$ mw: 400.13

TOXICITY DATA with REFERENCE
skn-mus TDLo:16 mg/kg:ETA EJCAAH 7,473,71

SAFETY PROFILE: Questionable carcinogen with experimental tumorigenic data. When heated to decomposition it emits toxic fumes of Br^-. See also BROMIDES.

BMX500 CAS:109-65-9 ***HR: 3***
1-BROMOBUTANE
DOT: UN 1126
mf: C_4H_9Br mw: 137.04

PROP: Colorless to pale straw-colored liquid. Mp: −112.4°, bp: 101.4°, flash p: 65°F (OC), d: 1.274 @ 25°/25°, autoign temp: 509°F, vap d: 4.72, lel: 2.8% @ 212°F, uel: 6.6% @ 212°F.

SYNS: BUTYL BROMIDE (DOT) ◇ BUTYL BROMIDE, NORMAL (DOT) ◇ N-BUTYL BROMIDE

TOXICITY DATA with REFERENCE
ihl-mam LC50:25800 mg/m^3 GTPZAB 18(4),55,74
ihl-rat LC50:237000 mg/m^3/30M FAVUAI 7,35,75
ipr-rat LD50:4450 mg/kg JPCEAO 320(1),133,78
ipr-mus LD50:6680 mg/kg JPCEAO 320(1),133,78
ipr-mam LD50:1424 mg/kg GTPZAB 18(4),55,74

CONSENSUS REPORTS: EPA Genetic Toxicology Program. Reported in EPA TSCA Inventory.

DOT Classification: Flammable Liquid; Label: Flammable Liquid.

SAFETY PROFILE: Moderately toxic by intraperitoneal route. Mildly toxic by inhalation. Dangerous fire hazard when exposed to heat, flame, or oxidizers. Violent reaction with bromobenzene + sodium above 30°C. Can react with oxidizing materials. To fight fire, use CO_2, dry chemical, mist or spray. See also BROMIDES.

BMX750 CAS:78-76-2 ***HR: 3***
2-BROMOBUTANE
DOT: UN 2339
mf: C_4H_9Br mw: 137.04

PROP: Colorless liquid, fp: < −50°, bp: 91.4°, flash p: 70°F, d: 1.257 @ 25°/25°.

SYNS: sec-BUTYL BROMIDE ◇ METHYLETHYLBROMOMETHANE

TOXICITY DATA with REFERENCE
ipr-mus TDLo:3000 mg/kg/8W-I:NEO CNREA8 35,1411,75

CONSENSUS REPORTS: EPA Genetic Toxicology Program. Reported in EPA TSCA Inventory.

DOT Classification: Flammable Liquid; Label: Flammable Liquid.

SAFETY PROFILE: Narcotic in high concentrations. Questionable carcinogen with experimental neoplastigenic data. See also BROMIDES and CHLORINATED HYDROCARBONS, ALIPHATIC. Dangerous fire hazard when exposed to heat or flame. When heated to decomposition it emits toxic fumes of Br^-; can react with oxidizing materials. To fight fire, use water, spray or mist, foam, CO_2, dry chemical.

BMX825 CAS:5162-44-7 ***HR: 3***
4-BROMO-1-BUTENE
mf: C_4H_7Br mw: 135.00

$$H_2C{=}CHCH_2CH_2Br$$

SAFETY PROFILE: A dangerous fire hazard (flash point < 1°C). Violent reaction with chloromethylphenylsilane + chloroplatinic acid. When heated to decomposition it emits toxic fumes of Br^-. See also BROMIDES.

BMY250 CAS:80-58-0 ***HR: 3***
2-BROMOBUTYRIC ACID
mf: $C_4H_7BrO_2$ mw: 167.02

PROP: Colorless, oily liquid; sol in alc and ether; sparingly sol in water. D: 1.54, bp: 181° @ 250 mm, mp: −4°.

SYN: α-BROMOBUTYRIC ACID

TOXICITY DATA with REFERENCE
orl-mus LD50:310 mg/kg JPETAB 86,336,46

CONSENSUS REPORTS: Reported in EPA TSCA Inventory.

SAFETY PROFILE: Poison by ingestion. Dangerous; when heated to decomposition it emits toxic fumes of Br^-. See also BROMIDES.

BMY500 CAS:5332-06-9 ***HR: 3***
4-BROMOBUTYRONITRILE
mf: C_4H_6BrN mw: 148.02

SYN: USAF DO-6

TOXICITY DATA with REFERENCE
ipr-mus LD50:100 mg/kg NTIS** AD277-689

CONSENSUS REPORTS: Reported in EPA TSCA Inventory. Cyanide and its compounds are on the Community Right-To-Know List.

SAFETY PROFILE: Poison by intraperitoneal route. When heated to decomposition it emits very toxic fumes of NO_x, CN^-, and Br^-. See also BROMIDES and NITRILES.

BMY800 CAS:83463-62-1 ***HR: 2***
BROMOCHLOROACETONITRILE
mf: $C_2HBrClN$ mw: 154.40

SYN: BROMOCHLOROMETHYL CYANIDE

TOXICITY DATA with REFERENCE
mmo-sat 1 nmol/plate ENMUDM 5,447,83
mma-sat 170 nmol/plate FAATDF 5,1065,85
dnd-hmn:lyms 2 μmol/L NTIS** PB84-246230
sce-ham:ovr 4200 nmol/L FAATDF 5,1065,85
orl-rat TDLo:825 mg/kg (7-21D post):REP TXCYAC 46,83,87
skn-mus TDLo:2400 mg/kg/2W-I:CAR FAATDF 5,1065,85

SAFETY PROFILE: Experimental reproductive data. Questionable carcinogen with experimental carcinogenic data. Mutation data reported. When heated to decomposition it emits toxic fumes of Br^-, Cl^-, and NO_x.

BMY825 CAS:25604-70-0 *HR: 3*
BROMOCHLOROACETYLENE
mf: C_2BrCl mw: 139.38

SAFETY PROFILE: An unstable high explosive. When heated to decomposition it emits toxic fumes of Cl^- and Br^-. See also ACETYLENE COMPOUNDS, BROMIDES, and CHLORIDES.

BMZ000 CAS:5579-85-1 *HR: 2*
6-BROMO-5-CHLORO-2-BENZOXAZOLINONE
mf: $C_7H_3BrClNO_2$ mw: 248.47

SYNS: BROMCHLORENONE ◇ 6-BROMO-5-CHLOROBENZOXAZOLONE ◇ NSC-24970

TOXICITY DATA with REFERENCE
orl-rat LD50:500 mg/kg MDCHAG 4(1),308,64
orl-mus LD50:871 mg/kg MDCHAG 4(1),308,64
skn-rbt LD50:3160 mg/kg HAZL** -,-,62

SAFETY PROFILE: Moderately toxic by ingestion and skin contact. When heated to decomposition it emits very toxic fumes of Cl^-, Br^-, and NO_x.

BNA000 CAS:758-24-7 *HR: 2*
2-BROMO-2-CHLORO-1,1-DIFLUOROETHYLENE
mf: C_2BrClF_2 mw: 177.38

SYNS: 1-BROMO-1-CHLORO-2,2-DIFLUOROETHENE ◇ 1-BROMO-1-CHLORO-2,2-DIFLUOROETHYLENE

TOXICITY DATA with REFERENCE
mma-sat 15 mmol/L ANESAV 51,424,79
oms-bcs 133 mol/L MUREAV 54,17,78
ihl-mus LC50:250 ppm/1H BJANAD 37,716,65

SAFETY PROFILE: Moderately toxic by inhalation. Mutation data reported. When heated to decomposition it emits very toxic fumes of Br^-, Cl^-, and F^-. See also CHLORINATED HYDROCARBONS, ALIPHATIC.

BNA250 CAS:353-59-3 *HR: 1*
BROMOCHLORODIFLUOROMETHANE
DOT: UN 1974
mf: $CBrClF_2$ mw: 165.37

PROP: Colorless gas.

SYNS: CHLORODIFLUOROBROMOMETHANE (DOT) ◇ CHLORODIFLUOROMONOBROMOMETHANE ◇ HALON 1211

TOXICITY DATA with REFERENCE
mmo-sat 10 pph MUREAV 142,187,85
mma-sat 5 pph MUREAV 142,187,85
ihl-rat LCLo:32 pph/15M FLCRAP 1,197,67

CONSENSUS REPORTS: Reported in EPA TSCA Inventory.

DOT Classification: Nonflammable Gas; Label: Nonflammable Gas.

SAFETY PROFILE: Mutation data reported. An asphyxiant. See also ARGON for description of inert gas asphyxiants. When heated to decomposition it emits very toxic fumes of Br^-, Cl^-, and F^-.

BNA325 CAS:126-06-7 *HR: 2*
3-BROMO-1-CHLORO-5,5-DIMETHYLHYDANTOIN
mf: $C_5H_6BrClN_2O_2$ mw: 241.49

SYN: 3-BROMO-1-CHLORO-5,5-DIMETHYL-2,4-IMIDAZOLIDINEDIONE

TOXICITY DATA with REFERENCE
skn-rbt 500 mg/24H SEV EPASR* 8EHQ-0181-0382
eye-rbt 100 mg/30S SEV EPASR* 8EHQ-0181-0382
orl-rat LD50:600 mg/kg EPASR* 8EHQ-0181-0382
orl-mus LD50:680 mg/kg EPASR* 8EHQ-0181-0382
skn-rbt LDLo:2 g/kg EPASR* 83HQ-0281-0382

CONSENSUS REPORTS: Reported in EPA TSCA Inventory.

SAFETY PROFILE: Moderately toxic by skin contact and ingestion. A severe eye and skin irritant. When heated to decomposition it emits toxic fumes of Cl^-, Br^-, and NO_x.

BNA500 CAS:796-13-4 *HR: D*
1-BROMO-1-(p-CHLOROPHENYL)-2,2-DIPHENYLETHYLENE
mf: $C_{20}H_{14}BrCl$ mw: 369.70

SYN: ETHYLENE,1-BROMO-1-(p-CHLOROPHENYL)-2,2-DIPHENYL-

TOXICITY DATA with REFERENCE
scu-rat TDLo:250 mg/kg (female 10D pre):REP JMCMAR 8,415,65

SAFETY PROFILE: Experimental reproductive effects. When heated to decomposition it emits toxic fumes of Br^- and Cl^-.

BNA750 CAS:41198-08-7 *HR: 3*
O-(4-BROMO-2-CHLOROPHENYL)-O-ETHYL-S-PROPYL PHOSPHOROTHIOATE
mf: $C_{11}H_{15}BrClO_3PS$ mw: 373.65

SYNS: CGA 15324 ◇ CURACRON ◇ POLYCRON ◇ PROFENOFOS ◇ SELECRON

TOXICITY DATA with REFERENCE
orl-rat LD50:400 mg/kg SPEADM 78-1,35,78
skn-rat LD50:300 mg/kg CIGET* -,-,77
orl-mus LD50:162 mg/kg TXAPA9 73,16,84
orl-mus LD50:298 mg/kg CIGET* -,-,77
orl-rbt LD50:700 mg/kg CIGET* -,-,77
skn-rbt LD50:192 mg/kg FMCHA2 -,C65,83
orl-ckn LD50:1900 µg/kg TXAPA9 73,16,84
skn-rbt LD50:472 mg/kg CIGET* -,-,77

SAFETY PROFILE: Poison by ingestion and skin contact. When heated to decomposition it emits very toxic SO_x, PO_x, Br^-, and Cl^-. See also ESTERS.

BNA825 CAS:109-70-6 ***HR: 2***
1-BROMO-3-CHLOROPROPANE
DOT: UN 2688
mf: C_3H_6BrCl mw: 157.45

SYNS: 3-BROMOPROPYL CHLORIDE ◇ 1,3-CHBP ◇ omega-CHLOROBROMOPROPANE ◇ 1-CHLORO-3-BROMOPROPANE (DOT) ◇ 3-CHLOROPROPYL BROMIDE ◇ TRIMETHYLENE BROMIDE CHLORIDE ◇ TRIMETHYLENE CHLOROBROMIDE

TOXICITY DATA with REFERENCE
orl-rat LD50:930 mg/kg TPKVAL 12,93,71
ihl-rat LC50:5668 mg/m^3 GTPZAB 19(9),36,75
orl-mus LD50:1290 mg/kg 85GMAT -,35,82
ihl-mus LCLo:7270 mg/m^3/2H 85GMAT -,35,82

CONSENSUS REPORTS: Reported in EPA TSCA Inventory.

DOT Classification: Poison B; Label: St. Andrews Cross.

SAFETY PROFILE: Moderately toxic by ingestion. When heated to decomposition it emits toxic fumes of Cl^- and Br^-. See also CHLORINATED HYDROCARBONS, ALIPHATIC, and BROMIDES.

BNB250 CAS:25614-03-3 ***HR: 3***
BROMOCRIPTINE
mf: $C_{32}H_{40}BrN_5O_5$ mw: 654.68

SYNS: BROMOCRIPTIN ◇ α-BROMOERGOCRIPTINE ◇ BROMOERGOCRYPTINE ◇ 2-BROMOERGOCRYPTINE ◇ 2-BROMO-α-ERGOKRYPTIN ◇ 2-BROMO-12'-HYDROXY-2'-(1-METHYLETHYL)-5'-α-(2-METHYLPROPYL)ERGOTAMIN-3',6',18-TRIONE ◇ CB-154

TOXICITY DATA with REFERENCE
oms-hmn:lym 100 μmol/L MUREAV 117,163,83
dna-rat-ipr 4 mg/kg CNREA8 36,2223,76
unr-wmn TDLo:2800 μg/kg (24-52D preg):TER BCPHBM 5,227,78
scu-rat TDLo:70 mg/kg (female 13-19D post):REP TJADAB 33,100C,86
orl-rat TDLo:7 g/kg/2Y-C:ETA,REP BMJOAE 2,1605,77
orl-wmn TDLo:6 mg/kg/60D-I:NOSE NEJMAG 306,178,82
ivn-rat LD50:72 mg/kg DRUGAY 17,313,78
ivn-rbt LD50:12 mg/kg USXXAM #3752814
unr-mus LD50:200 mg/kg BBIADT 43,1305,84

CONSENSUS REPORTS: EPA Genetic Toxicology Program.

SAFETY PROFILE: Poison by intravenous and possibly other routes data. Human teratogenic effects by an unspecified route: developmental abnormalities of the respiratory system, musculoskeletal system, urogenital system, craniofacial area and body wall. Human systemic effects by ingestion including: olfaction changes. An experimental teratogen. Other experimental reproductive effects. Human mutation data reported. Questionable carcinogen with experimental tumorigenic data. When heated to decomposition it emits very toxic fumes such as Br^- and NO_x.

BNB325 CAS:22260-51-1 ***HR: 3***
BROMOCRIPTINE MESILATE
mf: $C_{32}H_{40BrN5}O_5•CH_4O_3S$ mw: 714.43

SYNS: 2-BROMO-α-ERGOCRYPTINE METHANESULFONATE ◇ 2-BROMO-α-ERGOKRYPTINE-MESILATE (GERMAN) ◇ CB-154 ◇ PARLODEL

TOXICITY DATA with REFERENCE
orl-wmn TDLo:1400 mg/kg (lactating female 1-14D post):REP JRPMAP 33,630,88
par-rat TDLo:35 mg/kg (female 14-20D post):TER GCENA5 39,118,79
orl-cld TDLo:375 μg/kg JOPDAB 105,838,84
orl-wmn TDLo:1 mg/kg/20D-I AJPSAO 143,935,85
orl-man TDLo:52 mg/kg/35W-I:CNS NEURAI 35,1193,85
ivn-rat LD50:10500 μg/kg YKYUA6 29,1231,78
orl-mus LD50:2502 mg/kg YKYUA6 30,809,79
ivn-mus LD50:189 mg/kg YKYUA6 30,809,79
ivn-rbt LD50:8200 μg/kg IYKEDH 10,232,79

SAFETY PROFILE: Poison by intravenous route. Moderately toxic by ingestion. Human systemic effects by ingestion: cerebral spinal fluid changes. An experimental teratogen. Other experimental reproductive effects. When heated to decomposition it emits toxic fumes of Br^-, SO_x, and NO_x.

BNB750 ***HR: 3***
1-BROMO-12-CYCLOTRIDECADIEN-4,8,10-TRIYNE
mf: $C_{13}H_9Br$ mw: 245.12

SAFETY PROFILE: Explodes @ 65° and decomposes @ 0° in the dark.

BNC750 CAS:59-14-3 ***HR: 2***
5-BROMO-2'-DEOXY URIDINE
mf: $C_9H_{11}BrN_2O_5$ mw: 307.13

SYNS: BDU ◇ 5-BDU ◇ BROMODEOXYURIDINE ◇ 5-BROMODEOXYURIDINE ◇ 5-BROMO-2-DEOXYURIDINE ◇ 5-BROMODESOXYURIDINE ◇ BROMOURACIL DEOXYRIBOSIDE ◇ 5-BROMOURACIL DEOXYRIBOSIDE ◇ 5-BROMOURACIL-2-DEOXYRIBOSIDE ◇ BROXURIDINE ◇ BRUDR ◇ BUDR ◇ 5-BUDR

TOXICITY DATA with REFERENCE
mnt-hmn:fbr 82 μmol/L MUREAV 4,353,67
cyt-hmn:leu 200 mg/L ECREAL 34,182,64
msc-hmn:fbr 15 mg/L CSHSAZ 29,151,65

ipr-rat TDLo:700 mg/kg (female 9-15D post):REP TAKHAA 30,636,71
ipr-rat TDLo:250 mg/kg (female 13D post):TER TJADAB 23,383,81
orl-rat LD50:8400 mg/kg IYKEDH 4,467,73
scu-rat LD50:3900 mg/kg TAKHAA 30,530,71
unr-rat LD50:2300 mg/kg NIIRDN 6,721,82
orl-mus LD50:9100 mg/kg TAKHAA 30,530,71
ipr-mus LD50:3050 mg/kg TAKHAA 30,530,71
scu-mus LD50:3500 mg/kg TAKHAA 30,530,71
ivn-mus LD50:2500 mg/kg TAKHAA 30,530,71

CONSENSUS REPORTS: Reported in EPA TSCA Inventory. EPA Genetic Toxicology Program.

SAFETY PROFILE: Moderately toxic by subcutaneous, intravenous, intraperitoneal, and possibly other routes. Mildly toxic by ingestion. Experimental teratogenic and reproductive effects. Human mutation data reported. When heated to decomposition it emits very toxic fumes of Br^- and NO_x.

BNC800 CAS:27312-17-0 ***HR: 1***
2-BROMO-1,5-DIAMINO-4,8-DIHYDROXYANTHRAQUINONE
mf: $C_{14}H_9BrN_2O_4$ mw: 349.16

SYNS: ANTHRAQUINONE,2-BROMO-1,5-DIAMINO-4,8-DIHYDROXY- ◇ MODR OSTACETOVA LR

TOXICITY DATA with REFERENCE
eye-rbt 500 mg/24H MLD 28ZPAK-,245,72

CONSENSUS REPORTS: Reported in EPA TSCA Inventory.

SAFETY PROFILE: An eye irritant. When heated to decomposition it emits toxic fumes of NO_x and Br^-.

BND250 CAS:65235-63-4 ***HR: 1***
2-BROMO-1,8-DIAMINO-4,5-DIHYDROXYANTHRAQUINONE
mf: $C_{14}H_9BrN_2O_4$ mw: 349.16

SYN: MODR OSTACETOVA LG (CZECH)

TOXICITY DATA with REFERENCE
skn-rbt 500 mg/24H MLD 28ZPAK -,244,72
eye-rbt 500 mg/24H MLD 28ZPAK -,244,72
orl-rat LD50:12500 mg/kg 28ZPAK -,244,72

CONSENSUS REPORTS: Reported in EPA TSCA Inventory.

SAFETY PROFILE: Mildly toxic by ingestion. A skin and eye irritant. When heated to decomposition it emits very toxic fumes of Br^- and NO_x.

BND325 CAS:23834-96-0 ***HR: 3***
BROMODIBORANE
mf: B_2BrH_5 mw: 106.56

SAFETY PROFILE: May ignite violently on exposure to air. When heated to decomposition it emits toxic fumes of Br^-. See also BORANES and BORON COMPOUNDS.

BND500 CAS:75-27-4 ***HR: 3***
BROMODICHLOROMETHANE
mf: $CHBrCl_2$ mw: 163.83

PROP: Colorless liquid. Bp: 89.2–90.6°, d: 1.971 @ 25°/25°.

SYNS: BDCM ◇ DICHLOROBROMOMETHANE ◇ NCI-C55243

TOXICITY DATA with REFERENCE
mmo-sat 50 μL/plate DHEFDK FDA-78-1046,78
sce-hmn:lym 400 μmol/L ENVRAL 32,72,83
sce-mus-orl 200 mg/kg/4D-I ENVRAL 32,72,83
orl-rat TDLo:25500 mg/kg/2Y-C:CAR NTPTR* NTP-TR-321,87
orl-mus TDLo:25500 mg/kg/2Y-C:CAR NTPTR* NTP-TR-321,87
orl-rat LD50:916 mg/kg TXAPA9 52,351,80
orl-mus LD50:450 mg/kg TXAPA9 44,213,78

CONSENSUS REPORTS: NTP Carcinogenesis Studies (gavage): Clear Evidence: rat, mouse NTPTR* NTP-TR-321,87. EPA Genetic Toxicology Program. Community Right-To-Know List. Reported in EPA TSCA Inventory.

SAFETY PROFILE: Suspected carcinogen with experimental carcinogenic data. Moderately toxic by ingestion. Human mutation data reported. When heated to decomposition it emits very toxic fumes of Br^- and Cl^-. See also CHLORINATED HYDROCARBONS, ALIPHATIC and BROMIDES.

BND750 CAS:18936-66-8 ***HR: 3***
o-(4-BROMO-2,5-DICHLOROPHENYL)-o-ETHYL PHENYLPHOSPHONOTHIOATE
mf: $C_{14}H_{12}BrCl_2O_2PS$ mw: 426.10

SYN: VELSICOL FCS-303

TOXICITY DATA with REFERENCE
orl-mus LD50:75 mg/kg JAFCAU 27,1197,79
orl-gpg LDLo:100 mg/kg JEENAI 61,1261,68
scu-gpg LDLo:100 mg/kg JEENAI 61,1261,68

SAFETY PROFILE: Poison by ingestion and subcutaneous routes. When heated to decomposition it emits very toxic fumes of SO_x, PO_x, Cl^-, and Br^-. See also ESTERS.

BNE250 CAS:53581-53-6 ***HR: 3***
dl-4-BROMO-2,5-DIMETHOXYAMPHETAMINE HYDROBROMIDE
mf: $C_{11}H_{16}BrO_2$•BrH mw: 341.10

SYN: dl-4-BROMO-2,5-DIMETHOXY-α-METHYLPHENETHYLAMINE HYDROBROMIDE

TOXICITY DATA with REFERENCE
ipr-rat LD50:50 mg/kg TXAPA9 45(1),49,78
ivn-mus LD50:80 mg/kg TXAPA9 45(1),49,78
ivn-dog LD50:6400 μg/kg TXAPA9 45(1),49,78
orl-mky LD50:2 mg/kg TXAPA9 45,49,78

SAFETY PROFILE: Poison by ingestion, intraperitoneal and intravenous routes. See also BROMIDES and various amphetamine entries. When heated to decomposition it emits very toxic fumes of Br^-.

BNE325 CAS:70277-99-5 ***HR: 3***
2-BROMO-3,5-DIMETHOXYANILINE
mf: $C_8H_{10}BrNO_2$ mw: 232.08

SAFETY PROFILE: May explode when heated. Upon decomposition it emits toxic fumes of Br^- and NO_x.

BNE500 CAS:66969-02-6 ***HR: 3***
2-BROMO-N,N-DIMETHYL-1-ADAMANATANEMETHANAMINE HYDROCHLORIDEHEMIHYDRATE
mf: $C_{13}H_{22}BrN$•ClH•1/2H_2O mw: 317.74

SYN: 2-BROMO-1-(N,N-DIMETHYLAMINOMETHYL)ADAMANTANE HYDROCHLORIDEHEMIHYDRATE

TOXICITY DATA with REFERENCE
orl-mus LD50:413 mg/kg JMCMAR 19,967,76
ipr-mus LD50:159 mg/kg JMCMAR 19,967,76

SAFETY PROFILE: Poison by ingestion and intraperitoneal routes. When heated to decomposition it emits very toxic fumes of Br^-, NO_x, and HCl.

BNE600 CAS:17576-88-4 ***HR: 2***
3'-BROMO-4-DIMETHYLAMINOAZOBENZENE
mf: $C_{14}H_{14}BrN_3$ mw: 304.22

SYNS: ANILINE,p-(m-BROMOPHENYLAZO)-N,N-DIMETHYL- ◇ BENZENAMINE, 4-((3-BROMOPHENYL)AZO)-N,N-DIMETHYL-(9CI) ◇ p-(m-BROMOPHENYLAZO)-N,N-DIMETHYLANILINE

TOXICITY DATA with REFERENCE
orl-rat TDLo:7980 mg/kg/25W-C:CAR CBINA8 53,107,85
orl-rat LD50:13 g/kg NEOLA4 27,237,80

SAFETY PROFILE: Questionable carcinogen with experimental carcinogenic data. Low oral toxicity. When heated to decomposition it emits toxic fumes of Br^- and NO_x.

BNE750 CAS:980-71-2 ***HR: 3***
2-(p-BROMO-α-(2-(DIMETHYLAMINO)ETHYL) BENZYL)PYRIDINE MALEATE (1:1)
mf: $C_{16}H_{19}BrN_2$•$C_4H_4O_4$ mw: 435.36

SYNS: 2-(p-BROMO-α-(2-DIMETHYLAMINO)ETHYL)BENZYL)PYRIDINE BIMALEATE ◇ BROMOPHENIRAMINE MALEATE ◇ PARABROMODYLAMINE MALEATE

TOXICITY DATA with REFERENCE
orl-rat LD50:318 mg/kg 29ZVAB -,19,69
ipr-rat LD50:76 mg/kg 29ZVAB -,19,69

CONSENSUS REPORTS: Reported in EPA TSCA Inventory.

SAFETY PROFILE: Poison by ingestion and intraperitoneal routes. When heated to decomposition it emits very toxic fumes of Br^- and NO_x.

BNF000 CAS:52583-02-5 ***HR: 3***
2-BROMO-1-(3-DIMETHYLAMINOPROPYL)ADAMANTANE HYDROCHLORIDE
mf: $C_{15}H_{26}BrN$•ClH mw: 336.79

SYN: 2-BROMO-N,N-DIMETHYL-1-ADAMANTANEPROPANAMINE HYDROCHLORIDE

TOXICITY DATA with REFERENCE
orl-mus LD50:400 mg/kg JMCMAR 17,602,74
ipr-mus LD50:150 mg/kg JMCMAR 17,602,74

SAFETY PROFILE: Poison by ingestion and intraperitoneal routes. When heated to decomposition it emits very toxic fumes of Br^-, NO_x, and HCl.

BNF250 CAS:586-77-6 ***HR: 3***
p-BROMO-N,N-DIMETHYL ANILINE
mf: $C_8H_{10}BrN$ mw: 200.08

SYN: 4-BROMODIMETHYLANILINE

SAFETY PROFILE: May explode if heated. When heated to decomposition it emits toxic fumes of Br- and NO_x.

BNF300 ***HR: 2***
3-BROMO-7,12-DIMETHYLBENZ(a)ANTHRACENE
mf: $C_{20}H_{15}Br$ mw: 335.26

SYN: 3-BROMO-DMBA

TOXICITY DATA with REFERENCE
mma-sat 5 μg/plate CRNGDP 4,1221,83
skn-mus TDLo:520 mg/kg/50W-I:ETA CRNGDP 4,1221,83

SAFETY PROFILE: Questionable carcinogen with experimental tumorigenic data. Mutation data reported. When heated to decomposition it emits toxic fumes of Br^-.

BNF310 ***HR: 2***
4-BROMO-7,12-DIMETHYLBENZ(a)ANTHRACENE
mf: $C_{20}H_{15}Br$ mw: 335.26

SYN: 4-BROMO-DMBA

TOXICITY DATA with REFERENCE
mma-sat 10 μg/plate CRNGDP 4,1221,83
skn-mus TDLo:520 mg/kg/50W-I:CAR CRNGDP 4,1221,83

SAFETY PROFILE: Questionable carcinogen with experimental carcinogenic data. Mutation data reported. When heated to decomposition it emits toxic fumes of Br^-.

BNF315 CAS:63018-63-3 ***HR: 2***
5-BROMO-9,10-DIMETHYL-1,2-BENZANTHRACENE
mf: $C_{20}H_{15}Br$ mw: 335.26

SYN: BENZ(a)ANTHRACENE,8-BROMO-7,12-DIMETHYL-

TOXICITY DATA with REFERENCE
scu-mus TDLo:80 mg/kg:ETA CNREA8 6,454,46

SAFETY PROFILE: Questionable carcinogen with experimental tumorigenic data. When heated to decomposition it emits toxic fumes of Br^-.

BNF750 CAS:1463-08-7 ***HR: 3***
p-BROMO-α,α-DIMETHYLPHENETHYLAMINE HYDROCHLORIDE
mf: $C_{10}H_{14}BrN{\cdot}ClH$ mw: 264.62

SYN: S 84

TOXICITY DATA with REFERENCE
ipr-rat LD50:172 mg/kg APTOA6 17,121,60
orl-mus LD50:325 mg/kg CHTPBA 6,453,71

SAFETY PROFILE: Poison by ingestion and intraperitoneal routes. When heated to decomposition it emits very toxic fumes of Br^-, HCl, and NO_x.

BNG125 CAS:65036-47-7 ***HR: 3***
6-BROMO-2,4-DINITROBENZENEDIAZONIUM HYDROGEN SULFATE
mf: $C_6H_3BrN_4O_8S$ mw: 371.08

SAFETY PROFILE: Solution in sulfuric acid is explosive. When heated to decomposition it emits toxic fumes of Br^-, SO_x, and NO_x. See also SULFATES.

BNG250 ***HR: 3***
3-BROMO-2,7-DINITRO-5-BENZO(b)-THIOPHENEDIAZONIUM-4-OLATE
mf: $C_8HBrN_4O_5S$ mw: 345.09

SAFETY PROFILE: An explosive. When heated to decomposition it emits toxic fumes of Br^-, SO_x, and NO_x.

BNG750 CAS:776-74-9 ***HR: 3***
BROMODIPHENYLMETHANE
DOT: UN 1770
mf: $C_{13}H_{11}Br$ mw: 247.15

PROP: Solid. Mp: 45°; bp: 193° @ 26 mm. Decomp in hot water; sol in alc; very sol in benzene.

SYNS: DIPHENYLMETHYL BROMIDE (DOT) ◇ DIPHENYL METHYL BROMIDE, solid (DOT) ◇ DIPHENYL METHYL BROMIDE, solution (DOT)

CONSENSUS REPORTS: Reported in EPA TSCA Inventory.

DOT Classification: Corrosive Material; Label: Corrosive, Solid or Solution.

SAFETY PROFILE: A corrosive poison. When heated to decomposition it emits toxic fumes of Br^-. See also BROMIDES.

BNH000 CAS:776-74-9 ***HR: 3***
BROMODIPHENYLMETHANE (solution)
DOT: UN 1770
mf: $C_{13}H_{11}Br$ mw: 247.15

SYN: DIPHENYL METHYL BROMIDE, solution (DOT)

CONSENSUS REPORTS: Reported in EPA TSCA Inventory.

DOT Classification: Corrosive Material; Label: Corrosive.

SAFETY PROFILE: A corrosive, irritating liquid. When heated to decomposition it emits toxic fumes of Br^-. See also BROMODIPHENYLMETHANE and BROMIDES.

BNH500 CAS:17372-87-1 ***HR: 3***
BROMOEOSINE
mf: $C_{20}H_8Br_4O_5{\cdot}2Na$ mw: 693.90

SYNS: AIZEN EOSINE GH ◇ BROMO ACID ◇ BROMOFLUORESCEIC ACID ◇ BROMO FLUORESCEIN ◇ BRONZE BROMO ◇ CERTIQUAL EOSINE ◇ C.I. 45380 ◇ D&C RED No. 22 ◇ DISODIUM EOSIN ◇ EOSINE ◇ EOSINE SODIUM SALT ◇ EOSINE YELLOWISH ◇ EOSIN GELBLICH (GERMAN) ◇ FENAZO EOSINE XG ◇ HIDACID DIBROMO FLUORESCEIN ◇ IRGALITE BRONZE RED CL ◇ PHLOXINE TONER B ◇ PHLOX RED TONER X-1354 ◇ PURE EOSINE YY ◇ 11445 RED ◇ SODIUM EOSINATE ◇ SYMULER EOSIN TONER ◇ 2,4,5,7-TETRABROMO-9-o-CARBOXYPHENYL-6-HYDROXY-3-ISOXANTHONE, DISODIUM SALT ◇ 2,4,5,7-TETRABROMO-3,6-FLUORANDIOL ◇ TETRABROMOFLUORESCEIN ◇ 2′,4′,5′,7′-TETRABROMOFLUORESCEIN DISODIUM SALT ◇ TETRABROMOFLUORESCEIN S ◇ TETRABROMOFLUORESCEIN SOLUBLE ◇ 2-(2,4,5,7-TETRABROMO-6-HYDROXY-3-OXO-3H-XANTHENE-9-YL)BENZOIC ACID, DISODIUM SALT ◇ TOYO EOSINE G ◇ 1903 YELLOW PINK

TOXICITY DATA with REFERENCE
dnr-bcs 2 mg/disc TRENAF 27,153,76
scu-rat TDLo:13 g/kg/1Y-I:ETA GANNA2 47,51,56
ipr-rat LDLo:500 mg/kg IJLEAG 2,257,34
scu-rat LDLo:1500 mg/kg GANNA2 46,367,55
orl-mus LD50:2344 mg/kg EAPHA6 24,125,81
ivn-mus LD50:550 mg/kg TXAPA9 44,225,78
ivn-rbt LDLo:300 mg/kg IJLEAG 2,257,34

CONSENSUS REPORTS: IARC Cancer Review: Animal Inadequate Evidence IMEMDT 15,183,77. EPA Genetic Toxicology Program. Reported in EPA TSCA Inventory.

SAFETY PROFILE: Poison by intravenous route. Moderately toxic by ingestion, subcutaneous, and intraperitoneal routes. Questionable carcinogen with experimental tumorigenic data. When heated to decomposition it emits very toxic fumes of Br^- and Na_2O. See also BROMIDES.

BNI000 CAS:3132-64-7 ***HR: 3***
3-BROMO-1,2-EPOXYPROPANE
DOT: UN 2558
mf: C_3H_5BrO mw: 136.99

PROP: Flash p: < 22°.

SYNS: EPIBROMHYDRIN ◇ EPIBROMOHYDRIN (DOT) ◇ EPIBROMOHYDRINE

TOXICITY DATA with REFERENCE
mmo-sat 5 μmol/plate JTEHD6 5,1149,79
mmo-esc 20 μmol/L ARTODN 46,277,80

CONSENSUS REPORTS: EPA Genetic Toxicology Program. Reported in EPA TSCA Inventory.

DOT Classification: Poison B; Label: Flammable Liquid and Poison.

SAFETY PROFILE: Poison by intraperitoneal route. Human mutation data reported. A dangerous fire hazard when exposed to heat or flame. When heated to decomposition it emits toxic fumes of Br^-. See also BROMIDES.

BNI500 CAS:540-51-2 ***HR: 3***
2-BROMO ETHANOL
mf: C_2H_5BrO mw: 124.98

SYNS: BE ◇ BROMOETHANOL ◇ ETHYLENEBROMOHYDRIN ◇ GLYCOL BROMOHYDRIN

TOXICITY DATA with REFERENCE
mmo-sat 10 μL/plate EVHPAZ 21,79,77
mma-sat 10 μL/plate EVHPAZ 21,79,77
dnr-esc 10 μmol/plate EVHPAZ 21,79,77
dnr-bcs 20 μL/disc AEMIDF 43,177,82
mmo-klp 15 mmol/L EXPEAM 25,85,69
orl-mus TDLo:43 g/kg/80W-C:ETA JACTDZ 2(2),246,83
ipr-mus TDLo:150 mg/kg/8W-I:NEO CNREA8 39,391,79
ipr-mus LDLo:80 mg/kg TXAPA9 23,288,72

CONSENSUS REPORTS: EPA Genetic Toxicology Program. Reported in EPA TSCA Inventory.

SAFETY PROFILE: Poison by intraperitoneal route. Questionable carcinogen with experimental neoplastigenic and tumorigenic data. Mutation data reported. When heated to decomposition it emits toxic fumes of Br^-. See also BROMIDES.

BNK000 CAS:77-65-6 ***HR: 3***
2-BROMO-2-ETHYLBUTYRYLUREA
mf: $C_7H_{13}BrN_2O_2$ mw: 237.13

SYNS: ADALIN ◇ ADDISOMNOL ◇ N-(AMINOCARBONYL)-2-BROMO-2-ETHYLBUTANAMIDE ◇ BROMACETOCARBAMIDE ◇ BROMADAL ◇ BROMADEL ◇ BROMODIETHYLACETYLCARBAMIDE ◇ BROMODIETHYLACETYLUREA ◇ (α-BROMO-α-ETHYLBUTYRYL)CARBAMIDE ◇ (α-BROMO-α-ETHYLBUTYRYL)UREA ◇ 1-BROMO-ETHYL-BUTYRYL-UREA ◇ 2-BROMO-2-ETHYLBUTYRLUREA ◇ CARBOMAL ◇ DIACID ◇ DORMITURIN ◇ FYDALIN ◇ HOGGAR ◇ KARBROMAL ◇ KARTRYL ◇ NCI-C03805 ◇ NENESIN ◇ NYCTAL ◇ PARKOSED ◇ PELIDORM ◇ PIANADALIN ◇ PLANADALIN ◇ TILDIN ◇ URADAL

TOXICITY DATA with REFERENCE
cyt-smc 10 mmol/tube HEREAY 33,457,47
orl-rat LD50:316 mg/kg NCILB* NIH-NCI-E-C-72-3252,73
ipr-rat LD50:427 mg/kg ITMZBJ 17,305,80
ivn-rat LD50:427 mg/kg ARTODN 40,211,78
unr-rat LDLo:350 mg/kg JPHAA3 23,788,34
orl-mus LD50:464 mg/kg NCILB* NIH-NCI-E-C-72-3252,73
orl-dog LD50:450 mg/kg MEIEDD 10,254,83
scu-dog LDLo:300 mg/kg HBAMAK 4,1292,35
orl-cat LDLo:350 mg/kg HBAMAK 4,1293,35
orl-rbt LDLo:600 mg/kg SAPHAO 28,193,13
scu-frg LDLo:1667 mg/kg HBAMAK 4,1293,35

CONSENSUS REPORTS: NCI Carcinogenesis Bioassay (feed); No Evidence: mouse, rat NCITR* NCI-CG-TR-173,79. Reported in EPA TSCA Inventory.

SAFETY PROFILE: Poison by ingestion, subcutaneous, and possibly other routes. Moderately toxic via intravenous and intraperitoneal routes. Mutation data reported. A sedative, hypnotic and central nervous system depressant. When heated to decomposition it emits very toxic fumes of NO_x and Br^-.

BNK100 ***HR: 2***
4'-BROMO-3'-ETHYL-4-DIMETHYLAMINO-AZOBENZENE
mf: $C_{16}H_{18}BrN_3$ mw: 332.28

SYNS: ANILINE,p-((4-BROMO-3-ETHYLPHENYL)AZO)-N,N-DIMETHYL- ◇ BENZENAMINE, N,N-DIMETHYL-4'-BROMO-3'-ETHYL-

4-(PHENYLAZO)- ◇ p-((4-BROMO-3-ETHYLPHENYL)AZO)-N,N-DIMETHYLANILINE

TOXICITY DATA with REFERENCE
orl-rat TDLo:14414 mg/kg/52W-C:CAR CBINA8 53,107,85

SAFETY PROFILE: Questionable carcinogen with experimental carcinogenic data. When heated to decomposition it emits toxic fumes of Br^- and NO_x.

BNK250 ***HR: 3***
2-BROMO ETHYL ETHYL ETHER
mf: C_4H_9BrO mw: 155

PROP: Liquid. Vap d: 5.25, flash p: 5°.

SAFETY PROFILE: An insecticide. A dangerous fire hazard when exposed to heat or flame. See also ETHERS and BROMIDES.

BNK275 ***HR: 2***
p-((3-BROMO-4-ETHYLPHENYL)AZO)-N,N-DIMETHYLANILINE
mf: $C_{16}H_{18}BrN_3$ mw: 332.28

SYNS: ANILINE,p-((3-BROMO-4-ETHYLPHENYL)AZO)-N,N-DIMETHYL- ◇ BENZENAMINE, N,N-DIMETHYL-3′-BROMO-4′-ETHYL-4-(PHENYLAZO)- ◇ 3′-BROMO-4′-ETHYL-4-DIMETHYLAMINOAZOBENZENE

TOXICITY DATA with REFERENCE
orl-rat TDLo:6930 mg/kg/25W-C:CAR CBINA8 53,107,85

SAFETY PROFILE: Questionable carcinogen with experimental carcinogenic data. When heated to decomposition it emits toxic fumes of Br^- and NO_x.

BNK325 CAS:2758-06-7 ***HR: 3***
(2-BROMOETHYL)TRIMETHYLAMMONIUM BROMIDE
mf: $C_5H_{13}BrN•Br$ mw: 247.01

TOXICITY DATA with REFERENCE
orl-rat LD50:190 mg/kg QJPPAL 20,81,47
scu-rat LD50:60 mg/kg QJPPAL 20,81,47
orl-mus LD50:450 mg/kg QJPPAL 20,81,47
ipr-mus LD50:55 mg/kg QJPPAL 20,81,47
scu-mus LD50:65 mg/kg QJPPAL 20,81,47
ims-mus LD50:60 mg/kg QJPPAL 20,81,47

SAFETY PROFILE: Poison by ingestion, subcutaneous, intramuscular, and intraperitoneal routes. When heated to decomposition it emits toxic fumes of Br^-, NH_3, and NO_x.

BNK350 CAS:2028-52-6 ***HR: 2***
2-BROMOETHYNYL-2-BUTANOL
mf: C_6H_9BrO mw: 177.06

SYNS: BASON ◇ BROMOACETYLENYLETHYLMETHYLCARBINOL ◇ BROMOETHYNYLETHYLMETHYLCARBINOL ◇ 1-BROMO-3-METHYLPENTIN-3-OL ◇ 1-BROMO-3-METHYL-1-PENTYN-3-OL

TOXICITY DATA with REFERENCE
scu-rat LD50:940 mg/kg THERAP 10,56,55
orl-mus LD50:532 mg/kg THERAP 10,56,55
ipr-mus LD50:725 mg/kg AIPTAK 112,463,57
scu-mus LD50:910 mg/kg JPETAB 109,268,53

SAFETY PROFILE: Moderately toxic by ingestion and other routes. When heated to decomposition it emits toxic fumes of Br^-. See also ALCOHOLS.

BNK500 CAS:1940-57-4 ***HR: 3***
9-BROMOFLUORENE
mf: $C_{13}H_9Br$ mw: 245.13

TOXICITY DATA with REFERENCE
skn-man 2500 μg/48H SEV BJDEAZ 80,491,68
skn-hmn 2500 μg/24H SEV CHINAG (40),2080,67
ivn-mus LD50:180 mg/kg CSLNX* NX#01610

SAFETY PROFILE: Poison by intravenous route. A severe skin irritant in humans. When heated to decomposition it emits very toxic fumes of Br^-. See also BROMIDES.

BNK700 CAS:548-26-5 ***HR: 2***
BROMOFLUORESCEIC ACID
mf: $C_{20}H_8Br_4O_5•2Na$ mw: 693.90

SYNS: AIZEN EOSINE GH ◇ BROMO ACID ◇ BROMO B ◇ BROMOEOSINE ◇ BROMO FLUORESCEIN ◇ BRONZE BROMO ◇ CERTIQUAL EOSINE ◇ C.I. 45380 ◇ C.I. ACID RED 87 ◇ EOSIN ◇ EOSINE B ◇ EOSINE FA ◇ EOSINE LAKE RED Y ◇ FENAZO EOSINE XG ◇ FLUORESCEIN, 2′,4′,5′,7′-TETRABROMO-, DISODIUM SALT ◇ HIDACID BROMO ACID REGULAR ◇ HIDACID DIBROMO FLUORESCEIN ◇ IRGALITE BRONZE RED CL ◇ PHLOXINE RED 20-7600

TOXICITY DATA with REFERENCE
scu-rat TDLo:13 g/kg/1Y-I:ETA GANNA2 47,51,56
scu-rat LDLo:1500 mg/kg GANNA2 46,367,55
ivn-mus LD50:550 mg/kg TXAPA9 44,225,78

SAFETY PROFILE: Moderately toxic by subcutaneous and intravenous routes. Questionable carcinogen with experimental tumorigenic data. When heated to decomposition it emits toxic fumes of Br^-.

BNL000 CAS:75-25-2 ***HR: 3***
BROMOFORM
DOT: UN 2515
mf: $CHBr_3$ mw: 252.75

PROP: Colorless liquid or hexagonal crystals. Mp: 6-7°, bp: 149.5°, flash p: none, d: 2.890 @ 20°/4°.

SYNS: BROMOFORME (FRENCH) ◇ BROMOFORMIO (ITALIAN) ◇ METHENYL TRIBROMIDE ◇ NCI-C55130 ◇ RCRA WASTE NUMBER

U225 ◇ TRIBROMMETHAAN (DUTCH) ◇ TRIBROMMETHAN (GERMAN) ◇ TRIBROMOMETAN (ITALIAN) ◇ TRIBROMOMETHANE

TOXICITY DATA with REFERENCE
sln-dmg-orl 3000 ppm ENMUDM 7,677,85
sce-hmn:lym 80 μmol/L ENVRAL 32,72,83
sce-ham:ovr 290 μg/L ENMUDM 7,1,85
ipr-mus TDLo:1100 mg/kg/8W-I:NEO CNREA8 37,2717,77
orl-hmn LDLo:143 mg/kg 34ZIAG -,141,69
orl-rat LD50:1147 mg/kg TXAPA9 52,351,80
ihl-rat LCLo:45 g/m^3/4H 85GMAT -,28,82
ipr-rat LD50:414 mg/kg TOLED5 15,251,83
orl-mus LD50:1400 mg/kg TXAPA9 44,213,78
scu-mus LD50:1820 mg/kg TXAPA9 4,354,62
scu-rbt LDLo:410 mg/kg AEXPBL 28,201,1891

CONSENSUS REPORTS: Reported in EPA TSCA Inventory. Community Right-To-Know List.

OSHA PEL: TWA 0.5 ppm (skin)
ACGIH TLV: TWA 0.5 ppm (skin)
DOT Classification: Poison B; Label: St. Andrews Cross.

SAFETY PROFILE: A human poison by ingestion. Moderately toxic by intraperitoneal and subcutaneous routes. Human mutation data reported. A lachrymator. Questionable carcinogen with experimental neoplastigenic data. It can damage the liver to a serious degree and cause death. It has anesthetic properties similar to those of chloroform, but is not sufficiently volatile for inhalation purposes and is far too toxic for human use. As a sedative and antitussive its medicinal application has resulted in numerous poisonings. Inhalation of small amounts causes irritation, provoking the flow of tears and saliva, and reddening of the face. Abuse can lead to addiction and serious consequences. Explosive reaction with crown ethers or potassium hydroxide. Violent reaction with acetone or bases. Incompatible with Li or NaK alloys. When heated to decomposition it emits highly toxic fumes of Br^-. See also BROMIDES.

BNL250 CAS:2104-96-3 ***HR: 2***
BROMOFOSMETHYL
mf: $C_8H_8BrCl_2O_3PS$ mw: 366.00

SYNS: BROFENE ◇ O-(4-BROM-2,5-DICHLOR-PHENYL)-O,O-DIMETHYL-MONOTHIOPHOSPHAT (GERMAN) ◇ O-(4-BROMO-2,5-DICLORO-FENIL)-O,O-DIMETIL-MONOTIOFOSFATO(ITALIAN) ◇ 4-BROMO-2,5-DICHLOROPHENYL DIMETHYL PHOSPHOROTHIONATE ◇ O-(4-BROOM-2,5-DICHLOOR-FENYL)-O,O-DIMETHYL-MONOTHIOFOSFAAT (DUTCH) ◇ BROMOFOS ◇ BROMOPHOS ◇ BRUOMOPHOS (RUSSIAN) ◇ CELA S 1942 ◇ O,O-DIMETHYL-O-(4-BROMO-2,5-DICHLOROPHENYL) PHOSPHOROTHIOATE ◇ O,O-DIMETHYL-O-(2,5-DICHLOR-4-BROMPHENYL)-THIONOPHOSPHAT (GERMAN) ◇ O,O-DIMETHYL-O-(2,5-DICHLORO-4-BROMOPHENYL)-PHOSPHOROTHIOATE ◇ O,O-DIMETHYL-O-(2,5-DICHLORO-4-BROMOPHENYL) THIOPHOSPHATE ◇ EL 400 ◇ ENT 27,162 ◇ MONSANTO CP 51969 ◇ NETAL ◇ NEXION ◇ NEXION 40 ◇ OMS-658 ◇ S 1942 ◇ THIOPHOSPHATE de O,O-DIMETHYLE et de O-4-BROMO-2,5-DICHLOROPHENYLE (FRENCH)

TOXICITY DATA with REFERENCE
cyt-mus-ipr 73200 μg/kg RTOPDW 6,416,86
orl-mus TDLo:293 mg/kg (female 6-12D post):REP RTOPDW 6,416,86
orl-rat LD50:1600 mg/kg TXAPA9 14,515,69
ipr-rat LDLo:1625 mg/kg ATXKA8 22,36,66
orl-mus LD50:2829 mg/kg 28ZEAL 5,29,76
ipr-mus LD50:1040 mg/kg ATXKA8 22,36,66
orl-cat LDLo:750 mg/kg ATXKA8 22,36,66
skn-rbt LD50:2181 mg/kg 28ZEAL 5,29,76
orl-gpg LD50:1500 mg/kg ATXKA8 22,36,66
skn-mam LD50:2820 mg/kg GTPZAB 21(7),34,77

CONSENSUS REPORTS: EPA Genetic Toxicology Program.

SAFETY PROFILE: Moderately toxic by ingestion, skin contact, intraperitoneal, and possibly other routes. Experimental reproductive effects. Mutation data reported. When heated to decomposition it emits very toxic fumes of SO_x, PO_x, Br^-, and Cl^-.

BNL275 CAS:23483-74-1 ***HR: 3***
BROMO(2-HYDROXYETHYL)MERCURY AMMONIA SALT

SYNS: 2-(BROMOMERCURI) ETHANOL-AMMONIA (1:0.8 moles) compound ◇ MERCURY, BROMO(2-HYDROXYETHYL)-, compd. with AMMONIA (1:0.8 moles)

TOXICITY DATA with REFERENCE
ivn-mus LD50:56 mg/kg CSLNX* NX#05832

OSHA PEL: (Transitional: 0.01 mg/10m^3) 0.01 mg(Hg)/m^3; CL 0.03 mg(Hg)/m^3 (skin)
ACGIH TLV: TWA 0.01 mg(Hg)/m^3; STEL 0.03 mg(Hg)/m^3

SAFETY PROFILE: Poison by intravenous route. When heated to decomposition it emits toxic fumes of NH_3, Hg, and Br^-.

BNL750 CAS:87-48-9 ***HR: 2***
5-BROMOINDOLE-2,3-DIONE
mf: $C_8H_4BrNO_2$ mw: 226.04

SYN: 5-BROMISATIN (CZECH)

TOXICITY DATA with REFERENCE
eye-rbt 5 mg/24H SEV 28ZPAK -,143,72
orl-rat LDLo:4 g/kg 28ZPAK -,143,72
orl-mus LD50:437 mg/kg RPTOAN 45,10,82
ipr-mus LD50:437 mg/kg PCJOAU 15,858,81

SAFETY PROFILE: Moderately toxic by ingestion and intraperitoneal routes. A severe eye irritant. When heated to decomposition it emits very toxic fumes of Br^- and NO_x.

BNM000 CAS:314-42-1 ***HR: 2***
5-BROMO-3-ISOPROPYL-6-METHYLURACIL
mf: $C_8H_{11}BrN_2O_2$ mw: 247.12

PROP: Crystals. Mp: 158°. Sol in abs alc.

SYNS: 5-BROM-3-ISOPROPYL-6-METHYL-URACIL(GERMAN) ◇ 5-BROMO-3-ISOPROPYL-6-METHYL, 2,4-PYRIMIDINEDIONE (FRENCH) ◇ 5-BROMO-3-ISOPROPYL-6-METIL-URACIL (ITALIAN) ◇ 5-BROOM-3-ISOPROPYL-6-METHYL-URACIL DUTCH) ◇ HERBICIDE 82 ◇ HYVAR ◇ ISOCIL ◇ ISOPROCIL (FRENCH) ◇ 3-ISOPROPYL-5-BROMO-6-METHYLURACIL ◇ LOROX

TOXICITY DATA with REFERENCE
orl-rat LD50:3400 mg/kg RREVAH 10,97,65
orl-mus LDLo:3750 mg/kg TXAPA9 23,288,72

SAFETY PROFILE: Moderately toxic by ingestion. See also BROMIDES. When heated to decomposition it emits very toxic fumes of Br^- and NO_x.

BNM250 CAS:478-84-2 ***HR: 3***
2-BROMO-d-LYSERGIC ACID DIETHYLAMIDE
mf: $C_{20}H_{26}BrN_3O$ mw: 404.40

SYNS: BOL ◇ BOL-148 ◇ d-2-BROM-DIETHYLAMIDE of LYSERGIC ACID ◇ BROM LSD ◇ BROMLYSERGAMIDE ◇ 2-BROM-d-LYSERGIC ACID DIETHYLAMINE ◇ 2-BROMO-9,10-DIDEHYDRO-N,N-DIETHYL-6-METHYLERGOLINE-8-β-CARBOXAMIDE ◇ BROMOLYSERGIDE ◇ 9,10-DIDEHYDRO-N,N-DIETHYL-2-BROMO-6-METHYLERGOLINE-8-β-CARBOXAMIDE ◇ USAF SZ-1

TOXICITY DATA with REFERENCE
scu-ham TDLo:2 μg/kg (8D preg):TER SCIEAS 158,265,67
scu-ham TDLo:2 μg/kg (8D preg):REP SCIEAS 158,265,67
orl-hmn TDLo:75 μg/kg:CNS PSYPAG 1,20,59
ipr-mus LD50:25 mg/kg NTIS** AD277-689
ivn-mus LD50:20 mg/kg 28ZSAT -,-,64
ivn-rbt LD50:6 mg/kg ANYAA9 66,668,57

SAFETY PROFILE: Poison by intraperitoneal and intravenous routes. Experimental teratogenic and reproductive effects. Human systemic effects by ingestion: dilation or the arteries or veins. Many lysergic acid derivatives have central nervous system effects. When heated to decomposition it emits very toxic fumes such as Br^- and NO_x. See other lysergic acid derivatives.

BNM750 ***HR: 3***
BROMOMETHANE mixed with DIBROMOETHANE
DOT: UN 1647

SYN: METHYL BROMIDE and ETHYLENE DIBROMIDE MIXTURE, liquid (DOT)

DOT Classification: Poison B; Label: Poison.
SAFETY PROFILE: A poison. See also BROMIDES. When heated to decomposition it emits toxic fumes of Br-.

BNN125 CAS:102433-83-0 ***HR: 3***
2-(5-BROMO-2-METHOXYBENZYLOXY)TRIETHYLAMINE
mf: $C_{14}H_{22}BrNO_2$ mw: 316.28

TOXICITY DATA with REFERENCE
orl-rat LD50:220 mg/kg JPETAB 121,210,57
ipr-rat LD50:80 mg/kg JPETAB 121,210,57
orl-mus LD50:248 mg/kg JPETAB 121,210,57
ipr-mus LD50:123 mg/kg JPETAB 121,210,57

SAFETY PROFILE: Poison by ingestion and intraperitoneal routes. When heated to decomposition it emits toxic fumes of Br^- and NO_x.

BNN250 CAS:59177-64-9 ***HR: 3***
2-BROMO-N-METHYL-1-ADAMANTANEETHYLAMINE MALEATE
mf: $C_{13}H_{22}BrN \cdot C_4H_4O_4$ mw: 388.35

TOXICITY DATA with REFERENCE
orl-mus LD50:621 mg/kg JMCMAR 19,967,76
ipr-mus LD50:155 mg/kg JMCMAR 19,967,76

SAFETY PROFILE: Poison by intraperitoneal route. Moderately toxic by ingestion. When heated to decomposition it emits very toxic fumes of Br^- and NO_x.

BNN500 CAS:59177-85-4 ***HR: 3***
2-BROMO-N-METHYL-1-ADAMANTANEMETHANAMINE HYDROCHLORIDE
mf: $C_{12}H_{20}BrN \cdot ClH$ mw: 294.70

TOXICITY DATA with REFERENCE
orl-mus LD50:295 mg/kg JMCMAR 19,967,76
ipr-mus LD50:295 mg/kg JMCMAR 19,967,76

SAFETY PROFILE: Poison by ingestion and intraperitoneal routes. When heated to decomposition it emits very toxic fumes of Br^-, NO_x, and HCl.

BNN550 CAS:128-93-8 ***HR: 1***
1-BROMO-4-(METHYLAMINO)ANTHRAQUINONE
mf: $C_{15}H_{10}BrNO_2$ mw: 316.17

SYNS: 9,10-ANTHRACENEDIONE, 1-BROMO-4-(METHYLAMINO)- ◇ ANTHRAQUINONE, 1-BROMO-4-(METHYLAMINO)- ◇ 1-METHYLAMINO-4-BROMANTHRACHINON ◇ 1-(METHYLAMINO)-4-BROMOANTHRAQUINONE

TOXICITY DATA with REFERENCE
eye-rbt 500 mg/24H MLD 85JCAE-,566,86

CONSENSUS REPORTS: Reported in EPA TSCA Inventory.

SAFETY PROFILE: An eye irritant. When heated to decomposition it emits toxic fumes of NO_x and Br^-.

BNO000 CAS:31897-92-4 ***HR: 3***
2-BROMO-1-(2-METHYLAMINOPROPYL)ADAMANTANE HYDROCHLORIDE
mf: $C_{14}H_{24}BrN{\cdot}ClH$ mw: 322.76

SYN: 1-(2-BROMO-1-ADAMANTYL)-N-METHYL-2-PROPYLAMINE HYDROCHLORIDE

TOXICITY DATA with REFERENCE
orl-mus LD50:300 mg/kg JMCMAR 17,602,74
ipr-mus LD50:150 mg/kg JMCMAR 17,602,74

SAFETY PROFILE: Poison by ingestion and intraperitoneal routes. When heated to decomposition it emits very toxic fumes of Br^-, NO_x, and HCl.

BNO250 CAS:31898-11-0 ***HR: 3***
3-BROMO-1-(2-METHYLAMINOPROPYL)ADAMANTANE HYDROCHLORIDE
mf: $C_{14}H_{24}BrN{\cdot}ClH$ mw: 322.76

SYN: 1-(3-BROMO-1-ADAMANTYL)-N-METHYL-2-PROPYLAMINE HYDROCHLORIDE

TOXICITY DATA with REFERENCE
orl-mus LD50:400 mg/kg JMCMAR 17,602,74
ipr-mus LD50:150 mg/kg JMCMAR 17,602,74

SAFETY PROFILE: Poison by ingestion and intraperitoneal routes. When heated to decomposition it emits very toxic fumes of Br^-, NO_x, and HCl.

BNO500 CAS:2417-77-8 ***HR: 3***
9-BROMOMETHYLANTHRACENE
mf: $C_{15}H_{11}Br$ mw: 271.17

SYN: ICR 506

TOXICITY DATA with REFERENCE
mma-sat 10 μg/plate PNASA6 72,5135,75
ivn-mus TDLo:1350 μg/kg:NEO CNREA8 36,2423,76
ivn-mus LDLo:2700 μg/kg CNREA8 36,2423,76

CONSENSUS REPORTS: EPA Genetic Toxicology Program.

SAFETY PROFILE: Deadly poison by intravenous route. Questionable carcinogen with experimental neoplastigenic data. Mutation data reported. When heated to decomposition it emits toxic fumes of Br^-. See also BROMIDES.

BNO750 CAS:24961-39-5 ***HR: 3***
7-BROMO METHYL BENZ(a)ANTHRACENE
mf: $C_{19}H_{13}Br$ mw: 321.23

SYNS: 7-BMBA ◇ ICR 498

TOXICITY DATA with REFERENCE
dnr-esc 1 mg/L PNASA6 79,534,82
dns-hmn:fbr 1 μmol/L NARHAD 7,1343,79
dnr-ham:ovr 800 nmol/L PNASA6 79,534,82
dnd-ham:ovr 100 nmol/L SCMGDN 10,183,84
sce-ham:ovr 400 nmol/L PNASA6 79,534,82
msc-ham:ovr 50 nmol/L PNASA6 79,534,82
scu-rat TDLo:27 mg/kg:CAR CRNGDP 2,103,81
skn-mus TD:10 mg/kg/40W-I:ETA CNREA8 43,2034,83
ivn-mus TDLo:800 μg/kg:NEO CNREA8 36,2423,76
scu-rat TD:80 mg/kg:CAR CRNGDP 2,103,81
ivn-mus LDLo:1600 μg/kg CNREA8 36,2423,76

CONSENSUS REPORTS: EPA Genetic Toxicology Program.

SAFETY PROFILE: A deadly poison by intravenous route. Questionable carcinogen with experimental carcinogenic, neoplastigenic, and tumorigenic data. Human mutation data reported. When heated to decomposition it emits toxic fumes of Br^-. See also BROMIDES.

BNP000 CAS:49852-85-9 ***HR: 3***
6-BROMOMETHYLBENZO(a)PYRENE
mf: $C_{21}H_{13}Br$ mw: 345.25

TOXICITY DATA with REFERENCE
dnd-mam:lum 100 mg/L CBINA8 47,111,83
scu-rat TDLo:100 mg/kg/40D-I:ETA JMCMAR 16,714,73

SAFETY PROFILE: Questionable carcinogen with experimental tumorigenic data. Mutation data reported. See also BROMIDES. When heated to decomposition it emits toxic fumes of Br^-.

BNP250 CAS:107-82-4 ***HR: 3***
1-BROMO-3-METHYL BUTANE
DOT: UN 2341
mf: $C_5H_{11}Br$ mw: 151.05

PROP: Colorless liquid. D: 1.210, mp: −112°, bp: 120-121°, flash p: 21°. Sltly sol in water; misc with alc and ether.

SYNS: ISOAMYL BROMIDE ◇ ISOPENTYL BROMIDE ◇ 3-METHYLBUTYL BROMIDE

TOXICITY DATA with REFERENCE
ipr-rat LD50:6150 mg/kg 85GMAT -,76,82
ipr-mus LD50:13750 mg/kg 85GMAT -,76,82
ihl-mam LD50:21300 mg/m^3 GPTZAB 18(4),55,74
ipr-mam LD50:480 mg/kg GTPZAB 18(4),55,74

CONSENSUS REPORTS: EPA Genetic Toxicology Program.

DOT Classification: Label: Flammable Liquid.

SAFETY PROFILE: Moderately toxic by intraperitoneal route. Dangerous fire hazard when exposed to heat or flame. When heated to decomposition it emits toxic fumes of Br^-. See also BROMIDES.

BNP750 CAS:496-67-3 ***HR: 2***
2-BROMO-3-METHYLBUTYRYLUREA
mf: $C_6H_{11}BrN_2O_2$ mw: 223.10

SYNS: ABROVAL ◇ ALLUVAL ◇ ALURAL ◇ N-(AMINOCARBONYL)-2-BROMO-3-METHYLBUTANAMIDE ◇ BROMARAL ◇ BROMCARBAMIDE ◇ BROMISOVAL ◇ BROMISOVALERYLUREA ◇ α-BROMISOVALERYLUREA ◇ BROMISOVALUM ◇ BROMIZOVAL ◇ BROMOCARBAMIDE ◇ α-BROMO-β-DIMETHYLPROPANOYLUREA ◇ α-BROMOISOVALERIC ACID UREIDE ◇ α-BROMOISOVALEROYLUREA ◇ (α-BROMOISOVALERYL)UREA ◇ BROMOVAL ◇ BROMOVALEROCARBAMIDE ◇ BROMOVALERYLUREA ◇ BROMOXIL ◇ BROMURAL ◇ BROMUVAN ◇ BROMVALERYLUREA ◇ BROMVALETONE ◇ BROMVALETONUM ◇ BROMVALUREA ◇ BROMYL ◇ BROVALIN ◇ BROVALUREA ◇ BROVARIN ◇ BVU ◇ CALMOTIN ◇ DIAGRABROMYL ◇ DIBROLUUR ◇ DORMIGENE ◇ ISOBROMYL ◇ ISOVAL ◇ MONOBROMOISOVALERYLUREA ◇ 2-MONOBROMOISOVALERYLUREA ◇ PIVADORM ◇ PIVADORN ◇ SOMNUROL ◇ UPIOL ◇ UVALERAL

TOXICITY DATA with REFERENCE
orl-wmn TDLo:400 mg/kg:CNS BMJOAE 1,1238,55
orl-rat LD50:1000 mg/kg FEPRA7 7,262,48
orl-mus LD50:2 g/kg OYYAA2 11,693,76
orl-cat LD50:450 mg/kg NIIRDN 6,738,82
orl-rbt LD50:1200 mg/kg MEIEDD 10,193,83

CONSENSUS REPORTS: Reported in EPA TSCA Inventory.

SAFETY PROFILE: Moderately toxic by ingestion. Human systemic effects by ingestion: nausea or vomiting, and coma. A sedative and hypnotic agent. When heated to decomposition it emits very toxic fumes of Br^- and NO_x.

BNP850 CAS:25855-92-9 ***HR: 2***
9-(BROMOMETHYL)-10-CHLOROANTHRACENE
mf: $C_{15}H_{10}BrCl$ mw: 305.61

SYN: 10-BROMOMETHYL-9-CHLOROANTHRACENE

TOXICITY DATA with REFERENCE
ivn-mus TDLo:3066 μg/kg:NEO CNREA8 40,782,80

SAFETY PROFILE: Questionable carcinogen with experimental neoplastigenic data. When heated to decomposition it emits toxic fumes of Br^- and Cl^-.

BNQ000 CAS:34346-99-1 ***HR: 3***
7-BROMOMETHYL-4-CHLOROBENZ(a)ANTHRACENE
mf: $C_{19}H_{12}BrCl$ mw: 355.67

SYN: 4-CHLORO-7-BROMOMETHYLBENZ(a)ANTHRACENE

TOXICITY DATA with REFERENCE
skn-mus TDLo:14 mg/kg:ETA EJCAAH 7,473,71

SAFETY PROFILE: Questionable carcinogen with experimental tumorigenic data by skin contact. When heated to decomposition it emits very toxic fumes of Br^- and Cl^-. See also CHLRORINATED HYDROCARBONS, AROMATIC.

BNQ100 ***HR: 2***
3'-BROMO-4'-METHYL-4-DIMETHYLAMINOAZOBENZENE
mf: $C_{15}H_{16}BrN_3$ mw: 318.25

SYNS: ANILINE,p-((3-BROMO-p-TOLYL)AZO)-N,N-DIMETHYL- ◇ BENZENAMINE, N,N-DIMETHYL-3'-BROMO-4'-METHYL-4-(PHENYLAZO)- ◇ p-((3-BROMO-p-TOLYL)AZO)-N,N-DIMETHYLANILINE

TOXICITY DATA with REFERENCE
orl-rat TDLo:13977 mg/kg/52W-C:CAR CBINA8 53,107,85

SAFETY PROFILE: Questionable carcinogen with experimental carcinogenic data. When heated to decomposition it emits toxic fumes of Br^- and NO_x.

BNQ110 ***HR: 2***
4'-BROMO-3'-METHYL-4-DIMETHYLAMINOAZOBENZENE
mf: $C_{15}H_{16}BrN_3$ mw: 318.25

SYNS: ANILINE,p-((4-BROMO-m-TOLYL)AZO)-N,N-DIMETHYL- ◇ BENZENAMINE, N,N-DIMETHYL-4'-BROMO-3'-METHYL-4-(PHENYLAZO)- ◇ p-((4-BROMO-m-TOLYL)AZO)-N,N-DIMETHYLANILINE

TOXICITY DATA with REFERENCE
orl-rat TDLo:9677 mg/kg/36W-C:CAR CBINA8 53,107,85

SAFETY PROFILE: Questionable carcinogen with experimental carcinogenic data. When heated to decomposition it emits toxic fumes of Br^- and NO_x.

BNQ250 CAS:34346-97-9 ***HR: 3***
7-BROMOMETHYL-6-FLUOROBENZ(a)ANTHRACENE
mf: $C_{19}H_{12}BrF$ mw: 339.22

SYN: 6-FLUORO-7-BROMOMETHYLBENZ(a)ANTHRACENE

TOXICITY DATA with REFERENCE
skn-mus TDLo:14 mg/kg:ETA EJCAAH 7,473,71

SAFETY PROFILE: Questionable carcinogen with experimental tumorigenic data by skin contact. When heated to decomposition it emits very toxic fumes of Br^- and F^-.

BNQ500 CAS:4437-18-7 ***HR: 3***
2-BROMO METHYL FURAN
mf: C_5H_5BrO mw: 272.36

OCH=CHCH=CCH$_2$Br (ring: O bonded to C)

SAFETY PROFILE: A very unstable explosive. When heated to decomposition it emits toxic fumes of Br^-. See also BROMIDES.

BNQ750 CAS:34346-96-8 *HR: 3*
7-BROMOMETHYL-1-METHYLBENZ(a)ANTHRACENE
mf: $C_{20}H_{15}Br$ mw: 335.26

SYN: 1-METHYL-7-BROMOMETHYLBENZ(a)ANTHRACENE

TOXICITY DATA with REFERENCE
skn-mus TDLo:13 mg/kg:ETA EJCAAH 7,473,71

SAFETY PROFILE: Questionable carcinogen with experimental tumorigenic data by skin contact. When heated to decomposition it emits toxic fumes of Br^-.

BNR000 CAS:16238-56-5 *HR: 3*
7-BROMO METHYL-12-METHYLBENZ(a)ANTHRACENE
mf: $C_{20}H_{15}Br$ mw: 335.26

SYN: ICR 502

TOXICITY DATA with REFERENCE
mmo-sat 300 ng/plate ENMUDM 6(Suppl 2),1,84
mmo-esc 1 μg/plate ENMUDM 6(Suppl 2),1,84
otr-rat:emb 270 μg/L JJIND8 67,1303,81
otr-mus:fbr 16 μg/L JJIND8 67,1303,81
dnd-mus:emb 500 nmol/L CALEDQ 7,103,79
scu-rat TDLo:6 mg/kg:CAR CRNGDP 2,103,81
skn-mus TDLo:8040 mg/kg:NEO JJIND8 61,135,78
scu-rat TD:188 mg/kg:CAR CRNGDP 2,103,81

CONSENSUS REPORTS: EPA Genetic Toxicology Program.

SAFETY PROFILE: Questionable carcinogen with experimental carcinogenic and neoplastigenic data. Mutation data reported. When heated to decomposition it emits toxic fumes of Br^-.

BNR250 CAS:59230-81-8 *HR: 3*
12-BROMOMETHYL-7-METHYLBENZ(a)ANTHRACENE
mf: $C_{20}F_{15}Br$ mw: 605.11

TOXICITY DATA with REFERENCE
skn-mus TDLo:8040 mg/kg:NEO JJIND8 61,135,78

SAFETY PROFILE: Questionable carcinogen with experimental neoplastigenic data. When heated to decomposition it emits very toxic fumes of F^- and Br^-.

BNR325 CAS:57846-03-4 *HR: 3*
2-BROMOMETHYL-5-METHYLFURAN
mf: C_6H_7BrO mw: 175.02

SAFETY PROFILE: May decompose violently above 70°C. When heated to decomposition it emits toxic fumes of Br^-.

BNR750 CAS:78-77-3 *HR: 3*
1-BROMO-2-METHYL PROPANE
DOT: UN 2342
mf: C_4H_9Br mw: 137.04

PROP: Flash p: 22°C.

SYNS: 1-BUTYL BROMIDE ◇ i-BUTYL BROMIDE ◇ ISOBUTYL BROMIDE

TOXICITY DATA with REFERENCE
ipr-mus TDLo:3000 mg/kg/8W-I:NEO CNREA8 35,1411,75
ipr-uns LD50:1660 mg/kg GTPZAB 18(4),55,74

CONSENSUS REPORTS: EPA Genetic Toxicology Program. Reported in EPA TSCA Inventory.

DOT Classification: IMO: Flammable or Combustible Liquid; Label: Flammable Liquid.

SAFETY PROFILE: Questionable carcinogen with experimental neoplastigenic data. Moderately toxic by intraperitoneal route. A dangerous fire hazard when exposed to heat or flame. When heated to decomposition it emits toxic fumes of Br^-. See also BROMIDES.

BNS000 *HR: 3*
2-BROMO-2-METHYL PROPANE
mf: C_4H_9Br mw: 137.04

PROP: Flash p: −18°C.

SAFETY PROFILE: A very dangerous fire hazard when exposed to heat or flame. When heated to decomposition it emits toxic fumes of Br^-. See also BROMIDES.

BNS750 CAS:6954-48-9 *HR: 3*
6-BROMO-1,2-NAPHTHOQUINONE
mf: $C_{10}H_5BrO_2$ mw: 237.06

SYN: BONAPHTHON

TOXICITY DATA with REFERENCE
orl-rat TDLo:700 mg/kg (female 12D post):TER FATOAO 41,109,78
orl-rat LD50:3900 mg/kg FATOAO 43,337,80
ipr-rat LD50:130 mg/kg FATOAO 39,628,76
orl-mus LD50:260 mg/kg FATOAO 39,628,76
ipr-mus LD50:9 mg/kg FATOAO 39,628,76
orl-gpg LD50:900 mg/kg FATOAO 39,628,76

SAFETY PROFILE: Poison by ingestion and intraperitoneal routes. An experimental teratogen. When heated to decomposition it emits toxic fumes of Br^-.

BNT000 CAS:30007-47-7 ***HR: 3***
5-BROMO-5-NITRO-m-DIOXANE
mf: $C_4H_6BrNO_4$ mw: 212.02

SYNS: 5-BROM-5-NITRO-1,3-DIOXAN (GERMAN) ◇ 5-BROMO-5-NITRO-1,3-DIOXANE

TOXICITY DATA with REFERENCE
skn-rat 2500 μg/24H FSASAX 78,269,76
skn-mus 2500 μg/24H FSASAX 78,269,76
orl-rat LD50:455 mg/kg FSASAX 78,269,76
ipr-rat LD50:31 mg/kg FSASAX 78,269,76
orl-mus LD50:590 mg/kg FSASAX 78,269,76
scu-dog LDLo:500 mg/kg FSASAX 78,269,76

CONSENSUS REPORTS: Reported in EPA TSCA Inventory.

SAFETY PROFILE: Poison by intraperitoneal route. Moderately toxic by ingestion and subcutaneous routes. A skin irritant. When heated to decomposition it emits very toxic fumes of Br^- and NO_x.

BNT250 CAS:52-51-7 ***HR: 3***
2-BROMO-2-NITRO-1,3-PROPANEDIOL
mf: $C_3H_6BrNO_4$ mw: 200.01

SYNS: 2-BROMO-2-NITROPANE-1,3-DIOL ◇ 2-BROMO-2-NITROPROPAN-1,3-DIOL ◇ β-BROMO-β-NITROTRIMETHYLENEGLYCOL ◇ BRONOCOT ◇ BRONOPOL ◇ BRONOSOL

TOXICITY DATA with REFERENCE
skn-hmn 10 mg MOD JSCCA5 29,3,78
skn-rbt 500 mg/24H MLD JEPTDQ 4(4),47,80
skn-rbt 80 mg MOD JEPTDQ 4(4),47,80
eye-rbt 5 mg JSCCA5 29,3,78
orl-rat LD50:180 mg/kg 28ZEAL 5,30,76
skn-rat LD50:3500 mg/kg IYKEDH 8,680,77
ipr-rat LD50:26 mg/kg JSCCA5 29,3,78
scu-rat LD50:170 mg/kg KSRNAM 8,1029,74
ivn-rat LD50:37400 μg/kg IYKEDH 8,680,77
orl-mus LD50:250 mg/kg FMCHA2 -,C37,83
skn-mus LD50:4750 mg/kg IYKEDH 8,680,77
ipr-mus LD50:15500 μg/kg KHFZAN 11(1),73,77
scu-mus LD50:116 mg/kg IYKEDH 8,680,77
ivn-mus LD50:48 mg/kg IYKEDH 8,680,77
orl-dog LD50:250 mg/kg 28ZEAL 5,30,76

CONSENSUS REPORTS: Reported in EPA TSCA Inventory.

SAFETY PROFILE: Poison by ingestion, subcutaneous, intravenous, and intraperitoneal routes. Moderately toxic by skin contact. An eye and human skin irritant. An antiseptic. When heated to decomposition it emits very toxic fumes of NO_x and Br^-.

BNT500 CAS:14173-58-1 ***HR: 3***
3-BROMO-4-NITROQUINOLINE-1-OXIDE
mf: $C_9H_5BrN_2O_3$ mw: 269.07

TOXICITY DATA with REFERENCE
cyt-omi 37 μmol GANNA2 60,155,69
scu-mus TDLo:60 mg/kg/I:ETA CPBTAL 17,544,69

SAFETY PROFILE: Questionable carcinogen with experimental tumorigenic data. Mutation data reported. When heated to decomposition it emits very toxic fumes of Br^- and NO_x.

BNU000 CAS:111-83-1 ***HR: 1***
1-BROMOOCTANE
mf: $C_8H_{17}Br$ mw: 193.16

SYN: n-OCTYL BROMIDE

TOXICITY DATA with REFERENCE
orl-rat LD50:5020 mg/kg AIHAAP 30,470,69
skn-rbt LD50:8944 mg/kg AIHAAP 30,470,69

CONSENSUS REPORTS: Reported in EPA TSCA Inventory.

SAFETY PROFILE: Mildly toxic by ingestion and skin contact. When heated to decomposition it emits toxic fumes of Br^-. See also BROMIDES.

BNU125 CAS:23753-67-5 ***HR: 3***
1-BROMOPENTABORANE (9)
mf: B_5BrH_8 mw: 142.02

SAFETY PROFILE: Ignites spontaneously in air. Explosive reaction with hexamine above 90°C. When heated to decomposition it emits toxic fumes of Br^-. See also BORANES and BORON COMPOUNDS.

BNU250 CAS:63867-64-1 ***HR: 3***
4-BROMO-1,2,2,6,6-PENTAMETHYLPIPERIDINE
mf: $C_{10}H_{19}BrN$ mw: 233.21

TOXICITY DATA with REFERENCE
orl-mus LD50:172 mg/kg NATUAS 184,1707,59
ivn-mus LD50:51 mg/kg NATUAS 184,1707,59

SAFETY PROFILE: Poison by ingestion and intravenous routes. When heated to decomposition it emits very toxic fumes of NO_x and Br^-.

BNU500 CAS:107-81-3 ***HR: 3***
2-BROMOPENTANE
DOT: UN 2343
mf: $C_5H_{11}Br$ mw: 151.07

PROP: Colorless to yellow liquid, strong odor, bp: 120°, fp: < −30°, d: 1.211 @ 25°/25°, flash p: 90°F.

TOXICITY DATA with REFERENCE
ipr-mus LD50:150 mg/kg NTIS** AD691-490

CONSENSUS REPORTS: Reported in EPA TSCA Inventory.

DOT Classification: Flammable Liquid; Label: Flammable Liquid.

SAFETY PROFILE: Poison by intraperitoneal route. A local irritant and narcotic in high concentration. Ingestion can cause liver damage. A dangerous fire hazard when exposed to heat or flame. When heated to decomposition it emits toxic fumes of Br^-. See also BROMIDES, and CHLORINATED HYDROCARBONS, ALIPHATIC.

BNU660 ***HR: 3***
4-(2-(5-BROMO-2-PENTYLOXYBENZYLOXY) ETHYL)MORPHOLINE
mf: $C_{18}H_{28}BrNO_3$ mw: 386.38

TOXICITY DATA with REFERENCE
orl-rat LD50:1200 mg/kg JPETAB 121,210,57
ipr-rat LD50:265 mg/kg JPETAB 121,210,57
orl-mus LD50:620 mg/kg JPETAB 121,210,57
ipr-mus LD50:400 mg/kg JPETAB 121,210,57

SAFETY PROFILE: Poison by intraperitoneal route. Moderately toxic by ingestion. When heated to decomposition it emits toxic fumes of Br^- and NO_x.

BNU700 ***HR: 3***
2-(5-BROMO-2-PENTYLOXYBENZYLOXY)TRIETHYLAMINE
mf: $C_{18}H_{30}BrNO_2$ mw: 372.40

TOXICITY DATA with REFERENCE
orl-rat LD50:320 mg/kg JPETAB 121,210,57
ipr-rat LD50:85 mg/kg JPETAB 121,210,57
orl-mus LD50:190 mg/kg JPETAB 121,210,57
ipr-mus LD50:100 mg/kg JPETAB 121,210,57

SAFETY PROFILE: Poison by ingestion and intraperitoneal routes. When heated to decomposition it emits toxic fumes of Br^- and NO_x.

BNU725 CAS:10457-90-6 ***HR: 3***
BROMOPERIDOL
mf: $C_{21}H_{23}BrFNO_2$ mw: 420.36

PROP: Off-white, amorphous or microcrystalline powder. Mp: 155-158°. Solubility in water: 0.09 mg/mL; in 0.1M tartaric, lactic, citric and acetic acids: about 10 mg/mL.

SYNS: AZURENE ◇ 4-(4-(p-BROMOPHENYL)-4-HYDROXYPIPERIDINO)-4'-FLUOROBUTYROPHENONE ◇ 4-(4-(4-BROMOPHENYL)-4-HYDROXYPIPERIDINO)-4'-FLUOROBUTYROPHENONE ◇ 4-(4-(p-BROMOPHENYL)-4-HYDROXYPIPERIDINOL)-4-FLUOROBUTYROPHENONE ◇ 4-(4-(4-BROMOPHENYL)-4-HYDROXY-1-PIPERIDINYL)-1-(4-FLUOROPHENYL)-1-BUTANONE ◇ BROMPERIDOL ◇ IMPROMEN ◇ R 11333 ◇ TESOPREL

TOXICITY DATA with REFERENCE
orl-rat TDLo:110 mg/kg (female 7-17D post):REP JTSCDR 9(Suppl 1),109,84
orl-rat TDLo:110 mg/kg (female 7-17D post):TER JTSCDR 9(Suppl 1),109,84
orl-rat LD50:359 mg/kg ARZNAD 24,45,74
ipr-rat LD50:323 mg/kg IYKEDH 16,1461,85
scu-rat LD50:84 mg/kg ARZNAD 24,45,74
ivn-rat LD50:10 mg/kg IYKEDH 16,1461,85
orl-mus LD50:174 mg/kg IYKEDH 16,1461,85
ipr-mus LD50:156 mg/kg IYKEDH 16,1461,85
scu-mus LD50:114 mg/kg ARZNAD 24,45,74
ivn-mus LD50:18900 μg/kg ARZNAD 24,45,74

SAFETY PROFILE: Poison by ingestion, subcutaneous, intravenous, and intraperitoneal routes. An experimental teratogen. Other experimental reproductive effects. When heated to decomposition it emits toxic fumes of F^-, Br^- and NO_x.

BNU750 CAS:106-41-2 ***HR: 3***
p-BROMOPHENOL
mf: C_6H_5BrO mw: 173.02

SYN: 4-BROMOPHENOL

TOXICITY DATA with REFERENCE
skn-mus TDLo:7200 mg/kg/18W-I:ETA CNREA8 19,413,59
orl-mus LD50:523 mg/kg GISAAA 44(12),19,79
ipr-mus LD50:411 mg/kg GISAAA 44(12),19,79

CONSENSUS REPORTS: Reported in EPA TSCA Inventory.

SAFETY PROFILE: Moderately toxic by ingestion and intraperitoneal routes. Questionable carcinogen with experimental tumorigenic data. See also BROMOPHENOLS. When heated to decomposition it emits toxic fumes of Br^-.

BNV000 CAS:95-56-7 ***HR: 2***
o-BROMOPHENOL

TOXICITY DATA with REFERENCE
scu-rat LD50: 1500 mg/kg 14CYAT 2,1406,63
scu-gpg LDLo:1500 mg/kg RMSRA6 16,449,1896
orl-mus LD50:652 mg/kg GISAAA 44(12),19,79
ipr-mus LD50:633 mg/kg GISAAA 44(12),19,79

CONSENSUS REPORTS: Reported in EPA TSCA Inventory.

SAFETY PROFILE: Moderately toxic by ingestion, intraperitoneal and subcutaneous routes. See also

BROMOPHENOLS. When heated to decomposition it emits toxic fumes of Br^-.

BNV250 ***HR: 2***
BROMO PHENOLS
mf: $HO(C_6H_4)Br$ mw: 173

PROP: (m-) Crystals; insol in water; sol in alc, ether, and alkalis. (p-) Crystals; sltly sol in water; sol in alc, ether, chloroform, and glacial acetic acid. (o-) Yellow to oily, red liquid; unpleasant odor; insol in water; sol in alc, ether, and chloroform. D: (p-) 1.840 (15°), 1.5875 (80°); (o-) 1.5. Mp: (m-) 33°; (p-) 64°; (o-) 6°. Bp: (m-) 236°; (p-) 238°; (o-) 194°.

SAFETY PROFILE: Moderately toxic by several routes. Dangerous in a fire. When heated to decomposition it emits toxic fumes of Br^-. See also BROMIDES.

BNV500 CAS:21466-07-9 ***HR: D***
BROMOPHENOPHOS
mf: $C_{12}H_7Br_4O_5P$ mw: 581.80

SYNS: ACEDIST ◇ (1,1′-BIPHENYL)-2,2′-DIOL, 3,3′,5,5′-TETRABROMO-, MONO(DIHYDROGEN PHOSPHATE) (9CI) ◇ 2,2′-BIPHENYLDIOL, 3,3′,5,5′-TETRABROMO-, MONO(DIHYDROGEN PHOSPHATE) ◇ BROMFENOFOS ◇ BROMFENPHOS ◇ BROMOFENOFOS ◇ BROMPHENPHOS ◇ PH 1882

TOXICITY DATA with REFERENCE
orl-rat TDLo:50 mg/kg (female 9D post):TER ARTODN 60,319,87
orl-rat TDLo:50 mg/kg (female 9D post):REP ARTODN 60,319,87

SAFETY PROFILE: An experimental teratogen. Other experimental reproductive effects. When heated to decomposition it emits toxic fumes of PO_x and Br^-.

BNV750 CAS:16532-79-9 ***HR: 3***
4-BROMOPHENYLACETONITRILE
DOT: UN 1694
mf: C_8H_6BrN mw: 196.06

SYNS: 4-BROMOBENZENEACETONITRILE ◇ p-BROMOBENZYL CYANIDE ◇ 4-BROMOBENZYLCYANIDE ◇ p-BROMOPHENYLACETONITRILE ◇ 2-(4-BROMOPHENYL)ACETONITRILE

TOXICITY DATA with REFERENCE
ivn-mus LD50:56 mg/kg CSLNX* NX#03252

CONSENSUS REPORTS: Reported in EPA TSCA Inventory. Cyanide and its compounds are on the Community Right-To-Know List.

DOT Classification: Poison B; Label: Poison.

SAFETY PROFILE: Poison by intravenous route. See also BROMIDES and NITRILES. When heated to decomposition it emits very toxic fumes of Br^-, NO_x, and CN^-.

BNV850 ***HR: D***
4-BROMOPHENYL CHLOROMETHYL SULFONE
mf: $C_7H_6BrClO_2S$ mw: 269.55

SYN: SULFONE, p-BROMOPHENYL CHLOROMETHYL

TOXICITY DATA with REFERENCE
orl-rat TDLo:2450 mg/kg (female 6-15D post):TER JTSCDR 11,125,86

SAFETY PROFILE: Experimental reproductive effects. When heated to decomposition it emits toxic fumes of SO_x, Br^-, and Cl^-.

BNW250 CAS:7239-21-6 ***HR: 2***
1-(4-BROMOPHENYL)-3,3-DIMETHYLTRIAZENE
mf: $C_8H_{10}BrN_3$ mw: 228.12

SYNS: 1-p-BROMFENYL-3,3-DIMETHYLTRIAZEN(CZECH) ◇ 4-BROMO-PDMT

TOXICITY DATA with REFERENCE
mma-sat 5 mmol/L MUREAV 36,1,76
sln-dmg-orl 100 μmol/L CBINA8 9,365,74
orl-rat LD50:423 mg/kg 28ZPAK -,98,72

SAFETY PROFILE: Moderately toxic by ingestion. Mutation data reported. When heated to decomposition it emits very toxic fumes of Br^- and NO_x.

BNW500 CAS:1808-12-4 ***HR: 3***
BROMO PHENYL HYDRAMINE HYDROCHLORIDE
mf: $C_{17}H_{20}BrNO•ClH$ mw: 370.75

SYNS: β-(p-BROMOBENZHYDRYLOXY)ETHYLDIMETHYLAMINE HYDROCHLORIDE ◇ 2-(4-BROMOBENZOHYDRYLOXY)ETHYLDIMETHYLAMINE HYDROCHLORIDE

TOXICITY DATA with REFERENCE
orl-rat LD50:602 mg/kg CLDND* 112,318,54
ivn-mus LD50:63 mg/kg CLDND*
ivn-dog LD50:21 mg/kg CLDND*

SAFETY PROFILE: Poison by intravenous route. Moderately toxic by ingestion. When heated to decomposition it emits very toxic fumes of Br^-, NO_x, and HCl.

BNW625 ***HR: D***
2-(p-BROMOPHENYL)IMIDAZO(2,1-a)ISOQUINOLINE
mf: $C_{23}H_{16}N_2$ mw: 320.41

TOXICITY DATA with REFERENCE
orl-ham TDLo:2 mg/kg (4-8D preg):REP ARZNAD 33,1222,83

SAFETY PROFILE: Experimental reproductive effects. When heated to decomposition it emits toxic fumes of NO_x.

BNW750 CAS:1470-37-7 ***HR: 3***
4-BROMO-2-PHENYL-1,3-INDANDIONE
mf: $C_{15}H_9BrO_2$ mw: 301.15

SYN: 4-BROMO-2-FENILINDAN-1,3-DIONE(ITALIAN)

TOXICITY DATA with REFERENCE
orl-rat LD50:745 mg/kg FRPSAX 31,403,76
orl-mus LD50:114 mg/kg FRPSAX 31,403,76
ipr-mus LDLo:160 mg/kg FRPSAX 31,315,76

SAFETY PROFILE: Poison by ingestion and intraperitoneal routes. When heated to decomposition it emits toxic fumes of Br^-. See also BROMIDES.

BNW825 CAS:1985-12-2 ***HR: 3***
p-BROMOPHENYL ISOTHIOCYANATE
mf: C_7H_4BrNS mw: 214.09

SYN: p-BROMOPHENYL ESTER ISOTHIOCYANIC ACID

TOXICITY DATA with REFERENCE
orl-rat LD50:400 mg/kg FCTXAV 5,741,67
ipr-rat LDLo:100 mg/kg ARZNAD 19,558,69
ipr-mus LDLo:100 mg/kg ARZNAD 21,121,71

SAFETY PROFILE: Poison by ingestion and intraperitoneal routes. When heated to decomposition it emits toxic fumes of SO_x, Br^-, and NO_x. See also THIOCYANATES and ESTERS.

BNX000 CAS:22480-64-4 ***HR: 3***
p-BROMO PHENYL LITHIUM
mf: C_6H_4BrLi mw: 162.95

SAFETY PROFILE: Explodes on exposure to oxygen. When heated to decomposition it emits toxic fumes of Br^-. See also LITHIUM COMPOUNDS.

BNX125 CAS:23139-02-8 ***HR: 3***
3-(p-BROMOPHENYL)-1-METHYL-1-NITROSOUREA
mf: $C_8H_8BrN_3O_2$ mw: 258.10

SYNS: 1-METHYL-3-(p-BROMOPHENYL)-1-NITROSOUREA ◇ 1-METHYL-3-(p-BROMPHENYL)-1-NITROSOHARNSTOFF (GERMAN) ◇ 1-METHYL-1-NITROSO-3-(p-BROMOPHENYL)UREA

TOXICITY DATA with REFERENCE
cyt-ham:lng 10 μmol/L IAPUDO 31,797,80
sce-ham:lng 10 μmol/L IAPUDO 31,797,80
orl-rat TDLo:774 mg/kg/88W-I:CAR ARGEAR 53,329,83
orl-rat TD:1080 mg/kg/30W-I:ETA IAPUDO 31,685,80

SAFETY PROFILE: Questionable carcinogen with experimental carcinogenic and tumorigenic data. Mutation data reported. When heated to decomposition it emits toxic fumes of Br^- and NO_x.

BNX250 CAS:60050-37-5 ***HR: 2***
2-(m-BROMOPHENYL)-N-(4-MORPHOLINOMETHYL)SUCCINIMIDE
mf: $C_{15}H_{17}BrN_2O_3$ mw: 353.25

TOXICITY DATA with REFERENCE
orl-mus LD50:3012 mg/kg ARZNAD 29,290,79
ipr-mus LD50:443 mg/kg EJMCA5 13,465,78

SAFETY PROFILE: Moderately toxic by ingestion and intraperitoneal routes. When heated to decomposition it emits very toxic fumes of Br^- and NO_x.

BNX750 CAS:106-94-5 ***HR: 3***
1-BROMOPROPANE
DOT: UN 2344
mf: C_3H_7Br mw: 123.01

PROP: Liquid. Mp: −110°, bp: 70.9°, d: 1.353 @ 20°/4°, autoign temp: 914°F, flash p: <22°, lel: 4.6%.

SYNS: 1-BROMOPROPANE (DOT) ◇ PROPYL BROMIDE

TOXICITY DATA with REFERENCE
orl-rat TDLo:2 g/kg (5D male):REP MUREAV 101,321,82
orl-rat LDLo:4000 mg/kg MUREAV 101,321,82
ihl-rat LC50:253000 mg/m^3/30M FAVUAI 7,35,75
ipr-rat LD50:2950 mg/kg 85GMAT -,102,82
ipr-mus LD50:2530 mg/kg 85GMAT -,102,82

CONSENSUS REPORTS: Reported in EPA TSCA Inventory.

DOT Classification: Flammable or Combustible Liquid; Label: Flammable Liquid.

SAFETY PROFILE: Moderately toxic by ingestion and intraperitoneal routes. Mildly toxic by inhalation. Experimental reproductive effects. Mutation data reported. Dangerous fire hazard when heated or exposed to flame or oxidizers. To fight fire, use water, foam, CO_2, dry chemical. When heated to decomposition it emits toxic fumes of Br^-. See also BROMIDES.

BNY000 CAS:75-26-3 ***HR: 3***
2-BROMOPROPANE
DOT: UN 2344
mf: C_3H_7Br mw: 122.98

PROP: Flash p: <14°.

DOT Classification: Flammable or Combustible Liquid; Label: Flammable Liquid

SAFETY PROFILE: Flammable liquid. Dangerous fire hazard. When heated to decomposition it emits toxic fumes of Br^-. See also BROMIDES.

BNY750 CAS:627-18-9 ***HR: D***
3-BROMOPROPANOL
mf: C_3H_7BrO mw: 139.01

SYNS: 3-BROMO-1-PROPANOL ◇ 3-HYDROXYPROPYL BROMIDE

TOXICITY DATA with REFERENCE
mmo-sat 177 μg/plate MUREAV 57,381,78
mma-sat 350 μg/plate MUREAV 57,381,78

SAFETY PROFILE: Mutation data reported. When heated to decomposition it emits toxic fumes of Br^-. See also BROMIDES.

BNZ000 CAS:598-31-2 ***HR: 3***
BROMO-2-PROPANONE
DOT: UN 1569
mf: C_3H_5BrO mw: 136.99

SYNS: ACETONYL BROMIDE ◇ ACETYL METHYL BROMIDE ◇ BROMOACETONE ◇ BROMOACETONE (DOT) ◇ BROMOACETONE, liquid (DOT) ◇ BROMOMETHYL METHYL KETONE ◇ 1-BROMO-2-PROPANONE ◇ MONOBROMOACETONE ◇ RCRA WASTE NUMBER P017

TOXICITY DATA with REFERENCE
ihl-hmn LCLo:572 ppm/10M NTIS** PB214-270

DOT Classification: Poison A; Label: Poison Gas; Poison B; Label: Flammable Liquid and Poison.

SAFETY PROFILE: A poisonous gas. Moderately toxic to humans by inhalation. When heated to decomposition it emits toxic fumes of Br^-. See also BROMIDES.

BOA000 CAS:590-14-7 ***HR: D***
1-BROMOPROPENE
mf: C_3H_5Br mw: 120.97

SYNS: 1-BROMO-1-PROPENE ◇ 1-PROPENYL BROMIDE

TOXICITY DATA with REFERENCE
mmo-sat 1 μmol/plate ENMUDM 2,59,80
mma-sat 1 μmol/plate ENMUDM 2,59,80

SAFETY PROFILE: Mutation data reported. When heated to decomposition it emits toxic fumes of Br^-. See also BROMIDES.

BOA250 CAS:557-93-7 ***HR: D***
2-BROMOPROPENE
mf: C_3H_5Br mw: 120.97

SYNS: 2-BROMOPROPYLENE ◇ ISOPROPYLENE BROMIDE ◇ α-METHYLVINYL BROMIDE

TOXICITY DATA with REFERENCE
mmo-sat 1 μmol/plate ENMUDM 2,59,80
mma-sat 1 μmol/plate ENMUDM 2,59,80

SAFETY PROFILE: Mutation data reported. When heated to decomposition it emits toxic fumes of Br^-. See also BROMIDES.

BOA750 CAS:42461-89-2 ***HR: 2***
5-(3-BROMO-1-PROPENYL)-1,3-BENZODIOXOLE
mf: $C_{10}H_9BrO_2$ mw: 241.10

SYNS: 3′-BROMOISOSAFROLE ◇ 1,2-(METHYLENEDIOXY)-4-(3-BROMO-1-PROPENYL)BENZENE

TOXICITY DATA with REFERENCE
scu-rat TDLo:359 mg/kg/10W-:ETA CNREA8 33,590,73

SAFETY PROFILE: Questionable carcinogen with experimental tumorigenic data. When heated to decomposition it emits toxic fumes such as Br^-. See also BROMIDES.

BOB000 CAS:598-72-1 ***HR: 2***
α-BROMOPROPIONIC ACID
mf: $C_3H_5BrO_2$ mw: 152.99

TOXICITY DATA with REFERENCE
orl-mus LD50:250 mg/kg JPETAB 86,336,46

CONSENSUS REPORTS: Reported in EPA TSCA Inventory.

SAFETY PROFILE: Poison by ingestion. When heated to decomposition it emits toxic fumes of Br^-.

BOB250 CAS:590-92-1 ***HR: 2***
3-BROMOPROPIONIC ACID
mf: $C_3H_5BrO_2$ mw: 152.99

SYN: β-BROMOPROPIONIC ACID

TOXICITY DATA with REFERENCE
mmo-sat 25 μg/plate DHEFDK FDA-78-1046,78
ipr-mus TDLo:580 mg/kg/8W-I:ETA CNREA8 39,391,79
ipr-mus LDLo:500 mg/kg CBCCT* 6,228,54

CONSENSUS REPORTS: Reported in EPA TSCA Inventory.

SAFETY PROFILE: Moderately toxic by intraperitoneal route. Questionable carcinogen with experimental tumorigenic data. Mutation data reported. When heated to decomposition it emits toxic fumes of Br^-. See also BROMIDES.

BOB500 CAS:2417-90-5 ***HR: 3***
3-BROMO PROPIONITRILE
mf: C_3H_4BrN mw: 133.99

SYN: USAF DO-51

TOXICITY DATA with REFERENCE
ipr-mus LD50:50 mg/kg NTIS** AD277-689
par-mus LDLo:80 mg/kg CBCCT* 7,692,55

CONSENSUS REPORTS: Reported in EPA TSCA Inventory. Cyanide and its compounds are on the Community Right-To-Know List.

SAFETY PROFILE: Poison by parenteral and intraperitoneal routes. See also NITRILES. When heated to decomposition it emits very toxic fumes of NO_x, CN^-, and Br^-.

BOC510 CAS:626-55-1 ***HR: 3***
3-BROMOPYRIDINE
mf: C_5H_4BrN mw: 158.00

HC=CHCH=CBrCH=N (ring)

SAFETY PROFILE: Mixture with acetic acid + hydrogen peroxide explodes when heated above 50°C. When heated to decomposition it emits toxic fumes of Br^- and NO_x.

BOD000 CAS:41287-72-3 ***HR: 3***
3-(2-(5-BROMO-2-PYRIDYLOXY)ETHYL) THIAZOLIDINE HYDROCHLORIDE
mf: $C_{10}H_{13}BrN_2OS•ClH$ mw: 325.68

SYN: 5-BROMO-2-(2-(3-THIAZOLIDINYL)ETHOXY)PYRIDINEHYDROCHLORIDE

TOXICITY DATA with REFERENCE
orl-mus LD50:400 mg/kg JMCMAR 16,319,73
ipr-mus LD50:150 mg/kg JMCMAR 16,319,73

SAFETY PROFILE: Poison by ingestion and intraperitoneal routes. When heated to decomposition it emits very toxic fumes of Br^-, HCl, NO_x, and SO_x.

BOD500 CAS:41287-56-3 ***HR: 3***
2-(6-(5-BROMO-2-PYRIDYL OXY)HEXYL)AMINOETHANE THIOL HYDROCHLORIDE
mf: $C_{13}H_{21}BrN_2OS•ClH$ mw: 369.79

TOXICITY DATA with REFERENCE
orl-mus LD50:350 mg/kg JMCMAR 16,319,73
ipr-mus LD50:140 mg/kg JMCMAR 16,319,73

SAFETY PROFILE: Poison by ingestion and intraperitoneal route. When heated to decomposition it emits very toxic Br^-, Cl^-, SO_x, and NO_x.

BOE500 CAS:89-55-4 ***HR: 3***
5-BROMOSALICYLIC ACID
mf: $C_7H_5BrO_3$ mw: 217.03

TOXICITY DATA with REFERENCE
ivn-mus LD50:100 mg/kg CSLNX* NX#04478

CONSENSUS REPORTS: Reported in EPA TSCA Inventory.

SAFETY PROFILE: Poison by intravenous route. When heated to decomposition it emits toxic fumes of Br^-.

BOE750 CAS:13465-73-1 ***HR: 3***
BROMOSILANE
mf: BrH_3Si mw: 111.02

SYNS: SILYL BROMIDE

DOT Classification: Forbidden

SAFETY PROFILE: Ignites spontaneously upon exposure to air. When heated to decomposition it emits toxic fumes of Br^-. See also SILANE.

BOF000 CAS:103-64-0 ***HR: 2***
β-BROMOSTYRENE
mf: C_8H_7Br mw: 183.06

SYNS: α-BROMO-β-PHENYLETHYLENE ◇ ω-BROMOSTYRENE ◇ BROMOSTYROL ◇ BROMOSTYROLENE ◇ β-BROMSTYROL ◇ HYACINTH BASE

TOXICITY DATA with REFERENCE
orl-rat LD50:1250 mg/kg FCTXAV 11,1043,73

CONSENSUS REPORTS: Reported in EPA TSCA Inventory.

SAFETY PROFILE: Moderately toxic by ingestion. When heated to decomposition it emits toxic fumes of Br^-. See also BROMIDES.

BOF250 CAS:32017-76-8 ***HR: D***
4-BROMOSTYRENE OXIDE
mf: C_8H_7BrO mw: 199.06

SYNS: 1-BROMO-4-(EPOXYETHYL)BENZENE ◇ (p-BROMOPHENYL) OXIRANE ◇ (4-BROMOPHENYLOXIRANE) (9CI) ◇ p-BROMOSTYRENE OXIDE ◇ p-BROMOSTYRENE-7,8-OXIDE ◇ 4'-BROMOSTYRENE OXIDE

TOXICITY DATA with REFERENCE
mmo-sat 10 μg/plate MUREAV 111,99,83
mmo-esc 2 mmol/L CMSHAF 7,737,78

CONSENSUS REPORTS: EPA Genetic Toxicology Program.

SAFETY PROFILE: Mutation data reported. When heated to decomposition it emits toxic fumes of Br^-.

BOF500 CAS:128-08-5 ***HR: 3***
N-BROMO SUCCINIMIDE
mf: $C_4H_4BrNO_2$ mw: 178.00

$CH_2CO•NBrCO•CH_2$ (ring)

PROP: White to pale buff, fine, crystalline powder with faint odor of bromine. Mp: 173-175°, d: 2.098.

SYNS: 1-BROMO-2,5-PYRROLIDINEDIONE ◇ N-BROMOSUCCIMIDE ◇ SUCCINBROMIMIDE ◇ SUCCINIBROMIMIDE

TOXICITY DATA with REFERENCE
ipr-mus LDLo:256 mg/kg CBCCT* 2,244,50

CONSENSUS REPORTS: Reported in EPA TSCA Inventory.

SAFETY PROFILE: Poison by intraperitoneal route. An irritating poison to skin, eyes, and mucous membranes. Reacts explosively with aniline, diallyl sulfide, and hydrazine hydrate. Explosive reaction with propiononitrile after heating to 105°C for 24 hours. Violent reaction with dibenzoyl peroxide + 4-toluic acid. When heated to decomposition it emits toxic fumes of Br^- and NO_x. See also BROMIDES and NITROGEN MONOXIDE.

BOF750 CAS:679-84-5 ***HR: 2***
3-BROMO-1,1,2,2-TETRAFLUOROPROPANE
mf: $C_3H_3BrF_4$ mw: 194.97

SYNS: FHD-3 ◇ HALOPROPANE

TOXICITY DATA with REFERENCE
ihl-hmn TCLo:40000 ppm:CNS ANESAV 25,600,64
ihl-hmn TCLo:4000 ppm/30M:CVS ANESAV 25,600,64

SAFETY PROFILE: Human central nervous system and cardiovascular system effects by inhalation. When heated to decomposition it emits very toxic fumes of F^- and Br^-.

BOG000 CAS:14008-53-8 ***HR: 3***
3-BROMOTETRAHYDROTHIOPHENE-1,1-DIOXIDE
mf: $C_4H_7BrO_2S$ mw: 199.08

SYN: TETRAHYDRO-3-BROMOTHIOPHENE-1,1-DIOXIDE

TOXICITY DATA with REFERENCE
orl-rat LD50:215 mg/kg AIPTAK 119,423,59
ipr-rat LD50:44 mg/kg AIPTAK 119,423,59
orl-mus LD50:121 mg/kg AIPTAK 119,423,59
ipr-mus LD50:59 mg/kg AIPTAK 119,423,59
ivn-mus LD50:25 mg/kg AIPTAK 119,423,59
ivn-dog LD50:29 mg/kg AIPTAK 119,423,59

SAFETY PROFILE: Poison by ingestion, intraperitoneal, and intravenous routes. When heated to decomposition it emits very toxic fumes of Br^- and SO_x.

BOG250 CAS:6926-40-5 ***HR: 3***
N-BROMO TETRAMETHYL GUANIDINE
mf: $C_5H_{12}BrN_3$ mw: 194.07

$(CH_3)_2NC(:NBr)N(CH_3)_2$

SAFETY PROFILE: An unstable material which explodes when heated above 50°C. When heated to decomposition it emits toxic fumes of Br^- and NO_x.

BOH750 CAS:75-62-7 ***HR: 3***
BROMOTRICHLOROMETHANE
mf: $CBrCl_3$ mw: 198.27

PROP: Colorless liquid; bp: 103.8-105.1°; d: 1.997 @ 25°/25°.

TOXICITY DATA with REFERENCE
dnd-mam:lym 1 mmol/L TOLED5 11,243,82
orl-rat LDLo:100 mg/kg IJMDAI 10,301,74
ipr-rat LD50:119 mg/kg FAATDF 2,161,82

CONSENSUS REPORTS: Reported in EPA TSCA Inventory.

SAFETY PROFILE: Poison by ingestion and intraperitoneal routes. Narcotic in high concentration. Mutation data reported. See also CHLOROFORM. Incompatible with ethylene. When heated to decomposition it emits very toxic fumes of Cl^- and Br^-.

BOI000 CAS:13749-37-6 ***HR: 2***
3-BROMO-1,1,1-TRICHLORO PROPANE
mf: $C_3H_4BrCl_3$ mw: 224.31

SAFETY PROFILE: A preparative hazard. When heated to decomposition it emits toxic fumes of Cl^- and Br^-. See also BROMIDES; and CHLORINATED HYDROCARBONS, ALIPHATIC.

BOI250 CAS:63041-00-9 ***HR: 2***
3-BROMOTRICYCLOQUINAZOLINE
mf: $C_{21}H_{11}BrN_4$ mw: 399.27

TOXICITY DATA with REFERENCE
skn-mus TDLo:1240 mg/kg/1Y-I:NEO BJCAAI 16,275,62

SAFETY PROFILE: Questionable carcinogen with experimental neoplastigenic data. When heated to decomposition it emits very toxic fumes of Br^- and NO_x.

BOI750 CAS:2767-54-6 ***HR: 3***
BROMOTRIETHYLSTANNANE
mf: $C_6H_{15}BrSn$ mw: 285.81

PROP: Colorless liquid. Sol in organic solvents. D: 1.630; mp: −13.5°; bp: 224°.

SYN: TRIETHYLSTANNIUM BROMIDE ◇ TRIETHYL TIN BROMIDE

TOXICITY DATA with REFERENCE
ipr-rat TDLo:6 mg/kg (1D post):REP TXAPA9 72,557,84
ihl-mus LCLo:1640 mg/m^3 NDRC** NDCrc-132,Feb,42

CONSENSUS REPORTS: Reported in EPA TSCA Inventory.

OSHA PEL: TWA 0.1 mg(Sn)/m^3 (skin)
ACGIH TLV: TWA 0.1 mg(Sn)/m^3 (skin) (Proposed: TWA 0.1 mg(Sn)/m^3; STEL 0.2 mg(Sn)/m^3 (skin))
NIOSH REL: (Organotin Compounds) TWA 0.1 mg(Sn)/m^3

SAFETY PROFILE: Moderately toxic by inhalation. Experimental reproductive effects. See also TIN COMPOUNDS and BROMIDES. When heated to decomposition it emits toxic fumes of Br^-.

BOJ000 CAS:598-73-2 ***HR: 3***
BROMO TRIFLUOROETHYLENE
DOT: UN 2419
mf: BrF_3C_2 mw: 160.94

PROP: Bp: $-3°$.

SYNS: BROMOTRIFLUOROETHENE ◇ TRIFLUOROBROMOETHYLENE ◇ TRIFLUOROVINYLBROMIDE

CONSENSUS REPORTS: Reported in EPA TSCA Inventory.

DOT Classification: Flammable Gas; Label: Flammable Gas.

SAFETY PROFILE: A poison. Flammable gas or liquid. Ignites spontaneously in air. Incompatible with powerful oxidizers, O_2. When heated to decomposition it emits highly toxic fumes of Br^-, F^-, and $COCF_2$.

BOJ500 CAS:401-78-5 ***HR: 2***
m-BROMO-α,α,α-TRIFLUOROTOLUENE
mf: $C_7H_4BrF_3$ mw: 225.02

SYNS: 3-BROMBENZOTRIFLUORID (CZECH) ◇ m-BROMOBENZOTRIFLUORIDE ◇ 3-BROMOBENZOTRIFLUORIDE ◇ 3-BROMOBENZYLTRIFLUORIDE ◇ m-BROMO(TRIFLUOROMETHYL)BENZENE ◇ 3-BROMOTRIFLUOROMETHYLBENZENE ◇ m-(TRIFLUOROMETHYL)BROMOBENZENE ◇ 3-(TRIFLUOROMETHYL)BROMOBENZENE ◇ m-(TRIFLUOROMETHYL PHENYL BROMIDE ◇ 3-(TRIFLUOROMETHYL)PHENYL BROMIDE

TOXICITY DATA with REFERENCE
skn-rbt 500 mg/24H MLD 28ZPAK -,32,72
eye-rbt 500 mg/24H MOD 28ZPAK -,32,72
orl-rat LD50:2870 mg/kg 28ZPAK -,32,72

CONSENSUS REPORTS: Reported in EPA TSCA Inventory.

SAFETY PROFILE: Moderately toxic by ingestion. A skin and eye irritant. See also FLUORIDES and BROMIDES. When heated to decomposition it emits very toxic fumes of Br^- and F^-.

BOJ750 CAS:392-83-6 ***HR: 2***
o-BROMO-α,α,α-TRIFLUOROTOLUENE
mf: $C_7H_4BrF_3$ mw: 225.02

SYNS: 2-BROMBENZOTRIFLUORID (CZECH) ◇ o-BROMOBENZOTRIFLUORIDE ◇ 2-BROMOBENZOTRIFLUORIDE ◇ o-BROMOBENZYLTRIFLUORIDE ◇ o-(TRIFLUOROMETHYL)BROMOBENZENE

TOXICITY DATA with REFERENCE
skn-rbt 500 mg/24H MOD 28ZPAK -,32,72
eye-rbt 500 mg/24H MOD 28ZPAK -,32,72
orl-rat LD50:2720 mg/kg 28ZPAK -,32,72

CONSENSUS REPORTS: Reported in EPA TSCA Inventory.

SAFETY PROFILE: Moderately toxic by ingestion. A skin and eye irritant. See also FLUORIDES and BROMIDES. When heated to decomposition it emits very toxic fumes of Br^- and F^-.

BOK250 CAS:3091-18-7 ***HR: 3***
BROMOTRIPENTYLSTANNANE
mf: $C_{15}H_{33}BrSn$ mw: 412.08

SYN: TRI-N-PENTYLTIN BROMIDE

TOXICITY DATA with REFERENCE
ivn-mus LD50:56 mg/kg CSLNX* NX#05775

OSHA PEL: TWA 0.1 mg(Sn)/m^3 (skin)
ACGIH TLV: TWA 0.1 mg(Sn)/m^3 (skin) (Proposed: TWA 0.1 mg(Sn)/m^3; STEL 0.2 mg(Sn)/m^3 (skin))
NIOSH REL: (Organotin Compounds) TWA 0.1 mg(Sn)/m^3

SAFETY PROFILE: Poison by intravenous route. See also BROMIDES and TIN COMPOUNDS. When heated to decomposition it emits toxic fumes of Br^-.

BOK500 CAS:1607-57-4 ***HR: D***
BROMOTRIPHENYLETHYLENE
mf: $C_{20}H_{15}Br$ mw: 335.26

SYNS: ETHYLENE, BROMOTRIPHENYL- ◇ STILBENE, α'-BROMO-α-PHENYL-

TOXICITY DATA with REFERENCE
scu-rat TDLo:300 μg/kg (female 3D pre):REP AIPTAK 151,475,64

SAFETY PROFILE: Experimental reproductive effects. When heated to decomposition it emits toxic fumes of Br^-.

BOK750 CAS:2767-61-5 ***HR: 3***
BROMOTRIPROPYLSTANNANE
mf: $C_9H_{21}BrSn$ mw: 327.90

SYN: TRI-N-PROPYLTIN BROMIDE

TOXICITY DATA with REFERENCE
ihl-mus LCLo:1650 mg/m^3 NDRC** NDCrc-132,Feb,42
ivn-mus LD50:3600 μg/kg CSLNX* NX#02334

OSHA PEL: TWA 0.1 mg(Sn)/m^3 (skin)
ACGIH TLV: TWA 0.1 mg(Sn)/m^3 (skin) (Proposed: TWA 0.1 mg(Sn)/m^3; STEL 0.2 mg(Sn)/m^3 (skin))
NIOSH REL: (Organotin Compounds) TWA 0.1 mg(Sn)/m^3

SAFETY PROFILE: Poison by intravenous route. Moderately toxic by inhalation. When heated to decomposition it emits toxic fumes of Br^-. See also BROMIDES and TIN COMPOUNDS.

BOL000 CAS:51-20-7 ***HR: 2***
5-BROMOURACIL
mf: $C_4H_3BrN_2O_2$ mw: 191.00

TOXICITY DATA with REFERENCE
mmo-esc 5000 ppm AGACBH 4,286,74
cyt-grh-ipr 10 mg IDZAAW 38,305,83
ipr-mus TDLo:400 mg/kg (female 12D post):REP TCMUD8 7,7,87
ipr-rat LD50:1700 mg/kg PSEBAA 93,124,56
ipr-mus LD50:1400 mg/kg PSEBAA 93,124,56

CONSENSUS REPORTS: Reported in EPA TSCA Inventory. EPA Genetic Toxicology Program.

SAFETY PROFILE: Moderately toxic by intraperitoneal route. Experimental reproductive effects. Mutation data reported. When heated to decomposition it emits very toxic fumes of Br^- and NO_x.

BOL250 CAS:584-93-0 ***HR: 3***
α-BROMOVALERIC ACID
mf: $C_5H_9BrO_2$ mw: 181.05

TOXICITY DATA with REFERENCE
orl-mus LD50:380 mg/kg JPETAB 86,336,46

CONSENSUS REPORTS: Reported in EPA TSCA Inventory.

SAFETY PROFILE: Poison by ingestion. See also BROMIDES. When heated to decomposition it emits toxic fumes of Br^-.

BOL300 CAS:69304-47-8 ***HR: D***
(E)-5-(2-BROMOVINYL)-2'-DEOXYURIDINE
mf: $C_{10}H_{13}BrN_2O_5$ mw: 321.16

SYN: trans-5-(2-BROMOVINYL)-2'-DEOXYURIDINE

TOXICITY DATA with REFERENCE
dni-hmn:fbr 24 mg/L AMACCQ 24,803,83
sce-hmn:lym 50 mg/L BMJOAE 283,817,81
sce-hmn:fbr 50 mg/L BMJOAE 283,817,81
sce-hmn:lng 50 mg/L MUREAV 117,317,83

SAFETY PROFILE: Human mutation data reported. When heated to decomposition it emits toxic fumes of Br^- and NO_x.

BOL310 CAS:22585-64-4 ***HR: 3***
BROMYL FLUORIDE
mf: $BrFO_2$ mw: 130.90

SAFETY PROFILE: Reacts explosively with water. When heated to decomposition it emits toxic fumes of F^- and Br^-.

BOL325 CAS:23233-88-7 ***HR: 3***
BROTIANIDE
mf: $C_{15}H_{10}Br_2ClNO_2S$ mw: 463.59

SYNS: BAY 4059 ◇ BAY-VA 4059 ◇ 2-(ACETYLOXY)-3-BROMO-N-(4-BROMOPHENYL)-5-CHLORO-BENZENECARBOTHIOAMIDE ◇ 2-BROMO-6-(N-(p-BROMOPHENYL)THIOCARBAMOYL)-4-CHLOROBENZOIC ACID ◇ 3,4'-DIBROMO-5-CHLOROTHIOSALICYLANILIDE ACETATE (ESTER) ◇ DIRIAN

TOXICITY DATA with REFERENCE
orl-rat LD50:3000 mg/kg APFRAD 33,273,75
orl-mus LD50:184 mg/kg APFRAD 33,273,75
orl-rbt LD50:50 mg/kg APFRAD 33,273,75
orl-dom LD50:40 mg/kg FAZMAE 17,108,73

SAFETY PROFILE: Poison by ingestion. When heated to decomposition it emits toxic fumes of Cl^-, Br^-, SO_x, and NO_x.

BOL500 CAS:41451-75-6 ***HR: 3***
BRUCEANTIN
mf: $C_{28}H_{36}O_{11}$ mw: 548.64

PROP: A quassinoid from the *Brucea antidysenterica* plant.

SYN: NSC-165563

TOXICITY DATA with REFERENCE
dni-hmn:hla 50 nmol/L FEPRA7 33,581,74
dni-mus:lym 15 μmol/L JPMSAE 68,883,79
oms-mus:lym 15 μmol/L JPMSAE 68,883,79
orl-mus LD50:7027 μg/kg NCISP* JAN86
ipr-mus LD50:2727 μg/kg NCISP* JAN86
scu-mus LD50:3359 μg/kg NCISP* JAN86
ivn-mus LD50:1950 μg/kg TXAPA9 41,192,77
ivn-dog LDLo:500 μg/kg TXAPA9 41,192,77

SAFETY PROFILE: A deadly poison by ingestion, subcutaneous, intravenous, and intraperitoneal routes. Human mutation data reported. When heated to decomposition it emits acrid smoke and irritating fumes.

BOL750 CAS:357-57-3 ***HR: 3***
BRUCINE
DOT: UN 1570
mf: $C_{23}H_{26}N_2O_4$ mw: 394.51

PROP: Monoclinic prisms. Mp: 178°. An alkaloid extracted from *Strychnos* seeds (WQCHM* 4,-,74).

SYNS: BRUCIN (GERMAN) ◇ BRUCINA (ITALIAN) ◇ (−)-BRUCINE

◇ BRUCINE, solid (DOT) ◇ BRUCINE ALKALOID ◇ DIMETHOXY STRYCHNINE (DOT) ◇ 2,3-DIMETHOXYSTRYCHNINE ◇ 10,11-DIMETHYSTRYCHNINE ◇ RCRA WASTE NUMBER P018

TOXICITY DATA with REFERENCE
ipr-rat LD50:91 mg/kg JPETAB 131,185,61
scu-mus LD50:60 mg/kg APSXAS 7,329,70
ivn-dog LDLo:8 mg/kg HBAMAK 4,1289,35
ivn-rbt LDLo:30 mg/kg NTIS** PB214-270
ivn-gpg LDLo:120 mg/kg NTIS** PB214-270
scu-pgn LDLo:58 mg/kg HBAMAK 4,1289,35

CONSENSUS REPORTS: Reported in EPA TSCA Inventory.

DOT Classification: Poison B; Label: Poison.

SAFETY PROFILE: A poison by subcutaneous, intravenous, and intraperitoneal routes. An alkaloid-like strychnine, but one-sixth as toxic. When heated to decomposition it emits toxic fumes of NO_x. See also STRYCHNINE.

BOM000 CAS:60723-51-5 ***HR: 3***
BRUCINE METHIODIDE
mf: $C_{23}H_{26}N_2O_4 \cdot CH_3I$ mw: 536.45

SYNS: BRUCINE IODOMETHYLATE ◇ BRUCINE IODOMETHYLE (FRENCH)

TOXICITY DATA with REFERENCE
ivn-mus LDLo:10 mg/kg CRSBAW 144,53,50
ivn-rbt LDLo:30 mg/kg CRSBAW 144,53,50
ivn-gpg LDLo:120 mg/kg CRSBAW 144,53,50

SAFETY PROFILE: A poison via intravenous route. See also BRUCINE. When heated to decomposition it emits very toxic fumes of NO_x and I^-.

BOM125 ***HR: 3***
BUCKTHORN

PROP: A shrub which grows to 6 feet with small elliptical leaves 1 to 2 inches long. It produces a berry which turns black when mature and has a pit. It grows wild in western Texas and New Mexico.

SYNS: COYOTILLO ◇ KARWINSKIA HUMBOLDTIANA ◇ TULLIDORA

SAFETY PROFILE: The berry contains poisonous anthracenones. Ingestion may result (over a period of weeks or months) in loss of function in the peripheral nervous system including respiratory paralysis and death.

BOM250 CAS:129-74-8 ***HR: 2***
BUCLIZINE DIHYDROCHLORIDE
mf: $C_{28}H_{33}ClN_2 \cdot 2ClH$ mw: 506.00

SYNS: BUCLODIN ◇ 1-(p-tert-BUTYLBENZYL)-4-(p-CHLORODIPHENYLMETHYL)PIPERAZINE DIHYDROCHLORIDE ◇ 1-(p-tert-BUTYLBENZYL-4-p-CHLORO-α-PHENYLBENZYL)PIPERAZINE DIHYDROCHLORIDE ◇ 1-(p-CHLOROBENZHYDRYL)-4-(p-tert-BUTYLBENZYL)DIETHYLENEDIAMINE DIHYDROCHLORIDE ◇ 1-p-CHLOROBENZHYDRYL-4-p-(tert)-BUTYLBENZYLPIPERAZINEDIHYDROCHLORIDE ◇ HISTABUTYZINE DIHYDROCHLORIDE ◇ LONGIFENE ◇ SOFTRAN ◇ UCB 4445 ◇ VIBAZINE

TOXICITY DATA with REFERENCE
orl-rat TDLo:360 mg/kg (10-15D preg):TER AJOGAH 95,109,66
orl-rat LD50:1000 mg/kg MEXPAG 4,145,61
ivn-rat LD50:500 mg/kg MEXPAG 4,145,61
orl-mus LD50:2100 mg/kg JAPMA8 43,653,54
ipr-mus LD50:430 mg/kg JAPMA8 43,653,54

CONSENSUS REPORTS: Reported in EPA TSCA Inventory.

SAFETY PROFILE: Moderately toxic by ingestion, intravenous, and intraperitoneal routes. An experimental teratogen. When heated to decomposition it emits very toxic fumes of NO_x and HCl.

BOM510 CAS:36556-75-9 ***HR: 3***
BUCUMOLOL HYDROCHLORIDE
mf: $C_{17}H_{23}NO_4 \cdot ClH$ mw: 341.87

SYNS: dl-BUCUMOLOL HYDROCHLORIDE ◇ 8-(3-tert-BUTYLAMINO-2-HYDROXY)PROPOXY-5-METHYLCOUMARINHYDROCHLORIDE ◇ CS 359

TOXICITY DATA with REFERENCE
orl-rat LD50:1259 mg/kg IYKEDH 13,349,82
ipr-rat LD50:74200 μg/kg IYKEDH 13,349,82
scu-rat LD50:302 mg/kg IYKEDH 13,349,82
ivn-rat LD50:32400 μg/kg IYKEDH 13,349,82
orl-mus LD50:676 mg/kg IYKEDH 13,349,82
ipr-mus LD50:59200 μg/kg IYKEDH 13,349,82
scu-mus LD50:82300 μg/kg IYKEDH 13,349,82
ivn-mus LD50:31600 μg/kg JJPAAZ 23,497,73

SAFETY PROFILE: Poison by subcutaneous, intravenous, and intraperitoneal routes. Moderately toxic by ingestion. When heated to decomposition it emits toxic fumes of NO_x and HCl.

BOM520 CAS:51333-22-3 ***HR: 3***
BUDESONIDE
mf: $C_{25}H_{34}O_6$ mw: 430.59

PROP: Crystals. Mp: 221-232° (decomp). It is a mixture of two isomers; the content of the S-isomer in the mixture varies between 40-51%.

SYNS: (11-β,16-α)-16,17-(BUTYLIDENEBIS(OXY))-11,21-DIHYDROXYPREGNA-1,4-DIENE-3,20-DIONE ◇ 16-α,17-α-BUTYLIDENEDIOXY-11-β,21-DIHYDROXY-1,4-PREGNADIENE-3,20-DIONE ◇ PREFERID ◇ PULMICORT ◇ RHINOCORT

TOXICITY DATA with REFERENCE
scu-rat TDLo:44 μg/kg (female 7-17D post):REP KSRNAM 19,5093,85
scu-rat TDLo:220 μg/kg (female 7-17D post):TER KSRNAM 19,5093,85
ipr-rat LD50:138 mg/kg KSRNAM 19,4377,85
scu-rat LD50:58400 μg/kg KSRNAM 19,4377,85
ivn-rat LD50:98900 μg/kg KSRNAM 19,4377,85
orl-mus LD50:4750 mg/kg KSRNAM 19,4377,85
ipr-mus LD50:179 mg/kg KSRNAM 19,4377,85
scu-mus LD50:53600 μg/kg KSRNAM 19,4377,85
ivn-mus LD50:124 mg/kg KSRNAM 19,4377,85
scu-dog LD50:173 mg/kg KSRNAM 19,4377,85

SAFETY PROFILE: Poison by subcutaneous, intravenous, and intraperitoneal routes. Moderately toxic by ingestion. An experimental teratogen. Other experimental reproductive effects. When heated to decomposition it emits acrid smoke and fumes.

BOM530 CAS:57982-78-2 ***HR: 3***
BUDIPINE
mf: $C_{21}H_{27}N$ mw: 293.49

SYNS: BUDIPIN (GERMAN) ◇ 1-(1,1-DIMETHYLETHYL)-4,4-DIPHENYLPIPERIDINE

TOXICITY DATA with REFERENCE
orl-rat LD50:165 mg/kg ARZNAD 32,85,82
ivn-rat LD50:28 mg/kg ARZNAD 32,85,82
orl-mus LD50:120 mg/kg ARZNAD 32,85,82
ivn-mus LD50:33 mg/kg ARZNAD 32,85,82

SAFETY PROFILE: Poison by ingestion and intravenous routes. When heated to decomposition it emits toxic fumes of NO_x.

BOM600 CAS:35543-24-9 ***HR: 3***
BUFLOMEDIL HYDROCHLORIDE
mf: $C_{17}H_{25}NO_4 \cdot ClH$ mw: 343.89

SYNS: A-48257 ◇ BUFEDIL ◇ BUFLOMEDIL ◇ CHLORHYDRATE de (TRIMETHOXY-2-4-6)PHENYL-(PYRROLIDINE-3)PROPYLACETONE (FRENCH) ◇ FONZYLANE ◇ LL 1656 ◇ LOFTYL ◇ 4-(1-PYRROLIDINYL)-1-(2,4,6-TRIMETHOXYPHENYL)-1-BUTANONEHYDROCHLORIDE

TOXICITY DATA with REFERENCE
orl-mus LD50:275 mg/kg THERAP 30,207,75
ivn-mus LD50:55 mg/kg THERAP 30,207,75
ims-mus LD50:250 mg/kg THERAP 30,207,75
orl-dog LDLo:500 mg/kg THERAP 30,207,75
ivn-dog LDLo:50 mg/kg THERAP 30,207,75

SAFETY PROFILE: Poison by ingestion, intramuscular, and intravenous routes. A vasodilator. When heated to decomposition it emits toxic fumes of NO_x and HCl.

BOM650 CAS:465-39-4 ***HR: 3***
BUFOGENIN
mf: $C_{24}H_{32}O_4$ mw: 384.56

SYNS: 14,15-β-EPOXY-3-β-HYDROXY-5-β-BUFA-20,22-DIENOLIDE ◇ 3-β-HYDROXY-14,15-β-EPOXY-5-β-BUFA-20,22-DIENOLIDE ◇ RESIBUFOGENIN

TOXICITY DATA with REFERENCE
ivn-rat LD50:2200 μg/kg NIIRDN 6,899,82
ivn-mus LD50:4250 μg/kg NIIRDN 6,899,82
ivn-cat LD50:5 mg/kg JPETAB 111,365,54

SAFETY PROFILE: Deadly poison by intravenous route. When heated to decomposition it emits acrid smoke and fumes. See also BUFOGENIN B.

BOM655 CAS:465-19-0 ***HR: 3***
BUFOGENIN B
mf: $C_{24}H_{34}O_5$ mw: 402.58

PROP: Elongated prisms from methanol. Begins to sinter at 195°, decomp @ 210-223°. Very sparingly sol in chloroform, methanol, acetone.

SYNS: DESACETYLBUFOTALIN ◇ 3-β,14,16-β-TRIHYDROXY-5-β-BUFA-20,22-DIENOLIDE

TOXICITY DATA with REFERENCE
ivn-rat LDLo:2940 μg/kg OYYAA2 5,973,71
idu-rat LDLo:28400 μg/kg OYYAA2 5,973,71
orl-mus LD50:24500 μg/kg OYYAA2 5,973,71
scu-mus LD50:6950 μg/kg OYYAA2 5,973,71
ivn-mus LD50:10 μg/kg CPBTAL 24,1714,76
ivn-dog LDLo:580 μg/kg OYYAA2 5,973,71

SAFETY PROFILE: Deadly poison by ingestion, subcutaneous, intravenous, and intraduodenal routes. When heated to decomposition it emits acrid smoke and fumes. See also BUFOGENIN.

BOM750 CAS:1190-53-0 ***HR: 3***
BUFORMIN HYDROCHLORIDE
mf: $C_6H_{15}N_5 \cdot ClH$ mw: 193.72

SYNS: BUFONAMIN ◇ DIABRIN ◇ INSULAMIN

TOXICITY DATA with REFERENCE
cyt-hmn:emb 1900 μg/L SNSHBT (20),574,80
orl-rat LD50:320 mg/kg ARZNAD 12,314,62
orl-mus LD50:380 mg/kg ARZNAD 12,314,62
ipr-mus LD50:148 mg/kg PLRCAT 6,117,74

SAFETY PROFILE: A poison by intraperitoneal and ingestion routes. Mutation data reported. When heated to decomposition it emits very toxic fumes of HCl and NO_x.

BON000 CAS:471-95-4 ***HR: 3***
BUFOTALINE
mf: $C_{26}H_{36}O_6$ mw: 444.62

SYNS: BUFOTALIN ◇ 3-β,14,16-β-TRIHYDROXY-5-β-BUFA-20,22-DIENOLIDE-16-ACETATE

TOXICITY DATA with REFERENCE
scu-mus LD50:400 μg/kg CTOXAO 4,331,71
orl-dog LDLo:980 μg/kg CRSBAW 152,571,58
ivn-mus LD50:4130 μg/kg CPBTAL 24,1714,76
ivn-dog LDLo:360 μg/kg CRSBAW 152,571,58
ivn-cat LD50:130 μg/kg 85ELDJ -,189,63

SAFETY PROFILE: A deadly poison by ingestion, subcutaneous, and intravenous routes. When heated to decomposition it emits acrid and irritating fumes.

BON250 ***HR: 2***
BULAN and PROLAN MIXTURE (2:1)

SYNS: 1,1-BIS(p-CHLOROPHENYL)-2-NITROPROPANE mixed with 1,1-BIS(p-CHLOROPHENYL)-2-NITROBUTANE(1:2) ◇ CS 708 ◇ DILAN ◇ ENT 18,066

TOXICITY DATA with REFERENCE
orl-rat LD50:475 mg/kg FMCHA2 -,D103,80
skn-rat LD50:5900 mg/kg CMEP** -,1,56
orl-mus LD50:1100 mg/kg FEPRA7 12,368,53
ipr-mus LD50:950 mg/kg FEPRA7 12,368,53
orl-mam LD50:1100 mg/kg PCOC** -,929,66

SAFETY PROFILE: Moderately toxic by ingestion and intraperitoneal routes. Mildly toxic by skin contact. When heated to decomposition it emits very toxic fumes of Cl^- and NO_x. See also individual components.

BON300 ***HR: 3***
BULKOSOL

TOXICITY DATA with REFERENCE
orl-rat LD50: 3740 mg/kg NIIRDN 6,205,82
ivn-rat LD50:101 mg/kg NIIRDN 6,205,82
orl-mus LD50:2450 mg/kg NIIRDN 6,205,82
scu-mus LD50:4300 mg/kg NIIRDN 6,205,82
ivn-mus LD50:76800 μg/kg NIIRDN 6,205,82

SAFETY PROFILE: Poison by intravenous route. Moderately toxic by ingestion.

BON325 CAS:28395-03-1 ***HR: 3***
BUMETANIDE
mf: $C_{17}H_{20}N_2O_5S$ mw: 364.45

PROP: Crystals from aq ethanol. Mp: 230-231°.

SYNS: 3-(AMINOSULFONYL)-5-(BUTYLAMINO)-4-PHENOXY-3-(AMINOSULFONYL)-5-(BUTYLAMINO)-4-PHENOXYBENZOICACID ◇ BUMEX ◇ BURINE ◇ BURINEX ◇ 3-(BUTYLAMINO)-4-PHENOXY-5-SULFAMOYLBENZOIC ACID ◇ FONTEGO ◇ FORDIURAN ◇ LIXIL ◇ LUNETORON ◇ PF 1593 ◇ RO 10-6338 ◇ SEGUREX

TOXICITY DATA with REFERENCE
ipr-rat LD50:1000 mg/kg ARZNAD 1,218,51
scu-rat LD50:22500 μg/kg OYYAA2 9,413,75
ivn-rat LD50:4 mg/kg ARZNAD 1,218,51
orl-mus LD50:156 mg/kg OYYAA2 9,413,75
scu-mus LD50:140 mg/kg ARZNAD 1,218,51
ivn-mus LD50:4900 μg/kg OYYAA2 9,413,75
ivn-rbt LD50:2400 μg/kg OYYAA2 9,413,75

SAFETY PROFILE: Poison by ingestion, subcutaneous, and intravenous routes. Moderately toxic by intraperitoneal routes. When heated to decomposition it emits toxic fumes of SO_x and NO_x.

BON350 ***HR: 3***
BUNAZOCINE HYDROCHLORIDE
mf: $C_{19}H_{27}N_5O_3{\cdot}ClH$ mw: 409.91

SYNS: 4-AMINO-2-(4-BUTYRYLHEXAHYDRO-1H-1,4-DIAZEPIN-1-YL)-6,7-DIMETHOXYQUINAZOLINE HYDROCHLORIDE ◇ BUNAZOSIN HYDROCHLORIDE

TOXICITY DATA with REFERENCE
orl-rat LD50:1280 mg/kg IYKEDH 16,866,85
scu-rat LD50:365 mg/kg IYKEDH 16,866,85
ivn-rat LD50:50 mg/kg IYKEDH 16,866,85
ims-rat LD50:152 mg/kg IYKEDH 16,866,85
orl-mus LD50:1201 mg/kg IYKEDH 16,866,85
scu-mus LD50:730 mg/kg IYKEDH 16,866,85
ivn-mus LD50:57 mg/kg IYKEDH 16,866,85
ims-mus LD50:660 mg/kg IYKEDH 16,866,85

SAFETY PROFILE: Poison by subcutaneous, intramuscular, intravenous, and intraperitoneal routes. Moderately toxic by ingestion. When heated to decomposition it emits toxic fumes of NO_x and HCl.

BON365 ***HR: 3***
BUNGARUS CAERULEUS VENOM

SYN: VENOM, SNAKE, BUNGARUS CAERULEUS

TOXICITY DATA with REFERENCE
ipr-mus LD50:8 μg/kg TOXIA6 14,451,76
scu-mus LD50:450 μg/kg TOXIA6 5,47,67
ivn-mus LD50:96 μg/kg IJMRAQ 60,512,72
ivn-dog LDLo:120 μg/kg 19DDA6 1,269,67
ivn-rbt LDLo:40 μg/kg TOXIA6 2,5,64
ivn-mam LD50:90 μg/kg CLPTAT 8,849,67

SAFETY PROFILE: Deadly poison by subcutaneous, intravenous, and intraperitoneal routes.

BON367 ***HR: 3***
BUNGARUS FASCIATUS VENOM

SYN: VENOM, SNAKE, BUNGARUS FASCIATUS

TOXICITY DATA with REFERENCE
ipr-mus LD50:150 μg/kg 85EGD4 5,161,78
scu-mus LD50:3580 μg/kg/ TOXIA6 5,47,67
ivn-mus LD50:170 μg/kg TOXIA6 21,681,83

SAFETY PROFILE: Deadly poison by subcutaneous, intravenous, and intraperitoneal routes.

BON370 ***HR: 3***
BUNGARUS MULTICINCTUS VENOM

SYN: VENOM, FORMOSAN BANDED KRAIT, BUNGARUS MULTICINCTUS

TOXICITY DATA with REFERENCE
ipr-mus LD50:25 μg/kg JOBIAO 48,714,60
scu-mus LD50:160 μg/kg TIHHAH 61,239,62
ivn-mus LD50:71 μg/kg TOXIA6 9,131,71
ivn-rbt LDLo:1 mg/kg TOXIA6 3,281,66

SAFETY PROFILE: Deadly poison by subcutaneous, intravenous, and intraperitoneal routes.

BON400 CAS:23093-74-5 ***HR: 3***
BUNITROLOL HYDROCHLORIDE
mf: $C_{14}H_{20}N_2O_2$•ClH mw: 284.82

SYNS: BETRILOL ◇ o-(3-tert-BUTYLAMINO-2-HYDROXYPROPOXY)BENZONITRILE HYDROCHLORIDE ◇ 2-(3-((1,1-DIMETHYLETHYL)AMINO)-2-HYDROXYPROPOXY)-BENZONITRILEHYDROCHLORIDE ◇ o-(2-HYDROXY-3-(tert-BUTYLAMINO)PROPOXY) BENZONITRILE HYDROCHLORIDE ◇ KO 1366-CL ◇ KOE 1366 CHLORIDE ◇ STRESSON

TOXICITY DATA with REFERENCE
orl-rat TDLo:1100 mg/kg (female 7-17D post):REP IYKEDH 12,12,81
ipr-rat TDLo:70 mg/kg (female 7-13D post):TER OYYAA2 11,779,76
orl-rat LD50:639 mg/kg OYYAA2 11,795,76
ipr-rat LD50:222 mg/kg IYKEDH 14,484,83
scu-rat LD50:902 mg/kg IYKEDH 14,484,83
ivn-rat LD50:69 mg/kg IYKEDH 14,484,83
orl-mus LD50:250 mg/kg IYKEDH 12,25,81
ipr-mus LD50:264 mg/kg IYKEDH 14,484,83
scu-mus LD50:542 mg/kg IYKEDH 14,484,83
ivn-mus LD50:264 mg/kg OYYAA2 9,457,75
orl-dog LD50:490 mg/kg IYKEDH 14,484,83
ivn-dog LD50:36 mg/kg IYKEDH 14,484,83

CONSENSUS REPORTS: Cyanide and its compounds are on the Community Right-To-Know List.

SAFETY PROFILE: Poison by ingestion, intravenous, and intraperitoneal routes. Moderately toxic by subcutaneous route. Experimental reproductive effects. An experimental teratogen. When heated to decomposition it emits toxic fumes of NO_x, CN^- and HCl. See also NITRILES.

BON750 CAS:27262-46-0 ***HR: 3***
BUPICAINE HYDROCHLORIDE (+)
mf: $C_{18}H_{28}N_2O$•ClH mw: 324.94

SYN: 1-BUTYL-2',6'-PIPECOLOXYLIDIDE HYDROCHLORIDE (+)

TOXICITY DATA with REFERENCE
scu-rat LD50:43 mg/kg AIPTAK 200,359,72
ivn-rat LD50:6 mg/kg AIPTAK 200,359,72
scu-mus LD50:58 mg/kg AIPTAK 200,359,72
ivn-mus LD50:7200 μg/kg AIPTAK 200,359,72
orl-rbt LD50:18 mg/kg AIPTAK 200,359,72
ivn-rbt LD50:3300 μg/kg AIPTAK 200,359,72
par-rbt LD50:185 mg/kg AIPTAK 200,359,72
itr-rbt LD50:12 mg/kg AIPTAK 200,359,72

SAFETY PROFILE: Poison by ingestion, subcutaneous, intravenous, parenteral, and intratracheal routes. When heated to decomposition it emits very toxic fumes of HCl and NO_x. See other bupicaine or bupivacaine entries.

BOO000 CAS:14252-80-3 ***HR: 3***
BUPICAINE HYDROCHLORIDE (±)
mf: $C_{18}H_{28}N_2O$•ClH mw: 324.94

SYNS: BUPIVACAINE HYDROCHLORIDE ◇ 1-BUTYL-2',6'-PIPECOLOXYLIDIDE (±) ◇ (±)-1-BUTYL-2',6'-PIPECOLOXYLIDIDE MONOHYDROCHLORIDE, MONOHYDRATE ◇ CARBOSTESIN ◇ LAC-43 ◇ MARCAIN ◇ 2-PIPERIDINECARBOXAMIDE,1-BUTYL-N-(2,6-DIMETHYLPHENYL)MONOHYDROCHLORIDEMONOHYDRATE

TOXICITY DATA with REFERENCE
scu-rat LD50:43 mg/kg AIPTAK 200,359,72
ivn-rat LD50:6 mg/kg AIPTAK 200,359,72
scu-mus LD50:59 mg/kg AIPTAK 200,359,72
ivn-mus LD50:6400 μg/kg AIPTAK 200,359,72
orl-rbt LD50:18 mg/kg AIPTAK 200,359,72
ivn-rbt LD50:3400 μg/kg AIPTAK 200,359,72
par-rbt LD50:48 mg/kg AIPTAK 200,359,72
itr-rbt LD50:11 mg/kg AIPTAK 200,359,72
ipr-gpg LD50:50 mg/kg NIIRDN 6,680,82

SAFETY PROFILE: Poison by ingestion, subcutaneous, intravenous, intraperitoneal, parenteral, and intratracheal routes. A local anesthetic. When heated to decomposition it emits very toxic fumes of HCl and NO_x. See other bupicaine or bupivacaine entries.

BOO250 CAS:27262-45-9 ***HR: 3***
d(+)-BUPIVACAINE
mf: $C_{18}H_{28}N_2O$ mw: 288.48

SYN: d-(+)-1-BUTYL-2',6'-PIPECOLOXYLIDIDE

TOXICITY DATA with REFERENCE
scu-rat LD50:38 mg/kg APTOA6 31,273,72
ivn-rat LD50:3800 μg/kg APTOA6 31,273,72
scu-mus LD50:30 mg/kg APTOA6 31,273,72
ivn-mus LD50:7900 μg/kg APTOA6 31,273,72
ivn-rbt LDLo:5500 μg/kg APTOA6 31,273,72
itr-rbt LD50:10 mg/kg ARZNAD 26,78,76

SAFETY PROFILE: Poison by subcutaneous, intratracheal and intravenous routes. When heated to decomposition it emits toxic fumes of NO_x. See other bupicaine or bupivacaine entries.

BOO500 CAS:27262-47-1 ***HR: 3***
l(–)-BUPIVACAINE
mf: $C_{18}H_{28}N_2O$ mw: 288.48

SYN: l-(–)-1-BUTYL-2',6'-PIPECOLOXYLIDIDE

TOXICITY DATA with REFERENCE
scu-rat LD50:52 mg/kg APTOA6 31,273,72
ivn-rat LD50:7200 μg/kg APTOA6 31,273,72
scu-mus LD50:100 mg/kg APTOA6 31,273,72
ivn-mus LD50:9600 μg/kg APTOA6 31,273,72
ivn-rbt LDLo:9700 μg/kg APTOA6 31,273,72
itr-rbt LD50:14 mg/kg ARZNAD 26,78,76

SAFETY PROFILE: Poison by subcutaneous, intratracheal, and intravenous routes. When heated to decomposition it emits toxic fumes of NO_x. See other bupicaine or bupivacaine entries.

BOO625 ***HR: 2***
BUPLEURUM MARGINATUM WALL. EX. DC., EXTRACT

PROP: Indian plant belonging to the family *Apiaceae* (IJEBA6 22,312,84).

SYN: BURPLEURUM FALCATUM LINN. VAR. MARGINATUM (WALL. EX. DC.) CL., EXTRACT

TOXICITY DATA with REFERENCE
orl-rat TDLo:150 mg/kg (12-14D preg):REP IJEBA6 22,312,84
ipr-mus LD50:1 g/kg IJEBA6 22,312,84

SAFETY PROFILE: Moderately toxic by intraperitoneal route. Experimental reproductive effects.

BOO630 CAS:53152-21-9 ***HR: 3***
BUPRENORPHINE HYDROCHLORIDE
mf: $C_{29}H_{41}NO_4$•ClH mw: 504.17

SYNS: M-6029 ◇ MR 56

TOXICITY DATA with REFERENCE
ims-rat TDLo:1350 μg/kg (17-22D preg/21D post):REP IYKEDH 13,532,82
ims-rat TDLo:550 μg/kg (female 7-17D post):TER IYKEDH 13,509,82
orl-man TDLo:2857 μg/kg BMJOAE 296,214,88
ivn-rat LD50:62 mg/kg IYKEDH 13,486,82
orl-mus LD50:800 mg/kg IYKEDH 13,486,82
ivn-mus LD50:72 mg/kg IYKEDH 13,486,82
ivn-dog LD50:79 mg/kg YKYUA6 35,1351,84

SAFETY PROFILE: Poison by intravenous route. Moderately toxic by ingestion. An experimental teratogen. Other experimental reproductive effects. When heated to decomposition it emits toxic fumes of NO_x and HCl.

BOO635 CAS:21564-17-0 ***HR: 2***
BUSAN 72A
mf: $C_9H_6N_2S_3$ mw: 238.35

SYNS: TCMTB ◇ 2-(THIOCYANOMETHYLTHIO)BENZOTHIAZOLE, 60% ◇ THIOCYANIC ACID, 2-(BENZOTHIAZOLYLTHIO)METHYL ESTER

TOXICITY DATA with REFERENCE
orl-rat LD50:1590 mg/kg BUCKL* TCMTB,81
skn-rbt LD50:10 g/kg BUCKL* TCMTB,81
orl-dck LD50:1310 mg/kg BUCKL* TCMTB,81

SAFETY PROFILE: Moderately toxic by ingestion. When heated to decomposition it emits toxic fumes of SO_x and NO_x. See also THIOCYANATES.

BOO650 CAS:8059-83-4 ***HR: 3***
BUSCOPAN COMPOSITUM
mf: $C_{21}H_{30}NO_4$•$C_{13}H_{17}N_3O_4S$•Br•Na mw: 774.81

SYNS: N-BUTYLSCOPOLAMMONIUM BROMIDE combined with SODIUM SULPYRINE (1:25) ◇ SB 502

TOXICITY DATA with REFERENCE
ims-rat TDLo:22500 mg/kg (female 30D pre):REP OYYAA2 9,615,75
orl-rat LD50:4700 mg/kg KSRNAM 7,54,73
ipr-rat LD50:1400 mg/kg KSRNAM 7,54,73
scu-rat LD50:2850 mg/kg KSRNAM 7,54,73
orl-mus LD50:4350 mg/kg KSRNAM 7,54,73
ipr-mus LD50:2050 mg/kg KSRNAM 7,54,73
scu-mus LD50:2300 mg/kg KSRNAM 7,54,73
ivn-mus LD50:390 mg/kg KSRNAM 7,54,73
ivn-rbt LD50:480 mg/kg KSRNAM 7,54,73

SAFETY PROFILE: Poison by intravenous route. Moderately toxic by subcutaneous and intraperitoneal routes. Mildly toxic by ingestion. Experimental reproductive effects. When heated to decomposition it emits toxic fumes of Br^-, SO_x, NH_3, NO_x, and Na_2O.

BOO700 ***HR: 3***
BUSHMAN'S POISON

PROP: An evergreen shrub or small tree native to Africa but also found in California, Florida, and Hawaii. They are used as ornamental shrubs in California and in greenhouses in the rest of the US. The plant has large leaves, fragrant flowers shaped like a flared tube, and a small, plum-like fruit which is red or purple-black when mature.

SYNS: ACOKANTHERA (VARIOUS SPECIES) ◇ A. LONGIFLORA ◇ A. OBLONGIFOLIA ◇ A. OPPOSITIFOLIA ◇ POISON BUSH ◇ POISON TREE ◇ WINTERSWEET

SAFETY PROFILE: The toxic agent is a cardiac glycoside similar to ouabain. It is found in all parts of the plant with the highest concentration in the seeds. The fruit of some species has low levels of toxin and is considered edible. Human systemic effects may include: nausea, vomiting, pain in the mouth and abdomen, cramps, diarrhea, slowed heartbeat and high blood potassium levels. Symptoms develop after a delay period which is dependent upon the dose. See also OUABAIN.

BOO750 CAS:149-16-6 ***HR: 3***
BUTACAINE
mf: $C_{18}H_{30}N_2O_2$ mw: 306.50

PROP: Colorless, odorless powder. Mp: 98-100°.

SYNS: 3-(p-AMINOBENZOXY)-1-DI-n-BUTYLAMINOPROPANE ◇ p-AMINOBENZOYLDIBUTYLAMINOPROPANOL ◇ BUTYN ◇ 3-(DIBUTYLAMINO)-1-PROPANOL-p-AMINOBENZOATE ◇ 3-DIBUTYLAMINOPROPYL-p-AMINOBENZOATE

TOXICITY DATA with REFERENCE
scu-rat LDLo:150 mg/kg JPETAB 24,167,25
ivn-rat LDLo:7500 mg/kg PHREA7 12,190,32
scu-mus LDLo:100 mg/kg JPETAB 24,167,25
ivn-mus LDLo:12 mg/kg JAPMA8 39,4,50
scu-dog LDLo:55 mg/kg PHREA7 12,190,32
scu-cat LDLo:30 mg/kg JPETAB 24,167,25
ivn-cat LDLo:15 mg/kg AJPHAP 68,110,24
scu-rbt LDLo:50 mg/kg JPETAB 24,167,25
ivn-rbt LDLo:12 mg/kg PHREA7 12,190,32
scu-gpg LDLo:45 mg/kg JPETAB 62,69,38

SAFETY PROFILE: A poison via subcutaneous and intravenous routes. A weak allergen. Combustible. When heated to decomposition it emits toxic fumes of NO_x. See also BUTACAINE SULFATE.

BOP000 CAS:149-15-5 ***HR: 3***
BUTACAINE SULFATE
mf: $C_{36}H_{60}N_4O_4 \cdot H_2O_4S$ mw: 711.08

SYNS: 3-(p-AMINOBENZOXY)-1-DI-n-BUTYLAMINOPROPANESULFATE ◇ p-AMINOBENZOYLDIBUTYLAMINOPROPANOL SULFATE ◇ BUTELLINE ◇ BUTYN SULFATE ◇ 3-(DIBUTYLAMINO)-1-PROPANOL-p-AMINOBENZOATE (ESTER) SULFATE (2:1) ◇ 3-DIBUTYLAMINO-1-PROPANOL-4-AMINOBENZOATE (ESTER) SULFATE (SALT) (2:1) ◇ DIBUTYLAMINOPROPYL-p-AMINOBENZOATE SULFATE ◇ 3'-DIBUTYLAMINOPROPYL-4-AMINOBENZOATE SULFATE

TOXICITY DATA with REFERENCE
scu-rat LDLo:197 mg/kg PHREA7 12,262,32
ipr-mus LD50:80 mg/kg BJPCAL 1,90,46
scu-mus LDLo:100 mg/kg JPETAB 24,167,25
ivn-mus LD50:12 mg/kg JAPMA8 40,373,51
orl-bwd LD50:100 mg/kg TXAPA9 21,315,72

SAFETY PROFILE: A poison by ingestion, subcutaneous, intravenous, and intraperitoneal routes. A topical anesthetic. See also SULFATES. When heated to decomposition it emits very toxic fumes of SO_x and NO_2.

BOP250 CAS:590-19-2 ***HR: 2***
1,2-BUTADIENE
mf: C_4H_6 mw: 54.10

$$H_2C{=}C{=}CHCH_3$$

PROP: Flash p: <0°.

SAFETY PROFILE: A dangerous fire hazard. When heated to decomposition it emits acrid smoke and fumes. See also 1,3-BUTADIENE.

BOP500 CAS:106-99-0 ***HR: 3***
1,3-BUTADIENE
mf: C_4H_6 mw: 54.10

$$H_2C{=}CHCH{=}CH_2$$

PROP: Colorless gas; mild aromatic odor. Very reactive. Bp: −4.5°, mp: −113°, fp: −108.9°, flash p: −105°F, lel: 2.0%, uel: 11.5%, d: 0.621 @ 2 0°/4°, autoign temp: 788°F, vap d: 1.87, vap press: 1840 mm @ 21°.

SYNS: BIETHYLENE ◇ BIVINYL ◇ BUTADIEEN (DUTCH) ◇ BUTA-1,3-DIEEN (DUTCH) ◇ BUTADIEN (POLISH) ◇ BUTA-1,3-DIEN (GERMAN) ◇ BUTA-1,3-DIENE ◇ α-Γ-BUTADIENE ◇ DIVINYL ◇ ERYTHRENE ◇ NCI-C50602 ◇ PYRROLYLENE ◇ VINYLETHYLENE

TOXICITY DATA with REFERENCE
mnt-mus:ihl 100 ppm/6H/2D-C ENMUDM 8(Suppl 6),18,86
msc-mus:lym 20 pph ENMUDM 8(Suppl 6),75,86
ihl-rat TCLo:8000 ppm/6H (6-15D preg):TER EPASR* 8EHQ-0382-0441
ihl-rat TCLo:625 ppm/6H/61W:CAR NTPTR* NTP-TR-288,84
ihl-mus TCLo:1250 ppm/6H/60W-I:CAR SCIEAS 227,548,85
ihl-rat TC:8000 ppm/6H/2Y-I:NEO AIHAAP 48,407,87
ihl-hmn TCLo:2000 ppm/7H:EYE JIHTAB 26,69,44
ihl-hmn TCLo:8000 ppm:EYE,PUL INMEAF 17,199,48
orl-rat LD50:5480 mg/kg 85JCAE -,14,86
ihl-rat LC50:285 g/m³/4H RPTOAN 31,162,68
ihl-mus LC50:270 g/m³/2H RPTOAN 31,162,68
ihl-rbt LCLo:25 pph/23M JIHTAB 26,69,44

CONSENSUS REPORTS: NTP Fifth Annual Report on Carcinogens. IARC Cancer Review: Group 2B IMEMDT 7,136,87; Human Inadequate Evidence IMEMDT 39,155,86; Animal Sufficient Evidence IMEMDT 39,155,86. NTP Carcinogenesis Studies (inhalation); Clear Evidence: mouse NTPTR* NTP-TR-288,84. Reported in EPA TSCA Inventory. Community Right-To-Know List.

OSHA PEL: TWA 1000 ppm
ACGIH TLV: TWA 10 ppm; Suspected Human Carcinogen
DFG MAK: Processing after polymerization and loading: 15 ppm; Others: 5 ppm; Animal Carcinogen, Suspected Human Carcinogen.
NIOSH REL: Reduce to lowest feasible level
DOT Classification: Flammable Gas; Label: Flammable Gas.

SAFETY PROFILE: Confirmed carcinogen with experimental carcinogenic and neoplastigenic data. An experimental teratogen. Mutation data reported. Inhalation of high concentrations can cause unconsciousness and death. Human systemic effects by inhalation: cough, hallucinations, distorted perceptions, changes in the visual field and other unspecified eye effects. The vapors are irritating to eyes and mucous membranes. If spilled on skin or clothing, it can cause burns or frost bite (due to rapid vaporization). Chronic systemic poisoning in humans has not been reported. Dangerous fire hazard when exposed to heat, flame, or powerful oxidizers. Upon exposure to air it forms explosive peroxides sensitive to heat, shock, or heating above 27°C. May decompose explosively when heated above 200°C/1.0 kbar. Explodes on contact with aluminum tetrahydroborate. Potentially explosive reaction with $NO_x + O_2$, ethanol + iodine + mercury oxide (at 35°C), ClO_2, crotonaldehyde (above 180°C), buten-3-yne (with heat and pressure). Reaction with sodium nitrite forms a spontaneously flammable product. Exothermic reaction with boron trifluoride etherate + phenol. To fight fire, stop flow of gas. When heated to decomposition it emits acrid smoke and fumes.

BOP750 CAS:30031-64-2 ***HR: 3***
l-BUTADIENE DIEPOXIDE
mf: $C_4H_6O_2$ mw: 86.10

SYNS: (S-(R*,R*))-2,2'-BIOXIRANE ◇ l-DIEPOXYBUTANE ◇ (2S,3S)-DIEPOXYBUTANE ◇ l-1,2:3,4-DIEPOXYBUTANE ◇ (2S,3S)-1,2:3,4-DIEPOXYBUTANE ◇ NSC-32606

TOXICITY DATA with REFERENCE
dnd-omi 5 mmol/L BBACAQ 228,400,71
mmo-ssp 31 mmol/L ADWMAX -,193,62
ipr-mus TDLo:110 mg/kg/4W:NEO JNCIAM 36,915,66
ipr-mus LD50:61 mg/kg NCISA* PH-43-63-1132

CONSENSUS REPORTS: IARC Cancer Review: Group 2B IMEMDT 7,56,87; Animal Sufficient Evidence IMEMDT 11,115,76. EPA Genetic Toxicology Program.

SAFETY PROFILE: Suspected carcinogen with experimental neoplastigenic data. Poison by intraperitoneal route. Mutation data reported. When heated to decomposition it emits acrid and irritating fumes.

BOQ250 ***HR: 3***
BUTADIENE PEROXIDE
mf: $C_4H_6O_2$ mw: 86.09

SAFETY PROFILE: A shock-sensitive explosive formed by the peroxidation of butadiene upon prolonged exposure to air. Potentially explosive polymerization reaction with butadiene. Tank monitoring and a purging system are recommended to prevent explosion on contact with air over a long period of time. Concentration in butadiene as measured by standard methods of determining hydroperoxides may be only 5% of the true concentration. When heated to decomposition it emits acrid smoke and fumes. See also PEROXIDES.

BOQ500 CAS:16719-32-7 ***HR: 3***
N-2,3-BUTADIENYL-N-METHYLBENZYLAMINE HYDROCHLORIDE
mf: $C_{12}H_{15}N{\bullet}ClH$ mw: 209.74

SYN: U-1247

TOXICITY DATA with REFERENCE
orl-mus LD50:339 mg/kg JPMSAE 57,430,68
ipr-mus LD50:156 mg/kg JPMSAE 57,430,68
ivn-mus LD50:32 mg/kg JPMSAE 57,430,68

SAFETY PROFILE: Poison by ingestion, intraperitoneal, and intravenous routes. When heated to decomposition it emits very toxic fumes as Cl^- and NO_x.

BOQ625 CAS:460-12-8 ***HR: 3***
1,3-BUTADIYNE
mf: C_4H_2 mw: 50.06

SAFETY PROFILE: A dangerous explosive. Polymerizes violently above 0°C. Arsenic pentafluoride catalyzes explosive polymerization. Reaction with silver nitrate forms a very explosive friction-sensitive product. When heated to decomposition it emits acrid smoke and fumes.

BOQ750 CAS:125-88-2 ***HR: 3***
BUTALBITAL SODIUM
mf: $C_{10}H_{14}N_2O_3{\bullet}Na$ mw: 233.25

SYNS: APROBARBITAL SODIUM ◇ APROBARBITONE SODIUM ◇ SODIUM-5-ALLYL-5-ISOPROPYLBARBITURATE

TOXICITY DATA with REFERENCE
ipr-mus LDLo:400 mg/kg NTIS** AD691-490
ipr-rat LD50:85 mg/kg APSCAX 18,204,49
scu-rat LDLo:125 mg/kg JACSAT 47,2236,25

SAFETY PROFILE: A poison via intraperitoneal and subcutaneous routes. When heated to decomposition it emits toxic fumes of NO_x and Na_2O. See also BARBITURATES.

BOR000 CAS:1142-70-7 ***HR: 3***
BUTALLYLONAL
mf: $C_{11}H_{15}BrN_2O_3$ mw: 303.19

SYNS: 5-(2-BROMOALLYL)-5-sec-BUTYLBARBITURICACID ◇ 5-(2'-BROMOALLYL)-5-(1'-METHYL-N-PROPYL)BARBITURIC ACID ◇ BUTYLALYLONAL ◇ 5-sec-BUTYL-5-(β-BROMOALLYL)BARBITURIC ACID ◇ PERNOCTON ◇ PERNOSTON ◇ 2,4,6(1H,3H,5H)-PYRIMIDINETRIONE,5-(2-BROMO-2-PROPENYL)-5-(1-METHYLPROPYL)-(9CI) ◇ SONBUTAL

TOXICITY DATA with REFERENCE
ipr-rat LDLo:65 mg/kg JPETAB 44,325,32
scu-rat LD50:90 mg/kg AEPPAE 152,341,30
scu-mus LDLo:150 mg/kg REDH** #3850
orl-rbt LDLo:350 mg/kg REDH** #3850
ipr-rbt LDLo:75 mg/kg JPETAB 41,465,31
scu-rbt LDLo:160 mg/kg REDH** #3850
ivn-rbt LDLo:70 mg/kg REDH** #3850
scu-frg LDLo:150 mg/kg PHREA7 19,472,39
orl-mam LDLo:350 mg/kg JPETAB 42,253,31

SAFETY PROFILE: Poison by ingestion, intravenous, intraperitoneal, and subcutaneous routes. A central nervous system depressant (hypnotic) by ingestion. When heated to decomposition it emits very toxic fumes of Br^- and NO_x. See also BARBITURATES and ALLYL COMPOUNDS.

BOR250 CAS:3486-86-0 ***HR: 3***
BUTALLYLONAL SODIUM
mf: $C_{11}H_{14}BrN_2O_3$•Na mw: 325.17

SYNS: sec-BUTYL-BROM-ALLYL BARBITURIC ACID SODIUM SALT ◇ SODIUM-5-(2-BROMOALLYL)-5-sec-BUTYLBARBITURATE

TOXICITY DATA with REFERENCE
orl-cat LD50:135 mg/kg JPETAB 88,260,46
orl-rbt LD50:375 mg/kg JPETAB 42,253,31
ipr-rbt LD50:75 mg/kg JPETAB 42,253,31

SAFETY PROFILE: Poison by ingestion and intraperitoneal routes. When heated to decomposition it emits very toxic fumes of Br^- and NO_x. See also BARBITURATES.

BOR350 CAS:18109-81-4 ***HR: 2***
BUTAMIRATE CITRATE
mf: $C_{18}H_{29}NO_3$•$C_6H_8O_7$ mw: 499.62

SYNS: ABBOTT 36581 ◇ ACODEEN ◇ BUTAMYRATE CITRATE ◇ 2-(2-(DIETHYLAMINO)ETHOXY)ETHYL-2-PHENYLBUTYRATE CITRATE ◇ α-ETHYLBENZENEACETIC ACID-2-(2-DIETHYLAMINO) ETHOXY)ETHYL ESTER CITRATE ◇ HH-197 ◇ PHENYL ACETIC ACID DIETHYLAMINOETHOXYETHANOL ESTER CITRATE ◇ 2-PHENYLBUTYRIC ACID 2-(2-DIETHYLAMINO)ETHOXY)ETHYL ESTER CITRATE ◇ SINCODEEN ◇ SINCODEX ◇ SINCODIN ◇ SINCODIX ◇ SINECOD

TOXICITY DATA with REFERENCE
orl-rat TDLo:60 mg/kg (9-14D preg):REP TOIZAG 17,524,70
orl-rat TDLo:60 mg/kg (9-14D preg):TER TOIZAG 17,524,70
orl-rat LD50:4164 mg/kg TOIZAG 18,115,71
scu-rat LD50:3638 mg/kg TOIZAG 18,115,71
orl-mus LD50:865 mg/kg TOIZAG 18,115,71

SAFETY PROFILE: Moderately toxic by ingestion and subcutaneous routes. Experimental reproductive effects. An experimental teratogen. When heated to decomposition it emits toxic fumes of NO_x. See also ESTERS.

BOR500 CAS:106-97-8 ***HR: 1***
BUTANE
DOT: UN 1011/UN 1075
mf: C_4H_{10} mw: 58.14

PROP: Colorless gas; faint disagreeable odor. Bp: −0.5°, fp: −138°, lel: 1.9%, uel: 8.5%, flash p: −76°F (CC), d: 0.599, autoign temp: 761°F, vap press: 2 atm @ 18.8°, vap d: 2.046.

SYNS: n-BUTANE (DOT) ◇ BUTANEN (DUTCH) ◇ BUTANI (ITALIAN) ◇ DIETHYL ◇ METHYLETHYLMETHANE

TOXICITY DATA with REFERENCE
ihl-rat LC50:658 g/m³/4H FATOAO 30,102,67
ihl-mus LC50:680 g/m³/2H FATOAO 30,102,67

CONSENSUS REPORTS: Reported in EPA TSCA Inventory.

OSHA PEL: TWA 800 ppm
ACGIH TLV: TWA 800 ppm
DFG MAK: 1000 ppm (2350 mg/m³)
DOT Classification: Flammable Gas; Label: Flammable Gas.

SAFETY PROFILE: Mildly toxic via inhalation. Causes drowsiness. An asphyxiant. Very dangerous fire hazard when exposed to heat, flame, or oxidizers. Highly explosive when exposed to flame, or when mixed with $[Ni(CO)_4 + O_2]$. To fight fire, stop flow of gas. When heated to decomposition it emits acrid smoke and fumes.

BOR750 CAS:590-88-5 ***HR: 2***
1,3-BUTANEDIAMINE
mf: $C_4H_{12}N_2$ mw: 88.18

PROP: Liquid. Bp: 142-150°, flash p: 125°F, d: 0.85, vap d: 3.04.

SYN: 1,3-DIAMINOBUTANE

TOXICITY DATA with REFERENCE
skn-rbt 10 mg/24H open SEV AMIHBC 4,119,51
eye-rbt 250 μg open SEV AMIHBC 4,119,51

orl-rat LD50:1350 mg/kg AMIHBC 4,119,51
skn-rbt LD50:430 mg/kg AMIHBC 4,119,51

SAFETY PROFILE: Moderately toxic by ingestion and skin contact. Severe skin and eye irritant. Moderate fire hazard when exposed to heat or flame. To fight fire, use alcohol foam, foam, CO_2, dry chemical. Incompatible with oxidizing materials. When heated to decomposition it emits toxic fumes of NO_x. See also 1,4-BUTANEDIAMINE and AMINES.

BOS000 CAS:110-60-1 ***HR: 3***
1,4-BUTANEDIAMINE
mf: $C_4H_{12}N_2$ mw: 88.18

SYNS: BUTYLENEDIAMINE ◇ 1,4-BUTYLENEDIAMINE ◇ 1,4-DIAMINOBUTANE ◇ PUTRESCIN ◇ PUTRESCINE ◇ TETRAMETHYLENEDIAMINE ◇ 1,4-TETRAMETHYLENEDIAMINE

TOXICITY DATA with REFERENCE
cyt-hmn:hla 2 mmol/L JCLLAX 78,217,71
dns-mus:lvr 2 mmol/L AMOKAG 33,149,79
dni-mus:ast 10 mmol/L AMOKAG 33,149,79
dni-mus:lvr 20 mmol/L AMOKAG 33,149,79
ipr-mus TDLo:314 mg/kg (12D preg):TER TJADAB 28,237,83
scu-rat LDLo:300 mg/kg ZEPTAT 17,59,15
orl-mus LDLo:1600 mg/kg AECTCV 14,111,85
orl-rbt LDLo:1600 mg/kg CRSBAW 83,481,20
scu-rbt LDLo:200 mg/kg ZEPTAT 17,59,15
ivn-rbt LDLo:80 mg/kg CRSBAW 83,481,20
rec-rbt LDLo:400 mg/kg CRSBAW 83,481,20

CONSENSUS REPORTS: Reported in EPA TSCA Inventory.

SAFETY PROFILE: Poison by subcutaneous, intravenous, and rectal routes. Moderately toxic by ingestion. An experimental teratogen. Human mutation data reported. When heated to decomposition it emits toxic fumes of NO_x. See also 1,3-BUTANEDIAMINE and AMINES.

BOS250 CAS:584-03-2 ***HR: 2***
1,2-BUTANEDIOL
mf: $C_4H_{10}O_2$ mw: 90.14

PROP: D: 1.0, vap d: 3.1, bp: 194°, flash p: 194°F.

SYN: 1,2-BUTYLENE GLYCOL

TOXICITY DATA with REFERENCE
orl-mus LD50:3720 mg/kg TXAPA9 49,385,79

CONSENSUS REPORTS: Reported in EPA TSCA Inventory.

SAFETY PROFILE: Moderately toxic by ingestion. Combustible when exposed to heat or flame. To fight fire, use alcohol foam. When heated to decomposition it emits acrid and irritating fumes.

BOS500 CAS:107-88-0 ***HR: 1***
1,3-BUTANEDIOL
mf: $C_4H_{10}O_2$ mw: 90.14

PROP: Viscous liquid. Bp: 207.5°, fp: $< -50°$, flash p: 250°F, d: 1.006 @ 20°/20°, autoign temp: 741°F, vap press: 0.06 mm @ 20°, vap d: 3.2.

SYNS: 1,3-BUTANDIOL (GERMAN) ◇ BUTANE-1,3-DIOL ◇ β-BUTYLENE GLYCOL ◇ 1,3-BUTYLENE GLYCOL (FCC) ◇ 1,3-DIHYDROXYBUTANE ◇ METHYLTRIMETHYLENE GLYCOL

TOXICITY DATA with REFERENCE
skn-rbt 500 mg/24H MLD 85JCAE -,207,86
eye-rbt 505 mg AJOPAA 29,1363,46
eye-rbt 500 mg/24H MLD 85JCAE -,207,86
orl-rat TDLo:42360 mg/kg (female 6-15D post):REP JACTDZ 5(4),189,86
orl-rat LD50:18610 mg/kg JIDHAN 23,259,41
scu-rat LD50:20 g/kg NPIRI* 1,14,74
orl-mus LD50:12980 mg/kg JAPMA8 45,669,56
orl-gpg LD50:11 g/kg JIHTAB 23,259,41

CONSENSUS REPORTS: Reported in EPA TSCA Inventory.

SAFETY PROFILE: Mildly toxic by ingestion and subcutaneous routes. An skin and eye irritant. See also ETHYLENE GLYCOL. Experimental reproductive effects. Combustible when exposed to heat or flame. Incompatible with oxidizing materials. To fight fire, use foam, alcohol foam, CO_2, dry chemical. When heated to decomposition it emits acrid smoke and irritating fumes.

BOS750 CAS:110-63-4 ***HR: 2***
1,4-BUTANEDIOL
mf: $C_4H_{10}O_2$ mw: 90.14

PROP: Nearly odorless, colorless, viscid liquid. Bp: 228°, fp: 20.9°, flash p: 250°F (OC), d: 1.0154 @ 25°/4°, vap d: 3.1.

SYNS: BUTANE-1,4-DIOL ◇ 1,4-BUTYLENE GLYCOL ◇ 1,4-DIHYDROXYBUTANE ◇ 1,4-TETRAMETHYLENE GLYCOL

TOXICITY DATA with REFERENCE
orl-rat LD50:1525 mg/kg HYSAAV 33,41,68
ipr-rat LD50:1370 mg/kg TXAPA9 25,461,73
orl-mus LD50:2062 mg/kg HYSAAV 33,41,68
ipr-mus LDLo:500 mg/kg CBCCT* 3,363,51
orl-rbt LD50:2531 mg/kg HYSAAV 33,41,68
orl-gpg LD50:1200 mg/kg HYSAAV 33,41,68

CONSENSUS REPORTS: Reported in EPA TSCA Inventory.

SAFETY PROFILE: Moderately toxic by ingestion and

intraperitoneal routes. Combustible when exposed to heat or flame. To fight fire, use alcohol foam, mist, foam, CO_2, dry chemical. Incompatible with oxidizing materials. When heated to decomposition it emits acrid smoke and fumes.

BOT000 CAS:513-85-9 ***HR: 1***
2,3-BUTANEDIOL
mf: $C_4H_{10}O_2$ mw: 90.14

PROP: Colorless liquid or solid. Bp: 180°, fp: 19°, flash p: 185°F (TOC), d: 1.0095 @ 20°/20°, autoign temp: 756°F, vap press: 0.17 mm @ 20°, vap d: 3.1.

SYNS: 2,3-BUTYLENE GLYCOL ◇ 2,3-DIHYDROXYBUTANE ◇ DIMETHYLENE GLYCOL

TOXICITY DATA with REFERENCE
orl-mus LD50:5462 mg/kg TXAPA9 49,385,79

CONSENSUS REPORTS: Reported in EPA TSCA Inventory.

SAFETY PROFILE: Mildly toxic by ingestion. See also ETHYLENE GLYCOL. Flammable when exposed to heat or flame. Incompatible with oxidizing materials. To fight fire, use alcohol foam, CO_2, dry chemical. When heated to decomposition it emits acrid smoke and fumes.

BOT200 CAS:4437-85-8 ***HR: 1***
1,2-BUTANEDIOL, CYCLIC CARBONATE
mf: $C_5H_8O_3$ mw: 116.13

SYNS: 1,2-BUTYLENE CARBONATE ◇ CARBONIC ACID, CYCLIC ETHYLETHYLENE ESTER ◇ 1,3-DIOXOLAN-2-ONE, 4-ETHYL-

TOXICITY DATA with REFERENCE
skn-rbt 500 mg MLD JACTDZ 1,12,90
eye-rbt 100 mg MLD JACTDZ 1,12,90

SAFETY PROFILE: A skin and eye irritant. When heated to decomposition it emits acrid smoke and irritating fumes.

BOT250 CAS:55-98-1 ***HR: 3***
1,4-BUTANEDIOL DIMETHYL SULFONATE
mf: $C_6H_{14}O_6S_2$ mw: 246.32

PROP: White crystals. Mp: 114-118°.

SYNS: 1,4-BIS(METHANESULFONOXY)BUTANE ◇ (1,4-BIS(METHANESULFONYLOXY)BUTANE) ◇ BISULFAN ◇ BISULPHANE- ◇ 1,4-BUTANEDIOL DIMETHANESULPHONATE ◇ BUZULFAN ◇ C.B. 2041 ◇ CITOSULFAN ◇ 1,4-DIMESYLOXYBUTANE ◇ 1,4-DIMETHANESULFONOXYBUTANE ◇ 1,4-DI(METHANESULFONYLOXY)BUTANE ◇ 1,4-DIMETHANESULPHONYLOXYBUTANE ◇ 1,4-DIMETHYLSULFONOXYBUTANE ◇ GT41 ◇ GT 2041 ◇ LEUCOSULFAN ◇ MABLIN ◇ METHANESULFONIC ACID TETRAMETHYLENE ESTER ◇ MIELUCIN ◇ MISULBAN ◇ MITOSTAN ◇ MYELOLEUKON ◇ MYLERAN ◇ NCI-C01592 ◇ NSC-750 ◇ SULPHABUTIN ◇ TETRAMETHYLENE BIS(METHANESULFONATE) ◇ TETRAMETHYLENE DIMETHANE SULFONATE ◇ X 149

TOXICITY DATA with REFERENCE
mmo-sat 333 μg/plate ENMUDM 8(Suppl 7),1,86
mma-esc 25 μg/plate TAKHAA 44,96,85
orl-wmn TDLo:17 mg/kg (4-36W preg):REP PEDIAU 25,85,60
orl-wmn TDLo:17 mg/kg (4-36W preg):TER PEDIAU 25,85,60
orl-man TDLo:5684 μg/kg/21W-C:CAR BMJOAE 2,1513,77
orl-wmn TD:1140 mg/kg/9Y-I:CAR AIMDAP 124,66,69
ivn-mus TDLo:48 mg/kg/42D-I:NEO CNREA8 25,20,65
ipr-mus TD:1200 mg/kg/8W-I:ETA CNREA8 33,3069,73
orl-wmn TDLo:798 mg/kg/8Y JAMAAP 238,1951,77
orl-man TDLo:8 mg/kg/2D-I LANCAO 2,1463,84
ipr-rat LD50:18 mg/kg BCPCA6 1,39,58
scu-rat LD50:22 mg/kg KSRNAM 5,1894,71
ivn-rat LD50:1800 μg/kg ARZNAD 20,1467,70
orl-mus LD50:110 mg/kg KSRNAM 5,1894,71
ipr-mus LD50:86 mg/kg KSRNAM 5,1894,71
scu-mus LD50:63 mg/kg KSRNAM 5,1894,71
ivn-dog LDLo:8 mg/kg CCSUBJ 2,203,65
ivn-mky LDLo:8 mg/kg CCSUBJ 2,203,65

CONSENSUS REPORTS: NTP Fifth Annual Report on Carcinogens. IARC Cancer Review: Group 1 IMEMDT 7,137,87; Animal Inadequate Evidence IMEMDT 4,247,74; Human Inadequate Evidence IMEMDT 4,247,74. EPA Genetic Toxicology Program.

SAFETY PROFILE: Confirmed carcinogen producing leukemia, kidney, and uterine tumors. Experimental neoplastigenic and tumorigenic data. Poison by ingestion, subcutaneous, intraperitoneal, intravenous, and possibly other routes. Ingestion by pregnant women can cause cancer of the reproductive system of the fetus including the uterus. Human teratogenic effects by ingestion and possibly other routes include developmental abnormalities of the eye, ear, craniofacial area including the nose and tongue, gastrointestinal system, endocrine system, urogenital system, and other unspecified areas. Other human reproductive effects by ingestion and possibly other routes include: impotence, changes in the uterus, cervix, and vagina, and menstrual cycle disorders. Experimental reproductive effects. Human mutation data reported. When heated to decomposition it emits toxic fumes of SO_x. See also SULFONATES.

BOT500 CAS:431-03-8 ***HR: 3***
2,3-BUTANEDIONE
DOT: UN 2346
mf: $C_4H_6O_2$ mw: 86.10

$CH_3CO{\bullet}CO{\bullet}CH_3$

PROP: Greenish-yellow liquid; strong odor. Bp: 88°, flash p: 80°F, d: 0.9904 @ 15°/15°, refr index: 1.393-

1.397, vap d: 3.00.Misc in alc, fixed oils, propylene glycol; sol in glycerin, water.

SYNS: BIACETYL ◇ DIACETYL (FCC) ◇ 2,3-DIKETOBUTANE ◇ DIMETHYL DIKETONE ◇ DIMETHYLGLYOXAL ◇ FEMA No. 2370

TOXICITY DATA with REFERENCE
skn-rbt 500 mg/24H MOD FCTXAV 17(Suppl.),695,79
mmo-sat 1 mg/plate MUREAV 67,367,79
oms-hmn:emb 20 mg/L BEXBAN 74,828,72
ipr-rat LD50:400 mg/kg FCTXAV 7,571,69
orl-gpg LD50:990 mg/kg FCTXAV 2,327,64
orl-mam LD50:720 mg/kg RPTOAN 48,186,85

CONSENSUS REPORTS: Reported in EPA TSCA Inventory.

DOT Classification: Flammable Liquid; Label: Flammable Liquid.

SAFETY PROFILE: A poison by intraperitoneal route. Moderately toxic by ingestion. A skin irritant. Human mutation data reported. Flammable liquid. Dangerous fire hazard when exposed to heat or flame. To fight fire, use alcohol foam, CO_2, dry chemical. When heated to decomposition it emits acrid smoke and fumes. See also KETONES.

BOU100 CAS:60548-62-1 ***HR: D***
1,4-BUTANEDIYL SULFAMATE
mf: $C_4H_{12}N_2O_6S_2$ mw: 248.30

SYNS: 1,4-BUTANEDIOL, BISSULFAMATE (ester) ◇ SULFAMIC ACID, TETRAMETHYLENE ESTER

TOXICITY DATA with REFERENCE
orl-rat TDLo:1389 mg/kg (male 14D pre):REP JMCMAR 24,901,81

SAFETY PROFILE: Experimental reproductive effects. When heated to decomposition it emits toxic fumes of NO_x and SO_x.

BOU250 CAS:1633-83-6 ***HR: 3***
BUTANE SULTONE
mf: $C_4H_8O_3S$ mw: 136.18

SYNS: BUTANESULFONE ◇ Δ-BUTANE SULTONE ◇ 1,4-BUTANESULTONE (MAK) ◇ 1,4-BUTYLENE SULFONE ◇ Δ-VALEROSULTONE

TOXICITY DATA with REFERENCE
mmo-sat 100 μg/plate JNCIAM 62,893,79
dnr-esc 10 μL/disc JNCIAM 62,873,79
hma-mus/sat 138 mg/kg JNCIAM 62,911,79
orl-rat TDLo:1300 mg/kg/1Y-I:ETA,REP ZEKBAI 75,69,70
scu-rat TDLo:2280 mg/kg/76W-I:ETA ZEKBAI 75,69,70
orl-rat LD50:500 mg/kg ZEKBAI 75,69,70
scu-rat LD50:350 mg/kg ZEKBAI 75,69,70
ivn-rat LD50:270 mg/kg ZEKBAI 75,69,70
ipr-mus LD50:138 mg/kg JJIND8 62,911,79

CONSENSUS REPORTS: EPA Genetic Toxicology Program. Reported in EPA TSCA Inventory.

DFG MAK: Suspected Carcinogen.

SAFETY PROFILE: Suspected carcinogen with experimental tumorigenic data. Poison by subcutaneous, intravenous, and intraperitoneal routes. Moderately toxic by ingestion. Experimental reproductive effects. Human mutation data reported. See also SULFONATES. When heated to decomposition it emits toxic fumes of SO_x.

BOU500 CAS:1703-58-8 ***HR: 2***
1,2,3,4-BUTANETETRACARBOXYLIC ACID
mf: $C_8H_{10}O_8$ mw: 234.18

SYN: BUTANETETRACARBOXYLIC ACID

TOXICITY DATA with REFERENCE
eye-rbt 100 μg SpiEW# 13FEB80
orl-rat LD50:1720 mg/kg SpiEW# 13FEB80
skn-rbt LDLo:8000 mg/kg SpiEW# 13FEB80

CONSENSUS REPORTS: Reported in EPA TSCA Inventory.

SAFETY PROFILE: Moderately toxic by ingestion. Mildly toxic by skin contact. An eye irritant. When heated to decomposition it emits acrid smoke and irritating fumes.

BOU550 CAS:36169-16-1 ***HR: 2***
1-BUTANETHIOL, TIN(2+) SALT
mf: $C_4H_{10}S\cdot 1/_2Sn$ mw: 149.54

SYN: ESTABEX S

TOXICITY DATA with REFERENCE
orl-mus LD50:690 mg/kg ERNFA7 11,424,66

OSHA PEL: TWA 2 mg(Sn)/m^3
ACGIH TLV: TWA 2 mg(Sn)/m^3

SAFETY PROFILE: Moderately toxic by ingestion. When heated to decomposition it emits toxic fumes of SO_x and Sn.

BOV000 CAS:96-48-0 ***HR: 3***
4-BUTANOLIDE
mf: $C_4H_6O_2$ mw: 86.10

PROP: Colorless liquid; mild caramel odor. Mp: −44°, bp: 206°, flash p: 209°F (OC), d: 1.124 @ 25°/4°, refr index: 1.434-1.454 @ 25°, vap d: 3.0.

SYNS: Γ-6480 ◇ Γ-BL ◇ BLO ◇ BLON ◇ BUTYRIC ACID LACTONE ◇ Γ-BUTYROLACTONE (FCC) ◇ α-BUTYROLACTONE ◇ BUTYRYL LACTONE ◇ 4-DEOXYTETRONIC ACID ◇ DIHYDRO-2(3H)-FURANONE ◇ FEMA No. 3291 ◇ 4-HYDROXYBUTANOIC ACID LACTONE ◇ Γ-HYDROXYBUTYRIC ACID CYCLIC ESTER ◇ 4-HYDROXYBUTYRIC ACID Γ-LACTONE ◇ Γ-HYDROXYBUTYROLACTONE ◇ NCI-C55878 ◇ TETRAHYDRO-2-FURANONE

TOXICITY DATA with REFERENCE
dnd-bcs 20 μL/disc PMRSDJ 1,175,81
otr-ham:kdy 25 mg/L PMRSDJ 1,638,81
orl-rat TDLo:500 mg/kg (female 6-15D post):TER PHTXA6 62,57,88
orl-rat TDLo:25 g/kg (20D male):REP ARANDR 10,239,83
skn-mus TDLo:50 g/kg/42W-I:ETA JNCIAM 31,41,63
orl-rat LD50:1540 mg/kg GTPZAB 31(1),49,87
ipr-rat LD50:1000 mg/kg AITEAT 13,70,65
orl-mus LD50:1720 mg/kg GTPZAB 31(1),49,87
ipr-mus LD50:1100 mg/kg AITEAT 13,70,65
ivn-rbt LDLo:500 mg/kg AITEAT 13,70,65

CONSENSUS REPORTS: IARC Cancer Review: Group 3 IMEMDT 7,56,87; Animal No Evidence IMEMDT 11,231,76. EPA Genetic Toxicology Program. Reported in EPA TSCA Inventory.

SAFETY PROFILE: Moderately toxic by ingestion, intravenous, and intraperitoneal routes. An experimental teratogen. Other experimental reproductive effects. Questionable carcinogen with experimental tumorigenic data by skin contact. Mutation data reported. Less acutely toxic than β-propiolactone. Combustible when exposed to heat or flame; can react with oxidizing materials. To fight fire, use foam, alcohol foam, CO_2, dry chemical. Potentially explosive reaction with butanol + 2,4-dichlorophenol + sodium hydroxide. When heated to decomposition it emits acrid and irritating fumes.

BOV625 CAS:4154-69-2 ***HR: 3***
2-BUTANONE OXIME HYDROCHLORIDE
mf: $C_4H_{10}ClNO$ mw: 123.58

SYN: (2-HYDROXYLIMINIOBUTANECHLORIDE)

SAFETY PROFILE: Decomposes violently above 50°C. When heated to decomposition it emits toxic fumes of Cl^- and NO_x.

BOV750 CAS:129-18-0 ***HR: 3***
BUTAZOLIDINE SODIUM
mf: $C_{19}H_{20}N_2O_2$•Na mw: 331.40

SYNS: 4-BUTYL-1,2-DIPHENYL-3,5-PYRAZOLIDINEDIONESODIUM SALT ◇ 3,5-DIOXO-1,2-DIPHENYL-4-N-BUTYLPYRAZOLIDIN SODIUM ◇ DIPHENYLDIOXOBUTYLPYRAZOLIDINE-BUTAZOLIDINE-SODIUM ◇ PHENYLBUTAZONE SODIUM ◇ SODIUM BUTAZOLIDINE ◇ SODIUM PHENYLBUTAZONE ◇ SODIUM SALT of PHENYLBUTAZONE

TOXICITY DATA with REFERENCE
orl-wmn LDLo:16 mg/kg AIMEAS 39,1096,53
scu-rat LD50:360 mg/kg ARZNAD 8,229,58
orl-mus LD50:476 mg/kg RPOBAR 2,314,70
ipr-mus LD50:169 mg/kg RPOBAR 2,314,70
ivn-rat LD50:113 mg/kg FRPSAX 13,922,58
scu-mus LD50:271 mg/kg FRPSAX 12,521,57
ivn-mus LD50:94 g/kg FRPSAX 13,922,58

SAFETY PROFILE: A human poison by ingestion. Human systemic effects by ingestion: respiratory system damage, agranulocytosis, and dermatitis. An experimental poison via subcutaneous, intravenous, and intraperitoneal routes. An anti-inflammatory drug. When heated to decomposition it emits toxic fumes of NO_x and Na_2O.

BOV800 ***HR: 3***
BUTEA FRONDOSA, seed extract

PROP: Indian plant belonging to the family *Leguminosae* (IJEBA6 11,43,73).

SYN: PALASH SEED EXTRACT

TOXICITY DATA with REFERENCE
orl-mus TDLo:50 mg/kg (female 1-5D post):REP IJPPAZ 13,239,69
orl-mus LD50:7500 mg/kg IJPPAZ 13,239,69
ipr-mus LD50:20 mg/kg IJEBA6 11,43,73

SAFETY PROFILE: Poison by intraperitoneal route. Experimental reproductive effects.

BOV825 CAS:5716-20-1 ***HR: 3***
BUTEDRIN
mf: $C_{24}H_{38}N_2O_4$•H_2O_4S mw: 516.72

SYNS: BAMETAN SULFATE ◇ BAMETHAN SULFATE ◇ BASCURAT ◇ BUPATOL ◇ BUTIBATOL ◇ α-((BUTYLAMINO)METHYL)-p-HYDROXYBENZYL ALCOHOL SULFATE ◇ BUTYLNORSYMPATOL ◇ CYCLATE ◇ ECLERIN ◇ GARMIAN ◇ PERIPHETOL ◇ ROTESAR ◇ VASCULAT ◇ VASCULIT ◇ VASCUNICOL ◇ VASKULAT

TOXICITY DATA with REFERENCE
ipr-mus LD50:210 mg/kg NIIRDN 6,585,82
scu-mus LD50:422 mg/kg NIIRDN 6,585,82
ivn-mus LD50:72 mg/kg NIIRND 6,585,82

SAFETY PROFILE: Poison by intravenous and intraperitoneal routes. Moderately toxic by subcutaneous route. A vasodilator. When heated to decomposition it emits toxic fumes of SO_x and NO_x. See also SULFATES.

BOW250 CAS:25167-67-3 ***HR: 3***
1-BUTENE
DOT: UN 1012
mf: C_4H_8 mw: 56.11

PROP: A colorless, flammable gas; sltly aromatic odor. Bp: −6.3°, fp: −185.3°, lel: 1.6%, uel: 9.3% flash p: −80° (−112°F), d: 0.668 @ 0°/1°, vap d: 1.93, vap press: 3480 mm @ 21°, autoign temp: 723°F.

SYNS: BUTYLENE ◇ α-BUTYLENE

CONSENSUS REPORTS: Reported in EPA TSCA Inventory.

DOT Classification: Flammable Gas; Label: Flammable Gas.

SAFETY PROFILE: A simple asphyxiant. Very dangerous fire hazard when exposed to heat, flame, or oxidizers. To fight fire, stop flow of gas. Moderately explosive when exposed to flame. Mixtures with aluminum tetrahydroborate explode after an induction period. When heated to decomposition it emits acrid smoke and fumes.

BOW500 ***HR: 3***
cis-2-BUTENE
mf: C_4H_8 mw: 56.11

PROP: Colorless, flammable gas; sltly aromatic odor. Bp: 1°, fp: −139°, flash p: −100°F, d: 0.627 @ 15.5°/15.5°, vap press: 1410 mm @ 21°, autoign temp: 615°F, lel: 1.7%, uel: 9.0%, vap d: 1.9.

SYNS: DIMETHYLETHYLENE ◇ PSEUDO-BUTYLENE

SAFETY PROFILE: A simple asphyxiant. Very dangerous fire hazard when exposed to heat or flame. Very likely to explode. Incompatible with oxidizing materials. To fight fire, stop flow of gas. When heated to decomposition it emits acrid smoke and fumes.

BOW750 ***HR: 3***
trans-2-BUTENE
mf: C_4H_8 mw: 56.11

PROP: A colorless, flammable gas; sltly aromatic odor. Bp: 2.5°, fp: −105.6°, flash p: −100°F, d: 0.613 @ 15.5°/15.5° vap d: 1.95, vap press: 1592 mm @ 21°, autoign temp: 615 F, lel: 1.8%, uel: 9.7%, vap d: 1.9.

CONSENSUS REPORTS: EPA Extremely Hazardous Substances List.

SAFETY PROFILE: A simple asphyxiant. Very dangerous fire hazard when exposed to heat or flame. Very likely to explode. To fight fire, stop flow of gas. Incompatible with oxidizing materials. When heated to decomposition it emits acrid smoke and fumes.

BOX250 CAS:10099-70-4 ***HR: 2***
2-BUTENEDIOIC ACID BIS(1-METHYLETHYL) ESTER
mf: $C_{10}H_{16}O_4$ mw: 200.26

TOXICITY DATA with REFERENCE
skn-rbt 10 mg/24H open MLD AMIHBC 10,61,54
eye-rbt 500 mg open AMIHBC 10,61,54
orl-rat LD50:2140 mg/kg AMIHBC 10,61,54

CONSENSUS REPORTS: Reported in EPA TSCA Inventory.

SAFETY PROFILE: Moderately toxic by ingestion. A skin and eye irritant. See also ESTERS. When heated to decomposition it emits acrid smoke and irritating fumes.

BOX500 CAS:109-75-1 ***HR: 3***
3-BUTENE NITRILE
mf: C_4H_5N mw: 67.10

PROP: Colorless liquid, onion-like odor. Bp: 116-119°, d: 0.8341 @ 20°/4°, mp: −87°.

SYNS: ALLYL CYANIDE ◇ ALLYLNITRILE ◇ 1-BUTENE-4-NITRILE ◇ β-BUTENONITRILE ◇ TL 350 ◇ VINYLACETONITRILE

TOXICITY DATA with REFERENCE
skn-rbt 10 mg/24H open MLD AIHAAP 23,95,62
orl-rat LD50:115 mg/kg AIHAAP 30,470,69
ihl-rat LCLo:500 ppm/4H AIHAAP 30,470,69
orl-mus LD50:66770 μg/kg ARTODN 55,47,84
skn-rbt LD50:1410 mg/kg AIHAAP 23,95,65
ihl-gpg LC50:2500 mg/m³/4H GISAAA 34(4),36,69
ihl-mus LCLo:1000 mg/m³/10M NDRC** NDCrc-132,Sept,42
skn-rbt LDLo:1410 mg/kg AIHAAP 30,470,69
ihl-gpg LC50:2500 mg/m³/4H GISAAA 34,36,69

CONSENSUS REPORTS: Reported in EPA TSCA Inventory. Cyanide and its compounds are on the Community Right-To-Know List.

SAFETY PROFILE: A poison by ingestion. Moderately toxic by inhalation and skin contact. A skin irritant. See also NITRILES. Dangerous; emits highly toxic fumes of NO_x and CN^- when heated to decomposition or on contact with acids or acid fumes. To fight fire, use alcohol foam, mist.

BOX750 CAS:106-88-7 ***HR: 2***
1-BUTENE OXIDE
mf: C_4H_8O mw: 72.12

PROP: Colorless liquid. D: 0.8312 @ 20°/20°, bp: 63°, flash p: 5°F, lel: 1.5%, uel: 18.3%. Sol in water; misc with most organic solvents.

SYNS: 1,2-BUTENE OXIDE ◇ 1,2-BUTYLENE OXIDE ◇ 1,2-EPOXYBUTANE ◇ ETHYL ETHYLENE OXIDE ◇ ETHYLOXIRANE ◇ NCI-C55527

TOXICITY DATA with REFERENCE
skn-rbt 500 mg/24H MLD 85JCAE-,770,86
eye-rbt 100 mg/24H MOD 85JCAE-,770,86
mmo-klp mmol/L MUREAV 89,269,81
trn-dmg-orl 5 pph ENMUDM 7,349,85
mma-ssp 1600 μmol/L TCMUD8 3,75,83
ihl-rbt TCLo:1000 ppm/7H (10-24D preg):REP NTIS** PB81-168510
ihl-rat TCLo:400 ppm/6H/5D/2Y-C:CAR NTPTR* NTP-TR-329,88
orl-rat LD50:500 mg/kg NTIS** PB81-168510

ihl-rat LCLo:4000 ppm/4H AIHAAP 23,95,62
skn-rbt LD50:2100 mg/kg AIHAAP 23,95,62

CONSENSUS REPORTS: NTP Carcinogenesis Studies (inhalation); Clear Evidence: rat; No Evidence: mouse NTPTR* NTP-TR-329,88. Community Right-To-Know List. EPA Genetic Toxicology Program. Reported in EPA TSCA Inventory.

SAFETY PROFILE: Moderately toxic by ingestion and skin contact. Mildly toxic by inhalation. Experimental reproductive effects. Questionable carcinogen with experimental carcinogenic data. Mutation data reported. Dangerous fire hazard when exposed to heat, flame, or powerful oxidizers. To fight fire, use dry chemical, water spray, mist or fog, alcohol foam. When heated to decomposition it emits acrid smoke and fumes.

BOX825 CAS:16187-15-8 ***HR: 3***
trans-2-BUTENE OZONIDE
mf: $C_4H_8O_3$ mw: 104.11

SYN: (3,5-DIMETHYL-1,2,4-TRIOXOLANE)

SAFETY PROFILE: May explode when heated. When heated to decomposition it emits acrid smoke and fumes. See also OZONE.

BOY000 CAS:6117-91-5 ***HR: 2***
2-BUTEN-1-OL
mf: C_4H_8O mw: 72.12

PROP: Colorless liquid. Mp: <30°, bp: 118°, flash p: 92°F, d: 0.8726 @ 0°/4°, vap d: 2.49.

SYNS: 2-BUTENOL ◇ 2-BUTENYL ALCOHOL ◇ CROTONYL ALCOHOL ◇ CROTYL ALCOHOL

TOXICITY DATA with REFERENCE
mmo-sat 10 μg/plate TCMUD8 1,259,80
orl-rat LD50:930 mg/kg AIHAAP 23,95,62
ihl-rat LCLo:2000 ppm/4H AIHAAP 23,95,62
skn-rbt LD50:1270 mg/kg AIHAAP 23,95,62

CONSENSUS REPORTS: Reported in EPA TSCA Inventory.

SAFETY PROFILE: Moderately toxic by ingestion and skin contact. Mildly toxic by inhalation. Mutation data reported. Dangerous fire hazard when exposed to heat or flame; can react with oxidizing materials. To fight fire, use alcohol foam, CO_2, dry chemical. When heated to decomposition it emits acrid smoke and fumes. See also ALCOHOLS.

BOY250 ***HR: 3***
1-BUTEN-3-ONE
mf: C_4H_6O mw: 70.10

PROP: Flash p: −7°C.

SAFETY PROFILE: A dangerous fire hazard. When heated to decomposition it emits acrid smoke and fumes. See also KETONES.

BOY500 CAS:78-94-4 ***HR: 3***
3-BUTEN-2-ONE
mf: C_4H_6O mw: 70.10
DOT: UN 1251

PROP: Colorless liquid, powerfully irritating odor. Bp: 81.4°, flash p: 20°F (CC), d: 0.8393 @ 25°/4°, vap d: 2.41.

SYNS: ACETYL ETHYLENE ◇ 3-BUTENE-2-ONE ◇ METHYLENE ACETONE ◇ METHYL-VINYL-CETONE (FRENCH) ◇ METHYLVINYLKETON (GERMAN) ◇ METHYL VINYL KETONE ◇ Γ-OXO-α-BUTYLENE ◇ VINYL METHYL KETONE

TOXICITY DATA with REFERENCE
mmo-sat 250 μmol/L MUREAV 93,305,82
mma-sat 250 μmol/L MUREAV 93,305,82
orl-rat LD50:31 mg/kg 85GMAT -,88,82
ihl-rat LC50:7 mg/m³/4H 85GMAT -,88,82
orl-mus LD50:33 mg/kg 85GMAT -,88,82
ipr-mus LD50:76 mg/kg ZolH## 23OCT75

CONSENSUS REPORTS: Reported in EPA TSCA Inventory. EPA Extremely Hazardous Substances List.

DOT Classification: Flammable Liquid; Label: Flammable Liquid.

SAFETY PROFILE: Poison by ingestion, inhalation, and intraperitoneal routes. A severe irritant to skin, eyes, and mucous membranes. A lachrymator. Mutation data reported. See also KETONES. Dangerous fire hazard when exposed to heat, flame, or oxidizers. To fight fire, use CO_2, dry chemical. When heated to decomposition it emits acrid smoke and fumes.

BPA250 CAS:14746-03-3 ***HR: 3***
2-BUTEN-1-YL DIAZOACETATE
mf: $C_6H_8N_2O_2$ mw: 140.14

$N_2CHCO{\bullet}OCH_2CH{=}CHCH_3$

SAFETY PROFILE: Potentially explosive. When heated to decomposition it emits toxic fumes of NO_x. See also other diazo compounds.

BPA500 CAS:2237-92-5 ***HR: 3***
5-(1-BUTENYL)-5-ETHYLBARBITURIC ACID
mf: $C_{10}H_{14}N_2O_3$ mw: 210.26

TOXICITY DATA with REFERENCE
orl-mus LD50:320 mg/kg JACSAT 62,1199,40
ipr-mus LD50:225 mg/kg JACSAT 62,1199,40

SAFETY PROFILE: Poison by ingestion and intraperitoneal routes. When heated to decomposition it emits toxic fumes of NO_x. See also BARBITURATES.

BPA750 CAS:67050-00-4 ***HR: 3***
5-(2-BUTENYL)-5-ETHYL-2-THIOBARBITURIC ACID SODIUM SALT
mf: $C_{10}H_{13}N_2O_2S{\cdot}Na$ mw: 248.30

TOXICITY DATA with REFERENCE
ipr-rat LD50:123 mg/kg JAPMA8 34,183,45
ivn-rbt LD50:53 mg/kg JAPMA8 34,183,45

SAFETY PROFILE: Poison by intraperitoneal and intravenous routes. When heated to decomposition it emits very toxic fumes of NO_x, Na_2O, and SO_x. See also BARBITURATES.

BPB500 CAS:67050-04-8 ***HR: 3***
5-(1-BUTENYL)-5-ISOPROPYLBARBITURIC ACID
mf: $C_{11}H_{16}N_2O_3$ mw: 224.29

TOXICITY DATA with REFERENCE
orl-mus LD50:300 mg/kg JACSAT 62,1199,40
ipr-mus LD50:250 mg/kg JACSAT 62,1199,40

SAFETY PROFILE: Poison by ingestion and intraperitoneal routes. When heated to decomposition it emits toxic fumes of NO_x. See also BARBITURATES.

BPC500 CAS:67050-11-7 ***HR: 3***
5-(2-BUTENYL)-5-(1-METHYLBUTYL)-2-THIOBARBITURIC ACID SODIUM SALT
mf: $C_{13}H_{19}N_2O_2S{\cdot}Na$ mw: 290.39

TOXICITY DATA with REFERENCE
ipr-rat LD50:341 mg/kg JAPMA8 34,183,45
ivn-rbt LD50:49 mg/kg JAPMA8 34,183,45

SAFETY PROFILE: Poison by intraperitoneal and intravenous routes. When heated to decomposition it emits very toxic fumes of SO_x, Na_2O, and NO_x. See also BARBITURATES.

BPC600 ***HR: 2***
3-(3-BUTENYLNITROSAMINO)-1-PROPANOL
mf: $C_7H_{14}N_2O_2$ mw: 158.23

SYN: BUTENYL(3-HYDROXYPROPYL)NITROSAMINE

TOXICITY DATA with REFERENCE
scu-ham TDLo:22800 mg/kg/76W-I:CAR CDPRD4 4,79,81

SAFETY PROFILE: Questionable carcinogen with experimental carcinogenic data. When heated to decomposition it emits toxic fumes of NO_x.

BPC750 ***HR: 2***
2-BUTENYLPHENOL (mixed isomers)
mf: $C_{10}H_{12}O$ mw: 148.22

TOXICITY DATA with REFERENCE
skn-rbt 10 mg/24H SEV AMIHBC 4,119,51
eye-rbt 50 μg SEV AMIHBC 4,119,51
orl-rat LD50:410 mg/kg AMIHBC 4,119,51

SAFETY PROFILE: Moderately toxic by ingestion. A severe skin and eye irritant. When heated to decomposition it emits acrid smoke and irritating fumes.

BPD000 CAS:54746-50-8 ***HR: 3***
3-BUTENYL-(2-PROPENYL)-N-NITROSAMINE
mf: $C_7H_{12}N_2O$ mw: 140.21

SYN: N-ALLYL-N-NITROSO-3-BUTENYLAMINE

TOXICITY DATA with REFERENCE
mmo-sat 250 μg/plate MUREAV 68,195,79
scu-ham TDLo:15300 mg/kg/51W-I:CAR CDPRD4 4,79,81

SAFETY PROFILE: Questionable carcinogen with experimental carcinogenic data. Mutation data reported. When heated to decomposition it emits acrid smoke and irritating fumes. See also NITROSAMINES.

BPE109 CAS:689-97-4 ***HR: 3***
BUTEN-3-YNE
mf: C_4H_4 mw: 52.08

$$HC{\equiv}CCH{=}CH_2$$

PROP: Flash p: < −5°, lel: 2%, uel: 100% d: 0.68 @ 1.7 atm, vap d: 1.8, bp: 11°.

SYN: VINYL ACETYLENE

SAFETY PROFILE: Forms explosive peroxides with air or oxygen. Very exothermic decomposition when heated. Reacts explosively when heated with 1,3-butadiene or oxygen. Reacts with silver nitrate to form the explosive silver buten-3-ynide. When heated to decomposition it emits acrid smoke and irritating fumes. See also ACETYLENE COMPOUNDS.

BPE250 ***HR: 3***
3-BUTEN-1-YNYL DIETHYL ALUMINUM
mf: $C_8H_{13}Al$ mw: 137.17

$$H_2C{=}CH{=}CAl(CH_2CH_3)_2$$

SAFETY PROFILE: Ignites spontaneously in air. When heated to decomposition it emits acrid smoke and fumes. See also ALUMINUM COMPOUNDS.

BPE500 ***HR: 3***
3-BUTEN-1-YNYL DIISOBUTYL ALUMINUM
mf: $C_{12}H_{21}Al$ mw: 189.3

SAFETY PROFILE: Ignites spontaneously in air. When heated to decomposition it emits acrid smoke and fumes. See also ALUMINUM COMPOUNDS.

BPE750 ***HR: 3***
2-BUTEN-1-YNYL TRIETHYL LEAD
mf: $C_{10}H_{18}Pb$ mw: 341.41

CONSENSUS REPORTS: Lead and its compounds are on the Community Right-To-Know List.

SAFETY PROFILE: Explodes when heated rapidly. See also LEAD COMPOUNDS.

BPF000 CAS:125-40-6 ***HR: 3***
BUTISOL
mf: $C_{10}H_{16}N_2O_3$ mw: 212.28

SYNS: BUTABARB ◇ BUTABARBITAL ◇ BUTABARBITONE ◇ BUTATAB ◇ BUTATAL ◇ BUTICAPS ◇ BUTRATE ◇ 5-sec-BUTYL-5-ETHYLBARBITURIC ACID ◇ 5-sec-BUTYL-5-ETHYLMALONYL UREA ◇ 5-ETHYL-5-(1-METHYLPROPYL)BARBITURATE ◇ 5-ETHYL-5-(1-METHYLPROPYL)BARBITURIC ACID ◇ 5-ETHYL-5-(1-METHYL-PROPYL)-2,4,6(1H,3H,5H)-PYRIMIDINETRIONE (9CI) ◇ MEDARSED ◇ NILOX ◇ SECBUBARBITAL ◇ SECBUTABARBITAL ◇ SECBUTO-BARBITONE ◇ UNICELLES

TOXICITY DATA with REFERENCE
ipr-rat LD50:70 mg/kg JFETAB 44,325,32
scu-rat LDLo:140 mg/kg JACSAT 52,2440,30
ipr-mus LDLo:200 mg/kg JACSAT 58,731,36
ivn-mus LD50:175 mg/kg AIPTAK 132,164,61
orl-rbt LD50:140 mg/kg JPETAB 44,325,32
ipr-rbt LD50:75 mg/kg JPETAB 44,325,32
scu-rbt LDLo:200 mg/kg JACSAT 45,243,23
ivn-rbt LDLo:90 mg/kg JPPGAR 30,364,32

SAFETY PROFILE: Poison by ingestion, intravenous, intraperitoneal, and subcutaneous routes. A central nervous system depressant. When heated to decomposition it emits toxic fumes of NO_x. See also BARBITURATES.

BPF250 CAS:143-81-7 ***HR: 3***
BUTISOL SODIUM
mf: $C_{10}H_{16}N_2O_3 \cdot Na$ mw: 235.27

SYNS: BUTABARBITAL SODIUM ◇ 5-sec-BUTYL-5-ETHYLBARBITURIC ACID SODIUM SALT ◇ 5-ETHYL-5-(1-METHYLPROPYL)BARBITURIC ACID SODIUM SALT ◇ 5-ETHYL-5-(1-METHYLPROPYL)-2,4,6(1H,3H,5H)-PYRIMIDINETRIONE MONOSODIUM SALT ◇ SECBUBARBITAL SODIUM ◇ SODIUM BUTABARBITAL ◇ SODIUM-5-sec-BUTYL-5-ETHYLBARBITURATE ◇ SODIUM-5-ETHYL-5-sec-BUTYLBARBITURATE ◇ SODIUM-5-ETHYL-5-(1-METHYLPROPYL) BARBITURATE

TOXICITY DATA with REFERENCE
orl-hmn LDLo:125 mg/kg CTOXAO 10,327,77
orl-hmn TDLo:120 mg/kg:CNS,PSY BMJOAE 1,144,77
orl-rat LD50:78 mg/kg JPETAB 81,254,44
ipr-rat LD50:70 mg/kg JPETAB 81,254,44
ivn-rat LD50:70 mg/kg JPETAB 81,254,44
ipr-mus LD50:247 mg/kg JPETAB 81,254,44
ivn-dog LD50:90 mg/kg JPETAB 81,254,44
orl-rbt LD50:194 mg/kg JPETAB 81,254,44
ipr-rbt LD50:95 mg/kg JPETAB 81,254,44
ivn-rbt LD50:91 mg/kg JPETAB 81,254,44

SAFETY PROFILE: An experimental poison by ingestion, intraperitoneal, and intravenous routes. Human central nervous system and psychotropic effects by ingestion. When heated to decomposition it emits toxic fumes of NO_x and Na_2O. See also BARBITURATES.

BPF500 CAS:77-28-1 ***HR: 3***
BUTOBARBITAL
mf: $C_{10}H_{16}N_2O_3$ mw: 212.28

SYNS: BUDORM ◇ BUTETHAL ◇ BUTOBARBITONE ◇ BUTO-BARBITURAL ◇ 5-BUTYL-5-ETHYLBARBITURIC ACID ◇ 5-BUTYL-5-ETHYL-2,4,6(1H,3H,5H)-PYRIMIDINETRIONE (9CI) ◇ 5-ETHYL-5-N-BUTYLBARBITURIC ACID ◇ ETOVAL ◇ HYPERBUTAL ◇ LONGAN-OCT ◇ MEONAL ◇ MONODORM ◇ NEONAL ◇ SONERILE ◇ SONERYL

TOXICITY DATA with REFERENCE
scu-rat TDLo:1300 mg/kg (7-19D preg):TER EXPEAM 33,499,77
ims-rat TDLo:28 mg/kg (7D male):REP IJEBA6 16,316,78
orl-wmn TDLo:166 mg/kg:CNS BMJOAE 1,1238,55
ipr-rat LDLo:135 mg/kg JPETAB 44,325,32
scu-rat LDLo:190 mg/kg JPETAB 26,371,25
orl-mus LDLo:100 mg/kg AIPTAK 92,305,53
ipr-mus LD50:320 mg/kg JPETAB 89,356,47
orl-cat LDLo:80 mg/kg JPETAB 33,43,28
orl-rbt LDLo:100 mg/kg JPETAB 44,337,32
ipr-rbt LDLo:115 mg/kg JPETAB 44,325,32
scu-rbt LDLo:100 mg/kg JACSAT 45,243,23
ivn-rbt LDLo:90 mg/kg JACSAT 57,1961,35

SAFETY PROFILE: A poison by ingestion, intraperitoneal, subcutaneous, and intravenous routes. Experimental teratogenic and reproductive effects. Human systemic effects by ingestion: changes in motor activity, coma, and nausea or vomiting. A central nervous system depressant. When heated to decomposition it emits toxic fumes of NO_x. See also BARBITURATES.

BPF750 CAS:35763-44-1 ***HR: 3***
BUTOBARBITAL SODIUM
mf: $C_{10}H_{15}N_2O_3 \cdot Na$ mw: 234.26

SYNS: BUTETHAL SODIUM ◇ BUTOBARBITIONE SODIUM ◇ 5-BUTYL-5-ETHYL-2,4,6(1H,3H,5H)-PYRIMIDINETRIONE MONOSODIUM SALT (9CI) ◇ SODIUM ETHYL-N-BUTYL BARBITURATE

TOXICITY DATA with REFERENCE
ipr-rat LD50:197 mg/kg JPETAB 81,254,44
scu-rat LDLo:190 mg/kg JACSAT 47,2236,25
ipr-mus LDLo:275 mg/kg JEPTAB 31,455,27

SAFETY PROFILE: Poison by intraperitoneal and subcutaneous routes. When heated to decomposition it emits toxic fumes of NO_x and Na_2O. See also BARBITURATES.

BPF825 CAS:32838-28-1 ***HR: 3***
BUTOCTAMIDE SEMISUCCINATE
mf: $C_{16}H_{29}NO_5$ mw: 315.46

SYNS: BUTANEDIOIC ACID MONO(3-((2-ETHYLHEXYL)AMINO)-1-METHYL-3-OXOPROPYL) ESTER (9CI) ◇ BUTOCTAMIDE HYDROGEN SUCCINATE ◇ N-(2-ETHYLHEXYL)-3-HYDROXYBUTYRAMIDE HYDROGEN SUCCINATE ◇ N-2-ETHYLHEXYL-β-OXYBUTYRAMIDE SEMISUCCINATE ◇ M-2H ◇ SUCCINIC ACID MONOESTER with N-(2-ETHYLHEXYL)-3-HYDROXYBUTYRAMIDE

TOXICITY DATA with REFERENCE
orl-mus TDLo:8 g/kg (7-14D preg):TER OYYAA2 8,1413,74
orl-rat LD50:12100 mg/kg YAKUD5 24,2029,82
ipr-rat LD50:635 mg/kg TOIZAG 18,648,71
scu-rat LD50:3350 mg/kg TOIZAG 18,648,71
orl-mus LD50:5600 mg/kg TOIZAG 18,648,71
ipr-mus LD50:473 mg/kg TOIZAG 18,648,71
scu-mus LD50:3730 mg/kg TOIZAG 18,648,71

SAFETY PROFILE: Poison by intraperitoneal route. Moderately toxic by ingestion and subcutaneous routes. An experimental teratogen. When heated to decomposition it emits toxic fumes of NO_x.

BPG000 CAS:126-22-7 ***HR: 3***
BUTONATE
mf: $C_8H_{14}Cl_3O_5P$ mw: 327.54

SYNS: BUTANOIC ACID 2,2,2-TRICHLORO-1-(DIMETHOXYPHOSPHINYL)ETHYL ESTER ◇ BUTILCHLOROFOS ◇ DIMETHOXY-2,2,2-TRICHLORO-1-N-BUTYRYLOXY-ETHYLPHOSPHINEOXIDE ◇ O,O-DIMETHYL-(1-BUTYRYLOXY-2,2,2-TRICHLOROETHYL) PHOSPHONATE ◇ O,O-DIMETHYL 2,2,2-TRICHLORO-1-(N-BUTYRYLOXY) ETHYLPHOSPHONATE ◇ ENT 20,852 ◇ F-139 ◇ T-113 ◇ TRIBUFON

TOXICITY DATA with REFERENCE
dnd-mus-ipr 200 mg/kg PCBPBS 6,101,76
orl-rat LD50:1100 mg/kg ARSIM* 20,6,66
skn-rat LD50:7000 mg/kg FMCHA2 -,C40,83
ipr-rat LD50:700 mg/kg ZHYGAM 25,512,79
scu-rat LD50:3000 mg/kg PAREAQ 11,636,59
orl-mus LD50:760 mg/kg ZHYGAM 25,512,79
skn-dog LD50:3080 mg/kg ZHYGAM 25,512,79

SAFETY PROFILE: Poison by ingestion. Moderately toxic by skin contact, intraperitoneal, subcutaneous, and possibly other routes. Mutation data reported. When heated to decomposition it emits highly toxic fumes of PO_x and Cl^-. See also ESTERS.

BPG250 CAS:6365-83-9 ***HR: 3***
BUTOPHEN
mf: $C_{10}H_{12}N_2O_5 \cdot H_3N$ mw: 257.28

SYNS: 2-sec-BUTYL-4,6-DINITROPHENOL AMMONIUM SALT ◇ CHEMOX SELECTIVE ◇ 4,6-DINITRO-2-sec.BUTYLFENOLATE AMMONY (CZECH) ◇ 4,6-DINITRO-o-sec-BUTYLPHENOL AMMONIUM SALT ◇ 4,6-DINITRO-2-sec-BUTYLPHENOL AMMONIUM SALT ◇ DINOSEB (AMINE) ◇ DNBP AMMONIUM SALT ◇ DOW SELECTIVE ◇ 2-(1-METHYL-N-PROPYL) 4,6-DINITROPHENOL AMMONIUM SALT ◇ SELECTIVE ◇ SINOX W

TOXICITY DATA with REFERENCE
eye-rbt 50 μg/24H SEV 28ZPAK -,108,72
orl-rat LD50:45 mg/kg 28ZPAK -,108,72
skn-rat LDLo:67 mg/kg BJIMAG 26,59,69

SAFETY PROFILE: A poison by ingestion and skin contact. A severe eye irritant. When heated to decomposition it emits very toxic fumes of NH_3 and NO_x.

BPG325 CAS:58786-99-5 ***HR: 3***
BUTORPHANOL TARTRATE
mf: $C_{21}H_{29}NO_2 \cdot C_4H_6O_6$ mw: 477.61

SYNS: BT ◇ STADOL ◇ TORATE ◇ TORBUTROL

TOXICITY DATA with REFERENCE
ivn-rat TDLo:11 mg/kg (female 7-17D post):REP IYKEDH 13,446,82
ivn-rat TDLo:55 mg/kg (female 7-17D post):TER IYKEDH 13,446,82
orl-rat LD50:315 mg/kg IYKEDH 13,145,82
ipr-rat LD50:127 mg/kg IYKEDH 13,145,82
scu-rat LD50:425 mg/kg IYKEDH 13,145,82
ivn-rat LD50:17 mg/kg DRUGAY 16,474,78
ims-rat LD50:255 mg/kg IYKEDH 13,145,82
orl-mus LD50:395 mg/kg DRUGAY 16,474,78
ipr-mus LD50:192 mg/kg IYKEDH 13,145,82
scu-mus LD50:299 mg/kg PBPSDY 2,19,79
ivn-mus LD50:36 mg/kg IYKEDH 13,145,82
ims-mus LD50:208 mg/kg IYKEDH 13,145,82

SAFETY PROFILE: Poison by ingestion, subcutaneous, intramuscular, intravenous, and intraperitoneal routes. An experimental teratogen. Other experimental reproductive effects. An analgesic. When heated to decomposition it emits toxic fumes of NO_x.

BPG500 CAS:3329-56-4 ***HR: 3***
BUTOXY ACETYLENE
mf: $C_6H_{10}O$ mw: 98.14

$$(C_4H_9)OC \equiv CH$$

SAFETY PROFILE: Explodes at 100°C when heated in a sealed container. When heated to decomposition it emits acrid smoke and fumes. See also ACETYLENE COMPOUNDS.

BPG750 CAS:60444-92-0 ***HR: 3***
2-N-BUTOXYBENZAMIDE
mf: $C_{11}H_{15}NO_2$ mw: 193.27

SYNS: H.P. 165 ◇ o-BUTOXYBENZAMIDE

TOXICITY DATA with REFERENCE
orl-mus LD50:1300 mg/kg JPPMAB 4,872,52
ipr-mus LD50:360 mg/kg JPPMAB 4,872,52

SAFETY PROFILE: Poison by intraperitoneal route. Moderately toxic by ingestion. When heated to decomposition it emits toxic fumes of NO_x. See also AMIDES.

BPH750 CAS:67032-45-5 ***HR: 3***
p-BUTOXYBENZOIC ACID-3-(2-METHYLPIPERIDINO)PROPYL ESTER HYDROCHLORIDE
mf: $C_{20}H_{31}NO_3 \cdot ClH$ mw: 369.98

SYN: C-10

TOXICITY DATA with REFERENCE
skn-rbt 1% MLD AIPTAK 137,410,62
eye-rbt 2500 ppm MLD AIPTAK 137,410,62
ipr-mus LD50:73600 μg/kg AIPTAK 137,410,62
scu-mus LD50:177 mg/kg JACSAT 68,2592,46
ivn-mus LD50:22 mg/kg JACSAT 68,2592,46

SAFETY PROFILE: Poison by intraperitoneal, subcutaneous and intravenous routes. A skin and eye irritant. When heated to decomposition it emits very toxic fumes of HCl and NO_x. See also ESTERS.

BPI125 CAS:29025-14-7 ***HR: 3***
BUTOXYBENZYL HYOSCYAMINE BROMIDE
mf: $C_{28}H_{38}NO_4 \cdot Br$ mw: 532.58

PROP: Crystals from ethanol-acetone. Mp: 166-168°. Also reported as white needles from isopropanol, mp: 158-160°. Freely sol in glacial acetic acid; sol in chloroform, DMF. Sparingly sol in ethanol; sltly sol in water, 0.1N HCl, 0.1N NaOH. Practically insol in acetone, ether, and benzene.

SYNS: BHB ◇ p-BUTOXYBENZYL HYOSCYAMINIUM BROMIDE ◇ (−)-8-(p-BUTOXYBENZYL)-3-α-HYDROXY-1-α-H,5-α-H-TROPANIUM BROMIDE TROPATE (ester) ◇ 1-(1-(p-n-BUTOXYBENZYL)HYOSCYAMINIUM) BROMIDE ◇ BUTROPIUM BROMIDE ◇ COLIOPAN

TOXICITY DATA with REFERENCE
ipr-rat TDLo:144 mg/kg (female 9-14D post):REP OYYAA2 8,319,74
ipr-rat TDLo:36 mg/kg (female 9-14D post):TER OYYAA2 8,319,74
ipr-rat LD50:113 mg/kg IYKEDH 5,106,74
ivn-rat LD50:21 mg/kg IYKEDH 5,106,74
orl-mus LD50:1500 mg/kg USXXAM #3696110
scu-mus LD50:370 mg/kg NIIRDN 6,355,82
ivn-mus LD50:6400 μg/kg IYKEDH 5,106,74
ims-mus LD50:285 mg/kg NIIRDN 6,355,82
ivn-rbt LD50:6800 μg/kg OYYAA2 8,285,74

SAFETY PROFILE: Poison by subcutaneous, intramuscular, intravenous, and intraperitoneal routes. Moderately toxic by ingestion. An experimental teratogen. Other experimental reproductive effects. When heated to decomposition it emits toxic fumes of Br^- and NO_x. See also BROMIDES.

BPI300 CAS:832-06-4 ***HR: 2***
2-BUTOXYCARBONYLMETHYLENE-4-OXOTHIAZOLIDONE
mf: $C_9H_{13}NO_3S$ mw: 215.29

SYNS: ACETIC ACID, (4-OXO-2-THIAZOLIDINYLIDENE)-, BUTYL ESTER (9CI) ◇ 2-(n-BUTYLOXYCARBONYLMETHYLENE)THIAZOLID-4-ONE ◇ ICI 43823

TOXICITY DATA with REFERENCE
orl-rat TDLo:193 g/kg/2Y-C:ETA EXMDA4 (145),289,68

SAFETY PROFILE: Questionable carcinogen with experimental tumorigenic data. When heated to decomposition it emits toxic fumes of NO_x and SO_x.

BPI500 CAS:38915-40-1 ***HR: D***
N-(2-BUTOXY-7-CHLOROBENZO(b)-1,5-NAPHTHYRIDIN-10-(YL)-N'-(2-CHLOROETHYL-1,3-PROPANEDIAMINE-N'-ETHYL-)
mf: $C_{23}H_{29}Cl_2N_4O$ mw: 448.2

SYN: ICR 355

TOXICITY DATA with REFERENCE
mmo-sat 500 ng/plate MUREAV 136,185,84
msc-ham:ovr 1 μmol/L CNREA8 39,4875,79

SAFETY PROFILE: Mutation data reported. When heated to decomposition it emits very toxic fumes of Cl^- and NO_x.

BPI625 CAS:58763-31-8 ***HR: 3***
4'-BUTOXY-3'-CHLORO-5'-METHYL-3-PIPERIDINO-PROPIOPHENONE HYDROCHLORIDE
mf: $C_{19}H_{28}ClNO_2 \cdot ClH$ mw: 374.39

SYNS: 1-(4-BUTOXY-3-CHLORO-5-METHYLPHENYL)-3-(1-PIPERIDINYL)1-PROPANONE HYDROCHLORIDE (9CI) ◇ β-PIPERIDINOAETHYL-(3-CHLOR-4-n-BUTOXY-5-METHYLPHENYL)KETONHYDROCHLORID (GERMAN)

TOXICITY DATA with REFERENCE
orl-mus LD50:475 mg/kg PHARAT 31,21,76
scu-mus LD50:1000 mg/kg PHARAT 31,21,76
ivn-mus LD50:43 mg/kg PHARAT 31,21,76

SAFETY PROFILE: Poison by intravenous route. Moderately toxic by ingestion and subcutaneous routes.

When heated to decomposition it emits toxic fumes of NO_x, Cl^-, and HCl.

BPI750 ***HR: 3***
4'-BUTOXY-2'-CHLORO-2-PYRROLIDINYL ACETANILIDE HYDROCHLORIDE
mf: $C_{16}H_{23}ClN_2O_2$•ClH mw: 347.32

SYN: C 3187

TOXICITY DATA with REFERENCE
eye-rbt 2% SEV ARZNAD 8,270,58
ipr-rat LD50:287 mg/kg ARZNAD 8,270,58
scu-mus LD50:550 mg/kg ARZNAD 8,270,58

SAFETY PROFILE: Poison by intraperitoneal route. Moderately toxic by subcutaneous route. A severe eye irritant. When heated to decomposition it emits very toxic fumes of Cl^-, NO_x, and HCl.

BPJ000 CAS:41296-95-1 ***HR: 3***
4'-BUTOXY-2-(DIETHYLAMINO)ACETANILIDE HYDROCHLORIDE
mf: $C_{16}H_{26}N_2O_2$•ClH mw: 314.90

SYN: C 3121

TOXICITY DATA with REFERENCE
eye-rbt 2% MOD ARZNAD 8,270,58
ipr-rat LD50:220 mg/kg ARZNAD 8,270,58
scu-mus LD50:695 mg/kg ARZNAD 8,270,58

SAFETY PROFILE: Poison by intraperitoneal route. Moderately toxic by subcutaneous route. An eye irritant. When heated to decomposition it emits very toxic fumes of NO_x and HCl.

BPJ250 CAS:77966-20-2 ***HR: 3***
2-BUTOXY-N-(2-(DIETHYLAMINO)ETHYL)-N-(2,6-XYLYL)CINCHONINAMIDE HYDROCHLORIDE
mf: $C_{28}H_{37}N_3O_2$•ClH mw: 484.14

SYN: 2-BUTOXY-N-(2-DIETHYLAMINO)ETHYL)-N-(2,6-XYLYL)-4-QUINOLINECARBOXAMIDE HYDROCHLORIDE

TOXICITY DATA with REFERENCE
eye-rbt 2% SEV ARZNAD 8,708,58
ipr-rat LD50:300 mg/kg ARZNAD 8,708,58
scu-mus-LD50:1175 mg/kg ARZNAD 8,708,58

SAFETY PROFILE: Poison by intraperitoneal route. Moderately toxic by subcutaneous route. A severe eye irritant. When heated to decomposition it emits very toxic fumes of HCl and NO_x.

BPJ500 CAS:78109-80-5 ***HR: 3***
o-BUTOXY-N-(5-(DIETHYLAMINO)-2-PENTYL)BENZAMIDE HYDROCHLORIDE
mf: $C_{20}H_{34}N_2O_2$•ClH mw: 371.02

SYNS: 2-BUTOXYBENZOESAEURE-4'-DIAETHYLAMINO-L'-METHYL-BUTYLAMID(1') HYDROCHLORID (GERMAN) ◇ D-649

TOXICITY DATA with REFERENCE
scu-mus LD50:130 mg/kg ARZNAD 10,743,60
ivn-mus LD50:15 mg/kg ARZNAD 10,743,60

SAFETY PROFILE: Poison by subcutaneous and intravenous routes. When heated to decomposition it emits very toxic fumes of HCl and NO_x.

BPJ750 CAS:78109-81-6 ***HR: 3***
o-BUTOXY-N-(3-(DIETHYLAMINO)PROPYL)BENZAMIDE HYDROCHLORIDE
mf: $C_{18}H_{30}N_2O_2$•ClH mw: 342.96

SYNS: 2-BUTOXYBENZOESAEURE-3'-DIAETHYLAMINOPROPYLAMID-(1') HYDROCHLORID (GERMAN) ◇ D-638

TOXICITY DATA with REFERENCE
ipr-mus LD50:75 mg/kg ARZNAD 10,743,60
scu-mus LD50:160 mg/kg ARZNAD 10,743,60
ivn-mus LD50:30 mg/kg ARZNAD 10,743,60

SAFETY PROFILE: Poison by intraperitoneal, subcutaneous, and intravenous routes. When heated to decomposition it emits very toxic fumes of HCl and NO_x.

BPJ850 CAS:111-76-2 ***HR: 3***
2-BUTOXYETHANOL
DOT: UN 2369
mf: $C_6H_{14}O_2$ mw: 118.20

PROP: Clear, mobile liquid; pleasant odor. Bp: 168.4-170.2°, fp: −74.8°, flash p: 160°F (COC), d: 0.9012 @ 20°/20°, vap press: 300 mm @ 140°.

SYNS: BUCS ◇ BUTOKSYETYLOWY ALKOHOL (POLISH) ◇ 2-BUTOSSI-ETANOLO (ITALIAN) ◇ 2-BUTOXY-AETHANOL (GERMAN) ◇ BUTOXYETHANOL ◇ 2-BUTOXY-1-ETHANOL ◇ n-BUTOXYETHANOL ◇ BUTYL CELLOSOLVE ◇ o-BUTYL ETHYLENE GLYCOL ◇ BUTYL GLYCOL ◇ BUTYLGLYCOL (FRENCH, GERMAN) ◇ BUTYL OXITOL ◇ DOWANOL EB ◇ EKTASOLVE EB ◇ ETHYLENE GLYCOL-n-BUTYL ETHER ◇ ETHYLENE GLYCOL MONOBUTYL ETHER (MAK, DOT) ◇ GAFCOL EB ◇ GLYCOL BUTYL ETHER ◇ GLYCOL ETHER EB ◇ GLYCOL ETHER EB ACETATE ◇ GLYCOL MONOBUTYL ETHER ◇ JEFFERSOL EB ◇ MONOBUTYL GLYCOL ETHER ◇ 3-OXA-1-HEPTANOL ◇ POLY-SOLV EB

TOXICITY DATA with REFERENCE
skn-rbt 500 mg open MLD UCDS**
ihl-rat TCLo:200 ppm/6H (female 6-15D post):REP EVHPAZ 57,47,84
ihl-rbt TCLo:100 ppm/6H (female 6-18D post):TER EVHPAZ 57,47,84
orl-wmn TDLo:600 mg/kg HUTODJ 7,187,88
ihl-hmn TCLo:195 ppm/8H:GIT AMIHAB 14,114,56
ihl-hmn TCLo:100 ppm:NOSE,EYE,CNS NPIRI* 1,50,74
orl-rat LD50:470 mg/kg DOWCC* MSD-46
ihl-rat LC50:2900 mg/m^3 GTPZAB 32(3),48,88
ipr-rat LD50:220 mg/kg 85GMAT -,67,82

ivn-rat LD50:340 mg/kg AMIHAB 14,114,56
ihl-mus LC50:700 ppm/7H JIHTAB 25,157,43
scu-mus LDLo:500 mg/kg JPETAB 42,355,31
orl-rbt LD50:300 mg/kg YKYUA6 32,1241,81
skn-gpg LD50:230 mg/kg TXAPA9 7,559,65

CONSENSUS REPORTS: Reported in EPA TSCA Inventory. Glycol ethers are on the Community Right-To-Know List.

OSHA PEL: (Transitional: TWA 50 ppm (skin)) TWA 25 ppm (skin)
ACGIH TLV: TWA 25 ppm (skin)
DFG MAK: 20 ppm (100 mg/m^3)
DOT Classification: Poison B; Label: St. Andrews Cross, Flammable Liquid.

SAFETY PROFILE: Poison by ingestion, skin contact, intraperitoneal, and intravenous routes. Moderately toxic via inhalation and subcutaneous routes. Human systemic effects by inhalation: nausea or vomiting, headache, nose tumors, unspecified eye effects. Experimental teratogenic and reproductive effects. A skin irritant. Flammable liquid when exposed to heat or flame. To fight fire, use foam, CO_2, and dry chemical. Incompatible with oxidizing materials, heat, and flame. When heated to decomposition it emits acrid smoke and irritating fumes.

BPK250 CAS:78-51-3 ***HR: 3***
2-BUTOXYETHANOL PHOSPHATE
mf: $C_{18}H_{39}O_7P$ mw: 398.54

PROP: Light-colored liquid, butyl-like odor. Mp: −70°; bp: 200-230° @ 4 mm, flash p: 435°F, d: 1.02 @ 20°/20°, vap press: 0.03 mm @ 150°, vap d: 13.8.

SYNS: KP 140 ◇ KRONITEX KP-140 ◇ PHOSFLEX T-BEP ◇ TBEP ◇ TRI(2-BUTOXYETHANOL PHOSPHATE) ◇ TRIBUTOXYETHYL PHOSPHATE ◇ TRI(2-BUTOXYETHYL) PHOSPHATE ◇ TRIBUTYL CELLOSOLVE PHOSPHATE ◇ TRIS(2-BUTOXYETHYL) ESTER PHOSPHORIC ACID ◇ TRIS(2-BUTOXYETHYL) PHOSPHATE

TOXICITY DATA with REFERENCE
orl-rat LD50:3000 mg/kg NPIRI* 2,93,75
ivn-mus LD50:180 mg/kg CSLNX* NX#00391
orl-gpg LD50:3000 mg/kg 29ZWAE -,336,68

CONSENSUS REPORTS: Reported in EPA TSCA Inventory.

SAFETY PROFILE: A poison by intravenous route. Moderately toxic by ingestion. Combustible when exposed to heat or flame. Dangerous; see also PHOSPHATES; can react with oxidizing materials. To fight fire, use water, foam, CO_2, dry chemical. When heated to decomposition it emits toxic fumes of PO_x.

BPK500 CAS:7251-90-3 ***HR: 2***
2-BUTOXYETHOXY ACRYLATE
mf: $C_9H_{16}O_4$ mw: 188.25

SYNS: BUTYL CELLOSOLVE ACRYLATE ◇ 2-PROPENOIC ACID 2-BUTOXYETHYL ESTER

TOXICITY DATA with REFERENCE
orl-rat LD50:6500 mg/kg AIHAAP 30,470,69
skn-rbt LD50:640 mg/kg AIHAAP 30,470,69

CONSENSUS REPORTS: Reported in EPA TSCA Inventory.

SAFETY PROFILE: Moderately toxic by skin contact. Mildly toxic by ingestion. See also ESTERS. When heated to decomposition it emits acrid smoke and irritating fumes.

BPK750 CAS:4413-13-2 ***HR: 2***
1-BUTOXY-2-ETHOXYETHANE
mf: $C_8H_{18}O_2$ mw: 146.26

SYN: 1-(2-ETHOXYETHOXY)-BUTANE

TOXICITY DATA with REFERENCE
skn-rbt 10 mg/24H open MLD AMIHBC 10,61,54
eye-rbt 20 mg open AMIHBC 10,61,54
orl-rat LD50:2830 mg/kg AMIHBC 10,61,54
skn-rbt LD50:2120 mg/kg AMIHBC 10,61,54

SAFETY PROFILE: Moderately toxic by ingestion and skin contact. Irritating to skin and eyes. When heated to decomposition it emits acrid smoke and irritating fumes.

BPL250 CAS:112-56-1 ***HR: 3***
2-(2-BUTOXY ETHOXY)ETHYL THIOCYANATE
mf: $C_9H_{17}NO_2S$ mw: 203.33

SYNS: 2-(2-(BUTOXY)ETHOXY)ETHYL THIOCYANIC ACID ESTER ◇ BUTOXYRHODANODIETHYL ETHER ◇ β-BUTOXY-β′-THIOCYANODIETHYL ETHER ◇ 2-BUTOXY-2′-THIOCYANODIETHYL ETHER ◇ 1-BUTOXY-2-(2-THIOCYANOETHOXY)ETHANE ◇ 1-BUTOXY-2-(2-THIOCYANATOETHYXY)ENTHANE ◇ BUTYL CARBITOL RHODANATE ◇ BUTYL CARBITOL THIOCYANATE ◇ ENT 6 ◇ ETHANOL-2-(2-BUTOXYETHOXY) THIOCYANATE ◇ LETHANE ◇ LETHANE 384 ◇ LETHANE 384 REGULAR

TOXICITY DATA with REFERENCE
orl-rat LD50:90 mg/kg FMCHA2 -,D180,80
skn-rat LD50:250 mg/kg WRPCA2 9,119,70
ipr-rat LD50:90 mg/kg INMEAF 11,-,42
scu-rat LD50:550 mg/kg INMEAF 11,-,42
scu-rat LDLo:500 mg/kg JIHTAB 18,310,36
scu-mus LDLo:200 mg/kg JIHTAB 18,310,36
ivn-mus LD50:56 mg/kg CSLNX* NX#02402
orl-dog LD50:30 mg/kg PCOC** -,657,66
scu-dog LD50:200 mg/kg INMEAF 11,-,42
orl-rbt LD50:35 mg/kg JPETAB 82,377,44
skn-rbt LD50:125 mg/kg SPEADM 78-1,20,78

SAFETY PROFILE: A poison by ingestion, skin contact, intraperitoneal, subcutaneous, and intravenous routes. Moderately toxic by an unspecified route. High concentrations can cause central nervous system depression. An insecticide. See also THIOCYANATES, ESTERS, and ETHERS. When heated to decomposition it emits very toxic fumes of SO_x, NO_x, and CN^-.

BPL500 CAS:124-16-3 ***HR: 2***
1-BUTOXY ETHOXY-2-PROPANOL
mf: $C_9H_{20}O_3$ mw: 176.29

PROP: D: 0.9310 @ 20°/20°, bp: 230.3°, fp: −90°, flash p: 250°F (OC). Sol in water.

SYN: 1-(2-BUTOXYETHOXY)-2-PROPANOL

TOXICITY DATA with REFERENCE
skn-rbt 485 mg open MLD UCDS** 12/29/71
eye-rbt 100 mg MOD 34ZIAG -,730,69
orl-rat LD50:4000 mg/kg AIHAAP 30,470,69
skn-rbt LD50:2830 mg/kg AIHAAP 30,470,69

CONSENSUS REPORTS: Reported in EPA TSCA Inventory.

SAFETY PROFILE: Moderately toxic by ingestion and skin contact. A skin and eye irritant. Combustible when exposed to heat or flame. To fight fire, use alcohol foam, dry chemical, spray, or mist. When heated to decomposition it emits acrid and irritating fumes.

BPL750 CAS:10043-18-2 ***HR: 2***
3-(2-BUTOXYETHOXY)PROPANOL
mf: $C_9H_{20}O_3$ mw: 176.29

TOXICITY DATA with REFERENCE
skn-rbt 10 mg/24H open MLD AMIHBC 10,61,54
eye-rbt 20 mg open SEV AMIHBC 10,61,54
orl-rat LD50:5160 mg/kg AMIHBC 10,61,54
skn-rbt LD50:3000 mg/kg AMIHBC 10,61,54

SAFETY PROFILE: Moderately toxic by skin contact. Mildly toxic by ingestion. A skin and severe eye irritant. When heated to decomposition it emits acrid smoke and irritating fumes.

BPM000 CAS:112-07-2 ***HR: 3***
2-BUTOXYETHYL ACETATE
mf: $C_8H_{16}O_3$ mw: 160.24

PROP: Colorless liquid; fruity odor. Bp: 192.3°, d: 0.9424 @ 20°/20°, fp: −63.5°, flash p: 190°F. Sol in hydrocarbons and organic solvents; insol in water.

SYNS: 2-BUTOXYETHANOL ACETATE ◇ 2-BUTOXYETHYL ESTER ACETIC ACID ◇ BUTYL CELLOSOLVE ACETATE ◇ EKTASOLVE EB ACETATE ◇ ETHYLENE GLYCOL MONOBUTYL ETHER ACETATE (MAK) ◇ GLYCOL MONOBUTYL ETHERACETATE

TOXICITY DATA with REFERENCE
skn-rbt 500 mg open MLD UCDS** 1/31/66
orl-rat LD50:2400 mg/kg TXAPA9 51,117,79
orl-mus LD50:3200 mg/kg KODAK* 21MAY71
skn-rbt LD50:1500 mg/kg TXAPA9 51,117,79

CONSENSUS REPORTS: Reported in EPA ISCA Inventory. Glycol ethers are on the Community Right-To-Know List.

DFG MAK: 20 ppm (135 mg/m^3)

SAFETY PROFILE: Moderately toxic by ingestion and skin contact. Mild skin irritant. Flammable when exposed to heat, flame, or oxidizers. To fight fire, use alcohol foam. When heated to decomposition it emits acrid smoke and irritating fumes. See also ESTERS.

BPM500 CAS:57629-90-0 ***HR: D***
(BUTOXYMETHYL)NITROSOMETHYLAMINE
mf: $C_6H_{14}N_2O_2$ mw: 146.22

SYN: N-(BUTYLOXY)METHYL-N-METHYLNITROSAMINE

TOXICITY DATA with REFERENCE
mmo-sat 3 μmol/plate MUREAV 49,187,78
mma-sat 1 μmol/plate MUREAV 49,187,78

SAFETY PROFILE: Mutation data reported. Many nitrosamines are carcinogens. When heated to decomposition it emits toxic fumes of NO_x. See also NITROSAMINES.

BPM750 CAS:27471-60-9 ***HR: 3***
2-(p-BUTOXYPHENOXY)-N-(2-(DIETHYLAMINO) ETHYL)-2,5'-DIETHOXYACETANILIDE MONOHYDROCHLORIDE
mf: $C_{28}H_{42}N_2O_5 \cdot ClH$ mw: 523.18

SYNS: ANP 3548 ◇ CHLORHYDRATE de N-(DIETHOXY-2,5-PHENYL)-N-DIETHYLAMINO-2-ETHYLBUTOXY-4-PHENOXYACETAMIDE ◇ N,N-DIETHYL-N'-(2,5-DIETHOXYPHENYL)-N'-(4-BUTOXYPHENOXYACETYL) ETHYLENEDIAMINE HCl ◇ FENOXEDIL ◇ FENOXEDIL HYDROCHLORIDE ◇ SUPLEXEDIL

TOXICITY DATA with REFERENCE
orl-rat LD50:2400 mg/kg EJMCA5 10,291,75
ipr-rat LD50:175 mg/kg EJMCA5 10,291,75
scu-rat LD50:2065 mg/kg EJMCA5 10,291,75
ivn-rat LD50:10 mg/kg EJMCA5 10,291,75
orl-mus LD50:750 mg/kg EJMCA5 10,286,75
ipr-mus LD50:82 mg/kg EJMCA5 10,291,75
scu-mus LD50:341 mg/kg EJMCA5 10,291,75
ivn-mus LD50:17 mg/kg USXXAM #3818021
orl-rbt LD50:815 mg/kg EJMCA5 10,291,75

SAFETY PROFILE: Poison by intraperitoneal, intravenous, and subcutaneous routes. Moderately toxic by ingestion. A vasodilator. When heated to decomposition it emits very toxic fumes of NO_x and HCl.

BPN000 CAS:27468-64-0 ***HR: 3***
2-(p-BUTOXYPHENOXY)-N-(2-(DIETHYLAMINO) ETHYL)-N-(2,4-DIMETHOXYPHENYL)ACETAMIDE HYDROCHLORIDE
mf: $C_{26}H_{38}N_2O_5 \cdot ClH$ mw: 495.12

TOXICITY DATA with REFERENCE
orl-mus LD50:400 mg/kg EJMCA5 10,286,75
ivn-mus LD50:40 mg/kg EJMCA5 10,286,75

SAFETY PROFILE: Poison by ingestion and intravenous routes. When heated to decomposition it emits very toxic fumes of NO_x and HCl.

BPN250 CAS:27468-66-2 ***HR: 3***
2-(p-BUTOXYPHENOXY)-N-(2-(DIETHYLAMINO)ETHYL)-N-(2,5-DIMETHOXYPHENYL) ACETAMIDE HYDROCHLORIDE
mf: $C_{26}H_{38}N_2O_5 \cdot ClH$ mw: 495.12

TOXICITY DATA with REFERENCE
orl-mus LD50:400 mg/kg EJMCA5 10,286,75
ivn-mus LD50:25 mg/kg EJMCA5 10,286,75

SAFETY PROFILE: Poison by ingestion and intravenous routes. When heated to decomposition it emits very toxic fumes of NO_x and HCl.

BPO250 CAS:27468-71-9 ***HR: 3***
2-(p-BUTOXYPHENOXY)-N-(2(DIMETHYLAMINO)ETHYL)-N-(2,6-DIMETHYLPHENYL) ACETAMIDE HYDROCHLORIDE
mf: $C_{24}H_{34}N_2O_3 \cdot ClH$ mw: 435.06

TOXICITY DATA with REFERENCE
orl-mus LD50:400 mg/kg EJMCA5 10,286,75
ivn-mus LD50:18 mg/kg EJMCA5 10,286,75

SAFETY PROFILE: Poison by ingestion and intravenous routes. When heated to decomposition it emits very toxic fumes of HCl and NO_x.

BPP250 CAS:3102-00-9 ***HR: 3***
3-n-BUTOXY-1-PHENOXY-2-PROPANOL
mf: $C_{13}H_{20}O_3$ mw: 224.33

SYNS: (3-n-BUTOXY-2-HYDROXYPROPYL)PHENYLETHER ◇ 1-BUTOXY-3-PHENOXY-2-PROPANOL ◇ FEBUPROL ◇ H-33 ◇ K-10033 ◇ VALBIL

TOXICITY DATA with REFERENCE
orl-rat LD50:2370 mg/kg DRFUD4 3,191,78
ipr-rat LD50:400 mg/kg DRFUD4 3,191,78
orl-mus LD50:3050 mg/kg DRFUD4 3,191,78
ipr-mus LD50:436 mg/kg DRFUD4 3,191,78
orl-dog LD50:500 mg/kg DRFUD4 3,191,78
ivn-dog LD50:150 mg/kg DRFUD4 3,191,78

SAFETY PROFILE: Poison by intraperitoneal and intravenous routes. Moderately toxic by ingestion. Stimulates the production of bile by the liver. See also ETHERS. When heated to decomposition it emits acrid smoke and irritating fumes.

BPP750 CAS:2438-72-4 ***HR: 2***
p-BUTOXY PHENYL ACETOHYDROXAMIC ACID
mf: $C_{12}H_{17}NO_3$ mw: 223.30

SYNS: BUFEXAMIC ACID ◇ 4-BUTOXYPHENYLACETOHYDROXAMIC ACID ◇ CP 1044 J3 ◇ DROXAROL ◇ DROXARYL ◇ FLOGICID ◇ FLOGOCID N PLASTIGEL ◇ J3 ◇ PARFENAC ◇ PARFENAL

TOXICITY DATA with REFERENCE
sce-ham:ovr 20 µmol/L PAACA3 21,126,80
orl-rat TDLo:5 g/kg (6-15D preg):TER ARZNAD 20,565,70
orl-rat LD50:3370 mg/kg NIIRDN 6,681,82
ipr-rat LD50:805 mg/kg YKYUA6 28,253,77
orl-mus LD50:8000 mg/kg JMCMAR 13,211,70
ipr-mus LD50:1195 mg/kg YKYUA6 28,253,77

CONSENSUS REPORTS: EPA Genetic Toxicology Program.

SAFETY PROFILE: Moderately toxic by ingestion and intraperitoneal routes. Experimental teratogenic effects. Mutation data reported. When heated to decomposition it emits toxic fumes of NO_x.

BPQ000 CAS:77372-68-0 ***HR: D***
4-N-BUTOXYPHENYLACETOHYDROXAMIC ACID-o-PROPIONATE ESTER
mf: $C_{15}H_{21}NO_4$ mw: 279.2

TOXICITY DATA with REFERENCE
mmo-sat 1 nmol/plate PAACA3 21,126,80
sce-ham:ovr 20 µmol/L PAACA3 21,126,80
sce-ham:ovr 80 µmol/L MUREAV 88,81,81

SAFETY PROFILE: Mutation data reported. See also ESTERS. When heated to decomposition it emits toxic fumes of NO_x.

BPQ250 CAS:76790-19-7 ***HR: D***
N-(p-BUTOXYPHENYL ACETYL)-o-FORMYLHYDROXYLAMINE
mf: $C_{13}H_{17}NO_4$ mw: 251.2

SYN: 4-N-BUTOXYPHENYLACETOHYDROXAMICACID-o-FORMATE ESTER

TOXICITY DATA with REFERENCE
mmo-sat 1 nmol/plate PAACA3 21,126,80
sce-ham:ovr 400 µmol/L/13H-C MUREAV 88,81,81

SAFETY PROFILE: Mutation data reported. See also ESTERS. When heated to decomposition it emits toxic fumes of NO_x.

BPR000 CAS:77791-53-8 ***HR: 3***
4'-BUTOXY-2-PIPERIDINOACETANILIDE HYDROCHLORIDE
mf: $C_{17}H_{26}N_2O_2$•ClH mw: 326.91

SYN: C 3125

TOXICITY DATA with REFERENCE
eye-rbt 2% SEV ARZNAD 8,407,58
ipr-rat LD50:200 mg/kg ARZNAD 8,407,58
scu-mus LD50:665 mg/kg ARZNAD 8,407,58

SAFETY PROFILE: Poison by intraperitoneal route. Moderately toxic by subcutaneous route. A severe eye irritant. When heated to decomposition it emits very toxic fumes of NO_x and HCl.

BPR250 ***HR: 3***
4-BUTOXY-3-(PIPERIDINO)PROPIOPHENONE HYDROCHLORIDE
mf: $C_{18}H_{27}NO_2$•ClH mw: 325.92

SYN: C 5422

TOXICITY DATA with REFERENCE
eye-rbt 2% SEV ARZNAD 8,708,58
ipr-rat LD50:33 mg/kg ARZNAD 8,708,58
scu-mus LD50:37 mg/kg ARZNAD 8,708,58

SAFETY PROFILE: Poison by intraperitoneal and subcutaneous routes. A severe eye irritant. When heated to decomposition it emits very toxic fumes of HCl and NO_x.

BPR500 CAS:536-43-6 ***HR: 3***
4'-BUTOXY-3-PIPERIDINO PROPIOPHENONE HYDROCHLORIDE
mf: $C_{18}H_{27}NO_2$•ClH mw: 325.92

SYNS: 1-(2-(4-BUTOXYBENZOYL)ETHYL)PIPERIDINEHYDROCHLORIDE ◇ 4-n-BUTOXY-β-(1-PIPERIDYL)PROPIOPHENONE HYDROCHLORIDE ◇ DICLONIA ◇ DYCLOCAINUM ◇ DYCLONE HYDROCHLORIDE ◇ DYCLONINE HYDROCLORIDE ◇ DYCLOTHANE ◇ P-267 ◇ S 154

TOXICITY DATA with REFERENCE
skn-rbt 1% MLD AIPTAK 137,410,62
eye-rbt 1% MLD AIPTAK 137,410,62
ipr-rat LD50:46 mg/kg JPETAB 115,419,55
scu-rat LD50:201 mg/kg JPETAB 115,413,55
orl-mus LDLo:100 mg/kg TXAPA9 2,616,60
ipr-mus LD50:52 mg/kg AIPTAK 137,410,62
scu-mus LD50:42 mg/kg ARZNAD 5,559,55
ivn-mus LD50:20 mg/kg JPETAB 115,419,55
orl-dog LDLo:40 mg/kg TXAPA9 2,616,60
ivn-dog LD50:9500 μg/kg JPETAB 115,419,55
orl-rbt LDLo:200 mg/kg TXAPA9 2,616,60

SAFETY PROFILE: A poison by ingestion, intraperitoneal, subcutaneous, and intravenous routes. A skin and eye irritant. When heated to decomposition it emits very toxic fumes of HCl and NO_x.

BPS000 CAS:7420-06-6 ***HR: 2***
3-BUTOXY PROPANOIC ACID
mf: $C_7H_{14}O_3$ mw: 146.21

SYN: 3-BUTOXYPROPIONIC ACID

TOXICITY DATA with REFERENCE
skn-rbt 10 mg/24H open MLD AMIHBC 10,61,54
eye-rbt 250 μg open SEV AMIHBC 10,61,54
orl-rat LD50:5190 mg/kg AMIHBC 10,61,54
skn-rbt LD50:630 mg/kg AMIHBC 10,61,54

SAFETY PROFILE: Moderately toxic by skin contact. Mildly toxic by ingestion. A skin and severe eye irritant. When heated to decomposition it emits acrid smoke and irritating fumes.

BPS250 CAS:5131-66-8 ***HR: 2***
1-BUTOXY-2-PROPANOL
mf: $C_7H_{16}O_2$ mw: 132.23

SYNS: PROPASOL SOLVENT B ◇ PROPYLENE GLYCOL-n-BUTYL ETHER

TOXICITY DATA with REFERENCE
orl-rat LD50:2200 mg/kg 14CYAT 2,1572,63
skn-rbt LD50:3100 mg/kg NPIRI* 1,102,74

CONSENSUS REPORTS: Reported in EPA TSCA Inventory. Glycol ethers are on the Community Right-To-Know List.

SAFETY PROFILE: Moderately toxic by ingestion and skin contact. When heated to decomposition it emits acrid smoke and irritating fumes. See also ETHERS.

BPS500 CAS:10215-33-5 ***HR: 3***
3-BUTOXY-1-PROPANOL
mf: $C_7H_{16}O_2$ mw: 132.23

SYN: PROPYLENE GLYCOL MONO-n-BUTYL ETHER

TOXICITY DATA with REFERENCE
skn-rbt 500 mg open MLD UCDS** 7/28/66
eye-rbt 15 mg SEV UCDS** 7/28/66
orl-rat LD50:5950 mg/kg AIHAAP 30,470,69
skn-rbt LD50:1590 mg/kg AIHAAP 30,470,69
ivn-mus LD50:320 mg/kg CSLNX* NX#02921

CONSENSUS REPORTS: Glycol ethers are on the Community Right-To-Know List.

SAFETY PROFILE: Poison by intravenous route. Moderately toxic by skin contact. Mildly toxic by ingestion. A mild skin and severe eye irritant. When heated to decomposition it emits acrid smoke and irritating fumes.

BPS750 CAS:63716-40-5 ***HR: 2***
n-BUTOXYPROPANOL (mixed isomers)
mf: $C_7H_{16}O_2$ mw: 132.23

SYN: BUTOXYPROPANOL (mixed isomers)

TOXICITY DATA with REFERENCE
skn-rbt 500 mg open MLD UCDS** 10/13/64
eye-rbt 15 mg SEV UCDS** 10/13/64
orl-rat LD50:2830 mg/kg UCDS** 10/13/64
skn-rbt LD50:3560 mg/kg TXAPA9 28,313,74

SAFETY PROFILE: Moderately toxic by ingestion and skin contact. A mild skin and severe eye irritant. When heated to decomposition it emits acrid smoke and irritating fumes.

BPT000 CAS:6959-71-3 ***HR: 2***
3-BUTOXYPROPIONITRILE
mf: $C_7H_{13}NO$ mw: 127.21

SYN: 3-BUTOXYPROPANENITRILE

TOXICITY DATA with REFERENCE
eye-rbt 500 mg AMIHBC 10,61,54
orl-rat LD50:7460 mg/kg AMIHBC 10,61,54
ipr-mus LDLo:500 mg/kg CBCCT* 9,135,57
skn-rbt LD50:8980 mg/kg AMIHBC 10,61,54

CONSENSUS REPORTS: Reported in EPA TSCA Inventory. Cyanide and its compounds are on the Community Right-To-Know List.

SAFETY PROFILE: Moderately toxic by intraperitoneal route. Mildly toxic by ingestion and skin contact. An eye irritant. When heated to decomposition it emits toxic fumes of NO_x and CN^-. See also NITRILES.

BPT250 ***HR: 3***
4'-BUTOXY-2-PYRROLIDINYLACETANILIDE HYDROCHLORIDE
mf: $C_{16}H_{24}N_2O_2 \cdot ClH$ mw: 312.88

SYN: C 3130

TOXICITY DATA with REFERENCE
eye-rbt 2% SEV ARZNAD 8,270,58
ipr-rat LD50:186 mg/kg ARZNAD 8,270,58
scu-mus LD50:545 mg/kg ARZNAD 8,270,58

SAFETY PROFILE: Poison by intraperitoneal route. Moderately toxic by subcutaneous route. A severe eye irritant. When heated to decomposition it emits very toxic fumes of NO_x and HCl.

BPT500 CAS:35941-65-2 ***HR: D***
BUTRIPTYLINE
mf: $C_{21}H_{27}N$ mw: 293.49

SYNS: 5H-DIBENZO(a,d)CYCLOHEPTENE-5-PROPANAMINE,10-11-DIHYDRO-N,N,β-TRIMETHYL-, (±)- ◇ d,l-10,11-DIHYDRO-N,N,β-TRIMETHYL-5H-DIBENZO(a,d)-CYCLOHEPTENE-5-PROPYLAMINE

TOXICITY DATA with REFERENCE
orl-rat TDLo:75 mg/kg (female 4-18D post):TER JNMDBO 2,271,71

SAFETY PROFILE: An experimental teratogen. When heated to decomposition it emits toxic fumes of NO_x.

BPT750 CAS:5585-73-9 ***HR: 3***
BUTRIPTYLINE HYDROCHLORIDE
mf: $C_{21}H_{27}N \cdot ClH$ mw: 329.95

SYNS: (±)-10,11-DIHYDRO-N,N,β-TRIMETHYL-5H-DIBENZO(a,d,) CYCLOHEPTENE-5-PROPANAMINE HCl ◇ (±)-10,11-DIHYDRO-N,N,β-TRIMETHYL-5H-DIBENZO(a,d)-CYCLOHEPTENE-5-PROPYLAMINE HCl ◇ EVADYNE ◇ AY-62014

TOXICITY DATA with REFERENCE
orl-rat LD50:700 mg/kg 27ZQAG -,62,72
ipr-rat LD50:150 mg/kg 27ZQAG -,62,72
orl-mus LD50:345 mg/kg 27ZQAG -,62,72
ipr-mus LD50:120 mg/kg 27ZQAG -,62,72
ivn-mus LD50:48 mg/kg 27ZQAG -,62,72

SAFETY PROFILE: Poison by ingestion, intraperitoneal, and intravenous routes. An antidepressant. When heated to decomposition it emits toxic fumes of HCl and NO_x.

BPU000 CAS:16227-10-4 ***HR: 3***
BUTRIZOL
mf: $C_6H_{11}N_3$ mw: 125.20

SYNS: BT ◇ 4-N-BUTYL-4H-1,2,4-TRIAZOLE ◇ 4-BUTYL-s-TRIAZOLE ◇ DITHANE R-24 ◇ INDAR ◇ RH-124

TOXICITY DATA with REFERENCE
orl-rat LD50:50 mg/kg 85ARAE 4,94,76/77
skn-rbt LD50:315 mg/kg FMCHA2 -,C131,83

SAFETY PROFILE: A poison by ingestion and skin contact. When heated to decomposition it emits toxic fumes of NO_x.

BPU500 CAS:91-49-6 ***HR: 3***
N-BUTYLACETANILIDE
mf: $C_{12}H_{17}NO$ mw: 191.30

PROP: Yellowish liquid, mp: 20.8°, bp: 280°, flash p: 286°F, vap d: 6.6, d: 0.992 @ 25°/25°.

TOXICITY DATA with REFERENCE
orl-mus LD50:800 mg/kg TXAPA9 19,20,71
orl-gpg LD50:300 mg/kg 28ZEAL 4,78,69

CONSENSUS REPORTS: Reported in EPA TSCA Inventory.

SAFETY PROFILE: Poison by ingestion. Combustible.

To fight fire, use CO_2, dry chemical. When heated to decomposition it emits toxic fumes of NO_x.

BPU750 CAS:123-86-4 ***HR: 3***
n-BUTYL ACETATE
DOT: UN 1123
mf: $C_6H_{12}O_2$ mw: 116.18

PROP: Colorless liquid; strong fruity odor. Bp: 126°, fp: −73.5°, ULC: 50-60, lel: 1.4%, uel: 7.5°, flash p: 72°F, d: 0.88 @ 20°/20°, refr index: 1.393-1.396, autoign temp: 797°F, vap press: 15 mm @ 25°. Misc with alc, ether, and propylene glycol; sltly sol in water.

SYNS: ACETATE de BUTYLE (FRENCH) ◇ ACETIC ACID n-BUTYL ESTER ◇ BUTILE (ACETATI di) (ITALIAN) ◇ BUTYLACETAT (GERMAN) ◇ BUTYL ACETATE ◇ 1-BUTYL ACETATE ◇ BUTYLACETATEN (DUTCH) ◇ BUTYLE (ACETATE de) (FRENCH) ◇ BUTYL ETHANOATE ◇ FEMA No. 2174 ◇ OCTAN n-BUTYLU (POLISH)

TOXICITY DATA with REFERENCE
eye-hmn 300 ppm JIHTAB 25,282,43
skn-rbt 500 mg/24H MOD FCTXAV 17,509,79
skn-rbt 500 mg/24H MLD 85JCAE -,355,86
eye-rbt 20 mg SEV AMIHBC 10,61,54
ihl-rat TCLo:1500 ppm/7H (female 7-16D post):TER NTIS** PB83-258038
ihl-hmn TCLo:200 ppm:NOSE,EYE,PUL JIHTAB 25,282,43
orl-rat LD50:13100 mg/kg 85GMAT -,28,82
ihl-rat LC50:2000 ppm/4H NPIRI* 1,7,74
orl-mus LD50:7060 mg/kg YKYUA6 32,1241,81
ihl-mus LC50:6 g/m^3/2H YKYUA6 32,1241,81
ipr-mus LD50:1230 mg/kg SCCUR* -,2,61
ihl-cat LCLo:68 g/m^3/72M AGGHAR 5,1,33
orl-rbt LD50:3200 mg/kg 85GMAT -,28,82
orl-gpg LDLo:4700 mg/kg FCTXAV 17,509,79
ihl-gpg LCLo:67 g/m^3/4H FCTXAV 17,515,79
ipr-gpg LDLo:1500 mg/kg AIHAAP 35,21,74

CONSENSUS REPORTS: Reported in EPA TSCA Inventory.

OSHA PEL: (Transitional: TWA 150 ppm) TWA 150 ppm; STEL 200 ppm
ACGIH TLV: TWA 150 ppm; STEL 200 ppm
DFG MAK: 200 ppm (950 mg/m^3)
DOT Classification: Flammable Liquid; Label: Flammable Liquid.

SAFETY PROFILE: Moderately toxic by intraperitoneal route. Mildly toxic by inhalation and ingestion. An experimental teratogen. A skin and severe eye irritant. Human systemic effects by inhalation: conjunctiva irritation, unspecified nasal and respiratory system effects. A mild allergen. High concentrations are irritating to eyes and respiratory tract and cause narcosis. Evidence of chronic systemic toxicity is inconclusive. See also ESTERS. Flammable liquid. Moderately explosive when exposed to flame. Ignites on contact with potassium-tert-butoxide. To fight fire, use alcohol foam, CO_2, dry chemical. When heated to decomposition it emits acrid and irritating fumes.

BPV000 CAS:105-46-4 ***HR: 3***
sec-BUTYL ACETATE
DOT: UN 1123
mf: $C_6H_{12}O_2$ mw: 116.18

PROP: Colorless liquid, mild odor. Bp: 112°, flash p: 18°, d: 0.862-0.866 @ 20°/20°, vap d: 4.00. lel: 1.3%, uel: 7.5%.

SYNS: ACETATE de BUTYLE SECONDAIRE (FRENCH) ◇ ACETIC ACID-2-BUTOXY ESTER ◇ ACETIC ACID-1-METHYLPROPYL ESTER (9CI) ◇ 2-BUTANOL ACETATE ◇ 2-BUTYL ACETATE ◇ sec-BUTYL ALCOHOL ACETATE

CONSENSUS REPORTS: Reported in EPA TSCA Inventory.

OSHA PEL: TWA 200 ppm
ACGIH TLV: TWA 200 ppm
DFG MAK: 200 ppm (950 mg/m^3)
DOT Classification: Flammable Liquid; Label: Flammable Liquid.

SAFETY PROFILE: An irritant and allergen. See also ESTERS. Flammable liquid. To fight fire, use alcohol foam, CO_2, dry chemical. When heated to decomposition it emits acrid and irritating fumes.

BPV100 CAS:540-88-5 ***HR: 3***
tert-BUTYL ACETATE
DOT: UN 1123
mf: $C_6H_{12}O_2$ mw: 116.18

SYNS: ACETIC ACID-tert-BUTYL ESTER ◇ ACETIC ACID-1,1-DIMETHYLETHYL ESTER ◇ TEXACO LEAD APPRECIATOR ◇ TLA

CONSENSUS REPORTS: Reported in EPA TSCA Inventory.

OSHA PEL: TWA 200 ppm
ACGIH TLV: TWA 200 ppm
DFG MAK: 200 ppm (950 mg/m^3)
DOT Classification: IMO: Flammable Liquid; Label: Flammable Liquid.

SAFETY PROFILE: Poison by inhalation and ingestion. Flammable. To fight fire, use alcohol foam, CO_2, dry chemical. When heated to decomposition it emits acrid smoke and irritating fumes.

BPV250 CAS:591-60-6 ***HR: 1***
BUTYL ACETOACETATE
mf: $C_8H_{14}O_3$ mw: 158.22

PROP: Bp: 214°, flash p: 185°F, d: 0.96, vap d: 5.55.

SYNS: ACETOACETIC ACID BUTYL ESTER ◇ 3-OXO-BUTANOIC ACID BUTYL ESTER

TOXICITY DATA with REFERENCE
eye-rbt 500 mg open AMIHBC 10,61,54
orl-rat LD50:11260 mg/kg AMIHBC 10,61,54

CONSENSUS REPORTS: Reported in EPA TSCA Inventory.

SAFETY PROFILE: Mildly toxic by ingestion. An eye irritant. See also ESTERS. Flammable. To fight fire, use alcohol foam, CO_2, dry chemical. When heated to decomposition it emits acrid and irritating fumes.

BPV325 CAS:56986-35-7 ***HR: 3***
N-BUTYL-N-(1-ACETOXYBUTYL)NITROSAMINE
mf: $C_{10}H_{20}N_2O_3$ mw: 216.32

SYNS: ACETIC ACID-1-(BUTYLNITROSOAMINO)BUTYL ESTER ◇ N-(α-ACETOXY)BUTYL-N-BUTYLNITROSAMINE ◇ 1-ACETOXY-N-NITROSODIBUTYLAMINE ◇ BABN ◇ 1-(BUTYLNITROSOAMINO) BUTYL ACETATE

TOXICITY DATA with REFERENCE
mmo-sat 50 nmol/plate CNREA8 40,162,80
mmo-esc 500 nmol/plate CNREA8 40,162,80
dnr-bcs 500 nmol/plate CNREA8 40,162,80
dns-rat:oth 10 μmol/L CBINA8 53,99,85
cyt-ham:fbr 125 mg/L/48H MUREAV 48,337,77
cyt-ham:lng 32 mg/L GMCRDC 27,95,81
scu-rat TDLo:82 mg/kg/10W-I:CAR IAPUDO 41,619,82
scu-rat TD:70500 μg/kg/10W-I:ETA GANNA2 73,687,82

CONSENSUS REPORTS: EPA Genetic Toxicology Program.

SAFETY PROFILE: Questionable carcinogen with experimental carcinogenic and tumorigenic data. Mutation data reported. When heated to decomposition it emits toxic fumes of NO_x. See also NITROSAMINES and ESTERS.

BPV500 CAS:56986-37-9 ***HR: D***
sec-BUTYL ACETOXYMETHYL NITROSAMINE
mf: $C_7H_{14}N_2O_3$ mw: 174.23

SYNS: ACETIC ACID (sec-BUTYLNITROSAMINOMETHYL) ESTER ◇ N-sec-BUTYL-N-(ACETOXYMETHYL)NITROSAMINE

TOXICITY DATA with REFERENCE
mmo-sat 1 μmol/plate GANNA2 66,457,75
cyt-ham:fbr 63 mg/L/48H MUREAV 48,337,77
dnr-bcs 500 nmol/plate GANNA2 66,457,75
mmo-esc 25 μmol/plate GANNA2 70,663,79

SAFETY PROFILE: Mutation data reported. Many nitrosamines are carcinogens. When heated to decomposition it emits toxic fumes of NO_x. See also NITROSAMINES.

BPW000 CAS:66409-97-0 ***HR: 3***
n-BUTYL-3,o-ACETYL-12-β-13-α-DIHYDROJERVINE
mf: $C_{33}H_{49}NO_4$ mw: 523.83

SYN: n-BUTYL-12-β-13-α-DIHYDROJERVINE-3-ACETATE

TOXICITY DATA with REFERENCE
orl-ham TDLo:170 mg/kg (7D preg):TER JAFCAU 26,564,78
orl-ham LDLo:170 mg/kg JAFCAU 26(3),564,78

SAFETY PROFILE: Poison by ingestion. An experimental teratogen. When heated to decomposition it emits toxic fumes of NO_x.

BPW100 CAS:141-32-2 ***HR: 2***
n-BUTYL ACRYLATE
DOT: UN 2348
mf: $C_7H_{12}O_2$ mw: 128.19

PROP: Water-white, extremely reactive monomer. Bp: 69° @ 50 mm, fp: −64.6°, flash p: 120°F (OC), d: 0.89 @ 25°/25°, vap press: 10 mm @ 35.5°, vap d: 4.42.

SYNS: ACRYLIC ACID BUTYL ESTER ◇ ACRYLIC ACID n-BUTYL ESTER (MAK) ◇ BUTYL ACRYLATE ◇ BUTYLACRYLATE, INHIBITED (DOT) ◇ BUTYL-2-PROPENOATE

TOXICITY DATA with REFERENCE
skn-rbt 10 mg/24H open MLD AMIHBC 4,119,51
skn-rbt 500 mg open MLD UCDS** 4/5/73
eye-rbt 50 mg MLD UCDS** 4/5/73
ihl-rat TCLo:135 ppm/6H (6-15D preg):REP FAATDF 3,443,83
orl-rat LD50:900 mg/kg 85GMAT -,28,82
ihl-rat LC50:2730 ppm/4H JTEHD6 16,811,85
skn-rat LDLo:1700 mg/kg PJPPAA 32,223,80
ipr-rat LD50:550 mg/kg AMPMAR 36,58,75
orl-mus LD50:7561 mg/kg TOLED5 11,125,82
ihl-mus LC50:7800 mg/m³/2H 85GMAT -,28,82
ipr-mus LD50:853 mg/kg JDREAF 51,526,72
skn-rbt LD50:2000 mg/kg TXAPA9 28,313,74

CONSENSUS REPORTS: IARC Cancer Review: Group 3 IMEMDT 7,56,87, Animal Inadequate Evidence IMEMDT 39,67,86. Reported in EPA TSCA Inventory. Community Right-To-Know List.

OSHA PEL: TWA 10 ppm
ACGIH TLV: TWA 10 ppm
DFG MAK: 10 ppm (55 mg/m³)
DOT Classification: Flammable or Combustible Liquid; Label: Flammable Liquid.

SAFETY PROFILE: Moderately toxic by ingestion, inhalation, skin contact, and intraperitoneal routes. Experimental reproductive effects. A skin and eye irritant. Flammable when exposed to heat or flame. To fight fire, use foam, CO_2, dry chemical. Incompatible with oxidizing materials. When heated to decomposition it emits acrid and irritating fumes. See also ESTERS.

BPW250 ***HR: 3***
tert-BUTYL-1-ADAMANTANE PEROXYCARBOXYLATE
mf: $C_{15}H_{24}O_3$ mw: 252.35

SAFETY PROFILE: Explodes on heating to 90-100°. When heated to decomposition it emits acrid smoke and fumes. See also PEROXIDES.

BPW500 CAS:71-36-3 ***HR: 3***
n-BUTYL ALCOHOL
DOT: UN 1120
mf: $C_4H_{10}O$ mw: 74.14

PROP: Colorless liquid; vinous odor. Bp: 117.5°, ULC: 40, lel: 1.4%, uel: 11.2%, fp: −88.9°, flash p: 95-100°F, d: 0.80978 @ 20°/4°, autoign temp: 689°F, vap press: 5.5 mm @ 20°, vap d: 2.55. Misc in alc, ether, and organic solvents; sltly sol in water.

SYNS: ALCOOL BUTYLIQUE (FRENCH) ◇ BUTANOL (FRENCH) ◇ 1-BUTANOL ◇ n-BUTANOL ◇ BUTAN-1-OL ◇ BUTANOL (DOT) ◇ BUTANOLEN (DUTCH) ◇ BUTANOLO (ITALIAN) ◇ BUTYL ALCOHOL (DOT) ◇ BUTYL HYDROXIDE ◇ BUTYLOWY ALKOHOL (POLISH) ◇ BUTYRIC or NORMAL PRIMARY BUTYL ALCOHOL ◇ CCS 203 ◇ FEMA No. 2178 ◇ 1-HYDROXYBUTANE ◇ METHYLOLPROPANE ◇ PROPYLCARBINOL ◇ PROPYLMETHANOL ◇ RCRA WASTE NUMBER U031

TOXICITY DATA with REFERENCE
eye-hmn 50 ppm JIHTAB 25,282,43
skn-rbt 405 mg/24H MOD BIOFX* 2-5/69
skn-rbt 20 mg/24H MOD 85JCAE -,193,86
eye-rbt 1620 μg SEV AJOPAA 29,1363,46
eye-rbt 2 mg/24H SEV 85JCAE -,193,86
cyt-smc 10 mmol/tube HEREAY 33,457,47
ihl-rat TCLo:8000 ppm/7H (female 1-22D post):TER,REP TJADAB 35,56A,87
ihl-hmn TCLo:25 ppm:IRR JIHTAB 25,282,43
orl-rat LD50:790 mg/kg SAMJAF 43,795,69
ihl-rat LC50:8000 ppm/4H NPIRI* 1,10,74
ivn-rat LD50:310 mg/kg EVHPAZ 61,321,85
ipr-mus LD50:603 mg/kg 85GMAT -,28,82
ivn-mus LD50:377 mg/kg AIPTAK 135,330,62
orl-rbt LDLo:4250 mg/kg JLCMAK 10,985,25
skn-rbt LD50:3400 mg/kg NPIRI* 1,10,74

CONSENSUS REPORTS: Community Right-To-Know List. EPA Genetic Toxicology Program. Reported in EPA TSCA Inventory.

OSHA PEL: (Transitional: TWA 100 ppm) CL 50 ppm (skin)
ACGIH TLV: CL 50 ppm (skin)
DFG MAK: 100 ppm (300 mg/m^3)
DOT Classification: Flammable or Combustible Liquid; Label: Flammable Liquid.

SAFETY PROFILE: A poison by intravenous route. Moderately toxic by skin contact, ingestion, subcutaneous, and intraperitoneal routes. Human systemic effects by inhalation: conjunctiva irritation, unspecified respiratory system, and nasal effects. Experimental reproductive effects. A severe skin and eye irritant. Though animal experiments have shown the butyl alcohols to possess toxic properties, they have produced few cases of poisoning in industry probably because of their low volatility. The use of normal butyl alcohol is reported to have resulted in irritation of the eyes, with corneal inflammation, slight headache and dizziness, slight irritation of the nose and throat, and dermatitis about the fingernails and along the side of the fingers. Keratitis has also been reported. Mutation data reported. See also ALCOHOLS. Flammable liquid. Moderately explosive when exposed to flame. Incompatible with Al, chromium trioxide, oxidizing materials. To fight fire, use water spray, alcohol foam, CO_2, dry chemical. When heated to decomposition it emits acrid smoke and fumes.

BPW750 CAS:78-92-2 ***HR: 3***
sec-BUTYL ALCOHOL
DOT: UN 1120
mf: $C_4H_{10}O$ mw: 74.14

PROP: Colorless liquid. Mp: −89°, bp: 99.5°, flash p: 14°, d: 0.808 @ 20°/4°, autoign temp: 763°F, vap press: 10 mm @ 20°, vap d: 2.55, lel: 1.7% @ 212°F, uel: 9.8% @ 212°F.

SYNS: ALCOOL BUTYLIQUE SECONDAIRE (FRENCH) ◇ BUTAN-2-OL ◇ sec-BUTANOL (DOT) ◇ 2-BUTANOL ◇ BUTANOL SECONDAIRE (FRENCH) ◇ 2-BUTYL ALCOHOL ◇ BUTYLENE HYDRATE ◇ CCS 301 ◇ ETHYLMETHYL CARBINOL ◇ 2-HYDROXYBUTANE ◇ METHYLETHYLCARBINOL ◇ S.B.A.

TOXICITY DATA with REFERENCE
skn-rbt 500 mg/24H MLD 85JCAE -,193,86
eye-rbt 16 mg open AMIHBC 10,61,54
eye-rbt 100 mg/24H MOD 85JCAE -,193,86
ihl-rat TCLo:7000 ppm/7H (female 1-22D post):TER,REP TJADAB 35,56A,87
orl-rat LD50:6480 mg/kg AMIHBC 10,61,54
ihl-rat LCLo:16000 ppm/4H AMIHBC 10,61,54
ipr-rat LD50:1193 mg/kg EVHPAZ 61,321,85
ivn-rat LD50:138 mg/kg EVHPAZ 61,321,85
ipr-mus LD50:771 mg/kg SCCUR* -,2,61
ivn-mus LD50:764 mg/kg AIPTAK 135,330,62

orl-rbt LD50:4893 mg/kg IMSUAI 41,31,72
ipr-rbt LD50:277 mg/kg EVHPAZ 61,321,85

CONSENSUS REPORTS: Community Right-To-Know List. Reported in EPA TSCA Inventory.

OSHA PEL: (Transitional: TWA 150 ppm) TWA 100 ppm
ACGIH TLV: TWA 100 ppm
DFG MAK: 100 ppm (300 mg/m^3)
DOT Classification: Flammable or Combustible Liquid; Label: Flammable Liquid.

SAFETY PROFILE: Poison by intravenous and intraperitoneal routes. Mildly toxic by ingestion. Experimental reproductive effects. A skin and eye irritant. See also n-BUTYL ALCOHOL and ALCOHOLS. Dangerous fire hazard when exposed to heat or flame. Auto-oxidizes to an explosive peroxide. Ignites on contact with chromium trioxide. To fight fire, use water spray, alcohol foam, CO_2, dry chemical. Incompatible with oxidizing materials. When heated to decomposition it emits acrid smoke and fumes.

BPX000 CAS:75-65-0 ***HR: 3***
tert-BUTYL ALCOHOL
DOT: UN 1120
mf: $C_4H_{10}O$ mw: 74.14

PROP: Colorless liquid or rhombic prisms or plates. Mp: 25.3°, bp: 82.8°, flash p: 50°F (CC), d: 0.7887 @ 20°/4°, autoign temp: 896°F, vap press: 40 mm @ 24.5°, vap d: 2.55, lel: 2.4%, uel: 8.0%.

SYNS: ALCOOL BUTYLIQUE TERTIAIRE (FRENCH) ◇ tert-BUTANOL ◇ BUTANOL TERTIAIRE (FRENCH) ◇ tert-BUTYL HYDROXIDE ◇ 1,1-DIMETHYLETHANOL ◇ 2-METHYL-2-PROPANOL ◇ NCI-C55367 ◇ TRIMETHYLCARBINOL

TOXICITY DATA with REFERENCE
orl-mus TDLo:103 g/kg (female 6-20D post):REP JPETAB 222,294,82
ihl-rat TCLo:5000 ppm/7H (female 1-22D post):TER TJADAB 35,56A,87
orl-rat LD50:3500 mg/kg SCIEAS 116,663,52
ipr-mus LD50:933 mg/kg SCCUR* -,2,61
ivn-mus LD50:1538 mg/kg AIPTAK 135,330,62
orl-rbt LD50:3559 mg/kg IMSUAI 41,31,72
par-frg LDLo:12 g/kg AIPTAK 50,296,35

CONSENSUS REPORTS: Community Right-To-Know List. Reported in EPA TSCA Inventory. EPA Genetic Toxicology Program.

OSHA PEL: (Transitional: TWA 100 ppm) TWA 100 ppm; STEL 150 ppm
ACGIH TLV: TWA 100 ppm; STEL 150 ppm
DFG MAK: 100 ppm (300 mg/m^3)
DOT Classification: Flammable Liquid; Label: Flammable Liquid.

SAFETY PROFILE: Moderately toxic by ingestion, intravenous, and intraperitoneal routes. An experimental teratogen. Other experimental reproductive effects. Dangerous fire hazard when exposed to heat or flame. Moderately explosive in the form of vapor when exposed to flame. Ignites on contact with potassium-sodium alloys. To fight fire, use alcohol foam, CO_2, dry chemical. Incompatible with oxidizing materials, H_2O_2. See also n-BUTYL ALCOHOL and ALCOHOLS.

BPX500 CAS:13449-22-4 ***HR: 3***
n-BUTYL AMIDO SULFURYL AZIDE
mf: $C_4H_{10}N_4O_2S$ mw: 178.21

$(C_4H_9)NHSO_2N_3$

SAFETY PROFILE: May explode when heated. When heated to decomposition it emits toxic fumes of SO_x and NO_x. See also AZIDES.

BPX750 CAS:109-73-9 ***HR: 3***
n-BUTYLAMINE
DOT: UN 1125
mf: $C_4H_{11}N$ mw: 73.16

PROP: Liquid, ammonia-like odor. Mp: −50°, bp: 77°, flash p: 10°F (OC), 10°F (CC), d: 0.74-0.76 @ 20°/20°, autoign temp: 594°F, vap d: 2.52, lel: 1.7%, uel: 9.8%.

SYNS: 1-AMINO-BUTAAN (DUTCH) ◇ 1-AMINOBUTAN (GERMAN) ◇ 1-AMINOBUTANE ◇ 1-BUTANAMINE ◇ n-BUTILAMINA (ITALIAN) ◇ n-BUTYLAMIN (GERMAN) ◇ MONO-n-BUTYLAMINE ◇ NORVALAMINE

TOXICITY DATA with REFERENCE
cyt-rat-orl 110 mg/kg ZKKOBW 86,47,76
skn-rbt 10 mg/24H open JIHTAB 26,269,44
skn-rbt 500 mg open SEV UCDS** 7/19/65
ipr-mus TDLo:800 mg/kg:ETA BCPCA6 2,168,59
orl-rat LD50:366 mg/kg TXAPA9 63,150,82
ihl-rat LCLo:4000 ppm/4H JIHTAB 31,343,49
par-rat LDLo:600 mg/kg JPETAB 20,435,23
orl-mus LD50:430 mg/kg GISAAA 40(11),21,75
ihl-mus LC50:800 mg/m^3/2H 85GMAT -,28,82
ipr-mus LD50:629 mg/kg JPETAB 88,82,46
ivn-mus LD50:198 mg/kg JPETAB 88,82,46
orl-gpg LD50:430 mg/kg 85GMAT -,28,82
skn-rbt LD50:850 mg/kg UCDS** 7/19/65
skn-gpg LD50:370 mg/kg JIHTAB 26,269,44

CONSENSUS REPORTS: Reported in EPA TSCA Inventory.

OSHA PEL: CL 5 ppm (skin)
ACGIH TLV: CL 5 ppm
DFG MAK: 5 ppm (15 mg/m^3)
DOT Classification: Flammable Liquid; Label: Flammable Liquid.

SAFETY PROFILE: Poison by ingestion, skin contact, and intravenous routes. Moderately toxic by inhalation, intraperitoneal, and parenteral routes. A severe skin irritant. Questionable carcinogen with experimental tumorigenic data. Mutation data reported. Dangerous fire hazard when exposed to heat, flame, or oxidizing materials. To fight fire, use alcohol foam, CO_2, dry chemical. Explodes on contact with perchloryl fluoride. When heated to decomposition it emits toxic fumes of NO_x. See also AMINES.

BPY000 CAS:13952-84-6 ***HR: 3***
sec-BUTYLAMINE
DOT: UN 1125
mf: $C_4H_{11}N$ mw: 73.16

PROP: Liquid. Mp: −104°, bp: 63°, flash p: 15°F, d: 0.724 @ 20°.

SYNS: 2-AB ◇ 2-AMINOBUTANE ◇ BUTAFUME ◇ 2-BUTANAMINE ◇ DECCOTANE ◇ FRUCOTE ◇ 1-METHYLPROPYLAMINE ◇ TUTANE

TOXICITY DATA with REFERENCE
orl-rat LD50:152 mg/kg TXAPA9 63,150,82
orl-dog LD50:225 mg/kg 28ZEAL 5,33,76
skn-rbt LD50:2500 mg/kg 28ZEAL 5,33,76

CONSENSUS REPORTS: Reported in EPA TSCA Inventory.

DFG MAK: 5 ppm (15 mg/m^3)
DOT Classification: Flammable Liquid; Label: Flammable Liquid.

SAFETY PROFILE: A poison by ingestion. A powerful irritant. See also n-BUTYL AMINE and AMINES. Moderately toxic by skin contact. Dangerous fire hazard when exposed to heat or flame. To fight fire, use alcohol foam, water spray or mist, dry chemical. Incompatible with oxidizing materials. When heated to decomposition it emits toxic fumes of NO_x.

BPY250 CAS:75-64-9 ***HR: 3***
tert-BUTYLAMINE
DOT: UN 1125
mf: $C_4H_{11}N$ mw: 73.16

PROP: Colorless liquid. Mp: −67.5°, bp: 44-46°, d: 0.700 @ 15°, lel: 1.7% @ 212°F, uel: 8.9% @ 212°F, vap d: 2.5, autoign temp: 716°F.

SYNS: 2-AMINOISOBUTANE ◇ 2-AMINO-2-METHYLPROPANE ◇ BUTYLAMINE, tertiary ◇ 1,1-DIMETHYLETHYLAMINE ◇ TRIMETHYLAMINOMETHANE

TOXICITY DATA with REFERENCE
orl-rat LD50:78 mg/kg TXAPA9 63,150,82
orl-mus LD50:900 mg/kg WQCHM* 4,-,74

CONSENSUS REPORTS: Reported in EPA TSCA Inventory.

DFG MAK: 5 ppm (15 mg/m^3)
DOT Classification: Flammable Liquid; Label: Flammable Liquid.

SAFETY PROFILE: Poison by ingestion. See also BUTYLAMINE and AMINES. Very dangerous fire hazard when exposed to heat or flame. Very exothermic reaction with 2,2-dibromo-1,3-dimethylcyclopropanoic acid. To fight fire, use alcohol foam. When heated to decomposition it emits toxic fumes of NO_x.

BPY500 CAS:77966-25-7 ***HR: 3***
2-(BUTYLAMINO)-p-ACETOPHENETIDIDE HYDROCHLORIDE
mf: $C_{14}H_{22}N_2O_2 \cdot ClH$ mw: 286.84

SYN: C 5414

TOXICITY DATA with REFERENCE
eye-rbt 2% MLD ARZNAD 8,407,58
ipr-rat LD50:220 mg/kg ARZNAD 8,407,58
scu-mus LD50:800 mg/kg ARZNAD 8,407,58

SAFETY PROFILE: Poison by intraperitoneal route. Moderately toxic by subcutaneous route. An eye irritant. When heated to decomposition it emits very toxic fumes of HCl and NO_x.

BPY625 CAS:78907-16-1 ***HR: 3***
3-(tert-BUTYLAMINO)ACETYLINDOLE HYDROCHLORIDE HYDRATE
mf: $C_{14}H_{18}N_2O \cdot ClH \cdot H_2O$ mw: 284.82

SYN: 3-((tert-BUTYLAMINO)ACETYL)INDOLEHYDROCHLORIDE HYDRATE

TOXICITY DATA with REFERENCE
orl-mus LD50:410 mg/kg PCJOAU 15,412,81
scu-mus LD50:275 μg/kg PCJOAU 15,412,81
ivn-mus LD50:90 mg/kg PCJOAU 15,412,81

SAFETY PROFILE: Poison by intravenous and subcutaneous routes. Moderately toxic by ingestion. When heated to decomposition it emits toxic fumes of NO_x and HCl.

BPZ000 CAS:94-25-7 ***HR: 3***
BUTYL-p-AMINOBENZOATE
mf: $C_{11}H_{15}NO_2$ mw: 193.27

PROP: Yellow, amorphous powder; mp: 55°.

SYNS: p-AMINOBENZOIC ACID BUTYL ESTER ◇ BUTAMBEN

TOXICITY DATA with REFERENCE
ipr-mus LD50:67 mg/kg JMCMAR 17,900,74

CONSENSUS REPORTS: Reported in EPA TSCA Inventory.

SAFETY PROFILE: Poison by intraperitoneal route. An allergen. See also ESTERS and AMINES. Combustible when exposed to heat or flame. When heated to decomposition it emits toxic fumes such as NO_x.

BQA000 CAS:16488-48-5 ***HR: 3***
p-BUTYLAMINOBENZOIC ACID-2-(DIETHYLAMINO)ETHYL ESTER MONOHYDROCHLORIDE
mf: $C_{17}H_{28}N_2O_2{\cdot}ClH$ mw: 328.93

SYNS: BENZOE-DIAETHYL (GERMAN) ◇ HYDROCHLORID SALZ desp-N-n-BUTYLAMINO-BENZOESAURE-DIAETHYLAMINOAETHYLESTERS (GERMAN)

TOXICITY DATA with REFERENCE
ipr-rat LD50:27 mg/kg ARZNAD 1,218,51
scu-rat LD50:22500 μg/kg OYYAA2 9,413,75
ivn-rat LD50:4 mg/kg ARZNAD 1,218,51
orl-mus LD50:156 mg/kg OYYAA2 9,413,75
scu-mus LD50:140 mg/kg ARZNAD 1,218,51
ivn-mus LD50:4900 μg/kg OYYAA2 9,413,75
ivn-rbt LD50:2400 μg/kg OYYAA2 9,413,75

SAFETY PROFILE: Poison by ingestion, subcutaneous, intravenous, and intraperitoneal routes. When heated to decomposition it emits toxic fumes of NO_x and HCl. See also AMINES and ESTERS.

BQA010 CAS:94-24-6 ***HR: 3***
p-(BUTYLAMINO)BENZOIC ACID-2-(DIMETHYLAMINO)ETHYL ESTER
mf: $C_{15}H_{24}N_2O_2$ mw: 264.41

SYNS: AMETHOCAINE ◇ ANETAIN ◇ p-BUTYLAMINOBENZOYL-2-DIMETHYLAMINOETHANOL ◇ CONTRALGIN ◇ DICAIN ◇ DICAINE ◇ DIKAIN ◇ DIMETHYLAMINOETHYL-p-BUTYL-AMINOBENZOATE ◇ 2-DIMETHYLAMINOETHYL-p-BUTYLAMINOBENZOATE ◇ FISSUCAIN ◇ INTERCAIN ◇ LANDOCAINE ◇ LAUDOCAINE ◇ MEDICAINE ◇ MEDIHALER-TETRACAINE ◇ MEETHOBALM ◇ METRASPRAY ◇ MUCAESTHIN ◇ NIPHANOID ◇ PANTOCAINE ◇ PONTOCAINE ◇ REXOCAINE ◇ TETRACAINE ◇ UROMUCAESTHIN

TOXICITY DATA with REFERENCE
dnd-esc 30 μmol/L MUREAV 89,95,81
dns-hmn:hla 1 μmol/L BCPCA6 14,205,65
par-man LDLo:1 mg/kg:CNS,PUL SAVEAB 10,50,39
ivn-rat LD50:6 mg/kg ARZNAD 8,539,58
ipr-mus LD50:20 mg/kg RPTOAN 35(3),114,72
scu-mus LD50:25 mg/kg PHTXA6 20,521,57
ivn-mus LD50:6 mg/kg EJMCA5 10,291,75
scu-rbt LDLo:20 mg/kg AEPPAE 160,53,31
ivn-rbt LDLo:6 mg/kg AEPPAE 160,53,31
par-rbt LD50:33500 μg/kg ARZNAD 26,78,76
itr-rbt LD50:6500 μg/kg ARZNAD 26,78,76
par-frg LDLo:200 mg/kg AEPPAE 168,447,32

SAFETY PROFILE: A human poison by parenteral route with systemic effects including: muscle contractions, coma, and cyanosis. A poison experimentally by intravenous, parenteral, intratracheal, intraperitoneal, and subcutaneous routes. Human mutation data reported. A local anesthetic. See also ESTERS. When heated to decomposition it emits toxic fumes of NO_x.

BQA500 CAS:77791-55-0 ***HR: 2***
2-(BUTYLAMINO)-2′-CHLOROACETANILIDE HYDROCHLORIDE
mf: $C_{12}H_{17}ClN_2O{\cdot}ClH$ mw: 277.22

SYN: C 5413

TOXICITY DATA with REFERENCE
eye-rbt 2% MLD ARZNAD 8,407,58
ipr-rat LD50:670 mg/kg ARZNAD 8,407,58
scu-mus LD50:1075 mg/kg ARZNAD 8,407,58

SAFETY PROFILE: Moderately toxic by intraperitoneal and subcutaneous routes. An eye irritant. When heated to decomposition it emits very toxic fumes of Cl^-, NO_x, and HCl.

BQA750 CAS:6027-28-7 ***HR: 3***
2-(BUTYLAMINO)-6′-CHLORO-o-ACETOTOLUIDIDE MONOHYDROCHLORIDE
mf: $C_{13}H_{19}ClN_2O{\cdot}ClH$ mw: 291.25

SYNS: BUTANILICAINE HYDROCHLORIDE ◇ 2-(BUTYLAMINO)-N-(2-CHLORO-6-METHYLPHENYL)ACETAMIDEHYDROCHLORIDE ◇ HOSTACAIN ◇ HOSTACAINE ◇ HOSTACAINE HYDROCHLORIDE ◇ HOSTACAIN HYDROCHLORIDE

TOXICITY DATA with REFERENCE
eye-rbt 2% MLD ARZNAD 8,407,58
ipr-rat LD50:259 mg/kg ARZNAD 8,407,58
ipr-mus LD50:363 mg/kg ARZNAD 8,407,58
scu-mus LD50:570 mg/kg ARZNAD 8,181,58

SAFETY PROFILE: Poison by intraperitoneal route. Moderately toxic by subcutaneous route. An eye irritant. When heated to decomposition it emits very toxic fumes of Cl^-, NO_x, and HCl.

BQB000 CAS:5915-41-3 ***HR: 2***
2-tert-BUTYLAMINO-4-CHLORO-6-ETHYLAMINO-s-TRIAZINE
mf: $C_9H_{16}ClN_5$ mw: 229.75

SYNS: 2-tert-BUTYLAMINO-4-AETHYLAMINO-6-CHLOR-1,3,5-TRIAZIN (GERMAN) ◇ GARDOPRIM ◇ GS 13529 ◇ PRIMATOL-M80 ◇ SORGOPRIM ◇ TERBUTHYLAZINE ◇ TURBULETHYLAZIN (GERMAN)

TOXICITY DATA with REFERENCE
orl-rat LD50:1845 mg/kg GUCHAZ 6,60,73
unr-rat LD50:2500 mg/kg 30ZDA9 -,437,71

CONSENSUS REPORTS: Reported in EPA TSCA Inventory.

SAFETY PROFILE: Moderately toxic by ingestion and possibly other routes. When heated to decomposition it emits very toxic fumes of Cl^- and NO_x.

BQB250 CAS:15148-80-8 ***HR: 3***
1-(tert-BUTYLAMINO)-3-(2-CHLORO-5-METHYLPHENOXY)-2-PROPANOL HYDROCHLORIDE
mf: $C_{14}H_{22}ClNO_2 \cdot ClH$ mw: 308.28

SYNS: BETADRENOL ◇ BETADRENOL HYDROCHLORIDE ◇ BUPRANOLOL HYDROCHLORIDE ◇ 1-(2-CHLORO-5-METHYLPHENOXY)-3-((1,1-DIMETHYLETHYL)AMINO)-2-PROPANOLHYDROCHLORIDE ◇ KL 255 ◇ (−)-KL 255 ◇ SKF 16805A

TOXICITY DATA with REFERENCE
orl-rat TDLo:240 mg/kg (9-14D preg):REP OYYAA2 7,76,73
orl-rat TDLo:900 mg/kg (female 9-14D post):TER OYYAA2 7,65,73
orl-rat LD50:518 mg/kg NIIRDN 6,682,82
ipr-rat LD50:96 mg/kg NIIRDN 6,682,82
scu-rat LD50:630 mg/kg OYYAA2 7,75,73
orl-mus LD50:329 mg/kg NIIRDN 6,682,82
scu-mus LD50:567 mg/kg OYYAA2 7,75,73
ivn-mus LD50:39 mg/kg NIIRDN 6,682,82
orl-dog LD50:438 mg/kg NIIRDN 6,682,82
orl-rbt LD50:895 mg/kg NIIRDN 6,682,82
ivn-rbt LD50:15300 μg/kg NIIRDN 6,682,82

SAFETY PROFILE: Poison by ingestion, intraperitoneal, and intravenous routes. Moderately toxic by subcutaneous route. An experimental teratogen. Other experimental reproductive effects. When heated to decomposition it emits very toxic fumes of NO_x and Cl^-.

BQB825 CAS:81994-68-5 ***HR: 3***
4-BUTYLAMINO-N-(2-(DIETHYLAMINO)ETHYL) PHTHALIMIDE HYDROCHLORIDE
mf: $C_{18}H_{27}N_3O_2 \cdot ClH$ mw: 353.94

SYN: 5-BUTYLAMINO-2-(2-DIETHYLAMINOETHYL)-1H-ISOINDOLE-1,3(2H)-DIONE HYDROCHLORIDE

TOXICITY DATA with REFERENCE
orl-rat LD50:580 mg/kg EJMCA5 16,59,81
ipr-rat LD50:66 mg/kg EJMCA5 16,59,81
scu-rat LD50:130 mg/kg EJMCA5 16,59,81
ivn-rat LD50:6200 μg/kg EJMCA5 16,59,81
orl-mus LD50:312 mg/kg EJMCA5 16,59,81
ipr-mus LD50:71 mg/kg EJMCA5 16,59,81
scu-mus LD50:67 mg/kg EJMCA5 16,59,81
ivn-mus LD50:3700 μg/kg EJMCA5 16,59,81

SAFETY PROFILE: Poison by ingestion, subcutaneous, intravenous, and intraperitoneal routes. When heated to decomposition it emits toxic fumes of HCl and NO_x.

BQC000 CAS:111-75-1 ***HR: 2***
2-BUTYLAMINOETHANOL
mf: $C_6H_{15}NO$ mw: 117.22

PROP: Liquid. Bp: 192°, flash p: 170°F (OC), d: 0.89, vap d: 4.03.

SYN: 2-n-BUYTLAMINOETHANOL

TOXICITY DATA with REFERENCE
skn-rbt 10 mg/24H open AMIHBC 10,61,54
eye-rbt 250 μg open SEV AMIHBC 10,61,54
orl-rat LD50:1150 mg/kg AMIHBC 10,61,54
ipr-rat LD50:840 mg/kg TXAPA9 12,486,68
orl-mam LD50:7100 mg/kg TXAPA9 8,344,66

CONSENSUS REPORTS: Reported in EPA TSCA Inventory.

SAFETY PROFILE: Moderately toxic by ingestion and intraperitoneal routes. A skin and severe eye irritant. See also AMINES. Combustible when exposed to heat or flame. To fight fire, use alcohol foam, foam, CO_2, dry chemical. Incompatible with oxidizing materials. When heated to decomposition it emits toxic fumes of NO_x.

BQC250 CAS:26259-45-0 ***HR: 2***
2-sec-BUTYLAMINO-4-ETHYLAMINO-6-METHOXY-s-TRIAZINE
mf: $C_{10}H_{19}N_5O$ mw: 225.34

SYNS: 2-sec-BUTYLAMINO-4-ETHYLAMINO-6-METHOXY-1,3,5-TRIAZINE ◇ ETAZIN ◇ ETAZINE ◇ GEIGY G.S. 14254 ◇ GS 15254 ◇ 2-METHOXY-4-sec-BUTYLAMINO-6-AETHYLAMINO-s-TRIAZIN(GERMAN) ◇ SUMITOL ◇ SUMITOL 80W

TOXICITY DATA with REFERENCE
eye-rbt 35 mg SEV CIGET* -,-,77
orl-rat LD50:1000 mg/kg FMCHA2 -,C224,83

CONSENSUS REPORTS: Reported in EPA TSCA Inventory.

SAFETY PROFILE: Moderately toxic by ingestion. A severe eye irritant. An herbicide. See also AMINES. When heated to decomposition it emits toxic fumes of NO_x.

BQC500 CAS:33693-04-8 ***HR: 2***
2-tert-BUTYLAMINO-4-ETHYLAMINO-6-METHOXY-s-TRIAZINE
mf: $C_{10}H_{19}N_5O$ mw: 225.34

SYNS: 2-tert-BUTYLAMINO-4-ETHYLAMINO-6-METHOXY-1,3,5-TRIAZINE ◇ CARAGARD ◇ GS 14259 ◇ 2-METHOXY-4-tert-BUTYLAMINO-6-AETHYLAMINO-s-TRIAZIN (GERMAN) ◇ TERBUMETON

TOXICITY DATA with REFERENCE
orl-rat LD50:483 mg/kg GUCHAZ 6,62,73

CONSENSUS REPORTS: Reported in EPA TSCA Inventory. EPA Genetic Toxicology Program.

SAFETY PROFILE: Moderately toxic by ingestion. An herbicide. See also AMINES. When heated to decomposition it emits toxic fumes of NO_x.

BQC750 CAS:886-50-0 ***HR: 2***
2-tert-BUTYLAMINO-4-ETHYLAMINO-6-METHYLMERCAPTO-s-TRIAZINE
mf: $C_{10}H_{19}N_5S$ mw: 241.40

SYNS: 4-AETHYLAMINO-2-tert-BUTYLAMINO-6-METHYLTHIO-s-TRIAZIN (GERMAN) ◇ 2-tert-BUTYLAMINO-4-ETHYLAMINO-6-METHYLTHIO-s-TRIAZINE ◇ 2-METHYLTHIO-4-ETHYLAMINO-6-tert-BUTYLAMINO-s-TRIAZINE

TOXICITY DATA with REFERENCE
skn-rbt 380 mg open MLD CIGET* -,-,77
eye-rbt 76 mg MOD CIGET* -,-,77
orl-rat LD50:2045 mg/kg PESTD5 17,351,76
ipr-rat LD50:699 mg/kg PESTD5 17,351,76
orl-mus LD50:3884 mg/kg PESTD5 17,351,76
ipr-mus LD50:554 mg/kg PESTD5 17,351,76
orl-ckn LD50:4000 mg/kg 31ZOAD 1,56,68
unr-mam LD50:2900 mg/kg 30ZDA9 -,438,71

CONSENSUS REPORTS: EPA Extremely Hazardous Substances List.

SAFETY PROFILE: Moderately toxic by ingestion, intraperitoneal and possibly other routes. A skin and eye irritant. An herbicide. When heated to decomposition it emits very toxic fumes of NO_x and SO_x. See also MERCAPTANS and AMINES.

BQD000 CAS:54340-62-4 ***HR: 3***
2-tert-BUTYLAMINO-1-(7-ETHYL-2-BENZOFURANYL)ETHANOL HYDROCHLORIDE
mf: $C_{16}H_{23}NO_2$•ClH mw: 297.86

SYNS: BUFURALOL ◇ 1-(7-ETHYLBENZOFURAN-2-YL)-2-tert-BUTYLAMINO-1-HYDROXYETHANEHYDROCHLORIDE

TOXICITY DATA with REFERENCE
orl-rat LD50:750 mg/kg ARZNAD 27,1410,77
scu-rat LD50:1400 mg/kg ARZNAD 27,1410,77
orl-mus LD50:177 mg/kg ARZNAD 27,1410,77
ipr-mus LD50:88 mg/kg ARZNAD 27,1410,77
ivn-mus LD50:30 mg/kg ARZNAD 27,1410,77

SAFETY PROFILE: Poison by ingestion, intraperitoneal, and intravenous routes. Moderately toxic by subcutaneous route. When heated to decomposition it emits very toxic fumes of Cl^- and NO_x. See also AMINES.

BQD125 ***HR: 3***
3-(2-(tert-BUTYLAMINO)ETHYL)-6-HYDROXYBENZYL ALCOHOL SULFATE (2:1)
mf: $C_{26}H_{42}N_2O_4$•O_4S mw: 542.76

TOXICITY DATA with REFERENCE
ipr-rat LD50:295 mg/kg IYKEDH 9,222,78
ivn-rat LD50:59 mg/kg IYKEDH 9,222,78
orl-mus LD50:4750 mg/kg IYKEDH 9,222,78
ipr-mus LD50:239 mg/kg IYKEDH 9,222,78
scu-mus LD50:737 mg/kg IYKEDH 9,222,78
ivn-mus LD50:49 mg/kg IYKEDH 9,222,78

SAFETY PROFILE: Poison by intravenous and intraperitoneal routes. Moderately toxic by subcutaneous route. When heated to decomposition it emits toxic fumes of SO_x and NO_x. See also SULFATES.

BQD250 CAS:3775-90-4 ***HR: 3***
tert-BUTYL AMINO ETHYL METHACRYLATE
mf: $C_{10}H_{19}NO_2$ mw: 185.30

PROP: Liquid; bp: 100-105°; d: 0.914. flash p: 205°F (OC).

SYNS: AGEFLEX FM-4 ◇ 2-(tert-BUTYLAMINO)ETHYL METHACRYLATE

TOXICITY DATA with REFERENCE
ipr-mus LD50:174 mg/kg JDREAF 51,526,72

CONSENSUS REPORTS: Reported in EPA TSCA Inventory.

SAFETY PROFILE: Poison by intraperitoneal route. See also ESTERS and AMINES. Combustible when exposed to heat or flame. To fight fire, use alcohol foam, water spray or mist, dry chemical. When heated to decomposition it emits toxic fumes of NO_x.

BQD500 CAS:34866-46-1 ***HR: 3***
(5-(2-(tert-BUTYLAMINO)-1-HYDROXYETHYL)-2-HYDROXYPHENYL)UREA HYDROCHLORIDE
mf: $C_{13}H_{21}N_3O_3$•ClH mw: 303.83

SYN: CARBUTEROL HYDROCHLORIDE

TOXICITY DATA with REFERENCE
ivn-rat LD50:87 mg/kg JPETAB 189,167,74
orl-mus LD50:3543 mg/kg JPETAB 189,167,74
ivn-mus LD50:37 mg/kg JPETAB 189,167,74
scu-gpg LD50:473 mg/kg JPETAB 189,167,74

SAFETY PROFILE: Poison by intravenous route. Moderately toxic by ingestion and subcutaneous routes. When heated to decomposition it emits very toxic fumes of HCl and NO_x. See also AMINES.

BQE000 CAS:68377-91-3 ***HR: 3***
(±)-2-(3′-tert-BUTYLAMINO-2′-HYDROXYPROPYLTHIO)-4-(5′-CARBAMOYL-2′-THIENYL) THIAZOLE HYDROCHLORIDE
mf: $C_{15}H_{21}N_3O_2S_3$•ClH mw: 408.03

SYN: S 596

TOXICITY DATA with REFERENCE
orl-rat LD50:86 mg/kg DRFUD4 4,442,79
orl-mus LD50:5000 mg/kg DRFUD4 4,442,79
ipr-mus LD50:360 mg/kg DRFUD4 4,442,79

SAFETY PROFILE: Poison by ingestion and intraperitoneal routes. When heated to decomposition it emits very toxic fumes of NO_x, SO_x, and HCl.

BQE250 CAS:56776-01-3 ***HR: 3***
α-(tert-BUTYLAMINO)METHYL-2-CHLOROBENZYL ALCOHOL HYDROCHLORIDE
mf: $C_{12}H_{18}ClNO$•ClH mw: 264.22

SYNS: α-((tert-BUTYLAMINO)METHYL)-o-CHLOROBENZYLALCOHOL HYDROCHLORIDE ◇ C 78 ◇ o-CHLORO-α-((tert-BUTYLAMINO) METHYL)BENZYLALCOHOL HYDROCHLORIDE ◇ 1-(o-CHLOROPHENYL)-2-tert-BUTYLAMINO ETHANOL HYDROCHLORIDE ◇ LOBUTEROL ◇ TOLUBUTEROL HYDROCHLORIDE

TOXICITY DATA with REFERENCE
orl-rat TDLo:825 mg/kg (female 7-17D post):REP KSRNAM 11,439,77
orl-rat LD50:780 mg/kg DRFUD4 1,217,76
ipr-rat LD50:104 mg/kg ARZNAD 25,1028,75
scu-rat LD50:349 mg/kg ARZNAD 25,1028,75
ivn-rat LD50:42 mg/kg YAKUD5 23,1107,81
orl-mus LD50:243 mg/kg DRFUD4 1,217,76
ipr-mus LD50:76 mg/kg ARZNAD 25,1028,75
scu-mus LD50:121 mg/kg IYKEDH 12,933,81
ivn-mus LD50:40 mg/kg ARZNAD 25,1028,75
orl-dog LD50:300 mg/kg ARZNAD 27,1439,77

SAFETY PROFILE: Poison by ingestion, intraperitoneal, subcutaneous, and intravenous routes. Experimental reproductive effects. A bronchodilator. When heated to decomposition it emits very toxic fumes of Cl^- and NO_x. See also AMINES.

BQF250 CAS:3703-79-5 ***HR: 3***
α-((BUTYLAMINO)METHYL)-p-HYDROXYBENZYL ALCOHOL
mf: $C_{12}H_{19}NO_2$ mw: 209.32

SYNS: BAMETHANE ◇ BUTEDRINE ◇ 2-BUTYLAMINO-1-p-HYDROXYPHENYLETHANOL ◇ α-((BUTYLAMINO)METHYL)-4-HYDROXYBENZENEMETHANOL ◇ BUTYL-NOR-SYMPATOL ◇ n-BUTYLNORSYMPATHOL ◇ n-BUTYLNORSYNEPHRINE ◇ BUTYL-S- YMPATHOL ◇ 1-(p-HYDROXYPHENYL)-2-BUTYLAMINOETHANOL ◇ 1-(4-HYDROXYPHENYL)-1-HYDROXY-2-BUTYLAMINOETHANE

TOXICITY DATA with REFERENCE
ivn-rat LD50:80 mg/kg RPOBAR 2,272,70
orl-mus LD50:562 mg/kg RPOBAR 2,272,70
ipr-mus LD50:150 mg/kg JPETAB 89,297,47
ivn-mus LD50:72 mg/kg RPOBAR 2,271,70

SAFETY PROFILE: Poison by intravenous and intraperitoneal routes. Moderately toxic by ingestion. A vasodilator. When heated to decomposition it emits toxic fumes of NO_x. See also ALCOHOLS and AMINES.

BQF500 CAS:18559-94-9 ***HR: 3***
α′-((tert-BUTYL AMINO)METHYL)-4-HYDROXY-m-XYLENE-α,α′-DIOL
mf: $C_{13}H_{21}NO_3$ mw: 239.35

SYNS: AEORLIN ◇ AH 3365 ◇ ALBUTEROL ◇ BRONCOVALEAS ◇ 2-(tert-BUTYLAMINO)-1-(4-HYDROXY-3-HYDROXYMETHYLPHENYL)ETHANOL ◇ α-1-((tert-BUTYLAMINO)METHYL)-4-HYDROXY-m-XYLENE-α,α-DIOL ◇ α-1-(((1,1-DIMETHYLETHYL)AMINO) METHYL)-4-HYDROXY-1,3-BENZENEDIMETHANOL ◇ 4-HYDROXY-3-HYDROXYMETHYL-α-((tert-BUTYLAMINO)METHYL)BENZYLALCOHOL ◇ PROVENTIL ◇ SALBUTAMOL ◇ SOLBUTAMOL ◇ SULTANOL ◇ VENETLIN ◇ VENTOLIN

TOXICITY DATA with REFERENCE
orl-wmn TDLo:160 μg/kg (1D pre):REP RDCNBM 5,31,81
ivn-rat TDLo:3600 μg/kg (16-20D preg):TER DPTHDL 4(Suppl 1),150,82
orl-chd TDLo:1850 μg/kg:CNS,CVS BMJOAE 282,1932,81
ihl-man TCLo:36 μg/kg/6H BMJOAE 292,1430,86
ivn-hmn TDLo:6 μg/kg:CVS BMJOAE 1,365,76
orl-rat LD50:660 mg/kg USXXAM #4026897
ipr-rat LD50:295 mg/kg IYKEDH 4,193,73
ivn-rat LD50:57100 μg/kg USXXAM #4026987
ipr-mus LD50:239 mg/kg IYKEDH 4,193,73
scu-mus LD50:737 mg/kg IYKEDH 4,193,73
ivn-mus LD50:48700 μg/kg IYKEDH 4,193,73

SAFETY PROFILE: A poison by intraperitoneal and intravenous routes. Moderately toxic by ingestion and subcutaneous routes. Human cardiovascular system effects by intravenous route including arrythmias, change in heart rate and plasma or blood volume. Human (child) behavioral and cardiac effects by ingestion including tremors, excitement, and change in heart rate. Human maternal effects of the uterus, cervix, and vagina by ingestion. An experimental teratogen. Other experimental reproductive effects. A bronchodilator. When heated to decomposition it emits toxic fumes of NO_x.

BQF750 CAS:86166-58-7 ***HR: 3***
1-(tert-BUTYLAMINO)3-(3-METHYL-2-NITROPHENOXY)-2-PROPANOL
mf: $C_{13}H_{22}N_2O_4$ mw: 282.38

SYNS: dl-1-(2-NITRO-3-EMTHYLPHENOXY)-3-tert-BUTYLAMINOPROPAN-2-OL ◇ ZAMI 1305 ◇ dl-ZAMI 1305

TOXICITY DATA with REFERENCE
dni-rat:lvr 14 mmol/L CBINA8 50,77,84
oms-rat:lvr 28 mmol/L CBINA8 50,77,84
oms-rat-ipr 300 mg/kg/6D CBINA8 52,203,84
dni-rat-ipr 100 mg/kg TOPADD 13,18,85
orl-rat TDLo:9 g/kg/26W-C:CAR JJIND8 68,669,82
orl-rat TD:18 g/kg/26W-C:CAR JJIND8 68,669,82

SAFETY PROFILE: Questionable carcinogen with experimental carcinogenic data. Mutation data reported. When heated to decomposition it emits toxic fumes of NO_x.

BQF825 CAS:102071-76-1 ***HR: 3***
2-(BUTYLAMINO)-2-METHYL-1-PROPANOL BENZOATE HYDROCHLORIDE
mf: $C_{15}H_{23}NO_2$•ClH mw: 285.85

SYN: 2-(BUTYLAMINO)-2-METHYL-1-PROPANOL BENZOATE (ester) HYDROCHLORIDE

TOXICITY DATA with REFERENCE
ipr-mus LD50:230 mg/kg AIPTAK 115,483,58
scu-mus LD50:305 mg/kg AIPTAK 115,483,58
ivn-mus LD50:21 mg/kg AIPTAK 115,483,58

SAFETY PROFILE: Poison by subcutaneous, intravenous, and intraperitoneal routes. When heated to decomposition it emits toxic fumes of NO_x and HCl.

BQG250 ***HR: 3***
2-(BUTYLAMINO)-N-METHYL-N-(1-(2,6-XYLYLOXY)-2-PROPYL) ACETAMIDE HYDROCHLORIDE

SYN: C 6259

TOXICITY DATA with REFERENCE
eye-rbt 2% SEV ARZNAD 9,70,59
scu-mus LD50:170 mg/kg ARZNAD 9,70,59

SAFETY PROFILE: Poison by subcutaneous route. A severe eye irritant. When heated to decomposition it emits very toxic fumes of NO_x and HCl.

BQG500 CAS:102585-37-5 ***HR: 3***
2-(sec-BUTYLAMINO)-N-METHYL-N-(1-(2,4-XYLYLOXY)-2-PROPYL)ACETAMIDE HYDROCHLORIDE
mf: $C_{18}H_{30}N_2O_2$•ClH mw: 342.96

SYN: C 6260

TOXICITY DATA with REFERENCE
eye-rbt 2% SEV ARZNAD 9,70,59
scu-mus LD50:180 mg/kg ARZNAD 9,70,59

SAFETY PROFILE: Poison by subcutaneous route. A severe eye irritant. When heated to decomposition it emits very toxic fumes of NO_x and HCl.

BQG750 CAS:102585-38-6 ***HR: 3***
2-(BUTYLAMINO)-N-(1-PHENOXY-2-PROPYL) ACETAMIDE HYDROCHLORIDE
mf: $C_{15}H_{24}N_2O_2$•ClH mw: 300.87

SYN: C 6257

TOXICITY DATA with REFERENCE
eye-rbt 2% MLD ARZNAD 9,70,59
scu-mus LD50:245 mg/kg ARZNAD 9,70,59

SAFETY PROFILE: Poison by subcutaneous route. An eye irritant. When heated to decomposition it emits very toxic fumes of HCl and NO_x.

BQG850 CAS:78907-15-0 ***HR: 3***
3-(tert-BUTYLAMINO)PROPIONYLINDOLE HYDROCHLORIDE HYDRATE
mf: $C_{15}H_{20}N_2O$•ClH•H_20 mw: 298.85

SYNS: 2-(tert-BUTYLAMINO)-1-(3-INDOLYL)-1-PROPANONEHYDROCHLORIDE HYDRATE ◇ 2-(tert-BUTYLAMINO)-1-(3-INDOLYL)-1-PROPANONE MONOHYDROCHLORIDE, MONOHYDRATE

TOXICITY DATA with REFERENCE
orl-mus LD50:515 μg/kg PCJOAU 15,412,81
scu-mus LD50:315 μg/kg PCJOAU 15,412,81
ivn-mus LD50:95 mg/kg PCJOAU 15,412,81

SAFETY PROFILE: Poison by subcutaneous and intravenous routes. Moderately toxic by ingestion. When heated to decomposition it emits toxic fumes of NO_x and HCl.

BQH250 CAS:528-97-2 ***HR: 3***
p-BUTYLAMINO SALICYLIC ACID-2-(DIETHYLAMINO)ETHYL ESTER HYDROCHLORIDE
mf: $C_{17}H_{28}N_2O_3$•ClH mw: 344.93

SYNS: BRONCHIOCAIN ◇ BRONCHOCAIN ◇ BRONCHOCAINE ◇ 4-(BUTYLAMINO)SALICYLIC ACID 2-(DIETHYLAMINO)ETHYL ESTER HYDROCHLORIDE ◇ 4-(BUTYLAMINO)-SALICYLIC ACID 2-(DIETHYLAMINO)ETHYL ESTER MONOHYDROCHLORIDE ◇ C 4208 ◇ HCl SALZ DES p,N,N-BUTYLAMINOSALICYLSAEUREDIAETHYLAMINOAETHYLESTER (GERMAN) ◇ PARAESIN ◇ PHENOCAINE ◇ S 650 ◇ SALICYL-DIAETHYL (GERMAN) ◇ WOFACAIN A

TOXICITY DATA with REFERENCE
eye-rbt 2% SEV ARZNAD 8,708,58
ipr-rat LD50:62 mg/kg ARZNAD 1,218,51
ipr-mus LD50:12 mg/kg ARZNAD 1,218,51
scu-mus LD50:120 mg/kg ARZNAD 8,708,58
ivn-mus LD50:16 mg/kg ARZNAD 1,218,51

SAFETY PROFILE: A poison via intraperitoneal, subcutaneous, and intravenous routes. A severe eye irritant. See also AMINES and ESTERS. When heated to decomposition, it emits very toxic fumes of NO_x and HCl.

BQH500 CAS:17284-75-2 ***HR: 3***
p-BUTYLAMINOSALICYLIC ACID-2-(DIMETHYLAMINO)ETHYL ESTER HYDROCHLORIDE
mf: $C_{15}H_{24}N_2O_3 \cdot ClH$ mw: 316.87

SYNS: C 4207 ◇ SALICYL-DIMETHYL (GERMAN)

TOXICITY DATA with REFERENCE
eye-rbt 2% SEV ARZNAD 8,708,58
ipr-rat LD50:90 mg/kg ARZNAD 1,218,51
ivn-rat LD50:12 mg/kg ARZNAD 1,218,51
scu-mus LD50:130 mg/kg ARZNAD 8,708,58
ivn-mus LD50:30 mg/kg ARZNAD 1,218,51

SAFETY PROFILE: Poison by subcutaneous, intraperitoneal, and intravenous routes. A severe eye irritant. See also AMINES and ESTERS. When heated to decomposition it emits very toxic fumes of NO_x and HCl.

BQH750 CAS:78308-37-9 ***HR: 3***
p-BUTYLAMINOSALICYLIC ACID-1-ETHYL-4-PIPERIDYL ESTER HYDROCHLORIDE
mf: $C_{18}H_{28}N_2O_3 \cdot ClH$ mw: 356.94

SYN: C 4211

TOXICITY DATA with REFERENCE
eye-rbt 2% SEV ARZNAD 8,708,58
scu-mus LD50:57 mg/kg ARZNAD 8,708,58

SAFETY PROFILE: Poison by subcutaneous route. A severe eye irritant. See also AMINES and ESTERS. When heated to decomposition it emits very toxic fumes of Cl^- and NO_x.

BQH800 CAS:7532-60-7 ***HR: 3***
1-(BUTYLAMINO)-3-p-TOLUIDINO-2-PROPANOL
mf: $C_{14}H_{24}N_2O$ mw: 236.40

SYN: 1-(BUTYLAMINO)-3-((4-METHYLPHENYL)AMINO)-2-PROPANOL (9CI)

TOXICITY DATA with REFERENCE
ipr-rat LDLo:7600 μg/kg JPETAB 107,250,53
orl-mus LDLo:20 mg/kg JPETAB 107,250,53
ipr-mus LD50:12400 μg/kg JPETAB 109,407,53
ivn-mus LDLo:3 mg/kg JPETAB 107,250,53

SAFETY PROFILE: Poison by ingestion, intravenous, and intraperitoneal routes. When heated to decomposition it emits toxic fumes of NO_x.

BQH850 CAS:1126-78-9 ***HR: 3***
N-BUTYLANILINE
DOT: UN 2738
mf: $C_{10}H_{15}N$ mw: 149.26

SYNS: N-(n-BUTYL)ANILINE ◇ N-n-BUTYLANILINE (DOT) ◇ N-BUTYLBENZENAMINE (9CI) ◇ 4-(PHENYLAMINO)BUTANE

TOXICITY DATA with REFERENCE
skn-rbt 10 mg/24H open SEV AMIHBC 10,61,54
eye-rbt 500 mg open AMIHBC 10,61,54
orl-rat LD50:1620 mg/kg AMIHBC 10,61,54
skn-rbt LD50:5990 mg/kg AMIHBC 10,61,54
unr-mam LD50:282 mg/kg GISAAA 48(6),22,83

CONSENSUS REPORTS: Reported in EPA TSCA Inventory.

DOT Classification: Poison B; Label: Poison.

SAFETY PROFILE: Poison by an unspecified route. Moderately toxic by skin contact and ingestion. A severe skin and eye irritant. When heated to decomposition it emits toxic fumes of NO_x. See also ANILINE DYES.

BQI000 CAS:25013-16-5 ***HR: 3***
BUTYLATED HYDROXYANISOLE
mf: $C_{11}H_{16}O_2$ mw: 180.27

PROP: White waxy solid; faint characteristic odor. Mp: 48-63°. Sol in alc and propylene glycol; insol in water.

SYNS: ANTRANCINE 12 ◇ BHA (FCC) ◇ BUTYLHYDROXYANISOLE ◇ tert-BUTYLHYDROXYANISOLE ◇ tert-BUTYL-4-HYDROXYANISOLE ◇ 2(3)-tert-BUTYL-4-HYDROXYANISOLE ◇ BUTYLOHYDROKSYANIZOL (POLISH) ◇ EMBANOX ◇ FEMA No. 2183 ◇ NIPANTIOX 1-F ◇ PREMERGE PLUS ◇ SUSTANE ◇ SUSTANE 1-F ◇ TENOX BHA ◇ VERTAC

TOXICITY DATA with REFERENCE
mmo-omi 12500 μg/L FMLED7 14,183,82
sce-ham:fbr 100 μmol/L JNCIAM 58,1635,77
orl-mus TDLo:12600 mg/kg (female 1-21D post):REP FEPRA7 31,596,72
orl-rat TDLo:728 g/kg/2Y-C:CAR GANNA2 73,332,82
orl-mus TDLo:874 g/kg/1Y-C:ETA JJCREP 77,1083,86
orl-ham TDLo:437 g/kg/1Y-C:CAR JJCREP 77,1083,86
orl-ham TD:202 g/kg/24W-C:NEO GANNA2 74,459,83
orl-rat LD50:2 g/kg TRENAF 22,231,70
ipr-rat LD50:881 mg/kg TOLED5 27,15,85
orl-mus LD50:1100 mg/kg TRENAF 22,231,70
orl-rbt LD50:2100 mg/kg JAOCA7 54,239,77
orl-rat LDLo:2200 mg/kg AFREAW 3,197,51
orl-mus LD50:2000 mg/kg AFREAW 3,197,51

CONSENSUS REPORTS: IARC Cancer Review: Group 2B IMEMDT 7,56,87; Animal Sufficient Evidence IMEMDT 40,123,86. Reported in EPA TSCA Inventory. EPA Genetic Toxicology Program.

SAFETY PROFILE: Suspected carcinogen with experimental carcinogenic, neoplastigenic, and tumorigenic data. Moderately toxic by ingestion and intraperitoneal routes. Experimental reproductive effects. Mutation data reported. When heated to decomposition it emits acrid and irritating fumes.

BQI010 CAS:88-32-4 ***HR: 3***
3-tert-BUTYLATED HYDROXYANISOLE
mf: $C_{11}H_{16}O_2$ mw: 180.27

SYNS: 3-tert-BHA ◇ 3-tert-BUTYL-4-METHOXYPHENOL

TOXICITY DATA with REFERENCE
orl-ham TDLo:27 g/kg/3W-C:ETA JJIND8 76,143,86

SAFETY PROFILE: Questionable carcinogen with experimental tumorigenic data. When heated to decomposition it emits acrid and irritating fumes.

BQI125 CAS:84928-98-3 ***HR: 3***
N-BUTYL-N-2-AZIDOETHYLNITRAMINE
mf: $C_6H_{13}N_5O_2$ mw: 187.20

$(C_4H_9)N(NO_2)C_2H_4N_3$

SAFETY PROFILE: An impact-sensitive explosive. When heated to decomposition it emits toxic fumes of NO_x. See also AZIDES.

BQI250 CAS:1070-19-5 ***HR: 3***
tert-BUTYL AZIDO FORMATE
mf: $C_5H_9N_3O_2$ mw: 143.17

$(CH_3)_3COCO{\bullet}N_3$

SYNS: tert-BUTOXY CARBONYL AZIDE ◇ tert-BUTYLOXYCARBONYL AZIDE ◇ CARBONAZIDIC ACID, 1,1-DIMETHYLETHYL ESTER

CONSENSUS REPORTS: Reported in EPA TSCA Inventory.

DOT Classification: Forbidden.

SAFETY PROFILE: An unstable shock- and heat-sensitive explosive. It may explode above 100°C and ignites at 143°C. When heated to decomposition it emits toxic fumes of NO_x. See also AZIDES.

BQI500 CAS:63018-64-4 ***HR: 3***
5-n-BUTYL-1,2-BENZANTHRACENE
mf: $C_{22}H_{20}$ mw: 284.42

SYN: 8-BUTYLBENZ(a)ANTHRACENE

TOXICITY DATA with REFERENCE
skn-mus TDLo:860 mg/kg/36W-I:ETA PRLBA4 129,439,40

SAFETY PROFILE: Questionable carcinogen with experimental tumorigenic data. When heated to decomposition it emits acrid smoke and irritating fumes.

BQI750 CAS:104-51-8 ***HR: 1***
n-BUTYLBENZENE
DOT: UN 2709
mf: $C_{10}H_{14}$ mw: 134.24

PROP: Colorless liquid. Mp: −81.2°, bp: 182.1°, fp: −88.2°, flash p: 160°F (TOC), d: 0.8601 @ 20°/4°, vap press: 1 mm @ 22.7°, autoign temp: 774°F, lel: 0.8%, uel: 5.8%, vap d: 4.6.

SYN: 1-PHENYLBUTANE

TOXICITY DATA with REFERENCE
orl-rat LDLo:5000 mg/kg AMIHAB 19,403,59

CONSENSUS REPORTS: Reported in EPA TSCA Inventory.

DOT Classification: Flammable or Combustible Liquid; Label: Flammable Liquid.

SAFETY PROFILE: Mildly toxic by ingestion. Flammable when exposed to heat or flame. To fight fire, use alcohol foam, CO_2, dry chemical. Incompatible with oxidizing materials. When heated to decomposition it emits acrid and irritating fumes.

BQJ000 CAS:135-98-8 ***HR: 2***
sec-BUTYLBENZENE
DOT: UN 2709
mf: $C_{10}H_{14}$ mw: 134.24

PROP: Colorless liquid. Mp: −82.7°, bp: 173.5°, fp: −75.8°, flash p: 126°F (TOC), d: 0.8621 @ 20°, vap press: 1 mm @ 18.6°, vap d: 4.62, autoign temp: 788°F, lel: 0.8%, uel: 6.9%.

SYN: 2-PHENYLBUTANE

TOXICITY DATA with REFERENCE
orl-rat LD50:2240 mg/kg TXAPA9 28,313,74

CONSENSUS REPORTS: Reported in EPA TSCA Inventory.

DOT Classification: Flammable or Combustible Liquid; Label: Flammable Liquid.

SAFETY PROFILE: Moderately toxic by ingestion. Flammable when exposed to heat or flame. To fight fire, use foam, CO_2, dry chemical, water spray or mist. Incompatible with oxidizing materials. When heated to decomposition it emits acrid smoke and fumes.

BQJ250 CAS:98-06-6 ***HR: 1***
tert-BUTYLBENZENE
DOT: UN 2709
mf: $C_{10}H_{14}$ mw: 134.24

PROP: Colorless liquid. Bp: 168.2°, fp: −58°, flash p: 140°F (TOC), d: 0.8665 @ 20°, vap press: 1 mm @ 13.0°, vap d: 4.62, autoign temp: 842°F, lel: 0.7% @ 212°F, uel: 5.7% @ 212°F.

SYNS: 2-METHYL-2-PHENYLPROPANE ◇ PSEUDOBUTYLBENZENE ◇ TRIMETHYLPHENYLMETHANE

TOXICITY DATA with REFERENCE
orl-rat LDLo:5000 mg/kg AMIHAB 19,403,59

CONSENSUS REPORTS: Reported in EPA TSCA Inventory.

DOT Classification: Flammable or Combustible Liquid; Label: Flammable Liquid.

SAFETY PROFILE: Mildly toxic by ingestion. Flammable when exposed to heat or flame. To fight fire, use foam, CO_2, dry chemical, water spray, fog, mist. Incompatible with oxidizing materials. When heated to decomposition it emits acrid smoke and fumes.

BQJ350 CAS:122-43-0 ***HR: 1***
BUTYLBENZENEACETATE
mf: $C_{12}H_{16}O_2$ mw: 192.28

SYNS: ACETIC ACID, PHENYL-, BUTYL ESTER ◇ BENZENEACETIC ACID, BUTYL ESTER (9CI) ◇ BUTYL PHENYLACETATE ◇ n-BUTYL PHENYLACETATE ◇ PHENYLETHANOIC ACID BUTYL ESTER

TOXICITY DATA with REFERENCE
skn-rbt 500 mg/24H MLD FCTOD7 21,657,83

CONSENSUS REPORTS: Reported in EPA TSCA Inventory.

SAFETY PROFILE: A skin irritant. When heated to decomposition it emits acrid smoke and irritating fumes.

BQJ500 CAS:583-03-9 ***HR: 3***
α-BUTYLBENZENEMETHANOL
mf: $C_{11}H_{16}O$ mw: 164.27

SYNS: α-BUTYLBENZYL ALCOHOL ◇ FENIPENTOL ◇ 1-HYDROXY-1-PHENYLPENTANE ◇ PANCORAL ◇ PC 1 ◇ PH BC ◇ PHENYLBUTYLCARBINOL ◇ 1-PHENYL-1-HYDROXYPENTANE ◇ PHENYLPENTANOL ◇ 1-PHENYLPENTANOL

TOXICITY DATA with REFERENCE
orl-rat TDLo:700 mg/kg (female 7-13D post):REP GNRIDX 5,357,71
orl-rat TDLo:3500 mg/kg (7-13D preg):TER GNRIDX 5,357,71
orl-rat LD50:5432 mg/kg IYKEDH 4,90,72
ipr-rat LD50:256 mg/kg NIIRDN 6,657,82
scu-rat LD50:6930 mg/kg IYKEDH 4,90,73
orl-mus LD50:2900 mg/kg OSDIAF 14,261,65
ipr-mus LD50:188 mg/kg NIIRDN 6,657,82
scu-mus LD50:3153 mg/kg IYKEDH 4,90,73

CONSENSUS REPORTS: Reported in EPA TSCA Inventory.

SAFETY PROFILE: Poison by intraperitoneal route. Moderately toxic by ingestion and subcutaneous routes. An experimental teratogen. Other experimental reproductive effects. Stimulates the production of bile by the liver. When heated to decomposition it emits acrid smoke and irritating fumes.

BQJ750 CAS:24425-13-6 ***HR: 3***
2-tert-BUTYLBENZIMIDAZOLE
mf: $C_{11}H_{14}N_2$ mw: 174.27

TOXICITY DATA with REFERENCE
mmo-sat 250 μg/plate CHIMAD 27,68,73
ivn-mus LD50:56 mg/kg CSLNX* NX#07472

SAFETY PROFILE: Poison by intravenous route. Mutation data reported. When heated to decomposition it emits toxic fumes of NO_x.

BQK000 CAS:14255-87-9 ***HR: 2***
5-BUTYL-2-BENZIMIDAZOLECARBAMIC ACID METHYL ESTER
mf: $C_{13}H_{17}N_3O_2$ mw: 247.33

SYNS: N-(BUTYL-5-BENZIMIDAZOLYL)-2-CARBAMATE de METHYLE (FRENCH) ◇ (4-BUTYL-1H-BENZIMIDAZOL-2-YL)-CARBAMIC ACID METHYL ESTER ◇ 5-BUTYL-2-(CARBOMETHOXYAMINO) BENZIMIDAZOLE ◇ HELMATAC ◇ METHYL-5-BUTYL-2-BENZIMIDAZOLECARBAMATE ◇ PARBENDAZOLE ◇ PBDZ ◇ SKF 29044 ◇ VERMINUM ◇ WORM GUARD

TOXICITY DATA with REFERENCE
oms-hmn:leu 1 mg/L THERAP 31,505,76
oms-dom:leu 1 mg/L THERAP 31,505,76
orl-dom TDLo:60 mg/kg (female 21D post):REP COVEAZ 64(Suppl 4),41,74
orl-rat TDLo:80 mg/kg (female 8-15D post):TER THERAP 31,505,76
orl-mus LD50:1700 mg/kg BSVMA8 77,379,75
orl-dom LDLo:660 mg/kg AUVJA2 46,297,70

SAFETY PROFILE: Moderately toxic by ingestion. Experimental teratogenic and reproductive effects. Human mutation data reported. An anthelminthic agent. When heated to decomposition it emits toxic fumes of NO_x. See also CARBAMATES.

BQK250 CAS:136-60-7 ***HR: 2***
BUTYL BENZOATE
mf: $C_{11}H_{14}O_2$ mw: 178.25

PROP: Liquid. Mp: −21.5°, bp: 250°, flash p: 225°F (OC), d: 1.0073 @ 20°/20°, vap press: <0.01 mm @ 20°, vap d: 6.15.

SYNS: ANTHRAPOLE AZ ◇ BENZOIC ACID-n-BUTYL ESTER ◇ n-BUTYL BENZOATE ◇ DAI CARI XBN

TOXICITY DATA with REFERENCE
skn-rbt 10 mg/24H open SEV AMIHBC 10,61,54
skn-rbt 500 mg open MOD UCDS** 10/15/58
eye-rbt 500 mg AMIHBC 10,61,54
orl-rat LD50:5140 mg/kg AMIHBC 10,61,54
skn-rbt LD50:4000 mg/kg NPIRI* 2,7,75

CONSENSUS REPORTS: Reported in EPA TSCA Inventory.

SAFETY PROFILE: Moderately toxic by skin contact. Mildly toxic by ingestion. Severe skin irritant and moderate eye irritant. Combustible when exposed to heat or flame; can react with oxidizing materials. To fight fire, use CO_2, dry chemical, water mist, fog, spray. When heated to decomposition it emits acrid and irritating fumes. See also ESTERS.

BQK500 CAS:98-73-7 ***HR: 2***
p-tert-BUTYL BENZOIC ACID
mf: $C_{11}H_{14}O_2$ mw: 178.25

PROP: Colorless, fine, crystalline powder. Mp: 166.3°, d: 1.142 @ 20°/4°.

SYN: TBBA

TOXICITY DATA with REFERENCE
skn-rat TDLo:2450 mg/kg (male 7W pre):REP JACTDZ 6(2),233,87
orl-rat LD50:700 mg/kg TSCAT* OTS0510267

CONSENSUS REPORTS: Reported in EPA TSCA Inventory.

SAFETY PROFILE: Moderately toxic by ingestion. Experimental reproductive effects. An irritant. Combustible when exposed to heat or flame. Incompatible with oxidizing materials. To fight fire, use foam, CO_2, dry chemical. When heated to decomposition it emits acrid smoke and irritating fumes.

BQK750 CAS:95-31-8 ***HR: 3***
N-tert-BUTYL-2-BENZOTHIAZOLESULFENAMIDE
mf: $C_{11}H_{14}N_2S$ mw: 206.33

SYNS: PENNAC TBBS ◇ VANNAX NS

TOXICITY DATA with REFERENCE
mma-mus:lym 40 mg/L ENMUDM 5,193,83
otr-mus:emb 35 mg/L ENMUDM 5,193,83
ivn-mus LD50:180 mg/kg CSLNX* NX#02241

CONSENSUS REPORTS: Reported in EPA TSCA Inventory.

SAFETY PROFILE: Poison by intravenous route. Mutation data reported. When heated to decomposition it emits very toxic fumes of NO_x and SO_x.

BQK850 CAS:61481-19-4 ***HR: 3***
tert-BUTYL-BICYCLOPHOSPHATE
mf: $C_9H_{20}O_2$ mw: 206.20

SYNS: 2-(tert-BUTYL)-2-(HYEROXYMETHYL)-1,3-PROPANEDIOL, CYCLIC PHOSPHATE (1:1) ◇ 4-tert-BUTYL-1-OXO-1-PHOSPHA-2,6,7-TRIOXABICYCLO(2.2.2)OCTANE ◇ 4-(tert-BUTYL)-2,6,7-TRIOXA-1-PHOSPHABICYCLO(2.2.2)OCTAN-1-ONE

TOXICITY DATA with REFERENCE
ipr-rat LD50:35 µg/kg TXAPA9 46,411,78
orl-mus LD50:45 µg/kg TXAPA9 46,411,78
ipr-mus LD50:35 µg/kg TXAPA9 46,411,78
ivn-mus LD50:120 µg/kg

SAFETY PROFILE: Poison by ingestion, intravenous, and intraperitoneal routes. When heated to decomposition it emits toxic fumes of PO_x. See also PHOSPHATES.

BQL000 CAS:1190-53-0 ***HR: 3***
N-BUTYLBIGUANIDE HYDROCHLORIDE
mf: $C_6H_{15}N_5•ClH$ mw: 193.72

SYNS: ANDERE ◇ BIFORON ◇ BIGUNAL ◇ BUFONAMIN ◇ BUFORMIN HYDROCHLORIDE ◇ BULBONIN ◇ 1-BUTYLBIGUANIDE HYDROCHLORIDE ◇ 1-BUTYLDIGUANIDE HYDROCHLORIDE ◇ N-BUTYLIMIDODICARBONIMIDIC DIAMIDE MONOHYDROCHLORIDE (9CI) ◇ DIABRIN ◇ DIBETOS ◇ GLIBUTIDE ◇ GLIPORAL ◇ INSULAMIN ◇ KREBON ◇ PANFORMIN ◇ SILUBIN ◇ SINDIATIL ◇ TIDEMOL ◇ ZIAVETINE

TOXICITY DATA with REFERENCE
orl-mus LD50:380 mg/kg ARZNAD 12,314,62
ipr-mus LD50:380 mg/kg JAJAAA 18,196,65
ivn-mus LD50:105 mg/kg ARZNAD 12,314,62

SAFETY PROFILE: A poison via ingestion, intravenous, and intraperitoneal routes. When heated to decomposition it emits very toxic fumes of HCl and NO_x.

BQL500 CAS:64037-56-5 ***HR: 3***
sec-BUTYLBIS(2-CHLOROETHYL)AMINE HYDROCHLORIDE
mf: $C_8H_{17}Cl_2N•ClH$ mw: 234.62

SYNS: sec-BUTYL-BIS(β-CHLOROETHYL)AMINE HYDROCHLORIDE ◇ N-sec-BUTYL-2,2′-DICHLORODIETHYLAMINE, HYDROCHLORIDE ◇ TL 524

TOXICITY DATA with REFERENCE
orl-rat LDLo:50 mg/kg NCNSA6 5,11,53
ipr-mus LD50:2800 µg/kg CANCAR 2,1055,49
scu-mus LDLo:2 mg/kg NDRC** No.9-4-1-9,43

SAFETY PROFILE: Poison by ingestion, intraperitoneal, and subcutaneous routes. When heated to decomposition it emits very toxic fumes of HCl and NO_x.

BQL750 CAS:64037-57-6 ***HR: 3***
tert-BUTYLBIS(β-CHLOROETHYL)AMINE HYDROCHLORIDE
mf: $C_8H_{17}Cl_2N•ClH$ mw: 234.62

SYNS: tert-BUTYLBIS(2-CHLOROETHYL)AMINEHYDROCHLORIDE ◇ N-tert-BUTYL-2,2′-DICHLORO-DIETHYLAMINE HYDROCHLORIDE ◇ TL 568

TOXICITY DATA with REFERENCE
orl-rat LDLo:75 mg/kg NCNSA6 5,11,53
ipr-rat LD50:3 mg/kg CPBTAL 8,99,60
ipr-mus LD50:1420 μg/kg CANCAR 2,1055,49
scu-mus LDLo:25 mg/kg NTIS** PB158-507

SAFETY PROFILE: Poison by ingestion, intraperitoneal, and subcutaneous routes. When heated to decomposition it emits very toxic fumes of HCl and NO_x. See also AMINES.

BQM000 CAS:102-79-4 ***HR: 2***
N-BUTYL-N,N-BIS(HYDROXY ETHYL)AMINE
mf: $C_8H_{19}NO_2$ mw: 161.28

PROP: Liquid. Bp: 262°, flash p: 245°F (OC), d: 0.97, vap d: 5.55.

SYNS: N-BUTYLDIETHANOLAMINE ◇ N-BUTYL-2,2′-IMINODIETHANOL

TOXICITY DATA with REFERENCE
skn-rbt 10 mg/24H open MLD AMIHBC 10,61,54
eye-rbt 750 μg open SEV AMIHBC 10,61,54
orl-rat LD50:4250 mg/kg AMIHBC 10,61,54

CONSENSUS REPORTS: Reported in EPA TSCA Inventory.

SAFETY PROFILE: Mildly toxic via ingestion. No chronic effects data. A skin and severe eye irritant. Combustible when exposed to heat or flame. To fight fire, use alcohol foam, foam, CO_2, dry chemical. Incompatible with oxidizing materials. When heated to decomposition it emits toxic fumes of NO_x. See also AMINES.

BQM250 CAS:507-19-7 ***HR: 3***
tert-BUTYL BROMIDE
DOT: UN 2342
mf: C_4H_9Br mw: 137.04

PROP: Colorless liquid. Mp: −20°, bp: 73.3°, fp: −18°, d: 1.215 @ 25°/25°.

SYNS: 2-BROMOISOBUTANE ◇ 2-BROMO-2-METHYLPROPANE (DOT) ◇ TRIMETHYLBROMOMETHANE

TOXICITY DATA with REFERENCE
ipr-mus TDLo:3000 mg/kg/8W-I:NEO CNREA8 35,1411,75
ipr-rat LD50:1250 mg/kg 85GMAT -,29,82
ipr-mus LD50:4400 mg/kg 85GMAT -,29,82

CONSENSUS REPORTS: EPA Genetic Toxicology Program. Reported in EPA TSCA Inventory.

DOT Classification: Flammable or Combustible Liquid; Label: Flammable Liquid.

SAFETY PROFILE: Moderately toxic by intraperitoneal route. Questionable carcinogen with experimental neoplastigenic data. When heated to decomposition it emits toxic fumes of Br^-. See also BROMIDES.

BQM309 CAS:1867-72-7 ***HR: 3***
N-tert-BUTYL-1,4-BUTANEDIAMINE DIHYDROCHLORIDE
mf: $C_8H_{20}N_2{\cdot}2ClH$ mw: 217.22

SYNS: N-tert-BUTYL-1,4-DIAMINOBUTANEDIHYDROCHLORIDE ◇ CI-505 ◇ DIBUTADIAMIN DIHYDROCHLORIDE

TOXICITY DATA with REFERENCE
orl-rat LD50:810 mg/kg AIPTAK 154,263,65
ipr-rat LD50:349 mg/kg AIPTAK 154,263,65
scu-rat LD50:390 mg/kg AIPTAK 154,263,65
ivn-rat LD50:186 mg/kg AIPTAK 154,263,65
orl-mus LD50:1280 mg/kg AIPTAK 154,263,65
ipr-mus LD50:418 mg/kg AIPTAK 154,263,65
scu-mus LD50:1000 mg/kg AIPTAK 154,263,65
ivn-mus LD50:88 mg/kg AIPTAK 154,263,65

SAFETY PROFILE: Poison by subcutaneous, intravenous, and intraperitoneal routes. Moderately toxic by ingestion. When heated to decomposition it emits toxic fumes of NO_x and HCl. See also AMINES.

BQM500 CAS:109-21-7 ***HR: 2***
n-BUTYL n-BUTANOATE
mf: $C_8H_{16}O_2$ mw: 144.24

PROP: Colorless liquid; pineapple odor. Bp: 166°, flash p: 128°F (OC), d: 0.67-0.871, refr index: 1.405, vap d: 5.0. Misc with alc, ether, vegetable oils; sltly sol in propylene glycol, water.

SYNS: BUTYL BUTYRATE (FCC) ◇ n-BUTYL BUTYRATE ◇ n-BUTYL n-BUTYRATE ◇ FEMA No. 2186

TOXICITY DATA with REFERENCE
skn-rbt 500 mg/24H MOD FCTXAV 17,521,79
ipr-rat LD50:2300 mg/kg FCTXAV 17,521,79
ipr-mus LD50:8900 mg/kg FCTXAV 17,521,79
orl-rbt LD50:9520 mg/kg IMSUAI 41,31,72

CONSENSUS REPORTS: Reported in EPA TSCA Inventory.

SAFETY PROFILE: Moderately toxic via intraperitoneal route. Mildly toxic by ingestion. Moderately irritating to eyes, skin, and mucous membranes by inhalation. Narcotic in high concentrations. Combustible liquid. To fight fire, use alcohol foam, foam, CO_2, dry chemical. Incompatible with oxidizing materials. When heated to decomposition it emits acrid and irritating fumes.

BQM750 CAS:63937-32-6 ***HR: 3***
BUTYL-2-BUTOXYCYCLOPROPANE-1-CARBOXYLATE
mf: $C_{12}H_{22}O_3$ mw: 214.34

SYN: 2-BUTOXY-CYCLOPROPANECARBOXYLIC ACID BUTYL ESTER

TOXICITY DATA with REFERENCE
orl-rat LD50:24 mg/kg TXAPA9 28,313,74
skn-rbt LD50:110 mg/kg TXAPA9 28,313,74

SAFETY PROFILE: Poison by ingestion and skin contact. When heated to decomposition it emits acrid smoke and irritating fumes.

BQN250 CAS:78329-87-0 ***HR: 3***
p-(N-BUTYL-2-(BUTYLAMINO)ACETAMIDO) BENZOIC ACID BUTYL ESTER HYDROCHLORIDE
mf: $C_{20}H_{32}N_2O_3$•ClH mw: 385.00

SYN: C 3192

TOXICITY DATA with REFERENCE
eye-rbt 2% SEV ARZNAD 8,609,58
ipr-rat LD50:260 mg/kg ARZNAD 8,609,58
scu-mus LD50:2825 mg/kg ARZNAD 8,609,58

SAFETY PROFILE: Poison by intraperitoneal route. Moderately toxic by subcutaneous route. See also ESTERS. A severe eye irritant. When heated to decomposition it emits very toxic fumes of HCl and NO_x.

BQN500 CAS:78218-43-6 ***HR: 3***
N-BUTYL-2-(BUTYLAMINO)-2',6'-PROPIONOXYLIDIDE HYDROCHLORIDE
mf: $C_{19}H_{32}N_2O$•ClH mw: 340.99

SYN: C 3160

TOXICITY DATA with REFERENCE
eye-rbt 2% SEV ARZNAD 8,609,58
ipr-rat LD50:68 mg/kg ARZNAD 8,609,58
scu-mus LD50:256 mg/kg ARZNAD 8,609,58

SAFETY PROFILE: Poison by intraperitoneal and subcutaneous routes. A severe eye irritant. When heated to decomposition it emits very toxic fumes of HCl and NO_x. See also AMINES.

BQP000 CAS:7492-70-8 ***HR: 1***
BUTYL BUTYROLLACTATE
mf: $C_{11}H_{20}O_4$ mw: 216.28

PROP: Colorless liquid; butter, creamlike odor. D: 0.970, refr index: 1.420, flash p: +212°F. Misc with alc, fixed oils; sol in propylene glycol; insol in water.

SYNS: BUTANOICACID-2-BUTOXY-1-METHYL-2-OXOETHYL ESTER (9CI) ◇ BUTYL BUTYRYL LACTATE ◇ BUTYRIC ACID ESTER with BUTYL LACTATE ◇ FEMA No. 2190 ◇ LACTIC ACID, BUTYL ESTER, BUTYRATE

TOXICITY DATA with REFERENCE
skn-rbt 500 mg/24H FCTXAV 17,241,79

CONSENSUS REPORTS: Reported in EPA TSCA Inventory.

SAFETY PROFILE: A skin irritant. See also ESTERS. Combustible liquid. When heated to decomposition it emits acrid smoke and irritating fumes.

BQP250 CAS:592-35-8 ***HR: 3***
BUTYL CARBAMATE
mf: $C_5H_{11}NO_2$ mw: 117.17

SYNS: CARBAMIC ACID, BUTYL ESTER ◇ USAF EL-101 ◇ USAF FO-1

TOXICITY DATA with REFERENCE
mmo-esc 5000 ppm/3H AMNTA4 85,119,51
ipr-ham TDLo:492 mg/kg (8D preg):TER CNREA8 27,1696,67
ipr-mus TDLo:1980 mg/kg/6D-C:NEO PSEBAA 132,422,69
ipr-mus LD50:200 mg/kg NTIS** AD277-689
scu-mus LD50:540 mg/kg AJEBAK 45,507,67

CONSENSUS REPORTS: Reported in EPA TSCA Inventory.

SAFETY PROFILE: A poison via intraperitoneal route. Moderately toxic via subcutaneous route. Experimental teratogenic effects. Questionable carcinogen with experimental neoplastigenic data. Mutation data reported. See also CARBAMATES. When heated to decomposition it emits toxic fumes of NO_x.

BQP500 CAS:124-17-4 ***HR: 2***
BUTYL CARBITOL ACETATE
mf: $C_{10}H_{20}O_4$ mw: 204.30

PROP: Colorless liquid. fp: −32.2°, bp: 247°, flash p: 240°F (OC), d: 0.981 @ 20°/20°, autoign temp: 570°F, vap press: 0.01 mm @ 20°.

SYNS: 2-(2-BUTOXYETHOXY)ETHANOL ACETATE ◇ 2-(2-BUTOXYETHOXY)ETHYL ACETATE ◇ DIETHYLENE GLYCOL BUTYL ETHER ACETATE ◇ DIGLYCOL MONOBUTYL ETHER ACETATE ◇ EKTASOLVE DB ACETATE ◇ GLYCOL ETHER DB ACEATATE

TOXICITY DATA with REFERENCE
skn-rbt 500 mg open MLD UCDS** 12/29/71
eye-rbt 500 mg AJOPAA 29,1363,46
orl-rat LD50:6500 mg/kg 28ZEAL 5,32,76
orl-mus LD50:6600 mg/kg JPETAB 93,26,48
orl-rbt LD50:2600 mg/kg JPETAB 82,377,44
skn-rbt LD50:14500 mg/kg NPIRI* 1,27,74
orl-gpg LD50:2340 mg/kg JIHTAB 23,259,41
orl-ckn LD50:5000 mg/kg JPETAB 93,26,48

CONSENSUS REPORTS: Reported in EPA TSCA Inventory. Glycol ethers are on the Community Right-To-Know List.

SAFETY PROFILE: Moderately toxic by ingestion. Mild skin and eye irritant. Combustible when exposed to heat or flame. To fight fire, use foam, CO_2, dry chemical. Incompatible with oxidizing materials; heat; flame. When heated to decomposition it emits acrid and irritating fumes.

BQP750 CAS:85-70-1 ***HR: 2***
BUTYL CARBOBUTOXYMETHYL PHTHALATE
mf: $C_{18}H_{24}O_6$ mw: 336.42

SYNS: BUTYL PHTHALATE BUTYL GLYCOLATE ◇ BUTYL PHTHALYL BUTYL GLYCOLATE ◇ DIBUTYL-o-(o-CARBOXYBENZOYL) GLYCOLATE ◇ DIBUTYL-o-CARBOXYBENZOYLOXYACETATE ◇ SANTICIZIER B-16

TOXICITY DATA with REFERENCE
eye-rbt 500 mg AJOPAA 29,1363,46
cyt-ham:fbr 125 mg/L/24H MUREAV 48,337,77
ipr-rat TDLo:689 mg/kg (5-15D preg):TER JPMSAE 61,51,72
ipr-rat TDLo:2296 mg/kg (5-15D preg):REP JPMSAE 61,51,72
orl-rat LD50:7 g/kg EVHPAZ 3,131,73
ipr-rat LD50:6889 mg/kg JPMSAE 61,51,72
orl-mus LD50:12567 mg/kg IPSTB3 3,93,76
ipr-mus LD50:6880 mg/kg JSCCA5 28,667,77

CONSENSUS REPORTS: Reported in EPA TSCA Inventory.

SAFETY PROFILE: Mildly toxic via intraperitoneal route. Experimental teratogenic and reproductive effects. Mutation data reported. An eye irritant. When heated to decomposition it emits acrid and irritating fumes.

BQQ250 CAS:38252-74-3 ***HR: 1***
N-BUTYL-(3-CARBOXY PROPYL)NITROSAMINE
mf: $C_8H_{16}N_2O_3$ mw: 188.26

SYNS: BCPN ◇ 4-(BUTYLNITROSOAMINO)BUTANOIC ACID ◇ N-NITROSO-N-BUTYL-N-(3-CARBOXYPROPYL)AMINE

TOXICITY DATA with REFERENCE
mmo-sat 10 μmol/plate CNREA8 37,399,77
dnd-rat-par 50 mg/kg CBINA8 29,291,80
orl-rat TDLo:3760 mg/kg/12W-C:CAR CRNGDP 4,617,83
orl-mus TDLo:7 g/kg/20W-C:ETA GANNA2 67,175,76

CONSENSUS REPORTS: IARC Cancer Review: Animal Limited Evidence IMEMDT 17,51,78. EPA Genetic Toxicology Program.

SAFETY PROFILE: Questionable carcinogen with experimental carcinogenic and tumorigenic data. Mutation data reported. When heated to decomposition it emits toxic fumes of NO_x. See also NITROSAMINES.

BQQ750 CAS:109-69-3 ***HR: 3***
n-BUTYL CHLORIDE
DOT: UN 1127
mf: C_4H_9Cl mw: 92.58

PROP: Colorless liquid. Mp: −123.1°, bp: 78°, lel: 1.9%, uel: 10.1%, flash p: 15°F (OC), d: 0.884, autoign temp: 860°F, vap d: 3.20.

SYNS: BUTYL CHLORIDE (DOT) ◇ 1-CHLOROBUTANE (DOT) ◇ CHLORURE de BUTYLE (FRENCH) ◇ NCI-C06155 ◇ N-PROPYLCARBINYL CHLORIDE

TOXICITY DATA with REFERENCE
skn-rbt 10 mg/24H open MLD AMIHBC 10,61,54
eye-rbt 500 mg open AMIHBC 10,61,54
msc-mus:lym 500 mg/L NTPTR* NTP-TR-312,86
orl-rat LD50:2670 mg/kg AMIHBC 10,61,54
ihl-rat LCLo:8000 ppm/4H AMIHBC 10,61,54
skn-rbt LDLo:20 g/kg 34ZIAG -,745,69

CONSENSUS REPORTS: NTP Carcinogenesis Studies (gavage); No Evidence: mouse, rat NTPTR* NTP-TR-312,86. EPA Genetic Toxicology Program. Reported in EPA TSCA Inventory.

DOT Classification: Flammable Liquid; Label: Flammable Liquid.

SAFETY PROFILE: Moderately toxic by ingestion. Mutation data reported. See CHLORINATED HYDROCARBONS, ALIPHATIC. Skin and eye irritant. Dangerous fire hazard when exposed to heat or flame. Moderately explosive when exposed to flame. When heated to decomposition it emits highly toxic fumes of phosgene and Cl^-. To fight fire, use foam, CO_2, dry chemical. Incompatible with oxidizing materials.

BQR000 CAS:507-20-0 ***HR: 3***
tert-BUTYL CHLORIDE
mf: C_4H_9Cl mw: 92.58

PROP: Flash p: 32°F, d: 0.87, vap d: 3.2, bp: 51°.

SYNS: 2-CHLOROISOBUTANE ◇ 2-CHLORO-2-METHYLPROPANE ◇ TRIMETHYLCHLOROMETHANE

TOXICITY DATA with REFERENCE
ipr-mus TDLo:3000 mg/kg/8W-I:NEO CNREA8 35,1411,75

CONSENSUS REPORTS: Reported in EPA TSCA Inventory.

SAFETY PROFILE: Questionable carcinogen with experimental neoplastigenic data. Dangerous fire hazard when exposed to heat, flame (sparks), and oxidizers. To fight fire, use water, spray, fog, alcohol foam, dry chemical. When heated to decomposition it emits toxic fumes of Cl^-. See also CHLORINATED HYDROCARBONS, ALIPHATIC.

BQR250 CAS:27778-80-9 ***HR: 3***
β-sec-BUTYL-3-CHLORO-N,N-DIMETHYL-4-ETHOXYPHENETHYLAMINE
mf: $C_{16}H_{26}ClNO$ mw: 283.88

TOXICITY DATA with REFERENCE
orl-rat LD50:400 mg/kg CHTPBA 6,453,71
ivn-mus LD50:30 mg/kg CHTPBA 6,453,71

SAFETY PROFILE: Poison by ingestion and intravenous routes. When heated to decomposition it emits very toxic fumes of Cl^- and NO_x. See also AMINES.

BQR750 CAS:27778-78-5 ***HR: 3***
β-sec-BUTYL-3-CHLORO-N,N-DIMETHYL-4-METHOXYPHENETHYLAMINE
mf: $C_{15}H_{24}ClNO$ mw: 269.85

TOXICITY DATA with REFERENCE
orl-rat LD50:400 mg/kg CHTPBA 6,453,71
ivn-mus LD50:37 mg/kg CHTPBA 6,453,71

SAFETY PROFILE: Poison by ingestion and intravenous routes. When heated to decomposition it emits very toxic fumes of Cl^- and NO_x. See also AMINES.

BQS000 CAS:33132-85-3 ***HR: 3***
β-sec-BUTYL-5-CHLORO-N,N-DIMETHYL-2-METHOXYPHENETHYLAMINE
mf: $C_{15}H_{24}ClNO$ mw: 269.85

TOXICITY DATA with REFERENCE
orl-mus LD50:115 mg/kg CHTPBA 6,453,71
ivn-mus LD50:25 mg/kg CHTPBA 6,453,71

SAFETY PROFILE: Poison by ingestion and intravenous routes. When heated to decomposition it emits very toxic fumes of Cl^- and NO_x. See also AMINES.

BQS250 CAS:33132-71-7 ***HR: 3***
β-sec-BUTYL-p-CHLORO-N,N-DIMETHYL-PHENETHYLAMINE
mf: $C_{14}H_{22}ClN$ mw: 239.82

TOXICITY DATA with REFERENCE
orl-mus LD50:145 mg/kg CHTPBA 6,453,71
ivn-mus LD50:40 mg/kg CHTPBA 6,453,71

SAFETY PROFILE: Poison by ingestion and intravenous routes. When heated to decomposition it emits very toxic fumes of Cl^- and NO_x. See also AMINES.

BQT000 CAS:29122-56-3 ***HR: 3***
β-sec-BUTYL-5-CHLORO-2-ETHOXY-N,N-DIISOPROPYLPHENETHYLAMINE
mf: $C_{20}H_{34}ClNO$ mw: 340.00

TOXICITY DATA with REFERENCE
orl-mus LD50:220 mg/kg CHTPBA 6,453,71
ivn-mus LD50:31 mg/kg CHTPBA 6,453,71

SAFETY PROFILE: Poison by ingestion and intravenous routes. When heated to decomposition it emits very toxic fumes of Cl^- and NO_x. See also AMINES.

BQT250 CAS:29122-60-9 ***HR: 3***
1-(β-sec-BUTYL-5-CHLORO-2-ETHOXYPHENETHYL)PIPERIDINE
mf: $C_{19}H_{30}ClNO$ mw: 323.95

TOXICITY DATA with REFERENCE
orl-rat LD50:400 mg/kg CHTPBA 6,453,71
ivn-mus LD50:27 mg/kg CHTPBA 6,453,71

SAFETY PROFILE: Poison by ingestion and intravenous routes. When heated to decomposition it emits very toxic fumes of NO_x and Cl^-.

BQT500 CAS:16224-33-2 ***HR: 2***
BUTYL (3-CHLORO-2-HYDROXYPROPYL) ETHER
mf: $C_7H_{15}ClO_2$ mw: 166.67

SYN: BUTYL-CHLORHYDRINETHER (CZECH)

TOXICITY DATA with REFERENCE
skn-rbt 500 mg/24H MOD 28ZPAK -,81,72
eye-rbt 250 μg/24H SEV 28ZPAK -,81,72
orl-rat LD50:3520 mg/kg 28ZPAK -,81,72

CONSENSUS REPORTS: Reported in EPA TSCA Inventory.

SAFETY PROFILE: Moderately toxic by ingestion. A skin and severe eye irritant. See also ETHERS. When heated to decomposition it emits toxic fumes of Cl^-.

BQT750 CAS:5902-51-2 ***HR: 1***
3-tert-BUTYL-5-CHLORO-6-METHYLURACIL
mf: $C_9H_{13}ClN_2O_2$ mw: 216.69

SYNS: 3-tert-BUTYL-5-CHLOR-6-METHYLURACIL (GERMAN) ◇ 5-CHLORO-3-tert-BUTYL-6-METHYLURACIL ◇ 5-CHLORO-3-(1,1-DIMETHYLETHYL)-6-METHYL-2,4(1H,3H)-PYRIMIDINEDIONE ◇ COMPOUNE 732 ◇ DU PONT 732 ◇ DU PONT HERBICIDE 732 ◇ EXPERIMENTAL HERBICIDE 732 ◇ SINBAR ◇ TERBACIL ◇ TURBSVIL

TOXICITY DATA with REFERENCE
orl-rat LD50:7500 mg/kg FMCHA2 -,D302,80
unk-mam LD50:5000 mg/kg 30ZDA9 -,421,71

SAFETY PROFILE: Mildly toxic by ingestion and possibly other routes. When heated to decomposition it emits very toxic fumes of Cl^- and NO_x.

BQU000 CAS:56139-33-4 ***HR: 3***
tert-BUTYL CHLOROPEROXYFORMATE
mf: $C_5H_9ClO_3$ mw: 152.58

$(CH_3)_3COOCO{\cdot}Cl$

SAFETY PROFILE: A storage hazard. May ignite or explode at room temperature. When heated to decomposition it emits toxic fumes of Cl^-. See also PEROXIDES, ORGANIC.

BQU500 CAS:5902-52-3 ***HR: 3***
o-(4-tert-BUTYL-2-CHLOROPHENYL)-o-METHYL PHOSPHORAMIDOTHIONATE
mf: $C_{11}H_{17}ClNO_2PS$ mw: 293.77

SYNS: DOWCO 109 ◇ METHYL-PHOSPHORAMIDOTHIOIC ACID o-(tert-BUTYL-2-CHLOROPHENYL)ESTER ◇ NARLENE

TOXICITY DATA with REFERENCE
orl-rat LD50:820 mg/kg TXAPA9 21,315,72
orl-bwd LD50:75 mg/kg TXAPA9 21,315,72

SAFETY PROFILE: Poison by ingestion. See also ESTERS. When heated to decomposition it emits very toxic fumes of SO_x, PO_x, NO_x, and Cl^-.

BQU750 CAS:67195-50-0 ***HR: 3***
tert-20-BUTYLCHOLANTHRENE
mf: $C_{24}H_{22}$ mw: 310.46

SYN: 3-tert-BUTYLCHOLANTHRENE

TOXICITY DATA with REFERENCE
scu-mus TDLo:600 mg/kg/39W-I:ETA JNCIAM 2,99,41

SAFETY PROFILE: Questionable carcinogen with experimental tumorigenic data. When heated to decomposition it emits acrid smoke and irritating fumes.

BQV000 CAS:1189-85-1 ***HR: 3***
tert-BUTYL CHROMATE
mf: $C_8H_{18}CrO_4$ mw: 230.26

$[(CH_3)_3CO]_2CrO_2$

SYN: CHROMIC ACID, DI-tert-BUTYL ESTER

CONSENSUS REPORTS: Chromium and its compounds are on the Community Right-To-Know List.

OSHA PEL: CL 0.1 $mg(CrO_3)/m^3$ (skin)
ACGIH TLV: CL 0.1 $mg(CrO_3)/m^3$ (skin)
NIOSH REL: (Chromium(VI)) CL 0.001 Mg(Cr(VI))/m^3

SAFETY PROFILE: A very flammable mixture. When heated to decomposition it emits acrid and irritating fumes. See CHROMIUM COMPOUNDS and ESTERS.

BQV250 CAS:7492-44-6 ***HR: 2***
α-BUTYLCINNAMALDEHYDE
mf: $C_{13}H_{16}O$ mw: 188.2

SYNS: BUTYL CINNAMIC ALDEHYDE ◇ α-BUTYLCINNAMIC ALDEHYDE ◇ α-n-BUTYL-β-PHENYLACROLEIN ◇ 2-(PHENYLMETHYLENE)HEXANAL

TOXICITY DATA with REFERENCE
skn-rbt 500 mg/24H SEV FCTXAV 18,649,80
orl-rat LD50:4400 mg/kg FCTXAV 18,649,80

SAFETY PROFILE: A severe skin irritant. Mildly toxic by ingestion. When heated to decomposition it emits acrid smoke and irritating fumes. See also ALDEHYDES.

BQV500 CAS:538-65-8 ***HR: 1***
n-BUTYL CINNAMATE
mf: $C_{13}H_{16}O_2$ mw: 204.27

SYNS: n-BUTYL PHENYLACRYLATE ◇ CINNAMIC ACID-n-BUTYL ESTER

TOXICITY DATA with REFERENCE
skn-rbt 500 mg/24H MOD FCTXAV 18,649,80

SAFETY PROFILE: A skin irritant. See also ESTERS. When heated to decomposition it emits acrid smoke and irritating fumes.

BQV750 CAS:2409-55-4 ***HR: 3***
2-tert-BUTYL-p-CRESOL
mf: $C_{11}H_{16}O$ mw: 164.27

PROP: Clear liquid, sol in organic solvents and aqueous potassium hydroxide. Fp: 23.1°, bp: 244°, d: 0.922, flash p: 116°F.

SYNS: 2-tert-BUTYL-p-KRESOL (CZECH) ◇ 2-tert-BUTYL-4-METHYLPHENOL

TOXICITY DATA with REFERENCE
skn-rbt 2 mg/24H SEV 85JCAE-,227,86
eye-rbt 50 μg/24H SEV 28ZPAK -,55,72
dni-hmn:lyms 25 μmol/L RCOCB8 54,133,86
orl-rat LD50:2390 mg/kg 28ZPAK -,55,72
orl-ham TDLo:84 g/kg/20W-C:NEO CRNGDP 7,1285,86
orl-mus LD50:700 mg/kg JAPMA8 38,366,49
ipr-mus LD50:144 mg/kg JMCMAR 18,868,75
ivn-mus LD50:10 mg/kg CSLNX* NX#03020
skn-rbt LD50:2200 mg/kg JAPMA8 38,366,49
orl-gpg LD50:1180 mg/kg TPKVAL 12,124,71

CONSENSUS REPORTS: Reported in EPA TSCA Inventory.

SAFETY PROFILE: A poison by intraperitoneal and intravenous routes. Moderately toxic by ingestion and skin contact. Questionable carcinogen with experimental neoplastigenic data. A severe skin and eye irritant. Mutation data reported. Flammable when exposed to heat, flame, or oxidizers. To fight fire, use alcohol foam, foam, water spray, fog, dry chemical. When heated to decomposition it emits acrid and irritating fumes.

BQW000 CAS:98-52-2 ***HR: 3***
4-tert-BUTYLCYCLOHEXANOL
mf: $C_{10}H_{20}O$ mw: 156.30

SYNS: PADARYL ◇ USAF DO-20

TOXICITY DATA with REFERENCE
orl-rat LD50:4200 mg/kg FCTXAV 12,807,74
ipr-mus LD50:50 mg/kg NTIS** AD277-689

CONSENSUS REPORTS: Reported in EPA TSCA Inventory.

SAFETY PROFILE: Poison by intraperitoneal route. Moderately toxic by ingestion. When heated to decomposition it emits acrid smoke and irritating fumes. See also ALCOHOLS.

BQW250 CAS:98-53-3 ***HR: 1***
p-tert-BUTYLCYCLOHEXANONE
mf: $C_{10}H_{18}O$ mw: 154.28

TOXICITY DATA with REFERENCE
orl-rat LD50:5000 mg/kg FCTXAV 13,681,75
skn-rbt LD50:5000 mg/kg FCTXAV 13,681,75

CONSENSUS REPORTS: Reported in EPA TSCA Inventory.

SAFETY PROFILE: Mildly toxic by ingestion and skin contact. When heated to decomposition it emits acrid smoke and irritating fumes. See also KETONES.

BQW500 CAS:32210-23-4 ***HR: 1***
p-tert-BUTYLCYCLOHEXYL ACETATE
mf: $C_{12}H_{22}O_2$ mw: 198.34

SYNS: 4-tert-BUTYLCYCLOHEXYL ACETATE ◇ 4-tert-BUTYLHEXAHYDROPHENYL ACETATE ◇ VERTENEX

TOXICITY DATA with REFERENCE
skn-rbt 500 mg/24H MOD FCTXAV 16,637,78
orl-rat LD50:5000 mg/kg FCTXAV 16,637,78

CONSENSUS REPORTS: Reported in EPA TSCA Inventory.

SAFETY PROFILE: Mildly toxic by ingestion. A skin irritant. When heated to decomposition it emits acrid smoke and irritating fumes. See also ESTERS.

BQW750 CAS:10108-56-2 ***HR: 3***
N-BUTYL CYCLOHEXYL AMINE
mf: $C_{10}H_{21}N$ mw: 155.32

PROP: Flash p: 200°F (OC), d: 0.8, bp: 210°.

TOXICITY DATA with REFERENCE
skn-rbt 100 μg/24H open AIHAAP 23,95,62
orl-rat LD50:330 mg/kg AIHAAP 23,95,62
skn-rbt LD50:530 mg/kg AIHAAP 23,95,62

SAFETY PROFILE: A poison by ingestion. Moderately toxic by skin contact. See also AMINES. A skin irritant. Combustible when exposed to heat or flame. To fight fire, use alcohol foam. When heated to decomposition it emits toxic fumes of NO_x.

BQW825 CAS:841-73-6 ***HR: 3***
5-BUTYL-1-CYCLOHEXYLBARBITURIC ACID
mf: $C_{14}H_{22}N_2O_3$ mw: 266.38

PROP: Needles from methanol. Mp: 84°, bp: 185-187°.

SYNS: BCP ◇ BUCOLOM ◇ BUCOLOME ◇ 5-BUTYL-1-CYCLOHEXYL-2,4,6(1H,3H,5H)-PYRIMIDINETRIONE ◇ 5-n-BUTYL-1-CYCLOHEXYL-2,4,6-TRIOXOPERHYDROPYRIMIDINE ◇ PARAMIDIN ◇ PARAMIDINE

TOXICITY DATA with REFERENCE
orl-rat TDLo:2800 mg/kg (7-13D preg):TER JJPAAZ 17,381,67
orl-rat TDLo:25200 mg/kg (male 36W pre):REP NYKZAU 63,105,67
orl-rat LD50:1115 mg/kg NIIRDN 6,675,82
ipr-rat LD50:455 mg/kg NIIRDN 6,675,82
orl-mus LD50:1550 mg/kg NIIRDN 6,675,82
ipr-mus LD50:550 mg/kg ARZNAD 17,1519,67

SAFETY PROFILE: Poison by subcutaneous or intravenous routes. Moderately toxic by ingestion and other routes. An experimental teratogen. Other experimental reproductive effects. When heated to decomposition it emits toxic fumes of NO_x. See also BARBITURATES.

BQX000 CAS:61925-70-0 ***HR: 3***
N-(4-tert-BUTYL CYCLOHEXYL)-3,3-DIPHENYL PROPYLAMINE HYDROCHLORIDE
mf: $C_{25}H_{25}N\cdot ClH$ mw: 375.97

SYN: MG 18037

TOXICITY DATA with REFERENCE
orl-rat LD50:2550 mg/kg ARZNAD 26,2127,76
ipr-rat LD50:137 mg/kg ARZNAD 26,2127,76
orl-mus LD50:1850 mg/kg ARZNAD 26,2127,76
ipr-mus LD50:98 mg/kg ARZNAD 26,2127,76

SAFETY PROFILE: A poison by intraperitoneal route. Moderately toxic by ingestion. When heated to decomposition it emits very toxic fumes of HCl and NO_x.

BQX250 CAS:89-19-0 ***HR: 1***
BUTYL DECYL PHTHALATE
mf: $C_{22}H_{34}O_4$ mw: 362.56

SYNS: DECYL BUTYL PHTHALATE ◇ PLASTICIZER BDP

TOXICITY DATA with REFERENCE
orl-rat LD50:21 g/kg AIHAAP 30,470,69
skn-rbt LD50:16 g/kg AIHAAP 30,470,69

SAFETY PROFILE: Mildly toxic by ingestion and skin contact. See also ESTERS. When heated to decomposition it emits acrid smoke and irritating fumes.

BQX750 ***HR: 3***
tert-BUTYL DIAZOACETATE
mf: $C_6H_{10}N_2O_2$ mw: 142.16

SAFETY PROFILE: May explode during vacuum distillation. When heated to decomposition it emits toxic fumes of NO_x.

BQY000 CAS:10457-58-6 ***HR: 3***
14-n-BUTYL DIBENZ(a,h)ACRIDINE
mf: $C_{25}H_{21}N$ mw: 335.47

SYN: 10-n-BUTYL-1,2,5,6-DIBENZACRIDINE(FRENCH)

TOXICITY DATA with REFERENCE
scu-mus TDLo:60 mg/kg/9W-I:ETA BAFEAG 42,186,55

SAFETY PROFILE: Questionable carcinogen with experimental tumorigenic data. When heated to decomposition it emits toxic fumes of NO_x. See also AROMATIC AMINES.

BQY250 CAS:2422-88-0 ***HR: 2***
n-BUTYL-2-DIBUTYLTHIOUREA
mf: $C_{13}H_{28}N_2S$ mw: 244.49

TOXICITY DATA with REFERENCE
orl-rat LD50:3000 mg/kg TNICS* 13,78,73
orl-mus LD50:4300 mg/kg TNICS* 13,78,73

CONSENSUS REPORTS: Reported in EPA TSCA Inventory.

SAFETY PROFILE: Moderately toxic by ingestion. When heated to decomposition it emits very toxic fumes of NO_x and SO_x.

BQY300 CAS:684-82-2 ***HR: 3***
sec-BUTYLDICHLOROARSINE
mf: $C_4H_9AsCl_2$ mw: 202.95

SYNS: ARSINE, sec-BUTYLDICHLORO- ◇ ARSONOUS DICHLORIDE, (1-METHYLPROPYL)-(9CI) ◇ sec-BUTYLDICHLORARSINE ◇ DICHLORO(1-METHYLPROPYL)ARSINE

TOXICITY DATA with REFERENCE
ihl-mus LC50:12 g/m^3/10M NTIS** PB158-508

OSHA PEL: TWA 0.5 mg(As)/m^3

SAFETY PROFILE: Poison by inhalation. When heated to decomposition it emits toxic fumes of As and Cl^-.

BQY500 CAS:14090-22-3 ***HR: 3***
BUTYLDICHLOROBORANE
mf: $C_4H_9BCl_2$ mw: 138.7

SAFETY PROFILE: Explosive reaction on contact with water. Ignites in air after a delay period. When heated to decomposition it emits toxic fumes of Cl^-. See also BORANES and BORON COMPOUNDS.

BQZ000 CAS:94-80-4 ***HR: 2***
BUTYL DICHLOROPHENOXYACETATE
mf: $C_{12}H_{14}Cl_2O_3$ mw: 277.16

SYNS: BUTYL 2,4-D ◇ BUTYL (2,4-DICHLOROPHENOXY)ACETATE ◇ 2,4-D BUTYL ESTER ◇ BUTYL ESTER 2,4-D ◇ (2,4-DICHLOROPHENOXY)ACETIC ACID, BUTYL ESTER ◇ ESSO HERBICIDE 10 ◇ FERNESTA ◇ LIRONOX ◇ SHELL 40

TOXICITY DATA with REFERENCE
orl-rat TDLo:100 μg/kg (female 10D post):TER GISAAA 44(4),70,79
scu-mus TDLo:414 mg/kg (female 6-14D post):REP NTIS** PB223-160
orl-rat LD50:600 mg/kg FAATDF 9,423,87
orl-mus LD50:425 mg/kg 85GMAT -,29,82
orl-cat LD50:780 mg/kg 85GMAT -,29,82

CONSENSUS REPORTS: IARC Cancer Review: Animal Inadequate Evidence IMEMDT 15,111,77

SAFETY PROFILE: Moderately toxic by ingestion. Experimental teratogenic and reproductive effects. Questionable carcinogen. An herbicide. See also ESTERS. When heated to decomposition it emits toxic fumes of Cl^-.

BRA250 CAS:555-37-3 ***HR: 3***
1-BUTYL-3-(3,4-DICHLOROPHENYL)-1-METHYLUREA
mf: $C_{12}H_{16}Cl_2N_2O$ mw: 275.20

PROP: White or colorless crystals, low sol in water or hydrocarbon solvents. Mp: 103°.

SYNS: N-BUTYL-N'-(3,4-DICHLOROPHENYL)-N-METHYLUREA ◇ 3-(3,4-DICHLORPHENYL)-1-N-BUTYL-HARNSTOFF (GERMAN) ◇ 3-(3,4-DICHLOROPHENYL)-1-METHYL-1-BUTYLUREA ◇ GRANUREX ◇ KLOBEN ◇ KLOBEN NEBURON ◇ NEBUREA ◇ NEBUREX ◇ NEBURON

TOXICITY DATA with REFERENCE
orl-rat LD50:11000 mg/kg 85ARAE 2,144,77
ivn-mus LD50:180 mg/kg CSLNX* NX#03862

SAFETY PROFILE: Poison by intravenous route. Mildly toxic by ingestion. See also CHLORIDES and NITROGEN MONOXIDE. When heated to decomposition it emits toxic fumes of Cl^- and NO_x.

BRA500 CAS:102489-47-4 ***HR: 3***
2-(BUTYL(2-(DIETHYLAMINO)ETHYL)AMINO)-6'-CHLORO-o-ACETOTOLUIDIDE HYDROCHLORIDE
mf: $C_{19}H_{32}ClN_3O \cdot ClH$ mw: 390.45

SYN: C 5388

TOXICITY DATA with REFERENCE
eye-rbt 2% MLD ARZNAD 9,167,59
ipr-rat LD50:31 mg/kg ARZNAD 9,167,59
scu-mus LD50:47 mg/kg ARZNAD 9,167,59

SAFETY PROFILE: Poison by intraperitoneal and subcutaneous routes. An eye irritant. When heated to decomposition it emits very toxic fumes of Cl^- and NO_x.

BRA550 CAS:17563-48-3 ***HR: 3***
n-BUTYLDIETHYLTIN IODIDE
mf: $C_8H_{19}ISn$ mw: 360.86

SYN: STANNANE, BUTYLDIETHYLIODO-

TOXICITY DATA with REFERENCE
ivn-mus LD50:7100 μg/kg CSLNX* NX#05977

OSHA PEL: TWA 0.1 mg(Sn)/m^3
ACGIH TLV: TWA 0.1 mg(Sn)/m^3; STEL 0.2 mg/m^3 (skin)
NIOSH REL: (Organotin compound): 10H TWA 0.1 mg(Sn)/m^3

SAFETY PROFILE: Poison by intravenous route. When heated to decomposition it emits toxic fumes of Sn and I^-.

BRA600 CAS:29149-32-4 ***HR: 3***
tert-BUTYLDIFLUOROPHOSPHINE
mf: $C_4H_9F_2P$ mw: 126.09

$(CH_3)_3CPF_2$

SAFETY PROFILE: Ignites spontaneously in air. When heated to decomposition it emits toxic fumes of F^- and PO_x. See also PHOSPHINE.

BRA625 CAS:692-13-7 ***HR: 3***
1-BUTYLDIGUANIDE
mf: $C_6H_{15}N_5$ mw: 157.26

PROP: Strong base. Very sol in water.

SYNS: BUFORMIN ◇ BUFORMINE ◇ BUTFORMIN ◇ BUTYLBIGUANIDE ◇ BUTYLDIGUANIDE ◇ DBV ◇ GLYBIGID ◇ H 224 ◇ W 37

TOXICITY DATA with REFERENCE
orl-mus LD50:300 mg/kg JMCMAR 24,1521,81
ipr-mus LD50:140 mg/kg JMCMAR 24,1521,81
scu-gpg LD50:18 mg/kg MEXPAG 8,237,63

SAFETY PROFILE: Poison by ingestion, subcutaneous, and intraperitoneal routes. When heated to decomposition it emits toxic fumes of NO_x.

BRB000 CAS:62018-91-1 ***HR: D***
N-BUTYL-N-(2,4-DIHYDROXYBUTYL)NITROSAMINE
mf: $C_8H_{18}N_2O_3$ mw: 190.28

SYN: 4-(BUTYLNITROSOAMINO)-1,3-BUTANEDIOL

TOXICITY DATA with REFERENCE
mma-sat 11 μmol/plate CNREA8 37,399,77

SAFETY PROFILE: Mutation data reported. Many nitrosamines are carcinogens. When heated to decomposition it emits toxic fumes of NO_x. See also NITROSAMINES.

BRB450 CAS:24596-39-2 ***HR: 2***
4'-n-BUTYL-4-DIMETHYLAMINOAZOBENZENE
mf: $C_{18}H_{23}N_3$ mw: 281.44

SYNS: ANILINE, p-((p-BUTYLPHENYL)AZO)-N,N-DIMETHYL- ◇ p-((p-BUTYLPHENYL)AZO)-N,N-DIMETHYLANILINE

TOXICITY DATA with REFERENCE
orl-rat TDLo:13 mg/kg/Y-C:ETA JNCIAM 27,663,61

SAFETY PROFILE: Questionable carcinogen with experimental tumorigenic data. When heated to decomposition it emits toxic fumes of NO_x.

BRB460 CAS:24596-41-6 ***HR: 2***
4'-tert-BUTYL-4-DIMETHYLAMINOAZOBENZENE
mf: $C_{18}H_{23}N_3$ mw: 281.44

SYNS: ANILINE, p-((p-(tert-BUTYL)PHENYL)AZO)-N,N-DIMETHYL- ◇ p-((p-tert-BUTYLPHENYL)AZO)-N,N-DIMETHYLANILINE

TOXICITY DATA with REFERENCE
orl-rat TDLo:12852 mg/kg/Y-C:ETA JNCIAM 27,663,61

SAFETY PROFILE: Questionable carcinogen with experimental tumorigenic data. When heated to decomposition it emits toxic fumes of NO_x.

BRB500 CAS:69745-66-0 ***HR: 3***
4-(1-sec-BUTYL-2-(DIMETHYLAMINO)ETHYL) PHENOL
mf: $C_{14}H_{23}NO$ mw: 221.38

TOXICITY DATA with REFERENCE
orl-mus LD50:375 mg/kg CHTPBA 6,453,71
ivn-mus LD50:73 mg/kg CHTPBA 6,453,71

SAFETY PROFILE: Poison by ingestion and intravenous routes. When heated to decomposition it emits toxic fumes of NO_x.

BRB750 CAS:33098-26-9 ***HR: 3***
2-(1-sec-BUTYL-2-(DIMETHYLAMINO)ETHYL) QUINOLINE
mf: $C_{17}H_{24}N_2$ mw: 256.43

TOXICITY DATA with REFERENCE
orl-mus LD50:60 mg/kg CHTPBA 6,453,71
ivn-mus LD50:20 mg/kg CHTPBA 6,453,71

SAFETY PROFILE: Poison by ingestion and intrave-

nous routes. When heated to decomposition it emits toxic fumes of NO_x.

BRC000 CAS:33098-27-0 ***HR: 3***
2-(1-sec-BUTYL-2-(DIMETHYLAMINO)ETHYL) QUINOXALINE
mf: $C_{16}H_{23}N_3$ mw: 257.42

TOXICITY DATA with REFERENCE
orl-mus LD50:102 mg/kg CHTPBA 6,453,71
ivn-mus LD50:35 mg/kg CHTPBA 6,453,71

SAFETY PROFILE: Poison by ingestion and intravenous routes. When heated to decomposition it emits toxic fumes of NO_x.

BRC250 CAS:34548-72-6 ***HR: 3***
2-(1-sec-BUTYL-2-(DIMETHYLAMINO)ETHYL) THIOPHENE
mf: $C_{12}H_{21}NS$ mw: 211.40

TOXICITY DATA with REFERENCE
orl-mus LD50:260 mg/kg CHTPBA 6,453,71
ivn-mus LD50:50 mg/kg CHTPBA 6,453,71

SAFETY PROFILE: Poison by ingestion and intravenous routes. When heated to decomposition it emits very toxic fumes of NO_x and SO_x.

BRC500 CAS:51003-83-9 ***HR: 3***
2-n-BUTYL-3-DIMETHYLAMINO-5,6-METHYLENEDIOXYINDENE HYDROCHLORIDE
mf: $C_{16}H_{21}NO_2$•ClH mw: 295.84

SYNS: 6-BUTYL-5-DIMETHYLAMINO-5H-INDENO(5,6-d)-1,3-DIOXOLE HYDROCHLORIDE ◇ bu-MDI

TOXICITY DATA with REFERENCE
ipr-rat LD50:240 mg/kg RCOCB8 26,85,79
ipr-mus LD50:185 mg/kg RCOCB8 26,85,79
ivn-mus LD50:32 mg/kg RCOCB8 26,85,79

SAFETY PROFILE: A poison by intraperitoneal and intravenous routes. When heated to decomposition it emits very toxic fumes of NO_x and HCl.

BRC750 CAS:6279-54-5 ***HR: 2***
BUTYL-3-((DIMETHYLAMINO)METHYL)-4-HYDROXYBENZOATE
mf: $C_{14}H_{21}NO_3$ mw: 251.36

TOXICITY DATA with REFERENCE
orl-mus LDLo:2000 mg/kg ARZNAD 11,85,61
scu-mus LD50:475 mg/kg ARZNAD 11,85,61

SAFETY PROFILE: Moderately toxic by ingestion and subcutaneous route. When heated to decomposition it emits toxic fumes of NO_x. See also ESTERS.

BRD000 CAS:5221-53-4 ***HR: 3***
5-BUTYL-2-(DIMETHYLAMINO)-6-METHYL-4(1H)-PYRIMIDINONE
mf: $C_{11}H_{19}N_3O$ mw: 209.33

SYNS: 5-n-BUTYL-2-DIMETHYLAMINO-4-HYDROXY-6-METHYLPYRIMIDINE ◇ 5-BUTYL-2-(DIMETHYLAMINO)-6-METHYL-4-PYRIMIDINOL ◇ DIMETHIRIMOL ◇ 2-DIMETHYLAMINO-4-HYDROXY-5-n-BUTYL-6-METHYLPYRIMIDINE ◇ 2-DIMETHYLAMINO-4-METHYL-5-n-BUTYL-6-HYDROXYPYRIMIDINE ◇ METHYRIMOL ◇ MILCURB ◇ PP 675

TOXICITY DATA with REFERENCE
orl-rat LD50:2350 mg/kg WRPCA2 9,119,70
ipr-rat LDLo:200 mg/kg NATUAS 219,1160,68
orl-mus LD50:800 mg/kg 28ZEAL 5,79,76

SAFETY PROFILE: Poison by intraperitoneal route. Moderately toxic by ingestion and possibly other routes. When heated to decomposition it emits toxic fumes of NO_x.

BRD500 CAS:27778-82-1 ***HR: 3***
β-sec-BUTYL-N,N-DIMETHYL-2-ETHOXY-5-FLUOROPHENETHYLAMINE
mf: $C_{16}H_{26}FNO$ mw: 267.43

TOXICITY DATA with REFERENCE
orl-rat LD50:285 mg/kg CHTPBA 6,453,71
ivn-mus LD50:11 mg/kg CHTPBA 6,453,71

SAFETY PROFILE: Poison by ingestion and intravenous routes. When heated to decomposition it emits very toxic fumes of F^- and NO_x. See also AMINES.

BRD750 CAS:27684-90-8 ***HR: 3***
β-sec-BUTYL-N,N-DIMETHYL-5-FLUORO-2-METHOXYPHENETHYLAMINE
mf: $C_{15}H_{24}FNO$ mw: 253.40

TOXICITY DATA with REFERENCE
orl-rat LD50:300 mg/kg CHTPBA 6,453,71
ivn-mus LD50:18 mg/kg CHTPBA 6,453,71

SAFETY PROFILE: Poison by ingestion and intravenous routes. When heated to decomposition it emits very toxic fumes of F^- and NO_x. See also AMINES.

BRE000 CAS:56654-53-6 ***HR: 2***
1-BUTYL-3,3-DIMETHYL-1-NITROSOUREA
mf: $C_7H_{15}N_3O_2$ mw: 173.25

TOXICITY DATA with REFERENCE
mmo-esc 4 mmol/L CPBTAL 34,5056,86
mma-esc 8 mmol/L CPBTAL 34,5056,86
orl-rat TDLo:3140 mg/kg/45W-C:ETA,REP JNCIAM 56,1177,76

SAFETY PROFILE: Questionable carcinogen with experimental tumorigenic data. Experimental reproductive effects. Mutation data reported. When heated to decomposition it emits toxic fumes of NO_x.

BRE250 CAS:33132-61-5 ***HR: 3***
β-sec-BUTYL-N,N-DIMETHYLPHENETHYLAMINE
mf: $C_{14}H_{23}N$ mw: 205.38

TOXICITY DATA with REFERENCE
orl-mus LD50:170 mg/kg CHTPBA 6,453,71
ivn-mus LD50:31 mg/kg CHTPBA 6,453,71

SAFETY PROFILE: Poison by ingestion and intravenous routes. When heated to decomposition it emits toxic fumes of NO_x. See also AMINES.

BRE255 ***HR: 3***
β-sec-BUTYL-N,N-DIMETHYLPHENETHYLAMINE HYDROCHLORIDE
mf: $C_{14}H_{23}N{\cdot}ClH$ mw: 241.84

SYNS: 1-DIMETHYLAMINO-2-PHENYL-3-METHYLPENTANEHYDROCHLORIDE ◇ Z-134

TOXICITY DATA with REFERENCE
orl-rat LD50:357 mg/kg JPETAB 117,451,56
ipr-rat LD50:93 mg/kg JPETAB 117,451,56
ims-rat LD50:119 mg/kg JPETAB 117,451,56
orl-mus LD50:237 mg/kg JPETAB 117,451,56
ipr-mus LD50:110 mg/kg JPETAB 117,451,56

SAFETY PROFILE: Poison by ingestion, intraperitoneal, and intramuscular routes. When heated to decomposition it emits toxic fumes of NO_x and HCl.

BRE500 CAS:88-85-7 ***HR: 3***
2-sec-BUTYL-4,6-DINITROPHENOL
mf: $C_{10}H_{12}N_2O_5$ mw: 240.24

PROP: Crystals. Vap d: 7.73.

SYNS: ARETIT ◇ BASANITE ◇ BNP 30 ◇ BUTAPHENE ◇ CALDON ◇ CHEMOX GENERAL ◇ CHEMOX P.E. ◇ DINITRO ◇ DINITRO-3 ◇ 4,6-DINITRO-2-sec.BUTYLFENOL (CZECH) ◇ 2,4-DINITRO-6-sec-BUTYLPHENOL ◇ 4,6-DINITRO-o-sec-BUTYLPHENOL ◇ 4,6-DINITRO-2-sec-BUTYLPHENOL ◇ DINITROBUTYLPHENOL ◇ 4,6-DINITRO-2-(1-METHYL-N-PROPYL)PHENOL ◇ 2,4-DINITRO-6-(1-METHYL-PROPYL)PHENOL (FRENCH) ◇ DINOSEB ◇ DINOSEBE (FRENCH) ◇ DN 289 ◇ DNBP ◇ DNOSBP ◇ DNSBP ◇ DOW GENERAL ◇ DOW GENERAL WEED KILLER ◇ DOW SELECTIVE WEED KILLER ◇ ELGETOL ◇ ELGETOL 318 ◇ ENT 1,122 ◇ GEBUTOX ◇ HEL-FIRE ◇ KILOSEB ◇ 6-(1-METHYL-PROPYL)-2,4-DINITROFENOL (DUTCH) ◇ 2-(1-METHYLPROPYL)-4,6-DINITROPHENOL ◇ 6-(1-METIL-PROPIL)-2,4-DINITRO-FENOLO (ITALIAN) ◇ NITROPONE C ◇ PHENOTAN ◇ PREMERGE ◇ PREMERGE 3 ◇ RCRA WASTE NUMBER P020 ◇ SINOX GENERAL ◇ SPARIC ◇ SPURGE ◇ SUBITEX ◇ UNICROP DNBP ◇ VERTAC DINITRO WEED KILLER ◇ VERTAC GENERAL WEED KILLER ◇ VERTAC SELECTIVE WEED KILLER

TOXICITY DATA with REFERENCE
eye-rbt 50 μg/24H SEV 28ZPAK -,108,72
mrc-smc 185 ppm MUREAV 21,83,73
orl-rat TDLo:820 mg/kg (male/60D pre/1-22D preg):REP TXAPA9 45,235,78
orl-rat TDLo:150 mg/kg (female 6-15D post):TER TJADAB 33,19A,86
orl-mus TDLo:764 mg/kg/78W-I:ETA NTIS** PB223-159
orl-rat LD50:25 mg/kg TXAPA9 7,353,65
skn-rat LD50:80 mg/kg WRPCA2 9,119,70
scu-rat LD50:20368 μg/kg JPPMAB 4,1062,52
orl-mus LD50:16 mg/kg 85GMAT -,61,82
ihl-cat LCLo:45 mg/m^3/3H 85GMAT -,61,82
skn-rbt LD50:80 mg/kg 31ZOAD 1,178,68

CONSENSUS REPORTS: EPA Genetic Toxicology Program. EPA Extremely Hazardous Substances List.

SAFETY PROFILE: A poison by ingestion, inhalation, skin contact, subcutaneous, and intraperitoneal routes. Experimental teratogenic and reproductive effects. A severe eye irritant. Questionable carcinogen with experimental tumorigenic data. Mutation data reported. An herbicide. When heated to decomposition it emits toxic fumes of NO_x.

BRE750 CAS:6420-47-9 ***HR: 3***
o-sec-BUTYL-4,6-DINITROPHENOLTRIETHANOLAMINE SALT
mf: $C_{16}H_{27}N_3O_8$ mw: 389.46

SYNS: 2-sec-BUTYL-4,6-DINITROPHENOL-2,2′,2″-NITRILOTRIETHANOL SALT ◇ DINITROBUTYLPHENOL-2,2′,2″-NITRILOTRIETHANOL SALT ◇ 2-(1-METHYL-N-PROPYL)-4,6-DINITROPHENOL TRIETHANOLAMINE SALT

TOXICITY DATA with REFERENCE
orl-rat LD50:37 mg/kg SPEADM 74-1,-,74
skn-rat LD50:80 mg/kg SPEADM 74-1,-,74

SAFETY PROFILE: Poison by ingestion and skin contact. When heated to decomposition it emits toxic fumes of NO_x.

BRF500 CAS:50-33-9 ***HR: 3***
4-BUTYL-1,2-DIPHENYL-3,5-DIOXO PYRAZOLIDINE
mf: $C_{19}H_{20}N_2O_2$ mw: 308.41

SYNS: ALINDOR ◇ ALQOVERIN ◇ ANERVAL ◇ ANTADOL ◇ ANUSPIRAMIN ◇ ARTIZIN ◇ ARTRIZONE ◇ ARTROPAN ◇ AZDID ◇ AZOLID ◇ BENZONE ◇ BETAZED ◇ BUSONE ◇ BUTACOMPREN ◇ BUTACOTE ◇ BUTALAN ◇ BUTALGINA ◇ BUTALIDON ◇ BUTAPIRAZOL ◇ BUTAPYRAZOLE ◇ BUTARECBON ◇ BUTARTRINA ◇ BUTAZINA ◇ BUTAZONA ◇ BUTAZONE ◇ BUTIDIONA ◇ BUTONE ◇ BUTOZ ◇ 4-BUTYL-1,2-DIPHENYLPYRAZOLIDINE-3,5-DIONE ◇ BUTYLPYRIN ◇ BUVETZONE ◇ BUZON ◇ DIGIBUTINA ◇ DIOSSIDONE ◇ 3,5-DIOXO-1,2-DIPHENYL-4-N-BUTYLPYRAZOLIDENE ◇ DIOZOL ◇ DIPHEBUZOL ◇ DIPHENYLBUTAZONE ◇ 1,2-DIPHENYL-4-BUTYL-3,5-DIOXOPYRAZOLIDINE ◇ ELMEDAL ◇ EQUI BUTE ◇ ERIBUTAZONE ◇ ESTEVE ◇ FENARTIL ◇ FENIBUTAZONA ◇ FENIBUTOL ◇ FENILBUTINE ◇ FENILIDINA ◇ FENOTONE ◇ FENYLBUTAZON ◇ FLEXAZONE ◇ INTALBUT ◇ IPSOFLAME ◇ LINGEL ◇ MALGESIC ◇ MEPHABUTAZONE

◇ MERIZONE ◇ NADOZONE ◇ NCI-C56531 ◇ NOVOPHENYL ◇ PHEBUZIN ◇ PHENBUTAZOL ◇ PHENOPYRINE ◇ PHENYLBUTAZON (GERMAN) ◇ PHENYLBUTAZONE ◇ PIRARREUMOL "B" ◇ PRAECIRHEUMIN ◇ PYRAZOLIDIN ◇ REUDO ◇ REUMASYL ◇ REUMAZOL ◇ REUPOLAR ◇ RUBATONE ◇ SCANBUTAZONE ◇ SHIGRODIN ◇ TAZONE ◇ TEVCODYNE ◇ THERAZONE ◇ TODALGIL ◇ UZONE ◇ WESCOZONE ◇ ZOLAPHEN ◇ ZOLIDINUM ◇ ZORANE

TOXICITY DATA with REFERENCE
eye-rbt 100 mg MOD CMROCX 4,17,76
oms-hmn:emb 20 mg/L BEXBAN 74,828,72
mnt-mus-ipr 50 mg/kg IJEBA6 18,869,80
orl-rat TDLo:1100 mg/kg (female 7-17D post):REP OYYAA2 20,377,80
orl-rbt TDLo:780 mg/kg (female 6-18D post):TER IYKEDH 10,149,79
orl-man TDLo:4368 mg/kg/4Y-C:CAR,BLD BMJOAE 1,744,64
mul-wmn TDLo:4200 mg/kg/77W-I:CAR BMJOAE 2,1569,61
orl-man TD:140 mg/kg/3W-C:CAR,BLD BMJOAE 2,1552,60
par-hmn LDLo:168 mg/kg/2W-I:SYS 27ZXA3 -,448,63
orl-man TDLo:17500 μg/kg/3W-I:KID AIMEAS 41,1075,54
unr-man TDLo:200 mg/kg/5W-I:CVS,KID BMJOAE 282,950,81
unr-wmn TDLo:40 mg/kg/4D-I:BLD,MET PGPKA8 4(5),48,59
orl-rat LD50:245 mg/kg AIPTAK 123,48,59
ipr-rat LD50:142 mg/kg FRPSAX 14,347,59
scu-rat LD50:230 mg/kg OYYAA2 6,1285,72
ivn-rat LD50:100 mg/kg ARZNAD 10,665,60
ims-rat LD50:220 mg/kg ARZNAD 10,665,60
orl-mus LD50:270 mg/kg BCFAAI 111,293,72
ipr-mus LD50:128 mg/kg PCJOAU 19,33,85
scu-mus LD50:230 mg/kg JPPMAB 7,1022,55
ivn-mus LD50:90 mg/kg ARZNAD 19,36,69
ims-mus LD50:430 mg/kg OYYAA2 13,97,77
orl-dog LD50:332 mg/kg OYYAA2 20,265,80
ivn-dog LD50:121 mg/kg AIPTAK 149,571,64
ivn-cat LD50:100 mg/kg ARZNAD 19,36,69
orl-rbt LD50:781 mg/kg OYYAA2 20,265,80
ivn-rbt LD50:146 mg/kg ARZNAD 10,129,60
orl-gpg LD50:250 mg/kg ARZNAD 19,1207,69
orl-ham LD50:1260 mg/kg ATSUDG 7,365,84

CONSENSUS REPORTS: IARC Cancer Review: Group 3 IMEMDT 7,316,87; Human Inadequate Evidence IMEMDT 13,183,77. EPA Genetic Toxicology Program. Reported in EPA TSCA Inventory.

SAFETY PROFILE: Suspected human carcinogen producing leukemia. A human poison by parenteral route. An experimental poison by ingestion, intraperitoneal, subcutaneous, intravenous and intramuscular routes. Human systemic effects by ingestion and possibly other routes: fever, blood pressure increase, other unspecified vascular effects, damage to kidney tubules and glomeruli, decreased urine volume, blood in the urine, reduction in the number of white blood cells, and agranulocytosis. Experimental teratogenic and reproductive effects. Human mutation data reported. An eye irritant. An anti-inflammatory agent. When heated to decomposition it emits toxic fumes of NO_x.

BRF550 CAS:20333-40-8 ***HR: 3***
BUTYL DISELENIDE
mf: $C_8H_{18}Se_2$ mw: 272.18

SYNS: DIBUTYL DISELENIDE ◇ DIBUTYLDISELENIUM ◇ DI-n-BUTYL-DISELINIDE ◇ DISELENIDE, DIBUTYL-(9CI)

TOXICITY DATA with REFERENCE
ivn-mus LD50:75 mg/kg CSLNX* NX#09252

OSHA PEL: TWA 0.2 mg(Se)/m^3
ACGIH TLV: TWA 0.2 mg(Se)/m^3

SAFETY PROFILE: Poison by intravenous route. When heated to decomposition it emits toxic fumes of Se.

BRG000 CAS:110-57-6 ***HR: 3***
2-BUTYLENE DICHLORIDE
mf: $C_4H_6Cl_2$ mw: 125.00

PROP: Colorless liquid. Mp: 1-3°, bp: 156°, d: 1.183 @ 25°/4°.

SYNS: 1,4-DICHLOROBUTENE-2 (trans) ◇ 1,4-DICHLORO-2-BUTENE

TOXICITY DATA with REFERENCE
ipr-mus TDLo:150 mg/kg/77W-I:ETA CNREA8 35,2553,75
scu-mus TDLo:150 mg/kg/77W-I:NEO CNREA8 35,2553,75
ihl-rat LC50:86 ppm/4H AIHAM* -,-,68

CONSENSUS REPORTS: IARC Cancer Review: Group 3 IMEMDT 7,56,87; Animal Inadequate Evidence IMEMDT 15,149,77. Reported in EPA TSCA Inventory. EPA Extremely Hazardous Substances List.

SAFETY PROFILE: A poison by inhalation. Questionable carcinogen with experimental neoplastigenic and tumorigenic data. When heated to decomposition it emits toxic fumes of Cl^-. See also CHLORINATED HYDROCARBONS, ALIPHATIC.

BRG500 CAS:19485-03-1 ***HR: 2***
1,3-BUTYLENE GLYCOL DIACRYLATE
mf: $C_{10}H_{14}O_4$ mw: 198.24

SYNS: ACRYLIC ACID-1-METHYLTRIMETHYLENE ESTER ◇ 1,3-BUTANEDIOL DIACRYLATE ◇ 1,3-BUTYLENE DIACRYLATE ◇ 2-PROPENOIC ACID-1-METHYL-13-PROPANEDIYL ESTER

TOXICITY DATA with REFERENCE
orl-rat LD50:3540 mg/kg TXAPA9 28,313,74
skn-rbt LD50:450 mg/kg TXAPA9 28,313,74

CONSENSUS REPORTS: Reported in EPA TSCA Inventory.

SAFETY PROFILE: Moderately toxic by ingestion and skin contact. See also ESTERS. When heated to decomposition it emits acrid smoke and irritating fumes.

BRH250 CAS:106-83-2 ***HR: 3***
BUTYL-9,10-EPOXYSTEARATE
mf: $C_{22}H_{42}O_3$ mw: 354.64

SYN: 9,10-EPOXYOCTADECANOIC ACID BUTYL ESTER

TOXICITY DATA with REFERENCE
unr-mus TDLo:24 g/kg:ETA RARSAM 3,193,63

CONSENSUS REPORTS: Reported in EPA TSCA Inventory.

SAFETY PROFILE: Questionable carcinogen with experimental tumorigenic data. See also ESTERS. When heated to decomposition it emits acrid smoke and irritating fumes.

BRH750 CAS:142-96-1 ***HR: 1***
n-BUTYL ETHER
DOT: UN 1149
mf: $C_8H_{18}O$ mw: 130.26

PROP: Colorless liquid. Mp: −95°, bp: 142°, flash p: 77°F, d: 0.769 @ 20°/20°, autoign temp: 382°F, vap d: 4.48, lel: 1.5%, uel: 7.6%.

SYNS: 1-BUTOXYBUTANE ◇ BUTYL ETHER (DOT) ◇ DI-n-BUTYL ETHER (DOT) ◇ DIBUTYL OXIDE ◇ ETHER BUTYLIQUE (FRENCH) ◇ 1,1′-OXYBIS(BUTANE)

TOXICITY DATA with REFERENCE
eye-hmn 200 ppm/15M JIHTAB 28,262,46
skn-rbt 380 mg open MLD UCDS** 7/8/71
eye-rbt 500 mg open AMIHBC 10,61,54
ihl-hmn TCLo:200 ppm:NOSE,EYE JIHTAB 28,262,46
orl-rat LD50:7400 mg/kg AMIHBC 10,61,54
ihl-rat LCLo:4000 ppm/4H AMIHBC 10,61,54
skn-rbt LD50:10 g/kg AMIHBC 10,61,54

CONSENSUS REPORTS: Reported in EPA TSCA Inventory.

DOT Classification: Flammable Liquid; Label: Flammable Liquid; IMO: Flammable or Combustible Liquid; Label: Flammable Liquid.

SAFETY PROFILE: Mildly toxic by inhalation, ingestion, and skin contact. Human systemic effects by inhalation: conjunctiva irritation and unspecified nasal effects. An experimental skin and human eye irritant. See also ETHERS. Dangerous fire hazard when exposed to heat, flame or oxidizers. Incompatible with NCl_3 and oxidizing materials. To fight fire, use alcohol foam, dry chemical. When heated to decomposition it emits acrid smoke and fumes.

BRI000 CAS:123-05-7 ***HR: 3***
BUTYL ETHYL ACETALDEHYDE
DOT: UN 1191
mf: $C_8H_{16}O$ mw: 128.24

PROP: Bp: 163.4°, flash p: 125°F (OC). autoign temp: 387°F, d: 0.8205, vap press: 1.8 mm @ 20°, vap d: 4.42.

SYNS: ETHYLBUTYLACETALDEHYDE ◇ α-ETHYLCAPROALDEHYDE ◇ 2-ETHYLHEXALDEHYDE ◇ ETHYLHEXALDEHYDE (DOT) ◇ 2-ETHYLHEXANAL ◇ β-PROPYL-α-ETHYLACROLEIN

TOXICITY DATA with REFERENCE
skn-rbt 425 mg open MLD UCDS** 7/21/65
skn-rbt 10 mg/24H open SEV AMIHBC 4,119,51
eye-rbt 500 mg open AMIHBC 4,119,51
orl-rat LDLo:3200 mg/kg KODAK* -,-,71
ihl-rat LCLo:4000 ppm/4H AMIHBC 4,119,51
ipr-rat LD50:500 mg/kg HYERDA 3,201,78
orl-mus LD50:3550 mg/kg 85GMAT -,103,82
skn-rbt LD50:5040 mg/kg AMIHBC 4,119,51

CONSENSUS REPORTS: Reported in EPA TSCA Inventory.

DOT Classification: Flammable or Combustible Liquid; Label: Flammable Liquid.

SAFETY PROFILE: Moderately toxic by ingestion and intraperitoneal routes. Mildly toxic by inhalation and skin contact. An eye and severe skin irritant. See also ALDEHYDES. Dangerous fire hazard; spontaneously flammable in air. To fight fire, use foam, CO_2, dry chemical, water spray, mist, fog. Incompatible with oxidizing materials. When heated to decomposition it emits acrid and irritating fumes.

BRI250 CAS:149-57-5 ***HR: 2***
BUTYL ETHYL ACETIC ACID
mf: $C_8H_{16}O_2$ mw: 144.24

PROP: Flash p: 260°F (OC).

SYNS: α-ETHYLCAPROIC ACID ◇ 2-ETHYLHEXANOIC ACID ◇ 2-ETHYLHEXOIC ACID

TOXICITY DATA with REFERENCE
skn-rbt 10 mg/24H open JIHTAB 26,269,44
skn-rbt 450 mg open MLD UCDS** 11/4/71
eye-rbt 4500 μg SEV AJOPAA 29,1363,46
orl-rat TDLo:1803 mg/kg (female 12D post):TER TJADAB 35,41,87
orl-rat LD50:3000 mg/kg JIHTAB 26,269,44
skn-rbt LD50:1260 mg/kg UCDS** 11/4/71

CONSENSUS REPORTS: Reported in EPA TSCA Inventory.

SAFETY PROFILE: Moderately toxic by ingestion and skin contact. An experimental teratogen. A skin and severe eye irritant. Combustible when exposed to heat or flame. When heated to decomposition, it emits acrid and irritating fumes.

BRI500 CAS:77966-77-9 ***HR: 3***
n-BUTYL-2-(ETHYLAMINO)-2',6'-ACETOXYLIDIDE HYDROCHLORIDE
mf: $C_{16}H_{26}N_2O{\cdot}ClH$ mw: 298.90

SYN: C 3164

TOXICITY DATA with REFERENCE
eye-rbt 2% MLD ARZNAD 8,609,58
ipr-rat LD50:72 mg/kg ARZNAD 8,609,58
scu-mus LD50:125 mg/kg ARZNAD 8,609,58

SAFETY PROFILE: Poison by intraperitoneal and subcutaneous routes. An eye irritant. When heated to decomposition it emits very toxic fumes of HCl and NO_x.

BRI750 CAS:23947-60-6 ***HR: 2***
5-n-BUTYL-2-ETHYLAMINO-4-HYDROXY-6-METHYL-PYRIMIDINE
mf: $C_{11}H_{19}N_3O$ mw: 209.33

SYNS: 5-BUTYL-2-(ETHYLAMINO)-6-METHYL-4(1H)-PYRIMIDINONE ◇ ETHIRIMOL ◇ 2-ETHYLAMINO-4-METHYL-5-n-BUTYL-6-HYDROXYPYRIMIDINE ◇ MILCURB ◇ MILCURB SUPER ◇ MILGO ◇ MILGO E ◇ MILSTEM ◇ MILSTEM SEED DRESSING ◇ NEW MILSTEM ◇ PP149

TOXICITY DATA with REFERENCE
mmo-smc 50 ppm RSTUDV 6,161,76
orl-rat LD50:4000 mg/kg 28ZEAL 5,106,76
unr-mus LD50:4 g/kg TGANAK 16(1),45,82
ivn-mus LD50:800 mg/kg CHINAG (42),1512,69
orl-cat LD50:1000 mg/kg CHINAG (42),1512,69
orl-rbt LD50:1000 mg/kg CHINAG (42),1512,69
orl-gpg LD50:500 mg/kg CHINAG (42),1512,69

SAFETY PROFILE: Moderately toxic by ingestion, intravenous, and possibly other routes. Mutation data reported. When heated to decomposition it emits toxic fumes of NO_x.

BRJ000 CAS:41483-43-6 ***HR: 2***
5-BUTYL-2-ETHYLAMINO-6-METHYLPYRIMIDIN-4-YL DIMETHYLSULPHAMATE
mf: $C_{13}H_{24}N_4O_3S$ mw: 316.47

SYNS: 2-AETHYLAMINO-5-BUTYL-4-YL-DIMETHYLSULFAMAT (GERMAN) ◇ NIMROD

TOXICITY DATA with REFERENCE
orl-rat LD50:4000 mg/kg 85ARAE 4,115,76
orl-mus LD50:4000 mg/kg 85DPAN -,-,71/76
orl-gpg LD50:4000 mg/kg 85DPAN -,-,71/76

SAFETY PROFILE: Moderately toxic by ingestion. When heated to decomposition it emits very toxic fumes of NO_x and SO_x. See also ESTERS.

BRJ125 CAS:13080-06-3 ***HR: 3***
BUTYLETHYLMALONIC ACID-2-(DIETHYLAMINO)ETHYL ETHYL ESTER
mf: $C_{17}H_{33N}O_4$ mw: 315.51

SYNS: BUTYLAETHYLMALONSAEURE-AETHYL-DIAETHYLAMINOAETHYL-DI-ESTER (GERMAN) ◇ BUTYLETHYL-PROPANEDIOIC ACID-2-(DIETHYLAMINO)ETHYL ETHYL ESTER (9CI) ◇ Sch 5712

TOXICITY DATA with REFERENCE
orl-rat LD50:638 mg/kg AEPPAE 237,264,59
ipr-rat LD50:225 mg/kg AEPPAE 237,264,59
orl-mus LD50:412 mg/kg AEPPAE 237,264,59
ipr-mus LD50:258 mg/kg AEPPAE 237,264,59

SAFETY PROFILE: Poison by intraperitoneal route. Moderately toxic by ingestion. When heated to decomposition it emits toxic fumes of NO_x. See also ESTERS.

BRJ250 CAS:67050-26-4 ***HR: 3***
5-sec-BUTYL-5-ETHYL-1-METHYLBARBITURIC ACID
mf: $C_{11}H_{18}N_2O_3$ mw: 226.31

SYN: N-METHYLBUTABARBITAL

TOXICITY DATA with REFERENCE
ipr-rat LDLo:120 mg/kg JACSAT 58,1358,36
ivn-mus LD50:75 mg/kg AIPTAK 132,164,61
ivn-rbt LDLo:85 mg/kg JACSAT 58,1354,36

SAFETY PROFILE: Poison by intraperitoneal and intravenous routes. When heated to decomposition it emits toxic fumes of NO_x. See also BARBITURATES.

BRJ325 CAS:67330-25-0 ***HR: 2***
BUTYL FLUFENAMATE
mf: $C_{18}H_{18}F_3NO_2$ mw: 337.37

SYNS: BUTYL-o-((m-(TRIFLUOROMETHYL)PHENYL)AMINO)BENZOATE ◇ BUTYL-2-((3-(TRIFLUOROMETHYL)PHENYL)AMINO)BENZOATE ◇ HF 264 ◇ N-(α,α,α-TRIFLUORO-m-TOLYL)ANTHRANILIC ACID BUTYL ESTER

TOXICITY DATA with REFERENCE
orl-rat TDLo:990 mg/kg (7-17D preg):REP KSRNAM 13,3288,79
orl-rbt TDLo:1560 mg/kg (female 6-18D post):TER AMBNAS 27,33,79
orl-rat LD50:510 mg/kg OYYAA2 18,845,79
ipr-rat LD50:4550 mg/kg OYYAA2 18,845,79
scu-rat LD50:7800 mg/kg OYYAA2 18,845,79
ivn-rat LD50:650 mg/kg OYYAA2 18,845,79

orl-mus LD50:3100 mg/kg OYYAA2 18,845,79
ipr-mus LD50:4100 mg/kg OYYAA2 18,845,79
ivn-mus LD50:610 mg/kg IYKEDH 14,297,83
ipr-dog LD50:1500 mg/kg OYYAA2 18,845,79
scu-dog LD50:9300 mg/kg OYYAA2 18,845,79
ipr-rbt LD50:11500 mg/kg OYYAA2 18,845,79

SAFETY PROFILE: Moderately toxic by ingestion and other routes. An experimental teratogen. Experimental reproductive effects. When heated to decomposition it emits toxic fumes of F^- and NO_x. See also ESTERS.

BRJ750 CAS:2425-74-3 ***HR: 3***
tert-BUTYL FORMAMIDE
mf: $C_5H_{11}NO$ mw: 101.17

TOXICITY DATA with REFERENCE
ivn-mus LD50:180 mg/kg CSLNX* NX#04680

CONSENSUS REPORTS: Reported in EPA TSCA Inventory.

SAFETY PROFILE: A poison by intravenous route. When heated to decomposition it emits toxic fumes of NO_x.

BRK000 CAS:592-84-7 ***HR: 3***
n-BUTYL FORMATE
DOT: UN 1128
mf: $C_5H_{10}O_2$ mw: 102.15

PROP: Colorless liquid. Mp: −90°, bp: 106.0°, flash p: 64°F (CC), d: 0.911, autoign temp: 612°F, vap press: 40 mm @ 31.6°, vap d: 3.52, lel: 1.7%, uel: 8%.

SYN: BUTYLESTER KYSELINY MRAVENCI ◇ BUTYL FORMATE (DOT)

TOXICITY DATA with REFERENCE
ihl-hmn TCLo:10418 ppm:EYE,CNS,PUL AMIHAB 20,517,59
ihl-cat LCLo:10418 ppm/70M AMIHAB 20,517,59
orl-rbt LD50:2656 mg/kg IMSUAI 41,31,72

CONSENSUS REPORTS: Reported in EPA TSCA Inventory.

DOT Classification: Flammable Liquid; Label: Flammable Liquid.

SAFETY PROFILE: Moderately toxic by ingestion. Mildly toxic by inhalation. Human systemic effects by inhalation: muscle contractions and spasticity, conjunctiva irritation, and unspecified respiratory changes. An irritant and narcotic in high concentrations. See also ESTERS, BUTYL ALCOHOL, and FORMIC ACID. Dangerous fire hazard when exposed to heat or flame. To fight fire, use alcohol foam, foam, CO_2, dry chemical. Incompatible with oxidizing materials. When heated to decomposition it emits acrid and irritating fumes.

BRK100 CAS:16120-70-0 ***HR: 2***
N-n-BUTYL-N-FORMYLHYDRAZINE
mf: $C_5H_{12}N_2O$ mw: 116.19

SYNS: BFH ◇ FORMIC ACID, 1-BUTYLHYDRAZIDE

TOXICITY DATA with REFERENCE
orl-mus TDLo:70 g/kg/84W-C:CAR CRNGDP 1,589,80

SAFETY PROFILE: Questionable carcinogen with experimental carcinogenic data. When heated to decomposition it emits toxic fumes of NO_x.

BRK250 CAS:64441-42-5 ***HR: 3***
1-BUTYL-3-(2-FUROYL)UREA
mf: $C_{10}H_{14}NO_3$ mw: 196.25

SYNS: n-BUTYL-N'-(2-FUROYL) ◇ N-FUROYL-N'-n-BUTYLHARNSTOFF (GERMAN)

TOXICITY DATA with REFERENCE
orl-mus LD50:730 mg/kg ARZNAD 10,686,60
ipr-mus LD50:230 mg/kg ARZNAD 10,686,60

SAFETY PROFILE: Poison by intraperitoneal route. Moderately toxic by ingestion. When heated to decomposition it emits toxic fumes of NO_x.

BRK750 CAS:2426-08-6 ***HR: 3***
n-BUTYL GLYCIDYL ETHER
mf: $C_7H_{14}O_2$ mw: 130.21

SYNS: AGEFLEX BGE ◇ BGE ◇ 2,3-EPOXYPROPYL BUTYL ETHER ◇ GLYCIDYL BUTYL ETHER

TOXICITY DATA with REFERENCE
skn-rbt 454 mg/3D MLD AMIHAB 14,250,56
skn-rbt 500 mg/24H MOD 28ZPAK -,135,72
eye-rbt 91 mg MLD AMIHAB 14,250,56
eye-rbt 750 μg/24H SEV 28ZPAK -,135,72
mmo-esc 20 μmol/L ARTODN 46,277,80
dnd-esc 1 μmol/L ARTODN 46,277,80
skn-mus TDLo:36 g/kg (24D male):TER MUREAV 124,225,83
orl-rat LD50:2050 mg/kg AIHAAP 23,95,62
ihl-rat LCLo:670 ppm AMIHAB 14,250,56
ipr-rat LD50:1140 mg/kg AMIHAB 14,250,56
orl-mus LD50:1520 mg/kg AMIHAB 14,250,56
ipr-mus LD50:700 mg/kg AMIHAB 14,250,56
skn-rbt LD50:2520 mg/kg AIHAAP 23,95,62

CONSENSUS REPORTS: Reported in EPA TSCA Inventory.

OSHA PEL: (Transitional: TWA 50 ppm) TWA 25 ppm
ACGIH TLV: TWA 25 ppm
DFG MAK: Suspected Carcinogen.
NIOSH REL: (Glycidyl Ethers) CL 30 mg/m^3/15M

SAFETY PROFILE: Moderately toxic by ingestion, skin contact, and intraperitoneal routes. Mildly toxic by inhalation. An experimental teratogen. Mutation data reported. A skin and severe eye irritant. See also ETHERS. When heated to decomposition it emits acrid and irritating fumes.

BRK900 CAS:626-82-4 ***HR: 1***
BUTYL HEXANOATE
mf: $C_{10}H_{20}O_2$ mw: 172.30

SYNS: BUTYL CAPROATE ◇ n-BUTYL HEXANOATE ◇ HEXANOIC ACID, BUTYL ESTER

TOXICITY DATA with REFERENCE
skn-rbt 500 mg/24H MLD FCTOD7 21,653,83

CONSENSUS REPORTS: Reported in EPA TSCA Inventory.

SAFETY PROFILE: A skin irritant. When heated to decomposition it emits acrid smoke and irritating fumes.

BRL500 CAS:56795-65-4 ***HR: 3***
n-BUTYLHYDRAZINE HYDROCHLORIDE
mf: $C_4H_{12}N_2•ClH$ mw: 124.64

TOXICITY DATA with REFERENCE
orl-mus TDLo:14/g/kg/8W-C:NEO EJCAAH 11,473,75

SAFETY PROFILE: Questionable carcinogen with experimental neoplastigenic data. When heated to decomposition it emits very toxic fumes of NO_x and HCl. See also HYDRAZINE.

BRL750 CAS:40711-41-9 ***HR: D***
BUTYLHYDRAZINE OXALATE
mf: $C_4H_{12}N_2•C_2H_2O_4$ mw: 178.22

SYN: BUTYLHYDRAZINEETHANEDIOATE

TOXICITY DATA with REFERENCE
mma-sat 1200 μg/plate NEZAAQ 33,474,78
mmo-sat 1200 μg/plate NEZARQ 33,474,78

SAFETY PROFILE: Mutation data reported. When heated decomposition it emits toxic fumes such as NO_x. See also HYDRAZINE.

BRM000 ***HR: 3***
O,O-tert-BUTYL HYDROGEN MONOPEROXY MALEATE
mf: $C_8H_{12}O_5$ mw: 188.18

SAFETY PROFILE: Slightly shock-sensitive. Commercial grade 95% dry is very hazardous. When heated to decomposition it emits acrid smoke and fumes. See also PEROXIDES, ORGANIC.

BRM250 CAS:75-91-2 ***HR: 3***
tert-BUTYLHYDROPEROXIDE
DOT: UN 2093/UN 2094
mf: $C_4H_{10}O_2$ mw: 90.14

CH_3COOH

PROP: Water-white liquid. Flash p: 80°F or above, fp: −35°, d: 0.860, vap d: 2.07. Sltly sol in water; very sol in esters and alc.

SYNS: terc. BUTYLHYDROPEROXID (CZECH) ◇ CADOX TBH ◇ 1,1-DIMETHYLETHYL HYDROPEROXIDE ◇ HYDROPEROXYDE de BUTYLE TERTIAIRE (FRENCH) ◇ 2-HYDROPEROXY-2-METHYLPROPANE ◇ PERBUTYL H ◇ TBHP-70 ◇ TRIGONOX A-75 (CZECH)

TOXICITY DATA with REFERENCE
skn-rbt 500 mg AIHAAP 19,205,58
skn-rbt 500 mg/24H SEV 28ZPAK -,39,72
eye-rbt 7 mg AIHAAP 19,205,58
eye-rbt 100 mg/24H MOD 28ZPAK -,39,72
eye-rbt 150 mg/1M rns SEV ZAARAM 8,25,58
mmo-sat 17 μg/plate ENMUDM 5(Suppl 1),3,83
mma-sat 17 μg/plate ENMUDM 5(Suppl 1),3,83
pic-esc 25 mg/L VIRLAX 99,257,79
orl-rat LD50:406 mg/kg AIHAAP 19,205,58
ihl-rat LC50:500 ppm/4H AIHAAP 19,205,58
skn-rat LD50:790 mg/kg BSPII* 1/75-19B
ipr-rat LD50:87 mg/kg AIHAAP 19,205,58
orl-mus LD50:710 mg/kg BSPII* 1/75-19B
ihl-mus LC50:350 ppm/4H AIHAAP 19,205,58

CONSENSUS REPORTS: EPA Genetic Toxicology Program. Reported in EPA TSCA Inventory.

DFG MAK: Moderate skin effects.
DOT Classification: Organic Peroxide; Label: Organic Peroxide (UN2093, UN2094); Forbidden (more than 90% water).

SAFETY PROFILE: A poison by ingestion and inhalation. A severe skin and eye irritant. Mutation data reported. At highest dosage levels, symptoms noted were severe depression, incoordination, and cyanosis. Death was due to respiratory arrest. Very dangerous fire hazard when exposed to heat or flame, or by spontaneous chemical reaction such as with reducing materials. Moderately explosive; may explode during distillation. Violent reaction with traces of acid. Concentrated solutions may ignite spontaneously on contact with molecular sieve. Mixtures with transition metal salts may react vigorously and release oxygen. Forms an unstable solution with 1,2-dichloroethane. To fight fire, use alcohol foam, CO_2, dry chemical. When heated to decomposition it emits acrid smoke and fumes. See also PEROXIDES, ORGANIC.

BRM500 CAS:1948-33-0 ***HR: 3***
tert-BUTYLHYDROQUINONE
mf: $C_{10}H_{14}O_2$ mw: 166.24

PROP: White crystalline solid; characteristic odor. Mp: 126.5-128.5°. Sol in alc, ether; insol in water.

SYNS: MONO-tert-BUTYLHYDROQUINONE ◇ MTBHQ ◇ SUSTANE ◇ TBHQ (FCC) ◇ TENOX TBHQ

TOXICITY DATA with REFERENCE
cyt-mus-ipr 200 mg/kg FCTOD7 22,459,84
orl-rat LD50:700 mg/kg JAOCA7 52,53,75
ipr-rat LD50:300 mg/kg JAOCA7 52,53,75
orl-mus LD50:1000 mg/kg KODAK* 21MAY71
ipr-mus LD50:144 mg/kg DCTODJ 7,335,84

CONSENSUS REPORTS: Reported in EPA TSCA Inventory.

SAFETY PROFILE: Poison by intraperitoneal route. Moderately toxic by ingestion. Mutation data reported. When heated to decomposition it emits acrid smoke and irritating fumes.

BRM750 CAS:21070-33-7 ***HR: 3***
6-BUTYL-4-HYDROXYAMINOQUINOLINE-1-OXIDE
mf: $C_{13}H_{16}N_2O_2$ mw: 232.31

TOXICITY DATA with REFERENCE
scu-mus TDLo:60 mg/kg/I:ETA CPBTAL 17,544,69

SAFETY PROFILE: Questionable carcinogen with experimental tumorigenic data. When heated to decomposition it emits toxic fumes of NO_x.

BRN000 CAS:121-00-6 ***HR: 3***
3-tert-BUTYL-4-HYDROXYANISOLE
mf: $C_{11}H_{16}O_2$ mw: 180.27

SYNS: 2-tert-BUTYL-4-METHOXYPHENOL ◇ 4-METHOXY-2-tert-BUTYLPHENOL

TOXICITY DATA with REFERENCE
orl-ham TDLo:168 g/kg/20W-C:NEO CRNGDP 7,1285,86
orl-rat LD50:2910 mg/kg PLRCAT 16,1041,84
ipr-rat LD50:32 mg/kg PLRCAT 16,1041,84
orl-mus LD50:1583 mg/kg PLRCAT 16,1041,84
ipr-mus LD50:29 mg/kg PLRCAT 16,1041,84

CONSENSUS REPORTS: Reported in EPA TSCA Inventory.

SAFETY PROFILE: Poison by intraperitoneal route. Moderately toxic by ingestion. Questionable carcinogen with experimental neoplastigenic data. When heated to decomposition it emits acrid smoke and irritating fumes.

BRN250 CAS:55621-29-9 ***HR: D***
n-BUTYL-N-(2-HYDROXYBUTYL)NITROSAMINE
mf: $C_8H_{18}N_2O_2$ mw: 174.28

SYN: 1-(BUTYLNITROSOAMINO)-2-BUTANOL

TOXICITY DATA with REFERENCE
mma-sat 4 μmol/plate CNREA8 37,399,77

SAFETY PROFILE: Mutation data reported. Many nitrosamines are carcinogens. When heated to decomposition it emits toxic fumes of NO_x. See also NITROSAMINES.

BRN500 CAS:40911-07-7 ***HR: D***
n-BUTYL-N-(3-HYDROXYBUTYL)NITROSAMINE
mf: $C_8H_{18}N_2O_2$ mw: 174.28

SYN: 4-(BUTYLNITROSOAMINO)-2-BUTANOL

TOXICITY DATA with REFERENCE
mmo-sat 18 μmol/plate CNREA8 37,399,77
mma-sat 4 μmol/plate CNREA8 37,399,77

SAFETY PROFILE: Mutation data reported. Many nitrosamines are carcinogens. When heated to decomposition it emits toxic fumes of NO_x. See also NITROSAMINES.

BRO000 CAS:51938-14-8 ***HR: 3***
BUTYL(2-HYDROXYETHYL)NITROSOAMINE
mf: $C_6H_{14}N_2O_2$ mw: 146.22

SYNS: BHEN ◇ 2-(BUTYLNITROSAMINO)ETHANOL

TOXICITY DATA with REFERENCE
mmo-sat 100 μg/plate MUREAV 56,219,78
mma-sat 5 μmol/plate CNREA8 37,399,77
orl-rat TDLo:4800 mg/kg/20W-C:ETA GANNA2 65,13,74

CONSENSUS REPORTS: EPA Genetic Toxicology Program.

SAFETY PROFILE: Questionable carcinogen with experimental tumorigenic data. Mutation data reported. When heated to decomposition it emits toxic fumes of NO_x. See also NITROSAMINES.

BRO250 CAS:78128-80-0 ***HR: 2***
3-BUTYL-4-HYDROXY-2(5H)FURANONE
mf: $C_8H_{12}O_3$ mw: 156.20

SYN: α-n-BUTYL-β-HYDROXY-Δ^{α},β-BUTENOLID (GERMAN)

TOXICITY DATA with REFERENCE
scu-mus LD50:1750 mg/kg ARZNAD 11,277,61
ivn-mus LD50:1187 mg/kg ARZNAD 11,277,61

SAFETY PROFILE: Moderately toxic by subcutaneous and intravenous routes. When heated to decomposition it emits acrid smoke and irritating fumes.

BRO500 CAS:38252-75-4 ***HR: D***
n-BUTYL-N-(2-HYDROXYL-3-CARBOXYPROPYL)NITROSAMINE
mf: $C_8H_{16}N_2O_4$ mw: 204.26

SYN: 4-(BUTYLNITROSOAMINO)-3-HYDROXYBUTYRIC ACID

TOXICITY DATA with REFERENCE
mmo-sat 24 μmol/plate CNREA8 37,399,77
mma-sat 24 μmol/plate CNREA8 37,399,77

SAFETY PROFILE: Mutation data reported. Many nitrosamines are carcinogens. When heated to decomposition it emits toxic fumes of NO_x. See also NITROSAMINES.

BRO750 CAS:67590-46-9 ***HR: 3***
2-(tert-BUTYL)-2-(HYDROXYMETHYL)-1,3-PROPANEDIOL, CYCLIC PHOSPHITE (1:1)
mf: $C_8H_{15}O_3P$ mw: 190.20

SYN: 4-(tert-BUTYL)-2,6,7-TRIOXA-1-PHOSPHABICYCLO(2.2.2)OCTANE

TOXICITY DATA with REFERENCE
ipr-mus LD50:40 μg/kg TXAPA9 47,287,79
ivn-mus LD50:210 μg/kg EJMCA5 13,207,78

SAFETY PROFILE: A deadly poison by intraperitoneal and intravenous routes. When heated to decomposition it emits toxic fumes of PO_x.

BRP250 CAS:9003-13-8 ***HR: 1***
α-BUTYL-omega-HYDROXYPOLY(OXY(METHYL-1,2-ETHANEDIYL))
mf: $(C_3H_6O)_n \cdot C_4H_{10}O$

SYNS: BUTOXYPOLYPROPYLENE GLYCOL ◇ BUTOXYPROPANEDIOL POLYMER ◇ CRAG FLY REPELLENT ◇ ENT 8286 ◇ EXP. MITICIDE No. 7 ◇ NEWPOL LB3000 ◇ OPSB ◇ POLY(OXYPROPYLENE) BUTYL ETHER ◇ POLYOXYPROPYLENE MONOBUTYL ETHER ◇ POLYPROPYLENE GLYCOL MONOBUTYL ETHER ◇ PPG-14 BUTYL ETHER ◇ PPG-16 BUTYL ETHER ◇ PPG-33 BUTYL ETHER ◇ STABILENE ◇ STABILENE FLY REPELLENT ◇ UCON LB-250 ◇ UCON LB 1145 ◇ UCON LB 1800X

TOXICITY DATA with REFERENCE
skn-rbt 500 mg open MLD UCDS** 5/23/68
orl-rat LD50:9100 mg/kg ARSIM* 20,6,66
orl-rbt LD50:23900 mg/kg SPEADM 78-1,53,78
skn-rbt LD50:21 g/kg UCDS** 1/16/58

CONSENSUS REPORTS: Reported in EPA TSCA Inventory. Glycol ethers are on the Community Right-To-Know List.

SAFETY PROFILE: Mildly toxic by ingestion. Mildly toxic by skin contact. A skin irritant. An insect repellant. When heated to decomposition it emits acrid smoke and irritating fumes.

BRP500 CAS:507-40-4 ***HR: 3***
tert-BUTYL HYPOCHLORITE
mf: C_4H_9OCl mw: 106.6

SAFETY PROFILE: A storage hazard. Ultraviolet light causes exothermic decomposition. Reacts violently with rubber. Reaction with sodium hydrogen cyanamide forms the explosive cyanonitrene. When heated to decomposition it emits toxic fumes of Cl^-. See also HYPOCHLORITES.

BRP750 CAS:85-60-9 ***HR: 1***
4,4'-BUTYLIDENEBIS(3-METHYL-6-tert-BUTYLPHENOL)
mf: $C_{26}H_{38}O_2$ mw: 382.64

SYNS: 1,1-BIS(2-METHYL-4-HYDROXY-5-tert-BUTYLPHENYL)BUTANE ◇ 4,4'-BUTYLIDENEBIS(6-tert-BUTYL-m-CRESOL) ◇ 4,4'-BUTYLIDENEBIS(6-tert-BUTYL-3-METHYLPHENYL)◇ SANTOWHITE POWDER ◇ SUMILIT BBM ◇ SWP (ANTIOXIDANT)

TOXICITY DATA with REFERENCE
orl-rat LDLo:17 g/kg RCTEA4 45(3),627,72

CONSENSUS REPORTS: Reported in EPA TSCA Inventory.

SAFETY PROFILE: Mildly toxic by ingestion. An antioxidant. When heated to decomposition it emits acrid smoke and irritating fumes.

BRQ000 CAS:3772-23-4 ***HR: 1***
6,6'-BUTYLIDENEBIS(2,4-XYLENOL)
mf: $C_{20}H_{22}O$ mw: 278.42

TOXICITY DATA with REFERENCE
eye-rbt 100 mg IHFCAY 6,1,67
orl-rat LD50:5500 mg/kg IHFCAY 6,1,67

SAFETY PROFILE: Mildly toxic by ingestion. An eye irritant. When heated to decomposition it emits smoke and acrid, irritating fumes.

BRQ100 CAS:551-08-6 ***HR: 2***
3-BUTYLIDENE PHTHALIDE
mf: $C_{12}H_{12}O_2$ mw: 188.24

SYNS: BUTYLIDENE PHTHALIDE ◇ n-BUTYLIDENE PHTHALIDE ◇ 1(3H)-ISOBENZOFURANONE, 3-BUTYLIDENE-(9CI) ◇ PHTHALIDE, 3-BUTYLIDENE-

TOXICITY DATA with REFERENCE
skn-rbt 500 mg/24H MLD FCTOD7 21,659,83
orl-rat LD50:1850 mg/kg FCTOD7 21,659,83

CONSENSUS REPORTS: Reported in EPA TSCA Inventory.

SAFETY PROFILE: Moderately toxic by ingestion. A skin irritant. When heated to decomposition it emits acrid smoke and irritating fumes.

BRQ250 CAS:542-69-8 ***HR: 3***
n-BUTYL IODIDE
mf: C_4H_9I mw: 184.03

SYN: 1-IODOBUTANE

TOXICITY DATA with REFERENCE
ipr-mus TDLo:480 mg/kg/8W-I:NEO CNREA8 35,1411,75
ihl-rat LC50:6100 mg/m^3/4H 34ZIAG -,756,69
ipr-rat LD50:692 mg/kg 85GMAT -,30,82
ipr-mus LD50:101 mg/kg 85GMAT -,30,82

CONSENSUS REPORTS: EPA Genetic Toxicology Program. Reported in EPA TSCA Inventory.

SAFETY PROFILE: A poison by intraperitoneal route. Moderately toxic by inhalation. Questionable carcinogen with experimental neoplastigenic data. See also IODIDES. When heated to decomposition it emits toxic fumes of I^-.

BRQ350 ***HR: 2***
BUTYL ISOBUTYRATE
mf: $C_8H_{16}O_2$ mw: 44.44

PROP: Colorless liquid; apple-pineapple odor. D: 0.859-0.864, refr index: 1.401, flash p: 113°F. Misc with alc, ether, fixed oils; insol in glycerin, propylene glycol, water @ 166°.

SYN: FEMA No. 2188

SAFETY PROFILE: Combustible liquid. When heated to decomposition it emits acrid smoke and irritating fumes.

BRQ500 CAS:111-36-4 ***HR: 3***
n-BUTYL ISOCYANATE
DOT: UN 2485
mf: C_5H_9NO mw: 99.15

PROP: Colorless liquid. Bp: 115°, d: 0.880 @ 20°/4°.

SYNS: BIC ◇ ISOCYANIC ACID, BUTYL ESTER

TOXICITY DATA with REFERENCE
orl-rat LD50:600 mg/kg GTPZAB 20(3),53,76
ihl-rat LC50:3000 mg/m^3 GTPZAB 20(3),53,76
orl-mus LD50:150 mg/kg GTPZAB 20(3),53,76
ihl-mus LC50:680 mg/m^3 GTPZAB 20(3),53,76
ivn-mus LD50:1 mg/kg CSLNX* NX#05701
orl-gpg LD50:250 mg/kg GTPZAB 20(3),53,76

CONSENSUS REPORTS: Reported in EPA TSCA Inventory.

DOT Classification: Flammable Liquid; Label: Flammable Liquid and Poison.

SAFETY PROFILE: A poison by ingestion and intravenous routes. Mildly toxic by inhalation. A powerful irritant to eyes, skin, and mucous membranes. Flammable liquid. See also CYANATES and NITROGEN MONOXIDE.

BRQ750 CAS:7188-38-7 ***HR: 2***
tert-BUTYL ISOCYANIDE
mf: C_5H_9N mw: 83.15

SYN: tert-BUTYLISONITRILE

TOXICITY DATA with REFERENCE
ihl-rat LC50:710 mg/m^3/4H ARTODN 33,241,75
ihl-mus LC50:377 mg/m^3/4H ARTODN 33,241,75

CONSENSUS REPORTS: Cyanide and its compounds are on the Community Right-To-Know List.

SAFETY PROFILE: Moderately toxic by inhalation. When heated to decomposition it emits toxic fumes of NO_x and CN^-.

BRQ800 CAS:73791-40-9 ***HR: 3***
BUTYL(ISOPROPYL)ARSINIC ACID
mf: $C_7H_{17}AsO_2$ mw: 208.16

SYNS: ARSINE OXIDE, BUTYLHYDROXYISOPROPYL- ◇ BUTYLHYDROXYISOPROPYLARSINE OXIDE

TOXICITY DATA with REFERENCE
ivn-mus LD50:56 mg/kg CSLNX* NX#05105

OSHA PEL: TWA 0.5 mg(As)/m^3

SAFETY PROFILE: Poison by intravenous route. When heated to decomposition it emits toxic fumes of As.

BRR250 CAS:30026-92-7 ***HR: 3***
tert-BUTYL ISOPROPYL BENZENE HYDROPEROXIDE
DOT: NA 2091
mf: $C_{13}H_{20}O_2$ mw: 208.33

PROP: Crystals.

SYN: tert-BUTYL ISOPROPYL BENZENE HYDROPEROXIDE (DOT)

DOT Classification: Label: Organic Peroxide.

SAFETY PROFILE: Powerful irritant. See also PEROXIDES, ORGANIC. Dangerous fire hazard when exposed to heat or flame or by chemical reaction. Incompatible with oxidizing or reducing materials. When heated to decomposition it emits acrid smoke and fumes.

BRR500 CAS:74926-97-9 ***HR: 3***
2-sec-BUTYL-6-ISOPROPYLPHENOL
mf: $C_{13}H_{20}O$ mw: 192.2

TOXICITY DATA with REFERENCE
ivn-mus LD50:50 mg/kg JMCMAR 23,1350,80
ivn-rbt LDLo:15 mg/kg JMCMAR 23,1350,80

SAFETY PROFILE: Poison by intravenous route. When heated to decomposition it emits acrid smoke and irritating fumes.

BRR600 CAS:138-22-7 ***HR: 3***
n-BUTYL LACTATE
mf: $C_7H_{14}O_3$ mw: 146.21

PROP: Liquid. Sltly sol in water; misc in alc and ether.

Mp: −43°, bp: 188°, flash p: 160°F (OC), d: 0.968, autoign temp: 720°F, vap d: 5.04, vap press: 0.4 mm @ 20°.

SYNS: BUTYL α-HYDROXYPROPIONATE ◇ BUTYL LACTATE ◇ 2-HYDROXYPROPANOIC ACID, BUTYL ESTER ◇ LACTIC ACID, BUTYL ESTER

TOXICITY DATA with REFERENCE
skn-rbt 500 mg/24H MOD FCTXAV 17,727,79
scu-rat LD50:12 g/kg NPIRI* 1,15,74
ipr-mus LDLo:200 mg/kg CBCCT* 7,690,55
scu-mus LD50:11000 mg/kg FCTXAV 17,727,79

CONSENSUS REPORTS: Reported in EPA TSCA Inventory.

OSHA PEL: TWA 5 ppm
ACGIH TLV: TWA 5 ppm

SAFETY PROFILE: Poison by intraperitoneal route. A skin irritant. Toxic concentration in air for humans is about 4 ppm. Flammable when exposed to heat or flame; can react with oxidizing materials. To fight fire, use alcohol foam, foam, CO_2, dry chemical. When heated to decomposition it emits acrid smoke and irritating fumes. See also ESTERS, BUTYL ALCOHOL, and LACTIC ACID.

BRR739 CAS:109-72-8 ***HR: 3***
BUTYL LITHIUM
mf: C_4H_9Li mw: 64.06

SAFETY PROFILE: Probably very toxic. Solutions of greater than 20% will ignite spontaneously in air. Ignites on contact with water or CO_2. May cause potentially explosive polymerization of styrene. Extremely flammable. To fight fire, use dry chemical; see special instructions of manufacturer. See also LITHIUM COMPOUNDS and BUTYL LITHIUM.

BRR750 CAS:594-19-4 ***HR: 3***
tert-BUTYL LITHIUM
mf: C_4H_9Li mw: 64.06

SAFETY PROFILE: Probably very toxic. Solutions in heptane may ignite spontaneously in air. Potentially violent reaction with 2,2,2,4,4,4-hexafluoro-1,3-dimethyl-1,3,2,4-diazadiphosphetidine. Extremely flammable. To fight fire, use dry chemical; see special instructions of manufacturer. See also LITHIUM COMPOUNDS and BUTYL LITHIUM.

BRR900 CAS:109-79-5 ***HR: 3***
n-BUTYL MERCAPTAN
DOT: UN 2347
mf: $C_4H_{10}S$ mw: 90.20

PROP: Colorless liquid, skunk-like odor. Mp: −116°, bp: 98°, d: 0.8365 @ 25°/4°, flash p: 35°F, vap d: 3.1.

SYNS: BUTANETHIOL ◇ n-BUTANETHIOL ◇ BUTYL MERCAPTAN ◇ BUTYL MERCAPTAN (DOT) ◇ NCI-C60866

TOXICITY DATA with REFERENCE
eye-rbt 83 mg AIHAAP 19,171,58
ihl-mus TCLo:68 ppm/6H (female 6-16D post):REP FAATDF 8,170,87
ihl-mus TCLo:68 ppm/6H (female 6-16D post):TER FAATDF 8,170,87
orl-rat LD50:1500 mg/kg AIHAAP 19,171,58
ihl-rat LC50:4020 ppm/4H AIHAAP 19,171,58
ipr-rat LD50:399 mg/kg AIHAAP 19,171,58
orl-mus LD50:3 g/kg 85JCAE -,982,86
ihl-mus LC50:2500 ppm/4H AIHAAP 19,171,58
orl-rat LD50:3800 mg/kg

CONSENSUS REPORTS: Reported in EPA TSCA Inventory.

OSHA PEL: (Transitional: TWA 10 ppm) TWA 0.5 ppm
ACGIH TLV: TWA 0.5 ppm
DFG MAK: 0.5 ppm (1.5 mg/m^3)
NIOSH REL: (n-Alkane Mono Thiols) CL 0.5 ppm/15M
DOT Classification: Flammable Liquid; Label: Flammable Liquid.

SAFETY PROFILE: Poison by intraperitoneal route. Moderately toxic by ingestion. An eye irritant. Dangerous fire hazard by exposure to heat, flame, sparks, or powerful oxidizers. Reacts violently with HNO_3. Incompatible with acids, acid fumes, oxidizing materials, heat, flame, and sparks. To fight fire, use alcohol foam. When heated to decomposition it emits toxic SO_x. See also MERCAPTANS.

BRS000 CAS:486-17-9 ***HR: 3***
p-BUTYLMERCAPTOBENZHYDRYL-β-DIMETHYLAMINOETHYLSULPHIDE
mf: $C_{21}H_{29}NS_2$ mw: 359.63

SYNS: 2-((p-(BUTYLTHIO)-α-PHENYLBENZYL)THIO)-N,N-DIMETHYLETHYLAMINE ◇ CAPTODIAME ◇ CAPTODIAMIN ◇ CAPTODIAMINE ◇ COVATIN ◇ COVATIX ◇ N 68 ◇ SUVREN

TOXICITY DATA with REFERENCE
orl-rat LD50:3800 mg/kg ARZNAD 8,154,58
ipr-rat LD50:343 mg/kg ARZNAD 8,154,58
orl-mus LD50:1630 mg/kg ARZNAD 8,154,58
ipr-mus LD50:116 mg/kg JPETAB 108,201,53
scu-mus LD50:1750 mg/kg AIPTAK 136,440,62

SAFETY PROFILE: Poison by intraperitoneal route. Moderately toxic by ingestion and subcutaneous routes. See also MERCAPTANS and SULFIDES. When heated to decomposition it emits very toxic fumes of NO_x and SO_x.

BRS250 CAS:6192-29-6 ***HR: 3***
BUTYLMERCAPTOMETHYLPENICILLIN
mf: $C_{14}H_{22}N_2O_4S_2$ mw: 346.50

SYNS: n-BUTYLTHIOMETHYLPENICILLIN ◇ PENICILLIN BT

TOXICITY DATA with REFERENCE
ice-mus LD50:101 mg/kg JLCMAK 24,126,49
ice-dog LD50:11500 mg/kg JLCMAK 24,126,49
isp-dog LD50:56 mg/kg JLCMAK 24,126,49
ice-rbt LD50:15600 mg/kg JLCMAK 24,126,49

SAFETY PROFILE: Poison by intracerebral and intraspinal routes. When heated to decomposition it emits very toxic fumes of NO_x and SO_x. See also MERCAPTANS and other penicillin entries.

BRS500 CAS:6165-01-1 ***HR: 3***
9-BUTYL-6-MERCAPTOPURINE
mf: $C_9H_{12}N_4S$ mw: 208.31

SYNS: 9-BUTYL-1,9-DIHYDRO-6H-PURINE-6-THIONE ◇ 9-BUTYL-6-MP ◇ 9-BUTYL-9H-PURINE-6-THIOL ◇ NSC 19488 ◇ SRI 753

TOXICITY DATA with REFERENCE
ipr-rat LDLo:300 mg/kg CPCHAO 18,307,62
ipr-mus LD50:270 mg/kg NCISP* JAN86

SAFETY PROFILE: Poison by intraperitoneal route. When heated to decomposition it emits very toxic fumes of NO_x and SO_x. See also MERCAPTANS.

BRS750 CAS:543-63-5 ***HR: 3***
n-BUTYLMERCURIC CHLORIDE
mf: C_4H_9ClHg mw: 293.17

SYN: BMC

TOXICITY DATA with REFERENCE
dnr-esc 2 mmol/L MJDHDW 28,F39,80
cyt-hmn:hla 1 mg/L JJEMAG 39,47,69
scu-rat LDLo:73 mg/kg JJEMAG 39,47,69

CONSENSUS REPORTS: Mercury and its compounds are on the Community Right-To-Know List.

OSHA PEL: (Transitional: CL 1 mg/10m^3) TWA 0.01 mg(Hg)/m^3; STEL 0.03 mg/m^3 (skin)
ACGIH TLV: TWA 0.01 mg/(Hg)/m^3; STEL 0.03 mg(Hg)/m^3
NIOSH REL: (Mercury, Inorganic) TWA 0.05 mg(Hg)/m^3

SAFETY PROFILE: A poison by subcutaneous route. Mutation data reported. See also MERCURY COMPOUNDS, ORGANIC, and CHLORIDES. When heated to decomposition it emits very toxic fumes of Cl^- and Hg.

BRT000 CAS:532-34-3 ***HR: 2***
n-BUTYL MESITYL OXIDE OXALATE
mf: $C_{12}H_{18}O_4$ mw: 226.30

PROP: Yellow to reddish liquid. Bp: 113°, d: 1.052-1.060 @ 25°/25°, flash p: 315°F.

SYNS: BMOO ◇ BUTOPYRONOXYL ◇ BUTYL-3,4-DIHYDRO-2,2-DIMETHYL-4-OXO-2H-PYRAN-6-CARBOXYLATE ◇ n-BUTYL ESTER of3,4-DIHYDRO-2,2-DIMETHYL-4-OXO-2H-PYRAN-6-CARBOXYLIC ACID ◇ n-BUTYLMESITYLOXID OXALATE ◇ 2-CARBO-n-BUTOXY-6,6-DIMETHYL-5,6-DIHYDRO-1,4-PYRONE ◇ 3,4-DIHYDRO-2,2-DIMETHYL-4-OXO-2H-PYRAN-6-CARBOXYLIC ACID-n-BUTYL ESTER ◇ DIHDYROPYRONE ◇ α,α-DIMETHYL-α'-CARBOBUTOXY-DIHYDRO-Γ-PYRONE ◇ 2,2-DIMETHYL-6-CARBOBUTOXY-2,3-DIHYDRO-4-PYRONE ◇ ENT 9 ◇ INDALONE

TOXICITY DATA with REFERENCE
orl-rat LD50:77848 mg/kg YKYUA6 32,605,81
orl-mus LD50:12250 mg/kg ARSIM* 20,13,66
orl-rbt LD50:5400 mg/kg JPETAB 93,26,48
orl-gpg LD50:3200 mg/kg JPETAB 93,26,48

CONSENSUS REPORTS: Reported in EPA TSCA Inventory.

SAFETY PROFILE: Moderately toxic by ingestion. Produces liver necrosis in experimental animals. A mild skin irritant. See also OXALATES and ESTERS. Combustible when exposed to heat or flame. When heated to decomposition it emits acrid and irritant fumes.

BRT250 CAS:1912-32-9 ***HR: D***
BUTYL MESYLATE
mf: $C_5H_{12}O_3S$ mw: 152.23

SYNS: BUTYL METHANESULFONATE ◇ n-BUTYL METHANESULFONATE

TOXICITY DATA with REFERENCE
mmo-sat 20 mmol/L CNREA8 38,1595,78
sln-dmg-par 10 mmol/L JOGNAU 54,146,56
msc-hmn:lym 1 mmol/L/24H MUREAV 54,193,78
dnd-mam:lym 10 mmol/L CRNGDP 5,621,84

CONSENSUS REPORTS: EPA Genetic Toxicology Program.

SAFETY PROFILE: Human mutation data reported. See also SULFONATES. When heated to decomposition it emits toxic fumes of SO_x.

BRT750 ***HR: D***
BUTYL METHOXYMETHYLNITROSAMINE
mf: $C_6H_{14}N_2O_2$ mw: 146.22

TOXICITY DATA with REFERENCE
cyt-ham:fbr 1 g/L/48H MUREAV 48,337,77

SAFETY PROFILE: Mutation data reported. Many nitrosamines are carcinogens. When heated to decomposition it emits toxic fumes of NO_x. See also NITROSAMINES.

BRU000 ***HR: D***
sec-BUTYL METHOXYMETHYLNITROSAMINE
mf: $C_6H_{14}N_2O_2$ mw: 146.22

SYN: N-METHOXYMETHYL-N-NITROSO-sec-BUTYLAMINE

TOXICITY DATA with REFERENCE
cyt-ham:fbr 2 g/L/48H MUREAV 48,337,77

SAFETY PROFILE: Mutation data reported. Many nitrosamines are carcinogens. When heated to decomposition it emits toxic fumes of NO_x. See also NITROSAMINES.

BRU250 CAS:5412-64-6 ***HR: 3***
n-BUTYL-α-METHYLBENZYLAMINE
mf: $C_{12}H_{19}N$ mw: 177.32

TOXICITY DATA with REFERENCE
skn-rbt 10 mg/24H open MLD AIHAAP 23,95,62
orl-rat LD50:360 mg/kg AIHAAP 23,95,62
skn-rbt LD50:570 mg/kg AIHAAP 23,95,62

SAFETY PROFILE: Poison by ingestion. Moderately toxic by skin contact. A skin irritant. When heated to decomposition it emits toxic fumes of NO_x. See also AMINES.

BRU500 CAS:83-66-9 ***HR: 3***
6-tert-BUTYL-3-METHYL-2,4-DINITRO ANISOLE
mf: $C_{12}H_{16}N_2O_5$ mw: 268.30

SYNS: 2,6-DINITRO-3-METHOXY-4-tert-BUTYLTOLUENE ◇ MUSK AMBRETTE

TOXICITY DATA with REFERENCE
skn-rbt 500 mg/24H MOD FCTXAV 13,681,75
mmo-sat 2 μmol/plate FCTOD7 21,707,83
mma-sat 100 μg/plate FCTOD7 24,27,86
sln-dmg-orl 10 mmol/L FCTOD7 21,707,83
orl-rat LD50:339 mg/kg FCTXAV 2,327,64

CONSENSUS REPORTS: Reported in EPA TSCA Inventory.

SAFETY PROFILE: A poison by ingestion. Mutation data reported. A skin irritant. When heated to decomposition it emits toxic fumes of NO_x. See also AROMATIC AMINES.

BRU750 CAS:2487-01-6 ***HR: 3***
2-tert-BUTYL-5-METHYL-4,6-DINITROPHENYL ACETATE
mf: $C_{13}H_{16}N_2O_6$ mw: 296.31

SYNS: ACETIC ACID-2-(tert-BUTYL)-4,6-DINITRO-m-TOLYL ESTER ◇ 6-(1,1-DIMETHYLETHYL)-3-METHYL-2,4-DINITROPHENYL ACETATE ◇ 2,4-DINITRO-3-METHYL-6-tert-BUTYLPHNYLACETAT (GERMAN) ◇ 2,4-DINITRO-3-METHYL-6-tert-BUTYLPHENYL ACETATE ◇ MC 1488 ◇ MEDINOTERB ACETATE ◇ P 1488

TOXICITY DATA with REFERENCE
orl-rat LD50:42 mg/kg FMCHA2 -,D191,80
skn-rat LD50:1300 mg/kg GUCHAZ 6,326,73
orl-mus LD50:90 mg/kg 85GYAZ -,75-71
orl-rbt LD50:80 mg/kg 28ZEAL 4,82,69
orl-gpg LD50:55 mg/kg 28ZEAL 4,82,69
skn-gpg LD50:7200 mg/kg 85GYAZ -,75,71
orl-ckn LD50:560 mg/kg 28ZEAL 5,144,76

SAFETY PROFILE: Poison by ingestion. Moderately toxic by skin contact. See also ESTERS. When heated to decomposition it emits toxic fumes of NO_x.

BRV000 CAS:100836-63-3 ***HR: 3***
tert-BUTYL-N-(3-METHYL-2-THIAZOLIDINYLIDENE)CARBAMATE
mf: $C_9H_{16}N_2O_2S$ mw: 216.33

TOXICITY DATA with REFERENCE
orl-mus LD50:306 mg/kg JMCMAR 23,773,80
ivn-mus LD50:68 mg/kg JMCMAR 23,773,80

SAFETY PROFILE: Poison by ingestion and intravenous routes. See also CARBAMATES. When heated to decomposition it emits very toxic fumes of NO_x and SO_x.

BRV100 CAS:1005-67-0 ***HR: 3***
4-BUTYLMORPHOLINE
mf: $C_8H_{17}NO$ mw: 143.26

PROP: Bp: 110-115°

SYNS: N-BUTYLMORPHOLINE ◇ N-(n-BUTYL)MORPHOLINE ◇ MORPHOLINE, 4-BUTYL-

TOXICITY DATA with REFERENCE
skn-rbt 500/24H MOD JACTDZ 1,13,90
eye-rbt 100 mg MOD JACTDZ 1,13,90
orl-rat LD50:338 mg/kg JACTDZ 1,13,90
skn-rbt LD50:1800 mg/kg JACTDZ 1,13,90

SAFETY PROFILE: Poison by ingestion. Moderately toxic and corrosive to skin. A skin and eye irritant. When heated to decomposition it emits toxic fumes of NO_x.

BRV325 CAS:928-45-0 ***HR: 3***
BUTYL NITRATE
mf: $C_4H_9NO_3$ mw: 119.12

SAFETY PROFILE: An explosive. Reacts explosively with Lewis acids (e.g., boron trifluoride; aluminum chloride; etc.). When heated to decomposition it emits toxic fumes of NO_x. See also NITRATES.

BRV500 CAS:544-16-1 ***HR: 3***
n-BUTYL NITRITE
DOT: UN 2351
mf: $C_4H_9NO_2$ mw: 103.14

PROP: Oily liquid, characteristic odor, misc in alc and ether. Bp: 75°, d: 0.9114 @ 0°/4°, vap d: 3.5, flash p: 10°.

SYNS: BUTYL NITRITE (DOT) ◇ NBN ◇ NCI-C56553 ◇ NITROUS ACID-n-BUTYL ESTER

TOXICITY DATA with REFERENCE
mmo-sat 1 mg/plate PSEBAA 157,688,78
ipr-mus LD50:169 mg/kg TXAPA9 48,A43,79
orl-man TDLo:153 mg/kg:BLD AIMEAS 92,570,80
orl-rat LD50:83 mg/kg JJATDK 1,30,81
ihl-rat LC50:918 ppm/1H FEPRA7 41,1538,82
orl-mus LD50:171 mg/kg RDSADO 3,233,82
ihl-mus LC50:567 ppm/1H FAATDF 1,448,81

CONSENSUS REPORTS: Reported in EPA TSCA Inventory.

DOT Classification: Flammable Liquid; Label: Flammable Liquid.

SAFETY PROFILE: A poison by ingestion and intraperitoneal routes. Mildly toxic by inhalation. An irritant. Human systemic effects by ingestion: methemoglobinemia-carboxhemoglobinemia. Resembles amyl nitrite in causing fall in blood pressure, headache, pulse throbbing, and weakness. Mutation data reported. Flammable when exposed to heat or flame or by spontaneous chemical reaction. When heated to decomposition it emits toxic fumes of NO_x. See also NITRITES, BUTYL ALCOHOL, and ESTERS.

BRV750 CAS:924-43-6 ***HR: 3***
sec-BUTYL NITRITE
mf: $C_4H_9NO_2$ mw: 103.14

PROP: Liquid. Bp: 68°, d: 0.8981 @ 0°/4°, vap d: 3.5.

SYNS: NITROUS ACID-sec-BUTYL ESTER ◇ NITROUS ACID-1-METHYL PROPYL ESTER

TOXICITY DATA with REFERENCE
mma-sat 1 mg/plate BSIBAC 56,816,80
ipr-mus LD50:496 mg/kg TXAPA9 48,A43,79
orl-mus LD50:423 mg/kg RCSADO 3,233,82
ihl-mus LD50:1753 ppm/1H FAATDF 1,448,81

CONSENSUS REPORTS: Reported in EPA TSCA Inventory.

SAFETY PROFILE: Moderately toxic by ingestion, inhalation, and intraperitoneal routes. Mutation data reported. Flammable when exposed to heat or flame or by spontaneous chemical reaction. An oxidizer. Potentially explosive. To fight fire, use water, spray, foam, dry chemical. When heated to decomposition it emits toxic fumes of NO_x. See also n-BUTYL NITRITE, NITRITES, and ESTERS.

BRV760 CAS:540-80-7 ***HR: 3***
tert-BUTYL NITRITE
mf: $C_4H_9NO_2$ mw: 103.14

PROP: Yellow liquid, agreeable odor. D: 0.8941, bp: 63°, n (20/D) 1.3687. Very sol in alc, ether, chloroform, carbon disulfide; sltly sol in water; practically insol in glycerol.

SYNS: α,α-DIMETHYLETHYL NITRITE ◇ NITROUS ACID-1,1-DIMETHYLETHYL ESTER

TOXICITY DATA with REFERENCE
orl-mus LD50:308 mg/kg RCSADO 3,233,82
ihl-mus LC50:10852 ppm/1H FAATDF 1,448,81
ipr-mus LD50:496 mg/kg TXAPA9 48,A43,79

SAFETY PROFILE: Poison by ingestion. Moderately toxic by intraperitoneal route. Jet propellant. When heated to decomposition it emits toxic fumes of NO_x. See also n-BUTYL NITRITE, NITRITES, and ESTERS.

BRW000 ***HR: 3***
tert-BUTYL NITROACETYLENE
mf: $C_6H_9NO_2$ mw: 127.14

SAFETY PROFILE: When ignited in absence of a solvent, the primary, secondary and tertiary amines explode. Incompatible with amines. See also ACETYLENE COMPOUNDS.

BRW250 ***HR: 3***
tert-BUTYL-p-NITRO PEROXY BENZOATE
mf: $C_{11}H_{13}NO_5$ mw: 239.2

SAFETY PROFILE: Explodes in contact with flame. When heated to decomposition it emits toxic fumes of NO_x. See also PEROXIDES, ORGANIC.

BRW500 CAS:71002-67-0 ***HR: 3***
BUTYL-p-NITROPHENYL ESTER of ETHYLPHOSPHONIC ACID
mf: $C_{12}H_{18}NO_5P$ mw: 287.28

SYN: ETHYLPHOSPHONIC ACID BUTYL-p-NITROPHENYL ESTER

TOXICITY DATA with REFERENCE
scu-mus LD50:1500 μg/kg RPTOAN 42,106,79
ivn-mus LD50:1300 μg/kg RPTOAN 42,106,79
scu-rat LD50:1500 μg/kg FATOAO 42(3),299,79
ivn-rat LD50:1300 μg/kg FATOAO 42(3),299,79

SAFETY PROFILE: Deadly poison by subcutaneous and intravenous routes. See also ESTERS. When heated to decomposition it emits very toxic fumes of PO_x and NO_x.

BRW750 CAS:21070-32-6 ***HR: 3***
6-BUTYL-4-NITROQUINOLINE-1-OXIDE
mf: $C_{13}H_{14}N_2O_3$ mw: 246.29

TOXICITY DATA with REFERENCE
dns-ham:oth 4 μmol/L NATUAS 229,416,71
dnd-mus:fbr 100 μmol/L CNREA8 35,521,75
scu-mus TDLo:60 mg/kg/I:ETA CPBTAL 17,544,69

CONSENSUS REPORTS: EPA Genetic Toxicology Program.

SAFETY PROFILE: Questionable carcinogen with experimental tumorigenic data. Mutation data reported. When heated to decomposition it emits toxic fumes of NO_x.

BRX000 CAS:52731-39-2 ***HR: D***
4-(BUTYLNITROSAMINO)BUTYL ACETATE
mf: $C_{10}H_{20}N_2O_3$ mw: 216.2

SYNS: BABN ◇ N-BUTYL-N-(1-ACETOXYBUTYL)NITROSAMINE

TOXICITY DATA with REFERENCE
mmo-sat 1 μmol/plate GANNA2 71,124,80
mmo-esc 1 μmol/plate GANNA2 71,124,80
dnr-bcs 500 nmol/plate GANNA2 66,457,75
msc-ham:lng 100 μmol/L GANNA2 72,531,81

SAFETY PROFILE: Mutation data reported. Many nitrosamines are carcinogens. When heated to decomposition it emits toxic fumes of NO_x. See also NITROSAMINES.

BRX250 CAS:62018-92-2 ***HR: D***
N-BUTYL-N-NITROSO-β-ALANINE
mf: $C_7H_{14}N_2O_3$ mw: 174.23

SYN: N-BUTYL-N-(2-CARBOXYETHYL)NITROSAMINE

TOXICITY DATA with REFERENCE
mma-sat 30 μmol/plate CNREA8 37,399,77

SAFETY PROFILE: Mutation data reported. Many nitrosamines are carcinogens. See also NITROSAMINES. When heated to decomposition it emits toxic fumes of NO_x.

BRX500 CAS:56986-36-8 ***HR: 3***
BUTYLNITROSOAMINOMETHYL ACETATE
mf: $C_7H_{14}N_2O_3$ mw: 174.23

SYNS: ACETOXYMETHYLBUTYLNITROSAMINE ◇ N-(ACETOXY)METHYL-N,N-BUTYLNITROSAMINE ◇ BAMN ◇ BUTYL ACETOXYMETHYLNITROSAMINE ◇ N-BUTYL-N-(ACETOXYMETHYL)NITROSAMINE ◇ N-NITROSO-N-(1-ACETOXYMETHYL)BUTYLAMINE

TOXICITY DATA with REFERENCE
mmo-sat 1 μmol/plate MUREAV 49,187,78
mmo-esc 1 μmol/plate GANNA2 71,124,80
dnr-bcs 500 nmol/plate GANNA2 66,457,75
dns-rat:oth 10 μmol/L CBINA8 53,99,85
dnd-mus:fbr 260 nmol/L GANNA2 73,565,82
cyt-ham:fbr 16 mg/L/24H MUREAV 48,337,77
msc-ham:lng 100 μmol/L GANNA2 75,531,81
orl-rat TDLo:555 mg/kg/90D-I:ETA ZKKOBW 91,317,78
scu-rat TDLo:50 mg/kg/10W-I:CAR JCROD7 104,13,82
scu-rat TD:66 mg/kg/10W-I:CAR IAPUDO 41,619,82
orl-rat LD50:1500 mg/kg ZKKOBW 91,317,78

CONSENSUS REPORTS: EPA Genetic Toxicology Program.

SAFETY PROFILE: Moderately toxic by ingestion. Questionable carcinogen with experimental carcinogenic and tumorigenic data. Mutation data reported. When heated to decomposition it emits toxic fumes of NO_x. See also NITROSAMINES.

BRX750 CAS:51938-13-7 ***HR: D***
3-(BUTYLNITROSOAMINO)-1-PROPANOL
mf: $C_7H_{16}N_2O_2$ mw: 160.25

SYN: BUTYL-(3-HYDROXYPROPYL)NITROSAMINE

TOXICITY DATA with REFERENCE
mmo-sat 100 μg/plate MUREAV 56,219,78

SAFETY PROFILE: Mutation data reported. Many nitrosamines are carcinogens. When heated to decomposition it emits toxic fumes of NO_x. See also NITROSAMINES.

BRY000 CAS:51938-15-9 ***HR: 3***
1-(BUTYLNITROSOAMINO)-2-PROPANONE
mf: $C_7H_{14}N_2O_2$ mw: 158.23

SYNS: BUTYL(2-OXOPROPYL)NITROSOAMINE ◇ N-NITROSO-1-BUTYLAMINO-2-PROPANONE ◇ N-NITROSO-(2-OXOPROPYL)-N-BUTYLAMINE

TOXICITY DATA with REFERENCE
mmo-sat 31 μmol/plate CNREA8 37,399,77
mma-sat 4 μmol/plate CNREA8 37,399,77
orl-rat TDLo:2000 mg/kg/13W-C:ETA GANNA2 65,13,74

CONSENSUS REPORTS: EPA Genetic Toxicology Program.

SAFETY PROFILE: Questionable carcinogen with experimental tumorigenic data. Mutation data reported.

When heated to decomposition it emits toxic fumes of NO_x. See also NITROSAMINES.

BRY250 CAS:16339-05-2 ***HR: 3***
N-BUTYL-N-NITROSO AMYL AMINE
mf: $C_9H_{20}N_2O$ mw: 172.31

SYNS: BUTYLAMYLNITROSAMIN (GERMAN) ◇ N-BUTYL-N-NITROSOPENTYLAMINE ◇ N-BUTYL-N-PENTYLINITROSAMINE ◇ N-NITRO- SO-N-BUTYLPENTYLAMINE ◇ N-NITROSO-N-BUTYL-N-PENTYLAMINE

TOXICITY DATA with REFERENCE
scu-mus TDLo:17 g/kg/21W-I:ETA ZEKBAI 69,103,67
scu-rat LD50:2500 mg/kg ZEKBAI 69,103,67

SAFETY PROFILE: Moderately toxic by subcutaneous route. Questionable carcinogen with experimental tumorigenic data. See also N-NITROSO COMPOUNDS and NITROSAMINES. When heated to decomposition it emits toxic fumes of NO_x.

BRY500 CAS:924-16-3 ***HR: 3***
n-BUTYL-N-NITROSO-1-BUTAMINE
mf: $C_8H_{18}N_2O$ mw: 158.28

PROP: Pale yellow liquid. Bp: 235°.

SYNS: DBN ◇ DBNA ◇ DI-n-BUTYLNITROSAMIN (GERMAN) ◇ DIBUTYLNITROSOAMINE ◇ DI-n-BUTYLNITROSAMINE ◇ N,N-DI-n-BUTYLNITROSAMINE ◇ N,N-DIBUTYLNITROSOAMINE ◇ NDBA ◇ N-NITROSODIBUTYLAMINE ◇ N-NITROSODI-n-BUTYLAMINE (MAK) ◇ RCRA WASTE NUMBER U172

TOXICITY DATA with REFERENCE
mma-esc 1 μmol/plate GANNA2 75,8,84
dnd-esc 100 nmol/tube CRNGDP 3,781,82
dns-hmn:hla 10 μmol/L CNREA8 38,2621,78
dnd-rat:lvr 100 μmol/L CNREA8 42,2592,82
bfa-rat/sat 158 mg/kg CRNGDP 6,967,85
hma-rat/smc 2912 mg/kg TCMUD8 3,41,83
scu-ham TDLo:30 mg/kg (female 15D post):REP ZEKBAI 86,69,76
orl-rat TDLo:1200 mg/kg (12D preg):TER BEXBAN 78,1308,74
orl-rat TDLo:140 mg/kg/4W-C:CAR CNREA8 46,6160,86
scu-rat TDLo:8 g/kg/20W-I:ETA XENOBH 3,271,73
orl-mus TDLo:1200 mg/kg/8W-I:NEO TXAPA9 82,19,86
scu-mus TDLo:800 mg/kg/40W-I:CAR EJCAAH 6,433,70
orl-ham TDLo:9 g/kg:ETA,TER PSEBAA 136,1007,71
scu-ham TDLo:240 mg/kg:CAR,TER ZEKBAI 86,69,76
orl-rat LD50:1200 mg/kg NATWAY 50,735,63
scu-rat LD50:1200 mg/kg XENOBH 3,271,73
orl-ham LD50:2150 mg/kg ZKKOBW 79,85,73
ipr-ham LD50:1200 mg/kg ZKKOBW 79,85,73
scu-ham LD50:561 mg/kg PSEBAA 136,168,71

CONSENSUS REPORTS: NTP Fifth Annual Report on Carcinogens. IARC Cancer Review: Group 2B IMEMDT 7,56,87; Animal Sufficient Evidence IMEMDT 28,151,82; IMEMDT 17,51,78; IMEMDT 4,197,74; Human Limited Evidence IMEMDT 17,51,78. Community Right-To-Know List. EPA Genetic Toxicology Program. Reported in EPA TSCA Inventory.

DFG MAK: Animal Carcinogen, Suspected Human Carcinogen.

SAFETY PROFILE: Confirmed carcinogen with experimental carcinogenic, tumorigenic, and neoplastigenic data. Moderately toxic by ingestion, subcutaneous and intraperitoneal routes. Experimental teratogenic effects. Human mutation data reported. When heated to decomposition it emits toxic fumes of NOx. See also NITROSAMINES.

BRY750 CAS:76206-38-7 ***HR: D***
N-BUTYL-N-NITROSOCARBAMIC ACID-1-NAPHTHYL ESTER
mf: $C_{15}H_{16}N_2O_3$ mw: 272.2

SYN: 1-NAPHTHYL-N-BUTYL-N-NITROSOCARBAMATE

TOXICITY DATA with REFERENCE
mmo-sat 10 nmol/plate ENMUDM 2,395,80
mrc-smc 5 nmol/plate ENMUDM 2,395,80

SAFETY PROFILE: Mutation data reported. Many N-nitroso compounds are carcinogens. See also N-NITROSO COMPOUNDS and CARBAMATES. When heated to decomposition it emits toxic fumes of NO_x.

BRZ000 CAS:6558-78-7 ***HR: 3***
N-BUTYL-N-NITROSO ETHYL CARBAMATE
mf: $C_7H_{14}N_2O_3$ mw: 174.23

SYNS: N-BUTYL-N-NITROSOURETHAN ◇ 1-BUTYL-1-NITROSOURETHAN ◇ TL 478

TOXICITY DATA with REFERENCE
mmo-bcs 5 g/L MUREAV 42,19,77
dnr-bcs 5 g/L MUREAV 42,19,77
cyt-ham:fbr 120 mg/L/48H MUREAV 48,337,77
cyt-ham:lng 35 mg/L GMCRDC 27,95,81
sce-ham:fbr 100 μmol/L JNCIAM 58,1635,77
orl-rat TDLo:500 mg/kg (20D preg):ETA,TER GANNA2 71,811,80
orl-rat TDLo:2240 mg/kg/8W-C:ETA GANNA2 65,227,74
scu-rat TDLo:150 mg/kg (15-21D preg):ETA,TER GANNA2 71,811,80
orl-mus TDLo:5300 mg/kg/20W-C:NEO GANNA2 67,231,76
orl-rat LD50:900 mg/kg GANNA2 65,227,74
ihl-mus LCLo:300 mg/m^3/10M NDRC** NDCrc-132,Nov,42

CONSENSUS REPORTS: EPA Genetic Toxicology Program.

SAFETY PROFILE: A poison by inhalation. Moderately toxic by ingestion. Experimental teratogenic data. Questionable carcinogen with experimental neoplastigenic, and tumorigenic data. Mutation data reported. See also N-nitroso compounds and CARBAMATES. When heated to decomposition it emits toxic fumes of NO_x.

BRZ200 CAS:17721-94-7 ***HR: 2***
4-tert-BUTYL-1-NITROSOPIPERIDINE
mf: $C_9H_{18}N_2O$ mw: 170.29

SYN: N-NITROSO-4-tert-BUTYLPIPERIDINE

TOXICITY DATA with REFERENCE
mma-sat 250 μg/plate MUREAV 111,135,83
orl-rat TDLo:4500 mg/kg/2Y-I:CAR CRNGDP 2,1045,81

SAFETY PROFILE: Questionable carcinogen with experimental carcinogenic data. Mutation data reported. When heated to decomposition it emits toxic fumes of NO_x.

BSA000 CAS:72505-66-9 ***HR: D***
2-BUTYL-3-NITROSOTHIAZOLIDINE
mf: $C_7H_{14}N_2OS$ mw: 174.29

SYN: N-NITROSO-n-BUTYLTHIAZOLIDINE

TOXICITY DATA with REFERENCE
mmo-sat 1 mg/L JAFCAU 28,62,80
mma-sat 1 mg/L JAFCAU 28,62,80

SAFETY PROFILE: Mutation data reported. Many N-nitroso compounds are carcinogens. When heated to decomposition it emits very toxic fumes of NO_x and SO_x. See also N-NITROSO COMPOUNDS.

BSA250 CAS:869-01-2 ***HR: 3***
n-BUTYLNITROSOUREA
mf: $C_5H_{11}N_3O_2$ mw: 145.19

SYNS: BNU ◇ BUTYLNITROSOHARNSTOFF (GERMAN) ◇ N-n-BUTYL-N-NITROSOUREA ◇ 1-BUTYL-1-NITROSOUREA ◇ N-NITROSOBUTYLUREA

TOXICITY DATA with REFERENCE
pic-esc 2 mg/L TCMUE9 1,91,84
sce-ham:fbr 500 μmol/L CNREA8 44,3270,84
orl-rat TDLo:4867 mg/kg/24W-I:CAR GANNA2 67,33,76
orl-rat TDLo:120 mg/kg (22D preg):ETA,TER ARGEAR 48,9,78
orl-rat TDLo:16512 mg/kg/50W-I:CAR,REP JCROD7 107,32,84
ipr-rat TDLo:300 mg/kg:ETA GANMAX 12,283,72
ipr-rat TDLo:120 mg/kg (22D preg):ETA,TER ARGEAR 48,9,78
orl-mus TDLo:2800 mg/kg/10W-C:CAR GANNA2 68,281,77
orl-rat LD50:400 mg/kg PPTCBY 2,73,72
scu-rat LD50:1200 mg/kg ZEKBAI 69,103,67

CONSENSUS REPORTS: EPA Genetic Toxicology Program.

SAFETY PROFILE: Suspected carcinogen with experimental carcinogenic and tumorigenic data. An poison by ingestion. Moderately toxic by subcutaneous route. Experimental teratogenic and reproductive effects. Mutation data reported. When heated to decomposition it emits toxic fumes of NO_x. See also NITROSAMINES.

BSA500 CAS:3913-02-8 ***HR: 1***
2-BUTYL-1-OCTANOL
mf: $C_{12}H_{26}O$ mw: 186.38

PROP: Liquid. Mp: −80°, flash p: 230°F(OC), bp: 253.3°, d: 0.8355 @ 20°/20°, vap d: 6.42.

SYN: 2-BUTYLOCTYL ALCOHOL

TOXICITY DATA with REFERENCE
skn-rbt 10 mg/24H open MLD AMIHBC 4,119,51
eye-rbt 500 mg open AMIHBC 4,119,51
orl-rat LD50:13 g/kg AMIHBC 4,119,51

CONSENSUS REPORTS: Reported in EPA TSCA Inventory.

SAFETY PROFILE: Mildly toxic by ingestion. A skin and eye irritant. See also ALCOHOLS. Combustible when exposed to heat or flame. Incompatible with oxidizing materials. To fight fire, use CO_2, dry chemical. When heated to decomposition it emits acrid and irritating fumes.

BSA750 CAS:10097-26-4 ***HR: 1***
2-BUTYLOCTYL ESTER METHACRYLIC ACID
mf: $C_{16}H_{30}O_2$ mw: 254.46

TOXICITY DATA with REFERENCE
skn-rbt 10 mg/24H open MLD AMIHBC 10,61,54
eye-rbt 500 mg open AMIHBC 10,61,54
orl-rat LD50:26 g/kg AMIHBC 10,61,54

SAFETY PROFILE: Mildly toxic by ingestion. A skin and eye irritant. See also ESTERS. When heated to decomposition it emits smoke and irritating fumes.

BSB000 ***HR: 1***
BUTYL OLEATE
mf: $C_{22}H_{42}O_2$ mw: 338.64

PROP: Liquid. Bp: 173°, flash p: 356°F(OC), d: 0.873, vap d: 11.3.

SYN: (Z)-9-OCTADECENOIC ACID BUTYL ESTER

TOXICITY DATA with REFERENCE
skn-rbt 500 mg/24H MOD FCTXAV 17,241,79

SAFETY PROFILE: A skin irritant. Combustible when exposed to heat or flame. To fight fire, use CO_2, dry chemical. Incompatible with oxidizing materials. When heated to decomposition it emits acrid smoke and irritating fumes. See also ESTERS, BUTYL ALCOHOL, and OLEIC ACID.

BSB500 CAS:61734-89-2 ***HR: 3***
N-BUTYL-N-(2-OXOBUTYL)NITROSAMINE
mf: $C_8H_{16}N_2O_2$ mw: 172.26

SYN: N-NITROSO-N-(2-OXOBUTYL)BUTYLAMINE

TOXICITY DATA with REFERENCE
mma-sat 4 μmol/plate CNREA8 37,399,77
orl-rat TDLo:69 g/kg/20W-C:ETA GANNA2 67,825,76

CONSENSUS REPORTS: EPA Genetic Toxicology Program.

SAFETY PROFILE: Questionable carcinogen with experimental tumorigenic data. Mutation data reported. When heated to decomposition it emits toxic fumes of NO_x. See also NITROSAMINES.

BSB750 CAS:61734-90-5 ***HR: 3***
N-BUTYL-N-(3-OXOBUTYL)NITROSAMINE
mf: $C_8H_{16}N_2O_2$ mw: 172.26

SYN: N-NITROSO-N-(3-OXOBUTYL)BUTYLAMINE

TOXICITY DATA with REFERENCE
mma-sat 4 μmol/plate CNREA8 37,399,77
orl-rat TDLo:69 g/kg/20W-C:ETA GANNA2 67,825,76

CONSENSUS REPORTS: EPA Genetic Toxicology Program.

SAFETY PROFILE: Questionable carcinogen with experimental tumorigenic data. Mutation data reported. When heated to decomposition it emits toxic fumes of NO_x. See also NITROSAMINES.

BSC000 CAS:94-26-8 ***HR: 3***
BUTYL PARABEN
mf: $C_{11}H_{14}O_3$ mw: 194.25

SYNS: BUTOBEN ◇ BUTYL CHEMOSEPT ◇ BUTYL-p-HYDROXYBENZOATE ◇ n-BUTYL PARAHYDROXYBENZOATE ◇ BUTYL PARASEPT ◇ BUTYL TEGOSEPT ◇ p-HYDROXYBENZOIC ACID BUTYL ESTER ◇ NIPABUYL ◇ PARASEPT ◇ SOLBROL B ◇ TEGOSEPT B

TOXICITY DATA with REFERENCE
skn-gpg 5%/48H MLD JSCCA5 28,357,77
ipr-mus LD50:230 mg/kg JSCCA5 28,357,77

CONSENSUS REPORTS: Reported in EPA TSCA Inventory.

SAFETY PROFILE: Poison by intraperitoneal route. A skin irritant. When heated to decomposition it emits acrid smoke and irritating fumes. See also ESTERS.

BSC250 CAS:107-71-1 ***HR: 3***
tert-BUTYL PERACETATE
DOT: UN 2095/UN 2096
mf: $C_6H_{12}O_3$ mw: 132.18

PROP: Clear, colorless, benzene solution; insol in water; sol in organic solvents. D: 0.923, vap press: 50 mm @ 26°, flash p: <80°F (COC).

SYNS: tert-BUTYL PEROXYACETATE ◇ tert-BUTYL PEROXYACETATE, more than 76% in solution (DOT) ◇ ETHANEPEROXOIC ACID-1,1-DIMETHYLETHYL ESTER ◇ LUPERSOL 70

TOXICITY DATA with REFERENCE
orl-rat LD50:675 mg/kg 85GMAT -,30,82
ihl-rat LC33:8200 mg/m^3/4H 85GMAT -,30,82
orl-mus LD50:632 mg/kg 85GMAT -,30,82
ihl-mus LC33:6 g/m^3/2H 85GMAT -,30,82

CONSENSUS REPORTS: Reported in EPA TSCA Inventory.

DFG MAK: Moderate skin irritant.
DOT Classification: Organic Peroxide; Label: Organic Peroxide; Forbidden (> 76% in solution).

SAFETY PROFILE: Moderately toxic by ingestion. Mildly toxic by inhalation. Moderate skin and eye irritant. A shock and heat-sensitive explosive. Dangerous fire hazard when exposed to heat, flame, reducing agents. To fight fire, use dry chemical, alcohol foam, spray and mist. When heated to decomposition it emits acrid smoke and fumes. See also PEROXIDES, ORGANIC, and ESTERS.

BSC500 CAS:614-45-9 ***HR: 3***
tert-BUTYL PERBENZOATE
DOT: UN 2890/UN 2097/UN 2098
mf: $C_{11}H_{14}O_3$ mw: 194.25

PROP: Colorless to slt yellow liquid, mild aromatic odor. Bp: 112° (decomp), flash p: 19°, fp: 8°, vap press: 0.33 mm @ 50°, d: 1.0. Insol in water; sol in organic solvents.

SYNS: tert-BUTYLPERBENZOAN (CZECH) ◇ tert-BUTYL PEROXY BENZOATE ◇ tert-BUTYL PEROXYBENZOATE, technical pure or in concentration of more than 75% (DOT) ◇ ESPEROX 10 ◇ NOVOX ◇ PERBENZOATE de BUTYLE TERTIAIRE (FRENCH) ◇ TRIGONOX C

TOXICITY DATA with REFERENCE
skn-rbt 500 mg/24H MLD 28ZPAK -,52,72
eye-rbt 100 mg/1M rns MLD ZAARAM 8,25,58
eye-rbt 500 mg/24H MLD 28ZPAK -,52,72
mma-sat 67 μg/plate ENMUDM 8(Suppl 7),52,72
unr-mus TDLo:311 mg/kg:ETA RARSAM 3,193,63

orl-rat LD50:1012 mg/kg 85GMAT -,30,82
orl-mus LD50:914 mg/kg 85GMAT -,30,82

CONSENSUS REPORTS: Reported in EPA TSCA Inventory.

DOT Classification: Organic Peroxide; Label: Organic Peroxide.

SAFETY PROFILE: Moderately toxic by ingestion. A skin and eye irritant. Questionable carcinogen with experimental tumorigenic data. Mutation data reported. See also PEROXIDES, ORGANIC. Potentially explosive when heated above 115°C. Explosive reaction on contact with organic matter or copper(I) bromide + limonene. When heated to decomposition it emits acrid smoke and fumes.

BSC750 CAS:110-05-4 ***HR: 3***
tert-BUTYL PEROXIDE
DOT: UN 2102
mf: $C_8H_{18}O_2$ mw: 146.26

PROP: Clear, water white liquid. Mp: −40°, bp: 80° @ 284 mm, flash p: 65°F (OC), d: 0.79, vap press: 19.51 mm @ 20°, vap d: 5.03.

SYNS: CADOX ◇ DI-tert-BUTYLPEROXID (GERMAN) ◇ DI-tert-BUTYL PEROXIDE (MAK) ◇ DI-tert-BUTYL PEROXYDE (DUTCH) ◇ DTBP ◇ PEROSSIDO di BUTILE TERZIARIO (ITALIAN) ◇ PEROXYDE de BUTYLE TERTIAIRE (FRENCH) ◇ (TRIBUTYL)PEROXIDE

TOXICITY DATA with REFERENCE
skn-rbt 500 mg AIHAAP 19,205,58
eye-rbt 500 mg/24H MLD 28ZPAK -,40,72
eye-rbt 200 mg/1M rns MLD ZAARAM 8,25,58
unr-mus TDLo:585 mg/kg:ETA RARSAM 3,193,63
orl-rat LC50:10200 mg/kg 28ZPAK -,40,72
ipr-rat LD50:3210 mg/kg AIHAAP 19,205,58
orl-mus LD50:20 g/kg FEPRA7 7,252,48

CONSENSUS REPORTS: Reported in EPA ISCA Inventory

DFG MAK: Mild skin irritant.

DOT Classification: Organic Peroxide; Label: Organic Peroxide, Flammable Liquid.

SAFETY PROFILE: Moderately toxic by intraperitoneal route. A powerful irritant by ingestion and inhalation. A mild skin and eye irritant. Questionable carcinogen with experimental tumorigenic data. Flammable liquid; see PEROXIDES, ORGANIC, for fire and explosion hazards. Warning: Water may not work to fight fire. When heated to decomposition it emits acrid smoke and fumes.

BSD000 CAS:19910-65-7 ***HR: 2***
sec-BUTYL PEROXYDICARBONATE
DOT: UN 2150/UN 2151
mf: $C_{10}H_{18}O_6$ mw: 234.28

SYNS: DI-sec-BUTYL PEROXYDICARBONATE ◇ DI-sec-BUTYL PEROXYDICARBONATE, not more than 52% in solution (DOT) ◇ DI-sec-BUTYL PEROXYDICARBONATE, technically pure (DOT)

TOXICITY DATA with REFERENCE
skn-rbt LD50:1200 mg/kg BSPII* 1/75-19B

CONSENSUS REPORTS: Reported in EPA TSCA Inventory.

DOT Classification: Organic Peroxide; Label: Organic Peroxide.

SAFETY PROFILE: Moderately toxic by skin contact. See also PEROXIDES, ORGANIC. When heated to decomposition it emits acrid smoke and irritating fumes.

BSD250 CAS:927-07-1 ***HR: 3***
tert-BUTYL PEROXYPIVALATE
DOT: UN 2110
mf: $C_9H_{18}O_3$ mw: 174.27

PROP: Colorless liquid. D: 0.854 @ 25°/25°, fp: < 19°, flash p: > 155°F (OC), rapid decomp @ 21°.Insol in water and ethylene glycol; sol in most organic solvents.

SYNS: ESPEROX 31M ◇ TRIGONOZ 25-C75

TOXICITY DATA with REFERENCE
orl-rat LD50:4300 mg/kg BSPII* 1/75-19B

CONSENSUS REPORTS: Reported in EPA TSCA Inventory.

DOT Classification: Organic Peroxide; Label:Organic Peroxide

SAFETY PROFILE: Mildly toxic by ingestion. Moderately flammable by heat, flame (sparks), oxidizers. Can explode on heating. To fight fire, use water, fog, mist, alcohol foam, dry chemical. When heated to decomposition it emits acrid smoke and fumes. See also PEROXIDES, ORGANIC.

BSD500 CAS:3180-09-4 ***HR: 3***
2-n-BUTYLPHENOL
DOT: UN 2228/UN 2229
mf: $C_{10}H_{14}O$ mw: 150.24

SYNS: o-BUTYLPHENOL, liquid (DOT) ◇ o-BUTYLPHENOL, solid (DOT)

TOXICITY DATA with REFERENCE
skn-mus TDLo:3800 mg/kg/12W-I:NEO CNREA8 19,413,59
orl-rat LD50:650 mg/kg JPETAB 53,218,35

DOT Classification: Poison B; Label: St. Andrews Cross.

SAFETY PROFILE: Questionable carcinogen with experimental neoplastigenic data. When heated to decomposition it emits acrid smoke and irritating fumes. See also PHENOL and other butyl phenols.

BSD750 CAS:1638-22-8 ***HR: 3***
4-n-BUTYLPHENOL
DOT: UN 2228/UN 2229
mf: $C_{10}H_{14}O$ mw: 150.24

TOXICITY DATA with REFERENCE
skn-mus TDLo:3840 mg/kg/12W-I:ETA CNREA8 19,413,59

CONSENSUS REPORTS: Reported in EPA TSCA Inventory.

DOT Classification: Poison B; Label: St. Andrews Cross.

SAFETY PROFILE: A poison. Questionable carcinogen with experimental tumorigenic data. When heated to decomposition it emits acrid smoke and irritating fumes. See also PHENOL and other butyl phenols.

BSE000 CAS:89-72-5 ***HR: 3***
o-sec-BUTYLPHENOL
mf: $C_{10}H_{14}O$ mw: 150.24

PROP: Colorless liquid. Bp: 226-228° @ 25 mm, fp: 12°, flash p: 225°F, d: 0.981 @ 25°/25°.

SYN: 2-sec.-BUTYLFENOL (CZECH)

TOXICITY DATA with REFERENCE
skn-rbt 500 mg/24H SEV 28ZPAK -,55,72
eye-rbt 50 μg/24H SEV 28ZPAK -,55,72
orl-rat LD50:2700 mg/kg 28ZPAK -,55,72
ipr-mus LD50:63 mg/kg JMCMAR 18,868,75
ivn-mus LD50:60 mg/kg JMCMAR 23,1350,80
orl-gpg LD50:600 mg/kg DTLVS* 4,58,80
skn-gpg LD50:600 mg/kg DTLVS* 4,58,80

CONSENSUS REPORTS: Reported in EPA TSCA Inventory.

OSHA PEL: TWA 5 ppm (skin)
ACGIH TLV: TWA 5 ppm (skin)

SAFETY PROFILE: A poison by intraperitoneal and intravenous routes. Moderately toxic by ingestion and skin contact. A severe skin and eye irritant. Combustible when exposed to heat or flame. To fight fire, use foam, spray, CO_2, dry chemical. When heated to decomposition it emits acrid and irritating fumes. See also PHENOL and other butyl phenols.

BSE250 CAS:99-71-8 ***HR: 3***
p-sec-BUTYL PHENOL
mf: $(CH_3CHC_2H_5)C_6H_4OH$ mw: 150.2

PROP: Nearly white flakes. Bp: 135.4-136.5° @ 25 mm, fp: 51°, flash p: 240°F, d: 0.963 @ 60°/60°.

SYN: 4-sec BUTYL PHENOL

TOXICITY DATA with REFERENCE
ipr-mus LD50:66 mg/kg JMCMAR 18,868,75
ivn-mus LD50:40 mg/kg JMCMAR 23,1350,80

CONSENSUS REPORTS: Reported in EPA TSCA Inventory.

SAFETY PROFILE: Poison by intravenous and intraperitoneal routes. Moderately toxic by ingestion. Combustible when exposed to heat or flame. When heated to decomposition it emits toxic fumes. To fight fire, use foam, CO_2, dry chemical. Incompatible with oxidizing materials. See also PHENOL and other butyl phenols.

BSE500 CAS:98-54-4 ***HR: 3***
4-tert-BUTYLPHENOL
mf: $C_{10}H_{14}O$ mw: 150.24

PROP: Crystals or practically white flakes. Bp: 238°, fp: 97°, d: 0.9081 @ 114°/4°, vap press: 1 mm @ 70.0°, vap d: 5.1.

SYNS: p-tert-BUTYLFENOL (CZECH) ◇ BUTYLPHEN ◇ p-tert-BUTYLPHENOL (MAK) ◇ 4-(1,1-DIMETHYLETHYL)PHENOL ◇ 1-HYDROXY-4-tert-BUTYLBENZENE ◇ UCAR BUTYLPHENOL 4-T

TOXICITY DATA with REFERENCE
skn-rbt 500 mg/4H MLD DCTODJ 11,43,88
skn-rbt 500 mg/24H MLD 85JCAE-,224,86
eye-rbt 10 mg SEV DCTODJ 11,43,88
eye-rbt 50 μg/24H SEV 85JCAE-,224,86
orl-ham TDLo:252 g/kg/20W-C:NEO CRNGDP 7,1285,86
orl-rat LD50:2951 mg/kg AIHAAP 30,470,69
ipr-mus LD50:78 mg/kg JMCMAR 18,868,75
skn-rbt LD50:2288 mg/kg AIHAAP 30,470,69
orl-mam LD50:1500 mg/kg GISAAA 45(10),16,80
skn-mam LD50:1580 mg/kg GISAAA 45(10),16,80

CONSENSUS REPORTS: Reported in EPA TSCA Inventory.

DFG MAK: 0.08 ppm (0.5 mg/m^3)

SAFETY PROFILE: Poison by intraperitoneal route. Moderately toxic by skin contact and ingestion. A skin and severe eye irritant. Questionable carcinogen with experimental neoplastigenic data. Combustible when exposed to heat or flame; can react with oxidizing materials. To fight fire, use foam, CO_2, dry chemical. When heated to decomposition it emits acrid and irritating fumes. See also PHENOL and other butyl phenols.

BSE750 CAS:56488-59-6 ***HR: 3***
4'-(3-(4'-tert-BUTYLPHENOXY)-2-HYDROXY-PROPOXY)BENZOIC ACID
mf: $C_{20}H_{24}O_5$ mw: 344.44

SYN: 4-(3-(4-(1,1-DIMETHYLETHYL)PHENOXY)-2-HYDROXYPROPOXY)BENZOIC ACID

TOXICITY DATA with REFERENCE
orl-rat LD50:2400 mg/kg DRFUD4 4,140,79
ipr-rat LD50:500 mg/kg DRFUD4 4,140,79
orl-mus LD50:2100 mg/kg DRFUD4 4,140,79
ipr-mus LD50:335 mg/kg DRFUD4 4,140,79
orl-rbt LD50:1800 mg/kg DRFUD4 4,140,79
orl-gpg LD50:320 mg/kg DRFUD4 4,140,79
orl-mam LD50:5000 mg/kg DRFUD4 4,140,79

SAFETY PROFILE: Poison by ingestion and intraperitoneal routes. When heated to decomposition it emits acrid smoke and irritating fumes.

BSF250 CAS:61005-12-7 ***HR: 3***
o-sec-BUTYLPHENYL CARBAMATE
mf: $C_{12}H_{17}NO_2$ mw: 207.30

TOXICITY DATA with REFERENCE
orl-rat LD50:410 mg/kg OYYAA2 3,74,69
orl-mus LD50:340 mg/kg OYYAA2 3,74,69

SAFETY PROFILE: A poison by ingestion. See also CARBAMATES. When heated to decomposition it emits toxic fumes of NO_x.

BSF750 CAS:1126-79-0 ***HR: 2***
BUTYL PHENYL ETHER
mf: $C_{10}H_{14}O$ mw: 150.24

PROP: Flash p: 180°F (OC), d: 0.9, vap d: 5.2, bp: 210°.

SYN: BUTOXYPHENYL

TOXICITY DATA with REFERENCE
orl-mus LD50:3200 mg/kg JPETAB 88,400,46

CONSENSUS REPORTS: Reported in EPA TSCA Inventory.

SAFETY PROFILE: Moderately toxic by ingestion. See also ETHERS. When heated to decomposition it emits acrid and irritating fumes.

BSG000 CAS:329-21-5 ***HR: 3***
S-p-tert-BUTYLPHENYL-o-ETHYL ETHYLPHOSPHONODITHIOATE
mf: $C_{14}H_{23}OPS_2$ mw: 302.46

SYNS: S-(4-(1,1-DIMETHYLETHYL)PHENYL)-o-ETHYLETHYLPHOSPHONODITHIOATE ◇ ENT 25,765 ◇ N 3051 ◇ STAUFFER N-3051

TOXICITY DATA with REFERENCE
orl-rat LD50:141 mg/kg ARSIM* 20,22,66
orl-ckn LD50:64 mg/kg TXAPA9 7,606,65

SAFETY PROFILE: Poison by ingestion. When heated to decomposition it emits very toxic fumes of SO_x and PO_x. See also ESTERS.

BSG100 CAS:85303-89-5 ***HR: D***
3-(o-BUTYLPHENYL)-5-(m-METHOXYPHENYL)-s-TRIAZOLE
mf: $C_{19}H_{21}N_3O$ mw: 307.43

SYN: s-TRIAZOLE,3-(o-BUTYLPHENYL)-5-(m-METHOXYPHENYL)-

TOXICITY DATA with REFERENCE
scu-ham TDLo:60 mg/kg (female 4-8D post):REP JMCMAR 26,1187,83

SAFETY PROFILE: Experimental reproductive effects. When heated to decomposition it emits toxic fumes of NO_x.

BSG250 CAS:673-19-8 ***HR: 3***
m-sec-BUTYLPHENYL-N-METHYLCARBAMATE
mf: $C_{12}H_{17}NO_2$ mw: 207.30

SYNS: 3-sec-BUTYLPHENYL-N-METHYLCARBAMATE ◇ CALIFORNIA CHEMICAL COMPANY RE5305 ◇ CHEVRON RE5305 ◇ ENT 27,039 ◇ H-28 ◇ m-(1-METHYLPROPYL)PHENYLMETHYLCARBAMATE ◇ RE 5305 (CALIFORNIA CHEMICAL)

TOXICITY DATA with REFERENCE
orl-rat LD50:10 mg/kg ARSIM* 20,7,66
orl-ckn LD50:14 mg/kg TXAPA9 11,49,67
orl-bwd LD50:4600 μg/kg TXAPA9 21,315,72

SAFETY PROFILE: Poison by ingestion. See also CARBAMATES. When heated to decomposition it emits toxic fumes of NO_x.

BSH000 ***HR: 3***
2-tert-BUTYL-3-PHENYL OXAZIRANE
mf: $C_{11}H_{15}NO$ mw: 177.05

SAFETY PROFILE: May explode in vacuum. When heated to decomposition it emits toxic fumes of NO_x.

BSH075 CAS:85303-88-4 ***HR: D***
3-(o-BUTYLPHENYL)-5-PHENYL-s-TRIAZOLE
mf: $C_{18}H_{19}N_3$ mw: 277.40

SYN: s-TRIAZOLE,3-(o-BUTYLPHENYL)-5-PHENYL-

TOXICITY DATA with REFERENCE
scu-ham TDLo:85 mg/kg (female 4-8D post):REP RDMIDP 4,237,82

SAFETY PROFILE: Experimental reproductive effects. When heated to decomposition it emits toxic fumes of NO_x.

BSH250 CAS:78-48-8 ***HR: 3***
BUTYL PHOSPHOROTRITHIOATE
mf: $C_{12}H_{27}OPS_3$ mw: 314.54

PROP: Liquid. Bp: 150° @ 0.3 mm. Insol in water; sol in aliphatic, aromatic, and chlorinated hydrocarbons.

SYNS: B-1,776 ◇ BUTIFOS ◇ BUTIPHOS ◇ CHEMAGRO 1,776 ◇ CHEMAGRO B-1776 ◇ DEF ◇ DEF DEFOLIANT ◇ DE-GREEN ◇ E-Z-OFF D ◇ FOS-FALL "A" ◇ ORTHO PHOSPHATE DEFOLIANT ◇ S,S,S-TRIBUTYL PHOSPHOROTRITHIOATE ◇ S,S,S-TRIBUTYL TRITHIOPHOSPHATE

TOXICITY DATA with REFERENCE
orl-rat TDLo:216 mg/kg (8W pre):REP MZUZA8 (2),48,80
orl-rat LD50:150 mg/kg TXAPA9 14,515,69
skn-rat LD50:168 mg/kg WRPCA2 9,119,70
ipr-rat LD50:210 mg/kg 34ZIAG -,199,69
orl-mus LD50:77 mg/kg 85JCAE -,1188,86
ihl-mus LCLo:3804 mg/m^3/1H 34ZIAG -,199,69
ipr-mus LD50:290 mg/kg 34ZIAG -,199,69
skn-rbt LD50:97 mg/kg 85GMAT -,44,82

CONSENSUS REPORTS: Reported in EPA TSCA Inventory.

SAFETY PROFILE: A poison by ingestion, skin contact, and intraperitoneal routes. Experimental reproductive effects. Animal experiments show an anti-cholinesterase effect. When heated to decomposition it emits toxic fumes of PO_x and SO_x. See also PARATHION, PHOSPHATES, ESTERS, and SULFATES.

BSH500 CAS:6066-49-5 ***HR: 2***
3-n-BUTYLPHTHALIDE
mf: $C_{12}H_{14}O_2$ mw: 190.26

SYNS: BUTYLPHTHALIDE ◇ 3-BUTYLPHTHALIDE

TOXICITY DATA with REFERENCE
skn-rbt 500 mg/24H MOD FCTXAV 17,241,79
orl-rat LD50:2450 mg/kg FCTXAV 17,241,79

CONSENSUS REPORTS: Reported in EPA TSCA Inventory.

SAFETY PROFILE: Moderately toxic by ingestion. A skin irritant. When heated to decomposition it emits acrid smoke and irritating fumes.

BSI000 CAS:536-69-6 ***HR: 3***
5-BUTYL PICOLINIC ACID
mf: $C_{10}H_{13}NO_2$ mw: 179.24

SYNS: 5-BUTYL-2-PYRIDINECARBOXYLIC ACID ◇ FUSARIC ACID ◇ FUSARINIC ACID

TOXICITY DATA with REFERENCE
orl-mus LD50:180 mg/kg JOPHDQ 6,922,83
ipr-mus LD50:75 mg/kg JOPHDQ 6,922,83
ivn-mus LD50:100 mg/kg 85ERAY 3,1873,78

CONSENSUS REPORTS: EPA Genetic Toxicology Program. Reported in EPA TSCA Inventory.

SAFETY PROFILE: A poison by ingestion, intraperitoneal, and intravenous routes. When heated to decomposition it emits toxic fumes of NO_x.

BSI250 CAS:2180-92-9 ***HR: 3***
1-BUTYL-2',6'-PIPECOLOXYLIDIDE
mf: $C_{18}H_{28}N_2O$ mw: 288.48

SYNS: BUPIVACAINE ◇ dl-BUPIVACAINE

TOXICITY DATA with REFERENCE
isp-wmn TDLo:2396 mg/kg (female 39W post):REP BJOGAS 88,407,81
icv-wmn TDLo:500 µg/kg (female 39W post):TER JPEMAO 12,75,84
ivn-hmn TDLo:4300 µg/kg:BPR,PSY AANEAB 21,521,77
scu-rat LD50:48 mg/kg APTOA6 31,273,72
ivn-rat LD50:5600 µg/kg APTOA6 31,273,72
ipr-mus LD50:58700 µg/kg TXAPA9 54,501,80
scu-mus LD50:53 mg/kg APTOA6 31,273,72
ivn-mus LD50:7300 µg/kg APTOA6 31,273,72
ivn-rbt LD50:1620 µg/kg AACRAT 64,209,85
par-rbt LD50:64 mg/kg ARZNAD 26,78,76
itr-rbt LD50:12500 µg/kg ARZNAD 26,78,76

SAFETY PROFILE: A poison by subcutaneous, intraperitoneal, intratracheal, parenteral, and intravenous routes. An experimental teratogen. Other experimental reproductive effects. Human systemic effects by intravenous route: changes in regional blood flow rates and euphoria. When heated to decomposition it emits toxic fumes of NO_x.

BSI750 CAS:78329-88-1 ***HR: 2***
p-(N-BUTYL-2-(PIPERIDINO)ACETAMIDO)BENZOIC ACID BUTYL ESTER HYDROCHLORIDE
mf: $C_{22}H_{34}N_2O_3$•ClH mw: 411.04

SYN: C 3181

TOXICITY DATA with REFERENCE
eye-rbt 2% SEV ARZNAD 8,609,58
ipr-rat LD50:480 mg/kg ARZNAD 8,609,58
scu-mus LD50:3750 mg/kg ARZNAD 8,609,58

SAFETY PROFILE: Moderately toxic by intraperitoneal and subcutaneous routes. A severe eye irritant. See also ESTERS. When heated to decomposition it emits very toxic fumes of NO_x and HCl.

BSJ500 CAS:590-01-2 ***HR: 3***
BUTYL PROPANOATE
DOT: UN 1914
mf: $C_7H_{14}O_2$ mw: 130.2

PROP: Water-white liquid, apple-like odor. Mp: −89.6°,

bp: 145.4°, flash p: 90°F, d: 0.875 @ 20°, autoign temp: 800°F, vap d: 4.49.

SYNS: BUTYL PROPIONATE ◇ n-BUTYL PROPIONATE ◇ PROPANOIC ACID BUTYLESTER (9CI)

TOXICITY DATA with REFERENCE
skn-rbt 500 mg/24H MOD FCTXAV 18,649,80
orl-rat LD50:5000 mg/kg FCTXAV 18,649,80

DOT Classification: Flammable or Combustible Liquid; Label:Flammable Liquid

SAFETY PROFILE: Mildly toxic by ingestion. A skin irritant. Dangerously flammable when exposed to heat or flame. To fight fire, use foam, CO_2, dry chemical. Incompatible with oxidizing materials. See also ESTERS, n-BUTYL ALCOHOL, and PROPIONIC ACID.

BSK000 CAS:98-29-3 ***HR: 3***
4-tert-BUTYLPYROCATECHOL
mf: $C_{10}H_{14}O_2$ mw: 166.24

PROP: Fp: 52°, flash p: 265°F, bp: 285°, d: 1.049 @ 60°/25°.

SYNS: 4-tert-BUTYLCATECHOL ◇ p-tert-BUTYLPYROCATECHOL ◇ 4-tert-BUTYLPYROKATECHIN (CZECH) ◇ 4-(1,1-DIMETHYLETHYL)-1,2-BENZENEDIOL ◇ SYNOX TBC

TOXICITY DATA with REFERENCE
skn-rbt 10 mg/24H open AMIHBC 10,61,54
skn-rbt 500 mg/24H SEV 28ZPAK -,57,72
eye-rbt 50 μg open SEV AMIHBC 10,61,54
skn-gpg 0.1%/3W MLD JIDEAE 55,190,70
skn-gpg 1%/3W MOD JIDEAE 55,190,70
orl-rat LD50:2820 mg/kg AMIHBC 10,61,54
ivn-mus LD50:32 mg/kg CSLNX* NX#07874
skn-rbt LD50:630 mg/kg AMIHBC 10,61,54

CONSENSUS REPORTS: Reported in EPA TSCA Inventory.

SAFETY PROFILE: A poison by intravenous route. Moderately toxic by ingestion and skin absorption. A severe skin and eye irritant. Combustible when exposed to heat or flame. To fight fire, use CO_2, dry chemical, fog, mist. When heated to decomposition it emits acrid and irritating fumes.

BSK250 CAS:767-10-2 ***HR: 3***
n-BUTYLPYRROLIDINE
mf: $C_8H_{17}N$ mw: 127.26

TOXICITY DATA with REFERENCE
orl-mus LD50:51 mg/kg INHEAO 4,63,66
skn-mus LD50:1000 mg/kg INHEAO 4,63,66
ipr-mus LD50:37 mg/kg INHEAO 4,63,66
scu-mus LD50:57 mg/kg INHEAO 4,63,66

SAFETY PROFILE: Poison by ingestion, intraperitoneal, and subcutaneous routes. Moderately toxic by skin contact. When heated to decomposition it emits toxic fumes of NO_x.

BSL250 CAS:2052-14-4 ***HR: 2***
n-BUTYL SALICYLATE
mf: $C_{11}H_{14}O_3$ mw: 194.25

SYNS: BUTYL-o-HYDROXYBENZOATE ◇ n-BUTYL-o-HYDROXYBENZOATE ◇ BUTYL SALICYLATE ◇ 2-HYDROXYBENZOIC ACID BUTYL ESTER

TOXICITY DATA with REFERENCE
orl-rat LD50:1700 mg/kg FCTXAV 16,637,78

CONSENSUS REPORTS: Reported in EPA TSCA Inventory.

SAFETY PROFILE: Moderately toxic by ingestion. When heated to decomposition it emits acrid smoke and irritating fumes. See also ESTERS.

BSL325 ***HR: 3***
n-BUTYLSCOPOLAMINE TANNATE

TOXICITY DATA with REFERENCE
orl-rat TDLo:164 g/kg (91D male):REP KSRNAM 7,442,73
ipr-rat LD50:343 mg/kg OYYAA2 5,599,71
ipr-mus LD50:146 mg/kg OYYAA2 5,599,71
scu-mus LD50:228 mg/kg OYYAA2 5,599,71

SAFETY PROFILE: Poison by subcutaneous and intraperitoneal routes. Experimental reproductive effects. When heated to decomposition it emits toxic fumes of NO_x. See also AMINES.

BSL450 CAS:52670-52-7 ***HR: 3***
17-BUTYLSPARTEIN
mf: $C_{19}H_{34}N_2$ mw: 290.55

SYN: 6-BUTYLDODECAHYDRO-7,14-METHANO-2H,6H-DIPYRIDO (1,2-a:1′,2′-e)(1,5)DIAZOCINE

TOXICITY DATA with REFERENCE
orl-mus LD50:1820 mg/kg ARZNAD 30,1497,80
ipr-mus LD50:160 mg/kg ARZNAD 30,1497,80
ivn-mus LD50:27300 μg/kg ARZNAD 30,1497,80

SAFETY PROFILE: Poison by intravenous and intraperitoneal routes. Moderately toxic by ingestion. When heated to decomposition it emits toxic fumes of NO_x.

BSL500 CAS:2273-43-0 ***HR: 3***
BUTYL STANNOIC ACID
mf: $C_4H_{10}O_2Sn$ mw: 208.83

SYN: BUTYLHYDROXYOXOSTANNANE

TOXICITY DATA with REFERENCE
ivn-mus LD50:180 mg/kg CSLNX* NX#03474

CONSENSUS REPORTS: Reported in EPA TSCA Inventory.

OSHA PEL: TWA 0.1 mg(Sn)/m^3 (skin)
ACGIH TLV: TWA 0.1 mg(Sn)/m^3 (skin) (Proposed: TWA 0.1 mg(Sn)/m^3; STEL 0.2 mg(Sn)/m^3 (skin))
NIOSH REL: (Organotin Compounds) TWA 0.1 mg(Sn)/m^3

SAFETY PROFILE: A poison by intravenous route. See also TIN COMPOUNDS. When heated to decomposition it emits acrid smoke and irritating fumes.

BSL750 CAS:63979-65-7 ***HR: 3***
n-BUTYL-k-STROPHANTHIDIN
mf: $C_{27}H_{38}O_7$ mw: 474.65

TOXICITY DATA with REFERENCE
ivn-cat LDLo:350 μg/kg AEPPAE 185,329,37
ivn-rat LDLo:500 μg/kg AEPPAE 185,329,37

SAFETY PROFILE: Deadly poison by intravenous route. When heated to decomposition it emits acrid smoke and irritating fumes.

BSM000 CAS:339-43-5 ***HR: 3***
1-BUTYL-3-SULFANILYL UREA
mf: $C_{11}H_{17}N_3O_3S$ mw: 271.37

SYNS: ALENTIN ◇ N-(4-AMINOBENZENESULFONYL)-N'-BUTYLUREA ◇ 4-AMINO-N-((BUTYLAMINO)CARBONYL)BENZENESULFONAMIDE ◇ AMINOPHENUROBUTANE ◇ BUCARBAN ◇ BUCROL ◇ BUKARBAN ◇ BURCOL ◇ BUTISULFINA ◇ N'-(BUTYLCARBAMOYL)SULFANILAMIDE ◇ N^1-(BUTYLCARBAMOYL)SULFANILAMIDE ◇ N-BUTYLSULFANILYLUREA ◇ CARBUTAMID ◇ CARBUTAMIDE ◇ CICLORAL ◇ DIABORAL ◇ EMEDAN ◇ GLUCIDORAL ◇ GLUCOFREN ◇ GLYBUTAMIDE ◇ INBUTON ◇ INVENOL ◇ NADISAN ◇ NADIZAN ◇ NORBORAL ◇ ORANIL ◇ ORANYL ◇ ORASULIN ◇ N^1-SULFANILYL-N^2-BUTYLCARBAMIDE ◇ N^1-SULFANILYL-N^2-BUTYLUREA ◇ N-SULFANILYL-N'BUTYLUREE (FRENCH) ◇ U 6987

TOXICITY DATA with REFERENCE
unr-rat TDLo:1 g/kg (female 1D post):TER AKGIAO 42(12),35,66
unr-mus TDLo:8800 mg/kg (female 5-15D post):REP BSVMA8 71,289,69
orl-rat LD50:7800 mg/kg FATOAO 25,93,62
ivn-rat LD50:980 mg/kg DIAEAZ 6,2,57
orl-mus LD50:2800 mg/kg FATOAO 25,93,62
ipr-mus LD50:250 mg/kg NTIS** AD691-490
scu-mus LD50:2640 mg/kg DIAEAZ 6,2,57

CONSENSUS REPORTS: Reported in EPA TSCA Inventory.

SAFETY PROFILE: A poison by intraperitoneal route. Moderately toxic by ingestion and subcutaneous routes. An experimental teratogen. Other experimental reproductive effects. When heated to decomposition it emits very toxic fumes of NO_x and SO_x.

BSM125 CAS:544-40-1 ***HR: 2***
BUTYL SULFIDE
mf: $C_8H_{18}S$ mw: 146.32

PROP: Liquid. Mp: −79.7°, d: 0.839, bp: 182°. Insol in water; very sol in alc and ether.

SYNS: BUTYL MONOSULFIDE ◇ n-BUTYL-SULFIDE ◇ BUTYLTHIOBUTANE ◇ n-DIBUTYL SULFIDE ◇ DI-n-BUTYLSULFIDE ◇ DIBUTYL SULPHIDE ◇ DIBUTYL THIOETHER ◇ 5-THIANONANE ◇ THIANONANE-5

TOXICITY DATA with REFERENCE
skn-rbt 500 mg/24H MOD FCTXAV 17,769,79
orl-rat LD50:2220 mg/kg FCTXAV 17,769,79
ihl-mus LCLo:1800 mg/m^3 FCTXAV 17,769,79

CONSENSUS REPORTS: Reported in EPA TSCA Inventory.

SAFETY PROFILE: Moderately toxic by ingestion. A skin irritant. When heated to decomposition it emits toxic fumes of SO_x. See also SULFIDES.

BSM250 CAS:64910-63-0 ***HR: 3***
1-BUTYLSULFONIMIDOCYCLOHEXAMETHYLENE
mf: $C_{10}H_{21}NO_2S$ mw: 219.38

SYN: N-CYCLOHEXYL-1-BUTANESULFONAMIDE

TOXICITY DATA with REFERENCE
skn-rbt 175 mg/14D MLD NTIS** AD-A022-909
orl-rat LD50:2816 mg/kg NTIS** AD-A022-909
ipr-rat LD50:1074 mg/kg NTIS** AD-A022-909
ivn-rat LDLo:225 mg/kg NTIS** AD-A022-909
orl-mus LD50:5400 mg/kg NTIS** AD-A022-909
skn-mus LD50:7560 mg/kg NTIS** AD-A022-909
scu-mus LD50:519 mg/kg NTIS** AD-A022-909

SAFETY PROFILE: Poison by intravenous route. Moderately toxic by ingestion, intraperitoneal, and subcutaneous routes. A skin irritant. When heated to decomposition it emits very toxic fumes of SO_x and NO_x.

BSM825 CAS:63906-57-0 ***HR: 3***
1-BUTYL THEOBROMINE
mf: $C_{11}H_{14}N_4O_2$ mw: 234.29

SYNS: 1-(2'-BUTENYL)THEOBROMINE ◇ 1-CROTYL THEOBROMINE

TOXICITY DATA with REFERENCE
orl-mus LD50:667 mg/kg JPETAB 116,343,56
ipr-mus LD50:230 mg/kg JPETAB 116,343,56
ivn-mus LD50:95 mg/kg JPETAB 86,113,46

SAFETY PROFILE: Poison by intravenous and intra-

peritoneal routes. Moderately toxic by ingestion. When heated to decomposition it emits toxic fumes of NO_x.

BSN000 CAS:34014-18-1 ***HR: 3***
1-(5-(tert-BUTYL)-1,3,4-THIADIAZOL-2-YL)-1,3-DIMETHYLUREA
mf: $C_9H_{16}N_4OS$ mw: 228.35

SYNS: BRULAN ◇ 1-(5-tert-BUTYL-1,3,4-THIADIAZOL-2-YL)-3-DIMETHYLHARNSTOFF (GERMAN) ◇ N-(5-(1,1-DIMETHYLAETHYL)-1,3,4-THIADIAZOL-2-YL)-N,N'-DIMETHYLHARNSTOFF (GERMAN) ◇ E-103 ◇ EI-103 ◇ EL-103 ◇ GRASLAN ◇ PERFMID ◇ PREFLAN ◇ PREFMID ◇ SPIKE ◇ TEBULAN ◇ TEBUTHIURON ◇ TIUROLAN

TOXICITY DATA with REFERENCE
orl-rat LD50:644 mg/kg FMCHA2 -,D286,80
orl-mus LD50:579 mg/kg 85DPAN -,-,71/76
orl-rbt LD50:286 mg/kg 85DPAN -,-,71/76

SAFETY PROFILE: Poison by ingestion. When heated to decomposition it emits very toxic fumes of SO_x and NO_x.

BSN325 CAS:2314-17-2 ***HR: 2***
2-BUTYLTHIOBENZOTHIAZOLE
mf: $C_{11}H_{13}NS_2$ mw: 223.37

SYN: BUTYLCAPTAX

TOXICITY DATA with REFERENCE
orl-rat LD50:1270 mg/kg 85GMAT -,29,82
unr-rat LD50:1300 mg/kg GISAAA 47(2),63,82
orl-mus LD50:1610 mg/kg 85GMAT -,29,82
orl-rbt LD50:2344 mg/kg 85GMAT -,29,82

SAFETY PROFILE: Moderately toxic by ingestion and possibly other routes. When heated to decomposition it emits toxic fumes of SO_x and NO_x.

BSN500 CAS:628-83-1 ***HR: 3***
n-BUTYL THIOCYANATE
mf: C_5H_9NS mw: 115.21

SYNS: n-BUTYL RHODANATE ◇ BUTYRHODANID (GERMAN) ◇ 1-THIOCYANOBUTANE

TOXICITY DATA with REFERENCE
orl-rat LDLo:250 mg/kg JIHTAB 18,310,36
scu-rat LDLo:70 mg/kg JIHTAB 18,310,36
scu-mus LDLo:130 mg/kg JIHTAB 18,310,36
scu-rbt LDLo:6 mg/kg AEPPAE 150,257,30

CONSENSUS REPORTS: Reported in EPA TSCA Inventory.

SAFETY PROFILE: A poison by ingestion and subcutaneous routes. When heated to decomposition it emits very toxic fumes of NO_x and SO_x. See also THIOCYANATES.

BSO000 CAS:13071-79-9 ***HR: 3***
S-((tert-BUTYLTHIO)METHYL)-O,O-DIETHYL-PHOSPHORODITHIOATE
mf: $C_9H_{21}O_2PS_3$ mw: 288.45

SYNS: AC 921000 ◇ COUNTER ◇ COUNTER 15G SOIL INSECTICIDE ◇ COUNTER 15G SOIL INSECTICIDE-NEMATICIDE ◇ S-(((1,1-DIMETHYLETHYL)THIO)METHYL)-O,O-DIETHYLPHOSPHORODITHIOATE ◇ PHOSPHORODITHIOIC ACID S-((tert-BUTYLTHIO)METHYL)-O,O-DIETHYL ESTER ◇ PHOSPHORODITHIOIC ACID S-(((1,1-DIMETHYLETHYL)THIO)METHYL)-O,O-DIETHYL ESTER ◇ TERBUFOS

TOXICITY DATA with REFERENCE
orl-rat LD50:1600 μg/kg MEIEDD 10,1310,83
orl-mus LD50:3500 μg/kg FMCHA2 -,C63,83
orl-dog LD50:4500 μg/kg FMCHA2 -,C63,83
skn-rbt LD50:1100 μg/kg/24H FMCHA2 -,C63,83
orl-qal LD50:15 mg/kg EESADV 8,551,84

CONSENSUS REPORTS: EPA Genetic Toxicology Program. EPA Extremely Hazardous Substances List.

SAFETY PROFILE: Deadly poison by ingestion and skin contact. An insecticide. When heated to decomposition it emits very toxic fumes of SO_x and PO_x. See also ESTERS.

BSO200 CAS:70303-47-8 ***HR: 3***
(BUTYLTHIO)TRIOCTYLSTANNANE
mf: $C_{28}H_{60}SSn$ mw: 547.63

SYNS: STANNANE, (BUTYLTHIO)TRIOCTYL- ◇ TRIOCTYL(BUTYLTHIO)STANNANE

TOXICITY DATA with REFERENCE
ipr-mus LD50:389 mg/kg RPTOAN 42,73,79

OSHA PEL: TWA 0.1 mg(Sn)/m^3
ACGIH TLV: TWA 0.1 mg(Sn)/m^3; STEL 0.2 mg/m^3 (skin)
NIOSH REL: (Organotin compound): 10H TWA 0.1 mg(Sn)/m^3

SAFETY PROFILE: Poison by intraperitoneal route. When heated to decomposition it emits toxic fumes of SO_x and Sn.

BSO500 CAS:1516-32-1 ***HR: 3***
n-BUTYL THIOUREA
mf: $C_5H_{12}N_2S$ mw: 132.25

SYN: USAF D-5

TOXICITY DATA with REFERENCE
orl-rat LD50:200 mg/kg JPETAB 90,260,47
ipr-mus LD50:300 mg/kg NTIS** AD277-689

CONSENSUS REPORTS: Reported in EPA TSCA Inventory.

SAFETY PROFILE: A poison by ingestion and intraperitoneal routes. When heated to decomposition it emits very toxic fumes of NO_x and SO_x.

BSO750 CAS:25151-00-2 ***HR: 3***
BUTYLTIN TRILAURATE
mf: $C_{40}H_{72}O_6Sn$ mw: 767.81

SYNS: BTT ◇ n-BUTYLTIN TRICHLORIDE ◇ BUTYLTIN TRI(DODECANOATE) ◇ BUTYLTRI(LAUROYLOXY)STANNANE ◇ MONOBUTYLTIN TRICHLORIDE ◇ MONOBUTYLTIN TRILAURATE

TOXICITY DATA with REFERENCE
orl-mus LD50:325 mg/kg GISAAA 41(5),10,76

OSHA PEL: TWA 0.1 mg(Sn)/m^3 (skin)
ACGIH TLV: TWA 0.1 mg(Sn)/m^3 (skin) (Proposed: TWA 0.1 mg(Sn)/m^3; STEL 0.2 mg(Sn)/m^3 (skin))
NIOSH REL: (Organotin Compounds) TWA 0.1 mg(Sn)/m^3

SAFETY PROFILE: Poison by ingestion. See also TIN COMPOUNDS. When heated to decomposition it emits acrid smoke and irritating fumes.

BSP000 CAS:73927-88-5 ***HR: 3***
n-BUTYLTIN TRIS(DIBUTYLDITHIOCARBAMATE)
mf: $C_{31}H_{63}N_3S_6Sn$ mw: 789.02

SYN: BUTYLTRIS(DIBUTYLDITHIOCARBAMATO)STANNANE

TOXICITY DATA with REFERENCE
ivn-mus LD50:180 mg/kg CSLNX* NX#02083

OSHA PEL: TWA 0.1 mg(Sn)/m^3 (skin)
ACGIH TLV: TWA 0.1 mg(Sn)/m^3 (skin) (Proposed: TWA 0.1 mg(Sn)/m^3; STEL 0.2 mg(Sn)/m^3 (skin))
NIOSH REL: (Organotin Compounds) TWA 0.1 mg(Sn)/m^3

SAFETY PROFILE: Poison by intravenous route. See also CARBAMATES and TIN COMPOUNDS. When heated to decomposition it emits very toxic fumes of NO_x and SO_x.

BSP250 CAS:5593-70-4 ***HR: 3***
BUTYL TITANATE
mf: $C_{16}H_{36}O_4{\cdot}Ti$ mw: 340.42

PROP: Colorless to light yellow liquid, odor of butanol. Mp: −55°, bp: 312°, flash p: 170°F, vap d: 11.5.

SYN: TETRABUTYLTITANATE (CZECH)

TOXICITY DATA with REFERENCE
orl-rat LD50:3122 mg/kg MarJV# 29MAR77
ivn-mus LD50:180 mg/kg CSLNX* NX#01650

CONSENSUS REPORTS: Reported in EPA TSCA Inventory.

SAFETY PROFILE: A poison by intravenous route. Moderately toxic by ingestion. See BUTYL ALCOHOL and TITANIUM COMPOUNDS. Flammable when exposed to heat or flame. To fight fire, use water, spray, foam, dry chemical. Incompatible with oxidizing materials. When heated to decomposition it emits acrid and irritating fumes.

BSP500 CAS:98-51-1 ***HR: 2***
p-tert-BUTYLTOLUENE
mf: $C_{11}H_{16}$ mw: 148.27

PROP: Colorless liquid.

SYNS: p-METHYL-tert-BUTYLBENZENE ◇ 1-METHYL-4-tert-BUTYLBENZENE ◇ TBT

TOXICITY DATA with REFERENCE
eye-hmn 5 ppm/2H AMIHBC 9,227,54
skn-rbt 500 mg/24H MLD AMIHBC 9,227,54
eye-rbt 100 mg AMIHBC 9,227,54
ihl-hmn TCLo:10 ppm/3M:GIT AMIHBC 9,227,54
ihl-hmn TCLo:20 ppm/5M:EYE,IRR,GIT 28ZRAQ -,156,60
orl-rat LD50:1500 mg/kg AMIHBC 9,227,54
ihl-rat LC50:1500 mg/m^3/4H AMIHBC 9,227,54
orl-mus LD50:900 mg/kg AMIHBC 9,227,54
ihl-mus LC50:248 ppm/2H AMIHBC 9,227,54
orl-rbt LD50:2000 mg/kg AMIHBC 9,227,54

CONSENSUS REPORTS: Reported in EPA TSCA Inventory.

OSHA PEL: (Transitional: TWA 10 ppm) TWA 10 ppm; STEL 20 ppm
ACGIH TLV: TWA 10 ppm; STEL 20 ppm.
DFG MAK: 10 ppm (60 mg/m^3)

SAFETY PROFILE: Moderately toxic by inhalation and ingestion. A skin and human eye irritant. Human systemic effects by inhalation: nausea or vomiting, conjunctiva irritation, unspecified effects on the sense of taste. Inhalation of vapors causes irritation of lungs and depression of central nervous system. Prolonged exposure may result in damage to liver and kidneys. Flammable when exposed to heat or flame. Incompatible with oxidizing materials. When heated to decomposition it emits acrid smoke and fumes.

BSP750 CAS:778-28-9 ***HR: 3***
n-BUTYL-p-TOLUENESULFONATE
mf: $C_{11}H_{16}O_3S$ mw: 228.33

SYNS: BUTYL-p-METHYLBENZENESULFONATE ◇ BUTYL-p-TOLUENESULFONATE ◇ BUTYL TOSYLATE ◇ 4-METHYL-BENZENESULFONIC ACID BUTYL ESTER (9CI)

TOXICITY DATA with REFERENCE
scu-rat LD50:5000 mg/kg ZEKBAI 74,241,70
ivn-mus LD50:320 mg/kg CSLNX* NX#01764

CONSENSUS REPORTS: Reported in EPA TSCA Inventory.

SAFETY PROFILE: Poison by intravenous route. Mildly toxic by subcutaneous route. See also SULFONATES. When heated to decomposition it emits toxic fumes of SO_x.

BSQ000 CAS:64-77-7 ***HR: 2***
1-BUTYL-3-(p-TOLYL SULFONYL)UREA
mf: $C_{12}H_{18}N_2O_3S$ mw: 270.38

SYNS: AGLICID ◇ ARKOZAL ◇ ARTOSIN ◇ ARTOZIN ◇ BUTAMID ◇ N-((BUTYLAMINO)CARBONYL)-4-METHYLBENZENESULFONAMIDE ◇ 1-BUTYL-3-(p-METHYLPHENYLSULFONYL)UREA ◇ n-BUTYL-N'-p-TOLUENESULFONYLUREA ◇ N-n-BUTYL-N'-TOSYLUREA ◇ 1-BUTYL-3-TOSYLUREA ◇ BZ 55 ◇ D 860 ◇ DIABEN ◇ DIABETA- MID ◇ DIABETOL ◇ DIABUTON ◇ DOLIPOL ◇ DRABET ◇ HLS 831 ◇ IPOGLICONE ◇ MOBENOL ◇ NCI-CO1763 ◇ ORABET ◇ ORALIN ◇ OREZAN ◇ ORINASE ◇ ORINAZ ◇ OTERBEN ◇ RASTINON ◇ SK-TOLBUTAMIDE ◇ N-(SULFONYL-p-METHYLBENZENE)-N'-N-BUTYLUREA ◇ TOLBUSAL ◇ TOLBUTAMID ◇ TOLBUTAMIDE ◇ 1-p-TOLUENESULFONYL-3-BUTYLUREA ◇ TOLUINA ◇ TOLUMID ◇ TOLUVAN ◇ N-(p-TOLYLSULFONYL)-N'-BUTYLCARBAMIDE ◇ 3-(p-TOLYL-4-SULFONYL)-1-BUTYLUREA ◇ TOLYLSULFONYLBUTYLUREA ◇ WILLBUTAMIDE

TOXICITY DATA with REFERENCE
sce-mus-orl 28600 μg/kg MUREAV 77,349,80
sce-ham-ipr 28600 μg/kg MUREAV 77,349,80
orl-wmn TDLo:2 mg/kg (1-13W preg):REP CMAJAX 87,193,62
orl-wmn TDLo:2 mg/kg (1-13W preg):TER CMAJAX 87,193,62
orl-wmn LDLo:1 g/kg:GIT:SYS ATXKA8 23,153,68
orl-rat LD50:2490 mg/kg PMDCAY 1,187,61
ipr-rat LD50:860 mg/kg FRPSAX 12,268,57
ivn-rat LD50:700 mg/kg PMDCAY 1,187,61
orl-mus LD50:490 mg/kg IJCREE 26,81,88
scu-mus LD50:980 mg/kg NATUAS 193,891,62
ivn-rat LD50:700 mg/kg PMDCAY 1,187,61
ipr-mus LD50:700 mg/kg PCJOAU 14,107,80
ivn-mus LD50:770 mg/kg PMDCAY 1,187,61

CONSENSUS REPORTS: NCI Carcinogenesis Bioassay (feed); No Evidence: mouse, rat NCITR* NCI-CG-TR-31,77. Reported in EPA TSCA Inventory. EPA Genetic Toxicology Program.

SAFETY PROFILE: Moderately toxic by ingestion and several other routes. A human teratogen. Human reproductive effects by ingestion and possibly other routes: stillbirth, developmental abnormalities of the cardiovascular (circulatory) system and urogenital system, and unspecified neonatal effects. Human systemic effects by ingestion: nausea or vomiting, hypoglycemia. Other experimental teratogenic and reproductive effects. Mutation data reported. Implicated in aplastic anemia. When heated to decomposition it emits very toxic fumes of NO_x and SO_x.

BSQ250 CAS:473-41-6 ***HR: D***
1-BUTYL-3-(p-TOLYLSULFONYL)UREA, SODIUM SALT
mf: $C_{12}H_{17}N_2O_3S{\cdot}Na$ mw: 292.36

SYN: TOLBUTAMIDE SODIUM SALT

TOXICITY DATA with REFERENCE
ipr-rat TDLo:250 mg/kg (female 1-20D post):TER DIAEAZ 13,527,64
ipr-mus TDLo:400 mg/kg (female 13D post):REP TJADAB 13,65,76
ipr-mus LD50:467 mg/kg TJADAB 13,65,76

SAFETY PROFILE: An experimental teratogen. Other experimental reproductive effects. See also SULFONATES. When heated to decomposition it emits very toxic fumes of SO_x, Na_2O and NO_x.

BSQ500 CAS:4872-26-8 ***HR: 3***
BUTYLTRICHLOROGERMANE
mf: $C_4H_9Cl_3Ge$ mw: 236.07

TOXICITY DATA with REFERENCE
ipr-rat LDLo:48 mg/kg CHDDAT 262,1302,66
ipr-mus LD50:190 mg/kg CHDDAT 262,1302,66

SAFETY PROFILE: Poison by intraperitoneal route. When heated to decomposition it emits very toxic fumes of Cl^-. See also GERMANIUM COMPOUNDS and CHLORIDES.

BSQ750 CAS:93-79-8 ***HR: 1***
BUTYL-2,4,5-TRICHLOROPHENOXYACETATE
mf: $C_{12}H_{13}Cl_3O_3$ mw: 311.60

SYNS: ARBORICID ◇ BUTYL-2,4,5-T ◇ BUTYLATE-2,4,5-T ◇ N-BUTYLESTER KYSELINI-2,4,5-TRICHLORFENOXYOCTOVE (CZECH) ◇ N-BUTYL (2,4,5-TRICHLOROPHENOXY)ACETATE ◇ FLOMORE ◇ KILEX 3 ◇ KRZEWOTOKS ◇ 2,4,5-T-N-BUTYL ESTER ◇ TORMONA ◇ 2,4,5-TRICHLOROPHENOXYACETIC ACID, BUTYL ESTER ◇ TRIOXONE ◇ U46KW

TOXICITY DATA with REFERENCE
skn-rbt 500 mg/24H MOD 28ZPAK -,85,72
eye-rbt 100 mg/24H MOD 28ZPAK -,85,72
cyt-dmg-orl 250 ppm/24H HEREAY 68,115,71
cyt-rat-orl 10 μg/kg GTPZAB 18(4),24,74
orl-rat TDLo:2 mg/kg (1-20D preg):REP GTPZAB 20(8),5,76
orl-mus TDLo:1246 mg/kg (female 12-15D post):TER AECTCV 6,33,77

CONSENSUS REPORTS: EPA Genetic Toxicology Program.

SAFETY PROFILE: Moderately toxic by ingestion. Experimental teratogenic and reproductive effects. A skin and eye irritant. Mutation data reported. See also ESTERS. When heated to decomposition it emits toxic fumes of Cl^-.

BSR000 CAS:7521-80-4 ***HR: 3***
BUTYL TRICHLORO SILANE
DOT: UN 1747
mf: $C_4H_9Cl_3Si$ mw: 191.57

PROP: Liquid, vap d: 6.4, flash p: 130°F (OC), d: 1.2.

CONSENSUS REPORTS: Reported in EPA TSCA Inventory.

DOT Classification: Corrosive Material; Label: Corrosive.

SAFETY PROFILE: A corrosive poison. See also CHLOROSILANE. Flammable by heat, flame (sparks), oxidizers. To fight fire, use water to blanket fire, fog, mist, dry chemical, alcohol foam. Reacts with water or steam to produce heat and toxic and corrosive fumes. When heated to decomposition it emits highly toxic fumes of Cl^-.

BSR250 CAS:1118-46-3 ***HR: 2***
BUTYL TRICHLORO STANNANE
mf: $C_4H_9Cl_3Sn$ mw: 282.17

SYN: CHLORID-N-BUTYLCINICITY (CZECH)

TOXICITY DATA with REFERENCE
skn-rbt 50 mg/24H SEV 28ZPAK -,225,72
eye-rbt 50 μg/24H SEV 28ZPAK -,225,72
orl-rat LD50:2140 mg/kg 28ZPAK -,225,72

CONSENSUS REPORTS: Reported in EPA TSCA Inventory.

OSHA PEL: TWA 0.1 mg(Sn)/m^3 (skin)
ACGIH TLV: TWA 0.1 mg(Sn)/m^3 (skin) (Proposed: TWA 0.1 mg(Sn)/m^3; STEL 0.2 mg(Sn)/m^3 (skin))
NIOSH REL: (Organotin Compounds) TWA 0.1 mg(Sn)/m^3

SAFETY PROFILE: Moderately toxic by ingestion. A severe skin and eye irritant. See also TIN COMPOUNDS. When heated to decomposition it emits toxic fumes of Cl^-.

BSR500 CAS:313-94-0 ***HR: 3***
3-tert-BUTYLTRICYCLOQUINAZOLINE
mf: $C_{25}H_{21}N_4$ mw: 377.50

TOXICITY DATA with REFERENCE
skn-mus TDLo:1200 mg/kg/50W-I:CAR BCPCA6 14,323,65

SAFETY PROFILE: Questionable carcinogen with experimental carcinogenic data. When heated to decomposition it emits toxic fumes of NO_x.

BSR825 CAS:73452-32-1 ***HR: 3***
N-tert-BUTYL-N-TRIMETHYLSILYLAMINOBORANE
mf: C_7H_2OBNSi mw: 154.99

$(C_4H_9)N(Si(CH_3)_3)BH_2$

SAFETY PROFILE: Ignites spontaneously on contact with air. When heated to decomposition it emits toxic fumes of NO_x. See also BORANES, BORON COMPOUNDS, and SILANE.

BSS000 CAS:25852-70-4 ***HR: 2***
BUTYLTRIS(ISOOCTYLOXYCARBONYLMETHYLTHIO)STANNANE
mf: $C_{34}H_{66}O_6S_3Sn$ mw: 785.87

SYN: BUTYLTRIS(2-ETHYLHEXYLOXYCARBONYLMETHYLTHIO) STANNANE

TOXICITY DATA with REFERENCE
unk-rat LD50:1063 mg/kg TIUSAD 107,1,76

CONSENSUS REPORTS: Reported in EPA TSCA Inventory.

OSHA PEL: TWA 0.1 mg(Sn)/m^3 (skin)
ACGIH TLV: TWA 0.1 mg(Sn)/m^3 (skin) (Proposed: TWA 0.1 mg(Sn)/m^3; STEL 0.2 mg(Sn)/m^3 (skin))
NIOSH REL: (Organotin Compounds) TWA 0.1 mg(Sn)/m^3

SAFETY PROFILE: Moderately toxic by an unspecified route. See also TIN COMPOUNDS. When heated to decomposition it emits toxic fumes of SO_x.

BSS100 CAS:109-42-2 ***HR: 1***
BUTYL 10-UNDECENOATE
mf: $C_{15}H_{28}O_2$ mw: 240.43

SYNS: BUTYL UNDECYLENATE ◇ 10-UNDECENOIC ACID, BUTYL ESTER

TOXICITY DATA with REFERENCE
skn-rbt 500 mg/24H MLD FCTXAV 17,729,79
orl-rat LD50:5000 mg/kg FCTXAV 17,729,79

CONSENSUS REPORTS: Reported in EPA TSCA Inventory.

SAFETY PROFILE: Mildly toxic by ingestion. A skin irritant. When heated to decomposition it emits acrid smoke and irritating fumes.

BSS250 CAS:592-31-4 ***HR: 2***
N-BUTYLUREA
mf: $C_5H_{12}N_2O$ mw: 116.19

SYN: NCI-CO2131

TOXICITY DATA with REFERENCE
cyt-rat-orl 100 mg/kg ZKKOBW 86,47,76
cyt-ham:fbr 4 g/L/48H MUREAV 48,337,77
par-mus LDLo:1627 mg/kg JPETAB 51,217,34

CONSENSUS REPORTS: Reported in EPA TSCA Inventory. EPA Genetic Toxicology Program.

SAFETY PROFILE: Moderately toxic by parenteral route. Mutation data reported. When heated to decomposition it emits toxic fumes of NO_x.

BSS500 ***HR: 3***
1-BUTYLUREA and SODIUM NITRITE (2:1)

TOXICITY DATA with REFERENCE
orl-rat TDLo:126 g/kg/33W-C:CAR IJCNAW 23,253,79
orl-rat TDLo:1350 mg/kg (13-21D preg):ETA,TER GANNA2 68,81,77
orl-mus TDLo:380 g/kg/42W-C:CAR IJCNAW 23,253,79
orl-rat TD:27500 mg/kg/50D-C:NEO ZAPPAN 121,61,77

SAFETY PROFILE: Suspected carcinogen with experimental carcinogenic, neoplastigenic, and tumorigenic data. An experimental teratogen. When heated to decomposition it emits toxic fumes of NO_x. See also NITRITES.

BST000 CAS:1879-09-0 ***HR: 3***
6-tert-BUTYL-2,4-XYLENOL
mf: $C_{12}H_{18}O$ mw: 178.30

SYNS: 6-tert-BUTYL-2,4-DIMETHYLPHENOL ◇ PRODOX 340

TOXICITY DATA with REFERENCE
orl-mus LD50:530 mg/kg JAPMA8 38,366,49
orl-rbt LDLo:55 mg/kg JAPMA8 38,366,49
skn-gpg LDLo:7100 mg/kg JAPMA8 38,366,49
orl-gpg LDLo:420 mg/kg JAPMA8 38,366,49

CONSENSUS REPORTS: Reported in EPA TSCA Inventory.

SAFETY PROFILE: Poison by ingestion. Mildly toxic by skin contact. When heated to decomposition it emits smoke and acrid, irritating fumes.

BST500 CAS:110-65-6 ***HR: 3***
2-BUTYNE-1,4-DIOL
DOT: UN 2716
mf: $C_4H_6O_2$ mw: 86.10

$HOCH_2C{=}CCH2OH$

PROP: Straw to amber crystals. Mp: 57.5°, bp: 194° @ 100 mm.

SYN: 1,4-BUTYNEDIOL (DOT)

TOXICITY DATA with REFERENCE
orl-rat LD50:104 mg/kg HYSAAV 33,41,68
ihl-rat LCLo:150 mg/m^3/2H 85GMAT -,30,82
orl-mus LD50:105 mg/kg HYSAAV 33,41,68
ihl-mus LCLo:150 mg/m^3/2H 85GMAT -,30,82
orl-rbt LD50:150 mg/kg HYSAAV 33,41,68
orl-gpg LD50:130 mg/kg HYSAAV 33,41,68
orl-bwd LD50:75 mg/kg AECTCV 12,355,83

CONSENSUS REPORTS: Reported in EPA TSCA Inventory.

DOT Classification: Flammable Solid; Label: Flammable Solid.

SAFETY PROFILE: A poison by ingestion. A skin sensitizer upon long or repeated contact. Moderately explosive. When heated to decomposition it emits acrid smoke and fumes and may explode. Explosive reaction with traces of alkalies, alkali earth hydroxides, halide salts, strong acids, mercury salts + strong acids. See also ACETYLENE COMPOUNDS.

BST750 ***HR: 3***
2-BUTYNE-1-THIOL
mf: C_4H_6S mw: 86.16

SAFETY PROFILE: Forms an explosive polymer on exposure to air. Store at −20° in the presence of a stabilizer under nitrogen. When heated to decomposition it emits toxic fumes of SO_x. See also ACETYLENE COMPOUNDS.

BST900 CAS:1606-83-3 ***HR: D***
1,1'-(2-BUTYNYLENEDIOXY)BIS(3-CHLORO-2-PROPANOL)
mf: $C_{10}H_{16}Cl_2O_4$ mw: 271.16

SYNS: 2-PROPANOL, 1,1'-(2-BUTYNYLENEDIOXY)BIS(3-CHLORO- ◇ U 27,151

TOXICITY DATA with REFERENCE
orl-rat TDLo:960 mg/kg (male 8D pre):REP JRPFA4 21,263,70

CONSENSUS REPORTS: Reported in EPA TSCA Inventory.

SAFETY PROFILE: Experimental reproductive effects. When heated to decomposition it emits toxic fumes of Cl^-.

BSU000 ***HR: 3***
3-BUTYN-1-YL-p-TOLUENE SULFONATE
mf: $C_{11}H_{12}O_3S$ mw: 224.18

SAFETY PROFILE: Explodes in vacuum at 0.65 mbar. May be safe in small amounts below 0.01 mbar. When heated to decomposition it emits toxic fumes of SO_x. See also SULFONATES and ACETYLENE COMPOUNDS.

BSU250 CAS:123-72-8 ***HR: 3***
n-BUTYRALDEHYDE
DOT: UN 1129
mf: C_4H_8O mw: 72.12

PROP: Colorless, mobile liquid; pungent, nutty odor. Mp: −100°, bp: 74.7°, flash p: 20°F (CC), (−6°), d: 0.902 @ 20°/4°, autoign temp: 446°F, lel: 2.5%, uel: 12.5%, vap d: 2.5, D: 0.797-0.802. Sol in water; misc with ether @ 74.8°.

SYNS: ALDEHYDE BUTYRIQUE (FRENCH) ◇ ALDEIDE BUTIRRICA (ITALIAN) ◇ BUTAL ◇ BUTALDEHYDE ◇ BUTALYDE ◇ BUTANAL ◇ n-BUTANAL (CZECH) ◇ BUTYRAL ◇ BUTYRALDEHYD (GERMAN) ◇ BUTYRALDEHYDE (CZECH) ◇ n-BUTYL ALDEHYDE ◇ BUTYRIC ALDEHYDE ◇ FEMA No. 2219 ◇ NCI-C56291

TOXICITY DATA with REFERENCE
skn-rbt 410 mg open MLD UCDS** 7/20/65
skn-rbt 500 mg/24H SEV 28ZPAK -,40,72
eye-rbt 75 μg open SEV AMIHBC 4,119,51
eye-rbt 20 mg/24H MOD 28ZPAK -,40,72
skn-gpg 100% MOD FCTXAV 17,731,79
spm-mus-ipr 30 mg/kg MUREAV 39,317,77
spm-mus-orl 15 g/kg/50D MUREAV 39,317,77
ihl-hmn TCLo:580 mg/m³:IMM BMJOAE 2,913,56
orl-rat LD50:2490 mg/kg 28ZPAK -,40,72
ihl-rat LCLo:8000 ppm/4H AMIHBC 4,119,51
ipr-rat LD50:800 mg/kg FCTXAV 17,731,79
scu-rat LDLo:10 g/kg ARZNAD 11,73,61
ihl-mus LC50:44610 mg/m³/2H 85GMAT -,30,82
ipr-mus LD50:1140 mg/kg FCTXAV 17,731,79
scu-mus LD50:2700 mg/kg APTOA6 6,299,50
skn-rbt LD50:3560 mg/kg UCDS** 7/20/67
ihl-mam LC50:64 g/m³ GTPZAB 12(7),16,68

CONSENSUS REPORTS: Community Right-To-Know List. Reported in EPA TSCA Inventory.

DOT Classification: Flammable Liquid; Label: Flammable Liquid.

SAFETY PROFILE: Moderately toxic by ingestion, inhalation, skin contact, intraperitoneal, and subcutaneous routes. Severe skin and eye irritant. Human immunological effects by inhalation: delayed hypersensitivity. See also ALDEHYDES. Highly flammable liquid. To fight fire, use foam, CO_2, dry chemical. Incompatible with oxidizing materials. Reacts vigorously with chlorosulfonic acid, HNO_3, oleum, H_2SO_4. When heated to decomposition it emits acrid smoke and fumes.

BSU500 CAS:110-69-0 ***HR: 3***
m-BUTYRALDEHYDE OXIME
DOT: UN 2840
mf: C_4H_9NO mw: 87.14

PROP: Liquid. Mp: −29.5°, bp: 152°, flash p: 136°F (CC), d: 0.923, vap d: 3.01.

SYNS: BUTANAL OXIME ◇ BUTYRALDOXIME (DOT) ◇ N-BUTYRALDOXIME ◇ SKINO #1 ◇ TROYKYD ANTI-SKIN BTO ◇ USAF AM-6

TOXICITY DATA with REFERENCE
ipr-mus LD50:200 mg/kg NTIS** AD277-689

CONSENSUS REPORTS: Reported in EPA TSCA Inventory.

DOT Classification: Flammable or Combustible Liquid; Label: Flammable Liquid.

SAFETY PROFILE: A poison by intraperitoneal route. Flammable when exposed to heat or flame. To fight fire, use alcohol foam, dry chemical. Highly explosive. Can explode during vacuum distillation. Incompatible with oxidizing materials, metallic impurities. When heated to decomposition it emits toxic fumes of NO_x. See also ALDEHYDES.

BSV250 CAS:29067-70-7 ***HR: 3***
2-(3-BUTYRAMIDO-2,4,6-TRIIODOPHENYL)PROPIONIC ACID
mf: $C_{13}H_{14}I_3NO_3$ mw: 612.98

TOXICITY DATA with REFERENCE
orl-mus LD50:100 mg/kg JMCMAR 13,559,70
ivn-mus LD50:300 mg/kg JMCMAR 13,559,70

SAFETY PROFILE: Poison by ingestion and intravenous routes. When heated to decomposition it emits very toxic fumes of NO_x and I^-.

BSV500 CAS:1129-50-6 ***HR: 2***
n-BUTYRANILIDE
mf: $C_{10}H_{13}NO$ mw: 163.24

TOXICITY DATA with REFERENCE
orl-mus LD50:1630 mg/kg TXAPA9 19,20,71

CONSENSUS REPORTS: Reported in EPA TSCA Inventory.

SAFETY PROFILE: Moderately toxic by ingestion. When heated to decomposition it emits toxic fumes of NO_x.

BSV750 CAS:2440-29-1 ***HR: 3***
(BUTYRATO)PHENYLMERCURY
mf: $C_{10}H_{12}HgO_2$ mw: 364.81

SYN: PHENYL(BUTYRATE)MERCURY

TOXICITY DATA with REFERENCE
ipr-mus LDLo:16 mg/kg CBCCT* 3,53,51

CONSENSUS REPORTS: Mercury and its compounds are on the Community Right-To-Know List.

OSHA PEL: (Transitional: CL 1 mg/10m^3) CL 0.1 mg(Hg)/m^3 (skin)
ACGIH TLV: TWA 0.1 mg(Hg)/m^3 (skin)
NIOSH REL: TWA 0.05 mg(Hg)/m^3

SAFETY PROFILE: Poison by intraperitoneal route. See also MERCURY COMPOUNDS. When heated to decomposition it emits toxic Hg vapors.

BSW000 CAS:107-92-6 ***HR: 2***
n-BUTYRIC ACID
DOT: UN 2820
mf: $C_4H_8O_2$ mw: 88.12

PROP: Colorless liquid; strong, rancid butter odor. Mp: −7.9°, bp: 163.5°, flash p: 161°F, fp: −5.5°, d: 0.9590 @ 20°/20°, refr index: 1.397, autoign temp: 846°F, vap press: 0.43 mm @ 20°, vap d: 3.04, lel: 2.0%, uel: 10.0%.

SYNS: BUTANOIC ACID ◇ BUTTERSAEURE (GERMAN) ◇ ETHYLACETIC ACID ◇ FEMA No. 2221 ◇ 1-PROPANECARBOXYLIC ACID ◇ PROPYLFORMIC ACID

TOXICITY DATA with REFERENCE
skn-rbt 10 mg/24H open SEV AMIHBC 10,61,54
skn-rbt 500 mg open MOD UCDS** 4/10/68
eye-rbt 250 μg open SEV AMIHBC 10,61,54
dnd-hmn:hla 3 mmol/L CELLB5 12,855,77
dni-hmn:lym 4 mmol/L HAONDL 2,381,84
orl-rat LD50:2940 mg/kg AMIHBC 4,119,51
orl-mus LDLo:500 mg/kg TPKVAL 4,19,62
ipr-mus LD50:3180 mg/kg JPPMAB 21,85,69
scu-mus LD50:3180 mg/kg JPPMAB 21,85,69
ivn-mus LD50:800 mg/kg APTOA6 18,141,61
skn-rbt LD50:530 mg/kg UCDS** 4/10/68

CONSENSUS REPORTS: Reported in EPA TSCA Inventory.

DOT Classification: Corrosive Material; Label: Corrosive.

SAFETY PROFILE: Moderately toxic by ingestion, skin contact, subcutaneous, intraperitoneal, and intravenous routes. Human mutation data reported. Severe skin and eye irritant. A corrosive material. Combustible liquid. Could react with oxidizing materials. Incandescent reaction with chromium trioxide above 100°. To fight fire, use alcohol foam, CO_2, dry chemical. When heated to decomposition it emits acrid smoke and irritating fumes.

BSW500 CAS:539-90-2 ***HR: 1***
BUTYRIC ACID ISOBUTYL ESTER
mf: $C_8H_{16}O_2$ mw: 144.24

PROP: Colorless liquid; apple-pineapple odor. D: 0.858-0863, refr index: 1.402. Sol in alc, fixed oils; sltly sol in water; insol in glycerin.

SYNS: FEMA No. 2187 ◇ ISOBUTYL BUTANOATE ◇ ISOBUTYL BUTYRATE (FCC) ◇ 2-METHYLPROPYL BUTYRATE

TOXICITY DATA with REFERENCE
skn-rbt 500 mg/24H MLD FCTXAV 17,833,79
idu-rbt LD50:9500 mg/kg FCTXAV 17(Suppl),695,79
orl-rbt LD50:9520 mg/kg IMSUAI 41,31,72

CONSENSUS REPORTS: Reported in EPA TSCA Inventory.

SAFETY PROFILE: Mildly toxic by ingestion and intraduodenal routes. A skin irritant. See also ESTERS. When heated to decomposition it emits acrid smoke and irritating fumes.

BSX000 CAS:3068-88-0 ***HR: 3***
β-BUTYROLACTONE
mf: $C_4H_6O_2$ mw: 86.10

SYNS: 3-HYDROXYBUTANOIC ACID-β-LACTONE ◇ HYDROXYBUTYRIC ACID LACTONE ◇ 3-HYDROXYBUTYRIC ACID LACTONE ◇ 4-METHYL-2-OXETANONE

TOXICITY DATA with REFERENCE
skn-rbt 500 mg open MOD UCDS** 1/20/66
dnd-mam:lym 10 mmol/L BBACAQ 138,611,67
oms-mam:lym 286 nmol/L CBINA8 34,323,81
orl-rat TDLo:31 g/kg/61W-I:ETA JNCIAM 37,825,66
scu-rat TDLo:38 g/kg/78W-I:CAR JNCIAM 39,1213,67
skn-mus TDLo:59 g/kg/49W-I:CAR JNCIAM 35,707,65
scu-mus TDLo:12 g/kg/30W-I:NEO JNCIAM 37,825,66
orl-rat LD50:17 g/kg AIHAAP 30,470,69

CONSENSUS REPORTS: IARC Cancer Review: Group 2B IMEMDT 7,56,87; Animal Sufficient Evidence IMEMDT 11,225,76. Reported in EPA TSCA Inventory.

SAFETY PROFILE: Suspected carcinogen with experimental carcinogenic, neoplastigenic, and tumorigenic data. Mildly toxic by ingestion. A moderate skin irritant. Mutation data reported. When heated to decomposition it emits acrid and irritating fumes. See also BUTYROLACTONE.

BSX250 CAS:109-74-0 ***HR: 3***
BUTYRONITRILE
DOT: UN 2411
mf: C_4H_7N mw: 69.12

PROP: Colorless liquid. D: 0.796 @ 15°, mp: −112.6°, bp: 117°, flash p: 79°F (OC). Sltly sol in water; sol in alc and ether.

SYNS: BUTANENITRILE ◇ n-BUTANENITRILE ◇ BUTYRIC ACID NITRILE ◇ BUTYRONITRILE (DOT) ◇ 1-CYANOPROPANE ◇ PROPYL CYANIDE

TOXICITY DATA with REFERENCE
skn-rbt 395 mg open MLD UCDS** 5/17/60
orl-rat LD50:140 mg/kg AIHAAP 23,95,62

ihl-rat LCLo:1000 ppm/4H AIHAAP 23,95,62
orl-mus LD50:27689 μg/kg NEZAAQ 39,423,84
ihl-mus LC50:249 ppm/1H CTOXAO 18,991,81
ipr-mus LD50:38 mg/kg TXAPA9 59,589,81
skn-rbt LD50:500 mg/kg AIHAAP 23,95,62
scu-rbt LDLo:10 mg/kg AIPTAK 5,161,1899
ivn-rbt LDLo:980 mg/kg COREAF 153,895,11
skn-gpg LDLo:100 mg/kg KODAK* 21MAY71
scu-gpg LDLo:100 mg/kg COREAF 153,895,11
scu-frg LDLo:3100 mg/kg AIPTAK 5,161,1899

CONSENSUS REPORTS: Reported in EPA TSCA Inventory. Cyanide and its compounds are on the Community Right-To-Know List.

NIOSH REL: (Nitriles) TWA 22 mg/m^3

DOT Classification: Flammable Liquid; Label: Flammable Liquid and Poison.

SAFETY PROFILE: A poison by ingestion, skin contact, intraperitoneal, and subcutaneous routes. Moderately toxic by inhalation. A skin irritant. Dangerous fire hazard when exposed to heat, flame, or oxidizers. To fight fire, use alcohol foam. When heated to decomposition it emits toxic fumes of NO_x and CN^-.

BSX325 CAS:34291-02-6 ***HR: 2***
BUTYROSIN A
mf: $C_{21}H_{41}N_5O_{12}$ mw: 555.67

SYNS: AMBUTYROSIN A ◇ AMBUYROSIN A ◇ BUTIROSIN A ◇ o-2,6-DIAMINO-2,6-DIDEOXY-α-d-GLUCOPYRANOSYL-(1-4)-o-(β-d-XYLOFURANOSYL-(1-5))-N[1]-(4-AMINO-2-HYDROXY-1-OXOBUTYL)-2-DEOXY-d-STREPTAMINE

TOXICITY DATA with REFERENCE
ipr-mus LD50:2198 mg/kg 85GDA2 1,145,80
scu-mus LD50:3050 mg/kg 85GDA2 1,145,80
ivn-mus LD50:50 mg/kg 38KLAC -,239,77

SAFETY PROFILE: Poison by intravenous route. Moderately toxic by subcutaneous and intraperitoneal routes. When heated to decomposition it emits toxic fumes of NO_x. See also AMINES.

BSX500 CAS:67557-56-6 ***HR: 3***
N-(1-BUTYROXYMETHYL)METHYLNITROSAMINE
mf: $C_6H_{12}N_2O_3$ mw: 160.20

SYNS: N-(1-BUTYROXYMETHYL)-N-NITROSOMETHYLAMINE ◇ N-NITROSO-N-(1-BUTYROXYMETHYL)METHYL AMINE

TOXICITY DATA with REFERENCE
mmo-sat 1 μmol/plate ARTODN 39,51,77
mma-sat 200 nmol/plate ARTODN 39,51,77
orl-rat TDLo:60 mg/kg/90D-I:ETA ZKKOBW 91,317,78
orl-rat LD50:800 mg/kg ZKKOBW 91,317,78

SAFETY PROFILE: Moderately toxic by ingestion. Questionable carcinogen with experimental tumorigenic data. Mutation data reported. When heated to decomposition it emits toxic fumes of NO_x. See also NITROSAMINES.

BSX750 CAS:37415-56-8 ***HR: 3***
12-o-BUTYROYL-PHORBOLDODECANOATE
mf: $C_{36}H_{57}O_8$ mw: 617.93

SYN: PHORBOL-12-o-BUTYROYL-13-DODECANOATE

TOXICITY DATA with REFERENCE
skn-mus 3 ng MLD 85CVA2 5,213,70
skn-mus TDLo:12 mg/kg/12W-I:ETA 85CVA2 5,213,70

SAFETY PROFILE: Questionable carcinogen with experimental tumorigenic data. A skin irritant. When heated to decomposition it emits acrid smoke and irritating fumes.

BSY000 CAS:10431-86-4 ***HR: 3***
1-n-BUTYRYLAZIRIDINE
mf: $C_6H_{11}NO$ mw: 113.18

SYNS: 1-BUTYRYLAZIRIDINE ◇ BUTYRYLETHYLENEIMINE ◇ BUTYRYLETHYLENIMINE ◇ 1-(1-OXOBUTYL)AZIRIDINE

TOXICITY DATA with REFERENCE
cyt-rat-ipr 50 mg/kg BJPCAL 9,306,54
scu-rat TDLo:225 mg/kg/26W-I:ETA BJPCAL 9,306,54
ipr-mus LD50:970 mg/kg NCISA* PH-43-63-1132

SAFETY PROFILE: Moderately toxic by intraperitoneal route. Questionable carcinogen with experimental tumorigenic data. Mutation data reported. When heated to decomposition it emits toxic fumes of NO_x.

BSY250 ***HR: 3***
BUTYRYL CHLORIDE
mf: C_4H_7ClO mw: 106.51

PROP: Clear, colorless liquid with sharp odor. Mp: −89°, bp: 101°, d: 1.028 @ 20°/20°, vap d: 3.67, flash p: <21°.

SAFETY PROFILE: A poisonous irritant to skin, eyes, and mucous membranes. A dangerous fire hazard when exposed to heat or flame. Reaction with water, steam, or oxidizing materials produces toxic and corrosive fumes. When heated to decomposition it emits highly toxic fumes of Cl^-. See also CHLORIDES.

BSY750 ***HR: 3***
BUTYRYL NITRATE
mf: $C_4H_7NO_4$ mw: 133.06

SAFETY PROFILE: Explodes when heated. Upon decomposition it emits toxic fumes of NO_x. See also NITRATES.

BSZ000 CAS:1063-55-4 ***HR: 3***
BUTYRYLPERAZINE DIMALEATE
mf: $C_{24}H_{31}N_3OS•2C_4H_4O_4$ mw: 641.80

SYNS: BAYER 1362 ◇ BUTAPERAZINE DIMALEATE ◇ 1-(10-(3-(4-METHYL-1-PIPERAZINYL)PROPYL)PHENOTHIAZIN-2-YL)-1-BUTANONE DIMALEATE ◇ RANDOLECTIL ◇ REPOISE MALEATE ◇ RIKER 595

TOXICITY DATA with REFERENCE
orl-rat LD50:264 mg/kg CLPTAT 10,428,69
ivn-rat LD50:63 mg/kg CLPTAT 10,428,69
orl-mus LD50:296 mg/kg CLPTAT 10,428,69
ivn-mus LD50:67 mg/kg 27ZQAG -,11,72

SAFETY PROFILE: Poison by ingestion and intravenous routes. When heated to decomposition it emits very toxic fumes of SO_x and NO_x. See also KETONES.

BTA000 CAS:17730-82-4 ***HR: 3***
1-BUTYRYL-4-(PHENYLALLYL)PIPERAZINE HYDROCHLORIDE
mf: $C_{17}H_{24}N_2O•ClH$ mw: 308.89

SYNS: AP-237 ◇ 1-BUTYRYL-4-CINNAMYLPIPERAZINE HYDROCHLORIDE ◇ 1-N-BUTYRYL-4-CINNAMYL PIPERAZINE HYDROCHLORIDE ◇ P 237

TOXICITY DATA with REFERENCE
scu-rat TDLo:140 mg/kg (female 7-13D post):REP OYYAA2 6,271,72
orl-mus TDLo:1260 mg/kg (female 7-13D post):TER OYYAA2 6,271,72
orl-rat LD50:545 mg/kg AIPTAK 213,28,75
scu-rat LD50:339 mg/kg JJPAAZ 20,287,70
ivn-rat LD50:70 mg/kg JJPAAZ 20,287,70
orl-mus LD50:710 mg/kg AIPTAK 213,28,75
scu-mus LD50:258 mg/kg AIPTAK 213,28,75
ivn-mus LD50:80800 μg/kg OYYAA2 6,173,72
orl-dog LDLo:260 mg/kg TXAPA9 25,443,73
orl-gpg LD50:700 mg/kg JJPAAZ 20,287,70
scu-gpg LDLo:400 mg/kg TXAPA9 25,443,73

SAFETY PROFILE: Poison by ingestion, intravenous, and subcutaneous routes. An experimental teratogen. Other experimental reproductive effects. When heated to decomposition it emits very toxic fumes of HCl and NO_x. See also ALLYL COMPOUNDS.

BTA125 CAS:60479-97-2 ***HR: 3***
BUTYRYLPROMAZINE MALEATE
mf: $C_{21}H_{26}N_2OS•C_4H_4O_4$ mw: 470.63

SYNS: 1613-CB ◇ 2-BUTYRYL-10-(3-DIMETHYLAMINOPROPYL) PHENOTHIAZINE MALEATE ◇ 1-(10-(3-(DIMETHYLAMINO)PROPYL) PHENOTHIAZIN-2-YL)-1-BUTANONEMALEATE

TOXICITY DATA with REFERENCE
orl-rat LD50:300 mg/kg AIPTAK 123,78,59
scu-mus LD50:190 mg/kg PSCBAY 2,17,63
ivn-mus LD50:73 mg/kg AIPTAK 123,78,59

SAFETY PROFILE: Poison by ingestion, subcutaneous, and intravenous routes. When heated to decomposition it emits toxic fumes of SO_x and NO_x.

BTA250 CAS:8065-36-9 ***HR: 3***
BUX-TEN

PROP: A low-melting, amber solid. Mp: 26.4°. Very sol in xylene, ethanol; nearly insol in water.

SYNS: BUFENCARB ◇ BUX ◇ METALKAMATE ◇ METHYLCARBAMIC ACID-m-(1-METHYL)BUTYL)PHENYL ESTER mixed with CARBAMIC ACID, METHYL-m-(1-ETHYLPROPYL)PHENYL ESTER (3:1) ◇ ORTHO 5353

TOXICITY DATA with REFERENCE
orl-rat LD50:85 mg/kg FMCHA2 -,C40,83
skn-rat LD50:242 mg/kg WRPCA2 9,119,70
skn-dog LD50:1400 mg/kg GUCHAZ 6,271,73
skn-rbt LD50:400 mg/kg GUCHAZ 6,271,73
orl-pgn LD50:23700 μg/kg ASTTA8 (680),157,79
orl-qal LD50:42100 μg/kg ASTTA8 (680),157,79
orl-bwd LD50:4210 μg/kg ASTTA8 (680),157,79

SAFETY PROFILE: A poison by ingestion and skin contact. See also CARBAMATES and ESTERS. When heated to decomposition it emits toxic fumes of NO_x.

BTA325 CAS:15351-05-0 ***HR: 3***
BUZEPIDE METHIODIDE
mf: $C_{23}H_{31}N_2O•I$ mw: 478.46

SYNS: DIFEXAMIDE METHIODIDE ◇ 2,2-DIPHENYL-4-N-HEXAMETHYLENIMINOBUTYRAMIDE METHIODIDE ◇ DIPHEXAMIDE METHIODIDE ◇ FI 6146 ◇ R 661 ◇ SPACTIN

TOXICITY DATA with REFERENCE
orl-rat TDLo:30 mg/kg (10-15D preg):REP KSRNAM 4,2515,70
orl-mus TDLo:1500 mg/kg (female 8-13D post):TER KSRNAM 4,2515,70
orl-rat LD50:2800 mg/kg KSRNAM 4,1850,70
ipr-rat LD50:145 mg/kg KSRNAM 4,1850,70
scu-rat LD50:1210 mg/kg KSRNAM 4,1850,70
ivn-rat LD50:29300 μg/kg KSRNAM 4,1850,70
ims-rat LD50:800 mg/kg KSRNAM 4,1850,70
orl-mus LD50:820 mg/kg KSRNAM 4,1850,70
ipr-mus LD50:94 mg/kg KSRNAM 4,1850,70
scu-mus LD50:229 mg/kg KSRNAM 4,1850,70
ivn-mus LD50:14 mg/kg KSRNAM 4,1850,70
ims-mus LD50:201 mg/kg KSRNAM 4,1850,70

SAFETY PROFILE: A poison by intraperitoneal, intravenous, intramuscular, and subcutaneous routes. Moderately toxic by ingestion. An experimental teratogen. Experimental reproductive effects. When heated to decomposition it emits toxic fumes of I^- and NO_x. See also IODIDES.

BTA500 CAS:60452-14-4 ***HR: 3***
BZL
mf: $C_{20}H_{27}N_5O_3 \cdot ClH$ mw: 421.98

SYN: 7-(β-DIETHYLAMINOETHYL)-8-(α-HYDROXYBENZYL) THEOPHYLLINE HYDROCHLORIDE

orl-rat LD50:400 mg/kg TXAPA9 7,291,65
ipr-rat LD50:107 mg/kg TXAPA9 7,291,65
ipr-mus LD50:115 mg/kg TXAPA9 7,291,65
ivn-mus LD50:30 mg/kg TXAPA9 7,291,65

SAFETY PROFILE: Poison by ingestion, intraperitoneal, and intravenous routes. When heated to decomposition it emits very toxic fumes of HCl and NO_x. See also THIOPHYLLINE.

C

CAB125 CAS:20064-38-4 ***HR: 3***
C-666
mf: $C_{18}H_{20}N_4O_3 \cdot 2ClH$ mw: 413.34

SYNS: N,N-DIMETHYL-N'-(1-NITRO-9-ACRIDINYL)-1,3-PROPANEDIAMINE-N-OXIDE, DIHYDROCHLORIDE (9CI) ◇ 1-NITRO-9-(3-DIMETHYLAMINOPROPYLAMINE)ACRIDINE-N^{10}-OXIDE DIHYDROCHLORIDE ◇ N^{10}-OXIDE-1-NITRO-9-(3-DIMETHYLAMINOPROPYLAMINO)-DIHYDROCHLORIDEACRIDINE

TOXICITY DATA with REFERENCE
orl-rat LD50:110 mg/kg MMDPA6 8,252,76
ivn-rat LD50:5400 μg/kg MMDPA6 8,252,76
orl-mus LD50:108 mg/kg MMDPA6 8,252,76
ivn-mus LD50:9 mg/kg MMDPA6 8,252,76
ivn-pgn LD50:9000 μg/kg AITEAT 28,777,80

SAFETY PROFILE: Poison by ingestion and intravenous routes. When heated to decomposition it emits toxic fumes of NO_x and HCl. See also AMINES.

CAB250 CAS:78265-91-5 ***HR: 3***
C 3206
mf: $C_{17}H_{25}ClN_2O_2 \cdot ClH$ mw: 361.35

SYN: 6'-CHLORO-2-PYRROLIDINYL-o-HEXANOTOLUIDIDEHYDROCHLORIDE

TOXICITY DATA with REFERENCE
eye-rbt 2% MOD ARZNAD 8,544,58
ipr-rat LD50:42 mg/kg ARZNAD 8,544,58
ipr-mus LD50:37 mg/kg ARZNAD 8,544,58
scu-mus LD50:70 mg/kg ARZNAD 8,544,58

SAFETY PROFILE: Poison by intraperitoneal and subcutaneous routes. An eye irritant. When heated to decomposition it emits very toxic fumes of Cl^- and NO_x.

CAB500 CAS:78265-89-1 ***HR: 3***
C 3207
mf: $C_{17}H_{27}ClN_2O \cdot ClH$ mw: 347.37

SYN: 2-(BUTYLAMINO)-6'-CHLORO-o-HEXANOTOLUIDIDE HYDROCHLORIDE

TOXICITY DATA with REFERENCE
eye-rbt 2% SEV ARZNAD 8,544,58
ipr-rat LD50:33 mg/kg ARZNAD 8,544,58
scu-mus LD50:210 mg/kg ARZNAD 8,544,58

SAFETY PROFILE: Poison by intraperitoneal and subcutaneous routes. A severe eye irritant. When heated to decomposition it emits very toxic fumes of Cl^- and NO_x.

CAB750 CAS:78265-90-4 ***HR: 3***
C 3208
mf: $C_{15}H_{23}ClN_2O \cdot ClH$ mw: 319.31

SYN: 6'-CHLORO-2-(ETHYLAMINO)-o-HEXANOTOLUIDIDEHYDROCHLORIDE

TOXICITY DATA with REFERENCE
eye-rbt 2% MOD ARZNAD 8,544,58
ipr-rat LD50:25 mg/kg ARZNAD 8,544,58
scu-mus LD50:62 mg/kg ARZNAD 8,544,58

SAFETY PROFILE: Poison by intraperitoneal and subcutaneous routes. An eye irritant. When heated to decomposition it emits very toxic fumes of Cl^- and NO_x.

CAC250 ***HR: 3***
CACODYL SULFIDE
mf: $((CH_3)_2As)_2S$ mw: 242

PROP: Oily liquid. Bp: 211°. Sltly sol in water.

SYN: DICACODYL SULFIDE

CONSENSUS REPORTS: Arsenic and its compounds are on the Community Right-To-Know List.

SAFETY PROFILE: Poison by most routes. See also ARSENIC COMPOUNDS and SULFIDES. Dangerous fire hazard when exposed to heat or by spontaneous chemical reaction, i.e., in air. Vigorous reaction with oxidizing materials. When heated to decomposition it emits toxic fumes of As.

CAC500 ***HR: 3***
CADIA DEL PERRO

PROP: Aqueous extract from the dried leaves of the plant (JNCIAM 46,1131,71).

SYNS: K. IXINA ◇ KRAMERIA IXINA

TOXICITY DATA with REFERENCE
scu-rat TDLo:300 mg/kg/1Y-I:NEO JNCIAM 46,1131,71
ims-rat TDLo:45 g/kg/1Y-I:ETA JNCIAM 46,1131,71
skn-ham TDLo:53950 mg/kg/65W-I:CAR JNCIAM 53,1259,74

SAFETY PROFILE: Questionable carcinogen with experimental carcinogenic, tumorigenic, and neoplastigenic data. When heated to decomposition it emits acrid smoke and fumes.

CAD000 CAS:7440-43-9 ***HR: 3***
CADMIUM
af: Cd aw: 112.40

PROP: Hexagonal crystals, silver-white, malleable metal. Mp: 320.9°, bp: 767 ± 2°, d: 8.642, vap press: 1 mm @ 394°.

SYNS: C.I. 77180 ◇ COLLOIDAL CADMIUM ◇ KADMIUM (GERMAN)

TOXICITY DATA with REFERENCE
cyt-ham:ovr 1 μmol/L CGCGBR 26,251,80
orl-rat TDLo:155 mg/kg (male 13W pre):REP BECTA6 20,96,78
orl-rat TDLo:21500 μg/kg (multi):TER ENVRAL 22,466,80
ihl-wmn TCLo:129 $\mu g/m^3$/20Y-C:CAR AJIMD8 10,153,86
ims-rat TDLo:40 mg/kg/4W-I:CAR JEPTDQ 1(1),51,77 ZHYGAM 31,224,85
ims-rat TD:70 mg/kg:ETA BJCAAI 18,124,64
ims-rat TD:45 mg/kg/4W-I:NEO NCIUS* PH-43-64-886,SEPT,71
ihl-man TCLo:88 $\mu g/m^3$/8.6Y:KID AEHLAU 28,147,74
ihl-hmn LCLo:39 mg/m^3/20M AIHAAP 31,180,70
unk-man LDLo:15 mg/kg 85DCAI 2,73,70
orl-rat LD50:225 mg/kg TXAPA9 41,667,77
ihl-rat LC50:25 mg/m^3/30M SAIGBL 16,212,74
ipr-rat LD50:4 mg/kg TXAPA9 41,667,77
scu-rat LD50:9 mg/kg TXAPA9 41,667,77
ivn-rat LD50:1800 μg/kg JJATDK 1,264,81
orl-mus LD50:890 mg/kg 41HTAH -,14,78
ihl-mus LCLo:170 mg/m^3 NTIS** PB158-508
orl-rbt LDLo:70 mg/kg AMPMAR 34,127,73
scu-rbt LDLo:6 mg/kg PROTA* -,-,55
ivn-rbt LDLo:5 mg/kg JOGBAS 35,693,28
ims-ham LDLo:25 mg/kg NCIUS* PH-43-64-886

CONSENSUS REPORTS: NTP Fifth Annual Report on Carcinogens. IARC Cancer Review: Group 2A IMEMDT 7,139,87; Animal Sufficient Evidence IMEMDT 11,39,76; IMEMDT 2,74,73. Cadmium and its compounds are on the Community Right-To-Know List. Reported in EPA TSCA Inventory. EPA Genetic Toxicology Program.

OSHA PEL: Fume: TWA 0.1 $mg(Cd)/m^3$; CL 0.6 $mg(Cd)/m^3$; Dust: TWA 0.2 $mg(Cd)/m^3$; CL 0.6 $mg(Cd)/m^3$
ACGIH TLV: Dust and Salts: TWA 0.05 $mg(Cd)/m^3$ (Proposed: TWA 0.01 $mg(Cd)/m^3$ (dust), Suspected Human Carcinogen; 0.002 $mg(Cd)/m^3$ (respirable dust), Suspected Human Carcinogen); BEI: 10 μg/g creatinine in urine; 10 μg/L in blood. (Proposed: 5 μg/g creatinine in urine; 5 μg/L in blood.)
DFG BAT: Blood 1.5 μg/dL; Urine 15 μg/dL. MAK: Suspected Carcinogen.
NIOSH REL: (Cadmium) Reduce to lowest feasible level

SAFETY PROFILE: Confirmed human carcinogen with experimental carcinogenic, tumorigenic, and neoplastigenic data. A human poison by inhalation and possibly other routes. Poison experimentally by ingestion, inhalation, intraperitoneal, subcutaneous, intramuscular, and intravenous routes. In humans inhalation causes an excess of protein in the urine. Experimental teratogenic and reproductive effects. Mutation data reported. The dust ignites spontaneously in air and is flammable and explosive when exposed to heat, flame, or by chemical reaction with oxidizing agents; metals; HN_3; Zn; Se; and Te. Explodes on contact with hydrazoic acid. Violent or explosive reaction when heated with ammonium nitrate. Vigorous reaction when heated with nitryl fluoride. When heated to a high temperature it emits toxic fumes of Cd. See also CADMIUM COMPOUNDS.

CAD250 CAS:543-90-8 ***HR: 3***
CADMIUM(II) ACETATE
mf: $C_2H_4O_2 \cdot 1/2Cd$ mw: 116.25

PROP: Monoclinic, colorless crystals; odor of acetic acid. Mp: 256°, bp: decomp, d: 2.341.

SYNS: ACETIC ACID, CADMIUM SALT ◇ BIS(ACETOXY)CADMIUM ◇ CADMIUM ACETATE (DOT) ◇ CADMIUM DIACETATE ◇ C.I. 77185

TOXICITY DATA with REFERENCE
cyt-hmn:lym 10 nmol/L MUREAV 85,236,81
otr-ham:emb 1 μmol/L CNREA8 39,193,79
dnd-ham:emb 1 μmol/L CNREA8 39,193,79
ipr-rat TDLo:2 mg/kg (female 20D post):REP BECTA6 23,25,79
ipr-rat TDLo:1 mg/kg (14D preg):TER BECTA6 23,25,79
ipr-mus LD50:14 mg/kg TXAPA9 49,41,79

CONSENSUS REPORTS: Reported in EPA TSCA Inventory. EPA Genetic Toxicology Program. Cadmium and its compounds are on the Community Right-To-Know List.

OSHA PEL: TWA 0.2 $mg(Cd)/m^3$; CL 0.6 $mg(Cd)/m^3$ (dust)
ACGIH TLV: TWA 0.05 $mg(Cd)/m^3$ (Proposed: TWA 0.01 $mg(Cd)/m^3$ (dust), Suspected Human Carcinogen; 0.002 $mg(Cd)/m^3$ (respirable dust), Suspected Human Carcinogen); BEI: 10 μg/g creatinine in urine; 10 μg/L in blood.
NIOSH REL: (Cadmium) Reduce to lowest feasible level

SAFETY PROFILE: Confirmed human carcinogen. Poison by intraperitoneal route. An experimental teratogen. Other experimental reproductive effects. Human mutation data reported. When heated to decomposition it emits toxic fumes of Cd. See also CADMIUM COMPOUNDS.

CAD275 CAS:5743-04-4 ***HR: 1***
CADMIUM ACETATE DIHYDRATE
mf: $C_4H_6O_4 \cdot Cd \cdot 2H_2O$ mw: 266.54

PROP: Crystals, becoming anhydrous at 130°; slt acetic acid odor. D: 2.01, 2.341 (anhydrous), mp: 255° (anhydrous). Sol in water and alc.

SYNS: ACETIC ACID, CADMIUM SALT, DIHYDRATE ◇ CADMIUM DIACETATE DIHYDRATE

TOXICITY DATA with REFERENCE
cyt-hmn:lyms 1 mg/L CYGEDX 12(3),46,78

ACGIH TLV: TWA 0.01 mg(Cd)/m^3; Suspected Carcinogen.

SAFETY PROFILE: Suspected carcinogen. Mutation data reported. When heated to decomposition it emits toxic fumes of Cd.

CAD325 CAS:22750-53-4 ***HR: 3***
CADMIUM AMIDE
mf: CdH_4N_2 mw: 144.46

SYN: CADMIUM DIAMIDE

CONSENSUS REPORTS: Cadmium compounds are on the Community Right-To-Know List.

OSHA PEL: TWA 0.2 mg(Cd)/m^3; CL 0.6 mg(Cd)/m^3 (dust)
ACGIH TLV: TWA 0.05 mg(Cd)/m^3 (Proposed: TWA 0.01 mg(Cd)/m^3 (dust), Suspected Human Carcinogen; 0.002 mg(Cd)/m^3 (respirable dust), Suspected Human Carcinogen); BEI: 10 μg/g creatinine in urine; 10 μg/L in blood.
NIOSH REL: (Cadmium) Reduce to lowest feasible level

SAFETY PROFILE: Confirmed human carcinogen. May explode if heated. Reacts violently with water. See also CADMIUM COMPOUNDS and AMIDES. When heated to decomposition it emits toxic fumes of Cd and NO_x.

CAD350 CAS:14215-29-3 ***HR: 3***
CADMIUM AZIDE
mf: CdN_6 mw: 196.45

$Cd(N_3)_2$

SYN: CADMIUM DIAZIDE

CONSENSUS REPORTS: Cadmium compounds are on the Community Right-To-Know List.

OSHA PEL: TWA 0.2 mg(Cd)/m^3; CL 0.6 mg(Cd)/m^3 (dust)
ACGIH TLV: TWA 0.05 mg(Cd)/m^3 (Proposed: TWA 0.01 mg(Cd)/m^3 (dust), Suspected Human Carcinogen; 0.002 mg(Cd)/m^3 (respirable dust), Suspected Human Carcinogen); BEI: 10 μg/g creatinine in urine; 10 μg/L in blood.
NIOSH REL: (Cadmium) Reduce to lowest feasible level

SAFETY PROFILE: Confirmed human carcinogen. The dry solid is an unstable heat- and friction-sensitive explosive. When heated to decomposition it emits toxic fumes of NO_x and Cd. See also CADMIUM COMPOUNDS and AZIDES.

CAD500 CAS:7495-93-4 ***HR: 3***
CADMIUM BIS(2-ETHYLHEXYL) PHOSPHITE
mf: $C_{32}H_{68}O_6P_2 \cdot Cd$ mw: 723.34

SYN: BIS(2-ETHYLHEXYL) ESTER PHOSPHORUS ACID CADMIUM SALT

TOXICITY DATA with REFERENCE
ipr-mus LDLo:250 mg/kg CBCCT* 7,790,55

CONSENSUS REPORTS: Cadmium and its compounds are on the Community Right-To-Know List.

OSHA PEL: TWA 0.2 mg(Cd)/m^3; CL 0.6 mg(Cd)/m^3 (dust)
ACGIH TLV: TWA 0.05 mg(Cd)/m^3 (Proposed: TWA 0.01 mg(Cd)/m^3 (dust), Suspected Human Carcinogen; 0.002 mg(Cd)/m^3 (respirable dust), Suspected Human Carcinogen); BEI: 10 μg/g creatinine in urine; 10 μg/L in blood.
NIOSH REL: (Cadmium) Reduce to lowest feasible level

SAFETY PROFILE: Confirmed human carcinogen. Poison by intraperitoneal route. See also CADMIUM COMPOUNDS. When heated to decomposition it emits toxic fumes of PO_x and Cd.

CAD600 CAS:7789-42-6 ***HR: 3***
CADMIUM BROMIDE
mf: Br_2Cd mw: 272.22

PROP: Pearly hexagonal crystals, hygroscopic. Mp: 566°, bp: 963°, d: 5.192. Sol in water and alc; moderately sol in acetone; sltly sol in ethanol.

SYN: CADMIUM DIBROMIDE

CONSENSUS REPORTS: Reported in EPA TSCA Inventory.

ACGIH TLV: TWA 0.01 mg(Cd)/m^3; Suspected Carcinogen.

SAFETY PROFILE: Suspected carcinogen. When heated to decomposition it emits toxic fumes of Cd and Br^-.

CAD750 CAS:2191-10-8 ***HR: 3***
CADMIUM CAPRYLATE
mf: $C_{16}H_{30}O_4 \cdot Cd$ mw: 398.86

SYN: OCTANOIC ACID, CADMIUM SALT (2:1)

TOXICITY DATA with REFERENCE
orl-rat LD50:950 mg/kg JHEMA2 18,144,74
itr-rat LDLo:10 mg/kg JHEMA2 18,144,74
orl-mus LD50:300 mg/kg JHEMA2 18,144,74

CONSENSUS REPORTS: Reported in EPA TSCA Inventory. Cadmium and its compounds are on the Community Right-To-Know List.

OSHA PEL: TWA 0.2 mg(Cd)/m^3; CL 0.6 mg(Cd)/m^3 (dust)
ACGIH TLV: TWA 0.05 mg(Cd)/m^3 (Proposed: TWA 0.01 mg(Cd)/m^3 (dust), Suspected Human Carcinogen; 0.002 mg(Cd)/m^3 (respirable dust), Suspected Human Carcinogen); BEI: 10 μg/g creatinine in urine; 10 μg/L in blood.
NIOSH REL: (Cadmium) Reduce to lowest feasible level

SAFETY PROFILE: Confirmed human carcinogen. Poison by ingestion and intratracheal routes. See also CADMIUM COMPOUNDS. When heated to decomposition it emits toxic fumes of Cd.

CAE000 ***HR: 3***
CADMIUM CHLORATE
mf: $CdCl_2O_6$ mw: 279.31

$Cd(ClP_3)_2$

PROP: Colorless, deliquescent prisms. Mp: 80°, d: 2.28 @ 18°.

CONSENSUS REPORTS: Cadmium and its compounds are on the Community Right-To-Know List.

OSHA PEL: TWA 0.2 mg(Cd)/m^3; CL 0.6 mg(Cd)/m^3 (dust)
ACGIH TLV: TWA 0.05 mg(Cd)/m^3 (Proposed: TWA 0.01 mg(Cd)/m^3 (dust), Suspected Human Carcinogen; 0.002 mg(Cd)/m^3 (respirable dust), Suspected Human Carcinogen); BEI: 10 μg/g creatinine in urine; 10 μg/L in blood.
NIOSH REL: (Cadmium) Reduce to lowest feasible level

SAFETY PROFILE: Confirmed human carcinogen. A powerful oxidizing agent. Flammable by chemical reaction with reducing agents. Moderate explosion hazard when shocked or exposed to heat. Violent or explosive reaction with sulfides (e.g., copper(II) sulfide (explodes); antimony(II) sulfide; arsenic(III) sulfide; tin(II) sulfide; tin(IV) sulfide. When heated to decomposition it emits toxic fumes of Cd and Cl^-. See also CHLORATES.

CAE250 CAS:10108-64-2 ***HR: 3***
CADMIUM CHLORIDE
mf: $CdCl_2$ mw: 183.30

PROP: Hexagonal, colorless crystals. Mp: 568°, d: 4.047 @ 25°, vap press: 10 mm @ 656°, bp: 960°.

SYNS: CADDY ◇ CADMIUM DICHLORIDE ◇ KADMIUMCHLORID (GERMAN) ◇ VI-CAD

TOXICITY DATA with REFERENCE
dni-hmn:hla 250 μmol/L MUREAV 92,427,82
cyt-ofs-mul 630 μg/L/4W-C BECTA6 36,199,86
orl-mus TDLo:6 mg/kg (female 15-19D post):REP JEPTDQ 1(3),187,78
ihl-rat TCLo:130 μg/m^3 (female 1-19D post):TER GTPZAB 31(8),25,87
ihl-rat TCLo:20 μg/m^3/23H/78W-C:CAR JJIND8 70,367,83
scu-rat TDLo:3666 μg/kg:CAR PAACA3 24,84,83
ims-rat TDLo:4500 μg/kg:ETA ARPAAQ 83,493,67
scu-rat TD:5499 μg/kg:NEO PSEBAA 115,653,64
orl-wmn LDLo:3 g/kg:BPR,GIT BMJOAE 292,1559,86
orl-rat LD50:88 mg/kg AFDOAQ 15,122,51
ipr-rat LD50:1800 μg/kg EVHPAZ 28,89,79
orl-mus LD50:60 mg/kg APTOA6 48,108,81
ihl-mus LC50:2300 mg/m^3 NTIS** PB158-508
ipr-mus LD50:9300 μg/kg NEZAAQ 32,472,77
scu-mus LD50:3200 μg/kg APTOA6 48,108,81
ivn-mus LD50:3500 μg/kg TXAPA9 53,510,80
ihl-dog LC90:420 mg/m^3/30M JIHTAB 29,302,47
ivn-dog LDLo:5 mg/kg EQSSDX 1,1,75
scu-cat LDLo:25 mg/kg EQSSDX 1,1,75
ivn-cat LDLo:5 mg/kg HBAMAK 4,1289,35
orl-rbt LDLo:70 mg/kg EQSSDX 1,1,75

CONSENSUS REPORTS: Animal Sufficient Evidence IMEMDT 11,39,76; IMEMDT 2,74,73. EPA Genetic Toxicology Program. Cadmium and its compounds are on the Community Right-To-Know List. Reported in EPA TSCA Inventory.

OSHA PEL: TWA 0.2 mg(Cd)/m^3; CL 0.6 mg(Cd)/m^3 (dust)
ACGIH TLV: TWA 0.05 mg(Cd)/m^3 (Proposed: TWA 0.01 mg(Cd)/m^3 (dust), Suspected Human Carcinogen; 0.002 mg(Cd)/m^3 (respirable dust), Suspected Human Carcinogen); BEI: 10 μg/g creatinine in urine; 10 μg/L in blood.
DFG MAK: Animal Carcinogen, Suspected Human Carcinogen.
NIOSH REL: (Cadmium) Reduce to lowest feasible level

SAFETY PROFILE: Confirmed human carcinogen with experimental carcinogenic and tumorigenic data. Poison by ingestion, inhalation, skin contact, intraperitoneal, subcutaneous, intravenous, and possibly other routes. Human systemic effects by ingestion:

blood pressure, acute pulmonary edema, hypermotility, diarrhea. Experimental teratogenic and reproductive effects. Human mutation data reported. Reacts violently with BrF_3 and K. When heated to decomposition it emits very toxic fumes of Cd and Cl^-. See also CADMIUM COMPOUNDS and CHLORIDES.

CAE375 CAS:72589-96-9 ***HR: 3***
CADMIUM CHLORIDE, DIHYDRATE
mf: $CdCl_2•2H_2O$ mw: 219.34

TOXICITY DATA with REFERENCE
orl-mus TDLo:15 mg/kg (5D male):TER AXVMAW 34,399,80
itt-mam TDLo:29 μg/kg (1D male):REP BECTA6 26,233,81
scu-rat TDLo:6580 μg/kg:ETA ARGEAR 36,119,70

CONSENSUS REPORTS: Cadmium and its compounds are on the Community Right-To-Know List.

OSHA PEL: TWA 0.2 mg(Cd)/m^3; CL 0.6 mg(Cd)/m^3 (dust)
ACGIH TLV: TWA 0.05 mg(Cd)/m^3 (Proposed: TWA 0.01 mg(Cd)/m^3 (dust), Suspected Human Carcinogen; 0.002 mg(Cd)/m^3 (respirable dust), Suspected Human Carcinogen); BEI: 10 μg/g creatinine in urine; 10 μg/L in blood.
DFG MAK: Animal Carcinogen, Suspected Human Carcinogen.
NIOSH REL: (Cadmium) Reduce to lowest feasible level

SAFETY PROFILE: Confirmed human carcinogen with experimental tumorigenic data. An experimental teratogen. Other experimental reproductive effects. When heated to decomposition it emits toxic fumes of Cl^- and Cd. See also CADMIUM CHLORIDE, CADMIUM COMPOUNDS, and CHLORIDES.

CAE425 CAS:7790-78-5 ***HR: 3***
CADMIUM CHLORIDE, HYDRATE (2:5)
mf: $CdCl_2•5/2H_2O$ mw: 228.35

TOXICITY DATA with REFERENCE
dni-hmn:lym 28 μmol/L IAAAAM 79,83,86
scu-rat TDLo:2 mg/kg (male 1D pre):REP CCPTAY 27,521,83
orl-mus LD50:194 mg/kg JTEHD6 22,35,87
ipr-mus LD50:4567 μg/kg TXAPA9 63,461,82

CONSENSUS REPORTS: Cadmium and its compounds are on the Community Right-To-Know List.

OSHA PEL: TWA 0.2 mg(Cd)/m^3; CL 0.6 mg(Cd)/m^3 (dust)
ACGIH TLV: TWA 0.05 mg(Cd)/m^3 (Proposed: TWA 0.01 mg(Cd)/m^3 (dust), Suspected Human Carcinogen; 0.002 mg(Cd)/m^3 (respirable dust), Suspected Human Carcinogen); BEI: 10 μg/g creatinine in urine; 10 μg/L in blood.
DFG MAK: Animal Carcinogen, Suspected Human Carcinogen.
NIOSH REL: (Cadmium) Reduce to lowest feasible level

SAFETY PROFILE: Confirmed human carcinogen. Poison by ingestion and intraperitoneal routes. Experimental reproductive effects. Human mutation data reported. When heated to decomposition it emits toxic fumes of Cl^- and Cd. See also CADMIUM CHLORIDE, CADMIUM COMPOUNDS, and CHLORIDES.

CAE500 CAS:35658-65-2 ***HR: 3***
CADMIUM CHLORIDE, MONOHYDRATE
mf: $CdCl_2•H_2O$ mw: 201.32

TOXICITY DATA with REFERENCE
orl-rat TDLo:179 mg/kg (male 1D pre):REP CALEDQ 36,307,87
orl-rat TDLo:65 mg/kg/2Y-C:ETA CALEDQ 9,191,80
scu-rat TDLo:4478 μg/kg:CAR CALEDQ 36,307,87

CONSENSUS REPORTS: Cadmium and its compounds are on the Community Right-To-Know List.

OSHA PEL: TWA 0.2 mg(Cd)/m^3; CL 0.6 mg(Cd)/m^3 (dust)
ACGIH TLV: TWA 0.05 mg(Cd)/m^3 (Proposed: TWA 0.01 mg(Cd)/m^3 (dust), Suspected Human Carcinogen; 0.002 mg(Cd)/m^3 (respirable dust), Suspected Human Carcinogen); BEI: 10 μg/g creatinine in urine; 10 μg/L in blood.
DFG MAK: Animal Carcinogen, Suspected Human Carcinogen.
NIOSH REL: (Cadmium) Reduce to lowest feasible level

SAFETY PROFILE: Confirmed human carcinogen with experimental carcinogenic and tumorigenic data. Experimental reproductive effects. When heated to decomposition it emits very toxic fumes of Cd and Cl^-. See also CADMIUM CHLORIDE, CADMIUM COMPOUNDS, and CHLORIDES.

CAE750 ***HR: 3***
CADMIUM COMPOUNDS

TOXICITY DATA with REFERENCE
ihl-hmn TCLo:1500 μg/m^3/14Y-I:CAR,PUL ANYAA9 271,273,76

CONSENSUS REPORTS: Cadmium and its compounds are on the Community Right-To-Know List.

OSHA PEL: Fume: TWA 0.1 $mg(Cd)/m^3$; CL 0.6 $mg(Cd)/m^3$; Dust: TWA 0.2 $mg(Cd)/m^3$; CL 0.6 $mg(Cd)/m^3$
ACGIH TLV: Dust and Salts: TWA 0.05 $mg(Cd)/m^3$ (Proposed: TWA 0.01 $mg(Cd)/m^3$ (dust), Suspected Human Carcinogen; 0.002 $mg(Cd)/m^3$ (respirable dust), Suspected Human Carcinogen); BEI: 10 $\mu g/g$ creatinine in urine; 10 $\mu g/L$ in blood.
DFG BAT: Blood 1.5 $\mu g/dL$; Urine 15 $\mu g/dL$. MAK: Suspected Carcinogen.
NIOSH REL: (Cadmium) Reduce to lowest feasible level

SAFETY PROFILE: Confirmed human carcinogen producing lung tumors. Poison by ingestion. The irritating and emetic action is so violent, however, that little of the cadmium has time to be absorbed and fatal poisoning rarely ensues. Experimental carcinogens and teratogens. Cases of human poisoning have been reported from ingestion of food or beverages prepared or stored in cadmium-plated containers. Inhalation of fumes or dusts affects the respiratory tract and the kidneys. Brief exposure to high concentrations may result in pulmonary edema and death. Fatal concentrations may be breathed without sufficient discomfort to warn a worker to leave the exposure. Cadmium oxide fumes can cause metal fume fever resembling that caused by zinc oxide fumes. When heated to decomposition they emit toxic fumes of Cd.

CAF500 ***HR: 3***
CADMIUM DICYANIDE
mf: C_2CdN_2 mw: 164.44

CONSENSUS REPORTS: Cadmium and its compounds and Cyanide and its compounds are on the Community Right-To-Know List.

OSHA PEL: TWA 0.2 $mg(Cd)/m^3$; CL 0.6 $mg(Cd)/m^3$ (dust)
ACGIH TLV: TWA 0.05 $mg(Cd)/m^3$ (Proposed: TWA 0.01 $mg(Cd)/m^3$ (dust), Suspected Human Carcinogen; 0.002 $mg(Cd)/m^3$ (respirable dust), Suspected Human Carcinogen); BEI: 10 $\mu g/g$ creatinine in urine; 10 $\mu g/L$ in blood.
NIOSH REL: (Cadmium) Reduce to lowest feasible level

SAFETY PROFILE: Confirmed human carcinogen. A poison. Incompatible with magnesium. When heated to decomposition it emits toxic fumes of Cd and CN^-. See also CADMIUM COMPOUNDS and CYANIDE.

CAF750 CAS:15954-91-3 ***HR: 3***
CADMIUM(II) EDTA COMPLEX

SYN: (ETHYLENEDINITRILO)TETRAACETIC ACID CADMIUM(II) COMPLEX

TOXICITY DATA with REFERENCE
ipr-mus LD50:7800 $\mu g(Cd)/kg$ PABIAQ 11,853,63

CONSENSUS REPORTS: Cadmium and its compounds are on the Community Right-To-Know List.

OSHA PEL: TWA 0.2 $mg(Cd)/m^3$; CL 0.6 $mg(Cd)/m^3$ (dust)
ACGIH TLV: TWA 0.05 $mg(Cd)/m^3$ (Proposed: TWA 0.01 $mg(Cd)/m^3$ (dust), Suspected Human Carcinogen; 0.002 $mg(Cd)/m^3$ (respirable dust), Suspected Human Carcinogen); BEI: 10 $\mu g/g$ creatinine in urine; 10 $\mu g/L$ in blood.
NIOSH REL: (Cadmium) Reduce to lowest feasible level

SAFETY PROFILE: Confirmed human carcinogen. Poison by intraperitoneal route. When heated to decomposition it emits toxic fumes of NO_x and Cd.

CAG000 CAS:14486-19-2 ***HR: 3***
CADMIUM FLUOBORATE
mf: B_2CdF_8 mw: 286.02

SYNS: CADMIUM FLUOROBORATE ◇ TL 1026

TOXICITY DATA with REFERENCE
orl-rat LDLo:250 mg/kg NCNSA6 5,27,53
ihl-mus LCLo:650 mg/m^3/10M NDRC** No.9-4-1-19,44

CONSENSUS REPORTS: Reported in EPA TSCA Inventory. Cadmium and its compounds are on the Community Right-To-Know List.

OSHA PEL: TWA 0.2 $mg(Cd)/m^3$; CL 0.6 $mg(Cd)/m^3$ (dust); 2.5 $mg(F)/m^3$
ACGIH TLV: TWA 0.05 $mg(Cd)/m^3$ (Proposed: TWA 0.01 $mg(Cd)/m^3$ (dust), Suspected Human Carcinogen; 0.002 $mg(Cd)/m^3$ (respirable dust), Suspected Human Carcinogen); BEI: 10 $\mu g/g$ creatinine in urine; 10 $\mu g/L$ in blood.
NIOSH REL: (Cadmium) Reduce to lowest feasible level

SAFETY PROFILE: Confirmed human carcinogen. Poison by ingestion and inhalation. See TETRAFLUOROBORATE. When heated to decomposition it emits very toxic fumes of Cd and F^-.

CAG250 CAS:7790-79-6 ***HR: 3***
CADMIUM FLUORIDE
mf: CdF_2 mw: 150.40

PROP: Cubic, white crystals. Mp: 1100°, bp: 1758°, d: 6.64, vap press: 1 mm @ 1112°.

SYN: CADMIUM FLUORURE (FRENCH)

TOXICITY DATA with REFERENCE
scu-frg LDLo:280 mg/kg CRSBAW 124,133,37

CONSENSUS REPORTS: Reported in EPA TSCA In-

ventory. Cadmium and its compounds are on the Community Right-To-Know List.

OSHA PEL: TWA 0.2 mg(Cd)/m^3; CL 0.6 mg(Cd)/m^3 (dust); 2.5 mg(F)/m^3
ACGIH TLV: TWA 0.05 mg(Cd)/m^3 (Proposed: TWA 0.01 mg(Cd)/m^3 (dust), Suspected Human Carcinogen; 0.002 mg(Cd)/m^3 (respirable dust), Suspected Human Carcinogen); BEI: 10 μg/g creatinine in urine; 10 μg/L in blood.
NIOSH REL: (Cadmium) Reduce to lowest feasible level

SAFETY PROFILE: Confirmed human carcinogen. Poison by subcutaneous route. Violent reaction with K. See also FLUORIDES and CADMIUM COMPOUNDS. When heated to decomposition it emits very toxic fumes of Cd and F^-.

CAG500 CAS:17010-21-8 ***HR: 3***
CADMIUM FLUOSILICATE
mf: CdF_6Si mw: 254.49

PROP: Hexagonal, colorless crystals.

SYN: TL 1070

TOXICITY DATA with REFERENCE
orl-rat LDLo:100 mg/kg NCNSA6 5,27,53
ihl-mus LCLo:670 mg/m^3/10M NDRC** No.9-4-1-19,44

CONSENSUS REPORTS: Cadmium and its compounds are on the Community Right-To-Know List.

OSHA PEL: TWA 0.2 mg(Cd)/m^3; CL 0.6 mg(Cd)/m^3 (dust); 2.5 mg(F)/m^3
ACGIH TLV: TWA 0.05 mg(Cd)/m^3 (Proposed: TWA 0.01 mg(Cd)/m^3 (dust), Suspected Human Carcinogen; 0.002 mg(Cd)/m^3 (respirable dust), Suspected Human Carcinogen); BEI: 10 μg/g creatinine in urine; 10 μg/L in blood.
NIOSH REL: (Cadmium) Reduce to lowest feasible level

SAFETY PROFILE: Confirmed human carcinogen. Poison by ingestion and inhalation. When heated to decomposition it emits very toxic fumes of Cd and F^-.

CAG750 CAS:16039-55-7 ***HR: 3***
CADMIUM LACTATE
mf: $C_6H_{10}O_6$•Cd mw: 290.56

PROP: Needles.

SYN: LACTIC ACID, CADMIUM SALT

CONSENSUS REPORTS: Cadmium and its compounds are on the Community Right-To-Know List.

OSHA PEL: TWA 0.2 mg(Cd)/m^3; CL 0.6 mg(Cd)/m^3 (dust)
ACGIH TLV: TWA 0.05 mg(Cd)/m^3 (Proposed: TWA 0.01 mg(Cd)/m^3 (dust), Suspected Human Carcinogen; 0.002 mg(Cd)/m^3 (respirable dust), Suspected Human Carcinogen); BEI: 10 μg/g creatinine in urine; 10 μg/L in blood.
NIOSH REL: (Cadmium) Reduce to lowest feasible level

SAFETY PROFILE: Confirmed human carcinogen. A poison. When heated to decomposition it emits toxic fumes of Cd. See also CADMIUM COMPOUNDS.

CAG775 CAS:2605-44-9 ***HR: 3***
CADMIUM LAURATE
mf: $C_{24}H_{46}O_4$•Cd mw: 511.10

SYNS: CADMIUM DILAURATE ◇ CADMIUM DODECANOATE ◇ DODECANOIC ACID, CADMIUM SALT (9CI) ◇ LAURIC ACID, CADMIUM SALT (2:1)

TOXICITY DATA with REFERENCE
orl-rat LD50:2370 mg/kg 41HTAH -,14,78
orl-mus LD50:1060 mg/kg 41HTAH -,14,78

CONSENSUS REPORTS: Reported in EPA TSCA Inventory.
ACGIH TLV: TWA 0.01 mg(Cd)/m^3; Suspected Carcinogen.

SAFETY PROFILE: Suspected carcinogen. Moderately toxic by ingestion. When heated to decomposition it emits toxic fumes of Cd.

CAH000 CAS:10325-94-7 ***HR: 3***
CADMIUM NITRATE
mf: CdN_2O_6 mw: 236.42

PROP: White, prismatic needles; hygroscopic. Mp: 350°.

SYNS: CADMIUM DINITRATE ◇ NITRIC ACID, CADMIUM SALT

TOXICITY DATA with REFERENCE
mrc-bcs 5 mmol/L MUREAV 77,109,80
unr-rat LD50:200 mg/kg GISAAA 50(3),57,85
orl-mus LD50:100 mg/kg 41HTAH -,14,78
ihl-mus LC50:3850 mg/m^3 NTIS** PB1580508

CONSENSUS REPORTS: Reported in EPA TSCA Inventory. EPA Genetic Toxicology Program. Cadmium and its compounds are on the Community Right-To-Know List.

OSHA PEL: TWA 0.2 mg(Cd)/m^3; CL 0.6 mg(Cd)/m^3 (dust)
ACGIH TLV: TWA 0.05 mg(Cd)/m^3 (Proposed: TWA 0.01 mg(Cd)/m^3 (dust), Suspected Human Carcinogen; 0.002 mg(Cd)/m^3 (respirable dust), Suspected Human Carcinogen); BEI: 10 μg/g creatinine in urine; 10 μg/L in blood.
NIOSH REL: (Cadmium) Reduce to lowest feasible level

SAFETY PROFILE: Confirmed human carcinogen. Poison by ingestion and possibly other routes. Moderately toxic by inhalation. Mutation data reported. When heated to decomposition it emits very toxic fumes of Cd and NO_x. See also CADMIUM COMPOUNDS and NITRATES.

CAH250 CAS:10022-68-1 ***HR: 3***
CADMIUM(II) NITRATE TETRAHYDRATE (1:2:4)
mf: $N_2O_6 \cdot Cd \cdot 4H_2O$ mw: 308.50

SYNS: DUSICNAN KADEMNATY (CZECH) ◇ NITRIC ACID, CADMIUM SALT, TETRAHYDRATE

TOXICITY DATA with REFERENCE
skn-rbt 500 mg/24H SEV 28ZPAK -,12,72
eye-rbt 20 mg/24H MOD 28ZPAK -,12,72
mmo-esc 6 μmol/L ENVRAL 26,279,85
orl-rat LD50:300 mg/kg 28ZPAK -,12,72

CONSENSUS REPORTS: Cadmium and its compounds are on the Community Right-To-Know List.

OSHA PEL: TWA 0.2 mg(Cd)/m^3; CL 0.6 mg(Cd)/m^3 (dust)
ACGIH TLV: TWA 0.05 mg(Cd)/m^3 (Proposed: TWA 0.01 mg(Cd)/m^3 (dust), Suspected Human Carcinogen; 0.002 mg(Cd)/m^3 (respirable dust), Suspected Human Carcinogen); BEI: 10 μg/g creatinine in urine; 10 μg/L in blood.
NIOSH REL: (Cadmium) Reduce to lowest feasible level

SAFETY PROFILE: Confirmed human carcinogen. Poison by ingestion. A severe skin and moderate eye irritant. Mutation data reported. See also CADMIUM COMPOUNDS, CADMIUM NITRATE, and NITRATES. When heated to decomposition it emits very toxic fumes of Cd and NO_x.

CAH500 CAS:1306-19-0 ***HR: 3***
CADMIUM OXIDE
mf: CdO mw: 128.40

PROP: (1) Amorphous, brown crystals; (2) cubic, brown crystals. Mp (1): <1426°, mp (2): decomp @ 950°, bp: 1559°, d (1): 6.95, d (2): 8.15, vap press: 1 mm @ 1000°.

SYNS: KADMU TLENEK (POLISH) ◇ NCI-C02551

TOXICITY DATA with REFERENCE
ihl-rat TCLo:23 μg/m^3/5H (15W pre/1-20D preg):REP TOLED5 22,53,84
ihl-rat TCLo:91 μg/m^3 (female 1-19D post):TER GTPZAB 31(8),25,87
scu-rat TDLo:90 mg/kg:NEO BJCAAI 20,190,66
ihl-hmn TCLo:8630 μg/m^3/5H YAKUD5 22,455,80
ihl-man TCLo:500 μg/m^3/5Y-I:NOSE,KID QJMEA7 38,425,69
ihl-man TCLo:40 μg/m^3:CVS,KID GISAAA 45(10)22,80
orl-rat LD50:72 mg/kg YAKUD5 22,455,80
ihl-rat LC50:780 mg/m^3/10M NTIS** PB158-508
ipr-rat LD50:12 mg/kg ZDKAA8 38(9),18,78
orl-mus LD50:72 mg/kg 41HTAH -,14,78
ihl-mus LC50:340 mg/m^3/10M NTIS** PB158-508
ihl-dog LC50:400 mg/m^3/10M YAKUD5 22,455,80
ihl-mky LC50:15 g/m^3/10M NTIS** PB158-508
ihl-rbt LC50:3 g/m^3/15M NTIS** PB158-508
ihl-gpg LC50:3 g/m^3/15M NTIS** PB158-508

CONSENSUS REPORTS: NTP Fifth Annual Report on Carcinogens. IARC Cancer Review: Group 2A IMEMDT 7,139,87; Human Inadequate Evidence IMEMDT 2,74,73; IMEMDT 11,39,76; Animal Sufficient Evidence IMEMDT 11,39,76; IMEMDT 2,74,73. Reported in EPA TSCA Inventory. EPA Extremely Hazardous Substances List. Cadmium and its compounds are on the Community Right-To-Know List.

OSHA PEL: TWA 0.2 mg(Cd)/m^3; CL 0.6 mg(Cd)/m^3 (dust)
ACGIH TLV: TWA 0.05 mg(Cd)/m^3 (Proposed: TWA 0.01 mg(Cd)/m^3 (dust), Suspected Human Carcinogen; 0.002 mg(Cd)/m^3 (respirable dust), Suspected Human Carcinogen); BEI: 10 μg/g creatinine in urine; 10 μg/L in blood.
DFG MAK: Suspected Carcinogen.
NIOSH REL: (Cadmium) Reduce to lowest feasible level

SAFETY PROFILE: Confirmed human carcinogen with experimental neoplastigenic data. Poison by ingestion, inhalation, and intraperitoneal routes. An experimental teratogen. Other experimental reproductive effects. Human systemic effects by inhalation include: change in the sense of smell, change in heart rate, blood pressure increase, an excess of protein in the urine and other kidney or bladder changes. Mixtures with magnesium explode when heated. When heated to decomposition it emits toxic fumes of Cd. See also CADMIUM COMPOUNDS.

CAH750 ***HR: 3***
CADMIUM OXIDE FUME
mf: CdO mw: 128.40

SYN: CADMIUM FUME

TOXICITY DATA with REFERENCE
ihl-hmn LCLo:2500 mg/m^3 JIHTAB 29,279,47
ihl-man TCLo:8630 μg/m^3/5H:PUL BJIMAG 23,292,66
ihl-rat LC50:500 mg/m^3/10M JIHTAB 29,279,47
ihl-mus LCLo:700 mg/m^3/10M JIHTAB 29,279,47
ihl-dog LC50:4000 mg/m^3/10M JIHTAB 29,279,47
ihl-mky LC50:15000 mg/m^3/10M JIHTAB 29,279,47
ihl-rbt LC50:2500 mg/m^3/10M JIHTAB 29,279,47
ihl-gpg LC50:3500 mg/m^3/10M JIHTAB 29,279,47

CONSENSUS REPORTS: Reported in EPA TSCA Inventory. Cadmium and its compounds are on the Community Right-To-Know List.

OSHA PEL: TWA 0.1 mg(Cd)/m^3; CL 0.3 mg(Cd)/m^3
ACGIH TLV: TWA 0.05 mg(Cd)/m^3 (Proposed: TWA 0.01 mg(Cd)/m^3 (dust), Suspected Human Carcinogen; 0.002 mg(Cd)/m^3 (respirable dust), Suspected Human Carcinogen); BEI: 10 μg/g creatinine in urine; 10 μg/L in blood.
NIOSH REL: (Cadmium) Reduce to lowest feasible level

SAFETY PROFILE: Confirmed human carcinogen. Poison by inhalation. Moderately toxic to humans by inhalation. Human pulmonary system effects by inhalation including: coughing, difficult breathing, and cyanosis. A strong irritant via inhalation. When heated to decomposition it emits toxic fumes of Cd. See also CADMIUM OXIDE and CADMIUM COMPOUNDS.

CAI000 CAS:13477-17-3 ***HR: 3***
CADMIUM PHOSPHATE
mf: $Cd_3O_8P_2 \cdot 4H_2O$ mw: 599.22

PROP: Amorphous or colorless crystals. Mp: 1500°.

SYN: TL 1182

TOXICITY DATA with REFERENCE
ihl-mus LCLo:650 mg/m^3/10M NDRC** No.9-4-1-19,44

CONSENSUS REPORTS: Reported in EPA TSCA Inventory. Cadmium and its compounds are on the Community Right-To-Know List.

OSHA PEL: TWA 0.2 mg(Cd)/m^3; CL 0.6 mg(Cd)/m^3 (dust)
ACGIH TLV: TWA 0.05 mg(Cd)/m^3 (Proposed: TWA 0.01 mg(Cd)/m^3 (dust), Suspected Human Carcinogen; 0.002 mg(Cd)/m^3 (respirable dust), Suspected Human Carcinogen); BEI: 10 μg/g creatinine in urine; 10 μg/L in blood.
NIOSH REL: (Cadmium) Reduce to lowest feasible level

SAFETY PROFILE: Confirmed human carcinogen. Poison by inhalation. When heated to decomposition it emits toxic fumes of Cd and PO_x. See CADMIUM COMPOUNDS and PHOSPHATES.

CAI125 CAS:12014-28-7 ***HR: 3***
CADMIUM PHOSPHIDE
mf: Cd_3P_2 mw: 399.18

CONSENSUS REPORTS: Cadmium compounds are on the Community Right-To-Know List.

OSHA PEL: Fume: TWA 0.1 mg(Cd)/m^3; CL 0.6 mg (Cd)/m^3; Dust: TWA 0.2 mg(Cd)/m^3; CL 0.6 mg (Cd)/m^3
ACGIH TLV: Dust and Salts: TWA 0.05 mg(Cd)/m^3 (Proposed: TWA 0.01 mg(Cd)/m^3 (dust), Suspected Human Carcinogen; 0.002 mg(Cd)/m^3 (respirable dust), Suspected Human Carcinogen); BEI: 10 μg/g creatinine in urine; 10 μg/L in blood.
DFG BAT: Blood 1.5 μg/dL; Urine 15 μg/dL. MAK: Suspected Carcinogen.
NIOSH REL: (Cadmium) Reduce to lowest feasible level

SAFETY PROFILE: Confirmed carcinogen. Explosive reaction with concentrated nitric acid. When heated to decomposition it emits toxic fumes of PO_x and Cd. See also CADMIUM COMPOUNDS and PHOSPHIDES.

CAI250 ***HR: 3***
CADMIUM PROPIONATE
mf: $C_6H_{10}CdO_5$ mw: 258.55

CONSENSUS REPORTS: Cadmium and its compounds are on the Community Right-To-Know List.

OSHA PEL: Fume: TWA 0.1 mg(Cd)/m^3; CL 0.6 mg(Cd)/m^3; Dust: TWA 0.2 mg(Cd)/m^3; CL 0.6 mg(Cd)/m^3
ACGIH TLV: Dust and Salts: TWA 0.05 mg(Cd)/m^3 (Proposed: TWA 0.01 mg(Cd)/m^3 (dust), Suspected Human Carcinogen; 0.002 mg(Cd)/m^3 (respirable dust), Suspected Human Carcinogen); BEI: 10 μg/g creatinine in urine; 10 μg/L in blood.
DFG BAT: Blood 1.5 μg/dL; Urine 15 μg/dL. MAK: Suspected Carcinogen.
NIOSH REL: (Cadmium) Reduce to lowest feasible level

SAFETY PROFILE: Confirmed carcinogen. The salt has exploded. Incompatible with 3-pentanone vapor. When heated to decomposition it emits toxic fumes of Cd. See also CADMIUM COMPOUNDS.

CAI350 CAS:18897-36-4 ***HR: 3***
CADMIUM 2-PYRIDINETHIONE
mf: $C_{10}H_8CdN_2O_2S_2$ mw: 364.72

SYNS: CADMIUM,BIS(1-HYDROXY-2(1H)-PYRIDINETHIONATO)- ◇ CADMIUM PT ◇ CdPT

TOXICITY DATA with REFERENCE
orl-rat LD50:240 mg/kg TOANDB 3,1,79
ivn-rbt LD50:1340 μg/kg TOANDB 3,1,79

ACGIH TLV: TWA 0.01 mg(Cd)/m^3; Suspected Carcinogen
NIOSH REL: (Cadmium): TWA reduce to lowest feasible level

SAFETY PROFILE: Suspected carcinogen. Poison by ingestion and intravenous routes. When heated to decomposition it emits toxic fumes of NO_x, SO_x, and Cd.

CAI400 CAS:19010-79-8 ***HR: 3***
CADMIUM SALICYLATE
mf: $C_{14}H_{10}CdO_6$ mw: 386.64

PROP: Monohydrate small needles or plates. Mp: 242°. Sltly sol in cold water, methanol, eth; Very sol in boiling water.

SYNS: BIS(2-HYDROXYBENZOATO-O^1,O^2-, (T-4)-CADMIUM (9CI) ◇ CADMIUM, BIS(SALICYLATO)-

TOXICITY DATA with REFERENCE
orl-rat LD50:1200 mg/kg 41HTAH -,14,78
orl-mus LD50:164 mg/kg 41HTAH -,14,78
ACGIH TLV: TWA 0.01 mg(Cd)/m^3; Suspected Carcinogen

SAFETY PROFILE: Poison by ingestion. When heated to decomposition it emits toxic fumes of Cd.

CAI500 ***HR: 3***
CADMIUM SELENIDE
mf: CdSe mw: 191.36

PROP: Preparative hazard.

CONSENSUS REPORTS: Cadmium and its compounds as well as Selenium and its compounds are on the Community Right-To-Know List.

OSHA PEL: Fume: TWA 0.1 mg(Cd)/m^3; CL 0.6 mg(Cd)/m^3; Dust: TWA 0.2 mg(Cd)/m^3; CL 0.6 mg(Cd)/m^3
ACGIH TLV: Dust and Salts: TWA 0.05 mg(Cd)/m^3 (Proposed: TWA 0.01 mg(Cd)/m^3 (dust), Suspected Human Carcinogen; 0.002 mg(Cd)/m^3 (respirable dust), Suspected Human Carcinogen); BEI: 10 μg/g creatinine in urine; 10 μg/L in blood.
DFG BAT: Blood 1.5 μg/dL; Urine 15 μg/dL. MAK: Suspected Carcinogen; 0.1 mg(Se)/m^3.
NIOSH REL: (Cadmium) Reduce to lowest feasible level

SAFETY PROFILE: Confirmed carcinogen. Selenium compounds are considered to be poisons. When heated to decomposition it emits toxic fumes of Cd and Se. See also CADMIUM COMPOUNDS and SELENIUM COMPOUNDS.

CAI750 CAS:141-00-4 ***HR: 3***
CADMIUM SUCCINATE
mf: $C_4H_4O_4$•Cd mw: 228.48

SYNS: CADMINATE ◇ SUCCINIC ACID, CADMIUM SALT (1:1)

TOXICITY DATA with REFERENCE
orl-rat LD50:660 mg/kg FMCHA2 -,D53,80
orl-mus LD50:312 mg/kg 28ZEAL 5,35,76
ipr-mus LD50:270 mg/kg AIPTAK 128,391,60

CONSENSUS REPORTS: Reported in EPA TSCA Inventory. Cadmium and its compounds are on the Community Right-To-Know List.

OSHA PEL: Fume: TWA 0.1 mg(Cd)/m^3; CL 0.6 mg(Cd)/m^3; Dust: TWA 0.2 mg(Cd)/m^3; CL 0.6 mg(Cd)/m^3
ACGIH TLV: Dust and Salts: TWA 0.05 mg(Cd)/m^3 (Proposed: TWA 0.01 mg(Cd)/m^3 (dust), Suspected Human Carcinogen; 0.002 mg(Cd)/m^3 (respirable dust), Suspected Human Carcinogen); BEI: 10 μg/g creatinine in urine; 10 μg/L in blood.
DFG BAT: Blood 1.5 μg/dL; Urine 15 μg/dL. MAK: Suspected Carcinogen.
NIOSH REL: (Cadmium) Reduce to lowest feasible level

SAFETY PROFILE: Confirmed carcinogen. Poison by ingestion and intraperitoneal routes. Moderately toxic by ingestion. When heated to decomposition it emits toxic fumes of Cd. See also CADMIUM COMPOUNDS.

CAJ000 CAS:10124-36-4 ***HR: 3***
CADMIUM SULFATE (1:1)
mf: O_4S•Cd mw: 208.46

PROP: Rhombic, white crystals. Mp: 1000°, d: 4.691.

SYNS: CADMIUM SULFATE ◇ CADMIUM SULPHATE ◇ SULFURIC ACID, CADMIUM(2+) SALT ◇ SULPHURIC ACID, CADMIUM SALT (1:1)

TOXICITY DATA with REFERENCE
mrc-bcs 5 mmol/L MUREAV 77,109,80
dnd-rat:lvr 30 μmol/L MUREAV 113,357,83
msc-mus:lym 150 μg/L JTEHD6 9,367,82
par-mus TDLo:28 μg/kg (female 12D post):TER IGSBAL 92,65,76
ipr-mus TDLo:5150 μg/kg (female 9D post):REP TJADAB 29,427,84
orl-rat LD50:280 mg/kg 41HTAH -,14,78
orl-mus LD50:88 mg/kg 41HTAH -,14,78
ipr-mus LD50:12760 μg/kg COREAF 256,1043,63
orl-dog LDLo:105 mg/kg EQSSDX 1,1,75
scu-dog LDLo:27 mg/kg EQSSDX 1,1,75
scu-frg LDLo:105 mg/kg HBAMAK 4,1317,35

CONSENSUS REPORTS: NTP Fifth Annual Report on Carcinogens. IARC Cancer Review: Group 2A IMEMDT 7,139,87; Animal Sufficient Evidence IMEMDT 11,39,76; IMEMDT 2,74,73. Reported in EPA TSCA

Inventory. EPA Genetic Toxicology Program. Cadmium and its compounds are on the Community Right-To-Know List.

OSHA PEL: TWA 0.2 mg(Cd)/m^3; CL 0.6 mg(Cd)/m^3 (dust)
ACGIH TLV: TWA 0.05 mg(Cd)/m^3 (Proposed: TWA 0.01 mg(Cd)/m^3 (dust), Suspected Human Carcinogen; 0.002 mg(Cd)/m^3 (respirable dust), Suspected Human Carcinogen); BEI: 10 μg/g creatinine in urine; 10 μg/L in blood.
DFG MAK: Suspected Carcinogen.
NIOSH REL: (Cadmium) Reduce to lowest feasible level

SAFETY PROFILE: Confirmed human carcinogen with experimental carcinogenic data. Poison by ingestion, subcutaneous, and intraperitoneal routes. Experimental teratogenic and reproductive effects. Mutation data reported. See also CADMIUM COMPOUNDS and SULFATES. When heated to decomposition it emits very toxic fumes of Cd and SO_x.

CAJ250 CAS:7790-84-3 ***HR: 3***
CADMIUM SULFATE (1:1) HYDRATE (3:8)
mf: O_4S•Cd•8/3H_2O mw: 256.51

SYNS: CADMIUM SULFATE OCTAHYDRATE ◇ SULFURIC ACID, CADMIUM SALT, HYDRATE

TOXICITY DATA with REFERENCE
dnd-esc 3 μmol/L JOBAAY 133,75,78
cyt-ham:fbr 10 μmol/L/1H MUREAV 40,125,76
ivn-ham TDLo:2 mg/kg (8D preg):REP TJADAB 21,181,80
ivn-ham TDLo:2 mg/kg (8D preg):TER TJADAB 21,181,80
scu-rat TDLo:60 mg/kg/2Y-I:NEO AOHYA3 16,111,73
scu-rat TD:15 mg/kg/2Y-I:ETA AOHYA3 16,111,73

CONSENSUS REPORTS: IARC Cancer Review: Animal Sufficient Evidence IMEMDT 2,74,73. Cadmium and its compounds are on the Community Right-To-Know List.

OSHA PEL: TWA 0.2 mg(Cd)/m^3; CL 0.6 mg(Cd)/m^3 (dust)
ACGIH TLV: TWA 0.05 mg(Cd)/m^3 (Proposed: TWA 0.01 mg(Cd)/m^3 (dust), Suspected Human Carcinogen; 0.002 mg(Cd)/m^3 (respirable dust), Suspected Human Carcinogen); BEI: 10 μg/g creatinine in urine; 10 μg/L in blood.
NIOSH REL: (Cadmium) Reduce to lowest feasible level

SAFETY PROFILE: Confirmed human carcinogen with experimental tumorigenic and neoplastigenic data. Experimental teratogenic and reproductive effects. Mutation data reported. See also CADMIUM SULFATE, CADMIUM COMPOUNDS, and SULFATES. When heated to decomposition it emits very toxic fumes of Cd and SO_x.

CAJ500 CAS:13477-21-9 ***HR: 3***
CADMIUM SULFATE TETRAHYDRATE
mf: O_4S•Cd•4H_2O mw: 280.54

SYN: SULFURIC ACID, CADMIUM SALT, TETRAHYDRATE

TOXICITY DATA with REFERENCE
scu-rat TDLo:20 mg/kg/10W-I:NEO BJCAAI 18,667,64

CONSENSUS REPORTS: Cadmium and its compounds are on the Community Right-To-Know List.

OSHA PEL: TWA 0.2 mg(Cd)/m^3; CL 0.6 mg(Cd)/m^3 (dust)
ACGIH TLV: TWA 0.05 mg(Cd)/m^3 (Proposed: TWA 0.01 mg(Cd)/m^3 (dust), Suspected Human Carcinogen; 0.002 mg(Cd)/m^3 (respirable dust), Suspected Human Carcinogen); BEI: 10 μg/g creatinine in urine; 10 μg/L in blood.
NIOSH REL: (Cadmium) Reduce to lowest feasible level

SAFETY PROFILE: Confirmed human carcinogen with experimental neoplastigenic data. See also CADMIUM COMPOUNDS and CADMIUM SULFATE OCTAHYDRATE. When heated to decomposition it emits very toxic fumes of Cd and SO_x.

CAJ750 CAS:1306-23-6 ***HR: 3***
CADMIUM SULFIDE
mf: CdS mw: 144.46

PROP: Hexagonal, yellow-orange crystals. Mp: 1750 @ 100 atm, bp: subl in N_2, d: 4.82.

SYNS: AURORA YELLOW ◇ CADMIUM GOLDEN 366 ◇ CADMIUM LEMON YELLOW 527 ◇ CADMIUM ORANGE ◇ CADMIUM PRIMROSE 819 ◇ CADMIUM SULPHIDE ◇ CADMIUM YELLOW ◇ CADMOPUR YELLOW ◇ CAPSEBON ◇ C.I. 77199 ◇ C.I. PIGMENT ORANGE 20 ◇ C.I. PIGMENT YELLOW 37 ◇ FERRO YELLOW ◇ GREENOCKITE ◇ NCI-C02711

TOXICITY DATA with REFERENCE
cyt-hmn:leu 62 μg/L PJACAW 48,133,72
otr-ham:emb 1 mg/L CNREA8 42,2757,82
dnd-ham:ovr 10 mg/L CRNGDP 3,657,82
scu-rat TDLo:90 mg/kg:CAR BJCAAI 20,190,66
ims-rat TDLo:120 mg/kg:ETA BJCAAI 20,190,66
orl-rat LD50:7080 mg/kg 41HTAH -,14,78
orl-mus LD50:1166 mg/kg 41HTAH -,14,78
ihl-mus LCLo:1350 mg/m^3 NTIS** PB158-508

CONSENSUS REPORTS: NTP Fifth Annual Report on Carcinogens. IARC Cancer Review: Group 2A IMEMDT 7,139,87; Animal Sufficient Evidence IMEMDT 11,39,76; IMEMDT 2,74,73. EPA Genetic Toxicology Program. Cadmium and its compounds are on the Community Right-To-Know List. Reported in EPA TSCA Inventory.

OSHA PEL: TWA 0.2 mg(Cd)/m^3; CL 0.6 mg(Cd)/m^3 (dust)
ACGIH TLV: TWA 0.05 mg(Cd)/m^3 (Proposed: TWA 0.01 mg(Cd)/m^3 (dust), Suspected Human Carcinogen; 0.002 mg(Cd)/m^3 (respirable dust), Suspected Human Carcinogen); BEI: 10 μg/g creatinine in urine; 10 μg/L in blood.
DFG MAK: Suspected Carcinogen.
NIOSH REL: (Cadmium) Reduce to lowest feasible level

SAFETY PROFILE: Confirmed human carcinogen with experimental carcinogenic and tumorigenic data. Moderately toxic by ingestion and inhalation. Human mutation data reported. See also CADMIUM COMPOUNDS and SULFIDES. When heated to decomposition it emits very toxic fumes of Cd and SO_x.

CAK000 ***HR: 3***
CADMIUM THERMOVACUUM AEROSOL
mf: Cd mw: 112.40

SYN: AEROSOL of THERMOVACUUM CADMIUM

TOXICITY DATA with REFERENCE
unr-rat LD50:1365 mg/kg GTPZAB 22(5),6,78
unr-mus LD50:815 mg/kg GTPZAB 22(5),6,78

CONSENSUS REPORTS: Cadmium and its compounds are on the Community Right-To-Know List.

OSHA PEL: TWA 0.2 mg(Cd)/m^3; CL 0.6 mg(Cd)/m^3 (dust)
ACGIH TLV: TWA 0.05 mg(Cd)/m^3 (Proposed: TWA 0.01 mg(Cd)/m^3 (dust), Suspected Human Carcinogen; 0.002 mg(Cd)/m^3 (respirable dust), Suspected Human Carcinogen); BEI: 10 μg/g creatinine in urine; 10 μg/L in blood.
NIOSH REL: (Cadmium) Reduce to lowest feasible level

SAFETY PROFILE: Confirmed human carcinogen. Moderately toxic by an unspecified route. See also CADMIUM and CADMIUM COMPOUNDS. When heated to decomposition it emits very toxic fumes of Cd.

CAK250 CAS:73419-42-8 ***HR: 3***
CADMIUM-THIONEIN
mf: $C_{18}H_{30}N_6O_4S_2{\cdot}Cd$ mw: 571.06

PROP: Cadmium(II) is bound to the protein thioneine from rat or rabbit liver. (BCPCA6 26,25,77).

TOXICITY DATA with REFERENCE
ivn-rat LD50:280 μg/kg BCPCA6 26,25,77

CONSENSUS REPORTS: Cadmium and its compounds are on the Community Right-To-Know List.

OSHA PEL: TWA 0.2 mg(Cd)/m^3; CL 0.6 mg(Cd)/m^3 (dust)
ACGIH TLV: TWA 0.05 mg(Cd)/m^3 (Proposed: TWA 0.01 mg(Cd)/m^3 (dust), Suspected Human Carcinogen; 0.002 mg(Cd)/m^3 (respirable dust), Suspected Human Carcinogen); BEI: 10 μg/g creatinine in urine; 10 μg/L in blood.
NIOSH REL: (Cadmium) Reduce to lowest feasible level

SAFETY PROFILE: Confirmed human carcinogen. Deadly poison by intravenous route. When heated to decomposition it emits very toxic fumes of NO_x, SO_x, and Cd. See also CADMIUM COMPOUNDS.

CAK275 CAS:64241-34-5 ***HR: 3***
CADRALAZINE
mf: $C_{12}H_{21}N_5O_3$ mw: 283.38

SYNS: ETHYL-6-(ETHYL(2-HYDROXYPROPYL)AMINO)-3-PYRIDAZINECARBAZATE ◇ ETHYL-2-(6(ETHYL(2-HYDROXYPROPYL)AMINO)-3-PYRIDAZINYL)HYDRAZINECARBOXYLATE ◇ 3-(6-(ETHYL-(2-HYDROXYPROPYL)AMINO)PYRIDAZIN-3-YL)CARBAZIC ACID ETHYL ESTER ◇ 2-(6-ETHYL(2-HYDROXYPROPYL)AMINO)-3-PYRIDAZINYL)-HYDRAZINECARBOXYLIC ACID ETHYL ESTER ◇ ISF 2469

TOXICITY DATA with REFERENCE
orl-rat TDLo:132 mg/kg (female 7-17D post):TER YACHDS 15,3913,87
orl-rat LD50:2060 mg/kg JCPCDT 3,455,81
ipr-rat LD50:440 mg/kg DRFUD4 7,382,82
ivn-rat LD50:269 mg/kg JCPCDT 3,455,81
orl-mus LD50:825 mg/kg DRFUD4 7,382,82
ipr-mus LD50:362 mg/kg DRFUD4 7,382,82
ivn-mus LD50:162 mg/kg DRFUD4 7,382,82
ivn-dog LD50:400 mg/kg JCPCDT 3,455,81

SAFETY PROFILE: Poison by intravenous and intraperitoneal routes. Moderately toxic by ingestion. Experimental reproductive effects. When heated to decomposition it emits toxic fumes of NO_x.

CAK325 ***HR: 2***
CAESALPINIA (VARIOUS SPECIES)

PROP: Several species of shrubs of various sizes producing yellow flowers with red filaments. *C. gilliesti* (Bird of Paradise) is a tall, thornless shrub (15 ft) with compound leaves and a fruit pod about 4 inches long containing 6 to 8 seeds. It is common as a cultivated plant in the southern United States from Florida to Arizona. *C. pulcherrima* (Dwarf poinciana) is a vertical shrub with some thorns. Seed pods are similar to *C. gilliesti*. It is common as a cultivated plant in the West Indies and frost-free regions of the United States. *C. bonduc* is a ground-hugging shrub with many thorns. The seed pods are about 3 inches long, covered with thorns and contain 2 large seeds. It is common in the West Indies. *C. vesicaria* is a small, thorny tree native to the West Indies except Puerto Rico.

SYNS: BARBADOS PRIDE ◇ BIRD of PARADISE ◇ BRASIL (CUBA) ◇ BRASILETTO (BAHAMAS) ◇ BRIER (BAHAMAS) ◇ CARZAZO (DOMINICAN REPUBLIC) ◇ C. BONDUC ◇ C. DRUMMONDII ◇ C. GILLIESII ◇ CLAVELLINA (PUERTO RICO) ◇ C. MEXICANA ◇ C. PULCHERRIMA ◇ C. VESICARIA ◇ DODDLE-DO (PUERTO RICO) ◇ DULDUL (PUERTO RICO) ◇ DWARF POINCIANA ◇ ESPIGA de AMOR (PUERTO RICO) ◇ FLOR de CAMARON (MEXICO) ◇ FLOWER FENCE ◇ FRANCILLADE (HAITI) ◇ GREY NICKER ◇ GUACALOTE AMARILLO (CUBA) ◇ GUACAMAYA (CUBA) ◇ HABA de SAN ANTONIO (PUERTO RICO) ◇ HORSE NICKER ◇ INDIAN SAVIN TREE (JAMAICA) ◇ MARAVILLA (MEXICO) ◇ MATO AZUL (PUERTO RICO) ◇ MATO de PLAYA (PUERTO RICO) ◇ 'OHAI-ALI'I (HAWAII) ◇ SPANISH CARNATION ◇ TABACHIN (MEXICO)

SAFETY PROFILE: The seeds usually contain toxic tannins except the immature seeds of *C. pulcherrima* and cooked seeds of *C. bonduc*. Ingestion of the seeds may cause persistent vomiting and diarrhea after a delay of 30 minutes to 6 hours.

CAK375 CAS:331-39-5 ***HR: 2***
CAFFEIC ACID
mf: $C_9H_8O_4$ mw: 180.17

PROP: Constituent of plants, probably occurs in plants only in conjugated forms, e.g., chlorogenic acid. Yellow crystals from concentrated aq solns. Monohydrate from dil solns. Decomp 223-225° (softens at 194°). Sparingly sol in cold water; freely sol in hot water and cold alc. Alkaline solns turn from yellow to orange.

SYNS: 3,4-DIHYDROXYBENZENEACRYLIC ACID ◇ 3,4-DIHYDROXYCINNAMIC ACID ◇ 3-(3,4-DIHYDROXYPHENYL)-2-PROPENOIC ACID (9CI)

TOXICITY DATA with REFERENCE
mrc-smc 300 mg/L MUREAV 135,109,84
cyt-ham:ovr 200 mg/L CALEDQ 14,251,81
ipr-rat TDLo:480 mg/kg (5-12D preg):TER TXAPA9 36,227,76
ipr-rat LDLo:1500 mg/kg TXAPA9 36,227,76

SAFETY PROFILE: Moderately toxic by intraperitoneal route. An experimental teratogen. Mutation data reported. When heated to decomposition it emits acrid smoke and fumes.

CAK500 CAS:58-08-2 ***HR: 3***
CAFFEINE
mf: $C_8H_{10}N_4O_2$ mw: 194.22

PROP: White, fleecy masses; odorless with bitter taste. Mp: 236.8°. Sol in water, alc, chloroform, ether.

SYNS: CAFFEIN ◇ COFFEIN (GERMAN) ◇ COFFEINE ◇ 3,7-DIHYDRO-1,3,7-TRIMETHYL-1H-PURINE-2,6-DIONE ◇ ELDIATRIC C ◇ FEMA No. 2224 ◇ GUARANINE ◇ KOFFEIN (GERMAN) ◇ METHYLTHEOBROMIDE ◇ 1-METHYLTHEOBROMINE ◇ 7-METHYLTHEOPHYLLINE ◇ NCI-C02733 ◇ NO-DOZ ◇ ORGANEX ◇ THEIN ◇ THEINE ◇ 1,3,7-TRIMETHYL-2,6-DIOXOPURINE ◇ 1,3,7-TRIMETHYLXANTHINE

TOXICITY DATA with REFERENCE
dns-hmn:oth 1 mmol/L BIOJAU 35,665,81
dni-hmn:oth 4 mmol/L BIOJAU 35,665,81
cyt-hmn:lym 100 μg/L/24H MUREAV 46,205,77
orl-wmn TDLo:3276 mg/kg (1-39W preg):REP POMDAS 62(3),64,77
orl-wmn TDLo:6750 mg/kg (1-39W preg):TER LANCAO 1,1415,81
orl-mus TDLo:30800 mg/kg/44W-C:CAR CNREA8 48,2078,88
orl-wmn TDLo:96 mg/kg/1D-I:PSY,GIT JOPDAB 105,493,84
orl-man TDLo:13 mg/kg:PSY AJPSAO 143,1320,86
orl-hmn LDLo:192 mg/kg JNDRAK 5,252,65
orl-cld LDLo:320 mg/kg FNSCA6 3,275,74
orl-wmn LDLo:1 g/kg:GIT BIATDR-,6,73
ivn-hmn TDLo:7 mg/kg:PSY APTOA6 15,331,59
orl-inf TDLo:14700 μg/kg:CNS CLBIAS 10,148,77
ivn-inf TDLo:68 mg/kg:PSY AJDCAI 134,495,80
ivn-wmn LDLo:57 mg/kg:CNS,BLD APTOA6 15,331,59
ims-inf TDLo:36 mg/kg:PSY AJDCAI 134,495,80
orl-rat LD50:192 mg/kg JNDRAK 5,252,65
ipr-rat LD50:260 mg/kg ZERNAL 15,64,76
scu-rat LD50:170 mg/kg JCPHB8 7,131,67
ivn-rat LD50:105 mg/kg JPETAB 82,89,44
rec-rat LD50:300 mg/kg JCPHB8 7,131,67
orl-mus LD50:127 mg/kg TXAPA9 44,1,78
ipr-mus LD50:168 mg/kg CPBTAL 22,1459,74
scu-mus LD50:270 mg/kg AEPPAE 241,182,61
ivn-mus LD50:62 mg/kg TOLED5 29,25,85
orl-dog LD50:140 mg/kg NIIRDN 6,174,82

CONSENSUS REPORTS: Reported in EPA TSCA Inventory. EPA Genetic Toxicology Program.

SAFETY PROFILE: A human poison by ingestion. An experimental poison by ingestion, subcutaneous, intraperitoneal, intramuscular, rectal, and intravenous routes. Human systemic effects by ingestion, intravenous, and intramuscular routes include: ataxia, blood pressure elevation, convulsions or effect on seizure threshold, diarrhea, distorted perceptions, hallucinations, hypermotility, muscle contraction or spasticity, somnolence (general depressed activity), nausea or vomiting, toxic psychosis, tremors. A human teratogen causing developmental abnormalities of the craniofacial and musculoskeletal systems, pregnancy termination (abortion) and stillbirth. Human maternal effects include an unspecified effect on labor or childbirth. Human mutation data reported. An experimental teratogen. Other experimental reproductive effects. Questionable carcinogen with experimental carcinogenic data. Large doses (above 1.0 gram) cause palpitation, excitement, insomnia, dizziness, headache, and vomiting. Continued excessive use of caffeine in tea or coffee may lead to digestive disturbances, constipation, palpitations, shortness of breath, and depressed mental states. It is also implicated in cardiac disorders under those conditions. When heated to decomposition it emits toxic fumes of NO_x.

CAK750 CAS:5743-18-0 ***HR: 3***
CAFFEINE HYDROBROMIDE
mf: $C_8H_{10}N_4O_2$•BrH mw: 275.14

SYNS: CAFFEINE BROMIDE ◇ 3,7-DIHYDRO-1,3,7-TRIMETHYL-1H-PURINE-2,6-DIONEMONOHYDROBROMIDE

TOXICITY DATA with REFERENCE
orl-rbt LDLo:400 mg/kg HBAMAK 4,1289,35
scu-rbt LDLo:150 mg/kg HBAMAK 4,1289,35
ivn-rbt LDLo:100 mg/kg HBAMAK 4,1289,35

SAFETY PROFILE: Poison by ingestion, subcutaneous, and intravenous routes. See also CAFFEINE and BROMIDES. When heated to decomposition it emits very toxic fumes of NO_x and HBr.

CAK800 CAS:8000-95-1 ***HR: 2***
CAFFEINE and SODIUM BENZOATE
mf: $C_8H_{10}N_4O_2$•$C_7H_5NaO_2$ mw: 338.33

SYN: SODIUM BENZOATE and CAFFEINE

TOXICITY DATA with REFERENCE
ipr-rat TDLo:20 mg/kg (1D male):REP SCIEAS 127,84,58
orl-rat LD50:860 mg/kg 85GMAT -,31,82
orl-mus LD50:800 mg/kg 85GMAT -,31,82
ipr-mus LD50:525 mg/kg JPETAB 116,343,56

SAFETY PROFILE: Moderately toxic by ingestion and intraperitoneal routes. Experimental reproductive effects. When heated to decomposition it emits toxic fumes of NO_x and Na_2O. See also CAFFEINE and SODIUM BENZOATE.

CAL000 CAS:470-82-6 ***HR: 3***
CAJEPUTOL
mf: $C_{10}H_{18}O$ mw: 154.28

PROP: Colorless liquid characteristic odor with pungent, cooling taste. D: 0.921-0.924, refr index: 1.455-1.460, flash p: 122°F. Sol in alc, fixed oils, glycerin, and propylene glycol.

SYNS: 1,8-CINEOL ◇ CINEOLE ◇ 1,8-CINEOLE ◇ 1,8-EPOXY-p-MENTHANE ◇ EUCALYPTOL (FCC) ◇ EUCALYPTOLE ◇ FEMA No. 2465 ◇ LIMONENE OXIDE ◇ NCI-C56575 ◇ 1,8-OXIDO-p-MENTHANE ◇ 1,3,3-TRIMETHYL-2-OXABICYCLO(2.2.2)OCTANE

TOXICITY DATA with REFERENCE
scu-rat TDLo:2 g/kg (19-22D preg):REP BCPCA6 22,543,73
orl-rat LD50:2480 mg/kg FCTXAV 2,327,64
scu-mus LD50:1070 mg/kg SIZSAR 3,73,52
ims-mus LD50:100 mg/kg JSICAZ 21,342,62
scu-dog LDLo:1500 mg/kg TFAKA4 1,134,55
ims-gpg LDLo:2250 mg/kg TFAKA4 1,134,55

CONSENSUS REPORTS: Reported in EPA TSCA Inventory.

SAFETY PROFILE: Poison by intramuscular route. Moderately toxic by ingestion and subcutaneous routes. Experimental reproductive effects. Combustible liquid. When heated to decomposition it emits acrid smoke and fumes. See also LIMONENE.

CAL075 CAS:60996-85-2 ***HR: 3***
CALACIDOL
mf: $C_{20}H_{39}N_2$•Cl mw: 343.06

SYNS: 1-(2-(DICYCLOHEXYLAMINO)ETHYL)-1-METHYL-PIPERIDINIUM CHLORIDE ◇ I.U. 7

TOXICITY DATA with REFERENCE
orl-mus LD50:892 mg/kg FRPSAX 16,773,61
ipr-mus LD50:125 mg/kg FRPSAX 16,773,61
scu-mus LD50:151 mg/kg FRPSAX 16,773,61

SAFETY PROFILE: Poison by subcutaneous and intraperitoneal routes. Moderately toxic by ingestion. When heated to decomposition it emits toxic fumes of NO_x and Cl^-.

CAL125 ***HR: 2***
CALADIUM

PROP: The various species of this genus have variegated, heart-shaped leaves. The leaf coloration may be green with white, orange or red. They are popular house plants and may be cultivated all year in subtropical gardens and in the summer in temperate zones.

SYNS: ANGEL WINGS ◇ CALADIO (PUERTO RICO) ◇ CANANGA ◇ CAPOTILLO (MEXICO) ◇ C. BICOLOR ◇ CORAZON de CABRITO (CUBA) ◇ COUER SAIGNANT (HAITI) ◇ ELEPHANT'S EAR ◇ HEART-OF-JESUS ◇ LAGRIMAS de MARIA ◇ MOTHER-IN-LAW PLANT ◇ PALETA de PINTOR (PUERTO RICO)

SAFETY PROFILE: The whole plant contains toxic calcium oxalate raphides. Chewing any part of the plant results in burning pain in the lips, mouth and throat, possibly followed by inflammation and blistering. Systemic effects are usually not seen because of the insolubility of calcium oxalate, however, ingestion may cause inflammation of the stomach and intestines. See also OXALATES.

CAL250 CAS:7440-70-2 ***HR: 3***
CALCIUM
DOT: NA 1401/UN 1401/UN 1855
af: Ca aw: 40.08

PROP: Silver-white, soft metal. Mp: 842°, bp: 1484°, d: 1.54 @ 20°, vap press: 10 mm @ 983°.

SYNS: CALCICAT ◇ CALCIUM, non-pyrophoric (DOT) ◇ CALCIUM, pyrophoric (DOT) ◇ CALCIUM, METAL, CRYSTALLINE (DOT) ◇ CALCIUM, METAL (DOT)

CONSENSUS REPORTS: Reported in EPA TSCA Inventory.

DOT Classification: flammable Solid; Label: Flammable Solid and Dangerous When Wet (NA1401); Flammable Solid; Label: Spontaneously Combustible (UN1855); Flammable Solid; Label: Danger When Wet (UN1401).

SAFETY PROFILE: See CALCIUM COMPOUNDS. Flammable when heated or in intimate contact with moisture or acids. Moderate explosion hazard in intimate contact with very powerful oxidizing agents. Reacts with moisture or acids to liberate large quantities of hydrogen; can develop explosive pressure in containers. To fight fire, use special mixtures of dry chemical. Violent reaction with water may evolve explosive hydrogen gas. Potentially explosive reaction with alkali metal hydroxides or carbonates; dinitrogen tetraoxide; lead chloride + heat; phosphorus(V) oxide + heat; sulfur + heat. Molten calcium reacts explosively with asbestos cement. Hypergolic reaction with chlorine fluorides (e.g., chlorine trifluoride; chlorine pentafluoride). Ignition on contact with halogens (e.g., fluorine; chlorine); sulfur + vanadium(V) oxide. Violent reaction with mercury (at 390°C); silicon (above 1050°C); sodium + mixed oxides + heat. Incompatible with air.

CAL500 CAS:64046-96-4 ***HR: 3***
CALCIUM ACETARSONE
mf: $C_8H_{10}AsNO_5{\cdot}7Ca$ mw: 555.67

SYN: N-ACETYL-4-HYDROXY-m-ARSANILIC ACID, CALCIUM SALT

TOXICITY DATA with REFERENCE
orl-cat LDLo:135 mg/kg PSEBAA 27,267,30

CONSENSUS REPORTS: Arsenic and its compounds are on the Community Right-To-Know List.

OSHA PEL: TWA 0.5 mg(As)/m^3
ACGIH TLV: TWA 0.2 mg(As)/m^3

SAFETY PROFILE: Poison by ingestion. See also ARSENIC COMPOUNDS and CALCIUM COMPOUNDS. When heated to decomposition it emits very toxic fumes of As and NO_x.

CAL750 CAS:62-54-4 ***HR: 3***
CALCIUM ACETATE
mf: $C_4H_6O_4{\cdot}Ca$ mw: 158.18

PROP: Fine white, bulky powder. Sol in water; sltly sol in alc.

SYNS: ACETATE of LIME ◇ BROWN ACETATE ◇ CALCIUM DIACETATE ◇ GRAY ACETATE ◇ LIME ACETATE ◇ LIME PYROLIGNITE ◇ SORBO-CALCIAN ◇ SORBO-CALCION ◇ TELTOZAN ◇ VINEGAR SALTS

TOXICITY DATA with REFERENCE
dns-rat-rat 1290 μmol/kg/5D-I CRNGDP 6,1819,85
ivn-rat LDLo:147 mg/kg JPETAB 71,1,41
ivn-mus LD50:52 mg/kg JLCMAK 29,809,44

CONSENSUS REPORTS: Reported in EPA TSCA Inventory.

SAFETY PROFILE: Poison by intravenous route. Mutation data reported. See also CALCIUM COMPOUNDS. When heated to decomposition it emits acrid smoke and fumes.

CAM000 CAS:5902-95-4 ***HR: 3***
CALCIUM ACID METHYL ARSONATE
mf: $C_2H_8As_2O_6{\cdot}Ca$ mw: 318.02

SYNS: CALAR ◇ CALCIUM ACID METHANEARSONATE ◇ CALCIUM HYDROGEN METHANEARSONATE ◇ CALCIUM METHANEARSONATE ◇ CAMA ◇ SUPER CRAB-E-RAD-CALAR ◇ SUPER DAL-E-RAD ◇ SUPER DAL-E-RAD-CALAR ◇ USAF AN-11

TOXICITY DATA with REFERENCE
ipr-mus LD50:500 mg/kg NTIS** AD414-344
unr-mam LD50:4000 mg/kg FMCHA2 -,C241,83

CONSENSUS REPORTS: Arsenic and its compounds are on the Community Right-To-Know List.

OSHA PEL: TWA 0.5 mg(As)/m^3
ACGIH TLV: TWA 0.2 mg(As)/m^3

SAFETY PROFILE: Moderately toxic by intraperitoneal and possibly other routes. Arsenic compounds are considered to be poisons. An herbicide. When heated to decomposition it emits toxic fumes of As. See also ARSENIC COMPOUNDS and CALCIUM COMPOUNDS.

CAM500 CAS:27152-57-4 ***HR: 3***
CALCIUM ARSENITE
DOT: NA 1574
mf: $As2O_6{\cdot}3Ca$ mw: 366.08

PROP: White, granular powder.

SYNS: ARSENIOUS ACID, CALCIUM SALT ◇ CALCIUM ARSENITE, solid (DOT) ◇ MONOCALCIUM ARSENITE

CONSENSUS REPORTS: NTP Fifth Annual Report on Carcinogens. Arsenic and its compounds are on the Community Right-To-Know List.

OSHA PEL: Cancer Hazard
ACGIH TLV: TWA 0.2 mg(As)/m^3
NIOSH REL: (Inorganic Arsenic) CL 0.002 mg(As)/m^3/15M
DOT Classification: Poison B; Label: Poison.

SAFETY PROFILE: Confirmed carcinogen. A poison by inhalation and ingestion. When heated to decomposition it emits toxic fumes of As. See also ARSENIC COMPOUNDS and CALCIUM COMPOUNDS.

CAM675 CAS:21059-46-1 ***HR: 2***
CALCIUM ASPARTATE
mf: $C_4H_7NO_4 \cdot 7Ca$ mw: 413.68

SYNS: ASPARAGINATE CALCIUM ◇ CALCIRETARD ◇ CALCIUM-l-ASPARTATE

TOXICITY DATA with REFERENCE
orl-mus LD50:10 g/kg NIIRDN 6,12,82
ipr-mus LD50:1059 mg/kg NIIRDN 6,12,82
ivn-mus LD50:646 mg/kg NIIRDN 6,12,82

SAFETY PROFILE: Moderately toxic by intraperitoneal and intravenous routes. When heated to decomposition it emits toxic fumes of NO_x. See also CALCIUM COMPOUNDS.

CAM680 ***HR: 2***
CALCIUM BENZOATE
mf: $C_{14}H_{10}O_4 \cdot 3H_2O$ mw: 374.26

PROP: Orthorhombic crystals or powder. D: 1.44. Sol in water.

SAFETY PROFILE: Combustible when exposed to heat or flame. When heated to decomposition it emits acrid smoke and irritating fumes.

CAM750 CAS:6485-34-3 ***HR: D***
CALCIUM-o-BENZOSULFIMIDE
mf: $C_{14}H_{10}N_2O_6S_2 \cdot Ca$ mw: 406.46

PROP: White, crystalline powder; odorless or faint aromatic odor; sol in water.

SYNS: 1,2-BENZISOTHIAZOL-3(2H)-ONE-1,1-DIOXIDE,CALCIUM SALT ◇ CALCIUM-o-BENZOSULPHIMIDE ◇ CALCIUM-2-BENZOSULPHIMIDE ◇ CALCIUM SACCHARIN ◇ CALCIUM SACCHARINA ◇ CALCIUM SACCHARINATE ◇ DARAMIN ◇ SACCHARIN CALCIUM ◇ SULPHOBENZOIC IMIDE CALCIUM SALT

TOXICITY DATA with REFERENCE
dns-rat:lvr 100 mg/L CNREA8 40,4541,80
cyt-ham:lng 8 g/L MUREAV 163,63,86

CONSENSUS REPORTS: Reported in EPA TSCA Inventory.

SAFETY PROFILE: Mutagenic data reported. When heated to decomposition it emits toxic fumes of SO_x and NO_x.

CAN000 CAS:13780-03-5 ***HR: 3***
CALCIUM BISULFITE (solution)
DOT: UN 1923/UN 2693

PROP: Colorless or sltly yellowish liquid, strong sulfur dioxide odor. D: 1.06.

SYNS: CALCIUM BISULFITE, solution (DOT) ◇ CALCIUM HYDROGEN SULFITE, solution (DOT)

CONSENSUS REPORTS: Reported in EPA TSCA Inventory.

DOT Classification: Corrosive Material; Label: Corrosive (NA2693); Flammable Solid; Label: Spontaneously Combustible.

SAFETY PROFILE: A poison via ingestion. Strong irritant via skin contact, ingestion, and inhalation. When heated to decomposition it emits toxic fumes of SO_x. See also SULFITES and SULFUROUS ACID.

CAN250 CAS:12007-56-6 ***HR: 2***
CALCIUM BORATE
mf: B_4CaO_7 mw: 195.32

PROP: Colorless, rhombic or long, flat plates. Mp: 1154°.

SYN: COLEMANITE

TOXICITY DATA with REFERENCE
orl-rat LD50:5600 mg/kg GTPZAB 25(6),53,81
orl-mus LD50:5900 mg/kg GTPZAB 25(6),53,81
ipr-mus LD50:3900 mg/kg GTPZAB 25(6),53,81

CONSENSUS REPORTS: Reported in EPA TSCA Inventory.

SAFETY PROFILE: Moderately toxic by intraperitoneal route. Mildly toxic by ingestion. See also CALCIUM COMPOUNDS and BORON COMPOUNDS.

CAN400 ***HR: 1***
CALCIUM BROMATE
mf: $Ca(BrO_3)_2 \cdot H_2O$ mw: 313.90

PROP: White crystalline powder. Very sol in water.

SAFETY PROFILE: A nuisance dust.

CAN750 CAS:75-20-7 ***HR: 3***
CALCIUM CARBIDE
DOT: UN 1402
mf: C_2Ca mw: 64.10

PROP: Rhombic, gray crystals. Mp: approx 2300°, d: 2.222.

SYN: CALCIUM ACETYLIDE

CONSENSUS REPORTS: Reported in EPA TSCA Inventory.

DOT Classification: Flammable Solid; Label: Flammable Solid and Dangerous When Wet.

SAFETY PROFILE: Reaction on contact with moisture forms explosive acetylene gas. Flammable on contact with moisture, acid or acid fumes; evolves heat or flammable vapors. Moderate explosion hazard. Incandescent reaction with Cl_2 (245°C); Br_2 (350°C); I_2 (305°C); HCl gas + heat; PbF_2; Mg + heat. Incompatible with Se; (KOH + Cl_2); $AgNO_3$; Na_2O_2; $SnCl_2$; S; water. Mixtures with iron(III) chloride; iron(III) oxide; tin(II) chloride are easily ignited and burn fiercely. Vigorous reaction with methanol after an induction period. Addition to silver nitrate solutions precipitates the dangerously explosive silver acetylide. Copper salt solutions behave similarly. See also CALCIUM HYDROXIDE and ACETYLENE.

CAO000 CAS:1317-65-3 ***HR: 2***
CALCIUM CARBONATE
mf: $CO_3 \cdot Ca$ mw: 100.09

PROP: White microcrystalline powder. Mp: 825° (α), 1339° (β) @ 102.5 atm; d: 2.7-2.95. Found in nature as the minerals limestone, marble, aragonite, calcite, and vaterite. Odorless, tasteless powder or crystals. Two crystalline forms are of commercial importance: Aragonite, orthorhombic, mp: 825° (decomp), d: 2.83, formed at temperatures above 30°; calcite, hexagonal-rhombohedral, mp: 1339° (102.5 atm), d: 2.711, formed at temperatures below 30°. At about 825° it decomposes into CaO and CO_2. Practically insol in water, alc; sol in dilute acids.

SYNS: AGRICULTURAL LIMESTONE ◇ AGSTONE ◇ ARAGONITE ◇ ATOMIT ◇ BELL MINE PULVERIZED LIMESTONE ◇ CALCITE ◇ CARBONIC ACID, CALCIUM SALT (1:1) ◇ CHALK ◇ DOLOMITE ◇ FRANKLIN ◇ LIMESTONE (FCC) ◇ LITHOGRAPHIC STONE ◇ MARBLE ◇ NATURAL CALCIUM CARBONATE ◇ PORTLAND STONE ◇ SOHNHOFEN STONE ◇ VATERITE

TOXICITY DATA with REFERENCE
skn-rbt 500 mg/24H MOD 28ZPAK -,267,72
eye-rbt 750 μg/24H SEV 28ZPAK -,267,72
orl-rat LD50:6450 mg/kg 28ZPAK -,267,72

CONSENSUS REPORTS: Reported in EPA TSCA Inventory.

OSHA PEL: Total Dust: 15 mg/m^3; Respirable Fraction: 5 mg/m^3
ACGIH TLV: TWA (nuisance particulate) 10 mg/m^3 of total dust (when toxic impurities are not present, e.g., quartz < 1%).

SAFETY PROFILE: A severe eye and moderate skin irritant. Ignites on contact with F_2. Incompatible with acids; alum; ammonium salts; (Mg + H_2). Calcium carbonate is a common air contaminant. See also CALCIUM COMPOUNDS.

CAO250 CAS:9049-05-2 ***HR: 2***
CALCIUM CARRAGHEENATE

PROP: A mixture of highly sulfated polygalactosides. It is extracted from seaweed (FAONAU 53A,398,74).

SYNS: ALGIN GUM ◇ CALCIUM CARAGEENIN ◇ CALCIUM CARRAGEENAN ◇ CARRAGEENAN, CALCIUM(II) SALT ◇ VISCARIN 402

TOXICITY DATA with REFERENCE
orl-rat TDLo:13 g/kg (14D pre-21D post):REP TXAPA9 50,267,79
orl-rat LD50:5140 mg/kg FAONAU 53A,386,74
orl-mus LD50:8710 mg/kg FAONAU 53A,398,74
orl-rbt LD50:2280 mg/kg FAONAU 53A,398,74
orl-ham LD50:6180 mg/kg FAONAU 53A,398,74

CONSENSUS REPORTS: Reported in EPA TSCA Inventory.

SAFETY PROFILE: Moderately toxic by ingestion. Experimental reproductive effects. When heated to decomposition it emits toxic fumes of SO_x. See also CALCIUM COMPOUNDS.

CAO500 CAS:10137-74-3 ***HR: 2***
CALCIUM CHLORATE
DOT: UN 1452/UN 2429
mf: $Cl_2O_6 \cdot Ca$ mw: 206.98

PROP: Monoclinic, white-yellowish, deliq crystals. Mp: loses H_2O @ >100°, d: 2.711.

SYNS: CALCIUM CHLORATE, aqueous solution (DOT) ◇ CHLORATE de CALCIUM (FRENCH)

TOXICITY DATA with REFERENCE
orl-rat LDLo:4500 mg/kg JPETAB 35,1,29
ipr-rat LDLo:625 mg/kg JPETAB 35,1,29

DOT Classification: Oxidizer; Label: Oxidizer.

SAFETY PROFILE: Moderately toxic by ingestion and intraperitoneal routes. A powerful oxidant. Incompatible with Al; As; C; Cu; charcoal; MnO_2; metal sulfides; S; dibasic organic acids; organic matter; P. When heated to decomposition it emits toxic fumes of Cl^-. See also CHLORATES for fire, disaster, and explosion hazards.

CAO750 CAS:10043-52-4 ***HR: 2***
CALCIUM CHLORIDE
mf: $CaCl_2$ mw: 110.98

PROP: Cubic, colorless, deliq crystals. Mp: 772°, bp: >1600°, d: 2.512 @ 25°. Sol in water and alc.

SYNS: CALCIUM CHLORIDE, anhydrous ◇ CALPLUS ◇ CALTAC ◇ DOWFLAKE ◇ LIQUIDOW ◇ PELADOW ◇ SNOMELT ◇ SUPERFLAKE ANHYDROUS

TOXICITY DATA with REFERENCE
dns-rat-ipr 2500 μmol/kg JOENAK 65,45,75
cyt-rat:ast 3500 mg/kg GANNA2 7,165,87
orl-rat TDLo:112 g/kg/20W-C:ETA AJCAA7 23,550,35
orl-rat LD50:1000 mg/kg CJCMAV 12,216,48
ipr-rat LD50:264 mg/kg OYYAA2 14,963,77
scu-rat LD50:2630 mg/kg OYYAA2 14,963,77

ivn-rat LDLo:161 mg/kg JLCMAK 15,35,29
ims-rat LD50:25 mg/kg EMSUA8 4,223,46
orl-mus LD50:1940 mg/kg OYYAA2 14,963,77
ipr-mus LD50:245 mg/kg ABMGAJ 6,447,61
scu-mus LD50:823 mg/kg OYYAA2 14,963,77
ivn-mus LD50:42 mg/kg TXAPA9 22,150,72
ipr-dog LDLo:110 mg/kg AVERAG 44,555,37
scu-dog LDLo:274 mg/kg HBAMAK 4,1316,35

CONSENSUS REPORTS: Reported in EPA TSCA Inventory. EPA Genetic Toxicology Program.

SAFETY PROFILE: Moderately toxic by ingestion. Poison by intravenous, intramuscular, intraperitoneal, and subcutaneous routes. Questionable carcinogen with experimental tumorigenic data. Mutation data reported. Reacts violently with (B_2O_3 + CaO); BrF_3. Reaction with zinc releases explosive hydrogen gas. Catalyzes exothermic polymerization of methyl vinyl ether. Exothermic reaction with water. When heated to decomposition it emits toxic fumes of Cl^-. See also CALCIUM COMPOUNDS and CHLORIDES.

CAP000 CAS:14674-72-7 ***HR: 3***
CALCIUM CHLORITE
DOT: UN 1453
mf: $CaCl_2O_4$ mw: 174.98

$Ca(ClO_2)_2$

PROP: White solid.

DOT Classification: Oxidizer; Label: Oxidizer.

SAFETY PROFILE: A strong oxidizer. Ignites on contact with potassium thiocyanate. Reaction with Cl_2 yields explosive ClO_2. When heated to decomposition it emits toxic fumes of Cl^-. See also CHLORITES and CALCIUM COMPOUNDS.

CAP250 CAS:85721-24-0 ***HR: 2***
CALCIUM-4-(p-CHLOROPHENYL)-2-PHENYL-5-THIAZOLEACETATE
mf: $C_{17}H_{11}ClNO_2S•Ca$ mw: 368.88

SYNS: CALCIUM-2-PHENYL-4-(p-CHLOROPHENYL)-5-THIAZOLE ACETATE ◇ 4-(p-CHLOROPHENYL)-2-PHENYL-5-THIAZOLEACETIC ACID CALCIUM SALT ◇ FENTIAZAC CALCIUM SALT

TOXICITY DATA with REFERENCE
orl-rat LD50:860 mg/kg CMROCX 6,53,79
orl-mus LD50:1353 mg/kg CMROCX 6,53,79

SAFETY PROFILE: Moderately toxic by ingestion. When heated to decomposition it emits very toxic fumes of SO_x, NO_x, and Cl^-. See also CALCIUM COMPOUNDS.

CAP500 CAS:13765-19-0 ***HR: 3***
CALCIUM CHROMATE
mf: $CrO_4•Ca$ mw: 156.08

PROP: Monoclinic prisms; yellow color.

SYNS: CALCIUM CHROMATE (VI) ◇ CALCIUM CHROME YELLOW ◇ CALCIUM CHROMIUM OXIDE ($CaCrO_4$) ◇ CALCIUM MONOCHROMATE ◇ CHROMIC ACID, CALCIUM SALT (1:1) ◇ C.I. 77223 ◇ C.I. PIGMENT YELLOW 33 ◇ GELBIN ◇ RCRA WASTE NUMBER U032 ◇ YELLOW ULTRAMARINE

TOXICITY DATA with REFERENCE
mmo-sat 50 nmol/plate CRNGDP 2,283,81
mma-esc 100 μg/plate ENMUDM 6(Suppl 2),1,84
otr-rat:emb 58 μg/L JJIND8 67,1303,81
dlt-mus-unr 40 mg/kg MUREAV 97,180,82
dnd-ham:ovr 25 μmol/L/1H-C PAACA3 24,74,83
sce-ham:ovr 100 μg/L MUREAV 156,219,85
unr-mus TDLo:1400 mg/kg (35W male):REP MUREAV 97,180,82
ims-rat TDLo:76 mg/kg/19W-I:NEO BJCAAI 23,172,69
itr-rat TDLo:163 mg/kg/130W-I:CAR EXPADD 30,129,86
imp-rat TDLo:8 mg/kg:CAR CRNGDP 7,831,86
imp-rat TD:50 mg/kg:ETA AEHLAU 5,445,62

CONSENSUS REPORTS: NTP Fifth Annual Report on Carcinogens. IARC Cancer Review: Group 1 IMEMDT 7,165,87; Animal Sufficient Evidence IMEMDT 2,100,73; IMEMDT 23,205,80; Human Sufficient Evidence IMEMDT 23,205,80. Reported in EPA TSCA Inventory. EPA Genetic Toxicology Program. Chromium and its compounds are on the Community Right-To-Know List.

OSHA PEL: CL 0.1 mg(CrO_3)/m^3
ACGIH TLV: TWA 0.001 mg(Cr)/m^3; Suspected Human Carcinogen
DFG TRK: 0.1 mg/m^3 calculated as CrO_3 in that portion of dust that can possibly be inhaled; 0.2 mg/m^3 arc-welding by hand; others 0.1 mg/m^3. Animal Carcinogen, Suspected Human Carcinogen.
NIOSH REL: (Chromium(VI)) TWA 0.001 mg(Cr(VI))/m^3

SAFETY PROFILE: Confirmed human carcinogen with experimental carcinogenic, neoplastigenic, and tumorigenic data. Experimental reproductive effects. Mutation data reported. A powerful oxidizer. Mixture with boron burns violently if ignited. See also CHROMIUM COMPOUNDS and CALCIUM COMPOUNDS.

CAP750 CAS:8012-75-7 ***HR: 3***
CALCIUM CHROMATE(VI) DIHYDRATE
mf: $CrO_4•Ca•2H_2O$ mw: 192.12

SYNS: CALCIUM CHROME YELLOW ◇ CHROMIC ACID, CALCIUM SALT (1:1), DIHYDRATE ◇ C.I. 77223 ◇ C.I. PIGMENT YELLOW 33 ◇ GELBIN YELLOW ULTRAMARINE ◇ PIGMENT YELLOW 33 ◇ STEINBUHL YELLOW

TOXICITY DATA with REFERENCE
otr-ham:kdy 250 mg/L CNREA8 35,1058,75
scu-mus TDLo:400 mg/kg:ETA AMIHAB 21,530,60
imp-mus TDLo:400 mg/kg:CAR AMIHAB 21,530,60
orl-rat LD50:327 mg/kg TXAPA9 42,417,77
imp-rat LDLo:112 mg/kg AMIHAB 21,530,60

CONSENSUS REPORTS: IARC Cancer Review: Animal Sufficient Evidence IMEMDT 2,100,72. Chromium and its compounds are on the Community Right-To-Know List.

OSHA PEL: CL 0.1 mg(CrO^3)/m^3
ACGIH TLV: TWA 0.05 mg(Cr)/m^3; Confirmed Human Carcinogen
NIOSH REL: (Chromium(VI)) TWA 0.001 mg(Cr(VI))/m^3

SAFETY PROFILE: Confirmed human carcinogen with experimental tumorigenic and carcinogenic data. Poison by ingestion and implant routes. Mutation data reported. A powerful oxidizer. See also CHROMIUM COMPOUNDS and CALCIUM COMPOUNDS.

CAQ000 ***HR: 1***
CALCIUM COMPOUNDS

SAFETY PROFILE: The fumes evolved by burning calcium in air are composed of calcium oxide (quick lime) which is an irritant to the skin, eyes, and mucous membranes. Generally speaking, calcium compounds should be considered toxic only when they contain toxic components (such as arsenic, etc.) or as calcium oxide or hydroxide. Calcium compounds are common air contaminants.

CAQ250 CAS:156-62-7 ***HR: 3***
CALCIUM CYANAMIDE
mf: CN_2•Ca mw: 80.11
DOT: UN 1403

PROP: Hexagonal, rhombohedral, colorless crystals. Mp: 1300°, subl > 1500°. Compound not hydrated; compound contains more than 0.1% calcium (FEREAC 41,15972,76).

SYNS: AERO-CYANAMID ◇ AERO CYANAMID GRANULAR ◇ AERO CYANAMID SPECIAL GRADE ◇ ALZODEF ◇ CALCIUM CARBIMIDE ◇ CALCIUM CYANAMID ◇ CCC ◇ CYANAMIDE ◇ CYANAMIDE CALCIQUE (FRENCH) ◇ CYANAMIDE, CALCIUM SALT (1:1) ◇ CYANAMID GRANULAR ◇ CYANAMID SPECIAL GRADE ◇ CY-L 500 ◇ LIME-NITROGEN (DOT) ◇ NCI-C02937 ◇ NITROGEN LIME ◇ NITROLIME ◇ USAF CY-2

TOXICITY DATA with REFERENCE
mmo-sat 1 mg/plate ENMUDM 5(Suppl 1),3,83
mma-sat 100 μg/plate ENMUDM 5(Suppl 1),3,83
orl-mus TDLo:170 g/kg/2Y-C:ETA NCITR* NCI-CG-TR-163,79
orl-hmn LDLo:571 mg/kg 34ZIAG -,149,69
orl-rat LD50:158 mg/kg NIIRDN 6,304,82
ihl-rat LCLo:86 mg/m^3/4H 85GMAT -,40,82
skn-rat LD50:84 mg/kg 85GMAT -,40,82
ivn-rat LD50:125 mg/kg NIIRDN 6,304,82
unr-rat LD50:1000 mg/kg GUCHAZ 6,73,73
orl-mus LD50:334 mg/kg NIIRDN 6,304,82
ipr-mus LD50:100 mg/kg NTIS** AD277-689
ivn-mus LD50:282 mg/kg NIIRDN 6,304,82
orl-cat LD50:100 mg/kg 85GMAT -,40,82
orl-rbt LD50:1400 mg/kg PCOC** -,174,66
skn-rbt LD50:590 mg/kg 37ASAA 7,291,79

CONSENSUS REPORTS: NCI Carcinogenesis Bioassay (feed); No Evidence: mouse, rat NCITR* NCI-CG-TR-163,79. Community Right-To-Know List. Reported in EPA TSCA Inventory.

OSHA PEL: TWA 0.5 mg/m^3
ACGIH TLV: TWA 0.5 mg/m^3
DFG MAK: 1 mg/m^3
DOT Classification: ORM-C; Label: None; IMO: Flammable Solid; Label: Dangerous When Wet.

SAFETY PROFILE: Poison by ingestion, inhalation, skin contact, intravenous, and intraperitoneal routes. Moderately toxic to humans by ingestion. Questionable carcinogen with experimental tumorigenic data. Mutation data reported. The fatal dose, by ingestion, is probably around 20 to 30 grams for an adult. It does not have a cyanide effect. Calcium cyanamide is not believed to have a cumulative action. Flammable. Reaction with water forms the explosive acetylene gas. When heated to decomposition it emits toxic fumes of NO_x and CN^-. See also CALCIUM COMPOUNDS; AMIDES; and CYANIDES.

CAQ500 CAS:592-01-8 ***HR: 3***
CALCIUM CYANIDE
DOT: UN 1575
mf: C_2CaN_2 mw: 92.12

PROP: Rhombohedral crystals or white powder. Mp: decomp > 350°.

SYNS: CALCIUM CYANIDE, solid (DOT) ◇ CALCYANIDE ◇ CYANOGAS ◇ CYANURE de CALCIUM (FRENCH) ◇ RCRA WASTE NUMBER P021

TOXICITY DATA with REFERENCE
orl-rat LD50:39 mg/kg AIHAAP 30,470,69

CONSENSUS REPORTS: Cyanide and its compounds are on the Community Right-To-Know List. Reported in EPA TSCA Inventory.

OSHA PEL: TWA 5 mg(CN)/m^3
ACGIH TLV: TWA 5 mg(CN)/m^3 (skin)
DFG MAK: 5 mg/m^3
NIOSH REL: (Cyanide) CL 5 mg(CN)/m^3/10M
DOT Classification: Poison B; Label: Poison.

SAFETY PROFILE: A deadly poison by ingestion and probably other routes. When heated to decomposition it emits toxic fumes of NO_x and CN^-. See also CALCIUM COMPOUNDS and CYANIDES.

CAQ750 CAS:592-01-8 ***HR: 3***
CALCIUM CYANIDE (mixture)
DOT: UN 1575

SYN: CALCIUM CYANIDE MIXTURE, solid (DOT)

CONSENSUS REPORTS: Cyanide and its compounds are on the Community Right-To-Know List.

OSHA PEL: TWA 5 $mg(CN)/m^3$
ACGIH TLV: TWA 5 $mg(CN)/m^3$ (skin)
DFG MAK: 5 mg/m^3
NIOSH REL: (Cyanide) CL 5 $mg(CN)/m^3$/10M
DOT Classification: Poison B; Label: Poison.

SAFETY PROFILE: A poison. See also CALCIUM CYANIDE, CYANIDES, and CALCIUM COMPOUNDS. When heated to decomposition it emits toxic fumes of NO_x and CN^-.

CAR000 CAS:139-06-0 ***HR: 3***
CALCIUM CYCLOHEXYLSULPHAMATE
mf: $C_{12}H_{24}N_2O_6S_2 \cdot Ca$ mw: 396.58

PROP: White, crystalline powder; almost odorless; freely sol in water; practically insol in alc, benzene, chloroform, and ether.

SYNS: CALCIUM CYCLAMATE ◇ CALCIUM CYCLOHEXANESULFAMATE ◇ CALCIUM CYCLOHEXANE SULPHAMATE ◇ CALCIUM CYCLOHEXYLSULFAMATE ◇ CYCLAMATE CALCIUM ◇ CYCLAMATE, CALCIUM SALT ◇ CYCLAN ◇ CYCLOHEXANESULFAMIC ACID, CALCIUM SALT ◇ CYCLOHEXYLSULPHAMIC ACID, CALCIUM SALT ◇ CYLAN ◇ DIETIL ◇ KALZIUMZYKLAMATE (GERMAN) ◇ SUCARYL CALCIUM

TOXICITY DATA with REFERENCE
sln-dmg-orl 5 mmol/L DRISAA 46,114,71
dni-hmn:lng 100 mg/L JCLBA3 47,30a,70
cyt-hmn:leu 250 mg/L SCIEAS 164,568,69
cyt-ham:fbr 10 mg/L MUREAV 39,1,76
cyt-ham:lng 100 mg/L HEREAY 70,271,72
cyt-grb-ipr 150 mg/kg CNJGA8 13,189,71
orl-rat TDLo:55 mg/kg (1-22D preg):REP AJCNAC 23,782,70
orl-rat TDLo:3465 g/kg/88W-C:NEO JNCIAM 49,751,72
scu-rat TDLo:45 g/kg/66W-I:ETA FCTXAV 9,463,71
orl-rat LDLo:10 mg/kg CLDND* 7,178,87

CONSENSUS REPORTS: Reported in EPA TSCA Inventory. EPA Genetic Toxicology Program.

SAFETY PROFILE: Poison by ingestion and intravenous routes. Experimental reproductive effects. Questionable carcinogen with experimental tumorigenic and neoplastigenic data. Human mutation data reported. When heated to decomposition it emits very toxic fumes of SO_x and NO_x. See also CALCIUM COMPOUNDS.

CAR375 CAS:7789-41-5 ***HR: 2***
CALCIUM DIBROMIDE
mf: Br_2Ca mw: 199.90

PROP: The N.F. grade is a hydrated salt, containing not less than 84% and not more than 94% $CaBr_2$. Odorless, deliquesc granules or rhombic crystals; sharp, saline taste. Becomes yellow on long exposure to air. Mp: 730° (anhydrous), d: (25/4) 3.353. When strongly heated in air, becomes alkaline due to loss of bromine and formation of lime. Very sol in water, methanol, ethanol; sol in acetone; practically insol in dioxane, chloroform, ether.

SYN: CALCIUM BROMIDE

TOXICITY DATA with REFERENCE
cyt-rat/ast 2300 mg/kg GANNA2 54,155,63
ipr-rat LD50:437 mg/kg OYYAA2 16,229,78
ipr-mus LD50:740 mg/kg OYYAA2 16,229,78
scu-mus LD50:1580 mg/kg OYYAA2 20,693,80

CONSENSUS REPORTS: Reported in EPA TSCA Inventory.

SAFETY PROFILE: Moderately toxic by intraperitoneal and subcutaneous routes. Mutation data reported. Incompatible with potassium. When heated to decomposition it emits toxic fumes of Br^-. See also BROMIDES and CALCIUM COMPOUNDS.

CAR750 CAS:12013-56-8 ***HR: 3***
CALCIUM DISILICIDE
mf: $CaSi_2$ mw: 96.25

SAFETY PROFILE: Mixture with CCl_4 is a friction-sensitive explosive. Ignites on close contact with alkali metal fluorides. Mixture with iron(III) oxide (silicon thermite) reacts violently when heated producing molten iron as with the normal thermite mixture. Mixtures with potassium nitrate are easily ignited and burn at a very high temperature. See also CALCIUM COMPOUNDS.

CAR800 CAS:12264-18-5 ***HR: 2***
CALCIUM EDTA COMPLEX
mf: $C_{10}H_{12}CaN_2O_8 \cdot 2H$ mw: 330.34

SYNS: ACETIC ACID, (ETHYLENEDINITRILO)TETRA-, CALCIUM (II) COMPLEX ◇ CALCIATE(2-), ((ETHYLENEDINITRILO)TETRAACETATO)-, DIHYDROGEN (8CI) ◇ VERSENE CA

TOXICITY DATA with REFERENCE
scu-rat TDLo:1344 mg/kg (female 11-15D post):TER TXAPA9 82,426,86

scu-rat TDLo:2016 mg/kg (female 11-15D post):REP TXAPA9 82,426,86
ipr-mus LD50:573 mg(Ca)/kg PABIAQ 11,853,63

SAFETY PROFILE: Moderately toxic by intraperitoneal route. An experimental teratogen. Other experimental reproductive effects. When heated to decomposition it emits toxic fumes of NO_x.

CAR875 CAS:32266-82-3 ***HR: 2***
CALCIUM-N-2-ETHYLHEXYL-β-OXYBUTYRAMIDE SEMISUCCINATE
mf: $C_{32}H_{58}N_2O_{10}•Ca$ mw: 671.00

SYN: M-2

TOXICITY DATA with REFERENCE
orl-mus TDLo:300 mg/kg (female 7-14D post):REP TOIZAG 18,88,71
orl-rat TDLo:3 g/kg (9-14D preg):TER TOIZAG 18,88,71
orl-rat LD50:5746 mg/kg TOIZAG 17,579,70
ipr-rat LD50:741 mg/kg TOIZAG 17,579,70
scu-rat LD50:3037 mg/kg TOIZAG 17,579,70
orl-mus LD50:2129 mg/kg TOIZAG 17,579,70
ipr-mus LD50:549 mg/kg TOIZAG 17,579,70
scu-mus LD50:1187 mg/kg TOIZAG 17,579,70

SAFETY PROFILE: Moderately toxic by intraperitoneal, subcutaneous, and ingsetion routes. An experimental teratogen. Experimental reproductive effects. When heated to decomposition it emits toxic fumes of NO_x. See also CALCIUM COMPOUNDS and ESTERS.

CAS000 CAS:7789-75-5 ***HR: 2***
CALCIUM FLUORIDE
mf: CaF_2 mw: 78.08

PROP: Cubic, colorless crystals; luminous with heat. Mp: 1360°, d: 3.180.

SYNS: ACID-SPAR ◇ CALCIUM DIFLUORIDE ◇ FLUORITE ◇ FLUORSPAR ◇ IRTRAN 3 ◇ LIPARITE ◇ MET-SPAR

TOXICITY DATA with REFERENCE
cyt-rat/ast 1 g/kg GANNA2 54,155,63
ipr-mus TDLo:3200 mg/kg (9D preg):REP DZZEA7 34,124,79
ipr-mus TDLo:67200 mg/kg (1-21D preg):TER DZZEA7 34,484,79
orl-rat LD50:4250 mg/kg VAMNAQ 32,28,77
ipr-mus LD50:2638 mg/kg DZZEA7 34,484,79

CONSENSUS REPORTS: Reported in EPA TSCA Inventory.

OSHA PEL: TWA 2.5 mg(F)/m^3
ACGIH TLV: TWA 2.5 mg(F)/m^3
NIOSH REL: (Inorganic Fluorides) TWA 2.5 mg(F)/m^3

SAFETY PROFILE: Moderately toxic by intraperitoneal route. Mildly toxic by ingestion. An experimental teratogen. Other experimental reproductive effects. Mutation data reported. See also FLUORIDES and CALCIUM COMPOUNDS. When heated to decomposition it emits toxic fumes of F^-.

CAS250 CAS:544-17-2 ***HR: 3***
CALCIUM FORMATE
mf: $C_2H_2O_4•Ca$ mw: 130.12

SYNS: FORMIC ACID, CALCIUM SALT ◇ MRAVENCAN VAPENATY (CZECH)

TOXICITY DATA with REFERENCE
eye-rbt 100 mg/24H MOD 28ZPAK -,9,72
orl-rat LD50:2650 mg/kg 28ZPAK -,9,72
orl-mus LD50:1920 mg/kg ZERNAL 9,332,69
ivn-mus LD50:154 mg/kg ZERNAL 9,332,69

CONSENSUS REPORTS: Reported in EPA TSCA Inventory.

SAFETY PROFILE: Poison by intravenous route. Moderately toxic by ingestion. An eye irritant. When heated to decomposition it emits acrid smoke and fumes. See also CALCIUM COMPOUNDS.

CAS750 CAS:299-28-5 ***HR: 2***
CALCIUM GLUCONATE
mf: $C_{12}H_{22}O_{14}•Ca$ mw: 430.42

PROP: White, fluffy powder or granules; odorless and tasteless. Sol in hot water; less sol in cold water; insol in alc, acetic acid, and other organic solvents. Mp: loses H_2O @ 120°.

SYN: GLUCONATE de CALCIUM (FRENCH)

TOXICITY DATA with REFERENCE
ims-inf TDLo:143 mg/kg:SKN,MET JAMAAP 129,347,45
ims-inf LDLo:10 g/kg JAMAAP 129,347,45
ivn-rat LD50:950 mg/kg NIIRDN 6,226,82
ipr-mus LD50:2200 mg/kg JDGRAX 15(1-2),121,84
scu-mus LD50:2890 mg/kg JAPMA8 45,47,56
ivn-mus LD50:1210 mg/kg JAPMA8 45,47,56
ipr-dog LDLo:1100 mg/kg AVERAG 44,555,37
ivn-gpg LDLo:1810 mg/kg AIPTAK 191,44,71

CONSENSUS REPORTS: Reported in EPA TSCA Inventory.

SAFETY PROFILE: Moderately toxic by subcutaneous, intraperitoneal, and intravenous routes. Human systemic effects in infants by intramuscular route: dermatitis and fever. When heated to decomposition it emits acrid smoke and fumes. See also CALCIUM COMPOUNDS.

CAS825 ***HR: 1***
CALCIUM HEXAMETAPHOSPHATE

SAFETY PROFILE: A nuisance dust.

CAT125 CAS:17097-76-6 ***HR: 2***
CALCIUM HOMOPANTOTHENATE
mf: $C_{20}H_{38}N_2O_6$•Ca mw: 442.68

SYNS: CALCIUM-d-HOMOPANTOTHENATE ◇ CALCIUM HOPANTENATE ◇ (R)-4-((2,4-DIHYDROXY-3,3-DIMETHYL-1-OXOBUTYL) AMINO)-BUTANOIC ACID CALCIUM SALT (2:1) ◇ HOPANTENATE CALCIUM ◇ PANTOGAM

TOXICITY DATA with REFERENCE
orl-rat LD50:13800 mg/kg NIIRDN 6,788,82
scu-rat LD50:5600 mg/kg NIIRDN 6,788,82
orl-mus LD50:6000 mg/kg NIIRDN 6,788,82
scu-mus LD50:2600 mg/kg NIIRDN 6,788,82

SAFETY PROFILE: Moderately toxic by subcutaneous route. Mildly toxic by ingestion. When heated to decomposition it emits toxic fumes of NO_x. See also CALCIUM COMPOUNDS.

CAT175 CAS:1990-07-4 ***HR: 2***
CALCIUM HOPANTENATE HEMIHYDRATE
mf: $C_{20}H_{36}N_2O_{10}$•Ca•1/2H_2O mw: 513.61

SYNS: CALCIUM-d-(+)-4-(2,4-DIHYDROXY-3,3-DIMETHYLBUTYRAMIDE)BUTYRATE HEMIHDYRATE ◇ HOPA ◇ HOPANTENATE CALCIUM HEMIHDYRATE

TOXICITY DATA with REFERENCE
orl-rat TDLo:405 mg/kg (17-22D preg/21D post):REP OYYAA2 19,1011,80
orl-rat LD50:13900 mg/kg IYKEDH 9,829,78
ipr-rat LD50:13500 mg/kg IYKEDH 9,829,78
scu-rat LD50:5600 mg/kg IYKEDH 9,829,78
orl-mus LD50:6000 mg/kg IYKEDH 9,829,78
ipr-mus LD50:850 mg/kg YAKUD5 20,259,78
scu-mus LD50:2600 mg/kg IYKEDH 9,829,78

SAFETY PROFILE: Moderately toxic by subcutaneous and intraperitoneal routes. Experimental reproductive effects. When heated to decomposition it emits toxic fumes of NO_x. See also CALCIUM COMPOUNDS.

CAT200 CAS:7789-78-8 ***HR: 3***
CALCIUM HYDRIDE
mf: CaH_2 mw: 60.24

SYN: CALCIUM DIHYDRIDE

SAFETY PROFILE: Explosive reaction on heating with tetrahydrofuran. Mixtures with potassium chlorate and other metal oxohalogenates (e.g., chlorates; bromates; and perchlorates) are heat- and friction-sensitive explosives. Vigorous or incandescent reaction on heating with halogens (chlorine; bromine; or iodine); manganese dioxide; and silver halides (e.g., silver fluoride; silver iodide). See also CALCIUM COMPOUNDS and HYDRIDES.

CAT225 CAS:1305-62-0 ***HR: 2***
CALCIUM HYDROXIDE
mf: CaH_2O_2 mw: 74.10

PROP: Rhombic, trigonal, colorless crystals or white power; sltly bitter taste. Mp: loses H_2O @ 580°, bp: decomp, d: 2.343. Sol in water and glycerin; insol in alc.

SYNS: BELL MINE ◇ CALCIUM HYDRATE ◇ HYDRATED LIME ◇ KEMIKAL ◇ LIME WATER ◇ SLAKED LIME

TOXICITY DATA with REFERENCE
eye-rbt 10 mg SEV TXAPA9 55,501,80
cyt-rat/ast 1200 mg/kg GANNA2 54,155,62
orl-rat LD50:7340 mg/kg AIHAAP 30,470,69
orl-mus LD50:7300 mg/kg YKYUA6 32,1477,81

CONSENSUS REPORTS: Reported in EPA TSCA Inventory.

OSHA PEL: TWA 5 mg/m^3
ACGIH TLV: TWA 5 mg/m^3

SAFETY PROFILE: Mildly toxic by ingestion. A severe eye irritant. A skin, mucous membrane and respiratory system irritant. Mutation data reported. Causes dermatitis. Dust is considered to be a significant industrial hazard. A common air contaminant. Violent reaction with maleic anhydride; nitroethane; nitromethane; nitroparaffins; nitropropane; phosphorus. Reaction with polychlorinated phenols + potassium nitrate forms extremely toxic products. See also CALCIUM COMPOUNDS.

CAT250 CAS:7789-79-9 ***HR: 3***
CALCIUM HYPOPHOSPHITE
mf: $CaH_4O_4P_2$ mw: 170.06

$Ca(OP(O)H_2)_2$

SYN: CALCIUM PHOSPHINATE

SAFETY PROFILE: Mixture with potassium chlorate is a friction-sensitive explosive. When heated to decomposition it emits toxic fumes of PO_x. See also CALCIUM COMPOUNDS.

CAT500 CAS:7789-80-2 ***HR: 1***
CALCIUM IODATE
mf: $Ca(IO_3)_2$•H_2O mw: 407.90

PROP: White powder. Sltly sol in water; insol in alc.

SAFETY PROFILE: A nuisance dust.

CAT700 CAS:819-17-0 ***HR: 3***
CALCIUM METHIONATE
mf: $CH_2O_6S_2 \cdot Ca$ mw: 214.23

SYN: METHANEDISULFONIC ACID, CALCIUM SALT (1:1)

TOXICITY DATA with REFERENCE
ivn-rat LD50:329 mg/kg JAPMA8 45,47,56
scu-mus LD50:1085 mg/kg JAPMA8 45,47,56
ivn-mus LD50:422 mg/kg JAPMA8 45,47,56

SAFETY PROFILE: Poison by intravenous route. Moderately toxic by subcutaneous route. When heated to decomposition it emits toxic fumes of SO_x. See also CALCIUM COMPOUNDS and SULFONATES.

CAT750 CAS:7789-82-4 ***HR: 3***
CALCIUM MOLYBDATE
mf: $MoO_4 \cdot Ca$ mw: 200.02

TOXICITY DATA with REFERENCE
orl-rat LD50:101 mg/kg 28ZLA8 -,214,61
ipr-rat LD50:208 mg/kg EQSSDX 1,1,75

CONSENSUS REPORTS: Reported in EPA TSCA Inventory.

OSHA PEL: (Transitional: TWA Total Dust: 15 mg/m^3; Respirable Fraction: 5 mg/m^3) TWA Total Dust: 10 mg/m^3; Respirable Fraction: 5 mg/m^3
ACGIH TLV: TWA 10 mg(Mo)/m^3

SAFETY PROFILE: Poison by ingestion and intraperitoneal routes. See also MOLYBDENUM and CALCIUM COMPOUNDS.

CAU000 CAS:10124-37-5 ***HR: 3***
CALCIUM(II) NITRATE (1:2)
DOT: UN 1454
mf: $N_2O_6 \cdot Ca$ mw: 164.10

SYN: CALCIUM NITRATE (DOT)

CONSENSUS REPORTS: Reported in EPA TSCA Inventory.

DOT Classification: Oxidizer; Label: Oxidizer.

SAFETY PROFILE: An irritant. A strong oxidant. Forms powerfully explosive mixtures with aluminum + ammonium nitrate + formamide + water, ammonium nitrate + hydrocarbon oils, ammonium nitrate + water-soluble fuels, and organic materials. When heated to decomposition it emits toxic fumes of NO_x. See also NITRATES and CALCIUM COMPOUNDS.

CAU250 CAS:13477-34-4 ***HR: 2***
CALCIUM(II) NITRATE TETRAHYDRATE (1:2:4)
mf: $N_2O_6 \cdot Ca \cdot 4H_2O$ mw: 236.18

PROP: Cubic, colorless, hygroscopic crystals. Mp: 561.0°, d: 2.36.

SYNS: DUSICNAN VAPENATY (CZECH) ◇ NITRIC ACID, CALCIUM SALT, TETRAHYDRATE

TOXICITY DATA with REFERENCE
skn-rbt 500 mg/24H MLD 28ZPAK -,9,72
eye-rbt 500 mg/24H MLD 28ZPAK -,9,72
orl-rat LD50:3900 mg/kg 28ZPAK -,9,72

SAFETY PROFILE: Moderately toxic by ingestion. A skin and eye irritant. See also CALCIUM NITRATE, CALCIUM COMPOUNDS, and NITRATES. When heated to decomposition it emits toxic fumes of NO_x.

CAU500 CAS:1305-78-8 ***HR: 3***
CALCIUM OXIDE
DOT: UN 1910
mf: CaO mw: 56.08

PROP: Cubic, white crystals. Mp: 2580°, d: 3.37, bp: 2850°. Sol in water and glycerin; insol in alc.

SYNS: BURNT LIME ◇ CALCIA ◇ CALX ◇ LIME ◇ LIME, BURNED ◇ LIME, UNSLAKED (DOT) ◇ OXYDE de CALCIUM (FRENCH) ◇ QUICKLIME (DOT) ◇ WAPNIOWY TLENEK (POLISH)

CONSENSUS REPORTS: Reported in EPA TSCA Inventory.

OSHA PEL: TWA 5 mg/m^3
ACGIH TLV: TWA 2 mg/m^3
DFG MAK: 5 mg/m^3
DOT Classification: ORM-B; Label: None.

SAFETY PROFILE: A caustic and irritating material. See also CALCIUM COMPOUNDS. A common air contaminant. A powerful caustic to living tissue. The powdered oxide may react explosively with water. Mixtures with ethanol may ignite if heated and thus can cause an air-vapor explosion. Violent reaction with (B_2O_3 + $CaCl_2$); interhalogens (e.g., BF_3; ClF_3); F_2; HF; P_2O_5 + heat; water. Incandescent reaction with liquid HF. Incompatible with phosphorus(V) oxide.

CAU750 CAS:137-08-6 ***HR: 2***
CALCIUM-d-PANTOTHENATE
mf: $C_{19}H_{34}N_2O_{10} \cdot Ca$ mw: 490.63

PROP: White, sltly hygroscopic powder; odorless; bitter taste. Mp: 170-172°, decomp @ 195-196°. Sol in water and glycerin; insol in alc, chloroform, and ether.

SYNS: CALCIUM d(+)-N-(α,Γ-DIHYDROXY-β,β-DIMETHYLBUTYRYL)-β-ALANINATE ◇ CALCIUM PANTHOTHENATE (FCC) ◇ CALCIUM PANTOTHENATE ◇ d-CALCIUM PANTOTHENATE ◇ CALPANATE ◇ DEXTRO CALCIUM PANTOTHENATE ◇ N-(2,4-DIHYDROXY-3,3-DIMETHYLBUTYRYL)-β-ALANINE CALCIUM ◇ PANCAL ◇ PANTHOJECT ◇ PANTHOLIN ◇ PANTOTHENATE CALCIUM

◇ PANTOTHENIC ACID, CALCIUM SALT ◇ (+)-PANTOTHENIC ACID, CALCIUM SALT ◇ VITAMIN B-5

TOXICITY DATA with REFERENCE
ipr-rat LD50:820 mg/kg PSEBAA 45,311,40
scu-rat LD50:3400 mg/kg PSEBAA 45,311,40
ivn-rat LD50:830 mg/kg NIIRDN 6,599,82
orl-mus LD50:10 g/kg NIIRDN 6,599,82
ipr-mus LD50:920 mg/kg PSEBAA 45,311,40
scu-mus LD50:2700 mg/kg PSEBAA 45,311,40
ivn-mus LD50:910 mg/kg PSEBAA 45,311,40

CONSENSUS REPORTS: Reported in EPA TSCA Inventory.

SAFETY PROFILE: Moderately toxic by intraperitoneal, subcutaneous, and intravenous routes. Mildly toxic by ingestion. A vitamin. See also CALCIUM COMPOUNDS. When heated to decomposition it emits toxic fumes of NO_x.

CAU780 ***HR: 1***
CALCIUM PANTOTHENATE, CALCIUM CHLORIDE DOUBLE SALT
mf: $C_{19}H_{34}N_2O_{10}•Ca_2Cl_2$ mw: 601.61

PROP: White, sltly hygroscopic powder; odorless with bitter taste. Sol in water and glycerin; insol in alc, chloroform, and ether.

SAFETY PROFILE: Moderately toxic by intraperitoneal, subcutaneous, and intravenous routes. Mildly toxic by ingestion. A vitamin. See also CALCIUM COMPOUNDS. When heated to decomposition it emits toxic fumes of NO_x.

CAV000 CAS:7563-42-0 ***HR: 3***
CALCIUM PENTOBARBITAL
mf: $C_{11}H_{18}N_2O_3•7Ca$ mw: 506.87

SYNS: CALCIUM NEMBUTAL ◇ INSOM-RAPIDO ◇ NEMBUTAL CALCIUM ◇ PENTOBARBITAL CALCIUM ◇ 2,4,6(1H,3H,5H)-PYRIMIDINETRIONE, 5-ETHYL-5-(1-METHYLBUTYL)-, CALCIUM SALT (9CI) ◇ RAVONA ◇ REPOCAL ◇ SCHLAFEN

TOXICITY DATA with REFERENCE
orl-dog LDLo:60 mg/kg CRAAA7 20,350,41
ivn-dog LDLo:70 mg/kg CRAAA7 20,350,41

SAFETY PROFILE: Poison by ingestion and intravenous routes. When heated to decomposition it emits toxic fumes of NO_x. See also BARBITURATES.

CAV250 CAS:10118-76-0 ***HR: 3***
CALCIUM PERMANGANATE
DOT: UN 1456
mf: $Mn_2O_8•Ca$ mw: 277.96

$Ca(MnO_4)_2$

PROP: Violet, deliquescent crystals. Mp: decomp, d: 2.4.

SYN: KALIUMPERMANGANAT (GERMAN)

TOXICITY DATA with REFERENCE
ivn-rbt LDLo:50 mg/kg TDBU** -,-,33

CONSENSUS REPORTS: Manganese and its compounds are on the Community Right-To-Know List.

OSHA PEL: CL 5 $mg(Mn)/m^3$
ACGIH TLV: TWA 5 $mg(Mn)/m^3$
DOT Classification: Oxidizer; Label: Oxidizer.

SAFETY PROFILE: Poison by intravenous route. See also CALCIUM COMPOUNDS, MANGANESE COMPOUNDS, and PERMANGANATES. A strong oxidant. May explode on contact with acetic acid or acetic anhydride. Ignites on contact with cellulose. Incompatible with hydrogen peroxide.

CAV500 CAS:1305-79-9 ***HR: 3***
CALCIUM PEROXIDE
DOT: UN 1457
mf: CaO_2 mw: 72.08

PROP: Yellow crystals or powder or white crystals, decomposes in air. Mp: decomp @ 275°. Insol in water; sol in acids, forming hydrogen peroxide.

SYNS: CALCIUM DIOXIDE ◇ CALCIUM SUPEROXIDE

CONSENSUS REPORTS: Reported in EPA TSCA Inventory.

DOT Classification: Oxidizer; Label: Oxidizer.

SAFETY PROFILE: Irritating in concentrated form. Will react with moisture to form slaked lime. Flammable if hot and mixed with finely divided combustible material. Mixtures with oxidizable materials can also be ignited by grinding and are explosion hazards. A strong alkali. An oxidizer. Mixtures with polysulfide polymers may ignite. See also CALCIUM COMPOUNDS, CALCIUM HYDROXIDE, and PEROXIDES, INORGANIC.

CAV750 ***HR: 3***
CALCIUM PEROXOCHROMATE
mf: $Ca_3Cr_2O_{12}$ mw: 416.23

CONSENSUS REPORTS: Chromium and its compounds are on the Community Right-To-Know List.

SAFETY PROFILE: An explosive. See also CHROMIUM COMPOUNDS, PEROXIDES, and CALCIUM COMPOUNDS.

CAW000 CAS:13235-16-0 ***HR: 3***
CALCIUM PEROXODISULPHATE
mf: CaO_8S_2 mw: 232.21

SAFETY PROFILE: A powerful shock-sensitive explosive. Upon decomposition it emits toxic fumes of SO_x. See also CALCIUM COMPOUNDS, PEROXIDES, and SULPHATES.

CAW100 CAS:7757-93-9 ***HR: 1***
CALCIUM PHOSPHATE, DIBASIC
mf: $CaHPO_4 \cdot 2H_2O$ mw: 172.09

PROP: White powder. Sol in dilute acid; insol in water, alc.

SYN: DICALCIUM PHOSPHATE

SAFETY PROFILE: Skin and eye irritant. A nuisance dust.

CAW110 CAS:7758-23-8 ***HR: 1***
CALCIUM PHOSPHATE, MONOBASIC
mf: $Ca(H_2PO_4)_2$ mw: 234.05

PROP: White crystals or granular powder. Sltly sol in water; insol in alc.

SYNS: ACID CALCIUM PHOSPHATE ◇ CALCIUM BIPHOSPHATE ◇ MONOCALCIUM PHOSPHATE

SAFETY PROFILE: A nuisance dust.

CAW120 CAS:12167-74-7 ***HR: 1***
CALCIUM PHOSPHATE, TRIBASIC
mf: $10CaO \cdot 3P_2O_5 \cdot H_2O$ mw: 1004.64

PROP: White powder. Sol in dilute HCl; insol in water, alc.

SYNS: PERCIPITATED CALCIUM PHOSPHATE ◇ TRICALCIUM PHOSPHATE

SAFETY PROFILE: Skin and eye irritant. A nuisance dust.

CAW250 CAS:1305-99-3 ***HR: 3***
CALCIUM PHOSPHIDE
DOT: UN 1360
mf: Ca_3P_2 mw: 182.18

PROP: Red crystals. Mp: >1600°, d: 2.238 @ 25°.

SYN: TRICALCIUM DIPHOSPHIDE

CONSENSUS REPORTS: Reported in EPA TSCA Inventory.

DOT Classification: Flammable Solid; Label: Flammable Solid and Dangerous When Wet.

SAFETY PROFILE: Highly toxic due to phosphide which in presence of moisture emits phosphine. The phosphine may ignite spontaneously in air. Incandescent reaction with oxygen at 300°C. Incompatible with dichlorine oxide. When heated to decomposition it emits toxic fumes of PO_x. See also CALCIUM COMPOUNDS and PHOSPHIDES.

CAW376 CAS:26016-98-8 ***HR: 2***
CALCIUM PHOSPHONOMYCIN HYDRATE
mf: $C_3H_5O_4P \cdot Ca \cdot H_2O$ mw: 194.15

SYNS: CALCIUM (–)-(1R,2S)-(1,2-EPOXYPROPYL)PHOSPHONATE HYDRATE ◇ CALCIUM FOSFOMYCIN HYDRATE ◇ FOM-Ca HYDRATE ◇ FOSFOMYCIN-Ca HYDRATE ◇ FOSFOMYCIN CALCIUM HYDRATE

TOXICITY DATA with REFERENCE
orl-rat TDLo:1540 mg/kg (female 7-17D post):REP JJANAX 32,546,79
orl-rat TDLo:11 g/kg (male 9W pre):TER JJANAX 33,613,80
ipr-rat LD50:1036 mg/kg DRUGAY 6,785,82
ipr-mus LD50:994 mg/kg IYKEDH 11,811,80

SAFETY PROFILE: Moderately toxic by intraperitoneal route. An experimental teratogen. Other experimental reproductive effects. When heated to decomposition it emits toxic fumes of PO_x. See also CALCIUM COMPOUNDS.

CAW450 CAS:7790-76-3 ***HR: 1***
CALCIUM PYROPHOSPHATE
mf: $Ca_2P_2O_7$ mw: 254.10

PROP: Fine white powder. Sol in dilute HCl; insol in water.

SAFETY PROFILE: A nuisance dust.

CAW500 CAS:9007-13-0 ***HR: 1***
CALCIUM RESINATE
DOT: UN 1313/UN 1314
mf: $Ca(C_{44}H_{62}O_4)_2$ mw: 1349.50

PROP: Yellowish-white, amorphous powder or lumps.

SYNS: CALCIUM RESINATE, fused (DOT) ◇ CALCIUM RESINATE, technically pure (DOT) ◇ LIMED ROSIN

CONSENSUS REPORTS: Reported in EPA TSCA Inventory.

DOT Classification: Flammable Solid; Label: Flammable Solid; Flammable Solid; Label: Flammable Solid, fused.

SAFETY PROFILE: Flammable solid when heated; can react with oxidizing materials. When heated to decomposition it emits acrid smoke and fumes. See also CALCIUM COMPOUNDS.

CAW850 CAS:1344-95-2 ***HR: 1***
CALCIUM SILICATE

PROP: Varying proportions of CaO and SiO_2. White powder. Insol in water.

OSHA PEL: Total Dust: 15 mg/m^3; Respirable Fraction: 5 mg/m^3
ACGIH TLV: TWA (nuisance particulate) 10 mg/m^3 of total dust (when toxic impurities are not present, e.g., quartz < 1%).

SAFETY PROFILE: A nuisance dust.

CAX000 CAS:12013-55-7 ***HR: 3***
CALCIUM SILICIDE
mf: CaSi mw: 68.17

SAFETY PROFILE: Reacts with acids to evolve self-igniting silane gases. Reacts violently with F_2. See also CALCIUM HYDROXIDE and SILANES.

CAX250 CAS:16925-39-6 ***HR: 3***
CALCIUM SILICOFLUORIDE
mf: CaF_6Si mw: 182.17

PROP: White, crystalline powder. D: 2.662 @ 17.5°.

SYNS: CALCIUM FLUOSILICATE ô CALCIUM HEXAFLUOROSILICATE

CONSENSUS REPORTS: Reported in EPA TSCA Inventory.

OSHA PEL: TWA 2.5 mg(F)/m^3
NIOSH REL: (Inorganic Fluorides) TWA 2.5 mg(F)/m^3

SAFETY PROFILE: Poison by ingestion and subcutaneous routes. See also HEXAFLUORO SILICATE(2) DIHYDROGEN and CALCIUM COMPOUNDS. When heated to decomposition it emits toxic fumes of F^-.

CAX260 CAS:23209-59-8 ***HR: 2***
CALCIUM SODIUM METAPHOSPHATE
mf: $HO_3P•Ca•Na$ mw: 143.05

SYN: METAPHOSPHORIC ACID, CALCIUM SODIUM SALT

TOXICITY DATA with REFERENCE
ipl-rat TDLo:200 mg/kg/2Y-C:ETA EPASR* 8EHQ-0386-0619

CONSENSUS REPORTS: Reported in EPA TSCA Inventory.

SAFETY PROFILE: Questionable carcinogen with experimental tumorigenic data. When heated to decomposition it emits toxic fumes of PO_x.

CAX500 CAS:7778-18-9 ***HR: 3***
CALCIUM SULFATE
mf: $CaSO_4$ mw: 136.14

PROP: Pure anhydrous, white powder or odorless crystals. D: 2.964; mp: 1450°.

SYNS: GYPSUM ◇ PLASTER of PARIS

OSHA PEL: Total Dust: 15 mg/m^3; Respirable Fraction: 5 mg/m^3
ACGIH TLV: TWA (nuisance particulate) 10 mg/m^3 of total dust (when toxic impurities are not present, e.g., quartz < 1%)7

SAFETY PROFILE: A nuisance dust.Reacts violently with aluminum when heated. Mixtures with diazomethane react exothermically and eventually explode. Mixtures with phosphorus ignite at high temperatures. When heated to decomposition it emits toxic fumes of SO_x. See also CALCIUM COMPOUNDS and SULFATES.

CAX750 CAS:10101-41-4 ***HR: 1***
CALCIUM(II) SULFATE DIHYDRATE (1:1:2)
mf: $O_4S•Ca•2H_2O$ mw: 172.18

PROP: Colorless crystals. D: 2.32, mp: 128° (loses $1.5H_2O$), bp: 163° (loses $2H_2O$).

SYNS: ALABASTER ◇ ANNALINE ◇ C.I. 77231 ◇ C.I. PIGMENT WHITE 25 ◇ GYPSUM ◇ GYPSUM STONE ◇ LAND PLASTER ◇ LIGHT SPAR ◇ MAGNESIA WHITE ◇ MINERAL WHITE ◇ NATIVE CALCIUM SULFATE ◇ PRECIPITATED CALCIUM SULFATE ◇ SATINITE ◇ SATIN SPAR ◇ SULFURIC ACID, CALCIUM(2+) SALT, DIHYDRATE ◇ TERRA ALBA

TOXICITY DATA with REFERENCE
ipr-rat TDLo:450 mg/kg/3W-I:CAR ZHPMAT 162,467,76
ihl-hmn TCLo:194 g/m^3/10Y-I:NOSE,PUL GTPZAB 11(10),23,67

OSHA PEL: Total Dust: 15 mg/m^3; Respirable Fraction: 5 mg/m^3
ACGIH TLV: TWA (nuisance particulate) 10 mg/m^3 of total dust (when toxic impurities are not present, e.g., quartz < 1%)

SAFETY PROFILE: Human systemic effects by inhalation: fibrosing alveolitis (growth of fibrous tissue in the lung); unspecified respiratory system effects and unspecified effects on the nose. Questionable carcinogen with experimental carcinogenic data. Long considered a nuisance dust (depending on silica content). When heated to decomposition it emits toxic fumes of SO_x. See also CALCIUM SULFATE, CALCIUM COMPOUNDS, and SULFATES.

CAY000 CAS:20548-54-3 ***HR: 3***
CALCIUM SULFIDE
mf: CaS mw: 72.14

PROP: Cubic, colorless crystals. Bp: decomp, d: 218 @ 15°.

SYNS: CALCIC LIVER of SULFUR ◇ HEPAR CALCIS ◇ OLDHAMITE

SAFETY PROFILE: A poison via inhalation. Reacts violently with chromyl chloride, lead dioxide, potassium chlorate (mild explosion), potassium nitrate (violent explosion). Incompatible with oxidants. When heated to decomposition it emits toxic fumes of SO_x. See also CALCIUM COMPOUNDS and SULFIDES.

CAY250 CAS:2092-16-2 ***HR: 3***
CALCIUM THIOCYANATE
mf: $C_2N_2S_2{\cdot}Ca$ mw: 156.24

PROP: White, deliquescent crystals.

SYNS: CALCIUM RHODANID (GERMAN) ◇ THIOCYANIC ACID, CALCIUM SALT (2:1)

TOXICITY DATA with REFERENCE
orl-mus LDLo:120 mg/kg AEPPAE 169,429,33
ivn-rbt LDLo:250 mg/kg AEPPAE 169,429,33

CONSENSUS REPORTS: Reported in EPA TSCA Inventory.

SAFETY PROFILE: Poison by ingestion and intravenous routes. See also THIOCYANATES and CALCIUM COMPOUNDS. When heated to decomposition it emits toxic fumes of NO_x and SO_x.

CAY500 CAS:12111-24-9 ***HR: 2***
CALCIUM TRISODIUM DIETHYLENE TRIAMINE PENTAACETATE
mf: $C_{14}H_{18}N_3O_{10}{\cdot}CaNa_3$ mw: 497.40

SYNS: Ba 2797 ◇ CALCIUM CHEL-330 ◇ CALCIUM-DTPA ◇ CALCIUM TRISODIUM CHEL 330 ◇ CALCIUM TRISODIUM DTPA ◇ CALCIUM TRISODIUM PENTETATE ◇ CALCIUM TRISODIUM SALT of DIETHYLENETRIAMINEPENTAACETIC ACID ◇ DIETHYLENETRIAMINE PENTAACETIC ACID, CALCIUM TRISODIUM SALT ◇ DITRIPENTAT ◇ DTPA CALCIUM TRISODIUM SALT ◇ PENTACIN ◇ PENTACINE ◇ PENTETATE TRISODIUM CALCIUM ◇ PENTHAMIL

TOXICITY DATA with REFERENCE
dni-rat-scu 4 mmol/kg BCPCA6 23,901,74
dni-rat:lvr 20 mmol/L BCPCA6 23,901,74
ipr-rat TDLo:659 mg/kg (9D preg):REP JANSAG 23,908,64
scu-rat TDLo:1790 mg/kg (9-13D preg):TER HOKBAQ 18,37,83
ivn-rat LD50:2512 mg/kg AAJRDX 142,619,84
ipr-mus LD50:3200 mg/kg REPMBN 10,391,62

CONSENSUS REPORTS: Reported in EPA TSCA Inventory.

SAFETY PROFILE: Moderately toxic by intravenous and intraperitoneal routes. Experimental teratogenic and reproductive effects. Mutation data reported. When heated to decomposition it emits toxic fumes of Na_2O and NO_x. See also CALCIUM COMPOUNDS.

CAY675 ***HR: 3***
CALCIUM VALPROATE
mf: $C_{16}H_{30}O_4{\cdot}Ca$ mw: 326.54

SYNS: DIPROPYLACETIC ACID CALCIUM SALT ◇ 2-PROPYLVALERIC ACID CALCIUM SALT (2:1) ◇ VALONTIN ◇ VALPROIC ACID CALCIUM SALT ◇ VALPROIC ACID HEMI-CALCIUM SALT

TOXICITY DATA with REFERENCE
orl-rat TDLo:6 g/kg (6-15D preg):TER FAATDF 3,121,83
orl-rat TDLo:6 g/kg (6-15D preg):REP FAATDF 3,121,83
ipr-rat LD50:375 mg/kg JNPHAG 2,313,71
idu-rat LD50:1065 mg/kg JNPHAG 2,313,71
ipr-mus LD50:320 mg/kg JNPHAG 2,313,71
idu-mus LD50:673 mg/kg JNPHAG 2,313,71

SAFETY PROFILE: Poison by intraperitoneal route. Moderately toxic by intraduodenal route. Experimental teratogenic and reproductive effects. When heated to decomposition it emits acrid smoke and fumes. See also PROPYL VALERIC ACID, SODIUM SALT, and CALCIUM COMPOUNDS.

CAY800 ***HR: 2***
CALLA

PROP: A commonly cultivated ornamental. The leaves are shaped like an arrowhead and are sometimes mottled with white. The lily-type flower may be white, green, pink, or yellow. It is grown outdoors in mild climates and indoors elsewhere.

SYNS: CALLA LILY ◇ LIRIO CALA (SPANISH) ◇ ZANTEDESCHIA AETHIOPICA

SAFETY PROFILE: The leaves contain poisonous crystals of calcium oxalate. Chewing the leaves results in burning pain in the lips, mouth and throat, possibly followed by inflammation and blistering. Systemic effects are usually not seen because of the insolubility of calcium oxalate. The sap can cause contact dermatitis. See also OXALATES.

CAY875 CAS:42839-36-1 ***HR: 3***
CALNEGYT
mf: $C_9H_{20}N_4{\cdot}H_2O_4S{\cdot}H_2O$ mw: 300.43

SYNS: EGYT 739 ◇ GUANAZODINE SULFATE MONOHYDRATE ◇ ((OCTAHYDRO-2-AZOCINYL)METHYL)GUANIDINE SULFATE HYDRATE ◇ SANEGYT

TOXICITY DATA with REFERENCE
orl-rat LD50:3550 mg/kg OYYAA2 14,235,77
ipr-rat LD50:970 mg/kg OYYAA2 14,235,77
ivn-rat LD50:136 mg/kg NYKZAU 72,837,76
ims-rat LD50:1080 mg/kg OYYAA2 14,235,77
orl-mus LD50:2450 mg/kg USXXAM #3856778
scu-mus LD50:700 mg/kg USXXAM #3856778
ivn-mus LD50:100 mg/kg OYYAA2 14,235,77
ims-mus LD50:1240 mg/kg OYYAA2 14,235,77

SAFETY PROFILE: Poison by intravenous route. Moderately toxic by ingestion, intraperitoneal, intramuscular, and subcutaneous routes. When heated to decomposition it emits toxic fumes of SO_x and NO_x. An antihypertensive agent. See also SULFATES.

CAY950 CAS:8065-83-6 ***HR: 3***
CALO-CLOR
mf: $Cl_2Hg_2 \cdot Cl_2Hg$ mw: 743.57

TOXICITY DATA with REFERENCE
orl-rat LD50:55200 μg/kg FMCHA2-,C56,89

ACGIH TLV: TWA 0.1 mg(Hg)/m^3 (skin)

SAFETY PROFILE: Poison by ingestion. When heated to decomposition it emits toxic fumes of Hg and Cl^-.

CAZ000 ***HR: 3***
CALOMEL and MAGNESIUM SULFATE (5:8)

SYN: MAGNESIUM SULFATE and CALOMEL (8:5)

TOXICITY DATA with REFERENCE
orl-mus TDLo:44 g/kg/69W-I:ETA CNREA8 28,2272,68

CONSENSUS REPORTS: Mercury and its compounds are on the Community Right-To-Know List.

SAFETY PROFILE: Questionable carcinogen with experimental tumorigenic data. See also MERCUROUS CHLORIDE, MERCURY COMPOUNDS, MAGNESIUM COMPOUNDS, and SULFATES. When heated to decomposition it emits very toxic fumes of Hg, Cl^- and SO_x.

CAZ075 ***HR: 1***
CALOTROPIS PROCERA (Ait.) R.Br., flower extract

PROP: Indian plant belonging to the family *Asclepiadaceae* JOETD7 22,211,88

SYNS: AK, flower extract ◇ AKRA, flower extract

TOXICITY DATA with REFERENCE
orl-grb TDLo:3 g/kg (male 15D pre):REP IJEBA6 17,859,79
orl-mus LDLo:3 g/kg JOETD7 22,211,88

SAFETY PROFILE: Slightly toxic by ingestion. Experimental reproductive effects. When heated to decomposition it emits acrid smoke and irritating fumes.

CAZ125 CAS:6874-80-2 ***HR: 3***
CALPURNINE
mf: $C_{20}H_{27}N_3O_3$ mw: 357.50

SYNS: HOE 933 ◇ 13-HYDROXYLUPANINE-2-PYRROLE CARBOXYLIC ACID ESTER ◇ (2S-(2-α,7-β,7A-β,14-β,14a-α))-1H-PYRROLE-2-CARBOXYLICACID-DODECAHYDRO-11-OXO-7,14-METHANO-2H,6H-DIPYRIDO(1,2-α:1′,2′-e)(1,5)DIAZOCIN-2-YL)ESTER

TOXICITY DATA with REFERENCE
orl-rat LD50:132 mg/kg DRFUD4 2,365,77
scu-rat LD50:41 mg/kg DRFUD4 2,365,77
ivn-rat LD50:3 mg/kg DRFUD4 2,365,77
orl-mus LD50:32 mg/kg DRFUD4 2,365,77
ivn-mus LD50:3100 μg/kg DRFUD4 2,365,77

SAFETY PROFILE: Poison by ingestion, subcutaneous, and intravenous routes. When heated to decomposition it emits toxic fumes of NO_x.

CBA000 CAS:9012-59-3 ***HR: 3***
CALVACIN

PROP: High molecular weight Glycopeptide from the giant puffball mushroom *Calvatia Gigantea* (CNREA8 23,1036,63).

TOXICITY DATA with REFERENCE
ipr-rat LD50:65 mg/kg CNREA8 23,1036,63
ipr-mus LD50:138 mg/kg CNREA8 23,1036,63
ivn-rbt LDLo:13 mg/kg CNREA8 23,1036,63

SAFETY PROFILE: Poison by intraperitoneal and intravenous routes. When heated to decomposition it emits acrid smoke and fumes. See also MUSHROOMS.

CBA075 ***HR: 3***
CALYCANTHINE, HYDROCHLORIDE
mf: $C_{22}H_{26}N_4 \cdot ClH$ mw: 382.98

TOXICITY DATA with REFERENCE
ivn-rat LD50:17160 μg/kg JAPMA8 31,513,42
ivn-mus LD50:43790 μg/kg JAPMA8 31,513,42
ivn-rbt LDLo:10 mg/kg JAPMA8 31,513,42

SAFETY PROFILE: Poison by intravenous route. When heated to decomposition it emits toxic fumes of NO_x and HCl.

CBA100 CAS:26097-80-3 ***HR: D***
CAMBENDAZOLE
mf: $C_{14}H_{14}N_4O_2S$ mw: 302.38

PROP: Odorless, white, crystalline solid. Mp: 238-240° (decomp). Sol in alc, dimethylformamide; sparingly sol in acetone; sltly sol in benzene; very sltly sol in 0.1M HCl. Practically insol in isooctane and water (0.02 mg/mL). Stable in acid and base in range of pH 1 to 12.

SYNS: BONLAM ◇ BOVIDAM ◇ CAMBENZOLE ◇ CAMBET ◇ CBDZ ◇ EQUIBEN ◇ ISOPROPYL-2-(4-TRIAZOLYL)-5-BENZIMIDAZOLE-CARBAMATE ◇ MK 905 ◇ NOE (FRENCH) ◇ NOVAZOLE ◇ NOVIBEN ◇ (THIAZOLYL-4)-2 BENZIMIDAZOLYL CARBAMATE-5 D'ISOPROPYLE (FRENCH) ◇ N-((THIAZOLYL-4)-2-BENZIMIDAZOLYL-5-CARBAMATE D'ISOPROPYLE (FRENCH) ◇ (2-(4-THIAZOLYL)-1H-BENZIMIDAZOL-5-YL)-CARBAMIC ACID 1-METHYLETHYL ESTER (9CI)

TOXICITY DATA with REFERENCE
oms-hmn:leu 1 mg/L THERAP 31,505,76
oms-dom:leu 1 mg/L THERAP 31,505,76
orl-dom TDLo:75 mg/kg (female 21D post):REP BSVMA8 77,197,75
orl-rat TDLo:80800 μg/kg (female 8-15D post):TER THERAP 31,505,76

SAFETY PROFILE: An experimental teratogen. Experimental reproductive effects. Human mutation data reported. When heated to decomposition it emits toxic fumes of SO_x and NO_x. See also CARBAMATES.

CBA125 CAS:2752-65-0 ***HR: 3***
CAMBOGIC ACID
mf: $C_{38}H_{44}O_8$ mw: 628.82

SYNS: β-GUTTIFERIN ◇ B''-GUTTIFERIN

TOXICITY DATA with REFERENCE
ipr-rat LD50:88 mg/kg IJEBA6 5,96,67
ivn-rat LD50:107 mg/kg IJEBA6 5,96,67
scu-mus LD50:354 mg/kg 85DGA2 8(1),331,82

SAFETY PROFILE: Poison by subcutaneous, intravenous, and intraperitoneal routes. When heated to decomposition it emits acrid smoke and fumes.

CBA375 CAS:54063-28-4 ***HR: 3***
CAMIVERINE
mf: $C_{19}H_{30}N_2O_2$ mw: 318.51

SYNS: ESTERE ISOAMILICO dell'ACIDO α-(N-(PIRROLIDINOETIL))-AMINOFENILACETICO (ITALIAN) ◇ FC 4/58 ◇ 2-PHENYL-N-(2-(1-PYRROLIDINYL)ETHYL)GLYCINE ISOPENTYL ESTER ◇ SANASPASMINA

TOXICITY DATA with REFERENCE
ipr-rat LD50:140 mg/kg FRPSAX 17,914,62
ivn-rat LD50:21 mg/kg FRPSAX 17,914,62
orl-mus LD50:920 mg/kg FRPSAX 17,914,62
ipr-mus LD50:175 mg/kg FRPSAX 17,914,62
ivn-mus LD50:28 mg/kg FRPSAX 17,914,62
ivn-rbt LD50:13 mg/kg FRPSAX 17,914,62

SAFETY PROFILE: Poison by intravenous and intraperitoneal routes. Moderately toxic by ingestion. When heated to decomposition it emits toxic fumes of NO_x. See also ESTERS.

CBA500 CAS:79-92-5 ***HR: 1***
CAMPHENE
DOT: UN 9011
mf: $C_{10}H_{16}$ mw: 136.26

PROP: Colorless cubic crystals; oily odor. Mp: 50-51°, bp: 159°, d: 0.842 @ 54°/4°, refr index: 1.452 @ 55°. Sol in alc; misc in fixed oils; insol in water.

SYN: FEMA No. 2229

TOXICITY DATA with REFERENCE
bfa-rat/sat 2500 mg/kg NUCADQ 1,10,79

CONSENSUS REPORTS: Reported in EPA TSCA Inventory.

DOT Classification: ORM-A; Label: None.

SAFETY PROFILE: Mutation data reported. Combustible; yields flammable vapors when heated and can react with oxidizing materials. To fight fire, use water spray, foam, fog, CO_2. When heated to decomposition it emits acrid smoke and irritating fumes.

CBA750 CAS:76-22-2 ***HR: 3***
CAMPHOR
DOT: UN 2717
mf: $C_{10}H_{16}O$ mw: 152.26

PROP: White, transparent, crystalline masses; penetrating odor; pungent, aromatic taste. Mp: 180°, bp: 204°, lel: 0.6%, uel: 3.5%, flash p: 150°F (CC), d: 0.992 @ 25°/4°, autoign temp: 871°F, vap d: 5.24.

SYNS: 2-BORNANONE ◇ 2-CAMPHANONE ◇ CAMPHOR, synthetic (ACGIH, DOT) ◇ CAMPHOR-natural ◇ FORMOSA CAMPHOR ◇ GUM CAMPHOR ◇ HUILE de CAMPHRE (FRENCH) ◇ JAPAN CAMPHOR ◇ KAMPFER (GERMAN) ◇ 2-KETO-1,7,7-TRIMETHYLNORCAMPHANE ◇ LAUREL CAMPHOR ◇ MATRICARIA CAMPHOR ◇ 2-OXOBORNANE ◇ 1,7,7-TRIMETHYLBICYCLO(2.2.1)-2-HEPTANONE ◇ 1,7,7-TRIMETHYLNORCAMPHOR

TOXICITY DATA with REFERENCE
cyt-smc 2 mmol/tube HEREAY 33,457,47
orl-inf LDLo:70 mg/kg AJPAA4 30,857,54
unk-man LDLo:29 mg/kg 85CDAI 2,73,70
ipr-rat LDLo:900 mg/kg JPETAB 65,275,39
scu-rat LD50:70 mg/kg CDGU** -,-,34
orl-mus LD50:1310 mg/kg SHGKA3 75,934,75
ihl-mus LCLo:400 mg/m^3/3H 85GMAT -,31,82
ipr-mus LD50:3000 mg/kg AJPAA4 30,857,54
scu-mus LDLo:200 mg/kg HDTU** -,-,33
orl-dog LDLo:800 mg/kg HBAMAK 4,1289,35
ipr-cat LDLo:400 mg/kg HBAMAK 4,1289,35
orl-rbt LDLo:2000 mg/kg AJPAA4 30,857,54
scu-frg LDLo:240 mg/kg AEXPBL 50,199,1903

CONSENSUS REPORTS: Reported in EPA TSCA Inventory.

OSHA PEL: TWA 2 mg/m^3
ACGIH TLV: TWA 2 ppm; STEL 3 ppm
DFG MAK: 2 ppm (13 mg/m^3)
DOT Classification: Flammable Solid; Label: Flammable Solid.

SAFETY PROFILE: A human poison by ingestion, and possibly other routes. An experimental poison by inhalation, subcutaneous and intraperitoneal routes. A local irritant. Ingestion causes nausea, vomiting, dizziness, excitation, and convulsions. Mutation data reported. Used as a topical anti-infective and anti-itching agent. Flammable when exposed to heat or flame; can react with oxidizing materials. Vapor is explosive when exposed to heat or flame or CrO_3. To fight fire, use foam, carbon dioxide, dry chemical. See also KETONES and other camphor entries.

CBA800 CAS:21368-68-3 ***HR: 2***
dl-CAMPHOR
mf: $C_{10}H_{16}O$ mw: 152.26

SYN: (±)-CAMPHOR

TOXICITY DATA with REFERENCE
ipr-rat LD50:956 mg/kg KHFZAN 16(7),108,82
scu-rat LD50:3040 mg/kg KHFZAN 16(7),108,82
ipr-mus LD50:884 mg/kg KHFZAN 16(7),108,82
scu-mus LD50:3020 mg/kg KHFZAN 16(7),108,82

CONSENSUS REPORTS: Reported in EPA TSCA Inventory.

SAFETY PROFILE: Moderately toxic by subcutaneous and intraperitoneal routes. When heated to decomposition it emits acrid smoke and fumes. See other camphor entries.

CBB000 CAS:464-48-2 ***HR: 3***
l-(−)-CAMPHOR
mf: $C_{10}H_{16}O$ mw: 152.26

SYN: l-CAMPHOR

TOXICITY DATA with REFERENCE
orl-rat LDLo:800 μg/kg JPETAB 1,445,09
ivn-mus LD50:320 mg/kg CSLNX* NX#02534

CONSENSUS REPORTS: Reported in EPA TSCA Inventory.

SAFETY PROFILE: Deadly poison by ingestion. Poison by intravenous route. When heated to decomposition it emits acrid smoke and irritating fumes. See also (IR,4R)-(+)-CAMPHOR and CAMPHOR.

CBB250 CAS:464-49-3 ***HR: 3***
(1R,4R)-(+)-CAMPHOR
mf: $C_{10}H_{16}O$ mw: 152.26

SYNS: ALCANFOR ◇ (+)-2-BORNANONE ◇ d-2-BORNANONE ◇ d-2-CAMPHANONE ◇ (+)-CAMPHOR ◇ d-CAMPHOR ◇ d-(+)-CAMPHOR ◇ CAMPHOR USP ◇ JAPANESE CAMPHOR ◇ (1R)-1,7,7-TRIMETHYLBICYCLO(2.2.1)HEPTAN-2-ONE

TOXICITY DATA with REFERENCE
skn-rbt 500 mg/24H MLD FCTXAV 16,665,78
scu-rat LDLo:1700 mg/kg FCTXAV 16,665,78
ipr-rat LDLo:3500 mg/kg FCTXAV 16,665,78
orl-mus LD50:1310 mg/kg FCTXAV 16,665,78
scu-mus LDLo:2200 mg/kg FCTXAV 16,665,78
ivn-mus LD90:525 mg/kg FCTXAV 16,665,78
ipr-cat LDLo:400 mg/kg FCTXAV 16,665,78

CONSENSUS REPORTS: Reported in EPA TSCA Inventory.

SAFETY PROFILE: Poison by intraperitoneal route. Moderately toxic by ingestion, subcutaneous, and intravenous routes. A skin irritant. When heated to decomposition it emits acrid and irritating fumes. See other camphor entries.

CBB375 CAS:8011-47-0 ***HR: 3***
CAMPHORATED OIL

SYN: CAMPHOR LINIMENT

TOXICITY DATA with REFERENCE
orl-wmn TDLo:240 mg/kg (40W preg):REP OBGNAS 25,255,65
orl-wmn TDLo:240 mg/kg (female 40W post):TER OBGNAS 25,255,65
orl-wmn TDLo:148 mg/kg:CNS,GIT PEDIAU 52,713,73
unr-wmn TDLo:900 mg/kg:CNS,GIT JFMAAQ 43,999,57

SAFETY PROFILE: Human systemic effects by ingestion and possibly other routes: nausea or vomiting, convulsions, muscle weakness, and coma. Human teratogenic effects by ingestion include these developmental abnormalities: extra embryonic structures, homeostasis, reduced viability and other neonatal effects. Other experimental reproductive effects. When heated to decomposition it emits acrid smoke and fumes. See other camphor entries.

CBB500 CAS:8008-51-3 ***HR: 3***
CAMPHOR OIL
DOT: UN 1130

PROP: Colorless or yellowish, oily, fragrant liquid. Bp: 175-200°, flash p: 117°F (CC), d: 0.875-0.900 @ 20°/20°. Insol in water; sol in chloroform, ether, oils, and in approx 3 vols alc. Found in the trees and bark of *Cinnamomum carphora sieb* (*Fam. Lauraceae*) and prepared by fractional distillation of crude camphor oil after the camphor has been crystallized out; a white, viscous liquid with cineole as the principal ingredient along with monoterpenes (FCTXAV 11,1011,73).

SYNS: CAMPHOR OIL, RECTIFIED ◇ CAMPHOR OIL WHITE ◇ CAMPHOR OIL YELLOW ◇ FORMOSA CAMPHOR OIL ◇ FORMOSE OIL of CAMPHOR ◇ JAPANESE CAMPHOR OIL ◇ JAPANESE, OIL of CAMPHOR ◇ LIGHT CAMPHOR OIL ◇ LIGHT OIL of CAMPHOR ◇ LIQUID CAMPHOR ◇ OIL of CAMPHOR RECTIFIED ◇ OIL CAMPHOR SASSAFRASSY ◇ OIL of CAMPHOR WHITE ◇ WHITE CAMPHOR OIL ◇ WHITE OIL of CAMPHOR

TOXICITY DATA with REFERENCE
skn-rbt 500 mg MLD FCTXAV 11,1047,73
orl-hmn TDLo:29 mg/kg:CNS,PUL 34ZIAG -,150,69

orl-chd LDLo:50 mg/kg 34ZIAG -,150,69
orl-rat LD50:3730 mg/kg FCTXAV 13,739,75

CONSENSUS REPORTS: Reported in EPA TSCA Inventory.

DOT Classification: Combustible Liquid; Label: None; Flammable or Combustible Liquid; Label: Flammable Liquid.

SAFETY PROFILE: A human poison by ingestion. Human systemic effects by ingestion: convulsions, tremors, and unspecified respiratory system effects. A skin irritant. Flammable when exposed to heat or flame; can react with oxidizing materials. To fight fire, use foam, CO_2, dry chemical, mist, fog. See also SAFROL and CAMPHOR.

CBB870 CAS:7689-03-4 ***HR: 3***
CAMPTOTHECINE
mf: $C_{20}H_{16}N_2O_4$ mw: 348.38

PROP: Pale yellow needles from methanol + acetonitrile. Decomp 264-267°. Does not form stable salts with acids.

SYNS: CAMPTOTHECIN ◇ 20(S)-CAMPTOTHECINE ◇ (S)-4-ETHYL-4-HYDROXY-1H-PYRANO(3′,4′:6,7)INDOLIZINO(1,2-b)QUINOLINE-3,14(4H,12H)-DIONE ◇ NSC 94600 ◇ NSC 100880 ◇ 21,22-SECOCAMPTOTHECIN-21-OIC ACID LACTONE

TOXICITY DATA with REFERENCE
dnd-omi 100 mg/L/30M NATUAS 248,226,74
dnd-hmn:hla 20 μmol/L CNREA8 33,2834,73
dni-hmn:hla 5 μmol/L HXPHAU 38(Pt 2),649,75
oms-hmn:hla 5 μmol/L HXPHAU 38(Pt 2),649,75
oms-mus:lym 1 mg/L BCPCA6 21,1977,72
dni-ckn:emb 500 μg/L CJBIAE 55,1180,77
ipr-mus LD50:64 mg/kg CNREA8 39,2204,79
ivn-mus LD50:38 mg/kg NCISP* JAN86

SAFETY PROFILE: Poison by intravenous and intraperitoneal routes. Human mutation data reported. When heated to decomposition it emits toxic fumes of NO_x. See also CAMPTOTHECIN, SODIUM SALT.

CBB875 CAS:25387-67-1 ***HR: 3***
CAMPTOTHECIN, SODIUM SALT
mf: $C_{20}H_{15}N_2O_4 \cdot Na$ mw: 370.36

SYN: NSC-100880

TOXICITY DATA with REFERENCE
ivn-hmn TDLo:2500 μg/kg/7D-I:BLD CCROBU 56,515,72
orl-mus LD50:27 mg/kg PMDCAY 9,1,73
ivn-mus LD50:57 mg/kg PMDCAY 9,1,73

SAFETY PROFILE: Poison by ingestion and intravenous routes. Human systemic effects by intravenous route: reduction in the number of white blood cells (leukopenia), reduction in the number of blood platelets (thrombocytopenia), and changes in blood cell count. When heated to decomposition it emits toxic fumes of NO_x and Na_2O.

CBC375 ***HR: 3***
CANDIDA ALBICANS GLYCOPROTEINS

PROP: Glycoprotein complex isolated from the cell walls of the 29-3-109 strain of *Candida albicans* 40YJAX –,35,76

TOXICITY DATA with REFERENCE
ivn-rat TDLo:15 mg/kg (8D preg):TER 40YJAX -,35,79
ivn-rat TDLo:15 mg/kg (8D preg):REP 40YJAX -,35,79
ivn-mus LD50:290 mg/kg TOXIA6 12,103,74

SAFETY PROFILE: Poison by intravenous route. An experimental teratogen. Other experimental reproductive effects. When heated to decomposition it emits toxic fumes of NO_x.

CBC500 CAS:1405-90-9 ***HR: 3***
CANDIDIN
mf: $C_{46}H_{75}NO_{17}$ mw: 914.22

SYN: CANDIDINE

TOXICITY DATA with REFERENCE
orl-mus LD50:100 mg/kg 85GDA2 2,288,80
ipr-mus LD50:7 mg/kg MEIEDD 10,240,83
scu-mus LD50:30 mg/kg MEIEDD 10,240,83
ivn-mus LD50:1500 μg/kg MEIEDD 10,240,83

SAFETY PROFILE: Poison by ingestion, subcutaneous, intraperitoneal, and intravenous routes. When heated to decomposition it emits toxic fumes of NO_x.

CBC750 CAS:1403-22-1 ***HR: 1***
CANDIDIN B

PROP: Isolated from *Streptomyces viridoflavus* (ANTCAO 4,455,54).

TOXICITY DATA with REFERENCE
orl-rat LD50:8560 mg/kg PHARAT 14,435,59

CONSENSUS REPORTS: Reported in EPA TSCA Inventory.

SAFETY PROFILE: Mildly toxic by ingestion. When heated to decomposition it emits acrid smoke and irritating fumes.

CBD250 CAS:64854-99-5 ***HR: 1***
CANDLETOXIN A
mf: $C_{35}H_{44}O_9$ mw: 608.8

TOXICITY DATA with REFERENCE
skn-mus 290 ng OPEN ARTODN 44,279,80

SAFETY PROFILE: A skin irritant. When heated to decomposition it emits acrid smoke and irritating fumes.

CBD500 CAS:64854-98-4 ***HR: 1***
CANDLETOXIN B
mf: $C_{33}H_{42}O_8$ mw: 566.8

TOXICITY DATA with REFERENCE
skn-mus 110 ng OPEN ARTODN 44,279,80

SAFETY PROFILE: A skin irritant. When heated to decomposition it emits acrid smoke and irritating fumes.

CBD599 CAS:13956-29-1 ***HR: 3***
CANNABIDIOL
mf: $C_{21}H_{30}O_2$ mw: 314.51

PROP: Pale yellow resin or crystals. Mp: 66-67°, bp: 187-190°, d: 1.040, n (20/D) 1.5404. Practically insol in water or 10% NaOH; sol in ethanol, methanol, ether, benzene, chloroform, and petr ether.

SYNS: (−)-CANNABIDIOL ◇ (−)-trans-CANNABIDIOL ◇ CBD ◇ (−)-trans-2-p-MENTHA-1,8-DIEN-3-YL-5-PENTYLRESORCINOL ◇ (1R-trans)-2-(3-METHYL-6-(1-METHYLETHENYL)-2-CYCLOHEXEN-1-YL)-5-PENTYL-1,3-BENZENEDIOL

TOXICITY DATA with REFERENCE
mnt-mus-ipr 50 mg/kg/5D-I PHMGBN 21,277,80
dni-mus-ipr 200 mg/kg RCOCB8 17,703,77
dni-mus:lng 33700 nmol/L CNREA8 36,95,76
dni-mus:bmr 489 µmol/L CNREA8 36,95,76
cyt-mus-ipr 50 mg/kg/5D-I PHMGBN 21,277,80
orl-mus TDLo:50 mg/kg (female 12D post):REP TJADAB 33,195,86
orl-mus TDLo:750 mg/kg (15D male):TER SCIEAS 216,315,82
ivn-mus LD50:50 mg/kg JMCMAR 18,213,75
ivn-mky LD50:212 mg/kg TXAPA9 58,118,81

CONSENSUS REPORTS: EPA Genetic Toxicology Program.

SAFETY PROFILE: Poison by intravenous route. An experimental teratogen. Other experimental reproductive effects. Mutation data reported. When heated to decomposition it emits acrid smoke and fumes.

CBD625 CAS:521-35-7 ***HR: 3***
CANNABINOL
mf: $C_{21}H_{26}O_2$ mw: 310.42

PROP: Leaflets from petr ether. Mp: 76-77°. Sublimes at 4 mm with a bath temp of 180-190°, bp: (0.05) 185°. Insol in water; sol in methanol, ethanol, and aq alkaline solns.

SYNS: 3-AMYL-1-HYDROXY-6,6,9-TRIMETHYL-6H-DIBENZO(b,d) PYRAN ◇ CBN ◇ 6,6,9-TRIMETHYL-3-PENTYL-6H-DIBENZO(b,d) PYRAN-1-OL

TOXICITY DATA with REFERENCE
dni-hmn:hla 10 µmol/L ANTRD4 3,211,83
mnt-mus-ipr 50 mg/kg/5D-I PHMGBN 21,277,80
dni-mus-ipr 200 mg/kg RCOCB8 17,703,77
dni-mus:lng 2300 nmol/L CNREA8 36,95,76
cyt-mus-ipr 50 mg/kg/5D-I PHMGBN 21,277,80
spm-mus-ipr 50 mg/kg/5D-C PHMGBN 18,143,79
orl-mus TDLo:50 mg/kg (female 12D post):REP TJADAB 33,195,86
orl-mus TDLo:50 mg/kg (female 12D post):TER TJADAB 33,195,86
ivn-rbt LDLo:126 mg/kg JPETAB 88,154,46

CONSENSUS REPORTS: EPA Genetic Toxicology Program.

SAFETY PROFILE: Poison by intravenous route. An experimental teratogen. Other experimental reproductive effects. Human mutation data reported. When heated to decomposition it emits acrid smoke and fumes. See also CANNABIS and TETRAHYDROCANNABINOL.

CBD750 CAS:8063-14-7 ***HR: 3***
CANNABIS

PROP: A greenish-black, resinous, bitter substance from *Cannabis sativa*.

SYNS: BHANG ◇ CANNABIS RESIN ◇ CHARAS ◇ CME ◇ GANJA ◇ HASACH ◇ HASHISH ◇ INDIAN CANNABIS ◇ INDIAN HEMP ◇ MARIHUANA ◇ MARIJUANA

TOXICITY DATA with REFERENCE
sln-dmg-orl 1 pph 48NTAS 7,101,81
dlt-dmg-orl 5000 ppm 48NTAS 7,101,81
oms-hmn:lym 500 mg/L JAINAA 24,71,75
cyt-hmn:lym 500 mg/L JAINAA 24,71,75
orl-rat TDLo:3 g/kg (female 2-21D post):REP NETOD7 1,285,79
ihl-rbt TCLo:1440 µg/kg (female 6-18D post):TER FAATDF 7,236,86
orl-hmn TDLo:60 mg/kg/20D:CVS BMJOAE 1,460,78
orl-rat LD50:1380 mg/kg TXAPA9 25,363,73
ipr-mus LDLo:5 g/kg NATUAS 228,134,70

CONSENSUS REPORTS: EPA Genetic Toxicology Program.

SAFETY PROFILE: Moderately toxic by ingestion. An experimental teratogen. Experimental reproductive effects. Human systemic effects by ingestion include: change in heart rate, change in cardiac resting or action potential, and blood pressure decrease. Human mutation data reported. An allergen. When ingested or inhaled as smoke, it can cause euphoria, delirium, hallucinations, drowsiness, weakness, and hyporeflexia. An overdose can cause coma and death. Dried material can burn; can react with oxidizing materials. When heated to decomposition it emits toxic fumes of NO_x. See also TETRAHYDROCANNABINOL and THC.

CBD760 ***HR: 2***
CANNABIS SMOKE RESIDUE

SYN: MARIJUANA, SMOKE RESIDUE

TOXICITY DATA with REFERENCE
dnd-esc 10 ppm MUREAV 89,95,81
scu-rat TDLo:11640 mg/kg/18D-I:ETA VHTODE 21(Suppl),148,79

SAFETY PROFILE: Questionable carcinogen with experimental tumorigenic data. Mutation data reported. When heated to decomposition it emits acrid smoke and irritating fumes.

CBE250 ***HR: 2***
CANTHARIDES
mf: $C_{10}H_{12}O_4$ mw: 196.15

PROP: Brown to black powder or scales. Mp: 218°, bp: subl @ 90°.

SYNS: BLISTERING BEETLES ◇ BLISTERING FLIES ◇ SPANISH FLY

SAFETY PROFILE: Strong irritant via skin contact, ingestion, inhalation, and contact with eyes. An allergen. Can cause conjunctivitis, keratitis, blepharitis, slight swelling of cornea and inflammation of iris. It is often mistakenly used as an aphrodisiac, but it is much too dangerous and irritating a material for this purpose. When heated to decomposition it emits acrid smoke and fumes.

CBE750 CAS:56-25-7 ***HR: 3***
CANTHARIDINE
mf: $C_{10}H_{12}O_4$ mw: 196.22

SYNS: CANTHARIDES CAMPHOR ◇ CANTHARIDIN ◇ CANTHARONE ◇ exo-1,2-cis-DIMETHYL-3,6-EPOXYHEXAHYDROPHTHALIC ANHYDRIDE ◇ 2,3-DIMETHYL-7-OXABICYCLO(2.2.1)HEPTANE-2,3-DICARBOXYLIC ANHYDRIDE ◇ HEXAHYDRO-3A,7A-DIMETHYL-4,7-EPOXYISOBENZOFURAN-1,3-DIONE

TOXICITY DATA with REFERENCE
skn-mus TDLo:25 mg/kg/14W-I:NEO BJCAAI 9,177,55
skn-mus TD:70 mg/kg/52W-I:ETA CNREA8 32,1463,72
orl-hmn LDLo:428 μg/kg 34ZIAG -,646,69
ipr-mus LD50:1 mg/kg JAFCAU 35,823,87
orl-dog LDLo:50 mg/kg FDWU** -,-,31

CONSENSUS REPORTS: IARC Cancer Review: Group 3 IMEMDT 7,56,87; Animal Limited Evidence IMEMDT 10,79,76. EPA Extremely Hazardous Substances List. Reported in EPA TSCA Inventory.

SAFETY PROFILE: A deadly human poison by ingestion. Questionable carcinogen with experimental tumorigenic and neoplastigenic data. See also CANTHARIDES. When heated to decomposition it emits acrid and irritating fumes.

CBF000 CAS:76-90-4 ***HR: 3***
CANTRIL
mf: $C_{21}H_{26}NO_3{\cdot}Br$ mw: 420.39

SYNS: BENZILIC ACID ester with 3-HYDROXY-1,1-DIMETHYLPIPERIDINIUM BROMIDE ◇ CANTIL ◇ GASTROPIDIL ◇ 3-HYDROXY-1,1-DIMETHYLPIPERIDINIUM BROMIDE BENZILATE ◇ 3-((HYDROXYDIPHENYLACETYL)OXY)-1,1-DIMETHYLPIPERIDINIUMBROMIDE ◇ JB 340 ◇ MEPENZOLATE ◇ MEPENZOLATE BROMIDE ◇ N-METHYL-3-PIPERIDYL BENZILATE METHOBROMIDE ◇ N-METHYL-3-PIPERIDYLDIPHENYLGLYCOLATE METHOBROMIDE ◇ 1-METHYL-3-PIPERIDYL ESTER METHOBROMIDE BENZILIC ACID ◇ TRANCOLON

TOXICITY DATA with REFERENCE
orl-rat LD50:742 mg/kg JOPDAB 69,663,66
scu-rat LD50:740 mg/kg TXAPA9 18,185,71
ivn-rat LD50:22 mg/kg 27ZIAQ -,148,73
orl-mus LD50:900 mg/kg 27ZIAQ -,-,65
ivn-mus LD50:9800 μg/kg 27ZIAQ -,-,65
ipr-rat LD50:158 mg/kg NIIRDN 6,358,82
ipr-mus LD50:88 mg/kg NIIRDN 6,358,82
scu-mus LD50:455 mg/kg NIIRDN 6,358,82

SAFETY PROFILE: Poison by intravenous and intraperitoneal routes. Moderately toxic by ingestion and subcutaneous routes. When heated to decomposition it emits very toxic fumes of Br^- and NO_x.

CBF250 CAS:302-22-7 ***HR: 3***
CAP
mf: $C_{23}H_{29}ClO_4$ mw: 404.97

SYNS: 17-ACETOXY-6-CHLORO-6-DEHYDROPROGESTERONE ◇ 17-α-ACETOXY-6-CHLORO-6-DEHYDROPROGESTERONE ◇ 17-α-ACETOXY-6-CHLORO-6,7-DEHYDROPROGESTERONE ◇ 17-α-ACETOXY-6-CHLORO-4,6-PREGNADIENE-3,20-DIONE ◇ 17-α-ACETOXY-6-CHLOROPREGNA-4,6-DIENE-3,20-DIONE ◇ 17-(ACETYLOXY)-6-CHLOROPREGNA-4,6-DIENE-3,20-DIONE ◇ CHLORMADINON ACETATE ◇ CHLORMADINONE ACETATE ◇ CHLORMADINONU (POLISH) ◇ 6-CHLORO-17-α-ACETOXY-4,6-PREGNADIENE-3,20-DIONE ◇ Δ^6-6-CHLORO-17-α-ACETOXYPROGESTERONE ◇ 6-CHLORO-Δ^6-17-ACETOXYPROGESTERONE ◇ 6-CHLORO-Δ^6-(17-α) ACETOXYPROGESTERONE ◇ 6-CHLORO-Δ^6-DEHYDRO-17-ACETOXYPROGESTERONE ◇ 6-CHLORO-6-DEHYDRO-17-α-ACETOXYPROGESTERONE ◇ 6-CHLORO-6-DEHYDRO-17-α-HYDROXYPROGESTERONE ACETATE ◇ 6-CHLORO-17-α-HYDROXYPREGNA-4,6-DIENE-3,20-DIONEACETATE ◇ 6-CHLORO-17-α-HYDROXY-Δ^6-PROGESTERONE ACETATE ◇ CHLOROMADINONE ACETATE ◇ 6-CHLORO-$\Delta^{4,6}$-PREGNADIENE-17-α-OL-3,20-DIONE-17-ACETATE ◇ 6-CHLORO-PREGNA-4,6-DIEN-17-α-OL-3,20-DIONE ACETATE ◇ CLORDION ◇ CMA ◇ C-QUENS ◇ 6-DEHYDRO-6-CHLORO-17-α-ACETOXYPROGESTERONE ◇ LORMIN ◇ LUTINYL ◇ NSC-92338 ◇ RS 1280 ◇ SKEDULE ◇ ST 155

TOXICITY DATA with REFERENCE
orl-wmn TDLo:1200 μg/kg (female 17W pre):REP FESTAS 17,49,66

unr-wmn TDLo:3640 μg/kg (female 52W pre):TER CCPTAY 3,45,71
orl-dog TDLo:182 mg/kg/2Y-C:ETA JAMAAP 219,1601,72
ipr-mus LD50:3 g/kg KSRNAM 11,571,77

CONSENSUS REPORTS: IARC Cancer Review: Animal Limited Evidence IMEMDT 21,365,79; Animal Sufficient Evidence IMEMDT 6,149,74.

SAFETY PROFILE: Suspected carcinogen with experimental carcinogenic and tumorigenic data. Moderately toxic by intraperitoneal route. Human maternal and reproductive effects by ingestion, intramuscular, and possibly other routes: ovary, uterus, cervix, vagina, and fallopian tube changes; menstrual cycle changes or disorders; changes in fertility; and other unspecified female effects. A human teratogen which causes developmental abnormalities of the endocrine system in the fetus. Experimental teratogenic and reproductive effects. An oral contraceptive. When heated to decomposition it emits toxic fumes of Cl^-.

CBF500 CAS:11002-18-9 ***HR: 3***
CAPACIDIN

PROP: Produced from *Streptomycete* isolated from field soil (ANTCAO 10,702,60).

TOXICITY DATA with REFERENCE
ipr-mus LD50:4160 μg/kg ANTCAO 10,702,60
scu-mus LD50:7400 μg/kg ANTCAO 10,702,60

SAFETY PROFILE: Poison by intraperitoneal and subcutaneous routes. When heated to decomposition it emits acrid smoke and fumes.

CBF625 CAS:27276-25-1 ***HR: 2***
CAPOBENATE
mf: $C_{16}H_{22}NO_6$•Na mw: 347.38

SYNS: CAPOBENATE SODIUM ◇ C-3 SODIUM SALT ◇ epsilon-(3,4,5-TRIMETHOXYBENZAMIDO)CAPROIC ACID SODIUM SALT ◇ epsilon-(3,4,5-TRIMETHOXYBENZAMIDO)CAPRONSAEURE NATRIUM (GERMAN) ◇ 6-((3,4,5-TRIMETHOXYBENZOYL)AMINO)HEXANOIC ACID SODIUM SALT

TOXICITY DATA with REFERENCE
ipr-rat LD50:2500 mg/kg USXXAM #3697563
orl-mus LD50:5 g/kg USXXAM #3697563
ipr-mus LD50:3 g/kg USXXAM #3697563
ivn-mus LD50:2500 mg/kg USXXAM #3697563

SAFETY PROFILE: Moderately toxic by intravenous and intraperitoneal routes. Mildly toxic by ingestion. Used as a cardiac anti-arrhythmic. When heated to decomposition it emits toxic fumes of NO_x and Na_2O.

CBF675 CAS:1405-36-3 ***HR: 3***
CAPREOMYCIN DISULFATE

SYNS: CAPROCIN ◇ OGOSTAL

TOXICITY DATA with REFERENCE
ipr-rat LD50:157 mg/kg ANYAA9 135,960,66
scu-rat LD50:1191 mg/kg ANYAA9 135,960,66
ivn-rat LD50:325 mg/kg ANYAA9 135,960,66
scu-mus LD50:514 mg/kg ANYAA9 135,960,66
ivn-mus LD50:250 mg/kg ANYAA9 135,960,66

SAFETY PROFILE: Poison by intravenous and intraperitoneal routes. Moderately toxic by subcutaneous route. When heated to decomposition it emits toxic fumes of SO_x. See also SULFATES.

CBF680 CAS:37280-35-6 ***HR: 3***
CAPREOMYCIN IA
mf: $C_{25}H_{44}N_{14}O_8$ mw: 668.83

SYNS: A-250-II ◇ ANTIBIOTIC 29275 ◇ ANTIBIOTIC A-250-II ◇ CAPROMYCIN ◇ CAPSTAT

TOXICITY DATA with REFERENCE
orl-mus LD50:10 g/kg 85GDA2 4(1),288,80
scu-mus LD50:514 mg/kg 85GDA2 4(1),288,80
ivn-mus LD50:250 mg/kg 85GDA2 4(1),288,80

SAFETY PROFILE: Poison by intravenous route. Moderately toxic by subcutaneous route. When heated to decomposition it emits toxic fumes of NO_x.

CBF700 CAS:105-60-2 ***HR: 3***
CAPROLACTAM
mf: $C_6H_{11}NO$ mw: 113.18

$$\overline{HN(CH_2)_5CO}$$

PROP: White crystals. Mp: 69°, vap press: 6 mm @ 120°.

SYNS: AMINOCAPROIC LACTAM ◇ 6-AMINOHEXANOIC ACID CYCLIC LACTAM ◇ 2-AZACYCLOHEPTANONE ◇ 6-CAPROLACTAM ◇ omega-CAPROLACTAM (MAK) ◇ CAPROLATTAME (FRENCH) ◇ CYCLOHEXANONE ISO-OXIME ◇ EPSYLON KAPROLAKTAM (POLISH) ◇ HEXAHYDRO-2-AZEPINONE ◇ HEXAHYDRO-2H-AZEPIN-2-ONE ◇ 6-HEXANELACTAM ◇ HEXANONE ISOXIME ◇ HEXANONISOXIM (GERMAN) ◇ 1,6-HEXOLACTAM ◇ e-KAPROLAKTAM (CZECH) ◇ 2-KETOHEXAMETHYLENIMINE ◇ NCI-C50646 ◇ 2-OXOHEXAMETHYLENIMINE ◇ 2-PERHYDROAZEPINONE

TOXICITY DATA with REFERENCE
skn-rbt 500 mg/24H MLD 28ZPAK -,149,72
eye-rbt 20 mg/24H MOD 28ZPAK -,149,72
slt-dmg-orl 5 mmol/L PMRSDJ 5,313,85
mmo-smc 100 mg/L PMRSDJ 5,271,85
cyt-hmn:lym 270 mg/L PMRSDJ 5,457,85
orl-rbt TDLo:3450 mg/kg (female 6-28D post):TER JJATDK 7,317,87
orl-rat TDLo:10 g/kg (female 6-15D post):REP JJATDK 7,317,87

ihl-hmn TCLo:100 ppm:PUL AIHAAP 34,384,73
orl-rat LD50:1210 mg/kg NTPTR* NTP-TR-214,82
ihl-rat LC50:300 mg/m^3/2H 85GMAT -,32,82
ihl-rat LC50:300 mg/m^3/2H 85GMAT -,32,82
orl-mus LD50:930 mg/kg GTPZAB 10(10),54,66
ihl-mus LC50:450 mg/m^3 GTPZAB 10(10),54,66
ipr-mus LD50:650 mg/kg JPMSAE 60,1058,71
scu-mus LDLo:750 mg/kg AEXPBL 50,199,1903
skn-rbt LDLo:1438 mg/kg AIHAAP 30,470,69
scu-frg LDLo:2800 mg/kg AEXPBL 50,199,1903

CONSENSUS REPORTS: IARC Cancer Review: Group 4 IMEMDT 7,56,87; Animal No Evidence IMEMDT 39,247,86. Reported in EPA TSCA Inventory.

OSHA PEL: Dust: 1 mg/m^3; STEL 3 mg/m^3; Vapor: 5 ppm; STEL 10 ppm
ACGIH TLV: Dust: 1 mg/m^3; STEL 3 mg/m^3; Vapor: 4.3 ppm; STEL 8.6 ppm; (Proposed: TWA Dust: 1 mg/m^3; 5 ppm (vapor and aerosol); Vapor: 5 ppm; STEL 10 ppm)
DFG MAK: 25 mg/m^3

SAFETY PROFILE: Moderately toxic by ingestion, skin contact, intraperitoneal, and subcutaneous routes. Human systemic effects by inhalation: cough. An experimental teratogen. Other experimental reproductive effects. Human mutation data reported. A skin and eye irritant. Potentially explosive reaction with acetic acid + dinitrogen trioxide. When heated to decomposition it emits toxic fumes of NO_x.

CBF710 CAS:52622-27-2 ***HR: 1***
CAPRYLIC/CAPRIC TRIGLYCERIDE

SYNS: CAPTEX 300 ◇ MIGLYOL 810 NEUTRAL OIL ◇ MIGLYOL 812 NEUTRAL OIL ◇ MYRITOL 318 ◇ NEOBEE M-5 ◇ NEOBEE O ◇ OCTANOIC/DECANOIC ACID TRIGLYCERIDE ◇ VEGETABLE OIL 1400

TOXICITY DATA with REFERENCE
eye-rbt 100 mg/24H MLD JEPTDQ 4(4),105,80

SAFETY PROFILE: An eye irritant. When heated to decomposition it emits acrid smoke and irritating fumes.

CBF725 CAS:5299-65-0 ***HR: 3***
4-CAPRYLMORPHOLINE
mf: $C_{14}H_{27}NO_2$ mw: 241.42

SYNS: AI3-18285 ◇ 4-DECANOYLMORPHOLINE ◇ MORPHOLINE, 4-DECANOYL- ◇ MORPHOLINE, 4-(1-OXODECYL)-(9CI)

TOXICITY DATA with REFERENCE
eye-rbt 100 mg MLD NTIS** AD-A002-053
ivn-mus LD50:18 mg/kg CSLNX* NX#08964

SAFETY PROFILE: Poison by intravenous route. An eye irritant. When heated to decomposition it emits toxic fumes of NO_x.

CBF750 CAS:404-86-4 ***HR: 3***
CAPSAICIN
mf: $C_{18}H_{27}NO_3$ mw 305.46

PROP: Monoclinic, rectangular plates and scales. Mp: 65°C, bp: 210-220°C. Freely soluble in ethanol, ether, benzene, chloroform; sltly soluble in carbon disulfide; insoluble in water. Highly volatile with a pungent odor.

SYNS: CAPSAICINE ◇ N-((4-HYDROXY-3-METHOXYPHENYL) METHYL)-8-METHYL-6-NONENAMIDE ◇ trans-N-((4-HYDROXY-3-METHOXYPHENYL)METHYL)-8-METHYL-6-NONEAMIDE ◇ trans-8-METHYL-N-VANILLYL-6-NONEAMIDE ◇ NCI-C56564

TOXICITY DATA with REFERENCE
mma-sat 10 μg/plate ENMUDM 7,881,85
mnt-mus-ipr 7500 μg/kg ENMUDM 7,881,85
dni-mus-ipr 1800 μg/kg ENMUDM 7,881,85
ipr-rat LD50:9500 μg/kg TOXIA6 18,215,80
orl-mus LD50:47200 μg/kg YAHOA3 25,191,81
ipr-mus LD50:6500 μg/kg TOXIA6 18,215,80
scu-mus LD50:9000 μg/kg TOXIA6 18,215,80
ivn-mus LD50:400 μg/kg YAHOA3 25,101,81
ims-mus LD50:7800 μg/kg TOXIA6 18,215,80
itr-mus LD50:1600 μg/kg TOXIA6 18,215,80
ipr-gpg LD50:1100 μg/kg TOXIA6 18,215,80

CONSENSUS REPORTS: Reported in EPA TSCA Inventory.

SAFETY PROFILE: Deadly poison by intravenous and intraperitoneal routes. Poison by ingestion, subcutaneous, intramuscular and intratracheal routes. Mutation data reported. Capsaicin produced erythema and burning without blistering the human skin. Capsicum is considered a moderate irritant to human skin and a strong irritant to gastric mucosa. Irritating to mucous membranes; produces severe gastritis and diarrhea. Intragastric infusion of capsaicin in humans increased the DNA content of the gastric aspirate. Capsaicin inhibits transplanted tumors in mice. Capsicum chiles fed to rats produced tumors in 15 of 26 animals. It is the component in peppers which makes them hot. When heated to decomposition it emits toxic fumes of NO_x.

CBF800 CAS:2425-06-1 ***HR: 3***
CAPTAFOL
mf: $C_{10}H_9Cl_4NO_2S$ mw: 349.06

SYNS: CAPTOFOL ◇ DIFOLATAN ◇ DIFOSAN ◇ FOLCID ◇ ORTHO 5865 ◇ SANSPOR ◇ SULFONIMIDE ◇ SULPHEIMIDE ◇ N-(1,1,2,2-TETRACHLORAETHYLTHIO)CYCLOHEX-4-EN-1,4-DIACARBOXIMID (GERMAN) ◇ N-(1,1,2,2-TETRACHLORAETHYLTHIO)TETRAHYDROPHTHALAMID (GERMAN) ◇ N-1,1,2,2-TETRACHLOROETHYL-MERCAPTO-4-CYCLOHEXENE-1,2-CARBOXIMIDE ◇ N-((1,1,2,2-TETRACHLOROETHYL)SULFENYL)-cis-4-CYCLOHEXENE-1,2-DICARBOXIMIDE ◇ N-(1,1,2,2-TETRACHLOROETHYLTHIO)-4-CYCLOHEXENE-1,2-DICARBOXIMIDE

TOXICITY DATA with REFERENCE
mmo-esc 50 μg/plate MUREAV 40,19,76
mma-esc 50 μg/plate MUREAV 116,185,83
mrc-bcs 100 ng/disc/24H MUREAV 40,19,76
dlt-rat-ipr 25 mg/kg/5D FCTXAV 10,353,72
cyt-ham:lng 10 μmol/L MUREAV 78,177,80
sce-ham:lng 2 μmol/L MUREAV 78,177,80
orl-rat TDLo:11 g/kg (multi) :REP TXAPA9 13,420,68
orl-ham TDLo:200 mg/kg (female 8D post):TER TXAPA9 16,24,70
orl-mus TDLo:60480 mg/kg/96W-C:CAR GANNA2 75,853,84
orl-rat LD50:2500 mg/kg WRPCA2 9,119,70
ipr-mus LDLo:3 mg/kg FCTXAV 13,55,75

CONSENSUS REPORTS: EPA Genetic Toxicology Program.

OSHA PEL: TWA 0.1 mg/m^3
ACGIH TLV: TWA 0.1 mg/m^3

SAFETY PROFILE: Poison by intraperitoneal route. Moderately toxic by ingestion. An experimental teratogen. Other experimental reproductive effects. Questionable carcinogen with experimental carcinogenic data. Mutation data reported. A fungicide. When heated to decomposition it emits very toxic fumes of Cl^-, NO_x, and SO_x.

CBF825 CAS:1892-80-4 ***HR: 3***
CAPTAGON HYDROCHLORIDE
mf: $C_{18}H_{23}N_5O_2 \cdot ClH$ mw: 377.92

SYNS: AMFETYLINE HYDROCHLORIDE ◇ BZT ◇ FENETHYLLINE HYDROCHLORIDE ◇ 7-(2-((α-METHYLPHENETHYL)AMINO) ETHYL)THEOPHYLLINE HYDROCHLORIDE ◇ 7-(2-(1-METHYL-2-PHENETHYLAMINO)ETHYL)THEOPHYLLINEHYDROCHLORIDE ◇ 7-(PHENYL-ISOPROPYL-AMINO-AETHYL)-THEOPHYLLIN-HYDROCHLORID (GERMAN)

TOXICITY DATA with REFERENCE
orl-rat LD50:100 mg/kg 27ZQAG -,230,72
ipr-rat LD50:57 mg/kg 27ZQAG -,230,72
scu-rat LD50:196 mg/kg 27ZQAG -,230,72
orl-mus LD50:347 mg/kg 27ZQAG -,230,72
ipr-mus LD50:347 mg/kg ARZNAD 8,190,58
scu-mus LD50:80 mg/kg 27ZQAG -,230,72
ivn-mus LD50:55 mg/kg 27ZQAG -,230,72

SAFETY PROFILE: Poison by ingestion, subcutaneous, intravenous, and intraperitoneal routes. When heated to decomposition it emits toxic fumes of NO_x and HCl. See also THEOPHYLLINE and other theophylline entries.

CBG000 CAS:133-06-2 ***HR: 3***
CAPTAN
mf: $C_9H_8Cl_3NO_2S$ mw: 300.59

PROP: Odorless crystals. Insol in water; sol in benzene and chloroform.

SYNS: AACAPTAN ◇ AGROSOL S ◇ AGROX 2-WAY and 3-WAY ◇ AMERCIDE ◇ BANGTON ◇ BEAN SEED PROTECTANT ◇ CAPTAF ◇ CAPTANCAPTENEET 26,538 ◇ CAPTANE ◇ CAPTAN-STREPTOMYCIN 7.5-0.1 POTATO SEED PIECE PROTECTANT ◇ CAPTEX ◇ ENT 26,538 ◇ ESSO FUNGICIDE 406 ◇ FLIT 406 ◇ FUNGUS BAN TYPE II ◇ GLYODEX 3722 ◇ GRANOX PPM ◇ GUSTAFSON CAPTAN 30-DD ◇ HEXACAP ◇ KAPTAN ◇ LE CAPTANE (FRENCH) ◇ MALIPUR ◇ MERPAN ◇ MICRO-CHECK 12 ◇ NCI-C00077 ◇ NERACID ◇ ORTHOCIDE ◇ OSOCIDE ◇ SR406 ◇ STAUFFER CAPTAN ◇ 3a,4,7,7a-TETRAHYDRO-N-(TRICHLOROMETHANESULPHENYL)PHTHALIMIDE ◇ 3a,4,7,7a-TETRAHYDRO-2-((TRICHLOROMETHYL)THIO)-1H-ISOINDOLE-1,3(2H)-DIONE ◇ 1,2,3,6-TETRAHYDRO-N-(TRICHLOROMETHYLTHIO)PHTHALIMIDE ◇ N-(TRICHLOR-METHYLTHIO)-PHTHALIMID (GERMAN) ◇ N-TRICHLOROMETHYLMERCAPTO-4-CYCLOHEXENE-1,2-DICARBOXIMIDE ◇ N-(TRICHLOROMETHYLMERCAPTO)-Δ^4-TETRAHYDROPHTHALIMIDE ◇ N-TRICHLOROMETHYLTHIOCYCLOHEX-4-ENE-1,2-DICARBOXIMIDE ◇ N-TRICHLOROMETHYLTHIO-cis-Δ^4-CYCLOHEXENE-1,2-DICARBOXIMIDE ◇ N-((TRICHLOROMETHYL)THIO)-4-CYCLOHEXENE-1,2-DICARBOXIMIDE ◇ TRICHLOROMETHYLTHIO-1,2,5,6-TETRAHYDROPHTHALAMIDE ◇ N-((TRICHLOROMETHYL)THIO) TETRAHYDROPHTHALIMIDE ◇ N-TRICHLOROMETHYLTHIO-3A,4,7,7A-TETRAHYDROPHTHALIMIDE ◇ VANCIDE 89 ◇ VANGARD K ◇ VANICIDE ◇ VONDCAPTAN

TOXICITY DATA with REFERENCE
mmo-sat 310 ng/plate MUREAV 130,79,84
cyt-hmn:lng 10 mg/L ANYAA9 160,344,69
sce-hmn:lym 30 μmol/L MUREAV 79,53,80
oms-ctl:lvr 1 mmol/L CBINA8 56,289,85
orl-mus TDLo:250 mg/kg (male 5D pre):REP TXAPA9 23,277,72
orl-ham TDLo:200 mg/kg (female 8D post):TER TXAPA9 16,24,70
orl-mus TDLo:1075 g/kg/80W-C:NEO NCITR* NCI-TR-15,77
orl-mus TD:540 g/kg/80W-C:ETA NCITR* NCI-TR-15,77
orl-hmn LDLo:1071 mg/kg 34ZIAG -,151,69
orl-rat LD50:9 g/kg ARSIM* 20,6,66
ihl-mus LC50:5000 mg/m^3/2H TXAPA9 45,320,78
ipr-mus LD50:30 mg/kg ZBPHA6 234,110,76

CONSENSUS REPORTS: IARC Cancer Review: Group 3 IMEMDT 7,56,87; Animal Limited Evidence IMEMDT 30,295,83. NCI Carcinogenesis Bioassay (feed); Clear Evidence: mouse NCITR* NCI-CG-TR-15,77; No Evidence: rat NCITR* NCI-CG-TR-15,77. EPA Genetic Toxicology Program. Community Right-To-Know List. Reported in EPA TSCA Inventory.

OSHA PEL: TWA 5 mg/m^3
ACGIH TLV: TWA 5 mg/m^3

SAFETY PROFILE: Poison by intraperitoneal route. Moderately toxic to humans by ingestion. Moderately toxic experimentally by ingestion and inhalation routes. Experimental teratogenic and reproductive effects. Questionable carcinogen with experimental tumorigenic and neoplastigenic data. Human mutation data reported. When heated to decomposition it emits toxic fumes of Cl^-, SO_x, and NO_x.

CBG075 CAS:81424-67-1 **HR: 3**
CARACEMIDE
mf: $C_6H_{11}N_3O_4$ mw: 189.20

SYNS: N-ACETYL-N-(METHYLCARBAMOYLOXY)-N-METHYLUREA ◇ N-((METHYLAMINO)CARBONYL)-N-(((METHYLAMINO)CARBONYL)OXY)ACETAMIDE ◇ NSC-253272

TOXICITY DATA with REFERENCE
orl-mus LD50:388 mg/kg NCISP* JAN86
ipr-mus LD50:167 mg/kg NCISP* JAN86
ivn-mus LD50:238 mg/kg NTIS** PB84-152032

SAFETY PROFILE: Poison by ingestion, intravenous, and intraperitoneal routes. When heated to decomposition it emits toxic fumes of NO_x.

CBG125 CAS:8028-89-5 **HR: D**
CARAMEL

PROP: Dark brown to black liquid or solid; burnt sugar odor, pleasant bitter taste. Sol in water (colloidal).

SYN: CARAMEL COLOR

TOXICITY DATA with REFERENCE
mma-sat 50 mg/plate FCTOD7 22,623,84
cyt-ham:fbr 8 g/L FCTOD7 22,623,84
cyt-ham:lng 6900 mg/L GMCRDC 27,95,81

CONSENSUS REPORTS: Reported in EPA TSCA Inventory.

SAFETY PROFILE: Mutation data reported. When heated to decomposition it emits acrid smoke and irritating fumes.

CBG250 CAS:125-86-0 **HR: 3**
CARAMIPHEN ETHANE DISULFONATE
mf: $C_{18}H_{27}NO_2 \cdot 1/2C_2H_6O_6S_2$ mw: 479.66

SYNS: BIS(1-(CARBO-β-DIETHYLAMINOETHOXY)-1-PHENYLCYCLOPENTANE)ETHANE DISULFONATE ◇ BIS(1-(2-DIETHYLAMINOETHOXYCARBONYL)-1-PHENYLCYCLOPENTANE)ETHANE DISULFONATE ◇ DIETHYLAMINOETHYL-1-PHENYLCYCLOPENTANE-1-CARBOXYLATE ETHANE DISULFONATE ◇ PARANIT ETHANE DISULFONATE ◇ 1-PHENYLCYCLOPENTANECARBOXYLIC ACID 1-DIETHYLAMINOETHYL ESTER, 1,2-ETHANE DISULFONATE ◇ SKF No. 769-J(2) ◇ TAORYL ◇ TORYN

TOXICITY DATA with REFERENCE
orl-rat LDLo:1400 mg/kg CLDND* -,363,72
orl-mus LD50:485 mg/kg 27ZQAG -,363,72
ipr-mus LD50:240 mg/kg 27ZQAG -,363,72
ivn-mus LD50:67 mg/kg 27ZQAG -,363,72
ivn-rbt LD50:12 mg/kg 27ZQAG -,363,72

SAFETY PROFILE: Poison by intraperitoneal and intravenous routes. Moderately toxic by ingestion. When heated to decomposition it emits very toxic fumes of SO_x and NO_x. See also SULFONATES.

CBG375 CAS:57554-34-4 **HR: 3**
CARAMIPHEN HYDROCHLORIDE
mf: $C_{18}H_{27}NOS \cdot ClH$ mw: 341.98

SYN: CARAMIFENE (ITALIAN)

TOXICITY DATA with REFERENCE
ims-rat LD50:1148 μg/kg BJPCBM 39,822,70
orl-mus LD50:180 mg/kg BCFAAI 111,293,72
ipr-mus LD50:339 mg/kg EJMCA5 10,262,75
ims-mus LD50:651 μg/kg BJPCBM 39,822,70
ims-gpg LD50:115 μg/kg BJPCBM 39,822,70

SAFETY PROFILE: Poison by ingestion, intramuscular, and intraperitoneal routes. When heated to decomposition it emits toxic fumes of SO_x, NO_x, and HCl.

CBG500 CAS:8000-42-8 **HR: 2**
CARAWAY OIL

PROP: The main constituent of caraway oil is 1-carvone; found in the fruits of *Carum carvi* L. (Fam. *Umbelliferae*). (FCTXAV 11,1011,73). Colorless liquid; odor and taste of caraway.

SYNS: KUEMMEL OIL (GERMAN) ◇ OIL of CARAWAY

TOXICITY DATA with REFERENCE
skn-rbt 500 mg/24H FCTXAV 11,1051,73
mmo-sat 5 μg/plate KEKHB8 (9),11,79
orl-rat LD50:3500 mg/kg FCTXAV 11,1051,73
skn-rbt LD50:1780 mg/kg FCTXAV 11,1051,73

CONSENSUS REPORTS: Reported in EPA TSCA Inventory.

SAFETY PROFILE: Moderately toxic by ingestion and skin contact. A skin irritant. Mutation data reported. When heated to decomposition it emits acrid smoke and irritating fumes. See also 1-6,8(9)-p-MENTHADIEN-2-ONE (1-CARVONE)

CBH250 CAS:51-83-2 **HR: 3**
CARBACHOL CHLORIDE
mf: $C_6H_{15}N_2O_2 \cdot Cl$ mw: 182.68

SYNS: 2-((AMINOCARBONYL)OXY)-N,N,N-TRIMETHYLETHANAMINIUM CHLORIDE ◇ CARBACHOL ◇ CARBACHOLIN ◇ CARBACHOLINE CHLORIDE ◇ CARBACOLINA ◇ CARBAMIC ACID, ESTER with CHOLINE CHLORIDE ◇ CARBAMINOCHOLINE CHLORIDE ◇ CARBAMINOYLCHOLINE CHLORIDE ◇ CARBAMIOTIN ◇ CARBAMOYLCHOLINE CHLORIDE ◇ Γ-CARBAMOYL CHOLINE CHLORIDE ◇ CARBAMYLCHOLINE CHLORIDE ◇ CARBOCHOL ◇ CARBOCHOLIN ◇ CARBYL ◇ CARCHOLIN ◇ CHOLINE CARBAMATE CHLORIDE ◇ CHOLINE CHLORINE CARBAMATE ◇ CHOLINE, CHLORIDE CARBAMATE(ESTER) ◇ COLEYTL ◇ DORYL (PHARMACEUTICAL) ◇ (2-HYDROXYETHYL)TRIMETHYL

AMMONIUM CHLORIDE CARBAMATE ◇ ISOPTO CARBACHOL ◇ JESTRYL ◇ LENTIN ◇ LENTINE (FRENCH) ◇ MIOSTAT ◇ MISTURA C ◇ MORYL ◇ P.V. CARBACHOL ◇ TL 457 ◇ VASOPERIF

TOXICITY DATA with REFERENCE
ivn-man TDLo:1428 ng/kg:CVS,GIT CRSBAW 113,79,33
ims-hmn TDLo:6 μg/kg:EYE,CVS,SKN SCALA9 36,1,33
ims-man TDLo:2857 ng/kg:CVS CRSBAW 113,79,33
orl-rat LD50:40 mg/kg JPETAB 58,337,36
ipr-rat LD50:2 mg/kg AIPTAK 149,560,64
scu-rat LD50:4 mg/kg JPETAB 58,337,36
ivn-rat LD50:100 μg/kg JPETAB 58,337,36
orl-mus LD50:15 mg/kg NIIRDN 6,182,82
ipr-mus LD50:370 μg/kg ATXKA8 29,39,72
scu-mus LD50:3 mg/kg JPETAB 58,337,36
ivn-mus LD50:300 μg/kg JPETAB 58,337,36
orl-dog LDLo:3 mg/kg AEPPAE 164,346,32

CONSENSUS REPORTS: EPA Extremely Hazardous Substances List. Reported in EPA TSCA Inventory.

SAFETY PROFILE: Deadly poison by subcutaneous, intravenous, and intraperitoneal routes. Poison by ingestion and possibly other routes. Human systemic effects by intravenous and intramuscular routes including: lowered blood pressure, venous dilation, nausea or vomiting, sweating and lacrimation (increased flow of tears). A cholinergic agent (parasympathetic nerve stimulant). When heated to decomposition it emits very toxic fumes of Cl^-, NH_3, and NO_x. See also CARBAMATES.

CBH500 CAS:5942-95-0 ***HR: 3***
CARBADIPIMIDINE
mf: $C_{28}H_{38}N_4O$ mw: 446.70

SYNS: CARPIPRAMINE ◇ 10,11-DIHYDRO-5-(3-(4-PIPERIDINO-4-CARBAMOYLPIPERIDINO)PROPYL-(b,f)AZEPINE

TOXICITY DATA with REFERENCE
orl-rat LD50:1025 mg/kg 27ZQAG -,63,72
ipr-rat LD50:76 mg/kg 27ZQAG -,63,72
ivn-rat LD50:37 mg/kg 28ZQAG -,63,72
orl-mus LD50:2180 mg/kg 27ZQAG -,63,72
ipr-mus LD50:136 mg/kg 27ZQAG -,38,72
ivn-mus LD50:28 mg/kg 28ZQAG -,63,72
ivn-rbt LD50:18 mg/kg 27ZQAG -,63,72

SAFETY PROFILE: Poison by intraperitoneal and intravenous route. Moderately toxic by ingestion. When heated to decomposition it emits very toxic fumes of NO_x.

CBH750 ***HR: 3***
CARBAMATES

PROP: Compounds based upon carbamic acid, NH_2COOH. Used only in the form of its numerous salts and derivatives.

SAFETY PROFILE: Many carbamates are poisons or moderately toxic, and some are carcinogenic, teratogenic, or mutagenic. They are used as insecticides, fungicides, herbicides, and as accelerators in the vulcanization of rubber. There is little data on persistence or breakdown in the environment.

The N-alkylcarbamates and thiocarbamates can react with nitrite under mildly acid conditions to form N-nitroso compounds. Nitrite is found in soils, in human saliva, and in cured meats. N-nitrosodimethylamine is formed by soil microorganisms from Thiram. Other N-nitroso compounds could similarly be formed from other carbamate pesticides. However, the extent of the reaction of carbamates and nitrite in man is not known. The N-nitrosodialkylamines formed from dialkylthiocarbamate pesticides and nitrite are potent animal carcinogens and mutagens. The N-nitroso derivatives of several N-alkylcarbamates produce cancers in experimental animals at small doses.

Carbaryl, semicarbazide hydrochloride, n-propyl carbamate, Maneb, Zineb, Ferbam, and Thiram are experimental teratogens.

Many of the carbamates have central nervous system effects. Carbaryl and Zectran are acetylcholinesterase inhibitors.

Ethylenethiourea, which produces thyroid carcinomas in rats and liver cell tumors in mice by ingestion, is formed from ethylenebisdithiocarbamates such as Maneb and Zineb by metabolic processes and cooking.

See also individual compounds, NITROSAMINES, and N-NITROSO COMPOUNDS.

CBI000 CAS:709-90-0 ***HR: 2***
CARBAMIC ACID-α-METHYLPHENETHYL ESTER
mf: $C_{10}H_{13}NO_2$ mw: 179.24

SYNS: BETAQUIL ◇ SA 217

TOXICITY DATA with REFERENCE
orl-rat LD50:1100 mg/kg ARZNAD 13,856,63
orl-mus LD50:800 mg/kg ARZNAD 13,856,63

SAFETY PROFILE: Moderately toxic by ingestion. See also ESTERS and CARBAMATES. When heated to decomposition it emits toxic fumes of NO_x.

CBI250 CAS:120-02-5 ***HR: 3***
4-CARBAMIDOPHENYL BIS(CARBOXYMETHYLTHIO)ARSENITE
mf: $C_{11}H_{13}AsN_2O_5S_2$ mw: 392.30

SYNS: 2,2'-((4-((AMINOCARBONYL)AMINO)PHENYL)ARSINIDENE) BIS(THIO)BISACETIC ACID ◇ BIS(CARBOXYMETHYLMERCAPTO)(p-UREIDOPHENYL)ARSINE ◇ BIS(CARBOXYMETHYLTHIO)(p-UREIDOPHENYL)ARSINE ◇ (p-CARBAMOYLAMINO)PHENYLARSINOBIS (2-THIO-ACETIC ACID) ◇ CC 914 ◇ C.C. No. 914 ◇ MERCAPTO-ACETIC ACID, DIESTER with DITHIO-p-UREIDOBENZENEARSONOUS

ACID ◇ PHENYL UREA-p-DI(CARBOXYMETHYL) THIOARSENITE ◇ THIOCARBARSONE ◇ (p-UREIDOPHENYLARSYLENEDITHIO) DIACETIC ACID

TOXICITY DATA with REFERENCE
orl-rat LD50:1000 mg/kg JPETAB 91,112,47
ipr-rat LD50:75 mg/kg JPETAB 91,112,47
ivn-rat LD50:29 mg/kg JPETAB 91,112,47
ipr-mus LD50:100 mg/kg JPETAB 91,112,47
ivn-mus LD50:43 mg/kg JPETAB 91,112,47
ivn-rbt LDLo:100 mg/kg JPETAB 91,112,47

CONSENSUS REPORTS: Arsenic and its compounds are on the Community Right-To-Know List.

OSHA PEL: TWA 0.5 mg(As)/m^3
ACGIH TLV: TWA 0.2 mg(As)/m^3

SAFETY PROFILE: Poison by intraperitoneal and intravenous routes. Moderately toxic by ingestion. See also ARSENIC COMPOUNDS, MERCAPTANS, and ESTERS. When heated to decomposition it emits very toxic fumes of As and SO_x.

CBI500 CAS:2490-89-3 ***HR: 3***
4-CARBAMIDOPHENYLOXOARSINE
mf: $C_7H_7AsN_2O_2$ mw: 226.08

SYNS: 1-(p-ARSENOPHENYL)UREA ◇ p-CARBAMIDOPHENYL ARSENOUS ACID ◇ p-CARBAMIDOPHENYL ARSENOUS OXIDE ◇ CARBARSONE OXIDE ◇ CHEMOTHERAPY CENTER No. 606

TOXICITY DATA with REFERENCE
eye-rbt 2 mg JPETAB 82,377,44
orl-rat LD50:510 mg/kg FEPRA7 5,162,46
ipr-rat LD50:55 mg/kg JPETAB 91,112,47
ivn-rat LD50:17 mg/kg JPETAB 91,112,47
ipr-mus LD50:59 mg/kg JPETAB 91,112,47
ivn-mus LD50:41 mg/kg JPETAB 91,112,47
ivn-rbt LDLo:20 mg/kg JPETAB 91,112,47

CONSENSUS REPORTS: Arsenic and its compounds are on the Community Right-To-Know List.

OSHA PEL: TWA 0.5 mg(As)/m^3

SAFETY PROFILE: Poison by intraperitoneal and intravenous routes. Moderately toxic by ingestion. An eye irritant. When heated to decomposition it emits very toxic fumes of As and NO_x. See also ARSENIC COMPOUNDS.

CBI675 CAS:21704-46-1 ***HR: 3***
CARBAMIMIDOTHIOIC ACID, ETHYL ESTER, MONO(DIETHYL PHOSPHATE)
mf: $C_4H_{11}O_4P \cdot C_3H_8N_2S$ mw: 258.31

SYN: S-ETHYLISOTHIURONIUM DIETHYL PHOSPHATE

TOXICITY DATA with REFERENCE
ipr-rat LD50:113 mg/kg FATOAO 43,212,80
orl-mus LD50:2380 mg/kg FATOAO 43,212,80
ipr-mus LD50:680 mg/kg FATOAO 43,212,80
scu-mus LD50:705 mg/kg FATOAO 43,212,80
ims-mus LD50:772 mg/kg FATOAO 43,212,80

SAFETY PROFILE: Poison by intraperitoneal route. Moderately toxic by ingestion, subcutaneous, and intramuscular routes. When heated to decomposition it emits toxic fumes of NO_x, PO_x, and SO_x. See also ESTERS and PHOSPHATES.

CBJ000 CAS:121-59-5 ***HR: 3***
N-CARBAMOYLARSANILIC ACID
mf: $C_7H_9AsN_2O_4$ mw: 260.10

PROP: White, nearly odorless powder; slt acid taste; sol in alc and water. Mp: 174°.

SYNS: AMABEVAN ◇ AMEBAN ◇ AMEBARSONE ◇ AMIBIARSON ◇ AMINARSON ◇ AMINARSONE ◇ AMINOARSON ◇ (4-((AMINOCARBONYL)AMINO)PHENYL)ARSONIC ACID ◇ ARSAMBIDE ◇ p-ARSONOPHENYLUREA ◇ p-CARBAMIDOBENZENEARSONIC ACID ◇ p-CARBAMINO PHENYL ARSONIC ACID ◇ CARBAMINOPHENYL-p-ARSONIC ACID ◇ 4-CARBAMYLAMINOPHENYLARSONIC ACID ◇ N-CARBAMYL ARSANILIC ACID ◇ CARBARSONE (USDA) ◇ CARBASONE ◇ FENARSONE ◇ HISTOCARB ◇ LEUCARSONE ◇ p-UREIDOBENZENEARSONIC ACID ◇ 4-UREIDO-1-PHENYLARSONIC ACID

TOXICITY DATA with REFERENCE
orl-rat TDLo:5000 mg/kg:ETA CNREA8 26,619,66
orl-rat LD50:510 mg/kg MEIEDD 10,246,83
ipr-rat LDLo:1000 mg/kg JPETAB 80,393,44
orl-cat LDLo:250 mg/kg PSEBAA 29,125,31
orl-rbt LDLo:200 mg/kg PSEBAA 29,125,31
orl-gpg LDLo:200 mg/kg PSEBAA 29,125,31

CONSENSUS REPORTS: Arsenic and its compounds are on the Community Right-To-Know List.

OSHA PEL: TWA 0.5 mg(As)/m^3
ACGIH TLV: TWA 0.2 mg(As)/m^3

SAFETY PROFILE: Poison by ingestion. Moderately toxic by intraperitoneal route. Questionable carcinogen with experimental tumorigenic data. See also ARSENIC COMPOUNDS. When heated to decomposition it emits very toxic fumes of As and NO_x.

CBJ750 CAS:618-25-7 ***HR: 3***
N-(CARBAMOYLMETHYL)ARSANILIC ACID
mf: $C_8H_{11}AsN_2O_4$ mw: 274.13

PROP: White, crystalline powder.

SYNS: (4-((2-AMINO-2-OXOETHYL)AMINO)PHENYL)ARSONIC ACID ◇ 4-ARSONOPHENYLGLYCINAMIDE ◇ p-((CARBAMOYLMETHYL)AMINO)-BENXENEARSONIC ACID ◇ SODIUM-N-PHENYLGLYCINAMIDE-p-ARSONATE ◇ TRYPARSAMIDE

TOXICITY DATA with REFERENCE
ivn-rat LDLo:2000 mg/kg JPETAB 63,122,38

ims-rat LDLo:250 mg/kg JPETAB 63,122,38
ivn-mus LD50:4 g/kg THERAP 2,28,47
orl-cat LDLo:200 mg/kg PSEBAA 29,125,31
orl-rbt LDLo:200 mg/kg PSEBAA 29,125,31
ivn-rbt LD50:700 mg/kg JPETAB 80,93,44
orl-gpg LDLo:150 mg/kg PSEBAA 29,125,31

CONSENSUS REPORTS: Arsenic and its compounds are on the Community Right-To-Know List.

OSHA PEL: TWA 0.5 mg(As)/m^3
ACGIH TLV: TWA 0.2 mg(As)/m^3

SAFETY PROFILE: Poison by ingestion and intramuscular route. Moderately toxic by intravenous route. See also ARSENIC COMPOUNDS. When heated to decomposition it emits very toxic fumes of As and NO_x.

CBK000 CAS:817-99-2 ***HR: 3***
N-(CARBAMOYLMETHYL)-2-DIAZOACETAMIDE
mf: $C_4H_6N_4O_2$ mw: 142.14

SYNS: N-(2-AMINO-2-OXOETHYL)-2-DIAZOACETAMIDE ◇ N-DIAZOACETILGLICINA-AMIDE (ITALIAN) ◇ DIAZOACETYL-GLYCINAMIDE ◇ N-(DIAZOACETYL)GLYCINAMIDE ◇ DIAZO-ACETYLGLYCINE AMIDE ◇ N-DIAZOACETYLGLYCINE AMIDE

TOXICITY DATA with REFERENCE
mmo-sat 10 μg/plate AMACCQ 6,655,74
mma-sat 10 μg/plate PNASA6 72,5135,75
dnd-rat-ipr 3700 μg/kg BSIBAC 57,414,81
dnd-mus:fbr 620 μmol/L TOLED5 1,115,77
dni-mus/ast 1500 mg/kg BCPCA6 23,289,74
ipr-mus TDLo:720 mg/kg/4D-I:CAR BSIBAC 45,227,69
ipr-mus LD50:2630 mg/kg ARZNAD 23,690,73

CONSENSUS REPORTS: EPA Genetic Toxicology Program.

SAFETY PROFILE: Moderately toxic by intraperitoneal route. Questionable carcinogen with experimental carcinogenic data. Mutation data reported. When heated to decomposition it emits toxic fumes of NO_x. See also AMIDES.

CBK125 CAS:475-08-1 ***HR: 3***
2-CARBAMOYL-2-NITROACETONITRILE
mf: $C_3H_3N_3O_3$ mw: 129.08

$$H_2NCO \cdot CH(NO_2)C \equiv N$$

CONSENSUS REPORTS: Cyanide and its compounds are on the Community Right-To-Know List.

SAFETY PROFILE: A heat-sensitive explosive. When heated to decomposition it emits toxic fumes of NO_x and CN^-. See also NITRILES.

CBK250 CAS:42242-72-8 ***HR: D***
N-(1-CARBAMOYL-4-(NITROSOCYANAMIDO) BUTYL)BENZAMIDE
mf: $C_{13}H_{15}N_5O_3$ mw: 289.33

SYNS: N-(4-BENZAMIDO-4-CARBAMOYLBUTYL)-N-NITROSOCYANAMIDE ◇ 4-BENZOYLAMIDO-4-CARBOXAMIDO-n(N-NITROSO)-BUTYLCYANAMIDE ◇ BENZOYL-l-ARGININEAMIDE, NITROSATED

TOXICITY DATA with REFERENCE
mmo-sat 156 μmol/L GANNA2 65,45,74
mmo-esc 50 nmol/plate MUREAV 49,9,78

SAFETY PROFILE: Mutation data reported. Many N-nitroso compounds are carcinogens. When heated to decomposition it emits toxic fumes of NO_x. See also N-NITROSO COMPOUNDS.

CBK500 CAS:533-06-2 ***HR: 3***
1-CARBAMOYLOXY-2-HYDROXY-3(o-METHYLPHENOXY)PROPANE
mf: $C_{11}H_{15}NO_4$ mw: 225.27

SYNS: 2-HYDROXY-3-o-TOLYLOXYPROPYL-1-CARBAMATE ◇ KIMAVOXYL ◇ MC 2303 ◇ MEPHENESIN CARBAMATE ◇ 3-(2-METHYLPHENOXY)-1,2-PROPANEDIOL 1-CARBAMATE ◇ SQ 2303 ◇ 3-o-TOLOXY-2-HYDROXYPROPYL-1-CARBAMATE ◇ 3-o-TOLOXY-1,2-PROPANEDIOL-1-CARBAMIC ACID ESTER ◇ TOLSERAM ◇ 3-o-TOLYLOXY-2-HYDROXYPROPYL-1-CARBAMATE

TOXICITY DATA with REFERENCE
orl-rat LD50:1050 mg/kg JPETAB 129,75,60
ipr-rat LD50:413 mg/kg JPETAB 129,75,60
orl-mus LD50:1050 mg/kg JPETAB 129,75,60
ipr-mus LD50:490 mg/kg JPETAB 129,75,60
orl-ham LD50:982 mg/kg JPETAB 129,75,60
ipr-ham LD50:385 mg/kg JPETAB 129,75,60

SAFETY PROFILE: Poison by intraperitoneal route. Moderately toxic by ingestion. A skeletal muscle relaxant. When heated to decomposition it emits toxic fumes of NO_x. See also CARBAMATES and ESTERS.

CBK750 CAS:64046-99-7 ***HR: 3***
N-(1-CARBAMOYLPROPYL)ARSANILIC ACID
mf: $C_{10}H_{15}AsN_2O_4$ mw: 302.19

SYNS: n-BUTARSAMIDE ◇ PHENYL-α-AMINO-n-BUTYRAMIDE-p-ARSONIC ACID

TOXICITY DATA with REFERENCE
orl-rbt LDLo:50 mg/kg PSEBAA 29,125,31
orl-gpg LDLo:100 mg/kg PSEBAA 29,125,31

CONSENSUS REPORTS: Arsenic and its compounds are on the Community Right-To-Know List.

OSHA PEL: TWA 0.5 mg(As)/m^3

SAFETY PROFILE: Poison by ingestion. See also AR-

SENIC COMPOUNDS. When heated to decomposition it emits very toxic fumes of As and NO_x.

CBL000 CAS:103-03-7 **HR: 3**
1-CARBAMYL-2-PHENYLHYDRAZINE
mf: $C_7H_9N_3O$ mw: 151.19

PROP: Crystals. Mp: 172°.

SYNS: CPH ◇ CRYOGENINE ◇ KRYOGENIN ◇ 2-PHENYLDIAZENECARBOXAMIDE ◇ 2-PHENYLHYDRAZIDE, CARBAMIC ACID ◇ 1-PHENYLHYDRAZINE CARBOXAMIDE ◇ 2-PHENYLHYDRAZINECARBOXAMIDE ◇ PHENYLSEMICARBAZIDE ◇ 1-PHENYLSEMICARBAZIDE

TOXICITY DATA with REFERENCE
dnd-esc 250 μg/well MUREAV 133,161,84
orl-mus TDLo:394 mg/kg/62W-C:NEO JNCIAM 52,241,74
ipr-rat LD50:55 mg/kg AIPTAK 150,220,64

CONSENSUS REPORTS: IARC Cancer Review: Group 3 IMEMDT 7,56,87; Animal Limited Evidence IMEMDT 12,177,76. Reported in EPA TSCA Inventory.

SAFETY PROFILE: Poison by intraperitoneal route. Questionable carcinogen with experimental neoplastigenic data. Mutation data reported. When heated to decomposition it emits toxic fumes of NO_x. See also CARBAMATES.

CBL500 CAS:16118-49-3 **HR: 2**
d-(−)-CARBANILIC ACID (1-ETHYLCARBAMOYL)ETHYL ESTER
mf: $C_{12}H_{16}N_2O_3$ mw: 236.30

SYNS: CARBETAMEX ◇ CARBETAMID (GERMAN) ◇ CARBETAMIDE ◇ d-N-ETHYLACETAMIDE CARBANILATE ◇ d-(−)-1-(ETHYLCARBAMOYL)ETHYL PHENYLCARBAMATE ◇ d-N-ETHYLLACTAMIDE CARBANILATE (ESTER) ◇ (R)-N-ETHYL-2-((PHENYLAMINO)CARBONYL)OXY)PROPANAMIDE ◇ LEGURAME ◇ 2-PHENYL-CARBAMOYLOXY-N-AETHYL-PROPIONAMID (GERMAN) ◇ (PHENYLCARBAMOYLOXY)-2-N-ETHYLPROPIONAMIDE ◇ N-PHENYL-1-(ETHYLCARBAMOYL-1)-ETHYLCARBAMATE, D ISOMER ◇ 11,561 RP

TOXICITY DATA with REFERENCE
orl-rat LD50:11000 mg/kg 85ARAE 2,83,77
orl-mus LD50:1200 mg/kg GUCHAZ 6,80,73
orl-dog LD50:900 mg/kg GUCHAZ 6,80,73

SAFETY PROFILE: Moderately toxic by ingestion. An herbicide. When heated to decomposition it emits toxic fumes of NO_x. See also CARBAMATES.

CBL750 CAS:101-99-5 **HR: 3**
CARBANILIC ACID ETHYL ESTER
mf: $C_9H_{11}NO_2$ mw: 165.21

PROP: Crystals. Mp: 53°, bp: 238° (slt decomp), d: 1.106.

SYNS: EPC (the plant regulator) ◇ ETHYL CARBANILATE ◇ ETHYL-N-PHENYLCARBAMATE ◇ EUPHORIN ◇ KEIMSTOP ◇ PHENYLETHYL CARBAMATE ◇ PHENYLURETHAN ◇ PHENYLURETHAN(E) ◇ N-PHENYLURETHANE

TOXICITY DATA with REFERENCE
cyt-smc 12 mmol/tube HEREAY 33,457,47
skn-mus TDLo:20 g/kg/2W-I:NEO BJCAAI 9,177,55
unr-rat LDLo:500 mg/kg BJPCAL 7,142,52
ipr-mus LD50:350 mg/kg HBTXAC 5,45,59
scu-mus LDLo:1 g/kg HDTU** -,-,33
ivn-mus LD50:400 mg/kg HBTXAC 5,45,59

CONSENSUS REPORTS: Reported in EPA TSCA Inventory.

SAFETY PROFILE: Poison by intraperitoneal and intravenous routes. Moderately toxic by subcutaneous and possibly other routes. Questionable carcinogen with experimental neoplastigenic data. Mutation data reported. When heated to decomposition it emits toxic fumes of NO_x. See also CARBAMATES.

CBM000 CAS:122-42-9 **HR: 3**
CARBANILIC ACID ISOPROPYL ESTER
mf: $C_{10}H_{13}NO_2$ mw: 179.24

PROP: A white, crystalline solid; sol in acetone and benzene. Mp: 90°.

SYNS: BAN-HOE ◇ BEET-KLEEN ◇ CHEM-HOE ◇ IFC ◇ IPPC ◇ ISOPROPIL-N-FENIL-CARBAMMATO (ITALIAN) ◇ ISOPROPYL CARBANILATE ◇ ISOPROPYL CARBANILIC ACID ESTER ◇ ISOPROPYL-N-FENYL-CARBAMAAT (DUTCH) ◇ ISOPROPYL-N-PHENYL-CARBAMAT (GERMAN) ◇ ISOPROPYL PHENYLCARBAMATE ◇ ISOPROPYL-N-PHENYLCARBAMATE ◇ o-ISOPROPYL-N-PHENYL CARBAMATE ◇ ISOPROPYL-N-PHENYLURETHAN (GERMAN) ◇ ORTHO GRASS KILLER ◇ N-PHENYLCARBAMATE D'ISOPROPYLE (FRENCH) ◇ PHENYLCARBAMIC ACID-1-METHYLETHEL ESTER ◇ N-PHENYL ISOPROPYL CARBAMATE ◇ PREMALOX ◇ PROFAM ◇ PROPHAM ◇ TRIHERBIDE ◇ TRIHERBIDE-IPC ◇ TUBERIT ◇ TUBERITE ◇ USAF D-9 ◇ Y 2

TOXICITY DATA with REFERENCE
cyt-omi 550 μmol/L JCLBA3 63,84,74
sce-hmn:lym 2 mg/L MUREAV 147,296,85
scu-mus TDLo:7650 mg/kg (6-14D preg):TER NTIS** Pb 223-160
orl-mus TDLo:6 g/kg/10W-I:NEO BJCAAI 12,355,58
orl-hmn LDLo:714 mg/kg CRSBAW 175,496,81
orl-rat LD50:1000 mg/kg RREVAH 10,97,65
ipr-rat LD50:600 mg/kg CRSBAW 175,496,81
orl-mus LD50:2160 mg/kg 85GMAT -,79,82
ipr-mus LD50:200 mg/kg NTIS** AD277-689
unr-mam LD50:1000 mg/kg 30ZDA9 -,199,71

CONSENSUS REPORTS: IARC Cancer Review: Group 3 IMEMDT 7,56,87; Animal Inadequate Evidence IMEMDT 12,189,76. Reported in EPA TSCA Inventory. EPA Genetic Toxicology Program.

SAFETY PROFILE: Poison by intraperitoneal route. Moderately toxic to humans by ingestion. Moderately toxic experimentally by ingestion and possibly other routes. An experimental teratogen. Human mutation data reported. Questionable carcinogen with experimental neoplastigenic and teratogenic data. An herbicide. When heated to decomposition it emits toxic fumes of NO_x. See also CARBAMATES.

CBM250 CAS:102-07-8 ***HR: 3***
CARBANILIDE
mf: $C_{13}H_{12}N_2O$ mw: 212.27

SYNS: N,N'-DIPHENYLUREA ◇ sym-DIPHENYLUREA ◇ 1,3-DIPHENYLUREA ◇ USAF EK-534

TOXICITY DATA with REFERENCE
orl-rat LDLo:500 mg/kg JPETAB 90,260,47
ipr-mus LD50:200 mg/kg NTIS** AD277-689

CONSENSUS REPORTS: Reported in EPA TSCA Inventory.

SAFETY PROFILE: Poison by intraperitoneal route. Moderately toxic by ingestion. When heated to decomposition it emits toxic fumes of NO_x.

CBM500 CAS:116-06-3 ***HR: 3***
CARBANOLATE
mf: $C_7H_{14}N_2O_2S$ mw: 190.29

PROP: A solid material.

SYNS: ALDECARB ◇ ALDICARB (USDA) ◇ ALDICARBE (FRENCH) ◇ AMBUSH ◇ ENT 27,093 ◇ 2-METHYL-2-(METHYLTHIO)PROPANAL-O-((METHYLAMINO)CARBONYL)OXIME ◇ 2-METHYL-2-(METHYL-THIO)PROPIONALDEHYDE OXIME ◇ 2-METHYL-2-(METHYLTHIO) PROPIONALDEHYDE-O-(METHYLCARBAMOYL)OXIME ◇ 2-METHYL-2-METHYLTHIO-PROPIONALDEHYD-O-(N-METHYL-CARBAMOYL)-OXIM (GERMAN) ◇ 2-METIL-2-TIOMETIL-PROPIO-NALDEID-O-(N-METIL-CARBAMOIL)-OSSIMA(ITALIAN) ◇ NCI-C08640 ◇ OMS-771 ◇ RCRA WASTE NUMBER P070 ◇ TEMIC ◇ TEMIK ◇ TEMIK G10 ◇ UC-21149

TOXICITY DATA with REFERENCE
sce-hmn:lym 10 mg/L MUREAV 138,175,84
otr-rat:emb 117 μg/plate JJATDK 1,190,81
orl-rat LD50:650 μg/kg TXAPA9 14,515,69
skn-rat LD50:2500 μg/kg TXAPA9 14,515,69
scu-rat LDLo:666 μg/kg TXAPA9 25,569,73
unk-rat LD50:930 μg/kg 30ZDA9 -,197,71
orl-mus LD50:300 μg/kg JAFCAU 18,793,70
skn-rbt LD50:1400 mg/kg GUCHAZ 6,4,73
skn-gpg LD50:2400 mg/kg 85DPAN -,-,71/76
orl-pgn LD50:3160 μg/kg ASTTA8 (680),157,79
orl-ckn LD50:8 mg/kg 85GYAZ -,62,71
orl-qal LD50:2 mg/kg EESADV 8,551,84
orl-dck LD50:3400 μg/kg TXAPA9 47,451,79
orl-bwd LD50:750 μg/kg ASTTA8 (680),157,79

CONSENSUS REPORTS: NCI Carcinogenesis Bioassay (feed); No Evidence: mouse, rat NCITR* NCI-CG-TR-136,79. Reported in EPA TSCA Inventory. EPA Extremely Hazardous Substances List.

SAFETY PROFILE: Deadly poison by ingestion, skin contact, subcutaneous and possibly other routes. Human mutation data reported. A powerful systemic poison. In 1985 over 150 people in California exhibited toxic effects from eating watermelons contaminated with aldicarb. When heated to decomposition it emits very toxic fumes of NO_x and SO_x.

CBM750 CAS:63-25-2 ***HR: 3***
CARBARYL
mf: $C_{12}H_{11}NO_2$ mw: 201.24

PROP: White crystals. Mp: 142°, d: 1.232 @ 20°/20°.

SYNS: CARBATOX-60 ◇ CRAG SEVIN ◇ ENT 23,969 ◇ EXPERIMENTAL INSECTICIDE 7744 ◇ METHYLCARBAMATE-1-NAPHTHALENOL ◇ METHYLCARBAMATE-1-NAPHTHOL ◇ KARBARYL (POLISH) ◇ N-METHYLCARBAMATE de 1-NAPHTYLE (FRENCH) ◇ METHYLCARBAMIC ACID-1-NAPHTHYL ESTER ◇ N-METHYL-1-NAFTYL-CARBAMAAT (DUTCH) ◇ N-METHYL-1-NAPHTHYL-CARBAMAT (GERMAN) ◇ N-METHYL-α-NAPHTHYLCARBAMATE ◇ N-METHYL-1-NAPHTHYL CARBAMATE ◇ N-METHYL-α-NAPHTHYLURETHAN ◇ N-METIL-1-NAFTIL-CARBAMMATO (ITALIAN) ◇ α-NAFTYL-N-METHYLKARBAMAT (CZECH) ◇ 1-NAPHTHOL-N-METHYLCARBAMATE ◇ 1-NAPHTHYL METHYLCARBAMATE ◇ α-NAPHTHYL N-METHYLCARBAMATE ◇ 1-NAPHTHYL-N-METHYLCARBAMATE ◇ SEVIN

TOXICITY DATA with REFERENCE
skn-rbt 12 mg/24H SEV JAFCAU 9,30,61
eye-rbt 500 mg/24H MOD 28ZPAK -,164,72
mmo-sat 250 μg/plate RPZHAW 30,81,79
mma-hmn:fbr 1 μmol/L MUREAV 42,161,77
dns-hmn:fbr 1 μmol/L MUREAV 42,161,77
cyt-hmn:emb 40 μg/kg ZDVKAP 20(4),14,77
orl-rat TDLo:27500 μg/kg (multi) :REP TXAPA9 21,390,72
orl-ham TDLo:250 mg/kg (female 8D post):TER TXAPA9 15,152,69
orl-rat TDLo:5640 mg/kg/94W-I:ETA VPITAR 29,71,70
imp-rat TDLo:80 mg/kg:CAR VPITAR 29,71,70
orl-man TDLo:500 mg/kg:PNS NEURAI 37,1229,87
orl-rat LD50:230 mg/kg TXAPA9 11,546,67
skn-rat LD50:4000 mg/kg 85DPAN -,-,71/76
ipr-rat LD50:64 mg/kg PSEBAA 114,509,63
scu-rat LD50:1400 mg/kg 34ZIAG-,528,69
ivn-rat LD50:41900 μg/kg BWHOA6 44,241,71
orl-mus LD50:128 mg/kg JPETAB 181,576,72
ipr-mus LD50:25 mg/kg TXAPA9 6,402,64
scu-mus LD50:6717 mg/kg TOIZAG 17,60,70
skn-rbt LD50:2000 mg/kg 85DPAN-,-,71/76

CONSENSUS REPORTS: IARC Cancer Review: Group 3 IMEMDT 7,56,87; Animal Inadequate Evidence IMEMDT 12,37,76. Community Right-To-Know List.

OSHA PEL: TWA 5 mg/m^3
ACGIH TLV: TWA 5 mg/m^3
DFG MAK: 5 mg/m^3
NIOSH REL: (Carbaryl) TWA 5 mg/m^3
DOT Classification: ORM-A; Label: None.

SAFETY PROFILE: Poison by ingestion, intravenous, intraperitoneal, and possibly other routes. Human systemic effects by ingestion: sensory change involving peripheral nerve, muscle weakness. Experimental teratogenic and reproductive effects. Questionable carcinogen with experimental carcinogenic and tumorigenic data. Human mutation data reported. An eye and severe skin irritant. Absorbed by all routes, although skin absorption is slow. No accumulation in tissue. Symptoms include blurred vision, headache, stomach ache, vomiting. Symptoms similar to but less severe than those due to parathion. A reversible cholinesterase inhibitor. See also CARBAMATES and ESTERS. When heated to decomposition it emits toxic fumes of NO_x.

CBM875 CAS:33060-69-4 ***HR: 2***
CARBAVINE
mf: $C_6H_9NO_2$ mw: 127.16

SYN: METHYL-3-BUTYN-2-OL CARBAMATE

TOXICITY DATA with REFERENCE
ipr-rat LDLo:700 mg/kg RPTOAN 33,191,70
unr-mus LD50:1500 mg/kg RPTOAN 33,191,70
ipr-rbt LDLo:700 mg/kg RPTOAN 33,191,70

SAFETY PROFILE: Moderately toxic by intraperitoneal and possibly other routes. When heated to decomposition it emits toxic fumes of NO_x. See also CARBAMATES.

CBN000 CAS:86-74-8 ***HR: 3***
CARBAZOLE
mf: $C_{12}H_9N$ mw: 167.22

PROP: White crystals. Mp: 244.8°, bp: 354.8°, d: 1.10 @ 18°/4°, vap press: 400 mm @ 323.0°.

SYNS: 9-AZAFLUORENE ◇ 9H-CARBAZOLE ◇ DIBENZOPYRROLE ◇ DIBENZO(b,d)PYRROLE ◇ DIPHENYLENEIMINE ◇ DIPHENYLENIMIDE ◇ DIPHENYLENIMINE ◇ USAF EK-600

TOXICITY DATA with REFERENCE
orl-rat LDLo:500 mg/kg JPETAB 90,260,47
ipr-mus LD50:200 mg/kg NTIS** AD277-689

CONSENSUS REPORTS: IARC Cancer Review: Group 3 IMEMDT 7,56,87; Animal Limited Evidence IMEMDT 32,239,83. Reported in EPA TSCA Inventory.

SAFETY PROFILE: Poison by intraperitoneal route. Questionable carcinogen. Moderately toxic by ingestion. A pesticide. When heated to decomposition it emits toxic fumes of NO_x.

CBN100 CAS:86-72-6 ***HR: 1***
4-(3-CARBAZOLYLAMINO)PHENOL
mf: $C_{18}H_{14}N_2O$ mw: 274.34

SYNS: CARBAZOLE, 3-(p-HYDROXYANILINO)- ◇ 3-(4'-HYDROXYFENYL)AMINOKARBAZOL ◇ PHENOL, 4-(3-CARBAZOLYLAMINO)- ◇ R-BASE

TOXICITY DATA with REFERENCE
skn-rbt 500 mg/24H MLD 85JCAE-,825,86
eye-rbt 100 mg/24H MOD 85JCAE-,825,86

CONSENSUS REPORTS: Reported in EPA TSCA Inventory.

SAFETY PROFILE: A skin and eye irritant. When heated to decomposition it emits toxic fumes of NO_x.

CBN375 ***HR: 3***
CARBENDAZIM and SODIUM NITRITE (5:1)

SYNS: METHYL-2-BENZIMIDAZOLE CARBAMATE and SODIUM NITRITE ◇ SODIUM NITRITE and CARBENDAZIM (1:5) ◇ SODIUM NITRITE and METHYL-2-BENZIMIDAZOLE CARBAMATE

TOXICITY DATA with REFERENCE
orl-mus TDLo:31 g/kg/26W-I:CAR IJCNAW 15,830,75
orl-mus TDLo:3000 mg/kg (7-14D preg):TER IJCNAW 17,742,76
orl-mus TD:88 g/kg/12W-I:CAR MGONAD 19,175,75

SAFETY PROFILE: Suspected carcinogen with experimental carcinogenic data. An experimental teratogen. When heated to decomposition it emits toxic fumes of Na_2O and NO_x. See also NITRITES and CARBAMATES.

CBN750 CAS:27025-49-6 ***HR: 2***
CARBENICILLIN PHENYL
mf: $C_{23}H_{22}N_2O_6S$ mw: 454.53

SYNS: CARBENICILLIN PHENYL ESTER ◇ CARFECILLIN

TOXICITY DATA with REFERENCE
ipr-rat LD50:980 mg/kg ANTBAL 25,513,80
orl-mus LD50:3924 mg/kg ANTBAL 25,513,80
ivn-mus LD50:728 mg/kg ANTBAL 25,513,80

SAFETY PROFILE: Moderately toxic by ingestion, intraperitoneal, and intravenous routes. When heated to decomposition it emits very toxic fumes of NO_x and SO_x. See also CARBENICILLIN PHENYL SODIUM and ESTERS.

CBO000 CAS:21649-57-0 **HR: 2**
CARBENICILLIN PHENYL SODIUM
mf: $C_{23}H_{21}N_2NaO_6S$ mw: 476.51

SYNS: BRL 3475 ◇ CARBOXYBENZYLPENICILLIN PHENYL ESTER SODIUM SALT ◇ CARFECILLIN SODIUM ◇ SODIUM-α-PHENOXYCARBONYLBENZYLPENICILLIN ◇ UTICILLIN

TOXICITY DATA with REFERENCE
ipr-rat LD50:572 mg/kg NIIRDN 6,186,82
scu-rat LD50:4530 mg/kg NIIRDN 6,186,82
ivn-rat LD50:710 mg/kg NIIRDN 6,186,82
orl-mus LD50:3040 mg/kg ANTBAL 23(5),450,78
ipr-mus LD50:942 mg/kg NIIRDN 6,186,82
scu-mus LD50:2010 mg/kg NIIRDN 6,186,82
ivn-mus LD50:717 mg/kg NIIRDN 6,186,82
ivn-dog LD50:625 mg/kg NIIRDN 6,186,82
orl-rbt LD50:10 g/kg NIIRDN 6,186,82
ivn-rbt LD50:625 mg/kg NIIRDN 6,186,82

SAFETY PROFILE: Moderately toxic by ingestion, subcutaneous, intravenous, and intraperitoneal routes. When heated to decomposition it emits very toxic fumes of NO_x, Na_2O, and SO_x. See also ESTERS and other penicillin entries.

CBO250 CAS:4800-94-6 **HR: 1**
CARBENICILLIN SODIUM
mf: $C_{17}H_{18}N_2O_6S{\cdot}2Na$ mw: 424.41

SYNS: ANABACTYL ◇ BRL-2064 ◇ CARBECIN ◇ CARBENICILLIN DISODIUM SALT ◇ CARBOXYBENZYLPENICILLIN SODIUM ◇ N-(2-CARBOXY-3,3-DIMETHYL-7-OXO-4-THIA-1-AZABICYCLO(3.2.0)HEPT-6-YL)-2-PHENYL-MALONAMIC ACID DISODIUM SALT ◇ CBPC ◇ CP-15-639-2 ◇ FUGACILLIN ◇ GEOPEN ◇ GRIPENIN ◇ MICROCILLIN ◇ NSC-111071 ◇ PIOPEN ◇ PYOPEN ◇ PYOPENE ◇ SODIUM CARBENICILLIN

TOXICITY DATA with REFERENCE
ipr-rat LD50:10 g/kg NIIRDN 6,187,82
ivn-rat LD50:6800 mg/kg NIIRDN 6,187,82
ipr-mus LD50:7600 mg/kg NIIRDN 6,187,82
scu-mus LD50:9 g/kg NIIRDN 6,187,82
ivn-mus LD50:4500 mg/kg NKRZAZ 23,572,75
ivn-mky LD50:9800 mg/kg TAKHAA 34,405,75

SAFETY PROFILE: Mildly toxic by subcutaneous, intravenous, and intraperitoneal routes. When heated to decomposition it emits very toxic fumes of NO_x, Na_2O, and SO_x. See also CARBENICILLIN PHENYL SODIUM.

CBO500 CAS:7421-40-1 **HR: 3**
CARBENOXALONE, DISODIUM SALT
mf: $C_{34}H_{48}O_7{\cdot}2Na$ mw: 614.80

SYNS: BIOGASTRONE ◇ BIORAL ◇ CARBENOXOLONE, DISODIUM SALT ◇ CARBENOXOLONE SODIUM ◇ 3-(3-CARBOXY-1-OXOPROPOXY)-11-OXOOLEAN-12-EN-29-OIC ACID, DISODIUM SALT (3-β,20-β) ◇ 3-o-(β-CARBOXYPROPIONYL)-11-OXO-18-β-OLEAN-12-EN-30-OIC ACID, DISODIUM SALT ◇ DUOGASTRONE ◇ GLYCYRRHETINIC ACID HYDROGEN SUCCINATE DISODIUM SALT ◇ 18-β-GLYCYRRHETINIC ACID HYDROGEN SUCCINATE DISODIUM SALT ◇ 3-β-HYDROXY-11-OXOOLEAN-12-EN-30-OIC ACID HYDROGEN SUCCINATE DISODIUM SALT ◇ NEOGEL ◇ PYROGASTRONE ◇ SANODIN ◇ SODIUM-3-β-HYDROXY-11-OXO-12-OLEANEN-30-OATE SODIUM SUCCINATE ◇ ULCUS-TABLINEN

TOXICITY DATA with REFERENCE
orl-dog TDLo:3 g/kg (30D male):REP OYYAA2 11,831,76
orl-hmn TDLo:120 mg/kg/6W:CVS,MET CMAJAX 117,1155,77
orl-rat LD50:2450 mg/kg OYYAA2 11,263,76
ipr-rat LD50:112 mg/kg IYKEDH 10,710,79
scu-rat LD50:1515 mg/kg IYKEDH 10,710,79
ipr-mus LD50:120 mg/kg 21NDAB -,6,68
ivn-mus LD50:198 mg/kg 21NDAB -,6,68
orl-dog LD50:3900 mg/kg IYKEDH 10,710,79
ipr-dog LD50:371 mg/kg IYKEDH 10,710,79
scu-dog LD50:1060 mg/kg OYYAA2 11,263,76
ivn-dog LD50:371 mg/kg OYYAA2 11,263,76
orl-rbt LD50:2 g/kg 21NDAB -,6,68

SAFETY PROFILE: Poison by intravenous and intraperitoneal routes. Moderately toxic by ingestion and subcutaneous routes. Human systemic effects by ingestion: blood pressure increase, change in blood potassium levels. Experimental reproductive effects. An anti-inflammatory agent used to treat gastric ulcers. When heated to decomposition it emits toxic fumes of Na_2O.

CBO625 CAS:1755-52-8 **HR: 2**
CARBESTROL
mf: $C_{17}H_{22}O_3$ mw: 274.39

SYNS: 3-ETHYL-4-(p-METHOXYPHENYL)-2-METHYL-3-CYCLOHEXENE-1-CARBOXYLIC ACID ◇ 2-METHYL-3-ETHYL-4-p-METHOXYPHENYL-Δ^3-CYCLOHEXENE CARBOXYLIC ACID ◇ NSC-19962 ◇ ORF 2166

TOXICITY DATA with REFERENCE
orl-rat TDLo:30 μg/kg (1D preg):REP JPETAB 167,105,69
orl-hmn TDLo:126 mg/kg/6W-I:CNS,GIT CCROBU 56,641,72

SAFETY PROFILE: Human systemic effects by ingestion: anorexia, diarrhea, and nausea or vomiting. Experimental reproductive effects. When heated to decomposition it emits acrid smoke and fumes.

CBO750 CAS:16322-14-8 **HR: 3**
1-CARBETHOXY-1,2-DIHYDROQUINOLINE
mf: $C_{12}H_{13}NO_2$ mw: 203.26

SYN: ETHYL-1(2H)-QUINOLINECARBOXYLATE

TOXICITY DATA with REFERENCE
orl-rat LD50:195 mg/kg 27ZQAG -,207,72
ivn-rat LD50:125 mg/kg 27ZQAG -,207,72
orl-mus LD50:160 mg/kg 27ZQAG -,207,72
ipr-mus LD50:32 mg/kg JMCMAR 14,49,71

ivn-mus LD50:135 mg/kg 27ZQAG -,207,72
orl-dog LD50:125 mg/kg 27ZQAG -,207,72
orl-rbt LD50:220 mg/kg 27ZQAG -,207,72

SAFETY PROFILE: Poison by ingestion, intravenous, and intraperitoneal routes. When heated to decomposition it emits toxic fumes of NO_x.

CBP250 CAS:21600-51-1 ***HR: 3***
1(4-CARBETHOXYPHENYL)-3,3-DIMETHYLTRIAZENE
mf: $C_{11}H_{15}N_3O_2$ mw: 221.29

SYNS: 1-(p-CARBOXYAETHYLPHENYL)-3,3-DIMETHYLTRIAZEN (GERMAN) ◇ 1-(p-ETHYLCARBOXYPHENYL)-3,3-DIMETHYLTRIAZENE

TOXICITY DATA with REFERENCE
sln-dmg-orl 1 mmol/L CBINA8 9,365,74
mrc-smc 10 mmol/L CBINA8 9,365,74
hma-mus/smc 1 mmol/L CBINA8 9,365,74
scu-rat TDLo:760 mg/kg/19W-I:CAR ARZNAD 23,800,73
ivn-rat TDLo:805 mg/kg/23W-I:CAR ARZNAD 23,800,73
scu-mus TDLo:760 mg/kg/19W-I:ETA ARZNAD 23,800,73
scu-rat LD50:450 mg/kg ARZNAD 23,800,73
ivn-rat LD50:150 mg/kg ARZNAD 23,800,73

CONSENSUS REPORTS: EPA Genetic Toxicology Program.

SAFETY PROFILE: Poison by intravenous route. Moderately toxic by subcutaneous route. Questionable carcinogen with experimental carcinogenic and tumorigenic data. Mutation data reported. When heated to decomposition it emits toxic fumes of NO_x. See also ESTERS.

CBP325 ***HR: 3***
2-(N-(4-CARBETHOXY-4-PHENYL)PIPERIDINO) PROPIOPHENONE HYDROCHLORIDE
mf: $C_{23}H_{27}NO_3 \cdot ClH$ mw: 401.97

SYNS: 1-(2-BENZOYLETHYL)-4-PHENYLISONIPECOTICACID ETHYL ESTER HYDROCHLORIDE ◇ R 951

TOXICITY DATA with REFERENCE
orl-rat LD50:145 mg/kg APPNAH 7,373,58
scu-rat LD50:360 mg/kg APPNAH 7,373,58
ivn-rat LD50:3300 μg/kg APPNAH 7,373,58
orl-mus LD50:610 mg/kg APPNAH 7,373,58
scu-mus LD50:215 mg/kg APPNAH 7,373,58
ivn-mus LD50:13800 μg/kg APPNAH 7,373,58

SAFETY PROFILE: Poison by ingestion, subcutaneous, and intravenous routes. When heated to decomposition it emits toxic fumes of NO_x and HCl. See also ESTERS.

CBQ125 CAS:3811-06-1 ***HR: 3***
CARBIDIUM ETHANESULFONATE
mf: $C_{23}H_{22}N_3O_2 \cdot C_2H_5O_3S$ mw: 481.61

SYNS: 3-AMINO-9-p-CARBETHOXYAMINOPHENYL-10-METHYLPHENANTHRIDINIUM ETHANESULPHONATE ◇ 2-AMINO-6-(p-CARBOXYAMINOPHENYL)-5-METHYLPHENANTHRIDINIUMETHANESULFONATE ETHYL ESTER ◇ 74C48 ◇ CARBIDIUM ETHANESULPHONATE

TOXICITY DATA with REFERENCE
ipr-mus LD50:40 mg/kg BJPCAL 5,287,50
scu-mus LD50:130 mg/kg BJPCAL 5,287,50
ivn-mus LD50:10 mg/kg BJPCAL 5,287,50
ivn-rbt LD50:10 mg/kg BJPCAL 5,287,50

SAFETY PROFILE: Poison by subcutaneous, intravenous, and intraperitoneal routes. When heated to decomposition it emits toxic fumes of SO_x and NO_x. See also SULFONATES and ESTERS.

CBQ500 CAS:28860-95-9 ***HR: 2***
CARBIDOPA
mf: $C_{10}H_{14}N_2O_4$ mw: 226.26

PROP: Crystals from hot water. Mp: 203-205° (decomp). dl-Form: Tan, fluffy crystals; mp: 206-208° (decomp).

SYNS: N-AMINOMETHYLDOPA ◇ (S)-α-HYDRAZINO-3,4-DIHYDROXY-α-METHYL-BENZENEPROPANOIC ACID (9CI) ◇ HYDRAZINO-α-METHYLDOPA ◇ LODOSIN ◇ LODOSYN ◇ l-α-METHYLDOPAHYDRAZINE ◇ MK 486

TOXICITY DATA with REFERENCE
mma-sat 1 mg/plate RCOCB8 49,415,85
orl-rat TDLo:2880 mg/kg (90D pre):REP TXAPA9 66,201,82
orl-rat TDLo:2100 mg/kg (1-21D preg):TER TXAPA9 38,251,76
ipr-rat LD50:2804 mg/kg YKYUA6 31,1127,80
ipr-mus LD50:468 mg/kg YKYUA6 31,1127,80

SAFETY PROFILE: Moderately toxic by intraperitoneal route. An experimental teratogen. Other experimental reproductive effects. Mutation data reported. An antihypertensive agent. When heated to decomposition it emits toxic fumes of NO_x.

CBQ529 ***HR: 3***
CARBIDOPA MONOHYDRATE
mf: $C_{10}H_{14}N_2O_4 \cdot H_2O$ mw: 244.28

SYNS: (−)-l-α-HYDRAZINO-3,4-DIHYDROXY-α-METHYLHYDROCINNAMIC ACID MONOHYDRATE ◇ S(−)-α-HYDRAZINO-3,4-DIHYDROXY-α-METHYLHYDROCINNAMIC ACID MONOHYDRATE

TOXICITY DATA with REFERENCE
orl-rat LD50:4810 mg/kg TXAPA9 29,181,74
ipr-rat LD50:352 mg/kg TXAPA9 29,181,74
scu-rat LD50:3428 mg/kg YKYUA6 31,237,80
orl-mus LD50:1750 mg/kg TXAPA9 29,181,74

ipr-mus LD50:148 mg/kg TXAPA9 29,181,74
scu-mus LD50:4955 mg/kg YKYUA6 31,237,80
ivn-mus LD50:519 mg/kg TXAPA9 29,181,74

SAFETY PROFILE: Poison by intraperitoneal route. Moderately toxic by ingestion, subcutaneous, and intravenous routes. When heated to decomposition it emits toxic fumes of NO_x.

CBQ575 ***HR: 3***
CARBINOXAMINE DIPHENYLDISULFONATE
mf: $C_{16}H_{19}ClN_2O \cdot C_{12}H_{10}O_6S_2$ mw: 605.16

SYNS: 2-(p-CHLORO-α-(2-(DIMETHYLAMINO)ETHOXY)BENZYL)-PYRIDINE DIPHENYLDISULFONATE ◇ CXA-DPS

TOXICITY DATA with REFERENCE
orl-mus LD50:630 mg/kg TOIZAG 15,367,68
ipr-mus LD50:220 mg/kg TOIZAG 15,367,68
scu-mus LD50:560 mg/kg TOIZAG 15,367,68

SAFETY PROFILE: Poison by intraperitoneal route. Moderately toxic by ingestion and subcutaneous routes. When heated to decomposition it emits toxic fumes of Cl^-, NO_x, and SO_x. See also SULFONATES and AMINES.

CBQ625 CAS:467-22-1 ***HR: 3***
CARBIPHENE HYDROCHLORIDE
mf: $C_{28}H_{34}N_2O_2 \cdot ClH$ mw: 467.10

SYNS: BANDOL ◇ 2-ETHOXY-N-METHYL-N-(2-(METHYLPHENETHYLAMINO)ETHYL)-2,2-DIPHENYLACETAMIDEHYDROCHLORIDE ◇ ETOMIDE HYDROCHLORIDE ◇ NSC-106959 ◇ SQ 10269

TOXICITY DATA with REFERENCE
orl-man LDLo:2 mg/kg JMCMAR 6,547,63
orl-mus LD50:370 mg/kg AIPTAK 154,484,65
ipr-mus LD50:190 mg/kg AIPTAK 154,484,65
ivn-mus LD50:40 mg/kg AIPTAK 154,484,65

SAFETY PROFILE: A human poison by ingestion. An experimental poison by ingestion, intravenous, and intraperitoneal routes. When heated to decomposition it emits toxic fumes of NO_x and HCl.

CBQ750 CAS:112-15-2 ***HR: 2***
CARBITOL ACETATE
mf: $C_8H_{16}O_4$ mw: 176.24

PROP: Liquid. Bp: 217.4°, fp: −25°, flash p: 230°F (OC), d: 1.0114 @ 20°/20°, vap press: 0.05 mm @ 20°, vap d: 6.07.

SYNS: DIETHYLENE GLYCOL MONOETHYL ETHER ACETATE ◇ DIGLYCOL MONOETHYL ETHER ACETATE ◇ EKTASOLVE de ACETATE ◇ 2-(2-ETHOXYETHOXY)ETHANOL ACETATE ◇ GLYCOL ETHER de ACETATE

TOXICITY DATA with REFERENCE
skn-rbt 500 mg open MLD UCDS** 7/20/65
eye-rbt 505 mg AJOPAA 29,1363,46
orl-rat LD50:11 g/kg UCDS** 7/20/65
skn-rbt LD50:15 g/kg UCDS** 7/20/65
orl-gpg LD50:3930 mg/kg JIHTAB 23,259,41

CONSENSUS REPORTS: Reported in EPA TSCA Inventory. Glycol ether compounds are on the Community Right-To-Know List.

SAFETY PROFILE: Moderately toxic by ingestion. A skin and eye irritant. See also GLYCOL ETHERS. Combustible when exposed to heat; can react with oxidizing materials. To fight fire, use alcohol foam, water, CO_2, dry chemical. When heated to decomposition it emits acrid smoke and fumes.

CBR000 CAS:111-90-0 ***HR: 2***
CARBITOL CELLOSOLVE
mf: $C_6H_{14}O_3$ mw: 134.20

PROP: Colorless liquid, mild pleasant odor. Bp: 201.9°, flash p: 201°F (OC), d: 0.9902 @ 20°/4°, vap d: 4.62.

SYNS: APV ◇ CARBITOL ◇ CARBITOL SOLVENT ◇ DIETHYLENE GLYCOL ETHYL ETHER ◇ DIETHYLENE GLYCOL MONOETHYL ETHER ◇ DIGLYCOL MONOETHYL ETHER ◇ DIOXITOL ◇ DOWANOL ◇ DOWANOL DE ◇ ETHOXY DIGLYCOL ◇ 2-(2-ETHOXYETHOXY) ETHANOL ◇ ETHYL CARBITOL ◇ ETHYL DIETHYLENE GLYCOL ◇ ETHYLENE DIGLYCOL MONOETHYL ETHER ◇ LOSUNGSMITTEL APV ◇ MONOETHYL ETHER of DIETHYLENE GLYCOL ◇ POLY-SOLV ◇ SOLVOSOL

TOXICITY DATA with REFERENCE
skn-rbt 500 mg/24H MLD JPETAB 82,377,44
eye-rbt 500 mg MOD UCDS** 11/22/68
eye-rbt 125 mg MLD ADSYAF 45,553,42
orl-mus TDLo:44 g/kg (7-14D preg):REP EVHPAZ 57,141,84
orl-rat LD50:5500 mg/kg JIDHAN 21,173,39
skn-rat LD50:6000 mg/kg JIHTAB 29,190,47
ipr-rat LD50:6310 mg/kg TXAPA9 21,454,72
ivn-rat LD50:2200 mg/kg ARZNAD 28,1571,78
skn-mus LD50:6000 mg/kg JIHTAB 29,190,47
ipr-mus LD50:3905 mg/kg FEPRA7 6,342,47
scu-mus LD50:5500 mg/kg JPETAB 65,89,39
ivn-dog LD50:3000 mg/kg JIHTAB 29,190,47
ivn-cat LDLo:1 g/kg ARZNAD 28,1571,78
orl-rbt LD50:3620 mg/kg JIHTAB 23,259,41
skn-rbt LD50:8500 mg/kg JIHTAB 29,325,47

CONSENSUS REPORTS: Reported in EPA TSCA Inventory. Glycol ether compounds are on the Community Right-To-Know List.

SAFETY PROFILE: Moderately toxic by ingestion, intravenous, intraperitoneal, and possibly other routes. Mildly toxic by skin contact. A skin and eye irritant. Experimental reproductive effects. Combustible when exposed to heat; can react with oxidizing materials. To fight fire, use alcohol foam, CO_2, dry chemical. When heated to decomposition it emits acrid smoke and irritating fumes.

CBR125 CAS:1138-80-3 ***HR: 3***
CARBOBENZOXYLGLYCINE
mf: $C_{10}H_{11}NO_4$ mw: 209.22

SYNS: BENZYLOXYCARBONYLGLYCINE ◇ N-BENZYLOXYCARBONYLGLYCINE ◇ N-CARBOBENZOYLGLYCINE ◇ CARBOBENZOYL GLYCINE ◇ CARBOBENZYLOXYGLYCINE ◇ N-CARBOBENZYLOXYGLYCINE ◇ (CBZ)GLY ◇ Z-GLY

TOXICITY DATA with REFERENCE
ipr-mus TDLo:280 mg/kg (28D preg):REP JPMSAE 70,60,81
ivg-mus LD50:380 mg/kg JPMSAE 68,696,79

CONSENSUS REPORTS: Reported in EPA TSCA Inventory.

SAFETY PROFILE: Poison by intravaginal route. Experimental reproductive effects. When heated to decomposition it emits toxic fumes of NO_x. See also ESTERS.

CBR175 CAS:64187-25-3 ***HR: 3***
N-CARBOBENZOXYGLYCINE-1,2-DIBROMOETHYL ESTER
mf: $C_{12}H_{13}Br_2NO_4$ mw: 395.08

TOXICITY DATA with REFERENCE
ipr-mus TDLo:280 mg/kg (28D pre):REP JPMSAE 70,60,81
ivg-mus LD50:148 mg/kg JPMSAE 68,696,79

SAFETY PROFILE: Poison by intraperitoneal and intravaginal routes. Experimental reproductive effects. When heated to decomposition it emits toxic fumes of Br^- and NO_x. See also ESTERS.

CBR200 CAS:64187-24-2 ***HR: 2***
N-CARBOBENZOXYGLYCINE VINYL ESTER
mf: $C_{12}H_{13}NO_4$ mw: 235.26

TOXICITY DATA with REFERENCE
ipr-mus TDLo:280 mg/kg (28D pre):REP JPMSAE 70,60,81
ipr-mus LD50:501 mg/kg JMCMAR 20,1584,77
ivg-mus LD50:500 mg/kg JPMSAE 68,696,79

SAFETY PROFILE: Moderately toxic by intraperitoneal and intravaginal routes. Experimental reproductive effects. When heated to decomposition it emits toxic fumes of NO_x. See also ESTERS.

CBR210 ***HR: 3***
N-CARBOBENZOXY-l-LEUCINE-1,2-DIBROMOETHYL ESTER
mf: $C_{16}H_{21}Br_2NO_4$ mw: 451.20

SYN: l-N-CARBOXYLEUCINE-N-BENZYL-1-(1,2-DIBROMOETHYL) ESTER

TOXICITY DATA with REFERENCE
ipr-mus TDLo:280 mg/kg (28D pre):REP JPMSAE 68,696,79
ivg-mus LD50:81 mg/kg JPMSAE 68,696,79

SAFETY PROFILE: Poison by intravaginal route. Experimental reproductive effects. When heated to decomposition it emits toxic fumes of Br^- and NO_x. See also ESTERS.

CBR215 CAS:64187-27-5 ***HR: 2***
N-CARBOBENZOXY-l-LEUCINE VINYL ESTER
mf: $C_{16}H_{21}NO_4$ mw: 291.38

SYNS: l-N-CARBOXYLEUCINE N-BENZYL 1-VINYL ESTER ◇ LEUCINE, N-CARBOXY-, N-BENZYL 1-VINYL ESTER ◇ l-LEUCINE, N-((PHENYLMETHOXY)CARBONYL)-, ETHENYL ESTER

TOXICITY DATA with REFERENCE
ivg-mus TDLo:280 mg/kg (female 28D pre):REP JPMSAE 68,696,79
ivg-mus LD50:500 mg/kg JPMSAE 68,696,79

SAFETY PROFILE: Moderately toxic by intravaginal route. Experimental reproductive effects. When heated to decomposition it emits toxic fumes of NO_x.

CBR220 CAS:1161-13-3 ***HR: 3***
CARBOBENZOXYPHENYLALANINE
mf: $C_{17}H_{17}NO_4$ mw: 299.35

SYNS: (BENZYLOXYCARBONYL)PHENYLALANINE ◇ CARBOBENZOXY-l-PHENYLALANINE ◇ N-CARBOBENZOXY-l-PHENYLALANINE ◇ l-N-CARBOXY-3-PHENYLALANINE-N-BENZYL ESTER

TOXICITY DATA with REFERENCE
ipr-mus TDLo:280 mg/kg (28D pre):REP JPMSAE 68,696,79
unr-mus LD50:251 mg/kg JPMSAE 67,1726,78
ivg-mus LD50:250 mg/kg JPMSAE 68,696,79

SAFETY PROFILE: Poison by intravaginal route. Experimental reproductive effects. When heated to decomposition it emits toxic fumes of NO_x.

CBR225 CAS:64187-43-5 ***HR: 3***
N-CARBOBENZOXY-l-PHENYLALANINE-1,2-DIBROMOETHYL ESTER
mf: $C_{19}H_{19}Br_2NO_4$ mw: 485.21

SYN: l-N-BENZYLOXYCARBONYL-3-PHENYLALANINE-1,2-DIBROMOETHYL ESTER

TOXICITY DATA with REFERENCE
ivg-mus TDLo:280 mg/kg (28D pre):REP JPMSAE 68,696,79
ipr-mus LD50:73 mg/kg JMCMAR 20,1578,77
ivg-mus LD50:74 mg/kg JPMSAE 68,696,79

SAFETY PROFILE: Poison by intraperitoneal and intravaginal routes. Experimental reproductive effects. When heated to decomposition it emits toxic fumes of Br^- and NO_x. See also ESTERS.

CBR235 CAS:64187-42-4 ***HR: 2***
N-CARBOBENZOXY-l-PHENYLALANINE VINYL ESTER
mf: $C_{19}H_{19}NO_4$ mw: 325.39

SYNS: ALANINE, N-BENZYLOXYCARBONYL-3-PHENYL-, VINYL ESTER, l- ◇ ALANINE, N-CARBOXY-3-PHENYL-, N-BENZYL 1-VINYL ESTER, l- ◇ N-BENZYLOXYCARBONYL-l-PHENYLALANINE VINYL ESTER ◇ l-N-CARBOXY-3-PHENYLALANINE N-BENZYL 1-VINYL ESTER ◇ l-PHENYLALANINE, N-((PHENYLMETHOXY)CARBONYL)-, ETHENYL ESTER ◇ N-((PHENYLMETHOXY)CARBONYL)-l-PHENYLALANINE ETHENYL ESTER

TOXICITY DATA with REFERENCE
ivg-mus TDLo:280 mg/kg (female 28D pre):REP JPMSAE 68,696,79
unr-mus LD50:2001 mg/kg JPMSAE 67,1726,78

SAFETY PROFILE: Moderately toxic. Experimental reproductive effects. When heated to decomposition it emits toxic fumes of NO_x.

CBR245 ***HR: 3***
N-CARBOBENZOXY-l-PROLINE-1,2-DIBROMOETHYL ESTER
mf: $C_{15}H_{17}Br_2NO_4$ mw: 435.15

SYN: 1,2-PYRROLIDINEDICARBOXYLIC ACID-1-BENZYL 2-(1,2-DIBROMOETHYL) ESTER

TOXICITY DATA with REFERENCE
ipr-mus TDLo:280 mg/kg (28D pre):REP JPMSAE 68,696,79
ivg-mus LD50:225 mg/kg JPMSAE 68,696,79

SAFETY PROFILE: Poison by intravaginal route. Experimental reproductive effects. When heated to decomposition it emits toxic fumes of Br^- and NO_x. See also ESTERS.

CBR247 ***HR: 2***
N-CARBOBENZOXY-l-PROLINE VINYL ESTER
mf: $C_{15}H_{17}NO_4$ mw: 275.33

SYN: 1,2-PYRROLIDINECARBOXYLIC ACID, 1-BENZYL-2-VINYL ESTER

TOXICITY DATA with REFERENCE
ivg-mus TDLo:280 mg/kg (female 28D pre):REP JPMSAE 68,696,79
ivg-mus LD50:500 mg/kg JPMSAE 68,696,79

SAFETY PROFILE: Moderately toxic by intravaginal route. Experimental reproductive effects. When heated to decomposition it emits toxic fumes of NO_x.

CBR250 CAS:1722-62-9 ***HR: 3***
CARBOCAINE HYDROCHLORIDE
mf: $C_{15}H_{22}N_2O•ClH$ mw: 282.85

SYNS: CHLOROCAIN ◇ N-(2,6-DIMETHYLPHENYL)-1-METHYL-2-PIPERIDINECARBOXAMIDE-MONOHYDROCHLORIDE ◇ MEAVERIN ◇ MEPIVACAINE HYDROCHLORIDE ◇ dl-MEPIVACAINE HYDROCHLORIDE ◇ MEPIVASTESIN ◇ 1-METHYL-2′,6′-PIPECOLOXYLIDIDE HYDROCHLORIDE ◇ dl-1-METHYL-2′,6′-PIPECOLOXYLIDIDE HYDROCHLORID ◇ (1-METHYL-dl-PIPERIDINE-2-CARBOXYLIC ACID)-2,6-DIMETHYLANILIDE HYDROCHLORIDE ◇ SCANDICAIN

TOXICITY DATA with REFERENCE
ipr-rat TDLo:160 mg/kg (1D male):REP JRPFA4 51,477,77
scu-mus LD50:260 mg/kg NIIRDN 6,846,82
ivn-mus LD50:35 mg/kg APTOA6 42,88,78
imp-mus LD50:260 mg/kg 29ZVAB -,69,69
scu-rbt LD50:110 mg/kg NIIRDN 6,846,82
ivn-rbt LD50:22 mg/kg 29ZVAB -,69,69
imp-rbt LD50:110 mg/kg 29ZVAB -,69,69
scu-gpg LD50:94 mg/kg NIIRDN 6,846,82
ivn-gpg LD50:20 mg/kg NIIRDN 6,846,82

CONSENSUS REPORTS: Reported in EPA TSCA Inventory.

SAFETY PROFILE: Poison by subcutaneous, intravenous, and implant routes. Experimental reproductive effects. An anesthetic. When heated to decomposition it emits very toxic fumes of HCl and NO_x.

CBR500 CAS:655-35-6 ***HR: 3***
CARBOCHROMENE HYDROCHLORIDE
mf: $C_{20}H_{27}NO_5•ClH$ mw: 397.94

SYNS: A-27053 ◇ AG 3 ◇ ANTIANGOR ◇ CARBOCROMENE ◇ CASSELLA 4489 ◇ CHROMONAR HYDROCHLORIDE ◇ 3-(β-DIETHYLAMINOETHYL)-4-METHYL-7-(CARBETHOXYMETHOXY)-COUMARIN HYDROCHLORIDE ◇ INTENKORDIN ◇ INTENSAIN ◇ INTENSAIN HYDROCHLORIDE ◇ KARBOKROMEN (RUSSIAN) ◇ NSC-110430

TOXICITY DATA with REFERENCE
orl-mus LD50:6300 mg/kg ARZNAD 13,243,63
ipr-mus LD50:528 mg/kg ARZNAD 13,243,63
ivn-mus LD50:34 mg/kg KHFZAN 9,57,75

SAFETY PROFILE: Poison by intravenous route. Moderately toxic by intraperitoneal route. Mildly toxic by ingestion. A coronary vasodilator. When heated to decomposition it emits very toxic fumes of NO_x and HCl.

CBR675 CAS:638-23-3 ***HR: 2***
CARBOCISTEINE
mf: $C_5H_9NO_4S$ mw: 179.21

PROP: l-Form: Mp: 204-207°. dl-Form: Spherical aggregates of needles.

SYNS: CARBOCIT ◇ CARBOCYSTEINE ◇ l-CARBOXYMETHYLCYSTEINE ◇ S-(CARBOXYMETHYL)CYSTEINE ◇ 3-(CARBOXYMETHYLTHIO)ALANINE ◇ l-3-((CARBOXYMETHYL)THIO)ALANINE ◇ FLUIFORT ◇ L.J. 206 ◇ LOVISCOL ◇ MUCICLAR ◇ MUCOCIS ◇ MUCODYNE ◇ MUCOLASE ◇ MUCOLEX ◇ MUCOPRONT ◇ PECTOX ◇ PULMOCLASE ◇ REOMUCIL ◇ RHINATHIOL ◇ RINATIOL ◇ THIODRIL ◇ TRANSBRONCHIN

TOXICITY DATA with REFERENCE
orl-rat TDLo:625 mg/kg (17-21D preg):REP KSRNAM 13,1311,79
ipr-rat LD50:7800 mg/kg NIIRDN 6,190,82
scu-rat LD50:10300 mg/kg NIIRDN 6,190,82
orl-mus LD50:8400 mg/kg NIIRDN 6,190,82
ipr-mus LD50:1433 mg/kg YKKZAJ 94,1419,74
scu-mus LD50:9 g/kg NIIRDN 6,190,82

CONSENSUS REPORTS: Reported in EPA TSCA Inventory.

SAFETY PROFILE: Moderately toxic by intraperitoneal route. Mildly toxic by ingestion. Experimental reproductive effects. When heated to decomposition it emits toxic fumes of SO_x and NO_x.

CBR750 CAS:5675-57-0 ***HR: 3***
2-CARBOETHOXY-1-METHYLVINYL-DIETHYLPHOSPHATE
mf: $C_{10}H_{19}O_6P$ mw: 266.26

SYNS: 2-ETHOXYCARBONYL-1-METHYLVINYL DIETHYL PHOSPHATE ◇ PHOSPHORIC ACID, DIETHYL ESTER, with ETHYL 3-HYDROXYCROTONATE

TOXICITY DATA with REFERENCE
orl-rat LD50:22 mg/kg 28ZEAL 4,89,69
orl-mus LD50:11 mg/kg 28ZEAL 4,89,69

SAFETY PROFILE: Poison by ingestion. When heated to decomposition it emits toxic fumes of PO_x. See also PHOSPHATES.

CBS000 CAS:14679-73-3 ***HR: 3***
N^1-CARBOETHOXY-N^2-PHTHALAZINO HYDRAZINE
mf: $C_{11}H_{12}N_4O_2$ mw: 232.27

SYNS: BT 621 ◇ CARBOETHOXYPHTHALAZINO HYDRAZINE ◇ 3-(1-PHTHALAZINYL)CARBAZIC ACID ETHYL ESTER ◇ TODRALAZINA (ITALIAN) ◇ TODRALAZINE

TOXICITY DATA with REFERENCE
ipr-rat LD50:337 mg/kg BJPCAL 32,104,68
ivn-rat LD50:110 mg/kg BJPCAL 32,104,68
ims-rat LD50:333 mg/kg BJPCAL 32,104,68
ipr-mus LD50:500 mg/kg FRPSAX 34,299,79
ivn-mus LD50:360 mg/kg BJPCAL 32,104,68
ims-mus LD50:417 mg/kg BJPCAL 32,104,68
orl-frg LD50:650 mg/kg BJPCAL 32,104,68
par-frg LD50:636 mg/kg BJPCAL 32,104,68

SAFETY PROFILE: Poison by intraperitoneal, intravenous, and intramuscular routes. Moderately toxic by ingestion and parenteral routes. When heated to decomposition it emits toxic fumes of NO_x.

CBS250 CAS:4425-78-9 ***HR: 3***
CARBOFLUORENE AMINO ESTER
mf: $C_{20}H_{23}NO_2$ mw: 309.44

SYNS: FLUORENE-9-CARBOXYLIC ACID-2-(DIETHYLAMINO)ETHYL ESTER ◇ PAVATRIN ◇ PAVATRINEAT

TOXICITY DATA with REFERENCE
orl-mus LD50:900 mg/kg CLDND* 91,103,47
ivn-rbt LD50:16 mg/kg CLDND* 91,103,47

SAFETY PROFILE: Poison by intravenous routes. Moderately toxic by ingestion. See also AMINES and ESTERS. When heated to decomposition it emits toxic fumes of NO_x.

CBS275 CAS:1563-66-2 ***HR: 3***
CARBOFURAN
DOT: NA 2757
mf: $C_{12}H_{15}NO_3$ mw: 221.28

PROP: White, crystalline solid; odorless. Mp: 105-152°, d: 1.180 @ 20°/20°, vap press: 2×10^{-5} mm @ 33°. Sltly sol in water.

SYNS: BAY 70143 ◇ CURATERR ◇ D 1221 ◇ 2,3-DIHYDRO-2,2-DIMETHYLBENZOFURANYL-7-N-METHYLCARBAMATE ◇ 2,3-DIHYDRO-2,2-DIMETHYL-7-BENZOFURANYLMETHYLCARBAMATE ◇ 2,2-DIMETHYL-7-COUMARANYL N-METHYLCARBAMATE ◇ 2,2-DIMETHYL-2,3-DIHYDROBENZOFURAN-7-YLESTER, METHYLCARBAMIC ACID ◇ 2,2-DIMETHYL-2,3-DIHYDRO-7-BENZOFURANYL-N-METHYLCARBAMATE ◇ ENT 27,164 ◇ FMC 10242 ◇ FURADAN ◇ FURODAN ◇ METHYL CARBAMIC ACID 2,3-DIHYDRO-2,2-DIMETHYL-7-BENZOFURANYL ESTER ◇ NIA 10242 ◇ NIAGRA 10242 ◇ YALTOX

TOXICITY DATA with REFERENCE
mmo-sat 10 mg/plate MUREAV 116,185,83
cyt-hmn:lym 100 mg/L TGANAK 18(1),17,84
sce-hmn:lym 5 mg/L MUREAV 147,296,85
orl-mus TDLo:210 μg/kg (female 1-21D post):REP JEPTDQ 2(2),357,78
orl-mus TDLo:110 mg/kg (female 6-16D post):TER JESEDU 20,373,85
orl-rat LD50:5 mg/kg PSSCBG 1,117,70
ihl-rat LC50:85 mg/m^3 JOCMA7 12,16,70
skn-rat LD50:120 mg/kg WRPCA2 9,119,70
orl-mus LD50:2 mg/kg JAFCAU 18,793,70
ivn-mus LD50:450 μg/kg CSLNX* NX#11280
orl-dog LD50:19 mg/kg JOCMA7 12,16,70
ihl-dog LC50:52 mg/m^3 JOCMA7 12,16,70
skn-rbt LD50:885 mg/kg GUCHAZ 6,81,73
ihl-gpg LC50:43 mg/m^3/4H TobJS# 9NOV73

CONSENSUS REPORTS: EPA Extremely Hazardous

Substances List. Reported in EPA TSCA Inventory. EPA Genetic Toxicology Program.

OSHA PEL: TWA 0.1 mg/m^3
ACGIH TLV: TWA 0.1 mg/m^3
DOT Classification: Poison B; Label: Poison; Poison B; Label: Poison, liquid.

SAFETY PROFILE: Poison by inhalation, ingestion, skin contact, and intravenous routes. Experimental teratogenic and reproductive effects. Human mutation data reported. When heated to decomposition it emits toxic fumes of NO_x. See also CARBAMATES.

CBS500 CAS:497-18-7 ***HR: 3***
CARBOHYDRAZIDE
mf: CH_6N_4O mw: 90.11

SYNS: 4-AMINOSEMICARBAZIDE ◇ CARBAZIC ACID HYDRAZIDE ◇ CARBAZIDE ◇ CARBAZIDE (DOT) ◇ CARBODIHYDRAZIDE ◇ CARBONIC ACID DIHYDRAZIDE ◇ CARBONIC DIHYDRAZIDE ◇ CARBONOHYDRAZIDE ◇ CARBONYLDIHYDRAZINE ◇ 1,3-DIAMINOUREA

TOXICITY DATA with REFERENCE
ivn-mus LD50:120 mg/kg JPETAB 122,110,58
ipr-mus LD50:167 mg/kg JMPCAS 4,259,61

CONSENSUS REPORTS: Reported in EPA TSCA Inventory.

DOT Classification: Forbidden.

SAFETY PROFILE: Poison by intravenous and intraperitoneal routes. Explodes when heated. Reacts with nitrous acid to form the explosive carbonic diazide. When heated to decomposition it emits toxic fumes of NO_x.

CBS750 CAS:63042-08-0 ***HR: 2***
4'-CARBOMETHOXY-2,3'-DIMETHYLAZOBENZENE
mf: $C_{16}H_{16}N_2O_3$ mw: 284.34

SYNS: 4'-CARBOMETHOXY-2,3'-DIMETHYLAZOBENZOL ◇ CARBONIC ACID METHYL-4-(o-TOLYLAZO)-o-TOLYL ESTER ◇ 2,3'-DIMETHYLAZOBENZENE-4'-METHYLCARBONATE

TOXICITY DATA with REFERENCE
orl-rat TDLo:27 g/kg/43W-C:ETA GANNA2 33,196,39

SAFETY PROFILE: Questionable with experimental tumorigenic data. When heated to decomposition it emits toxic fumes of NO_x.

CBT125 CAS:25147-05-1 ***HR: 3***
N-CARBOMETHOXYMETHYLIMINOPHOSPHORYL CHLORIDE
mf: $C_2H_3Cl_3NO_2P$ mw: 310.38

SAFETY PROFILE: Violent or explosive spontaneous decomposition. Upon decompositon it emits toxic fumes of Cl^-, PO_x and NO_x. See also CHLORIDES.

CBT175 CAS:89022-11-7 ***HR: 2***
2'-CARBOMETHOXYPHENYL 4-GUANIDINOBENZOATE
mf: $C_{16}H_{15}N_3O_4$ mw: 313.34

SYNS: 4-((AMINOIMINOMETHYL)AMINO)BENZOIC ACID 2-(METHOXYCARBONYL)PHENYL ESTER ◇ BENZOIC ACID, 4-((AMINOIMINOMETHYL)AMINO)-,2-(METHOXYCARBONYL)PHENYL ESTER ◇ SALICYLIC ACID, METHYL ESTER, ESTER with p-GUANIDINOBENZOIC ACID

TOXICITY DATA with REFERENCE
ivg-rbt TDLo:100 μg/kg (female 1D pre):REP CCPTAY 32,183,85
ipr-mus LD50:750 mg/kg JMCMAR 29,514,86

SAFETY PROFILE: Moderately toxic by intraperitoneal route. Experimental reproductive effects. When heated to decomposition it emits toxic fumes of NO_x.

CBT250 CAS:4564-87-8 ***HR: 3***
CARBOMYCIN
mf: $C_{42}H_{67}NO_{16}$ mw: 842.10

SYNS: CARBOMYCIN A ◇ DELTAMYCIN A ◇ 9-DEOXY-12,13-EPOXY-9-OXOLEUCOMYCIN V 3-ACETATE 4^B-(3-METHYLBUTANOATE) ◇ M-4209 ◇ MAGNAMYCIN ◇ MAGNAMYCIN A

TOXICITY DATA with REFERENCE
scu-mus LD50:295 mg/kg ANTCAO 3,55,53
ivn-mus LD50:550 mg/kg MEIEDD 10,250,83
ims-mus LD50:1000 mg/kg ANTCAO 3,55,53
ivn-rbt LD50:700 mg/kg ANTCAO 3,55,53

SAFETY PROFILE: Poison by subcutaneous route. Moderately toxic by intravenous and intramuscular routes. When heated to decomposition it emits toxic fumes of NO_x.

CBT500 CAS:7440-44-0 ***HR: 1***
CARBON
af: C aw: 12.01

PROP: Black crystals, powder or diamond form. Mp: 3652-3697° (subl), bp: approx 4200°, d (amorp): 1.8-2.1, d (graphite): 2.25, d (diamond): 3.51, vap press: 1 mm @ 3586°.

SYNS: BLACK PEARLS ◇ CARBONE (ITALIAN) ◇ CHARCOAL BLACK ◇ C.I. 77266 ◇ COLUMBIAN CARBON ◇ GRAPHITE (MAK) ◇ GRAPHITE, NATURAL (ACGIH) ◇ GRAPHITE, SYNTHETIC ◇ PURIFIED CHARCOAL

TOXICITY DATA with REFERENCE
scu-rat TDLo:167 mg/kg (8D preg):REP TJADAB 4,327,71
ivn-mus LD50:440 mg/kg TXAPA9 24,497,73

CONSENSUS REPORTS: Reported in EPA TSCA Inventory.

OSHA PEL: (Natural graphite) (Transitional: TWA 50 mppcf) TWA 2.5 mg/m^3; (Synthetic graphite) (Transitional: TWA Total Dust: 15 mg/m^3; Respirable Fraction: 5 mg/m^3) TWA Total Dust: 10 mg/m^3; Respirable Fraction: 5 mg/m^3
ACGIH TLV: TWA 2 mg/m^3 (respirable dust)
DFG MAK: 6 mg/m^3
DOT Classification: Flammable Solid; Label: Spontaneously Combustible

SAFETY PROFILE: Moderately toxic by intravenous route. Experimental reproductive effects. It can cause a dust irritation, particularly to the eyes and mucous membranes. See also CARBON BLACK, SOOT. Combustible when exposed to heat. Dust is explosive when exposed to heat or flame or oxides, peroxides, oxosalts, halogens, interhalogens, O_2, (NH_4NO_3 + heat), (NH_4ClO_4 @ 240°), bromates, $Ca(OCl)_2$, chlorates, (Cl_2 + $Cr(OCl)_2$), ClO, iodates, IO_5, $(Pb(NO_3)_2$, $HgNO_3$, HNO_3, (oils + air), (K + air), Na_2S, $Zn(NO_3)_2$. Incompatible with air, metals, oxidants, unsaturated oils.

CBT750 CAS:1333-86-4 ***HR: 1***
CARBON BLACK

PROP: A generic term applied to a family of high-purity colloidal carbons commercially produced by carefully controlled pyrolysis of gaseous or liquid hydrocarbons. Carbon blacks, including commercial colloidal carbons such as furnace blacks, lamp blacks and acetylene blacks, usually contain less than several tenths percent of extractable organic matter and less than one percent ash.

SYNS: ACETYLENE BLACK ◇ CHANNEL BLACK ◇ FURNACE BLACK ◇ LAMP BLACK

CONSENSUS REPORTS: IARC Cancer Review: Group 3 IMEMDT 7,142,87; Human Inadequate Evidence IMEMDT 33,35,84; Animal Inadequate Evidence IMEMDT 33,35,84

OSHA PEL: TWA 3.5 mg/m^3
ACGIH TLV: TWA 3.5 mg/m^3
NIOSH REL: (Carbon Black) TWA 3.5 mg/m^3

SAFETY PROFILE: Mildly toxic by ingestion, inhalation, and skin contact. Questionable carcinogen. See also CARBON. A nuisance dust in high concentrations. While it is true that the tiny particulates of carbon black contain some molecules of carcinogenic materials, the carcinogens are apparently held tightly and are not eluted by hot or cold water, gastric juices, or blood plasma.

CBU250 CAS:124-38-9 ***HR: 1***
CARBON DIOXIDE
DOT: UN 1013/UN 1845/UN 2187
mf: CO_2 mw: 44.01

PROP: Colorless, odorless gas. Mp: sublimes @ −78.5° (−56.6° @ 5.2 atm), vap d: 1.53.

SYNS: ANHYDRIDE CARBONIQUE (FRENCH) ◇ CARBONIC ACID GAS ◇ CARBONIC ANHYDRIDE ◇ KOHLENDIOXYD (GERMAN) ◇ KOHLENSAURE (GERMAN)

TOXICITY DATA with REFERENCE
ihl-rat TCLo:6 pph/24H (10D preg):REP CIRUAL 8,1218,60
ihl-rat TCLo:6 pph/24H (10D preg):TER CIRUAL 8,1218,60
ihl-hmn LCLo:9 pph/5M TABIA2 3,231,33
ihl-mam LCLo:90000 ppm/5M AEPPAE 138,65,28

CONSENSUS REPORTS: Reported in EPA TSCA Inventory.

OSHA PEL: (Transitional: TWA 5000 ppm) TWA 10,000 ppm; STEL 30,000 ppm
ACGIH TLV: TWA 5000 ppm; STEL 30,000 ppm
DFG MAK: 5000 ppm (9000 mg/m^3)
NIOSH REL: (Carbon Dioxide) TWA 10000 ppm; CL 30000 ppm/10M
DOT Classification: Nonflammable Gas; Label: Nonflammable Gas (UN1013, UN2187); ORM-A; Label: None (UN1845).

SAFETY PROFILE: An asphyxiant. See discussion of simple asphyxiants under ARGON. Experimental teratogenic and reproductive effects. Contact of carbon dioxide snow with the skin can cause burns. Dusts of magnesium, zirconium, titanium, and some magnesium-aluminum alloys ignite and then explode in CO_2 atmospheres. Dusts of aluminum, chromium, and manganese ignite and then explode when heated in CO_2. Several bulk metals will burn in CO_2. Reacts vigorously with (Al + Na_2O_2); Cs_2O; $Mg(C_2H_5)_2$; Li; (Mg + Na_2O_2); K; KHC; Na; Na_2C_2; NaK; Ti. CO_2 fire extinguishers can produce highly incendiary sparks of 5-15 mJ at 10-20 KV by electrostatic discharge. Incompatible with acrylaldehyde, aziridine, metal acetylides, sodium peroxide.

CBU500 CAS:124-38-9 ***HR: 1***
CARBON DIOXIDE (liquefied)
DOT: UN 1013/UN 1845/UN 2187

SYNS: CARBON DIOXIDE, liquefied (DOT) ◇ CARBON DIOXIDE, refrigerated liquid (DOT)

CONSENSUS REPORTS: Reported in EPA TSCA Inventory.

NIOSH REL: TWA 10000 ppm; CL 30000 ppm/10M
DOT Classification: Nonflammable Gas; Label: Nonflammable Gas.

SAFETY PROFILE: See CARBON DIOXIDE.

CBU750 CAS:124-38-9 ***HR: 2***
CARBON DIOXIDE (solid)

PROP: White snow-like solid. D: 1.35.

SYNS: CARBON DIOXIDE, solid (DOT) ◇ CARBON ICE (DOT) ◇ DRY ICE (DOT)

CONSENSUS REPORTS: Reported in EPA TSCA Inventory.

DOT Classification: ORM-A; Label: None.

SAFETY PROFILE: See CARBON DIOXIDE. The solid is very cold and can cause frostbite if in contact with the skin.

CBV000 CAS:53569-62-3 ***HR: 2***
CARBON DIOXIDE mixed with NITROUS OXIDE
DOT: UN 1015
mf: $CO_2 \cdot N_2O$ mw:88.03

SYNS: CARBON DIOXIDE, mixture with NITROGEN OXIDE (N_2O) ◇ CARBON DIOXIDE-NITROUS OXIDE mixture (DOT)

NIOSH REL: (Carbon Dioxide) TWA 10000 ppm; CL 30000 ppm/10M; (N_2O as Anesthetic Agent) TWA 25 ppm/1H
DOT Classification: Nonflammable Gas; Label: Nonflammable Gas.

SAFETY PROFILE: See components as listed. An anesthetic mixture. Combustible. An oxidizing mixture. Can react with reducing materials.

CBV250 CAS:8063-77-2 ***HR: 1***
CARBON DIOXIDE mixed with OXYGEN
DOT: UN 1014
SYN: CARBON DIOXIDE-OXYGEN mixture (DOT)

NIOSH REL: (Carbon Dioxide) TWA 10000 ppm; CL 30000 ppm/10M
DOT Classification: Nonflammable Gas; Label: Nonflammable Gas.

SAFETY PROFILE: Possible asphyxiant.

CBV500 CAS:75-15-0 ***HR: 3***
CARBON DISULFIDE
DOT: UN 1131
mf: CS_2 mw: 76.13

PROP: Clear, colorless liquid; nearly odorless when pure. Mp: −110.8°, bp: 46.5°, lel: 1.3%, uel: 50%, flash p: −22°F (CC), d: 1.261 @ 20°/20°, autoign temp: 257°F, vap press: 400 mm @ 28°, vap d: 2.64.

SYNS: CARBON BISULFIDE (DOT) ◇ CARBON BISULPHIDE ◇ CARBON DISULPHIDE ◇ CARBONE (SUFURE de) (FRENCH) ◇ CARBONIO (SOLFURO di) (ITALIAN) ◇ CARBON SULFIDE ◇ CARBON SULPHIDE (DOT) ◇ DITHIOCARBONIC ANHYDRIDE ◇ KOHLENDISULFID (SCHWEFELKOHLENSTOFF) (GERMAN) ◇ KOOLSTOFDISULFIDE (ZWAVELKOOLSTOF) (DUTCH) ◇ NCI-C04591 ◇ RCRA WASTE NUMBER P022 ◇ SCHWEFELKOHLENSTOFF (GERMAN) ◇ SOLFURO di CARBONIO (ITALIAN) ◇ SULPHOCARBONIC ANHYDRIDE ◇ WEEVILTOX ◇ WEGLA DWUSIARCZEK (POLISH)

TOXICITY DATA with REFERENCE
mmo-sat 100 μL/plate NIOSH* 5AUG77
sce-hmn:lym 10200 μg/L BCTKAG 14,115,81
ihl-man TCLo:40 mg/m^3 (91W male):REP MELAAD 60,566,69
ihl-rat TCLo:100 mg/m^3/8H (1-21D preg):TER TJADAB 14,374,76
ihl-hmn LCLo:4000 ppm/30M 29ZWAE -,118,68
ihl-hmn LCLo:2000 ppm/5M TABIA2 3,231,33
unr-man LDLo:186 mg/kg 85DCAI 2,73,70
orl-rat LD50:3188 mg/kg GISAAA 31(1),13,66
ihl-rat LC50:25 g/m^3/2H 85GMAT -,32,82
orl-mus LD50:2780 mg/kg GISAAA 31(1),13,66
ihl-mus LC50:10 g/m^3/2H 85GMAT -,32,82 GISAAA 31(1),13,66
orl-gpg LD50:2125 mg/kg GISAAA 31(1),13,66
ipr-gpg LDLo:400 mg/kg AIHAAP 35,21,74

CONSENSUS REPORTS: Reported in EPA TSCA Inventory. EPA Genetic Toxicology Program. Community Right-To-Know List. EPA Extremely Hazardous Substances List.

OSHA PEL: (Transitional: TWA 20 ppm; CL 30 ppm; PK 100 ppm/30 min) TWA 4 ppm; STEL 12 (skin)
ACGIH TLV: TWA 10 ppm (skin); BEI: 5 mg(2-thiothiazolidine-4-carboxylic acid (TTCA))/g creatinine in urine.
DFG MAK: 10 ppm (30 mg/m^3); BAT: 8 mg/L of 4-thio-4-thiazolidine carboxylic acid (TTCA) at end of shift.
NIOSH REL: (Carbon Disulfide) TWA 1 ppm; CL 10 ppm/15M
DOT Classification: Flammable Liquid; Label: Flammable Liquid; IMO: Flammable Liquid; Label: Flammable Liquid, Poison.

SAFETY PROFILE: A human poison by unspecified route. Mildly toxic to humans by inhalation. An experimental poison by intraperitoneal route. Human reproductive effects on spermatogenesis by inhalation. Experimental teratogenic and reproductive effects. Human mutation data reported. The main toxic effect is on the central nervous system, acting as a narcotic and anesthetic in acute poisoning with death following from respiratory failure. In chronic poisoning, the effect on the nervous system is one of central and peripheral damage which may be permanent if the damage has been severe.

Flammable liquid. A dangerous fire hazard when exposed to heat, flame, sparks, friction, or oxidizing materials. Severe explosion hazard when exposed to heat or flame. Ignition and potentially explosive reaction when heated in contact with rust or iron. Mixtures with sodium or potassium-sodium alloys are powerful, shock-sensitive explosives. Explodes on contact with permanganic acid. Potentially explosive reaction with nitrogen oxide; chlorine (catalyzed by iron). Mixtures with dinitrogen tetraoxide are heat-, spark- and shock-sensitive explosives. Reacts with metal azides to produce shock- and heat-sensitive, explosive metal azidodithioformates. Aluminum powder ignites in CS_2 vapor. The vapor ignites on contact with fluorine. Reacts violently with azides; CsN_3; ClO; ethylamine diamine; ethylene imine; $Pb(N_3)_2$; LiN_3; (H_2SO_4 + permanganates); KN_3; RbN_3; NaN_3; phenylcopper-triphenylphosphine complexes. Incompatible with air, metals, oxidants. To fight fire, use water, CO_2, dry chemical, fog, mist. When heated to decomposition it emits highly toxic fumes of SO_x.

CBV750 CAS:64057-79-0 ***HR: 3***
CARBONIC ACID BIS(2-METHYLALLYL) ESTER
mf: $C_9H_{14}O_3$ mw: 170.23

TOXICITY DATA with REFERENCE
ivn-mus LD50:250 mg/kg CBCCT* 6,139,54

CONSENSUS REPORTS: Reported in EPA TSCA Inventory.

SAFETY PROFILE: Poison by intravenous route. See also ALLYL COMPOUNDS and ESTERS. When heated to decomposition it emits acrid smoke and irritating fumes.

CBW000 CAS:973-21-7 ***HR: 3***
CARBONIC ACID-2-sec-BUTYL-4,6-DINITROPHENYL ISOPROPYL ESTER
mf: $C_{14}H_{18}N_2O_7$ mw: 326.34

SYNS: ACREX ◇ 2-sec-BUTYL-4,5-DINITROPHENOL ISOPROPYL CARBONATE ◇ 2-sec-BUTYL-4,6-DINITROPHENYL ISOPROPYL CARBONATE ◇ DESSIN ◇ 2,4-DINITRO-6-sek.BUTYL-ISOPROPYLPHENYL-CARBONAT (GERMAN) ◇ 2,4-DINITRO-6-sec-BUTYLPHENYL ISOPROPYL CARBONATE ◇ DINOBUTON ◇ DINOFEN ◇ DRAWINOL ◇ DS 18302 ◇ ENT 27,244 ◇ ISOPHEN ◇ ISOPHEN (pesticide) ◇ ISOPROPYL-2,4-DINITRO-6-SEC-BUTYLPHENYL CARBONATE ◇ ISOPROPYL-2-(1-METHYL-N-PROPYL)-4,6-DINITROPHENYLCARBONATE ◇ KASEBON ◇ MC 1053 ◇ 1-METHYLETHYL-2-(1-ETHYLPROPYL)-4,6-DINITROPHENYL CARBONATE ◇ 1-METHYLETHYL-2-(1-METHYLPROPYL)-4,5-DINITROPHENYLESTER CARBONIC ACID ◇ 2-(1-METHYL-2-PROPYL)-4,6-DINITROPHENYLISOPROPYLCARBONATE ◇ SYTASOL ◇ TALAN ◇ UC 19786 ◇ UNION CARBIDE 19786

TOXICITY DATA with REFERENCE
cyt-mus-unr 25 mg/kg TGANAK 14(6),41,80
cyt-mus-orl 25 mg/kg CYGEDX 14(6),38,80
orl-rat LD50:59 mg/kg TXAPA9 14,515,69
ihl-rat LCLo:4850 $\mu g/m^3$/4H 85GYAZ -,72,71
skn-rat LDLo:1500 mg/kg TXAPA9 14,515,69
unk-rat LD50:140 mg/kg 30ZDA9 -,100,71
orl-mus LD50:170 mg/kg GTPZAB 19(9),55,75
ipr-mus LD50:125 mg/kg BCPCA6 18,1389,69
unk-mus LD50:2540 mg/kg 30ZDA9 -,100,71
skn-rbt LD50:3200 mg/kg FMCHA2 -,C82,83
orl-ckn LD50:150 mg/kg GUCHAZ 6,224,73

SAFETY PROFILE: Poison by ingestion, inhalation, intraperitoneal, and possibly other routes. Moderately toxic by skin contact. Mutation data reported. A miticide. See also ESTERS. When heated to decomposition it emits toxic fumes of NO_x.

CBW400 CAS:2463-45-8 ***HR: 3***
CARBONIC ACID, CYCLIC 3-CHLOROPROPYLENE ESTER
mf: $C_4H_5ClO_3$ mw: 136.54

SYN: 1,3-DIOXOLAN-2-ONE,4-(CHLOROMETHYL)-

TOXICITY DATA with REFERENCE
orl-rat TDLo:70 mg/kg (male 14D pre):REP CCPTAY 9,451,74
orl-rat LD50:80 mg/kg CCPTAY 9,451,74

CONSENSUS REPORTS: Reported in EPA TSCA Inventory.

SAFETY PROFILE: Poison by ingestion. Experimental reproductive effects. When heated to decomposition it emits toxic fumes of Cl^-.

CBW500 CAS:108-32-7 ***HR: 1***
CARBONIC ACID CYCLIC PROPYLENE ESTER
mf: $C_4H_6O_3$ mw: 102.10

PROP: A clear liquid. Bp: 242.1°, fp: −48.8°, flash p: 275°F (OC), d: 1.2069 @ 20°/20°, vap press: 0.03 mm @ 20°.

SYNS: CYCLIC METHYLETHYLENE CARBONATE ◇ CYCLIC PROPYLENE CARBONATE ◇ CYCLIC-1,2-PROPYLENE CARBONATE ◇ 1-METHYLETHYLENE CARBONATE ◇ 1,2-PROPANEDIOL CARBONATE ◇ 1,2-PROPANEDIOL CYCLIC CARBONATE ◇ 1,2-PROPANEDIYL CARBONATE ◇ 1,2-PROPYLENE CARBONATE ◇ PROPYLENE GLYCOL CYCLIC CARBONATE

TOXICITY DATA with REFERENCE
skn-hmn 100 mg/3D-I MOD 85DKA8 -,127,77
eye-rbt 60 mg MOD UCDS** 4/25/58
orl-rat LD50:29 g/kg UCDS** 4/25/58

CONSENSUS REPORTS: Reported in EPA TSCA Inventory.

SAFETY PROFILE: Mildly toxic by ingestion. A human skin irritant. An eye irritant. See also ESTERS. Combustible when exposed to heat or flame. To fight fire, use alcohol foam. Can react with oxidizing materials. When heated to decomposition it emits acrid smoke and irritating fumes. See also ESTERS.

CBW750 CAS:630-08-0 ***HR: 3***
CARBON MONOXIDE
DOT: UN 1016/NA 9202
mf: CO mw: 28.01

PROP: Colorless, odorless gas. Mp: −207°, bp: −191.3°, lel: 12.5%, uel: 74.2%, d: (gas) 1.250 g/L @ 0°, (liquid) 0.793, autoign temp: 1128°F.

SYNS: CARBONE (OXYDE de) (FRENCH) ◇ CARBONIC OXIDE ◇ CARBONIO (OSSIDO di) (ITALIAN) ◇ CARBON MONOXIDE, CRYOGENIC liquid (DOT) ◇ ON OXIDE (CO) ◇ EXHAUST GAS ◇ FLUE GAS ◇ KOHLENMONOXID (GERMAN) ◇ KOHLENOXYD (GERMAN) ◇ KOOLMONOXYDE (DUTCH) ◇ OXYDE de CARBONE (FRENCH) ◇ WEGLA TLENEK (POLISH)

TOXICITY DATA with REFERENCE
ihl-mus TCLo:65 ppm/24H (female 7-18D post):REP TJADAB 29(2),8B,84
ihl-mus TCLo:8 pph/1H (female 8D post):TER FPNJAG 11,301,58
ihl-hmn TCLo:600 mg/m^3/10M GTPZAB 31(4),34,87
ihl-man LCLo:4000 ppm/30M 29ZWAE -,207,68
ihl-man TCLo:650 ppm/45M:CNS,BLD AIHAAP 34,212,73
ihl-hmn LCLo:5000 ppm/5M TABIA2 3,231,33
ihl-rat LC50:1807 ppm/4H TXAPA9 17,752,70
ihl-mus LC50:2444 ppm/4H TXAPA9 17,752,70
ihl-dog LCLo:4000 ppm/46M HBAMAK 4,1360,35
ihl-rbt LCLo:4000 ppm HBAMAK 4,1360,35
ihl-gpg LC50:5718 ppm/4H TXAPA9 17,752,70
ihl-mam LCLo:5000 ppm/5M AEPPAE 138,65,28
ihl-bwd LD50:1334 ppm AECTCV 12,355,83

CONSENSUS REPORTS: Reported in EPA TSCA Inventory.

OSHA PEL: (Transitional: TWA 50 ppm) TWA 35; CL 200 ppm
ACGIH TLV: TWA 50 ppm; STEL 400 ppm (Proposed: 25 ppm); BEI: less than 8% carboxyhemoglobin in blood at end of shift; less than 40 ppm CO in end-exhaled air at end of shift. (Proposed: less than 3.5% carboxyhemoglobin in blood at end of shift; less than 20 ppm CO in end-exhaled air at end of shift.)
DFG MAK: 30 ppm (33 mg/m^3); BAT: 5% in blood at end of shift.
NIOSH REL: (Carbon Monoxide) TWA 35 ppm; CL 200 ppm
DOT Classification: Flammable Gas; Label: Flammable Gas; Flammable Gas; Label: Flammable Gas and Poison Gas.

SAFETY PROFILE: Mildly toxic by inhalation in humans but has caused many fatalities. Experimental teratogenic and reproductive effects. Human systemic effects by inhalation: changes in psychophysiological tests and methemoglobinemia-carboxhemoglobinemia. Can cause asphyxiations by preventing hemoglobin from binding oxygen. After being removed from exposure, the half-life of its elimination from the blood is one hour. Chronic exposure effects can occur at lower concentrations. A common air contaminant. Acute cases of poisoning resulting from brief exposures to high concentrations seldom result in any permanent disability if recovery takes place. Chronic effects as the result of repeated exposure to lower concentrations have been described, particularly in the Scandinavian literature. Auditory disturbances and contraction of the visual fields have been demonstrated. Glycosuria does occur, and heart irregularities have been reported. Other workers have found that where the poisoning has been relatively long and severe, cerebral congestion and edema may occur, resulting in long-lasting mental or nervous damage. Repeated exposure to low concentration of the gas, up to 100 ppm in air, is generally believed to cause no signs of poisoning or permanent damage. Industrially, sequelae are rare, as exposure, though often severe, is usually brief. It is a common air contaminant.

A dangerous fire hazard when exposed to flame. Severe explosion hazard when exposed to heat or flame. Violent or explosive reaction on contact with bromine trifluoride, bromine pentafluoride, chlorine dioxide, or peroxodisulfuryl difluoride. Mixture of liquid CO with liquid O_2 is explosive. Reacts with sodium or potassium to form explosive products sensitive to shock, heat, or contact with water. Mixture with copper powder + copper(II) perchlorate + water forms an explosive complex. Mixture of liquid CO with liquid dinitrogen oxide is a rocket propellant combination. Ignites on warming with iodine heptafluoride. Ignites on contact with cesium oxide + water. Potentially explosive reaction with iron(III) oxide between 0-150°C. Exothermic reaction with ClF_3; (Li + H_2O); NF_3; OF_2; (K + O_2); Ag_2O; (Na + NH_3). To fight fire, stop flow of gas.

CBX109 CAS:1885-14-9 ***HR: 3***
CARBONOCHLORIDIC ACID PHENYL ESTER
DOT: UN 2746
mf: $C_7H_5ClO_2$ mw: 156.57

SYNS: CHLOROFORMIC ACID PHENYL ESTER ◇ FENYLESTER KYSELINY CHLORMRAVENCI (CZECH) ◇ PHENYL CHLOROCARBONATE ◇ PHENYL CHLOROFORMATE ◇ PHENYLCHLOROFORMATE (DOT)

TOXICITY DATA with REFERENCE
skn-rbt 500 mg/24H MLD 28ZPAK -,163,72
eye-rbt 50 μg/24H SEV 28ZPAK -,163,72
orl-rat LD50:1410 mg/kg AIHAAP 30,470,69
ihl-rat LCLo:44 ppm/4H AIHAAP 30,470,69
skn-rbt LD50:3970 mg/kg AIHAAP 30,470,69

CONSENSUS REPORTS: Reported in EPA TSCA Inventory.

DOT Classification: Poison B; Label: Corrosive and Poison.

SAFETY PROFILE: Poison by inhalation. Moderately toxic by ingestion and skin contact. A corrosive skin and eye irritant. See also ESTERS. When heated to decomposition it emits toxic fumes of Cl^-.

CBX250 ***HR: 3***
CARBON REMOVER (liquid)
DOT: UN 1132

PROP: Flash p: <80°F.

DOT Classification: Flammable Liquid; Label: Flammable Liquid.

SAFETY PROFILE: Dangerous fire hazard when exposed to heat or flame; can react with oxidizing materials. To fight fire, use CO_2, dry chemical.

CBX750 CAS:558-13-4 ***HR: 3***
CARBON TETRABROMIDE
DOT: UN 2516
mf: CBr_4 mw: 331.65

PROP: Colorless, monoclinic tablets. Mp: (α) 48.4°, (β) 90.1°, bp: 189.5°, d: 3.42, vap press: 40 mm @ 96.3°.

SYNS: CARBON BROMIDE ◇ TETRABROMIDE METHANE ◇ TETRABROMOMETHANE

TOXICITY DATA with REFERENCE
orl-rat LDLo:1000 mg/kg 14CYAT 2,1270,63
scu-mus LD50:298 mg/kg TXAPA9 4,354,62
ivn-mus LD50:56 mg/kg CSLNX* NX#01612

CONSENSUS REPORTS: Reported in EPA TSCA Inventory.

OSHA PEL: TWA 0.1 ppm; STEL 0.3 ppm
ACGIH TLV: TWA 0.1 ppm; STEL 0.3 ppm
DOT Classification: Poison B; Label: St. Andrews Cross.

SAFETY PROFILE: Poison by subcutaneous and intravenous routes. Moderately toxic by ingestion. Narcotic in high concentration. Mixture with Li particles is an impact-sensitive explosive. Explodes on contact with hexacylcohexyldilead. When heated to decomposition it emits toxic fumes of Br^-. See also CHLORINATED HYDROCARBONS, ALIPHATIC.

CBY000 CAS:56-23-5 ***HR: 3***
CARBON TETRACHLORIDE
DOT: UN 1846
mf: CCl_4 mw: 153.81

PROP: Colorless liquid; heavy, ethereal odor. Mp: −22.6°, bp: 76.8°, fp: −22.9°, flash p: none, d: 1.597 @ 20°, vap press: 100 mm @ 23.0°.

SYNS: BENZINOFORM ◇ CARBONA ◇ CARBON CHLORIDE ◇ CARBON TET ◇ CZTEROCHLOREK WEGLA (POLISH) ◇ ENT 4,705 ◇ FASCIOLIN ◇ FLUKOIDS ◇ METHANE TETRACHLORIDE ◇ NECATORINA ◇ NECATORINE ◇ PERCHLOROMETHANE ◇ R 10 ◇ RCRA WASTE NUMBER U211 ◇ TETRACHLOORKOOLSTOF (DUTCH) ◇ TETRACHLOORMETAAN ◇ TETRACHLORKOHLENSTOFF, TETRA (GERMAN) ◇ TETRACHLORMETHAN (GERMAN) ◇ TETRACHLOROCARBON ◇ TETRACHLOROMETHANE ◇ TETRACHLORURE de CARBONE (FRENCH) ◇ TETRACLOROMETANO (ITALIAN) ◇ TETRACLORURO di CARBONIO (ITALIAN) ◇ TETRAFINOL ◇ TETRAFORM ◇ TETRASOL ◇ UNIVERM ◇ VERMOESTRICID

TOXICITY DATA with REFERENCE
skn-rbt 4 mg MLD XEURAQ MDDC-1715
skn-rbt 500 mg/24H MLD 85JCAE-,91,86
eye-rbt 2200 μg/30S MLD XEURAQ MDDC-1715
eye-rbt 500 mg/24H MLD 85JCAE-,91,86
mmo-sat 20 uL/L EJMBA2 18,213,83
mmo-asn 5000 ppm MUREAV 147,288,85
ihl-rat TCLo:250 ppm/8H (female 10-15D post):REP DABBBA 32,2021,71
orl-rat TDLo:3 g/kg (14D preg):TER BEXBAN 82,1262,76
scu-rat TDLo:15600 mg/kg/12W-I:ETA JJIND8 38,891,67
orl-mus TDLo:4400 mg/kg/19W-I:NEO JJIND8 20,431,58
scu-rat TD:182 g/kg/70W-I:CAR JJIND8 44,419,70
ihl-hmn TCLo:20 ppm:GIT 85CYAB 2,136,59
orl-wmn TDLo:1800 mg/kg:EYE,CNS TXMDAX 69,86,73
orl-man TDLo:1700 mg/kg:CNS,PUL,GIT SAMJAF 49,635,75
ihl-hmn LCLo:1000 ppm PCOC** -,198,66
ihl-hmn TCLo:45 ppm/3D:CNS,GIT LANCAO 1,360,60
ihl-hmn TCLo:317 ppm/30M:GIT JAMAAP 103,962,34
ihl-hmn LCLo:5 pph/5M TABIA2 3,231,33
unk-man LDLo:93 mg/kg 85DCAI 2,73,70
orl-rat LD50:2350 mg/kg ARTODN 54,275,83
ihl-rat LC50:8000 ppm/4H NPIRI* 1,16,74
skn-rat LD50:5070 mg/kg SPEADM 78-1,16,78
ipr-rat LD50:1500 mg/kg XEURAQ MDDC-1715
orl-mus LD50:8263 mg/kg JPPMAB 3,169,51
ihl-mus LC50:9526 ppm/8H JIDHAN 29,382,47
ipr-mus LD50:572 mg/kg PHMCAA 10,172,68
orl-dog LDLo:1000 mg/kg QJPPAL 7,205,34
ihl-dog LCLo:14620 ppm/8H NIHBAZ 191,1,49
ipr-dog LD50:1500 mg/kg TXAPA9 10,119,67
ivn-dog LDLo:125 mg/kg QJPPAL 7,205,34
ihl-cat LCLo:38110 ppm/2H HBAMAK 4,1405,35
scu-cat LDLo:300 mg/kg JPETAB 63,153,38

CONSENSUS REPORTS: NTP Fifth Annual Report on Carcinogens. IARC Cancer Review: Group 2B IMEMDT 7,143,87; Animal Sufficient Evidence IMEMDT 20,371,79; IMEMDT 1,53,72; Human Inadequate Evidence IMEMDT 1,53,72; Human Limited Evidence IMEMDT 20,371,79. Community Right-To-Know List. EPA Genetic Toxicology Program. Reported in EPA TSCA Inventory.

OSHA PEL: (Transitional: TWA 10 ppm; CL 25 ppm; PK 200 ppm/5 min) TWA 2 ppm
ACGIH TLV: TWA 5 ppm; STEL 30 (skin); Suspected Human Carcinogen
DFG MAK: 10 ppm (65 mg/m^3); BEI: 1.6 mL/m^3 in alveolar air 1 hour after exposure; Suspected Carcinogen.
NIOSH REL: (Carbon Tetrachloride) CL 2 ppm/60M
DOT Classification: ORM-A; Label: None; Poison B; Label: Poison.

SAFETY PROFILE: Confirmed carcinogen with experimental carcinogenic, neoplastigenic, and tumorigenic data. A human poison by ingestion and possibly other routes. Poison by subcutaneous and intravenous routes. Mildly toxic by inhalation. Human systemic effects by inhalation and ingestion: nausea or vomiting, pupillary constriction, coma, antipsychotic effects, tremors, somnolence, anorexia, unspecified respiratory system and gastrointestinal system effects. Experimental teratogenic and reproductive effects. An eye and skin irritant. Damages liver, kidneys, and lungs. Mutation data reported. A narcotic. Individual susceptibility varies widely. Contact dermatitis can result from skin contact.

Carbon tetrachloride has a narcotic action resembling that of chloroform, though not as strong. Following exposure to high concentrations, the victim may become unconscious, and if exposure is not terminated, death can follow from respiratory failure. The after-effects following recovery from narcosis are more serious than those of delayed chloroform poisoning, usually taking the form of damage to the kidneys, liver, and lungs. Exposure to lower concentrations, insufficient to produce unconsciousness, usually results in severe gastrointestinal upset and may progress to serious kidney and hepatic damage. The kidney lesion is an acute nephrosis; the liver involvement consists of an acute degeneration of the central portions of the lobules. When recovery takes place, there may be no permanent disability. Marked variation in individual susceptibility to carbon tetrachloride exists; some persons appear to be unaffected by exposures which seriously poison their fellow workers. Alcoholism and previous liver and kidney damage seem to render the individual more susceptible. Concentrations on the order of 1000 to 1500 ppm are sufficient to cause symptoms if exposure continues for several hours. Repeated daily exposure to such concentration may result in poisoning.

Though the common form of poisoning following industrial exposure is usually one of gastrointestinal upset, which may be followed by renal damage, other cases have been reported in which the central nervous system has been effected resulting in the production of polyneuritis, narrowing of the visual fields, and other neurological changes. Prolonged exposure to small amounts of carbon tetrachloride has also been reported as causing cirrhosis of the liver.

Locally, a dermatitis may be produced following long or repeated contact with the liquid. The skin oils are removed and the skin becomes red, cracked and dry. The effect of carbon tetrachloride on the eyes either as a vapor or as a liquid, is one of irritation with lacrimation and burning.

Industrial poisoning is usually acute with malaise, headache, nausea, dizziness, and confusion which may be followed by stupor and sometimes loss of consciousness. Symptoms of liver and kidney damage may follow later with development of dark urine, sometimes jaundice and liver enlargement, followed by scanty urine, albumenuria and renal casts; uremia may develop and cause death. Where exposure has been less acute, the symptoms are usually headache, dizziness, nausea, vomiting, epigastric distress, loss of appetite, and fatigue. Visual disturbances (blind spots, spots before the eyes, a visual "haze" and restriction of the visual fields), secondary anemia, and occasionally a slight jaundice may occur. Dermatitis may be noticed on the exposed parts.

Forms impact-sensitive explosive mixtures with particulates of many metals, e.g., aluminum (when ball milled or heated to 152° in a closed container); barium (bulk metal also reacts violently); beryllium; potassium (200 times more shock-sensitive than mercury fulminate); potassium-sodium alloy (more sensitive than potassium); lithium; sodium; zinc (burns readily). Also forms explosive mixtures with chlorine trifluoride; calcium hypochlorite (heat sensitive); calcium disilicide (friction and pressure sensitive); triethyldialuminum trichloride (heat sensitive); decaborane(14) (impact sensitive); dinitrogen tetraoxide. Violent or explosive reaction on contact with fluorine. Forms explosive mixtures with ethylene between 25-105° and 30-80 bar. Potentially explosive reaction on contact with boranes. 9:1 mixtures of methanol and CCl_4 react exothermically with aluminum, magnesium, or zinc. Potentially dangerous reaction with dimethyl formamide; 1,2,3,4,5,6-hexachlorocyclohexane; or dimethylacetamide when iron is present as a catalyst. CCl_4 has caused explosions when used as a fire extinguisher on wax and uranium fires. Incompatible with aluminum trichloride, dibenzoyl peroxide, potassium-tert-butoxide. Vigorous exothermic reaction with allyl alcohol; $Al(C_2H_5)_3$; (benzoyl peroxide + C_2H_4); BrF_3; diborane; disilane; liquid O_2; Pu; ($AgClO_4$ + HCl); potassium-tert-butoxide; tetraethylenepentamine; tetrasilane; trisilane; Zr. When heated to decomposition it emits toxic fumes of Cl^- and phosgene. It has been banned from household use by the FDA. See also CHLORINATED HYDROCARBONS, ALIPHATIC.

CBY250 CAS:75-73-0 ***HR: 2***
CARBON TETRAFLUORIDE
DOT: UN 1982
mf: CF_4 mw: 88.01

PROP: Colorless gas. Mp: −184°, bp: −127.7°, d: 1.96 @ −184°.

SYNS: ARCTON O ◇ CARBON FLUORIDE ◇ FC 14 ◇ FREON 14 ◇ HALOCARBON 14 ◇ HALON 14 ◇ PERFLUOROMETHANE ◇ R 14 ◇ R 14 (refrigerant) ◇ TETRAFLUOROMETHANE (DOT)

TOXICITY DATA with REFERENCE
ihl-rat LCLo:895000 ppm/15M MRLR** No.23,50

CONSENSUS REPORTS: Reported in EPA TSCA Inventory.

DOT Classification: Nonflammable Gas; Label: Nonflammable Gas.

SAFETY PROFILE: Mildly toxic by inhalation. Less chronically toxic than carbon tetrachloride. Violent reaction with Al. When heated to decomposition it emits toxic fumes of F^-. See also FLUORIDES.

CBY500 CAS:507-25-5 ***HR: 3***
CARBON TETRAIODIDE
mf: CI_4 mw: 519.61

PROP: Octahedral, red crystals. Mp: 171° decomp, d: 4.32.

SYNS: CARBON IODIDE ◇ TETRAIODOMETHANE

TOXICITY DATA with REFERENCE
ivn-mus LD50:178 mg/kg CSLNX* NX#02298

CONSENSUS REPORTS: Reported in EPA TSCA Inventory.

SAFETY PROFILE: Poison by intravenous route. See also IODOFORM. Explodes on contact with bromine trifluoride. Mixtures with lithium particles are impact-sensitive explosives. Vigorous reaction with BrF_2. When heated to decomposition it emits toxic fumes of I^-.

CBY750 CAS:75-46-7 ***HR: 2***
CARBON TRIFLUORIDE
DOT: UN 1984
mf: CHF_3 mw: 70.02

PROP: Colorless, odorless gas. Mp: −163°, bp: −82.2°, d: 1.52 (liquid) @ −100°.

SYNS: ARCTON ◇ FLUOROFORM ◇ FLUORYL ◇ FREON 23 ◇ FREON F-23 ◇ GENETRON-23 ◇ HALOCARBON 23 ◇ METHYL TRIFLUORIDE ◇ R 23 ◇ TRIFLUOROMETHANE (DOT)

TOXICITY DATA with REFERENCE
sln-dmg-ihl 98 pph/10M ENVRAL 7,275,74

CONSENSUS REPORTS: EPA Genetic Toxicology Program. Reported in EPA TSCA Inventory.

DOT Classification: Nonflammable Gas; Label: Nonflammable Gas.

SAFETY PROFILE: Narcotic in high concentration. A mild respiratory irritant. Mutation data reported. See also FLUORIDES. When heated to decomposition it emits toxic fumes of F^-.

CCA000 CAS:14435-92-8 ***HR: 3***
CARBONYL DIAZIDE
mf: CN_6O mw: 112.05

$O{:}C(N_3)_2$

SYNS: CARBONIC DIAZIDE ◇ CARBONYL AZIDE

SAFETY PROFILE: A very dangerous high explosive. May explode violently in ice water or on exposure to light. When heated to decomposition it emits toxic fumes of CO and NO_x. See also AZIDES and CARBONYLS.

CCA125 CAS:6470-09-3 ***HR: 2***
CARBONYL DIISOTHIOCYANATE
mf: $C_3N_2OS_2$ mw: 144.17

$O{:}C(N{=}C{=}S)_2$

SAFETY PROFILE: A strong Lewis acid. It reacts explosively with dimethyl sulfoxide. When heated to decomposition it emits toxic fumes of CO, SO_x, NO_x, and CN^-. See also CARBONYLS and THIOCYANATES.

CCA500 CAS:353-50-4 ***HR: 3***
CARBONYL FLUORIDE
DOT: UN 2417
mf: CF_2O mw: 66.01

PROP: Colorless gas; pungent. Hygroscopic, mp: −114°, bp: −83°, d: 1.139 @ −114°.

SYNS: CARBON DIFLUORIDE OXIDE ◇ CARBON FLUORIDE OXIDE ◇ CARBONIC DIFLUORIDE ◇ CARBON OXYFLUORIDE ◇ CARBONYL DIFLUORIDE ◇ DIFLUOROFORMALDEHYDE ◇ FLUOPHOSGENE ◇ FLUOROFORMYL FLUORIDE ◇ FLUOROPHOSGENE ◇ RCRA WASTE NUMBER U033

TOXICITY DATA with REFERENCE
ihl-rat LC50:360 ppm/1H AIHAAP 29,41,68

CONSENSUS REPORTS: Reported in EPA TSCA Inventory.

OSHA PEL: TWA 2 ppm; STEL 5 ppm
ACGIH TLV: TWA 2 ppm; STEL 5 ppm
DOT Classification: Poison A; Label: Poison Gas.

SAFETY PROFILE: A poison. Moderately toxic by inhalation. A powerful irritant. Hydrolyzes instantly to form HF on contact with moisture. See also CARBONYLS, HYDROFLUORIC ACID, and FLUORINE. Incompatible with hexafluoroisopropylideneamino-lithium. When heated to decomposition it emits toxic fumes of CO and F^-. See CARBON MONOXIDE for fire and explosion hazard.

CCB250 ***HR: 3***
CARBONYL LITHIUM
mf: CLiO mw: 34.95

SAFETY PROFILE: Explodes on contact with water. When heated to decomposition it emits toxic fumes of CO. See also CARBONYLS and LITHIUM COMPOUNDS.

CCB500 ***HR: 3***
CARBONYL POTASSIUM
mf: CKO mw: 67.11

SAFETY PROFILE: Explodes on heating in air or contact with water. Incompatible with oxygen. When heated to decomposition it emits toxic fumes of CO and K_2O. See also CARBONYLS.

CCB609 ***HR: 3***
CARBONYLS

PROP: The (CO) group with a metal (M). They may exist as dimeric acetylene derivatives (MOC≡COM) or as salts of hexahydroxybenzene.

SAFETY PROFILE: Most carbonyls are highly toxic. The toxicity of carbonyls depends in part, but not always entirely, on their ready decomposition which releases carbon monoxide. Symptoms are due in part to carbon monoxide and in part to the direct irritating action of the carbonyl. See specific carbonyl in question. Many carbonyl metals ignite spontaneously in air, some with a delay period. Others are moderate fire and explosion hazards when exposed to heat or flame. Carbonyls of alkali metals are potentially explosive. Hypergolic reaction with dinitrogen tetraoxide. They react with water or steam to produce toxic and flammable vapors; can react vigorously with oxidizing materials. When heated to decomposition they emit highly toxic fumes of carbon monoxide. See also CARBON MONOXIDE and POWDERED METALS.

CCB750 ***HR: 3***
CARBONYL SODIUM
mf: CNaO mw: 51.00

SAFETY PROFILE: Incompatible with water. Explodes when heated in air @ 90°C. When heated to decomposition it emits toxic fumes of CO and Na_2O. See also CARBONYLS and SODIUM COMPOUNDS.

CCC000 CAS:463-58-1 ***HR: 3***
CARBONYL SULFIDE
DOT: UN 2204
mf: COS mw: 60.07

PROP: Gas or liquid. Mp: −138°, bp: 49.9°, lel: 12%, uel: 28.5%, d: liq 1.24 @ −87°, vap d: 2.1.

SYNS: CARBON OXIDE SULFIDE ◇ CARBON OXYSULFIDE ◇ CARBONYL SULFIDE-^{32}S ◇ OXYCARBON SULFIDE

TOXICITY DATA with REFERENCE
ipr-rat LD50:23 mg/kg TXAPA9 55,198,80
ihl-mus LCLo:1200 ppm/35M BDCGAS 76,299,43

CONSENSUS REPORTS: Community Right-To-Know List. Reported in EPA TSCA Inventory.

DOT Classification: Poison A; Label: Poison Gas and Flammable Gas.

SAFETY PROFILE: Poison by intraperitoneal route. Mildly toxic by inhalation. Narcotic in high concentration. An irritant. May liberate highly toxic hydrogen sulfide upon decomposition. A very dangerous fire hazard and moderate explosion hazard when exposed to heat or flame. Can react vigorously with oxidizing materials. To fight fire, stop flow of gas or use CO_2, dry chemical, or water spray. When heated to decomposition it emits toxic fumes of CO. See also CARBONYLS and SULFIDES.

CCC075 CAS:41575-94-4 ***HR: 3***
CARBOPLATIN
mf: $C_6H_{12}N_2O_4Pt$ mw: 371.29

SYNS: CBDCA ◇ cis-(1,1-CYCLOBUTANEDICARBOSYLATO) DIAMMINEPLATINUM(II) ◇ 1,1-CYCLOBUTANEDICARBOXYLATE DIAMMINE PLATINUM(II) ◇ cis-(1,1-CYCLOBUTANEDICARBOXYLATO)DIAMMINEPLATINUM(II) ◇ DIAMMINE(1,1-CYCLOBUTANEDICARBOXYLATO)PLATINUM (II) ◇ cis-DIAMMINE(1,1-CYCLOBUTANEDICARBOXYLATO)PLATINUM(II) ◇ JM 8 ◇ NSC-241240

TOXICITY DATA with REFERENCE
mmo-esc 300 μmol/L MUREAV 173,13,86
dnd-mus:leu 200 μmol/L CNREA8 45,4043,85
mnt-ham:lng 8250 nmol/L NEOLA4 31,655,84
sce-ham:lng 8250 nmol/L NEOLA4 31,655,84
ivn-rat LD50:61 mg/kg JJIND8 67,201,81
ipr-mus LD50:150 mg/kg EJCODS 20,1087,84
ivn-mus LD50:89360 μg/kg NCISP* JAN86
unr-mus LD50:180 mg/kg RRCRBU 48,12,74

SAFETY PROFILE: Poison by intravenous, intraperitoneal, and possibly other routes. Mutation data reported. When heated to decomposition it emits toxic fumes of NO_x. See also AMINES and PLATINUM COMPOUNDS.

CCC100 CAS:35700-23-3 ***HR: 3***
CARBOPROST
mf: $C_{21}H_{36}O_5$ mw: 368.57

SYNS: 15-M3-PGF2-α ◇ METHYL-PGF2-α ◇ 15(S)-15-METHYL PGF2-α ◇ 15-METHYLPROSTAGLANDIN F2-α ◇ 15(S)-METHYLPROSTAGLANDIN F2-α ◇ 15(S)-15-METHYL-PROSTAGLANDIN F2-α ◇ (15S)-15-METHYLPROSTAGLANDIN F2-α ◇ PROSTIN ◇ (5Z,9-α,11-α,13E,15S)-9,11,15-TRIHYDROXY-15-METHYLPROSTA-5,13-DIEN-1-OIC ACID ◇ U 32921E

TOXICITY DATA with REFERENCE
ims-wmn TDLo:12 μg/kg (30D preg):REP PRGLBA 14,785,77

SAFETY PROFILE: In humans very small amounts cause abortion by intramuscular, intravaginal and intraplacental routes. Experimental reproductive effects. When heated to decomposition it emits acrid smoke and fumes.

CCC110 CAS:58551-69-2 ***HR: 3***
CARBOPROST TROMETHAMINE
mf: $C_{21}H_{36}O_5 \cdot C_4H_{11}NO_3$ mw: 489.73

SYNS: 15(2)15-METHYL PGF2-α TROMETHAMINE SALT ◇ 15(S)15-METHYL PROSTAGLANDIN F2-α TROMETHAMINE ◇ 9,11,15-TRIHYDROXY-15-METHYL-PROSTA-5,13-DIEN-1-OIC ACID, (5Z,9-α,11-α,13E,15S)-compd. with 2-AMINO-2-(HYDROXYMETHYL)-1,3-PROPANEDIOL (1:1)

TOXICITY DATA with REFERENCE
ims-wmn TDLo:30 μg/kg (14W preg):REP CCPTAY 11,533,75
scu-rat TDLo:300 μg/kg (female 9-11D post):TER APTRDI 4,157,78
ivn-rat LD50:25100 μg/kg APTRDI 4,157,78
ivn-mus LD50:131 mg/kg APTRDI 4,157,78

SAFETY PROFILE: Poison by intravenous route. In humans, very small amounts cause abortion by intramuscular route. An experimental teratogen. Other experimental reproductive effects. When heated to decomposition it emits toxic fumes of NO_x.

CCC250 CAS:59-31-4 ***HR: 3***
CARBOSTYRIL
mf: C_9H_7NO mw: 145.17

PROP: White crystals or powder. Mp: 76°, bp: 267°. Very sltly sol in water.

SYN: 2-QUINOLINOL

TOXICITY DATA with REFERENCE
ipr-mus LD50:150 mg/kg NTIS** AD607-952

CONSENSUS REPORTS: Reported in EPA TSCA Inventory.

SAFETY PROFILE: Poison by intraperitoneal route. A fungicide. A central nervous system stimulant. When heated to decomposition it emits toxic fumes of NO_x.

CCC325 CAS:33330-91-5 ***HR: 3***
1-p-CARBOXAMIDOPHENYL)-3,3-DIMETHYLTRIAZINE
mf: $C_9H_{12}N_4O$ mw: 192.25

SYNS: 1-(4'-CARBOXYLAMIDOPHENYL)-3,3-DIMETHYLTRIAZINE ◇ CB 10286 ◇ p-(3,3-DIMETHYLTRIAZENO)BENZAMIDE ◇ p-(3,3-DIMETHYL-1-TRIAZENYL)BENZAMIDE ◇ 4-(3,3-DIMETHYL-1-TRIAZENYL)BENZAMIDE ◇ 1-(4'-KARBOXYLAMIDOFENYL)-3,3-DIMETHYLTRIAZENU (CZECH) ◇ 1-(4'-KARBOXYLAMIDOPHENYL)-3,3-DIMETHYLTRIAZEN (GERMAN)

TOXICITY DATA with REFERENCE
mma-sat 91 μmol/L JMCMAR 22,473,79
orl-rat LD50:54 mg/kg CKFRAY 27,384,78
ipr-mus LD50:356 mg/kg CTRRDO 62,721,78

SAFETY PROFILE: Poison by ingestion and intraperitoneal routes. Mutation data reported. When heated to decomposition it emits toxic fumes of NO_x. See also AMIDES.

CCC500 CAS:5234-68-4 ***HR: 3***
CARBOXINE
mf: $C_{12}H_{13}NO_2S$ mw: 235.32

SYNS: 5-CARBOXANILIDO-2,3-DIHYDRO-6-METHYL-1,4-OXATHIIN ◇ CARBOXIN (USDA) ◇ D 735 ◇ DCMO ◇ 2,3-DIHYDRO-5-CARBOXANILIDO-6-METHYL-1,4-OXATHIIN ◇ 5,6-DIHYDRO-2-METHYL-3-CARBOXANILIDO-1,4-OXATHIIN (GERMAN) ◇ 2,3-DIHYDRO-6-METHYL-1,4-OXATHIIN-5-CARBOXANILIDE ◇ 5,6-DIHYDRO-2-METHYL-1,4-OXATHIIN-3-CARBOXANILIDE ◇ 5,6-DIHYDRO-2-METHYL-N-PHENYL-1,4-OXATHIIN-3-CARBOXAMIDE ◇ F 735 ◇ FLO PRO V SEED PROTECTANT ◇ VITAVAX

TOXICITY DATA with REFERENCE
orl-rat LD50:430 mg/kg GTPZAB 23(2),55,79
skn-rat LD50:1050 mg/kg GTPZAB 23(2),55,79
orl-mus LD50:3200 mg/kg GTPZAB 23(2),55,79
orl-ckn LD50:24 g/kg VETNAL 54(6),85,78
unk-mam LD50:3200 mg/kg 30ZDA9 -,423,71
orl-bwd LD50:42200 μg/kg AECTCV 12,355,83

SAFETY PROFILE: Poison by ingestion. Moderately toxic by skin contact and possibly other routes. When heated to decomposition it emits very toxic fumes of NO_x and SO_x.

CCC750 CAS:141-82-2 ***HR: 3***
CARBOXYACETIC ACID
mf: $C_3H_4O_4$ mw: 104.07

SYNS: DICARBOXYMETHANE ◇ METHANEDICARBOXYLIC ACID ◇ PROPANEDIOIC ACID ◇ USAF EK-695

TOXICITY DATA with REFERENCE
skn-rbt 500 mg/24H MLD BIOFX* 22-3/71

eye-rbt 100 mg SEV BIOFX* 22-3/71
orl-rat LD50:1310 mg/kg BIOFX* 22-3/71
orl-mus LD50:4000 mg/kg BIJOAK 34,1196,40
ipr-mus LD50:300 mg/kg NTIS** AD277-689

CONSENSUS REPORTS: Reported in EPA TSCA Inventory.

SAFETY PROFILE: Poison by intraperitoneal route. Moderately toxic by ingestion. A skin and severe eye irritant. When heated to decomposition it emits acrid smoke and irritating fumes.

CCD625 CAS:56743-33-0 ***HR: 3***
CARBOXYBENZENESULFONYL AZIDE
mf: $C_7H_5N_3O_4S$ mw: 227.19

$HOCO•C_6H_4SO_2N_3$

SAFETY PROFILE: Decomposes explosively at 120°C. When heated to decomposition it emits toxic fumes of SO_x and NO_x. See also AZIDES.

CCD750 CAS:69365-73-7 ***HR: 3***
N-(2-CARBOXYCAPROYL)HYDRAZOBENZENE CALCIUM SALT HEMIHYDRATE
mf: $C_{38}H_{42}N_4O_6•Ca•1/2H_2O$ mw: 699.93

SYNS: BUMADIZON CALCIUM SALT HEMIHYDRATE ◇ BUTYL-MALONIC ACID MONO(1,2-DIPHENYLHYDRAZIDE) CALCIUM SALT HEMIHYDRATE ◇ BUTYL-MALONSAEURE-MONO-(1,2-DIPHENYL-HYDRAZID)-CALCIUM-SEMIHYDRAT (German) ◇ BUTYLPROPA-NEDIOIC ACID MONO(1,2-DIPHENYLHYDRAZIDE) CALCIUM SALT HEMIHYDRATE ◇ α-CARBOXYCAPROYL-N,N'-DIPHENYLHYDRAZ-INE CALCIUM SALT HEMIHYDRATE ◇ EUMOTOL ◇ RHEUMATOL

TOXICITY DATA with REFERENCE
orl-rat LD50:1250 mg/kg ARZNAD 23,1215,73
ivn-mus LD50:263 mg/kg ARZNAD 23,1215,73
orl-mus LD50:2500 mg/kg ARZNAD 23,1215,73
ivn-mus LD50:258 mg/kg ARZNAD 23,1215,73

SAFETY PROFILE: Poison by intravenous route. Moderately toxic by ingestion. An analgesic, antipyretic, and antirheumatic. When heated to decomposition it emits toxic fumes of NO_x.

CCE000 CAS:148-78-7 ***HR: 3***
p-CARBOXYCARBANILIC ACID-4-BIS(2-CHLOROETHYLAMINO)PHENYL ESTER
mf: $C_{18}H_{18}Cl_2N_2O_4$ mw: 397.28

SYN: p-(N,N-DI-2-CHLOROETHYLAMINO)PHENYL-N-(p-CARBOXY-PHENYL)CARBAMATE

TOXICITY DATA with REFERENCE
ipr-rat LD50:56 mg/kg JMCMAR 8,167,65
ipr-mus LD50:290 mg/kg JMCMAR 8,167,65

SAFETY PROFILE: Poison by intraperitoneal route. See also CARBAMATES. When heated to decomposition it emits very toxic fumes of Cl^- and NO_x.

CCE250 CAS:22788-18-7 ***HR: D***
CARBOXYCYCLOPHOSPHAMIDE
mf: $C_7H_{15}Cl_2N_2O_4P$ mw: 366.33

SYNS: 3-((AMINO(BIS(2-CHLOROETHYL)AMINO)PHOSPHINYL) OXY)PROPANOIC ACID ◇ N,N-BIS(2-CHLOROETHYL)PHOSPHORO-DIAMIDATE HYDRACRYLIC ACID ◇ CARBOXYPHOSPHAMIDE

TOXICITY DATA with REFERENCE
mma-sat 125 μg/plate MUREAV 129,47,84
mma-esc 20 mmol/L JTEHD6 3,637,77
dni-hmn:lym 800 μmol/L AGACBH 4,117,74
sce-hmn:lym 100 mmol/L MUREAV 129,47,84

SAFETY PROFILE: Human mutation data reported. When heated to decomposition it emits very toxic fumes of PO_x, NO_x, and Cl^-.

CCE500 CAS:493-52-7 ***HR: 3***
2-CARBOXY-4'-(DIMETHYLAMINO)AZOBENZENE
mf: $C_{15}H_{15}N_3O_2$ mw: 269.33

PROP: Shiny violet crystals.

SYNS: C.I. 13020 ◇ C.I. ACID RED 2 ◇ p-(DIMETHYLAMINO)AZO-BENZENE-o-CARBOXYLIC ACID ◇ 4'-DIMETHYLAMINOAZOBEN-ZENE-2-CARBOXYLIC ACID ◇ o-((p-(DIMETHYLAMINO)PHENYL)AZO) BENZOIC ACID ◇ 2-((4-DIMETHYLAMINO)PHENYLAZO)BENZOIC ACID ◇ METHYL RED

TOXICITY DATA with REFERENCE
mma-sat 50 μg/plate MUREAV 56,249,78
dnr-bcs 2 mg/disc TRENAF 27,153,76
dns-rat:lvr 10 μmol/L CNREA8 46,1654,86
orl-rat TDLo:12 g/kg/57W-C:ETA BJCAAI 9,310,55

CONSENSUS REPORTS: IARC Cancer Review: Group 3 IMEMDT 7,56,87; Animal Inadequate Evidence IMEMDT 8,161,75. Reported in EPA TSCA Inventory. EPA Genetic Toxicology Program.

SAFETY PROFILE: Questionable carcinogen with experimental tumorigenic data. Mutation data reported. When heated to decomposition it emits toxic fumes of NO_x.

CCE750 CAS:20691-84-3 ***HR: 3***
3'-CARBOXY-4-DIMETHYLAMINOAZOBENZENE
mf: $C_{15}H_{15}N_3O_2$ mw: 269.33

SYN: 3-((p-(DIMETHYLAMINO)PHENYL)AZO)BENZOICACID

TOXICITY DATA with REFERENCE
mma-sat 1 μmol/plate CRNGDP 1,121,80
dns-rat:lvr 10 μmol/L CNREA8 46,1654,86
orl-rat TDLo:12 g/kg/57W-C:ETA BJCAAI 9,310,55
orl-rat LD50:3757 mg/kg NEOLA4 27,237,80

SAFETY PROFILE: Moderately toxic by ingestion. Questionable carcinogen with experimental tumorigenic data. Mutation data reported. When heated to decomposition it emits toxic fumes of NO_x.

CCF125 CAS:12758-40-6 ***HR: 2***
CARBOXYETHYLGERMANIUM SESQUIOXIDE
mf: $C_6H_{10}Ge_2O_7$ mw: 339.34

SYNS: BIS-β-CARBOXYETHYLGERMANIUM SESQUIOXIDE ◇ 2-CARBOXYETHYLGERMASESQUIOXANE ◇ 3,3'-(DIOXODIGERMOXANYLENE) DIPROPANOIC ACID ◇ DIPROPANOIC ACID GERMANIUM SEQUIOXIDE ◇ Ge 132 ◇ GERMANATE(2-), BIS(2-CARBOXYLATOETHYL)TRIOXODI-, DIHYDROGEN (9CI) ◇ 3,3'-(GERMANOIC ANHYDRIDE) DIPROPANOIC ACID

TOXICITY DATA with REFERENCE
ipr-rat TDLo:94 g/kg (male 90D pre):REP OYYAA2 20,271,80
orl-rat LD50:9500 mg/kg SIGZAL 46,227,86
ipr-rat LD50:3200 mg/kg SIGZAL 46,227,86
scu-rat LD50:16300 mg/kg SIGZAL 46,227,86
ivn-rat LD50:3200 mg/kg DRFUD4 5,545,80
orl-mus LD50:11400 mg/kg SIGZAL 46,227,86
ivn-mus LD50:2110 mg/kg DRFUD4 5,548,80

CONSENSUS REPORTS: Reported in EPA TSCA Inventory.

SAFETY PROFILE: Moderately toxic by intravenous route. Experimental reproductive effects. See also GERMANIUM COMPOUNDS.

CCF250 CAS:4033-46-9 ***HR: 3***
3-((2-CARBOXYETHYL)THIO)ALANINE
mf: $C_6H_{11}NO_4S$ mw: 193.24

SYN: S-2-CARBOXYETHYL-l-CYSTEINE

TOXICITY DATA with REFERENCE
scu-rat TDLo:520 mg/kg/52W-I:ETA BJCAAI 15,85,61

SAFETY PROFILE: Questionable carcinogen with experimental tumorigenic data. When heated to decomposition it emits very toxic fumes of SO_x and NO_x. See also AMINES.

CCF500 CAS:63907-33-5 ***HR: 3***
(3-(4-(CARBOXYLATOMETHOXY)PHENYL)-2-HYDROXYPROPYL)HYDROXY-MERCURATE(1-), SODIUM
mf: $C_{11}H_{13}HgO_5 \cdot Na$ mw: 448.82

SYN: (p-(2-HYDROXY-3-HYDROXYMERCURI)PROPYL)PHENOXY)ACETIC ACID, SODIUM SALT

TOXICITY DATA with REFERENCE
ivn-rbt LDLo:7 mg/kg JPETAB 41,21,31

CONSENSUS REPORTS: Mercury and its compounds are on the Community Right-To-Know List.

OSHA PEL: (Transitional: CL 1 mg(Hg)/10m^3) CL 0.1 mg(Hg)/m^3 (skin)
ACGIH TLV: TWA 0.1 mg(Hg)/m^3 (skin)
NIOSH REL: (Inorganic Mercury) TWA 0.05 mg(Hg)/m^3

SAFETY PROFILE: Poison by intravenous route. See also MERCURY COMPOUNDS. When heated to decomposition it emits toxic fumes of Hg and Na_2O.

CCF750 CAS:13442-14-3 ***HR: 3***
6-CARBOXYL-4-HYDROXYLAMINOQUINOLINE-1-OXIDE
mf: $C_{10}H_8N_2O_4$ mw: 220.20

SYN: 4-(HYDROXYAMINO)-6-QUINOLINECARBOXYLICACID-1-OXIDE

TOXICITY DATA with REFERENCE
scu-mus TDLo:120 mg/kg/50D-I:ETA BCPCA6 16,631,67

SAFETY PROFILE: Questionable carcinogen with experimental tumorigenic data. When heated to decomposition it emits toxic fumes of NO_x.

CCG000 CAS:1425-67-8 ***HR: 3***
6-CARBOXYL-4-NITROQUINOLINE-1-OXIDE
mf: $C_{10}H_6N_2O_5$ mw: 234.18

SYNS: 6-CARBOXY-4-NITROQUINOLINE-1-OXIDE ◇ 4-NITROQUINOLINE-6-CARBOXYLIC ACID-1-OXIDE ◇ 4-NITRO-6-QUINOLINE-CARBOXYLIC ACID-1-OXIDE

TOXICITY DATA with REFERENCE
mmo-esc 500 μg/plate CNREA8 32,2369,72
mrc-esc 500 μg/well CNREA8 32,2369,72
mmo-smc 100 mg/L IGSBAL 85,127,72
dnd-mus:fbr 100 μmol/L CNREA8 35,521,75
dns-ham:oth 4 μmol/L NATUAS 229,416,71
dnd-mam:lym 5 mg BIPMAA 4,409,66
scu-rat TDLo:90 mg/kg/20W-I:ETA GANNA2 58,397,67

CONSENSUS REPORTS: EPA Genetic Toxicology Program.

SAFETY PROFILE: Questionable carcinogen with experimental tumorigenic data. Mutation data reported. When heated to decomposition it emits toxic fumes of NO_x.

CCG250 CAS:64038-41-1 ***HR: 2***
2-(CARBOXY-METHOXY)BENZALDEHYDE SODIUM SALT
mf: $C_9H_7O_4 \cdot Na$ mw: 202.15

TOXICITY DATA with REFERENCE
orl-mus LD50:4200 mg/kg FEPRA7 19,24,60
ipr-mus LD50:1900 mg/kg FEPRA7 19,24,60

SAFETY PROFILE: Moderately toxic by intraperitoneal

route. Mildly toxic by ingestion. See also ALDEHYDES. When heated to decomposition it emits toxic fumes of Na_2O.

CCG500 CAS:36568-91-9 ***HR: 3***
(4-(CARBOXY METHOXY)-3-CHLOROPHENYL) (5,5-DIETHYL-2,4,6(1H,3H,5H)-PYRIMIDINETRIONATO-O)^2-MERCURY, MONOSODIUM SALT
mf: $C_{16}H_{18}ClHgN_2O_6{\cdot}Na$ mw: 593.39

SYNS: MERBAPHEN ◇ NOVASUROL

TOXICITY DATA with REFERENCE
ivn-rbt LDLo:20 mg/kg JPETAB 41,21,31

CONSENSUS REPORTS: Mercury and its compounds are on the Community Right-To-Know List.

OSHA PEL: (Transitional: CL 1 mg/10m^3) CL 0.1 mg(Hg)/m^3 (skin)
ACGIH TLV: TWA 0.1 mg(Hg)/m^3 (skin)
NIOSH REL: (Inorganic Mercury) TWA 0.05 mg(Hg)/m^3

SAFETY PROFILE: Poison by intravenous route. See also MERCURY COMPOUNDS. When heated to decomposition it emits very toxic fumes of Cl^-, NO_x, and Hg.

CCH000 CAS:9086-60-6 ***HR: 3***
CARBOXYMETHYLCELLULOSE NORDIC

SYNS: AMMONIUM CARBOXYMETHYL CELLULOSE ◇ CARBOXYMETHYL CELLULOSE, AMMONIUM SALT

TOXICITY DATA with REFERENCE
scu-rat TDLo:6600 mg/kg/73W-I:NEO RCBIAS 20,701,61

CONSENSUS REPORTS: Reported in EPA TSCA Inventory.

SAFETY PROFILE: Questionable carcinogen with experimental neoplastigenic data. When heated to decomposition it emits toxic fumes of NO_x and NH_3. See also CARBOXYMETHYLCELLULOSE.

CCH125 CAS:2387-59-9 ***HR: 2***
S-CARBOXYMETHYLCYSTEINE
mf: $C_5H_9NO_4S$ mw: 179.21

SYNS: AHR-3053 ◇ 3-((CARBOXYMETHYL)THIO)ALANINE ◇ LJ 206 ◇ S-CMC

TOXICITY DATA with REFERENCE
ipr-rat LD50:7800 mg/kg OYYAA2 14,567,77
scu-rat LD50:10300 mg/kg OYYAA2 14,567,77
ipr-mus LD50:2980 mg/kg IYKEDH 12,668,81
scu-mus LD50:9000 mg/kg IYKEDH 12,668,81

SAFETY PROFILE: Moderately toxic by intraperitoneal route. Mildly toxic by subcutaneous route. When heated to decomposition it emits toxic fumes of SO_x and NO_x. See also AMINES.

CCH199 ***HR: 2***
2-CARBOXYMETHYLISOTHIOURONIUM CHLORIDE
mf: $C_3H_7ClN_2O_2S$ mw: 170.61

$HOCO{\cdot}CH_2SC(:N^-H_2)NH_2Cl^-$

SYN: CARBOXYMETHYL CARBAMIMONIOTHIOATE CHLORIDE

SAFETY PROFILE: Reaction with chlorine may form the dangerously explosive nitrogen trichloride. When heated to decomposition it emits toxic fumes of Cl^-, SO_x, and NO_x. See also CHLORIDES.

CCH250 CAS:63938-93-2 ***HR: 3***
2-(CARBOXYMETHYLMERCAPTO)PHENYLSTIBONIC ACID
mf: $C_8H_9O_5SSb$ mw: 338.98

SYNS: 2-(CARBOXYMETHYLMERCAPTO)PHENYL-STIBONSAEURE (GERMAN) ◇ RO 2-1160 ◇ ((2-STIBONOPHENYL)THIO)ACETIC ACID

TOXICITY DATA with REFERENCE
orl-rat LD50:5000 mg/kg ARZNAD 4,116,54
orl-mus LD50:5000 mg/kg AIPTAK 85,100,51
scu-mus LD50:2520 mg/kg AIPTAK 85,100,51
ivn-mus LD50:965 mg/kg AIPTAK 85,100,51
ivn-rbt LD50:186 mg/kg AIPTAK 85,100,51
ipr-gpg LD50:350 mg/kg AIPTAK 85,100,51
ipr-ham LD50:550 mg/kg AIPTAK 85,100,51

CONSENSUS REPORTS: Antimony and its compounds are on the Community Right-To-Know List.

OSHA PEL: TWA 0.5 mg(Sb)/m^3
ACGIH TLV: TWA 0.5 mg(Sb)/m^3
NIOSH REL: (Antimony) TWA: 0.5 mg(Sb)/m^3

SAFETY PROFILE: Poison by intravenous and intraperitoneal routes. Moderately toxic by subcutaneous route. Mildly toxic by ingestion. When heated to decomposition it emits very toxic fumes of antimony and SO_x. See also ANTIMONY COMPOUNDS and MERCAPTANS.

CCH500 ***HR: 3***
CARBOXYMETHYLNITROSOUREA
mf: $C_3H_5N_3O_4$ mw: 147.11

TOXICITY DATA with REFERENCE
orl-rat TDLo:4 g/kg/74W-I:NEO JJIND8 62,1523,79
ipr-rat LD50:210 mg/kg JJIND8 62,1523,79

SAFETY PROFILE: Poison by intraperitoneal route. Questionable carcinogen with experimental neoplastigenic data. When heated to decomposition it emits toxic fumes of NO_x. See also N-NITROSO COMPOUNDS.

CCH750 CAS:6295-57-4 ***HR: 3***
2-CARBOXYMETHYLTHIOBENZOTHIAZOLE
mf: $C_9H_7NO_2S_2$ mw: 225.29

SYN: S-2-BENZOTHIAZOLYLTHIOGLYCOLICACID

TOXICITY DATA with REFERENCE
ipr-mus LD50:200 mg/kg NTIS** AD277-689

CONSENSUS REPORTS: Reported in EPA TSCA Inventory.

SAFETY PROFILE: Poison by intraperitoneal route. When heated to decomposition it emits very toxic fumes of NO_x and SO_x.

CCH800 ***HR: 2***
5-CARBOXYMETHYL-3-p-TOLYL-THIAZOLIDINE-2,4-DIONE-2-ACETOPHENONE HYDRAZONE
mf: $C_{20}H_{16}N_3O_4S$ mw: 394.45

SYN: 5-KARBOKSIMETIL-3-p-TOLIL-TIAZOLIDIN-2,4-DION-2-ACETOFENONHIDRAZON (CZECH)

TOXICITY DATA with REFERENCE
orl-rat LD50:600 mg/kg ZDVEA7 39(Suppl 1),20,70
ipr-rat LD50:2160 mg/kg ZDVEA7 39(Suppl 1),20,70
orl-mus LD50:820 mg/kg ZDVEA7 39(Suppl 1),20,70
ipr-mus LD50:1870 mg/kg ZDVEA7 39(Suppl 1),20,70

SAFETY PROFILE: Moderately toxic by ingestion and intraperitoneal routes. When heated to decomposition it emits toxic fumes of SO_x and NO_x.

CCI250 CAS:62-23-7 ***HR: 2***
1-CARBOXY-4-NITROBENZENE
mf: $C_7H_5NO_4$ mw: 167.13

PROP: Crystals. Mp: 242.4°, bp: sublimes, d: 1.550 @ 32°/4°.

SYNS: KYSELINA-p-NITROBENZOOVA (CZECH) ◇ p-NITROBENZOIC ACID ◇ 4-NITROBENZOIC ACID ◇ 4-NITRODRACYLIC ACID

TOXICITY DATA with REFERENCE
eye-rbt 20 mg/24H MOD 28ZPAK -,129,72
mmo-sat 100 ug/plate MUREAV 137,71,84
mma sat 10 μmol/plate MUREAV 58,11,78
bfa-rat/sat 400 mg/kg/4D PNASA6 72,4607,75
orl-rat LD50:1960 mg/kg CRSBAW 160,1097,66
ipr-rat LD50:1210 mg/kg CRSBAW 160,1097,66
par-rat LD50:1960 mg/kg CRSBAW 160,1097,66
ipr-mus LD50:880 mg/kg CRSBAW 160,1097,66
ivn-mus LD50:770 mg/kg CRSBAW 160,1097,66
par-mus LD50:1470 mg/kg CRSBAW 160,1097,66

CONSENSUS REPORTS: Reported in EPA TSCA Inventory. EPA Genetic Toxicology Program.

SAFETY PROFILE: Moderately toxic by ingestion, intravenous, parenteral, and intraperitoneal routes. An eye irritant. Mutation data reported. When heated to decomposition it emits toxic fumes of NO_x. See also NITRO COMPOUNDS of AROMATIC HYDROCARBONS.

CCI500 CAS:41956-77-8 ***HR: 2***
2-(5-CARBOXYPENTYL)-4-THIAZOLIDONE
mf: $C_9H_{15}NO_3S$ mw: 217.31

SYNS: ACIDOMYCIN ◇ ACTITHIAZIC ACID ◇ CINNAMONIN ◇ 1-MYCOBACIDIN ◇ 1-4-OXO-2-THIAZOLIDINEHEXANOIC ACID ◇ 4-THIAZOLIDONE-2-CAPROIC ACID ◇ epsilon-(2-(4-THIAZOLIDONE)) HEXANOIC ACID

TOXICITY DATA with REFERENCE
scu-mus LD50:20 g/kg PHBUA9 1,84,53
ivn-mus LD50:3500 mg/kg PHBUA9 1,84,53

SAFETY PROFILE: Moderately toxic by intravenous route. Mildly toxic by subcutaneous route. When heated to decomposition it emits very toxic fumes of NO_x and SO_x.

CCI550 CAS:1197-16-6 ***HR: 3***
p-CARBOXY PHENYLARSENOXIDE
mf: $C_7H_5AsO_3$ mw: 212.04

SYNS: ARSINE, OXO(4-CARBOXY)PHENYL- ◇ BENZOIC ACID, 4-ARSENOSO-

TOXICITY DATA with REFERENCE
ivn-rbt LD50:2800 μg/kg JPETAB 80,93,44

OSHA PEL: TWA 0.5 mg(As)/m^3

SAFETY PROFILE: Poison by intravenous route. When heated to decomposition it emits toxic fumes of As.

CCJ000 CAS:64050-44-8 ***HR: 2***
4'-CARBOXYPHENYLMETHANESULFONANILIDE, SODIUM SALT
mf: $C_{14}H_{12}NO_4S{\cdot}Na$ mw: 313.32

TOXICITY DATA with REFERENCE
orl-mus LD50:2450 mg/kg JPETAB 91,263,47
scu-mus LD50:1650 mg/kg JPETAB 91,263,47
ivn-mus LD50:1300 mg/kg JPETAB 91,263,47
ivn-dog LD50:1693 mg/kg JPETAB 91,263,47
ivn-rbt LD50:1419 mg/kg JPETAB 91,263,47

SAFETY PROFILE: Moderately toxic by ingestion, subcutaneous, and intravenous routes. When heated to decomposition it emits very toxic fumes of SO_x, Na_2O, and NO_x.

CCJ350 CAS:65296-81-3 ***HR: 3***
4-CARBOXYPHTHALATO(1,2-DIAMINOCYCLOHEXANE)PLATINUM(II)
mf: $C_{15}H_{18}N_2O_6Pt$ mw: 517.44

SYNS: (CYCLOHEXANE-1,2-DIAMMINE)(4-CARBOXYPHTHLATO)PLATINUM(II) ◇ NSC 271674

TOXICITY DATA with REFERENCE
ivn-rat LD50:84 mg/kg JJIND8 67,201,81
ipr-mus LD50:46100 μg/kg NCISP* JAN86
ivn-mus LD50:40130 μg/kg NCISP* JAN86

SAFETY PROFILE: Poison by intravenous and intraperitoneal routes. When heated to decomposition it emits toxic fumes of NO_x. See also PLATINUM COMPOUNDS.

CCJ375 ***HR: 2***
3-CARBOXYPROPYL(2-PROPENYL)NITROSAMINE
mf: $C_7H_{12}N_2O_3$ mw: 172.21

SYN: 4-(ALLYLNITROSOAMINO)BUTRIC ACID

TOXICITY DATA with REFERENCE
scu-ham TDLo:23100 mg/kg/77W-I:CAR CDPRD4 4,79,81

SAFETY PROFILE: Questionable carcinogen with experimental carcinogenic data. When heated to decomposition it emits toxic fumes of NO_x.

CCJ400 ***HR: 2***
CARBOXY VINYL POLYMER

PROP: A finely-divided white powder which disperses in water to yield a low viscosity acid solution. When neutralized, the solution is changed into a clear, stable gel (AIPTAK 114,258,58).

SYN: CP

TOXICITY DATA with REFERENCE
orl-rat LD50:4000 mg/kg AIPTAK 114,258,58
orl-mus LD50:4300 mg/kg AIPTAK 114,258,58
orl-gpg LD50:2000 mg/kg AIPTAK 114,258,58

SAFETY PROFILE: Moderately toxic by ingestion. When heated to decomposition it emits acrid smoke and fumes. See also POLYMERS, SOLUBLE.

CCJ500 CAS:19477-24-8 ***HR: 3***
CARCINOLIPIN
mf: $C_{44}H_{78}O_2$ mw: 639.22

SYNS: CHOLESTERYL-14-METHYLHEXADECANOATE ◇ 3-β-14-METHYLHEXADECANOATE-CHOLEST-5-EN-3-OL

TOXICITY DATA with REFERENCE
scu-mus TDLo:720 mg/kg/(14-21D preg):ETA,TER NEOLA4 20,347,73

SAFETY PROFILE: Questionable carcinogen with experimental tumorigenic data. An experimental teratogen. When heated to decomposition it emits acrid smoke and irritating fumes.

CCJ625 CAS:8000-66-6 ***HR: D***
CARDAMON OIL

PROP: From the seed of *Elettaria acrdamomun* (L.) Maton (Fam. *Zingiberazeae*). Colorless liquid; aromatic penetrating odor of cardamom, pungent taste. Misc with alc.

SYNS: CARDAMON ◇ OIL of CARDAMON

TOXICITY DATA with REFERENCE
mmo-sat 2500 ng/plate KEKHB8 (9),11,79
mmo-esc 2500 ng/plate KEKHB8 (9),11,79
dnr-bcs 19 mg/disc SKEZAP 25,378,84

CONSENSUS REPORTS: Reported in EPA TSCA Inventory.

SAFETY PROFILE: Mutation data reported. When heated to decomposition it emits acrid smoke and fumes.

CCJ825 ***HR: 3***
CARDINAL FLOWER

PROP: Annual weeds with distinctive flowers. They have 2 small petals opposed by 3 large petals and may be blue, pink, white, red or yellow. The various species grow wild across the United States. Indian tobacco is cultivated as a drug plant. Cardinalis is grown as an ornamental.

SYNS: ASTHMA WEED ◇ BLADDERPOD LOBELIA ◇ BLUE CARDINAL FLOWER ◇ CARDENAL de MACETA (MEXICO) ◇ EMETIC WEED ◇ EYE BRIGHT ◇ GAG ROOT ◇ GREAT BLUE LOBELIA ◇ HIGH BELIA ◇ HOG PHYSIC ◇ INDIAN PINK ◇ INDIAN TOBACCO ◇ KINNIKINNIK ◇ LOBELIA INFLATA ◇ LOBELIA SIPHILITICA ◇ LOBELLOA CARDINALIS ◇ LOUISIANA LOBELIA ◇ LOW BELIA ◇ PUKE WEED ◇ RED LOBELIA ◇ SCARLET LOBELIA ◇ WILD TOBACCO

SAFETY PROFILE: The whole plant contains the poisonous lobeline and related alkaloids. Poisonings are most common when the plant is used in home medicine. The leaves are sold for use in tea and tobacco as a psychoactive ingredient. Ingestion of the leaves may cause nausea, vomiting, sensory disturbances, dizziness, and convulsions. See also LOBELINE.

CCK000 CAS:3599-32-4 ***HR: 3***
CARDIO-GREEN
mf: $C_{43}H_{48}N_2O_6S_2 \cdot Na$ mw: 776.04

SYN: ICG

TOXICITY DATA with REFERENCE
ipr-rat LD50:700 mg/kg TXAPA9 24,37,73

ipr-mus LD50:400 mg/kg TXAPA9 24,37,73
ivn-mus LD50:60 mg/kg TXAPA9 44,225,78

CONSENSUS REPORTS: Reported in EPA TSCA Inventory.

SAFETY PROFILE: Poison by intraperitoneal and intravenous routes. When heated to decomposition it emits very toxic fumes of SO_x, Na_2O, and NO_x.

CCK125 CAS:87-33-2 ***HR: 2***
CARDIS
mf: $C_6H_8N_2O_8$ mw: 236.16

PROP: Hard, colorless crystals. Mp: 70°. Sparingly sol in water. Freely sol in organic solvents, such as acetone, alc, and ether.

SYNS: ASTRIDINE ◇ CARDIO ◇ CARVANIL ◇ CARVASIN ◇ CEDOCARD ◇ CLAODICAL ◇ COROSORBIDE ◇ COROVLISS ◇ 1,4:3,6-DIANHYDROSORBITOL-2,5-DINITRATE ◇ DINITROSORBIDE ◇ DISORLON ◇ DURANITRAT ◇ EURECOR ◇ FLINDIX ◇ GLENTONIN-RETARD ◇ HARRICAL ◇ IBD ◇ ISDIN ◇ ISO-BID ◇ ISOKET ◇ ISOMACK ◇ ISO-PUREN ◇ ISORBID ◇ ISORDIL ◇ ISORDIL TEMBIDS ◇ ISOSORBIDE DINITRATE ◇ ISOSTENASE ◇ ISOTRATE ◇ KORODIL ◇ LANGORAN ◇ LASERDIL ◇ MAYCOR ◇ MONOCLAIR ◇ MYOREXON ◇ NITROSORBID ◇ NITROSORBIDE ◇ NITROSORBON ◇ NOSIM ◇ RESOIDAN ◇ RIFLOC RETARD ◇ RIGEDAL ◇ SORBANGIL ◇ SORBID ◇ SORBIDE NITRATE ◇ SORBIDILAT ◇ SORBIDINITRATE ◇ SORBISLO ◇ SORBITRATE ◇ SORBONIT ◇ SORQUAD ◇ SORQUAT ◇ VASCARDIN ◇ VASORBATE ◇ VASOTRATE

TOXICITY DATA with REFERENCE
cyt-mus:mmr 1 mmol/L/48H-C JTSCDR 5,141,80
ivn-rat TDLo:260 mg/kg (6-22D preg/20D post):REP KSRNAM 19,5021,85
orl-rat LD50:747 mg/kg YAKUD5 26,309,84
ipr-rat LD50:620 mg/kg NIIRDN 6,72,82
scu-rat LD50:1237 mg/kg YACHDS 10,2109,82
orl-mus LD50:1050 mg/kg NIIRDN 6,72,82
ipr-mus LD50:960 mg/kg NIIRDN 6,72,82
scu-mus LD50:1050 mg/kg NIIRDN 6,72,82
ims-mus LD50:1080 mg/kg NIIRDN 6,72,82

CONSENSUS REPORTS: Reported in EPA TSCA Inventory.

SAFETY PROFILE: Moderately toxic by ingestion, intraperitoneal, intramuscular, and subcutaneous routes. Experimental reproductive effects. Mutation data reported. When heated to decomposition it emits toxic fumes of NO_x. A coronary vasodilator. See also NITRATES.

CCK250 CAS:959-24-0 ***HR: 3***
β-CARDONE
mf: $C_{12}H_{20}N_2O_3S \cdot ClH$ mw: 308.86

SYNS: 4'-(1-HYDROXY-2-(ISOPROPYLAMINO)ETHYL)METHANESULFOANILIDEHYDROCHLORIDE ◇ 4'-(1-HYDROXY-2-ISOPROPYLAMINO) ETHYL)METHANESULFONANILIDE MONOHYDROCHLORIDE ◇ 4-(2-ISOPROPYLAMINE-1-HYDROXYETHYL)METHANESULFOANILIDE HYDROCHLORIDE ◇ 4-(2-ISOPROPYLAMINO-1-HYDROXYAETHYL) METHANESULFONALID HYDROCHLORID (GERMAN) ◇ ISOPROPYLAMINOHYDROXYETHYLMETHANESULFONALIDE HYDROCHLORIDE ◇ N-ISOPROPYL-β-(4-METHAN- ESULFONAMIDOPHENYL) ETHANOLAMINE HYDROCHLORIDE ◇ MEAD JOHNSON 1999 ◇ MJ 1999 ◇ MJ 1999 HYDROCHLORIDE ◇ SOTACOR ◇ SOTALEX ◇ SOTALOL ◇ SOTALOL HYDROCHLORIDE

TOXICITY DATA with REFERENCE
orl-man LDLo:45714 μg/kg ARTODN 43,221,80
orl-rat LD50:3450 mg/kg JPETAB 149,161,65
ipr-rat LD50:680 mg/kg JPETAB 149,161,65
orl-mus LD50:2600 mg/kg JPETAB 149,161,65
ipr-mus LD50:670 mg/kg JPETAB 149,161,65
ivn-mus LD50:166 mg/kg ARZNAD 27,1022,77
ipr-dog LD50:330 mg/kg JPETAB 149,161,65
orl-rbt LD50:1000 mg/kg JPETAB 149,161,65

SAFETY PROFILE: A human poison by ingestion. Poison experimentally by intravenous and intraperitoneal routes. Moderately toxic by ingestion. Human systemic effects by ingestion: excitement, dyspnea, and convulsions. When heated to decomposition it emits very toxic fumes of HCl, SO_x, and NO_x.

CCK500 CAS:13466-78-9 ***HR: 1***
3-CARENE
mf: $C_{10}H_{16}$ mw: 136.26

PROP: Colorless, mobile liquid; found in many volatile oils (such as Swedish and Finnish turpentine oils, galanga root oil and in German pine needle oils such as those from *Pinus pumilio* and *Pinus sylvestris*) and isolated from turpentine fractions (FCTXAV 11,1011,73).

SYNS: Δ^3-CARENE ◇ S-3-CARENE ◇ ISODIPRENE ◇ 3,7,7-TRIMETHYLBICYCLO(4.1.0)-3-HEPTENE ◇ 3,7,7-TRIMETHYL-3-NORCARENE ◇ 4,7,7-TRIMETHYL-3-NORCARENE

TOXICITY DATA with REFERENCE
skn-rbt 500 mg/24H FCTXAV 11,1053,73
orl-rat LD50:4800 mg/kg FCTXAV 11,1053,73

CONSENSUS REPORTS: Reported in EPA TSCA Inventory.

SAFETY PROFILE: Mildly toxic by ingestion. A skin irritant. When heated to decomposition it emits acrid smoke and fumes.

CCK550 CAS:33605-67-3 ***HR: D***
CARGUTOCIN
mf: $C_{42}H_{65}N_{11}O_{12}$ mw: 916.18

SYNS: DEAMINO-DICARBA-(GLY^7)-OXYTOCIN ◇ STATOCIN ◇ Y 5350

TOXICITY DATA with REFERENCE
ivn-rat TDLo:27 iu/kg (17-22D preg/21D post):REP IYKEDH 10,41,79
ivn-rat TDLo:11 iu/kg (female 7-17D post):TER IYKEDH 10,26,79

SAFETY PROFILE: An experimental teratogen. Other experimental reproductive effects. When heated to decomposition it emits toxic fumes of NO_x. See also OXYTOCIN.

CCK575 CAS:42583-55-1 ***HR: D***
CARMETIZIDE
mf: $C_{10}H_{12}ClN_3O_6S_2$ mw: 369.82

SYN: DU-5747

TOXICITY DATA with REFERENCE
orl-rat TDLo:6 g/kg (9-14D preg):TER OYYAA2 7,267,73

SAFETY PROFILE: An experimental teratogen. When heated to decomposition it emits toxic fumes of Cl^-, SO_x, and NO_x. See also ESTERS.

CCK625 CAS:50935-04-1 ***HR: 3***
CARMINOMYCIN I
mf: $C_{26}H_{27}NO_{10}$ mw: 513.54

SYNS: CARMINOMICIN I ◇ CARUBICIN ◇ NSC-180024

TOXICITY DATA with REFERENCE
dni-mus:leu 390 nmol/L JANTAJ 34,1596,81
oms-mus:leu 490 nmol/L JANTAJ 34,1596,81
orl-mus LD50:7300 μg/kg ANTBAL 19,57,74
ipr-mus LD50:1100 μg/kg ANTBAL 29,666,84
scu-mus LD50:3800 μg/kg ANTBAL 19,57,74
ivn-mus LD50:3700 μg/kg ANTBAL 19,57,74

SAFETY PROFILE: Deadly poison by ingestion, subcutaneous, intravenous, and intraperitoneal routes. Mutation data reported. When heated to decomposition it emits toxic fumes of NO_x.

CCK630 CAS:61422-45-5 ***HR: 3***
CARMOFUR
mf: $C_{11}H_{16}FN_3O_3$ mw: 257.30

PROP: White crystals from ethanol. Mp: 110-111°.

SYNS: 2,4-DIOXO-5-FLUORO-N-HEXYL-3,4-DIHYDRO-1(2H)-PYRIMIDINECARBOXAMIDME ◇ 2,4-DIOXO-5-FLUORO-N-HEXYL-1,2,3,4-TETRAHYDRO-1-PYRIMIDINECARBOXAMIDE ◇ 5-FLUORO-1-HEXYLCARBAMOYL-URACIL ◇ HCFU ◇ 1-HEXYLCARBAMOYL-5-FLUOROURACIL ◇ MIFUROL ◇ 1,2,3,4-TETRAHYDRO-2,4-DIOXO-5-FLUORO-N-HEXYL-1-PYRIMIDINECARBOXAMIDE ◇ YAMAFUL

TOXICITY DATA with REFERENCE
dnr-bcs 40 μg/plate TAKHAA 44,96,85
cyt-mus-orl 400 mg/kg OYYAA2 19,363,80
orl-rat TDLo:550 mg/kg (female 7-17D post):REP KSRNAM 14,1373,80
orl-rat TDLo:275 mg/kg (7-17D preg):TER KSRNAM 14,1373,80
orl-wmn TDLo:1152 mg/kg/14W-I:CNS,PSY JNRYA9 234,365,87
orl-man TDLo:1749 mg/kg/29W-I JNRYA9 234,365,87
orl-rat LD50:268 mg/kg NIIRDN 6,191,82
ipr-rat LD50:93 mg/kg NIIRDN 6,191,82
scu-rat LD50:260 mg/kg NIIRDN 6,191,82
orl-mus LD50:1260 mg/kg NIIRDN 6,191,82
ipr-mus LD50:96 mg/kg NIIRDN 6,191,82
scu-mus LD50:532 mg/kg NIIRDN 6,191,82
orl-dog LD50:65 mg/kg NIIRDN 6,191,82
orl-rbt LD50:55 mg/kg NIIRDN 6,191,82

SAFETY PROFILE: Poison by ingestion, subcutaneous, and intraperitoneal routes. Human systemic effects by ingestion: encephalitis, hallucinations, distorted perceptions, ataxia. Experimental reproductive effects. An experimental teratogen. Mutation data reported. When heated to decomposition it emits toxic fumes of F^- and NO_x.

CCK650 CAS:56-99-5 ***HR: 2***
CARNITINE CHLORIDE
mf: $C_7H_{16}NO_3{\cdot}Cl$ mw: 197.69

SYNS: (3-CARBOXY-2-HYDROXYPROPYL)TRIMETHYLAMMONIUM CHLORIDE ◇ 3-CARBOXY-2-HYDROXY-N,N,N-TRIMETHYL-1-PROPANAMINIUM CHLORIDE (9CI)

TOXICITY DATA with REFERENCE
orl-mus LD50:6690 mg/kg NIIRDN 6,135,82
scu-mus LD50:4030 mg/kg NIIRDN 6,135,82
ivn-mus LD50:1150 mg/kg NIIRDN 6,135,82

SAFETY PROFILE: Moderately toxic by intravenous route. Mildly toxic by ingestion and subcutaneous routes. When heated to decomposition it emits toxic fumes of Cl^-, NH_3, and NO_x. See also CHLORIDES.

CCK660 CAS:6645-46-1 ***HR: 2***
l-CARNITINE HYDROCHLORIDE
mf: $C_7H_{15}NO_3{\cdot}Cl$ mw: 196.68

SYNS: AMMONIUM, (3-CARBOXY-2-HYDROXYPROPYL)TRIMETHYL-, CHLORIDE, (-)- ◇ (-)-(3-CARBOXY-2-HYDROXYPROPYL)TRIMETHYLAMMONIUM CHLORIDE ◇ l-(3-CARBOXY-2-HYDROXYPROPYL) TRIMETHYLAMMONIUM CHLORIDE ◇ (R)-3-CARBOXY-2-HYDROXY-N,N,N-TRIMETHYL-1-PROPANAMINIUM CHLORIDE ◇ l-CARNITINE CHLORIDE ◇ (R)-CARNITINE HYDROCHLORIDE ◇ LC-80 ◇ 1-PROPANAMINIUM, 3-CARBOXY-2-HYDROXY-N,N,N-TRIMETHYL-, CHLORIDE, (R)- (9CI)

TOXICITY DATA with REFERENCE
orl-rat TDLo:33 g/kg (female 7-17D post):REP IYKEDH 19,465,88
orl-rat TDLo:33 g/kg (female 7-17D post):TER IYKEDH 19,465,88
orl-rat LD50:6890 mg/kg IYKEDH 19,191,88

ipr-rat LD50:1920 mg/kg IYKEDH 19,191,88
ivn-rat LD50:1440 mg/kg IYKEDH 19,191,88
orl-mus LD50:8 g/kg IYKEDH 19,446,88
ipr-mus LD50:1690 mg/kg IYKEDH 19,446,88
scu-mus LD50:4320 mg/kg IYKEDH 19,446,88
ivn-dog LD50:2272 mg/kg IYKEDH 19,238,88

SAFETY PROFILE: Moderately toxic by intraperitoneal and intravenous routes. An experimental teratogen. Other experimental reproductive effects. When heated to decomposition it emits toxic fumes of NO_x and Cl^-.

CCK665 CAS:305-84-0 ***HR: 1***
CARNOSINE
mf: $C_9H_{14}N_4O_3$ mw: 226.27

SYNS: β-ALANYL-l-HISTIDINE ◇ l-CARNOSINE ◇ l-HISTIDINE, N-β-ALANYL- ◇ IGNOTINE ◇ KARNOZZN ◇ N-2-M

TOXICITY DATA with REFERENCE
ipr-rat TDLo:21 mg/kg (female 8-14D post):TER OYYAA2 8,1219,74
ipr-rat TDLo:21 mg/kg (female 8-14D post):REP OYYAA2 8,1219,74
ipr-mus LD50:9087 mg/kg USXXAM #4446149

CONSENSUS REPORTS: Reported in EPA TSCA Inventory.

SAFETY PROFILE: Mildly toxic by intraperitoneal route. An experimental teratogen. Other experimental reproductive effects. When heated to decomposition it emits toxic fumes of NO_x.

CCK675 ***HR: 2***
CAROLINA ALLSPICE

PROP: A large shrub (to 12 feet tall) with large (2- to 3-inch), fruity smelling, brownish-red or purple flowers. The fruit is fig-shaped and contains large glossy brown seeds. Various species are native to the eastern states from Pennsylvania, through northern Florida to Alabama and in California.

SYNS: AMERICAN ALLSPICE ◇ BUBBIE BLOSSOMS ◇ BUBBY BUSH ◇ CALYCANTH ◇ CALYCANTHUS (VARIOUS SPECIES) ◇ C. FERTILIS ◇ C. FLORIDUS ◇ C. OCCIDENTALIS ◇ PINEAPPLE SHRUB ◇ SPICEBUSH ◇ STRAWBERRY BUSH ◇ SWEET BETTIE ◇ SWEET SHRUB

SAFETY PROFILE: The seeds contain the toxin calycanthin and some related alkaloids. No human poisonings have been reported, but ingestion of the seeds could cause symptoms similar to strychnine poisoning: convulsions, weak contractions of the heart, and low blood pressure.

CCK775 CAS:7075-03-8 ***HR: 3***
CARPIPRAMINE DIHYDROCHLORIDE
mf: $C_{28}H_{38}N_4O{\cdot}2ClH$ mw: 519.62

SYNS: CARPIPRAMINE HYDROCHLORIDE ◇ DEFEKTON

TOXICITY DATA with REFERENCE
orl-rat LD50:1025 mg/kg NIIRDN 6,185,82
ipr-rat LD50:76 mg/kg NIIRDN 6,185,82
ivn-rat LD50:37 mg/kg NIIRDN 6,185,82
orl-mus LD50:2180 mg/kg NIIRDN 6,185,82
ipr-mus LD50:136 mg/kg NIIRDN 6,185,82
ivn-mus LD50:28200 μg/kg NIIRDN 6,185,82

SAFETY PROFILE: Poison by intravenous and intraperitoneal routes. Moderately toxic by ingestion. When heated to decomposition it emits toxic fumes of NO_x and HCl. See also CARPIPRAMINE DIHYDROCHLORIDE MONOHYDRATE and AMINES.

CCK780 ***HR: 3***
CARPIPRAMINE DIHYDROCHLORIDE MONOHYDRATE
mf: $C_{28}H_{38}N_4O{\cdot}2ClH{\cdot}H_2O$ mw: 537.64

SYNS: DEFEKTON ◇ PRAZINIL ◇ PZ 1511

TOXICITY DATA with REFERENCE
orl-rat LD50:1025 mg/kg MEIEDD 10,260,83
ipr-rat LD50:76 mg/kg MEIEDD 10,260,83
ivn-rat LD50:37 mg/kg MEIEDD 10,260,83
orl-mus LD50:2180 mg/kg MEIEDD 10,260,83
ipr-mus LD50:136 mg/kg MEIEDD 10,260,83
ivn-mus LD50:28200 μg/kg MEIEDD 10,260,83

SAFETY PROFILE: Poison by intravenous and intraperitoneal routes. Moderately toxic by ingestion. When heated to decomposition it emits toxic fumes of NO_x and HCl. A psychotropic agent. See other carpipramine entries.

CCK790 CAS:100482-23-3 ***HR: 3***
CARPIPRAMINE MALEATE
mf: $C_{28}H_{38}N_4O{\cdot}C_4H_4O_4$ mw: 562.78

SYN: CARBADIPIMIDINE MALEATE

TOXICITY DATA with REFERENCE
ipr-rat LD50:169 mg/kg NIIRDN 6,185,82
orl-mus LD50:2055 mg/kg NIIRDN 6,185,82
ipr-mus LD50:147 mg/kg NIIRDN 6,185,82

SAFETY PROFILE: Poison by intraperitoneal route. Moderately toxic by ingestion. When heated to decomposition it emits toxic fumes of NO_x. See other carpipramine entries.

CCK800 CAS:53716-49-7 ***HR: 3***
CARPROFEN
mf: $C_{15}H_{12}ClNO_2$ mw: 273.72

PROP: Crystals from chloroform. Mp: 197-198°.

SYNS: dl-6-CHLORO-α-METHYLCARBAZOLE-2-ACETIC ACID ◇ IMADYL ◇ RIMADYL

TOXICITY DATA with REFERENCE
orl-rat TDLo:2360 mg/kg (9W male/2W pre-3W post):REP TXAPA9 56,376,80
orl-rat LD50:74 mg/kg OYYAA2 14,251,77
orl-mus LD50:186 mg/kg MDACAP 18,170,82

SAFETY PROFILE: Poison by ingestion. Experimental reproductive effects. When heated to decomposition it emits toxic fumes of Cl^- and NO_x.

CCL109 CAS:23734-06-7 ***HR: 3***
CARQUEJOL
mf: $C_{10}H_{14}O$ mw: 150.24

SYN: (1S-cis)-5-METHYLENE-6-(1-METHYLETHENYL)-2-CYCLOHEXEN-1-OL

TOXICITY DATA with REFERENCE
ipr-rat LD50:410 mg/kg APFRAD 18,715,60
orl-mus LD50:1800 mg/kg APFRAD 18,715,60
ipr-mus LD50:456 mg/kg APFRAD 18,715,60
ipr-dog LDLo:250 mg/kg APFRAD 18,715,60

SAFETY PROFILE: Poison by intraperitoneal route. Moderately toxic by ingestion. When heated to decomposition it emits acrid smoke and fumes.

CCL250 CAS:9000-07-1 ***HR: 3***
CARRAGEEN

PROP: A sulfated polysaccharide. Dried plant of seaweed *Chondrus crispus, Chondrus ocellatus, Eucheuma cottonil, Eucheuma spinosum, Gigartina acicularis, Gigartina pistillata, Gigartina radula, Gigartina stellata*. Yellow-white when powdered. Sol in water @ 80°; insol in organic solvents. Dried, bleached *Chondrus crispus* containing salts of sulfated polygalactose esters.

SYNS: 3,6-ANHYDRO-d-GALACTAN ◇ AUBYGEL GS ◇ AUBYGUM DM ◇ BURTONITE-V-40-E ◇ CARASTAY ◇ CARASTAY G ◇ CARRAGEENAN (FCC) ◇ CARRAGEENAN GUM ◇ CARRAGHEANIN ◇ CARRAGHEEN ◇ CARRAGHEENAN ◇ CHONDRUS ◇ CHONDRUS EXTRACT ◇ COLLOID 775 ◇ COREINE ◇ EUCHEUMA SPINOSUM GUM ◇ FLANOGEN ELA ◇ GALOZONE ◇ GELCARIN ◇ GELCARIN HMR ◇ GELOZONE ◇ GENU ◇ GENUGEL ◇ GENUGEL CJ ◇ GENUGOL RLV ◇ GENUVISCO J ◇ GUM CARRAGEENAN ◇ GUM CHON 2 ◇ GUM CHROND ◇ IRISH GUM ◇ IRISH MOSS EXTRACT ◇ IRISH MOSS GELOSE ◇ KILLEEN ◇ LYGOMME CDS ◇ PEARLPUSS ◇ PELLUGEL ◇ PENCOGEL ◇ PIG-WRACK ◇ SATIAGEL GS 350 ◇ SATIAGUM 3 ◇ SATIAGUM STANDARD ◇ SEAKEM CARRAGEENIN ◇ SEATREM ◇ SELF ROCK MOSS ◇ VISCARIN

TOXICITY DATA with REFERENCE
orl-rat TDLo:2100 g/kg/40W-C:ETA CNREA8 38,4427,78
scu-rat TDLo:525 mg/kg/21W-I:NEO 13BYAH -,83,62
ivn-rbt LDLo:5 mg/kg JPPMAB 17,647,65
ivn-gpg LDLo:20 mg/kg NATUAS 202,401,64

CONSENSUS REPORTS: IARC Cancer Review: Group 3 IMEMDT 7,56,87; Animal Limited Evidence IMEMDT 10,181,76. Reported in EPA TSCA Inventory.

SAFETY PROFILE: Poison by intravenous route. Questionable carcinogen with experimental neoplastigenic and tumorigenic data. When heated to decomposition it emits acrid smoke and fumes.

CCL350 CAS:11114-20-8 ***HR: 2***
kappa-CARRAGEENAN

SYNS: kappa-CARRAGEEN ◇ kappa-CARRAGEENIN ◇ SATIAGEL GS 350

TOXICITY DATA with REFERENCE
par-rat TDLo:320 mg/kg:ETA OYYAA2 32,711,86
orl-rbt LDLo:3 mg/kg JPPMAB 17,647,65

SAFETY PROFILE: Poison by ingestion. Questionable carcinogen with experimental tumorigenic data. When heated to decomposition it emits acrid smoke and irritating fumes.

CCL500 ***HR: 3***
CARRAGEENAN, DEGRADED

PROP: Carrageenan derived from *Eucheuma spinosum*, degraded by acid hydrolysis; average molecular weight 20,000-40,000 (CALEDQ 4,171,78).

TOXICITY DATA with REFERENCE
orl-rat TDLo:360 g/kg/9W-C:CAR CALEDQ 14,267,81
orl-rat TD:1700 g/kg/52W-C:CAR CALEDQ 4,171,78
orl-rat TD:3116 g/kg/77W-C:NEO PPTCBY 9,127,79
orl-rat TD:2250 g/kg/64W-C:ETA PPTCBY 9,127,79

CONSENSUS REPORTS: IARC Cancer Review: Animal Sufficient Evidence IMEMDT 31,79,83.

SAFETY PROFILE: Confirmed carcinogen with experimental carcinogenic, neoplastigenic, and tumorigenic data. See also CARRAGEEN. When heated to decomposition it emits toxic fumes of SO_x.

CCL750 CAS:8015-88-1 ***HR: 1***
CARROT SEED OIL

PROP: Distilled from the seeds of *Daucus carota* L. (Fam. *Umbelliferae*). (FCTXAV 14,659,76). Light yellow to amber liquid; aromatic odor. Sol in fixed oils, mineral oil; insol in glycerin, propylene glycol.

TOXICITY DATA with REFERENCE
skn-rbt 500 mg/24H MLD FCTXAV 14,659,76
skn-gpg 500 mg/24H MLD FCTXAV 14,659,76

CONSENSUS REPORTS: Reported in EPA TSCA Inventory.

SAFETY PROFILE: A skin irritant. When heated to decomposition it emits acrid smoke and irritating fumes.

CCL800 CAS:51781-06-7 ***HR: 3***
CARTEOLOL
mf: $C_{16}H_{24}N_2O_3$ mw: 292.42

SYN: 5-(3-((1,1-DIMETHYLETHYL)AMINO)-2-HYDROXYPROPOXY)-3,4-DIHYDRO-2(1H)-QUINOLINONE

TOXICITY DATA with REFERENCE
orl-dog LD50:830 mg/kg OYYAA2 19,323,80
orl-rbt LD50:740 mg/kg OYYAA2 19,323,80
ivn-rbt LD50:112 mg/kg OYYAA2 19,323,80

SAFETY PROFILE: Poison by intravenous route. Moderately toxic by ingestion. When heated to decomposition it emits toxic fumes of NO_x.

CCM000 CAS:499-75-2 ***HR: 3***
CARVACROL
mf: $C_{10}H_{14}O$ mw: 150.24

PROP: Colorless to pale yellow liquid; spicy thymol odor. D: 0.974-0.980, refr index: 1.521-1.526, flash p: 212.°F. Sol in alc, ether; insol in water.

SYNS: 2-p-CYMENOL ◇ FEMA No. 2245 ◇ 2-HYDROXY-p-CYMENE ◇ ISOPROPYL-o-CRESOL ◇ 5-ISOPROPYL-2-METHYLPHENOL ◇ ISOTHYMOL ◇ 2-METHYL-5-ISOPROPYLPHENOL ◇ o-THYMOL

TOXICITY DATA with REFERENCE
skn-rbt 500 mg/24H SEV FCTXAV 17(suppl)695,79
orl-rat LD50:810 mg/kg FCTXAV 2,327,64
orl-cat LDLo:100 mg/kg HBTXAC 5,46,59
orl-rbt LDLo:100 mg/kg AEPPAE 161,196,31
skn-rbt LDLo:2700 mg/kg JAPMA8 38,366,49
scu-rbt LDLo:1000 mg/kg HBTXAC 5,46,59
scu-frg LDLo:75 mg/kg HBTXAC 5,46,59

CONSENSUS REPORTS: Reported in EPA TSCA Inventory.

SAFETY PROFILE: Poison by ingestion and subcutaneous route. Moderately toxic by skin contact. A severe skin irritant. Combustible liquid. When heated to decomposition it emits acrid smoke and irritating fumes.

CCM100 CAS:2244-16-8 ***HR: 3***
d-CARVONE
mf: $C_{10}H_{14}O$ mw: 150.24

PROP: Colorless liquid; caraway odor. D: 0.956-0.960, refr index: 1.96-1.499. Sol in propylene glycol, fixed oils; misc in alc; insol in glycerin.

SYNS: (+)-CARVONE ◇ d(+)-CARVONE ◇ (S)-CARVONE ◇ (S)-(+)-CARVONE ◇ FEMA No. 2249 ◇ d-p-MENTHA-6,8,(9)-DIEN-2-ONE ◇ d-1-METHYL-4-ISOPROPENYL-6-CYCLOHEXEN-2-ONE ◇ (S)-2-METHYL-5-(1-METHYLETHENYL)-2-CYCLOHEXEN-1-ONE

TOXICITY DATA with REFERENCE
skn-rbt 500 mg/24H MLD FCTXAV 16,673,78
orl-rat LD50:3710 μg/kg FCTXAV 16,673,78
skn-rbt LD50:4 mg/kg FCTXAV 16,673,78

CONSENSUS REPORTS: Reported in EPA TSCA Inventory.

SAFETY PROFILE: Poison by ingestion and skin contact. A skin irritant. When heated to decomposition it emits acrid smoke and irritating fumes.

CCM120 CAS:6485-40-1 ***HR: 2***
l(−)-CARVONE
mf: $C_{10}H_{14}O$ mw: 150.22

PROP: Colorless liquid; spearmint odor. D: 0.956-0.960, refr index: 1.495-1.499. Sol in propylene glycol, fixed oils; misc in alc; insol in glycerin.

SYNS: (−)-CARVONE ◇ 1-CARVONE ◇ (R)-CARVONE ◇ FEMA No. 2249 ◇ 1-6,8(9)-p-MENTHADIEN-2-ONE ◇ (R)-(−)-p-MENTHA-6,8-DIEN-2-ONE ◇ 1-1-METHYL-4-ISOPROPENYL-6-CYCLOHEXEN-2-ONE ◇ (R)-2-METHYL-5-(1-METHYLETHENYL)-2-CYCLOHEXEN-1-ONE(9CI)

TOXICITY DATA with REFERENCE
orl-rat LD50:1640 mg/kg FCTXAV 11,1057,73
orl-gpg LD50:766 mg/kg FCTXAV 11,1057,73

SAFETY PROFILE: Moderately toxic by ingestion. When heated to decomposition it emits acrid smoke and irritating fumes.

CCM750 CAS:97-42-7 ***HR: 1***
1-CARVYL ACETATE
mf: $C_{12}H_{18}O_2$ mw: 194.30

SYNS: 1-p-MENTHA-6(8,9)-DIEN-2-YL ACETATE ◇ 2-METHYL-5-(1-METHYLETHENYL)-2-CYCLOHEXEN-1-OLACETATE

TOXICITY DATA with REFERENCE
skn-rbt 500 mg/24H MLD FCTXAV 16,637,78

CONSENSUS REPORTS: Reported in EPA TSCA Inventory.

SAFETY PROFILE: A skin irritant. When heated to decomposition it emits acrid smoke and fumes. See also ESTERS.

CCN000 CAS:87-44-5 ***HR: 1***
CARYOPHYLLENE
mf: $C_{15}H_{26}$ mw: 206.41

PROP: Found in oil of clove, cinnamon leaves, and copaiba balsam and in minor quantities in various other essential oils, especially lavender; prepared by isolation from clove leaf oil, clove stem oil, cinnamon leaf oil or pine oil fractions (FCTXAV 11,1011,73). Colorless to sltly yellow oily liquid; clove odor. D: 0.897-0.910, refr index: 1.498-1.504, flash p: 206°F. Sol in alc, ether; insol in water.

SYNS: β-CARYOPHYLLENE (FCC) ◇ FEMA No. 2252 ◇ 8-METHYLENE-4,11,11-(TRIMETHYL)BICYCLO(7.2.0)UNDEC-4-ENE

TOXICITY DATA with REFERENCE
skn-rbt 500 mg/24H FCTXAV 11,1059,73

CONSENSUS REPORTS: Reported in EPA TSCA Inventory.

SAFETY PROFILE: A skin irritant. Combustible liquid. When heated to decomposition it emits acrid smoke and irritating fumes.

CCN100 CAS:1139-30-6 ***HR: 1***
β-CARYOPHYLLENE EPOXIDE
mf: $C_{15}H_{24}O$ mw: 220.39

PROP: Mp: 61-62°.

SYNS: CARYOPHYLENE OXIDE ◇ CARYOPHYLLENE EPOXIDE ◇ CARYOPHYLLENE OXIDE ◇ (-)-CARYOPHYLLENE OXIDE ◇ β-CARYOPHYLLENE OXIDE ◇ EPOXYCARYOPHYLLENE ◇ (-)-EPOXYDIHYDROCARYOPHYLLENE ◇ 5-OXATRICYCLO(8.2.0.$0^{4,6}$) DODECANE, 4,12,12-TRIMETHYL-9-METHYLENE-, (1R,4R,6R,10S)- ◇ 4,11,11-TRIMETHYL-8-METHYLENE-5-OXATRICYCLO(8.2.0.0(4,6)) DODECANE

TOXICITY DATA with REFERENCE
skn-rbt 500 mg/24H MOD FCTOD7 21,661,83

CONSENSUS REPORTS: Reported in EPA TSCA Inventory.

SAFETY PROFILE: A skin irritant. When heated to decomposition it emits acrid smoke and irritating fumes.

CCN250 CAS:1403-27-6 ***HR: 3***
CARZINOCIDIN

PROP: An antitumor substance from *Streptomyces sahachiroi*.

SYNS: CARCINOCIDIN

TOXICITY DATA with REFERENCE
ipr-mus LD50:43500 μg/kg JAJAAA 9,9,56
scu-mus LD50:20 mg/kg JAJAAA 9,9,56
ivn-mus LD50:4700 μg/kg JAJAAA 9,6,56

SAFETY PROFILE: Poison by subcutaneous, intravenous, and intraperitoneal routes. When heated to decomposition it emits acrid smoke and irritating fumes.

CCN500 CAS:1403-28-7 ***HR: 3***
CARZINOPHILIN

SYNS: CARDINOPHILLIN ◇ CARDINOPHYLLIN

TOXICITY DATA with REFERENCE
mmo-esc 10 mg/disc ANYAA9 76,475,58
scu-rat TDLo:50 μg/kg (female 6-10D post):TER OSDIAF 14,107,65
scu-rat TDLo:50 μg/kg (female 6-10D post):REP OSDIAF 14,107,65
ipr-mus LD50:8000 unit/kg JAJAAA 13,27,60
scu-mus LD50:3 mg/kg 85GDA2 6,300,81
ivn-mus LD50:500 μg/kg 85GDA2 6,300,81

SAFETY PROFILE: Poison by subcutaneous and intravenous routes. An experimental teratogen. Other experimental reproductive effects. Mutation data reported. When heated to decomposition it emits acrid smoke and irritating fumes.

CCN750 CAS:1403-29-8 ***HR: 3***
CARZINOPHILIN A
mf: $C_{31}H_{33}N_3O_{12}$ mw: 639.67

PROP: Active fraction of antitumor substance *Carzinophilin* obtained from *Streptomyces sahachiroi*.

TOXICITY DATA with REFERENCE
mmo-esc 500 μg/disc APMBAY 6,23,58
ivn-mus LD50:15 μg/kg 85ERAY 2,1356,78

SAFETY PROFILE: Deadly poison by intravenous route. Mutation data reported.

CCO000 CAS:11002-21-4 ***HR: 3***
CARZINOSTATININ

TOXICITY DATA with REFERENCE
ipr-mus LD50:137 mg/kg JAJAAA 15,53,63
scu-mus LD50:115 mg/kg JAJAAA 15,53,63
ivn-mus LD50:283 mg/kg JAJAAA 15,53,63

SAFETY PROFILE: Poison by subcutaneous, intravenous, and intraperitoneal routes.

CCO675 CAS:33445-03-3 ***HR: 3***
CASSAINE HYDROCHLORIDE
mf: $C_{24}H_{39}NO_4$•ClH mw: 442.10

SYN: (E)-7-OXO-3-β-HYDROXY-14-α-METHYL-8-β-PODOCARPANE-Δ^{13}-α-ACETIC ACID-2-(DIMETHYLAMINO)ETHYL ESTER HYDROCHLORIDE

TOXICITY DATA with REFERENCE
ivn-dog LDLo:400 μg/kg JMCMAR 10,582,67
ivn-cat LDLo:806 μg/kg JPHAA3 27,9,38
ivn-gpg LDLo:2640 μg/kg APSXAS 13,35,76

SAFETY PROFILE: Deadly poison by intravenous route. When heated to decomposition it emits toxic fumes of NO_x and HCl.

CCO680 ***HR: 3***
CASSAVA

PROP: A bushy shrub up to 9 feet tall which grows long, tuberous roots. The alternate leaves have 3 to 7 lobes. It is cultivated for food in the United States Gulf Coast states, Hawaii, Guam, and the West Indies.

SYNS: JUCA ◇ MANIHOT ESCULENTA ◇ MANIOC ◇ MANIOKA ◇ SWEET POTATO PLANT ◇ TAPIOCA ◇ YUCA ◇ YUCA BRAVA

SAFETY PROFILE: The leaves, and especially the tubers, contain the cyanogenetic glycosides linamarin and lotaustralin. Cyanogenetic glycosides release cyanide when exposed to stomach acid. Ingestion may cause after a delay period of several hours: abdominal pain, vomiting, lack of muscle control, coma, and convulsions. See also CYANIDE.

CCO700 CAS:92456-72-9 ***HR: D***
CASSAVA, MANIHOT UTILISSIMA

PROP: Cassava tubers form the staple diet for may people in Nigeria and other tropical countries (BNEOBV 36,233,79).

SYNS: CASSAVA ◇ CASSAVA MEAL ◇ CASSAVA POWDER ◇ GARI ◇ MANIHOT UTILISSIMA ◇ MILLED CASSAVA POWDER

TOXICITY DATA with REFERENCE
orl-rat TDLo:650 g/kg (14D pre/1-22D preg):REP BNEOBV 36,233,79
orl-rat TDLo:375 g/kg (female 1-15D post):TER TJADAB 24,289,81

SAFETY PROFILE: An experimental teratogen. Other experimental reproductive effects.

CCO750 CAS:8007-80-5 ***HR: 3***
CASSIA OIL

PROP: Chief constituent is cinnamic aldehyde, found in the leaves and twigs of *Cinnamomum cassia blume* (FCTXAV 13,91,75). Yellow liquid; cinnamon odor, spicy burning taste. Sol in fixed oils, propylene glycol; insol in glycerin, mineral oil.

SYNS: ARTIFICIAL CINNAMON OIL ◇ CINNAMON BARK OIL ◇ CINNAMON BARK OIL, CEYLON TYPE (FCC) ◇ CINNAMON OIL ◇ KASSIA OEL (GERMAN) ◇ OIL of CASSIA ◇ OIL of CHINESE CINNAMON ◇ OIL of CINNAMON ◇ OIL of CINNAMON, CEYLON ◇ OILS, CINNAMON

TOXICITY DATA with REFERENCE
skn-hmn 100% FCTXAV 13,109,75
skn-mus 100% MLD FCTXAV 13,109,75
skn-rbt 500 mg/24H SEV FCTXAV 13,91,75
dnr-bcs 600 μg/disc TOFOD5 8,91,85
orl-rat LD50:2800 mg/kg FCTXAV 13,91,75
orl-mus LD50:2670 mg/kg TOFOD5 8,91,85
ipr-mus LD50:500 mg/kg PHMCAA 3,62,61
skn-rbt LD50:320 mg/kg FCTXAV 13,91,75

CONSENSUS REPORTS: Reported in EPA TSCA Inventory.

SAFETY PROFILE: Poison by skin contact. Moderately toxic by ingestion and intraperitoneal routes. A human skin irritant. Mutation data reported. See also CINNAMALDEHYDE and ALDEHYDES. When heated to decomposition it emits acrid smoke and irritating fumes.

CCO800 ***HR: 3***
CASSIA TORA Linn., leaf extract

PROP: Indian plant belonging to the family *Leguminosae* INDRBA 15,49,78).

TOXICITY DATA with REFERENCE
orl-mus LDLo:200 mg/kg INDRBA 15,49,78
ipr-mus LDLo:100 mg/kg INDRBA 15,49,78
ivn-mus LDLo:20 mg/kg INDRBA 15,49,78

SAFETY PROFILE: Poison by ingestion, intravenous, and intraperitoneal routes.

CCP000 ***HR: 3***
CASTOR BEAN
DOT: UN 2969

PROP: An annual which may grow higher than 15 feet. The large, lobed leaves may be 3 feet across. The spiny seed pods grow in clusters and contain plump seeds that are white with brown or black mottling. The seeds have a pleasant taste.

SYNS: AFRICAN COFFEE TREE ◇ CASTOR BEANS (DOT) ◇ CASTOR FLAKE (DOT) ◇ CASTOR MEAL (DOT) ◇ CASTOR OIL PLANT ◇ CASTOR POMACE (DOT) ◇ HIGUERETA (CUBA, PUERTO RICO) ◇ HIGUERILLA (MEXICO) ◇ KOLI (HAWAII) ◇ LA'AU-'AILA (HAWAII) ◇ MAN'S MOTHERWORT ◇ MEXICO WEED ◇ PA'AILA (HAWAII) ◇ PALMA CHRISTI (HAITI) ◇ RICIN (HAITI) ◇ RICINO (PUERTO RICO) ◇ RICINUS COMMUNIS ◇ STEADFAST ◇ WONDER TREE

TOXICITY DATA with REFERENCE
orl-chd LDLo:500 μg/kg 34ZIAG -,158,69

DOT Classification: ORM-C; Label: None.

SAFETY PROFILE: Deadly poison by ingestion in humans. The seeds contain the deadly poison ricin, a plant lectin (toxalbumin) which inhibits protein synthesis in the intestinal wall. Ingestion of the seeds can cause after a delay period of several hours: nausea, vomiting, diarrhea and intestinal dysfunction. There may be massive fluid and electrolyte loss. Ingestion of as few as 2 seeds could be fatal. A potent allergen. When heated to decomposition it emits toxic fumes of NO_x. See also RICIN.

CCP250 CAS:8001-79-4 ***HR: 2***
CASTOR OIL

PROP: From seeds of *Ricinus communis* L. (Fam. *Euphorbiaceae*). A colorless to pale yellow, viscous liquid; bland taste, characteristic odor. Mp: −12°, bp: 313°, flash p: 445°F (CC), d: 0.96, autoign temp: 840°F. Sol in alc; misc in abs alc, glacial acetic acid, chloroform, and ether.

SYNS: AROMATIC CASTOR OIL ◇ CASTOR OIL AROMATIC ◇ COSMETOL ◇ CRYSTAL O ◇ GOLD BOND ◇ NCI-C55163 ◇ NEOLOID ◇ OIL of PALMA CHRISTI ◇ PHORBYOL ◇ RICINUS OIL ◇ RICIRUS OIL ◇ TANGANTANGAN OIL

TOXICITY DATA with REFERENCE
eye-rbt 500 mg AJOPAA 29,1363,46

CONSENSUS REPORTS: Reported in EPA TSCA Inventory.

SAFETY PROFILE: Moderately toxic by ingestion. An allergen. An eye irritant. Combustible when exposed to heat. Spontaneous heating may occur. To fight fire, use CO_2, dry chemical, fog, mist. See also CASTOR BEAN.

CCP500 CAS:535-89-7 ***HR: 3***
CASTRIX
mf: $C_7H_{10}ClN_3$ mw: 171.65

PROP: Sltly water-sol crystals.

SYNS: 2-CHLOOR-4-DIMETHYLAMINO-6-METHYL-PYRIMIDINE (DUTCH) ◇ 2-CHLOR-4-DIMETHYLAMINO-6-METHYLPYRIMIDIN (GERMAN) ◇ 2-CHLORO-4-DIMETHYLAMINO-6-METHYL-PYRIMIDINE ◇ 2-CHLORO-4-METHYL-6-DIMETHYLAMINOPYRIMIDINE ◇ 2-CLORO-4-DIMETILAMINO-6-METIL-PIRIMIDINA (ITALIAN) ◇ CRIMIDIN (GERMAN) ◇ CRIMIDINA (ITALIAN) ◇ CRIMIDINE ◇ W 491

TOXICITY DATA with REFERENCE
orl-rat LD50:1250 μg/kg GUCHAZ 6,139,73
ipr-rat LD50:1 mg/kg JAPMA8 27,307,48
orl-mus LD50:1200 μg/kg MEIEDD 11,405,89
ipr-mus LD50:420 μg/kg JAPMA8 37,307,48
ipr-dog LD50:500 μg/kg JAPMA8 37,307,48
orl-rbt LD50:5 mg/kg 28ZEAL 5,59,76
ipr-rbt LD50:5 mg/kg JAPMA8 37,307,48
orl-gpg LD50:2660 μg/kg PCOC** -,202,66
ipr-gpg LD50:2660 μg/kg JAPMA8 37,307,48

CONSENSUS REPORTS: EPA Extremely Hazardous Substances List.

SAFETY PROFILE: Deadly poison by ingestion and intraperitoneal routes. Can cause central nervous system damage and convulsions. Intensely poisonous to mammals. A pesticide. When heated to decomposition it emits very toxic fumes of Cl^- and NO_x.

CCP675 CAS:3758-54-1 ***HR: 2***
CATANAC SP ANTISTATIC AGENT
mf: $C_{25}H_{53}N_2O_2 \cdot H_2O_4P$ mw:510.79

SYNS: CATANAC SP ◇ CATIONIC SP ◇ (2-HYDROXYETHYL)DIMETHYL(3-STEARAMIDOPROPYL)-AMMOIUM PHOSPHATE (1:1) (SALT)

TOXICITY DATA with REFERENCE
eye-rbt 100 mg SEV 34ZIAG -,158,69
orl-rat LD50:2835 mg/kg 34ZIAG -,158,69

CONSENSUS REPORTS: Reported in EPA TSCA Inventory.

SAFETY PROFILE: Moderately toxic by ingestion. A severe eye irritant. When heated to decomposition it emits toxic fumes of NH_3, NO_x, and PO_x.

CCP750 CAS:595-57-3 ***HR: D***
CATATOXIC STEROID No. 1
mf: $C_{22}H_{31}FO_5 \cdot K$ mw: 433.63

SYNS: CS-1 ◇ 9-α-FLUORO-11-β,17-DIHYDROXY-3-OXO-4-ANDROSTENE-17-α-PROPIONIC ACID POTASSIUM ◇ SC 11927 ◇ SU 11927

TOXICITY DATA with REFERENCE
orl-rat TDLo:1 g/kg (female 3-7D post):REP JRPFA4 37,257,74
orl-rat TDLo:500 mg/kg (female 3-7D post):TER JRPFA4 37,257,74

SAFETY PROFILE: An experimental teratogen. Other experimental reproductive effects. When heated to decomposition it emits toxic fumes of F^- and K_2O.

CCP800 CAS:100786-01-4 ***HR: D***
CATECHIN

SYNS: CATECHU ◇ C.I. NATURAL BROWN 3 ◇ CUTCH (DYE) ◇ GAMBIER ◇ KATHA

TOXICITY DATA with REFERENCE
sce-mus-ipr 3 mg/kg CALEDQ 36,189,87
dlt-mus-orl 200 mg/kg CALEDQ 36,189,87
orl-mus TDLo:200 mg/kg (male 1D pre):REP CALEDQ 36,189,87

SAFETY PROFILE: Experimental reproductive effects. Mutation data reported. When heated to decomposition it emits acrid smoke and irritating fumes.

CCP850 CAS:120-80-9 ***HR: 3***
CATECHOL
mf: $C_6H_6O_2$ mw: 110.12

PROP: Colorless crystals. Mp: 105°, bp: 246°, flash p: 261°F (CC), d: 1.341 @ 15°, vap press: 10 mm @ 118.3°, vap d: 3.79. Sol in water, chloroform, and benzene; very sol in alc and ether.

SYNS: o-BENZENEDIOL ◇ 1,2-BENZENEDIOL ◇ CATECHIN ◇ C.I. 76500 ◇ C.I. OXIDATION BASE 26 ◇ o-DIHYDROXYBENZENE ◇ 1,2-DIHYDROXYBENZENE ◇ o-DIOXYBENZENE ◇ o-DIPHENOL ◇ DURAFUR DEVELOPER C ◇ FOURAMINE PCH ◇ FOURRINE 68 ◇ o-HYDROQUINONE ◇ o-HYDROXYPHENOL ◇ 2-HYDROXYPHENOL ◇ NCI-C55856 ◇ OXYPHENIC ACID ◇ PELAGOL GREY C ◇ o-PHENYLENEDIOL ◇ PYROCATECHIN ◇ PYROCATECHINIC ACID ◇ PYROCATECHOL ◇ PYROCATECHUIC ACID

TOXICITY DATA with REFERENCE
mrc-smc 300 mg/L MUREAV 135,109,84
dni-hmn:hla 200 μmol/L MUREAV 92,427,82
dns-rat-orl 1 g/kg JJIND8 74,1283,85
scu-rat TDLo:5 mg/kg (1D pre):REP ENDOAO 57,466,55
orl-rat LD50:260 mg/kg AFREAW 3,197,51
scu-rat LDLo:110 mg/kg AIPTAK 176,193,68
orl-mus LD50:260 mg/kg AFREAW 3,197,51
ipr-mus LD50:175 mg/kg BEXBAN 61,291,66
scu-mus LD50:247 mg/kg INHEAO 5,143,67
ivn-dog LDLo:40 mg/kg HBTXAC 1,62,56
skn-rbt LD50:800 mg/kg AIHAAP 37,596,76
ipr-gpg LDLo:150 mg/kg HBTXAC 1,62,55
par-frg LDLo:160 mg/kg AEPPAE 166,437,32

CONSENSUS REPORTS: IARC Cancer Review: Group 3 IMEMDT 7,56,87; Animal Inadequate Evidence IMEMDT 15,155,77. EPA Extremely Hazardous Substances List. Reported in EPA TSCA Inventory. EPA Genetic Toxicology Program.

OSHA PEL: TWA 5 ppm (skin)
ACGIH TLV: TWA 5 ppm

SAFETY PROFILE: Poison by ingestion, subcutaneous, intraperitoneal, intravenous, and parenteral routes. Moderately toxic by skin contact. Experimental reproductive effects. Can cause dermatitis on skin contact. An allergen. Human mutation data reported. Questionable carcinogen. Systemic effects similar to phenol. Combustible when exposed to heat or flame; can react vigorously with oxidizing materials. Hypergolic reaction with concentrated nitric acid. To fight fire, use water, CO_2, dry chemical. When heated to decomposition it emits acrid smoke and irritating fumes. See also PHENOL.

CCP875 CAS:154-23-4 ***HR: 2***
d-CATECHOL
mf: $C_{15}H_{14}O_6$ mw: 290.29

PROP: dl-Form: Needles from water + acetic acid. Mp: 212-216°. Sltly sol in cold water, ether; sol in hot water, alc, glacial acetic acid, acetone. Practically insol in benzene, chloroform, petr ether. Hydrated d-form: Needles from water + acetic acid. Mp: 93-96° (175-177° when anhydrous). Hydrated l-form: Needles from water + acetic acid. Mp: 93-96° (175-177° when anhydrous).

SYNS: CATECHIN ◇ (+)-CATECHIN ◇ d-CATECHIN ◇ d-(+)-CATECHIN ◇ CATECHIN (FLAVAN) ◇ CATECHINIC ACID ◇ CATECHOL ◇ (+)-CATECHOL ◇ CATECHOL (FLAVAN) ◇ CATECHUIC ACID ◇ CATERGEN ◇ CIANIDANOL ◇ KB-53

TOXICITY DATA with REFERENCE
oms-hmn:lym 5 μmol/L CNREA8 45,2471,85
sce-hmn:lym 5 μmol/L CNREA8 45,2471,85
orl-rat TDLo:12150 mg/kg (17-22D preg/21D post):REP OYYAA2 24,509,82
orl-rat TDLo:55 g/kg (7-17D preg):TER OYYAA2 24,495,82
ipr-rat LD50:1084 mg/kg OYYAA2 24,361,82
ipr-mus LD50:1 g/kg PLMEAA 42,75,81

SAFETY PROFILE: Moderately toxic by intraperitoneal route. An experimental teratogen. Other experimental reproductive effects. Human mutation data reported. When heated to decomposition it emits acrid smoke and fumes.

CCP900 CAS:2050-46-6 ***HR: 1***
CATECHOL DIETHYL ETHER
mf: $C_{10}H_{14}O_2$ mw: 166.24

PROP: Crystals. Mp: 43-45°, bp: 219°, d: 1.0.

SYNS: BENZENE, o-DIETHOXY- ◇ BENZENE, 1,2-DIETHOXY-(9CI) ◇ o-DIETHOXYBENZENE ◇ 1,2-DIETHOXYBENZENE

TOXICITY DATA with REFERENCE
eye-rbt 100 mg MLD FCTOD7 20,573,82
eye-rbt 100 mg/30S RNS MLD FCTOD7 20,573,82

CONSENSUS REPORTS: Reported in EPA TSCA Inventory.

SAFETY PROFILE: An eye irritant. When heated to decomposition it emits acrid smoke and irritating fumes.

CCQ125 ***HR: 3***
CAULOPHYLLUM THALICTROIDES, glycoside extract

PROP: Crystalline glycoside isolated from *Caulophyllum thalictroides* blue cohosh) (JAPMA8 43,16,54).

TOXICITY DATA with REFERENCE
eye-rbt 5000 ppm/1M JAPMA8 43,16,54
ivn-rat LDLo:20300 μg/kg JAPMA8 43,16,54
ivn-mus LD50:11800 μg/kg JAPMA8 43,16,54

SAFETY PROFILE: Poison by intravenous route. An eye irritant. When heated to decomposition it emits acrid smoke and fumes.

CCQ200 CAS:72558-82-8 ***HR: D***
CAZ PENTAHYDRATE
mf: $C_{22}H_{22}N_6O_7S_2 \cdot 5H_2O$ mw: 636.72

SYNS: (6R-(6-α,7-β(Z)))-1-((7-(((2-AMINO-4-THIAZOLYL)((1-CABOXY-1-METHYLETHOXY)IMINO)ACETYL)AMINO)-2-CARBOXY-8-OXO-5-THIA-1-AZABICYCLO(4.2.0)OCT-2-EN-3-YL)METHYL)-PYRIDINIUM HYDROXIDE, inner salt, PENTAHYDRATE ◇ CEFTAZIDIME ◇ CEFTAZIDIME PENTAHYDRATE ◇ FORTAM ◇ FORTAZ ◇ GLAZIDIM ◇ GR 20263 PENTAHYDRATE ◇ SN 401 PENTAHYDRATE ◇ TAZICEF ◇ TAZIDIME

TOXICITY DATA with REFERENCE
ivn-rat TDLo:13500 mg/kg (17-22D preg/21D post):REP NKRZAZ 31(Suppl 3),968,83
ivn-rat TDLo:11 g/kg (female 7-17D post):TER NKRZAZ 31(Suppl 3),940,83

SAFETY PROFILE: Experimental reproductive effects. An experimental teratogen. When heated to decomposition it emits toxic fumes of NO_x and SO_x.

CCQ500 CAS:8007-20-3 ***HR: 2***
CEDAR LEAF OIL

PROP: Constituent is d-α-thujone, found in leaves of *Thuja occidentalis* L. (Fam. *Cupressaaceae*) (FCTXAV 12,807,74). Yellowish, volatile oil; strong sage odor. D: 0.910-0.920. Sol in fixed oils, mineral oil, propylene glycol; insol in glycerin.

SYNS: OIL of ARBOR VITAE ◇ OIL of CEDAR LEAF ◇ OIL of THUJA ◇ OIL of WHITE CEDAR ◇ OILS, CEDAR LEAF ◇ OIL THUJA ◇ THUJA OIL ◇ WHITE CEDAR OIL

TOXICITY DATA with REFERENCE
skn-rbt 500 mg/24H MOD FCTXAV 12,807,74
orl-rat LD50:830 mg/kg FCTXAV 12,807,74

CONSENSUS REPORTS: Reported in EPA TSCA Inventory.

SAFETY PROFILE: Moderately toxic by ingestion. A skin irritant. Ingestion of large quantities causes hypertension, bradycardia, tachypnea, convulsions, death. When heated to decomposition it emits acrid smoke and fumes. See also ABSINTHIUM (THUJONE).

CCQ750 CAS:8023-85-6 ***HR: 1***
CEDARWOOD OIL ATLAS

PROP: From *Cedrus atlantica*, contains α- and β-atalantone (FCTXAV 14,659,76).

SYNS: CEDARWOOD OIL MOROCCAN ◇ CEDRUS ATLANTICA OIL

TOXICITY DATA with REFERENCE
skn-rbt 500 mg/24H MLD FCTXAV 14,659,76

CONSENSUS REPORTS: Reported in EPA TSCA Inventory.

SAFETY PROFILE: A skin irritant. When heated to decomposition it emits acrid smoke and irritating fumes.

CCR000 CAS:8000-27-9 ***HR: 1***
CEDARWOOD OIL (VIRGINIA)

PROP: Colorless or sltly yellow, viscid liquid. Composition: Cedrene and cedrol. D: 0.940-0.950 @ 20°/20°. From steam distillation of the wood of *Juniperus virginiana L*. The main constituents are cedrene, thujopsene and cedrol (FCTXAV 12,807,74).

SYNS: OIL CEDAR ◇ RED CEDARWOOD OIL

TOXICITY DATA with REFERENCE
skn-rbt 500 mg/24H MOD FCTXAV 12,807,74

CONSENSUS REPORTS: Reported in EPA TSCA Inventory.

SAFETY PROFILE: A skin irritant and allergen. Combustible when exposed to heat or flame. When heated to decomposition it emits acrid smoke and irritating fumes.

CCR250 CAS:77-54-3 ***HR: 1***
8-β-H-CEDRAN-8-OL ACETATE
mf: $C_{17}H_{28}O_2$ mw: 264.45

SYNS: ACETIC ACID, CEDROL ESTER ◇ CEDRANYL ACETATE ◇ CEDRYL ACETATE ◇ OCTAHYDRO-3,6,8,8-TETRAMETHYL-1H-3a,7-METHANOAZULEN-6-OL ACETATE

TOXICITY DATA with REFERENCE
skn-rbt 500 mg/24H MOD FCTXAV 12,847,74

CONSENSUS REPORTS: Reported in EPA TSCA Inventory.

SAFETY PROFILE: A skin irritant. When heated to decomposition it emits acrid smoke and irritating fumes.

CCR500 CAS:469-61-4 ***HR: 1***
α-CEDRENE
mf: $C_{15}H_{24}$ mw: 204.39

SYN: CEDR-8-ENE

TOXICITY DATA with REFERENCE
skn-rbt 500 mg/24H MLD FCTXAV 16,637,78

CONSENSUS REPORTS: Reported in EPA TSCA Inventory.

SAFETY PROFILE: A skin irritant. When heated to decomposition it emits acrid smoke and irritating fumes.

CCR510 CAS:29597-36-2 ***HR: 1***
CEDR-8-ENE EPOXIDE

SYNS: ANDRANE ◇ CEDRANE, 8,9-EPOXIDE

TOXICITY DATA with REFERENCE
skn-rbt 500 mg/24H MOD FCTXAV 18,663,80

CONSENSUS REPORTS: Reported in EPA TSCA Inventory.

SAFETY PROFILE: A skin irritant. When heated to decomposition it emits acrid smoke and irritating fumes.

CCR524 CAS:39900-38-4 ***HR: 1***
CEDROL FORMATE
mf: $C_{16}H_{26}O_2$ mw: 250.42

SYNS: CEDRYL FORMATE ◇ 1H-3-α-7-METHANOAZULEN-6-OL, OCTAHYDRO-3,6,8,8-TETRAMETHYL-, FORMATE, (3R-(3-α-3a-β,6-α-7-β,8aα-)-

TOXICITY DATA with REFERENCE
skn-rbt 500 mg/24H MOD FCTOD7 20,647,82

CONSENSUS REPORTS: Reported in EPA TSCA Inventory.

SAFETY PROFILE: A skin irritant. When heated to decomposition it emits acrid smoke and irritating fumes.

CCR525 CAS:67874-81-1 ***HR: 1***
CEDROL METHYL ETHER
mf: $C_{16}H_{28}O$ mw: 236.44

SYNS: CEDRAMBER ◇ 1H-3a,7-METHANOAZULENE, OCTAHYDRO-6-METHOXY-3,6,8,8-TETRAMETHYL-,(3R-(3-α-3a-β, 6-α-7-β,8aα-)- ◇ METHYL CEDRYL ETHER

TOXICITY DATA with REFERENCE
skn-rbt 500 mg/24H MOD FCTXAV 17,747,79

CONSENSUS REPORTS: Reported in EPA TSCA Inventory.

SAFETY PROFILE: A skin irritant. When heated to decomposition it emits acrid smoke and irritating fumes.

CCR850 CAS:70356-03-5 ***HR: 2***
CEFACLOR HYDRATE
mf: $C_{15}H_{14}ClN_3O_4S{\bullet}H_2O$ mw: 385.85

PROP: Crystalline solid. Sol in water; practically insol in methanol, chloroform, benzene.

SYNS: ALFATIL ◇ CECLOR ◇ CEFACLOR ◇ DISTACLOR ◇ LILLY 99638 HYDRATE ◇ PANACEF ◇ PANORAL ◇ PANORAL HYDRATE

TOXICITY DATA with REFERENCE
ipr-rat LD50:1259 mg/kg IYKEDH 13,637,82
scu-rat LD50:4838 mg/kg IYKEDH 13,637,82
ipr-mus LD50:1227 mg/kg IYKEDH 13,637,82
scu-mus LD50:4180 mg/kg IYKEDH 13,637,82

SAFETY PROFILE: Moderately toxic by intraperitoneal route. Mildly toxic by subcutaneous route. When heated to decomposition it emits toxic fumes of Cl^-, SO_x, and NO_x.

CCR875 CAS:3254-89-5 ***HR: 3***
CEFADOL
mf: $C_{21}H_{27}NO{\bullet}ClH$ mw: 345.95

SYNS: CELMIDOL ◇ DEPHENIDOL HYDROCHLORIDE ◇ DIFENIDOL HYDROCHLORIDE ◇ DIFENIDOLIN ◇ α,α-DIPHENYL-1-PIPERIDINEBUTANOL HYDROCHLORIDE ◇ MANIOL ◇ MECALMIN ◇ PINERORO ◇ SATANOLON ◇ TENESDOL ◇ WANSAR ◇ YESDOL

TOXICITY DATA with REFERENCE
orl-rat TDLo:4950 mg/kg (11D pre):REP GNRIDX 5,430,71
orl-rat LD50:515 mg/kg IYKEDH 4,193,73
ipr-rat LD50:82 mg/kg IYKEDH 4,193,73
scu-rat LD50:670 mg/kg IYKEDH 4,193,73
ivn-rat LD50:29 mg/kg IYKEDH 4,193,73
ims-rat LD50:635 mg/kg IYKEDH 4,193,73
orl-mus LD50:400 mg/kg IYKEDH 4,193,73
ipr-mus LD50:105 mg/kg IYKEDH 4,193,73
scu-mus LD50:163 mg/kg IYKEDH 4,193,73
ivn-mus LD50:37 mg/kg IYKEDH 4,193,73

SAFETY PROFILE: Poison by ingestion, subcutaneous, intravenous, and intraperitoneal routes. Moderately toxic by intramuscular route. Experimental reproductive effects. When heated to decomposition it emits toxic fumes of NO_x and HCl.

CCR890 CAS:3577-01-3 ***HR: 2***
CEFALOGLYCIN
mf: $C_{18}H_{19}N_3O_6S$ mw: 405.46

PROP: Dihydrate, Kafocin. Crystalline powder. Mp: 223-250° (decomp).

SYNS: 7-(d-α-AMINOPHENYL-ACETAMIDO)CEPHALOSPORANIC ACID ◇ CEPHALOGLYCIN ◇ CEPHALOGLYCINE ◇ d-CEPHALOGLYCINE ◇ CEPHAOGLYCIN ACID ◇ KAFOCIN ◇ KEFGLYCIN ◇ LILLY 39435

TOXICITY DATA with REFERENCE
orl-rat TDLo:1200 mg/kg (7-12D preg):REP NKRZAZ 18,39,70
ipr-rat LD50:1300 mg/kg NKRZAZ 18,22,70
scu-rat LD50:2800 mg/kg NKRZAZ 18,22,70
ipr-mus LD50:1030 mg/kg NKRZAZ 18,22,70
scu-mus LD50:3700 mg/kg NKRZAZ 18,22,70

SAFETY PROFILE: Moderately toxic by subcutaneous and intraperitoneal routes. Experimental reproductive effects. When heated to decomposition it emits toxic fumes of SO_x and NO_x.

CCR925 CAS:30034-03-8 ***HR: 2***
CEFAMANDOLE SODIUM
mf: $C_{18}H_{17}N_6O_5S_2{\bullet}Na$ mw: 484.52

SYN: SODIUM CEFAMANDOLE

TOXICITY DATA with REFERENCE
ivn-rat TDLo:5500 mg/kg (female 7-17D post):REP NKRZAZ 27(Suppl 5),658,79
ipr-rbt TDLo:195 mg/kg (female 6-18D post):TER NKRZAZ 27(Suppl 5),658,79
scu-rat LD50:12100 mg/kg YAKUD5 26,115,84
ivn-rat LD50:4410 mg/kg YAKUD5 26,115,84
scu-mus LD50:10300 mg/kg YAKUD5 26,115,84
ivn-mus LD50:4460 mg/kg YAKUD5 26,115,84

SAFETY PROFILE: Moderately toxic by intravenous route. An experimental teratogen. Experimental reproductive effects. When heated to decomposition it emits toxic fumes of SO_x, NO_x, and Na_2O.

CCR950 CAS:64485-93-4 ***HR: 2***
CEFATOXIME SODIUM
mf: $C_{16}H_{16}N_5O_7S_2{\bullet}Na$ mw: 477.48

SYNS: (6R-(6-α,7-β(Z)))-3-((ACETYLOXY)METHYL)-7-(((2-AMINO-4-THIAZOLYL)(METHOXYIMIO)ACETYL)AMINO)-8-OXO-5-THIA-1-AZABICYCLO(4,2,0)OCT-2-ENE-2-CARBOXYLIC ACID, SODIUM SALT ◇ CEFOTAXIME SODIUM ◇ CTX ◇ HR 756 ◇ RU 24756 ◇ SODIUM-7-(2-(2-AMINO-4-THIAZOLYL)-2-METHOXYIMINOACETAMIDO) CEPHALOSPORANATE

TOXICITY DATA with REFERENCE
ivn-rbt TDLo:325 mg/kg (6-18D preg):TER OYYAA2 21,375,81
ivn-cld TDLo:1800 mg/kg/18D-I:BLD DICPBB 17,739,83
orl-rat LD50:20 g/kg NKRZAZ 28(Suppl 1),98,80
ipr-rat LD50:10 g/kg NKRZAZ 28(Suppl 1),98,80
scu-rat LD50:18400 mg/kg NKRZAZ 28(Suppl 1),98,80
ivn-rat LD50:7000 mg/kg NIIRDN 6,APP-8,82
ims-rat LD50:2000 mg/kg NKRZAZ 28(Suppl 1),98,80
orl-mus LD50:20 g/kg NKRZAZ 28(Suppl 1),98,80
ipr-mus LD50:10 g/kg NIIRDN 6,APP-8,82
scu-mus LD50:12950 mg/kg NIIRDN 6,APP-8,82
ivn-mus LD50:8350 mg/kg NKRZAZ 28(Suppl 1),98,80
ivn-rbt LD50:1880 mg/kg NKRZAZ 28(Suppl 1),98,80

SAFETY PROFILE: Moderately toxic by intravenous and intramuscular routes. Mildly toxic by ingestion and intraperitoneal routes. Human systemic effects by intravenous route: agranulocytosis. An experimental teratogen. When heated to decomposition it emits toxic fumes of SO_x, NO_x and Na_2O.

CCS250 CAS:27164-46-1 ***HR: 2***
CEFAZOLIN SODIUM SALT
mf: $C_{14}H_{13}N_8O_4S_3$•Na mw: 476.52

SYNS: ACEF ◇ ANCEF ◇ ATIRIN ◇ BIAZOLINA ◇ CEFACIDAL ◇ CEFAMEDIN ◇ CEFAMEZIN ◇ CEFAZIL ◇ CEFAZINA ◇ CEFAZOLIN ◇ CEFAZOLINE SODIUM ◇ CEZ SODIUM ◇ ELZOGRAM ◇ FIRMACEF ◇ GRAMAXIN ◇ KEFZOL ◇ LIVICLINA ◇ MONOSODIUM CEFAZOLIN ◇ SKF 41588 ◇ SODIUM CEFAZOLIN ◇ SODIUM CEPHAZOLIN ◇ SODIUM CEZ ◇ TOTACEF ◇ ZOLICEF

TOXICITY DATA with REFERENCE
ivn-rat TDLo:5500 mg/kg (female 7-17D post):REP NKRZAZ 28(Suppl 7),1119,80
ivn-rat TDLo:5500 mg/kg (7-17D preg):TER NKRZAZ 28(Suppl 7),1119,80
ivn-wmn TDLo:660 mg/kg/11D-I NPRNAY 45,72,87
ims-hmn TDLo:14 mg/kg/D:GIT,SKN JMGZAI 8(8),10,71
scu-rat LD50:7400 mg/kg NIIRDN 6,404,82
scu-rat LD50:10 g/kg MEIEDD 10,269,83
ivn-rat LD50:2760 mg/kg NKRZAZ 35(Suppl 1),207,87
ipr-mus LD50:6200 mg/kg NIIRDN 6,404,82
scu-mus LD50:7600 mg/kg JIDIAQ 128,S379,73
ivn-mus LD50:3900 mg/kg JIDIAQ 128,S379,73
scu-dog LD50:4 g/kg NKRZAZ 18,528,70
ivn-dog LD50:2200 mg/kg ARZNAD 29,424,79
ivn-rbt LD50:2500 mg/kg ARZNAD 29,424,79

SAFETY PROFILE: Moderately toxic by subcutaneous and intravenous routes. Mildly toxic by intraperitoneal route. Human systemic effects by intramuscular route: changes in structure or function of the salivary glands, nausea or vomiting, and allergic dermatitis. An experimental teratogen. Other experimental reproductive effects. When heated to decomposition it emits very toxic fumes of NO_x, Na_2O, and SO_x.

CCS300 CAS:75738-58-8 ***HR: 2***
CEFMENOXIME HEMIHYDROCHLORIDE
mf: $C_{16}H_{17}N_9O_5S_3$•1/2ClH mw: 529.77

SYNS: AB 50912 HEMIHYDROCHLORIDE ◇ SCE 1365 HYDROCHLORIDE

TOXICITY DATA with REFERENCE
ipr-rat TDLo:10500 mg/kg (35D male):REP TAKHAA 42,104,83
scu-rat LD50:13150 mg/kg JJANAX 25,2615,82
ivn-rat LD50:2680 mg/kg JJANAX 35,2615,82
orl-mus LD50:17540 mg/kg IYKEDH 14,297,83
scu-mus LD50:11830 mg/kg JJANAX 35,2615,82
ivn-mus LD50:7830 mg/kg YAKUD5 22,1605,80

SAFETY PROFILE: Moderately toxic by intravenous route. Mildly toxic by ingestion. Experimental reproductive effects. When heated to decomposition it emits toxic fumes of SO_x, NO_x, and HCl.

CCS350 CAS:56796-20-4 ***HR: 1***
CEFMETAZOLE
mf: $C_{15}H_{17}N_7O_5S_2$ mw: 471.57

SYNS: CS 1170 ◇ SKF 83088

TOXICITY DATA with REFERENCE
ivn-dog TDLo:9 g/kg (18-35D preg):TER JZKEDZ 6,289,80
ipr-mus LD50:10233 mg/kg SKKNAJ 31,49,79
scu-mus LD50:12190 mg/kg SKKNAJ 31,49,79
ivn-mus LD50:8690 mg/kg SKKNAJ 31,49,79

SAFETY PROFILE: Mildly toxic by subcutaneous, intravenous, and intraperitoneal routes. An experimental teratogen. When heated to decomposition it emits toxic fumes of SO_x and NO_x.

CCS360 ***HR: 2***
CEFMETAZOLE SODIUM
mf: $C_{15}H_{16}N_7O_5S_3$•Na mw: 493.55

SYNS: CMZ SODIUM ◇ CS 1170 SODIUM ◇ SKF 83088 SODIUM

TOXICITY DATA with REFERENCE
ivn-mus TDLo:5 g/kg (female 6-15D post):REP SKKNAJ 30,148,78
ivn-rat TDLo:5500 mg/kg (female 7-17D post):TER SKKNAJ 30,148,78
orl-rat LD50:3204 mg/kg JOPHDQ 8,633,85
orl-mus LD50:3228 mg/kg JOPHDQ 8,633,85

ipr-mus LD50:10233 mg/kg SKKNAJ 30,112,78
scu-mus LD50:12190 mg/kg SKKNAJ 30,112,78
ivn-mus LD50:8690 mg/kg SKKNAJ 30,112,78

SAFETY PROFILE: Moderately toxic by ingestion. Mildly toxic by subcutaneous, intravenous, and intraperitoneal routes. An experimental teratogen. Other experimental reproductive effects. When heated to decomposition it emits toxic fumes of SO_x, NO_x, and Na_2O.

CCS365 ***HR: 1***
CEFMINOX
mf: $C_{16}H_{20}N_7O_7S3 \cdot Na$ mw: 541.60

SYNS: (6R-(6-α,7-α))-7-(((2-AMINO-2-CARBOXYETHYL)THIO)ACETYL)AMINO)-7-METHOXY-3-(((1-METHYL-1H-TETRAZOL-5-YL)THIO)METHYL)-8-OXO-5-THIA-1-AZABICYCLO(4.2.0)OCT-2-ENE-2-CARBOXYLIC ACID MONOSODIUM SALT ◇ CEPHAMYCIN ◇ MEICELIN ◇ MT-141

TOXICITY DATA with REFERENCE
unr-rat TDLo:36 g/kg (male 36D pre):REP TOLED5 31(Suppl),66,86
ipr-rat LD50:8550 mg/kg JJANAX 37,847,84
ivn-rat LD50:5700 mg/kg JJANAX 37,847,84
ims-rat LD50:9600 mg/kg JJANAX 37,847,84
ivn-mus LD50:5200 mg/kg JJANAX 37,847,84
ims-mus LD50:8200 mg/kg JJANAX 37,847,84

SAFETY PROFILE: Mildly toxic by intravenous, intramuscular, and intraperitoneal routes. Experimental reproductive effects. When heated to decomposition it emits toxic fumes of SO_x, NO_x, and Na_2O.

CCS369 CAS:62893-20-3 ***HR: 2***
CEFOPERAZONE SODIUM
mf: $C_{25}H_{27}N_9O_8S_2 \cdot Na$ mw: 668.72

SYNS: CPZ ◇ T-1551

TOXICITY DATA with REFERENCE
unr-man TDLo:229 mg/kg/4D-I:BLD AIMEAS 102,721,85
ivn-rat LD50:4260 mg/kg NKRZAZ 28(Suppl 6),179,80
ipr-mus LD50:8200 mg/kg NKRZAZ 28(Suppl 6),179,80
scu-mus LDLo:15 g/kg NKRZAZ 28(Suppl 6),179,80
ivn-mus LD50:3840 mg/kg NKRZAZ 28(Suppl 6),179,80
ivn-dog LDLo:6 g/kg NKRZAZ 28(Suppl 6),179,80

SAFETY PROFILE: Moderately toxic by intravenous routes. Mildly toxic by subcutaneous and intraperitoneal routes. Human systemic effects by an unspecified route: change in clotting factors. When heated to decomposition it emits toxic fumes of SO_x, NO_x, and Na_2O.

CCS371 CAS:74356-00-6 ***HR: 1***
CEFOTAN
mf: $C_{17}H_{17}N_7O_8S_4 \cdot 2Na$ mw: 621.63

SYNS: (6R-cis)-7-(((4-(2-AMINO-1-CARBOXY-2-OXOETHYL)-1,3-DITHIETAN-2-YL)CARBONYL)AMINO)-7-METHOXY-3-(((1-METHYL-1H-TETRAZOL-5-YL)THIO)METHYL)-8-OXO-5-THIA-1-AZABICYCLO(4.2.0)OCT-2-ENE-2-CARBOXYLIC ACID MONOSODIUM SALT ◇ CEFOTETAN DISODIUM SALT ◇ ICI 156834 DISODIUM ◇ YM 09330

TOXICITY DATA with REFERENCE
ivn-rat TDLo:2700 mg/kg (17-22D preg/21D post):REP OYYAA2 23,767,82
ipr-rat LD50:8250 mg/kg NKRZAZ 30(Suppl 1),212,82
ivn-rat LD50:6790 mg/kg NKRZAZ 30(Suppl 1),212,82
ipr-mus LD50:8120 mg/kg NKRZAZ 30(Suppl 1),212,82
ivn-mus LD50:4990 mg/kg NKRZAZ 30(Suppl 1),212,82

SAFETY PROFILE: Mildly toxic by intravenous and intraperitoneal routes. Experimental reproductive effects. When heated to decomposition it emits toxic fumes of SO_x, NO_x, and Na_2O.

CCS373 CAS:69712-56-7 ***HR: 1***
CEFOTETAN
mf: $C_{17}H_{17}N_7O_8S_4$ mw: 575.65

SYNS: CTT ◇ 5-THIA-1-AZABICYCLO(4.2.0)OCT-2-ENE-2-CARBOXYLIC ACID, 7-(((4-(2-AMINO-1-CARBOXY-2-OXOETHYLIDENE)-1,3-DITHIETAN-2-YL)CARBONYL)AMINO)-7-METHOXY-3-(((1-METHYL-1H-TETRAZOL-5-YL)THIO)METHYL)-8-OXO-, (6R-(6-α-7-α))- ◇ YM 09330

TOXICITY DATA with REFERENCE
ivn-rat TDLo:1100 mg/kg (female 7-17D post):REP NKRZAZ 30(Suppl 1),278,82
ivn-rat LD50:5 g/kg 43MKAT 1,273,80

SAFETY PROFILE: Slightly toxic by intravenous route. Experimental reproductive effects. When heated to decomposition it emits toxic fumes of NO_x and SO_x.

CCS375 CAS:66309-69-1 ***HR: 2***
CEFOTIAM DIHYDROCHLORIDE
mf: $C_{18}H_{23}N_9O_4S_2 \cdot 2ClH$ mw: 598.60

SYN: CEFOTIAM HYDROCHLORIDE

TOXICITY DATA with REFERENCE
ivn-rat LD50:3680 mg/kg IYKEDH 12,668,81
scu-mus LD50:7800 mg/kg NIIRDN 6,411,82
ivn-mus LD50:3840 mg/kg IYKEDH 12,668,81

SAFETY PROFILE: Moderately toxic by intravenous route. Mildly toxic by subcutaneous route. When heated to decomposition it emits toxic fumes of SO_x, NO_x, and HCl.

CCS500 CAS:35607-66-0 ***HR: 2***
CEFOXITIN
mf: $C_{16}H_{17}N_3O_7S_2$ mw: 427.48

SYNS: CEPHOXITIN ◇ CFX ◇ REPHOXITIN

TOXICITY DATA with REFERENCE
ivn-wmn TDLo:75 mg/kg/18H-I:BLD AIMEAS 92,874,80

ivn-rat LD50:8580 mg/kg NKRZAZ 26(Suppl 1),150,78
scu-mus LD50:9250 mg/kg NKRZAZ 26(Suppl 1),150,78
ivn-mus LD50:4970 mg/kg NKRZAZ 26(Suppl 1),150,78

SAFETY PROFILE: Mildly toxic by subcutaneous and intravenous routes. Human systemic effects by intravenous route: reduction in the white blood cell count. When heated to decomposition it emits very toxic fumes of NO_x and SO_x. See also AMINES.

CCS510 CAS:33564-30-6 ***HR: 1***
CEFOXOTIN SODIUM
mf: $C_{16}H_{16}N_3O_7S_2 \cdot Na$ mw: 449.46

SYNS: (6R-cis)-3-(((AMINOCARBONYL)OXY)METHYL)-7-METHYOXY-8-OXO-7-((2-THIENYLACETYL)AMINO)-5-THIA-1-AZABICYCLO(4.2.0)OCT-2-ENE-2-CARBOXYLIC ACID MONOSODIUM SALT ◇ CEFOXITIN SODIUM SALT ◇ CENOMYCIN ◇ MEFOXIN ◇ MEFOXITIN ◇ MERXIN ◇ MONOSODIUM CEROXITIN

TOXICITY DATA with REFERENCE
ivn-rat LD50:8580 mg/kg IYKEDH 11,181,80
scu-mus LD50:9250 mg/kg YKYUA6 31,629,80
ivn-mus LD50:4970 mg/kg IYKEDH 11,181,80
ivn-dog LD50:10000 mg/kg NIIRDN 6,410,82

SAFETY PROFILE: Mildly toxic by subcutaneous and intravenous routes. When heated to decomposition it emits toxic fumes of SO_x, NO_x, and Na_2O. See also CEFOXITIN.

CCS525 CAS:85287-61-2 ***HR: 2***
CEFPIMIZOLE SODIUM
mf: $C_{28}H_{26}N_6O_{10}S_2 \cdot Na$ mw: 693.66

SYNS: AC 1370 ◇ AC 1370 SODIUM ◇ (6R-(6-α,7-β(R*)))-1-((2-CARBOXY-7-(((((5-CARBOXY-1H-IMIDAZOL-4-YL)CARBONYL)AMINO)PHENYLACETYLE)AMINO)-8-OXO-5-THIA-1-AZABICYCLO(4.2.0)OCT-2-EN-3-YL)METHYL)-4-(2-SULFOETHYL)-PYRIDINIUM HYDROXODIE, inner salt, MONOSODIUM SALT ◇ U 631963

TOXICITY DATA with REFERENCE
scu-rat LD50:11500 mg/kg TOLED5 23,135,84
ivn-rat LD50:3500 mg/kg TOLED5 23,135,84
scu-mus LD50:6800 mg/kg TOLED5 23,135,84
ivn-mus LD50:2700 mg/kg TOLED5 23,135,84

SAFETY PROFILE: Moderately toxic by intravenous route. Mildly toxic by subcutaneous route. When heated to decomposition it emits toxic fumes of NO_x, SO_x, and Na_2O.

CCS530 CAS:51762-05-1 ***HR: 1***
CEFROXADIN
mf: $C_{16}H_{19}N_3O_5S$ mw: 365.44

PROP: Internal salt. Mp: 170° (decomp).

SYNS: (6R-(6-α,7-β(R*)))-7-((AMINO-1,4-CYCLOHEXADIEN-1-YLACETAL)AMINO)-3-METHYL-8-OXO-5-THIA-1-AZABICYCLO(4.2.0)OCT-2-ENE-2-CARBOXYLIC ACID ◇ 7-(D-2-AMINO-2-(1,4-CYCLOHEXADIENYL)ACETAMIDE)-3-METHOXY-3-CEPHEM-4-CARBYLIC ACID ◇ ANTIBIOTIC CGP 9000 ◇ CEFROXADINE ◇ CGP 9000 ◇ CXD ◇ ORASPOR

TOXICITY DATA with REFERENCE
orl-rat TDLo:2700 mg/kg (female 17-22D post):REP OYYAA2 19,615,80
orl-rat TDLo:27 g/kg (17-22D preg/21D post):TER OYYAA2 19,615,80
ipr-rat LD50:6 g/kg NKRZAZ 28(Suppl 3),98,80
ipr-mus LD50:7090 mg/kg JANTAJ 29,653,76
orl-rbt LD50:10 g/kg NIIRDN 6,APP-11,82

SAFETY PROFILE: Mildly toxic by ingestion and intraperitoneal routes. An experimental teratogen. Other experimental reproductive effects. When heated to decomposition it emits toxic fumes of SO_x and NO_x.

CCS535 ***HR: 2***
CEFROXADIN DIHYDRATE
mf: $C_{16}H_{19}N_3O_5S \cdot 2H_2O$ mw: 401.48

SYNS: (6R-(6-α,7-β(R*)))-7-((AMINO-1,4-CYCLOHEXADIEN-1-YLACETAL)AMINO)-3-METHYL-8-OXO-5-THIA-1-AZABICYCLO(4.2.0)OCT-2-ENE-2-CARBOXYLIC ACID DIHYDRATE ◇ CGP-9000 DIHYDRATE

TOXICITY DATA with REFERENCE
orl-rat TDLo:11 g/kg (7-17D preg):REP IYKEDH 10,802,79
ipr-rat LD50:3320 mg/kg IYKEDH 13,349,82

SAFETY PROFILE: Moderately toxic by intraperitoneal route. Experimental reproductive effects. When heated to decomposition it emits toxic fumes of SO_x and NO_x. See also CEFROXADIN.

CCS550 CAS:52152-93-9 ***HR: 2***
CEFSULODIN SODIUM
mf: $C_{22}H_{19}N_4O_8S_2 \cdot Na$ mw: 554.56

SYNS: (6R-(6-α,7-β(R*)))-4-(AMINOCARBONYL)-1-((2-CARBOXY-8-OXO-7-((PHENYLSULFOACETYL)AMINO)-5-THIO-1-AZABICYCLO(4.2.0)OCT-2-EN-3-YL)METHYL)-PYRIDINIUM HYDROXIDE, inner salt, MONOSODIUM SALT ◇ CGP 71743 ◇ SCE 129

TOXICITY DATA with REFERENCE
ipr-rat LD50:3030 mg/kg IYKEDH 12,668,81
scu-rat LD50:5550 mg/kg IYKEDH 12,668,81
ivn-rat LD50:3150 mg/kg IYKEDH 12,668,81
ims-rat LD50:5530 mg/kg IYKEDH 12,668,81
ipr-mus LD50:6350 mg/kg IYKEDH 12,668,81
scu-mus LD50:6940 mg/kg IYKEDH 12,668,81
ivn-mus LD50:3780 mg/kg NIIRDN 6,412,82
ims-mus LD50:3800 mg/kg IYKEDH 12,668,81

SAFETY PROFILE: Moderately toxic by intramuscular, intravenous, and intraperitoneal routes. Mildly toxic by subcutaneous route. When heated to decomposition it emits toxic fumes of NO_x, SO_x, and Na_2O.

CCS560 CAS:41136-22-5 ***HR: 2***
CEFTEZOLE SODIUM
mf: $C_{13}H_{11}N_8O_4S_3$•Na mw: 462.49

TOXICITY DATA with REFERENCE
scu-rat TDLo:56400 mg/kg (26W male):REP NKRZAZ 24,671,76
ivn-rat LD50:3800 mg/kg NIIRDN 6,413,82
ipr-mus LD50:8900 mg/kg NIIRDN 6,413,82
ivn-mus LD50:4700 mg/kg NIIRDN 6,413,82

SAFETY PROFILE: Moderately toxic by intravenous route. Mildly toxic by intraperitoneal route. Experimental reproductive effects. When heated to decomposition it emits toxic fumes of SO_x, NO_x and Na_2O.

CCS600 CAS:55268-75-2 ***HR: 1***
CEFUROXIM
mf: $C_{16}H_{16}N_4O_8S$ mw: 424.42

PROP: White, crystalline solid.

SYNS: (6R-(6-α,7-β(Z)))-3-(((AMINOCARBONYL)OXY)METHYL)-7-((2-FURANYL(METHYOXYIMINO)ACETYL)AMINO)-8-OXO-5-THIA-1-AZABICYCLO(4.2.0)OCT-2-ENE-2-CARBOXYLIC ACID ◇ CEFUROXIME ◇ CEPHUROXIME ◇ CXM ◇ ZINACEF

TOXICITY DATA with REFERENCE
scu-rat TDLo:8800 mg/kg (7-17D preg):REP NKRZAZ 27(Suppl 6),245,79
scu-rat TDLo:4400 mg/kg (female 7-17D post):TER NKRZAZ 27(Suppl 6),245,79
ivn-man TDLo:64 mg/kg/16H-I LANCAO 1,965,84
ivn-mus LD50:10400 mg/kg DRUGAY 17,233,79

SAFETY PROFILE: Mildly toxic by intravenous route. An experimental teratogen. Other experimental reproductive effects. When heated to decomposition it emits toxic fumes of SO_x and NO_x.

CCS625 CAS:64544-07-6 ***HR: 3***
CEFUROXIME AXETIL
mf: $C_{20}H_{22}N_4O_{10}S$ mw: 510.52

SYNS: CXM-AX ◇ SN 407 ◇ 5-THIA-1-AZABICYCLO(4.2.0)OCT-2-ENE-2-CARBOXYLIC ACID, 3-(((AMINOCARBONYL)OXY)METHYL)-7-((2-FURANYL(METHOXYIMINO)ACETYL)AMINO)-8-OXO-, 1-(ACETYLOXY)ETHYL ESTER, (6R-(6-α-7-β (Z)))-

TOXICITY DATA with REFERENCE
orl-rat TDLo:26 g/kg (female 17-22D post):REP NKRZAZ 34(Suppl 5),251,86
ipr-rat LD50:950 mg/kg NKRZAZ 34(Suppl 5),64,86
scu-rat LD50:2500 mg/kg NKRZAZ 34(Suppl 5),64,86
ipr-mus LD50:510 mg/kg NKRZAZ 34(Suppl 5),64,86
scu-mus LD50:1840 mg/kg NKRZAZ 34(Suppl 5),68,86
orl-rbt LD50:200 mg/kg NKRZAZ 34(Suppl 5),64,86

SAFETY PROFILE: Poison by ingestion. Moderately toxic by intraperitoneal and subcutaneous routes. Experimental reproductive effects. When heated to decomposition it emits toxic fumes of NO_x and SO_x.

CCS635 CAS:82219-81-6 ***HR: 2***
CEFZONAME SODIUM
mf: $C_{16}H_{14}N_7O_5S_4$•Na mw: 535.60

SYNS: CL 251931 SODIUM SALT ◇ CZON ◇ L-105 ◇ 5-THIA-1-AZABICYCLO(4.2.0)OCT-2-ENE-2-CARBOXYLIC ACID, 7-(((2-AMINO-4-THIAZOLYL)(METHOXYIMINO)ACETYL)AMINO)-8-OXO-3-((1,2,3-THIADIAZOL-5-YLTHIO)METHYL)-, SODIUM SALT, (6R-(6-α-7-β(Z)))-

TOXICITY DATA with REFERENCE
ipr-rat TDLo:233 g/kg (female 26W pre):REP YACHDS 14,4013,86
ivn-rat LD50:4222 mg/kg NKRZAZ 34(Suppl 3),96,86
ipr-mus LD50:6424 mg/kg NKRZAZ 34(Suppl 3),96,86
scu-mus LD50:8 g/kg NKRZAZ 34(Suppl 3),96,86
ivn-mus LD50:4117 mg/kg NKRZAZ 34(Suppl 3),96,86
ivn-dog LD50:2500 mg/kg NKRZAZ 34(Suppl 3),96,86

SAFETY PROFILE: Moderately toxic by intravenous route. Experimental reproductive effects. When heated to decomposition it emits toxic fumes of NO_x and SO_x.

CCS650 ***HR: 3***
CELANDINE

PROP: A low (1 to 3 feet) herb which produces small yellow flowers from March to August. Most of the plant is covered with fine white hairs and its sap is a red-orange color. It grows in wet soil in the region bounded by Georgia, Missouri, British Columbia, and Nova Scotia.

SYNS: CHELIDONIUM MAJUS L. ◇ ELON WORT ◇ FELONWORT ◇ SWALLOW WORT ◇ TETTERWORT ◇ WORT-WEED

SAFETY PROFILE: The whole plant contains poisonous isoquinoline alkaloids some of which are adrenergic blockers. Ingestion (rare because of the unpleasant taste) can cause headache and sleepiness within 14 hours, followed by fever, vomiting, diarrhea, coma and circulatory collapse within 6 hours.

CCS675 CAS:8064-08-2 ***HR: D***
CELESTAN-DEPOT
mf: $C_{24}H_{31}FO_6$•$C_{22}H_{30}FO_8P$•2Na mw: 953.02

SYNS: BETAMETHASONE ACETATE mixed with BETAMETHASONE SODIUM PHOSPHATE ◇ BETAMETHASONE SODIUM PHOSPHATE mixed with BETAMETHASONE ACETATE ◇ CELESTONE CHRONODOSE ◇ CELESTONE SOLOSPAN ◇ CELESTONE SOLUSPAN ◇ 21-(PHOSPHONOOXY)PREGNA-1,4-DIENE-3,20-DIONE DISODIUM SALT

TOXICITY DATA with REFERENCE
ims-rbt TDLo:800 μg/kg (female 24-26D post):REP AJOGAH 136,234,80
scu-rat TDLo:4200 μg/kg (female 12-13D post):TER TJADAB 23,15,81

SAFETY PROFILE: An experimental teratogen. Experimental reproductive effects. When heated to decomposition it emits toxic fumes of F^-, PO_x, and Na_2O.

CCT250 CAS:9005-81-6 ***HR: 3***
CELLOPHANE
mf: $(C_6H_{10}O_5)_n$

SYN: VISKING CELLOPHANE

TOXICITY DATA with REFERENCE
imp-rat TDLo:18 mg/kg:ETA CNREA8 15,333,55

CONSENSUS REPORTS: Reported in EPA TSCA Inventory.

SAFETY PROFILE: Questionable carcinogen with experimental tumorigenic data by implant. See also POLYMERS. When heated to decomposition it emits acrid smoke and irritating fumes.

CCT825 ***HR: 1***
CELLRYL

PROP: Protein-free extract of calf blood which comprises various kinds of amino acids, peptides, nucleosides, electrolytes, and unidentified organic substances; exerts healing effect on experimentally induced ulcer and wound (UsuT## 29JUN79).

TOXICITY DATA with REFERENCE
scu-rat TDLo:100 g/kg (8-17D preg):REP YACHDS 4,1114,76
scu-rat TDLo:25 g/kg (8-17D preg):TER YACHDS 4,1114,76
ivn-mus LD50:43 g/kg YACHDS 4,74,76

SAFETY PROFILE: Mildly toxic by intravenous route. An experimental teratogen. Other experimental reproductive effects. When heated to decomposition it emits toxic fumes of NO_x.

CCT900 ***HR: 2***
CELLULASE AP3

TOXICITY DATA with REFERENCE
ipr-rat LD50:2650 mg/kg KSRNAM 8,3751,74
scu-rat LD50:11920 mg/kg KSRNAM 8,3751,74
orl-mus LD50:30900 mg/kg KSRNAM 8,3751,74
ipr-mus LD50:3660 mg/kg KSRNAM 8,3751,74
scu-mus LD50:6710 mg/kg KSRNAM 8,3751,74

SAFETY PROFILE: Moderately toxic by intraperitoneal route. Mildly toxic by ingestion and subcutaneous routes. When heated to decomposition it emits toxic fumes of NO_x.

CCU000 CAS:8050-88-2 ***HR: 1***
"CELLULOID"
DOT: UN 2000/UN 2002

PROP: Clear or colored cellulose nitrate. D: 1.35-1.60.

SYNS: CELLULOID, in blocks, rods, rolls, sheets, tubes (DOT) ◇ CELLULOID SCRAP (DOT)

DOT Classification: Flammable Solid; Label: Flammable Solid; Flammable Solid; Label: Spontaneously Combustible (Scrap).

SAFETY PROFILE: Mildly toxic. Flammable when exposed to heat or flame. Can react with oxidizing materials. When heated to decomposition it emits acrid smoke and fumes. See also NITRATES.

CCU050 CAS:9004-38-0 ***HR: D***
CELLULOSE ACETATE MONOPHTHALATE

SYNS: ACETYL PHTHALYL CELLULOSE ◇ CAP-WAKO ◇ CELLACETATE ◇ CELLULOSE, ACETATE HYDROGEN 1,2-BENZENEDICARBOXYLATE (9CI) ◇ CELLULOSE, ACETATE PHTHALATE ◇ CELLULOSE ACETOPHTHALATE ◇ CELLULOSE ACETYLPHTHALATE

TOXICITY DATA with REFERENCE
orl-mus TDLo:3 g/kg (female 7-12D post):TER IYKEDH 6,49,75

CONSENSUS REPORTS: Reported in EPA TSCA Inventory.

SAFETY PROFILE: An experimental teratogen. When heated to decomposition it emits acrid smoke and irritating fumes.

CCU100 ***HR: 1***
CELLULOSE, MICROCRYSTALLINE

PROP: Fine white crystalline powder from treatment of α-cellulose with mineral acids. Insol in water, most organic solvents.

SYN: CELLULOSE GEL

SAFETY PROFILE: A nuisance dust. When heated to decomposition it emits acrid smoke and irritating fumes.

CCU150 CAS:9004-34-6 ***HR: 1***
CELLULOSE, POWDERED

PROP: Fine white fibrous particles from treatment of bleached cellulose from wood or cotton. Insol in water and most organic solvents.

OSHA PEL: Total Dust: 15 mg/m^3; Respirable Fraction: 5 mg/m^3
ACGIH TLV: TWA (nuisance particulate) 10 mg/m^3 of total dust (when toxic impurities are not present, e.g., quartz $<$ 1%).

SAFETY PROFILE: A nuisance dust. When heated to decomposition it emits acrid smoke and irritating fumes.

CCU250 CAS:9004-70-0 ***HR: 3***
CELLULOSE TETRANITRATE
DOT: UN 0340/UN 0341/UN 0342/UN 0343/UN 1324/UN 2059/UN 2060/UN 2556/UN 2557/NA 1324/NA 1325/NA 2059/NA 2555/NA 2556
mf: $C_{12}H_{16}(ONO_2)_4O_6$ mw: 504.3

PROP: White, amorphous solid. D: 1.66, flash p: 55°F.

SYNS: CELLOIDIN ◇ CELLULOSE NITRATE ◇ COLLODION COTTON ◇ COLLOXYLIN ◇ GUNCOTTON ◇ NITROCELLULOSE ◇ NITROCOTTON ◇ PYRALIN ◇ PYROXYLIN ◇ PYROXYLIN PLASTICS (DOT) ◇ PYROXYLIN PLASTIC SCRAP (DOT) ◇ SOLUBLE GUN COTTON ◇ XYLOIDIN

CONSENSUS REPORTS: Reported in EPA TSCA Inventory.

DOT Classification: Flammable Solid; Label: Flammable Solid (NA1324, NA1325, NA2555); Class A Explosive; Label: Explosive A, (dry, UN0340, UN0341); Flammable Solid; Label: Flammable Solid (UN1324, UN2555, UN2556,UN2557); Class B Explosive; Label: Explosive B (UN0342, UN0343); Flammable or Combustible Liquid.; Label: Flammable Liquid (UN2556, UN2060).
DOT Classification: Flammable Solid; Label: Flammable Solid.

SAFETY PROFILE: Flammable solid. Highly dangerous fire hazard in the dry state when exposed to heat, flame, or powerful oxidizers. When wet with 35% of denatured ethanol it is about as hazardous as ethanol alone or gasoline. Dry cellulose tetranitrate burns rapidly with intense heat and ignites easily. Moderately dangerous explosion hazard. To fight fire, use copious volumes of water; alcohol foam. CO_2 is effective in extinguishing fires of nitrocellulose solvents. See also EXPLOSIVES, HIGH.

CCV000 ***HR: 3***
CEMENT, leather
DOT: NA 1133

DOT Classification: Flammable Liquid; Label: Flammable Liquid.

SAFETY PROFILE: Flammable when exposed to heat or flame.

CCV250 ***HR: 3***
CEMENT (liquid)
DOT: NA 1133/UN 1133

SYN: CEMENT, adhesive (DOT)

DOT Classification: Combustible Liquid; Label: None; Flammable Liquid; Label: Flammable Liquid; IMO: Flammable or Combustible Liquid; Label: Flammable Liquid.

SAFETY PROFILE: Combustible when exposed to heat or flame.

CCV750 ***HR: 2***
CEMENT (pyroxylin)
DOT: NA 1133

SYN: CEMENT, PYROXYLIN (DOT)

DOT Classification: Flammable Liquid; Label: Flammable Liquid.

SAFETY PROFILE: Dangerous fire hazard when exposed to heat or flame; can react with oxidizing materials.

CCW000 ***HR: 3***
CEMENT (roofing liquid)
DOT: NA 1133

SYN: CEMENT, ROOFING, liquid (DOT)

DOT Classification: Flammable Liquid; Label: Flammable Liquid.

SAFETY PROFILE: Dangerous fire hazard when exposed to heat or flame; can react with oxidizing materials.

CCW250 ***HR: 3***
CEMENT (rubber)
DOT: NA 1133

PROP: Flash p: 50° F or less.

SYN: CEMENT, RUBBER (DOT)

DOT Classification: Flammable Liquid; Label: Flammable Liquid.

SAFETY PROFILE: May contain benzene or other toxic solvents. See specific constituent. Dangerous fire hazard when exposed to heat or flame; can react with oxidizing materials.

CCW375 CAS:82636-28-0 ***HR: 3***
CENTBUCRIDINE HYDROCHLORIDE
mf: $C_{17}H_{22}N \cdot 7ClH$ mw: 495.62

SYN: 1,2,3,4-TETRAHYDRO-4-(N-BUTYLAMINO)ACRIDINEHYDROCHLORIDE

TOXICITY DATA with REFERENCE
scu-mus TDLo:400 mg/kg (6-15D preg):TER IJEBA6 20,337,82
scu-mus TDLo:400 mg/kg (6-15D preg):REP IJEBA6 20,337,82
scu-rat LD50:45 mg/kg IJEBA6 20,330,82
ipr-mus LD50:25 mg/kg INJPD2 19,44,87
scu-mus LD50:26 mg/kg IJEBA6 20,330,82
scu-mky LD50:10500 μg/kg IJEBA6 20,330,82

SAFETY PROFILE: Poison by subcutaneous and intraperitoneal routes. Experimental reproductive effects. When heated to decomposition it emits toxic fumes of NO_x and HCl.

CCW500 CAS:41510-23-0 ***HR: 3***
CENTBUTINDOLE
mf: $C_{24}H_{26}FN_3O$ mw: 391.53

SYN: 1,2,3,4,6,7,12A-OCTAHYDRO-2-(1-(p-FLUOROPHENYL)-1-OXO-4-BUTYL)-PYRAZINO(2,1:6,1)PYRIDO(3,4-B)INDOLE

TOXICITY DATA with REFERENCE
orl-rat LD50:700 mg/kg DRFUD4 3,803,78
ipr-mus LD50:180 mg/kg DRFUD4 3,803,78

SAFETY PROFILE: Poison by intraperitoneal route. Moderately toxic by ingestion. When heated to decomposition it emits very toxic fumes of F^- and NO_x.

CCW725 CAS:31477-60-8 ***HR: D***
CENTCHROMAN
mf: $C_{30}H_{35}NO_3$ mw: 457.66

SYNS: COMPOUND 67/20 ◇ trans-2,2-DIMETHYL-3-PHENYL-4-(p-(β-PYRROLIDINOETHOXY)PHENYL)-7-METHOXYCHROMAN ◇ 3,4-trans-2,2-DIMETHYL-3-PHENYL-4-(p-(β-PYRROLIDINOETHOXY)PHENYL)-7-METHOXYCHROMAN ◇ trans-1-(2-(p-(7-METHOXY-2,2-DIMETHYL-3-PHENYL-4-CHROMANYL)PHENOXY)ETHYL)PYRROLIDINE

TOXICITY DATA with REFERENCE
orl-man TDLo:6 mg/kg (14D male):REP IJEBA6 15,1177,77
orl-rat TDLo:1250 μg/kg (female 8D post):TER IJEBA6 15,1144,77

CONSENSUS REPORTS: EPA Genetic Toxicology Program.

SAFETY PROFILE: Human reproductive effects by ingestion: impaired spermatogenesis. Experimental reproductive effects. When heated to decomposition it emits toxic fumes of NO_x.

CCW750 CAS:51023-56-4 ***HR: 3***
CENTCHROMAN HYDROCHLORIDE
mf: $C_{30}H_{35}NO_2 \cdot ClH$ mw: 478.12

SYNS: 67/20CDRI ◇ 3,4-trans-2,2-DIMETHYL-3-PHENYL-4-p-(β-PYRROLIDINOETHOXY)PHENYL-7-METHOXYCHROMANHCl

TOXICITY DATA with REFERENCE
scu-rat TDLo:1250 μg/kg (9D preg):REP CCPTAY 9,279,74
orl-rat LDLo:1600 mg/kg IJEBA6 15,1159,77
ipr-mus LD50:400 mg/kg IJEBA6 15,1159,77

SAFETY PROFILE: Poison by intraperitoneal route. Moderately toxic by ingestion. Experimental reproductive effects. When heated to decomposition it emits very toxic fumes of NO_x and HCl.

CCW800 CAS:98459-16-6 ***HR: 3***
CENTPHENAQUIN
mf: $C_{24}H_{27}N_3 \cdot 2ClH$ mw: 429.41

SYN: 7,8,9,10-TETRAHYDRO-11-(4-PHENYL-1-PIPERAZINYL)-6H-CYCLOHEPTA(b)QUINOLINEDIHYDROCHLORIDE

TOXICITY DATA with REFERENCE
ipr-rat LD50:493 mg/kg IJEBA6 23,214,85
ipr-mus LD50:494 mg/kg IJEBA6 23,214,85
ivn-mus LD50:56 mg/kg IJEBA6 23,214,85

SAFETY PROFILE: Poison by intravenous route. Moderately toxic by intraperitoneal route. When heated to decomposition it emits toxic fumes of NO_x and HCl.

CCW925 ***HR: 3***
CENTRUROIDES SUFFUSUS SUFFUSUS VENOM

SYNS: C. SUFFUSUS SUFFUSUS VENOM ◇ VENOM, SCORPION, CENTRUROIDES SUFFUSUS SUFFUSUS

TOXICITY DATA with REFERENCE
ipr-mus LD50:78 μg/kg TOXIA6 22,308,84
ice-mus LD50:1600 ng/kg TOXIA6 22,308,84
unr-mus LD50:25 μg/kg TOXIA6 20,9,82

SAFETY PROFILE: Deadly poison by intraperitoneal, intracerebral, and possibly other routes.

CCX000 CAS:123-03-5 ***HR: 3***
CEPACOL CHLORIDE
mf: $C_{21}H_{38}N \cdot Cl$ mw: 340.05

SYNS: ACETOQUAT CPC ◇ AKTIVEX ◇ AMMONYX CPC ◇ BIOSEPT ◇ CEEPRYN ◇ CEEPRYN CHLORIDE ◇ CEPRIM ◇ CETAMIUM ◇ CETYLPYRIDINIUM CHLORIDE ◇ N-CETYLPYRIDINIUM CHLORIDE ◇ 1-CETYLPYRIDINIUM CHLORIDE ◇ DOBENDAN ◇ HEXADECYLPYRIDINIUM CHLORIDE ◇ n-HEXADECYLPYRIDINIUM CHLORIDE ◇ 1-HEXADECYLPYRIDINIUM CHLORIDE ◇ INTEXSAN CPC ◇ PRISTACIN ◇ PYRISEPT ◇ QUATERNARIO CPC

TOXICITY DATA with REFERENCE
skn-rbt 50 mg/24H MOD 33NFA8 -,2,75
eye-rbt 1% ARZNAD 18,137,68
eye-rbt 100 mg JPMSAE 59,188,70
orl-rat LD50:200 mg/kg SDSTBT 5R,24,72
ipr-rat LD50:6 mg/kg JAPMA8 35,89,46
scu-rat LD50:250 mg/kg JAPMA8 35,89,46
ivn-rat LD50:30 mg/kg AFDOAQ 18,43,54
orl-mus LD50:108 mg/kg PSEBAA 120,511,65
ipr-mus LD50:10 mg/kg JMCMAR 23,469,80
orl-rbt LD50:400 mg/kg PCOC** -,208,66
skn-rbt LDLo:2 g/kg JPMSAE 59,188,70
ivn-rbt LD50:36 mg/kg PCOC** -,208,66

CONSENSUS REPORTS: Reported in EPA TSCA Inventory.

SAFETY PROFILE: Poison by ingestion, intraperitoneal, subcutaneous, and intravenous routes. Moderately toxic by skin contact. A skin and eye irritant. When heated to decomposition it emits very toxic fumes of NO_x and Cl^-.

CCX125 CAS:5853-29-2 ***HR: 3***
CEPHAELINE HYDROCHLORIDE
mf: $C_{28}H_{38}N_2O_4 \cdot 2ClH$ mw: 539.60

SYN: (−)-CEPHAELINE DIHYDROCHLORIDE

TOXICITY DATA with REFERENCE
ipr-rat LD50:10 mg/kg JPETAB 104,421,52
orl-mus LD50:74970 μg/kg NCISP* JAN86
ipr-mus LD50:20530 μg/kg NCISP* JAN86

SAFETY PROFILE: Poison by ingestion and intraperitoneal routes. When heated to decomposition it emits toxic fumes of NO_x and HCl.

CCX175 CAS:11005-92-8 ***HR: 3***
CEPHALOMYCIN

TOXICITY DATA with REFERENCE
orl-mus LD50:1000 mg/kg 85GDA2 4(2),235,80
ipr-mus LD50:55 mg/kg 85FZAT -,204,67
scu-mus LD50:161 mg/kg 85FZAT -,204,67
ivn-mus LD50:31 mg/kg 85ERAY 2,1237,78

SAFETY PROFILE: Poison by subcutaneous, intravenous, and intraperitoneal routes. Moderately toxic by ingestion.

CCX250 CAS:153-61-7 ***HR: 3***
CEPHALOTHIN
mf: $C_{16}H_{16}N_2O_6S_2$ mw: 396.46

SYNS: CEFALOTIN ◇ CEPHALOTIN ◇ CET ◇ CT ◇ 7-(2-THIENYLACETAMIDO)CEPHALOSPORANIC ACID ◇ 7-(THIOPHENE-2-ACETAMIDO)CEPHALOSPORANIC ACID

TOXICITY DATA with REFERENCE
pic-omi 25 μg/plate ZMMPAO 231,369,75
scu-mus TDLo:18 g/kg (9-11D preg):TER SEIJBO 16,250,76
ipr-rat LD50:4296 mg/kg ANTBAL 26(1),44,82
scu-rat LDLo:10 g/kg NKRZAZ 27(Suppl 6),124,79
par-rat LD50:23 mg/kg AACHAX -,863,65
ivn-mus LD50:4990 mg/kg NKRZAZ 26(Suppl 1),150,78
ims-mus LD50:7 g/kg BYYADW 3,220,78
ice-mus LD50:81 mg/kg AACHAX -,863,65

SAFETY PROFILE: Poison by parenteral and intracerebral routes. Moderately toxic by intravenous route. Mildly toxic by subcutaneous and intraperitoneal routes. An experimental teratogen. Mutation data reported. See also ESTERS. When heated to decomposition it emits very toxic fumes of NO_x and SO_x.

CCX300 CAS:34444-01-4 ***HR: D***
CEPHAMANDOLE
mf: $C_{18}H_{18}N_6O_5S_2$ mw: 462.54

SYNS: CEFADOLE ◇ CEFAMANDOL ◇ CEFAMANDOLE ◇ 1-CEFAMANDOLE ◇ CEPHADOLE ◇ MANDOKEF ◇ 5-THIA-1-AZABICYCLO(4.2.0)OCT-2-ENE-2-CARBOXYLIC ACID, 7-((HYDROXYPHENYLACETYL)AMINO)-3-(((1-METHYL-1H-TETRAZOL-5-YL)THIO)METHYL)-8-OXO-, (6R-(6-α-7-β(R*)))-

TOXICITY DATA with REFERENCE
scu-rat TDLo:30 g/kg (male 30D pre):REP FAATDF 8,280,87

SAFETY PROFILE: Experimental reproductive effects. When heated to decomposition it emits toxic fumes of NO_x and SO_x.

CCX500 CAS:21593-23-7 ***HR: 2***
CEPHAPIRIN
mf: $C_{17}H_{17}N_3O_6S_2$ mw: 423.49

SYNS: CEFAPIRIN (GERMAN) ◇ 3-(HYDROXYMETHYL)-8-OXO-7-(2-(4-PYRIDYLTHIO)ACETAMIDO)-5-THIA-1-AZABICYCLO(4.2.0)OCT-2-ENE-2-CARBOXYLIC ACID, ACETATE (ESTER)

TOXICITY DATA with REFERENCE
ipr-rat TDLo:68250 mg/kg (female 13W pre):REP TOIZAG 21,211,74
ivn-man TDLo:514 μg/kg/9D-I:SYS DICPBB 19,553,85
orl-rat LD50:16356 mg/kg TOIZAG 21,279,74
ipr-rat LD50:7850 mg/kg TOIZAG 21,279,74
orl-mus LD50:26088 mg/kg TOIZAG 21,279,74
ipr-mus LD50:8899 mg/kg TOIZAG 21,279,74
scu-mus LD50:13556 mg/kg TOIZAG 21,279,74

SAFETY PROFILE: Moderately toxic by intraperitoneal route. Human systemic effects by intravenous route: jaundice. Experimental reproductive effects. When heated to decomposition it emits very toxic fumes of NO_x and SO_x.

CCX550 CAS:481-49-2 ***HR: 3***
CEPHARANTHINE
mf: $C_{37}H_{38}N_2O_6$ mw: 606.77

PROP: From tubers of *Stephania cephalantha Hayata*, and *Stephania sasahii Hayata, Menispermaceae*. Yellow powder. Mp: 145-155°. Obtained by drying solvated needles from acetone + benzene. Soluble in the usual organic solvents except petr ether.

SYNS: CEPHARANTHIN ◇ 6′,12′-DIMETHOXY-2,2′-DIMETHYL-6,7-(METHYLENEBIS(OXY))OXYACANTHAN

TOXICITY DATA with REFERENCE
ipr-rat TDLo:2400 mg/kg (female 30D pre):REP KSRNAM 16,3855,82
orl-rat LD50:2 g/kg KSRNAM 16,3855,82
scu-rat LD50:100 mg/kg KSRNAM 16,3855,82
ivn-rat LD50:57 mg/kg KSRNAM 16,3855,82

orl-mus LD50:1900 mg/kg KSRNAM 16,3855,82
ipr-mus LD50:125 mg/kg CPBTAL 24,2413,76
scu-mus LD50:100 mg/kg KSRNAM 16,3855,82
ivn-mus LD50:43500 μg/kg KSRNAM 16,3855,82

SAFETY PROFILE: Poison by subcutaneous, intravenous, and intraperitoneal routes. Moderately toxic by ingestion. Experimental reproductive effects. When heated to decomposition it emits toxic fumes of NO_x.

CCX600 CAS:67055-59-8 ***HR: 2***
CEPHEDRINE
mf: $C_{13}H_{18}N_2O$ mw: 218.33

SYNS: CEFEDRIN ◇ 3-((1-HYDROXY-1-PHENYL-2-PROPYL) METHYLAMINO)PROPIONITRILE ◇ 1-PHENYL-2-(METHYL-(β-CYANOETHYL) AMINO)PROPAN-1-OL

TOXICITY DATA with REFERENCE
orl-rat LD50:940 mg/kg PCJOAU 14,773,80
scu-rat LD50:600 mg/kg PCJOAU 14,773,80
orl-mus LD50:465 mg/kg PCJOAU 14,773,80
scu-mus LD50:410 mg/kg PCJOAU 14,773,80
orl-gpg LD50:1200 mg/kg FATOAO 41,345,78

CONSENSUS REPORTS: Cyanide and its compounds are on the Community Right-To-Know List.

SAFETY PROFILE: Moderately toxic by ingestion and subcutaneous routes. When heated to decomposition it emits toxic fumes of NO_x and CN^-. See also NITRILES.

CCX620 ***HR: 3***
CERASTES CERASTES VENOM

SYNS: C. CERASTES VENOM ◇ VENOM, SNAKE, CERASTES CERASTES

TOXICITY DATA with REFERENCE
ipr-mus LD50:1080 μg/kg TOXIA6 18,384,80
ivn-mus LD50:375 μg/kg TOXIA6 14,146,76
ivn-dog LDLo:100 μg/kg TOXIA6 6,221,69

SAFETY PROFILE: Deadly poison by intravenous and intraperitoneal routes.

CCX625 CAS:11005-70-2 ***HR: 3***
CERBEROSIDE
mf: $C_{42}H_{66}O_{18}$ mw: 859.08

SYNS: CERBEROSID (GERMAN) ◇ CERBROSIDE ◇ THEVETIN B

TOXICITY DATA with REFERENCE
ivn-cat LD50:810 μg/kg 85ELDJ -,189,63
unr-cat LDLo:636 μg/kg 85ELDJ 134,63
ivn-gpg LDLo:3539 μg/kg AEPPAE 252,314,66

SAFETY PROFILE: Deadly poison by intravenous and possibly other routes. When heated to decomposition it emits acrid smoke and fumes.

CCX725 CAS:55467-31-7 ***HR: 3***
CEREXIN A
mf: $C_{63}H_{103}N_{15}O_{19}$ mw: 1374.81

SYN: ANTIBIOTIC 60-6

TOXICITY DATA with REFERENCE
ipr-mus LD50:50 mg/kg 85GDA2 4(1),261,80
scu-mus LD50:500 mg/kg 85GDA2 4(1),261,80
ivn-mus LD50:25 mg/kg 85GDA2 4(1),261,80

SAFETY PROFILE: Poison by intravenous and intraperitoneal routes. Moderately toxic by subcutaneous route. When heated to decomposition it emits toxic fumes of NO_x.

CCY000 CAS:1306-38-3 ***HR: 2***
CERIC OXIDE
mf: CeO_2 mw: 172.12

SYN: CERIUM DIOXIDE

TOXICITY DATA with REFERENCE
orl-rat LDLo:1000 mg/kg CURL** 35,25,60

CONSENSUS REPORTS: Reported in EPA TSCA Inventory.

SAFETY PROFILE: Moderately toxic by ingestion. See also CERIUM COMPOUNDS.

CCY250 CAS:7440-45-1 ***HR: 3***
CERIUM
DOT: UN 1333
af: Ce aw: 140.13

PROP: Cubic or hexagonal, steel gray crystals. Mp: 815°, bp: 3257°, d: (cubic form): 6.90, hexagonal form 6.75.

CONSENSUS REPORTS: Reported in EPA TSCA inventory.

DOT Classification: Flammable Solid; Label: Flammable Solid

SAFETY PROFILE: Cerium resembles aluminum in its pharmacological action as well as in its chemical properties. The insoluble salts such as the oxalates are stated to be nontoxic even in large doses. It is used to prevent vomiting in pregnancy. The average dose is from 0.05 to 0.5 gram.

The effect on the central nervous system of the rare-earth metals following inhalation may preclude welding operations with these materials to any large extent. Cerium is stated to produce polycythemia but is useless in the treatment of anemia owing to its toxic effects. The salts of cerium increase the blood coagulation rate. See also RARE EARTHS. A strong reducing agent. Moderate fire hazard; ignites spontaneously in air at 150-180°.

Moderate explosion hazard in the form of dust when exposed to flame. The metal or its alloys spark with friction. Many alloys are pyrophoric in air. See also IRON DUST. Explosive reaction with zinc. Very exothermic reaction with antimony or bismuth. Ignites when heated in atmospheres of CO_2 + N_2; Cl_2; or Br_2. Violent reaction when heated with phosphorus (400°C); silicon (1400°C).

CCY500 CAS:537-00-8 ***HR: 3***
CERIUM ACETATE
mf: $C_6H_9O_6{\cdot}Ce$ mw: 317.27

SYNS: CERIUM TRIACETATE ◇ CEROUS ACETATE

TOXICITY DATA with REFERENCE
ivn-hmn TDLo:2 mg/kg:CNS JCINAO 21,447,42

CONSENSUS REPORTS: Reported in EPA TSCA Inventory.

SAFETY PROFILE: Human central nervous system effects. See also CERIUM COMPOUNDS. When heated to decomposition it emits acrid and irritating fumes.

CCY699 ***HR: 3***
CERIUM AZIDE
mf: CeN_9 mw:266.18

$Ce(N_3)_3$

SAFETY PROFILE: An explosive. Upon decomposition it emits toxic fumes of NO_x. See also CERIUM COMPOUNDS and AZIDES.

CCY750 CAS:7790-86-5 ***HR: 3***
CERIUM CHLORIDE
mf: $CeCl_3$ mw: 246.47

PROP: Colorless, deliquescent crystals. Mp: 848°, bp: 1727°, d: 3.92.

SYNS: CERIUM(III) CHLORIDE ◇ CERIUM TRICHLORIDE ◇ CEROUS CHLORIDE

TOXICITY DATA with REFERENCE
orl-rat LD50:2111 mg/kg EQSSDX 1,1,75
scu-rat LDLo:4000 mg/kg AEXPBL 100,230,23
ivn-rat LD50:5096 μg/kg APYPAY 32,205,81
orl-mus LD50:5277 mg/kg EQSSDX 1,1,75
ipr-mus LD50:172 mg/kg COREAF 256,1043,63
scu-mus LDLo:4000 mg/kg AEPPAE 188,465,38
ivn-dog LDLo:60 mg/kg HBAMAK 4,1289,35
ipr-gpg LD50:56 mg/kg AMIHAB 15,9,57
scu-gpg LDLo:2 g/kg AEXPBL 72,228,13
scu-frg LDLo:211 mg/kg EQSSDX 1,1,75

CONSENSUS REPORTS: Reported in EPA TSCA Inventory. EPA Genetic Toxicology Program.

SAFETY PROFILE: Poison by intravenous, intraperitoneal, and subcutaneous routes. Moderately toxic by ingestion. See also CERIUM COMPOUNDS. When heated to decomposition it emits toxic fumes of Cl^-.

CCZ000 CAS:512-24-3 ***HR: 3***
CERIUM CITRATE
mf: $C_6H_8O_7{\cdot}Ce$ mw: 332.26

SYNS: CERIUM(III) CITRATE ◇ CEROUS CITRATE ◇ 2-HYDROXY-1,2,3-PROPANETRISCARBOXYLIC ACID CERIUM(3+) SALT (1:1) (9CI)

TOXICITY DATA with REFERENCE
scu-mus TDLo:190 mg/kg (12D preg):REP JTEHD6 10,449,82
ipr-mus LD50:149 mg/kg AEHLAU 5,437,62
ipr-gpg LD50:83 mg/kg AEHLAU 5,437,62

SAFETY PROFILE: Poison by intraperitoneal route. Experimental reproductive effects. See also CERIUM COMPOUNDS. When heated to decomposition it emits acrid and irritating fumes.

CDA250 ***HR: 2***
CERIUM COMPOUNDS

PROP: Compounds of cerium and the other rare earth elements are generally of low toxicity. The greatest exposures are likely to be during manufacture of cerium. Exposed workers have experienced sensitivity to heat, itching, and skin lesions. Large doses to experimental animals have caused writhing, ataxia (loss of muscle coordination), labored respiration, sedation, hypotension, and death by cardiovascular collapse. The chloride, bromide, nitrate, bromate, and perchlorate salts are water soluble and thus are more likely to cause systemic effects when ingested. The sulfates, iodides, and iodates are less water soluble. Oxides, oxalates, sulfides, carbonates, fluorides, and phosphates are insoluble. The salts of cerium increase the blood coagulation rate. Cerium tartrate has been found to produce a direct injurious action on the hearts of small animals. Cerium oxalate has been used to suppress motion sickness and to suppress vomiting during pregnancy (by ingestion of 1 gram/24 hours). The toxicity of cerium compounds may be taken to be that of cerium, except when the anion has a toxicity of its own. See also CERIUM and RARE EARTHS.

CDA500 CAS:15158-67-5 ***HR: 3***
CERIUM EDETATE

TOXICITY DATA with REFERENCE
ipr-mus LD50:37.6 mg/kg AEHLAU 5,437,62
ipr-gpg LD50:129 mg/kg AEHLAU 5,437,62

SAFETY PROFILE: Poison by intraperitoneal route. See also CERIUM COMPOUNDS. When heated to decomposition it emits acrid smoke and irritating fumes.

CDA750 CAS:7758-88-5 ***HR: 3***
CERIUM FLUORIDE
mf: CeF_3 mw: 197.12

PROP: White, hexagonal crystals. D: 6.16, mp: 1430°, bp: 2327°. Insol in water; sol in H_2SO_4.

SYNS: CERIUM FLUORURE (FRENCH) ◇ CERIUM TRIFLUORIDE ◇ CEROUS FLUORIDE

CONSENSUS REPORTS: Reported in EPA TSCA Inventory.

OSHA PEL: TWA 2.5 mg(F)/m^3
ACGIH TLV: TWA 2.5 mg(F)/m^3
NIOSH REL: (Inorganic Fluorides) TWA 2.5 mg(F)/m^3.

SAFETY PROFILE: A poison. See FLUORIDES and CERIUM COMPOUNDS. When heated to decomposition it emits toxic fumes of F^-.

CDB000 CAS:10108-73-3 ***HR: 3***
CERIUM(III) NITRATE
mf: $N_3O_9{\cdot}Ce$ mw: 326.15

SYNS: CERIUM NITRATE ◇ CERIUM(3+) NITRATE ◇ CERIUM TRINITRATE ◇ CEROUS NITRATE ◇ DUSICNAN CERITY (CZECH) ◇ NITRIC ACID, CERIUM(3+) SALT (8CI, 9CI)

TOXICITY DATA with REFERENCE
itt-rat TDLo:26092 μg/kg (1D male):REP JRPFA4 7,21,64
orl-rat LD50:3154 mg/kg EQSSDX 1,1,75
ipr-rat LD50:216 mg/kg EQSSDX 1,1,75
ivn-rat LD50:37 mg/kg EQSSDX 1,1,75

CONSENSUS REPORTS: Reported in EPA TSCA Inventory.

SAFETY PROFILE: Poison by intravenous and intraperitoneal routes. Moderately toxic by ingestion. Experimental reproductive effects. See also CERIUM COMPOUNDS and NITRATES. When heated to decomposition it emits toxic fumes of NO_x.

CDB250 CAS:10294-41-4 ***HR: 3***
CERIUM(III) NITRATE, HEXAHYDRATE (1:3:6)
mf: $N_3O_9{\cdot}Ce{\cdot}6H_2O$ mw: 434.27

SYNS: CERIUM NITRATE, HEXAHYDRATE ◇ CERIUM TRINITRATE HEXAHYDRATE ◇ CEROUS NITRATE HEXAHYDRATE ◇ NITRIC ACID, CERIUM(3+) SALT, HEXAHYDRATE

TOXICITY DATA with REFERENCE
orl-rat LD50:4200 mg/kg TXAPA9 5,750,63
ipr-rat LD50:290 mg/kg TXAPA9 5,750,63
ivn-rat LD50:4 mg/kg TXAPA9 5,750,63
ipr-mus LD50:470 mg/kg TXAPA9 5,750,63

SAFETY PROFILE: Poison by intraperitoneal and intravenous routes. Moderately toxic by ingestion. See also CERIUM COMPOUNDS and NITRATES. When heated to decomposition it emits toxic fumes of NO_x.

CDB325 CAS:25764-08-3 ***HR: 3***
CERIUM NITRIDE
mf: CeN mw: 154.13

SAFETY PROFILE: Reaction with water or dilute acids may cause ignition and the release of toxic ammonia gas and explosive hydrogen gas. When heated to decomposition it emits toxic fumes of NO_x. See also CERIUM COMPOUNDS and NITRIDES.

CDB400 CAS:13590-82-4 ***HR: D***
CERIUM(IV) SULFATE
mf: $O_8S_2{\cdot}Ce$ mw: 332.24

SYNS: CERIC DISULFATE ◇ CERIC SULFATE ◇ CERIC SULPHATE ◇ CERIUM DISULFATE ◇ CERIUM SULFATE ◇ CERIUM(4+) SULFATE ◇ SULFURIC ACID, CERIUM SALT (2:1)

TOXICITY DATA with REFERENCE
itt-rat TDLo:26902 μg/kg (male 1D pre):REP JRPFA4 7,21,64

CONSENSUS REPORTS: Reported in EPA TSCA Inventory.

SAFETY PROFILE: Experimental reproductive effects. When heated to decomposition it emits acrid smoke and irritating fumes.

CDB500 ***HR: 3***
CERIUM(III) TETRAHYDROALUMINATE
mf: Al_3CeH_{12} mw: 236.46

PROP: Decomp @ −80°C.

SAFETY PROFILE: A dangerous fire hazard. Ignites spontaneously in air. Unstable. See also CERIUM COMPOUNDS and ALUMINUM COMPOUNDS.

CDB750 ***HR: 3***
CERIUM TRIHYDRIDE
mf: CeH_3 mw: 143.14

SAFETY PROFILE: May ignite spontaneously in moist air. It is stable in dry air. See also CERIUM COMPOUNDS and HYDRIDES.

CDB760 CAS:8054-43-1 ***HR: 1***
CERNILTON

SYN: CN 009

TOXICITY DATA with REFERENCE
orl-rat TDLo:576 g/kg (male 26W pre):REP TOIZAG 15,201,68
ipr-rat LD50:6660 mg/kg TOIZAG 15,201,68
orl-mus LD50:27610 mg/kg TOIZAG 15,201,68
ipr-mus LD50:6940 mg/kg TOIZAG 15,201,68
scu-mus LD50:13060 mg/kg TOIZAG 15,201,68

SAFETY PROFILE: Mildly toxic. Experimental reproductive effects. When heated to decomposition it emits acrid smoke and irritating fumes.

CDB770 CAS:106440-54-4 ***HR: 1***
CERNITIN GBX

SYN: GBX

TOXICITY DATA with REFERENCE
orl-rat TDLo:700 g/kg (male 35D pre):REP TOIZAG 15,201,68
ipr-rat LD50:3310 mg/kg TOIZAG 15,201,68
orl-mus LD50:52250 mg/kg TOIZAG 15,201,68
ipr-mus LD50:1720 mg/kg TOIZAG 15,201,68
scu-mus LD50:26130 mg/kg TOIZAG 15,201,68

SAFETY PROFILE: Moderately toxic by intraperitoneal route. Experimental reproductive effects. When heated to decomposition it emits acrid smoke and irritating fumes.

CDB772 CAS:106440-55-5 ***HR: 1***
CERNITIN T-60

TOXICITY DATA with REFERENCE
orl-rat TDLo:420 g/kg (male 35D pre):REP TOIZAG 15,201,68
ipr-rat LD50:7580 mg/kg TOIZAG 15,201,68
orl-mus LD50:27750 mg/kg TOIZAG 15,201,68
ipr-mus LD50:8310 mg/kg TOIZAG 15,201,68
scu-mus LD50:9470 mg/kg TOIZAG 15,201,68

SAFETY PROFILE: Mildly toxic. Experimental reproductive effects. When heated to decomposition it emits acrid smoke and irritating fumes.

CDB775 CAS:64318-79-2 ***HR: 3***
CERVAGEM
mf: $C_{23}H_{38}O_5$ mw: 394.61

SYNS: 16,16-DIMETHYL-trans-Δ^2-PGE1 METHYL ESTER ◇ 16,16-DIMETHYL-trans-Δ^2-PROSTAGLANDIN E1 METHYL ESTER ◇ GEMEPROST ◇ ONO 802 ◇ PREGLANDIN

TOXICITY DATA with REFERENCE
ivg-wmn TDLo:20 μg/kg (10W preg):REP CCPTAY 27,51,83
ivg-rat TDLo:438 μg/kg (female 1-7D post):TER GEIRDK 14,809,82
orl-rat LD50:56500 μg/kg GEIRDK 14,188,82
scu-rat LD50:22600 μg/kg GEIRDK 14,188,82
ivn-rat LD50:28600 μg/kg GEIRDK 14,188,82
ivg-rat LD50:32500 μg/kg GEIRDK 14,188,82
orl-mus LD50:59 mg/kg GEIRDK 14,188,82
scu-mus LD50:32500 μg/kg GEIRDK 14,188,82
ivn-mus LD50:29500 μg/kg GEIRDK 14,188,82
ivg-mus LD50:36 mg/kg GEIRDK 14,188,82

SAFETY PROFILE: Poison by ingestion, subcutaneous, intravenous, and intravaginal routes. Human reproductive effects by intravaginal route: abortion, changes in the uterus, cervix, and vagina. An experimental teratogen. Other experimental reproductive effects. When heated to decomposition it emits acrid smoke and fumes. See other prostaglandin entries.

CDB800 ***HR: 3***
CESALIN

SYN: NCS 110435

TOXICITY DATA with REFERENCE
ipr-rat LD50:1838 μg/kg NCISP* JAN86
scu-rat LD50:7777 μg/kg NCISP* JAN86
ipr-mus LD50:2286 μg/kg NCISP* JAN86

SAFETY PROFILE: Deadly poison by subcutaneous and intraperitoneal routes.

CDC000 CAS:7440-46-2 ***HR: 3***
CESIUM
af: Cs aw: 132.91
DOT: UN 1383/UN 1407

PROP: Hexagonal crystals, silver-white, ductile metal or possibly a silvery liquid. Mp: 28.5°, bp: 705°, d: 1.873, vap press: 1 mm @ 279°.

SYNS: CESIUM-133 ◇ CESIUM METAL (DOT) ◇ CESIUM, POWDERED (DOT)

CONSENSUS REPORTS: Reported in EPA TSCA Inventory.

DOT Classification: Flammable Solid; Label: Flammable Solid and Danger When Wet (UN1407); Flammable Solid; Label: Spontaneously Combustible (UN1383).

SAFETY PROFILE: Cesium is quite similar to potassium in its elemental state. It has been shown, however, to have pronounced physiological action in experimentation with animals. Hyper-irritability, including marked spasms, has been shown to follow the administration of cesium in amounts equal to the potassium content of the diet. It has been found that replacing the potassium in the diet of rats with cesium caused death after 10-17 days. Ignites spontaneously in air. Violent reaction with water, moisture, or steam releases hydrogen gas which explodes. Violent reaction with acids, halogens, and other oxidizing materials. Incandescent reaction with non-metals (e.g., sulfur, phosphorus). See also SODIUM.

CDC125 CAS:22750-56-7 ***HR: 3***
CESIUM ACETYLIDE
mf: C_2Cs_2 mw: 289.83

SAFETY PROFILE: Explosive reaction on contact with nitric acid. Ignition on contact with fluorine, chlorine, bromine, iodine, and hydrogen chloride. Vigorous or incandescent reaction on heating with iron(III) choride, boron, or silicon. See also CESIUM and ACETYLIDES.

CDC250 CAS:22205-57-8 ***HR: 3***
CESIUM AMIDE
mf: CsH_2N mw: 148.93

SAFETY PROFILE: Incandescent reaction on contact with air. Incompatible with water. When heated to decomposition it emits toxic fumes of NO_x. See also CESIUM and AMIDES.

CDC375 CAS:61136-62-7 ***HR: 3***
CESIUM ARSENATE
mf: $AsO_4 \cdot 3Cs$ mw: 537.65

SYN: ARSENIC ACID, TRICESIUM SALT

TOXICITY DATA with REFERENCE
ihl-rat TCLo:430 $\mu g/m^3$/24H (1-22D preg):TER VAMNAQ (8),10,78
orl-mus LD50:116 mg/kg VAMNAQ (8),10,78

CONSENSUS REPORTS: Arsenic and its compounds are on the Community Right-To-Know List.

OSHA: Cancer Hazard

SAFETY PROFILE: Poison by ingestion. Experimental teratogenic effects by inhalation. When heated to decomposition it emits toxic fumes of As. See also ARSENIC COMPOUNDS and CESIUM.

CDC500 CAS:7787-69-1 ***HR: 2***
CESIUM BROMIDE
mf: BrCs mw: 212.82

TOXICITY DATA with REFERENCE
ipr-rat LD50:1400 mg/kg AIHOAX 1,637,50

CONSENSUS REPORTS: Reported in EPA TSCA Inventory.

SAFETY PROFILE: Moderately toxic by intraperitoneal route. See also CESIUM and BROMIDES. When heated to decomposition it emits toxic fumes of Br^-.

CDC699 ***HR: 1***
CESIUM BROMOXENATE
mf: $BrCsO_3Xe$ mw: 392.10

SAFETY PROFILE: Solution in water is extremely unstable. When heated to decomposition it emits toxic fumes of Br^-. See also CESIUM and BROMIDES.

CDC750 CAS:534-17-8 ***HR: 2***
CESIUM CARBONATE
mf: $CO_3 \cdot 2Cs$ mw: 325.83

SYNS: CARBONIC ACID, DICESIUM SALT ◇ DICESIUM CARBONATE

TOXICITY DATA with REFERENCE
mrc-bcs 5 mol/L MUREAV 77,109,80
orl-rat LD50:2333 mg/kg VAMNAQ (8),10,78
orl-mus LD50:2170 mg/kg VAMNAQ (8),10,78

CONSENSUS REPORTS: EPA Genetic Toxicology Program. Reported in EPA TSCA Inventory.

SAFETY PROFILE: Moderately toxic by ingestion. Mutation data reported. When heated to decomposition it emits acrid smoke and fumes. See also CESIUM.

CDD000 CAS:7647-17-8 ***HR: 2***
CESIUM CHLORIDE
mf: ClCs mw: 168.36

PROP: D: 3.99, mp: 646°, bp: 1303°.

SYNS: CESIUM MONOCHLORIDE ◇ DICESIUM DICHLORIDE ◇ TRICESIUM TRICHLORIDE

TOXICITY DATA with REFERENCE
mrc-bcs 5 mol/L MUREAV 77,109,80
sln-smc 20 mmol/L MUTAEX 1,21,86
orl-rat LD50:2600 mg/kg VAMNAQ (8),10,78
ipr-rat LD50:1500 mg/kg AIHOAX 1,637,50
orl-mus LD50:2306 mg/kg VAMNAQ (8),10,78
ipr-mus LD50:1849 mg/kg COREAF 256,1043,63

CONSENSUS REPORTS: Reported in EPA TSCA Inventory. EPA Genetic Toxicology Program.

SAFETY PROFILE: Moderately toxic by ingestion and intraperitoneal routes. Mutation data reported. Reacts violently with BF_3. See also CESIUM. When heated to decomposition it emits toxic fumes of Cl^-.

CDD250 CAS:26283-13-6 ***HR: 3***
CESIUM CHLOROXENATE
mf: $ClCsO_3Xe$ mw: 347.66

SAFETY PROFILE: Explodes at 205°C in vacuum. When heated to decomposition it emits toxic fumes of Cl^-. See also CESIUM and CHLORIDES.

CDD325 CAS:71250-00-5 ***HR: 3***
CESIUM CYANOTRIDECAHYDRODECABORATE (2-)
mf: $CH_{13}B_{10}Cs_2N$ mw: 413.03

CONSENSUS REPORTS: Cyanide compounds are on the Community Right-To-Know List.

SAFETY PROFILE: A poison. Violent reaction with concentrated hydrochloric acid. When heated to decomposition it emits toxic fumes of NO_x. See also CESIUM, CYANIDE, and BORON COMPOUNDS.

CDD500 CAS:13400-13-0 ***HR: 3***
CESIUM FLUORIDE
mf: CsF mw: 151.91

SYNS: CESIUM MONOFLUORIDE ◇ DICESIUM DIFLUORIDE ◇ TRICESIUM TRIFLUORIDE

CONSENSUS REPORTS: Reported in EPA TSCA Inventory.

OSHA PEL: TWA 2.5 mg(F)m^3
ACGIH TLV: TWA 2.5 mg(F)/m^3
NIOSH REL: (Inorganic Fluorides) TWA 2.5 mg(F)/m^3.

SAFETY PROFILE: A poison. See also FLUORIDES and CESIUM COMPOUNDS. Incompatible with benzenediazonium tetrafluoroborate and difluoroamine. When heated to decomposition it emits toxic fumes of F^-.

CDD625 CAS:12079-66-2 ***HR: 3***
CESIUM GRAPHITE
mf: C_8Cs mw: 228.99

SAFETY PROFILE: Explodes on contact with water. Ignites spontaneously in air. See also CESIUM.

CDD750 CAS:21351-79-1 ***HR: 3***
CESIUM HYDROXIDE
DOT: UN 2681/UN 2682
mf: CsHO mw: 149.92

PROP: Colorless to yellowish, very deliquescent crystals. Mp: 272.3°, d: 3.675.

SYNS: CESIUM HYDRATE ◇ CESIUM HYDROXIDE DIMER ◇ CESIUM HYDROXIDE, solid (DOT) ◇ CESIUM HYDROXIDE, solution (DOT)

TOXICITY DATA with REFERENCE
skn-rbt 5 mg/24H MLD TXAPA9 32,239,75
eye-rbt 5 mg/5M rns SEV TXAPA9 32,239,75
orl-rat LD50:570 mg/kg GTPZAB 21(1),29,77
ipr-rat LD50:100 mg/kg AIHOAX 1,637,50
orl-mus LD50:800 mg/kg 20PKA3 -,-,67

CONSENSUS REPORTS: Reported in EPA TSCA Inventory.

OSHA PEL: TWA 2 mg/m^3
ACGIH TLV: TWA 2 mg/m^3
DOT Classification: Corrosive; IMO: Corrosive Material; Label: Corrosive.

SAFETY PROFILE: Poison by intraperitoneal route. Moderately toxic by ingestion. A powerful caustic. A corrosive skin and eye irritant. See also CESIUM.

CDE000 CAS:7789-17-5 ***HR: 2***
CESIUM IODIDE
mf: CsI mw: 259.81

TOXICITY DATA with REFERENCE
orl-rat LD50:2386 mg/kg NIOSH* TR-74,1,72
ipr-rat LD50:1400 mg/kg AIHOAX 1,637,50

CONSENSUS REPORTS: Reported in EPA TSCA Inventory.

SAFETY PROFILE: Moderately toxic by ingestion and intraperitoneal route. See also CESIUM and IODIDES. When heated to decomposition, it emits toxic fumes of I^-.

CDE125 CAS:12430-27-2 ***HR: 3***
CESIUM LITHIUM TRIDECAHYDRONONABORATE
mf: $H_{13}B_9CsLi$ mw: 250.24

SAFETY PROFILE: Ignites spontaneously in air. See also CESIUM, LITHIUM COMPOUNDS, and BORON COMPOUNDS.

CDE250 CAS:7789-18-6 ***HR: 2***
CESIUM(I) NITRATE (1:1)
mf: NO_3•Cs mw: 194.92
DOT: UN 1451

PROP: Colorless, hexagonal or cubic, glittering crystalline powder. Mp: 414°, bp: decomp, d: 3.685, 2.71 @ 500° (liq).

SYNS: CESIUM NITRATE (DOT) ◇ NITRIC ACID, CESIUM SALT

TOXICITY DATA with REFERENCE
mrc-bcs 5 mol/L MUREAV 77,109,80
orl-rat LD50:2390 mg/kg VAMNAQ (8),10,78
ipr-rat LD50:1200 mg/kg AIHOAX 1,637,50
orl-mus LD50:2300 mg/kg VAMNAQ (8),10,78

CONSENSUS REPORTS: Reported in EPA TSCA Inventory. EPA Genetic Toxicology Program.

DOT Classification: Oxidizer; Label: Oxidizer.

SAFETY PROFILE: Moderately toxic by ingestion and intraperitoneal routes. Mutation data reported. When heated to decomposition it emits toxic fumes of NO_x. See also CESIUM and NITRATES.

CDE325 CAS:20281-00-9 ***HR: 3***
CESIUM OXIDE
mf: Cs_2O mw: 297.81

SAFETY PROFILE: Ignition or incandescent reaction on contact with water; ethanol; moisture + carbon monoxide or carbon dioxide; sulfur dioxide + heat; or halogens (fluorine; chlorine; or iodine) above 150°C. See also CESIUM.

CDE400 CAS:78937-12-9 ***HR: 3***
CESIUM PENTACARBONYLVANADATE (3-)
mf: $C_5Cs_3O_5V$ mw: 589.71

SAFETY PROFILE: Ignites spontaneously in air or when scratched under a non-reactive gas. Explodes on contact with water or alcohols. When heated to decomposition it emits toxic fumes of VO_x. See also CESIUM and VANADIUM COMPOUNDS.

CDE500 CAS:10294-54-9 ***HR: 2***
CESIUM SULFATE
mf: Cs_2O_4S mw: 361.88

SYNS: DICESIUM SULFATE ◇ SULFURIC ACID, DICESIUM SALT

TOXICITY DATA with REFERENCE
mrc-bcs 5 mol/L MUREAV 77,109,80
orl-rat LD50:2830 mg/kg VAMNAQ (8),10,78
orl-mus LD50:3180 mg/kg VAMNAQ (8),10,78

CONSENSUS REPORTS: Reported in EPA TSCA Inventory. EPA Genetic Toxicology Program.

SAFETY PROFILE: Moderately toxic by ingestion. Mutation data reported. See also CESIUM COMPOUNDS and SULFATES. When heated to decomposition it emits toxic fumes of SO_x.

CDF000 CAS:12053-67-7 ***HR: 3***
CESIUM TRIOXIDE ("OZONATE")
mf: CsO_3 mw: 180.91

SYN: CESIUM OZONIDE

SAFETY PROFILE: Violent reaction with water. See also CESIUM.

CDF250 CAS:29144-42-1 ***HR: 3***
CETOCYLINE
mf: $C_{22}H_{21}NO$ mw: 315.44

SYNS: β-CHELOCARDIN ◇ 2-DECARBOXAMIDO-2-ACETYL-4-DESDIMETHYLAMINO-4-AMINO-9-METHYL-5A,6-ANHYDROTETRACYCLINE

TOXICITY DATA with REFERENCE
orl-mus LD50:2500 μg/kg 85ERAY 1,534,78
ipr-mus LD50:140 mg/kg 85ERAY 1,534,78
ivn-mus LD50:88 mg/kg 85ERAY 1,534,78

SAFETY PROFILE: Poison by ingestion, intraperitoneal, and intravenous routes. When heated to decomposition it emits toxic fumes of NO_x.

CDF375 CAS:34675-84-8 ***HR: 3***
CETRAXATE
mf: $C_{17}H_{23}NO_4$ mw: 305.41

PROP: Crystals from methanol, melts over a range of 200-280°.

SYNS: trans-4-(((4-(AMINOMETHYL)CYCLOHEXYL)CARBONYL)OXY)BENZENEPROPANOIC ACID ◇ trans-p-HYDROXY HYDROCINNAMIC ACID-4-(AMINOMETHYL)CYCLOHEXANE CARBOXYLATE

TOXICITY DATA with REFERENCE
ipr-rat LD50:716 mg/kg OYYAA2 19,323,80
scu-rat LD50:1503 mg/kg OYYAA2 19,323,80
ivn-rat LD50:345 mg/kg OYYAA2 19,323,80
ipr-mus LD50:1520 mg/kg OYYAA2 19,323,80
scu-mus LD50:4310 mg/kg OYYAA2 19,323,80
ivn-mus LD50:681 mg/kg OYYAA2 19,323,80

SAFETY PROFILE: Poison by intravenous route. Moderately toxic by intraperitoneal and subcutaneous routes. When heated to decomposition it emits toxic fumes of NO_x.

CDF380 CAS:27724-96-5 ***HR: 3***
CETRAXATE HYDROCHLORIDE
mf: $C_{17}H_{23}NO_4 \cdot ClH$ mw: 341.87

SYNS: trans-4-(((4-(AMINOMETHYL)CYCLOHEXYL)CARBONYL)OXY)-BENZENEPROPANOIC ACID HYDROCHLORIDE ◇ 4'-(2-CARBOXYETHYL)PHENYL-trans-4-AMINOMETHYLCYCLOHEXANECARBOXYLATE HYDROCHLORIDE ◇ CV 1006

TOXICITY DATA with REFERENCE
ipr-rat LD50:716 mg/kg OYYAA2 12,265,76
scu-rat LD50:1415 mg/kg IYKEDH 10,710,79
ivn-rat LD50:298 mg/kg IYKEDH 10,710,79
ipr-mus LD50:1520 mg/kg OYYAA2 12,265,76
scu-mus LD50:4210 mg/kg IYKEDH 10,710,79
ivn-mus LD50:666 mg/kg IYKEDH 10,710,79

SAFETY PROFILE: Poison by intravenous route. Moderately toxic by subcutaneous and intraperitoneal routes. When heated to decomposition it emits toxic fumes of NO_x and HCl.

CDF400 CAS:3151-59-5 ***HR: 3***
CETYLAMINE HYDROFLUORIDE
mf: $C_{16}H_{35}N \cdot FH$ mw: 261.53

SYNS: CETYLAMINE-HF ◇ CETYLAMINHYDROFLUORID (GERMAN) ◇ GA 242 ◇ HEPTAFLUR ◇ HEXADECYLAMINE HYDROFLUORIDE ◇ 1-HEXADECANAMINE HYDROFLUORIDE (9CI) ◇ SKF 2208K

TOXICITY DATA with REFERENCE
orl-mus TDLo:40 mg/kg (9D preg):REP DZZEA7 32,861,77
orl-mus TDLo:40 mg/kg (9D preg):TER DZZEA7 32,861,77
ipr-mus LD50:45246 μg/kg DZZEA7 35,1070,80

OSHA PEL: TWA 2.5 mg(F)/m^3
ACGIH TLV: TWA 2.5 mg(F)/m^3

SAFETY PROFILE: Poison by intraperitoneal route. An experimental teratogen. Other experimental reproductive effects. When heated to decomposition it emits toxic fumes of NO_x and HF.

CDF500 CAS:13316-70-6 **HR: 3**
CETYLDIETHYLETHYLAMMONIUM BROMIDE
mf: $C_{22}H_{48}N \cdot Br$ mw: 406.62

SYNS: CETYLTRIETHYLAMMONIUM BROMIDE ◇ HEXADECYLTRIETHYLAMMONIUM BROMIDE ◇ TRIETHYLHEXADECYLAMMONIUM BROMIDE

TOXICITY DATA with REFERENCE
eye-rbt 1% SEV JAPMA8 38,428,49
orl-mus TDLo:900 mg/kg (female 8-12D post):REP TCMUD8 7,7,87
orl-mus LD50:60 mg/kg JAPMA8 38,428,49
ivn-mus LD50:50 mg/kg JAPMA8 38,428,49

SAFETY PROFILE: Poison by ingestion and intravenous routes. Experimental reproductive effects. A severe eye irritant. When heated to decomposition it emits very toxic fumes of Br^-, NH_3 and NO_x. See also BROMIDES.

CDF750 CAS:6004-24-6 **HR: 3**
CETYLPYRIDINIUM CHLORIDE MONOHYDRATE
mf: $C_{21}H_{38}N \cdot Cl \cdot H_2O$ mw: 358.07

SYNS: CEEPRYN ◇ CEPACOL ◇ 1-HEXADECYLPYRIDINIUM CHLORIDE MONOHYDRATE

TOXICITY DATA with REFERENCE
ipr-rat LDLo:15 mg/kg JPETAB 74,401,42
ipr-mus LDLo:3 mg/kg JPETAB 74,401,42
ivn-dog LDLo:100 mg/kg JPETAB 74,401,42
orl-rbt LDLo:400 mg/kg JPETAB 74,401,42
ipr-rbt LDLo:5 mg/kg JPETAB 74,401,42
scu-rbt LDLo:200 mg/kg JPETAB 74,401,42
ivn-rbt LDLo:20 mg/kg JPETAB 74,401,42
ipr-gpg LDLo:5 mg/kg JPETAB 74,401,42

SAFETY PROFILE: Poison by ingestion, intraperitoneal, intravenous, and subcutaneous routes. When heated to decomposition it emits very toxic fumes of Cl^- and NO_x.

CDG000 CAS:62-59-9 **HR: 3**
CEVADINE
mf: $C_{32}H_{49}NO_9$ mw: 591.82

SYNS: CEVADENE ◇ CEVADIN ◇ (Z)-3-((Z)-2-METHYLCROTONATE)4,9-EPOXYCEVANE-3-β,4-β,12,14,16-β,17,20-HEPTOL ◇ VERATRINE ◇ VERATRINE (CRYSTALLIZED)

TOXICITY DATA with REFERENCE
ipr-mus LD50:3500 μg/kg PSEBAA 76,847,51
scu-mus LD50:4900 μg/kg JPETAB 113,89,55
ivn-mus LD50:1 mg/kg PHREA7 26,383,46
scu-gpg LDLo:1 mg/kg PHREA7 26,383,46

SAFETY PROFILE: Poison by intravenous, subcutaneous, and intraperitoneal routes. When heated to decomposition it emits toxic fumes of NO_x.

CDG250 CAS:5205-82-3 **HR: 3**
CG 201
mf: $C_{22}H_{28}NO_3 \cdot CH_3O_4S$ mw: 465.61

PROP: Crystals from petr ether. Mp: 134-135°.

SYNS: ACABEL ◇ BENZILIC ACID ester with 2-(HYDROXYMETHYL)-1,1-DIMETHYLPIPERIDINIUM METHYL SULFATE ◇ BENZILSAEURE-(N,N-DIMETHYL-2-HYDROXYMETHYL-PIPERIDINIUM)-ESTER-METHYLSULFAT (GERMAN) ◇ BEVONIUM METHYL SULFATE ◇ BEVONIUM METILSULFATE ◇ ESTER d'ACIDE BENZILIQUE et DU-1-METHYLSULFATE de 1,1-DIMETHYL-(2-HYDROXY-METHYL) PIPERIDINIUM ◇ ESTER del ACIDO BENCILICO del-1,1-DIMETIL-2-OXIMETIL-PIPERIDINIO-METILSULFATO (SPANISH) ◇ 2-((HYDROXYDIPHENYLACETYL)OXY)METHYL)-1,1-DIMETHYLPIPERIDINIUM METHYL SULFATE (SALT) ◇ 2-(HYDROXYMETHYL)-1,1-DIMETHYLPIPERIDINIUM METHYL SULFATE BENZILATE ◇ L-99 ◇ α-PHENYLMANDELIC ACID-N,N-DIMETHYLPIPERIDINIUM-2-METHYL ESTER METHYLSULFATE ◇ PIRIBENZIL METHYL SULFATE

TOXICITY DATA with REFERENCE
orl-rat LD50:5080 mg/kg ARZNAD 16,901,66
scu-rat LD50:2400 mg/kg ARZNAD 16,901,66
ivn-rat LD50:26 mg/kg ARZNAD 16,901,66
scu-mus LD50:436 mg/kg ARZNAD 16,901,66
orl-mus LD50:1360 mg/kg ARZNAD 16,901,66
ivn-mus LD50:17400 μg/kg ARZNAD 16,901,66
orl-dog LD50:1000 mg/kg ARZNAD 16,901,66
orl-rbt LD50:1000 mg/kg ARZNAD 16,901,66
ims-rbt LD50:500 mg/kg ARZNAD 16,901,66
orl-gpg LD50:3860 mg/kg ARZNAD 16,901,66
scu-gpg LD50:182 mg/kg ARZNAD 16,901,66

SAFETY PROFILE: Poison by subcutaneous and intravenous routes. Moderately toxic by ingestion and intramuscular routes. When heated to decomposition it emits toxic fumes of NO_x and SO_x. See also ESTERS.

CDG300 **HR: 3**
CGS 10787B
mf: $C_{15}H_{13}N_3O_2 \cdot C_6H_{15}NO_3$ mw: 416.53

SYN: α-CYANO-1-METHYL-β-OXO-PYRROLE-2-PROPIONANILIDE compd. with 2,2',2''-NITRILOTRIETHANOL

TOXICITY DATA with REFERENCE
orl-rat LD50:1071 mg/kg JACTDZ 2(2),249,83
ivn-rat LD50:243 mg/kg JACTDZ 2(2),249,83
orl-mus LD50:674 mg/kg JACTDZ 2(2),249,83
ivn-mus LD50:144 mg/kg JACTDZ 2(2),249,83

CONSENSUS REPORTS: Cyanide and its compounds are on the Community Right-To-Know List.

SAFETY PROFILE: Poison by intravenous route. Moderately toxic by ingestion. When heated to decomposition it emits toxic fumes of NO_x and CN^-.

CDG500 CAS:20562-03-2 ***HR: 3***
α-CHACONINE

TOXICITY DATA with REFERENCE
ipr-rat TDLo:20 mg/kg (female 5-12D post):TER TXAPA9 36,227,76
ipr-rat TDLo:40 mg/kg (female 5-12D post):REP TXAPA9 36,227,76
ipr-rat LD50:84 mg/kg TXAPA9 36,227,76
ipr-mus LD50:19 mg/kg TOLED5 3,349,79
ipr-rbt LDLo:50 mg/kg RCOCB8 12,657,75

SAFETY PROFILE: Poison by intraperitoneal route. An experimental teratogen. Other experimental reproductive effects.

CDG750 CAS:50335-03-0 ***HR: 3***
CHAETOGLOBOSIN A
mf: $C_{32}H_{36}N_2O_5$ mw: 528.70

PROP: Isolated from cultures of *Chaetonium globosum* (JJEMAG 48,105,78).

TOXICITY DATA with REFERENCE
ipr-mus TDLo:700 μg/kg (8D preg):REP MAIKD3 (10),17,80
orl-rat LD50:400 mg/kg JJEMAG 48,105,78
ipr-rat LDLo:2000 μg/kg JJEMAG 48,105,78
orl-mus LD50:400 mg/kg JJEMAG 48,105,78
scu-mus LD50:6500 μg/kg JJEMAG 48,105,78

SAFETY PROFILE: Poison by ingestion, intraperitoneal, and subcutaneous routes. Experimental reproductive effects. When heated to decomposition it emits toxic fumes of NO_x.

CDH000 CAS:94-41-7 ***HR: 3***
CHALCONE
mf: $C_{15}H_{12}O$ mw: 208.27

SYNS: 2-BENZALACETOPHENONE ◇ 1-BENZOYL-1-PHENYLETHENE ◇ β-BENZOYLSTYRENE ◇ 2-BENZYLIDENEACETOPHENONE ◇ CINNAMOPHENONE ◇ 1,3-DIPHENYL-1-PROPEN-3-ONE ◇ 3-PHENYLACRYLOPHENONE ◇ β-PHENYLACRYLOPHENONE ◇ 1-PHENYL-2-BENZOYLETHYLENE ◇ PHENYL STYRYL KETONE

TOXICITY DATA with REFERENCE
ivn-mus LD50:56 mg/kg CSLNX* NX#04476

CONSENSUS REPORTS: Reported in EPA TSCA Inventory.

SAFETY PROFILE: Poison by intravenous route. See also KETONES. When heated to decomposition it emits acrid smoke and irritating fumes.

CDH125 ***HR: 2***
CHALICE VINE

PROP: Climbing or erect vines with large yellow or creamy yellow trumpet-shaped flowers and elongated berries. They are native to Mexico and the subtropical areas of the United States, and are cultivated in Florida, Hawaii, and the West Indies.

SYNS: BEJUCO DO PEO (PUERTO RICO) ◇ CHAMICO BEJUCO (CUBA) ◇ CUP-OF-GOLD ◇ PALO GUACO (CUBA) ◇ SILVER CUP ◇ SOLANDRA (VARIOUS SPECIES) ◇ TRUMPET PLANT

SAFETY PROFILE: All parts of the plant including the nectar contain poisonous atropine alkaloids. Ingestion of any part of the plant can cause rapid heartbeat, fever, blurred vision, dilated pupils, excitement, headache, delirium, and hallucinations. See also ATROPINE.

CDH250 CAS:520-36-5 ***HR: 1***
CHAMOMILE
mf: $C_{15}H_{10}O_5$ mw: 270.25

PROP: Blue liquid, turning brownish-yellow. Composed of amyl and butyl esters of angelic and tiglic acids, butyric acid, etc. D: 0.905-0.915 @ 15°/15°.

SYNS: APIGENIN ◇ APIGENINE ◇ APIGENOL ◇ C.I. NATURAL YELLOW 1 ◇ 5,7-DIHYDROXY-2-(4-HYDROXYPHENYL)-4H-1-BENZOPYRAN-4-ONE ◇ 2-(p-HYDROXYPHENYL)-5,7-DIHYDROXYCHROMONE ◇ PELARGIDENON 1449 ◇ 4′,5,7-TRIHYDROXYFLAVONE ◇ VERSULIN

TOXICITY DATA with REFERENCE
mmo-sat 100 μg/plate BCSTB5 5,1489,77
mma-sat 100 μg/plate BCSTB5 5,1489,77

SAFETY PROFILE: Mutation data reported. A mild allergen. When heated to decomposition it emits acrid smoke and irritating fumes. See also ESTERS.

CDH500 CAS:8002-66-2 ***HR: 1***
CHAMOMILE OIL

PROP: By steam distillation of the flowers and stalks of *Matrilaria chamomilla* L. (FCTXAV 12,807,74). Blue-yellowish-brown liquid; strong odor and bitter aromatic taste. Composed of amyl and butyl esters of angelic, tiglic acids, and butyric acid. D: 0.905-0.915 @ 15°/15°. Sol in fixed oils, propylene glycol; insol in mineral oil, glycerin.

SYNS: CAMOMILE OIL GERMAN ◇ CHAMOMILE-GERMAN OIL ◇ GERMAN CHAMOMILE OIL ◇ HUNGARIAN CHAMOMILE OIL

TOXICITY DATA with REFERENCE
skn-rbt 500 mg/24H MOD FCTXAV 12,851,74

CONSENSUS REPORTS: Reported in EPA TSCA Inventory.

SAFETY PROFILE: A mild allergen. A skin irritant. See also ESTERS. When heated to decomposition it emits acrid and irritating fumes.

CDH750 CAS:8015-92-7 ***HR: 1***
CHAMOMILE OIL (ROMAN)

PROP: Obtained by the steam distillation of the dried flowers of *Anthemis nobilis* L. (FCTXAV 12,807,74). Blue liquid, turning brownish-yellow; strong aromatic odor. Composition: Amyl and butyl esters of angelic and tiglic acids, butyric acid, etc. D: 0.905-0.915 @ 15°/15°. Sol in fixed oils, mineral oil, propylene glycol; insol in glycerin.

SYN: CAMOMILE OIL, ENGLISH TYPE (FCC)

TOXICITY DATA with REFERENCE
skn-rbt 500 mg/24H MOD FCTXAV 12,853,74

CONSENSUS REPORTS: Reported in EPA TSCA Inventory.

SAFETY PROFILE: A mild allergen. A skin irritant. See also ESTERS. Combustible when heated. When heated to decomposition it emits acrid smoke and irritating fumes.

CDI000 CAS:64365-11-3 ***HR: 1***
CHARCOAL, ACTIVATED (DOT)
DOT: UN 1361/UN 1362
af: C aw: 12.01

PROP: Black porous solid, coarse granules or powder. Insol in water, organic solvents.

SYNS: ACTIVATED CARBON ◇ CARBON, ACTIVATED ◇ CARBORAFFIN ◇ CARBORAFINE ◇ KARBORAFIN ◇ NUCHAR 722

DOT Classification: Flammable Solid; Label: Spontaneously Combustible.

SAFETY PROFILE: It can cause a dust irritation, particularly to the eyes and mucous membranes. Combustible when exposed to heat. Dust is flammable and explosive when exposed to heat or flame or oxides.

CDI250 CAS:16291-96-6 ***HR: 2***
CHARCOAL (BRIQUETTES)
DOT: NA 1361

PROP: Black amorphous solid. Composition: carbon + impurities. Mw: 12.0, mp: >3500°, bp: 4200°, d: 3.51.

SYN: CHARCOAL

CONSENSUS REPORTS: Reported in EPA TSCA Inventory.

DOT Classification: Flammable Solid; Label: Flammable Solid.

SAFETY PROFILE: Carbon itself has no toxic action, but it contains impurities that may be toxic. Fire hazard: reacts with liquid air; $Ba(ClO_3)_2$; BrF_5; ClO; $Ca(ClO_3)_2$; ClF_2; F_2; H_2O_2; $Mg(ClO_3)_2$; (O_2 + wood); perchlorates; peroxides; (P + air); K + $KClO_3$; KNO_3; RuO_4; $AgNO_3$; $NaClO_3$; (AgCl + NaO_2); S; (S + $NaNO_3$); $Zn(ClO_3)_2$. Heats spontaneously, particularly when wet, freshly calcined, or tightly packed, and it can ignite and burn. Slight explosion hazard when exposed to heat or flame. To fight fire, use water, mist, foam or dry chemical. When heated to decomposition it emits acrid smoke and fumes.

CDI500 CAS:16291-96-6 ***HR: 2***
CHARCOAL SCREENINGS, MADE from "PINON" WOOD (DOT)
DOT: NA 1361

DOT Classification: Flammable Solid; Label: Flammable Solid.

SAFETY PROFILE: A flammable solid. See also CHARCOAL (BRIQUETTES).

CDJ000 ***HR: 2***
CHARCOAL (SHELL)
DOT: NA 1361

SYN: CHARCOAL, SHELL (DOT)

DOT Classification: Flammable Solid; Label: Flammable Solid.

SAFETY PROFILE: A flammable solid. See also CHARCOAL (BRIQUETTES).

CDJ500 CAS:16291-96-6 ***HR: 2***
CHARCOAL (wood, ground, crushed, granulated or pulverized)
DOT: NA 1361

DOT Classification: Flammable Solid; Label: Flammable Solid.

SAFETY PROFILE: A flammable solid. See also CHARCOAL (BRIQUETTES).

CDJ750 ***HR: 2***
CHARCOAL (wood, lump)

DOT Classification: Flammable Solid; Label: Flammable Solid.

SAFETY PROFILE: A flammable solid. See also CHARCOAL (BRIQUETTES).

CDK000 CAS:16291-96-6 ***HR: 2***
CHARCOAL WOOD SCREENINGS, OTHER THAN "PINON" WOOD SCREENINGS (DOT)

DOT Classification: Flammable Solid; Label: Flammable Solid.

SAFETY PROFILE: A flammable solid. See also CHARCOAL (BRIQUETTES).

CDK250 CAS:6377-18-0 ***HR: 3***
CHARTREUSIN
mf: $C_{32}H_{32}O_{14}$ mw: 640.64

PROP: Antibiotic substances produced by *Streptomyces chartreusis* from soil (JACSAT 75,4011,53).

SYNS: ANTIBIOTIC X-465A ◇ LAMBDAMYCIN ◇ NSC 5159 ◇ U-7257

TOXICITY DATA with REFERENCE
mmo-sat 200 μg/plate ABCHA6 44(4),919,80
pic-esc 150 ng/plate CNREA8 43,2819,83
dni-hmn:oth 5 mg/L RCOCB8 34,173,81
oms-hmn:oth 5 mg/L RCOCB8 34,173,81
ipr-mus LD50:300 mg/kg CNCRA6 30,9,63
scu-mus LD50:500 mg/kg 85GDA2 6,335,81

SAFETY PROFILE: Poison by intraperitoneal route. Moderately toxic by subcutaneous route. Human mutation data reported. When heated to decomposition it emits acrid smoke and irritating fumes.

CDK500 CAS:1393-72-2 ***HR: 3***
CHARTREUSIN, SODIUM SALT
mf: $C_{18}H_{17}O_8 \cdot Na$ mw: 384.34

SYNS: ANTIBIOTIC X465A SODIUM SALT ◇ X465A SODIUM SALT

TOXICITY DATA with REFERENCE
ipr-mus LD50:600 mg/kg 85ERAY 1,768,78
ivn-mus LD50:250 mg/kg JACSAT 75,4011,53

SAFETY PROFILE: Poison by intravenous route. Moderately toxic by intraperitoneal route. See also CHARTREUSIN. When heated to decomposition it emits toxic fumes of Na_2O.

CDK750 CAS:8001-74-9 ***HR: 1***
CHAULMOOGRA OIL

PROP: Active ingredients are glycerides of a series of unsaturated fatty acids possessing a 5-carbon ring (27ZTAP 3,33,69).

SYN: HYDNOCARPUS OIL

TOXICITY DATA with REFERENCE
orl-cat LDLo:4300 mg/kg JPETAB 24,221,25
orl-rbt LDLo:5 g/kg JPETAB 24,221,25

SAFETY PROFILE: Mildly toxic by ingestion. When heated to decomposition it emits acrid smoke and irritating fumes.

CDL000 CAS:476-32-4 ***HR: 3***
CHELIDONINE
mf: $C_{20}H_{19}NO_5$ mw: 353.40

PROP: White, crystalline powder. Mp: 135-136°.

TOXICITY DATA with REFERENCE
scu-rat LDLo:300 mg/kg JAMAAP 75,1324,20
ivn-mus LD50:35 mg/kg FEPRA7 5,163,46
scu-mus LDLo:300 mg/kg JAMAAP 75,1324,20
scu-rbt LDLo:300 mg/kg JAMAAP 75,1324,20
scu-gpg LDLo:300 mg/kg JAMAAP 75,1324,20
scu-frg LDLo:300 mg/kg JAMAAP 75,1324,20

SAFETY PROFILE: Poison by intravenous and subcutaneous routes. A central nervous system depressant causing sleepiness, depression, slowing of the pulse, and, in large doses, coma and circulatory failures. Combustible when exposed to heat or flame. When heated to decomposition it emits toxic fumes of NO_x.

CDL325 CAS:474-25-9 ***HR: 3***
CHENODESOXYCHOLIC ACID
mf: $C_{24}H_{40}O_4$ mw: 392.64

PROP: Needles from ethyl acetate + heptane. Mp: 119°. Freely sol in methanol, alc, acetone, acetic acid; more sol in ether and ethyl acetate than deoxycholic acid. Practically insol in water, petr ether, benzene. Forms beautiful crystalline salts of Na, K, and Ba. While the acid is tasteless, the Na salt tastes slightly sweet at first, then bitter.

SYNS: ANTHROPODEOXYCHOLIC ACID ◇ ANTHROPODESOXYCHOLIC ACID ◇ ANTHROPODODESOXYCHOLIC ACID ◇ CDC ◇ CDCA ◇ CHENDAL ◇ CHENDOL ◇ CHENIC ACID ◇ CHENIX ◇ CHENOCEDON ◇ CHENODEOXYCHOLIC ACID ◇ CHENODESOXYCHOLSAEURE (GERMAN) ◇ CHENODEX ◇ CHENODIOL ◇ CHENOFALK ◇ CHENOSAURE ◇ CHENOSSIL ◇ CHOLANORM ◇ 3-α,7-α-DIHYDROXYCHOLANIC ACID ◇ 3-α,7-α-DIHYDROXY-5-β-CHOLAN-24-OIC ACID ◇ FLUIBIL ◇ GALLODESOXYCHOLIC ACID ◇ HEKBILIN ◇ KEBILIS ◇ ULMENIDE

TOXICITY DATA with REFERENCE
mmo-sat 20 mg/L MUREAV 158,45,85
sln-smc 100 mg/L CRNGDP 5,447,84
orl-rat TDLo:3480 mg/kg (female 15-22D post):REP NTIS** PB81-127581
orl-rat TDLo:100 mg/kg (male 1D pre):TER NTIS** PB81-127581
orl-wmn TDLo:24 g/kg/5Y-C:CAR CLONEA 7,245,81
orl-rat LD50:4000 mg/kg IYKEDH 13,1128,82
ipr-rat LD50:105 mg/kg OYYAA2 15,915,78
ivn-rat LD50:86 mg/kg OYYAA2 15,915,78
orl-mus LD50:3000 mg/kg IYKEDH 13,1128,82
ipr-mus LD50:48500 μg/kg IYKEDH 13,1128,82
ivn-mus LD50:100 mg/kg ARZNAD 20,323,70

SAFETY PROFILE: Poison by intravenous and intraperitoneal routes. Moderately toxic by ingestion. An experimental teratogen. Experimental reproductive effects. Questionable human carcinogen producing liver tumors. Mutation data reported. When heated to decomposition it emits acrid smoke and fumes.

CDL375 CAS:2646-38-0 ***HR: 3***
CHENODESOXYCHOLIC ACID SODIUM SALT
mf: $C_{24}H_{39}O_4 \cdot Na$ mw: 414.62

SYNS: CHENODEOXYCHOLIC ACID SODIUM SALT ◇ SODIUM CHENODEOXYCHOLATE ◇ SODIUM CHENODESOXYCHOLATE

TOXICITY DATA with REFERENCE
ivn-rat LD50:100 mg/kg KSRANM 11,2499,77
scu-mus LD50:1450 mg/kg KSRNAM 11,2499,77
ivn-mus LD50:114 mg/kg KSRNAM 11,2499,77

SAFETY PROFILE: Poison by intravenous route. Moderately toxic by subcutaneous route. When heated to decomposition it emits toxic fumes of Na_2O. See also CHENODESOXYCHOLIC ACID.

CDL500 CAS:8006-99-3 ***HR: 3***
CHENOPODIUM OIL

PROP: American wormseed. Ingredients are ascaridol, cymene, camphor and saponins (27ZTAP 3,33,69). Colorless or pale yellow liquid, characteristic disagreeable odor and taste. Composition: 60-70% ascaridol. D: 0.950-0.980 @ 25°/25°. Insol in water; sol in 8 vols 70% alc; sltly sol in glacial acetic acid. Keep well closed, cool, and protected from light.

SYNS: OIL of AMERICAN WORMSEED ◇ OIL of CHENOPODIUM

TOXICITY DATA with REFERENCE
skn-mus 100% FCTXAV 14,713,76
skn-rbt 500 mg/24H MLD FCTXAV 14,713,76
skn-pig 100% FCTXAV 14,713,76
orl-rat LD50:255 mg/kg FCTXAV 14,713,76
skn-rbt LD50:415 mg/kg FCTXAV 14,713,76

SAFETY PROFILE: Poison by ingestion. Moderately toxic by skin contact. A skin irritant. See also ASCARIDOL, CAMPHOR, SAPONIN, and CYMENE. When heated to decomposition it emits acrid smoke and irritating fumes.

CDL750 ***HR: 3***
CHERRY BARK OAK

PROP: Tannin containing fraction of bark used (JNCIAM 57,207,76)

SYNS: QUERCUS FALCATA PAGODAEFOLIA ◇ TANNIN from CHERRY BARK OAK

TOXICITY DATA with REFERENCE
scu-rat TDLo:720 mg/kg/45W-I:NEO JNCIAM 57,207,76

SAFETY PROFILE: Questionable carcinogen with experimental neoplastigenic data. See also TANNIN. When heated to decomposition it emits acrid and irritating fumes.

CDM000 ***HR: 3***
CHERRY LAUREL OIL

PROP: Volatile oil from leaves of *Prunus laurocerasus L., Rosacene*. Pale yellow liquid, odor and taste similar to oil of bitter almond. D: 1.054-1.066 @ 20°/20°. Sltly sol in water; sol in 2 vols 70% alc, benzene, chloroform, and ether.

SAFETY PROFILE: Very poisonous. Hydrogen cyanide component is responsible for highly toxic properties. Keep well closed, cool, and protected from light. See also CYANIDE. When heated to decomposition it emits toxic fumes of CN^-.

CDM250 CAS:1401-55-4 ***HR: 3***
CHESTNUT TANNIN

SYNS: CASTANEA SATIVA MILL TANNIN ◇ TANNIN from CHESTNUT

TOXICITY DATA with REFERENCE
scu-mus TDLo:750 mg/kg/12W-I:ETA BJCAAI 14,147,60
ipr-mus LD50:150 mg/kg JPPMAB 9,98,57
scu-mus LD50:140 mg/kg JPPMAB 9,98,57
ivn-mus LD50:50 mg/kg JPPMAB 9,98,57
ims-mus LD50:120 mg/kg JPPMAB 9,98,57

SAFETY PROFILE: Poison by subcutaneous, intramuscular, intravenous, and intraperitoneal routes. Questionable carcinogen with experimental tumorigenic data. See also TANNIN. When heated to decomposition it emits acrid and irritating fumes.

CDM325 ***HR: 3***
CHINABERRY

PROP: A tree which may grow to 50 feet. The leaves are compound with 2-inch long serrated leaflets. The flowers are purple, fragrant, and form clusters. The yellow berries remain on the tree after the leaves fall. They grow wild and are cultivated in the US coastal states from Virginia to Texas, Hawaii, Guam, and the West Indies.

SYNS: AFRICAN LILAC TREE ◇ ALELAILA (PUERTO RICO) ◇ ARBOL DEL QUITASOL (CUBA) ◇ BEAD TREE ◇ CHINA TREE ◇ FALSE SYCAMORE ◇ HOG BUSH ◇ 'INIA (HAWAII) ◇ INDIAN LILAC ◇ JAPANESE BEAD TREE ◇ LILAILA ◇ LILAS (HAITI, DOMINICAN REPUBLIC) ◇ MELIA AZEDARACH ◇ PARADISE TREE ◇ PARAISO (MEXICO) ◇ PASILLA (PUERTO RICO) ◇ PERSIAN LILAC ◇ PRIDE of CHINA ◇ PRIDE of INDIA ◇ SYRIAN BEAD TREE ◇ TEXAS UMBRELLA TREE ◇ WEST INDIAN LILAC ◇ WHITE CEDAR

SAFETY PROFILE: The fruit and bark contain tetranortriterpene neurotoxins and gastroenteric toxins. In some areas the fruit is edible. Ingestion may cause, after an indefinite delay period, poor muscle coordination, confusion, stupor, intense gastritis, vomiting, diarrhea, difficult breathing, convulsions, and partial to complete paralysis.

CDM500 CAS:71392-29-5 ***HR: 3***
CHINOIN-127
mf: $C_{11}H_{17}N_3O_2$ mw: 223.31

SYN: 1,6-DIMETHYL-4-OXO-1,6,7,8,9,9a-HEXAHYDRO-4H-PYRIDO (1,2-a)PYRIMIDINE-3-CARBOXAMIDE

TOXICITY DATA with REFERENCE
orl-rat LD50:370 mg/kg ARZNAD 29,766,79
scu-rat LD50:280 mg/kg ARZNAD 29,266,79
ivn-rat LD50:210 mg/kg ARZNAD 29,766,79
orl-mus LD50:360 mg/kg ARZNAD 29,766,79

SAFETY PROFILE: Poison by ingestion, subcutaneous, and intravenous routes. When heated to decomposition it emits toxic fumes of NO_x.

CDM575 CAS:88338-63-0 ***HR: 3***
CHINOIN-170
mf: $C_{11}H_{12}N_6O_3$ mw: 276.29

SYNS: 3,7-DIHYDRO-1,3-DIMETHYL-7-((5-METHYL-1,2,4-OXADIAZOL-3-YL)METHYL)-1H-PURINE-2,6-DIONE ◇ 3-((1,3-DIMETHYLXANTHIN-7-YL)METHYL)-5-METHYL-1,2,4-OXADIAZOLE

TOXICITY DATA with REFERENCE
orl-rat LD50:1549 mg/kg DRFUD4 10,624,85
ivn-rat LD50:397 mg/kg DRFUD4 10,624,85
orl-mus LD50:1243 mg/kg DRFUD4 10,624,85
ivn-mus LD50:417 mg/kg DRFUD4 10,624,85
ims-mus LD50:721 mg/kg DRFUD4 10,624,85

SAFETY PROFILE: Poison by intravenous route. Moderately toxic by ingestion and intramuscular routes. When heated to decomposition it emits toxic fumes of NO_x.

CDM625 CAS:41024-90-2 ***HR: 3***
CHIRAL BINAPHTHOL
mf: $C_{20}H_{14}O_2$ mw: 286.33

SYN: (±)-1,1-BI-2-NAPHTHOL

TOXICITY DATA with REFERENCE
eye-rbt 50 mg MOD DCTODJ 8,451,85
orl-rat LD50:113 mg/kg DCTODJ 8,451,85
ipr-rat LD50:20 mg/kg DCTODJ 8,451,85
ipr-mus LD50:6 mg/kg DCTODJ 8,451,85

SAFETY PROFILE: Poison by ingestion and intraperitoneal routes. An eye irritant. When heated to decomposition it emits acrid smoke and fumes. See also various napthol entries.

CDM700 ***HR: 3***
CHIRONEX FLECKERI TOXIN

SYN: TOXIN, JELLYFISH, CHIRONEX FLECKERI

TOXICITY DATA with REFERENCE
ivn-mus LDLo:1 mg/kg BJPCBM 35,510,69
par-mus LDLo:167 μg/kg TOXIA6 9,145,71
unr-mam LD50:5 mg/kg CLPTAT 8,849,67

SAFETY PROFILE: Poison by intravenous, parenteral, and possibly other routes.

CDM750 CAS:1398-61-4 ***HR: 3***
CHITIN
mf: $C_{30}H_{50}N_4O_{19}$ mw: 770.84

SYN: CHITINA (ITALIAN)

TOXICITY DATA with REFERENCE
ivn-rat LD50:50 mg/kg BSIBAC 44,1685,68

CONSENSUS REPORTS: Reported in EPA TSCA Inventory.

SAFETY PROFILE: Poison by intravenous route. When heated to decomposition it emits toxic fumes of NO_x.

CDN000 CAS:53-19-0 ***HR: 3***
CHLODITHANE
mf: $C_{14}H_{10}Cl_4$ mw: 320.04

SYNS: CHLODITAN ◇ 1-CHLORO-2-(2,2-DICHLORO-1-(4-CHLOROPHENYL)ETHYL)BENZENE ◇ 2-(o-CHLOROPHENYL)-2-(p-CHLOROPHENYL)-1,1-DICHLOROETHANE ◇ o,p′-DDD ◇ 2,4′-DDD ◇ 1,1-DICHLORO-2,2-BIS(2,4′-DICHLOROPHENYL)ETHANE ◇ 1,1-DICHLORO-2-(o-CHLOROPHENYL)-2-(p-CHLOROPHENYL)ETHANE ◇ o,p′-DICHLORODIPHENYLDICHLOROETHANE ◇ 2,4′-DICHLOROPHENYLDICHLOROETHANE ◇ MITOTANE ◇ NCI-C04933 ◇ NSC 38721 ◇ o,p-TDE ◇ o,p′-TDE

TOXICITY DATA with REFERENCE
cyt-rat:oth 10 μg/L 34LXAP -,555,76
orl-man TDLo:16 g/kg (male 15W pre):REP JTCTDW 25,463,87
scu-mus TDLo:900 mg/kg (female 6-14D post):TER NTIS** PB223-160
orl-rat TDLo:10 g/kg/52W-C:ETA BAFEAG 52,89,65
ipr-rat TDLo:2500 mg/kg/7W-I:ETA,TER CANCAR 40(Suppl 4),1935,77
orl-man TDLo:17 g/kg/35W:CNS CANCAR 42,2177,78
orl-wmn TDLo:800 mg/kg/4D:SKN CANCAR 42,2177,78
orl-wmn TDLo:11 g/kg/15W:BLD,CVS CANCAR 42,2177,78
orl-wmn TDLo:14 g/kg/22W:GIT,CNS CANCAR 42,2177,78

CONSENSUS REPORTS: NCI Carcinogenesis Studies (ipr); Equivocal Evidence: mouse, rat CANCAR 40,1935,77. EPA Genetic Toxicology Program.

SAFETY PROFILE: Human systemic effects by ingestion: somnolence, blood pressure depression, diarrhea, nausea or vomiting, normocytic anemia (decrease in the number of red blood cells), and pigmented or nucleated red blood cells. Experimental teratogenic and reproductive effects. Questionable carcinogen with experimental carcinogenic and tumorigenic data. Mutation data reported. When heated to decomposition it emits toxic fumes of Cl^-.

CDN200 CAS:78-95-5 ***HR: 3***
CHLORACETONE
DOT: UN 1695
mf: C_3H_5ClO mw: 92.53

PROP: Colorless liquid, pungent odor. Mp: −44.5°, bp: 119°, d: 1.162.

SYNS: ACETONYL CHLORIDE ◇ A-STOFF ◇ CHLORACETONE ◇ CHLOROACETONE ◇ CHLOROACETONE, stabilized (DOT) ◇ CHLOROPROPANONE ◇ 1-CHLORO-2-PROPANONE ◇ MONOCHLORACETONE ◇ MONOCHLOROACETONE ◇ MONOCHLOROACETONE, inhibited (DOT) ◇ MONOCHLOROACETONE, stabilized (DOT) ◇ MONOCHLOROACETONE, unstabilized (DOT) ◇ TONITE

TOXICITY DATA with REFERENCE
sln-oin-dmg-ihl 100 pph/6M PREBA3 62,284,46/47
ihl-hmn LCLo:605 ppm/10M NTIS** PB214-270
orl-rat LD50:100 mg/kg AIHAAP 47,375,86
ihl-rat LC50:262 ppm/1H AIHAAP 47,375,86
skn-rat LDLo:100 mg/kg KODAK* 21MAY71
ipr-rat LD50:80 mg/kg OYYAA2 33,695,87
orl-mus LD50:127 mg/kg AIHAAP 47,375,86
ipr-mus LD50:92 mg/kg OYYAA2 33,695,87
skn-rbt LD50:141 mg/kg AIHAAP 47,375,86

CONSENSUS REPORTS: Reported in EPA TSCA Inventory.

ACGIH TLV: CL 1 ppm (skin)
DOT Classification: Forbidden, unstabilized; Irritating Material; LABEL: Irritant; Poison B; LABEL: Poison, stabilized

SAFETY PROFILE: Poison by inhalation, ingestion, and skin contact. Mutation data reported. A lachrymator poison gas. See also CHLORINATED HYDROCARBONS, ALIPHATIC; ACETONE. Flammable when exposed to heat or flame, or oxidizers. Old material can explode. When heated to decomposition it emits highly toxic fumes.

CDN500 CAS:107-14-2 ***HR: 3***
CHLORACETONITRILE
DOT: UN 2668
mf: C_2H_2ClN mw: 75.50

SYNS: CHLOROACETONITRILE (DOT) ◇ α-CHLOROACETONITRILE ◇ 2-CHLOROACETONITRILE ◇ CHLOROMETHYL CYANIDE ◇ MONOCHLOROACETONITRILE ◇ MONOCHLOROMETHYL CYANIDE ◇ USAF KF-5

TOXICITY DATA with REFERENCE
skn-rbt 14 mg/24H open MLD AIHAAP 23,95,62
dnd-hmn:lym 50 μmol/L AIHAAP 23,95,62
sce-ham:ovr 79100 nmol/L FAATDF 5,1065,85
skn-mus TDLo:4800 mg/kg/2W-I:CAR FAATDF 5,1065,85
orl-rat LD50:220 mg/kg AIHAAP 23,95,62
ihl-rat LCLo:250 ppm/4H AIHAAP 23,95,62
orl-mus LD50:139 mg/kg ARTODN 55,47,84
ipr-mus LD50:100 mg/kg NTIS** AD277-689
skn-rbt LDLo:71 mg/kg AIHAAP 23,95,62

CONSENSUS REPORTS: Reported in EPA TSCA Inventory. Cyanide and its compounds are on the Community Right-To-Know List.

DOT Classification: Poison B; Label: Flammable Liquid and Poison.

SAFETY PROFILE: Poison by ingestion, skin contact, and intraperitoneal route. Moderately toxic by inhalation. A skin irritant. Human mutation data reported. Questionable carcinogen with experimental tumorigenic data. See also NITRILES. When heated to decomposition it emits very toxic fumes of Cl^-, NO_x, and CN^-.

CDO000 CAS:302-17-0 ***HR: 3***
CHLORAL HYDRATE
mf: $C_2HCl_3O{\cdot}H_2O$ mw: 165.40

PROP: Transparent, colorless crystals; aromatic, penetrating, sltly acrid odor and sltly bitter, caustic taste. Mp: 52°, bp: 97.5°, d: 1.9.

SYNS: AQUACHLORAL ◇ Bi 3411 ◇ CHLORALDURAT ◇ DORMAL ◇ FELSULES ◇ HYDRAL ◇ HYDRAL de CHLORAL ◇ KESSODRATE ◇ LORINAL ◇ NOCTEC ◇ NORTEC ◇ NYCOTON ◇ PHALDRONE ◇ RECTULES ◇ SK-CHORAL HYDRATE ◇ SOMNI SED ◇ SOMNOS ◇ SONTEC ◇ TOSYL ◇ TRAWOTOX ◇ TRICHLORACETALDEHYD-HYDRAT (GERMAN) ◇ TRICHLOROACETALDEHYDE HYDRATE ◇ TRICHLOROACETALDEHYDE MONOHYDRATE ◇ 2,2,2-TRICHLORO-1,1-ETHANEDIOL

TOXICITY DATA with REFERENCE
mrc-smc 15 mmol/L MUREAV 141,19,84
sce-hmn:lym 54 mg/L AGTQAH 24,105,81
orl-mus TDLo:13 g/kg (female 3W pre-3W post):REP NETOD7 6,137,84
orl-mus TDLo:10 mg/kg:CAR CDPRD4 9,279,86
skn-mus TDLo:960 mg/kg/1W-I:ETA BJCAAI 9,177,55
orl-wmn TDLo:465 mg/kg:EYE,BPR AMSVAZ 223,269,88
orl-hmn LDLo:4 mg/kg PHBTH* 3,132,67
orl-hmn TDLo:300 mg/kg:CNS,CVS BMJOAE 2,994,77
unr-cld TDLo:48 mg/kg/6H-I:BPR AACRAT 65,691,86
unk-man LDLo:103 mg/kg 85DCAI 2,73,70
unr-inf TDLo:160 mg/kg/3D-I:CVS AACRAT 65,691,86
rec-cld TDLo:97 mg/kg:CVS AACRAT 65,691,86
orl-rat LD50:479 mg/kg TXAPA9 18,185,71
skn-rat LD50:3030 mg/kg 85JFAN A065,84
ipr-rat LD50:472 mg/kg JAPMA8 41,100,52
par-rat LD50:710 mg/kg NIIRDN 6,784,82
orl-mus LD50:1100 mg/kg JPETAB 106,444,52
ipr-mus LD50:580 mg/kg ARZNAD 10,665,60

scu-mus LDLo:800 mg/kg AEPPAE 166,437,32
orl-dog LDLo:1000 mg/kg JPETAB 78,340,43
orl-cat LDLo:400 mg/kg JPETAB 78,340,43

CONSENSUS REPORTS: Reported in EPA TSCA Inventory. EPA Genetic Toxicology Program.

SAFETY PROFILE: A human poison by ingestion and possibly other routes. Poison experimentally by ingestion, intravenous, and rectal routes. Moderately toxic by subcutaneous, parenteral, and intraperitoneal routes. Experimental reproductive effects. Human systemic effects by ingestion: general anesthetic; cardiac arrythmias, blood pressure depression, eye effects, coma, pulse rate increase, arrhythmias. Human mutation data reported. Questionable carcinogen with experimental carcinogenic and tumorigenic data by skin contact. A sedative, anesthetic, and narcotic. Combustible when exposed to heat or flame. When heated to decomposition it emits toxic fumes of Cl^-.

CDO250 CAS:95-06-7 ***HR: 3***
2-CHLORALLYL DIETHYLDITHIOCARBAMATE
mf: $C_8H_{14}ClNS_2$ mw: 223.80

PROP: Amber liquid. Bp: 129° @ 1 mm.

SYNS: CDEC ◇ CHLORALLYL DIETHYLDITHIOCARBAMATE ◇ 2-CHLOROALLYL DIETHYLDITHIOCARBAMATE ◇ 2-CHLOROALLYL-N,N-DIETHYLDITHIOCARBAMATE ◇ 2-CHLORO-2-PROPENE-1-THIOL DIETHYLDITHIOCARBAMATE ◇ 2-CHLORO-2-PROPENYL DIETHYLCARBAMODITHIOATE ◇ CP 4572 ◇ DIETHYLCARBAMODITHIOIC ACID 2-CHLORO-2-PROPENYL ESTER ◇ DIETHYLDITHIOCARBAMIC ACID-2-CHLOROALLYL ESTER ◇ NCI-C00453 ◇ SULFALLATE ◇ THIOALLATE ◇ VEGADEX ◇ VEGADEX SUPER

TOXICITY DATA with REFERENCE
mmo-sat 10 uL/plate PMRSDJ 2,87,81
mma-sat 10 uL/plate PMRSDJ 2,87,81
mmo-omi 1100 μg/plate PMRSDJ 2,87,81
mmo-asn 20 uL/plate JSFAAE 32,826,81
orl-rat TDLo:6825 mg/kg/78W-C:CAR NCITR* NCI-CG-TR-115,78
orl-mus TDLo:59 g/kg/78W-C:CAR NCITR* NCI-CG-TR-115,78
orl-rat LD50:850 mg/kg RREVAH 10,97,65
skn-rbt LD50:2200 mg/kg 85DPAN -,-,71/76

CONSENSUS REPORTS: NTP Fifth Annual Report on Carcinogens. IARC Cancer Review: Group 2B IMEMDT 7,56,87; Animal Sufficient Evidence IMEMDT 30,283,83. NCI Carcinogenesis Bioassay (feed); Clear Evidence: mouse, rat NCITR* NCI-CG-TR-115,78. EPA Genetic Toxicology Program.

SAFETY PROFILE: Confirmed with experimental carcinogenic data. Moderately toxic by ingestion and skin contact. Mutation data reported. An herbicide. When heated to decomposition it emits very toxic fumes of Cl^-, NO_x, and SO_x. See also ALLYL COMPOUNDS; CARBAMATES; and ESTERS.

CDO500 CAS:305-03-3 ***HR: 3***
CHLORAMBUCIL
mf: $C_{14}H_{19}Cl_2NO_2$ mw: 304.24

SYNS: AMBOCHLORIN ◇ AMBOCLORIN ◇ 4-(BIS(2-CHLOROETHYL)AMINO)BENZENEBUTANOIC ACID ◇ Γ-(p-BIS(2-CHLOROETHYL)AMINOPHENYL)BUTYRIC ACID ◇ 4-(p-(BIS(2-CHLOROETHYL)AMINO)PHENYL)BUTYRIC ACID ◇ 4-(p-BIS(β-CHLOROETHYL)AMINOPHENYL)BUTYRIC ACID ◇ CB 1348 ◇ CHLORAMINOPHEN ◇ CHLORAMINOPHENE ◇ CHLOROAMBUCIL ◇ CHLOROBUTIN ◇ CHLOROBUTINE ◇ N,N-DI-2-CHLOROETHYL-Γ-p-AMINOPHENYLBUTYRIC ACID ◇ p-(N,N-DI-2-CHLOROETHYL)AMINOPHENYL BUTYRIC ACID ◇ p-N,N-DI-(β-CHLOROETHYL)AMINOPHENYL BUTYRIC ACID ◇ Γ-(p-DI(2-CHLOROETHYL)AMINOPHENYL)BUTYRIC ACID ◇ ECLORIL ◇ ELCORIL ◇ LEUKERAN ◇ LEUKERSAN ◇ LEUKORAN ◇ LINFOLIZIN ◇ LINFOLYSIN ◇ NCI-C03485 ◇ NSC-3088 ◇ PHENYLBUTYRIC ACID NITROGEN MUSTARD ◇ RCRA WASTE NUMBER U035

TOXICITY DATA with REFERENCE
mma-sat 100 μg/plate ENMUDM 8(Suppl 7),1,86
dns-hmn:lym 1 μmol/L JTEHD6 6,1059,80
sce-hmn:lym 150 μg/L MUREAV 143,225,85
sln-dmg-orl 2000 ppm ENMUDM 7,677,85
trn-dmg-orl 3000 ppm ENMUDM 7,677,85
orl-wmn TDLo:13 mg/kg (56D pre):REP JAMAAP 191,444,65
orl-wmn TDLo:5160 μg/kg (33-75D preg):TER JAMAAP 186,74,63
orl-man TDLo:84 mg/kg/2.5Y-C:CAR,BLD AMSVAZ 199,373,76
orl-wmn TDLo:101 mg/kg/82W-C:CAR,BLD GYNOA3 6,115,78
unr-wmn TDLo:161 mg/kg/3Y-I:CAR,BLD SJHAAQ 13,179,74
ipr-rat TDLo:120 mg/kg/26W-I:CAR RRCRBU 52,1,75
skn-mus TDLo:108 mg/kg/9W-I:NEO BJCAAI 10,363,56
ipr-mus TDLo:18 mg/kg/4W:CAR JNCIAM 36,915,66
orl-man TD:59 mg/kg/96W-C:CAR,BLD NPMDAD 10,1717,81
orl-wmn TD:200 mg/kg/6Y-C:CAR,BLD ACHAAH 62,283,80
orl-wmn TD:307 mg/kg/7Y-C:CAR,BLD AIMDAP 134,728,74
orl-wmn TD:141 mg/kg/5Y-I:CAR,BLD NEJMAG 304,441,81
orl-hmn TD:180 mg/kg/3Y-I:CAR,BLD AIMDAP 137,355,77
orl-wmn TD:135 mg/kg/4Y-C:CAR,BLD AIMDAP 137,355,77
orl-hmn TD:84 mg/kg/3Y-C:CAR,BLD ACCBAT 38,228,83
orl-wmn TD:70 mg/kg/94W-C:CAR,BLD ACCBAT 38,228,83

orl-chd TD:108 mg/kg/77W-C:CAR,BLD AFPEAM 36,592,79
orl-wmn TDLo:82600 μg/kg:PUL CANCAR 41,455,78
orl-rat LD50:76 mg/kg FCTOD7 22,665,84
ipr-rat LD50:14 mg/kg BCPCA6 13,969,64
scu-rat LDLo:32 mg/kg EJCAAH 13,873,77
orl-mus LD50:101 mg/kg NCISP* JAN86
ipr-mus LD50:30 mg/kg PHMGBN 11,231,74
scu-mus LD50:115 mg/kg NCISP* JAN86
ivn-dog LDLo:3 mg/kg CCSUBJ 2,201,65
ivn-mky LDLo:3 mg/kg CCSUBJ 2,201,65

CONSENSUS REPORTS: NTP Fifth Annual Report on Carcinogens. IARC Cancer Review: Group 1 IMEMDT 7,144,87; Human Inadequate Evidence IMEMDT 9,125,75; Human Limited Evidence IMEMDT 26,115,81; Animal Limited Evidence IMEMDT 26,115,81; Animal Sufficient Evidence IMEMDT 9,125,75. EPA Genetic Toxicology Program.

SAFETY PROFILE: Confirmed carcinogen producing leukemia. Experimental carcinogenic and neoplastigenic data. Poison by ingestion, intravenous, intraperitoneal, and subcutaneous routes. Human respiratory system effects by ingestion: cough, dyspnea, and interstitial fibrosis. Human reproductive effects by ingestion and possibly other routes: changes in spermatogenesis; menstrual cycle changes or disorders; and teratogenic effects of the fetal urogenital system. Experimental teratogenic and reproductive effects. Human mutation data reported. An anti-neoplastic agent. When heated to decomposition it emits very toxic fumes of Cl^- and NO_x.

CDO625 CAS:1030-06-4 ***HR: 3***
CHLORAMBUCIL SODIUM SALT
mf: $C_{14}H_{19}Cl_2NO_2$•Na mw: 327.23

SYNS: CB 1348 SODIUM SALT ◇ p-(DI-2-CHLOROETHYLAMINE) PHENYL BUTYRIC ACID SODIUM SALT ◇ SODIUM CHLORAMBUCIL

TOXICITY DATA with REFERENCE
sln-oin-dmg-par 2 mmol/L GENRA8 1,173,60
ipr-mus TDLo:10 mg/kg (12D preg):REP BJPCAL 11,437,56
unr-mus LD50:45 mg/kg BJPCAL 11,437,56

CONSENSUS REPORTS: EPA Genetic Toxicology Program.

SAFETY PROFILE: Poison by an unspecified route. Experimental reproductive effects. Mutation data reported. When heated to decomposition it emits toxic fumes of Cl^-, NO_x, and Na_2O. See also CHLORAMBUCIL.

CDO750 CAS:10599-90-3 ***HR: 3***
CHLORAMIDE
mf: ClH_2N mw: 51.48

SYNS: CHLORAMINE ◇ CHLORAMINE (inorganic compound) ◇ CHLOROAMINE ◇ MONOCHLORAMIDE ◇ MONOCHLORAMINE ◇ MONOCHLOROAMINE ◇ MONOCHLOROAMMONIA

TOXICITY DATA with REFERENCE
mmo-bcs 18 μmol/L SCIEAS 192,1141,76
dnr-bcs 18 μmol/L SCIEAS 192,1141,76

CONSENSUS REPORTS: Reported in EPA TSCA Inventory.

SAFETY PROFILE: Mutation data reported. The dry material decomposes violently at −50°C. When heated to decomposition it emits very toxic fumes of NO_x, NH_3, and Cl^-.

CDP000 CAS:127-65-1 ***HR: 3***
CHLORAMINE T
mf: $C_7H_8ClNO_2S$•Na mw: 228.66

SYNS: ACTI-CHLORE ◇ AKTIVIN ◇ ANEXOL ◇ BENZENESULFONAMIDE, N-CHLORO-4-METHYL-, SODIUM SALT (9CI) ◇ BERKENDYL ◇ CHLORALONE ◇ CHLORASAN ◇ CHLORASEPTINE ◇ CHLORAZAN ◇ CHLORAZENE ◇ CHLORAZONE ◇ CHLOROZONE ◇ CHLORSEPTOL ◇ CLORINA ◇ CLOROSAN ◇ DESINFECT ◇ EUCLORINA ◇ GANSIL ◇ GYNECLORINA ◇ HALAMID ◇ HELIOGEN ◇ KLORAMIN ◇ KLORAMINE-T ◇ MULTICHLOR ◇ SODIUM CHLORAMINE T ◇ SODIUM p-TOLUENESULFONYLCHLORAMIDE ◇ SODIUM TOSYLCHLORAMIDE ◇ TAMPULES ◇ TOCHLORINE ◇ TOLAMINE ◇ TOSYLCHLORAMIDE SODIUM

TOXICITY DATA with REFERENCE
cyt-hmn:lyms 100 ppm/24H ARMCAH 21,409,70
par-mus LDLo:300 mg/kg JPETAB 14,259,20
ivn-rbt LDLo:25 mg/kg JPETAB 14,259,20
scu-gpg LDLo:900 mg/kg JPETAB 14,259,20
par-frg LDLo:200 mg/kg JPETAB 14,259,20

CONSENSUS REPORTS: Reported in EPA TSCA Inventory.

SAFETY PROFILE: Poison by parenteral and intravenous routes. Human mutagenic data reported. When heated to decomposition it emits toxic fumes of Cl^-, SO_x, Na_2O, and NO_x. See also SULFONATES and CHLORIDES.

CDP250 CAS:56-75-7 ***HR: 3***
CHLORAMPHENICOL
mf: $C_{11}H_{12}Cl_2N_2O_5$ mw: 323.15

PROP: Crystalline. Mp: 151°. Sltly sol in water.

SYNS: ALFICETYN ◇ AMBOFEN ◇ AMPHENICOL ◇ AMPHICOL ◇ AMSECLOR ◇ ANACETIN ◇ AQUAMYCETIN ◇ AUSTRACIL ◇ AUSTRACOL ◇ BIOCETIN ◇ BIOPHENICOL ◇ CAF ◇ CAM ◇ CAP ◇ CATILAN ◇ CHEMICETIN ◇ CHEMICETINA ◇ CHLOMIN ◇ CHLOMYCOL ◇ CHLORAMEX ◇ CHLORAMFICIN ◇ CHLORAMFILIN ◇ d-CHLORAMPHENICOL ◇ d-threo-CHLORAMPHENICOL ◇ CHLORAMSAAR ◇ CHLORASOL ◇ CHLORA-TABS ◇ CHLORICOL ◇ CHLORNITROMYCIN ◇ CHLOROCAPS ◇ CHLOROCID ◇ CHLOROCIDIN C TETRAN ◇ CHLOROCOL ◇ CHLOROJECT L

◇ CHLOROMAX ◇ CHLOROMYCETIN ◇ CHLORONITRIN ◇ CHLOROPTIC ◇ CHLOROVULES ◇ CIDOCETINE ◇ CIPLAMYCETIN ◇ CLORAMIDINA ◇ CLOROAMFENICOLO (ITALIAN) ◇ CLOROMISAN ◇ CLOROSINTEX ◇ COMYCETIN ◇ CPH ◇ CYLPHENICOL ◇ DESPHEN ◇ DETREOMYCINE ◇ DEXTROMYCETIN ◇ d-(−)-threo-2-DICHLOROACETAMIDO-1-p-NITROPHENYL-1,3-PROPANEDIOL ◇ d-threo-N-DICHLOROACETYL-1-p-NITROPHENYL-2-AMINO-1,3-PROPANEDIOL ◇ d-(−)-threo-2,2-DICHLORO-N-(β-HYDROXY-α-(HYDROXYMETHYL))-p-NITROPHENETHYLACETAMIDE ◇ d-(−)-2,2-DICHLORO-N-(β-HYDROXY-α-(HYDROXYMETHYL)-p-NITROPHENYLETHYL)ACETAMIDE ◇ d-threo-N-(1,1'-DIHYDROXY-1-p-NITROPHENYLISOPROPYL)DICHLOROACETAMIDE ◇ DOCTAMICINA ◇ ECONOCHLOR ◇ EMBACETIN ◇ EMETREN ◇ ENICOL ◇ ENTEROMYCETIN ◇ ERBAPLAST ◇ ERTILEN ◇ FARMICETINA ◇ FENICOL ◇ GLOBENICOL ◇ GLOROUS ◇ HALOMYCETIN ◇ HORTFENICOL ◇ I 337A ◇ INTRAMYCETIN ◇ ISMICETINA ◇ ISOPHENICOL ◇ ISOPTO FENICOL ◇ KAMAVER ◇ KEMICETINE ◇ LEUKOMYAN ◇ LEVOMYCETIN ◇ LOROMISIN ◇ MASTIPHEN ◇ MEDIAMYCETINE ◇ MICOCHLORINE ◇ MICROCETINA ◇ MYCHEL ◇ MYCINOL ◇ NCI-C55709 ◇ d-(−)-threo-1-p-NITROPHENYL-2-DICHLORACETAMIDO-1,3-PROPANEDIOL ◇ d-threo-1-(p-NITROPHENYL)-2-(DICHLOROACETYLAMINO)-1,3-PROPANE DIOL ◇ NORIMYCIN V ◇ NOVOCHLOROCAP ◇ NOVOMYCETIN ◇ NOVOPHENICOL ◇ NSC 3069 ◇ OFTALENT ◇ OLEOMYCETIN ◇ OPTHOCHLOR ◇ OTOPHEN ◇ PANTOVERNIL ◇ PARAXIN ◇ PETNAMYCETIN ◇ QUEMICETINA ◇ RIVOMYCIN ◇ ROMPHENIL ◇ SEPTICOL ◇ SINTOMICETINA ◇ STANOMYCETIN ◇ SYNTHOMYCINE ◇ TEVCOCIN ◇ TIFOMYCINE ◇ TREOMICETINA ◇ U-6062 ◇ UNIMYCETIN ◇ VETICOL

TOXICITY DATA with REFERENCE
dni-hmn:bmr 1500 μmo/L 46GFA5 -,17,81
cyt-hmn:lym 500 mg/L HUMAA7 7,305,69
orl-mus TDLo:175 mg/kg (female 15-21D post):REP NEPHBW 13,233,74
ivn-rat TDLo:2 g/kg (female 10-14D post):TER AEMBAP 27,291,72
orl-wmn TDLo:300 mg/kg/60W-I:CAR,BLD NEJMAG 277,1003,67
ipr-mus TDLo:2500 mg/kg/5W-I:ETA,BLD CNREA8 41,3478,81
orl-wmn TD:1680 mg/kg/6W-I:CAR,BLD NEJMAG 277,1003,67
orl-man TD:434 mg/kg/W-C:CAR,BLD ACHAAH 66,267,81
orl-inf TDLo:440 mg/kg:CNS,GIT,MET JAMAAP 234,149,75
orl-wmn LDLo:400 mg/kg JAMAAP 234,149,75
unr-chd TDLo:250 mg/kg/10D:BLD,LIV CPEDAM 14,499,75
ims-inf TDLo:250 mg/kg/2D:CVS NEJMAG 262,787,60
orl-rat TDLo:2500 mg/kg FRPSAX 10,3,55
ipr-rat LD50:1811 mg/kg TXAPA9 18,185,71
scu-rat LD50:5 g/kg TXAPA9 9,445,66
ivn-rat LD50:171 mg/kg JCINAO 28,943,49
orl-mus LD50:1500 mg/kg ARZNAD 5,1,55
ipr-mus LD50:1100 mg/kg DIPHAH 14,21,62
scu-mus LD50:400 mg/kg 85ERAY 1,493,78
ivn-mus LD50:110 mg/kg JCINAO 28,943,49
ivn-dog LDLo:150 mg/kg JOBAAY 55,425,48
ivn-rbt LD50:117 mg/kg JCINAO 28,943,49
orl-gpg LD50:500 mg/kg FRPSAX 10,3,55
ivn-gpg LD50:560 mg/kg FRPSAX 9,21,54

CONSENSUS REPORTS: IARC Cancer Review: Group 2B IMEMDT 7,145,87; Human Limited Evidence IMEMDT 10,85,76. Reported in EPA TSCA Inventory. EPA Genetic Toxicology Program.

SAFETY PROFILE: Suspected human carcinogen producing leukemia, aplastic anemia, and other bone marrow changes. Experimental tumorigenic data. Poison by intravenous and subcutaneous routes. Moderately toxic by ingestion and intraperitoneal routes. Human systemic effects by an unknown route: changes in plasma or blood volume, unspecified liver effects, and hemorrhaging. Experimental teratogenic and reproductive effects. Human mutation data reported. An antibiotic. When heated to decomposition it emits very toxic fumes of NO_x and Cl^-. See also other chloramphenicol entries.

CDP325 CAS:134-90-7 ***HR: D***
l(+)-threo-CHLORAMPHENICOL
mf: $C_{23}H_{22}N_2O_5$ mw: 406.47

TOXICITY DATA with REFERENCE
mmo-sat 600 μmol/L BBRCA9 78,151,77
dnd-sat 5 mmol/L BBRCA9 78,151,77
dnd-esc 500 μmol/L BBRCA9 78,151,77

CONSENSUS REPORTS: EPA Genetic Toxicology Program.

SAFETY PROFILE: Mutation data reported. When heated to decomposition it emits toxic fumes of NO_x. See also other chloramphenicol entries.

CDP500 CAS:982-57-0 ***HR: 2***
CHLORAMPHENICOL MONOSUCCINATE SODIUM SALT
mf: $C_{15}H_{15}Cl_2N_2O_8 \cdot Na$ mw: 445.21

SYNS: CHLORAMPHENICOL SODIUM MONOSUCCINATE ◇ CHLORAMPHENICOL SODIUM SUCCINATE ◇ CHLORAMPHENICOL SUCCINATE SODIUM ◇ CHLORAMPHENICOL-SUKZINAT-NATRIUM (GERMAN) ◇ PROTOPHENICOL ◇ SODIUM CHLORAMPHENICOL SUCCINATE

TOXICITY DATA with REFERENCE
scu-rat TDLo:6 g/kg (female 11-14D post):TER DEGEA3 31,1181,76
ivn-rbt TDLo:1200 mg/kg (female 7-12D post):REP VHAGAS 71,623,77
ivn-inf LDLo:135 mg/kg/3D-I JOPDAB 103,485,83
ipr-rat LD50:1400 mg/kg NIIRDN 6,248,82
ivn-rat LD50:1500 mg/kg NIIRDN 6,248,82

SAFETY PROFILE: Moderately toxic by intravenous and intraperitoneal routes. An experimental teratogen.

Other experimental reproductive effects. When heated to decomposition it emits very toxic fumes of Cl^-, Na_2O, and NO_x. See also other chloramphenicol entries.

CDP700 CAS:530-43-8 ***HR: 2***
CHLORAMPHENICOL PALMITATE
mf: $C_{27}H_{42}Cl_2N_2O_6$ mw: 561.61

SYNS: CAP-P ◇ CAP-PALMITATE ◇ CHLORAMPHENICOL MONOPALMITATE ◇ DETREOPAL ◇ α-ESTER PALMITIC ACID with D-threo-(−)-2,2-DICHLORO-N-(β-HYDROXY-α-(HYDROXYMETHYL)-p-NITROPHENETHYL)ACETAMIDE

TOXICITY DATA with REFERENCE
orl-rbt TDLo:1200 mg/kg (7-12D preg):TER VHAGAS 71,623,77
orl-rbt TDLo:1200 mg/kg (7-12D preg):REP VHAGAS 71,623,77
orl-mus LD50:2640 mg/kg NIIRDN 6,248,82

SAFETY PROFILE: Moderately toxic by oral route. An experimental teratogen. Other experimental reproductive effects. An antibiotic. When heated to decomposition it emits very toxic fumes of NO_x and Cl^-. See also other chloramphenicol entries.

CDP725 CAS:3544-94-3 ***HR: 2***
CHLORAMPHENICOL SUCCINATE
mf: $C_{15}H_{16}Cl_2N_2O_8$ mw: 423.23

SYNS: CHLORAMPHENICOL ACID SUCCINATE ◇ CHLORAMPHENICOL HEMISUCCINATE ◇ CHLORAMPHENICOL HYDROGEN SUCCINATE ◇ CHLORAMPHENICOL MONOSUCCINATE ◇ CHLOROMYCETIN SUCCINATE ◇ CHRONICIN FOAM ◇ CPSA ◇ KEMICETINE SUCCINATE ◇ LEVOMYCETIN HEMISUCCINATE ◇ LEVOMYCETIN SUCCINATE ◇ PARAXIN SUCCINATE ◇ SUCCINATO de CLORANFENICOL (SPANISH) ◇ SUCCINIC ACID-α-MONOESTER with d-threo-(−)-2,2-DICHLORO-N-(β-HYDROXY-α-(HYDROXYMETHYL)-p-NITROPHENETHYL)ACETAMIDE

TOXICITY DATA with REFERENCE
scu-rat TDLo:750 mg/kg (7-21D preg):REP SCIEAS 213,238,81
orl-rat LD50:8300 mg/kg KSRNAM 4,135,70
scu-rat LD50:2400 mg/kg THERAP 22,1405,67
ivn-rat LD50:1720 mg/kg THERAP 22,1405,67
orl-mus LD50:11 g/kg KSRNAM 4,135,70
ipr-mus LD50:1400 mg/kg FRPSAX 9,21,54
scu-mus LD50:4200 mg/kg KSRNAM 4,135,70
ivn-mus LD50:1500 mg/kg FRPSAX 9,21,54
ivn-gpg LD50:1000 mg/kg FRPSAX 9,21,54

CONSENSUS REPORTS: EPA Genetic Toxicology Program.

SAFETY PROFILE: Moderately toxic by subcutaneous, intravenous, and intraperitoneal routes. Experimental reproductive effects. When heated to decomposition it emits toxic fumes of Cl^- and NO_x. See also other chloramphenicol entries.

CDP750 CAS:20856-57-9 ***HR: 3***
CHLORANIFORMETHANE
mf: $C_9H_7Cl_5N_2O$ mw: 336.43

SYNS: BAY 79770 ◇ CHLORANIFORMETHAN ◇ 1-(3,4-DICHLORANILINO)-1-FORMYLAMINO-2,2,2-TRICHLORAETHAN (GERMAN) ◇ N-FORMYL-N'-(3',4'-DICHLORPHENYL)-2,2,2-TRICHLORACETALDEHYDAM (GERMAN) ◇ IMUGAN ◇ MILFARON ◇ N-(2,2,2-TRICHLORO-1-(3,4-DICHLOROANILINO)ETHYLFORMAMIDE

TOXICITY DATA with REFERENCE
orl-rat LD50:2500 mg/kg 85ARAE 4,129,76/77
orl-gpg LD50:250 mg/kg 85DPAN -,-,71/76

SAFETY PROFILE: Poison by ingestion. A pesticide. When heated to decomposition it emits very toxic fumes of Cl^- and NO_x. See also ALDEHYDES.

CDQ000 ***HR: 3***
CHLORATES

PROP: Chlorates are a combination of a metal or hydrogen and $^-ClO_3$ monovalent radical. They are crystalline and somewhat deliquescent.

SAFETY PROFILE: The principal toxic effects of chlorates are the production of methemoglobin in the blood and destruction of red blood corpuscles. The latter may lead to irritation of the kidneys. Damage to heart muscle has been reported.

Dangerous fire hazard in contact with flammable matter. When contaminated with oxidizable materials, they are particularly sensitive to friction, heat, and shock. They are powerful oxidizing agents and can undergo violent reactions with reducing materials. Dangerous explosion hazard when shocked, exposed to heat, or rubbed, particularly when contaminated with sugar; charcoal; shellac; sulfur; starch; sawdust; sulfuric acid; ammonium compounds; cyanides; phosphorous or antimony sulfide; Al; (metals + acids); As_2S_3; CaH_2; MnO_2; metal sulfides; organic acids; powdered metals; Hg_3P_4; PHI_4; SCN; (S + Cu); Se; NaH_2PO_2; SrH; SO_2. Chlorates when mixed with combustible materials may form explosive mixtures. For instance, potassium chlorate, when mixed with sulfur or with other combustible substances explodes on friction. Pure chlorates which have been spilled on the floor, or mixed with small amounts of impurities, become very sensitive to shock and friction. Water is considered the best agent for fighting fires involving chlorates. When heated to decomposition they can emit toxic fumes of Cl^- and explode.

CDQ250 CAS:57109-90-7 ***HR: 3***
CHLORAZEPATE DIPOTASSIUM
mf: $C_{16}H_{10}ClN_2O_3 \cdot K \cdot HKO$ mw: 408.94

SYNS: AB 35616 ◇ ABBOTT-35616 ◇ AH 3232 ◇ BELSEREN ◇ BIPOTASSIUM CHLORAZEPATE ◇ CB 4306 ◇ CHLORAZEPAM ◇ CLORAZEPATE DIPOTASSIUM ◇ DIPOTASSIUM CHLORAZEPATE

◇ DIPOTASSIUM CLORAZEPATE ◇ MENDON ◇ NEVRACTEN ◇ POTASSIUM 7-CHLORO-2,3-DIHYDRO-2-OXO-5-PHENYL-1H-1,4-BENZODIAZEPINE-3-CARBOXYLATE KOH ◇ TENCILAN ◇ TRANSENE ◇ TRANSILIUM ◇ TRANXENE ◇ TRANXILEN ◇ TRANXILENE ◇ TRANXILIUM

TOXICITY DATA with REFERENCE
sln-asn 1 mg/L MUREAV 26,159,74
unr-wmn TDLo:18 mg/kg (25-39W preg):REP THERAP 36,305,81
orl-rat LD50:880 mg/kg IYKEDH 10,710,79
ipr-rat LD50:31200 μg/kg IYKEDH 10,710,79
scu-rat LD50:1478 mg/kg NIIRDN 6,245,82
ivn-rat LD50:279 mg/kg NIIRDN 6,245,82
orl-mus LD50:700 mg/kg ARZNAD 20,123,70
ipr-mus LD50:290 mg/kg ARZNAD 20,123,70
scu-mus LD50:443 mg/kg IYKEDH 10,710,79
ivn-mus LD50:157 mg/kg IYKEDH 10,710,79

SAFETY PROFILE: Poison by intravenous and intraperitoneal routes. Moderately toxic by ingestion and subcutaneous routes. Experimental reproductive effects. Mutation data reported. A tranquilizer. When heated to decomposition it emits very toxic fumes of Cl^-, NO_x and K_2O.

CDQ325 CAS:580-48-3 ***HR: 2***
CHLORAZINE
mf: $C_{11}H_{20}ClN_5$ mw: 257.81

SYNS: 2-CHLORO-4,6-BIS(DIETHYLAMINO)-s-TRIAZINE ◇ 6-CHLORO-N,N,N',N'-TETRAETHYL-1,3,5-TRIAZINE-2,4-DIAMINE

TOXICITY DATA with REFERENCE
orl-rat LD50:850 mg/kg FMCHA2 -,C50,83
unr-rat LD50:3500 mg/kg 30ZDA9 -,420,71
orl-mus LD50:743 mg/kg 85GMAT -,35,82

SAFETY PROFILE: Moderately toxic by ingestion and possibly other routes. When heated to decomposition it emits toxic fumes of Cl^- and NO_x.

CDQ500 CAS:5576-62-5 ***HR: 3***
CHLORBENZOXYETHAMINE DIHYDROCHLORIDE
mf: $C_{27}H_{31}ClN_2O•2ClH$ mw: 507.97

SYNS: ANTIULCERA MASTER ◇ CHLORBENZOSAMINE DIHYDROCHLORIDE ◇ CHLORBENZOXAMINE DIHYDROCHLORIDE ◇ GASTOMAX ◇ LIBRATAR ◇ U.C.B. 1474

TOXICITY DATA with REFERENCE
orl-rat LD50:3350 mg/kg AIPTAK 118,167,59
ivn-rat LD50:66 mg/kg AIPTAK 118,167,59
orl-mus LD50:1400 mg/kg AIPTAK 118,167,59

SAFETY PROFILE: Poison by intravenous route. Moderately toxic by ingestion. When heated to decomposition it emits very toxic fumes of NO_x and Cl^-.

CDQ750 CAS:95-25-0 ***HR: 3***
5-CHLORBENZOZAZOLIN-2-ON
mf: $C_7H_4ClNO_2$ mw: 169.57

SYNS: BIOMIORAN ◇ 5-CHLORBENZOXAZOLIN-2-ON ◇ 5-CHLOROBENZOXAZOLIDONE ◇ 5-CHLORO-2-BENZOXAZOLINONE ◇ 6-CHLORO-2-BENZOXAZOLINONE ◇ 5-CHLOROBENZOXAZOL-2-ONE ◇ 5-CHLORO-3(H)-2-BENZOXAZOLONE ◇ CHLOROXAZONE ◇ CHLORZOXAZONE ◇ MYOFLEXINE ◇ PARAFLEX ◇ SOLAXIN ◇ USAF MA-10

TOXICITY DATA with REFERENCE
orl-rat LD50:763 mg/kg JPETAB 129,75,60
ipr-rat LD50:150 mg/kg JPETAB 129,75,60
orl-mus LD50:440 mg/kg ARZNAD 17,242,67
ipr-mus LD50:50 mg/kg NTIS** AD277-689
scu-mus LD50:170 mg/kg APTOA6 19,247,62
orl-ham LD50:662 mg/kg JPETAB 129,75,60
ipr-ham LD50:166 mg/kg JPETAB 129,75,60
ipr-mam LD50:550 mg/kg CHTPBA 6,65,71

CONSENSUS REPORTS: Reported in EPA TSCA Inventory.

SAFETY PROFILE: Poison by intraperitoneal and subcutaneous routes. Moderately toxic by ingestion. A skeletal muscle relaxant. When heated to decomposition it emits very toxic fumes of Cl^- and NO_x.

CDR000 CAS:129-71-5 ***HR: 3***
CHLORCYCLIZINE DIHYDROCHLORIDE
mf: $C_{18}H_{21}ClN_2•2ClH$ mw: 373.78

SYNS: AH 289 ◇ 1-(4-CHLOROBENZHYDRYL)-4-METHYLPIPERAZINE DIHYDROCHLORIDE ◇ 1-(p-CHLORO-α-PHENYLBENZYL)-4-METHYL-PIPERAZINE DIHYDROCHLORIDE ◇ DI-PARALENE-2-HYDROCHLORIDE ◇ HISTANTINE DIHYDROCHLORIDE ◇ N-METHYL-N'-(4-CHLOROBENZHYDRYL)PIPERAZINEDIHYDROCHLORIDE ◇ PERAZIL ◇ PERAZIL DIHYDROCHLORIDE ◇ TRIHISTAN

TOXICITY DATA with REFERENCE
ipr-rat LD50:100 mg/kg AIPTAK 80,378,49
orl-mus LDLo:150 mg/kg AIPTAK 80,378,49
ipr-mus LD50:137 mg/kg JPETAB 96,388,49
scu-mus LDLo:150 mg/kg AIPTAK 80,378,49
ivn-mus LD50:35 mg/kg AIPTAK 80,378,49
ipr-dog LDLo:100 mg/kg AIPTAK 80,378,49
ipr-cat LD50:75 mg/kg AIPTAK 80,378,49
ipr-gpg LD50:100 mg/kg AIPTAK 80,378,49

SAFETY PROFILE: Poison by ingestion, subcutaneous, intravenous, and intraperitoneal routes. An antihistamine. When heated to decomposition it emits very toxic fumes of Cl^- and NO_x. See also CHLORCYCLIZINE DIHYDROCHLORIDE.

CDR250 CAS:14362-31-3 ***HR: 3***
CHLORCYCLIZINE HYDROCHLORIDE
mf: $C_{18}H_{21}ClN_2•xClH$ mw: 556.08

SYNS: AH-289 HYDROCHLORIDE ◇ CHLORCYCLIZINIUM CHLO-

RIDE ◇ 1-(p-CHLOROBENZHYDRYL)-4-METHYLPIPERAZINE HYDROCHLORIDE ◇ DIPARALENE HYDROCHLORIDE ◇ ERAMIDE

TOXICITY DATA with REFERENCE
ims-mus TDLo:200 mg/kg (female 12-13D post):TER AIPTAK 194,168,71
ims-mus TDLo:200 mg/kg (female 12-13D post):REP AIPTAK 194,168,71
ipr-rat LD50:100 mg/kg CLDND*
orl-mus LD50:300 mg/kg CLDND*
ipr-mus LD50:100 mg/kg CLDND*
scu-mus LD50:200 mg/kg CLDND*
ivn-mus LD50:35 mg/kg CLDND*
ipr-dog LD50:125 mg/kg CLDND*
ipr-cat LD50:100 mg/kg CLDND*
ipr-gpg LD50:100 mg/kg CLDND*

SAFETY PROFILE: Poison by ingestion, subcutaneous, intravenous, and intraperitoneal routes. An experimental teratogen. Other experimental reproductive effects. When heated to decomposition it emits very toxic fumes of HCl and NO_x. See also CHLORCYCLIZINE DIHYDRACHLORIDE.

CDR500 CAS:894-56-4 ***HR: 3***
CHLORCYCLIZINE HYDROCHLORIDE A
mf: $C_{18}H_{21}ClN_2 \cdot ClH$ mw: 337.32

SYNS: CHLOROCYCLIZINE HYDROCHLORIDE ◇ 1-(p-CHLORO-α-PHENYLBENZYL)-4-METHYLPIPERAZINE HYDROCHLORIDE ◇ PERAZIL

TOXICITY DATA with REFERENCE
orl-rat TDLo:650 mg/kg (10-22D preg):REP VAAZA2 39,59,82
orl-rat TDLo:500 mg/kg (6-15D preg):TER TXCYAC 8,87,77
ipr-rat LD50:100 mg/kg 27ZQAG -,213,72
orl-mus LD50:300 mg/kg 27ZQAG -,213,72
ipr-mus LD50:137 mg/kg JPETAB 96,388,49
scu-mus LD50:200 mg/kg 27ZQAG -,213,72
ivn-mus LD50:50800 μg/kg TXAPA9 1,454,56
ipr-dog LD50:125 mg/kg 27ZQAG -,213,72
ipr-cat LD50:75 mg/kg 27ZQAG -,213,72
ipr-gpg LD50:100 mg/kg 27ZQAG -,213,72

SAFETY PROFILE: Poison by ingestion, subcutaneous, intravenous, and intraperitoneal routes. An experimental teratogen. Other experimental reproductive effects. When heated to decomposition it emits very toxic fumes of HCl and NO_x.

CDR550 CAS:963-03-1 ***HR: 2***
CHLORCYCLOHEXAMIDE
mf: $C_{13}H_{17}ClN_2O_3S$ mw: 316.83

SYNS: BENZENESULFONAMIDE, 4-CHLORO-N-((CYCLOHEXYLAMINO)CARBONYL)- ◇ CHLORHEXAMIDE ◇ CHLOROCYCLAMIDE-R ◇ 1-((p-CHLOROPHENYL)SULFONYL)-3-CYCLOHEXYLUREA ◇ K 694 ◇ ORADIAN ◇ UREA, 1-((p-CHLOROPHENYL)SULFONYL)-3-CYCLOHEXYL-

TOXICITY DATA with REFERENCE
orl-rat TDLo:1 g/kg (female 10D post):TER PROEAS 14,89,68
orl-rat LD50:1525 mg/kg FATOAO 25,93,62
orl-mus LD50:1525 mg/kg FATOAO 25,93,62

SAFETY PROFILE: Moderately toxic by ingestion. Experimental teratogenic effects. When heated to decomposition it emits toxic fumes of SO_x, NO_x, and Cl^-.

CDR575 CAS:5566-34-7 ***HR: 2***
trans-CHLORDAN
mf: $C_{10}H_6Cl_8$ mw: 409.76

SYNS: γ-CHLORDAN ◇ Γ(trans)-CHLORDANE ◇ 4,7-METHANOINDAN, 2,2,4,5,6,7,8,8-OCTACHLORO-3a,4,7,7a-TETRAHYDRO- ◇ 4,7-METHANO-1H-INDENE, 2,2,4,5,6,7,8,8-OCTACHLORO-2,3,3a,4,7,7a-HEXAHYDRO- (9CI) ◇ 2,2,4,5,6,7,8,8-OCTACHLORO-3a,4,7,7a-TETRAHYDRO-4,7-METHANOINDAN

TOXICITY DATA with REFERENCE
ipr-mus TDLo:100 mg/kg (female 1D post):REP BECTA6 17,559,77
orl-rat LD50:500 mg/kg NTIS** PB85-143766

SAFETY PROFILE: Moderately toxic by ingestion. Experimental reproductive effects. When heated to decomposition it emits toxic fumes of Cl^-.

CDR675 CAS:5103-71-9 ***HR: 3***
α-CHLORDAN
mf: $C_{10}H_6Cl_8$ mw: 409.76

SYNS: cis-CHLORDAN ◇ α-CHLORDANE ◇ α(cis)-CHLORDANE ◇ cis-CHLORDANE

TOXICITY DATA with REFERENCE
ipr-mus TDLo:50 mg/kg (1D preg):REP BECTA6 17,559,77
orl-rat LD50:500 mg/kg NTIS** PB85-143766
orl-mus LD50:125 mg/kg JAFCAU 21,1113,73
ipr-mus LDLo:290 mg/kg TXAPA9 23,288,72

SAFETY PROFILE: Poison by ingestion and intraperitoneal routes. Experimental reproductive effects. When heated to decomposition it emits toxic fumes of Cl^-. See also CHLORDANE.

CDR750 CAS:57-74-9 ***HR: 3***
CHLORDANE
DOT: UN 2762
mf: $C_{10}H_6Cl_8$ mw: 409.76

PROP: Colorless to amber; odorless, viscous liquid. Bp: 175°, d: 1.57-1.63 @ 15.5°/15.5°.

SYNS: ASPON-CHLORDANE ◇ BELT ◇ CD 68 ◇ CHLOORDAAN (DUTCH) ◇ CHLORDAN ◇ Γ-CHLORDAN ◇ CHLORDANE, liquid (DOT) ◇ CHLORINDAN ◇ CHLOR KIL ◇ CHLORODANE

◇ CHLORTOX ◇ CLORDAN (ITALIAN) ◇ CORODANE ◇ CORTILANNEU ◇ DICHLOROCHLORDENE ◇ DOWCHLOR ◇ ENT 9,932 ◇ ENT 25,552-X ◇ HCS 3260 ◇ KYPCHLOR ◇ M 140 ◇ M 410 ◇ NCI-C00099 ◇ NIRAN ◇ 1,2,4,5,6,7,8,8-OCTACHLOOR-3a,4,7,7a-TETRAHYDRO-4,7-endo-METHANO-INDAAN (DUTCH) ◇ OCTACHLOR ◇ OCTACHLORODIHYDRODICYCLOPENTADIENE ◇ 1,2,4,5,6,7,8,8-OCTACHLORO-2,3,3a,4,7,7a-HEXAHYDRO-4,7-METHANOINDENE ◇ 1,2,4,5,6,7,8,8-OCTACHLORO-2,3,3a,4,7,7a-HEXAHYDRO-4,7-METHANO-1H-INDENE ◇ 1,2,4,5,6,7,8,8-OCTACHLORO-3a,4,7,7a-HEXAHYDRO-4,7-METHYLENE INDANE ◇ OCTACHLORO-4,7-METHANOHYDROINDANE ◇ OCTACHLORO-4, 7-METHANOTETRAHYDROINDANE ◇ 1,2,4,5,6,7,8,8-OCTACHLORO-4,7-METHANO-3a,4,7,7a-TETRAHYDROINDANE ◇ 1,2,4,5,6,7,8,8-OCTACHLORO-3a,4,7,7a-TETRAHYDRO-4,7-METHANOINDAN ◇ 1,2,4,5,6,7,8,8-OCTACHLORO-3a,4,7,7a-TETRAHYDRO-4,7-METHANOINDANE ◇ 1,2,4,5,6,7,10,10-OCTACHLORO-4,7,8,9-TETRAHYDRO-4,7-METHYLENEINDANE ◇ 1,2,4,5,6,7,8,8-OCTACHLOR-3a,4,7,7a-TETRAHYDRO-4,7-endo-METHANO-INDAN (GERMAN) ◇ OCTA-KLOR ◇ OKTATERR ◇ ORTHO-KLOR ◇ 1,2,4,5,6,7,8,8-OTTOCHLORO-3A,4,7,7A-TETRAIDRO-4,7-endo-METANO-INDANO(ITALIAN) ◇ RCRA WASTE NUMBER U036 ◇ SD 5532 ◇ SHELL SD-5532 ◇ SYNKLOR ◇ TAT CHLOR 4 ◇ TOPICHLOR 20 ◇ TOPICLOR ◇ TOPICLOR 20 ◇ TOXICHLOR ◇ VELSICOL 1068

TOXICITY DATA with REFERENCE

sce-ofs-mul 54 pmol/L MUREAV 118,61,83
sce-hmn:lym 10 μmol/L ARTODN 52,221,83
orl-mus TDLo:7 mg/kg (15-21D preg):REP BJPCBM 49,311,73
orl-mus TDLo:3360 μg/kg (1-21D preg):TER JEPTDQ 2(2),357,78
orl-mus TDLo:2020 mg/kg/80W-C:CAR NCITR* NCI-CG-TR-8,77
orl-mus TD:3780 mg/kg/80W-C:CAR NCITR* NIC-CG-TR-8,77
orl-man TDLo:3071 μg/kg JTCTDW 20,291,83
orl-hmn LDLo:29 mg/kg:LIV CMEP** -,1,56
orl-wmn LDLo:120 μg/kg:CNS,GIT CMEP** -,1,56
skn-hmn LDLo:428 mg/kg:CNS 34ZIAG -,648,69
unr-man LDLo:118 mg/kg 85DCAI 2,73,70
orl-rat LD50:200 mg/kg ARZNAD 17,614,67
skn-rat LD50:690 mg/kg JAVMA4 157,1835,70
ipr-rat LD50:343 mg/kg TXAPA9 32,443,75
orl-mus LD50:145 mg/kg ARSIM* 20,19,66
ipr-mus LDLo:240 mg/kg TXAPA9 23,288,72
ivn-mus LD50:100 mg/kg CSLNX* NX#04876
ihl-cat LC50:100 mg/m^3/4H GTPZAB 8(4),30,64
orl-rbt LD50:100 mg/kg PCOC** -,226,66
skn-rbt LD50:780 mg/kg 85DPAN -,-,71/76
ivn-rbt LDLo:10 mg/kg AIHOAX 1,13,50

CONSENSUS REPORTS: IARC Cancer Review: Group 3 IMEMDT 7,146,87; Human Inadequate Evidence IMEMDT 20,45,79; Animal Sufficient Evidence IMEMDT 20,45,79. NCI Carcinogenesis Bioassay (feed); Clear Evidence: mouse NCITR* NCI-CG-TR-8,77; No Evidence: rat NCITR* NCI-CG-TR-8,77. EPA Genetic Toxicology Program. Community Right-To-Know List. EPA Extremely Hazardous Substances List.

OSHA PEL: TWA 0.5 mg/m^3 (skin)
ACGIH TLV: TWA 0.5 mg/m^3 (skin)
DFG MAK: Suspected Carcinogen.
DOT Classification: Combustible Liquid; Label: None; Flammable Liquid; Label: Flammable Liquid.

SAFETY PROFILE: Suspected carcinogen with experimental carcinogenic data. Poison to humans by ingestion and possibly other routes. An experimental poison by ingestion, inhalation, intravenous, and intraperitoneal routes. Moderately toxic by skin contact. Human systemic effects by ingestion or skin contact: tremors, convulsions, excitement, ataxia (loss of muscle coordination), and gastritis. Experimental teratogenic and reproductive effects. Human mutation data reported. Combustible liquid. It is no longer permitted for use as a termiticide in homes.

A central nervous system stimulant whose exact mode of action is unknown, but it may involve microsomal enzyme stimulation. Animals poisoned by this and related compounds show an extremely marked loss of appetite and neurological symptoms. The fatal dose to humans is unknown. It has been estimated to be between 6 to 60 grams (0.2 to 2 ounces). One person receiving an accidental skin application of 25% solution (amounting to something over 30 grams of technical chlordane) developed symptoms within about 40 minutes and died, apparently of respiratory failure, before medical attention was obtained. In two patients, death followed exposure to low ingestion doses of chlordane (2-4 grams). On microscopic examination, both patients showed severe chronic fatty degeneration of the liver, characteristic of chronic alcoholism. Although these two fatalities cannot be attributed exclusively to chlordane, they are entirely consistent with previous observations that the toxicity of other chlorinated hydrocarbons is much enhanced in the presence of chronic liver damage. The dangerous chronic dose in humans is unknown.

Experimental animals exposed to repeated small doses exhibit hyperexcitability, tremors, and convulsions, and those which survive long enough show marked anorexia and loss of weight. Symptoms in animals frequently occur within an hour of the administration of a large dose, but death often is delayed for several days depending on the dosage and route of administration. In any event, symptoms are of longer duration with chlordane than with DDT under similar conditions.

Laboratory analyses on poisoned animals are essentially normal, except that the insecticide is found in tissues by means of bioassay. A method for specific, quantitative chemical analysis for chlordane is now available using small amounts of subcutaneous fat. Chronically poisoned animals show degenerative changes in the liver and kidney tubules.

When heated to decomposition it emits toxic fumes of Cl^-.

CDS000 CAS:115-28-6 ***HR: 1***
CHLORENDIC ACID
mf: $C_9H_4Cl_6O_4$ mw:388.83

SYNS: 1,4,5,6,7,7-HEXACHLORO-5-NORBORNENE-2,3-DICARBOXYLIC ACID ◇ KYSELINA 3,6-ENDOMETHYLEN-3,4,5,6,7,7-HEXACHLOR-Δ^4-TETRAHYDROFTALOVA (CZECH) ◇ KYSELINA HET (CZECH) ◇ NCI-C55072

TOXICITY DATA with REFERENCE
skn-rbt 500 mg/24H MLD 28ZPAK -,92,72
eye-rbt 250 μg/24H SEV 28ZPAK -,92,72
msc-mus:lyms 1700 mg/L NTPTR* NTP-TR-304,87
orl-rat TDLo:45063 mg/kg/2Y-C:CAR NTPTR* NTP-TR-304,87
orl-mus TDLo:108 g/kg/2Y-C:CAR NTPTR* NTP-TR-304,87

CONSENSUS REPORTS: NTP Fifth Annual Report on Carcinogens. NTP Carcinogenesis Studies (feed): Clear Evidence: mouse,rat NTPTR* NTP-TR-304,87. Reported in EPA TSCA Inventory.

SAFETY PROFILE: Confirmed carcinogen with experimental carcinogenic data. A severe eye and mild skin irritant. When heated to decomposition it emits toxic fumes of Cl^-.

CDS100 CAS:6889-41-4 ***HR: 3***
CHLORENDIC IMIDE
mf: $C_9H_3Cl_6NO_2$ mw: 369.83

SYNS: 1,4,5,6,7,7-HEXACHLOROBICYCLO(2.2.1)HEPTENE-2,3-DICARBOXIMIDE ◇ 1,4,5,6,7,7-HEXACHLORO-5-NORBORNENE-2,3-DICARBOXIMIDE

TOXICITY DATA with REFERENCE
orl-rat LD50:2300 mg/kg GTPZAB 26(4),49,82
orl-mus LD50:2400 mg/kg GTPZAB 26(4),49,82
ivn-mus LD50:320 mg/kg CSLNX* NX#00789

SAFETY PROFILE: Poison by intravenous route. Moderately toxic by ingestion. When heated to decomposition it emits toxic fumes of Cl^- and NO_x.

CDS125 CAS:16672-87-0 ***HR: 2***
CHLORETHEPHON
mf: $C_2H_6ClO_3P$ mw: 144.50

PROP: Very hygroscopic needles from benzene. Mp: 74-75°. Freely sol in water, methanol, acetone, ethylene glycol, propylene glycol; sltly sol in benzene, toluene; practically insol in petr ether.

SYNS: AMCHEM 68-250 ◇ BROMOFLOR ◇ CAMPOSAN ◇ CEP ◇ 2-CEPA ◇ CEPHA ◇ CEPHA 10LS ◇ 2-CHLORAETHYL-PHOSPHONSAEURE (GERMAN) ◇ 2-CHLORETHYLPHOSPHONIC ACID ◇ 2-CHLOROETHANEPHOSPHONIC ACID ◇ ETHEFON ◇ ETHEL ◇ ETHEPHON ◇ ETHEVERSE ◇ ETHREL ◇ FLORDIMEX ◇ FLOREL ◇ G 996 ◇ KAMPOSAN ◇ ROLL-FRUCT ◇ TOMATHREL

TOXICITY DATA with REFERENCE
orl-rat LD50:3400 mg/kg ZKMAAX 20,274,80
orl-mus LD50:2850 mg/kg PHARAT 32,181,77
orl-rbt LD50:5000 mg/kg GISAAA 48(8),79,83
skn-rbt LD50:5730 mg/kg 85DPAN -,-,71/76
orl-gpg LD50:4200 mg/kg GISAAA 48(8),79,83
unr-mam LD50:4200 mg/kg 30ZDA9 -,384,71

CONSENSUS REPORTS: EPA Genetic Toxicology Program.

SAFETY PROFILE: Moderately toxic by ingestion. Mildly toxic by skin contact. A plant growth regulator. Caution: Spray formulations are quite acidic, about pH 1.0. May be irritating to exposed skin and eyes, or if inhaled. When heated to decomposition it emits toxic fumes of Cl^- and PO_x.

CDS250 CAS:132-89-8 ***HR: 2***
CHLORETHYLBENZMETHOXAZONE
mf: $C_{10}H_{10}ClNO_2$ mw: 211.66

SYNS: 2-(2-CHLOROETHYL)-3-AZA-4-CHROMANONE ◇ 2-(2-CHLOROETHYL)-2,3-DIHYDRO-4H-1,3-BENZOXAZIN-4-ONE ◇ 2-(2-CHLOROETHYL)-2,3-DIHYDRO-4-OXO-1,3-BENZOXAZINE ◇ 2-(β-CHLOROETHYL)-2,3-DIHYDRO-4-OXO(BENZO-1,3-OXAZINE) ◇ 4-OXO-2-(β-CHLOROETHYL)-2,3-DIHYDROBENZO-1,3-OXAZINE

TOXICITY DATA with REFERENCE
ipr-rat LD50:730 mg/kg BMJOAE 1,36,60
orl-rat LD50:10 g/kg ARZNAD 7,651,57

SAFETY PROFILE: Moderately toxic by intraperitoneal route. Mildly toxic by ingestion. When heated to decomposition it emits very toxic fumes of Cl^- and NO_x.

CDS275 CAS:21267-72-1 ***HR: 3***
CHLORETIN
mf: $C_{12}H_{12}ClNO$ mw: 221.70

SYNS: 2903 H ◇ 2-CHLORO-N-(1-METHYL-2-PROPYNYL)-ACETANILIDE (8CI) ◇ 2-CHLORO-N-(1-METHYL-2-PROPYNYL)-N-PHENYLACETAMIDE ◇ BASAMAIZE ◇ BAS-290-H ◇ BAS 2900H ◇ BAS 2903H ◇ BUTISAN ◇ BUTISANE ◇ CHLORESSIGSAEURE-N-ISOBUTINYLANILID (GERMAN) ◇ 2-CHLORO-N-(1-METHYL-2-PROPYNYL)ACETANILIDE ◇ PRYNACHLOR

TOXICITY DATA with REFERENCE
orl-rat LD50:1170 mg/kg GUCHAZ 6,440,73
orl-mus LD50:150 mg/kg GUCHAZ 6,440,73
skn-rbt LD50:1926 mg/kg GUCHAZ 6,440,73

CONSENSUS REPORTS: EPA Genetic Toxicology Program.

SAFETY PROFILE: Poison by ingestion. Moderately toxic by skin contact. When heated to decomposition it emits toxic fumes of Cl^- and NO_x.

CDS500 CAS:2274-74-0 ***HR: 2***
CHLORFENSULFIDE
mf: $C_{12}H_6Cl_4N_2S$ mw: 352.06

SYNS: CHLORFENSULFID (GERMAN) ◇ 4-CHLOROPHENYL-2,4,5-TRICHLOROPHENYLAZOSULFIDE ◇ 4-CHLORPHENYL-2',4',5'-TRICHLORPHENYLAZOSULFID (GERMAN) ◇ CPAS ◇ MICASIN ◇ 2,4,5-TRICHLOROBENZENEDIAZO p-CHLOROPHENYL SULFIDE ◇ 2,4,5-TRICHLOROPHENYLAZO-4'-CHLOROPHENYL-SULFIDE

TOXICITY DATA with REFERENCE
orl-rat LD50:4000 mg/kg 85DPAN -,-,71/76
unr-mus LD50:3000 mg/kg 30ZDA9 -,281,71

SAFETY PROFILE: Moderately toxic by ingestion and possibly other routes. See also SULFIDES. When heated to decomposition it emits very toxic fumes of Cl^-, and NO_x, and SO_x.

CDS750 CAS:470-90-6 ***HR: 3***
CHLORFENVINFOS
mf: $C_{12}H_{14}Cl_3O_4P$ mw: 359.58

SYNS: APACHLOR ◇ BIRLANE ◇ C-10015 ◇ CFV ◇ CGA 26351 ◇ CHLOFENVINPHOS ◇ O-2-CHLOOR-1-(2,4-DICHLOOR-FENYL)-VINYL-O,O-DIETHYLFOSFAAT (DUTCH) ◇ O-2-CHLOR-1-(2,4-DICHLOR-PHENYL)-VINYL-O,O-DIAETHYLPHOSPHAT(GERMAN) ◇ CHLORFENVINFOS ◇ CHLORFENVINPHOS ◇ 2-CHLORO-1-(2,4-DICHLOROPHENYL)VINYL DIETHYL PHOSPHATE ◇ β-2-CHLORO-1-(2',4'-DICHLOROPHENYL) VINYL DIETHYLPHOSPHATE ◇ CHLOROFENVINPHOS ◇ CHLORPHENVINFOS ◇ CHLORPHENVINPHOS ◇ O-2-CLORO-1-(2,4-DICLORO-FENIL)-VINYL-O,O-DIETILFOSFATO (ITALIAN) ◇ COMPOUND 4072 ◇ CVP ◇ DERMATON ◇ O,O-DIAETHYL-O-1-(4,5-DICHLORPHENYL)-2-CHLOR-VINYL-PHOSPHAT (GERMAN) ◇ 2,4-DICHLORO-α-(CHLOROMETHYLENE)BENZYL ALCOHOL DIETHYL PHOSPHATE ◇ O,O-DIETHYL-O-(2-CHLORO-1-(2',4'-DICHLOROPHENYL)VINYL) PHOSPHATE ◇ ENT 24,969 ◇ GC 4072 ◇ OMS 1328 ◇ PHOSPHATE de O,O-DIETHYLE et de O-2-CHLORO-1-(2,4-DICHLOROPHENYL) VINYLE (FRENCH) ◇ SAPECRON ◇ SHELL 4072 ◇ STELADONE ◇ SUPONA ◇ SUPONE ◇ UNITOX ◇ VINYLPHATE

TOXICITY DATA with REFERENCE
mmo-sat 500 μg/plate MUREAV 116,185,83
skn-hmn TDLo:10 mg/kg:BLD,BIO IMSUAI 38,25,69
orl-rat LD50:10 mg/kg FMCHA2 -,C52,83
skn-rat LD50:26400 μg/kg APYPAY 32,507,81
ipr-rat LD50:8500 μg/kg VETRAX 77,1140,65
scu-rat LD50:7 mg/kg APYPAY 32,507,81
ivn-rat LD50:6600 μg/kg TXAPA9 17,323,70
orl-mus LD50:117 mg/kg GUCHAZ 6,100,73
ipr-mus LD50:87 mg/kg JPPMAB 19,612,67
scu-mus LD50:339 mg/kg JPPMAB 19,612,67
ivn-mus LD50:87 mg/kg JPPMAB 19,612,67
orl-dog LD50:1200 mg/kg 85GYAZ -,18,71
ivn-dog LD50:51 mg/kg TXAPA9 17,323,70

CONSENSUS REPORTS: EPA Extremely Hazardous Substances List.

SAFETY PROFILE: Poison by ingestion, skin contact, intraperitoneal, subcutaneous, and intravenous routes. Human systemic effects by skin contact: unspecified blood system effects. Mutation data reported. A cholinesterase inhibitor. An insecticide. See also PARATHION. When heated to decomposition it emits very toxic fumes of Cl^- and PO_x.

CDT000 CAS:2536-31-4 ***HR: 2***
CHLORFLURENOL METHYL ESTER
mf: $C_{15}H_{11}ClO_3$ mw: 274.71

SYNS: CF 125 ◇ CHLORFLURECOL ◇ CHLORFLURECOL-METHYL ◇ CHLORFLURECOL-METHYL ESTER ◇ CHLORFLURENOL ◇ 2-CHLOR-9-HYDROXYFLUOREN-CARBONSAEURE-(9)-METHYLESTER (GERMAN) ◇ CHLOROFLURENOL-METHYL ESTER ◇ 2-CHLORO-9-HYDROXY-9-METHYLCARBOXYLATEFLUORENE ◇ CURBISET ◇ IT 3456 ◇ MAINTAIN A ◇ MAINTAIN CF125 ◇ METHYL-2-CHLORO-9-HYDROXYFLUORENE-9-CARBOXYLATE ◇ MORPHACTIN ◇ MULTIPROP

TOXICITY DATA with REFERENCE
orl-rat LD50:3100 mg/kg GUCHAZ 6,102,73

CONSENSUS REPORTS: Reported in EPA TSCA Inventory.

SAFETY PROFILE: Moderately toxic by ingestion. An herbicide. When heated to decomposition it emits very toxic fumes of Cl^-. See also ESTERS.

CDT125 CAS:56-95-1 ***HR: 3***
CHLORHEXIDINE DIACETATE
mf: $C_{22}H_{30}Cl_2N_{10}{\bullet}2C_2H_4O_2$ mw: 625.64

SYNS: 1,6-BIS(5-(p-CHLOROPHENYL)BIGUANIDINO)HEXANE DIACETATE ◇ CHLORHEXIDINE ACETATE ◇ 10,040 DIACETATE ◇ 1,6-DI(4'-CHLOROPHENYLDIGUANIDINO)HEXANE DIACETATE ◇ 1,1'-HEXAMETHYLENEBIS(5-(p-CHLOROPHENYL)BIGUANIDE DIACETATE ◇ HIBITANE DIACETATE

TOXICITY DATA with REFERENCE
skn-rbt 500 mg/24H MLD JACTDZ 4(6),309,85
orl-mus LD50:2 g/kg BJPCAL 9,192,54
ipr-mus LD50:38 mg/kg BJPCAL 9,192,54
scu-mus LD50:325 mg/kg BJPCAL 9,192,54
ivn-mus LD50:25 mg/kg BJPCAL 9,192,54

SAFETY PROFILE: Poison by subcutaneous, intravenous and intraperitoneal routes. Moderately toxic by ingestion. A skin irritant. When heated to decomposition it emits toxic fumes of Cl^- and NO_x.

CDT250 CAS:18472-51-0 ***HR: 3***
CHLORHEXIDINE DIGLUCONATE
mf: $C_{22}H_{30}Cl_2N_{10}{\bullet}2C_6H_{12}O_7$ mw: 897.88

SYNS: 1,6-BIS(5-(p-CHLOROPHENYL)BIGUANDINO)HEXANE DIGLUCONATE ◇ 1,1'-HEXAMETHYLENEBIS(5-(p-CHLOROPHENYL)BIGUANIDEDIGLUCONATE

TOXICITY DATA with REFERENCE
orl-mus LD50:1800 mg/kg JPRSB6 8(suppl.12),55,73
ivn-mus LD50:22 mg/kg JPRSB6 8(suppl.12),55,73

CONSENSUS REPORTS: Reported in EPA TSCA Inventory.

SAFETY PROFILE: Poison by intravenous route. Moderately toxic by ingestion. When heated to decomposition it emits very toxic fumes of Cl^- and NO_x.

CDT500 CAS:14007-07-9 ***HR: 3***
CHLORHEXIDINE GLUCONATE
mf: $C_{22}H_{30}Cl_2N_{10}{\cdot}7C_6H_{12}O_7$ mw: 1878.78

SYN: CHLORHEXIDIN GLUKONATU (CZECH)

TOXICITY DATA with REFERENCE
orl-mus LD50:1800 mg/kg CESTAT 74(5),392,74
ivn-mus LD50:22 mg/kg CESTAT 74(5),392,74

SAFETY PROFILE: Poison by intravenous route. Moderately toxic by ingestion. When heated to decomposition it emits very toxic fumes of NO_x and Cl^-.

CDT750 CAS:96-24-2 ***HR: 3***
CHLORHYDRIN
DOT: UN 2689
mf: $C_3H_7ClO_2$ mw: 110.55

$ClCH_2CHOHCH_2$

PROP: Colorless liquid. Bp: 213° decomp, d: 1.326.

SYNS: α-CHLORHYDRIN ◇ CHLORODEOXYGLYCEROL ◇ 1-CHLORO-2,3-DIHYDROXYPROPANE ◇ 3-CHLORO-1,2-DIHYDROXYPROPANE ◇ α-CHLOROHYDRIN ◇ 1-CHLOROPROPANE-2,3-DIOL ◇ 1-CHLORO-2,3-PROPANEDIOL ◇ 3-CHLOROPROPANE-1,2-DIOL ◇ 3-CHLORO-1,2-PROPANEDIOL ◇ 3-CHLOROPROPYLENE GYLCOL ◇ β,β′-DIHYDROXYISOPROPYL CHLORIDE ◇ 2,3-DIHYDROXYPROPYL CHLORIDE ◇ EPIBLOC ◇ GLYCERIN-α-MONOCHLORHYDRIN ◇ GLYCEROL CHLOROHYDRIN ◇ GLYCEROL-α-CHLOROHYDRIN ◇ GLYCEROL-α-MONOCHLOROHYDRIN (DOT) ◇ GLYCERYL-α-CHLOROHYDRIN ◇ MONOCHLORHYDRIN ◇ MONOCHLOROHYDRIN ◇ α-MONOCHLOROHYDRIN ◇ U-5897

TOXICITY DATA with REFERENCE
eye-rbt 27 mg JRPFA4 24,267,71
mmo-ssp 100 mmol/L MUREAV 118,213,83
mma-ssp 300 mmol/L MUREAV 118,213,83
spm-rat-orl 600 mg/kg/24D-C CUSCAM 44,193,75
msc-mus:lym 10 mmol/L PAACA3 21,74,80
orl-mky TDLo:1260 mg/kg (male 42D pre):REP JRPFA4 21,275,70
orl-rat TDLo:34580 mg/kg/72W-C:ETA JJIND8 67,75,81
orl-rat TDLo:26 mg/kg IPCLBZ 24,20,82
ihl-rat LCLo:125 ppm/4H JIHTAB 31,343,49
ipr-rat LDLo:10 mg/kg NCNSA6 5,9,53
orl-mus LD50:160 mg/kg AMIHAB 14,250,56
ipr-mus LD50:73 mg/kg JMCMAR 18,116,75
orl-bwd LD50:23700 μg/kg AECTCV 12,355,83

CONSENSUS REPORTS: Reported in EPA TSCA Inventory. EPA Genetic Toxicology Program.

DOT Classification: Poison B; Label: St. Andrews Cross.

SAFETY PROFILE: Poison by ingestion and intraperitoneal routes. Moderately toxic by inhalation. Experimental reproductive effects. An eye irritant. Questionable carcinogen with experimental tumorigenic data. Mutation data reported. A chemosterilant for rodents. Combustible when exposed to heat or flame. Reaction with perchloric acid forms a sensitive explosive product more powerful than glyceryl nitrate. When heated to decomposition it emits toxic fumes of Cl^-.

CDU000 CAS:7790-93-4 ***HR: 3***
CHLORIC ACID
DOT: UN 2626/NA 2626
mf: $ClHO_3$ mw: 84.46

PROP: Colorless solution. Mp: $< -20°$, bp: decomp @ 40°, d: 1.282 @ 14.2°.

SYN: CHLORIC ACID, solution, containing not more than 10% acid (DOT)

CONSENSUS REPORTS: Reported in EPA TSCA Inventory.

DOT Classification: Oxidizer; Label: Oxidizer and Poison (NA2626); Oxidizer; Label: Oxidizer (UN2626)

SAFETY PROFILE: A poison. A strong irritant by ingestion and inhalation. Dangerous fire hazard; ignites organic matter upon contact. A very powerful oxidizing agent. Violent or explosive reaction with oxidizable materials. Aqueous solutions decompose explosively during evaporation. Solutions greater than 40% are unstable. Reacts violently with NH_3; Sb; Sb_2S_3; As_2S_3; Bi; CuS; PHI_4; SnS_2; SnS. Reaction with cellulose causes ignition after a delay period. Dangerous reaction with metal sulfides and metal chlorides (e.g., incandescent reaction with antimony trisulfide; arsenic trisulfide; tin(II)sulfide; tin(IV) sulfide; explosion on contact with copper sulfide). Reaction with metals (e.g., antimony; bismuth; iron) forms explosive products. When heated to decomposition it emits toxic fumes of Cl^-. See also CHLORATES and CHLORINE.

CDU250 ***HR: D***
CHLORIDES

SAFETY PROFILE: Varies widely. Sodium chloride (table salt) has very low toxicity, while carbonyl chloride (phosgene) is lethal in small doses. Therefore, see specific entries. When heated to decomposition or on contact with acids or acid fumes, they evolve highly toxic chloride fumes. Some organic chlorides decompose to yield phosgene.

CDU325 ***HR: 3***
CHLORIERTES CAMPHEN

PROP: Consists of a mixture of chlorinated camphene with 67-69% chlorine (85GYAZ -,50,71).

TOXICITY DATA with REFERENCE
orl-rat LD50:60 mg/kg 85GYAZ -,50,71
skn-rat LD50:1 g/kg 85GYAZ -,50,71
skn-rbt LD50:250 mg/kg 85GYAZ -,50,71

SAFETY PROFILE: Poison by ingestion and skin contact. When heated to decomposition it emits toxic fumes of Cl^-. See also CAMPHENE.

CDU750 CAS:303-49-1 ***HR: 3***
CHLORIMIPRAMINE
mf: $C_{19}H_{23}ClN_2$ mw: 314.89

SYNS: ANAFRANIL ◇ 3-CHLORO-5-(3-(DIMETHYLAMINO)PROPYL)-10,11-DIHYDRO-5H-DIBENZ(b,f)AZEPINE ◇ 3-CHLOROIMIPRAMINE ◇ CIM ◇ CLOMIPRAMINE ◇ MONOCHLORIMIPRAMINE

TOXICITY DATA with REFERENCE
ivn-wmn TDLo:50 mg/kg (38-47D preg):TER TJADAB 24(1),42A,81
orl-rat TDLo:420 mg/kg (lactating female 21D post):REP PSCHDL 56,93,78
orl-man TDLo:357 μg/kg:GIT JCPYDR 2,215,82
orl-wmn TDLo:10 mg/kg/5D-I:CVS BJMOAE 1,406,71
ivn-wmn TDLo:3400 μg/kg/47M-I:CNS,CVS BMJOAE 3,698,72
orl-rat LD50:613 mg/kg TXCYAC 24,335,82
ipr-rat LD50:149 mg/kg TXCYAC 24,335,82
orl-mus LD50:380 mg/kg GWXXBX #2618152
ipr-mus LD50:150 mg/kg JMCMAR 21,448,78
ivn-mus LD50:27 mg/kg APSXAS 13,485,76

SAFETY PROFILE: Poison by ingestion, intraperitoneal, and intravenous routes. Human systemic effects by ingestion and intravenous routes: convulsions, heart damage, and blood pressure increase, nausea or vomiting. Human teratogenic effects on the cardiovascular system of the fetus. Experimental reproductive effects. When heated to decomposition it emits very toxic fumes of Cl^- and NO_x.

CDV000 CAS:17321-77-6 ***HR: 3***
CHLORIMIPRAMINE HYDROCHLORIDE
mf: $C_{19}H_{23}ClN_2•ClH$ mw: 351.35

SYNS: ANAFRANIL ◇ ANAPHRANIL ◇ 3-CHLORO-10,11-DIHYDRO-N,N-DIMETHYL-5H-DIBENZ(b,f)AZEPINE-5-PROPANAMINEMONOHYDROCHLORIDE ◇ 3-CHLORO-5-(3-(DIMETHYLAMINO)PROPYL)-10,11-DIHYDRO-5H-DIBENZ(b,f)AZEPINEMONOHYDROCHLORIDE ◇ 3-CHLOROIMIPRAMINE HYDROCHLORIDE ◇ CHLOROIMIPRAMINE MONOHYDROCHLORIDE ◇ CLOMIPRAMINE HYDROCHLORIDE ◇ G 34586

TOXICITY DATA with REFERENCE
sln-dmg-orl 200 mg SOGEBZ 7,1042,71
scu-mus TDLo:110 mg/kg (9D preg):REP DGDFA5 22,61,80
orl-wmn TDLo:360 mg/kg/13W-I:BPR JCLPDE 46,290,85
orl-rat LD50:1150 mg/kg NIIRDN 6,243,82
ipr-rat LD50:135 mg/kg NIIRDN 6,243,82
scu-rat LD50:1750 mg/kg NIIRDN 6,243,82
ivn-rat LD50:26 mg/kg NIIRDN 6,243,82
orl-mus LD50:470 mg/kg NIIRDN 6,243,82
ipr-mus LD50:90 mg/kg NIIRDN 6,243,82
scu-mus LD50:400 mg/kg KSRNAM 4,2105,70
ivn-mus LD50:26 mg/kg NIIRDN 6,243,82
orl-dog LD50:383 mg/kg KSRNAM 4,2105,70
ivn-dog LD50:32 mg/kg KSRNAM 4,2105,70
orl-rbt LD50:792 mg/kg KSRNAM 4,2105,70
scu-rbt LD50:217 mg/kg KSRNAM 4,2105,70
ivn-rbt LD50:12 mg/kg KSRNAM 4,2105,70

SAFETY PROFILE: Poison by ingestion, subcutaneous, intravenous, and intraperitoneal routes. Human systemic effects by ingestion: pulse rate increase, no fall in blood pressure. Experimental reproductive effects. Mutation data reported. When heated to decomposition it emits very toxic fumes of Cl^- and NO_x. See also CHLORIMIPRAMINE.

CDV100 CAS:8001-35-2 ***HR: 3***
CHLORINATED CAMPHENE
DOT: NA 2761
mf: $C_{10}H_{10}Cl_8$ mw: 413.80

PROP: Yellow, waxy solid; pleasant piney odor. Mp: 65-90°. Almost insol in water; very sol in aromatic hydrocarbons.

SYNS: AGRICIDE MAGGOT KILLER (F) ◇ ALLTEX ◇ ALLTOX ◇ ATTAC 6 ◇ ATTAC 6-3 ◇ CAMPHECHLOR ◇ CAMPHOCHLOR ◇ CAMPHOCLOR ◇ CAMPHOFENE HUILEUX ◇ CHEM-PHENE ◇ CHLOROCAMPHENE ◇ CLOR CHEM T-590 ◇ COMPOUND 3956 ◇ CRESTOXO ◇ CRISTOXO 90 ◇ ENT 9,735 ◇ ESTONOX ◇ FASCO-TERPENE ◇ GENIPHENE ◇ GY-PHENE ◇ HERCULES 3956 ◇ HERCULES TOXAPHENE ◇ KAMFOCHLOR ◇ M 5055 ◇ MELIPAX ◇ MOTOX ◇ NCI-C00259 ◇ OCTACHLOROCAMPHENE ◇ PCC ◇ PENPHENE ◇ PHENACIDE ◇ PHENATOX ◇ POLYCHLORCAMPHENE ◇ POLYCHLORINATED CAMPHENES ◇ POLYCHLOROCAMPHENE ◇ RCRA WASTE NUMBER P123 ◇ STROBANE-T-90 ◇ SYNTHETIC 3956 ◇ TOXADUST ◇ TOXAFEEN (DUTCH) ◇ TOXAKIL ◇ TOXAPHEN (GERMAN) ◇ TOXAPHENE ◇ TOXON 63 ◇ TOXYPHEN ◇ VERTAC 90% ◇ VERTAC TOXAPHENE 90

TOXICITY DATA with REFERENCE
skn-mam 500 mg MOD JAMAAP 149,1135,52
mmo-sat 100 μg/plate ENMUDM 8(Suppl 7),1,86
mma-sat 500 μg/plate SCIEAS 205,591,79
sce-hmn:lym 10 μmol/L ARTODN 52,221,83
otr-mus:emb 12400 μg/L PMRSDJ 5,659,85

orl-rat TDLo:900 μg/kg (female 5-22D post):REP AECTCV 9,247,80
orl-rat TDLo:250 mg/kg (female 7-16D post):TER BECTA6 15,660,76
orl-rat TDLo:30 g/kg/80W-C:ETA NCITR* NCI-CG-TR-37,79
orl-mus TDLo:6600 mg/kg/80W-C:CAR NCITR* NCI-CG-TR-37,79
orl-mus TD:13 g/kg/80W-C:CAR NCITR* NCI-CG-TR-37,79
orl-hmn LDLo:28 mg/kg:CNS 34ZIAG -,598,69
orl-man LDLo:29 mg/kg CMEP** -,-,56
skn-hmn TDLo:657 mg/kg:SKN CMEP** -,1,56
unr-man LDLo:44 mg/kg 85DCAI 2,73,70
orl-rat LD50:50 mg/kg ARZNAD 17,614,67
skn-rat LD50:600 mg/kg SPEADM 74-1,-,74
ipr-rat LDLo:70 mg/kg BECTA6 19,47,78
unr-rat LD50:240 mg/kg CYGEDX 8(1),23,74
orl-mus LD50:112 mg/kg SPEADM 74-1,-,74
ihl-mus LCLo:2000 mg/m^3/2H JAMAAP 149,1135,52
ipr-mus LD50:47 mg/kg JAFCAU 25,1394,77
unr-mus LD50:45 mg/kg CYGEDX 8(1),23,74
orl-dog LD50:15 mg/kg SPEADM 74-1,-,74
orl-rbt LDLo:780 mg/kg GUCHAZ 6,507,73
skn-rbt LD50:1025 mg/kg JEENAI 46,702,53
orl-gpg LD50:250 mg/kg 85DPAN -,-,71/76
orl-ham LD50:200 mg/kg TXAPA9 48,A192,79
orl-dck LD50:31 mg/kg DOEAAH 35,25,79

CONSENSUS REPORTS: NTP Fifth Annual Report on Carcinogens. IARC Cancer Review: Group 2B IMEMDT 7,56,87; Human Limited Evidence IMEMDT 20,327,79; Animal Sufficient Evidence IMEMDT 20,327,79. NCI Carcinogenesis Bioassay (feed); Clear Evidence: mouse, rat NCITR* NCI-CG-TR-37,79.

OSHA PEL: (Transitional: TWA 0.5 mg/m^3 (skin)) TWA 0.5 mg/m^3; STEL 1 mg/m^3 (skin)
ACGIH TLV: TWA 0.5 mg/m^3; STEL 1 mg/m^3 (skin)
DFG MAK: 0.5 mg/m^3
DOT Classification: ORM-A; Label: None.

SAFETY PROFILE: Confirmed carcinogen with experimental carcinogenic and tumorigenic data. Human poison by ingestion and possibly other routes. Experimental poison by ingestion, intraperitoneal, and possibly other routes. Moderately toxic experimentally by inhalation and skin contact. May be a human carcinogenic. Human systemic effects by ingestion and skin contact: somnolence, convulsions or effect on seizure threshold, coma and allergic skin dermatitis. A skin irritant; absorbed through the skin. Experimental teratogenic and reproductive effects. Human mutation data reported. Liver injury has been reported. Lethal amounts of toxaphene can enter the body through the mouth, lungs, and skin. Systemic absorption of the insecticide is increased by the presence of digestible oils, and liquid preparations of the insecticide which penetrate the skin more readily than do dusts and wettable powders.

A toxic mixture of organochlorine pesticides stored to some extent in body fat. It resembles chlordane and, to some extent, camphor in its physiological action. It causes diffuse stimulation of the brain and spinal cord resulting in generalized convulsions of a tonic or clonic character. Death usually results from respiratory failure. Detoxification appears to occur in the liver. The lethal ingestion dose for man is estimated to be 2-7 grams, a toxicity of about four times that of DDT. At least seven human deaths have been reported due to toxaphene, all in children. Two families have been made ill by eating vegetables containing a large residue of toxaphene. When heated to decomposition it emits toxic fumes of Cl^-.

CDV125 ***HR: 3***
CHLORINATED DIBENZO DIOXINS

PROP: See individual entries for physical properties.

SYNS: 1-CHLORODIBENZO-p-DIOXIN ◇ DIBENZO-p-DIOXIN ◇ 1,2,3,8-TETRACHLORODIBENZO-p-DIOXIN ◇ 1,2,4-TRICHLORO DIBENZO-p-DIOXIN

SAFETY PROFILE: The chlorinated dibenzo dioxins are not manufactured on a commercial basis, but some are present as impurities in herbicide and fungicide formulations, such as 2,4,5-T, the pentachlorophenols, and hexachlorphene (from trichlorophenol). The chlorinated dibenzo dioxins include some with anitbacterial, flameproofing, insecticidal, and fungicidal actions. Their acute toxicity ranges from moderate to high and some are carcinogens, mutagens, and teratogens. They tend to accumulate in living organisms. When heated to decomposition they emit toxic fumes of Cl^-. See also SPECIFIC COMPOUNDS.

CDV175 CAS:31242-93-0 ***HR: 3***
CHLORINATED DIPHENYL OXIDE
mf: $C_{12}H_4Cl_6O$ mw: 376.86

PROP: Light yellow, very viscous liquid. Bp: 230-260° @ 8 mm, d: 1.60 @ 20°/60°, autoign temp: 1148°F, vap d: 13.0.

SYNS: BENZENE, 1,1'-OXYBIS-, HEXACHLORO derivatives (9CI) ◇ ETHER, HEXACHLOROPHENYL ◇ HEXACHLORODIPHENYL ETHER ◇ HEXACHLORO DIPHENYL OXIDE ◇ PHENYL ETHER, HEXACHLORO deriv. (8CI) ◇ TRICHLORO DIPHENYL ETHER ◇ TRICHLORO DIPHENYL OXIDE

TOXICITY DATA with REFERENCE
orl-gpg LDLo:50 mg/kg 14CYAT 2,1707,63

OSHA PEL: TWA 0.5 mg/m^3
ACGIH TLV: TWA 0.5 mg/m^3
DFG MAK: 0.5 mg/m^3

SAFETY PROFILE: Poison by ingestion and probably by inhalation. Combustible when exposed to heat, flame, or oxidizing materials. To fight fire, use water spray, fog, foam, dry chemical, CO_2. When heated to decomposition it emits toxic fumes of Cl^-. See also ETHERS and ALDRIN (a closely related compound).

CDV250 ***HR: 2***
CHLORINATED HYDROCARBONS, ALIPHATIC

SYN: ALIPHATIC CHLORINATED HYDROCARBONS ◇ CHLORINATED HC, ALIPHATIC

SAFETY PROFILE: Suspected carcinogen with experimental tumors of the liver, lung, skin, and blood-forming tissues. The substitution of a chlorine (or other halogen) atom for a hydrogen greatly increases the anesthetic action of the aliphatic hydrocarbons and increases the range of their systemic effects. In many cases, the chlorine derivative is quite toxic. In general, the unsaturated chlorine derivatives are more narcotic but less toxic than the saturated derivatives. In the saturated group, the narcotic effect is proportional to the number of chlorine atoms. This relationship is not true for toxicity.

In dealing with these chlorinated hydrocarbons, it must be remembered that a toxic action may result from repeated exposure to concentrations which are too low to produce a narcotic effect, and which, consequently, are too low to give warning of danger. Individual susceptibility varies widely. Certain workmen may be seriously affected by concentrations that seem to have no effect on fellow employees in the same exposure.

In general reactivity decreases with greater substitution of halogen for hydrogen atoms. Halogenated (e.g., fluorine, chlorine, or bromine containing) acetylene compounds are unstable and should be treated as explosives. Lightly substituted haloalkanes are highly flammable and can react with divalent light metals to form dangerously reactive products. Lightly substituted haloalkenes are highly flammable, peroxidizable, and may polymerize violently. When heated to decomposition they emit highly toxic fumes of phosgene. They may react violently with Al, liquid O_2, K, and Na.

CDV500 ***HR: 3***
CHLORINATED HYDROCARBONS, AROMATIC

SYN: CHLORINATED HC AROMATIC

SAFETY PROFILE: In most instances, it is difficult to predict the toxicity of these compounds. However, in the case of most aromatic chlorine compounds, their toxicity is usually no greater, and frequently is less, than that of the corresponding aromatic hydrocarbons with the notable exception of naphthalene and the various biphenyls. They can react with oxidizing materials. React violently with Al, liquid O_2, K, or Na. When heated to decomposition it emits toxic fumes of Cl^-.

CDV575 ***HR: 3***
CHLORINATED NAPHTHALENES

SAFETY PROFILE: Questionable carcinogens whith can cause tumors of the liver. Severe irritants by ingestion, inhalation, and skin contact. The action of the chlorinated naphthalenes on the body is quite similar to that of the chlorinated biphenyls. The chief effects being the production of chloracne of the skin, and systemically an acute yellow atrophy of the liver. When heated to decomposition they emit toxic fumes of Cl^-.

CDV625 CAS:56641-03-3 ***HR: 3***
CHLORINATED POLYETHER POLYURETHAN

PROP: Polymer formed from toluene diisocyanate and 1,4-butanediol and cured with 4,4′-methylenebis(o-chloroaniline) (CNREA8 36,3973,76).
mf: $(C_{13}H_{12}Cl_2N_2{\cdot}C_9H_6N_2O_2{\cdot}(C_4H_8O)_nH_2O)_x$

SYNS: OSTAMER ◇ POLYURETHANE Y-238 ◇ Y-238

TOXICITY DATA with REFERENCE
imp-rat TDLo:20 mg/kg:ETA CNREA8 36,3973,76

CONSENSUS REPORTS: IARC Cancer Review: Animal Sufficient Evidence IMEMDT 19,303,79.

SAFETY PROFILE: Confirmed carcinogen with experimental tumorigenic data. When heated to decomposition it emits toxic fumes of Cl^- and NO_x.

CDV700 CAS:145-94-8 ***HR: 3***
CHLORINDANOL
mf: C_9H_9ClO mw: 168.63

PROP: Needles from petr ether. Mp: 91-93°.

SYNS: CLORINDANOL ◇ 7-CHLORO-4-INDANOL ◇ 2,3-DIHYDRO-7-CHLORO-1H-INDEN-4-OL (9CI) ◇ LANESTA

TOXICITY DATA with REFERENCE
orl-mus LD50:920 mg/kg JAPMA8 48,212,59
ipr-mus LD50:96 mg/kg JAPMA8 48,212,59
ivn-mus LD50:48 mg/kg JAPMA8 48,212,59

SAFETY PROFILE: Poison by intravenous and intraperitoneal routes. Moderately toxic by ingestion. When heated to decomposition it emits toxic fumes of Cl^-.

CDV750 CAS:7782-50-5 ***HR: 3***
CHLORINE
DOT: UN 1017
mf: Cl_2 mw: 70.90

PROP: Greenish-yellow gas, liquid, or rhombic crystals. Mp: −101°, bp: −34.5°, d: (liquid) 1.47 @ 0° (3.65

atm), vap press: 4800 mm @ 20°, vap d: 2.49. Sol in water.

SYNS: BERTHOLITE ◇ CHLOOR (DUTCH) ◇ CHLOR (GERMAN) ◇ CHLORE (FRENCH) ◇ CHLORINE MOL. ◇ CLORO (ITALIAN) ◇ MOLECULAR CHLORINE

TOXICITY DATA with REFERENCE
cyt-hmn:lym 20 ppm CBINA8 6,375,73
spm-mus-orl 20 mg/kg/5D-C ENMUDM 7,201,85
ihl-hmn LCLo:2530 mg/m^3/30M:PUL 28ZOAH -,150,37
ihl-hmn LCLo:500 ppm/5M TABIA2 3,231,33
ihl-rat LC50:293 ppm/1H NTIS** PB214-270
ihl-mus LC50:137 ppm/1H NTIS** PB214-270
ihl-dog LCLo:800 ppm/30M JPETAB 14,65,19
ihl-cat LCLo:660 ppm/4H AHYGAJ 7,233,1887
ihl-rbt LDLo:660 ppm/4H AHYGAJ 7,233,1887

CONSENSUS REPORTS: Reported in EPA TSCA Inventory. Community Right-To-Know List. EPA Extremely Hazardous Substances List.

OSHA PEL: (Transitional: TWA CL 1 ppm) TWA 0.5 ppm; STEL 1 ppm
ACGIH TLV: TWA 0.5 ppm; STEL 1 ppm
DFG MAK: 0.5 ppm (1.5 mg/m^3)
NIOSH REL: (Chlorine) CL 0.5 ppm/15M
DOT Classification: Nonflammable Gas; Label: Nonflammable Gas and Poison; Poison A; Label: Poison Gas.

SAFETY PROFILE: Moderately toxic to humans by inhalation. Very irritating by inhalation. Human mutation data reported. Human respiratory system effects by inhalation: changes in the trachea or bronchi, emphysema, chronic pulmonary edema or congestion. A strong irritant to eyes and mucous membranes.

Chlorine is extremely irritating to the mucous membranes of the eyes and the respiratory tract at 3 ppm. Combines with moisture to liberate O_2 and forms HCl. Both these substances, if present in quantity, cause inflammation of the tissues with which they come in contact. A concentration of 3.5 ppm produces a detectable odor; 15 ppm causes immediate irritation of the throat. Concentrations of 50 ppm are dangerous for even short exposures; 1000 ppm may be fatal, even when exposure is brief. Because of its intensely irritating properties, severe industrial exposure seldom occurs, as the worker is forced to leave the exposure area before he can be seriously affected. In cases where this is impossible, the initial irritation of the eyes and mucous membranes of the nose and throat is followed by coughing, a feeling of suffocation, and later, pain and a feeling of constriction in the chest. If exposure has been severe, pulmonary edema may follow with rales being heard over the chest. It is a common air contaminant.

Explodes on contact with acetylene + heat or UV light, air + ethylene, molten aluminum, ammonia, amidosulfuric acid, antimony trichloride + tetramethyl silane (at 100°), benzene + light, biuret, bromine pentafluoride + heat, tert-butanol, butyl rubber + naphtha, carbon disulfide + iron catalyst, chlorinated pyridine + iron powder, 3-chloropropyne, cobalt(II) chloride + methanol, diborane, dibutyl phthalate (at 118°), dichloro (methyl)arsine (in a sealed container), diethyl ether, dimethyl phosphoramidiate, dioxygen difluoride, disilyl oxide, 4,4′-dithiodimorpholine, ethane over activated carbon (at 350°), fluorine + sparks, gasoline, glycerol (above 70° in a sealed container), hexachlorodisilane (above 300°), hydrocarbon oils or waxes, iron(III) chloride + monomers (e.g., styrene), methane over mercury oxide, methanol, methanol + tetrapyridine cobalt(II) chloride, naphtha + sodium hydroxide, nitrogen triiodide, oxygen difluoride, white phosphorus (in liquid Cl_2), phosphorus compounds, polypropylene + zinc oxide, propane (at 300°), silicones when heated in a sealed container [e.g., polydimethyl siloxane (above 88°), polymethyl trifluoropropylsiloxane (above 68°)], stibine, synthetic rubber (in liquid Cl_2), tetraselenium tetranitride, trimethyl thionophosphate. Explosive products are formed on reaction with alkylthiouronium salts, amidosulfuric acid, acidic ammonium chloride solutions, aziridine, bis(2,4-dinitrophenyl) disulfide, cyanuric acid, phenyl magnesium bromide. Mixtures with ethylene are explosives initiated by light, heat, or by the presence or mercury, mercury oxide, silver oxide, lead oxide (at 100°). Mixtures with hydrogen are explosives initiated by sparks, light, heating to over 280°, or the presence of yellow mercuric oxide or nitrogen trichloride. Mixtures with hydrogen and other gases (e.g., air, hydrogen chloride, oxygen) are also explosive.

Ignition or explosive reaction with metals (e.g., aluminum, antimony powder, bismuth powder, brass, calcium powder, copper, germanium, iron, manganese, potassium, tin, vanadium powder). Reaction with some metals requires moist Cl_2 or heat. Ignites with diethyl zinc (on contact), polyisobutylene (at 130°), metal acetylides, metal carbides, metal hydrides (e.g., potassium hydride, sodium hydride, copper hydride), metal phosphides (e.g., copper(II) phosphide), methane + oxygen, hydrazine, hydroxylamine, calcium nitride, non-metals (e.g., boron, active carbon, silicon, phosphorus), non-metal hydrides (e.g., arsine, phosphine, silane), steel (above 200° or as low as 50° when impurities are present), sulfides (e.g., arsenic disulfide, boron trisulfide, mercuric sulfide), trialkyl boranes.

Violent reaction with alcohols, N-aryl sulfinamides, dimethyl formamide, polychlorobiphenyl, sodium hydroxide, hydrochloric acid + dinitroanilines. Incandescent reaction when warmed with cesium oxide (above 150°), tellurium, arsenic, tungsten dioxide. Potentially dangerous reaction with hydrocarbons + Lewis acids releases toxic and reactive HCl gas.

Can react to cause fires or explosions upon contact with turpentine, illuminating gas, polypropylene, rubber, sulfamic acid, $As_2(CH_3)_4$, UC_2, acetaldehyde, alcohols, alkylisothiourea salts, alkyl phosphines, Al, Sb, As, AsS_2, AsH_3, Ba_3P_2, C_6H_6, Bi, B, BPI_2, B_2S_3, brass BrF_5, Ca, (CaC_2 + KOH), $Ca(ClO_2)_2$, Ca_3N_2 Ca_3P_2, C, CS_2, Cs, $CsHC_2$, Co_2O, Cs_3N, (C + $Cr(OCl)_2$), CuH_2, CuC_2, dialklyl phosphines, diborane, dibutyl phthalate, $Zn(C_2H_5)_2$, C_2H_6, C_2H_4, ethylene imine, $C_2H_5PH_2$, F_2, Ge, glycerol, $(NH_2)_2$, (H_2O + KOH), I_2, hydroxylamine, Fe, FeC_2, Li, Li_2C_2, Li_6C_2, Mg, Mg_2P_3, Mn, Mn_3P_2, HgO, HgS, Hg, Hg_3P_2, CH_4, Nb, NI_3, OF_2, H_2SiO, (OF_2 + Cu), PH_3, P, $P(SNC)_3$, P_2O_3, PCB's, K, KHC_2, KH, Ru, $RuHC_2$, Si, SiH_2, Ag_2O, Na, $NaHC_2$, Na_2C_2, SnF_2, SbH_3, Sr_3P, Te, Th, Sn, WO_2, U, V, Zn, ZrC_2.

CDW000 CAS:13973-88-1 ***HR: 3***
CHLORINE AZIDE
mf: ClN_3 mw: 77.47

PROP: An explosive gas.

SYN: CHLOR(O)AZIDE

DOT Classification: Forbidden

SAFETY PROFILE: Strong irritant by inhalation. An extremely unstable explosive. Reacts with liquid ammonia to form an explosive liquid. Explosive reaction with 1,3-butadiene, C_2H_6, C_2H_4, CH_4, C_3H_8, phophorus, silver azide, sodium. Reacts with water or steam to produce toxic and corrosive fumes of HCl. Has been used as an initiator in chemical gas lasers. When heated to decomposition it emits toxic fumes of Cl^- and NO_x. See also CHLORINE and AZIDES.

CDW450 CAS:10049-04-4 ***HR: 3***
CHLORINE DIOXIDE
mf: ClO_2 mw: 67.45

PROP: Red-yellow gas or orange-red crystals. Mp: −59°, bp: 9.9° @ 731 mm (explodes), d: 3.09 g/L @ 11°.

SYNS: ALCIDE ◇ ANTHIUM DIOXCIDE ◇ CHLORINE DIOXIDE, not hydrated (DOT) ◇ CHLORINE OXIDE ◇ CHLORINE(IV) OXIDE ◇ CHLORINE PEROXIDE ◇ CHLOROPEROXYL ◇ CHLORYL RADICAL ◇ DOXCIDE 50

TOXICITY DATA with REFERENCE
eye-rbt 100 mg MLD JJATDK 2,160,82
mma-sat 400 μg/plate FCTOD7 22,623,84
orl-rat TDLo:570 mg/kg (female 14D pre-21D post):REP NTIS** PB85-245983
orl-rat LD50:292 mg/kg JJATDK 2,160,82
ihl-rat LCLo:500 ppm/15M TXAPA9 27,527,74
unr-rat LD50:140 mg/kg GISAAA 36(11),18,71

CONSENSUS REPORTS: Reported in EPA TSCA Inventory. Community Right-To-Know List.

OSHA PEL: (Transitional: TWA 0.1 ppm) TWA 0.1 ppm; STEL 0.3 ppm
ACGIH TLV: TWA 0.1 ppm; STEL 0.3 ppm
DFG MAK: 0.1 ppm (0.3 mg/m^3)
DOT Classification: Forbidden (not hydrated); Oxidizer; Label: Oxidizer and Poison (hydrated, frozen).

SAFETY PROFILE: Moderately toxic by inhalation. Experimental reproductive effects. Mutation data reported. An eye irritant. A powerful explosive sensitive to spark, impact, sunlight, or heating rapidly to 100°C. A powerful oxidizer. Concentrations of greater than 10% in air are explosive. Explodes on mixing with carbon monoxide, hydrocarbons (e.g., butadiene, ethane, ethylene, methane, propane), fluoramines (e.g., difluoramine, trifluoramine). Mixtures with hydrogen explode with sparking or or contact with platinum. Explodes on contact with mercury, potassium hydroxide, phosphorus pentachloride + chlorine. Ignites or explodes on contact with non-metals (e.g., phosphorus, sulfur, sugar). Reacts violently with F_2, NHF_2. Reacts with water or steam to produce toxic and corrosive fumes of HCl. When heated to decomposition it emits toxic fumes of Cl^-. See also CHLORINE.

CDW500 CAS:12133-60-7 ***HR: 3***
CHLORINE DIOXYGEN TRIFLUORIDE
mf: ClF_3O_2 mw: 124.45

PROP: Powerful oxidant, low stability.

SAFETY PROFILE: A poison. A very powerful oxidant. Explosive reaction with organic materials. When heated to decomposition it emits toxic fumes of F^- and Cl^-. See also CHLORINE and FLUORIDES.

CDX000 CAS:14545-72-3 ***HR: 3***
CHLORINE NITRATE
mf: $ClNO_3$ mw: 97.46

SYN: NITRYL HYPOCHLORITE

SAFETY PROFILE: Explosive reaction with metals, metal chlorides, alcohols, ethers, and most organic materials. When heated to decomposition it emits toxic fumes of Cl^- and NO_x. See also CHLORINE and NITRATES.

CDX250 CAS:13637-63-3 ***HR: 3***
CHLORINE PENTAFLUORIDE
mf: ClF_5 mw: 130.45
DOT: UN 2548

SYNS: CHLORINE FLUORIDE (ClF_5) ◇ CHLORINE PENTAFLUORIDE (DOT)

TOXICITY DATA with REFERENCE
ihl-rat LC50:122 ppm/1H AIHAAP 33,661,72
ihl-mus LC50:57 ppm/1H AIHAAP 33,661,72

ihl-dog LC50:122 ppm/1H AIHAAP 33,661,72
ihl-mky LC50:173 ppm/1H AIHAAP 33,661,72

OSHA PEL: TWA 2.5 mg(F)/m^3
ACGIH TLV: TWA 2.5 mg(F)/m^3
NIOSH REL: (Inorganic Fluorides) TWA 2.5 mg(F)/m^3
DOT Classification: Poison A; Label: Poison Gas, Oxidizer, Corrosive.

SAFETY PROFILE: Poison by inhalation. A corrosive material. Vigorous reaction in contact with water or anhydrous nitric acid. Violent reaction on contact with metals. When heated to decomposition it emits very toxic fumes of Cl^- and F^-. See also CHLORINE, FLUORINE, FLUORIDES, and CHLORINE. TRIFLUORIDE.

CDX500 CAS:27218-16-2 ***HR: 3***
CHLORINE PERCHLORATE
mf: Cl_2O_4 mw: 134.91

SAFETY PROFILE: A shock-sensitive explosive. Potentially explosive reaction with chlorotrifluoroethylene, perfluoroalkyl iodides (e.g., perfluoromethyl iodide, 1,2-diiodoperfluoroethane, 1,3-diiodoperfluoropropane). When heated to decomposition it emits toxic fumes of Cl^-. See also CHLORINE and PERCHLORATES.

CDX750 CAS:7790-91-2 ***HR: 3***
CHLORINE TRIFLUORIDE
DOT: UN 1749
mf: ClF_3 mw: 92.45

PROP: Colorless gas to yellow liquid, sweet odor, mp: −83°, bp: 11.8°, d: 1.77 @ 13°.

SYNS: CHLORINE FLUORIDE ◇ CHLOROTRIFLUORIDE ◇ TRIFLUORURE de CHLORE (FRENCH)

TOXICITY DATA with REFERENCE
eye-rat 21 ppm/12H-I AMIHAB 12,515,55
eye-dog 21 ppm/12H-I AMIHAB 12,515,55
ihl-hmn LCLo:50 ppm 34ZIAG -,651,66
ihl-rat LCLo:400 ppm/30M TXAPA9 27,527,74
ihl-mus LC50:178 ppm/1H AMRL** TR-70-55/70
ibl-mky LC50:230 ppm/1H AMRL** TR-70-77/70

CONSENSUS REPORTS: Reported in EPA TSCA Inventory.

OSHA PEL: CL 0.1 ppm
ACGIH TLV: CL 0.1 ppm
DFG MAK: 0.1 ppm (0.4 mg/m^3)
DOT Classification: Oxidizer; Label: Oxidizer and Poison; Poison A; Label: Poison Gas, Oxidizer, Corrosive.

SAFETY PROFILE: Human poison by inhalation. An eye irritant. See also FLUORIDES; CHLORINE; and FLUORINE. Spontaneously flammable. A powerful oxidant which may react violently with oxidizable materials. A rocket propellant.

Explosive reaction with water, bis(trifluoromethyl)sulfide or -disulfide, polychlorotrifluoroethylene, trifluoromethanesulfenyl chloride, and other hydrogen containing materials (e.g., ammonia, coal gas, hydrogen, hydrogen sulfide, methane, acetic acid, benzene, ether, cotton, paper, wood). Forms shock-sensitive explosive mixtures with highly chlorinated compounds (e.g., carbon tetrachloride), nitroaryl compounds (e.g., trinitrotoluene, hexanitrobiphenyl, hexanitrodiphenyl amine, hexanitrodiphenyl sulfide, hexanitrodiphenyl ether). Reaction with ammonium fluoride or ammonium hydrogen fluoride forms explosive gaseous products.

Ignition on contact with boron-containing materials, iodine, finely divided refractory materials (e.g., asbestos, glass wool, sand, tungsten carbide), fluorinated polymers (with flowing trifluoride).

Violent reaction with acids (e.g., nitric or sulfuric), chromium trioxide, ruthenium, selenium tetrafluoride (above 106°C), metals, metal oxides, metal salts, nonmetals, non-metal salts, organic matter, glass wool, acetic acid, Al, Sb, As, Cu, Ir, Fe, Pb, Mg, Mo, Os, P, K, Rh, Se, Si, Ag, Na, S, Te, Sn, W, Zn, oxides, CO, graphite, HgI_2, HNO_3, K_2CO_3, KI, rubber, AgN_3, $AgNO_3$, NaOH, V_2P_5, WO_3. Incompatible with fuels, nitro compounds. When heated to decomposition or in reaction with water or steam it emits toxic fumes of F^- and Cl^-.

CDX800 CAS:65597-24-2 ***HR: 3***
CHLORINE(1)TRIFLUOROMETHANESULFONATE
mf: $CClF_3O_3S$ mw: 184.52

SAFETY PROFILE: Explodes on contact with easily oxidizable materials (e.g., organic materials). When heated to decomposition it emits toxic fumes of F^-, Cl^- and SO_x. See also SULFONATES.

CDY000 CAS:69-27-2 ***HR: 3***
CHLORISONDAMINE CHLORIDE
mf: $C_{14}H_{20}Cl_4N_2{\cdot}2Cl$ mw: 429.06

SYNS: CHLORISONDAMINE ◇ CHLORISONDAMINE DIMETHOCHLORIDE ◇ N-((2-DIMETHYLAMMONIUM)ETHYL)-4,5,6,7-TETRACHLOROISOINDOLINIUM DIMETHOCHLORIDE ◇ ECOLID ◇ ECOLID CHLORIDE ◇ HISINDAMONE A ◇ ISOINDOLINE, 2-(2-DIMETHYLAMINOETHYL)-4,5,6,7-TETRACHLORODIMETHOCHLORIDE ◇ SU 3088 ◇ 4,5,6,7-TETRACHLORO-2-(2-DIMETHYLAMINOETHYL)-ISOINDOLINE DIMETHOCHLORIDE

TOXICITY DATA with REFERENCE
orl-rat LD50:300 mg/kg SKNEA7 10,15,60
ivn-rat LD50:28 mg/kg JPETAB 115,172,55
orl-mus LD50:380 mg/kg BCFAAI 103,490,64
ipr-mus LD50:62 mg/kg AIPTAK 155,69,65

scu-mus LD50:240 mg/kg FRPSAX 20,482,65
ivn-mus LD50:28 mg/kg CSLNX* NX#00058

SAFETY PROFILE: Poison by ingestion, subcutaneous, intravenous and intraperitoneal routes. When heated to decomposition it emits very toxic fumes of Cl^-, NH_3 and NO_x.

CDY100 CAS:2281-78-9 ***HR: 3***
CHLORISOPROPAMIDE
mf: $C_{10}H_{13}ClN_2O_3S$ mw: 276.76

SYNS: BENZENESULFONAMIDE,4-CHLORO-N-(((1-METHYLETHYL)AMINO)CARBONYL)- (9CI) ◇ UREA, 1-((p-CHLOROPHENYL) SULFONYL)-3-ISOPROPYL-

TOXICITY DATA with REFERENCE
orl-rat TDLo:1 g/kg (female 10D post):TER PROEAS 14,89,68
ipr-mus LD50:153 mg/kg FRZKAP (6),26,87

SAFETY PROFILE: Poison by intraperitoneal route. An experimental teratogen. When heated to decomposition it emits toxic fumes of SO_x, NO_x, and Cl^-.

CDY250 ***HR: 3***
CHLORITES

SAFETY PROFILE: Many chlorite salts are heat- and impact-sensitive explosives. The metal salts are powerful oxidants. They are much less stable than the analogous chlorates. React violently with NH_3, organic matter, or metals. See individual chlorites.

CDY275 CAS:1961-77-9 ***HR: 2***
CHLORMADINON
mf: $C_{21}H_{27}ClO_3$ mw: 362.93

SYNS: CHLORMADINONE ◇ 6-CHLORO-17-HYDROXYPREGNA-4,6-DIENE-3,20-DIONE

TOXICITY DATA with REFERENCE
orl-wmn TDLo:200 μg/kg (20D pre):REP FESTAS 16,158,65
scu-rbt TDLo:300 mg/kg (female 1-3D post):TER FESTAS 20,211,69

SAFETY PROFILE: Human reproductive effects by ingestion: changes in the uterus, cervix, vagina, and female fertility. An experimental teratogen. Other experimental reproductive effects. When heated to decomposition it emits toxic fumes of Cl^-.

CDY299 CAS:24934-91-6 ***HR: 3***
CHLORMEPHOS
mf: $C_5H_{12}ClO_2PS_2$ mw: 234.70

PROP: Bp: 81-85°C @ 0.1 mm Hg, d: 1.260, vap. press: 0.0056 (mm Hg) @ 30°C, sltly water sol (60 mg/L @ 20C).

SYNS: DOTAN ◇ MC 2188 ◇ S-(CHLOROMETHYL)-O,O-DIETHYL PHOSPHORODITHIOATE ◇ S-CHLOROMETHYL-O,O-DIETHYL PHOSPHOROTHIOLOTHIOATE ◇ S-CHLOROMETHYL-O,O-DIETHYL PHOSPHOROTHIOLOTHIONATE ◇ S-(CHLOROMETHYL)-O,O-DIMETHYL PHOSPHORODITHIOIC ACID, ESTER

TOXICITY DATA with REFERENCE
orl-rat LD50:7 mg/kg FMCHA2 -,C53,83
skn-rat LD50:27 mg/kg GUCHZA 6,103,73

CONSENSUS REPORTS: EPA Extremely Hazardous Substances List.

SAFETY PROFILE: Poison by inhalation, ingestion and skin contact. Symptoms may include nausea, vomiting, abdominal cramps, diarrhea, excessive salivation, headache, giddiness, weakness, muscle twitching, difficult breathing, blurring or dimness of vision, and loss of muscle coordination. Death may occur from failure of the respiratory center, paralysis of the respiratory muscles, or intense bronchoconstriction. An organophosphorus pesticide. Combustible. For small fires, use dry chemical, carbon dioxide, water spray, or foam. For large fires, use water spray, fog, or foam. When heated to decomposition it emits toxic fumes of Cl^-, PO_x, and SO_x.

CDY325 CAS:3689-76-7 ***HR: 3***
CHLORMIDAZOLE
mf: $C_{15}H_{13}ClN_2$ mw: 256.75

PROP: Bp: 240-242°.

SYNS: 1-p-CHLORBENZYL-2-METHYL-BENZIMIDAZOL(GERMAN) ◇ 1-((4-CHLOROPHENYL)METHYL)-2-METHYL-1H-BENZIMIDAZOLE (9CI) ◇ CLOMIDAZOLE ◇ DIAMYCELINE ◇ FUNGO-POLYCID ◇ FUTRICAN ◇ H115 ◇ MYCO-POLYCID

TOXICITY DATA with REFERENCE
orl-rat LD50:2200 mg/kg ARZNAD 9,489,59
ipr-rat LD50:58 mg/kg ARZNAD 9,489,59
orl-mus LD50:200 mg/kg ARZNAD 9,489,59
ipr-mus LD50:90 mg/kg ARZNAD 9,489,59
scu-mus LD50:250 mg/kg ARZNAD 9,489,59

SAFETY PROFILE: Poison by ingestion, subcutaneous, and intraperitoneal routes. When heated to decomposition it emits toxic fumes of Cl^- and NO_x.

CDY500 CAS:107-20-0 ***HR: 3***
CHLOROACETALDEHYDE
DOT: UN 2232
mf: C_2H_3ClO mw: 78.50

PROP: Clear, colorless liquid; pungent odor. Bp: 90.0°-100.1° (40% soln), fp: −16.3° (40% soln), flash p: 190°F, d: 1.19 @ 25°/25° (40% soln), vap press: 100 mm @ 45° (40% soln).

SYNS: 2-CHLOROACETALDEHYDE ◇ CHLOROACETALDEHYDE MONOMER ◇ 2-CHLORO-1-ETHANAL ◇ MONOCHLOROACETALDEHYDE ◇ RCRA WASTE NUMBER P023

TOXICITY DATA with REFERENCE
mmo-omi 250 μL/plate CBINA8 30,9,80
mmo-omi 1 mmol/L MUREAV 73,1,80
mmo-asn 30 μL/plate CBINA8 30,9,80
dnd-sat:tes 3 g CRNGDP 3,663,82
orl-rat LD50:75 mg/kg JPMSAE 61,19,72
ipr-rat LD50:6 mg/kg JPMSAE 61,19,72
orl-mus LD50:69 mg/kg JPMSAE 61,19,72
ipr-mus LD50:6 mg/kg JPMSAE 61,19,72
skn-rbt LD50:224 mg/kg JPMSAE 61,19,72
ipr-rbt LD50:4640 mg/kg JPMSAE 61,19,72
ipr-gpg LD50:2 mg/kg JPMSAE 61,19,72

CONSENSUS REPORTS: Reported in EPA TSCA Inventory. EPA Genetic Toxicology Program.

OSHA PEL: CL 1 ppm
ACGIH TLV: CL 1 ppm
DFG MAK: 1 ppm (3 mg/m^3)
DOT Classification: Poison B; Label: Poison.

SAFETY PROFILE: Poison by ingestion, skin contact, and intraperitoneal routes. Mutation data reported. Combustible when exposed to heat or flame. Reacts with oxidizing materials. To fight fire, use water, foam, CO_2, dry chemical. When heated to decomposition it emits toxic fumes of Cl^-. See also ALDEHYDES and CHLORIDES.

CDY825 CAS:598-49-2 ***HR: 3***
N-CHLOROACETAMIDE
mf: C_2H_4ClNO mw: 93.51

SAFETY PROFILE: Solutions of the acetamide may explode during drying or concentration operations. When heated to decomposition it emits toxic fumes of Cl^- and NO_x. See also 2-CHLORO ACETAMIDE.

CDY850 CAS:79-07-2 ***HR: 3***
2-CHLORO ACETAMIDE
mf: C_2H_4ClNO mw: 93.52

PROP: Crystals, moderately water-sol, mp: 120°, bp: 225° (decomp).

SYNS: CHLORACETAMID (GERMAN) ◇ CHLOROACETAMIDE ◇ α-CHLOROACETAMIDE ◇ 2-CHLOROETHANAMIDE ◇ USAF DO-29

TOXICITY DATA with REFERENCE
orl-rat LD50:70 mg/kg ARZNAD 19,1073,69
ipr-mus LD50:100 mg/kg NTIS** AD277-689
ivn-mus LD50:180 mg/kg CSLNX* NX#03764

CONSENSUS REPORTS: Reported in EPA TSCA Inventory.

SAFETY PROFILE: Poison by ingestion, intravenous, and intraperitoneal routes. When heated to decomposition it emits very toxic Cl^- and NO_x. See also N-CHLOROACETAMIDE.

CDZ000 CAS:3272-96-6 ***HR: 3***
CHLOROACETAMIDE OXIME
mf: $C_2H_5ClN_2O$ mw: 108.53

SAFETY PROFILE: An unstable heat-sensitive explosive. When heated to decomposition it emits toxic fumes of Cl^- and NO_x. See also EXPLOSIVES.

CDZ100 CAS:539-03-7 ***HR: 3***
4'-CHLOROACETANILIDE
mf: C_8H_8ClNO mw: 169.62

SYNS: ACETIC-4-CHLOROANILIDE ◇ N-(4-CHLOROPHENYL)ACETAMIDE

TOXICITY DATA with REFERENCE
mmo-sat 2500 μg/plate PCBPBS 10,174,79
ipr-rat LD50:245 mg/kg JAPMA8 48,204,59
ipr-mus LD50:730 mg/kg 49RQAC 1,140,82

CONSENSUS REPORTS: Reported in EPA TSCA Inventory.

SAFETY PROFILE: Poison by intraperitoneal route. Mutation data reported. When heated to decomposition it emits toxic fumes of Cl^- and NO_x.

CEA000 CAS:79-11-8 ***HR: 3***
CHLOROACETIC ACID
mf: $C_2H_3ClO_2$ mw: 94.50
DOT: UN 1750/UN 1751

PROP: Colorless crystals. Mp: (α) 63°, (β) 56°, (τ) 50°, bp: 189°, flash p: 259°F, d: 1.58 @ 20°/20°, vap d: 3.26.

SYNS: ACIDE CHLORACETIQUE (FRENCH) ◇ ACIDE MONOCHLORACETIQUE (FRENCH) ◇ ACIDOMONOCLOROACETICO (ITALIAN) ◇ CHLORACETIC ACID ◇ α-CHLOROACETIC ACID ◇ CHLOROACETIC ACID, liquid (DOT) ◇ CHLOROACETIC ACID, solid (DOT) ◇ CHLOROETHANOIC ACID ◇ MCA ◇ MONOCHLOORAZIJNZUUR (DUTCH) ◇ MONOCHLORACETIC ACID ◇ MONOCHLORESSIGSAEURE (GERMAN) ◇ MONOCHLOROACETIC ACID ◇ MONOCHLORO-7ETHANOIC ACID ◇ NCI-C60231

TOXICITY DATA with REFERENCE
mma-mus:lym 548 mg/L MUREAV 97,49,82
scu-mus TDLo:100 mg/kg:ETA NTIS** PB223-159
ihl-rat LC50:180 mg/m^3 GTPZAB 18(9),32,74
scu-rat LD50:5 mg/kg TXAPA9 22,303,72
ivn-rat LD50:55 mg/kg GTPZAB 18(9),32,74
orl-mus LD50:165 mg/kg JPETAB 86,336,46
scu-mus LD50:250 mg/kg AIPTAK 116,154,58

CONSENSUS REPORTS: Reported in EPA TSCA Inventory. EPA Genetic Toxicology Program. EPA Extremely Hazardous Substances List. Community Right-To-Know List.

DOT Classification: Corrosive Material; Label: Corrosive, liquid, solution or solid.

SAFETY PROFILE: Poison by ingestion, inhalation, subcutaneous, and intravenous route. A corrosive skin, eye, and mucous membrane irritant. Questionable carcinogen with experimental tumorigenic data. Mutation data reported. Combustible liquid when exposed to heat or flame. To fight fire, use water spray, fog, mist, dry chemical, foam. When heated to decomposition it emits toxic fumes of Cl^-. See also CHLORIDES.

CEA750 CAS:532-27-4 ***HR: 3***
α-CHLOROACETOPHENONE
DOT: UN 1697
mf: C_8H_7ClO mw: 154.60

SYNS: CAF ◇ CAP ◇ CHEMICAL MACE ◇ 1-CHLOROACETOPHENONE ◇ omega-CHLOROACETOPHENONE ◇ CHLOROACETOPHENONE, gas, liquid or solid (DOT) ◇ CHLOROMETHYL PHENYL KETONE ◇ 2-CHLORO-1-PHENYLETHANONE ◇ CN ◇ MACE (lachrymator) ◇ NCI-C55107 ◇ PHENACYL CHLORIDE ◇ PHENYLCHLOROMETHYLKETONE

TOXICITY DATA with REFERENCE
skn-rat 12%/6H open MOD ARTODN 40,75,78
skn-rbt 5 mg/24H MLD TXAPA9 17,295,70
skn-rbt 12%/6H open MOD ARTODN 40,75,78
eye-rbt 1 mg MLD TXAPA9 17,295,70
eye-rbt 3 mg SEV TXAPA9 17,295,70
skn-gpg 12%/6H open MOD ARTODN 40,75,78
skn-mus TDLo:2400 mg/kg/27W-I:NEO BJCAAI 7,482,53
ihl-hmn LCLo:159 mg/m³/20M 34ZIAG -,163,69
ihl-hmn TCLo:93 mg/m³/3M:EYE AIHAAP 23,199,62
ihl-hmn TCLo:20 mg/m³:EYE BJIMAG 29,298,72
orl-rat LD50:127 mg/kg ARTODN 40,75,78
ihl-rat LCLo:417 mg/m³/15M ARTODN 40,75,78
ipr-rat LD50:36 mg/kg ARTODN 40,75,78
ivn-rat LD50:41 mg/kg ARTODN 40,75,78
orl-mus LD50:139 mg/kg NTIS** AD837-111
ihl-mus LCLo:600 mg/m³/15M ARTODN 40,75,78
ipr-mus LD50:60 mg/kg NTIS** AD837-111
ivn-mus LD50:81 mg/kg ARTODN 40,75,78
orl-rbt LD50:118 mg/kg ARTODN 40,75,78
ihl-rbt LCLo:465 mg/m³/20M ARTODN 40,75,78
ivn-rbt LD50:30 mg/kg ARTODN 40,75,78
orl-gpg LD50:158 mg/kg ARTODN 40,75,78
ihl-gpg LCLo:490 mg/m³/30M ARTODN 40,75,78
ipr-gpg LD50:17 mg/kg ARTODN 40,75,78

CONSENSUS REPORTS: Reported in EPA TSCA Inventory. Community Right-To-Know List.

OSHA PEL: TWA 0.05 ppm
ACGIH TLV: TWA 0.05 ppm
DOT Classification: Irritating Material; Label: Irritant; Poison B; Label: Poison.

SAFETY PROFILE: A human poison by inhalation. An experimental poison by ingestion, inhalation, intraperitoneal, and intravenous routes. Human systemic effects by inhalation: lacrimation, conjunctiva irritation, and unspecified eye effects, cough, and dyspnea. A severe eye and moderate skin irritant. Questionable carcinogen with experimental neoplastigenic data by skin contact. A riot control agent. When heated to decomposition it emits toxic fumes of Cl^-. See also KETONES.

CEB000 CAS:2142-68-9 ***HR: 2***
2'-CHLOROACETOPHENONE
mf: C_8H_7ClO mw: 154.60

TOXICITY DATA with REFERENCE
orl-rat LD50:1820 mg/kg CroHP# 02SEPT80
orl-mus LD50:880 mg/kg CroHP# 02SEPT80

CONSENSUS REPORTS: Reported in EPA TSCA Inventory.

SAFETY PROFILE: Moderately toxic by ingestion. See also 2-CHLOROACETOPHENONE. When heated to decomposition it emits toxic fumes of Cl^-.

CEB250 CAS:99-91-2 ***HR: 3***
p-CHLOROACETOPHENONE
mf: C_8H_7ClO mw: 154.60

PROP: Pale straw-colored liquid or white crystals; fragrant, non-persistent odor. Mp: 56°, bp: 237-247°, fp: 59°, d: 1.19 @ 25°/25°, vap press: 0.012 mm @ 0°, vap d: 5.2, flash p: 244°F.

SYNS: 4'-CHLOROACETOPHENONE ◇ 4-CHLOROACETOPHENONE ◇ 1-(4-CHLOROPHENYL)ETHANONE ◇ USAF DO-1

TOXICITY DATA with REFERENCE
ihl-hmn TCLo:1 mg/m³/1M:NOSE,EYE 27ZZA9 8,192,57
orl-mus LD50:1207 mg/kg PHARAT 31,317,76
ipr-mus LD50:100 mg/kg NTIS** AD277-689

CONSENSUS REPORTS: Reported in EPA TSCA Inventory.

SAFETY PROFILE: Poison by intraperitoneal route. Moderately toxic by ingestion. A powerful irritant and lachrymator. Human systemic effects by inhalation: lachrimation and unspecified effects on the eye and sense of smell. Combustible when exposed to heat or flame. To fight fire, use water, foam, alcohol foam, dry chemical. When heated to decomposition or on contact with water or steam it emits toxic fumes of Cl^-. See also CHLORIDES and 2-CHLOROACETOPHENONE.

CEB500 CAS:16634-82-5 ***HR: 3***
2-CHLORO-4-ACETOTOLUIDIDE
mf: $C_9H_{10}ClNO$ mw: 183.65

SYNS: 2-CHLOROACETO-p-TOLUIDIDE ◇ N-(CHLOROACETYL)-p-TOLUIDINE ◇ 2-CHLORO-4'-METHYLACETANILIDE ◇ p-METHYL-CHLOROACETANILIDE ◇ 4-METHYL-α-CHLOROACETANILIDE ◇ p-TOLUENE CHLOROMETHYLAMIDE

TOXICITY DATA with REFERENCE
orl-rat LDLo:2740 mg/kg TXAPA9 18,517,71
ipr-rat LD50:452 mg/kg TXAPA9 18,517,71
ivn-rat LD50:60 mg/kg TXAPA9 18,517,71
ivn-mus LD50:320 mg/kg CSLNX* NX#02137

SAFETY PROFILE: Poison by intravenous route. Moderately toxic by ingestion and intraperitoneal routes. When heated to decomposition it emits very toxic fumes of Cl^- and NO_x.

CEB750 CAS:7149-79-3 ***HR: 3***
3'-CHLORO-p-ACETOTOLUIDIDE
mf: $C_9H_{10}ClNO$ mw: 183.65

TOXICITY DATA with REFERENCE
orl-bwd LD50:1300 μg/kg TXAPA9 21,315,72

CONSENSUS REPORTS: Reported in EPA TSCA Inventory.

SAFETY PROFILE: A deadly poison by ingestion. When heated to decomposition it emits very toxic fumes of Cl^- and NO_x.

CEB875 ***HR: D***
6-αoCHLORO-17-α-ACETOXYPROGESTERONE
mf: $C_{23}H_{31}ClO_4$ mw: 406.99

SYN: 6-α-CHLORO-17-α-HYDROXYPREGN-4-ENE-3,20-DIONE ACETATE

TOXICITY DATA with REFERENCE
orl-rbt TDLo:313 μg/kg (1D pre):REP ACEDAB 73,17,63

SAFETY PROFILE: Experimental reproductive effects. When heated to decomposition it emits toxic fumes of Cl^-.

CEC000 CAS:140-49-8 ***HR: 2***
4'-CHLOROACETYL ACETANILIDE
mf: $C_{10}H_{10}ClNO_2$ mw: 211.66

SYNS: p-ACETAMIDOPHENACYL CHLORIDE ◇ p-(ACETYLAMINO)PHENACYL CHLORIDE ◇ 4'-(CHLOROACETYL)ACETANILIDE ◇ NCI-C03770

TOXICITY DATA with REFERENCE
mmo-sat 3300 ng/plate ENMUDM 7(Suppl 5),1,85
mma-sat 10 μg/plate ENMUDM 7(Suppl 5),1,85
orl-rat LD50:2150 mg/kg NCILB* NIH-NCI-E-C-72-3252
orl-mus LD50:1470 mg/kg NCILB* NIH-NCI-E-C-72-3252

CONSENSUS REPORTS: NCI Carcinogenesis Bioassay (feed); No Evidence: mouse, rat NCITR* NCI-CG-TR-177,79. Reported in EPA TSCA Inventory.

SAFETY PROFILE: Moderately toxic by ingestion. Mutation data reported. When heated to decomposition it emits very toxic fumes of Cl^- and NO_x. See also CHLORIDES.

CEC250 CAS:79-04-9 ***HR: 3***
CHLOROACETYL CHLORIDE
DOT: UN 1752
mf: $C_2H_2Cl_2O$ mw: 112.94

PROP: Water-white or sltly yellow liquid. Bp: 105-106°, fp: −22.5°, flash p: none, d: 1.495 @ 0°.

SYNS: CHLORACETYL CHLORIDE ◇ CHLOROACETIC ACID CHLORIDE ◇ CHLOROACETIC CHLORIDE ◇ CHLORURE de CHLORACETYLE (FRENCH) ◇ MONOCHLOROACETYL CHLORIDE

TOXICITY DATA with REFERENCE
orl-rat LD50:120 mg/kg DTLVS* 4,84,80
ihl-rat LCLo:1000 ppm/4H 34ZIAG -,607,69
ivn-mus LD50:32 mg/kg CSLNX* NX#04538

CONSENSUS REPORTS: Reported in EPA TSCA Inventory.

OSHA PEL: TWA 0.05 ppm
ACGIH TLV: TWA 0.05 ppm; STEL: 0.15 ppm
DOT Classification: Corrosive Material; Label; Corrosive.

SAFETY PROFILE: Poison by ingestion and intravenous routes. Mildly toxic by inhalation. Corrosive. A lacrymator. When heated to decomposition it emits toxic fumes of Cl^-.

CEC500 CAS:593-63-5 ***HR: 3***
CHLOROACETYLENE
mf: C_2HCl mw: 60.47

SAFETY PROFILE: Explodes or ignites on contact with air. Its reactivity and volatility make it extremely dangerous. Probably a strong irritant by inhalation. When heated to decomposition it emits toxic fumes of Cl^-. See also ACETYLENE COMPOUNDS.

CED500 CAS:61462-73-5 ***HR: 3***
4'-(2-CHLORO-9-ACRIDINYLAMINO) METHANESULFONANILIDE
mf: $C_{20}H_{16}ClN_3O_2S$ mw: 397.90

SYN: N-(4-((2-CHLORO-9-ACRIDINYL)AMINO)PHENYL)METHANESULFONAMIDE

TOXICITY DATA with REFERENCE
mmo-sat 25 μmol/L JMCMAR 23,269,80
ipr-mus LD10:200 mg/kg JMCMAR 23,269,80

SAFETY PROFILE: Poison by intraperitoneal route. Mutation data reported. When heated to decomposition it emits very toxic fumes of Cl^-, NO_x, and SO_x.

CED750 CAS:61417-08-1 ***HR: 3***
4'-(3-CHLORO-9-ACRIDINYLAMINO) METHANESULFONANILIDE
mf: $C_{20}H_{16}ClN_3O_2S$ mw: 397.90

TOXICITY DATA with REFERENCE
mmo-sat 23 µmol/L JMCMAR 23,269,80
ipr-mus LD10:120 mg/kg JMCMAR 23,269,80

SAFETY PROFILE: Poison by intraperitoneal route. Mutation data reported. When heated to decomposition it emits very toxic fumes of Cl^-, NO_x, and SO_x.

CEE500 CAS:598-79-8 ***HR: 3***
2-CHLOROACRYLIC ACID
mf: $C_3H_3ClO_2$ mw: 106.51

SYNS: CHLOROACRYLIC ACID ◇ α-CHLOROACRYLIC ACID

TOXICITY DATA with REFERENCE
ivn-mus LD50:42 mg/kg CSLNX* NX#07795

CONSENSUS REPORTS: Reported in EPA TSCA Inventory.

SAFETY PROFILE: Poison by intravenous route. When heated to decomposition it emits toxic fumes of Cl^-.

CEE750 CAS:920-37-6 ***HR: 3***
2-CHLOROACRYLONITRILE
mf: C_4H_2ClN mw: 87.51

PROP: Flash p: 46.4°F.

SYNS: CHLOROACRYLONITRILE ◇ α-CHLOROACRYLONITRILE

TOXICITY DATA with REFERENCE
ivn-mus LD50:100 mg/kg CSLNX* NX#02164

CONSENSUS REPORTS: Reported in EPA TSCA Inventory. Cyanide and its compounds are on the Community Right-To-Know List.

SAFETY PROFILE: Poison by intravenous route. A powerful irritant. A dangerous fire hazard when exposed to heat or flame. To fight fire, use water, dry chemical, CO_2, foam. When heated to decomposition it emits very toxic fumes of Cl^-, NO_x, and CN^-. See also NITRILES.

CEF100 CAS:146-77-0 ***HR: 3***
2-CHLOROADENOSINE
mf: $C_{10}H_{12}ClN_5O_4$ mw: 301.72

SYNS: ADENOSINE, 2-CHLORO- ◇ Cl-ADO

TOXICITY DATA with REFERENCE
ipr-mus TDLo:603 µg/kg (female 11-12D post):TER TJADAB 25(2),35A,82
ivn-mus LDLo:50 mg/kg AIPTAK 118,95,59

SAFETY PROFILE: Poison by intravenous route. An experimental teratogen. When heated to decomposition it emits toxic fumes of NO_x and Cl^-.

CEF125 ***HR: 3***
2-CHLOROADENOSINE-5'-SULFAMATE
mf: $C_{10}H_{13}ClN_6O_6S$ mw: 380.80

PROP: Isolated from a subspecies of *Streptomyces rishiriensis* (JANTAJ 35,939,82).

SYN: 5'-SULFAMOYL-2-CHLOROADENOSINE

TOXICITY DATA with REFERENCE
dni-esc 10 mg/L JANTAJ 35,939,82
oms-esc 10 mg/L JANTAJ 35,939,82
orl-mus LD50:3400 µg/kg JANTAJ 35,939,82
ipr-mus LD50:200 µg/kg JANTAJ 35,939,82

SAFETY PROFILE: Deadly poison by ingestion and intraperitoneal routes. Mutation data reported. When heated to decomposition it emits toxic fumes of Cl^-, SO_x, and NO_x.

CEF250 CAS:5976-47-6 ***HR: 3***
β-CHLORO ALLYL ALCOHOL
mf: C_3H_5ClO mw: 92.53

SYN: 2-CHLORO-2-PROPEN-1-OL

TOXICITY DATA with REFERENCE
skn-rbt 500 mg SEV SCCUR* -,2,61
mmo-sat 1 nmol/plate JAFCAU 28,880,80
mma-sat 1 nmol/plate JAFCAU 28,880,80
orl-mus LDLo:63 mg/kg SCCUR* -,2,61
ihl-mus LCLo:198 ppm/6H SCCUR* -,2,61
ivn-mus LD50:56 mg/kg CSLNX* NX#02539
skn-rbt LDLo:25 mg/kg SCCUR* -,2,61

SAFETY PROFILE: Poison by ingestion, skin contact, and intravenous routes. Moderately tox by inhalation. A severe skin irritant. Mutation data reported. When heated to decomposition it emits toxic fumes of Cl^-. See also pi-CHLORO ALLYL ALCOHOL, ALLYL COMPOUNDS, CHLORIDES, and ALCOHOLS.

CEF500 CAS:29560-84-7 ***HR: 3***
pi-CHLORO ALLYL ALCOHOL
mf: C_3H_5ClO mw: 92.53

SYN: 3-CHLORO-2-PROPEN-1-OL

TOXICITY DATA with REFERENCE
skn-rbt 500 mg SEV SCCUR* -,2,61
orl-rat LD50:102 mg/kg SCCUR* -,2,61

ihl-rat LC50:370 ppm/1H SCCUR* -,2,61
orl-mus LD50:175 mg/kg SCCUR* -,2,61
ihl-mus LC50:540 ppm/1H SCCUR* -,2,61
skn-mus LDLo:170 mg/kg SCCUR* -,2,61

SAFETY PROFILE: Poison by ingestion and skin contact. Moderately toxic by inhalation. A severe skin irritant. When heated to decomposition it emits toxic fumes of Cl^-. See also β-CHLORO ALLYL ALCOHOL, ALLYL COMPOUNDS, CHLORIDES, and ALCOHOLS.

CEG550 CAS:4080-31-3 ***HR: 3***
1-(3-CHLOROALLYL)-3,5,7-TRIAZA-1-AZONIAADAMANTANE CHLORIDE
mf: $C_9H_{16}ClN_4 \cdot Cl$ mw: 251.19

SYNS: DOWCO 184 ◇ DOWICIDE Q ◇ DOWICIL 75 ◇ DOWICIL 100 ◇ QUATERNIUM 15 ◇ 3,5,7-TRIAZA-1-AZONIAADAMANTANE, 1-(3-CHLOROALLYL)-, CHLORIDE

TOXICITY DATA with REFERENCE
skn-rbt 500 mg/24H MLD JACTDZ 5(3),61,86
mma-sat 333 μg/plate EMMUEG 11(Suppl 12),1,88
orl-rat TDLo:250 mg/kg (female 6-15D post):TER JACTDZ 5(3),61,86
orl-rat LD50:500 mg/kg PCOC** -,455,66
orl-rbt LD50:78500 μg/kg JACTDZ 5(3),61,86
skn-rbt LD50:565 mg/kg JACTDZ 5(3),61,86

CONSENSUS REPORTS: Reported in EPA TSCA Inventory.

SAFETY PROFILE: Poison by ingestion. Moderately toxic by skin contact. Experimental teratogenic effects. A skin irritant. Mutation data reported. When heated to decomposition it emits toxic fumes of NO_x and Cl^-.

CEG600 CAS:615-66-7 ***HR: 1***
3-CHLORO-4-AMINOANILINE
mf: $C_6H_7ClN_2$ mw: 142.60

$Cl(H_2N)C_6H_3NH_2$

SYNS: 3-CHLOR-p-FENYLENDIAMIN (CZECH) ◇ 2-CHLORO-1,4-BENZENEDIAMINE ◇ o-CHLORO-p-PHENYLENEDIAMINE ◇ 2-CHLORO-p-PHENYLENEDIAMINE ◇ C.I. 76065 ◇ URSOL BROWN O

TOXICITY DATA with REFERENCE
eye-rbt 20 mg/24H MOD 28ZPAK -,96,72
orl-rat TDLo:4 g/kg (6-15D preg):TER FCTOD7 22,147,84
orl-rat TDLo:4 g/kg (6-15D preg):REP FCTOD7 22,147,84

CONSENSUS REPORTS: Reported in EPA TSCA Inventory.

SAFETY PROFILE: An experimental teratogen. Other experimental reproductive effects. An eye irritant. Decomposes explosively at 165°C/33 mbar. When heated to decomposition it emits toxic fumes of Cl^- and NO_x. See also AROMATIC AMINES.

CEG625 CAS:6219-71-2 ***HR: 3***
3-CHLORO-4-AMINOANILINE SULFATE
mf: $C_6H_5ClN_2 \cdot H_2O_4S$ mw: 238.66

SYNS: 2-CHLORO-1,4-BENZENEDIAMINE SULFATE ◇ 2-CHLORO-p-PHENYLENEDIAMINE SULFATE ◇ C.I. 76066 ◇ C.I. OXIDATION BASE 13A ◇ 2-Cl-P-PD ◇ FOURRINE 81 ◇ FOURRINE SO ◇ NCI-C03316 ◇ RENAL SO

TOXICITY DATA with REFERENCE
mmo-sat 333 μg/plate ENMUDM 5(Suppl 1),3,83
mma-sat 100 μg/plate ENMUDM 5(Suppl 1),3,83
orl-rat TDLo:41 g/kg/77W-C:ETA CRNGDP 1,495,80

CONSENSUS REPORTS: NCI Carcinogenesis Bioassay (Feed); Results Negative: mouse, rat NCITR* NCI-CG-TR-113,78

SAFETY PROFILE: Questionable carcinogen with experimental tumorigenic data. Mutation data reported. When heated to decomposition it emits toxic fumes of Cl^-, SO_x, and NO_x. See also AROMATIC AMINES and SULFATES.

CEG750 CAS:2457-76-3 ***HR: 2***
2-CHLORO-4-AMINOBENZOIC ACID
mf: $C_7H_6ClNO_2$ mw: 171.59

SYNS: 4-AMINO-2-CHLOROBENZOIC ACID ◇ USAF NB-1

TOXICITY DATA with REFERENCE
mmo-esc 500 mg/L JGMIAN 18,543,58
ipr-mus LDLo:500 mg/kg NTIS** AD277-689

CONSENSUS REPORTS: Reported in EPA TSCA Inventory.

SAFETY PROFILE: Moderately toxic by intraperitoneal route. Mutation data reported. When heated to decomposition it emits very toxic fumes of Cl^- and NO_x. See also AROMATIC AMINES.

CEH000 CAS:5730-85-8 ***HR: 3***
3-CHLORO-4-AMINODIPHENYL
mf: $C_{12}H_{10}ClN$ mw: 203.68

SYN: 3-CHLOROBIPHENYLAMINE

TOXICITY DATA with REFERENCE
scu-rat TDLo:7000 mg/kg/W-I:ETA BMBUAQ 14,141,58

SAFETY PROFILE: Questionable carcinogen with experimental tumorigenic data. When heated to decomposition it emits very toxic fumes of Cl^- and NO_x.

CEH125 CAS:101-79-1 ***HR: 3***
4-CHLORO-4′-AMINODIPHENYL ETHER
mf: $C_{12}H_{10}ClNO$ mw: 219.68

SYNS: 4′-CHLORO-4-AMINOBIPHENYL ETHER ◇ p-(p-CHLOROPHENOXY)ANILINE ◇ 4-(4-CHLOROPHENOXY)ANILINE ◇ 4-(4-CHLOROPHENOXY)-BENZENAMINE (9CI)

TOXICITY DATA with REFERENCE
mma-sat 100 μg/plate CBINA8 44,133,83
orl-rat TDLo:38 g/kg/78W-C:CAR JEPTDQ 2(2),325,78
orl-mus TDLo:150 g/kg/78W-C:NEO JEPTDQ 2(2),325,78
orl-mus TD:302 g/kg/78W-C:ETA JEPTDQ 2(2),325,78

CONSENSUS REPORTS: Reported in EPA TSCA Inventory.

SAFETY PROFILE: Questionable carcinogen with experimental carcinogenic, neoplastigenic, and tumorigenic data. Mutation data reported. When heated to decomposition it emits toxic fumes of Cl^- and NO_x. See also ETHERS.

CEH250 CAS:95-85-2 ***HR: 3***
p-CHLORO-o-AMINOPHENOL
mf: C_6H_6ClNO mw: 143.58
DOT: UN 2673

SYN: 2-AMINO-4-CHLOROPHENOL (DOT)

TOXICITY DATA with REFERENCE
orl-rat LD50:690 mg/kg GTPZAB 25(8),50,81
orl-mus LD50:1030 mg/kg GTPZAB 25(8),50,81

CONSENSUS REPORTS: Reported in EPA TSCA Inventory.

DOT Classification: Poison B; Label: Poison.

SAFETY PROFILE: A poison. Moderately toxic by ingestion. When heated to decomposition it emits very toxic fumes of Cl^- and NO_x. See also AROMATIC AMINES and CHLORIDES.

CEH500 CAS:2047-30-5 ***HR: 3***
CHLOROAMITRIPTYLINE HYDROCHLORIDE
mf: $C_{20}H_{22}ClN{\cdot}ClH$ mw: 348.34

SYNS: CHLORPROHEPTADIEN ◇ CHLORPROHEPTADIENE HYDROCHLORIDE

TOXICITY DATA with REFERENCE
orl-mus LD50:210 mg/kg AIPTAK 144,481,63
ivn-mus LD50:43 mg/kg AIPTAK 144,481,63

SAFETY PROFILE: Poison by ingestion and intravenous routes. When heated to decomposition it emits very toxic fumes of Cl^- and NO_x.

CEH670 CAS:95-51-2 ***HR: 3***
2-CHLOROANILINE
DOT: UN 2018/UN 2019
mf: C_6H_6ClN mw: 127.58

PROP: Liquid. Bp: 208.84°, mp: −1.94°, d: 1.2114, n (20/D) 1.5895. Practically insol in water; sol in most organic solvents, also in acids.

SYNS: 1-AMINO-2-CHLOROBENZENE ◇ o-CHLORANILINE ◇ o-CHLOROANILINE ◇ o-CHLOROANILINE, liquid (DOT) ◇ o-CHLOROANILINE, solid (DOT) ◇ 2-CHLORO-BENZENAMINE (9CI) ◇ FAST YELLOW GC BASE

TOXICITY DATA with REFERENCE
dnr-esc 500 μg/L JJIND8 62,873,79
mmo-asn 200 mg/L CJMIAZ 16,369,70
orl-mus LD50:256 mg/kg GTPZAB 13(5),29,69
skn-cat LD50:222 mg/kg GTPZAB 13(5),29,69
scu-cat LDLo:310 mg/kg AGBAAM 110,12,33

CONSENSUS REPORTS: EPA Genetic Toxicology Program. Reported in EPA TSCA Inventory.

DOT Classification: Poison B; Label: Poison.

SAFETY PROFILE: Poison by skin contact, ingestion, and subcutaneous routes. Mutation data reported. When heated to decomposition it emits toxic fumes of Cl^- and NO_x. See also ANILINE DYES.

CEH675 CAS:108-42-9 ***HR: 3***
3-CHLOROANILINE
DOT: UN 2018/UN 2019
mf: C_6H_6ClN mw: 127.58

PROP: Liquid. Bp: 230.5°, mp: −10.4°, d: 1.2150, n (20/D) 1.5931. Practically insol in water; sol in most common organic solvents.

SYNS: m-AMINOCHLOROBENZENE ◇ 1-AMINO-3-CHLOROBENZENE ◇ 3-CHLOORANILINEN (DUTCH) ◇ m-CHLORANILINE ◇ m-CHLOROANILINE ◇ 3-CHLOROANILINE (ITALIAN) ◇ m-CHLOROANILINE, liquid (DOT) ◇ m-CHLOROANILINE, solid (DOT) ◇ 3-CHLOROBENZENAMINE ◇ m-CHLOROPHENYLAMINE ◇ 3-CHLOROPHENYLAMINE ◇ FAST ORANGE GC BASE ◇ ORANGE GC BASE

TOXICITY DATA with REFERENCE
mmo-asn 200 mg/L CJMIAZ 16,369,70
orl-rat LD50:256 mg/kg GISAAA 31(12),6,66
skn-rat LD50:250 mg/kg 85GMAT-,34,82
orl-mus LD50:334 mg/kg GTPZAB 13(5),29,69
ihl-mus LC50:550 mg/m³/4H 85GMAT-,34,82
ivn-dog LDLo:50 mg/kg AEPPAE 244,387,63
skn-cat LD50:223 mg/kg GTPZAB 13(5),29,69
scu-cat LDLo:125 mg/kg AHBAAM 110,12,33
orl-gpg LD50:250 mg/kg GISAAA 31(12),6,66

CONSENSUS REPORTS: EPA Genetic Toxicology Program. Reported in EPA TSCA Inventory.

DOT Classification: Poison B; Label: Poison.

SAFETY PROFILE: Poison by ingestion, skin contact, subcutaneous, and intravenous routes. Mutation data reported. When heated to decomposition it emits toxic fumes of Cl^- and NO_x. See also ANILINE DYES.

CEH680 CAS:106-47-8 **HR: 3**
4-CHLOROANILINE
DOT: UN 2018/UN 2019
mf: C_6H_6ClN mw: 127.58

PROP: Orthorhombic crystals from alc or petr ether. Mp: 72.5°, bp: 232°, d: 1.169. Sol in hot water; freely sol in alc, ether, acetone, carbon disulfide.

SYNS: 1-AMINO-4-CHLOROBENZENE ◇ 4-CHLORANILIN (CZECH) ◇ p-CHLORANILINE ◇ p-CHLOROANILINE ◇ p-CHLOROANILINE, liquid (DOT) ◇ p-CHLOROANILINE, solid (DOT) ◇ 4-CHLOROBENZENAMINE ◇ 4-CHLORO BENZENEAMINE ◇ 4-CHLOROPHENYLAMINE ◇ NCI-C02039 ◇ RCRA WASTE NUMBER P024

TOXICITY DATA with REFERENCE
skn-rbt 500 mg/24H MLD 28ZPAK -,96,72
eye-rbt 250 μg/24H SEV 28ZPAK -,96,72
mma-sat 100 μg/plate ENMUDM 7(Suppl 5),1,85
otr-rat:emb 14500 ng/plate JJATDK 1,190,81
dns-rat:lvr 5 mg/L MUREAV 97,359,82
orl-rat TDLo:18200 mg/kg/2Y-C:NEO FCTOD7 25,619,87
orl-rat TDLo:14 g/kg/78W-C:ETA NCITR* NCI-CG-TR-189,79
orl-rat LD50:310 mg/kg AIHAAP 23,95,62
skn-rat LD50:3200 mg/kg AGGHAR 15,447,57
ipr-rat LD50:420 mg/kg AGGHAR 15,447,57
orl-mus LD50:100 mg/kg NCILB* NCI-E-C-72-3252,73
ihl-mus LC12:250 mg/m^3/6H 85GMAT -,34,82
ivn-dog LDLo:100 mg/kg AEPPAE 244,387,63
skn-cat LD50:239 mg/kg GTPZAB 13(5),29,69
scu-cat LDLo:125 mg/kg AHBAAM 110,12,33
skn-rbt LD50:360 mg/kg AIHAAP 23,95,62
orl-gpg LD50:350 mg/kg 85GMAT -,34,82

CONSENSUS REPORTS: EPA Genetic Toxicology Program. Reported in EPA TSCA Inventory.

DOT Classification: Poison B; Label: Poison.

SAFETY PROFILE: Poison by ingestion, inhalation, skin contact, subcutaneous, and intravenous routes. Moderately toxic by inhalation and intraperitoneal routes. A skin and severe eye irritant. Questionable carcinogen with experimental neoplastigenic and tumorigenic data. Mutation data reported. When heated to decomposition it emits toxic fumes of Cl^- and NO_x. See also ANILINE DYES.

CEH700 CAS:10238-21-8 **HR: 1**
1-((p-(2-(CHLORO-o-ANISAMIDO)ETHYL)PHENYL)SULFONYL)-3-CYCLOHEXYL UREA
mf: $C_{23}H_{28}ClN_3O_5S$ mw: 494.05

SYNS: 5-CHLORO-N-(2-(4-((((CYCLOHEXYLAMINO)CARBONYL)AMINO)SULFONYL)PHENYL)ETHYL)-2-METHOXYBENZAMIDE ◇ N-(4-(2-(5-CHLORO-2-METHOXYBENZAMIDO)ETHYL) PHENYLSULFONYL)-N'-CYCLOHEXYLUREA ◇ DAONIL ◇ DIABETA ◇ EUGLUCAN ◇ EUGLUCON ◇ EUGLUCON 5 ◇ EUGLYKON ◇ GILEMAL ◇ GLIBENCLAMIDE ◇ GLYBENZCYCLAMIDE ◇ GLYBURIDE ◇ HB 419 ◇ HD 419 ◇ MANINIL ◇ MICRONASE ◇ U 26452 ◇ UR 606

TOXICITY DATA with REFERENCE
ipr-rat TDLo:4 mg/kg (5-9D preg):REP ARZNAD 21,846,71
scu-rat TDLo:1800 mg/kg (9-14D preg):TER SEIJBO 17,31,77
orl-wmn TDLo:147 μg/kg DICPBB 18,142,84
ipr-rat LD50:3750 mg/kg ARZNAD 19,1413,69
orl-mus LD50:3250 mg/kg ARZNAD 16,1640,66
ipr-mus LD50:5900 mg/kg OYYAA2 4,247,70
scu-mus LD50:20 g/kg OYYAA2 4,247,70

SAFETY PROFILE: Mildly toxic by ingestion, intraperitoneal, and subcutaneous routes. An experimental teratogen. Other experimental reproductive effects. When heated to decomposition it emits very toxic fumes of SO_x, NO_x, and Cl^-.

CEH750 CAS:5345-54-0 **HR: 2**
3-CHLOROANISIDINE
mf: C_7H_8ClNO mw: 157.61

SYNS: 3-CHLORO-4-METHOXY-BENZENAMINE (9CI) ◇ OCPA ◇ ORTHOCHLOROPARANISIDINE

TOXICITY DATA with REFERENCE
skn-rbt 500 mg/24H MLD TOERD9 2,77,79
eye-rbt 100 mg MLD TOERD9 2,77,79
orl-rat LD50:550 mg/kg TOERD9 2,77,79
ipr-rat LD50:510 mg/kg TOERD9 2,77,79
orl-mus LD50:650 mg/kg TOERD9 2,77,79
ipr-mus LD50:670 mg/kg TOERD9 2,77,79

SAFETY PROFILE: Moderately toxic by ingestion and intraperitoneal routes. A skin and eye irritant. When heated to decomposition it emits toxic fumes of Cl^- and NO_x. See also AROMATIC AMINES.

CEI000 CAS:82-44-0 **HR: 3**
1-CHLOROANTHRAQUINONE
mf: $C_{14}H_8ClO_2$ mw: 243.67

SYNS: 1-CHLORANTHRACHINON (CZECH) ◇ 1-CHLORO-9,10-ANTHRACENEDIONE ◇ α-CHLOROANTHRAQUINONE ◇ 1-CHLORO-9,10-ANTHRAQUINONE ◇ α-MONOCHLOROANTHRAQUINONE

TOXICITY DATA with REFERENCE
skn-rbt 500 mg/24H MLD 28ZPAK -,86,72
eye-rbt 100 mg/24H MOD 28ZPAK -,86,72
orl-rat LD50:15100 mg/kg 28ZPAK -,86,72
ivn-mus LD50:56 mg/kg CSLNX* NX#03287

CONSENSUS REPORTS: Reported in EPA TSCA Inventory.

SAFETY PROFILE: Poison by intravenous route. Mildly toxic by ingestion. A skin and eye irritant. When

heated to decomposition it emits very toxic fumes of Cl^- and NO_x.

CEI250 CAS:300-88-9 ***HR: 3***
CHLOROARSENOL
mf: $C_7H_{14}AsClO_3 \cdot H_3N$ mw: 273.62

SYNS: AMMONIUM CHLOROHEPTENE ARSONATE ◇ ARSION ◇ CHLORARSENOL ◇ (2-CHLORO-1-HEPTENYL)ARSONIC ACID MONOAMMONIUM SALT ◇ SOLARSON ◇ SOLASON

TOXICITY DATA with REFERENCE
cyt-hmn:hla 1000 ppm/6H IDZAAW 40,135,65
cyt-hmn:lng 1000 ppm/12H IDZAAW 40,135,65
scu-rbt LDLo:1 g/kg HBAMAK 4,1289,35
ivn-rbt LDLo:1 g/kg HBAMAK 4,1289,35

CONSENSUS REPORTS: Arsenic and its compounds are on the Community Right-To-Know List.

OSHA PEL: TWA 0.5 mg(As)/m^3
ACGIH TLV: TWA 0.2 mg(As)/m^3

SAFETY PROFILE: Arsenic compounds are generally poisons. Moderately toxic by subcutaneous and intravenous routes. Human mutation data reported. When heated to decomposition it emits very toxic fumes of As, Cl^-, NO_x, and NH_3. See also ARSENIC COMPOUNDS.

CEI325 CAS:25167-31-1 ***HR: 3***
1-CHLOROAZIRIDINE
mf: C_2H_4ClN mw: 77.51

CH_2NClCH_2 (ring)

SAFETY PROFILE: A dangerous storage hazard, it may explode at room temperature. When heated to decomposition it emits toxic fumes of Cl^- and NO_x. See also various aziridine compounds.

CEI500 CAS:89-98-5 ***HR: 3***
o-CHLOROBENZALDEHYDE
mf: C_7H_5ClO mw: 140.57

SYNS: o-CHLOORBENZALDEHYDE (DUTCH) ◇ 2-CHLOORBENZALDEHYDE (DUTCH) ◇ 2-CHLORBENZALDEHYD (GERMAN) ◇ 2-CHLOROBENZALDEHYDE ◇ 2-CLOROBENZALDEIDE (ITALIAN) ◇ o-CHLOROBENZENECARBOXALDEHYDE ◇ USAF M-7

TOXICITY DATA with REFERENCE
ipr-mus LD50:10 mg/kg NTIS** AD277-689
ivn-rbt LD50:8500 μg/kg PJPPAA 31,563,79

CONSENSUS REPORTS: Reported in EPA TSCA Inventory.

SAFETY PROFILE: Poison by intraperitoneal and intravenous routes. When heated to decomposition it emits toxic fumes of Cl^-. See also ALDEHYDES and CHLORIDES.

CEJ000 CAS:20268-52-4 ***HR: 3***
10-CHLORO-1,2-BENZANTHRACENE
mf: $C_{18}H_{11}Cl$ mw: 262.74

SYN: 7-CHLOROBENZ(a)ANTHRACENE

TOXICITY DATA with REFERENCE
mma-sat 20 μg/plate MUREAV 155,91,85
skn-mus TDLo:530 mg/kg/22W-I:ETA COREAF 226,1852,48

SAFETY PROFILE: Questionable carcinogen with experimental tumorigenic data. Mutation data reported. When heated to decomposition it emits toxic fumes of Cl^-.

CEJ125 CAS:108-90-7 ***HR: 3***
CHLOROBENZENE
DOT: UN 1134
mf: C_6H_5Cl mw: 112.56

PROP: Clear, colorless liquid. Bp: 131.7°, lel: 1.3%, uel: 7.1%, @ 150°, mp: −45°, flash p: 85°F (CC), d: 1.113 @ 15.5°/15.5°, autoign temp: 1180°F, vap press: 10 mm @ 22.2°, vap d: 3.88.

SYNS: BENZENE CHLORIDE ◇ CHLOORBENZEEN (DUTCH) ◇ CHLORBENZENE ◇ CHLORBENZOL ◇ CHLOROBENZEN (POLISH) ◇ CHLOROBENZOL (DOT) ◇ CLOROBENZENE (ITALIAN) ◇ MCB ◇ MONOCHLOORBENZEEN (DUTCH) ◇ MONOCHLORBENZENE ◇ MONOCHLORBENZOL (GERMAN) ◇ MONOCHLOROBENZENE ◇ MONOCLOROBENZENE (ITALIAN) ◇ NCI-C54886 ◇ PHENYL CHLORIDE ◇ RCRA WASTE NUMBER U037

TOXICITY DATA with REFERENCE
mrc-smc 1000 ppm NTIS** PB84-138973
ihl-rbt TCLo:590 ppm/6H (6-18D preg):REP TXAPA9 76,365,84
ihl-rat TCLo:210 ppm/6H (6-15D preg):TER TXAPA9 76,365,84
orl-rat LD50:2290 mg/kg 38MKAJ 2B,3603,81
ipr-rat LDLo:7400 mg/kg RMSRA6 16,449,1896
scu-rat LDLo:7000 mg/kg RMSRA6 16,449,1896
orl-mus LD50:2300 mg/kg 85GMAT -,34,82
ihl-mus LCLo:15 g/m^3 GISAAA 20(8),19,55
ipr-mus LD50:515 mg/kg PHMCAA 10,172,68
orl-rbt LD50:2830 mg/kg 14CYAT 2,1394,63
ipr-gpg LDLo:4100 mg/kg RMSRA6 16,449,1896

CONSENSUS REPORTS: NTP Carcinogenesis Studies (gavage); Some Evidence: rat NTPTR* NTP-TR-261,85; No Evidence: mouse NTPTR* NTP-TR-261,85. Reported in EPA TSCA Inventory. Community Right-To-Know List.

OSHA PEL: TWA 75 ppm
ACGIH TLV: TWA 10 ppm
DFG MAK: 50 ppm (230 mg/m^3)
DOT Classification: Flammable or Combustible Liquid; Label: Flammable Liquid.

SAFETY PROFILE: Moderately toxic by ingestion and intraperitoneal routes. Experimental teratogenic and reproductive effects. Mutation data reported. Strong narcotic with slight irritant qualities. Dichlorobenzols are strongly narcotic. Little is known of the effects of repeated exposures at lower concentrations, but it may cause kidney and liver damage. The industrial illnesses reported may possibly be due to nitrobenzol. Dangerous fire hazard when exposed to heat or flame. Moderate explosion hazard when exposed to heat or flame. Potentially explosive reaction with powdered sodium or phosphorus trichloride + sodium. Violent reaction with $AgClO_4$ or dimethyl sulfoxide. Reacts vigorously with oxidizers. See also CHLORINATED HYDROCARBONS, AROMATIC. To fight fire, use foam, CO_2, dry chemical, water to blanket fire. Associated with EPA Superfund sites.

CEJ250 CAS:17333-84-5 ***HR: 3***
m-CHLOROBENZENEDIAZONIUM SALTS
mf: $C_6H_4ClN_2X$

$$C_6H_4ClN_2^+X^-$$

SAFETY PROFILE: Violent explosion on contact with sodium disulfide. Reaction with potassium O,O-diphenylphosphorodithioates; o-alkyldithiocarbonate (xanthate) solutions; thiophenoxide solutions (e.g., sodium-2-chlorothiophenoxide; potassium thiophenoxide); various sulfides or derivatives (e.g., hydrogen sulfide; sodium hydrogen sulfide; sodium mono-, di- or poly-sulfides) forms highly explosive products. When heated to decomposition it emits toxic fumes of Cl^- and NO_x.

CEJ500 CAS:17333-83-4 ***HR: 3***
o-CHLOROBENZENEDIAZONIUM SALTS
mf: $C_6H_4ClN_2X$

$$C_6H_4ClN_2^+X^-$$

SYN: 2-CHLOROBENZENEDIAZONIUMSALTS

SAFETY PROFILE: Reaction with potassium O,O-diphenylphosphorodithioates; o-alkyldithiocarbonate (xanthate) solutions; thiophenoxide solutions (e.g., sodium-2-chlorothiophenoxide; potassium thiophenoxide); various sulfides or derivatives (e.g., hydrogen sulfide; sodium hydrogen sulfide; sodium mono-, di- or polysulfides) forms highly explosive products. Incompatible with potassium-2-chlorothiophenolate. When heated to decomposition it emits toxic fumes of Cl^- and NO_x.

CEK000 CAS:98-64-6 ***HR: 3***
p-CHLOROBENZENESULFONAMIDE
mf: $C_6H_6ClNO_2S$ mw: 191.64

SYN: USAF MA-3

TOXICITY DATA with REFERENCE
orl-rat LDLo:500 mg/kg NCNSA6 5,20,53
ipr-mus LD50:200 mg/kg NTIS** AD277-689

CONSENSUS REPORTS: Reported in EPA TSCA Inventory.

SAFETY PROFILE: Poison by intraperitoneal route. Moderately toxic by ingestion. When heated to decomposition it emits very toxic fumes of SO_x, NO_x, and Cl^-.

CEK250 CAS:5138-90-9 ***HR: 1***
p-CHLOROBENZENESULFONIC ACID, SODIUM SALT
mf: $C_6H_4ClO_3S{\cdot}Na$ mw: 214.60

SYN: p-CHLORBENZENSULFONAN SODNY (CZECH)

TOXICITY DATA with REFERENCE
skn-rbt 500 mg/24H MLD 28ZPAK -,178,72
eye-rbt 100 mg/24H MOD 28ZPAK -,178,72
orl-rat LD50:10 g/kg 28ZPAK -,178,72

CONSENSUS REPORTS: Reported in EPA TSCA Inventory.

SAFETY PROFILE: Mildly toxic by ingestion. A skin and eye irritant. When heated to decomposition it emits very toxic fumes of SO_x, Na_2O, and Cl^-. See also SULFONATES.

CEK375 CAS:98-60-2 ***HR: 3***
p-CHLOROBENZENESULFONYL CHLORIDE
mf: $C_6H_4Cl_2O_2S$ mw: 211.06

SYNS: p-CHLORBENZENESULFOCHLORID (CZECH) ◇ CHLORID KYSELINY-p-CHLORBENSULFONOVE (CZECH)

TOXICITY DATA with REFERENCE
skn-rbt 500 mg/24H MOD 28ZPAK -,198,72
eye-rbt 50 μg/24H SEV 28ZPAK -,198,72
orl-rat LD50:4250 mg/kg 28ZPAK -,198,72
ipr-mus LDLo:250 mg/kg CBCCT* 7,774,55

CONSENSUS REPORTS: Reported in EPA TSCA Inventory.

SAFETY PROFILE: Poison by intraperitoneal route. Mildly toxic by ingestion. A severe eye and moderate skin irritant. When heated to decomposition it emits toxic fumes of Cl^- and SO_x. See also CHLORIDES.

CEK425 CAS:106-54-7 ***HR: 3***
4-CHLOROBENZENETHIOL
mf: C_6H_5ClS mw: 144.62

SYNS: p-CHLORO-PHENYL MERCAPTAN ◇ p-CHLOROTHIOPHENOL ◇ 4-CHLOROTHIOPHENOL ◇ p-CHLORTHIOFENOL (CZECH)

TOXICITY DATA with REFERENCE
skn-rbt 500 mg/24H MOD 28ZPAK -,167,72

eye-rbt 50 μg/24H SEV 28ZPAK -,167,72
skn-mus TDLo:8000 mg/kg/20W-I:ETA CNREA8 19,413,59
orl-rat LD50:500 mg/kg 28ZPAK -,167,72
ipr-mus LD50:75 mg/kg NTIS** AD691-490

CONSENSUS REPORTS: Chlorophenol compounds are on the Community Right-To-Know List. Reported in EPA TSCA Inventory.

SAFETY PROFILE: Poison by intraperitoneal route. Moderately toxic by ingestion. A severe eye and moderate skin irritant. Questionable carcinogen with experimental tumorigenic data by skin contact. When heated to decomposition it emits toxic fumes of Cl^- and SO_x. See also CHLOROPHENOLS and MERCAPTANS

CEK500 CAS:21248-01-1 ***HR: 3***
6-CHLOROBENZENO(a)PYRENE
mf: $C_{20}H_{11}Cl$ mw: 286.69

TOXICITY DATA with REFERENCE
unk-mus TDLo:80 mg/kg/8D-I:ETA BEBMAE 88(11),592,79

SAFETY PROFILE: Questionable carcinogen with experimental tumorigenic data. When heated to decomposition it emits toxic fumes of Cl^-. See also CHLORINATED HYDROCARBONS, AROMATIC; and PYRENE.

CEK875 CAS:55981-23-2 ***HR: 3***
N[1]-(4'-CHLOROBENZHYDRYL)-N[4]-SPIROMORPHOLINO-PIPERAZINIUM CHLORIDE HYDROCHLORIDE
mf: $C_{21}H_{26}ClN_2O{\bullet}Cl{\bullet}ClH$ mw: 429.85

SYNS: 9-((4-CHLOROPHENYL)PHENYLMETHYL)3-OXA-9-AZA-6-AZONIASPIRO(5.5)UNDECANE CHLORIDE HCl ◇ CRC 7001

TOXICITY DATA with REFERENCE
ipr-rat LD50:40500 μg/kg APJUA8 28,21,78
ivn-rat LD50:2400 μg/kg APJUA8 28,21,78
ims-rat LD50:320 mg/kg APJUA8 28,21,78
ipr-mus LD50:51300 μg/kg APJUA8 28,21,78
ims-mus LD50:281 mg/kg APJUA8 28,21,78
orl-dog LD50:315 mg/kg APJUA8 28,21,78

SAFETY PROFILE: Poison by ingestion, intramuscular, intravenous, and intraperitoneal routes. When heated to decomposition it emits toxic fumes of NO_x and Cl^-.

CEL000 CAS:32226-65-6 ***HR: 3***
2-CHLOROBENZO(e)(1)BENZOTHIOPYRANO (4,3-b) INDOLE
mf: $C_{19}H_{10}ClNS$ mw: 319.81

TOXICITY DATA with REFERENCE
scu-mus TDLo:378 mg/kg/27W-I:NEO JNCIAM 46,1257,71

SAFETY PROFILE: Questionable carcinogen with experimental neoplastigenic data. When heated to decomposition it emits very toxic fumes of Cl^-, NO_x, and SO_x.

CEL250 CAS:118-91-2 ***HR: 2***
2-CHLOROBENZOIC ACID
mf: $C_7H_5ClO_2$ mw: 156.57

SYNS: 2-CBA ◇ o-CHLOROBENZOIC ACID ◇ KYSELINA o-CHLORBENZOOVA (CZECH)

TOXICITY DATA with REFERENCE
skn-rbt 500 mg/24H MLD 28ZPAK -,91,72
eye-rbt 20 mg/24H MOD 28ZPAK -,91,72
orl-rat LD50:6460 mg/kg 28ZPAK -,91,72
ipr-rat LD50:2300 mg/kg BCFAAI 112,53,73

CONSENSUS REPORTS: Reported in EPA TSCA Inventory.

SAFETY PROFILE: Moderately toxic by intraperitoneal route. Mildly toxic by ingestion. An eye and skin irritant. When heated to decomposition it emits toxic fumes of Cl^-.

CEL500 CAS:7250-60-4 ***HR: 3***
o-CHLOROBENZOIC ACID NICKEL(II) SALT
mf: $C_{14}H_8Cl_2O_4{\bullet}Ni$ mw: 369.83

TOXICITY DATA with REFERENCE
ipr-mus LDLo:125 mg/kg CBCCT* 4,317,52

CONSENSUS REPORTS: Nickel and its compounds are on the Community Right-To-Know List.

NIOSH REL: (Inorganic Nickel) TWA: 0.015 mg(Ni)/m^3

SAFETY PROFILE: Poison by intraperitoneal route. See also NICKEL COMPOUNDS. When heated to decomposition it emits toxic fumes of Cl^-.

CEL750 CAS:81-45-8 ***HR: 2***
CHLOROBENZONE
mf: $C_{21}H_{12}ClNO_3$ mw: 361.79

SYNS: 1BA-4-XA (RUSSIAN) ◇ 1-BENZOYLAMINO-2-CHLOROANTHRAQUINONE ◇ N-(4-CHLORO-9,10-DIHYDRO-9,10-DIOXO-1-ANTHRACENYL) BENZAMIDE (9CI)

TOXICITY DATA with REFERENCE
ipr-rat LD50:2000 mg/kg GTPZAB 21(12),27,77

CONSENSUS REPORTS: Reported in EPA TSCA Inventory.

SAFETY PROFILE: Moderately toxic by intraperitoneal route. When heated to decomposition it emits very toxic fumes of Cl^- and NO_x.

CEM000 CAS:873-32-5 ***HR: 3***
o-CHLOROBENZONITRILE
mf: C_7H_4ClN mw: 137.57

PROP: (o- and p-). Crystals.

SYNS: o-CHLORBENZONITRIL (CZECH) ◇ NITRIL KYSELINY-o-CHLORBENZOOVE (CZECH)

TOXICITY DATA with REFERENCE
eye-rbt 100 mg/24H MOD 28ZPAK -,160,72
orl-rat LD50:435 mg/kg 28ZPAK -,160,72
ipr-mus LD50:150 mg/kg NTIS** AD691-490

CONSENSUS REPORTS: Reported in EPA TSCA Inventory. Cyanide and its compounds are on the Community Right-To-Know List.

SAFETY PROFILE: Poison by intraperitoneal route. Moderately toxic by ingestion. An eye irritant. When heated to decomposition or on contact with water, steam, acid, or acid fumes it emits toxic fumes of Cl^-, and CN^-. See also NITRILES.

CEM250 CAS:623-03-0 ***HR: 3***
p-CHLOROBENZONITRILE
mf: C_7H_4ClN mw: 137.57

SYN: NITRIL KYSELINY p-CHLORBENZOOVE (CZECH)

TOXICITY DATA with REFERENCE
eye-rbt 100 mg/24H MOD 28ZPAK -,161,72
orl-rat LD50:887 mg/kg 28ZPAK -,161,72
ipr-mus LD50:150 mg/kg NTIS** AD691-490

CONSENSUS REPORTS: Reported in EPA TSCA Inventory. Cyanide and its Compounds are on the Community Right-To-Know List.

SAFETY PROFILE: Poison by intraperitoneal route. Moderately toxic by ingestion. An eye irritant. When heated to decomposition it emits very toxic fumes of NO_x, Cl^-, and CN^-. See also NITRILES.

CEM500 CAS:615-20-3 ***HR: 3***
2-CHLOROBENZOTHIAZOLE
mf: C_7H_4C1NS mw: 169.63

SYN: USAF EK-2784

TOXICITY DATA with REFERENCE
orl-rat LDLo:250 mg/kg NCNSA6 5,24,53
ipr-mus LD50:200 mg/kg NTIS** AD277-689

CONSENSUS REPORTS: Reported in EPA TSCA Inventory.

SAFETY PROFILE: Poison by ingestion and intraperitoneal routes. When heated to decomposition it emits very toxic fumes of SO_x, NO_x, and Cl^-.

CEM625 ***HR: 3***
1-CHLOROBENZOTRIAZOL
mf: $C_6H_4ClN_3$ mw: 153.57

$C_6H_4N(Cl)N{=}N$ (ring)

SAFETY PROFILE: May ignite spontaneously. When heated to decomposition it emits toxic fumes of Cl^- and NO_x.

CEM825 CAS:98-56-6 ***HR: 1***
p-CHLOROBENZOTRIFLUORIDE
DOT: UN 2234
mf: $C_7H_4ClF_3$ mw: 180.56

SYNS: (p-CHLOROPHENYL)TRIFLUOROMETHANE ◇ p-CHLOROTRIFLUOROMETHYLBENZENE ◇ 4-CHLOROTRIFLUOROMETHYLBENZENE ◇ 1-CHLORO-4-(TRIMETHYL)-BENZENE (9CI) ◇ α,α,α-TRIFLUORO-4-CHLOROTOLUENE ◇ p-TRIFLUOROMETHYLPHENYL CHLORIDE ◇ p-(TRIFLUOROMETHYL)CHLOROBENZENE

TOXICITY DATA with REFERENCE
dns-hmn:emb 1 g/L AISSAW 18,123,82
orl-rat LD50:13 g/kg GTPZAB 28(5),49,84
ihl-rat LC50:22 g/m^3 GTPZAB 28(5),49,84
orl-mus LD50:11500 mg/kg GTPZAB 28(5),49,84
ihl-mus LC50:20 g/m^3 GTPZAB 28(5),49,84

CONSENSUS REPORTS: Reported in EPA TSCA Inventory.

DOT Classification: Flammable or Combustible Liquid; Label: Flammable Liquid.

SAFETY PROFILE: Mildly toxic by ingestion and inhalation. Human mutation data reported. Flammable. Strongly exothermic reaction with sodium dimethylsulfinate. When heated to decomposition it emits toxic fumes of F^- and Cl^-. See also CHLORINATED HYDROCARBONS, AROMATIC; and FLUORIDES.

CEO000 CAS:14848-01-2 ***HR: 3***
p-CHLOROBENZOYL AZIDE
mf: $C_7H_4ClN_3O$ mw: 195.59

$ClC_6H_4CO{\bullet}N_3$

SAFETY PROFILE: An unstable explosive. When heated to decomposition it emits toxic fumes of Cl^- and NO_x. See also AZIDES.

CEP000 CAS:103-17-3 ***HR: 2***
p-CHLOROBENZYL-p-CHLOROPHENYL SULFIDE
mf: $C_{13}H_{10}Cl_2S$ mw: 269.19

PROP: Crystals, almond-like odor. Mp: 75-76°, d: 1.4210 @ 25°/4°, vap press: 1.21 × 10^{-5} mm @ 30°. Insol in water, sol in most organic solvents.

SYNS: CHLOORBENZIDE (DUTCH) ◇ (4-CHLOOR-BENZYL)-(4-CHLOOR-FENYL)-SULFIDE (DUTCH) ◇ CHLORBENSID (GERMAN) ◇ CHLORBENSIDE ◇ CHLORBENXIDE ◇ CHLORBENZIDE ◇ (4-CHLOR-BENZYL)-(4-CHLOR-PHENYL)-SULFID (GERMAN) ◇ p-CHLOROBENZYL-p-CHLOROPHENYL SULPHIDE ◇ 4-CHLOROBENZYL-4-CHLOROPHENYL SULPHIDE ◇ 1-CHLORO-4-(((4-CHLOROPHENYL)METHYL)THIO)BENZENE ◇ CHLOROCIDE ◇ CHLOROPARACIDE ◇ 4-CHLOROPHENYL-4'-CHLOROBENZYL SULFIDE ◇ CHLOROSULFACIDE ◇ CHLORPARACIDE ◇ CHLORSULPHACIDE ◇ (4-CLORO-BENZIL)-(4-CLORO-FENIL)-SOLFURO (ITALIAN) ◇ p,p'-DICHLORODIPHENYL SULFIDE ◇ ENT 20,696 ◇ HRS 860 ◇ METOX ◇ MITOX ◇ RD 2195 ◇ SULFURE de 4-CHLOROBENZYLE et de 4-CHLOROPHENYLE (FRENCH)

TOXICITY DATA with REFERENCE
orl-rat LD50:2000 mg/kg WRPCA2 9,119,70
unr-mus LD50:3000 mg/kg 30ZDA9 -,242,71

SAFETY PROFILE: Moderately toxic by ingestion and possibly other routes. Has caused liver and kidney injury and skin irritation in experimental animals. When heated to decomposition it emits toxic fumes of Cl^- and SO_x. See also CHLORIDES and SULFIDES.

CEP250 CAS:78393-38-1 ***HR: 3***
1-(4-CHLOROBENZYL)-3-(6-CHLORO-o-TOLYL)-1-(2-PYRROLIDINYLETHYL) UREA HYDROCHLORIDE
mf: $C_{21}H_{25}Cl_2N_3O{\bullet}ClH$ mw: 442.85

SYN: C 5323

TOXICITY DATA with REFERENCE
eye-rbt 2% MLD ARZNAD 8,664,58
ipr-rat LD50:105 mg/kg ARZNAD 8,664,58
scu-mus LD50:205 mg/kg ARZNAD 8,664,58

SAFETY PROFILE: Poison by intraperitoneal and subcutaneous routes. An eye irritant. When heated to decomposition it emits very toxic fumes of NO_x and Cl^-.

CEP675 CAS:13835-15-9 ***HR: 3***
4-(p-CHLOROBENZYL)-2-(2-DIMETHYLAMINO)ETHYL)-1(2H)-PHTHALAZINONE HYDROCHLORIDE
mf: $C_{19}H_{20}ClN_3O{\bullet}ClH$ mw: 378.33

SYN: HL 2197

TOXICITY DATA with REFERENCE
orl-mus LD50:300 mg/kg ARZNAD 8,219,58
ipr-mus LD50:157 mg/kg ARZNAD 7,678,57
scu-mus LD50:100 mg/kg ARZNAD 8,219,58
ivn-mus LD50:75 mg/kg ARZNAD 8,219,58

SAFETY PROFILE: Poison by ingestion, subcutaneous, intravenous, and intraperitoneal routes. When heated to decomposition it emits toxic fumes of NO_x and Cl^-.

CEQ500 CAS:3309-77-1 ***HR: 3***
p-CHLOROBENZYL-3-HYDROXYCROTONATE DIMETHYL PHOSPHATE
mf: $C_{13}H_{16}ClO_6P$ mw: 334.71

SYNS: BAS 4239 ◇ 1-(p-CHLOROBENZYLOXYCARBONYL)-1-PROPEN-2-YL-DIMETHYLPHOSPHATE ◇ DIMETHYL PHOSPHATE-3-HYDROXY-CROTONIC ACID, p-CHLOROBENZYL ESTER ◇ ENT 24716 ◇ PHOSPHORIC ACID, DIMETHYL ESTER with p-CHLOROBENZYL-3-HYDROXYCROTONATE ◇ SD 4,239 ◇ SHELL SD 4,239

TOXICITY DATA with REFERENCE
orl-rat LD50:139 mg/kg 28ZEAL 4,98,69
orl-mus LD50:212 mg/kg 28ZEAL 4,98,69

SAFETY PROFILE: Poison by ingestion. When heated to decomposition it emits very toxic fumes of PO_x and Cl^-. See also ESTERS and PHOSPHATES.

CEQ600 CAS:2698-41-1 ***HR: 3***
o-CHLOROBENZYLIDENE MALONONITRILE
mf: $C_{10}H_5ClN_2$ mw: 188.62

PROP: White crystals. Mp: 95°, bp: 313°.

SYNS: 2-CHLOROBENZAL MALONONITRILE ◇ o-CHLOROBENZAL MALONONITRILE ◇ o-CHLOROBENZYLIDENE MALONITRILE ◇ 2-CHLOROBENZYLIDENE MALONONITRILE ◇ 2-CHLOROBMN ◇ CS ◇ β,β-DICYANO-o-CHLOROSTYRENE ◇ NCI-C55118 ◇ PROPANEDINITRILE((2-CHLOROPHENYL)METHYLENE) ◇ USAF KF-11

TOXICITY DATA with REFERENCE
skn-hmn 10 mg/1H MLD BJDEAZ 90,657,74
eye-man 5 mg/m³/20S SEV MMEDA9 134,219,69
eye-man 624 ng APTOA6 35,412,74
skn-rat 12%/6H open MLD ARTODN 40,75,78
skn-rbt 12%/6H open MLD ARTODN 40,75,78
eye-rbt 5 mg TXAPA9 4,656,62
eye-rbt 1 mg MLD TXAPA9 17,295,70
eye-rbt 1150 ng APTOA6 35,412,74
skn-gpg 12%/6H open MLD ARTODN 40,75,78
eye-gpg 429 ng APTOA6 35,412,74
mmo-sat 100 μg/plate ARTODN 49,15,81
mma-sat 100 μg/plate ARTODN 49,15,81
ihl-rat TCLo:6 mg/m³/5M (female 6-15D post):TER 37QLAZ -,79,77
ihl-hmn TCLo:1500 μg/m³/90M:EYE,PUL AEHLAU 6,366,63
orl-rat LD50:178 mg/kg AEHLAU 24,449,72
ihl-rat LCLo:1806 mg/m³/45M ARTODN 40,75,78
ipr-rat LD50:48 mg/kg TXAPA9 24,45,73
ivn-rat LD50:28 mg/kg ARTODN 40,75,78
orl-mus LD50:282 mg/kg NTIS** AD837-111
ihl-mus LCLo:2753 mg/m³/20M ARTODN 40,75,78
ipr-mus LD50:32320 μg/kg TOLED5 8,73,81
ivn-mus LD50:47700 μg/kg TXAPA9 25,111,73
orl-rbt LD50:143 mg/kg ARTODN 40,75,78
ihl-rbt LCLo:1802 mg/m³/10M ARTODN 40,75,78
ivn-rbt LDLo:8 mg/kg TXAPA9 4,656,62

orl-gpg LD50:212 mg/kg ARTODN 40,75,78
ihl-gpg LCLo:2326 mg/m^3/10M ARTODN 40,75,78
ipr-gpg LD50:73 mg/kg ARTODN 40,75,78

CONSENSUS REPORTS: Reported in EPA TSCA Inventory. Cyanide and its compounds are on the Community Right-To-Know List.

OSHA PEL: (Transitional: TWA 0.05 ppm) CL 0.05 ppm (skin)
ACGIH TLV: CL 0.05 ppm (skin)
OSHA PEL: TWA 0.05 ppm

SAFETY PROFILE: Poison by ingestion, intraperitoneal, and intravenous routes. Moderately toxic by inhalation. Human systemic effects by inhalation: conjuntiva irritation, cough, and unspecified respiratory system effects. A human skin and eye irritant. Human exposure data suggest relatively low systematic toxicity, but intense irritation of eyes, skin, and mucous membranes. Mutation data reported. A tear-gas used for riot control. When heated to decomposition it emits very toxic fumes of Cl^-, NO_x, and CN^-. See also NITRILES.

CEQ625 CAS:50264-86-3 ***HR: 3***
1-p-CHLOROBENZYL-1H-INDAZOLE-3-CARBOXYLIC ACID
mf: $C_{15}H_{11}ClN_2O_2$ mw: 286.73

SYNS: AF 1312/TS ◇ 1-(4-CHLOROBENZYL)-1H-INDAZOLE-3-CARBOXYLIC ACID ◇ 1-((4-CHLOROPHENYL)METHYL)-1H-INDOLE-3-CARBOXYLIC ACID

TOXICITY DATA with REFERENCE
spm-rat-orl 200 mg/kg EXMPA6 23,288,75
spm-mky-orl 14400 mg/kg/24W-I EXMPA6 23,357,75
orl-rat TDLo:1 g/kg (female 6-15D post):TER ARTODN 5,197,82
orl-rat TDLo:19 g/kg (63D male):REP JRPFA4 50,159,77
orl-rat LD50:1140 mg/kg DRFUD4 3,87,78
ipr-rat LD50:58 mg/kg EXMPA6 23,288,75

SAFETY PROFILE: Poison by intraperitoneal route. Moderately toxic by ingestion. An experimental teratogen. Other experimental reproductive effects. Mutation data reported. When heated to decomposition it emits toxic fumes of Cl^- and NO_x.

CEQ750 CAS:3694-45-9 ***HR: 3***
4-CHLOROBENZYL ISOTHIOCYANATE
mf: C_8H_6ClNS mw: 183.66

SYN: ISOTHIOCYANIC ACID,-p-CHLOROBENZYL ESTER

TOXICITY DATA with REFERENCE
ipr-rat LDLo:100 mg/kg ARZNAD 16,870,66
ipr-mus LDLo:100 mg/kg ARZNAD 21,121,71
ivn-mus LD50:56 mg/kg CSLNX* NX#02757

SAFETY PROFILE: Poison by intraperitoneal and intravenous routes. See also ESTERS and THIOCYANATES. When heated to decomposition it emits very toxic fumes of SO_x, Cl^- and CN^-.

CER250 ***HR: 2***
2-(p-CHLOROBENZYL(2-(PYRROLIDINYL) ETHYL)AMINO)-o-ACETOTOLUIDIDE DIHYDROCHLORIDE
mf: $C_{22}H_{28}ClN_3O{\bullet}2ClH$ mw: 458.90

SYN: C 5352.

TOXICITY DATA with REFERENCE
eye-rbt 2% SEV ARZNAD 9,167,59
scu-mus LD50:550 mg/kg ARZNAD 9,167,59

SAFETY PROFILE: Moderately toxic by subcutaneous route. A severe eye irritant. When heated to decomposition it emits very toxic fumes of Cl^- and NO_x.

CER825 CAS:72985-56-9 ***HR: 3***
2-CHLORO-1,1-BIS(FLUOROOXY) TRIFLUOROETHANE
mf: $C_2ClF_5O_2$ mw: 186.47

SAFETY PROFILE: An extremely unstable explosive. Upon decomposition it emits toxic fumes of F^- and Cl^-. See also EXPLOSIVES.

CES250 CAS:64037-53-2 ***HR: 3***
1-CHLORO-3-BROMO-BUTENE-1
mf: C_4H_6BrCl mw: 169.46

TOXICITY DATA with REFERENCE
orl-rat LD50:74 mg/kg SCCUR* -,3,61
orl-mus LD50:56500 μg/kg SCCUR* -,3,61

SAFETY PROFILE: Poison by ingestion. When heated to decomposition it emits very toxic fumes of Br^- and Cl^-. See also CHLORINATED HYDROCARBONS, ALIPHATIC; and BROMIDES.

CES500 CAS:107-04-0 ***HR: 3***
1-CHLORO-2-BROMOETHANE
mf: C_2H_4BrCl mw: 143.42

PROP: Colorless, volatile liquid; sweet chloroform-like odor. Bp: 106.1°, fp: −18.4°, flash p: none, d: 1.7272 @ 25°/4°, vap press 40 mm @ 29.7°, vap d: 4.94.

SYNS: sym-CHLOROBROMOETHANE ◇ ETHYLENE CHLOROBROMIDE

TOXICITY DATA with REFERENCE
mmo-sat 1 mmol/L CRNGDP 2,499,81
mma-sat 10 μmol/plate EVHPAZ 21,79,77
dnr-esc 10 μL/plate EVHPAZ 21,79,77
dnd-mus-ipr 500 μmol/kg CRNGDP 4,1491,83
msc-ham:ovr 200 μmol/L MUREAV 90,183,81
orl-rat LD50:64 mg/kg 28ZEAL 4,73,69

CONSENSUS REPORTS: Reported in EPA TSCA Inventory. EPA Genetic Toxicology Program.

SAFETY PROFILE: Poison by ingestion and probably inhalation routes. An irritant to the skin, eyes, and mucous membranes. May cause injury to liver and kidneys. Mutation data reported. When heated to decomposition it emits toxic fumes of Cl^- and Br^-. See also CHLORINATED HYDROCARBONS, ALIPHATIC; and BROMIDES.

CES650 CAS:74-97-5 ***HR: 3***
CHLOROBROMOMETHANE
DOT: UN 1887
mf: CH_2BrCl mw: 129.39

PROP: Clear, colorless liquid; sweet odor. Bp: 67.8°, fp: −88°, flash p: none, d: 1.930 @ 25°/25°, vap d: 4.46.

SYNS: BROMOCHLOROMETHANE ◇ HALON 1011 ◇ METHYLENE CHLOROBROMIDE ◇ MIL-B-4394-B ◇ MONO-CHLORO-MONO-BROMO-METHANE

TOXICITY DATA with REFERENCE
mmo-sat 10 μL/plate DHEFDK FDA-78-1046,78
orl-rat LD50:5000 mg/kg 34ZIAG -,390,69
ihl-rat LCLo:28800 ppm/15M MRLR** #113,52
orl-mus LD50:4300 mg/kg JIHTAB 29,382,47
ihl-mus LC50:15850 mg/m^3/8H JIHTAB 29,382,47

CONSENSUS REPORTS: Reported in EPA TSCA Inventory.

OSHA PEL: TWA 200 ppm
ACGIH TLV: TWA 200 ppm
DFG MAK: 200 ppm (1050 mg/m^3)
DOT Classification: Poison B; Label: St. Andrews Cross.

SAFETY PROFILE: A poison. Mildly toxic by ingestion and inhalation. Mutation data reported. This material has a narcotic action of moderate intensity, although of prolonged duration. Animals exposed for several weeks to 1000 ppm had blood bromide levels as high as 350 mg/100 gram. Therefore, until further data are available, it should be considered at least as toxic as carbon tetrachloride and more than minimal exposure to its vapors should be avoided. Dangerous; when heated to decomposition it emits highly toxic fumes of Br^- and Cl^-. See also BROMIDES and CHLORINATED HYDROCARBONS, ALIPHATIC.

CES750 CAS:13360-45-7 ***HR: 2***
1-(3-CHLORO-4-BROMOPHENYL)-3-METHYL-3-METHOXYUREA
mf: $C_9H_{10}BrClN_2O_2$ mw: 293.57

SYNS: BROMEX ◇ N-(4-BROMO-3-CHLOROPHENYL)-N'-METHOXY-N'-METHYLUREA ◇ N'-(4-BROMO-3-CHLOROPHENYL)-N-METHOXY-N-METHYLUREA ◇ 3-(4-BROMO-3-CHLOROPHENYL)-1-METHOXY-1-METHYLUREA ◇ C-6313 ◇ CHLORBROMURON ◇ CHLOROBRUMURON ◇ CIBA 6313 ◇ MALORAN

TOXICITY DATA with REFERENCE
eye-rbt 50 mg MOD CIGET* -,-,77
orl-rat LD50:2150 mg/kg PHARAT 37,370,82

SAFETY PROFILE: Moderately toxic by ingestion. An eye irritant. An herbicide. When heated to decomposition it emits very toxic fumes of NO_x, Cl^-, and Br^-.

CET000 CAS:73926-87-1 ***HR: 3***
trans-CHLORO(2-(3-BROMOPROPIONAMIDO) CYCLOHEXYL)MERCURY
mf: $C_9H_{15}BrClHgNO$ mw: 469.20

SYNS: 3-BROMO-N-(2-CHLOROMERCURICYCLOHEXYL)PROPIONAMIDE ◇ MERCURY (E)-CHLORO(2-(3-BROMOPROPIONAMIDO)CYCLOHEXYL)

TOXICITY DATA with REFERENCE
ivn-mus LD50:32 mg/kg CSLNX* NX#04830

CONSENSUS REPORTS: Mercury and its compounds are on the Community Right-To-Know List.

OSHA PEL: (Transitional: CL 1 mg/10m^3) TWA 0.01 mg(Hg)/m^3; STEL 0.03 mg/m^3 (skin)
ACGIH TLV: TWA 0.1 mg(Hg)/m^3 (skin)
NIOSH REL: (Inorganic Mercury) TWA: 0.05 mg(Hg)/m^3

SAFETY PROFILE: Poison by intravenous route. See also MERCURY COMPOUNDS. When heated to decomposition it emits very toxic fumes of Br^-, Cl^-, NO_x, and Hg.

CET250 CAS:627-22-5 ***HR: 3***
1-CHLOROBUTADIENE
mf: C_4H_5Cl mw: 88.54

PROP: Colorless liquid. Bp: 59.4°, d: 0.9583, flash p: −4°F, lel: 4.0%, uel: 20.0%, vap d: 3.0.

SYN: 1-CHLORO-1,3-BUTADIENE

TOXICITY DATA with REFERENCE
mmo-sat 8 pph/4H ARTODN 41,249,79
mma-sat 8 pph/4H ARTODN 41,249,79
sln-dmg-orl 5700 μmol/L/3D-I 35WYAM -,63,76

CONSENSUS REPORTS: EPA Genetic Toxicology Program.

SAFETY PROFILE: Mutation data reported. Probably a poison by ingestion, subcutaneous, and intravenous routes. Dangerous fire hazard when exposed to heat or flame. To fight fire, use alcohol foam. When heated to decomposition it emits toxic fumes of Cl^-. See also CHLORINATED HYDROCARBONS, ALIPHATIC.

CEU000 ***HR: 3***
1-CHLOROBUTANE
mf: C_4H_9Cl mw: 92.57

PROP: Flash p: 10.4°F; lel: 1.9%; uel: 10.1%.

SAFETY PROFILE: A dangerous fire hazard when exposed to heat or flame. To fight fire, use water, water spray, fog, mist, dry chemical, alcohol foam. Reaction with divalent metals may form dangerously reactive products. When heated to decomposition it emits toxic fumes of Cl^-. See also 2-CHLOROBUTANE and CHLORINATED HYDROCARBONS, ALIPHATIC.

CEU250 CAS:78-86-4 ***HR: 3***
2-CHLOROBUTANE
DOT: UN 1127
mf: C_4H_9Cl mw: 92.58

PROP: Flash p: 14°F, d: 0.87, vap d: 3.2, bp: 68.50.

SYN: sec-BUTYL CHLORIDE

TOXICITY DATA with REFERENCE
ipr-mus TDLo:3240 mg/kg/8W-I:NEO CNREA8 35,1411,75
orl-rat LD50:17460 mg/kg AIHAAP 30,470,69
ihl-rat LCLo:8000 ppm/4H AIHAAP 30,470,69
skn-rbt LD50:20 g/kg AIHAAP 30,470,69

CONSENSUS REPORTS: Reported in EPA TSCA Inventory.

DOT Classification: Flammable Liquid; Label: Flammable Liquid.

SAFETY PROFILE: Mildly toxic by ingestion, inhalation, and skin contact. Questionable carcinogen with experimental neoplastigenic data. Dangerous fire hazard when exposed to heat, open flame (sparks), or oxidizers. To fight fire, use water, water spray, fog, mist, dry chemical, alcohol foam. When heated to decomposition it emits toxic fumes of Cl^-. See also CHLORINATED HYDROCARBONS, ALIPHATIC.

CEU500 CAS:928-51-8 ***HR: 3***
4-CHLORO-1-BUTANOL
mf: C_4H_9ClO mw: 108.58

SYNS: 4-CHLORBUTAN-1-OL (GERMAN) ◇ 4-CHLORO-1-BUTANEOL ◇ 4-CHLOROBUTANOL ◇ TETRAMETHYLENE CHLOROHYDRIN

TOXICITY DATA with REFERENCE
mmo-sat 20 μmol/plate MUREAV 90,91,81
ipr-mus TDLo:3650 mg/kg/8W-I:NEO CNREA8 39,391,79
orl-mus LD50:990 mg/kg ZHYGAM 26,17,80

CONSENSUS REPORTS: Reported in EPA TSCA Inventory.

SAFETY PROFILE: Moderately toxic by ingestion. Questionable carcinogen with experimental neoplastigenic data. Mutation data reported. When heated to decomposition it emits toxic fumes of Cl^-. See also CHLORIDES and ALCOHOLS.

CEU750 CAS:616-27-3 ***HR: 3***
1-CHLORO-2-BUTANONE
mf: C_4H_7ClO mw: 106.55

SAFETY PROFILE: A dangerously unstable explosive; may explode spontaneously. Upon decomposition it emits toxic fumes of Cl^-.

CEU825 CAS:591-97-9 ***HR: 1***
1-CHLORO-2-BUTENE
mf: C_4H_7Cl mw: 90.56

PROP: trans-Isomer: Liquid. Bp: 84.8°, n (20/D) 1.4350, n (25/D) 1.4327, d: 0.9295. cis-Isomer: Liquid. Bp: 84.1°, n (20/D) 1.4390, d: 0.9426.

TOXICITY DATA with REFERENCE
mmo-sat 5 μmol/plate BCPCA6 29,2611,80
mma-sat 5 μmol/plate BCPCA6 29,2611,80
dns-hmn:hla 100 μmol/L CALEDQ 20,263,83

SAFETY PROFILE: Human mutation data reported. An eye and mucous membrane irritant. When heated to decomposition it emits toxic fumes of Cl^-. See also CHLORINATED HYDROCARBONS, ALIPHATIC.

CEV000 CAS:4461-41-0 ***HR: 2***
2-CHLORO-2-BUTENE
mf: C_4H_7Cl mw: 90.55

$CH_3CH{=}CClCH_3$

PROP: Flash p: −13°F; lel: 2.3%; uel: 9.3%.

SAFETY PROFILE: A dangerous fire hazard when exposed to heat or flame. When heated to decomposition it emits toxic fumes of Cl^-. See also CHLORINATED HYDROCARBONS, ALIPHATIC.

CEV250 CAS:563-52-0 ***HR: 3***
3-CHLORO-1-BUTENE
mf: C_4H_7Cl mw: 90.55

$H_2C{=}CHCHClCH_3$

PROP: dl-Form: Liquid. Bp: 63.9-64.2°, d: 0.9001, n (20/D) 1.4150. d(−)-Form: Liquid. Bp: −5°. l(+)-Form: Liquid. Flash p.: −16.6°F.

TOXICITY DATA with REFERENCE
mmo-sat 10 μmol/plate BCPCA6 29,2611,80
mma-sat 10 μmol/plate BCPCA6 29,2611,80
dns-hmn:hla 500 μmol/L CALEDQ 20,263,83

SAFETY PROFILE: Human mutation data reported.

Irritates eyes, respiratory passages. A dangerous fire hazard when exposed to heat or flame. When heated to decomposition it emits toxic fumes of Cl^-. See also CHLORINATED HYDROCARBONS, ALIPHATIC.

CEV500 CAS:7119-27-9 ***HR: 3***
1-CHLORO-1-BUTEN-3-ONE
mf: C_4H_5ClO mw: 104.54

$ClCH{=}CHCO{\bullet}CH_3$

SAFETY PROFILE: An unstable explosive. Upon decomposition it emits toxic fumes of Cl^-.

CEV750 CAS:55477-20-8 ***HR: 3***
4-CHLORO-2-(tert-BUTYLAMINO)-6-(4-METHYL-PIPERAZINO)-5-METHYLTHIOPYRIMIDINE
mf: $C_{14}H_{24}ClN_5S$ mw: 329.94

TOXICITY DATA with REFERENCE
orl-mus LD50:600 mg/kg JMCMAR 18,553,75
ivn-mus LD50:67 mg/kg JMCMAR 18,553,75

SAFETY PROFILE: Poison by intravenous route. Moderately toxic by ingestion. When heated to decomposition it emits very toxic fumes of SO_x, NO_x, and Cl^-.

CEV800 CAS:13280-07-4 ***HR: 3***
4-CHLORO-2-BUTYNOL
mf: C_4H_5ClO mw: 104.54

$ClCH_2C{\equiv}CCH_2OH$

SAFETY PROFILE: A heat-sensitive explosive. When heated to decomposition it emits toxic fumes of Cl^-.

CEW000 CAS:1951-12-8 ***HR: 3***
β-CHLOROBUTYRIC ACID
mf: $C_4H_7ClO_2$ mw: 122.56

SYNS: 3-CHLOROBUTANOIC ACID ◇ 3-CHLOROBUTYRIC ACID

TOXICITY DATA with REFERENCE
ipr-mus TDLo:1180 mg/kg/8W-I:ETA CNREA8 39,391,79

CONSENSUS REPORTS: Reported in EPA TSCA Inventory.

SAFETY PROFILE: Questionable carcinogen with experimental tumorigenic data. When heated to decomposition it emits toxic fumes of Cl^-. See also CHLORIDES.

CEW500 CAS:101-27-9 ***HR: 3***
m-CHLORO CARBANILIC ACID-4-CHLORO-2-BUTYNYL ESTER
mf: $C_{11}H_9Cl_2NO_2$ mw: 258.11

SYNS: A-980 ◇ BARBAMATE ◇ BARBAN ◇ BARBANE ◇ 2-BUTYNYL-4-CHLORO-m-CHLOROCARBANILATE ◇ C-847 ◇ CARBIN ◇ CARBYNE ◇ CARYNE ◇ CBN ◇ (4-CHLOOR-BUT-2-YN-YL)-N-(3-CHLOOR-FENYL)-CARBAMAAT (DUTCH) ◇ (4-CHLOR-BUT-2-IN-YL)-N-(3-CHLOR-PHENYL)-CARBAMAT (GERMAN) ◇ CHLORINAT ◇ CHLORO-2-BUTYNYL-m-CHLOROCARBAMATE ◇ 4-CHLOROBUT-2-YNYL-m-CHLOROCARBANILATE ◇ 4-CHLORO-2-BUTYNYL-m-CHLOROCARBANILATE ◇ 4-CHLOROBUT-2-YNYL-3-CHLOROPHENYLCARBAMATE ◇ 4-CHLORO-2-BUTYNYL-N-(3-CHLOROPHENYL)CARBAMATE ◇ N-(3-CHLORO PHENYL) CARBAMATE de 4-CHLORO 2-BUTYNYLE (FRENCH) ◇ (3-CHLOROPHENYL) CARBAMIC ACID 4-CHLORO-2-BUTYNYL ESTER ◇ (4-CLORO-BUT-2-IN-IL)-N-(3-CLORO-FENIL)-CARBAMMATO (ITALIAN) ◇ CS-847 ◇ FISONS B25 ◇ NEOBAN ◇ S-847

TOXICITY DATA with REFERENCE
mrc-bcs 20 μg/disc/24H MUREAV 40,19,76
orl-rat LD50:600 mg/kg RREVAH 10,97,65
ihl-rat LD50:527 mg/kg EQSFAP 3,618,75
orl-mus LD50:322 mg/kg FATOAO 35,356,72
orl-rbt LD50:600 mg/kg PAREAQ 14,225,62
skn-rbt LD50:23000 mg/kg 85DPAN -,-,71/76
orl-gpg LD50:240 mg/kg PAREAQ 14,225,62
unr-mam LD50:240 mg/kg 30ZDA9 -,202,71

CONSENSUS REPORTS: EPA Genetic Toxicology Program.

SAFETY PROFILE: Poison by ingestion, inhalation, and possibly other routes. Mildly toxic by skin contact. Mutation data reported. An herbicide. See also CARBAMATES and ESTERS. When heated to decomposition it emits very toxic fumes of Cl^- and NO_x.

CEX250 CAS:1967-16-4 ***HR: 3***
m-CHLOROCARBANILIC ACID-1-METHYL-2-PROPYNYL ESTER
mf: $C_{11}H_{10}ClNO_2$ mw: 223.67

SYNS: BICP ◇ BIPC (the herbicide) ◇ BUTYN-1-OL-3-ESTER of m-CHLOROPHENYLCARBAMIC ACID ◇ 3-BUTYNYL-m-CHLOROCARBANILATE ◇ 1-BUTYN-3-YL-m-CHLOROPHENYL-CARBAMATE ◇ CHLORBUFAM ◇ CHLORBUPHAM ◇ CHLOROBUFAM ◇ 3-CHLOROPHENYLCARBAMIC ACID-1-METHYLPROPYNYL ESTER ◇ 3-CHLORPHENYL-CARBAMIDSAURE-BUTIN-(1)-YL(3)-ESTER (GERMAN) ◇ GRISEMIN ◇ GRISIN ◇ IEM-1-15 ◇ ISOBUTINYL-N-(3-CHLORPHENYL)-CARBAMAT (GERMAN) ◇ 1-METHYL-2-PROPYNYL-m-CHLOROCARBANILATE ◇ 1-METHYL-2-PROPYNYL-m-CHLOROPHENYLCARBAMATE ◇ 1-METHYLPROPYNYL 3-CHLOROPHENYLCARMATE ◇ 1-METHYLPROPYNYL ESTER of 3-CHLOROPHENYLCARBAMIC ACID

TOXICITY DATA with REFERENCE
orl-rat LD50:2500 mg/kg RREVAH 10,97,65
ipr-mus LD50:250 mg/kg 85DPAN -,-,71/76
unr-mus LD50:15500 μg/kg 85GDA2 1,262,80

SAFETY PROFILE: Poison by intraperitoneal and possibly other routes. Moderately toxic by ingestion. A pesticide. See also CARBAMATES and ESTERS. When heated to decomposition it emits very toxic fumes of Cl^- and NO_x.

CEX255 CAS:52716-12-8 ***HR: 3***
N-(CHLOROCARBONYLOXY)TRIMETHYLUREA
mf: $C_5H_9ClN_2O_3$ mw: 180.59

$ClCO{\bullet}ONCH_3CO{\bullet}N(CH_3)_2$

SAFETY PROFILE: An unstable explosive. When heated to decomposition it emits toxic fumes of Cl^- and NO_x. See also EXPLOSIVES.

CEX275 CAS:5665-94-1 ***HR: 3***
5-CHLOROCARVACROL
mf: $C_{10}H_{13}ClO$ mw: 184.68

SYNS: CARVASEPT ◇ CHLORCARVACROL ◇ MCIT ◇ MONOCHLOROISOTHYMOL

TOXICITY DATA with REFERENCE
ipr-mus LDLo:250 mg/kg CBCCT* 5,338,53
scu-mus LDLo:1500 mg/kg AEPPAE 161,196,31
orl-cat LDLo:1000 mg/kg AEPPAE 161,196,31
orl-rbt LDLo:750 mg/kg AEPPAE 161,196,31
scu-frg LDLo:400 mg/kg AEPPAE 161,196,31

SAFETY PROFILE: Poison by subcutaneous and intraperitoneal routes. Moderately toxic by ingestion. When heated to decomposition it emits toxic fumes of Cl^-.

CEX500 CAS:77966-38-2 ***HR: 3***
6'-CHLORO-2-(p-CHLOROBENZYL(2-(DIETHYLAMINO)ETHYL)AMINO)-o-ACETOLUIDIDE DIHYDROCHLORIDE
mf: $C_{22}H_{29}Cl_2N_3O{\bullet}2ClH$ mw: 495.36

SYN: C 5364

TOXICITY DATA with REFERENCE
eye-rbt 2% SEV ARZNAD 9,167,59
ipr-rat LD50:103 mg/kg ARZNAD 9,167,59
scu-mus LD50:240 mg/kg ARZNAD 9,167,59

SAFETY PROFILE: Poison by intraperitoneal and subcutaneous routes. A severe eye irritant. When heated to decomposition it emits very toxic fumes of Cl^- and NO_x.

CEX750 ***HR: 3***
6'-CHLORO-2-(p-CHLOROBENZYL(2-(PYRROLIDINYL)ETHYL)AMINO)-o-ACETOTOLUIDIDE DIHYDROCHLORIDE
mf: $C_{22}H_{27}Cl_2N_3O{\bullet}2ClH$ mw: 493.34

SYN: C 5343

TOXICITY DATA with REFERENCE
eye-rbt 2% SEV ARZNAD 9,167,59
ipr-rat LD50:91 mg/kg ARZNAD 9,167,59
scu-mus LD50:225 mg/kg ARZNAD 9,167,59

SAFETY PROFILE: Poison by intraperitoneal and subcutaneous routes. A severe eye irritant. When heated to decomposition it emits very toxic fumes of Cl^- and NO_x.

CEY250 CAS:78110-37-9 ***HR: 3***
6-CHLORO-9-((2-((2-CHLOROETHYL)AMINO)ETHYL)AMINO)-2-METHOXYACRIDINE 2-HYDROCHLORIDE SESQUIHYDRATE
mf: $C_{18}H_{19}Cl_2N_3O{\bullet}2ClH{\bullet}3/2H_2O$ mw: 464.21

TOXICITY DATA with REFERENCE
mmo-sat 5 μg/plate JMCMAR 15,739,72
ipr-mus LD20:19 mg/kg JMCMAR 15,739,72

SAFETY PROFILE: Poison by intraperitoneal route. Mutation data reported. When heated to decomposition it emits very toxic fumes of NO_x and Cl^-.

CFA250 ***HR: 3***
6-CHLORO-9-(3-(2-CHLOROETHYL)MERCAPTOPROPYLAMINO)-2-METHOXYACRIDINE HYDROCHLORIDE
mf: $C_{19}H_{20}Cl_2N_2OS{\bullet}ClH$ mw: 431.83

SYNS: ICR 342 ◇ 2-METHOXY-6-CHLORO-9-(3-(2-CHLOROETHYL) MERCAPTO PROPYLAMINO) ACRIDINE HYDROCHLORIDE

TOXICITY DATA with REFERENCE
ivn-mus TDLo:6500 μg/kg:NEO CNREA8 36,2423,76
ivn-mus LDLo:12 mg/kg CNREA8 36,2423,76

SAFETY PROFILE: Poison by intravenous route. Questionable carcinogen with experimental neoplastigenic data. When heated to decomposition it emits very toxic fumes of Cl^-, SO_x, and NO_x. See also MERCAPTANS.

CFA500 CAS:126-85-2 ***HR: 3***
2-CHLORO-N-(2-CHLOROETHYL)-N-METHYL ETHANAMINE-N-OXIDE
mf: $C_5H_{11}Cl_2NO$ mw: 172.07

SYNS: 2,2'-DICHLORO-N-METHYLDIETHYLAMINE-N-OXIDE ◇ DIETHYLAMINE, 2,2'-DICHLORO-N-METHYL-, OXIDE ◇ HN_2 AMINE OXIDE ◇ HN_2 OXIDE MUSTARD ◇ MBAO ◇ MECHLORETHAMINE OXIDE ◇ METHYL-BIS(β-CHLOROETHYL)AMINE OXIDE ◇ METHYLBIS(β-CHLOROETHYL)AMINE N-OXIDE ◇ N-METHYL-DI-2-CHLOROETHYLAMINE-N-OXIDE ◇ MITOMEN ◇ MITOMIN ◇ NITROGEN MUSTARD OXIDE ◇ NITROGEN MUSTARD-N-OXIDE ◇ NITROMIN ◇ NMO ◇ N-OXYD-LOST ◇ N-OXYD-MUSTARD ◇ NSC-10107 ◇ OXY-NH2

TOXICITY DATA with REFERENCE
mmo-esc 50 μg/plate TAKHAA 44,96,85
dlt-mus-ipr-20 mg/kg MUREAV 26,285,74
ipr-rat TDLo:20 mg/kg (male 1D pre):REP 85GUAJ -,37,66
ivn-rat TDLo:218 mg/kg/1Y-I:CAR ARZNAD 20,1461,70
skn-mus TDLo:204 g/kg/17W-I:ETA GANNA2 57,295,66

ipr-mus LD50:100 mg/kg NTIS** PB158-507
ivn-mus LD50:80 mg/kg JAJAAA 13,19,60

SAFETY PROFILE: Poison by intravenous and intraperitoneal routes. Experimental reproductive effects. Questionable carcinogen with experimental carcinogenic and tumorigenic data. Mutation data reported. When heated to decomposition it emits toxic fumes of Cl^- and NO_x.

CFA750 CAS:302-70-5 ***HR: 3***
2-CHLORO-N-(2-CHLOROETHYL)-N-METHYL-ETHANAMINE-N-OXIDE HYDROCHLORIDE
mf: $C_5H_{11}Cl_2NO{\cdot}ClH$ mw: 208.53

SYNS: CHLORMETHINE-N-OXIDE HYDROCHLORIDE ◇ 2,2'-DICHLORO-N-METHYLDIETHYLAMINE N-OXIDE HYDROCHLORIDE ◇ HN_2 OXIDE HYDROCHLORIDE ◇ MBAO HYDROCHLORIDE ◇ MECHLORETHAMINE OXIDE HYDROCHLORIDE ◇ METHYL-BIS-(β-CHLORAETHYL)-AMIN-N-OXYD-HYDROCHLORID(GERMAN) ◇ METHYLBIS(β-CHLOROETHYL)AMINE-N-OXIDE HYDROCHLORIDE ◇ N-METHYLBIS(2-CHLOROETHYL)AMINE-N-OXIDE HYDROCHLORIDE ◇ N-METHYL-2,2'-DICHLORODIETHYLAMINE-N-OXIDE HYDROCHLORIDE ◇ METHYLDI(2-CHLOROETHYL)AMINE-N-OXIDE HYDROCHLORIDE ◇ MITOMEN ◇ MUSTRON ◇ NITROGEN MUSTARD OXIDE ◇ NITROGEN MUSTARD-N-OXIDE ◇ NITROGEN MUSTARD-N-OXIDE HYDROCHLORIDE ◇ NITROMIM ◇ NITROMIN HYDROCHLORIDE ◇ N-OXYD-LOST ◇ NSC-10107 ◇ OSSIAMINA ◇ OSSICHLORIN ◇ OXYAMINE ◇ SK-598 ◇ XA 2

TOXICITY DATA with REFERENCE
mmo-sat 500 μg/plate URLRA5 7,119,79
pic-esc 200 mg/L ARMKA7 51,9,65
cyt-slw-par 1670 μL ZEVBA5 89,216,58
bfa-rat/sat 250 mg/kg URLRA5 7,119,79
mul-rat TDLo:1300 mg/kg/93W-I:ETA ZEKBAI 62,112,57
unr-man TDLo:2 mg/kg:CNS,BLD CCROBU 50,219,66
orl-rat LD50:60500 μg/kg NIIRDN 6,256,82
ipr-rat LD50:79500 μg/kg NIIRDN 6,256,82
scu-rat LD50:76100 μg/kg NIIRDN 6,256,82
ipr-mus LD50:125 mg/kg NCISP* JAN86
ivn-dog LDLo:4 mg/kg CCSUBJ 2,201,65
ivn-mky LDLo:8 mg/kg CCSUBJ 2,201,65

CONSENSUS REPORTS: IARC Cancer Review: Group 2B IMEMDT 7,56,87; Animal Sufficient Evidence IMEMDT 9,209,75; EPA Genetic Toxicology Program.

SAFETY PROFILE: Suspected carcinogen with experimental tumorigenic data. Poison by ingestion, subcutaneous, intravenous, and intraperitoneal routes. Human systemic effects by an unspecified route: convulsions and unspecified changes in bone marrow. Mutation data reported. An antineoplastic agent. When heated to decomposition it emits toxic fumes of Cl^- and NO_x.

CFB500 CAS:19996-03-3 ***HR: 3***
9-CHLORO-10-CHLOROMETHYL ANTHRACENE
mf: $C_{15}H_{10}Cl_2$ mw: 261.15

SYNS: 10-CHLOROMETHYL-9-CHLOROANTHRACENE ◇ ICR 486

TOXICITY DATA with REFERENCE
mma-sat 1 μg/plate PNASA6 72,5135,75
pic-esc 50 ng/plate CNREA8 43,2819,83
ivn-mus TDLo:2600 μg/kg:NEO CNREA8 36,2423,76

CONSENSUS REPORTS: EPA Genetic Toxicology Program.

SAFETY PROFILE: Questionable carcinogen with experimental neoplastigenic data. Mutation data reported. When heated to decomposition it emits toxic fumes of Cl^-. See also CHLORINATED HYDROCARBONS, AROMATIC; and ANTHRACENE.

CFB750 CAS:7205-90-5 ***HR: 2***
1-CHLORO-4-(CHLOROMETHYLTHIO)BENZENE
mf: $C_7H_6Cl_2S$ mw: 193.09

SYN: p-CHLORFENYLMERKAPTOMETHYLCHLORID(CZECH)

TOXICITY DATA with REFERENCE
skn-rbt 500 mg/24H MOD 28ZPAK -,172,72
eye-rbt 50 μg/24H SEV 28ZPAK -,172,72
orl-rat LD50:1030 mg/kg 28ZPAK -,172,72

CONSENSUS REPORTS: Reported in EPA TSCA Inventory.

SAFETY PROFILE: Moderately toxic by ingestion. A skin and severe eye irritant. When heated to decomposition it emits very toxic fumes of SO_x and Cl^-. See also MERCAPTANS.

CFB825 CAS:23784-96-5 ***HR: 3***
2-CHLORO-5-CHLOROMETHYLTHIOPHENE
mf: $C_5H_4Cl_2S$ mw: 167.05

$CH{=}C(Cl)SC(CH_2Cl){=}CH$ (ring)

SAFETY PROFILE: A storage hazard, it may decompose explosively at room temperature. When heated to decomposition it emits toxic fumes of Cl^- and SO_x.

CFC000 CAS:1018-71-9 ***HR: 3***
3-CHLORO-4-(3-CHLORO-2-NITROPHENYL) PYRROLE
mf: $C_{10}H_6Cl_2N_2O_2$ mw: 257.08

SYNS: 3-CHLORO-4-(2'-NITRO-3'-CHLOROPHENYL)PYRROLE ◇ NSC-107654 ◇ PN ◇ PYROACE ◇ PYROLLNITRIN ◇ PYRROLNITRIN

TOXICITY DATA with REFERENCE
ipr-rat LD50:68 mg/kg 85ERAY 3,1479,78
orl-mus LD50:1000 mg/kg 85ERAY 3,1479,78
ipr-mus LD50:680 mg/kg MEIEDD 10,1157,83

scu-mus LD50:2000 mg/kg 85GDA2 5,44,81
ipr-rbt LD50:105 mg/kg 85ERAY 3,1479,78

SAFETY PROFILE: Poison by intraperitoneal route. Moderately toxic by ingestion and subcutaneous routes. When heated to decomposition it emits very toxic fumes of Cl^- and NO_x.

CFC100 CAS:57808-65-8 ***HR: 3***
N-(5-CHLORO-4-((4-CHLOROPHENYL) CYANOMETHYL)-2-METHYLPHENYL)-2-HYDROXY-3, 5-DIIODOBENZAMIDE
mf: $C_{22}H_{14}Cl_2I_2N_2O_2$ mw: 663.08

SYNS: BENZAMIDE, N-(5-CHLORO-4-((4-CHLOROPHENYL)CYANOMETHYL)-2-METHYLPHENYL)-2-HYDROXY-3,5- DIIODO- ◇ CLOSANTEL ◇ R 31520

TOXICITY DATA with REFERENCE
orl-rat TDLo:3600 mg/kg (male 90D pre):REP DCTODJ 8,101,85
orl-rat LD50:262 mg/kg DCTODJ 8,101,85
ims-rat LD50:28400 μg/kg DCTODJ 8,101,85
orl-mus LD50:331 mg/kg DCTODJ 8,101,85
ims-mus LD50:56800 μg/kg DCTODJ 8,101,85

SAFETY PROFILE: Poison by ingestion and intramuscular routes. Experimental reproductive effects. When heated to decomposition it emits toxic fumes of NO_x, I^-, and Cl^-.

CFC250 CAS:846-49-1 ***HR: 2***
7-CHLORO-5-(o-CHLOROPHENYL)-1,3-DIHYDRO-3-HYDROXY-2H-1,4-BENZODIAZEPIN-2-ONE
mf: $C_{15}H_{10}Cl_2N_2O_2$ mw: 321.17

SYNS: ALMAZINE ◇ ATIVAN ◇ 7-CHLORO-5-(2-CHLOROPHENYL)-1,3-DIHYDRO-3-HYDROXY-2H-1,4-BENZODIAZEPIN-2-ONE◇ 7-CHLORO-5-(2-CHLOROPHENYL)-3-HYDROXY-1H-1,4-BENZODIAZEPIN-2(3H)-ONE ◇ EMOTIVAL ◇ LORAX ◇ LORAZEPAM ◇ LORSILAN ◇ PSICOPAX ◇ TAVOR ◇ TEMESTA ◇ WY 4036 ◇ WYPAX

TOXICITY DATA with REFERENCE
orl-wmn TDLo:2420 μg/kg (36-39W preg/2D post):REP BMJOAE 282,1106,81
orl-rat TDLo:2800 μg/kg (9-15D preg):TER JZKEDZ 1,25,75
orl-chd TDLo:71 μg/kg:CNS TOLED5 2,109,78
orl-wmn TDLo:380 μg/kg/19D-I:BLD BMJOAE 296,1332,88
orl-hmn TDLo:21 μg/kg CLPTAT 39,526,86
orl-rat LD50:4500 mg/kg IYKEDH 8,680,77
ipr-rat LD50:870 mg/kg PHMGBN 10,345,73
orl-mus LD50:1850 mg/kg IYKEDH 8,680,77
ipr-mus LD50:1810 mg/kg JZKEDZ 1,5,74

SAFETY PROFILE: Moderately toxic by ingestion and intraperitoneal routes. Human systemic effects by ingestion: general anesthetic, hallucinations or distorted perceptions, loss of muscle control (ataxia), aplastic anemia, bone marrow changes. Human reproductive effects by ingestion: unspecified neonatal effects. An experimental teratogen. Other experimental reproductive effects. An anticonvulsant and tranquilizer. When heated to decomposition it emits very toxic fumes of Cl^- and NO_x. See also DIAZEPAM.

CFC500 CAS:24283-57-6 ***HR: 3***
5-CHLORO-3-(4-CHLOROPHENYL)-4'-FLUORO-2'-METHYLSALICYLANILIDE
mf: $C_{20}H_{14}Cl_2FNO_2$ mw: 390.25

SYNS: CP 48985 ◇ 4',5-DICHLORO-N-(4-FLUORO-2-METHYLPHENYL)-2-HYDROXY-(1,1'-BIPHENYL)-3-CARBOXAMIDE ◇ ENT 27,349 ◇ MONSANTO CP-48985 ◇ NSC 190947

TOXICITY DATA with REFERENCE
orl-rat LD50:2510 mg/kg ARSIM* 20,16,66
ipr-mus LD50:151 mg/kg BCPCA6 18,1389,69

SAFETY PROFILE: Poison by intraperitoneal route. Moderately toxic by ingestion. When heated to decomposition it emits very toxic fumes of Cl^-, F^-, and NO_x.

CFC750 CAS:14437-17-3 ***HR: 2***
2-CHLORO-3-(4-CHLOROPHENYL)METHYLPROPIONATE
mf: $C_{10}H_{10}Cl_2O_2$ mw: 233.10

SYNS: BAY 70533 ◇ BAYER 70533 ◇ BIDISIN ◇ CHLORFENPROP-METHYL ◇ 2-CHLORO-3-(4-CHLOROPHENYL)PROPIONIC ACID METHYL ESTER ◇ CHLORPHENPROP-METHYL ◇ 3-(4-CHLORPHENYL)-2-CHLORPROPIONSAEUREMETHYLESTER(GERMAN) ◇ METHACHLORPHENPROP ◇ METHYL-2-CHLORO-3-(4-CHLOROPHENYL)PROPIONATE ◇ METHYL α,4-DICHLOROPHENYLPROPANOATE ◇ W5769

TOXICITY DATA with REFERENCE
orl-rat LD50:1072 mg/kg WRPCA2 9,119,70
orl-mus LD50:1000 mg/kg 85DPAN -,-,71,76
orl-rbt LD50:500 mg/kg 28ZEAL 5,44,76
skn-rbt LD50:756 mg/kg WRPCA2 9,119,70
orl-gpg LD50:500 mg/kg 28ZEAL 5,44,76

SAFETY PROFILE: Moderately toxic by ingestion and skin contact. An herbicide. See also ESTERS. When heated to decomposition it emits toxic fumes of Cl^-.

CFD250 CAS:5980-86-9 ***HR: 3***
CHLORO(2-CHLOROVINYL)MERCURY
mf: $C_2H_2Cl_2Hg$ mw: 297.53

SYN: (2-CHLOROVINYL)MERCURICCHLORIDE

TOXICITY DATA with REFERENCE
ivn-mus LD50:56200 μg/kg CSLNX* NX#05831

CONSENSUS REPORTS: Mercury and its compounds are on the Community Right-To-Know List.

OSHA PEL: (Transitional: CL 1 mg/10m^3) CL 0.1 mg(Hg)/m^3 (skin)
ACGIH TLV: TWA 0.1 mg(Hg)/m^3
NIOSH REL: (Mercury, Inorganic) TWA 0.05 mg(Hg)/m^3

SAFETY PROFILE: Poison by intravenous route. See also MERCURY COMPOUNDS. When heated to decomposition it emits very toxic fumes of Cl^- and Hg.

CFE000 CAS:1570-64-5 ***HR: 3***
4-CHLORO-o-CRESOL
mf: C_7H_7ClO mw: 142.59

TOXICITY DATA with REFERENCE
orl-mus LD50:1320 mg/kg PHARAT 30,147,75
ivn-mus LD50:56 mg/kg CSLNX* NX#03270

CONSENSUS REPORTS: Reported in EPA TSCA Inventory. Chlorophenol compounds are on the Community Right-To-Know List.

SAFETY PROFILE: Poison by intravenous route. Moderately toxic by ingestion. When heated to decomposition it emits toxic fumes of Cl^- and phosgene. See also CRESOL and CHLOROPHENOLS.

CFE250 CAS:59-50-7 ***HR: 3***
4-CHLORO-m-CRESOL
mf: C_7H_7ClO mw: 142.59

PROP: Odorless crystals (when pure). Somewhat sol in water, very sol in organic solvents. Mp: 66°, bp: 235°.

SYNS: APTAL ◇ BAKTOL ◇ BAKTOLAN ◇ CANDASETPIC ◇ p-CHLOR-m-CRESOL ◇ CHLOROCRESOL ◇ p-CHLOROCRESOL ◇ p-CHLORO-m-CRESOL ◇ 6-CHLORO-m-CRESOL ◇ 2-CHLORO-HYDROXYTOLUENE ◇ 6-CHLORO-3-HYDROXYTOLUENE ◇ 4-CHLORO-3-METHYLPHENOL ◇ 3-METHYL-4-CHLOROPHENOL ◇ OTTAFACT ◇ PARMETOL ◇ PAROL ◇ PCMC ◇ PREVENTOL CMK ◇ RASCHIT ◇ RASEN-ANICON ◇ RCRA WASTE NUMBER U039

TOXICITY DATA with REFERENCE
orl-rat LD50:1830 mg/kg SchP## 04APR86
scu-rat LD50:400 mg/kg QJPPAL 12,212,39
ipr-mus LDLo:30 mg/kg QJPPAL 12,212,39
scu-mus LDLo:200 mg/kg QJPPAL 12,212,39

CONSENSUS REPORTS: Reported in EPA TSCA Inventory. Chlorophenol compounds are on the Community Right-To-Know List.

SAFETY PROFILE: Poison by intraperitoneal and subcutaneous routes. Moderately toxic by ingestion. An allergen. Incompatible with sodium hydroxide. When heated to decomposition it emits toxic fumes of Cl^- and phosgene. See also CRESOL and CHLOROPHENOLS.

CFE750 CAS:2003-31-8 ***HR: 3***
CHLOROCYANOACETYLENE
mf: C_3ClN mw: 85.49

$$ClC \equiv CC \equiv N$$

SYN: CHLOROPROPYNENITRILE

CONSENSUS REPORTS: Cyanide and its compounds are on the Community Right-To-Know List.

SAFETY PROFILE: Ignites spontaneously in air. A dangerous storage hazard, it may explode in a sealed container. When heated to decomposition it emits toxic fumes of Cl^-, NO_x, and CN^-. See also ACETYLENE COMPOUNDS, CYANIDE, NITRILES, and CHLORINATED HYDROCARBONS, ALIPHATIC.

CFF100 ***HR: 3***
2-CHLORO-α-CYANO-6-METHYLERGOLINE-8-PROPIONAMIDE
mf: $C_{19}H_{21}ClN_4O$ mw: 356.89

SYN: ERGOLINE-8-PROPIONAMIDE, 2-CHLORO-α-CYANO-6-METHYL-

TOXICITY DATA with REFERENCE
orl-rat TDLo:8 mg/kg (female 5D post):REP ARZNAD 33,1094,83
orl-mus LD50:12500 μg/kg ARZNAD 33,1094,83

SAFETY PROFILE: Poison by ingestion. Experimental reproductive effects. When heated to decomposition it emits toxic fumes of NO_x and Cl^-.

CFF250 CAS:15271-41-7 ***HR: 3***
3-CHLORO-6-CYANO-2-NORBORNANONE-o-(METHYLCARBAMOYL)OXIME
mf: $C_{10}H_{12}ClN_3O_2$ mw: 241.70

SYNS: endo-3-CHLORO-exo-6-CYANO-2-NORBORNANONE-o-(METHYLCARBAMOYL)OXIME ◇ 2-exo-CHLORO-6-endo-CYANO-2-NORBORNANONE-o-(METHYLCARBAMOYL)OXIME2-CARBONITRILE ◇ 3-CHLORO-6-CYANONORBORNANONE-2-OXIME-o,N-METHYL-CARBAMATE ◇ 5-CHLORO-6-((((METHYLAMINO)CARBONYL)OXY)IMINO)BICYCLO(2.2.1)HEPTANE ◇ exo-5-CHLORO-6-OXO-endo-2-NORBORNANECARBONITRILE-o-(METHYLCARBAMOYL)OXIME ◇ COMPOUND UC-20047 A ◇ ENT 25,962 ◇ TRANID ◇ UC 20047 ◇ UC 26089 ◇ UC 20,047A ◇ UNION CARBIDE UC 20047

TOXICITY DATA with REFERENCE
orl-rat LD50:19 mg/kg TXAPA9 14,515,69
skn-rat LD50:303 mg/kg WRPCA2 9,119,70
unk-rat LD50:26 mg/kg 30ZDA9 -,198,71
skn-rbt LDLo:303 mg/kg BESAAT 12,161,66

CONSENSUS REPORTS: EPA Extremely Hazardous Substances List. Cyanide and its compounds are on the Community Right-To-Know List.

SAFETY PROFILE: Poison by ingestion, skin contact,

and possibly other routes. A pesticide. When heated to decomposition it emits very toxic fumes of Cl^- and NO_x. See also CARBAMATES.

CFF500 CAS:82-93-9 ***HR: 3***
CHLOROCYCLINE
mf: $C_{18}H_{21}ClN_2$ mw: 300.86

SYNS: CHLORCYCLINE ◇ CHLORCYCLIZINE ◇ 1-(4-CHLOROBENZHYDRYL)-4-METHYLPIPERAZINE ◇ CHLOROCYCLIZINE ◇ 1-(p-CHLORO-α-PHENYLBENZYL)-4-METHYLPIPERAZINE ◇ DI-PARALEN ◇ DIPARALENE ◇ HISTANTIN ◇ HISTANTINE

TOXICITY DATA with REFERENCE
dni-rat-ipr 50 mg/kg JPETAB 171,109,70
orl-rat TDLo:330 mg/kg (female 7-17D post):TER ARZNAD 31,1225,81
orl-uns TDLo:240 mg/kg (female 18-25D post):REP AOBIAR 13,1281,68

SAFETY PROFILE: Unspecified human reproductive effects. Experimental teratogenic and reproductive effects. Mutation data reported. When heated to decomposition it emits very toxic fumes of Cl^- and NO_x.

CFG250 CAS:822-87-7 ***HR: 2***
α-CHLOROCYCLOHEXANONE
mf: C_6H_9ClO mw: 132.60

SYN: 2-CHLOROCYCLOHEXANONE

TOXICITY DATA with REFERENCE
mmo-sat 370 nmol/plate CBINA8 45,305,83
ipr-mus LD50:830 mg/kg COREAF 254,2683,62

CONSENSUS REPORTS: Reported in EPA TSCA Inventory.

SAFETY PROFILE: Moderately toxic by intraperitoneal route. Mutation data reported. When heated to decomposition it emits toxic fumes of Cl^-. See also KETONES and CHLORIDES.

CFG500 CAS:14737-08-7 ***HR: 2***
4-CHLORO-4-CYCLOHEXENE-1,2-DICARBOXYLIC ANHYDRIDE
mf: $C_8H_7ClO_3$ mw: 186.60

SYNS: ANHYDRIDE KYSELINY 4-CHLOR-1,2,3,6-TETRAHYDROFTA-LOVE (CZECH) ◇ 4-CHLORTETRAHYDROFTALANHYDRID (CZECH)

TOXICITY DATA with REFERENCE
skn-rbt 500 mg/24H MLD 28ZPAK -,140,72
eye-rbt 20 mg/24H SEV 28ZPAK -,140,72
orl-rat LD50:3390 mg/kg 28ZPAK -,140,72

SAFETY PROFILE: Moderately toxic by ingestion. A severe eye and mild skin irritant. When heated to decomposition it emits toxic fumes of Cl^-. See also ANHYDRIDES.

CFG750 CAS:10379-14-3 ***HR: 2***
7-CHLORO-5-(CYCLOHEXEN-1-YL)-1,3-DIHYDRO-1-METHYL-2H-1,4-BENZODIAZEPIN-2-ONE
mf: $C_{16}H_{17}ClN_2O$ mw: 288.80

SYNS: CB 4261 ◇ 4361 CB ◇ 7-CHLORO-5-(1-CYCLOHEXENYL)-1-METHYL-2-OXO-2,3-DIHYDRO-1H-(1,4)-BENZO(f)DIAZEPINE ◇ CLINOXAN ◇ MUSARIL ◇ MYOLASTAN ◇ TETRAZEPAM

TOXICITY DATA with REFERENCE
orl-mus LD50:2000 mg/kg CHTPBA 2,254,67
ipr-mus LD50:415 mg/kg 27ZQAG -,171,72

SAFETY PROFILE: Moderately toxic by ingestion and intraperitoneal routes. A tranquilizer and muscle relaxant. When heated to decomposition it emits very toxic fumes of Cl^- and NO_x. See also DIAZEPAM and KETONES.

CFH000 CAS:77966-40-6 ***HR: 3***
6'-CHLORO-2-(CYCLOHEXYLAMINO)-o-ACETOTOLUIDIDE HYDROCHLORIDE
mf: $C_{15}H_{21}ClN_2O \cdot ClH$ mw: 317.29

SYN: C 3115

TOXICITY DATA with REFERENCE
eye-rbt 2% MLD ARZNAD 8,407,58
ipr-rat LD50:83 mg/kg ARZNAD 8,407,58
scu-mus LD50:122 mg/kg ARZNAD 8,407,58

SAFETY PROFILE: Poison by intraperitoneal and subcutaneous routes. An eye irritant. When heated to decomposition it emits very toxic fumes of Cl^- and NO_x.

CFH500 CAS:13909-11-0 ***HR: 3***
cis-3-(2-CHLOROCYCLOHEXYL)-1-(2-CHLOROETHYL)-1-NITROSOUREA
mf: $C_9H_{15}Cl_2N_3O_2$ mw: 268.17

SYNS: cis-N'-(2-CHLOROCYCLOHEXYL)-N-(2-CHLOROETHYL)-N-NITROSOUREA ◇ (Z)-3-(2-CHLOROCYCLOHEXYL)-1-(2-CHLOROETHYL)-1-NITROSOUREA ◇ NSC 84954

TOXICITY DATA with REFERENCE
ipr-mus LD50:52100 μg/kg NCISP* JAN86

SAFETY PROFILE: Poison by intraperitoneal route. Many N-nitroso compounds are carcinogens. When heated to decomposition it emits very toxic fumes of Cl^- and NO_x. See also N-NITROSO COMPOUNDS.

CFH750 CAS:13909-12-1 ***HR: 3***
trans-3-(2-CHLOROCYCLOHEXYL)-1-(2-CHLOROETHYL)-1-NITROSOUREA
mf: $C_9H_{15}Cl_2N_3O_2$ mw: 268.17

SYNS: trans-N'-(2-CHLOROCYCLOHEXYL)-N-(2-CHLOROETHYL)-N-NITROSOUREA ◇ NSC 88104 ◇ SRI 2656

TOXICITY DATA with REFERENCE
orl-mus LD50:51100 μg/kg NCISP* JAN86
ipr-mus LD50:62230 μg/kg NCISP* JAN86
scu-mus LD50:74600 μg/kg NCISP* JAN86

SAFETY PROFILE: Poison by ingestion, subcutaneous, and intraperitoneal routes. Many N-nitroso compounds are carcinogens. When heated to decomposition it emits very toxic fumes of NO_x and Cl^-. See also N-NITROSO COMPOUNDS.

CFH825 CAS:28968-07-2 *HR: 3*
6-CHLORO-5-CYCLOHEXYL-1-INDANCARBOXYLIC ACID
mf: $C_{16}H_{19}ClO_2$ mw: 278.80

PROP: Colorless crystals from petr ether. Mp: 150.5-152.5°. (S)-(+)-Form: Colorless needles from petr ether. Mp: 135-136°. (R)-(−)-Form: Colorless crystals from petr ether. Mp: 134-135°.

SYNS: BRITAI ◇ (±)-6-CHLORO-5-CYCLOHEXYL-2,3-DIHYDRO-1H-INDENE-1-CARBOXYLIC ACID (9CI) ◇ (±)-6-CHLORO-5-CYCLOHEXYLINDAN-1-CARBOXYLIC ACID ◇ (±)-6-CHLORO-5-CYCLOHEXYL-1-INDANCARBOXYLIC ACID ◇ CLIDANAC ◇ (±)-2,3-DIHYDRO-6-CHLORO-5-CYCLOHEXYL-1H-INDENE-1-CARBOXYLIC ACID ◇ INDANAL ◇ TAI-284 ◇ (±)-TAI 284 ◇ dl-TAI 284

TOXICITY DATA with REFERENCE
orl-mus TDLo:810 mg/kg (7-15D preg):TER SEIJBO 17,177,77
orl-rat TDLo:90 mg/kg (30D male):REP OYYAA2 7,333,73
orl-rat LD50:41 mg/kg JMCMAR 15,1297,72
ipr-rat LD50:50 mg/kg OYYAA2 7,333,73
scu-rat LD50:60 mg/kg OYYAA2 7,333,73
ivn-rat LD50:45 mg/kg OYYAA2 7,333,73
orl-mus LD50:750 mg/kg OYYAA2 7,333,73
ipr-mus LD50:500 mg/kg OYYAA2 7,333,73
scu-mus LD50:800 mg/kg OYYAA2 7,333,73
ivn-mus LD50:150 mg/kg OYYAA2 7,333,73
orl-rbt LD50:250 mg/kg OYYAA2 7,333,73
orl-gpg LD50:400 mg/kg OYYAA2 7,333,73

SAFETY PROFILE: Poison by ingestion, subcutaneous, intravenous, and intraperitoneal routes. Experimental teratogenic and reproductive effects. A non-steroidal anti-inflammatory agent. When heated to decomposition it emits toxic fumes of Cl^-.

CFI000 CAS:77966-41-7 *HR: 3*
6′-CHLORO-2-(N-CYCLOHEXYL-N-METHYLAMINO)- o-ACETOTOLUIDIDE HYDROCHLORIDE
mf: $C_{16}H_{23}ClN_2O{\cdot}ClH$ mw: 331.32

SYN: C 3120

TOXICITY DATA with REFERENCE
eye-rbt 2% SEV ARZNAD 8,407,58
ipr-rat LD50:105 mg/kg ARZNAD 8,407,58
scu-mus LD50:337 mg/kg ARZNAD 8,407,58

SAFETY PROFILE: Poison by intraperitoneal and subcutaneous routes. A severe eye irritant. When heated to decomposition it emits very toxic fumes of Cl^- and NO_x.

CFI250 *HR: 2*
CHLOROCYCLOPENTANE
mf: C_5H_9Cl mw: 104.58

$ClCHCH_2CH_2CH_2CH_2$ (ring)

PROP: Flash p: 60.8°F.

SAFETY PROFILE: A dangerous fire hazard when exposed to heat or flame. When heated to decomposition it emits toxic fumes of Cl^-. See also CHLORINATED HYDROCARBONS, ALIPHATIC.

CFI500 CAS:694-28-0 *HR: 3*
α-CHLOROCYCLOPENTANONE
mf: C_5H_7ClO mw: 118.57

SYN: 2-CHLOROCYCLOPENTANONE

TOXICITY DATA with REFERENCE
ipr-mus LD50:310 mg/kg COREAF 254,2683,62

CONSENSUS REPORTS: Reported in EPA TSCA Inventory.

SAFETY PROFILE: Poison by intraperitoneal route. When heated to decomposition it emits toxic fumes of Cl^-. See also KETONES and CHLORIDES.

CFI625 CAS:96-40-2 *HR: 3*
3-CHLOROCYCLOPENTENE
mf: C_5H_7Cl mw: 102.56

$CH{=}CHC_2H_4CHCl$ (ring)

SAFETY PROFILE: A dangerous storage hazard; it may decompose explosively at room temperature. When heated to decomposition it emits toxic fumes of Cl^-. See also CHLORINATED HYDROCARBONS, ALIPHATIC.

CFI750 CAS:13347-42-7 *HR: 2*
4-CHLORO-2-CYCLOPENTYL PHENOL
mf: $C_{11}H_{13}ClO$ mw: 196.69

SYN: DOWICIDE 9

TOXICITY DATA with REFERENCE
skn-hmn 25 mg MLD DOWCC* Apr.55
eye-rbt 100 mg SEV DOWCC* Apr.55
orl-rat LDLo:420 mg/kg DOWCC*

orl-rbt LDLo:420 mg/kg DOWCC*
skn-rbt LD50:850 mg/kg DOWCC*
orl-gpg LDLo:420 mg/kg DOWCC*
orl-ckn LDLo:420 mg/kg DOWCC*

CONSENSUS REPORTS: Chlorophenol compounds are on the Community Right-To-Know List.

SAFETY PROFILE: Moderately toxic by ingestion and skin contact. A human skin irritant. A severe eye irritant. When heated to decomposition it emits toxic fumes of Cl^-. See also CHLOROPHENOLS.

CFJ000 CAS:63007-70-5 ***HR: 2***
3-CHLORO-4-CYCLO-PROPYLMETHOXY-PHENYLACETIC ACID LYSINE SALT (d,l)
mf: $C_{12}H_{13}ClO_3 \cdot C_6H_{14}N_2O_2$ mw: 386.92

SYNS: 2-(3-CHLORO-4-CYCLOPROPYLMETHOXYPHENYL)ACETIC ACID LYSINE SALT (d,l) ◇ ISF 2508

TOXICITY DATA with REFERENCE
orl-rat LD50:895 mg/kg FRPSAX 32,286,77
ipr-rat LD50:429 mg/kg FRPSAX 32,286,77

SAFETY PROFILE: Moderately toxic by ingestion and intraperitoneal routes. When heated to decomposition it emits very toxic fumes of NO_x and Cl^-.

CFJ375 ***HR: D***
6-CHLORO-6-DEOXYGLUCOSE
mf: $C_6H_{11}ClO_5$ mw: 198.62

SYNS: 6CDG ◇ 6-CHLORO-6-DEOXY-d-GLUCOSE

TOXICITY DATA with REFERENCE
orl-rat TDLo:168 mg/kg (male 14D pre):REP JRPFA4 63,67,81

SAFETY PROFILE: Experimental reproductive effects. When heated to decomposition it emits toxic fumes of Cl^-.

CFJ750 CAS:50-90-8 ***HR: 2***
5-CHLORO-2'-DEOXYURIDINE
mf: $C_9H_{11}ClN_2O_5$ mw: 262.67

SYNS: 5-CHLORODEOXYURIDINE ◇ CLUDR

TOXICITY DATA with REFERENCE
msc-hmn:lym 100 μmol/L LIFSAK 19,563,76
cyt-ham:ovr 10 μmol/L BLFSBY 29A,69,84
sce-ham:ovr 10 μmol/L MUREAV 91,395,81
msc-ham:ovr 500 nmol/L ENMUDM 4,301,82
ipr-rat LD50:2000 mg/kg ADTEAS 3,181,68

SAFETY PROFILE: Moderately toxic by intraperitoneal route. Human mutation data reported. When heated to decomposition it emits very toxic fumes of Cl^- and NO_x.

CFK000 CAS:93-71-0 ***HR: 3***
2-CHLORO-N,N-DIALLYLACETAMIDE
mf: $C_8H_{12}ClNO$ mw: 173.66

PROP: Amber liquid. Bp: 74° @ 0.3 mm. Sltly sol in water; sol in alc, hexane, and xylene.

SYNS: ALIDOCHLOR ◇ ALLIDOCHLOR ◇ CDAA ◇ CDAAT ◇ α-CHLORO-N,N-DIALLYLACETAMIDE ◇ 2-CHLORO-N,N-DI-2-PROPENYLACETAMIDE ◇ CP 6,343 ◇ DIALLYLCHLOROACETAMIDE ◇ N,N-DIALLYLCHLOROACETAMIDE ◇ N,N-DIALLYL-α-CHLOROACETAMIDE ◇ N,N-DIALLYL-2-CHLOROACETAMIDE ◇ NCI-C04035 ◇ RADOX ◇ RANDOX ◇ RANTOX T

TOXICITY DATA with REFERENCE
orl-rat LD50:700 mg/kg RREVAH 10,97,65
skn-rat LD50:360 mg/kg WRPCA2 9,119,70

CONSENSUS REPORTS: Reported in EPA TSCA Inventory.

SAFETY PROFILE: Poison by skin contact. Moderately toxic by ingestion. An herbicide. When heated to decomposition it emits very toxic fumes of Cl^- and NO_x. See also ALLYL COMPOUNDS.

CFK125 CAS:95-83-0 ***HR: 3***
4-CHLORO-1,2-DIAMINOBENZENE
mf: $C_6H_7ClN_2$ mw: 142.60

SYNS: p-CHLORO-o-PHENYLENEDIAMINE ◇ 4-CHLORO-o-PHENYLENEDIAMINE ◇ 4-CHLORO-1,2-PHENYLENEDIAMINE ◇ 4-Cl-o-PD ◇ 1,2-DIAMINO-4-CHLOROBENZENE ◇ 3,4-DIAMINOCHLOROBENZENE ◇ 3,4-DIAMINO-1-CHLOROBENZENE ◇ NCI-C03292 ◇ URSOL OLIVE 6G

TOXICITY DATA with REFERENCE
mma-sat 10 μg/plate ENMUDM 7(Suppl 5),1,85
mma-esc 1 mg/plate ENMUDM 7(Suppl 5),1,85
dnd-hmn:fbr 50 μmol/L MUREAV 127,107,84
cyt-mus-ipr 100 mg/kg ENMUDM 8(Suppl 6),53,86
orl-rat TDLo:135 g/kg/77W-C:CAR CRNGDP 1,495,80
orl-mus TDLo:324 g/kg/77W-C:CAR CRNGDP 1,495,80
orl-mus TD :40500 g/kg/96W-C:NEO CRNGDP 1,495,80

CONSENSUS REPORTS: NTP Fifth Annual Report on Carcinogens. IARC Cancer Review: Group 2B IMEMDT 7,56,87; Human Limited Evidence IMEMDT 27,81,82; Animal Sufficient Evidence IMEMDT 27,81,82. NCI Carcinogenesis Bioassay (feed); Clear Evidence: mouse, rat NCITR* NCI-CG-TR-63,78. Reported in EPA TSCA Inventory.

SAFETY PROFILE: Confirmed with experimental carcinogenic and neoplastigenic data. Human mutation data reported. When heated to decomposition it emits toxic fumes of Cl^- and NO_x. See also AROMATIC AMINES.

CFK325 CAS:17927-57-0 ***HR: 3***
CHLORODIBORANE
mf: B_2ClH_5 mw: 62.11

SAFETY PROFILE: A dangerous fire hazard. It ignites spontaneously in air and is a gas above −11°C. When heated to decomposition it emits toxic fumes of Cl^-. See also BORANES.

CFK500 CAS:124-48-1 ***HR: 2***
CHLORODIBROMOMETHANE
mf: $CHBr_2Cl$ mw: 208.29

PROP: Colorless to pale yellow, heavy liquid. Bp: 118-122°, fp: < −20°, d: 2.440 @ 25°/25°.

SYNS: CDBM ◇ DIBROMOCHLOROMETHANE ◇ NCI-C55254

TOXICITY DATA with REFERENCE
mmo-sat 10 uL/plate DHEFDK FDA-78-1046,78
mrc-smc 100 uL/L MUREAV 155,53,85
sce-hmn:lym 400 μmol/L ENVRAL 32,72,83
sce-mus-orl 100 mg/kg/4D-I ENVRAL 32,72,83
orl-rat LD50:848 mg/kg TXAPA9 52,351,80
orl-mus LD50:800 mg/kg TXAPA9 44,213,78

CONSENSUS REPORTS: NTP Carcinogenesis Studies (gavage); Some Evidence: mouse NTPTR* NTP-TR-282,86; No Evidence: rat NTPTR* NTP-TR-282,85; Reported in EPA TSCA Inventory.

SAFETY PROFILE: Moderately toxic by ingestion. Human mutation data reported. Compounds of this type are generally irritating and narcotic. See also BROMOFORM and CHLOROFORM. When heated to decomposition it emits toxic fumes of Cl^- and Br^-.

CFK750 CAS:63869-01-2 ***HR: 3***
CHLORO(DIBUTOXYPHOSPHINYL)MERCURY
mf: $C_8H_{18}ClHgO_3P$ mw: 429.27

SYN: (DIBUTOXYPHOSPHINYL)MERCURY CHLORIDE

TOXICITY DATA with REFERENCE
ipr-mus LDLo:31 mg/kg CBCCT* 8,104,56

CONSENSUS REPORTS: Mercury and its compounds are on the Community Right-To-Know List.

OSHA PEL: (Transitional: CL 1 mg/10m³) CL 0.1 mg(Hg)/m³ (skin)
ACGIH TLV: TWA 0.1 mg(Hg)/m³ (skin)
NIOSH REL: (Inorganic Mercury) TWA: 0.05 mg(Hg)/m³

SAFETY PROFILE: Poison by intraperitoneal route. When heated to decomposition it emits very toxic fumes of PO_x, Hg, and Cl^-. See also MERCURY COMPOUNDS and CHLORIDES.

CFL000 CAS:77966-42-8 ***HR: 3***
6'-CHLORO-2-(DIBUTYLAMINO)-o-ACETOTOLUIDIDE, HYDROCHLORIDE
mf: $C_{17}H_{27}ClN_2O{\cdot}ClH$ mw: 347.37

SYN: C 3072

TOXICITY DATA with REFERENCE
eye-rbt 2% SEV ARZNAD 8,407,58
ipr-rat LD50:250 mg/kg ARZNAD 8,407,58
scu-mus LD50:2295 mg/kg ARZNAD 8,407,58

SAFETY PROFILE: Poison by intraperitoneal route. Moderately toxic by subcutaneous route. A severe eye irritant. When heated to decomposition it emits very toxic fumes of Cl^- and NO_x.

CFL200 CAS:68786-66-3 ***HR: 3***
6-CHLORO-5-(2,3-DICHLOROPHENOXY)-2-METHYLTHIO-BENZIMIDAZOLE
mf: $C_{14}H_9Cl_3N_2OS$ mw: 359.66

SYNS: 1H-BENZIMIDAZOLE, 5-CHLORO-6-(2,3-DICHLOROPHENOXY)-2-(METHYLTHIO)- ◇ CGA 89317 ◇ 5-CHLORO-6-(2,3-DICHLOROPHENOXY)-2-(METHYLTHIO)-1H-BENZIMIDAZOLE ◇ FASINEX ◇ TRICLABENDAZOLE

TOXICITY DATA with REFERENCE
orl-rat TDLo:800 mg/kg (female 8-15D post):TER TXCYAC 43,283,87
orl-rbt LD50:206 mg/kg MDACAP 21,227,85

SAFETY PROFILE: Poison by ingestion. Experimental teratogenic effects. When heated to decomposition it emits toxic fumes of SO_x, NO_x, and Cl^-.

CFL500 CAS:77791-58-3 ***HR: 2***
2'-CHLORO-2-(DIETHYLAMINO)ACETANILIDE HYDROCHLORIDE
mf: $C_{12}H_{17}ClN_2O{\cdot}ClH$ mw: 277.22

SYN: C 3070

TOXICITY DATA with REFERENCE
eye-rbt 2% MOD ARZNAD 8,270,58
ipr-rat LD50:550 mg/kg ARZNAD 8,270,58
ipr-mus LD50:600 mg/kg ARZNAD 8,270,58
scu-mus LD50:2150 mg/kg ARZNAD 8,270,58

SAFETY PROFILE: Moderately toxic by intraperitoneal and subcutaneous routes. An eye irritant. When heated to decomposition it emits very toxic fumes of Cl^- and NO_x.

CFL750 CAS:55489-49-1 ***HR: 3***
3'-CHLORO-2-(DIETHYLAMINO)ACETANILIDE HYDROCHLORIDE
mf: $C_{12}H_{17}ClN_2O{\cdot}ClH$ mw: 277.22

SYN: C 3191

TOXICITY DATA with REFERENCE
eye-rbt 2% MLD ARZNAD 8,170,58
ipr-rat LD50:370 mg/kg ARZNAD 8,170,58
ipr-mus LD50:175 mg/kg JAPMA8 49,80,60
scu-mus LD50:375 mg/kg JAPMA8 49,80,60

SAFETY PROFILE: Poison by subcutaneous and intraperitoneal routes. An eye irritant. When heated to decomposition it emits very toxic fumes of Cl^-, and NO_x.

CFM000 CAS:74816-28-7 ***HR: 3***
4'-CHLORO-2-(DIETHYLAMINO)ACETANILIDE HYDROCHLORIDE
mf: $C_{12}H_{17}ClN_2O{\cdot}ClH$ mw: 277.22

SYN: C 3061

TOXICITY DATA with REFERENCE
eye-rbt 2% MLD ARZNAD 8,170,58
ipr-rat LD50:318 mg/kg ARZNAD 8,170,58
ipr-mus LD50:375 mg/kg ARZNAD 8,170,58
scu-mus LD50:800 mg/kg ARZNAD 8,170,58

SAFETY PROFILE: Poison by intraperitoneal route. Moderately toxic by subcutaneous route. An eye irritant. When heated to decomposition it emits very toxic fumes of Cl^- and NO_x.

CFM250 CAS:77966-46-2 ***HR: 3***
3'-CHLORO-2-(DIETHYLAMINO)-o-ACETOTOLUIDIDE HYDROCHLORIDE
mf: $C_{13}H_{19}ClN_2O{\cdot}ClH$ mw: 291.25

SYNS: C 3037 ◇ 2'-CHLORO-2-(DIETHYLAMINO)-2'-METHYLACETANILIDE HYDROCHLORIDE

TOXICITY DATA with REFERENCE
eye-rbt 2% MLD ARZNAD 8,270,58
ipr-rat LD50:245 mg/kg ARZNAD 8,270,58
ipr-mus LD50:200 mg/kg JAPMA8 49,80,60
scu-mus LD50:375 mg/kg JAPMA8 49,80,60

SAFETY PROFILE: Poison by subcutaneous and intraperitoneal routes. An eye irritant. When heated to decomposition it emits very toxic fumes of Cl^- and NO_x.

CFM500 CAS:77966-47-3 ***HR: 3***
4'-CHLORO-2-(DIETHYLAMINO)-o-ACETOTOLUIDIDE HYDROCHLORIDE
mf: $C_{13}H_{19}ClN_2O{\cdot}ClH$ mw: 291.25

SYNS: C 3101 ◇ 4'-CHLORO-2-(DIETHYLAMINO)-2'-METHYLACETANILIDE HYDROCHLORIDE

TOXICITY DATA with REFERENCE
eye-rbt 2% MLD ARZNAD 8,270,58
ipr-rat LD50:350 mg/kg ARZNAD 8,270,58
scu-mus LD50:860 mg/kg ARZNAD 8,270,58

SAFETY PROFILE: Poison by intraperitoneal route. Moderately toxic by subcutaneous route. An eye irritant. When heated to decomposition it emits very toxic fumes of Cl^- and NO_x.

CFM750 CAS:77966-48-4 ***HR: 3***
5'-CHLORO-2-(DIETHYLAMINO)-o-ACETOTOLUIDIDE HYDROCHLORIDE
mf: $C_{13}H_{19}ClN_2O{\cdot}ClH$ mw: 291.25

SYNS: C 3152 ◇ 3'-CHLORO-2-(DIETHYLAMINO)-6'-METHYLACETANILIDE HYDROCHLORIDE

TOXICITY DATA with REFERENCE
eye-rbt 2% MLD ARZNAD 8,270,58
ipr-rat LD50:261 mg/kg ARZNAD 8,270,58
scu-mus LD50:1125 mg/kg ARZNAD 8,270,58

SAFETY PROFILE: Poison by intraperitoneal route. Moderately toxic by subcutaneous route. An eye irritant. When heated to decomposition it emits very toxic fumes of Cl^- and NO_x.

CFN000 CAS:77966-49-5 ***HR: 3***
6'-CHLORO-2-(DIETHYLAMINO)-m-ACETOTOLUIDIDE HYDROCHLORIDE
mf: $C_{13}H_{19}ClN_2O{\cdot}ClH$ mw: 291.25

SYNS: C 3201 ◇ 2'-CHLORO-2-(DIETHYLAMINO)-5'-METHYLACETANILIDE HYDROCHLORIDE

TOXICITY DATA with REFERENCE
eye-rbt 2% MOD ARZNAD 8,270,58
ipr-rat LD50:370 mg/kg ARZNAD 8,270,58
scu-mus LD50:2450 mg/kg ARZNAD 8,270,58

SAFETY PROFILE: Poison by intraperitoneal route. Moderately toxic by subcutaneous route. An eye irritant. When heated to decomposition it emits very toxic fumes of Cl^- and NO_x.

CFN500 CAS:77985-16-1 ***HR: 3***
6'-CHLORO-3-(DIETHYLAMINO)-o-BUTYROTOLUIDIDE HYDROCHLORIDE
mf: $C_{15}H_{23}ClN_2O{\cdot}ClH$ mw: 319.31

SYN: C 5126

TOXICITY DATA with REFERENCE
ipr-rat LD50:35 mg/kg ARZNAD 8,544,58
ipr-mus LD50:30 mg/kg ARZNAD 8,544,58
scu-mus LD50:40 mg/kg ARZNAD 8,544,58

SAFETY PROFILE: Poison by intraperitoneal and subcutaneous routes. When heated to decomposition it emits very toxic fumes of NO_x and Cl^-.

CFN750 ***HR: 2***
6'-CHLORO-2-(2-DIETHYLAMINO)ETHOXY) ACETANILIDE HYDROCHLORIDE
mf: $C_{14}H_{21}ClN_2O_2{\cdot}ClH$ mw: 321.28

SYN: C 7239

TOXICITY DATA with REFERENCE
ipr-rat LD50:590 mg/kg ARZNAD 9,683,59
scu-mus LD50:2050 mg/kg ARZNAD 9,683,59

SAFETY PROFILE: Moderately toxic by intraperitoneal and subcutaneous routes. When heated to decomposition it emits very toxic fumes of Cl^- and NO_x.

CFO000 CAS:102489-48-5 ***HR: 3***
6'-CHLORO-2-(2-(DIETHYLAMINO)ETHOXY)-o-ACETOTOLUIDIDE HYDROCHLORIDE
mf: $C_{15}H_{23}ClN_2O_2 \cdot ClH$ mw: 335.31

SYN: C 3068

TOXICITY DATA with REFERENCE
eye-rbt 2% MLD ARZNAD 9,113,59
ipr-rat LD50:175 mg/kg ARZNAD 9,113,59
scu-mus LD50:465 mg/kg ARZNAD 9,113,59

SAFETY PROFILE: Poison by intraperitoneal route. Moderately toxic by subcutaneous route. An eye irritant. When heated to decomposition it emits very toxic fumes of Cl^- and NO_x.

CFO250 CAS:7432-27-1 ***HR: 3***
5-CHLORO-2-(2-(DIETHYLAMINO)ETHOXY) BENZANILIDE
mf: $C_{19}H_{23}ClN_2O_2$ mw: 346.89

SYN: o-DIAETHYLAMINOAETHOXY-5-CHLOR-BENZANILID (GERMAN)

TOXICITY DATA with REFERENCE
scu-mus TDLo:20 mg/kg (1-19D preg):REP ARZNAD 18,658,68
scu-mus TDLo:20 mg/kg (1-19D preg):TER ARZNAD 18,658,68
orl-mus LD50:480 mg/kg ARZNAD 16,1127,66
scu-mus LD50:720 mg/kg ARZNAD 16,1127,66
ivn-mus LD50:49 mg/kg ARZNAD 16,1127,66

SAFETY PROFILE: Poison by intravenous route. Moderately toxic by ingestion and subcutaneous routes. An experimental teratogen. Other experimental reproductive effects. When heated to decomposition it emits very toxic fumes of Cl^- and NO_x.

CFO750 CAS:56287-41-3 ***HR: 3***
5-CHLORO-2-(2-(2-(DIETHYLAMINO)ETHOXY)ETHYL)-2-METHYL-1,3-BENZODIOXOLE
mf: $C_{16}H_{24}ClNO_3$ mw: 313.86

SYNS: 2-(2-(5-CHLORO-2-METHYL-1,3-BENZODIOXOL-2-YL) ETHOXY)-N,N-DIETHYLETHANAMINE ◇ LR-529 ◇ 2-METHYL-5-CHLORO-2-(N,N-DIETHYLAMINOETHOXYETHYL)-1,3-BENZODIOXOLE

TOXICITY DATA with REFERENCE
ivn-rat LD50:18 mg/kg DRFUD4 3,379,78
ipr-mus LD50:111 mg/kg DRFUD4 3,379,78

SAFETY PROFILE: Poison by intravenous and intraperitoneal routes. When heated to decomposition it emits very toxic fumes of Cl^- and NO_x.

CFP000 CAS:102489-49-6 ***HR: 3***
6'-CHLORO-2-(2-(DIETHYLAMINO)ETHYL) AMINO-o-ACETOTOLUIDIDE HYDROCHLORIDE
mf: $C_{15}H_{24}ClN_3O \cdot ClH$ mw: 334.33

SYN: C 3173

TOXICITY DATA with REFERENCE
ipr-rat LD50:305 mg/kg ARZNAD 9,167,59
scu-mus LD50:1010 mg/kg ARZNAD 9,167,59

SAFETY PROFILE: Poison by intraperitoneal route. Moderately toxic by subcutaneous route. When heated to decomposition it emits very toxic fumes of Cl^- and NO_x.

CFP250 CAS:52400-77-8 ***HR: 3***
5-CHLORO-2-(2-(2-(DIETHYLAMINO) ETHYLAMINO)ETHYL)-2-METHYL-1,3-BENZODIOXOLE DIHYDROCHLORIDE
mf: $C_{16}H_{25}ClN_2O_2 \cdot 2ClH$ mw: 385.80

TOXICITY DATA with REFERENCE
ivn-rat LD50:35 mg/kg EJMCA5 12,413,77
ipr-mus LD50:132 mg/kg EJMCA5 12,413,77

SAFETY PROFILE: Poison by intravenous and intraperitoneal routes. When heated to decomposition it emits very toxic fumes of NO_x and Cl^-.

CFP750 CAS:43047-59-2 ***HR: 3***
8-CHLORO-2-(2-(DIETHYLAMINO)ETHYL-2H-(1)-BENZOTHIOPYRANO(4,3,2-cd)INDAZOLE-5-METHANOL MONOMETHANE SULFONATE
mf: $C_{20}H_{22}ClN_3OS \cdot CH_4O_3S$ mw: 484.07

SYN: IA-4

TOXICITY DATA with REFERENCE
mmo-sat 465 nmol/plate JPETAB 200,1,77
mma-sat 465 nmol/plate JPETAB 200,1,77
mmo-nsc 20 μmol/L JTEHD6 1,271,75
bfa-mus/sat 100 mg/kg JPETAB 200,1,77
hma-mus/sat 100 mg/kg JPETAB 200,1,77
ivn-mus LD50:131 mg/kg JPETAB 186,402,73
ims-mus LD50:1560 mg/kg JPETAB 186,402,73

CONSENSUS REPORTS: EPA Genetic Toxicology Program.

SAFETY PROFILE: Poison by intravenous route. Moderately toxic by intramuscular route. Mutation data reported. When heated to decomposition it emits very toxic fumes of SO_x, Cl^-, and NO_x. See also SULFONATES.

CFQ000 CAS:54484-91-2 ***HR: 2***
8-CHLORO-2-(2-(DIETHYLAMINO)ETHYL)-2H-(1)BENZOTHIOPYRANO(4,3,2-cd)INDAZOLE-5-METHANOL-N-OXIDE
mf: $C_{20}H_{22}ClN_3O_2S$ mw: 403.96

SYN: IA-4 N-OXIDE

TOXICITY DATA with REFERENCE
mmo-sat 492 nmol/plate JPETAB 200,1,77
mma-sat 492 nmol/plate JPETAB 200,1,77
sln-dmg-par 2800 μmol/L MUREAV 82,111,81
bfa-mus/sat 200 mg/kg CNREA8 38,4478,78
hma-mus/sat 100 mg/kg JPETAB 200,1,77
ims-mus LD50:3500 mg/kg JPETAB 200,1,77

CONSENSUS REPORTS: EPA Genetic Toxicology Program.

SAFETY PROFILE: Moderately toxic by intramuscular route. Mutation data reported. When heated to decomposition it emits very toxic fumes of SO_x, NO_x, and Cl^-.

CFQ250 CAS:101651-60-9 ***HR: 3***
2′-CHLORO-2-((2-(DIETHYLAMINO)ETHYL)ETHYLAMINO)ACETANILIDE DIHYDROCHLORIDE
mf: $C_{16}H_{26}ClN_3O \cdot 2ClH$ mw: 384.82

SYN: C 5412

TOXICITY DATA with REFERENCE
eye-rbt 2% MLD ARZNAD 9,262,59
ipr-rat LD50:100 mg/kg ARZNAD 9,262,59
scu-mus LD50:460 mg/kg ARZNAD 9,262,59

SAFETY PROFILE: Poison by intraperitoneal route. Moderately toxic by subcutaneous route. An eye irritant. When heated to decomposition it emits very toxic fumes of Cl^- and NO_x.

CFQ500 CAS:102489-50-9 ***HR: 3***
6′-CHLORO-2-((2-(DIETHYLAMINO)ETHYL)ETHYLAMINO)-o-ACETOTOLUIDIDE HYDROCHLORIDE
mf: $C_{17}H_{28}ClN_3O \cdot ClH$ mw: 362.39

SYN: C 3253

TOXICITY DATA with REFERENCE
eye-rbt 2% MLD ARZNAD 9,167,59
ipr-rat LD50:55 mg/kg ARZNAD 9,167,59
scu-mus LD50:210 mg/kg ARZNAD 9,167,59

SAFETY PROFILE: Poison by subcutaneous and intraperitoneal routes. An eye irritant. When heated to decomposition it emits very toxic fumes of Cl^- and NO_x.

CFQ750 CAS:102489-51-0 ***HR: 3***
6′-CHLORO-2-((2-(DIETHYLAMINO)ETHYL)ISOPROPYLAMINO)-o-ACETOTOLUIDIDE HYDROCHLORIDE
mf: $C_{18}H_{30}ClN_3O \cdot ClH$ mw: 376.42

SYN: C 5384

TOXICITY DATA with REFERENCE
eye-rbt 2% MLD ARZNAD 9,167,59
ipr-rat LD50:22 mg/kg ARZNAD 9,167,59
scu-mus LD50:60 mg/kg ARZNAD 9,167,59

SAFETY PROFILE: Poison by subcutaneous and intraperitoneal routes. An eye irritant. When heated to decomposition it emits very toxic fumes of Cl^- and NO_x.

CFR000 CAS:55489-49-1 ***HR: 3***
2′-CHLORO-2-(2-(DIETHYLAMINO)ETHYL)METHYLAMINOACETANILIDE DIHYDROCHLORIDE
mf: $C_{15}H_{24}ClN_3O \cdot 2ClH$ mw: 370.79

SYN: C 5366

TOXICITY DATA with REFERENCE
ipr-rat LD50:160 mg/kg ARZNAD 9,262,59
scu-mus LD50:1040 mg/kg ARZNAD 9,262,59

SAFETY PROFILE: Poison by intraperitoneal route. Moderately toxic by subcutaneous route. When heated to decomposition it emits very toxic fumes of Cl^- and NO_x.

CFR250 CAS:77791-57-2 ***HR: 3***
4′-CHLORO-2-(2-(DIETHYLAMINO)ETHYL)METHYLAMINOACETANILIDE DIHYDROCHLORIDE
mf: $C_{15}H_{24}ClN_3O \cdot 2ClH$ mw: 370.79

SYN: C 5400

TOXICITY DATA with REFERENCE
eye-rbt 2% MLD ARZNAD 8,262,59
ipr-rat LD50:166 mg/kg ARZNAD 9,262,59
scu-mus LD50:435 mg/kg ARZNAD 9,262,59

SAFETY PROFILE: Poison by intraperitoneal route. Moderately toxic by subcutaneous route. An eye irritant. When heated to decomposition it emits very toxic fumes of Cl^- and NO_x.

CFR500 CAS:77984-94-2 ***HR: 3***
3′-CHLORO-2-(2-(DIETHYLAMINO)ETHYL) METHYLAMINO-o-ACETOTOLUIDIDE DIHYDROCHLORIDE
mf: $C_{16}H_{26}ClN_3O \cdot 2ClH$ mw: 384.82

SYN: C 5397

TOXICITY DATA with REFERENCE
ipr-rat LD50:72 mg/kg ARZNAD 9,262,59
scu-mus LD50:500 mg/kg ARZNAD 9,262,59

SAFETY PROFILE: Poison by intraperitoneal route. Moderately toxic by subcutaneous route. When heated to decomposition it emits very toxic fumes of Cl^- and NO_x.

CFR750 CAS:77966-43-9 ***HR: 3***
4′-CHLORO-2-(2-(DIETHYLAMINO)ETHYL) METHYLAMINO-o-ACETOTOLUIDIDE DIHYDROCHLORIDE
mf: $C_{16}H_{26}ClN_3O \cdot 2ClH$ mw: 384.82

SYN: C 5401

TOXICITY DATA with REFERENCE
eye-rbt 2% MLD ARZNAD 9,262,59
ipr-rat LD50:135 mg/kg ARZNAD 9,262,59
scu-mus LD50:350 mg/kg ARZNAD 9,262,59

SAFETY PROFILE: Poison by subcutaneous and intraperitoneal routes. An eye irritant. When heated to decomposition it emits very toxic fumes of Cl^- and NO_x.

CFS000 CAS:77966-44-0 ***HR: 3***
5′-CHLORO-2-(2-(DIETHYLAMINO)ETHYL) METHYLAMINO-o-ACETOTOLUIDIDE DIHYDROCHLORIDE
mf: $C_{16}H_{26}ClN_3O \cdot 2ClH$ mw: 384.82

SYN: C 5402

TOXICITY DATA with REFERENCE
ipr-rat LD50:114 mg/kg ARZNAD 9,262,59
scu-mus LD50:340 mg/kg ARZNAD 9,262,59

SAFETY PROFILE: Poison by subcutaneous and intraperitoneal routes. When heated to decomposition it emits very toxic fumes of Cl^- and NO_x.

CFS250 CAS:77966-45-1 ***HR: 3***
6′-CHLORO-2-(2-(DIETHYLAMINO)ETHYL) METHYLAMINO-m-ACETOTOLUIDIDE DIHYDROCHLORIDE
mf: $C_{16}H_{26}ClN_3O \cdot 2ClH$ mw: 384.82

SYN: C 5398

TOXICITY DATA with REFERENCE
eye-rbt 2% MLD ARZNAD 9,262,59
ipr-rat LD50:104 mg/kg ARZNAD 9,262,59
scu-mus LD50:750 mg/kg ARZNAD 9,262,59

SAFETY PROFILE: Poison by intraperitoneal route. Moderately toxic by subcutaneous route. An eye irritant. When heated to decomposition it emits very toxic fumes of Cl^- and NO_x.

CFS500 CAS:102489-52-1 ***HR: 3***
6′-CHLORO-2-((2-(DIETHYLAMINO)ETHYL) METHYLAMINO)-o-ACETOTOLUIDIDE HYDROCHLORIDE
mf: $C_{16}H_{26}ClN_3O \cdot ClH$ mw: 348.36

SYN: C 3249

TOXICITY DATA with REFERENCE
eye-rbt 2% MLD ARZNAD 9,167,59
ipr-rat LD50:110 mg/kg ARZNAD 9,167,59
scu-mus LD50:460 mg/kg ARZNAD 9,167,59

SAFETY PROFILE: Poison by intraperitoneal route. Moderately toxic by subcutaneous route. An eye irritant. When heated to decomposition emits very toxic fumes of Cl^- and NO_x.

CFS750 CAS:102489-53-2 ***HR: 3***
6′-CHLORO-2-((2-(DIETHYLAMINO)ETHYL)OCTYLAMINO)-o-ACETOTOLUIDIDE HYDROCHLORIDE
mf: $C_{23}H_{40}ClN_3O \cdot ClH$ mw: 446.57

SYN: C 5347

TOXICITY DATA with REFERENCE
eye-rbt 2% SEV ARZNAD 9,167,59
ipr-rat LD50:98 mg/kg ARZNAD 9,167,59
scu-mus LD50:260 mg/kg ARZNAD 9,167,59

SAFETY PROFILE: Poison by intraperitoneal and subcutaneous routes. A severe eye irritant. When heated to decomposition it emits very toxic fumes of Cl^- and NO_x.

CFT000 CAS:102489-54-3 ***HR: 3***
6′-CHLORO-2-((2-(DIETHYLAMINO)ETHYL)(2-PHENOXYETHYL)AMINO)-o-ACETOTOLUIDIDE HYDROCHLORIDE
mf: $C_{23}H_{32}ClN_3O_2 \cdot ClH$ mw: 454.49

SYN: C 5290

TOXICITY DATA with REFERENCE
eye-rbt 2% SEV ARZNAD 9,113,59
scu-mus LD50:92 mg/kg ARZNAD 9,113,59

SAFETY PROFILE: Poison by subcutaneous route. A severe eye irritant. When heated to decomposition it emits very toxic fumes of Cl^- and NO_x.

CFT250 CAS:102489-55-4 ***HR: 3***
6′-CHLORO-2-((2-(DIETHYLAMINO)ETHYL)PROPYLAMINO)-o-ACETOTOLUIDIDE HYDROCHLORIDE
mf: $C_{18}H_{30}ClN_3O{\bullet}ClH$ mw: 376.42

SYN: C 5385

TOXICITY DATA with REFERENCE
eye-rbt 2% MLD ARZNAD 9,167,59
ipr-rat LD50:28 mg/kg ARZNAD 9,167,59
scu-mus LD50:74 mg/kg ARZNAD 9,167,59

SAFETY PROFILE: Poison by intraperitoneal and subcutaneous routes. An eye irritant. When heated to decomposition it emits very toxic fumes of Cl^- and NO_x.

CFT500 CAS:101651-61-0 ***HR: 2***
2′-CHLORO-2-(2-(DIETHYLAMINO) ETHYLTHIO)ACETANILIDE HYDROCHLORIDE
mf: $C_{14}H_{21}ClN_2OS{\bullet}ClH$ mw: 337.34

SYN: C 4920

TOXICITY DATA with REFERENCE
eye-rbt 2% MLD ARZNAD 9,683,59
ipr-rat LD50:490 mg/kg ARZNAD 9,683,59
scu-mus LD50:1750 mg/kg ARZNAD 9,683,59

SAFETY PROFILE: Moderately toxic by intraperitoneal and subcutaneous routes. An eye irritant. When heated to decomposition it emits very toxic fumes of Cl^-, SO_x, and NO_x.

CFT750 CAS:102489-56-5 ***HR: 3***
6′-CHLORO-2-(2-(DIETHYLAMINO)ETHYLTHIO)-o-ACETOTOLUIDIDE HYDROCHLORIDE
mf: $C_{15}H_{23}ClN_2OS{\bullet}ClH$ mw: 351.37

SYN: C 4926

TOXICITY DATA with REFERENCE
eye-rbt 2% MLD ARZNAD 9,683,59
ipr-rat LD50:118 mg/kg ARZNAD 9,683,59
scu-mus LD50:350 mg/kg ARZNAD 9,683,59

SAFETY PROFILE: Poison by intraperitoneal and subcutaneous routes. An eye irritant. When heated to decomposition it emits very toxic fumes of Cl^-, SO_x, and NO_x.

CFU000 CAS:101651-90-5 ***HR: 3***
7-CHLORO-10-(3-(DIETHYLAMINO)-2-HYDROXYPROPYL)ISOALLOXAZINE SULFATE
mf: $C_{17}H_{20}ClN_5O_3{\bullet}H_2O_4S$ mw: 475.95

TOXICITY DATA with REFERENCE
ipr-rat LD50:50 mg/kg CMTRAG 2,96,61
scu-mus LD50:48 mg/kg CMTRAG 2,96,61
ivn-mus LD50:132 mg/kg CMTRAG 2,96,61
ims-mus LD50:54 mg/kg CMTRAG 2,96,61

SAFETY PROFILE: Poison by intraperitoneal, subcutaneous, intravenous, and intramuscular routes. When heated to decomposition it emits very toxic fumes of SO_x, NO_x, and Cl^-.

CFU250 CAS:74816-32-3 ***HR: 3***
4′-CHLORO-2-(DIETHYLAMINO)-N-METHYLACETANILIDE HYDROCHLORIDE
mf: $C_{13}H_{19}ClN_2O{\bullet}ClH$ mw: 291.25

SYN: C 3049

TOXICITY DATA with REFERENCE
eye-rbt 2% MLD ARZNAD 8,609,58
ipr-rat LD50:220 mg/kg ARZNAD 8,609,58
scu-mus LD50:350 mg/kg ARZNAD 8,609,58

SAFETY PROFILE: Poison by intraperitoneal and subcutaneous routes. An eye irritant. When heated to decomposition it emits very toxic fumes of NO_x and Cl^-.

CFU500 CAS:77966-51-9 ***HR: 3***
6′-CHLORO-2-(DIETHYLAMINO)-N-METHYL-o-ACETOTOLUIDIDE HYDROCHLORIDE
mf: $C_{14}H_{21}ClN_2O{\bullet}ClH$ mw: 305.28

SYN: V 316

TOXICITY DATA with REFERENCE
ipr-rat LD50:136 mg/kg ARZNAD 8,609,58
scu-mus LD50:305 mg/kg ARZNAD 8,609,58

SAFETY PROFILE: Poison by intraperitoneal and subcutaneous routes. When heated to decomposition it emits very toxic fumes of Cl^- and NO_x.

CFU750 CAS:69-05-6 ***HR: 3***
6-CHLORO-9-((4-(DIETHYLAMINO)-1-METHYLBUTYL)AMINO)-2-METHOXYACRIDINE DIHYDROCHLORIDE
mf: $C_{23}H_{30}ClN_3O{\bullet}2ClH$ mw: 472.93

SYNS: ACRICHINE ◇ ARICHIN ◇ ATABRINE DIHYDROCHLORIDE ◇ ATABRINE HYDROCHLORIDE ◇ CHEMIOCHIN ◇ CHINACRIN HYDROCHLORIDE ◇ 2-CHLORO-5-(omega-DIETHYLAMINO-α-METHYLBUTYLAMINO)-7-METHOXYACRIDINE DIHYDROCHLORIDE ◇ 3-CHLORO-9-(4′-DIETHYLAMINO-1′-METHYLBUTYLAMINO)-7-METHOXYACRIDINE DIHYDROCHLORIDE ◇ 3-CHLORO-7-METHOXY-9-(1-METHYL-4-DIETHYLAMINOBUTYLAMINO)ACRIDINEDIHYDROCHLORIDE ◇ CRINODORA ◇ DIAL ◇ ERION ◇ ITALCHIN ◇ MALARICIDA ◇ MECRYL ◇ MEPACRINE DIHYDROCHLORIDE ◇ MEPACRINE HYDROCHLORIDE ◇ METHOQUINE ◇ 2-METHOXY-6-CHLORO-9-(4-DIETHYLAMINO-1-METHYLBUTYLAMINO) ACRIDINEDIHYDROCHLORIDE ◇ METOQUINE ◇ PALACRIN ◇ PENTILEN ◇ QUINACRINE DIHYDROCHLORIDE ◇ QUINACRINE HYDROCHLORIDE ◇ 866 R.P. ◇ SN 390

TOXICITY DATA with REFERENCE
mma-sat 1 mg/plate MUREAV 22,295,74
sln-dmg-orl 5 mmol/L MUREAV 158,177,85
icv-wmn TDLo:20 mg/kg (1D pre):REP CCPTAY 14,75,76
orl-man TDLo:34 mg/kg/8D-I SMJOAV 75,359,82
orl-wmn TDLo:18 mg/kg/3D-I SMJOAV 75,359,82
orl-rat LD50:660 mg/kg JPETAB 91,157,47
ivn-rat LD50:29 mg/kg JPETAB 91,157,47
iut-rat LD50:100 mg/kg IJEBA6 1074,78
ipr-mus LD50:189 mg/kg JPETAB 91,133,47
scu-mus LD50:212 mg/kg ABEMAV 1,317,41
ivn-mus LD50:38 mg/kg JPETAB 91,157,47
ivn-dog LDLo:20 mg/kg JAPMA8 34,20,45
orl-cat LDLo:200 mg/kg AEPPAE 170,328,33

CONSENSUS REPORTS: EPA Genetic Toxicology Program.

SAFETY PROFILE: Poison by ingestion, subcutaneous, intrauterine, intravenous, and intraperitoneal routes. Human reproductive effects by intrauterine and intracervical routes: changes in fertility and unspecified effects on the uterus, cervix and vagina. Experimental reproductive effects. Mutation data reported. Used as a treatment for parasitic worms. When heated to decomposition it emits very toxic fumes of Cl^- and NO_x.

CFV250 CAS:63673-37-0 ***HR: 3***
4-CHLORO-2-DIETHYLAMINO-6-(4-METHYLPIPERAZINO)-5-METHYLTHIOPYRIMIDINE
mf: $C_{14}H_{24}ClN_4S$ mw: 329.94

TOXICITY DATA with REFERENCE
orl-mus LD50:600 mg/kg JMCMAR 18,553,75
ivn-mus LD50:123 mg/kg JMCMAR 18,553,75

SAFETY PROFILE: Poison by intravenous route. Moderately toxic by ingestion. When heated to decomposition it emits very toxic fumes of Cl^-, NO_x, and SO_x.

CFW000 CAS:77966-53-1 ***HR: 2***
2'-CHLORO-2-(DIETHYLAMINO)-5'-TRIFLUOROMETHYLACETANILIDE HYDROCHLORIDE
mf: $C_{13}H_{16}ClF_3N_2O{\bullet}ClH$ mw: 345.22

SYN: C 3078

TOXICITY DATA with REFERENCE
eye-rbt 2% SEV ARZNAD 8,270,58
ipr-rat LD50:590 mg/kg ARZNAD 8,270,58
scu-mus LD50:4250 mg/kg ARZNAD 8,270,58

SAFETY PROFILE: Moderately toxic by intraperitoneal route. Mildly toxic by subcutaneous route. A severe eye irritant. When heated to decomposition it emits very toxic fumes of Cl^-, F^-, and NO_x.

CFW250 CAS:77966-52-0 ***HR: 3***
4'-CHLORO-2-(DIETHYLAMINO)-3'-TRIFLUOROMETHYLACETANILIDE HYDROCHLORIDE
mf: $C_{13}H_{16}ClF_3N_2O{\bullet}ClH$ mw: 345.22

SYN: C 3074

TOXICITY DATA with REFERENCE
eye-rbt 2% SEV ARZNAD 8,270,58
ipr-rat LD50:300 mg/kg ARZNAD 8,270,58
scu-mus LD50:1175 mg/kg ARZNAD 8,270,58

SAFETY PROFILE: Poison by intraperitoneal route. Moderately toxic by subcutaneous route. A severe eye irritant. See also FLUORIDES. When heated to decomposition it emits very toxic fumes of Cl^-, F^-, and NO_x.

CFW625 CAS:5314-83-0 ***HR: 3***
CHLORODIETHYLBORANE
mf: $C_4H_{10}BCl$ mw: 104.39

SAFETY PROFILE: Ignites spontaneously in air. When heated to decomposition it emits toxic fumes of Cl^-. See also BORANES.

CFW750 CAS:23184-66-9 ***HR: 2***
2-CHLORO-2',6'-DIETHYL-N-(BUTOXYMETHYL) ACETANILIDE
mf: $C_{17}H_{26}ClNO_2$ mw: 311.89

PROP: Amber liquid, oily. Sltly sol in H_2O.

SYNS: BUTACHLOR ◇ BUTANEX ◇ N-BUTOXYMETHYL-2-CHLORO-2',6'-DIETHYLACETANILIDE ◇ N-(BUTOXYMETHYL)-2-CHLORO-N-(2,6-DIETHYLPHENYL)ACETAMIDE ◇ CP 53619 ◇ LAMBAST ◇ MACHETE ◇ MACHETE (herbicide) ◇ MACHETTE

TOXICITY DATA with REFERENCE
mma-sat 500 μg/plate MUREAV 116,185,83
mmo-omi 4 mg/L JASIAB 104,571,85
orl-rat LD50:1740 mg/kg EKMMA8 13,123,74
skn-rbt LD50:4080 mg/kg FMCHA2 -,C39,83

SAFETY PROFILE: Moderately toxic by ingestion. Mildly toxic by skin contact. Mutation data reported. An herbicide. When heated to decomposition it emits very toxic fumes of Cl^- and NO_x.

CFW800 CAS:14215-58-8 ***HR: D***
CHLORODIETHYLENETRIAMINE PLATINUM (II) CHLORIDE
mf: $C_4H_{13}ClN_3Pt{\bullet}Cl$ mw: 369.19

TOXICITY DATA with REFERENCE
mmo-sat 100 μg/plate MUREAV 48,139,77
dnd-esc 20 μmol/L MUREAV 93,327,82
dni-esc 40 μmol/L MUREAV 93,327,82

SAFETY PROFILE: Mutation data reported. When heated to decomposition it emits toxic fumes of Cl^- and NO_x. See also PLATINUM COMPOUNDS and CHLORIDES.

CFX000 CAS:15972-60-8 ***HR: 2***
2-CHLORO-2',6'-DIETHYL-N-(METHOXYMETHYL)ACETANILIDE
mf: $C_{14}H_{20}ClNO_2$ mw: 269.80

SYNS: ALACHLOR (USDA) ◇ ALANEX ◇ ALOCHLOR ◇ CHLORESSIGSAEURE-N-(METHOXYMETHYL)-2,6-DIAETHYLANILID (GERMAN) ◇ 2-CHLORO-N-(2,6-DIETHYLPHENYL)-N-(METHOXYMETHYL)ACETAMIDE ◇ CP 50144 ◇ LASSO ◇ LAZO ◇ METACHLOR ◇ METHACHLOR ◇ PILLARZO

TOXICITY DATA with REFERENCE
mmo-omi 90 mg/L JASIAB 104,571,85
mrc-smc 33 μg/plate MUREAV 136,233,84
cyt-hmn:lym 4 mg/L MUREAV 116,341,83
cyt-rat-ipr 1250 μg/kg MUREAV 116,341,83
orl-rat LD50:1200 mg/kg WRPCA2 9,119,70
orl-mus LD50:462 mg/kg GTPZAB 21(12),30,77
skn-rbt LD50:3500 mg/kg GUCHAZ 6,3,73
orl-mam LD50:3000 mg/kg GUCHAZ 6,3,73
unk-mam LD50:1200 mg/kg 30ZDA9 -,126,71

CONSENSUS REPORTS: EPA Genetic Toxicology Program.

SAFETY PROFILE: Moderately toxic by ingestion, skin contact, and possibly other routes. Human mutation data reported. When heated to decomposition it emits very toxic fumes of Cl^- and NO_x.

CFX125 CAS:68674-44-2 ***HR: 3***
CHLORODIFLUOROACETYL HYPOCHLORITE
mf: $C_2Cl_2F_2O_2$ mw: 164.92

SAFETY PROFILE: An unstable explosive above 22°C. When heated to decomposition it emits toxic fumes of F^- and Cl^-. See also HYPOCHLORITES.

CFX250 CAS:75-68-3 ***HR: 1***
1-CHLORO-1,1-DIFLUOROETHANE
DOT: UN 2517
mf: $C_2H_3ClF_2$ mw: 100.50

PROP: Gas. Mp: −131°, bp: −9.5°, d: 1.19, lel: 9.0%; uel: 14.8%.

SYNS: CHLORODIFLUOROETHANE (DOT) ◇ α-CHLOROETHYLIDENE FLUORIDE ◇ 1,1-DIFLUORO-1-CHLOROETHANE ◇ DIFLUOROMONOCHLOROETHANE (DOT) ◇ FC142b ◇ FLUOROCARBON FC142b ◇ FREON 142 ◇ FREON 142b ◇ GENETRON 101 ◇ GENETRON 142b

TOXICITY DATA with REFERENCE
mma-sat 50 pph/24H TXAPA9 72,15,84
ihl-rat LC50:2050 g/m³/4H 85GMAT -,53,82
ihl-mus LC50:1758 g/m³/2H 85GMAT -,53,82

CONSENSUS REPORTS: Reported in EPA TSCA Inventory.

DOT Classification: Flammable Gas; Label: Flammable Gas.

SAFETY PROFILE: Very mildly toxic by inhalation. Mutation data reported. A very dangerous fire hazard when exposed to heat, flame, or oxidizing materials. To fight fire, stop flow of gas. Can react vigorously with oxidizing materials. When heated to decomposition it emits toxic fumes of F^- and Cl^-.

CFX500 CAS:75-45-6 ***HR: 1***
CHLORODIFLUOROMETHANE
DOT: UN 1018
mf: $CHClF_2$ mw: 86.47

PROP: Gas. D: 3.87 air @ 0°, mp: −146°, bp: −40.8°, autoign temp: 1170°F.

SYNS: ALGOFRENE TYPE 6 ◇ ARCTON 4 ◇ DIFLUOROCHLOROMETHANE ◇ DIFLUOROMONOCHLOROMETHANE ◇ ELECTRO-CF 22 ◇ ESKIMON 22 ◇ F 22 ◇ FLUOROCARBON-22 ◇ FREON ◇ FREON 22 ◇ FRIGEN ◇ GENETRON 22 ◇ ISCEON 22 ◇ ISOTRON 22 ◇ MONOCHLORODIFLUOROMETHANE ◇ PROPELLANT 22 ◇ R 22 (DOT) ◇ REFRIGERANT 22 ◇ UCON 22/HALOCARBON 22

TOXICITY DATA with REFERENCE
mmo-sat 33 pph/24H-C TOLED5 2,1,78
mma-sat 33 pph/24H-C TOLED5 2,1,78
ihl-rat TCLo:50000 ppm/5H (56D male):REP FAATDF 1,266,81
ihl-rat LC50:35 pph/15M HUTODJ 1,239,82
ihl-mus LC50:28 pph/20M TXAPA9 59,64,81
ihl-dog LCLo:70 pph TXAPA9 2,363,60

CONSENSUS REPORTS: IARC Cancer Review: Group 3 IMEMDT 7,149,87; Human Inadequate Evidence IMEMDT 41,237,86; Animal Limited Evidence IMEMDT 41,237,86. Reported in EPA TSCA Inventory. EPA Genetic Toxicology Program.

OSHA PEL: TWA 1000 ppm
ACGIH TLV: TWA 1000 ppm
DFG MAK: 500 ppm (1800 mg/m³)
DOT Classification: Nonflammable Gas; Label: Nonflammable Gas.

SAFETY PROFILE: Mildly toxic by inhalation. Experimental reproductive effects. Mutation data reported. An asphyxiant in high concentrations. At elevated pressures, 50% mixtures with air are combustible although ignition is difficult. When heated to decomposition it emits toxic fumes of F^- and Cl^-. See also CHLORINATED HYDROCARBONS, ALIPHATIC, and FLUORIDES.

CFX625 CAS:59034-34-3 ***HR: 3***
1-CHLORO-3,3-DIFLUORO-2-METHOXYCYCLOPROPENE
mf: $C_4H_3ClF_2O$ mw: 140.52

$ClC{=}C(OCH_3)CF_2$

SAFETY PROFILE: Explosive reaction on contact with water or methanol. When heated to decomposition it emits toxic fumes of F^- and Cl^-. See also CHLORINATED HYDROCARBONS, ALIPHATIC.

CFY000 CAS:58-93-5 ***HR: 3***
6-CHLORO-3,4-DIHYDRO-2H-1,2,4-BENZOTHIADIAZINE-7-SULFONAMIDE- 1,1-DIOXIDE
mf: $C_7H_8ClN_3O_4S_2$ mw: 297.75

SYNS: AQUARILLS ◇ AQUARIUS ◇ BREMIL ◇ 6-CHLORO-3,4-DIHYDRO-7-SULFAMOYL-2H-1,2,4-BENZOTHIADIAZINE-1,1-DIOXIDE ◇ 6-CHLORO-7-SULFAMOYL-3,4-DIHYDRO-2H-1,2,4-BENZOTHIADIAZINE-1,1-DIOXIDE ◇ CHLOROSULTHIADIL ◇ CHLORSULFONAMIDO DIHYDROBENZOTHIADIAZINE DIOXIDE ◇ CHLORZIDE ◇ CIDREX ◇ DICHLOROSAL ◇ DICHLOTIAZID ◇ DICHLOTRIDE ◇ DICLOTRIDE ◇ 3,4-DIHYDRO-6-CHLORO-7-SULFAMYL-1,2,4-BENZOTHIADIAZINE-1,1-DIOXIDE ◇ DIHYDROCHLOROTHIAZID ◇ DIHYDROCHLOROTHIAZIDE ◇ 3,4-DIHYDROCHLOROTHIAZIDE ◇ DIHYDROXYCHLOROTHIAZIDUM ◇ DIREMA ◇ DISALUNIL ◇ DRENOL ◇ DYAZIDE ◇ ESIDREX ◇ ESIDRIX ◇ FLUVIN ◇ HCTZ ◇ HCZ ◇ HIDRIL ◇ HIDROCHLORTIAZID ◇ HIDRORONOL ◇ HIDROTIAZIDA ◇ HYDRO-AQUIL ◇ HYDROCHLORTHIAZID ◇ HYDRODIURETIC ◇ HYDRO-DIURIL ◇ HYDROSALURIC ◇ HYDROTHIDE ◇ HYPOTHIAZIDE ◇ IDROTIAZIDE ◇ IVAUGAN ◇ JEN-DIRIL ◇ MASCHITT ◇ MEGA-DIURIL ◇ NCI-C55925 ◇ NEFRIX ◇ NEO-CODEMA ◇ NEOFLUMEN ◇ ORETIC ◇ PANURIN ◇ RO-HYDRAZIDE ◇ SU 5879 ◇ THIARETIC ◇ THIURETIC ◇ THLARETIC ◇ URODIAZIN ◇ VETIDREX ◇ ZIDE

TOXICITY DATA with REFERENCE
dnd-esc 5 mg/L MUREAV 89,95,81
cyt-ham:lng 500 mg/L/48H GMCRDC 27,95,81
orl-rat TDLo:582 g/kg (26W pre):REP IYKEDH 11,294,80
orl-wmn TDLo:2 mg/kg/12H-I:SYS SMJOAV 76,1363,83
orl-wmn TDLo:500 μg/kg:PUL,GIT DICPBB 18,238,84
ipr-rat LD50:234 mg/kg 27ZIAQ -,124,73
scu-rat LD50:1270 mg/kg 27ZIAQ -,124,73
ivn-rat LD50:990 mg/kg JPETAB 140,249,63
orl-mus LD50:1175 mg/kg FRZKAP (1),44,83
ipr-mus LD50:578 mg/kg 27ZIAQ -,77,65
scu-mus LD50:1470 mg/kg 27ZIAQ -,124,73
ivn-mus LD50:590 mg/kg JPETAB 134,273,61
ivn-dog LD50:250 mg/kg 27ZIAQ -,124,73
ivn-rbt LD50:461 mg/kg 27ZIAQ -,124,73

CONSENSUS REPORTS: Reported in EPA TSCA Inventory. EPA Genetic Toxicology Program.

SAFETY PROFILE: Poison by intraperitoneal and intravenous routes. Moderately toxic by ingestion and subcutaneous routes. Human systemic effects by ingestion: sodium level changes, chlorine level changes, acute pulmonary edema, nausea or vomiting. Experimental reproductive effects. Mutation data reported. A diuretic. When heated to decomposition it emits very toxic fumes of SO_x, Cl^-, and NO_x.

CFY250 CAS:36104-80-0 ***HR: 2***
7-CHLORO-1,3-DIHYDRO-3-(N,N-DIMETHYLCARBAMOYL)-1-METHYL-5-PHENYL-2H-1,4-BENZODIAZEPIN-2-ONE
mf: $C_{19}H_{18}ClN_3O_3$ mw: 371.85

SYNS: ALBEGO ◇ B 5333 ◇ CAMAZEPAM ◇ 7-CHLORO-1,3-DIHYDRO-3-HYDROXY-1-METHYL-5-PHENYL-1,4-BENZODIAZEPIN-2-ONE DIMETHYLCARBAMATE ◇ SB 5833

TOXICITY DATA with REFERENCE
orl-mus LD50:970 mg/kg DRFUD4 1,458,76
ipr-mus LD50:800 mg/kg DRFUD4 1,458,76

SAFETY PROFILE: Moderately toxic by ingestion and intraperitoneal routes. A tranquilizer. See also CARBAMATES and DIAZEPAM. When heated to decomposition it emits very toxic fumes of Cl^- and NO_x.

CFY500 CAS:4700-56-5 ***HR: 3***
7-CHLORO-1,3-DIHYDRO-3-HEMISUCCINYLOXY-2H-1,4-BENZODIAZEPIN-2-ONE
mf: $C_{19}H_{15}ClN_2O_5$ mw: 386.81

SYNS: BUTANEDIOIC ACID MONO(7-CHLORO-2,3-DIHYDRO-2-OXO-5-PHENYL-1H-1,4-BENZODIAZEPIN-3-YL)ESTER ◇ NULANS ◇ SAS 538 ◇ SUCCINIC ACID MONOESTER with 7-CHLORO-1,3-DIHYDRO-3-HYDROXY-5-PHENYL-2H-1,4-BENZODIAZEPIN-2-ONE

TOXICITY DATA with REFERENCE
orl-mus LD50:1148 mg/kg FRPPAO 22,506,67
ipr-mus LD50:375 mg/kg FRPPAO 22,506,67
ivn-mus LD50:285 mg/kg FRPPAO 22,506,67

SAFETY PROFILE: Poison by intraperitoneal and intravenous routes. Moderately toxic by ingestion. See also DIAZEPAM and ESTERS. When heated to decomposition it emits very toxic fumes of Cl^- and NO_x.

CFY750 CAS:846-50-4 ***HR: 3***
7-CHLORO-1,3-DIHYDRO-3-HYDROXY-1-METHYL-5-PHENYL-2H-1,4-BENZODIAZEPIN-2-ONE
mf: $C_{16}H_{13}ClN_2O_2$ mw: 300.76

SYNS: CEREPAX ◇ CRISONAR ◇ 1,3-DIHYDRO-7-CHLORO-3-HYDROXY-1-METHYL-5-PHENYL-2H-1,4-BENZODIAZEPIN-2-ONE ◇ ER 115 ◇ EUHYPNOS ◇ HYDROXYDIAZEPAM ◇ 3-HYDROXY-DIAZEPAM ◇ K3917 ◇ LEVANXENE ◇ LEVANXOL ◇ MABERTIN ◇ METHYLOXAZEPAM ◇ N-METHYLOXAZEPAM ◇ NORMISON ◇ OXYDIAZEPAM ◇ PLANUM ◇ REMESTAN ◇ RESTORIL ◇ RO 5-5345 ◇ SIGNOPAM ◇ TEMAZEPAM ◇ WY 2917 ◇ WY 3917

TOXICITY DATA with REFERENCE
orl-hmn TDLo:15430 μg/kg/12D-I:CNS,GIT ARZNAD 22,93,72
orl-rat LD50:2000 mg/kg DRUGAY 21,321,81
ipr-rat LD50:600 mg/kg DRUGAY 21,321,81
orl-mus LD50:370 mg/kg EJPHAZ 4,467,68
ipr-mus LD50:85 mg/kg AIPTAK 185,135,70
orl-dog LD50:3620 mg/kg DRUGAY 21,321,81

SAFETY PROFILE: Poison by ingestion and intraperitoneal routes. Human systemic effects by ingestion: muscle weakness and changes in the structure or function of the salivary glands. A tranquilizer. When heated to decomposition it emits toxic fumes of Cl^- and NO_x. See also DIAZEPAM.

CFZ000 CAS:604-75-1 ***HR: 2***
7-CHLORO-1,3-DIHYDRO-3-HYDROXY-5-PHENYL-2H-1,4-BENZODIAZEPINE-2-ONE
mf: $C_{15}H_{11}ClN_2O_2$ mw: 286.73

SYNS: ADUMBRAN ◇ ANSIOLISINA ◇ ANSIOXACEPAM ◇ ANXIOLIT ◇ APLAKIL ◇ ASTRESS ◇ BONARE ◇ 7-CHLORO-3-HYDROXY-5-PHENYL-1,3-DIHYDRO-2H-1,4-BENZODIAZEPIN-2-ONE ◇ ENIDREL ◇ HILONG ◇ ISODIN ◇ LIMBIAL ◇ NESONTIL ◇ NOCTAZEPAM ◇ NOTARAL ◇ OX ◇ OXAZEPAM ◇ PACIENX ◇ PRAXITEN ◇ PROPAX ◇ PSICOPAX ◇ QUEN ◇ QUILIBREX ◇ RO 5-6789 ◇ RONDAR ◇ SERAX ◇ SERENAL ◇ SERENID ◇ SERENID-D ◇ SEREPAX ◇ SERESTA ◇ SERPAX ◇ SIGACALM ◇ SOBRIL ◇ TAZEPAM ◇ TRANQUO-BUSCOPAN-WIRKSTOFF ◇ VABEN ◇ WY-3498 ◇ Z10-TR

TOXICITY DATA with REFERENCE
mma-sat 5200 pmol/plate CNREA8 38,4478,78
orl-mus TDLo:2520 mg/kg (1-21D preg/21D post):REP PLRCAT 9,325,77
orl-mus TDLo:65 g/kg/52W-C:NEO RCOCB8 8,481,74
orl-chd TDLo:8 mg/kg:CNS JAMAAP 196,662,66
ipr-rat LD50:1535 mg/kg PHMGBN 10,345,73
ipr-mus LD50:767 mg/kg JMCMAR 11,777,68

CONSENSUS REPORTS: IARC Cancer Review: Group 3 IMEMDT 7,56,87; Animal Limited Evidence IMEMDT 13,57,77

SAFETY PROFILE: Moderately toxic by intraperitoneal route. Human (child) systemic effects by ingestion: somnolence, changes in REM sleep, and loss of muscle control (ataxia). Experimental reproductive effects. Questionable carcinogen with experimental neoplastigenic data. Mutation data reported. Used to treat anxiety and tension. When heated to decomposition it emits very toxic fumes of NO_x and HCl. See also DIAZEPAM.

CGA000 CAS:2898-12-6 ***HR: 3***
7-CHLORO-2,3-DIHYDRO-1-METHYL-5-PHENYL-1H-1,4-BENZODIAZEPINE
mf: $C_{16}H_{15}ClN_2$ mw: 270.78

SYNS: ANSILAN ◇ DIEPIN ◇ 2,3-DIHYDRO-7-CHLORO-1-METHYL-5-PHENYL-1H-1,4-BENZODIAZEPINE ◇ ELBRUS ◇ ESMAIL ◇ MEDAZEPAM ◇ MEDAZEPOL ◇ MEGASEDAN ◇ MEZEPAN ◇ NARSIS ◇ NOBRIUM ◇ PAZITAL ◇ PSIQUIM ◇ RESMIT ◇ RUDOTEL ◇ SERENIUM ◇ SIMAN ◇ TRANQUILAX

TOXICITY DATA with REFERENCE
orl-rat TDLo:860 mg/kg (1-22D preg/21D post):REP PSYPAG 41,113,75
orl-rat LD50:900 mg/kg 26RAAN -,35,73
orl-mus LD50:475 mg/kg ARZNAD 24,2029,74
ipr-mus LD50:360 mg/kg AIPTAK 185,135,70

SAFETY PROFILE: Poison by intraperitoneal route. Moderately toxic by ingestion. Experimental reproductive effects. A tranquilizer. When heated to decomposition it emits very toxic fumes of Cl^- and NO_x. See also DIAZEPAM.

CGA500 CAS:1088-11-5 ***HR: 3***
7-CHLORO-1,3-DIHYDRO-5-PHENYL-2H-1,4-BENZODIAZEPIN-2-ONE
mf: $C_{15}H_{11}ClN_2O$ mw: 270.73

SYNS: A-101 ◇ CALMDAY ◇ DEALKYLPRAZEPAM ◇ DEMETHYL-DIAZEPAM ◇ N-DEMETHYLDIAZEPAM ◇ 1-DEMETHYLDIAZEPAM ◇ N-DEOXYDEMOXAPAM ◇ DESALKYLPRAZEPAM ◇ N-DESCYCLO-PROPYLMETHYLPRAZEPAM ◇ DESMETHYLDIAZEPAM ◇ N-DES-METHYLDIAZEPAM ◇ DMDZ ◇ MADAR ◇ NDD ◇ NORDIAZEPAM ◇ NORPRAZEPAM ◇ RO 5-2180 ◇ STILNY

TOXICITY DATA with REFERENCE
mma-sat 57 nmol/plate CNREA8 38,4478,78
orl-mus TDLo:400 mg/kg/(14D preg):TER TXAPA9 25,453,73
orl-mus LD50:670 mg/kg EJPHAZ 4,467,68
ipr-mus LD50:290 mg/kg EJPHAZ 16,311,71

SAFETY PROFILE: Poison by intraperitoneal route. Moderately toxic by ingestion. An experimental teratogen. Mutation data reported. A tranquilizer. When heated to decomposition it emits very toxic fumes of NO_x and Cl^-. See also DIAZEPAM.

CGB000 CAS:55299-24-6 ***HR: 2***
7-CHLORO-1,3-DIHYDRO-5-PHENYL-1-TRIMETHYLSILYL-2H-1,4-BENZODIAZEPIN-2-ONE
mf: $C_{18}H_{18}ClOSi$ mw: 313.90

SYNS: ST 720 (FRENCH) ◇ TRIMETHYL SILYL-1-CHLORO-7-DIHYDRO-1,3-PHENYL-5,2H-BENZODIAZEPINE-1,4-ONE-2(FRENCH)

TOXICITY DATA with REFERENCE
orl-mus LD50:1000 mg/kg APFRAD 36,621,78
ipr-mus LD50:600 mg/kg APFRAD 36,621,78

SAFETY PROFILE: Moderately toxic by ingestion and intraperitoneal routes. When heated to decomposition it emits toxic fumes of Cl^-. See also DIAZEPAM.

CGB250 CAS:14437-41-3 ***HR: 2***
4é-CHLORO-3,5-DIIODOSALICYLANILIDE ACETATE
mf: $C_{15}H_{10}ClI_2NO_3$ mw: 541.51

SYNS: ACETOXY-4′-CHLORO-3,5-DIIODOBENZANILIDE◇ 2-(ACETYLOXY)-N-(4-CHLOROPHENYL)-3,5-DIIODOBENZAMIDE◇ C.I. 633 ◇ CLIOXANIDE ◇ CN 59,567 ◇ SYD 230 ◇ TREMERAD

TOXICITY DATA with REFERENCE
ipr-mus LD50:720 mg/kg AUVJA2 46,297,70
orl-dom LD50:414 mg/kg AUVJA2 46,297,70

CONSENSUS REPORTS: Reported in EPA TSCA Inventory.

SAFETY PROFILE: Moderately toxic by ingestion and intraperitoneal routes. Used in treatment against parasitic worms. When heated to decomposition it emits very toxic fumes of Cl^-, I^-, and NO_x.

CGB500 CAS:1779-25-5 ***HR: 1***
CHLORO DIISOBUTYL ALUMINUM
mf: $C_8H_{18}AlCl$ mw: 176.69

SYNS: ALLUMINIO DIISOBUTIL-MONOCLORURO (ITALIAN) ◇ BIS(ISOBUTYL)ALUMINUM CHLORIDE ◇ CHLOROBIS(2-METHYLPROPYL)ALUMINUM ◇ DIISOBUTYLALUMINUM CHLORIDE ◇ DIISOBUTYLALUMINUM MONOCHLORIDE ◇ DIISOBUTYLCHLOROALUMINUM

TOXICITY DATA with REFERENCE
ihl-mus LDLo:680 g/kg/15M MELAAD 58,290,67

CONSENSUS REPORTS: Reported in EPA TSCA Inventory.

ACGIH TLV: TWA 2 mg(Al)/m^3

SAFETY PROFILE: Mildly toxic by inhalation. See also ALUMINUM COMPOUNDS and CHLORIDES. Ignites spontaneously in air. When heated to decomposition it emits toxic fumes of Cl^-.

CGB750 CAS:63869-02-3 ***HR: 3***
CHLORO(DIISOPROPOXYPHOSPHINYL) MERCURY
mf: $C_6H_{14}ClHgO_3P$ mw: 401.21

TOXICITY DATA with REFERENCE
ipr-mus LDLo:16 mg/kg CBCCT* 8,104,56

CONSENSUS REPORTS: Mercury and its compounds are on the Community Right-To-Know List.

OSHA PEL: (Transitional: CL 1 mg/10m^3) CL 0.1 mg(Hg)/m^3 (skin)
ACGIH TLV: TWA 0.1 mg/(Hg)/m^3 (skin)
NIOSH REL: (Inorganic Mercury) TWA: 0.05 mg(Hg)/m^3

SAFETY PROFILE: Poison by intraperitoneal route. See also MERCURY COMPOUNDS. When heated to decomposition it emits very toxic fumes of PO_x, Cl^-, and Hg.

CGC000 CAS:77966-54-2 ***HR: 3***
6′-CHLORO-2-(DIISOPROPYLAMINO)-o-ACETOTOLUIDIDE HYDROCHLORIDE
mf: $C_{15}H_{23}ClN_2O•ClH$ mw: 319.31

TOXICITY DATA with REFERENCE
eye-rbt 2% MLD ARZNAD 8,407,58
ipr-rat LD50:100 mg/kg ARZNAD 8,407,58
scu-mus LD50:515 mg/kg ARZNAD 8,407,58

SAFETY PROFILE: Poison by intraperitoneal route. Moderately toxic by subcutaneous route. An eye irritant. When heated to decomposition it emits very toxic fumes of Cl^- and NO_x.

CGC200 CAS:1585-74-6 ***HR: 3***
N-CHLORODIMETHYLAMINE
mf: C_2H_6ClN mw: 79.53

SAFETY PROFILE: Products of reaction with antimony chlorides (e.g. antimony trichloride and antimony pentachloride) are dangerous heat- and shock-sensitive explosives which may explode at room temperature. When heated to decomposition it emits toxic fumes of Cl^- and NO_x. See also AMINES.

CGD000 CAS:77966-55-3 ***HR: 3***
6′-CHLORO-2-(DIMETHYLAMINO)-o-ACETOTOLUIDIDE HYDROCHLORIDE
mf: $C_{11}H_{15}ClN_2O•ClH$ mw: 263.19

SYN: V 252

TOXICITY DATA with REFERENCE
ipr-rat LD50:218 mg/kg ARZNAD 8,407,58
ipr-mus LD50:243 mg/kg ARZNAD 8,407,58
scu-mus LD50:600 mg/kg ARZNAD 8,407,58

SAFETY PROFILE: Poison by intraperitoneal route. Moderately toxic by subcutaneous route. When heated to decomposition it emits very toxic fumes of Cl^- and NO_x.

CGD250 CAS:2491-76-1 ***HR: 3***
p-CHLORO DIMETHYLAMINOAZOBENZENE
mf: $C_{14}H_{14}ClN_3$ mw: 259.76

SYNS: 4'-CHLORO-4-DIMETHYLAMINOAZOBENZENE ◇ N,N-DIMETHYL-p-((p-CHLOROPHENYL)AZO)ANILINE

TOXICITY DATA with REFERENCE
scu-mus TDLo:500 mg/kg (9D preg):TER OFAJAE 36,195,60
orl-rat TDLo:6100 mg/kg/21W-C:NEO JEMEAV 87,139,48
scu-mus LDLo:500 mg/kg OFAJAE 36,195,60

SAFETY PROFILE: Moderately toxic by subcutaneous route. Questionable carcinogen with experimental neoplastigenic data. Experimental teratogenic effects. When heated to decomposition it emits very toxic fumes of Cl^- and NO_x.

CGD399 ***HR: 3***
β-CHLORODIMETHYLAMINO DIBORANE
mf: $C_2H_{10}B_2ClN$ mw: 105.18

$ClHB:H_2:BHN(CH_3)_2$

SYN: B-CHLORO-N,N-DIMETHYLAMINODIBORANE

SAFETY PROFILE: Ignites spontaneously in air. When heated to decomposition it emits toxic fumes of Cl^- and NO_x. See also BORANES, CHLORIDES, and AMINES.

CGD500 CAS:3505-38-2 ***HR: 3***
2-(p-CHLORO-α-(2-(DIMETHYLAMINO)ETHOXY) BENZYL)PYRIDINE BIMALEATE
mf: $C_{16}H_{19}ClN_2O•C_4H_4O_4$ mw: 406.90

SYNS: ALLERGEFON MALEATE ◇ CARBINOXAMINE MALEATE ◇ p-CARBINOXAMINE MALEATE ◇ 2-(p-CHLORO-α-(2-(DIMETHYLAMINO)ETHOXY)BENZYL)PYRIDINE MALEATE ◇ 2-((4-CHLOROPHENYL)-2-PYRIDINYLMETHOXY)-N,N-DIMETHYLETHANAMINE-(Z)-2-BUTENEDIOATE (1:1) ◇ CLISTIN ◇ CLISTIN MALEATE ◇ CLISTINE MALEATE

TOXICITY DATA with REFERENCE
orl-mus LD50:162 mg/kg CLDND* 15,367,68
scu-mus LD50:350 mg/kg TOIZAG 15,367,68
ivn-mus LD50:32 mg/kg CLDND*
ivn-dog LDLo:36 mg/kg CLDND*
orl-gpg LD50:411 mg/kg CLDND*
scu-gpg LD50:120 mg/kg APFRAD 20,463,62

CONSENSUS REPORTS: Reported in EPA TSCA Inventory.

SAFETY PROFILE: Poison by ingestion, subcutaneous, and intravenous routes. When heated to decomposition it emits very toxic fumes of Cl^- and NO_x.

CGD750 CAS:101651-62-1 ***HR: 3***
2'-CHLORO-2-((2-(DIMETHYLAMINO)ETHYL) ETHYLAMINO)ACETANILIDE DIHYDROCHLORIDE
mf: $C_{14}H_{22}ClN_3O•2ClH$ mw: 356.76

SYN: C 5417

TOXICITY DATA with REFERENCE
eye-rbt 2% MLD ARZNAD 9,262,59
ipr-rat LD50:148 mg/kg ARZNAD 9,262,59
scu-mus LD50:575 mg/kg ARZNAD 9,262,59

SAFETY PROFILE: Poison by intraperitoneal route. Moderately toxic by subcutaneous route. An eye irritant. When heated to decomposition it emits very toxic fumes of Cl^- and NO_x.

CGE000 CAS:101651-94-9 ***HR: 3***
7-CHLORO-10-(2-(DIMETHYLAMINO)ETHYL) ISOALLOXAZINE SULFATE
mf: $C_{14}H_{14}ClN_5O_2•H_2O_4S$ mw: 417.86

TOXICITY DATA with REFERENCE
orl-mus LD50:1900 mg/kg CMTRAG 2,96,61
scu-mus LD50:38 mg/kg CMTRAG 2,96,61
ivn-mus LD50:60 mg/kg CMTRAG 2,96,61

SAFETY PROFILE: Poison by subcutaneous and intravenous routes. Moderately toxic by ingestion. See also SULFATES. When heated to decomposition it emits very toxic fumes of SO_x, Cl^-, and NO_x.

CGE250 CAS:95770-03-9 ***HR: 2***
2'-CHLORO-2-(2-(DIMETHYLAMINO)ETHYLTHIO) ACETANILIDE HYDROCHLORIDE
mf: $C_{12}H_{17}ClN_2OS•ClH$ mw: 309.28

SYN: C 5501

TOXICITY DATA with REFERENCE
eye-rbt 2% MOD ARZNAD 9,683,59
ipr-rat LD50:750 mg/kg ARZNAD 9,683,59
scu-mus LD50:1400 mg/kg ARZNAD 9,683,59

SAFETY PROFILE: Moderately toxic by intraperitoneal and subcutaneous routes. An eye irritant. When heated to decomposition it emits very toxic fumes of Cl^-, NO_x, and SO_x.

CGE500 CAS:100620-36-8 ***HR: 3***
6'-CHLORO-2-(2-(DIMETHYLAMINO) ETHYLTHIO)-o-ACETOTOLUIDIDE
mf: $C_{13}H_{19}ClN_2OS$ mw: 286.85

SYN: C 5458

TOXICITY DATA with REFERENCE
ipr-rat LD50:140 mg/kg ARZNAD 9,683,59
scu-mus LD50:460 mg/kg ARZNAD 9,683,59

SAFETY PROFILE: Poison by intraperitoneal route. Moderately toxic by subcutaneous route. When heated to decomposition it emits very toxic fumes of SO_x, NO_x, and Cl^-.

CGE750 CAS:77966-56-4 ***HR: 3***
6'-CHLORO-2-(DIMETHYLAMINO)-N-METHYL-o-ACETOTOLUIDIDE HYDROCHLORIDE
mf: $C_{12}H_{17}ClN_2O•ClH$ mw: 277.22

SYN: C 3133

TOXICITY DATA with REFERENCE
ipr-rat LD50:390 mg/kg ARZNAD 8,609,58
scu-mus LD50:445 mg/kg ARZNAD 8,609,58

SAFETY PROFILE: Poison by intraperitoneal route. Moderately toxic by subcutaneous route. When heated to decomposition it emits very toxic fumes of Cl^- and NO_x.

CGF000 CAS:19986-35-7 ***HR: 3***
5-CHLORO-3-(DIMETHYLAMINOMETHYL)-2-BENZOXAZOLINONE
mf: $C_{10}H_{11}ClN_2O_2$ mw: 226.68

TOXICITY DATA with REFERENCE
orl-mus LD50:1500 mg/kg MDCHAG 4(1),308,64
ipr-mus LD50:400 mg/kg MDCHAG 4(1),308,64

SAFETY PROFILE: Poison by intraperitoneal route. Moderately toxic by ingestion. When heated to decomposition it emits very toxic fumes of Cl^- and NO_x.

CGF250 CAS:101651-96-1 ***HR: 3***
7-CHLORO-10-(4-(DIMETHYLAMINO)-1-METHYLBUTYL)ISOALLOXAZINE SULFATE
mf: $C_{19}H_{24}ClN_5O_2•H_2O_4S$ mw: 488.01

TOXICITY DATA with REFERENCE
ipr-rat LD50:55 mg/kg CMTRAG 2,96,61
scu-mus LD50:120 mg/kg CMTRAG 2,96,61
ivn-mus LD50:28 mg/kg CMTRAG 2,96,61

SAFETY PROFILE: Poison by intraperitoneal, subcutaneous, and intravenous routes. See also SULFATES. When heated to decomposition it emits very toxic fumes of SO_x, NO_x and Cl^-.

CGF500 CAS:78218-37-8 ***HR: 3***
6'-CHLORO-3-(DIMETHYLAMINO)-o-PROPIONOTOLUIDIDE HYDROCHLORIDE
mf: $C_{12}H_{17}ClN_2O•ClH$ mw: 277.22

TOXICITY DATA with REFERENCE
ipr-rat LD50:114 mg/kg ARZNAD 8,544,58
ipr-mus LD50:175 mg/kg ARZNAD 8,544,58
scu-mus LD50:445 mg/kg ARZNAD 8,544,58

SAFETY PROFILE: Poison by intraperitoneal route. Moderately toxic by subcutaneous route. When heated to decomposition it emits very toxic fumes of NO_x, and Cl^-.

CGG500 CAS:28907-45-1 ***HR: 3***
7-CHLORO-10-(3-DIMETHYLAMINOPROPYL)-BENZO-(b)(1,8)-5(10H)-NAPHTHAPYRIDONE HYDROCHLORIDE
mf: $C_{17}H_{18}ClN_3O•ClH$ mw: 352.29

SYNS: C 45 ◇ C 45 (pharmaceutical) ◇ CHLOROWODORKU 10-Γ-DWUMETYLOAMINOPROPYLO-7-CHLOROBENZO(b)-(1,8)-NAFTYRYDONU-5 (POLISH) ◇ IFC-45

TOXICITY DATA with REFERENCE
orl-rat LD50:500 mg/kg PJPPAA 27,503,75
ipr-rat LD50:105 mg/kg PJPPAA 27,503,75
ivn-rat LD50:33 mg/kg DRFUD4 3,303,78
orl-mus LD50:200 mg/kg DRFUD4 3,303,78
ipr-mus LD50:106 mg/kg PJPPAA 27,503,75
ivn-mus LD50:33 mg/kg DRFUD4 3,303,78
ivn-rbt LD50:12 mg/kg PJPPAA 27,503,75

SAFETY PROFILE: Poison by ingestion, intravenous, and intraperitoneal routes. When heated to decomposition it emits very toxic fumes of Cl^- and NO_x.

CGG600 CAS:14051-55-9 ***HR: 3***
3-CHLORO-5-(3-(DIMETHYLAMINO)PROPYL)-10,11-DIHYDRO-5H-DIBENZ(b,f)AZEPINE HYDROCHLORIDE
mf: $C_{19}H_{23}ClN_2•ClH$ mw: 351.35

TOXICITY DATA with REFERENCE
orl-rat LD50:1150 mg/kg IYKEDH 4,193,73
ipr-rat LD50:135 mg/kg IYKEDH 4,193,73
scu-rat LD50:1750 mg/kg IYKEDH 4,193,73
ivn-rat LD50:26 mg/kg IYKEDH 4,193,73
orl-mus LD50:470 mg/kg IYKEDH 4,193,73
ipr-mus LD50:90 mg/kg IYKEDH 4,193,73
scu-mus LD50:400 mg/kg IYKEDH 4,193,73
ivn-mus LD50:26 mg/kg IYKEDH 4,193,73

SAFETY PROFILE: Poison by subcutaneous, intravenous, and intraperitoneal routes. Moderately toxic by ingestion. When heated to decomposition it emits toxic fumes of NO_x and HCl.

CGG750 CAS:101651-97-2 ***HR: 3***
7-CHLORO-10-(3-(DIMETHYLAMINO)PROPYL)ISOALLOXAZINE HYDROCHLORIDE
mf: $C_{15}H_{16}ClN_5O_2•ClH$ mw: 370.27

TOXICITY DATA with REFERENCE
orl-mus LD50:1250 mg/kg CMTRAG 2,96,61
ipr-mus LD50:70 mg/kg CMTRAG 2,96,61
scu-mus LD50:24 mg/kg CMTRAG 2,96,61
ivn-mus LD50:60 mg/kg CMTRAG 2,96,61

SAFETY PROFILE: Poison by intraperitoneal, subcutaneous, and intravenous routes. Moderately toxic by ingestion. When heated to decomposition it emits very toxic fumes of Cl^- and NO_x.

CGH250 CAS:63019-52-3 ***HR: 3***
9-CHLORO-8,12-DIMETHYLBENZ(a)ACRIDINE
mf: $C_{19}H_{14}ClN$ mw: 291.79

SYNS: 2-CHLORO-1,10-DIMETHYL-5,6-BENZACRIDINE(FRENCH) ◇ 1,10-DIMETHYL-2-CHLORO-5,6-BENZACRIDINE ◇ 8,12-DIMETHYL-9-CHLOROBENZ(a)ACRIDINE

TOXICITY DATA with REFERENCE
skn-mus TDLo:500 mg/kg/41W-I:ETA AICCA6 11,736,55

SAFETY PROFILE: Questionable carcinogen with experimental tumorigenic data by skin contact. When heated to decomposition it emits very toxic fumes of NO_x and Cl^-.

CGH500 CAS:64050-23-3 ***HR: 3***
10-CHLORO-6,9-DIMETHYL-5,10-DIHYDRO-3,4-BENZOPHENARSAZINE
mf: $C_{18}H_{15}AsClN$ mw: 355.71

SYN: 12-CHLORO-7,12-DIHYDRO-8,11-DIMETHYLBENZO(a)PHENARSAZINE

TOXICITY DATA with REFERENCE
skn-mus TDLo:380 mg/kg/16W-I:ETA CRSBAW 145,1451,51

CONSENSUS REPORTS: Arsenic and its compounds are on the Community Right-To-Know List.

OSHA PEL: TWA 0.5 mg(As)/m^3

SAFETY PROFILE: Questionable carcinogen with experimental tumorigenic data by skin contact. When heated to decomposition it emits very toxic fumes of As, Cl^-, and NO_x. See also ARSENIC COMPOUNDS.

CGH675 CAS:10140-91-7 ***HR: 3***
p-CHLORO-5,10-DIMETHYL-2,4-DIOXA-p-THIONO-3-PHOSPHABICYCLO(4.4.0)DECANE

TOXICITY DATA with REFERENCE
skn-rbt 10 mg/24H open MLD AIHAAP 23,95,62
orl-rat LD50:110 mg/kg AIHAAP 23,95,62
skn-rbt LD50:200 mg/kg AIHAAP 23,95,62

SAFETY PROFILE: Poison by ingestion and skin contact. When heated to decomposition it emits toxic fumes of Cl^-, PO_x, and SO_x.

CGI125 CAS:26096-99-1 ***HR: 3***
N-(p-CHLORO-α,α-DIMETHYLPHENETHYL)-2-(DIETHLAMINO)PROPIONAMIDE HYDROCHLORIDE
mf: $C_{17}H_{27}ClN_2O•ClH$ mw: 347.37

SYN: N-(2-(4-CHLOROPHENYL)-1,1-DIMETHYLETHYL)-2-(DIETHYLAMINO)-PROPANAMIDEHYDROCHLORIDE

TOXICITY DATA with REFERENCE
orl-mus LD50:690 mg/kg APSXAS 15,87,78
ipr-mus LD50:220 mg/kg APSXAS 15,87,78
ivn-mus LD50:35 mg/kg APSXAS 15,87,78

SAFETY PROFILE: Poison by intravenous and intraperitoneal routes. Moderately toxic by ingestion. When heated to decomposition it emits toxic fumes of Cl^- and NO_x.

CGI500 CAS:671-04-5 ***HR: 3***
2-CHLORO-4,5-DIMETHYLPHENYL METHYLCARBAMATE
mf: $C_{10}H_{12}ClNO_2$ mw: 213.68

SYNS: BANOL ◇ BANOL TUCO SOK ◇ CARBANOLATE ◇ 2-CHLORO-4,5-DIMETHYLPHENOL, METHYL CARBAMATE ◇ (2-CHLORO-4,5-DIMETHYL)PHENYL ESTER, CARBAMIC ACID ◇ CHLOROXYLAM ◇ 6-CHLORO-3,4-XYLENYL N-METHYLCARBAMATE ◇ 2-CHLORO-4,5-XYLYL ESTER, CARBAMIC ACID ◇ 6-CHLORO-3,4-XYLYL N-METHYLCARBAMATE ◇ OMS-174 ◇ U 12927 ◇ U-17004 ◇ UPJOHN U-12,927

TOXICITY DATA with REFERENCE
orl-rat LD50:30 mg/kg WRPCA2 9,119,70
ipr-rat LD50:11200 μg/kg BWHOA6 44(1-3),241,71
ivn-rat LD50:3 mg/kg BJIMAG 22,317,65
ims-rat LD50:24 mg/kg BJIMAG 22,317,65
unk-rat LD50:293 mg/kg 30ZDA9 -,190,71
orl-mus LD50:300 mg/kg ARSIM* 20,26,66
orl-pgn LD50:4200 μg/kg TXAPA9 21,315,72
orl-qal LD50:7500 μg/kg ASTTA8 (680),157,79
orl-dck LD50:2400 μg/kg TXAPA9 21,315,72
orl-bwd LD50:1780 μg/kg ASTTA8 (680),157,79

CONSENSUS REPORTS: Chlorophenol compounds are on the Community Right-To-Know List.

SAFETY PROFILE: Poison by ingestion, intraperitoneal, intravenous, intramuscular, and possibly other routes. See also CARBAMATES; CHLOROPHENOLS; and ESTERS. A pesticide. When heated to decomposition it emits very toxic fumes of Cl^- and NO_x.

CGI625 CAS:811-62-1 ***HR: 3***
CHLORODIMETHYLPHOSPHINE
mf: C_2H_6ClP mw: 96.50

SAFETY PROFILE: Ignites spontaneously in air. When heated to decomposition it emits toxic fumes of Cl^- and PO_x. See also PHOSPHINE.

CGI750 CAS:77966-93-9 ***HR: 3***
6'-CHLORO-2-(2,6-DIMETHYLPIPERIDINO)-o-ACETOTOLUIDIDE HYDROCHLORIDE
mf: $C_{16}H_{23}ClN_2O{\bullet}ClH$ mw: 331.32

SYN: V 375

TOXICITY DATA with REFERENCE
eye-rbt 2% MLD ARZNAD 8,407,58
ipr-rat LD50:72 mg/kg ARZNAD 8,407,58
ipr-mus LD50:77 mg/kg ARZNAD 8,407,58
scu-mus LD50:180 mg/kg ARZNAD 8,407,58

SAFETY PROFILE: Poison by intraperitoneal and subcutaneous routes. An eye irritant. When heated to decomposition it emits very toxic fumes of Cl^- and NO_x.

CGJ000 CAS:102504-64-3 ***HR: 3***
6'-CHLORO-3-(2,6-DIMETHYLPIPERIDINO)-o-PROPIONOTOLUIDIDE HYDROCHLORIDE
mf: $C_{16}H_{25}ClN_2O{\bullet}ClH$ mw: 333.34

SYN: C 3140

TOXICITY DATA with REFERENCE
eye-rbt 2% MLD ARZNAD 8,544,58
ipr-rat LD50:33 mg/kg ARZNAD 8,544,58
scu-mus LD50:25 mg/kg ARZNAD 8,544,58

SAFETY PROFILE: Poison by intraperitoneal and subcutaneous routes. An eye irritant. When heated to decomposition it emits very toxic fumes of Cl^- and NO_x.

CGJ250 CAS:24358-29-0 ***HR: 3***
2-CHLORO-5-(3,5-DIMETHYLPIPERIDINO SULPHONYL)BENZOIC ACID
mf: $C_{14}H_{18}ClNO_4S$ mw: 331.84

SYN: TIBRIC ACID

TOXICITY DATA with REFERENCE
dni-mus:oth 500 µmol/L CNREA8 40,36,80
orl-rat TDLo:39 g/kg/71W-C:CAR NATUAS 283,397,80

SAFETY PROFILE: Questionable carcinogen with experimental carcinogenic data. Mutation data reported. When heated to decomposition it emits very toxic fumes of SO_x, NO_x, and Cl^-.

CGK500 CAS:63020-91-7 ***HR: 3***
2'-CHLORO-N,N-DIMETHYL-4-STILBENAMINE
mf: $C_{16}H_{16}ClN$ mw: 257.78

SYNS: 2'-CHLORO-4-DIMETHYLAMINOSTILBENE ◇ 2'-CHLORO-4-STILBENYL-N,N-DIMETHYLAMINE

TOXICITY DATA with REFERENCE
orl-rat TDLo:625 mg/kg/60W-C:ETA ABMGAJ 9,87,62
ipr-rat TDLo:300 mg/kg/8W-I:CAR BJCAAI 10,123,56

SAFETY PROFILE: Questionable carcinogen with experimental carcinogenic and tumorigenic data. When heated to decomposition it emits very toxic fumes of Cl^- and NO_x.

CGK750 CAS:63040-27-7 ***HR: 3***
3'-CHLORO-N,N-DIMETHYL-4-STILBENAMINE
mf: $C_{16}H_{16}ClN$ mw: 257.78

SYNS: 3'-CHLORO-N,N-DIMETHYLAMINOSTIBEN (GERMAN) ◇ 3'-CHLORO-4-DIMETHYLAMINOSTILBENE ◇ 3'-CHLORO-4-STILBENYL-N,N-DIMETHYLAMINE

TOXICITY DATA with REFERENCE
orl-rat TDLo:440 mg/kg/42W-C:ETA ABMGAJ 9,87,62

SAFETY PROFILE: Questionable carcinogen with experimental tumorigenic data. When heated to decomposition it emits very toxic fumes of Cl^- and NO_x.

CGL000 CAS:7378-50-9 ***HR: 3***
4'-CHLORO-N,N-DIMETHYL-4-STILBENAMINE
mf: $C_{16}H_{16}ClN$ mw: 257.78

SYNS: 4'-CHLORO-N,N-DIMETHYLAMINOSTIBEN (GERMAN) ◇ 4'-CHLORO-4-DIMETHYLAMINOSTILBENE ◇ 4'-CHLORO-4-STILBENYL-N,N-DIMETHYLAMINE

TOXICITY DATA with REFERENCE
orl-rat TDLo:625 mg/kg/59W-C:ETA ABMGAJ 9,87,62

SAFETY PROFILE: Questionable carcinogen with experimental tumorigenic data. When heated to decomposition it emits very toxic fumes of Cl^- and NO_x.

CGL125 CAS:72040-09-6 ***HR: 3***
N-CHLORO-4,5-DIMETHYLTRIAZOLE
mf: $C_4H_6ClN_3$ mw: 131.56

ClNN=NC(CH_3)=CCH$_3$ (ring)

SAFETY PROFILE: The solid triazole and its concentrated solutions are unstable and may decompose violently at room temperature. When heated to decomposition it emits toxic fumes of Cl^- and NO_x.

CGL250 CAS:7287-36-7 ***HR: 2***
4'-CHLORO-2,2-DIMETHYLVALERANILIDE
mf: $C_{13}H_{18}ClNO$ mw: 239.77

SYNS: N-(4-CHLOROPHENYL)-2,2-DIMETHYLPENTANAMIDE ◇ N-(4-CHLOROPHENYL)-2,2-DIMETHYLVALEROAMIDE ◇ N-(4-CHLORPHENYL)-2,2-DIMETHYLPENTAMID (GERMAN) ◇ N-(4-CHLOR-PHENYL)-2,2-DIMETHYL-VALERIANSAEUREAMID (GERMAN) ◇ D-90-A ◇ MONALIDE ◇ POTABLAN ◇ SCHERING-35830 ◇ SN 35830

TOXICITY DATA with REFERENCE
orl-rat LD50:2600 mg/kg 85ARAE 2,220,77
skn-rbt LD50:2600 mg/kg 31ZOAD 1,307,68
unk-mam LD50:4000 mg/kg 30ZDA9 -,126,71

CONSENSUS REPORTS: EPA Genetic Toxicology Program.

SAFETY PROFILE: Moderately toxic by ingestion, skin contact, and possibly other routes. A pesticide. When heated to decomposition it emits very toxic fumes of Cl^- and NO_x.

CGL325 CAS:3531-19-9 ***HR: 3***
2-CHLORO-4,6-DINITROANILINE
mf: $C_6H_4ClN_3O_4$ mw: 217.57

$Cl(NO_2)_2C_6H_2NH_2$

SAFETY PROFILE: An explosive sensitive to heat or an initiating charge. Solution in nitrosylsulfuric acid explodes between 50-160°C depending on the concentration. When heated to decomposition it emits toxic fumes of Cl^- and NO_x. See also EXPLOSIVES and ANILINE DYES.

CGL500 CAS:5388-62-5 ***HR: 3***
4-CHLORO-2,6-DINITROANILINE
mf: $C_6H_4ClN_3O_4$ mw: 217.57

$Cl(O_2N)_2C_6H_2NH_2$

SAFETY PROFILE: Solution in nitrosylsulfuric acid may explode when heated. When heated to decomposition it emits toxic fumes of Cl^- and NO_x. See also 2-CHLORO-4,6-DINITROANILINE; ANILINE; and ANILINE DYES.

CGL750 CAS:25567-67-3 ***HR: 3***
CHLORODINITROBENZENE
DOT: UN 1577
mf: $C_6H_3ClN_2O_4$ mw: 202.56

SYNS: CHLORODINITROBENZENE (DOT) ◇ CHLORODINITRO BENZENE (mixed isomers) ◇ DINITROCHLOROBENZENE ◇ DINITROCHLOROBENZENE (DOT)

TOXICITY DATA with REFERENCE
dnd-mus-ipr 60 mg/kg BSIBAC 56,1680,80
dnd-ham:lvr 5 μmol/L MUREAV 131,215,84
dns-ham:lvr 5 μmol/L MUREAV 131,215,84

DOT Classification: Poison B; Label: Poison.

SAFETY PROFILE: A poison. Mutation data reported. Potentially explosive. When heated to decomposition it emits very toxic fumes of Cl^- and NO_x. See also other chloro-dinitrobenzenes.

CGM000 CAS:97-00-7 ***HR: 3***
1-CHLORO-2,4-DINITROBENZENE
mf: $C_6H_3ClN_2O_4$ mw: 202.56

PROP: Yellow rhombic crystals, insol in water. mp(α): 53.4°, mp(β): 43°, mp(Γ): 27°, bp: 315°, lel: 2.0%, uel: 22%, flash p: 382°F (CC), d(α): 1.687 @ 22°, d(β):1.680 @ 20°/4°, vap d: 6.98.

SYNS: 1-CHLOOR-2,4-DINITROBENZEEN (DUTCH) ◇ 1-CHLOR-2,4-DINITROBENZENE ◇ 4-CHLORO-1,3-DINITROBENZENE ◇ 6-CHLORO-1,3-DINITROBENZENE ◇ 1-CHLORO-2,4-DINITROBENZOL (GERMAN) ◇ 1-CLORO-2,4-DINITROBENZENE (ITALIAN) ◇ 2,4-DINITROCHLOROBENZENE ◇ 1,3-DINITRO-4-CHLOROBENZENE ◇ 2,4-DINITRO-1-CHLOROBENZENE ◇ DINITROCHLOROBENZOL ◇ DINITROCHLOROBENZOL (DOT) ◇ DNCB

TOXICITY DATA with REFERENCE
skn-hmn 30 μg CODEDG 2,247,76
skn-rbt 100 μg/24H open AIHAAP 23,95,62
mmo-sat 3 μg/plate ARDEAC 121,348,85
mma-sat 50 μg/plate ADREDL 266,315,79
dnd-mus-ipr 30 mg/kg MUREAV 116,239,83
otr-ham:kdy 10 mg/L ARTODN 45,307,80
orl-rat LD50:1070 mg/kg AIHAAP 23,95,62
ipr-rat LD50:280 mg/kg AGGHAR 17,217,59
skn-rbt LD50:130 mg/kg AIHAAP 23,95,62

CONSENSUS REPORTS: Reported in EPA TSCA Inventory.

DOT Classification: Poison B; Label: Poison.

SAFETY PROFILE: Poison by by skin contact and intraperitoneal routes. Moderately toxic by by ingestion. A human skin irritant. Acts as a primary irritant as well as a sensitizer of skin. An allergen. Mutation data reported. Combustible when exposed to heat or flame. A moderate explosion hazard when exposed to flame, sparks, heated to 150°, or when shocked in a sealed container. Explosive reaction with ammonia at 170°C/40 bar. To fight fire use CO_2, dry chemical. Reacts violently with hydrazine sulfate or hydrazine hydrate. See also NITRO COMPOUNDS of AROMATIC HYDROCARBONS.

CGM199 ***HR: 3***
4-CHLORO-2,5-DINITROBENZENE DIAZONIUM-6-OXIDE
mf: $C_6HClN_4O_5$ mw: 244.55

SAFETY PROFILE: A very shock-sensitive explosive solid. When heated to decomposition it emits toxic fumes of Cl^- and NO_x. See also NITRO COMPOUNDS of AROMATIC HYDROCARBONS.

CGM375 CAS:63886-82-8 ***HR: 3***
p-CHLORO-2,4-DIOXA-5-ETHYL-p-THIONO-3-PHOSPHABICYCLO(4.4.0)DECANE

TOXICITY DATA with REFERENCE
skn-rbt 10 mg/24H open MLD AIHAAP 23,95,62
orl-rat LD50:140 mg/kg AIHAAP 23,95,62
skn-rbt LD50:400 mg/kg AIHAAP 23,95,62

SAFETY PROFILE: Poison by ingestion and skin contact. A skin irritant. When heated to decomposition it emits toxic fumes of Cl^-, PO_x, and SO_x.

CGM400 CAS:2921-31-5 ***HR: 3***
p-CHLORO-2,4-DIOXA-5-METHYL-p-THIONO-3-PHOSPHABICYCLO(4.4.0)DECANE
mf: $C_8H_{14}ClO_2PS$ mw: 240.70

SYNS: ENT 23,970 ◇ UC 8305 ◇ UNION CARBIDE UC-8305

TOXICITY DATA with REFERENCE
skn-rbt 10 mg/24H open MLD AIHAAP 23,95,62
orl-rat LD50:120 mg/kg ARSIM* 20,25,66
skn-rbt LD50:360 mg/kg AIHAAP 23,95,62
orl-ckn LD50:26 mg/kg TXAPA9 11,49,67

SAFETY PROFILE: Poison by ingestion and skin contact. A skin irritant. When heated to decomposition it emits toxic fumes of Cl^-, SO_x, and PO_x.

CGM450 CAS:3367-31-5 ***HR: 2***
CHLORO((3-(2,4-DIOXO-5-IMIDAZOLIDINYL)-2-METHOXY)PROPYL) MERCURY
mf: $C_7H_{11}ClHgN_2O_3$ mw: 407.24

TOXICITY DATA with REFERENCE
orl-mus LD50:3320 mg/kg JMPCAS 5,168,62

ACGIH TLV: TWA 0.1 mg(Hg)/m^3 (skin)
NIOSH REL: (Mercury, Inorganic): 8H TWA 0.05 mg(Hg)/m^3

SAFETY PROFILE: Moderately toxic by ingestion. When heated to decomposition it emits toxic fumes of NO_x, Hg, and Cl^-.

CGM500 CAS:3861-99-2 ***HR: 2***
6-CHLORO-1,3-DIOXO-5-ISOINDOLINESULFONAMIDE
mf: $C_8H_5ClN_2O_4S$ mw: 260.66

SYNS: 4-CHLORO-5-SULPHAMOYLPHTHALIMIDE ◇ 1H-ISOINDOLE-5-SULFONAMIDE, 6-CHLORO-2,3-DIHYDRO-1,3-DIOXO- ◇ 5-ISOINDOLINESULFONAMIDE, 6-CHLORO-1,3-DIOXO-

TOXICITY DATA with REFERENCE
skn-rbt 500 mg MLD FCTOD7 20,573,82
eye-rbt 100 mg SEV FCTOD7 20,573,82
eye-rbt 100 mg/4S RNS SEV FCTOD7 20,573,82

SAFETY PROFILE: A severe eye and mild skin irritant. When heated to decomposition it emits toxic fumes of NO_x, SO_x, and Cl^-.

CGM750 CAS:2051-60-7 ***HR: 2***
2-CHLORODIPHENYL
mf: $C_{12}H_9Cl$ mw: 188.66

SYNS: 2-CHLORO-1,1'-BIPHENYL ◇ o-CHLORODIPHENYL

TOXICITY DATA with REFERENCE
orl-mam LDLo:2500 mg/kg JIDHAN 13,87,31

SAFETY PROFILE: Moderately toxic by ingestion. When heated to decomposition it emits toxic fumes of Cl^-.

CGN000 CAS:712-48-1 ***HR: 3***
CHLORODIPHENYLARSINE
DOT: UN 1699
mf: $C_{12}H_{10}AsCl$ mw: 264.59

PROP: Colorless crystals when pure, technical product is dark brown liquid. Bp: 333° (decomp), fp: 44°, d: 1.363 @ 40° (solid): 1.358 @ 45° (liquid), vap press: 0.00049 mm @ 20°, vap d: 9.15.

SYNS: BLUE CROSS ◇ CLARK I ◇ DA ◇ DIPHENYLARSINOUS CHLORIDE ◇ DIPHENYLCHLOORARSINE (DUTCH) ◇ DIPHENYLCHLOROARSINE (DOT) ◇ SNEEZING GAS

TOXICITY DATA with REFERENCE
ihl-hmn LCLo:55 ppm/30M NTIS** PB214-270
ihl-dog LCLo:31 ppm/50M ZGEMAZ 13,523,21
ihl-cat LCLo:6 ppm/12M ZGEMAZ 13,523,21
scu-cat LDLo:500 μg/kg ZGEMAZ 13,523,21

CONSENSUS REPORTS: Arsenic and its compounds are on the Community Right-To-Know List.

OSHA PEL: TWA 0.5 mg(As)/m^3
DOT Classification: Poison B; Label: Poison.

SAFETY PROFILE: A human poison by inhalation. Poison experimentally by inhalation and skin contact. A powerfully irritating military poison. Exposure yields cold-like symptoms, plus headache, vomiting and nausea. A nonpersistent gas. Decontamination is by use of chlorine or caustic soda in confined spaces. When heated to decomposition it emits toxic fumes of As and Cl^-. See also ARSENIC COMPOUNDS.

CGN250 CAS:77966-58-6 ***HR: 3***
6'-CHLORO-2-(DIPROPYLAMINO)-o-ACETOTOLUIDIDE HYDROCHLORIDE
mf: $C_{15}H_{23}ClN_2O{\bullet}ClH$ mw: 319.31

SYN: C 3071

TOXICITY DATA with REFERENCE
eye-rbt 2% SEV ARZNAD 8,407,58
ipr-rat LD50:260 mg/kg ARZNAD 8,407,58
ipr-mus LD50:245 mg/kg ARZNAD 8,407,58
scu-mus LD50:1780 mg/kg ARZNAD 8,407,58

SAFETY PROFILE: Poison by intraperitoneal route. Moderately toxic by subcutaneous route. A severe eye irritant. When heated to decomposition it emits very toxic fumes of Cl^- and NO_x.

CGN325 CAS:22086-53-9 ***HR: 3***
CHLORODIPROPYLBORANE
mf: $C_6H_{14}BCl$ mw: 132.44

SAFETY PROFILE: Ignites spontaneously in air. When heated to decomposition it emits toxic fumes of Cl^-. See also BORANES and CHLORIDES.

CGO125 CAS:1622-32-8 ***HR: 3***
2-CHLOROETHANE SULFOCHLORIDE
mf: $C_2H_4Cl_2O_2S$ mw: 163.02

SYNS: β-CHLOROETHANESULFONYL CHLORIDE ◇ 2-CHLOROETHANESULFONYL CHLORIDE ◇ 2-CHLOROETHYLSULFONYL CHLORIDE

TOXICITY DATA with REFERENCE
orl-rat LD50:240 mg/kg GTPZAB 15(6),59,71
ihl-rat LC50:420 mg/m³/4H GTPZAB 15(6),59,71
ihl-mus LC50:250 mg/m³/4H GTPZAB 15(6),59,71

CONSENSUS REPORTS: EPA Extremely Hazardous Substances List.

SAFETY PROFILE: Poison by inhalation and ingestion. When heated to decomposition it emits toxic fumes of Cl^- and SO_x.

CGO500 CAS:115-96-8 ***HR: 3***
2-CHLOROETHANOL PHOSPHATE
mf: $C_6H_{12}Cl_3O_4P$ mw: 285.50

PROP: Flash p: 421°F (COC), boiling range: 210-220° @ 20 mm d: 1.425 @ 20°/20°, autoign temp: 1115°F, vap press: 0.5 mm @ 145°.

SYNS: CELLUFLEX ◇ FYROL CEF ◇ NCI-C60128 ◇ NIAX FLAME RETARDANT 3 CF ◇ TRICHLORETHYL PHOSPHATE ◇ TRI-β-CHLOROETHYL PHOSPHATE ◇ TRI(2-CHLOROETHYL)PHOSPHATE ◇ TRIS(2-CHLOROETHYL)ESTER PHOSPHORIC ACID ◇ TRIS(β-CHLOROETHYL) PHOSPHATE ◇ TRIS(2-CHLOROETHYL) PHOSPHATE

TOXICITY DATA with REFERENCE
skn-rbt 10 mg/24H open MLD AMIHBC 4,119,51
eye-rbt 500 mg open AMIHBC 4,119,51
mmo-sat 333 μg/plate ENMUDM 5(Suppl 1),3,83
mma-sat 1 μmol/plate MUREAV 66,373,79
mnt-ham-ipr 125 mg/kg EJCODS 18,1337,82
otr-ham:emb 400 mg/L EJCODS 18,1337,82
sce-ham:lng 700 mg/L EJCODS 18,1337,82
orl-rat TDLo:1800 mg/kg (7-15D preg):REP ESKHA5 101,55,83
orl-rat LD50:1230 mg/kg BECTA6 17,720,77
ipr-mus LDLo:250 mg/kg CBCCT* 7,396,55

CONSENSUS REPORTS: Reported in EPA TSCA Inventory. EPA Genetic Toxicology Program.

SAFETY PROFILE: Poison by intraperitoneal route. Moderately toxic by ingestion. Experimental reproductive effects. A skin and eye irritant. Combustible when exposed to heat or flame. When heated to decomposition it emits very toxic fumes of PO_x and Cl^-. See also PHOSPHATES, CHLORIDES, and ESTERS.

CGP125 CAS:689-98-5 ***HR: 3***
2-CHLOROETHYLAMINE
mf: C_2H_6ClN mw: 79.53

SAFETY PROFILE: Unstable and may polymerize explosively. When heated to decomposition it emits toxic fumes of Cl^- and NO_x. See also AMINES and CHLORIDES.

CGP250 CAS:870-24-6 ***HR: 3***
2-CHLOROETHYLAMINE HYDROCHLORIDE
mf: $C_2H_7Cl_2N$ mw: 115.99

SYN: 2-CHLOROETHYLAMMONIUM CHLORIDE

SAFETY PROFILE: Explosive reaction with concentrated alkali above 50°C. When heated to decomposition it emits toxic fumes of Cl^-, NH_3, and NO_x.

CGP375 CAS:110335-28-9 ***HR: 3***
3'-CHLORO-2-ETHYLAMINO-o-ACETOTOLUIDIDE HYDROCHLORIDE
mf: $C_{11}H_{15}ClN_2O{\cdot}ClH$ mw: 263.19

TOXICITY DATA with REFERENCE
eye-rbt 100 mg MOD JAPMA8 49,80,60
ipr-mus LD50:125 mg/kg JAPMA8 49,80,60
scu-mus LD50:300 mg/kg JAPMA8 49,80,60

SAFETY PROFILE: Poison by subcutaneous and intraperitoneal routes. An eye irritant. When heated to decomposition it emits toxic fumes of NO_x and Cl^-.

CGP380 CAS:77966-59-7 ***HR: 3***
6'-CHLORO-2-(ETHYLAMINO)-o-ACETOTOLUIDIDE HYDROCHLORIDE
mf: $C_{11}H_{15}ClN_2O{\cdot}ClH$ mw: 263.19

SYNS: C 3063 ◇ 2'-CHLORO-2-(ETHYLAMINO)-6'-METHYLACETANILIDE, HYDROCHLORIDE

TOXICITY DATA with REFERENCE
ipr-rat LD50:235 mg/kg ARZNAD 8,407,58
ipr-mus LD50:270 mg/kg ARZNAD 8,407,58
scu-mus LD50:655 mg/kg ARZNAD 8,407,58

SAFETY PROFILE: Poison by intraperitoneal route. Moderately toxic by subcutaneous route. When heated to decomposition it emits toxic fumes of NO_x and Cl^-.

CGP500 CAS:38915-49-0 ***HR: 3***
7-((2-((2-CHLOROETHYL)AMINO)ETHYL) AMINO)BENZ(c)ACRIDINE DIHYDROCHLORIDE HYDRATE
mf: $C_{21}H_{20}ClN_3 \cdot 2ClH \cdot H_2O$ mw: 440.83

TOXICITY DATA with REFERENCE
mmo-sat 5 μg/plate JMCMAR 15,739,72
ipr-mus LD20:170 mg/kg JMCMAR 15,739,72

SAFETY PROFILE: Poison by intraperitoneal route. Mutation data reported. When heated to decomposition it emits very toxic fumes of Cl^- and NO_x.

CGQ250 CAS:40713-31-3 ***HR: 3***
2-CHLOROETHYLAMINOETHYL DEHYDROABIETATE HYDROCHLORIDE
mf: $C_{24}H_{36}ClNO_2 \cdot ClH$ mw: 442.52

SYN: DEHYDRO-ABIETIC ACID-2-(2-(CHLOROETHYL)AMINO)ETHYL ESTER

TOXICITY DATA with REFERENCE
ipr-rat LD50:400 mg/kg PCJOAU 6,647,72
ipr-mus LD50:400 mg/kg PCJOAU 6,647,72

SAFETY PROFILE: Poison by intraperitoneal route. See also ESTERS. When heated to decomposition it emits very toxic fumes of Cl^- and NO_x.

CGQ280 CAS:56538-00-2 ***HR: 3***
N-(2-CHLOROETHYL)AMINOMETHYL-4-HYDROXYNITROBENZENE
mf: $C_9H_{11}ClN_2O_3$ mw: 230.67

SYNS: 2-(((2-CHLOROETHYL)AMINO)METHYL)-4-NITROPHENOL ◇ PHENOL, 2-(((2-CHLOROETHYL)AMINO)METHYL)-4-NITRO-

TOXICITY DATA with REFERENCE
orl-rat TDLo:35 mg/kg (female 1-7D post):REP IJMRAQ 66,987,77
orl-rat LD50:25 mg/kg IJMRAQ 66,987,77

SAFETY PROFILE: Poison by ingestion. Experimental reproductive effects. When heated to decomposition it emits toxic fumes of NO_x and Cl^-.

CGQ400 CAS:56538-01-3 ***HR: 3***
N-(2-CHLOROETHYL)AMINOMETHYL-4-METHOXYNITROBENZENE
mf: $C_{10}H_{13}ClN_2O_3$ mw: 244.70

SYNS: BENZYLAMINE,N-(2-CHLOROETHYL)-2-METHOXY-5-NITRO- ◇ N-(2-CHLOROETHYL)-2-METHOXY-5-NITROBENZYLAMINE

TOXICITY DATA with REFERENCE
orl-rat TDLo:35 mg/kg (female 1-7D post):REP IJMRAQ 66,987,77
orl-rat LD50:28 mg/kg IJMRAQ 66,987,77

SAFETY PROFILE: Poison by ingestion. Experimental reproductive effects. When heated to decomposition it emits toxic fumes of NO_x and Cl^-.

CGQ500 CAS:21715-46-8 ***HR: 3***
6-CHLORO-2-ETHYLAMINO-4-METHYL-4-PHENYL-4H-3,1-BENZOXAZINE
mf: $C_{17}H_{17}ClN_2O$ mw: 300.81

SYNS: 2-AETHYLAMINO-6-CHLOR-4-METHYL-4-PHENYL-4H-3,1-BENZOXAZIN (GERMAN) ◇ 6-CHLORO-N-ETHYL-4-METHYL-4-PHENYL-4H-3,1-BENZOXAZIN-2-AMINE ◇ ETIFOXIN ◇ ETIFOXINE ◇ HOE 36801

TOXICITY DATA with REFERENCE
orl-rat LD50:1500 mg/kg THERAP 27,325,72
ipr-rat LD50:292 mg/kg THERAP 27,325,72
ivn-rat LD50:55 mg/kg THERAP 27,325,72
orl-mus LD50:1388 mg/kg THERAP 27,325,72
ipr-mus LD50:450 mg/kg THERAP 27,325,72
ivn-mus LD50:120 mg/kg THERAP 27,325,72
ivn-gpg LDLo:133 mg/kg THERAP 27,325,72

SAFETY PROFILE: Poison by intravenous and intraperitoneal routes. Moderately toxic by ingestion. A tranquilizer. When heated to decomposition it emits very toxic fumes of Cl^- and NO_x.

CGR000 CAS:38914-96-4 ***HR: 2***
9-((3-((2-CHLOROETHYL)AMINO)PROPYL) AMINO)ACRIDINE DIHYDROCHLORIDE HYDRATE
mf: $C_{18}H_{20}ClN_3 \cdot 2ClH \cdot H_2O$ mw: 404.80

TOXICITY DATA with REFERENCE
mmo-sat 5 μg/plate JMCMAR 15,739,72
ipr-mus LD20:243 mg/kg JMCMAR 15,739,72

SAFETY PROFILE: Moderately toxic by intraperitoneal route. Mutation data reported. When heated to decomposition it emits very toxic fumes of Cl^- and NO_x.

CGR250 CAS:38915-50-3 ***HR: 3***
7-((3-((2-CHLOROETHYL)AMINO)PROPYL) AMINO)BENZ(c)ACRIDINE DIHYDROCHLORIDE SESQUIHYDRATE
mf: $C_{22}H_{22}ClN_3 \cdot 2ClH \cdot 3/2H_2O$ mw: 463.87

TOXICITY DATA with REFERENCE
mmo-sat 5 μg/plate JMCMAR 15,739,72
ipr-mus LD20:70 mg/kg JMCMAR 15,739,72

SAFETY PROFILE: Poison by intraperitoneal route. Mutation data reported. When heated to decomposition it emits very toxic fumes of NO_x and Cl^-.

CGR500 CAS:38915-61-6 ***HR: 3***
7-((3-((2-CHLOROETHYL)AMINO)PROPYL) AMINO)BENZO(b)(1,10)PHENANTHROLINE DIHYDROCHLORIDE
mf: $C_{21}H_{21}ClN_4 \cdot 2ClH$ mw: 437.83

SYN: ICR 395

TOXICITY DATA with REFERENCE
mmo-sat 500 ng/plate MUREAV 136,185,84
ipr-mus LD20:18 mg/kg JMCMAR 15,739,72

SAFETY PROFILE: Poison by intraperitoneal route. Mutation data reported. When heated to decomposition it emits very toxic fumes of Cl^- and NO_x.

CGR750 CAS:38915-61-6 ***HR: 3***
7-((3-((2-CHLOROETHYL)AMINO)PROPYL) AMINO)BENZO(b)(1,8)PHENANTHROLINE DIHYDROCHLORIDE HYDRATE
mf: $C_{21}H_{21}ClN_4 \cdot 2ClH \cdot H_2O$ mw: 455.85

TOXICITY DATA with REFERENCE
mmo-sat 5 μg/plate JMCMAR 15,739,72
ipr-mus LD20:68 mg/kg JMCMAR 15,739,72

SAFETY PROFILE: Poison by intraperitoneal route. Mutation data reported. When heated to decomposition it emits very toxic fumes of Cl^- and NO_x.

CGS500 CAS:36167-69-8 ***HR: D***
10-((2-CHLOROETHYLAMINO)PROPYLAMINO)-2-METHOXY-7-CHLOROBENZO(b)-(1,5)-NAPHTHYRIDINE
mf: $C_{18}H_{20}Cl_2N_4O$ mw: 379.32

SYNS: N-(2-CHLOROETHYL)-N'-(7-CHLORO-2-METHOXYBENZO(b)-1,5-NAPHTHYRIDIN-10-YL)-1,3-PROPANEDIAMINE ◇ ICR 372

TOXICITY DATA with REFERENCE
mmo-sat 500 ng/plate MUREAV 136,185,84
mmo-esc 5 μg/plate GENTAE 78,823,74
msc-ham:ovr 1 μmol/L CNREA8 38,506,78

CONSENSUS REPORTS: EPA Genetic Toxicology Program.

SAFETY PROFILE: Mutation data reported. When heated to decomposition it emits very toxic fumes of NO_x and Cl^-.

CGS750 CAS:38925-90-5 ***HR: 3***
4-((3-((2-CHLOROETHYL)AMINO)PROPYL) AMINO)-6-METHOXYQUINOLINE HYDROCHLORIDE
mf: $C_{15}H_{20}ClN_3O \cdot 2ClH$ mw: 366.75

SYN: ICR 180

TOXICITY DATA with REFERENCE
mmo-sat 5 μg/plate JMCMAR 15,739,72
ipr-mus LD20:91 mg/kg JMCMAR 15,739,72

SAFETY PROFILE: Poison by intraperitoneal route. Mutation data reported. When heated to decomposition it emits very toxic fumes of Cl^- and NO_x.

CGT000 CAS:102584-97-4 ***HR: 3***
6'-CHLORO-2-(ETHYLAMINO)-o-VALEROTOLUIDIDE HYDROCHLORIDE
mf: $C_{14}H_{21}ClN_2O \cdot ClH$ mw: 305.28

SYN: C 3205

TOXICITY DATA with REFERENCE
eye-rbt 2% MLD ARZNAD 8,544,58
ipr-rat LD50:35 mg/kg ARZNAD 8,544,58
scu-mus LD50:65 mg/kg ARZNAD 8,544,58

SAFETY PROFILE: Poison by intraperitoneal and subcutaneous routes. An eye irritant. When heated to decomposition it emits very toxic fumes of Cl^- and NO_x.

CGU000 CAS:2114-18-3 ***HR: 2***
β-CHLOROETHYL CARBAMATE
mf: $C_3H_6ClNO_2$ mw: 123.55

TOXICITY DATA with REFERENCE
ipr-mus LD50:1000 mg/kg JNCIAM 8,99,47

CONSENSUS REPORTS: Reported in EPA TSCA Inventory.

SAFETY PROFILE: Moderately toxic by intraperitoneal route. See also CARBAMATES. When heated to decomposition it emits very toxic fumes of Cl^- and NO_x.

CGU199 CAS:627-11-2 ***HR: 3***
CHLOROETHYL CHLOROFORMATE
mf: $C_3H_4Cl_2O_2$ mw: 142.97

PROP: Bp: 155.7°, d: 1.3847, insoluble in water.

SYNS: CARBONOCHLORIDIC ACID-2-CHLOROETHYL ESTER ◇ (2-CHLOROETHOXY)CARBONYL CHLORIDE ◇ 2-CHLOROETHYL CHLOROCARBONATE ◇ 2-CHLOROETHYL CHLOROFORMATE ◇ β-CHLOROETHYL CHLOROFORMATE ◇ CHLOROFORMIC ACID-2-CHLOROETHYL ESTER ◇ TL 207

TOXICITY DATA with REFERENCE
ihl-mus LCLo:200 mg/m^3/10M NDRC** NCDrc-132,June,42

CONSENSUS REPORTS: EPA Extremely Hazardous Substances List. EPA Genetic Toxicology Program. Reported in EPA TSCA Inventory.

SAFETY PROFILE: Poison by inhalation. May also cause fatalities by ingestion or skin contact. May cause burns to the eyes and skin. When heated to decomposition it emits toxic fumes of Cl^-.

CGV000 CAS:13909-14-3 ***HR: 3***
1-(2-CHLOROETHYL)-3-CYCLODODECYL-1-NITROSOUREA
mf: $C_{15}H_{28}ClN_3O_2$ mw: 317.91

SYNS: N-(2-CHLOROETHYL)-N'-CYCLODODECYL-N-NITROSOUREA ◇ NSC 91726

TOXICITY DATA with REFERENCE
ipr-mus LD50:1297 mg/kg NCISP* JAN86

SAFETY PROFILE: Poison by intraperitoneal route. Many N-nitroso compounds are carcinogens. When heated to decomposition it emits very toxic fumes of Cl^- and NO_x. See also N-NITROSO COMPOUNDS.

CGV250 CAS:13010-47-4 ***HR: 3***
1-(2-CHLOROETHYL)-3-CYCLOHEXYL-1-NITROSOUREA
mf: $C_9H_{16}ClN_3O_2$ mw: 233.73

SYNS: BELUSTINE ◇ CCNU ◇ CECENU ◇ CEENU ◇ CHLOROETHYLCYCLOHEXYLNITROSOUREA ◇ N-(2-CHLOROETHYL)-N'-CYCLOHEXYL-N-NITROSOUREA ◇ ((CHLORO-2-ETHYL)-1-CYCLOHEXYL-3-NITROSOUREA ◇ CINU ◇ (CLORO-2-ETIL)-1-CICLOESIL-3-NITROSOUREA (ITALIAN) ◇ ICIG 1109 ◇ LOMUSTINE ◇ NCI-C04740 ◇ NSC-79037 ◇ RB 1509 ◇ SRI 2200

TOXICITY DATA with REFERENCE
mma-sat 100 nmol/plate JJIND8 65,149,80
mmo-esc 500 μmol/L GANNA2 71,674,80
dnd-hmn:emb 100 μmol/L CNREA8 44,1352,84
dns-hmn:lym 10 mg/L FRPSAX 36,947,81
dni-hmn:leu 1 μmol/L BBACAQ 425,463,76
sce-hmn:lym 50 μmol/L CGCYDF 9,261,83
msc-ham:lng 10 μmol/L CNREA8 40,2719,80
ivn-rbt TDLo:39 mg/kg (female 6-18D post):REP TXAPA9 34,456,75
ipr-rat TDLo:8 mg/kg (9-12D preg):TER TXAPA9 34,456,75
orl-rat TDLo:50 mg/kg:ETA PTBIAN 35,139,87
orl-hmn TDLo:30 mg/kg:CNS,GIT CCYPBY 3,33,72
orl-hmn TDLo:3 mg/kg:GIT,BLD CTRRDO 60,709,76
orl-rat LD50:70 mg/kg TXAPA9 21,405,72
ipr-rat LD50:50350 μg/kg NCISP* JAN86
orl-mus LD50:38 mg/kg TXAPA9 21,405,72
ipr-mus LD50:53 mg/kg TXAPA9 21,405,72
scu-mus LD50:54 mg/kg TXAPA9 21,405,72
ivn-mus LD10:40 mg/kg ANBCB3 23,64,78
orl-dog LDLo:10 mg/kg ACRSAJ 16,273,72
ivn-dog LDLo:5 mg/kg CCYPBY 4(3),13,73

CONSENSUS REPORTS: NTP Fifth Annual Report on Carcinogens. IARC Cancer Review: Group 2A IMEMDT 7,150,87; Human Limited Evidence IMEMDT 26,137,81; Animal Sufficient Evidence IMEMDT 26,137,81. NCI Carcinogenesis Studies (ipr); Clear Evidence: mouse CANCAR 40,1935,77; No Evidence: rat CANCAR 40,1935,77. EPA Genetic Toxicology Program.

SAFETY PROFILE: Confirmed carcinogen with experimental carcinogenic and tumorigenic data. Poison by ingestion, intraperitoneal, subcutaneous, intravenous, and possibly other routes. Human systemic effects by ingestion: anorexia, nausea or vomiting, leukopenia (decrease in the white blood cell count), and thrombocytopenia (decrease in the number of blood platelets). Experimental teratogenic and reproductive effects. Human mutation data reported. When heated to decomposition it emits very toxic fumes of Cl^- and NO_x. See also N-NITROSO COMPOUNDS.

CGV275 CAS:30077-45-3 ***HR: 1***
β-CHLOROETHYLDICHLOROARSINE
mf: $C_2H_4AsCl_3$ mw: 209.33

SYNS: ARSINE, (2-CHLOROETHYL)DICHLORO- ◇ β-CHLOROETHYLDICHLORARSINE

TOXICITY DATA with REFERENCE
ihl-mus LC50:13 g/m^3/10M NTIS** PB158-508

OSHA PEL: TWA 0.5 mg(As)/m^3

SAFETY PROFILE: Toxic by inhalation. When heated to decomposition it emits toxic fumes of As and Cl^-.

CGV500 CAS:100-35-6 ***HR: 3***
N-(2-CHLORO ETHYL)DIETHYLAMINE
mf: $C_6H_{14}ClN$ mw: 135.66

SYNS: (2-CHLOROETHYL)DIETHYLAMINE ◇ β-CHLOROTRIETHYLAMINE ◇ 2-CHLOROTRIETHYLAMINE ◇ 2-(DIETHYLAMINO) CHLOROETHANE ◇ DIETHYLAMINOETHYL CHLORIDE ◇ β-(DIETHYLAMINO)ETHYL CHLORIDE ◇ N-DIETHYLAMINOETHYL CHLORIDE ◇ 2-(DIETHYLAMINO)ETHYL CHLORIDE ◇ DIETHYL(2-CHLOROETHYL)AMINE

TOXICITY DATA with REFERENCE
skn-rbt 10 mg/24H MOD AMIHBC 4,119,51
eye-rbt 2 mg SEV AMIHBC 4,119,51
dni-mus-ivg 5000 ppm JIDEAE 62,378,74
orl-rat LD50:17 mg/kg AMIHBC 4,119,51
unr-rat LD50:30 mg/kg PHBUA9 1,297,53
skn-rbt LD50:300 mg/kg AMIHBC 4,119,51

CONSENSUS REPORTS: EPA Genetic Toxicology Program.

SAFETY PROFILE: Poison by ingestion, skin contact and possibly other routes. A severe eye and moderate skin irritant. When heated to decomposition it emits very toxic fumes of Cl^- and NO_x. See also AMINES and CHLORIDES.

CGW000 CAS:107-99-3 ***HR: 3***
N-(2-CHLOROETHYL)DIMETHYLAMINE
mf: $C_4H_{10}ClN$ mw: 107.60

PROP: Liquid. Vap d: 3.72.

SYNS: CHLORO(DIMETHYLAMINO)ETHANE ◇ β-CHLOROETHYLDIMETHYLAMINE ◇ (2-CHLOROETHYL) DIMETHYLAMINE ◇ DIMETHYLAMINOETHYL CHLORIDE ◇ β-(DIMETHYLAMINO) ETHYL CHLORIDE ◇ 2-DIMETHYLAMINOETHYLCHLORIDE ◇ DIMETHYL(2-CHLOROETHYL)AMINE ◇ HN 1 ◇ NITROGEN HALF MUSTARD

TOXICITY DATA with REFERENCE
mmo-sat 100 nmol/plate ARTODN 56,267,85
mma-sat 100 nmol/plate ARTODN 56,267,85
mmo-esc 100 nmol/plate ARTODN 56,267,85
mma-esc 100 nmol/plate ARTODN 56,267,85
oms-omi 2 mol EXPEAM 29,1344,73
mmo-smc 20 mmol/L GENTAE 92,83,79
dni-mus-ivg 2 pph JIDEAE 62,378,74
unr-rat LD50:30 mg/kg PHBUA9 1,297,53

CONSENSUS REPORTS: EPA Genetic Toxicology Program. Reported in EPA TSCA Inventory.

SAFETY PROFILE: Poison by an unspecified route. A systemic irritant. Mutation data reported. When heated to decomposition it emits highly toxic fumes of Cl^- and NO_x.

CGW100 ***HR: 2***
3'-CHLORO-4'-ETHYL-4-DIMETHYLAMINOAZOBENZENE
mf: $C_{16}H_{18}ClN_3$ mw: 287.82

SYNS: BENZENAMINE, N,N-DIMETHYL-3'-CHLORO-4'-ETHYL-4-(PHENYLAZO)- ◇ p-((3-CHLORO-4-ETHYLPHENYL)AZO)-N,N-DIMETHYLANILINE

TOXICITY DATA with REFERENCE
orl-rat TDLo:4515 mg/kg/25W-C:CAR CBINA8 53,107,85

SAFETY PROFILE: Questionable carcinogen with experimental carcinogenic data. When heated to decomposition it emits toxic fumes of Cl^- and NO_x.

CGW105 ***HR: 2***
4'-CHLORO-3'-ETHYL-4-DIMETHYLAMINOAZOBENZENE
mf: $C_{16}H_{18}ClN_3$ mw: 287.82

SYNS: BENZENAMINE, N,N-DIMETHYL-4'-CHLORO-3'-ETHYL-4-(PHENYLAZO)- ◇ p-((4-CHLORO-3-ETHYLPHENYL)AZO)-N,N-DIMETHYLANILINE

TOXICITY DATA with REFERENCE
orl-rat TDLo:15725 mg/kg/52W-C:CAR CBINA8 53,107,85

SAFETY PROFILE: Questionable carcinogen with experimental carcinogenic data. When heated to decomposition it emits toxic fumes of Cl^- and NO_x.

CGW250 CAS:13909-02-9 ***HR: 3***
1-(2-CHLOROETHYL)-3-(2,6-DIOXO-3-PIPERIDYL)-1-NITROSOUREA
mf: $C_8H_{10}ClN_4O_4$ mw: 261.67

SYNS: N-(2-CHLOROETHYL)-N'-(2,6-DIOXO-3-PIPERIDINYL)-N-NITROSOUREA ◇ NSC-95466 ◇ PCNU

TOXICITY DATA with REFERENCE
mmo-sat 200 μg/plate TCMUD8 5,319,85
dnd-mus:leu 200 μmol/L PAACA3 24,249,83
dni-mus/oth 16 mg/kg INSSDM 19,85,81
oms-mus/leu 16 mg/kg INSSDM 19,85,81
orl-mus LD50:35700 μg/kg NTIS** PB282-250
ipr-mus LD50:15210 μg/kg NCISP* JAN86
ivn-mus LD50:22 mg/kg NTIS** PB282-250
ivn-dog LDLo:3 mg/kg NTIS** PB282-250
ivn-mky LDLo:10 mg/kg NTIS** PB282-250

SAFETY PROFILE: Poison by ingestion, intravenous, and intraperitoneal routes. Many N-nitroso compounds are carcinogens. Mutation data reported. When heated to decomposition it emits very toxic fumes of Cl^- and NO_x. See also N-NITROSO COMPOUNDS.

CGW750 CAS:63019-51-2 ***HR: 3***
4-CHLORO-6-ETHYLENEIMINO-2-PHENYLPYRIMIDINE
mf: $C_{12}H_{10}ClN_3$ mw: 231.70

SYN: 6-(1-AZIRIDINYL)-4-CHLORO-2-PHENYLPYRIMIDINE

TOXICITY DATA with REFERENCE
scu-rat TDLo:960 mg/kg/16W-I:NEO BJPCAL 9,306,54

SAFETY PROFILE: Questionable carcinogen with experimental neoplastigenic data. When heated to decomposition it emits very toxic fumes of Cl^- and NO_x.

CGX000 CAS:7763-77-1 ***HR: 3***
CHLOROETHYLENE OXIDE
mf: C_2H_3ClO mw: 78.50

SYNS: CHLOROEPOXYETHANE ◇ CHLOROOXIRANE ◇ MONOCHLOROETHYLENE OXIDE

TOXICITY DATA with REFERENCE
mma-sat 400 μmol/L MUREAV 58,217,78
mmo-esc 500 μmol/L MUREAV 152,147,85
mrc-smc 1 mmol/L TOERD9 3,131,81
dns-rat-ivn 5 g/kg CBINA8 17,239,77
msc-ham:lng 6 μmol/L IJCNAW 16,639,75
scu-mus TDLo:128 mg/kg/42W-I:NEO CNREA8 40,352,80

CONSENSUS REPORTS: EPA Genetic Toxicology Program.

SAFETY PROFILE: Questionable carcinogen with experimental neoplastigenic data. Mutation data reported.

When heated to decomposition it emits very toxic fumes of Cl^-. See also CHLORIDES.

CGX250 CAS:4535-87-9 ***HR: 3***
N-(2-CHLOROETHYL)ETHANAMINE HYDROCHLORIDE
mf: $C_4H_{10}ClN{\cdot}ClH$ mw: 144.06

SYNS: N-(2-CHLOROETHYL)ETHYLAMINEHYDROCHLORIDE ◇ ETHYL-β-CHLOROETHYLAMINE HYDROCHLORIDE

TOXICITY DATA with REFERENCE
ipr-rat LD50:400 mg/kg ARZNAD 11,143,61
ipr-mus LD50:1120 mg/kg CANCAR 2,1055,49
scu-mus LD50:1 g/kg NTIS** PB158-507
ivn-mus LD50:100 mg/kg JPETAB 91,224,47

SAFETY PROFILE: Poison by intravenous and intraperitoneal routes. Moderately toxic by subcutaneous route. When heated to decomposition it emits toxic fumes of Cl^- and NO_x.

CGX325 CAS:56538-02-4 ***HR: 3***
N-(2-CHLOROETHYL)-2-ETHOXY-5-NITROBENZYLAMINE
mf: $C_{11}H_{15}ClN_2O_3$ mw: 258.73

SYNS: BENZENEMETHANAMINE-N-(2-CHLOROETHYL)-2-ETHOXY-5-NITRO ◇ N-(2-CHLOROETHYL)AMINOETHYL-4-ETHOXYNITROBENZENE

TOXICITY DATA with REFERENCE
orl-rat TDLo:20 mg/kg (female 6-9D post):REP IJMRAQ 66,987,77
orl-rat LD50:25 mg/kg IJMRAQ 66,987,77

SAFETY PROFILE: Poison by ingestion. Experimental reproductive effects. When heated to decomposition it emits toxic fumes of Cl^- and NO_x.

CGX500 CAS:38915-22-9 ***HR: 3***
2-((2-CHLOROETHYL)ETHYLAMINO)-N-(3-((6-CHLORO-2-METHOXY-9-ACRIDINYL)AMINO)PROPYL)PROPYL)ACETAMIDE DIHYDROCHLORIDE, HEMIHYDRATE
mf: $C_{23}H_{28}Cl_2N_4O_2{\cdot}2ClH{\cdot}1/2H_2O$ mw: 545.38

SYN: ICR 290

TOXICITY DATA with REFERENCE
mmo-sat 5 μg/plate JMCMAR 15,739,72
ipr-mus LD20:6 mg/kg JMCMAR 15,739,72

SAFETY PROFILE: Poison by intraperitoneal route. Mutation data reported. When heated to decomposition it emits very toxic fumes of Cl^- and NO_x.

CGX625 CAS:51775-17-8 ***HR: 3***
2-(N-(2-CHLOROETHYL)-N-ETHYLAMINOMETHYL)-1,4-BENZODIOXAN HYDROCHLORIDE
mf: $C_{13}H_{18}ClNO_2{\cdot}ClH$ mw: 292.23

SYNS: CHLORHYDRATE de (N-ETHYL,N,β-CHLORETHYL)AMINOMETHYLBENZODIOXANE (FRENCH) ◇ N-(2-CHLOROETHYL)-N-ETHYL-1,4-BENZODIOXAN-2-METHYLAMINEHYDROCHLORIDE ◇ 3718 RP

TOXICITY DATA with REFERENCE
scu-rat LDLo:40 mg/kg AIPTAK 95,285,53
scu-mus LDLo:40 mg/kg AIPTAK 95,285,53
ivn-mus LD50:56 mg/kg CSLNX* NX#07473
ivn-dog LDLo:15 mg/kg AIPTAK 95,285,53
scu-rbt LDLo:30 mg/kg AIPTAK 95,285,53
ivn-rbt LDLo:15 mg/kg AIPTAK 95,285,53
scu-frg LDLo:50 mg/kg AIPTAK 95,285,53

SAFETY PROFILE: Poison by subcutaneous and intravenous routes. When heated to decomposition it emits toxic fumes of NO_x and Cl^-.

CGX750 CAS:38914-97-5 ***HR: 3***
9-((3-((2-CHLOROETHYL)ETHYLAMINO)PROPYL)AMINO)ACRIDINE DIHYDROCHLORIDE
mf: $C_{20}H_{24}ClN_3{\cdot}2ClH$ mw: 414.84

SYNS: N′-9-ACRIDINYL-N-(2-CHLOROETHYL)-N-ETHYL-1,3-PROPANEDIAMINE DIHYDROCHLORIDE ◇ ICR 217

TOXICITY DATA with REFERENCE
mmo-sat 500 ng/plate MUREAV 136,185,84
msc-ham:ovr 1 μmol/L CNREA8 39,4875,79
ipr-mus LD20:2 mg/kg JMCMAR 15,739,72

SAFETY PROFILE: Poison by intraperitoneal route. Mutation data reported. When heated to decomposition it emits very toxic fumes of Cl^- and NO_x.

CGY000 CAS:39013-93-9 ***HR: 3***
7-((3-((2-CHLOROETHYL)ETHYLAMINO)PROPYL)AMINO)BENZO(b) (1,10)PHENANTHROLINE DIHYDROCHLORIDE
mf: $C_{23}H_{25}ClN_4{\cdot}2ClH$ mw: 465.89

SYNS: 7-((3-((2-CHLOROETHYL)ETHYLAMINO)PROPYL)AMINO)BENZO(b)(1,10)PHENANTHROLINE 2HCl ◇ ICR 368

TOXICITY DATA with REFERENCE
mmo-sat 500 ng/plate MUREAV 136,185,84
msc-ham:ovr 250 nmol/L CNREA8 39,4875,79
ipr-mus LD20:1 mg/kg JMCMAR 15,739,72

SAFETY PROFILE: Poison by intraperitoneal route. Mutation data reported. When heated to decomposition it emits very toxic fumes of Cl^- and NO_x.

CGY500 CAS:78218-16-3 ***HR: 3***
9-((3-((2-CHLOROETHYL)ETHYLAMINO)PROPYL)AMINO)-4-METHOXYACRIDINE DIHYDROCHLORIDE HEMIHYDRATE
mf: $C_{21}H_{26}ClN_3O \cdot 2ClH \cdot 1/2H_2O$ mw: 453.84

SYNS: 9-((3-((2-CHLOROETHYL)ETHYLAMINO)PROPYL)AMINO)-4-METHOXYACRIDINE 2HCl HEMIHYDRATE ◇ N-(2-CHLOROETHYL)-N-ETHYL-N'-(4-METHOXY-9-ACRIDINYL)-1,3-PROPANEDIAMINEDIHYDROCHLORIDE HEMIHYDRATE

TOXICITY DATA with REFERENCE
mmo-sat 5 μg/plate JMCMAR 15,739,72
ipr-mus LD20:1 mg/kg JMCMAR 15,739,72

SAFETY PROFILE: Poison by intraperitoneal route. Mutation data reported. When heated to decomposition it emits very toxic fumes of Cl^- and NO_x.

CGY750 CAS:693-07-2 ***HR: 3***
CHLOROETHYL ETHYL SULFIDE
mf: C_4H_9ClS mw: 124.64

SYNS: 2-CHLOROETHYL ETHYL SULFIDE ◇ 2-CHLOROETHYL ETHYL THIOETHER ◇ 1-CHLORO-2-(ETHYLTHIO)ETHANE ◇ ETHYL-β-CHLOROETHYL SULFIDE ◇ ETHYL-2-CHLOROETHYL SULFIDE ◇ β-ETHYLMERKAPTOETHYLCHLORID (CZECH) ◇ 2-(ETHYLTHIO) CHLOROETHANE ◇ 2-ETHYLTHIOETHYL CHLORIDE ◇ HALF-MUSTARD GAS ◇ h-MG

TOXICITY DATA with REFERENCE
skn-rbt 500 mg/24H SEV 28ZPAK -,170,72
eye-rbt 250 μg/24H SEV 28ZPAK -,170,72
mmo-esc 500 μmol/L MUREAV 28,257,75
dnd-esc 500 μmol/L MUREAV 28,257,75
slt-dmg-par 5 mmol/L MUREAV 13,19,71
sln-dmg-par 5 mmol/L CNREA8 32,550,72
orl-rat LD50:252 mg/kg 28ZPAK -,170,72
scu-mus LDLo:25 mg/kg NTIS** PB158-507

CONSENSUS REPORTS: Reported in EPA TSCA Inventory. EPA Genetic Toxicology Program.

SAFETY PROFILE: Poison by ingestion and subcutaneous routes. Mutation data reported. A severe skin and eye irritant. See also ETHERS and SULFIDES. When heated to decomposition it emits very toxic fumes of Cl^- and SO_x.

CGY825 CAS:1537-62-8 ***HR: 3***
2-CHLOROETHYL FLUOROACETATE
mf: $C_4H_5ClFO_2$ mw: 140.55

SYNS: β-CHLOROETHYL FLUOROACETATE ◇ TL 671

TOXICITY DATA with REFERENCE
ihl-rat LC50:200 mg/m³/10M NTIS** PB158-508
ihl-mus LC50:700 mg/m³/10M NTIS** PB158-508
ihl-rbt LC50:100 mg/m³/10M NTIS** PB158-508
ihl-gpg LC50:150 mg/m³/10M NTIS** PB158-508

SAFETY PROFILE: Poison by inhalation. When heated to decomposition it emits toxic fumes of F^- and Cl^-.

CGZ000 CAS:371-28-8 ***HR: 3***
2-CHLOROETHYL-γ-FLUOROBUTYRATE
mf: $C_6H_{10}ClFO_2$ mw: 168.61

SYN: 4-FLUORO-BUTYRIC ACID-2-CHLOROETHYL ESTER

TOXICITY DATA with REFERENCE
ihl-mus LC50:54 mg/m³/10M NTIS** PB158-508
ihl-gpg LC50:100 mg/m³/10M NTIS** PB158-508

SAFETY PROFILE: Poison by inhalation. See also ESTERS. When heated to decomposition it emits very toxic fumes of Cl^- and F^-.

CHA000 CAS:58484-07-4 ***HR: 3***
1-(2-CHLOROETHYL)-3-(β-d-GLUCOPYRANOSYL)-1-NITROSOUREA
mf: $C_9H_{16}ClN_3O_7$ mw: 313.73

SYNS: GANU ◇ NSC D 254157

TOXICITY DATA with REFERENCE
dnd-mus:leu 100 μmol/L INSSDM 19,49,81
dni-mus/leu 10 mg/kg CNREA8 37,783,77
ivn-rat LD50:40 mg/kg GANNA2 68,247,77
ipr-mus LD50:15 mg/kg CNREA8 37,783,77
ipr-mus LDLo:10 mg/kg CNREA8 37,2615,77
ivn-mus LD10:10 mg/kg ANBCB3 23,64,78

SAFETY PROFILE: Poison by intravenous and intraperitoneal routes. Many N-nitroso compounds are carcinogens. Mutation data reported. When heated to decomposition it emits very toxic fumes of NO_x and Cl^-. See also N-NITROSO COMPOUNDS.

CHA250 CAS:60784-48-7 ***HR: 3***
1-(2-CHLOROETHYL)-3-(4-HYDROXYBUTYL)-1-NITROSOUREA
mf: $C_7H_{14}ClN_3$ mw: 175.69

TOXICITY DATA with REFERENCE
mrc-smc 1 mmol/L/16H MUREAV 42,45,77
ipr-rat LD50:32 mg/kg EJCAAH 13,937,77

SAFETY PROFILE: Poison by intraperitoneal route. Mutation data reported. Many N-nitroso compounds are carcinogens. When heated to decomposition it emits very toxic fumes of NO_x and Cl^-. See also N-NITROSO COMPOUNDS.

CHA500 CAS:52049-26-0 ***HR: 3***
1-(2-CHLOROETHYL)-3-(cis-4-HYDROXYCYCLOHEXYL)-1-NITROSOUREA
mf: $C_9H_{16}ClN_3O_3$ mw: 249.73

SYNS: cis-4-HYDROXY-CCNU ◇ cis-N-(2-CHLOROETHYL)-N'-(4-HYDROXYCYCLOHEXYL)-N-NITROSOUREA ◇ cis-4-OH-CCNU ◇ NSC 239724

TOXICITY DATA with REFERENCE
ipr-mus LD50:60180 μg/kg NCISP* JAN86
ivn-mus LD10:25 mg/kg ANBCB3 23,64,78
ipl-mus LDLo:33 mg/kg JMCMAR 18,634,75

SAFETY PROFILE: Poison by intravenous, implant and intraperitoneal routes. Many N-nitroso compounds are carcinogens. When heated to decomposition it emits very toxic fumes of Cl^- and NO_x. See also N-NITROSO COMPOUNDS.

CHA750 CAS:58494-43-2 ***HR: 3***
1-(2-CHLOROETHYL)-3-(trans-2-HYDROXY-CYCLOHEXYL)-1-NITROSOUREA
mf: $C_9H_{16}ClN_3O_4$ mw: 265.73

SYNS: N-(2-CHLOROETHYL)-N'-(trans-2-HYDROXYCYCLOHEXYL)-N-NITROSOUREA ◇ NSC 253947 ◇ trans-2-OH-CCNU ◇ trans-N-(2-CHLOROETHYL)-N'-(2-HYDROXYCYCLOHEXYL)-N-NITROSOUREA

TOXICITY DATA with REFERENCE
mmo-sat 100 nmol/plate JJIND8 65,149,80
mma-sat 100 nmol/plate JJIND8 65,149,80
ipr-mus LD50:90140 μg/kg NCISP* JAN86

SAFETY PROFILE: Poison by intraperitoneal route. Mutation data reported. Many N-nitroso compounds are carcinogens. When heated to decomposition it emits very toxic fumes of Cl^- and NO_x. See also N-NITROSO COMPOUNDS.

CHB000 CAS:56239-24-8 ***HR: 3***
1-(2-CHLOROETHYL)-3-(trans-4-HYDROXY-CYCLOHEXYL)-1-NITROSOUREA
mf: $C_9H_{16}ClN_3O_3$ mw: 249.73

SYNS: N-(2-CHLOROETHYL)-N'-(trans-4-HYDROXYCYCLOHEXYL)-N-NITROSOUREA ◇ trans-N-(2-CHLOROETHYL)-N'-(4-HYDROXYCYCLOHEXYL)-N-NITROSOUREA ◇ trans-4-HYDROXY-CCNU ◇ NSC-239717 ◇ trans-4-OH CCNU

TOXICITY DATA with REFERENCE
mmo-sat 100 nmol/plate JJIND8 65,149,80
mma-sat 100 nmol/plate JJIND8 65,149,80
dnd-hmn:emb 100 μmol/L CNREA8 44,1352,84
ipr-mus LD50:52420 μg/kg NCISP* JAN86
ivn-mus LD10:12 mg/kg ANBCB3 23,64,78
ipl-mus LDLo:35 mg/kg JMCMAR 18,634,75

SAFETY PROFILE: Poison by implant, intravenous and intraperitoneal routes. Human mutation data reported. Many N-nitroso compounds are carcinogens. When heated to decomposition it emits very toxic fumes of NO_x and Cl^-. See also N-NITROSO COMPOUNDS.

CHB250 CAS:101651-98-3 ***HR: 3***
7-CHLORO-10-(3-(N-ETHYL-N-(2-HYDROXY-ETHYL)AMINO)PROPYL)ISOALLOXAZINE HYDROCHLORIDE
mf: $C_{17}H_{20}ClN_5O_3 \cdot ClH$ mw: 414.33

TOXICITY DATA with REFERENCE
ipr-rat LD50:75 mg/kg CMTRAG 2,96,61
orl-mus LD50:5000 mg/kg CMTRAG 2,96,61
ipr-mus LD50:90 mg/kg CMTRAG 2,96,61
scu-mus LD50:145 mg/kg CMTRAG 2,96,61
ivn-mus LD50:60 mg/kg CMTRAG 2,96,61
ims-mus LD50:90 mg/kg CMTRAG 2,96,61

SAFETY PROFILE: Poison by intraperitoneal, subcutaneous, intravenous, and intramuscular routes. Mildly toxic by ingestion. When heated to decomposition it emits very toxic fumes of Cl^- and NO_x.

CHB750 CAS:60784-46-5 ***HR: 3***
1-(2-CHLOROETHYL)-3-(2-HYDROXYETHYL)-1-NITROSOUREA
mf: $C_5H_{10}ClN_3O_3$ mw: 195.63

SYNS: CNU-ETHANOL ◇ HECNU ◇ 1-(2-HYDROXYETHYL)-3-(2-CHLOROETHYL)-3-NITROSOUREA ◇ HYDROXYETHYL CNU ◇ NSC 294895

TOXICITY DATA with REFERENCE
sln-dmg-orl 5 mmol/L DRISAA 52,20,77
sln-dmg-par 5 mmol/L DRISAA 52,20,77
mrc-smc 1 mmol/L/16H MUREAV 42,45,77
dnd-rat-ipr 100 μmol/kg CNREA8 44,514,84
cyt-mus:lng 1 mg/L/1H MUREAV 44,87,77
ivn-rat TDLo:16 mg/kg/60W-I:ETA DTESD7 8,273,80
ipr-rat LD50:25300 μg/kg ONCOBS 37,177,80
ipr-mus LD50:25 mg/kg INSSDM 19,123,81

CONSENSUS REPORTS: EPA Genetic Toxicology Program.

SAFETY PROFILE: Questionable carcinogen with experimental tumorigenic data. Poison by intraperitoneal route. Mutation data reported. Many N-nitroso compounds are carcinogens. When heated to decomposition it emits very toxic fumes of Cl^- and NO_x. See also N-NITROSO COMPOUNDS.

CHC000 CAS:693-30-1 ***HR: 3***
2-CHLOROETHYL-2-HYDROXYETHYL SULFIDE
mf: C_4H_9ClOS mw: 140.64

SYNS: β-CHLOROETHYL-β-HYDROXYETHYL SULFIDE ◇ 2-((2-CHLOROETHYL)THIO)ETHANOL ◇ HALF MUSTARD GAS ◇ HALF SULFUR MUSTARD ◇ 2-HYDROXYETHYL-2-CHLOROETHYL SULFIDE ◇ MUSTARD CHLOROHYDRIN ◇ SULFUR HALF-MUSTARD

TOXICITY DATA with REFERENCE
dni-hmn:hla 1500 mg/L IUSMDJ 9,41,79
oms-hmn:hla 1500 mg/L IUSMDJ 9,41,79

dnd-rat:lvr 100 μmol/L BIJOAK 80,496,61
dnd-mus/ast 4 mg/kg BIJOAK 80,496,61
ims-rat LD50:500 μg/kg NTIS** PB158-507
skn-mus LD50:600 mg/kg JPETAB 93,1,48
ivn-mus LD50:35 mg/kg JPETAB 93,1,48

SAFETY PROFILE: Poison by intravenous and intramuscular routes. Moderately toxic by skin contact. Human mutation data reported. When heated to decomposition it emits very toxic fumes of Cl^- and SO_x. See also SULFIDES.

CHC250 CAS:60784-47-6 ***HR: 3***
3-(2-CHLOROETHYL)-1-(3-HYDROXYPROPYL)-3-NITROSOUREA
mf: $C_6H_{12}ClN_3O_3$ mw: 209.66

SYN: 1-(3-HYDROXYPROPYL)-CNU

TOXICITY DATA with REFERENCE
mrc-smc 1 mmol/L/16H MUREAV 42,45,77
ipr-rat LDLo:16 mg/kg JNCIAM 60,345,78

SAFETY PROFILE: Poison by intraperitoneal route. Mutation data reported. Many N-nitroso compounds are carcinogens. When heated to decomposition it emits very toxic fumes of Cl^- and NO_x. See also N-NITROSO COMPOUNDS.

CHC500 CAS:107-27-7 ***HR: 3***
CHLOROETHYL MERCURY
mf: C_2H_5ClHg mw: 265.11

PROP: Silvery, irridescent leaflets. Mp: 192.5°.

SYNS: CERESAN ◇ EMC ◇ ETHYLMERCURIC CHLORIDE ◇ ETHYLMERCURY CHLORIDE ◇ GANOZAN ◇ GRANOSAN

TOXICITY DATA with REFERENCE
mmo-esc 20 nmol/L MJDHDW 28,F39,80
dnr-esc 3 mmol/L MJDHDW 28,F39,80
orl-rat TDLo:5600 μg/kg (28D male):REP TIVSAI 46,157,73
orl-rat TDLo:9 mg/kg (14D pre/1-22D preg):TER TIVSAI 46,157,73
orl-rat LD50:40 mg/kg TAGTER 9,25,74
skn-rat LD50:200 mg/kg PHJOAV 185,361,60
scu-rat LD50:66 mg/kg JJEMAG 39,47,69
orl-mus LD50:56 mg/kg 85GMAT -,68,82
ihl-mus LC50:5 mg/m^3 85JCAE -,1199,86
ipr-mus LD50:16 mg/kg OCRAAH 3,137,68

CONSENSUS REPORTS: Mercury and its compounds are on the Community Right-To-Know List.

OSHA PEL: (Transitional: CL 1 mg/10m^3) TWA 0.01 mg(Hg)/m^3; STEL 0.03 mg/m^3 (skin)
ACGIH TLV: TWA 0.01 mg(Hg)/m^3; STEL 0.03 mg(Hg)/m^3

SAFETY PROFILE: Poison by ingestion, inhalation, skin contact, subcutaneous, and intraperitoneal routes. An experimental teratogen. Other experimental reproductive effects. Human mutation data reported. See also MERCURY COMPOUNDS, ORGANIC. When heated to decomposition it emits very toxic fumes of Cl^- and Hg.

CHC675 CAS:1888-94-4 ***HR: 3***
CHLOROETHYL METHACRYLATE
mf: $C_6H_9ClO_2$ mw: 148.60

SYNS: β-CHLOROETHYL METHACRYLATE ◇ 2-CHLOROETHYL METHACRYLATE

TOXICITY DATA with REFERENCE
orl-rat LD50:200 mg/kg 85GMAT -,36,82
ihl-rat LC50:550 mg/m^3/4H 85GMAT -,36,82
ihl-mus LC50:700 mg/m^3/2H 85GMAT -,36,82

SAFETY PROFILE: Poison by inhalation and ingestion. When heated to decomposition it emits toxic fumes of Cl^-. See also ESTERS.

CHC750 CAS:3570-58-9 ***HR: 3***
2-CHLOROETHYL METHANESULFONATE
mf: $C_3H_7ClO_3S$ mw: 158.61

SYNS: CB 1506 ◇ β-CHLOROETHYLMETHANESULFONATE ◇ CHLOROETHYL METHANESULPHONATE ◇ CHLOROMETHANE SULFONATE d'ETHYLE (FRENCH) ◇ METHANESULFONIC ACID CHLOROETHYL ESTER ◇ NSC 18016

TOXICITY DATA with REFERENCE
dnd-dmg-par 10500 μmol/L CNREA8 30,195,70
sln-dmg-par 15 mmol/L ANYAA9 68,731,58
mmo-nsc 100 mmol/L MGBUA3 17,5,60
mmo-ssp 8 mmol/L ADWMAX -,193,62
ipr-rat TDLo:250 mg/kg (5D male):REP 85GUAJ -,55,66
ipr-rat LD50:135 mg/kg CNCRA6 9,56,60
ivn-rat LD50:143 mg/kg CNCRA6 9,56,60
ipr-mus LD50:182 mg/kg CNCRA6 9,56,60
ivn-mus LD50:182 mg/kg CNCRA6 9,56,60
ivn-dog LDLo:32 mg/kg CCSUBJ 2,203,65
ivn-mky LD50:127 mg/kg CNCRA6 9,56,60
orl-qal LD50:208 mg/kg JRPFA4 48,271,76

CONSENSUS REPORTS: EPA Genetic Toxicology Program.

SAFETY PROFILE: Poison by ingestion, intravenous, and intraperitoneal routes. Experimental reproductive effects. Mutation data reported. See also SULFONATES. When heated to decomposition it emits very toxic fumes of Cl^- and SO_x.

CHD250 CAS:13909-09-6 ***HR: 3***
1-(2-CHLOROETHYL)-3-(4-METHYL-CYCLO-HEXYL)-1-NITROSOUREA
mf: $C_{10}H_{18}ClN_3O_2$ mw: 247.76

SYNS: 1-(2-CHLOROETHYL)-3-(trans-4-METHYL-CYCLOHEXYL)-1-NITROSOUREA ◇ N-(2-CHLOROETHYL)-N'-(trans-4-METHYL-CYCLOHEXYL)-N-NITROSOUREA ◇ ME-CCNU ◇ METHYL-CCNU ◇ trans-METHYL-CCNU ◇ METHYL-LOMUSTINE ◇ NCI-C04955 ◇ NSC-95441 ◇ SEMUSTINE

TOXICITY DATA with REFERENCE
skn-rbt 5 mg/24H rns TXCYAC 14,117,79
mmo-sat 100 nmol/plate JJIND8 65,149,80
mma-sat 100 nmol/plate JJIND8 65,149,80
dnd-esc 50 μmol/L MUREAV 89,95,81
dni-mus:oth 10 μmol/L CNREA8 43,5837,83
orl-hmn TDLo:22 mg/kg/60W-C:CAR NEJMAG 309,1079,83
ipr-rat TDLo:30 mg/kg/7W-I:CAR CANCAR 40(Suppl 4),1935,77
ivn-rat TDLo:32 mg/kg/60W-I:ETA DTESD7 8,273,80
orl-hmn TDLo:90 mg/kg:BLD,GIT CCYPBY 4,257,73
orl-hmn TDLo:6 mg/kg:GIT,BLD CTRRDO 60,709,76
orl-chd LDLo:5550 mg/kg/30W-I NEJMAG 300,1200,79
orl-chd TDLo:37950 mg/kg/2Y-I NEJMAG 300,1200,79
orl-mus LD50:49900 μg/kg NTIS** PB269-473
ipr-mus LD10:37 mg/kg CNREA8 34,194,74
unr-mus LD50:117 mg/kg NCISP* JAN86
orl-dog LDLo:25 mg/kg ACRSAJ 16,273,72
ivn-dog LDLo:14 mg/kg CTRRDO 60,1559,76
orl-mky LDLo:100 mg/kg ACRSAJ 16,273,72
ivn-mky LDLo:45 mg/kg CTRRDO 60,1559,76

CONSENSUS REPORTS: NCI Carcinogenesis Studies (ipr); Clear Evidence: rat CANCAR 40,1935,77; No Evidence: mouse CANCAR 40,1935,77.

SAFETY PROFILE: Suspected human carcinogen producing leukemia. Experimental carcinogenic and tumorigenic data. Poison by ingestion, intraperitoneal, intravenous, and other possibly other routes. Mutation data reported. Human systemic effects by ingestion: nausea or vomiting, damage to kidney tubules and glomeruli, and hematuria (blood in the urine). When heated to decomposition it emits very toxic fumes of Cl^- and NO_x. See also N-NITROSO COMPOUNDS.

CHD500 CAS:61137-63-1 ***HR: 3***
trans-1-(2-CHLOROETHYL)-3-(3-METHYLCYCLO-HEXYL)-1-NITROSOUREA
mf: $C_{10}H_{18}ClN_3O_2$ mw: 247.76

SYNS: Me-CCNU ◇ trans-N-(2-CHLOROETHYL)-N'-(3-METHYLCYCLOHEXYL)-N-NITROSOUREA

TOXICITY DATA with REFERENCE
msc-ham:ovr 4 mg/L CNREA8 35,460,75
ipr-mus LD50:71460 μg/kg NCISP* JAN86

SAFETY PROFILE: Poison by intraperitoneal route. Mutation data reported. Many N-nitroso compounds are carcinogens. When heated to decomposition it emits very toxic fumes of Cl^- and NO_x. See also N-NITROSO COMPOUNDS.

CHD675 CAS:1755-67-5 ***HR: 3***
5-(2-CHLOROETHYL)-4-METHYLOXAZOLE-1,2-ETHANEDISULFONATE (2:1)
mf: $C_{12}H_{16}Cl_2O_2 \cdot C_2H_6O_6S_2$ mw: 453.38

TOXICITY DATA with REFERENCE
orl-mus LD50:380 mg/kg APSXAS 4,269,67
ipr-mus LD50:315 mg/kg APSXAS 4,269,67
scu-mus LD50:380 mg/kg APSXAS 4,269,67
ivn-mus LD50:250 mg/kg APSXAS 4,269,67

SAFETY PROFILE: Poison by ingestion, subcutaneous, intravenous, and intraperitoneal routes. When heated to decomposition it emits toxic fumes of Cl^- and SO_x. See also SULFONATES.

CHD700 CAS:5959-98-8 ***HR: 3***
3-(2-CHLOROETHYL)-2-METHYLPYRIDINE HYDROCHLORIDE
mf: $C_8H_{10}ClN \cdot ClH$ mw: 192.10

SYN: 3-(2-CHLOROETHYL)-2-PICOLINEHYDROCHLORIDE

TOXICITY DATA with REFERENCE
orl-mus LD50:1300 mg/kg APSXAS 4,269,67
ipr-mus LD50:350 mg/kg APSXAS 4,269,67
ivn-mus LD50:150 mg/kg APSXAS 4,269,67

SAFETY PROFILE: Poison by intravenous and intraperitoneal routes. Moderately toxic by ingestion. When heated to decomposition it emits toxic fumes of Cl^- and NO_x.

CHD750 CAS:533-45-9 ***HR: 3***
5-(2-CHLOROETHYL)-4-METHYLTHIAZOLE
mf: C_6H_8ClNS mw: 161.66

SYNS: CHLORETHIAZOL ◇ CHLORMETHIAZOLE ◇ CHLORO-S.C.T.Z. ◇ CLOMETHIAZOLE ◇ CLOMETHIAZOLUM ◇ DISTRANEURIN ◇ EMINEURINA ◇ HEMINEVRIN ◇ 4-METHYL-5-(β-CHLOROETHYL)THIAZOLE ◇ SCTZ ◇ SOMNEVRIN ◇ WY 1485

TOXICITY DATA with REFERENCE
orl-mus LD50:2110 mg/kg JMCMAR 7,167,64
ipr-mus LD50:190 mg/kg APSXAS 8,39,71
ivn-mus LD50:94 mg/kg APSXAS 7,423,70

SAFETY PROFILE: Poison by intravenous and intraperitoneal routes. Moderately toxic by ingestion. When heated to decomposition it emits very toxic fumes of Cl^-, NO_x, and SO_x.

CHD800 ***HR: 2***
5-(2-CHLOROETHYL)-4-METHYLTHIAZOLE ETHANE DISULFONATE
mf: $C_6H_8ClNS \cdot C_2H_6O_6S_2$ mw: 351.86

TOXICITY DATA with REFERENCE
orl-mus LD50:835 mg/kg APSXAS 4,269,67
ipr-mus LD50:590 mg/kg APSXAS 4,269,67
scu-mus LD50:620 mg/kg APSXAS 4,269,67
ivn-mus LD50:620 mg/kg APSXAS 4,269,67

SAFETY PROFILE: Moderately toxic by ingestion, subcutaneous, intravenous and intraperitoneal routes. When heated to decomposition it emits toxic fumes of Cl^-, SO_x, and NO_x. See also SULFONATES.

CHD875 CAS:3240-94-6 ***HR: 3***
4-(2-CHLOROETHYL)MORPHOLINE
mf: $C_6H_{12}ClNO$ mw: 149.64

SYN: TL 401

TOXICITY DATA with REFERENCE
dni-mus ivg 1 pph JIDEAE 62,378,74
unr-rat LD50:70 mg/kg PHBUA9 1,297,53
ihl-mus LCLo:370 mg/m^3/10M NDRC** NDCrc-132,Nov,42

SAFETY PROFILE: Poison by inhalation and possibly other routes. Mutation data reported. When heated to decomposition it emits toxic fumes of Cl^- and NO_x.

CHE000 CAS:3647-69-6 ***HR: 3***
4-(2-CHLOROETHYL)MORPHOLINE HYDROCHLORIDE
mf: $C_6H_{12}ClNO \cdot ClH$ mw: 186.10

TOXICITY DATA with REFERENCE
mmo-sat 21500 nmol/L ENMUDM 3,11,81
mmo-esc 21500 nmol/L ENMUDM 3,11,81
dns-rat:lvr 100 μmol/L ENMUDM 3,11,81
msc-mus:lym 44 μmol/L ENMUDM 3,33,81
ipr-mus LD50:150 mg/kg (1D male):REP CANCAR 2,1075,49
ipr-mus LD50:161 mg/kg CANCAR 2,1075,49

CONSENSUS REPORTS: Reported in EPA TSCA Inventory.

SAFETY PROFILE: Poison by intraperitoneal route. Experimental reproductive effects. Mutation data reported. When heated to decomposition it emits very toxic fumes of Cl^- and NO_x.

CHE250 CAS:64057-51-8 ***HR: D***
N-(β-CHLOROETHYL)-N-NITROSOACETAMIDE
mf: $C_4H_7ClN_2O_2$ mw: 150.58

SYN: N-(2-CHLOROETHYL)-N-NITROSOACETAMIDE

TOXICITY DATA with REFERENCE
mmo-sat 1 nmol/plate CNREA8 43,175,83
mma-sat 2 nmol/plate CNREA8 39,1328,79
pic-esc 2 mg/L CNREA8 39,1328,79
dnd-mus:leu 500 nmol/L CNREA8 43,175,83

SAFETY PROFILE: Mutation data reported. Many N-nitroso compounds are carcinogens. When heated to decomposition it emits very toxic fumes of Cl^- and NO_x. See also N-NITROSO COMPOUNDS.

CHE325 CAS:60784-40-9 ***HR: 3***
N-(2-CHLOROETHYL)-N-NITROSOCARBOMOYL AZIDE
mf: $C_3H_4ClN_5O_2$ mw: 177.55

$ClC_2H_4N(N{:}O)CO \cdot N_3$

SAFETY PROFILE: A potentially explosive material. Many N-nitroso compounds are experimental carcinogens. When heated to decomposition it emits toxic fumes of Cl^- and NO_x. See also N-NITROSO COMPOUNDS and AZIDES.

CHE500 CAS:13909-13-2 ***HR: 3***
1-(2-CHLOROETHYL)-1-NITROSO-3-(2-NORBORNYL)UREA
mf: $C_{10}H_{16}ClN_3O_2$ mw: 245.74

SYNS: N'-BICYCLO(2.2.1)HEPT-2-YL-N-(2-CHLOROETHYL)-N-NITROSOUREA ◇ 1-(2-CHLOROETHYL)-3-(2-NORBORNYL)-1-NITROSOUREA ◇ NSC-88106 ◇ SRI 2638

TOXICITY DATA with REFERENCE
orl-rat LD50:83 mg/kg NCIMR* -,372,68
orl-mus LD50:56 mg/kg NCIMR* -,372,68
ipr-mus LD50:54760 μg/kg NCISP* JAN86

SAFETY PROFILE: Poison by ingestion and intraperitoneal routes. Many N-nitroso compounds are carcinogens. When heated to decomposition it emits very toxic fumes of Cl^- and NO_x. See also N-NITROSO COMPOUNDS.

CHE750 CAS:2365-30-2 ***HR: 3***
1-(2-CHLOROETHYL)-1-NITROSOUREA
mf: $C_3H_6ClN_3O_2$ mw: 151.57

SYNS: N-(2-CHLOROETHYL)-N-NITROSOUREA ◇ CNU ◇ MP 655 ◇ N-NITROSO-2-CHLOROETHYLUREA ◇ 1-NITROSO-1-(2-CHLOROETHYL)UREA ◇ NSC 47547 ◇ SKI 28404

TOXICITY DATA with REFERENCE
mmo-sat 1 μg/plate MUREAV 68,1,79
mma-sat 20 nmol/plate CNREA8 39,1328,79
dnr-esc 1 mmol/L CNREA8 45,6471,85
dnd-hmn:ovr 50 μmol/L INSSDM 19,33,81
dnd-mus:leu 10 μmol/L CNREA8 43,175,83
msc-ham:lng 10 μmol/L CNREA8 40,2719,80

ipr-mus LD50:4368 μg/kg NCISP* JAN86
ivn-mus LD10:4700 μg/kg ANBCB3 23,64,78

CONSENSUS REPORTS: EPA Genetic Toxicology Program.

SAFETY PROFILE: Poison by intraperitoneal and intravenous routes. Human mutation data reported. Many N-nitroso compounds are carcinogens. When heated to decomposition it emits very toxic fumes of Cl^- and NO_x. See also N-NITROSO COMPOUNDS.

CHF000 CAS:33073-60-8 ***HR: 3***
trans-4-(3-(2-CHLOROETHYL)-3-NITROSOUREIDOCYCLOHEXANE CARBOXYLIC ACID ETHYL ESTER
mf: $C_{12}H_{20}ClN_3O_4$ mw: 305.80

SYNS: 4-((((2-CHLOROETHYL)NITROSOAMINO)CARBONYL) AMINO)CYCLOHEXANE CARBOXYLIC ACID, ETHYL ESTER ◇ NSC 103548

TOXICITY DATA with REFERENCE
ipr-mus LD50:24150 μg/kg NCISP* JAN86

SAFETY PROFILE: Poison by intraperitoneal route. Many N-nitroso compounds are carcinogens. When heated to decomposition it emits very toxic fumes of Cl^- and NO_x. See also N-NITROSO COMPOUNDS.

CHF250 CAS:61866-12-4 ***HR: 3***
2-(3-(2-CHLOROETHYL)3-NITROSOUREIDO)ETHYL METHANE SULFONATE
mf: $C_5H_{10}ClN_3O_5S$ mw: 259.69

SYNS: 2-(3-(2-CHLORATHYL)-3-NITROSOUREIDO)ATHYLMETHANSULFONAT(GERMAN) ◇ CNUEMS ◇ ETHYLMETHANESULFONATO-CNU ◇ NSC-294896

TOXICITY DATA with REFERENCE
mrc-smc 500 μmol/L/16H MUREAV 42,45,77
ipr-rat LD50:23 mg/kg VDGIA2 85,1293,79
ivn-rat LD50:13800 μg/kg ONCOBS 37,177,80

SAFETY PROFILE: Poison by intraperitoneal and intravenous routes. Mutation data reported. Many N-nitroso compounds are carcinogens. When heated to decomposition it emits very toxic fumes of Cl^-, NO_x, and SO_x. See also N-NITROSO COMPOUNDS, SULFONATES, and ESTERS.

CHF500 CAS:6296-45-3 ***HR: 3***
2-CHLOROETHYL-N-NITROSOURETHANE
mf: $C_5H_9ClN_2O_3$ mw: 180.61

SYNS: N-(2-CHLOROETHYL)-N-NITROSOETHYLCARBAMATE◇ N-(β-CHLOROETHYL)-N-NITROSOURETHAN◇ ETHYL-N-(β-CHLOROETHYL)-N-NITROSOCARBAMATE◇ TL 154

TOXICITY DATA with REFERENCE
mmo-sat 1 nmol/plate CNREA8 43,175,83
dnd-mus:leu 500 nmol/L CNREA8 43,175,83
orl-rat TDLo:6 mg/kg:ETA CNREA8 31,573,71
orl-rat LDLo:10 mg/kg CNREA8 31,573,71
ihl-rat LCLo:330 mg/m³/10M NDRC** NDCrc-132,Apr,42
ipr-rat LDLo:6500 μg/kg CNREA8 31,573,71
ihl-dog LCLo:330 mg/m³/10M NDRC** NDCrc-132,Apr,42
ihl-cat LCLo:330 mg/m³/10M NDRC** NDCrc-132,Apr,42

SAFETY PROFILE: Poison by inhalation, ingestion, and intraperitoneal routes. Questionable carcinogen with experimental tumorigenic data. Mutation data reported. Many N-nitroso compounds are carcinogens. See also CARBAMATES and N-NITROSO COMPOUNDS. When heated to decomposition it emits very toxic fumes of Cl^- and NO_x.

CHG000 CAS:113-18-8 ***HR: 3***
1-CHLORO-3-ETHYL-1-PENTEN-4-YN-3-OL
mf: C_7H_9ClO mw: 144.61

SYNS: A 71 ◇ AETHYL-CHLORVYNOL ◇ ALVINOL ◇ ARVYNOL ◇ β-CHLOROVINYL ETHYLETHYNYL CARBINOL ◇ 3-(β-CHLOROVINYL)-1-PENTYN-3-OL ◇ ETCHLORVINOLO ◇ ETHCHLOROVYNOL ◇ ETHCHLORVINYL ◇ ETHCLORVYNOL ◇ ETHOCHLORVYNOL ◇ ETHYL-β-CHLOROVINYLETHYNYL CARBINOL ◇ ETHYLCHLORVYNOL ◇ NORMONSON ◇ NORMOSAN ◇ NORMOSON ◇ NOSTEL ◇ PLACIDIL ◇ PLACIDYL ◇ ROERIDORM ◇ SERENIL ◇ SERENSIL

TOXICITY DATA with REFERENCE
unr-wmn TDLo:643 mg/kg (26-39W preg):REP PEDIAU 52,714,73
orl-rat TDLo:400 mg/kg (1-20D preg):TER EESADV 5,494,81
orl-wmn TDLo:10 mg/kg:CNS BMJOAE 2,1610,62
orl-wmn TDLo:15 mg/kg/2D:BLD AIMEAS 77,73,72
scu-rat LD50:200 mg/kg 27ZQAG -,420,72
orl-mus LD50:290 mg/kg JPETAB 114,326,55
ipr-mus LD50:275 mg/kg 27ZQAG -,420,72
scu-mus LD50:240 mg/kg JPETAB 114,326,55
ivn-dog LD50:55 mg/kg 27ZQAG -,420,72
orl-bwd LD50:42 mg/kg TXAPA9 21,315,72

SAFETY PROFILE: Poison by ingestion, subcutaneous, intraperitoneal, and intravenous routes. Human systemic effects by ingestion: general anesthesia and thrombocytopenia (reduction in the number of blood platelets). Human effects on newborn by an unspecified route: drug dependency and apgar score (condition of newborn). Experimental teratogenic and reproductive effects. When heated to decomposition it emits toxic fumes of Cl^-.

CHG250 CAS:73816-74-7 ***HR: 1***
4-(4-CHLORO-6-ETHYLPHENYLAMINO-2-s-TRIAZINYLAMINO-5-HYDROXY-6-(4-METHYL- 2-SULFOPHENYLAZO)-2,7-NAPHTHALENE DISULFONIC ACID TRISODIUM SALT
mf: $C_{28}H_{21}ClN_7O_{10}S_3$•3Na mw: 816.16

SYN: CERVEN BRILANTNI OSTAZINOVA H-3B (CZECH)

TOXICITY DATA with REFERENCE
eye-rbt 20 mg/24H MOD 28ZPAK -,234,72
orl-rat LD50:6730 mg/kg 28ZPAK -,234,72

SAFETY PROFILE: Mildly toxic by ingestion. An eye irritant. When heated to decomposition it emits very toxic fumes of Cl^-, NO_x, Na_2O, and SO_x.

CHG375 CAS:10419-79-1 ***HR: 2***
(2-CHLOROETHYL)PHOSPHONIC ACID DIETHYL ESTER
mf: $C_6H_{14}ClO_3P$ mw: 200.62

TOXICITY DATA with REFERENCE
orl-rat LD50:1 g/kg GISAAA 48(8),79,83
orl-rbt LD50:2 g/kg GISAAA 48(8),79,83
orl-gpg LD50:1450 mg/kg GISAAA 48(8),79,83

SAFETY PROFILE: Moderately toxic by ingestion. When heated to decomposition it emits toxic fumes of Cl^- and PO_x. See also ESTERS.

CHG400 CAS:23510-39-6 ***HR: 2***
(2-CHLOROETHYL)PHOSPHONIC ACID MONOETHYL ESTER
mf: $C_4H_{10}ClO_3P$ mw: 172.56

TOXICITY DATA with REFERENCE
orl-rat LD50:1250 mg/kg GISAAA 48(8),79,83
orl-rbt LD50:2 g/kg GISAAA 48(8),79,83
orl-gpg LD50:1800 mg/kg GISAAA 48(8),79,83

SAFETY PROFILE: Moderately toxic by ingestion. When heated to decomposition it emits toxic fumes of Cl^- and PO_x. See also ESTERS.

CHG500 CAS:2008-75-5 ***HR: 3***
1-(2-CHLOROETHYL)PIPERIDINE HYDROCHLORIDE
mf: $C_7H_{14}ClN$•ClH mw: 184.13

SYNS: β-CHLOROETHYLPIPERIDINE HYDROCHLORIDE ◇ PIPERIDINOETHYL CHLORIDE, HYDROCHLORIDE

TOXICITY DATA with REFERENCE
mmo-sat 5400 nmol/L ENMUDM 3,11,81
mmo-esc 5400 nmol/L ENMUDM 3,11,81
dns-rat:lvr 500 μmol/L ENMUDM 3,11,81
msc-mus:lym 44 μmol/L ENMUDM 3,33,81
ipr-mus LD50:93 mg/kg JPETAB 94,249,48
scu-mus LD50:125 mg/kg JPETAB 97,25,49
ivn-mus LD50:56 mg/kg CSLNX* NX#07326

SAFETY PROFILE: Poison by subcutaneous, intravenous, and intraperitoneal routes. Mutation data reported. When heated to decomposition it emits very toxic fumes of Cl^- and NO_x.

CHG750 CAS:101651-64-3 ***HR: 3***
2'-CHLORO-2-(ETHYL(2-PIPERIDINOETHYL)AMINO)ACETANILIDE DIHYDROCHLORIDE
mf: $C_{17}H_{26}ClN_3O$•2ClH mw: 396.83

SYN: C 5410

TOXICITY DATA with REFERENCE
eye-rbt 2% MLD ARZNAD 9,262,59
ipr-rat LD50:89 mg/kg ARZNAD 9,262,59
scu-mus LD50:400 mg/kg ARZNAD 9,262,59

SAFETY PROFILE: Poison by intraperitoneal and subcutaneous routes. An eye irritant. When heated to decomposition it emits very toxic fumes of Cl^- and NO_x.

CHH000 CAS:38915-59-2 ***HR: 3***
7-(3-(2-CHLOROETHYL-n-PROPYLAMINO)PROPYLAMINO)BENZO(b)(1,10)-PHENATHROLINE HYDROCHLORIDE
mf: $C_{24}H_{27}ClN_4$•3ClH mw: 516.38

SYN: ICR 394

TOXICITY DATA with REFERENCE
mmo-sat 5 μg/plate JMCMAR 15,739,72
ivn-mus TDLo:1550 μg/kg:NEO CNREA8 36,2423,76
ipr-mus LD20:1 mg/kg JMCMAR 15,739,72

SAFETY PROFILE: Poison by intraperitoneal route. Questionable carcinogen with experimental neoplastigenic data. Mutation data reported. When heated to decomposition it emits very toxic fumes of Cl^- and NO_x.

CHH125 CAS:10140-94-0 ***HR: 3***
2-CHLORO-5-ETHYL-4-PROPYL-2-THIONO-1,3,2-DIOXAPHOSPHORINANE
mf: $C_8H_{16}ClO_2PS$ mw: 242.72

SYN: PHOSPHOROCHLORIDOTHIOIC ACID, cyclic O,O-(2-ETHYL-1-PROPYLTRIMETHYLENE) ESTER

TOXICITY DATA with REFERENCE
skn-rbt 10 mg/24H open SEV AIHAAP 23,95,62
orl-rat LD50:300 mg/kg AIHAAP 23,95,62
skn-rbt LD50:1000 mg/kg AIHAAP 23,95,62

SAFETY PROFILE: Poison by ingestion. Moderately toxic by skin contact. A severe skin irritant. When heated to decomposition it emits toxic fumes of Cl^-, PO_x, and SO_x. See also ESTERS.

CHI125 CAS:80-41-1 ***HR: 2***
2-CHLOROETHYL TOSYLATE
mf: $C_9H_{11}ClO_3S$ mw: 234.71

SYNS: 2-CHLORO-ETHANOL-4-METHYLBENZENESULFONATE (9CI) ◇ 2-CHLORO-ETHANOL-p-TOLUENESULFONATE (8CI) ◇ β-CHLOROETHYLESTER KYSELINY-p-TOLUENSULFONOVE (CZECH)

TOXICITY DATA with REFERENCE
eye-rbt 500 mg/24H MLD 28ZPAK -,197,72
orl-rat LD50:498 mg/kg 28ZPAK -,197,72
ipr-mus LDLo:512 mg/kg CBCCT* 2,303,50

CONSENSUS REPORTS: Reported in EPA TSCA Inventory.

SAFETY PROFILE: Moderately toxic by ingestion and intraperitoneal routes. An eye irritant. When heated to decomposition it emits toxic fumes of Cl^- and SO_x. See also SULFONATES.

CHI250 CAS:110-75-8 ***HR: 3***
2-CHLOROETHYL VINYL ETHER
mf: C_4H_7ClO mw: 106.56

PROP: Liquid. Bp: 109° @ 740 mm, d: 1.0525, flash p: 80°F (OC), mp: −70.3°.

SYNS: 2-CHLORETHYL VINYL ETHER ◇ (2-CHLOROETHOXY)ETHENE ◇ RCRA WASTE NUMBER U042 ◇ VINYL-β-CHLOROETHYL ETHER ◇ VINYL-2-CHLOROETHYL ETHER

TOXICITY DATA with REFERENCE
skn-rbt 525 mg open SEV UCDS** 11/15/71
eye-rbt 500 mg open JIHTAB 31,60,49
orl-rat LD50:250 mg/kg JIHTAB 31,60,49
ihl-rat LCLo:250 ppm/4H JIHTAB 31,343,49
skn-rbt LD50:3354 mg/kg JIHTAB 31,60,49

CONSENSUS REPORTS: Reported in EPA TSCA Inventory.

SAFETY PROFILE: Poison by ingestion. Moderately toxic by inhalation and skin contact. A severe eye and skin irritant. See also ETHERS. Dangerous fire hazard when exposed to heat, flame, or oxidizers. Potentially explosive. May form dangerous peroxides on exposure to air. To fight fire, use alcohol foam, dry chemical. When heated to decomposition it emits toxic fumes of Cl^-. See also CHLORIDES and ETHERS.

CHI750 ***HR: 3***
CHLOROETHYNYL NORGESTREL mixed with MESTRANOL (20:1)

SYNS: MESTRANOL mixed with CHLOROETHYNYL NORGESTREL (1:20) ◇ WY-4355 mixed with MESTRANOL (20:1)

TOXICITY DATA with REFERENCE
orl-dog TDLo:463 mg/kg/90W-I:CAR JJIND8 65,137,80
orl-dog TD:423 mg/kg/4Y-I:ETA JTEHD6 3,179,77

SAFETY PROFILE: Questionable carcinogen with experimental carcinogenic and tumorigenic data. When heated to decomposition it emits very toxic fumes of Cl^- and NO_x.

CHI825 CAS:5096-17-3 ***HR: 3***
N-(7-CHLORO-2-FLUORENYL)ACETAMIDE
mf: $C_{15}H_{11}ClNO$ mw: 256.72

SYN: N-2-(7-CHLORO)FLUORENYLACETAMIDE

TOXICITY DATA with REFERENCE
orl-rat TDLo:1265 mg/kg/11W-C:ETA JNCIAM 24,149,60
orl-rat LDLo:5000 mg/kg CNREA8 26,619,66

SAFETY PROFILE: Questionable carcinogen with experimental tumorigenic data. Mildly toxic by ingestion. When heated to decomposition it emits toxic fumes of Cl^- and NO_x.

CHI900 CAS:593-70-4 ***HR: 3***
CHLOROFLUOROMETHANE
mf: CH_2ClF mw: 68.48

SYNS: CFC 31 ◇ FC 31 ◇ FREON 31 ◇ MONOCHLOROMONOFLUOROMETHANE ◇ R 31 ◇ R 31 (refrigerant)

TOXICITY DATA with REFERENCE
mmo-sat 5 pph MUREAV 118,277,83
mma-sat 5 pph MUREAV 118,277,83
otr-ham:kdy 100 μmol/L TXAPA9 72,15,84
msc-ham:ovr 10 pph EVSRBT 25,91,82
orl-rat TDLo:78 g/kg/1Y-I:CAR TXAPA9 72,15,84
ihl-mky LCLo:1000 ppm/4H TXAPA9 48,A109,79

CONSENSUS REPORTS: IARC Cancer Review: Group 3 IMEMDT 7,56,87; Animal Limited Evidence IMEMDT 41,229,86.

DFG MAK: Animal Carcinogen, Suspected Human Carcinogen.

SAFETY PROFILE: Confirmed carcinogen with experimental carcinogenic data. Moderately toxic by inhalation. Mutation data reported. When heated to decomposition it emits very toxic fumes of Cl^- and F^-. See also CHLORINATED HYDROCARBONS, ALIPHATIC, and FLUORIDES.

CHJ000 CAS:34214-51-2 ***HR: 2***
3-(2-CHLORO-6-FLUOROPHENYL)-5-METHYL-4-ISOXAZOLYLPENICILLIN SODIUM MONOHYDRATE
mf: $C_{19}H_{17}ClFN_3O_5S•Na•H_2O$ mw: 494.91

SYNS: 6-(3-(2-CHLORO-6-FLUOROPHENYL)-5-METHYL-4-ISOXAZOLECARBOXAMIDO)PENICILLANIC ACID SODIUM SALT ◇ CULPEN ◇ FLOXACILLIN SODIUM MONOHYDRATE ◇ FLOXAPEN ◇ FLUCLOXACILLIN SODIUM MONOHYDRATE ◇ STAPHYLEX

TOXICITY DATA with REFERENCE
orl-mus LD50:3800 mg/kg MEIEDD 10,589,83
scu-mus LD50:2200 mg/kg MEIEDD 10,589,83

SAFETY PROFILE: Moderately toxic by ingestion and subcutaneous routes. When heated to decomposition it emits very toxic fumes of Cl^-, F^-, NO_x, Na_2O and SO_x. See other penilillin entries.

CHJ250 CAS:6186-91-0 ***HR: 3***
3-CHLORO-2-FLUOROPROPENE
mf: C_3H_4ClF mw: 94.52

SYN: 3-CHLORO-2-FLUORO-1-PROPENE

TOXICITY DATA with REFERENCE
orl-rat LD50:280 mg/kg AIHAAP 23,95,62
ihl-rat LCLo:1000 ppm/4H AIHAAP 23,95,62
skn-rbt LD50:200 mg/kg AIHAAP 23,95,62

SAFETY PROFILE: Poison by ingestion and skin contact. Mildly toxic by inhalation. When heated to decomposition it emits very toxic fumes of F^- and Cl^-. See also CHLORINATED HYDROCARBONS, ALIPHATIC, and FLUORIDES.

CHJ500 CAS:67-66-3 ***HR: 3***
CHLOROFORM
DOT: UN 1888
mf: $CHCl_3$ mw: 119.37

PROP: Colorless liquid; heavy, ethereal odor. Mp: −63.5°, bp: 61.26°, fp: −63.5°, flash p: none, d: 1.49845 @ 15°, vap press: 100 mm @ 10.4°, vap d: 4.12.

SYNS: CHLOROFORME (FRENCH) ◇ CLOROFORMIO (ITALIAN) ◇ FORMYL TRICHLORIDE ◇ METHANE TRICHLORIDE ◇ METHENYL TRICHLORIDE ◇ METHYL TRICHLORIDE ◇ NCI-C02686 ◇ R 20 (refrigerant) ◇ RCRA WASTE NUMBER U044 ◇ TCM ◇ TRICHLORMETHAAN (DUTCH) ◇ TRICHLORMETHAN (CZECH) ◇ TRICHLOROFORM ◇ TRICHLOROMETHANE ◇ TRICLOROMETANO (ITALIAN)

TOXICITY DATA with REFERENCE
skn-rbt 10 mg/24H open MLD AIHAAP 23,95,62
skn-rbt 500 mg/24H MLD 28ZPAK -,27,72
eye-rbt 148 mg AIHAAP 37,697,76
eye-rbt 20 mg/24H MOD 28ZPAK -,27,72
sce-hmn:lym 10 mmol/L ENVRAL 32,72,83
dns-mus-ipr 50 mg/kg TOLED5 21,357,84
orl-mus TDLo:2177 mg/kg (male 3W pre):REP NETOD7 1,199,79
ihl-rat TCLo:20100 μg/m³/1H (female 7-14D post):TER NTIS** PB277-077
orl-rat TDLo:13832 mg/kg/2Y-C:CAR FAATDF 5,760,85
orl-mus TDLo:127 g/kg/92W-I:CAR NCITR* NCI-CG-TR-0,76
orl-rat TD:98 g/kg/78W-I:NEO NCITR* NCI-CG-TR-0,76
orl-mus TD:24752 mg/kg/2Y-C:ETA FAATDF 5,760,85
ihl-hmn TCLo:10 mg/m³/1Y:CNS,GIT IRGGAJ 24,127,67
ihl-hmn LCLo:25000 ppm/5M TABIA2 3,231,33
ihl-hmn TCLo:5000 mg/m³/7M:CNS AHBAAM 116,131,36
unr-man LDLo:546 mg/kg 85DCAI 2,73,70
orl-rat LD50:908 mg/kg JPFCD2 17,205,82
ihl-rat LC50:47702 mg/m³/4H ENVRAL 40,411,86
orl-mus LD50:36 mg/kg ATSUDG 2,371,79
ihl-mus LC50:28 g/m³ PCOC** -,230,66
ipr-mus LD50:623 mg/kg AGGHAR 18,109,60
scu-mus LD50:704 mg/kg JPETAB 123,224,58
orl-dog LDLo:1000 mg/kg QJPPAL 7,205,34
ihl-dog LC50:100 g/m³ PCOC** -,230,66
ipr-dog LD50:1000 mg/kg TXAPA9 10,119,67
ivn-dog LDLo:75 mg/kg QJPPAL 7,205,34
ihl-cat LCLo:35000 mg/m³/4H AHBAAM 116,131,36
orl-rbt LDLo:500 mg/kg AEXPBL 97,86,23

CONSENSUS REPORTS: NTP Fifth Annual Report on Carcinogens. IARC Cancer Review: Group 2B IMEMDT 7,152,87; Animal Limited Evidence IMEMDT 1,61,72; Human Limited Evidence IMEMDT 20,401,79; Animal Sufficient Evidence IMEMDT 20,401,79. NCI Carcinogenesis Bioassay (gavage); Clear Evidence: mouse, rat NCITR* NCI-CG-TR,1976. EPA Genetic Toxicology Program. EPA Extremely Hazardous Substances List. Community Right-To-Know List. Reported in EPA TSCA Inventory.

OSHA PEL: (Transitional: CL 50 ppm) TWA 2 ppm
ACGIH TLV: TWA 10 ppm; Suspected Human Carcinogen
DFG MAK: Suspected Carcinogen.
NIOSH REL: (Waste Anesthetic Gases and Vapors) CL 2 ppm/1H; (Chloroform) CL 2 ppm/60M
DOT Classification: ORM-A; Label: None; IMO: Poison B; Label: Poison.

SAFETY PROFILE: Confirmed carcinogen with experimental carcinogenic, neoplastigenic, and tumorigenic data. A human poison by ingestion and inhalation. An experimental poison by ingestion and intravenous routes. Moderately toxic experimentally by intraperitoneal and subcutaneous routes. Human systemic effects by inhalation: hallucinations and distorted perceptions, nausea, vomiting, and other unspecified gastrointestinal effects. Human mutation data reported. Experimental teratogenic and reproductive effects.

Inhalation of the concentrated vapor causes dilation of the pupils with reduced reaction to light, as well as reduced intraocular pressure (experimental). In the initial stages there is a feeling of warmth of the face and body, then an irritation of the mucous membranes, conjunctiva, and skin; followed by excitation, loss of reflexes, sensation, and consciousness. Prolonged inhalation will bring on paralysis accompanied by cardiac respiratory failure and finally death.

Chloroform has been widely used as an anesthetic.

However, due to its toxic effects, this use is being abandoned. Concentrations of 68,000-82,000 ppm in air can kill most animals in a few minutes. 14,000 ppm may cause death after an exposure of from 30 to 60 minutes. 5,000–6,000 ppm can be tolerated by animals for 1 hour without serious disturbances. The maximum concentration tolerated for several hours or for prolonged exposure with slight symptoms is 2,000-2,500 ppm. Prolonged administration as an anesthetic may lead to such serious effects as profound toxemia and damage to the liver, heart, and kidneys. Experimental prolonged but light anesthesia in dogs produces a typical hepatitis.

Explosive reaction with sodium + methanol or sodium methoxide + methanol. Mixtures with sodium or potassium are impact sensitive explosives. Reacts violently with acetone + alkali (e.g., sodium hydroxide, potassium hydroxide, or calcium hydroxide), Al, disilane, Li, Mg, methanol + alkali, nitrogen tetroxide, perchloric acid + phosphorus pentoxide, potassium-tert-butoxide, sodium methylate, NaK. Incompatible with dinitrogen tetraoxide, fluorine, metals, or triisopropylphosphine. Nonflammable. When heated to decomposition it emits toxic fumes of Cl^-.

See also CHLORINATED HYDROCARBONS, ALIPHATIC.

CHJ599 CAS:29671-92-9 ***HR: 3***
CHLOROFORMAMIDINIUM CHLORIDE
mf: CH_4Cl2N_2 mw: 102.95

ClC(:NH)$N^-H_3Cl^-$

SAFETY PROFILE: Reaction with perchloric acid forms highly explosive products. When heated to decomposition it emits toxic fumes of Cl^- and NO_x. See also CHLORIDES.

CHJ625 CAS:75524-40-2 ***HR: 3***
CHLOROFORMAMIDINIUM NITRATE
mf: $CH_3ClN_3O_3$ mw: 140.51

SAFETY PROFILE: A powerful explosive and a strong oxidant. Mixtures with wet magnesium powder, powdered aluminum, or powdered iron ignite and then explode. Reacts violently with ammonia or amines. When heated to decomposition it emits toxic fumes of Cl^- and NO_x. See also NITRATES and EXPLOSIVES.

CHJ750 CAS:54-31-9 ***HR: 3***
4-CHLORO-N-FURFURYL-5-SULFAMOYLANTHRANILIC ACID
mf: $C_{12}H_{11}ClN_2O_5S$ mw: 330.76

SYNS: AISEMIDE ◇ ALUZINE ◇ 5-(AMINOSULFONYL)-4-CHLORO-2-((2-FURNAYLMETHYL)AMINO)BENZOIC ACID ◇ BERONALD ◇ CHLOR-N-(2-FURYLMETHYL)-5-SULFAMYLANTHRANILSAEURE (GERMAN) ◇ 4-CHLORO-N-(2-FURYLMETHYL)-5-SULFAMOYLANTHRANILIC ACID ◇ DESDEMIN ◇ DIURAL ◇ DRYPTAL ◇ ERROLON ◇ EUTENSIN ◇ FRUSEMIDE ◇ FRUSEMIN ◇ FRUSID ◇ FULSIX ◇ FULUVAMIDE ◇ FURANTHRIL ◇ FURANTHRYL ◇ FURANTRIL ◇ FURESIS ◇ FUROSEDON ◇ FUROSEMID ◇ FUROSEMIDE ◇ FUROSEMIDE "MITA" ◇ FURSEMID ◇ FURSEMIDE ◇ FUSID ◇ HYDRO-RAPID ◇ KATLEX ◇ LASEX ◇ LASIX ◇ LB 502 ◇ LOWPSTRON ◇ MACASIROOL ◇ NICOROL ◇ NCI-C55936 ◇ PREFEMIN ◇ PROFEMIN ◇ RADONNA ◇ ROSEMIDE ◇ SALIX ◇ SEGURIL ◇ TRANSIT ◇ TROFURIT ◇ UREX ◇ UROSEMIDE

TOXICITY DATA with REFERENCE
cyt-hmn:leu 200 mg/L/24H MUREAV 66,69,79
cyt-ham:lng 1 g/L ATSUDG (4),41,80
orl-rat TDLo:300 mg/kg (female 16D post):TER TJADAB 34,452,86
orl-mus TDLo:12500 mg/kg (female 6-15D post):REP SEIJBO 24,111,84
orl-wmn TDLo:120 mg/kg/21W-I:SYS JRSMD9 79,239,86
ivn-man TDLo:29 mg/kg:EAR,KID NEJMAG 282,1413,70
ivn-hmn TDLo:1300 μg/kg:CVS AIMEAS 103,1,85
ivn-inf TDLo:1 mg/kg/4H-I:SYS ADCHAK 59,907,84
ivn-wmn TDLo:2500 μg/kg/2M-C:BPR ICMED9 12,54,86
orl-rat LD50:2600 mg/kg TXAPA9 18,185,71
ipr-rat LD50:800 mg/kg APPHAX 42,199,85
ivn-rat LD50:800 mg/kg NIIRDN 6,725,82
orl-mus LD50:2200 mg/kg PCJOAU 19,706,85
ivn-mus LD50:308 mg/kg ARZNAD 14,44,64
orl-dog LD50:2000 mg/kg NIIRDN 6,725,82
orl-rbt LD50:800 mg/kg NIIRDN 6,725,82
ivn-rbt LD50:400 mg/kg NIIRDN 6,725,82

CONSENSUS REPORTS: EPA Genetic Toxicology Program.

SAFETY PROFILE: Poison by intravenous route. Moderately toxic by ingestion and intraperitoneal routes. Human systemic effects by intravenous route: change in the sensitivity of the ear to sound, tinnitus, unspecified effects on the heart, constriction of the arteries, and a decrease in urine volume, interstitial nephritis, metabolic alkalosis, pulse rate decrease, fall in BP. Ingestion can damage the liver. Experimental teratogenic and reproductive effects. Human mutation data reported. When heated to decomposition it emits very toxic fumes of Cl^-, NO_x, and SO_x.

CHK000 CAS:5857-37-4 ***HR: 3***
CHLORO(2-FURYL)MERCURY
mf: C_4H_3ClHgO mw: 303.11

SYNS: CHLORO-2-FURANYL MERCURY ◇ 2-CHLOROMERCURIFURAN ◇ 2-FURYLMERCURIC CHLORIDE ◇ 2-FURYLMERCURY CHLORIDE

TOXICITY DATA with REFERENCE
ipr-mus LDLo:20 mg/kg HBTXAC 5,81,59
ivn-mus LD50:56 mg/kg CSLNX* NX#03269

CONSENSUS REPORTS: Mercury and its compounds are on the Community Right-To-Know List.

OSHA PEL: (Transitional: CL 1 mg/10m^3) CL 0.1 mg(Hg)/m^3 (skin)
ACGIH TLV: TWA 0.1 mg(Hg)/m^3 (skin)
NIOSH REL: (Inorganic Mercury) TWA: 0.05 mg(Hg)/m^3

SAFETY PROFILE: Poison by intravenous and intraperitoneal routes. See also MERCURY COMPOUNDS, ORGANIC. When heated to decomposition it emits very toxic fumes of Hg and Cl^-.

CHK125 CAS:102489-58-7 ***HR: 3***
6'-CHLORO-2-(2-FURYLMETHYL)AMINO-o-ACETOTOLUIDIDE HYDROCHLORIDE
mf: $C_{14}H_{15}ClN_2O_2•ClH$ mw: 315.22

SYN: C 3211

TOXICITY DATA with REFERENCE
eye-rbt 2% MLD ARZNAD 8,407,58
ipr-rat LD50:330 mg/kg ARZNAD 8,407,58
scu-mus LD50:2350 mg/kg ARZNAD 8,407,58

SAFETY PROFILE: Poison by intraperitoneal route. Moderately toxic by subcutaneous route. When heated to decomposition it emits toxic fumes of NO_x and Cl^-.

CHK175 CAS:327-97-9 ***HR: 2***
CHLOROGENIC ACID
mf: $C_{16}H_{18}O_9$ mw: 354.34

SYNS: 3-CAFFEOYLQUINIC ACID ◇ 3-o-CAFFEOYLQUINIC ACID

TOXICITY DATA with REFERENCE
mrc-smc 1 g/L MUREAV 135,109,84
cyt-ham:ovr 250 mg/L MUREAV 111,209,83
ipr-rat TDLo:40 mg/kg (5-12D preg):TER TXAPA9 36,227,76
ipr-rat LDLo:4000 mg/kg TXAPA9 36,337,76

SAFETY PROFILE: Moderately toxic by intraperitoneal route. An experimental teratogen. Mutation data reported. When heated to decomposition it emits acrid smoke and fumes.

CHK250 CAS:13637-65-5 ***HR: 3***
CHLOROGERMANE
mf: $ClGeH_3$ mw: 111.07

SAFETY PROFILE: Reaction with ammonia forms heat-sensitive explosive products. When heated to decomposition it emits toxic fumes of Cl^-. See also GERMANIUM COMPOUNDS.

CHK750 CAS:400-44-2 ***HR: 3***
2-CHLORO-1,1,1,4,4,4-HEXAFLUOROBUTENE-2
mf: C_4HClF_6 mw: 198.50

SYNS: CHFB ◇ 1,1,1,4,5,5-HEXAFLUORO-2-CHLORO-2-BUTENE

TOXICITY DATA with REFERENCE
ihl-hmn TCLo:10 ppm/1H:PUL CENEAR 44,6,66
ihl-rat LC50:3 ppm/6H 34ZIAG -,310,69

SAFETY PROFILE: Poison by inhalation. Human respiratory system effects by inhalation. When heated to decomposition it emits very toxic fumes of F^- and Cl^-. See also CHLORINATED HYDROCARBONS, ALIPHATIC, and FLUORIDES.

CHK825 CAS:73803-48-2 ***HR: D***
endo-4-CHLORO-N-(HEXAHYDRO-4,7-METHANO-ISOINDOL-2-YL)-3-SULFAMOYLBENZAMIDE

SYNS: TDS ◇ N-(4-AZA-endo-TRICYCLO(5.2.1.5^{2,6})-DECAN-4-YL)-4-CHLORO-3-SULFAMOYLBENZAMIDE

TOXICITY DATA with REFERENCE
orl-rat TDLo:5 g/kg (8-17D preg):REP YKRYAH 12,651,79
orl-rat TDLo:40 g/kg (8-17D preg):TER YKRYAH 12,651,79

SAFETY PROFILE: An experimental teratogen. Other experimental reproductive effects. When heated to decomposition it emits toxic fumes of Cl^-, SO_x and NO_x.

CHL000 CAS:73926-88-2 ***HR: 3***
trans-CHLORO(2-HEXANAMIDOCYCLOHEXYL) MERCURY
mf: $C_{12}H_{22}ClHgNO$ mw: 432.39

SYNS: CHLORO(2-HEXANAMIDOCYCLOHEXYL)MERCURY,(E)- ◇ N-(2-CHLOROMERCURICYCLOHEXYL) HEXANAMIDE, (E)-

TOXICITY DATA with REFERENCE
ivn-mus LD50:56 mg/kg CSLNX* NX#04829

CONSENSUS REPORTS: Mercury and its compounds are on the Community Right-To-Know List.

OSHA PEL: (Transitional: CL 1 mg/10m^3) CL 0.1 mg(Hg)/m^3 (skin)
ACGIH TLV: TWA 0.1 mg(Hg)/m^3 (skin)
NIOSH REL: (Inorganic Mercury) TWA: 0.05 mg(Hg)/m^3

SAFETY PROFILE: Poison by intravenous route. See also MERCURY COMPOUNDS, ORGANIC. When heated to decomposition it emits very toxic fumes of Cl^-, NO_x, and Hg.

CHL250 CAS:13654-91-6 ***HR: 3***
CHLOROHEXYL ISOCYANATE
mf: $C_7H_{10}ClNO$ mw: 159.63

SYN: ISOCYANIC ACID-6-CHLOROHEXYL ESTER

TOXICITY DATA with REFERENCE
ihl-rat LCLo:60 mg/m^3/4H GTPZAB 12(10),40,68
ihl-mus LC50:69 mg/m^3/2H GTPZAB 12(10),40,68

SAFETY PROFILE: Poison by inhalation. See also THIOCYANATES and ESTERS. When heated to decomposition it emits very toxic fumes of Cl^- and NO_x.

CHL500 CAS:18979-94-7 ***HR: 3***
4-CHLORO-2-HEXYLPHENOL
mf: $C_{12}H_{17}ClO$ mw: 212.74

SYN: 2-HEXYL-4-CHLOROPHENOL

TOXICITY DATA with REFERENCE
skn-mus TDLo:8400 mg/kg/21W-I:NEO CNREA8 19,413,59

CONSENSUS REPORTS: Chlorophenol compounds are on the Community Right-To-Know List.

SAFETY PROFILE: Questionable carcinogen with experimental neoplastigenic data by skin contact. When heated to decomposition it emits toxic fumes of Cl^- and NO_x. See also CHLOROPHENOLS.

CHL875 CAS:52340-46-2 ***HR: D***
dl-α-CHLOROHYDRIN
mf: $C_3H_7ClO_2$ mw: 110.55

SYNS: (±)-3-CHLORO-1,2-PROPANEDIOL ◇ dl-3-CHLORO-1,2-PROPANEDIOL ◇ (±)-2,3-DIHYDROXYCHLOROPROPANE

TOXICITY DATA with REFERENCE
orl-rat TDLo:50 mg/kg (male 1D pre):REP CCPTAY 13,639,76

SAFETY PROFILE: Experimental reproductive effects. When heated to decomposition it emits toxic fumes of Cl^-.

CHM000 CAS:615-67-8 ***HR: 3***
CHLOROHYDROQUINONE
mf: $C_6H_5ClO_2$ mw: 144.56

TOXICITY DATA with REFERENCE
scu-rat TDLo:5 mg/kg (female 1D pre):REP ENDOAO 57,466,55
orl-rat LDLo:200 mg/kg KODAK* 21MAY71
skn-rat LDLo:500 mg/kg KODAK* 21MAY71
ipr-rat LDLo:100 mg/kg KODAK* 21MAY71

CONSENSUS REPORTS: Reported in EPA TSCA Inventory.

SAFETY PROFILE: Poison by ingestion and intraperitoneal routes. Moderately toxic by skin contact. Experimental reproductive effects. When heated to decomposition it emits toxic fumes of Cl^-. See also CHLORIDES.

CHM500 CAS:13442-11-0 ***HR: 3***
5-CHLORO-4-(HYDROXYAMINO)QUINOLINE-1-OXIDE
mf: $C_9H_7ClN_2O_2$ mw: 210.63

TOXICITY DATA with REFERENCE
scu-mus TDLo:120 mg/kg/50D-I:ETA BCPCA6 16,631,67

SAFETY PROFILE: Questionable carcinogen with experimental tumorigenic data. When heated to decomposition it emits very toxic fumes of Cl^- and NO_x.

CHM750 CAS:14076-05-2 ***HR: 3***
6-CHLORO-4-(HYDROXYAMINO)QUINOLINE-1-OXIDE
mf: $C_9H_7ClN_2O_2$ mw: 210.63

TOXICITY DATA with REFERENCE
scu-mus TDLo:120 mg/kg/50D-I:ETA BCPCA6 16,631,67

SAFETY PROFILE: Questionable carcinogen with experimental tumorigenic data. When heated to decomposition it emits very toxic fumes of Cl^- and NO_x.

CHN000 CAS:13442-12-1 ***HR: 3***
7-CHLORO-4-(HYDROXYAMINO)QUINOLINE-1-OXIDE
mf: $C_9H_7ClN_2O_2$ mw: 210.63

TOXICITY DATA with REFERENCE
scu-mus TDLo:120 mg/kg/50D-I:ETA BCPCA6 16,631,67

SAFETY PROFILE: Questionable carcinogen with experimental tumorigenic data. When heated to decomposition it emits very toxic fumes of Cl^- and NO_x.

CHN500 CAS:92-04-6 ***HR: 2***
3-CHLORO-4-HYDROXYBIPHENYL
mf: $C_{12}H_9ClO$ mw: 204.66

PROP: White flakes. Bp: 322° (decomp), fp: 74.2°, flash p: 345°F, d: <1, mp: 80°.

SYNS: 3-CHLOR-4-HYDROXYBIFENYL (CZECH) ◇ 3-CHLORO-4-HYDROXYDIPHENYL ◇ 2-CHLORO-4-PHENYLPHENOL ◇ DOWICIDE 4 ◇ 4-PHENYL-2-CHLOROPHENOL

TOXICITY DATA with REFERENCE
skn-rbt 500 mg/24H MLD 28ZPAK -,82,72
eye-rbt 50 μg/24H SEV 28ZPAK -,82,72
orl-rat LD50:4220 mg/kg 28ZPAK -,82,72

CONSENSUS REPORTS: Reported in EPA TSCA Inventory. Chlorophenol compounds are on the Community Right-To-Know List.

SAFETY PROFILE: Mildly toxic by ingestion. A severe eye and mild skin irritant. A pesticide. Combustible when exposed to heat or flame. To fight fire, use alcohol foam, CO_2, dry chemical. When heated to decomposition it emits toxic fumes of Cl^-. See also CHLOROPHENOLS.

CHN750 CAS:24579-91-7 ***HR: 2***
CHLORO(2-HYDROXY-3,5-DINITROPHENYL) MERCURY
mf: $C_6H_3ClHgN_2O_5$ mw: 419.15

SYN: 2-(CHLOROMERCURI)-4,6-DINITROPHENOL

TOXICITY DATA with REFERENCE
orl-rat LDLo:500 mg/kg NCNSA6 5,37,53

CONSENSUS REPORTS: Mercury and its compounds are on the Community Right-To-Know List.

OSHA PEL: (Transitional: CL 1 mg/10m³) CL 0.1 mg(Hg)/m³ (skin)
ACGIH TLV: TWA 0.1 mg(Hg)/m³ (skin)
NIOSH REL: (Inorganic Mercury) TWA: 0.05 mg(Hg)/m³

SAFETY PROFILE: Moderately toxic by ingestion. See also MERCURY COMPOUNDS, ORGANIC. When heated to decomposition it emits very toxic fumes of Cl^-, NO_x, and Hg.

CHO125 CAS:94-87-1 ***HR: 3***
2-CHLORO-N-(2-HYDROXYETHYL)ANILINE
mf: $C_8H_{10}ClNO$ mw: 171.63

$ClC_6H_4NHC_2H_4OH$

SAFETY PROFILE: Potentially explosive decomposition above 210°C, catalyzed by the presence of mild steel. This reaction has caused a violent explosion during an industrial scale distillation. Upon decomposition it emits toxic fumes of Cl^-, HCl, and NO_x. Decomposition also produces primary amines, ethylene, methane, carbon monoxide, and carbon dioxide. See also ANILINE DYES.

CHO250 CAS:55477-27-5 ***HR: 3***
4-CHLORO-6-(2-HYDROXYETHYLPIPERAZINO-2-METHYLAMINO-5-METHYLTHIOPYRIMIDINE
mf: $C_{12}H_{20}ClN_5OS$ mw: 317.88

TOXICITY DATA with REFERENCE
orl-mus LD50:525 mg/kg JMCMAR 18,553,75
ivn-mus LD50:124 mg/kg JMCMAR 18,553,75

SAFETY PROFILE: Poison by intravenous route. Moderately toxic by ingestion. When heated to decomposition it emits very toxic fumes of Cl^-, NO_x, and SO_x.

CHO750 CAS:538-04-5 ***HR: 3***
2-CHLORO-4-(HYDROXY MERCURI)PHENOL
mf: $C_6H_5ClHgO_2$ mw: 345.15

PROP: Insol solid. Contains 20% mercury (27ZTAP 3,36,69).

SYNS: (3-CHLORO-4-HYDROXYPHENYL)HYDROXYMERCURY ◇ SEMESAN

CONSENSUS REPORTS: Mercury and its compounds and chlorophenol compounds are on the Community Right-To-Know List.

ACGIH TLV: TWA 0.1 mg(Hg)/m³ (skin)
NIOSH REL: TWA 0.05 mg(Hg)/m³

SAFETY PROFILE: Poison by ingestion, inhalation, and intravenous routes. See also MERCURY COMPOUNDS and CHLOROPHENOLS. When heated to decomposition it emits very toxic fumes of Cl^- and Hg.

CHP250 CAS:303-47-9 ***HR: 3***
(−)-N-((5-CHLORO-8-HYDROXY-3-METHYL-1-OXO-7-ISOCHROMANYL)CARBONYL)-3-PHENYLALANINE
mf: $C_{29}H_{18}ClNO_6$ mw: 403.84

PROP: Crystals. Mp: 169°.

SYNS: (R)N-((5-CHLORO-3,4-DIHYDRO-8-HYDROXY-3-METHYL-1-OXO-1H-2-BENZOPYRAN-7-YL)PHENYLALANINE◇ NCI-C56586 ◇ OCHRATOXIN A

TOXICITY DATA with REFERENCE
cyt-mky:kdy 20 mg/L TXAPA9 32,198,75
orl-mus TDLo:3750 μg/kg (female 15-17D post):REP TJADAB 27,293,83
orl-mus TDLo:3 mg/kg (female 15D post):TER TXAPA9 57,127,81
orl-mus TDLo:2216 mg/kg/44W-C:CAR GANNA2 69,599,78
orl-mus TD:1478 mg/kg/44W-C:NEO MAIKD3 (18),15,83
orl-mus TD:1478 mg/kg/44W-C:ETA GANRAE 30,1445,84
orl-mus TD:3504 mg/kg/2Y-C:CAR JJIND8 75,733,85
orl-rat LD50:20 mg/kg FCTXAV 6,479,68
ipr-rat LD50:12600 μg/kg ARCVBP 5(2),233,74
ivn-rat LD50:12750 μg/kg ARCVBP 5(2),233,74
orl-mus LD50:46 mg/kg TOLED5 25,1,85
ipr-mus LD50:22 mg/kg APTOA6 2,109,46
ivn-mus LD50:25710 μg/kg ARCVBP 5(2),233,74
orl-dog LD50:200 μg/kg CRTXB2 2,499,74
orl-pig LD50:1 mg/kg CRTXB2 2,499,74
orl-ckn LD50:3300 μg/kg APMBAY 21,492,71

CONSENSUS REPORTS: IARC Cancer Review: Group 3 IMEMDT 7,271,87; Animal Limited Evidence IMEMDT 31,191,83; Animal Inadequate Evidence IMEMDT 10,191,76; Human Inadequate Evidence IMEMDT 31,191,83.

SAFETY PROFILE: Poison by ingestion, intraperitoneal, intravenous, and subcutaneous routes. Experimental teratogenic and reproductive effects. Questionable carcinogen with experimental carcinogenic, neoplastigenic, tumorigenic data. Mutation data reported. When heated to decomposition it emits very toxic fumes of Cl^- and NO_x.

CHP375 ***HR: D***
6-CHLORO-17-α-HYDROXY-16-α-METHYL-PREGNA-4,6-DIENE-3,20-DIONE
mf: $C_{24}H_{31}ClO_4$ mw: 419.00

SYN: 6-CHLORO-16-α-METHYL-Δ(6)-DEHYDRO-17-α-ACETOXY-PROGESTERONE

TOXICITY DATA with REFERENCE
scu-rbt TDLo:10 μg/kg (female 1D pre):REP ACEDAB 73,3,63

SAFETY PROFILE: Experimental reproductive effects. When heated to decomposition it emits toxic fumes of Cl^-.

CHP500 CAS:5160-02-1 ***HR: 3***
5-CHLORO-2-((2-HYDROXY-1-NAPHTHYL)AZO)-p-TOLUENE SULFONIC ACID, BARIUM SALT
mf: $C_{17}H_{12}ClN_2O_4S{\cdot}1/2Ba$ mw: 444.49

SYNS: BRIGHT RED ◇ BRILLIANT RED ◇ BRILLIANT SCARLET ◇ BRILLIANT TONER Z ◇ BRONZE RED RO ◇ BRONZE SCARLET ◇ 5-CHLORO-2-((2-HYDROXY-1-NAPHTHALENYL)AZO)-4-METHYLBENZENE SULFONIC ACID, BARIUM SALT (2:1) ◇ 5-CHLORO-2-((2-HYDROXY-1-NAPHTHALENYL)AZO)-4-METHYLBENZENE SULPHONIC ACID, BARIUM SALT ◇ 1-(4-CHLORO-o-SULFO-5-TOLYLAZO)-2-NAPHTHOL,BARIUM SALT ◇ C.I. PIGMENT RED ◇ COSMETIC CORAL RED KO BLUISH ◇ DAINICHI LAKE RED C ◇ D&C RED No. 9 ◇ DESERT RED ◇ ELJON LAKE RED C ◇ HAMILTON RED ◇ HELIO RED TONER LCLL ◇ IRGALITE RED CBN ◇ ISOL LAKE RED LCS 12527 ◇ LAKE RED C ◇ LATEXOL SCARLET R ◇ LD RUBBER RED 16913 ◇ LUTETIA RED CLN ◇ MICROTEX LAKE RED CR ◇ MOHICAN RED A-8008 ◇ NCI-C53792 ◇ No. 3 CONC. SCARLET ◇ PARIDINE RED LCL ◇ PIGMENT RED CD ◇ POTOMAC RED ◇ RECOLITE RED LAKE C ◇ 1860 RED ◇ RED SCARLET ◇ SANYO LAKE RED C ◇ SEGNALE RED LC ◇ SICO LAKE RED 2L ◇ SUPEROL RED C RT-265 ◇ SYMULER LAKE RED C ◇ TERMOSOLIDO RED LCG ◇ TEXAN RED TONER D ◇ TONER LAKE RED C ◇ TRANSPARENT BRONZE SCARLET ◇ VULCAFIX SCARLET R ◇ VULCAN RED LC ◇ VULCOL FAST RED L ◇ WAYNE RED X-2486

TOXICITY DATA with REFERENCE
mmo-sat 1 mg/plate SCIEAS 236,933,87
orl-rat TDLo:130 g/kg/2Y-C:CAR NTPTR* NTP-TR-225,82
orl-rat TD:109 g/kg/2Y-C:NEO FCTOD7 25,619,87

CONSENSUS REPORTS: IARC Cancer Review: Group 3 IMEMDT 7,56,87; Animal Inadequate Evidence IMEMDT 8,107,75; NTP Carcinogenesis Bioassay (feed); Clear Evidence; rat NTPTR* NTP-TR-225,82; No Evidence: mouse NTPTR* NTP-TR-225,82. Reported in EPA TSCA Inventory.

SAFETY PROFILE: Questionable carcinogen with experimental carcinogenic and tumorigenic data. Mutation data reported. When heated to decomposition it emits very toxic fumes of SO_x, NO_x, and Cl^-. See also SULFONATES.

CHP750 CAS:3124-93-4 ***HR: 3***
21-CHLORO-17-HYDROXY-19-NOR-17-α-PREGNA-4,9-DIEN-20-YN-3-ONE
mf: $C_{20}H_{23}ClO_2$ mw: 330.88

SYNS: 17-α-CHLOROETHINYL-17-β-HYDROXYESTRA-4,9-DIEN-3-ONE ◇ 17-α-CHLOROETHYNYL-17-β-HYDROXY-19-NOR-4,9-ANDROSTADIEN-3-ONE ◇ 17-α-CHLOROETHYNLY-19-NOR-4,9-ANDROSTADIEN-17-β-OL-3-ONE ◇ ETHYNERONE ◇ MK 665

TOXICITY DATA with REFERENCE
orl-dog TDLo:1008 mg/kg/4Y-I:ETA JTEHD6 3,179,77

SAFETY PROFILE: Questionable carcinogen with experimental tumorigenic data. When heated to decomposition it emits toxic fumes of Cl^-.

CHQ000 CAS:101652-00-0 ***HR: 3***
7-CHLORO-10-(2-HYDROXY-3-PIPERIDINO-PROPYL)ISOALLOXAZINE SULFATE
mf: $C_{18}H_{20}ClN_5O_3{\cdot}H_2O_4S$ mw: 487.96

TOXICITY DATA with REFERENCE
ipr-rat LD50:38 mg/kg CMTRAG 2,96,61
scu-mus LD50:60 mg/kg CMTRAG 2,96,61
ivn-mus LD50:115 mg/kg CMTRAG 2,96,61

SAFETY PROFILE: Poison by intraperitoneal, subcutaneous and intravenous routes. See also SULFATES. When heated to decomposition it emits very toxic fumes of SO_x, Cl^-, and NO_x.

CHQ250 ***HR: 3***
3-CHLORO-2-HYDROXYPROPYL PERCHLORATE
mf: $C_3H_6Cl_2O_5$ mw: 192.99

$$ClCH_2CH(OH)CH_2OClO_3$$

SAFETY PROFILE: Explodes violently when shaken. Upon decomposition it emits toxic fumes of Cl^-. See also PERCHLORATES.

CHQ500 CAS:637-61-6 ***HR: 3***
4-CHLOROIMINO-2,5-CYCLOHEXADIENE-1-ONE
mf: C_6H_4ClNO mw: 141.56

$$O{:}C_6H_4{:}NCl$$

TOXICITY DATA with REFERENCE
ipr-mus LD50:12 mg/kg JMCMAR 21,11,78

CONSENSUS REPORTS: Reported in EPA TSCA Inventory.

SAFETY PROFILE: Poison by intraperitoneal route.

Explodes on heating. Upon decomposition it emits toxic fumes of Cl^-.

CHQ750 CAS:537-45-1 *HR: 3*
4-CHLOROIMINO-2,6-DIBROMO-2,5-CYCLOHEXADIENE-1-ONE
mf: $C_6H_2Br_2ClNO$ mw: 299.36

HC=CBrCO•CBr=CHC:NCl

TOXICITY DATA with REFERENCE
ipr-mus LD50:63 mg/kg JMCMAR 21,11,78

CONSENSUS REPORTS: Reported in EPA TSCA Inventory.

SAFETY PROFILE: Poison by intraperitoneal route. A storage hazard. May explode at room temperature. Explodes when heated above 50°C. When heated to decomposition it emits very toxic fumes of Br^-, Cl^-, and NO_x.

CHR000 CAS:101-38-2 *HR: 3*
4-CHLOROIMINO-2,6-DICHLORO-2,5-CYCLOHEXADIENE-1-ONE
mf: $C_6H_2Cl_3NO$ mw: 210.44

HC=CClCO•CCl=CHC:NCl

SYNS: 2,6-DICHLOROQUINONE CHLOROIMIDE ◇ N,2,6-TRICHLORO-p-BENZOQUINONEIMINE

TOXICITY DATA with REFERENCE
ipr-mus LD50:20 mg/kg JMCMAR 21,11,78
ivn-mus LD50:56 mg/kg CSLNX* NX#00254

CONSENSUS REPORTS: Reported in EPA TSCA Inventory.

SAFETY PROFILE: Poison by intraperitoneal and intravenous routes. A storage hazard. It may explode at room temperature. When heated to decomposition it emits very toxic fumes of Cl^- and NO_x.

CHR325 CAS:25604-71-1 *HR: 3*
CHLOROIODOACETYLENE
mf: C_2ClI mw: 186.38

SYN: CHLOROIODOETHYNE

SAFETY PROFILE: A very unstable, explosive material. When heated to decomposition it emits toxic fumes of Cl^- and I^-. See also ACETYLENE COMPOUNDS and EXPLOSIVES.

CHR400 CAS:109-71-7 *HR: 3*
3-CHLORO-1-IODOPROPYNE
mf: C_3H_2ClI mw: 200.41

SAFETY PROFILE: Reacts explosively with air when heated to 47°C. When heated to decomposition it emits toxic fumes of Cl^- and I^-. See also CHLORINATED HYDROCARBONS, ALIPHATIC.

CHR500 CAS:130-26-7 *HR: 3*
5-CHLORO-7-IODO-8-QUINOLINOL
mf: C_9H_5ClINO mw: 305.50

SYNS: ALCHLOQUIN ◇ AMEBIL ◇ AMOENOL ◇ BACTOL ◇ BARQUINOL ◇ BUDOFORM ◇ CHINOFORM ◇ 5-CHLOR-7-JOD-8-8HYDROXY-CHINOLIN (GERMAN) ◇ 5-CHLORO-8-HYDROXY-7-IODOQUINOLINE ◇ 5-CHLORO-7-IODO-8-HYDROXYQUINOLINE ◇ CHLOROIODOQUINE ◇ CLIOQUINOL ◇ CLIQUINOL ◇ ECZECIDIN ◇ EMAFORM ◇ ENTERO-BIO FORM ◇ ENTEROQUINOL ◇ ENTEROSEPTOL ◇ ENTERO-VIOFORM ◇ ENTEROZOL ◇ ENTERUM LOCORTEN ◇ ENTROKIN ◇ HI-ENTEROL ◇ HYDRIODIDE-ENTROL ◇ IODENTEROL ◇ IODOCHLORHYDROXYQUINOL ◇ IODOCHLORHYDROXYQUINOLINE ◇ 7-IODO-5-CHLORO-8-HYDROXYQUINOLINE ◇ 7-IODO-5-CHLOROXINE ◇ IODOENTEROL ◇ NIOFORM ◇ QUINAMBICIDE ◇ ROMETIN ◇ VIOFORM ◇ VIOFORM N.N.R.

TOXICITY DATA with REFERENCE
sln-asn 1 g/L MUREAV 26,159,74
dnd-hmn:hla 40 μmol/L ANYAA9 284 525,77
orl-rat TDLo:528 mg/kg (7-17D preg):REP OYYAA2 14,211,77
orl-rat TDLo:1320 mg/kg (7-17D preg):TER OYYAA2 14,211,77
orl-wmn TDLo:11 g/kg/36W:EYE LANCAO 1,1015,72
orl-hmn TDLo:1400 mg/kg/20D-I:BRN JJMCAQ 24,195,71
ipr-rat LD50:3400 mg/kg OYYAA2 14,75,77
orl-mus LD50:69 mg/kg ATSUDG 2,371,79
orl-cat LD50:400 mg/kg AJTMAQ 24,29,44
orl-rbt LDLo:250 mg/kg JAMAAP 100,1658,33
orl-gpg LDLo:175 mg/kg AJTMAQ 24,29,44

CONSENSUS REPORTS: Reported in EPA TSCA Inventory. EPA Genetic Toxicology Program.

SAFETY PROFILE: Poison by ingestion. Moderately toxic by intraperitoneal route. Human systemic effects by ingestion: change in central nervous system electrical function, optic nerve damage, and changes in vision. Experimental teratogenic and reproductive effects. Human mutation data reported. When heated to decomposition it emits very toxic fumes of Cl^-, I^-, and NO_x.

CHR700 CAS:109651-74-3 *HR: 3*
3′-CHLORO-2-ISOBUTYLAMINO-p-ACETOTOLUIDIDE HYDROCHLORIDE
mf: $C_{13}H_{19}ClN_2O•ClH$ mw: 291.25

TOXICITY DATA with REFERENCE
eye-rbt 100 mg MOD JAPMA8 49,80,60
ipr-mus LD50:75 mg/kg JAPMA8 49,80,60
scu-mus LD50:225 mg/kg JAPMA8 49,80,60

SAFETY PROFILE: Poison by subcutaneous and intraperitoneal routes. An eye irritant. When heated to decomposition it emits toxic fumes of NO_x and Cl^-.

CHR750 CAS:102489-59-8 ***HR: 3***
6′-CHLORO-2-(ISOBUTYLAMINO)-o-ACETOTOLUIDIDE HYDROCHLORIDE
mf: $C_{13}H_{19}ClN_2O•ClH$ mw: 291.25

SYN: C 3156

TOXICITY DATA with REFERENCE
ipr-rat LD50:345 mg/kg ARZNAD 8,407,58
ipr-mus LD50:390 mg/kg ARZNAD 8,407,58
scu-mus LD50:700 mg/kg ARZNAD 8,407,58

SAFETY PROFILE: Poison by intraperitoneal route. Moderately toxic by subcutaneous route. When heated to decomposition it emits very toxic fumes of Cl^- and NO_x.

CHR850 CAS:109509-25-3 ***HR: 3***
3′-CHLORO-3-ISOBUTYLAMINO-o-PROPIONOTOLUIDIDE HYDROCHLORIDE
mf: $C_{14}H_{21}ClN_2O•ClH$ mw: 305.28

TOXICITY DATA with REFERENCE
eye-rbt 100 mg MOD JAPMA8 49,80,60
ipr-mus LD50:325 mg/kg JAPMA8 49,80,60
scu-mus LD50:1500 mg/kg JAPMA8 49,80,60

SAFETY PROFILE: Poison by intraperitoneal route. Moderately toxic by subcutaneous route. An eye irritant. When heated to decomposition it emits toxic fumes of NO_x and HCl. See also CHLORINATED HYDROCARBONS, AROMATIC.

CHS250 CAS:4288-84-0 ***HR: D***
1-CHLORO-2-ISOPROPOXY-2-PROPANOL
mf: $C_6H_{13}ClO_2$ mw: 152.64

SYNS: 1-CHLORO-3-(PENTYLOXY)-2-PROPANOL ◇ 2-PROPANOL, 1-CHLORO-3-ISOPROPOXY- ◇ U 25,352

TOXICITY DATA with REFERENCE
orl-rat TDLo:960 mg/kg (male 8D pre):REP JRPFA4 21,263,70

CONSENSUS REPORTS: Reported in EPA TSCA Inventory.

SAFETY PROFILE: Experimental reproductive effects. When heated to decomposition it emits toxic fumes of Cl^-.

CHS500 CAS:1918-16-7 ***HR: 3***
2-CHLORO-N-ISOPROPYLACETANILIDE
mf: $C_{11}H_{14}ClNO$ mw: 211.71

SYNS: BEXTON ◇ CHLORESSIGSAEURE-N-ISOPROPYLANILID (GERMAN) ◇ α-CHLORO-N-ISOPROPYLACETANILIDE ◇ 2-CHLORO-N-ISOPROPYL-N-PHENYLACETAMIDE ◇ 2-CHLORO-N-(1-METHYLETHYL)-N-PHENYLACETAMIDE ◇ CP 31393 ◇ N-ISOPROPYL-α-CHLOROACETANILIDE ◇ N-ISOPROPYL-2-CHLOROACETANILIDE ◇ NITICID ◇ PROPACHLOR ◇ PROPACHLORE ◇ RAMROD ◇ SATECID

TOXICITY DATA with REFERENCE
cyt-mus-unr 10 mg/kg TGANAK 14(6),41,80
cyt-mus-orl 10 mg/kg CYGEDX 14(6),38,80
orl-rat LD50:710 mg/kg 85ARAE 2,63,77
orl-mus LD50:290 mg/kg EKMMA8 13,123,74
orl-rbt LD50:710 mg/kg KHZDAN 17,90,74
skn-rbt LD50:380 mg/kg WRPCA2 9,119,70
orl-dck LD50:512 mg/kg DOEAAH 35,25,79

CONSENSUS REPORTS: EPA Genetic Toxicology Program.

SAFETY PROFILE: Poison by ingestion and skin contact. Mutation data reported. A selective herbicide. When heated to decomposition it emits very toxic fumes of Cl^- and NO_x.

CHS750 CAS:77966-61-1 ***HR: 3***
6′-CHLORO-2-(ISOPROPYLAMINO)-o-ACETOTOLUIDIDE HYDROCHLORIDE
mf: $C_{12}H_{17}ClN_2O•ClH$ mw: 277.22

SYNS: C 3059 ◇ 2′-CHLORO-2-(ISOPROPYLAMINO)-6′-METHYLACETANILIDE HYDROCHLORIDE

TOXICITY DATA with REFERENCE
ipr-rat LD50:330 mg/kg ARZNAD 8,407,58
ipr-mus LD50:275 mg/kg ARZNAD 8,407,58
scu-mus LD50:730 mg/kg ARZNAD 8,407,58

SAFETY PROFILE: Poison by intraperitoneal route. Moderately toxic by subcutaneous route. When heated to decomposition it emits very toxic fumes of NO_x and Cl^-.

CHT500 CAS:20794-96-1 ***HR: 3***
β-CHLORO-N-ISOPROPYL-2-NAPHTHALENEETHYLAMINE HYDROCHLORIDE
mf: $C_{15}H_{18}ClN•ClH$ mw: 284.25

SYNS: 2-(α-CHLORO-β-ISOPROPYLAMINE)ETHYLNAPHTHALENE HYDROCHLORIDE ◇ ICI 42464

TOXICITY DATA with REFERENCE
orl-rat TDLo:19800 mg/kg/38W-C:CAR PSDTAP 10,183,69
orl-mus TDLo:18 g/kg/11W-C:ETA NATUAS 207,594,65

CONSENSUS REPORTS: EPA Genetic Toxicology Program.

SAFETY PROFILE: Questionable carcinogen with experimental carcinogenic and tumorigenic data. When heated to decomposition it emits very toxic fumes of Cl^- and NO_x.

CHU000 CAS:33965-80-9 ***HR: 3***
3-CHLORO-LACTONITRILE
mf: C_3H_4ClNO mw: 105.53

$ClCH_2CH(CN)OH$

SYN: 2-CHLORO-1-CYANOETHANOL

TOXICITY DATA with REFERENCE
ipr-mus LDLo:4 mg/kg CBCCT* 2,241,50
ivn-dog LD50:7 mg/kg CBCCT* 2,299,50
skn-rbt LDLo:25 mg/kg CBCCT* 2,299,50
ivn-rbt LDLo:10 mg/kg CBCCT* 2,299,50

CONSENSUS REPORTS: Cyanide and its compounds are on the Community Right-To-Know List.

SAFETY PROFILE: Poison by skin contact, intravenous, and intraperitoneal routes. Heating above 110°C may cause explosive decomposition to 2-chloroacetaldehyde and HCN. When heated to decomposition it emits very toxic fumes of Cl^-, CN^-, and NO_x. See also NITRILES.

CHU500 CAS:59-85-8 ***HR: 3***
p-CHLOROMERCURIC BENZOIC ACID
mf: $C_7H_5ClHgO_2$ mw: 357.16

SYNS: (p-CARBOXYPHENYL)CHLOROMERCURY ◇ p-(CHLOROMERCURI)BENZOIC ACID ◇ USAF D-3

TOXICITY DATA with REFERENCE
ipr-mus LD50:25 mg/kg NTIS** AD277-689

CONSENSUS REPORTS: Reported in EPA TSCA Inventory. Mercury and its compounds are on the Community Right-To-Know List.

OSHA PEL: (Transitional: CL 1 mg/10m³) CL 0.1 mg(Hg)/m³ (skin)
ACGIH TLV: TWA 0.1 mg(Hg)/m³ (skin)
NIOSH REL: (Inorganic Mercury) TWA: 0.05 mg(Hg)/m³

SAFETY PROFILE: Poison by intraperitoneal route. See also MERCURY COMPOUNDS. When heated to decomposition it emits very toxic fumes of Cl^- and Hg.

CHU750 CAS:73940-90-6 ***HR: 3***
N-(CHLOROMERCURI)FORMANILIDE
mf: $C_7H_6ClHgNO$ mw: 356.18

SYN: CHLORO(N-PHENYLFORMAMIDO)MERCURY

TOXICITY DATA with REFERENCE
ivn-mus LD50:32 mg/kg CSLNX* NX#05982

CONSENSUS REPORTS: Mercury and its compounds are on the Community Right-To-Know List.

OSHA PEL: (Transitional: CL 1 mg/10m³) CL 0.1 mg(Hg)/m³ (skin)
ACGIH TLV: TWA 0.1 mg(Hg)/m³ (skin)
NIOSH REL: (Inorganic Mercury) TWA: 0.05 mg(Hg)/m³

SAFETY PROFILE: Poison by intravenous route. See also MERCURY COMPOUNDS. When heated to decomposition it emits very toxic fumes of Cl^-, NO_x, and Hg.

CHV250 CAS:3477-28-9 ***HR: 3***
3-(3-CHLOROMERCURI-2-METHOXY-1-PROPYL)-5,5-DIMETHYLHYDANTOIN
mf: $C_9H_{15}ClN_2O_3Hg$ mw: 435.21

SYN: CHLORO((3-(5,5-DIMETHYL-2,4-DIOXO-3-IMIDAZOLIDINYL)-2-METHOXY)PROPYL)MERCURY

TOXICITY DATA with REFERENCE
orl-mus LD50:346 mg/kg JMPCAS 5,168,62

CONSENSUS REPORTS: Mercury and its compounds are on the Community Right-To-Know List.

NIOSH REL: (Inorganic Mercury) TWA: 0.05 mg(Hg)/m³

SAFETY PROFILE: Poison by ingestion. See also MERCURY COMPOUNDS. When heated to decomposition it emits very toxic fumes of Cl^-, NO_x and Hg.

CHV500 CAS:3367-32-6 ***HR: 2***
1-(3-CHLOROMERCURI-2-METHOXY)PROPYLHYDANTOIN
mf: $C_7H_{10}ClHgN_2O_3$ mw: 406.23

SYNS: 1-(3-CHLOROMERCURI-2-METHOXY-1-PROPYL)-HYDANTOIN ◇ 1-(3-(CHLOROMERCURY)-2-METHOXYPROPYL)HYDANTOIN

TOXICITY DATA with REFERENCE
orl-mus LD50:1580 mg/kg AIPTAK 149,415,64

CONSENSUS REPORTS: Mercury and its compounds are on the Community Right-To-Know List.

OSHA PEL: (Transitional: CL 1 mg/10m³) CL 0.1 mg(Hg)/m³ (skin)
ACGIH TLV: TWA 0.1 mg(Hg)/m³ (skin)
NIOSH REL: (Inorganic Mercury) TWA: 0.05 mg(Hg)/m³

SAFETY PROFILE: Moderately toxic by ingestion. See also MERCURY COMPOUNDS. When heated to decomposition it emits very toxic fumes of Cl^-, NO_x and Hg.

CHV750 CAS:3367-29-1 ***HR: 3***
3-(3-CHLOROMERCURI-2-METHOXY-1-PROPYL) HYDANTOIN
mf: $C_7H_{11}ClHgN_2O_3$ mw: 407.24

SYN: CHLORO((3-(2,4-DIOXO-3-IMIDAZOLIDINYL)-2-METHOXY) PROPYL)MERCURY

TOXICITY DATA with REFERENCE
orl-mus LD50:358 mg/kg JMPCAS 5,168,62

CONSENSUS REPORTS: Mercury and its compounds are on the Community Right-To-Know List.

OSHA PEL: (Transitional: CL 1 mg/10m^3) CL 0.1 mg(Hg)/m^3 (skin)
ACGIH TLV: TWA 0.1 mg(Hg)/m^3 (skin)
NIOSH REL: (Inorganic Mercury) TWA: 0.05 mg(Hg)/m^3

SAFETY PROFILE: Poison by ingestion. See also MERCURY COMPOUNDS. When heated to decomposition it emits very toxic fumes of Cl^-, NO_x, and Hg.

CHW000 CAS:67465-39-8 ***HR: 3***
1-(3-CHLOROMERCURI-2-METHOXY-1-PROPYL)-3-METHYLHYDANTOIN
mf: $C_8H_{13}ClHgN_2O_3$ mw: 421.27

SYN: CHLORO((3-(2,4-DIOXO-3-METHYL-1-IMIDAZOLIDINYL)-2-METHOXY)PROPYL)MERCURY

TOXICITY DATA with REFERENCE
orl-mus LD50:298 mg/kg JMPCAS 5,168,62

CONSENSUS REPORTS: Mercury and its compounds are on the Community Right-To-Know List.

OSHA PEL: (Transitional: CL 1 mg/10m^3) CL 0.1 mg(Hg)/m^3 (skin)
ACGIH TLV: TWA 0.1 mg(Hg)/m^3 (skin)
NIOSH REL: (Inorganic Mercury) TWA: 0.05 mg(Hg)/m^3

SAFETY PROFILE: Poison by ingestion. See also MERCURY COMPOUNDS. When heated to decomposition it emits very toxic fumes of Cl^-, Hg, and NO_x.

CHW250 CAS:3367-28-0 ***HR: 3***
3-(3-CHLOROMERCURI-2-METHOXY-1-PROPYL)-1-METHYLHYDANTOIN
mf: $C_8H_{13}ClHgN_2O_3$ mw: 421.27

SYNS: CHLORO((3-(2,4-DIOXO-1-METHYL-3-IMIDAZOLIDINYL)-2-METHOXY)PROPYL)MERCURY ◇ 3-(3-(CHLOROMERCURI)-2-METHOXYPROPYL)-1-METHYLHYDANTOIN

TOXICITY DATA with REFERENCE
orl-mus LD50:264 mg/kg JMPCAS 5,168,62

CONSENSUS REPORTS: Mercury and its compounds are on the Community Right-To-Know List.

OSHA PEL: (Transitional: CL 1 mg/10m^3) CL 0.1 mg(Hg)/m^3 (skin)
ACGIH TLV: TWA 0.1 mg(Hg)/m^3
NIOSH REL: (Inorganic Mercury) TWA: 0.05 mg(Hg)/m^3

SAFETY PROFILE: Poison by ingestion. See also MERCURY COMPOUNDS. When heated to decomposition it emits very toxic fumes of Cl^-, NO_x, and Hg.

CHW500 CAS:3367-30-4 ***HR: 3***
5-(3-CHLOROMERCURI-2-METHOXY-1-PROPYL)-3-METHYLHYDANTOIN
mf: $C_7H_{11}ClHgN_2O_3$ mw: 407.24

SYN: CHLORO((3-(2,4-DIOXO-3-METHYL-5-IMIDAZOLIDINYL)-2-METHOXY)PROPYL)MERCURY

TOXICITY DATA with REFERENCE
orl-mus LD50:715 mg/kg JMPCAS 5,168,62

CONSENSUS REPORTS: Mercury and its compounds are on the Community Right-To-Know List.

OSHA PEL: (Transitional: CL 1 mg/10m^3) CL 0.1 mg(Hg)/m^3 (skin)
ACGIH TLV: TWA 0.1 mg(Hg)/m^3 (skin)
NIOSH REL: (Inorganic Mercury) TWA: 0.05 mg(Hg)/m^3

SAFETY PROFILE: A poison. Moderately toxic by ingestion. See also MERCURY COMPOUNDS. When heated to decomposition it emits very toxic fumes of Cl^-, NO_x, and Hg.

CHW675 CAS:90-03-9 ***HR: 3***
o-CHLOROMERCURIPHENOL
mf: C_6H_5ClHgO mw: 329.15

SYNS: CHLORO(o-HYDROXYPHENYL)MERCURY ◇ o-HYDROXYPHENYLMERCURIC CHLORIDE ◇ MERCUFENOL CHLORIDE ◇ MYRINGACAINE DROPS ◇ SALICRESIN FLUID ◇ U-7743

TOXICITY DATA with REFERENCE
orl-rat LDLo:100 mg/kg NCNSA6 5,36,53
ipr-rat LDLo:25 mg/kg NCNSA6 5,36,53
scu-mus LD50:36 mg/kg HBTXAC 5,114,59
ivn-mus LD50:23 mg/kg HBTXAC 5,114,59

CONSENSUS REPORTS: Reported in EPA TSCA Inventory. Mercury and its compounds as well as Chlorophenol compounds are on the Community Right-To-Know List.

OSHA PEL: (Transitional: CL 1 mg/10m^3) CL 0.1 mg(Hg)/m^3 (skin)
ACGIH TLV: TWA 0.1 mg(Hg)/m^3 (skin)
NIOSH REL: (Mercury, Inorganic) TWA 0.05 mg(Hg)/m^3

SAFETY PROFILE: Poison by ingestion, subcutaneous, intravenous, and intraperitoneal routes. An antiseptic. See also MERCURY COMPOUNDS, ORGANIC and CHLOROPHENOLS. When heated to decomposition it emits toxic fumes of Cl^- and Hg.

CHW750 CAS:623-07-4 ***HR: 3***
p-CHLOROMERCURIPHENOL
mf: C_6H_5ClHgO mw: 329.15

SYNS: CHLORO(p-HYDROXYPHENYL)MERCURY ◇ p-(CHLOROMERCURI)PHENOL

TOXICITY DATA with REFERENCE
ipr-rat LDLo:50 mg/kg NCNSA6 5,36,53

CONSENSUS REPORTS: Reported in EPA TSCA Inventory. Mercury and its compounds as well as chlorophenol compounds are on the Community Right-To-Know List.

OSHA PEL: (Transitional: CL 1 mg/10m^3) CL 0.1 mg(Hg)/m^3 (skin)
ACGIH TLV: TWA 0.1 mg(Hg)/m^3 (skin)
NIOSH REL: (Inorganic Mercury) TWA: 0.05 mg(Hg)/m^3

SAFETY PROFILE: Poison by intraperitoneal route. When heated to decomposition it emits very toxic fumes of Cl^- and Hg. See also MERCURY COMPOUNDS, ORGANIC, and CHLOROPHENOLS.

CHX250 CAS:62-37-3 ***HR: 3***
CHLOROMERODRIN
mf: $C_5H_{11}ClHgN_2O_2$ mw: 367.22

SYNS: (3-((AMINOCARBONYL)AMINO)-2-METHOXYPROPYL) CHLOROMERCURY ◇ CHLORMEROPRIN ◇ (3-(CHLOROMERCURI)-2-METHOXYPROPYL)UREA ◇ 1-(3-(CHLOROMERCURI)-2-METHOXY-PROPYL)UREA ◇ CHLOROMERIDIN ◇ CHLOROMERODRIN ◇ DIURONE ◇ HG-203 ◇ KATONIL ◇ MERCLORAN ◇ MERCORAL ◇ MERILID ◇ (2-METHOXYPROPYL)UREA, MERCURY COMPLEX ◇ NEOHYDRIN ◇ ORICUR ◇ PERCAPYL ◇ PROMERAN

TOXICITY DATA with REFERENCE
orl-rat LDLo:82 mg/kg TXAPA9 18,185,71
orl-mus LD50:215 mg/kg AIPTAK 143,181,63
ipr-mus LDLo:63 mg/kg CBCCT* 5,144,53

CONSENSUS REPORTS: Mercury and its compounds are on the Community Right-To-Know List.

OSHA PEL: (Transitional: CL 1 mg/10m^3) CL 0.1 mg(Hg)/m^3 (skin)
ACGIH TLV: TWA 0.1 mg(Hg)/m^3 (skin)
NIOSH REL: (Inorganic Mercury) TWA: 0.05 mg(Hg)/m^3

SAFETY PROFILE: Poison by ingestion and intraperitoneal routes. A diuretic. See also MERCURY COMPOUNDS. When heated to decomposition it emits very toxic fumes of Cl^-, NO_x, and Hg.

CHX750 ***HR: 3***
CHLOROMETHANE mixed with DICHLOROMETHANE
DOT: UN 1912

SYN: METHYL CHLORIDE-METHYLENE CHLORIDE MIXTURE (DOT)

DOT Classification: Flammable Gas; Label: Flammable Gas.

SAFETY PROFILE: Flammable when exposed to heat or flame. See also CHLOROMETHANE and DICHLOROMETHANE. When heated to decomposition it emits toxic fumes of Cl^-.

CHY000 CAS:3518-65-8 ***HR: 3***
CHLOROMETHANE SULFONYL CHLORIDE
mf: $CH_2Cl_2O_2S$ mw: 148.99

SYNS: CHLORID KYSELINY CHLORMETHANSULFONOVE (CZECH) ◇ CHLORMETHANSULFOCHLORID (CZECH)

TOXICITY DATA with REFERENCE
skn-rbt 500 mg/24H SEV 28ZPAK -,198,72
eye-rbt 50 μg/24H SEV 28ZPAK -,198,72
orl-rat LD50:372 mg/kg 28ZPAK -,198,72

SAFETY PROFILE: Poison by ingestion. A severe skin and eye irritant. When heated to decomposition it emits very toxic fumes of Cl^- and SO_x.

CHY250 CAS:148-65-2 ***HR: 3***
CHLOROMETHAPYRILENE
mf: $C_{14}H_{18}ClN_3S$ mw: 295.86

SYNS: CHLOROPYRILENE ◇ CHLOROTHEN ◇ 2-((5-CHLORO-2-THENYL)(2-DIMETHYLAMINOETHYL)AMINO)PYRIDINE ◇ CHLOROTHENYLPYRAMINE ◇ N,N-DIMETHYL-N'-(2-PYRIDYL)-N'-(5-CHLORO-2-THENYL)ETHYLENEDIAMINE ◇ ETHYLENEDIAMINE, N-(5-CHLORO-2-THENYL)-N',N'-DIMETHYL-N-2-PYRIDYL- ◇ NCI-C60559 ◇ PYRITHEN ◇ TAGATHEN ◇ 2-THENYLAMINE, 5-CHLORO-N-(2-(DIMETHYLAMINO)ETHYL)-N-2-PYRIDYL-

TOXICITY DATA with REFERENCE
ipr-mus LD50:105 mg/kg JPETAB 96,388,49

SAFETY PROFILE: Poison by intraperitoneal route. When heated to decomposition it emits very toxic fumes of Cl^-, NO_x, and SO_x.

CHY750 CAS:63074-03-3 ***HR: D***
2-((3-((6-CHLORO-2-METHOXY-9-ACRIDINYL) AMINO))PROPYL)ETHYLAMINOETHANOL DIHYDROCHLORIDE
mf: $C_{21}H_{26}ClN_3O_2 \cdot 2ClH$ mw: 460.87

SYN: ICR 170-OH

TOXICITY DATA with REFERENCE
mmo-sat 100 nmol/plate CRNGDP 3,187,82

mma-sat 100 nmol/plate CRNGDP 3,187,82
msc-ham:ovr 1 µmol/L CNREA8 39,4875,79

SAFETY PROFILE: Mutation data reported. When heated to decomposition it emits very toxic fumes of Cl^- and NO_x.

CIA000 CAS:116-80-3 ***HR: 2***
N-(5-CHLORO-4-METHOXYANTHRAQUINONYL) BENZAMIDE
mf: $C_{22}H_{14}ClNO_4$ mw: 391.82

SYN: 1-BENZOYLAMINO-4-METHOXY-5-CHLORANTHRACHINON (CZECH)

TOXICITY DATA with REFERENCE
eye-rbt 500 mg/24H SEV 28ZPAK -,90,72

SAFETY PROFILE: A severe eye irritant. When heated to decomposition it emits very toxic fumes of Cl^- and NO_x.

CIB500 CAS:5185-84-2 ***HR: 3***
CHLORO(trans-2-METHOXYCYCLOOCTYL) MERCURY
mf: $C_9H_{17}ClHgO$ mw: 377.30

TOXICITY DATA with REFERENCE
ivn-mus LD50:14 mg/kg CSLNX* NX#02812

CONSENSUS REPORTS: Mercury and its compounds are on the Community Right-To-Know List.

OSHA PEL: (Transitional: CL 1 mg/10m^3) CL 0.1 mg(Hg)/m^3 (skin)
ACGIH TLV: TWA 0.1 mg(Hg)/m^3 (skin)
NIOSH REL: (Inorganic Mercury) TWA: 0.05 mg(Hg)/m^3

SAFETY PROFILE: Poison by intravenous route. See also MERCURY COMPOUNDS. When heated to decomposition it emits very toxic fumes of Cl^- and Hg.

CIB625 CAS:4222-27-9 ***HR: 3***
3-CHLORO-3-METHOXYDIAZIRINE
mf: $C_2H_3ClN_2O$ mw: 106.51

SAFETY PROFILE: The liquid is a dangerously unstable explosive. Upon decomposition it emits toxic fumes of Cl^- and NO_x. See also EXPLOSIVES.

CIB700 CAS:91-38-3 ***HR: 3***
4-CHLORO-N-(p-METHOXYPHENYL)ANTHRANILIC ACID
mf: $C_{14}H_{12}ClNO_3$ mw: 277.72

SYN: 5-CHLORO-4-METHOXYDIPHENYLAMINE-2-CARBOXYLIC ACID

TOXICITY DATA with REFERENCE
orl-mus LD50:450 mg/kg QJPPAL 21,10,48
ipr-mus LD50:150 mg/kg QJPPAL 21,10,48
scu-mus LD50:250 mg/kg QJPPAL 21,10,48

SAFETY PROFILE: Poison by subcutaneous and intraperitoneal routes. Moderately toxic by ingestion. When heated to decomposition it emits toxic fumes of Cl^- and NO_x.

CIB725 CAS:4222-26-8 ***HR: 3***
CHLORO-(4-METHOXYPHENYL)DIAZIRINE
mf: $C_8H_7ClN_2O$ mw: 182.61

$CH_3OC_6H_4(Cl)CN{=}N$ (ring)

SAFETY PROFILE: Explodes at room temperature. When heated to decomposition it emits toxic fumes of Cl^- and NO_x.

CIC000 CAS:73926-89-3 ***HR: 3***
CHLORO(2-(3-METHOXYPROPIONAMIDO) CYCLOHEXYL)MERCURY
mf: $C_{10}H_{18}ClHgNO_2$ mw: 420.33

SYN: N-(2-CHLOROMERCURICYCLOHEXYL)PROPIONAMIDE

TOXICITY DATA with REFERENCE
ivn-mus LD50:18 mg/kg CSLNX* NX#04831

CONSENSUS REPORTS: Mercury and its compounds are on the Community Right-To-Know List.

OSHA PEL: (Transitional: CL 1 mg/10m^3) CL 0.1 mg(Hg)/m^3 (skin)
ACGIH TLV: TWA 0.1 mg(Hg)/m^3 (skin)
NIOSH REL: (Inorganic Mercury) TWA: 0.05 mg(Hg)/m^3

SAFETY PROFILE: Poison by intravenous route. See also MERCURY COMPOUNDS. When heated to decomposition it emits very toxic fumes of Cl^-, Hg, and NO_x.

CIC500 CAS:99999-42-5 ***HR: 3***
7-CHLORO-8-METHOXY-10-(2-PYRROLIDINYLETHYL)ISOALLOXAZINE ACETATE
mf: $C_{17}H_{18}ClN_5O_3 \cdot C_2H_4O_2$ mw: 435.91

TOXICITY DATA with REFERENCE
ipr-rat LD50:155 mg/kg CMTRAG 2,96,61
scu-mus LD50:1000 mg/kg CMTRAG 2,96,61
ivn-mus LD50:105 mg/kg CMTRAG 2,96,61

SAFETY PROFILE: Poison by intraperitoneal and intravenous routes. Moderately toxic by subcutaneous route. When heated to decomposition it emits very toxic fumes of Cl^- and NO_x.

CID000 CAS:59177-62-7 **HR: 3**
2-CHLORO-N-METHYL-1-ADAMANTANE METHANAMINE HYDROCHLORIDE
mf: $C_{12}H_{20}ClN{\cdot}ClH$ mw: 250.24

TOXICITY DATA with REFERENCE
orl-mus LD50:300 mg/kg JMCMAR 19,967,76
ipr-mus LD50:150 mg/kg JMCMAR 19,967,76

SAFETY PROFILE: Poison by ingestion and intraperitoneal routes. When heated to decomposition it emits very toxic fumes of Cl^- and NO_x.

CID250 CAS:77966-62-2 **HR: 3**
6'-CHLORO-2-(METHYLAMINO)-o-ACETOTOLUIDIDE HYDROCHLORIDE
mf: $C_{10}H_{13}ClN_2O{\cdot}ClH$ mw: 249.16

SYNS: C 3167 ◇ 2'-CHLORO-6'-METHYL-2-(METHYLAMINO)ACETANILIDE HYDROCHLORIDE

TOXICITY DATA with REFERENCE
eye-rbt 2% MLD ARZNAD 8,407,58
ipr-rat LD50:330 mg/kg ARZNAD 8,407,58
ipr-mus LD50:305 mg/kg ARZNAD 8,407,58
scu-mus LD50:775 mg/kg ARZNAD 8,407,58

SAFETY PROFILE: Poison by intraperitoneal route. Moderately toxic by subcutaneous route. An eye irritant. When heated to decomposition it emits very toxic fumes of Cl^- and NO_x.

CID825 CAS:27683-73-4 **HR: 3**
4'-CHLORO-2-((METHYLAMINO)METHYL) BENZHYDROL HYDROCHLORIDE
mf: $C_{15}H_{16}ClNO{\cdot}ClH$ mw: 298.23

SYNS: α-(4-CHLOROPHENYL)-2-((METHYLAMINO)METHYL)-BENZENEMETHANOL HYDROCHLORIDE (9CI) ◇ PR-F 36 Cl

TOXICITY DATA with REFERENCE
orl-rat LD50:560 mg/kg AIPTAK 211,253,74
ipr-rat LD50:115 mg/kg AIPTAK 211,253,74
scu-rat LD50:328 mg/kg AIPTAK 211,253,74
orl-mus LD50:320 mg/kg AIPTAK 211,253,74
ipr-mus LD50:54 mg/kg AIPTAK 211,253,74
scu-mus LD50:83 mg/kg AIPTAK 211,253,74
ivn-mus LD50:53 mg/kg AIPTAK 211,253,74

SAFETY PROFILE: Poison by ingestion, subcutaneous, intravenous, and intraperitoneal routes. When heated to decomposition it emits toxic fumes of NO_x and Cl^-.

CIE250 CAS:102504-65-4 **HR: 3**
6'-CHLORO-2-(METHYLAMINO)-o-PROPIONOTOLUIDIDE HYDROCHLORIDE
mf: $C_{11}H_{15}ClN_2O{\cdot}ClH$ mw: 263.19

SYN: C 3158

TOXICITY DATA with REFERENCE
eye-rbt 2% MLD ARZNAD 8,544,58
ipr-rat LD50:183 mg/kg ARZNAD 8,544,58
scu-mus LD50:340 mg/kg ARZNAD 8,544,58

SAFETY PROFILE: Poison by intraperitoneal and subcutaneous routes. An eye irritant. When heated to decomposition it emits very toxic fumes of NO_x and Cl^-.

CIE500 CAS:78218-38-9 **HR: 3**
6'-CHLORO-3-(METHYLAMINO)-o-PROPIONOTOLUIDIDE HYDROCHLORIDE
mf: $C_{11}H_{15}ClN_2O{\cdot}ClH$ mw: 263.19

SYN: C 3162

TOXICITY DATA with REFERENCE
eye-rbt 2% MLD ARZNAD 8,544,58
ipr-rat LD50:365 mg/kg ARZNAD 8,544,58
scu-mus LD50:700 mg/kg ARZNAD 8,544,58

SAFETY PROFILE: Poison by intraperitoneal route. Moderately toxic by subcutaneous route. An eye irritant. When heated to decomposition it emits very toxic fumes of Cl^- and NO_x.

CIF000 CAS:52583-06-9 **HR: 3**
2-CHLORO-1-(2-METHYLAMINOPROPYL)-3,5,7-TRIMETHYLADAMANTANE HYDROCHLORIDE
mf: $C_{17}H_{30}ClN{\cdot}ClH$ mw: 320.39

SYN: 1-(2-CHLORO-5,7-DIMETHYL-3-METHYL-1-ADAMANTYL)-N-METHYL-2-PROPYL AMINE HYDROCHLORIDE

TOXICITY DATA with REFERENCE
orl-mus LD50:400 mg/kg JMCMAR 17,602,74
ipr-mus LD50:100 mg/kg JMCMAR 17,602,74

SAFETY PROFILE: Poison by ingestion and intraperitoneal routes. When heated to decomposition it emits very toxic fumes of Cl^- and NO_x.

CIF250 CAS:1199-85-5 **HR: 3**
p-CHLORO-N-METHYLAMPHETAMINE
mf: $C_{10}H_{14}ClN$ mw: 183.70

SYNS: p-CHLORO-N-α-DIMETHYLPHENETHYLAMINE ◇ d-1-p-CHLORO-METHYLAMPHETAMINE (FRENCH) ◇ CMA ◇ pCMA ◇ RO 4-6861 ◇ S-33

TOXICITY DATA with REFERENCE
orl-rat LD50:110 mg/kg THERAP 26,219,71
scu-rat LD50:55 mg/kg AIPTAK 159,442,66
ivn-rat LD50:52 mg/kg THERAP 26,219,71
orl-mus LD50:100 mg/kg THERAP 26,219,71
ipr-mus LD50:20 mg/kg ISYAM* -,729,70
ivn-mus LD50:50 mg/kg THERAP 26,219,71

SAFETY PROFILE: Poison by ingestion, intravenous, intraperitoneal, and subcutaneous routes. When heated to decomposition it emits very toxic fumes of Cl^- and NO_x. See also BENZEDRINE and other amphetamine entries.

CIF750 CAS:73637-11-3 ***HR: 2***
2-CHLOROMETHYL-p-ANISALDEHYDE
mf: $C_9H_9ClO_2$ mw: 184.63

SYN: o-CHLOROMETHYLANISALDEHYDE

TOXICITY DATA with REFERENCE
skn-hmn 200 mg SEV CHMBAY 7,490,71

SAFETY PROFILE: A severe human skin irritant. See also ALDEHYDES. When heated to decomposition it emits toxic fumes of Cl^-.

CIG000 CAS:63018-67-7 ***HR: 3***
5-CHLORO-10-METHYL-1,2-BENZANTHRACENE
mf: $C_{19}H_{13}Cl$ mw: 276.77

SYN: 8-CHLORO-7-METHYLBENZ(a)ANTHRACENE

TOXICITY DATA with REFERENCE
scu-mus TDLo:80 mg/kg:ETA JNCIAM 1,303,40

SAFETY PROFILE: Questionable carcinogen with experimental tumorigenic data. When heated to decomposition it emits toxic fumes of Cl^-. See also CHLORINATED HYDROCARBONS, AROMATIC.

CIG250 CAS:6325-54-8 ***HR: 3***
7-CHLOROMETHYL BENZ(a)ANTHRACENE
mf: $C_{19}H_{13}Cl$ mw: 276.77

SYN: ICR 451

TOXICITY DATA with REFERENCE
mma-sat 1 μg/plate PNASA6 72,5135,75
ivn-mus TDLo:700 μg/kg:NEO CNREA8 36,2423,76
ivn-mus LDLo:1384 μg/kg CNREA8 36,2423,76

SAFETY PROFILE: Poison by intravenous route. Questionable carcinogen with experimental neoplastigenic data. Mutation data reported. When heated to decomposition it emits toxic Cl^-. See also CHLORINATED HYDROCARBONS, AROMATIC.

CIG500 CAS:6366-24-1 ***HR: 3***
7-CHLORO-10-METHYL-1,2-BENZANTHRACENE
mf: $C_{19}H_{13}Cl$ mw: 276.77

SYN: 10-CHLORO-7-METHYLBENZ(a)ANTHRACENE

TOXICITY DATA with REFERENCE
scu-mus TDLo:80 mg/kg:ETA JNCIAM 1,303,40

SAFETY PROFILE: Questionable carcinogen with experimental tumorigenic data. When heated to decomposition it emits toxic fumes of Cl^-. See also CHLORINATED HYDROCARBONS, AROMATIC.

CIG750 CAS:27165-08-8 ***HR: 3***
4-CHLORO-2-METHYLBENZENEDIAZONIUM SALTS
mf: $C_7H_6ClN_2^+X^-$

SAFETY PROFILE: Reaction with sulfides (e.g., hydrogen sulfide, ammonium sulfide sodium hydrogen sulfide, disodium sulfide, or disodium polysulfide), forms explosive products. When heated to decomposition it emits toxic fumes of Cl^- and NO_x.

CIH000 CAS:49852-84-8 ***HR: 3***
6-CHLOROMETHYL BENZO(a)PYRENE
mf: $C_{21}H_{13}Cl$ mw: 300.79

TOXICITY DATA with REFERENCE
mmo-sat 750 ng/plate CBINA8 56,101,85
scu-rat TDLo:100 mg/kg/40D-I:CAR JMCMAR 16,714,73
scu-rat TD:2256 mg/kg/60D-I:NEO CBINA8 29,159,80

SAFETY PROFILE: Questionable carcinogen with experimental carcinogenic and neoplastigenic data. Mutation data reported. When heated to decomposition it emits very toxic fumes of Cl^-. See also CHLORINATED HYDROCARBONS, AROMATIC.

CIH825 CAS:1667-11-4 ***HR: D***
4-CHLOROMETHYLBIPHENYL
mf: $C_{13}H_{11}Cl$ mw: 202.69

SYNS: 4-CMB ◇ p-PHENYLBENZYL CHLORIDE ◇ 4-PHENYLBENZYL CHLORIDE

TOXICITY DATA with REFERENCE
dnd-hmn:fbr 200 μmol/L MUREAV 145,209,85
dnd-hmn:oth 200 μmol/L MUREAV 145,209,85
ipr-mus TDLo:250 mg/kg (5D male):REP MUREAV 100,345,82

SAFETY PROFILE: Experimental reproductive effects. Human mutation data reported. When heated to decomposition it emits toxic fumes of Cl^-. See also CHLORINATED HYDROCARBONS, AROMATIC.

CIH900 CAS:65313-33-9 ***HR: 3***
CHLOROMETHYL BISMUTHINE
mf: C_2H_6BiCl mw: 274.50

SAFETY PROFILE: Ignites spontaneously in air. When heated to decomposition it emits toxic fumes of Cl^- and Bi. See also BISMUTH COMPOUNDS.

CII000 CAS:107-84-6 ***HR: 3***
1-CHLORO-3-METHYLBUTANE
mf: $C_5H_{11}Cl$ mw: 106.60

PROP: Flash p: 16°, mp: −104°; bp: 99°, d: 0.8704 @ 20°/4°, lel: 1.5%, uel: 7.4%.

SAFETY PROFILE: Very dangerous fire hazard when exposed to heat, flame, or powerful oxidizers. Reaction with divalent metals may form reactive products. When heated to decomposition it emits toxic fumes of Cl^-. See also CHLORINATED HYDROCARBONS, ALIPHATIC.

CII250 CAS:594-36-5 ***HR: 2***
2-CHLORO-2-METHYLBUTANE
mf: $C_5H_{11}Cl$ mw: 106.60

PROP: Flash p: 16°, d: 0.8650 @ 20°/4°, mp: −73.7°, bp: 85°, lel: 1.5%, uel: 7.4%.

SAFETY PROFILE: Very dangerous fire hazard when exposed to heat, flame, or powerful oxidizers. Reaction with divalent metals may form very reactive products. When heated to decomposition it emits toxic fumes of Cl^-. See also CHLORINATED HYDROCARBONS, ALIPHATIC.

CIJ250 CAS:20228-97-1 ***HR: 3***
2-CHLORO-6-METHYLCARBANILIC ACID-2-(DIETHYLAMINO)ETHYL ESTER, HYDROCHLORIDE
mf: $C_{14}H_{21}ClN_2O_2{\bullet}ClH$ mw: 321.28

SYN: C 3069

TOXICITY DATA with REFERENCE
eye-rbt 2% MLD ARZNAD 8,664,58
ipr-rat LD50:55 mg/kg ARZNAD 8,664,58
scu-mus LD50:112 mg/kg ARZNAD 8,664,58

SAFETY PROFILE: Poison by intraperitoneal and subcutaneous routes. An eye irritant. See also ESTERS. When heated to decomposition it emits very toxic fumes of Cl^- and NO_x.

CIK250 CAS:33531-34-9 ***HR: 3***
2-CHLORO-6-METHYLCARBANILIC ACID-N-METHYL-4-PIPERIDINYL ESTER
mf: $C_{14}H_{19}ClN_2O_2$ mw: 282.80

TOXICITY DATA with REFERENCE
scu-mus LD50:63 mg/kg JMCMAR 14,710,71
ivn-mus LD50:14 mg/kg JMCMAR 14,710,71

SAFETY PROFILE: Poison by subcutaneous and intravenous routes. See also ESTERS. When heated to decomposition it emits very toxic fumes of Cl^- and NO_x.

CIK500 CAS:77944-89-9 ***HR: 3***
2-CHLORO-6-METHYLCARBANILIC ACID-2-(PYRROLIDINYL)ETHYL ESTER HYDROCHLORIDE
mf: $C_{14}H_{19}ClN_2O_2{\bullet}ClH$ mw: 319.26

SYNS: C 3067 ◇ 2-(PYRROLIDINYL)ETHYL-2-CHLORO-6-METHYLCARBANILATE HYDROCHLORIDE

TOXICITY DATA with REFERENCE
eye-rbt 2% MLD ARZNAD 8,664,58
ipr-rat LD50:72 mg/kg ARZNAD 8,664,58
scu-mus LD50:160 mg/kg ARZNAD 8,664,58
ivn-mus LD50:36 mg/kg ARZNAD 10,475,60

SAFETY PROFILE: Poison by intraperitoneal, subcutaneous, and intravenous routes. An eye irritant. See also ESTERS. When heated to decomposition it emits very toxic fumes of NO_x and Cl^-.

CIK750 CAS:321-54-0 ***HR: 3***
3-CHLORO-4-METHYL-7-COUMARINYL DIETHYLPHOSPHATE
mf: $C_{14}H_{16}ClO_6P$ mw: 346.72

SYNS: COROXON ◇ COUMAPHOS-O-ANALOG ◇ COUMAPHOS OXYGEN ANALOG (USDA) ◇ O,O-DI(2-CHLOROETHYL)-7-(3-CHLORO-4-METHYLCOUMARINYL)PHOSPHATE ◇ O,O-DIETHYL-O-(3-CHLORO-4-METHYLCOUMARIN-7-YL)PHOSPHATE ◇ DIETHYL-3-CHLORO-4-METHYL-7-COUMARINYL PHOSPHATE ◇ PHOSPHORIC ACID, DIETHYL ESTER, with 3-CHLORO-7-HYDROXY-4-METHYLCOUMARIN

TOXICITY DATA with REFERENCE
orl-ckn LD50:2200 μg/kg BCPCA6 16,1183,67

SAFETY PROFILE: Deadly poison by ingestion. When heated to decomposition it emits very toxic fumes of PO_x and Cl^-. See also ESTERS and PHOSPHATES.

CIK825 CAS:4222-21-3 ***HR: 3***
3-CHLORO-3-METHYLDIAZIRINE
mf: $C_2H_3ClN_2$ mw: 90.51

SAFETY PROFILE: A powerful, extremely shock-sensitive explosive. Upon decomposition it emits toxic fumes of Cl^- and NO_x. See also EXPLOSIVES.

CIL000 CAS:102129-02-2 ***HR: 3***
o-CHLORO-2-(METHYL(2-(DIETHYLAMINO) ETHYL)AMINO)PROPIONANILIDE DIHYDROCHLORIDE
mf: $C_{16}H_{26}ClN_3O{\bullet}2ClH$ mw: 389.54

SYN: C 5405

TOXICITY DATA with REFERENCE
ipr-rat LD50:100 mg/kg ARZNAD 9,262,59
scu-mus LD50:700 mg/kg ARZNAD 9,262,59

SAFETY PROFILE: Poison by intraperitoneal route. Moderately toxic by subcutaneous route. When heated to decomposition it emits very toxic fumes of Cl^- and NO_x.

CIL500 CAS:29053-27-8 ***HR: 2***
7-CHLORO-2-METHYL-3,3a-DIHYDRO-2H,9H-ISOXAZOLO(3,2-b)(1,3)BENZOXAZIN-9-ONE
mf: $C_{11}H_9ClNO_3$ mw: 238.66

SYNS: 3,3a-DIHYDRO-7-CHLORO-2-METHYL-2H,9H-ISOXAZOLO(3,2-b)(1,3)BENZOXAZIN-9-ONE ◇ MESECLAZONE ◇ W-2395

TOXICITY DATA with REFERENCE
orl-rat LD50:1160 mg/kg TXAPA9 33,147,75
orl-mus LD50:2250 mg/kg TXAPA9 33,147,75

SAFETY PROFILE: Moderately toxic by ingestion. When heated to decomposition it emits very toxic fumes of NO_x and Cl^-.

CIL700 CAS:63951-11-1 ***HR: 3***
3'-CHLORO-4'-METHYL-4-DIMETHYLAMINO-AZOBENZENE
mf: $C_{15}H_{16}ClN_3$ mw: 273.79

SYNS: p-((3-CHLORO-p-TOLYL)AZO)-N,N-DIMETHYLANILINE ◇ N,N-DIMETHYL-3'-CHLORO-4'-METHYL-4-(PHENYLAZO)-BENZENAMINE

TOXICITY DATA with REFERENCE
mma-sat 250 nmol/plate CNREA8 46,1654,86
dns-rat:lvr 10 μmol/L CNREA8 46,1654,86
orl-rat TDLo:14062 mg/kg/36W-C:CAR CBINA8 53,107,85
orl-rat TD:9774 mg/kg/34W-I:ETA CNREA8 30,1520,70

SAFETY PROFILE: Questionable carcinogen with experimental carcinogenic and tumorigenic data. Mutation data reported. When heated to decomposition it emits toxic fumes of Cl^- and NO_x.

CIL710 CAS:17010-59-2 ***HR: 2***
4'-CHLORO-3'-METHYL-4-DIMETHYLAMINO-AZOBENZENE
mf: $C_{15}H_{16}ClN_3$ mw: 273.79

SYNS: ANILINE, N,N-DIMETHYL-p-(4'-CHLORO-3'-METHYLPHENYLAZO)- ◇ p-((4-CHLORO-m-TOLYL)AZO)-N,N-DIMETHYLANILINE ◇ N,N-DIMETHYL-p-((4-CHLORO-m-TOLYL)AZO)ANILINE

TOXICITY DATA with REFERENCE
orl-rat TDLo:3927 mg/kg/17W-C:CAR CBINA8 53,107,85
orl-rat TD:1027 mg/kg/50D-I:ETA CNREA8 30,1520,70

SAFETY PROFILE: Questionable carcinogen with experimental carcinogenic and tumorigenic data. When heated to decomposition it emits toxic fumes of Cl^- and NO_x.

CIL750 CAS:101651-65-4 ***HR: 3***
o-CHLORO-2-(METHYL(2-(DIMETHYLAMINO)ETHYL)AMINO)ACETANILIDE DIHYDROCHLORIDE
mf: $C_{13}H_{20}ClN_3O \cdot 2ClH$ mw: 342.73

SYN: C 5415

TOXICITY DATA with REFERENCE
ipr-rat LD50:240 mg/kg ARZNAD 9,262,59
scu-mus LD50:1280 mg/kg ARZNAD 9,262,59

SAFETY PROFILE: Poison by intraperitoneal route. Moderately toxic by subcutaneous route. When heated to decomposition it emits very toxic fumes of Cl^- and NO_x.

CIL775 CAS:73639-62-0 ***HR: 2***
4-(CHLOROMETHYL)-2,2-DIMETHYL-1,3-DIOXA-2-SILACYCLOPENTANE
mf: $C_5H_{11}ClO_2Si$ mw: 166.70

SYNS: 2,2-DIMETHYL-4-(CHLOROMETHYL)-1,3-DIOXA-2-SILACYCLOPENTANE ◇ 1,3-DIOXA-2-SILACYCLOPENTANE, 4-(CHLOROMETHYL)-2,2-DIMETHYL- ◇ SOC

TOXICITY DATA with REFERENCE
ipr-rat TDLo:525 mg/kg (male 21D pre):REP CMBID4 29,299,83
ipr-rat LD50:500 mg/kg CMBID4 29,299,83

SAFETY PROFILE: Moderately toxic by intraperitoneal route. Experimental reproductive effects. When heated to decomposition it emits toxic fumes of Cl^-.

CIL800 CAS:4362-40-7 ***HR: 3***
4-(CHLOROMETHYL)-2,2-DIMETHYL-1,3-DIOXOLANE
mf: $C_6H_{11}ClO_2$ mw: 150.62

TOXICITY DATA with REFERENCE
orl-rat TDLo:140 mg/kg (14D male):REP CCPTAY 9,451,74
orl-rat LD50:115 mg/kg CCPTAY 9,451,74

CONSENSUS REPORTS: Reported in EPA TSCA Inventory.

SAFETY PROFILE: Poison by ingestion. Experimental reproductive effects. When heated to decomposition it emits toxic fumes of Cl^-.

CIL850 CAS:53460-80-3 ***HR: 3***
1-(4-CHLOROMETHYL-1,3-DIOXOLAN-2-YL)-2-PROPANONE
mf: $C_7H_{11}ClO_3$ mw: 178.63

SYN: 2-PROPANONE,1-(4-(CHLOROMETHYL)-1,3-DIOXOLAN-2-YL)-

TOXICITY DATA with REFERENCE
orl-rat TDLo:140 mg/kg (male 14D pre):REP CCPTAY 9,451,74
orl-rat LD50:120 mg/kg CCPTAY 9,451,74

SAFETY PROFILE: Poison by ingestion. Experimental reproductive effects. When heated to decomposition it emits toxic fumes of Cl^-.

CIL900 CAS:869-50-1 ***HR: 3***
1-CHLOROMETHYL-1,2-ETHANEDIOL DIACETATE
mf: $C_7H_{11}ClO_4$ mw: 194.63

SYNS: ACETIC ACID, 3-CHLOROPROPYLENE ESTER ◇ CHLORODEOXYGLYCEROL DIACETATE ◇ α-CHLOROHYDRIN DIACETATE ◇ 1-CHLORO-2,3-PROPANEDIOL DIACETATE ◇ 1,2-DIACETOXY-3-CHLOROPROPANE ◇ 1,2-PROPANEDIOL, 3-CHLORO-, DIACETATE

TOXICITY DATA with REFERENCE
orl-rat TDLo:77500 μg/kg (male 31D pre):REP JRPFA4 30,117,72
ipr-mus LD50:340 mg/kg JMCMAR 20,644,77

SAFETY PROFILE: Poison by intraperitoneal route. Experimental reproductive effects. When heated to decomposition it emits toxic fumes of Cl^-.

CIM000 CAS:3188-13-4 ***HR: 3***
CHLOROMETHYL ETHYL ETHER
DOT: UN 2354
mf: C_3H_7ClO mw: 94.54

PROP: Flash p: < −2.2°F.

SYNS: CHLOROMETHOXY ETHANE ◇ ETHOXY CHLOROMETHANE ◇ ETHOXY METHYL CHLORIDE

CONSENSUS REPORTS: Reported in EPA TSCA Inventory.

DOT Classification: Flammable Liquid and Poison; Label: Flammable Liquid and Poison.

SAFETY PROFILE: A poison by inhalation and ingestion. A very dangerous fire and explosion hazard when exposed to heat or flame. See also ETHERS.

CIM300 ***HR: 3***
3-CHLOROMETHYLFURAN
mf: C_5H_5ClO mw: 116.55

SAFETY PROFILE: A storage hazard. It may explode spontaneously at room temperature. Violent reaction with lithium aluminum hydride, ethylacetate. When heated to decomposition it emits toxic fumes of Cl^-.

CIM399 ***HR: 3***
N-CHLORO-4-METHYL-2-IMIDAZOLINONE
mf: $C_4H_5ClN_2O$ mw: 132.55

ClNCO•NHCCH$_3$=CH (ring)

SAFETY PROFILE: Explodes spontaneously at room temperature. Upon decomposition it emits toxic fumes of Cl^- and NO_x.

CIM500 ***HR: 3***
4-CHLORO-1-METHYLIMIDAZOLIUM NITRATE
mf: $C_4H_6ClN_3O_3$ mw: 179.57

ClC=CHN$^+$(CH$_3$)CH=NNO$_3$$^-$ (ring)

SAFETY PROFILE: Thermally unstable and can decompose violently. Upon decomposition it emits toxic fumes of Cl^- and NO_x.

CIN000 CAS:16781-80-9 ***HR: 3***
α-(CHLOROMETHYL)-5-IODO-2-METHYL-4-NITROIMIDAZOLE-2-ETHANOL
mf: $C_7H_9ClIN_3O_3$ mw: 345.54

SYN: 1-(3-CHLORO-2-HYDROXYPROPYL)-5-IODO-2-METHYL-4-NITROIMIDAZOLE

TOXICITY DATA with REFERENCE
orl-mus LD50:901 mg/kg JMCMAR 17,1019,74
ipr-mus LD50:354 mg/kg JMCMAR 17,1019,74

SAFETY PROFILE: Poison by intraperitoneal route. Moderately toxic by ingestion. When heated to decomposition it emits very toxic fumes of I^-, Cl^-, and NO_x.

CIN500 CAS:25148-26-9 ***HR: 3***
10-CHLOROMETHYL-9-METHYLANTHRACENE
mf: $C_{16}H_{13}Cl$ mw: 240.74

SYN: ICR 433

TOXICITY DATA with REFERENCE
mma-sat 5 μg/plate PNASA6 72,5135,75
ivn-mus TDLo:2400 μg/kg:NEO CNREA8 36,2423,76

SAFETY PROFILE: Questionable carcinogen with experimental neoplastigenic data. Mutation data reported. When heated to decomposition it emits toxic fumes of Cl^-. See also CHLORINATED HYDROCARBONS, AROMATIC.

CIN750 CAS:13345-62-5 ***HR: 3***
7-CHLOROMETHYL-12-METHYL BENZ(a)ANTHRACENE
mf: $C_{20}H_{15}Cl$ mw: 290.80

SYN: IRC 453

TOXICITY DATA with REFERENCE
mma-sat 1 μg/plate PNASA6 72,5135,75
scu-rat TDLo:150 mg/kg/39D-I:NEO CNREA8 31,1951,71
ivn-mus LDLo:2 mg/kg CNREA8 36,2423,76

CONSENSUS REPORTS: EPA Genetic Toxicology Program.

SAFETY PROFILE: Poison by intravenous route. Questionable carcinogen with experimental neoplastigenic data. Mutation data reported. When heated to decomposition it emits toxic Cl^-. See also CHLORINATED HYDROCARBONS, AROMATIC.

CIO000 CAS:2212-10-4 ***HR: 2***
CHLOROMETHYLMETHYLDIETHOXY SILANE
mf: $C_6H_{15}ClO_2Si$ mw: 182.75

SYN: CHLORMETHYL-METHYL-DIETHOXYSILAN(CZECH)

TOXICITY DATA with REFERENCE
skn-rbt 500 mg/24H MOD 28ZPAK -,218,72
eye-rbt 100 mg/24H SEV 28ZPAK -,218,72
orl-rat LD50:1300 mg/kg 28ZPAK -,218,72
ihl-rat LCLo:2270 ppm/4H 28ZPAK -,218,72

CONSENSUS REPORTS: Reported in EPA TSCA Inventory.

SAFETY PROFILE: Moderately toxic by ingestion and inhalation. A skin and severe eye irritant. When heated to decomposition it emits toxic fumes of Cl^-. See also SILANE.

CIO250 CAS:107-30-2 ***HR: 3***
CHLOROMETHYL METHYL ETHER
DOT: UN 1239
mf: C_2H_5ClO mw: 80.52

$ClCH_2OCH_3$

PROP: Flash p: <73.4°F.

SYNS: CHLORDIMETHYLETHER (CZECH) ◇ CMME ◇ DIMETHYLCHLOROETHER ◇ ETHER METHYLIQUE MONOCHLORE (FRENCH) ◇ METHYLCHLOROMETHYL ETHER ◇ METHYLCHLOROMETHYL ETHER (DOT) ◇ METHYL CHLOROMETHYL ETHER, anhydrous (DOT) ◇ MONOCHLORODIMETHYL ETHER (MAK) ◇ RCRA WASTE NUMBER U046

TOXICITY DATA with REFERENCE
dni-hmn:lym 5 mL/L CALEDQ 13,213,81
ihl-rat TCLo:1 ppm/6H/72W:ETA AEHLAU 30,70,75
scu-mus TDLo:312 mg/kg/26W-I:NEO JNCIAM 46,143,71
orl-rat LD50:817 mg/kg 28ZPAK -,84,72
ihl-rat LC50:55 ppm/7H AEHLAU 30,61,75
ihl-mus LC50:1030 mg/m^3/2H 85GMAT -,89,82
ihl-ham LC50:65 ppm/7H AEHLAU 30,61,75

CONSENSUS REPORTS: NTP Fifth Annual Report on Carcinogens. IARC Cancer Review: Group 1 IMEMDT 7,131,87; Animal Sufficient Evidence IMEMDT 4,239,74; Human Limited Evidence IMEMDT 4,239,74. EPA Genetic Toxicology Program. Reported in EPA TSCA Inventory. Community Right-To-Know List. EPA Extremely Hazardous Substances List.

OSHA: Cancer Suspect Agent
ACGIH TLV: Suspected Human Carcinogen.
DFG MAK: Human Carcinogen.
NIOSH REL: (Methyl Chloromethyl Ether) TWA use 29 CFR 1910.1006
DOT Classification: Flammable Liquid; Label: Flammable Liquid, Poison, anhydrous; IMO: Flammable Liquid; Label: Flammable Liquid.

SAFETY PROFILE: Confirmed human carcinogen with experimental carcinogenic, tumorigenic, and neoplastigenic data. Poison by inhalation. Moderately toxic by ingestion. Human mutation data reported. A very dangerous fire hazard when exposed to heat or flame. To fight fire, use alcohol foam, water, CO_2, or dry chemical. Reaction with divalent metals forms a very reactive product. When heated to decomposition it emits toxic fumes of Cl^-. See also ETHERS and CHLORINATED HYDROCARBONS, ALIPHATIC.

CIO275 CAS:52157-57-0 ***HR: 3***
2-CHLOROMETHYL-5-METHYLFURAN
mf: C_6H_7ClO mw: 130.57

$OC(CH_3)=CHCH=CH_2Cl$

SAFETY PROFILE: A storage hazard. It may explode spontaneously at room temperature. Less stable than 2-chloromethylfuran. When heated to decomposition it emits toxic fumes of Cl^-. See also CHLORINATED HYDROCARBONS, AROMATIC.

CIO375 CAS:36236-73-4 ***HR: D***
4-(CHLOROMETHYL)-2-METHYL-2-PENTYL-1,3-DIOXOLANE
mf: $C_{10}H_{19}ClO_2$ mw: 206.74

SYN: AY-22,352

TOXICITY DATA with REFERENCE
orl-rat TDLo:50 mg/kg (20D male):REP JRPFA4 30,117,72

SAFETY PROFILE: Experimental reproductive effects. When heated to decomposition it emits toxic fumes of Cl^-.

CIO500 CAS:67293-64-5 ***HR: 3***
2-CHLORO-10-((2-METHYL-3-(4-METHYL-1-PIPERAZINYL)PROPYL-PHENOTHIAZINE
mf: $C_{21}H_{26}ClN_3S$ mw: 388.01

SYN: 6710 RP

TOXICITY DATA with REFERENCE
orl-mus LD50:430 mg/kg CRSBAW 152,1371,58
ipr-mus LD50:120 mg/kg CRSBAW 152,1371,58
scu-mus LD50:420 mg/kg CRSBAW 152,1371,58
ivn-mus LD50:95 mg/kg CRSBAW 152,1371,58

SAFETY PROFILE: Poison by intraperitoneal and in-

travenous routes. Moderately toxic by ingestion and subcutaneous routes. When heated to decomposition it emits very toxic fumes of Cl^-, NO_x, and SO_x.

CIO750 CAS:50308-83-3 ***HR: 3***
7-CHLORO-1-METHYL-4-(p-((1-METHYLPYRIDINIUM-4-YL)AMINO)PHENYL)CARBAMOYL)ANILINO)QUINOLINIUM DIBROMIDE
mf: $C_{29}H_{26}ClN_5O{\cdot}2Br$ mw: 655.87

TOXICITY DATA with REFERENCE
dnd-mus:lym 1 μmol/L JMCMAR 22,134,79
ipr-mus LD10:39 mg/kg JMCMAR 22,134,79

SAFETY PROFILE: Poison by intraperitoneal route. Mutation data reported. When heated to decomposition it emits very toxic fumes of Br^-, NO_x, and Cl^-.

CIP000 CAS:50308-82-2 ***HR: 3***
6-CHLORO-1-METHYL-4-(p-((p-((1-METHYLPYRIDINIUM-4-YL)AMINO)PHENYL)CARBAMOYL)ANILINO)QUINOLINIUM, DI-p-TOLUENESULFONATE
mf: $C_{29}H_{26}ClN_5O{\cdot}2C_7H_7O_3S$ mw: 838.45

TOXICITY DATA with REFERENCE
dnd-mus:lym 790 nmol/L JMCMAR 22,134,79
ipr-mus LD10:40 mg/kg JMCMAR 22,134,79

SAFETY PROFILE: Poison by intraperitoneal route. Mutation data reported. When heated to decomposition it emits very toxic fumes of SO_x, NO_x and Cl^-.

CIP250 CAS:50308-84-4 ***HR: 3***
8-CHLORO-1-METHYL-4-(p-((p-((1-METHYLPYRIDINIUM-4-YL)AMINO)PHENYL)CARBAMOYL)ANILINO)QUINOLINIUM) DI-p-TOLUENESULFONATE
mf: $C_{29}H_{26}ClN_5O{\cdot}2C_7H_7O_3S$ mw: 838.45

TOXICITY DATA with REFERENCE
dnd-mus:lym 1100 nmol/L JMCMAR 22,134,79
ipr-mus LD10:55 mg/kg JMCMAR 22,134,79

SAFETY PROFILE: Poison by intraperitoneal route. Mutation data reported. See also SULFONATES. When heated to decomposition it emits very toxic fumes of SO_x, NO_x, and Cl^-.

CIP500 CAS:3688-85-5 ***HR: 2***
4-CHLORO-N-METHYL-3-(METHYLSULFAMOYL)BENZAMIDE
mf: $C_9H_{11}ClN_2O_3S$ mw: 262.73

SYNS: 4-CHLORO-N-METHYL-3-((METHYLAMINO)SULFONYL)BENZAMIDE ◇ C.I. 456 ◇ CN-36337 ◇ D 1593 ◇ DIAPAMIDE ◇ THIAMIZIDE ◇ TIAM!ZID ◇ TIAMIZIDE ◇ VECTREN

TOXICITY DATA with REFERENCE
orl-rat LD50:1400 mg/kg JNDRAK 3,302,63
orl-mus LD50:2580 mg/kg JNDRAK 3,302,63
ipr-mus LD50:520 mg/kg JNDRAK 3,302,63

SAFETY PROFILE: Moderately toxic by ingestion and intraperitoneal routes. When heated to decomposition it emits very toxic SO_x, NO_x, and Cl^-.

CIP750 CAS:86-52-2 ***HR: 2***
1-CHLOROMETHYL NAPHTHALENE
mf: $C_{11}H_9Cl$ mw: 176.65

PROP: Oily liquid, volatile with steam. Flash p: 270°F (OC), autoign temp: 1036°F, d: 1.19382 @ 20°/4°, mp: –20°, bp: 259°. Sol in benzene, petr ether, alc; insol in water.

SYN: α-CHLOROMETHYLNAPHTHALENE

TOXICITY DATA with REFERENCE
orl-rat LD50:890 mg/kg IHFCAY 6,1,67
skn-rbt LD50:2000 mg/kg IHFCAY 6,1,67

CONSENSUS REPORTS: Reported in EPA TSCA Inventory.

SAFETY PROFILE: Moderately toxic ingestion and skin contact. See also CHLORINATED HYDROCARBONS, AROMATIC. Combustible when exposed to heat or flame. To fight fire, use dry chemical, spray or mist, CO_2. When heated to decomposition it emits toxic fumes of Cl^-.

CIQ000 CAS:64059-42-3 ***HR: 2***
2-(8-CHLOROMETHYL-1-NAPHTHYLTHIO)ACETIC ACID
mf: $C_{13}H_{11}ClO_2S$ mw: 266.75

SYN: KYSELINA-S-(8-CHLORMETHYL-1-NAFTYL)THIOGLYKOLOVA (CZECH)

TOXICITY DATA with REFERENCE
eye-rbt 5 mg/24H SEV 28ZPAK -,173,72
orl-rat LDLo:2500 mg/kg 28ZPAK -,173,72

SAFETY PROFILE: Moderately toxic by ingestion. A severe eye irritant. When heated to decomposition it emits very toxic fumes of Cl^- and SO_x.

CIQ250 CAS:67292-88-0 ***HR: 3***
α-(CHLOROMETHYL)-2-NITROIMIDAZOLE-2-ETHANOL
mf: $C_6H_8ClN_3O_3$ mw: 205.62

SYN: 1-(3-CHLORO-2-HYDROXYPROPYL)-2-NITROIMIDAZOLE

TOXICITY DATA with REFERENCE
orl-mus LD50:330 mg/kg JMCMAR 17,1019,74
ipr-mus LD50:158 mg/kg JMCMAR 17,1019,74

SAFETY PROFILE: Poison by ingestion and intraperi-

toneal routes. When heated to decomposition it emits very toxic fumes of Cl^- and NO_x.

CIQ400 CAS:53460-81-4 ***HR: 3***
4-(CHLOROMETHYL)-2-(o-NITROPHENYL)-1,3-DIOXOLANE
mf: $C_{10}H_{10}ClNO_4$ mw: 243.66

SYN: 1,3-DIOXOLANE,4-(CHLOROMETHYL)-2-(o-NITROPHENYL)-

TOXICITY DATA with REFERENCE
orl-rat TDLo:280 mg/kg (male 14D pre):REP CCPTAY 9,451,74
orl-rat LD50:65 mg/kg CCPTAY 9,451,74

SAFETY PROFILE: Poison by ingestion. Experimental reproductive effects. When heated to decomposition it emits toxic fumes of NO_x and Cl^-.

CIQ500 CAS:16339-16-5 ***HR: 3***
2-CHLORO-N-METHYL-N-NITROSOETHYLAMINE
mf: $C_3H_7ClN_2O$ mw: 122.57

SYNS: 2-CHLORO-2-METHYL-N-NITROSOETHANAMINE ◇ METHYL-2-CHLORAETHYLNITROSAMIN (GERMAN) ◇ METHYL(2-CHLOROETHYL)NITROSAMINE ◇ N-NITROSOMETHYL-2-CHLOROETHYLAMINE

TOXICITY DATA with REFERENCE
orl-rat TDLo:111 mg/kg/53W-C:ETA ZEKBAI 69,103,67
orl-rat LD50:22 mg/kg ZEKBAI 69,103,67
ivn-rat LD50:22 mg/kg ZEKBAI 69,103,67
unr-mam LD50:22 mg/kg GMCRDC 17,107,75

SAFETY PROFILE: Poison by ingestion, intravenous, and possibly other routes. Questionable carcinogen with experimental tumorigenic data. Many nitrosamine compounds are carcinogens. When heated to decomposition it emits very toxic fumes of Cl^- and NO_x. See also NITROSAMINES.

CIQ625 CAS:25480-76-6 ***HR: 3***
N-CHLORO-5-METHYL-2-OXAZOLIDINONE
mf: $C_4H_6ClNO_2$ mw: 135.55

$OCO{\bullet}NClCH_2CHCH_3$ (ring)

SAFETY PROFILE: Potentially explosive above 160°C. When heated to decomposition it emits toxic fumes of Cl^- and NO_x.

CIR000 ***HR: 2***
4-CHLORO-2-METHYLPHENOL
mf: C_7H_7ClO mw: 142.59

$Cl(CH_3)C_6H_3OH$

CONSENSUS REPORTS: Chlorophenol compounds are on the Community Right-To-Know List.

SAFETY PROFILE: Very exothermic reaction with concentrated sodium hydroxide; releases explosive fumes. When heated to decomposition it emits toxic fumes of Cl^-. See also CHLOROPHENOLS.

CIR250 CAS:94-74-6 ***HR: 3***
(4-CHLORO-2-METHYLPHENOXY)ACETIC ACID
mf: $C_9H_9ClO_3$ mw: 200.63

SYNS: AGRITOX ◇ AGROXONE ◇ ANICON KOMBI ◇ ANICON M ◇ BH MCPA ◇ BORDERMASTER ◇ BROMINAL M & PLUS ◇ B-SELEKTONON M ◇ CHIPTOX ◇ 4-CHLORO-o-CRESOXYACETIC ACID ◇ 4-CHLORO-o-TOLOXYACETIC ACID ◇ ((4-CHLORO-o-TOLYL)OXY)ACETIC ACID ◇ CHWASTOX ◇ CORNOX-M ◇ DED-WEED ◇ DICOPUR-M ◇ DICOTEX ◇ DOW MCP AMINE WEED KILLER ◇ EMCEPAN ◇ EMPAL ◇ HEDAPUR M 52 ◇ HERBICIDE M ◇ HORMOTUHO ◇ 4K-2M ◇ KILSEM ◇ KREZONE ◇ LEGUMEX DB ◇ LEUNA M ◇ LEYSPRAY ◇ LINORMONE ◇ M 40 ◇ 2M-4C ◇ MCP ◇ MCPA ◇ MEPHANAC ◇ METAXON ◇ METHOXONE ◇ 2-METHYL-4-CHLOROPHENOXYACETIC ACID ◇ 2-METHYL-4-CHLORPHENOXYESSIGSAEURE (GERMAN) ◇ 2M-4KH ◇ NETAZOL ◇ OKULTIN M ◇ PHENOXYLENE SUPER ◇ RAPHONE ◇ RAZOL DOCK KILLER ◇ RHOMENE ◇ RHONOX ◇ SEPPIC MMD ◇ SHAMROX ◇ SOVIET TECHNICAL HERBICIDE 2M-4C ◇ TRASAN ◇ U 46 M-FLUID ◇ USTINEX ◇ VACATE ◇ VERDONE ◇ VESAKONTUHO MCPA ◇ WEEDAR MCPA CONCENTRATE ◇ WEEDONE MCPA ESTER ◇ WEED-RHAP ◇ ZELAN

TOXICITY DATA with REFERENCE
sln-dmg-orl 5 mmol/L EXPEAM 30,621,74
mmo-smc 30 μmol/L/3H MUREAV 60,291,79
dns-mus-orl 200 mg/kg MUREAV 55,197,78
hma-mus/sat 200 mg/kg ECBUDQ 27,182,78
sce-ham:ovr 10 μmol/L/1H CRNGDP 5,703,84
orl-mus TDLo:1 g/kg (6-15D preg):TER ARZNAD 33,1479,83
orl-mus TDLo:3 g/kg (6-15D preg):REP ARZNAD 33,1479,83
orl-man LDLo:814 mg/kg:CNS,CVS BMJOAE 2,629,65
orl-rat LD50:700 mg/kg AJVRAH 15,622,54
orl-mus LD50:439 mg/kg RPZHAW 31,373,80
scu-mus LDLo:28 mg/kg PCOC** -,711,66
ivn-mus LD50:28 mg/kg PCOC** -,711,66

CONSENSUS REPORTS: IARC Cancer Review: Group 2B IMEMDT 7,156,87; Animal Inadequate Evidence IMEMDT 30,255,83; Human Inadequate Evidence IMEMDT 30,255,83; Human Limited Evidence IMEMDT 41,357,86. Reported in EPA TSCA Inventory. EPA Genetic Toxicology Program.

SAFETY PROFILE: Suspected carcinogen. Poison by subcutaneous and intravenous routes. Moderately toxic by ingestion. Human systemic effects by ingestion: blood pressure decrease and coma. Experimental teratogenic and reproductive effects. Mutation data reported. An herbicide. When heated to decomposition it emits toxic fumes of Cl^-.

CIR325 CAS:16484-77-8 ***HR: 2***
2-(4-CHLORO-2-METHYLPHENOXY)PROPANOIC ACID (R) (9CI)
mf: $C_{10}H_{11}ClO_3$ mw: 214.66

SYN: 2M-4XP

TOXICITY DATA with REFERENCE
orl-mus TDLo:3 g/kg (6-15D preg):TER ARZNAD 33,1479,83
orl-mus TDLo:5 g/kg (6-15D preg):REP ARZNAD 33,1479,83
orl-rat LD50:1050 mg/kg FMCHA2 -,C185,89
ipr-rat LD50:680 mg/kg ZDBEA9 23(11),83,77

SAFETY PROFILE: Moderately toxic by ingestion and intraperitoneal route. An experimental teratogen. Experimental reproductive effects. When heated to decomposition it emits toxic fumes of Cl^-.

CIR500 CAS:93-65-2 ***HR: 3***
4-CHLORO-2-METHYLPHENOXY-α-PROPIONIC ACID
mf: $C_{10}H_{11}ClO_3$ mw: 214.66

SYNS: ACIDE2-(4-CHLORO-2-METHYL-PHENOXY)PROPIONIQUE (FRENCH) ◇ ACIDO 2-(4-CLORO-2-METIL-FENOSSI)-PROPIONICO (ITALIAN) ◇ BH MECOPROP ◇ CHIPCO TURF HERBICIDE MCPP ◇ 2-(4-CHLOOR-2-METHYL-FENOXY)-PROPIONZUUR (DUTCH) ◇ 2-(4-CHLOR-2-METHYL-PHENOXY)-PROPIONSAEURE (GERMAN) ◇ 2-(4-CHLORO-2-METHYLPHENOXY)PROPIONIC ACID ◇ (+)-α-(4-CHLORO-2-METHYLPHENOXY) PROPIONIC ACID ◇ 2-(4-CHLORO-PHENOXY-2-METHYL)PROPIONIC ACID ◇ 2-(p-CHLORO-o-TOLYLOXY)PROPIONIC ACID ◇ CMPP ◇ COMPITOX ◇ FBC CMPP ◇ HEDONAL MCPP ◇ ISO-CORNOX ◇ KILPROP ◇ LIRANOX ◇ 2M-4CP ◇ MCPP ◇ 2-MCPP ◇ MCPP 2,4-D ◇ MCPP-D-4 ◇ MCPP-K-4 ◇ MECOMEC ◇ MECOPEOP ◇ MECOPER ◇ MECOPEX ◇ MECOPROP ◇ MECOTURF ◇ MECPROP ◇ MEPRO ◇ METHOXONE ◇ α-(2-METHYL-4-CHLOROPHENOXY)PROPIONIC ACID ◇ 2-(2-METHYL-4-CHLOROPHENOXY)PROPIONIC ACID ◇ 2-METHYL-4-CHLOROPHENOXY-α-PROPIONIC ACID ◇ 2-(2-METHYL-4- CHLORPHENOXY)-PROPIONSAEURE (GERMAN) ◇ 2M 4KHP ◇ N.B. MECOPROP ◇ PROPAL ◇ PROPONEX-PLUS ◇ RANKOTEX ◇ RUNCATEX ◇ RD 4593 ◇ U 46 ◇ U 46 KV-ESTER ◇ U 46 KV-FLUID ◇ VI-PAR ◇ VI-PEX

TOXICITY DATA with REFERENCE
mrc-smc 742 ppm MUREAV 21,83,73
dns-mus-orl 100 mg/kg MUREAV 55,197,78
orl-mus TDLo:3 g/kg (female 6-15D post):TER ARZNAD 33,1479,83
orl-mus TDLo:7 g/kg (6-15D preg):REP ARZNAD 33,1479,83
orl-rat LD50:650 mg/kg WRPCA2 9,119,70
ipr-rat LD50:402 mg/kg TXCYAC 3,349,75
orl-mus LD50:369 mg/kg RPZHAW 31,373,80
skn-rbt LD50:900 mg/kg PCOC** -,683,66

CONSENSUS REPORTS: IARC Cancer Review: Group 2B IMEMDT 7,156,87; Human Limited Evidence IMEMDT 41,357,86. EPA Genetic Toxicology Program. Reported in EPA TSCA Inventory.

SAFETY PROFILE: Suspected carcinogen. Poison by ingestion. Moderately toxic by skin contact and intraperitoneal routes. Experimental teratogenic and reproductive effects. Mutation data reported. An herbicide. When heated to decomposition it emits toxic fumes of Cl^-.

CIR750 CAS:22316-47-8 ***HR: 3***
7-CHLORO-1-METHYL-5-PHENYL-1H-1,5-BENZODIAZEPINE-2,4(3H,5H)-DIONE
mf: $C_{16}H_{13}ClN_2O_2$ mw: 300.76

SYNS: CHLOREPIN ◇ CLOBAZAM ◇ CLOREPIN ◇ FRISIUM ◇ H-4723 ◇ HR 376 ◇ LM-2717 ◇ 1-PHENYL-5-METHYL-8-CHLORO-1,2,4,5-TETRAHYDRO-2,4-DIOXO-3H-1,5-BENZODIAZEPINE ◇ RU-4723 ◇ URBANYL

TOXICITY DATA with REFERENCE
orl-rat TDLo:6500 mg/kg (female 17-21D post):REP OYYAA2 25,917,83
orl-rbt TDLo:650 mg/kg (female 7-17D post):TER OYYAA2 25,1055,83
orl-man TDLo:104 mg/kg/1Y-C:CNS,GIT BMJOAE 282,1931,81
orl-rat LD50:6 g/kg MDACAP 16,9,80
ipr-rat LD50:740 mg/kg BCPHBM 7,33S,79
scu-rat LD50:8700 mg/kg OYYAA2 25,663,83
orl-mus LD50:580 mg/kg ARZNAD 35,133,85
ipr-mus LD50:289 mg/kg BCPHBM 7,33S,79
scu-mus LD50:2250 mg/kg BCPHBM 7,33S,79
orl-rbt LD50:320 mg/kg BCPHBM 7,33S,79
orl-gpg LD50:109 mg/kg BCPHBM 7,33S,79

SAFETY PROFILE: Poison by ingestion and intraperitoneal routes. Moderately toxic by subcutaneous route. Human systemic effects by ingestion: wakefulness, withdrawal, nausea and vomiting. An experimental teratogen. Other experimental reproductive effects. A tranquilizer. When heated to decomposition it emits very toxic fumes of NO_x and Cl^-. See also DIAZEPAM.

CIS000 CAS:562-09-4 ***HR: 3***
2-((p-CHLORO-α-METHYL-α-PHENYLBENZYL) OXY)-N,N-DIMETHYLAMINE HYDROCHLORIDE
mf: $C_{18}H_{22}ClNO{\cdot}ClH$ mw: 340.32

SYNS: 2-(α-(p-CHLOROPHENYL)-α-METHYLBENZYLOXY)-N,N-DIMETHYLAMINE ◇ (1-(p-CHLOROPHENYL)-1-PHENYL)ETHYL (β-DIMETHYLAMINOETHYL) ETHER HYDROCHLORIDE ◇ 2-(1-(4-CHLOROPHENYL)-1-PHENYLETHOXY)-N,N-DIMETHYLETHANAMINE HYDROCHLORIDE ◇ CHLORPHENOXAMINE HYDROCHLORIDE ◇ CONTRISTAMINE HYDROCHLORIDE ◇ β-DIMETHYLAMINO-ETHYL (p-CHLORO-α-METHYLBENZHYDRYL) ETHER HYDROCHLORIDE ◇ PHENOXENE HYDROCHLORIDE ◇ SUBSTANZ NR. 1766 (GERMAN) ◇ SYSTRAL

TOXICITY DATA with REFERENCE
orl-rat LD50:1000 mg/kg 29ZVAB -,31,69

orl-mus LD50:345 mg/kg CLDND* 2,83,60
scu-mus LD50:159 mg/kg CLDND* 4,638,62
scu-gpg LD50:140 mg/kg ARZNAD 4,189,54

SAFETY PROFILE: Poison by ingestion and subcutaneous routes. Moderately toxic by ingestion. When heated to decomposition it emits very toxic fumes of Cl^- and NO_x.

CIS250 CAS:15545-48-9 ***HR: 2***
N-(3-CHLORO-4-METHYLPHENYL)-N',N'-DIMETHYLUREA
mf: $C_{10}H_{13}ClN_2O$ mw: 212.70

SYNS: C 2242 ◇ 3-(3-CHLOR-4-METHYLPHENYL)-1,1-DIMETHYLHARNSTOFF (GERMAN) ◇ 3-(3-CHLORO-4-METHYLPHENYL)-1,1-DIMETHYL-UREA ◇ CHLOROTOLURON ◇ CHLORTOLURON ◇ CLORTOKEM ◇ DICURAN

TOXICITY DATA with REFERENCE
mma-sat 1 μg/plate MUREAV 58,353,78
dni-mus-orl 200 mg/kg MUREAV 58,353,78
orl-rat TDLo:2 g/kg (4D preg):REP KHZDAN 22,362,79
orl-rat LD50:5800 mg/kg KHZDAN 22,362,79
ihl-rat LC50:1300 mg/m³ 85DPAN -,-,71/76

SAFETY PROFILE: Moderately toxic by inhalation. Mildly toxic by ingestion. Experimental reproductive effects. Mutation data reported. A pesticide. When heated to decomposition it emits very toxic fumes of Cl^- and NO_x.

CIS325 CAS:65039-20-5 ***HR: 3***
2-CHLORO-5-METHYLPHENYLHYDROXYLAMINE
mf: C_7H_8ClNO mw: 157.60

$Cl(CH_3)C_6H_3NHOH$

SAFETY PROFILE: Explodes when heated above 120°C. When heated to decomposition it emits toxic fumes of Cl^- and NO_x. See also AROMATIC AMINES.

CIS625 CAS:1631-82-9 ***HR: 2***
CHLOROMETHYLPHENYLSILANE
mf: C_7H_9ClSi mw: 156.69

SAFETY PROFILE: Vigorous reaction above 100°C with 4-bromobutene + chloroplatinic acid. When heated to decomposition it emits toxic fumes of Cl^-. See also SILANE.

CIS750 CAS:2058-52-8 ***HR: 3***
2-CHLORO-11-(4-METHYLPIPERAZINO) DIBENZO(b,f)(1,4)THIAZEPINE
mf: $C_{18}H_{18}ClN_3S$ mw: 343.90

SYNS: 2-CHLORO-11-(4-METHYL-1-PIPERAZINYL)DIBENZO (b,f)(1,4)THIAZEPINE ◇ DIBENZOTHIAZEPINE

TOXICITY DATA with REFERENCE
orl-rat TDLo:72 mg/kg (female 9-14D post):REP OYYAA2 4,305,70
orl-rat TDLo:6 mg/kg (9-14D preg):TER OYYAA2 4,305,70
orl-rat LD50:280 mg/kg DRUGAY 6,234,82
orl-mus LD50:270 mg/kg ARZNAD 15,841,65
orl-gpg LD50:150 mg/kg TXAPA9 14,657,69

SAFETY PROFILE: Poison by ingestion. An experimental teratogen. Other experimental reproductive effects. When heated to decomposition it emits very toxic fumes of Cl^-, NO_x, and SO_x.

CIT000 CAS:4956-31-4 ***HR: 3***
2-CHLORO-11-(4'-METHYL)PIPERAZINO-DIBENZO(b,f)(1,4)THIAZEPINE HYDROCHLORIDE
mf: $C_{18}H_{18}ClN_3S•ClH$ mw: 380.36

SYNS: 2-CHLORO-11-(4-METHYL-1-PIPERAZINYL)DIBENZO(b,f)(1,4)THIAZEPINE HYDROCHLORIDE ◇ HF-2159 HYDROCHLORIDE

TOXICITY DATA with REFERENCE
orl-rat LD50:430 mg/kg IJNEAQ 4,375,65
ivn-rat LD50:42 mg/kg IJNEAQ 4,375,65
orl-mus LD50:270 mg/kg IJNEAQ 4,375,65
ivn-mus LD50:46 mg/kg IJNEAQ 4,375,65
orl-gpg LD50:154 mg/kg IJNEAQ 4,375,65

SAFETY PROFILE: Poison by ingestion and intravenous routes. When heated to decomposition it emits very toxic fumes of Cl^-, NO_x, and SO_x.

CIT625 CAS:59943-31-6 ***HR: 3***
7-CHLORO-3-(4-METHYL-1-PIPERAZINYL)-4H-1,2,4-BENZOTHIADIAZINE-1,1-DIOXIDE
mf: $C_{12}H_{15}ClN_4O_2S$ mw: 314.82

SYN: DU-717

TOXICITY DATA with REFERENCE
ivn-rat LD50:551 mg/kg IYKEDH 11,294,80
ivn-mus LD50:242 mg/kg IYKEDH 11,294,80
ivn-dog LD50:400 mg/kg IYKEDH 11,294,80

SAFETY PROFILE: Poison by intravenous route. When heated to decomposition it emits toxic fumes of Cl^-, SO_x, and NO_x.

CIT750 CAS:77966-63-3 ***HR: 3***
6'-CHLORO-2-(2-METHYLPIPERIDINO)-o-ACETOTOLUIDIDE HYDROCHLORIDE
mf: $C_{15}H_{21}ClN_2O•ClH$ mw: 317.29

SYN: V 315

TOXICITY DATA with REFERENCE
eye-rbt 2% MLD ARZNAD 8,407,58
ipr-rat LD50:77 mg/kg ARZNAD 8,407,58

ipr-mus LD50:110 mg/kg ARZNAD 8,407,58
scu-mus LD50:139 mg/kg ARZNAD 8,407,58

SAFETY PROFILE: Poison by intraperitoneal and subcutaneous routes. An eye irritant. When heated to decomposition it emits very toxic fumes of Cl^- and NO_x.

CIU000 CAS:101651-66-5 ***HR: 3***
o-CHLORO-2-(METHYL(2-(PIPERIDINO)ETHYL) AMINO)ACETANILIDE DIHYDROCHLORIDE
mf: $C_{16}H_{24}ClN_3O{\bullet}2ClH$ mw: 382.80

SYN: C 5406

TOXICITY DATA with REFERENCE
ipr-rat LD50:124 mg/kg ARZNAD 9,262,59
scu-mus LD50:590 mg/kg ARZNAD 9,262,59

SAFETY PROFILE: Poison by intraperitoneal route. Moderately toxic by subcutaneous route. When heated to decomposition it emits very toxic fumes of Cl^- and NO_x.

CIU250 CAS:78218-42-5 ***HR: 3***
6'-CHLORO-3-(2-METHYLPIPERIDINO)-o-PROPIONOTOLUIDIDE HYDROCHLORIDE
mf: $C_{16}H_{23}ClN_2O{\bullet}ClH$ mw: 331.32

SYN: C 3139

TOXICITY DATA with REFERENCE
ipr-rat LD50:39 mg/kg ARZNAD 8,544,58
scu-mus LD50:36 mg/kg ARZNAD 8,544,58

SAFETY PROFILE: Poison by intraperitoneal and subcutaneous routes. When heated to decomposition it emits very toxic fumes of Cl^- and NO_x.

CIU325 CAS:58763-27-2 ***HR: 3***
3'-CHLORO-5'-METHYL-3-PIPERIDINO-4-PROPOXY-PROPIOPHENONE HYDROCHLORIDE
mf: $C_{18}H_{26}ClNO_2{\bullet}ClH$ mw: 360.36

SYNS: 1-(3-CHLORO-5-METHYL-4-PROPOXYPHENYL)-3-(1-PIPERIDINYL)1-PROPANONE HYDROCHLORIDE (9CI) ◇ β-PIPERIDINOAETHYL-(3-CHLOR-4-PROPOXY-5-METHYLPHENYL)-KETONHYDROCHLORID (GERMAN)

TOXICITY DATA with REFERENCE
orl-mus LD50:460 mg/kg PHARAT 31,21,76
scu-mus LD50:330 mg/kg PHARAT 31,21,76
ivn-mus LD50:50 mg/kg PHARAT 31,21,76

SAFETY PROFILE: Poison by subcutaneous and intravenous routes. Moderately toxic by ingestion. When heated to decomposition it emits toxic fumes of NO_x and Cl^-.

CIU500 CAS:513-36-0 ***HR: 3***
1-CHLORO-2-METHYLPROPANE
mf: C_4H_9Cl mw: 92.57

PROP: Flash p: 21.2°F, lel: 2.0%, uel: 8.7%, d: 0.8829 @ 15°, mp: -130.3°, bp: 68.9°.

SYN: ISOBUTYL CHLORIDE

SAFETY PROFILE: A poison by ingestion and inhalation. A very dangerous fire and explosion hazard when exposed to heat or flame. When heated to decomposition it emits toxic fumes of Cl^-. See also CHLORINATED HYDROCARBONS, ALIPHATIC.

CIU750 CAS:563-47-3 ***HR: 3***
3-CHLORO-2-METHYLPROPENE
DOT: UN 2554
mf: C_4H_7Cl mw: 90.56

$$H_2C{=}C(CH_3)CH_2Cl$$

PROP: Colorless, volatile liquid, disagreeable odor. Bp: 72.17°, lel: 2.3%, uel: 9.3%, fp: $< -80°$, d: 0.9257 @ 20°/4°, vap press: 101.7 mm @ 20°, vap d: 3.12, flash p: −10°. Misc in alc and ether.

SYNS: 3-CHLOR-2-METHYL-PROP-1-EN (GERMAN) ◇ Γ-CHLOROISOBUTYLENE ◇ 3-CHLORO-2-METHYL-1-PROPENE ◇ CHLORURE de METHALLYLE (FRENCH) ◇ 3-CLORO-2-METIL-PROP-1-ENE (ITALIAN) ◇ CLORURO di METALLILE (ITALIAN) ◇ ISOBUTENYL CHLORIDE ◇ METHALLYL CHLORIDE ◇ α-METHALLYL CHLORIDE ◇ 2-METHYL-ALLYLCHLORID (GERMAN) ◇ β-METHYLALLYL CHLORIDE ◇ 2-METHYLALLYL CHLORIDE ◇ METHYL ALLYL CHLORIDE (DOT) ◇ NCI-C54820

TOXICITY DATA with REFERENCE
mmo-sat 6 μmol/plate BCPCA6 29,2611,80
dns-hmn:hla 1 mmol/L CALEDQ 20,263,83
orl-rat TDLo:77250 mg/kg/2Y-I:CAR NTPTR* NTP-TR-300,86
orl-rat TDLo:38625 mg/kg/2Y-I:ETA PAACA3 26,95,85
orl-mus TDLo:51500 mg/kg/2Y-I:CAR NTPTR* NTP-TR-300,86
orl-rat TD:77250 mg/kg/2Y-I:NEO,REP NTPTR* NTP-TR-300,86
ihl-hmn LCLo:22000 ppm/10M 34ZIAG -,377,69
ihl-rat LC50:34000 mg/m³/30M FAVUAI 7,35,75

CONSENSUS REPORTS: NTP Fifth Annual Report on Carcinogens. NTP Carcinogenesis Studies (gavage); Clear Evidence: mouse, rat NTPTR* NTP-TR-300,86. Reported in EPA TSCA Inventory.

DOT Classification: IMO: Flammable Liquid; Label: Flammable Liquid.

SAFETY PROFILE: Confirmed carcinogen with experimental carcinogenic, neoplastigenic, and tumorigenic data. Mildly toxic to humans by inhalation. Experimental reproductive effects. An irritant. Human mutation data reported. Very dangerous fire hazard when exposed to heat, flame, or oxidizers. Moderately explosive when exposed to heat or flame. Can react vigorously with oxidizing materials. To fight fire, use alcohol foam, CO_2, dry chemical. When heated to decomposition it emits toxic fumes of Cl^-. See also CHLORINATED HYDROCARBONS, ALIPHATIC; and ALLYL COMPOUNDS.

CIV000 CAS:6959-48-4 ***HR: 3***
3-(CHLOROMETHYL) PYRIDINE HYDROCHLORIDE
mf: $C_6H_6ClN \cdot ClH$ mw: 164.04

SYN: NCI-C03838

TOXICITY DATA with REFERENCE
mmo-sat 333 µg/plate IARCCD 27,283,80
mma-sat 33300 ng/plate ENMUDM 7(Suppl 5),1,85
mma-esc 1 mg/plate ENMUDM 7(Suppl 5),1,85
otr-rat:emb 640 ng/plate JJATDK 1,190,81
orl-rat TDLo:37 g/kg/83W-I:CAR NCITR* NCI-CG-TR-95,78
orl-mus TDLo:49 g/kg/81W-I:CAR NCITR* NCI-CG-TR-95,78
orl-rat LD50:316 mg/kg NCILB* NIH-NCI-E-C-72-3252
orl-mus LD50:316 mg/kg NCILB* NIH-NCI-E-C-72-3252

CONSENSUS REPORTS: NCI Carcinogenesis Bioassay (gavage); Clear Evidence: mouse, rat NCITR* NCI-CG-TR-95,78. EPA Genetic Toxicology Program.

SAFETY PROFILE: Suspected carcinogen with experimental carcinogenic data. Poison by ingestion. Mutation data reported. When heated to decomposition it emits very toxic fumes of NO_x and Cl^-.

CIW250 ***HR: 3***
5'-CHLORO-2-(METHYL(2-(PYRROLIDINYL) ETHYL)AMINO)-O-ACETOTOLUIDIDE DIHYDROCHLORIDE
mf: $C_{16}H_{24}ClN_3O \cdot 2ClH$ mw: 382.80

SYNS: C 5420

TOXICITY DATA with REFERENCE
eye-rbt 2% MLD ARZNAD 9,262,59
ipr-rat LD50:118 mg/kg ARZNAD 9,262,59
scu-mus LD50:385 mg/kg ARZNAD 9,262,59

SAFETY PROFILE: Poison by intraperitoneal and subcutaneous routes. An eye irritant. When heated to decomposition it emits very toxic fumes of Cl^- and NO_x.

CIX000 CAS:102129-03-3 ***HR: 3***
o-CHLORO-2-(METHYL(2-(PYRROLIDINYL) ETHYL)AMINO)PROPIONANILIDE DIHYDROCHLORIDE
mf: $C_{16}H_{24}ClN_3O \cdot 2ClH$ mw: 382.80

SYN: C 5407

TOXICITY DATA with REFERENCE
eye-rbt 2% MLD ARZNAD 9,262,59
ipr-rat LD50:112 mg/kg ARZNAD 9,262,59
scu-mus LD50:855 mg/kg ARZNAD 9,262,59

SAFETY PROFILE: Poison by intraperitoneal route. Moderately toxic by subcutaneous route. An eye irritant. When heated to decomposition it emits very toxic fumes of NO_x and Cl^-.

CIX750 CAS:3003-84-7 ***HR: 3***
2-CHLOROMETHYL TETRAHYDROFURAN
mf: C_5H_9ClO mw: 120.59

TOXICITY DATA with REFERENCE
ipr-mus LDLo:250 mg/kg CBCCT* 4,319,52

CONSENSUS REPORTS: Reported in EPA TSCA Inventory.

SAFETY PROFILE: Poison by intraperitoneal route. When heated to decomposition it emits toxic fumes of Cl^-.

CIY250 CAS:617-88-9 ***HR: 3***
2-CHLOROMETHYLTHIOPHENE
mf: C_5H_5ClS mw: 132.61

SCH=CHCH=CH_2Cl (ring bond from S to CH)

SAFETY PROFILE: Flammable when exposed to heat or flame. A storage hazard. It decomposes at room temperature to release hydrogen chloride, and may explode in a sealed container. Highly explosive when shocked, exposed to heat, or by spontaneous chemical reaction. Can react vigorously with oxidizing materials. See also THIOPHENE.

CIY325 CAS:1558-25-4 ***HR: 3***
(CHLOROMETHYL)TRICHLOROSILANE
mf: CH_2Cl_4Si mw: 183.92

SYNS: CHLOROMETHYL(TRICHLORO)SILANE ◇ TRICHLORO(CHLOROMETHYL)SILANE (9CI)

TOXICITY DATA with REFERENCE
orl-mus LDLo:100 mg/kg 85GMAT -,37,82
ihl-mus LD50:30 mg/m^3/2H 85GMAT -,37,82
skn-mus LDLo:100 mg/kg 85GMAT -,37,82
ipr-mus LDLo:100 mg/kg 85GMAT -,37,82

SAFETY PROFILE: Poison by inhalation, skin contact, ingestion, and intraperitoneal routes. When heated to decomposition it emits toxic fumes of Cl^-. See also SILANE.

CIY500 CAS:15267-95-5 ***HR: 2***
(CHLOROMETHYL)TRIETHOXYSILANE
mf: $C_7H_{17}ClO_3Si$ mw: 212.78

SYN: CHLORMETHYL-TRIETHOXYSILAN (CZECH)

TOXICITY DATA with REFERENCE
skn-rbt 500 mg/24H MOD 28ZPAK -,219,72
eye-rbt 500 mg/24H MOD 28ZPAK -,219,72
orl-rat LD50:2400 mg/kg 28ZPAK -,219,72

CONSENSUS REPORTS: Reported in EPA TSCA Inventory.

SAFETY PROFILE: Moderately toxic by ingestion. A skin and eye irritant. When heated to decomposition it emits toxic fumes of Cl^-.

CIY750 CAS:5926-26-1 ***HR: 2***
CHLOROMETHYL TRIMETHOXYSILANE
mf: $C_4H_{11}ClO_3Si$ mw: 170.69

SYN: CHLOROMETHYL-TRIMETHOXYSILAN (CZECH)

TOXICITY DATA with REFERENCE
skn-rbt 500 mg/24H MOD 28ZPAK -,217,72
eye-rbt 250 μg/24H SEV 28ZPAK -,217,72
orl-rat LDLo:500 mg/kg 28ZPAK -,217,72
ihl-rat LCLo:3600 ppm/30M 28ZPAK -,217,72

SAFETY PROFILE: Moderately toxic by ingestion. Mildly toxic by inhalation. A skin and severe eye irritant. When heated to decomposition it emits toxic fumes of Cl^-.

CIY899 ***HR: 3***
N-CHLORO-3-MORPHOLINONE
mf: $C_4H_6ClNO_2$ mw: 135.55

$ClNCH_2CO{\bullet}OCH_2CH_2$ (ring)

SAFETY PROFILE: May explode when heated to 115°C. When heated to decomposition it emits toxic fumes of Cl^- and NO_x.

CIZ000 CAS:90-13-1 ***HR: 2***
1-CHLORONAPHTHALENE
mf: $C_{10}H_7Cl$ mw: 162.62

PROP: Insol in water, sol in alc, benzene. D: 1.1938 @ 20°/4°, mp: −2.3°, bp: 259°, flash p: 121°.

SYNS: α-CHLORNAPHTHALENE ◇ α-CHLORONAPHTHALENE

TOXICITY DATA with REFERENCE
orl-rat LD50:1540 mg/kg NTIS** PB214-270
orl-mus LD50:1091 mg/kg NTIS** PB214-270
orl-gpg LD50:2000 mg/kg GISAAA 47(11),78,82

CONSENSUS REPORTS: Reported in EPA TSCA Inventory.

SAFETY PROFILE: Moderately toxic by ingestion. When heated to decomposition it emits toxic fumes of Cl^-. See also CHLORINATED HYDROCARBONS, AROMATIC.

CJA000 CAS:91-58-7 ***HR: 2***
2-CHLORONAPHTHALENE
mf: $C_{10}H_7Cl$ mw: 162.62

PROP: D: 1.377 @ 71°, mp: 59.5°, bp: 206°. Insol in water; sol in alc, benzene, chloroform, ether, CS_2

SYNS: β-CHLORONAPHTHALENE ◇ RCRA WASTE NUMBER U047

TOXICITY DATA with REFERENCE
orl-rat LD50:2078 mg/kg NTIS** PB214-270
orl-mus LD50:886 mg/kg NTIS** PB214-270

CONSENSUS REPORTS: Reported in EPA TSCA Inventory.

SAFETY PROFILE: Moderately toxic by by ingestion. When heated to decomposition it emits toxic fumes of Cl^-. See also CHLORINATED HYDROCARBONS, AROMATIC.

CJA100 CAS:2675-77-6 ***HR: 1***
CHLORONEB
mf: $C_8H_8Cl_2O_2$ mw: 207.06

SYNS: CHLORONEBE (FRENCH) ◇ DEMOSAN ◇ 1,4-DICHLORO-2,5-DIMETHOXYBENZENE ◇ SOIL FUNGICIDE 1823 ◇ TERSAN-SP

TOXICITY DATA with REFERENCE
mmo-asn 24 μmol/L PHYTAJ 66,217,76
sln-asn 48 μmol/L EVHPAZ 31,81,79
orl-rat LD50:11 g/kg 85ARAE 4,82,76/77

CONSENSUS REPORTS: EPA Genetic Toxicology Program.

SAFETY PROFILE: Very mildly toxic by ingestion. Mutation data reported. When heated to decomposition it emits toxic fumes of Cl^-.

CJA150 CAS:59483-61-3 ***HR: 3***
N-CHLORO-4-NITROANILINE
mf: $C_6H_5ClN_2O_2$ mw: 172.57

SAFETY PROFILE: Decomposes explosively at room temperature. Upon decomposition it emits toxic fumes of Cl^- and NO_x. See also ANILINE DYES and NITRO COMPOUNDS of AROMATIC HYDROCARBONS.

CJA175 CAS:121-87-9 ***HR: 3***
2-CHLORO-4-NITROANILINE
mf: $C_6H_5ClN_2O_2$ mw: 172.58

PROP: Mp: 109°. Sltly sol in water; very sol in alc, ether.

SYNS: 1-AMINO-2-CHLORO-4-NITROBENZENE ◇ o-CHLORO-p-NITROANILINE ◇ 4-NITRO-2-CHLOROANILINE ◇ OCPNA

TOXICITY DATA with REFERENCE
orl-rat LD50:6430 mg/kg GTPZAB 25(8),50,81
orl-mus LD50:1250 mg/kg GTPZAB 25(8),50,81

ipr-mus LDLo:500 mg/kg CBCCT* 5,337,53
ivn-mus LDLo:50 mg/kg CBCCT* 6,138,54

CONSENSUS REPORTS: Reported in EPA TSCA Inventory.

SAFETY PROFILE: Poison by intravenous route. Moderately toxic by ingestion and intraperitoneal routes. When heated to decomposition it emits toxic fumes of Cl^- and NO_x. See also ANILINE DYES.

CJA185 CAS:635-22-3 ***HR: 3***
4-CHLORO-3-NITROANILINE
mf: $C_6H_5ClN_2O_2$ mw: 172.58

SYN: ANILINE, 4-CHLORO-3-NITRO-

TOXICITY DATA with REFERENCE
eye-rbt 100 mg MLD EPASR* 8EHQ-0882-0452
skn-gpg 500 mg/24H MLD EPASR* 8EHQ-0882-0452
orl-rat TDLo:5850 mg/kg (male 90D pre):REP FAATDF 6,551,86
orl-rat LD50:400 mg/kg EPASR* 8EHQ-0882-0452
ipr-rat LD50:200 mg/kg EPASR* 8EHQ-0882-0452
orl-mus LD50:800 mg/kg EPASR* 8EHQ-0882-0452
ipr-mus LD50:200 mg/kg EPASR* 8EHQ-0882-0452
orl-brd LD50:100 mg/kg TXAPA9 21,315,72

CONSENSUS REPORTS: Reported in EPA TSCA Inventory.

SAFETY PROFILE: Poison by ingestion and intraperitoneal routes. Experimental reproductive effects. Skin and eye irritant. When heated to decomposition it emits toxic fumes of NO_x and Cl^-.

CJA200 CAS:4920-79-0 ***HR: 2***
2-CHLORO-4-NITRO-ANISOLE
mf: $C_7H_6ClNO_3$ mw: 187.59

SYNS: 2-CHLORO-1-METHOXY-4-NITROBENZENE(9CI) ◇ CHLORONITROANISOLE ◇ OCNA ◇ o-CHLORO-p-NITROANISOLE

TOXICITY DATA with REFERENCE
skn-rbt 500 mg/24H MLD TOERD9 2,77,79
eye-rbt 100 mg MLD TOERD9 2,77,79
orl-rat LD50:1180 mg/kg TOERD9 2,77,79
ipr-rat TDLo:445 mg/kg TOERD9 2,77,79
orl-mus LD50:1550 mg/kg TOERD9 2,77,79
ipr-mus LD50:815 mg/kg TOERD9 2,77,79

SAFETY PROFILE: Moderately toxic by ingestion and intraperitoneal routes. An eye and skin irritant. When heated to decomposition it emits toxic fumes of Cl^- and NO_x. See also NITRO COMPOUNDS of AROMATIC HYDROCARBONS.

CJA250 CAS:129-40-8 ***HR: 1***
1-CHLORO-5-NITROANTHRAQUINONE
mf: $C_{14}H_6ClNO_4$ mw: 287.66

SYNS: 1-CHLOR-5-NITROANTHRACHINON (CZECH) ◇ 1-CHLORO-5-NITRO-9,10-ANTHRACENEDIONE ◇ 5-CHLORO-1-NITROANTHRAQUINONE ◇ 1-NITRO-5-CHLOROANTHRAQUINONE

TOXICITY DATA with REFERENCE
eye-rbt 500 mg/24H MLD 28ZPAK -,86,72

CONSENSUS REPORTS: Reported in EPA TSCA Inventory.

SAFETY PROFILE: An eye irritant. When heated to decomposition it emits very toxic fumes of Cl^- and NO_x. See also NITRO COMPOUNDS of AROMATIC HYDROCARBONS.

CJA950 CAS:25167-93-5 ***HR: D***
CHLORONITROBENZENE
mf: $C_6H_4ClNO_2$ mw: 157.56

SYNS: MONONITROCHLOROBENZENE ◇ NITROCHLOROBENZENE

TOXICITY DATA with REFERENCE
dnd-mus-ipr 60 mg/kg ARTODN (5),355,82
dnd-ham:lvr 5 μmol/L MUREAV 131,215,84
dns-ham:lvr 5 μmol/L MUREAV 131,215,84

SAFETY PROFILE: Mutation data reported. When heated to decomposition it emits toxic fumes of Cl^- and NO_x. See also other chloronitrobenzene entries and NITRO COMPOUNDS of AROMATIC HYDROCARBONS.

CJB250 CAS:121-73-3 ***HR: 3***
1-CHLORO-3-NITROBENZENE
DOT: UN 1578
mf: $C_6H_4ClNO_2$ mw: 157.56

PROP: Yellowish crystals. Mp: 46°, flash p: 103°, bp: 236°, d: 1.534 @ 20°/4°.

SYNS: CHLORO-m-NITROBENZENE ◇ m-CHLORONITROBENZENE ◇ m-CHLORONITROBENZENE (DOT) ◇ m-NITROCHLOROBENZENE ◇ m-NITROCHLOROBENZENE, solid (DOT)

TOXICITY DATA with REFERENCE
orl-rat LD50:470 mg/kg GTPZAB 25(8),50,81
orl-mus LD50:380 mg/kg GTPZAB 25(8),50,81

CONSENSUS REPORTS: Reported in EPA TSCA Inventory.

DOT Classification: Poison B; Label: Poison.

SAFETY PROFILE: Poison by ingestion and inhalation. It forms methemoglobin in the body and gives rise to cyanosis and blood changes. Its effects are cumulative and analogous to those of nitrobenzene. The para compound is thought to be somewhat less toxic than the ortho compound. Chemically, it is probably converted in the body to chloroaniline, which is also poisonous. In industry, it is the dust of this material that is most often the source of intoxication. Dangerous fire hazard when exposed to heat or flame. It can react with oxidizing materials. When heated to decomposition it emits toxic fumes of Cl^-, NO_x, and phosgene. See also other chloronitrobenzene entries and NITRO COMPOUNDS of AROMATIC HYDROCARBONS.

CJB750 CAS:88-73-3 ***HR: 3***
CHLORO-o-NITROBENZENE
DOT: UN 1578
mf: $C_6H_4ClNO_2$ mw: 157.56

PROP: Yellow crystals. Mp: 32-33°, bp: 245-246°, d: 1.348, flash p: 123°.

SYNS: o-CHLORONITROBENZENE ◇ o-CHLORONITROBENZENE (DOT) ◇ 1-CHLORO-2-NITROBENZENE ◇ 2-CHLORONITROBENZENE ◇ 2-CHLORO-1-NITROBENZENE ◇ o-NITROCHLOROBENZENE ◇ o-NITROCHLOROBENZENE, liquid (DOT) ◇ ONCB

TOXICITY DATA with REFERENCE
mmo-sat 205 µg/plate MUREAV 116,217,83
mma-sat 100 µg/plate ENMUDM 5(Suppl 1),3,83
orl-rat TDLo:22 g/kg/78W-C:NEO JEPTDQ 2(2),325,78
orl-mus TDLo:140 g/kg/78W-C:CAR JEPTDQ 2(2),325,78
orl-mus TD:280 g/kg/78W-C:CAR JEPTDQ 2(2),325,78
orl-rat LD50:288 mg/kg NTIS** PB214-270
orl-mus LD50:135 mg/kg NTIS** PB214-270
orl-rbt LD50:280 mg/kg 85GMAT -,92,82

CONSENSUS REPORTS: Reported in EPA TSCA Inventory.

DOT Classification: Poison B; Label: Poison.

SAFETY PROFILE: Poison by ingestion and probably inhalation. Questionable carcinogen with experimental carcinogenic and neoplastigenic data. Combustible when exposed to heat or flame. To fight fire, use water, foam. Potentially explosive reaction with ammonia at 160°C/30 bar. When heated to decomposition it emits toxic fumes of Cl^-, NO_x, and phosgene. See also other chloronitrobenzene entries and NITRO COMPOUNDS of AROMATIC HYDROCARBONS.

CJB825 CAS:96-73-1 ***HR: 3***
2-CHLORO-5-NITROBENZENESULFONIC ACID
mf: $C_6H_4ClNO_5S$ mw: 237.61

$Cl(NO_2)C_6H_3SO_2OH$

SAFETY PROFILE: Decomposes violently at 150°C. When heated to decomposition it emits toxic fumes of Cl^-, SO_x and NO_x. See also SULFONATES and NITRO COMPOUNDS of AROMATIC HYDROCARBONS.

CJC000 CAS:4515-30-4 ***HR: 2***
2-CHLORO-3,5-NITROBENZENESULFONIC ACID, SODIUM SALT
mf: $C_6H_2ClN_2O_7S{\cdot}Na$ mw: 304.60

SYN: 2,4-DINITROCHLORBENZEN-6-SULFONAN SODNY (CZECH)

TOXICITY DATA with REFERENCE
skn-rbt 500 mg/24H MLD 28ZPAK -,181,72
eye-rbt 20 mg/24H MOD 28ZPAK -,181,72
orl-rat LD50:1640 mg/kg 28ZPAK -,181,72

SAFETY PROFILE: Moderately toxic by ingestion. A skin and eye irritant. When heated to decomposition it emits very toxic fumes of Cl^-, Na_2O, NO_x, and SO_x. See also SULFONATES.

CJC250 CAS:99-60-5 ***HR: 2***
2-CHLORO-4-NITROBENZOIC ACID
mf: $C_7H_4ClNO_4$ mw: 201.57

PROP: Mp: 141°. Sol in hot water and hot benzene.

SYN: KYSELINA-2-CHLORO-4-NITROBENZOOVA (CZECH)

TOXICITY DATA with REFERENCE
eye-rbt 250 µg/24H SEV 28ZPAK -,91,72
orl-rat LDLo:5070 mg/kg 28ZPAK -,91,72

CONSENSUS REPORTS: Reported in EPA TSCA Inventory.

SAFETY PROFILE: Mildly toxic by ingestion. A severe eye irritant. When heated to decomposition it emits very toxic fumes of Cl^- and NO_x.

CJC500 CAS:96-99-1 ***HR: 3***
4-CHLORO-3-NITROBENZOIC ACID
mf: $C_7H_4ClNO_4$ mw: 201.57

PROP: Sltly sol in alc, sol in hot water. D: 1.645 @ 18°, mp: 183°.

SYN: KYSELINA 4-CHLORO-3-NITROBENZOOVA (CZECH)

TOXICITY DATA with REFERENCE
orl-rat LD50:3150 mg/kg MarJV# 29MAR77
orl-bwd LD50:75 mg/kg TXAPA9 21,315,72

CONSENSUS REPORTS: Reported in EPA TSCA Inventory.

SAFETY PROFILE: Poison by ingestion. When heated to decomposition it emits very toxic fumes of NO_x and Cl^-.

CJC515 CAS:6307-82-0 ***HR: 1***
2-CHLORO-5-NITROBENZOIC ACID METHYL ESTER
mf: $C_8H_6ClNO_4$ mw: 215.60

SYNS: BENZOIC ACID, 2-CHLORO-5-NITRO-, METHYL ESTER ◇ METHYLESTER KYSELINY 2-CHLOR-5-NITROBENZOOVE

TOXICITY DATA with REFERENCE
skn-rbt 500 mg/24H MLD 85JCAE-,591,86
eye-rbt 500 mg/24H MLD 85JCAE-,591,86
orl-rat LD50:5360 mg/kg 85JCAE-,591,86

SAFETY PROFILE: Mildly toxic by ingestion. A skin and eye irritant. When heated to decomposition it emits toxic fumes of NO_x and Cl^-.

CJC549 ***HR: 3***
2-CHLORO-5-NITROBENZYL ALCOHOL
mf: $C_7H_6ClNO_3$ mw: 187.58

$O_2N(Cl)C_6H_3CH_2OH$

PROP: Mp: 79°.

SAFETY PROFILE: Decomposes exothermically at 211°C. When heated to decomposition it emits toxic fumes of Cl^- and NO_x. See also NITRO COMPOUNDS of AROMATIC HYDROCARBONS.

CJC600 CAS:56433-01-3 ***HR: 3***
6-CHLORO-2-NITROBENZYL BROMIDE
mf: $C_7H_5BrClNO_2$ mw: 250.48

SAFETY PROFILE: Potentially dangerous exothermic reaction when heated above 190°C in a sealed container. When heated to decomposition it emits toxic fumes of Cl^-, Br^-, and NO_x. See also NITRO COMPOUNDS of AROMATIC HYDROCARBONS and BROMIDES.

CJC610 CAS:50274-95-8 ***HR: 3***
2-CHLORO-4-NITROBENZYL CHLORIDE
mf: $C_7H_5Cl_2NO_2$ mw: 206.03

SAFETY PROFILE: Potentially dangerous exothermic reaction when heated above 190°C in a sealed container. When heated to decomposition it emits toxic fumes of Cl^- and NO_x. See also NITRO COMPOUNDS of AROMATIC HYDROCARBONS.

CJC625 CAS:938-71-6 ***HR: 3***
4-CHLORO-2-NITROBENZYL CHLORIDE
mf: $C_7H_5Cl_2NO_2$ mw: 206.03

SAFETY PROFILE: Potentially dangerous exothermic reaction when heated above 190°C in a sealed container. When heated to decomposition it emits toxic fumes of Cl^- and NO_x. See also NITRO COMPOUNDS of AROMATIC HYDROCARBONS.

CJC750 CAS:22236-53-9 ***HR: 3***
2-CHLORO-2-NITROBUTANE
mf: $C_4H_8ClNO_2$ mw: 137.58

TOXICITY DATA with REFERENCE
orl-mus LD50:970 mg/kg HYSAAV 34(10-12),429,69
ihl-mus LC50:135 g/m³/3H HYSAAV 34(10-12),429,69
scu-mus LD50:250 mg/kg HYSAAV 30,169,65

SAFETY PROFILE: Poison by subcutaneous route. Moderately toxic by ingestion. Very mildly toxic by inhalation. When heated to decomposition it emits very toxic fumes of Cl^- and NO_x.

CJC800 CAS:598-92-5 ***HR: 3***
1-CHLORO-1-NITROETHANE
mf: $C_2H_4ClNO_2$ mw: 109.52

PROP: Sol in alc. D: 1.258 @ 20°/20°, bp: 125°, flash p: 56°.

TOXICITY DATA with REFERENCE
orl-mus LD50:620 mg/kg HYSAAV 34(10-12),429,69
ihl-mus LC50:21 g/m³/3H HYSAAV 34(10-12),429,69
ipr-mus LD50:560 mg/kg KHFZAN 10(6),53,76
scu-mus LD50:185 mg/kg HYSAAV 34(10-12),429,69
orl-rbt LDLo:100 mg/kg JIHTAB 22,315,40

CONSENSUS REPORTS: Reported in EPA TSCA Inventory.

SAFETY PROFILE: Poison by ingestion and subcutaneous route. Moderately toxic by intraperitoneal route. Very mildly toxic by inhalation. When heated to decomposition it emits toxic fumes of Cl^- and NO_x.

CJD250 CAS:619-08-9 ***HR: 3***
2-CHLORO-4-NITROPHENOL
mf: $C_6H_4ClNO_3$ mw: 173.56

PROP: Mp: 106°

TOXICITY DATA with REFERENCE
orl-rat LD50:900 mg/kg 34ZIAG -,169,69
ipr-mus LD50:80 mg/kg JMCMAR 18,868,75

CONSENSUS REPORTS: Reported in EPA TSCA Inventory. Chlorophenol compounds are on the Community Right-To-Know List.

SAFETY PROFILE: Poison by intraperitoneal route. Moderately toxic by ingestion. When heated to decomposition it emits very toxic fumes of Cl^- and NO_x. See also CHLOROPHENOLS and NITRO COMPOUNDS of AROMATIC HYDROCARBONS.

CJD500 CAS:2814-77-9 ***HR: D***
1-((2-CHLORO-4-NITROPHENYL)AZO)-2-NAPHTHOL
mf: $C_{16}H_{10}ClN_3O_3$ mw: 327.74

SYNS: ADC PERMANENT RED TONER R ◇ AMERICAN VERMILION ◇ BLAZING RED ◇ CARNELIO RED R ◇ CHLORPARANITRANILINE RED ◇ C.I. 12085 ◇ C.I. PIGMENT RED 4 ◇ DAINICHI PERMA-

NENT RED RX ◇ D&C RED No. 36 ◇ DUPLEX PERMATON RED L 20-7022 ◇ FASTONA RED R ◇ FAST ORANGE 3R ◇ FLAME TONES ◇ FLAMING RED ◇ GRAPHTAL RED RL ◇ IRGALITE RED PRR ◇ ISOL FAST RED R ◇ KROMON RED R ◇ LATEXOL RED J ◇ LINCOLN RED 1002 ◇ LUTETIA FAST ORANGE 3R ◇ MONOLITE FAST RED G ◇ No. 1 FORTHFAST RED R ◇ ORALITH RED ◇ PERMANENT RED TONER R ◇ PERMANSA RED ◇ PERMATON RED XL 20-7015 ◇ PIGMENT RED 4 ◇ PYROTONE RED TONER RA-5520 ◇ 12094 RED ◇ RED No. 228 ◇ RUBBER RED R EXTRA ◇ SEGNALE LIGHT RED PRG ◇ SILOPOL RED G ◇ SYTON FAST RED R ◇ TANAGER RED X-761 ◇ TIGER ORANGE ◇ VERSAL FAST RED R ◇ VULCAFOR ORANGE R ◇ VULCAN RED R

TOXICITY DATA with REFERENCE
mmo-sat 100 μg/plate JJIND8 64,665,80
mma-sat 5 μg/plate ESKGA2 29,212,83

CONSENSUS REPORTS: Reported in EPA TSCA Inventory. EPA Genetic Toxicology Program.

SAFETY PROFILE: Mutation data reported. When heated to decomposition it emits very toxic fumes of Cl^- and NO_x. See also NITRO COMPOUNDS of AROMATIC HYDROCARBONS.

CJD600 CAS:135-12-6 ***HR: 1***
4-CHLORO-2-NITROPHENYL p-CHLOROPHENYL ETHER
mf: $C_{12}H_7Cl_2NO_3$ mw: 284.10

SYNS: BENZENE,4-CHLORO-1-(4-CHLOROPHENOXY)-2-NITRO- ◇ 4,4'-DICHLOR-2-NITRODIFENYLETHER ◇ ETHER, 4-CHLOROPHENYL (4'-CHLORO-2'-NITRO)PHENYL

TOXICITY DATA with REFERENCE
skn-rbt 500 mg/24H MLD 28ZPAK-,84,72
eye-rbt 100 mg/24H MOD 28ZPAK-,84,72

CONSENSUS REPORTS: Reported in EPA TSCA Inventory.

SAFETY PROFILE: A skin and eye irritant. When heated to decomposition it emits toxic fumes of NO_x and Cl^-.

CJD625 CAS:64046-47-5 ***HR: 1***
2-CHLORO-5-NITROPHENYL ESTER ACETIC ACID
mf: $C_8H_6ClNO_4$ mw: 215.60

SYN: METHYLESTERKYSELINY-2-CHLOR-5-NITROBENZOOVE (CZECH)

TOXICITY DATA with REFERENCE
skn-rbt 500 mg/24H MLD 28ZPAK -,92,72
eye-rbt 500 mg/24H MLD 28ZPAK -,92,72
orl-rat LD50:5360 mg/kg 28ZPAK -,92,72

SAFETY PROFILE: Mildly toxic by ingestion. An eye and skin irritant. When heated to decomposition it emits toxic fumes of Cl^- and NO_x. See also ESTERS.

CJD650 CAS:328-04-1 ***HR: 3***
o-(2-CHLORO-4-NITROPHENYL)-o-ISOPROPYL ETHYLPHOSPHONOTHIOATE
mf: $C_{11}H_{15}ClNO_4PS$ mw: 323.75

SYNS: ENT 25,755 ◇ N 2404 ◇ STAUFFER N-2404

TOXICITY DATA with REFERENCE
orl-rat LD50:32 mg/kg ARSIM* 20,22,66
orl-mus LDLo:62 mg/kg AECTCV 14,111,85
orl-ckn LD50:4 mg/kg TXAPA9 7,606,65

SAFETY PROFILE: Poison by ingestion. When heated to decomposition it emits toxic fumes of Cl^-, NO_x, PO_x, and SO_x.

CJD750 CAS:2425-66-3 ***HR: 3***
CHLORONITROPROPANE
mf: $C_3H_6ClNO_2$ mw: 123.55

SYNS: CHLORONITROPROPAN (POLISH) ◇ 1-CHLORO-2-NITROPROPANE

TOXICITY DATA with REFERENCE
orl-rat LD50:197 mg/kg PCOC** -,647,66
ihl-rat LC50:1070 mg/m^3 GUCHAZ 6,111,73
skn-rat LD50:362 mg/kg 31ZOAD 1,89,68
orl-mus LD50:105 mg/kg GUCHAZ 6,111,73
skn-rbt LD50:362 mg/kg WRPCA2 9,119,70

SAFETY PROFILE: Poison by ingestion and skin contact. Moderately toxic by inhalation. When heated to decomposition it emits very toxic fumes of Cl^- and NO_x. See also other chloronitropropane entries.

CJE000 CAS:600-25-9 ***HR: 3***
1-CHLORO-1-NITROPROPANE
mf: $C_3H_6ClNO_2$ mw: 123.55

PROP: Liquid. Bp: 139.5°, flash p: 144°F (OC), d: 1.209 @ 20°/20°, vap d: 4.26.

SYNS: CHLORONITROPROPANE ◇ KORAX ◇ LANSTAN

TOXICITY DATA with REFERENCE
orl-rbt LDLo:50 mg/kg JIHTAB 27,95,45
unr-rat LD50:197 mg/kg FMCHA2 -,C54,83
orl-mus LD50:510 mg/kg HYSAAV 34(10-12),429,69
ihl-mus LC50:66 g/m^3/3H HYSAAV 34(10-12),429,69
scu-mus LD50:165 mg/kg HYSAAV 34(10-12),429,69
ihl-rbt LCLo:2000 mg/m^3/6H JIHTAB 27,95,45
ihl-gpg LCLo:580 mg/m^3/1H JIHTAB 27,95,45

OSHA PEL: (Transitional: TWA 20 ppm) TWA 2 ppm
ACGIH TLV: TWA 2 ppm
DFG MAK: 20 ppm (100 mg/m^3)

SAFETY PROFILE: Poison by ingestion, subcutaneous, and possibly other routes. Moderately toxic by inhalation. Causes injury to kidneys, liver, and cardiovascular system. Combustible when exposed to heat, flame (sparks), and oxidizers. Moderately explosive when exposed to heat. To fight fire, use alcohol foam, water, CO_2, or dry chemical. When heated to decomposition it emits toxic fumes of Cl^- and NO_x. See also other chloropropane entries and CHLORIDES.

CJE250 CAS:594-71-8 ***HR: 3***
2-CHLORO-2-NITROPROPANE
mf: $C_3H_6ClNO_2$ mw: 123.55

PROP: Liquid. Bp: 132°, flash p: 135°F (OC), d: 1.193 @ 20°/20°, vap d: 4.26.

TOXICITY DATA with REFERENCE
orl-rbt LDLo:500 mg/kg JIHTAB 27,95,45
orl-mus LD50:580 mg/kg HYSAAV 34(10-12),429,69
ihl-mus LC50:55 g/m³/3H HYSAAV 34(10-12),429,69
scu-mus LD50:195 mg/kg HYSAAV 34(10-12),429,69

SAFETY PROFILE: Poison by subcutaneous route. Moderately toxic by ingestion. Mildly toxic by inhalation. Flammable when exposed to heat, flame, or oxidizers. Explodes on rapid heating. When heated to decomposition it emits toxic fumes of Cl^- and NO_x. See also other chloronitropropane entries and CHLORIDES.

CJE500 CAS:14100-52-8 ***HR: 3***
3-CHLORO-4-NITROQUINOLINE-1-OXIDE
mf: $C_9H_5ClN_2O_3$ mw: 224.61

TOXICITY DATA with REFERENCE
cyt-omi 36 μmol/L GANNA2 60,155,69
scu-mus TDLo:347 mg/kg/I:ETA CPBTAL 17,544,59

SAFETY PROFILE: Questionable carcinogen with experimental tumorigenic data. Mutation data reported. When heated to decomposition it emits very toxic fumes of Cl^- and NO_x.

CJE750 CAS:14076-19-8 ***HR: 3***
5-CHLORO-4-NITROQUINOLINE-1-OXIDE
mf: $C_9H_5ClN_2O_3$ mw: 224.61

TOXICITY DATA with REFERENCE
scu-mus TDLo:120 mg/kg/50D-I:ETA BCPCA6 16,631,67

SAFETY PROFILE: Questionable carcinogen with experimental tumorigenic data. When heated to decomposition it emits very toxic fumes of Cl^- and NO_x.

CJF000 CAS:3741-12-6 ***HR: 3***
6-CHLORO-4-NITROQUINOLINE-1-OXIDE
mf: $C_9H_5ClN_2O_3$ mw: 224.61

TOXICITY DATA with REFERENCE
cyt-hmn:lvr 5260 nmol/L JNCIAM 47,367,71
dnd-mus:fbr 10 μmol/L CNREA8 35,521,75
otr-ham:emb 4 μmol/L PJACAW 42,1211,66
dns-ham:oth 500 μmol/L NATUAS 229,416,71
dnd-mam:lym 5 mg BIPMAA 4,409,66
skn-rat TDLo:38 mg/kg/17W-I:ETA GANNA2 53,167,62

CONSENSUS REPORTS: EPA Genetic Toxicology Program.

SAFETY PROFILE: Questionable carcinogen with experimental tumorigenic data by skin contact and other routes. Mutation data reported. When heated to decomposition it emits very toxic fumes of Cl^- and NO_x.

CJF250 CAS:14753-14-1 ***HR: 3***
7-CHLORO-4-NITROQUINOLINE-1-OXIDE
mf: $C_9H_5ClN_2O_3$ mw: 224.61

TOXICITY DATA with REFERENCE
pic-esc 94 mg/L EXPEAM 24,1245,68
cyt-omi 44 μmol/L GANNA2 60,155,69
mmo-smc 2 mg/L IGSBAL 85,127,72
scu-mus TDLo:120 mg/kg/50D-I:ETA BCPCA6 16,631,67

SAFETY PROFILE: Questionable carcinogen with experimental tumorigenic data. Mutation data reported. When heated to decomposition it emits very toxic fumes of Cl^- and NO_x.

CJF500 CAS:6505-75-5 ***HR: 3***
3'-CHLORO-5-NITROSALICYLANILIDE
mf: $C_{13}H_9ClN_2O_4$ mw: 292.69

SYN: USAF BV-8

TOXICITY DATA with REFERENCE
orl-mus LDLo:710 mg/kg AECTCV 14,111,85
ipr-mus LD50:15 mg/kg NTIS** AD610-947
ivn-mus LD50:100 mg/kg CSLNX* NX#00478

SAFETY PROFILE: Poison by intraperitoneal and intravenous routes. Moderately toxic by ingestion. When heated to decomposition it emits very toxic fumes of Cl^- and NO_x.

CJF825 CAS:695-64-7 ***HR: 3***
1-CHLORO-1-NITROSOCYCLOHEXANE
mf: $C_6H_{10}ClNO$ mw: 147.60

SAFETY PROFILE: May explode during vacuum distillation. When heated to decomposition it emits toxic fumes of Cl^- and NO_x.

CJG000 CAS:6866-10-0 ***HR: 3***
2-CHLORO-1-NITROSO-2-PHENYLPROPANE
mf: $C_9H_{10}ClNO$ mw: 183.64

$CH_3CClC_6H_5CH_2N{:}O$

SAFETY PROFILE: A storage hazard. The dry material can decompose vigorously in a sealed container. When heated to decomposition it emits toxic fumes of Cl^- and NO_x.

CJG375 CAS:65445-60-5 ***HR: 3***
3-CHLORONITROSOPIPERIDINE
mf: $C_5H_9ClN_2O$ mw: 148.61

SYNS: 3-CHLORO-1-NITROSOPIPERIDINE ◇ N-NITROSO-3-CHLOROPIPERIDINE

TOXICITY DATA with REFERENCE
mma-sat 1 μg/plate MUREAV 56,131,77
sln-dmg-orl 5 mmol/L/24H MUREAV 67,27,79
orl-rat TDLo:300 mg/kg/30W-C:ETA CNREA8 40,3325,80

SAFETY PROFILE: Questionable carcinogen with experimental tumorigenic data. Mutation data reported. Many N-nitroso compounds are carcinogens. When heated to decomposition it emits toxic fumes of Cl^- and NO_x. See also N-NITROSO COMPOUNDS.

CJG500 CAS:65445-61-6 ***HR: 3***
4-CHLORONITROSOPIPERIDINE
mf: $C_5H_9ClN_2O$ mw: 148.61

SYNS: 4-CHLORO-1-NITROSOPIPERIDINE ◇ N-NITROSO-4-CHLOROPIPERIDINE

TOXICITY DATA with REFERENCE
mma-sat 1 μmol/plate MUREAV 56,131,77
sln-dmg-orl 5 mmol/L/24H MUREAV 67,27,79
sce-hmn:lym 1 mmol/L TCMUE9 1,129,84
orl-rat TDLo:300 mg/kg/30W-C:ETA CNREA8 40,3325,80

CONSENSUS REPORTS: EPA Genetic Toxicology Program.

SAFETY PROFILE: Questionable carcinogen with experimental tumorigenic data. Human mutation data reported. Many N-nitroso compounds are carcinogens. When heated to decomposition it emits very toxic fumes of Cl^- and NO_x. See also N-NITROSO COMPOUNDS.

CJG750 CAS:10140-97-3 ***HR: 3***
α-CHLORO-p-NITROSTYRENE
mf: $C_8H_6ClNO_2$ mw: 183.60

TOXICITY DATA with REFERENCE
orl-rat LDLo:710 mg/kg AIHAAP 23,95,62
skn-rbt LD50:390 mg/kg AIHAAP 23,95,62

SAFETY PROFILE: Poison by skin contact. Moderately toxic by ingestion. When heated to decomposition it emits very toxic fumes of Cl^- and NO_x.

CJG800 CAS:121-86-8 ***HR: 1***
2-CHLORO-4-NITROTOLUENE
mf: $C_7H_6ClNO_2$ mw: 171.59

SYN: TOLUENE, 2-CHLORO-4-NITRO-

TOXICITY DATA with REFERENCE
skn-rbt 500 mg/24H MLD 85JCAE-,603,86
eye-rbt 500 mg/24H MLD 85JCAE-,603,86

CONSENSUS REPORTS: Reported in EPA TSCA Inventory.

SAFETY PROFILE: A skin and eye irritant. When heated to decomposition it emits toxic fumes of NO_x and Cl^-.

CJH250 ***HR: 3***
CHLORONIUM PERCHLORATE
mf: $Cl_2H_2O_4$ mw: 136.92

SAFETY PROFILE: May explode spontaneously. Upon decomposition it emits toxic fumes of Cl^-. See also PERCHLORATES and CHLORIDES.

CJH500 CAS:1120-10-1 ***HR: 2***
9-CHLORONONANOIC ACID
mf: $C_9H_{17}ClO_2$ mw: 192.71

SYNS: CHLOROPERALGONIC ACID ◇ CHLORPERALGONIC

TOXICITY DATA with REFERENCE
orl-rat LD50:3000 mg/kg GISAAA 27(10),12,62
orl-mus LD50:3000 mg/kg GISAAA 27(10),12,62
orl-mam LD50:3000 mg/kg GISAAA 27(10),12,62

SAFETY PROFILE: Moderately toxic by ingestion. When heated to decomposition it emits toxic fumes of Cl^-.

CJH750 CAS:35941-71-0 ***HR: 3***
4-((5-CHLORO-2-OXO-3(2H)-BENZOTHIAZOLYL) ACETYL)-1-PIPERAZINEETHANOL
mf: $C_{15}H_{18}ClN_3O_3S•ClH$ mw: 392.33

SYNS: 5-CHLORO-3-(4-(2-HYDROXYETHYL)-1-PIPERAZINYL)CARBONYLMETHYL-2-BENZOTHIAZOLINONE ◇ FK 1160 ◇ NTA-194 ◇ SOLANTAL ◇ TIARAMIDE HYDROCHLORIDE

TOXICITY DATA with REFERENCE
eye-rbt 200 mg rns MLD ARZNAD 22,724,72
orl-rat TDLo:6 g/kg (9-14D preg):TER ARZNAD 23,504,73
orl-rat LD50:2300 mg/kg ARZNAD 32,225,82
ipr-rat LD50:530 mg/kg NIIRDN 6,451,82
scu-rat LD50:930 mg/kg ARZNAD 23,504,73
ivn-rat LD50:203 mg/kg ARZNAD 23,504,73
orl-mus LD50:564 mg/kg ARZNAD 23,504,73
ipr-mus LD50:298 mg/kg ARZNAD 23,504,73
scu-mus LD50:375 mg/kg ARZNAD 23,504,73
ivn-mus LD50:178 mg/kg ARZNAD 23,504,73

SAFETY PROFILE: Poison by subcutaneous, intravenous, and intraperitoneal routes. Moderately toxic by ingestion. An experimental teratogen. An eye irritant.

When heated to decomposition it emits very toxic fumes of Cl^-, NO_x, and SO_x.

CJI000 ***HR: 3***
CHLOROPARAFFIN XP-470

SYN: XP-470

TOXICITY DATA with REFERENCE
scu-rat TDLo:90 g/kg/52W-I:CAR GISAAA 44(7),68,79

SAFETY PROFILE: Questionable carcinogen with experimental carcinogenic data. When heated to decomposition it emits toxic fumes of Cl^-.

CJI100 CAS:7203-90-9 ***HR: 3***
CHLORO-PDMT
mf: $C_8H_{10}ClN_3$ mw: 183.66

SYNS: 1-p-CHLORFENYL-3,3-DIMETHYLTRIAZEN (CZECH) ◇ 1-(p-CHLOROPHENYL)-3,3-DIMETHYL-TRIAZENE ◇ 1-(4-CHLOROPHENYL)-3,3-DIMETHYLTRIAZENE ◇ 1-(p-CHLOR-PHENYL)-3,3-DIMETHYL-TRIAZEN (GERMAN)

TOXICITY DATA with REFERENCE
mma-sat 5 mmol/L MUREAV 36,1,76
sln-dmg-orl 100 μmol/L CBINA8 9,365,74
mnt-mus-ipr 50 mg/kg/24H MUREAV 56,319,78
scu-rat TDLo:3200 mg/kg/70W-I:CAR ZKKOBW 81,285,74
scu-rat TD:300 mg/kg:NEO ZKKOBW 81,285,74
orl-rat LD50:362 mg/kg 28ZPAK -,98,72
scu-rat LD50:330 mg/kg ZKKOBW 81,285,74

CONSENSUS REPORTS: EPA Genetic Toxicology Program.

SAFETY PROFILE: Poison by ingestion and subcutaneous routes. Questionable carcinogen with experimental carcinogenic and neoplastigenic data. Mutation data reported. When heated to decomposition it emits toxic fumes of Cl^- and NO_x.

CJI250 CAS:6984-99-2 ***HR: 3***
CHLOROPENTAFLUOROACETONE HYDRATE
mf: $C_6Cl_2F_{10}O_2 \cdot 5H_2O$ mw: 455.06

TOXICITY DATA with REFERENCE
orl-rat LD50:85 mg/kg TXAPA9 7,592,65
skn-rat LD50:81 mg/kg TXAPA9 7,592,65

SAFETY PROFILE: Poison by ingestion and skin contact. When heated to decomposition it emits very toxic fumes of Cl^- and F^-.

CJI500 CAS:76-15-3 ***HR: 1***
CHLOROPENTAFLUOROETHANE
DOT: UN 1020
mf: C_2ClF_5 mw: 154.47

PROP: Colorless gas. Bp: −39.3°, mp: −38°, d: 1.5678 @ -42°. Insol in water; sol in alc and ether.

SYNS: F-115 ◇ FLUOROCARBON-115 ◇ FREON 115 ◇ GENETRON 115 ◇ HALOCARBON 115 ◇ MONOCHLOROPENTAFLUOROETHANE (DOT)

CONSENSUS REPORTS: Reported in EPA TSCA Inventory.

OSHA PEL: TWA 1000 ppm
ACGIH TLV: TWA 1000 ppm
DOT Classification: Nonflammable Gas; Label: Nonflammable Gas.

SAFETY PROFILE: Mildly toxic by inhalation. A nonflammable gas. When heated to decomposition it emits toxic fumes of F^- and Cl^-.

CJI609 ***HR: 3***
CHLOROPEPTIDE

PROP: Chlorine-containing cyclic pentapeptide from metabolites of *Penicillium islandicum Sopp.* (JTSCDR 2,349,77).

TOXICITY DATA with REFERENCE
orl-rat LD50:5 mg/kg JTSCDR 2,349,77
scu-rat LD50:400 μg/kg JTSCDR 2,349,77
orl-mus LD50:7 mg/kg JTSCDR 2,349,77
ipr-mus LD50:450 μg/kg JTSCDR 2,349,77
scu-mus LD50:500 μg/kg JTSCDR 2,349,77
ivn-mus LD50:400 μg/kg JTSCDR 2,349,77
scu-gpg LD50:500 mg/kg JTSCDR 2,349,77

SAFETY PROFILE: Poison by subcutaneous, intravenous, and intraperitoneal routes. Mildly toxic by ingestion. When heated to decomposition it emits toxic fumes of Cl^- and NO_x.

CJI750 CAS:937-14-4 ***HR: 3***
3-CHLOROPEROXYBENZOIC ACID
DOT: UN 2755
mf: $C_7H_5ClO_3$ mw: 172.57

$ClC_6H_4CO{:}OOH$

PROP: Mp: 94°.

SYNS: 3-CHLORO-BENZENECARBOPEROXOIC ACID (9CI) ◇ m-CHLOROBENZOYL HYDROPEROXIDE ◇ m-CHLOROPERBENZOIC ACID ◇ 3-CHLOROPERBENZOIC ACID ◇ m-CHLOROPEROXYBENZOIC ACID

TOXICITY DATA with REFERENCE
skn-mus TDLo:21 g/kg/52W-I:ETA JNCIAM 55,1359,75

CONSENSUS REPORTS: Reported in EPA TSCA Inventory.

SAFETY PROFILE: Questionable carcinogen with experimental tumorigenic data. Reaction with 2-bromo-4-

methyl pyridine forms a violently unstable product. When heated to decomposition it emits toxic fumes of Cl^-. See also PEROXIDES and CHLORIDES.

CJI809 CAS:32755-26-3 ***HR: 3***
CHLOROPEROXYTRIFLUOROMETHANE
mf: $CClF_3O_2$ mw: 98.46

SAFETY PROFILE: Initiates explosive polymerization of tetrafluoroethylene. When heated to decomposition it emits toxic fumes of F^- and Cl^-. See also PEROXIDES.

CJJ000 CAS:3691-35-8 ***HR: 3***
CHLOROPHACINONE
mf: $C_{23}H_{15}ClO_3$ mw: 374.83

SYNS: AFNOR ◇ CAID ◇ CHLOORFACINON (DUTCH) ◇ 2(2-(4-CHLOOR-FENYL-2-FENYL)-ACETYL)-INDAAN-1,3-DION(DUTCH) ◇ CHLORFACINON (GERMAN) ◇ 2-(α-p-CHLOROPHENYLACETYL)INDANE-1,3-DIONE ◇ 2-((p-CHLOROPHENYL)PHENYLACETYL)-1,3-INDANDIONE ◇ 2(2-(4-CHLOROPHENYL)-2-PHENYLACETYL)INDAN-1,3-DIONE ◇ 2-((4-CHLOROPHENYL)PHENYLACETYL)-1H-INDENE-1,3(2H)-DIONE ◇ CHLORPHACINON (ITALIAN) ◇ 2(2-(4-CHLOR-PHENYL-2-PHENYL)ACETYL)INDAN-1,3-DION(GERMAN) ◇ ((4-CHLORPHENYL)-1-PHENYL)-ACETYL-1,3-INDANDION (GERMAN) ◇ 1-(4-CHLORPHENYL)-1-PHENYL-ACETYL-INDAN-1,3-DION (GERMAN) ◇ 2(2-(4-CLORO-FENIL-2-FENIL)-ACETIL)INDAN-1,3-DIONE (ITALIAN) ◇ DELTA ◇ DRAT ◇ LIPHADIONE ◇ LM 91 ◇ MICROZUL ◇ MURIOL ◇ 2-(2-PHENYL-2-(4-CHLOROPHENYL)ACETYL)-1,3-INDANDIONE ◇ QUICK ◇ RAMUCIDE ◇ RANAC ◇ RATOMET ◇ RAVIAC ◇ ROZOL ◇ TOPITOX

TOXICITY DATA with REFERENCE
orl-rat LD50:2100 μg/kg GUCHAZ 6,112,73
orl-mus LD50:1060 μg/kg TXAPA9 25,42,73
orl-rbt LD50:50 mg/kg GUCHAZ 6,112,73
skn-rbt LD50:200 mg/kg GUCHAZ 6,112,73
orl-dck LD50:100 mg/kg GUCHAZ 6,112,73
orl-mam LD50:7500 μg/kg GUCHAZ 6,112,73
orl-brd LD50:430 mg/kg GUCHAZ 6,112,73

CONSENSUS REPORTS: EPA Extremely Hazardous Substances List.

SAFETY PROFILE: Poison by ingestion and skin contact. A pesticide. When heated to decomposition it emits toxic fumes of Cl^-.

CJJ250 CAS:6164-98-3 ***HR: 3***
CHLOROPHENAMIDINE
mf: $C_{10}H_{13}ClN_2$ mw: 196.70

SYNS: ACARON ◇ BERMAT ◇ C 8514 ◇ CARZOL ◇ CDM ◇ CHLORDIMEFORM ◇ CHLORFENAMIDINE ◇ N'-(4-CHLORO-2-METHYLPHENYL)-N,N-DIMETHYLMETHANIMIDAMIDE◇ CHLOROPHENAMADIN ◇ N'-(4-CHLORO-o-TOLYL)-N,N-DIMETHYLFORMAMIDINE ◇ CHLORPHENAMIDINE ◇ N'-(4-CHLOR-o-TOLYL)-N,N-DIMETHYLFORMAMIDIN (GERMAN) ◇ CIBA 8514 ◇ N,N-DIMETHYL-N'-(2-METHYL-4-CHLOROPHENYL)-FORMAMIDINE ◇ N,N-DIMETHYL-N'-(2-METHYL-4-CHLORPHENYL)-FORMADIN (GERMAN) ◇ ENT 27,335 ◇ ENT 27,567 ◇ EP-333 ◇ FUNDAL ◇ FUNDAL 500 ◇ FUNDEX ◇ GALECRON ◇ N'-(2-METHYL-4-CHLOROPHENYL)-N,N-DIMETHYLFORMAMIDINE ◇ N'-(2-METHYL-4-CHLORPHENYL)-FORMAMIDIN-HYDROCHLORID(GERMAN) ◇ NSC 190935 ◇ RS 141 ◇ SCHERING 36268 ◇ SN 36268 ◇ SPANON ◇ SPANONE

TOXICITY DATA with REFERENCE
skn-rbt 500 mg open MLD CIGET* 6/2/75
eye-rbt 100 mg MLD CIGET* 6/2/75
mmo-smc 5 ppm RSTUDV 6,161,76
dni-hmn:hla 1 mmol/L BECTA6 11,184,74
oms-hmn:hla 1 mmol/L BECTA6 11,184,74
orl-rat TDLo:1800 μg/kg (5-22D preg):REP BECTA6 20,760,78
orl-mus TDLo:6552 mg/kg/78W-C:CAR CHYCDW 19,154,85
orl-rat LD50:160 mg/kg KSKZAN 16(2),59,78
skn-rat LD50:263 mg/kg FAATDF 7,299,86
ipr-rat LD50:90 mg/kg PSDTAP 15,239,74
orl-mus LD50:160 mg/kg SPEADM 78-1,-,78
skn-mus LD50:225 mg/kg SPEADM 78-1,21,78
ipr-mus LD50:71 mg/kg OYYAA2 1,78,67
orl-rbt LD50:625 mg/kg 28ZEAL 5,43,76
skn-rbt LD50:640 mg/kg 85DPAN -,-,71/76

CONSENSUS REPORTS: IARC Cancer Review: Group 3 IMEMDT 7,56,87. EPA Genetic Toxicology Program.

SAFETY PROFILE: Poison by ingestion, skin contact, and intraperitoneal routes. Experimental reproductive effects. Human mutation data reported. An eye and skin irritant. Questionable carcinogen with experimental carcinogenic data. When heated to decomposition it emits very toxic fumes of NO_x and Cl^-.

CJJ500 CAS:19750-95-9 ***HR: 3***
CHLOROPHENAMIDINE HYDROCHLORIDE
mf: $C_{10}H_{13}ClN_2 \cdot ClH$ mw: 233.16

SYNS: CHLORDIMEFORM HYDROCHLORIDE ◇ N'-(4-CHLORO-2-METHYLPHENYL)-METHANIMIDAMIDEMONOHYDROCHLORIDE ◇ N'-(4-CHLORO-o-TOLYL)-N,N-DIMETHYLFORMAMIDINE HYDROCHLORIDE ◇ N,N-DIMETHYL-N'-(2-METHYL-4-CHLOROPHENYL)-FORMAMIDINE HYDROCHLORIDE ◇ ENT 27,567 ◇ EP 333 ◇ FUNDAL SP ◇ GALECRON SP ◇ MORTON EP 333 ◇ NOR-AM EP 333 ◇ NSC 195102 ◇ SCHERING 36268

TOXICITY DATA with REFERENCE
orl-rat LD50:225 mg/kg SPEADM 78-1,-,78
skn-rat LD50:4000 mg/kg SPEADM 78-1,22,78
orl-mus LD50:290 mg/kg BESAAT 15,103,69
orl-rbt LD50:625 mg/kg BESAAT 15,103,69

SAFETY PROFILE: Poison by ingestion. Moderately toxic by skin contact. A pesticide. When heated to decomposition it emits very toxic fumes of NO_x and Cl^-.

CJK000 CAS:2598-25-6 ***HR: 3***
1-(p-CHLOROPHENETHYL)HYDRAZINE HYDROGEN SULFATE
mf: $C_8H_{11}ClN_2 \cdot H_2O_4S$ mw: 268.74

SYNS: p-CHLORO-β-PHENYLETHYLHYDRAZINE DIHYDROGEN SULFATE ◇ WL 28

TOXICITY DATA with REFERENCE
scu-mus TDLo:360 mg/kg (1-6D preg):REP JOENAK 30,205,64
orl-mus LD50:175 mg/kg JMPCAS 5,221,62
scu-mus LD50:250 mg/kg JOENAK 30,205,64

SAFETY PROFILE: Poison by ingestion and subcutaneous routes. Experimental reproductive effects. When heated to decomposition it emits very toxic fumes of Cl^-, NO_x, and SO_x. See also HYDRAZINE and SULFATES.

CJK100 CAS:2598-75-6 ***HR: 2***
2-(o-CHLOROPHENETHYL)-3-THIOSEMICARBAZIDE
mf: $C_9H_{12}ClN_3S$ mw: 229.75

SYNS: SEMICARBAZIDE,2-(o-CHLOROPHENETHYL)-3-THIO- ◇ WL 34

TOXICITY DATA with REFERENCE
scu-mus TDLo:1500 mg/kg (female 1-6D post):REP JOENAK 30,205,64
scu-mus LD50:500 mg/kg JOENAK 30,205,64

SAFETY PROFILE: Moderately toxic by subcutaneous route. Experimental reproductive effects. When heated to decomposition it emits toxic fumes of SO_x, NO_x, and Cl^-.

CJK250 CAS:95-57-8 ***HR: 3***
2-CHLOROPHENOL
DOT: UN 2020/UN 2021
mf: C_6H_5ClO mw: 128.56

PROP: Light amber liquid. Bp: 174.5°, fp: 7°, d: 1.256 @ 25°/25°, flash p: 147°F, vap press: 1 mm @ 12.1°. Sltly water sol; very sol in alc, ether, and alkali.

SYNS: o-CHLOROPHENOL ◇ o-CHLOROPHENOL, liquid (DOT) ◇ o-CHLOROPHENOL, solid (DOT) ◇ o-CHLORPHENOL (GERMAN) ◇ RCRA WASTE NUMBER U048

TOXICITY DATA with REFERENCE
sln-ham:lng 800 umol/L MUREAV 182,135,87
orl-rat TDLo:4550 mg/kg (70 pre/1-21D preg):REP EVHPAZ 46,137,82
skn-mus TDLo:4800 mg/kg/12W-I:ETA CNREA8 19,413,59
orl-rat LD50:670 mg/kg FEPRA7 2,76,43
ipr-rat LD50:230 mg/kg BJPCAL 13,20,58
scu-rat LD50:950 mg/kg FEPRA7 2,76,43
orl-mus LD50:345 mg/kg TOLED5 29 39,85
ipr-mus LD50:235 mg/kg JMCMAR 18,868,75
scu-rbt LDLo:950 mg/kg HBAMAK 4,1361,35
ivn-rbt LDLo:120 mg/kg HBIXAC 5,112,59

CONSENSUS REPORTS: Reported in EPA TSCA Inventory. Chlorophenol compounds are on the Community Right-To-Know List.

DOT Classification: Poison B; Label: St. Andrews Cross.

SAFETY PROFILE: Poison by ingestion, intraperitoneal, and intravenous routes. Experimental reproductive effects. Questionable carcinogen with experimental tumorigenic data. Mutation data reported. Flammable when exposed to heat, flame, or oxidizers. To fight fire, use alcohol foam. When heated to decomposition it emits toxic fumes of Cl^-. See also CHLOROPHENOLS and CHLORIDES.

CJK500 CAS:108-43-0 ***HR: 3***
3-CHLOROPHENOL
DOT: UN 2020/UN 2021
mf: C_6H_5ClO mw: 128.56

PROP: Crystals. Mp: 33.5°, bp: 214°, d: 1.245 @ 45°/4°, vap press: 1 mm @ 44.2°, flash p: >112°.

SYNS: m-CHLOROPHENOL ◇ m-CHLOROPHENOL, liquid (DOT) ◇ m-CHLOROPHENOL, solid (DOT)

TOXICITY DATA with REFERENCE
skn-mus TDLo:6000 mg/kg/15W-I:ETA CNREA8 19,413,59
orl-rat LD50:570 mg/kg FEPRA7 2,76,43
ipr-rat LD50:355 mg/kg BJPCAL 13,20,58
scu-rat LD50:1390 mg/kg FEPRA7 2,76,43
orl-mus LD50:521 mg/kg TOLED5 29,39,85

CONSENSUS REPORTS: Reported in EPA TSCA Inventory. Chlorophenol compounds are on the Community Right-To-Know List.

DOT Classification: Poison B; Label: St. Andrews Cross.

SAFETY PROFILE: Poison by intraperitoneal route. Moderately toxic by ingestion and subcutaneous routes. Questionable carcinogen with experimental tumorigenic data by skin contact. When heated to decomposition it emits toxic fumes of Cl^-. See also CHLOROPHENOLS.

CJK750 CAS:106-48-9 ***HR: 3***
4-CHLOROPHENOL
mf: C_6H_5ClO mw: 128.56

DOT: UN 2020/UN 2021

PROP: Needle-like, white to straw colored crystals; unpleasant odor. fp: 42.8°, flash p: 250°F, d: 1.246 @ 60°/25°, vap press: 1 mm @ 49.8°, mp: 43.5°, bp: 220°, sltly water sol; very sol in alc, chloroform, and ether.

SYNS: p-CHLORFENOL (CZECH) ◇ p-CHLOROPHENOL ◇ p-CHLOROPHENOL, liquid (DOT) ◇ p-CHLOROPHENOL, solid (DOT)

TOXICITY DATA with REFERENCE
skn-rbt 500 mg/24H SEV 28ZPAK -,78,72
eye-rbt 250 μg/24H SEV 28ZPAK -,78,72
mmo-sat 200 μg/plate PCBPBS 10,174,79
orl-rat LD50:261 mg/kg 28ZPAK -,78,72
ipr-rat LD50:281 mg/kg BJPCAL 13,20,58
scu-rat LD50:1030 mg/kg FEPRA7 2,76,43
orl-mus LD50:1373 mg/kg TOLED5 29,39,85
ipr-mus LD50:332 mg/kg JMCMAR 18,868,75
orl-mam LD50:500 mg/kg GISAAA 45(10),16,80
skn-mam LD50:1000 mg/kg GISAAA 45(10),16,80

CONSENSUS REPORTS: Reported in EPA TSCA Inventory. Chlorophenol compounds are on the Community Right-To-Know List.

DOT Classification: IMO: Poison B; Label: St. Andrews Cross.

SAFETY PROFILE: Poison by ingestion and intraperitoneal routes. Moderately toxic by skin contact and subcutaneous routes. A severe skin and eye irritant. Mutation data reported. Combustible when exposed to heat or flame. To fight fire, use water, spray, mist, fog, foam, dry chemical. When heated to decomposition it emits toxic fumes of Cl^-. See also CHLOROPHENOLS and CHLORIDES.

CJL000 ***HR: 3***
CHLOROPHENOLS

CONSENSUS REPORTS: Chlorophenol compounds are on the Community Right-To-Know List.

SAFETY PROFILE: Many are suspected experimental carcinogens. Most are strong eye and skin irritants. They are systemic irritants by inhalation, ingestion, and skin contact. Generally mutagenic.

Trichlorophenols are generally poisons and may be carcinogens. They may contain 2,3,7,8-tetrachlorodibenzo-p-dioxin (TCDD) as a contaminant. Some trichlorophenols are used as herbicides (e.g., 2,4,5-T and silvex). Human exposure may cause chloracne, liver dysfunction, muscle weakness, and prophyria.

Pentachlorophenol is a poison by several routes. Human exposure causes increased respiration, fever, tachycardia, muscle weakness, and cardiac failure. Many toxic effects are due to impurities in commercial grade material. A teratogen and mutagen. Pentachlorophenol and 2,4,6-trichlorophenol may interfere with mitochondrial oxidative phosphorylation. When heated to decomposition they emit toxic fumes of Cl^-. See also specific compounds, PHENOL, and CHLORIDES.

CJL409 CAS:19142-68-8 ***HR: 3***
2-(2-(4-(2-((2-CHLORO-10-PHENOTHIAZINYL) METHYL)PROPYL)-1-PIPERAZINYL) ETHOXY)ETHANOL
mf: $C_{24}H_{32}ClN_3O_2S$ mw: 462.10
SYN: UCB 2493

TOXICITY DATA with REFERENCE
orl-rat LD50:400 mg/kg ANPBAZ 61,669,61
ivn-rat LD50:50 mg/kg ANPBAZ 61,669,61

SAFETY PROFILE: Poison by ingestion and intravenous routes. When heated to decomposition it emits very toxic fumes of SO_x, NO_x, and Cl^-.

CJL500 CAS:84-04-8 ***HR: 3***
1-(3-(3-CHLOROPHENOTHIAZIN-10-YL)PROPYL)-ISONIPECOTAMIDE
mf: $C_{21}H_{24}ClN_3OS$ mw: 401.99

SYNS: MOMETINE ◇ MORNIDINE ◇ NAUSIDOL ◇ NOMETINE ◇ PIPAMAZINE ◇ SC 8016 ◇ SC 9387

TOXICITY DATA with REFERENCE
orl-chd TDLo:500 μg/kg:CNS 34ZIAG -,478,69
orl-rat LD50:620 mg/kg 27ZQAG -,39,72
orl-mus LD50:370 mg/kg 27ZQAG -,39,72
ipr-mus LD50:80 mg/kg 27ZQAG -,39,72

SAFETY PROFILE: Poison by ingestion and intraperitoneal routes. Human (child) systemic effects by ingestion: somnolence, sleep, and coma. When heated to decomposition it emits very toxic fumes of SO_x, NO_x, and Cl^-.

CJM250 CAS:58-39-9 ***HR: 3***
4-(3-(2-CHLOROPHENOTHIAZIN-10-YL)PROPYL)-1-PIPERAZINEETHANOL
mf: $C_{21}H_{26}ClN_3OS$ mw: 404.01

SYNS: 2-CHLORO-10-3-(1-(2-HYDROXYETHYL)-4-PIPERAZINYL) PROPYL PHENOTHIAZINE ◇ DECENTAN ◇ PERAZIN ◇ ETAPERAZINE ◇ ETHAPERAZINE ◇ FENTAZIN ◇ 1-(2-HYDROXYETHYL)-4-(3-(2-CHLORO-10-PHENOTHIAZINYL)PROPYL)PIPERAZINE◇ Γ-(4-(β-HYDROXYETHYL)PIPERAZIN-1-YL)PROPYL-2-CHLOROPHENOTHIAZINE ◇ 1′,1-(2-IDROSSIETIL)4-(3- (2-CLORO-10-FENOTIAZIL)PROPILPIPERAZINA (ITALIAN) ◇ PERFENAZINA (ITALIAN) ◇ PERPHENAZIN ◇ PERPHENAZINE ◇ TRIFARON ◇ TRILAFON

TOXICITY DATA with REFERENCE
cyt-hmn:leu 270 mg/kg BMJOAE 3,634,69
orl-mus TDLo:150 mg/kg (female 10-12D post):TER TOIZAG 28,621,81
unr-mus TDLo:4 mg/kg (female 1D pre):REP 85GRAA -,57,65

ims-hmn TDLo:71428 ng/kg:CNS BMJOAE 3,867,67
orl-rat LD50:318 mg/kg TXAPA9 21,315,72
ipr-rat LD50:146 mg/kg ARZNAD 24,917,74
ivn-rat LD50:34 mg/kg FRPPAO 26,585,71
orl-mus LD50:120 mg/kg 27ZIAQ -,-,65
ipr-mus LD50:64 mg/kg JPMSAE 59,976,70
ivn-mus LD50:19 mg/kg ARZNAD 11,932,61
ivn-dog LD50:51 mg/kg ARZNAD 10,638,60

SAFETY PROFILE: Poison by ingestion, intravenous, subcutaneous, intraperitoneal, and intramuscular routes. Human systemic effects by intramuscular route: muscle spasms. Experimental teratogenic and reproductive effects. Human mutation data reported. When heated to decomposition it emits very toxic fumes of SO_x, NO_x, and Cl^-.

CJM750 CAS:2865-70-5 ***HR: 3***
10-CHLOROPHENOXARSINE
mf: $C_{12}H_8AsClO$ mw: 278.57

SYNS: 10-CHLORO-10H-PHENOXARSINE ◇ 10-CHLOROPHENOXYARSINE ◇ DID 95

TOXICITY DATA with REFERENCE
orl-mus LDLo:42 mg/kg AECTCV 14,111,85
orl-bwd LD50:100 mg/kg TXAPA9 21,315,72

CONSENSUS REPORTS: Reported in EPA TSCA Inventory. Arsenic and its compounds are on the Community Right-To-Know List.

OSHA PEL: TWA 0.5 mg(As)/m^3

SAFETY PROFILE: Poison by ingestion. See also ARSINE and ARSENIC COMPOUNDS. When heated to decomposition it emits very toxic fumes of As and Cl^-.

CJN000 CAS:122-88-3 ***HR: 2***
p-CHLOROPHENOXYACETIC ACID
mf: $C_8H_7ClO_3$ mw: 186.60

PROP: Mp: 159°.

SYNS: (4-CHLOROPHENOXY)ACETIC ACID ◇ 4-CP ◇ CPA ◇ MARKS 4-CPA ◇ PCPA ◇ SURE-SET ◇ TOMATO HOLD ◇ TOMATO FIX CONCENTRATE ◇ TOMATOTONE

TOXICITY DATA with REFERENCE
dns-hmn:oth 10 mmol/L ZYDXDM 13(2),74,84
orl-rat LD50:850 mg/kg RREVAH 10,97,65
ipr-mus LD50:680 mg/kg CHTPBA 5,211,70

CONSENSUS REPORTS: Reported in EPA TSCA Inventory.

SAFETY PROFILE: Moderately toxic by ingestion and intraperitoneal routes. Human mutation data reported. When heated to decomposition it emits toxic fumes of Cl^-.

CJN250 CAS:3544-35-2 ***HR: 3***
p-CHLOROPHENOXYACETIC ACID-2-ISOPROPYLHYDRAZIDE
mf: $C_{11}H_{15}ClN_2O_2$ mw: 242.73

SYNS: 1-(p-CHLOROPHENOXYACETYL)-2-ISOPROPYLHYDRAZINE ◇ IPROCLOZIDE ◇ ISOPROPILIDRAZIDE dell'ac. p-CLOROFENOSSIACETICO (ITALIAN) ◇ PC 603 ◇ PU 603 ◇ SINDERESIN ◇ SOG-4 ◇ SURSUM

TOXICITY DATA with REFERENCE
orl-rat TDLo:600 mg/kg (15D preg):REP AOGNAX 71,153,66
orl-rat LD50:550 mg/kg 27ZQAG -,394,72
orl-mus LD50:278 mg/kg 27ZQAG -,394,72
ipr-mus LD50:219 mg/kg PCJOAU 11,887,77

SAFETY PROFILE: Poison by ingestion and intraperitoneal routes. Experimental reproductive effects. When heated to decomposition it emits very toxic fumes of Cl^- and NO_x. See also HYDRAZINE.

CJN750 CAS:1223-36-5 ***HR: 3***
2-(p-CHLOROPHENOXY)-N-(2-(DIETHYLAMINO)ETHYL)ACETAMIDE
mf: $C_{14}H_{21}ClN_2O_2$ mw: 284.82

SYNS: AMICHLOPHENE ◇ ANP 246 ◇ CHLOFEXAMIDE ◇ 2-(4-CHLOROPHENOXY)-N-(2-(DIETHYLAMINO)ETHYL)ACETAMIDE ◇ CLOFEXAMIDE ◇ IEM 455 ◇ NP 246

TOXICITY DATA with REFERENCE
orl-rat LD50:1040 mg/kg OYYAA2 17,935,79
orl-mus LD50:1200 μg/kg CHTPBA 1,444,66
ivn-mus LD50:215 mg/kg ARZNAD 14,556,64
ivn-rbt LDLo:90 mg/kg ARZNAD 14,556,64

SAFETY PROFILE: Poison by ingestion and intravenous routes. When heated to decomposition it emits very toxic fumes of Cl^- and NO_x.

CJO250 CAS:43121-43-3 ***HR: 3***
1-(4-CHLOROPHENOXY)-3,3-DIMETHYL-1-(1,2,4-TRIAZOL-1-YL)-2-BUTAN-2-ONE
mf: $C_{14}H_{16}ClN_3O_2$ mw: 293.78

SYNS: AMIRAL ◇ BAY 6681 F ◇ BAYLETON ◇ BAY-MEB-6447 ◇ 1-((tert-BUTYLCARBONYL-4-CHLOROPHENOXY)METHYL)-1H-1,2,4-TRIAZOLE ◇ 1-(4-CHLOROPHENOXY)-3,3-DIMETHYL-1-(1H-1,2,4-TRIAZOL-1-YL)-2-BUTANONE ◇ MEB 6447 ◇ TRIADIMEFON

TOXICITY DATA with REFERENCE
orl-rat LD50:400 mg/kg FMCHA2 -,C27,83
orl-mus LD50:363 mg/kg MEIEDD 10,1372,83
orl-rbt LD50:500 mg/kg 85DPAN -,-,71/76

SAFETY PROFILE: Poison by ingestion. When heated to decomposition it emits very toxic fumes of Cl^- and NO_x. See also KETONES.

CJO500 CAS:1892-43-9 ***HR: 2***
2-(p-CHLOROPHENOXY)ETHANOL
mf: $C_8H_9ClO_2$ mw: 172.62

SYNS: 2-(4′-CHLORFENOXY)ETHANOL (CZECH) ◇ p-CHLORFENYLMONOGLYKOLETHER (CZECH) ◇ 2-(4-CHLOROPHENOXY)ETHANOL

TOXICITY DATA with REFERENCE
skn-rbt 10 mg/24H open MLD AIHAAP 23,95,62
skn-rbt 500 mg/24H MOD 28ZPAK -,81,72
eye-rbt 250 μg/24H SEV 28ZPAK -,81,72
orl-rat LD50:1490 mg/kg 28ZPAK -,81,72
skn-rbt LD50:500 mg/kg AIHAAP 23,95,62

CONSENSUS REPORTS: Reported in EPA TSCA Inventory.

SAFETY PROFILE: Moderately toxic by ingestion and skin contact. A severe eye and moderate skin irritant. When heated to decomposition it emits very toxic fumes of Cl^-.

CJP250 CAS:69782-24-7 ***HR: 3***
(2-(p-CHLOROPHENOXY)ETHYL)HYDRAZINE HYDROCHLORIDE
mf: $C_8H_{11}ClN_2O\cdot ClH$ mw: 223.12

TOXICITY DATA with REFERENCE
orl-mus LD50:300 mg/kg JMCMAR 6,63,63
ipr-mus LD50:300 mg/kg JMCMAR 6,63,63

SAFETY PROFILE: Poison by ingestion and intraperitoneal routes. When heated to decomposition it emits very toxic fumes of Cl^- and NO_x. See also HYDRAZINE.

CJP500 CAS:2598-73-4 ***HR: 3***
1-(2-(o-CHLOROPHENOXY)ETHYL)HYDRAZINE HYDROGEN SULFATE
mf: $C_8H_{11}ClN_2O\cdot H_2O_4S$ mw: 284.74

SYNS: HYDRAZINE, 1-(2-(o-CHLOROPHENOXY)ETHYL)-, HYDROGEN SULFATE (1:1) ◇ HYDRAZINE, 1-(2-(o-CHLOROPHENOXY) ETHYL)-, SULFATE (1:1) ◇ WEG 147

TOXICITY DATA with REFERENCE
scu-mus TDLo:180 mg/kg (female 1-6D post):REP JOENAK 30,205,64
scu-mus LD50:150 mg/kg JOENAK 30,205,64

SAFETY PROFILE: Poison by subcutaneous route. Experimental reproductive effects. When heated to decomposition it emits toxic fumes of SO_x, NO_x, and Cl^-.

CJP750 CAS:26129-32-8 ***HR: 3***
2-(4-(4-CHLOROPHENOXY)PHENOXY)PROPIONIC ACID
mf: $C_{15}H_{13}ClO_4$ mw: 292.73

SYNS: ACIDO FENOFIBRICO ◇ 2-(4-(4′-CHLORPHENOXY)-PHENOXY)-PROPIONSAEURE (GERMAN) ◇ FENOFIBRIC ACID ◇ HCG-004

TOXICITY DATA with REFERENCE
orl-rat LD50:1200 mg/kg DRFUD4 4,326,79
ivn-rat LD50:190 mg/kg DRFUD4 4,326,79
ipr-mus LD50:500 mg/kg NPMDAD 9,3737,80

SAFETY PROFILE: Poison by intravenous route. Moderately toxic by ingestion and intraperitoneal routes. When heated to decomposition it emits toxic fumes of Cl^-.

CJQ000 CAS:1982-47-4 ***HR: 3***
3-(p-(p-CHLOROPHENOXY)PHENYL)-1,1-DIMETHYLUREA
mf: $C_{15}H_{15}ClN_2O_2$ mw: 290.77

SYNS: C 1983 ◇ 3-(4-(4-CHLOOR-FENOXY)-FENOXY)-FENYL)-1,1-DIMETHYLUREUM (DUTCH) ◇ 3-(4-(4-CHLORO-FENOSSIL)-1,1-DIMETIL-UREA (ITALIAN) ◇ N′-4-(4-CHLOROPHENOXY)PHENYL-N,N-DIMETHYLUREA ◇ 1-(4-(4-CHLORO-PHENOXY)PHENYL)-3,3-D′METHYLUREE (FRENCH) ◇ CHLOROXIFENIDIM ◇ CHLOROXURON ◇ 3-(4-(4-CHLOR-PHENOXY)-PHENYL)-1,1-DIMETHYL-HARNSTOFF (GERMAN) ◇ CIBA 1983 ◇ NOREX ◇ TENORAN

TOXICITY DATA with REFERENCE
orl-rat LD50:3700 mg/kg WRPCA2 9,119,70
unk-rat LD50:1000 mg/kg 30ZDA9 -,232,71
orl-dog LD50:10 mg/kg 28ZEAL 4,114,69

CONSENSUS REPORTS: EPA Genetic Toxicology Program. EPA Extremely Hazardous Substances List.

SAFETY PROFILE: Poison by ingestion. Moderately toxic by an unspecified route. An herbicide. When heated to decomposition it emits very toxic fumes of Cl^- and NO_x.

CJQ250 CAS:886-74-8 ***HR: 3***
3-(p-CHLOROPHENOXY)-1,2-PROPANEDIOL-1-CARBAMATE
mf: $C_{10}H_{12}ClNO_4$ mw: 245.68

SYNS: 3-(p-CHLOROPHENOXY)-2-HYDROXYPROPYLCARBAMATE ◇ 3-(4-CHLOROPHENOXY)-1,2-PROPANEDIOL-1-CARBAMATE ◇ CHLORPHENESIN CARBAMATE ◇ MAOLATE ◇ U-19,646

TOXICITY DATA with REFERENCE
orl-dog TDLo:40 g/kg (14W pre):REP OYYAA2 14,27,77
orl-hmn TDLo:23 mg/kg/D:CNS 34ZIAG -,171,69
orl-rat LD50:632 mg/kg IYKEDH 10,710,70
ipr-rat LD50:354 mg/kg 27ZQAG -,386,72
scu-rat LD50:952 mg/kg OYYAA2 13,659,77
ivn-rat LD50:237 mg/kg YKYUA6 31,363,80
orl-mus LD50:741 mg/kg IYKEDH 10,710,79
ipr-mus LD50:452 mg/kg IYKEDH 10,710,79
scu-mus LD50:798 mg/kg IYKEDH 10,710,79
ivn-mus LD50:239 mg/kg 27ZQAG -,386,72

SAFETY PROFILE: Poison by intraperitoneal and intravenous routes. Moderately toxic by ingestion and subcutaneous routes. Human systemic effects by ingestion:

somnolence, hallucinations or distorted perceptions, and muscle weakness. Experimental reproductive effects. When heated to decomposition it emits very toxic fumes of Cl^- and NO_x. See also CARBAMATES.

CJQ500 CAS:102585-42-2 ***HR: 3***
N-(1-(o-CHLOROPHENOXY)-2-PROPYL)-2-(DIETHYLAMINO)-N-ETHYLACETAMIDE HYDROCHLORIDE
mf: $C_{17}H_{27}ClN_2O_2{\bullet}ClH$ mw: 363.37

SYN: C 2096

TOXICITY DATA with REFERENCE
eye-rbt 2% MLD ARZNAD 9,70,59
scu-mus LD50:200 mg/kg ARZNAD 9,70,59

SAFETY PROFILE: Poison by subcutaneous route. An eye irritant. When heated to decomposition it emits very toxic fumes of Cl^- and NO_x.

CJQ750 CAS:102585-43-3 ***HR: 3***
N-(1-(o-CHLOROPHENOXY)-2-PROPYL)-2-(DIETHYLAMINO)-N-METHYLACETAMIDE HYDROCHLORIDE
mf: $C_{16}H_{25}ClN_2O_2{\bullet}ClH$ mw: 349.34

SYN: C 2095

TOXICITY DATA with REFERENCE
eye-rbt 2% MLD ARZNAD 9,70,59
scu-mus LD50:330 mg/kg ARZNAD 9,70,59

SAFETY PROFILE: Poison by subcutaneous route. An eye irritant. When heated to decomposition it emits very toxic fumes of Cl^- and NO_x.

CJR125 CAS:1991-78-2 ***HR: D***
4-CHLOROPHENYLALANINE
mf: $C_9H_{10}ClNO_2$ mw: 199.65

SYNS: p-CHLORO-dl-PHENYLALANINE ◇ p-CHLOROPHENYLALANINE ◇ 3-(p-CHLOROPHENYL)ALANINE ◇ PCPA

TOXICITY DATA with REFERENCE
scu-rat TDLo:300 mg/kg (female 18-21D post):REP FEPRA7 36,1034,77
orl-rat TDLo:1896 mg/kg (15-20D preg):TER BIREBV 11,280,74

SAFETY PROFILE: An experimental teratogen. Experimental reproductive effects. When heated to decomposition it emits toxic fumes of Cl^- and NO_x.

CJR250 CAS:3647-19-6 ***HR: 3***
N-(3-CHLOROPHENYL)-1-AZIRIDINECARBOXAMIDE
mf: $C_9H_9ClN_2O$ mw: 196.65

SYNS: 1-(1-AZIRIDINYL)-N-(m-CHLOROPHENYL)FORMAMIDE◇ 3-CHLOROPHENYL-N-CARBAMOYLAZIRIDINE

TOXICITY DATA with REFERENCE
ipr-mus TDLo:120 mg/kg/4W-I:NEO CNREA8 29,2184,69
ivn-mus LD50:180 mg/kg CSLNX* NX#03943

SAFETY PROFILE: Poison by intravenous route. Questionable carcinogen with experimental neoplastigenic data. When heated to decomposition it emits very toxic fumes of Cl^- and NO_x.

CJR500 CAS:80-38-6 ***HR: 2***
4-CHLOROPHENYL BENZENESULFONATE
mf: $C_{12}H_9ClO_3S$ mw: 268.72

PROP: Colorless crystals. Mp: 62°. Insol in water; sol in organic solvents.

SYNS: ARACID ◇ BENZENESULFONATE de 4-CHLOROPHENYLE (FRENCH) ◇ BENZENESULFONIC ACID, 4-CHLOROPHENYL ESTER ◇ (4-CHLOOR-FENYL)-BENZEEN-SULFONAAT (DUTCH) ◇ p-CHLOROFENYLESTER KYSELINY BENZENSULFONOVE (CZECH) ◇ p-CHLOROPHENYL BENZENESULFONATE ◇ p-CHLOROPHENYL BENZENESULPHONATE ◇ 4-CHLOROPHENYL BENZENESULPHONATE ◇ (4-CHLOR-PHENYL)-BENZOLSULFONAT (GERMAN) ◇ (4-CLORO-FENIL)-BENZOL-SOLFONATO (ITALIAN) ◇ CPB ◇ CPBS ◇ ENT 4,585 ◇ FENIZON (FRENCH) ◇ FENSON ◇ GC 928 ◇ MURVESCO ◇ PCBS ◇ PCI ◇ PCPBS ◇ TRIFENSON

TOXICITY DATA with REFERENCE
skn-rbt 500 mg/24H MLD 28ZPAK -,197,72
eye-rbt 100 mg/24H MOD 28ZPAK -,197,72
orl-rat LD50:1350 mg/kg ARSIM* 20,10,66
unk-mam LD50:1300 mg/kg 30ZDA9 -,274,71

SAFETY PROFILE: Moderately toxic by ingestion and possibly other routes. An eye and skin irritant. See also ESTERS and SULFONATES. An acaricide. When heated to decomposition it emits toxic fumes of Cl^- and SO_x.

CJR809 CAS:1982-36-1 ***HR: 3***
1-(p-CHLORO-α-PHENYLBENZYL)HEXAHYDRO-4-METHYL-1H-1,4-DIAZEPINE DIHYDROCHLORIDE
mf: $C_{19}H_{23}ClN_2{\bullet}2ClH$ mw: 387.81

SYNS: HOMOCHLORCYCLIZINE DIHYDROCHLORIDE ◇ HOMOCHLOROCYCLIZINE DIHYDROCHLORIDE ◇ SA 97 DIHYDROCHLORIDE

TOXICITY DATA with REFERENCE
orl-rat LD50:490 mg/kg JOALAS 31,237,60
ipr-rat LD50:80 mg/kg JOALAS 31,237,60
ivn-rat LD50:36 mg/kg JOALAS 31,237,60
orl-mus LD50:390 mg/kg JOALAS 31,237,60
ipr-mus LD50:125 mg/kg JOALAS 31,237,60
scu-mus LD50:135 mg/kg JOALAS 31,237,60
ivn-mus LD50:47 mg/kg JOALAS 31,237,60
ipr-dog LD50:50 mg/kg JOALAS 31,237,60

SAFETY PROFILE: Poison by ingestion, subcutaneous, intravenous, and intraperitoneal routes. When

heated to decomposition it emits very toxic fumes of Cl^- and NO_x.

CJR909 CAS:68-88-2 ***HR: 3***
1-(p-CHLORO-α-PHENYLBENZYL)-4-(2-((2-HYDROXYETHOXY)ETHYL)PIPERAZINE
mf: $C_{21}H_{27}ClN_2O_2$ mw: 374.95

SYNS: ATARA ◇ ATARAX ◇ ATARAXOID ◇ ATARAZOID ◇ ATAZINA ◇ ATERAX ◇ 1-(p-CHLOROBENZHYDRYL)-4-(2-(2-HYDROXYETHOXY)ETHYL)DIETHYLENEDIAMINE ◇ 1-(p-CHLOROBENZHYDRYL)-4-(2-(2-HYDROXYETHOXY)ETHYL)PIPERAZINE ◇ N-(4-CHLOROBENZHYDRYL)-N'-(HYDROXYETHOXYETHYL) PIPERAZINE ◇ 1-(p-CHLORODIPHENYLMETHYL)-4-(2-(2-HYDROXYETHOXY)ETHYL)PIPERAZINE ◇ 2-(2-(4-(p-CHLORO-α-PHENYLBENZYL)-1-PIPERAZINYL)ETHOXY)ETHANOL ◇ DEINAIT ◇ EQUIPOISE ◇ FENAROL ◇ HYCHOTINE ◇ HYDROXINE ◇ HYDROXYCINE ◇ HYDROXYZINE ◇ IDROSSIZINA ◇ NEO-CALMA ◇ NEUROZINA ◇ NP 212 ◇ PAMAZONE ◇ PARENTERAL ◇ PAXISTIL ◇ PLACIDOL ◇ PLAXIDOL ◇ TRAN-Q ◇ TRAQUIZINE ◇ UCB 492 ◇ U.CB 4492 ◇ VESPARAZ-WIRKSTOFF

TOXICITY DATA with REFERENCE
orl-dog TDLo:1500 mg/kg (female 1-60D post):REP 28QFAD -,233,73
orl-rat TDLo:400 mg/kg (13-16D preg):TER TXAPA9 17,67,70
orl-rat LD50:840 mg/kg CHTPBA 3,210,68
ipr-rat LD50:160 mg/kg CHTPBA 3,210,68
ivn-rat LD50:45 mg/kg ANPBAZ 61,669,61
orl-mus LD50:480 mg/kg AANEAB 7,87,63
ipr-mus LD50:81300 μg/kg DPHFAK 23,281,71
ivn-mus LD50:137 mg/kg 27ZQAG -,237,72

SAFETY PROFILE: Poison by intravenous and intraperitoneal routes. Moderately toxic by ingestion. Experimental teratogenic and reproductive effects. When heated to decomposition it emits very toxic fumes of Cl^- and NO_x.

CJR959 ***HR: 3***
2-(α-(p-CHLOROPHENYL)BENZYLOXY)-N,N-DIMETHYLETHYLAMINE HYDROCHLORIDE
mf: $C_{17}H_{20}ClNO{\cdot}ClH$ mw: 326.29

SYN: SUBSTANZ NR. 1602 (GERMAN)

TOXICITY DATA with REFERENCE
ivn-mus LD50:46 mg/kg ARZNAD 4,189,54
scu-gpg LD50:94 mg/kg ARZNAD 4,189,54

SAFETY PROFILE: Poison by intravenous and subcutaneous routes. When heated to decomposition it emits very toxic fumes of NO_x and Cl^-.

CJT125 CAS:15842-89-4 ***HR: 3***
1-(4-CHLOROPHENYL)BIGUANIDINIUM HYDROGEN DICHROMATE
mf: $C_8H_{12}ClCr_2N_5O_7$ mw: 429.66

$ClC_6H_4NHC(:NH)NHC(:NH)N^+H_3HCr_2O_7^-$

CONSENSUS REPORTS: Chromium compounds are on the Community Right-To-Know List.

SAFETY PROFILE: Decomposes violently above 130°C. When heated to decomposition it emits toxic fumes of Cl^- and NO_x. See also CHROMIUM COMPOUNDS.

CJT750 CAS:80-33-1 ***HR: 3***
4-CHLOROPHENYL-4-CHLOROBENZENESULFONATE
mf: $C_{12}H_8Cl_2O_3S$ mw: 303.16

SYNS: ACARICYDOL E 20 ◇ C-854 ◇ C 1,006 ◇ CCS ◇ CHLOORFENSON (DUTCH) ◇ (4-CHLOOR-FENYL)-4-CHLOOR-BENZEEN-SULFONAAT (DUTCH) ◇ CHLOREFENIZON (FRENCH) ◇ CHLORFENSON ◇ CHLORFENSONE ◇ 4-CHLOROBENZENESULFONATE de 4-CHLOROPHENYLE (FRENCH) ◇ p-CHLOROBENZENESULFONIC ACID-p-CHLOROPHENYL ESTER ◇ CHLOROFENIZON ◇ p-CHLOROPHENYL-p-CHLOROBENZENE SULFONATE ◇ 4-CHLOROPHENYL-4-CHLOROBENZENESULPHONATE ◇ 4-CHLORPHENYL-4'-CHLORBENZOLSULFONAT (GERMAN) ◇ (4-CHLOR-PHENYL)-4-CHLOR-BENZOL-SULFONATE (GERMAN) ◇ (4-CLORO-FENIL)-4-CLORO-VENZOL-SOLFONATO (ITALIAN) ◇ COROTRAN ◇ CPCBS ◇ D 854 ◇ DIFENSON ◇ ENT 16,358 ◇ EPHIRSULPHONATE ◇ ESTER SULFONATE ◇ ESTONMITE ◇ ETHERSULFONATE ◇ GENITE 883 ◇ K 6451 ◇ LETHALAIRE G-58 ◇ MITICIDE K-101 ◇ NIAGARATRAN ◇ ONEX ◇ ORTHOTRAN ◇ OTRACID ◇ OVATRAN ◇ OVEX ◇ OVOCHLOR ◇ OVOTOX ◇ OVOTRAN ◇ PCPCBS ◇ SAPPILAN ◇ SAPPIRAN ◇ TRICHLORFENSON (OBS.)

TOXICITY DATA with REFERENCE
orl-mus TDLo:115 g/kg/78W-I:ETA NTIS** PB223-159
orl-rat LD50:2000 mg/kg GUCHAZ 6,381,73
unr-rat LD50:2050 mg/kg DABBBA 32,4116,72
orl-mus LD50:2000 mg/kg FMCHA2 -,C167,83
orl-rbt LD50:5660 mg/kg PCOC** -,841,66
orl-gpg LD50:640 mg/kg PCOC** -,841,66
orl-ckn LD50:3780 mg/kg PCOC** -,841,66

SAFETY PROFILE: Moderately toxic by ingestion and possibly other routes. Questionable carcinogen with experimental tumorigenic data. A pesticide. When heated to decomposition it emits very toxic fumes of Cl^- and SO_x. See also SULFONATES and CHLORINATED HYDROCARBONS, AROMATIC.

CJU125 CAS:17710-62-2 ***HR: 2***
p-CHLOROPHENYL-N-(4'-CHLOROPHENYL)THIOCARBAMATE
mf: $C_{13}H_9Cl_2NOS$ mw: 298.19

SYN: p-CHLOROTHIOCARBANILICACID-o-(p-CHLOROPHENYL) ESTER

TOXICITY DATA with REFERENCE
orl-mus LD50:2161 mg/kg YKKZAJ 88,465,68
ipr-mus LD50:426 mg/kg YKKZAJ 88,465,68
scu-mus LD50:462 mg/kg YKKZAJ 88,465,68

SAFETY PROFILE: Moderately toxic by ingestion, subcutaneous, and intraperitoneal routes. When heated to decomposition it emits toxic fumes of Cl^-, SO_x, and NO_x. See also CARBAMATES and ESTERS.

CJU250 CAS:120-32-1 ***HR: 3***
4-CHLORO-α-PHENYL-o-CRESOL
mf: $C_{13}H_{11}ClO$ mw: 218.69

PROP: Nearly colorless flakes. Mp: 49°, bp: 175° @ 5 mm, d: 1.2 @ 55°/25°.

SYNS: o-BENZYL-p-CHLOROPHENOL ◇ 2-BENZYL-4-CHLOROPHENOL ◇ BIO-CLAVE ◇ 5-CHLORO-2-HYDROXYDIPHENYLMETHANE ◇ CHLOROPHENE ◇ 4-CHLORO-2-(PHENYLMETHYL)PHENOL ◇ CLOROPHENE ◇ KETOLIN-H ◇ NCI-C61201 ◇ NEOSABENYL ◇ SANTOPHEN ◇ SANTOPHEN I GERMICIDE ◇ SENTIPHENE

TOXICITY DATA with REFERENCE
skn-mus TDLo:14 g/kg/34W-I:ETA CNREA8 19,413,59
orl-rat LD50:1700 mg/kg JPMSAE 63,1068,74

CONSENSUS REPORTS: Reported in EPA TSCA Inventory. Chlorophenol compounds are on the Community Right-To-Know List.

SAFETY PROFILE: Moderately toxic by ingestion. Questionable carcinogen with experimental tumorigenic data. When heated to decomposition it emits toxic fumes of Cl^-. See also CHLOROPHENOLS.

CJV250 CAS:35367-38-5 ***HR: 2***
1-(4-CHLOROPHENYL)-3-(2,6-DIFLUOROBENZOYL)UREA
mf: $C_{14}H_9ClF_2N_2O_2$ mw: 310.70

SYNS: N-(((4-CHLOROPHENYL)AMINO)CARBONYL)-2,6-DIFLUOROBENZAMIDE ◇ DIFLUBENZURON ◇ DIFLURON ◇ DIMILIN ◇ DU 112307 ◇ ENT 29,054 ◇ OMS 1804 ◇ PDD 6040I ◇ PH 60-40 ◇ PHILIPS-DUPHAR PH 60-40 ◇ TH 6040 ◇ THOMPSON-HAYWARD TH6040

TOXICITY DATA with REFERENCE
mmo-sat 1 mg/plate PCBPBS 10,174,79
cyt-mus-unr 500 mg/kg TGANAK 16(1),45,82
orl-mus LD50:4640 mg/kg SPEADM 78-1,-,78
unr-mus LD50:4600 mg/kg TGANAK 16(1),45,82
skn-rbt LD50:2000 mg/kg SPEADM 74-1,-,74

CONSENSUS REPORTS: Reported in EPA TSCA Inventory. EPA Genetic Toxicology Program.

SAFETY PROFILE: Moderately toxic by skin contact. Mildly toxic by ingestion and possibly other routes. Mutation data reported. When heated to decomposition it emits very toxic fumes of Cl^-, F^-, and NO_x.

CJW500 CAS:54708-51-9 ***HR: 3***
1-(m-CHLOROPHENYL)-3-N,N-DIMETHYLCARBAMOYL-5-METHOXYPYRAZOLE
mf: $C_{13}H_{14}ClN_3O_2$ mw: 279.75

SYNS: 1-(3-CHLOROPHENYL)-3,N,N-DIMETHYLCARBAMOYL-5-METHOXYPYRAZOLE ◇ PZ 177

TOXICITY DATA with REFERENCE
orl-rat TDLo:2750 mg/kg (7-17D preg):TER IYKEDH 9,538,78
orl-rbt TDLo:5200 mg/kg (female 6-18D post):REP IYKEDH 9,558,78
orl-rat LD50:790 mg/kg NYKZAU 72,31,76
ipr-rat LD50:235 mg/kg NYKZAU 72,31,76
scu-rat LD50:1100 mg/kg NYKZAU 72,31,76
orl-mus LD50:955 mg/kg NYKZAU 72,31,76
ipr-mus LD50:490 mg/kg NYKZAU 72,31,76
scu-mus LD50:1190 mg/kg NYKZAU 72,31,76
orl-rbt LD50:1650 mg/kg IYKEDH 8,494,77
ipr-rbt LD50:1040 mg/kg IYKEDH 8,494,77

SAFETY PROFILE: Poison by intraperitoneal route. Moderately toxic by ingestion and subcutaneous routes. An experimental teratogen. Other experimental reproductive effects. An analgesic and anti-inflammatory agent. When heated to decomposition it emits very toxic fumes of NO_x and Cl^-.

CJX000 CAS:60719-83-7 ***HR: 3***
2-(4-CHLOROPHENYL)-1,1-DIMETHYLETHYL 2-AMINOPROPANOATE HYDROCHLORIDE
mf: $C_{13}H_{18}ClNO_2{\bullet}ClH$ mw: 292.23

SYN: p-CHLORO-α,α-DIMETHYL-2-AMINOPROPIONATE-PHENETHYL ALCOHOL HYDROCHLORIDE

TOXICITY DATA with REFERENCE
orl-mus LD50:901 mg/kg JMCMAR 21,448,78
ipr-mus LD50:200 mg/kg JMCMAR 21,448,78
ivn-mus LD50:54 mg/kg JMCMAR 21,448,78

SAFETY PROFILE: Poison by intraperitoneal and intravenous routes. Moderately toxic by ingestion. When heated to decomposition it emits very toxic fumes of Cl^- and NO_x.

CJX750 CAS:150-68-5 ***HR: 3***
3-(p-CHLOROPHENYL)-1,1-DIMETHYLUREA
mf: $C_9H_{11}ClN_2O$ mw: 198.67

PROP: Crystals; slight odor. Mp: 171°, vap press: 0.002 mm @ 100°. Nearly water insol.

SYNS: 3-(4-CHLOOR-FENYL)-1,1-DIMETHYLUREUM(DUTCH) ◇ CHLORFENIDIM ◇ N-(p-CHLOROPHENYL)-N′,N′-DIMETHYLUREA ◇ N′-(4-CHLOROPHENYL)-N,N-DIMETHYLUREA ◇ 1-(p-CHLOROPHENYL)-3,3-DIMETHYLUREA ◇ 3-(4-CHLOROPHENYL)-1,1-DIMETHYLUREA ◇ 1-(4-CHLORO PHENYL)-3,3-DIMETHYLUREE (FRENCH) ◇ 3-(4-CHLOR-PHENYL)-1,1-DIMETHYL-HARNSTOFF (GERMAN) ◇ 3-(4-CLORO-FENIL)-1,1-DIMETIL-UREA (ITALIAN) ◇ CMU ◇ N,N-DIMETHYL-N′-(4-CHLOROPHENYL)UREA ◇ 1,1-DIMETHYL-3-(p-CHLOROPHENYL)UREA ◇ HERBICIDES, MONURON ◇ KARMEX MONURON HERBICIDE ◇ KARMEX W. MONURON HERBICIDE ◇ LIROBETAREX ◇ MONUREX ◇ MONURON ◇ MONUROX

◇ MONURUON ◇ MONUURON ◇ NCI-C02846 ◇ TELVAR ◇ TELVAR MONURON WEEDKILLER ◇ USAF P-8 ◇ USAF XR-41

TOXICITY DATA with REFERENCE
mma-sat 1 μg/plate MUREAV 58,353,78
dni-mus-orl 200 mg/kg MUREAV 58,353,78
otr-ham:emb 5 mg/L CRNGDP 4,291,83
orl-mus TDLo:1935 mg/kg (6-15D preg):TER NTIS** PB223-160
orl-mus TDLo:1935 mg/kg (6-15D preg):REP NTIS** PB223-160
orl-rat TDLo:46350 mg/kg/2Y-C:CAR NTPTR* NTP-TR 266,88
orl-mus TDLo:4320 mg/kg/16W-I:CAR VOONAW 16(10),51,70
orl-rat LD50:1053 mg/kg FAATDF 7,299,86
unk-rat LD50:3600 mg/kg JPFCD2 B15,929,80
ipr-mus LD50:1000 mg/kg NTIS** AD277-689
orl-gpg LDLo:670 mg/kg PAREAQ 14,225,62
unr-mam LD50:3500 mg/kg 30ZDA9 -,231,71

CONSENSUS REPORTS: IARC Cancer Review: Group 3 IMEMDT 7,56,87; Animal Sufficient Evidence IMEMDT 12,167,76. Reported in EPA TSCA Inventory. EPA Genetic Toxicology Program.

SAFETY PROFILE: Moderately toxic by ingestion, intraperitoneal, and possibly other routes. Experimental teratogenic and reproductive effects. Questionable carcinogen with experimental carcinogenic data. Mutation data reported. An herbicide. When heated to decomposition it emits very toxic fumes of NO_x, and Cl^-.

CJY000 CAS:140-41-0 ***HR: 2***
3-(p-CHLOROPHENYL)-1,1-DIMETHYLUREA TRICHLOROACETATE
mf: $C_2HCl_3O_2 \cdot C_9H_{11}ClN_2O$ mw: 362.05

SYNS: 3-(p-CHLOROPHENYL)-1,1-DIMETHYLUREA compounded with TRICHLOROACETIC ACID (1:1) ◇ GC-2996 ◇ MONURON-TCA ◇ TRICHLOROACETIC ACID compounded with N'-(4-CHLOROPHENYL)-N,N-DIMETHYLUREA (1:1) ◇ UROX 379 ◇ XORU-OX

TOXICITY DATA with REFERENCE
orl-rat LD50:2300 mg/kg 28ZEAL 4,292,69
scu-rbt LD50:1000 mg/kg FMCHA2 -,C249,83

SAFETY PROFILE: Moderately toxic by ingestion and subcutaneous routes. When heated to decomposition it emits very toxic fumes of Cl^- and NO_x. See also 3-(p-CHLOROPHENYL)-1,1-DIMETHYLUREA and TRICHLOROACETIC ACID.

CJY120 CAS:5131-60-2 ***HR: 3***
4-CHLORO-m-PHENYLENEDIAMINE
mf: $C_6H_7ClN_2$ mw: 142.60

PROP: Mp: 90°.

SYNS: C.I. 76027 ◇ 4-CHLORO-1,3-BENZENEDIAMINE ◇ 1-CHLORO-2,4-DIAMINOBENZENE ◇ 4-CHLOROPHENE-1,3-DIAMINE ◇ 4-CHLOROPHENYLENE-1,3-DIAMINE ◇ 4-CHLORO-1,3-PHENYLENEDIAMINE ◇ 4-Cl-M-PD ◇ NCI-C03305

TOXICITY DATA with REFERENCE
mmo-sat 1 mg/plate ENMUDM 7(Suppl 5),1,85
mma-sat 10 μg/plate ENMUDM 7(Suppl 5),1,85
orl-rat TDLo:164 g/kg/78W-C:CAR NCITR* NCI-CG-TR-85,78
orl-mus TDLo:648 g/kg/77W-C:CAR CRNGDP 1,495,80
orl-mus TD:917 g/kg/78W-C:ETA NCITR* NCI-CG-TR-85,78
orl-rat TD:108 g/kg/77W-C:NEO CRNGDP 1,495,80
orl-rat TD:1092 g/kg/78W-C:CAR,REP IARC** 27,81,82
orl-rat TD:2184 g/kg/78W-C:CAR,REP IARC** 27,81,82

CONSENSUS REPORTS: IARC Cancer Review: Group 3 IMEMDT 7,56,87; Animal Inadequate Evidence IMEMDT 27,81,82. NCI Carcinogenesis Bioassay (feed); Clear Evidence: mouse, rat NCITR* NCI-CG-TR-85,78. Reported in EPA TSCA Inventory.

SAFETY PROFILE: Suspected carcinogen with experimental carcinogenic, neoplastigenic, and tumorigenic data. Experimental reproductive effects. Mutation data reported. When heated to decomposition it emits toxic fumes of Cl^- and NO_x. See also AROMATIC AMINES.

CJY250 CAS:61583-30-0 ***HR: D***
(4-CHLORO-o-PHENYLENEDIAMMINE) DICHLOROPLATINUM(II)
mf: $C_6H_7Cl_3N_2Pt$ mw: 408.59

TOXICITY DATA with REFERENCE
mmo-sat 100 nmol/L JMCMAR 23,459,80
mma-sat 100 nmol/L JMCMAR 23,459,80
dnd-omi 1 mL BICMBE 60,901,78
dnd-sal:spr 1 mL BICMBE 60,901,78

SAFETY PROFILE: Mutation data reported. See also PLATINUM COMPOUNDS. When heated to decomposition it emits very toxic fumes of Cl^- and NO_x.

CJZ000 CAS:155-00-0 ***HR: 3***
o-CHLORO-β-PHENYLETHYLHYDRAZINE DIHYDROGEN SULPHATE
mf: $C_8H_{11}ClN_2 \cdot H_2O_4S$ mw: 268.74

TOXICITY DATA with REFERENCE
scu-mus TDLo:240 mg/kg (female 1-6D post):REP JOENAK 49,635,71
scu-mus LD50:182 mg/kg JOENAK 30,205,64

SAFETY PROFILE: Poison by acute subcutaneous route. Experimental reproductive effects. See also SULFATES. When heated to decomposition it emits very toxic fumes of SO_x, Cl^-, and NO_x.

CKA000 CAS:33671-46-4 ***HR: 2***
5-(2-CHLOROPHENYL)-7-ETHYL-1-METHYL-1,3-DIHYDRO-2H-THIENO(2,3-e)(1,4)DIAZEPIN-2-ONE
mf: $C_{16}H_{15}ClN_2OS$ mw: 318.84

SYNS: 5-(o-CHLOROPHENYL)-7-ETHYL-1,3-DIHYDRO-1-METHYL-2H-THIENO(2,3-e)-1,4-DIAZEPIN-2-ONE ◇ CLOTIAZEPAM ◇ RISE ◇ TRECALMO ◇ Y 6047

TOXICITY DATA with REFERENCE
orl-rat TDLo:7 g/kg (35D pre):REP KSRNAM 6,2228,72
orl-mus TDLo:600 mg/kg (7-12D preg):TER KSRNAM 6,2264,72
orl-rat LD50:1461 mg/kg NIIRDN 6,233,82
ipr-rat LD50:682 mg/kg NIIRDN 6,233,82
orl-mus LD50:636 mg/kg JMCMAR 16,214,73
ipr-mus LD50:440 mg/kg JMCMAR 16,214,73
scu-mus LD50:2837 mg/kg NIIRDN 6,233,82

SAFETY PROFILE: Moderately toxic by ingestion, subcutaneous, and intraperitoneal routes. An experimental teratogen. Other experimental reproductive effects. A tranquilizer. When heated to decomposition it emits very toxic fumes of SO_x, NO_x, and Cl^-. See also DIAZEPAM.

CKA500 CAS:13822-05-4 ***HR: 3***
1-(p-CHLOROPHENYL)-1,2,3,4,5,6-HEXAHYDRO-2,5-BENZODIAZOCINO DIHYDROCHLORIDE
mf: $C_{16}H_{17}ClN_2 \cdot 2ClH$ mw: 345.72

SYNS: 1-(p-CHLOROPHENYL)-1,2,3,4,5,6-HEXAHYDRO-2,5-BENZODIAZOCINE, DIHYDROCHLORIDE ◇ WY 5244

TOXICITY DATA with REFERENCE
orl-rat LD50:100 mg/kg TXAPA9 18,185,71
orl-bwd LD50:100 mg/kg TXAPA9 21,315,72

SAFETY PROFILE: Poison by ingestion. When heated to decomposition it emits very toxic fumes of NO_x and Cl^-.

CKA575 CAS:21905-40-8 ***HR: 3***
(m-CHLOROPHENYL)HYDROXY(β-HYDROXYPHENETHYL)ARSINE OXIDE
mf: $C_{14}H_{14}AsClO_3$ mw: 340.65

SYNS: ARSINE OXIDE, (m-CHLOROPHENYL)HYDROXY(β-HYDROXYPHENETHYL)- ◇ 2-PHENYL-2-HYDROXYETHYL, m-CHLOROPHENYL ARSINIC ACID

TOXICITY DATA with REFERENCE
ivn-mus LD50:100 mg/kg CSLNX* NX#06919

OSHA PEL: TWA 0.5 mg(As)/m^3

SAFETY PROFILE: Poison by intravenous route. When heated to decomposition it emits toxic fumes of As and Cl^-.

CKA625 ***HR: D***
2-(p-CHLOROPHENYL)IMIDAZO(2,1-a)ISOQUINOLINE
mf: $C_{17}H_{11}ClN_2$ mw: 278.75

SYN: 2-(p-CHLOROPHENYL)IMIDAZO(2,1-b)ISOQUINOLINE

TOXICITY DATA with REFERENCE
orl-ham TDLo:20 mg/kg (4-8D preg):REP ARZNAD 33,1222,83

SAFETY PROFILE: Experimental reproductive effects. When heated to decomposition it emits toxic fumes of Cl^- and NO_x.

CKA630 ***HR: D***
2-(4-CHLOROPHENYL)IMIDAZO(5,1-a)ISOQUINOLINE
mf: $C_{17}H_{12}ClN_2$ mw: 279.76

SYN: L 12236

TOXICITY DATA with REFERENCE
scu-rat TDLo:10 mg/kg (1-5D preg):REP JAPRAN 23,295,82

SAFETY PROFILE: Experimental reproductive effects. When heated to decomposition it emits toxic fumes of Cl^- and NO_x.

CKA750 CAS:2909-38-8 ***HR: 3***
m-CHLOROPHENYL ISOCYANATE
mf: C_7H_4ClNO mw: 153.57

PROP: Water-white liquid, sol in organic solvents. Mp: −4°, bp: 101° @ 30 mm, flash p: 215°F (COC).

SYN: ISOCYANIC ACID-m-CHLOROPHENYL ESTER

TOXICITY DATA with REFERENCE
ihl-mam LD50:63 mg/kg GTPZAB 11(4),23,67

CONSENSUS REPORTS: Reported in EPA TSCA Inventory.

SAFETY PROFILE: Poison by inhalation and ingestion. Combustible when exposed to heat or flame. When heated to decomposition it emits toxic fumes of Cl^-, CN^-, and NO_x. See also ESTERS and THIOCYANATES.

CKB000 CAS:104-12-1 ***HR: 3***
p-CHLOROPHENYL ISOCYANATE
mf: C_7H_4ClNO mw: 153.57

PROP: White solid, sol in organic solvents. Mp: 31°, bp: 204°, flash p: 230°F.

SYNS: p-CHLORFENYLISOKYANAT (CZECH) ◇ ISOCYANIC ACID-p-CHLOROPHENYL ESTER ◇ PCPI

TOXICITY DATA with REFERENCE
skn-rbt 500 mg/24H MOD 28ZPAK -,165,72
eye-rbt 250 μg/24H SEV 28ZPAK -,165,72
ihl-man TCLo:800 $\mu g/m^3$/1M HYSAAV 31(7-9),481,66
orl-rat LD50:4710 mg/kg 28ZPAK -,165,72
orl-mus LD50:530 mg/kg GTPZAB 11(4),23,67
ihl-mus LCLo:40 mg/m^3 HYSAAV 31(7-9),481,66

CONSENSUS REPORTS: Reported in EPA TSCA Inventory.

SAFETY PROFILE: Poison by inhalation. Moderately toxic by ingestion. Unspecified human systemic effects. A severe eye and moderate skin irritant. Combustible when exposed to heat or flame. Dangerous, can explode on distillation. When heated to decomposition it emits toxic fumes of Cl^-, CN^-, and NO_x.

CKB250 CAS:500-92-5 ***HR: 3***
1-(p-CHLOROPHENYL)-5-ISOPROPYLBIGUANIDE
mf: $C_{11}H_{16}ClN_5$ mw: 253.77

PROP: White powder. Mp: 244°.

SYNS: BIGUMAL ◇ CHLORGUANIDE ◇ CHLOROGUANIDE ◇ PALUDRINE ◇ PROGUANIL

TOXICITY DATA with REFERENCE
orl-rat TDLo:700 mg/kg (7D pre):REP MEXPAG 10,361,64
ipr-rat LD50:15 mg/kg 14XBAV -,367,64
ivn-rat LDLo:40 mg/kg CLDND* 4,14,49
orl-rbt LD50:150 mg/kg CLDND* 4,14,49
ipr-rbt LD50:50 mg/kg CLDND* 4,14,49
orl-ckn LDLo:400 mg/kg CLDND* 4,14,49
ivn-ckn LDLo:60 mg/kg CLDND* 4,14,49

SAFETY PROFILE: Poison by ingestion, intravenous, and intraperitoneal routes. Experimental reproductive effects. When heated to decomposition it emits toxic fumes of Cl^- and NO_x.

CKB500 CAS:637-32-1 ***HR: 3***
1-(p-CHLOROPHENYL)-5-ISOPROPYLBIGUANIDE HYDROCHLORIDE
mf: $C_{11}H_{16}ClN_5$•ClH mw: 290.23

SYNS: CHLORGUANIDE HYDROCHLORIDE ◇ CHLOROGUANIDE HYDROCHLORIDE ◇ CHLOROGUANIDINE HYDROCHLORIDE ◇ N-4-CHLOROPHENYL-N^5-ISOPROPYLDIGUANIDE HYDROCHLORIDE ◇ DIGUANYL ◇ DRINUPAL HYDROCHLORIDE ◇ GUANATOL HYDROCHLORIDE ◇ M 4888 ◇ PALUDRINE HYDROCHLORIDE ◇ PALUSIL HYDROCHLORIDE ◇ PROGUANIL HYDROCHLORIDE ◇ SN 12,837 ◇ TIRIAN HYDROCHLORIDE ◇ 3359 RP

TOXICITY DATA with REFERENCE
orl-mus TDLo:168 mg/kg (7D pre/1-7D preg):REP ANTCAO 12,671,62
orl-rat LD50:58 mg/kg JPETAB 91,157,47
ivn-rat LD50:33 mg/kg JPETAB 91,157,47
orl-mus LD50:27 mg/kg JPETAB 91,157,47
ivn-mus LD50:23 mg/kg JPETAB 91,157,47
orl-rbt LD50:172 mg/kg BJPCAL 2,181,47
ivn-rbt LD50:44850 μg/kg JPETAB 91,157,47
ivn-gpg LD50:39510 μg/kg JPETAB 91,157,47
orl-ckn LD50:400 mg/kg BJPCAL 2,181,47

SAFETY PROFILE: Poison by ingestion and intravenous routes. Experimental reproductive effects. An antimalarial drug. When heated to decomposition it emits very toxic fumes of Cl^- and NO_x.

CKC000 CAS:101-21-3 ***HR: 3***
N-3-CHLOROPHENYLISOPROPYLCARBAMATE
mf: $C_{10}H_{12}ClNO_2$ mw: 213.68

PROP: Light brown, crystalline solid; faint characteristic odor. Mp: 41°, bp: 247° (decomp).

SYNS: BEET-KLEEN ◇ BUD-NIP ◇ N-(3-CHLOOR-FENYL)-ISOPROPYL CARBAMAAT (DUTCH) ◇ CHLOR-IFC ◇ CHLOR-IPC ◇ m-CHLOROCARBANILIC ACID, ISOPROPYL ESTER ◇ 3-CHLOROCARBANILIC ACID, ISOPROPYL ESTER ◇ N-(3-CHLORO PHENYL) CARBAMATE D'ISOPROPYLE (FRENCH) ◇ N-(3-CHLOROPHENYL) CARBAMIC ACID, ISOPROPYL ESTER ◇ (3-CHLOROPHENYL)CARBAMIC ACID, 1-METHYLETHYL ESTER ◇ CHLOROPROPHAM ◇ N-(3-CHLOR-PHENYL)-ISOPROPYL-CARBAMAT (GERMAN) ◇ CHLORPROPHAM ◇ CHLORPROPHAME (FRENCH) ◇ CICP ◇ CI-IPC ◇ CIPC ◇ N-(3-CLORO-FENIL)-ISOPROPIL-CARBAMMATO (ITALIAN) ◇ ELBANIL ◇ ENT 18,060 ◇ FASCO WY-HOE ◇ FURLOE ◇ FURLOE 4EC ◇ ISOPROPYL-m-CHLOROCARBANILATE ◇ ISOPROPYL-3-CHLOROCARBANILATE ◇ ISOPROPYL-3-CHLOROPHENYLCARBAMATE ◇ ISOPROPYL-N-(3-CHLOROPHENYL)CARBAMATE ◇ o-ISOPROPYL-N-(3-CHLOROPHENYL)CARBAMATE ◇ ISOPROPYL-N-(3-CHLORPHENYL)-CARBAMAT (GERMAN) ◇ JACK WILSON CHLORO 51 (oil) ◇ LIRO CIPC ◇ METOXON ◇ NEXOVAL ◇ PREVENOL ◇ PREVENOL 56 ◇ PREVENTOL ◇ PREVENTOL 56 ◇ PREWEED ◇ SPROUT NIP ◇ SPROUT-NIP EC ◇ SPUD-NIC ◇ SPUD-NIE ◇ STOPGERME-S ◇ TATERPEX ◇ TRIHERBICIDE CIPC ◇ UNICROP CIPC ◇ Y 3

TOXICITY DATA with REFERENCE
mmo-smc 100 mg/L PMRSDJ 1,414,81
dns-hmn:fbr 4 mg/L PMRSDJ 1,528,81
scu-mus TDLo:9 g/kg (6-14D preg):TER NTIS** PB223-160
orl-mus TDLo:600 mg/kg:NEO BJCAAI 12,355,58
orl-rat LD50:1200 mg/kg AEHLAU 19,621,69
ipr-rat LD50:700 mg/kg CRSBAW 175,496,81
orl-mus LD50:6500 mg/kg CRSBAW 175,496,81
ipr-mus LD50:2600 mg/kg CRSBAW 175,496,81
orl-rbt LD50:5000 mg/kg PAREAQ 14,225,62

CONSENSUS REPORTS: IARC Cancer Review: Group 3 IMEMDT 7,56,87; Animal Inadequate Evidence IMEMDT 12,55,76. EPA Genetic Toxicology Program.

SAFETY PROFILE: Moderately toxic by ingestion, intraperitoneal, and possibly other routes. Questionable carcinogen with experimental neoplastigenic and teratogenic data. Human mutation data reported. An herbicide. When heated to decomposition it emits highly toxic fumes of Cl^-, NO_x, and phosgene. See also CARBAMATES.

CKC325 CAS:14774-78-8 ***HR: 3***
4-CHLOROPHENYLLITHIUM
mf: C_6H_4ClLi mw: 118.49

SAFETY PROFILE: Explodes on contact with oxygen. When heated to decomposition it emits toxic fumes of Cl^-. See also LITHIUM COMPOUNDS.

CKC500 CAS:1631-29-4 ***HR: 3***
N-(p-CHLOROPHENYL)MALEIMIDE
mf: $C_{10}H_6ClNO_2$ mw: 207.62

TOXICITY DATA with REFERENCE
skn-rbt 500 mg MLD SCCUR* -,7,61
orl-rat LD50:580 mg/kg SCCUR* -,7,61
orl-mus LD50:290 mg/kg SCCUR* -,7,61
orl-rbt LDLo:350 mg/kg SCCUR* -,7,61

SAFETY PROFILE: Poison by ingestion. A skin irritant. When heated to decomposition it emits very toxic fumes of Cl^- and NO_x.

CKD250 CAS:54708-68-8 ***HR: 2***
1-(3-CHLOROPHENYL-5-METHOXY-N-METHYL-1H-PYRAZOLE-3-CARBOXAMIDE
mf: $C_{12}H_{12}ClN_3O_2$ mw: 265.72

SYNS: PC 222 ◇ PZ 222

TOXICITY DATA with REFERENCE
orl-rat LD50:1540 mg/kg NYKZAU 72,31,76
ipr-rat LD50:410 mg/kg NYKZAU 72,31,76
orl-mus LD50:1540 mg/kg NYKZAU 72,31,76
ipr-mus LD50:910 mg/kg NYKZAU 72,31,76
orl-rbt LD50:4800 mg/kg NYKZAU 72,31,76

SAFETY PROFILE: Moderately toxic by ingestion and intraperitoneal routes. When heated to decomposition it emits very toxic fumes of Cl^- and NO_x.

CKD500 CAS:1746-81-2 ***HR: 3***
3-(4-CHLOROPHENYL)-1-METHOXY-1-METHYLUREA
mf: $C_9H_{11}ClN_2O_2$ mw: 214.67

SYNS: AFESIN ◇ ARESIN ◇ AREZIN ◇ AREZINE ◇ ARRESIN ◇ 3-(4-CHLORPHENYL)-1-METHOXY-1-METHYLHARNSTOFF(GERMAN) ◇ N-(4-CHLOROPHENYL)-N'-METHOXY-N-METHYLUREA ◇ N'-(4-CHLOROPHENYL)-N-METHOXY-N-METHYLUREA ◇ HOE 2747 ◇ MONOLINURON ◇ PREMALIN

TOXICITY DATA with REFERENCE
orl-mus TDLo:8 g/kg (6-15D preg):TER ARTODN 38,261,77
orl-mus TDLo:8 g/kg (6-15D preg):REP ARTODN 38,261,77
orl-rat LD50:1800 mg/kg WRPCA2 9,119,70
orl-dog LD50:500 mg/kg 28ZEAL 5,158,76

CONSENSUS REPORTS: Reported in EPA TSCA Inventory.

SAFETY PROFILE: Moderately toxic by ingestion. Experimental teratogenic and reproductive effects. When heated to decomposition it emits very toxic fumes of Cl^- and NO_x.

CKD750 CAS:1867-66-9 ***HR: 3***
2-(o-CHLOROPHENYL)-2-(METHYLAMINO)CYCLOHEXANONE HYDROCHLORIDE
mf: $C_{13}H_{16}ClNO•ClH$ mw: 274.21

SYNS: CI 581 ◇ CL 369 ◇ CN-52,372-2 ◇ KETAJECT ◇ KETALAR ◇ KETAMINE ◇ KETAMINE HYDROCHLORIDE ◇ KETANEST ◇ KETASET ◇ KETAVET ◇ KETOLAR ◇ VETALAR

TOXICITY DATA with REFERENCE
ipr-rat TDLo:1700 mg/kg (1-17D preg):TER REANBJ 26,137,79
unr-man TDLo:1 mg/kg:CNS CSLNX* NX#03551
ivn-hmn TDLo:2 mg/kg:CNS BMJOAE 2,943,76
mul-inf TDLo:11587 μg/kg:PUL BJANAD 58,573,86
orl-rat LD50:447 mg/kg NIIRDN 6,264,82
ipr-rat LD50:224 mg/kg TXAPA9 18,185,71
ivn-rat LD50:58880 μg/kg NIIRDN 6,264,82
orl-mus LD50:617 mg/kg NIIRDN 6,164,82
ivn-mus LD50:224 mg/kg NIIRDN 6,262,82
ivn-mus LD50:55900 μg/kg NIIRDN 6,264,82
ims-gpg LD50:361 mg/kg VHTODE 24,410,82
ivn-ckn LD50:67900 μg/kg AJVRAH 45,531,84

CONSENSUS REPORTS: EPA Genetic Toxicology Program.

SAFETY PROFILE: Poison by intramuscular, intraperitoneal, and intravenous routes. Moderately toxic by ingestion. Human systemic effects by intravenous and possibly other routes: analgesia, coma, hallucinations and distorted perceptions, dyspnea. An experimental teratogen. An anesthetic. When heated to decomposition it emits very toxic fumes of Cl^- and NO_x.

CKD800 CAS:73791-42-1 ***HR: 3***
(4-CHLOROPHENYL)METHYLARSINIC ACID
mf: $C_7H_8AsClO_2$ mw: 234.52

SYNS: ARSINE OXIDE, (p-CHLOROPHENYL)HYDROXYMETHYL- ◇ (p-CHLOROPHENYL)HYDROXYMETHYLARSINE OXIDE

TOXICITY DATA with REFERENCE
ivn-mus LD50:100 mg/kg CSLNX* NX#01207

OSHA PEL: TWA 0.5 mg(As)/m^3

SAFETY PROFILE: Poison by intravenous route. When heated to decomposition it emits toxic fumes of As and Cl^-.

CKE000 CAS:511-46-6 ***HR: 3***
2-(α-(p-CHLOROPHENYL)-α-METHYLBENZYLOXY)- N,N-DIETHYLETHYLAMINE
mf: $C_{20}H_{26}ClNO$ mw: 331.92

SYNS: ETHYLAMINE, 2-(α-(p-CHLOROPHENYL)-α-METHYLBENZYLOXY)-N,N-DIETHYL ◇ SUBSTANZ NR. 1925 (GERMAN)

TOXICITY DATA with REFERENCE
ivn-mus LD50:32 mg/kg ARZNAD 4,189,54
scu-gpg LD50:120 mg/kg ARZNAD 4,189,54

SAFETY PROFILE: Poison by intravenous and subcutaneous routes. See also AMINES. When heated to decomposition it emits very toxic fumes of Cl^- and NO_x.

CKE250 CAS:102584-42-9 ***HR: 3***
1-(2-(α-(p-CHLOROPHENYL)-α-METHYLBENZYLOXY)ETHYL PYRROLIDINE
mf: $C_{20}H_{24}ClNO$ mw: 329.90

SYN: SUBSTANZ NR. 2135

TOXICITY DATA with REFERENCE
scu-mus LD50:90 mg/kg ARZNAD 4,189,54
ivn-mus LD50:50 mg/kg ARZNAD 4,189,54

SAFETY PROFILE: Poison by subcutaneous and intravenous routes. When heated to decomposition it emits very toxic fumes of Cl^- and NO_x.

CKE750 CAS:79-93-6 ***HR: 3***
2-p-CHLOROPHENYL-3-METHYL-2,3-BUTANEDIOL
mf: $C_{11}H_{15}ClO_2$ mw: 214.71

SYNS: ACALMID ◇ ACALO ◇ ALTERTON ◇ ATADIOL ◇ B586 ◇ 2-(4-CHLOROPHENYL)-3-METHYL-2,3-BUTANEDIOL ◇ ENT 15,208 ◇ FELIXYN ◇ FENAGLICODOLO ◇ PAUSITAL ◇ PHENAGLYCODOL ◇ PHENGLYKODOL ◇ REMIN ◇ SEDAPSIN ◇ SINFORIL ◇ STESIL ◇ ULTRAN ◇ USAF EL-44

TOXICITY DATA with REFERENCE
orl-rat LD50:832 mg/kg TXAPA9 21,315,72
orl-mus LD50:514 mg/kg 27ZQAG -,401,72
ipr-mus LD50:200 mg/kg NTIS** AD277-689
scu-mus LD50:358 mg/kg 27ZQAG -,401,72
ivn-mus LD50:254 mg/kg 27ZQAG -,401,72
orl-bwd LD50:32 mg/kg TXAPA9 21,315,72

SAFETY PROFILE: Poison by ingestion, intraperitoneal, subcutaneous, and intravenous routes. A tranquilizer. When heated to decomposition it emits toxic fumes of Cl^-.

CKF000 CAS:3942-54-9 ***HR: 3***
o-CHLOROPHENYL METHYLCARBAMATE
mf: $C_8H_8ClNO_2$ mw: 185.62

SYNS: 2-CHLOROPHENYL-N-METHYLCARBAMATE ◇ CPMC ◇ ETROFOL ◇ HOPCIDE

TOXICITY DATA with REFERENCE
orl-rat LD50:648 mg/kg FMCHA2 -,D132,80
orl-mus LD50:150 mg/kg GUCHAZ 6,115,73

CONSENSUS REPORTS: EPA Genetic Toxicology Program. Reported in EPA TSCA Inventory.

SAFETY PROFILE: A poison by ingestion and possibly other routes. An insecticide. See also CARBAMATES. When heated to decomposition it emits very toxic fumes of Cl^- and NO_x.

CKF500 CAS:80-77-3 ***HR: 3***
2-(4-CHLOROPHENYL)-3-METHYL-4-METATHIAZANONE-1,1-DIOXIDE
mf: $C_{11}H_{12}ClNO_3S$ mw: 273.75

SYNS: BANABIN ◇ BANABIN-SINTYAL ◇ BISINA ◇ CHLORMETHAZANONE ◇ CHLORMETHAZONE ◇ CHLORMEZANONE ◇ 2-(p-CHLOROPHENYL)TETRAHYDRO-3-METHYL-4H-1,3-THIAZIN-4-ONE 1,1-DIOXIDE ◇ 2-(p-CHLORPHENYL)-3-METHYL-1,3-PERHYDROTHIAZIN-4-ON-1,1-DIOXIDE ◇ CLORILAX ◇ CLORMETAZANONE ◇ CLORMETHAZON ◇ DICHLOROMETHAZANONE ◇ FENAROL ◇ LOBAK ◇ MIORILAX ◇ MIO-SED ◇ MUSKEL ◇ MUSKEL-TRANCOPAL ◇ PHENAROL ◇ REXAN ◇ RILANSYL ◇ RILAQUIL ◇ RILASSOL ◇ RILAX ◇ RILLASOL ◇ SUPOTRAN ◇ SUPROTAN ◇ TANAFOL ◇ TETRAHYDRO-2-(p-CHLOROPHENYL)-3-METHYL-4H-1,3-THIAZIN-4-ONE-1,1-DIOXIDE ◇ TRANCOPAL

TOXICITY DATA with REFERENCE
orl-man TDLo:157 mg/kg:EYE,CNS BMJOAE 292,732,86
orl-rat LD50:605 mg/kg AIPTAK 130,280,61
ipr-rat LD50:370 mg/kg ARZNAD 17,242,67
orl-mus LD50:600 mg/kg OYYAA2 9,601,75
ipr-mus LD50:570 mg/kg ARZNAD 17,242,67
scu-mus LD50:322 mg/kg APTOA6 19,247,62
orl-dog LD50:500 mg/kg TXAPA9 1,168,59
ipr-dog LD50:500 mg/kg TXAPA9 1,168,59
ipr-gpg LD50:600 mg/kg TXAPA9 1,168,59

SAFETY PROFILE: Poison by intraperitoneal and subcutaneous routes. Moderately toxic by ingestion. Human systemic effects by ingestion: dilation of the pupils, ataxia (loss of muscle coordination), and coma. When heated to decomposition it emits very toxic fumes of SO_x, NO_x, and Cl^-.

CKF750 CAS:3766-60-7 ***HR: 2***
3-(p-CHLOROPHENYL)-1-METHYL-1-(1-METHYL-2-PROPYNYL)UREA
mf: $C_{12}H_{13}ClN_2O$ mw: 236.72

SYNS: ARISAN ◇ BUTURON ◇ BUTYRON ◇ N'-(4-

CHLOROPHENYL)-N-ISOBUTINYL-N-METHYLUREA ◇ N'-(4-CHLOROPHENYL)-N-METHYL-N-(1-METHYL-2-PROPYNYL)-UREA ◇ N-(4-CHLORPHENYL)-N'-METHYL-N'-ISOBUTINYLHARNSTOFF (GERMAN) ◇ 3-(4-CHLORPHENYL)-1-METHYL-1-ISOBUTINYL-HARNSTOFF (GERMAN) ◇ EPTAPUR ◇ H 95

TOXICITY DATA with REFERENCE
orl-mus TDLo:700 mg/kg (12-13D preg):REP ARTODN 38,261,77
orl-mus TDLo:3 g/kg (6-15D preg):TER ARTODN 38,261,77
orl-rat LD50:1791 mg/kg ARTODN 38,261,77
ipr-mus LD50:500 mg/kg 85DPAN -,-,71/76

CONSENSUS REPORTS: EPA Genetic Toxicology Program.

SAFETY PROFILE: Moderately toxic by ingestion and intraperitoneal routes. Experimental teratogenic and reproductive effects. An herbicide. When heated to decomposition it emits very toxic fumes of Cl^- and NO_x.

CKG000 CAS:15687-18-0 ***HR: 2***
2-(p-CHLOROPHENYL)-4-METHYLPENTANE-2,4-DIOL
mf: $C_{12}H_{17}ClO_2$ mw: 228.74

SYNS: 2-(p-CHLOROPHENYL)-4-METHYL-2,4-PENTANEDIOL ◇ FENPENTADIOL ◇ 2-METHYL-4-(p-CHLOROPHENYL)-2,4-PENTANEDIOL ◇ RD 292 ◇ TREDUM

TOXICITY DATA with REFERENCE
orl-rat LD50:1200 mg/kg ARZNAD 21,9,71
orl-mus LD50:940 mg/kg ARZNAD 21,9,71

SAFETY PROFILE: Moderately toxic by ingestion. A tranquilizer and an analeptic agent (stimulant). When heated to decomposition it emits toxic fumes of Cl^-.

CKG500 CAS:123-09-1 ***HR: 3***
p-CHLOROPHENYL METHYL SULFIDE
mf: C_7H_7ClS mw: 159.56

PROP: D: 1.222, bp: 169°.

SYNS: p-CHLOROTHIOANISOLE ◇ 4-CHLOROTHIOANISOLE ◇ METHYL-p-CHLOROPHENYL SULFIDE ◇ METHYL-4-CHLOROPHENYL SULFIDE

TOXICITY DATA with REFERENCE
eye-rbt 100 mg MLD NTIS** AD-A082-824
orl-rat LD50:479 mg/kg NTIS** AD-A082-824
skn-rat LDLo:5630 mg/kg NTIS** AD-A082-824
orl-mus LD50:672 mg/kg NTIS** AD-A082-824
orl-rat LD50:400 mg/kg TOLED5 1000 (Sp 1ssI),32,80

SAFETY PROFILE: Poison by ingestion. Mildly toxic by skin contact. An eye irritant. When heated to decomposition it emits very toxic fumes of Cl^- and SO_x. See also CHLORINATED HYDROCARBONS, AROMATIC; and SULFIDES.

CKG750 CAS:98-57-7 ***HR: 3***
p-CHLOROPHENYL METHYL SULFONE
mf: $C_7H_7ClO_2S$ mw: 191.56

SYNS: 4-CHLOROPHENYL METHYL SULFONE ◇ METHYL-4-CHLOROPHENYL SULFONE

TOXICITY DATA with REFERENCE
skn-rbt 500 mg/24H MLD NTIS** AD-A082-824
orl-rat LD50:400 mg/kg NTIS** AD-A082-824
orl-mus LD50:606 mg/kg NTIS** AD-A082-824

SAFETY PROFILE: Poison by ingestion. A skin irritant. When heated to decomposition it emits very toxic fumes of Cl^- and SO_x.

CKH000 CAS:934-73-6 ***HR: 2***
p-CHLOROPHENYL METHYL SULFOXIDE
mf: C_7H_7ClOS mw: 175.56

SYNS: 4-CHLOROPHENYL METHYL SULFOXIDE ◇ METHYL-4-CHLOROPHENYL SULFOXIDE

TOXICITY DATA with REFERENCE
skn-rbt 500 mg/24H SEV NTIS** AD-A082-824
eye-rbt 100 mg SEV NTIS** AD-A082-824
orl-rat LD50:463 mg/kg NTIS** AD-A082-824
orl-mus LD50:440 mg/kg NTIS** AD-A082-824

SAFETY PROFILE: Moderately toxic by ingestion. A severe eye and skin irritant. When heated to decomposition it emits very toxic fumes of Cl^- and SO_x.

CKI000 CAS:3818-90-4 ***HR: 3***
1-p-CHLOROPHENYL PENTYL SUCCINATE
mf: $C_{15}H_{19}ClO_4$ mw: 298.79

SYNS: AF 425 ◇ α-BUTYL-p-CHLOROBENZYL ESTER of SUCCINIC ACID ◇ SUCCINATO ACIDO DI 1-p-CLOROFENILPENTILE (ITALIAN) ◇ SUCCINIC ACID-α-BUTYL-p-CHLOROBENZYL ESTER

TOXICITY DATA with REFERENCE
orl-rat LD50:1130 mg/kg BCFAAI 100,504,61
ipr-mus LD50:259 mg/kg BCFAAI 100,855,61
ivn-mus LD50:76 mg/kg BCFAAI 100,504,61

SAFETY PROFILE: Poison by intravenous and intraperitoneal routes. Moderately toxic by ingestion. When heated to decomposition it emits toxic fumes of Cl^-.

CKI175 CAS:102071-30-7 ***HR: 3***
α-(p-CHLOROPHENYL)-α-PHENYL-2-PIPERIDINEMETHANOL HYDROCHLORIDE
mf: $C_{18}H_{20}ClNO \cdot ClH$ mw: 338.30

SYN: α-(p-CLOROFENIL)-α-FENIL-2-PIPERIDILMETANOLO CLORIDRATO (ITALIAN)

TOXICITY DATA with REFERENCE
orl-rat LD50:400 mg/kg FRPSAX 12,853,57
orl-mus LD50:145 mg/kg FRPSAX 12,853,57
ipr-mus LD50:78 mg/kg FRPSAX 12,853,57

SAFETY PROFILE: Poison by ingestion and intraperitoneal routes. When heated to decomposition it emits toxic fumes of NO_x and Cl^-.

CKI250 ***HR: 3***
1-(p-CHLOROPHENYL)-1-PHENYL-2-PROPYN-1-OL CARBAMATE
mf: $C_{16}H_{12}ClNO_2$ mw: 285.74

SYNS: 4-CHLORO-α-ETHYNYL-α-PHENYLBENZENEMETHANOL CARBAMATE ◇ 1-(4-CHLOROPHENYL)-1-PHENYL-2-PROPYNYL ESTER CARBAMIC ACID

TOXICITY DATA with REFERENCE
orl-rat TDLo:4040 mg/kg/42W-C:CAR TXAPA9 21,414,72
orl-rat TD:14 g/kg/77W-C:ETA JJIND8 71,211,83
ipr-mus LD50:347 mg/kg JMCMAR 11,1155,68

SAFETY PROFILE: Poison by intraperitoneal route. Questionable carcinogen with experimental carcinogenic and tumorigenic data. When heated to decomposition it emits toxic fumes of Cl^- and NO_x. See also CARBAMATES; and CHLORINATED HYDROCARBONS, AROMATIC.

CKI500 CAS:10473-70-8 ***HR: 3***
1-(4-CHLOROPHENYL)-1-PHENYL-2-PROPYNYL CARBAMATE
mf: $C_{16}H_{12}ClNO_2$ mw: 285.74

SYN: CARBAMIC ACID, 1-(4-CHLOROPHENYL)-1-PHENYL-2-PROPYNYL ESTER ◇ 4-CHLORO-α-ETHYNYL-α-PHENYLBENZENEMETHANOL CARBAMATE

TOXICITY DATA with REFERENCE
orl-rat TDLo:4040 mg/kg/42W-C:CAR TXAPA9 21,414,72
orl-rat TD:14 g/kg/77W-C:ETA JJIND8 71,211,83
ipr-mus LD50:347 mg/kg JMCMAR 11,1155,68

SAFETY PROFILE: Questionable carcinogen with experimental carcinogenic and tumorigenic data. See also CARBAMATES. When heated to decomposition it emits very toxic fumes of Cl^- and NO_x.

CKI600 CAS:31301-20-9 ***HR: D***
erythro-3-(p-CHLOROPHENYL)-2-PHENYL-4-(2-(1-PYRROLIDINYL)ETHOXY)BUTYROPHENONE
mf: $C_{28}H_{30}ClNO_2$ mw: 448.04

SYN: BUTYROPHENONE,3-(p-CHLOROPHENYL)-2-PHENYL-4′-(2-(1-PYRROLIDINYL)ETHOXY)-, erythro-

TOXICITY DATA with REFERENCE
orl-rat TDLo:2500 μg/kg (female 1-5D post):REP CCPTAY 2,199,70

SAFETY PROFILE: Experimental reproductive effects. When heated to decomposition it emits toxic fumes of NO_x, and Cl^-.

CKI625 CAS:80-00-2 ***HR: 2***
p-CHLOROPHENYL PHENYL SULFONE
mf: $C_{12}H_9ClO_2S$ mw: 252.72

PROP: Crystals, slt aromatic odor, no taste, insol in water. Mp: 90-94°.

SYNS: 4-CHLORODIPHENYL SULFONE ◇ R-CHLORODIPHENYL SULPHONE ◇ p-CHLOROPHENYL PHENYL SULPHONE ◇ 1-CHLORO-4-(PHENYLSULFONYL)BENZENE ◇ COMPOUND R-242 ◇ ENT 17,941 ◇ p-MONOCHLOROPHENYL PHENYL SULFONE ◇ R-242 ◇ R-242-B ◇ SULFENONE ◇ SULPHENONE

TOXICITY DATA with REFERENCE
orl-rat LD50:1400 mg/kg GUCHAZ 5,95,68
ipr-rat LDLo:500 mg/kg JAFCAU 3,836,55
orl-mus LD50:2700 mg/kg JAFCAU 3,836,55
ipr-mus LD50:1000 mg/kg JAFCAU 3,836,55

CONSENSUS REPORTS: Reported in EPA TSCA Inventory.

SAFETY PROFILE: Moderately toxic by ingestion and intraperitoneal routes. An acaricide. When heated to decomposition and it emits very toxic fumes of Cl^- and SO_x.

CKI750 CAS:18046-21-4 ***HR: 3***
4-(p-CHLOROPHENYL)-2-PHENYL-5-THIAZOLEACETIC ACID
mf: $C_{17}H_{12}ClNO_2S$ mw: 329.81

SYNS: BR 700 ◇ CH 800 ◇ 4-(p-CHLOROPHENYL)-2-PHENYLTHIAZOLE-5-ACETIC ACID ◇ 4-(4-CHLOROPHENYL)-2-PHENYL)-5-THIAZOLEACETIC ACID (9CI) ◇ DONOREST ◇ FENTIAZAC ◇ FLOGENE ◇ NORVEDAN

TOXICITY DATA with REFERENCE
orl-rat TDLo:1350 mg/kg (female 17-22D post):REP KSRNAM 13,1929,79
orl-rat TDLo:525 mg/kg (14D pre/1-7D preg):TER KSRNAM 13,1929,79
orl-rat LD50:409 mg/kg NIIRDN 6,APP-17,82
ipr-rat LD50:325 mg/kg NIIRDN 6,APP-17,82
scu-rat LD50:543 mg/kg IYKEDH 13,637,82
ivn-rat LD50:160 mg/kg KSRNAM 13,1895,79
orl-mus LD50:520 mg/kg NIIRDN 6,APP-17,82
ipr-mus LD50:385 mg/kg IYKEDH 13,637,82
scu-mus LD50:655 mg/kg IYKEDH 13,637,82

ivn-mus LD50:161 mg/kg KSRNAM 13,1895,79
ivn-dog LD50:103 mg/kg KSRNAM 13,1895,79
orl-rbt LD50:625 mg/kg KSRNAM 13,1895,79
ivn-rbt LD50:71 mg/kg KSRNAM 13,1895,79

SAFETY PROFILE: Poison by intravenous and intraperitoneal routes. Moderately toxic by ingestion and subcutaneous routes. An experimental teratogen. Other experimental reproductive effects. An anti-inflammatory agent. When heated to decomposition it emits very toxic fumes of SO_x, NO_x, and Cl^-.

CKI825 CAS:4415-51-4 ***HR: 3***
4-(p-CHLOROPHENYL)-1-PIPERAZINE-ETHANOL-2-PYRIDINEACRYLATE(ESTER)
mf: $C_{20}H_{22}ClN_3O_2$ mw: 371.90

SYN: 1-(4-CHLOROPHENYL)-4-(2-(3-(2-PYRIDYL)ACRYLOXY)ETHYL)-PIPERAZINE

TOXICITY DATA with REFERENCE
orl-rat LD50:478 mg/kg TXAPA9 10,444,67
ipr-rat LD50:311 mg/kg JPMSAE 55,1105,66
orl-mus LD50:529 mg/kg JPMSAE 55,1105,66
ipr-mus LD50:280 mg/kg JPMSAE 55,290,66

SAFETY PROFILE: Poison by intraperitoneal route. Moderately toxic by ingestion. When heated to decomposition it emits toxic fumes of Cl^- and NO_x.

CKJ000 CAS:25332-39-2 ***HR: 3***
2-(3-(4-(3-CHLOROPHENYL)-1-PIPERAZINYL)PROPYL)-1,2,4-TRIZOLO(4,3-a)PYRIDIN-3(2H)-ONE HYDROCHLORIDE
mf: $C_{19}H_{22}ClN_5O \cdot ClH$ mw: 408.37

SYNS: AF 1161 ◇ 2-(3-(4-(m-CHLOROPHENYL)-1-PIPERAZINYL)PROPYL)-s-TRIAZOLE-(4,3-α)-PIRIDIN-3(WH)-ONE HCl ◇ DESYREL ◇ MOLIPAXIN ◇ PRAGMAZONE ◇ THOMBRAN ◇ TOMBRAN ◇ TRAZODONE HYDROCHLORIDE ◇ TRITTICO

TOXICITY DATA with REFERENCE
orl-man TDLo:159 mg/kg (male 7W pre):REP AJPSAO 140,1256,83
orl-man TDLo:112 mg/kg (male 6W pre):REP JCLPDE 45,232,84
orl-wmn TDLo:750 μg/kg AJPSAO 141,434,84
orl-wmn TDLo:7500 μg/kg/5D-I AJPSAO 140,642,83
orl-man TDLo:46 mg/kg/8D-I JCPYDR 6,117,86
orl-man TDLo:667 μg/kg:CVS AJPSAO 141,1472,84
orl-rat LD50:690 mg/kg MPPPBK 9,76,74
ipr-rat LD50:178 mg/kg MPPPBK 9,76,74
ivn-rat LD50:91 mg/kg MPPPBK 9,76,74
orl-mus LD50:610 mg/kg MPPPBK 9,76,74
ivn-mus LD50:91 mg/kg MPPPBK 9,76,74
orl-dog LD50:500 mg/kg MPPPBK 9,76,74
ivn-mky LD50:25 mg/kg PBPSDY 3,94,81
orl-rbt LD50:560 mg/kg MPPPBK 9,76,74
ivn-rbt LD50:52 mg/kg MPPPBK 9,76,74

SAFETY PROFILE: Poison by intravenous route. Moderately toxic by ingestion. Human systemic effects by ingestion: cardiomyopathy including infarction. Experimental reproductive effects. When heated to decomposition it emits very toxic fumes of Cl^- and NO_x. A tranquilizer and hypotensive agent. See also KETONES.

CKJ750 CAS:13037-55-3 ***HR: 3***
N-(4-CHLOROPHENYL) RHODANINE
mf: $C_9H_6ClNOS_2$ mw: 243.73

SYN: USAF T-1

TOXICITY DATA with REFERENCE
unk-rat LDLo:200 mg/kg ARZNAD 16,1092,66
ipr-mus LD50:200 mg/kg NTIS** AD277-689

SAFETY PROFILE: Poison by intraperitoneal and possibly other routes. When heated to decomposition it emits very toxic fumes of Cl^-, NO_x, and SO_x.

CKJ825 CAS:20064-40-8 ***HR: 3***
5-((p-CHLOROPHENYL)SULFONYL)-3-METHYL-1,2,4-THIADIAZOLE
mf: $C_9H_7ClN_2O_2S_2$ mw: 274.75

TOXICITY DATA with REFERENCE
eye-rbt 200 mg/1M YKKZAJ 88,1437,68
ivn-rat LD50:132 mg/kg YKKZAJ 88,1437,68
orl-mus LD50:3 g/kg YKKZAJ 88,1437,68
ipr-mus LD50:347 mg/kg YKKZAJ 88,1437,68
scu-mus LD50:646 mg/kg YKKZAJ 88,1437,68
ivn-mus LD50:151 mg/kg YKKZAJ 88,1437,68

SAFETY PROFILE: Poison by intravenous and intraperitoneal routes. Moderately toxic by ingestion and subcutaneous routes. An eye irritant. When heated to decomposition it emits toxic fumes of Cl^-, SO_x, and NO_x.

CKK000 CAS:94-20-2 ***HR: 3***
1-(p-CHLOROPHENYLSULFONYL)-3-PROPYLUREA
mf: $C_{10}H_{13}ClN_2O_3S$ mw: 276.76

SYNS: ADIABEN ◇ ASUCROL ◇ CATANIL ◇ 1-(p-CHLOROBENZENESULFONYL)-3-PROPYLUREA ◇ N-(p-CHLOROBENZENESULFONYL)-N-PROPYLUREA ◇ CHLORODIABINA ◇ CHLORONASE ◇ 1-p-CHLOROPHENYL-3-(PROPYLSULFONYL)UREA ◇ CHLOROPROPAMIDE ◇ 4-CHLORO-4-((PROPYLAMINO)CARBONYL)BENZENESULFONAMIDE

◇ CHLORPROPAMID ◇ CHLORPROPAMIDE ◇ CLORPROPAMIDE (ITALIAN) ◇ DIABARIL ◇ DIABECHLOR ◇ DIABENAL ◇ DIABENESE ◇ DIABENEZA ◇ DIABETORAL ◇ DIABET-PAGES ◇ DIABINESE ◇ GLISEMA ◇ MELDIAN ◇ MELITASE ◇ MELLINESE ◇ NCI-CO1752 ◇ ORADIAN ◇ N-PROPYL-N'-(p-CHLOROBENZENE-SULFONYL)UREA ◇ 1-PROPYL-3-(p-CHLOROBENZENE-SULFONYL)UREA ◇ N-PROPYL-N'-p-CHLORPHENYLSULFONYLCARBAMIDE ◇ STABINOL ◇ U-3818

TOXICITY DATA with REFERENCE
sce-mus-orl 177 mg/kg MUREAV 77,349,80
sce-ham-ipr 7100 μg/kg MUREAV 77,349,80
orl-wmn TDLo:840 mg/kg (9-39W preg):REP ADCHAK 49,283,74
orl-wmn TDLo:884 mg/kg (1-35W preg):TER ADCHAK 49,283,74
orl-wmn TDLo:25 mg/kg/5D-I:BLD DICPBB 18,981,84
orl-wmn TDLo:300 mg/kg:CNS CTOXAO 11,13,77
orl-man TDLo:83 mg/kg/25D-I:GIT AJGAAR 80,381,85
orl-rat LD50:2150 mg/kg FATOAO 25,93,62
ipr-rat LD50:580 mg/kg TXAPA9 18,185,71
ivn-rat LD50:590 mg/kg PMDCAY 1,187,61
orl-mus LD50:1546 mg/kg AIPTAK 153,379,65
ivn-mus LD50:500 mg/kg PMDCAY 1,187,61

CONSENSUS REPORTS: NCI Carcinogenesis Bioassay (feed); No Evidence: mouse, rat NCITR* NCI-CG-TR-45,78. EPA Genetic Toxicology Program.

SAFETY PROFILE: Moderately toxic by ingestion, intraperitoneal, and intravenous routes. Human systemic effects by ingestion: hemolysis with or without anemia, ulceration or bleeding from large intestine, cholestatic jaundice. Human teratogenic effects by ingestion: fetal death. Human reproductive effects by ingestion and possibly other routes: stillbirth, reduced viability and abnormal characteristics of newborn at birth. An experimental teratogen. Experimental reproductive effects. Mutation data reported. An antidiabetic agent. When heated to decomposition it emits very toxic fumes of Cl^-, NO_x, and SO_x.

CKK099 ***HR: 3***
N-CHLORO-5-PHENYLTETRAZOLE
mf: $C_7H_6ClN_3$ mw: 167.60

SAFETY PROFILE: May explode when heated. When heated to decomposition it emits toxic fumes of Cl^- and NO_x.

CKK250 CAS:17969-20-9 ***HR: 3***
2-(p-CHLOROPHENYL)-4-THIAZOLE ACETIC ACID
mf: $C_{11}H_8ClNO_2S$ mw: 253.71

SYNS: ACIDEp-CHLOROPHENYL-2-THIAZOLE-ACETIQUE-4 (FRENCH) ◇ ACIDO FENCLOZICO (ITALIAN) ◇ ACIDUM FENCLOZICUM ◇ 2-(4-CHLOROPHENYL)-4-THIAZOLEACETIC ACID (9CI) ◇ 2-(p-CHLOROPHENYL)THIAZOL-4-YLACETIC ACID ◇ FENCLOZIC ACID ◇ ICI 54,450 ◇ MIALEX ◇ MYALEX

TOXICITY DATA with REFERENCE
orl-rat TDLo:50 mg/kg (20-21D preg):REP NATUAS 240,37,72
orl-rat LD50:850 mg/kg NATUAS 221,582,69
ivn-rat LD50:300 mg/kg NATUAS 221,582,69
orl-mus LD50:400 mg/kg FRPSAX 40,875,85
ipr-mus LD50:250 mg/kg FRPSAX 40,875,85
ivn-mus LD50:250 mg/kg NATUAS 221,582,69

SAFETY PROFILE: Poison by ingestion, intraperitoneal, and intravenous routes. Experimental reproductive effects. An anti-inflammatory agent. When heated to decomposition it emits very toxic fumes of Cl^-, NO_x, and SO_x.

CKK500 CAS:15446-08-9 ***HR: 2***
4-(p-CHLOROPHENYL)THIO-1-BUTANOL
mf: $C_9H_{13}ClOS$ mw: 204.73

SYN: 4-(p-CHLORPHENYLTHIO)-BUTANOL(GERMAN)

TOXICITY DATA with REFERENCE
orl-rat TDLo:18200 mg/kg (91D male):REP ARZNAD 29,1382,79
ipr-rat LD50:885 mg/kg ARZNAD 29,1141,79
orl-mus LD50:1145 mg/kg ARZNAD 19,1141,79
ipr-mus LD50:880 mg/kg ARZNAD 29,1141,79

SAFETY PROFILE: Moderately toxic by ingestion and intraperitoneal routes. Experimental reproductive effects. When heated to decomposition it emits very toxic fumes of Cl^- and SO_x.

CKL000 CAS:5344-82-1 ***HR: 3***
2-CHLOROPHENYL THIOUREA
mf: $C_7H_7ClN_2S$ mw: 186.67

SYN: RCRA WASTE NUMBER P026

TOXICITY DATA with REFERENCE
orl-rat LD50:4600 μg/kg JPETAB 90,260,47

CONSENSUS REPORTS: EPA Extremely Hazardous Substances List. Reported in EPA TSCA Inventory.

SAFETY PROFILE: Poison by ingestion. When heated to decomposition it emits very toxic fumes of Cl^-, NO_x, and SO_x.

CKL100 CAS:69095-74-5 ***HR: D***
5-(o-CHLOROPHENYL)-3-(o-TOLYL)-s-TRIAZOLE
mf: $C_{15}H_{12}ClN_3$ mw: 269.75

SYN: s-TRIAZOLE,5-(o-CHLOROPHENYL)-3-(o-TOLYL)-

TOXICITY DATA with REFERENCE
scu-ham TDLo:75 mg/kg (female 4-8D post):REP JMCMAR 26,1187,83

SAFETY PROFILE: Experimental reproductive effects. When heated to decomposition it emits toxic fumes of NO_x and Cl^-.

CKL250 CAS:29975-16-4 ***HR: 3***
8-CHLORO-6-PHENYL-4H-s-TRIAZOLO(4,3-a)(1,4) BENZODIAZEPINE
mf: $C_{16}H_{11}ClN_4$ mw: 294.76

SYNS: 8-CHLORO-6-PHENYL-4H-(1,2,4)TRIAZOLO(4,3-a)(1,4-BENZODIAZEPINE (9CI) ◇ D-40TA ◇ ESILGAN ◇ ESTAZOLAM ◇ EURODIN ◇ JULODIN ◇ NUCTALON

TOXICITY DATA with REFERENCE
orl-rat TDLo:300 mg/kg (30D pre):REP TAKHAA 32,158,73
orl-rat LD50:2500 mg/kg NIIRDN 6,103,82
ipr-rat LD50:339 mg/kg IYKEDH 6,530,75
scu-rat LD50:3580 mg/kg IYKEDH 6,530,75
ims-rat LD50:868 mg/kg IYKEDH 6,530,75
orl-mus LD50:600 mg/kg CCCCAK 48,2395,83
ipr-mus LD50:438 mg/kg JJPAAZ 21,497,71
orl-rbt LD50:300 mg/kg TAKHAA 32,152,73

SAFETY PROFILE: Poison by ingestion and intraperitoneal routes. Moderately toxic by intramuscular and subcutaneous routes. Experimental reproductive effects. An hypnotic and sedative agent. When heated to decomposition it emits very toxic fumes of Cl^- and NO_x. See also DIAZEPAM.

CKL325 CAS:66535-86-2 ***HR: 3***
2-(p-CHLOROPHENYL)-s-TRIAZOLO(5,1-a)ISO QUINOLINE
mf: $C_{16}H_{10}ClN_3$ mw: 279.74

PROP: Crystals. Mp: 238-240°.

SYNS: CANOCENTA ◇ 2-(4-CHLOROPHENYL)-(1,2,4)TRIAZOLO(5,1-a)ISOQUINOLINE (9CI) ◇ DL 717-IT ◇ L 12717 ◇ LOTRIFEN

TOXICITY DATA with REFERENCE
orl-ham TDLo:40 mg/kg (female 4-8D post):REP RDMIDP 4,237,82
ipr-mus LD50:1 g/kg MDACAP 23,75,87
ims-dog LDLo:75 mg/kg MDACAP 23,75,87

SAFETY PROFILE: Poison by intramuscular route. Experimental reproductive effects. When heated to decomposition it emits toxic fumes of Cl^- and NO_x.

CKL500 CAS:8072-20-6 ***HR: 2***
4-CHLOROPHENYL-2,4,5-TRICHLORO-PHENYLAZOSULFIDE mixed with 1,1-BIS(4-CHLOROPHENYL)ETHANOL
mf: $C_{14}H_{12}Cl_2O \cdot C_{12}H_6Cl_4N_2S$ mw: 619.22

SYNS: ANILIX ◇ BCPE mixed with SPAS ◇ CHLOROFENSULPHIDE ◇ CPAS mixed with BCPE ◇ DANINON ◇ MICASIN ◇ MILBEX ◇ MILSAR ◇ MISASIN

TOXICITY DATA with REFERENCE
orl-mus TDLo:1200 mg/kg (1-12D preg):TER OYYAA2 2,355,68
orl-rat LD50:1856 mg/kg OYYAA2 2,161,68
orl-mus LD50:2000 mg/kg OYYAA2 2,355,68

SAFETY PROFILE: Moderately toxic by ingestion. An experimental teratogen. A miticide. When heated to decomposition it emits very toxic fumes of Cl^-, SO_x, and NO_x. See also SULFIDES.

CKL750 CAS:2227-13-6 ***HR: 2***
p-CHLOROPHENYL-2,4,5-TRICHLOROPHENYL SULFIDE
mf: $C_{12}H_6Cl_4S$ mw: 324.04

SYNS: ANIMERT ◇ ANIMERT V-10 ◇ ANIMERT V-101 ◇ ANIMERT V-10K ◇ 4-CHLOROPHENYL 2,4,5-TRICHLOROPHENYL SULFIDE ◇ ENT 27,115 ◇ PHILIPS-DUPHAR V-101 ◇ 3,4,6,R'-TETRACHLORDIPHENYLSULFID (GERMAN) ◇ 2,4,4',5-TETRACHLORODIPHENYL SULFIDE ◇ 2,4,5,4'-TETRACHLORODIPHENYL SULFIDE ◇ TETRASUL ◇ 1,2,4-TRICHLORO-5-((4-CHLOROPHENYL)THIO)-BENZENE

TOXICITY DATA with REFERENCE
orl-rat TDLo:18 g/kg (17W male):REP TXCYAC 1,63,73
orl-rat LD50:3960 mg/kg BESAAT 15,129,69
orl-mus LD50:5010 mg/kg GUCHAZ 6,495,73
orl-rbt LDLo:1350 mg/kg TXCYAC 1,63,73
skn-rbt LD50:2000 mg/kg SPEADM 78-1,55,78
orl-gpg LD50:500 mg/kg TXCYAC 1,63,73
skn-gpg LD50:8200 mg/kg 85DPAN -,-,71/76
ipr-gpg LD50:550 mg/kg TXCYAC 1,63,73

SAFETY PROFILE: Moderately toxic by ingestion, skin contact, and intraperitoneal routes. Experimental reproductive effects. A pesticide. When heated to decomposition it emits very toxic fumes of Cl^- and SO_x. See also SULFIDES.

CKM000 CAS:116-29-0 ***HR: 2***
p-CHLOROPHENYL-2,4,5-TRICHLOROPHENYL SULFONE
mf: $C_{12}H_6Cl_4O_2S$ mw: 356.04

PROP: Crystals. Mp: 147°. Nearly water-insol.

SYNS: AKARITOX ◇ AREDION ◇ 4-CHLOROPHENYL-2,4,5-TRICHLOROPHENYL SULFONE ◇ p-CHLOROPHENYL-2,4,5-TRICHLOROPHENYL SULPHONE ◇ DUPHAR ◇ ENT 23,737 ◇ FMC 5488 ◇ MITION ◇ NIA 5488 ◇ POLACARITOX ◇ ROZTOZOL ◇ SULFONE-2,4,4′,5-TETRACHLORODIPHENYL ◇ TEDION ◇ TEDION V-18 ◇ 2,4,4′,5-TETRACHLOOR-DIFENYL-SULFON (DUTCH) ◇ 2,4,4′,5-TETRACHLOR-DIPHENYL-SULFON (GERMAN) ◇ 2,4,4′,5-TETRACHLORODIPHENYL SULFONE ◇ 2,4,5,4′-TETRACHLORODIPHENYLSULPHONE ◇ 2,4,4′,5-TETRACLORO- DIFENIL-SOLFONE (ITALIAN) ◇ TETRADICHLONE ◇ TETRADIFON ◇ TETRADIPHON ◇ TETRAFIDON ◇ 1,2,4-TRICHLORO-5-((4-CHLOROPHENYL)SULFONYL)-BENZENE ◇ V-18

TOXICITY DATA with REFERENCE
scu-mus TDLo:1953 mg/kg (6-14D preg):TER NTIS** PB223,-160
orl-rat LD50:566 mg/kg WRPCA2 9,119,70
orl-dog LD50:2000 mg/kg SPEADM 78-1,55,78
skn-rbt LD50:10 g/kg 28ZEAL 5,220,76

CONSENSUS REPORTS: EPA Genetic Toxicology Program.

SAFETY PROFILE: Moderately toxic by ingestion. Mildly toxic by skin contact. An experimental teratogen. Used to control worms in crops. When heated to decomposition it emits highly toxic fumes of Cl^- and SO_x.

CKM250 CAS:26571-79-9 ***HR: 3***
CHLOROPHENYLTRICHLOROSILANE
DOT: UN 1753
mf: $C_6H_4Cl_4Si$ mw: 245.99

PROP: Colorless to pale yellow liquid, readily hydrolyzed by moisture with the liberation of HCl (a mixture of 3 isomers). Bp: 230°, d: 1.439 @ 25°/25°, flash p: 255°F (COC).

DOT Classification: Corrosive Material; Label: Corrosive.

SAFETY PROFILE: A poison irritant by ingestion and inhalation. A corrosive irritant to the skin, eyes, and mucous membranes. Combustible when exposed to heat or flame. In contact with water it readily hydrolyzes to HCl and evolves heat. When heated to decomposition it emits toxic fumes of Cl^-. See also CHLOROSILANES.

CKM750 CAS:29025-67-0 ***HR: 3***
1-(p-CHLOROPHENYL)-2,8,9-TRIOXA-5-AZA-1-SILABICYCLO(3.3.3) UNDECANE
mf: $C_{12}H_{16}ClNO_3Si$ mw: 285.83

SYNS: p-CHLORFENYLSILATRAN (CZECH) ◇ 1-(p-CHLOROPHENYL)SILATRANE

TOXICITY DATA with REFERENCE
orl-rat LD50:520 mg/kg MarJV# 29MAR77
orl-mus LDLo:8 mg/kg AECTCV 14,111,85
ipr-mus LD50:1050 mg/kg RCRVAB 38(12),975,69

SAFETY PROFILE: Poison by ingestion. Moderately toxic by intraperitoneal route. When heated to decomposition it emits very toxic fumes of Cl^- and NO_x.

CKN000 CAS:1406-65-1 ***HR: 3***
CHLOROPHYLL

PROP: Dark green solution.

SYNS: BIOPHYLL ◇ CHLOROPHYL, GREEN ◇ C.I. 1956 ◇ DAROTOL ◇ DEODOPHYLL ◇ E 140 ◇ GREEN CHLOROPHYL ◇ L-GRUEN No. 1 (GERMAN) ◇ No. 1249 ◇ No. 1403 ◇ No. 75810

TOXICITY DATA with REFERENCE
ipr-mus LD50:400 mg/kg ARZNAD 4,19,54
ivn-mus LD50:285 mg/kg ARZNAD 4,19,54
ivn-gpg LD50:85 mg/kg ARZNAD 7,357,57

SAFETY PROFILE: Poison by intravenous and intraperitoneal routes. When heated to decomposition it emits toxic fumes of NO_x.

CKN250 CAS:11006-92-1 ***HR: 3***
CHLOROPHYLLIN

TOXICITY DATA with REFERENCE
ipr-mus LD50:400 mg/kg ARZNAD 4,19,54
ivn-mus LD50:285 mg/kg ARZNAD 4,19,54
ivn-gpg LDLo:80 mg/kg ARZNAD 4,19,54

SAFETY PROFILE: Poison by intraperitoneal and intravenous routes.

CKN375 CAS:4684-94-0 ***HR: 2***
6-CHLOROPICOLINIC ACID
mf: $C_6H_4ClNO_2$ mw: 157.56

SYNS: 6-CHLORO-2-PYRIDINECARBOXYLIC ACID ◇ 6-CPA

TOXICITY DATA with REFERENCE
orl-rat LD50:2180 mg/kg DOEAAH 32,22,76
unr-rat LD50:1487 mg/kg GISAAA 48(4),52,83
unr-mus LD50:777 mg/kg GISAAA 48(4),52,83

SAFETY PROFILE: Moderately toxic by ingestion and possibly other routes. When heated to decomposition it emits toxic fumes of Cl^- and NO_x.

CKN500 CAS:76-06-2 ***HR: 3***
CHLOROPICRIN
DOT: UN 1580/NA 1583/NA 1955/NA 2929/UN 1583
mf: CCl_3NO_2 mw: 164.37

PROP: Sltly oily, colorless liquid. D: 1.651 @ 22.8°/4°,

mp: −64°, bp: 112.3 @ 766 mm, vap press: 40 mm @ 33.80, vap d: 6.69. Sol in water, alc, and ether.

SYNS: ACQUINITE ◇ CHLOORPIKRINE (DUTCH) ◇ CHLOR-O-PIC ◇ CHLOROPICRIN, liquid (DOT) ◇ CHLOROPICRIN, ABSORBED (DOT) ◇ CHLOROPICRINE (FRENCH) ◇ CHLORPIKRIN (GERMAN) ◇ CLOROPICRINA (ITALIAN) ◇ DOJYOPICRIN ◇ DOLOCHLOR ◇ LARVACIDE ◇ MICROLYSIN ◇ NCI-C00533 ◇ NITROCHLOROFORM ◇ NITROTRICHLOROMETHANE ◇ PIC-CLOR ◇ PICFUME ◇ PICRIDE ◇ PROFUME A ◇ PS ◇ TRICHLOORNITROMETHAAN (DUTCH) ◇ TRICHLORNITROMETHAN (GERMAN) ◇ TRICHLORONITROMETHANE ◇ TRI-CLOR ◇ TRICLORO-NITRO-METANO (ITALIAN)

TOXICITY DATA with REFERENCE
mma-sat 50 μg/plate MUREAV 116,185,83
orl-mus TDLo:26 g/kg/78W-I:ETA NCITR* NCI-CG-TR-65,78
ihl-hmn TCLo:2 mg/m³:EYE,PUL ZGEMAZ 106,296,39
ihl-hmn TCLo:2000 mg/m³/10M 34ZIAG -,169,69
orl-rat LD50:250 mg/kg DOWCC* -,-,72
ihl-mus LC50:1600 mg/m³/10M NDRC** NDCrc-132, an,42
ipr-mus LD50:25 mg/kg KHFZAN 10(6),53,75
ihl-cat LCLo:800 mg/m³/20M ZGEMAZ 106,296,39
ihl-rbt LCLo:800 mg/m³/20M ZGEMAZ 106,296,39
ihl-gpg LCLo:800 mg/m³/20M ZGEMAZ 106,296,39
ivn-gpg LD50:4200 μg/kg AIPTAK 114,146,58

CONSENSUS REPORTS: NCI Carcinogenesis Bioassay (gavage); No Evidence: mouse NCITR* NCI-GC-TR-65,78. Reported in EPA TSCA Inventory.

OSHA PEL: TWA 0.1 ppm
ACGIH TLV: TWA 0.1 ppm
DFG MAK: 0.1 ppm (0.7 mg/m³)
DOT Classification: Poison B; Label: Poison (NA1583, UN1580, UN1583); Poison B; Label: Poison and Flammable Liquid (NA2929); Poison B; Label: St. Andrews Cross (UN1583)
DOT Classification: Poison A; Label: Poison and Nonflammable Gas (NA1955)

SAFETY PROFILE: Poison by ingestion, intravenous, and intraperitoneal routes. Moderately toxic by inhalation. Human systemic effects by inhalation: lacrimation, conjunctiva irritation, and pulmonary changes. Mutation data reported. A powerful irritant that affects all body surfaces. It causes lachrymation, vomiting, bronchitis, pulmonary edema, irritation to gastrointestinal and respiratory tracts. Questionable carcinogen with experimental tumorigenic data. An additional toxic effect is its reaction with SH-groups in hemoglobin thus interfering with oxygen transport. Photochemical transformation of chloropicrin into phosgene (carboxy chloride, $COCl_2$) has been reported. A concentration of 1 ppm causes a smarting pain in the eyes and therefore in itself constitutes a good warning of exposure. Inhalation causes vomiting, probably due to swallowing saliva in which small amounts of chloropicrin have dissolved. Its primary lethal effect is to produce lung injury and it is a difficult gas to protect oneself against because it is chemically inert and does not react with the usual chemicals used in gas masks. Four ppm is sufficient to render a worker unfit for action and 20 ppm, when breathed from 1 to 2 minutes, causes definite bronchial or pulmonary lesions. Industrially it is used as a warning agent in commercial fumigants. It is more toxic than chlorine but less so than phosgene.

Above a critical volume it can be shocked into detonation. Mixtures with 3-bromopropyne are shock- and heat-sensitive explosives. Violent reaction with aniline + heat, alcoholic sodium hydroxide, sodium methoxide, and propargyl bromide. When heated to decomposition it emits very toxic fumes of Cl^- and NO_x.

Used for insect and rodent control in grain elevators and bins and as a soil fumigant and fungicide. See also NITRO COMPOUNDS.

CKN510 CAS:76-06-2 ***HR: 3***
CHLOROPICRIN MIXTURE (flammable)
DOT: NA 2929
mf: CCl_3NO_2 mw: 164.37

SYNS: CHLOROPICRINE ◇ TRICHLORONITROMETHANE (flammable mixture)

TOXICITY DATA with REFERENCE
ihl-cat LCLo:800 mg/m³/20M PEMNDP 8,168,87
ihl-rbt LCLo:800 mg/m³/20M PEMNDP 8,168,87
ihl-gpg LCLo:800 mg/m³/20M PEMNDP 8,168,87

CONSENSUS REPORTS: Reported in EPA TSCA Inventory.

DOT Classification: Poison B; Label: Poison and Flammable Liquid.

SAFETY PROFILE: A poison. Moderately toxic by inhalation. When heated to decomposition it emits very toxic fumes of NO_x and Cl^-. See also TRICHLORONITROMETHANE.

CKN675 CAS:2156-71-0 ***HR: 3***
1-CHLOROPIPERIDINE
mf: $C_5H_{10}ClN$ mw: 119.61

$ClN(CH_2)_4CH_2$ (ring)

SYN: N-CHLOROPIPERIDINE

TOXICITY DATA with REFERENCE
mmo-sat 32 μg/plate JEPTDQ 4(2-3),345,80
bfa-mus/sat 50 mg/kg JEPTDQ 4(2-3),345,80
oms-ham:ovr 1200 μg/L JEPOEC 6,241,85
cyt-ham:ovr 400 μg/L JEPOEC 6,241,85
orl-mus TDLo:2800 mg/kg (female 7D pre):REP JACTDZ 2(2),209,83
ipr-mus LD50:100 mg/kg JEPTDQ 4(2-3),345,80

SAFETY PROFILE: Poison by intraperitoneal route. Mutation data reported. Experimental reproductive effects. Can undergo rapid spontaneous decomposition. When heated to decomposition it emits toxic fumes of Cl^- and NO_x.

CKN750 CAS:77966-67-7 ***HR: 3***
6'-CHLORO-2-PIPERIDINO-o-ACETOTOLUIDIDE HYDROCHLORIDE
mf: $C_{14}H_{19}ClN_2O{\cdot}ClH$ mw: 303.26

SYN: V 255

TOXICITY DATA with REFERENCE
eye-rbt 2% MLD ARZNAD 8,407,58
ipr-rat LD50:161 mg/kg ARZNAD 8,407,58
ipr-mus LD50:170 mg/kg ARZNAD 8,407,58
scu-mus LD50:265 mg/kg ARZNAD 8,407,58

SAFETY PROFILE: Poison by intraperitoneal and subcutaneous routes. An eye irritant. When heated to decomposition it emits very toxic fumes of Cl^- and NO_x.

CKO000 CAS:78218-40-3 ***HR: 3***
6'-CHLORO-3-(PIPERIDINO)-o-PROPIONOTOLUIDIDE HYDROCHLORIDE
mf: $C_{15}H_{21}ClN_2O{\cdot}ClH$ mw: 317.29

SYN: C 3138

TOXICITY DATA with REFERENCE
ipr-rat LD50:58 mg/kg ARZNAD 8,544,58
scu-mus LD50:67 mg/kg ARZNAD 8,544,58

SAFETY PROFILE: Poison by intraperitoneal and subcutaneous routes. When heated to decomposition it emits very toxic fumes of NO_x and Cl^-.

CKO750 CAS:16941-12-1 ***HR: 3***
CHLOROPLATINIC ACID
DOT: UN 2507
mf: $Cl_6Pt{\cdot}2H$ mw: 409.81

PROP: Brownish-yellow, very deliq, crystalline mass. D: 2.431, mp: 60°. Easily sol in water and alc.

SYNS: CHLOROPLATINIC(IV) ACID ◇ DIHYDROGEN HEXACHLOROPLATINATE ◇ DIHYDROGEN HEXACHLOROPLATINATE(2-) ◇ HEXACHLOROPLATINIC ACID ◇ HEXACHLOROPLATINIC(IV) ACID ◇ HEXACHLOROPLATININIC(4+) ACID, HYDROGEN- ◇ HYDROGEN HEXACHLOROPLATINATE(4+) ◇ PLATINIC CHLORIDE

TOXICITY DATA with REFERENCE
mrc-bcs 10 mmol/L MUREAV 77,109,80
ivn-rat LD50:49 mg/kg NTIS** PB291-731
ipr-mus LD50:61 mg/kg COREAF 256,1043,63

CONSENSUS REPORTS: Reported in EPA TSCA Inventory. EPA Genetic Toxicology Program.

OSHA PEL: TWA 0.002 mg(Pt)/m^3
ACGIH TLV: TWA 0.002 mg(Pt)/m^3
DOT Classification: ORM-B; Label: None; Corrosive Material; Label: Corrosive

SAFETY PROFILE: Poison by intravenous and intraperitoneal routes. Mutation data reported. See PLATINUM COMPOUNDS and CHLORIDES. Incompatible with BrF_3. When heated to decomposition it emits toxic fumes of Cl^-.

CKP000 CAS:16941-12-1 ***HR: 3***
CHLOROPLATINIC ACID (technical)

PROP: Red-brown, deliq prisms. Mp: 60°, d: 2.431.

SYN: CHLOROPLATINIC ACID, solid (DOT)

CONSENSUS REPORTS: Reported in EPA TSCA Inventory.

DOT Classification: ORM-B; Label: None.

SAFETY PROFILE: See CHLOROPLATINIC ACID, PLATINUM COMPOUNDS, and CHLORIDES. When heated to decomposition it emits toxic fumes of Cl^-.

CKP250 CAS:50-53-3 ***HR: 3***
CHLOROPROMAZINE
mf: $C_{17}H_{19}ClN_2S$ mw: 318.89

SYNS: AMINASINE ◇ AMINAZIN ◇ AMINAZINE ◇ AMPLIACTIL ◇ AMPLICITIL ◇ 2-CHLORO-10-(3-(DIMETHYLAMINO)PROPYL)PHENOTHIAZINE ◇ CHLORO-3-(DIMETHYLAMINO-3-PROPYL)-10 PHENOTHIAZINE (FRENCH) ◇ CHLORPROMAZIN ◇ CHLORPROMAZINE ◇ CHLOR-PZ ◇ CLOPROMAZINA (ITALIAN) ◇ 2-CLORO-10-(3-DIMETILAMINOPROPIL)FENOTIAZINA (ITALIAN) ◇ CPZ ◇ FENACTIL ◇ FENAKTYL ◇ FRACTION AB ◇ HIBANIL ◇ HIBERNAL ◇ LARGACTIL ◇ LARGACTILOTHIAZINE ◇ LARGACTYL ◇ MEGAPHEN ◇ NOVOMAZINA ◇ PHENACTYL ◇ PHENATHYL ◇ PLEGOMAZIN ◇ PRAZIL ◇ PROMACTIL ◇ PROMAZIL ◇ PROPAPHENIN ◇ PROZIL ◇ PROZIN ◇ SKF-2601 ◇ THORAZINE ◇ TORAZINA ◇ WINTERMIN

TOXICITY DATA with REFERENCE
unr-wmn TDLo:540 mg/kg (1-39W preg):REP AGPSA3 2,606,60
ims-rat TDLo:270 mg/kg (female 18-22D post):TER TJADAB 26,21,82
orl-inf TDLo:20 mg/kg:CVS,CNS AJDCAI 130,507,76
orl-wmn TDLo:200 μg/kg:CNS,LIV NYSJAM 57,1922,57
orl-hmn TDLo:8570 μg/kg/12D-I:CNS ARZNAD 22,93,72
orl-rat LD50:142 mg/kg ARZNAD 18,261,68
ihl-rat LC50:209 mg/m^3/2H TPKVAL 10,73,68
ihl-rat LC50:209 mg/m^3/2H TPKVAL 10,73,68
ipr-rat LD50:58 mg/kg 27ZTAP 3,69,69
scu-rat LD50:75 mg/kg NYKZAU 56,377,60
ivn-rat LD50:23 mg/kg FRPPAO 26,585,71
orl-mus LD50:135 mg/kg AIPTAK 118,358,59
ihl-mus LC50:209 mg/m^3/2H TPKVAL 10,73,68

ipr-mus LD50:14 mg/kg FRPSAX 14,269,59
scu-mus LD50:33 mg/kg ARZNAD 11,932,61
ivn-mus LD50:16 mg/kg ARZNAD 11,932,61

CONSENSUS REPORTS: EPA Genetic Toxicology Program.

SAFETY PROFILE: A human poison by an unspecified route. Poison experimentally by ingestion, inhalation, intravenous, intraperitoneal, and subcutaneous routes. Human systemic effects by ingestion: decrease in blood pressure, somnolence, sleep, wakefulness, tremors, muscle weakness, and jaundice. Human teratogenic effects by an unspecified route: stillbirth, physical abnormalities, and other neonatal effects. Human reproductive effects by an unspecified route: terminates pregnancy. Experimental reproductive effects. An experimental teratogen. Has been implicated in aplastic anemia. Mutation data reported. When heated to decomposition it emits very toxic fumes of Cl^-, NO_x, and SO_x.

CKP500 CAS:69-09-0 ***HR: 3***
CHLOROPROMAZINE HYDROCHLORIDE
mf: $C_{17}H_{19}ClN_2S{\cdot}ClH$ mw: 355.35

SYNS: AMINAZIN MONOHYDROCHLORIDE ◇ AMPLIACTIL MONOHYDROCHLORIDE ◇ CHLORACTIL ◇ CHLORAZIN ◇ 2-CHLORO-10-(3-DIMETHYLAMINOPROPYL)PHENOTHIAZINE MONOHYDROCHLORIDE ◇ CHLOROPROMAZINE MONOHYDROCHLORIDE ◇ CPZ ◇ 10-(3-DIMETHYLAMINOPROPYL)-2-CHLOROPHENOTHIAZINE MONOHYDROCHLORIDE ◇ HEBANIL ◇ HIBANIL ◇ HIBERNAL ◇ HYBERNAL ◇ KLORPROMAN ◇ KLORPROMEX ◇ LARGACTIL MONOHYDROCHLORIDE ◇ LARGAKTYL ◇ MEGAPHEN ◇ NCI-CO5210 ◇ NEURAZINE ◇ NORCOZINE ◇ PHENOTHIAZINE HYDROCHLORIDE ◇ PLEGOMAZIN ◇ PROMACID ◇ PROMAPAR ◇ PROPAPHEN ◇ PROPAPHENIN HYDROCHLORIDE ◇ PSYCHOZINE ◇ 4560 RP HYDROCHLORIDE ◇ SONAZINE ◇ TAROCTYL ◇ THORAZINE ◇ THORAZINE HYDROCHLORIDE ◇ TORAZINA ◇ TRANZINE ◇ UNITENSEN

TOXICITY DATA with REFERENCE
orl-rat TDLo:300 mg/kg (female 6-20D post):REP TJADAB 37,185,88
scu-mus TDLo:10 mg/kg (female 3D post):TER JRPFA4 76,527,86
orl-wmn TDLo:6 mg/kg PGMJAO 60,564,84
orl-wmn TDLo:35 g/kg/16Y-I BIPCBF 18,1441,83
ivn-man TDLo:1786 μg/kg/2D-I BIPCBF 18,1441,83
ivn-wmn TDLo:822 μg/kg AEMED3 17,380,88
orl-rat LD50:145 mg/kg JPETAB 148,151,65
ihl-rat LC50:40 mg/m^3/2H 85JCAE -,1106,86
ipr-rat LD50:62 mg/kg PSYPAG 12,142,68
scu-rat LD50:140 mg/kg 27ZMA4 2,-,67
ivn-rat LD50:26 mg/kg FAZMAE 5,269,63
orl-mus LD50:135 mg/kg CPBTAL 24,1179,76
ihl-mus LC50:40 mg/m^3/2H 85JCAE -,1106,86
ipr-mus LD50:115 mg/kg AIPTAK 134,328,61
ivn-mus LD50:20 mg/kg ARZNAD 21,808,71

CONSENSUS REPORTS: Reported in EPA TSCA Inventory.

SAFETY PROFILE: Poison by ingestion, intraperitoneal, intravenous, and subcutaneous routes. An experimental teratogen. Experimental reproductive effects. An anti-emetic and antipsychotic drug. When heated to decomposition it emits very toxic fumes of Cl^-, NO_x, and SO_x. See also CHLORPROMAZINE.

CKP700 CAS:683-50-1 ***HR: 2***
2-CHLOROPROPANAL
mf: C_3H_5ClO mw: 92.53

SYNS: α-CHLOROPROPANAL ◇ 3-CHLOROPROPANAL ◇ α-CHLOROPROPIONALDEHYDE ◇ 2-CHLOROPROPIONALDEHYDE ◇ α-CHLOROPROPYLALDEHYDE ◇ PROPANAL, 2-CHLORO-(9CI)

TOXICITY DATA with REFERENCE
orl-mus TDLo:3560 mg/kg/89W-I:NEO JJIND8 63,1433,79

SAFETY PROFILE: Questionable carcinogen with experimental neoplastigenic data. When heated to decomposition it emits toxic fumes of Cl^-.

CKP750 CAS:540-54-5 ***HR: 3***
1-CHLOROPROPANE
DOT: UN 1278
mf: C_3H_7Cl mw: 78.55

PROP: Colorless liquid, chloroform-like odor. Mp: −122.8°, bp: 47.2°, lel: 2.6%, uel: 11.1%, flash p: <0°F, d: 0.890, vap d: 2.71, autoign temp: 968°F.

SYN: N-PROPYL CHLORIDE

CONSENSUS REPORTS: Reported in EPA TSCA Inventory.

DOT Classification: Flammable Liquid; Label: Flammable Liquid.

SAFETY PROFILE: A moderately poisonous irritant to skin, eyes, and mucous membranes. Narcotic in high concentrations. Dangerous fire hazard when exposed to heat, flame, or oxidizers. Moderately explosive when exposed to flame. Keep away from heat and open flame; can react vigorously with oxidizing materials. To fight fire, use CO_2, dry chemical. When heated to decomposition it emits toxic fumes of Cl^-. See also CHLORINATED HYDROCARBONS, ALIPHATIC.

CKQ000 CAS:75-29-6 ***HR: 3***
2-CHLOROPROPANE
DOT: UN 2356
mf: C_3H_7Cl mw: 78.54

$CH_3CHClCH_3$

PROP: Flash p.: −25.6°F; lel: 2.8%; uel: 10.7%.

SYN: ISOPROPYL CHLORIDE

DOT Classification: Flammable Liquid; Label: Flammable Liquid

SAFETY PROFILE: A very dangerous fire hazard when exposed to heat, flame, or oxidizers. When heated to decomposition it emits toxic fumes of Cl^-. See also 1-CHLOROPROPANE.

CKQ250 CAS:1331-07-3 ***HR: 3***
CHLOROPROPANE DIOL-1,3
mf: $C_3H_7ClO_2$ mw: 110.55

SYNS: CHLORO-1,3-PROPANEDIOL

TOXICITY DATA with REFERENCE
skn-rbt 10 mg/24H JIHTAB 30,63,48
eye-rbt 20 mg JIHTAB 30,63,48
orl-rat LD50:110 mg/kg JIHTAB 30,63,48
ihl-rat LCLo:1000 ppm/1H JIHTAB 30,63,48
skn-rbt LD50:450 mg/kg JIHTAB 30,63,48

SAFETY PROFILE: Poison by ingestion. Moderately toxic by skin contact. Mildly toxic by inhalation. A skin and eye irritant. Reaction with 70% perchloric acid forms a dangerously unstable explosive product. When heated to decomposition it emits toxic fumes of Cl^-.

CKQ500 CAS:3477-94-9 ***HR: 2***
3-CHLORO-1,2-PROPANEDIOL 1-BENZOATE
mf: $C_{10}H_{11}ClO_3$ mw: 214.66

SYNS: BENZOIC ACID, 3-CHLORO-2-HYDROXYPROPYL ESTER ◇ 1-BENZOYLOXY-3-CHLOROPROPAN-2-OL ◇ 1,2-PROPANEDIOL, 3-CHLORO-, 1-BENZOATE ◇ U 27,574

TOXICITY DATA with REFERENCE
orl-rat TDLo:960 mg/kg (male 8D pre):REP JRPFA4 21,263,70
ipr-mus LD50:450 mg/kg JMCMAR 20,644,77

SAFETY PROFILE: Moderately toxic by intraperitoneal route. Experimental reproductive effects. When heated to decomposition it emits toxic fumes of Cl^-.

CKQ750 CAS:15121-11-6 ***HR: 3***
CHLOROPROPANEDIOL CYCLIC SULFITE
mf: $C_3H_5ClO_3S$ mw: 156.69

SYN: 1-CHLOROMETHYLETHYLENE GLYCOL CYCLIC SULFITE

TOXICITY DATA with REFERENCE
orl-rat TDLo:112 mg/kg (male 14D pre):REP CCPTAY 9,451,74
ivn-mus LD50:178 mg/kg CSLNX* NX#02154

CONSENSUS REPORTS: Reported in EPA TSCA Inventory.

SAFETY PROFILE: Poison by intravenous route. See also SULFITES. Experimental reproductive effects. When heated to decomposition it emits very toxic fumes of Cl^- and SO_x.

CKR000 CAS:51034-39-0 ***HR: 2***
3-CHLORO-1-PROPANESULFONIC ACID, MONOSODIUM SALT 2-CHLOROPROPANAL
mf: $C_3H_6ClO_3S{\cdot}Na$ mw: 180.59

TOXICITY DATA with REFERENCE
ipr-mus LD50:750 mg/kg NTIS** AD691-490

CONSENSUS REPORTS: Reported in EPA TSCA Inventory.

SAFETY PROFILE: Moderately toxic by intraperitoneal route. See also SULFONATES. When heated to decomposition it emits very toxic fumes of SO_x, Na_2O, and Cl^-.

CKR500 CAS:78-89-7 ***HR: 3***
2-CHLORO-1-PROPANOL
DOT: UN 2611
mf: C_3H_7ClO mw: 94.55

PROP: Colorless liquid, mild non-residual odor. Bp: 133.5°, flash p: 125°F (CC), d: 1.103 @ 20°, vap d: 3.26.

SYNS: 2-CHLOROPROPYL ALCOHOL ◇ PROPYLENECHLOROHYDRIN

TOXICITY DATA with REFERENCE
skn-rbt 500 mg open MLD UCDS** 7/23/71
eye-rbt 2230 µg SEV AJOPAA 29,1363,46
orl-rat LD50:218 mg/kg FAONAU 53A,359,74
ihl-rat LCLo:500 ppm/4H AIHAAP 30,470,69
orl-dog LDLo:200 mg/kg FAONAU 53A,359,74
skn-rbt LD50:529 mg/kg AIHAAP 30,470,69
orl-gpg LD50:720 mg/kg JIHTAB 23,259,41

CONSENSUS REPORTS: Reported in EPA TSCA Inventory.

DOT Classification: Poison B; Label: Flammable Liquid and Poison.

SAFETY PROFILE: Poison by ingestion. Moderately toxic by inhalation and skin contact. A skin and severe eye irritant. Flammable when exposed to heat, flame, or powerful oxidizers. To fight fire, use alcohol foam, CO_2, dry chemical. When heated to decomposition it emits toxic fumes of Cl^-.

CKR750 CAS:127-00-4 ***HR: D***
1-CHLORO-2-PROPANOL with 2-CHLORO-1-PROPANOL
mf: C_3H_7ClO mw: 94.55

TOXICITY DATA with REFERENCE
mmo-sat 40 μmol/plate FCTXAV 18,115,80
mma-sat 1100 μg/plate MUREAV 30,303,75

SAFETY PROFILE: Mutation data reported. When heated to decomposition it emits toxic fumes of Cl^-. See also individual components.

CKS000 CAS:557-98-2 ***HR: 3***
2-CHLORO-1-PROPENE
DOT: UN 2456
mf: C_3H_5Cl mw: 76.53

PROP: Colorless liquid. Bp: 22.65°, fp: −137.4°, d: 0.918 @ 9°, flash p: −4°. lel: 4.5%; uel: 16%.

SYN: 2-CHLOROPROPENE (DOT)

TOXICITY DATA with REFERENCE
mma-sat 100 μmol/plate BCPCA6 29,2611,80
ihl-mus LC50:267 g/m^3 UCPHAQ 2,39,41

CONSENSUS REPORTS: Reported in EPA TSCA Inventory.

DOT Classification: Flammable Liquid; Label: Flammable Liquid.

SAFETY PROFILE: Mildy toxic by inhalation. Mutation data reported. Very dangerous fire hazard when exposed to heat, flame, sparks, or powerful oxidizers. To fight fire, use water, spray, mist, fog, dry chemical, alcohol foam. When heated to decomposition it emits toxic fumes of Cl^-. See also CHLORIDES.

CKS099 CAS:21947-75-1 ***HR: 2***
cis-1-CHLOROPROPENE OXIDE
mf: C_3H_5ClO mw: 90.51

SYNS: cis-1-CHLORO-1,2-EPOXYPROPANE ◇ cis-2-CHLORO-3-METHYLOXIRANE ◇ cis-CPO ◇ OXIRANE, 2-CHLORO-3-METHYL-, cis-(9CI) ◇ PROPANE, 1-CHLORO-1,2-EPOXY-, (Z)-

TOXICITY DATA with REFERENCE
mmo-sat 550 μmol/L MUREAV 101,115,82
mmo-esc 1130 μmol/L MUREAV 101,115,82
dnr-esc 110 μmol/L MUREAV 101,115,82
otr-ham:emb 110 μmol/L JJIND8 69,531,82
skn-mus TDLo:400 mg/kg/62W-I:CAR CNREA8 43,159,83
scu-mus TDLo:40 mg/kg/69W-I:CAR CNREA8 43,159,83

SAFETY PROFILE: Questionable carcinogen with experimental carcinogenic data. Mutation data reported. When heated to decomposition it emits toxic fumes of Cl^-.

CKS100 CAS:21947-76-2 ***HR: 2***
trans-1-CHLOROPROPENE OXIDE
mf: C_3H_5ClO mw: 92.53

SYNS: trans-1-CHLORO-1,2-EPOXYPROPANE ◇ trans-2-CHLORO-3-METHYLOXIRANE ◇ trans-CPO ◇ OXIRANE, 2-CHLORO-3-METHYL-, trans-(9CI) ◇ PROPANE, 1-CHLORO-1,2-EPOXY-, (E)-

TOXICITY DATA with REFERENCE
mmo-sat 550 μmol/L MUREAV 101,115,82
mmo-esc 2250 μmol/L MUREAV 101,115,82
dnr-esc 110 μmol/L MUREAV 101,115,82
otr-ham:emb 550 μmol/L JJIND8 69,531,82
skn-mus TDLo:400 mg/kg/73W-I:CAR CNREA8 43,159,83
scu-mus TDLo:40 mg/kg/60W-I:CAR CNREA8 43,159,83

SAFETY PROFILE: Questionable carcinogen with experimental carcinogenic data. Mutation data reported. When heated to decomposition it emits toxic fumes of Cl^-.

CKS325 CAS:62861-56-7 ***HR: 3***
2-CHLORO-2-PROPENYL TRIFLUOROMETHANE SULFONATE
mf: $C_4H_4ClF_3O_3S$ mw: 224.58

$$H_2C{=}CClCH_2OSO_2CF_3$$

SAFETY PROFILE: A dangerous storage hazard. Store in a vented container at −78°C. Reacts violently with aprotic solvents (e.g., DMF and DMSO). When heated to decomposition it emits toxic fumes of F^-, Cl^-, and SO_x. See also SULFONATES.

CKS500 CAS:107-94-8 ***HR: 3***
3-CHLOROPROPIONIC ACID
mf: $C_3H_5ClO_2$ mw: 108.53

SYNS: β-CHLOROPROPIONIC ACID ◇ β-MONOCHLOROPROPIONIC ACID

TOXICITY DATA with REFERENCE
ipr-mus TDLo:730 mg/kg/4W-I:NEO CNREA8 39,391,79
skn-mus LDLo:1040 mg/kg CNREA8 28,653,68

CONSENSUS REPORTS: Reported in EPA TSCA Inventory.

SAFETY PROFILE: Moderately toxic by skin contact. Questionable carcinogen with experimental neoplastigenic data. Mutation data reported. When heated to decomposition it emits toxic fumes of Cl^-. See also α-CHLOROPROPIONIC ACID.

CKS750 CAS:598-78-7 ***HR: 2***
α-CHLOROPROPIONIC ACID
mf: $C_3H_5ClO_2$ mw: 108.53

PROP: Sol in water. D: 1.260-1.268 @ 20°, bp: 183-187°, flash p: 225°F.

TOXICITY DATA with REFERENCE
orl-mus LD50:980 mg/kg JPETAB 86,336,46
orl-rat LD50:500 mg/kg DTLVS* 4,95,80

CONSENSUS REPORTS: Reported in EPA TSCA Inventory.

SAFETY PROFILE: Moderately toxic by ingestion. Combustible when exposed to heat or flame. To fight fire, use water, foam, alcohol foam. When heated to decomposition it emits toxic fumes of Cl^-. See also 3-CHLOROPROPIONIC ACID.

CKT000 CAS:17639-93-9 ***HR: 3***
2-CHLOROPROPIONIC ACID METHYL ESTER
DOT: UN 2933
mf: $C_4H_7ClO_2$ mw: 122.56

SYN: METHYL-2-CHLOROPROPIONATE(DOT)

TOXICITY DATA with REFERENCE
ipr-mus LDLo:250 mg/kg CBCCT* 6,228,54

CONSENSUS REPORTS: Reported in EPA TSCA Inventory.

DOT Classification: Flammable or Combustible Liquid; Label: Flammable Liquid.

SAFETY PROFILE: Poison by intraperitoneal route. See also ESTERS. Flammable when exposed to heat of flame. When heated to decomposition it emits toxic fumes of Cl^-.

CKT100 CAS:16987-02-3 ***HR: 2***
2-CHLOROPROPIONIC ACID SODIUM SALT
mf: $C_3H_4ClO_2 \cdot Na$ mw: 130.51

SYNS: 2-CHLOROPROPANOIC ACID SODIUM SALT ◇ 2-CHLOROPROPIONATE SODIUM SALT ◇ α-CHLOROPROPIONIC ACID SODIUM SALT ◇ PROPANOIC ACID, 2-CHLORO-, SODIUM SALT ◇ SODIUM-2-CHLOROPROPIONATE

TOXICITY DATA with REFERENCE
orl-mus TDLo:44 g/kg (male 84D pre):REP JPETAB 222,501,82
orl-mus LD50:1671 mg/kg JPETAB 222,501,82

SAFETY PROFILE: Moderately toxic by ingestion. Experimental reproductive effects. When heated to decomposition it emits toxic fumes of Cl^-.

CKT250 CAS:542-76-7 ***HR: 3***
3-CHLOROPROPIONITRILE
mf: C_3H_4ClN mw: 89.53

PROP: Colorless liquid. Mp: −51°, bp: 176° decomp, flash p: 168°F (CC), d: 1.1363 @ 25°, vap press: 6 mm @ 50°, vap d: 3.09.

SYNS: 3-CHLOROPROPANENITRILE ◇ 3-CHLOROPROPANONITRILE ◇ β-CHLOROPROPIONITRILE ◇ RCRA WASTE NUMBER P027 ◇ USAF A-8798

TOXICITY DATA with REFERENCE
orl-rat LD50:100 mg/kg 14CYAT 2,2025,62
orl-mus LD50:9 mg/kg 14CYAT 2,2025,62
ipr-mus LD50:100 mg/kg NTIS** AD277-689
ipr-mus LDLo:25 mg/kg NTIS** AD277-689
ipr-mus LD50:100 mg/kg NTIS** AD277-689
ivn-mus LD50:56 mg/kg CSLNX* NX#01996

CONSENSUS REPORTS: EPA Extremely Hazardous Substances List. Cyanide and its compounds are on the Community Right-To-Know List. Reported in EPA TSCA Inventory.

SAFETY PROFILE: Poison by ingestion, intravenous, and intraperitoneal routes. Flammable in its liquid form when exposed to heat or flame. To fight fire, use alcohol foam, water, foam, CO_2, or dry chemical. When heated to decomposition it emits very toxic fumes of Cl^-, CN^-, and NO_x. See also NITRILES.

CKT500 CAS:6285-05-8 ***HR: 3***
p-CHLOROPROPIOPHENONE
mf: C_9H_9ClO mw: 168.63

SYN: USAF EK-5296

TOXICITY DATA with REFERENCE
ipr-mus LD50:200 mg/kg NTIS** AD277-689
ivn-mus LD50:100 mg/kg CSLNX* NX#04477

CONSENSUS REPORTS: Reported in EPA TSCA Inventory.

SAFETY PROFILE: Poison by intraperitoneal and intravenous routes. When heated to decomposition it emits toxic fumes of Cl^-.

CKT750 CAS:77966-68-8 ***HR: 3***
6′-CHLORO-2-(PROPYLAMINO)-o-ACETOTOLUIDIDE HYDROCHLORIDE
mf: $C_{12}H_{17}ClN_2O \cdot ClH$ mw: 277.22

SYNS: C 3058 ◇ 2′-CHLORO-6′-METHYL-2-(PROPYLAMINO)ACETANILIDE HYDROCHLORIDE

TOXICITY DATA with REFERENCE
ipr-rat LD50:400 mg/kg ARZNAD 8,407,58
ipr-mus LD50:400 mg/kg ARZNAD 8,407,58
scu-mus LD50:1125 mg/kg ARZNAD 8,407,58

SAFETY PROFILE: Poison by intraperitoneal route. Moderately toxic by subcutaneous route. When heated to decomposition it emits very toxic fumes of Cl^- and NO_x.

CKU000 CAS:77985-17-2 ***HR: 3***
6′-CHLORO-2-(PROPYLAMINO)-o-BUTYROTOLUIDIDE HYDROCHLORIDE
mf: $C_{14}H_{21}ClN_2O \cdot ClH$ mw: 305.28

SYN: C 3189

TOXICITY DATA with REFERENCE
eye-rbt 2% MLD ARZNAD 8,544,58
ipr-rat LD50:34 mg/kg ARZNAD 8,544,58
scu-mus LD50:57 mg/kg ARZNAD 8,544,58

SAFETY PROFILE: Poison by intraperitoneal and subcutaneous routes. An eye irritant. When heated to decomposition it emits very toxic fumes of Cl^- and NO_x.

CKU250 CAS:77846-96-9 ***HR: 3***
9-((2-((2-CHLOROPROPYL)AMINO)ETHYL) AMINO)-2-METHOXYACRIDINE DIHYDROCHLORIDE HEMIHYDRATE
mf: $C_{19}H_{22}ClN_3O \cdot 2ClH \cdot 1/2H_2O$ mw: 425.79

TOXICITY DATA with REFERENCE
mmo-sat 5 μg/plate JMCMAR 15,739,72
ipr-mus LD20:64 mg/kg JMCMAR 15,739,72

SAFETY PROFILE: Poison by intraperitoneal route. Mutation data reported. When heated to decomposition it emits very toxic fumes of Cl^- and NO_x.

CKU625 CAS:2612-33-1 ***HR: 3***
1-CHLORO-2,3-PROPYLENE DINITRATE
mf: $C_3H_5ClN_2O_6$ mw: 200.54

$ClCH_2CH(ONO_2)CH_2ONO_2$

SAFETY PROFILE: A viscous liquid explosive. Upon decomposition it emits toxic fumes of Cl^- and NO_x. See also NITRATES and EXPLOSIVES.

CKU750 CAS:3569-57-1 ***HR: 3***
3-CHLOROPROPYL-n-OCTYLSULFOXIDE
mf: $C_{11}H_{23}ClOS$ mw: 238.85

SYN: MGK REPELLENT 1,207

TOXICITY DATA with REFERENCE
orl-rat LD50:5660 mg/kg PCOC** -,759,66
skn-rbt LD50:8 mg/kg PCOC** -,759,66

CONSENSUS REPORTS: EPA Extremely Hazardous Substances List.

SAFETY PROFILE: Poison by skin contact. Mildly toxic by ingestion. When heated to decomposition it emits very toxic fumes of Cl^- and SO_x.

CKV250 ***HR: 2***
1-CHLORO-2-PROPYNE
mf: C_3H_2Cl mw: 74.51

PROP: Flash p: < 59°F.

SAFETY PROFILE: A dangerous fire and explosion hazard when exposed to heat or flame. Incompatible with ammonia. When heated to decomposition it emits toxic fumes of Cl^-. See also ACETYLENE COMPOUNDS and CHLORINATED HYDROCARBONS, ALIPHATIC.

CKV275 CAS:624-65-7 ***HR: 3***
3-CHLOROPROPYNE
mf: C_3H_3Cl mw: 74.51

SYN: PROPARGYL CHLORIDE

SAFETY PROFILE: A pressure-sensitive explosive. Reacts explosively with ammonia in a closed container. When heated to decomposition it emits toxic fumes of Cl^-. See also ACETYLENE COMPOUNDS; CHLORINATED HYDROCARBONS, ALIPHATIC; and EXPLOSIVES.

CKV500 CAS:87-42-3 ***HR: 3***
6-CHLOROPURINE
mf: $C_5H_3ClN_4$ mw: 154.57

SYNS: 6-CHLORO-9H-PURINE ◇ 6-CHLORO-1H-PURINE (9CI) ◇ ClP ◇ NSC 744 ◇ SK 6048

TOXICITY DATA with REFERENCE
mmo-sat 10 μL/plate ANYAA9 76,475,58
mmo-esc 10 μL/disc ANYAA9 76,475,58
mmo-omi 500 mg/L SOGEBZ 6,1509,70
hma-mus/sat 25 mg/kg MUREAV 26,455,74
ipr-rat TDLo:100 mg/kg (7D preg):REP JRPFA4 4,291,62
ipr-rat LD50:400 mg/kg ADTEAS 3,181,68
orl-mus LD50:720 mg/kg NCISP* JAN86
ipr-mus LD50:132 mg/kg NCISP* JAN86
scu-mus LD50:514 mg/kg NCISP* JAN86
unr-mus LD50:230 mg/kg PMDCAY 7,69,70

SAFETY PROFILE: Poison by intraperitoneal and possibly other routes. Moderately toxic by ingestion and subcutaneous routes. Experimental reproductive effects. Mutation data reported. When heated to decomposition it emits very toxic fumes of Cl^- and NO_x.

CKV625 CAS:59-32-5 ***HR: 3***
CHLOROPYRAMINE
mf: $C_{16}H_{20}ClN_3$ mw: 289.84

PROP: Light yellow, viscous, oily liquid; pungent odor. Bp: 154-155°.

SYNS: ALLERGAN ◇ AVAPENA ◇ 2-((p-CHLOROBENZYL)(2-(DIMETHYLAMINO)ETHYL)AMINO)PYRIDINE ◇ N-(p-CHLOROBENZYL)-N',N'-DIMETHYL-N-(2-PYRIDYL)ETHYLENEDIAMINE ◇ p-CHLORBENZYL-α-PYRIDYL-DIMETHYL-AETHYLENDIAMIN (GERMAN) ◇ CHLORONEOANTERGAN ◇ N-((4-CHLOROPHENYL) METHYL)-N',N'-DIMETHYL-N-2-PYRIDINYL-1,2-ETHANEDIAMINE (9CI) ◇ CHLOROPYRIBENZAMINE ◇ HALOPYRAMINE ◇ SUPRASTIN ◇ SYNOPEN ◇ SYNOPEN R ◇ SYNPEN

TOXICITY DATA with REFERENCE
orl-rat LD50:920 mg/kg ARZNAD 25,1723,75
ipr-rat LD50:104 mg/kg ARZNAD 25,1723,75

ivn-rat LD50:32500 μg/kg ARZNAD 25,1723,75
orl-mus LD50:354 mg/kg ARZNAD 25,1723,75
ipr-mus LD50:79200 μg/kg ARZNAD 25,1723,75
ivn-mus LD50:24100 μg/kg ARZNAD 25,1723,75
ipr-gpg LD50:108 mg/kg ARZNAD 25,1723,75
scu-gpg LD50:142 mg/kg ARZNAD 7,131,57

SAFETY PROFILE: Poison by ingestion, subcutaneous, intravenous, and intraperitoneal routes. When heated to decomposition it emits toxic fumes of Cl^- and NO_x.

CKW000 CAS:109-09-1 ***HR: 3***
2-CHLOROPYRIDINE
DOT: UN 2822
mf: C_5H_4ClN mw: 113.55

PROP: Colorless, oily liquid. Bp: 170°, d: 1.205 @ 25°, vap press: 1 mm @ 13.3°, vap d: 3.93.

SYNS: o-CHLOROPYRIDINE ◇ α-CHLOROPYRIDINE

TOXICITY DATA with REFERENCE
ihl-rat LCLo:100 ppm/4H TXAPA9 11,361,67
orl-mus LD50:110 mg/kg TXAPA9 11,361,67
ipr-mus LD50:130 mg/kg TXAPA9 11,361,67
skn-rbt LD50:64 mg/kg TXAPA9 11,361,67
ipr-rbt LD50:48 mg/kg TXAPA9 11,361,67
orl-bwd LD50:1 g/kg AECTCV 12,355,83

CONSENSUS REPORTS: Reported in EPA TSCA Inventory.

DOT Classification: Poison B; Label: Poison.

SAFETY PROFILE: Poison by ingestion, inhalation, skin contact, and intraperitoneal routes. Combustible when exposed to heat or flame. Can react with oxidizing materials. When heated to decomposition it emits very toxic fumes of Cl^-, NO_x, and phosgene.

CKW250 CAS:626-60-8 ***HR: 3***
3-CHLOROPYRIDINE
mf: C_5H_4ClN mw: 113.55

SYN: m-CHLOROPYRIDINE

TOXICITY DATA with REFERENCE
ipr-mus LD50:235 mg/kg TXAPA9 11,361,67
orl-bwd LD50:750 mg/kg AECTCV 12,355,83

CONSENSUS REPORTS: Reported in EPA TSCA Inventory.

SAFETY PROFILE: Poison by intraperitoneal route. Moderately toxic by ingestion. When heated to decomposition it emits very toxic fumes of Cl^- and NO_x.

CKW325 CAS:2402-95-1 ***HR: 3***
2-CHLOROPYRIDINE-N-OXIDE
mf: C_5H_4ClNO mw: 129.55

CH=CHCH=CHCCl=N:O (ring)

SAFETY PROFILE: Decomposes violently when heated above 90°C. Upon decomposition it emits toxic fumes of Cl^- and NO_x.

CKW500 CAS:5428-90-0 ***HR: 3***
CHLORO-3-PYRIDYLMERCURY
mf: C_5H_4ClHgN mw: 314.14

SYN: 3-(CHLOROMERCURI)PYRIDINE

TOXICITY DATA with REFERENCE
ipr-mus LDLo:16 mg/kg CBCCT* 3,128,51

CONSENSUS REPORTS: Mercury and its compounds are on the Community Right-To-Know List.

OSHA PEL: (Transitional: CL 1 mg/10m^3) CL 0.1 mg(Hg)/m^3 (skin)
ACGIH TLV: TWA 0.1 mg(Hg)/m^3 (skin)
NIOSH REL: (Mercury, Inorganic) TWA 0.05 mg(Hg)/m^3

SAFETY PROFILE: Poison by intraperitoneal route. See also MERCURY COMPOUNDS. When heated to decomposition it emits very toxic fumes of Cl^-, Hg, and NO_x.

CLB250 CAS:78110-10-8 ***HR: 3***
6'-CHLORO-2-(PYRROLIDINYL)-o-DIACETOTOLUIDIDE HYDROCHLORIDE
mf: $C_{15}H_{19}ClN_2O_2$•ClH mw: 294.81

SYNS: N-ACETYL-N-(2-CHLORO-6-METHYLPHENYL)-1-PYRROLIDINEACETAMIDE MONOHYDROCHLORIDE ◇ C 3199

TOXICITY DATA with REFERENCE
eye-rbt 2% MLD ARZNAD 8,609,58
ipr-rat LD50:180 mg/kg ARZNAD 8,609,58
ipr-mus LD50:120 mg/kg ARZNAD 8,609,58
scu-mus LD50:325 mg/kg ARZNAD 8,609,58

SAFETY PROFILE: Poison by intraperitoneal and subcutaneous routes. An eye irritant. When heated to decomposition it emits toxic fumes of Cl^- and NO_x.

CLC100 ***HR: 3***
2'-CHLORO-2-PYRROLIDINYL-5'-TRIFLUOROMETHYLACETANILIDE HYDROCHLORIDE
mf: $C_{13}H_{14}ClF_3N_2O$•ClH mw: 343.20

SYNS: C 3078 ◇ 6'-CHLORO-2-PYRROLIDINYL-α,α,α-TRIFLUORO-m-ACETOTOLUIDINE, HYDROCHLORIDE

TOXICITY DATA with REFERENCE
eye-rbt 2% MOD ARZNAD 8,270,58

ipr-rat LD50:157 mg/kg ARZNAD 8,270,58
scu-mus LD50:232 mg/kg ARZNAD 8,270,58

SAFETY PROFILE: Poison by subcutaneous and intraperitoneal routes. An eye irritant. When heated to decomposition it emits toxic fumes of F^-, NO_x, and Cl^-.

CLC125 ***HR: 3***
4'-CHLORO-2-PYRROLIDINYL-3'-TRIFLUOROMETHYLACETANILIDE HYDROCHLORIDE
mf: $C_{13}H_{14}ClF_3N_2O{\cdot}ClH$ mw: 343.20

SYNS: C 3073 ◇ 4'-CHLORO-2-PYRROLIDINYL-α,α,α-TRIFLUORO-m-ACETOTOLUIDIDE, HYDROCHLORIDE

TOXICITY DATA with REFERENCE
eye-rbt 2% SEV ARZNAD 8,270,58
ipr-rat LD50:174 mg/kg ARZNAD 8,270,58
scu-mus LD50:460 mg/kg ARZNAD 8,270,58

SAFETY PROFILE: Poison by intraperitoneal route. Moderately toxic by subcutaneous route. A severe eye irritant. When heated to decomposition it emits toxic fumes of F^-, NO_x and Cl^-.

CLC500 CAS:72-80-0 ***HR: 3***
CHLOROQUINALDOL
mf: $C_{10}H_7Cl_2NO$ mw: 228.08

SYNS: CHLORQUINALDOL ◇ 5,7-DICHLORO-8-HYDROXY-QUINALDINE ◇ 5,7-DICHLORO-2-METHYL-8-HYDROXYQUINOLINE ◇ 5,7-DICHLORO-2-METHYL-8-QUINOLINOL ◇ 5,7-DICHLORO-8-QUINALDINOL ◇ HYDROXYDICHLOROQUINALDINE

TOXICITY DATA with REFERENCE
orl-rat LD50:660 mg/kg 29ZVAB -,33,69
orl-dog LD50:2250 mg/kg 29ZVAB -,33,69
orl-rbt LD50:160 mg/kg 29ZVAB -,33,69

SAFETY PROFILE: Poison by ingestion. When heated to decomposition it emits very toxic fumes of Cl^- and NO_x.

CLC750 CAS:16064-14-5 ***HR: 3***
6-CHLORO-4-QUINAZOLINONE
mf: $C_8H_5ClN_2O$ mw: 180.60

SYN: 6-CHLORO-4(3H)-QUINAZOLINONE

TOXICITY DATA with REFERENCE
orl-mus LD50:404 mg/kg ARZNAD 12,1204,62
ipr-mus LD50:340 mg/kg ARZNAD 12,1204,62

SAFETY PROFILE: Poison by intraperitoneal route. Moderately toxic by ingestion. When heated to decomposition it emits very toxic fumes of Cl^- and NO_x.

CLD000 CAS:54-05-7 ***HR: 3***
CHLOROQUINE
mf: $C_{18}H_{26}ClN_3$ mw: 319.92

SYNS: AMOKIN ◇ ARALEN ◇ ARTHROCHIN ◇ AVLOCLOR ◇ BEMACO ◇ BEMAPHATE ◇ BEMASULPH ◇ CHEMOCHIN ◇ CHINGAMIN ◇ CHLORAQUINE ◇ CHLOROCHIN ◇ 7-CHLORO-4-(4-DIETHYLAMINO-1-METHYLBUTYLAMINO)QUINOLINE ◇ CHLOROQUINIUM ◇ N^4-(7-CHLORO-4-QUINOLINYL)-N^1,N^1-DIETHYL-1,4-PENTANEDIAMINE ◇ CIDANCHIN ◇ CLOROCHINA ◇ COCARTRIT ◇ DELAGIL ◇ DICHINALEX ◇ ELESTOL ◇ GONTOCHIN ◇ HELIOPAR ◇ IMAGON ◇ IROQUINE ◇ KLOROKIN ◇ LAPAQUIN ◇ MALAQUIN ◇ MALAREN ◇ MALAREX ◇ MESYLITH ◇ NEOCHIN ◇ NIVACHINE ◇ NIVAQUINE B ◇ QUINACHLOR ◇ QUINAGAMINE ◇ QUINERCYL ◇ QUINILON ◇ QUINOSCAN ◇ RESOCHIN ◇ RESOQUINA ◇ RESOQUINE ◇ REUMACHLOR ◇ REUMAQUIN ◇ ROQUINE ◇ RP 3377 ◇ SANOQUIN ◇ SENAQUIN ◇ SILBESAN ◇ SIRAGAN ◇ SN 6718 ◇ SN 7618 ◇ SOLPRINA ◇ SOPAQUIN ◇ TANAKAN ◇ TRESOCHIN ◇ TROCHIN ◇ W 7618 ◇ WIN 244

TOXICITY DATA with REFERENCE
mmo-sat 100 μmol/L AMACCQ 9,77,76
cyt-hmn:lym 100 mg/L BEXBAN 82,1095,76
unr-wmn TDLo:1155 mg/kg (6-39W preg):REP JOPDAB 69,1150,66
unr-wmn TDLo:1155 mg/kg (6-39W preg):TER JOPDAB 69,1150,66
orl-wmn LDLo:110 mg/kg:CVS,GIT NEJMAG 318,1,88
orl-wmn TDLo:3600 mg/kg/3Y TGMEAJ 32,216,80
orl-man LDLo:86 mg/kg:CVS,GIT NEJMAG 318,1,88
orl-hmn LDLo:20 mg/kg JETOAS 6,86,73
orl-rat LD50:330 mg/kg JTCTDW 20,271,83
ipr-rat LD50:102 mg/kg PHMGBN 13,401,75
orl-mus LD50:311 mg/kg OYYAA2 7,753,73
ipr-mus LD50:66 mg/kg ARZNAD 32,1219,82
scu-mus LD50:150 mg/kg JETOAS 6,86,73
ivn-mus LD50:21600 μg/kg CYLPDN 4,69,83
ims-mus LD50:71 mg/kg CYLPDN 4,69,83
scu-rbt LD50:75 mg/kg JETOAS 6,86,73
ivn-rbt LD50:8 mg/kg JETOAS 6,86,73
ims-rbt LDLo:80 mg/kg YHHPAL 15,630,80

CONSENSUS REPORTS: IARC Cancer Review: Group 3 IMEMDT 7,56,87; Animal Inadequate Evidence IMEMDT 13,47,77. EPA Genetic Toxicology Program.

SAFETY PROFILE: Poison by ingestion, intraperitoneal, intravenous, intramuscular, and subcutaneous routes. Human systemic effects by ingestion: heart rate changes, nausea or vomiting. Human teratogenic effects by an unspecified route include developmental abnormalities of the urogenital system, eyes and ears, other unspecified areas, and postnatal effects. Human reproductive effects by an unspecified route: terminates pregnancy. Human mutation data reported. Questionable carcinogen. An antimalarial agent. When heated to decomposition it emits very toxic fumes of Cl^- and NO_x.

CLD250 CAS:50-63-5 ***HR: 3***
CHLOROQUINE DIPHOSPHATE
mf: $C_{18}H_{26}ClN_3{\cdot}2H_3O_4P$ mw: 515.92

SYNS: ALERMINE ◇ ARALEN DIPHOSPHATE ◇ ARALEN PHOSPHATE ◇ ARECHIN ◇ AROCLOR 54 ◇ AVLOCLOR ◇ BEMAPHATE ◇ CHINGAMIN ◇ 7-CHLOR-4-(4-(DIAETHYLAMINO)-1-METHYLBUTYLAMINO)-CHINOLINDIPHOSPHAT (GERMAN) ◇ 7-CHLORO-4-((4'-DIETHYLAMINO-1-METHYLBUTYL)AMINO)QUINOLINEDIPHOSPHATE ◇ CHLOROIN ◇ 2-(p-CHLORO-α- (2-DIMETHYLAMINO) ETHYL)BENZYL)PYRIDINE MALEATE (1:1) ◇ CHLOROQUINE PHOSPHATE ◇ CHLOR-TRIMETON ◇ CQ ◇ DELAGIL ◇ GONTOCHIN PHOSPHATE ◇ HISTASPAN ◇ H-STADUR ◇ KHINGAMIN ◇ NOSCOSED ◇ RESOCHIN ◇ RESOCHIN DIPHOSPHATE ◇ RESOQUINE ◇ SANOQUIN ◇ TANAKAN ◇ TELDRIN ◇ TELODRON

TOXICITY DATA with REFERENCE
mmo-sat 100 mg/L MUREAV 68,41,79
scu-rat TDLo:475 mg/kg (female 10-14D post):REP VAAZA2 39,59,82
orl-rat TDLo:750 mg/kg (female 9D post):TER RPTOAN 30,114,67
orl-cld LDLo:250 mg/kg ATXKA8 23,204,68
orl-wmn TDLo:167 mg/kg:CVS,PUL JTCTDW 19,1067,82/83
orl-man LDLo:179 mg/kg ATXKA8 23,204,68
orl-rat LDLo:600 mg/kg 85GLAQ 1,390,46
orl-mus LD50:500 mg/kg TMPRAD 30,308,79
ipr-mus LD50:68 mg/kg 85GLAQ 1,390,46
scu-mus LD60:200 mg/kg ATMPA2 74,393,80
ivn-brd LD50:64500 μg/kg ARZNAD 20,1775,70

SAFETY PROFILE: A human poison by ingestion. Poison by intravenous, subcutaneous, and intraperitoneal routes. Human systemic effects by ingestion: EKG changes, blood pressure lowering, respiratory depression. An experimental teratogen. Experimental reproductive effects. Mutation data reported. When heated to decomposition it emits very toxic fumes of Cl^-, NO_x, and PO_x. See also CHLOROQUINE.

CLD500 CAS:4213-44-9 ***HR: 3***
CHLOROQUINE MUSTARD
mf: $C_{18}H_{24}Cl_3N_3{\cdot}2ClH$ mw: 461.72

SYNS: 4-((4-(BIS(2-CHLOROETHYL)AMINO)-1-METHYLBUTYL) AMINO-7-CHLOROQUINOLINE, DIHYDROCHLORIDE ◇ ICR-25A ◇ NSC-17118

TOXICITY DATA with REFERENCE
dnd-mus:lvr 70 μmol/L CNREA8 21,1124,61
dnd-mus:oth 70 μmol/L CNREA8 21,1124,61
ipr-mus TDLo:8 mg/kg/4W:CAR JNCIAM 36,915,66
ipr-rat LD10:1100 μg/kg CCROBU 17,63,62
ivn-dog LDLo:200 μg/kg CCSUBJ 2,202,65
ivn-mky LDLo:410 μg/kg CCSUBJ 2,202,65

SAFETY PROFILE: A deadly poison by intraperitoneal and intravenous routes. Questionable carcinogen with experimental carcinogenic data. Mutation data reported. When heated to decomposition it emits very toxic fumes of Cl^- and NO_x. See also CHLOROQUINE.

CLD750 CAS:95-88-5 ***HR: 3***
4-CHLORORESORCINOL
mf: $C_6H_5ClO_2$ mw: 144.56

TOXICITY DATA with REFERENCE
eye-rbt 5% MLD JAPMA8 46,185,57
orl-rat TDLo:2 g/kg (6-15D preg):REP JACTDZ 2(4),325,83
orl-rat LD50:369 mg/kg FCTXAV 15,607,77
ipr-mus LD50:195 mg/kg JAPMA8 46,185,57

CONSENSUS REPORTS: Reported in EPA TSCA Inventory. EPA Genetic Toxicology Program. Chlorophenols are on the Community Right-To-Know List.

SAFETY PROFILE: Poison by ingestion and intraperitoneal routes. Experimental reproductive effects. An eye irritant. A hair dye component. When heated to decomposition it emits toxic fumes of Cl^-. See also CHLOROPHENOLS and RESORCINOL.

CLE250 ***HR: 3***
CHLOROSILANES

PROP: Compounds of Si, Cl, and H where the total number of atoms of Cl and H add up to 4. SiH_xCl_{4-x}.
SAFETY PROFILE: Poison by ingestion and inhalation, and a poisonous irritant to skin, eyes, and mucous membranes. Toxicity is based on HCl which is formed upon hydrolysis of a chlorosilane. Self-ignites in air. With a little ammonia, it forms a self-igniting product. They react with water or steam to produce heat and toxic and corrosive fumes of HCl. When heated to decomposition they emit highly toxic fumes of Cl^-.

CLE500 CAS:73928-01-5 ***HR: 3***
3-CHLORO-4-STILBENAMINE
mf: $C_{12}H_{12}ClN$ mw: 205.70

SYN: 3-CHLORO-4-AMINOSTILBENE

TOXICITY DATA with REFERENCE
scu-rat TDLo:200 mg/kg/W-I:ETA BMBUAQ 14,141,58

SAFETY PROFILE: Questionable carcinogen with experimental tumorigenic data. When heated to decomposition it emits very toxic fumes of Cl^- and NO_x.

CLE750 CAS:2039-87-4 ***HR: 1***
o-CHLOROSTYRENE
mf: C_8H_7Cl mw: 138.60

TOXICITY DATA with REFERENCE
skn-rbt 10 mg/24H JIHTAB 30,63,48
eye-rbt 500 mg JIHTAB 30,63,48
orl-rat LD50:5200 mg/kg JIHTAB 30,63,48
skn-rbt LD50:20 g/kg JIHTAB 30,63,48

CONSENSUS REPORTS: Reported in EPA TSCA Inventory.

OSHA PEL: TWA 50 ppm; STEL: 75 ppm
ACGIH TLV: TWA 50 ppm; STEL: 75 ppm

SAFETY PROFILE: Mildly toxic by ingestion and skin contact. A skin and eye irritant. When heated to decomposition it emits toxic fumes of Cl^-. See also CHLORINATED HYDROCARBONS, AROMATIC.

CLF000 CAS:20697-04-5 ***HR: D***
3-CHLOROSTYRENE OXIDE
mf: C_8H_7ClO mw: 154.60

SYNS: (m-CHLOROPHENYL)OXIRANE ◇ (3-CHLOROPHENYL)OXIRANE (9CI) ◇ m-CHLOROSTYRENE OXIDE

TOXICITY DATA with REFERENCE
mmo-sat 2060 μmol/L MUREAV 58,159,78
mmo-esc 2 mmol/L CMSHAF 7,737,78
msc-ham:lng 200 μmol/L CMSHAF 8,369,79

SAFETY PROFILE: Mutation data reported. When heated to decomposition it emits toxic fumes of Cl^-.

CLF325 CAS:14293-44-8 ***HR: 3***
4-CHLORO-5-SULFAMOYL-2',6'-SALICYLOXYLIDIDE
mf: $C_{15}H_{15}ClN_2O_4S$ mw: 354.83

PROP: Crystals from methanol-water. Mp: 256°.

SYNS: 5-(AMINOSULFONYL)-4-CHLORO-N-(2,6-DIMETHYLPHENYL)-2-HYDROXY BENZAMIDE (9CI) ◇ AQUAPHOR ◇ BE 1293 ◇ BEI-1293 ◇ 4-CHLOR-5-SULFAMOYL-2',6'-SALICYLOXYLIDID (GERMAN) ◇ DIUREXAN ◇ XIPAMID ◇ XIPAMIDE

TOXICITY DATA with REFERENCE
orl-rat LD50:1640 mg/kg ARZNAD 25,245,75
ipr-rat LD50:320 mg/kg ARZNAD 25,245,75
orl-mus LD50:1810 mg/kg ARZNAD 25,245,75
ipr-mus LD50:520 mg/kg ARZNAD 25,245,75
scu-mus LD50:1480 mg/kg ARZNAD 25,245,75

SAFETY PROFILE: Poison by intraperitoneal route. Moderately toxic by ingestion and subcutaneous routes. When heated to decomposition it emits toxic fumes of Cl^-, SO_x, and NO_x.

CLF500 CAS:25081-01-0 ***HR: 3***
N-CHLOROSULFINYLIMIDE
mf: ClNOS mw: 97.52

O:S:NCl

SAFETY PROFILE: May explode if melted in a sealed container. Reacts with chlorine fluoride to form an explosive, powerfully oxidizing product. When heated to decomposition it emits toxic fumes of Cl^-, SO_x, and NO_x.

CLG000 CAS:73926-94-0 ***HR: 2***
4-CHLORO-4'-(6-SULFO-2H-NAPHTHO(1,2-d)TRIAZOL-2-YL)-2,2'-STILBENEDISULFONIC ACID TRISODIUM SALT
mf: $C_{24}H_{13}ClN_3O_9S_3 \cdot 3Na$ mw: 688.00

SYN: 2-(4''-CHLOR-4'-STILBYL)NAFTOTRIAZOL-6,2',2''-TRISULFONAN SODNY (CZECH)

TOXICITY DATA with REFERENCE
eye-rbt 5 mg/24H SEV 28ZPAK -,250,72
orl-rat LD50:19900 mg/kg 28ZPAK -,250,72

SAFETY PROFILE: Mildly toxic by ingestion. A severe eye irritant. When heated to decomposition it emits very toxic fumes of Cl^-, NO_x, Na_2O, and SO_x.

CLG200 ***HR: 2***
5-(CHLOROSULFONYL)-2,4-DICHLOROBENZOIC ACID
mf: $C_7H_4Cl_3O_4S$ mw: 290.52

SYN: 2,4-DICHLORO-5-CHLOROSULPHONYLBENZOICACID

TOXICITY DATA with REFERENCE
skn-rbt 500 mg MLD FCTOD7 20,563,82
eye-rbt 100 mg SEV FCTOD7 20,573,82
eye-rbt 100 mg/4S rns SEV FCTOD7 20,573,82

SAFETY PROFILE: A skin and severe eye irritant. When heated to decomposition it emits toxic fumes of Cl^- and SO_x.

CLG250 CAS:1189-71-5 ***HR: 3***
CHLOROSULFONYLISOCYANATE
mf: $CClNO_3S$ mw: 141.54

ClSO2N:C:O

PROP: Fp: −43°; bp: 107°.

SAFETY PROFILE: A very strong irritant. Reacts violently with water. When heated to decomposition it emits toxic fumes of Cl^-, SO_x, and NO_x.

CLG500 CAS:7790-94-5 ***HR: 3***
CHLOROSULFURIC ACID
DOT: UN 1754
mf: $ClHO_3S$ mw: 116.52

PROP: Clear to cloudy, colorless to pale yellow liquid; sharp odor. Mp: −80°, bp: 151.0°, d: 1.766 @ 18°, vap press: 1 mm @ 32°, vap d: 4.02.

SYNS: CHLOROSULFONIC ACID (DOT) ◇ MONOCHLOROSULFURIC ACID ◇ SULFONIC ACID, MONOCHLORIDE ◇ SULFURIC CHLOROHYDRIN

CONSENSUS REPORTS: Reported in EPA TSCA Inventory.

DOT Classification: Corrosive Material; Label: Corrosive.

SAFETY PROFILE: A poison irritant. See also SULFURIC ACID. Chlorosulfonic acid is corrosive, can cause severe acid burns and is very irritating to the eyes, lungs, and mucous membranes. It can cause acute toxic effects either in the liquid or vapor state. Inhalation of concentrated vapor may cause loss of consciousness with serious damage to lung tissue. Contact of liquid with the eyes can cause severe burns if not immediately and completely removed. It also causes severe skin burns due to its highly corrosive action. Upon ingestion it will irritate the mouth, esophagus, and stomach to a serious degree and on contact with skin cause dermatitis. It may cause conjunctivitis even in the vapor form. If spilled on a person, remove all contaminated clothing, wash contaminated skin with a lot of water, followed by baking soda solution. Irrigate eyes with warm water for 15 minutes. Consult a physician.

Vent stored drums two times per month to control pressure of hydrogen produced by action of acid on metal of drum. Decomposes explosively on contact with water, alcohol, or acids. Explosive reaction with phosphorus. Violent reaction with silver nitrate. Potentially violent reaction with sulfuric acid or diphenyl ether. Incompatible with acetic acid, acetic anhydride, acetonitrile, acrolein, acrylic acid, acrylonitrile, allyl alcohol, allyl chloride, 2-amino ethanol, ammonium hydroxide, aniline, n-butyraldehyde, creosote oil, cresol, cumene, dichloroethyl ether, diethylene glycol monomethyl ether, diisobutylene, diisopropylether, epichloro hydrin, ethyl acetate, ethyl acrylate, ethylene chlorohydrin, ethylene cyanohydrin, ethylene diamine, ethylene glycol, ethylene glycol monoethyl ether acetate, ethylene imine, glyoxal, HCl, HF, H_2O_2, isoprene, mesityl oxide, metal powders, methyl ethyl ketone, HNO_3, 2-nitropropane, β-propiolactone, propylene oxide, pyridene, NaOH, sulfolane, styrene monomer, vinyl acetate, vinylidene chloride, water, organic matter, combustibles. Dangerous. To fight fire, avoid water, use dry chemicals. When heated to decomposition it emits toxic fumes of Cl^- and SO_x. See SULFURIC ACID, HYDROCHLORIC ACID, and SULFONATES.

CLG750 ***HR: 3***
CHLOROSULFURIC ACID, mixed with SULFUR DIOXIDE

SYN: CHLOROSULFONIC ACID-SULFUR DIOXIDE MIXTURE (DOT)

DOT Classification: Corrosive Material; Label: Corrosive.

SAFETY PROFILE: A poison. A corrosive material which is highly irritating to skin, eyes, and mucous membranes. See also CHLOROSULFURIC ACID and SULFUR DIOXIDE. When heated to decomposition it emits very toxic fumes of Cl^- and SO_x.

CLG825 ***HR: 2***
4-CHLORO-5-SULPHAMOYLPHTHALIMIDE
mf: $C_8H_5ClN_2O_4$ mw: 228.60

SYN: 6-CHLORO-1,3-DIOXO-5-ISOINDOLINESULFONAMIDE

TOXICITY DATA with REFERENCE
skn-rbt 500 mg MLD FCTOD7 20,573,82
eye-rbt 100 mg SEV FCTOD7 20,573,82
eye-rbt 100 mg/4S rns SEV FCTOD7 20,573,82

SAFETY PROFILE: A severe eye and mild skin irritant. When heated to decomposition it emits toxic fumes of Cl^- and NO_x.

CLH000 CAS:63938-10-3 ***HR: 1***
CHLOROTETRAFLUOROETHANE
DOT: UN 1021
mf: C_2HClF_4 mw: 136.48

PROP: Colorless gas.

SYN: MONOCHLOROTETRAFLUOROETHANE (DOT)

DOT Classification: Nonflammable Gas; Label: Nonflammable Gas.

SAFETY PROFILE: Probably acts as a simple asphyxiant. See also CHLORINATED HYDROCARBONS, ALIPHATIC; and FLUORIDES. When heated to decomposition it emits highly toxic fumes of F^- and Cl^-.

CLH500 CAS:6926-39-2 ***HR: 3***
N-CHLOROTETRAMETHYLGUANIDINE
mf: $C_5H_{12}ClN_3$ mw: 149.63

SAFETY PROFILE: An unstable material (even at 0°C) which explodes if heated above 50°C. Upon decomposition it emits toxic fumes of Cl^- and NO_x.

CLH625 CAS:4113-57-9 ***HR: 3***
5-CHLORO-1,2,3-THIADIAZOLE
mf: C_2HClN_2S mw: 120.56

SAFETY PROFILE: A heat- and impact-sensitive explosive. When heated to decomposition it emits toxic fumes of Cl^-, SO_x, and NO_x. See also EXPLOSIVES.

CLH750 CAS:58-94-6 ***HR: 2***
CHLOROTHIAZIDE
mf: $C_7H_6ClN_3O_4S_2$ mw: 295.73

SYNS: ALURENE ◇ CHLORIAZID ◇ 6-CHLORO-2H-1,2,4-BENZOTHIADIAZINE-7-SULFONAMIDE-1,1-DIOXIDE ◇ 6-CHLORO-7-SULFAMOYL-2H-1,2,4-BENZOTHIADIAZINE-1,1-DIOXIDE ◇ CHLOROTHIAZID ◇ CHLORSAL ◇ CHLORTHIAZIDE ◇ CHLORURIT ◇ CHLOTRIDE ◇ CLOTRIDE ◇ CT ◇ DIURESAL ◇ DIURIL ◇ DIURILIX ◇ DIURITE ◇ DIUTRID ◇ FLUMEN ◇ MINZIL

◇ NEO-DEMA ◇ SALISAN ◇ SALUNIL ◇ SALURETIL ◇ SALURIC ◇ SK-CHLOROTHIAZIDE ◇ THIAZIDE ◇ URINEX ◇ WARDUZIDE ◇ YADALAN

TOXICITY DATA with REFERENCE
orl-rat TDLo:12450 mg/kg (8-22D preg):REP JCINAO 41,710,62
orl-rat LD50:10 g/kg YAKUD5 21,775,79
ipr-rat LD50:1386 mg/kg 27ZIAQ -,77,73
orl-mus LD50:8000 mg/kg AIPTAK 118,467,59
ipr-mus LD50:1400 mg/kg JPETAB 134,273,61
ivn-mus LD50:940 mg/kg JPETAB 134,273,61
ivn-dog LD50:1000 mg/kg 27ZIAQ -,77,73

CONSENSUS REPORTS: Reported in EPA TSCA Inventory. EPA Genetic Toxicology Program.

SAFETY PROFILE: Moderately toxic by intraperitoneal and intravenous routes. Mildly toxic by ingestion. Experimental reproductive effects. Has been implicated in aplastic anemia. When heated to decomposition it emits very toxic fumes of SO_x, NO_x and Cl^-.

CLJ750 CAS:2812-73-9 ***HR: 3***
CHLOROTHIOFORMIC ACID ETHYL ESTER
DOT: UN 2826
mf: C_3H_5ClOS mw: 124.59

SYN: ETHYL CHLOROTHIOFORMATE (DOT)

DOT Classification: Corrosive Material; Label: Corrosive, Flammable Liquid.

SAFETY PROFILE: Probably a poison by inhalation and ingestion. A corrosive irritant to skin, eyes, and mucous membranes. See also ESTERS and CHLORIDES. Flammable when exposed to heat or flame. When heated to decomposition it emits very toxic fumes of Cl^- and SO_x.

CLJ875 CAS:21923-23-9 ***HR: 3***
CHLORTHIOPHOS
mf: $C_{11}H_{15}Cl_2O_3PS_2$ mw: 361.25

SYNS: CELAMERCK S-2957 ◇ CELA S-2957 ◇ CELATHION ◇ CM S 2957 ◇ O-(DICHLORO(METHYLTHIO)PHENYL) O,O-DIETHYL PHOSPHOROTHIOATE (3 isomers) ◇ O,O-DIETHYL-O-2,4,5-DICHLORO-(METHYLTHIO)PHENYL THIONOPHOSPHATE ◇ ENT 27,635 ◇ NSC 195164 ◇ OMS 1342 ◇ S 2957

TOXICITY DATA with REFERENCE
orl-rat LD50:7800 μg/kg FMCHA2 -,C56,8
skn-rat LD50:58 mg/kg AHRTAN 27,3,76
orl-mus LD50:141 mg/kg AHRTAN 27,3,76
orl-rbt LD50:20 mg/kg AHRTAN 27,3,76
skn-rbt LD50:48 mg/kg AHRTAN 27,3,76
orl-gpg LD50:58 mg/kg AHRTAN 27,3,76
orl-ckn LD50:45 mg/kg AHRTAN 27,3,76
orl-qal LD50:45 mg/kg AHRTAN 27,3,76

CONSENSUS REPORTS: EPA Extremely Hazardous Substances List.

SAFETY PROFILE: Poison by ingestion and skin contact. When heated to decomposition it emits toxic fumes of Cl^-, PO_x, and SO_x.

CLK100 CAS:95-49-8 ***HR: 3***
o-CHLOROTOLUENE
DOT: UN 2238
mf: C_7H_7Cl mw: 126.59

PROP: Liquid. Bp: 158.97°, d: (20/4) 1.0826, mp: −35.59°. Volatile with steam. Sltly sol in water; freely sol in alc, benzene, chloroform, ether.

SYNS: 2-CHLORO-1-METHYLBENZENE (9CI) ◇ 2-CHLOROTOLUENE ◇ HALSO 99 ◇ 1-METHYL-2-CHLOROBENZENE ◇ 2-METHYLCHLOROBENZENE ◇ o-TOLYL CHLORIDE

TOXICITY DATA with REFERENCE
ihl-rat LCLo:17500 ppm DTLVS* 4,95,80
unr-rat LD50:5700 mg/kg GISAAA 45(12),64,80
unr-mus LD50:4400 mg/kg GISAAA 45(12),64,80
unr-gpg LD50:3000 mg/kg GISAAA 46(2),14,81

CONSENSUS REPORTS: Reported in EPA TSCA Inventory.

OSHA PEL: TWA 50 ppm
ACGIH TLV: TWA 50 ppm
DOT Classification: Flammable or Combustible Liquid; Label: Flammable Liquid.

SAFETY PROFILE: Moderately toxic by unspecified routes. Flammable when exposed to heat or flame. When heated to decomposition it emits toxic fumes of Cl^-. See also TOLYL CHLORIDE, and CHLORINATED HYDROCARBONS, AROMATIC.

CLK200 CAS:87-60-5 ***HR: 3***
3-CHLORO-o-TOLUIDINE
mf: C_7H_8ClN mw: 141.61

SYNS: 1-AMINO-2-CHLORO-6-METHYLBENZENE ◇ 1-AMINO-3-CHLORO-2-METHYLBENZENE ◇ 2-AMINO-6-CHLOROTOLUENE ◇ AZOIC DIAZO COMPONENT 46 ◇ 3-CHLORO-2-METHYLANILINE ◇ 3-CHLOR-2-TOLUIDIN (CZECH) ◇ FAST SCARLET TR BASE ◇ SCARLET TR BASE

TOXICITY DATA with REFERENCE
dni-mus-orl 200 mg/kg MUREAV 46,305,77
orl-rat LD50:574 mg/kg MarJV# 29MAR77
orl-bwd LD50:237 mg/kg AECTCV 12,355,83

CONSENSUS REPORTS: Reported in EPA TSCA Inventory.

SAFETY PROFILE: Poison by ingestion. Mutation data reported. When heated to decomposition it emits toxic fumes of Cl^- and NO_x. See also other chloro toluidine entries.

CLK210 CAS:615-65-6 ***HR: 3***
2-CHLORO-p-TOLUIDINE
mf: C_7H_8ClN mw: 141.61

SYN: 2-CHLOR-4-TOLUIDIN (CZECH)

TOXICITY DATA with REFERENCE
skn-rbt 500 mg/24H SEV 28ZPAK -,97,72
eye-rbt 250 μg/24H SEV 28ZPAK -,97,72
mma-sat 1 μmol/plate MUREAV 77,317,80
orl-rat LD50:367 mg/kg 28ZPAK -,97,72

CONSENSUS REPORTS: Reported in EPA TSCA Inventory.

SAFETY PROFILE: Poison by ingestion. A severe eye and skin irritant. Mutation data reported. When heated to decomposition it emits toxic fumes of Cl^- and NO_x. See also other chloro toluidine entries.

CLK215 CAS:95-74-9 ***HR: 3***
3-CHLORO-p-TOLUIDINE
mf: C_7H_8ClN mw: 141.61

SYNS: 1-AMINO-3-CHLORO-4-METHYLBENZENE ◇ 4-AMINO-2-CHLOROTOLUENE ◇ 2-CHLORO-4-AMINOTOLUENE ◇ 3-CHLORO-4-METHYLANILINE ◇ CPT ◇ DKC 1347 ◇ DRC 1339 ◇ NCI-C02040

TOXICITY DATA with REFERENCE
dni-mus-orl 200 mg/kg MUREAV 46,305,77
orl-rat LD50:1500 mg/kg TXAPA9 21,315,72
ipr-rat LD50:325 mg/kg TXAPA9 18,517,71
ivn-rat LD50:48 mg/kg TXAPA9 18,517,71
orl-mus LD50:316 mg/kg NCILB* NCI-E-C-72-3252,73
orl-pgn LD50:13 mg/kg TXAPA9 21,315,72
orl-qal LD50:1 mg/kg AECTCV 12,355,83
orl-bwd LD50:2400 μg/kg TXAPA9 21,315,72

CONSENSUS REPORTS: Reported in EPA TSCA Inventory. NCI Carcinogenogenesis Bioassay (Feed); Results Negative: Mouse, Rat NCITR* NCI-CG-TR-145,78

SAFETY PROFILE: Poison by ingestion, intravenous, and intraperitoneal routes. Mutation data reported. When heated to decomposition it emits toxic fumes of Cl^- and NO_x. See also other chlorotoluidine entries.

CLK220 CAS:95-69-2 ***HR: 3***
4-CHLORO-o-TOLUIDINE
mf: C_7H_8ClN mw: 141.61

SYNS: AMARTHOL FAST RED TR BASE ◇ 2-AMINO-5-CHLOROTOLUENE ◇ AZOENE FAST RED TR BASE ◇ AZOGENE FAST RED TR ◇ AZOIC DIAZO COMPONENT 11 BASE ◇ BRENTAMINE FAST RED TR BASE ◇ 5-CHLORO-2-AMINOTOLUENE ◇ 4-CHLORO-2-METHYLANILINE ◇ 4-CHLORO-6-METHYLANILINE ◇ 4-CHLORO-2-METHYLBENZENEAMINE ◇ 4-CHLORO-2-TOLUIDINE ◇ DAITO RED BASE TR ◇ DEVAL RED K ◇ DEVAL RED TR ◇ DIAZO FAST RED TRA ◇ FAST RED BASE TR ◇ FAST RED 5CT BASE ◇ FAST RED TR ◇ FAST RED TR11 ◇ FAST RED TR BASE ◇ FAST RED TRO BASE ◇ KAKO RED TR BASE ◇ KAMBAMINE RED TR ◇ 2-METHYL-4-CHLOROANILINE ◇ MITSUI RED TR BASE ◇ RED BASE CIBA IX ◇ RED BASE IRGA IX ◇ RED BASE NTR ◇ RED TR BASE ◇ SANYO FAST RED TR BASE ◇ TULABASE FAST RED TR

TOXICITY DATA with REFERENCE
mmo-sat 400 μg/plate JPFCD2 19,95,84
dnr-sat 250 mg/disc JPFCD2 19,95,84
dnr-esc 2 g/disc JPFCD2 19,95,84
oms-hmn:hla 1 mmol/L BECTA6 11,184,74
slt-mus-orl 12 g/kg/3D-I MUREAV 135,219,84
dnd-ham:lng 3 mmol/L/2H MUREAV 77,317,80
scu-cat LDLo:310 mg/kg AHBAAM 110,12,33
orl-bwd LD50:75 mg/kg AECTCV 12,355,83

CONSENSUS REPORTS: IARC Cancer Review: Group 2A IMEMDT 7,56,87; Human Inadequate Evidence IMEMDT 16,277,78; Animal Sufficient Evidence IMEMDT 30,61,83. Reported in EPA TSCA Inventory.

DFG MAK: Human Carcinogen.

SAFETY PROFILE: Confirmed carcinogen. Poison by ingestion and subcutaneous routes. Human mutation data reported. In the presence of Copper(II) chloride catalyst decomposition occurs above 239°C. When heated to decomposition it emits toxic fumes of Cl^- and NO_x. See also other chloro toluidine entries.

CLK225 CAS:95-79-4 ***HR: 3***
5-CHLORO-o-TOLUIDINE
mf: C_7H_8ClN mw: 141.61

PROP: Solid. Bp: 241°, mp: 29°.

SYNS: ACCO FAST RED KB BASE ◇ 1-AMINO-3-CHLORO-6-METHYLBENZENE ◇ 2-AMINO-4-CHLOROTOLUENE ◇ ANSIBASE RED KB ◇ AZOENE FAST RED KB BASE ◇ AZOIC DIAZO COMPONENT 32 ◇ 4-CHLORO-2-AMINOTOLUENE ◇ 3-CHLORO-6-METHYLANILINE ◇ 5-CHLORO-2-METHYLANILINE ◇ FAST RED KB AMINE ◇ FAST RED KB BASE ◇ FAST RED KB SALT ◇ FAST RED KB SALT SUPRA ◇ FAST RED KBS SALT ◇ GENAZO RED KB SOLN ◇ HILTONIL FAST RED KB BASE ◇ LAKE RED KB BASE ◇ METROGEN RED FORMER KB SOLN ◇ NAPHTHOSOL FAST RED KB BASE ◇ NCI-C02051 ◇ PHARMAZOID RED KB ◇ RED KB BASE ◇ SPECTROLENE RED KB ◇ STABLE RED KB BASE

TOXICITY DATA with REFERENCE
dni-mus-orl 200 mg/kg MUREAV 46,305,77
orl-rat TDLo:164 g/kg/78W-C:ETA NCITR* NCI-CG-TR-187,79
orl-mus TDLo:131 g/kg/78W-C:CAR NCITR* NCI-CG-TR-187,79
orl-mus TD:262 g/kg/78W-C:CAR NCITR* NCI-CG-TR-187,79
orl-rat LD50:464 mg/kg NCILB* NIH-NCI-E-C-72-3252

CONSENSUS REPORTS: NTP Carcinogenesis Bioassay (feed): Clear Evidence: mouse NCITR* NCI-TR-

187,79; (feed): Inadequate Studies: rat NCITR* NCI-TR-187,79. Reported in EPA TSCA Inventory. EPA Genetic Toxicology Program.

DFG MAK: Suspected Carcinogen.

SAFETY PROFILE: Suspected carcinogen with experimental carcinogenic and tumorigenic data. Moderately toxic by ingestion. When heated to decomposition it emits very toxic fumes of Cl^- and NO_x. See also AROMATIC AMINES.

CLK230 CAS:7745-89-3 ***HR: 3***
3-CHLORO-p-TOLUIDINE HYDROCHLORIDE
mf: $C_7H_8ClN \cdot ClH$ mw: 178.07

SYNS: CTH ◇ DRC-1,339 ◇ 4-METHYL-3-CHLOROANILINE HYDROCHLORIDE ◇ STARLICIDE

TOXICITY DATA with REFERENCE
orl-rat LD50:655 mg/kg TXAPA9 18,517,71
ipr-mus LD50:338 mg/kg TXAPA9 29,135,74
orl-pgn LD50:18 mg/kg TXAPA9 21,315,72
orl-ckn LD50:4 mg/kg PCOC** -,457,66
ipr-ckn LDLo:100 mg/kg TXAPA9 22,458,72
orl-dck LD50:18 mg/kg TXAPA9 21,315,72
orl-bwd LD50:2400 μg/kg TXAPA9 21,315,72

SAFETY PROFILE: Poison by ingestion and intraperitoneal routes. When heated to decomposition it emits toxic fumes of NO_x and Cl^-. See also other chloro toluidine entries.

CLK235 CAS:3165-93-3 ***HR: 3***
4-CHLORO-2-TOLUIDINE HYDROCHLORIDE
DOT: UN 1579
mf: $C_7H_8ClN \cdot ClH$ mw: 178.07

SYNS: AMARTHOL FAST RED TR BASE ◇ AMARTHOL FAST RED TR SALT ◇ 2-AMINO-5-CHLOROTOLUENE HYDROCHLORIDE ◇ AZANIL RED SALT TRD ◇ AZOENE FAST RED TR SALT ◇ AZOGENE FAST RED TR ◇ AZOIC DIAZO COMPONENT 11 BASE ◇ BRENTAMINE FAST RED TR SALT ◇ CHLORHYDRATE de 4-CHLOROORTHOTOLUIDINE (FRENCH) ◇ 5-CHLORO-2-AMINOTOLUENE HYDROCHLORIDE ◇ 4-CHLORO-2-METHYLANILINE HYDROCHLORIDE ◇ 4-CHLORO-6-METHYLANILINE HYDROCHLORIDE ◇ 4-CHLORO-2-METHYLBENZENEAMINE HYDROCHLORIDE ◇ 4-CHLORO-o-TOLUIDINE HYDROCHLORIDE ◇ 4-CHLORO-o-TOLUIDINE HYDROCHLORIDE (DOT) ◇ C.I. 37085 ◇ C.I. AZOIC DIAZO COMPONENT 11 ◇ DAITO RED SALT TR ◇ DEVOL RED K ◇ DEVOL RED TA SALT ◇ DEVOL RED TR ◇ DIAZO FAST RED TR ◇ DIAZO FAST RED TRA ◇ FAST RED 5CT SALT ◇ FAST RED SALT TR ◇ FAST RED SALT TRA ◇ FAST RED SALT TRN ◇ FAST RED TR SALT ◇ HINDASOL RED TR SALT ◇ KROMON GREEN B ◇ 2-METHYL-4-CHLOROANILINE HYDROCHLORIDE ◇ NATASOL FAST RED TR SALT ◇ NCI-C02368 ◇ NEUTROSEL RED TRVA ◇ OFNA-PERL SALT RRA ◇ RCRA WASTE NUMBER U049 ◇ RED BASE CIBA IX ◇ RED BASE IRGA IX ◇ RED SALT CIBA IX ◇ RED SALT IRGA IX ◇ RED TRS SALT ◇ SANYO FAST RED SALT TR

TOXICITY DATA with REFERENCE
orl-mus TDLo:49 g/kg/78W-C:CAR JEPTDQ 2(2),325,78
orl-mus TD:104 g/kg/99W-C:CAR NCITR* NCI-CG-TR-165,78
ipr-rat LD50:560 mg/kg NCIBR* NCI-E-68-1311,73
ipr-mus LD50:680 mg/kg NCIBR* NCI-E-68-1311,73

CONSENSUS REPORTS: IARC Cancer Review: Group 2A IMEMDT 48,123,90; Animal Inadequate Evidence, Human Inadequate Evidence IMEMDT 16,277,78. NCI Carcinogenesis Bioassay (Feed); Clear Evidence: Mouse; No Evidence: Rat NCITR* NCI-CG-TR-165,79. Reported in EPA TSCA Inventory.

DOT Classification: Poison B; Label: Poison; DOT-IMO: Poison B; Label: St. Andrews Cross.

SAFETY PROFILE: Suspected carcinogen with experimental carcinogenic data. Moderately toxic by intraperitoneal route. When heated to decomposition it emits toxic fumes of NO_x and Cl^-. See also other chloro toluidine entries.

CLK325 CAS:13710-19-5 ***HR: 3***
N-(3-CHLORO-o-TOLYL)ANTHRANILIC ACID
mf: $C_{14}H_{12}ClNO_2$ mw: 261.72

PROP: Crystals from abs ethanol. Mp: 207-207.5°.

SYNS: N-(3-CHLORO-2-METHYLPHENYL)ANTHRANILIC ACID ◇ CLOTAM ◇ GEA 6414 ◇ N-(2-METHYL-3-CHLOROPHENYL)ANTHRANILIC ACID ◇ TOLFENAMIC ACID

TOXICITY DATA with REFERENCE
orl-rat TDLo:648 mg/kg (female 17-22D post):REP TOIZAG 29,889,83
orl-rat TDLo:1 mg/kg (21D preg):TER OYYAA2 27,117,84
orl-rat LD50:225 mg/kg TOIZAG 28,99,81
ipr-rat LD50:238 mg/kg TOIZAG 29,851,83
scu-rat LD50:246 mg/kg IYKEDH 14,838,83
orl-mus LD50:280 mg/kg IYKEDH 14,838,83
ipr-mus LD50:185 mg/kg IYKEDH 14,838,83
scu-mus LD50:267 mg/kg IYKEDH 14,838,83

SAFETY PROFILE: Poison by ingestion, subcutaneous, and intraperitoneal routes. An experimental teratogen. Experimental reproductive effects. When heated to decomposition it emits toxic fumes of Cl^- and NO_x.

CLK500 CAS:78371-90-1 ***HR: 3***
1-(6-CHLORO-o-TOLYL)-3-CYCLOHEXYL-3-(2-(DIETHYLAMINO)ETHYL)UREA HYDROCHLORIDE
mf: $C_{20}H_{32}ClN_3O \cdot ClH$ mw: 402.46

TOXICITY DATA with REFERENCE
ipr-rat LD50:41 mg/kg ARZNAD 8,664,58
scu-mus LD50:87 mg/kg ARZNAD 8,664,58

SAFETY PROFILE: Poison by intraperitoneal and subcutaneous routes. When heated to decomposition it emits very toxic fumes of Cl^- and NO_x.

CLK750 CAS:78371-91-2 ***HR: 3***
1-(6-CHLORO-o-TOLYL)-3-(3-(DIBUTYLAMINO) PROPYL)UREA HYDROCHLORIDE
mf: $C_{19}H_{32}ClN_3O•ClH$ mw: 390.45

TOXICITY DATA with REFERENCE
eye-rbt 2% MOD ARZNAD 8,664,58
ipr-rat LD50:275 mg/kg ARZNAD 8,664,58
scu-mus LD50:450 mg/kg ARZNAD 8,664,58

SAFETY PROFILE: Poison by intraperitoneal route. Moderately toxic by subcutaneous route. An eye irritant. When heated to decomposition it emits very toxic fumes of Cl^- and NO_x.

CLL000 CAS:78371-93-4 ***HR: 3***
1-(6-CHLORO-o-TOLYL)-3-(2-(DIETHYLAMINO) ETHYL)-3-METHYLUREA
mf: $C_{15}H_{24}ClN_3O$ mw: 297.87

SYN: C 3247

TOXICITY DATA with REFERENCE
eye-rbt 2% MLD ARZNAD 8,664,58
ipr-rat LD50:108 mg/kg ARZNAD 8,664,58
scu-mus LD50:262 mg/kg ARZNAD 8,664,58

SAFETY PROFILE: Poison by intraperitoneal and subcutaneous routes. An eye irritant. When heated to decomposition it emits very toxic fumes of Cl^- and NO_x.

CLL250 CAS:78371-92-3 ***HR: 3***
1-(6-CHLORO-o-TOLYL)-3-(2-(DIETHYLAMINO)ETHYL)UREA HYDROCHLORIDE
mf: $C_{14}H_{22}ClN_3O•ClH$ mw: 320.30

SYN: C 3182

TOXICITY DATA with REFERENCE
eye-rbt 2% MLD ARZNAD 8,664,58
ipr-rat LD50:212 mg/kg ARZNAD 8,664,58
scu-mus LD50:500 mg/kg ARZNAD 8,664,58

SAFETY PROFILE: Poison by intraperitoneal route. Moderately toxic by subcutaneous route. An eye irritant. When heated to decomposition it emits very toxic fumes of Cl^- and NO_x.

CLL500 CAS:78371-95-6 ***HR: 3***
1-(6-CHLORO-o-TOLYL)-3-(2-(DIETHYLAMINO) ETHYL)-3-(2,6-XYLYL)UREA HYDROCHLORIDE
mf: $C_{22}H_{30}ClN_3O•ClH$ mw: 424.46

SYN: C 3184

TOXICITY DATA with REFERENCE
eye-rbt 2% MLD ARZNAD 8,664,58
ipr-rat LD50:84 mg/kg ARZNAD 8,664,58
scu-mus LD50:75 mg/kg ARZNAD 8,664,58

SAFETY PROFILE: Poison by intraperitoneal and subcutaneous routes. An eye irritant. When heated to decomposition it emits very toxic fumes of Cl^- and NO_x.

CLL750 CAS:78371-94-5 ***HR: 3***
1-(6-CHLORO-o-TOLYL)-1-(2-(DIETHYLAMINO) ETHYL)-3-(2,6-XYLYL)UREA HYDROCHLORIDE
mf: $C_{22}H_{30}ClN_3O•ClH$ mw: 424.46

SYN: C 3186

TOXICITY DATA with REFERENCE
ipr-rat LD50:62 mg/kg ARZNAD 8,664,58
scu-mus LD50:90 mg/kg ARZNAD 8,664,58

SAFETY PROFILE: Poison by intraperitoneal and subcutaneous routes. When heated to decomposition it emits very toxic fumes of Cl^- and NO_x.

CLM000 CAS:78371-96-7 ***HR: 3***
1-(6-CHLORO-o-TOLYL)-3-(3-(DIETHYLAMINO) PROPYL)UREA
mf: $C_{15}H_{24}ClN_3O$ mw: 297.87

SYN: C 3214

TOXICITY DATA with REFERENCE
eye-rbt 2% MLD ARZNAD 8,664,58
ipr-rat LD50:275 mg/kg ARZNAD 8,664,58
scu-mus LD50:450 mg/kg ARZNAD 8,664,58

SAFETY PROFILE: Poison by intraperitoneal route. Moderately toxic by subcutaneous route. An eye irritant. When heated to decomposition it emits very toxic fumes of Cl^- and NO_x.

CLM250 CAS:78371-98-9 ***HR: 3***
1-(6-CHLORO-o-TOLYL)-3-(2-(DIMETHYLAMINO)ETHYL)-3-ISOPROPYLUREA HYDROCHLORIDE
mf: $C_{15}H_{24}ClN_3O•ClH$ mw: 334.33

SYN: C 3246

TOXICITY DATA with REFERENCE
eye-rbt 2% MLD ARZNAD 8,664,58
ipr-rat LD50:22 mg/kg ARZNAD 8,664,58
scu-mus LD50:30 mg/kg ARZNAD 8,664,58

SAFETY PROFILE: Poison by intraperitoneal and subcutaneous routes. An eye irritant. When heated to decomposition it emits very toxic fumes of Cl^- and NO_x.

CLM500 CAS:78371-97-8 ***HR: 3***
1-(6-CHLORO-o-TOLYL)-3-(2-(DIMETHYLAMINO)ETHYL)UREA HYDROCHLORIDE
mf: $C_{12}H_{18}ClN_3O{\cdot}ClH$ mw: 292.24

SYN: C 3213

TOXICITY DATA with REFERENCE
eye-rbt 2% MLD ARZNAD 8,664,58
ipr-rat LD50:362 mg/kg ARZNAD 8,664,58
scu-mus LD50:1025 mg/kg ARZNAD 8,664,58

SAFETY PROFILE: Poison by intraperitoneal route. Moderately toxic by subcutaneous route. An eye irritant. When heated to decomposition it emits very toxic fumes of NO_x and Cl^-.

CLM750 CAS:78371-99-0 ***HR: 3***
1-(6-CHLORO-o-TOLYL)-3-(3-(DIMETHYLAMINO)PROPYL)UREA HYDROCHLORIDE
mf: $C_{13}H_{20}ClN_3O{\cdot}ClH$ mw: 306.27

SYN: C 3229

TOXICITY DATA with REFERENCE
ipr-rat LD50:300 mg/kg ARZNAD 8,664,58
scu-mus LD50:1375 mg/kg ARZNAD 8,664,58

SAFETY PROFILE: Poison by intraperitoneal route. Moderately toxic by subcutaneous route. When heated to decomposition it emits very toxic fumes of Cl^- and NO_x.

CLN000 CAS:78372-00-6 ***HR: 3***
1-(6-CHLORO-o-TOLYL)-3-(4-METHOXYBENZYL)-3-(2-PIPERIDINOETHYL)UREA
mf: $C_{23}H_{30}ClN_3O_2$ mw: 416.01

SYN: C 5320

TOXICITY DATA with REFERENCE
eye-rbt 2% MLD ARZNAD 8,664,58
ipr-rat LD50:70 mg/kg ARZNAD 8,664,58
scu-mus LD50:130 mg/kg ARZNAD 8,664,58

SAFETY PROFILE: Poison by intraperitoneal and subcutaneous routes. An eye irritant. When heated to decomposition it emits very toxic fumes of Cl^- and NO_x.

CLN250 CAS:78393-39-2 ***HR: 3***
1-(6-CHLORO-o-TOLYL)-3-(4-METHOXYBENZYL)-3-(2-(PYRROLIDINYL)ETHYL)UREA HYDROCHLORIDE
mf: $C_{22}H_{28}ClN_3O_2{\cdot}ClH$ mw: 438.44

SYN: C 5324

TOXICITY DATA with REFERENCE
ipr-rat LD50:72 mg/kg ARZNAD 8,664,58
scu-mus LD50:145 mg/kg ARZNAD 8,664,58

SAFETY PROFILE: Poison by intraperitoneal and subcutaneous routes. When heated to decomposition it emits very toxic fumes of Cl^- and NO_x.

CLN325 CAS:75318-76-2 ***HR: D***
3-(4-CHLORO-o-TOLYL)-5-(m-METHOXYPHENYL)-s-TRIAZOLE
mf: $C_{16}H_{14}ClN_3O$ mw: 299.78

TOXICITY DATA with REFERENCE
scu-ham TDLo:200 μg/kg (female 4-8D post):REP RDMIDP 4,237,82

SAFETY PROFILE: Experimental reproductive effects. When heated to decomposition it emits toxic fumes of Cl^- and NO_x.

CLN500 CAS:78372-01-7 ***HR: 3***
1-(4-CHLORO-o-TOLYL)-3-(p-METHYLBENZYL)-3-(2-PYRROLIDINYLETHYL)UREA HYDROCHLORIDE
mf: $C_{22}H_{28}ClN_3O{\cdot}ClH$ mw: 422.44

SYN: C 5326

TOXICITY DATA with REFERENCE
eye-rbt 2% MLD ARZNAD 8,664,58
ipr-rat LD50:85 mg/kg ARZNAD 8,664,58
scu-mus LD50:175 mg/kg ARZNAD 8,664,58

SAFETY PROFILE: Poison by intraperitoneal and subcutaneous routes. An eye irritant. When heated to decomposition it emits very toxic fumes of Cl^- and NO_x.

CLN750 CAS:94-81-5 ***HR: 2***
4-((4-CHLORO-o-TOLYL)OXY)BUTYRIC ACID
mf: $C_{11}H_{13}ClO_3$ mw: 228.69

SYNS: BEXANE ◇ BEXONE ◇ CAN-TROL ◇ 4-(4-CHLOR-2-METHYLPHENOXY)-BUETTERSAEURE (GERMAN) ◇ 4-(4-CHLOR-2-METHYLPHENOXY)-BUTTERSAEURE (GERMAN) ◇ 4-(4-CHLORO-2-METHYLPHENOXY)BUTANOIC ACID ◇ Γ-(4-CHLORO-2-METHYLPHENOXY)BUTYRIC ACID ◇ 4-(4-CHLORO-2-METHYLPHENOXY)BUTYRIC ACID ◇ (4-CHLORO-o-TOLYLOXY)BUTYRIC ACID ◇ LEGUMEX ◇ 4-(MCB) ◇ MCPB ◇ MCP-BUTYRIC ◇ 2-METHYL-4-CHLOROPHENOXYBUTYRIC ACID ◇ Γ-2-METHYL-4-CHLOROPHENOXYBUTYRIC ACID ◇ 4-(2-METHYL-4-CHLOROPHENOXY)BUTYRIC ACID ◇ 4-(2-METHYL-4-CHLORPHENOXY)-BUTTERSAEURE (GERMAN) ◇ PDQ ◇ THISTROL ◇ TRIFOLEX ◇ TRITROL ◇ TROPOTOX ◇ TROTOX ◇ U46 MCPB

TOXICITY DATA with REFERENCE
sln-dmg-orl 4400 μmol/L EXPEAM 30,621,74
mrc-smc 13500 μmol/L IARCCD 10,161,74
orl-rat LD50:680 mg/kg WRPCA2 4,36,65
orl-mus LD50:800 mg/kg FMCHA2 -,C43,83

CONSENSUS REPORTS: EPA Genetic Toxicology Program.

SAFETY PROFILE: Moderately toxic by ingestion. Mutation data reported. An herbicide. When heated to decomposition it emits toxic fumes of Cl^-.

CLO000 CAS:6062-26-6 ***HR: 2***
(4-CHLORO-o-TOLYLOXY)BUTYRIC ACID SODIUM SALT
mf: $C_{11}H_{12}ClO_3$•Na mw: 250.67

SYNS: CANTROL ◇ 4-(4-CHLOR-2-METHYL-PHENOXY)-BUTTERSAEURE NATRIUMSALZ (GERMAN) ◇ CHLOROMETHYLPHENOXYBUTYRIC ACID SODIUM SALT ◇ 4-(4-CHLORO-2-METHYLPHENOXY)BUTYRIC ACID SODIUM SALT ◇ 4-(4-CHLORO-2-METHYLPHENOXY)BUTANOIC ACID, SODIUM SALT ◇ M&B 3046 ◇ MCPB ◇ 4-(MCPD) ◇ 4-(2-METHYL-4-CHLOROPHENOXY)BUTYRIC ACID, SODIUM SALT ◇ THISTROL ◇ TROPOTOX

TOXICITY DATA with REFERENCE
orl-rat LD50:700 mg/kg PCOC** -,715,66
skn-rat LD50:1000 mg/kg WRPCA2 9,119,70
orl-mus LD50:700 mg/kg GUCHAZ 6,108,73

SAFETY PROFILE: Moderately toxic by ingestion and skin contact. A pesticide. When heated to decomposition it emits toxic fumes of Cl^- and Na_2O.

CLO200 CAS:1929-86-8 ***HR: 2***
2-((4-CHLORO-o-TOLYL)OXY)PROPIONIC ACID POTASSIUM SALT
mf: $Cl_0H_{10}ClO_3$•K mw: 168.10

SYNS: GORDON'S MECOMEC ◇ HEDONAL MCPP ◇ MCPP POTASSIUM SALT ◇ MECOPEX ◇ MECOPROP POTASSIUM SALT ◇ METHOXONE M ◇ PROPANOIC ACID, 2-(4-CHLORO-2-METHYLPHENOXY)-, POTASSIUM SALT (9CI) ◇ PROPIONIC ACID, 2-((4-CHLORO-o-TOLYL)OXY)-, POTASSIUM SALT ◇ SYS 67MPROP ◇ VI-PEX

TOXICITY DATA with REFERENCE
orl-rat TDLo:520 mg/kg (female 4-18D post):REP TJADAB 33,11A,86
orl-rat TDLo:1320 mg/kg (female 4-18D post):TER TJADAB 33,11A,86
orl-rat LD50:930 mg/kg FMCHA2 -,C184,89

SAFETY PROFILE: Moderately toxic by ingestion. An experimental teratogen. Other experimental reproductive effects. When heated to decomposition it emits toxic fumes of Cl^-.

CLO500 CAS:78372-02-8 ***HR: 3***
1-(6-CHLORO-o-TOLYL)-3-(2-PYRROLIDINYLETHYL)UREA HYDROCHLORIDE
mf: $C_{14}H_{20}ClN_3O$•ClH mw: 318.28

SYN: C 3193

TOXICITY DATA with REFERENCE
eye-rbt 2% MLD ARZNAD 8,664,58
ipr-rat LD50:210 mg/kg ARZNAD 8,664,58
scu-mus LD50:550 mg/kg ARZNAD 8,664,58

SAFETY PROFILE: Poison by intraperitoneal route. Moderately toxic by subcutaneous route. An eye irritant. When heated to decomposition it emits very toxic fumes of Cl^- and NO_x.

CLO750 CAS:569-57-3 ***HR: 3***
CHLOROTRIANISENE
mf: $C_{23}H_{21}ClO_3$ mw: 380.89

SYNS: ANISENE ◇ CHLORESTROLO ◇ 1,1′,1″-(1-CHLORO-1-ETHENYL-2-YLIDENE)-TRIS(4-METHOXYBENZENE) ◇ CHLOROTRIANIZEN ◇ CHLOROTRISIN ◇ CHLOROTRIS(p-METHOXYPHENYL)ETHYLENE ◇ CHLORTRIANISEN ◇ CLORESTROLO ◇ CLOROTRISIN ◇ CTA ◇ HORMONISENE ◇ KHLORTRIANIZEN ◇ MERBENTUL ◇ METACE ◇ NSC-10108 ◇ RIANIL ◇ TACE ◇ TACE-FN ◇ TRI-p-ANISYLCHLOROETHYLENE ◇ TRIS(p-METHOXYPHENYL)CHLOROETHYLENE

TOXICITY DATA with REFERENCE
orl-wmn TDLo:48 mg/kg (14W pre):REP OBGNAS 8,399,56
orl-rat TDLo:37 mg/kg/2Y-C:ETA TXAPA9 11,489,67

CONSENSUS REPORTS: IARC Cancer Review: Animal Inadequate Evidence IMEMDT 21,139,79; Human Limited Evidence IMEMDT 21,139,79.

SAFETY PROFILE: Suspected human carcinogen with experimental tumorigenic data. Human reproductive effects by ingestion: changes in fertility. Used in cancer treatment. When heated to decomposition it emits very toxic fumes of Cl^-.

CLP000 CAS:3151-41-5 ***HR: 3***
CHLOROTRIBENZYLSTANNANE
mf: $C_{21}H_{21}ClSn$ mw: 427.56

SYNS: CHLORID TRIBENZYLCINICITY (CZECH) ◇ TRIBENZYLCHLOROSTANNANE ◇ TRIBENZYLTIN CHLORIDE

TOXICITY DATA with REFERENCE
skn-rbt 500 mg/24H MLD 28ZPAK -,232,72
eye-rbt 5 mg/24H SEV 28ZPAK -,232,72
orl-rat LD50:175 mg/kg 28ZPAK -,232,72

OSHA PEL: TWA 0.1 mg(Sn)/m^3 (skin)
ACGIH TLV: TWA 0.1 mg(Sn)/m^3 (skin) (Proposed: TWA 0.1 mg(Sn)/m^3; STEL 0.2 mg(Sn)/m^3 (skin))
NIOSH REL: (Organotin Compounds) TWA 0.1 mg(Sn)/m^3

SAFETY PROFILE: Poison by ingestion. A skin and severe eye irritant. See also TIN COMPOUNDS. When heated to decomposition it emits toxic fumes of Cl^-.

CLP250 CAS:2117-36-4 ***HR: 2***
CHLOROTRIBUTYLGERMANIUM
mf: $C_{12}H_{27}ClGe$ mw: 279.43

TOXICITY DATA with REFERENCE
ipr-rat LDLo:1970 mg/kg CHDDAT 262,1302,66
ipr-mus LDLo:1280 mg/kg CHDDAT 262,1302,66

CONSENSUS REPORTS: Reported in EPA TSCA Inventory.

SAFETY PROFILE: Moderately toxic by intraperitoneal route. When heated to decomposition it emits very toxic fumes of Cl^-. See also GERMANIUM COMPOUNDS.

CLP500 CAS:1461-22-9 ***HR: 3***
CHLOROTRIBUTYLSTANNANE
mf: $C_{12}H_{27}ClSn$ mw: 325.53

SYNS: CHLORID TRI-n-BUTYLCINICITY (CZECH) ◇ TRIBUTYLCHLOROSTANNANE ◇ TRI-n-BUTYLTIN CHLORIDE ◇ TRI-n-BUTYLZINN-CHLORID (GERMAN)

TOXICITY DATA with REFERENCE
eye-rbt 50 μg/24H SEV 28ZPAK -,231,72
orl-rat LD50:129 mg/kg 28ZPAK -,231,72
orl-mus LD50:117 mg/kg ATXKA8 23,283,68
orl-rbt LDLo:30 mg/kg SAIGBL 15,3,73
skn-rbt LDLo:70 mg/kg SAIGBL 15,3,73

CONSENSUS REPORTS: Reported in EPA TSCA Inventory.

OSHA PEL: TWA 0.1 mg(Sn)/m^3 (skin)
ACGIH TLV: TWA 0.1 mg(Sn)/m^3 (skin) (Proposed: TWA 0.1 mg(Sn)/m^3; STEL 0.2 mg(Sn)/m^3 (skin))
NIOSH REL: (Organotin Compounds) TWA 0.1 mg(Sn)/m^3

SAFETY PROFILE: Poison by ingestion and skin contact. A severe eye irritant. Tributyl tin compounds are extremely toxic to marine life. See also TIN COMPOUNDS. When heated to decomposition it emits toxic fumes of Cl^-.

CLP625 ***HR: 3***
3-CHLORO-3-TRICHLOROMETHYLDIAZIRINE
mf: $C_2Cl_4N_2$ mw: 193.85

SAFETY PROFILE: An extremely shock-sensitive explosive. Upon decomposition it emits toxic fumes of Cl^- and NO_x. See also EXPLOSIVES.

CLP750 CAS:1929-82-4 ***HR: 3***
2-CHLORO-6-(TRICHLOROMETHYL)PYRIDINE
mf: $C_6H_3Cl_4N$ mw: 230.90

SYNS: DOWCO-163 ◇ NITRAPYRIN (ACGIH) ◇ N-SERVE NITROGEN STABILIZER

TOXICITY DATA with REFERENCE
orl-rat LD50:940 mg/kg PCOC** -,819,66
orl-mus LD50:710 mg/kg GUCHAZ 6,122,73
orl-rbt LD50:500 mg/kg FMCHA2 -,C171,83
skn-rbt LD50:850 mg/kg PCOC** -,819,66
orl-ckn LD50:235 mg/kg 28ZEAL 5,166,76

CONSENSUS REPORTS: NCI Carcinogenesis Studies (ipr); Clear Evidence: mouse, rat RRCRBU 52,1,75. Reported in EPA TSCA Inventory.

OSHA PEL: Total Dust: 15 mg/m^3; Respirable Fraction: 5 mg/m^3
ACGIH TLV: TWA 10 mg/m^3; STEL 20 mg/m^3

SAFETY PROFILE: Poison by ingestion. Moderately toxic by skin contact. When heated to decomposition it emits very toxic fumes of Cl^- and NO_x.

CLQ250 CAS:869-24-9 ***HR: 3***
2-CHLOROTRIETHYLAMINE HYDROCHLORIDE
mf: $C_6H_{14}ClN \cdot ClH$ mw: 172.12

SYNS: β-CHLOROETHYLDIETHYLAMINE HYDROCHLORIDE ◇ (2-CHLOROETHYL)DIETHYLAMINE HYDROCHLORIDE ◇ DIETHYLAMINOETHYL CHLORIDE HYDROCHLORIDE ◇ β-DIETHYLAMINOETHYL CHLORIDE HYDROCHLORIDE ◇ DIETHYL-β-CHLOROETHYLAMINEHYDROCHLORIDE

TOXICITY DATA with REFERENCE
mmo-sat 500 μmol/L FNMUDM 3,11,81
mmo-esc 1 μmol/L JPPMAB 31,67P,79
dns-rat:lvr 100 μmol/L ENMUDM 3,11,81
msc-mus:lym 22 μmol/L ENMUDM 3,33,81
ipr-rat LD50:30 mg/kg CPBTAL 8,807,60
orl-mus LDLo:320 mg/kg AECTCV 14,111,85
ipr-mus LD50:71 mg/kg JPETAB 94,249,48
scu-mus LD50:100 mg/kg JPETAB 91,224,47
ivn-mus LD50:100 mg/kg JPETAB 91,224,47
ivn-rbt LDLo:40 mg/kg JPETAB 91,224,47
orl-bwd LD50:42 mg/kg TXAPA9 21,315,72

CONSENSUS REPORTS: Reported in EPA TSCA Inventory. EPA Genetic Toxicology Program.

SAFETY PROFILE: Poison by ingestion, intraperitoneal, subcutaneous, and intravenous routes. Mutation data reported. When heated to decomposition it emits very toxic fumes of Cl^- and NO_x. See also AMINES and CHLORIDES.

CLQ500 CAS:15529-90-5 ***HR: 3***
CHLORO(TRIETHYLPHOSPHINE)GOLD
mf: $C_6H_{15}AuClP$ mw: 350.60

SYNS: SK&F 36914 ◇ TRIETHYLPHOSPHINEAUROUS CHLORIDE

TOXICITY DATA with REFERENCE
dni-hmn:oth 37500 nmol/L BCPCA6 34,3243,85

orl-rbt TDLo:46280 μg/kg (6-18D preg):TER VTPHAK 15(Suppl 5),97,78
orl-rbt TDLo:92560 μg/kg (6-18D preg):REP VTPHAK 15(Suppl 5),97,78
orl-rat LD50:79 mg/kg VTPHAK 15(Suppl 5),1,78
orl-mus LD50:68 mg/kg VTPHAK 15(Suppl 5),1,78

SAFETY PROFILE: Poison by ingestion. Experimental teratogenic and reproductive effects. Human mutation data reported. When heated to decomposition it emits very toxic fumes of Cl^- and PO_x. See also PHOSPHINE and GOLD.

CLQ750 CAS:79-38-9 ***HR: 3***
CHLOROTRIFLUOROETHYLENE
DOT: UN 1082
mf: C_2ClF_3 mw: 116.47

PROP: A gas. Lel: 24%, uel: 40.3%, flash p: −18°F.

SYNS: 1-CHLORO-1,2,2-TRIFLUOROETHYLENE ◇ 2-CHLORO-1,1,2-TRIFLUOROETHYLENE ◇ CHLORTRIFLUORAETHYLEN (GERMAN) ◇ CTFE ◇ DAIFLON ◇ FLUOROPLAST 3 ◇ GENETRON 1113 ◇ MONOCHLOROTRIFLUOROETHYLENE ◇ TRIFLUOROCHLOROETHYLENE (DOT) ◇ 1,1,2-TRIFLUORO-2-CHLOROETHYLENE ◇ TRIFLUOROMONOCHLOROETHYLENE ◇ TRIFLUOROVINYL CHLORIDE ◇ TRITHENE

TOXICITY DATA with REFERENCE
ihl-rat LC50:1000 ppm/4H FLCRAP 1,197,67
orl-mus LD50:268 mg/kg ABMGAJ 21,377,68
ihl-mus LC50:3000 ppm/7H ABMGAJ 21,377,68
ipr-mus LD50:175 mg/kg ABMGAJ 21,377,68
ihl-gpg LC50:4300 mg/m³/4H GTPZAB 21(5),36,77

CONSENSUS REPORTS: Reported in EPA TSCA Inventory.

DOT Classification: Flammable Gas; Label: Flammable Gas.

SAFETY PROFILE: Poison by ingestion and intraperitoneal routes. Moderately toxic by inhalation. Very dangerous fire hazard when exposed to heat, flames (sparks), or oxidizers. To fight fire, stop flow of gas. Violent reaction when mixed with ($Br_2 + O_2$) or (ClF_3 + water). Potentially explosive polymerization reaction with ethylene. Incompatible with 1,1-dichloroethylene; oxygen. When heated to decomposition it emits toxic fumes of F^- and Cl^-. See also CHLORINATED HYDROCARBONS, ALIPHATIC; and FLUORIDES.

CLR000 CAS:425-87-6 ***HR: 3***
2-CHLORO-1,1,2-TRIFLUOROETHYL METHYL ETHER
mf: $C_3H_4ClF_3O$ mw: 148.52

TOXICITY DATA with REFERENCE
eye-rbt 2 mg open SEV AMIHBC 4,119,51
orl-rat LD50:5130 mg/kg AMIHBC 4,119,51
skn-rbt LD50:200 mg/kg AMIHBC 4,119,51

SAFETY PROFILE: Poison by skin contact. Mildly toxic by ingestion. Severe eye irritant. See also ETHERS. When heated to decomposition it emits very toxic fumes of Cl^- and F^-. See also ETHERS, CHLORIDES, and FLUORIDES.

CLR250 CAS:75-72-9 ***HR: 1***
CHLOROTRIFLUOROMETHANE
DOT: UN 1022
mf: $CClF_3$ mw: 104.46

PROP: Colorless gas, ethereal odor. Mp: −181°, bp: −80°.

SYNS: ARCTON 3 ◇ F 13 ◇ FREON 13 ◇ GENETRON 13 ◇ HALOCARBON 13/UCON 13 ◇ MONOCHLOROTRIFLUOROMETHANE (DOT) ◇ R 13 ◇ TRIFLUOROCHLOROMETHANE (DOT) ◇ TRIFLUOROMETHYL CHLORIDE ◇ TRIFLUOROMONOCHLOROCARBON

CONSENSUS REPORTS: Reported in EPA TSCA Inventory.

DOT Classification: Nonflammable Gas; Label: Nonflammable Gas.

SAFETY PROFILE: A mild irritant. Narcotic in high concentrations. Reacts violently with Al. When heated to decomposition it emits highly toxic fumes of F^- and Cl^-.

CLR825 CAS:58911-30-1 ***HR: 3***
3-CHLORO-3-TRIFLUOROMETHYLDIAZIRINE
mf: $C_2ClF_3N_2$ mw: 144.48

SAFETY PROFILE: Potentially explosive. When heated to decomposition it emits toxic fumes of F^-, Cl^-, and NO_x.

CLS000 CAS:6294-93-5 ***HR: 3***
4-CHLORO-3-TRIFLUOROMETHYLPHENOL
mf: $C_7H_4ClF_3O$ mw: 196.56

SYN: p-CHLORO-m-TRIFLUOROMETHYLPHENOL

TOXICITY DATA with REFERENCE
orl-mus LD50:630 mg/kg 11FYAN 3,84,63
ivn-mus LD50:64 mg/kg 11FYAN 3,84,63

CONSENSUS REPORTS: Chlorophenols on the Community Right-To-Know List.

SAFETY PROFILE: Poison by intravenous route. Moderately toxic by ingestion. When heated to decomposition it emits very toxic fumes of Cl^- and F^-. See also CHLOROPHENOLS.

CLS125 CAS:25238-02-2 ***HR: 3***
2-CHLORO-N,N,Né-TRIFLUOROPROPIONAMIDINE
mf: $C_3H_4ClF_3N_2$ mw: 160.53

SAFETY PROFILE: A shock-sensitive explosive. When heated to decomposition it emits toxic fumes of F^-, Cl^-, and NO_x. See also EXPLOSIVES.

CLS250 CAS:17230-87-4 ***HR: 3***
4-(4-(4-CHLORO-α,α,α-TRIFLUORO-m-TOLYL)-4-HYDROXYPIPERIDINO)BUTYROPHENONE-4-FLUOROHYDROCHLORIDE
mf: $C_{22}H_{22}ClF_4NO_2 \cdot ClH$ mw: 480.36

SYNS: CLOFLUPEROL HYDROCHLORIDE ◇ R 9298 ◇ SEPERIDOL ◇ SEPEROL

TOXICITY DATA with REFERENCE
orl-rat LD50:195 mg/kg 27ZQAG -,186,72
scu-rat LD50:69 mg/kg 27ZQAG -,186,72
ivn-rat LD50:17 mg/kg 27ZQAG -,186,72
scu-mus LD50:47 mg/kg 27ZQAG -,186,72
ivn-mus LD50:19 mg/kg 27ZQAG -,186,72

SAFETY PROFILE: Poison by ingestion, subcutaneous, and intravenous routes. When heated to decomposition it emits very toxic fumes of Cl^-, F^-, and NO_x.

CLS500 CAS:7342-38-3 ***HR: 3***
CHLORO(TRIISOBUTYL)STANNANE
mf: $C_{12}H_{27}ClSn$ mw: 325.53

PROP: Solid. D: 1.1290 @ 34°, mp: 30.2°, bp: 174° @ 13 mm.

SYN: TRIISOBUTYLTIN CHLORIDE

TOXICITY DATA with REFERENCE
ivn-mus LD50:5 mg/kg CSLNX* NX#05523

OSHA PEL: TWA 0.1 mg(Sn)/m^3 (skin)
ACGIH TLV: TWA 0.1 mg(Sn)/m^3 (skin) (Proposed: TWA 0.1 mg(Sn)/m^3; STEL 0.2 mg(Sn)/m^3 (skin))
NIOSH REL: (Organotin Compounds) TWA 0.1 mg(Sn)/m^3

SAFETY PROFILE: Poison by intravenous route. Tributyl tin compounds are very toxic to marine life. See also TIN COMPOUNDS. When heated to decomposition it emits toxic fumes of Cl^-.

CLS750 ***HR: 3***
CHLOROTRIMETHYLSILANE
mf: C_3H_9ClSi mw: 108.64

PROP: Flash p: −4°F.

SAFETY PROFILE: Reacts violently with water or hexafluoroisopropylideneaminolithium. A very dangerous fire hazard when exposed to heat or flame. When heated to decomposition it emits toxic fumes of Cl^-. See also CHLOROSILANES.

CLT000 CAS:1066-45-1 ***HR: 3***
CHLOROTRIMETHYLSTANNANE
mf: C_3H_9ClSn mw: 199.26

PROP: Mp: 37°.

SYNS: CHLOROTRIMETHYLTIN ◇ TRIMETHYLCHLOROSTANNANE ◇ TRIMETHYLCHLOROTIN ◇ TRIMETHYLSTANNYL CHLORIDE ◇ TRIMETHYLTIN CHLORIDE

TOXICITY DATA with REFERENCE
dni-rbt:oth 10 μg/L JTEHD6 16,229,85
orl-rat TDLo:9066 μg/kg (14D pre-21D post):REP NTOTDY 4,539,82
orl-rat LD50:12600 μg/kg AJPAA4 97,59,79
ipr-rat LD50:7450 μg/kg NETOD7 4,127,82
ivn-mus LD50:1800 μg/kg CSLNX* NX#02983

CONSENSUS REPORTS: EPA Extremely Hazardous Substances List. Reported in EPA TSCA Inventory.

OSHA PEL: TWA 0.1 mg(Sn)/m^3 (skin)
ACGIH TLV: TWA 0.1 mg(Sn)/m^3 (skin) (Proposed: TWA 0.1 mg(Sn)/m^3; STEL 0.2 mg(Sn)/m^3 (skin))
NIOSH REL: (Organotin Compounds) TWA 0.1 mg(Sn)/m^3

SAFETY PROFILE: A deadly poison by intravenous route. Experimental reproductive effects. See also TIN COMPOUNDS. When heated to decomposition it emits toxic fumes of Cl^-.

CLT250 CAS:1943-16-4 ***HR: 3***
CHLOROTRINITROMETHANE
mf: $CClN_3O_6$ mw: 185.49

TOXICITY DATA with REFERENCE
ipr-mus LD50:29300 μg/kg KHFZAN 10(6),53,76
ipr-mam LDLo:500 mg/kg COREAF 171,1396,20
ihl-mam LCLo:5 g/m^3 COREAF 171,1396,20

SAFETY PROFILE: Poison by intraperitoneal route. Mildly toxic by inhalation. Potentially explosive. When heated to decomposition it emits very toxic fumes of Cl^- and NO_x.

CLT500 CAS:76-83-5 ***HR: 3***
CHLOROTRIPHENYLMETHANE
mf: $C_{19}H_{15}Cl$ mw: 278.79

SYN: TRITYL CHLORIDE

TOXICITY DATA with REFERENCE
ivn-mus LD50:180 mg/kg CSLNX* NX#04021

CONSENSUS REPORTS: Reported in EPA TSCA Inventory.

SAFETY PROFILE: Poison by intravenous route. When heated to decomposition it emits toxic fumes of Cl^-. See also CHLORINATED HYDROCARBONS, AROMATIC.

CLU000 CAS:639-58-7 ***HR: 3***
CHLOROTRIPHENYLSTANNANE
mf: $C_{18}H_{15}ClSn$ mw: 385.47

PROP: Colorless crystals, insol in water, sol in organic solvents. Mp: 106°, bp: 240° @ 13.5 mm.

SYNS: AQUATIN ◇ BRESTANOL ◇ CHLOROTRIPHENYLTIN ◇ FENTIN CHLORIDE ◇ GC 8993 ◇ GENERAL CHEMICALS 8993 ◇ HOE 2872 ◇ LS 4442 ◇ TINMATE ◇ TPTC ◇ TRIPHENYLCHLOROSTANNANE ◇ TRIPHENYLCHLOROTIN ◇ TRIPHENYLTIN CHLORIDE

TOXICITY DATA with REFERENCE
orl-rat TDLo:380 mg/kg (19D male):REP JEENAI 61,32,68
orl-rat LD50:135 mg/kg FMCHA2 -,C245,83
orl-mus LD50:18 mg/kg FMCHA2 -,C245,83
ivn-mus LD50:18 mg/kg CSLNX* NX#01649

CONSENSUS REPORTS: Reported in EPA TSCA Inventory. EPA Extremely Hazardous Substances List.

OSHA PEL: TWA 0.1 mg(Sn)/m^3 (skin)
ACGIH TLV: TWA 0.1 mg(Sn)/m^3 (skin) (Proposed: TWA 0.1 mg(Sn)/m^3; STEL 0.2 mg(Sn)/m^3 (skin))
NIOSH REL: (Organotin Compounds) TWA 0.1 mg(Sn)/m^3

SAFETY PROFILE: Poison by ingestion and intravenous routes. Experimental reproductive effects. An insect chemosterilant. See also TIN COMPOUNDS. When heated to decomposition it emits toxic fumes of Cl^-.

CLU250 CAS:2279-76-7 ***HR: 3***
CHLOROTRIPROPYLSTANNANE
mf: $C_9H_{21}ClSn$ mw: 283.44

PROP: Colorless liquid. D: 1.2678 @ 28°, mp: −23.5°. Sol in organic solvents.

SYNS: TRIPROPYLTIN CHLORIDE ◇ TRI-n-PROPYLTIN CHLORIDE

TOXICITY DATA with REFERENCE
ivn-mus LD50:4 mg/kg CSLNX* NX#02220

OSHA PEL: TWA 0.1 mg(Sn)/m^3 (skin)
ACGIH TLV: TWA 0.1 mg(Sn)/m^3 (skin) (Proposed: TWA 0.1 mg(Sn)/m^3; STEL 0.2 mg(Sn)/m^3 (skin))
NIOSH REL: (Organotin Compounds) TWA 0.1 mg(Sn)/m^3

SAFETY PROFILE: Poison by intravenous route. See also TIN COMPOUNDS. When heated to decomposition it emits toxic fumes of Cl^-.

CLU500 CAS:10008-90-9 ***HR: 3***
CHLORO(TRIVINYL)STANNANE
mf: C_6H_9ClSn mw: 235.29

SYN: TRIVINYLTIN CHLORIDE

TOXICITY DATA with REFERENCE
ivn-mus LD50:40 mg/kg CSLNX* NX#05524

OSHA PEL: TWA 0.1 mg(Sn)/m^3 (skin)
ACGIH TLV: TWA 0.1 mg(Sn)/m^3 (skin) (Proposed: TWA 0.1 mg(Sn)/m^3; STEL 0.2 mg(Sn)/m^3 (skin))
NIOSH REL: TWA (Organotin Compounds) 0.1 mg(Sn)/m^3

SAFETY PROFILE: Poison by intravenous route. See also TIN COMPOUNDS and CHLORIDES. When heated to decomposition it emits toxic fumes of Cl^-.

CLV000 CAS:541-25-3 ***HR: 3***
CHLOROVINYLARSINE DICHLORIDE
mf: $C_2H_2AsCl_3$ mw: 207.31

PROP: Liquid, faint odor of geranium. Bp: 190° decomp, fp: −13°, d: 1.888 @ 20°/4°, vap press: 0.4 mm @ 20°, vap d: 7.15.

SYNS: (2-CHLOROETHENYL) ARSONOUS DICHLORIDE ◇ β-CHLOROVINYLBICHLOROARSINE ◇ 2-CHLOROVINYLDICHLOROARSINE ◇ (2-CHLOROVINYL)DICHLOROARSINE ◇ DICHLORO(2-CHLOROVINYL) ARSINE ◇ LEWISITE ◇ LEWISITE (ARSENIC COMPOUND)

TOXICITY DATA with REFERENCE
skn-hmn 95 μg NTIS** PB158-508
orl-rat TDLo:20 mg/kg (female 6-15D post):TER NTIS** DE88-008303
ihl-hmn LCLo:6 ppm/30M NTIS** PB214-270
skn-hmn LDLo:38 mg/kg 27ZZA9 8,192,57
ihl-rat LC50:580 mg/m^3/1H NTIS** PB158-508
skn-rat LD50:15 mg/kg NTIS** PB158-508
scu-rat LD50:1 mg/kg JPBAA7 58,411,46
ihl-mus LC50:500 mg/m^3/9M NTIS** PB158-508
skn-mus LD50:15 mg/kg NTIS** PB158-508
ihl-dog LC50:1400 mg/m^3/15M NTIS** PB158-508
skn-dog LD50:15 mg/kg JPBAA7 58,411,46
scu-dog LD50:2 mg/kg JPBAA7 58,411,46
ivn-dog LD50:2 mg/kg NTIS** PB158-508
ihl-cat LC50:30 g/m^3/30M NTIS** PB158-508
ihl-rbt LC50:1200 mg/m^3/8M NTIS** PB158-508

CONSENSUS REPORTS: Reported in EPA TSCA Inventory. EPA Genetic Toxicology Program. EPA Extremely Hazardous Substances List. Arsenic and its compounds are on the Community Right-To-Know List.

SAFETY PROFILE: A human poison by inhalation and skin contact. Poison experimentally by inhalation, skin contact, subcutaneous, intraperitoneal, and intravenous routes. An experimental teratogen. A blistering type military poison. Lewisite is absorbed through skin, as little as 2 mL on the skin can cause death. Has a delayed action similar to distilled mustard gas. This gas exhibits a systemic poisoning effect on humans. When heated to decomposition it emits toxic fumes of Cl^- and As. See also ARSENIC COMPOUNDS.

CLV250 CAS:64049-11-2 ***HR: 3***
(2-CHLOROVINYL)DIETHOXYARSINE
mf: $C_6H_{12}AsClO_2$ mw: 226.55

TOXICITY DATA with REFERENCE
ihl-mus LCLo:500 mg/m^3 NDRC** -,7,43
skn-mus LDLo:80 mg/kg NDRC** -,24,42

CONSENSUS REPORTS: Arsenic and its compounds are on the Community Right-To-Know List.

SAFETY PROFILE: Poison by skin contact. Moderately toxic by inhalation. See also ARSENIC COMPOUNDS. When heated to decomposition it emits very toxic fumes of As and Cl^-.

CLV375 CAS:311-47-7 ***HR: 3***
2-CHLOROVINYL DIETHYL PHOSPHATE
mf: $C_6H_{12}ClO_4P$ mw: 214.60

SYNS: COMPOUND 1836 ◇ DIETHYL-2-CHLOROVINYL PHOSPHATE ◇ O,O-DIETHYL-O-(2-CHLOROVINYL) PHOSPHATE ◇ OS 1836 ◇ SD 1836 ◇ SHELL OS 1836

TOXICITY DATA with REFERENCE
orl-rat LD50:10 mg/kg AMIHBC 9,45,54
ihl-rat LC50:22 ppm AMIHBC 9,45,54
ipr-rat LD50:9 mg/kg AMIHBC 9,45,54
orl-mus LD50:32 mg/kg PAREAQ 11,636,59
orl-rbt LD50: 3mg/kg AMIHBC 9,45,54
skn-rbt LD50:18 mg/kg AMIHBC 9,45,54

SAFETY PROFILE: Poison by inhalation, skin contact, ingestion, and intraperitoneal routes. When heated to decomposition it emits toxic fumes of Cl^- and PO_x.

CLV500 CAS:556-97-8 ***HR: D***
1-CHLORO-3,5-XYLENE
mf: C_8H_9Cl mw: 140.62

SYNS: 5-CHLORO-m-XYLENE ◇ 5-CHLORO-1,3-XYLENE ◇ m-XYLENE, 5-CHLORO-

TOXICITY DATA with REFERENCE
scu-rat TDLo:20 mg/kg (female 4D pre):REP JSICAZ 19,264,60

SAFETY PROFILE: Experimental reproductive effects. When heated to decomposition it emits toxic fumes of Cl^-.

CLW000 CAS:88-04-0 ***HR: 3***
4-CHLORO-3,5-XYLENOL
mf: C_8H_9ClO mw: 156.62

PROP: Crystals, phenolic odor, sltly water-sol. Mp: 115.5°, bp: 246°.

SYNS: BENZYTOL ◇ 4-CHLORO-3,5-DIMETHYLPHENOL ◇ CHLORO-XYLENOL ◇ p-CHLORO-m-XYLENOL ◇ DESSON ◇ DETTOL ◇ ESPADOL ◇ HUSEPT EXTRA ◇ OTTASEPT ◇ OTTASEPT EXTRA ◇ PCMX ◇ RBA 777

TOXICITY DATA with REFERENCE
orl-rat TDLo:17100 mg/kg (1-19D preg):TER AOISDR (45),100,83
orl-rat LD50:3830 mg/kg JACTDZ 4(5),147,85
orl-mus LDLo:1600 mg/kg AECTCV 14,111,85
ipr-mus LD50:115 mg/kg JAPMA8 41,595,52

CONSENSUS REPORTS: Reported in EPA TSCA Inventory. Chlorophenols are on the Community Right-To-Know List.

SAFETY PROFILE: Poison by intraperitoneal route. Moderately toxic by ingestion. An experimental teratogen. An antimicrobial agent. See also CHLOROPHENOLS; and CHLORINATED HYDROCARBONS, AROMATIC. When heated to decomposition it emits toxic fumes of Cl^-.

CLW250 CAS:50892-23-4 ***HR: 3***
(4-CHLORO-6-(2,3-XYLIDINO)-2-PYRIMIDINYLTHIO)ACETIC ACID
mf: $C_{14}H_{14}ClN_3O_2S$ mw: 323.82

SYNS: ((4-CHLORO-6-((2,3-DIMETHYLPHENYL)AMINO)-2-PYRIMIDINYL)THIO)ACETIC ACID ◇ WY-14,643

TOXICITY DATA with REFERENCE
dns-rat:lvr 1 mmol/L CALEDQ 24,147,84
dni-mus:oth 100 μmol/L CNREA8 40,36,80
orl-rat TDLo:29 g/kg/69W-C:CAR CNREA8 39,152,79
orl-mus TDLo:37 g/kg/62W-C:CAR CNREA8 39,152,79
orl-rat TD:27 g/kg/64W-C:ETA CALEDQ 32,33,86

SAFETY PROFILE: Suspected carcinogen with experimental carcinogenic and tumorigenic data. Mutation data reported. When heated to decomposition it emits very toxic fumes of Cl^-, NO_x, and SO_x.

CLW500 CAS:65089-17-0 ***HR: 3***
2-((4-CHLORO-6-(2,3-XYLIDINO)-2-PYRIMIDINYL)THIO)-N-(2-HYDROXYETHYL)ACETAMIDE
mf: $C_{16}H_{19}ClN_4O_2S$ mw: 366.90

SYNS: BR-931 ◇ PIRINIXIL

TOXICITY DATA with REFERENCE
dns-rat:lvr 100 μmol/L CALEDQ 24,147,84
dni-mus:oth 25 μmol/L CNREA8 40,36,80

orl-rat TDLo:17 g/kg/81W-C:CAR NATUAS 283,397,80
orl-mus TDLo:137 g/kg/81W-C:CAR NATUAS 283,397,80

SAFETY PROFILE: Suspected carcinogen with experimental carcinogenic data. Mutation data reported. When heated to decomposition it emits very toxic fumes of Cl^-, NO_x, and SO_x.

CLW625 CAS:30544-72-0 ***HR: 3***
4-(p-CHLORO-N-2,6-XYLYLBENZAMIDO) BUTYRIC ACID
mf: $C_{19}H_{29}ClNO_3$ mw: 345.85

SYNS: B 66347 ◇ B 67347 ◇ N-(p-CHLORBENZOYL)-Γ-(2,6-DIMETHYLANILINO)-BUTTERSAEURE(GERMAN)

TOXICITY DATA with REFERENCE
orl-rat LD50:1900 mg/kg GWXXBX #1917036
ivn-rat LD50:300 mg/kg GWXXBX #1917036
orl-mus LD50:710 mg/kg GWXXBX #1917036
ipr-mus LD50:305 mg/kg GWXXBX #1917036

SAFETY PROFILE: Poison by intravenous and intraperitoneal routes. Moderately toxic by ingestion. When heated to decomposition it emits toxic fumes of Cl^- and NO_x.

CLX000 CAS:54749-90-5 ***HR: 3***
CHLOROZOTOCIN
mf: $C_9H_{16}ClN_3O_7$ mw: 313.73

SYNS: 1-(2-CHLOROETHYL)-3-(d-GLUCOPYRANOS-2-YL)-1-NITROSOUREA ◇ 2-((((2-CHLOROETHYL)NITROSOAMINO)CARBONYL)AMINO)-2-DEOXY-d-GLUCOPYRANOSE◇ 2-((((2-CHLOROETHYL)NITROSOAMINO)CARBONYL)AMINO)-2-DEOXY-d-GLUCOSE ◇ 2-(3-(2-CHLOROETHYL)-3-NITROSOUREIDO)-2-DEOXY-d-GLUCOSOPYRANOSE ◇ 2-(3-(2-CHLOROETHYL)-3-NITROSOUREIDO)-d-GLUCO-PYRANOSE ◇ CHLZ ◇ CZT ◇ DCNU ◇ NSC 178248 ◇ NSC D 254157

TOXICITY DATA with REFERENCE
mmo-sat 100 nmol/plate JJIND8 65,149,80
mma-sat 100 mg/L/1H MUREAV 40,281,76
dnd-rat-ipr 100 μmol/kg CNREA8 44,514,84
sce-rat:oth 1 μmol/L CNREA8 43,473,83
dnd-mam:lym 10 mmol/L CNREA8 44,1887,84
ipr-rat TDLo:34 mg/kg/85W-I:CAR CALEDQ 8,133,79
ivn-rat TDLo:16 mg/kg/60W-I:ETA DTESD7 8,273,80
ivn-man TDLo:500 mg/kg:CNS,GIT,BLD CANCAR 46,2365,80
ipr-rat LD50:28 mg/kg CALEDQ 8,133,79
ivn-rat LD50:22500 μg/kg ONCOBS 37,177,80
ipr-mus LD50:35 mg/kg INSSDM 19,123,81
scu-mus LD50:66230 μg/kg NCISP* JAN86
ivn-mus LD10:15 mg/kg GANNA2 71,686,80

CONSENSUS REPORTS: EPA Genetic Toxicology Program.

SAFETY PROFILE: Poison by subcutaneous, intravenous, and intraperitoneal routes. Human systemic effects by intravenous route: anorexia, nausea or vomiting, and thrombocytopenia (decrease in the number of blood platelets). Questionable carcinogen with experimental carcinogenic and tumorigenic data. Mutation data reported. When heated to decomposition it emits very toxic fumes of Cl^- and NO_x. See also NITROSAMINES.

CLX250 CAS:633-59-0 ***HR: 3***
CHLORPERPHENTHIXENE DIHYDROCHLORIDE
mf: $C_{22}H_{25}ClN_2OS•2ClH$ mw: 473.92

SYNS: AY 62021 ◇ 4-(3-(2-CHLOROTHIOXANTHEN-9-YLIDENE)PROPYL)-1-PIPERAZINEETHANOLDIHYDROCHLORIDE ◇ CHLORPENTHIXOL DIHYDROCHLORIDE ◇ CIATYL ◇ CLOPENTHIXOL DIHYDROCHLORIDE ◇ CLOPIXOL ◇ N-746 ◇ SORDENAC ◇ SORDINOL

TOXICITY DATA with REFERENCE
orl-rat LD50:660 mg/kg 27ZQAG -,67,72
ipr-rat LD50:105 mg/kg 27ZQAG -,67,72
ivn-rat LD50:125 mg/kg 27ZQAG -,67,72
orl-mus LD50:560 mg/kg 27ZQAG -,67,72
ipr-mus LD50:222 mg/kg 27ZQAG-,67,72
ivn-mus LD50:111 mg/kg USXXAM #3996211

SAFETY PROFILE: Poison by intraperitoneal and intravenous routes. Moderately toxic by ingestion. When heated to decomposition it emits very toxic fumes of SO_x, NO_x, and Cl^-.

CLY250 CAS:461-78-9 ***HR: 3***
CHLORPHENTERMINE
mf: $C_{10}H_{14}ClN$ mw: 183.70

SYNS: p-CHLORO-α,α-DIMETHYLPHENETHYLAMINE ◇ CHLOROPHENTERMINE ◇ β-(p-CHLOROPHENYL)-α,α-DIMETHYLETHYLAMINE ◇ 1-(p-CHLOROPHENYL)-2-METHYL-2-AMINOPROPANE ◇ CHLORPHENTERAMINE

TOXICITY DATA with REFERENCE
dns-rat-unr 300 mg/kg/5D-C 40QBA3 -,459,78
oms-rat-unr 300 mg/kg/5D-C 40QBA3 -,459,78
scu-rat TDLo:480 mg/kg (10-12D preg):TER VAAZA2 12,295,73
orl-man TDLo:5357 μg/kg:ANS,CVS THERAP 34,205,79
orl-rat LD50:250 mg/kg NYKZAU 65(6),218S,69
orl-mus LD50:270 mg/kg APPHAX 26,598,69
ipr-mus LD50:150 mg/kg APSXAS 15,87,78
scu-mus LD50:260 mg/kg APPHAX 26,598,69
ivn-mus LD50:56 mg/kg CSLNX* NX#00697

SAFETY PROFILE: Poison by ingestion, intraperitoneal, subcutaneous, and intravenous routes. Human systemic effects by ingestion: blood pressure elevation and sympathetic nervous system stimulation. An experimental teratogen. Mutation data reported. An anorectic drug which diminishes the appetite. When heated to decomposition it emits very toxic fumes of Cl^- and NO_x.

CLY500 CAS:52-86-8 ***HR: 3***
γ-(4-(p-CHLORPHENYL)-4-HYDROXPIPERIDINO)-p-FLUORBUTYROPHENONE
mf: $C_{21}H_{23}ClFNO_2$ mw: 375.90

SYNS: ALDO ◇ ALOPERIDIN ◇ ALOPERIDOLO ◇ BROTOPON ◇ 4-(4-(4-CHLOROPHENYL)-4-HYDROXY-1-PIPERIDINYL)-1-(4-FLUORO PHENYL)-1-BUTANONE ◇ EINALON S ◇ EUKYSTOL ◇ 1-(3-p-FLUOROBENZOYLPROPYL)-4-p-CHLOROPHENYL-4-HYDROXYPIPERIDINE ◇ 4′-FLUORO-4-(4-HYDROXY-4-(4′-CHLOROPHENYL) PIPERIDINO)BUTYROPHENONE ◇ GALOPERIDOL ◇ HALDOL ◇ HALOPERIDOL ◇ HALOSTEN ◇ 4-(4-HYDROXY-4′-CHLORO-4-PHENYLPIPERIDINO)-4′-FLUOROBUTYROPHENONE ◇ KESELAN ◇ LEALGIN COMPOSITUM ◇ LINTON ◇ PELUCES ◇ PERNOX ◇ R 1625 ◇ SERENACE ◇ SERNAS ◇ SERNEL ◇ ULCOLIND ◇ ULIOLIND ◇ VESALIUM

TOXICITY DATA with REFERENCE
mma-sat 100 nmol/plate CRNGDP 3,223,82
cyt-hmn:fbr 10 g/L AMBUCH 6,42,79
orl-wmn TDLo:14700 μg/kg (1-7W preg):REP JAMAAP 231,62,75
orl-wmn TDLo:14700 μg/kg (1-7W preg):TER JAMAAP 231,62,75
ipr-mus TDLo:25 mg/kg/5D-C:CAR CRNGDP 3,223,82
ipr-mus TD:50 mg/kg/10D-C:CAR CRNGDP 3,223,82
orl-man TDLo:9 mg/kg/30W-I:PNS BIPCBF 22,111,87
orl-man TDLo:480 μg/kg/6D-I AJPSAO 142,389,85
orl-cld TDLo:72 μg/kg AJPSAO 143,1176,85
orl-wmn TDLo:100 μg/kg/10D-I:BLD JAGSAF 35,248,87
orl-hmn TDLo:71 μg/kg JCPYDR 5,120,85
unr-cld TDLo:375 μg/kg/3D LANCAO 2,479,80
unr-hmn TDLo:9800 μg/kg/28D ARZNAD 32,911,82
unr-man TDLo:500 μg/kg/5D-I JAMAAP 250,485,83
mul-man TDLo:343 μg/kg SMJOAV 76,546,83
mul-man TDLo:1 mg/kg/1D-I JCPYDR 3,338,83
orl-rat LD50:128 mg/kg ARZNAD 24,45,74
ipr-rat LD50:27 mg/kg 27ZQAG -,190,72
scu-rat LD50:60 mg/kg NIIRDN 6,594,82
ivn-rat LD50:15 mg/kg NIIRDN 6,594,82
orl-mus LD50:71 mg/kg FRPSAX 31,442,76
ipr-mus LD50:30 mg/kg BCFAAI 111,293,72
scu-mus LD50:41 mg/kg OYYAA2 1,74,67
ivn-mus LD50:13 mg/kg ARZNAD 11,932,61

CONSENSUS REPORTS: EPA Genetic Toxicology Program.

SAFETY PROFILE: Poison by ingestion, intraperitoneal, intravenous, and subcutaneous routes. Human systemic effects by ingestion: fasciculations, agranulocytosis. A human teratogen by ingestion which causes developmental abnormalites of the musculoskeletal and cardiovascular (circulatory) systems, and abnormal conditions of newborn at birth. Human mutation data reported. An experimental teratogen. Experimental reproductive effects. Questionable carcinogen with experimental carcinogenic data. A tranquilizer used in the treatment of schizophrenia and agitated psychoses. When heated to decomposition it emits very toxic fumes of F^-, Cl^-, and NO_x.

CLY600 CAS:77-36-1 ***HR: 1***
CHLORPHTHALIDOLONE
mf: $C_{14}H_{11}ClN_2O_4S$ mw: 338.78

SYNS: BENZENESULFONAMIDE,2-CHLORO-5-(2,3-DIHYDRO-1-HYDROXY-3-OXO-1H-ISOINDOL-1-YL)-(9CI) ◇ BENZENESULFONAMIDE, 2-CHLORO-5-(1-HYDROXY-3-OXO-1-ISOINDOLINYL)-◇ CHLOROTHALIDONE ◇ CHLORPHTHALIDONE ◇ CHLORTALIDONE ◇ CHLORTHALIDON ◇ CHLORTHALIDONE ◇ G 33182 ◇ HYGROTON ◇ IGROTON ◇ ISOREN ◇ NATRIURAN ◇ ORADIL ◇ OXODOLIN ◇ PHTHALAMODINE ◇ PHTHALAMUDINE ◇ RENON ◇ SALURETIN ◇ ZAMBESIL

TOXICITY DATA with REFERENCE
orl-man TDLo:30 mg/kg (male 3W pre):REP BMJOAE 281,714,80
orl-hmn TDLo:2587 μg/kg/8D-I:EYE,GIT JAMAAP 258,484,87
orl-man TDLo:5714 μg/kg/4D-I:BPR,GLN JAMAAP 220,1592,72
orl-wmn TDLo:12 mg/kg/6D-I:SYS SMJOAV 79,629,86

SAFETY PROFILE: Human systemic effects by ingestion: BP elevation, hyperglycemia, sodium and chlorine level changes, headache, nausea or vomiting. Experimental reproductive effects. When heated to decomposition it emits toxic fumes of SO_x, NO_x, and Cl^-.

CLY750 CAS:84-01-5 ***HR: 3***
CHLORPROETHAZINE
mf: $C_{19}H_{23}ClN_2S$ mw: 346.95

SYNS: 2-CHLORO-10-(3-DIETHYLAMINOPROPYL)PHENOTHIAZINE ◇ NEURIPLEGE ◇ RP 4909

TOXICITY DATA with REFERENCE
orl-mus LD50:300 mg/kg PSCBAY 2,17,63
ipr-mus LD50:90 mg/kg PSCBAY 2,17,63
scu-mus LD50:325 mg/kg PSCBAY 2,17,63
ivn-mus LD50:80 mg/kg PSCBAY 2,17,63

SAFETY PROFILE: Poison by ingestion, intraperitoneal, subcutaneous, and intravenous route. When heated to decomposition it emits very toxic fumes of Cl^-, NO_x, and SO_x.

CLZ000 CAS:4611-02-3 ***HR: 3***
CHLORPROETHAZINE HYDROCHLORIDE
mf: $C_{19}H_{23}ClN_2S \cdot ClH$ mw: 383.41

SYN: 2-CHLORO-10-(3′-DIETHYLAMINOPROPYL)PHENOTHIAZINE HYDROCHLORIDE

TOXICITY DATA with REFERENCE
orl-mus LD50:300 mg/kg 27ZQAG -,14,72
ipr-mus LD50:90 mg/kg 27ZQAG -,14,72
scu-mus LD50:325 mg/kg 27ZQAG -,14,72
ivn-mus LD50:80 mg/kg 27ZQAG -,14,72

SAFETY PROFILE: Poison by ingestion, intravenous, intraperitoneal, and subcutaneous routes. When heated to decomposition it emits very toxic fumes of Cl^-, NO_x, and SO_x.

CMA000 CAS:5490-31-3 ***HR: 3***
CHLORPROHEPTATRIEN
mf: $C_{20}H_{20}ClN{\cdot}ClH$ mw: 346.32

TOXICITY DATA with REFERENCE
orl-mus LD50:250 mg/kg AIPTAK 144,481,63
ivn-mus LD50:28500 μg/kg AIPTAK 144,481,63

SAFETY PROFILE: Poison by ingestion and intravenous routes. When heated to decomposition it emits very toxic fumes of Cl^- and NO_x.

CMA100 CAS:2921-88-2 ***HR: 3***
CHLORPYRIFOS
DOT: NA 2783
mf: $C_9H_{11}Cl_3NO_3PS$ mw: 350.59

SYNS: BRODAN ◇ O,O-DIAETHYL-O-3,5,6-TRICHLOR-2-PYRIDYLMONOTHIOPHOSPHAT (GERMAN) ◇ O,O-DIETHYL-O-3,5,6-TRICHLORO-2-PYRIDYL PHOSPHOROTHIOATE ◇ DOWCO 179 ◇ DURSBAN ◇ DURSBAN F ◇ ENT 27,311 ◇ ERADEX ◇ LORSBAN ◇ OMS-0971 ◇ PYRINEX ◇ 3,5,6-TRICHLORO-2-PYRIDINOL-O-ESTER with O,O-DIETHYL PHOSPHOROTHIOATE

TOXICITY DATA with REFERENCE
cyt-dmg-orl 50 ppb/3S ENMUDM 5,835,83
orl-mus TDLo:250 mg/kg (female 6-15D post):TER TXAPA9 54,31,80
orl-man TDLo:300 mg/kg:PNS ARTODN 59,176,86
orl-rat LD50:82 mg/kg TXAPA9 14,515,69
ihl-rat LD50:78 mg/kg BECTA6 19,113,78
skn-rat LD50:202 mg/kg TXAPA9 14,515,69
orl-mus LD50:60 mg/kg JESEDU 13,11,78
ihl-mus LD50:94 mg/kg BECTA6 19,113,78
ipr-mus LD50:192 mg/kg TXAPA9 65,144,82
orl-rbt LD50:1000 mg/kg SPEADM 78-1,45,78
skn-rbt LD50:2000 mg/kg GUCHAZ 6,203,73

CONSENSUS REPORTS: EPA Genetic Toxicology Program.

OSHA PEL: TWA 0.2 mg/m^3 (skin)
ACGIH TLV: TWA 0.2 mg/m^3 (skin)
DOT Classification: ORM-A; Label: None.

SAFETY PROFILE: Poison by ingestion, intraperitoneal, skin contact, and inhalation routes. Human systemic effects by ingestion: paresthesia, muscle weakness, coma. An experimental teratogen. Mutation data reported. When heated to decomposition it emits very toxic fumes of Cl^-, NO_x, PO_x, and SO_x.

CMA250 CAS:5598-13-0 ***HR: 2***
CHLORPYRIFOS-METHYL
mf: $C_7H_7Cl_3NO_3PS$ mw: 322.53

SYNS: O,O-DIMETHYL-O-(3,5,6-TRICHLORO-2-PYRIDYL)PHOSPHOROTHIOATE ◇ DOWCO 217 ◇ DURSBAN METHYL ◇ ENT 27,520 ◇ METHYL CHLORPYRIFOS ◇ METHYL DURSBAN ◇ NOLTRAN ◇ NSC 60380 ◇ OMS-1155 ◇ RELDAN ◇ ZERTELL

TOXICITY DATA with REFERENCE
skn-rbt 500 mg/24H MLD TXAPA9 21,369,72
orl-mus TDLo:2184 mg/kg (male 26W pre):REP HOEKAN 23,71,73
orl-rat LD50:1828 mg/kg HOEKAN 23,57,73
skn-rat LD50:3713 mg/kg YKYUA6 35,1315,84
orl-mus LD50:2032 mg/kg HOEKAN 23,57,73
ipr-mus LD50:2325 mg/kg TXAPA9 65,144,82
scu-mus LD50:23800 mg/kg YKYUA6 30,409,79
orl-rbt LD50:2000 mg/kg BESAAT 15,123,69

SAFETY PROFILE: Moderately toxic by ingestion, intraperitoneal, and skin contact. A skin irritant. Experimental reproductive effects. When heated to decomposition it emits very toxic fumes of Cl^-, NO_x, PO_x, and SO_x. A pesticide.

CMA500 CAS:3495-42-9 ***HR: 3***
CHLORQUINOX
mf: $C_8H_2Cl_4N_2$ mw: 267.92

SYNS: LUCEL ◇ 5,6,7,8-TETRACHLOROQUINOXALINE

TOXICITY DATA with REFERENCE
orl-rat LD50:6400 mg/kg 28ZEAL 5,50,76
orl-rbt LD50:3000 mg/kg 28ZEAL 5,50,76
orl-brd LD50:400 mg/kg 28ZEAL 5,50,76

SAFETY PROFILE: Poison by ingestion. When heated to decomposition it emits very toxic fumes of Cl^- and NO_x.

CMA600 CAS:97919-22-7 ***HR: 3***
CHLORSULFAQUINOXALINE
mf: $C_{14}H_{11}ClN_4O_2S$ mw: 334.80

SYNS: 4-AMINO-N-(5-CHLORO-2-QUINOXALINYL)BENZENESULFONAMIDE ◇ BENZENESULFONAMIDE, 4-AMINO-N-(5-CHLORO-2-QUINOXALINYL)- ◇ NSC-339004

TOXICITY DATA with REFERENCE
ivn-rat TDLo:600 mg/kg (male 1D pre):REP NTIS** PB87-128658
ivn-rat LDLo:600 mg/kg NTIS** PB87-128658
ivn-mus LD50:607 mg/kg NTIS** PB87-128658
ivn-dog LDLo:12 mg/kg NTIS** PB87-128658

SAFETY PROFILE: Poison by intravenous route. Experimental reproductive effects. When heated to decomposition it emits toxic fumes of SO_x, NO_x, and Cl^-.

CMA750 CAS:57-62-5 ***HR: 3***
CHLORTETRACYCLINE
mf: $C_{22}H_{23}ClN_2O_8$ mw: 478.92

PROP: Golden yellow crystals. Mp: 168-169°. Sltly sol in water; very sol in aq soln pH 7.65; freely sol in the "cellosolves," dioxane, "Carbitol;" sol in methanol, ethanol, butanol, acetone, ethyl acetate, and benzene; insol in ether and petroleum ether.

SYNS: ACRONIZE ◇ AUREOCINA ◇ AUREOMYCIN ◇ AUREOMYCIN A-377 ◇ AUREOMYKOIN ◇ BIOMITSIN ◇ BIOMYCIN ◇ 7-CHLORO-4-(DIMETHYLAMINO)-1,4,4a,5,5a,6,11,12a-OCTAHYDRO-2-NAPHTHACENECARBOXAMIDE ◇ 7-CHLOROTETRACYCLINE ◇ CHRYSOMYKINE ◇ CTC ◇ DUOMYCIN ◇ FLAMYCIN

TOXICITY DATA with REFERENCE
scu-mus TDLo:372 mg/kg (1-6D preg):REP ASPHAK 23,481,69
orl-rat LDLo:3 g/kg JAFCAU 17,497,69
ipr-rat LDLo:335 mg/kg CLDND*
orl-dog LDLo:750 mg/kg AAGAAW -,595,60
ivn-dog LD50:150 mg/kg HBTXAC 5,52,59
ipr-gpg LDLo:1800 mg/kg ANTBAL 20,793,75
ivn-gpg LDLo:100 mg/kg ANYAA9 51,182,48

CONSENSUS REPORTS: Reported in EPA TSCA Inventory.

SAFETY PROFILE: Poison by intravenous and intraperitoneal routes. Moderately toxic by ingestion. Experimental reproductive effects. When heated to decomposition it emits toxic fumes of Cl^- and NO_x. See also TETRACYCLINE.

CMB000 CAS:64-72-2 ***HR: 3***
CHLORTETRACYCLINE HYDROCHLORIDE
mf: $C_{22}H_{23}ClN_2O_8 \cdot ClH$ mw: 515.38

SYNS: AUREOCICLINA ◇ AUREOCYCLINE ◇ AUREOMYCIN HYDROCHLORIDE ◇ AUXEOMYCIN ◇ CHLOROTETRACYCLINE HYDROCHLORIDE ◇ CLOROTETRACICLINA CLORIDRATO (ITALIAN) ◇ ISPHAMYCIN ◇ NSC-13252 ◇ U-6780

TOXICITY DATA with REFERENCE
scu-mus TDLo:1125 mg/kg (10-18D preg):REP CRSBAW 161,300,67
orl-rat LD50:10300 mg/kg TXAPA9 18,185,71
ivn-rat LD50:100 mg/kg BCFAAI 102,660,63
orl-mus LD50:2740 mg/kg FRPSAX 10,197,55
ipr-mus LD50:197 mg/kg RPOBAR 2,278,70
scu-mus LDLo:1000 mg/kg ANYAA9 51,254,48
ivn-mus LD50:101 mg/kg RPOBAR 2,278,70

SAFETY PROFILE: Poison by intraperitoneal and intravenous routes. Moderately toxic by ingestion and subcutaneous routes. Experimental reproductive effects. When heated to decomposition it emits very toxic fumes of Cl^- and NO_x. See also TETRACYCLINE.

CMB125 CAS:14008-79-8 ***HR: 3***
CHLORTROPBENZYL
mf: $C_{21}H_{21}ClNO \cdot ClH$ mw: 378.37

SYNS: 3-α-((p-CHLORO-α-PHENYLBENZYL)OXY)-1-α-H,5-α-H-TROPANE HYDROCHLORIDE ◇ FC-1 ◇ SL-6057 ◇ TROPINE-4-CHLOROBENZHYDRYL ETHER HYDROCHLORIDE ◇ WY 2149

TOXICITY DATA with REFERENCE
orl-rat LD50:364 mg/kg JPETAB 114,192,55
ipr-rat LD50:58 mg/kg JPETAB 114,192,55
orl-mus LD50:174 mg/kg JPETAB 114,192,55
ipr-mus LD50:32 mg/kg JPETAB 114,192,55
ivn-dog LDLo:28 mg/kg JPETAB 114,192,55

SAFETY PROFILE: Poison by ingestion, intravenous, and intraperitoneal routes. When heated to decomposition it emits toxic fumes of NO_x and HCl.

CMB250 ***HR: 3***
CHLORYL HYPOFLUORITE
mf: $ClFO_3$ mw: 102.45

SAFETY PROFILE: An explosive. Upon decomposition it emits toxic fumes of F^- and Cl^-. See also CHLORIDES and FLUORIDES.

CMB500 CAS:12442-63-6 ***HR: 3***
CHLORYL PERCHLORATE
mf: Cl_2O_6 mw: 166.91

SAFETY PROFILE: Probably a poison and irritant due to its reactivity. A very powerful oxidant. Explodes when heated or on contact with water or thionyl chloride. Violent or explosive reaction with organic matter (e.g., ethanol; stopcock grease; wood). The least explosive of the chlorine oxide compounds. When heated to decomposition it emits toxic fumes of Cl^-. See also CHLORIDES and PERCHLORATES.

CMB675 CAS:35317-79-4 ***HR: 2***
CHLOTAZOLE
mf: $C_5H_5Cl_3N_2OS$ mw: 247.53

SYNS: KHLOTAZOL ◇ 2,2,2-TRICHLORO-1-(2-THIAZOLYLAMINO)ETHANOL

TOXICITY DATA with REFERENCE
orl-rat LD50:2500 mg/kg PCJOAU 17,519,83
orl-mus LD50:1 g/kg RPTOAN 46,213,83
ipr-mus LD50:708 mg/kg FRXXBL #2400361
orl-cat LD50:700 mg/kg PCJOAU 17,519,83
orl-rbt LD50:700 mg/kg PCJOAU 17,519,83

SAFETY PROFILE: Moderately toxic by ingestion and

intraperitoneal routes. When heated to decomposition it emits toxic fumes of Cl^-, SO_x, and NO_x.

CMB750 CAS:12236-46-3 ***HR: 3***
CHOCOLATE BROWN FB

SYNS: 11660 BROWN ◇ C.I. FOOD BROWN 2

TOXICITY DATA with REFERENCE
ipr-rat LDLo:250 mg/kg FCTXAV 5,159,67
ipr-mus LD50:210 mg/kg FCTXAV 5,159,67

SAFETY PROFILE: Poison by intraperitoneal route.

CMC000 CAS:479-23-2 ***HR: 3***
CHOLANTHRENE
mf: $C_{20}H_{14}$ mw: 254.34

SYNS: BENZ(j)ACEANTHRYLENE ◇ 1,2-DIHYDRO-BENZ(j) ACEANTHRYLENE ◇ 7,8-DIMETHYLENEBENZ(a)ANTHRACENE

TOXICITY DATA with REFERENCE
mma-sat 3 μg/plate MUREAV 119,259,83
msc-ham:lng 10 mg/L CNREA8 44,4993,84
skn-mus TDLo:170 mg/kg/7W-I:ETA PRLBA4 123,343,37
scu-mus TDLo:40 mg/kg:CAR CNREA8 1,695,41

SAFETY PROFILE: Questionable carcinogen with experimental carcinogenic and tumorigenic data. Mutation data reported. When heated to decomposition it emits acrid smoke and irritating fumes.

CMC750 CAS:67-97-0 ***HR: 3***
CHOLECALCIFEROL
mf: $C_{27}H_{44}O$ mw: 384.71

PROP: White crystals; odorless. Insol in water; sol in alc, chloroform, and fatty oils.

SYNS: COLECALCIFEROL ◇ 7-DEHYDROCHOLESTROL, ACTIVATED ◇ DELSTEROL ◇ DEPARAL ◇ D3-VIGANTOL ◇ OLEOVITAMIN D3 ◇ RICKETON ◇ 9,10-SECOCHOLESTA-5,7,10(19)-TRIEN-3-β-OL ◇ TRIVITAN ◇ VIGORSAN ◇ VITAMIN D3 ◇ VITINC DAN-DEE-3

TOXICITY DATA with REFERENCE
scu-rat TDLo:90 mg/kg (12-20D preg):TER FOMOAJ 29,333,70
orl-inf TDLo:39 mg/kg/34W-I BMJOAE 295,1173,87
orl-rat LD50:42 mg/kg TXAPA9 43,125,78
orl-dog LD50:80 mg/kg JAVMA4 193,211,88

CONSENSUS REPORTS: Reported in EPA TSCA Inventory.

SAFETY PROFILE: Poison by ingestion. An experimental teratogen. When heated to decomposition it emits acrid smoke and irritating fumes.

CMC800 ***HR: 3***
CHOLERA ENTEROTOXIN

SYNS: CHOLERA ENTERO-EXOTOXIN ◇ CHOLERAGEN ◇ ENTERO-EXOTOXIN, CHOLERA ◇ ENTEROTOXIN, CHOLERA

TOXICITY DATA with REFERENCE
ivn-mus TDLo:8 μg/kg (female 4-6D post):REP IMLCAV 1,223,72
ivn-mus LD50:260 μg/kg IMLCAV 1,223,72
ivn-mky LDLo:10 μg/kg TOXIA6 18,309,80
ivn-rbt LDLo:100 μg/kg TOXIA6 19,701,81

SAFETY PROFILE: Poison by intravenous route. Experimental reproductive effects. When heated to decomposition it emits acrid smoke and irritating fumes.

CMD000 CAS:80-99-9 ***HR: 3***
5-α-CHOLEST-7-EN-3-β-OL
mf: $C_{27}H_{46}O$ mw: 386.73

SYNS: Δ^7-CHOLESTENOL ◇ 7-CHOLESTEN-3-β-OL ◇ CHOLESTERIN (GERMAN) ◇ LATHOSTEROL

TOXICITY DATA with REFERENCE
scu-mus TDLo:800 mg/kg/4W-I:ETA NATWAY 60,525,73

SAFETY PROFILE: Questionable carcinogen with experimental tumorigenic data. When heated to decomposition it emits acrid smoke and irritating fumes.

CMD250 CAS:3328-25-4 ***HR: 3***
CHOLEST-6-EN-3-β-OL-5-α-HYDROPEROXIDE
mf: $C_{27}H_{46}O_3$ mw: 418.73

SYNS: Δ^6-CHOLESTEN-3-β-OL-5-α-HYDROPEROXIDE ◇ Δ^6-CHOLESTEN-3-β-OL-5-α-HYDROPEROXYD (GERMAN) ◇ CHOLESTEROL-5-α-HYDROPEROXIDE

TOXICITY DATA with REFERENCE
ipr-mus TDLo:200 mg/kg:NEO STRAAA 124,626,64

SAFETY PROFILE: Questionable carcinogen with experimental neoplastigenic data. When heated to decomposition it emits acrid smoke and irritating fumes.

CMD500 CAS:601-54-7 ***HR: 3***
CHOLEST-5-EN-3-ONE
mf: $C_{27}H_{44}O$ mw: 384.71

SYNS: CHOLESTENONE ◇ $\Delta^{(5)}$-CHOLESTENONE ◇ 5-CHOLESTEN-3-ONE ◇ CHOLESTERONE

TOXICITY DATA with REFERENCE
scu-mus TDLo:760 mg/kg/16W:CAR CNREA8 6,403,46

SAFETY PROFILE: Questionable carcinogen with experimental carcinogenic data. When heated to decomposition it emits acrid smoke and irritating fumes.

CMD750 CAS:57-88-5 ***HR: 3***
CHOLESTEROL
mf: $C_{27}H_{46}O$ mw: 386.73

PROP: White or faint yellow, pearly leaflets. Mp: 148.5°, bp: 360° decomp.

SYNS: CHOLEST-5-EN-3-β-OL ◇ Δ^5-CHOLESTEN-3-β-OL ◇ 5-CHOLESTEN-3-β-OL ◇ 5:6-CHOLESTEN-3-β-OL ◇ CHOLESTERIN ◇ CHOLESTEROL BASE H ◇ CHOLESTERYL ALCOHOL ◇ CHOLESTRIN ◇ CHOLESTROL ◇ CORDULAN ◇ DUSOLINE ◇ DUSORAN ◇ DYTHOL ◇ HYDROCERIN ◇ 3-β-HYDROXYCHOLEST-5-ENE ◇ KATHRO ◇ LANOL ◇ NIMCO CHOLESTEROL BASE H ◇ PROVITAMIN D ◇ SUPER HARTOLAN ◇ TEGOLAN

TOXICITY DATA with REFERENCE
mmo-sat 500 μg/plate FCTOD7 20,35,82
dnd-mus:oth 1 μmol/L CJBBDU 62,94,84
orl-rbt TDLo:6 g/kg (1-31D preg):REP JONUAI 102,1681,72
orl-rbt TDLo:6 g/kg (1-31D preg):TER JONUAI 102,1681,72
ipr-rat TDLo:800 mg/kg/43W-I:ETA EMSUA8 3,95,45
scu-mus TDLo:15 g/kg/47W-I:CAR NATUAS 160,270,47

CONSENSUS REPORTS: IARC Cancer Review: Group 3 IMEMDT 7,161,87; Human Inadequate Evidence IMEMDT 31,95,83; Animal Inadequate Evidence IMEMDT 10,99,76. Reported in EPA TSCA Inventory. EPA Genetic Toxicology Program.

SAFETY PROFILE: Experimental teratogenic and reproductive effects. Questionable carcinogen with experimental carcinogenic and tumorigenic data. Mutation data reported. Used in pharmaceutical and dermal preparations as an emulsifying agent. When heated to decomposition it emits acrid smoke and irritating fumes.

CME000 CAS:63019-46-5 ***HR: 3***
CHOLESTEROL ISOHEPTYLATE
mf: $C_{34}H_{58}O_2$ mw: 498.92

SYN: CHOLESTEROL-5-METHYL-1-HEXANOATE

TOXICITY DATA with REFERENCE
scu-mus TDLo:600 mg/kg/72W-I:CAR JNCIAM 19,977,57

SAFETY PROFILE: Questionable carcinogen with experimental carcinogenic data. When heated to decomposition it emits acrid smoke and irritating fumes.

CME250 CAS:3546-10-9 ***HR: 3***
CHOLESTERYL-p-BIS(2-CHLOROETHYL)AMINO PHENYLACETATE
mf: $C_{39}H_{59}Cl_2NO_2$ mw: 644.89

SYNS: (p-BIS(2-CHLOROETHYL)AMINO)PHENYL)ACETATECHOLESTEROL ◇ (p-(BIS(2-CHLOROETHYL)AMINO)PHENYL)ACETIC ACID CHOLESTEROL ESTER ◇ (4-(BIS(2-CHLOROETHYL) AMINO)PHENYL)ACETIC ACID CHOLESTERYL ESTER ◇ 5-CHOLESTEN-3-β-OL3-(p-(BIS(2-CHLOROETHYL)AMINO)PHENYL)ACETATE ◇ FENESTERIN ◇ FENESTRIN ◇ NCI-C01558 ◇ NSC 104469 ◇ PHENESTERINE ◇ PHENESTRIN

TOXICITY DATA with REFERENCE
orl-rat TDLo:780 mg/kg/52W-I:CAR NCITR* NCI-CG-TR-60,78
orl-mus TDLo:1092 mg/kg/1Y-I:CAR NCITR* NCI-CG-TR-60,78
ipr-mus TDLo:2400 mg/kg/8W-I:NEO CNREA8 33,3069,73

CONSENSUS REPORTS: NCI Carcinogenesis Bioassay (gavage); Clear Evidence: mouse, rat NCITR* NCI-CG-TR-60,78.

SAFETY PROFILE: Suspected carcinogen with experimental carcinogenic and neoplastigenic data. When heated to decomposition it emits very toxic fumes of Cl^- and NO_x.

CME400 CAS:11041-12-6 ***HR: 2***
CHOLESTYRAMINE

SYNS: CHOLESTYRAMINE CHLORIDE ◇ CHOLESTYRAMINE RESIN ◇ COLESTYRAMIN ◇ CUEMID ◇ QUANTALAN ◇ QUESTRAN

TOXICITY DATA with REFERENCE
orl-rat TDLo:6 g/kg (17-22D post):REP KSRNAM 16,2078,82
orl-rat TDLo:42 g/(9 W male pre/female 2 W pre/1-7D post):TER KSRNAM 16,2040,82
orl-man TDLo:112 g/kg/W-C:CAR NEJMAG 301,1007,79
orl-inf TDLo:4 g/kg/2D-I:SYS AJDCAI 141,479,87
orl-cld TDLo:46 g/kg/39W-I:NOSE CMAJAX 134,609,86

CONSENSUS REPORTS: Reported in EPA TSCA Inventory.

SAFETY PROFILE: Questionable human carcinogen producing colon tumors. An experimental teratogen. Other experimental reproductive effects. Toxic effects by ingestion: acidosis and nose bleeds. When heated to decomposition it emits acrid smoke and irritating fumes.

CME675 CAS:27959-26-8 ***HR: 1***
CHOLEXAMIN
mf: $C_{34}H_{32}N_4O_9$ mw: 640.70

PROP: Crystals from dilute acetic acid; practically odorless and tasteless. Mp: 177-180°. Sltly sol in water, ethanol, and ether.

SYNS: CHOLEXAMINE ◇ 2-HYDROXYCYCLOHEXANE-1,1,3,3-TETRAMETHANOL TETRAESTER with NICOTINIC ACID ◇ K 31 (pharmaceutical) ◇ NICOMOL ◇ 3-PYRIDINECARBOXYLIC ACID, (2-HYDROXY-1,3-CYCLOHEXANEDIYLIDENE)TETRAKIS(METHYLENE) ESTER ◇ 2,2,6,6-TETRAKIS(NICOTINOYLOXYMETHYL) CYCLOHEXANOL ◇ TETRANICOTINIC ACID-2-HYDROXYCYCLOHEXA-1,1,3,3-TETRAMETHYLESTER

TOXICITY DATA with REFERENCE
orl-rat TDLo:10500 mg/kg (female 35D pre):REP OYYAA2 14,755,77
orl-rat LD50:10 g/kg OYYAA2 14,741,77

SAFETY PROFILE: Mildly toxic by ingestion. Experimental reproductive effects. When heated to decomposition it emits toxic fumes of NO_x. An anticholesteremic agent which reduces the blood cholesterol level. See also ESTERS.

CME750 CAS:81-25-4 ***HR: 3***
CHOLIC ACID
mf: $C_{24}H_{40}O_5$ mw: 408.64

PROP: The most abundant bile acid; the monohydrate crystallizes in plates from dilute acetic acid; sol in glacial acetic acid, acetone, and alc. Sltly sol in chloroform, practically insol in water and benzene.

SYNS: CHOLALIN ◇ CHOLSAEURE (GERMAN) ◇ COLALIN ◇ 3-α,7-α,12-α-TRIHYDROXY-5-β-CHOLAN-24-OIC ACID ◇ 3,7,12-TRIHYDROXY-CHOLAN-24-OIC ACID (3-α,5-β,7-α,12-α) ◇ 3-α,7-α,12-α-TRIHYDROXYCHOLANSAEURE (GERMAN)

TOXICITY DATA with REFERENCE
mmo-sat 50 mg/L MUREAV 158,45,85
sln-smc 400 mg/L CRNGDP 5,447,84
ipr-mus LD50:330 mg/kg ARZNAD 20,323,70

CONSENSUS REPORTS: Reported in EPA TSCA Inventory.

SAFETY PROFILE: Poison by intraperitoneal route. Mutation data reported. When heated to decomposition it emits acrid smoke and irritating fumes.

CMF000 CAS:62-49-7 ***HR: 3***
CHOLINE
mf: $C_5H_{14}NO$ mw: 104.20

SYNS: BILINEURINE ◇ CHOLINE ION ◇ (2-HDYROXYETHYL) TRIMETHYLAMMONIUM ◇ 2-HYDROXY-N,N,N-TRIMETHYL-ETHANAMINIUM

TOXICITY DATA with REFERENCE
ipr-rat LD50:400 mg/kg TXAPA9 12,486,68
orl-cat LD50:6640 mg/kg TXAPA9 12,486,68
scu-cat LDLo:150 mg/kg HBAMAK 4,1289,35
scu-rbt LDLo:800 mg/kg CRSBAW 83,481,20
ivn-rbt LDLo:70 mg/kg CRSBAW 83,481,20
rec-rbt LDLo:460 mg/kg CRSBAW 83,481,20
scu-frg LDLo:1500 mg/kg HBAMAK 4,1289,35

CONSENSUS REPORTS: Reported in EPA TSCA Inventory.

SAFETY PROFILE: Poison by intraperitoneal, subcutaneous, and intravenous routes. Moderately toxic by rectal route. Mildly toxic by ingestion. When heated to decomposition it emits toxic fumes of NO_x and NH_3.

CMF250 CAS:51-84-3 ***HR: 3***
CHOLINE ACETATE (ESTER)
mf: $C_7H_{16}NO_2$ mw: 146.24

SYNS: ACECOLINE ◇ ACETYLCHOLINE ◇ ACETYL CHOLINE ION ◇ 2-(ACETYLOXY)-N,N,N-TRIMETHYLETHANAMINIUM ◇ ACH ◇ ARTEROCOLINE ◇ CHOLINE ACETATE ◇ OVISOT

TOXICITY DATA with REFERENCE
scu-rat LD50:250 mg/kg 27ZIAQ -,29,73
ivn-rat LD50:22 mg/kg JPETAB 58,337,36
orl-mus LD50:3000 mg/kg JPETAB 58,337,36
ipr-mus LD50:170 mg/kg AIPTAK 192,88,71
scu-mus LD50:170 mg/kg JPETAB 58,337,36
ivn-mus LD50:11 mg/kg ATXKA8 29,39,72
ivn-rbt LD50:300 μg/kg 27ZIAQ -,39,73

SAFETY PROFILE: Poison by subcutaneous, intravenous, and intraperitoneal routes. Moderately toxic by ingestion. When heated to decomposition it emits toxic fumes of NO_x. See also CHOLINE and ESTERS.

CMF350 CAS:987-78-0 ***HR: 1***
CHOLINE CYTIDINE DIPHOSPHATE
mf: $C_{14}H_{26}N_4O_{11}P_2$ mw: 488.38

SYNS: CHOLINE, HYDROXIDE, 5'-ESTER with CYTIDINE 5'-(TRIHYDROGEN PYROPHOSPHATE), inner salt ◇ CDP-CHOLIN ◇ CDP-CHOLINE ◇ CDP-COLINA ◇ CEREB ◇ CHOLINE 5'-CYTIDINE DIPHOSPHATE ◇ CITICHOLINE ◇ CITICOLINE ◇ CITIDIN DIFOSFATO de COLINA ◇ CITIDOLINE ◇ COLITE ◇ CYTIDINDIPHOSPHOCHOLIN ◇ CYTIDINE CHOLINE DIPHOSPHATE ◇ CYTIDINE 5'-(CHOLINE DIPHOSPHATE) ◇ CYTIDINE DIPHOSPHATE CHOLINE ◇ CYTIDINE 5'-DIPHOSPHATE CHOLINE ◇ CYTIDINE DIPHOSPHATE CHOLINE ESTER ◇ CYTIDINE DIPHOSPHATE CHOLIN ESTER ◇ CYTIDINE DIPHOSPHOCHOLINE ◇ CYTIDINE 5'-DIPHOSPHOCHOLINE ◇ CYTIDINE DIPHOSPHORYLCHOLINE ◇ CYTIDOLINE ◇ ENSIGN ◇ NICHOLIN ◇ NICOLIN ◇ NITICOLIN ◇ RECOFNAN ◇ RECOGNAN ◇ SOMAZINA ◇ SUNCHOLIN

TOXICITY DATA with REFERENCE
ipr-rat TDLo:7500 mg/kg (male 30D pre):REP OYYAA2 20,109,80
orl-rat LD50:18 g/kg DRUGAY 6,322,82
ipr-rat LD50:5344 mg/kg OYYAA2 20,109,80
scu-rat LD50:8218 mg/kg OYYAA2 20,109,80
ivn-rat LD50:2973 mg/kg OYYAA2 20,109,80
orl-mus LD50:12500 mg/kg DRUGAY 6,322,82
ipr-mus LD50:5393 mg/kg OYYAA2 20,109,80
scu-mus LD50:5800 mg/kg DRUGAY 6,322,82
ivn-mus LD50:4600 mg/kg ARZNAD 33,1033,83

SAFETY PROFILE: Moderately toxic by intravenous route. Experimental reproductive effects. When heated to decomposition it emits toxic fumes of NO_x and PO_x.

CMF400 CAS:999-81-5 ***HR: 2***
CHOLINE DICHLORIDE
mf: $C_5H_{13}ClN \cdot Cl$ mw: 158.09

SYNS: ANTYWYLEGACZ ◇ CCC PLANT GROWTH REGULANT

◇ CE CE CE ◇ 2-CHLORAETHYL-TRIMETHYLAMMONIUMCHLORID ◇ CHLORCHOLINCHLORID ◇ CHLORCHOLINE CHLORIDE ◇ CHLORMEQUAT ◇ CHLORMEQUAT CHLORIDE ◇ CHLOROCHOLINE CHLORIDE ◇ (β-CHLOROETHYL)TRIMETHYLAMMONIUM CHLORIDE ◇ (2-CHLOROETHYL)TRIMETHYLAMMONIUM CHLORIDE ◇ 2-CHLORO-N,N,N-TRIMETHYLETHANAMINIUM CHLORIDE ◇ 60-CS-16 ◇ CYCLOCEL ◇ CYCOCEL ◇ CYCOCEL-EXTRA ◇ CYCOGAN ◇ CYCOGAN EXTRA ◇ CYOCEL ◇ EI 38,555 ◇ ETHANAMINIUM, 2-CHLORO-N,N,N-TRIMETHYL-, CHLORIDE (9CI) ◇ HICO CCC ◇ HORMOCEL-2CCC ◇ INCRECEL ◇ LIHOCIN ◇ NCI-C02960 ◇ RETACEL ◇ STABILAN ◇ TRIMETHYL-β-CHLORETHYLAMMONIUMCHLORID ◇ TUR

TOXICITY DATA with REFERENCE
skn-rbt 500 mg/24H MLD 85JCAE-,616,86
dni-mus-ivg 5 pph JIDEAE 62,378,74
orl-mus TDLo:7100 mg/kg/78W-I:NEO NTIS** PB223-159
orl-hmn LDLo:10 mg/kg:PUL AXVMAW 31,527,77
ivn-hmn LDLo:1 mg/kg:PUL AXVMAW 31,527,77
skn-rat LD50:4000 mg/kg FMCHA2-,C53,83
ipr-rat LD50:64 mg/kg ABMGAJ 33,89,74
ivn-rat LD50:12500 μg/kg AXVMAW 31,527,77
unr-rat LD50:780 mg/kg VINIT* #4758-80
orl-mus LD50:54 mg/kg ABMGAJ 27,663,71
ipr-mus LD50:62 mg/kg ABMGAJ 27,663,71
ivn-mus LD50:7 mg/kg AXVMAW 31,527,77
unr-mus LD50:560 mg/kg VINIT* #4758-80
orl-dog LD50:50 mg/kg AXVMAW 24,1049,70
orl-cat LD50:7 mg/kg AXVMAW 24,1049,70
ivn-rbt LDLo:4 mg/kg AXVMAW 31,527,77

CONSENSUS REPORTS: NTP Carcinogenesis Bioassay (feed): No Evidence: mouse, rat NCITR* NCI-TR-158,79. Reported in EPA TSCA Inventory.

SAFETY PROFILE: Human poison by ingestion and intravenous routes. Moderately toxic by skin contact. Human systemic effects: respiratory depression. Questionable carcinogen with experimental neoplastigenic data. Mutation data reported. When heated to decomposition it emits toxic fumes of Cl^-.

CMF500 CAS:4499-40-5 ***HR: 3***
CHOLINE, with THEOPHYLLINE (1:1)
mf: $C_7H_7N_4O_2{\cdot}C_5H_{14}NO$ mw: 283.38

SYNS: CHOLEDYL ◇ CHOLEGYL ◇ CHOLINE THEOPHYLLINATE ◇ CHOLINE THEOPHYLLINE SALT ◇ CHOLINOPHYLLINE ◇ FILORAL ◇ (2-HYDROXYETHYL)TRIMETHYLAMMONIUM with THEOPHYLLINE ◇ 2-HYDROXY-N,N,N-TRIMETHYLETHANAMINIUM SALT with 3,7-DIHYDRO-1,3-DIMETHYLPURINE-2,6-DIONE ◇ OXTRIMETHYLLINE ◇ OXTRIPHYLLINE ◇ SOLIPHYLLINE ◇ TEOFILCOLINA ◇ TEOKOLIN ◇ THEOPHYLLINE CHOLINATE ◇ THEOPHYLLINE SALT of CHOLINE ◇ THEOXYLLINE ◇ THIOPHYLLINE CHOLINATE ◇ THIOPHYLLINE with CHOLLINE

TOXICITY DATA with REFERENCE
orl-inf TDLo:113 mg/kg/31D:END,KID ADCHAK 53,757,78
orl-wmn TDLo:420 mg/kg:GIT,CNS SMJOAV 71,965,78
orl-rat LD50:600 mg/kg NIIRDN 6,278,82
ipr-rat LD50:185 mg/kg NIIRDN 6,278,82
ims-rat LD50:240 mg/kg NIIRDN 6,278,82
orl-mus LD50:770 mg/kg CLDND*
ivn-mus LD50:112 mg/kg CLDND*
ims-mus LD50:360 mg/kg CLDND*
orl-gpg LD50:210 mg/kg CLDND*
ivn-gpg LD50:118 mg/kg CLDND*
ims-gpg LD50:185 mg/kg CLDND*

SAFETY PROFILE: Poison by ingestion, intravenous, intraperitoneal, and intramuscular routes. Human systemic effects by ingestion: tremors and excitement, nausea or vomiting, hyperglycemia and changes in urine composition. When heated to decomposition it emits toxic fumes of NO_x and NH_3. See also THEOPHYLLINE and CHOLINE.

CMF750 CAS:67-48-1 ***HR: 3***
CHOLINE HYDROCHLORIDE
mf: $C_5H_{14}NO{\cdot}Cl$ mw: 139.65

PROP: Colorless to white hygroscopic crystals; slt odor of trimethylamine. Sol in water and alc.

SYNS: BIOCOLINA ◇ CHLORIDE de CHOLINE (FRENCH) ◇ CHOLINE CHLORHYDRATE ◇ CHOLINE CHLORIDE (FCC) ◇ CHOLINIUM CHLORIDE ◇ HEPACHOLINE ◇ (2-HYDROXYETHYL) TRIMETHYLAMMONIUM CHLORIDE ◇ LIPOTRIL

TOXICITY DATA with REFERENCE
cyt-ham:ovr 500 μg/L ENMUDM 7,1,85
sce-ham:ovr 500 μg/L ENMUDM 7,1,85
orl-rat LD50:3400 mg/kg PSEBAA 58,87,45
ipr-rat LD50:400 mg/kg TXAPA9 12,486,68
orl-mus LD50:3900 mg/kg ARZNAD 33,1016,83
ipr-mus LD50:320 mg/kg PSEBAA 51,281,42
scu-mus LDLo:735 mg/kg JPETAB 6,477,14/15
ivn-mus LD50:53 mg/kg ARZNAD 33,1016,83
ivn-dog LDLo:5 mg/kg HBAMAK 4,1289,35
ivn-cat LDLo:25 mg/kg HBAMAK 4,1289,35
ipr-rbt LDLo:500 mg/kg JIDIAQ 42,473,28
scu-rbt LDLo:1 g/kg PSEBAA 51,281,42
ivn-rbt LDLo:1100 μg/kg PSEBAA 51,281,42

CONSENSUS REPORTS: Reported in EPA TSCA Inventory.

SAFETY PROFILE: Poison by intraperitoneal and intravenous routes. Moderately toxic experimentally by ingestion and subcutaneous routes. Mutation data reported. A lipotropic agent which induces the reduction in fats contained in the liver. When heated to decomposition it emits toxic fumes of Cl^-, SO_x, and NO_x. See also CHOLINE.

CMG000 CAS:2016-36-6 ***HR: 2***
CHOLINE SALICYLATE
mf: $C_{12}H_{19}NO_4$ mw: 241.32

SYNS: ACTASAL ◇ ARRET ◇ ARTHROPAN ◇ ARTROBIONE ◇ CHOLINE SALICYLATE B ◇ CHOLINE, SALICYLATE (SALT) ◇ CHOLINE SALICYLIC ACID SALT ◇ (2-HYDROXYETHYL)TRIMETHYLAMMONIUM SALICYLATE ◇ 2-HYDROXY-N,N,N-TRIMETHYLETHANAMINIUM SALT with 2-HYDROXYBENZOIC ACID (1:1) ◇ MUNDISAL ◇ SALICOL ◇ SALICYLIC ACID CHOLINE SALT ◇ SYRAP

TOXICITY DATA with REFERENCE
orl-rat LD50:1530 mg/kg MEIEDD 10,313,83
orl-mus LD50:2690 NIIRDN 6,291,82
ipr-mus LD50:410 mg/kg NIIRDN 6,291,82
scu-mus LD50:1 g/kg NIIRDN 6,291,82

CONSENSUS REPORTS: Reported in EPA TSCA Inventory.

SAFETY PROFILE: Moderately toxic by ingestion, intraperitoneal, and subcutaneous routes. An analgesic and antipyretic. When heated to decomposition it emits toxic fumes of NO_x and NH_3. See also CHOLINE.

CMG250 CAS:306-40-1 ***HR: 3***
CHOLINE SUCCINATE (2:1) (ESTER)
mf: $C_{14}H_{30}N_2O_4$ mw: 290.46

SYNS: ANECTINE ◇ CHOLINE SUCCINATE (ester) ◇ DIACETYLCHOLINE ◇ DICHOLINE SUCCINATE ◇ 2,2'-((1,4-DIOXO-1,4-BUTANEDIYL)BIS(OXY))BIS(N,N,N-TRIMETHYLETHANAMINIUM ◇ DITILIN ◇ DITILINE ◇ QUELICIN ◇ SUCCINIC ACID DIESTER with CHOLINE ◇ SUCCINOCHOLINE ◇ SUCCINOYLCHOLINE ◇ SUCCINYLBISCHOLINE ◇ SUCCINYLDICHOLINE ◇ SUXAMETHONIUM ◇ SUXEMETHONIUM

TOXICITY DATA with REFERENCE
ivn-hmn TDLo:1430 μg/kg:PUL ANATAE 21,27,72
orl-mus LD50:125 mg/kg 27ZIAQ -,-,65
ipr-mus LD50:2140 μg/kg AIPTAK 152,277,64
scu-mus LD50:7500 μg/kg ARZNAD 15,126,65
ivn-mus LD50:280 μg/kg RCOCB8 1,141,70
ivn-dog LDLo:300 μg/kg AIPTAK 88,1,51
ivn-rbt LD50:800 μg/kg AIPTAK 88,1,51

SAFETY PROFILE: Poison by ingestion, intraperitoneal, intravenous,and subcutaneous routes. Human systemic effects by intravenous route: changes in the trachea or bronchi. When heated to decomposition it emits toxic fumes of NO_x.

CMG675 CAS:9002-61-3 ***HR: D***
CHORIONIC GONADOTROPIN

PROP: Long, thin rods or needles from 60% alc. Freely sol in water; sol in aq glycerol and glycols; insol in the anhydrous organic solvents.

SYNS: AMBINON ◇ ANTUITRIN S ◇ APL ◇ APL (hormone) ◇ APOIDINA ◇ CHORIGON ◇ CHORIGONADOTROPIN ◇ CHORIGONIN ◇ CHORIONIC GONADOTROPHIN ◇ CHORIONIC GONADOTROPIC HORMONE ◇ CHORULON ◇ CORIANTIN ◇ FOLLUTEIN ◇ GONABION ◇ GONADEX ◇ HCG ◇ HUMAN CHORIONIC GONADOTROPIN ◇ KOROTRIN ◇ PHYSEX ◇ PRAEDYN ◇ PREGNYL ◇ PRIMOGONYL ◇ RANDONOS ◇ SYNAPHORIN

TOXICITY DATA with REFERENCE
spm-nml-ipr 18 mg/kg/2D ENDKAC 67,167,76
unr-wmn TDLo:36 mg/kg (6D pre):REP ACENA7 78,332,75
ipr-ham TDLo:60 mg/kg (female 4-6D post):TER ENDOAO 83,217,68

SAFETY PROFILE: Human reproductive effects by an unspecified route: changes in female fertility. An experimental teratogen. Other reproductive effects with experimental animals. Mutation data reported.

CMG700 ***HR: 3***
CHRISTMAS ROSE

PROP: An evergreen perennial herb which grows to 2 feet. The white or pink-white flower has 5 petals and is 2 to 3 inches across. It is native to Europe and grows wild in the northern United States and Canada.

SYNS: HELLEBORE ◇ HELLEBORUS NIGER

SAFETY PROFILE: The whole plant contains the poisons hellebrin, helleborin and helleborein (cardiac glycosides), and the direct irritants saponin and protoanemonin. Ingestion may cause mouth and abdominal pain, nausea, vomiting, and diarrhea. Cardiac glycosides may cause death due to their effect on heart function. See also DIGITALIS and SAPONIN.

CMG750 CAS:1836-22-2 ***HR: D***
CHROMACID FAST RED 3B
mf: $C_{17}H_{12}N_2O_9S_2 \cdot 3Na$ mw: 521.40

SYNS: C.I. 16105 ◇ ERIO CHROME RED PE ◇ EXT. D&C RED No. 2

TOXICITY DATA with REFERENCE
mmo-sat 500 μg/plate MUREAV 56,249,78

CONSENSUS REPORTS: Reported in EPA TSCA Inventory.

SAFETY PROFILE: Mutation data reported. When heated to decomposition it emits very toxic fumes of NO_x, Na_2O, and SO_x.

CMG800 CAS:15005-90-0 ***HR: 3***
CHROMALUM HEXAHYDRATE
mf: $Cr_2O_{12}S_3 \cdot 6H_2O$ mw: 500.30

SYN: CHROMIUM(III) SULFATE, HEXAHYDRATE (2:3:6)

TOXICITY DATA with REFERENCE
ivn-rat LDLo:144 mg/kg AJPHAP 209,489,65

OSHA PEL: TWA 0.5 mg(Cr)/m^3
ACGIH TLV: TWA 0.5 mg(Cr)/m^3

SAFETY PROFILE: Poison by intravenous route. When heated to decomposition it emits toxic fumes of SO_x and Cr.

CMG850 CAS:7788-99-0 ***HR: 2***
CHROME ALUM (DODECAHYDRATE)
mf: $CrKO_8S_2•12H_2O$ mw: 499.41

SYNS: CHROME ALUM ◇ POTASSIUM CHROMIUM ALUM ◇ SULFURIC ACID, CHROMIUM(3+)POTASSIUM SALT(2:1:1), DODECAHYDRATE

TOXICITY DATA with REFERENCE
dnr-esc 125 μg/well MUREAV 133,161,84
scu-rat TDLo:135 mg/kg:ETA PBPHAW 14,47,78
ivn-rat LD50:112 mg/kg EQSFAP 1,1,75

OSHA PEL: TWA 0.5 mg(Cr)/m^3
ACGIH TLV: TWA 0.5 mg(Cr)/m^3

SAFETY PROFILE: Poison by intravenous route. Questionable carcinogen with experimental tumorigenic data. Mutation data reported. When heated to decomposition it emits toxic fumes of Cr^-.

CMH000 CAS:1066-30-4 ***HR: 3***
CHROMIC ACETATE
mf: $C_6H_9O_6•Cr$ mw: 229.15

PROP: Gray, green powder or bluish-green pasty mass.

SYNS: CHROMIC ACETATE(III) ◇ CHROMIUM ACETATE ◇ CHROMIUM(III) ACETATE ◇ CHROMIUM TRIACETATE

TOXICITY DATA with REFERENCE
mmo-esc 16 mmol/L MUREAV 58,175,78
cyt-hmn:leu 16 μmol/L MUREAV 58,175,78
imp-rat TDLo:1000 mg/kg/56W-I:ETA AEHLAU 5,445,62
ivn-mus LDLo:2290 mg/kg EQSSDX 1,1,75
ivn-rbt LDLo:1604 mg/kg EQSSDX 1,1,75
ivn-frg LDLo:6185 mg/kg AIPTAK 62,330,39

CONSENSUS REPORTS: NTP Fifth Annual Report on Carcinogens. IARC Cancer Review: Group 3 IMEMDT 7,165,87; Animal Inadequate Evidence IMEMDT 2,100,73; IMEMDT 23,205,80. Chromium and its compounds are on the Community Right-To-Know List. Reported in EPA TSCA Inventory.

OSHA PEL: TWA 0.5 mg(Cr)/m^3
ACGIH TLV: TWA 0.5 mg(Cr)/m^3
DOT Classification: ORM-E; Label: None.

SAFETY PROFILE: Confirmed carcinogen with experimental tumorigenic data. Moderately toxic by intravenous route. Human mutation data reported. See also CHROMIUM COMPOUNDS. When heated to decomposition it emits acrid smoke and irritating fumes.

CMH250 CAS:7738-94-5 ***HR: 3***
CHROMIC ACID
mf: CrH_2O_4 mw: 118.02

SYNS: ACIDE CHROMIQUE (FRENCH) ◇ CHROMIC(VI) ACID

TOXICITY DATA with REFERENCE
mmo-sat 80 μg/plate MUREAV 54,139,78
dnr-smc 1200 nmol/L CNJGA8 24,771,82
dnr-ssp 1200 nmol/L CNJGA8 24,771,82
scu-dog LDLo:320 mg/kg EQSSDX 1,1,75

CONSENSUS REPORTS: Reported in EPA TSCA Inventory. Chromium and its compounds are on the Community Right-To-Know List.

OSHA PEL: (Transitional: CL 1 mg/10m^3) CL 0.1 mg(CrO_3)/m^3
ACGIH TLV: TWA 0.05 mg(Cr)/m^3, Confirmed Human Carcinogen.
DFG MAK: Animal Carcinogen, Suspected Human Carcinogen.
NIOSH REL: (Chromium(VI)) TWA 0.025 mg(Cr(VI))/m^3; CL 0.05/15M

SAFETY PROFILE: Poison by subcutaneous route. Mutation data reported. A powerful oxidizer. A storage hazard; it may burst a sealed container due to carbon dioxide release. Potentially explosive reactions with oxidizable materials. May ignite on contact with acetone or alcohols. When heated to decomposition it emits acrid smoke and irritating fumes. See also CHROMIUM COMPOUNDS.

CMH500 ***HR: 3***
CHROMIC ACID (mixture)
DOT: NA 1463
mf: CrO_3 mw: 99.98

PROP: Dark red crystals. Mp: 196°, d: 2.70, decomp @ 250° to Cr_2O_3 + O_2, a powerful oxidizer. Water-sol.

SYN: CHROMIC ACID MIXTURE, DRY (DOT)

CONSENSUS REPORTS: Chromium and its compounds are on the Community Right-To-Know List.

OSHA PEL: (Transitional: CL 1 mg/10m^3) CL 0.1 mg(CrO_3)/m^3
ACGIH TLV: TWA 0.05 mg(Cr)/m^3, Confirmed Human Carcinogen.
DFG MAK: Animal Carcinogen, Suspected Human Carcinogen.
NIOSH REL: TWA 0.025 mg(Cr(VI))/m^3; CL 0.05/15M
DOT Classification: Oxidizer; Label: Oxidizer.

SAFETY PROFILE: A poison. A powerful irritant of

skin, eyes, and mucous membranes. Can cause a dermatitis, bronchoasthma, "chrome holes," damage to the eyes. Dangerously reactive. Incompatible with acetic acid, acetic anhydride, tetrahydronaphthalene, acetone, alcohols, alkali metals, ammonia, arsenic, bromine penta fluoride, butyric acid, n,n-dimethylformamide, hydrogen sulfide, peroxyformic acid, phosphorus, potassium hexacyanoferrate, pyridine, selenium, sodium, sulfur and many other materials. See also CHROMIC ACID, CHROMIUM COMPOUNDS, and CHROMATES.

CMH750 CAS:1308-14-1 ***HR: 3***
CHROMIC ACID (solution)
DOT: UN 1755
mf: CrH_3O_3 mw: 103.03

SYN: CHROMIC(III) HYDROXIDE

CONSENSUS REPORTS: Reported in EPA TSCA Inventory. Chromium and its compounds are on the Community Right-To-Know List.

OSHA PEL: (Transitional: CL 1 $mg/10m^3$) CL 0.1 $mg(CrO_3)/m^3$
ACGIH TLV: TWA 0.05 $mg(Cr)/m^3$
DFG MAK: Animal Carcinogen, Suspected Human Carcinogen.
NIOSH REL: (Chromium(VI)) TWA 0.025 mg(Cr(VI)) $/m^3$; CL 0.05 mg/m^3/15M
DOT Classification: Corrosive Material; Label: Corrosive

SAFETY PROFILE: A poison by many routes. Dangerously reactive. See also CHROMIC ACID and CHROMIUM COMPOUNDS.

CMI250 CAS:24613-89-6 ***HR: 3***
CHROMIC CHROMATE
mf: $Cr_3O_{12}\cdot 2Cr$ mw: 452.00

SYNS: CHROMIC ACID, CHROMIUM(3+) SALT (3:2) ◇ CHROMIUM CHROMATE (MAK)

TOXICITY DATA with REFERENCE
imp-rat TDLo:112 mg/kg:NEO AIHAAP 20,274,59

CONSENSUS REPORTS: IARC Cancer Review: Animal Sufficient Evidence IMEMDT 2,100,73. Reported in EPA TSCA Inventory. Chromium and its compounds are on the Community Right-To-Know List.

OSHA PEL: (Transitional: CL 1 $mg/10m^3$) CL 0.1 $mg(CrO_3)/m^3$
ACGIH TLV: TWA 0.05 $mg(Cr)/m^3$
DFG MAK: Animal Carcinogen, Suspected Human Carcinogen.
NIOSH REL: (Chromium(VI)) TWA 0.001 mg(Cr(VI)) $/ m^3$

SAFETY PROFILE: Confirmed carcinogen with experimental carcinogenic and neoplastigenic data. Very powerful oxidizer. See also CHROMIUM COMPOUNDS.

CMI500 CAS:1308-31-2 ***HR: 3***
CHROMITE (mineral)
mf: Cr_2FeO_4 mw: 223.85

SYNS: CHROME ORE ◇ CHROMITE ◇ CHROMITE ORE ◇ IRON CHROMITE

TOXICITY DATA with REFERENCE
mma-sat 2 mg/plate CRNGDP 3,1331,82
cyt-hmn:oth 500 mg/L BJCAAI 44,219,81
sce-ham:ovr 10 mg/L CRNGDP 3,1331,82

CONSENSUS REPORTS: IARC Cancer Review: Group 3 IMEMDT 7,165,87; Animal Inadequate Evidence IMEMDT 23,205,80. Chromium and its compounds are on the Community Right-To-Know List.

OSHA PEL: TWA 0.5 $mg(Cr)/m^3$
ACGIH TLV: TWA 0.05 mg/m^3 (ore processing); Confirmed Human Carcinogen (ore processing)

SAFETY PROFILE: Confirmed human carcinogen during ore processing. Human mutation data reported. See also CHROMIUM COMPOUNDS and IRON.

CMI750 CAS:7440-47-3 ***HR: 3***
CHROMIUM
af: Cr aw: 52.00

SYN: CHROME

TOXICITY DATA with REFERENCE
ivn-rat TDLo:2160 μg/kg/6W-I:ETA JNCIAM 16,447,55
orl-hmn LDLo:71 mg/kg:GIT 34ZIAG -,176,69

CONSENSUS REPORTS: NTP Fifth Annual Report on Carcinogens. IARC Cancer Review: Group 3 IMEMDT 7,165,87; Animal Inadequate Evidence IMEMDT 23,205,80. Chromium and its compounds are on the Community Right-To-Know List. Reported in EPA TSCA Inventory.

OSHA PEL: TWA 1 mg/m^3
ACGIH TLV: TWA 0.5 $(Cr)mg/m^3$

SAFETY PROFILE: Confirmed human carcinogen with experimental tumorigenic data. Human poison by ingestion with gastrointestinal effects. Powder will explode spontaneously in air. Ignites and is potentially explosive in atmospheres of carbon dioxide. Violent or explosive reaction when heated with ammonium nitrate. May ignite or react violently with bromine pentafluoride. Incandescent reaction with nitrogen oxide or sulfur dioxide. Incompatible with oxidants. See also CHROMIUM COMPOUNDS.

CMJ000 CAS:628-52-4 ***HR: 3***
CHROMIUM ACETATE HYDRATE
mf: $C_4H_6O_4{\cdot}Cr{\cdot}H_2O$ mw: 188.12

PROP: Red crystals.

SYNS: ACETIC ACID, CHROMIUM (2+) SALT (8CI, 9CI) ◇ CHROMIUM(2+) ACETATE ◇ CHROMIUM(II) ACETATE ◇ CHROMIUM DIACETATE ◇ CHROMOUS ACETATE ◇ CHROMOUS ACETATE MONOHYDRATE

TOXICITY DATA with REFERENCE
orl-rat LD50:11260 mg/kg AIHAAP 30,470,69

CONSENSUS REPORTS: Reported in EPA TSCA Inventory. Chromium and its compounds are on the Community Right-To-Know List.

OSHA PEL: TWA 0.5 mg(Cr)/m^3
ACGIH TLV: TWA 0.5 mg(Cr)/m^3

SAFETY PROFILE: Mildly toxic by ingestion. The anhydrous acetate ignites spontaneously in air. See also CHROMIUM COMPOUNDS. When heated to decomposition it emits acrid smoke and irritating fumes.

CMJ100 CAS:29689-14-3 ***HR: 3***
CHROMIUM CARBONATE
mf: $CH_2O_3{\cdot}xCr$ mw: 426.03

SYNS: BASIC CHROMIUM CARBONATE ◇ CARBONIC ACID, CHROMIUM SALT

CONSENSUS REPORTS: NTP Fifth Annual Report on Carcinogens. Reported in EPA TSCA Inventory.

SAFETY PROFILE: Confirmed carcinogen. When heated to decomposition it emits toxic fumes of Cr.

CMJ250 CAS:10025-73-7 ***HR: 3***
CHROMIUM CHLORIDE
mf: Cl_3Cr mw: 158.35
PROP: Bp: 1300° (subl).

SYNS: CHROMIC CHLORIDE ◇ CHROMIUM(III) CHLORIDE (1:3) ◇ CHROMIUM CHLORIDE, anhydrous ◇ CHROMIUM TRICHLORIDE ◇ C.I. 77295 ◇ PURATRONIC CHROMIUM CHLORIDE ◇ TRICHLOROCHROMIUM

TOXICITY DATA with REFERENCE
cyt-hmn:oth 500 mg/L BJCAAI 44,219,81
sce-ham:lng 39 mg/L CRNGDP 4,605,83
ipr-mus TDLo:59500 μg/kg (female 8D post):TER JTSCDR 1(2),1,76
ipr-mus TDLo:59500 μg/kg (female 9D post):REP JTSCDR 1(2),1,76
orl-rat LD50:1870 mg/kg YAKUD5 22,291,80
ihl-mus LC50:31500 μg/m^3/2H 85GMAT -,39,82
ipr-mus LD50:434 mg/kg COREAF 256,1043,63
skn-gpg LDLo:202 mg/kg AEHLAU 11,201,65*
ipr-gpg LDLo:200 mg/kg AEHLAU 11,201,65

CONSENSUS REPORTS: IARC Cancer Review: Group 3; IMEMDT 49,49,90; Human Inadequate Evidence IMEMDT 49,49,90; Animal Inadequate Evidence IMEMDT 49,49,90. Reported in EPA TSCA Inventory. EPA Genetic Toxicology Program. Chromium and its compounds are on the Community Right-To-Know List. EPA Extremely Hazardous Substances List.

OSHA PEL: TWA 0.5 mg(Cr)/m^3
ACGIH TLV: TWA 0.5 mg(Cr)/m^3

SAFETY PROFILE: Poison by skin contact, inhalation, and intraperitoneal routes. Experimental teratogenic and reproductive effects. Human mutation data reported. Reacts violently with lithium under nitrogen atmosphere. When heated to decomposition it emits toxic fumes of Cl^-.

CMJ500 ***HR: 3***
CHROMIUM COMPOUNDS

CONSENSUS REPORTS: Chromium and its compounds are on the Community Right-To-Know List.

SAFETY PROFILE: Chromate salts are suspected human carcinogens producing tumors of the lungs, nasal cavity, and paranasal sinus. Chromic acid and its salts have a corrosive action on the skin and mucous membranes. The lesions are confined to the exposed parts, affecting chiefly the skin of the hands and forearms and the mucous membranes of the nasal septum. The characteristic lesion is a deep, penetrating ulcer, which, for the most part, does not tend to suppurate, and which is slow in healing. Small ulcers, about the size of a matchhead, may be found, chiefly around the base of the nails, on the knuckles, dorsum of the hands and forearms. These ulcers tend to be clean and progress slowly. They are frequently painless, even though quite deep. They heal slowly and leave scars. On the mucous membranes of the nasal septum, the ulcers are usually accompanied by purulent discharge and crusting. If exposure continues, perforation of the nasal septum may result but produces no deformity of the nose. Hexavalent compounds are more toxic than the trivalent. Eczematous dermatitis due to trivalent chromium compounds has been reported.

CMJ600 CAS:13548-38-4 ***HR: 3***
CHROMIUM(III) NITRATE
mf: CrN_3O_9 mw: 238.03
DOT: UN 2720

SYNS: CHROMIC NITRATE ◇ CHROMIUM NITRATE ◇ CHROMIUM (3+) NITRATE ◇ CHROMIUM NITRATE (DOT) ◇ CHROMIUM TRINITRATE ◇ NITRIC ACID, CHROMIUM (3+) SALT

TOXICITY DATA with REFERENCE
dnr-bcs 160 mmol/L MUREAV 58,175,78
orl-rat LD50:3250 mg/kg YAKUD5 22,291,80

orl-mus LD50:2976 mg/kg SAIGBL 20,590,78
ipr-mus LD50:110 mg/kg SAIGBL 20,590,78
scu-mus LD50:3232 mg/kg SAIGBL 20,590,78

CONSENSUS REPORTS: IARC Cancer Review: Group 3 IMEMDT 49,49,90; Human Inadequate Evidence IMEMDT 49,49,90; Animal Inadequate Evidence IMEMDT 49,49,90. Reported in EPA TSCA Inventory.

OSHA PEL: TWA 0.5 mg(Cr)/m^3
ACGIH TLV: TWA 0.5 mg(Cr)/m^3
DOT Classification: Oxidizer; Label: Oxidizer

SAFETY PROFILE: Poison by intraperitoneal route. Moderately toxic by subcutaneous and ingestion routes. Mutation data reported. Questionable carcinogen. When heated to decomposition it emits toxic fumes of NO_x and Cr.

CMJ850 CAS:24094-93-7 ***HR: 3***
CHROMIUM NITRIDE
mf: CrN mw: 66.00

CONSENSUS REPORTS: Chromium compounds are on the Community Right-To-Know List.

SAFETY PROFILE: Mixture with potassium nitrate ignites when heated. When heated to decomposition it emits toxic fumes of NO_x. See also CHROMIUM COMPOUNDS and NITRIDES.

CMJ900 CAS:1308-38-9 ***HR: 3***
CHROMIUM(III) OXIDE (2:3)
mf: Cr_2O_3 mw: 152.00

SYNS: ANADOMIS GREEN ◇ ANIDRIDE CROMIQUE (FRENCH) ◇ CASALIS GREEN ◇ CHROME GREEN ◇ CHROME OCHER ◇ CHROME OXIDE ◇ CHROME OXIDE GREEN ◇ CHROMIA ◇ CHROMIC ACID ◇ CHROMIC ACID GREEN ◇ CHROMIC OXIDE ◇ CHROMIUM OXIDE ◇ CHROMIUM(III) OXIDE ◇ CHROMIUM(3+) OXIDE ◇ CHROMIUM SESQUIOXIDE ◇ CHROMIUM(3+) TRIOXIDE ◇ C.I. 77288 ◇ C.I. No. 77278 ◇ C.I. PIGMENT GREEN 17 ◇ DICHROMIUM TRIOXIDE ◇ 11661 GREEN ◇ GREEN CHROME OXIDE ◇ GREEN CHROMIC OXIDE ◇ GREEN CINNABAR ◇ GREEN ROUGE ◇ GUIGNER'S GREEN ◇ LEAF GREEN ◇ LEVANOX GREEN GA ◇ OIL GREEN ◇ OXIDE of CHROMIUM ◇ ULTRAMARINE GREEN

TOXICITY DATA with REFERENCE
mmo-sat 1 mmol/L TOLED5 8,195,81
dnr-sat 50 mmol/L TOLED5 7,439,81
dnd-esc 5 mmol/L CNREA8 40,2455,80
sce-ham:lng 34 mg/L CRNGDP 4,605,83
ipr-rat TDLo:90 mg/kg:ETA VOONAW 13(11),57,67

CONSENSUS REPORTS: NTP Fifth Annual Report on Carcinogens. IARC Cancer Review: Group 3 IMEMDT 7,165,87; Animal Inadequate Evidence IMEMDT 23,205,80. Reported in EPA TSCA Inventory. Chromium and its compounds are on the Community Right-To-Know List.

OSHA PEL: TWA 0.5 mg(Cr)/m^3
ACGIH TLV: TWA 0.5 mg(Cr)/m^3
DFG MAK: Suspected Carcinogen.

SAFETY PROFILE: Confirmed carcinogen with experimental tumorigenic data. Mutation data reported. Probably a severe eye, skin, and mucous membrane irritant. A powerful oxidizer. Reacts violently with CLF_3. See also CHROMIUM COMPOUNDS.

CMJ910 CAS:12018-40-5 ***HR: 2***
CHROMIUM OXIDE, aerosols
mf: Cr_5O_{12} mw: 452.00

TOXICITY DATA with REFERENCE
ihl-rat TCLo:143 μg/m^3/78W-I:ETA TXCYAC 42,219,86

SAFETY PROFILE: Questionable carcinogen with experimental tumorigenic data. When heated to decomposition it emits toxic fumes of Cr^-.

CMK000 CAS:1333-82-0 ***HR: 3***
CHROMIUM(VI) OXIDE (1:3)
DOT: UN 1463/NA 1463/UN 1755
mf: CrO_3 mw: 100.00

PROP: Red, rhombic, deliquescent crystals. D: 2.70, mp: 196°, bp: decomp, sol: 61.7 g/100 cc @ 0°, 67.45 g/100 cc @ 100°.

SYNS: ANHYDRIDE CHROMIQUE (FRENCH) ◇ ANIDRIDE CROMICA (ITALIAN) ◇ CHROME (TRIOXYDE de) (FRENCH) ◇ CHROMIC ACID ◇ CHROMIC(VI) ACID ◇ CHROMIC ACID, solid (DOT) ◇ CHROMIC ACID, solution (DOT) ◇ CHROMIC ANHYDRIDE (DOT) ◇ CHROMIC TRIOXIDE (DOT) ◇ CHROMIUM OXIDE ◇ CHROMIUM(VI) OXIDE ◇ CHROMIUM TRIOXIDE ◇ CHROMIUM(6+) TRIOXIDE ◇ CHROMIUM TRIOXIDE, anhydrous (DOT) ◇ CHROMO (TRIOSSIDO di) (ITALIAN) ◇ CHROMSAEUREANHYDRID (GERMAN) ◇ CHROMTRIOXID (GERMAN) ◇ CHROOMTRIOXYDE (DUTCH) ◇ CHROOMZUURANHYDRIDE (DUTCH) ◇ MONOCHROMIUM OXIDE) ◇ MONOCHROMIUM TRIOXIDE ◇ PURATRONIC CHROMIUM TRIOXIDE

TOXICITY DATA with REFERENCE
mmo-sat 1 mmol/L TOLED5 8,195,81
cyt-hmn:leu 2 mg/L MUREAV 58,175,78
scu-mus TDLo:20 mg/kg (8D preg):TER SEIJBO 19,171,79
ivn-ham TDLo:7500 μg/kg (female 8D post):REP ENVRAL 16,101,78
ihl-ham TCLo:110 μg/m^3:CAR,NOSE,PUL AGGHAR 13,528,55
imp-rat TDLo:125 mg/kg:CAR AIHAAP 20,274,59
ihl-mus TCLo:3480 μg/m^3/2H/1Y-I:ETA SAIGBL 29,17,87
ihl-hmn TCLo:110 μg/m^3 YAKUD5 22,291,80
orl-rat LD50:80 mg/kg TRENAF 27(2),119,76
orl-mus LD50:127 mg/kg CHYCDW 14,86,80
ipr-mus LD50:14 mg/kg NEZAAQ 34,193,79

scu-mus LDLo:20 mg/kg SEIJBO 19,171,79
scu-dog LDLo:330 mg/kg 27ZWAY 3.3,1521,-

CONSENSUS REPORTS: NTP Fifth Annual Report on Carcinogens. IARC Cancer Review: Group 1 IMEMDT 7,165,87; Animal Sufficient Evidence IMEMDT 23,205,80. EPA Genetic Toxicology Program. Chromium and its compounds are on the Community Right-To-Know List. Reported in EPA TSCA Inventory.

OSHA PEL: CL 0.1 mg(CrO_3)/m^3
ACGIH TLV: TWA 0.05 mg(Cr)/m^3; Confirmed Human Carcinogen
DFG MAK: 0.1 mg/m^3, Suspected Carcinogen.
NIOSH REL: (Chromium(VI)) TWA 0.025 mg(Cr(VI))/m^3; CL 0.05/15M
DOT Classification: Oxidizer; Label: Oxidizer (NA1463); Corrosive Material; Label: Corrosive (UN1755); Oxidizer; Label: Oxidizer, Corrosive (UN1463).

SAFETY PROFILE: Confirmed human carcinogen producing nasal and lung tumors. Experimental carcinogenic and tumorigenic data. Poison by ingestion, intraperitoneal, and subcutaneous routes. Experimental teratogenic and reproductive effects. Human mutation data reported. Corrosive. Probably a severe eye, skin, and mucous membrane irritant. See also CHROMIUM COMPOUNDS.

A powerful oxidizer. Explosive reaction with acetaldehyde, acetic acid + heat, acetic anhydride + heat, benzaldehyde, benzene, benzylthylaniline, butyraldehyde, 1,3-dimethylhexahydropyrimidone, diethyl ether, ethylacetate, isopropylacetate, methyl dioxane, pelargonic acid, pentyl acetate, phosphorus + heat, propionaldehyde and other organic materials or solvents. Forms a friction- and heat-sensitive explosive mixture with potassium hexacyanoferrate. Ignites on contact with alcohols, acetic anhydride + tetrahydronaphthalene, acetone, butanol, chromium(II) sulfide, cyclohexanol, dimethyl formamide, ethanol, ethylene glycol, methanol, 2-propanol, pyridine. Violent reaction with acetic anhydride + 3-methylphenol (above 75°C), acetylene, bromine pentafluoride, glycerol, hexamethylphosphoramide, peroxyformic acid, selenium, sodium amide. Incandescent reaction with alkali metals (e.g., sodium, potassium), ammonia, arsenic, butyric acid (above 100°C), chlorine trifluoride, hydrogen sulfide + heat, sodium + heat, and sulfur. Incompatible with N,N-dimethylformamide.

CMK275 CAS:14884-42-5 ***HR: 3***
CHROMIUM PENTAFLUORIDE
mf: CrF_5 mw: 146.99

CONSENSUS REPORTS: Chromium compounds are on the Community Right-To-Know List.

SAFETY PROFILE: Undergoes violent redox and halogen exchange reactions. Mixtures with phosphorus trichloride react violently on slight heating. When heated to decomposition it emits toxic fumes of F^-. See also CHROMIUM COMPOUNDS.

CMK300 CAS:7789-04-0 ***HR: 3***
CHROMIUM PHOSPHATE
mf: $Cr{\bullet}H_3O_4P$ mw: 150.00

SYNS: ARNAUDON'S GREEN ◇ ARNAUDON'S GREEN (HEMIHEPTAHYDRATE) ◇ CHROMIC PHOSPHATE ◇ CHROMIUM MONOPHOSPHATE ◇ CHROMIUM ORTHOPHOSPHATE ◇ PHOSPHORIC ACID CHROMIUM (III) SALT ◇ PHOSPHORIC ACID, CHROMIUM(3+) SALT (1:1) ◇ PLESSY'S GREEN (HEMIHEPTAHYDRATE)

CONSENSUS REPORTS: NTP Fifth Annual Report on Carcinogens. IARC Cancer Review: Group 3 IMEMDT 49,49,90; Human Inadequate Evidence IMEMDT 49,49,90; IARC Cancer Review: Animal Inadequate Evidence IMEMDT 49,49,90. Reported in EPA TSCA Inventory.

OSHA PEL: TWA 0.5 mg(Cr)/m^3
ACGIH TLV: TWA 0.5 mg(Cr)/m^3

SAFETY PROFILE: Confirmed carcinogen. When heated to decomposition it emits toxic fumes of PO_x and Cr.

CMK400 CAS:37224-57-0 ***HR: 2***
CHROMIUM POTASSIUM ZINC OXIDE

SYNS: POTASSIUM ZINC CHROMATE ◇ ZINC POTASSIUM CHROMATE

TOXICITY DATA with REFERENCE
imp-rat TDLo:10746 μg/kg:CAR BJIMAG 43,243,86

CONSENSUS REPORTS: IARC Cancer Review: Human Sufficient Evidence IMEMDT 23,205,80; Animal Sufficient Evidence IMEMDT 2,100,73. Chromium and its compounds, as well as zinc and its compounds, are on the Community Right-To-Know List.

OSHA PEL: (Transitional: 1 mg(CrO_3)/10m^3) CL 0.1 mg(CrO_3)/m^3
ACGIH TLV: TWA 0.01 mg(Cr)/M^3; Confirmed Human Carcinogen
DFG MAK: Human Carcinogen.
NIOSH REL: (Chromium (VI)) TWA 0.001 mg(Cr(VI))/m^3

SAFETY PROFILE: Confirmed carcinogen with experimental carcinogenic data. When heated to decomposition it emits toxic fumes of Cr^- and Zn^-.

CMK425 CAS:10031-37-5 ***HR: 3***
CHROMIUM SULFATE, PENTADECAHYDRATE
mf: $O_{12}S_3{\bullet}2Cr{\bullet}15H_2O$ mw: 662.38

SYNS: SULFURIC ACID, CHROMIUM(3+) SALT (3:2), PENTADECAHYDRATE ◇ WOOL MORDANT

TOXICITY DATA with REFERENCE
ipr-mus LD50:258 mg/kg CRNGDP 4,1535,83

ACGIH TLV: TWA 0.5 mg(Cr)/m^3

SAFETY PROFILE: Poison by intraperitoneal route. When heated to decomposition it emits toxic fumes of SO_x and Cr.

CMK450 CAS:10060-12-5 ***HR: 3***
CHROMIUM TRICHLORIDE HEXAHYDRATE
mf: $Cl_3Cr•6HO_2$ mw: 356.41

SYNS: CHLORID CHROMITY HEXAHYDRAT ◇ CHROMIC CHLORIDE HEXAHYDRATE ◇ CHROMIUM CHLORIDE, HEXAHYDRATE (8CI,9CI) ◇ CHROMIUM(III) CHLORIDE, HEXAHYDRATE (1:3:6) ◇ CHROMIUM SESQUICHLORIDE ◇ HEXAAQUACHROMIUM CHLORIDE ◇ HEXAAQUACHROMIUM (III) CHLORIDE

TOXICITY DATA with REFERENCE
oth-hmn:oth 500 mg/L BJCAAI 44,219,81
cyt-hmn:leu 400 mg/L SAIGBL 18,136,76
dnd-ham:kdy 500 mg/L CBINA8 37,309,81
sce-ham:fbr 32 mg/L MUREAV 104,141,82
ipr-mus TDLo:134 mg/kg (female 8D post):TER TXCYAC 26,257,83
orl-rat LD50:1790 mg/kg SinJF# 29MAR77
ipr-mus LD50:285 mg/kg TXAPA9 63,461,82
ivn-mus LDLo:1602 mg/kg AIPTAK 62,330,39
ivn-rbt LDLo:576 mg/kg AIPTAK 62,330,39
ivn-frg LDLo:374 mg/kg AIPTAK 62,330,39

ACGIH TLV: TWA 0.5 mg(Cr)/m^3
OSHA PEL: TWA 0.5 mg(Cr)/m^3

SAFETY PROFILE: Poison by intraperitoneal and intravenous routes. Moderately toxic by ingestion. An experimental teratogen. Human mutation data reported. When heated to decomposition it emits acrid smoke and toxic fumes.

CMK500 CAS:15930-94-6 ***HR: 3***
CHROMIUM(6+)ZINC OXIDE HYDRATE (1:2:6:1)
mf: $CrO_4•H_2O_2•Zn_2•H_2O$ mw: 298.78

SYNS: BUTTERCUP YELLOW ◇ CHROMIC ACID, ZINC SALT (1:2) ◇ ZINC CHROMATE HYDROXIDE ◇ ZINC CHROMATE(VI) HYDROXIDE ◇ ZINC HYDROXYCHROMATE ◇ ZINC YELLOW

TOXICITY DATA with REFERENCE
sce-ham:ovr 100 μg/L MUREAV 156,219,85

CONSENSUS REPORTS: IARC Cancer Review: Human Sufficient Evidence IMEMDT 23,205,80; Animal Sufficient Evidence IMEMDT 2,100,73. Chromium and its compounds, as well as zinc and its compounds, are on the Community Right-To-Know List.

OSHA PEL: (Transitional: 1 mg(CrO_3)/10m^3) CL 0.1 mg(CrO_3)/m^3
ACGIH TLV: TWA 0.01 mg(Cr)/M^3; Confirmed Human Carcinogen
DFG MAK: Human Carcinogen.
NIOSH REL: (Chromium (VI)) TWA 0.001 mg(Cr(VI))/m^3

SAFETY PROFILE: Confirmed human carcinogen. Mutation data reported. When heated to decomposition it emits toxic fumes of ZnO. See also CHROMIUM and ZINC COMPOUNDS.

CMK650 CAS:7059-24-7 ***HR: 3***
CHROMOMYCIN A3
mf: $C_{57}H_{82}O_{26}$ mw: 1183.39

SYNS: ABURAMYCIN B ◇ 3B-o-(4-o-ACETYL-2,6-DIDEOXY-3-C-METHYL-α-l-ARABINOHEXOPYRANOSYL)-7-METHYL-OLIVOMYCIN D ◇ ANTIBIOTIC B 599 ◇ CHROMOMYSIN A_3 ◇ NSC-58514 ◇ TOYOMYCIN

TOXICITY DATA with REFERENCE
dnr-bcs 800 ng/plate TAKHAA 44,96,85
dnd-hmn:hla 400 μg/L CNREA8 45,2813,85
msc-hmn:hla 8 μg/L CNREA8 45,2813,85
msc-mus:emb 25 μg/L CNREA8 45,2813,85
scu-rat TDLo:125 μg/kg (6-10D preg):TER OSDIAF 14,107,65
scu-rat TDLo:125 μg/kg (6-10D preg):REP OSDIAF 14,107,65
ipr-rat LDLo:250 μg/kg 85ERAY 2,1401,78
orl-mus LD50:1431 μg/kg NCISP* JAN86
ipr-mus LD50:800 μg/kg JAJAAA 16,22,63
scu-mus LD50:2800 μg/kg NIIRDN 6,245,82
ivn-mus LD50:1 mg/kg JAJAAA 16,22,63
ipr-dog LDLo:250 μg/kg 85ERAY 2,1401,78
ivn-dog LDLo:200 μg/kg TXAPA9 27,259,74
ivn-mky LDLo:330 μg/kg TXAPA9 27,259,74
ipr-cat LDLo:250 μg/kg 85ERAY 2,1041,78

SAFETY PROFILE: A deadly poison by ingestion, subcutaneous, intravenous, and intraperitoneal routes. An experimental teratogen. Experimental reproductive effects. Human mutation data reported. When heated to decomposition it emits acrid smoke and fumes.

CMK750 CAS:12622-79-6 ***HR: 3***
CHROMOMYCIN SODIUM

PROP: Produced by a strain of *Actinomyces olivoreticuli* (85ERAY 2,1322,78).

SYN: OLIVOMYCIN, SODIUM SALT

TOXICITY DATA with REFERENCE
ipr-rat LDLo:1 mg/kg ANTBAL 7,53,62
ivn-rat LDLo:1 mg/kg ANTBAL 7,53,62
orl-mus LDLo:250 mg/kg ANTBAL 7,53,62

ipr-mus LD50:12700 μg/kg ANTBAL 7,53,62
scu-mus LD50:15600 μg/kg ANTBAL 7,53,62
ivn-mus LD50:138 mg/kg 85ERAY 2,1322,78
ivn-dog LDLo:300 μg/kg ANTBAL 7,53,62
ivn-rbt LDLo:2500 μg/kg ANTBAL 7,53,62
ipr-gpg LDLo:2 mg/kg ANTBAL 7,53,62

SAFETY PROFILE: Poison by ingestion, intraperitoneal, intravenous, and subcutaneous routes. When heated to decomposition it emits toxic fumes including Na_2O.

CML000 CAS:14259-67-7 ***HR: 3***
CHROMYL AZIDE CHLORIDE
mf: $ClCrN_3O_2$ mw: 161.47

$CrO_2(N_3)Cl$

CONSENSUS REPORTS: Chromium and its compounds are on the Community Right-To-Know List.

SAFETY PROFILE: An exlosive. When heated to decomposition it emits toxic fumes of Cl^- and NO_x. See also CHROMIUM COMPOUNDS, AZIDES, and CHLORIDES.

CML125 CAS:14977-61-8 ***HR: 3***
CHROMYL CHLORIDE
DOT: UN 1758
mf: Cl_2CrO_2 mw: 154.90

PROP: Dark red liquid, musty burning odor. Mp: −96.5°, bp: 115.7°, d: 1.9145 @ 25°/4°, vap press: 20 mm @ 20°.

SYNS: CHLORURE de CHROMYLE (FRENCH) ◇ CHROMIC OXYCHLORIDE ◇ CHROMIUM CHLORIDE OXIDE ◇ CHROMIUM DICHLORIDE DIOXIDE ◇ CHROMIUM DIOXIDE DICHLORIDE ◇ CHROMIUM(VI) DIOXYCHLORIDE ◇ CHROMIUM OXYCHLORIDE ◇ CHROMOXYCHLORID (GERMAN) ◇ CHROMYLCHLORID (GERMAN) ◇ CHROOMOXYLCHLORIDE (DUTCH) ◇ CROMILE, CLORURO di (ITALIAN) ◇ CROMO, OSSICLORURO di (ITALIAN) ◇ DICHLORODIOXOCHROMIUM ◇ DIOXODICHLOROCHROMIUM ◇ OXYCHLORURE CHROMIQUE (FRENCH)

TOXICITY DATA with REFERENCE
mmo-sat 50 μg/plate CRNGDP 1,583,80
mma-sat 100 μg/plate CRNGDP 1,583,80

CONSENSUS REPORTS: Reported in EPA TSCA Inventory. Chromium and its compounds are on the Community Right-To-Know List.

OSHA PEL: TWA 0.5 mg(Cr)/m^3
ACGIH TLV: TWA 0.025 ppm
DFG MAK: Suspected Carcinogen.
NIOSH REL: (Chromium(VI)) TWA 0.001 mg(Cr(VI))/m^3
DOT Classification: Corrosive Material; Label: Corrosive.

SAFETY PROFILE: Probably a poison by various routes. Mutation data reported. Corrosive. A strong irritant. Hydrolyzes to form chromic and hydrochloric acids. A strong oxidizer and chlorinating agent. Violent reaction with water. Reacts violently with alcohol, ether, acetone, turpentine. Ignites or explodes on contact with non-metal halides (e.g., disulfur dichloride, phosphorus trichloride, and phosphorus tribromide), non-metal hydrides (e.g., hydrogen sulfide, and hydrogen phosphide), flowers of sulfur, moist phosphorus, sodium azide, and urea. During preparation can violently explode. Incompatible with ammonia, disulfur dichloride, organic solvents, phosphorus, phosphorus trichloride, sodium azide, and sulfur. When heated to decomposition it emits toxic fumes of Cl^-. See also CHROMIUM COMPOUNDS.

CML325 CAS:16017-38-2 ***HR: 3***
CHROMYL NITRATE
mf: CrN_2O_8 mw: 208.00

CONSENSUS REPORTS: Chromium compounds are on the Community Right-To-Know List.

SAFETY PROFILE: A powerful oxidant and nitrating agent. Ignites on contact with many organic materials (e.g., hydrocarbons, orgnic solvents, paper, rubber and wood). When heated to decomposition it emits toxic fumes of NO_x. See also CHROMIUM COMPOUNDS.

CML500 ***HR: 3***
CHROMYL PERCHLORATE
mf: Cl_2CrO_{10} mw: 282.90

$CrO_2(ClO_4)_2$

CONSENSUS REPORTS: Chromium and its compounds are on the Community Right-To-Know List.

SAFETY PROFILE: A powerful oxidant. Explodes when heated above 80°C. Ignites on contact with organic solvents. When heated to decomposition it emits toxic fumes of Cl^-. See also CHROMIUM COMPOUNDS and PERCHLORATES.

CML750 CAS:491-59-8 ***HR: 3***
CHRYSAROBIN
mf: $C_{15}H_{12}O_3$ mw: 240.27

PROP: Brownish to orange-yellow crystals.

SYNS: CHRYSOPHANIC ACID ANTHRANOL ◇ 3-METHYL-1,8,9-ANTHRACENETRIOL ◇ 3-METHYLANTHRALIN ◇ 1,8,9-TRIHYDROXY-3-METHYLANTHRACENE

TOXICITY DATA with REFERENCE
mmo-sat 100 μg/plate BCSTB5 5,1489,77
mma-sat 100 μg/plate BCSTB5 5,1489,77

mmo-smc 1000 ppm/16H ADVEA4 51,45,71
ipr-mus LDLo:4 mg/kg CBCCT* 2,241,50

SAFETY PROFILE: Poison by intraperitoneal route. Mutation data reported. An irritant and an allergen. Combustible when exposed to heat or flame. When heated to decomposition it emits acrid smoke and fumes.

CML800 CAS:2642-98-0 ***HR: 3***
6-CHRYSENAMINE
mf: $C_{18}H_{13}N$ mw: 243.32

PROP: Leaflets from alc. Mp: 210-211°. Sltly sol in alc, benzene, ethyl acetate.

SYNS: 6-AMC ◇ 6-AMINOCHRYSENE ◇ CHRYSENEX ◇ CHRYSONEX

TOXICITY DATA with REFERENCE
mmo-sat 2500 ng/plate CNREA8 44,3408,84
mma-sat 500 ng/plate MUREAV 155,7,85
dnr-bcs 20 uL/disc MUREAV 97,1,82
dns-rat:lvr 500 nmol/L ENMUDM 3,11,81
msc-ham:ovr 50 mg/L JTEHD6 13,531,84
skn-mus TDLo:1100 g/kg/39W-I:CAR EJCAAH 11,327,75

CONSENSUS REPORTS: EPA Genetic Toxicology Program.

SAFETY PROFILE: Questionable carcinogen with experimental carcinogenic data by skin contact. Mutation data reported. When heated to decomposition it emits toxic fumes of NO_x. See also AROMATIC AMINES and CHRYSENE.

CML810 CAS:218-01-9 ***HR: 3***
CHRYSENE
mf: $C_{18}H_{12}$ mw: 228.30

PROP: Occurs in coal tar. Is formed during distillation of coal, in very small amount during distillation or pyrolysis of many fats and oils. Orthorhombic bipyramidal plates from benzene. D: 1.274, mp: 254°. Sublimes easily in vacuum, bp: 448°. Sltly sol in alc, ether, carbon bisulfide, and glacial acetic acid; moderately sol in boiling benzene; insol in water. Chrysene is generally only sltly sol in cold organic solvents, but fairly sol in these solvents when hot, including glacial acetic acid.

SYNS: 1,2-BENZOPHENANTHRENE ◇ BENZO(a)PHENANTHRENE ◇ 1,2-BENZPHENANTHRENE ◇ BENZ(a)PHENANTHRENE ◇ 1,2,5,6-DIBENZONAPHTHALENE ◇ RCRA WASTE NUMBER U050

TOXICITY DATA with REFERENCE
mma-sat 5 μg/plate MUREAV 156,61,85
msc-hmn:lym 6 μmol/L DTESD7 10,227,82
msc-hmn:oth 12 μmol/L MUREAV 130,127,84
skn-mus TDLo:3600 μg/kg:NEO CNREA8 38,1831,78
scu-mus TDLo:200 mg/kg:ETA CNREA8 15,632,55

CONSENSUS REPORTS: IARC Cancer Review: Group 1 IMEMDT 7,56,87; Animal Limited Evidence IMEMDT 32,247,83; Animal Sufficient Evidence IMEMDT 3,159,73. EPA Genetic Toxicology Program. Reported in EPA TSCA Inventory.

OSHA PEL: 0.2 mg/m^3
ACGIH TLV: Suspected Human Carcinogen
DFG MAK: Animal Carcinogen, Suspected Human Carcinogen.
NIOSH REL: (Chrysene) To be controlled as a carcinogen.

SAFETY PROFILE: Confirmed carcinogen with experimental carcinogenic, neoplastigenic, and tumorigenic data by skin contact. Human mutation data reported. When heated to decomposition it emits acrid smoke and fumes.

CML815 CAS:15131-84-7 ***HR: 3***
CHRYSENE-5,6-EPOXIDE
mf: $C_{18}H_{12}O$ mw: 244.30

SYNS: CHRYSENE-K-REGION EPOXIDE ◇ CHRYSENE-5,6-OXIDE ◇ 5,6-EPOXY-5,6-DIHYDROCHRYSENE

TOXICITY DATA with REFERENCE
mma-sat 50 μg/plate PNASA6 72,5135,75
dnr-esc 100 μmol/L ZKKOBW 92,157,78
dns-esc 100 μmol/L ZKKOBW 92,157,78
scu-mus TDLo:400 mg/kg/10W-I:ETA IJCNAW 2,500,67

CONSENSUS REPORTS: EPA Genetic Toxicology Program.

SAFETY PROFILE: Questionable carcinogen with experimental tumorigenic data. Mutation data reported. When heated to decomposition it emits acrid smoke and fumes. See also CHRYSENE.

CML820 CAS:63339-68-4 ***HR: 2***
CHUANGHSINMYCIN
mf: $C_{12}H_{11}NO_2S$ mw: 233.30

SYNS: CHUANGXINMYCIN ◇ cis-(−)-3,5-DIHYDRO-3-METHYL-2H-THIOPYRANO(4,3,2-cd)INDOLE-2-CARBOXYLICACID

TOXICITY DATA with REFERENCE
orl-mus LD50:1770 mg/kg 85GDA2 5,133,81
ipr-mus LD50:875 mg/kg 85GDA2 5,133,81
ivn-mus LD50:600 mg/kg 85GDA2 5,133,81

SAFETY PROFILE: Moderately toxic by ingestion, intravenous, and intraperitoneal routes. When heated to decomposition it emits toxic fumes of SO_x and NO_x.

CML822 CAS:102419-73-8 ***HR: 2***
CHUANGHSINMYCIN SODIUM
mf: $C_{12}H_{10}NO_2•Na$ mw: 223.22

SYN: CHUANGXIMYCIN SODIUM

TOXICITY DATA with REFERENCE
orl-mus LD50:1770 mg/kg JANTAJ 32,79-23,79
ipr-mus LD50:875 mg/kg JANTAJ 32,79-23,79
ivn-mus LD50:600 mg/kg JANTAJ 32,79-23,79

SAFETY PROFILE: Moderately toxic by ingestion, intravenous, and intraperitoneal routes. When heated to decomposition it emits toxic fumes of NO_x and Na_2O.

CML825 CAS:58812-37-6 ***HR: 3***
CHUANLIANSU
mf: $C_{30}H_{38}O_{11}$ mw: 574.68

SYN: TOOSENDANIN

TOXICITY DATA with REFERENCE
ipr-rat LD50:98 mg/kg CTYAD8 13,317,82
orl-mus LD50:244 mg/kg CTYAD8 13,317,82
ipr-mus LD50:13800 μg/kg CTYAD8 13,317,82
scu-mus LD50:14300 μg/kg CTYAD8 13,317,82
ivn-mus LD50:14600 μg/kg CTYAD8 13,317,82
ivn-rbt LD50:4200 μg/kg CTYAD8 13,317,82

SAFETY PROFILE: Poison by ingestion, subcutaneous, intravenous and intraperitoneal routes. When heated to decomposition it emits acrid smoke and fumes.

CML835 CAS:37106-97-1 ***HR: 2***
CHYMEX
mf: $C_{23}H_{20}N_2O_5$ mw: 404.45

PROP: Crystals from methanol/water. Mp: 240-242°.

SYNS: BENTIROMIDE ◇ (S)-p-(α-BENZAMIDO-p-HYDROXYHYDROCINNAMAMIDO)BENZOIC ACID ◇ 4-((2-(BENZOYLAMINO)-3-(4-HYDROXYPHENYL)-1-OXOPROPYL)AMINO)BENZOIC ACID ◇ (S)-4-((2-BENZOYLAMINO)-3-(4-HYDROXYPHENYL)-1-OXOPROPYL)AMINO)BENZOIC ACID ◇ p-((N-BENZOYL-l-TYROSIN)AMIDO)BENZOICACID ◇ BENZOYLTYROSYL-p-AMINOBENZOIC ACID ◇ N-BENZOYL-l-TYROSYL-p-AMINOBENZOIC ACID ◇ BTPABA ◇ E-2663 ◇ PFT

TOXICITY DATA with REFERENCE
ipr-rat LD50:2000 mg/kg IYKEDH 11,181,80
ivn-rat LD50:485 mg/kg IYKEDH 11,181,80
ipr-mus LD50:1650 mg/kg IYKEDH 11,181,80
ivn-mus LD50:1020 mg/kg IYKEDH 11,181,80

SAFETY PROFILE: Moderately toxic by intravenous and intraperitoneal routes. When heated to decomposition it emits toxic fumes of NO_x.

CML850 CAS:9004-07-3 ***HR: 3***
α-CHYMOTRYPSIN

SYNS: ALPHA CHYMAR ◇ ALPHA-CHYMAR OPHTH ◇ AVAZYME ◇ CHYMAR ◇ CHYMOTEST ◇ CHYMOTRYPSIN A ◇ CHYMOTRYPSIN B ◇ E.C. 3.4.4.5 ◇ E.C. 3.4.4.6 ◇ E.C. 3.4.21.1 ◇ ENZEON ◇ QUIMAR ◇ QUIMOTRASE

TOXICITY DATA with REFERENCE
ipr-rat LD50:65100 μg/kg JAPMA8 39,42,50
scu-rat LD50:250 mg/kg NIIRDN 6,202,82
ivn-rat LD50:84 mg/kg NIIRDN 6,202,82
ims-rat LD50:116 mg/kg NIIRDN 6,202,82
scu-mus LD50:185 mg/kg NIIRDN 6,202,82
ivn-mus LD50:89 mg/kg NIIRDN 6,202,82
ims-mus LD50:104 mg/kg NIIRDN 6,202,82
ivn-rbt LD50:24000 units/kg AIPTAK 106,164,56
ivn-gpg LDLo:50000 units/kg AIPTAK 106,164,56

CONSENSUS REPORTS: Reported in EPA TSCA Inventory.

SAFETY PROFILE: Poison by subcutaneous, intramuscular, intravenous and intraperitoneal routes. When heated to decomposition it emits toxic fumes of NO_x.

CML890 ***HR: 3***
CI-914
mf: $C_{13}H_{12}N_4O \cdot ClH$ mw: 276.75

SYN: 4,5-DIHYDRO-6-(4-1H-IMIDAZOL-1-YL)PHENYL)-3(2H)-PYRIDAZINONE HYDROCHLORIDE

TOXICITY DATA with REFERENCE
orl-rat LD50:125 mg/kg TOXID9 4,7,84
ivn-rat LD50:98 mg/kg TOXID9 4,6,84
orl-mus LD50:250 mg/kg TOXID9 4,7,84
ivn-mus LD50:226 mg;kg TOXID9 4,6,84

SAFETY PROFILE: Poison by ingestion and intravenous routes. When heated to decomposition it emits toxic fumes of NO_x and HCl.

CMM000 CAS:6441-77-6 ***HR: 2***
C.I. 45405
mf: $C_{20}H_6Br_4Cl_2O_5 \cdot 2K$ mw: 795.00

SYNS: C.I. ACID RED 98 ◇ 4.6-DICHLORO-2′,4′,5′,7′-TETRABROMO-FLUORESCEIN DIPOTASSIUM SALT ◇ FOOD DYE RED No. 104 ◇ PHLOXIN ◇ PHLOXINE ◇ PHLOXINE K ◇ 2′,4′,5′,7′-TETRABROMO-4,7-DICHLORO-FLUORESCEIN DIPOTASSIUM SALT ◇ 2,4,5,7-TETRABROMO-12,15-DICHLOROFLUORESCEIN, DIPOTASSIUM SALT ◇ TOYO ACID PHLOXINE

TOXICITY DATA with REFERENCE
dnr-esc 1 mg/disc MUREAV 88,1,81
dnd-esc 1 mg/disc FCTXAV 18,215,80
dni-hmn:leu 500 mg/L NEZAAQ 30,574,75
cyt-hmn:leu 1 mg/L NEZAAQ 30,574,75
cyt-hmn:fbr 1 mg/L NEZAAQ 30,574,75
orl-rat TDLo:75 g/kg (1-20D preg):REP SKEZAP 16,34,75
orl-rat TDLo:75 g/kg (1-20D preg):TER SKEZAP 16,34,75
orl-rat LD50:2870 mg/kg SKEZAP 16,34,75

CONSENSUS REPORTS: EPA Genetic Toxicology Program.

SAFETY PROFILE: Moderately toxic by ingestion. An

experimental teratogen. Other experimental reproductive effects. Human mutation data reported. When heated to decomposition it emits toxic fumes of Cl^-, Br^-, and K_2O.

CMM400 CAS:12219-87-3 ***HR: 1***
C.I. ACID GREEN 40

SYNS: ACID GREEN 40 ◇ LANASYN GREEN BL ◇ NYLOSAN GREEN F-BL ◇ SILK FAST GREEN B ◇ XYLENE ACID MILLING GREEN BL ◇ ZELEN KYSELA 40

TOXICITY DATA with REFERENCE
eye-rbt 500 mg/24H MLD 85JCAE-,1330,86
orl-rat LD50:8680 mg/kg 85JCAE-,1330,86

SAFETY PROFILE: Mildly toxic by ingestion. An eye irritant. When heated to decomposition it emits acrid smoke and irritating fumes.

CMM750 CAS:2391-30-2 ***HR: 3***
C.I. ACID YELLOW 7
mf: $C_{19}H_{13}N_2O_5S \cdot Na$ mw: 404.39

SYNS: BRILLIANT SULFAFLAVINE ◇ C.I. 56205 ◇ 2,3-DIHYDRO-6-AMINO-1,3-DIOXO-2-(p-TOLYL)1H-BENZ(de)ISOQUINOLINE-5-SULFONIC ACID, MONOSODIUM SALT ◇ FENAZO YELLOW XX ◇ SOLAR PURE YELLOW 8G

TOXICITY DATA with REFERENCE
ivn-mus LD50:110 mg/kg TXAPA9 44,225,78

CONSENSUS REPORTS: Reported in EPA TSCA Inventory.

SAFETY PROFILE: Poison by intravenous route. When heated to decomposition it emits very toxic fumes of SO_x, Na_2O, and NO_x.

CMM800 CAS:68650-43-1 ***HR: D***
CICHORIUM INTYBUS, ETHANOL EXTRACT

TOXICITY DATA with REFERENCE
orl-rat TDLo:600 mg/kg (female 14-16D post):REP IJEBA6 14,623,76

CONSENSUS REPORTS: Reported in EPA TSCA Inventory.

SAFETY PROFILE: Experimental reproductive effects. When heated to decomposition it emits acrid smoke and irritating fumes.

CMN000 CAS:505-75-9 ***HR: 3***
CICUTOXIN
mf: $C_{17}H_{22}O_2$ mw: 258.39

SYN: 8,10,12-HEPTADECATRIENE-4,6-DIYNE-1,14-DIOL,(E,E,E)-(-)-

TOXICITY DATA with REFERENCE
ipr-mus LD50:48300 μg/kg NYKZAU 52,29S,56
orl-cat LDLo:7 mg/kg HBAMAK 4,1289,35
ivn-cat LDLo:5360 μg/kg NYKZAU 52,29S,56
unr-frg LDLo:61 mg/kg HBAMAK 4,1289,35

SAFETY PROFILE: Poison by ingestion, intravenous, and possibly other routes. When heated to decomposition it emits acrid smoke and irritating fumes.

CMN125 CAS:64440-87-5 ***HR: 3***
CIDEFERRON

TOXICITY DATA with REFERENCE
ivn-rat LD50:385 mg/kg NIIRDN 6,293,82
ivn-mus LD50:403 mg/kg NIIRDN 6,923,82
ivn-rbt LD50:280 mg/kg NIRIDN 6,923,82

SAFETY PROFILE: Poison by intravenous route.

CMN750 CAS:2610-05-1 ***HR: D***
C.I. DIRECT BLUE 1, TETRASODIUM SALT
mf: $C_{34}H_{28}N_6O_{16}S_4 \cdot 4Na$ mw: 996.88

SYNS: AIREDALE BLUE FFD ◇ AMANIL SKY BLUE 6B ◇ ATLANTIC RESIN FAST BLUE ◇ BELAMINE SKY BLUE FF ◇ CALCODUR RESIN FAST BLUE ◇ CHICAGO BLUE 6B ◇ CHLORAZOL SKY BLUE FF ◇ CHROME LEATHER SKY BLUE ◇ C.I. 24410 ◇ C.I. DIRECT BLUE 1 ◇ DIACOTTON SKY BLUE 6B ◇ DIPHENYL BRILLIANT BLUE FF ◇ DIRECT BRILLIANT BLUE FF ◇ ENIANIL BRILLIANT BLUE FF ◇ FENAMIN SKY BLUE 3F ◇ HISPAMIN SKY BLUE 6B ◇ KAYAKU DIRECT SKY BLUE 6B ◇ LUMICREASE BLUE 4GL ◇ NAPHTAMINE SKY BLUE DD ◇ NCI-C61109 ◇ PONTAMINE SKY BLUE ◇ PYRAZOL FAST BRILLIANT BLUE VP ◇ SHIKISO DIRECT SKY BLUE 6B ◇ TERTRODIRECT BLUE FF ◇ VONDACEL BLUE FF

TOXICITY DATA with REFERENCE
mmo-sat 100 nmol/plate MUREAV 136,33,84
mma-sat 100 nmol/plate MUREAV 97,103,82
ipr-rat TDLo:200 mg/kg (8D preg):TER PSEBAA 127,215,68
ipr-rat TDLo:140 mg/kg (8D preg):REP PSEBAA 127,215,68

CONSENSUS REPORTS: Reported in EPA TSCA Inventory.

SAFETY PROFILE: An experimental teratogen. Other experimental reproductive effects. Mutation data reported. When heated to decomposition it emits very toxic fumes of NO_x, Na_2O, and SO_x.

CMO000 CAS:2602-46-2 ***HR: 3***
C.I. DIRECT BLUE 6, TETRASODIUM SALT
mf: $C_{32}H_{20}N_6O_{14}S_4 \cdot 4Na$ mw: 932.78

PROP: A dye.

SYNS: AIREDALE BLUE 2BD ◇ AIZEN DIRECT BLUE 2BH ◇ AMANIL BLUE 2BX ◇ ATLANTIC BLUE 2B ◇ ATUL DIRECT BLUE 2B ◇ AZOCARD BLUE 2B ◇ AZOMINE BLUE 2B ◇ BELAMINE BLUE 2B ◇ BENCIDAL BLUE 2B ◇ BENZANIL BLUE 2B ◇ BENZO BLUE GS ◇ BLUE 2B ◇ BRASILAMINA BLUE 2B ◇ CALCOMINE BLUE 2B ◇ CHLORAMINE BLUE 2B ◇ CHLORAZOL BLUE B ◇ CHROME

LEATHER BLUE 2B ◇ C.I. 22610 ◇ CRESOTINE BLUE 2B ◇ DIACOTTON BLUE BB ◇ DIAMINE BLUE 2B ◇ DIAPHTAMINE BLUE BB ◇ DIAZINE BLUE 2B ◇ DIAZOL BLUE 2B ◇ DIPHENYL BLUE 2B ◇ DIRECT BLUE 6 ◇ ENIANIL BLUE 2BN ◇ FENAMIN BLUE 2B ◇ FIXANOL BLUE 2B ◇ HISPAMIN BLUE 2B ◇ INDIGO BLUE 2B ◇ KAYAKU DIRECT ◇ MITSUI DIRECT BLUE 2BN ◇ NAPHTAMINE BLUE 2B ◇ NB2B ◇ NCI-C54579 ◇ NIAGARA BLUE 2B ◇ NIPPON BLUE BB ◇ PARAMINE BLUE 2B ◇ PHENAMINE BLUE BB ◇ PHENO BLUE 2B ◇ PONTAMINE BLUE BB ◇ SODIUM DIPHENYL-4,4′-BIS-AZO-2″-8″-AMINO-1″-NAPHTHOL-3″,6″ DISULPHONATE ◇ TERTRODIRECT BLUE 2B ◇ VONDACEL BLUE 2B

TOXICITY DATA with REFERENCE
mma-sat 100 nmol/plate MUREAV 136,33,94
dnd-rat-ipr 61200 μg/kg TXCYAC 32,315,84
ipr-rat TDLo:200 mg/kg (8D preg):TER PSEBAA 127,215,68
ipr-rat TDLo:140 mg/kg (8D preg):REP PSEBAA 127,215,68
orl-rat TDLo:5250 mg/kg/5W-C:CAR NCITR* NCI-CG-TR-108,78
scu-rat TDLo:750 mg/kg/27W-I:ETA BJEPA5 38,291,57
orl-rat TD:2310 mg/kg/4W-C:NEO NCITR* NCI-CG-TR-108,78

CONSENSUS REPORTS: NTP Fifth Annual Report on Carcinogens. IARC Cancer Review: Human Limited Evidence IMEMDT 29,311,82; Animal Sufficient Evidence IMEMDT 29,311,82. NCI Carcinogenesis Bioassay (feed); Clear Evidence: rat NCITR* NCI-CG-TR-108,78; No Evidence: mouse NCITR* NCI-CG-TR-108,78. Reported in EPA TSCA Inventory. Community Right-To-Know List.

SAFETY PROFILE: Confirmed carcinogen with experimental carcinogenic, neoplastigenic, and tumorigenic data. Experimental teratogenic and reproductive effects. Mutation data reported. When heated to decomposition it emits very toxic fumes of NO_x, Na_2O, and SO_x.

CMO250 CAS:72-57-1 ***HR: 3***
C.I. DIRECT BLUE 14, TETRASODIUM SALT
mf: $C_{34}H_{28}N_6O_{14}S_4 \cdot 4Na$ mw: 964.88

SYNS: AMANIL SKY BLUE ◇ AMIDINE BLUE 4B ◇ AZIDINE BLUE 3B ◇ AZURRO DIRETTO 3B ◇ BENCIDAL BLUE 3B ◇ BENZAMINE BLUE ◇ BENZO BLUE ◇ BLEU DIAMINE ◇ BLUE EMB ◇ BRASILAMINA BLUE 3B ◇ CENTRALINE BLUE 3B ◇ CHLORAMINE BLUE ◇ CHLORAZOL BLUE 3B ◇ CHROME LEATHER BLUE 3B ◇ C.I. 23850 ◇ C.I. DIRECT BLUE 14 ◇ CONGOBLAU 3B ◇ CONGO BLUE ◇ CRESOTINE BLUE 3B ◇ DIAMINE BLUE 3B ◇ DIANILBLAU ◇ DIANIL BLUE ◇ DIAZINE BLUE 3B ◇ DIPHENYL BLUE 3B ◇ DIRECT BLUE 14 ◇ HISPAMIN BLUE 3BX ◇ NAPHTAMINE BLUE 2B ◇ NAPHTHYLAMINE BLUE ◇ NCI-C61289 ◇ NIAGARA BLUE ◇ ORION BLUE 3B ◇ PARAMINE BLUE 3B ◇ PARKIBLEU ◇ PARKIPAN ◇ PONTAMINE BLUE 3BX ◇ PYRAZOL BLUE 3B ◇ PYROTROPBLAU ◇ RCRA WASTE NUMBER U236 ◇ RENOLBLAU 3B ◇ SODIUM DITOLYLDIAZOBIS-8-AMINO-1-NAPHTHOL-3,6-DISULFONATE ◇ SODIUM DITOLYLDIAZOBIS-8-AMINO-1-NAPHTHOL-3,6-DISULPHONATE ◇ TB ◇ TRIANOL DIRECT BLUE 3B ◇ TRIPAN BLUE ◇ TRYPANBLAU (GERMAN) ◇ TRYPAN BLUE ◇ TRYPAN BLUE SODIUM SALT

TOXICITY DATA with REFERENCE
mma-sat 250 μg/plate MUREAV 56,249,78
dnd-esc 20 μmol/L MUREAV 89,95,81
dns-rat:lvr 10 μmol/L MUREAV 136,255,84
bfa-rat/sat 500 mg/kg MUREAV 156,131,85
dns-ham:lvr 100 μmol/L MUREAV 136,255,84
scu-rat TDLo:10 mg/kg (female 7D post):REP ESKHA5 (103),75,85
scu-ham TDLo:20 μg/kg (female 8D post):TER LIFSAK 8,525,69
scu-rat TDLo:630 mg/kg/43W-I:CAR BJEPA5 33,524,52
par-rat TDLo:250 mg/kg/10W-I:ETA CANCAR 5,792,52
scu-rat TD:7500 mg/kg/86W-I:NEO LAINAW 12,1221,63
ipr-rat LDLo:300 mg/kg PSEBAA 31,825,34
scu-rat LDLo:300 mg/kg PSEBAA 31,825,34
ivn-rat LDLo:300 mg/kg PSEBAA 31,825,34
ipr-mus LDLo:350 mg/kg PSEBAA 31,825,34
scu-mus LD50:267 mg/kg NNAPBA 267,31,70
ivn-mus LD50:328 mg/kg TXAPA9 23,537,72
ipr-rbt LDLo:400 mg/kg PSEBAA 31,825,34
ivn-rbt LDLo:100 mg/kg PSEBAA 31,825,34
ipr-gpg LDLo:250 mg/kg PSEBAA 31,825,34
scu-gpg LDLo:300 mg/kg PSEBAA 31,825,34

CONSENSUS REPORTS: IARC Cancer Review: Group 2B IMEMDT 7,56,87; Animal Sufficient Evidence IMEMDT 8,267,75. EPA Genetic Toxicology Program. Reported in EPA TSCA Inventory.

SAFETY PROFILE: Suspected carcinogen with experimental carcinogenic, neoplastigenic, and tumorigenic data. Poison by intraperitoneal, intravenous, and subcutaneous routes. Experimental teratogenic and reproductive effects. Mutation data reported. When heated to decomposition it emits very toxic fumes of NO_x, Na_2O, and SO_x.

CMO500 CAS:2429-73-4 ***HR: D***
C.I. DIRECT BLUE 15, TETRASODIUM SALT
mf: $C_{34}H_{28}N_6O_{16}S_4 \cdot 4Na$ mw: 996.88

SYNS: AIREDALE BLUE D ◇ AIZEN DIRECT SKY BLUE 5BH ◇ BELAMINE SKY BLUE A ◇ BENZO SKY BLUE A-CF ◇ CHLORAMINE SKY BLUE 4B ◇ C.I. 24400 ◇ DIACOTTON SKY BLUE 5B ◇ DIAPHTAMINE PURE BLUE ◇ DIPHENYL SKY BLUE 6B ◇ DIRECT PURE BLUE ◇ ENIANIL PURE BLUE AN ◇ FENAMIN SKY BLUE ◇ HISPAMIN SKY BLUE 3B ◇ NAPHTAMINE BLUE 10G ◇ NCI-C61290 ◇ NIAGARA SKY BLUE ◇ PHENAMINE SKY BLUE A ◇ SHIKISO DIRECT SKY BLUE 5B ◇ SKY BLUE 5B ◇ TERTRODIRECT BLUE F ◇ VONDACEL BLUE HH

TOXICITY DATA with REFERENCE
mmo-sat 500 μg/plate VHTODE 22,413,80
mma-sat 300 nmol/plate MUREAV 116,305,83

ipr-rat TDLo:140 mg/kg (8D preg):TER TJADAB 2,85,69
ipr-rat TDLo:70 mg/kg (8D preg):REP PSEBAA 127,215,68

CONSENSUS REPORTS: Reported in EPA TSCA Inventory.

SAFETY PROFILE: An experimental teratogen. Other experimental reproductive effects. Mutation data reported. When heated to decomposition it emits very toxic fumes of NO_x, Na_2O, and SO_x.

CMO750 CAS:16071-86-6 ***HR: 3***
C.I. DIRECT BROWN
mf: $C_{31}H_{20}N_6O_9S{\cdot}Cu{\cdot}2Na$ mw: 762.15

SYNS: AIZEN PRIMULA BROWN BRLH ◇ AMANIL SUPRA BROWN LBL ◇ ATLANTIC RESIN FAST BROWN BRL ◇ BENZAMIL SUPRA BROWN BRLL ◇ CALCODUR BROWN BRL ◇ CHLORAMINE FAST BROWN BRL ◇ CHROME LEATHER BROWN BRLL ◇ C.I. 30145 ◇ DERMA FAST BROWN W-GL ◇ DIPHENYL FAST BROWN BRL ◇ DIRECT BROWN 95 ◇ NCI-C54568 ◇ SATURN BROWN LBR ◇ SOLAR BROWN PL ◇ TETRAMINE FAST BROWN BRS

TOXICITY DATA with REFERENCE
mmo-sat 100 nmol/plate MUREAV 136,147,84
mma-sat 30 nmol/plate MUREAV 136,147,84
dns-rat-orl 100 mg/kg MUREAV 136,147,84
orl-rat TDLo:2625 mg/kg/5W-C:NEO NCITR* NCI-CG-TR-108,78

CONSENSUS REPORTS: IARC Cancer Review: Animal Limited Evidence IMEMDT 29,321,82; Human Limited Evidence IMEMDT 29,321,82; NCI Carcinogenesis Bioassay (feed); Clear Evidence: rat NCITR* NCI-CG-TR-108,78; No Evidence: mouse NCITR* NCI-CG-TR-108,78. Reported in EPA TSCA Inventory. Community Right-To-Know List.

SAFETY PROFILE: Suspected carcinogen with experimental carcinogenic and neoplastigenic data. Mutation data reported. When heated to decomposition it emits very toxic fumes of Na_2O, SO_x, and NO_x.

CMP000 CAS:2586-60-9 ***HR: 3***
C.I. DIRECT VIOLET 1, DISODIUM SALT
mf: $C_{32}H_{22}N_6O_8S_2{\cdot}2Na$ mw: 728.70

SYNS: AIREDALE VIOLET ND ◇ AMANIL FAST VIOLET N ◇ ATLANTIC VIOLET N ◇ ATUL DIRECT VIOLET N ◇ AZOCARD VIOLET N ◇ BENCIDAL FAST VIOLET N ◇ BENZANIL VIOLET N ◇ BENZO VIOLET N ◇ BRASILAMINA VIOLET 3R ◇ CALCOMINE VIOLET N ◇ CHLORAZOL VIOLET N ◇ C.I. 22570 ◇ COTTON VIOLET R ◇ DIAMINE VIOLET N ◇ DIAPHTAMINE VIOLET N ◇ DIAZINE VIOLET N ◇ DIAZOL VIOLET N ◇ DIRECT FAST VIOLET N ◇ DIRECT VIOLET C ◇ ERIE VIOLET 3R ◇ FIXANOL VIOLET N ◇ HISPAMIN VIOLET 3R ◇ JAPANOL VIOLET J ◇ NAPHTAMINE VIOLET N ◇ PARAMINE FAST VIOLET N ◇ PHENO VIOLET N ◇ PONTAMINE VIOLET N ◇ TERTRODIRECT VIOLET N ◇ TRISULFON VIOLET N

TOXICITY DATA with REFERENCE
orl-rat TDLo:1125 g/kg/71W-C:ETA VOONAW 23(7),72,77

SAFETY PROFILE: Questionable carcinogen with experimental tumorigenic data. When heated to decomposition it emits very toxic fumes of NO_x, Na_2O, and SO_x.

CMP250 CAS:8062-14-4 ***HR: 2***
C.I. FOOD BROWN 1
mf: $C_{18}H_{14}N_6O_6S_2{\cdot}C_{13}H_{13}N_4O_3S{\cdot}3Na$ mw: 848.83

SYNS: 1545 BROWN ◇ BROWN FK ◇ 4,4'-((4,6-DIAMINO-m-PHENYLENE)BIS(AZO))DIBENZENESULFONIC ACID, DISODIUM SALT mixed with p-((4,6-DIAMINO-m-TOLYL)AZO)-BENZENESULFONIC ACID, SODIUM SALT ◇ EEC SERIAL No. 124 ◇ GOLDEN BROWN RK-FQ

TOXICITY DATA with REFERENCE
mmo-esc 1 g/L FCTXAV 18,215,80
mma-esc 10 g/L FCTXAV 18,215,80
orl-rat LDLo:4000 mg/kg FCTXAV 6,1,68
ipr-rat LD50:750 mg/kg FCTXAV 6,737,68
ipr-mus LDLo:1500 mg/kg FCTXAV 6,1,68

CONSENSUS REPORTS: EPA Genetic Toxicology Program.

SAFETY PROFILE: Moderately toxic by ingestion and intraperitoneal routes. Mutation data reported. When heated to decomposition it emits very toxic fumes of NO_x, Na_2O, and SO_x.

CMP500 CAS:4553-89-3 ***HR: 3***
C.I. FOOD BROWN 3, DISODIUM SALT
mf: $C_{27}H_{20}N_4O_9S_2{\cdot}2Na$ mw: 654.61

SYNS: C.I. 20285 ◇ 2,4-DIHYDROXY-3,5-DI(4-SULPHO-1-NAPHTHYLAZO)BENZYL ALCOHOL, DISODIUM SALT

TOXICITY DATA with REFERENCE
ipr-rat LD50:375 mg/kg FCTXAV 4,143,66
ipr-mus LD50:220 mg/kg FCTXAV 4,143,66

SAFETY PROFILE: Poison by intraperitoneal route. When heated to decomposition it emits very toxic fumes of NO_x, Na_2O, and SO_x.

CMP800 ***HR: 3***
CIGARETTE REFINED TAR

SYNS: CIGARETTE TAR ◇ COLOMBIAN BLACK TOBACCO CIGARETTE REFINED TAR ◇ TAR, from tobacco ◇ TOBACCO REFINED TAR ◇ TOBACCO TAR ◇ U.S. BLENDED LIGHT TOBACCO CIGARETTE REFINED TAR

TOXICITY DATA with REFERENCE
mmo-sat 1 mg/L GANNA2 69,85,78
otr-ham:emb 1 mg/L GANNA2 69,85,78
otr-ham:lng 10 mg/L BJCAAI 25,574,71
cyt-ham:lng 10 mg/L BJCAAI 25,574,71
skn-mus TDLo:118 g/kg/52W-I:CAR CANCAR 21,376,68

ivg-mus TDLo:1320 mg/kg/44W-I:CAR,REP JNCIAM 23,1,59
skn-mus TD:235 g/kg/52W-I:CAR CANCAR 21,376,68

SAFETY PROFILE: Suspected carcinogen with experimental carcinogenic data. Experimental reproductive effects. Mutation data reported. See also TOBACCO and NICOTINE.

CMP825 CAS:73963-72-1 ***HR: 1***
CILOSTAZOL
mf: $C_{20}H_{27}N_5O_2$ mw: 369.52

SYNS: 6-(4-(1-CYCLOHEXYL-1H-TETRAZOL-5-YL)BUTOXY)-3,4-DIHYDRO-2(1H)-QUINOLINONE ◇ 3,4-DIHYDRO-6-(4-(1-CYCLOHEXYL-1H-TETRAZOL-5-YL)BUTOXY)-2(1H)-QUINOLINONE ◇ OPC-13013

TOXICITY DATA with REFERENCE
orl-rat TDLo:1650 mg/kg (female 17-22D post):REP IYKEDH 16,1073,85
orl-rat TDLo:11 g/kg (female 7-17D post):TER IYKEDH 16,1053,85
orl-man TDLo:1248 μg/kg:CNS ARZNAD 35,1173,85

SAFETY PROFILE: Human systemic effects by ingestion: headache. An experimental teratogen. Experimental reproductive effects. A vasodilator. When heated to decomposition it emits toxic fumes of NO_x.

CMP875 ***HR: D***
CIMETIDINE mixed with SODIUM NITRITE (4:1)
mf: $C_{10}H_{16}N_6S{\bullet}NNaO_2$ mw: 321.38

SYN: SODIUM NITRITE mixed with CIMETIDINE (1:4)

TOXICITY DATA with REFERENCE
mmo-sat 39 μg/plate LANCAO 2,993,81
mma-sat 80 μmol/plate CRNGDP 1,925,80
dnr-esc 2500 μg/well LANCAO 3,993,81

SAFETY PROFILE: Mutation data reported. When heated to decomposition it emits toxic fumes of SO_x, NO_x, CN^-, and Na_2O. See also TAGAMET and SODIUM NITRITE.

CMP885 CAS:539-35-5 ***HR: 2***
CINAMONIN
mf: $C_9H_{15}NO_3S$ mw: 217.31

SYNS: ACIDOMYCIN ◇ ACTITHIAZIC ACID ◇ MYCOBACIDIN

TOXICITY DATA with REFERENCE
ipr-mus LD50:2500 mg/kg TDKNAF 14,60,55
scu-mus LD50:20 g/kg 85GDA2 4(1),143,80
ivn-mus LD50:1500 mg/kg 85GDA2 4(1),143,80

SAFETY PROFILE: Moderately toxic intravenous and intraperitoneal routes. Mildly toxic by subcutaneous routes. When heated to decomposition it emits toxic fumes of SO_x and NO_x.

CMP900 CAS:54-84-2 ***HR: 3***
CINANSERIN HYDROCHLORIDE
mf: $C_{20}H_{24}N_2OS{\bullet}ClH$ mw: 376.98

SYNS: n-(2-((3-(DIMETHYLAMINO)PROPYL)THIO)PHENYL)-3-PHENYL-2-PROPENAMIDE MONOHYDROCHLORIDE ◇ 2′-((3-(DIMETHYLAMINO)PROPYL)THIO)CINNAMANILIDEHYDROCHLORIDE ◇ MAPTC ◇ SQ 10,643

TOXICITY DATA with REFERENCE
orl-rat LD50:1500 mg/kg AIPTAK 152,132,64
orl-mus LD50:480 mg/kg AIPTAK 152,132,64
ivn-mus LD50:35 mg/kg AIPTAK 152,132,64
ivn-dog LDLo:8 mg/kg AIPTAK 152,132,64

SAFETY PROFILE: Poison by intravenous route. Moderately toxic by ingestion. When heated to decomposition it emits toxic fumes of SO_x, NO_x, and HCl.

CMP925 CAS:118-10-5 ***HR: 3***
d-CINCHONINE
mf: $C_{19}H_{22}N_2O$ mw: 294.43

PROP: Occurs in most varieties of cinchona bark, especially in bark of *Cinchona micrantha R. & P., Rubiaceae*. Prisms, needles from alcohol or ether. Mp: about 265°; begins to sublime at 220°. One gram dissolves in 60 mL alc, 25 mL boiling alc, 110 mL chloroform, 500 mL ether. Practically insol in water.

SYNS: α-4-QUINOLYL-5-VINYL-2-QUINUCLIDINEMETHANOL ◇ α-(5-VINYL-2-QUINOLYL)-2-QUINUCLIDINEMETHANOL

TOXICITY DATA with REFERENCE
ipr-rat LD50:152 mg/kg APTOA6 4,265,48
scu-mus LDLo:400 mg/kg AEPPAE 205,129,48
scu-frg LDLo:200 mg/kg AEPPAE 205,129,48

CONSENSUS REPORTS: Reported in EPA TSCA Inventory.

SAFETY PROFILE: Poison by intraperitoneal and subcutaneous routes. When heated to decomposition it emits toxic fumes of NO_x.

CMP950 CAS:20168-99-4 ***HR: 3***
CINDOMET
mf: $C_{21}H_{19}NO_4$ mw: 349.39

PROP: Yellow crystals from acetone. Loses water. Mp: 170-172°.

SYNS: CINMETACIN ◇ CINMETHACIN ◇ 1-CINNAMOYL-5-METHOXY-2-METHYLINDOLE-3-ACETIC ACID ◇ 1-CINNAMOYL-2-METHOXY-5-METHOXY-3-INDOLYLACETIC ACID ◇ INDOLACIN ◇ 5-METHOXY-2-METHYL-1-(1-OXO-3-PHENYL-2-PROPENYL)-1H-INDOLE-3-ACETIC ACID (9CI)

TOXICITY DATA with REFERENCE
orl-rat LD50:1020 mg/kg ARZNAD 23,1690,73
ipr-rat LD50:590 mg/kg ARZNAD 23,1690,73

orl-mus LD50:750 mg/kg ARZNAD 23,1690,73
ipr-msu LD50:360 mg/kg ARZNAD 23,1690,73

SAFETY PROFILE: Poison by intraperitoneal route. Moderately toxic by ingestion. An anti-inflammatory agent. When heated to decomposition it emits toxic fumes of NO_x.

CMP969 CAS:104-55-2 *HR: 3*
CINNAMALDEHYDE
mf: C_9H_8O mw: 132.17

PROP: Found in Ceylon and Chinese cinnamon oils. Yellowish, oily liquid; strong odor of cinnamon. D: 1.048-1.052, mp: −7.5°, bp: 246.0° (some decomp), d: 1.048-1.052 @ 25°/25°, refr index: 1.619-1.623, flash p: 248°F. Very sltly sol in water; misc with alc, ether, chloroform, fixed oils.

SYNS: BENZYLIDENEACETALDEHYDE ◇ CASSIA ALDEHYDE ◇ CINNAMAL ◇ CINNAMYL ALDEHYDE ◇ CINNIMIC ALDEHYDE ◇ FEMA No. 2286 ◇ NCI-C56111 ◇ PHENYLACROLEIN ◇ 3-PHENYLACROLEIN ◇ 3-PHENYLPROPENAL ◇ 3-PHENYL-2-PROPENAL ◇ ZIMTALDEHYDE

TOXICITY DATA with REFERENCE
skn-hmn 40 mg/48H SEV FCTXAV 17,253,79
mma-sat 500 μg/plate FCTOD7 22,623,84
sln-dmg-par 2 pph ENMUDM 7,677,85
dni-mus:leu 31500 μg/L DCTODJ 6,521,83
oms-mus:leu 31500 μg/L DCTODJ 6,521,83
cyt-ham:fbr 15 mg/L FCTOD7 22,623,84
orl-rat LD50:2220 mg/kg FCTXAV 2,327,64
orl-mus LD50:2225 mg/kg YKKZAJ 92,135,72
ipr-mus LD50:610 mg/kg YKKZAJ 92,135,72
ivn-mus LD50:75 mg/kg CSLNX* NX#07571
par-mus LDLo:200 mg/kg CBCCT* 7,687,55
orl-gpg LD50:1160 mg/kg FCTXAV 2,327,64

CONSENSUS REPORTS: Reported in EPA TSCA Inventory.

SAFETY PROFILE: Poison by intravenous and parenteral routes. Moderately toxic by ingestion and intraperitoneal routes. A severe human skin irritant. Mutation data reported. Combustible liquid. May ignite after a delay period in contact with NaOH. When heated to decomposition it emits acrid smoke and fumes. See also ALDEHYDES.

CMP975 CAS:621-82-9 *HR: 3*
CINNAMIC ACID
mf: $C_9H_8O_2$ mw: 148.17

PROP: Occurs free and partly esterified in storax, balsam Peru or Tolu, oil of cinnamon, coca leaves. White monoclinic crystals; honey floral odor. D: (4/4) 1.2475, mp: 133°, bp: 300°, flash p: +212°F. One gram dissolves in about 2000 mL water at 25° (more sol in hot water), in 6 mL alc, 5 mL methanol, 12 mL chloroform. Freely sol in benzene, ether, acetone, glacial acetic acid, carbon disulfide, fixed oils.

SYNS: FEMA No. 2288 ◇ PHENYLACRYLIC ACID ◇ tert-β-PHENYLACRYLIC ACID ◇ 3-PHENYLACRYLIC ACID ◇ 3-PHENYLPROPENOIC ACID ◇ 3-PHENYL-2-PROPENOIC ACID ◇ ZIMTSAEURE (GERMAN)

TOXICITY DATA with REFERENCE
skn-rbt 500 mg/24H MLD FCTXAV 16,687,78
orl-rat LD50:2500 mg/kg FCTXAV 16,687,78
ipr-rat LD50:1600 mg/kg BCFAAI 112,53,73
orl-mus LD50:5 g/kg GISAAA 46(1),94,81
ipr-mus LD50:160 mg/kg FCTXAV 16,687,78
ivn-mus LD50:380 mg/kg ARZNAD 19,617,69

CONSENSUS REPORTS: Reported in EPA TSCA Inventory.

SAFETY PROFILE: Poison by intravenous and intraperitoneal routes. Moderately toxic by ingestion. A skin irritant. Combustible liquid. When heated to decomposition it emits acrid smoke and fumes.

CMQ000 CAS:63938-16-9 *HR: 3*
CINNAMIC ACID, NICKEL(II) SALT
mf: $C_{18}H_{14}O_4{\cdot}Ni$ mw: 353.03

TOXICITY DATA with REFERENCE
par-mus LDLo:40 mg/kg CBCCT* 7,687,55

CONSENSUS REPORTS: Nickel and its compounds are on the Community Right-To-Know List.
NIOSH REL: (Nickel, Inorganic) TWA 0.015 mg(Ni)/m^3

SAFETY PROFILE: Poison by parenteral route. See also NICKEL COMPOUNDS. When heated to decomposition it emits acrid smoke and irritating fumes.

CMQ475 CAS:3669-32-7 *HR: 3*
CINNAMOHYDROXAMIC ACID
mf: $C_9H_9NO_2$ mw: 163.19

SYNS: CINNAMOYLHYDROXAMIC ACID ◇ N-HYDROXY-3-PHENYL-2-PROPENAMIDE (9CI)

TOXICITY DATA with REFERENCE
orl-rat LD50:2140 mg/kg TXAPA9 28,313,74
orl-mus LD50:1350 mg/kg BJPCAL 26,41,66
scu-mus LD50:86 mg/kg BJPCAL 26,41,66

SAFETY PROFILE: Poison by subcutaneous route. Moderately toxic by ingestion. When heated to decomposition it emits toxic fumes of NO_x.

CMQ500 CAS:4360-47-8 *HR: 3*
CINNAMONITRILE
mf: C_9H_7N mw: 129.17

SYN: CINNAMYL NITRILE

TOXICITY DATA with REFERENCE
skn-rbt 500 mg/24H MOD FCTXAV 14,659,76
orl-rat LD50:4150 mg/kg FCTXAV 14,659,76
ivn-rbt LDLo:34 mg/kg COREAF 153,895,11
scu-gpg LDLo:130 mg/kg COREAF 153,895,11

CONSENSUS REPORTS: Cyanide and its compounds are on the Community Right-To-Know List. Reported in EPA TSCA Inventory.

SAFETY PROFILE: Poison by subcutaneous and intravenous routes. Mildly toxic by ingestion. A skin irritant. See also NITRILES. When heated to decomposition it emits toxic fumes of NO_x and CN^-.

CMQ625 CAS:751-01-9 ***HR: 3***
14-CINNAMOYLOXYCODEINONE
mf: $C_{27}H_{25}NO_5$ mw: 443.53

TOXICITY DATA with REFERENCE
orl-mus LD50:1100 mg/kg JPPMAB 17,759,65
scu-mus LD50:530 mg/kg JPPMAB 17,759,65
ivn-mus LD50:31 mg/kg JPPMAB 17,759,65

SAFETY PROFILE: Poison by intravenous route. Moderately toxic by ingestion and subcutaneous route. When heated to decomposition it emits toxic fumes of NO_x.

CMQ725 CAS:1405-39-6 ***HR: 3***
CINNAMYCIN

SYNS: ANTIBIOTIC NSC-71936 ◇ NSC-71936

TOXICITY DATA with REFERENCE
ipr-mus LD50:2 mg/kg 85GDA2 4(1),249,80
scu-mus LD50:400 mg/kg 85GDA2 4(1),249,80
ivn-mus LD50:2500 μg/kg 85GDA2 4(1),249,80

SAFETY PROFILE: Poison by subcutaneous, intraperitoneal, and intravenous routes.

CMQ730 CAS:103-54-8 ***HR: 2***
CINNAMYL ACETATE
mf: $C_{11}H_{12}O_2$ mw: 176.23

PROP: Colorless liquid; sweet floral odor. D: 1.047-1.051, refr index: 1.539-1.543, flash p: 244°F. Misc with chloroform, ether, fixed oils; insol in glycerin, water @ 264°.

SYNS: ACETIC ACID, CINNAMYL ESTER ◇ FEMA No. 2293 ◇ Γ-PHENYLALLYL ACETATE ◇ 3-PHENYL-2-PROPEN-1-YL ACETATE

TOXICITY DATA with REFERENCE
orl-rat LD50:3300 mg/kg FCTXAV 11,1065,73
orl-mus LD50:4750 mg/kg VPITAR 33(5),48,74
ipr-mus LD50:1200 mg/kg PHMCAA 3,62,61
orl-gpg LD50:4750 mg/kg VPITAR 33(5),48,74

CONSENSUS REPORTS: Reported in EPA TSCA Inventory.

SAFETY PROFILE: Moderately toxic by ingestion and intraperitoneal routes. Combustible liquid. When heated to decomposition it emits acrid smoke and fumes. See also ALLYL COMPOUNDS.

CMQ740 CAS:104-54-1 ***HR: 2***
CINNAMYL ALCOHOL
mf: $C_9H_{10}O$ mw: 134.19

PROP: Occurs (in the esterified form) in storax and in balsam Peru, cinnamon leaves, hyacinth oil. Needles or crystalline mass; odor of hyacinth. Mp: 33°, d: 1.0397, bp: 250.0°, n (20/D) 1.58190. Sol in water, glycerol, and propylene glycol; freely sol in alc, ether, other common organic solvents.

SYNS: CINNAMIC ALCOHOL ◇ CINNAMYL ALCOHOL, SYNTHETIC ◇ FEMA No. 2294 ◇ Γ-PHENYLALLYL ALCOHOL ◇ 3-PHENYLALLYL ALCOHOL ◇ 3-PHENYL-2-PROPEN-1-OL ◇ STYRONE ◇ STYRYL CARBINOL

TOXICITY DATA with REFERENCE
skn-rbt 500 mg/24H MOD FCTXAV 12,855,74
orl-rat LD50:2000 mg/kg FCTXAV 12,855,74
orl-mus LD50:2675 mg/kg VPITAR 33(5),48,74
orl-gpg LD50:2675 mg/kg VPITAR 33(5),48,74

CONSENSUS REPORTS: Reported in EPA TSCA Inventory.

SAFETY PROFILE: Moderately toxic by ingestion. A skin irritant. When heated to decomposition it emits acrid smoke and fumes. See also ALCOHOLS and ALLYL COMPOUNDS.

CMQ750 CAS:5320-75-2 ***HR: 2***
CINNAMYL BENZOATE
mf: $C_{16}H_{14}O_2$ mw: 238.30

SYNS: BENZOIC ACID, CINNAMYL ESTER ◇ CINNAMYL ALCOHOL, BENZOATE

TOXICITY DATA with REFERENCE
skn-rbt 500 mg/24H MOD FCTXAV 14,659,76
orl-rat LD50:4 g/kg FCTXAV 14,659,76

CONSENSUS REPORTS: Reported in EPA TSCA Inventory.

SAFETY PROFILE: Moderately toxic by ingestion. A skin irritant. See also ESTERS. When heated to decomposition it emits acrid smoke and irritating fumes.

CMQ800 CAS:103-61-7 ***HR: 1***
CINNAMYL BUTYRATE
mf: $C_{13}H_{16}O_2$ mw: 204.29

SYNS: BUTYNOIC ACID, 3-PHENYL-2-PROPENYL ESTER ◇ BUTYRIC ACID, CINNAMYL ESTER ◇ PHENYLPROPENYL n-BUTYRATE

TOXICITY DATA with REFERENCE
skn-rbt 500 mg/24H FCTXAV 16,691,78

CONSENSUS REPORTS: Reported in EPA TSCA Inventory.

SAFETY PROFILE: A skin irritant. When heated to decomposition it emits acrid smoke and irritating fumes.

CMR100 CAS:298-57-7 ***HR: 3***
1-CINNAMYL-4-(DIPHENYLMETHYL)PIPERAZINE
mf: $C_{26}H_{28}N_2$ mw: 368.56

SYNS: trans-1-CINNAMYL-(4-DIPHENYLMETHYL)PIPERAZINE ◇ CINNARIZINE ◇ DIMITRON ◇ DIMITRONAL ◇ FOLCODAL ◇ GLANIL ◇ LABYRIN ◇ LAZETA ◇ MARISAN ◇ 516 MD ◇ MIDRONAL ◇ MITRONAL ◇ PIPERAZINE, 1-CINNAMYL-4-(DIPHENYLMETHYL)- ◇ PIPERAZINE, 1-(DIPHENYLMETHYL)-4-(3-PHENYL-2-PROPENYL)-(9CI) ◇ R 516 ◇ R 1575 ◇ SEPAN ◇ STUGERON ◇ STUTGERON ◇ STUTGIN ◇ TOLIMAN

TOXICITY DATA with REFERENCE
orl-rat TDLo:3750 mg/kg (female 30D pre):REP KSRNAM 16,1748,82
orl-wmn TDLo:252 mg/kg/12W-I:SKN BJDEAZ 112,607,85
ipr-rat LD50:1050 mg/kg MDACAP 21,443,85
ivn-rat LD50:24 mg/kg KSRNAM 16,1735,82
ipr-mus LD50:730 mg/kg MDACAP 21,443,85
ivn-mus LD50:22 mg/kg KSRNAM 16,1735,82

SAFETY PROFILE: Poison by intravenous route. Moderately toxic by intraperitoneal route. Experimental reproductive effects. Human systemic effects by ingestion: allergic dermatitis. When heated to decomposition it emits toxic fumes of NO_x.

CMR250 CAS:64043-53-4 ***HR: 3***
d-CINNAMYLEPHEDRINE HYDROCHLORIDE
mf: $C_{19}H_{23}NO•ClH$ mw: 317.89

SYN: CINNAMYLEPHEDRINE HYDROCHLORIDE, DEXTRO

TOXICITY DATA with REFERENCE
idr-hmn TDLo:143 ng/kg:SKN JPETAB 76,295,42
scu-mus LD50:75 mg/kg JPETAB 76,295,42

SAFETY PROFILE: Poison by subcutaneous route. Human systemic effects by very small amounts administered intradermally: unspecified skin effects. When heated to decomposition it emits very toxic fumes of HCl and NO_x.

CMR500 CAS:104-65-4 ***HR: 2***
CINNAMYL FORMATE
mf: $C_{10}H_{10}O_2$ mw: 162.20

PROP: Colorless liquid; balsamic odor. D: 1.077-1.082, refr index: 1.550-1.556, flash p: 212°F. Misc with alc, chloroform, ether, fixed oils; insol in water @ 250°.

SYNS: CINNAMYL ALCOHOL, FORMATE ◇ CINNAMYL METHANOATE ◇ FEMA No. 2299 ◇ FORMIC ACID, CINNAMYL ESTER ◇ 3-PHENYL-2-PROPEN-1-YL FORMATE

TOXICITY DATA with REFERENCE
orl-rat LD50:2900 mg/kg FCTXAV 14,659,76

CONSENSUS REPORTS: Reported in EPA TSCA Inventory.

SAFETY PROFILE: Moderately toxic by ingestion. See also ESTERS. Combustible liquid. When heated to decomposition it emits acrid smoke and irritating fumes.

CMR750 CAS:103-59-3 ***HR: 1***
CINNAMYL ISOBUTYRATE
mf: $C_{13}H_{16}O_2$ mw: 204.29

SYNS: ISOBUTYRIC ACID, CINNAMYL ESTER ◇ 2-METHYL-PROPANOIC ACID-3-PHENYL-2-PROPENYL ESTER

TOXICITY DATA with REFERENCE
skn-rbt 500 mg/24H MLD FCTXAV 17,509,79

CONSENSUS REPORTS: Reported in EPA TSCA Inventory.

SAFETY PROFILE: A skin irritant. When heated to decomposition it emits acrid smoke and irritating fumes. See also ESTERS.

CMR800 ***HR: 1***
CINNAMYL ISOVALERATE
mf: $C_{14}H_{18}O_2$ mw: 218.30

PROP: Colorless to sltly yellow liquid; spicy, floral, fruity odor. D: 0.991-0.996, refr index: 1.518-1.524, flash p: +212°F. Misc in alc, chloroform, ether, most oils; insol in glycerin, propylene glycol, and water @ 313°.

SYN: FEMA No. 2302

SAFETY PROFILE: Combustible liquid. When heated to decomposition it emits acrid smoke and irritating fumes.

CMR850 CAS:103-56-0 ***HR: 2***
CINNAMYL PROPIONATE
mf: $C_{12}H_{14}O_2$ mw: 190.24

PROP: Colorless to pale yellow liquid; spicy, fruity, balsamic odor. D: 1.029-1.033, refr index: 1.523-1.537, flash p: +212°F. Misc in alc, chloroform, ether, most oils; insol in glycerin, propylene glycol, and water @ 289°.

SYN: FEMA No. 2301 ◇ 3-PHENYL-2-PROPENYL PROPIONATE ◇ 3-

PHENYL-2-PROPEN-1-YL PROPIONATE ◇ PROPIONIC ACID, CINNAMYL ESTER

TOXICITY DATA with REFERENCE
skn-rbt 500 mg/24H MLD FCTXAV 12,859,74
orl-rat LD50:3400 mg/kg FCTXAV 12,859,74

CONSENSUS REPORTS: Reported in EPA TSCA Inventory.

SAFETY PROFILE: Moderately toxic by ingestion. A skin irritant. Combustible liquid. When heated to decomposition it emits acrid smoke and irritating fumes.

CMS125 CAS:60763-49-7 ***HR: 2***
CINNARIZINE CLOFIBRATE
mf: $C_{26}H_{28}N_2{\cdot}C_{10}H_{11}ClO_3$ mw: 583.22

SYNS: CLOFIBRATO de CINARIZINA (SPANISH) ◇ LM-16

TOXICITY DATA with REFERENCE
orl-rat LD50:6800 mg/kg DRFUD4 3,572,78
ipr-rat LD50:2000 mg/kg DRFUD4 3,572,78
orl-mus LD50:4300 mg/kg DRFUD4 3,572,78

SAFETY PROFILE: Moderately toxic by ingestion and intraperitoneal routes. When heated to decomposition it emits toxic fumes of Cl^- and NO_x.

CMS200 CAS:28657-80-9 ***HR: 2***
CINOXACIN
mf: $C_{12}H_{10}N_2O_5$ mw: 262.24

PROP: Light tan crystals. Mp: 261-262° (decomp). Sol in most polar organic solvents.

SYNS: CINOBAC ◇ CINX ◇ COMPOUND 64716 ◇ 1-ETHYL-1,4-DIHYDRO-4-OXO(1,3)DIOXOLO(4,5-g)CINNOLINE-3-CARBOXYLIC ACID ◇ 1-ETHYL-6,7-METHYLENEDIOXY-4(1H)-OXOCINNOLINE-3-CARBOXYLIC ACID

TOXICITY DATA with REFERENCE
orl-rat TDLo:1100 mg/kg (female 7-17D post):TER NKRZAZ 28,484,80
orl-rat TDLo:1050 mg/kg (female 14D pre):REP NKRZAZ 28,484,80
orl-rat LD50:4160 mg/kg NKRZAZ 28(Suppl 4),406,80
scu-rat LD50:1380 mg/kg IYKEDH 14,297,83
ivn-rat LD50:860 mg/kg IYKEDH 14,297,83
orl-mus LD50:2330 mg/kg IYKEDH 14,297,83
scu-mus LD50:900 mg/kg IYKEDH 14,297,83
ivn-mus LD50:850 mg/kg IYKEDH 14,297,83

SAFETY PROFILE: Moderately toxic by ingestion, subcutaneous and intravenous routes. An experimental teratogen. Other experimental reproductive effects. When heated to decomposition it emits toxic fumes of NO_x.

CMS205 CAS:62865-26-3 ***HR: 3***
C.I. PIGMENT YELLOW 35

SYNS: B-3-Zh ◇ CADMIUM GOLDEN ◇ CADMIUM LEMON ◇ CADMIUM PRIMROSE ◇ CADMIUM SULFIDE mixed with ZINC SULFIDE (1:1) ◇ C.I. 77205

TOXICITY DATA with REFERENCE
ACGIH TLV: TWA 0.01 $mg(Cd)/m^3$; Suspected Carcinogen.
NIOSH REL: (Cadmium): TWA reduce to lowest feasible level

SAFETY PROFILE: Suspected carcinogen. When heated to decomposition it emits toxic fumes of Cd.

CMS210 CAS:52214-84-3 ***HR: 2***
CIPROFIBRATE
mf: $C_{13}H_{14}Cl_2O_3$ mw: 289.17

SYNS: 2-(p-(2,2-DICHLOROCYCLOPROPYL)PHENOXY)-2-METHYL PROPIONIC ACID ◇ PROPANOIC ACID, 2-(4-(2,2-DICHLOROCYCLOPROPYL)PHENOXY)-2-METHYL- ◇ WIN 35833

TOXICITY DATA with REFERENCE
otr-rat-orl 3650 mg/kg/1Y C CNREA8 46,4601,86
orl-rat TDLo:280 mg/kg (15-21D post):REP TXCYAC 12,1,79
orl-rat TDLo:6390 mg/kg/60W-C:ETA CALEDQ 38,65,87

SAFETY PROFILE: Questionable carcinogen with experimental tumorigenic data. Experimental reproductive data. Mutation data reported. When heated to decomposition it emits toxic fumes of Cl^-.

CMS225 CAS:9008-54-2 ***HR: 3***
CIRCULIN
mf: $C_{56}H_{96}N_{12}O_{13}$ mw: 1145.64

SYN: POLYPEPTIN

TOXICITY DATA with REFERENCE
ivn-rat LD50:20 mg/kg JCINAO 28,1032,49
ims-rat LD50:23 mg/kg JCINAO 28,1032,49
ipr-mus LD50:15 mg/kg JOBAAY 56,749,48
scu-mus LD50:77 mg/kg JCINAO 29,1032,49
ivn-mus LD50:10 mg/kg JCINAO 28,1032,49

SAFETY PROFILE: Poison by subcutaneous, intramuscular, intravenous, and intraperitoneal routes. When heated to decomposition it emits toxic fumes of NO_x.

CMS227 CAS:12225-26-2 ***HR: 1***
C.I. REACTIVE BLACK 8

SYNS: CERN REAKTIVNI 8 ◇ CIBACRON BLACK B-D ◇ HELAKTYN BLACK DN ◇ OSTAZIN BLACK H-N ◇ PROCION BLACK H-N ◇ REACTIVE BLACK 8

TOXICITY DATA with REFERENCE
skn-rbt 500 mg/24H MLD 85JCAE-,1293,86

eye-rbt 500 mg/24H MLD 85JCAE-,1293,86
orl-rat LD50:9120 mg/kg 85JCAE-,1293,86

SAFETY PROFILE: Mildly toxic by ingestion. A skin and eye irritant. When heated to decomposition it emits acrid smoke and irritating fumes.

CMS230 CAS:58128-20-4 ***HR: 2***
C.I. REACTIVE YELLOW 73

SYNS: POLACTINE G YELLOW ◇ POLAKTYN YELLOW G ◇ YELLOW POLAKTIN G ◇ ZOLCIENI POLACTYNOWEJ G (POLISH)

TOXICITY DATA with REFERENCE
eye-rbt 20 mg MLD MEPAAX 30,157,79
orl-rat LD50:8200 mg/kg MEPAAX 30,157,79
ipr-rat LD50:1330 mg/kg MEPAAX 30,157,79

SAFETY PROFILE: Moderately toxic by intraperitoneal route. Mildly toxic by ingestion. An eye irritant.

CMS232 CAS:11056-12-5 ***HR: 3***
CIROLEMYCIN

SYN: U 12241

TOXICITY DATA with REFERENCE
dnd-sat 217 μg ABBIA4 120,292,67
dnd-esc 5 mg/L MUREAV 89,95,81
dnd-omi 217 μg ABBIA4 120,292,67
dnd-sal:spr 100 mmol ABBIA4 120,292,67
dnd-mam:lym 217 μg ABBIA4 120,292,67
ipr-mus LD50:5600 μg/kg 85GDA2 3,161,80
scu-mus LD50:1250 μg/kg 85GDA2 3,161,80

SAFETY PROFILE: Poison by subcutaneous and intraperitoneal routes. Mutation data reported.

CMS237 CAS:81098-60-4 ***HR: 2***
CISAPRIDE
mf: $C_{23}H_{29}ClFN_3O_4 \cdot H_2O$ mw: 484.02

SYNS: BENZAMIDE,4-AMINO-5-CHLORO-N-(1-(3-(4-FLUOROPHENOXY)PROPYL)-3-METHOXY-4-PIPERIDINYL)-2-METHOXY-, MONOHYDRATE, cis- ◇ R 51619

TOXICITY DATA with REFERENCE
orl-rat TDLo:14600 mg/kg (female 1Y pre):REP OYYAA2 35,85,88
orl-rat LD50:4166 mg/kg IYKEDH 19,599,88
ipr-rat LD50:3435 mg/kg IYKEDH 19,599,88
ivn-rat LD50:27400 μg/kg IYKEDH 19,599,88
orl-mus LD50:8715 mg/kg IYKEDH 19,599,88
ivn-mus LD50:32200 μg/kg IYKEDH 19,599,88

SAFETY PROFILE: Moderately toxic by intraperitoneal and intravenous routes. Mildly toxic by ingestion. Experimental reproductive effects. When heated to decomposition it emits toxic fumes of NO_x, F^-, and Cl^-.

CMS241 CAS:33204-76-1 ***HR: D***
CISOBITAN
mf: $C_{18}H_{28}O_4Si_4$ mw: 420.82

SYNS: CYCLIC-2,6-cis-DIPHENYLHEXAMETHYLCYCLOTETRASILOXANE ◇ (Z)-2,6-DIPHENYL-2,4,4,6,8,8-HEXAMETHYL-CYCLOTETRASILOXANE ◇ cis-2,6-DIPHENYL-2,4,4,6,8,8-HEXAMETHYLCYCLOTETRASILOXANE ◇ 2,6-cis-DIPHENYLHEXAMETHYLCYCLOTETRASILOXANE ◇ cis-2,2,4,6,6,8-HEXAMETHYL-4,8-DIPHENYL-CYCLOTETRASILOXANE ◇ KABI 1774 ◇ QUADROSILAN

TOXICITY DATA with REFERENCE
orl-rat TDLo:1 mg/kg (female 1D post):REP BIREBV 7,260,72

SAFETY PROFILE: Experimental reproductive effects. When heated to decomposition it emits acrid smoke and fumes.

CMS242 CAS:1229-55-6 ***HR: 1***
C.I. SOLVENT RED
mf: $C_{17}H_{14}N_2O_2$ mw: 278.33

SYNS: C.I. 12150 ◇ OIL PINK ◇ SOLVENT RED 1

TOXICITY DATA with REFERENCE
eye-rbt 100 mg NTIS** AD-A172-758

CONSENSUS REPORTS: Reported in EPA TSCA Inventory.

SAFETY PROFILE: An eye irritant. When heated to decomposition it emits toxic fumes of NO_x.

CMS248 ***HR: 3***
CISTANCHE TUBULOSA Wight (extract)

PROP: Indian plant belonging to the family *Orobanchaceae* (IJEBA6 18,594,80).

SYN: PHELIPAEA CALOTROPIDIS Walp., extract

TOXICITY DATA with REFERENCE
orl-rat TDLo:150 mg/kg (12-14D preg):REP IJEBA6 18,594,80
ipr-mus LD50:250 mg/kg IJEBA6 18,594,80

SAFETY PROFILE: Poison by intraperitoneal route. Experimental reproductive effects.

CMS250 CAS:1786-81-8 ***HR: 3***
CITANEST HYDROCHLORIDE
mf: $C_{13}H_{20}N_2O \cdot ClH$ mw: 256.81

SYNS: CITANEST ◇ L-67 HYDROCHLORIDE ◇ N-(2-METHYLPHENYL)-2-(PROPYLAMINO)-PROPANAMIDE MONOHYDROCHLORIDE ◇ PRILOCAINE HYDROCHLORIDE ◇ PROPITOCAINE HYDROCHLORIDE ◇ α-PROPYLAMINE-2-METHYL-PROPIONANILIDEHYDROCHLORIDE ◇ 2-(PROPYLAMINO)-o-PROPIONOTOLUIDIDE HYDROCHLORIDE

TOXICITY DATA with REFERENCE
ipr-rat LD50:148 mg/kg NIIRDN 6,692,82

scu-rat LD50:790 mg/kg NIIRDN 6,692,82
ivn-rat LD50:56600 μg/kg NIIRDN 6,692,82
ipr-mus LD50:30 mg/kg 29ZVAB -,99,69
scu-mus LD50:632 mg/kg NIIRDN 6,692,82
ivn-mus LD50:55 mg/kg NIIRDN 6,692,82

SAFETY PROFILE: Poison by intraperitoneal and intravenous routes. Moderately toxic by subcutaneous route. When heated to decomposition it emits very toxic fumes of HCl and NO_x.

CMS320 CAS:498-23-7 ***HR: 2***
CITRACONIC ACID
mf: $C_5H_6O_4$ mw: 130.11

PROP: Obtained by carefully heating citric acid at about 175°. Hygroscopic, monoclinic crystals; characteristic odor. D: 1.62, mp: about 90° with decomp. Freely sol in water, alc, ether; sltly sol in chloroform; insol in benzene or petr ether.

SYNS: (Z)-2-METHYL-2-BUTENEDIOIC ACID (9CI) ◇ cis-METHYLBUTENEDIOIC ACID ◇ 2-METHYL-2-BUTENEDIOIC ACID ◇ METHYLMALEIC ACID

TOXICITY DATA with REFERENCE
orl-rat LD50:1320 mg/kg FCTXAV 2,327,64
orl-mus LD50:2260 mg/kg FCTXAV 2,327,64
orl-gpg LD50:1350 mg/kg FCTXAV 2,327,64

SAFETY PROFILE: Moderately toxic by ingestion. When heated to decomposition it emits acrid smoke and fumes.

CMS322 CAS:616-02-4 ***HR: 3***
CITRACONIC ANHYDRIDE
mf: $C_5H_4O_3$ mw: 112.09

SYNS: CITRACONIC ACID ANHYDRIDE ◇ 3-METHYL-2,5-FURANDIONE ◇ METHYLMALEIC ANHYDRIDE ◇ α-METHYLMALEIC ANHYDRIDE ◇ 2-METHYLMALEIC ANHYDRIDE ◇ 3-METHYLMALEIC ANHYDRIDE ◇ MONOMETHYLMALEIC ANHYDRIDE

TOXICITY DATA with REFERENCE
skn-rbt 10 mg/24H open JIHTAB 26,269,44
orl-rat LD50:2600 mg/kg JIHTAB 26,269,44
skn-rbt LD50:218 mg/kg SHELL*
skn-gpg LD50:1247 mg/kg JIHTAB 26,269,44

CONSENSUS REPORTS: Reported in EPA TSCA Inventory.

SAFETY PROFILE: Poison by skin contact. Moderately toxic by ingestion. A skin irritant. When heated to decomposition it emits acrid smoke and fumes. See also ANHYDRIDES.

CMS324 CAS:66408-78-4 ***HR: 1***
CITRAL ETHYLENE GLYCOL ACETAL
mf: $C_{12}H_{20}O_2$ mw: 196.32

SYNS: CITRACETAL ◇ 1,3-DIOXOLANE, 2-(2,6-DIMETHYL-1,5-HEPTADIENYL)-

TOXICITY DATA with REFERENCE
skn-rbt 500 mg/24H MOD FCTXAV 17,749,79

CONSENSUS REPORTS: Reported in EPA TSCA Inventory.

SAFETY PROFILE: A skin irritant. When heated to decomposition it emits acrid smoke and irritating fumes.

CMS325 ***HR: 2***
CITRAL METHYLANTHRANILATE, SCHIFF'S BASE
mf: $C_{18}H_{23}NO_2$ mw: 285.42

SYN: N-(3,7-DIMETHYL-2,6-OCTADIENYLIDENE)ANTHRANILIC ACID METHYL ESTER

TOXICITY DATA with REFERENCE
skn-rbt 500 mg/24H MOD FCTOD7 20(Suppl),651,82
orl-rat LD50:3800 mg/kg FCTOD7 20(Suppl),651,82
skn-rbt LD50:2500 mg/kg FCTOD7 20(Suppl),651,82

SAFETY PROFILE: Moderately toxic by skin contact and ingestion. A skin irritant. When heated to decomposition it emits toxic fumes of NO_x.

CMS500 CAS:25425-12-1 ***HR: 3***
CITREOVIRIDIN
mf: $C_{23}H_{30}O_6$ mw: 402.53

SYNS: CITREOVIRIDINE ◇ 4-METHOXY-5-METHYL-6-(7-METHYL-8-(TETRAHYDRO-3,4-DIHYDROXY-2,4,5-TRIMETHYL-2-FURANYL)-1,3,5,7-OCTATETRAENYL)-2H-PYRAN-2-ONE

TOXICITY DATA with REFERENCE
orl-rat TDLo:60 mg/kg (female 8-11D post):TER FCTOD7 24,1315,86
orl-rat TDLo:60 mg/kg (female 8-11D post):REP FCTOD7 24,1315,86
scu-rat LD50:3600 μg/kg JJEMAG 42,91,72
orl-mus LD50:29 mg/kg JJEMAG 42,91,72
ipr-mus LD50:7200 μg/kg JJEMAG 42,91,72
scu-mus LD50:9600 μg/kg RCOCB8 59,31,88

CONSENSUS REPORTS: EPA Genetic Toxicology Program.

SAFETY PROFILE: Poison by ingestion, subcutaneous, and intraperitoneal routes. When heated to decomposition it emits acrid smoke and irritating fumes.

CMS750 CAS:77-92-9 ***HR: 3***
CITRIC ACID
mf: $C_6H_8O_7$ mw: 192.14

PROP: Colorless, odorless crystals (crystals are monoclinic holohedra and crystallize from hot concentrated aq soln); acid taste. Mp: 153° (anhyd form), bp: decomp; d: 1.665, flash p: +212°F. Sol in water, alc, ether.

SYNS: ACILETTEN ◇ CITRETTEN ◇ CITRIC ACID, anhydrous ◇ CITRO ◇ FEMA No. 2306 ◇ 2-HYDROXY-1,2,3-PROPANETRICARBOXYLIC ACID ◇ β-HYDROXYTRICARBALLYLIC ACID ◇ KYSELINA CITRONOVA (CZECH)

TOXICITY DATA with REFERENCE
skn-rbt 500 mg/24H MOD 28ZPAK -,105,72
eye-rbt 750 μg/24H SEV 28ZPAK -,105,72
orl-rat LD50:6730 mg/kg 28ZPAK -,105,72
ipr-rat LD50:883 mg/kg JPETAB 94,65,48
scu-rat LD50:5500 mg/kg TAKHAA 30,25,71
orl-mus LD50:5040 mg/kg TAKHAA 30,25,71
ipr-mus LD50:961 mg/kg JPETAB 94,65,48
scu-mus LD50:2700 mg/kg TAKHAA 30,25,71
ivn-mus LD50:42 mg/kg JPETAB 94,65,48
orl-rbt LDLo:7000 mg/kg IECHAD 15,628,23
ivn-rbt LD50:330 mg/kg JPETAB 94,65,48

CONSENSUS REPORTS: Reported in EPA TSCA Inventory.

SAFETY PROFILE: Poison by intravenous route. Moderately toxic by subcutaneous and intraperitoneal routes. Mildly toxic by ingestion. A severe eye and moderate skin irritant. An irritating organic acid, some allergenic properties. Combustible liquid. Potentially explosive reaction with metal nitrates. When heated to decomposition it emits acrid smoke and fumes.

CMS775 CAS:518-75-2 ***HR: 3***
CITRININ
mf: $C_{13}H_{14}O_5$ mw: 250.27

SYNS: ANTIMYCIN ◇ 3H-2-BENZOPYRAN-7-CARBOXYLIC ACID, 4,6-DIHYDRO-8-HYDROXY-3,4,5-TRIMETHYL-6-OXO-,(3R-trans)- ◇ (3R,4S)-4,6-DIHYDRO-8-HYDROXY-3,4,5-TRIMETHYL-6-OXO-3H-2-BENZOPYRAN-7-CARBOXYLIC ACID

TOXICITY DATA with REFERENCE
skn-gpg 40 mg/24H SEV JANCA2 57,1121,74
dnd-esc 100 mg/L AEMIDF 52,1273,86
pic-esc 300 mg/L AEMIDF 52,1273,86
dnr-bcs 20 μg/disc CNREA8 36,445,76
ipr-mus TDLo:30 mg/kg (10D post):TER FCTXAV 14,175,76
scu-rat TDLo:35 mg/kg (10D post):REP TXCYAC 25,151,82
orl-rat TDLo:25200 mg/kg/60W-C:NEO CALEDQ 17,281,83
orl-rat TD:13 g/kg/32W-C:ETA FGIGDO 5,77,81
ipr-rat LD50:67 mg/kg JPETAB 88,173,46
scu-rat LD50:67 mg/kg JPETAB 88,173,46
orl-mus LD50:112 mg/kg TXAPA9 37,139,76
ipr-mus LD50:35 mg/kg JPETAB 88,173,46
scu-mus LD50:73 mg/kg FCTXAV 15,29,77
orl-rbt LD50:134 mg/kg FCTOD7 21,487,83
ipr-rbt LD50:50 mg/kg FCTOD7 21,487,83
ivn-rbt LD50:19 mg/kg JPETAB 88,173,46
scu-gpg LD50:37 mg/kg JPETAB 88,173,46
orl-ham LD50:75 mg/kg FCTXAV 16,355,78
ipr-ham LD50:66 mg/kg FCTXAV 16,355,79

CONSENSUS REPORTS: IARC Cancer Review: Group 3 IMEMDT 7,56,87, Animal Limited Evidence IMEMDT 40,67,86

SAFETY PROFILE: Poison by ingestion and other routes. An experimental teratogen. Other experimental reproductive effects. A severe skin irritant. Questionable carcinogen with experimental neoplastigenic and tumorigenic data. Mutation data reported. When heated to decomposition it emits acrid smoke and irritating fumes.

CMS845 CAS:106-23-0 ***HR: 2***
CITRONELLAL
mf: $C_{10}H_{18}O$ mw: 154.25

PROP: Colorless to sltly yellow liquid; intense lemon-citronnella-rose odor. D: 0.850-0.860, refr index: 1.446-1.456, flash p: 170°F. Sol in alc, most oils; sltly sol in propylene glycol; insol glycerin and water.

SYNS: 3,7-DIMETHYL-6-OCTENAL ◇ FEMA No. 2307

SAFETY PROFILE: Combustible liquid. When heated to decomposition it emits acrid smoke and irritating fumes.

CMS850 CAS:107-75-5 ***HR: 1***
CITRONELLAL HYDRATE
mf: $C_{10}H_{20}O_2$ mw: 172.30

PROP: Colorless liquid; sweet, floral, lily odor. D: 0.918-0.923, refr index: 1.447-1.450, flash p: +212°F. Sol in fixed oils and propylene glycol; insol in glycerin.

SYNS: CYCLALIA ◇ CYCLOSIA ◇ 3,7-DIMETHYL-7-HYDROXYOCTANAL ◇ FEMA No. 2583 ◇ FIXOL ◇ HYDROXYCITRONELLAL (FCC) ◇ 7-HYDROXYCITRONELLAL ◇ 7-HYDROXY-3,7-DIMETHYLOCTAN-1-AL ◇ 7-HYDROXY-3,7-DIMETHYL OCTANAL ◇ LAURINE ◇ LILYL ALDEHYDE ◇ MUSUET SYNTHETIC ◇ MUSUETTINE PRINCIPLE ◇ PHIXIA

TOXICITY DATA with REFERENCE
skn-rbt 500 mg/24H FCTXAV 12,921,74

CONSENSUS REPORTS: Reported in EPA TSCA Inventory.

SAFETY PROFILE: A skin irritant. Combustible liquid. When heated to decomposition it emits acrid smoke and irritating fumes. See also ALDEHYDES.

CMT000 CAS:8000-29-1 ***HR: 2***
CITRONELLA OIL

SYNS: ESSENTIAL OIL of CYMBOPOGON NARDUS ◇ OIL of CITRONELLA ◇ OILS, CINTONELLA ◇ ZITRONELL OEL (GERMAN)

TOXICITY DATA with REFERENCE
skn-rbt 500 mg/24H FCTXAV 11,1067,73
eye-rbt 100 mg AJOPAA 29,1363,46
dnr-bcs 5 uL/disc TOFOD5 8,91,85
orl-rat LD50:7200 mg/kg PHARAT 14,435,59
ipr-rat LD50:713 mg/kg IJEBA6 9,515,71
orl-mus LD50:4600 mg/kg TOFOD5 8,91,85
skn-rbt LD50:4700 mg/kg FCTXAV 11,1067,73

CONSENSUS REPORTS: Reported in EPA TSCA Inventory.

SAFETY PROFILE: Moderately toxic by intraperitoneal route. Mildly toxic by ingestion and skin contact. Mutation data reported. A skin and eye irritant. When heated to decomposition it emits acrid smoke and fumes.

CMT050 CAS:2436-90-0 ***HR: 1***
CITRONELLENE
mf: $C_{10}H_{18}$ mw: 138.28

PROP: Bp: 154-155°, d: 0.757 @ 20°/4°.

SYNS: DIHYDROMYRCENE ◇ 3,7-DIMETHYL-1,6-OCTADIENE ◇ 1,6-OCTADIENE, 3,7-DIMETHYL-

TOXICITY DATA with REFERENCE
skn-rbt 500 mg/24H MOD FCTOD7 21,845,83

CONSENSUS REPORTS: Reported in EPA TSCA Inventory.

SAFETY PROFILE: A skin irritant. When heated to decomposition it emits acrid smoke and irritating fumes.

CMT125 CAS:502-47-6 ***HR: 2***
CITRONELLIC ACID
mf: $C_{10}H_{18}O_2$ mw: 170.28

SYN: 3,7-DIMETHYL-6-OCTENOIC ACID

TOXICITY DATA with REFERENCE
skn-rbt 500 mg/24H MOD FCTOD7 20(Suppl),653,82
orl-rat LD50:2610 mg/kg FCTOD7 20(Suppl),653,82
skn-rbt LD50:450 mg/kg FCTOD7 20(Suppl),653,82

CONSENSUS REPORTS: Reported in EPA TSCA Inventory.

SAFETY PROFILE: Moderately toxic by ingestion and skin contact. A skin irritant. Reaction with ozone produces an explosive product. When heated to decomposition it emits acrid smoke and fumes.

CMT250 CAS:106-22-9 ***HR: 3***
CITRONELLOL
mf: $C_{10}H_{20}O$ mw: 156.30

PROP: Colorless oily liquid; rose odor. D: 0.850-0.860, refr index: 1.454-1.462, flash p: 215°F. Sol in fixed oils, propylene glycol; sltly sol in water; insol in glycerin @ 225°.

SYNS: CEPHROL ◇ 2,6-DIMETHYL-2-OCTEN-8-OL ◇ 3,7-DIMETHYL-6-OCTEN-1-OL ◇ FEMA No. 2309 ◇ FEMA No. 2980 ◇ RHODINOL ◇ RODINOL

TOXICITY DATA with REFERENCE
orl-rat LD50:3450 mg/kg FCTXAV 13,757,75
ivn-mus LDLo:100 mg/kg CBCCT* 5,139,53
ims-mus LD50:4000 mg/kg JSICAZ 21,342,62
skn-rbt LD50:2650 mg/kg FCTXAV 13,757,75

CONSENSUS REPORTS: Reported in EPA TSCA Inventory.

SAFETY PROFILE: Poison by intravenous route. Moderately toxic by ingestion, skin contact, and intramuscular routes. Combustible liquid. When heated to decomposition it emits acrid smoke and irritating fumes. See also ALCOHOLS.

CMT500 CAS:68039-38-3 ***HR: 1***
CITRONELLYL-2-BUTENOATE
mf: $C_{14}H_{24}O_2$ mw: 224.38

SYNS: 2-BUTENOIC ACID-3,7-DIMETHYL-6-OCTENYL ESTER ◇ CITRONELLYL-α-CROTONATE ◇ 3,7-DIMETHYL-6-OCTEN-1-OL CROTONATE

TOXICITY DATA with REFERENCE
skn-rbt 500 mg/24H MOD FCTXAV 14,725,76

CONSENSUS REPORTS: Reported in EPA TSCA Inventory.

SAFETY PROFILE: A skin irritant. See also ESTERS. When heated to decomposition it emits acrid smoke and irritating fumes.

CMT600 ***HR: 1***
CITRONELLYL BUTYRATE
mf: $C_{14}H_{26}O_2$ mw: 226.36

PROP: Colorless liquid; strong, fruity-rosy odor. D: 0.873-0.883; refr index: 1.444-1.448, flash p: +212°F. Misc in alc, ether, chloroform, most oils; insol water @ 245°.

SYNS: 3.7-DIMETHYL-6-OCTEN-1-YL BUTYRATE ◇ FEMA No. 2312

SAFETY PROFILE: Combustible liquid. When heated to decomposition it emits acrid smoke and irritating fumes.

CMT750 CAS:105-85-1 ***HR: 1***
CITRONELLYL FORMATE
mf: $C_{11}H_{20}O_2$ mw: 184.31

PROP: Colorless liquid; strong, fruity odor. D: 0.890-0.93, refr index: 1.443-1.452, flash p: 197°F. Sol in alc, fixed oils; sltly sol in propylene glycol; insol in glycerin, water @ 235°.

SYNS: 3,7-DIMETHYL-6-OCTEN-1-OL FORMATE ◇ 2,6-DIMETHYL-2-OCTEN-8-YL FORMATE ◇ 3,7-DIMETHYL-6-OCTEN-1-YL FORMATE ◇ FEMA No. 2314 ◇ FORMIC ACID, CITRONELLYL ESTER ◇ FORMIC ACID-3,7-DIMETHYL-6-OCTEN-1-YL ESTER

TOXICITY DATA with REFERENCE
skn-hmn 20 mg/48H MLD FCTXAV 11,1073,73
skn-rbt 500 mg/24H FCTXAV 11,1073,73
orl-rat LD50:8400 mg/kg FCTXAV 11,1011,73

CONSENSUS REPORTS: Reported in EPA TSCA Inventory.

SAFETY PROFILE: Mildly toxic by ingestion. A human skin irritant. Combustible liquid. When heated to decomposition it emits acrid smoke and irritating fumes. See also ESTERS and FORMIC ACID.

CMT900 ***HR: 1***
CITRONELLYL ISOBUTYRATE
mf: $C_{14}H_{26}O_2$ mw: 226.36

PROP: Colorless liquid; rosy-fruity odor. D: 0.870-0.880, refr index: 1.440-1.448, flash p: +212°F. Misc in alc, chloroform, ether, most oils; insol in water @ 249°.

SYNS: 3,7-DIMETHYL-6-OCTEN-1-YL ISOBUTYRATE ◇ FEMA No. 2313

SAFETY PROFILE: Combustible liquid. When heated to decomposition it emits acrid smoke and irritating fumes.

CMU000 CAS:51566-62-2 ***HR: 1***
CITRONELLYL NITRILE

SYN: 3,7-DIMETHYL-6-OCTENENITRILE

TOXICITY DATA with REFERENCE
skn-rbt 500 mg/24H MLD FCTXAV 17,509,79
orl-rat LD50:5300 mg/kg FCTXAV 17,509,79

CONSENSUS REPORTS: Cyanide and its compounds are on the Community Right-To-Know List.

SAFETY PROFILE: Mildly toxic by ingestion. A skin irritant. See also NITRILES. When heated to decomposition it emits toxic fumes of NO_x and CN^-.

CMU050 CAS:139-70-8 ***HR: 1***
CITRONELLYL PHENYLACETATE
mf: $C_{18}H_{26}O_2$ mw: 274.44

SYNS: ACETIC ACID, PHENYL-, 3,7-DIMETHYL-6-OCTENYL ESTER ◇ BENZENEACETIC ACID, 3,7-DIMETHYL-6-OCTENYL ESTER (9CI) ◇ 3,7-DIMETHYL-6-OCTEN-1-YL PHENYLACETATE

TOXICITY DATA with REFERENCE
skn-rbt 500 mg/24H MLD FCTOD7 20,657,82

CONSENSUS REPORTS: Reported in EPA TSCA Inventory.

SAFETY PROFILE: A skin irritant. When heated to decomposition it emits acrid smoke and irritating fumes.

CMU100 ***HR: 1***
CITRONELLYL PROPIONATE
mf: $C_{13}H_{24}O_2$ mw: 212.33

PROP: Colorless liquid; fruity-rosy odor. D: 0.877-0.886, refr index: 1.443-1.449, flash p: +212°F. Misc in alc, most oils; insol in water @ 242°.

SYN: FEMA No. 2316

SAFETY PROFILE: Combustible liquid. When heated to decomposition it emits acrid smoke and irritating fumes.

CMU300 ***HR: 1***
CITRUS HYSTRIX DC., fruit peel extract

PROP: Thailand plant belonging to the family *Rutaceae* JOETD7 13,105,85

TOXICITY DATA with REFERENCE
orl-rat TDLo:16 g/kg (female 15-22D post):TER JOETD7 13,105,85
orl-rat TDLo:25 g/kg (female 8-12D post):REP JOETD7 13,105,85
orl-rat LDLo:100 g/kg JOETD7 13,105,85

SAFETY PROFILE: Slightly toxic by ingestion. An experimental teratogen. Other experimental reproductive effects. When heated to decomposition it emits acrid smoke and irritating fumes.

CMU475 CAS:2278-50-4 ***HR: 1***
C.I. VAT BLACK 8
mf: $C_{45}H_{19}N_3O_4$ mw: 665.67

SYNS: BENZADONE GREY M ◇ 1H-BENZ(6,7)INDAZOLO(2,3,4-fgh)NAPHTH(2″,3″:6′,7′)INDOLO(3′,2′:5,6)ANTHR A(2,1,9-mna) ACRIDINE-5,8,13,25-TETRAONE ◇ CALEDON GREY M ◇ CERN KYPOVA 8 ◇ C.I. 71000 ◇ INDANTHREN GREY M ◇ INDANTHREN GREY MG ◇ MIKETHRENE GREY M ◇ MIKETHRENE GREY MG ◇ NIHONTHRENE GREY M ◇ OSTANTHREN GREY M ◇ PARADONE GREY M ◇ PARADONE GREY MG ◇ SED OSTANTHRENOVA M ◇ VAT GRAY S ◇ VAT GREY S

TOXICITY DATA with REFERENCE
eye-rbt 500 mg/24H MLD 85JCAE-,1324,86
orl-rat LD50:11200 mg/kg 85JCAE-,1324,86
ipr-mus LD50:3 g/kg GISAAA 51(1),78,86

CONSENSUS REPORTS: Reported in EPA TSCA Inventory.

SAFETY PROFILE: Mildly toxic by intraperitoneal and ingestion routes. An eye irritant. When heated to decomposition it emits toxic fumes of NO_x.

CMU500 CAS:6424-76-6 ***HR: 1***
C.I. VAT BLUE 16
mf: $C_{36}H_{18}O_4$ mw: 514.54

SYNS: CALCOLOID NAVY BLUE 2GC ◇ CALEDON DARK BLUE G ◇ CALEDON PRINTING NAVY G ◇ CARBANTHRENE NAVY BLUE G ◇ C.I. 71200 ◇ 16,17-ETHYLENEDIOXYVIOLANTHRONE ◇ INDANTHRENE NAVY BLUE G ◇ MIKETHRENE MARINE BLUE G ◇ MODR NAMORNICKA OSTANTHRENOVA G (CZECH) ◇ NIHONTHRENE NAVY BLUE G ◇ PALANTHRENE NAVY BLUE G ◇ PARADONE NAVY BLUE G ◇ ROMANTRENE NAVY BLUE FG

TOXICITY DATA with REFERENCE
eye-rbt 500 mg/24H MLD 28ZPAK -,243,72

CONSENSUS REPORTS: Reported in EPA TSCA Inventory.

SAFETY PROFILE: An eye irritant. When heated to decomposition it emits acrid smoke and irritating fumes.

CMU750 CAS:6373-20-2 ***HR: 1***
C.I. VAT BLUE 22
mf: $C_{34}H_{12}Cl_4O_2$ mw: 594.26

SYNS: C.I. 59815 (CZECH) ◇ C.I. 59820 ◇ INDANTHREN NAVY BLUE TRR ◇ MODR NAMORNICKA OSTANTHRENOVA RA (CZECH) ◇ PALANTHRENE NAVY BLUE RB

TOXICITY DATA with REFERENCE
eye-rbt 500 mg/24H MLD 28ZPAK -,244,72

CONSENSUS REPORTS: Reported in EPA TSCA Inventory.

SAFETY PROFILE: An eye irritant. When heated to decomposition it emits toxic fumes of Cl^-.

CMU800 CAS:6247-46-7 ***HR: 1***
C.I. VAT BROWN 25
mf: $C_{43}H_{25}N_3O_7$ mw: 695.71

SYNS: CARBANTHRENE RED BROWN 5R ◇ C.I. 69020 ◇ 16H-DINAPHTHO(2,3-a:2',3'-i)CARBAZOLE-5,10,15,17-TETRAONE,6,9-DIBENZAMIDO-1-METHOXY- ◇ FENANTHRENE RED BROWN 5R ◇ HNED CERVENAVA OSTANTHRENOVA 5 RF ◇ HNED KYPOVA 25 ◇ INDANTHRENE RED BROWN 5RF ◇ INDANTHRENE REDDISH BROWN 5RF ◇ INDANTHREN RED BROWN 5RF ◇ MIKETHRENE RED BROWN 5RF ◇ OSTANTHREN REDDISH BROWN 5RF

TOXICITY DATA with REFERENCE
eye-rbt 500 mg/24H MLD 85JCAE-,1325,86

SAFETY PROFILE: An eye irritant. When heated to decomposition it emits toxic fumes of NO_x.

CMU850 CAS:542-46-1 ***HR: 1***
cis-CIVETONE
mf: $C_{17}H_{30}O$ mw: 250.47

PROP: Crystals; strong musty odor. Mp: 31-32°, bp: 342°.

SYNS: CIVETONE ◇ 9-CYCLOHEPTADECEN-1-ONE ◇ 9-CYCLOHEPTADECEN-1-ONE,(Z)-(8CI,9CI)

TOXICITY DATA with REFERENCE
skn-rbt 500 mg/24H MOD FCTXAV 14,727,76

CONSENSUS REPORTS: Reported in EPA TSCA Inventory.

SAFETY PROFILE: A skin irritant. When heated to decomposition it emits acrid smoke and irritating fumes.

CMU875 CAS:76541-72-5 ***HR: 2***
CLANICLOR
mf: $C_{11}H_{17}ClO_7P_2$ mw: 358.67

SYNS: (4-CHLOROPHENYL)(DIMETHOXYPHOSPHINYL)METHYL PHOSPHORIC ACID DIMETHYL ESTER ◇ DIMETHYL-α-(DIMETHOXYPHOSPHINYL)-p-CHLOROBENZYL PHOSPHATE ◇ SR-202

TOXICITY DATA with REFERENCE
orl-rat LD50:3200 mg/kg DRFUD4 7,271,82
ipr-rat LD50:580 mg/kg DRFUD4 7,271,82
orl-mus LD50:2 g/kg DRFUD4 7,271,82
ipr-mus LD50:500 mg/kg DRFUD4 7,271,82

SAFETY PROFILE: Moderately toxic by ingestion and intraperitoneal routes. When heated to decomposition it emits toxic fumes of Cl^- and PO_x.

CMV000 CAS:149-29-1 ***HR: 3***
CLAVACIN
mf: $C_7H_6O_4$ mw: 154.13

PROP: Colorless crystals. Mp: 111°.

SYNS: CLAIRFORMIN ◇ 2,4-DIHYDROXY-2H-PYRAN-Δ-3(6H),α-ACETIC ACID-3,4-LACTONE ◇ (2,4-DIHYDROXY-2H-PYRAN-3(6H)-YLIDENE)ACETIC ACID-3,4-LACTONE ◇ EXPANSIN ◇ GIGANTIN ◇ 4-HYDROXY-4H-FURO(3,2-C)PYRAN-2(6H)-ONE ◇ LEUCOPIN ◇ MYCOIN ◇ PATULIN ◇ PENATIN ◇ PENICIDIN ◇ TERCININ ◇ TERININ

TOXICITY DATA with REFERENCE
dni-hmn:lym 50 mg/L FCTOD7 20,893,82
oms-hmn:hla 3200 mg/L FCTOD7 20,893,82
orl-rat TDLo:135 mg/kg (5W male/5W pre-20D preg):REP JTEHD6 2,713,77
orl-rat TDLo:135 mg/kg (MGN):TER JTEHD6 2,713,77
scu-rat TDLo:232 mg/kg/58W-I:NEO BJCAAI 15,85,61
orl-rat LD50:27790 μg/kg ARCVBP 8,41,77
ipr-rat LD50:4590 μg/kg ARCVBP 8,41,77
ivn-rat LD50:8570 μg/kg ARCVBP 8,41,77
orl-mus LD50:17 mg/kg TXAPA9 45,275,78
ipr-mus LD50:5 mg/kg APTOA6 2,109,46

scu-mus LD50:10 mg/kg FCTXAV 18,181,80
ivn-mus LD50:5 mg/kg 85GDA2 5,439,81
ice-mus LD50:570 μg/kg TXCYAC 13,91,79

CONSENSUS REPORTS: IARC Cancer Review: Group 3 IMEMDT 7,56,87; Animal Limited Evidence IMEMDT 10,205,76; Animal Inadequate Evidence IMEMDT 40,83,86. EPA Genetic Toxicology Program.

SAFETY PROFILE: Poison by ingestion, subcutaneous, intracerebral, intraperitoneal, intravenous, and possibly other routes. An experimental teratogen. Other experimental reproductive effects. Human mutation data reported. Questionable carcinogen with experimental neoplastigenic and teratogenic data. An antimicrobial agent. When heated to decomposition it emits acrid smoke and irritating fumes.

CMV250 CAS:57943-81-4 ***HR: 2***
CLAVULANIC ACID SODIUM SALT

SYNS: ANTIBIOTIC MM 14151 ◇ MM 14151

TOXICITY DATA with REFERENCE
scu-mus LD50:4500 mg/kg JANTAJ 30,77-27,77
ipr-mus LD50:4000 mg/kg JANTAJ 30,77-27,77

SAFETY PROFILE: Moderately toxic by intraperitoneal route. Mildly toxic by subcutaneous route. When heated to decomposition it emits toxic fumes of NO_x and Na_2O.

CMV325 CAS:57645-91-7 ***HR: 3***
CLEBOPRIDE MALATE
mf: $C_{20}H_{24}ClN_3O_2 \cdot C_4H_6O_5$ mw: 508.02

SYNS: 4-AMINO-5-CHLORO-2-METHOXY-N-(1-BENZYL-4-PIPERIDYL)BENZAMIDE MALATE ◇ CLEBOPRIDE HYDROGEN MALATE ◇ LAS

TOXICITY DATA with REFERENCE
orl-rat TDLo:12500 μg/kg (female 17-20D post):REP KSRNAM 16,5661,82
orl-rat LD50:2540 mg/kg OYYAA2 25,803,83
ipr-rat LD50:155 mg/kg OYYAA2 25,803,83
scu-rat LD50:4850 mg/kg OYYAA2 25,803,83
ivn-rat LD50:39 mg/kg OYYAA2 25,803,83
ims-rat LD50:1450 mg/kg OYYAA2 25,803,83
orl-mus LD50:490 mg/kg OYYAA2 25,803,83
ipr-mus LD50:145 mg/kg OYYAA2 25,803,83
scu-mus LD50:305 mg/kg OYYAA2 25,803,83
ivn-mus LD50:51 mg/kg OYYAA2 25,803,83
ims-mus LD50:260 mg/kg OYYAA2 25,803,83

SAFETY PROFILE: Poison by subcutaneous, intramuscular, intravenous and intraperitoneal routes. Moderately toxic by ingestion. Experimental reproductive effects. When heated to decomposition it emits toxic fumes of Cl^- and NO_x.

CMV375 CAS:25047-48-7 ***HR: 3***
CLEISTANTHIN A
mf: $C_{28}H_{28}O_{11}$ mw: 540.52

SYNS: CIBA GO.4350 ◇ DIPHYLLIN-3,4-o-DIMETHYL XYLOSIDE ◇ SHENG BAI XIN (CHINESE)

TOXICITY DATA with REFERENCE
orl-rat LD50:12500 μg/kg PHMGBN 4,347,70
ipr-rat LD50:2630 μg/kg PHMGBN 4,347,70
ivn-rat LD50:2580 μg/kg PHMGBN 4,347,70
orl-mus LD50:38110 μg/kg PHMGBN 4,347,70
orl-mky LDLo:40 mg/kg PHMGBN 4,347,70
ivn-mky LDLo:1 mg/kg PHMGBN 4,347,70
orl-cat LDLo:100 mg/kg PHMGBN 4,347,70
ipr-cat LDLo:5 mg/kg PHMGBN 4,347,70
orl-rbt LDLo:50 mg/kg PHMGBN 4,347,70
ipr-rbt LDLo:2500 μg/kg PHMGBN 4,347,70

SAFETY PROFILE: Poison by ingestion, intravenous and intraperitoneal routes. When heated to decomposition it emits acrid smoke and fumes.

CMV390 ***HR: 3***
CLEMATIS

PROP: A perennial climbing herb. The leaves resemble those of poison ivy. It is native to Canada and the northern United States, and is available commercially throughout the United States.

SYNS: BLUEBELL ◇ BLUE JESSAMINE ◇ CABELLOS de ANGEL (CUBA, PUERTO RICO) ◇ CABEZA de VIEJO (MEXICO) ◇ CASCARITA ◇ CLEMATITE AUX GEAUX (CANADA) ◇ CURL FLOWER ◇ CURLY HEADS ◇ DEVILS HAIR ◇ DEVILS THREAD ◇ FLAMULA (CUBA) ◇ HEADACHE WEED ◇ HERBE AUX GEAUX (CANADA) ◇ LEATHER FLOWER ◇ LIANE BON GARCON (HAITI) ◇ PIPE STEM ◇ SUGAR-BOWLS ◇ TRAVELERS JOY ◇ VASE FLOWER ◇ VASE VINE ◇ VIRGIN'S BOWER ◇ YERBA de PORDIOSEROS (CUBA)

SAFETY PROFILE: The whole plant contains the poison protoanemonin which is a direct irritant and drying agent on the skin and mucous membranes. Ingestion causes pain and blistering of the mouth possibly followed by vomiting, diarrhea, cramps, bloody and painful urination, dizziness, fainting, confusion, and convulsions.

CMV400 CAS:1163-36-6 ***HR: 3***
CLEMIZOLE HDYROCHLORIDE
mf: $C_{19}H_{20}ClN_3 \cdot ClH$ mw: 362.33

SYNS: ALLERCUR ◇ ALLERCURE HYDROCHLORIDE ◇ 1-(p-CHLOROBENZIL)-2-PIRROLIDIL-METIL-BENZIMIDAZOLO CLORIDATO (ITALIAN) ◇ 1-(p-CHLOROBENZYL)-2-(1-PYRROLIDINYLMETHYL)BENZIMIDAZOLE HYDROCHLORIDE ◇ 1-p-CHLOROBENZYL-PYRROLIDYL-METHYLENE-BENZIMIDAZOLEHYDROCHLORIDE ◇ P 48 ◇ REACTROL

TOXICITY DATA with REFERENCE
orl-rat LD50:1950 mg/kg JAPMA8 49,18,60

ivn-rat LD50:74 mg/kg JAPMA8 49,18,60
orl-mus LD50:837 mg/kg JAPMA8 49,18,60
ipr-mus LD50:290 mg/kg FRPSAX 14,194,59
ivn-mus LD50:75 mg/kg JAPMA8 49,18,60
ivn-gpg LD50:26 mg/kg AIPTAK 113,313,58

SAFETY PROFILE: Poison by intravenous and intraperitoneal routes. Moderately toxic by ingestion. When heated to decomposition it emits toxic fumes of NO_x and HCl.

CMV475 CAS:1532-19-0 ***HR: 3***
CLEP
mf: $C_7H_7Cl_2N \cdot ClH$ mw: 212.51

SYNS: 2-(α,β-DICHLORETHYL)PYRIDINE HYDROCHLORIDE ◇ 2-(1,2-DICHLOROETHYL)PYRIDINEHYDROCHLORIDE

TOXICITY DATA with REFERENCE
orl-rat TDLo:5940 mg/kg/84W-C:CAR GANMAX 3,51,66
skn-mus TDLo:19 g/kg/37W-I:ETA PAACA3 9,28,68
mul-mus TDLo:2640 mg/kg/37W-I:NEO GANMAX 3,51,66

SAFETY PROFILE: Questionable carcinogen with experimental carcinogenic, neoplastigenic, and tumorigenic data. When heated to decomposition it emits very toxic fumes of Cl^- and NO_x.

CMV500 CAS:34148-01-1 ***HR: 3***
CLIDANAC
mf: $C_{16}H_{19}ClO_2$ mw: 278.80

SYNS: BRITAI ◇ 6-CHLORO-5-CYCLOHEXYL-2,3-DIHYDRO-1H-INDENE-1-CARBOXYLIC ACID (9CI) ◇ 6-CHLORO-5-CYCLOHEXYL-1-INDANCARBOXYLIC ACID ◇ INDANAL ◇ TAI 284

TOXICITY DATA with REFERENCE
orl-rat LD50:41 mg/kg JMCMAR 15,1297,72
scu-rat LD50:62700 μg/kg IYKEDH 12,668,81
ivn-rat LD50:88400 μg/kg IYKEDH 12,668,81
orl-mus LD50:825 mg/kg NIIRDN 6,217,82
scu-mus LD50:860 mg/kg IYKEDH 12,668,81
ivn-mus LD50:212 mg/kg IYKEDH 12,668,81
orl-rbt LD50:380 mg/kg NIIRDN 6,217,82

SAFETY PROFILE: Poison by ingestion, subcutaneous, and intravenous routes. An anti-inflammatory and antipyretic agent. When heated to decomposition it emits toxic fumes of Cl^-.

CMV675 CAS:18323-44-9 ***HR: 3***
CLINDAMYCIN
mf: $C_{18}H_{33}ClN_2O_5S$ mw: 425.04

PROP: Yellow, amorphous solid.

SYNS: 7(S)-CHLORO-7-DEOXYLINCOMYCIN ◇ CLEOCIN ◇ CLINDAMYCINE (FRENCH) ◇ DALACIN C ◇ SOBELIN ◇ U-21,251

TOXICITY DATA with REFERENCE
scu-rat LD50:2618 mg/kg TXAPA9 18,185,71
orl-rbt LDLo:1 mg/kg RMVEAG 156,915,80
orl-ham LDLo:1 mg/kg ARZNAD 34,794,84

SAFETY PROFILE: A poison by ingestion. Moderately toxic by subcutaneous route. When heated to decomposition it emits toxic fumes of Cl^-, SO_x, and NO_x.

CMV680 ***HR: 2***
CLINDAMYCIN-2-PALMITATE MONOHYDROCHLORIDE
mf: $C_{34}H_{63}ClN_2O_6S \cdot ClH$ mw: 699.96

TOXICITY DATA with REFERENCE
orl-rat TDLo:2450 mg/kg (9-15D preg):TER KSRNAM 7,1724,73
orl-mus TDLo:2450 mg/kg (7-13D preg):REP KSRNAM 7,1724,73
scu-rat LD50:2 g/kg TXAPA9 21,516,72
orl-mus LD50:1956 mg/kg TXAPA9 21,516,72

SAFETY PROFILE: Moderately toxic by ingestion and subcutaneous routes. An experimental teratogen. Experimental reproductive effects. When heated to decomposition it emits toxic fumes of SO_x, NO_x, and Cl^-.

CMV690 CAS:24729-96-2 ***HR: 3***
CLINDAMYCIN-2-PHOSPHATE
mf: $C_{18}H_{34}ClN_2O_8PS$ mw: 505.02

SYNS: 7(S)-CHLORO-7-DEOXYLINCOMYCIN-2-PHOSPHATE ◇ CLEOCIN PHOSPHATE ◇ CLINDAMYCIN PHOSPHATE ◇ SOBELIN ◇ U-28,508

TOXICITY DATA with REFERENCE
ipr-rat TDLo:700 mg/kg (9-15D preg):REP KSRNAM 7,1697,73
ipr-rat TDLo:700 mg/kg (9-15D preg):TER KSRNAM 7,1697,73
ivn-wmn TDLo:12 mg/kg:BPR SMJOAV 75,768,82
orl-rat LD50:1832 mg/kg TXAPA9 27,308,74
ipr-rat LD50:745 mg/kg IYKEDH 14,484,83
scu-rat LD50:3861 mg/kg IYKEDH 14,484,83
ivn-rat LD50:321 mg/kg IYKEDH 14,484,83
orl-mus LD50:2539 mg/kg IYKEDH 14,484,83
ipr-mus LD50:784 mg/kg DECRDP 3,79,77
scu-mus LD50:1036 mg/kg IYKEDH 14,484,83
ivn-mus LD50:820 mg/kg IYKEDH 14,484,83
ims-mus LD50:1100 mg/kg IYKEDH 14,484,83

SAFETY PROFILE: Poison by intravenous route. Moderately toxic by ingestion, intramuscular, and subcutaneous routes. Human systemic effects by intravenous route: pulse rate increase, blood pressure lowering. An experimental teratogen. Experimental reproductive effects. When heated to decomposition it emits toxic fumes of Cl^-, PO_x, SO_x, and NO_x.

CMV700 CAS:30299-08-2 ***HR: 3***
CLINOFIBRATE
mf: $C_{29}H_{36}O_6$ mw: 468.64

PROP: Off-white powder. Mp: 143-146° (decomp). Sol in methanol, ethanol, acetone, chloroform, glacial acetic acid. Sltly sol in CCl_4. Practically insol in water.

SYNS: 2,2'-CYCLOHEXANE-1,1-DIYLBIS(p-PHENYLENEOXY)BIS(2-METHYLBUTYRIC ACID) ◇ 2,2'-(CYCLOHEXYLIDENEBIS(4,1-PHENYLENEOXY)BIS(2-METHYLBUTANOIC ACID) ◇ 2,2'-(4,4'-CYCLOHEXYLIDENEDIPHENOXY)-2,2'-DIMETHYLDIBUTYRIC ACID ◇ LIPOCLIN ◇ S 8527

TOXICITY DATA with REFERENCE
ipr-rat LD50:205 mg/kg IYKEDH 12,933,81
scu-rat LD50:1930 mg/kg NIIRDN 6,219,82
orl-mus LD50:1600 mg/kg NIIRDN 6,219,82
ipr-mus LD50:255 mg/kg NIIRDN 6,219,82
scu-mus LD50:350 mg/kg NIIRDN 6,219,82
ivn-mus LD50:150 mg/kg IYKEDH 12,933,81

SAFETY PROFILE: Poison by subcutaneous, intravenous, and intraperitoneal routes. Moderately toxic by ingestion. When heated to decomposition it emits acrid smoke and fumes.

CMV850 CAS:12173-10-3 ***HR: 2***
CLINOPTILOLITE

SYN: KLINOSORB

TOXICITY DATA with REFERENCE
ipl-rat TDLo:240 mg/kg/13W-I:ETA GTPZAB 30(5),29,86

SAFETY PROFILE: Questionable carcinogen with experimental tumorigenic data. When heated to decomposition it emits acrid smoke and irritating fumes.

CMV950 CAS:33979-15-6 ***HR: 2***
CLIVORINE
mf: $C_{21}H_{28}NO_7$ mw: 406.50

SYN: SENECIONANIUM, 12-(ACETYLOXY)-14,15,20,21-TETRADEHYDRO-15,20-DIHYDRO-8-HYDROXY-4-METHYL-11,16-DIOXO-, (8-xi,12-β,14-Z)-

TOXICITY DATA with REFERENCE
mma-sat 1 mg/plate MUREAV 68,211,79
slt-slw-par 10 μg KIKNAJ (30),70,79
dns-rat:lvr 2 μmol/L CNREA8 45,3125,85
dns-ham:lvr 2 μmol/L CNREA8 45,3125,85
orl-rat TDLo:1700 mg/kg/49W-C:CAR CALEDQ 10,117,80

SAFETY PROFILE: Questionable carcinogen with experimental carcinogenic data. Mutation data reported. When heated to decomposition it emits toxic fumes of NO_x.

CMW000 CAS:1622-61-3 ***HR: 1***
CLOAZEPAM
mf: $C_{15}H_{10}ClN_3O_3$ mw: 315.73

SYNS: 5-(o-CHLOROPHENYL)-1,3-DIHYDRO-7-NITRO-2H-1,4-BENZODIAZEPIN-2-ONE ◇ 5-(o-CHLOROPHENYL)-7-NITRO-1H-1,4-BENZODIAZEPIN-2(3H)-ONE ◇ CLOAZEPAM ◇ CLONAZEPAM ◇ RIVOTRIL ◇ RO 4-8180 ◇ RO 5-4023

TOXICITY DATA with REFERENCE
scu-mus TDLo:8 mg/kg (female 9D post):TER DGDFA5 22,61,80
orl-rbt TDLo:24 mg/kg (female 6-13D post):REP YACHDS 5,2457,77
orl-man TDLo:21 mg/kg/26W-I:PNS PGMJAO 63,311,87
ipr-rat LD50:14200 mg/kg IYKEDH 11,811,80
orl-mus LD50:2 g/kg PJPPAA 35,77,83
ipr-mus LD50:13300 mg/kg IYKEDH 11,811,80

CONSENSUS REPORTS: EPA Genetic Toxicology Program.

SAFETY PROFILE: Mildly toxic by ingestion. Human systemic effects by ingestion: fasciculations. Experimental teratogenic and reproductive effects. An anticonvulsant. When heated to decomposition it emits very toxic fumes of Cl^- and NO_x. See also DIAZEPAM.

CMW250 CAS:2726-03-6 ***HR: 3***
CLOBENZEPAM HYDROCHLORIDE
mf: $C_{17}H_{18}ClN_3O \cdot ClH$ mw: 352.29

SYNS: 5,10-DIHYDRO-7-CHLOR-10-(2-(DIMETHYLAMINO)ETHYL)-11H-DIBENZO(b,e)(1,4)DIAZEPIN-11-ONE HCl ◇ TARPAN

TOXICITY DATA with REFERENCE
orl-rat LD50:242 mg/kg MEXPAG 6,205,62
ivn-rat LD50:26 mg/kg MEXPAG 6,205,62
orl-mus LD50:324 mg/kg 27ZQAG -,66,72
ipr-mus LD50:75 mg/kg JMCMAR 14,153,71
ivn-mus LD50:28500 μg/kg MEXPAG 6,205,62
orl-gpg LD50:262 mg/kg MEXPAG 6,205,62

SAFETY PROFILE: Poison by ingestion, intravenous, and intraperitoneal routes. An antihistamine. When heated to decomposition it emits very toxic fumes of Cl^- and NO_x. See also DIAZEPAM.

CMW300 CAS:25122-46-7 ***HR: 3***
CLOBETASOL PROPIONATE
mf: $C_{25}H_{32}ClFO_5$ mw: 467.02

SYNS: 21-CHLORO-9-FLUORO-11-β,17-DIHYDROXY-16-β-METHYLPREGNA-1,4-DIENE-3,20-DIONE-17-PROPIONATE ◇ CLOBETASOL-17-PROPIONATE

TOXICITY DATA with REFERENCE
skn-rat TDLo:17500 mg/kg (35D male):REP OYYAA2 27,1217,84
ipr-rat LD50:351 mg/kg NIIRDN 6,241,82
scu-rat LD50:366 mg/kg IYKEDH 9,1066,78

ipr-mus LD50:118 mg/kg NIIRDN 6,241,82
scu-mus LD50:81700 μg/kg NIIRDN 6,241,82

SAFETY PROFILE: Poison by subcutaneous and intraperitoneal routes. Experimental reproductive effects. When heated to decomposition it emits toxic fumes of F^- and Cl^-.

CMW400 CAS:25122-57-0 ***HR: 2***
CLOBETASONE BUTYRATE
mf: $C_{26}H_{32}ClFO_5$ mw: 479.03

SYNS: 21-CHLORO-9-FLUORO-17-HYDROXY-16β-METHYL PREGNA-1,4-DIENE-3,11,20-TRIONE BUTYRATE ◇ CLOBETASONE-17-BUTYRATE ◇ EUMOVATE ◇ MOLIVATE ◇ SN 203

TOXICITY DATA with REFERENCE
scu-rat TDLo:3 mg/kg (female 17-22D post):REP KSRNAM 14,359,80
scu-rat TDLo:3300 μg/kg (female 7-17D post):TER KSRNAM 14,343,80
ipr-rat LD50:1510 mg/kg JTSCDR 5,45,80
ipr-mus LD50:500 mg/kg IYKEDH 14,838,83

SAFETY PROFILE: Moderately toxic by intraperitoneal route. An experimental teratogen. Other experimental reproductive effects. When heated to decomposition it emits toxic fumes of F^- and Cl^-.

CMW459 CAS:14860-49-2 ***HR: 3***
CLOBUTINOL
mf: $C_{14}H_{22}ClNO$ mw: 255.82

PROP: Bp: 179-180°.

SYNS: p-CHLORO-α-(2-(DIMETHYLAMINO)-1-METHYLETHYL)-α-METHYL-PHENETHYL ALCOHOL ◇ 1-(4-CHLOROPHENYL)-2,3-DIMETHYL-4-DIMETHYLAMINO-2-BUTANOL ◇ p-CLORO-α-(2-DIMETILAMINO)-1-METILETIL)-α-METIL FENETIL ALCOOL (ITALIAN)

TOXICITY DATA with REFERENCE
orl-rat LD50:802 mg/kg IYKEDH 6,119,75
ipr-rat LD50:165 mg/kg IYKEDH 6,119,75
scu-rat TDLo:775 mg/kg IYKEDH 6,119,75
ivn-rat LD50:63 mg/kg IYKEDH 6,119,75
orl-mus LD50:334 mg/kg IYKEDH 6,119,75
ipr-mus LD50:128 mg/kg IYKEDH 6,119,75
scu-mus LD50:262 mg/kg OYYAA2 8,1067,74
ivn-mus LD50:53 mg/kg IYKEDH 6,119,75
ivn-dog LD50:45300 μg/kg OYYAA2 8,1067,74

SAFETY PROFILE: Poison by ingestion, subcutaneous, intravenous, and intraperitoneal routes. When heated to decomposition it emits very toxic fumes of Cl^- and NO_x. See also ALCOHOLS.

CMW500 CAS:1215-83-4 ***HR: 3***
CLOBUTINOL HYDROCHLORIDE
mf: $C_{14}H_{22}ClNO{\bullet}ClH$ mw: 292.28

SYNS: BIOTERTUSSIN ◇ 2-(p-CHLOROBENZYL)-3-DIMETHYLAMINOMETHYL-2-BUTANOL HYDROCHLORIDE ◇ 4-CHLORO-α-(2-(DIMETHYLAMINO)-1-METHYLETHYL)-α-METHYL-BENZENEETHANOL HYDROCHLORIDE ◇ p-CHLORO-α-(2-DIMETHYLAMINO-1-METHYLETHYL)-α-METHYLPHENETHYL ALCOHOL HYDROCHLORIDE ◇ 1-p-CHLOROPHENYL-2,3-DIMETHYL-4-DIMETHYLAMINO-2-BUTANOL HYDROCHLORIDE ◇ p-CLORO-α-(2-DIMETILAMINO)-1-METILETIL)-α-METIL FENETIL ALCOOL CLORIDRATO (ITALIAN) ◇ KAT 256 ◇ PERTOXIL ◇ SILOMAT ◇ SILONIST

TOXICITY DATA with REFERENCE
orl-rat LD50:802 mg/kg NIIRDN 6,241,82
ipr-rat LD50:151 mg/kg NIIRDN 6,241,82
scu-rat LD50:702 mg/kg NIIRDN 6,241,82
ivn-rat LD50:63 mg/kg NIIRDN 6,241,82
orl-mus LD50:334 mg/kg NIIRDN 6,241,82
ipr-mus LD50:125 mg/kg NIIRDN 6,241,82
scu-mus LD50:262 mg/kg OYYAA2 4,961,70
ivn-mus LD50:40900 μg/kg NIIRDN 6,241,82
ivn-dog LD50:45300 μg/kg OYYAA2 4,961,70

SAFETY PROFILE: Poison by ingestion, subcutaneous, intravenous, and intraperitoneal routes. An antitussive. When heated to decomposition it emits toxic fumes of Cl^- and NO_x. See also ALCOHOLS.

CMW550 CAS:77174-66-4 ***HR: 2***
CLOCONAZOLE HYDROCHLORIDE
mf: $C_{18}H_{15}ClN_2O{\bullet}HCl$ mw: 347.26

SYNS: 1-(1-(2-((3-CHLOROBENZYL)OXY)PHENYL)VINYL)-1H-IMIDAZOLE HYDROCHLORIDE ◇ 1-(1-(2-((3-CHLOROPHENYL)METHOXY)PHENYL)ETHENYL)-1H-IMIDAZOLE HYDROCHLORIDE ◇ 1-(1-(2-((3-CHLOROPHENYL)METHOXY)PHENYL)ETHENYL)-1H-IMIDAZOLE MONOHYDROCHLORIDE (9CI) ◇ 710674-S

TOXICITY DATA with REFERENCE
scu-rat TDLo:1050 mg/kg (female 14D pre):REP KSRNAM 18,4917,84
scu-rat TDLo:1050 mg/kg (female 14D pre):TER KSRNAM 18,4917,84
orl-rat LD50:2 g/kg DRFUD4 10,451,85
scu-rat LD50:4700 mg/kg KSRNAM 18,4811,84
orl-mus LD50:1350 mg/kg KSRNAM 18,4811,84
ipr-mus LD50:94 mg/kg KSRNAM 18,4811,84

SAFETY PROFILE: Moderately toxic by ingestion and subcutaneous routes. An experimental teratogen. Other experimental reproductive effects. When heated to decomposition it emits toxic fumes of Cl^- and NO_x.

CMW600 CAS:4755-59-3 ***HR: 3***
CLODAZONE
mf: $C_{18}H_{20}ClN_3O$ mw: 329.86

SYN: AW-14-2446

TOXICITY DATA with REFERENCE
orl-rat LD50:2000 mg/kg INPHB6 1,214,68
ipr-rat LD50:142 mg/kg INPHB6 1,214,68

ivn-rat LD50:77 mg/kg INPHB6 1,214,68
orl-mus LD50:715 mg/kg INPHB6 1,214,68
ivn-mus LD50:90 mg/kg INPHB6 1,214,68
orl-gpg LD50:505 mg/kg INPHB6 1,214,68

SAFETY PROFILE: Poison by intravenous and intraperitoneal routes. Moderately toxic by ingestion. When heated to decomposition it emits toxic fumes of Cl^- and NO_x.

CMW700 CAS:511-13-7 ***HR: 3***
CLOFEDANOL HYDROCHLORIDE
mf: $C_{17}H_{20}ClNO•ClH$ mw: 326.29

SYNS: BAYER-186 ◇ BAYER B-186 ◇ CHLOPHEDIANOL HYDROCHLORIDE ◇ 2-CHLORO-α-(2-(DIMETHYLAMINO)-ETHYL)BENZHYDROL HYDROCHLORIDE ◇ 2-CHLORO-α-(2-(DIMETHYLAMINO) ETHYL)-α-PHENYL-BENZENEMETHANOL HYDROCHLORIDE (9CI) ◇ 1-o-CHLOROPHENYL-1-PHENYL-3-DIMETHYLAMINO-1-PROPANOL HYDROCHLORIDE ◇ CHLORPHEDIANOL HYDROCHLORIDE ◇ CLOPHEDIANOL HYDROCHLORIDE ◇ COLDRIN ◇ DETIGON ◇ DETIGON-BAYER ◇ α-(2-DIMETHYLAMINOETHYL)-o-CHLOROBENZHYDROL HYDROCHLORIDE ◇ REFUGAL ◇ SK 74 ◇ SL 501 ◇ ULO ◇ ULONE

TOXICITY DATA with REFERENCE
orl-mus TDLo:192 mg/kg (female 7-12D post):REP GNRIDX 7,177,73
orl-rat TDLo:192 mg/kg (female 9-14D post):TER GNRIDX 7,177,73
orl-rat LD50:422 mg/kg 27ZKAW -,-,69
ipr-rat LD50:295 mg/kg IYKEDH 12,933,81
scu-rat LD50:440 mg/kg GNRIDX 7,177,73
ivn-rat LD50:53 mg/kg GNRIDX 7,177,73
ims-rat LD50:268 mg/kg GNRIDX 7,177,73
orl-mus LD50:284 mg/kg THERAP 15,93,60
ipr-mus LD50:130 mg/kg GNRIDX 7,177,73
scu-mus LD50:144 mg/kg THERAP 15,93,60
ivn-mus LD50:42 mg/kg THERAP 15,93,60
orl-dog LD50:84 mg/kg YKYUA6 32,1281,81
scu-mam LD50:186 mg/kg 27ZQAG -,369,72

SAFETY PROFILE: Poison by ingestion, subcutaneous, intramuscular, intravenous, and intraperitoneal routes. Experimental reproductive effects. An experimental teratogen. When heated to decomposition it emits toxic fumes of NO_x and HCl.

CMW750 CAS:17449-96-6 ***HR: 3***
CLOFEXAMIDE PHENYLBUTAZONE
mf: $C_{14}H_{21}ClN_2O_2•C_{19}H_{20}N_2O_2$ mw: 593.23

SYNS: 2-(p-CHLOROPHENOXY)-N-(2-(DIETHYLAMINO)ETHYL) ACETAMIDE COMPOUND with 4-BUTYL-1,2-DIPHENYL-3,5-PYRAZOLIDINEDIONE (1:1) ◇ CLOFEXAMIDE-PHENYLBUTAZONE MIXTURE ◇ CLOFEZON ◇ CLOFEZONE ◇ PERCLUSON ◇ PERCLUSONE

TOXICITY DATA with REFERENCE
orl-mus TDLo:1600 mg/kg (female 7-14D post):TER OYYAA2 18,235,79
orl-rat TDLo:1920 mg/kg (8-15D preg):REP OYYAA2 18,235,79
orl-rat LD50:1950 mg/kg OYYAA2 15,41,78
ipr-rat LD50:1500 mg/kg NIIRDN 6,237,82
orl-mus LD50:1700 mg/kg OYYAA2 17,935,79
ipr-mus LD50:1650 mg/kg NIIRDN 6,237,82
orl-rbt LD50:390 mg/kg CHTPBA 3,53,68
orl-gpg LD50:720 mg/kg CHTPBA 3,53,68

SAFETY PROFILE: Poison by ingestion. Moderately toxic by intraperitoneal route. An experimental teratogen. Other experimental reproductive effects. When heated to decomposition it emits very toxic fumes of Cl^- and NO_x.

CMX000 CAS:882-09-7 ***HR: 3***
CLOFIBRIC ACID
mf: $C_{10}H_{11}ClO_3$ mw: 214.66

SYNS: (p-CHLOROPHENOXY)DIMETHYL-ACETIC ACID ◇ α-(p-CHLOROPHENOXY)ISOBUTYRIC ACID ◇ 2-(4-CHLOROPHENOXY)-2-METHYLPROPANOIC ACID ◇ 2-(p-CHLOROPHENOXY)-2-METHYLPROPIONIC ACID ◇ CHLOROPHIBRINIC ACID ◇ CLOFIBRINIC ACID ◇ CLOFIBRINSAEURE (GERMAN)

TOXICITY DATA with REFERENCE
unk-hmn TDLo:260 mg/kg (91D):LIV ATHSBL 36,159,80
orl-rat LD50:897 mg/kg ARZNAD 30,2023,80
scu-rat LD50:120 mg/kg PHARAT 22,167,67
orl-mus LD50:1170 mg/kg ARZNAD 27,1173,77
ipr-mus LD50:290 mg/kg PHARAT 22,167,67
scu-mus LD50:683 mg/kg RPTOAN 33,150,70

SAFETY PROFILE: Poison by intraperitoneal and subcutaneous routes. Moderately toxic by ingestion. Liver damage in humans by an unspecified route. When heated to decomposition it emits toxic fumes of Cl^-.

CMX500 CAS:911-45-5 ***HR: 3***
CLOMIPHENE
mf: $C_{26}H_{28}ClNO$ mw: 406.00

SYNS: CHLOMAPHENE ◇ CHLORAMIFENE ◇ CHLORAMIPHENE ◇ 2-(4-(2-CHLORO-1,2-DIPHENYLETHENYL)PHENOXY)-N,N-DIETHYLETHANAMINE ◇ 2-(p-(β-CHLORO-α-PHENYLSTYRYL) PHENOXY)-TRIETHYLAMINE ◇ CISCLOMIPHENE ◇ CLOMEPHENE B ◇ CLOMIFENE ◇ 1-(p-(β-DIETHYLAMINOETHOXY)PHENYL)-1,2-DIPHENYLCHLOROETHYLENE

TOXICITY DATA with REFERENCE
unr-wmn TDLo:5 mg/kg (5D pre):REP UMJOAJ 45,59,76
unr-wmn TDLo:120 mg/kg (15D pre):TER UMJOAJ 45,59,76
orl-wmn TDLo:15 mg/kg/13W-I:CAR,SPN,SKN LANCAO 2,1176,77
orl-mus TDLo:552 mg/kg/69W-I:ETA PEXTAR 11,440,69
orl-wmn TDLo:5 mg/kg/5D-I:SKN BMJOAE 292,380,86
orl-mus LD50:1700 mg/kg FEPRA7 20,419,61
ipr-mus LD50:390 mg/kg FEPRA7 20,419,61

SAFETY PROFILE: Poison by intraperitoneal route. Moderately toxic by ingestion. Experimental teratogenic data. Human systemic effects by ingestion: dermatitis. Human reproductive effects by unspecified routes: death of fetus, stillbirth, and poor viability. Human teratogenic effects by unspecified routes include developmental abnormalities of the central nervous system, cardiovascular system, gastrointestinal system, and urogenital system. Other experimental teratogenic and reproductive effects. Questionable human carcinogen producing spinal cord and skin tumors. When heated to decomposition it emits very toxic fumes of Cl^- and NO_x.

CMX700 CAS:50-41-9 ***HR: 3***
racemic-CLOMIPHENE CITRATE
mf: $C_{26}H_{28}ClNO{\cdot}C_6H_8O_7$ mw: 598.14

SYNS: CHLORAMIPHENE ◇ CHLORAMIPHENE CITRATE ◇ 2-CHLORO-1-(p-(β-DIETHYLAMINOETHOXY)PHENYL)-1,2-DIPHENYLETHYLENE ◇ 2-(p-(2-CHLORO-1,2-DIPHENYL VINYL)PHENOXY)TRIETHYLAMINE CITRATE (1:1) ◇ CLOMID ◇ CLOMIFEN CITRATE ◇ CLOMIFENO ◇ CLOMIPHENE CITRATE ◇ CLOMIPHENE DIHYDROGEN CITRATE ◇ CLOMIPHENE-R ◇ CLOMIPHINE ◇ CLOMIVID ◇ CLOMPHID ◇ 1-(p-(β-DIETHYLAMINO ETHOXY)PHENYL)-1,2-DIPHENYL-2-CHLOROETHYLENE CITRATE ◇ DYNERIC ◇ GENOZYM ◇ IKACLOMIN ◇ MER-41 ◇ MRL 41 ◇ NSC 35770 ◇ OMIFIN

TOXICITY DATA with REFERENCE
dnd-esc 25 mg/L MUREAV 165,57,86
dni-esc 50 mg/L MUREAV 165,57,86
orl-man TDLo:43 mg/kg (60D male):REP JCEMAZ 29,638,69
unr-wmn TDLo:15 mg/kg (15D pre):TER LANCAO 2,1107,81
scu-rat TDLo:2500 μg/kg:ETA JSTBBK 12,47,80
orl-rat LD50:5750 mg/kg TXAPA9 9,44,66
ipr-rat LD50:530 mg/kg TXAPA9 9,44,66
orl-mus LD50:1700 mg/kg OYYAA2 3,187,69

CONSENSUS REPORTS: IARC Cancer Review: Group 3 IMEMDT 7,172,87; Human Inadequate Evidence IMEMDT 21,551,79; Animal Inadequate Evidence IMEMDT 21,551,79. EPA Genetic Toxicology Program.

SAFETY PROFILE: Moderately toxic by ingestion and intraperitoneal routes. Human reproductive effects by ingestion: changes in spermatogenesis and effects on testes, epididymis, and sperm duct. Human teratogenic effects by an unspecified route: developmental abnormalities of the eye and ear. Experimental reproductive effects. Questionable carcinogen with experimental tumorigenic data. Used as a drug to induce ovulation and for the treatment of oligospermia. When heated to decomposition it emits very toxic fumes of Cl^- and NO_x.

CMX750 CAS:7599-79-3 ***HR: D***
trans-CLOMIPHENE CITRATE
mf: $C_{26}H_{28}ClNO{\cdot}C_6H_8O_7$ mw: 598.14

SYNS: (E)-2-(p-(2-CHLORO-1,2-DIPHENYLVINYL)PHENOXY)TRIETHYLAMINE, CITRATE ◇ trans-CLOMIFENE CITRATE

TOXICITY DATA with REFERENCE
ims-rbt TDLo:90 mg/kg (21-26D preg):REP TXAPA9 23,339,72
ims-rbt TDLo:90 mg/kg (21-26D preg):TER TXAPA9 23,339,72

SAFETY PROFILE: An experimental teratogen. Other experimental reproductive effects. When heated to decomposition it emits very toxic fumes of Cl^- and NO_x.

CMX760 CAS:4205-91-8 ***HR: 3***
CLONIDINE HYDROCHLORIDE
mf: $C_9H_9Cl_2N_3{\cdot}ClH$ mw: 266.57

SYNS: CATAPRES ◇ CATAPRESAN ◇ 2-(2,6-DICHLOROANILINO)-2-IMIDAZOLINE HYDROCHLORIDE ◇ 2,6-DICHLORO-N-2-IMIDAZOLIDINYLIDENE-BENZENAMINE HYDROCHLORIDE ◇ 2-(2,6-DICHLOROPHENYLAMINO)-2-IMIDAZOLINHYDROCHLORID (GERMAN) ◇ 2-((2,6-DICHLOROPHENYL)IMINO)IMIDAZOLIDINE MONOHYDROCHLORIDE ◇ DIXARIT ◇ HEMITON ◇ ISOGLAUCON ◇ ST-155

TOXICITY DATA with REFERENCE
scu-uns TDLo:2080 μg/kg (female 8-20D post):REP TJADAB 31,10B,85
ivn-dom TDLo:5 μg/kg (female 1D post):TER ANESAV 67,A449,87
orl-inf TDLo:390 μg/kg PEDIAU 72,500,83
orl-wmn TDLo:126 μg/kg/4W-I:GIT BMJOAE 292,174,86
orl-man TDLo:69 μg/kg:CNS,CVS LANCAO 2,694,76
orl-chd TDLo:70 μg/kg:CNS,CVS,PUL AJDCAI 137,171,83
orl-rat LD50:126 mg/kg IYKEDH 18,366,87
ipr-rat LD50:100 mg/kg IYKEDH 9,829,78
scu-rat LD50:77 mg/kg IYKEDH 18,366,87
ivn-rat LD50:29 mg/kg ARZNAD 16,1038,66
orl-mus LD50:135 mg/kg IYKEDH 18,366,87
ipr-mus LD50:100 mg/kg IYKEDH 9,829,78
scu-mus LD50:59 mg/kg YKKZAJ 95,966,75
ivn-mus LD50:17600 μg/kg NIIRDN 6,235,82
orl-dog LD50:30 mg/kg PBPSDY 1,67,77
orl-rbt LD50:80 mg/kg ARZNAD 16,1038,66

SAFETY PROFILE: Poison by ingestion, subcutaneous, intravenous, and intraperitoneal routes. Human systemic effects by ingestion: somnolence, effects on sleep, reduced blood pressure, increase in pulse rate, respiratory system effects and gastrointestinal system effects. An experimental teratogen. Other experimental reproductive effects. When heated to decomposition it emits toxic fumes of NO_x and Cl^-.

CMX770 CAS:17737-65-4 ***HR: 3***
CLONIXIC ACID
mf: $C_{13}H_{11}ClN_2O_2$ mw: 262.71

SYNS: CHLONIXIN ◇ CLONIXIN ◇ CLONIXINE ◇ 2-(2′-METHYL-3′-CHLORO)ANILINONICOTINIC ACID ◇ SCH 10304

TOXICITY DATA with REFERENCE
orl-rat TDLo:280 mg/kg (female 9-15D post):REP KSRNAM 9,453,75
orl-mus TDLo:560 mg/kg (female 7-13D post):TER KSRNAM 9,453,75
orl-rat LD50:335 mg/kg OYYAA2 7,655,73
ipr-rat LD50:148 mg/kg OYYAA2 7,655,73
scu-rat LD50:325 mg/kg OYYAA2 7,655,73
orl-mus LD50:400 mg/kg OYYAA2 7,655,73
ipr-mus LD50:198 mg/kg OYYAA2 7,655,73
scu-mus LD50:263 mg/kg OYYAA2 7,655,73

SAFETY PROFILE: Poison by ingestion, subcutaneous, and intraperitoneal routes. An experimental teratogen. Other experimental reproductive effects. When heated to decomposition it emits toxic fumes of Cl^- and NO_x.

CMX800 CAS:3703-76-2 ***HR: 3***
CLOPERASTINE
mf: $C_{20}H_{24}ClNO$ mw: 329.90

PROP: Bp: 172-174°.

SYN: 1-(2-((4-CHLOROPHENYL)PHENYLMETHOXY)ETHYL)PIPERIDINE (9CI)

TOXICITY DATA with REFERENCE
orl-rat LD50:2325 mg/kg NIIRDN 6,241,82
ipr-rat LD50:120 mg/kg NIIRDN 6,241,82
orl-mus LD50:600 mg/kg NIIRDN 6,241,82
ipr-mus LD50:96 mg/kg NIIRDN 6,241,82

SAFETY PROFILE: Poison by intraperitoneal route. Moderately toxic by ingestion. When heated to decomposition it emits toxic fumes of Cl^- and NO_x.

CMX820 CAS:22199-30-0 ***HR: 3***
CLOPERASTINE HYDROCHLORIDE
mf: $C_{20}H_{24}ClNO•ClH$ mw: 366.36

SYNS: CLOPERASTINA CLORIDRATO (ITALIAN) ◇ 1-(2-(p-CLORO-α-FENILBENZILOSSI)ETIL)PIPERIDINA CLORIDRATO (ITALIAN)

TOXICITY DATA with REFERENCE
orl-rat LD50:1986 mg/kg BCFAAI 122,384,83
ipr-rat LD50:150 mg/kg BCFAAI 122,384,83
orl-mus LD50:553 mg/kg BCFAAI 122,384,83
ipr-mus LD50:140 mg/kg BCFAAI 122,384,83

SAFETY PROFILE: Poison by intraperitoneal route. Moderately toxic by ingestion. When heated to decomposition it emits toxic fumes of NO_x and HCl.

CMX840 CAS:525-26-8 ***HR: 2***
CLOPERIDONE HYDROCHLORIDE
mf: $C_{21}H_{23}ClN_4O_2•ClH$ mw: 435.39

SYNS: 3-(3-(4-(m-CHLOROPHENYL)-1-PIPERAZINYL)PROPYL)-2,4(1H,3H)-QUINAZOLINEDIONE HYDROCHLORIDE ◇ MA 1337

TOXICITY DATA with REFERENCE
orl-rat LD50:7650 mg/kg JPETAB 148,151,65
ipr-rat LD50:3290 mg/kg JPETAB 148,151,65
ipr-mus LD50:2610 mg/kg JPETAB 148,151,65

SAFETY PROFILE: Moderately toxic by intraperitoneal route. Mildly toxic by ingestion. When heated to decomposition it emits toxic fumes of NO_x and Cl^-.

CMX845 CAS:55600-34-5 ***HR: 2***
CLOPHEN A-30

TOXICITY DATA with REFERENCE
orl-rat TDLo:4992 mg/kg/2Y-C:NEO TXAPA9 75,278,84

SAFETY PROFILE: Questionable carcinogen with experimental neoplastigenic data. When heated to decomposition it emits acrid smoke and irritating fumes.

CMX850 CAS:2971-90-6 ***HR: 1***
CLOPIDOL
mf: $C_7H_7Cl_2NO$ mw: 192.05

SYNS: COCCIDIOSTAT C ◇ COYDEN ◇ 3,5-DICHLORO-2,6-DIMETHYL-4-PYRIDINOL ◇ LERBEK ◇ METHYLCHLOROPINDOL ◇ METHYLCHLORPINDOL ◇ METILCLORPINDOL

TOXICITY DATA with REFERENCE
orl-rat LD50:18 g/kg MEIEDD 10,341,83

OSHA PEL: Total Dust: 15 mg/m^3; Respirable Fraction: 5 mg/m^3
ACGIH TLV: TWA 10 mg/m^3

SAFETY PROFILE: A nuisance dust. When heated to decomposition it emits very toxic fumes of Cl^- and NO_x.

CMX860 CAS:60086-22-8 ***HR: 3***
CLOPIPAZAN MESYLATE
mf: $C_{19}H_{18}ClNO•CH_4O_3S$ mw: 407.94

SYNS: 4(2-CHLORO-9H-XANTHEN-9-YLIDENE)-1-METHYLPIPERIDINE METHANESULFONATE ◇ SKF-69,634
orl-hmn TDLo:9140 μg/kg/28D DRFUD4 6,15,81
orl-rat LD50:319 mg/kg DRFUD4 6,15,81
ipr-rat LD50:80 mg/kg DRFUD4 6,15,81
ipr-mus LD50:57 mg/kg DRFUD4 6,15,81

SAFETY PROFILE: Poison by ingestion and intraperitoneal routes. Unspecified human systemic effects by ingestion. When heated to decomposition it emits toxic fumes of Cl^-, NO_x, and SO_x.

CMX880 CAS:40665-92-7 ***HR: D***
CLOPROSTENOL
mf: $C_{22}H_{29}ClO_6$ mw: 424.96

SYNS: ESTRUMATE ◇ ICI 80996 ◇ racemic-ICI 80,996 ◇ I.C.I. LTD. COMPOUND NUMBER 80996

TOXICITY DATA with REFERENCE
ims-pig TDLo:2917 ng/kg (female 16W post):REP JASIAB 89,267,77

SAFETY PROFILE: Experimental reproductive effects. When heated to decomposition it emits toxic fumes of Cl^-.

CMX920 CAS:34255-03-3 ***HR: 3***
CLOQUINOZINE TARTRATE
mf: $C_{16}H_{22}ClNC_4H_6O_6$ mw: 413.94

SYNS: 3-(p-CHLOROBENZYL)OCTAHYDRO-QUINOLIZINETARTRATE (1:1) ◇ 3-(p-CHLOROBENZYL)QUINOLIZIDINE TARTRATE ◇ QB-1

TOXICITY DATA with REFERENCE
ivn-rat LD50:45 mg/kg JJPAAZ 16,353,66
scu-mus LD50:430 mg/kg JJPAAZ 16,353,66
ivn-mus LD50:31800 μg/kg JJPAAZ 16,353,66

SAFETY PROFILE: Poison by intravenous route. Moderately toxic by subcutaneous route. When heated to decomposition it emits toxic fumes of Cl^- and NO_x.

CMY000 CAS:17780-75-5 ***HR: 3***
CLORGYLINE HYDROCHLORIDE
mf: $C_{13}H_{15}Cl_2NO \cdot ClH$ mw: 308.65

SYN: N-METHYL-N-PROPARGYL-3-(2,4-DICHLOROPHENOXY)PROPYLAMINE HYDROCHLORIDE

TOXICITY DATA with REFERENCE
orl-rat LD50:210 mg/kg BCPCA6 17,1285,68
ivn-rat LD50:62 mg/kg BCPCA6 17,1285,68
orl-mus LD50:430 mg/kg BCPCA6 17,1285,68
ipr-mus LD50:350 mg/kg JMCMAR 21,56,78
scu-mus LD50:400 mg/kg BCPCA6 17,1285,68
ivn-mus LD50:94 mg/kg BCPCA6 17,1285,68

SAFETY PROFILE: Poison by ingestion, intraperitoneal, intravenous, and subcutaneous routes. When heated to decomposition it emits very toxic fumes of HCl and NO_x.

CMY075 CAS:8000-34-8 ***HR: 2***
CLOVE BUD OIL

SYNS: NELKEN OEL (GERMAN) ◇ OIL of CLOVE ◇ OILS, CLOVE

TOXICITY DATA with REFERENCE
dnr-bcs 30 uL/disc TOFOD5 8,91,85
orl-rat LD50:2650 mg/kg FCTXAV 13,761,75
skn-rbt LD50:5000 mg/kg FCTXAV 13,761,75

CONSENSUS REPORTS: EPA Genetic Toxicology Program. Reported in EPA TSCA Inventory.

SAFETY PROFILE: Moderately toxic by ingestion. Mildly toxic by skin contact. Mutation data reported. When heated to decomposition it emits acrid smoke and fumes.

CMY100 CAS:8015-97-2 ***HR: 2***
CLOVE LEAF OIL MADAGASCAR

PROP: From steam distillation of leaves of *Eugenis caryophyllata* Thunberg (*Eugenia aromatica* L. Baill.) (Fam. *Myrtaceae*). Pale yellow liquid. Ref. index: 1.527-1.538 ZBJ000 20°. Sol in propylene glycol, fixed oils; insol in glycerin, mineral oil.

SYNS: CLOVE LEAF OIL ◇ OILS, CLOVE LEAF

TOXICITY DATA with REFERENCE
skn-mus 100 % FCTXAV 16,695,78
skn-rbt 500 mg/24H SEV FCTXAV 16,695,78
skn-pig 100 % FCTXAV 16,695,78
orl-rat LD50:1370 mg/kg FCTXAV 16,695,78
skn-rbt LD50:1200 mg/kg FCTXAV 16,695,78

CONSENSUS REPORTS: Reported in EPA TSCA Inventory.

SAFETY PROFILE: Moderately toxic by ingestion and skin contact. A severe skin irritant. When heated to decomposition it emits acrid smoke and fumes.

CMY125 CAS:24166-13-0 ***HR: 2***
CLOXAZOLAZEPAM
mf: $C_{17}H_{14}Cl_2N_2O_2$ mw: 349.23

PROP: Crystals. Mp: 202-204° (decomp). Freely sol in glacial acetic acid; sparingly sol in chloroform; sltly sol in acetone, dehydrated ethanol, ethyl acetate, and benzene. Practically insol in water.

SYNS: CLOXAZOLAM ◇ CS 370 ◇ ENADEL ◇ LUBALIX ◇ MT 14-411 ◇ OLCADIL ◇ SEPAZON ◇ TOLESTAN

TOXICITY DATA with REFERENCE
orl-mus TDLo:1800 mg/kg (7-12D preg):TER SKKNAJ 23,180,71
orl-rat LD50:1780 mg/kg IYKEDH 5,106,74
orl-mus LD50:2630 mg/kg IYKEDH 5,106,74

SAFETY PROFILE: Moderately toxic by ingestion. An experimental teratogen. When heated to decomposition it emits toxic fumes of Cl^- and NO_x. A minor tranquilizer. See also DIAZEPAM.

CMY135 CAS:15311-77-0 ***HR: 3***
CLOXYPENDYL
mf: $C_{20}H_{25}ClN_4OS\cdot 2ClH$ mw: 477.92

SYNS: 2-(4-(3-(3-CHLORO-10H-PYRIDO(3,2-b)-1,4-BENZOTHIAZINE-10-YL)PROPYL)-1-PIPERAZINYLETHANOL ◇ 4-(3-(3-CHLORO-10H-PYRIDO(3,2-b)(1,4)-BENZOTHIAZIN-10-YL)PROPYL)-1-PIPERAZINE ETHANOL ◇ 2-(4-(3-(3-CHLORO-10H-PYRIDO(3,2-b)(1,4)BENZO-THIAZIN-1-OYL)PROPYL)-1-PIPERAZINYL)ETHANOL

TOXICITY DATA with REFERENCE
orl-rat LD50:657 mg/kg ARZNAD 18,435,68
ipr-rat LD50:280 mg/kg ARZNAD 18,435,68
orl-mus LD50:610 mg/kg ARZNAD 18,435,68
ipr-mus LD50:166 mg/kg ARZNAD 18,435,68
scu-mus LD50:683 mg/kg ARZNAD 18,435,68
ivn-mus LD50:91 mg/kg ARZNAD 18,435,68

SAFETY PROFILE: Poison by intravenous and intraperitoneal routes. Moderately toxic by ingestion and subcutaneous routes. When heated to decomposition it emits toxic fumes of NO_x, SO_x, and HCl.

CMY250 CAS:5786-21-0 ***HR: 3***
CLOZAPINE
mf: $C_{18}H_{19}ClN_4$ mw: 326.86

SYNS: 8-CHLORO-11-(4-METHYL-1-PIPERAZINYL)-5H-DIBENZO (b,e)(1,4)DIAZEPINE ◇ CLOZAPIN ◇ HF-1854 ◇ IPROX ◇ LEPONEX ◇ LEPOTEX ◇ W-801

TOXICITY DATA with REFERENCE
sln dmg-orl 2 mg/2D SOGEBZ 11,718,75
cyt-hmn:lym 10 mg/L HUGEDQ 38,77,77
orl-rat TDLo:1080 mg/kg (16-22D preg/20D post):REP FRPPAO 26,585,71
orl-mus TDLo:48 mg/kg (female 7-12D post):TER KSRNAM 7,696,73
orl-hmn TDLo:6428 μg/kg/3D-I:CNS ARZNAD 22,919,72
orl-hmn TDLo:2143 μg/kg/3D-I:CNS ARZNAD 22,919,72
orl-hmn TDLo:5 mg/kg/7D-I:CVS ARZNAD 22,919,72
orl-rat LD50:251 mg/kg KSRNAM 7,667,73
scu-rat LD50:240 mg/kg KSRNAM 7,667,73
ivn-rat LD50:41600 μg/kg KSRNAM 7,667,73
ims-rat LD50:210 mg/kg FRPPAO 26,585,71
orl-mus LD50:150 mg/kg JMCMAR 23,878,80
ipr-mus LD50:98 mg/kg ARZNAD 32,668,82
scu-mus LD50:194 mg/kg KSRNAM 7,667,73
ivn-mus LD50:36500 μg/kg KSRNAM 7,667,73
orl-dog LD50:145 mg/kg FRPPAO 26,585,71
orl-gpg LD50:510 mg/kg FRPPAO 26,585,71

SAFETY PROFILE: Poison by ingestion, subcutaneous, intramuscular, intravenous, and intraperitoneal routes. Human systemic effects by ingestion: somnolence, hallucinations or distorted perceptions, and increased pulse rate. An experimental teratogen. Other experimental reproductive effects. Human mutation data reported. Used as a sedative. When heated to decomposition it emits toxic fumes of Cl^- and NO_x. See also DIAZEPAM.

CMY325 CAS:9001-13-2 ***HR: 1***
COAGULASE

SYNS: HEMOCOAGULASE ◇ HEPTOCOAGULASE ◇ PLASMA COAGULASE ◇ PLASMOCOAGULASE ◇ REPTILASE S ◇ RP-093 ◇ STAPHYLOCOAGULASE ◇ THROMBIN COAGULASE

TOXICITY DATA with REFERENCE
scu-rat TDLo:24000 units/kg (9-14D preg):TER KSRNAM 9,2337,75
scu-rat TDLo:24000 units/kg (9-14D preg):REP KSRNAM 9,2337,75
ivn-rat LD50:160200 units/kg KSRNAM 9,2304,75
scu-mus LD50:107 g/kg TOIZAG 15,383,68
ivn-mus LD50:53700 mg/kg TOIZAG 15,383,68

SAFETY PROFILE: An experimental teratogen. Other experimental reproductive effects.

CMY625 ***HR: 3***
COAL CONVERSION MATERIALS, SRC-II HEAVY DISTILLATE

SYN: SRC-II HEAVY DISTILLATE

TOXICITY DATA with REFERENCE
skn-mus TDLo:2870 mg/kg/2Y-I:CAR NTIS** CONF-801143
skn-mus TD:285 g/kg/2Y-I:CAR NTIS** CONF-801143

SAFETY PROFILE: Suspected carcinogen with experimental carcinogenic data by skin contact.

CMY635 ***HR: 3***
COAL DUST

PROP: Black powder or dust.

SYN: ANTHRACITE PARTICLES ◇ COAL FACINGS ◇ COAL, GROUND BITUMINOUS (DOT) ◇ COAL-MILLED ◇ COAL SLAG-MILLED ◇ SEA COAL

TOXICITY DATA with REFERENCE
ihl-rat TCLo:6600 $\mu g/m^3$/6H/86W-I:ETA AIHAAP 42,382,81

OSHA PEL: (Transitional: Respirable Quartz Fraction less than 5% SiO_2: TWA 2.4 mg/m^3; Respirable Quartz Fraction greater than or equal to 5% SiO_2: 10 mg/m^3) Respirable Quartz Fraction less than 5% SiO_2: TWA 2 mg/m^3; Respirable Quartz Fraction greater than or equal to 5% SiO_2: 0.1 mg/m^3
ACGIH TLV: TWA 2 mg/m^3
DOT Classification: Flammable Solid; Label: Flammable Solid.

SAFETY PROFILE: Questionable carcinogen with ex-

perimental tumorigenic data. Variable toxicity depending upon SiO_2 content. See also SILICA. Moderately flammable when exposed to heat, flame, or chemical reaction with oxidizers. Slightly explosive when exposed to flame.

CMY650 CAS:68131-74-8 ***HR: D***
COAL FLY ASH

SYNS: ASHES (residues) ◇ COAL ASH

TOXICITY DATA with REFERENCE
mmo-sat 1800 μg/plate BECTA6 32,179,84
mma-sat 1800 μg/plate BECTA6 32,179,84
mma-omi 535 mg/L SCIEAS 211,180,81

CONSENSUS REPORTS: Reported in EPA TSCA Inventory.

SAFETY PROFILE: Mutation data reported.

CMY800 CAS:8007-45-2 ***HR: 3***
COAL TAR
DOT: UN 1999

SYNS: CARBO-CORT ◇ CRUDE COAL TAR ◇ ESTAR ◇ IMPERVOTAR ◇ LAV ◇ LAVATAR ◇ PIXALBOL ◇ PIX CARBONIS ◇ POLYTAR BATH ◇ SUPERTAH ◇ SYNTAR ◇ TAR ◇ TAR, COAL ◇ TAR, liquid (DOT) ◇ ZETAR

TOXICITY DATA with REFERENCE
skn-hmn 15 μg/3D-I MLD 85DKA8 -,127,77
skn-rbt 5%/3H MLD SCPHA4 43,11,75
mmo-sat 5 μg/plate NTIS** PB84-138973
mma-esc 50 μg/plate NTIS** PB84-138973
orl-mus TDLo:12 g/kg/30W-C:ETA AJCAA7 26,552,36
skn-mus TDLo:64 g/kg/36W-I:CAR AMIHBC 4,299,51

CONSENSUS REPORTS: NTP Fifth Annual Report on Carcinogens. IARC Cancer Review: Group 1 IMEMDT 7,175,87; Animal Sufficient Evidence IMEMDT 34,65,84; IMEMDT 35,83,85; IMEMDT 3,22,73; Human Sufficient Evidence IMEMDT 34,65,84; IMEMDT 3,22,73; Human Limited Evidence IMEMDT 35,83,85. Reported in EPA TSCA Inventory.

OSHA PEL: TWA 0.2 mg/m^3; Carcinogen
DFG MAK: Human Carcinogen.
NIOSH REL: (Coal Tar Products) TWA 0.1 mg/m^3
DOT Classification: Flammable or Combustible Liquid; Label: Flammable Liquid.

SAFETY PROFILE: Confirmed human carcinogen with experimental carcinogenic and tumorigenic data. Mutation data reported. A human and experimental skin irritant. When heated to decomposition it emits acrid smoke and irritating fumes.

CMY805 CAS:8007-45-2 ***HR: 2***
COAL TAR, AEROSOL

TOXICITY DATA with REFERENCE
ihl-mus TCLo:22 g/m^3/55W-I:CAR JNCIAM 39,175,67

OSHA PEL: TWA 0.2 mg/m^3
NIOSH REL: TWA 0.1 mg/m^3 CHE fraction

CONSENSUS REPORTS: Reported in EPA TSCA Inventory.

SAFETY PROFILE: Questionable carcinogen with experimental carcinogenic data. When heated to decomposition it emits acrid smoke and irritating fumes.

CMY825 CAS:8001-58-9 ***HR: 3***
COAL TAR CREOSOTE
DOT: UN 1136

SYNS: AWPA #1 ◇ BRICK OIL ◇ COAL TAR OIL ◇ COAL TAR OIL (DOT) ◇ CREOSOTE ◇ CREOSOTE, from COAL TAR ◇ CREOSOTE OIL ◇ CREOSOTE P1 ◇ CREOSOTUM ◇ CRESYLIC CREOSOTE ◇ HEAVY OIL ◇ LIQUID PITCH OIL ◇ NAPHTHALENE OIL ◇ PRESERV-O-SOTE ◇ RCRA WASTE NUMBER U051 ◇ TAR OIL ◇ WASH OIL

TOXICITY DATA with REFERENCE
mma-sat 20 μg/plate MUREAV 119,21,83
bfa-rat/sat 250 mg/kg IAPUDO 59,279,84
orl-rat TDLo:52416 mg/kg (female 91D pre):REP OYYAA2 21,899,81
skn-mus TDLo:99 g/kg/33W-I:CAR FAATDF 7,228,86
orl-rat LD50:725 mg/kg TXAPA9 6,378,64
orl-mus LD50:433 mg/kg OYYAA2 21,899,81
orl-dog LDLo:600 mg/kg 14CYAT 2,1394,63
orl-cat LDLo:600 mg/kg 14CYAT 2,1394,63
orl-rbt LDLo:600 mg/kg 14CYAT 2,1394,63

CONSENSUS REPORTS: NTP Fifth Annual Report on Carcinogens. IARC Cancer Review: Group 2A IMEMDT 7,177,87; Animal Sufficient Evidence, Human Limited Evidence IMEMDT 35,83,85; Animal Sufficient Evicence IMEMDT 3,22,73. Reported in EPA TSCA Inventory.

NIOSH REL: (Coal Tar Products) TWA 0.1 mg/m^3 CHE fraction
DOT Classification: Flammable or Combustible Liquid; Label: Flammable Liquid.

SAFETY PROFILE: Confirmed carcinogen with experimental carcinogenic data. Moderately toxic by ingestion. Experimental reproductive effects. Mutation data reported. When heated to decomposition it emits acrid smoke and fumes.

CMZ100 CAS:65996-93-2 ***HR: 3***
COAL TAR PITCH VOLATILES

SYNS: PITCH ◇ PITCH, COAL TAR

TOXICITY DATA with REFERENCE
skn-mus TDLo:36 g/kg/18W-I:CAR AJIMD8 2,59,81
skn-mus TD:4200 mg/kg/31W-I:NEO TXAPA9 18,41,71
skn-mus TD:82 g/kg/52W-I:CAR HYSAAV 33(5),180,68

CONSENSUS REPORTS: IARC Cancer Review: Group 1 IMEMDT 7,174,87; Animal Sufficient Evidence, Human Sufficient Evidence IMEMDT 35,83,85; Human Sufficient Evidence IMEMDT 3,22,73. Reported in EPA TSCA Inventory.

OSHA PEL: TWA 0.2 mg/m^3; Carcinogen
ACGIH TLV: TWA 0.2 mg/m^3 (volatile), Confirmed Human Carcinogen
NIOSH REL: (Coal Tar Products) TWA 0.1 mg/m^3 CHE fraction

SAFETY PROFILE: Confirmed carcinogen with experimental carcinogenic and neoplastigenic data by skin contact. When heated to decomposition it emits acrid smoke and fumes.

CNA250 CAS:7440-48-4 ***HR: 3***
COBALT
af: Co aw: 58.93

PROP: Gray, hard, magnetic, ductile, somewhat malleable metal. Exists in two allotropic forms. At room temperature, the hexagonal form is more stable than the cubic form; both forms can exist at room temperature. Stable in air or toward water at ordinary temperatures. D 8.92, mp 1493°, bp about 3100°, Brinell hardness: 125, latent heat of fusion 62 cal/g, latent heat of vaporization 1500 cal/g, specific heat (15-100°): 0.1056 cal/g/°C. Readily sol in dil HNO_3; very slowly attacked by HCl or cold H_2SO_4. The hydrated salts of cobalt are red, and the sol salts form red solns which become blue on adding concd HCl.

SYNS: AQUACAT ◇ C.I. 77320 ◇ COBALT-59 ◇ KOBALT (GERMAN, POLISH) ◇ NCI-C60311 ◇ SUPER COBALT

TOXICITY DATA with REFERENCE
ims-rat TDLo:126 mg/kg:NEO NATUAS 173,822,54
imp-rbt TDLo:75 mg/kg:ETA ZEKBAI 52,425,42
orl-rat LDLo:1500 mg/kg 15CYAT 2,1026,63
ipr-rat LDLo:250 mg/kg EQSSDX 1,1,75
ivn-rat LDLo:100 mg/kg EQSSDX 1,1,75
itr-rat LDLo:25 mg/kg NTIS** AEC-TR-6710
ipr-mus LDLo:100 mg/kg EQSSDX 1,1,75
orl-rbt LDLo:750 mg/kg AIPTAK 62,347,39
ivn-rbt LDLo:100 mg/kg EQSSDX 1,1,75

CONSENSUS REPORTS: Reported in EPA TSCA Inventory. Cobalt and its compounds are on the Community Right-To-Know List.

OSHA PEL: (Transitional: TWA 0.1 mg/m^3) TWA 0.05 mg/m^3
ACGIH TLV: (metal, dust, and fume) TWA 0.05 mg(Co)/m^3
DFG TRK: 0.5 mg/m^3 calculated as cobalt in that portion of dust that can possibly be inhaled in the production of cobalt powder and catalysts; hard metal (tungsten carbide) and magnet production (processing of powder, machine pressing, and mechanical processing of unsintered articles); others 0.1 mg/m^3 calculated as cobalt in that portion of dust that can possibly be inhaled. Animal Carcinogen, Suspected Human Carcinogen.
NIOSH REL: (Cobalt): insufficient evidence for recommending limit

SAFETY PROFILE: Confirmed carcinogen with experimental neoplastigenic and tumorigenic data. Poison by intravenous, intratracheal, and intraperitoneal routes. Moderately toxic by ingestion. Inhalation of the dust may cause pulmonary damage. The powder may cause dermatitis. Ingestion of soluble salts produces nausea and vomiting by local irritation. Powdered cobalt ignites spontaneously in air. Flammable when exposed to heat or flame. Explosive reaction with hydrazinium nitrate, ammonium nitrate + heat, and 1,3,4,7-tetramethylisoindole (at 390°C). Ignites on contact with bromine pentafluoride. Incandescent reaction with acetylene or nitryl fluoride. See also COBALT COMPOUNDS.

CNA500 CAS:6147-53-1 ***HR: 2***
COBALT ACETATE TETRAHYDRATE
mf: $C_4H_6O_4{\cdot}Co{\cdot}4H_2O$ mw: 249.11

SYNS: ACETIC ACID, COBALT(2+) SALT, TETRAHYDRATE ◇ COBALT DIACETATE TETRAHYDRATE ◇ COBALTOUS ACETATE TETRAHYDRATE ◇ OCTAN KOBALTNATY (CZECH)

TOXICITY DATA with REFERENCE
skn-rbt 500 mg/24H MOD 28ZPAK -,21,72
eye-rbt 500 mg/24H MLD 28ZPAK -,21,72
cyt-hmn:lym 600 µg/L CYGEDX 12(3),46,78
orl-rat LD50:708 mg/kg FCTOD7 20,311,82

CONSENSUS REPORTS: Cobalt and its compounds are on the Community Right-To-Know List.

SAFETY PROFILE: Moderately toxic by ingestion. A skin and eye irritant. Human mutation data reported. See also COBALT COMPOUNDS. When heated to decomposition it emits acrid smoke and irritating fumes.

CNA750 CAS:11114-92-4 ***HR: 3***
COBALT ALLOY, Co,Cr

SYNS: CHROMIUM-COBALT ALLOY ◇ COBALT-CHROMIUM ALLOY ◇ DIN 2.4602 ◇ DIN 2.4964 ◇ HASTELLOY C ◇ HAYNES STELLITE 21 ◇ HEV-4 ◇ VITALLIUM ◇ ZIMALLOY

TOXICITY DATA with REFERENCE
ims-rat TDLo:140 mg/kg:ETA LANCAO 1,564,71

CONSENSUS REPORTS: NTP Fifth Annual Report on Carcinogens. IARC Cancer Review: Animal Limited Evidence IMEMDT 23,205,80. Cobalt and its compounds, as well as chromium and its compounds, are on the Community Right-To-Know List.

OSHA PEL: TWA 1 mg(Cr)/m^3; 0.1 mg(Co)/m^3 (fume and dust)
ACGIH TLV: TWA 0.5 mg(Cr)/m^3
NIOSH REL: (Cobalt) Insufficient evidence for recommending limit.

SAFETY PROFILE: Confirmed carcinogen with experimental tumorigenic data. Violent reaction with molten Li. See also COBALT COMPOUNDS and CHROMIUM COMPOUNDS.

CNB000 ***HR: 3***
COBALT(III) AMIDE
mf: CoH_6N_3 mw: 107.00

$Co(NH_2)_3$

CONSENSUS REPORTS: Cobalt and its compounds are on the Community Right-To-Know List.

SAFETY PROFILE: Powdered material will spontaneously explode in air. When heated it converts to cobalt(III) nitride which ignites spontaneously in air. When heated to decomposition it emits toxic fumes of NO_x. See also COBALT COMPOUNDS and AMIDES.

CNB099 ***HR: 3***
COBALT(II) AZIDE
mf: CoN_6 mw: 142.97

$Co(N_3)_2$

CONSENSUS REPORTS: Cobalt and its compounds are on the Community Right-To-Know List.

SAFETY PROFILE: Explodes when heated to 200°C. When heated to decomposition it emits toxic fumes of NO_x. See also COBALT COMPOUNDS and AZIDES.

CNB250 CAS:7789-43-7 ***HR: D***
COBALT(II) BROMIDE
mf: Br_2Co mw: 218.75

CONSENSUS REPORTS: Cobalt and its compounds are on the Community Right-To-Know List.

SAFETY PROFILE: Exothermic reaction when heated with sodium. When heated to decomposition it emits toxic fumes of Br^-. See also COBALT COMPOUNDS and BROMIDES.

CNB500 CAS:10210-68-1 ***HR: 3***
COBALT CARBONYL
mf: $C_8Co_2O_8$ mw: 341.94

$(OC)_3Co:(CO)_2:Co(CO)_3$

PROP: Orange platelets. D: 1.87, mp: 51°, decomp above 52°. Decomp on exposure to air. Insol in water; sol in organic solvents.

SYNS: COBALT OCTACARBONYL ◇ COBALT TETRACARBONYL ◇ COBALT TETRACARBONYL DIMER ◇ DI-mu-CARBONYLHEXACARBONYLDICOBALT ◇ DICOBALT CARBONYL ◇ DICOBALT OCTACARBONYL ◇ OCTACARBONYLDICOBALT

TOXICITY DATA with REFERENCE
orl-rat LD50:754 mg/kg MEIEDD 10,448,83
ihl-rat LC16:15200 μg/kg/2H 85GMAT -,49,82
orl-mus LD50:378 mg/kg MEIEDD 10,448,83
ihl-mus LC50:27 mg/m^3/2H 85GMAT -,49,82

CONSENSUS REPORTS: Reported in EPA TSCA Inventory. Cobalt and its compounds are on the Community Right-To-Know List. EPA Extremely Hazardous Substances List.

OSHA PEL: TWA 0.1 mg(Co)/m^3
ACGIH TLV: TWA 0.1 mg(Co)/m^3

SAFETY PROFILE: Poison by ingestion and inhalation. Decomposes in air to form a product which ignites spontaneously in air. When heated to decomposition it emits acrid smoke and fumes. See also CARBONYLS and COBALT COMPOUNDS.

CNB599 CAS:7646-79-9 ***HR: 3***
COBALT(II) CHLORIDE
mf: Cl_2Co mw: 129.83

PROP: Blue powder. Mp: 724°, bp: 1049°, d: 3.348.

SYNS: COBALT DICHLORIDE ◇ COBALT MURIATE ◇ COBALTOUS CHLORIDE ◇ COBALTOUS DICHLORIDE ◇ KOBALT CHLORID (GERMAN)

TOXICITY DATA with REFERENCE
mrc-bcs 50 mmol/L MUREAV 77,109,80
mmo-smc 100 mmol/L CPBTAL 33,1571,85
mrc-smc:3 g/L MUREAV 155,117,85
dni-hmn:hla 1 mmol/L MUREAV 92,427,82
dns-ham:emb 200 μmol/L MUREAV 131,173,84
msc-ham:lng 200 μmol/L MUREAV 68,259,79
ipr-rat TDLo:30 g/kg (15-16D preg):TER TJADAB 29(3),23A,84
ipr-rat TDLo:30 g/kg (15-16D preg):REP TJADAB 29(3),23A,84
scu-rat TDLo:400 mg/kg:CAR LBANAX 11,43,77
orl-chd TDLo:48 mg/kg:CNS,END,MET JAMAAP 157,117,55
orl-chd LDLo:1500 mg/kg 34ZIAG -,182,69
orl-rat LD50:80 mg/kg HYSAAV 36,277,71
ivn-rat LD50:20 mg/kg AIPTAK 143,219,63

orl-mus LD50:80 mg/kg HYSAAV 36,277,71
ipr-mus LD50:49 mg/kg AEPPAE 244,17,62
scu-mus LDLo:100 mg/kg 27ZWAY 3.2,1444,-
ivn-dog LDLo:36 mg/kg HBAMAK 4,1289,35
orl-rbt LDLo:1272 mg/kg SMSJAR 26,131,1826
scu-rbt LDLo:200 mg/kg HBAMAK 4,1289,35
orl-gpg LD50:55 mg/kg HYSAAV 36,277,71
skn-gpg LDLo:165 mg/kg AEHLAU 11,201,65
ipr-gpg LDLo:165 mg/kg AEHLAU 11,201,65

CONSENSUS REPORTS: Reported in EPA TSCA Inventory. EPA Genetic Toxicology Program. Cobalt and its compounds are on the Community Right-To-Know List.

SAFETY PROFILE: Poison experimentally by ingestion, skin contact, intraperitoneal, intravenous, and subcutaneous routes. Moderately toxic to humans by ingestion. Human systemic effects by ingestion: anorexia, goiter (increased thyroid size), and weight loss. Experimental teratogenic and reproductive effects. Human mutation data reported. Questionable carcinogen with experimental carcinogenic data. Incompatible with metals (e.g., sodium and potassium). See also COBALT. When heated to decomposition it emits toxic fumes of Cl^-.

CNB750 ***HR: 3***
COBALT(III) CHLORIDE
mf: Cl_3Co mw: 165.29

CONSENSUS REPORTS: Cobalt and its compounds are on the Community Right-To-Know List.

SAFETY PROFILE: Ignites on contact with lithium. Incompatible with pentacarbonyl iron and zinc. When heated to decomposition it emits toxic fumes of Cl^-. See also COBALT COMPOUNDS and CHLORIDES.

CNB800 CAS:7791-13-1 ***HR: 3***
COBALT(II) CHLORIDE HEXAHYDRATE
mf: $Cl_2Co\bullet 6H_2O$ mw: 237.95

SYNS: COBALT CHLORIDE, HEXAHYDRATE (8CI, 9CI) ◇ COBALT(2+) CHLORIDE HEXAHYDRATE ◇ COBALT DICHLORIDE HEXAHYDRATE ◇ COBALTOUS CHLORIDE, HEXAHYDRATE

TOXICITY DATA with REFERENCE
ivn-mus TDLo:47590 mg/kg (8D preg):TER ENVRAL 33,47,84
orl-rat LD50:766 mg/kg FCTOD7 20,311,82
ipr-rat LD50:35 mg/kg JAPYAA 32,315,72
scu-rat LDLo:121 mg/kg EQSSDX 1,1,75
ipr-mus LD50:90 mg/kg AEPPAE 244,17,62
scu-mus LDLo:100 mg/kg EQSSDX 1,1,75
ivn-dog LDLo:30300 μg/kg EQSSDX 1,1,75
scu-rbt LDLo:200 mg/kg EQSSDX 1,1,75
ivn-rbt LDLo:25400 μg/kg EQSSDX 1,1,75

CONSENSUS REPORTS: Cobalt and its compounds are on the Community Right-To-Know List.

SAFETY PROFILE: Poison by subcutaneous, intravenous, and intraperitoneal routes. Moderately toxic by ingestion. An experimental teratogen. Experimental reproductive effects. When heated to decomposition it emits toxic fumes of Cl^-. See also COBALT and CHLORIDES.

CNB850 ***HR: 3***
COBALT COMPOUNDS

CONSENSUS REPORTS: Cobalt and its compounds are on the Community Right-To-Know List.

DFG TRK: 0.5 mg/m^3 calculated as cobalt in that portion of dust that can possibly be inhaled in the production of cobalt powder and catalysts; hard metal (tungsten carbide) and magnet production (processing of powder, machine pressing, and mechanical processing of unsintered articles); others 0.1 mg/m^3 calculated as cobalt in that portion of dust that can possibly be inhaled. Animal Carcinogen, Suspected Human Carcinogen.

SAFETY PROFILE: Confirmed carcinogen with experimental neoplastigenic and tumorigenic data. Cobalt has a low toxicity by ingestion. Ingestion of soluble salts produces nausea and vomiting by local irritation. In animals, administration of cobalt salts produces an increase in the total red cell mass of the blood. In humans, a single case of poisoning with liver and kidney damage has been attributed to cobalt. Locally, cobalt has been shown to produce dermatitis and investigators have been able to demonstrate a hypersensitivity of the skin to cobalt. There have been reports of hematologic, digestive, and pulmonary changes in humans. See also specific compounds.

CNC000 CAS:71-48-7 ***HR: 3***
COBALT DIACETATE
mf: $C_4H_6O_4\bullet Co$ mw: 177.03

SYNS: ACETIC ACID, COBALT(2+) SALT ◇ COBALT ACETATE ◇ COBALT(2+) ACETATE ◇ COBALT(II) ACETATE ◇ COBALTOUS DIACETATE

TOXICITY DATA with REFERENCE
otr-ham:emb 200 μmol/L CNREA8 39,193,79
dnd-ham:emb 200 μmol/L CNREA8 39,193,79
orl-rat LD50:503 mg/kg FCTOD7 20,311,82
ivn-mus LD50:31 mg/kg BJPCAL 23,455,64
ivn-rbt LD50:25 mg/kg BJPCAL 23,455,64

CONSENSUS REPORTS: Cobalt and its compounds are on the Community Right-To-Know List. Reported in EPA TSCA Inventory. EPA Genetic Toxicology Program.

SAFETY PROFILE: Poison by intravenous route. Moderately toxic by ingestion. Mutation data reported. See also COBALT COMPOUNDS. When heated to decomposition it emits acrid smoke and irritating fumes.

CNC100 CAS:10026-17-2 ***HR: 3***
COBALT(II) FLUORIDE
mf: CoF_2 mw: 96.93

PROP: Small light brown hexagonal crystals. Reacts with water.

SYNS: COBALT DIFLUORIDE ◇ COBALTOUS FLUORIDE

TOXICITY DATA with REFERENCE
orl-rat LD50:150 mg/kg FCTOD7 20,311,82

CONSENSUS REPORTS: Reported in EPA TSCA Inventory.

OSHA PEL: TWA 2.5 mg(F)/m^3
ACGIH TLV: TWA 2.5 mg(F)/m^3
NIOSH REL: (Fluorides, Inorganic): 10H TWA 2.5 mg(F)/m^3

SAFETY PROFILE: Poison by ingestion. When heated to decomposition it emits toxic fumes of Co and F^-.

CNC230 CAS:16842-03-8 ***HR: 3***
COBALT HYDROCARBONYL
mf: C_4HCoO_4 mw: 171.98

TOXICITY DATA with REFERENCE
ihl-rat LC50:165 mg/m^3/30M 34ZIAG -,182,69

CONSENSUS REPORTS: Cobalt and its compounds are on the Community Right-To-Know List.

OSHA PEL: TWA 0.1 mg(Co)/m^3
ACGIH TLV: TWA 0.1 mg(Co)/m^3

SAFETY PROFILE: Poison by inhalation. See also COBALT COMPOUNDS.

CNC250 CAS:13762-14-6 ***HR: D***
COBALT MOLYBDATE
mf: $CoMoO_4$ mw: 218.87

SYNS: COBALT(2+) MOLYBDATE ◇ COBALT MOLYBDENUM OXIDE ◇ COBALTOUS MOLYBDATE ◇ HT-400 E 1/8"

TOXICITY DATA with REFERENCE
otr-ham:emb 250 μmol/L CNREA8 39,193,79
dnd-ham:emb 6100 μmol/L CNREA8 39,193,79

CONSENSUS REPORTS: Reported in EPA TSCA Inventory. EPA Genetic Toxicology Program. Cobalt and its compounds are on the Community Right-To-Know List.

OSHA PEL: TWA 5 mg(Mo)/m^3
ACGIH TLV: TWA 5 mg(Mo)/m^3
NIOSH REL: (Cobalt) Insufficient evidence for recommending limit.

SAFETY PROFILE: Mutation data reported. See also COBALT COMPOUNDS and MOLYBDENUM COMPOUNDS.

CNC500 CAS:10141-05-6 ***HR: 3***
COBALT(II) NITRATE
mf: CoN_2O_6 mw: 182.95

PROP: Mp: 55°, d: 1.87.

SYNS: COBALT DINITRATE ◇ COBALTOUS NITRATE ◇ NITRIC ACID, COBALT (2+) SALT

TOXICITY DATA with REFERENCE
itt-rat TDLo:14636 μg/kg (1D male):REP JRPFA4 7,21,64
scu-rbt TDLo:4530 μg/kg/5D-C:ETA COREAF 236,1387,53
orl-rat LD50:434 mg/kg FCTOD7 20,311,82
orl-rbt LDLo:250 mg/kg EQSSDX 1,1,75
scu-rbt LDLo:75 mg/kg EQSSDX 1,1,75
ims-pgn LDLo:50 mg/kg HBAMAK 4,1289,35
scu-frg LDLo:150 mg/kg HBAMAK 4,1289,35

CONSENSUS REPORTS: Reported in EPA TSCA Inventory. Cobalt and its compounds are on the Community Right-To-Know List.

SAFETY PROFILE: Poison by ingestion, intramuscular, and subcutaneous routes. Experimental reproductive effects. Questionable carcinogen with experimental tumorigenic data. Used in animal feed. Explosive reaction with ammonium hexacyanoferrate(II) at 220°C. Potentially explosive reaction with carbon. When heated to decomposition it emits toxic fumes of NO_x. See also COBALT COMPOUNDS and NITRATES.

CNC750 CAS:12139-70-7 ***HR: 3***
COBALT(II) NITRIDE
mf: CoN mw: 72.94

CONSENSUS REPORTS: Cobalt and its compounds are on the Community Right-To-Know List.

SAFETY PROFILE: Powder will spontaneously explode in air. When heated to decomposition it emits toxic fumes of NO_x. See also COBALT COMPOUNDS and NITRIDES.

CND000 CAS:63919-21-1 ***HR: 3***
COBALT NITROPRUSSIDE

SYN: COBALTNITROSOPENTACYANOFERRATE(3)

TOXICITY DATA with REFERENCE
orl-rat LD50:147 mg/kg AIPTAK 172,487,68
ipr-rat LD50:15 mg/kg AIPTAK 172,487,68
orl-mus LD50:74 mg/kg AIPTAK 172,487,68
ipr-mus LD50:10700 μg/kg AIPTAK 172,487,68
ivn-rbt LDLo:9400 μg/kg AIPTAK 172,487,68

CONSENSUS REPORTS: Cobalt and its compounds, as well as cyanide and its compounds, are on the Community Right-To-Know List.

SAFETY PROFILE: Poison by ingestion, intravenous, and intraperitoneal routes. When heated to decomposition it emits very toxic fumes of CN^- and NO_x. See also COBALT COMPOUNDS and CYANIDE.

CND125 CAS:1307-96-6 ***HR: 3***
COBALT(II) OXIDE
mf: CoO mw: 74.93

PROP: Powder, or cubic or hexagonal crystals. Color varies from olive green to red, depending on the particle size, but the commercial material is usually dark grey and contains about 76% Co. Mp: about 1935°, d: 5.7 to 6.7. Practically insol in water; sol in acids or alkalies. Easily reduced to Co by C or CO. Reacts at high temperatures with silica, alumina, and zinc oxide to form pigments.

SYNS: C.I. 77322 ◇ C.I. PIGMENT BLACK 13 ◇ COBALT BLACK ◇ COBALT MONOOXIDE ◇ COBALT MONOXIDE ◇ COBALTOUS OXIDE ◇ COBALT OXIDE ◇ COBALT(2+) OXIDE ◇ MONOCOBALT OXIDE ◇ ZAFFRE

TOXICITY DATA with REFERENCE
ims-rat TDLo:135 mg/kg:CAR CNREA8 22,152,62
ims-rat TD:90 mg/kg:ETA CNREA8 22,158,62
orl-rat LD50:202 mg/kg FCTOD7 20,311,82
itr-rat LDLo:50 mg/kg NTIS** AEC-TR-6710
scu-mus LD50:125 mg/kg ZVKOA6 19,186,74
ims-mus LDLo:800 mg/kg CNREA8 22,152,62
orl-dog LDLo:89 mg/kg EQSSDX 1,1,75

CONSENSUS REPORTS: Cobalt and its compounds are on the Community Right-To-Know List. Reported in EPA TSCA Inventory.

SAFETY PROFILE: Poison by ingestion, subcutaneous, and intratracheal routes. Moderately toxic by intramuscular route. Questionable carcinogen with experimental carcinogenic and tumorigenic data. Violent reaction with hydrogen peroxide. See also COBALT. Note: The commercial oxides are usually not definite chemical compounds but mixtures of the cobalt oxides.

CND825 CAS:1308-04-9 ***HR: 2***
COBALT(III) OXIDE
mf: Co_2O_3 mw: 165.86

CONSENSUS REPORTS: Cobalt compounds are on the Community Right-To-Know List.

SAFETY PROFILE: Violent reaction with hydrogen peroxide. The oxide increases the sensitivity of nitroalkanes (e.g. nitromethane, nitroethane, and 1-nitropropane) to heat or detonation. See also COBALT COMPOUNDS.

CNE000 ***HR: 2***
COBALT RESINATE, precipitated
mf: $Co(C_{44}H_{62}O_4)_2$ mw: 1368.81

PROP: Brown-red powder.

CONSENSUS REPORTS: Cobalt and its compounds are on the Community Right-To-Know List.

DOT Classification: Flammable Solid; Label: Flammable Solid.

SAFETY PROFILE: A dangerous fire hazard when exposed to heat, flame, oxidizers or air. Ignites spontaneously in air. See also COBALT COMPOUNDS. When heated to decomposition it emits acrid smoke and irritating fumes.

CNE125 CAS:10124-43-3 ***HR: 3***
COBALT(II) SULFATE (1:1)
mf: $O_4S{\cdot}Co$ mw: 154.99

PROP: Red to lavender dimorphic, orthorhombic crystals. D: 3.71. Stable to 708°. Dissolves slowly in boiling water.

SYNS: COBALTOUS SULFATE ◇ COBALT SULFATE ◇ COBALT SULFATE (1:1) ◇ COBALT (2+) SULFATE ◇ COBALT(II) SULPHATE ◇ SULFURIC ACID, COBALT(2+) SALT (1:1)

TOXICITY DATA with REFERENCE
orl-rat LD50:424 mg/kg FCTOD7 20,311,82
ipr-mus LD50:143 mg/kg COREAF 256,1043,63
ivn-dog LDLo:20 mg/kg HBAMAK 4,1289,35
orl-rbt LDLo:1800 mg/kg HBAMAK 4,1289,35

CONSENSUS REPORTS: Cobalt and its compounds are on the Community Right-To-Know List. EPA Genetic Toxicology Program. Reported in EPA TSCA Inventory.

SAFETY PROFILE: Poison by intravenous and intraperitoneal routes. Moderately toxic by ingestion. When heated to decomposition it emits toxic fumes of SO_x. See also COBALT COMPOUNDS.

CNE200 CAS:1317-42-6 ***HR: 3***
COBALT(II) SULFIDE
mf: CoS mw: 90.99

PROP: Exists in two forms. α-CoS: black, amorphous powder. Sol in HCl. β-CoS: grey powder or reddish-silver octahedral crystals. Mp: above 1100°, d: 5.45. Practically insol in water; sol in acids.

SYNS: COBALT MONOSULFIDE ◇ COBALTOUS SULFIDE ◇ COBALT SULFIDE ◇ COBALT SULFIDE (amorphous)

TOXICITY DATA with REFERENCE
otr-ham:emb 1 mg/L CNREA8 42,2757,82
dnd-ham:ovr 10 mg/L CRNGDP 3,657,82
ims-rat TDLo:180 mg/kg:ETA CNREA8 22,158,62

CONSENSUS REPORTS: Reported in EPA TSCA Inventory. Cobalt and its compounds are on the Community Right-To-Know List.

SAFETY PROFILE: Questionable carcinogen with ex-

perimental tumorigenic data. Mutation data reported. If dried at 300°C it ignites spontaneously in air. See also COBALT COMPOUNDS and SULFIDES.

CNE250 CAS:10026-18-3 ***HR: 3***
COBALT TRIFLUORIDE
mf: CoF_3 mw: 226.47

CONSENSUS REPORTS: Cobalt and its compounds are on the Community Right-To-Know List.

SAFETY PROFILE: A powerful fluorinating agent and oxidizer. Violent reaction with hydrocarbons or water. Very exothermic reaction when warmed with silicon. When heated to decomposition it emits toxic fumes of F^-. See also COBALT COMPOUNDS and FLUORIDES.

CNE375 CAS:24699-40-9 ***HR: 3***
COBEN
mf: $C_{19}H_{25}N_3$•ClH mw: 331.93

SYNS: 1-(2-(PHENYL(2-PYRIDYLMETHYL)AMINO)ETHYL)PIPERIDINE HYDROCHLORIDE ◇ N-PHENYL-N-(2-PYRIDYLMETHYL)-2-PIPERIDINOETHYLAMINE HYDROCHLORIDE ◇ N-(2-PICOLYL)-N-PHENYL-N-(2-PIPERIDINOETHYL)AMINE HYDROCHLORIDE ◇ PICOPERIDAMINE HYDROCHLORIDE ◇ PICOPERINE HYDROCHLORIDE ◇ N-(2-PIPERIDINOETHYL)-N-(2-PYRIDYLMETHYL)ANILINE HYDROCHLORIDE ◇ 1-(2-(N-(2-PYRIDYLMETHYL)ANILINO)ETHYL)PIPERIDINEHYDROCHLORIDE ◇ N-(2-PYRIDYL-METHYL)-N-PHENYL-N-2-(PIPERIDINOETHYL)AMINEHYDROCHLORIDE ◇ TAT-3 HYDROCHLORIDE

TOXICITY DATA with REFERENCE
orl-mus TDLo:180 mg/kg (8-13D preg):TER TAKHAA 29,297,70
orl-rat LD50:740 mg/kg KSRNAM 4,403,70
ipr-rat LD50:69 mg/kg KSRNAM 4,403,70
scu-rat LD50:480 mg/kg KSRNAM 4,403,70
ivn-rat LD50:11500 μg/kg KSRNAM 4,403,70
orl-mus LD50:210 mg/kg KSRNAM 4,403,70
ipr-mus LD50:55500 μg/kg KSRNAM 4,403,70
scu-mus LD50:142 mg/kg KSRNAM 4,403,70
ivn-mus LD50:10 mg/kg KSRNAM 4,403,70
orl-dog LDLo:100 mg/kg ARZNAD 19,1916,69
ivn-dog LDLo:25 mg/kg ARZNAD 19,1916,69

SAFETY PROFILE: Poison by ingestion, subcutaneous, intravenous, and intraperitoneal routes. An experimental teratogen. When heated to decomposition it emits toxic fumes of NO_x and HCl.

CNE500 CAS:29091-05-2 ***HR: 2***
COBEXO
mf: $C_{11}H_{13}F_3N_4O_4$ mw: 322.28

SYNS: COBEX ◇ N^3,N^3-DIETHYL-2,4-DINITRO-6-(TRIFLUOROMETHYL)-1,3-BENZENEDIAMINE◇ N^4,N^4-DIETHYL-α,α,α-TRIFLUORO-3,5-DINITRO-TOLUENE-2,4-DIAMINE◇ DINITRAMINE ◇ DINITROAMINE ◇ USB-3584

TOXICITY DATA with REFERENCE
orl-rat LD50:3000 mg/kg 28ZEAL 5,80,76
skn-rbt LD50:2000 mg/kg FMCHA2 -,D106,80

SAFETY PROFILE: Moderately toxic by ingestion and skin contact. An herbicide. When heated to decomposition it emits very toxic fumes of F^- and NO_x. See also FLUORIDES and NITRO COMPOUNDS of AROMATIC HYDROCARBONS.

CNE750 CAS:50-36-2 ***HR: 3***
COCAINE
mf: $C_{17}H_{21}NO_4$ mw: 303.39

PROP: Colorless to white crystals. Mp: 98°, bp: 187-188°. Volatile, especially above 90°. Soluble in alcohol, chloroform, ether, oil turpentine, olive oil, liquid petrolatum, acetone, ethyl acetate, carbon disulfide. Sparingly soluble in water.

SYNS: BENZOYLMETHYLECGONINE ◇ BERNICE ◇ BERNIES ◇ BURESE ◇ 2-β-CARBOMETHOXY-3-β-BENZOXYTROPANE ◇ "C" CARRIE ◇ CECIL ◇ CHOLLY ◇ (−)-COCAINE ◇ β-COCAINE ◇ l-COCAINE ◇ COKE ◇ CORINE ◇ ECGONINE, METHYL ESTER, BENZOATE (ESTER) ◇ ERITROXILINA ◇ ERYTROXYLIN ◇ GIRL ◇ GOLD DUST ◇ HAPPY DUST ◇ 3-β-HYDROXY-1-α-H,5-α-H-TROPANE-2-β-CARBOXYLIC ACID METHYL ESTER, BENZOATE ◇ KOKAIN ◇ KOKAN ◇ KOKAYEEN ◇ METHYL-3-β-HYDROXY-1-α-H,5-α-H-TROPANE-2-β-CARBOXYLATE BENZOATE (ESTER) ◇ NEUROCAINE ◇ STAR DUST ◇ 2-β-TROPANECARBOXYLIC ACID, 3-β-HYDROXY-, METHYL ESTER, BENZOATE (ESTER) ◇ 3-TROPANYLBENZOATE-2-CARBOXYLIC ACID METHYL ESTER

TOXICITY DATA with REFERENCE
eye-rbt 16% MLD JAPMA8 42,685,53
orl-man LDLo:7353 μg/kg 85DCAI 2,73,70
orl-hmn TDLo:714 mg/kg:CNS JAMAAP 238,1391,77
unk-hmn LDLo:286 μg/kg 34ZIAG -,183,69
ipr-rat LD50:70 mg/kg JLCMAK 15,731,30
scu-rat LD50:250 mg/kg JLCMAK 15,731,30
ivn-rat LD50:17500 μg/kg JLCMAK 15,731,30
orl-mus LD50:99 mg/kg ARZNAD 16,1275,66
ipr-mus LD50:75 mg/kg JATOD3 4,19,80
scu-mus LDLo:125 mg/kg APBOAI 12,189,66
ivn-mus LD50:30 mg/kg RPTOAN 35(3),114,72
scu-dog LDLo:3500 μg/kg BDHU** -,-,36
ivn-dog LD50:13 mg/kg AIPTAK 235,328,78
orl-rbt LDLo:126 mg/kg 27ZIAQ -,78,73
scu-rbt LDLo:50 mg/kg AEPPAE 160,53,31
ivn-rbt LD50:17 mg/kg JLCMAK 15,731,30

SAFETY PROFILE: A human poison by ingestion and possibly other routes. Poison experimentally by ingestion, intraperitoneal, intravenous, subcutaneous, and parenteral routes. Human central nervous system effects by ingestion and possibly other routes: general anesthesia, hallucinations or distorted perceptions, and convulsions. An eye irritant. A widely abused, controlled substance. Abuse leads to habituation or addiction. In medicine, it is used as a local narcotic anesthetic applied topically to mucous membranes. The free base is soluble in fats and thus is used for ointments and oily solutions. For water-soluble applications, the sulfate or hydrochloride is used. See also ESTERS. When heated to decomposition it emits highly toxic fumes.

CNF000 CAS:53-21-4 *HR: 3*
COCAINE HYDROCHLORIDE
mf: $C_{17}H_{21}NO_4 \cdot ClH$ mw: 339.85

PROP: Mp: 195°. Soluble in water, alc, chloroform, glycerol, acetone, ether, and oils. Decomposes when heated.

SYNS: COCAIN-CHLORHYDRAT (GERMAN) ◇ (−)-COCAINE HYDROCHLORIDE ◇ l-COCAINE HYDROCHLORIDE ◇ COCAINE CHLORIDE ◇ COCAINE MURIATE ◇ 3-β-HYDROXY-1-α-H,5-α-H-TROPANE-2-β-CARBOXYLIC ACID METHYL ESTER, BENZOATE (ESTER), HYDROCHLORIDE ◇ SAL de MERCK

TOXICITY DATA with REFERENCE
eye-rbt 2% MLD ARZNAD 8,181,58
cyt-ham:lng 830 mg/L GMCRDC 27,95,81
scu-rat TDLo:1080 mg/kg (male 72D pre):REP JOAND3 10,17,89
scu-rat TDLo:910 mg/kg (female 7-19D post):TER NRTXDN 10,51,88
ipr-rat LD50:78 mg/kg ARZNAD 8,181,58
orl-mus LD50:96 mg/kg RPOBAR 2,279,70
ipr-mus LD50:68 mg/kg AIPTAK 189,198,71
scu-mus LD50:30 mg/kg ARZNAD 8,181,58
ivn-mus LD50:15 mg/kg AIPTAK 105,221,56
ivn-dog LD50:21 mg/kg AIPTAK 235,328,78
scu-rbt LDLo:100 mg/kg JPETAB 47,255,33
scu-gpg LDLo:60 mg/kg JPETAB 47,255,33

SAFETY PROFILE: Poison by ingestion, intravenous, intraperitoneal, and subcutaneous routes. An experimental teratogen. Other experimental reproductive effects. An eye irritant. Mutation data reported. A widely abused, controlled substance. Abuse leads to habituation or addiction. In medicine, it is used as a local anesthetic and central nervous system stimulant. Incompatible with calomel; mercuric oxide; silver nitrate. When heated to decomposition it emits very toxic fumes of NO_x and HCl. See also COCAINE.

CNF109 CAS:4611-05-6 *HR: 3*
COCHLIOBOLIN
mf: $C_{25}H_{36}O_4$ mw: 400.61

SYNS: COCHLIOBOLIN A ◇ OPHIOBOLIN ◇ OPHIOBOLIN A

TOXICITY DATA with REFERENCE
orl-mus LD50:238 mg/kg 85GDA2 6,143,81
ipr-mus LD50:21 mg/kg 85GDA2 6,143,81
scu-mus LD50:73 mg/kg 85GDA2 6,143,81
ivn-mus LD50:12 mg/kg 85GDA2 6,143,81

SAFETY PROFILE: Poison by ingestion, subcutaneous, intravenous, and intraperitoneal routes. When heated to decomposition it emits acrid smoke and fumes.

CNF159 CAS:11051-88-0 *HR: 3*
COCHLIODINOL
mf: $C_{32}H_{30}N_2O_4$ mw: 506.64

SYNS: 3,6-BIS(5-(3-METHYL-2-BUTENYL)INDOL-3-YL)-2,5-DIHYDROXY-p-BENZOQUINONE ◇ 2,5-DIHYDROXY-3,6-BIS(5-(3-METHYL-2-BUTENYL)-1H-INDOL-3-YL)-2,5-CYCLOHEXADIENE-1,4-DIONE

TOXICITY DATA with REFERENCE
orl-mus LD50:221 mg/kg 85GDA2 3,310,80
ipr-mus LD50:41 mg/kg 85GDA2 3,310,80
scu-mus LD50:147 mg/kg 85GDA2 3,310,80

SAFETY PROFILE: Poison by ingestion, subcutaneous, and intraperitoneal routes. When heated to decomposition it emits toxic fumes of NO_x.

CNF175 CAS:61789-30-8 *HR: D*
COCOA FATTY ACIDS, POTASSIUM SALTS

SYN: SOAP

TOXICITY DATA with REFERENCE
skn-mus TDLo:6 g/kg (female 2-13D post):REP TXCYAC 4,171,75
skn-mus TDLo:600 mg/kg (female 2-13D post):TER TXCYAC 4,171,75

CONSENSUS REPORTS: Reported in EPA TSCA Inventory.

SAFETY PROFILE: An experimental teratogen. Other experimental reproductive effects. When heated to decomposition it emits acrid smoke and irritating fumes.

CNF250 CAS:104-61-0 *HR: 2*
COCONUT ALDEHYDE
mf: $C_9H_{16}O_2$ mw: 156.25

PROP: Colorless to sltly yellow liquid; coconut odor. D: 0.958-0.966, refr index: 1.446-1.450, flash p: +212°F. Sol in alc, fixed oils, propylene glycol; insol in water.

SYNS: ALDEHYDE C-18 ◇ Γ-N-AMYLBUTYROLACTONE ◇ FEMA No. 2781 ◇ 4-HYDROXYNONANOIC ACID, Γ-LACTONE ◇ Γ-NONALACTONE (FCC) ◇ 1,4-NONALOLIDE ◇ PRUNOLIDE

TOXICITY DATA with REFERENCE
skn-rbt 500 mg/24H MLD FCTXAV 13,681,75
orl-rat LD50:6600 mg/kg FCTXAV 13,681,75
orl-gpg LD50:3440 mg/kg FCTXAV 2,327,64

CONSENSUS REPORTS: Reported in EPA TSCA Inventory.

SAFETY PROFILE: Moderately toxic by ingestion. A skin irritant. Combustible liquid. When heated to decomposition it emits acrid smoke and irritating fumes. See also ALDEHYDES.

CNF325 CAS:61788-90-7 ***HR: 1***
COCONUT DIMETHYL AMINE OXIDE

TOXICITY DATA with REFERENCE
skn-hmn 2500 μg/24H MOD AKEDAX 235,180,69
skn-rbt 10 mg MLD JSCCA5 22,411,71
skn-rbt 230 mg/5W open MLD JCSSA5 22,411,71
skn-gpg 115 mg/5W open MLD JSCCA5 22,411,71

CONSENSUS REPORTS: Reported in EPA TSCA Inventory.

SAFETY PROFILE: A human skin irritant. When heated to decomposition it emits toxic fumes of NO_x. See also AMINES.

CNF500 CAS:76-57-3 ***HR: 3***
CODEINE
mf: $C_{18}H_{21}NO_3$ mw: 299.40

PROP: Crystals from H_2O or dil alc. Mp: 155°, d: 1.32 @ 20°/4°.

SYNS: METHYLMORPHINE ◇ N-METHYL-NORDOCEINE ◇ MORPHINE-3-METHYL ETHER ◇ MORPHINE MONOMETHYL ETHER

TOXICITY DATA with REFERENCE
orl-rat TDLo:350 mg/kg (6-15D preg):TER ARZNAD 26,551,76
orl-rat TDLo:1200 mg/kg (6-15D preg):REP ARZNAD 26,551,76
unr-man LDLo:12 mg/kg 85DCAI 2,73,70
orl-rat LD50:427 mg/kg JMCMAR 16,782,73
ipr-rat LD50:130 mg/kg ARZNAD 26,551,76
scu-rat LD50:229 mg/kg ARZNAD 24,600,74
ivn-rat LD50:75 mg/kg JPPMAB 25,929,73
orl-mus LD50:250 mg/kg MDCHAG 5,318,65
ipr-mus LD50:60 mg/kg APFRAD 8,261,50
scu-mus LD50:84100 μg/kg ARZNAD 24,600,74
ivn-mus LD50:54 mg/kg ARZNAD 16,617,66
ims-mus LD50:290 mg/kg OYYAA2 13,97,77
ivn-dog LD50:69 mg/kg CPBTAL 7,372,59
ivn-rbt LD50:34 mg/kg EXPEAM 18,446,62

SAFETY PROFILE: A human poison by an unspecified route. An experimental poison by ingestion, intraperitoneal, intravenous, intramuscular, and subcutaneous routes. Human reproductive effects. An experimental teratogen. Other experimental reproductive effects. An addictive drug. Flammable when exposed to heat or flame. To fight fire, use alcohol foam. When heated to decomposition it emits toxic fumes of NO_x. See also MORPHINE and ETHERS.

CNF750 CAS:1422-07-7 ***HR: 3***
CODEINE HYDROCHLORIDE
mf: $C_{18}H_{21}NO_3{\cdot}ClH$ mw: 335.86

TOXICITY DATA with REFERENCE
orl-rat LD50:750 mg/kg ARZNAD 21,719,71
scu-rat LD50:215 mg/kg ARZNAD 21,727,71
orl-mus LD50:365 mg/kg BJPCAL 16,209,61
ipr-mus LD50:115 mg/kg ARZNAD 25,873,75
scu-mus LD50:270 mg/kg JMCMAR 19,1054,76
ivn-mus LD50:67 mg/kg BJPCAL 16,209,61
scu-rbt LDLo:36 mg/kg JPETAB 66,182,39
par-rbt LDLo:24 mg/kg JPETAB 66,182,39

SAFETY PROFILE: Poison by ingestion, subcutaneous, intravenous, intraperitoneal and parenteral routes. An addictive drug. See also CODEINE. When heated to decomposition it emits very toxic fumes of NO_x and HCl.

CNG000 CAS:63732-63-8 ***HR: 3***
CODEINE METHOCHLORIDE
mf: $C_{18}H_{21}NO_3{\cdot}CH_3Cl$ mw: 349.89

TOXICITY DATA with REFERENCE
scu-mus LDLo:1000 mg/kg JPETAB 49,319,33
scu-rbt LDLo:400 mg/kg JPETAB 49,319,33
par-frg LDLo:2500 mg/kg JPETAB 49,319,33

SAFETY PROFILE: Poison by subcutaneous route. Moderately toxic by parenteral route. See also CODEINE. An addictive drug. When heated to decomposition it emits very toxic fumes of NO_x and Cl^-.

CNG250 CAS:3688-66-2 ***HR: 3***
CODEINE NICOTINATE (ESTER)
mf: $C_{24}H_{24}N_2O_4$ mw: 404.50

SYNS: LYOPECT ◇ NICOCODINE ◇ NICOTINIC ACID-7,8-DIDEHYDRO-4,5-α-EPOXY-3-METHOXY-17-METHYLMORPHINAN-6α-YL ESTER ◇ NICOTINIC ACID, ESTER with CODEINE ◇ RC 146

TOXICITY DATA with REFERENCE
ipr-rat LD50:37 mg/kg AIPTAK 143,466,63
orl-rat LD50:840 mg/kg AIPTAK 143,466,63
orl-mus LD50:280 mg/kg AIPTAK 143,466,63
ipr-mus LD50:48 mg/kg AIPTAK 143,466,63

SAFETY PROFILE: Poison by ingestion and intraperitoneal routes. See also CODEINE. An addictive drug. When heated to decomposition it emits toxic fumes of NO_x.

CNG500 CAS:52-28-8 ***HR: 3***
CODEINE PHOSPHATE
mf: $C_{18}H_{21}NO_3{\cdot}H_3O_4P$ mw: 397.40

TOXICITY DATA with REFERENCE
scu-mus TDLo:110 mg/kg (9D preg):TER DGDFA5 22,61,80
orl-wmn TDLo:3600 μg/kg/2D-I:EYE,BPR CPHADV 5,15,86
orl-man TDLo:5143 μg/kg/5D-I CPHADV 5,15,86
orl-rat LD50:266 mg/kg JPETAB 128,384,60
ipr-rat LD50:140 mg/kg APTOA6 22,241,65
scu-rat LD50:312 mg/kg JPETAB 154,161,66
ivn-rat LD50:54 mg/kg AIPTAK 103,200,55
ims-rat LD50:208 mg/kg AIPTAK 190,124,71
orl-mus LD50:237 mg/kg JPETAB 128,384,60
ipr-mus LD50:110 mg/kg ARZNAD 19,1916,69
scu-mus LD50:120 mg/kg JPPMAB 16,174,64
ivn-mus LD50:62 mg/kg JJPAAZ 17,538,67
ims-mus LD50:191 mg/kg AIPTAK 190,124,71
ivn-dog LD50:978800 μg/kg OYYAA2 4,961,70
orl-rbt LDLo:100 mg/kg HBTXAC 1,72,55

CONSENSUS REPORTS: EPA Genetic Toxicology Program.

SAFETY PROFILE: Poison by ingestion, subcutaneous, intravenous, intramuscular, and intraperitoneal routes. Human systemic effects by ingestion: miosis (pupillary constriction), sleep disturbance, and blood pressure lowering. An experimental teratogen. Used as an analgesic. See also CODEINE. When heated to decomposition it emits very toxic fumes of NO_x and PO_x.

CNG675 CAS:5913-76-8 ***HR: 3***
CODEINE PHOSPHATE SESQUIHYDRATE
mf: $C_{18}H_{21}NO_3 \cdot H_3O_4P \cdot 3/2H_2O$ mw: 424.38

SYN: 7,8-DIDEHYDRO-4,5-α-EPOXY-3-METHOXY-17-METHYLMORPHINAN-6-α-OL PHOSPHATE SESQUIHYDRATE (3:3:2)

TOXICITY DATA with REFERENCE
orl-rat LD50:209 mg/kg KSRNAM 5,1011,71
ipr-rat LD50:38 mg/kg KSRNAM 5,1011,71
scu-rat LD50:81 mg/kg KSRNAM 5,1011,71
orl-mus LD50:290 mg/kg KSRNAM 5,1787,71
ipr-mus LD50:83 mg/kg KSRNAM 5,1011,71
scu-mus LD50:191 mg/kg YKKZAJ 81,740,61
ivn-mus LD50:70 mg/kg KSRNAM 5,1787,71
ivn-dog LD50:97800 μg/kg YKKZAJ 81,740,61
orl-rbt LDLo:100 mg/kg MEIEDD 10,350,83

SAFETY PROFILE: Poison by ingestion, subcutaneous, intravenous, and intraperitoneal routes. When heated to decomposition it emits toxic fumes of NO_x and PO_x. A narcotic analgesic and antitussive. See also CODEINE and CODEINE PHOSPHATE.

CNG750 CAS:1420-53-7 ***HR: 3***
CODEINE SULFATE
mf: $C_{36}H_{42}N_2O_6 \cdot O_4S$ mw: 694.86

TOXICITY DATA with REFERENCE
cyt-ofs-ipr 100 mg/L BEXBBO 16,425,80
scu-mus TDLo:100 mg/kg (9D preg):REP JPMSAE 66,1727,77
scu-mus TDLo:100 mg/kg (9D preg):TER JPMSAE 66,1727,77
orl-rat LD50:430 mg/kg AIPTAK 123,48,59
ipr-rat LD50:107 mg/kg TXCYAC 14,217,79
scu-rat LD50:332 mg/kg JPETAB 134,332,61
ivn-rat LD50:55 mg/kg JPETAB 134,332,61
orl-mus LD50:395 mg/kg JPETAB 134,332,61
ipr-mus LD50:159 mg/kg AIPTAK 136,333,62
scu-mus LD50:183 mg/kg JPETAB 134,332,61
ivn-mus LD50:68 mg/kg JPETAB 134,332,61

SAFETY PROFILE: Poison by ingestion, intraperitoneal, intravenous, and subcutaneous routes. An experimental teratogen. Other experimental reproductive effects. Mutation data reported. An addictive, narcotic drug. When heated to decomposition it emits very toxic fumes of NO_x and SO_x. See also CODEINE.

CNG775 ***HR: D***
COFFEE

TOXICITY DATA with REFERENCE
mmo-esc 12 mg/plate MUREAV 116,179,83
pic-esc 12 mg/plate MUREAV 116,179,83
orl-rat TDLo:1340 g/kg (female 91D pre-21D post):REP TXAPA9 58,171,81
orl-rat TDLo:2811 g/kg (35D pre/1-22D preg):TER TXAPA9 44,1,78

SAFETY PROFILE: An experimental teratogen. Experimental reproductive effects. Mutation data reported. A stimulant. See also CAFFEINE.

CNG825 ***HR: 3***
COFFEE SENNA

PROP: An annual herb (up to 3 feet tall) with smooth compound leaves and yellow flowers. The seed pod is about 4.5 inches long and thin with dark olive green seeds. It is common, particularly along highways and coastal areas, from Virginia to Texas, Hawaii and Guam.

SYNS: 'AUKO'I (HAWAII) ◇ BICHE PRIETO (MEXICO) ◇ CASSIA OCCIDENTALIS ◇ DANDELION (JAMAICA) ◇ HEDIONDA (PUERTO RICO) ◇ MIKIPALAOA (HAWAII) ◇ PISS-A-BED (JAMAICA) ◇ POIS PUANTE (HAITI) ◇ STINKING WEED ◇ STYPTIC WEED ◇ WILD COFFEE ◇ YERBA HEDIONDA (CUBA)

SAFETY PROFILE: The whole plant contains the irritant chrysarobin, the cathartic emodin and toxalbumin. Reported poisonings are the result of ingesting raw seeds. Roasted seeds are used as a coffee substitute. The raw seeds have a strong laxative action in man. Chronic ingestion may cause damage to the muscles, kidneys, liver and lungs, and may result in death. See also 3-METHYL-1,8,9-ANTHRACENETRIOL, 6-METHYL-1,3,8-TRIHYDROXYANTHRAQUINONE, and ABRIN.

CNG830 CAS:64-86-8 ***HR: 3***
COLCHICINE
mf: $C_{22}H_{25}NO_6$ mw: 399.48

PROP: Pale yellow scales or powder. Mp: 142-150°. Crystals from ethyl acetate, pale yellow needles. Mp: 157°. One gram dissolves in 22 mL water, 220 mL ether, 100 mL benzene; freely sol in alcohol or chloroform; practically insol in petr ether.

SYNS: 7-ACETAMIDO-6,7-DIHYDRO-1,2,3,10-TETRAMETHOXY-BENZO(a)HEPTALEN-9(5H)-ONE ◇ N-ACETYL TRIMETHYL-COLCHICINIC ACID METHYLETHER ◇ COLCHICIN (GERMAN) ◇ COLCHICINA (ITALIAN) ◇ 7-α-H-COLCHICINE ◇ COLCHINEOS ◇ COLCHISOL ◇ COLCIN ◇ COLSALOID ◇ CONDYLON ◇ NSC 757 ◇ N-(5,6,7,9-TETRAHYDRO-1,2,3,10-TETRAMETHOXY-9-OX-OBENZO(α)HEPTALEN-7-YL)-ACETAMIDE

TOXICITY DATA with REFERENCE
eye-rbt 1%/3D SEV AJOPAA 31,837,48
dnd-hmn:fbr 1 mg/L BBACAQ 824,117,85
sln-mus-ipr 200 μg/kg ENMUDM 8(Suppl 6),51,86
scu-rat TDLo:1200 μg/kg (18-20D preg):REP DEPBA5 9,119,76
scu-rat TDLo:1200 μg/kg (18-20D preg):TER DEPBA5 9,119,76
orl-wmn TDLo:320 μg/kg:PNS LANCAO 2,1271,87
orl-hmn LDLo:86 μg/kg:PUL,GIT,MET 34ZIAG -,184,69
orl-man TDLo:12514 μg/kg/2Y-I NEJMAG 316,1562,87
orl-man LDLo:11 mg/kg:PUL,SYS CMAJAX 138,335,88
ivn-wmn LDLo:360 μg/kg/6D:BLD,KID NPMDAD 9,1587,80
ivn-man LDLo:143 μg/kg/5D-I:BLD NEJMAG 309,310,83
par-hmn TDLo:710 μg/kg:CNS,CVS,GIT AIMDAP 137,394,77
ipr-rat LD50:6100 μg/kg TXAPA9 24,37,73
scu-rat LDLo:4 mg/kg AIPTAK 84,257,50
ivn-rat LD50:1600 μg/kg TXAPA9 13,50,68
ice-rat LDLo:1 mg/kg AIPTAK 109,386,57
par-rat LDLo:1000 μg/kg FRPSAX 15,533,60
orl-mus LD50:5886 μg/kg NCISP* JAN86
ipr-mus LD50:2 mg/kg BAFEAG 42,308,55
scu-mus LD50:1200 μg/kg NIIRDN 6,280,82
ivn-mus LD50:1700 μg/kg NIIRDN 6,280,82
ims-mus LD50:1197 μg/kg JMCMAR 24,257,50
orl-dog LDLo:125 μg/kg 85DZAJ -,315,68
scu-dog LDLo:571 μg/kg HBAMAK 4,1337,35

CONSENSUS REPORTS: EPA Extremely Hazardous Substances List. EPA Genetic Toxicology Program. Reported in EPA TSCA Inventory.

SAFETY PROFILE: A human poison by ingestion and intravenous routes. Poison experimentally by ingestion, subcutaneous, intramuscular, parenteral, intraperitoneal, intracerebral, and possibly other routes. Human systemic effects: aplastic anemia, blood pressure depression, body temperature decrease, changes in kidney tubules, dyspnea, flaccid paralysis without anesthesia, gastrointestinal effects, kidney damage and hemorrhaging, muscle contraction or spasticity, muscle weakness, nausea or vomiting, respiratory stimulation, and somnolence. An experimental teratogen. Experimental reproductive effects. A severe eye irritant. Human mutation data reported. Inhibits the formation of microtubules and thus impairs cell division. When heated to decomposition it emits toxic fumes of NO_x.

CNG835 CAS:41826-92-0 ***HR: 3***
COLIBIL
mf: $C_{16}H_{22}O_6$ mw: 310.38

PROP: Colorless needles from aq ethanol or plates from aq acetone. Mp: 150-151°. Stable to heat, humidity, indoor diffused sunlight.

SYNS: CHOLIBIL ◇ SUPACAL ◇ TREPIBUTONE ◇ 3-(2,4,5-TRI-ETHOXYBENZOYL)PROPIONIC ACID ◇ 2,4,5-TRIETHOXY-Γ-OX-OBENZENEBUTANOIC ACID

TOXICITY DATA with REFERENCE
orl-rat TDLo:9100 mg/kg (female 4W pre):REP TAKHAA 36,263,77
orl-rat LD50:2450 mg/kg IYKEDH 11,811,80
ipr-rat LD50:410 mg/kg IYKEDH 11,811,80
scu-rat LD50:570 mg/kg IYKEDH 11,811,80
ivn-rat LD50:350 mg/kg IYKEDH 11,811,80
orl-mus LD50:1340 mg/kg IYKEDH 11,811,80
ipr-mus LD50:510 mg/kg YAKUD5 22,1501,80
scu-mus LD50:625 mg/kg YAKUD5 22,1501,80
ivn-mus LD50:500 mg/kg IYKEDH 11,811,80

SAFETY PROFILE: Poison by intravenous route. Moderately toxic by ingestion, subcutaneous, and intraperitoneal routes. Experimental reproductive effects. When heated to decomposition it emits acrid fumes and smoke.

CNH000 CAS:9004-70-0 ***HR: 3***
COLLODION
$C_{12}H_{16}O_6(NO_3)_4C_{13}H_{17}O_7(NO_3)_3$ mw: 975

PROP: Soln of nitrated cellulose in ether + alc. Flash p: <0°F.

CONSENSUS REPORTS: Reported in EPA TSCA Inventory.

DOT Classification: Flammable Liquid; Label: Flammable Liquid.

SAFETY PROFILE: Very dangerous fire hazard when

exposed to heat or flame. To fight fire, use alcohol foam. When heated to decomposition it emits toxic fumes of NO_x.

CNH125 CAS:26591-12-8 ***HR: 2***
COLORFIX
mf: $(C_2H_4N_4 \cdot CH_2O)_x$

SYNS: CARFLOC D 1000 ◇ CETRAMIN ◇ DICYANDIAMIDE-FORMALDEHYDE ADDUCT ◇ DICYANDIAMIDE-FORMALDEHYDE POLYMER ◇ DICYANDIAMIDE-FORMALDEHYDE RESIN ◇ DRASIL 507 ◇ FIXATIVE IS ◇ FIXER IS ◇ NEOFIX FP ◇ NIKAFLOC D 1000 ◇ NONISOLD ◇ PARAMEL DC ◇ PERMINAL FC-P ◇ RESIN T ◇ SATILAN PLZ ◇ SATILAN RL 2 ◇ SUMISET D ◇ SYNTEFIX ◇ US 2 ◇ WARCO F 71

TOXICITY DATA with REFERENCE
skn-rbt 500 mg/24H MLD 28ZPAK -,305,72
eye-rbt 100 mg/24H MOD 28ZPAK -,305,72
orl-rat LD50:1750 mg/kg 28ZPAK -,305,72

SAFETY PROFILE: Moderately toxic by ingestion. An eye and skin irritant. When heated to decomposition it emits toxic fumes of NO_x and CN^-. See also ALDEHYDES.

CNH250 ***HR: 3***
COLTSFOOT

PROP: It is an herb of the tribe *Senecione* and from the family *Compositae* (GANNA2 67,125,76).

SYNS: KAN-TO-KA (JAPANESE) ◇ TUSSILAGO FARFARA L

TOXICITY DATA with REFERENCE
orl-rat TDLo:4800 g/kg/77W-C:CAR GANNA2 67,125,76

SAFETY PROFILE: Questionable carcinogen with experimental carcinogenic data.

CNH275 ***HR: 3***
COMBRETODENDRON AFRICANUM (Welw), extract

PROP: Plant belonging to the family *Lecythidaceae* (THERAP 22,325,67).

TOXICITY DATA with REFERENCE
scu-rat TDLo:1 g/kg (14D preg):TER THERAP 22,325,67
scu-mus TDLo:1 g/kg (female 1D post):REP THERAP 22,325,67
ipr-rat LD50:1 g/kg THERAP 22,325,67
scu-rat LD50:5 g/kg THERAP 22,325,67
ipr-mus LD50:950 mg/kg THERAP 22,325,67
ivn-rbt LD50:250 mg/kg THERAP 22,325,67

SAFETY PROFILE: Poison by intravenous route. Moderately toxic by intraperitoneal route. An experimental teratogen. Other experimental reproductive effects.

CNH300 CAS:67110-84-3 ***HR: 3***
COMPOUND 20-438
mf: $C_{22}H_{27}N \cdot ClH$ mw: 341.96

SYN: (4aRS,5RS,9bRS)-2,3,4,4a,5,9b-HEXAHYDRO-2-ETHYL-7-METHYL-5-p-TOLYL-1H-INDENO(1,2-c)PYRIDINEHYDROCHLORIDE

TOXICITY DATA with REFERENCE
orl-mus TDLo:10 mg/kg (male 1D pre):REP MUREAV 66,113,79
orl-mus LDLo:400 mg/kg MUREAV 66,113,79

SAFETY PROFILE: Poison by ingestion. Experimental reproductive effects. When heated to decomposition it emits toxic fumes of NO_x.

CNH375 CAS:4091-50-3 ***HR: 3***
COMPOUND 48/80
mf: $C_{10}H_{15}NO$ mw: 165.26

TOXICITY DATA with REFERENCE
ipr-mus TDLo:1200 mg/kg (1-6D preg):REP JRPFA4 6,179,63
ivn-rat LD50:2400 μg/kg AIPTAK 120,353,59
ivn-mus LD50:1950 μg/kg AIPTAK 120,353,59
ivn-rbt LD50:1500 μg/kg AIPTAK 120,353,59
ivn-gpg LD50:1290 μg/kg AIPTAK 120,353,59

SAFETY PROFILE: Poison by intravenous route. Experimental reproductive effects. When heated to decomposition it emits toxic fumes of NO_x. See also AMINES.

CNH500 CAS:27114-11-0 ***HR: 3***
COMPOUND 69/183
mf: $C_{22}H_{25}FN_2O \cdot 2ClH$ mw: 425.41

SYNS: 3-(Γ-(p-FLUOROBENZOYL)PROPYL)-2,3,4,4a,5,6-HEXAHYDRO-1(H)-PYRAZINO(1,2A)QUINOLINE HCl ◇ 4'-FLURO-4-(1,2,4,4a,5,6-HEXAHYDRO-3H-PYRANZINO(1,2-A)QUINOLIN-3-YL)-BUTYROPHENONE 2HCl

TOXICITY DATA with REFERENCE
orl-rat LD50:800 mg/kg DRFUD4 4,185,79
ipr-rat LD50:161 mg/kg ARZNAD 28,1641,78
orl-mus LD50:1 g/kg DRFUD4 4,185,79
ipr-mus LD50:300 mg/kg JMCMAR 13,516,70
ivn-mus LD50:95 mg/kg ARZNAD 28,1641,78

SAFETY PROFILE: Poison by intraperitoneal and intravenous routes. Moderately toxic by ingestion. When heated to decomposition it emits very toxic fumes of F^-, NO_x, and HCl.

CNH525 CAS:74886-24-1 ***HR: D***
COMPOUND 78/702
mf: $C_{21}H_{26}O_2$ mw: 310.47

SYNS: 1,2-DIETHYL-1,3-BIS-(p-METHOXYPHENYL)-1-PROPENE◇ 3-(p-METHOXYBENZYL)-4-(p-METHOXYPHENYL)-3-HEXENE

TOXICITY DATA with REFERENCE
orl-rat TDLo:1500 μg/kg (2D preg):REP IJEBA6 18,557,80

SAFETY PROFILE: Experimental reproductive effects. When heated to decomposition it emits acrid smoke and fumes.

CNH550 ***HR: 3***
COMPOUND 14045 METHIODIDE
mf: $C_{20}H_{32}NO \cdot I$ mw: 429.43

SYN: 1-(3-CYCLOHEXYL-3-HYDROXY-3-PHENYLPROPYL)-1-METHYL-PYRROLIDINIUM IODIDE

TOXICITY DATA with REFERENCE
orl-mus LD50:515 mg/kg JAPMA8 43,408,54
ipr-mus LD50:86 mg/kg JAPMA8 43,408,54
ivn-mus LD50:12510 μg/kg JAPMA8 43,408,54
scu-dog LDLo:100 mg/kg JAPMA8 43,408,54

SAFETY PROFILE: Poison by subcutaneous, intravenous, and intraperitoneal routes. Moderately toxic by ingestion. When heated to decomposition it emits toxic fumes of I^- and NO_x. See also IODIDES.

CNH625 CAS:11028-71-0 ***HR: 3***
CONCANAVALIN A

PROP: The most extensively investigated member of the lectin family of plant proteins. Unlike most lectins, it lacks covalently bound carbohydrate and therefore is not a glycoprotein.

SYNS: CON A ◇ NSC 143504 ◇ RICIN-TOXIN CON A

TOXICITY DATA with REFERENCE
cyt-oin:emb 100 mg/L DKBSAS 223,316,75
cyt-mnl:emb 25 mg/L EXPEAM 23,1568,76
cyt-nml:emb 25 mg/L EXPEAM 32,1568,76
dns-mus-ivn 400 μg/kg JJMCAQ 36,43,83
dns-mus:oth 1 mg/L JJIND8 65,1321,80
ivn-mus TDLo:5 mg/kg (female 7D post):TER NDKIA2 32,230,81
par-ham TDLo:800 μg/kg (female 3D post):REP CCPTAY 35,507,87
ipr-mus LD50:41500 μg/kg NCISP* JAN86
ivn-mus LD50:50 mg/kg ARZNAD 30,759,80

SAFETY PROFILE: Poison by intravenous and intraperitoneal routes. An experimental teratogen. Other experimental reproductive effects. Mutation data reported. When heated to decomposition it emits toxic fumes of NO_x.

CNH650 CAS:66231-56-9 ***HR: 2***
CONCTASE C

TOXICITY DATA with REFERENCE
ipr-rat LD50:1050 mg/kg OYYAA2 19,503,80
scu-rat LD50:2660 mg/kg OYYAA2 19,503,80
ipr-mus LD50:735 mg/kg OYYAA2 19,503,80
scu-mus LD50:850 mg/kg OYYAA2 19,503,80

SAFETY PROFILE: Moderately toxic by subcutaneous and intraperitoneal routes.

CNH660 ***HR: 3***
CONESSINE HYDROCHLORIDE
mf: $C_{24}H_{40}N_2 \cdot ClH$ mw: 393.12

SYNS: CHLORHYDRATE de CONESSINE (FRENCH) ◇ 3-β-(DIMETHYLAMINO)CON-5-ENINEHYDROCHLORIDE

TOXICITY DATA with REFERENCE
ipr-mus LDLo:126 mg/kg APFRAD 7,549,49
scu-mus LDLo:150 mg/kg APFRAD 7,549,49
ivn-gpg LDLo:105 mg/kg APFRAD 7,549,49
ivn-frg LDLo:100 mg/kg

SAFETY PROFILE: Poison by subcutaneous, intravenous, and intraperitoneal routes. When heated to decomposition it emits toxic fumes of NO_x and HCl.

CNH730 CAS:1604-01-9 ***HR: 3***
Γ-CONICEIN
mf: $C_8H_{15}N$ mw: 125.24

PROP: Alkaline liquid, mousy odor. Bp: 171°, volatile with steam, d: 0.8753, n (16/D) 1.4661. Sltly sol in water; freely sol in alc, chloroform, and ether.

SYNS: Γ-CONICEINE ◇ 2,3,4,5-TETRAHYDRO-6-PROPYL-PYRIDINE

TOXICITY DATA with REFERENCE
orl-pig TDLo:17 mg/kg (female 30-45D post):TER AJVRAH 46,1368,85
orl-mus LD50:12 mg/kg JPPMAB 15,1,63
scu-mus LD50:12 mg/kg JPPMAB 15,1,63
ivn-mus LD50:2600 μg/kg JPPMAB 15,1,63

SAFETY PROFILE: Poison by ingestion, subcutaneous, and intravenous routes. An experimental teratogen. When heated to decomposition it emits toxic fumes of NO_x.

CNH750 ***HR: 2***
CONIUM MACULATUM

PROP: Colorless, oily liquid with mousy odor. Bp: 166.5°, fp: −2.5°, d: 0.844-0.848 @ 20°/4°. The toxic component of poison hemlock (CTOXAO 12,49,78).

SYN: KURDUMANA, root extract

TOXICITY DATA with REFERENCE
ipr-rat LD50:750 mg/kg IJEBA6 16,228,78

SAFETY PROFILE: Moderately toxic by intraperitoneal route. Combustible when exposed to heat or flame.

CNH775 CAS:8001-64-7 ***HR: 3***
CONVALLARIN

TOXICITY DATA with REFERENCE
scu-mus LDLo:70 mg/kg 27ZWAY E,1,78,-
scu-rbt LDLo:10 mg/kg HBAMAK 4,1289,35
ivn-rbt LDLo:4 mg/kg HBAMAK 4,1289,35
orl-pgn LDLo:100 mg/kg HBAMAK 4,1289,35
scu-pgn LDLo:3 mg/kg HBAMAK 4,1289,35
orl-frg LDLo:200 mg/kg HBAMAK 4,1289,35
scu-frg LDLo:15 mg/kg HBAMAK 4,1289,35

SAFETY PROFILE: Poison by ingestion, subcutaneous, and intravenous routes.

CNH780 CAS:508-75-8 ***HR: 3***
CONVALLATOXIN
mf: $C_{29}H_{42}O_{10}$ mw: 550.71

PROP: Prisms from methanol + ether. Mp: 235-242°. Sol in alc, acetone; sltly sol in chloroform, ethyl acetate, and water; practically insol in ether, petr ether.

SYNS: CONVALLAOTOXIN ◇ CONVALLATON ◇ CONVALLATOXOSIDE ◇ CONVALLOTOXIN ◇ CORGLYCON ◇ CORGLYCONE ◇ CORGLYKON ◇ KORGLYKON ◇ MANNOPYRANOSIDE, STROPHANTHIDIN-3 6-DEOXY-, α-l- ◇ RHAMNOSIDE, STROPHANTHIDIN-3, α-l- ◇ STROPHANTHIDIN-3-(6-DEOXY-α-l-MANNOPYRANOSIDE) ◇ STROPHANTHIDIN-α-l-RHAMNOSIDE

TOXICITY DATA with REFERENCE
ivn-rat LD50:15 mg/kg ARZNAD 11,848,61
ipr-mus LD50:10 mg/kg AIPTAK 155,165,65
scu-mus LD50:15 mg/kg FATOAO 44,342,81
ivn-mus LD50:1 mg/kg CSLNX* NX#00462
ivn-mky LDLo:90 μg/kg ARZNAD 13,412,63
ipr-cat LD50:200 μg/kg AIPTAK 155,165,65
ivn-cat LD50:76 mg/kg ARZNAD 13,220,63
idu-cat LDLo:1290 μg/kg ARZNAD 20,229,70
ivn-pig LDLo:50 μg/kg ARZNAD 20,229,70
idu-pig LDLo:3170 μg/kg ARZNAD 20,229,70

SAFETY PROFILE: Poison by subcutaneous, intravenous, intraduodenal, intraperitoneal, and possibly other routes. When heated to decomposition it emits acrid smoke and fumes.

CNH785 CAS:3253-62-1 ***HR: 3***
CONVALLATOXOL
mf: $C_{29}H_{44}O_{10}$ mw: 552.73

SYNS: CONVALLOTOXOL ◇ CONVALOTOXOL ◇ COVALLATOXOL ◇ α-l-STROPHANTHIDOL-3,6-DEOXY-MANNOPYRANOSIDE

TOXICITY DATA with REFERENCE
ivn-cat LD50:87 μg/kg JPETAB 111,365,54
ivn-rat LD50:56 mg/kg AIPTAK 155,165,65
ipr-mus LD50:30 mg/kg AIPTAK 155,165,65
ipr-cat LD50:130 μg/kg AIPTAK 155,165,65

SAFETY PROFILE: Poison by intravenous and intraperitoneal routes. When heated to decomposition it emits acrid smoke and fumes.

CNH789 ***HR: 3***
COONTIE

PROP: The coontie looks like a low palm with a short trunk and a few 2-foot long pinnate leaves. It produces seed cones. It grows on the southeastern coast of Georgia, Florida, the Bahamas, the northwestern coast of Jamaica, the Dominican Republic and Puerto Rico.

SYNS: BAY BUSH (BAHAMAS) ◇ COMPTIE ◇ FLORIDA ARROWROOT ◇ GUAYIGA (DOMINICAN REPUBLIC) ◇ MARUNGUEY (PUERTO RICO) ◇ PALMITA de JARDIN (PUERTO RICO) ◇ SAGO CYCAS ◇ SEMINOLE BREAD ◇ YUGILLA (CUBA) ◇ ZAMIA PUMILA

SAFETY PROFILE: The roots and trunk contain the poison cycasin. Ingestion of these plant parts may cause persistent vomiting, diarrhea, colic, depression, and muscular paralysis. Washing the grated root with water renders it edible. See also CYCASIN.

CNH792 CAS:8001-61-4 ***HR: 2***
COPAIBA OIL

PROP: From steam distillation of South American *Copaifera* L. (Fam. *Leguminosae*) balsam. Colorless to yellow liquid; characteristic odor, aromatic, slightly bitter taste. D: 0.880-0.907; ref. index: 1.493-1.500 @ 20°. Sol in alc, fixed oils, mineral oil.

SYNS: BALSAM CAPTIVI ◇ BALSAMS, COPAIBA ◇ COPAIBA BALSAM ◇ COPAIBA OLEORESIN ◇ JESUIT'S BALSAM

TOXICITY DATA with REFERENCE
orl-rat LD50:5 g/kg FCTXAV 14,687,76

CONSENSUS REPORTS: Reported in EPA TSCA Inventory.

SAFETY PROFILE: Mildly toxic by ingestion. Large doses cause vomiting and diarrhea. Can also cause dermatitis and kidney damage. When heated to decomposition it emits acrid smoke and irritating fumes.

CNH800 CAS:11078-23-2 ***HR: 2***
COPIAMYCIN
mf: $C_{55}H_{97}NO_{22}$ mw: 1124.53

SYN: NSC-110326

TOXICITY DATA with REFERENCE
orl-mus LD50:1030 mg/kg 85GDA2 7,53,81
ipr-mus LD50:1200 mg/kg 85GDA2 7,53,81
scu-mus LD50:1050 mg/kg 85GDA2 7,53,81

SAFETY PROFILE: Moderately toxic by ingestion, subcutaneous, and intraperitoneal routes. When heated to decomposition it emits toxic fumes of NO_x.

CNI000 CAS:7440-50-8 ***HR: 3***
COPPER
af: Cu aw: 63.54

PROP: A metal with a distinct reddish color. Mp: 1083°, bp: 2324°, d: 8.92, vap press: 1 mm @ 1628°.

SYNS: ALLBRI NATURAL COPPER ◇ ANAC 110 ◇ ARWOOD COPPER ◇ BRONZE POWDER ◇ CDA 101 ◇ CDA 102 ◇ CDA 110 ◇ CDA 122 ◇ C.I. 77400 ◇ C.I. PIGMENT METAL 2 ◇ COPPER-AIRBORNE ◇ COPPER BRONZE ◇ COPPER-MILLED ◇ COPPER SLAG-AIRBORNE ◇ COPPER SLAG-MILLED ◇ 1721 GOLD ◇ GOLD BRONZE ◇ KAFAR COPPER ◇ M1 (COPPER) ◇ M2 (COPPER) ◇ OFHC Cu ◇ RANEY COPPER

TOXICITY DATA with REFERENCE
orl-rat TDLo:152 mg/kg (22W pre):TER GISAAA 45(3),8,80
iut-rat TDLo:250 μg/kg (female 1D pre):REP IJEBA6 19,1124,81
ipl-rat TDLo:100 mg/kg:ETA AIHAAP 41,836,80
orl-hmn TDLo:120 μg/kg:GIT PHRPA6 73,910,58

CONSENSUS REPORTS: Reported in EPA TSCA Inventory. Copper and its compounds are on the Community Right-To-Know List.

OSHA PEL: TWA (dust, mist) 1 mg(Cu)/m^3; (fume) 0.1 mg/m^3
ACGIH TLV: TWA (dust, mist) 1 mg(Cu)/m^3; (fume) 0.2 mg/m^3
DFG MAK: (dust) 1 mg/m^3; (fume) 0.1 mg/m^3

SAFETY PROFILE: Questionable carcinogen with experimental tumorigenic data. Experimental teratogenic and reproductive effects. Human systemic effects by ingestion: nausea and vomiting. See also COPPER COMPOUNDS. Liquid copper explodes on contact with water. Potentially explosive reaction with actylenic compounds; 3-bromopropyne; ethylene oxide; lead azide; and ammonium nitrate. Ignites on contact with chlorine; chlorine trifluoride; fluorine (above 121°); and hydrazinium nitrate (above 70°). Reacts violently with C_2H_2; bromates; chlorates; iodates; (Cl_2 + OF_2); dimethyl sulfoxide + trichloroacetic acid; ethylene oxide; H_2O_2; hydrazine mononitrate; hydrazoic acid; H_2S + air, $Pb(N_3)_2$; K_2O_2; NaN_3; Na_2O_2; sulfuric acid. Incandescent reaction with potassium dioxide. Incompatible with 1-bromo-2-propyne.

CNI250 CAS:142-71-2 ***HR: 3***
COPPER ACETATE
mf: $C_4H_6O_4$•Cu mw: 181.64

PROP: Greenish-blue powder or small crystals.

SYNS: ACETATE de CUIVRE (FRENCH) ◇ ACETIC ACID, CUPRIC SALT ◇ COPPER(2+) ACETATE ◇ COPPER(II) ACETATE ◇ COPPER DIACETATE ◇ COPPER(2+) DIACETATE ◇ CRYSTALLIZED VERDIGRIS ◇ CRYSTALS of VENUS ◇ CUPRIC ACETATE ◇ CUPRIC DIACETATE ◇ NEUTRAL VERDIGRIS ◇ OCTAN MEDNATY (CZECH)

TOXICITY DATA with REFERENCE
scu-rat TDLo:40 mg/kg (7-10D preg):REP CRSBAW 166,1237,72
orl-rat LD50:595 mg/kg MarJV# 29MAR77
scu-rat LD50:350 mg/kg PMDCAY 15,211,77
ipr-mus LD50:2500 μg/kg BCPCA6 30,771,81

CONSENSUS REPORTS: Reported in EPA TSCA Inventory. Copper and its compounds are on the Community Right-To-Know List.

ACGIH TLV: TWA 1 mg(Cu)/m^3

SAFETY PROFILE: Poison by subcutaneous and intraperitoneal routes. Moderately toxic by ingestion. Experimental reproductive effects. When heated to decomposition it emits acrid smoke and irritating fumes. See also COPPER COMPOUNDS.

CNI325 CAS:6046-93-1 ***HR: 3***
COPPER(II) ACETATE MONOHYDRATE
mf: $C_4H_6O_4$•Cu•H_2O mw: 199.66

SYNS: COPPER(2+) ACETATE, MONOHYDRATE ◇ COPPER DIACETATE MONOHYDRATE ◇ CUPRIC ACETATE MONOHYDRATE

TOXICITY DATA with REFERENCE
orl-rat LD50:710 mg/kg AIHAAP 30,470,69
orl-mus LDLo:1600 mg/kg AECTCV 14,111,85
ipr-mam LD50:5 mg/kg JANSAG 55,337,82

CONSENSUS REPORTS: Copper and its compounds are on the Community Right-To-Know List.

SAFETY PROFILE: Poison by intraperitoneal route. Moderately toxic by ingestion. See also COPPER COMPOUNDS

CNI500 CAS:12540-13-5 ***HR: 3***
COPPER(II) ACETYLIDE
mf: C_2Cu mw: 87.56

PROP: A black or brown solid.

CONSENSUS REPORTS: Copper and its compounds are on the Community Right-To-Know List.

ACGIH TLV: TWA 1 mg(Cu)/m^3

DOT Classification: Forbidden

SAFETY PROFILE: Ignites and then explodes when heated to 100°C. Much more sensitive to impact, friction, and heat than copper(I) acetylide (the red-brown form). See also COPPER COMPOUNDS and ACETYLIDES.

CNI600 CAS:11133-98-5 ***HR: 3***
COPPER ALLOY, Cu, Be

SYN: COPPER-BERYLLIUM ALLOY

CONSENSUS REPORTS: IARC Cancer Review: Animal Inadequate Evidence IMEMDT 23,143,80. Copper and its compounds, as well as beryllium and its compounds, are on the Community Right-To-Know List.

OSHA PEL: (Transitional: TWA 0.002 mg(Be)/m^3; CL 0.005; Pk 0.025/30M/8H) TWA 0.002 mg(Be)/m^3; STEL 0.005 mg(Be)/m^3/30M; CL 0.025 mg(Be)/m^3
ACGIH TLV: TWA 0.002 mg(Be)/m^3, Suspected Human Carcinogen.
NIOSH REL: (Beryllium) CL not to exceed 0.0005 mg(Be)/m^3

SAFETY PROFILE: Confirmed carcinogen. Cases of berylliosis have been reported from exposure to so called low beryllium alloys. See also BERYLLIUM COMPOUNDS and COPPER COMPOUNDS. When heated to decomposition it emits very toxic fumes of BeO.

CNI900 **HR: 3**
COPPER ARSENATE HYDROXIDE
mf: $AsCu_2HO_5$ mw: 283.01

PROP: A green solid.

SYNS: COPPER ARSENATE (BASIC) ◇ CUPROUS ARSENATE, BASIC

CONSENSUS REPORTS: Copper and its compounds as well as arsenic and its compounds are on the Community Right-To-Know List.

NIOSH REL: CL 2 μg/m^3/15M

SAFETY PROFILE: A poison by various routes. See also ARSENIC COMPOUNDS and COPPER COMPOUNDS. When heated to decomposition it emits toxic fumes of As.

CNJ325 **HR: D**
COPPER ASCORBATE
mf: $C_6H_8O_6{\cdot}CuO_4S$ mw: 335.74

TOXICITY DATA with REFERENCE
mmo-sat 15 mmol/L NATUAS 260,722,76
dnd-hmn:oth 5 μmol/L CRNGDP 4,583,83
dns-hmn:fbr 1 mmol/L NATUAS 260,722,76

CONSENSUS REPORTS: Copper and its compounds are on the Community Right-To-Know List.

SAFETY PROFILE: Human mutation data reported. When heated to decomposition it emits toxic fumes of SO_x. See also COPPER COMPOUNDS.

CNJ500 CAS:14215-30-6 **HR: 3**
COPPER(II) AZIDE
mf: CuN_6 mw: 147.58

$Cu(N_3)_2$

CONSENSUS REPORTS: Copper and its compounds are on the Community Right-To-Know List.

SAFETY PROFILE: An unstable explosive sensitive to friction and impact. It may explode spontaneously. When heated to decomposition it emits toxic fumes of NO_x. See also COPPER COMPOUNDS and AZIDES.

CNJ750 CAS:12069-69-1 **HR: 3**
COPPER(II) CARBONATE HYDROXIDE (2:1:2)
mf: $CO_3{\cdot}H_2O_2{\cdot}2Cu$ mw: 221.11

PROP: Green powder. Mp: decomp @ 200°, d: 4.0.

SYNS: BASIC COPPER CARBONATE ◇ BASIC CUPRIC CARBONATE ◇ (CARBONATO)DIHYDROXYDICOPPER ◇ CHESTNUT COMPOUND ◇ COPPER CARBONATE HYDROXIDE ◇ CUPRIC CARBONATE ◇ DICOPPER DIHYDROXYCARBONATE ◇ KOP KARB ◇ KUPFERCARBONAT (GERMAN) ◇ MALACHITE

TOXICITY DATA with REFERENCE
orl-rat LD50:159 mg/kg EQSSDX 1,1,75
orl-rbt LD50:159 mg/kg GUCHAZ 6,128,73
orl-pgn LDLo:1000 mg/kg AUVJA2 16,147,40
orl-dck LDLo:900 mg/kg AUVJA2 16,147,40
orl-mam LD50:625 mg/kg FMCHA2 -,C60,83
orl-brd LD50:900 mg/kg PCOC** -,258,66

CONSENSUS REPORTS: Copper and its compounds are on the Community Right-To-Know List. Reported in EPA TSCA Inventory.

SAFETY PROFILE: Poison by ingestion. When heated to decomposition it emits acrid smoke and fumes. See also COPPER COMPOUNDS.

CNJ900 CAS:26506-47-8 **HR: 2**
COPPER CHLORATE
DOT: UN 2721
mf: $HClO_3{\cdot}xCu$ mw: 529.24

SYNS: CHLORIC ACID, COPPER SALT ◇ COPPER CHLORATE (DOT)

ACGIH TLV: TWA 1 mg(Cu)/m^3
DOT Classification: Oxidizer; Label: Oxidizer

SAFETY PROFILE: An oxidizer. When heated to decomposition it emits toxic fumes of Cl^-.

CNK250 CAS:7758-89-6 **HR: 3**
COPPER(I) CHLORIDE
mf: ClCu mw: 98.99

PROP: Cubic, white crystals. Mp: 422°, bp: 1366°, d: 3.53, vap press: 1 mm @ 546°.

SYNS: CHLORID MEDNY (CZECH) ◇ COPPER MONOCHLORIDE ◇ CUPROUS CHLORIDE ◇ CUPROUS DICHLORIDE ◇ DICOPPER DICHLORIDE

TOXICITY DATA with REFERENCE
cyt-rat/ast 120 mg/kg GANNA2 54,155,63
orl-rat LD50:140 mg/kg EQSSDX 1,1,75
scu-gpg LD50:100 mg/kg EQSSDX 1,1,75

CONSENSUS REPORTS: Copper and its compounds are on the Community Right-To-Know List. Reported in EPA TSCA Inventory. EPA Genetic Toxicology Program.

SAFETY PROFILE: Poison by ingestion and subcutaneous routes. Mutation data reported. Reacts violently with potassium or with lithium nitride + heat. When heated to decomposition it emits toxic fumes of Cl^-. See also CHLORIDES and COPPER COMPOUNDS.

CNK500 CAS:1344-67-8 ***HR: 3***
COPPER(II) CHLORIDE (1:2)
DOT: UN 2802
mf: Cl_2Cu mw: 134.44

PROP: Yellowish-brown, hygroscopic powder. Mp: 498°, d: 3.054.

SYNS: COPPER CHLORIDE (DOT) ◇ CUPRIC CHLORIDE ◇ KIRTICOPPER

TOXICITY DATA with REFERENCE
dnd-omi 2 mmol/L SCIEAS 198,513,77
mmo-smc 100 μmol/L CPBTAL 33,1571,85
dnd-mam:lym 2 mmol/L SCIEAS 198,513,77
orl-rat LD50:140 mg/kg AFDOAQ 15,122,51
orl-mus LD50:190 mg/kg FOREAE 7,313,42
ipr-mus LD50:7400 μg/kg AEPPAE 244,17,62
ivn-mus LD50:17500 μg/kg EJMCA5 19,425,84
orl-gpg LD50:31 mg/kg FOREAE 7,313,42
scu-gpg LDLo:100 mg/kg HBTXAC 1,76,56

CONSENSUS REPORTS: Copper and its compounds are on the Community Right-To-Know List. Reported in EPA TSCA Inventory.

DOT Classification: ORM-B; Label: None.

SAFETY PROFILE: Poison by ingestion, subcutaneous, intravenous, and intraperitoneal routes. Mutation data reported. See also COPPER COMPOUNDS and CHLORIDES. Can react violently with K and Na. When heated to decomposition it emits toxic fumes of Cl^-.

CNK559 CAS:1332-40-7 ***HR: 2***
COPPER CHLORIDE, mixed with COPPER OXIDE, HYDRATE
mf: $Cl_2Cu_4H_6O_6$ mw: 427.12

SYNS: AGRIZAN ◇ BASF-GRUNKUPFER ◇ BASIC COPPER CHLORIDE ◇ BLITOX ◇ BLITOX 50 ◇ BLUE COPPER ◇ BLUE COPPER-50 ◇ CHEMOCIN ◇ CHEMPAR ◇ COBOX ◇ COLLOIDOX ◇ COPPER CHLORIDE, BASIC ◇ COPPER CHLORIDE OXIDE ◇ COPPER OC FUNGICIDE ◇ COPPER OXYCHLORIDE ◇ COPPERSAN ◇ COP-TOX ◇ CU-56 ◇ KUPFEROXYCHLORID (GERMAN)

TOXICITY DATA with REFERENCE
orl-rat LD50:700 mg/kg PHJOAV 185,361,60

CONSENSUS REPORTS: Copper and its compounds are on the Community Right-To-Know List.

SAFETY PROFILE: Moderately toxic by ingestion. A pesticide. When heated to decomposition it emits toxic fumes of Cl^-. See also COPPER COMPOUNDS and individual components.

CNK599 ***HR: 3***
COPPER(I) CHLOROACETYLIDE
mf: $C_4Cl_2Cu_2$ mw: 246.04

CONSENSUS REPORTS: Copper and its compounds are on the Community Right-To-Know List.

SAFETY PROFILE: An explosive. When heated to decomposition it emits toxic fumes of Cl^-. See also COPPER COMPOUNDS, CHLORIDES, and ACETYLIDES.

CNK609 ***HR: 3***
COPPER CHROMATE
mf: $CrCuO_4$ mw: 179.54

CONSENSUS REPORTS: Copper and its compounds as well as chromium and its compounds are on the Community Right-To-Know List.

SAFETY PROFILE: Ignites on contact with hydrogen sulfide gas. See also COPPER COMPOUNDS and CHROMIUM COMPOUNDS.

CNK625 CAS:866-82-0 ***HR: 2***
COPPER(I) CITRATE
mf: $C_6H_4O_7 \cdot 2Cu$ mw: 315.18

SYNS: CITRIC ACID, COPPER(2+) SALT (8CI) ◇ COPPER CITRATE ◇ COPPER SALT 2-HYDROXY-1,2,3-PROPANETRICARBOXYLIC ACID (1:2) ◇ CUPRIC CITRATE ◇ CUPROCITROL ◇ 2-HYDROXY-1,2,3-PROPANETRICARBOXYLIC ACID COPPER(2+) SALT (1:2) (9CI)

TOXICITY DATA with REFERENCE
ivn-ham TDLo:1800 μg/kg (8D preg):REP BIREBV 11,97,74
ipr-ham TDLo:2700 μg/kg (8D preg):TER TJADAB 21,89,80
orl-rat LD50:1580 mg/kg MarJV# 29MAR77

CONSENSUS REPORTS: Copper and its compounds are on the Community Right-To-Know List. Reported in EPA TSCA Inventory.

SAFETY PROFILE: Moderately toxic by ingestion. Experimental reproductive effects. An experimental teratogen. See also COPPER COMPOUNDS.

CNK700 CAS:55158-44-6 ***HR: 3***
COPPER-COBALT-BERYLLIUM

SYNS: COPPER ALLOY, Cu, Be, Co ◇ BERYLLIUM-COPPER-COBALT ALLOY

CONSENSUS REPORTS: IARC Cancer Review: Animal Inadequate Evidence IMEMDT 23,143,80. Copper, cobalt, and beryllium and their compounds are on the Community Right-To-Know List.

OSHA PEL: (Transitional: TWA 0.002 mg(Be)/m^3; CL 0.005; Pk 0.025/30M/8H) TWA 0.002 mg(Be)/m^3; STEL 0.005 mg(Be)/m^3/30M; CL 0.025 mg(Be)/m^3
ACGIH TLV: TWA 0.002 mg(Be)/m^3, Suspected Human Carcinogen
NIOSH REL: (Beryllium) CL not to exceed 0.0005 mg(Be)/m^3; (Cobalt) Insufficient evidence for recommending limit.

SAFETY PROFILE: Confirmed carcinogen. See COPPER, BERYLLIUM and COBALT COMPOUNDS. When heated to decomposition it emits very toxic fumes of BeO.

CNK750 **HR: 3**
COPPER COMPOUNDS

CONSENSUS REPORTS: Copper and its compounds are on the Community Right-To-Know List.

SAFETY PROFILE: As the sublimed oxide, copper may be responsible for one form of metal fume fever. In animals, inhalation of copper dust has caused hemolysis of the red blood cells, deposition of hemofuscin in the liver and pancreas, and injury to the lung cells; injection of the dust has caused cirrhosis of the liver and pancreas, and a condition closely resembling hemochromatosis, or bronzed diabetes. However, considerable trial exposure to copper compounds has not resulted in such disease. As regards local effect, copper chloride and sulfate have been reported as causing irritation of the skin and conjunctivae which may be on an allergic basis. Cuprous oxide is irritating to the eyes and upper respiratory tract. Discoloration of the skin is often seen in persons handling copper, but this does not indicate any actual injury. There is an excess of cancer cases in the copper smelting industry. In humans the ingestion of a large quantity of copper sulfate has caused vomiting, gastric pain, dizziness, exhaustion, anemia, cramps, convulsions, shock, coma and death. Symptoms attributed to damage to the nervous system and kidney have been recorded, jaundice has been observed and, in some cases, the liver has been enlarged. Deaths have been reported to have occurred following the ingestion of as little as 27 grams of the salt, while other victims have recovered after having taken up to 120 grams. Many copper-containing compounds are used as fungicides. Many copper salts form highly unstable acetylides. Those formed in basic solutions from (Cu^- salts + C_2H_2) are less stable than those formed from Cu^{Σ} salts. (copper salts + hydrazine) react strongly, and with nitro-methane are explosive.

CNL000 CAS:544-92-3 **HR: 3**
COPPER CYANIDE
mf: CCuN mw: 89.56

PROP: Monoclinic, white prisms. Mp: 473° in N_2, bp: decomp, d: 2.92.

SYNS: CUPRICIN ◇ CUPROUS CYANIDE ◇ RCRA WASTE NUMBER P029

CONSENSUS REPORTS: Reported in EPA TSCA Inventory. Cyanide and its compounds, as well as copper and its compounds, are on the Community Right-To-Know List.

ACGIH TLV: TWA 1 mg(Cu)/m^3
DOT Classification: Poison B; Label: Poison.

SAFETY PROFILE: A poison. Reacts violently with magnesium. When heated to decomposition it emits very toxic CN^- and NO_x. See also CYANIDES and COPPER COMPOUNDS.

CNL250 CAS:14763-77-0 **HR: 3**
COPPER(II) CYANIDE
DOT: UN 1587
mf: C_2CuN_2 mw: 115.58

PROP: Yellowish-green powder. Mp: decomp before melting.

SYNS: COPPER CYANIDE (DOT) ◇ COPPER CYNANAMIDE ◇ CUPRIC CYANIDE (DOT) ◇ CYANURE de CUIVRE (FRENCH)

TOXICITY DATA with REFERENCE
ipr-rat LDLo:50 mg/kg NCNSA6 5,27,53

CONSENSUS REPORTS: Copper and its compounds, as well as cyanide and its compounds, are on the Community Right-To-Know List.

ACGIH TLV: TWA 1 mg(Cu)/m^3
DOT Classification: Poison B; Label: Poison.

SAFETY PROFILE: Poison by intraperitoneal route. See also CYANIDES and COPPER COMPOUNDS. Incompatible with magnesium. When heated to decomposition it emits toxic fumes of NO_x and CN^-.

CNL500 CAS:137-29-1 **HR: 3**
COPPER DIMETHYLDITHIOCARBAMATE
mf: $C_6H_{12}N_2S_4$•Cu mw: 303.98

SYNS: COMPOUND-4018 ◇ CUMATE ◇ DIMETHYLDITHIOCARBAMIC ACID COPPER SALT ◇ WOLFEN

TOXICITY DATA with REFERENCE
mmo-sat 2 mg/plate MUREAV 68,313,79
mma-sat 1 mg/plate MUREAV 68,313,79
ipr-rat LDLo:25 mg/kg NCNSA6 5,14,53
ipr-mus LDLo:8 mg/kg CBCCT* 4,227,52

CONSENSUS REPORTS: Reported in EPA TSCA In-

ventory. Copper and its compounds are on the Community Right-To-Know List.

SAFETY PROFILE: Poison by intraperitoneal route. Mutation data reported. See also COPPER COMPOUNDS and CARBAMATES. When heated to decomposition it emits very toxic fumes of NO_x and SO_x.

CNL625 CAS:32061-49-7 ***HR: 3***
COPPER(II)-1,3-DI(5-TETRAZOLYL)TRIAZENIDE
mf: $C_4H_4CuN_{22}$ mw: 423.77

$Cu(\text{-}NN{=}NN{=}CN{=}NNHC{=}NN{=}NNH)_2$

CONSENSUS REPORTS: Copper compounds are on the Community Right-To-Know List.

SAFETY PROFILE: A heat- and friction-sensitive explosive. Upon decomposition it emits toxic fumes of NO_x. See also COPPER COMPOUNDS and EXPLOSIVES.

CNL750 CAS:54453-03-1 ***HR: 3***
COPPER EDTA COMPLEX

SYN: (ETHYLENEDINITRILO)TETRAACETIC ACID COPPER(II) COMPLEX

TOXICITY DATA with REFERENCE
ipr-mus LD50:2090 μg(Cu)/kg PABIAQ 11,853,63

CONSENSUS REPORTS: Copper and its compounds are on the Community Right-To-Know List. Reported in EPA TSCA Inventory.

SAFETY PROFILE: Poison by intraperitoneal route. See also COPPER COMPOUNDS. When heated to decomposition it emits toxic fumes of NO_x.

CNM000 ***HR: 3***
COPPER FUME
af: Cu aw: 63.54

SYN: MIEDZ (POLISH)

TOXICITY DATA with REFERENCE
ihl-hmn TCLo:1 mg/m^3:IRR DTLVS* 3,59,71

CONSENSUS REPORTS: Copper and its compounds are on the Community Right-To-Know List.

OSHA PEL: Fume: TWA 0.1 mg/m^3
ACGIH TLV: Fume: TWA 0.2 mg/m^3
DFG MAK: Fume: 0.1 mg/m^3

SAFETY PROFILE: Human systemic irritant by inhalation. See also COPPER COMPOUNDS.

CNM250 ***HR: 3***
COPPER(I) HYDRIDE
mf: Cu_2H_2 mw: 129.11

PROP: Red-brown crystals. Mp: decomp @ 60°, d: 6.38.

CONSENSUS REPORTS: Copper and its compounds are on the Community Right-To-Know List.

SAFETY PROFILE: Ignites spontaneously in air. Ignites on contact with fluorine; bromine; or iodine. See also COPPER COMPOUNDS and HYDRIDES.

CNM500 CAS:20427-59-2 ***HR: 3***
COPPER HYDROXIDE
mf: $H_2O_2 \cdot Cu$ mw: 97.56

PROP: Blue, gelatinous or amorphous powder. D: 3.368.

SYNS: COMAC ◇ COPPER DIHYDROXIDE ◇ COPPER(2+) HYDROXIDE ◇ CUPRAVIT BLAU ◇ CUPRAVIT BLUE ◇ CUPRIC HYDROXIDE ◇ KOCIDE ◇ KUPRABLAU ◇ PARASOL

TOXICITY DATA with REFERENCE
orl-hmn LDLo:200 mg/kg 28ZLA8 2,-,69
unr-rat LD50:1000 mg/kg RMCHA2 -,C61,83

CONSENSUS REPORTS: Copper and its compounds are on the Community Right-To-Know List. Reported in EPA TSCA Inventory.

SAFETY PROFILE: A human poison by ingestion. An experimental poison by an unspecified route. See also COPPER COMPOUNDS.

CNM750 CAS:3251-23-8 ***HR: 2***
COPPER(II) NITRATE
DOT: UN 1479
mf: CuN_2O_6 mw: 187.55

PROP: Large, blue-green, deliquescent, orthorhombic crystals. D: 2.047, sublimes at 150-225°, mp: 255-256°. Sol in water, ethyl acetate, dioxane; dissolves in and reacts vigorously with ether.

SYNS: COPPER DINITRATE ◇ COPPER(2+) NITRATE ◇ CUPRIC DINITRATE ◇ CUPRIC NITRATE (DOT)

TOXICITY DATA with REFERENCE
skn-rbt 500 mg SEV FCTOD7 20,563,82
eye-rbt 100 mg SEV FCTOD7 20,573,82
eye-rbt 100 mg/4S rns SEV FCTOD7 20,573,82
orl-rat LD50:940 mg/kg EQSSDX 1,1,75

CONSENSUS REPORTS: Copper and its compounds are on the Community Right-To-Know List. Reported in EPA TSCA Inventory.

ACGIH TLV: TWA 1 mg(Cu)/m^3
DOT Classification: Oxidizer; Label: Oxidizer.

SAFETY PROFILE: Moderately toxic by ingestion. A severe eye and skin irritant. Potentially explosive reaction above 220°C with ammonium or potassium hexacyanoferrate(II). Reaction with ammonia + potassium amide gives explosive product. Violent reaction with acetic anhydride. May ignite on prolonged contact with paper. Concentrated solutions may ignite in contact with tin foil. Used as a fungicide, herbicide, and as a catalyst component in solid rocket fuel. When heated to decomposition it emits toxic fumes of NO_x. See also COPPER COMPOUNDS and NITRATES.

CNN000 CAS:10031-43-3 ***HR: 2***
COPPER(II) NITRATE, TRIHYDRATE (1:2:3)
mf: $N_2O_6 \cdot Cu \cdot 3H_2O$ mw: 241.62

PROP: Blue, deliquescent crystals. Mp: 114.5°, d: 2.047.

SYNS: COPPER DINITRATE TRIHYDRATE ◇ CUPRIC NITRATE TRIHYDRATE ◇ GERHARDITE ◇ NITRIC ACID COPPER(2+) SALT TRIHYDRATE

TOXICITY DATA with REFERENCE
cyt-rat/ast 600 mg/kg GANNA2 54,155,63
orl-rat LD50:940 mg/kg AIHAAP 30,470,69

CONSENSUS REPORTS: Copper and its compounds are on the Community Right-To-Know List.

SAFETY PROFILE: Moderately toxic by ingestion. Mutation data reported. Can ignite on prolonged contact with paper. Can explode when finely mixed with potassium ferrocyanide. Concentrated solutions may ignite in contact with tin foil. See also COPPER COMPOUNDS and COPPER NITRATE. When heated to decomposition it emits toxic fumes of NO_x.

CNN250 ***HR: 3***
COPPER(I) NITRIDE
mf: Cu_3N mw: 123.63

PROP: Dark-green powder. Mp: decomp @ 300°, d: 5.84 @ 25°/4°.

CONSENSUS REPORTS: Copper and its compounds are on the Community Right-To-Know List.

SAFETY PROFILE: Explodes on heating in air. When mixed with HNO_3 (conc) explodes violently. See also COPPER COMPOUNDS and NITRIDES.

CNN399 ***HR: 3***
COPPER 1,3,5-OCTATRIEN-7-YNIDE
mf: C_8H_7Cu mw: 166.69

$$H(CH{=}CH)_3C{\equiv}CCu$$

CONSENSUS REPORTS: Copper and its compounds are on the Community Right-To-Know List.

SAFETY PROFILE: Explodes when heated in air. When heated to decomposition it emits acrid smoke and fumes. See also COPPER COMPOUNDS and ACETYLENE COMPOUNDS.

CNN500 CAS:10290-12-7 ***HR: 3***
COPPER ORTHOARSENITE
DOT: UN 1586
mf: $AsCuHO_3$ mw: 187.47

PROP: Yellowish-green powder. Mp: decomp.

SYNS: ACID COPPER ARSENITE ◇ AIR-FLO GREEN ◇ ARSONIC ACID, COPPER(2+) SALT (1:1) (9CI) ◇ COPPER ARSENITE, solid (DOT) ◇ CUPRIC ARSENITE ◇ CUPRIC GREEN ◇ SCHEELES GREEN ◇ SCHEELE'S MINERAL ◇ SWEDISH GREEN

CONSENSUS REPORTS: Arsenic and its compounds, as well as copper and its compounds, are on the Community Right-To-Know List.

OSHA PEL: Cancer Hazard
ACGIH TLV: TWA 0.2 mg(As)/m^3
NIOSH REL: (Arsenic, Inorganic) CL 0.002 mg(As)/m^3/15M
DOT Classification: Poison B; Label: Poison.

SAFETY PROFILE: Poison. When heated to decomposition it emits toxic fumes of As. See also ARSENIC COMPOUNDS and COPPER COMPOUNDS.

CNN750 CAS:53421-36-6 ***HR: 2***
COPPER(I) OXALATE
mf: $C_2Cu_2O_4$ mw: 183.10

PROP: Solid, light bluish-green powder.

SYN: CUPRIC OXALATE

CONSENSUS REPORTS: Copper and its compounds are on the Community Right-To-Know List.

SAFETY PROFILE: Explodes weakly when heated slightly. See also OXALATES and COPPER COMPOUNDS.

CNO000 CAS:1317-39-1 ***HR: 2***
COPPER(I) OXIDE
mf: Cu_2O mw: 143.08

PROP: Octahedral, cubic red crystals. Mp: 1235°, bp: $-O_2$ @ 1800°, d: 6.0.

SYNS: BROWN COPPER OXIDE ◇ C.I. 77402 ◇ COPOX ◇ COPPER NORDOX ◇ COPPER SARDEX ◇ CUPPER OXIDE (RUSSIAN) ◇ CUPROUS OXIDE ◇ DICOPPER MONOXIDE ◇ FUNGIMAR ◇ KUPFEROXYDUL (GERMAN) ◇ OLEOCUIVRE ◇ OLEO NORDOX ◇ RED COPPER OXIDE ◇ YELLOW CUPROCIDE

TOXICITY DATA with REFERENCE
ihl-rat TCLo:11 $\mu g/m^3$/24H (14W male):REP GISAAA 41(6),8,76
orl-rat LD50:470 mg/kg AIHAAP 30,470,69

CONSENSUS REPORTS: Reported in EPA TSCA In-

ventory. Copper and its compounds are on the Community Right-To-Know List.

SAFETY PROFILE: Moderately toxic by ingestion. Experimental reproductive effects. A fungicide. Violent, potentially explosive reaction with concentrated peroxyformic acid. Violent reaction when heated with aluminum. See also COPPER COMPOUNDS.

CNO250 CAS:1317-38-0 ***HR: 2***
COPPER(II) OXIDE
mf: CuO mw: 79.54

PROP: Fine black powder. Bp: decomp @ 1026°, d: 6.4.

SYNS: CUPRIC OXIDE ◇ PARAMELACONITE

CONSENSUS REPORTS: Copper and its compounds are on the Community Right-To-Know List.

SAFETY PROFILE: Used as fungicide. Also a trace mineral added to animal feeds. Explodes when heated with powdered aluminum; anilinium perchlorate; hydrogen; magnesium; phthalic anhydride. Ignites on contact with dichloromethylsilane; hydrogen sulfide; hydrogen trisulfide. Incandescent reaction when heated with boron; rubidium acetylide (at 350°C); potassium; sodium; phospham. Reacts violently with $CsHC_2$; hydrazine; PN_2H; Ti; Zr. Incompatible with metals and reductants (e.g., hydroxylamine or hydrazine). See also COPPER COMPOUNDS.

CNO325 CAS:15061-57-1 ***HR: 3***
COPPER(I) PERCHLORATE
mf: $Cu_2Cl_2O_8$ mw: 325.99

CONSENSUS REPORTS: Copper compounds are on the Community Right-To-Know List.

SAFETY PROFILE: Mixtures with 1,4-oxathiane, carbon monoxide, or alkenes (e.g. ethylene, allene, or 1,3-butadiene) form heat-sensitive explosive complexes. See also COPPER COMPOUNDS and PERCHLORATES.

CNO350 CAS:10294-46-9 ***HR: 3***
COPPER(II) PERCHLORATE
mf: $CuCl_2O_8$ mw: 262.45

CONSENSUS REPORTS: Copper compounds are on the Community Right-To-Know List.

SAFETY PROFILE: Forms explosive complexes with polyfunctional amines. See also COPPER COMPOUNDS and PERCHLORATES.

CNO500 CAS:17031-32-2 ***HR: 3***
COPPER(II) PERCHLORATE, DIHYDRATE
mf: Cl_2O_8•Cu•$2H_2O$ mw: 298.48

SYNS: CUPRIC DIPERCHLORATE TETRAHYDRATE ◇ PERCHLORIC ACID, COPPER(II) SALT, DIHYDRATE (8CI, 9CI)

TOXICITY DATA with REFERENCE
ipr-rat LD50:29 mg/kg JAFCAU 14,512,66

CONSENSUS REPORTS: Copper and its compounds are on the Community Right-To-Know List.

SAFETY PROFILE: Poison by intraperitoneal route. See also COPPER COMPOUNDS and COPPER(II) PERCHLORATE. When heated to decomposition it emits toxic fumes of Cl^-.

CNO750 CAS:34461-68-2 ***HR: 3***
COPPER(II) PHOSPHINATE
mf: $CuH_2O_4P_2$ mw: 191.50

$Cu(OPHOH)_2$

CONSENSUS REPORTS: Copper and its compounds are on the Community Right-To-Know List.

SAFETY PROFILE: Explodes when heated to about 90°C or on impact. When heated to decomposition it emits toxic fumes of PO_x. See also COPPER COMPOUNDS.

CNP000 CAS:13991-87-2 ***HR: 3***
COPPER SORBATE
mf: $C_{12}H_{14}O_4$•Cu mw: 285.80

SYN: SORBIC ACID, COPPER SALT

TOXICITY DATA with REFERENCE
orl-mus LDLo:125 mg/kg CHDDAT 266,1080,68
ipr-mus LDLo:15 mg/kg CHDDAT 266,1080,68

CONSENSUS REPORTS: Copper and its compounds are on the Community Right-To-Know List.

SAFETY PROFILE: Poison by ingestion and intraperitoneal routes. See also COPPER COMPOUNDS. When heated to decomposition it emits acrid smoke and irritating fumes.

CNP250 CAS:7758-98-7 ***HR: 3***
COPPER(II) SULFATE (1:1)
DOT: NA 9109
mf: O_4S•Cu mw: 159.60

PROP: Blue crystals or blue, crystalline granules or powder. D: 2.284.

SYNS: BCS COPPER FUNGICIDE ◇ BLUE COPPER ◇ BLUE STONE ◇ BLUE VITRIOL ◇ COPPER MONOSULFATE ◇ COPPER SULFATE ◇ CP BASIC SULFATE ◇ CUPRIC SULFATE ◇ KUPPERSULFAT (GERMAN) ◇ ROMAN VITRIOL ◇ SULFATE de CUIVRE (FRENCH) ◇ SULFURIC ACID, COPPER(2+) SALT (1:1) ◇ TNCS 53 ◇ TRINAGLE

TOXICITY DATA with REFERENCE
dni-mus-ipr 20 g/kg ARGEAR 51,605,81

dns-ham:emb 200 μmol/L MUREAV 131,173,84
ivn-mus TDLo:3200 μg/kg (female 8D post):TER WRABDT 186,297,79
ipr-rat TDLo:7500 μg/kg (3D preg):REP BECTA6 25,702,80
par-ckn TDLo:10 mg/kg:ETA BEXBAN 9,519,40
orl-man LDLo:857 mg/kg:GIT ATXKA8 17,20,58
orl-cld TDLo:150 mg/kg:SYS,BLD AJDCAI 131,149,77
orl-hmn LDLo:50 mg/kg JAMAAP 235,801,76
orl-hmn TDLo:11 mg/kg:GIT LANCAO 2,700,60
orl-rat LD50:300 mg/kg 36SBA8 1,507,77
scu-rat LD50:43 mg/kg PESTD5 16,252,75
unr-rat LD50:520 mg/kg GTPZAB 26(6),21,82
ipr-mus LD50:18 mg/kg COREAF 256,1043,63
scu-mus LDLo:500 μg/kg TJIZAF 48,313,78
ivn-mus LDLo:50 mg/kg HBTXAC 1,76,56
ivn-rbt LD50:10 mg/kg JIDHAN 31,301,49

CONSENSUS REPORTS: Copper and its compounds are on the Community Right-To-Know List. Reported in EPA TSCA Inventory. EPA Genetic Toxicology Program.

ACGIH TLV: TWA 1 mg(Cu)/m^3
DOT Classification: ORM-E; label: None.

SAFETY PROFILE: A human poison by ingestion. An experimental poison by ingestion, subcutaneous, parenteral, intravenous, and intraperitoneal routes. Human systemic effects by ingestion: gastritis, diarrhea, nausea or vomiting, damage to kidney tubules, and hemolysis. Questionable carcinogen with experimental tumorigenic data. An experimental teratogen. Other experimental reproductive effects. Mutation data reported. Reacts violently with hydroxylamine; magnesium. See also COPPER COMPOUNDS and SULFATES. When heated to decomposition it emits toxic fumes of SO_x.

CNP500 CAS:7758-99-8 ***HR: 3***
COPPER(II) SULFATE PENTAHYDRATE (1:1:5)
mf: $O_4S{\cdot}Cu{\cdot}5H_2O$ mw: 249.70

PROP: Mp: $-4H_2O$ @ 110°.

SYNS: BLUE COPPERRAS ◇ BLUESTONE ◇ BLUE VITRIOL ◇ COPPERFINE-ZINC ◇ CSP ◇ CUPRIC SULFATE PENTAHYDRATE ◇ KUPFERSULFAT-PENTAHYDRAT (GERMAN) ◇ KUPFERVITRIOL (GERMAN) ◇ ROMAN VITRIOL ◇ SALZBURG VITRIOL ◇ SULFURIC ACID, COPPER(2+) SALT, PENTAHYDRATE ◇ TRIANGLE

TOXICITY DATA with REFERENCE
dni-hmn:lym 76 μmol/L IAAAAM 79,83,86
cyt-rat/ast 300 mg/kg GANNA2 54,155,63
orl-hmn TDLo:272 mg/kg:LIV,KID,BLD IPRAA8 18,807,65
orl-hmn LDLo:1088 mg/kg IPRAA8 18,807,65
unk-man LDLo:221 mg/kg 85DCAI 2,73,70
orl-rat LD50:300 mg/kg 85ARAE 2,182,77
ipr-mus LD50:33 mg/kg BCPCA6 14,289,65
orl-dog LDLo:60 mg/kg HBAMAK 4,1289,35
scu-gpg LDLo:62 mg/kg BMJOAE 2,217,13
orl-pgn LDLo:1000 mg/kg AUVJA2 16,147,40
orl-dom LDLo:5 mg/kg JCVPAR 82,47,72
orl-mam LD50:470 mg/kg FMCHA2 -,C62,83
ipr-mam LD50:7500 μg/kg JANSAG 55,337,82
orl-bwd LDLo:300 mg/kg AUVJA2 16,147,40

CONSENSUS REPORTS: Copper and its compounds are on the Community Right-To-Know List.

SAFETY PROFILE: A human poison by an unspecified route. Moderately toxic to humans by ingestion. An experimental poison by ingestion, subcutaneous, and intraperitoneal routes. Human systemic effects by ingestion: jaundice, unspecified urinary system effects, and hemolysis. Human mutation data reported. Used widely as a fungicide. When heated to decomposition it emits toxic fumes of SO_x. See also COPPER COMPOUNDS and COPPER(II) SULFATE.

CNP750 CAS:22205-45-4 ***HR: D***
COPPER(I) SULFIDE
mf: Cu_2S mw: 159.14

PROP: Rhombic, black crystals. Mp: 1100°, d: 5.6.

SYNS: COPPER SULFIDE ◇ CUPRASULFIDE ◇ CUPROUS SULFIDE ◇ DICOPPER MONOSULFIDE ◇ DICOPPER SULFIDE

TOXICITY DATA with REFERENCE
otr-ham:emb 380 μmol/L GNREA8 39,193,79
dnd-ham:emb 380 μmol/L CNREA8 39,193,79

CONSENSUS REPORTS: Copper and its compounds are on the Community Right-To-Know List. Reported in EPA TSCA Inventory. EPA Genetic Toxicology Program.

SAFETY PROFILE: Mutation data reported. See also SULFIDES and COPPER COMPOUNDS. When heated to decomposition it emits toxic fumes of SO_x.

CNQ000 CAS:1317-40-4 ***HR: 3***
COPPER (II) SULFIDE
mf: CuS mw: 95.60

PROP: Found in nature as the mineral covellite, or indigo copper (blue, hexagonal or monoclinic crystals). Black powder or crystals. Mp: transition @ 103°; bp: decomp @ 220°; d: 4.6. Practically insol in water, alc, dil acids, alkalies; sol in KCN soln, NH_4OH, hot HNO_3.

SYNS: C.I. 77450 ◇ C.I. PIGMENT BLUE 34 ◇ COPPER BLUE ◇ COPPER MONOSULFIDE ◇ COPPER SULFIDE ◇ COPPER(2+) SULFIDE ◇ CUPRIC SULFIDE ◇ HORACE VERNET'S BLUE ◇ MONOCOPPER MONOSULFIDE ◇ OIL BLUE

TOXICITY DATA with REFERENCE
cyt-rat/ast 150 mg/kg GANNA2 54,155,63

otr-ham:emb 5 mg/L CNREA8 42,2757,82
dnd-ham:ovr 10 mg/L CRNGDP 3,657,82

CONSENSUS REPORTS: Copper and its compounds are on the Community Right-To-Know List. Reported in EPA TSCA Inventory.

SAFETY PROFILE: Mutation data reported. Explodes on contact with magnesium chlorate, zinc chlorate, cadmium chlorate, or concentrated solutions of chloric acid. Can react violently with H_2O_2, NH_4MgNO_3 + water, $Zn(ClO_3)_2$. See also COPPER COMPOUNDS and SULFIDES.

CNQ250 CAS:62126-20-9 ***HR: 3***
COPPER(I) TETRAHYDROALUMINATE
mf: $AlCuH_4$ mw: 94.55

PROP: Unstable, decomp @ −70°C.

CONSENSUS REPORTS: Copper and its compounds are on the Community Right-To-Know List.

SAFETY PROFILE: Ignites spontaneously in air. Unstable. See also COPPER COMPOUNDS and ALUMINUM COMPOUNDS.

CNQ375 CAS:25267-55-4 ***HR: 3***
COPPER-2,4,5-TRICHLOROPHENOLATE
mf: $C_{12}H_6Cl_6O_2$•Cu mw: 458.42

SYNS: COPPER TRICHLOROPHENOLATE ◇ CTCP ◇ TRIKHLORFENOLYAT MEDI (RUSSIAN)

TOXICITY DATA with REFERENCE
orl-rat LD50:5500 mg/kg 85GMAT -,40,82
unr-rat LDLo:100 mg/kg KSKZAN 18(5),55,80
unr-mus LDLo:100 mg/kg KSKZAN 18(5),55,80
orl-rbt LD50:1537 mg/kg 85GMAT -,40,82
ihl-gpg LCLo:200 mg/m^3/1H 85GMAT -,40,82

CONSENSUS REPORTS: Copper and its compounds are on the Community Right-To-Know List.

SAFETY PROFILE: Poison by inhalation and possibly other routes. Moderately toxic by ingestion. When heated to decomposition it emits toxic fumes of Cl^-. See also COPPER COMPOUNDS and CHLOROPHENOLS.

CNQ500 ***HR: 2***
COPPER-ZINC ALLOYS
mf: Cu-Zn mw: 128.91

CONSENSUS REPORTS: Copper and its compounds, as well as zinc and its compounds, are on the Community Right-To-Know List.

SAFETY PROFILE: Potentially explosive reaction with alkyl bromides. Violent reaction with diiodomethane + ether. See also COPPER COMPOUNDS and ZINC COMPOUNDS.

CNQ750 ***HR: 3***
COPPER-ZINC CHROMATE COMPLEX

SYNS: CHROMIC ACID, COPPER-ZINC-COMPLEX ◇ CRAG FUNGICIDE 658 ◇ ZINC-COPPER CHROMATE COMPLEX

CONSENSUS REPORTS: Copper, zinc, chromium and their compounds are on the Community Right-To-Know List.

OSHA PEL: (Transitional: 1 mg(CrO_3)/10m^3) CL 0.1 mg(CrO_3)/m^3
ACGIH TLV: TWA 0.01 mg(Cr)/M^3; Confirmed Human Carcinogen
DFG MAK: Human Carcinogen.
NIOSH REL: (Chromium (VI)) TWA 0.001 mg(Cr(VI))/m^3

SAFETY PROFILE: A poison. See also CHROMIUM COMPOUNDS, ZINC COMPOUNDS, and COPPER COMPOUNDS. A fungicide. When heated to decomposition it emits toxic fumes of ZnO.

CNR000 CAS:8001-31-8 ***HR: 3***
COPRA (OIL)
DOT: UN 1363

PROP: From the kernel of the fruit of the coconut palm *Cocos nucifera*. Fatty solid or liquid; sweet, nutty taste. Mp: 21-27°.

SYNS: COCONUT BUTTER ◇ COCONUT MEAL PELLETS, containing 6-13% moisture and no more than 10% residual fat (DOT) ◇ COCONUT OIL (FCC) ◇ COCONUT PALM OIL ◇ COPRA (DOT) ◇ COPRA PELLETS (DOT) ◇ FREE COCONUT OIL

CONSENSUS REPORTS: Reported in EPA TSCA Inventory.

DOT Classification: ORM-C; Label: None; Flammable Solid; Label: None.

SAFETY PROFILE: Flammable solid when exposed to heat or flame. May spontaneously heat and ignite if stored wet and hot.

CNR125 CAS:314-35-2 ***HR: 3***
CORAFIL
mf: $C_{13}H_{21}N_5O_2$ mw: 279.39

PROP: Waxy solid. Mp: 75°. Very sol in water, acetone; sltly sol in ethanol, ether.

SYNS: CAMPHOPHYLINE ◇ DIAETHYLAMINOAETHYL-THEOPHYLLIN (GERMAN) ◇ 7-(2-(DIETHYLAMINO)ETHYL)-3,7-DIHYDRO-1,3-DIMETHYL-1H-PURINE-2,6-DIONE (9CI) ◇ DIETHYLAMINOETHYL THEOPHYLLINE ◇ 7-(2-DIETHYLAMINO)ETHYLTHEOPHYLLINE ◇ 7-(DIETHYLAMINOETHYL)THEOPHYLLINE ◇ ETAMINOPHYLLINE ◇ ETAMIPHYLLIN ◇ ETAMIPHYLLINE

◇ MILLIPHYLLINE ◇ MILLOPHYLLINE ◇ PAREPHYLLIN ◇ QUERYL ◇ R-3588 ◇ SOLUFILINA ◇ SOLUPHYLINE

TOXICITY DATA with REFERENCE
orl-mus LD50:1237 mg/kg JPETAB 116,343,56
ipr-mus LD50:254 mg/kg JPETAB 116,343,56
scu-mus LD50:183 mg/kg ARZNAD 4,649,54

SAFETY PROFILE: Poison by subcutaneous and intraperitoneal routes. Moderately toxic by ingestion. When heated to decomposition it emits toxic fumes of NO_x.

CNR135 ***HR: 3***
CORAL PLANT

PROP: The various species of Jatropha range in size from shrubs to small trees. Most are perennials. They bear a 3-sided seed capsule, and each section carries one seed. They are commonly cultivated, and most grow wild in tropical areas.

SYNS: BARBADOS NUT ◇ BELLYACHE BUSH ◇ CORAL VEGETAL (CUBA) ◇ CUIPU (MEXICO) ◇ FRAILECILLO (CUBA) ◇ GOUT STALK ◇ HIGUERETA CIMARRONA ◇ JATROPHA CATHARTICA ◇ JATROPHA CURCAS ◇ JATROPHA GOSSYPIIFOLIA ◇ JATROPHA INTEGERRIMA ◇ JATROPHA MACRORHIZA ◇ JATROPHA MULTIFIDA ◇ JATROPHA PODAGRICA ◇ JICAMILLA (MEXICO, TEXAS) ◇ MALA MUJER (MEXICO) ◇ PEREGRINA (PUERTO RICO, CUBA, US) ◇ PHYSIC NUT ◇ PINON (PUERTO RICO, DOMINICAN REPUBLIC) ◇ ROSE-FLOWERED JATROPHA ◇ SPICY JATROPHA ◇ TARTAGO (PUERTO RICO) ◇ TAUTUBA (PUERTO RICO) ◇ TINAJA (PUERTO RICO) ◇ TUATUA (PUERTO RICO)

SAFETY PROFILE: The seeds contain the poison jatrophin (curcin) a toxalbumin which inhibits protein synthesis in the intestinal wall. Ingestion of only one seed can rapidly cause severe nausea, vomiting and diarrhea with resulting fluid and electrolyte loss. See also ABRIN.

CNR150 ***HR: 3***
CORAL SNAKE VENOM

SYNS: M.F. FULVIUS VENOM ◇ M. FULVIUS FULVIUS VENOM ◇ MICRURUS FULVIUS FULVIUS VENOM

TOXICITY DATA with REFERENCE
ipr-mus LD50:538 μg/kg TOXIA6 13,139,75
ivn-mus LD50:225 μg/kg TOXIA6 13,139,75
ims-mus LD50:200 μg/kg BIJOAK 193,899,81

SAFETY PROFILE: Deadly poison by intramuscular, intravenous, and intraperitoneal routes.

CNR250 CAS:61503-59-1 ***HR: 3***
CORALYNE SULFOACETATE
mf: $C_{22}H_{22}NO_4•C_2H_3O_5S$ mw: 503.56

SYNS: NSC 154890 ◇ 2,3,10,11-TETRAMETHOXY-8-METHYLDIBENZO(a,g)QUINOLIZINIUM SALT with SULFOACETIC ACID (1:1)

TOXICITY DATA with REFERENCE
ipr-mus LD50:173 mg/kg NCISP* JAN86
ivn-mus LD50:221 mg/kg TXAPA9 37,165,76
ivn-dog LDLo:37 mg/kg TXAPA9 37,165,76

SAFETY PROFILE: Poison by intravenous and intraperitoneal routes. When heated to decomposition it emits very toxic fumes of SO_x and NO_x.

CNR500 CAS:6452-71-7 ***HR: 3***
CORETAL
mf: $C_{15}H_{23}NO_3$ mw: 265.39

SYNS: 1-(o-ALLYLOXY)PHENOXY)-3-(ISOPROPYLAMINO)-2-PROPANOL ◇ 1-(ISOPROPYLAMINO)-2-HYDROXY-3-(o-(ALLYLOXY)PHENOXY)PROPANE ◇ OXPRENOLOL ◇ 1-((1-METHYLETHYL)AMINO)-3-(2-(2-PROPENYLOXY)PHENOXY)-2-PROPANOL

TOXICITY DATA with REFERENCE
orl-hmn TDLo:50 mg/kg:CVS,PUL BMJOAE 1,776,77
orl-rat LD50:730 mg/kg ARZNAD 18,164,68
scu-rat LD50:940 mg/kg ARZNAD 18,164,68
ivn-rat LD50:33 mg/kg ARZNAD 18,164,68
ipr-mus LD50:170 mg/kg PJPPAA 25,145,73
scu-mus LD50:245 mg/kg ARZNAD 18,164,68
ivn-mus LD50:20 mg/kg ARZNAD 27,1022,77
orl-cat LD90:200 mg/kg ARZNAD 18,164,68
ivn-rbt LD90:20 mg/kg ARZNAD 18,164,68

SAFETY PROFILE: Poison by ingestion, subcutaneous, intravenous, and intraperitoneal routes. Human systemic effects by ingestion: decreased pulse rate and blood pressure, and unspecified changes in the cardiovascular and respiratory systems. When heated to decomposition it emits toxic fumes of NO_x. See also ALLYL COMPOUNDS.

CNR675 CAS:42200-33-9 ***HR: 3***
CORGARD
mf: $C_{17}H_{27}NO_4$ mw: 309.45

PROP: Crystalline powder. Mp: 124-136°. Freely sol in alc, propylene glycol; sltly sol in chloroform. Insol in acetone, benzene, ether, and hexane.

SYNS: ANABET ◇ 1-(tert-BUTYLAMINO)-3-((5,6,7,8-TETRAHYDRO-cis-6,7-DIHYDROXY-1-NAPHTHYL)OXY)-2-PROPANOL◇ 5-(3-((1,1-DIMETHYLETHYL)AMINO)-2-HYDROXYPROPOXY)-1,2,3,4-TETRAHYDRO-2,3-NAPHTHALENEDIOL◇ NADOLOL ◇ SOLGOL ◇ SQ 11725 ◇ 2,3-cis-1,2,3,4-TETRAHYDRO-5-((2-HYDROXY-3-tert-BUTYLAMINO)PROPOXY)-2,3-NAPHTHALENEDIOL

TOXICITY DATA with REFERENCE
orl-rat TDLo:550 mg/kg (7-17D preg):REP YACHDS 11,5119,83
orl-rat TDLo:550 mg/kg (7-17D preg):TER YACHDS 11,5119,83
orl-man TDLo:571 μg/kg:EYE AIMEAS 97,454,82
orl-rat LD50:5300 mg/kg MEIEDD 10,909,83
ipr-rat LD50:322 mg/kg IYKEDH 16,1461,85

ivn-rat LD50:59200 μg/kg IYKEDH 16,1461,85
orl-mus LD50:3800 mg/kg YACHDS 9,527,81
ipr-mus LD50:298 mg/kg IYKEDH 16,1461,85
ivn-mus LD50:47100 μg/kg IYKEDH 16,1461,85

SAFETY PROFILE: Poison by intravenous and intraperitoneal routes. Moderately toxic by ingestion. Human systemic effects by ingestion: visual field changes. An experimental teratogen. Other experimental reproductive effects. When heated to decomposition it emits toxic fumes of NO_x. Used to reduce cardiac arrhythmias, lower blood pressure and treat angina pectoris.

CNR725 CAS:2571-86-0 ***HR: 3***
CORIAMYRTIN
mf: $C_{15}H_{18}O_5$ mw: 278.33

PROP: Bitter, monoclinic prisms. Mp: 229-230°. Sltly sol in water, cold alc; freely sol in hot alc and in ether.

SYNS: CORIAMYRTINE ◇ CORIAMYRTIONE

TOXICITY DATA with REFERENCE
scu-rat LDLo:1 mg/kg JPETAB 57,410,36
ivn-rat LDLo:700 μg/kg JPETAB 57,410,36
ipr-mus LD50:3 mg/kg JMCMAR 11,729,68
scu-mus LD50:3234 μg/kg JAPMA8 29,2,40
ivn-mus LDLo:1 mg/kg JPETAB 57,410,36
unr-mus LD50:1 mg/kg MEIEDD 10,361,83
unr-cat LDLo:237 μg/kg JPETAB 57,361,36
scu-rbt LD50:930 μg/kg JAPMA8 29,2,40
ivn-rbt LD50:371 μg/kg JAPMA8 29,2,40
scu-gpg LDLo:2439 μg/kg JPETAB 57,361,36
scu-frg LDLo:10 mg/kg JPETAB 57,410,36
unr-frg LD50:5800 μg/kg MEIEDD 10,361,83

SAFETY PROFILE: Poison by subcutaneous, intravenous, intraperitoneal, and possibly other routes. When heated to decomposition it emits acrid smoke and fumes.

CNR735 CAS:8008-52-4 ***HR: 2***
CORIANDER OIL

PROP: From steam distillation of ripe fruit of *Coriandrum sativum* L. (Fam. *Umbelliferae*). Colorless liquid; characteristic odor and taste. D: 0.863-0.875, refr index: 1.462 @ 20°.

SYNS: OIL of CORIANDER ◇ OILS, CORIANDER

TOXICITY DATA with REFERENCE
skn-rbt 500 mg/24H FCTXAV 11,1077,73
dnr-bcs 10 mg/disc TOFOD5 8,91,85
orl-rat LD50:4130 mg/kg FCTXAV 11,1077,73
orl-mus LD50:3520 mg/kg TOFOD5 8,91,85

CONSENSUS REPORTS: Reported in EPA TSCA Inventory.

SAFETY PROFILE: Moderately toxic by ingestion. Mutation data reported. A skin irritant. When heated to decomposition it emits acrid smoke and fumes.

CNR740 ***HR: 3***
CORIARI MYRTIFOLIA

PROP: Small shrubs or trees which produce small green flowers and purple-black berries. They are grown as ornamentals in the southern US and California.

SAFETY PROFILE: The fruit contains the poison coriamyrtin. Ingestion may cause convulsions similar to those produced by picrotoxin. See also PICROTOXIN.

CNR750 CAS:20153-98-4 ***HR: 3***
CORMELIAN
mf: $C_{31}H_{44}N_2O_{10}$•2ClH mw: 677.69

SYNS: ASTA C ◇ N,N'-(BIS-ω-HYDROXYPROPYL)HOMOPIPERAZINE 3,4,5-TRIMETHOXYBENZOATE DIHYDROCHLORIDE ◇ N,N'-BIS(3-(3,4,5-TRIMETHOXYBENZOYLOXY)PROPYL)HOMOPIPERAZINE DIHYDROCHLORIDE ◇ 1,4-BIS(3-(3,4,5-TRIMETH OXYBENZOYLOXY)-PROPYL)PERHYDRO-1,4-DIAZEPINE DIHYDROCHLORIDE ◇ COMELIAN ◇ DILAZEP DIHYDROCHLORIDE

TOXICITY DATA with REFERENCE
orl-mus TDLo:6 g/kg (female 7-12D post):TER KSRNAM 8,2084,74
orl-rat TDLo:30 g/kg (female 30D pre):REP KSRNAM 8,2050,74
ipr-rat LD50:85 mg/kg ARZNAD 22,667,72
scu-rat LD50:369 mg/kg KSRNAM 8,2044,74
ivn-rat LD50:13700 μg/kg KSRNAM 8,2044,74
ivn-rat LD50:2860 mg/kg KSRNAM 8,2044,74
ipr-mus LD50:161 mg/kg ARZNAD 22,667,72
scu-mus LD50:154 mg/kg ARZNAD 22,667,72
ivn-mus LD50:16800 μg/kg KSRNAM 8,2044,74
ivn-dog LD50:11 mg/kg ARZNAD 22,667,72

SAFETY PROFILE: Poison by subcutaneous, intravenous, and intraperitoneal routes. Moderately toxic by ingestion. An experimental teratogen. Other experimental reproductive effects. Used as a coronary vasodilator and as an anti-anginal agent. When heated to decomposition it emits very toxic fumes of HCl and NO_x.

CNR825 ***HR: 3***
CORMELIAN-DIGOTAB
mf: $C_{31}H_{44}N_2O_{10}$•2ClH•$C_{43}H_{66}O_{15}$ mw: 1500.78

SYNS: ASTA CD 072 ◇ DILAZEP/β-ACETYLDIGOXIN

TOXICITY DATA with REFERENCE
ipr-rat LD50:83200 μg/kg ARZNAD 24,1914,74
orl-mus LD50:1980 mg/kg ARZNAD 24,1914,74
ipr-mus LD50:142 mg/kg ARZNAD 24,1914,74
orl-dog LD50:422 μg/kg ARZNAD 24,1914,74

SAFETY PROFILE: Poison by intraperitoneal route. Moderately toxic by ingestion. When heated to decomposition it emits toxic fumes of NO_x and HCl. See also CORMELIAN.

CNS000 CAS:8001-30-7 ***HR: 1***
CORN OIL

PROP: Light yellow, clear, oily liquid; faint characteristic odor. Mp: −10°, flash p: 490°F (CC), d: 0.92, autoign temp: 740°F. From wet milling of *Zea mays* (85DIA2 2,70,77).

TOXICITY DATA with REFERENCE
skn-hmn 300 mg/3D-I MLD 85DKA8 -,127,77
orl-rat TDLo:12500 mg/kg (15-19D preg):TER TJADAB 27,75A,83

CONSENSUS REPORTS: Reported in EPA TSCA Inventory.

SAFETY PROFILE: Human skin irritant. An experimental teratogen. May be an allergen. Combustible liquid when exposed to heat or flame. Dangerous spontaneous heating may occur during storage if leaks impregnate rags, waste, etc. To fight fire, use CO_2, dry chemical.

CNS200 CAS:4503-12-2 ***HR: 2***
CORONARIDINE HYDROCHLORIDE
mf: $C_{21}H_{26}N_2O_2$•ClH mw: 374.95

SYNS: (-)-CORONARIDINE MONOHYDROCHLORIDE ◇ IBOGAMINE-18-CARBOXYLIC ACID, METHYL ESTER, HYDROCHLORIDE ◇ IBOGAMINE-18-CARBOXYLIC ACID, METHYL ESTER, MONOHYDROCHLORIDE (9CI) ◇ 6,9-METHANO-8H-PYRIDO(1′,2′:1,2)AZEPINO(4,5-b)INDOLE-6(6aH)-CARBOXYLICACID, 7,8,9,10,12,13-HEXAHYDRO-6-ETHYL-13a-HYDROXY-, METHYL ESTER, HYDROCHLORIDE

TOXICITY DATA with REFERENCE
orl-rat TDLo:50 mg/kg (female 10D pre):REP JPMSAE 62,1199,73
orl-mus LDLo:500 mg/kg JPMSAE 52,598,63
ipr-cat LDLo:20 mg/kg JPMSAE 52,598,63
ivn-cat LDLo:10 mg/kg JPMSAE 52,598,63

SAFETY PROFILE: Moderately toxic by ingestion, intraperitoneal, and intravenous routes. Experimental reproductive effects. When heated to decomposition it emits toxic fumes of NO_x and HCl.

CNS250 CAS:191-07-1 ***HR: 3***
CORONENE
mf: $C_{24}H_{12}$ mw: 300.36

SYN: HEXABENZOBENZENE

TOXICITY DATA with REFERENCE
mma-sat 1 μg/plate/48H FCTXAV 17,141,79
dnd-mam:lym 20 mg BIPMAA 4,409,66
skn-mus TDLo:20 mg/kg/W-I:ETA NCIMAV 28,173,68

CONSENSUS REPORTS: IARC Cancer Review: Group 3 IMEMDT 7,56,87; Animal Inadequate Evidence IMEMDT 32,263,83.

SAFETY PROFILE: Questionable carcinogen with experimental tumorigenic data. Mutation data reported. A polycyclic hydrocarbon air pollutant. When heated to decomposition it emits acrid smoke and irritating fumes.

CNS625 CAS:50-22-6 ***HR: D***
CORTICOSTERONE
mf: $C_{21}H_{30}O_4$ mw: 346.51

PROP: Trigonal plates from acetone. Mp: 180-182°. Insol in water; sol in usual organic solvents.

SYNS: COMPOUND B ◇ CORTICOSTERON ◇ 17-DEOXYCORTISOL ◇ 11-β,21-DIHYDROXYPREGN-3,20-DIONE ◇ (11-β)-11,21-DIHYDROXY-PREGN-4-ENE-3,20-DIONE (9CI) ◇ 11-β,21-DIHYDROXY-PREGN-4-ENE-3,20-DIONE ◇ 11,12-DIHYDROXYPROGESTERONE ◇ 11-β,21-DIHYDROXYPROGESTERONE ◇ 11-HYDROXYCORTI OALDOSTERONE ◇ KENDALL'S COMPOUND B ◇ 4-PREGNENE-11-α,21-DIOL-3,20-DIONE ◇ REICHSTEIN'S SUBSTANCE H

TOXICITY DATA with REFERENCE
dni-rat:oth 100 nmol/L ECREAL 161,29,85
dni-ckn:emb 1500 pmol/L NATUAS 257,804,75
ipr-rat TDLo:14 mg/kg (19-20D preg):REP BNEOBV 36,1,79
scu-rat TDLo:525 mg/kg (12-18D preg):TER ACEDAB 173,44,73

SAFETY PROFILE: An experimental teratogen. Experimental reproductive effects. Mutation data reported. A steroid. When heated to decomposition it emits acrid smoke and fumes. See also CORTSOL.

CNS650 CAS:1173-26-8 ***HR: D***
CORTICOSTERONE ACETATE
mf: $C_{23}H_{32}O_5$ mw: 388.55

SYNS: 21-ACETATECORTICOSTERONE ◇ (11-β)-21-(ACETYLOXY)-11-HYDROXY-PREGN-4-ENE-3,20-DIONE (9CI) ◇ CORT A ◇ 11-β,21-DI-HYDROXY-PREGN-4-ENE-3,20-DIONE-21-ACETATE ◇ 11-β,21-DIHYDROXYPREGN-4-ENE-3,20-DIONE ACETATE ◇ SKF 5654

TOXICITY DATA with REFERENCE
scu-mus TDLo:400 mg/kg (female 11-14D post):TER TJADAB 6,313,72
scu-mus TDLo:800 mg/kg (female 11-14D post):REP TJADAB 6,313,72

SAFETY PROFILE: An experimental teratogen. Experimental reproductive effects. A steroid. When heated to decomposition it emits acrid smoke and fumes. See also CORTISOL.

CNS750 CAS:50-23-7 ***HR: 3***
CORTISOL
mf: $C_{21}H_{30}N_{40}O_5S$ mw: 954.97

SYNS: AEROSEB-HC ◇ ANTI-INFLAMMATORY HORMONE

◇ BARSEB HC ◇ CETACORT ◇ COBADEX ◇ COMPOUND F ◇ CORTDOME ◇ CORTISOL ALCOHOL ◇ CORTISPRAY ◇ DERMACORT ◇ EF CORLIN ◇ GENACORT ◇ HC ◇ HEB-CORT ◇ HIDRO-COLISONA ◇ 11-β-HYDROCORTISONE ◇ HYDROCORTISONE FREE ALCOHOL ◇ HYDROCORTISYL ◇ HYDROCORTONE ◇ 17-HYDROXYCORTICOSTERONE ◇ 11-β-HYDROXYCORTISONE ◇ HYTONE LOTION ◇ KENDALL'S COMPOUND F ◇ NSC 10483 ◇ OPTEF ◇ PERMICORT ◇ 4-PREGNENE-11-β,17-α,21-TRIOL 3,20-DIONE ◇ REICHSTEIN'S SUBSTANCE M ◇ SCHEROSON F ◇ 11-β,17,21-TRIHYDROXYPREGN-4-ENE-3,20-DIONE ◇ 11-β,17-α-21-TRIHYDROXY-4-PREGNENE-3,20-DIONE

TOXICITY DATA with REFERENCE
dns-rat-ipr 20 mg/kg BEXBAN 94,1511,82
dni-rat:mmr 1 mg/L AMBPBZ 85,57,77
cyt-mus-ipr 50 mg/kg PSEBAA 171,109,82
dni-gpg:lng 1 mg/L PSEBAA 171,109,82
dns-ckn:emb 10 mg/L ITCSAF 20,172,84
ipr-rat TDLo:80 mg/kg (female 14-15D post):REP BBRCA9 75,125,77
ims-mus TDLo:200 mg/kg (female 12D post):TER JCGBDF 7,341,87
ipr-rat LD50:150 mg/kg JJPAAZ 21,377,71
scu-rat LD50:449 mg/kg TXAPA9 8,250,66

CONSENSUS REPORTS: Reported in EPA TSCA Inventory.

SAFETY PROFILE: Poison by intraperitoneal route. Moderately toxic by subcutaneous route. An experimental teratogen. Other experimental reproductive effects. Mutation data reported. A steroid. When heated to decomposition it emits very toxic fumes of SO_x and NO_x. See also CORTISONE.

CNS800 CAS:53-06-5 ***HR: 3***
CORTISONE
mf: $C_{21}H_{28}O_5$ mw: 360.49

PROP: Rhombohedral platelets from 95% alc. Mp: 220-224° (some decomp). Fairly sol in cold methanol, ethanol, and acetone; much less sol in ether, benzene, and chloroform; sltly sol in water (28 mg/100 mL at 25°).

SYNS: ADRENALEX ◇ ADRESON ◇ COMPOUND E ◇ CORLIN ◇ CORTADREN ◇ CORTISAL ◇ CORTISATE ◇ CORTISTAL ◇ CORTIVITE ◇ CORTOGEN ◇ CORTONE ◇ 11-DEHYDRO-17-HYDROXYCORTICOSTERONE ◇ 17α,21-DIHYDROXY-4-PREGNENE-3,11,20-TRIONE ◇ 17-HYDROXY-11-DEHYDROCORTICOSTERONE ◇ 17-α-HYDROXY-11-DEHYDROCORTICOSTERONE ◇ 17α,21-HYDROXYPREGN-4-ENE-3,11,20-TRIONE ◇ INCORTIN ◇ KE ◇ KENDALL'S COMPOUND E ◇ PREGN-4-EN-17α,21-DIOL-3,11,20-TRIONE ◇ 4-PREGNENE-17α,21-DIOL-3,11,20-17,21-DIHYDROXY- ◇ Δ^4-PREGNENE-17α,21-DIOL-3,11,20-TRIONE ◇ REICHSTEIN'S SUBSTANCE FA ◇ SCHEROSON ◇ WINTERSTEINER'S COMPOUND F

TOXICITY DATA with REFERENCE
dns-rat-scu 168 mg/kg/3D-C ENDOAO 94,1637,74
dnd-mus:lvr 100 μmol/L ENZYAS 41,183,71
orl-wmn TDLo:160 mg/kg (16W pre/1-39W preg):REP OBGNAS 1,276,53
orl-wmn TDLo:90 mg/kg (90D preg):TER AJOGAH 65,237,53

SAFETY PROFILE: Human teratogenic effects by ingestion and possibly other routes: developmental abnormalities of the eye and ear, craniofacial area, musculoskeletal system, cardiovascular system and other neonatal measures or effects. Other experimental teratogenic effects. Additional reproductive effects in experimental animals. Mutation data reported. When heated to decomposition it emits acrid smoke and fumes. See also CORTISOL.

CNS825 CAS:50-04-4 ***HR: 3***
CORTISONE-21-ACETATE
mf: $C_{23}H_{30}O_6$ mw: 402.53

SYNS: ACETATE CORTISONE ◇ 21-ACETOXY-17,α-HYDROXYPREGN-4-ENE-3,11,20-TRIONE ◇ 21-ACETOXY-17,α-HYDROXY-3,11,20-TRIKETOPREGNENE-4 ◇ 21-(ACETYLOXY)-17-HYDROXY-PREGN-4-ENE-3,11,20-TRIONE (9CI) ◇ ADRESON ◇ ARTRIONA ◇ BIOCORT ACETATE ◇ COMPOUND E ACETATE ◇ CORTADREN ◇ CORTELAN ◇ CORTISAL ◇ CORTISATE ◇ CORTISONE ACETATE ◇ CORTISONE MONOACETATE ◇ CORTISTAB ◇ CORTISYL ◇ CORTIVITE ◇ CORTOGEN ◇ CORTOGEN ACETATE ◇ CORTONE ◇ CORTONE ACETATE ◇ 11-DEHYDRO-17-HYDROXYCORTICOSTERONE ACETATE ◇ 11-DEHYDRO-17-HYDROXYCORTICOSTERONE-21-ACETATE ◇ 17,21-DIHYDROXY-PREGN-4-ENE-3,11,20-TRIONE ACETATE ◇ 17,21-DIHYDROXY-PREGN-4-ENE-3,11,20-TRIONE 21-ACETATE ◇ INCORTIN ◇ IRISONE ACETATE ◇ 4-PREGNENE-17,α,21-DIOL-3,11,20-TRIONE 21-ACETATE ◇ RICORTEX ◇ SCHEROSON

TOXICITY DATA with REFERENCE
mmo-bcs 5 g/L MUREAV 42,19,77
dnr-bcs 5 g/L MUREAV 42,19,77
ims-rat TDLo:300 mg/kg (female 12-15D post):REP ACPAAN 8,217,67
ims-dog TDLo:37500 μg/kg (female 20-34D post):TER JZKEDZ 6,37,80
ipr-mus TDLo:35 mg/kg:ETA EXPEAM 33,1640,77

SAFETY PROFILE: Questionable carcinogen with experimental tumorigenic data. An experimental teratogen. Experimental reproductive effects. Mutation data reported. When heated to decomposition it emits acrid smoke and fumes. See also CORTISON.

CNT250 ***HR: 3***
CORUNDUM FUME

PROP: Half finely divided alumina, half silica (JPBAA7 69,81,55).

TOXICITY DATA with REFERENCE
itr-rat LDLo:90 mg/kg JPBAA7 69,81,55

SAFETY PROFILE: Poison by intratracheal route. See also ALUMINUM OXIDE (2:3) and SILICA.

CNT325 ***HR: 3***
CORYDALOID

PROP: Extracted from *Corydalis bulbosa dc* (KSRNAM 5,1129,71).

TOXICITY DATA with REFERENCE
orl-rat TDLo:60 mg/kg (multi) :REP KSRNAM 5,1153,71
orl-rat LD50:4100 mg/kg KSRNAM 5,1129,71
ipr-rat LD50:175 mg/kg KSRNAM 5,1129,71
scu-rat LD50:220 mg/kg KSRNAM 5,1129,71
orl-mus LD50:917 mg/kg KSRNAM 5,1129,71
ipr-mus LD50:205 mg/kg KSRNAM 5,1129,71
scu-mus LD50:155 mg/kg KSRNAM 5,1129,71

SAFETY PROFILE: Poison by subcutaneous and intraperitoneal routes. Moderately toxic by ingestion. Experimental reproductive effects.

CNT625 CAS:82-54-2 ***HR: 3***
COTARNIN
mf: $C_{12}H_{15}NO_4$ mw: 237.28

PROP: Small needles from benzene. Decomp @ 132-133°. Sol in alc, chloroform, ether, benzene; sltly sol in water; sol in dilute acids, in ammonia or sodium carbonate soln, but only sltly sol in potassium hydroxide soln. Aq or alcoholic solns are yellow.

SYNS: COTARNINE ◇ 5,6,7,8-TETRAHYDRO-4-METHOXY-6-METHYL-1,2-DIOXOLO-(4,5-g)ISOQUINOLIN-5-OL(9CI)

TOXICITY DATA with REFERENCE
unr-dog LDLo:200 mg/kg HBAMAK 4,1289,35
unr-rbt LDLo:180 mg/kg HBAMAK 4,1289,35
unr-frg LDLo:40 mg/kg HBAMAK 4,1289,35

SAFETY PROFILE: Poison by an unspecified route. When heated to decomposition it emits toxic fumes of NO_x.

CNT750 ***HR: 2***
COTTON DUST

TOXICITY DATA with REFERENCE
ihl-hmn TCLo:10000 mg/m^3/10Y:PUL BJIMAG 17,1,60
ipr-rat LD50:2 g/kg JDGRAX 5(1),93,73
ipr-gpg LD50:4 g/kg JDGRAX 5(1),93,73

OSHA PEL: TWA 1 mg/m^3 (raw dust); 0.2 mg/m^3 (yarn manufacturing); 0.75 mg/m^3 (slashing and weaving); 0.5 mg/m^3 (other operations)
ACGIH TLV: TWA 0.2 mg/m^3 (raw dust)
DFG MAK: 1.5 mg/m^3 (raw cotton)
NIOSH REL: (Cotton Dust) CL 0.200 mg/m^3 lint-free

SAFETY PROFILE: Human pulmonary effects. Causes a mild febrile condition of the lungs resembling metal fume fever. Coarser grades of cotton contain more dust than the finer varieties, and therefore constitute a greater hazard. It is considered an inert dust and indeed it is, within the meaning of the term. However, it can cause some illness, due to the allergens or fungi in the cotton or on the dust. Workers in processing rooms may develop conjunctivitis or blepharitis from the burned products of the gassing of the double yarn. It is a mild allergen. Inhalation may produce bronchial asthma, sneezing and eczema in sensitized persons. Moderate fire and explosion hazard when exposed to heat or flame; can react with oxidizing materials.

CNU000 CAS:8001-29-4 ***HR: 3***
COTTONSEED OIL (unhydrogenated)

PROP: Oily, pale yellow, nearly odorless liquid from seeds of species of *Gossypium hirsutum*. Flash p: 486°F (CC), fp: 0-5°, d: 0.915-0.921 @ 25°/25°, autoign temp: 650°F.

SYNS: DEODORIZED WINTERIZED COTTONSEED OIL ◇ NCI-C50168

TOXICITY DATA with REFERENCE
ipr-rat TDLo:2256 mg/kg (5-15D preg):TER JDREAF 51,1632,72
orl-mus TDLo:2940 g/kg/35W-C:ETA LPDSAP 17,115,82

SAFETY PROFILE: Questionable carcinogen with experimental tumorigenic data. Experimental teratogenic effects. An allergen. Combustible liquid when exposed to heat or flame. However, if allowed to impregnate rags or oily waste, it can become a dangerous hazard due to spontaneous heating. To fight fire, use CO_2, dry chemical.

CNU750 CAS:56-72-4 ***HR: 3***
COUMAPHOS
DOT: UN 2783
mf: $C_{14}H_{16}ClO_5PS$ mw: 362.78

SYNS: AGRIDIP ◇ ASUNTHOL ◇ BAYER 21/199 ◇ BAYMIX 50 ◇ 3-CHLORO-7-HYDROXY-4-METHYL-COUMARIN-O,O-DIETHYLPHOSPHOROTHIOATE ◇ 3-CHLORO-7-HYDROXY-4-METHYL-COUMARIN-O-ESTER with O,O-DIETHYL PHOSPHOROTHIOATE ◇ O-3-CHLORO-4-METHYL-7-COUMARINYL-O,O-DIETHYL PHOSPHOROTHIOATE ◇ 3-CHLORO-4-METHYL-7-COUMARINYL DIETHYL PHOSPHOROTHIOATE ◇ 3-CHLORO-4-METHYL-7-HYDROXYCOUMARIN DIETHYL THIOPHOSPHORIC ACID ESTER ◇ 3-CHLORO-4-METHYLUMBELLIFERONE-O-ESTER with O,O-DIETHYL PHOSPHOROTHIOATE ◇ CUMAFOS (DUTCH) ◇ O,O-DIAETHYL-O-(3-CHLOR-4-METHYL-CUMARIN-7-YL)-MONOTHIOPHOSPHAT (GERMAN) ◇ O,O-DIETHYL-O-(3-CHLOOR-4-METHYL-CUMARIN-7-YL)MONOTHIOFOSFAAT (DUTCH) ◇ O,O-DIETHYL-O-(3-CHLORO-4-METHYL-7-COUMARINYL)PHOSPHOROTHIOATE ◇ O,O-DIETHYL-O-(3-CHLORO-4-METHYLCOUMARINYL-7) THIOPHOSPHATE ◇ O,O-DIETHYL-O-(3-CHLORO-4-METHYL-2-OXO-2H-BENZOPYRAN-7-YL)PHOSPHOROTHIOATE ◇ O,O-DIETHYL-3-CHLORO-4-METHYL-7-UMBELLIFERONE THIOPHOSPHATE ◇ O,O-DIETHYL-O-(3-CHLORO-4-METHYLUMBELLIFERYL)PHOSPHOROTHIOATE ◇ DIETHYL-3-CHLORO-4-METHYLUMBELLIFERYL THIONOPHOSPHATE

◇ DIETHYL THIOPHOSPHORIC ACIDESTER of 3-CHLORO-4-METHYL-7-HYDROXYCOUMARIN ◇ O,O-DIETIL-O-(3-CLORO-4-METIL-CUMARIN-7-IL-MONOTIOFOSFATO) (ITALIAN) ◇ DIOLICE ◇ ENT 17,956 ◇ MELDONE ◇ NCI-C08662 ◇ THIOPHOSPHATE de O,O-DIETHYLE et de O-(3-CHLORO-4-METHYL-7-COUMARINYLE) (FRENCH)

TOXICITY DATA with REFERENCE
otr-rat:emb 1400 ng/plate JJATDK 1,190,81
orl-rat LD50:13 mg/kg DOEAAH 35,25,79
ihl-rat LC50:303 mg/m3 VHTODE 24,87,82
skn-rat LD50:860 mg/kg WRPCA2 9,119,70
ipr-rat LD50:7500 μg/kg PSEBAA 129,699,68
orl-mus LD50:28 mg/kg HYSAAV 33(12),334,68
ipr-mus LD50:200 mg/kg PCOC** -,282,66
orl-rbt LD50:80 mg/kg HYSAAV 33(12),334,68
skn-rbt LD50:500 mg/kg VHTODE 24,87,82

CONSENSUS REPORTS: NCI Carcinogenesis Bioassay (feed); No Evidence: mouse, rat NCITR* NCI-CG-TR-96,79. EPA Extremely Hazardous Substances List.

DOT Classification: Poison B; Label: Poison; Poison B; Label: Poison, liquid.

SAFETY PROFILE: Poison by ingestion, skin contact, inhalation, and intraperitoneal routes. Mutation data reported. When heated to decomposition, it emits very toxic fumes of SO_x, PO_x, and Cl^-. See also COUMARIN.

CNU825 CAS:7400-08-0 ***HR: D***
4-COUMARIC ACID
mf: $C_9H_8O_3$ mw: 164.17

PROP: Needles. Mp: 210-213°. Crystallizes in anhydrous form from conc hot aq soln, but as monohydrate from dilute aq soln on slow cooling. Sltly sol in cold water; sol in hot water, alc, and ether. Practically insol in benzene and ligroin.

SYNS: p-COUMARIC ACID ◇ p-CUMARIC ACID ◇ p-HYDROXYCINNAMIC ACID ◇ 4-HYDROXYCINNAMIC ACID ◇ 4'-HYDROXYCINNAMIC ACID ◇ p-HYDROXYPHENYLACRYLIC ACID ◇ β-(4-HYDROXYPHENYL)ACRYLIC ACID ◇ 3-(4-HYDROXYPHENYL)-2-PROPENOIC ACID

TOXICITY DATA with REFERENCE
orl-mus TDLo:50 mg/kg (female 6D post):REP CCPTAY 20,49,79

CONSENSUS REPORTS: Reported in EPA TSCA Inventory.

SAFETY PROFILE: Experimental reproductive effects. When heated to decomposition it emits acrid smoke and fumes. See also COUMARIN.

CNV000 CAS:91-64-5 ***HR: 3***
COUMARIN
mf: $C_9H_6O_2$ mw: 146.15

PROP: Crystals; fragrant, pleasant odor; burning taste. Mp: 70°, bp: 291.0°, vap press: 1 mm @ 106.0°.

SYNS: 2H-1-BENZOPYRAN-2-ONE ◇ 1,2-BENZOPYRONE ◇ cis-o-COUMARINIC ACID LACTONE ◇ COUMARINIC ANHYDRIDE ◇ o-HYDROXYCINNAMIC ACID LACTONE ◇ o-HYDROXYZIMTSAURE-LACTON (GERMAN) ◇ NCI-C60297 ◇ 2-OXO-1,2-BENZOPYRAN ◇ RATTEX ◇ TONKA BEAN CAMPHOR

TOXICITY DATA with REFERENCE
mma-sat 1 mg/plate ENMUDM 5(Suppl 1),3,83
dnd-mam:Lym 20 mmol/L PNASA6 48,686,62
orl-mus TDLo:3600 mg/kg (6-17D preg):TER ARZNAD 17,97,67
orl-rat TD:200 g/kg/2Y-C:ETA TXCYAC 1,93,73
orl-rat LD50:293 mg/kg FCTXAV 12,385,74
orl-mus LD50:196 mg/kg YKKZAJ 83,1124,63
ipr-mus LD50:220 mg/kg ARZNAD 15,897,65
scu-mus LD50:242 mg/kg YKKZAJ 83,1124,63
orl-gpg LD50:202 mg/kg FCTXAV 2,327,64

CONSENSUS REPORTS: IARC Cancer Review: Group 3 IMEMDT 7,56,87; Animal Limited Evidence IMEMDT 10,113,76. EPA Genetic Toxicology Program. Reported in EPA TSCA Inventory.

SAFETY PROFILE: Poison by ingestion, intraperitoneal, and subcutaneous routes. Questionable carcinogen with experimental tumorigenic data. Experimental teratogenic effects. Mutation data reported. Combustible when exposed to heat or flame. When heated to decomposition it emits acrid smoke and fumes. See also KETONES and ANHYDRIDES.

CNV500 CAS:4434-05-3 ***HR: 3***
COUMERMYCIN AL
mf: $C_{55}H_{59}N_5O_{20}$ mw: 1110.19

SYNS: COUMAMYCIN ◇ NOTOMYCIN A1

TOXICITY DATA with REFERENCE
mmo-esc 200 mg/L NATUAS 271,385,78
orl-mus LD50:2000 mg/kg 85ERAY 1,147,78
ipr-mus LD50:159 mg/kg 85ERAY 1,147,78
scu-mus LD50:250 mg/kg AACHAX -,786,65
ivn-mus LD50:25 mg/kg 85ERAY 1,147,78
ims-mus LD50:500 mg/kg 85ERAY 1,147,78

SAFETY PROFILE: Poison by subcutaneous, intramuscular, intravenous, and intraperitoneal routes. Moderately toxic by ingestion. Mutation data reported. When heated to decomposition it emits toxic fumes of NO_x.

CNV750 CAS:8065-91-6 ***HR: 3***
C-QUENS
mf: $C_{23}H_{29}ClO_4 \cdot C_{21}H_{26}O_2$ mw: 715.44

SYNS: ACONCEN ◇ CHLORMADINONE ACETATE mixed with MESTRANOL ◇ 6-CHLORO-6-DEHYDRO-17-α-ACETOXYPROGESTERONE mixed with MESTRENOL ◇ LUTESTRAL (FRENCH) ◇ MESTRANOL mixed with CHLORMADINONE ACETATE ◇ MESTRANOL mixed with 6-CHLORO-6-DEHYDRO-17-α-ACETOXYPROGESTERONE ◇ MESTRENOL mixed with 6-CHLORO-6-DEHYDRO-17-α-ACETOXYPROGESTERONE

TOXICITY DATA with REFERENCE
orl-wmn TDLo:1302 μg/kg (21D pre):REP CCPTAY 1,389,70
orl-wmn TDLo:60480 μg/kg/6Y-I:NEO SURGAZ 77,137,75
ipr-rat LD50:5 g/kg OYYAA2 4,217,70
scu-rat LD50:5 g/kg OYYAA2 4,217,70

SAFETY PROFILE: Questionable human carcinogen producing liver tumors. Human reproductive effects by ingestion: changes in the uterus, cervix, vagina, and fertility. Other experimental reproductive effects. Mildly toxic by subcutaneous and intraperitoneal routes. A steroid. When heated to decomposition it emits toxic fumes of Cl^-.

CNW000 CAS:136-78-7 ***HR: 2***
CRAG HERBICIDE
mf: $C_8H_7Cl_2O_5S{\cdot}Na$ mw: 309.10

SYNS: CRAG HERBICIDE 1 ◇ CRAG SESONE ◇ 2,4-DES-Na ◇ 2,4-DES-NATRIUM (GERMAN) ◇ 2-(2,4-DICHLOROPHENOXY)ETHANOL HYDROGEN SULFATE SODIUM SALT ◇ 2,4-DICHLOROPHENOXYETHYL SULFATE, SODIUM SALT ◇ DISUL ◇ DISUL-Na ◇ DISUL-SODIUM ◇ NATRIUM-2,4-DICHLORPHENOXYATHYLSULFAT (GERMAN) ◇ SES ◇ SESONE (ACGIH) ◇ SODIUM-2-(2,4-DICHLOROPHENOXY)ETHYL SULFATE ◇ SODIUM-2,4-DICHLOROPHENOXYETHYL SULPHATE ◇ SODIUM-2,4-DICHLOROPHENYL CELLOSOLVE SULFATE

TOXICITY DATA with REFERENCE
orl-rat LD50:730 mg/kg RREVAH 10,97,65
orl-mam LD50:1230 mg/kg FMCHA2 -,C212,83

OSHA PEL: (Transitional: TWA Total Dust: 15 mg/m^3; Respirable Fraction: 5 mg/m^3) TWA Total Dust: 10 mg/m^3; Respirable Fraction: 5 mg/m^3
ACGIH TLV: TWA 10 mg/m^3

SAFETY PROFILE: Moderately toxic by ingestion. Strong solutions are skin irritants. An herbicide. When heated to decomposition it emits very toxic fumes of Cl^-, Na_2O, and SO_x.

CNW100 CAS:11005-94-0 ***HR: 3***
CRANOMYCIN

TOXICITY DATA with REFERENCE
orl-mus LD50:8600 μg/kg 85FZAT -,232,67
ipr-mus LD50:760 μg/kg 85FZAT -,232,67
ivn-mus LD50:840 μg/kg 85FZAT -,232,67

SAFETY PROFILE: Poison by ingestion, intravenous, and intraperitoneal routes.

CNW105 CAS:101516-88-5 ***HR: 3***
CRANOMYCIN HYDROCHLORIDE

TOXICITY DATA with REFERENCE
orl-mus LD50:8600 mg/kg 85ERAY 1,453,78
ipr-mus LD50:760 μg/kg 85ERAY 1,453,78
ivn-mus LD50:840 μg/kg 85ERAY 1,453,78

SAFETY PROFILE: Poison by ingestion, intravenous, and intraperitoneal routes.

CNW125 CAS:55769-64-7 ***HR: 3***
CRAVITEN
mf: $C_{32}H_{48}N_2O_{10}{\cdot}2ClH$ mw: 693.74

SYN: BUTOBENDINE DIHYDROCHLORIDE

TOXICITY DATA with REFERENCE
orl-rat TDLo:1500 mg/kg (6-15D preg):TER PJPPAA 32,893,80
orl-rat LD50:1500 mg/kg DRFUD4 5,610,80
ipr-rat LD50:142 mg/kg PJPPAA 32,823,80
ivn-rat LD50:15800 μg/kg MEIEDD 10,212,83
orl-mus LD50:4500 mg/kg DRFUD4 5,610,80
ipr-mus LD50:550 mg/kg PJPPAA 32,823,80
ivn-mus LD50:18 mg/kg DRFUD4 5,610,80
ivn-rbt LD50:5100 μg/kg PJPPAA 32,823,80

SAFETY PROFILE: Poison by intravenous and intraperitoneal routes. Moderately toxic by ingestion. An experimental teratogen. When heated to decomposition it emits toxic fumes of NO_x and HCl. See also ESTERS.

CNW500 CAS:1319-77-3 ***HR: 3***
CRESOL
DOT: UN 2022/UN 2076
mf: C_7H_8O mw: 108.15

PROP: Description (U.S.P. XVI): mixture of isomeric cresols obtained from coal tar, colorless or yellowish to brown-yellow or pinkish liquid, phenolic odor. Mp: 10.9-35.5°, bp: 191-203°, flash p: 178°F, d: 1.030-1.038 @ 25°/25°, vap press: 1 mm @ 38-53°, vap d: 3.72.

SYNS: ACEDE CRESYLIQUE (FRENCH) ◇ BACILLOL ◇ CRESOLI (ITALIAN) ◇ CRESYLIC ACID ◇ HYDROXYTOLUOLE (GERMAN) ◇ KRESOLE (GERMAN) ◇ KRESOLEN (DUTCH) ◇ KREZOL (POLISH) ◇ RCRA WASTE NUMBER U052 ◇ TEKRESOL ◇ ar-TOLUENOL

TOXICITY DATA with REFERENCE
orl-rat LD50:1454 mg/kg NTIS** PB214-270
orl-mus LD50:760 mg/kg KSGZA3 36,932,82
skn-rbt LD50:2000 mg/kg TXAPA9 42,417,77

CONSENSUS REPORTS: Community Right-To-Know List.

OSHA PEL: TWA 5 ppm (skin)
ACGIH TLV: TWA 5 ppm
DFG MAK: (all isomers) 5 ppm (22 mg/m^3)
NIOSH REL: (Cresol) TWA 10 mg/m^3
DOT Classification: Corrosive Material; Label: Corrosive; Poison B; Label: Poison.

SAFETY PROFILE: Moderately toxic by ingestion and skin contact. Corrosive to skin and mucous membranes. Systemic poisoning has rarely been reported, but it is possible that absorption may result in damage to the kidneys, liver, and nervous system. The main hazard accompanying its use in industry lies in severe chemical burns and dermatitis. Flammable when exposed to heat or flame; can react vigorously with oxidizing materials. Slightly explosive in the form of vapor when exposed to heat or flame. Explosive Range: 1.35% @ 300°F. Reacts violently with HNO_3, oleum, or chlorosulfonic acid. When heated to decomposition it emits highly toxic and irritating fumes. To fight fire, use foam, CO_2, dry chemical. See also other cresol entries PHENOL.

CNW750 CAS:108-39-4 ***HR: 3***
m-CRESOL
DOT: UN 2076
mf: C_7H_8O mw: 108.15

PROP: Colorless to yellowish liquid, phenolic odor. Mp: 10.9° bp: 202.8°, lel: 1.1% @ 302°F, flash p: 202°F, d: 1.034 @ 20°/4°, autoign temp: 1038°F, vap press: 1 mm @ 52.0°, vap d: 3.72.

SYNS: 3-CRESOL ◇ m-CRESYLIC ACID ◇ 1-HYDROXY-3-METHYLBENZENE ◇ m-HYDROXYTOLUENE ◇ m-KRESOL ◇ m-METHYLPHENOL ◇ 3-METHYLPHENOL ◇ m-OXYTOLUENE ◇ RCRA WASTE NUMBER U052 ◇ m-TOLUOL

TOXICITY DATA with REFERENCE
skn-rbt 517 mg/24H SEV BIOFX* 3-5/69
eye-rbt 103 mg SEV BIOFX* 3-5/69
dni-hmn:hla 10 μmol/L/4H BECTA6 32,220,84
scu-rbt TDLo:134 g/kg/(6-18D preg):TER OYYAA2 16,1191,78
skn-mus TDLo:2280 mg/kg/20W-I:NEO CNREA8 19,413,59
orl-rat LD50:242 mg/kg BIOFX* 3-5/69
skn-rat LD50:1100 mg/kg GTPZAB 18(2),58,74
scu-rat LDLo:900 mg/kg HBTXAC 5,56,59
orl-mus LD50:828 mg/kg GTPZAB 18,58,74
ipr-mus LD50:168 mg/kg HBTXAC 5,56,59
scu-mus LDLo:450 mg/kg HBAMAK 4,1361,35
ivn-dog LDLo:150 mg/kg HBTXAC 5,56,59
scu-cat LDLo:180 mg/kg JPETAB 80,233,44
orl-rbt LDLo:1400 mg/kg JPETAB 80,233,44
skn-rbt LD50:2050 mg/kg BIOFX* 3-5/69
ivn-rbt LDLo:280 mg/kg JPETAB 80,233,44
ipr-gpg LDLo:100 mg/kg HBAMAK 4,1361,35

CONSENSUS REPORTS: Community Right-To-Know List. Reported in EPA TSCA Inventory. EPA Genetic Toxicology Program.

OSHA PEL: TWA 5 ppm (skin)
ACGIH TLV: TWA 5 ppm
NIOSH REL: (Cresol) TWA 10 mg/m^3
DOT Classification: Poison B; Label: Poison.

SAFETY PROFILE: Poison by ingestion, intravenous, intraperitoneal, and subcutaneous routes. Moderately toxic by skin contact. Severe eye and skin irritant. An experimental teratogen. Human mutation data reported. Questionable carcinogen with experimental neoplastigenic data. Flammable when exposed to heat or flame. Moderately explosive in the form of vapor when exposed to heat or flame. See also other cresol entries and PHENOL.

CNX000 CAS:95-48-7 ***HR: 3***
o-CRESOL
DOT: UN 2076
mf: C_7H_8O mw: 108.15

PROP: Crystals or liquid darkening with exposure to air and light. Mp: 30.8°, bp: 190.8°, flash p: 178°F, d: 1.047 @ 20°/4°, autoign temp: 1110°F, vap press: 1 mm @ 38.2°, vap d: 3.72, lel: 1.4% @ 300°F.

SYNS: 2-CRESOL ◇ o-CRESYLIC ACID ◇ 1-HYDROXY-2-METHYLBENZENE ◇ o-HYDROXYTOLUENE ◇ o-KRESOL (GERMAN) ◇ o-METHYLPHENOL ◇ 2-METHYLPHENOL ◇ ORTHOCRESOL ◇ o-OXYTOLUENE ◇ RCRA WASTE NUMBER U052 ◇ o-TOLUOL

TOXICITY DATA with REFERENCE
skn-rbt 524 mg/24H SEV BIOFX* 4-5/69
eye-rbt 105 mg SEV BIOFX* 4-5/69
sce-hmn:fbr 8 mmol/L MUREAV 137,51,84
skn-mus TDLo:4800 mg/kg/12W-I:NEO CNREA8 19,413,59
orl-rat LD50:121 mg/kg BIOFX* 4-5/69
skn-rat LD50:620 mg/kg GTPZAB 18(2),58,74
scu-rat LDLo:65 mg/kg RMSRA6 15,561,1895
orl-mus LD50:344 mg/kg GTPZAB 18,58,74
ihl-mus LC50:179 mg/m^3/2H 85GMAT -,40,82
skn-mus LD50:620 mg/kg 85GMAT -,40,82
ipr-mus LDLo:200 mg/kg RBPMAZ 22,1,52
scu-mus LDLo:410 mg/kg ZHINAV 64,113,1909
ivn-dog LDLo:80 mg/kg HBTXAC 5,56,59
scu-cat LDLo:55 mg/kg JPETAB 80,233,44

CONSENSUS REPORTS: EPA Extremely Hazardous Substances List. Community Right-To-Know List. EPA Genetic Toxicology Program. Reported in EPA TSCA Inventory.

OSHA PEL: TWA 5 ppm (skin)
ACGIH TLV: TWA 5 ppm
NIOSH REL: (Cresol) TWA 10 mg/m^3
DOT Classification: Poison B; Label: Poison.

SAFETY PROFILE: Poison by ingestion, inhalation, subcutaneous, intravenous, and intraperitoneal routes. Moderately toxic by skin contact. A severe eye and skin irritant. Human mutation data reported. Questionable carcinogen with experimental neoplastigenic data. Flammable when exposed to heat, flame, or oxidants. To fight fire, water may be used to blanket fire; foam, fog, mist, dry chemical. See also other cresol entries and PHENOL.

CNX250 CAS:106-44-5 ***HR: 3***
p-CRESOL
DOT: UN 2076
mf: C_7H_8O mw: 108.15

PROP: Found in a score of essential oils, including ylang-ylang and oil of jasmine (FCTXAV 12,385,74) Crystals, phenolic odor. Mp: 35.5°, bp: 201.8°, lel: 1.1% @ 302°F, flash p: 202°F, d: 1.0341 @ 20°/4°, autoign temp: 1038°F, vap press 1 mm @ 53.0°, vap d: 3.72.

SYNS: 4-CRESOL ◇ p-CRESYLIC ACID ◇ 1-HYDROXY-4-METHYLBENZENE ◇ p-HYDROXYTOLUENE ◇ 4-HYDROXYTOLUENE ◇ p-KRESOL ◇ 1-METHYL-4-HYDROXYBENZENE ◇ p-METHYLPHENOL ◇ 4-METHYLPHENOL ◇ p-OXYTOLUENE ◇ PARAMETHYL PHENOL ◇ RCRA WASTE NUMBER U052 ◇ p-TOLUOL ◇ p-TOLYL ALCOHOL

TOXICITY DATA with REFERENCE
skn-rbt 517 mg/24H SEV BIOFX* 5-5/69
eye-rbt 103 mg SEV BIOFX* 5-5/69
skn-mus TDLo:2280 mg/kg/20W-I:NEO CNREA8 19,413,59
orl-rat LD50:207 mg/kg BIOFX* 5-5/69
skn-rat LD50:750 mg/kg GTPZAB 18,58,74
scu-rat LDLo:500 mg/kg HBTXAC 5,58,59
unr-rat LD50:1440 mg/kg GTPPAF 8,145,72
orl-mus LD50:344 mg/kg GTPZAB 18,58,74
ipr-mus LD50:25 mg/kg HBTXAC 5,58,59
scu-mus LDLo:150 mg/kg HBAMAK 4,1361,35
unk-mus LD50:160 mg/kg BJCAAI 6,160,52
scu-cat LDLo:80 mg/kg JPETAB 80,233,44
orl-rbt LDLo:620 mg/kg JPETAB 80,233,44
skn-rbt LD50:301 mg/kg BIOFX* 5-5/69
scu-rbt LDLo:300 mg/kg HBAMAK 4,1361,35
ivn-rbt LDLo:180 mg/kg JPETAB 80,233,44
scu-gpg LDLo:200 mg/kg HBTXAC 5,58,59
scu-frg LDLo:150 mg/kg HBAMAK 4,1361,35

CONSENSUS REPORTS: Community Right-To-Know List. Reported in EPA TSCA Inventory. EPA Genetic Toxicology Program.

OSHA PEL: TWA 5 ppm (skin)
ACGIH TLV: TWA 5 ppm
NIOSH REL: (Cresol) TWA 10 mg/m^3
DOT Classification: Poison B; Label: Poison.

SAFETY PROFILE: Poison by ingestion, skin contact, subcutaneous, intravenous, and intraperitoneal routes. A severe skin and eye irritant. Questionable carcinogen with experimental neoplastigenic data by itself and with 7,12-dimethyl benz(a)anthracene. Combustible when exposed to heat or flame. Moderately explosive in the form of vapor when exposed to heat or flame. To fight fire, use CO_2, dry chemical, alcohol foam. See also other cresol entries and PHENOL.

CNX625 CAS:83-40-9 ***HR: 3***
2,3-CRESOTIC ACID
mf: $C_8H_8O_3$ mw: 152.16

PROP: White to sltly reddish, odorless crystals. Mp: 165-166°, volatile with steam. Sltly sol in cold water, more sol in hot water; sol in chloroform, alc, ether, alkali hydroxides.

SYNS: ACIDO ORTOCRESOTINICO (ITALIAN) ◇ ACIDO 3-OSSI-5-METIL-BENZOICO (ITALIAN) ◇ CRESOTIC ACID ◇ o-CRESOTIC ACID ◇ CRESOTINIC ACID ◇ β-CRESOTINIC ACID ◇ o-CRESOTINIC ACID ◇ 2,3-CRESOTINIC ACID ◇ HOMOSALICYLIC ACID ◇ 2-HYDROXY-3-METHYL-BENZOIC ACID (9CI) ◇ 3-METHYLSALYCILIC ACID ◇ 3-MS

TOXICITY DATA with REFERENCE
orl-rat LD50:445 mg/kg BIALAY 10,270,63
orl-mus LD50:1 g/kg APSXAS 7,289,70
ivn-mus LD50:345 mg/kg FRPSAX 20,506,65

CONSENSUS REPORTS: Reported in EPA TSCA Inventory.

SAFETY PROFILE: Poison by intravenous route. Moderately toxic by ingestion. When heated to decomposition it emits acrid smoke and fumes.

CNX700 ***HR: 2***
CRESSA CRETICA Linn., extract

PROP: Indian plant belonging to the family *Convolvulaceae* IJEBA6 22,312,84

TOXICITY DATA with REFERENCE
orl-ham TDLo:500 mg/kg (female 1-5D post):REP IJEBA6 22,312,84
ipr-mus LD50:681 mg/kg IJEBA6 22,312,84

SAFETY PROFILE: Moderately toxic by intraperitoneal route. Experimental reproductive effects. When heated to decomposition it emits acrid smoke and irritating fumes.

CNX800 ***HR: 3***
CROCUS

PROP: A type of lilly which grows from a bulb and produces tubular purple or white flowers. They are grown only as houseplants or outdoor ornamentals.

SYNS: AUTUMN CROCUS ◇ COLCHICUM AUTUMNALE ◇ COLCHICUM SPECIOSUM ◇ COLCHICUM VERNUM ◇ FALL CROCUS ◇ MEADOW SAFFRON ◇ MYSTERIA ◇ VELLORITA (CUBA) ◇ WONDER BULB

SAFETY PROFILE: The whole plant contains the poison colchicine. Ingestion of any part of the plant causes a burning pain in the mouth, intense thirst, nausea, vomiting, abdominal cramps, severe diarrhea, and sometimes kidney damage. Colchicine is excreted slowly so the effects may persist for some time. See also COLCHICINE.

CNX825 CAS:15826-37-6 ***HR: 2***
CROMOGLYCATE DISODIUM
mf: $C_{23}H_{14}O_{11}$•2Na mw: 512.35

SYNS: AARANE ◇ AARARRE ◇ CROMOGLYCATE ◇ CROMOLYN SODIUM ◇ CROMOLYN SODIUM SALT ◇ DISODIUM CHROMOGLYCATE ◇ DISODIUM CROMOGLICATE ◇ DISODIUM CROMOGLYCATE ◇ DISODIUM-5,5′-((2-HYDROXYTRIMETHYLENE)DIOXY)-BIS(4-OXO-4H-1-BENZOPYRAN-2-CARBOXYLATE)◇ FPL 670 ◇ FRENASMA ◇ INOSTRAL ◇ INTAL ◇ LOMUDAL ◇ LOMUDAS ◇ NALCROM ◇ NASMIL ◇ RYNACROM ◇ SODIUM CROMOGLYCATE ◇ SODIUM CROMOLYN

TOXICITY DATA with REFERENCE
orl-wmn TDLo:96 mg/kg/6D-I:SKN BMJOAE 289,470,84
orl-hmn TDLo:34 mg/kg/4W:PUL BMJOAE 2,916,76
scu-rat LD50:6 g/kg NIIRDN 6,244,82
ipr-mus LD50:4100 mg/kg NIIRDN 6,244,82
scu-mus LD50:4400 mg/kg NIIRDN 6,244,82
ivn-mus LD50:3300 mg/kg NIIRDN 6,244,82
ivn-rbt LD50:2 g/kg KSRNAM 4,189,70

SAFETY PROFILE: Moderately toxic by intravenous route. Mildly toxic by subcutaneous and intraperitoneal routes. Human systemic effects by ingestion: allergic dermatitis, respiratory depression and other lung effects. When heated to decomposition it emits toxic fumes of Na_2O.

CNX827 ***HR: D***
CROTALARIA JUNCEA Linn., seed extract

PROP: Indian plant belonging to the family *Fabaceae* (IJEBA6 22,312,84).

SYN: SANAI, seed extract

TOXICITY DATA with REFERENCE
orl-ham TDLo:500 mg/kg (1-5D preg):REP IJEBA6 22,312,84

SAFETY PROFILE: Experimental reproductive effects.

CNX830 ***HR: 3***
CROTALUS CERASTES VENOM

SYN: VENOM, SNAKE, CROTALUS CERASTES

TOXICITY DATA with REFERENCE
ipr-mus LD50:2080 μg/kg 14FHAR -,409,63
ivn-mus LD50:2600 μg/kg 14FHAR -,409,63
ipr-mam LD50:4 mg/kg CLPTAT 8,849,67

SAFETY PROFILE: A deadly poison by intravenous and intraperitoneal routes.

CNX835 ***HR: 3***
CROTALUS DURISSUS DURISSUS VENOM

SYN: VENOM, COSTA RICAN SNAKE, CROTALUS (CROTALUS) DURISSUS DURISSUS

TOXICITY DATA with REFERENCE
ipr-mus LD50:850 μg/kg AJTHAB 21,360,72
ivn-mus LD50:1938 μg/kg AJTHAB 21,360,72

SAFETY PROFILE: Poison by intravenous and intraperitoneal routes.

CNY000 ***HR: 3***
CROTALUS DURISSUS TERRIFICUS VENOM

SYN: VENOM, SNAKE, CROTALUS DURISSUS TERRIFICUS

TOXICITY DATA with REFERENCE
ipr-mus LD50:400 μg/kg TOXIA6 9,131,71
scu-mus LD50:600 μg/kg TOXIA6 23,825,85
ivn-mus LD50:47 μg/kg TOXIA6 23,361 85
ims-mus LD50:1400 μg/kg TOXIA6 15,129,77
unr-mus LD50:200 μg/kg TOXIA6 19,473,81
ims-dog LDLo:1500 μg/kg COREAF 260,5408,65
ipr-mam LD50:300 μg/kg CLPTAT 8,849,67

SAFETY PROFILE: A deadly poison by subcutaneous, intramuscular, intravenous, intraperitoneal, and possibly other routes.

CNY300 ***HR: 3***
CROTALUS HORRIDUS VENOM

SYN: VENOM, SNAKE, CROTALUS HORRIDUS

TOXICITY DATA with REFERENCE
ipr-mus LD50:2940 μg/kg 14FHAR -,409,63
scu-mus LD50:24900 μg/kg AJTMAQ 31,489,51
ivn-mus LD50:2630 μg/kg 14FHAR -,409,63
scu-dog LDLo:60 mg/kg AJPHAP 173,535,53
ivn-dog LDLo:500 μg/kg AJPHAP 173,535,53

SAFETY PROFILE: Poison by subcutaneous, intravenous, and intraperitoneal routes.

CNY325 ***HR: 3***
CROTALUS RUBER RUBER VENOM

SYN: VENOM, SNAKE, CROTALUS RUBER RUBER

TOXICITY DATA with REFERENCE
ipr-mus LD50:4650 μg/kg 14FHAR -,409,63
ivn-mus LD50:3700 μg/kg 14FHAR -,409,63
ims-mus LD50:11500 μg/kg TOXIA6 23,769,85
ipr-mam LD50:6690 μg/kg CLPTAT 8,849,67
ivn-mam LD50:3700 μg/kg CLPTAT 8,849,67

SAFETY PROFILE: Poison by intramuscular, intravenous, and intraperitoneal routes.

CNY339 ***HR: 3***
CROTALUS SCUTULATUS SCUTULATUS (CALIF) VENOM

SYN: VENOM, SNAKE, CROTALUS SCUTULATUS SCUTULATUS (CALIFORNIA)

TOXICITY DATA with REFERENCE
ivn-rat LD50:1 mg/kg 85EGD4 5,217,78
ipr-mus LD50:240 μg/kg TOXIA6 15,129,77
ivn-mus LD50:150 μg/kg TOXIA6 15,129,77
ims-mus LD50:700 μg/kg TOXIA6 15,129,77

SAFETY PROFILE: Poison by intramuscular, intravenous, and intraperitoneal routes.

CNY350 ***HR: 3***
CROTALUS SCUTULATUS SCUTULATUS VENOM

SYN: VENOM, SNAKE, CROTALUS SCUTULATUS SCUTULATUS

TOXICITY DATA with REFERENCE
ivn-rat LD50:1 mg/kg 85EGD4 -,211,78
ipr-mus LD50:110 μg/kg TOXIA6 16,81,78
scu-mus LD50:320 μg/kg TOXIA6 23,825,85
ivn-mus LD50:30 μg/kg TOXIA6 23,11,85
ims-mus LD50:700 μg/kg TOXIA6 15,129,77
unr-mus LD50:200 μg/kg TOXIA6 19,473,81

SAFETY PROFILE: A deadly poison by subcutaneous, intramuscular, intravenous, intraperitoneal and possibly other routes.

CNY375 ***HR: 3***
CROTALUS SCUTULATUS VENOM

SYN: VENOM, SNAKE, CROTALUS SCUTULATUS SCUTULATUS

TOXICITY DATA with REFERENCE
ipr-mus LD50:178 μg/kg TOXIA6 9,131,71
ivn-mus LD50:178 μg/kg TOXIA6 9,131,71
ipr-mam LD50:230 μg/kg CLPTAT 8,849,67
ivn-mam LD50:210 μg/kg CLPTAT 8,849,67

SAFETY PROFILE: A deadly poison by intravenous and intraperitoneal routes.

CNY750 ***HR: 3***
CROTALUS VIRIDIS CONCOLOR VENOM

SYN: VENOM, MIDGET FADED RATTLESNAKE, CROTALUS VIRIDIS CONCOLOR

TOXICITY DATA with REFERENCE
ipr-mus LD50:250 μg/kg TOXIA6 15,129,77
ivn-mus LD50:45 μg/kg TOXIA6 23,361,85
ims-mus LD50:1200 μg/kg TOXIA6 15,129,77
unr-mus LD50:200 μg/kg TOXIA6 19,473,81

SAFETY PROFILE: Deadly poison by intramuscular, intravenous, intraperitoneal, and possibly other routes.

CNZ000 ***HR: 3***
CROTALUS VIRIDUS CERERUS VENOM

SYN: VENOM, MIDGET FADED RATTLESNAKE, CROTALUS VIRIDUS CERBERUS

TOXICITY DATA with REFERENCE
ipr-mus LD50:2500 μg/kg TOXIA6 15,129,77
ims-mus LD50:6 mg/kg TOXIA6 15,129,77

SAFETY PROFILE: Poison by intramuscular and intraperitoneal routes.

COA000 ***HR: 3***
CROTALUS VIRIDIS HELLERI VENOM

SYN: VENOM, MIDGET FADED RATTLESNAKE, CROTALUS VIRIDIS HELLERI

TOXICITY DATA with REFERENCE
ipr-mus LD50:1560 μg/kg PAASAH 44,145,56
scu-mus LD50:3560 μg/kg PAASAH 44,145,56
ivn-mus LD50:580 μg/kg TOXIA6 16,431,78
ims-mus LD50:5100 μg/kg TOXIA6 15,129,77
ivn-dog LD50:50 μg/kg PWPSA8 16,58,73
ivn-cat LD50:1 g/kg PWPSA8 16,58,73
ipr-mam LD50:1600 μg/kg CLPTAT 8,849,67
ivn-mam LD50:1290 μg/kg CLPTAT 8,849,67

SAFETY PROFILE: A deadly poison by subcutaneous, intramuscular, intravenous, and intraperitoneal routes.

COA250 ***HR: 3***
CROTALUS VIRIDIS LUTOSUS VENOM

SYN: VENOM, MIDGET FADED RATTLESNAKE, CROTALUS VIRIDIS LUTOSUS

TOXICITY DATA with REFERENCE
ipr-mus LD50:2100 μg/kg TOXIA6 15,129,77
ims-mus LD50:4200 μg/kg TOXIA6 15,129,77

SAFETY PROFILE: Poison by intramuscular and intraperitoneal routes.

COA500 ***HR: 3***
CROTALUS VIRIDIS OREGANUS VENOM

SYN: VENOM, MIDGET FADED RATTLESNAKE, CROTALUS VIRIDIS OREGANUS

TOXICITY DATA with REFERENCE
ipr-mus LD50:3200 μg/kg TOXIA6 15,129,77
ims-mus LD50:3600 μg/kg TOXIA6 15,129,77

SAFETY PROFILE: A deadly poison by intramuscular and intraperitoneal routes.

COA750 ***HR: 3***
CROTALUS VIRIDIS VENOM

SYN: VENOM, MIDGET FADED RATTLESNAKE, CROTALUS VIRIDIS VIRIDIS

TOXICITY DATA with REFERENCE
ipr-mus LD50:2 mg/kg TOXIA6 15,129,77
scu-mus LD50:41 mg/kg TOXIA6 16,431,78

SAFETY PROFILE: Poison by subcutaneous and intraperitoneal routes.

COB000 CAS:21284-11-7 ***HR: 2***
CROTOCIN
mf: $C_{19}H_{24}O_5$ mw: 332.43

SYNS: ANTIBIOTIC T ◇ 7-β,8-β:12,13-DIEPOXY-TRICHOTHEC-9-EN-6-β-OL ◇ (4-β(Z),7-β,8-β)-7,8:12,13-DIEPOXY-TRICHOTHEC-9-EN-4-OL 2-BUTENOATE

TOXICITY DATA with REFERENCE
skn-gpg 332 ng MLD FAATDF 4(2, Pt 2),5124,84
mma-sat 100 μg/plate CNREA8 38,536,78
orl-mus LD50:1000 mg/kg 85GDA2 6,183,81
ipr-mus LD50:810 mg/kg 85ERAY 3,2043,78
ivn-mus LD50:700 mg/kg 85GDA2 6,183,81

SAFETY PROFILE: Moderately toxic by ingestion, intravenous, and intraperitoneal routes. A skin irritant. Mutation data reported. When heated to decomposition it emits acrid smoke and fumes.

COB250 CAS:4170-30-3 ***HR: 3***
CROTONALDEHYDE
DOT: UN 1143
mf: C_4H_6O mw: 70.09

PROP: Water-white, mobile liquid; pungent suffocating odor. Bp: 104°, fp: −76.0°, lel: 2.1%, uel: 15.5%, flash p: 55°F, d: 0.853 @ 20°/20°, vap d: 2.41, autoign temp: 405°F.

SYNS: 2-BUTENAL ◇ CROTONIC ALDEHYDE ◇ KROTONALDEHYD (CZECH) ◇ β-METHYL ACROLEIN ◇ RCRA WASTE NUMBER U053

TOXICITY DATA with REFERENCE
mmo-sat 250 μmol/L ENMUDM 7(Suppl 3),56,85
sln-dmg-par 3500 ppm ENMUDM 7,677,85
trn-dmg-par 3500 ppm ENMUDM 7,677,85
dnd-mam:lym 21500 mg/L/16H CNREA8 44,990,84
orl-rat TDLo:2664 mg/kg/2Y-C:CAR CNREA8 46,1285,86
orl-rat LD50:206 mg/kg GTPZAB 26(8),53,82
ihl-rat LC50:200 mg/m^3/2H GTPZAB 26(8),53,82
orl-mus LD50:104 mg/kg GTPZAB 26(8),53,82
ihl-mus LD50:580 mg/m^3/2H GTPZAB 26(8),53,82

CONSENSUS REPORTS: EPA Extremely Hazardous Substances List. Reported in EPA TSCA Inventory.

OSHA PEL: TWA 2 ppm
DFG MAK: Suspected Carcinogen.
DOT Classification: Flammable Liquid; Label: Flammable Liquid and Poison.

SAFETY PROFILE: Suspected carcinogen with experimental carcinogenic data. Poison by ingestion and inhalation. Mutation data reported. An eye, skin, and mucous membrane irritant. A lachrymating material which can cause corneal burns and is very dangerous to the eyes. Caution: Keep away from heat and open flame. Keep container closed. Use with adequate ventilation. Extremely irritating to eyes, skin, mucous membranes. When necessary, the lacrimatory effect of the vapors may be counteracted by ammonia fumes. Dangerous fire hazard when exposed to heat or flame; can react with oxidizing materials. To fight fire, use alcohol foam, CO_2, dry chemical. Reacts violently with 1,3-butadiene. Violent hypergolic reaction with concentrated nitric acid. When heated to decomposition it emits acrid smoke and fumes. See also ALDEHYDES.

COB260 CAS:123-73-9 ***HR: 3***
(E)-CROTONALDEHYDE
mf: C_4H_6O mw: 70.10

PROP: Water-white, mobile liquid; pungent, suffocating odor. Bp: 104°, fp: −76.0°, lel: 2.1%, uel: 15.5%, flash p: 55°F, d: 0.853 @ 20°/20°, vap d: 2.41, autoign temp: 450°F.

SYNS: ALDEHYDE CROTONIQUE (FRENCH) ◇ trans-2-BUTENAL ◇ (E)-2-BUTENAL ◇ CROTONAL ◇ CROTONALDEHYDE ◇ CROTONIC ALDEHYDE ◇ 1,2-ETHANEDIOL DIPROPANOATE (9CI) ◇ ETHYLENE GLYCOL DIPROPIONATE (8CI) ◇ ETHYLENE PROPIONATE ◇ β-METHYL ACROLEIN ◇ NCI-C56279 ◇ PROPYLENE ALDEHYDE ◇ RCRA WASTE NUMBER U053 ◇ TOPANEL

TOXICITY DATA with REFERENCE
eye-hmn 45 ppm AIHAAP 28,561,67
skn-rbt 500 mg open MLD UCDS** 4/21/67
spm-mus-ipr 30 mg/kg MUREAV 39-317-77
ihl-hmn TCLo:12 mg/m^3/10M:IRR JAMAAP 165,1908,57
orl-rat LD50:300 mg/kg JIHTAB 26,269,44
ihl-rat LC50:4000 mg/m^3/30M APTOA6 6,299,50
scu-rat LD50:140 mg/kg APTOA6 6,299,50
orl-mus LD50:240 mg/kg BIJOAK 34,1196,40

ipr-mus LD50:160 mg/kg ZolH## 23OCT75
scu-mus LD50:160 mg/kg APTOA6 6,299,50
skn-rbt LD50:380 mg/kg UCDS** 4/21/67
skn-gpg LD50:1331 mg/kg AEXPBL 18,218,1884

CONSENSUS REPORTS: Reported in EPA TSCA Inventory.

OSHA PEL: TWA 2 ppm
ACGIH TLV: TWA 2 ppm
DFG MAK: Suspected Carcinogen.

SAFETY PROFILE: A poison by ingestion, subcutaneous, and intraperitoneal routes. Mutation data reported. A lachrymating material which is very dangerous to the eyes. Human respiratory system irritant by inhalation. Can cause corneal burns and is irritating to the skin. In case of contact, immediately flush the skin or eyes with water for at least 15 minutes and get medical attention. See also ALDEHYDES. Dangerous fire hazard when exposed to heat or flame. To fight fire, use alcohol foam, CO_2, dry chemical. Incompatible with 1,3-butadiene and oxidizing materials. When heated to decomposition it emits acrid smoke and fumes.

COB500 CAS:3724-65-0 ***HR: 3***
CROTONIC ACID
DOT: UN 2823
mf: $C_4H_6O_2$ mw: 86.10

PROP: Colorless, needle-like crystals. Bp: 185°, mp: 72°, flash p: 190°F (COC), d: 1.018 @ 15°/4°, vap press: 0.19 mm @ 20°, vap d: 2.97.

SYNS: α-BUTENOIC ACID ◇ 2-BUTENOIC ACID ◇ CROTONIC ACID, solid ◇ α-CROTONIC ACID ◇ 3-METHYLACRYLIC ACID ◇ β-METHYLACRYLIC ACID

TOXICITY DATA with REFERENCE
skn-rbt 10 mg/24H open JIHTAB 26,269,44
orl-rat LD50:1000 mg/kg JIHTAB 26,269,44
ipr-rat LD50:100 mg/kg 34ZIAG -,190,69
orl-mus LD50:4800 mg/kg BIJOAK 34,1196,40
ipr-mus LD50:3590 mg/kg JPPMAB 21,85,69
scu-mus LD50:3590 mg/kg JPPMAB 21,85,69
skn-gpg LD50:616 mg/kg JIHTAB 26,269,44
ipr-gpg LD50:60 mg/kg 34ZIAG -,190,69

CONSENSUS REPORTS: Reported in EPA TSCA Inventory.

DOT Classification: Corrosive Material; Label: Corrosive.

SAFETY PROFILE: Poison by intraperitoneal route. Moderately toxic by ingestion, skin contact, and subcutaneous routes. A powerful corrosive and irritant. Flammable when exposed to heat or flame; can react with oxidizing materials. To fight fire, use alcohol Foam, CO_2, dry chemical. When heated to decomposition it emits acrid smoke and irritating fumes.

COB750 CAS:623-70-1 ***HR: 3***
α-CROTONIC ACID ETHYL ESTER
DOT: UN 1862
mf: $C_6H_{10}O_2$ mw: 114.16

PROP: Colorless, monoclinic prisms or water-white liquid; pungent odor. Mp: 45° (solid). Bp: 209° (solid), 139° (liquid), flash p: 36.0°F, d: 0.9207 @ 20°/20°, vap d: 3.93.

SYNS: CROTONATE d'ETHYLE (FRENCH) ◇ (E)-CROTONIC ACID, ETHYL ESTER ◇ ETHYLCROTONATE

TOXICITY DATA with REFERENCE
skn-rbt 10 mg/24H open JIHTAB 26,269,44
eye-rbt 5 mg SEV AJOPAA 29,1363,46
orl-rat LD50:3000 mg/kg JIHTAB 26,269,44

CONSENSUS REPORTS: Reported in EPA TSCA Inventory.

DOT Classification: Flammable Liquid; Label: Flammable Liquid.

SAFETY PROFILE: Moderately toxic by ingestion and probably by inhalation. A skin, mucous membrane, and severe eye irritant. Very dangerous fire hazard when exposed to heat or flame; can react vigorously with oxidizing materials. To fight fire, use foam, CO_2, or dry chemical. See also ESTERS. When heated to decomposition it emits acrid smoke and fumes.

COB825 CAS:623-43-8 ***HR: 2***
(E)-CROTONIC ACID METHYL ESTER
mf: $C_5H_8O_2$ mw: 100.13

SYNS: trans-2-BUTENOIC ACID METHYL ESTER ◇ METHYL trans-2-BUTENOATE ◇ METHYL CROTONATE ◇ METHYL α-CROTONATE ◇ METHYL E-CROTONATE ◇ METHYL trans-CROTONATE

TOXICITY DATA with REFERENCE
skn-rbt 500 mg/24H MOD FCTXAV 17,865,79
orl-mus LD50:1600 mg/kg FCTXAV 17,865,79
skn-gpg LD50:10 g/kg FCTXAV 17,865,79

CONSENSUS REPORTS: Reported in EPA TSCA Inventory.

SAFETY PROFILE: Moderately toxic by ingestion. Mildly toxic by skin contact. A skin irritant. When heated to decomposition it emits acrid smoke and fumes. See also ESTERS.

COB900 CAS:623-68-7 ***HR: 2***
CROTONIC ANHYDRIDE
mf: $C_8H_{10}O_3$ mw: 154.18

SYNS: ANHYDRID KYSELINY KROTONOVE ◇ 2-BUTENOIC ACID, ANHYDRIDE (9CI) ◇ CROTONIC ACID ANHYDRIDE

TOXICITY DATA with REFERENCE
skn-rbt 10 mg/24H open MLD AIHAAP 23,95,62
orl-rat LD50:2830 mg/kg AIHAAP 23,95,62

CONSENSUS REPORTS: Reported in EPA TSCA Inventory.

SAFETY PROFILE: Moderately toxic by ingestion. A skin irritant. When heated to decomposition it emits acrid smoke and irritating fumes.

COC250 CAS:8001-28-3 ***HR: 3***
CROTON OIL

PROP: Oil from the seeds of *Croton tiglium* (BJCAAI 10,72,56). Brownish-yellow, viscid oil; slt offensive odor. Composition: croton resin, glycerides of fatty acids and crotin. D: 0.935 @ 25°/25°.

SYNS: CROTONOEL (GERMAN) ◇ CROTON RESIN ◇ CROTON TIGLIUM L. OIL ◇ OLEUM TIGLII ◇ OLIO DI CROTON (ITALIAN)

TOXICITY DATA with REFERENCE
skn-mus 190 ng MLD CNREA8 28,2338,68
skn-gpg 400 μg/20H JIDEAE 24,35,55
mma-sat 2500 μg/plate BJCAAI 37,873,78
dni-hmn:fbr 10 ppm CNREA8 35,1392,75
dni-hmn:lym 50 ppm BBRCA9 45,630,71
skn-mus TDLo:2 mg/kg/27W-I:NEO BJCAAI 7,482,53
skn-mus TD:285 mg/kg/42W-I:ETA BJCAAI 10,72,56
ipr-mus LDLo:1 mg/kg TXAPA9 23,288,72
ipr-frg LD50:60 mg/kg ZEKBAI 65,325,63
par-frg LD50:60 mg/kg ZEKBAI 65,325,63
unk-frg LD50:60 mg/kg ZEKBAI 65,325,63

CONSENSUS REPORTS: Reported in EPA TSCA Inventory.

SAFETY PROFILE: Poison by parenteral, intraperitoneal, and possibly other routes. A skin and eye irritant. An allergen. Human mutation data reported. Questionable carcinogen with experimental neoplastigenic and tumorigenic data by skin contact. When heated to decomposition it emits toxic fumes.

COC500 CAS:503-17-3 ***HR: 3***
CROTONYLENE
DOT: UN 1144
mf: CH_3CCCH_3 mw: 54.09

PROP: Liquid. Bp: 27°, flash p: < −4°F, lel: 1.4%, d: 0.688 @ 25°, vap d: 1.91.

SYNS: 2-BUTYNE ◇ DIMETHYLACETYLENE

CONSENSUS REPORTS: Reported in EPA TSCA Inventory.

DOT Classification: Flammable Liquid; Label: Flammable Liquid.

SAFETY PROFILE: A simple asphyxiant. Very dangerous fire hazard when exposed to heat or flame; can react with oxidizing materials. Moderately explosive in the form of vapor when exposed to heat or flame. To fight fire, use foam, CO_2, dry chemicals. See also ACETYLENE COMPOUNDS and ARGON (for a description of simple asphyxiants).

COC750 ***HR: 2***
/0 α-(N-CROTONYL-N-ETHYL)AMINO-BUTYRIC ACID mixed with BUTRIC ACID, α-(N-CROTONYL-N-PROPYL)AMINO

PROP: Equal parts of dimethylamines of N-crotonyl-α-ethylaminobutyric acid and of N-crotonyl-α-propylaminobutyric acid. (JPETAB 128,176,60)
SYN: DCB

TOXICITY DATA with REFERENCE
ipr-mus LD50:698 mg/kg JPETAB 128,176,60
scu-mus LD50:800 mg/kg JPETAB 128,176,60

SAFETY PROFILE: Moderately by subcutaneous and intraperitoneal routes. When heated to decomposition it emits toxic fumes of NO_x.

COC825 CAS:57449-30-6 ***HR: D***
CROTONYLOXYMETHYL-4,5,6-TRIHYDROOXYCYCLOHEX-2-ENONE
mf: $C_{11}H_{14}O_6$ mw: 242.25

TOXICITY DATA with REFERENCE
dni-mus:leu 6500 μg/L JANTAJ 35,1222,82
oms-mus:leu 25 mg/L JANTAJ 35,1222,82
oms-mus:leu 20 mg/L JANTAJ 35,1222,82

SAFETY PROFILE: Mutation data reported. When heated to decomposition it emits acrid smoke and fumes. See also ESTERS.

COC875 CAS:9007-40-3 ***HR: 3***
CROTOXIN

TOXICITY DATA with REFERENCE
ipr-mus LD50:50 mg/kg PNASA6 68,1560,71
scu-mus LD50:400 μg/kg BBACAQ 5,98,50
ivn-mus LD50:108 μg/kg NNAPBA 270,274,71

SAFETY PROFILE: Poison by subcutaneous, intravenous, and intraperitoneal routes.

COD000 CAS:7700-17-6 ***HR: 3***
CROTOXYPHOS
mf: $C_{14}H_{19}O_6P$ mw: 314.30

SYNS: CIODRIN ◇ CIODRIN VINYL PHOSPHATE ◇ CIOVAP

◇ CYODRIN ◇ DECROTOX ◇ (E)-3-((DIMETHOXYPHOSPHINYL)OXY)-2-BUTENOIC ACID 1-PHENYLETHYL ESTER (9CI) ◇ O,O-DIMETHYL-O-(1-METHYL-2-CARBOXY-α-PHENYLETHYL)VINYL PHOSPHATE ◇ DIMETHYL-cis-1-METHYL-2-(1-PHENYLETHOXYCARBONYL) VINYL PHOSPHATE ◇ DIMETHYL PHOSPHATE of α-METHYL-BENZYL-3-HYDROXY-cis-CROTONATE ◇ DUO-KILL ◇ ENT 24,717 ◇ (E)-3-HYDROXY-CROTONIC ACID α-METHYLBENZYL ESTER, DIMETHYL PHOSPHATE ◇ 1-METHYLBENZYL-3-(DIMETHOXY-PHOSPHINYLOXO) ISOCROTONATE ◇ α-METHYL BENZYL-3-(DIMETHOXY-PHOSPHINYLOXY)-cis-CROTONATE ◇ α-METHYL-BENZYL-3-HYDROXY-CROTONATE DIMETHYL PHOSPHATE ◇ PANTOZOL 1 ◇ cis-2-(1-PHENYLETHOXY)CARBONYL-1-METHYLVINYL DIMETHYLPHOSPHATE ◇ SD 4294 ◇ SHELL SD 4294 ◇ VOLFAZOL

TOXICITY DATA with REFERENCE
mma-smc 5000 ppm NTIS** PB80-133226
msc-mus:lym 180 mg/L NTIS** PB84-138973
orl-rat LD50:74 mg/kg WRPCA2 9,119,70
skn-rat LD50:202 mg/kg WRPCA2 9,119,70
scu-rat LD50:47 mg/kg BJPCBM 40,124,70
orl-mus LD50:90 mg/kg PCOC** -,244,66
ipr-mus LD50:71 mg/kg JPPMAB 19,612,67
scu-mus LD50:15 mg/kg JPPMAB 19,612,67
ivn-mus LD50:4500 μg/kg JPPMAB 19,612,67
skn-rbt LD50:385 mg/kg PCOC** -,244,66

CONSENSUS REPORTS: Reported in EPA TSCA Inventory.

SAFETY PROFILE: Poison by ingestion, skin contact, subcutaneous, intravenous, and intraperitoneal routes. Mutation data reported. An insecticide. When heated to decomposition it emits highly toxic PO_x.

COD100 CAS:10141-07-8 ***HR: 2***
CROTYLIDENE DICROTONATE
mf: $C_{12}H_{16}O_4$ mw: 224.28

SYNS: 2-BUTENOIC ACID, 2-BUTENYLIDENE ESTER, (E,E,E)-(8CI,9CI) ◇ 2-BUTENYLIDENE CROTONATE ◇ CROTONIC ACID, 2-BUTENYLIDENE ESTER

TOXICITY DATA with REFERENCE
skn-rbt 10 mg/24H open MLD AIHAAP 23,95,62
orl-rat LD50:2590 mg/kg AIHAAP 23,95,62

SAFETY PROFILE: Moderately toxic by ingestion. A skin irritant. When heated to decomposition it emits acrid smoke and irritating fumes.

COD475 CAS:294-93-9 ***HR: 2***
12-CROWN-4
mf: $C_8H_{16}O_4$ mw: 176.24

SYNS: EOCT ◇ ETHYLENE OXIDE CYCLIC TETRAMER ◇ 1,4,7,10-TETRAOXACYCLODODECANE

TOXICITY DATA with REFERENCE
skn-rbt 100 mg/24H MLD DCTODJ 8,451,85
eye-rbt 50 mg MLD DCTODJ 8,451,85
ihl-rat TCLo:1 ppm/7H (15D male):REP TXAPA9 27,342,74
orl-rat LD50:2830 mg/kg DCTODJ 1,339,78
ipr-rat LD50:1550 mg/kg DCTODJ 8,451,85
orl-mus LD50:3150 mg/kg TXAPA9 44,263,78
ipr-mus LD50:1290 mg/kg DCTODJ 8,451,85

SAFETY PROFILE: Moderately toxic by ingestion and intraperitoneal routes. Experimental reproductive effects. An eye and skin irritant.

COD500 CAS:17455-13-9 ***HR: 2***
18-CROWN-6
mf: $C_{12}H_{24}O_6$ mw: 264.36

$OC_2H_4(OC_2H_4)_4OCH_2CH_2$

SYN: 1,4,7,10,13,16-HEXANOXACYCLOOCTADECANE

TOXICITY DATA with REFERENCE
skn-rbt 100 mg/24H MLD DCTODJ 8,451,85
eye-rbt 50 mg MOD DCTODJ 8,451,85
orl-rat LD50:1390 mg/kg DCTODJ 1,339,78
ipr-rat LD50:830 mg/kg DCTODJ 8,451,85
orl-mus LD50:705 mg/kg TXAPA9 44,263,78
ipr-mus LD50:464 mg/kg DCTODJ 8,451,85

CONSENSUS REPORTS: Reported in EPA TSCA Inventory.

SAFETY PROFILE: Moderately toxic by ingestion and intraperitoneal routes. A skin and eye irritant. When heated to decomposition it emits acrid smoke and irritating fumes.

COD575 CAS:14187-32-7 ***HR: 2***
CROWN 18
mf: $C_{20}H_{24}O_6$ mw: 360.41

SYNS: DIBENZO-18-CROWN-6 ◇ 6,7,9,10,17,18,20,21-OCTAHYDRODIBENZO(b,k)(1,4,7,10,13,16)HEXAOXYCYCLOOCTADECIN

TOXICITY DATA with REFERENCE
skn-rbt 100 mg/24H MLD DCTODJ 8,451,85
eye-rbt 50 mg MOD DCTODJ 8,451,85
orl-rat LD50:2600 mg/kg DCTODJ 8,451,85
ipr-rat LD50:1880 mg/kg DCTODJ 8,451,85
ipr-mus LD50:920 mg/kg DCTODJ 8,451,85

SAFETY PROFILE: Moderately toxic by ingestion and intraperitoneal routes. An eye and skin irritant. When heated to decomposition it emits acrid smoke and fumes.

COD675 ***HR: 3***
CROWN FLOWER

PROP: Large shrubs which produce seeds and pods similar to milkweed. The crown flower grows to 15 feet. The small crown flower grows to 6 feet. Crown flowers are cultivated in south Florida and Hawaii. The small crown flower is a weed in the West Indies.

SYNS: ALGODON de SEDA (CUBA, PUERTO RICO) ◇ CALOTROPIS (VARIOUS SPECIES) ◇ C. GIGANTEA ◇ C. PROCERA ◇ FRENCH JASMINE ◇ GIANT MILKWOOD ◇ MUDAR ◇ PUA KALAUNU (HAWAII) ◇ TULA (PUERTO RICO)

SAFETY PROFILE: The whole plant contains cardiac glycosides. The sap also contains calcium oxalate and an allergan and can produce severe inflammation on contact with the eye. Chewing any part of the plant results in burning pain in the lips, mouth, and throat, possibly followed by inflammation and blistering. Large amounts are seldom swallowed due to the bitter taste. Ingestion may cause inflammation of the stomach and intestines and cardiac arrythimias. See also DIGITALIS and OXALATES.

COD725 ***HR: 2***
CRUDE OIL, synthetic

TOXICITY DATA with REFERENCE
skn-mus TDLo:339 g/kg/2Y-I:CAR FAATDF 7,228,86

SAFETY PROFILE: Questionable carcinogen with experimental carcinogenic data. When heated to decomposition it emits acrid smoke and irritating fumes.

COD750 CAS:68308-34-9 ***HR: 3***
CRUDE SHALE OILS
DOT: UN 1288

SYNS: BLUE OIL ◇ GREEN OIL ◇ RAW SHALE OIL ◇ SHALE OIL (DOT) ◇ UNFINISHED LUBRICATING OIL

TOXICITY DATA with REFERENCE
skn-rbt 500 mg/72H AIHAAP 40,460,79
mmo-sat 1 mg/plate 45KQAH -,173,80
cyt-mus-ipr 1500 mg/kg ENMUDM 4,639,82
sce-mus-ipr 1500 mg/kg ENMUDM 4,408,82
mma-ham:ovr 1 g/L 45KQAH -,173,80
slt-ham:ovr 1 g/L NTIS** CONF-800680-3
ipr-mus TDLo:200 mg/kg (1D male):REP NTIS** BNL-51002
skn-mus TDLo:20700 mg/kg/69W-I:CAR EVHPAZ 38,149,81
skn-mus TD:26208 mg/kg/2Y-I:CAR NTIS** CONF-801143
skn-mus TD:28800 mg/kg/30W-I:ETA NTIS** CONF-790334-3
orl-rat LD50:8 g/kg AIHAAP 40,460,79
orl-mus LD50:11 g/kg NTIS** Conf. 800680-1
ipr-mus LD50:4300 mg/kg NTIS** Conf. 800680-1
skn-rbt LD50:5 g/kg NTIS** BNL-51002-273,79

CONSENSUS REPORTS: IARC Cancer Review: Group 1 IMEMDT 7,339,87; Human Sufficient Evidence IMEMDT 35,161,85; Animal Limited Evidence IMEMDT 35,161,85; Animal Sufficient Evidence IMEMDT 3,22,73.

DOT Classification: Flammable or Combustible Liquid; Label: Flammable Liquid.

SAFETY PROFILE: Confirmed human carcinogen with experimental carcinogenic, neoplastigenic, and tumorigenic data. Mildly toxic by ingestion, skin contact, and intraperitoneal routes. A skin irritant. Experimental reproductive effects. Mutation data reported. Flammable when exposed to heat and flame. When heated to decomposition it emits acrid smoke and fumes.

COD850 CAS:299-86-5 ***HR: 3***
CRUFORMATE
mf: $C_{12}H_{19}ClNO_3P$ mw: 291.74

SYNS: AMIDOFOS ◇ AMIDOPHOS ◇ o-(4-terz.-BUTIL-2-CLORO-FENIL)-o-METIL-FOSFORAMMIDE (ITALIAN) ◇ o-(4-tert BUTYL-2-CHLOOR-FENYL)-o-METHYL-FOSFORZUUR-N-METHYL-AMIDE (DUTCH) ◇ 4-tert-BUTYL-2-CHLORO PHENYL METHYL METHYL PHOSPHORAMIDATE ◇ 4-tert.-BUTYL 2-CHLOROPHENYL METHYLPHOSPHORAMIDATE de METHYLE (FRENCH) ◇ o-(4-tert-BUTYL-2-CHLOR-PHENYL)-o-METHYL-PHOSPHORSAEURE-N-METHYL AMID (GERMAN) ◇ CRUFOMATE ◇ CRUFOMATE A ◇ DOWCC 132 ◇ ENT 25,602-X ◇ o-METHYL-o-2-CHLORO-4-tert-BUTYLPHENYL-N-METHYLAMIDOPHOSPHATE ◇ MONTREL ◇ RUELENE ◇ RUELENE DRENCH ◇ RUELENE 25E ◇ RULENE

TOXICITY DATA with REFERENCE
skn-mus TDLo:100 mg/kg (female 2D pre):REP JESEDU 16,141,81
orl-rat LD50:460 mg/kg TXAPA9 14,515,69
ihl-rat LCLo:12 mg/m^3/4H 85GMAR -,29,82
unk-rat LD50:770 mg/kg 30ZDA9 -,317,71
orl-rbt LD50:400 mg/kg 28ZEAL 5,59,76
skn-rbt LD50:2000 mg/kg SPEADM 78-1,37,78
orl-gpg LD50:1000 mg/kg FCTXAV 6,185,68
orl-bwd LD50:100 mg/kg TXAPA9 21,315,72
orl-mam LD50:251 mg/kg VETNAL 54(12),94,78

OSHA PEL: TWA 5 mg/m^3
ACGIH TLV: TWA 5 mg/m^3

SAFETY PROFILE: A poison by ingestion and inhalation. Moderately toxic via skin contact and possibly other routes. Experimental reproductive effects. When heated to decomposition it emits very toxic fumes of PO_x, NO_x, and Cl^-.

COE000 CAS:1309-32-6 ***HR: 3***
CRYPTOHALITE
DOT: UN 2854
mf: $F_6Si•2H_4N$ mw: 178.19

PROP: Mp: subl, d: 2.01.

SYNS: AMMONIUM FLUOSILICATE ◇ AMMONIUM HEXAFLUOROSILICATE ◇ AMMONIUM SILICOFLUORIDE (DOT)

◇ DIAMMONIUM HEXAFLUOROSILICATE ◇ FLUOSILICATE de AMMONIUM (FRENCH)

TOXICITY DATA with REFERENCE
orl-rat LDLo:100 mg/kg NCNSA6 5,27,53
orl-mus LD50:70 mg/kg GISAAA 53(11),80,88
orl-gpg LDLo:150 mg/kg CRSBAW 124,133,37
scu-frg LDLo:224 mg/kg CRSBAW 124,133,37

OSHA PEL: TWA 2.5 $mg(F)/m^3$
NIOSH REL: (Fluorides, Inorganic) TWA 2.5 mg(F) $/m^3$

DOT Classification: ORM-B; Label: None; Poison B; Label: St. Andrews Cross.

SAFETY PROFILE: Poison by ingestion and subcutaneous routes. See also HEXAFLUROSILICATE (2-) DIHYDROGEN and FLUORIDES. When heated to decomposition it emits very toxic fumes of F^-, NH_3, and NO_x.

COE100 CAS:37226-23-6 ***HR: 3***
CRYSTALLOMYCIN

TOXICITY DATA with REFERENCE
orl-mus LD50:1500 mg/kg 85ERAY 1,393,78
ipr-mus LD50:109 mg/kg ANTBAL 4(4),63,59
scu-mus LD50:220 mg/kg ANTBAL 4(4),63,59
ivn-mus LD50:124 mg/kg ANTBAL 4(4),63,59
scu-rbt LDLo:200 mg/kg ANTBAL 4(4),63,59
ivn-rbt LDLo:50 mg/kg ANTBAL 4(4),63,59
scu-gpg LDLo:125 mg/kg ANTBAL 4(4),63,59

SAFETY PROFILE: Poison by subcutaneous, intravenous, and intraperitoneal routes. Moderately toxic by ingestion.

COE125 CAS:10380-77-5 ***HR: 3***
CT 3318
mf: $C_{31}H_{42}N_2O_3 \cdot 2I$ mw: 744.55

SYNS: DIIODOMETHYLATE de la BIS(PIPERIDINOMETHYL-COUMARANYL-5)CETONE (FRENCH) ◇ DIIODOMETILATO del BISPIPERIDINOMETILCUMARANIL-5-CHETONE(ITALIAN)

TOXICITY DATA with REFERENCE
ivn-rat LD50:1500 μg/kg AIPTAK 120,53,59
scu-mus LD50:1100 μg/kg AIPTAK 106,395,56
ivn-dog LDLo:400 μg/kg AIPTAK 106,395,56
scu-cat LDLo:700 μg/kg AIPTAK 106,395,56
ivn-rbt LDLo:250 μg/kg AIPTAK 106,395,56
scu-gpg LDLo:400 μg/kg AIPTAK 106,395,56
ivn-ckn LD50:1 mg/kg AIPTAK 120,53,59

SAFETY PROFILE: A deadly poison by subcutaneous and intravenous routes. When heated to decomposition it emits toxic fumes of NO_x and I^-.

COE250 CAS:18444-66-1 ***HR: 3***
CUCURBITACIN E
mf: $C_{32}H_{44}O_8$ mw: 556.76

SYNS: CUCURBITACINE-E ◇ α-ELATERIN

TOXICITY DATA with REFERENCE
orl-mus LD50:340 mg/kg CHTPBA 5,205,70

SAFETY PROFILE: Poison by ingestion. When heated to decomposition it emits acrid smoke and irritating fumes.

COE500 CAS:122-03-2 ***HR: 2***
CUMALDEHYDE
mf: $C_{10}H_{12}O$ mw: 148.22

PROP: Found in at least 50 essential oils such as cumin, eucalyptus species, cinnamon, boldo and rue, and as main constituent of oil of *Pectis papposa harn* and *gray* (FCTXAV 12,385,74). Colorless to pale yellow liquid; pungent odor of cumin. D: 0.976-0.980, refr index: 1.529-1.534, flash p: 199°F. Sol in alc, ether; insol in water.

SYNS: p-CUMIC ALDEHYDE ◇ CUMINALDEHYDE ◇ CUMINIC ALDEHYDE (FCC) ◇ CUMINYL ALDEHYDE ◇ FEMA No. 2341 ◇ p-ISOPROPYLBENZALDEHYDE ◇ 4-ISOPROPYLBENZALDEHYDE ◇ p-ISOPROPYLBENZENECARBOXALDEHYDE ◇ 4-(1-METHYLETHYL)-BENZALDEHYDE (9CI)

TOXICITY DATA with REFERENCE
skn-rbt 500 mg/24H FCTXAV 12,395,74
orl-rat LD50:1390 mg/kg FCTXAV 2,327,64
orl-mus LD50:2400 mg/kg BIJOAK 34,1196,40
skn-rbt LD50:2800 mg/kg FCTXAV 12,395,74

CONSENSUS REPORTS: Reported in EPA TSCA Inventory.

SAFETY PROFILE: Moderately toxic by ingestion and skin contact. A skin irritant. Combustible liquid. When heated to decomposition it emits acrid smoke and irritating fumes. See also ALDEHYDES.

COE750 CAS:98-82-8 ***HR: 2***
CUMENE
DOT: UN 1918
mf: C_9H_{12} mw: 120.21

PROP: Colorless liquid. Mp: −96.0°, bp: 152°, flash p: 111°F, d: 0.864 @ 20°/4°, vap press: 10 mm @ 38.3°, autoign temp: 795°F, lel: 0.9%, uel: 6.5%, vap d: 4.1.

SYNS: BENZENE ISOPROPYL ◇ CUMEEN (DUTCH) ◇ CUM ◇ 2-FENILPROPANO (ITALIAN) ◇ 2-FENYL-PROPAAN (DUTCH) ◇ ISOPROPYLBENZEEN (DUTCH) ◇ ISOPROPILBENZENE (ITALIAN) ◇ ISOPROPYL BENZENE ◇ ISOPROPYLBENZOL ◇ ISOPROPYL-BENZOL (GERMAN) ◇ 2-PHENYLPROPANE ◇ RCRA WASTE NUMBER U055

TOXICITY DATA with REFERENCE
skn-rbt 10 mg/24H open MLD AMIHBC 4,119,51
eye-rbt 86 mg MLD AMIHAB 14,387,56
ihl-hmn TCLo:200 ppm:NOSE,CNS,PUL TGNCDL 2,39,61
orl-rat LD50:1400 mg/kg AMIHAB 14,387,56
ihl-rat LC50:8000 ppm/4H AMIHBC 4,119,51
ihl-mus LC50:24700 mg/m^3/2H 85GMAT -,78,82
skn-rbt LD50:12300 mg/kg AMIHBC 4,119,51

CONSENSUS REPORTS: Community Right-To-Know List. Reported in EPA TSCA Inventory. EPA Genetic Toxicology Program.

OSHA PEL: TWA 50 ppm (skin)
ACGIH TLV: TWA 50 ppm (skin)
DFG MAK: 50 ppm (245 mg/m^3)
DOT Classification: Flammable or Combustible Liquid; Label: Flammable Liquid.

SAFETY PROFILE: Moderately toxic by ingestion. Mildly toxic by inhalation and skin contact. Human systemic effects by inhalation: an antipsychotic, unspecified changes in the sense of smell and respiratory system. An eye and skin irritant. Potential narcotic action. Central nervous system depressant. There is no apparent difference between the toxicity of natural cumene or that derived from petroleum. See also BENZENE and TOLUENE. Flammable when exposed to heat or flame; can react with oxidizing materials. Violent reaction with HNO_3; oleum; chlorosulfonic acid. To fight fire, use foam, CO_2, dry chemical.

COF000 CAS:93-53-8 ***HR: 2***
CUMENE ALDEHYDE
mf: $C_9H_{10}O$ mw: 134.19

PROP: Colorless liquid; floral odor. D: 0.998-1.006, refr index: 1.515-1.520, flash p: 156°F. Sol in fixed oils; sltly sol in propylene glycol; insol in glycerin.

SYNS: FEMA No. 2886 ◇ α-FORMYLETHYLBENZENE ◇ HYACINTHAL ◇ HYDRATROP ALDEHYDE ◇ HYDRATROPIC ALDEHYDE ◇ α-METHYL PHENYLACETALDEHYDE ◇ α-METHYL-α-TOLUIC ALDEHYDE ◇ 2-PHENYLPROPANAL ◇ α-PHENYLPROPIONALDEHYDE ◇ 2-PHENYLPROPIONALDEHYDE (FCC)

TOXICITY DATA with REFERENCE
orl-rat LD50:2800 mg/kg FCTXAV 2,327,64

CONSENSUS REPORTS: Reported in EPA TSCA Inventory.

SAFETY PROFILE: Moderately toxic by ingestion. Combustible liquid. When heated to decomposition it emits acrid smoke and irritating fumes. See also ALDEHYDES.

COF250 CAS:64-00-6 ***HR: 3***
m-CUMENOL METHYLCARBAMATE
mf: $C_{11}H_{15}NO_2$ mw: 193.27

SYNS: COMPOUND 10854 ◇ m-CUMENYL METHYLCARBAMATE ◇ ENT 25,500 ◇ ENT 25,543 ◇ HERCULES 5727 ◇ HIP ◇ m-ISOPROPYLPHENOL-N-METHYLCARBAMATE ◇ m-ISOPROPYLPHENYL METHYLCARBAMATE ◇ m-ISOPROPYLPHENYL-N-METHYLCARBAMATE ◇ 3-ISOPROPYLPHENYL METHYLCARBAMATE ◇ N-METHYL-m-ISOPROPYLPHENYL CARBAMATE ◇ N-METHYL-3-ISOPROPYLPHENYL CARBAMATE ◇ OMS-15 ◇ UC 10854 ◇ UNION CARBIDE UC-10,854

TOXICITY DATA with REFERENCE
orl-rat LD50:29 mg/kg TXAPA9 21,315,72
skn-rat LD50:113 mg/kg 31ZOAD 1,263,68
ipr-rat LD50:14200 μg/kg BWHOA6 44(1-3),241,71
ivn-rat LD50:3150 μg/kg BJIMAG 22,317,65
ims-rat LD50:14 mg/kg BJIMAG 22,317,65
unk-rat LD50:41 mg/kg 30ZDA9 -,193,71
orl-mus LD50:16 mg/kg JAFCAU 18,793,70
ipr-mus LDLo:6 mg/kg TXAPA9 6,402,64
ivn-mus LD50:1410 μg/kg CSLNX* NX#02085
ims-dog LDLo:13 mg/kg BJIMAG 22,317,65
skn-rbt LD50:40 mg/kg FMCHA2 -,D323,80
orl-gpg LD50:10 mg/kg 31ZOAD 1,263,68
orl-ckn LD50:12 mg/kg TXAPA9 11,49,67
orl-bwd LD50:3200 μg/kg TXAPA9 21,315,72

CONSENSUS REPORTS: EPA Extremely Hazardous Substances List.

SAFETY PROFILE: Poison by ingestion, skin contact, intraperitoneal, intravenous, intramuscular, and possibly other routes. A pesticide. See also CARBAMATES. When heated to decomposition it emits toxic fumes of NO_x.

COF325 CAS:8014-13-9 ***HR: 2***
CUMIN OIL

PROP: From steam distillation of *Cuminum cyminum* L. Light yellow to brown liquid; strong odor. D: 0.905-0.925, refr index: 1.501 @ 20°. Sol in fixed oils, mineral oil; very sol in glycerin, propylene glycol.

SYNS: CUMMIN ◇ OILS, CUMIN

TOXICITY DATA with REFERENCE
skn-rbt 500 mg/24H MOD FCTXAV 12,869,74
mmo-sat 100 μL/plate NUCADQ 1,10,79
dnr-bcs 10 mg/disc TOFOD5 8,91,85
bfa-rat/sat 2500 mg/kg NUCADQ 1,10,79
orl-rat LD50:2500 mg/kg FCTXAV 12,869,74
skn-rbt LD50:3560 mg/kg FCTXAV 12,869,74

CONSENSUS REPORTS: Reported in EPA TSCA Inventory.

SAFETY PROFILE: Moderately toxic by ingestion and skin contact. A skin irritant. Mutation data reported.

When heated to decomposition it emits acrid smoke and irritating fumes.

COF350 CAS:479-13-0 ***HR: D***
CUMOESTEROL
mf: $C_{15}H_8O_5$ mw: 268.23

PROP: Crystals. Mp: 385° (sublimes @ 325°). Sltly sol in alkaline water, methanol, chloroform, ether, carbon tet, and benzene. Insol in acidified water and pet ether.

SYNS: 6H-BENZOFURO(3,2-c)(1)BENZOPYRAN-6-ONE,3,9-DIHDYROXY- ◇ COUMESTROL ◇ CUMOSTROL

TOXICITY DATA with REFERENCE
dns-mus-scu 400 mg/kg CNREA8 48,14,88
orl-mus TDLo:35 g/kg (female 14D pre):TER PSEBAA 167,237,81
orl-mus TDLo:35 g/kg (female 14D pre):REP PSEBAA 167,237,81

CONSENSUS REPORTS: Reported in EPA TSCA Inventory.

SAFETY PROFILE: An experimental teratogen. Other experimental reproductive effects. Mutation data reported. When heated to decomposition it emits acrid smoke and irritating fumes.

COF500 CAS:12002-03-8 ***HR: 3***
CUPRIC ACETOARSENITE
DOT: UN 1585
mf: $C_4H_6As_6Cu_4O_{16}$ mw: 1013.78

PROP: Emerald green powder.

SYNS: (ACETATO)(TRIMETAARSENITO)DICOPPER ◇ ACETOARSENITE de CUIVRE (FRENCH) ◇ BASLE GREEN ◇ C.I. 77410 ◇ C.I. PIGMENT GREEN 21 (9CI) ◇ COPPER ACETOARSENITE (DOT) ◇ COPPER ACETOARSENITE, solid (DOT) ◇ EMERALD GREEN ◇ ENT 884 ◇ FRENCH GREEN ◇ GENUINE PARIS GREEN ◇ IMPERIAL GREEN ◇ KING'S GREEN ◇ MEADOW GREEN ◇ MINERAL GREEN ◇ MITIS GREEN ◇ MOSS GREEN ◇ MOUNTAIN GREEN ◇ NEUWIED GREEN ◇ NEW GREEN ◇ ORTHO P-G BAIT ◇ PARIS GREEN ◇ PARROT GREEN ◇ PATENT GREEN ◇ POWDER GREEN ◇ SCHWEINFURTERGRUN ◇ SCHWEINFURT GREEN ◇ SOWBUG & CUTWORM BAIT ◇ SWEDISH GREEN ◇ VIENNA GREEN

TOXICITY DATA with REFERENCE
orl-rat LD50:22 mg/kg PCOC** -,254,66
unk-mam LD50:18 mg/kg 30ZDA9 -,388,71

CONSENSUS REPORTS: Arsenic and its compounds as well as copper and its compounds are on the Community Right-To-Know List.

OSHA PEL: TWA 0.5 mg(As)/m^3
DOT Classification: Poison B; Label: Poison.

SAFETY PROFILE: Poison by ingestion and and possibly other routes. An insecticide. When heated to decomposition it emits very toxic fumes of As. See also ARSENIC COMPOUNDS and COPPER COMPOUNDS.

COF675 CAS:370-81-0 ***HR: D***
CUPRIZONE
mf: $C_{14}H_{22}N_4O_2$ mw: 278.40

SYNS: BISCYCLOHEXANONE OXALDIHYDRAZONE ◇ CUPRIZANE ◇ ETHANEDIOIC ACID BIS(CYCLOHEXYLIDENE HYDRAZIDE)

TOXICITY DATA with REFERENCE
cyt-rat-orl 14 g/kg/14D-C EJBCAI 47,313,74
orl-mus TDLo:24 g/kg (3-22D preg):REP TXAPA9 8,512,66

CONSENSUS REPORTS: Reported in EPA TSCA Inventory.

SAFETY PROFILE: Experimental reproductive effects. Mutation data reported. When heated to decomposition it emits toxic fumes of NO_x. See also OXALATES.

COF750 CAS:8063-06-7 ***HR: 3***
CURARE

PROP: Brown, brittle, resinous mass. Rendering of an Indian name given to the unstandardized extracts derived mainly from the bark of various species of *Strychnos* and *Chondodendron* (12VSA5 9,347,76).

SYNS: INTOCOSTRINE ◇ OURARI ◇ URARI ◇ WOORALI ◇ WOORARI ◇ WOURARA

TOXICITY DATA with REFERENCE
unr-man LDLo:735 μg/kg 85DCAI 2,73,70
ipr-mus LD50:3200 μg/kg AIPTAK 153,308,65
scu-mus LD50:670 μg/kg PSEBAA 76,536,51
ivn-mus LD50:140 μg/kg PSEBAA 118,756,65
ivn-dog LD50:1200 μg/kg JPETAB 82,266,44
orl-rbt LDLo:270 mg/kg AEXPBL 61,283,09
scu-rbt LDLo:2700 μg/kg AEXPBL 61,283,09

SAFETY PROFILE: A deadly human poison by an unspecified route. A deadly experimental poison by ingestion, intraperitoneal, intravenous, and subcutaneous routes. When heated to decomposition it emits highly toxic fumes.

COF825 CAS:22260-42-0 ***HR: 3***
CURARINE
mf: $C_{38}H_{44}N_2O_6$ mw: 624.84

SYNS: CURARIN ◇ (+)-CURARINE ◇ (1′-α)-7′,12′-DIHYDROXY-6,6′-DIMETHOXY-2,2,2′,2′-TETRAMETHYLTUBOCURARANIUM(9CI) ◇ 2,2′-DIMETHYL-BEBEERINIUM (8CI)

TOXICITY DATA with REFERENCE
scu-mus LDLo:380 μg/kg HBAMAK 4,1289,35
scu-dog LDLo:340 μg/kg FDWU** -,-,31
scu-cat LDLo:340 μg/kg HBAMAK 4,1289,35
orl-rbt LDLo:334 μg/kg AEXPBL 61,283,09

scu-rbt LDLo:340 μg/kg HBAMAK 4,1289,35
ivn-rbt LDLo:80 μg/kg HBAMAK 4,1289,35
scu-gpg LDLo:90 μg/kg HBAMAK 4,1289,35
scu-pgn LDLo:618 μg/kg FDWU** -,-,31
ims-pgn LDLo:618 μg/kg HBAMAK 4,1289,35
scu-frg LDLo:16800 μg/kg FDWU** -,-,31

SAFETY PROFILE: A deadly poison by ingestion, subcutaneous, intramuscular and intravenous routes. When heated to decomposition it emits toxic fumes of NO_x.

COF850 ***HR: 2***
CURCUMA LONGA Linn., rhizome extract

PROP: Indian plant belonging to the family *Zingiberaceae* (IJEBA6 6,232,68).

SYNS: HALDI RHIZOME EXTRACT ◇ TURMERIC rhizome extract

TOXICITY DATA with REFERENCE
orl-rat TDLo:700 mg/kg (1-7D preg):REP IJEBA6 16,1077,78
orl-rat LD50:12200 mg/kg IJMRAQ 59,1289,71
ipr-mus LD50:430 mg/kg IJMRAQ 64,601,76

SAFETY PROFILE: Moderately toxic by intraperitoneal route. Mildly toxic by ingestion. Experimental reproductive effects.

COG000 CAS:8024-37-1 ***HR: 2***
CURCUMIN

SYNS: C.I. 75300 ◇ CURCUMA OIL ◇ CURCUMINE ◇ NCI-C60015 ◇ TURMERIC OIL ◇ TURMERIC OLEORESIN

TOXICITY DATA with REFERENCE
skn-rbt 500 mg/24H MLD FCTOD7 21,839,83
cyt-ham:lng 20 mg/L GMCRDC 27,95,81
ipr-mus LD50:1500 mg/kg IJMRAQ 64,601,76

CONSENSUS REPORTS: Reported in EPA TSCA Inventory.

SAFETY PROFILE: Moderately toxic by intraperitoneal route. A skin irritant. Mutation data reported. When heated to decomposition it emits acrid smoke and irritating fumes.

COG250 CAS:29929-77-9 ***HR: 3***
CURETARD
mf: $(C_{12}H_{14}N_2O)_n$

SYNS: N-NITROSO-2,2,4-TRIMETHYL-1,2-DIHYDROQUINOLINE,POLYMER ◇ 1-NITROSO-2,2,4-TRIMETHYL-1,2,-DIHYDRO-QUINOLINE (POLYMER) ◇ 1-NITROSO-2,2,4-TRIMETHYL-1(2H)QUINOLINE, POLYMER

TOXICITY DATA with REFERENCE
ipr-rat TDLo:2600 mg/kg/26W-I:NEO EJCAAH 4,233,68
scu-rat TDLo:2500 mg/kg/20W-I:CAR BJCAAI 23,408,69
ipr-rat TD:2500 mg/kg/20W-I:ETA BJCAAI 23,408,69

SAFETY PROFILE: Questionable carcinogen with experimental carcinogenic, tumorigenic, and neoplastigenic data. See also N-NITROSO COMPOUNDS and POLYMERS. When heated to decomposition it emits toxic fumes of NO_x.

COG500 CAS:62355-03-7 ***HR: 3***
CUSCOHYGRINE BIS(METHYL BENZENESULFONATE)
mf: $C_{15}H_{30}N_2O•C_{12}H_{10}O_6S_2$ mw: 568.81

SYNS: 1,3-BIS(N,N-DIMETHYL-2-PYRROLIDINIUM)PROPANE DIBENZENESULFONATE ◇ 2,2'-(2-OXO-TRIMETHYLENE)BIS(1,1-DIMETHYLPYRROLIDINIUM)DIBENZENESUFLONATE

TOXICITY DATA with REFERENCE
ipr-mus LD50:117 mg/kg PCJOAU 11,478,77
ipr-cat LD50:117 mg/kg PCJOAU 11,135,77

SAFETY PROFILE: Poison by intraperitoneal route. See also SULFONATES. When heated to decomposition it emits very toxic fumes of NO_x and SO_x.

COG750 CAS:62209-22-7 ***HR: 3***
CUSCOHYGRINE DIMETHYLIODIDE
mf: $C_{15}H_{30}N_2O•I_2$ mw: 508.27

SYNS: 1,3-BIS(N,N-DIMETHYL-2-PYRROLIDINIUM)PROPANONE DIIODIDE ◇ CUSCOHYGRIN DIMETHIODIDE ◇ 2,2'-(2-OXOTRIMETHYLENE)BIS(1,1-DIMETHYLPYRROLIDINIUM)DIIODIDE

TOXICITY DATA with REFERENCE
ipr-mus LD50:111 mg/kg PCJOAU 11,478,77
ipr-cat LD50:111 mg/kg PCJOAU 11,135,77

SAFETY PROFILE: Poison by intraperitoneal route. See also IODIDES. When heated to decomposition it emits very toxic fumes of NO_x and I^-.

COH000 ***HR: 3***
CUTTING OILS

SAFETY PROFILE: Often carcinogenic. The cause of "cutting oil" dermatitis is generally due to an insoluble oil. However it can occasionally be caused by a soluble oil. Many have looked for a causative factor other than the oil itself. Bacteria have frequently been blamed, although insoluble oils are usually sterile while the soluble oils may contain bacteria. The metal slivers which occur in these oils after use have also been blamed as well as the sulfur, chlorine, and inhibitors which they contain. The oil itself can plug the pores and form boils. Combustible when exposed to heat or flame. See also MINERAL OILS.

COH250 CAS:140-87-4 ***HR: 3***
CYANACETIC ACID HYDRAZIDE
mf: $C_3H_5N_3O$ mw: 99.11

PROP: Mp: 115°, a solid.

SYNS: AB-42 ◇ ARMAZAL ◇ CIANAZIL ◇ CYACETACID ◇ CYACETACIDE ◇ CYACETAZID ◇ CYACETAZIDE ◇ CYANACETHYDRAZIDE ◇ CYANACETIC ACID HYDRAZIDE ◇ CYANACETOHYDRAZIDE ◇ CYANACETYLHYDRAZIDE ◇ CYANAZIDE ◇ CYANIZIDE ◇ CYANOACETHYDRAZIDE ◇ CYANOACETIC ACID HYDRAZIDE ◇ CYANOACETOHYDRAZIDE ◇ α-CYANOACETOHYDRAZIDE ◇ CYANOACETYLHYDRAZIDE ◇ CYANOETHYDRAZIDE ◇ CYAZID ◇ CYAZIDE ◇ DICTYCIDE ◇ DICTYZIDE ◇ HELMOX ◇ HIDACIAN ◇ HIDACIANN ◇ KYANACETHYDRAZID ◇ LEANDIN ◇ MACKREAZID ◇ MALONITRILE HYDRAZIDE ◇ MALONONITRILE HYDRAZIDE ◇ NEOHYDRAZID ◇ REACID ◇ REAZID ◇ REAZIDE ◇ TSIAZID ◇ USAF KF-18

TOXICITY DATA with REFERENCE
orl-mus LD50:250 mg/kg CLDND* AD277-689
scu-mus LD50:228 mg/kg ABMGAJ 21,635,68

CONSENSUS REPORTS: Cyanide and its compounds are on the Community Right To Know List.

SAFETY PROFILE: Poison by ingestion and intraperitoneal route. When heated to decomposition it emits toxic fumes of NO_x and CN^-. See also NITRILES.

COH500 CAS:420-04-2 ***HR: 3***
CYANAMIDE
mf: CH_2N_2 mw: 42.05

PROP: Deliquescent crystals. Mp: 45°, bp: 260°, flash p: 285°F, d: 1.282, vap d: 1.45.

SYNS: AMIDOCYANOGEN ◇ CARBAMONITRILE ◇ CARBIMIDE ◇ CYANOAMINE ◇ CYANOGENAMIDE ◇ CYANOGEN NITRIDE ◇ HYDROGEN CYANAMIDE ◇ USAF EK-1995

TOXICITY DATA with REFERENCE
orl-rat TDLo:2600 mg/kg (male 70D pre):REP PHTXA6 61,20,87
orl-rat LD50:125 mg/kg MEIEDD 10,383,83
ihl-rat LCLo:86 mg/m³/4H 85GMAT -,40,82
skn-rat LD50:84 mg/kg 85GMAT -,40,82
ipr-rat LDLo:200 mg/kg PSEBAA 54,254,43
ipr-mus LD50:200 mg/kg NTIS** AD277-689
skn-rbt LD50:590 mg/kg 34ZIAG -,190,69

CONSENSUS REPORTS: Reported in EPA TSCA Inventory. Cyanide and its compounds are on the Community Right-To-Know List.

OSHA PEL: TWA 2 mg/m³
ACGIH TLV: TWA 2 mg/m³

SAFETY PROFILE: Poison by ingestion, inhalation, and intraperitoneal route. Moderately toxic by skin contact. Experimental reproductive effects. Combustible when exposed to heat or flame. To fight fire, use CO_2, dry chemical. Thermally unstable. Contact with moisture (water), acids, or alkalies may cause a violent reaction above 40°. Concentrated aqueous solutions may undergo explosive polymerization. Mixture with 1,2-phenylenediamine salts may cause explosive polymerization. When heated to decomposition or on contact with acid or acid fumes, it emits toxic fumes of CN^- and NO_x. See also CYANIDE and AMIDES.

COH750 ***HR: 3***
CYANATES

SAFETY PROFILE: Variable. See individual entry. When heated to decomposition, or on contact with acid or acid fumes, they emit toxic fumes of CN^-.

COI000 CAS:4027-17-2 ***HR: 3***
CYANATOTRIBUTYLSTANNANE
mf: $C_{13}H_{27}NOSn$ mw: 332.10

SYN: TRIBUTYLTIN CYANATE

TOXICITY DATA with REFERENCE
ivn-mus LD50:6300 μg/kg CSLNX* NX#02224

CONSENSUS REPORTS: Cyanide and its compounds are on the Community Right-To-Know List.

OSHA PEL: TWA 0.1 mg(Sn)/m³ (skin)
ACGIH TLV: TWA 0.1 mg(Sn)/m³ (skin) (Proposed: TWA 0.1 mg(Sn)/m³; STEL 0.2 mg(Sn)/m³ (skin))
NIOSH REL: (Organotin Compounds) TWA 0.1 mg(Sn)/m³

SAFETY PROFILE: Poison by intravenous route. Tributyl tin compounds are very toxic to marine life. See also CYANATES and TIN COMPOUNDS. When heated to decomposition it emits toxic fumes of NO_x and CN^-.

COI125 ***HR: 3***
CYANEA CAPILLATA TOXIN

SYN: TOXIN, JELLYFISH, SCYPHOZOAN, CYANEA CAPILLATA

TOXICITY DATA with REFERENCE
ivn-rat LD50:800 μg/kg TOXIA6 15,3,77
ipr-mus LD50:300 μg/kg TOXIA6 15,3,77
ivn-mus LD50:300 μg/kg TOXIA6 15,3,77
ivn-rbt LD50:300 μg/kg TOXIA6 15,3,77
ivn-ckn LD50:2 mg/kg TOXIA6 15,3,77

SAFETY PROFILE: A deadly poison by intravenous and intraperitoneal routes.

COI250 CAS:917-61-3 ***HR: 3***
CYANIC ACID, SODIUM SALT
mf: CNO•Na mw: 65.01

SYNS: CYANSAN ◇ SAN-CYAN ◇ SODIUM ISOCYANATE ◇ WEECON ◇ ZASSOL

TOXICITY DATA with REFERENCE
orl-hmn TDLo:5400 mg/kg/24W:EYE,MET AROPAW 94,927,76
orl-rat LD50:1500 mg/kg JPETAB 185,653,73

ims-rat LD50:310 mg/kg BJPCAL 1,186,46
orl-mus LDLo:4 mg/kg APFRAD 19,740,61
ipr-mus LD50:260 mg/kg JPETAB 185,653,73

CONSENSUS REPORTS: Cyanide and its compounds are on the Community Right-To-Know List. Reported in EPA TSCA Inventory.

SAFETY PROFILE: Poison by ingestion, intraperitoneal, and intramuscular routes. Human systemic effects by ingestion: weight loss, changes in the visual field, and other eye effects. See also CYANATES. When heated to decomposition it emits very toxic fumes of CN^- and Na_2O.

COI500 CAS:57-12-5 ***HR: 3***
CYANIDE
mf: CN^- mw: 26.02

SYNS: CARBON NITRIDE ION (CN^{1-}) ◇ CYANIDE ANION ◇ CYANURE (FRENCH) ◇ ISOCYANIDE ◇ RCRA WASTE NUMBER P030

TOXICITY DATA with REFERENCE
ipr-mus LD50:3 mg/kg NATUAS 228,1315,70

CONSENSUS REPORTS: Cyanide and its compounds are on the Community Right-To-Know List.

OSHA PEL: TWA 5 mg(CN)/m^3
ACGIH TLV: TWA 5 mg/m^3 (skin)
DFG MAK: 5 mg/m^3
NIOSH REL: (Cyanide) TWA CL 5 mg/m^3/10M

SAFETY PROFILE: Very poisonous by most routes. Cyanide directly stimulates the chemoreceptors of the carotid and aortic bodies with a resultant hyperpnea (increase in the depth and rate of respiration). Cardiac irregularities are often noted, but the heart invariably outlasts the respirations. Death is due to respiratory arrest of central origin. It can occur within seconds or minutes of the inhalation of high concentrations of HCN gas. Because of slower absorption, death may be more delayed after the ingestion of cyanide salts, but the critical events still occur within the first hour. Two other sources of cyanide have been responsible for human poisoning: the naturally occurring amygdalin and the drug nitroprusside.

Amygdalin is a cyanogenic glycoside found in apricot, peach, and similar fruit pits and in sweet almonds (Sayre and Kaymakcalan, 1941). It is a chemical combination of glucose, benzaldehyde, and cyanide from which the latter can be released by the action of β-glucosidase or emulsion. Although these enzymes are not found in mammalian tissues, the human intestinal microflora appears to possess these or similar enzymes capable of effecting cyanide release resulting in human poisoning. For this reason, amygdalin may be as much as 40 times more toxic by the oral route as compared with intravenous injection. Amygdalin is the major ingredient of Laetrile, and this alleged anticancer drug has also been responsible for human cyanide poisoning.

An ethical drug that may also cause cyanide poisoning in overdose is the potent vascular smooth muscle relaxant sodium nitroprusside. Although nitroprusside is related chemically to ferricyanide, unlike the latter it penetrates into erythrocytes and reacts with hemoglobin to release its cyanide (Smith and Kruszyna, 1974). Fortunately, the therapeutic margin for nitroprusside appears to be quite large.

Cyanide is commonly found in certain rat and pest poisons, silver and metal polishes, photographic solutions, and fumigating products. Compounds such as potassium cyanide can also be readily purchased from chemical stores. Cyanide is readily absorbed from all routes, including the skin, mucous membranes, and by inhalation, although alkali salts of cyanide are toxic only when ingested. Death may occur with ingestion of even small amounts of sodium or potassium cyanide and can occur within minutes or hours depending on route of exposure. Inhalation of toxic fumes represents a potentially rapidly fatal type of exposure. A blood cyanide level of greater than 0.2 μg/mL is considered toxic. Lethal cases have usually had levels above 1 μg/mL.

Clinically, cyanide poisoning is reported to produce a bitter, almond odor on the breath of the patient; however, only a small proportion of the population is genetically able to discern this characteristic odor. Typically, cyanide has a bitter, burning taste, and following poisoning, symptoms of salivation, nausea without vomiting, anxiety, confusion, vertigo, giddiness, lower jaw stiffness, convulsions, opisthotonos, paralysis, coma, cardiac arrhythmias, and transient respiratory stimulation followed by respiratory failure may occur. Bradycardia is a common finding, but in most cases heartbeat usually outlasts respiration (Wexler et al., 1947). A prolonged expiratory phase is considered to be characteristic of cyanide poisoning. (Casarett and Doull's, "Toxicology, The Basic Science of Poisons" 2nd ed. Doull, Klaassen and Amdur (eds). Macmillan Pub. Co. Inc. New York, NY).

The volatile cyanides resemble HCN physiologically, inhibiting tissue oxidation and causing death through asphyxia. Cyanogen is probably as toxic as HCN; the nitriles are generally considered somewhat less toxic, probably because of their lower volatility. The non-volatile cyanide salts appear to be relatively non-toxic systemically, so long as they are not ingested and care is taken to prevent the formation of HCN. Workers, such as electroplaters and picklers who are daily exposed to cyanide solutions may develop a "cyanide" rash, characterized by itching, and by macular, papular, and vesicular eruptions. Frequently there is secondary infection. Exposure to small amounts of cyanide compounds over long periods of time is reported to cause loss of appetite, headache, weakness, nausea, dizziness, and symptoms of irritation of the upper respiratory tract and eyes. See also specific cyanide compounds.

Flammable by chemical reaction with heat, moisture, acid. Many cyanides evolve HCN rather easily. This is a flammable gas and is highly toxic. Carbon dioxide from the air is sufficiently acidic to liberate HCN from cyanide solutions. Reaction with hypochlorite solutions may be violent at pH 10.0 - 10.3. Explodes if melted with nitrites or chlorates at about 450°. Violent reaction with F_2; Mg; nitrates; HNO_3; nitrites. Metal cyanides are easily oxidized and may be thermally unstable. N-cyano derivatives may be reactive or unstable. Many organic nitriles can be very reactive under the right conditions. When heated to decomposition or on contact with acid, acid fumes, water or steam, it emits toxic and flammable vapors of CN^-. See also HYDROCYANIC ACID.

COI750 CAS:528-58-5 ***HR: 3***
CYANIDOL
mf: $C_{15}H_{11}O_6$ mw: 287.26

SYNS: 2-(3,4-DIHYDROXYPHENYL)-3,5,7-TRIHYDROXY-BENZOPYRYLIUM ACID ANION ◇ 3,3′,4′,5,7-PENTAHYDROXYFLAVYLIUM ACID ANION

TOXICITY DATA with REFERENCE
ipr-rat LD50:2350 mg/kg CHTPBA 2,33,67
ivn-rat LD50:240 mg/kg CHTPBA 2,33,67
ipr-mus LD50:4110 mg/kg CHTPBA 2,33,67
ivn-mus LD50:840 mg/kg CHTPBA 2,33,67

SAFETY PROFILE: Poison by intravenous route. Moderately toxic by intraperitoneal route. When heated to decomposition it emits acrid smoke and irritating fumes.

COJ250 CAS:107-91-5 ***HR: 3***
2-CYANOACETAMIDE
mf: $C_3H_4N_2O$ mw: 84.09

PROP: White powder. Mp: 119°, bp: decomp.

SYNS: CYANACETAMIDE ◇ CYANOACETAMIDE ◇ CYANOIMINOACETIC ACID ◇ MALONAMIDE NITRILE ◇ MALONAMONITRILE ◇ NITRILOMALONAMIDE ◇ USAF KF-14

TOXICITY DATA with REFERENCE
orl-rat LD50:7230 mg/kg 28ZPAK -,162,72
orl-mus LD50:1680 mg/kg KHZDAN 9,50,66
ipr-mus LD50:200 mg/kg NTIS** AD691-689

CONSENSUS REPORTS: Cyanide and its compounds are on the Community Right-To-Know List. Reported in EPA TSCA Inventory.

SAFETY PROFILE: Poison by intraperitoneal route. Moderately toxic by ingestion. See also NITRILES. When heated to decomposition it emits toxic fumes of NO_x and CN^-.

COJ500 CAS:372-09-8 ***HR: 3***
CYANOACETIC ACID
mf: $C_3H_3NO_2$ mw: 85.07

PROP: A solid. Mp: 66°, bp: 108° @ 15 mm.

SYNS: ACIDE CYANACETIQUE (FRENCH) ◇ CAA ◇ CYANESSIGSAEURE (GERMAN) ◇ MALONIC MONONITRILE ◇ MONOCYANOACETIC ACID ◇ USAF KF-17

TOXICITY DATA with REFERENCE
orl-rat LD50:1500 mg/kg LONZA# 12JAN81
ipr-mus LD50:200 mg/kg NTIS** AD691-490
scu-rbt LDLo:2000 mg/kg AIPTAK 5,161,1899
scu-frg LDLo:2000 mg/kg AIPTAK 5,161,1899

CONSENSUS REPORTS: Reported in EPA TSCA Inventory. Cyanide and its compounds are on the Community Right-To-Know List.

SAFETY PROFILE: Poison by intraperitoneal route. Moderately toxic by ingestion and subcutaneous routes. See also NITRILES. Mixture with furfuryl alcohol explodes when heated. When heated to decomposition it emits toxic fumes of NO_x and CN^-.

COJ625 CAS:16130-58-8 ***HR: 3***
CYANOACETYL CHLORIDE
mf: C_3H_2ClNO mw: 103.51

CONSENSUS REPORTS: Cyanide compounds are on the Community Right-To-Know List.

SAFETY PROFILE: A poison. A storage hazard. It may explode at room temperature. Upon decomposition it emits toxic fumes of Cl^-, NO_x, and CN^-. See also CYANIDE and CHLORIDES.

COJ750 CAS:6629-04-5 ***HR: 3***
N-CYANOACETYL ETHYL CARBAMATE
mf: $C_6H_8N_2O_3$ mw: 156.16

TOXICITY DATA with REFERENCE
ipr-mus TDLo:2400 mg/kg/4W-I:NEO CNREA8 29,2184,69

CONSENSUS REPORTS: Cyanide and its compounds are on the Community Right-To-Know List.

SAFETY PROFILE: Questionable carcinogen with experimental neoplastigenic data. See also CARBAMATES and CYANIDES. When heated to decomposition it emits toxic fumes of NO_x.

COK125 CAS:873-74-5 ***HR: 3***
4-CYANOANILINE
mf: $C_7H_6N_2$ mw: 118.15

SYNS: p-AMINOBENZONITRILE (8CI) ◇ 4-AMINOBENZONITRILE (9CI) ◇ p-CYANOANILINE

TOXICITY DATA with REFERENCE
orl-mus LD50:650 mg/kg APFRAD 41,391,83
ipr-mus LD50:155 mg/kg JMCMAR 17,900,74
orl-qal LD50:23700 μg/kg AECTCV 12,355,83
orl-bwd LD50:23700 μg/kg AECTCV 12,355,83

CONSENSUS REPORTS: EPA Genetic Toxicology Program. Reported in EPA TSCA Inventory. Cyanide and its compounds are on the Community Right-To-Know List.

SAFETY PROFILE: Poison by ingestion and intraperitoneal routes. When heated to decomposition it emits toxic fumes of NO_x and CN^-. See also CYANIDES and ANILINE DYES.

COK250 CAS:105-07-7 ***HR: 3***
p-CYANOBENZALDEHYDE
mf: C_8H_5NO mw: 131.14

SYNS: 4-CYANOBENZALDEHYDE ◇ p-CYANOBENZENE CARBOXALDEHYDE ◇ p-FORMYLBENZONITRILE ◇ 4-FORMYLBENZONITRILE ◇ TEREPHTHALALDEHYDONITRILE ◇ USAF KF-1

TOXICITY DATA with REFERENCE
ipr-mus LD50:100 mg/kg NTIS** AD277-689

CONSENSUS REPORTS: Cyanide and its compounds are on the Community Right-To-Know List. Reported in EPA TSCA Inventory.

SAFETY PROFILE: Poison by intraperitoneal route. See also NITRILES and ALDEHYDES. When heated to decomposition it emits toxic fumes of NO_x and CN^-.

COK500 CAS:7476-08-6 ***HR: 3***
10-CYANO-1,2-BENZANTHRACENE
mf: $C_{19}H_{11}N$ mw: 253.31

SYN: 7-CYANOBENZ(a) ANTHRACENE

TOXICITY DATA with REFERENCE
skn-mus TDLo:1370 mg/kg/57W-I:ETA PRLBA4 131,170,42

CONSENSUS REPORTS: Cyanide and its compounds are on the Community Right-To-Know List.

SAFETY PROFILE: Questionable carcinogen with experimental tumorigenic data. When heated to decomposition it emits toxic fumes of NO_x and CN^-. See also CYANIDE.

COK659 CAS:60633-76-3 ***HR: 3***
CYANOBORANE OLIGOMER
mf: $(CH_2BN)_n$

CONSENSUS REPORTS: Cyanide compounds are on the Community Right-To-Know List.

SAFETY PROFILE: A heat- and shock-sensitive explosive. When heated to decomposition it emits toxic fumes of NO_x and CN^-. See also CYANIDE; BORANES; and EXPLOSIVES.

COK750 ***HR: 1***
N-CYANO-2-BROMOETHYLBUTYLAMINE
mf: $C_7H_{13}BrN$ mw: 185.10

CONSENSUS REPORTS: Cyanide and its compounds are on the Community Right-To-Know List.

SAFETY PROFILE: Decomposes exothermically at 160°C. When heated to decomposition it emits toxic fumes of Br^-, NO_x and CN^-. See also CYANIDE and BROMIDES.

COL000 ***HR: 3***
N-CYANO-2-BROMOETHYLCYCLOHEXYLAMINE
mf: $C_9H_{15}BrN$ mw: 217.14

CONSENSUS REPORTS: Cyanide and its compounds are on the Community Right-To-Know List.

SAFETY PROFILE: A poison. May explode when heated to 160°C. When heated to decomposition it emits toxic fumes of Br^-, NO_x, and CN^-. See also CYANIDE, AMINES and BROMIDES.

COL125 CAS:82423-05-0 ***HR: 3***
CYANOCYCLINE A
mf: $C_{22}H_{26}N_4O_5$ mw: 426.52

TOXICITY DATA with REFERENCE
dni-esc 2 mg/L JANTAJ 36,1228,83
oms-esc 2 mg/L JANTAJ 26,1228,83
dni-mus:ast 1 mg/L JANTAJ 36,1228,83
oms-mus:ast 100 μg/L JANTAJ 36,1228,83
ipr-mus LD50:10 mg/kg JANTAJ 35,771,82

CONSENSUS REPORTS: Cyanide and its compounds are on the Community Right-To-Know List.

SAFETY PROFILE: Poison by intraperitoneal route. Mutation data reported. When heated to decomposition it emits toxic fumes of NO_x and CN^-. See also CYANIDE.

COL250 CAS:1769-99-9 ***HR: 3***
5-CYANO-10,11-DIHYDRO-5-(3-DIMETHYLAMINOPROPYL)-5H-DIBENZO (a,d)CYCLOHEPTENE HYDROCHLORIDE
mf: $C_{21}H_{24}N_2$•ClH mw: 340.93

SYNS: 10,11-DIHYDRO-5-CYANO-N,N-DIMETHYL-5H-DIBENZO (a,d)CYCLOHEPTENE-5-PROPYLAMINE HCl ◇ UCB 4208

TOXICITY DATA with REFERENCE
orl-rat LD50:1250 mg/kg 27ZQAG -,91,72

ivn-rat LD50:28 mg/kg 27ZQAG -,91,72
ims-rat LD50:620 mg/kg 27ZQAG -,91,72
ipr-mus LD50:120 mg/kg JMCMAR 6,251,63
ivn-mus LD50:56 mg/kg CSLNX* NX#01197

CONSENSUS REPORTS: Cyanide and its compounds are on the Community Right-To-Know List.

SAFETY PROFILE: Poison by intravenous and intraperitoneal routes. Moderately toxic by ingestion and intramuscular routes. When heated to decomposition it emits toxic fumes of NO_x, CN^-, and HCl. See also CYANIDE.

COL750 CAS:683-45-4 ***HR: 3***
CYANODIMETHYLARSINE
mf: C_3H_6AsN mw: 131.02

$N{\equiv}CAs(CH_3)_2$

SYN: DIMETHYLCYANOARSINE

TOXICITY DATA with REFERENCE
orl-rat LD50:50 mg/kg NCNSA6 5,13,53
ihl-mus LCLo:400 mg/m^3/15M ZGEMAZ 13,523,21
ihl-dog LCLo:900 mg/m^3/8M ZGEMAZ 13,523,21
ihl-cat LCLo:100 mg/m^3/20M ZGEMAZ 13,523,21

CONSENSUS REPORTS: Cyanide and its compounds, as well as arsenic and its compounds, are on the Community Right-To-Know List.

OSHA PEL: TWA 0.5 mg(As)/m^3

SAFETY PROFILE: Poison by ingestion and inhalation. Ignites spontaneously in air. See also ARSENIC COMPOUNDS and CYANIDES. When heated to decomposition it emits very toxic fumes of As, CN^-, and NO_x.

COM000 CAS:63018-68-8 ***HR: 3***
5-CYANO-9,10-DIMETHYL-1,2-BENZANTHRACENE
mf: $C_{21}H_{15}N$ mw: 281.37

SYN: 7,12-DIMETHYLBENZ(a)ANTHRACENE-8-CARBONITRILE

TOXICITY DATA with REFERENCE
scu-mus TDLo:80 mg/kg:ETA CNREA8 6,454,46

CONSENSUS REPORTS: Cyanide and its compounds are on the Community Right-To-Know List.

SAFETY PROFILE: Questionable carcinogen with experimental tumorigenic data. When heated to decomposition it emits toxic fumes of NO_x and CN^-. See also CYANIDE.

COM075 ***HR: 3***
α-CYANO-2,6-DIMETHYLERGOLINE-8-PROPIONAMIDE
mf: $C_{20}H_{24}N_4O$ mw: 336.48

SYN: ERGOLINE-8-PROPIONAMIDE, α-CYANO-2,6-DIMETHYL-

TOXICITY DATA with REFERENCE
orl-rat TDLo:4 mg/kg (female 5D post):REP ARZNAD 33,1094,83
orl-mus LD50:200 mg/kg ARZNAD 33,1094,83

SAFETY PROFILE: Poison by ingestion. Experimental reproductive effects. When heated to decomposition it emits toxic fumes of NO_x.

COM125 CAS:7790-03-6 ***HR: 2***
2-(2-CYANOETHOXY)ETHYL ESTER ACRYLIC ACID
mf: $C_8H_{11}NO_3$ mw: 169.20

TOXICITY DATA with REFERENCE
skn-rbt 10 mg/24H open MLD AIHAAP 23,95,62
orl-rat LD50:1120 mg/kg AIHAAP 23,95,62
skn-rbt LD50:750 mg/kg AIHAAP 23,95,62

CONSENSUS REPORTS: Cyanide and its compounds are on the Community Right-To-Know List.

SAFETY PROFILE: Moderately toxic by ingestion and skin contact. A skin irritant. When heated to decomposition it emits toxic fumes of NO_x and CN^-. See also CYANIDE and ESTERS.

COM250 CAS:10141-15-8 ***HR: 2***
4-CYANOETHOXY-2-METHYL-2-PENTANOL
mf: $C_9H_{17}NO_2$ mw: 171.27

TOXICITY DATA with REFERENCE
skn-rbt 10 mg/24H open MLD AIHAAP 23,95,62
orl-rat LDLo:3200 mg/kg AIHAAP 23,95,62
skn-rbt LDLo:1500 mg/kg AIHAAP 23,95,62

CONSENSUS REPORTS: Cyanide and its compounds are on the Community Right-To-Know List.

SAFETY PROFILE: Moderately toxic by ingestion and skin contact. A skin irritant. See also NITRILES. When heated to decomposition it emits toxic fumes of NO_x and CN^-.

COM500 CAS:65216-94-6 ***HR: 3***
N-(CYANOETHYL)DIETHYLENETRIAMINE

TOXICITY DATA with REFERENCE
skn-rbt 500 mg MLD SCCUR* -,7,61
orl-rat LD50:4550 mg/kg AMIHAB 17,129,58
ipr-rat LD50:261 mg/kg AMIHAB 17,129,58
ipr-mus LD50:222 mg/kg AMIHAB 17,129,58

CONSENSUS REPORTS: Cyanide and its compounds are on the Community Right-To-Know List.

SAFETY PROFILE: Poison by intraperitoneal route. Mildly toxic by ingestion. A skin irritant. When heated to decomposition it emits toxic fumes of NO_x and CN^-. See also CYANIDE.

COM750 CAS:1001-58-7 ***HR: 3***
β-CYANOETHYLMERCAPTAN
mf: C_3H_5NS mw: 87.15

TOXICITY DATA with REFERENCE
ipr-mus LD50:100 mg/kg NTIS** AD691-490

CONSENSUS REPORTS: Reported in EPA TSCA Inventory. Cyanide and its compounds are on the Community Right-To-Know List.

SAFETY PROFILE: Poison by intraperitoneal route. When heated to decomposition it emits very toxic fumes of SO_x and NO_x.

CON000 CAS:1071-22-3 ***HR: 2***
2-CYANOETHYLTRICHLOROSILANE
mf: $C_3H_4Cl_3NSi$ mw: 188.52
SYN: β-CYANOETHYLTRICHLOROSILANE

TOXICITY DATA with REFERENCE
skn-rbt 100 μg/24H open AIHAAP 23,95,62
orl-rat LD50:2000 mg/kg AIHAAP 23,95,62

CONSENSUS REPORTS: Cyanide and its compounds are on the Community Right-To-Know List. Reported in EPA TSCA Inventory.

SAFETY PROFILE: Moderately toxic by ingestion. A skin irritant. When heated to decomposition it emits very toxic fumes of NO_x and Cl^-. See also CYANIDE and SILANE.

CON250 CAS:919-31-3 ***HR: 1***
(2-CYANOETHYL)TRIETHOXYSILANE
mf: $C_9H_{19}NO_3Si$ mw: 217.38
SYN: β-CYANOETHYLTRIETHOXYSILANE

TOXICITY DATA with REFERENCE
skn-rbt 10 mg/24H open MLD AIHAAP 23,95,62
orl-rat LD50:5630 mg/kg AIHAAP 23,95,62
skn-rbt LD50:5950 mg/kg AIHAAP 23,95,62

CONSENSUS REPORTS: Reported in EPA TSCA Inventory. Cyanide and its compounds are on the Community Right-To-Know List.

SAFETY PROFILE: Mildly toxic by ingestion and skin contact. A skin irritant. See also SILANES and CYANIDE. When heated to decomposition it emits toxic fumes of NO_x and CN^-.

CON300 CAS:13067-93-1 ***HR: 3***
CYANOFENPHOS
mf: $C_{15}H_{14}NO_2PS$ mw: 303.33

PROP: Crystalline solid. Mp: 83°, n (25/D) 1.5839. Solubility in water at 30°: 6 ppm. Moderately sol in ketones and aromatic solvents. Vap. press at 25°: 0.0000132 mm Hg.

SYNS: B-10094 ◇ CP 19699 ◇ CYANOPHENPHOS ◇ o-p-CYANOPHENYL-o-ETHYL PHENYLPHOSPHONOTHIOATE ◇ o-(4-CYANOPHENYL)-o-ETHYL PHENYLPHOSPHONOTHIOATE ◇ CYP ◇ ENT 25,832 ◇ ENT 25,832-a ◇ o-ETHYL-o-4-CYANOPHENYL PHENYLPHOSPHOROTHIOATE ◇ o-ETHYLPHENYLPHOSPHONOTHIOATE-o-ESTER with p-HYDROXYBENZONITRILE ◇ EXPERIMENTAL INSECTICIDE S-4087 ◇ MONSANTO CP-19699 ◇ PHENYLPHOSPHONOTHIOIC ACID-o-ETHYL ESTER-o-ESTER with p-HYDROXYBENZONITRILE ◇ PHOSPHOROTHIOIC ACID-o-(4-CYANOPHENYL)-o-ETHYL PHENYL ESTER ◇ S 4087 ◇ STAUFFER B-10094 ◇ SURECIDE ◇ UPJOHN U-32714

TOXICITY DATA with REFERENCE
orl-rat LD50:79 mg/kg ARSIM* 20,15,66
orl-mus LD50:43700 μg/kg FMCHA2 -,C224,83
scu-mus LD50:122 mg/kg GUCHAZ 6,257,73
orl-gpg LDLo:50 mg/kg JEENAI 61,1261,68
scu-gpg LDLo:100 mg/kg JEENAI 61,1261,68
orl-ckn LD50:20 mg/kg TXAPA9 11,49,67

CONSENSUS REPORTS: Cyanide and its compounds are on the Community Right-To-Know List.

SAFETY PROFILE: Poison by ingestion and subcutaneous routes. Toxic to fish and bees. When heated to decomposition it emits toxic fumes of CN^-, PO_x, and SO_x. See also CYANIDE.

CON500 CAS:353-18-4 ***HR: 3***
2-CYANO-2'-FLUORODIETHYL ETHER
mf: C_5H_8FNO mw: 117.14
SYNS: 2-CYANOETHYL-2'-FLUOROETHYLETHER ◇ 2-FLUORO-2'-CYANODIETHYL ETHER

TOXICITY DATA with REFERENCE
ipr-mus LD50:10 mg/kg 30ZFAF -,140,59
scu-mus LD50:10 mg/kg JCSOA9 -,2774,49

CONSENSUS REPORTS: Cyanide and its compounds are on the Community Right To Know List.

SAFETY PROFILE: Poison by intraperitoneal and subcutaneous routes. See also CYANIDE and ETHERS. When heated to decomposition it emits toxic F^-, NO_x, and CN^-.

CON825 CAS:4474-17-3 ***HR: 3***
CYANOFORMYL CHLORIDE
mf: C_2ClNO mw: 89.48

CONSENSUS REPORTS: Cyanide compounds are on the Community Right-To-Know List.

SAFETY PROFILE: A poison. Violent reaction with water. When heated to decomposition it emits toxic fumes of Cl^-, CN^-, and NO_x. See also CYANIDE and CHLORIDES.

COO000 CAS:460-19-5 ***HR: 3***
CYANOGEN
DOT: UN 1026
mf: C_2N_2 mw: 52.04

$$N{\equiv}CC{\equiv}N$$

PROP: Colorless gas, pungent odor. Mp: −34.4°, bp: −21.0°, d: 0.866 @ 17°/4°, lel: 6.6%, uel: 32%, vap d: 1.8.

SYNS: CARBON NITRIDE ◇ CYANOGENE (FRENCH) ◇ CYANOGEN GAS (DOT) ◇ DICYANOGEN ◇ ETHANEDINITRILE ◇ NITRILOACETONITRILE ◇ OXALIC ACID DINITRILE ◇ OXALONITRILE ◇ OXALYL CYANIDE ◇ PRUSSITE ◇ RCRA WASTE NUMBER P031

TOXICITY DATA with REFERENCE
eye-hmn 16 ppm/6M AIHAAP 21,121,60
ihl-hmn TCLo:16 ppm:EYE,NOSE AIHAAP 21,121,60
ihl-rat LC50:350 ppm/1H AIHAAP 21,121,60
unk-dog LDLo:15 mg/kg AIPTAK 3,77,1897
scu-rbt LDLo:13 mg/kg AIPTAK 3,77,1897
scu-pgn LDLo:9 mg/kg AIPTAK 3,77,1897
scu-frg LDLo:43 mg/kg AIPTAK 3,77,1897

CONSENSUS REPORTS: Reported in EPA TSCA Inventory. Cyanide and its compounds are on the Community Right-To-Know List.

OSHA PEL: TWA 10 ppm
ACGIH TLV: TWA 10 ppm
DFG MAK: 10 ppm (22 mg/m^3)
DOT Classification: Poison A; Label: Flammable Gas and Poison Gas.

SAFETY PROFILE: A poison by subcutaneous and possibly other routes. Moderately toxic by inhalation. Human systemic effects by inhalation: damage to the olfactory nerves, and irritation of the conjunctiva. A systemic irritant by inhalation and subcutaneous routes. A human eye irritant. Very dangerous fire hazard when exposed to heat, flames (sparks), or oxidizers. To fight fire, stop flow of gas. Potentially explosive reaction with powerful oxidants (e.g., dichlorine oxide; fluorine; oxygen; ozone). When heated to decomposition or on contact with acid, acid fumes, water or steam, will react to produce highly toxic fumes of NO_x and CN^-. See also other cyanogen entries and CYANIDE.

COO250 CAS:764-05-6 ***HR: 3***
CYANOGEN AZIDE
mf: CN_4 mw: 68.04

CONSENSUS REPORTS: Cyanide and its compounds are on the Community Right-To-Know List.

SAFETY PROFILE: A poison. Explodes violently with mild mechanical, thermal, or electrical shock. May spontaneously explode in storage even when cooled to −20 °C. Reacts with 10% sodium hydroxide to form the violently explosive 5-azidotetrazolide. See also other cyanogen entries; CYANIDE; and AZIDES.

COO500 CAS:506-68-3 ***HR: 3***
CYANOGEN BROMIDE
DOT: UN 1889
mf: CBrN mw: 105.93

PROP: Colorless needles. Mp: 52°, bp: 61.6°, d: 2.015 @ 20°/4°, vap press: 100 mm @ 22.6°.

SYNS: BROMINE CYANIDE ◇ BROMOCYAN ◇ BROMOCYANOGEN ◇ BROMURE de CYANOGEN (FRENCH) ◇ CAMPILIT ◇ CYANOBROMIDE ◇ CYANOGEN MONOBROMIDE ◇ RCRA WASTE NUMBER U246 ◇ TL 822

TOXICITY DATA with REFERENCE
ihl-hmn LCLo:92 ppm/10M NTIS** PB214-270
ihl-mus LCLo:500 mg/m^3/10M NDRC** No.9-4-1-9,43

CONSENSUS REPORTS: Cyanide and its compounds are on the Community Right-To-Know List. EPA Extremely Hazardous Substances List. Reported in EPA TSCA Inventory.

DOT Classification: Poison B; Label: Poison, Corrosive

SAFETY PROFILE: A human and experimental poison by inhalation. Corrosive. When heated to decomposition it emits very toxic fumes of CN^- and Br^-. Possibly unstable. See also other cyanogen entries; CYANIDE; and BROMIDES.

COO750 CAS:506-77-4 ***HR: 3***
CYANOGEN CHLORIDE
DOT: UN 1589
mf: CClN mw: 61.47

PROP: Colorless liquid or gas; lachrymatory and irritating odor. Mp: −6.5°, bp: 13.1°, d: 1.218 @ 4°/4°, vap press: 1010 mm @ 20°, vap d: 1.98.

SYNS: CHLORCYAN ◇ CHLORINE CYANIDE ◇ CHLOROCYAN ◇ CHLOROCYANIDE ◇ CHLOROCYANOGEN ◇ CHLORURE de CYANOGENE (FRENCH) ◇ CYANOGEN CHLORIDE, containing less than 0.9% water (DOT) ◇ CYANOGEN CHLORIDE, inhibited (DOT) ◇ RCRA WASTE NUMBER P033

TOXICITY DATA with REFERENCE
eye-hmn 100 mg/m^3/2M SEV BJEPA5 33,241,46
ihl-hmn TCLo:10 mg/m^3:EYE WHOTAC -,31,70
ihl-man TCLo:2 g/m^3:SKN NTIS** PB158-508
ihl-rat LC50:5400 mg/m^3/3M NTIS** PB158-508
ihl-mus LC50:3 g/m^3/30S NTIS** PB158-508

scu-mus LDLo:39 mg/kg 27ZWAY 1.1,779,-
ihl-dog LC50:3800 mg/m^3/1M NTIS** PB158-508
scu-dog LDLo:5 mg/kg HBAMAK 4,1341,35
ihl-mky LC50:4400 mg/m^3/1M NTIS** PB158-508
orl-cat LD50:6 mg/kg NTIS** PB158-508
ihl-cat LC50:6 g/m^3/1M NTIS** PB158-508
ihl-rbt LC50:6 g/m^3/7M NTIS** PB158-508
scu-rbt LDLo:20 mg/kg HBAMAK 4,1341,35
ihl-gpg LC50:5500 mg/m^3/2M NTIS** PB158-508
scu-pgn LDLo:8700 μg/kg HBAMAK 4,1341,35
ihl-dom LC50:3600 mg/m^3/2M NTIS** PB158-508

CONSENSUS REPORTS: Cyanide and its compounds are on the Community Right-To-Know List. Reported in EPA TSCA Inventory.

OSHA PEL: CL 0.3 ppm
ACGIH TLV: CL 0.3 ppm
DOT Classification: Poison A; Label: Flammable Gas and Poison Gas.

SAFETY PROFILE: Poison by ingestion, subcutaneous, and possibly other routes. Toxic by inhalation. Human systemic effects by inhalation: lacrimation, conjunctiva irritation, and chronic pulmonary edema or congestion. A primary irritant. A severe human eye irritant. An insecticide. Flammable when exposed to heat or flame. When heated to decomposition or on contact with water or steam, it will react to produce highly toxic and corrosive fumes of Cl^-, CN^-, and NO_x. See also other cyanogen entries, CYANIDES, and CHLORIDES.

COO825 CAS:1495-50-7 ***HR: 3***
CYANOGEN FLUORIDE
mf: CFN mw: 45.02

CONSENSUS REPORTS: Cyanide and its compounds are on the Community Right-To-Know List.

SAFETY PROFILE: A poison. Explosive polymerization is catalyzed by hydrogen fluoride. When heated to decomposition it emits toxic fumes of F^- and NO_x. See also other cyanogen entries, CYANIDE, and FLUORIDES.

COP000 CAS:506-78-5 ***HR: 3***
CYANOGEN IODIDE
mf: CIN mw: 152.92

PROP: Colorless solid. Mp: 146.5°, vap press: 1 mm @ 25.2°.

SYNS: IODINE CYANIDE ◇ JODCYAN ◇ NCI

TOXICITY DATA with REFERENCE
scu-rat LDLo:44 mg/kg 27ZWAY 1.1,779,-
scu-mus LDLo:27 mg/kg HBAMAK 4,1289,35
scu-dog LDLo:19 mg/kg HBAMAK 4,1289,35
orl-cat LDLo:18 mg/kg HBAMAK 4,1289,35
scu-cat LDLo:20 mg/kg HBAMAK 4,1289,35
scu-rbt LDLo:360 mg/kg HBAMAK 4,1289,35
scu-frg LDLo:110 mg/kg HBAMAK 4,1289,35

CONSENSUS REPORTS: Cyanide and its compounds are on the Community Right-To-Know List. EPA Extremely Hazardous Substances List. Reported in EPA TSCA Inventory.

SAFETY PROFILE: A poison by ingestion and subcutaneous route. Violent reaction with P. See other cyanogen entries; CYANIDES and IODIDES. When heated to decomposition it emits very toxic fumes of NO_x, CN^-, and I^-.

COP125 CAS:461-58-5 ***HR: 3***
CYANOGUANIDINE
mf: $C_2H_4N_4$ mw: 84.08

SYN: DICYANODIAMIDE

CONSENSUS REPORTS: Cyanide compounds are on the Community Right-To-Know List.

SAFETY PROFILE: Mixtures with ammonium nitrate, potassium chlorate, and related compounds are powerful explosives. When heated to decomposition it emits toxic fumes of CN^- and NO_x. See also CYANIDE and AMIDES.

COP400 CAS:6071-81-4 ***HR: 3***
S-1-CYANO-2-HYDROXY-3-BUTENE
mf: C_5H_7NO mw: 97.13

SYNS: S-3-HYDROXY-4-PENTENONITRILE ◇ 4-PENTENONITRILE, 3-HYDROXY-, S-

TOXICITY DATA with REFERENCE
scu-rat TDLo:150 mg/kg (female 8D post):TER FCTXAV 18,159,80
scu-rat LD50:200 mg/kg FCTXAV 18,159,80
unr-mus LD50:170 mg/kg JAFCAU 17,483,69

SAFETY PROFILE: Poison by subcutaneous route. An experimental teratogen. When heated to decomposition it emits toxic fumes of NO_x.

COP500 CAS:31065-88-0 ***HR: 3***
CYANOHYDROXYMERCURY
mf: CHHgNO mw: 243.62

TOXICITY DATA with REFERENCE
scu-mus LDLo:10 mg/kg MOLAAF 73,751,39

CONSENSUS REPORTS: Cyanide and its compounds as well as mercury and its compounds are on the Community Right-To-Know List.

OSHA PEL: (Transitional: CL 1 mg/10m^3) CL 0.1 mg (Hg)/m^3 (skin)
ACGIH TLV: TWA 0.1 mg(Hg)/m^3 (skin)
NIOSH REL: (Mercury, Inorganic) TWA 0.05 mg(Hg)/m^3

SAFETY PROFILE: Poison by subcutaneous route. See also MERCURY COMPOUNDS and CYANIDE. When heated to decomposition it emits very toxic fumes of Hg, NO_x, and CN^-.

COP600 ***HR: 3***
α-CYANO-6-ISOBUTYLERGOLINE-8-PROPIONAMIDE
mf: $C_{22}H_{28}N_4O$ mw: 364.54

SYN: ERGOLINE-8-PROPIONAMIDE, α-CYANO-6-ISOBUTYL-

TOXICITY DATA with REFERENCE
orl-rat TDLo:24 mg/kg (female 5D post):REP ARZNAD 33,1094,83
orl-mus LD50:200 mg/kg ARZNAD 33,1094,83

SAFETY PROFILE: Poison by ingestion. Experimental reproductive effects. When heated to decomposition it emits toxic fumes of NO_x.

COP750 CAS:1001-55-4 ***HR: 3***
CYANOMETHYL ACETATE
mf: $C_4H_5NO_2$ mw: 99.10

PROP: Colorless liquid. Mp: −22.5°, bp: 200°, d: 1.123 @ 15°.

SYN: GLYCOLONITRILE ACETATE

TOXICITY DATA with REFERENCE
skn-rbt 10 mg/24H open JIHTAB 30,63,48
eye-rbt 20 mg open SEV JIHTAB 30,63,48
orl-rat LD50:32 mg/kg JIHTAB 30,63,48
ihl-rat LCLo:16 ppm/4H JIHTAB 31,343,49
skn-rbt LD50:43 mg/kg JIHTAB 30,63,48

CONSENSUS REPORTS: Cyanide and its compounds are on the Community Right-To-Know List.

SAFETY PROFILE: A poison by skin contact, ingestion, and inhalation. A skin and severe eye irritant. When heated to decomposition it emits very toxic fumes of NO_x and CN^-. See also NITRILES.

COP759 ***HR: 3***
5-CYANO-5-METHYLTETRAZOLE
mf: $C_3H_3N_5$ mw: 109.09

N=NNCH$_3$N=CCN (ring: N to C)

CONSENSUS REPORTS: Cyanide compounds are on the Community Right-To-Know List.

SAFETY PROFILE: Explosive reaction with aluminum hydride. When heated to decomposition it emits toxic fumes of CN^- and NO_x. See also CYANIDE.

COP765 CAS:88254-07-3 ***HR: 2***
CYANOMORPHOLINOADRIAMYCIN
mf: $C_{32}H_{34}N_2O_{12}$ mw: 638.68

SYNS: 3′-DEAMINO-3′-(3-CYANO-4-MORPHOLINYL)DOXORUBICIN ◇ MRA-CN ◇ 5,12-NAPHTHACENEDIONE, 7,8,9,10-TETRAHYDRO-10-((3-(3-CYANOMORPHOLINO)-2,3,6-TRIDEOXY-α-L-lyxo-HEXO-PYRANOSYL)OXY)-8-(HYDROXYACETYL)-1-METHOXY-6,8,11-TRIHYDR OXY-, (8s-cis)-

TOXICITY DATA with REFERENCE
mma-sat 500 ng/plate CNREA8 44,5599,84
dni-hmn:leu 2200 pmol/L CNREA8 46,4041,86
ivn-rat TDLo:10 μg/kg:CAR CBTOE2 3,17,87

SAFETY PROFILE: Questionable carcinogen with experimental carcinogenic data. Mutation data reported. When heated to decomposition it emits toxic fumes of NO_x.

COP775 CAS:1884-64-6 ***HR: 3***
CYANONITRENE
mf: CN_2 mw: 40.02

CONSENSUS REPORTS: Cyanide compounds are on the Community Right-To-Know List.

SAFETY PROFILE: An explosive. When heated to decomposition it emits toxic fumes of CN^- and NO_x. See also CYANIDE.

COQ325 CAS:68597-10-4 ***HR: 3***
2-CYANO-4-NITROBENZENEDIAZONIUM HYDROGEN SULFATE
mf: $C_7H_4N_4O_6S$ mw: 272.19

CONSENSUS REPORTS: Cyanide compounds are on the Community Right-To-Know List.

SAFETY PROFILE: Reacts violently with sulfuric acid when heated. When heated to decomposition it emits toxic fumes of NO_x, CN^-, and SO_x. See also CYANIDE, SULFATES, and NITRO COMPOUNDS of AROMATIC HYDROCARBONS.

COQ375 CAS:56092-91-2 ***HR: 3***
3-(3-CYANO-1,2,4-OXADIAZOL-5-YL)-4-CYANO FURAZAN-2-(5-) OXIDE
mf: $C_6N_6O_3$ mw: 204.10

CONSENSUS REPORTS: Cyanide compounds are on the Community Right-To-Know List.

SAFETY PROFILE: Reacts explosively with hydrazines and other nitrogenous bases (e.g., mono- or dimethyl hydrazine, piperidine, piperazine, and diethylamine).

When heated to decomposition it emits toxic fumes of CN^- and NO_x. See also CYANIDE.

COQ399 CAS:2636-26-2 ***HR: 3***
CYANOPHOS
mf: $C_9H_{10}NO_3PS$ mw: 243.23

PROP: Yellow to reddish-yellow transparent liquid. Bp: 119-120° (decomp), mp: 14-15°, n (32.5/D) 1.5404. Very sol in methanol, ethanol, acetone, chloroform. Sparingly sol in n-hexane, kerosene; sltly sol in water. Rapid decomp under alkaline conditions and upon exposure to light.

SYNS: BAY 34727 ◇ BAYER 34727 ◇ CIAFOS ◇ O-p-CYANOPHENYL O,O-DIMETHYL PHOSPHOROTHIOATE ◇ O-(4-CYANOPHENYL) O,O-DIMETHYL PHOSPHOROTHIOATE ◇ CYANOX ◇ CYAP ◇ O,O-DIMETHYL-O-(4-CYANO-PHENYL)-MONOTHIOPHOSPHAT(GERMAN) ◇ O,O-DIMETHYL-O-p-CYANOPHENYL PHOSPHOROTHIOATE ◇ O,O-DIMETHYL-O-4-CYANOPHENYL PHOSPHOROTHIOATE ◇ O,O-DIMETHYL-O-4-CYANOPHENYL THIOPHOSPHATE ◇ ENT 25,675 ◇ MAY & BAKER S-4084 ◇ PHOSPHOROTHIOIC ACID o-(4-CYANOPHENYL)-9,9-DIMETHYL ESTER ◇ S 4084 ◇ SUMITOMO S 4084 ◇ SUNITOMO S 4084

TOXICITY DATA with REFERENCE
orl-rat LD50:25 mg/kg SPEADM 78-1,26,78
skn-rat LD50:800 mg/kg 28ZEAL 5,62,76
orl-mus LD50:324 mg/kg GISAAA 48(9),76,83
ipr-mus LD50:880 mg/kg MEIEDD 10,322,83
orl-gpg LD50:324 mg/kg GISAAA 48(9),76,83
orl-ckn LD50:24 mg/kg TXAPA9 11,49,67
skn-mam LD50:2010 mg/kg GTPZAB 21(7),34,77

CONSENSUS REPORTS: Cyanide and its compounds are on the Community Right-To-Know List. On EPA Extremely Hazardous Substances List.

SAFETY PROFILE: Poison by ingestion. Moderately toxic by skin contact and intraperitoneal routes. An insecticide and cholinesterase inhibitor. See also PARATHION. When heated to decomposition it emits toxic fumes of NO_x, PO_x, CN^-, and SO_x.

COQ750 CAS:627-26-9 ***HR: 3***
1-CYANOPROPENE
mf: C_4H_5N mw: 67.09

$N{\equiv}CCH{=}CHCH_3$

PROP: Flash p: 60.8°F.

SYN: 2-BUTENENITRILE

CONSENSUS REPORTS: Cyanide and its compounds are on the Community Right-To-Know List.

SAFETY PROFILE: A poison. Very reactive. A very dangerous fire hazard when exposed to heat or flame. When heated to decomposition it emits toxic fumes of CN^-. See also CYANIDE.

COR325 CAS:1190-16-5 ***HR: 3***
3-CYANOPROPYLDICHLOROMETHYLSILANE
mf: $C_5H_9Cl_2NSi$ mw: 182.14

SYN: 4-(DICHLOROMETHYLSILYL)BUTYRONITRILE

TOXICITY DATA with REFERENCE
skn-rbt 100 μg/24H open AIHAAP 23,95,62
skn-rbt 5 mg/24H SEV 85JCAE-,1225,86
eye-rbt 750 μg/24H SEV 85JCAE-,1225,86
orl-rat LD50:2830 mg/kg AIHAAP 23,95,62
ihl-mus LCLo:770 mg/m^3/2H 85JCAE-,1225,86
skn-rbt LD50:1490 mg/kg AIHAAP 23,95,62

CONSENSUS REPORTS: Cyanide and its compounds are on the Community Right-To-Know List. Reported in EPA TSCA Inventory.

SAFETY PROFILE: Moderately toxic by inhalation, ingestion, and skin contact. A severe skin and eye irritant. When heated to decomposition it emits toxic fumes of Cl^-, NO_x and CN^-. See also NITRILES and SILANE.

COR500 CAS:1067-99-8 ***HR: 2***
(3-CYANOPROPYL)DIETHOXY(METHYL) SILANE
mf: $C_9H_{19}NO_2Si$ mw: 201.38

SYNS: BUTYRONITRILE,4-(DIETHOXYMETHYLSILYL)- ◇ DIETHOXY-3-KYANPROPYL-METHYLSILAN ◇ SILANE, (3-CYANOPROPYL)DIETHOXY(METHYL)-

TOXICITY DATA with REFERENCE
skn-rbt 10 mg/24H open MLD AIHAAP 23,95,62
orl-rat LD50:3730 mg/kg AIHAAP 23,95,62

CONSENSUS REPORTS: Reported in EPA TSCA Inventory.

SAFETY PROFILE: Moderately toxic by ingestion. A skin irritant. When heated to decomposition it emits toxic fumes of NO_x and Si.

COR750 CAS:40561-27-1 ***HR: 3***
2-CYANO-2-PROPYL NITRATE
mf: $C_4H_6N_2O_3$ mw: 130.10

$(CH_3)_2C(CN)ONO_2$

CONSENSUS REPORTS: Cyanide compounds are on the Community Right-To-Know List.

SAFETY PROFILE: An impact-sensitive explosive. When heated to decomposition it emits toxic fumes of NO_x and CN^-. See also CYANIDE and NITRATES.

COS500 CAS:60560-33-0 ***HR: 2***
2-CYANO-3-(4-PYRIDYL)-1-(1,2,3,TRIMETHYLPROPYL)GUANIDINE
mf: $C_{13}H_{19}N_5$ mw: 245.37

SYNS: P 1134 ◇ PINACIDIL ◇ PND

TOXICITY DATA with REFERENCE
orl-rat LD50:570 mg/kg JMCMAR 21,773,78
orl-mus LD50:490 mg/kg NYKZAU 86,341,85

CONSENSUS REPORTS: Cyanide and its compounds are on the Community Right-To-Know List.

SAFETY PROFILE: Moderately toxic by ingestion. When heated to decomposition it emits very toxic fumes of NO_x and CN^-. See also CYANIDE.

COS750 CAS:41427-34-3 ***HR: 3***
2-CYANO-4-STILBENAMINE
mf: $C_{15}H_{12}N_2$ mw: 220.29

SYNS: 4-AMINO-2-STILBENECARBONITRILE ◇ 2-CYANO-4-AMINOSTILBENE

TOXICITY DATA with REFERENCE
scu-rat TDLo:2700 mg/kg/30W-I:NEO TXAPA9 5,344,63

CONSENSUS REPORTS: Cyanide and its compounds are on the Community Right-To-Know List.

SAFETY PROFILE: Questionable carcinogen with experimental neoplastigenic data. See also NITRILES. When heated to decomposition it emits toxic fumes of NO_x and CN^-.

COS800 CAS:1067-47-6 ***HR: 1***
(3-CYANOPROPYL) TRIETHOXYSILANE
mf: $C_{10}H_{21}NO_3Si$ mw: 231.41

SYNS: BUTYRONITRILE, 4-(TRIETHOXYSILYL)- ◇ SILANE, (3-CYANOPROPYL)TRIETHOXY- ◇ SILANE, TRIETHOXY(3-CYANOPROPYL)- ◇ TRIETHOXY-3-KYANPROPYLSILAN

TOXICITY DATA with REFERENCE
skn-rbt 100 μg/24H open AIHAAP 23,95,62
orl-rat LD50:4920 mg/kg AIHAAP 23,95,62

CONSENSUS REPORTS: Reported in EPA TSCA Inventory.

SAFETY PROFILE: Mildly toxic by ingestion. A skin irritant. When heated to decomposition it emits toxic fumes of NO_x and Si.

COS825 CAS:70247-32-4 ***HR: 3***
2-(5-CYANOTETRAZOLE)PENTAMMINECOBALT(III) PERCHLORATE
mf: $C_2H_{16}Cl_3CoN_{10}O_{12}$ mw: 537.50

CONSENSUS REPORTS: Cyanide compounds and cobalt compounds are on the Community Right-To-Know List.

SAFETY PROFILE: A relatively insensitive explosive. When heated to decomposition it emits toxic fumes of Cl^-, NO_x and CN^-. See also CYANIDE, PERCHLORATES, and COBALT COMPOUNDS.

COS899 CAS:63833-98-7 ***HR: 3***
CYANOTRIMEPRAZINE MALEATE
mf: $C_{19}H_{21}N_3S \cdot C_4H_4O_4$ mw: 439.57

SYNS: CIANATIL MALEATE ◇ CYAMEMAZINE MALEATE ◇ CYAMEPROMAZINE MALEATE ◇ CYANO-3-(DIMETHYLAMINO-3-METHYL-2-PROPYL)-10-PHENOTHIAZINE MALEATE ◇ 10-(3-(DIMETHYLAMINO)-2-METHYLPROPYL)PHENOTHIAZINE-2-CARBONITRILE MALEATE ◇ KYAMEPROMAZINE MALEATE ◇ 7204 RP

TOXICITY DATA with REFERENCE
orl-mus LD50:640 mg/kg CRSBAW 155,1029,61
ipr-mus LD50:210 mg/kg CRSBAW 155,1029,61
scu-mus LD50:690 mg/kg CRSBAW 155,1029,61
ivn-mus LD50:90 mg/kg CRSBAW 155,1029,61

CONSENSUS REPORTS: Cyanide and its compounds are on the Community Right-To-Know List.

SAFETY PROFILE: Poison by intravenous and intraperitoneal routes. Moderately toxic by ingestion and subcutaneous routes. See also NITRILES. When heated to decomposition it emits very toxic fumes of CN^-, NO_x, and SO_x.

COS909 CAS:4248-66-2 ***HR: D***
CYANOTRIMETHYLANDROSTENOLONE
mf: $C_{23}H_{33}NO_2$ mw: 355.57

SYNS: CYANOKETONE ◇ 2-α-CYANO-4,4,17-α-TRIMETHYLANDROST-5-EN-17-β-OL-3-ONE

TOXICITY DATA with REFERENCE
ims-rat TDLo:60 mg/kg (female 19D post):REP PSEBAA 121,757,66
ims-rat TDLo:720 mg/kg (15-20D preg):TER ENDOAO 87,432,70

CONSENSUS REPORTS: Cyanide and its compounds are on the Community Right-To-Know List.

SAFETY PROFILE: Experimental reproductive effects. An experimental teratogen. A steroid. When heated to decomposition it emits toxic fumes of CN^- and NO_x. See also CYANIDES.

COT000 CAS:16176-02-6 ***HR: 3***
2-CYANO-1,2,3-TRIS(DIFLUOROAMINO)PROPANE
mf: $C_4H_4F_6N_4$ mw: 222.10

$$F_2NC(CN)(CH_2NF_2)_2$$

CONSENSUS REPORTS: Cyanide and its compounds are on the Community Right-To-Know List.

SAFETY PROFILE: A shock-sensitive explosive. When heated to decomposition it emits toxic fumes of F^-, NO_x, and CN^-. See also CYANIDE and FLUORIDES.

COT500 ***HR: 3***
CYCAD HUSK

PROP: The active substance in the cycad meal is aglycone of Cycasin, a methylazoxymethanol (JNCIAM 41,605,68).

SYN: CYCAS CIRCINALIS HUSK

TOXICITY DATA with REFERENCE
orl-rat TDLo:28 g/kg/10D-C:ETA JNCIAM 41,605,68

SAFETY PROFILE: Questionable carcinogen with experimental tumorigenic data. When heated to decomposition it emits toxic and irritating fumes.

COT750 ***HR: 3***
CYCAD MEAL

PROP: Obtained from the nut of *Cycas circinalis L.* (FEPRA7 23,1384,64).

SYN: CYCAD NUT, aqueous extract

TOXICITY DATA with REFERENCE
orl-rat TDLo:21 g/kg/3W-C:CAR FEPRA7 23,1383,64
skn-mus TDLo:68 g/kg/6D-C:ETA FEPRA7 23,1383,64
orl-gpg TDLo:30 g/kg/44W-I:NEO FEPRA7 23,1384,64

SAFETY PROFILE: Questionable carcinogen with experimental carcinogenic, neoplastigenic, and tumorigenic data. When heated to decomposition it emits acrid smoke and fumes.

COU000 CAS:14901-08-7 ***HR: 3***
CYCASIN
mf: $C_8H_{16}N_2O_7$ mw: 252.26

SYNS: CYCAS REVOLUTA GLUCOSIDE ◇ CYKAZINE ◇ β-d-GLUCOSYLOXYAZOXYMETHANE ◇ METHYLAZOXYMETHANOL GLUCOSIDE ◇ METHYLAZOXYMETHANOL-β-d-GLUCOSIDE ◇ (METHYL-ONN-AZOXY)METHYL-β-d-GLUCOPYRANOSIDE

TOXICITY DATA with REFERENCE
mma-sat 10 μmol/plate CNREA8 39,3780,79
dnd-rat-orl 56 mg/kg MUREAV 54,39,78
orl-rat TDLo:100 mg/kg:ETA JCROD7 100,231,81
orl-rat TDLo:230 mg/kg (2-5D preg):TER JNCIAM 38,233,67
orl-ham TDLo:150 mg/kg:CAR CNREA8 31,283,71
orl-rat TD:21 g/kg/3W-C:CAR FEPRA7 23,1386,64
orl-rat LD50:270 mg/kg GANNA2 62,353,71
orl-mus LD50:500 mg/kg FEPRA7 31,1493,72
orl-rbt LDLo:30 mg/kg FEPRA7 31,1493,72
orl-gpg LDLo:20 mg/kg FEPRA7 31,1493,72
orl-ham LDLo:250 mg/kg FEPRA7 31,1493,72

CONSENSUS REPORTS: EPA Genetic Toxicology Program. IARC Cancer Review: Group 2B IMEMDT 7,56,87; Human Inadequate Evidence IMEMDT 10,121,76; Animal Sufficient Evidence IMEMDT 10,121,76; IMEMDT 1,157,72.

SAFETY PROFILE: Confirmed carcinogen with experimental carcinogenic and tumorigenic data. A poison by ingestion. An experimental teratogen. Mutation data reported. When heated to decomposition it emits toxic fumes of NO_x.

COU250 CAS:532-76-3 ***HR: 3***
CYCLAINE HYDROCHLORIDE
mf: $C_{16}H_{23}NO_2•ClH$ mw: 297.86

SYNS: CYCLAINE ◇ 1-(CYCLOHEXYLAMINO)-2-PROPANOL BENZOATE (ESTER) HYDROCHLORIDE ◇ D 109 ◇ HEXYLCAINE HYDROCHLORIDE

TOXICITY DATA with REFERENCE
orl-mus LD50:1080 mg/kg 29ZVAB -,57,69
scu-mus LD50:1080 mg/kg CLDND* 5,683,71
scu-rbt LD50:164 mg/kg CLDND*-,57,69
scu-gpg LD50:166 mg/kg JPETAB 93,388,48
ivn-rbt LD50:14 mg/kg 29ZVAB -,57,69

SAFETY PROFILE: Poison by intravenous and subcutaneous routes. Moderately toxic by ingestion. A local anesthetic. When heated to decomposition it emits very toxic fumes of HCl and NO_x.

COU500 CAS:103-95-7 ***HR: 2***
CYCLAMEN ALDEHYDE
mf: $C_{13}H_{18}O$ mw: 190.31

PROP: Colorless liquid; strong, floral odor. D: 0.946-0.952, refr index: 1.503-1.508. Sol in fixed oils; insol in propylene glycol, glycerin.

SYNS: ALDEHYDE B ◇ CYCLAMAL ◇ FEMA No. 2743 ◇ p-ISOPROPYL-α-METHYLHYDROCINNAMIC ALDEHYDE ◇ p-ISOPROPYL-α-METHYLPHENYLPROPYL ALDEHYDE ◇ α-METHYL-p-ISOPROPYLHYDROCINNAMALDEHYDE ◇ 2-METHYL-3-(p-ISOPROPYLPHENYL)PROPIONALDEHYDE

TOXICITY DATA with REFERENCE
skn-hmn 15 mg/48H MLD FCTXAV 12,385,74
orl-rat LD50:3810 mg/kg FCTXAV 2,327,64

CONSENSUS REPORTS: Reported in EPA TSCA Inventory.

SAFETY PROFILE: Moderately toxic by ingestion. A human skin irritant. See also ALDEHYDES. When heated to decomposition it emits acrid smoke and irritating fumes.

COU510 CAS:7149-24-8 ***HR: 1***
CYCLAMEN ALDEHYDE DIETHYL ACETAL
mf: $C_{17}H_{28}O_2$ mw: 264.45

SYNS: HYDROCINNAMALDEHYDE, p-ISOPROPYL-α-METHYL-, DI-

ETHYL ACETAL ◇ α-METHYL-p-ISOPROPYL HYDROCINNAMIC ALDEHYDE DIETHYL ACETAL

TOXICITY DATA with REFERENCE
skn-rbt 500 mg/24H MOD FCTXAV 14,731,76

CONSENSUS REPORTS: Reported in EPA TSCA Inventory.

SAFETY PROFILE: A skin irritant. When heated to decomposition it emits acrid smoke and irritating fumes.

COU525 CAS:29886-96-2 ***HR: 1***
CYCLAMEN ALDEHYDE DIMETHYL ACETAL
mf: $C_{15}H_{24}O_2$ mw: 236.39

SYN: HYDROCINNAMALDEHYDE, p-ISOPROPYL-α-METHYL-, DIMETHYL ACETAL ◇ α-METHYL-p-ISOPROPYLHYDROCINNAMIC ALDEHYDE DIMETHYL ACETAL ◇ PROPIONALDEHYDE, 3-(p-ISOPROPYLBENZYL)-, DIMETHYL ACETAL

TOXICITY DATA with REFERENCE
skn-rbt 500 mg/24H MOD FCTOD7 20,659,82

SAFETY PROFILE: A skin irritant. When heated to decomposition it emits acrid smoke and irritating fumes.

COV125 ***HR: 3***
CYCLAMIDOMYCIN
mf: $C_7H_{10}N_{20}$ mw: 374.37

SYNS: DESDANINE ◇ PYRACRYMYCIN-1 ◇ 1-PYRROLINE-2-ACRYLAMIDE

TOXICITY DATA with REFERENCE
orl-mus LD50:240 mg/kg 85GDA2 5,80,81
ipr-mus LD50:150 mg/kg 85GDA2 5,80,81
scu-mus LD50:150 mg/kg 85GDA2 5,80,81
ivn-mus LD50:125 mg/kg 85GDA2 5,80,81

SAFETY PROFILE: Poison by ingestion, subcutaneous, intravenous, and intraperitoneal routes. When heated to decomposition it emits toxic fumes of NO_x.

COV500 CAS:3572-80-3 ***HR: 3***
CYCLAZOCINE
mf: $C_{18}H_{25}NO$ mw: 271.44

SYNS: 2-CYCLOPROPYLMETHYL-5,9-DIMETHYL-2'-HYDROXY-6,7-BENEOMORPHAN ◇ 3-CYCLOPROPYLMETHYL-6(eq),11(ax)-DIMETHYL-2,6-METHANO-3-BENZAZOCIN-8-OL ◇ 3-(CYCLOPROPYLMETHYL)1-1,2,3,4,5,6-HEXAHYDRO-6,11-DIMETHYL-2,6-METHANO-3-BENZAZOCIN-8-OL ◇ 2-CYCLOPROPYLMETHYL-2'-HYDROXY-5,9-DIMETHYL-6,7-BENZOMORPHAN ◇ NIH 7981 ◇ NSC-107429 ◇ WIN 20740

TOXICITY DATA with REFERENCE
orl-rbt TDLo:13 mg/kg (6-18D preg):REP TXAPA9 31,534,75
scu-rat LD50:310 mg/kg JPETAB 143,141,64
ivn-rat LD50:32 mg/kg AIPTAK 165,112,67
ipr-mus LDLo:10 mg/kg RCOCB8 17,255,77
scu-mus LD50:153 mg/kg AIPTAK 241,79,79
ivn-mus LD50:28 mg/kg AIPTAK 165,112,67

SAFETY PROFILE: A poison by subcutaneous, intravenous, and intraperitoneal routes. Experimental reproductive effects. Used in the treatment of narcotic addiction. When heated to decomposition it emits toxic fumes of NO_x.

COV625 CAS:362-74-3 ***HR: D***
CYCLIC AMP DIBUTYRATE
mf: $C_{18}H_{24}N_5O_8P$ mw: 469.44

SYNS: 3',5'-CYCLIC AMP DIBUTYRATE ◇ CYCLIC AMP N^6,2'-DIBUTYRYL cAMP ◇ CYCLIC DIBUTYRYL AMP ◇ DIBUYTYRYL cAMP ◇ N^6,$O^{2\prime}$-DIBUTYRYL cAMP ◇ N^6,2'-o-DIBUTYRYL cAMP ◇ DIBUTYRYL CYCLIC AMP ◇ N^6,$O^{2\prime}$-DIBUYTYRL CYCLIC AMP ◇ N^6,2'-o-DIBUTYRYL CYCLIC AMP ◇ DIBUTYRYL-3',5'-CYCLIC AMP ◇ DIBUTYRYL CYCLIC-3',5'-AMP

TOXICITY DATA with REFERENCE
oms-hmn:oth 10 μmol/L JIDEAE 65,52,75
oms-rat:oth 1 mmol/L INOPAO 13,210,74
dni-mus:oth 5 μmol/L CNREA8 43,3514,83

SAFETY PROFILE: Human mutation data reported. When heated to decomposition it emits toxic fumes of PO_x and NO_x. See also AMIDES.

COV750 CAS:3741-38-6 ***HR: 3***
CYCLIC ETHYLENE SULFITE
mf: $C_2H_4O_3S$ mw: 108.12

SYNS: 1,3,2-DIOXATHIOLANE-2-OXIDE (9CI) ◇ ETHYLENE SULFITE ◇ 1,2-ETHYLENE SULFITE ◇ GLYCOL SULFITE

TOXICITY DATA with REFERENCE
ipr-mus TDLo:768 mg/kg/64W-I:ETA JNCIAM 53,695,74
ipr-mus LD50:250 mg/kg CBCCT* 5,341,53

CONSENSUS REPORTS: Reported in EPA TSCA Inventory.

SAFETY PROFILE: Poison by intraperitoneal route. Questionable carcinogen with experimental tumorigenic data. When heated to decomposition it emits toxic fumes of SO_x. See also SULFITES.

COV825 ***HR: 3***
CYCLOBARBITAL-SALICYLAMIDE COMPLEX
mf: $C_{12}H_{16}N_2O_3 \cdot C_7H_7NO_2$ mw: 373.45

TOXICITY DATA with REFERENCE
ipr-rat LD50:300 mg/kg KSRNAM 4,2536,70
scu-rat LD50:1780 mg/kg KSRNAM 4,2536,70
ipr-mus LD50:580 mg/kg KSRNAM 4,2536,70
scu-mus LD50:2300 mg/kg KSRNAM 4,2536,70

SAFETY PROFILE: Poison by intraperitoneal route. Moderately toxic by subcutaneous route. When heated to decomposition it emits toxic fumes of NO_x. See also

TETRAHYDROPHENOBARBITAL, BARBITURATES, and SALICYLAMIDE.

COW000 CAS:287-23-0 ***HR: 1***
CYCLOBUTANE
DOT: UN 2601
mf: C_4H_8 mw: 56.12

PROP: A gas. Mp: −50°, bp: 12.9°, flash p: <50°F(CC), d: 0.708 @ 11°, vap d: 1.93, lel: 1.8%.

SYN: TETRAMETHYLENE

DOT Classification: Flammable Gas; Label: Flammable Gas

SAFETY PROFILE: May be a simple asphyxiant. See also CYCLOHEXANE. Very dangerous fire hazard when exposed to heat or flame; can react with oxidizing materials. To fight fire, stop flow of gas; CO_2, dry chemicals, or water spray. When heated to decomposition it emits acrid smoke and fumes.

COW250 ***HR: 1***
CYCLOBUTENE
mf: C_4H_6 mw: 54.09

PROP: Gas. Bp: 2.4°; d: 0.733 @ 0°/4°; flash p: <15°F.

SYN: CYCLOBUTYLENE

SAFETY PROFILE: May be a simple asphyxiant. Dangerous fire hazard when exposed to heat or flame; can react with oxidizing materials. When heated to decomposition it emits acrid smoke and fumes.

COW500 CAS:60550-91-6 ***HR: D***
CYCLOBUTYL-N-(2-FLUORENYL)FORMAMIDE
mf: $C_{18}H_{17}NO$ mw: 263.36

SYNS: CYCLOBUTANECARBOXAMIDE-N-(2-FLUORENYL)◇ N-FLUORENYLCYCLOBUTANECARBOXAMIDE

TOXICITY DATA with REFERENCE
mma-sat 100 ng/plate BBRCA9 71,1201,76

SAFETY PROFILE: Mutation data reported. When heated to decomposition it emits toxic fumes of NO_x.

COW675 CAS:77287-90-2 ***HR: 3***
17-CYCLOBUTYLMETHYL-3-HYDROXY-6-METHYLENE-8-β-METHYLMORPHINAN
mf: $C_{22}H_{29}NO•CH_4O_3S$ mw: 419.63

SYNS: (8-β)-17-(CYCLOBUTYLMETHYL)-6-METHYLENEMORPHINAN-3-OL METHANESULFONATE ◇ TR5379M

TOXICITY DATA with REFERENCE
orl-rat LD50:750 mg/kg FAATDF 3,478,83
ivn-rat LD50:22300 μg/kg FAATDF 3,478,83
orl-mus LD50:365 mg/kg FAATDF 3,478,83
ivn-mus LD50:35 mg/kg FAATDF 3,478,83

SAFETY PROFILE: Poison by ingestion and intravenous routes. When heated to decomposition it emits toxic fumes of SO_x and NO_x. A narcotic antagonist.

COW700 CAS:512-16-3 ***HR: 2***
CYCLOBUTYROL
mf: $C_{10}H_{18}O_3$ mw: 186.28

PROP: Colorless crystals from ether-petr ether. Mp: 81-82°. Sltly sol in water, petr ether. Very sol in alcohols, acetone, dioxane, chloroform, ether.

SYNS: 1-CYCLOHEXANOL-α-BUTYRIC ACID ◇ α-ETHYL-1-HYDROXYCYCLOHEXANEACETIC ACID ◇ HEBUCOL ◇ α-(1-HYDROXYCYCLOHEXYL)BUTYRIC ACID ◇ 1-HYDROXY-α-ETHYLCYCLOHEXYLACETIC ACID ◇ JL 130

TOXICITY DATA with REFERENCE
orl-rat LD50:4820 mg/kg NIIRDN 6,313,82
scu-rat LD50:3230 mg/kg NIIRDN 6,313,82
ivn-rat LD50:1760 mg/kg NIIRDN 6,313,82
scu-mus LD50:4200 mg/kg NIIRDN 6,313,82
ivn-mus LD50:2900 mg/kg NIIRDN 6,313,82
ivn-rbt LD50:1920 mg/kg NIIRDN 6,313,82

CONSENSUS REPORTS: Reported in EPA TSCA Inventory.

SAFETY PROFILE: Moderately toxic by intravenous and subcutaneous routes. Mildly toxic by ingestion. When heated to decomposition it emits acrid smoke and fumes.

COW750 CAS:12663-46-6 ***HR: 3***
CYCLOCHLOROTINE
mf: $C_{24}H_{30}Cl_2N_5O_7$ mw: 571.49

PROP: White needles. Mp: 251°, decomp. Chlorine containing peptide produced by *P. islandicum* (85CVA2 5,177,70).

SYN: ISLANDITOXIN

TOXICITY DATA with REFERENCE
orl-mus TDLo:362 mg/kg/32W-C:ETA FCTXAV 10,193,72
orl-mus LD50:6550 μg/kg FCTXAV 10,193,72
ipr-mus LD50:330 μg/kg CTOXAO 17,45,80
scu-mus LD50:475 μg/kg FCTXAV 10,193,72
ivn-mus LD50:335 μg/kg FCTXAV 10,193,72

CONSENSUS REPORTS: IARC Cancer Review: Group 3 IMEMDT 7,56,87; Animal Limited Evidence IMEMDT 10,139,76.

SAFETY PROFILE: A deadly poison by ingestion, subcutaneous, intraperitoneal, and intravenous routes. Questionable carcinogen with experimental tumorigenic data. When heated to decomposition it emits very toxic fumes of Cl^- and NO_x.

COW780 CAS:72117-72-7 ***HR: 1***
α-CYCLOCITRYLIDENE-4-METHYLBUTAN-3-ONE
mf: $C_{15}H_{24}O$ mw: 220.39

SYNS: DIMETHYLIONONE ◇ 1,3-DIMETHYL-α-IONONE ◇ 1-PENTEN-3-ONE,1-(2,6,6-TRIMETHYL-2-CYCLOHEXEN-1-YL)-2-METHYL-

TOXICITY DATA with REFERENCE
skn-rbt 500 mg/24H MOD FCTXAV 16,717,78

CONSENSUS REPORTS: Reported in EPA TSCA Inventory.

SAFETY PROFILE: A skin irritant. When heated to decomposition it emits acrid smoke and irritating fumes.

COW825 CAS:51022-69-6 ***HR: 3***
CYCLOCORT
mf: $C_{28}H_{35}FO_7$ mw: 502.63

SYNS: (11-β,16-α)-21-(ACETYLOXY)-16,17-(CYCLOPENTYLIDENEBIS(OXY))-9-FLUORO-11-HYDROXYPREGNA-1,4-DIENE-3,20-DIONE ◇ AMCINONIDE ◇ CL-34699 ◇ PENTICORT

TOXICITY DATA with REFERENCE
ipr-rat LD50:243 mg/kg IYKEDH 13,637,82
scu-rat LD50:145 mg/kg IYKEDH 13,637,82
ipr-mus LD50:896 mg/kg IYKEDH 13,637,82
scu-mus LD50:143 mg/kg IYKEDH 13,637,82

SAFETY PROFILE: Poison by subcutaneous and intraperitoneal routes. A steroid. When heated to decomposition it emits toxic fumes of F^-.

COW875 CAS:31698-14-3 ***HR: 2***
CYCLOCYTIDINE
mf: $C_9H_{11}N_3O_4$ mw: 225.23

PROP: A cytostatic agent and intermediate in the synthesis of cytarabine.

SYNS: ANCITABINE ◇ ANCYTABINE ◇ 2,2'-ANHYDROARABINOSYLCYTOSINE ◇ ANHYDROARA C ◇ ANHYDROCYTIDINE ◇ 2,2'-ANHYDROCYTIDINE ◇ o-2,2'-CYCLOCYTIDINE ◇ 2,2'-CYCLOCYTIDINE ◇ 2,2'-o-CYCLOCYTIDINE

TOXICITY DATA with REFERENCE
ipr-rat LD50:1700 mg/kg EKFMA7 9,31,80
ivn-rat LD50:820 mg/kg IYKEDH 7,108,76
orl-mus LD50:3400 mg/kg EKFMA7 9,31,80
ipr-mus LD50:1600 mg/kg EKFMA7 9,31,80
scu-mus LD50:4050 mg/kg IYKEDH 7,108,76
ivn-mus LD50:800 mg/kg IYKEDH 7,108,76

SAFETY PROFILE: Moderately toxic by ingestion, intravenous, and intraperitoneal routes. Mildly toxic by subcutaneous route. When heated to decomposition it emits toxic fumes of NO_x.

COW900 CAS:10212-25-6 ***HR: 3***
CYCLOCYTIDINE HYDROCHLORIDE
mf: $C_9H_{11}N_3O_4$•ClH mw: 261.69

SYNS: ALEXAN ◇ ANCITABINE HYDROCHLORIDE ◇ 2,2'-ANHYDROARABINOSYLCYTOSINE HYDROCHLORIDE ◇ 2,2'-ANHYDRO-1-β-d-ARABINOFURANOSYLCYTOSINEHYDROCHLORIDE ◇ 2,2'-ANHYDROCYTARABINE HYDROCHLORIDE ◇ 2,2'-ANHYDROCYTIDINE HYDROCHLORIDE ◇ 1-β-d-ARABINOFURANOSYL-2,2'-ANHYDRO-CYTOSINE HYDROCHLORIDE ◇ CYCLO-CMP HYDROCHLORIDE ◇ CYCLOCYTIDINE ◇ 2,2'-CYCLOCYTIDINE HYDROCHLORIDE ◇ 2,2'-o-CYCLOCYTIDINE HYDROCHLORIDE ◇ o-2,2'-CYCLOCYTIDINE MONOHYDROCHLORIDE ◇ NSC-145668

TOXICITY DATA with REFERENCE
sce-hmn:lym 100 ng/L MUREAV 53,215,78
dni-mus:leu 10 mg/L CPBTAL 20,2286,72
ipr-rat TDLo:800 mg/kg (female 8-15D post):REP OYYAA2 8,1681,74
ipr-rat TDLo:800 mg/kg (female 8-15D post):TER OYYAA2 8,1681,74
scu-wmn TDLo:120 mg/kg:CVS,GIT,BLD CTRRDO 62,455,78
ipr-rat LD50:3800 mg/kg YAKUD5 21,359,79
ivn-rat LD50:820 mg/kg YAKUD5 21,359,79
ipr-mus LD50:2528 mg/kg NCISP* JAN86
ivn-mus LD50:800 mg/kg NIIRDN 6,55,82
ivn-dog LD50:344 mg/kg OYYAA2 8,353,74
ivn-mky LD50:1045 mg/kg OYYAA2 8,353,74

SAFETY PROFILE: Poison by intravenous route. Moderately toxic by intraperitoneal route. Human systemic effects by subcutaneous route: blood pressure depression, nausea or vomiting, and changes in bone marrow. An experimental teratogen. Other experimental reproductive effects. Human mutation data reported. When heated to decomposition it emits toxic fumes of NO_x and HCl.

COW925 CAS:7585-39-9 ***HR: 3***
β-CYCLODEXTRIN
mf: $C_{42}H_{70}O_{35}$ mw: 1135.12

SYNS: α-CYCLOAMYLOSE ◇ CYCLOHEPTAAMYLOSE ◇ β-CYCLOHEPTAAMYLOSE ◇ CYCLOHEPTAGLUCOSAN ◇ CYCLOMALTOHEPTAOSE ◇ CYCLOPHEPTAGLUCAN ◇ β-DEXTRIN ◇ SCHAARDINGER α-DEXTRIN

TOXICITY DATA with REFERENCE
orl-rat LD50:18800 mg/kg OYYAA2 26,287,83
ipr-rat LD50:356 mg/kg 48THAM 1,109,82
scu-rat LD50:3700 mg/kg OYYAA2 26,287,83
ivn-rat LD50:1008 mg/kg AJPAA4 83,367,76
ipr-mus LD50:330 mg/kg 48THAM 1,109,82
scu-mus LD50:412 mg/kg 48THAM 1,109,82

SAFETY PROFILE: Poison by intraperitoneal route. Moderately toxic by subcutaneous and intravenous routes. Mildly toxic by ingestion. When heated to decomposition it emits acrid smoke and fumes. See also DEXTRINS.

COW935 ***HR: 2***
cis,trans,trans-CYCLODODECA-1,5,9-TRIENE
mf: $C_{12}H_{18}$ mw: 162.30

SYNS: CDT ◇ 1,5,9-CYCLODODECATRIENE (Z,E,E)

TOXICITY DATA with REFERENCE
skn-mus 100%/12D open SEV BJIMAG 25,75,68
skn-rbt 2670 mg SEV BJIMAG 25,75,68
skn-rbt 20 g/31D-I open SEV BJIMAG 25,75,68
skn-gpg 10 g/31D-I open SEV BJIMAG 25,75,68
eye-rbt 89 mg MLD BJIMAG 25,75,68

SAFETY PROFILE: An eye and severe skin irritant. When heated to decomposition it emits acrid smoke and fumes.

COX000 CAS:31717-87-0 ***HR: 3***
CYCLODODECYL-2,6-DIMETHYLMORPHOLINE ACETATE
mf: $C_{18}H_{36}NO \cdot C_2H_4O_2$ mw: 342.61

SYNS: N-CYCLODODECYL-2,6-DIMETHYLMORPHOLINIUMACETATE ◇ CYCLOMORPH ◇ DODEMORFE (FRANCE)

TOXICITY DATA with REFERENCE
orl-rat LD50:4180 mg/kg FMCHA2 -,D115,80
ipr-mus LD50:320 mg/kg GUCHAZ 6,244,73

SAFETY PROFILE: Poison by intraperitoneal route. Mildly toxic by ingestion. When heated to decomposition it emits toxic fumes of NO_x.

COX250 CAS:64058-30-6 ***HR: 3***
p-N-CYCLO-ETHYLENEUREIDOAZOBENZENE
mf: $C_{15}H_{14}N_4O$ mw: 266.33

SYNS: AZOBENZEN (CZECH) ◇ 4-N-CYCLOETHYLENEUREIDOAZOBENZENE

TOXICITY DATA with REFERENCE
scu-rat TDLo:1140 mg/kg/15W-I:ETA BJPCAL 9,306,54

SAFETY PROFILE: Questionable carcinogen with experimental tumorigenic data. When heated to decomposition it emits toxic fumes of NO_x.

COX325 CAS:152-53-4 ***HR: 3***
CYCLOGUANIL HYDROCHLORIDE
mf: $C_{11}H_{14}ClN_5 \cdot ClH$ mw: 288.21

SYNS: 1-(4-CHLOROPHENYL)-1,6-DIHYDRO-6,6-DIMETHYL-1,3,5-TRIAZINE-2,4-DIAMINE MONOHYDROCHLORIDE ◇ 4,6-DIAMINO-1-(p-CHLOROPHENYL)-1,2-DIHYDRO-2,2-DIMETHYL-s-TRIAZINE MONOHYDROCHLORIDE

TOXICITY DATA with REFERENCE
ipr-mus LD50:54 mg/kg TXAPA9 18,487,71
scu-mus LD50:220 mg/kg ATMPA2 74,393,80
ivn-dog LDLo:24 mg/kg TXAPA9 18,487,71

SAFETY PROFILE: Poison by subcutaneous, intravenous, and intraperitoneal routes. When heated to decomposition it emits toxic fumes of NO_x and Cl^-.

COX400 CAS:516-21-2 ***HR: 3***
CYCLOGUANYL
mf: $C_{11}H_{14}ClN_5$ mw: 251.75

SYNS: 4-AMINO-6-p-CHLOROANILINO-1,2-DWUHYDRO-2,2-DWUMETHYLO-1,3,5-TROJAZYNA ◇ CHLORGUANIDE TRIAZINE ◇ CGT ◇ 1-(p-CHLOROPHENYL)-4,6-DIAMINO-2,2-DIMETHYL-1,2-DIHYDRO-s-TRIAZINE ◇ 1-p-CHLOROPHENYL-1,2-DIHYDRO-2,2-DIMETHYL-4,6-DIAMINO-s-TRIAZINE ◇ CYCLOGUANIL ◇ WR 5473 ◇ s-TRIAZINE, 1,2-DIHYDRO-1-(p-CHLOROPHENYL)-4,6-DIAMINO-2,2-DIMETHYL-

TOXICITY DATA with REFERENCE
mmo-omi 100 μg/plate AACHAX 12,84,77
orl-rat TDLo:50 mg/kg (female 1D post):TER BEBMAE 77(6),56,74
ipr-rat LD50:98 mg/kg DIPHAH 10,81,58

SAFETY PROFILE: Poison by intraperitoneal route. An experimental teratogen. Mutation data reported. When heated to decomposition it emits toxic fumes of NO_x.

COX500 CAS:291-64-5 ***HR: 3***
CYCLOHEPTANE
DOT: UN 2241
mf: C_7H_{14} mw: 98.19

PROP: An oil. Mp: −12°, bp: 117°, flash p: 59°F, d: 0.8099 @ 20°/4°, vap d: 3.3.

SYN: SUBERANE

DOT Classification: Flammable Liquid; Label: Flammable Liquid

SAFETY PROFILE: Dangerous fire hazard when exposed to heat or flame; can react with oxidizing materials. To fight fire, use foam, CO_2, dry chemicals. See also CYCLOHEXANE.

COY000 CAS:544-25-2 ***HR: 3***
1,3,5-CYCLOHEPTATRIENE
DOT: UN 2603
mf: C_7H_8 mw: 92.14

$$\overline{CH{=}CH(CH{=}CH)_2CH_2}$$

PROP: Flash p: 39.2°F.

SYNS: CYCLOHEPTATRIENE (DOT) ◇ TROPILIDENE

TOXICITY DATA with REFERENCE
cyt-rat:lvr 100 mg/L MUREAV 155,57,85

orl-rat LD50:57 mg/kg AOHYA3 10,123,67
skn-rat LD50:442 mg/kg AOHYA3 10,123,67
orl-mus LD50:171 mg/kg AOHYA3 10,123,67

DOT Classification: Flammable Liquid; Label: Flammable Liquid and Poison.

SAFETY PROFILE: Poison by ingestion and skin contact. Mutation data reported. A very dangerous fire hazard when exposed to heat, flame, or oxidizers. Potentially violent reaction with nitrogen monoxide. When heated to decomposition it emits acrid smoke and fumes.

COY100 CAS:12125-77-8 ***HR: 3***
CYCLOHEPTATRIENE MOLYBDENUM TRICARBONYL
mf: $C_{10}H_8MoO_3$ mw: 272.12

PROP: Mp: 100-101° decomp.

SYN: MOLYBDENUM,TRICARBONYL(1,3,5-CYCLOHEPTATRIENE)-

TOXICITY DATA with REFERENCE
ivn-mus LD50:32 mg/kg CSLNX* NX#04764

OSHA PEL: (Transitional: TWA Total Dust: 15 mg/m^3; Respirable Fraction: 5 mg/m^3) TWA Total Dust: 10 mg/m^3; Respirable Fraction: 5 mg/m^3
ACGIH TLV: TWA 10 mg(Mo)/m^3

SAFETY PROFILE: Poison by intravenous route. A skin and eye irritant. When heated to decomposition it emits toxic fumes of Mo.

COY250 CAS:628-92-2 ***HR: 2***
CYCLOHEPTENE
DOT: UN 2242
mf: C_7H_{12} mw: 96.174

$CH{=}CH(CH_2)_4CH_2$ (ring)

PROP: Flash p: < 73.4°F.

SAFETY PROFILE: A dangerous fire hazard when exposed to heat, flame, or oxidizers. When heated to decomposition it emits acrid smoke and fumes.

COY500 CAS:509-86-4 ***HR: 3***
CYCLOHEPTENYL ETHYLBARBITURIC ACID
mf: $C_{13}H_{18}N_2O_3$ mw: 250.33

SYNS: 5-(1-CYCLOHEPTEN-1-YL)-5-ETHYLBARBITURICACID ◇ CYCLOHEPTENYLETHYLMALONYLUREA ◇ 5-(1-CYCLOHEPTEN-1-YL)-5-ETHYL-2,4,6(1H,3H,5H)-PYRIMIDINETRIONE (9CI) ◇ 5-ETHYL-5-(1′-CYCLOHEPTENYL)-BARBITURIC ACID ◇ G 475 ◇ HEPTABARB ◇ HEPTABARBITAL ◇ HEPTABARBITONE ◇ HEPTABARBUM ◇ HEPTADORM ◇ HEPTAMAL ◇ MEDAPAN ◇ MEDOMIN ◇ MEDOMINE

TOXICITY DATA with REFERENCE
ipr-rat LD50:220 mg/kg JPETAB 93,101,48
ipr-mus LD50:210 mg/kg ARZNAD 9,360,59
ivn-dog LD50:105 mg/kg JPETAB 93,101,48
ivn-rbt LD50:119 mg/kg JPETAB 93,101,48

SAFETY PROFILE: A poison by intraperitoneal and intravenous routes. Psychotropic effects by ingestion. When heated to decomposition it emits toxic fumes of NO_x. See also BARBITURATES.

CPA500 CAS:592-57-4 ***HR: 3***
1,3-CYCLOHEXADIENE
mf: C_6H_8 mw: 80.13

$CH_2(CH{=}CH)_2CH_2$ (ring)

PROP: Flash p: < 73.4°F.

SAFETY PROFILE: A dangerous fire hazard when exposed to heat, flame, or oxidizers. It forms explosive polymeric oxides on exposure to air. When heated to decomposition it emits acrid smoke and fumes.

CPA750 CAS:628-41-1 ***HR: 3***
1,4-CYCLOHEXADIENE
mf: C_6H_8 mw: 80.13

$CH{=}CHCH_2CH{=}CHCH_2$ (ring)

PROP: Flash p: 12.2°F.

SAFETY PROFILE: A very dangerous fire hazard when exposed to heat, flame, or oxidizers. When heated to decomposition it emits acrid smoke and fumes.

CPA775 CAS:4998-76-9 ***HR: 3***
CYCLOHEXANAMINE HYDROCHLORIDE
mf: $C_6H_{13}N{\cdot}ClH$ mw: 135.66

SYNS: AMINOCYCLOHEXANE HYDROCHLORIDE ◇ AMINOHEXAHYDROBENZENE HYDROCHLORIDE ◇ HEXAHYDROANILINE HYDROCHLORIDE

TOXICITY DATA with REFERENCE
spm-rat-orl 32400 mg/kg/90D TXCYAC 8,143,77
orl-rat TDLo:1 g/kg (female 6-15D post):TER TOLED5 17,137,83
orl-mus TDLo:1 g/kg (female 6-15D post):REP TOLED5 17,137,83
orl-rat LD50:720 mg/kg GISAAA 51(9),80,86
ipr-rat LDLo:350 mg/kg LIFSAK 8,843,69
orl-mus LD50:760 mg/kg GISAAA 51(9),80,86
ipr-mus LD50:300 mg/kg TXAPA9 22,465,72

CONSENSUS REPORTS: EPA Genetic Toxicology Program. Reported in EPA TSCA Inventory.

SAFETY PROFILE: Poison by intraperitoneal route. Moderately toxic by ingestion. An experimental teratogen. Other experimental reproductive effects. Mutation data reported. When heated to decomposition it emits toxic fumes of NO_x and HCl.

CPB000 CAS:110-82-7 ***HR: 3***
CYCLOHEXANE
DOT: UN 1145
mf: C_6H_{12} mw: 84.18

PROP: Colorless, mobile liquid; pungent odor. Mp: 6.5°, bp: 80.7°, fp: 4.6°, flash: p: 1.4°F, ULC: 90-95, lel: 1.3%, uel: 8.4%, d: 0.7791 @ 20°/4°, autoign temp: 473°F, vap press: 100 mm @ 60.8°, vap d: 2.90.

SYNS: CICLOESANO (ITALIAN) ◇ CYCLOHEXAAN (DUTCH) ◇ CYCLOHEXAN (GERMAN) ◇ CYKLOHEKSAN (POLISH) ◇ HEXAHYDROBENZENE ◇ HEXAMETHYLENE ◇ HEXANAPHTHENE ◇ RCRA WASTE NUMBER U056

TOXICITY DATA with REFERENCE
skn-rbt 1548 mg/2D-I JIHTAB 25,199,43
dnd-esc 10 μmol/L MUREAV 89,95,81
orl-rat LD50:29820 mg/kg JIHTAB 25,415,43
orl-mus LD50:813 mg/kg NPIRI* 1,17,74
orl-rbt LDLo:5500 mg/kg JIHTAB 25,199,43
ivn-rbt LDLo:77 mg/kg JPMRAB 3,1,28

CONSENSUS REPORTS: Community Right-To-Know List.

OSHA PEL: TWA 300 ppm
ACGIH TLV: TWA 300 ppm
DFG MAK: 300 ppm (1050 mg/m^3)
DOT Classification: Flammable Liquid; Label: Flammable Liquid.

SAFETY PROFILE: Poison by intravenous route. Moderately toxic by ingestion. A systemic irritant by inhalation and ingestion. A skin irritant. Mutation data reported. Flammable liquid. Dangerous fire hazard when exposed to heat or flame; can react with oxidizing materials. Moderate explosion hazard in the form of vapor when exposed to flame. When mixed hot with liquid dinitrogen tetraoxide an explosion resulted. To fight fire, use foam, CO_2, dry chemical, spray, fog. When heated to decomposition it emits acrid smoke and fumes.

CPB050 CAS:1122-56-1 ***HR: 1***
CYCLOHEXANECARBOXAMIDE
mf: $C_7H_{13}NO$ mw: 127.21

PROP: Mp: 186-188°. Hygroscopic.

SYNS: CYCLOHEXAMETHYLENE CARBAMIDE ◇ CYCLOHEXANAMIDE ◇ CYCLOHEXANEFORMAMIDE ◇ CYCLOHEXYLCARBOXAMIDE ◇ CYCLOHEXYL CARBOXYAMIDE ◇ HEXAHYDROBENZOIC ACID AMIDE

TOXICITY DATA with REFERENCE
skn-rbt 500 mg/24H MOD 33NFA8 -,2,75

CONSENSUS REPORTS: Reported in EPA TSCA Inventory.

SAFETY PROFILE: A skin irritant. When heated to decomposition it emits toxic fumes of NO_x.

CPB100 CAS:694-83-7 ***HR: 1***
1,2-CYCLOHEXANEDIAMINE
mf: $C_6H_{14}N_2$ mw: 114.22

PROP: Bp: 92-93 @ 18 mm, d: 0.931, flash p: 167° F.

SYN: 1,2-DIAMINOCYCLOHEXANE

TOXICITY DATA with REFERENCE
skn-rbt 500 mg/24H MOD JACTDZ 1,8,90
orl-rat LDLo:1 g/kg JACTDZ 1,8,90

SAFETY PROFILE: Slightly toxic by ingestion. A skin irritant. A combustible liquid. When heated to decomposition it emits toxic fumes of NO_x.

CPB500 CAS:4442-85-7 ***HR: 3***
CYCLOHEXANEETHYLAMINE
mf: $C_8H_{17}N$ mw: 127.26

SYN: WIN 5522-2

TOXICITY DATA with REFERENCE
ipr-mus LDLo:140 mg/kg JPETAB 106,341,52
ivn-mus LD50:44 mg/kg JPETAB 105,336,52

SAFETY PROFILE: Poison by intraperitoneal and intravenous routes. When heated to decomposition it emits toxic fumes of NO_x. See also AMINES.

CPB625 CAS:1569-69-3 ***HR: 2***
CYCLOHEXANETHIOL
mf: $C_6H_{12}S$ mw: 116.24

SYN: CYKLOHEXANTHIOL ◇ CYKLOHEXYLMERKATPAN (CZECH)

TOXICITY DATA with REFERENCE
skn-rbt 500 mg/24H SEV 28ZPAK -,167,72
eye-rbt 500 g/24H MLD 28ZPAK -,167,72

CONSENSUS REPORTS: Reported in EPA TSCA Inventory.

NIOSH REL: (Cyclohexanethiol) CL 0.5 ppm/15M

SAFETY PROFILE: An eye and severe skin irritant. When heated to decomposition it emits toxic fumes of SO_x. See also MERCAPTANS.

CPB650 CAS:3570-93-2 ***HR: 3***
1,2,3-CYCLOHEXANETRIONE TRIOXIME
mf: $C_6H_9N_3O_3$ mw: 171.16

HON:CCH₂C(:NOH)CH₂C(:NOH)CH₂ (ring: first C bonded to last CH₂)

SAFETY PROFILE: Explodes violently when heated to

155°C. Potentially explosive reaction with sulfinyl chloride. When heated to decomposition it emits toxic fumes of NO_x.

CPB750 CAS:108-93-0 ***HR: 3***
CYCLOHEXANOL
mf: $C_6H_{12}O$ mw: 100.16

PROP: Colorless needles or viscous liquid; hygroscopic, camphor-like odor. Mp: 24°, bp: 161.5°, flash p: 154°F (CC), d: 0.9449 @ 25°/4°, vap press: 1 mm @ 21.0°, vap d: 3.45, autoign temp: 572°F.

SYNS: ADRONAL ◇ ANOL ◇ CICLOESANOLO (ITALIAN) ◇ CYCLOHEXYL ALCOHOL ◇ CYKLOHEKSANOL (POLISH) ◇ HEXAHYDROPHENOL ◇ HEXALIN ◇ HYDRALIN ◇ HYDROPHENOL ◇ HYDROXYCYCLOHEXANE ◇ NAXOL

TOXICITY DATA with REFERENCE
eye-hmn 100 ppm JIHTAB 25,282,43
skn-rbt 14600 μg/24H open MLD AIHAAP 23,95,62
eye-rbt 2 mg SEV AJOPAA 29,1363,46
cyt-hmn:leu 100 μmol/L DBTEAD 19,215,71
dnd-mam:lym 150 mmol/L PNASA6 48,686,62
scu-rat TDLo:315 mg/kg (21D male):REP IJEBA6 17,1305,79
ihl-hmn TCLo:75 ppm:NOSE,EYE,PUL JIHTAB 25,282,43
orl-rat LD50:2060 mg/kg MDZEAK 8,244,67
scu-mus LD50:2480 mg/kg REMBA8 5,7,67
ivn-mus LD50:272 mg/kg AIPTAK 135,342,62
ims-mus LD50:1000 mg/kg JSICAZ 21,342,62
orl-rbt LDLo:2200 mg/kg JIHTAB 25,199,43
skn-rbt LDLo:12 g/kg JIHTAB 25,199,43
ipr-rbt LDLo:1420 mg/kg JPMRAB 3,1,28

CONSENSUS REPORTS: EPA Genetic Toxicology Program. Reported in EPA TSCA Inventory.

OSHA PEL: TWA 50 ppm (skin)
ACGIH TLV: TWA 50 ppm (skin)
DFG MAK: 50 ppm (200 mg/m^3)

SAFETY PROFILE: Poison by intravenous route. Moderately toxic by ingestion, subcutaneous, and intramuscular routes. Mildly toxic by skin contact. Human systemic effects by inhalation: conjunctiva irritation, and changes in the olfactory and respiratory systems. Has caused damage to kidneys, liver, and blood vessels in experimental animals. Experimental reproductive effects. Human mutation data reported. A severe eye irritant. Narcotic-like action. Has caused liver, kidney, vascular injury in experimental animals. Flammable when exposed to heat or flame; can react with oxidizing materials. Ignites on contact with chromium trioxide. Violent reaction with HNO_3. Incompatible with oxidants. To fight fire, use alcohol foam, foam, CO_2, dry chemical. When heated to decomposition it emits acrid smoke and fumes. See also ALCOHOLS.

CPC000 CAS:108-94-1 ***HR: 2***
CYCLOHEXANONE
DOT: UN 1915
mf: $C_6H_{10}O$ mw: 98.16

PROP: Colorless liquid; acetone-like odor. Mp: −45.0°, bp: 115.6°. ULC: 35-40, lel: 1.1% @ 100°, flash p: 111°F, d: 0.9478 @ 20°/4°, autoign temp: 788°F, vap press: 10 mm @ 38.7°, vap d: 3.4.

SYNS: CICLOESANONE (ITALIAN) ◇ CYCLOHEXANON (DUTCH) ◇ CYKLOHEKSANON (POLISH) ◇ HEXANON ◇ KETOHEXAMETHYLENE ◇ NADONE ◇ NCI-C55005 ◇ PIMELIC KETONE ◇ RCRA WASTE NUMBER U057 ◇ SEXTONE

TOXICITY DATA with REFERENCE
eye-hmn 75 ppm JIHTAB 25,282,43
skn-rbt 500 mg open MLD UCDS**
eye-rbt 4740 μg SEV AJOPAA 29,1363,46
mma-sat 20 μL/L EJMBA2 18,213,83
mmo-bcs 200 μL/L EJMBA2 18,213,83
sce-ham:ovr 7500 μL/L ENMUDM 7(Suppl 3),60,85
orl-mus TDLo:11 g/kg (female 8-12D post):REP TCMUD8 6,361,86
ihl-hmn TCLo:75 ppm:NOSE,EYE,PUL JIHTAB 25,282,43
orl-rat LD50:1535 mg/kg AIHAAP 30,470,69
ihl-rat LC50:8000 ppm/4H NPIRI* 1,18,74
scu-rat LD50:2170 mg/kg JIHTAB 25,415,43
orl-mus LD50:1400 mg/kg NTIS** AD-A066-307
ipr-mus LD50:1350 mg/kg COREAF 254,2245,62
scu-mus LDLo:1300 mg/kg AEXPBL 50,199,1903
ivn-dog LDLo:630 mg/kg 14CYAT 2,1719,63
orl-rbt LDLo:1600 mg/kg JIHTAB 25,199,43
skn-rbt LD50:948 mg/kg AIHAAP 30,470,69

CONSENSUS REPORTS: Reported in EPA TSCA Inventory.

OSHA PEL: (Transitional: TWA 50 ppm) TWA 25 ppm (skin)
ACGIH TLV: TWA 25 ppm (skin)
DFG MAK: 50 ppm (200 mg/m^3)
NIOSH REL: (Ketone (Cyclohexanone)) TWA 100 mg/m^3
DOT Classification: Flammable or Combustible Liquid; Label: Flammable Liquid.

SAFETY PROFILE: Moderately toxic by ingestion, inhalation, subcutaneous, intravenous, and intraperitoneal routes. A skin and severe eye irritant. Human systemic effects by inhalation: changes in the sense of smell, conjunctiva irritation, and unspecified respiratory system changes. Human irritant by inhalation. Mild narcotic properties have also been ascribed to it. Human mutation data reported. Experimental reproductive effects. Flammable when exposed to heat or flame; can react vigorously with oxidizing materials. Slight explosion hazard in its vapor form, when exposed to flame. Explosive reaction with nitric acid at 75°C. Reaction with hydrogen peroxide + nitric acid forms an explosive peroxide. To fight fire, use alcohol foam, dry chemical, or CO_2. When heated to decomposition it emits acrid smoke and irritating fumes. See also KETONES and CYCLOHEXANE.

CPC250 ***HR: 2***
CYCLOHEXANONE-Δ
mf: C_6H_8O mw: 96.12

PROP: Liquid. Bp: 155.5°, flash p: 93°F (CC), vap d: 3.31, vap press: 4 mm @ 20°.

SAFETY PROFILE: Skin contact can cause a dermatitis. Irritating to eyes, skin, and mucous membranes. Can damage the liver and kidneys. Dangerous fire hazard when exposed to flame and heat; can react with oxidizing materials. To fight fire, use CO_2, dry chemical.

CPC500 ***HR: 3***
CYCLOHEXANONE PEROXIDE and BIS (1-HYDROXYCYCLOHEXYL)PEROXIDE MIXTURE

SYN: 1-((1-HYDROPEROXYCYCLOHEXYL)DIOXY) CYCLOHEXENOL with BIS(1-HYDROXYCYCLOHEXYL)PEROXIDE

DOT Classification: Organic Peroxide; Label: Organic Peroxide.

SAFETY PROFILE: A powerful oxidizer. When heated to decomposition it emits acrid smoke and irritating fumes. See PEROXIDES, ORGANIC.

CPC579 CAS:110-83-8 ***HR: 3***
CYCLOHEXENE
DOT: UN 2256
mf: C_6H_{10} mw: 82.15

PROP: Colorless liquid. Bp: 83°, fp: −103.7°, flash p: < 21.2°F, d: 0.8102 @ 20°/4°, vap press: 160 mm @ 38°, autoign temp: 590°F, vap d: 2.8, lel: 1.2%.

SYNS: BENZENETETRAHYDRIDE ◇ CYKLOHEKSEN (POLISH) ◇ 1,2,3,4-TETRAHYDROBENZENE

CONSENSUS REPORTS: Reported in EPA TSCA Inventory.

OSHA PEL: 300 ppm
ACGIH TLV: 300 ppm
DFG MAK: 300 ppm (1015 mg/m^3)
DOT Classification: Flammable Liquid.

SAFETY PROFILE: Moderately toxic by inhalation and ingestion. A very dangerous fire hazard when exposed to flame; can react with oxidizers. Dangerous; keep away from heat and open flame. To fight fire, use foam, CO_2, dry chemical.

CPC625 CAS:100-45-8 ***HR: 2***
3-CYCLOHEXENE-1-CARBONITRILE
mf: C_7H_9N mw: 107.17

SYN: 3-CYCLOHENENYL CYANIDE

TOXICITY DATA with REFERENCE
eye-rbt 500 mg open AMIHBC 10,61,54
orl-rat LD50:460 mg/kg AMIHBC 10,61,54
skn-rbt LD50:9460 mg/kg AMIHBC 10,61,54

CONSENSUS REPORTS: Cyanide and its compounds are on the Community Right-To-Know List.

SAFETY PROFILE: Moderately toxic by ingestion. An eye irritant. When heated to decomposition it emits toxic fumes of NO_x and CN^-. See also NITRILES.

CPC650 CAS:4771-80-6 ***HR: 2***
3-CYCLOHEXENE-1-CARBOXYLIC ACID
mf: $C_7H_{10}O_2$ mw: 126.17

TOXICITY DATA with REFERENCE
skn-rbt 10 mg/24H open SEV AMIHBC 10,61,54
eye-rbt 250 μg open SEV AMIHBC 10,61,54
orl-rat LD50:4260 mg/kg AMIHBC 10,61,54
skn-rbt LD50:1000 mg/kg AMIHBC 10,61,54

SAFETY PROFILE: Moderately toxic by skin contact. Mildly toxic by ingestion. A severe eye and skin irritant. When heated to decomposition it emits acrid smoke and fumes.

CPD000 CAS:286-20-4 ***HR: 3***
CYCLOHEXENE OXIDE
mf: $C_6H_{10}O$ mw: 98.16

PROP: Clear liquid. Bp: 129.5°, flash p: 81°F, d: 0.9678 @ 25°/4°, vap d: 3.5.

SYNS: CCHO ◇ CYCLOHEXANE OXIDE ◇ CYCLOHEXENE EPOXIDE ◇ CYCLOHEXENE-1-OXIDE ◇ 1,2-CYCLOHEXENE OXIDE ◇ CYCLOHEXYLENE OXIDE ◇ 1,2-EPOXYCYCLOHEXANE ◇ 7-OXABICYCLO(4.1.0)HEPTANE ◇ TETRAMETHYLENEOXIRANE

TOXICITY DATA with REFERENCE
mmo-klp 5 mmol/L MUREAV 89,269,81
mmo-sat 10 μmol/plate BCPCA6 29,1068,80
mma-sat 1 mg/plate MUREAV 58,217,78
msc-ham:lng 5 mmol/L CBINA8 51,77,84
unk-mus TDLo:79 mg/kg:ETA RARSAM 3,193,63
orl-rat LD50:1090 mg/kg AIHAAP 30,470,69
ihl-rat LCLo:2000 ppm/4H AIHAAP 30,470,69
ipr-rat LD50:549 mg/kg TXAPA9 52,422,80

ims-mus LD50:1000 mg/kg JSICAZ 21,342,62
skn-rbt LD50:630 mg/kg AIHAAP 30,470,69

CONSENSUS REPORTS: Reported in EPA TSCA Inventory. EPA Genetic Toxicology Program.

SAFETY PROFILE: Moderately toxic by ingestion, skin contact, intraperitoneal, and intramuscular routes. Mildly toxic by inhalation. Questionable carcinogen with experimental tumorigenic data. Mutation data reported. A dangerous fire hazard when exposed to heat or flame. When heated to decomposition it emits acrid smoke and irritant fumes.

CPD250 CAS:930-68-7 ***HR: 3***
2-CYCLOHEXEN-1-ONE
mf: C_6H_8O mw: 96.14

SYN: CYCLOHEXENONE

TOXICITY DATA with REFERENCE
eye-rbt 98 mg AIHAAP 33,338,72
orl-rat LD50:220 mg/kg AIHAAP 33,338,72
ihl-rat LC50:250 ppm/4H AIHAAP 33,338,72
ipr-mus LD50:170 mg/kg ZolH## 23OCT75
skn-rbt LD50:70 mg/kg AIHAAP 33,338,72

CONSENSUS REPORTS: Reported in EPA TSCA Inventory.

SAFETY PROFILE: A poison by ingestion, inhalation, intraperitoneal, and skin routes. When heated to decomposition it emits acrid smoke and irritant fumes. See also KETONES.

CPD500 CAS:19143-00-1 ***HR: 3***
S-2-((4-CYCLOHEXEN-3-YLBUTYL)AMINO)ETHYL THIOSULFATE
mf: $C_{12}H_{23}NO_3S_2$ mw: 293.48

SYN: 2-((4-CYCLOHEXEN-3-YLBUTYL)AMINO)ETHANETHIOLHYDROGEN SULFATE (ESTER)

TOXICITY DATA with REFERENCE
orl-mus LD50:900 mg/kg JMCMAR 11,1190,68
ipr-mus LD50:75 mg/kg JMCMAR 11,1190,68

SAFETY PROFILE: Poison by intraperitoneal route. Moderately toxic by ingestion. See also THIOSULFATES. When heated to decomposition it emits very toxic fumes of NO_x and SO_x.

CPD625 CAS:77251-47-9 ***HR: 2***
1-(2-CYCLOHEXEN-1-YLCARBONYL)-2-METHYLPIPERIDINE
mf: $C_{13}H_{21}NO$ mw: 207.35

SYN: AI3-37220

TOXICITY DATA with REFERENCE
skn-rbt 500 mg/24H MLD NTIS** AD-A087-646
eye-rbt 100 mg/24H MLD NTIS** AD-A087-646
orl-rat LDLo:1270 mg/kg NTIS** AD-A087-646

SAFETY PROFILE: Moderately toxic by ingestion. A skin and eye irritant. When heated to decomposition it emits toxic fumes of NO_x.

CPD750 CAS:100-40-3 ***HR: 3***
CYCLOHEXENYLETHYLENE
mf: C_8H_{12} mw: 108.20

PROP: Liquid. Bp: 128°, fp: −109°, flash p: 60°F (TOC), d: 0.832 @ 20°/4°, autoign temp: 517°F, vap press: 25.8 mm @ 38°, vap d: 3.76.

SYNS: BUTADIENE DIMER ◇ 4-ETHENYL-1-CYCLOHEXENE ◇ NCI-C54999 ◇ 1,2,3,4-TETRAHYDROSTYRENE ◇ 1-VINYLCYCLOHEXENE-3 ◇ 1-VINYLCYCLOHEX-3-ENE ◇ 4-VINYLCYCLOHEXENE-1 ◇ 4-VINYL-1-CYCLOHEXENE

TOXICITY DATA with REFERENCE
orl-mus TDLo:103 g/kg/2Y-I:CAR,REP JTEHD6 21,507,87
skn-mus TDLo:16 g/kg/54W-I:ETA JNCIAM 31,41,63
orl-mus TD:103 g/kg/2Y-I:NEO,REP NTPTR* NTP-TR-303,86
orl-mus TD:206 g/kg/2Y-I:CAR NTPTR* NTP-TR-303,86
orl-rat LD50:2563 mg/kg AIHAAP 30,470,69
ihl-rat LCLo:8000 ppm/4H AIHAAP 30,470,69
ihl-mus LC50:27000 mg/m^3 IARC** 11,-,76
skn-rbt LD50:16640 mg/kg AIHAAP 30,470,69

CONSENSUS REPORTS: IARC Cancer Review: Group 3 IMEMDT 7,56,87; Animal Inadequate Evidence IMEMDT 11,277,76; Animal Limited Evidence IMEMDT 39,181,86. NTP Carcinogenesis Studies (gavage); Clear Evidence: mouse NTPTR* NTP-TR-303,86; Inadequate Studies: rat NTPTR* NTP-TR-303,86. Reported in EPA TSCA Inventory.

ACGIH TLV: (Proposed: TWA 0.1 ppm; Suspected Human Carcinogen)

SAFETY PROFILE: Moderately toxic by ingestion and inhalation. Mildly toxic by skin contact. Questionable carcinogen with experimental carcinogenic, neoplastigenic, and tumorigenic data. Experimental reproductive effects. Dangerous fire hazard when exposed to heat, flame or oxidizers. Can react with oxidizers. To fight fire, use foam, CO_2, dry chemical.

CPE125 CAS:4845-05-0 ***HR: 3***
2-CYCLOHEXENYL HYDROPEROXIDE
mf: $C_6H_{10}O_2$ mw: 114.14

SAFETY PROFILE: A heat-sensitive explosive. When heated to decomposition it emits acrid smoke and fumes. See also PEROXIDES and EXPLOSIVES.

CPE500 CAS:10137-69-6 ***HR: 2***
CYCLOHEXENYL TRICHLOROSILANE
DOT: UN 1762
mf: $C_6H_9Cl_3Si$ mw: 215.59

PROP: Colorless, fuming liquid; HCl odor. Bp: 202°; d: 1.263 @ 25°/25°; flash p: 200°F (COC).

TOXICITY DATA with REFERENCE
skn-rbt 100 μg/24H open AIHAAP 23,95,62
orl-rat LD50:2830 mg/kg AIHAAP 23,95,62
skn-rbt LD50:630 mg/kg AIHAAP 23,95,62

CONSENSUS REPORTS: Reported in EPA TSCA Inventory.

DOT Classification: Corrosive Material; Label: Corrosive.

SAFETY PROFILE: Moderately toxic by ingestion and skin contact. A skin irritant. A corrosive material. It fumes in moist air releasing HCl. Combustible when exposed to heat or flame. When heated to decomposition it emits toxic fumes of Cl^-. See also CHLOROSILANES.

CPE750 CAS:66-81-9 ***HR: 3***
CYCLOHEXIMIDE
mf: $C_{15}H_{23}NO_4$ mw: 281.39

PROP: Crystals. Mp: 116°. Moderately sol in water; sol in chloroform, ether, and acetone.

SYNS: ACTI-AID ◇ ACTIDIONE ◇ ACTIDIONE TGF ◇ ACTIDONE ◇ ACTISPRAY ◇ 3-(2-(3,5-DIMETHYL-2-OXOCYCLOHEXYL)-2-HYDROXYETHYL)GLUTARIMIDE ◇ HIZAROCIN ◇ KAKEN ◇ NARAMYCIN ◇ NEOCYCLOHEXIMIDE ◇ NSC-185 ◇ U-4527

TOXICITY DATA with REFERENCE
skn-rbt 5 mg/24H rns TXCYAC 14,117,49
skn-rbt 1%/24H MOD NTIS** PB-274-414
dni-hmn:oth 100 mg/L BBACAQ 696,15,82
dni-hmn:oth 300 nmol/L CNREA8 44,2421,84
ipr-rat TDLo:1 mg/kg (female 18D post):REP BJEPA5 51,361,70
ipr-mus TDLo:30 mg/kg (female 9D post):TER TJADAB 25,345,82
orl-rat LD50:2 mg/kg UPJOH* 2(6),-,71
ipr-rat LD50:3700 μg/kg JPETAB 136,400,62
scu-rat LD50:2500 μg/kg ANTCAO 10,682,60
ivn-rat LD50:2 mg/kg ANTCAO 10,682,60
orl-mus LD50:133 mg/kg UPJOH* 2(6),-,71
ipr-mus LD50:100 mg/kg CNCRA6 30,9,63
scu-mus LD50:160 mg/kg UPJOH* 2(6),-,71
ivn-mus LD50:150 mg/kg JACSAT 69,474,47
orl-dog LD50:65 mg/kg PCOC** -,292,66
orl-mky LD50:60 mg/kg GUCHAZ 6,146,73

CONSENSUS REPORTS: EPA Extremely Hazardous Substances List. EPA Genetic Toxicology Program.

SAFETY PROFILE: A poison by ingestion, subcutaneous, intraperitoneal and intravenous routes. An experimental teratogen. Other experimental reproductive effects. Human mutation data reported. A skin irritant. A pesticide. When heated to decomposition it emits toxic fumes of NO_x.

CPF000 CAS:622-45-7 ***HR: 2***
CYCLOHEXYL ACETATE
DOT: UN 2243
mf: $C_8H_{14}O_2$ mw: 142.22

PROP: Pale yellow liquid; fruity odor. Bp: 177°, d: 0.996, vap d: 4.9, flash p: 136°F, autoign temp: 633°F.

SYNS: CYCLOHEXANOL ACETATE ◇ CYCLOHEXANOLAZETAT (GERMAN) ◇ CYCLOHEXANYL ACETATE

TOXICITY DATA with REFERENCE
skn-rbt 500 mg/24H MOD FCTXAV 17(suppl),695,79
ihl-hmn TCLo:3000 mg/m³/45M:IRR AHYGAJ 78,260,13
orl-rat LD50:6730 mg/kg TXAPA9 28,313,74
scu-cat LDLo:606 mg/kg AHYGAJ 78,260,13
skn-rbt LD50:10 g/kg TXAPA9 28,313,74

CONSENSUS REPORTS: Reported in EPA TSCA Inventory.

DOT Classification: Flammable or Combustible Liquid; Label: Flammable Liquid.

SAFETY PROFILE: Moderately toxic by subcutaneous route. Mildly toxic by ingestion and skin contact. Human systemic effects by inhalation: conjunctiva irritation and unspecified respiratory system changes. A systemic irritant to humans. Flammable when exposed to heat or flame. When heated to decomposition it emits acrid smoke and irritating fumes.

CPF500 CAS:108-91-8 ***HR: 3***
CYCLOHEXYLAMINE
DOT: UN 2357
mf: $C_6H_{13}N$ mw: 99.20

PROP: Liquid; strong, fishy odor. Mp: −17.7°, bp: 134.5°, flash p: 69.8°F, d: 0.865 @ 25°/25°, autoign temp: 560°F, vap d. 3.42.

SYNS: AMINOCYCLOHEXANE ◇ AMINOHEXAHYDROBENZENE ◇ CHA ◇ CYCLOHEXANAMINE ◇ HEXAHYDROANILINE ◇ HEXAHYDROBENZENAMINE

TOXICITY DATA with REFERENCE
skn-hmn 125 mg/48H SEV AMIHBC 5,311,52
cyt-hmn:leu 10 μmol/L/5H MUREAV 39,1,76
cyt-ham:fbr 10 mg/L MUREAV 39,1,76
dni-hmn:hla 100 μg/L INHEAO 9,188,71
orl-mus TDLo:600 mg/kg (female 6-11D post):TER SEIJBO 11,51,71

ipr-rat TDLo:300 mg/kg (male 1D pre):REP FCTXAV 10,29,72
orl-rat LD50:156 mg/kg SKEZAP 14,542,73
ihl-rat LC50:7500 mg/m^3 GTPZAB 7(11),51,63
orl-mus LD50:224 mg/kg 85GMAT -,41,82
ihl-mus LC50:1070 mg/m^3 GTPZAB 7(11),51,63
scu-mus LD50:1150 mg/kg VOONAW 4,659,58
skn-rbt LD50:277 mg/kg AIHAAP 30,470,69
par-rbt LDLo:500 mg/kg IECHAD 29,1247,37
ipr-mam LD50:200 mg/kg AMIHBC 5,311,52

CONSENSUS REPORTS: IARC Cancer Review: Group 3 IMEMDT 7,178,87; Animal No Evidence IMEMDT 22,55,80. EPA Extremely Hazardous Substances List. EPA Genetic Toxicology Program. Reported in EPA TSCA Inventory.

OSHA PEL: TWA 10 ppm
ACGIH TLV: TWA 10 ppm
DFG MAK: 10 ppm (40 mg/m^3)
DOT Classification: Flammable Liquid; Label: Flammable Liquid, Corrosive; Flammable or Combustible Liquid; Label: Flammable, Corrosive.

SAFETY PROFILE: A poison by ingestion, skin contact, and intraperitoneal routes. Moderately toxic by subcutaneous and parenteral routes. Experimental teratogenic and reproductive effects. Severe human skin irritant. Can cause dermatitis and convulsions. Human mutation data reported. Questionable carcinogen. Flammable liquid. Dangerous fire hazard when exposed to heat, flame, or oxidizers. To fight fire, use alcohol foam, CO_2, dry chemical. When heated to decomposition it emits toxic fumes of NO_x.

CPF750 CAS:19834-02-7 *HR: 3*
CYCLOHEXYLAMINE SULFATE
mf: $C_6H_{13}N \cdot H_2O_4S$ mw: 197.28

SYNS: CHA-SULFATE ◇ CHS ◇ CYCLOHEXAMINE SULFATE

TOXICITY DATA with REFERENCE
orl-mus TDLo:42 g/kg (multi) :REP TXCYAC 8,285,77
orl-mus TDLo:42 g/kg (multi) :TER TXCYAC 8,285,77
orl-rat TDLo:11 g/kg/2Y-C:ETA SCIEAS 167,1131,70

SAFETY PROFILE: Questionable carcinogen with experimental tumorigenic data. An experimental teratogen. Other experimental reproductive effects. When heated to decomposition it emits very toxic fumes of SO_x and NO_x. See also AMINES and SULFATES.

CPG000 CAS:58695-41-3 *HR: 2*
CYCLOHEXYLAMINO ACETIC ACID
mf: $C_8H_{15}NO_2$ mw: 157.24

SYNS: CYKLOHEXYLAMINACETAT (CZECH) ◇ OCTAN CYKLOHEXYLAMINU (CZECH)

TOXICITY DATA with REFERENCE
skn-rbt 500 mg/24H SEV 28ZPAK -,64,72
eye-rbt 500 mg/24H MLD 28ZPAK -,64,72
orl-rat LD50:2120 mg/kg 28ZPAK -,64,72

SAFETY PROFILE: Moderately toxic by ingestion. An eye and severe skin irritant. When heated to decomposition it emits toxic fumes of NO_x.

CPG125 CAS:2842-38-8 *HR: 1*
2-(CYCLOHEXYLAMINO)ETHANOL
mf: $C_8H_{17}NO$ mw: 143.26

SYNS: ABROMEEN E-25 ◇ N-(2-HYDROXYETHYL)CYCLOHEXYLAMINE

TOXICITY DATA with REFERENCE
scu-mus TDLo:900 mg/kg (6-14D preg):TER NTIS** PB223-160
orl-rat LD50:38300 mg/kg 34ZIAG -,61,69

SAFETY PROFILE: Very mildly toxic by ingestion. An experimental teratogen. An eye irritant. When heated to decomposition it emits toxic fumes of NO_x.

CPG250 CAS:65210-28-8 *HR: 3*
2-(2-(CYCLOHEXYLAMINO)ETHYL-2-METHYL-1,3-BENZODIOXOLE HYDROCHLORIDE
mf: $C_{16}H_{23}NO_2 \cdot ClH$ mw: 297.86

TOXICITY DATA with REFERENCE
ivn-rat LD50:20 mg/kg EJMCA5 12,413,77
ipr-mus LD50:79 mg/kg EJMCA5 12,413,77

SAFETY PROFILE: Poison by intravenous and intraperitoneal routes. When heated to decomposition it emits very toxic fumes of HCl and NO_x.

CPG500 CAS:57281-35-3 *HR: 3*
4-(CYCLOHEXYLAMINO)-1-(NAPHTHALENYLOXY)-2-BUTANOL
mf: $C_{20}H_{27}NO_2$ mw: 313.48

SYN: CHINOIN 103

TOXICITY DATA with REFERENCE
orl-mus LD50:178 mg/kg DRFUD4 4,12,79
ivn-mus LD50:50 mg/kg DRFUD4 4,12,79

SAFETY PROFILE: Poison by ingestion and intravenous routes. When heated to decomposition it emits toxic fumes of NO_x.

CPG625 CAS:64011-62-7 *HR: 3*
dl-1-CYCLOHEXYL-2-AMINOPROPANE HYDROCHLORIDE
mf: $C_9H_{19}N \cdot ClH$ mw: 177.75

SYN: (±)-α-METHYLCYCLOHEXANEETHYLAMINE HYDROCHLORIDE

TOXICITY DATA with REFERENCE
ipr-rat LD50:65 mg/kg JPETAB 100,267,50
ipr-rbt LDLo:100 mg/kg JPETAB 100,267,50
ipr-gpg LDLo:50 mg/kg JPETAB 100,267,50

SAFETY PROFILE: Poison by intraperitoneal route. When heated to decomposition it emits toxic fumes of NO_x and HCl.

CPH250 CAS:34961-28-9 ***HR: 2***
CYCLOHEXYLAMMONIUM FORMATE
mf: $CH_2O_2 \cdot C_6H_{13}N$ mw: 145.23

TOXICITY DATA with REFERENCE
skn-hmn 250 mg/48H SEV AMIHBC 5,311,52
skn-rbt 500 mg MOD AMIHBC 5,311,52
ipr-mam LD50:580 mg/kg AMIHBC 5,311,52

SAFETY PROFILE: Moderately toxic by intraperitoneal route. A severe human skin irritant. When heated to decomposition it emits toxic fumes of NO_x and NH_3. See also FORMIC ACID and CYCLOHEXYLAMINE.

CPH500 CAS:15860-21-6 ***HR: 2***
CYCLOHEXYLAMMONIUM STEARATE
mf: $C_{18}H_{36}O_2 \cdot C_6H_{13}N$ mw: 383.74

SYN: STEARIC ACID with CYCLOHEXYLAMINE (1:1)

TOXICITY DATA with REFERENCE
skn-hmn 500 mg/48H SEV AMIHBC 5,311,52
skn-rbt 500 mg MLD AMIHBC 5,311,52
ipr-mam LD50:4 g/kg AMIHBC 5,311,52

SAFETY PROFILE: Moderately toxic by intraperitoneal route. A severe human skin irritant. When heated to decomposition it emits toxic fumes of NO_x and NH_3. See also STEARIC ACID and CYCLOHEXYLAMINE.

CPI000 CAS:13311-57-4 ***HR: 3***
N-CYCLOHEXYL-1-AZIRIDINECARBOXAMIDE
mf: $C_9H_{16}N_2O$ mw: 168.27

SYNS: CYCLOHEXYL-N-CARBAMOYLAZIRIDINE ◇ N-CYCLOHEXYL-N-CARBAMOYLAZIRIDINE

TOXICITY DATA with REFERENCE
ipr-mus TDLo:240 mg/kg/4W-I:NEO CNREA8 29,2184,69

SAFETY PROFILE: Questionable carcinogen with experimental neoplastigenic data. When heated to decomposition it emits toxic fumes of NO_x.

CPI250 CAS:95-33-0 ***HR: 3***
N-CYCLOHEXYL-2-BENZOTHIAZOLESULFENAMIDE
mf: $C_{13}H_{16}N_2S_2$ mw: 264.43

PROP: Light tan or buff powder. Mp: 94°, d: 1.27 @ 25°.

SYNS: DURAX ◇ PENNAC CBS ◇ SANTOCURE ◇ SULFENAMIDE TS

TOXICITY DATA with REFERENCE
par-rat TDLo:400 mg/kg (4-11D preg):TER BEXBAN 93,107,82
par-rat TDLo:400 mg/kg (4-11D preg):REP BEXBAN 93,107,82
orl-mus TDLo:76 g/kg/78W-I:ETA NTIS** PB223-159
ivn-mus LD50:32 mg/kg CSLNX* NX#02243

CONSENSUS REPORTS: Reported in EPA TSCA Inventory.

SAFETY PROFILE: A poison by intravenous route. Questionable carcinogen with experimental tumorigenic data. An experimental teratogen. Experimental reproductive effects. When heated to decomposition it emits toxic fumes of SO_x and NO_x.

CPI350 CAS:63441-20-3 ***HR: 1***
1-(CYCLOHEXYLCARBONYL)-3-METHYLPIPERIDINE
mf: $C_{13}H_{23}NO$ mw: 209.37

SYNS: AI3-36537 ◇ PIPERIDINE, 1-(CYCLOHEXYLCARBONYL)-3-METHYL-

TOXICITY DATA with REFERENCE
eye-rbt 100 mg/24H MOD AEHA** 51-029-76

SAFETY PROFILE: An eye irritant. When heated to decomposition it emits toxic fumes of NO_x.

CPJ000 CAS:32808-51-8 ***HR: 3***
4-(4-CYCLOHEXYL-3-CHLOROPHENYL)-4-OXOBUTYRIC ACID
mf: $C_{16}H_{19}ClO_3$ mw: 294.80

SYNS: l'ACIDE BUCLOXIQUE (FRENCH) ◇ BENZENEBUTANOIC ACID, 3-CHLORO-4-CYCLOHEXYL-α-OXO- ◇ BUCLOSINSAEURE (GERMAN) ◇ BUCLOXIC ACID ◇ BUCLOXONIC ACID ◇ 804 CB ◇ 3-(3-CHLORO-4-CYCLOHEXYLBENZOYL)PROPIONIC ACID ◇ 3-CHLORO-4-CYCLOHEXYL-α-OXOBENZENEBUTANOIC ACID ◇ 4-(3-CHLORO-4-CYCLOHEXYLPHENYL)-4-OXO-BUTYRIC ACID ◇ ESFAR

TOXICITY DATA with REFERENCE
orl-mus TDLo:975 mg/kg (6-18D preg):TER ARZNAD 24,1398,74
orl-rbt TDLo:550 mg/kg (7-28D preg):REP ARZNAD 24,1398,74
orl-rat LD50:120 mg/kg ARZNAD 24,1364,74
ipr-rat LD50:195 mg/kg ARZNAD 24,1398,74
orl-mus LD50:852 mg/kg ARZNAD 24,1398,74
ipr-mus LD50:1100 mg/kg ARZNAD 24,1364,74

SAFETY PROFILE: Poison by intraperitoneal and ingestion routes. Experimental teratogenic and reproductive effects. An anti-inflammatory agent. When heated to decomposition it emits toxic fumes of Cl^-.

CPJ250 CAS:32808-53-0 ***HR: 3***
4-(4-CYCLOHEXYL-3-CHLOROPHENYL)-4-OXOBUTYRIC ACID CALCIUM SALT
mf: $C_{32}H_{36}Cl_2O_6{\bullet}Ca$ mw: 627.66

SYNS: ACIDE BUCLOXIQUE CALCIUM (FRENCH) ◇ l'ACIDE (CYCLOHEXYL-4, CHLORO-3, PHENYL)-4,OXO-4, BUTYRIQUE CALCIUM (FRENCH) ◇ BUCLOXIC ACID CALCIUM ◇ BUCLOXIC ACID CALCIUM SALT ◇ BUCLOXINSAEURE KALZIUM (FERMAN) ◇ BUCLOXONIC ACID CALCIUM SALT ◇ CALCIUM BUCLOXATE ◇ CALCIUM ESFAR ◇ CB 804 CALCIUM ◇ 4-(3-CHLOR-4-CYCLOHEXYL-PHENYL)-4-OXO-BUTTERSAEURE KALZIUM (GERMAN) ◇ 3-(3-CHLORO-4-CYCLOHEXYLBENZOYL)PROPIONIC ACID CALCIUM SALT ◇ 3-CHLORO-4-CYCLOHEXYL-α-OXO-BENZENEBUTANOIC ACID ◇ 3-CHLORO-4-CYCLOHEXYL-α-OXOBENZENEBUTANOIC ACID CALCIUM SALT ◇ 4-(3-CHLORO-4-CYCLOHEXYLPHENYL)-4-OXOBUTYRIC ACID CALCIUM SALT ◇ ESFAR CALCIUM

TOXICITY DATA with REFERENCE
orl-rat LD50:175 mg/kg ARZNAD 24,1364,74
ipr-rat LD50:200 mg/kg ARZNAD 24,1364,74
orl-mus LD50:1700 mg/kg ARZNAD 24,1360,74
ipr-mus LD50:1700 mg/kg ARZNAD 24,1364,74

SAFETY PROFILE: Poison by ingestion and intraperitoneal routes. Used as an anti-inflammatory agent. When heated to decomposition it emits toxic fumes of Cl^-.

CPJ500 CAS:92-64-8 ***HR: 3***
CYCLOHEXYLCYANOETHYLETHANOLAMINE
mf: $C_{11}H_{14}N_2O$ mw: 190.27

SYNS: 2-(N-(2-CYANOETHYL)-N-CYCLOHEXYL)AMINO-ETHANOL ◇ N-(β-CYANOETHYL)-N-(β-HYDROXYETHY)-ANILINE ◇ N-β-HYDROXYETHYL-N-β-KYANETHYLANILIN (CZECH)

TOXICITY DATA with REFERENCE
eye-rbt 500 mg/24H MLD 28ZPAK -,162,73
orl-rat LD50:3410 mg/kg 28ZPAK -,162,72
ipr-mus LD50:200 mg/kg NTIS** AD691-490

CONSENSUS REPORTS: Cyanide and its compounds are on the Community Right-To-Know List.

SAFETY PROFILE: Poison by intraperitoneal route. Moderately toxic by ingestion. An eye irritant. When heated to decomposition it emits toxic fumes of NO_x and CN^-. See also NITRILES.

CPK000 ***HR: 2***
N-CYCLOHEXYL-N-DIETHYLTHIOCARBONYL SULFONAMIDE

SYN: THIOPENTEX

TOXICITY DATA with REFERENCE
skn-hmn 250 mg/48H MOD AMIHBC 5,311,52
skn-rbt 500 mg MLD AMIHBC 5,311,52
ipr-mam LD50:1200 mg/kg AMIHBC 5,311,52

SAFETY PROFILE: Moderately toxic by intraperitoneal route. A human skin irritant. When heated to decomposition it emits very toxic fumes of SO_x.

CPK500 CAS:131-89-5 ***HR: 3***
2-CYCLOHEXYL-4,6-DINITROPHENOL
DOT: NA 9026
mf: $C_{12}H_{14}N_2O_5$ mw: 266.28

PROP: Crystals.

SYNS: 6-CICLOESIL-2,4-DINITR-FENOLO (ITALIAN) ◇ 2-CYCLOHEXYL-4,6-DINITROFENOL (DUTCH) ◇ 6-CYCLOHEXYL-2,4-DINITROPHENOL ◇ DINEX ◇ DINITROCYCLOHEXYLPHENOL ◇ DINITRO-o-CYCLOHEXYLPHENOL ◇ 2,4-DINITRO-6-CYCLOHEXYLPHENOL ◇ 4,6-DINITRO-o-CYCLOHEXYLPHENOL ◇ DINITROCYCLOHEXYLPHENOL (DOT) ◇ DN DRY MIX No. 1 ◇ DN DUST No. 12 ◇ DNOCHP ◇ DOWSPRAY 17 ◇ DRY MIX No. 1 ◇ ENT 157 ◇ PEDINEX (FRENCH) ◇ RCRA WASTE NUMBER P034 ◇ SN 46

TOXICITY DATA with REFERENCE
skn-rbt 105 mg/9D-I MOD JIHTAB 30,10,48
orl-rat LD50:65 mg/kg ARSIM* 20,9,66
orl-mus LD50:50 mg/kg 85DPAN -,-,71/76
ipr-mus LD50:25 mg/kg BCPCA6 18,1389,69
scu-mus LDLo:30 mg/kg UCPHAQ 1,151,39
ivn-dog LDLo:8 mg/kg AIPTAK 50,20,35
orl-rbt LDLo:100 mg/kg UCPHAQ 1,151,39
scu-rbt LDLo:40 mg/kg UCPHAQ 1,151,39
orl-gpg LD50:50 mg/kg PCOC** -,417,66
skn-gpg LDLo:1000 mg/kg PCOC** -,417,66
scu-gpg LDLo:20 mg/kg PCOC** -,417,66

DOT Classification: ORM-A; Label: None.

SAFETY PROFILE: A poison by ingestion, intraperitoneal, intravenous, subcutaneous, and possibly other routes. Moderately toxic by skin contact. A skin irritant. Fire hazard. See also NITRATES and PHENOLS. Can react with oxidizers. When heated to decomposition it emits toxic fumes of NO_x.

CPK625 CAS:52694-54-9 ***HR: 3***
(+)-1-CYCLOHEXYL-4-(1,2-DIPHENYLETHYL) PIPERAZINE DIHYDROCHLORIDE
mf: $C_{24}H_{32}N_3{\bullet}2ClH$ mw: 421.50

SYN: (S)-1-CYCLOHEXYL-4-(1,2-DIPHENYLETHYL)-PIPERAZINEDIHYDROCHLORIDE

TOXICITY DATA with REFERENCE
ivn-rat LD50:8 mg/kg AIPTAK 221,105,76
orl-mus LD50:274 mg/kg AIPTAK 221,105,76

scu-mus LD50:320 mg/kg AIPTAK 221,105,76
ivn-mus LD50:18500 μg/kg AIPTAK 221,105,76

SAFETY PROFILE: Poison by ingestion, subcutaneous, and intravenous routes. When heated to decomposition it emits toxic fumes of NO_x and HCl.

CPL100 CAS:10328-51-5 ***HR: 3***
N,N′-(1,4-CYCLOHEXYLENEDIMETHYLENE)BIS(2-(1-AZIRIDINYL)ACETAMIDE)
mf: $C_{16}H_{28}N_4O_2$ mw: 308.48

SYNS: ACETAMIDE, N,N′-(1,4-CYCLOHEXYLENEDIMETHYLENE) BIS(2-(1-AZIRIDINYL)- ◇ 1-AZIRIDINEACETAMIDE, N,N′-(1,4-CYCLOHEXYLENEDIMETHYLENE)BIS- ◇ N,N′-BIS-AZIRIDINYLACETYL-1,4-CYCLOHEXYLDIMETHYLENEDIAMINE

TOXICITY DATA with REFERENCE
cyt-rat-orl 300 μg/kg MUREAV 31,115,75
ipr-mus TDLo:50 mg/kg (male 5D pre):REP EXPEAM 24,924,68
orl-mus LD50:71 mg/kg EXPEAM 24,924,68
ipr-mus LD50:45 mg/kg EXPEAM 24,924,68

SAFETY PROFILE: Poison by ingestion and intraperitoneal routes. Experimental reproductive effects. Mutation data reported. When heated to decomposition it emits toxic fumes of NO_x.

CPL250 CAS:4442-79-9 ***HR: 2***
2-CYCLOHEXYLETHANOL
mf: $C_8H_{16}O$ mw: 128.24

SYNS: CYCLOHEXYLETHYL ALCOHOL ◇ HEXAHYDROPHENYLETHYL ALCOHOL

TOXICITY DATA with REFERENCE
orl-rat LD50:940 mg/kg FCTXAV 13,785,75
skn-rbt LD50:1220 mg/kg FCTXAV 13,785,75

CONSENSUS REPORTS: Reported in EPA TSCA Inventory.

SAFETY PROFILE: Moderately toxic by ingestion and skin contact. When heated to decomposition it emits acrid smoke and irritating fumes. See also ALCOHOLS.

CPL750 CAS:13908-93-5 ***HR: 3***
CYCLOHEXYL FLUOROETHYL NITROSOUREA
mf: $C_9H_{16}FN_3O_2$ mw: 217.28

SYNS: CFNU ◇ 3-CYCLOHEXYL-1-(2-FLUOROETHYL)-1-NITROSOUREA ◇ N′-CYCLOHEXYL-N-(2-FLUOROETHYL)-N-NITROSOUREA ◇ FCNU ◇ 1-FLUOROETHYL-3-CYCLOHEXYL-1-NITROSOUREA ◇ NSC 87974 ◇ SRI 2619

TOXICITY DATA with REFERENCE
orl-rat LD50:18500 μg/kg TXAPA9 10,397,67
ivn-rat LD50:12 mg/kg TXAPA9 10,397,67
orl-mus LD50:111 mg/kg TXAPA9 10,397,67
ipr-mus LD10:34 mg/kg CNREA8 34,194,74
scu-mus LD50:25210 μg/kg NCISP* JAN86
ivn-mus LD50:51 mg/kg TXAPA9 10,397,67

SAFETY PROFILE: Poison by ingestion, subcutaneous, intravenous, and intraperitoneal routes. Many N-nitroso compounds are carcinogens. When heated to decomposition it emits very toxic fumes of F^- and NO_x. See also N-NITROSO COMPOUNDS.

CPM250 CAS:78128-81-1 ***HR: 3***
3-CYCLOHEXYL-4-HYDROXY-2(5H)FURANONE
mf: $C_{10}H_{14}O_3$ mw: 182.24

SYN: α-CYCLOHEXYL-β-HYDROXY-Δ^{α},β-BUTENOLID (GERMAN)

TOXICITY DATA with REFERENCE
scu-mus LD50:416 mg/kg ARZNAD 11,277,61
ivn-mus LD50:155 mg/kg ARZNAD 11,277,61

SAFETY PROFILE: Poison by intravenous route. Moderately toxic by subcutaneous route. When heated to decomposition it emits acrid smoke and irritating fumes. See also KETONES.

CPM750 CAS:6856-43-5 ***HR: 3***
1-(3-CYCLOHEXYL-3-HYDROXY-3-PHENYL-PROPYL)- 1-METHYL-PIPERIDINIUM IODIDE
mf: $C_{20}H_{31}NO•CH_3I$ mw: 443.46

SYNS: 1-CYCLOHEXYL-1-PHENYL-3-PIPERIDINO-PROPANOL, METHYLIODIDE ◇ α-CYCLOHEXYL-α-(2-(PIPERIDINO)ETHYL)-BENZYLALCOHOL METHYLIODIDE ◇ WIN 1593

TOXICITY DATA with REFERENCE
orl-mus LD50:2520 mg/kg JPETAB 110,282,54
ivn-mus LD50:12 mg/kg JPETAB 110,282,54

SAFETY PROFILE: Poison by intravenous route. Moderately toxic by ingestion. See also IODIDES. When heated to decomposition it emits very toxic fumes of I^- and NO_x.

CPN500 CAS:3173-53-3 ***HR: 3***
CYCLOHEXYL ISOCYANATE
DOT: UN 2488
mf: $C_7H_{11}NO$ mw: 125.19

SYNS: ISOCYANIC ACID, CYCLOHEXYL ESTER ◇ NSC 87419

TOXICITY DATA with REFERENCE
mmo-sat 150 μg/plate ABCHA6 44,3017,80
ipr-mus LD50:13 mg/kg NCISP* JAN86
ivn-mus LD50:18 mg/kg CSLNX* NX#04502

CONSENSUS REPORTS: Reported in EPA TSCA Inventory.

DOT Classification: Poison B; Label: Flammable Liquid and Poison.

SAFETY PROFILE: Poison by intravenous and intraperitoneal routes. Mutation data reported. Flammable when exposed to heat or flame. When heated to decomposition it emits toxic fumes of NO_x. See also CYANATES and ESTERS.

CPN750 CAS:3687-61-4 ***HR: 3***
2-(N-CYCLOHEXYL-N-ISOPROPYLAMINOMETHYL)-1,3,4-OXADIAZOLE
mf: $C_{15}H_{21}N_3O_2$ mw: 275.39

SYNS: AF 594 ◇ 5-(2-(DIETHYLAMINO)ETHYL)-3-(p-METHOXYPHENYL)-1,2,4-OXADIAZOLE ◇ N,N-DIETHYL-3-(4-METHOXYPHENYL)-1,2,4-OXADIAZOLE-5-ETHANAMINE ◇ 3-p-METHOXYPHENYL-5-DIETHYLAMINOETHYL-1,2,4-OXADIAZOLE ◇ MEXOLAMINE ◇ R 1067

TOXICITY DATA with REFERENCE
orl-rat LD50:1899 mg/kg ARZNAD 12,539,62
ipr-rat LD50:288 mg/kg ARZNAD 12,539,62
orl-mus LD50:722 mg/kg ARZNAD 12,539,62
ipr-mus LD50:331 mg/kg ARZNAD 12,539,62
scu-mus LD50:691 mg/kg ARZNAD 12,539,62
ivn-mus LD50:83 mg/kg ARZNAD 12,539,62
orl-gpg LD50:700 mg/kg ARZNAD 12,539,62
ipr-gpg LD50:233 mg/kg ARZNAD 12,539,62

SAFETY PROFILE: Poison by intraperitoneal and intravenous routes. Moderately toxic by ingestion and subcutaneous routes. When heated to decomposition it emits toxic fumes of NO_x.

CPO500 CAS:4388-82-3 ***HR: 3***
1,1-CYCLOHEXYL-2-METHYLAMINOPROPANE-5,5-PHENYLETHYLBARBITURATE
mf: $C_{12}H_{12}N_2O_3 \cdot C_{10}H_{21}N$ mw: 387.58

SYNS: BARBEXACLONE ◇ BARBEXACLONUM ◇ BARBITURIC ACID,1-(1-(1-CYCLOHEXYL-N-METHYL-2-PROPANAMINE)-5-ETHYL-5-PHENYL ◇ CHP-PHENOBARBITALAT (GERMAN) ◇ MALIASIN

TOXICITY DATA with REFERENCE
orl-rat LD50:306 mg/kg ARZNAD 13,613,63
orl-mus LD50:334 mg/kg ARZNAD 13,613,63

SAFETY PROFILE: Poison by ingestion. Human reproductive effects. When heated to decomposition it emits toxic fumes of NO_x. See also BARBITURATES.

CPP000 CAS:59182-63-7 ***HR: 3***
N-(2-CYCLOHEXYL-1-METHYLETHYL)-3,3-DIPHENYLPROPYLAMINE HYDROCHLORIDE
mf: $C_{24}H_{33}N \cdot ClH$ mw: 372.04

SYNS: DROPRENILAMINE HYDROCHLORIDE ◇ MG 8926 ◇ VALCOR

TOXICITY DATA with REFERENCE
orl-rat LD50:1550 mg/kg ARZNAD 26,2127,76
ipr-rat LD50:65 mg/kg ARZNAD 26,2127,76
orl-mus LD50:2850 mg/kg ARZNAD 26,2127,76
ipr-mus LD50:68 mg/kg ARZNAD 26,2127,76

SAFETY PROFILE: Poison by intraperitoneal route. Moderately toxic by ingestion. When heated to decomposition it emits very toxic fumes of HCl and NO_x.

CPP750 CAS:70907-61-8 ***HR: 3***
1-(2-CYCLOHEXYLPHENOXY)-1-(2-IMIDAZOLINYL)ETHANE HYDROCHLORIDE
mf: $C_{17}H_{24}N_2O \cdot ClH$ mw: 308.89

SYN: MG 18512

TOXICITY DATA with REFERENCE
orl-mus LD50:1650 μg/kg ARZNAD 29,729,79
ipr-mus LD50:300 μg/kg ARZNAD 29,729,79

SAFETY PROFILE: A deadly poison by ingestion and intraperitoneal routes. When heated to decomposition it emits very toxic fumes of HCl and NO_x.

CPQ250 CAS:77-37-2 ***HR: 3***
1-CYCLOHEXYL-1-PHENYL-3-PYRROLIDINO-1-PROPANOL
mf: $C_{19}H_{29}NO$ mw: 287.49

SYNS: 1-CYCLOHEXYL-1-PHENYL-3-(1-PYRROLIDINYL)-1-PROPANOL ◇ ELORINE ◇ KEMADRINE ◇ LERGINE ◇ METANIN ◇ OSNERVAN ◇ PROCIDLIDINA ◇ PROCYCLIDINE ◇ PROCYKLIDIN ◇ PROSYKLIDIN ◇ SPAMOL ◇ TRICICLIDINA ◇ TRICILOID ◇ TRICOLOID ◇ TRICYCLAMOL ◇ VAGOSIN

TOXICITY DATA with REFERENCE
ipr-mus LD50:131 mg/kg 27ZQAG -,291,72
ivn-mus LD50:60 mg/kg 27ZQAG -,291,72

SAFETY PROFILE: A poison by intraperitoneal and intravenous routes. When heated to decomposition it emits toxic fumes of NO_x.

CPQ275 CAS:6837-24-7 ***HR: 3***
1-CYCLOHEXYL-2-PYRROLIDINONE
mf: $C_{10}H_{17}NO$ mw: 167.28

SYNS: N-CYCLOHEXYLPYRROLIDINONE ◇ N-CYCLOHEXYLPYRROLIDONE ◇ 2-PYRROLIDINONE, 1-CYCLOHEXYL-

TOXICITY DATA with REFERENCE
skn-rbt 500 mg/24H SEV FCTOD7 26,475,88
eye-rbt 100 mg SEV FCTOD7 26,475,88
orl-rat LD50:370 mg/kg FCTOD7 26,475,88
ihl-rat LC50:120 ppm/1H FAATDF 4,587,84
skn-rbt LD50:1600 mg/kg FAATDF 4,587,84

CONSENSUS REPORTS: Reported in EPA TSCA Inventory.

SAFETY PROFILE: Poison by ingestion. Moderately toxic by inhalation and skin contact. A severe skin and eye irritant. When heated to decomposition it emits toxic fumes of NO_x.

CPQ625 CAS:100-88-9 ***HR: 3***
N-CYCLOHEXYLSULPHAMIC ACID
mf: $C_6H_{13}NO_3S$ mw: 179.26

PROP: Crystals; sweet-sour taste. Mp: 169-170°. Fairly strong acid. Very sparingly soluble in water. Slowly hydrolyzed by hot water.

SYNS: CYCLAMATE ◇ CYCLAMIC ACID ◇ CYCLOHEXANESULPHAMIC ACID ◇ CYCLOHEXYLAMIDOSULPHURIC ACID ◇ CYCLOHEXYLAMINESULPHONIC ACID ◇ CYCLOHEXYLSULFAMIC ACID (9CI) ◇ CYCLOHEXYLSULPHAMIC ACID ◇ HEXAMIC ACID ◇ SUCARYL ◇ SUCARYL ACID

TOXICITY DATA with REFERENCE
orl-man TDLo:22 g/kg/77W-C:CAR,KID JOURAA 118,258,77
orl-man TD:131 g/kg/5Y-C:CAR,KID JOURAA 118,258,77
orl-man TD:164 g/kg/6Y-C:CAR,KID JOURAA 118,258,77
orl-rat LD50:12 g/kg AJMSA9 225,551,53
ivn-rat LD50:4 g/kg AJMSA9 225,551,53
orl-mus LD50:10 g/kg AJMSA9 225,551,53
ivn-mus LD50:180 mg/kg CSLNX* NX#01774

CONSENSUS REPORTS: Reported in EPA TSCA Inventory.

SAFETY PROFILE: Suspected human carcinogen producing bladder tumors. Poison by intravenous route. Mildly toxic by ingestion. When heated to decomposition it emits toxic fumes of SO_x and NO_x.

CPQ650 CAS:29396-39-2 ***HR: 3***
3-CYCLOHEXYLSYDNONE IMINE MONOHYDROCHLORIDE
mf: $C_8H_{13}N_3O{\bullet}ClH$ mw: 203.70

SYN: N-(ZYKLOHEXYL)-SYDNONIMIN HYDROCHLORID (GERMAN)

TOXICITY DATA with REFERENCE
orl-mus LD50:206 mg/kg ABMGAJ 14,369,65
ipr-mus LD50:63 mg/kg OYYAA2 2,280,68
ivn-mus LD50:70 mg/kg JMCMAR 14,1013,71

SAFETY PROFILE: Poison by ingestion, intravenous and intraperitoneal routes. When heated to decomposition, it emits toxic fumes of NO_x and HCl.

CPR000 CAS:664-95-9 ***HR: 2***
1-CYCLOHEXYL-3-p-TOLYSULFONYLUREA
mf: $C_{14}H_{20}N_2O_3S$ mw: 296.42

PROP: Fine, white crystals.

SYNS: 1-CICLOESIL-3-p-TOLILSOLFONILUREA(ITALIAN) ◇ CYCHLORAL ◇ CYCLAMID ◇ CYCLAMIDE ◇ 1-CYCLOHEXYL-3-p-TOLUENESULFONYLUREA ◇ DIABORAL ◇ GLICOSIL ◇ GLYCYCLAMIDE ◇ N-(4-METHYLBENZENESULFONYL)-N'-CYCLOHEXYLUREA ◇ TOLCYCLAMIDE ◇ TOLHEXAMIDE ◇ 1-(p-TOLYLSULFONYL)-3-CYCLOHEXYLUREA

TOXICITY DATA with REFERENCE
orl-rat TDLo:800 mg/kg (9D preg):TER FATOAO 28,616,65
ipr-rat LD50:870 mg/kg FRPSAX 12,268,57
ipr-mus LD50:1150 mg/kg RPOBAR 2,280,70

SAFETY PROFILE: Moderately toxic by intraperitoneal route. An experimental teratogen. When heated to decomposition it emits toxic fumes of SO_x and NO_x.

CPR250 CAS:98-12-4 ***HR: 3***
CYCLOHEXYLTRICHLOROSILANE
DOT: UN 1763
mf: $C_6H_{11}Cl_3Si$ mw: 217.61

CONSENSUS REPORTS: Reported in EPA TSCA Inventory.

DOT Classification: Corrosive Material; Label: Corrosive.

SAFETY PROFILE: A highly toxic and corrosive material. When heated to decomposition it emits toxic fumes of Cl^-. See also CHLOROSILANES.

CPR500 CAS:16607-80-0 ***HR: 3***
1-CYCLOHEXYLTRIMETHYLAMINE
mf: $C_9H_{19}N$ mw: 141.29

SYN: N,N-DIMETHYL-N-CYCLOHEXYLMETHYLAMINE

TOXICITY DATA with REFERENCE
orl-rat LD50:1230 mg/kg AIHAAP 30,470,69
skn-rbt LD50:210 mg/kg AIHAAP 30,470,69

SAFETY PROFILE: Poison by skin contact. Moderately toxic by ingestion. See also AMINES. When heated to decomposition it emits toxic fumes of NO_x.

CPR750 CAS:742-20-1 ***HR: 3***
CYCLOMETHIAZIDE
mf: $C_{13}H_{18}ClN_3O_4S_2$ mw: 379.91

SYNS: CYCLOPENTHIAZIDE ◇ 3-CYCLOPENTYLMETHYL HYDROCHLOROTHIAZIDE DERIV ◇ NAVIDREX ◇ NAVIDRIX ◇ SALIMED ◇ SALIMID ◇ SU 8341 ◇ TSIKLOMETIAZID

TOXICITY DATA with REFERENCE
ivn-rat LD50:142 mg/kg AIPTAK 131,325,61
ivn-mus LD50:232 mg/kg MEIEDD 10,394,83

SAFETY PROFILE: Poison by intravenous route. An antihypertensive agent. When heated to decomposition it emits very toxic fumes of SO_x, NO_x, and Cl^-.

CPR800 CAS:121-82-4 ***HR: 3***
CYCLONITE
DOT: UN 0072/UN 0118
mf: $C_3H_6N_6O_6$ mw: 222.15

PROP: White, crystalline powder. Mp: 202°.

SYNS: CYCLOTRIMETHYLENENITRAMINE ◇ CYCLOTRIMETHYLENETRINITRAMINE ◇ CYCLOTRIMETHYLENETRINITRAMINE, containing at least 10%-25% water (DOT) ◇ CYCLOTRIMETHYLENETRINITRAMINE, desensitized (DOT) ◇ ESAIDRO-1,3,5-TRINITRO-1,3,5-TRIAZINA (ITALIAN) ◇ HEKSOGEN (POLISH) ◇ HEXAHYDRO-1,3,5-TRINITRO-1,3,5-TRIAZIN (GERMAN) ◇ HEXAHYDRO-1,3,5-TRINITRO-s-TRIAZINE ◇ HEXAHYDRO-1,3,5-TRINITRO-1,3,5-TRIAZINE ◇ HEXOGEEN (DUTCH) ◇ HEXOGEN (explosive) ◇ HEXOGEN 5W ◇ HEXOLITE ◇ HEXOLITE, dry or containing, by weight, less than 15% water (DOT) ◇ PBX(AF) 108 ◇ RDX ◇ TRIMETHYLEENTRINITRAMINE (DUTCH) ◇ TRIMETHYLENETRINITRAMINE ◇ sym-TRIMETHYLENETRINITRAMINE ◇ TRINITROCYCLOTRIMETHYLENE TRIAMINE ◇ 1,3,5-TRINITRO-1,3,5-TRIAZACYCLOHEXANE

TOXICITY DATA with REFERENCE
orl-rat TDLo:3 g/kg (13W male/13W pre-22D preg):REP NTIS** AD-A092-531
orl-rat TDLo:20 mg/kg (female 6-15D post):TER NTIS** AD-A166-249
orl-rat LD50:100 mg/kg TXAPA9 39,531,77
ipr-rat LDLo:10 mg/kg EATR** EB-TR-73040
ivn-rat LDLo:18 mg/kg EATR** EB-TR-73040
orl-mus LD50:59 mg/kg NTIS** AD-A092-531
ivn-mus LD50:19 mg/kg EATR** EB-TR-73040
orl-cat LDLo:100 mg/kg FATOAO 7,43,44
orl-rbt LDLo:500 mg/kg FATOAO 7,43,44
ivn-gpg LD50:25 mg/kg EATR** EB-TR-73040

CONSENSUS REPORTS: Reported in EPA TSCA Inventory.

OSHA PEL: TWA 1.5 mg/m³ (skin)
ACGIH TLV: TWA 1.5 mg/m³ (skin)
DOT Classification: Class A Explosive; Label: Explosive A, Corrosive.

SAFETY PROFILE: Poison by ingestion, intraperitoneal, and intravenous routes. An experimental teratogen. Other experimental reproductive effects. A corrosive irritant to skin, eyes, and mucous membranes. Cases of epileptiform convulsions have been reported from exposure. It is one of the most powerful high explosives in use today. Has more shattering power than TNT and is often mixed with TNT as a bursting charge for aerial bombs, mines, and torpedoes. It is easily initiated by mercury fulminate which may be used as a booster. When heated to decomposition it emits toxic fumes of NO_x. See also AMINES, NITRATES, and EXPLOSIVES, HIGH.

CPR825 CAS:1552-12-1 ***HR: 2***
cis,cis-CYCLOOCTA-1,5-DIENE
mf: C_8H_{12} mw: 108.20

SYNS: COD ◇ 1,5-CYCLOOCTADIENE (Z,Z)

TOXICITY DATA with REFERENCE
skn-mus 100%/12D open SEV BJIMAG 25,75,68
skn-rbt 2640 mg SEV BJIMAG 25,75,68
skn-rbt 20 g/31D-I open SEV BJIMAG 25,75,68
eye-rbt 88 mg MLD BJIMAG 25,75,68
skn-gpg 10 g/31D-I open SEV BJIMAG 25,75,68

SAFETY PROFILE: An eye and severe skin irritant. When heated to decomposition it emits acrid smoke and fumes.

CPR840 CAS:12245-39-5 ***HR: 3***
(1,5-CYCLOOCTADIENE)(2,4-PENTANEDIONATO)RHODIUM
mf: $C_{13}H_{19}O_2Rh$ mw: 310.23

SYNS: ACETYLACETONATE-1,5-CYCLOOCTADIENERHODIUM ◇ RHODIUM, ((1,2,5,6-eta)-1,5-CYCLOOCTADIENE)(2,4-PENTANEDIONATO-O,O')- ◇ RHODIUM, (1,5-CYCLOOCTADIENE)(2,4-PENTANEDIONATO)-

TOXICITY DATA with REFERENCE
ipr-mus LD50:34 mg/kg CBINA8 45,1,83

OSHA PEL: TWA 0.1 mg(Rh)/m³
ACGIH TLV: TWA 1 mg(Rh)/m³

SAFETY PROFILE: Poison by intraperitoneal route. When heated to decomposition it emits toxic fumes of Rh.

CPS000 CAS:115-25-3 ***HR: 1***
CYCLOOCTAFLUOROBUTANE
DOT: UN 1976
mf: C_4F_8 mw: 200.03

PROP: Colorless, odorless gas. Bp: −6.04°, mp: −41.4°, d (liquid): 1.513 @ −70°F.

SYNS: FC-C 318 ◇ FREON C-318 ◇ HALOCARBON C-138 ◇ OCTAFLUOROCYCLOBUTANE (DOT) ◇ PERFLUOROCYCLOBUTANE ◇ PROPELLANT C318 ◇ R-C 318

TOXICITY DATA with REFERENCE
sln-dmg-ihl 99 pph/10M ENVRAL 7,275,74

CONSENSUS REPORTS: EPA Genetic Toxicology Program. Reported in EPA TSCA Inventory.

DOT Classification: Nonflammable Gas; Label: Nonflammable Gas.

SAFETY PROFILE: Mildly toxic by ingestion and inhalation. Can cause slight transient effects at high concentrations. No anesthesia or central nervous system effects. Nonflammable Gas. Mutation data reported. When heated to decomposition it emits highly toxic fumes of F^-.

CPS250 CAS:502-49-8 *HR: 2*
CYCLOOCTANONE
mf: $C_8H_{14}O$ mw: 126.22

TOXICITY DATA with REFERENCE
ipr-mus LD50:740 mg/kg COREAF 254,2245,62

CONSENSUS REPORTS: Reported in EPA TSCA Inventory.

SAFETY PROFILE: Moderately toxic by intraperitoneal route. When heated to decomposition it emits acrid smoke and irritating fumes. See also KETONES.

CPS500 CAS:629-20-9 *HR: 3*
1,3,5,7-CYCLOOCTATETRAENE
DOT: UN 2538
mf: C_8H_8 mw: 104.15

CH=CH(CH=CH)$_2$CH=CH (ring)

PROP: Liquid. Mp: −7°, bp: 140.6°, fp: −4.7°, vap press: 7.9 mm @ 25°, flash p: <71.6°F.

DOT Classification: Flammable Liquid; Label: Flammable Liquid

SAFETY PROFILE: May be a simple asphyxiant. A dangerous fire hazard when exposed to heat or flame; can react with oxidizing materials. To fight fire, use spray, mist, fog, foam, dry chemicals. Reaction with oxygen gives explosive peroxide byproducts. When heated to decomposition it emits acrid smoke and fumes.

CPT000 CAS:2163-69-1 *HR: 3*
3-CYCLOOCTYL-1,1-DIMETHYLUREA
mf: $C_{11}H_{22}N_2O$ mw: 198.35

SYNS: ALIPUR-O ◇ 3-CYCLOOCTYL-1,1-DIMETHYLHARNSTOFF (GERMAN) ◇ N-CYCLOOCTYL-N',N'-DIMETHYLUREA ◇ CYCLOURON ◇ CYCLURON

TOXICITY DATA with REFERENCE
orl-rat LD50:1500 mg/kg PCOC** -,294,66
ihl-rat LD50:1125 mg/kg EQSFAP 3,618,75
ipr-mus LD50:300 mg/kg 85DPAN -,-,71/76
unk-mus LD50:300 mg/kg 30ZDA9 -,229,71
orl-mam LD50:2600 mg/kg 85GYAZ -,81,71

SAFETY PROFILE: Poison by intraperitoneal and possibly other routes. Moderately toxic by ingestion and inhalation. A pesticide. When heated to decomposition it emits toxic fumes of NO_x.

CPT750 CAS:4449-51-8 *HR: 3*
CYCLOPAMINE
mf: $C_{27}H_{41}NO_2$ mw: 411.69

PROP: Derived from *Veratrum californicum* (TJADAB 3,175,70).

SYNS: ALKALOID V ◇ 11-DEOXOJERVINE

TOXICITY DATA with REFERENCE
orl-rbt TDLo:116 mg/kg (female 7-8D post):TER PSEBAA 136,1174,71
orl-ham TDLo:250 mg/kg (female 7D post):REP PSEBAA 149,302,75
orl-mus LDLo:180 mg/kg PSEBAA 149,302,75
orl-ham LDLo:170 mg/kg PSEBAA 149,302,75

SAFETY PROFILE: A poison by ingestion. An experimental teratogen. Other experimental reproductive effects. When heated to decomposition it emits toxic fumes of NO_x.

CPU000 CAS:202-98-2 *HR: 3*
4H-CYCLOPENTA(def)CHRYSENE
mf: $C_{19}H_{12}$ mw: 240.31

SYN: 4,5-METHYLENECHRYSENE

TOXICITY DATA with REFERENCE
scu-mus TDLo:80 mg/kg:ETA CNREA8 3,606,43

SAFETY PROFILE: Questionable carcinogen with experimental tumorigenic data. When heated to decomposition it emits acrid smoke and irritating fumes. See other chrysene compounds.

CPU250 CAS:502-72-7 *HR: 1*
CYCLOPENTADECANONE
mf: $C_{15}H_{28}O$ mw: 224.43

SYN: NORMUSCONE

TOXICITY DATA with REFERENCE
skn-rbt 500 mg/24H MLD FCTXAV 14,659,76

CONSENSUS REPORTS: Reported in EPA TSCA Inventory.

SAFETY PROFILE: A skin irritant. When heated to decomposition it emits acrid smoke and irritating fumes. See also KETONES.

CPU500 CAS:542-92-7 *HR: 3*
1,3-CYCLOPENTADIENE
mf: C_5H_6 mw: 66.11

PROP: Colorless liquid. Mp: −85°, bp: 42.5°, d: 0.80475 @ 19°/4°. flash p: 77°F.

SYNS: CYCLOPENTADIENE ◇ PENTOLE ◇ PYROPENTYLENE ◇ R-PENTINE

CONSENSUS REPORTS: Reported in EPA TSCA Inventory.

OSHA PEL: TWA 75 ppm
ACGIH TLV: TWA 75 ppm
DFG MAK: 75 ppm (200 mg/m^3)

SAFETY PROFILE: Probably moderately toxic by inhalation. A dangerous fire hazard when exposed to heat or flame; can react with oxidizing materials. Moderate explosion hazard in the form of gas when exposed to heat or by chemical reaction. It decomposes violently at high temperatures and pressures. Dimerization is highly exothermic. Explosive reaction with fuming nitric acid, dinitrogen tetroxide, sulfuric acid. Reaction with nitrogen oxide + oxygen forms an explosive product. Reaction with oxygen forms a flame-sensitive explosive product. Ignites on contact with oxygen + ozone. Reacts vigorously on contact with potassium hydroxide. Incompatible with oxides of nitrogen, sulfuric acid. When heated to decomposition it emits acrid smoke and fumes.

CPU750 CAS:21254-73-9 ***HR: 3***
CYCLOPENTADIENYL GOLD(I)
mf: C_5H_5Au mw: 262.06

SAFETY PROFILE: Ignites with friction or low heat. Will burn easily. When heated to decomposition it emits acrid smoke and fumes. See also GOLD.

CPV000 CAS:12079-65-1 ***HR: 3***
CYCLOPENTADIENYLMANGANESE TRICARBONYL
mf: $C_8H_5MnO_3$ mw: 204.07

SYNS: MANGANESE CYCLOPENTADIENYL TRICARBONYL ◇ MCT

TOXICITY DATA with REFERENCE
orl-rat LDLo:20 mg/kg HYSAAV 30,40,65
ihl-rat LCLo:120 mg/m^3/2H HYSAAV 30,40,65
ipr-rat LD50:14 mg/kg TXCYAC 34,341,85
orl-mus LD50:150 mg/kg GISAAA 28(4),29,63
ivn-mus LD50:710 μg/kg CSLNX* NX#11285

CONSENSUS REPORTS: Reported in EPA TSCA Inventory. Manganese and its compounds are on the Community Right-To-Know List.

OSHA PEL: TWA 0.1 mg(Mn)/m^3 (skin)
ACGIH TLV: TWA 0.1 mg(Mn)/m^3

SAFETY PROFILE: A poison by ingestion, inhalation, intraperitoneal, and intravenous routes. A mild narcotic which can damage kidneys. When heated to decomposition it emits acrid smoke and irritating fumes. See also MANGANESE COMPOUNDS and CARBON MONOXIDE.

CPV250 ***HR: 3***
CYCLOPENTADIENYL SILVER PERCHLORATE
mf: $C_5H_5AgClO_4$ mw: 272.42

CONSENSUS REPORTS: Silver and its compounds are on the Community Right-To-Know List.

SAFETY PROFILE: Explodes on heating. When heated to decomposition it emits toxic fumes of Cl^-. See also SILVER COMPOUNDS.

CPV500 CAS:4984-82-1 ***HR: 3***
CYCLOPENTADIENYL SODIUM
mf: C_5H_5Na mw: 88.08

CH=CHCHNaCH=CH (ring)

SAFETY PROFILE: Ignites spontaneously in air. Evaporation of a solution leaves a pyrophoric residue. Mixture with lead(II)nitrate may be explosive above 100°C. When heated to decomposition it emits toxic fumes of Na_2O.

CPV609 CAS:538-02-3 ***HR: 3***
CYCLOPENTAMINE HYDROCHLORIDE
mf: $C_9H_{19}N•ClH$ mw: 177.75

SYNS: CLOPANE HYDROCHLORIDE ◇ N,α-DIMETHYLCYCLOPENTANEETHYLAMINEHYDROCHLORIDE

TOXICITY DATA with REFERENCE
orl-rat LD50:169 mg/kg 28ZVAB -,37,69
orl-mus LD50:169 mg/kg JPETAB 93,423,48
ivn-mus LD50:41600 μg/kg JPETAB 93,423,48

SAFETY PROFILE: Poison by ingestion and intravenous routes. When heated to decomposition it emits toxic fumes of NO_x and HCl. See also AMINES.

CPV750 CAS:287-92-3 ***HR: 1***
CYCLOPENTANE
DOT: UN 1146
mf: C_5H_{10} mw: 70.15

PROP: Colorless liquid. Bp: 49.3°, fp: −93.7°, flash p: 19.4°F, autoign temp: 716°F, d: 0.745 @ 20°/4°, vap press: 400 mm @ 31.0°, vap d: 2.42.

SYN: PENTAMETHYLENE

TOXICITY DATA with REFERENCE
ihl-mus LCLo:110000 mg/m^3 AEPPAE 149,116,30

CONSENSUS REPORTS: Reported in EPA TSCA Inventory.

OSHA PEL: TWA 600 ppm
ACGIH TLV: TWA 600 ppm
DOT Classification: Flammable Liquid; Label: Flammable Liquid.

SAFETY PROFILE: Mildly toxic by ingestion and inhalation. High concentrations have narcotic action. A very dangerous fire hazard when exposed to heat or flame; can react with oxidizers. To fight fire, use foam, CO_2, dry chemical. When heated to decomposition it emits acrid smoke and fumes.

CPW250 CAS:35944-73-1 ***HR: 3***
1,3-CYCLOPENTANEDISULFONYL DIFLUORIDE
mf: $C_5H_8F_2O_4S_2$ mw: 234.25

SYN: PHILIPS 2133

TOXICITY DATA with REFERENCE
orl-rat LD50:7900 μg/kg TXAPA9 21,315,72
orl-mus LDLo:94 mg/kg AECTCV 14,111,85
orl-bwd LD50:1300 μg/kg TXAPA9 21,315,72

SAFETY PROFILE: A deadly poison by ingestion. See also FLUORIDES. When heated to decomposition it emits very toxic fumes of F^- and SO_x.

CPW325 CAS:54573-23-8 ***HR: 3***
4,5-CYCLOPENTANOFURAZAN-N-OXIDE
mf: $C_5H_6N_2O_2$ mw: 126.11

SYN: TRIMETHYLENEFUROXAN

SAFETY PROFILE: Decomposes explosively at 150°C. Upon decomposition it emits toxic fumes of NO_x.

CPW500 CAS:120-92-3 ***HR: 3***
CYCLOPENTANONE
DOT: UN 2245
mf: C_5H_8O mw: 84.13

O:C(CH$_2$)$_3$CH$_2$ (ring)

PROP: Liquid. Mp: −58.2°, bp: 130.6°, flash p: 79°F, d: 0.9509 @ 18°/4°, vap d: 2.3.

SYNS: ADIPIC KETONE ◇ DUMASIN ◇ KETOCYCLOPENTANE ◇ KETOPENTAMETHYLENE

TOXICITY DATA with REFERENCE
skn-rbt 500 mg/24H FCTXAV 17,241,79
skn-rbt 500 mg MLD FCTOD7 20,573,82
eye-rbt 100 mg SEV FCTOD7 20,573,82
eye-rbt 100 mg/4S rns SEV FCTOD7 20,573,82
ipr-mus LD50:1950 mg/kg COREAF 254,2245,62
scu-mus LDLo:2600 mg/kg AEXPBL 50,199,1903
scu-frg LDLo:3000 mg/kg AEXPBL 50,199,1903

CONSENSUS REPORTS: Reported in EPA TSCA Inventory.

DOT Classification: Flammable or Combustible Liquid; Label: Flammable Liquid.

SAFETY PROFILE: Moderately toxic by intraperitoneal and subcutaneous routes. A skin and severe eye irritant. Dangerous fire hazard when exposed to heat or flame; can react with oxidizers. To fight fire, use alcohol foam, foam, CO_2, dry chemical. Potentially explosive reaction with hydrogen peroxide + nitric acid. When heated to decomposition it emits acrid smoke and fumes. See also KETONES.

CPW750 CAS:1192-28-5 ***HR: 2***
CYCLOPENTANONE OXIME
mf: C_5H_9NO mw: 99.13

CH$_2$(CH$_2$)$_3$C=NOH (ring)

PROP: Solid, mp: 57.5°.

TOXICITY DATA with REFERENCE
unk-mus LD50:1200 mg/kg PCJOAU 12,227,78

CONSENSUS REPORTS: Reported in EPA TSCA Inventory.

SAFETY PROFILE: Moderately toxic by an unspecified route. Violent reaction when heated with 85% sulfuric acid. When heated to decomposition it emits toxic fumes of NO_x.

CPX250 CAS:203-64-5 ***HR: 3***
4H-CYCLOPENTA(def)PHENANTHRENE
mf: $C_{15}H_{10}$ mw: 190.25

SYN: CYCLOPENTAPHENANTHRENE

TOXICITY DATA with REFERENCE
orl-rat TDLo:3000 mg/kg:ETA CNREA8 26,619,66

SAFETY PROFILE: Questionable carcinogen with experimental tumorigenic data. When heated to decomposition it emits acrid smoke and irritating fumes.

CPX500 CAS:27208-37-3 ***HR: 3***
CYCLOPENTA(cd)PYRENE
mf: $C_{18}H_{10}$ mw: 226.28

SYNS: ACEPYRENE ◇ ACEPYRYLENE ◇ CYCLOPENTENO(c,d)PYRENE

TOXICITY DATA with REFERENCE
mma-hmn:lym 4 mg/L JJIND8 63,309,79
msc-hmn:lym 50 nmol/L MUREAV 128,221,84
otr-mus:fbr 300 μg/L CNREA8 40,4482,80
otr-mus:emb 300 μg/L EVSRBT 22,445,81
msc-mus:lym 1200 μg/L CNREA8 40,4482,80
msc-ham:lng 300 μg/L CRNGDP 3,763,82
skn-mus TDLo:9051 μg/kg:NEO CNREA8 40,642,80
scu-mus TDLo:260 mg/kg:ETA HLSCAE 9,32,72

CONSENSUS REPORTS: IARC Cancer Review: Group 3 IMEMDT 7,56,87; Animal Limited Evidence IMEMDT 32,269,83. EPA Genetic Toxicology Program.

SAFETY PROFILE: Questionable carcinogen with experimental neoplastigenic and tumorigenic data. Human mutation data reported. When heated to decomposition it emits acrid smoke and irritating fumes.

CPX625 CAS:73473-54-8 ***HR: D***
CYCLOPENTA(cd)PYRENE-3,4-OXIDE
mf: $C_{18}H_{10}O$ mw: 242.28

SYN: CPP-3,4-OXIDE

TOXICITY DATA with REFERENCE
mmo-sat 100 ng/plate CNREA8 40,3940,80
otr-mus:fbr 3 mg/L CNREA8 40,4482,80
otr-mus:emb 300 μg/L EVSRBT 22,445,81

SAFETY PROFILE: Mutation data reported. When heated to decomposition it emits acrid smoke and fumes.

CPX750 CAS:142-29-0 ***HR: 3***
CYCLOPENTENE
DOT: UN 2246
mf: C_5H_8 mw: 68.13

PROP: Liquid. Mp: −93.3°, bp: 44.242°, fp: −135.2°, flash p: −20°F, d: 0.77199 @ 20°.

TOXICITY DATA with REFERENCE
orl-rat LD50:1656 mg/kg AIHAAP 30,470,69
skn-rbt LD50:1231 mg/kg AIHAAP 30,470,69

CONSENSUS REPORTS: Reported in EPA TSCA Inventory.

DOT Classification: Flammable Liquid; Label: Flammable Liquid.

SAFETY PROFILE: Moderately toxic by ingestion and skin contact. A very dangerous fire hazard when exposed to flame or heat; can react with oxidizing materials. Keep away from heat and open flame. To fight fire, use foam, CO_2, dry chemical.

CPY000 CAS:3212-60-0 ***HR: 3***
2-CYCLOPENTENE-1-OL
mf: C_5H_8O mw: 84.13

SYN: 1-CYCLOPENTEN-3-OL

TOXICITY DATA with REFERENCE
skn-rbt 10 mg/24H open MLD AIHAAP 23,95,62
orl-rat LD50:470 mg/kg AIHAAP 30,470,69
ihl-rat LCLo:1000 ppm/4H AIHAAP 30,470,69
skn-rbt LDLo:180 mg/kg AIHAAP 23,95,62

SAFETY PROFILE: Poison by skin contact. Moderately toxic by ingestion and inhalation. A skin irritant. When heated to decomposition it emits acrid smoke and irritating fumes.

CPY500 CAS:7129-91-1 ***HR: 3***
1,2-CYCLOPENTENO-5,10-ACEANTHRENE
mf: $C_{19}H_{16}$ mw: 244.35

SYN: 2,7,8,9-TETRAHYDRO-1H-CYCLOPENT(j)ACEANTHRYLENE

TOXICITY DATA with REFERENCE
scu-mus TDLo:160 mg/kg/43W-I:NEO JNCIAM 2,99,41

SAFETY PROFILE: Questionable carcinogen with experimental neoplastigenic data. When heated to decomposition it emits acrid smoke and irritating fumes.

CPY750 CAS:7099-43-6 ***HR: 3***
5:6-CYCLOPENTENO-1:2-BENZANTHRACENE
mf: $C_{21}H_{18}$ mw: 270.39

SYN: 2,3-DIHYDRO-1H-BENZO(a)CYCLOPENT(b)ANTHRACENE

TOXICITY DATA with REFERENCE
skn-mus TDLo:480 mg/kg/20W-I:ETA PRLBA4 111,485,32

SAFETY PROFILE: Questionable carcinogen with experimental tumorigenic data. When heated to decomposition it emits acrid smoke and irritating fumes.

CPZ125 CAS:5870-29-1 ***HR: 3***
CYCLOPENTOLATE HYDROCHLORIDE
mf: $C_{17}H_{25}NO_3$•ClH mw: 327.89

SYNS: CYCLOGYL ◇ β-DIMETHYLAMINOETHYL (1-HYDROXYCYCLOPENTYL)PHENYLACETATE HYDROCHLORIDE ◇ 2-(DIMETHYLAMINO)ETHYL 1-HYDROXY-α-PHENYLCYCLOPENTANEACETATE HYDROCHLORIDE

TOXICITY DATA with REFERENCE
scu-chd TDLo:40 μg/kg:CNS AROPAW 87,634,72
orl-mus LD50:960 mg/kg NIIRDN 6,314,82
ipr-mus LD50:314 mg/kg JPETAB 106,141,52
ivn-mus LD50:84 mg/kg JPETAB 106,141,52

SAFETY PROFILE: Poison by intravenous and intraperitoneal routes. Moderately toxic by ingestion. Human systemic effects by subcutaneous route: convulsions. When heated to decomposition it emits toxic fumes of NO_x and HCl. See also ESTERS.

CQA000 CAS:1003-03-8 ***HR: 1***
CYCLOPENTYLAMINE
mf: $C_5H_{11}N$ mw: 85.15

PROP: Flash p: 55.4°F.

SAFETY PROFILE: A dangerous fire hazard when exposed to heat or flame. When heated to decomposition it emits toxic fumes of NO_x. See also AMINES.

CQB250 CAS:40202-39-9 ***HR: 3***
2-CYCLOPENTYL-4,6-DINITROPHENOL
mf: $C_{11}H_{12}N_2O_5$ mw: 252.25

SYN: DINITROCYCLOPENTYLPHENOL

TOXICITY DATA with REFERENCE
ivn-dog LDLo:10 mg/kg AIPTAK 50,20,35
ivn-pgn LDLo:5 mg/kg AIPTAK 50,20,35

SAFETY PROFILE: Poison by intravenous route. When heated to decomposition it emits toxic fumes of NO_x. See also NITRO COMPOUNDS of AROMATIC HYDROCARBONS and PHENOLS.

CQB275 CAS:10137-73-2 ***HR: 2***
CYCLOPENTYL ETHER
mf: $C_{10}H_{18}O$ mw: 154.28

SYN: ETHER, DICYCLOPENTYL

TOXICITY DATA with REFERENCE
skn-rbt 10 mg/24H open MLD AIHAAP 23,95,62
orl-rat LD50:470 mg/kg AIHAAP 23,95,62
ihl-rat LCLo:250 ppm/4H AIHAAP 23,95,62
skn-rbt LD50:1410 mg/kg AIHAAP 23,95,62

SAFETY PROFILE: Moderately toxic by ingestion, inhalation, and skin contact. A skin irritant. When heated to decomposition it emits acrid smoke and irritating fumes.

CQC250 CAS:21208-99-1 ***HR: 3***
S-2-((5-CYCLOPENTYLPENTYL)AMINO)ETHYL THIOSULFATE
mf: $C_{12}H_{25}NO_3S_2$ mw: 295.50

TOXICITY DATA with REFERENCE
orl-mus LD50:800 mg/kg JMCMAR 11,1190,68
ipr-mus LD50:25 mg/kg JMCMAR 11,1190,68

SAFETY PROFILE: Poison by intraperitoneal route. Moderately toxic by ingestion. When heated to decomposition it emits very toxic fumes of NO_x and SO_x. See also THIOSULFATES.

CQC500 CAS:6055-19-2 ***HR: 3***
CYCLOPHOSPHAMIDE HYDRATE
mf: $C_7H_{15}Cl_2N_2O_2P{\bullet}H_2O$ mw: 279.13

SYNS: 1-BIS(2-CHLOROETHYL)AMINO-1-OXO-2-AZA-5-OXAPHOSPHORIDINE MONOHYDRATE ◇ 2-(BIS(2-CHLOROETHYL)AMINO)-1-OXA-3-AZA-2-PHOSPHOCYCLOHEXANE 2-OXIDE MONOHYDRATE ◇ (BIS(CHLORO-2-ETHYL)AMINO)-2-TETRAHYDRO-3,4,5,6-OXAZAPHOSPHORINE-1,3,2-OXIDE-2-MONOHYDRATE ◇ BIS(2-CHLOROETHYL)PHOSPHORAMIDE CYCLIC PROPANOLAMIDE ESTER MONOHYDRATE ◇ N,N-BIS(β-CHLOROETHYL)-N',O-PROPYLENEPHOSPHORIC ACID ESTER AMINE MONOHYDRATE ◇ N,N-BIS(2-CHLOROETHYL)TETRAHYDRO-2H-1,3,2-OXAPHOSPHORIN-2-AMINE-2-OXIDE MONOHYDRATE ◇ N,N-BIS(β-CHLOROETHYL)-N',O-TRIMETHYLENEPHOSPHORIC ACID ESTER DIAMIDE MONOHYDRATE ◇ CB-4564 ◇ CLAFEN ◇ CYCLIC N',O-PROPYLENE ESTER of N,N-BIS(2-CHLOROETHYL)PHOSPHORODIAMIDIC ACID MONOHYDRATE ◇ CYCLOPHOSPHAMIDE MONOHYDRATE ◇ CYCLOPHOSPHAMIDUM ◇ CYCLOPHOSPHAN ◇ CYCLOPHOSPHANE ◇ CYCLOPHOSPHANUM ◇ CYTOPHOSPHAN ◇ CYTOXAN ◇ 2-(DI(2-CHLOROETHYL)AMINO)-1-OXA-3-AZA-2-PHOSPHACYCLOHEXANE-2-OXIDE MONOHYDRATE ◇ N,N-DI(2-CHLOROETHYL)AMINO-N,O-PROPYLENE PHOSPHORIC ACID ESTER DIAMIDE MONOHYDRATE ◇ ENDOXANA ◇ ENDOXAN-ASTA ◇ ENDOXAN MONOHYDRATE ◇ ENDOXAN R ◇ ENDUXAN ◇ GENOXAL ◇ MITOXAN ◇ NSC 26271 ◇ PROCYTOX ◇ SEMDOXAN ◇ SENDOXAN ◇ SENDUXAN

TOXICITY DATA with REFERENCE
bfa-mus/smc 500 mg/kg EVSRBT 24,893,81
ivn-rat TDLo:27500 μg/kg (7-17D preg):REP KSRNAM 16,517,82
ivn-rat TDLo:27500 μg/kg (7-17D preg):TER KSRNAM 16,517,82
orl-rat LD50:94 mg/kg TXAPA9 4,324,62
ipr-rat LD50:121 mg/kg FRMBAZ 18,409,70
orl-mus LD50:350 mg/kg TXAPA9 4,324,62
ivn-mus LD50:275 mg/kg TXAPA9 4,324,62
orl-dog LD50:44 mg/kg TXAPA9 4,324,62

CONSENSUS REPORTS: IARC Cancer Review: Animal Sufficient Evidence IMEMDT 9,135,75; Human Limited Evidence IMEMDT 9,135,75; Human Sufficient Evidence IMEMDT 26,165,81.

SAFETY PROFILE: Confirmed human carcinogen. Poison by ingestion and intravenous routes. Experimental reproductive effects. Mutation data reported. When heated to decomposition it emits toxic fumes of Cl^-, PO_x, and NO_x.

CQC600 ***HR: 3***
CYCLOPHOSPHAMIDE and MNU (1:2)
mf: $C_7H_{15}Cl_2N_2O_2P{\bullet}C_4H_5N_6O_2$ mw: 430.26

PROP: A combination of these two drugs is used in chemotherapy to combat far advanced malignant tumors. (ZKKOBW 89,311,77)

SYN: MNU and CYCLOPHOSPHAMIDE (2:1)

TOXICITY DATA with REFERENCE
ivn-hmn TDLo:48 mg/kg/28D-I:GIT,BLD ZKKOBW 89,311,77

SAFETY PROFILE: Human systemic effects by intravenous route: nausea or vomiting and bone marrow changes. When heated to decomposition it emits very toxic fumes of Cl^-, NO_x, and PO_x.

CQC650 CAS:50-18-0 ***HR: 3***
CYCLOPHOSPHORAMIDE
mf: $C_7H_{15}Cl_2N_2O_2P$ mw: 261.11

PROP: Crystals. Mp: 41-45°. Water-sol; sltly sol in organic solvents.

SYNS: ASTA ◇ ASTA B518 ◇ B 518 ◇ N,N-BIS-(β-CHLORAETHYL)-N',O-PROPYLEN-PHOSPHORSAEURE-ESTER-DIAMID (GERMAN) ◇ 2-(BIS(2-CHLOROETHYL)AMINO)-2H-1,3,2-OXAAZAPHOSPHORINE2-OXIDE ◇ N,N-BIS(2-CHLOROETHYL)-N'-(3-HYDROXYPROPYL) PHOSPHORODIAMIDIC ACID intramol. ESTER ◇ BIS(2-CHLOROETHYL)PHOSPHORAMIDE-CYCLIC PROPANOLAMIDE ESTER ◇ N,N-BIS(2-CHLOROETHYL)-N',O-PROPYLENEPHOSPHORIC ACID ESTER DIAMIDE ◇ N,N-BIS(2-CHLOROETHYL)TETRAHYDRO-2H-1,3,2-OXAPHOSPHORIN-2-AMINE-2-OXIDE ◇ N,N-BIS(β-CHLOROETHYL)-

N',O-TRIMETHYLENEPHOSPHORIC ACID ESTER DIAMIDE ◇ CB 4564 ◇ CLAFEN ◇ CLAPHENE ◇ CP ◇ CPA ◇ CTX ◇ CY ◇ CYCLOPHOSPHAMIDE ◇ CYCLOPHOSPHAMIDUM ◇ CYCLOPHOSPHAN ◇ CYCLOSTIN ◇ CYTOPHOSPHAN ◇ CYTOXAN ◇ N,N-DI(2-CHLOROETHYL)-N,o-PROPYLENE-PHOSPHORIC ACID ESTER DIAMIDE ◇ ENDOXAN ◇ ENDOXANAL ◇ GENOXAL ◇ HEXADRIN ◇ MITOXAN ◇ NCI-C04900 ◇ NEOSAR ◇ NSC 26271 ◇ 2-H-1,3,2-OXAZAPHOSPHORINANE ◇ PROCYTOX ◇ RCRA WASTE NUMBER U058 ◇ SEMDOXAN ◇ SENDUXAN ◇ SK 20501 ◇ ZYKLOPHOSPHAMID (GERMAN)

TOXICITY DATA with REFERENCE
sce-hmn:oth 500 mg/L ENMUDM 7(Suppl 3),26,85
sce-hmn:fbr 10 μmol/L CNREA8 45,3626,85
mma-ham:lng 10 mg/L MUREAV 157,189,85
orl-wmn TDLo:60 mg/kg (60D preg):REP NEJMAG 289,2259,73
mul-wmn TDLo:215 mg/kg (43-70D preg):TER AIMEAS 74,87,71
orl-wmn TDLo:1890 mg/kg/3Y-I:CAR,BLD JHMJAX 142,211,78
orl-man TDLo:2310 mg/kg/4.5Y-C:CAR,GIT BMJOAE 280,524,80
unr-man TDLo:857 mg/kg/3Y-C:CAR,BLD JCPAAK 26,649,73
unr-wmn TDLo:1050 mg/kg/69W-C:CAR,BLD JCPAAK 26,649,73
orl-rat TDLo:475 mg/kg/100W-I:CAR IJCNAW 23,706,79
ivn-rat TDLo:676 mg/kg/1Y-I:CAR ARZNAD 20,1461,70
ipr-mus TDLo:1950 mg/kg/26W-I:NEO RRCRBU 52,1,75
scu-mus TDLo:1352 mg/kg/1Y-I:CAR,TER ARZNAD 20,1461,70
orl-wmn TD:2700 mg/kg/6Y-C:CAR,KID URGABW 17,105,78
orl-man TD:1078 mg/kg/3Y-C:CAR,KID RIHYAC 32,1073,78
orl-man TD:1800 mg/kg/6Y-C:CAR,KID JOURAA 126,544,81
orl-hmn TD:920 mg/kg/3Y-C:CAR,KID AIMEAS 91,221,79
orl-man TD:1190 mg/kg/4Y-I:CAR,BLD MEDIAV 58,32,79
orl-wmn TD:1760 mg/kg/4Y-C:CAR,KID SJRHAT 12,73,83
orl-wmn TDLo:45 mg/kg:KID ARHEAW 15,530,72
orl-hmn TDLo:20 mg/kg:GIT,KID,SKN ARHEAW 12,663,69
orl-cld TDLo:2500 μg/kg AJDCAI 140,1094,86
orl-man TDLo:56 mg/kg/4W-I:SYS SMJOAV 78,222,85
orl-hmn TDLo:20 mg/kg:GIT,SYS,SKN ARHEAW 12,663,69
orl-wmn LDLo:16 mg/kg/4D-I:BLD AJMSA9 254,48,67
ivn-wmn TDLo:60 mg/kg/9W-I AIMDAP 145,548,85
mul-man LDLo:45 mg/kg/26W-I:BLD AJMSA9 254,48,67
ipr-rat LD50:40 mg/kg CPCHAO 18,307,62
scu-rat LD50:144 mg/kg KSRNAM 16,431,82
ivn-rat LD50:148 mg/kg KSRNAM 16,431,82
orl-mus LD50:137 mg/kg RPTOAN 36,240,73
ipr-mus LD50:159 μg/kg CTRRDO 61,103,77
scu-mus LD50:200 mg/kg ASBDD9 2,95,79
ivn-mus LD50:425 mg/kg ATSUDG 7,90,84
par-mus LD50:315 mg/kg TRPLAU 13,316,72
ipr-dog LDLo:50 mg/kg KSRNAM 16,431,82
ivn-dog LDLo:11 mg/kg CCSUBJ 2,191,65
ivn-mky LDLo:45 mg/kg CCSUBJ 2,191,65

CONSENSUS REPORTS: NTP Fifth Annual Report on Carcinogens. IARC Cancer Review: Group 1 IMEMDT 7,182,87; Human Sufficient Evidence IMEMDT 26,165,81; Animal Sufficient Evidence IMEMDT 26,165,81; IMEMDT 9,135,75; Human Limited Evidence IMEMDT 9,135,75. NCI Carcinogenesis Studies (ipr); Clear Evidence: mouse, rat RRCRBU 52,1,75. EPA Genetic Toxicology Program.

SAFETY PROFILE: Confirmed human carcinogen producing leukemia, Hodgkin's disease, gastrointestinal, and bladder tumors. Experimental carcinogenic, neoplastigenic, and teratogenic data. A human poison by ingestion and many other routes. Human systemic effects: kidney changes (hepatic dysfunction), leukopenia (reduced white blood cell count), nausea and alopecia (loss of hair), liver changes. Human reproductive and teratogenic effects by multiple routes: spermatogenesis, testical changes, epididymis and sperm duct changes, menstrual cycle changes; fetal developmental abnormalities of the craniofacial area, musculoskeletal and cardiovascular systems. Experimental reproductive effects. Human mutation data reported. A powerful skin irritant. Used as an immunosuppressive agent in nonmalignant diseases. When heated to decomposition it emits highly toxic fumes of PO_x, NO_x, and Cl^-.

CQD000 CAS:18172-33-3 ***HR: 3***
α-CYCLOPIAZONIC ACID
mf: $C_{20}H_{20}N_2O_3$ mw: 336.42

SYN: CYCLOPIAZONIC ACID

TOXICITY DATA with REFERENCE
mma-sat 1 μmol/plate AEMIDF 47,1355,84
orl-rat TDLo:20 mg/kg (8-11D preg):REP JTEHD6 14,585,84
orl-rat TDLo:4 mg/kg (8-11D preg):TER JTEHD6 14,585,84
orl-rat LD50:36 mg/kg TXAPA9 18,114,71
ipr-rat LD50:2 mg/kg TXAPA9 18,114,71
ipr-mus LD50:13 mg/kg FCTOD7 23,831,85

CONSENSUS REPORTS: EPA Genetic Toxicology Program.

SAFETY PROFILE: Poison by ingestion and intraperitoneal routes. An experimental teratogen. Experimental reproductive effects. Mutation data reported. When heated to decomposition it emits toxic fumes of NO_x.

CQD250 CAS:39071-30-2 *HR: 3*
5H-CYCLOPROPA(3,4)BENZ(1,2-e)AZULEN-5-ONE,1,1a-α,1b-β,4,4a,7a-α,7b,8,9,9a-DECAHYDRO-4a-α,7b-α,9a-α-TRIHYDROXY-3-HYDROXYMETHYL-1,6,8-α-TRIMETHYL-1-ACETOXYMETHYL-,9a-(2-METHYLBUT-2-ENOATE)
mf: $C_{27}H_{36}O_8$ mw: 488.63

TOXICITY DATA with REFERENCE
skn-mus 44 ng MLD 85CVA2 5,213,70
skn-mus TDLo:225 mg/kg/36W-I:ETA 85CVA2 5,213,70

SAFETY PROFILE: Questionable carcinogen with experimental tumorigenic data. A skin irritant. When heated to decomposition it emits acrid smoke and irritating fumes.

CQD750 CAS:75-19-4 *HR: 3*
CYCLOPROPANE
DOT: UN 1027
mf: C_3H_6 mw: 42.09

$CH_2CH_2CH_2$ (ring)

PROP: Colorless gas. Mp: −126.6°, bp: −33.5°, lel: 2.4%, uel: 10.4%, d: 1.879 g/L @ 0°, autoign temp: 932°F. A minor constituent of MAPP gas.

SYNS: CYCLOPROPANE, liquefied (DOT) ◇ TRIMETHYLENE

TOXICITY DATA with REFERENCE
cyt-ckn-ihl 20 pph/3H ANESAV 34,157,71

CONSENSUS REPORTS: IARC Cancer Review: Animal No Adequate Data IMEMDT 7,93,87. Reported in EPA TSCA Inventory.

DOT Classification: Flammable Gas; Label: Flammable Gas.

SAFETY PROFILE: Mutation data reported. Questionable carcinogen. High concentrations are narcotic. Human reproductive effects. Very dangerous fire hazard when exposed to heat or flame; can react with oxidizing materials. Explosion Hazard: Moderate in the form of vapor when exposed to heat or flame. To fight fire, stop flow of gas, then use CO_2, dry chemical or water spray. When heated to decomposition it emits acrid smoke and fumes.

CQE250 CAS:765-30-0 *HR: 2*
CYCLOPROPYLAMINE
mf: C_3H_7N mw: 57.10

$CH_2CH_2CHNH_2$ (ring)

PROP: Flash p: 33.8°F.

SAFETY PROFILE: A very dangerous fire hazard when exposed to heat or flame. When heated to decomposition it emits toxic fumes of NO_x. See also AMINES.

CQE325 CAS:31431-43-3 *HR: D*
5-(CYCLOPROPYLCARBONYL)-2-BENZIMIDAZOLECARBAMIC ACID METHYL ESTER
mf: $C_{13}H_{13}N_3O_3$ mw: 259.29

PROP: Crystals from acetic acid. Mp: 250.5°.

SYNS: CYCLOBENDAZOLE ◇ HAPTOCIL

TOXICITY DATA with REFERENCE
oms-hmn:leu 1 mg/L THERAP 31,505,76
oms-hmn:oth 2 mg/L THERAP 31,505,76
orl-rat TDLo:68800 μg/kg (female 8-15D post):TER THERAP 31,505,76

SAFETY PROFILE: An experimental teratogen. Human mutation data reported. When heated to decomposition it emits toxic fumes of NO_x. See also CARBAMATES and ESTERS.

CQE750 CAS:540-47-6 *HR: 2*
CYCLOPROPYL METHYL ETHER
mf: C_4H_8O mw: 72.11

PROP: Liquid. Mp: −119°, bp: 44.7°, d: 0.786 @ 25°/4°, flash p: <50°F.

SYNS: CYPRONIC ETHER ◇ METHOXYCYCLOPROPANE

SAFETY PROFILE: A very dangerous fire hazard when exposed to heat or flame. When heated to decomposition it emits acrid smoke and fumes. Can form unstable explosive peroxides. See also ETHERS and PEROXIDES.

CQF059 CAS:765-43-5 *HR: 2*
CYCLOPROPYL METHYL KETONE
mf: C_5H_8O mw: 84.119

$H_2CCH_2CHCO \cdot CH_3$ (ring)

PROP: Flash p: 55.4°F.

SAFETY PROFILE: A dangerous fire hazard when exposed to heat or flame. When heated to decomposition it emits acrid smoke and fumes. See also KETONES.

CQF079 CAS:42281-59-4 *HR: 3*
(−)-17-CYCLOPROPYLMETHYLMORPHINAN-3,4-DIOL
mf: $C_{20}H_{27}NO_2$ mw: 313.48

SYNS: (L)-BC-2605 ◇ BRISTOL LABORATORIES BC 2605 ◇ 1-N-CYCLOPROPYLMETHYL-3,14-DIHYDROXYMORPHINAN ◇ (−)-3,14-DIHYDROXY-N-(CYCLOBUTYLMETHYL)MORPHINAN ◇ OXILORPHAN

TOXICITY DATA with REFERENCE
orl-hmn TDLo:14 μg/kg:CNS,GIT,KID JCPCBR 16(4),183,76
scu-hmn TDLo:26 μg/kg:CNS DRFUD4 2,746,77
scu-mus LD50:315 mg/kg JPETAB 193,23,75
ivn-mus LD50:32 mg/kg JPETAB 193,23,75

SAFETY PROFILE: Poison by subcutaneous and intravenous routes. Human systemic effects by ingestion and subcutaneous routes: somnolence, tremors, ataxia (loss of muscle coordination), hypermotility, diarrhea, and changes in the kidney, ureter, or bladder. When heated to decomposition it emits toxic fumes of NO_x.

CQF099 CAS:16590-41-3 ***HR: 2***
N-CYCLOPROPYLMETHYLNOROXYMORPHONE
mf: $C_{20}H_{23}NO_4$ mw: 341.44

SYNS: (5-α)-17-(CYCLOPROPYLMETHYL-4,5-EPOXY-3,14-DIHYDROXY-MORPHINAN-6-ONE (9CI) ◇ N-CYCLOPROPYLMETHYL-14-HYDROXYDIHYDROMORPHINONE ◇ EN 1639 ◇ EN 1939 ◇ NALTREXONE ◇ UM-792

TOXICITY DATA with REFERENCE
sce-hmn:lym 1 g/L ENMUDM 1,180,79
cyt-hmn:lym 1 g/L ENMUDM 1,180,79
sln-dmg-orl 10 g/L/24H MUREAV 66,129,79
sln-dmg-par 10 g/L MUREAV 66,129,79
scu-mus TDLo:21 mg/kg (female 10-19D post):REP DEPBA5 21,283,88
scu-mus LD50:551 mg/kg ANYAA9 281,321,76

CONSENSUS REPORTS: EPA Genetic Toxicology Program.

SAFETY PROFILE: Moderately toxic by subcutaneous route. Experimental reproductive effects. Human mutation data reported. When heated to decomposition it emits toxic fumes of NO_x. See also MORPHINE and KETONES.

CQF125 CAS:33453-19-9 ***HR: 2***
1-CYCLOPROPYLMETHYL-4-PHENYL-6-CHLORO-2(1H)-QUINAZOLINONE
mf: $C_{18}H_{15}ClN_2O$ mw: 310.80

SYNS: 6-CHLORO-1-(CYCLOPROPYLMETHYL)-4-PHENYL-2(1H)-QUINAZOLINONE ◇ SL-512

TOXICITY DATA with REFERENCE
orl-rat LD50:2 g/kg ARZNAD 23,1266,73
ipr-rat LD50:790 mg/kg ARZNAD 23,1266,73
orl-mus LD50:1800 mg/kg ARZNAD 23,1266,73
ipr-mus LD50:660 mg/kg ARZNAD 23,1266,73

SAFETY PROFILE: Moderately toxic by ingestion and intraperitoneal route. When heated to decomposition it emits toxic fumes of Cl^- and NO_x.

CQG250 CAS:26399-36-0 ***HR: 2***
N-(CYCLOPROPYLMETHYL)-α,α,α-TRIFLUORO-2,6-DINITRO-N-PROPYL-p-TOLUIDINE
mf: $C_{14}H_{16}F_3N_3O_4$ mw: 347.33

SYNS: CGA 10832 ◇ ER5461 ◇ GA-10832 ◇ PREGARD ◇ PROFLURALIN ◇ TOLBAN

TOXICITY DATA with REFERENCE
skn-rbt 218 mg open MLD CIGET* -,-,77
eye-rbt 44 mg SEV CIGET* -,-,77
orl-rat LD50:1808 mg/kg FMCHA2 -,D310,80
ihl-rat LCLo:3970 mg/m^3 CIGET* -,-,77
skn-rbt LD50:13754 mg/kg CIGET* -,-,77

SAFETY PROFILE: Moderately toxic by ingestion and inhalation. Mildly toxic by skin contact. A skin and severe eye irritant. An herbicide. See also FLUORIDES. When heated to decomposition it emits very toxic fumes of F^- and NO_x.

CQG750 CAS:4163-15-9 ***HR: 3***
CYCLORPHAN
mf: $C_{20}H_{27}NO$ mw: 297.48

SYNS: 17-(CYCLOPROPYLMETHYL)MORPHINAN-3-OL ◇ (−)-3-HYDROXY-N-CYCLOPROPYLMETHYLMORPHINAN

TOXICITY DATA with REFERENCE
ivn-rat LD50:23 mg/kg AIPTAK 165,112,67
scu-mus LD50:215 mg/kg AIPTAK 165,112,67
ivn-mus LD50:24 mg/kg AIPTAK 165,112,67

SAFETY PROFILE: Poison by intravenous and subcutaneous routes. When heated to decomposition it emits toxic fumes of NO_x.

CQH000 CAS:68-41-7 ***HR: 3***
CYCLOSERINE
mf: $C_3H_6N_2O_2$ mw: 102.11

PROP: Produced by *Streptomyces orchidaceus* (ANTCAO 6,360,56).

SYNS: d-R-AMINO-3-ISOSSAZOLIDONE (ITALIAN) ◇ d-4-AMINO-3-ISOXAZOLIDINONE ◇ d-4-AMINO-3-ISOXAZOLIDONE ◇ CICLOSERINA (ITALIAN) ◇ CYCLOMYCIN ◇ CYCLO-d-SERINE ◇ E-733-A ◇ FARMISERINE ◇ I-1431 ◇ JN-21 ◇ K-300 ◇ MIROSERINA ◇ NOVOSERIN ◇ ORIENTOMYCIN ◇ d-OXAMICINA (ITALIAN) ◇ d-OXAMYCIN ◇ OXYMYCIN ◇ PA 94 ◇ RO-1-9213 ◇ SEROMYCIN ◇ TISOMYCIN ◇ WASSERINA

TOXICITY DATA with REFERENCE
orl-hmn TDLo:560 mg/kg/4W-I:CNS DICHAK 29,241,56
orl-wmn TDLo:60 mg/kg:CNS BMJOAE 1,907,65
unr-wmn TDLo:40 mg/kg/2D-I:CNS TUBEAS 38,297,57
unr-man TDLo:64 mg/kg/4D-I:CNS ABANAE 3,148,55/56
orl-mus LD50:5290 mg/kg 85ERAY 2,906,78
ipr-mus LD50:180 mg/kg 85FZAT -,238,67
scu-mus LD50:1400 mg/kg YKYUA6 31,1085,80
ivn-mus LD50:560 mg/kg YKYUA6 31,1085,80

scu-dog LDLo:2000 mg/kg ANTCAO 6,708,56
scu-mky LDLo:4000 mg/kg ANTCAO 6,708,56

CONSENSUS REPORTS: EPA Genetic Toxicology Program.

SAFETY PROFILE: Poison by intraperitoneal route. Moderately toxic by intravenous and subcutaneous routes. Mildly toxic by ingestion. Human systemic effects by ingestion and possibly other routes: wakefulness, sleep, altered sleep time, hallucinations, distorted perceptions, tremors, convulsions, and coma. An antibiotic used in the treatment of human pulmonary tuberculosis. When heated to decomposition it emits toxic fumes of NO_x.

CQH100 CAS:59865-13-3 ***HR: 2***
CYCLOSPORIN A
mf: $C_{62}H_{111}N_{11}O_{12}$ mw: 1202.84

SYNS: ANTIBIOTIC S 7481F1 ◇ CICLOSPORIN ◇ CYCLOSPORIN ◇ CYCLOSPORINE ◇ CYCLOSPORINE A ◇ OL 27-400 ◇ S 7481F1 ◇ SANDIMMUN ◇ SANDIMMUNE

TOXICITY DATA with REFERENCE
sce-hmn:lyms 1 mg/L IGAYAY 134,403,85
ims-rbt TDLo:210 mg/kg (14D pre):REP INJFA3 29,218,84
orl-man TDLo:259 mg/kg/2W-C:CAR CEDEDE 8,159,83
orl-wmn TDLo:62500 μg/kg/5D-I:SYS LANCAO 1,1221,86
orl-man TDLo:20 mg/kg/2D-I:BLD LANCAO 2,1092,86
unr-man TDLo:30 mg/kg/4D-I:SYS AIMEAS 107,786,87
orl-rat LD50:1489 mg/kg IYKEDH 17,365,86
ipr-rat LD50:147 mg/kg IYKEDH 17,365,86
scu-rat LD50:286 mg/kg IYKEDH 17,365,86
ivn-rat LD50:24 mg/kg IYKEDH 17,365,86
orl-mus LD50:2803 mg/kg IYKEDH 17,365,86
ivn-mus LD50:96 mg/kg IYKEDH 17,365,86
ivn-rbt LD50:10 mg/kg TOPADD 14,73,86

SAFETY PROFILE: Questionable human carcinogen producing Hodgkin's disease. Experimental carcinogenic data. Experimental reproductive effects. Poison by intraperitoneal and intravenous routes. Moderately toxic by ingestion. Human systemic effects by ingestion: increased body temperature, cyanosis. Mutation data reported. When heated to decomposition it emits toxic fumes of NO_x.

CQH250 CAS:2691-41-0 ***HR: 3***
CYCLOTETRAMETHYLENE TETRANITRAMINE
DOT: UN 0226
mf: $C_4H_8N_8O_8$ mw: 296.20

$$O_2NN[CH_2N(NO_2)]_3CH_2$$

SYNS: CYCLOTETRAMETHYLENE TETRANITRAMINE, dry (DOT) ◇ HMX (DOT) ◇ beta HMY ◇ HW 4 ◇ LX 14-0 ◇ OCTOGEN ◇ OKTOGEN ◇ TETRAMETHYLENETETRANITRAMINE ◇ 1,3,5,7-TETRANITROPERHYDRO-1,3,5,7-TETRAZOCINE

TOXICITY DATA with REFERENCE
orl-mus LD50:1500 mg/kg GISAAA 40(11),17,75
ivn-dog LDLo:40 mg/kg EATR** EB-TR-73040
orl-gpg LD50:300 mg/kg GISAAA 40(11),17,75
ivn-gpg LD50:28 mg/kg EATR** EB-TR-73040

CONSENSUS REPORTS: Reported in EPA TSCA Inventory.

DOT Classification: Forbidden, Dry; Class A Explosive; Label: Explosive A, wet.

SAFETY PROFILE: A poison by ingestion and intravenous routes. An explosive. Decomposes violently at 279°C. When heated to decomposition it emits toxic fumes of NO_x. See also EXPLOSIVES, HIGH.

CQH325 CAS:860-79-7 ***HR: 3***
CYCLOVIROBUXINE D
mf: $C_{26}H_{46}N_2O$ mw: 402.74

SYNS: BEBUXINE ◇ CYCLOVIROBUXIN D ◇ CYCLOVIROBUXINE

TOXICITY DATA with REFERENCE
orl-mus LD50:293 mg/kg CYLPDN 3,101,82
ipr-mus LD50:9200 μg/kg CYLPDN 3,101,82
ivn-mus LD50:8900 μg/kg CYLPDN 3,101,82

SAFETY PROFILE: Poison by ingestion, intravenous, and intraperitoneal routes. When heated to decomposition it emits toxic fumes of NO_x.

CQH500 CAS:126-02-3 ***HR: 3***
CYCRIMINE HYDROCHLORIDE
mf: $C_{19}H_{29}NO \cdot ClH$ mw: 323.95

SYNS: COMPOUND 8958 ◇ α-CYCLOPENTYL-α-PHENYL-1-PIPERIDINEPROPANOL HYDROCHLORIDE ◇ PAGITANE HYDROCHLORIDE ◇ 1-PHENYL-1-CYCLOPENTYL-3-PIPERIDINO-1-PROPANOL HYDROCHLORIDE

TOXICITY DATA with REFERENCE
orl-rat LD50:628 mg/kg 27ZQAG -,218,72
orl-mus LD50:349 mg/kg 27ZQAG -,218,72
ipr-mus LD50:250 mg/kg NTIS** AD691-490
ivn-mus LD50:50 mg/kg 27ZQAG -,218,72

SAFETY PROFILE: Poison by ingestion, intraperitoneal, and intravenous routes. When heated to decomposition it emits very toxic fumes of HCl and NO_x.

CQH625 CAS:7199-29-3 ***HR: 2***
CYHEPTAMIDE
mf: $C_{16}H_{15}NO$ mw: 237.29

PROP: Long needles from acetonitrile. Mp: 193-194°.

Sol in chloroform; sparingly sol in methanol, acetone; sltly sol in ethanol, ether. Practically insol in water.

SYNS: AY 8682 ◇ BS 7029 ◇ CYHEPTAMINE ◇ DIBENZO(a,d)CYCLOHEPTADIENE-5-CARBOXAMIDE ◇ DIBENZO(a,d)(1,4)-CYCLOHEPTADIENE-5-CARBOXAMIDE ◇ 10,11-DIHYDRO-5H-DIBENZO(a,d)CYCLOHEPTENE-5-CARBOXAMIDE ◇ ICI 51426

TOXICITY DATA with REFERENCE
orl-rat LD50:2400 mg/kg 27ZQAG -,68,72
ipr-rat LD50:2000 mg/kg 27ZQAG -,68,72
orl-mus LD50:1830 mg/kg 27ZQAG -,68,72
ipr-mus LD50:630 mg/kg JMCMAR 7,88,64

SAFETY PROFILE: Moderately toxic by ingestion and intraperitoneal routes. An anticonvulsant. When heated to decomposition it emits toxic fumes of NO_x.

CQH650 CAS:13121-70-5 ***HR: 3***
CYHEXATIN
mf: $C_{18}H_{34}OSn$ mw: 385.21

SYNS: DOWCO-213 ◇ ENT 27,395-X ◇ M 3180 ◇ PLICTRAN ◇ PLYCTRAN ◇ TCTH ◇ TRICYCLOHEXYLHYDROXYSTANNANE ◇ TRICYCLOHEXYLHYDROXYTIN ◇ TRICYCLOHEXYLTIN HYDROXIDE ◇ TRICYCLOHEXYLZINNHYDROXID (GERMAN)

TOXICITY DATA with REFERENCE
orl-rat TDLo:108 mg/kg (MGN):REP EQSFAP 4,80,75
orl-rat LD50:180 mg/kg KSKZAN 16(2),59,78
ihl-rat LC50:244 mg/m^3 GISAAA 49(2),74,84
skn-rat LD50:446 mg/kg FAATDF 7,299,86
ipr-rat LD50:13 mg/kg DOWCC* 47(7),80,82
orl-rbt LD50:458 mg/kg GISAAA 47(7),80,82
skn-rbt LD50:2422 mg/kg GISAAA 47(7),80,82
orl-gpg LD50:780 mg/kg TRIPA7 -,1,73
orl-ckn LD50:654 mg/kg TRIPA7 -,1,73
orl-dom LDLo:150 mg/kg TXAPA9 31,66,75

OSHA PEL: TWA 0.1 mg(Sn)/m^3; TWA 5 mg/m^3
ACGIH TLV: TWA 5 mg/m^3; TWA 0.1 mg(Sn)/m^3; STEL 0.2 mg/m^3 (skin)
NIOSH REL: (Organotin Compounds) TWA 0.1 mg(Sn)/m^3.

SAFETY PROFILE: Poison by ingestion, inhalation, and intraperitoneal routes. Moderately toxic by skin contact. Experimental reproductive effects. When heated to decomposition it emits acrid smoke and irritating fumes. See also TIN COMPOUNDS.

CQH750 CAS:508-77-0 ***HR: 3***
CYMARIN
mf: $C_{30}H_{44}O_9$ mw: 548.74

SYNS: CYMARINE ◇ 3-β-(β-d-CYMAROSYLOXY)-5,14-DIHYDROXY-19-OXO-5-β-CARD-20(22)-ENOLIDE ◇ STROPHANTHIDIN-d-CYMAROSID (GERMAN) ◇ K-STROPHANTHIN-α

TOXICITY DATA with REFERENCE
ivn-rat LD50:20 mg/kg AIPTAK 155,165,65
ipr-mus LD50:12 mg/kg AIPTAK 155,165,65
ivn-cat LDLo:95 μg/kg MEIEDD 10,397,83
unr-cat LDLo:110 μg/kg AIPTAK 148,471,64

SAFETY PROFILE: Poison by intravenous, intraperitoneal, and possibly other routes. Used as a cardiotonic. When heated to decomposition it emits acrid smoke and fumes.

CQI000 CAS:99-87-6 ***HR: 2***
p-CYMENE
DOT: UN 2046
mf: $C_{10}H_{14}$ mw: 134.24

PROP: Colorless to pale yellow liquid; odorless. Mp: −68.2°, bp: 176°, lel: 0.7%, @ 100°, ULC: 30-35, flash p: 117°F (CC), d: 0.853, refr index: 1.489, autoign temp: 817°F, vap d: 4.62, vap press: 1 mm @ 17.3°, flash p: (technical) 127°F, uel (technical): 5.6%. Found in nearly 100 volatile oils including lemongrass, sage, thyme, coriander, star anise, and cinnamon (FCTXAV 12,385,74). Sol in alc, ether, acetone, benzene.

SYNS: CAMPHOGEN ◇ CYMENE ◇ CYMOL ◇ DOLCYMENE ◇ FEMA No. 2356 ◇ 4-ISOPROPYL-1-METHYLBENZENE ◇ p-ISOPROPYLTOLUENE ◇ p-METHYL-CUMENE ◇ p-METHYLISOPROPYL BENZENE ◇ 1-METHYL-4-ISOPROPYLBENZENE ◇ PARACYMENE ◇ PARACYMOL

TOXICITY DATA with REFERENCE
skn-rbt 500 mg/24H MOD FCTXAV 12,401,74
cyt-smc 200 μmol/tube HEREAY 33,457,47
orl-rat LD50:4750 mg/kg FCTXAV 2,327,64

CONSENSUS REPORTS: Reported in EPA TSCA Inventory.

DOT Classification: Flammable or Combustible Liquid; Label: Flammable Liquid.

SAFETY PROFILE: Mildly toxic by ingestion. Humans sustain central nervous system effects at low doses. Mutation data reported. A skin irritant. Flammable or combustible liquid. Explosion Hazard: Slight in the form of vapor. To fight fire, use foam, CO_2, dry chemical. When heated to decomposition it emits acrid smoke and fumes.

CQI250 CAS:536-60-7 ***HR: 2***
p-CYMEN-7-OL
mf: $C_{10}H_{14}O$ mw: 150.24

SYNS: CUMIC ALCOHOL ◇ CUMINIC ALCOHOL ◇ CUMINOL ◇ CUMINYL ALCOHOL ◇ CUMYL ALCOHOL ◇ p-ISOPROPYLBENZYL ALCOHOL

TOXICITY DATA with REFERENCE
skn-rbt 500 mg/24H MOD FCTXAV 12,871,74

orl-rat LD50:1020 mg/kg FCTXAV 12,871,74
skn-rbt LD50:2500 mg/kg FCTXAV 12,871,74

CONSENSUS REPORTS: Reported in EPA TSCA Inventory.

SAFETY PROFILE: Moderately toxic by ingestion and skin contact. A skin irritant. When heated to decomposition it emits acrid smoke and irritating fumes. See also ALCOHOLS.

CQI500 CAS:2631-37-0 ***HR: 3***
m-CYM-5-YL METHYLCARBAMATE
mf: $C_{12}H_{17}NO_2$ mw: 207.30

SYNS: CARBAMULT ◇ ENT 27,300 ◇ ENT 27,300-A ◇ EP 316 ◇ METHYLCARBAMIC ACID-m-CYM-5-YL ESTER ◇ 3-METHYL-5-ISOPROPYLPHENYL-N-METHYLCARBAMATE ◇ (3-METHYL-5-ISOPROPYLPHENYL)-N-METHYLCARBAMAT (GERMAN) ◇ 3-METHYL-5-(1-METHYLETHYL)PHENOLMETHYLCARBAMATE ◇ MINACIDE ◇ MORTON EP-316 ◇ PROMECARB ◇ SCHERING 34615 ◇ UC 9880 ◇ UNION CARBIDE UC-9880

TOXICITY DATA with REFERENCE
orl-rat LD50:60 mg/kg MEIEDD 10,1122,83
skn-rat LD50:450 mg/kg WRPCA2 9,119,70
ipr-rat LD50:27200 μg/mg BWHOA6 44(1-3),241,71
ivn-rat LD50:5 mg/kg BJIMAG 22,317,65
ims-rat LD50:44 mg/kg BJIMAG 22,317,65
orl-mus LD50:16 mg/kg BESAAT 15,131,69
orl-gpg LDLo:25 mg/kg JEENAI 61(5),1261,68
scu-gpg LDLo:25 mg/kg JEENAI 61(5),1261,68

CONSENSUS REPORTS: EPA Extremely Hazardous Substances List.

SAFETY PROFILE: Poison by ingestion, intraperitoneal, intravenous, intramuscular, and subcutaneous routes. Moderately toxic by skin contact. An insecticide. When heated to decomposition it emits toxic fumes of NO_x. See also CARBAMATES.

CQI750 CAS:22936-86-3 ***HR: 2***
CYPRAZINE
mf: $C_9H_{14}ClN_5$ mw: 227.73

SYNS: 6-CHLOR-N-CYCLOPROPYL-N′-(1-METHYLETHYL)-1,3,5-TRIZAINE-2,4-DIAMINE ◇ 2-CHLORO-4-CYCLOPROPYLAMINO-6-ISOPROPYLAMINO-sec-TRIAZINE ◇ 2-CHLORO-4-CYCLOPROPYL-AMINO-6-ISOPROPYLAMINO-1,3,5-TRIAZINE ◇ OUTFOX ◇ S-6115 ◇ S-9115

TOXICITY DATA with REFERENCE
orl-rat LD50:1200 mg/kg FMCHA2 -,D227,80
skn-rbt LD50:7500 mg/kg GUCHAZ 6,147,73

SAFETY PROFILE: Moderately toxic by ingestion. Mildly toxic by skin contact. An herbicide. When heated to decomposition it emits very toxic fumes of Cl^- and NO_x.

CQJ000 CAS:8013-86-3 ***HR: 1***
CYPRESS OIL

PROP: The constituents include furfural, d-α-pinene, d-camphene, cymene, d-terpineol, l-cadinene, sylvestrene, cypress camphor and cedrol (FCTXAV 16,637,78).

TOXICITY DATA with REFERENCE
skn-rbt 500 mg/24H MOD FCTXAV 16,699,78

CONSENSUS REPORTS: Reported in EPA TSCA Inventory.

SAFETY PROFILE: A skin irritant. When heated to decomposition it emits acrid smoke and irritating fumes. See also individual components.

CQJ250 CAS:2759-71-9 ***HR: 3***
CYPROMID
mf: $C_{10}H_9Cl_2NO$ mw: 230.10

SYNS: CIPROMID ◇ CLOBBER ◇ 3,4′-DICHLOROCYCLOPROPANE-CARBOXANILIDE ◇ N-(3,4-DICHLOROPHENYL)CYCLOPROPANE-CARBOXAMIDE

TOXICITY DATA with REFERENCE
orl-rat LD50:215 mg/kg WRPCA2 9,119,70
orl-rbt LD50:3028 mg/kg 28ZEAL 5,64,76
skn-rbt LD50:3038 mg/kg WRPCA2 7,135,68

SAFETY PROFILE: A poison by ingestion. Moderately toxic by skin contact. An herbicide. When heated to decomposition it emits very toxic fumes of HCl and NO_x.

CQJ500 CAS:427-51-0 ***HR: 3***
CYPROSTERONE ACETATE
mf: $C_{24}H_{29}ClO_4$ mw: 416.98

SYNS: 17-α-ACETOXY-6-CHLORO-1-α,2-α-METHYLENEPREGNA-4,6-DIENE-3,20-DIONE ◇ 6-CHLORO-1,2-α-METHYLENE-6-DEHYDRO-17-α-HYDROXYPROGESTERONE ACETATE ◇ 6-CHLORO-Δ^6-1,2-α-METHYLENE-17-α-HYDROXYPROGESTERONE ACETATE ◇ 6-CHLORO-1,2-α-METHYLENE-17-α-HYDROXY-Δ^6-PROGESTERONE ACETATE ◇ CPA ◇ CYPROTERONE ACETATE ◇ CYPROTERON-R ACETATE ◇ 1,2-α-METHYLENE-6-CHLORO-Δ^6-17-α-HYDROXYPROGESTERONE ACETATE ◇ 1,2-α-METHYLENE-6-CHLORO-PREGNA-4,6-DIENE-3,20-DIONE 17-α-ACETATE ◇ 1,2-α-METHYLENE-6-CHLORO-$\Delta^{-4,6}$-PREGNADIENE-17-α-OL-3,20-DIONE 17-α-ACETATE ◇ -α-METHYLENE-6-CHLORO-$\Delta^{-4,6}$-PREGNADIENE-17-α-OL-3,20-DIONE ACETATE ◇ NSC-81430 ◇ PREGNA-4,6-DIENE-3,20-DIONE, 6-CHLORO-17-HYDROXY-1-α,2-α-METHYLENE-, ACETATE ◇ SH 714

TOXICITY DATA with REFERENCE
dns-rat-orl 40 mg/kg CBINA8 31,287,80
orl-man TDLo:320 mg/kg (16W male):REP JUPFA4 32,365,78
scu-mus TDLo:30 mg/kg (female 2D post):TER TJADAB 25,27,82
orl-rat TLDo:27 g/kg/78W:ETA LANCAO 2,688,76
ipr-rat LD50:565 mg/kg IYKEDH 13,349,82
ipr-mus LD50:3300 mg/kg NIIRDN 6,APP-4,82

CONSENSUS REPORTS: EPA Genetic Toxicology Program.

SAFETY PROFILE: Questionable carcinogen with experimental tumorigenic and teratogenic data. Moderately toxic by intraperitoneal route. Human reproductive effects by ingestion and possibly other routes: abnormal spermatogenesis, changes in the testes, epididymis, and sperm duct, impotence, and other paternal effects. Experimental reproductive effects. Mutation data reported. Used as a drug to arrest precocious puberty in children and hirsutism in women. A steroid. When heated to decomposition it emits toxic fumes of Cl^-.

CQJ750 CAS:56-17-7 ***HR: 2***
CYSTAMINE DIHYDROCHLORIDE
mf: $C_4H_{12}N_2S_2$•2ClH mw: 225.22

SYNS: AED ◇ 2-AMINOETHYL DISULFIDE DIHYDROCHLORIDE ◇ 2,2'-DITHIO-BIS-(ETHYLAMINE) DIHYDROCHLORIDE ◇ USAF CB-34

TOXICITY DATA with REFERENCE
orl-rat TDLo:1250 mg/kg (13-17D preg):REP JEEMAF 8,94,60
ipr-mus LD50:405 mg/kg ARZNAD 21,284,71

CONSENSUS REPORTS: Reported in EPA TSCA Inventory.

SAFETY PROFILE: Moderately toxic by intraperitoneal route. Experimental reproductive effects. When heated to decomposition it emits very toxic fumes of HCl, SO_x, and NO_x. See also SULFIDES.

CQK000 CAS:52-90-4 ***HR: 2***
l-CYSTEINE
mf: $C_3H_7NO_2S$ mw: 121.17

PROP: An amino acid derived from cystine, occurring naturally in the l-form, which will be considered here. Colorless crystals; sol in water, ammonium hydroxide, and acetic acid; insol in ether, acetone, benzene, carbon disulfide, and carbon tetrachloride.

SYNS: CYSTEIN ◇ CYSTEINE ◇ l-(+)-CYSTEINE ◇ HALF-CYSTEINE ◇ HALF-CYSTINE ◇ β-MERCAPTOALANINE ◇ THIOSERINE

TOXICITY DATA with REFERENCE
mmo-sat 60 μmol/plate BCPCA6 34,3725,85
dns-hmn:fbr 1 mmol/L CALEDO 5,199,78
orl-mus TDLo:3600 mg/kg (female 7-12D post):REP TOIZAG 24,667,77
orl-rat LD50:1890 mg/kg AGACBH 4,125,74
ipr-rat LD50:1620 mg/kg OYYAA2 7,1251,73
scu-rat LD50:1550 mg/kg OYYAA2 7,1251,73
orl-mus LD50:660 mg/kg ARTODN 41,79,78
ipr-mus LD50:1400 mg/kg OYYAA2 7,1251,73
scu-mus LD50:1360 mg/kg OYYAA2 7,1251,73

CONSENSUS REPORTS: Reported in EPA TSCA Inventory. EPA Genetic Toxicology Program.

SAFETY PROFILE: Moderately toxic by ingestion, intraperitoneal, and subcutaneous routes. Experimental reproductive effects. Human mutation data reported. When heated to decomposition it emits very toxic fumes of SO_x and NO_x.

CQK100 CAS:51025-94-6 ***HR: 2***
CYSTEINE-GERMANIC ACID
mf: $C_3H_{11}GeNO_6S$ mw: 261.80

SYNS: (l-CYSTEINE)TETRAHYDROXYGERMANIUM ◇ DB ◇ GERMANIUM, (l-CYSTEINE)TETRAHYDROXY-

TOXICITY DATA with REFERENCE
scu-rat TDLo:70 mg/kg (female 7-13D post):TER YIKUAO 22,107,73
orl-rat LD50:3400 mg/kg TOIZAG 20,180,73
ipr-rat LD50:1090 mg/kg TOIZAG 20,180,73
scu-rat LD50:1200 mg/kg TOIZAG 20,180,73
orl-mus LD50:3320 mg/kg TOIZAG 20,180,73
ipr-mus LD50:2350 mg/kg TOIZAG 20,180,73
scu-mus LD50:2160 mg/kg TOIZAG 20,180,73

SAFETY PROFILE: Moderately toxic by ingestion, intraperitoneal, and subcutaneous routes. An experimental teratogen. When heated to decomposition it emits toxic fumes of NO_x and SO_x.

CQK125 CAS:58100-26-8 ***HR: 2***
CYSTEINE HYDRAZIDE
mf: $C_3H_9N_3OS$ mw: 135.18

SYN: l-HYDRAZIDE CYSTEINE

TOXICITY DATA with REFERENCE
orl-rat LD50:1050 mg/kg AITEAT 27,733,79
ipr-rat LD50:430 mg/kg AITEAT 27,733,79
orl-mus LD50:1010 mg/kg AITEAT 27,733,79
ipr-mus LD50:525 mg/kg AITEAT 27,733,79

SAFETY PROFILE: Moderately toxic by ingestion and intraperitoneal routes. When heated to decomposition it emits toxic fumes of SO_x and NO_x. See also l-CYSTEINE.

CQK250 CAS:52-89-1 ***HR: 2***
l-CYSTEIN HYDROCHLORIDE
mf: $C_3H_7NO_2S$•ClH mw: 157.63

PROP: White crystalline powder; characteristic acetic taste. Mp: 175° (decomp). Sol in water, alc.

SYNS: CYSTEINE CHLORHYDRATE ◇ CYSTEINE HYDROCHLORIDE ◇ l-CYSTEINE HYDROCHLORIDE ◇ l-CYSTEINE MONOHYDROCHLORIDE (FCC)

TOXICITY DATA with REFERENCE
mma-sat 20 mg/plate FCTOD7 22,623,84

cyt-ham:fbr 2 g/L FCTOD7 22,623,84
ipr-mus LD50:1250 mg/kg NTIS** AD691-490
ivn-mus LD50:771 mg/kg JJANAX 38,137,85
unk-mus LD50:3 g/kg BJCAAI 6,160,52

CONSENSUS REPORTS: Reported in EPA TSCA Inventory.

SAFETY PROFILE: Moderately toxic by intraperitoneal, intravenous, and possibly other routes. Mutation data reported. When heated to decomposition it emits very toxic fumes of NO_x, SO_x, and Cl^-.

CQK325 CAS:56-89-3 ***HR: D***
l-CYSTINE
mf: $C_6H_{12}N_2O_4S_2$ mw: 240.30

PROP: Naturally occurring levorotatory form. Colorless to white hexagonal tablets from water. Decomp 260-261°. Sltly sol in water and alc. d-Cystine: Crystals. Sltly sol in water. dl-Cystine, the synthetic racemic form: Crystals. Sltly sol in water. meso-Cystine, the internally compensated form: Crystals. Sltly sol in water.

SYNS: CYSTEINE DISULFIDE ◇ CYSTIN ◇ (−)-CYSTINE ◇ CYSTINE ACID ◇ DICYSTEINE ◇ β,β'-DITHIODIALANINE ◇ GELUCYSTINE ◇ OXIDIZED l-CYSTEINE

TOXICITY DATA with REFERENCE
orl-rat TDLo:9300 mg/kg (93D male):REP OYYAA2 15,199,78
orl-rat LDLo:25 g/kg OYYAA2 15,199,78

SAFETY PROFILE: Experimental reproductive effects. When heated to decomposition it emits toxic fumes of PO_x and SO_x.

CQK500 CAS:53317-25-2 ***HR: 3***
l-CYSTINE-BIS(N,N-β-CHLOROETHYL)HYDRAZIDEHYDROBROMIDE
mf: $C_{14}H_{28}Cl_4N_6O_2S_2•2BrH$ mw: 680.24

SYN: CYDRIN

TOXICITY DATA with REFERENCE
ipr-rat LD50:47 mg/kg NEOLA4 24,401,77
ipr-mus LD50:71 mg/kg NEOLA4 24,401,77
scu-mus LD50:76 mg/kg NEOLA4 24,401,77

SAFETY PROFILE: Poison by intraperitoneal and subcutaneous routes. When heated to decomposition it emits very toxic fumes of HBr, SO_x, NO_x, and Cl^-.

CQK600 CAS:21739-91-3 ***HR: 3***
CYTEMBENA
mf: $C_{11}H_8BrO_4•Na$ mw: 307.09

SYNS: ACRYLIC ACID, 3-p-ANISOYL-3-BROMO-, SODIUM SALT, (E)- ◇ (E)-3-p-ANISOYL-3-BROMOACRYLIC ACID SODIUM SALT ◇ 2-BUTENOIC ACID, 3-BROMO-4-(4-METHOXYPHENYL)-4-OXO-, SODIUM SALT, (E)- (9CI) ◇ MBBA ◇ NCI-C50737 ◇ NSC-104801 ◇ SODNA SUL KYSELINY cis-β-4-METHOXYBENZOYL-β-BROMAKRYLOVE

TOXICITY DATA with REFERENCE
mmo-sat 100 μg/plate SCIEAS 236,933,87
msc-mus:lyms 25 mg/L SCIEAS 236,933,87
ipr-rat LD:14 mg/kg/2Y-I:CAR,REP NTPTR* NTP-TR-207,81
ipr-rat TDLo:7 mg/kg/2Y-I:CAR,REP NTPTR* NTP-TR-207,81
ipr-rat LD50:155 mg/kg CKFRAY 29,106,80
scu-rat LD50:155 mg/kg CKFRAY 29,106,80
ivn-rat LD50:245 mg/kg CKFRAY 29,106,80
ipr-mus LD50:50 mg/kg CKFRAY 29,106,80
scu-mus LD50:52 mg/kg CKFRAY 29,106,80
ivn-mus LD50:98 mg/kg CKFRAY 29,106,80

CONSENSUS REPORTS: NTP Carcinogenesis Bioassay (ipr): Clear Evidence: rat NTPTR* NTP-TR-207,81; NTP Carcinogenesis Bioassay (ipr): No Evidence: mouse NTPTR* NTP-TR-207,81

SAFETY PROFILE: Poison by intraperitoneal, subcutaneous, and intravenous routes. Questionable carcinogen with experimental carcinogenic data. Experimental reproductive effects. Mutation data reported. When heated to decomposition it emits toxic fumes of NaO_2 and Br^-.

CQL250 CAS:115-93-5 ***HR: 3***
CYTHIOATE
mf: $C_8H_{12}NO_5PS_2$ mw: 297.30

SYNS: O,O-DIMETHYL-O,p-SULFAMOYLPHENYLPHOSPHOROTHIOATE ◇ ENT 25,640 ◇ p-HYDROXYBENZENESULFONAMIDE-O-ESTER with O,O-DIMETHYL PHOSPHOROTHIOATE

TOXICITY DATA with REFERENCE
orl-rat LD50:160 mg/kg FMCHA2 -,D88,80
orl-mus LD50:38 mg/kg BESAAT 15,116,69

CONSENSUS REPORTS: Reported in EPA TSCA Inventory.

SAFETY PROFILE: Poison by ingestion. An insecticide. When heated to decomposition it emits very toxic fumes of NO_x, PO_x, and SO_x.

CQL500 CAS:485-35-8 ***HR: 3***
CYTISINE
mf: $C_{11}H_{14}N_2O$ mw: 190.27

SYNS: BAPTITOXIN ◇ BAPTITOXINE ◇ CYSTISINE ◇ CYTITONE ◇ 1,2,3,4,5,6-HEXAHYDRO-1,5-METHANO-8H-PYRIDO(1,2-A)(1,5) DIAZOCIN-8-ONE ◇ SOPHORINE ◇ ULEXINE

TOXICITY DATA with REFERENCE
orl-mus LD50:101 mg/kg BJPCBM 35,161,69
ipr-mus LD50:9400 μg/kg BJPCBM 35,161,69

ivn-mus LD50:1730 μg/kg BJPCBM 35,161,69
inv-cat LD50:400 μg/kg ITOBA9 (2),104,78

CONSENSUS REPORTS: Reported in EPA TSCA Inventory.

SAFETY PROFILE: Poison by ingestion, intravenous and intraperitoneal routes. A toxin found in some plants. When heated to decomposition it emits toxic fumes of NO_x.

CQL750 CAS:6047-01-4 ***HR: 3***
CYTISINE HYDROCHLORIDE
mf: $C_{11}H_{14}N_2O \cdot ClH$ mw: 226.73

SYN: (−)7R:9S-CYTISINE HYDROCHLORIDE

TOXICITY DATA with REFERENCE
orl-mus LD50:101 mg/kg BJPCBM 35,161,69
ipr-mus LD50:9400 μg/kg BJPCBM 35,161,69
ivn-mus LD50:1730 μg/kg BJPCBM 35,161,69

SAFETY PROFILE: Poison by ingestion, intraperitoneal, and intravenous routes. When heated to decomposition it emits very toxic fumes of HCl and NO_x.

CQM125 CAS:14930-96-2 ***HR: 3***
CYTOCHALASIN B
mf: $C_{29}H_{37}NO_5$ mw: 479.67

SYN: PHOMIN

TOXICITY DATA with REFERENCE
dni-hmn:oth 1 mg/L CNREA8 45,311,85
dni-hmn:hla 1 μmol/L MUREAV 92,427,82
cyt-hmn:fbr 1 mg/L JCLBA3 89,194,81
cyt-hmn:oth 1 mg/L JNCIAM 52,653,74
cyt-mus:mmr 1 mg/L ITCSAF 19,58,83
ipr-ham TDLo:5 mg/kg (female 8D post):TER TJADAB 22,59,80
ipr-rat LD50:11 mg/kg TOXIA6 17,137,79
ipr-mus LD50:30 mg/kg FEPRA7 38,438,79

CONSENSUS REPORTS: EPA Genetic Toxicology Program.

SAFETY PROFILE: Poison by intraperitoneal route. An experimental teratogen. Human mutation data reported. When heated to decomposition it emits toxic fumes of NO_x.

CQM250 CAS:36011-19-5 ***HR: 3***
CYTOCHALASIN E
mf: $C_{28}H_{32}NO_7$ mw: 494.62

PROP: Food storage mold metabolite of *Aspergillus clavatus* (TXAPA9 32,135,75).

TOXICITY DATA with REFERENCE
ipr-mus TDLo:3900 μg/kg (female 7-9D post):TER TJADAB 25,11,82
ipr-mus LD50:2700 μg/kg FEPRA7 38,438,79
ipr-gpg LD50:500 μg/kg JJEMAG 48,105,78

SAFETY PROFILE: A poison by intraperitoneal route. An experimental teratogen. When heated to decomposition it emits toxic fumes of NO_x.

CQM325 CAS:9007-43-6 ***HR: D***
CYTOCHROME C

PROP: Reduced form crystallizes as separate needles; oxidized form as rosettes. Mol wt about 13,000. Cytochrome c2: Needles changing to squares. Mol wt about 13,000. Cytochrome c3: Needles. Mol wt 11,300.

SYNS: CROMOCI ◇ CYTOREST ◇ FERRICYTOCHROME C ◇ FERROCYTOCHROME C ◇ HEMATIN-PROTEIN ◇ HORSE-CYTOCHROME C ◇ HORSE HEART CYTOCHROME C ◇ LANDRAX ◇ MYOHEMATIN ◇ NITROSYLFERRICYTOCHROME C

TOXICITY DATA with REFERENCE
mmo-sat 20 μg/plate ABCHA6 45,327,81
mma-sat 20 μg/plate ABCHA6 45,327,81
orl-rat TDLo:250 mg/kg (1-22D preg):REP AJANA2 110,29,62

CONSENSUS REPORTS: Reported in EPA TSCA Inventory.

SAFETY PROFILE: Experimental reproductive effects. Mutation data reported. When heated to decomposition it emits toxic fumes of NO_x.

CQM500 CAS:65-46-3 ***HR: 2***
CYTOSINE RIBOSIDE
mf: $C_9H_{13}N_3O_5$ mw: 243.25

SYNS: 4-AMINO-1-β-d-RIBOFURANOSYL-2(1H)-PYRIMIDINONE ◇ CYTIDINE ◇ 1-β-RIBOFURANOSYLCYTOSINE

TOXICITY DATA with REFERENCE
pic-esc 1 g/L ZAPOAK 12,583,72
oms-hmn:oth 100 μmol/L JIDEAE 65,52,75
dnd-mam:lym 150 μmol/L PNASA6 48,686,62
ipr-mus LD50:2700 mg/kg RPTOAN 40,66,77

CONSENSUS REPORTS: Reported in EPA TSCA Inventory.

SAFETY PROFILE: Moderately toxic by intraperitoneal route. Human mutation data reported. When heated to decomposition it emits toxic fumes of NO_x.

CQM750 CAS:3543-75-7 ***HR: 3***
CYTOSTASAN
mf: $C_{16}H_{21}Cl_2N_3O_2 \cdot ClH$ mw: 394.76

SYNS: IMET 3393 ◇ Γ(1-METHYL-5-BIS(β-CHLORAETHYL)AMINOBENZIMIDAZOLYL)BUTTERSAEUREHYDROCHLORID(GERMAN) ◇ Γ-(1-METHYL-5-BIS(β-CHLOROAETHYL) AMINOBENZIMIDAZOYL)BUTTERSAUERHYDROCHLORID(GERMAN)

TOXICITY DATA with REFERENCE
ipr-mus TDLo:70 mg/kg (7D preg):REP ZPPLBF 110,1067,71
ipr-mus TDLo:70 mg/kg (7D preg):TER ZPPLBF 110,1067,71
orl-mus TDLo:250 mg/kg/4D-I:CAR ARGEAR 43,16,74
ipr-mus TDLo:50 mg/kg/4D-I:CAR ARGEAR 43,16,74
orl-rat LD50:200 mg/kg ATSUDG 8,504,85
ivn-rat LD50:40 mg/kg ATSUDG 8,504,85
orl-mus LD50:250 mg/kg ARGEAR 43,16,74
ipr-mus LD50:100 mg/kg ARGEAR 43,16,74
ivn-mus LD50:80 mg/kg ATSUDG 8,504,85

SAFETY PROFILE: Questionable carcinogen with experimental carcinogenic and teratogenic data. A poison by ingestion, intravenous, and intraperitoneal routes. Experimental reproductive effects. When heated to decomposition it emits very toxic fumes of HCl and NO_x.

CQN000 CAS:4465-94-5 ***HR: 3***
CYTOXAL ALCOHOL
mf: $C_7H_{17}Cl_2N_2O_3P{\bullet}C_6H_{13}N$ mw: 378.33

SYNS: 2-(BIS(2-CHLOROETHYL)AMINO)TETRAHYDROOXAZAPHOSPHORINE CYCLOHEXYLAMINE SALT ◇ N,N-BIS(2-CHLOROETHYL)-N′-(3-HYDROXYPROPYL)PHOSPHORODIAMIDATE,CYCLOHEXYLAMMONIUM SALT ◇ N,N-BIS(2-CHLOROETHYL)-N′-3-PHOSPHORODIAMIDIC ACID HYDROXYLPROPYLCYCLOHEXYLAMINE SALT ◇ CYTOXYL ALCOHOL CYCLOHEXYLAMMONIUM SALT ◇ NCI-C04922 ◇ NSC-52695

TOXICITY DATA with REFERENCE
mmo-sat 100 μg/plate NTPTB* JAN82
mma-sat 100 μg/plate NTPTB* JAN82
dni-hmn:lym 500 μmol/L AGACBH 4,117,74
ipr-mus TDLo:29 mg/kg (11D preg):TER TJADAB 4,141,71
ipr-mus TDLo:290 mg/kg (11D preg):REP TJADAB 4,141,71
ipr-rat TDLo:2900 mg/kg/26W-I:CAR RRCRBU 52,1,75
ipr-mus TDLo:3900 mg/kg/26W-I:ETA CANCAR 40S,1935,77
orl-mus LD50:1618 mg/kg NCISP* JAN86
scu-mus LD50:966 mg/kg NCISP* JAN86
ivn-mus LD50:400 mg/kg NCISP* JAN86

CONSENSUS REPORTS: NCI Carcinogenesis Studies (ipr); Clear Evidence: mouse, rat RRCRBU 52,1,75.

SAFETY PROFILE: Suspected carcinogen with experimental carcinogenic and tumorigenic data. Poison by intravenous route. Moderately toxic by ingestion and subcutaneous routes. Experimental teratogenic and reproductive effects. Human mutation data reported. When heated to decomposition it emits very toxic fumes of NO_x, NH_3, PO_x, and Cl^-. See also ALCOHOLS.

CQN125 CAS:3308-51-8 ***HR: D***
CYTOXYL AMINE
mf: $C_5H_7Cl_2N_2O_3P$ mw: 245.01

SYN: N,N-BIS(2-CHLOROETHYL)-o-(3-AMINOPROPYL)PHOSPHORAMIDATE, ZWITTERION

TOXICITY DATA with REFERENCE
dni-hmn:lym 500 μmol/L AGACBH 4,117,74
cyt-hmn:leu 50 μmol/L BLOOAW 27,816,66
ipr-mus TDLo:107 mg/kg (11D preg):TER TJADAB 4,141,71
ipr-mus TDLo:270 mg/kg (11D preg):REP TJADAB 4,141,71

SAFETY PROFILE: Human mutation data reported. An experimental teratogen. Other experimental reproductive effects. When heated to decomposition it emits toxic fumes of NO_x, Cl^-, and PO_x.

D

DAA800 CAS:94-75-7 ***HR: 3***
2,4-D
DOT: NA 2765
mf: $C_8H_6Cl_2O_3$ mw: 221.04

PROP: White powder. Mp: 141°; bp: 160° @ 0.4 mm; vap d: 7.63.

SYNS: ACIDE-2,4-DICHLORO PHENOXYACETIQUE (FRENCH) ◇ ACIDO (2,4-DICLORO-FENOSSI)-ACETICO (ITALIAN) ◇ AGROTECT ◇ AMIDOX ◇ AMOXONE ◇ AQUA-KLEEN ◇ BH 2,4-D ◇ CHIPCO TURF HERBICIDE "D" ◇ CHLOROXONE ◇ CROP RIDER ◇ CROTILIN ◇ D 50 ◇ DACAMINE ◇ 2,4-D ACID ◇ DEBROUSSAILLANT 600 ◇ DECAMINE ◇ DED-WEED ◇ DED-WEED LV-69 ◇ DESORMONE ◇ (2,4-DICHLOOR-FENOXY)-AZIJNZUUR (DUTCH) ◇ DICHLOROPHENOXYACETIC ACID ◇ 2,4-DICHLOROPHENOXYACETIC ACID (DOT) ◇ 2,4-DICHLORPHENOXYACETIC ACID ◇ (2,4-DICHLOR-PHENOXY)-ESSIGSAEURE (GERMAN) ◇ DICOPUR ◇ DICOTOX ◇ DINOXOL ◇ DMA-4 ◇ DORMONE ◇ 2,4-DWU-CHLOROFENOKSYOCTOSY KWAS (POLISH) ◇ EMULSAMINE BK ◇ EMULSAMINE E-3 ◇ ENT 8,538 ◇ ENVERT 171 ◇ ENVERT DT ◇ ESTERON ◇ ESTERON 99 ◇ ESTERON 76 BE ◇ ESTERON BRUSH KILLER ◇ ESTERON 99 CONCENTRATE ◇ ESTERONE FOUR ◇ ESTERON 44 WEED KILLER ◇ FARMCO ◇ FERNESTA ◇ FERNIMINE ◇ FERNOXONE ◇ FOREDEX 75 ◇ FORMOLA 40 ◇ HEDONAL (The herbicide) ◇ HERBIDAL ◇ IPANER ◇ KROTILINE ◇ LAWN-KEEP ◇ MACRONDRAY ◇ MIRACLE ◇ MONOSAN ◇ MOXONE ◇ NETAGRONE 600 ◇ NSC 423 ◇ PENNAMINE ◇ PHENOX ◇ PIELIK ◇ PLANOTOX ◇ PLANTGARD ◇ RCRA WASTE NUMBER U240 ◇ RHODIA ◇ SALVO ◇ SPRITZ-HORMIN/2,4-D ◇ SUPER D WEEDONE ◇ SUPERORMONE CONCENTRE ◇ TRANSAMINE ◇ TRIBUTON ◇ TRINOXOL ◇ U 46 ◇ U 46DP ◇ U-5043 ◇ VERGEMASTER ◇ VERTON D ◇ VIDON 638 ◇ VISKO-RHAP DRIFT HERBICIDES ◇ WEED-AG-BAR ◇ WEEDAR-64 ◇ WEED-B-GON ◇ WEEDEZ WONDER BAR ◇ WEEDONE LV4 ◇ WEED TOX ◇ WEEDTROL

TOXICITY DATA with REFERENCE
skn-rbt 500 mg/24H MLD 28ZPAK -,279,72
eye-rbt 750 μg/24H SEV 28ZPAK -,279,72
sce-hmn:lym 10 mg/L JOHEA8 73,224,82
dni-ham:ovr 1 mmol/L TOLED5 29,137,85
orl-rat TDLo:500 mg/kg (female 6-15D post):REP FCTXAV 9,801,71
orl-mus TDLo:707 mg/kg (female 11-14D post):TER AECTCV 6,33,77
orl-hmn LDLo:80 mg/kg:GIT,CNS ARPAAQ 94,270,72
orl-man LDLo:93 mg/kg:CNS PAREAQ 14,225,52
orl-rat LD50:370 mg/kg FMCHA2 -,C68,83
skn-rat LD50:1500 mg/kg WRPCA2 9,119,70
ipr-rat LDLo:666 mg/kg JIHTAB 29,85,47
orl-mus LD50:368 mg/kg AJVRAH 15,622,54
ipr-mus LDLo:125 mg/kg TXAPA9 23,288,72
orl-dog LD50:100 mg/kg AEHLAU 7,202,63
orl-rbt LDLo:800 mg/kg AMPMAR 12,26,51
skn-rbt LD50:1400 mg/kg AFDOAQ 16,3,52

CONSENSUS REPORTS: IARC Cancer Review: Group 2B IMEMDT 7,156,87; Human Limited Evidence IMEMDT 41,357,86; Animal Inadequate Evidence IMEMDT 15,111,77; Human Inadequate Evidence IMEMDT 15,111,77. EPA Genetic Toxicology Program. Reported in EPA TSCA Inventory. Community Right-To-Know List.

OSHA PEL: TWA 10 mg/m^3
ACGIH TLV: TWA 10 mg/m^3
DFG MAK: 10 mg/m^3
DOT Classification: ORM-A; Label: None.

SAFETY PROFILE: Suspected human carcinogen. Experimental teratogenic and reproductive effects. Poison by ingestion, intravenous, and intraperitoneal routes. Moderately toxic by skin contact. Human systemic effects by ingestion: somnolence, convulsions, coma, and nausea or vomiting. Can cause liver and kidney injury. A skin and severe eye irritant. Human mutation data reported. When heated to decomposition it emits toxic fumes of Cl^-.

DAB000 CAS:8015-35-8 ***HR: D***
2,4-D and 2,4,5-T (2:1)
mf: $C_8H_6Cl_2O_3 \cdot C_8H_5Cl_3O_3$ mw: 476.52

SYN: HORMOSLYR 64

TOXICITY DATA with REFERENCE
scu-mus TDLo:990 mg/kg (6-14D preg):TER APTOA6 32,408,73
scu-mus TDLo:450 mg/kg (6-14D preg):REP APTOA6 32,408,73

SAFETY PROFILE: An experimental teratogen. Other experimental reproductive effects. When heated to decomposition it emits toxic fumes of Cl^-. See also 2,4-DICHLOROPHENOXYACETIC ACID and 2,4,5-TRICHLOROPHENOXYACETIC ACID.

DAB020 CAS:2307-55-3 ***HR: 3***
2,4-D AMMONIUM SALT
mf: $C_8H_6Cl_2O_3 \cdot H_3N$ mw: 238.08

TOXICITY DATA with REFERENCE
orl-rat TDLo:100 μg/kg (female 9D post):TER GISAAA 44(4),70,79

skn-mus TDLo:1300 mg/kg/86W-I:ETA VPITAR 33(5),83,74
ipr-mus LDLo:250 mg/kg JIDHAN 29,85,47
ipr-mus LDLo:250 mg/kg JIDHAN 29,85,47

CONSENSUS REPORTS: IARC Cancer Review: Animal Inadequate Evidence IMEMDT 15,111,77

SAFETY PROFILE: A poison by intraperitoneal route. Questionable carcinogen with experimental tumorigenic data. An experimental teratogen. When heated to decomposition it emits very toxic fumes of Cl^-, NO_x, and NH_3.

DAB200 CAS:33400-47-4 ***HR: 3***
D-10,242
mf: $C_{22}H_{24}N_4O_2$•ClH mw: 412.96

SYN: 2-AMINO-6-((1,2-DIPHENYLETHYL)AMINO)-3-PYRIDINE-CARBAMIC ACID ETHYL ESTER, MONOHYDROCHLORIDE

TOXICITY DATA with REFERENCE
orl-rat LD50:163 mg/kg DRFUD4 7,801,82
orl-mus LD50:295 mg/kg DRFUD4 7,801,82
orl-dog LD50:40 mg/kg DRFUD4 7,801,82

SAFETY PROFILE: Poison by ingestion. When heated to decomposition it emits toxic fumes of NO_x and HCl. See also CARBAMATES.

DAB400 CAS:39196-18-4 ***HR: 3***
DACAMOX
mf: $C_9H_{18}N_2O_2S$ mw: 218.35

SYNS: DIAMOND SHAMROCK DS-15647 ◇ 3,3-DIMETHYL-1-(METHYLTHIO)-2-BUTANONE-o-((METHYLAMINO)CARBONYL) OXIME ◇ DS-15647 ◇ ENT 27,851 ◇ RCRA WASTE NUMBER P045 ◇ THIOFANOX

TOXICITY DATA with REFERENCE
orl-rat LD50:8500 μg/kg 85ARAE 1,45,77
skn-rbt LD50:39 mg/kg SPEADM 78-1,60,78

CONSENSUS REPORTS: Reported in EPA TSCA Inventory. EPA Extremely Hazardous Substances List.

SAFETY PROFILE: Poison by ingestion and skin contact. When heated to decomposition it emits very toxic fumes of NO_x and SO_x.

DAB600 CAS:4342-03-4 ***HR: 3***
DACARBAZINE
mf: $C_6H_{10}N_6O$ mw: 182.22

SYNS: DETICENE ◇ DIC ◇ (DIMETHYLTRIAZENO)IMIDAZOLE-CARBOXAMIDE ◇ 4-(DIMETHYLTRIAZENO)IMIDAZOLE-5-CARBOXAMIDE ◇ 4-(3,3-DIMETHYL-1-TRIAZENO)IMIDAZOLE-5-CARBOXAMIDE ◇ 4-(5)-(3,3-DIMETHYL-1-TRIAZENO) IMIDAZOLE-5(4)-CARBOXAMIDE ◇ 5-(DIMETHYLTRIAZENO)IMIDAZOLE-4-CARBOXAMIDE ◇ 5-(3,3-DIMETHYLTRIAZENO)IMIDAZOLE-4-CARBOXAMIDE ◇ 5-(3,3-DIMETHYL-1-TRIAZENO)IMIDAZOLE-4-CARBOXAMIDE ◇ 5-(3,3-DIMETHYL-1-TRIAZENYL)-1H-IMIDAZOLE-4-CARBOXAMIDE ◇ DTIC ◇ DTIC-DOME ◇ NCI-C04717 ◇ NSC-45388

TOXICITY DATA with REFERENCE
mma-sat 100 μg/plate CRNGDP 3,467,82
sce-ham:ovr 200 mg/L CNREA8 43,577,83
ipr-rat TDLo:50 mg/kg (12D preg):TER CNREA8 33,2231,73
orl-rat TDLo:1730 mg/kg/15W-C:CAR JNCIAM 54,951,75
ipr-rat TDLo:25 mg/kg (20D preg):ETA,TER ARGEAR 50,3-06,80
ipr-rat TDLo:3900 mg/kg/26W-I:CAR RRCRBU 52,1,75
ivn-hmn TDLo:3500 μg/kg:GIT,BLD,BIO CCROBU 57,83,73
orl-rat LD50:2147 mg/kg UACHDS 9,3105,81
ipr-rat LD50:350 mg/kg ARGEAR 50,3-06,80
ivn-rat LD50:411 mg/kg YACHDS 9,3105,81
orl-mus LD50:2032 mg/kg YACHDS 9,3105,81
ipr-mus LD50:567 mg/kg CTRRDO 62,721,78
par-ham LD10:250 mg/kg JSONAU 15,355,80

CONSENSUS REPORTS: NTP Fifth Annual Report on Carcinogens. IARC Cancer Review: Group 2B IMEMDT 7,184,87; Human Limited Evidence IMEMDT 26,203,81; Animal Sufficient Evidence IMEMDT 26,203,81. NCI Carcinogenesis Studies (ipr); Clear Evidence: mouse, rat RRCRBU 52,1,75. EPA Genetic Toxicology Program.

SAFETY PROFILE: Confirmed carcinogen with experimental carcinogenic and tumorigenic data. Poison by intraperitoneal and parenteral routes. Moderately toxic by ingestion and intravenous routes. Experimental teratogenic effects. Human systemic effects by intravenous route: nausea or vomiting, luekopenia (reduced white blood cell count), and changes in dehydrogenase enzymatic activity. Mutation data reported. When heated to decomposition it emits toxic fumes of NO_x.

DAB700 ***HR: 2***
DAFFODIL

PROP: Bulb-producing plants with leaves that emerge directly from the bulb which looks much like an onion. A leafless stem carries one or more white or yellow flowers which have a trumpet-shaped section growing from the center of a flat corona. They are native to Europe and northern Africa, and are cultivated as ornamentals in the United States.

SYNS: JONQUIL ◇ NARCISO (CUBA, MEXICO) ◇ NARCISSUS ◇ NARCISSUS POETICUS ◇ NARCISSUS PSEUDONARCISSUS ◇ PACIENCIA

SAFETY PROFILE: The bulbs contain the poisonous lycorine and related alkaloids. They are sometimes mistaken for onions. Ingestion of large amounts can cause nausea, persistent vomiting, and diarrhea.

DAB750 CAS:18067-13-5 ***HR: 3***
DAIPIN
mf: $C_{18}H_{24}NO_4 \cdot CH_3O_4S$ mw: 429.53

SYNS: DD 234 ◇ ESPASMO GASIUM ◇ N-METHYLHYOSCINE METHYL SULFATE ◇ N-METHYLSCOPOLAMINE METHOSULFATE ◇ METHYLSCOPOLAMINE METHYL SULFATE ◇ N-METHYLSCOPOLAMINE METHYL SULFATE ◇ METHYLSCOPOLAMMONIUM METHYLSULFATE ◇ SANDRIX

TOXICITY DATA with REFERENCE
scu-rat TDLo:45500 μg/kg (91D male):REP KSRNAM 7,192,73
orl-rat LD50:5590 mg/kg IYKEDH 4,90,73
ipr-rat LD50:212 mg/kg IYKEDH 4,90,73
scu-rat LD50:1340 mg/kg IYKEDH 4,90,73
ivn-rat LD50:56800 μg/kg IYKEDH 4,90,73
orl-mus LD50:3010 mg/kg IYKEDH 4,90,73
ipr-mus LD50:116 mg/kg IYKEDH 4,90,73
scu-mus LD50:558 mg/kg IYKEDH 4,90,73
ivn-mus LD50:36900 μg/kg IYKEDH 4,90,73

SAFETY PROFILE: Poison by intravenous and intraperitoneal routes. Moderately toxic by ingestion and subcutaneous routes. Experimental reproductive effects. When heated to decomposition it emits toxic fumes of SO_x and NO_x.

DAB800 CAS:1172-18-5 ***HR: 3***
DALMANE
mf: $C_{21}H_{23}ClFN_3O \cdot 2ClH$ mw: 460.84

SYNS: BENOZIL ◇ DALMADORM ◇ DALMADORM HYDROCHLORIDE ◇ DALMATE ◇ DORMODOR ◇ FELISON ◇ FLURAZEPAN DIHYDROCHLORIDE ◇ FLURAZEPAM HYDROCHLORIDE ◇ ID 480 DIHYDROCHLORIDE ◇ INSUMIN ◇ LUNIPAK ◇ NSC-78559 ◇ RO 5-6901 ◇ SOMLAN

TOXICITY DATA with REFERENCE
orl-rat LD50:978 mg/kg NIIRDN 6,7-06,82
ipr-rat LD50:179 mg/kg NIIRDN 6,7-06,82
scu-rat LD50:859 mg/kg NIIRDN 6,7-06,82
ivn-rat LD50:40500 μg/kg NIIRDN 6,7-06,82
orl-mus LD50:660 mg/kg 26RAAN -,47,73
ipr-mus LD50:201 mg/kg NIIRDN 6,7-06,82
scu-mus LD50:440 mg/kg OYYAA2 14,637,77
ivn-mus LD50:66900 μg/kg NIIRDN 6,7-06,82
orl-rbt LD50:568 mg/kg AIPTAK 178,216,69

SAFETY PROFILE: Poison by intravenous and intraperitoneal routes. Moderately toxic by ingestion and subcutaneous routes. Habituating and possibly addictive. An hypnotic and sedative. When heated to decomposition it emits very toxic fumes of F^-, NO_x, and Cl^-.

DAB820 CAS:101052-67-9 ***HR: 3***
DAMPA D

PROP: Contains 3% dibenzthion, 0.3% dichlorothiocyanoaniline, 1% diphenylhydramine (NIIRDN 6,455, 82).

TOXICITY DATA with REFERENCE
ipr-rat LD50:390 mg/kg NIIRDN 6,445,82
orl-mus LD50:1880 mg/kg NIIRDN 6,445,82
ipr-mus LD50:380 mg/kg NIIRDN 6,445,82
scu-mus LD50:940 mg/kg NIIRDN 6,445,82

SAFETY PROFILE: Poison by intraperitoneal route. Moderately toxic by ingestion and subcutaneous routes. A topical antibacterial agent. When heated to decomposition it emits toxic fumes of SO_x, NO_x, and CN^-. See also BENZHYDRYL, THIOCYANATES, and individual components.

DAB825 CAS:39515-41-8 ***HR: 3***
DANITOL
mf: $C_{22}H_{23}NO_3$ mw: 349.43

PROP: Synthetic pyrethroid insecticide with repellent and contact activity. Pale yellow oil, n (26/D) 1.5283.

SYNS: α-CYANO-3-PHENOXYBENZYL2,2,3,3-TETRAMETHYL-1-CYCLOPROPANECARBOXYLATE ◇ FENPROPANAGE ◇ FENPROPATHRIN ◇ GENPROPATHRIN ◇ MEOTHRIN ◇ RODY ◇ S 32-06 ◇ SD 417-06 ◇ WL 417-06

TOXICITY DATA with REFERENCE
orl-rat LD50:49 mg/kg FMCHA2 -,C70,83
ivn-rat LDLo:2500 μg/kg ARTODN 45,325,80
skn-rbt LD50:2000 mg/kg FMCHA2 -,C70,83

CONSENSUS REPORTS: Cyanide and its compounds are on the Community Right-To-Know List.

SAFETY PROFILE: Poison by ingestion and intravenous routes. Moderately toxic by skin contact. When heated to decomposition it emits toxic fumes of NO_x and CN^-. See also CYANIDE and ESTERS.

DAB830 CAS:17230-88-5 ***HR: 3***
DANOCRINE
mf: $C_{22}H_{27}NO_2$ mw: 337.50

PROP: Crystals from acetone. Mp: 224.4-226.8°.

SYNS: CHRONOGYN ◇ CYCLOMEN ◇ DANAZOL ◇ DANOL ◇ LADOGAL ◇ 17-α-2,4-PREGNADIEN-20-YNO(2,3-d)ISOXAZOL-17-OL ◇ 17-α-PREGNA-2,4-DIEN-20-YNO(2,3-d)ISOXAZOL-17-OL ◇ 17-α-PREGN-4-EN-20-YNO(2,3-d)ISOXAZOL-17-OL ◇ WIN 17757 ◇ WINOBANIN

TOXICITY DATA with REFERENCE
orl-wmn TDLo:16 μg/kg (female 5D post):REP CCPTAY 27,39,83
orl-wmn TDLo:2920 mg/kg/2Y-C:CAR JSONAU 28,114,85

orl-wmn TDLo:120 mg/kg/10D-I:SKN AIMDAP 145,2251,85
ipr-mus LD50:6770 mg/kg IYKEDH 14,484,83

SAFETY PROFILE: Human systemic effects by ingestion: allergic dermatitis. Human male reproductive effects by ingestion: changes in spermatogenesis, impotence, and other unspecified effects. Human female reproductive effects by ingestion: menstrual cycle changes or disorders, changes in fertility, and other unspecified effects. Other experimental reproductive effects. Questionable human carcinogen producing liver tumors. When heated to decomposition it emits toxic fumes of NO_x.

DAB840 CAS:14663-23-1 ***HR: 2***
DANTRIUM HEMIHEPTAHYDRATE
mf: $C_{14}H_9N_4O_5$•Na mw: 336.26

SYNS: DANTOROLENE SODIUM ◇ DANTRIUM ◇ DANTROLENE SODIUM ◇ F-400 ◇ 2,4-IMIDAZOLIDINEDIONE, 1-(((5-(4-NITROPHENYL)-2-FURANYL)METHYLENE)AMINO)-, SODIUM SALT ◇ 1-((5-(p-NITROPHENYL)FURFURYLIDENE)AMINO)HYDANTOIN SODIUM

TOXICITY DATA with REFERENCE
orl-rat TDLo:500 mg/kg (14D pre/7-17D preg):REP KSRNAM 11,2218,77
orl-rat TDLo:500 mg/kg (14D pre/7-17D preg):TER KSRNAM 11,2218,77
orl-rat LD50:7431 mg/kg IYKEDH 12,668,81
ipr-rat LD50:413 mg/kg IYKEDH 12,668,81
orl-mus LD50:1188 mg/kg JPMSAE 69,327,80
ipr-mus LD50:534 mg/kg IYKEDH 12,668,81

SAFETY PROFILE: Moderately toxic by intraperitoneal and ingestion routes. An experimental teratogen. Other experimental reproductive effects. When heated to decomposition it emits toxic fumes of NO_x and Na_2O.

DAB845 CAS:7261-97-4 ***HR: 2***
DANTROLENE
mf: $C_{14}H_{10}N_4O_5$ mw: 314.28

SYN: HYDANTOIN, 1-((5-(p-NITROPHENYL)FURFURYLIDENE) AMINO)-

TOXICITY DATA with REFERENCE
unr-man TDLo:9386 mg/kg/3Y-C:CAR PGMJAO 56,261,80

SAFETY PROFILE: Questionable human carcinogen producing Hodgkin's disease. When heated to decomposition it emits toxic fumes of NO_x.

DAB850 CAS:28164-88-7 ***HR: 3***
DAPHNETOXIN
mf: $C_{27}H_{30}O_8$ mw: 482.57

SYN: ORTHOBENZOIC ACID, cyclic 7,8,10a-ESTER with 5,6-EPOXY-4,5,6,6a,7,8,9,10,10a,10b-DECAHYDRO-3a,4,7,8,10a-PENTAHYDROXY-5-(HYDROXYMETHYL)-8-ISOPROPENYL-2,10-DIMETHYLBENZ(e)AZULEN-3(3aH)-ONE

TOXICITY DATA with REFERENCE
skn-hmn 5 μg/24H SEV TOXIA6 19,841,81
ipr-mus LD50:1100 μg/kg SCYYDZ 9(2),48,89

SAFETY PROFILE: Poison by intraperitoneal route. A severe skin irritant. When heated to decomposition it emits acrid smoke and irritating fumes.

DAB875 CAS:71-81-8 ***HR: 3***
DARBID
mf: $C_{23}H_{33}N_2O$•I mw: 480.48

PROP: Crystals or amorphous powder. Mp: 198-201° (decomp). The methiodide is freely sol in boiling water, methanol, ethanol, chloroform; practically insol in ether.

SYNS: 2,2-DIPHENYL-4-DIISOPROPYLAMINOBUTYRAMIDE METHIODIDE ◇ DIPRAMID ◇ DIPRAMIDE ◇ ISAMID ◇ ISOPROPAMIDE ◇ ISOPROPAMIDE IODIDE ◇ MARYGIN-M ◇ 5579 MD ◇ PIACCAMIDE ◇ PRIAMIDE ◇ PRIAZIMIDE ◇ R 79 ◇ SANULCIN ◇ SKF 4740 ◇ TYRIMIDE

TOXICITY DATA with REFERENCE
orl-mus LD50:1600 mg/kg AIPTAK 103,120,55
ipr-mus LD50:120 mg/kg AIPTAK 103,120,55
ivn-mus LD50:12779 mg/kg AIPTAK 103,100,55

SAFETY PROFILE: Poison by intraperitoneal route. Moderately toxic by ingestion. When heated to decomposition it emits toxic fumes of I^- and NO_x.

DAB879 CAS:469-62-5 ***HR: 3***
DARVON
mf: $C_{22}H_{29}NO_2$ mw: 339.52

SYNS: DEXTROPROPOXYPHENE ◇ α-(+)-4-DIMETHYLAMINO-1,2-DIPHENYL-3-METHYL-2-BUTANOL PROPIONATE ESTER ◇ DOLENE ◇ DOLOXENE ◇ (+)-PROPOXYPHENE ◇ d-PROPOXYPHENE ◇ PROXAGESIC ◇ SK 65

TOXICITY DATA with REFERENCE
orl-hmn TDLo:650 mg/kg:PUL,CNS,CVS AMSVAZ 200,241,76
orl-rat LD50:135 mg/kg AIPTAK 178,446,69
ipr-rat LD50:50 mg/kg AIPTAK 178,446,69
scu-rat LD50:79100 μg/kg ARZNAD 24,600,74
orl-mus LD50:300 mg/kg JMCMAR 15,295,72
ipr-mus LD50:110 mg/kg AIPTAK 178,446,69
scu-mus LD50:113 mg/kg ARZNAD 24,600,74
ivn-mus LD50:25 mg/kg JPETAB 134,154,61

SAFETY PROFILE: Poison by ingestion, intraperitoneal, subcutaneous, and intravenous routes. Human systemic effects by ingestion: change in cardiac rate, respiratory depression and coma. When heated to decomposition it emits toxic fumes of NO_x.

DAB880 CAS:26570-10-5 ***HR: 3***
DARVON-N
mf: $C_{10}H_8O_3S{\cdot}C_{22}H_{29}NO_2{\cdot}H_2O$ mw: 565.78

SYNS: d-4-DIMETHYLAMINO-3-METHYL-1,2-DIPHENYL-2-PROPIONOXYBUTANENAPHTHALENE-2-SULPHONATEHYDRATE ◇ d-PROPOXYPHENE NAPSYLATE HYDRATE ◇ S-9700

TOXICITY DATA with REFERENCE
orl-rat LD50:485 mg/kg KSRNAM 5,1011,71
ipr-rat LD50:56 mg/kg KSRNAM 5,1011,71
scu-rat LD50:169 mg/kg KSRNAM 5,1011,71
orl-mus LD50:973 mg/kg OYYAA2 5,359,71
ipr-mus LD50:171 mg/kg KSRNAM 5,2043,71
scu-mus LD50:749 mg/kg KSRNAM 5,1011,71

SAFETY PROFILE: Poison by ingestion, subcutaneous, and intraperitoneal routes. When heated to decomposition it emits toxic fumes of SO_x and NO_x. See also d-PROPOXYPHENE HYDROCHLORIDE and SULFONATES.

DAB925 CAS:41093-93-0 ***HR: D***
DATURALACTONE
mf: $C_{28}H_{38}O_7$ mw: 486.66

TOXICITY DATA with REFERENCE
orl-rat TDLo:175 mg/kg (1-7D preg):REP IJEBA6 16,419,78

SAFETY PROFILE: Experimental reproductive effects. When heated to decomposition it emits acrid smoke and fumes.

DAC000 CAS:20830-81-3 ***HR: 3***
DAUNOMYCIN
mf: $C_{27}H_{29}NO_{10}$ mw: 527.57

PROP: Thin, red needles. Mp: 190° (decomp). Isolated from cultures of a *Streptomyces* (CNREA8 32,1029,72).

SYNS: ACETYLADRIAMYCIN ◇ CERUBIDIN ◇ DAUNAMYCIN ◇ DAUNORUBICIN ◇ DAUNORUBICINE ◇ DM ◇ FI6339 ◇ LEUKAEMOMYCIN C ◇ NCI-C04693 ◇ NSC-82151 ◇ RCRA WASTE NUMBER U059 ◇ RP 13057 ◇ 13,057 R.P. ◇ RUBIDOMYCIN ◇ RUBIDOMYCINE ◇ RUBOMYCIN C ◇ RUBOMYCIN C 1 ◇ STREPTOMYCES PEUCETIUS

TOXICITY DATA with REFERENCE
mmo-sat 1 μg/plate ENMUDM 7,129,85
pic-sat 800 ng/plate MUREAV 110,243,83
dni-hmn:oth 30 nmol/L CNREA8 44,2421,84
ipr-rat TDLo:8 mg/kg (7-14D preg):TER CRSBAW 163,1299,69
ipr-rat TDLo:16 mg/kg (7-14D preg):REP CRSBAW 163,1299,69
ipr-rat TDLo:2200 μg/kg/7W-I:ETA CANCAR 40,1935,77
ivn-rat TDLo:6250 μg/kg:CAR CNREA8 38,1444,78
scu-mus TDLo:15 g/kg/12W-I:CAR RRCRBU 20,73,69
ivn-rat TD:5 mg/kg:NEO,REP CNREA8 32,1029,72
scu-mus TD:15 mg/kg/12W-I:NEO ARZNAD 17,948,67
orl-hmn LDLo:6 mg/kg 34ZIAG -,521,69
ipr-rat LD50:20 mg/kg ADTEAS 3,181,68
ivn-rat LD50:13 mg/kg NCINS* -,304,67
orl-mus LD50:205 mg/kg YKYUA6 25,573,74
orl-mus LD50:205 mg/kg YKYUA6 25,573,74
ipr-mus LD50:2500 μg/kg NCINS* -,304,67
scu-mus LD50:16 mg/kg COREAF 257,1813,63
ivn-mus LD50:18 mg/kg 85FZAT -,243,67

CONSENSUS REPORTS: IARC Cancer Review: Group 2B IMEMDT 7,56,87; Animal Sufficient Evidence IMEMDT 10,145,76. NCI Carcinogenesis Studies (ipr); Clear Evidence: rat CANCAR 40,1935,77; No Evidence: mouse CANCAR 40,1935,77. EPA Genetic Toxicology Program.

SAFETY PROFILE: Suspected carcinogen with experimental carcinogenic, neoplastigenic, and tumorigenic data. Human poison by ingestion. Experimental poison by subcutaneous, intravenous, and intraperitoneal routes. Experimental teratogenic and reproductive effects. Human mutation data reported. When heated to decomposition it emits toxic fumes of NO_x. See also DAUNOMYCIN HYDROCHLORIDE.

DAC200 CAS:23541-50-6 ***HR: 3***
DAUNOMYCIN HYDROCHLORIDE
mf: $C_{27}H_{29}NO_{10}{\cdot}ClH$ mw: 564.03

SYNS: CERUBIDINE ◇ DAUNOBLASTIN ◇ DAUNOBLASTINA ◇ DAUNOMYCIN CHLOROHYDRATE ◇ DAUNORUBICIN HYDROCHLORIDE ◇ NDC 0082-4155 ◇ NSC-82151 ◇ ONDENA ◇ RP 13057 HYDROCHLORIDE ◇ RUBIDOMYCIN ◇ RUBIDOMYCIN HYDROCHLORIDE ◇ RUBOMYCIN C

TOXICITY DATA with REFERENCE
mmo-sat 83 ng/plate CNREA8 38,2148,78
mmo-asn 250 mg/L MUREAV 97,293,82
orl-rat LD50:290 mg/kg NIIRDN 6,437,82
ipr-rat LD50:14300 μg/kg NIIRDN 6,437,82
scu-rat LD50:33200 μg/kg NIIRDN 6,437,82
ivn-rat LD50:14300 μg/kg NIIRDN 6,437,82
orl-mus LD50:205 mg/kg NIIRDN 6,437,82
ipr-mus LD50:3050 μg/kg NIIRDN 6,437,82
scu-mus LD50:28800 μg/kg NIIRDN 6,437,82

CONSENSUS REPORTS: EPA Genetic Toxicology Program.

SAFETY PROFILE: Poison by ingestion, intraperitoneal, subcutaneous, and intravenous routes. Mutation data reported. A antineoplastic agent. When heated to decomposition it emits very toxic fumes of NO_x and HCl. See also DAUNOMYCIN.

DAC300 CAS:28008-55-1 ***HR: 3***
DAUNOMYCINOL
mf: $C_{27}H_{31}NO_{10}$ mw: 529.59

SYNS: ANTIBIOTIC 20-798RP ◇ DAUNORUBICINOL ◇ DIHYDRO-DAUNOMYCIN ◇ 13-DIHYDRODAUNOMYCIN ◇ 13-DIHYDRO-DAUNORUBICIN ◇ DUBORIMYCIN ◇ 1-HYDROXY-13-DIHYDRO-DAUNOMYCIN ◇ LEUKAEMOMYCIN D

TOXICITY DATA with REFERENCE
pic-esc 1400 μg/L MUREAV 77,197,80
dni-mus:leu 7600 nmol/L JANTAJ 34,1596,81
oms-mus:leu 3 μmol/L JANTAJ 34,1596,81
ipr-mus LD50:6500 μg/kg 85GDA2 3,122,80

SAFETY PROFILE: Poison by intraperitoneal route. Mutation data reported. When heated to decomposition it emits toxic fumes of NO_x. See also DAUNOMYCIN.

DAC400 CAS:8016-03-3 ***HR: 1***
DAVANA OIL

PROP: Distilled from *Artemisia pallens* (FCTXAV 14,659,76).

TOXICITY DATA with REFERENCE
skn-rbt 500 mg/24H MOD FCTXAV 14,737,76

CONSENSUS REPORTS: Reported in EPA TSCA Inventory.

SAFETY PROFILE: A moderate skin irritant. When heated to decomposition it emits acrid smoke and irritating fumes.

DAC450 CAS:12011-76-6 ***HR: 2***
DAWSONITE
mf: CH_2AlO_5•Na mw: 144.00

SYN: CRYSTALLINE DEHYDROXY SODIUM ALUMINUM, CARBONATE

TOXICITY DATA with REFERENCE
imp-rat TDLo:200 mg/kg:NEO JJIND8 67,965,81

SAFETY PROFILE: Questionable carcinogen with experimental neoplastigenic data.

DAC500 ***HR: 2***
DAY BLOOMING JESSAMINE

PROP: Evergreen shrubs with smooth-edged, oval leaves and clusters of white, tubular flowers. The day blooming jessamine has flowers that are fragrant during the day with black berries. The night blooming jessamine has flowers that are fragrant at night with white berries. They are native to the West Indies and grow wild and are under cultivation in the United States (Florida, Texas) and Guam.

SYNS: 'ALA-AUMOE (HAWAII) ◇ C. DIURNUM ◇ C. NOCTURNUM ◇ CESTRUM (VARIOUS SPECIES) ◇ CHINESE INKBERRY ◇ DAMA de DIA (PUERTO RICO) ◇ DAMA de NOCHE (PUERTO RICO) ◇ GALAN de DIA (CUBA) ◇ GALAN de NOCHE (CUBA) ◇ HUELE de NOCHE (MEXICO) ◇ JASMIN de NUIT (HAITI) ◇ KUPAOA (HAWAII) ◇ LILAS de NUIT (HAITI) ◇ MAKAHALA (HAWAII) ◇ NIGHT BLOOMING JESSAMINE ◇ ONAONA-IAPANA (HAWAII)

SAFETY PROFILE: The berries and sap contain toxic saponins and nicotine. Ingestion of these plant parts may result in inflammation of the stomach and intestines.

DAC800 CAS:33857-26-0 ***HR: 3***
DCDD
mf: $C_{12}H_6Cl_2O_2$ mw: 253.08

PROP: Colorless crystals. Mp: 210°.

SYNS: 2,7-DICHLORODIBENZODIOXIN ◇ 2,7-DICHLORODI-BENZO(b,e)(1,4)DIOXIN ◇ 2,7-DICHLORODIBENZO-p-DIOXIN ◇ NCI-C03667

TOXICITY DATA with REFERENCE
eye-rbt 2 mg MLD EVHPAZ 5,87,73
orl-rat TDLo:5 mg/kg (6-15D preg):TER ADCSAJ 120,70,73
orl-mus TDLo:378 g/kg/90W-C:ETA NCITR* NCI-CG-TR-123,79

CONSENSUS REPORTS: IARC Cancer Review: Animal Inadequate Evidence IMEMDT 15,41,77. NCI Carcinogenesis Bioassay (feed); Clear Evidence: mouse NCITR* NCI-CG-TR-123,79; No Evidence: rat NCITR* NCI-CG-TR-123,79.

SAFETY PROFILE: An eye irritant. Experimental teratogenic data. Questionable carcinogen with experimental tumorigenic data. When heated to decomposition it emits toxic fumes of Cl^-.

DAC975 CAS:66826-72-0 ***HR: 3***
cis-DCPO
mf: $C_3H_4Cl_2O$ mw: 126.97

SYNS: cis-2-CHLORO-3-(CHLOROMETHYL)OXIRANE ◇ cis-1,3-DICHLORO-1,2-EPOXYPROPANE ◇ cis-1,3-DICHLOROPROPENE OXIDE

TOXICITY DATA with REFERENCE
otr-ham:emb 5 μmol/L JJIND8 69,531,82
skn-mus TDLo:400 mg/kg/53W-I:CAR CNREA8 43,159,83
scu-mus TDLo:20 mg/kg/71W-I:CAR CNREA8 43,159,83

SAFETY PROFILE: Questionable carcinogen with experimental carcinogenic data. Mutation data reported. When heated to decomposition it emits toxic fumes of Cl^-.

DAD000 CAS:14600-07-8 ***HR: 2***
2,4-D CROTYL ESTER
mf: $C_{12}H_{12}Cl_2O_3$ mw: 275.14

SYN: CROTYL-2,4-DICHLOROPHENOXYACETATE

TOXICITY DATA with REFERENCE
ihl-rat LCLo:520 mg/m^3/4H HYSAAV 31,383,66
unr-rat LDLo:452 mg/kg HYSAAV 31,383,66
unr-mus LDLo:580 mg/kg HYSAAV 31,383,66

SAFETY PROFILE: Moderately toxic by inhalation and possibly other routes. When heated to decomposition it emits toxic fumes of Cl^-. See also ESTERS.

DAD040 CAS:61848-70-2 ***HR: 3***
cis-DDCP
mf: $C_6H_4Cl_2N_2Pt$ mw: 370.11

SYNS: 1,2-DIAMINOCYCLOHEXANEPLATINUM(II)CHLORIDE ◇ DICHLORO(1,2-CYCLOHEXANEDIAMINE)PLATINUM ◇ DICHLORO(1,2-DIAMINOCYCLOHEXANE)PLATINUM ◇ DICHLORO (1,2-DIAMINOCYCLOHEXANE)PLATINUM(II) ◇ cis-DICHLORO-1,2-DIAMINOCYCLOHEXANE PLATINUM(II) ◇ NSC 194814 ◇ PT 155

TOXICITY DATA with REFERENCE
mmo-sat 20 nmol/plate CNREA8 41,4368,81
dnd-sat 10 mg/L/20H-C CNREA8 41,4368,81
oms-bcs 9900 nmol/L/3H-C CNREA8 41,4368,81
ipr-mus TDLo:20530 μg/kg/10W-I:NEO CNREA8 41,4368,81
ipr-mus LD50:40130 μg/kg NCISP* JAN86

SAFETY PROFILE: Poison by intraperitoneal route. Questionable carcinogen with experimental neoplastigenic data. Mutation data reported. When heated to decomposition it emits toxic fumes of Cl^- and NO_x. See also PLATINUM COMPOUNDS.

DAD050 ***HR: 3***
trans(+)-DDCP
mf: $C_6H_{14}Cl_2N_2Pt$ mw: 380.176

SYNS: (SP-4-2)-trans(+)-DICHLORO(1,2-CYCLOHEXANEDIAMINE-N,N')- (9CI) ◇ XX 212

TOXICITY DATA with REFERENCE
mmo-sat 20 nmol/plate CNREA8 41,4368,81
dnd-sat 10 mg/L/20H-C CNREA8 41,4368,81
oms-bcs 20 μmol/L/3H-C CNREA8 41,4368,81
ipr-mus TDLo:20530 μg/kg/10W-I:NEO CNREA8 41,4368,81

SAFETY PROFILE: Questionable carcinogen with experimental neoplastigenic data. Mutation data reported. When heated to decomposition it emits toxic fumes of Cl^- and NO_x. See also DDCP and PLATINUM COMPOUNDS.

DAD075 CAS:61848-66-6 ***HR: 3***
trans(−)-DDCP
mf: $C_6H_{14}Cl_2N_2Pt$ mw: 380.21

SYN: trans(−)-DICHLORO-1,2-DIAMINOCYCLOHEXANE-PLATINUM(II)

TOXICITY DATA with REFERENCE
mmo-sat 20 nmol/plate CNREA8 41,4368,81
dnd-sat 10 mg/L/20H-C CNREA8 41,4368,81
oms-bcs 13 μmol/L/3H-C CNREA8 41,4368,81
ipr-mus TDLo:20530 μg/kg/10W-I:NEO CNREA8 41,4368,81

SAFETY PROFILE: Questionable carcinogen with experimental neoplastigenic data. Mutation data reported. When heated to decomposition it emits toxic fumes of Cl^- and NO_x. See also DDCP and PLATINUM COMPOUNDS.

DAD200 CAS:50-29-3 ***HR: 3***
DDT
DOT: NA 2761
mf: $C_{14}H_9Cl_5$ mw: 354.48

PROP: Colorless crystals or white to sltly off-white powder. Odorless or with slight aromatic odor. Mp: 108.5-109°.

SYNS: AGRITAN ◇ ANOFEX ◇ ARKOTINE ◇ AZOTOX ◇ α,α-BIS(p-CHLOROPHENYL)-β,β,β-TRICHLORETHANE ◇ 1,1-BIS-(p-CHLOROPHENYL)-2,2,2-TRICHLOROETHANE ◇ 2,2-BIS(p-CHLOROPHENYL)-1,1,1-TRICHLOROETHANE ◇ BOSAN SUPRA ◇ BOVIDERMOL ◇ CHLOROPHENOTHAN ◇ CHLOROPHENOTHANE ◇ CHLOROPHENOTOXUM ◇ CITOX ◇ CLOFENOTANE ◇ p,p'-DDT ◇ DEDELO ◇ DEOVAL ◇ DETOX ◇ DETOXAN ◇ DIBOVAN ◇ DICHLORODIPHENYLTRICHLOROETHANE ◇ DICHLORODIPHENYLTRICHLOROETHANE (DOT) ◇ p,p'-DICHLORODIPHENYLTRICHLOROETHANE ◇ 4,4'-DICHLORODIPHENYLTRICHLOROETHANE ◇ DICOPHANE ◇ DIDIGAM ◇ DIDIMAC ◇ DIPHENYLTRICHLOROETHANE ◇ DODAT ◇ DYKOL ◇ ENT 1,506 ◇ ESTONATE ◇ GENITOX ◇ GESAFID ◇ GESAPON ◇ GESAREX ◇ GESAROL ◇ GUESAPON ◇ GUESAROL ◇ GYRON ◇ HAVERO-EXTRA ◇ HILDIT ◇ IVORAN ◇ IXODEX ◇ KOPSOL ◇ MICRO DDT 75 ◇ MUTOXIN ◇ NCI-C00464 ◇ NEOCID ◇ PARACHLOROCIDUM ◇ PEB1 ◇ PENTACHLORIN ◇ PENTECH ◇ PPZEIDAN ◇ R50 ◇ RCRA WASTE NUMBER U061 ◇ RUKSEAM ◇ SANTOBANE ◇ TECH DDT ◇ 1,1,1-TRICHLOOR-2,2-BIS(4-CHLOOR FENYL)-ETHAAN (DUTCH) ◇ 1,1,1-TRICHLOR-2,2-BIS(4-CHLOR-PHENYL)-AETHAN (GERMAN) ◇ TRICHLOROBIS (4-CHLOROPHENYL) ETHANE ◇ 1,1,1-TRICHLORO-2,2-DI(4-CHLOROPHENYL)-ETHANE ◇ 1,1'-(2,2,2-TRICHLOROETHYLIDENE)BIS(4-CHLOROBENZENE) ◇ 1,1,1-TRICLORO-2,2-BIS(4-CLORO-FENIL)-ETANO (ITALIAN) ◇ ZEIDANE ◇ ZERDANE

TOXICITY DATA with REFERENCE
dni-dmg:oth 250 ppb EXPEAM 41,745,85
cyt-hmn:lym 200 μg/L/72H MUREAV 40,131,76
dnd-mus:ast 15 μmol/L MUREAV 89,95,81
orl-mus TDLo:504 mg/kg (lactating female 21D post):REP ENPBBC 4,189,74

scu-mus TDLo:418 mg/kg (female 6-14D post):TER NTIS** PB223-160
orl-rat TDLo:1225 mg/kg/7W-C:CAR TUMOAB 61,113,75
orl-mus TDLo:24 mg/kg (MGN):NEO,TER JNCIAM 51,983,73
orl-mus TDLo:73 mg/kg/26W-C:CAR FCTXAV 7,215,69
scu-mus TDLo:370 mg/kg/80W-I:NEO IJCNAW 19,725,77
orl-ham TDLo:21280 mg/kg/38W-I:ETA,TER PSEBAA 134,113,70
orl-rat TD:8100 mg/kg/2Y-C:ETA TXAPA9 11,88,67
orl-mus TD:3408 mg/kg (MGN):NEO,TER IJCNAW 11,688,73
orl-inf LDLo:150 mg/kg BMJOAE 2,845,45
orl-man TDLo:6 mg/kg:CNS,GIT,SKN CMEP** -,1,56
orl-hmn TDLo:16 mg/kg:CNS CMEP** -,1,56
orl-hmn LDLo:500 mg/kg:CNS,CVS,PUL MEIEDD 10,409,83
orl-hmn TDLo:5 mg/kg:CNS PHARAT 2,268,47
unr-man LDLo:221 mg/kg 85DCAI 2,73,70
orl-rat LD50:87 mg/kg DOEAAH 35,25,79
skn-rat LD50:1931 mg/kg SPEADM 74-1,-,74
ipr-rat LD50:9100 μg/kg PESTD5 17,351,76
scu-rat LD50:1500 mg/kg BMJOAE 1,865,45
ivn-rat LDLo:30 mg/kg JPETAB 86,213,46
ivn-rat LD50:68 mg/kg ANTBAL 14,316,69
orl-mus LD50:135 mg/kg FEPRA7 12,368,53
ipr-mus LD50:32 mg/kg PESTD5 17,351,76
ivn-mus LD50:68500 μg/kg ANTBAL 14,316,69
skn-rbt LD50:300 mg/kg BMJOAE 1,865,45

CONSENSUS REPORTS: NTP Fifth Annual Report on Carcinogens. IARC Cancer Review: Group 2B IMEMDT 7,186,87; Animal Sufficient Evidence IMEMDT 5,83,74; Human Inadequate Evidence IMEMDT 5,83,74. NCI Carcinogenesis Bioassay (feed); No Evidence: mouse, rat NCITR* NCI-CG-TR-131,78. Reported in EPA TSCA Inventory. EPA Genetic Toxicology Program.

OSHA PEL: TWA 1 mg/m^3 (skin)
ACGIH TLV: TWA 1 mg/m^3
NIOSH REL: (DDT) TWA 0.5 mg/m^3; avoid skin contact
DFG MAK: 1 mg/m^3
DOT Classification: ORM-A; Label: None.

SAFETY PROFILE: Confirmed carcinogen with experimental carcinogenic, neoplastigenic, tumorigenic, and teratogenic data. Human poison by ingestion. Experimental poison by ingestion, skin contact, subcutaneous, intravenous, and intraperitoneal routes. Experimental reproductive effects. Human systemic effects by ingestion: anesthetic, convulsions, headache, analgesia, cardiac arrythmias, nausea or vomiting, sweating and unspecified pulmonary changes. Human mutation data reported. An insecticide. When heated to decomposition it emits toxic fumes of Cl^-. See also CHLORINATED HYDROCARBONS, AROMATIC.

A dose of 20 grams has proved highly dangerous though not fatal to a human. This dose was taken by 5 persons who vomited an unknown portion of the material and even so recovered only incompletely after 5 weeks. Smaller doses produced less important symptoms with relatively rapid recovery. Experimental ingestion of 1.5 grams resulted in great discomfort and moderate neurological changes including paraesthesia, tremor, moderate ataxia, exaggeration of part of the reflexes, headache, and fatigue. Vomiting followed only after 11 hours. Recovery was complete on the following day. The fatal dose of DDT for humans is not known. Judging from the literature, no one has ever been killed by DDT in the absence of other insecticides and/or a variety of toxic solvents. However, these common solvent formulations are highly fatal when taken in small doses, partly because of the toxicity of the solvent, and perhaps because of the increased absorbability of the DDT; several fatal cases in humans have been reported. Little is known of the hazard of chronic DDT poisoning. Human volunteers have ingested up to 35 mg/day for 21 months with no ill effects.

DDT and some of its degradation products, particularly DDE, are stored in fat. This storage effect leads to a concentration of DDT at higher levels of the food chain. DDT stored in the fat is at least largely inactive since a greater total dose may be stored in an experimental animal than is sufficient as a lethal dose for that same animal if given at one time. A study based on 75 human cases reported an average of 5.3 ppm of DDT stored in the fat. A higher content of DDT and its derivatives (up to 434 ppm of DDE and 648 ppm of DDT) was found in workers who had very extensive exposure. Without exception, the samples were taken from persons who were either asymptomatic or suffering from some disease completely unrelated to DDT. Careful hospital examination of workers who had been very extensively exposed and who had volunteered for examination revealed no abnormality which could be attributed to DDT. Much higher levels have been found in humans than have been observed in the fat of experimental animals which were apparently asymptomatic. DDT stored in the fat is eliminated only very gradually when further dosage is discontinued. However, weight loss can speed the release of this stored DDT (and DDE) into the blood. After a single dose, the secretion of DDT in the milk and its excretion in the urine reach their height within a day or two and continue at a lower level thereafter.

DAD500 CAS:72732-50-4 ***HR: 3***
DEACETYLDEMETHYLTHYMOXAMINE
mf: $C_{13}H_{21}NO_2$ mw: 223.35

SYNS: (2-(4-HYDROXY-2-ISOPROPYL-5-METHYLPHENOXY) ETHYL)METHYLAMINE ◇ 2-METHYL-4-(2-(METHYLAMINO) ETHOXY)-5-ISOPROPYL-PHENOL ◇ 2-METHYL-4-(2-(METHYLAMINO)ETHOXY)-5-(1-METHYLETHYL)-PHENOL

TOXICITY DATA with REFERENCE
orl-mus LD50:340 mg/kg KSRNAM 16,1147,82
ipr-mus LD50:84 mg/kg KSRNAM 16,1147,82
ivn-mus LD50:28 mg/kg KSRNAM 16,1147,82

SAFETY PROFILE: Poison by ingestion, intravenous, and intraperitoneal routes. When heated to decomposition it emits toxic fumes of NO_x.

DAD600 CAS:34114-98-2 ***HR: 3***
DEACETYL-HT-2 TOXIN
mf: $C_{19}H_{30}O_7$ mw: 370.49

SYNS: 12,13-EPOXY-3-α,4-β,8-α,15-TETRAHYDROXYTRICHOTHEC-9-ENE-8-ISOVALERATE ◇ TOXIN T-2 TRIOL ◇ 3-α,4-β,15-TRIHYDROXY-8,α-(3-METHYLBUTYRYLOXY)-12,13-EPOXYTRICHOTHEC-9-ENE

TOXICITY DATA with REFERENCE
ipr-mus LD50:108 mg/kg AEMIDF 46,120,83
orl-ckn LD50:30180 μg/kg AEMIDF 35,636,78

SAFETY PROFILE: Poison by ingestion and intraperitoneal routes. When heated to decomposition it emits acrid smoke and fumes.

DAD650 CAS:19855-39-1 ***HR: 3***
DEACETYLLANATOSIDE B
mf: $C_{47}H_{74}O_{19}$ mw: 943.21

SYNS: DEACETYL-LANATOSIDE B (8CI) ◇ DESACETYL-LANTOSID B (GERMAN) ◇ DESACETYLLANATOSIDE B ◇ GLUCOGITOXIN ◇ PURPUREA B ◇ PURPUREA GLYCOSIDE B ◇ PURPUREA-GLYKOSID B (GERMAN)

TOXICITY DATA with REFERENCE
ivn-cat LD50:370 μg/kg 85ELDJ -,193,63
orl-frg LD50:8600 μg/kg JJPAAZ 9,91,60
scu-frg LD50:940 μg/kg JJPAAZ 9,91,60

SAFETY PROFILE: Poison by ingestion, subcutaneous, and intravenous routes. When heated to decomposition it emits acrid smoke and fumes.

DAD800 CAS:36069-46-2 ***HR: D***
DEACETYLMULDAMINE
mf: $C_{27}H_{46}NO_2$ mw: 416.74

TOXICITY DATA with REFERENCE
orl-ham TDLo:150 mg/kg (7D preg):REP JAFCAU 26,561,78
orl-ham TDLo:150 mg/kg (7D preg):TER JAFCAU 26,561,78

SAFETY PROFILE: An experimental teratogen. Other experimental reproductive effects. When heated to decomposition it emits toxic fumes of NO_x.

DAD850 CAS:35231-36-8 ***HR: 3***
DEACETYLTHYMOXAMINE
mf: $C_{14}H_{23}NO$ mw: 221.38

SYNS: DESACETYLTHYMOXAMINE ◇ 4-(2-(DIMETHYLAMINO) ETHOXY)-5-ISOPROPYL-2-METHYLPHENOL ◇ 4-(2-(DIMETHYLAMINO)ETHOXY)-2-METHYL-5-(1-METHYLETHYL)PHENOL ◇ (2-(4-HYDROXY-2-ISOPROPYL-5-METHYLPHENOXY)ETHYL) DIMETHYLAMINE

TOXICITY DATA with REFERENCE
orl-mus LD50:335 mg/kg KSRNAM 16,1147,82
ipr-mus LD50:82 mg/kg KSRNAM 16,1147,82
scu-mus LD50:120 mg/kg KSRNAM 16,1147,82
ivn-mus LD50:29 mg/kg KSRNAM 16,1147,82

SAFETY PROFILE: Poison by ingestion, subcutaneous, intravenous, and intraperitoneal routes. When heated to decomposition it emits toxic fumes of NO_x.

DAD880 ***HR: 3***
DEADLY NIGHTSHADE

PROP: A heavily-branched perennial plant about 3 feet tall with 6-inch oval leaves and red sap. The flowers are about 1-inch long and grow from the leaf joints. The berries are about 0.5-inch in diameter and purple to black in color. It is native to Eurasia and North Africa, and is cultivated in the United States.

SYNS: ATROPA BELLADONNA ◇ BELLADONNA ◇ BLACK NIGHTSHADE ◇ NIGHTSHADE ◇ SLEEPING NIGHTSHADE

SAFETY PROFILE: The whole plant contains various belladonna alkaloids including atropine. Human systemic effects by ingestion: difficulty in speaking and swallowing, rapid heartbeat, fever, dilation of the pupil, blurred vision, hallucinations, urinary retention and delirium. See also ATROPINE and BELLADONNA.

DAE100 ***HR: 3***
DEATH CAMAS

PROP: Bulb-producing plants with 1.5-foot long grassy leaves. It produces yellow or green-white flowers at the end of a leafless stalk. They grow wild throughout Canada and the United States, except southern Florida and Hawaii.

SYNS: ALKALI GRASS ◇ HOG'S POTATO ◇ MYSTER GRASS ◇ POISON SEGO ◇ SAND CORN ◇ SOAP PLANT ◇ SQUIRREL FOOD ◇ WATER LILY ◇ WILD ONION ◇ ZIGADENUS (VARIOUS SPECIES) ◇ ZYGADENUS (VARIOUS SPECIES)

SAFETY PROFILE: All parts of the plant contain the poisons zygadenine, zygacine, germidines and protoveratridine. Human systemic effects by ingestion: mouth pain, persistent vomiting, headache, dizziness, slowed heartbeat, low blood pressure, and convulsions.

DAE200 CAS:2862-16-0 ***HR: 3***
7-DEAZAINOSINE
mf: $C_{11}H_{13}N_3O_5$ mw: 267.27

SYNS: DEAMINOHYDROXYTUBERCINDIN ◇ 7-β-d-RIBOFURANOSYL-7H-PYRROLO(2,3-d)PYRIMIDIN-4-OL

TOXICITY DATA with REFERENCE
orl-rat LD50:26 mg/kg CNREA8 29,116,69
ipr-rat LD50:25 mg/kg CNREA8 29,116,69
scu-rat LD50:24 mg/kg CNREA8 29,116,69
ipr-mus LD50:30 mg/kg CNREA8 29,116,69
orl-dog LDLo:48 mg/kg CNREA8 29,116,69
ivn-dog LDLo:48 mg/kg CNREA8 29,116,69

SAFETY PROFILE: Poison by ingestion, intraperitoneal, subcutaneous, and intravenous routes. When heated to decomposition it emits toxic fumes of NO_x.

DAE400 CAS:17702-41-9 ***HR: 3***
DECABORANE
DOT: UN 1868
mf: $B_{10}H_{14}$ mw: 122.24

PROP: Colorless needles. Mp: 99.7°, d: 0.94. (solid), d: 0.78 (liquid @ 100°), vap press: 19 mm @ 100°.

SYN: DECABORANE(14)

TOXICITY DATA with REFERENCE
orl-rat LD50:64 mg/kg MLSR** #8,51
ihl-rat LC50:46 ppm/4H AMIHAB 17,362,58
skn-rat LD50:740 mg/kg AMIHAB 11,132,55
ipr-rat LD50:23 mg/kg AMIHBC 8,335,53
orl-mus LD50:41 mg/kg AMIHAB 11,132,55
ihl-mus LC50:12 ppm/4H NTIS** AD224-0-06
ipr-mus LD50:33 mg/kg AMIHAB 11,132,55
skn-rbt LD50:71 mg/kg AMIHAB 11,132,55
ipr-rbt LD50:28 mg/kg AMIHBC 8,335,53

CONSENSUS REPORTS: Reported in EPA TSCA Inventory. EPA Extremely Hazardous Substances List.

OSHA PEL: (Transitional: TWA 0.05 ppm (skin)) TWA 0.05 ppm; STEL 0.15 ppm (skin)
ACGIH TLV: TWA 0.05 ppm; STEL 0.15 ppm (skin)
DFG MAK: 0.05 ppm (0.3 mg/m^3)
DOT Classification: Flammable Solid; Label: Flammable Solid and Poison.

SAFETY PROFILE: Poison by inhalation, ingestion, skin contact, and intraperitoneal routes. Ignites in O_2 at 100°C. Forms impact-sensitive explosive mixtures with ethers (e.g., dioxane) and halocarbons (e.g., carbon tetrachloride). Incompatible with dimethyl sulfoxide. When heated to decomposition it emits toxic fumes of boron oxides. See also BORON COMPOUNDS and BORANES.

DAE425 CAS:2227-17-0 ***HR: 2***
DECACHLOROBI-2,4-CYCLOPENTADIEN-1-YL
mf: $C_{10}Cl_{10}$ mw: 474.60

PROP: Yellow prisms from petr ether. Mp: 121.5-122°. Stable to alkali.

SYNS: BIS(PENTACHLOR-2,4-CYCLOPENTADIEN-1-YL) ◇ BIS(PENTACHLOROCYCLOPENTADIENYL) ◇ BIS(PENTACHLORO-2,4-CYCLOPENTADIEN-1-YL) ◇ DECACHLOR ◇ 1,1′,2,2′,3,3′,4,4′,5,5′-DECACHLOROBI-2,4-CYCLOPENTADIEN-1-YL ◇ DIENOCHLOR ◇ ENT 25,718 ◇ HOOKER HRS-16 ◇ HOOKER HRS 1654 ◇ HRS-16 ◇ HRS 16A ◇ HRS 1654 ◇ PENTAC ◇ PENTAC WP

TOXICITY DATA with REFERENCE
mmo-sat 10 μg/plate MUREAV 116,185,83
mma-sat 500 μg/plate MUREAV 116,185,83
orl-rat LD50:1200 mg/kg ARSIM* 20,13,66

SAFETY PROFILE: Moderately toxic by ingestion. Mutation data reported. See also CHLORINATED HYDROCARBONS, ALIPHATIC.

DAE450 CAS:25152-84-5 ***HR: 1***
trans,trans-2,4-DECADIENAL
mf: $C_{10}H_{16}O$ mw: 152.23

PROP: Yellow liquid; chicken fat odor. D: 0.806-0.876, refr index: 1.514-1.516, flash p: +212°F. Sol in alc, fixed oils; insol in water @ 104°.

SYNS: FEMA No. 3135 ◇ HEPTENYL ACROLEIN

SAFETY PROFILE: Combustible liquid. When heated to decomposition it emits acrid smoke and irritating fumes.

DAE500 CAS:26660-76-4 ***HR: 3***
DECADONIUM DIIODIDE
mf: $C_{34}H_{62}N_2•2I$ mw: 752.78

SYNS: 1,10-BIS(N-METHYL-N-(1′-ADAMANTYL)AMINO)DECANE DIIODOMETHYLATE ◇ N,N′-DECAMETHYLENEBIS((1-ADAMANTYL) DIMETHYLAMMONIUM, DIIODIDE ◇ 1,10-(N-METHYL-N-(1′-ADAMANTYL)AMINO)DECANEDIIODOMETHYLATE

TOXICITY DATA with REFERENCE
ivn-mus LD50:780 μg/kg RPTOAN 33,185,70
ivn-cat LDLo:100 mg/kg RPTOAN 33,185,70
ivn-rbt LDLo:50 mg/kg RPTOAN 33,185,70

SAFETY PROFILE: Poison by intravenous route.

When heated to decomposition it emits toxic fumes of I^- and NO_x.

DAE525 CAS:2392-39-4 ***HR: 3***
DECADRON PHOSPHATE
mf: $C_{22}H_{28}FO_8P \cdot 2Na$ mw: 516.45

SYNS: DEXACORT ◇ DEXADRESON ◇ DEXAGRO ◇ DEXAMETHASONE DISODIUM PHOSPHATE ◇ DEXAMETHASONE SODIUM PHOSPHATE ◇ DEXAMETHAZONE SODIUM PHOSPHATE ◇ DISODIUM DEXAMETHASONE PHOSPHATE ◇ 9-FLUORO-11-β,17,21-TRIHYDROXY-16-α-METHYLPREGNA-1,4-DIENE-3,20-DIONE-21-(DIHYDROGEN PHOSPHATE) DISODIUM SALT ◇ MEGACORT ◇ SODIUM DEXAMETHASONE PHOSPHATE ◇ SOLDESAM ◇ SOLU-DECADRON ◇ SPERSADOX ◇ TURBINAIRE

TOXICITY DATA with REFERENCE
ipr-rat TDLo:400 μg/kg (female 19-20D post):REP PEDIAU 65,287,80
orl-mus TDLo:10 mg/kg (female 6-15D post):TER BCFAAI 119,391,80
ivn-inf TDLo:1500 μg/kg/3D-I:CNS LANCAO 1,632,87
ivn-wmn TDLo:320 mg/kg:GIT LANCAO 1,1035,86
ivn-rat LD50:159 mg/kg KSRNAM 19,3961,85
orl-mus LD50:1800 mg/kg ATSUDG 7,90,84
ipr-mus LD50:550 mg/kg ATSUDG 7,90,84
ivn-mus LD50:112 mg/kg KSRNAM 19,3961,85

CONSENSUS REPORTS: Reported in EPA TSCA Inventory.

SAFETY PROFILE: Poison by intravenous route. Moderately toxic by ingestion and intraperitoneal routes. Human systemic effects by intravenous route: CNS and gastrointestinal changes. An experimental teratogen. Other experimental reproductive effects. When heated to decomposition it emits toxic fumes of F^-, PO_x, and Na_2O.

DAE600 CAS:15652-38-7 ***HR: 3***
DECAFENTIN
mf: $C_{28}H_{36}P \cdot C_{18}H_{15}BrClSn$ mw: 868.99

SYNS: A-36 ◇ CELA A-36 ◇ DECYLTRIPHENYLPHOSPHONIUM BROMOCHLOROTRIPHENYLSTANNATE ◇ (DECYL-TRIPHENYL-PHOSPHONIUM)-TRIPHENYL-BROM-CHLOR-STANNAT(GERMAN) ◇ STANNOPLUS ◇ STANNORAM ◇ STANNPLOUS

TOXICITY DATA with REFERENCE
orl-rat LD50:700 mg/kg FMCHA2 -,D287,80
orl-mus LD50:550 mg/kg BESAAT 15,120,69
skn-rbt LD50:310 mg/kg BESAAT 15,120,69

OSHA PEL: TWA 0.1 mg(Sn)/m^3 (skin)
ACGIH TLV: TWA 0.1 mg(Sn)/m^3 (skin) (Proposed: TWA 0.1 mg(Sn)/m^3; STEL 0.2 mg(Sn)/m^3 (skin))
NIOSH REL: (Organotin Compounds) TWA 0.1 mg(Sn)/m^3

SAFETY PROFILE: Poison by skin contact. Moderately toxic by ingestion. When heated to decomposition it emits very toxic fumes of PO_x, Br^-, and Cl^-. A pesticide. See also TIN COMPOUNDS.

DAE625 CAS:41409-50-1 ***HR: 3***
DECAFLUOROBUTYRAMIDINE
mf: $C_4F_{10}N_2$ mw: 266.04

$C_3F_7C(:NF)NF_2$

SAFETY PROFILE: A shock-sensitive explosive. Upon decomposition it emits toxic fumes of F^- and NO_x.

DAE695 CAS:19590-85-3 ***HR: 3***
cis-N-(DECAHYDRO-2-METHYL-5-ISOQUINOLYL)-3,4,5-TRIMETHOXYBENZAMIDE
mf: $C_{20}H_{30}N_2O_4$ mw: 362.52

SYNS: M-30 ◇ cis-5,8,10-H-5-(3,4,5-TRIMETHOXYBENZAMIDO)-2-METHYL DECAHYDROISOQUINOLINE

TOXICITY DATA with REFERENCE
ipr-mus LD50:278 mg/kg DECRDP 10,197,84
ivn-dog LDLo:40 mg/kg DECRDP 10,197,84
ivn-rbt LDLo:31600 μg/kg DECRDP 10,197,84

SAFETY PROFILE: Poison by intravenous and intraperitoneal routes. When heated to decomposition it emits toxic fumes of NO_x.

DAE700 CAS:27460-73-7 ***HR: 3***
trans-N-(DECAHYDRO-2-METHYL-5-ISOQUINOLYL)-3,4,5-TRIMETHOXYBENZAMIDE
mf: $C_{20}H_{30}N_2O_4$ mw: 362.52

SYNS: M-32 ◇ (E)-N-(2-METHYLDECAHYDROISOQUINOL-5-YL)-3,4,5-TRIMETHOXY-BENZAMIDE ◇ trans-N-(2-METHYLDECAHYDROISOQUINOL-5-YL)-3,4,5-TRIMETHOXYBENZAMIDE ◇ 5-(3,4,5-TRIMETHOXYBENZAMIDO)-2-METHYL-trans-DECAHYDROISOQUINOLINE ◇ trans-9,10-t-5-H-5-(3,4,5-TRIMETHOXYBENZAMIDO)-2-METHYL DECAHYDROISOQUINOLINE

TOXICITY DATA with REFERENCE
ipr-mus LD50:221 mg/kg JMCMAR 23,2-06,80
ivn-dog LDLo:40 mg/kg DECRDP 10,197,84
ivn-rbt LDLo:17800 μg/kg DECRDP 10,197,84

SAFETY PROFILE: Poison by intravenous and intraperitoneal routes. When heated to decomposition it emits toxic fumes of NO_x.

DAE800 CAS:91-17-8 ***HR: 3***
DECAHYDRONAPHTHALENE
DOT: UN 1147
mf: $C_{10}H_{18}$ mw: 138.28

PROP: Water-white liquid. Mp (cis): −43.3°, mp (trans): −30.7°, bp: (cis): 195.6°, bp: (trans) 187.3°, flash p: 136°F, (CC), autoign temp: 482°F, vap press: (cis) 1 mm @ 22.5°, (trans) 10 mm @ 47.2°, d: (cis)

0.8963 @ 20°/4°, vap d: 4.76, lel: 0.7% @ 212°F, uel: 4.9% @ 212°F.

SYNS: BICYCLO(4.4.0)DECANE ◇ DEC ◇ DECALIN ◇ DECALIN (DOT) ◇ DECALIN SOLVENT ◇ DE-KALIN ◇ DEKALINA (POLISH) ◇ NAPHTHALANE ◇ NAPHTHANE ◇ PERHYDRONAPHTHALENE

TOXICITY DATA with REFERENCE
skn-rbt 10 mg/24H open MLD AMIHBC 4,119,51
eye-rbt 500 mg open AMIHBC 4,119,51
ihl-rat TCLo:5 ppm/24H/90D-C:NEO FAATDF 5,785,85
ihl-mus TCLo:50 ppm/24H/90D-C:CAR FAATDF 5,785,85
ihl-hmn TCLo:100 ppm:NOSE,EYE,PUL TGNCDL 2,40,61
orl-rat LD50:4170 mg/kg AMIHBC 4,119,51
ihl-rat LCLo:500 ppm/4H AMIHBC 4,119,51
ihl-mus LCLo:993 ppm/4H NTIS** AD-A-062-138
skn-rbt LD50:5900 mg/kg AMIHBC 4,119,51
ihl-gpg LCLo:319 ppm/8H NTIS** AD-A086-341

CONSENSUS REPORTS: Reported in EPA TSCA Inventory.

DOT Classification: Combustible Liquid; Label: None; Flammable or Combustible Liquid; Label: Flammable Liquid.

SAFETY PROFILE: Moderately toxic by inhalation and ingestion. Questionable carcinogen with experimental carcinogenic and neoplastigenic data. Mildly toxic by skin contact. Human systemic effects by inhalation: conjuctiva irritation, unspecified olfactory and pulmonary system changes. Can cause kidney damage. Mutation data reported. A skin and eye irritant. Flammable when exposed to heat or flame, can react with oxidizing materials. To fight fire, use foam, CO_2, dry chemical. When heated to decomposition it emits acrid smoke and fumes.

DAF000 CAS:825-51-4 ***HR: 1***
DECAHYDRO-2-NAPHTHALENOL
mf: $C_{10}H_{18}O$ mw: 154.28

PROP: D: 0.996, bp: 109° @ 14mm, flash p: > 112°.

SYNS: DECAHYDRONAPHTHALEN-2-OL ◇ DECAHYDRO-NAPTHOL-2 ◇ DECAHYDRO-β-NAPHTHOL ◇ trans-DECAHYDRO-β-NAPHTHOL ◇ 2-DECALINOL ◇ 2-DECALOL ◇ 2-HYDROXYDECALIN

TOXICITY DATA with REFERENCE
skn-rbt 500 mg/24H MOD FCTXAV 12,873,74

CONSENSUS REPORTS: Reported in EPA TSCA Inventory.

SAFETY PROFILE: A moderate skin irritant. When heated to decomposition it emits acrid smoke and irritating fumes.

DAF100 CAS:10519-11-6 ***HR: 1***
DECAHYDRO-β-NAPHTHYL ACETATE
mf: $C_{12}H_{20}O_2$ mw: 196.32

SYN: 2-NAPHTHOL, DECAHYDRO-, ACETATE

TOXICITY DATA with REFERENCE
skn-rbt 500 mg/24H MOD FCTXAV 17,755,79

CONSENSUS REPORTS: Reported in EPA TSCA Inventory.

SAFETY PROFILE: A skin irritant. When heated to decomposition it emits acrid smoke and irritating fumes.

DAF150 CAS:10519-12-7 ***HR: 1***
DECAHYDRO-β-NAPHTHYL FORMATE
mf: $C_{11}H_{18}O_2$ mw: 182.29

SYNS: DECALINYL FORMATE ◇ 2-NAPHTHALENOL, DECAHYDRO-, FORMATE ◇ 2-NAPHTHOL, DECAHYDRO-, FORMATE ◇ SANTALOZONE

TOXICITY DATA with REFERENCE
skn-rbt 500 mg/24H MOD FCTXAV 17,757,79

CONSENSUS REPORTS: Reported in EPA TSCA Inventory.

SAFETY PROFILE: A skin irritant. When heated to decomposition it emits acrid smoke and irritating fumes.

DAF200 CAS:705-86-2 ***HR: 1***
Δ-DECALACTONE
mf: $C_{10}H_{18}O_2$ mw: 170.28

PROP: Colorless liquid; coconut, fruity odor, butterlike on dilution. Refr index: 1.456-1.459. Very sol in alc and propylene glycol; insol in water @ 281°.

SYNS: AMYL-Δ-VALEROLACTONE ◇ DECANOLIDE-1,5 ◇ FEMA No. 2361

TOXICITY DATA with REFERENCE
skn-rbt 500 mg/24H MLD FCTXAV 14,659,76
eye-rbt 100 mg MLD NTIS** AD-A053-896

CONSENSUS REPORTS: Reported in EPA TSCA Inventory.

SAFETY PROFILE: A skin and eye irritant. When heated to decomposition it emits acrid smoke and irritating fumes.

DAF300 CAS:52918-63-5 ***HR: 3***
DECAMETHRINE
mf: $C_{22}H_{19}Br_2NO_3$ mw: 505.24

PROP: Crystals. Mp: 98-101°, sol in ethanol, acetone, and dioxane; insol in water.

SYNS: BUTOFLIN ◇ BUTOX ◇ DECAMETHRIN ◇ DECIS ◇ DEKAMETRIN (HUNGARIAN) ◇ DELTAMETHRIN ◇ ESBECYTHRIN ◇ JMC 45498 ◇ K-OTHRIN ◇ NRDC 161 ◇ RU 22974

TOXICITY DATA with REFERENCE
cyt-ofs-mul 100 nL/L JFIBA9 26,13,85
orl-rat TDLo:70 mg/kg (7-20D preg):REP JEPTDQ 2(3),751,79
orl-mus TDLo:30 mg/kg (7-16D preg):TER JEPTDQ 2(3),751,79
orl-rat LD50:30 mg/kg FAATDF 7,299,86
ihl-rat LC50:785 mg/m^3/2H JEPTDQ 2(3),751,79
ivn-rat LD50:2526 mg/kg PCBPBS 30,79,88
orl-mus LD50:3450 mg/kg IJTEDP 6,127,84
ice-mus LD50:26100 μg/kg PCBPBS 24,200,85
ivn-dog LD50:3440 μg/kg IJTEDP 6,127,84

CONSENSUS REPORTS: EPA Genetic Toxicology Program.

SAFETY PROFILE: Poison by ingestion, inhalation, intravenous and intracerebral routes. An experimental teratogen. Other experimental reproductive effects. Mutation data reported. When heated to decomposition it emits toxic fumes of Br^-, CN^-, and NO_x. See also ESTERS.

DAF450 CAS:63884-28-6 ***HR: 3***
1,1'-DECAMETHYLENEBIS(1-METHYLPIPERIDINIUM IODIDE)
mf: $C_{22}H_{46}N_2{\cdot}2I$ mw: 592.50

TOXICITY DATA with REFERENCE
scu-mus LD50:3 mg/kg YKKZAJ 74,1267,54
ivn-mus LD50:20 mg/kg BJPCAL 10,124,55
ivn-rbt LDLo:3 mg/kg YKKZAJ 74,911,54

SAFETY PROFILE: Poison by subcutaneous and intravenous routes. When heated to decomposition it emits toxic fumes of NO_x and I^-. See also IODIDES.

DAF600 CAS:541-22-0 ***HR: 3***
DECAMETHYLENEBIS(TRIMETHYLAMMONIUM BROMIDE)
mf: $C_{16}H_{38}N_2{\cdot}2Br$ mw: 418.38

SYNS: DECACURAN ◇ DECAMETHONIUM ◇ DECAMETHONIUM BROMIDE ◇ DECAMETHONIUM DIBROMIDE ◇ DECAMETHYLENE-1,10-BISTRIMETHYLAMMONIUM DIBROMIDE ◇ N,N,N,N',N',N'-HEXAMETHYL-1,10-DECANEDIAMINIUM DIBROMIDE ◇ SYNCURINE

TOXICITY DATA with REFERENCE
ipr-rat LD50:2900 μg/kg TXAPA9 14,67,69
orl-mus LD50:190 mg/kg JPETAB 118,395,56
ipr-mus LD50:900 μg/kg JAPMA8 45,792,56
scu-mus LD50:4 mg/kg JKKZAJ 74,911,54
ivn-mus LD50:630 μg/kg AIPTAK 105,221,56
ivn-cat LD50:35 μg/kg JPETAB 118,395,56
ivn-rbt LD50:125 μg/kg JPETAB 118,395,56

CONSENSUS REPORTS: Reported in EPA TSCA Inventory.

SAFETY PROFILE: Deadly poison by ingestion, intraperitoneal, subcutaneous, and intravenous routes. When heated to decomposition it emits very toxic fumes of NO_x, NH_3, and Br^-. See also BROMIDES.

DAF800 CAS:1420-40-2 ***HR: 3***
DECAMETHYLENEBIS(TRIMETHYLAMMONIUM DIIODIDE)
mf: $C_{16}H_{38}N_2{\cdot}2I$ mw: 512.36

SYNS: DECAMETHIONIUM IODIDE ◇ DECAMETHONIUM DIIODIDE ◇ DECAMETHONIUM IODIDE ◇ DECAMETHYLENEBIS (TRIMETHYLAMMONIUM IODIDE) ◇ EULISSIN A ◇ EULIXINE ◇ N,N,N,N^1,N^1,N^1-HEXAMETHYL-1,10-DECANEDIAMINIUM DIIODIDE ◇ PROCURAN

TOXICITY DATA with REFERENCE
orl-rat LD50:85 mg/kg PSEBAA 120,511,65
ipr-rat LD50:1550 μg/kg TXAPA9 36,585,76
ivn-rat LD50:3020 μg/kg BJPCAL 4,381,49
orl-mus LD50:115 mg/kg PSEBAA 120,511,65
scu-mus LD50:5 mg/kg YKKZAJ 74,911,54
ivn-mus LD50:838 μg/kg BJPCAL 4,381,49
ivn-rbt LD50:200 μg/kg RISSAF 12,158,49
ivn-ckn LD50:80 μg/kg AIPTAK 122,152,59

SAFETY PROFILE: Poison by ingestion, intraperitoneal, intravenous, and subcutaneous routes. When heated to decomposition it emits very toxic fumes of NO_x, NH_3 and I^-.

DAG000 CAS:112-31-2 ***HR: 2***
1-DECANAL
mf: $C_{10}H_{20}O$ mw: 156.30

PROP: Found in over 50 sources including citrus oils, citronella, and lemongrass (FCTXAV 11,477,73). Colorless to light yellow liquid; floral, fatty odor. D: 0.830 @ 15°/4°, bp: 208°, flash p: 185°F. Sol in 80% alc, fixed oils, volatile oils, and mineral oils; insol in water and glycerol.

SYNS: ALDEHYDE C10 ◇ C-10 ALDEHYDE ◇ CAPRALDEHYDE ◇ 1-DECYL ALDEHYDE ◇ FEMA No. 2362

TOXICITY DATA with REFERENCE
skn-rbt 14372 μg/24H open SEV AIHAAP 23,95,62
skn-rbt 500 mg/24H MLD FCTXAV 11,1079,73
orl-rat LD50:3730 mg/kg FCTXAV 11,477,73
skn-rbt LD50:5040 mg/kg FCTXAV 11,477,73

CONSENSUS REPORTS: Reported in EPA TSCA Inventory.

SAFETY PROFILE: Moderately toxic by ingestion. Mildly toxic by skin contact. A severe skin irritant. Combustible liquid. When heated to decomposition it emits acrid smoke and irritating fumes. See also ALDEHYDES.

DAG200 CAS:112-31-2 ***HR: 2***
1-DECANAL (mixed isomers)

SYN: FEMA No. 2362

TOXICITY DATA with REFERENCE
skn-rbt 14372 μg/24H open SEV AIHAAP 23,95,62
orl-rat LD50:3730 mg/kg AIHAAP 23,95,62

CONSENSUS REPORTS: Reported in EPA TSCA Inventory.

SAFETY PROFILE: Moderately toxic by ingestion. A severe skin irritant. See also 1-DECANAL. When heated to decomposition it emits acrid smoke and fumes.

DAG400 CAS:124-18-5 ***HR: 3***
DECANE
DOT: UN 2247
mf: $C_{10}H_{22}$ mw: 142.29

PROP: Liquid. Mp: −29.7°, bp: 174.1°, lel: 0.8%, uel: 5.4%, flash p: 115°F (CC), d: 0.730 @ 20°/4°, autoign temp: 410°F, vap press: 1 mm @ 16.5°, vap d: 4.90.

SYN: n-DECANE (DOT)

TOXICITY DATA with REFERENCE
skn-mus TDLo:25 g/kg/52W-I:ETA TXAPA9 9,70,66
ihl-mus LC50:72300 mg/m³/2H GTPZAB 26(8),53,82

CONSENSUS REPORTS: Reported in EPA TSCA Inventory.

DOT Classification: Flammable or Combustible Liquid; Label: Flammable Liquid.

SAFETY PROFILE: Questionable carcinogen with experimental tumorigenic data. A simple asphyxiant. Narcotic in high concentrations. Flammable when exposed to heat or flame. Can react with oxidizing materials. Moderately explosive in its vapor form. To fight fire, use foam, CO_2, dry chemical. Emitted from modern building materials. (CENEAR 69,22,91) See also ARGON for discussion of asphyxiants.

DAG600 CAS:2016-57-1 ***HR: 3***
1-DECANEAMINE
mf: $C_{10}H_{23}N$ mw: 157.34

PROP: Liquid. Mp: 17°, bp: 95° @ 10 mm, flash p: 210°F, d: 0.79 @ 20°, vap d: 5.5.

SYNS: 1-AMINODECANE ◇ DECYLAMINE

TOXICITY DATA with REFERENCE
skn-rbt 100 μg/24H open AIHAAP 23,95,62
orl-rat LD50:280 mg/kg AIHAAP 23,95,62
skn-rbt LD50:350 mg/kg AIHAAP 23,95,62

CONSENSUS REPORTS: Reported in EPA TSCA Inventory.

SAFETY PROFILE: Poison by ingestion and skin contact. A skin irritant. Flammable when exposed to heat or flame; can react with oxidizing materials. To fight fire, use alcohol foam, foam, dry chemical. When heated to decomposition it emits toxic fumes of NO_x. See AMINES; and AMINES, FATTY.

DAH400 CAS:334-48-5 ***HR: 3***
DECANOIC ACID
mf: $C_{10}H_{20}O_2$ mw: 172.30

PROP: White crystals; unpleasant odor. D: 0.8782 @ 50°/4°, bp: 270°, mp: 31.4°. Sol in most organic solvents and in dilute nitric acid; insol in water.

SYNS: CAPRIC ACID ◇ n-CAPRIC ACID ◇ CAPRINIC ACID ◇ CAPRYNIC ACID ◇ n-DECANOIC ACID ◇ n-DECOIC ACID ◇ DECYLIC ACID ◇ n-DECYLIC ACID ◇ HEXACID 1095 ◇ NEO-FAT 10 ◇ 1-NONANECARBOXYLIC ACID

TOXICITY DATA with REFERENCE
skn-rbt 500 mg/24H MOD FCTXAV 17,735,79
sln-smc 14500 ppb ANYAA9 407,186,83
ivn-mus LD50:129 mg/kg APTOA6 18,141,61

CONSENSUS REPORTS: Reported in EPA TSCA Inventory.

SAFETY PROFILE: Poison by intravenous route. Mutation data reported. A moderate skin irritant. When heated to decomposition it emits acrid smoke and irritating fumes.

DAH450 CAS:10024-58-5 ***HR: 1***
DECANOIC ACID, DIESTER with TRIETHYLENE GLYCOL (mixed isomers)
mf: $C_{26}H_{50}O_6$ mw: 458.76

SYN: DIDECANOYLTRIETHYLENE GLYCOL ESTER (mixed isomers)

TOXICITY DATA with REFERENCE
skn-rbt 10 mg/24H open MLD AIHAAP 23,95,62
orl-rat LD50:7460 mg/kg AIHAAP 23,95,62
skn-rbt LD50:11200 mg/kg AIHAAP 23,95,62

CONSENSUS REPORTS: Reported in EPA TSCA Inventory.

SAFETY PROFILE: Mildly toxic by ingestion and skin contact. When heated to decomposition it emits acrid smoke and irritating fumes.

DAI000 CAS:26909-37-5 ***HR: 3***
10-DECARBAMOYLMITOMYCIN C
mf: $C_{14}H_{17}N_3O_4$ mw: 291.34

SYNS: DCMC ◇ DECARBAMOYLMITOMYCIN C ◇ DECARBAMYLMITOMYCIN C

TOXICITY DATA with REFERENCE
mmo-sat 2500 ng/plate MUREAV 149,485,85

sce-hmn:lym 100 nmol/L MUREAV 149,485,85
sce-ham:lng 100 μg/L CNREA8 44,3270,84
ivn-mus LD50:33800 μg/kg YKKZAJ 92,1218,72

SAFETY PROFILE: Poison by intravenous route. Human mutation data reported. When heated to decomposition it emits toxic fumes of NO_x.

DAI200 CAS:5053-08-7 ***HR: 3***
DECASPIRIDE HYDROCHLORIDE
mf: $C_{15}H_{20}N_2O_2$•ClH mw: 296.83

SYNS: CHLORHYDRATE de PHENETHYL-8-OXA-1-DIAZA-3,8-SPIRO(4,5)DECANONE-2 (FRENCH) ◇ ESPIRAN ◇ 8-N-FENETIL-1-OXA-2-OXO-3,8-DIAZASPIRO-(4,5)-DECANO CLORIDRATO (ITALIAN) ◇ FENSPIRIDE ◇ FENSPIRIDE HYDROCHLORIDE ◇ JP 428 HYDROCHLORIDE ◇ NAT-333 HYDROCHLORIDE ◇ NDR-5998A HYDROCHLORIDE ◇ 8-(2-PHENYLETHYL)-1-OXA-3,8-DIAZASPIRO(4.5) DECAN-2-ONE HYDROCHLORIDE ◇ PHENETHYL-8-OXA-1-DIAZA-3,8-SPIRO(4,5)DECANONE-2-HYDROCHLORIDE ◇ PNEUMOREL ◇ RESPIRIDE ◇ TEGENCIA HYDROCHLORIDE ◇ VIARESPAN HYDROCHLORIDE

TOXICITY DATA with REFERENCE
orl-rat LD50:437 mg/kg ARZNAD 19,1263,69
ivn-rat LD50:122 mg/kg AIPTAK 193,111,71
orl-mus LD50:250 mg/kg AIPTAK 193,111,71
ipr-mus LD50:230 mg/kg ARZNAD 19,1263,69
ivn-mus LD50:106 mg/kg MEIEDD 10,575,83
ivn-dog LD50:74 mg/kg BCFAAI 117,343,78
orl-gpg LD50:260 mg/kg AIPTAK 193,111,71
ipr-gpg LD50:210 mg/kg AIPTAK 193,111,71

SAFETY PROFILE: Poison by ingestion, intraperitoneal, and intravenous routes. A bronchodilator and antiadrenergic agent. When heated to decomposition it emits very toxic fumes of NO_x and HCl.

DAI350 CAS:3913-71-1 ***HR: 2***
2-DECENAL
mf: $C_{10}H_{18}O$ mw: 154.28

PROP: Sltly yellow liquid; orange odor. D: 0.836-0.846, refr index: 1.452-1.457. Sol in alc, fixed oils; insol in water.

SYNS: trans-2-DECEN-1-AL ◇ DECENALDEHYDE ◇ FEMA No. 2366

TOXICITY DATA with REFERENCE
skn-rbt 500 mg/24H SEV FCTXAV 17,761,79
orl-rat LD50:5000 mg/kg FCTXAV 17,761,79
skn-rbt LD50:3400 mg/kg FCTXAV 17,761,79

CONSENSUS REPORTS: Reported in EPA TSCA Inventory.

SAFETY PROFILE: Moderately toxic by skin contact. Mildly toxic by ingestion. A severe skin irritant. When heated to decomposition it emits acrid smoke and fumes. See also ALDEHYDES.

DAI360 CAS:21662-09-9 ***HR: 1***
cis-4-DECENAL
mf: $C_{10}H_{18}O$ mw: 154.28

PROP: Colorless to sltly yellow liquid; fatty, orangelike odor. D: 0.847, refr index: 1.442-1.444, Sol in alc, fixed oils; insol in water.

SYNS: cis-4-DECEN-1-AL (FCC) ◇ FEMA No. 3264

TOXICITY DATA with REFERENCE
skn-gpg 100%/24H MOD FCTOD7 20,663,82

CONSENSUS REPORTS: Reported in EPA TSCA Inventory.

SAFETY PROFILE: A skin irritant. When heated to decomposition it emits acrid smoke and irritating fumes.

DAI400 CAS:13019-22-2 ***HR: 1***
9-DECEN-1-OL
mf: $C_{10}H_{20}O$ mw: 156.30

PROP: D: 0.875, bp: 236°, flash p: 104°.

SYNS: omega-DECENOL ◇ 1-DECEN-10-OL ◇ DECYLENIC ALCOHOL

TOXICITY DATA with REFERENCE
skn-rbt 100% FCTXAV 12,405,74

CONSENSUS REPORTS: Reported in EPA TSCA Inventory.

SAFETY PROFILE: A skin irritant. When heated to decomposition it emits acrid smoke and irritating fumes. See also ALCOHOLS.

DAI450 CAS:50816-18-7 ***HR: 1***
9-DECENYL ACETATE
mf: $C_{12}H_{22}O_2$ mw: 198.34

SYNS: ACETIC ACID, 9-DECENYL ESTER ◇ 9-DECEN-1-OL, ACETATE ◇ DECENYL ACETATE

TOXICITY DATA with REFERENCE
skn-rbt 500 mg/24H MLD FCTOD7 20,665,82

CONSENSUS REPORTS: Reported in EPA TSCA Inventory.

SAFETY PROFILE: A skin irritant. When heated to decomposition it emits acrid smoke and irritating fumes.

DAI460 CAS:13560-89-9 ***HR: 3***
DECHLORANE PLUS
mf: $C_{18}H_{12}Cl_{12}$ mw: 653.70

PROP: Colorless crystals. Mp: > 325°. Soluble in o-dichlorobenzene.

SYNS: DECHLORANE 605 ◇ DECHLORANE PLUS 515 ◇ DECHLORANE PLUS 2520

TOXICITY DATA with REFERENCE
orl-rat LD50:25 g/kg LitL## -06MAY85
ihl-rat LC50:2250 mg/m^3 LitL## -06MAY85
skn-rbt LD50:8 g/kg LitL## -06MAY85

SAFETY PROFILE: Poison by inhalation. Mildly toxic by skin contact and ingestion. When heated to decomposition it emits toxic fumes of Cl^-. See also CHLORINATED HYDROCARBONS, AROMATIC.

DAI475 ***HR: 3***
DECLINAX
mf: $C_{10}H_{13}N_3$•BrH mw: 256.18

SYNS: DEBRISOQUIN HYDROBROMIDE ◇ 3,4-DIHYDRO-2(1H) ISOQUINOLINECARBOXYAMIDINE HYDROBROMIDE ◇ RO 5-3307

TOXICITY DATA with REFERENCE
orl-mus LD50:242 mg/kg CTCEA9 6,299,64
ipr-mus LD50:150 mg/kg CTCEA9 6,299,64
scu-mus LD50:163 mg/kg CTCEA9 6,299,64
ivn-mus LD50:40500 μg/kg CTCEA9 6,299,64

SAFETY PROFILE: Poison by ingestion, subcutaneous, intravenous, and intraperitoneal routes. When heated to decomposition it emits toxic fumes of NO_x and HBr.

DAI485 CAS:64-73-3 ***HR: 3***
DECLOMYCIN HYDROCHLORIDE
mf: $C_{21}H_{21}ClN_2O_8$•ClH mw: 501.35

SYNS: 7-CHLORO-6-DEMETHYLTETRACYCLINEHYDROCHLORIDE ◇ CHLORTETRIN ◇ DEMECLOCYCLINE HYDROCHLORIDE ◇ DEMETHYLCHLOROTETRACYCLINE HYDROCHLORIDE ◇ DEMETHYLCHLORTETRACYCLINE HYDROCHLORIDE ◇ DEMETRACICLINA ◇ DETRAVIS ◇ LEDERMYCIN HYDROCHLORIDE ◇ MECICLIN ◇ MEXOCINE

TOXICITY DATA with REFERENCE
unr-wmn TDLo:48 μg/kg (28W preg):REP PEDIAU 34,423,64
orl-man TDLo:69 mg/kg/4D-I:SYS AJKDDP 5,270,85
orl-rat LD50:2372 mg/kg TXAPA9 18,185,71
ivn-mus LD50:275 mg/kg AISMAE 43,143,62

SAFETY PROFILE: Poison by intravenous route. Moderately toxic by ingestion. Human systemic effects by ingestion: depressed renal function tests, urine composition changes, weight loss or decreased weight gain. Human female reproductive effects by an unreported route: delayed effects on newborn. An antibacterial agent. Experimental reproductive effects. When heated to decomposition it emits toxic fumes of NO_x and Cl^-.

DAI500 CAS:2156-96-9 ***HR: 2***
DECYL ACRYLATE
mf: $C_{13}H_{24}O_2$ mw: 212.37

SYN: n-DECYL ACRYLATE

TOXICITY DATA with REFERENCE
skn-rbt 10 mg/24H open SEV AIHAAP 23,95,62
orl-rat LD50:6460 mg/kg AIHAAP 23,95,62
skn-rbt LD50:6300 mg/kg AIHAAP 23,95,62

CONSENSUS REPORTS: Reported in EPA TSCA Inventory.

SAFETY PROFILE: Mildly toxic by ingestion and skin contact. A severe skin irritant. When heated to decomposition it emits smoke and acrid fumes.

DAI600 CAS:112-30-1 ***HR: 3***
DECYL ALCOHOL
mf: $C_{10}H_{22}O$ mw: 158.32

PROP: Found in sweet orange and a few other essential oils (FCTXAV 11,95,73). Colorless, viscous, refractive liquid; floral fruity odor. Mp: 7°, bp: 232.9°, flash p: 180°F (OC), d: 0.8297 @ 20°/4°, refr index: 1.435-1.439, vap press: 1 mm @ 69.5°, vap d: 5.3. Sol in alc, ether, mineral oil, propylene glycol, fixed oils; insol in glycerin water @ 233°.

SYNS: AGENT 504 ◇ ALCOHOL C-10 ◇ ANTAK ◇ C 10 ALCOHOL ◇ CAPRIC ALCOHOL ◇ CAPRINIC ALCOHOL ◇ DECANAL DIMETHYL ACETAL ◇ DECANOL ◇ n-DECANOL ◇ 1-DECANOL (FCC) ◇ n-DECATYL ALCOHOL ◇ n-DECYL ALCOHOL ◇ DECYLIC ALCOHOL ◇ DYTOL S-91 ◇ EPAL 10 ◇ FEMA No. 2365 ◇ LOROL 22 ◇ NONYLCARBINOL ◇ PRIMARY DECYL ALCOHOL ◇ ROYALTAC ◇ SIPOL L10

TOXICITY DATA with REFERENCE
skn-hmn 75 mg/3D-I SEV 85DKA8 -,127,77
skn-rbt 2600 mg/kg/24H MOD AIHAAP 34,493,73
eye-rbt 83 mg SEV AIHAAP 34,493,73
skn-mus TDLo:12 g/kg/25W-I:ETA TXAPA9 9,70,66
orl-rat LD50:4720 mg/kg AIHAAP 34,493,73
orl-mus LD50:6500 mg/kg FMCHA2 -,C208,83
ihl-mus LC50:4 g/m^3/2H 85GMAT -,42,82
skn-rbt LD50:3560 mg/kg FCTXAV 11,95,73

CONSENSUS REPORTS: Reported in EPA TSCA Inventory.

SAFETY PROFILE: Moderately toxic by skin contact. Mildly toxic by ingestion and inhalation. A severe human skin and eye irritant. Questionable carcinogen with experimental tumorigenic data. Combustible when exposed to heat or flame; can react with oxidizing materials. To fight fire, use foam, CO_2, dry chemical. When heated to decomposition it emits acrid smoke and irritating fumes. See also ALCOHOLS.

DAI800 CAS:66988-15-6 ***HR: 2***
DECYL ALCOHOL (mixed isomers)
mf: $C_{10}H_{22}O$ mw: 158.32

SYNS: DECANOL (mixed isomers) ◇ FAIR 85 ◇ SPROUT-OFF ◇ TOBACCO SUCKER CONTROL AGENT 148

TOXICITY DATA with REFERENCE
skn-rbt 10 mg/24H open SEV AMIHBC 4,119,51
eye-rbt 500 mg open AMIHBC 4,119,51
orl-rat LD50:9800 mg/kg AMIHBC 4,119,51
skn-rbt LD50:3560 mg/kg AMIHBC 4,119,51

SAFETY PROFILE: Moderately toxic by skin contact. Mildly toxic by ingestion. An eye and severe skin irritant. When heated to decomposition it emits acrid smoke and irritating fumes. See also ALCOHOLS and DECYL ALCOHOL.

DAJ000 CAS:1322-98-1 ***HR: 3***
DECYL BENZENE SODIUM SULFONATE
mf: $C_{16}H_{25}O_3S{\cdot}Na$ mw: 320.46

SYNS: SODIUM DECYLBENZENESULFONAMIDE ◇ SODIUM DECYLBENZENESULFONATE

TOXICITY DATA with REFERENCE
eye-rbt 450 mg SEV AROPAW 40,668,48
eye-rbt 1% SEV JAPMA8 38,428,49
orl-mus LD50:2000 mg/kg PSTGAW 3,1,45
ivn-mus LD50:115 mg/kg JAPMA8 38,428,49

CONSENSUS REPORTS: Reported in EPA TSCA Inventory.

SAFETY PROFILE: Poison by intravenous route. Moderately toxic by ingestion. A severe eye irritant. When heated to decomposition it emits toxic fumes of SO_x. See also SULFONATES.

DAJ200 CAS:28519-06-4 ***HR: 1***
DECYL CHLORIDE (mixed isomers)
mf: $C_{10}H_{21}Cl$ mw: 176.76

SYN: CHLORODECANE

TOXICITY DATA with REFERENCE
skn-rbt 500 mg MLD 34ZIAG -,745,69
orl-rat LD50:45300 mg/kg AIHAAP 23,95,62
skn-rbt LD50:5660 mg/kg AIHAAP 23,95,62

SAFETY PROFILE: Mildly toxic by skin contact and ingestion. A mild skin irritant. When heated to decomposition it emits toxic fumes of Cl^-. See also CHLORINATED HYDROCARBONS, ALIPHATIC.

DAJ300 CAS:23489-03-4 ***HR: 3***
1-DECYL-1-ETHYLPIPERIDINIUM BROMIDE
mf: $C_{17}H_{36}N{\cdot}Br$ mw: 334.45

TOXICITY DATA with REFERENCE
orl-mus LD50:202 mg/kg PSDTAP 15,331,74
ipr-mus LD50:34964 μg/kg PSDTAP 15,331,74
ivn-mus LD50:4293 μg/kg PSDTAP 15,331,74

SAFETY PROFILE: Poison by ingestion, intravenous and intraperitoneal routes. When heated to decomposition it emits toxic fumes of NO_x and Br^-.

DAJ400 CAS:14817-09-5 ***HR: 2***
4-n-DECYLOXY-3,5-DIMETHOXYBENZOIC ACID AMIDE
mf: $C_{19}H_{31}NO_4$ mw: 337.51

SYNS: DECIMEMIDE ◇ 4-(DECYLOXY)-3,5-DIMETHOXYBENZAMIDE ◇ DENEGYT ◇ EGYT-1050

TOXICITY DATA with REFERENCE
orl-rat LD50:1650 mg/kg 27ZQAG -,377,72
orl-mus LD50:2950 mg/kg 27ZQAG -,377,72

SAFETY PROFILE: Moderately toxic by ingestion. An anticonvulsant. When heated to decomposition it emits toxic fumes of NO_x.

DAJ800 CAS:8024-14-4 ***HR: 2***
DEERTONGUE INCOLORE

PROP: Found in leaves of *Liatris odoratissima*; contains coumarin (12VSA5 9,5315,76).

SYNS: DEER'S TONGUE ◇ LIATRIS ◇ LIATRIX OLEORESIN ◇ VANILLA PLANT

TOXICITY DATA with REFERENCE
orl-rat LD50:730 mg/kg FCTXAV 14,743,76
skn-rbt LD50:3670 mg/kg FCTXAV 14,743,76

SAFETY PROFILE: Moderately toxic by ingestion and skin contact. When heated to decomposition it emits toxic fumes of NO_x. See also COUMARIN.

DAK000 CAS:56283-74-0 ***HR: 3***
16-DEETHYL-3-o-DEMETHYL-16-METHYL-3-o-(1-OXOPROPYL)MONENSIN
mf: $C_{37}H_{62}O_{12}$ mw: 698.99

SYN: LAIDLOMYCIN

TOXICITY DATA with REFERENCE
ipr-mus LD50:5 mg/kg 85ERAY 1,814,78
scu-mus LD50:2500 μg/kg 85ERAY 1,814,78
ivn-mus LD50:1 mg/kg 85ERAY 1,814,78

SAFETY PROFILE: Poison by intraperitoneal, subcutaneous, and intravenous routes. When heated to decomposition it emits acrid smoke and irritating fumes.

DAK200 CAS:70-51-9 ***HR: 3***
DEFEROXAMINE
mf: $C_{25}H_{48}N_6O_8$ mw: 560.79

SYNS: 30-AMINO-3,14,25-TRIHYDROXY-3,9,14,20,25-PENTAAZATRIACONTANE-2,10,13,21,24-PENTAONE ◇ N-BENZOYLFERRIOXAMINE B ◇ DEFEROXAMINUM ◇ DEFERRIOXAMINE ◇ DEFERRIOXAMINE B ◇ DESFERAL ◇ DESFERRAL ◇ DESFERRIN ◇ DESFERRIOXAMINE ◇ DESFERRIOXAMINE B ◇ Df B ◇ DFO ◇ DFOA ◇ DFOM ◇ NSC-52760

TOXICITY DATA with REFERENCE
scu-rat LD50:12240 mg/kg IYKEDH 6,119,75
ivn-rat LD50:329 mg/kg TXAPA9 18,185,71
orl-mus LD50:4329 mg/kg ARZNAD 17,748,67
ipr-mus LD50:1680 mg/kg KSRNAM 4,99,70
scu-mus LD50:1450 mg/kg IYKEDH 6,119,75

SAFETY PROFILE: Poison by intravenous route. Moderately toxic by intraperitoneal and subcutaneous routes. Mildly toxic by ingestion. When heated to decomposition it emits toxic fumes of NO_x.

DAK300 CAS:138-14-7 ***HR: 3***
DEFEROXAMINE MESYLATE
mf: $C_{25}H_{48}N_6O_8 \cdot CH_4O_3S$ mw: 656.90

SYNS: DEFEROXAMINE MESILATE ◇ DEFEROXAMINE METHANESULFONATE ◇ DESFERAL METHANESULFONATE ◇ DESFERRIOXAMINE B MESYLATE ◇ DESFERRIOXAMINE B METHANESULFONATE ◇ DESFERRIOXAMINE METHANESULFONATE

TOXICITY DATA with REFERENCE
orl-rat LD50:17300 mg/kg NIIRDN 6,496,82
scu-rat LD50:5500 mg/kg NIIRDN 6,496,82
ivn-rat LD50:330 mg/kg NIIRDN 6,496,82
orl-mus LD50:15200 mg/kg NIIRDN 6,496,82
ipr-mus LD50:1240 mg/kg OYYAA2 7,1181,73
scu-mus LD50:1280 mg/kg OYYAA2 7,1181,73
ivn-mus LD50:273 mg/kg NIIRDN 6,496,82

SAFETY PROFILE: Poison by intravenous route. Moderately toxic by subcutaneous and intraperitoneal routes. Mildly toxic by ingestion. When heated to decomposition it emits toxic fumes of NO_x and SO_x.

DAK325 ***HR: D***
DEGLYCOSYLATED HUMAN CHORIONIC GONADOTROPIN

SYNS: DEGLYCOSYLATED HCG ◇ DG-HCG ◇ HUMAN CHORIOGONADOTROPIN, DEGLYCOSYLATED

TOXICITY DATA with REFERENCE
scu-rat TDLo:1 mg/kg (13-16D preg):REP ENDOAO 113,195,83
scu-rat TDLo:600 μg/kg (8-11D preg):TER ENDOAO 113,195,83

SAFETY PROFILE: Experimental teratogenic and reproductive effects.

DAK400 CAS:1740-19-8 ***HR: 2***
DEHYDROABIETIC ACID
mf: $C_{21}H_{30}O_2$ mw: 314.51

SYNS: DHA ◇ 13-ISOPROPYLPODOCARPA-8,11,13-TRIEN-15-OIC ACID

TOXICITY DATA with REFERENCE
orl-rat LD50:1710 mg/kg BECTA6 18,42,77

CONSENSUS REPORTS: Reported in EPA TSCA Inventory.

SAFETY PROFILE: Moderately toxic by ingestion. When heated to decomposition it emits acrid smoke and irritating fumes.

DAK600 CAS:434-16-2 ***HR: 3***
7-DEHYDROCHOLESTEROL
mf: $C_{27}H_{44}O$ mw: 384.71

SYNS: (3-β)CHOLESTA-5,7-DIEN-3-OL ◇ 5,7-CHOLESTADIEN-3-β-OL ◇ Δ^7-CHOLESTEROL ◇ $\Delta^{5,7}$-CHOLESTEROL ◇ DEHYDROCHOLESTERIN (GERMAN) ◇ 7-DEHYDROCHOLESTERIN ◇ DEHYDROCHOLESTEROL ◇ 7,8-DIDEHYDROCHOLESTEROL ◇ PROVITAMIN D_3

TOXICITY DATA with REFERENCE
scu-mus TDLo:800 mg/kg/4W-I:ETA NATWAY 60,525,73

CONSENSUS REPORTS: Reported in EPA TSCA Inventory.

SAFETY PROFILE: Questionable carcinogen with experimental tumorigenic data. When heated to decomposition it emits acrid smoke and irritating fumes.

DAK800 CAS:1059-86-5 ***HR: 3***
7-DEHYDROCHOLESTEROL ACETATE
mf: $C_{29}H_{46}O_2$ mw: 426.75

SYNS: CHOLESTA-5,7-DIEN-3-β-OL ACETATE ◇ 7-DEHYDROCHOLESTERYL ACETATE

TOXICITY DATA with REFERENCE
scu-mus TDLo:200 mg/kg/90D-I:ETA NATUAS 209,1026,66

SAFETY PROFILE: Questionable carcinogen with experimental tumorigenic data. When heated to decomposition it emits acrid smoke and irritating fumes.

DAL000 CAS:81-23-2 ***HR: 2***
DEHYDROCHOLIC ACID
mf: $C_{24}H_{34}O_5$ mw: 402.58

SYNS: ACIDE DEHYDROCHOLIQUE (FRENCH) ◇ ACOLEN ◇ BILIDREN ◇ BILOSTAT ◇ CHOLAGON ◇ CHOLAN DH ◇ CHOLIMED ◇ CHOLOGON ◇ CHOLOLIN ◇ DECHOLIN ◇ DEHYCHOL ◇ DEHYDROCHOLSAEURE (GERMAN) ◇ DEHYSTOLIN ◇ DEIDROCOLICO VITA ◇ DHC ◇ DIDOCOL ◇ DIDROCOLO ◇ DILABIL ◇ DILAHIL ◇ DRENOBYL ◇ EREBILE ◇ FELACRINOS ◇ HYKOLEX ◇ KETOCHOL ◇ KETOCHOLANIC ACID ◇ NOVOCOLIN ◇ OXYCHOLIN ◇ PROCHOLON ◇ SANOCHOLEN ◇ TRIKETOCHOLANIC ACID ◇ 3,7,12-TRIKETOCHOLANIC ACID ◇ 3,7,12-TRIOXOCHOLANIC ACID ◇ 3,7,12-TRIOXO-5-β-CHOLAN-24-OIC ACID

TOXICITY DATA with REFERENCE
orl-rat LD50:4 g/kg AIPTAK 116,154,58
ivn-rat LD50:750 mg/kg NIIRDN 6,495,82
ims-rat LD50:1500 mg/kg AEPPAE 222,244,54
orl-mus LD50:3100 mg/kg ARZNAD 12,857,62

scu-mus LD50:1620 mg/kg AIPTAK 116,154,58
ivn-mus LD50:1492 JJPAAZ 22,235,72

CONSENSUS REPORTS: Reported in EPA TSCA Inventory.

SAFETY PROFILE: Moderately toxic by ingestion, intravenous, intramuscular, and subcutaneous routes. When heated to decomposition it emits acrid smoke and fumes.

DAL030 ***HR: 3***
DEHYDROEPIANDROSTERONE SODIUM SULFATE DIHYDRATE
mf: $C_{19}H_{27}O_5S \cdot Na \cdot 2H_20$ mw: 426.55

SYNS: 3-β-HYDROXYANDROST-5-EN-17-ONE ESTER with SODIUM SULFATE DIHYDRATE ◇ PRASTERONE SODIUM SULFATE DIHYDRATE ◇ SODIUM ANDROST-5-EN-17-ONE-3-β-YL SULFATE DIHYDRATE

TOXICITY DATA with REFERENCE
ivn-wmn TDLo:4 mg/kg:GIT,SKN JMGZAI 18(5),10,81
ipr-rat LD50:523 mg/kg JMGZAI 18(5),10,81
scu-rat LD50:1005 mg/kg JMGZAI 18(5),10,81
ivn-rat LD50:468 mg/kg JMGZAI 18(5),10,81
ipr-mus LD50:460 mg/kg JMGZAI 18(5),10,81
scu-mus LD50:899 mg/kg JMGZAI 18(5),10,81
ivn-mus LD50:274 mg/kg JMGZAI 18(5),10,81

SAFETY PROFILE: Poison by intravenous route. Moderately toxic by intraperitoneal and subcutaneous routes. Human systemic effects by intravenous route: dermatitis, nausea, and vomiting. A steroid. When heated to decomposition it emits toxic fumes of SO_x and Na_2O.

DAL040 ***HR: 3***
DEHYDROEPIANDROSTERONE SULFATE SODIUM
mf: $C_{19}H_{27}O_5S \cdot Na$ mw: 390.51

SYNS: DHA-S SODIUM ◇ MYLIS ◇ PRASTERONE SODIUM SULFATE ◇ SODIUM DEHYDROEPIANDROSTERONE SULFATE

TOXICITY DATA with REFERENCE
ipr-rat TDLo:500 mg/kg (8-17D preg):REP OYYAA2 12,201,76
ipr-mus TDLo:100 mg/kg (female 7-16D post):TER OYYAA2 12,201,76
ivn-wmn TDLo:4 mg/kg:GIT,SKN JMGZAI 18(5),10,81
ipr-rat LD50:523 mg/kg KSRNAM 10,1852,76
scu-rat LD50:1005 mg/kg KSRNAM 10,1852,76
ivn-rat LD50:468 mg/kg KSRNAM 10,1852,76
ipr-mus LD50:460 mg/kg KSRNAM 10,1852,76
scu-mus LD50:899 mg/kg KSRNAM 10,1852,76
ivn-mus LD50:274 mg/kg KSRNAM 10,1852,76

SAFETY PROFILE: Poison by intravenous route. Moderately toxic by intraperitoneal and subcutaneous routes. Experimental teratogenic and reproductive effects. Human systemic effects by intravenous route: nausea or vomiting, dermatitis. A steroid. When heated to decomposition it emits toxic fumes of SO_x and Na_2O.

DAL060 CAS:26400-24-8 ***HR: 3***
DEHYDROHELIOTRIDINE
mf: $C_8H_{11}NO_2$ mw: 153.20

SYN: 3,8-DIDEHYDRO-HELIOTRIDINE

TOXICITY DATA with REFERENCE
dni-mus:kdy 5 mg/L CBINA8 4,421,71/72
dnd-dom:kdy 10 mg/L CBINA8 10,133,75
dns-dom:kdy 10 μmol/L CBINA8 13,243,76
dni-dom:kdy 40 mg/L CBINA8 10,133,75
oms-dom:kdy 10 mg/L CBINA8 10,133,75
ipr-rat TDLo:60 mg/kg (14D preg):TER JPTLAS 131,339,80
ipr-rat LDLo:92 mg/kg CBINA8 12,299,76

SAFETY PROFILE: Poison by intraperitoneal route. An experimental teratogen. Mutation data reported. When heated to decomposition it emits toxic fumes of NO_x.

DAL100 CAS:23107-11-1 ***HR: 3***
DEHYDROHELIOTRINE
mf: $C_{16}H_{25}NO_5$ mw: 311.42

TOXICITY DATA with REFERENCE
ipr-rat TDLo:30 mg/kg (female 14D post):TER JPTLAS 131,339,80
ipr-rat LDLo:62 mg/kg CBINA8 12,299,76

SAFETY PROFILE: Poison by intraperitoneal route. An experimental teratogen. When heated to decomposition it emits toxic fumes of NO_x.

DAL200 CAS:3343-10-0 ***HR: 3***
1,2-DEHYDRO-3-METHYLCHOLANTHRENE
mf: $C_{21}H_{14}$ mw: 266.35

SYNS: DEHYDRO-3-METHYLCHOLANTHRENE ◇ 3-METHYLBENZ (j)ACEANTHRYLENE ◇ 3-METHYLCHOLANTHRYLENE ◇ 20-METHYLCHOLANTHRYLENE

TOXICITY DATA with REFERENCE
skn-mus TDLo:85 mg/kg/20W-I:CAR CBINA8 22,69,78
scu-mus TDLo:120 mg/kg/6W-I:NEO IJCNAW 2,505,67

SAFETY PROFILE: Questionable carcinogen with experimental carcinogenic and neoplastigenic data. When heated to decomposition it emits acrid smoke and irritating fumes.

DAL300 CAS:72-63-9 ***HR: 2***
1-DEHYDRO-17-α-METHYLTESTOSTERONE
mf: $C_{20}H_{28}O_2$ mw: 300.42

PROP: Crystals from acetone + ether. Mp: 163-164°.

SYNS: ABIROL ◇ ANABOLIN ◇ CIBA 17309 BA ◇ COMPOUND 17309 ◇ CREIN ◇ DANABOL ◇ DEHYDROMETHYLTESTERONE ◇ A1-DEHYDROMETHYLTESTERONE ◇ DIANABOL ◇ DIANABOLE ◇ GEABOL ◇ 17-β-HYDROXY-17-α-METHYLANDROSTRA-1,4-DIEN-3-ONE ◇ MA ◇ METANABOL ◇ METANDIENON ◇ METANDIENONE ◇ METANDIENONUM ◇ METANDROSTENOLON ◇ METANDROSTENOLONE ◇ METASTENOL ◇ METHANDIENONE ◇ METHANDROLONE ◇ METHANDROSTENOLONE ◇ 17-α-METHYL-17-β-HYDROXY-1,4-ANDROSTADIEN-3-ONE ◇ Δ'-17-METHYLTESTOSTERONE ◇ Δ(1)-17-α-METHYLTESTOSTERONE ◇ NEROBOL ◇ NEROBOLETTES ◇ NSC-42722 ◇ PROTOBOLIN ◇ STENOLON ◇ STENOLONE

TOXICITY DATA with REFERENCE
dlt-mus-orl 200 mg/kg/10D-I VPITAR 38(4),63,80
orl-man TDLo:13 mg/kg (60D male):REP CCPTAY 15,151,77
orl-man TDLo:561 mg/kg/7Y-C:CAR LANCAO 2,1273,72
ipr-rat LD50:425 mg/kg PCJOAU 20,143,86

CONSENSUS REPORTS: EPA Genetic Toxicology Program.

SAFETY PROFILE: Moderately toxic by intraperitoneal route. Human reproductive effects by ingestion route: changes in spermatogenesis. Other experimental reproductive effects. Questionable human carcinogen producing liver tumors. Mutation data reported. When heated to decomposition it emits acrid smoke and irritating fumes. See also TESTOSTERONE.

DAL350 CAS:23291-96-5 ***HR: 2***
DEHYDROMONOCROTALINE
mf: $C_{16}H_{21}NO_6$ mw: 323.38

SYNS: MONOCROTALINE, 3,8-DIDEHYDRO- ◇ 20-NORCROTALANAN-11,15-DIONE,3,8-DIDEHYDRO-14,19-DIHYDRO-12,13-DIHYDROXY-, (13-α-14-α)-

TOXICITY DATA with REFERENCE
skn-mus TDLo:1504 mg/kg/47W-I:ETA CALEDQ 17,61,82
ipr-rat LDLo:16 mg/kg CBINA8 12,299,76

SAFETY PROFILE: Poison by intraperitoneal route. Questionable carcinogen with experimental tumorigenic data. When heated to decomposition it emits toxic fumes of NO_x.

DAL400 CAS:23107-12-2 ***HR: 3***
DEHYDRORETRONECINE
mf: $C_8H_{11}NO_2$ mw: 153.20

SYNS: 3,8-DIDEHYDRORETRONECINE ◇ (R)-2,3-DIHYDRO-1-HYDROXY-1H-PYRROLIZINE-7-METHANOL

TOXICITY DATA with REFERENCE
mmo-sat 500 μg/plate MUREAV 149,485,85
sce-hmn:lym 1 μmol/L MUREAV 149,485,85
dni-rat:lvr 100 μmol/L CBINA8 30,325,80
oms-rat:lvr 20 μmol/L CBINA8 30,325,80
otr-ham:kdy 250 μg/L CRNGDP 1,161,80
scu-rat TDLo:350 mg/kg/1Y-I:NEO JNCIAM 56,787,76
skn-mus TDLo:120 mg/kg/28W-I:CAR JJIND8 61,85,78
scu-mus TDLo:80 mg/kg/4W-I:CAR JJIND8 61,85,78
ipr-rat LD50:122 mg/kg CBINA8 12,299,76

CONSENSUS REPORTS: IARC Cancer Review: Animal Sufficient Evidence IMEMDT 10,333,76.

SAFETY PROFILE: Confirmed carcinogen with experimental carcinogenic and neoplastigenic data. Poison by intraperitoneal route. Human mutation data reported. When heated to decomposition it emits toxic fumes of NO_x.

DAL600 CAS:84-17-3 ***HR: 3***
DEHYDROSTILBESTROL
mf: $C_{18}H_{18}O_2$ mw: 266.36

SYNS: 3,4-BIS(p-HYDROXYPHENYL)-2,4-HEXADIENE ◇ 3,4-BIS(4-HYDROXYPHENYL)-2,4-HEXADIENE ◇ CYCLADIENE ◇ DIENESTROL ◇ DIENOESTROL ◇ β-DIENOESTROL ◇ DIENOL ◇ 4,4'-(1,2-DIETHYLIDENE-1,2-ETHANEDIYL)BISPHENOL ◇ p,p'-(DIETHYLIDENEETHYLENE)DIPHENOL ◇ 4,4'-(DIETHYLIDENEETHYLENE) DIPHENOL ◇ DINOVEX ◇ DI(p-OXYPHENYL)-2,4-HEXADIENE ◇ DV ◇ ESTRAGARD ◇ ESTRODIENOL ◇ ESTRORAL ◇ FOLLIDIENE ◇ FOLLORMON ◇ GYNEFOLLIN ◇ HORMOFEMIN ◇ 4,4'-HYDROXY-Γ,Δ-DIPHENYL-β,Δ-HEXADIENE ◇ ISODIENESTROL ◇ OESTRASID ◇ OESTRODIENE ◇ OESTRODIENOL ◇ OESTRORAL ◇ PARA-DIEN ◇ RESTROL ◇ RETALON ◇ SEXADIEN ◇ SYNESTROL ◇ TESERENE ◇ WILLNESTROL

TOXICITY DATA with REFERENCE
sce-hmn:fbr 5 nmol/L NATUAS 281,392,79
dns-ham:emb 3 mg/L CNREA8 44,184,84
orl-rat TDLo:40 μg/kg (3D preg):REP JRPFA4 13,101,67

CONSENSUS REPORTS: IARC Cancer Review: Animal Inadequate Evidence IMEMDT 21,161,79; Human Limited Evidence IMEMDT 21,161,79

SAFETY PROFILE: Suspected human carcinogen. Human mutation data reported. Experimental reproductive effects. Used as a drug for the treatment of postmenopausal symptoms. When heated to decomposition it emits acrid smoke and irritating fumes.

DAM000 CAS:60504-95-2 ***HR: 3***
DEISOVALERYL BLASTMYCIN
mf: $C_{21}H_{28}N_2O_8$ mw: 436.51

PROP: Produced by *Streptomyces sp.* 5140-Al (JANTAJ 29,804,76).

SYNS: N-(7-BUTYL-4,9-DIMETHYL-2,6-DIOXO-8-HYDROXY-1,5-DIOXONAN-3-YL)-3-FORMAMIDOSALICYLAMIDE ◇ N-(7-BUTYL-8-HYDROXY-4,9-DIMETHYL-2,6-DIOXO-1,5-DIOXONAN-3-YL)-3-FORMAMIDOSALICYLAMIDE

TOXICITY DATA with REFERENCE
ipr-mus LD50:25 mg/kg JANTAJ 29,804,76
ivn-mus LD50:15 mg/kg JANTAJ 29,804,76

SAFETY PROFILE: Poison by intraperitoneal and intravenous routes. When heated to decomposition it emits toxic fumes of NO_x.

DAM300 CAS:24356-94-3 ***HR: D***
DELADROXONE
mf: $C_{29}H_{36}O_4$ mw: 448.65

PROP: Crystals from 95% ethanol. Mp: 150-151°. Stable to boiling mineral acids.

SYNS: ALGESTONE ACETOPHENIDE ◇ ALPHASONE ACETOPHENIDE ◇ BOVITROL ◇ 16-α,17-DIHYDROXYPREGN-4-ENE-3,20-DIONE CYCLIC ACETAL with ACETOPHENONE ◇ DIHYDROXYPROGESTERONE ACETOPHENIDE

TOXICITY DATA with REFERENCE
orl-rat TDLo:2500 mg/kg (20D pre):REP FESTAS 15,63,64
scu-rat TDLo:300 mg/kg (female 14-19D post):TER ENDOAO 71,448,62

SAFETY PROFILE: An experimental teratogen. Other experimental reproductive effects. A steroid. When heated to decomposition it emits acrid smoke and fumes. See other progesterone entries.

DAM315 CAS:83435-67-0 ***HR: 3***
DELAPRIL HYDROCHLORIDE
mf: $C_{26}H_{32}N_2O_5 \cdot ClH$ mw: 489.06

SYNS: CV 3317 ◇ N-(N-((S)-1-ETHOXYCARBONYL-3-PHENYLPROPYL)-l-ALANYL)-N-(INDAN-2-YL)GLYCINEHYDROCHLORIDE ◇ GLYCINE, N-(2,3-DIHYDRO-1H-INDEN-2-YL)-N-(N-(1-(ETHOXYCARBONYL)-3-PHENYLPROPYL)-l-ALANYL)-, MONOHYDROCHLORIDE, (S)-

TOXICITY DATA with REFERENCE
orl-rat TDLo:8700 mg/kg (female 15-22D post):REP YACHDS 15(Suppl 1),203,87
orl-rat LD50:8260 mg/kg YACHDS 15(Suppl 1),203,87
ipr-rat LD50:208 mg/kg YACHDS 15(Suppl 1),203,87
scu-rat LD50:5900 mg/kg YACHDS 15(Suppl 1),203,87
orl-mus LD50:3120 mg/kg YACHDS 15(Suppl 1),203,87
ipr-mus LD50:164 mg/kg YACHDS 15(Suppl 1),203,87
scu-mus LD50:2340 mg/kg YACHDS 15(Suppl 1),203,87

SAFETY PROFILE: Poison by intraperitoneal route. Moderately toxic by ingestion and subcutaneous routes. Experimental reproductive effects. When heated to decomposition it emits toxic fumes of NO_x and HCl.

DAM325 ***HR: 3***
DELATESTRYL and DEPO-PROVERA
mf: $C_{24}H_{34}O_4 \cdot C_{26}H_{41}O_3$ mw: 788.25

SYNS: DELATESTRYL and DEPO-MEDROXYPROGESTERONE ACETATE ◇ DEPO-MEDROXYPROGESTERONE ACETATE and DELATESTRYL ◇ DEPO-MEDROXYPROGESTERONE ACETATE and TESTOSTERONE ENANTHATE ◇ DEPO-PROVERA and DELATESTRYL ◇ DEPO-PROVERA and TESTOSTERONE ENANTHATE ◇ TESTOSTERONE ENANTHATE and DEPO-MEDROXYPROGESTERONE ACETATE ◇ TESTOSTERONE ENANTHATE and DEPO-PROVERA

TOXICITY DATA with REFERENCE
ims-man TDLo:15 mg/kg (3D male):REP CCPTAY 15,627,77

SAFETY PROFILE: Human male reproductive effects by intramuscular route: spermatogenesis, impotence. A steroid. When heated to decomposition it emits acrid smoke and fumes. See other testosterone entries.

DAM400 CAS:528-53-0 ***HR: 3***
DELPHINIDOL
mf: $C_{15}H_{11}O_7$ mw: 303.26

SYNS: 3,3′,4′,5,5′,7-HEXAHYDROXYFLAVYLIUM ACID ANION ◇ 3,4,7-TRIHYDROXY-2-(3,4,5-TRIHYDROXYPHENYL)BENZOPYRYLIUM, ACID ANION

TOXICITY DATA with REFERENCE
ipr-rat LD50:2350 mg/kg CHTPBA 2,33,67
ivn-rat LD50:240 mg/kg CHTPBA 2,33,67
ipr-mus LD50:4110 mg/kg CHTPBA 2,33,67
ivn-mus LD50:840 mg/kg CHTPBA 2,33,67

SAFETY PROFILE: Poison by intravenous route. Moderately toxic by intraperitoneal route. When heated to decomposition it emits acrid smoke and irritating fumes.

DAM600 CAS:57-42-1 ***HR: 3***
DEMAROL
mf: $C_{15}H_{21}NO_2$ mw: 247.37

SYNS: DEMEROL ◇ DOLCONTRAL ◇ DOLOSAL ◇ DOLSIN ◇ ETHYL-1-METHYL-4-PHENYLISONIPECOTATE ◇ ETHYL-1-METHYL-4-PHENYLPIPERIDINE-4-CARBOXYLATE ◇ ISONIPECAINE ◇ LIDOL ◇ MEPERIDINE ◇ N-METHYL-4-PHENYL-4-CARBETHOXYPIPERIDINE ◇ 1-METHYL-4-PHENYLISONIPECOTIC ACID, ETHYL ESTER ◇ 1-METHYL-4-PHENYL-PIPERIDIN-4-CARBON-SAEURE-AETHYLESTER-HYDROCHLORID (GERMAN) ◇ 1-METHYL-4-PHENYLPIPERIDINE-4-CARBOXYLIC ACID ETHYL ESTER ◇ NEMEROL ◇ PETHIDINETER ◇ PETHIDOINE ◇ PHETIDINE ◇ PIPERSAL ◇ PIRIDOSAL

TOXICITY DATA with REFERENCE
ims-wmn TDLo:2 mg/kg (39W preg):REP AJOGAH 64,1368,52
unr-man LDLo:15 mg/kg 85DCAI 2,73,70
unr-man TDLo:57 mg/kg:CNS NEJMAG 312,509,85
orl-rat LD50:162 mg/kg AIPTAK 180,155,69
ipr-rat LD50:87 mg/kg JAPMA8 47,323,58
scu-rat LD50:113 mg/kg ARZNAD 24,600,74
ivn-rat LD50:29 mg/kg DDREDK 1,83,81
idu-rat LD50:90 mg/kg AIPTAK 180,155,69
orl-mus LD50:200 mg/kg AIPTAK 135,376,62

ipr-mus LD50:150 mg/kg AIPTAK 135,376,62
par-mus LD50:178 mg/kg JMCMAR 11,889,68

SAFETY PROFILE: A human poison by an unspecified route. Poison experimentally by ingestion, subcutaneous, intravenous, intradermal, parenteral, and intraperitoneal routes. Human systemic effects by unspecified route: changes in sleep patterns and muscle weakness. Human reproductive effects by intramuscular route: effects on measurements and viability of newborn. A pharmaceutical pain killer. When heated to decomposition it emits toxic fumes of NO_x. See also ESTERS.

DAM700 CAS:50-13-5 ***HR: 3***
DEMEROL HYDROCHLORIDE
mf: $C_{15}H_{21}NO_2{\cdot}ClH$ mw: 283.83

SYNS: ALGIL ◇ ALODAN (GEROT) ◇ ANTIDUROL ◇ CENTRALGIN ◇ CHLORBICYCLENE (FRENCH) ◇ DEMEROL ◇ DISPADOL ◇ DOLANTAL ◇ DOLANTIN ◇ DOLANTIN HYDROCHLORIDE ◇ DOLANTOL ◇ DOLAREN ◇ DOLARGAN ◇ DOLCONTRAL ◇ DOLENAL ◇ DOLENOL ◇ DOLESTINE ◇ DOLIN ◇ DOLOGAL ◇ DOLONEURINE ◇ DOLOPETHIN ◇ DOLOSAL ◇ DOLVANOL ◇ ENDOLAT ◇ ETHYL-1-METHYL-4-PHENYLISONIPECOTATE HYDROCHLORIDE ◇ ETHYL-1-METHYL-4-PHENYLPIPERIDINE-4-CARBOXYLATE HYDROCHLORIDE ◇ ETHYL-1-METHYL-4-PHENYLPIPERIDYL-4-CARBOXYLATE HYDROCHLORIDE ◇ ISONIPECAINE HYDROCHLORIDE ◇ LIDOL ◇ LYDOL ◇ MEFEDINA ◇ MEPADIN ◇ MEPERIDINE HYDROCHLORIDE ◇ MEPHEDINE ◇ 1-METHYL-4-CARBETHOXY-4-PHENYLPIPERIDINE HYDROCHLORIDE ◇ N-METHYL-4-PHENYL-4-CARBETHOXYPIPERIDINE HYDROCHLORIDE ◇ 1-METHYL-4-PHENYL-4-CARBOETHOXYPIPERIDINEHYDROCHLORIDE ◇ 1-METHYL-4-PHENYLISONIPECOTIC ACID ETHYL ESTER HYDROCHLORIDE ◇ OPERIDINE ◇ PANTALGINE ◇ PENTANTIN ◇ PETANTIN HYDROCHLORIDE ◇ PETHIDINE CHLORIDE ◇ PETIDIN ◇ PIRIDOSAL ◇ S 140 ◇ SAUTERALGYL ◇ SPASMEDAL ◇ SPASMODOLIN ◇ SYNELAUDINE ◇ WY 554

TOXICITY DATA with REFERENCE
mnt-mus-ipr 8 mg/kg IJMRAQ 75,112,82
ivn-wmn TDLo:735 μg/kg (female 39W post):TER JPEMAO 16,23,88
ims-wmn TDLo:43 mg/kg/2D-I AEMED3 14,1007,85
ims-wmn TDLo:13500 μg/kg/3D-I AJPSAO 144,1062,87
orl-rat LD50:170 mg/kg JPETAB 103,147,51
ipr-rat LD50:65 mg/kg TXCYAC 14,217,79
scu-rat LD50:175 mg/kg ARZNAD 3,238,53
ivn-rat LD50:17 mg/kg AIPTAK 115,213,58
orl-mus LD50:178 mg/kg NIIRDN 6,755,82
ipr-mus LD50:120 mg/kg JPETAB 119,26,57
orl-mus LD50:178 mg/kg ARZNAD 28,164,78
ivn-dog LD50:68 mg/kg AIPTAK 149,571,64
par-frg LD50:515 mg/kg JPETAB 103,147,51

SAFETY PROFILE: Poison by ingestion, subcutaneous, intravenous, and intraperitoneal routes. Moderately toxic by parenteral route. Experimental teratogenic effects. Mutation data reported. An analgesic. When heated to decomposition it emits very toxic fumes of HCl and NO_x.

DAM800 CAS:64314-52-9 ***HR: D***
4-DEMETHOXYADRIAMYCIN
mf: $C_{26}H_{27}NO_{10}$ mw: 513.54

TOXICITY DATA with REFERENCE
mmo-sat 1 μg/plate ENMUDM 5,473,83
dns-rat:lvr 2 mg/L CNREA8 44,5599,84
dnd-mus:fbr 1800 nmol/L CNREA8 37,4523,77

SAFETY PROFILE: Mutation data reported. When heated to decomposition it emits toxic fumes of NO_x.

DAN000 CAS:58957-92-9 ***HR: 3***
4-DEMETHOXYDAUNOMYCIN
mf: $C_{26}H_{27}NO_9$ mw: 497.54

SYNS: 4-DEMETHOXYDAUNORUBICIN ◇ NSC-256439

TOXICITY DATA with REFERENCE
mmo-sat 2 μg/plate ENMUDM 7,129,85
dns-rat:lvr 2 mg/L CNREA8 44,5599,84
dnd-mus:fbr 1800 nmol/L CNREA8 37,4523,77
dnd-mam:lym 200 nmol/L BBRCA9 69,744,76
orl-mus LD50:16 mg/kg CTRRDO 61,893,77
ivn-mus LD50:4 mg/kg CTRRDO 61,893,77

SAFETY PROFILE: Poison by ingestion and intravenous routes. Mutation data reported. When heated to decomposition it emits toxic fumes of NO_x. See also DAUNOMYCIN.

DAN200 ***HR: 3***
N-DEMETHYLACLACINOMYCIN A
mf: $C_{41}H_{51}NO_{13}$ mw: 765.93

TOXICITY DATA with REFERENCE
mma-sat 1 nmol/plate CNREA8 38,1782,78
ipr-mus LD50:46 mg/kg JANTAJ 31,78-90,78

SAFETY PROFILE: Poison by intraperitoneal route. Mutation data reported. When heated to decomposition it emits toxic fumes of NO_x.

DAN300 CAS:7336-36-9 ***HR: 3***
2-DEMETHYLCOLCHICINE
mf: $C_{21}H_{24}NO_6$ mw: 386.46

SYN: O^2-DEMETHYLCOLCHICINE

TOXICITY DATA with REFERENCE
ice-rat LDLo:75 μg/kg AIPTAK 109,386,57
orl-mus LD50:19 mg/kg AIPTAK 94,453,53
ipr-mus LD50:16 mg/kg AIPTAK 109,386,57
ims-mus LD50:16577 μg/kg JMCMAR 24,257,81

SAFETY PROFILE: Poison by ingestion, intramuscular, intracerebral, and intraperitoneal routes. When heated to decomposition it emits toxic fumes of NO_x. See also CHOLCHICINE.

DAN375 CAS:477-29-2 ***HR: 3***
3-DEMETHYLCOLCHICINE GLUCOSIDE
mf: $C_{27}H_{33}NO_{11}$ mw: 547.61

SYN: COLCHICOSIDE

TOXICITY DATA with REFERENCE
ice-rat LDLo:75 μg/kg AIPTAK 109,386,57
par-rat LDLo:75 μg/kg FRPSAX 15,533,60
orl-mus LD50:84 mg/kg AIPTAK 94,386,57
ipr-mus LD50:280 mg/kg AIPTAK 109,386,57
ivn-mus LD50:320 mg/kg AIPTAK 94,453,53

SAFETY PROFILE: Poison by ingestion, parenteral, intracerebral, intravenous, and intraperitoneal routes. When heated to decomposition it emits toxic fumes of NO_x. See also COLCHICINE.

DAO500 CAS:298-03-3 ***HR: 3***
DEMETON
mf: $C_8H_{19}O_3PS_2$ mw: 258.36

SYNS: BAYER 8169 ◇ DEMETON-O ◇ DIAETHYLTHIO-PHOSPHORSAEUREESTER des AETHYLTHIOGLYKOL (GERMAN) ◇ O,O-DIETHYL-O-(2-ETHTHIOETHYL)PHOSPHOROTHIOATE ◇ DIETHYL 2-ETHTHIOETHYL THIONOPHOSPHATE ◇ O,O-DI-ETHYL-O-2-(ETHYLTHIO)ETHYL PHOSPHOROTHIOATE ◇ O,O-DI-ETHYL-2-ETHYLTHIO ETHYL PHOSPHOROTHIOATE ◇ DIETHYL-2-(ETHYLTHIO(ETHYL PHOSPHOROTHIONATE) ◇ DI-SEPTON ◇ E-1059 ◇ ETHYLTHIOMETON ◇ MERCAPTOFOS (RUSSIAN) ◇ THIODEMETON ◇ THIOLMECAPTOPHOS

TOXICITY DATA with REFERENCE
mmo-esc 200 μg/plate NTIS** PB80-133226
mma-esc 200 μg/plate NTIS** PB80-133226
dnr-bcs 5 μg/disc NTIS** PB80-133226
mma-smc 1000 ppm NTIS** PB80-133226
dns-hmn:fbr 100 mg/L NTIS** PB80-133226
cyt-ham-ipr 2 mg/kg ARTODN 58,152,85
orl-rat LD50:7500 μg/kg GUCHAZ 6,161,73
scu-mus LD50:15 mg/kg AEPPAE 217,144,53
ipr-ham LD50:10 mg/kg ARTODN 58,152,85

SAFETY PROFILE: Poison by ingestion, subcutaneous, and intraperitoneal routes. Human mutation data reported. An insecticide. When heated to decomposition it emits toxic fumes of PO_x and SO_x. See also DEMETON-O + DEMETON-S and other demeton entries.

DAO600 CAS:8065-48-3 ***HR: 3***
DEMETON-O + DEMETON-S
mf: $C_8H_{19}O_3PS_2 \cdot C_8H_{19}O_3PS_2$ mw: 516.72

PROP: A light brown liquid, sulfur compound odor.

SYNS: BAY 10756 ◇ BAYER 8169 ◇ DEMETON ◇ DEMOX ◇ DIETH-OXY THIOPHOSPHORIC ACID ESTER of 2-ETHYLMERCAP-TOETHANOL ◇ O,O-DIETHYL 2-ETHYLMERCAPTOETHYL THIO-PHOSPHATE ◇ O,O-DIETHYL O(and S)-2-(ETHYLTHIO)ETHYL PHOSPHOROTHIOATE MIXTURE ◇ E 1059 ◇ ENT 17,295 ◇ MERCAPTOPHOS ◇ PHOSPHOROTHIOIC ACID-O,O-DIETHYL-O-(2-(ETHYLTHIO)ETHYL) ESTER, mixed with O,O-DIETHYL S-(2-(ETHYLTHIO)ETHYL) ESTER (7:3) ◇ SYSTEMOX ◇ SYSTOX ◇ ULV

TOXICITY DATA with REFERENCE
sce-hmn:lym 10 mg/L MUREAV 88,307,81
sce-ham:lng 10 mg/L MUREAV 88,307,81
ipr-mus TDLo:10 mg/kg/(7-11D preg):TER TXAPA9 24,324,73
orl-hmn LDLo:171 μg/kg CMEP** -,1,56
orl-man TDLo:144 mg/kg/24D-I TXAPA9 14,603,69
orl-rat LD50:1700 μg/kg TXAPA9 21,315,72
ihl-rat LCLo:15 mg/m³/4H 85GMAT -,52,82
skn-rat LD50:8200 μg/kg TXAPA9 2,88,60
ipr-rat LD50:2500 μg/kg APCRAW 4,117,61
ivn-rat LD50:1750 μg/kg BLLIAX 38,151,58
ims-rat LDLo:3 mg/kg 13ZGAF -,220,62
orl-mus LD50:7850 μg/kg BLLIAX 38,151,58
ipr-mus LD50:4 mg/kg PSEBAA 129,699,68
ivn-mus LD50:3900 μg/kg BLLIAX 38,151,58
ims-dog LD50:3650 μg/kg 13ZGAF -,221,62
ihl-cat LCLo:15 mg/m³/4H 85GMAT -,52,82
ims-cat LD50:3900 μg/kg 13ZGAF -,220,62
skn-rbt LD50:24 mg/kg SPEADM 74-1,-,74

CONSENSUS REPORTS: EPA Genetic Toxicology Program. EPA Extremely Hazardous Substances List.

OSHA PEL: TWA 0.1 mg/m³ (skin)
ACGIH TLV: TWA 0.01 ppm (skin)
DFG MAK: 0.01 ppm (0.1 mg/m³)

SAFETY PROFILE: A deadly human poison by ingestion. Poison experimentally by ingestion, inhalation, skin contact, intramuscular, intravenous, subcutaneous, and intraperitoneal routes. An experimental teratogen. Human mutation data reported. An insecticide which inhibits cholinesterase in humans and animals and thus causes the buildup of acetylcholine. Doses are cumulative. If illness occurs, it is acute in nature whether caused by a single large dose or by repeated exposure. Persons poisoned with demeton may be expected to show the following symptoms: headache, giddiness, blurred vision, weakness, nausea, diarrhea, and discomfort in the chest. When heated to decomposition it emits very toxic fumes of PO_x and SO_x. See also PARATHION and other demeton entries.

DAO800 CAS:867-27-6 ***HR: 3***
DEMETON-O-METHYL
mf: $C_6H_{15}O_3PS_2$ mw: 230.30

SYNS: BAY 15203 ◇ DEMETON-O-METILE (ITALIAN) ◇ O,O-DIMETHYL-O-(2-AETHYLTHIO-AETHYLMONOTHIOPHOSPHAT (GERMAN) ◇ O,O-DIMETHYL-O-ETHYLMERCAPTOETHYL THIO-PHOSPHATE ◇ O,O-DIMETHYL 2-ETHYLMERCAPTOETHYL THIO-PHOSPHATE, THIONO ISOMER ◇ O,O-DIMETHYL-O-(2-ETHYL-THIO-ETHYL)-MONOTHIOFOSFAAT (DUTCH) ◇ O,O-DIMETHYL-O-2-(ETHYLTHIO)ETHYL PHOSPHOROTHIOATE ◇ O,O-DIMETIL-O-(2-ETILTIO-ETIL)-MONOTIOFOSFATO (ITALIAN) ◇ ENT 18,862

◇ β-ETHYLMERCAPTOETHYL DIMETHYL THIONOPHOSPHATE ◇ O-(2-(ETHYLTHIO)ETHYL)-O,O-DIMETHYL PHOSPHOROTHIOATE ◇ 2-(ETHYLTHIO)ETHYL DIMETHYL PHOSPHOROTHIONATE ◇ METHYL-DEMETON-O ◇ O-METHYLDEMETON ◇ METHYL-LCISTOX ◇ METHYLMERCAPTOPHOS ◇ METHYLSYSTOX ◇ THIOPHOSPHATE de O,O-DIMETHYLE et de O-2-ETHYLTHIO-ETHYLE (FRENCH)

TOXICITY DATA with REFERENCE
ihl-rat TCLo:500 μg/m^3/24H (female 1-22D post):REP HYSAAV 36(1),34,71
ihl-rat TCLo:200 ng/m^3/(1-22D preg):TER HYSAAV 36(1),34,71
orl-rat LD50:75 mg/kg FATOAO 22,559,59
ivn-rat LD50:216 mg/kg BIJOAK 67,187,57
orl-mus LD50:46 mg/kg FATOAO 22,559,59
orl-cat LDLo:30 mg/kg FATOAO 22,559,59
ihl-cat LCLo:20 mg/m^3 GISAAA 28(3),21,63
skn-rbt LDLo:75 mg/kg FATOAO 22,559,59

SAFETY PROFILE: Poison by ingestion, skin contact, inhalation, and intravenous routes. Experimental teratogenic and reproductive effects. When heated to decomposition it emits very toxic fumes of PO_x and SO_x. See also DEMETON-O + DEMETON-S and other demeton entries.

DAP000 CAS:301-12-2 ***HR: 3***
DEMETON-O-METHYL SULFOXIDE
mf: $C_6H_{15}O_3PS_2$ mw: 230.30

SYNS: BAY 21097 ◇ DEMETON-S-METHYL-SULFOXID (GERMAN) ◇ DEMETON-S-METHYL SULFOXIDE ◇ DEMETON-METHYL SULPHOXIDE ◇ O,O-DIMETHYL-S-(2-AETHYLSULFINYL-AETHYL)-THIOLPHOSPHAT (GERMAN) ◇ O,O-DIMETHYL-S-(2-ETHTHIONYLETHYL) PHOSPHOROTHIOATE ◇ DIMETHYL-S-(2-ETHTHIONYLETHYL) THIOPHOSPHATE ◇ O,O-DIMETHYL-S-(2-ETHYLSULFINYL-ETHYL)-MONOTHIOFOSFAAT(DUTCH) ◇ O,O-DIMETHYL-S-(2-(ETHYLSULFINYL)ETHYL) PHOSPHOROTHIOATE ◇ O,O-DIMETHYL-S-(2-ETHYLSULFINYL)ETHYL THIOPHOSPHATE ◇ O,O-DIMETHYL-S-ETHYLSULPHINYLETHYL PHOSPHOROTHIOLATE ◇ O,O-DIMETHYL-S-(3-OXO-3-THIA-PENTYL)-MONOTHIOPHOSPHAT (GERMAN) ◇ O,O-DIMETIL-S-(2-ETIL-SOLFINIL-ETIL)-MONOTIOFOSFATO (ITALIAN) ◇ ENT 24,964 ◇ S-(2-(ETHYLSULFINYL)ETHYL)-O,O-DIMETHYL PHOSPHOROTHIOATE ◇ ISOMETHYLSYSTOX SULFOXIDE ◇ METAISOSYSTOX-SULFOXIDE ◇ METASYSTEMOX ◇ METASYSTOX-R ◇ METHYL DEMETON-O-SULFOXIDE ◇ METILMERCAPTOFOSOKSID ◇ OXYDEMETONMETHYL ◇ OXYDEMETON-METILE (ITALIAN) ◇ R 2170 ◇ THIOPHOSPHATE de O,O-DIMETHYLE et de S-2-ETHYLSULFINYLETHYLE (FRENCH)

TOXICITY DATA with REFERENCE
mmo-sat 50 μg/plate MUREAV 124,97,83
sce-hmn:lym 20 mg/L MUREAV 124,97,83
mnt-mus-ipr 10 mg/kg MUREAV 124,97,83
orl-rat LD50:30 mg/kg AEPPAE 234,352,58
skn-rat LD50:100 mg/kg WRPCA2 9,119,70
ipr-rat LD50:20 mg/kg GUCHAZ 6,385,73
ivn-rat LD50:47 mg/kg BIJOAK 67,187,57
orl-mus LD50:30 mg/kg SPEADM 78-1,29,78
ipr-mus LD50:8 mg/kg TXAPA9 4,621,62
orl-gpg LD50:120 mg/kg PCOC** -,721,66
orl-pgn LD50:15 mg/kg TXAPA9 20,57,71

CONSENSUS REPORTS: EPA Genetic Toxicology Program.

SAFETY PROFILE: Poison by ingestion, skin contact, intravenous, and intraperitoneal routes. Human mutation data reported. When heated to decomposition it emits very toxic fumes of PO_x and SO_x. See also other demeton entries.

DAP200 CAS:126-75-0 ***HR: 3***
DEMETON-S
mf: $C_8H_{19}O_3PS_2$ mw: 258.36

SYNS: O,O-DIAETHYL-S-(2-AETHYLTHIO-AETHYL)-MONOTHIOPHOSPHAT (GERMAN) ◇ DIAETHYLTHIOPHOSPHORSAEUREESTER des AETHYLTHIOGLYKOL (GERMAN) ◇ DIETHYL-S-(2-ETHIOETHYL) THIOPHOSPHATE ◇ O,O-DIETHYL-S-(2-ETHTHIOETHYL)PHOSPHOROTHIOATE ◇ O,O-DIETHYL-S-ETHYL-2-ETHYLMERCAPTOPHOSPHOROTHIOLATE ◇ O,O-DIETHYL-S-(2-ETHYLTHIO-ETHYL)-MONOTHIOFOSFAAT (DUTCH) ◇ O,O-DIETHYL-S-2-(ETHYLTHIO) ETHYL PHOSPHOROTHIOATE ◇ O,O-DIETHYL-S-(2-(ETHYLTHIO) ETHYL) PHOSPHOROTHIOLATE (USDA) ◇ O,O-DIETIL-S-(2-ETILTIO-ETIL)-MONOTIOFOSFATO (ITALIAN) ◇ O,O-DIETYL-S-2-ETYLMERKAPTOETYLTIOFOSFAT (CZECH) ◇ 2-(ETHYLTHIO)-ETHANETHIOL S-ESTER with O,O-DIETHYL PHOSPHOROTHIOATE ◇ ISODEMETON ◇ IZOSYSTOX (CZECH) ◇ PO-SYSTOX ◇ THIOLDEMETON ◇ THIOL SYSTOX ◇ THIOPHOSPHATE de O,O-DIETHYLE et de S-(2-ETHYLTHIO-ETHYLE) (FRENCH)

TOXICITY DATA with REFERENCE
orl-rat LD50:1500 μg/kg AEPPAE 217,144,53
ipr-rat LD50:1500 μg/kg AMIHAB 13,606,56
ipr-mus LD50:1850 μg/kg BLLIAX 38,151,58
scu-mus LD50:6 mg/kg AEPPAE 217,144,53
ipr-gpg LD50:5500 μg/kg AMIHAB 13,606,56

SAFETY PROFILE: Poison by ingestion, intraperitoneal, and subcutaneous routes. When heated to decomposition it emits very toxic fumes of PO_x and SO_x. See also DEMETON-O + DEMETON-S and other demeton entries.

DAP400 CAS:919-86-8 ***HR: 3***
DEMETON-S-METHYL
mf: $C_6H_{15}O_3PS_2$ mw: 230.30

SYNS: BAY 18436 ◇ BAYER 25/154 ◇ DEMETON-S-METILE (ITALIAN) ◇ O,O-DIMETHYL-S-(2-AETHYLTHIO-AETHYL)-MONOTHIOPHOSPHAT (GERMAN) ◇ O,O-DIMETHYL-S-(2-ETHTHIOETHYL) PHOSPHOROTHIOATE ◇ DIMETHYL-S-(2-ETHTHIOETHYL)THIOPHOSPHATE ◇ O,O-DIMETHYL-S-ETHYLMERCAPTOETHYL THIOPHOSPHATE ◇ O,O-DIMETHYL-S-ETHYLMERCAPTOETHYL THIOPHOSPHATE, THIOLO ISOMER ◇ O,O-DIMETHYL-S-(2-ETHYLTHIO-ETHYL)-MONOTHIOFOSFAAT (DUTCH) ◇ O,O-DIMETHYL-S-(2-(ETHYLTHIO)ETHYL)PHOSPHOROTHIOATE ◇ O,O-DIMETHYL-S-(3-THIA-PENTYL)-MONOTHIOPHOSPHAT (GERMAN) ◇ O,O-DIMETIL-S-(2-ETILITIO-ETIL)-MONOTIOFOSFATO (ITALIAN) ◇ DURATOX ◇ S-(2-(ETHYLTHIO)ETHYL)-O,O-DIMETHYL PHOSPHOROTHIOATE

◇ S-(2-(ETHYLTHIO)ETHYL)DIMETHYL PHOSPHOROTHIOLATE ◇ S-(2-(ETHYLTHIO)ETHYL)-O,O-DIMETHYLTHIOPHOSPHATE ◇ ISOMETASYSTOX ◇ ISOMETHYLSYSTOX ◇ METAISOSEPTOX ◇ METAISOSYSTOX ◇ METASYSTOX FORTE ◇ METHYL DEMETON THIOESTER ◇ METHYL ISOSYSTOX ◇ METHYL-MERCAPTOFOS TEOLOVY ◇ THIOPHOSPHATE de O,O-DIMETHYLE et de S-2-ETHYLTHIOETHYLE (FRENCH)

TOXICITY DATA with REFERENCE
mmo-sat 5 μL/plate MUREAV 28,405,75
sln-dmg-orl 80 ppm MUREAV 28,405,75
orl-rat LD50:60 mg/kg PAREAQ 11,636,59
ihl-rat LC50:500 mg/m^3/4H 85DPAN -,-,71/76
skn-rat LD50:85 mg/kg PHJOAV 185,361,60
ipr-rat LD50:7500 μg/kg TXAPA9 4,621,62
ivn-rat LDLo:54 mg/kg BCPCA6 6,244,61
orl-gpg LD50:110 mg/kg TXAPA9 4,621,62
ipr-gpg LD50:12500 μg/kg TXAPA9 4,621,62

CONSENSUS REPORTS: Reported in EPA TSCA Inventory. EPA Extremely Hazardous Substances List.

SAFETY PROFILE: Poison by ingestion, inhalation, skin contact, intraperitoneal, and intravenous routes. Mutation data reported. An insecticide. When heated to decomposition it emits very toxic fumes of PO_x and SO_x. See also DEMETON-O + DEMETON-S and other demeton entries.

DAP600 CAS:17040-19-6 ***HR: 3***
DEMETON-S-METHYL-SULPHONE
mf: $C_6H_{15}O_5PS_2$ mw: 262.30

SYNS: BAYER 20315 ◇ DEMETON-S-METHYLSULFON (GERMAN) ◇ DEMETON-S-METHYLSULFONE ◇ O,O-DIMETHYL-S-(2-AETHYL-SULFONYL-AETHYL)-THIOLPHOSPHAT (GERMAN) ◇ O,O-DIMETHYL-S-(2-ETHSULFONYLETHYL)PHOSPHOROTHIOATE ◇ DIMETHYL-S-(2-ETHSULFONYLETHYL)THIOPHOSPHATE ◇ O,O-DIMETHYL-S-ETHYL-2-SULFONYLETHYL PHOSPHOROTHIOLATE ◇ O,O-DIMETHYL-S-ETHYLSULPHONYLETHYL PHOSPHOROTHIOLATE ◇ DIOXYDEMETON-S-METHYL ◇ E 158 ◇ ISOMETASYSTOX SULFONE ◇ ISOMETHYLSYSTOX SULFONE ◇ M 3/158 ◇ METAISOSYSTOX-SOLFON 20 315

TOXICITY DATA with REFERENCE
mmo-esc 5 μL/plate MUREAV 28,405,75
orl-rat LD50:40 mg/kg PAREAQ 11,636,59
ihl-rat LC50:195 mg/m^3/4H GUCHAZ 6,165,73
skn-rat LD50:500 mg/kg GUCHAZ 6,165,73
ipr-rat LD50:21 mg/kg GUCHAZ 6,165,73
ivn-rat LD50:22 mg/kg BIJOAK 67,187,57
orl-rbt LD50:50 mg/kg 85GYAZ -,20,71
orl-gpg LD50:120 mg/kg TXAPA9 4,621,62
ipr-gpg LD50:85 mg/kg TXAPA9 4,621,62

SAFETY PROFILE: Poison by ingestion, inhalation, intraperitoneal, and intravenous routes. Moderately toxic by skin contact. Mutation data reported. An insecticide. When heated to decomposition it emits very toxic fumes of PO_x and SO_x. See also DEMETON-O + DEMETON-S and other demeton entries.

DAP700 CAS:2955-38-6 ***HR: 3***
DEMETRIN
mf: $C_{19}H_{17}ClN_2O$ mw: 324.83

PROP: Crystals from methanol. Mp: 145-146°.

SYNS: CENTRAX ◇ 7-CHLORO-1-(CYCLOPROPYLMETHYL)-1,3-DIHYDRO-5-PHENYL-2H-1,4-BENZODIAZEPIN-2-ONE ◇ 7-CHLORO-1-CYCLOPROPYLMETHYL-5-PHENYL-1H-1,4-BENZODIAZEPIN-2(3H)-ONE ◇ K-373 ◇ LYSANXIA ◇ PRAZEPAM ◇ SEDAPRAN ◇ SETTIMA ◇ TREPIDAN ◇ VERSTRAN

TOXICITY DATA with REFERENCE
orl-mus TDLo:504 mg/kg (female 1-21D post):REP PLRCAT 9,325,77
orl-rat TDLo:10 g/kg (female 8-17D post):TER OYYAA2 15,797,78
orl-rat TDLo:5284 mg/kg/2Y-C:ETA TXAPA9 57,39,81
orl-rat TD:55 g/kg/2Y-C TXAPA9 57,39,81
orl-mus LD50:2300 mg/kg JKXXAF #79-92631
ipr-mus LD50:1020 mg/kg OYYAA2 15,241,78

SAFETY PROFILE: Moderately toxic by ingestion and intraperitoneal routes. Experimental teratogenic and reproductive effects. Questionable carcinogen with experimental tumorigenic data. Note: This is a controlled substance (depressant) listed in the U.S. Code of Federal Regulations, Title 21 Part 1308.14 (1985). A tranquilizer and muscle relaxant. When heated to decomposition it emits toxic fumes of Cl^- and NO_x.

DAP800 CAS:62-97-5 ***HR: 3***
DEMOTIL
mf: $C_{20}H_{24}N•CH_3O_4S$ mw: 389.55

SYNS: N,N-DIMETHYL-4-PIPERIDYLIDENE-1,1-DIPHENYLMETHANE METHYLSULFATE ◇ DIPHEMANIL ◇ DIPHEMANIL METHYLSULFATE ◇ DIPHENATIL ◇ DIPHENMANIL METHYLSULFATE ◇ DIPHENMETHANIL ◇ DIPHENMETHANIL METHYLSULFATE ◇ 4-(DIPHENYLMETHYLENE)-1,1-DIMETHYLPIPERIDINIUM METHYLSULFATE ◇ NIVELONA ◇ p-(α-PHENYLBENZYLIDENE)-1,1-DIMETHYLPIPERIDINIUM METHYL SULFATE ◇ PRANTAL ◇ PRANTAL METHYLSULFATE ◇ VAGOPHEMANIL ◇ VAGOPHEMANIL METHYL SULFATE ◇ VARITON

TOXICITY DATA with REFERENCE
orl-rat LD50:1107 mg/kg PSEBAA 78,576,51
ivn-rat LD50:5 mg/kg OYYAA2 23,461,82
orl-mus LD50:317 mg/kg PSEBAA 78,576,51
ipr-mus LD50:47 mg/kg PSEBAA 78,576,51
ivn-mus LD50:4012 μg/kg AIPTAK 103,100,55
ivn-dog LD50:42 mg/kg PSEBAA 78,576,51
orl-gpg LD50:404 mg/kg PSEBAA 78,576,51

SAFETY PROFILE: Poison by ingestion, intraperitoneal, and intravenous routes. An anticholinergic agent.

When heated to decomposition it emits very toxic fumes of SO_x and NO_x.

DAP810 CAS:8075-78-3 ***HR: D***
DEMULEN
mf: $C_{24}H_{32}O_4 \cdot C_{20}H_{24}O_2$ mw: 681.00

SYNS: BISECURIN II ◇ CONOVA 30 ◇ ETHYNODIOL DIACETATE mixture with ETHYNYLESTRADIOL ◇ ETHYNYLESTRADIOL mixture with ETHYNODIOL DIACETATE ◇ OVAMIN 30 ◇ OVULEN 30 ◇ SC 11800 EE

TOXICITY DATA with REFERENCE
orl-wmn TDLo:420 μg/kg (20D pre):REP CCPTAY 26,23,82

SAFETY PROFILE: Human reproductive effects by ingestion: menstrual cycle changes or disorders. Other experimental reproductive effects. When heated to decomposition it emits acrid smoke and fumes.

DAP815 ***HR: 3***
DENDROASPIS ANGUSTICEPS VENOM

SYN: VENOM, SNAKE, DENDROASPIS ANGUSTICEPS

TOXICITY DATA with REFERENCE
ipr-mus LD50:2800 μg/kg 19DDA6 1,283,67
scu-mus LD50:3320 μg/kg 19DDA6 1,223,67
ivn-mus LD50:381 μg/kg TOXIA6 2,5,64
ivn-mam LD50:450 μg/kg CLPTAT 8,849,67

SAFETY PROFILE: Poison by subcutaneous, intravenous, and intraperitoneal routes.

DAP820 ***HR: 3***
DENDROASPIS JAMESONI VENOM

SYNS: D. JAMESONI VENOM ◇ VENOM, SNAKE, DENDROASPIS JAMESONI

TOXICITY DATA with REFERENCE
ipr-mus LD50:440 μg/kg TOXIA6 13,295,75
scu-mus LD50:1020 μg/kg 19DDA6 1,223,67
ivn-mus LD50:840 μg/kg 19DDA6 1,223,67

SAFETY PROFILE: Poison by subcutaneous, intravenous, and intraperitoneal routes.

DAP825 CAS:35306-34-4 ***HR: 3***
DENDROBINE HYDROCHLORIDE
mf: $C_{16}H_{25}NO_2 \cdot ClH$ mw: 299.88

SYN: DENDROBAN-12-ONE HYDROCHLORIDE

TOXICITY DATA with REFERENCE
ivn-rat LDLo:20 mg/kg JPETAB 55,319,35
ivn-mus LDLo:20 mg/kg JPETAB 55,319,35
ivn-rbt LDLo:17 mg/kg JPETAB 55,319,35
ivn-gpg LDLo:22 mg/kg JPETAB 55,319,35

SAFETY PROFILE: Poison by intravenous route. When heated to decomposition it emits toxic fumes of NO_x and HCl.

DAP840 ***HR: 2***
DENDROCALAMUS MEMBRANACEUS Munro, extract excluding roots

PROP: Indian plant belonging to the family *Poaceae* IJEBA6 22,312,84

TOXICITY DATA with REFERENCE
orl-ham TDLo:500 mg/kg (female 1-5D post):REP IJEBA6 22,312,84
ipr-mus LD50:825 mg/kg IJEBA6 22,312,84

SAFETY PROFILE: Moderately toxic by intraperitoneal route. Experimental reproductive effects. When heated to decomposition it emits acrid smoke and irritating fumes.

DAP850 CAS:71771-90-9 ***HR: 2***
DENOPAMINE
mf: $C_{18}H_{23}NO_4$ mw: 317.42

SYNS: BENZENEMETHANOL, α-(((2-(3,4-DIMETHOXYPHENYL) ETHYL)AMINO)METHYL)-4-HYDROXY-, (R)- ◇ (-)-(R)-1-(p-HYDROXY-PHENYL)-2-((3,4-DIMETHOXYPHENETHYL)AMINO)ETHANOL ◇ TA 064

TOXICITY DATA with REFERENCE
orl-rat TDLo:810 mg/kg (female 17-21D post):REP OYYAA2 32,769,86
orl-rbt TDLo:1300 mg/kg (female 6-18D post):TER OYYAA2 32,769,86
orl-rat LD50:9369 mg/kg OYYAA2 32,751,86
ipr-rat LD50:1785 mg/kg OYYAA2 32,751,86

SAFETY PROFILE: Moderately toxic by intraperitoneal route. Experimental reproductive effects. When heated to decomposition it emits toxic fumes of NO_x.

DAP875 CAS:26166-37-0 ***HR: 3***
DENUDATINE
mf: $C_{22}H_{33}NO_2$ mw: 343.56

SYN: 16,17-DIDEHYDRO-21-ETHYL-4-METHYL-7,20-CYCLOATIDANE-11-β,15-β-DIOL

TOXICITY DATA with REFERENCE
ivn-rat LD50:130 mg/kg CYLPDN 3,104,82
orl-mus LD50:290 mg/kg CYLPDN 3,104,82
ivn-mus LD50:128 mg/kg CYLPDN 3,104,82

SAFETY PROFILE: Poison by ingestion and intravenous routes. When heated to decomposition it emits toxic fumes of NO_x.

DAP880 CAS:77234-90-3 ***HR: 3***
DENZIMOL HYDROCHLORIDE
mf: $C_{19}H_{20}N_2O \cdot ClH$ mw: 328.87

SYNS: α-(p-PHENETHYLPHENYL)-1-IMIDAZOLEETHANOL MONOHYDROCHLORIDE ◇ N-(β-(4-(β-PHENYLETHYL)PHENYL)-β-HYDROXYETHYL)IMIDAZOLE HYDROCHLORIDE ◇ REC 15-1533

TOXICITY DATA with REFERENCE
ipr-rat LD50:332 mg/kg ARZNAD 33,1155,83
ivn-rat LD50:22 mg/kg ARZNAD 33,1155,83
orl-mus LD50:434 mg/kg JMCMAR 24,727,81
ipr-mus LD50:246 mg/kg ARZNAD 33,1168,83
ivn-mus LD50:46 mg/kg ARZNAD 33,1168,83
ivn-rbt LD50:22400 μg/kg ARZNAD 33,1168,83

SAFETY PROFILE: Poison by intravenous and intraperitoneal routes. Moderately toxic by ingestion. When heated to decomposition it emits toxic fumes of NO_x and HCl.

DAP900 CAS:8044-51-7 ***HR: 1***
DEOBASE

SYNS: DEODORIZED KEROSENE ◇ DEODORIZED KEROSINE

TOXICITY DATA with REFERENCE
skn-rbt 500 mg/24H MLD FCTOD7 20(Suppl),699,82
eye-rbt 100 mg/24H MLD FCTOD7 20(Suppl),699,82
orl-rat LD50:45 g/kg AMIHBC 2,420,50
orl-rbt LD50:8840 mg/kg AMIHBC 2,420,50

SAFETY PROFILE: Mildly toxic by ingestion. An eye and skin irritant. See also KEROSENE.

DAQ000 CAS:73825-59-9 ***HR: 3***
11-DEOXO-12-β,13-α-DIHYDRO-11-α-HYDROXYJERVINE
mf: $C_{26}H_{40}NO_3$ mw: 414.67

SYN: 12-β,13-α-DIHYDROJERVINE-11-α-OL

TOXICITY DATA with REFERENCE
orl-ham TDLo:113 mg/kg (7D preg):REP JAFCAU 26,561,78
orl-ham TDLo:150 mg/kg (7D preg):TER JAFCAU 26,561,78
orl-ham LDLo:150 mg/kg JAFCAU 26,561,78

SAFETY PROFILE: Poison by ingestion. Experimental teratogenic and reproductive effects. When heated to decomposition it emits toxic fumes of NO_x. See also JERVINE.

DAQ002 CAS:51340-26-2 ***HR: 3***
11-DEOXO-12-β,13-α-DIHYDRO-11-β-HYDROXYJERVINE
mf: $C_{26}H_{40}NO_3$ mw: 414.67

SYNS: 12-β,13-α-DIHYDRO-11-DEOXO-11-β-HYDROXYJERVINE ◇ 12-β,13-α-DIHYDROJERVINE-11-β-OL ◇ JERVINE, 11-DEOXO-12-β,13-α-DIHYDRO-11-β-HYDROXY-

TOXICITY DATA with REFERENCE
orl-ham TDLo:150 mg/kg (female 7D post):REP 41CIAR -,409,78
orl-ham LDLo:150 mg/kg JAFCAU 26,561,78

SAFETY PROFILE: Poison by ingestion. Experimental reproductive effects. When heated to decomposition it emits toxic fumes of NO_x.

DAQ100 ***HR: 3***
9-DEOXO-16,16-DIMETHYL-9-METHYLENE-PGE77
mf: $C_{23}H_{38}O_4$ mw: 378.61

SYNS: 9-DEOXO-16,16-DIMETHYL-9-METHYLENEPROSTAGLANDIN E2 ◇ 7-(5,5-DIMETHYL-3-HYDROXY-2-(3-HYDROXY-1-OCTENYL)CYCLOPENTYL)-5-HEPTENOIC ACID

TOXICITY DATA with REFERENCE
ivg-wmn TDLo:800 μg/kg (13W preg):REP CCPTAY 22,153,80

SAFETY PROFILE: Human reproductive effects by intravaginal route: changes in the uterus, cervix, and vagina; terminates pregnancy. A steroid. When heated to decomposition it emits acrid smoke and fumes. See various prostaglandins.

DAQ110 CAS:85559-57-5 ***HR: 3***
11-DEOXOGLYCYRRHETINIC ACID HYDROGEN MALEATE SODIUM SALT
mf: $C_{34}H_{51}O_6 \cdot 7Na$

TOXICITY DATA with REFERENCE
ipr-rat LD50:67800 μg/kg EPXXDW #69380
scu-rat LD50:98300 μg/kg EPXXDW #69380
ivn-rat LD50:54200 μg/kg EPXXDW #69380

SAFETY PROFILE: Poison by subcutaneous, intravenous, and intraperitoneal routes. When heated to decomposition it emits toxic fumes of Na_2O.

DAQ125 CAS:14410-98-1 ***HR: D***
11-DEOXOJERVINE-4-EN-3-ONE
mf: $C_{27}H_{41}NO_2$ mw: 411.69

SYN: 3′,6′,10,11b-TETRAMETHYL-2,3′a,4′,5,5′,6,6′,6a,6b,7,7′,7′a,8,11,11a,11b-HEXADECAHYDRO-SPIRO(9H-BENZO(a)FLUORENE-9,2′(3′H)-FURO(3,2-b)PYRIDIN)-3(1H)-ONE

TOXICITY DATA with REFERENCE
orl-ham TDLo:15 mg/kg (7D preg):REP 41CIAR -,409,78
orl-ham TDLo:38 mg/kg (7D preg):TER 41CIAR -,409,78

SAFETY PROFILE: Experimental teratogenic and reproductive effects. When heated to decomposition it emits toxic fumes of NO_x. See also JERVINE.

DAQ200 CAS:958-09-8 ***HR: D***
DEOXYADENOSINE
mf: $C_{10}H_{13}N_5O_3$ mw: 251.28

SYNS: ADENINE DEOXYRIBONUCLEOSIDE ◇ ADENINE DEOXYRIBOSE ◇ ADENYLDEOXYRIBOSIDE ◇ 2'-DEOXYADENOSINE ◇ DESOXYADENOSINE

TOXICITY DATA with REFERENCE
cyt-hmn:lym 6 mmol/L CHROAU 90,239,84
dni-ham:lng 1 mmol/L BICMBE 64,809,82
msc-ham:lng 1 mmol/L BICMBE 64,809,82

CONSENSUS REPORTS: EPA Genetic Toxicology Program. Reported in EPA TSCA Inventory.

SAFETY PROFILE: Human mutation data reported. When heated to decomposition it emits toxic fumes of NO_x.

DAQ225 CAS:73-03-0 ***HR: D***
3'-DEOXYADENOSINE
mf: $C_{10}H_{13}N_5O_3$ mw: 251.28

PROP: Needles from ethanol, n-butanol, n-propanol, or water. Mp: 225-226°, pH aq soln: 7.1.

SYNS: CORDYCEPIN ◇ CORDYCEPINE ◇ 9-CORDYCEPOSIDOADENINE

TOXICITY DATA with REFERENCE
oms-nml:lng 100 μmol/L CARYAB 28,301,75
dnd-hmn:fbr 500 μmol/L ENMUDM 7,267,85
dni-hmn:hla 20 mg/L JANTAJ 35,119,82
ipr-mus TDLo:80 mg/kg (7D preg):REP JOENAK 81,351,79

CONSENSUS REPORTS: EPA Genetic Toxicology Program.

SAFETY PROFILE: Experimental reproductive effects. Human mutation data reported. When heated to decomposition it emits toxic fumes of NO_x.

DAQ400 CAS:83-44-3 ***HR: 3***
DEOXYCHOLATIC ACID
mf: $C_{24}H_{40}O_4$ mw: 392.64

PROP: A white crystalline powder. Mp: 178°. Sol in alc, acetone; sltly sol in ether and chloroform; insol in water.

SYNS: CHOLEIC ACID ◇ CHOLEREBIC ◇ CHOLOREBIC ◇ DEGALOL ◇ DEOXYCHOLIC ACID (FCC) ◇ 7-α-DEOXYCHOLIC ACID ◇ DESOXYCHOLIC ACID ◇ DESOXYCHOLSAEURE (GERMAN) ◇ 3,12-DIHYDROXYCHOLANIC ACID ◇ 3-α,12-α-DIHYDROXYCHOLANIC ACID ◇ 3-α,12-α-DIHYDROXY-5-β-CHOLAN-24-OIC ACID ◇ 3-α,12-α-DIHYDROXY-5-β-CHOLANOIC ACID ◇ 3-α,12-α-DIHYDROXYCHOLANSAEURE (GERMAN) ◇ DROXOLAN ◇ 17-β-(1-METHYL-3-CARBOXYPROPYL)-ETIOCHOLANE-3-α,12-α-DIOL ◇ PYROCHOL ◇ SEPTOCHOL

TOXICITY DATA with REFERENCE
mmo-sat 20 mg/L MUREAV 158,45,85
sln-smc 100 mg/L CRNGDP 5,447,84
otr-ham:emb 7250 μg/L TOLED5 9,177,81
skn-mus TDLo:2700 mg/kg/10W-I:ETA BJCAAI 10,363,56
orl-rat LD50:1 g/kg NAIZAM 33,71,82
orl-mus LD50:1 g/kg NAIZAM 33,71,82
ipr-mus LD50:130 mg/kg ARZNAD 20,323,70
ivn-rbt LDLo:1000 mg/kg ZGEMAZ 52,779,26

CONSENSUS REPORTS: Reported in EPA TSCA Inventory.

SAFETY PROFILE: Poison by intraperitoneal route. Moderately toxic by ingestion and intravenous routes. Questionable carcinogen with experimental tumorigenic data. Mutation data reported. When heated to decomposition it emits acrid smoke and irritating fumes.

DAQ600 CAS:64-85-7 ***HR: 2***
11-DEOXYCORTICOSTERONE
mf: $C_{21}H_{30}O_3$ mw: 330.51

SYNS: CORTEXONE ◇ DESOXYCORTICOSTERONE ◇ DESOXYCORTONE ◇ 21-HYDROXYPREGN-4-ENE-3,20-DIONE ◇ 21-HYDROXYPROGESTERONE ◇ 4-PREGNEN-21-OL-3,20-DIONE

TOXICITY DATA with REFERENCE
ims-rat TDLo:14500 μg/kg (7D pre/1-22D preg):REP JCINAO 41,710,62
unr-mus LD50:1000 mg/kg JMCMAR 7,673,64

SAFETY PROFILE: Moderately toxic by unspecified route. Experimental reproductive effects. A steroid. When heated to decomposition it emits acrid smoke and irritating fumes.

DAQ800 CAS:56-47-3 ***HR: 3***
11-DEOXYCORTICOSTERONE ACETATE
mf: $C_{23}H_{32}O_4$ mw: 372.55

SYNS: 21-ACETOXY-3,20-DIKETOPREGN-4-ENE ◇ CORTACET ◇ CORTATE ◇ CORTENIL ◇ CORTESAN ◇ CORTEXONE ACETATE ◇ CORTIFAR ◇ CORTIGEN ◇ CORTINAQ ◇ CORTIRON ◇ CORTIVIS ◇ CORTIXYL ◇ DCA ◇ DECORTIN ◇ DECORTON ◇ DECOSTERONE ◇ DECOSTRATE ◇ 11-DEOXYCORTICOSTERONE-21-ACETATE ◇ DEOXYCORTONE ACETATE ◇ DESCORTERONE ◇ DESCOTONE ◇ DESOXYCORTICOSTERONE ACETATE ◇ DESOXYCORTONE ACETATE ◇ DOCA ◇ DOCA ACETATE ◇ DOC-AC ◇ DOC ACETATE ◇ DORCOSTRIN ◇ DOXO ◇ 21-HYDROXYPREGN-4-ENE-3,20-DIONE-21-ACETATE ◇ KRINOCORTS ◇ OCRITEN ◇ ORGANON'S DOCA ACETATE ◇ PERCORTEN ◇ PERCOTOL ◇ 4-PREGNENE-3,20-DIONE-21-OL ACETATE ◇ PRIMOCORT ◇ PRIMOCORTAN ◇ SINCORTEX ◇ STERAQ ◇ SYNCORT ◇ SYNCORTA ◇ SYNCORTYL ◇ UNIDOCAN

TOXICITY DATA with REFERENCE
par-rat TDLo:316 mg/kg (female 56D pre):REP PSEBAA 120,238,65
scu-mus TDLo:520 mg/kg/13W-I:ETA PSEBAA 83,14,53

SAFETY PROFILE: Questionable carcinogen with experimental tumorigenic data. Experimental reproductive effects. A steroid. When heated to decomposition it emits acrid smoke and irritating fumes.

DAQ850 CAS:951-77-9 ***HR: D***
DEOXYCYTIDINE
mf: $C_9H_{13}N_3O_4$ mw: 227.25

SYNS: CYTOSINE DEOXYRIBOSIDE ◇ dCYD ◇ 2′-DEOXYCYTIDINE ◇ DEOXYRIBONUCLEOSIDE CYTOSINE ◇ DEOXYRIBOSE CYTIDINE ◇ DESOXYCYTIDIN (GERMAN)

TOXICITY DATA with REFERENCE
dns-mus:leu 3 μmol/L CALEDQ 19,147,83
dni-ham:lng 1 mmol/L BICMBE 64,809,82
ipr-mus TDLo:800 mg/kg (9D preg):REP DZZEA7 32,861,77

CONSENSUS REPORTS: EPA Genetic Toxicology Program. Reported in EPA TSCA Inventory.

SAFETY PROFILE: Experimental reproductive effects. Mutation data reported. When heated to decomposition it emits toxic fumes of NO_x.

DAR000 CAS:55297-96-6 ***HR: 3***
14-DEOXY-14-((2-DIETHYLAMINOETHYL) MERCAPTOACETOXY)-MUTILIN HYDROGEN FUMARATE
mf: $C_{28}H_{47}NO_4S•C_4H_4O_4$ mw: 609.90

SYNS: SQ 22947 ◇ TIAMUTIN

TOXICITY DATA with REFERENCE
orl-mus LD50:841 mg/kg AMACCQ 7,507,75
scu-mus LD50:521 mg/kg AMACCQ 7,507,75
orl-ckn LD50:1860 mg/kg AMACCQ 7,507,75
ims-ckn LD50:270 mg/kg AMACCQ 7,507,75
orl-trk LD50:1345 mg/kg AMACCQ 7,507,75

SAFETY PROFILE: Poison by intramuscular route. Moderately toxic by ingestion and subcutaneous routes. When heated to decomposition it emits very toxic fumes of NO_x and SO_x.

DAR100 CAS:4298-16-2 ***HR: 3***
dl-DEOXYEPHEDRINE HYDROCHLORIDE
mf: $C_{10}H_{15}N•ClH$ mw: 185.72

SYNS: dl-DESOXYEPHEDRINE HYDROCHLORIDE ◇ dl-METHAMPHETAMINE HYDROCHLORIDE ◇ dl-N-METHYL-β-PHENYLISOPROPYLAMINE HYDROCHLORIDE

TOXICITY DATA with REFERENCE
ipr-rat LD50:54480 μg/kg JAPMA8 37,223,48
orl-mus LD50:143 mg/kg JAPMA8 37,223,48
ipr-mus LD50:61160 μg/kg JAPMA8 37,223,48

SAFETY PROFILE: Poison by ingestion and intraperitoneal routes. A powerful central nerous system stimulant. When heated to decomposition it emits toxic fumes of NO_x and HCl. See also various amphetamines.

DAR150 CAS:447-25-6 ***HR: 3***
6-DEOXY-6-FLUOROGLUCOSE
mf: $C_6H_{11}FO_5$ mw: 182.17

TOXICITY DATA with REFERENCE
orl-rat TDLo:120 mg/kg (1D male):REP CCPTAY 25,535,82
orl-rat LDLo:240 mg/kg CCPTAY 25,535,82
orl-mus LDLo:480 mg/kg CCPTAY 25,535,82

SAFETY PROFILE: Poison by ingestion. Experimental reproductive effects. When heated to decomposition it emits toxic fumes of F^-.

DAR200 CAS:25526-93-6 ***HR: D***
3′-DEOXY-3′-FLUOROTHYMIDINE
mf: $C_{10}H_{13}FN_2O_4$ mw: 244.25

SYN: 3′-FLUORO-3′-DEOXYTHYMIDINE

TOXICITY DATA with REFERENCE
cyt-mus:lym 10 mmol/L/12H MUREAV 40,101,76
cyt-mus:ast 10 μmol/L PHARAT 32,178,77

SAFETY PROFILE: Mutation data reported. When heated to decomposition it emits very toxic fumes of F^- and NO_x. See also FLUORIDES.

DAR400 CAS:50-91-9 ***HR: 3***
2′-DEOXY-5-FLUOROURIDINE
mf: $C_9H_{11}FN_2O_5$ mw: 246.22

SYNS: DEOXYFLUOROURIDINE ◇ 1-β-d-2′-DEOXYRIBOFURANOSYL-5-FLUROURACIL ◇ FDUR ◇ FLOXURIDIN ◇ FLOXURIDINE ◇ FLUORODEOXYURIDINE ◇ β-5-FLUORO-2′-DEOXYURIDINE ◇ 5-FLUORODEOXYURIDINE ◇ 5-FLUORO-2-DEOXYURIDINE ◇ 5-FLUORO-2′-DEOXYURIDINE ◇ 5-FLUOROURACIL DEOXYRIBOSIDE ◇ 5-FLUOROURACIL-2′-DEOXYRIBOSIDE ◇ FLUORURIDINE DEOXYRIBOSE ◇ FUDR ◇ 5-FUDR ◇ NSC-27640 ◇ RO 5-0360

TOXICITY DATA with REFERENCE
dnd-hmn:leu 500 nmol/L CNREA8 43,5145,83
dni-hmn:oth 1 μmol/L CNREA8 42,3005,82
oms-hmn:oth 1 μmol/L CNREA8 42,3005,82
ipr-mus TDLo:10 mg/kg (female 16D post):REP TJADAB 29(2),5B,84
unr-mus TDLo:45 mg/kg (12D preg):TER AJANA2 137,87,73
ivn-hmn TDLo:5 mg/kg/14D-C:GIT CANCAR 34,972,74
par-wmn TDLo:173 mg/kg/82W-I:SKN JNMAAE 79,669,87
orl-rat LD50:215 mg/kg NCILB* NIH-NCI-E-C-72-3252,73
ipr-rat LD50:1600 mg/kg CPCHAO 18,307,62
orl-mus LD50:147 mg/kg NCILB* NIH-NCI-E-C-72-3252,73

CONSENSUS REPORTS: EPA Genetic Toxicology Program.

SAFETY PROFILE: Poison by ingestion. Moderately toxic by intraperitoneal route. An experimental teratogen. Other experimental reproductive effects. Human systemic effects: hypermotility, diarrhea, nausea, vomiting and other gastrointestinal effects, allergic dermatitis, and bone marrow changes. Human mutation data reported. When heated to decomposition it emits very toxic fumes of F^- and NO_x.

DAR600 CAS:154-17-6 ***HR: 3***
2-DEOXYGLUCOSE
mf: $C_6H_{12}O_5$ mw: 164.18

PROP: Crystals from acetone or butanone. Mp: 142-144°. α-Form: Crystals from isopropanol. Mp: 134-136°.

SYNS: 2-DEOXY-d-ARABINO-HEXOSE ◇ 2-DEOXY-3-ARABINO-HEXOSE ◇ d-2-DEOXYGLUCOSE ◇ 2-DEOXY-d-GLUCOSE ◇ 2-DESOXY-d-GLUCOSE (FRENCH) ◇ 2-DG ◇ NSC 15193

TOXICITY DATA with REFERENCE
orl-rat TDLo:4 g/kg (7-14D preg):TER 85DJA5 -,95,71
orl-rat TDLo:2 g/kg (7-8D preg):REP 85DJA5 -,95,71
scu-rat LD50:250 mg/kg APFRAD 39,327,81
ipr-rat LD50:2000 mg/kg JPPMAB 17,814,65

CONSENSUS REPORTS: Reported in EPA TSCA Inventory.

SAFETY PROFILE: Poison by subcutaneous route. Moderately toxic by intraperitoneal route. An experimental teratogen. Other experimental reproductive effects. When heated to decomposition it emits acrid smoke and fumes.

DAR800 CAS:961-07-9 ***HR: D***
2'-DEOXYGUANOSINE
mf: $C_{10}H_{13}N_5O_4$ mw: 267.28

SYNS: DEOXYGUANOSINE ◇ GUANINE DEOXYRIBOSIDE

TOXICITY DATA with REFERENCE
dni-ham:lng 1 mmol/L BICMBE 64,809,82
cyt-ham:fbr 500 μmol/L CYTOAN 49,667,84
sce-ham:fbr 500 μmol/L CYTOAN 49,667,84
dnd-mam:lym 50 mmol/L PNASA6, 48,686,62

CONSENSUS REPORTS: Reported in EPA TSCA Inventory. EPA Genetic Toxicology Program.

SAFETY PROFILE: Mutation data reported. When heated to decomposition it emits toxic fumes of NO_x.

DAS000 CAS:54-42-2 ***HR: 3***
2'-DEOXY-5-IODOURIDINE
mf: $C_9H_{11}IN_2O_5$ mw: 354.12

SYNS: ALLERGAN 211 ◇ DENDRID ◇ 1-(2-DEOXY-β-d-RIBOFURANOSYL)-5-IODOURACIL ◇ 1-β-d-2'-DEOXYRIBOFURANOSYL-5-IODOURACIL ◇ EMANIL ◇ HERPESIL ◇ HERPIDU ◇ HERPLEX ◇ HERPLEX LIQUIFILM ◇ IDEXUR ◇ IDOXENE ◇ IDOXURIDIN ◇ IDOXURIDINE ◇ IDU ◇ IDUCHER ◇ IDULEA ◇ IDUOCULOS ◇ IDUR ◇ IDURIDIN ◇ 5-IODODEOXYURIDINE ◇ 5-IODO-2'-DEOXYURIDINE ◇ 5-IODOURACIL DEOXYRIBOSIDE ◇ IUDR ◇ 5-IUDR ◇ JODDEOXIURIDIN ◇ KERECID ◇ NSC 39661 ◇ OPHTHALMADINE ◇ SK&F 14287 ◇ STOXIL ◇ SYNMIOL

TOXICITY DATA with REFERENCE
sce-hmn:lym 50 mg/L BMJOAE 283,817,81
sce-hmn:fbr 50 mg/L BMJOAE 283,817,81
msc-hmn:lym 100 μmol/L LIFSAK 19,563,76
msc-ham:lng 1 mg/L CRNGDP 6,1207,85
dni-rbt:kdy 1 mg/L JMCMAR 24,390,81
scu-mus TDLo:600 mg/kg (female 16-18D post):REP TJADAB 11,103,75
ocu-rbt TDLo:1350 mg/kg (female 6-18D post):TER AROPAW 93,46,75
ipr-mus TDLo:100 mg/kg/8W-I:CAR EXPEAM 29,1132,73
ipr-rat LD50:4000 mg/kg ADTEAS 3,181,68
ipr-mus LD50:1000 mg/kg JJIND8 62,911,79

CONSENSUS REPORTS: Reported in EPA TSCA Inventory. EPA Genetic Toxicology Program.

SAFETY PROFILE: Moderately toxic by intraperitoneal route. Experimental teratogenic and reproductive effects. Questionable carcinogen with experimental carcinogenic data. Human mutation data reported. When heated to decomposition it emits very toxic fumes of I^- and NO_x.

DAS400 CAS:10356-92-0 ***HR: 3***
1-DEOXY-1-(N-NITROSOMETHYLAMINO)-d-GLUCITOL
mf: $C_7H_{16}N_2O_6$ mw: 224.25

SYNS: 1-DEOXY-1-(METHYLNITROSAMINO)-d-GLUCITOL ◇ 1-N-METHYL-N-NITROSAMINO-1-DEOXY-d-GLUCITOLE ◇ 1-(N-METHYL-N-NITROSOAMINO)-1-DEOXY-d-GLUCITOL ◇ 1-N-METHYL-N-NITROSOAMINO-1-DESOXY-d-GLUCIT(GERMAN)

TOXICITY DATA with REFERENCE
orl-rat TDLo:2500 mg/kg/25W-I:ETA ZEKBAI 75,296,71

SAFETY PROFILE: Questionable carcinogen with experimental tumorigenic data. Many N-nitroso compounds are carcinogens. When heated to decomposition it emits toxic fumes of NO_x. See also NITROSAMINES.

DAS500 CAS:62422-00-8 ***HR: D***
7-DEOXYNOGALAROL
mf: $C_{29}H_{31}NO_{11}$ mw: 569.61

TOXICITY DATA with REFERENCE
dni-hmn:oth 2040 nmol/L HXPHAU 38(Pt 2),623,75
oms-hmn:oth 2040 nmol/L HXPHAU 38(Pt 2),623,75
dnd-mam:lym 12 μmol/L CBINA8 36,1,81

SAFETY PROFILE: Human mutation data reported.

When heated to decomposition it emits toxic fumes of NO_x.

DAS600 CAS:37636-51-4 **HR: 3**
3'-DEOXYPAROMOMYCIN I
mf: $C_{23}H_{45}N_5O_{13}$ mw: 599.73

SYNS: LIVIDOMYCIN B ◇ QUINTOMYCIN D

TOXICITY DATA with REFERENCE
orl-mus LD50:10 g/kg YKYUA6 31,1085,80
scu-mus LD50:1245 μg/kg YKYUA6 31,1085,80
ivn-mus LD50:123 mg/kg 85ERAY 1,682,78
ims-mus LD50:1343 μg/kg YKYUA6 31,1085,80

SAFETY PROFILE: Poison by intravenous, intramuscular, and subcutaneous routes. Mildly toxic by ingestion. When heated to decomposition it emits toxic fumes of NO_x.

DAS800 CAS:56530-49-5 **HR: 1**
12-DEOXY-PHORBOL-20-ACETATE-13-DODECANOATE
mf: $C_{34}H_{52}O_7$ mw: 572.86

SYN: 12-DEOXY-PHORBOL-13-DODECANOATE-20-ACETATE

TOXICITY DATA with REFERENCE
skn-mus 500 ng/4H APTOA6 37,250,75
skn-mus 2500 ng/24H APTOA6 37,250,75

SAFETY PROFILE: A skin irritant. When heated to decomposition it emits acrid smoke and irritating fumes.

DAT000 CAS:25090-71-5 **HR: 1**
12-DEOXYPHORBOL-20-ACETATE-13-ISOBUTYRATE
mf: $C_{26}H_{36}O_7$ mw: 460.62

SYN: 12-DEOXYPHORBOL-13-ISOBUTYRATE-20-ACETATE

TOXICITY DATA with REFERENCE
skn-mus 500 ng/4H APTOA6 37,250,75
skn-mus 6400 ng/24H APTOA6 37,250,75

SAFETY PROFILE: A skin irritant. When heated to decomposition it emits acrid smoke and irritating fumes.

DAT200 CAS:25090-73-7 **HR: 1**
12-DEOXY-PHORBOL-20-ACETATE-13-(2-METHYLBUTYRATE)
mf: $C_{27}H_{38}O_7$ mw: 474.65

SYN: 12-DEOXY-PHORBOL-13-α-METHYLBUTYRATE-20-ACETATE

TOXICITY DATA with REFERENCE
skn-mus 390 ng OPEN ARTODN 44,279,80
skn-mus 2800 ng/24H APTOA6 37,250,75

SAFETY PROFILE: A skin irritant. When heated to decomposition it emits acrid smoke and irritating fumes.

DAT400 CAS:56602-09-6 **HR: 1**
12-DEOXY-PHORBOL-20-ACETATE-13-OCTENOATE
mf: $C_{30}H_{42}O_7$ mw: 514.72

SYN: 12-DEOXY-PHORBOL-13-OCTENOATE-20-ACETATE

TOXICITY DATA with REFERENCE
skn-mus 1800 ng/4H APTOA6 37,250,75
skn-mus 4000 ng/24H APTOA6 37,250,75

SAFETY PROFILE: A skin irritant. When heated to decomposition it emits acrid smoke and irritating fumes.

DAT600 CAS:25090-72-6 **HR: 1**
12-DEOXY-PHORBOL-20-ACETATE-13-TIGLATE
mf: $C_{27}H_{36}O_7$ mw: 472.63

SYN: 12-DEOXYPHORBOL-13-TIGLATE-20-ACETATE

TOXICITY DATA with REFERENCE
skn-mus 1800 ng/4H APTOA6 37,250,75
skn-mus 7800 ng/24H APTOA6 37,250,75

SAFETY PROFILE: A skin irritant. When heated to decomposition it emits acrid smoke and irritating fumes.

DAT800 CAS:69883-99-4 **HR: 1**
12-DEOXYPHORBOL-13-(4-ACETOXY-PHENYLACETATE)-20-ACETATE
mf: $C_{32}H_{38}O_9$ mw: 566.4

TOXICITY DATA with REFERENCE
skn-mus 130 ng OPEN ARTODN 44,279,80

SAFETY PROFILE: A skin irritant. When heated to decomposition it emits acrid smoke and irritating fumes.

DAU000 CAS:65700-60-9 **HR: 1**
12-DEOXYPHORBOL-13-ANGELATE
mf: $C_{25}H_{34}O_6$ mw: 430.59

TOXICITY DATA with REFERENCE
skn-mus 720 ng OPEN ARTODN 44,279,80

SAFETY PROFILE: A skin irritant. When heated to decomposition it emits acrid smoke and irritating fumes.

DAU200 CAS:65700-59-6 **HR: 3**
12-DEOXYPHORBOL-13-ANGELATE-20-ACETATE
mf: $C_{27}H_{36}O_7$ mw: 472.63

TOXICITY DATA with REFERENCE
skn-mus 3 μg OPEN ARTODN 44,279,80
skn-mus TDLo:180 mg/kg/24W-I:ETA EXPEAM 30,1438,74

SAFETY PROFILE: Questionable carcinogen with experimental tumorigenic data. A skin irritant. When heated to decomposition it emits acrid smoke and irritating fumes.

DAU400 CAS:56726-04-6 ***HR: 1***
12-DEOXY-PHORBOL-13-DECDIENOATE-20-ACETATE
mf: $C_{32}H_{44}O_7$ mw: 540.76

SYN: 12-DEOXY-PHORBOL-20-ACETATE-13-DECDIENOATE

TOXICITY DATA with REFERENCE
skn-mus 1100 ng/4H APTOA6 37,250,75
skn-mus 3500 ng/24H APTOA6 37,250,75

SAFETY PROFILE: A skin irritant. When heated to decomposition it emits acrid smoke and irritating fumes.

DAU600 CAS:56530-47-3 ***HR: 1***
12-DEOXY-PHORBOL-13-DODECANOATE
mf: $C_{32}H_{50}O_6$ mw: 530.82

TOXICITY DATA with REFERENCE
skn-mus 110 ng/4H APTOA6 37,250,75
skn-mus 280 ng/24H APTOA6 37,250,75

SAFETY PROFILE: A skin irritant. When heated to decomposition it emits acrid smoke and irritating fumes.

DAW200 CAS:28152-97-8 ***HR: 1***
12-DEOXY-PHORBOL-13-α-METHYLBUTYRATE
mf: $C_{25}H_{36}O_6$ mw: 432.61

TOXICITY DATA with REFERENCE
skn-mus 400 ng/4H APTOA6 37,250,75
skn-mus 1600 ng/24H APTOA6 37,250,75

SAFETY PROFILE: A skin irritant. When heated to decomposition it emits acrid smoke and irritating fumes. See also ESTERS.

DAY600 CAS:28152-96-7 ***HR: 1***
12-DEOXY-PHORBOL-13-TIGLATE
mf: $C_{25}H_{34}O_6$ mw: 430.59

TOXICITY DATA with REFERENCE
skn-mus 300 ng/4H APTOA6 37,250,75
skn-mus 600 ng/24H APTOA6 37,250,75

SAFETY PROFILE: A skin irritant. When heated to decomposition it emits acrid smoke and irritating fumes.

DAY800 CAS:61-67-6 ***HR: D***
4-DEOXYPYRIDOXAL
mf: $C_8H_{11}NO_2$ mw: 153.20

SYNS: DEOXYPYRIDOXINE ◇ 4-DEOXYPYRIDOXINE ◇ 4-DEOXYPYRIDOXOL ◇ DESOXYPYRIDOXINE ◇ 4,6-DIMETHYL-5-HYDROXY-3-PYRIDINEMETHANOL

TOXICITY DATA with REFERENCE
mmo-esc 500 mg/L/1H CRSUBM 3,69,55
orl-rat TDLo:400 mg/kg (16D pre):REP PSEBAA 68,274,48

SAFETY PROFILE: Experimental reproductive effects. Mutation data reported. When heated to decomposition it emits toxic fumes of NO_x.

DAY825 CAS:148-51-6 ***HR: 3***
4-DEOXYPYRIDOXOL HYDROCHLORIDE
mf: $C_8H_{11}NO_2$•ClH mw: 189.66

PROP: Crystals from alcohol + ether + acetone. Mp: 257°. Sol in water and alc.

SYNS: DESOXYPYRIDOXIME HYDROCHLORIDE ◇ 5-HYDROXY-4,6-DIMETHYL-3-PYRIDINEMETHANOLHYDROCHLORIDE ◇ NSC 3063

TOXICITY DATA with REFERENCE
orl-rat TDLo:11500 μg/kg (1-20D preg):TER SCIEAS 169,1329,70
orl-rat TDLo:11500 μg/kg (1-20D preg):REP SCIEAS 169,1329,70
ipr-mus LD50:150 mg/kg NTIS** AD691-490
orl-ckn LD50:1570 mg/kg JMCMAR 17,1235,74

SAFETY PROFILE: Poison by intraperitoneal route. Moderately toxic by ingestion. Experimental teratogenic and reproductive effects. When heated to decomposition it emits toxic fumes of NO_x and HCl.

DAY835 CAS:60504-57-6 ***HR: 3***
1-DEOXYPYRROMYCIN
mf: $C_{30}H_{35}NO_{10}$ mw: 569.66

SYNS: AKLAVIN ◇ ANTIBIOTIC MA 144T1

TOXICITY DATA with REFERENCE
dni-mus:leu 630 nmol/L JANTAJ 34,1596,81
oms-mus:leu 280 nmol/L JANTAJ 34,1596,81
ipr-mus LDLo:100 mg/kg 85FZAT -,111,67
ivn-mus LDLo:150 mg/kg 85FZAT -,111,67

CONSENSUS REPORTS: EPA Genetic Toxicology Program.

SAFETY PROFILE: Poison by intraperitoneal and intravenous routes. Mutation data reported. When heated to decomposition it emits toxic fumes of NO_x.

DAZ000 CAS:9007-49-2 ***HR: D***
DEOXYRIBONUCLEIC ACID

SYNS: CALF THYMUS DNA ◇ DNA

TOXICITY DATA with REFERENCE
dlt-dmg-par 1000 ppm FOBLAN 20,271,74
spm-dmg-par 1000 ppm FOBLAN 20,271,74
dns-nml-par 30 ng DEBIAO 43,189,75

SAFETY PROFILE: Mutation data reported. When heated to decomposition it emits toxic fumes of PO_x and NO_x.

DAZ100 CAS:30517-65-8 ***HR: D***
6-DEOXYVERSICOLORIN A
mf: $C_{18}H_{10}O_6$ mw: 322.28

SYN: (3aS-cis)-3a,12a-DIHYDRO-4,6-DIHYDROXY-ANTHRA(2,3-b) FURO(3,2-d)FURAN-5,10-DIONE(9CI)

TOXICITY DATA with REFERENCE
mmo-sat 500 nmol/plate MUREAV 143,121,85
mma-sat 500 nmol/plate MUREAV 143,121,85
dns-rat:lvr 500 nmol/L MUREAV 143,121,85

SAFETY PROFILE: Mutation data reported. When heated to decomposition it emits acrid smoke and fumes.

DAZ110 CAS:21987-62-2 ***HR: D***
DEPHOSPHATE BROMOFENOFOS
mf: $C_{12}H_6Br_4O_2$ mw: 501.82

SYNS: (1,1′-BIPHENYL)-2,2′-DIOL, 3,3′,5,5′-TETRABROMO- ◇ 2,2′-BIPHENYLDIOL, 3,3′,5,5′-TETRABROMO- ◇ 3,3′,5,5′-TETRABROMO-2,2′-BIPHENYLDIOL

TOXICITY DATA with REFERENCE
orl-rat TDLo:50 mg/kg (female 10D post):TER ARTODN 60,325,87
orl-rat TDLo:50 mg/kg (female 10D post):REP ARTODN 60,325,87

SAFETY PROFILE: An experimental teratogen. Other experimental reproductive effects. When heated to decomposition it emits toxic fumes of Br^-.

DAZ115 CAS:313-06-4 ***HR: D***
DEPOFEMIN
mf: $C_{26}H_{36}O_3$ mw: 396.62

PROP: Crystals from benzene + petr ether. Mp: 151-152°. Sol in ether, methanol, benzene, chloroform, peanut oil, cottonseed oil, corn oil, and sesame oil. The limit of solubility in the oils is about 400 mg/mL.

SYNS: DEPOESTRADIOL ◇ DEPOESTRADIOL CYPIONATE ◇ ECP ◇ ESTRADEP ◇ ESTRADIOL CYCLOPENTYLPROPIONATE ◇ ESTRADIOL-17-CYCLOPENTYLPROPIONATE ◇ ESTRADIOL-17-β-CYCLOPENTYLPROPIONATE ◇ ESTRADIOL-CYPIONATE ◇ ESTRADIOL-17-CYPIONATE ◇ ESTRADIOL-17-β-CYPIONATE ◇ (17-β)-ESTRA-1,3,5(10)-TRIENE-3,17-DIOL 17-CYCLOPENTANEPROPANOATE (9CI)

TOXICITY DATA with REFERENCE
orl-rbt TDLo:690 μg/kg (female 6-8D post):TER FESTAS 16,281,65
orl-ham TDLo:24 mg/kg (female 1-3D post):REP FESTAS 16,281,65

CONSENSUS REPORTS: Reported in EPA TSCA Inventory.

SAFETY PROFILE: An experimental teratogen. Other experimental reproductive effects. A steroid. When heated to decomposition it emits acrid smoke and fumes.

DAZ117 CAS:53-36-1 ***HR: 3***
DEPO-MEDRATE
mf: $C_{24}H_{32}O_6$ mw: 416.56

SYNS: DEPO-MEDROL ◇ DEPO-MEDRONE ◇ DEPO-METHYLPREDNISOLONE ◇ DEPO-METHYLPREDNISOLONE ACETATE ◇ DEPOT-MEDROL ◇ MEDROL ACETATE ◇ METHYLPREDNISOLONE ACETATE ◇ METHYLPREDNISOLONE 21-ACETATE ◇ 6-METHYLPREDNISOLONE ACETATE ◇ 6-α-METHYLPREDNISOLONE ACETATE ◇ MPA ◇ 11-β,17,21-TRIHYDROXY-6-α-METHYL-PREGNA-1,4-DIENE-3,20-DIONE 21-ACETATE ◇ U 8210 ◇ URBASON CRYSTAL SUSPENSION

TOXICITY DATA with REFERENCE
ims-rbt TDLo:1800 μg/kg (female 7-18D post):REP ANREAK 193,598,79
ims-rbt TDLo:1200 μg/kg (female 7-18D post):TER ANREAK 193,598,79
scu-rat LD50:265 mg/kg JTSCDR 10(Suppl 1),1,85
ipr-mus LD50:2145 mg/kg NIIRDN 6,833,82
scu-mus LD50:1320 mg/kg JTSCDR 10(Suppl 1,),1,85

SAFETY PROFILE: Poison by subcutaneous route. Moderately toxic by intraperitoneal route. An experimental teratogen. Other experimental reproductive effects. A steroid. When heated to decomposition it yields acrid smoke and fumes.

DAZ118 ***HR: 3***
(±)-DEPRENIL HYDROCHLORIDE
mf: $C_{13}H_{17}N{\bullet}ClH$ mw: 223.77

SYNS: (±)-N,α-DIMETHYL-N-2-PROPYNYLPHENETHYLAMINE HYDROCHLORIDE ◇ (±)-E-250 ◇ (±)-PHENYLISOPROPYLMETHYLPROPYNYLAMINE HYDROCHLORIDE

TOXICITY DATA with REFERENCE
scu-rat LD50:218 mg/kg APACAB 32,377,67
ivn-rat LD50:75 mg/kg APACAB 32,377,67

SAFETY PROFILE: Poison by subcutaneous, and intravenous routes. When heated to decomposition it emits toxic fumes of NO_x and HCl. See also AMINES.

DAZ120 CAS:4528-52-3 ***HR: 3***
(+)-DEPRENIL HYDROCHLORIDE
mf: $C_{13}H_{17}N{\bullet}ClH$ mw: 223.77

SYNS: (+)-DEPRENYL HYDROCHLORIDE ◇ (+)-N,α-DIMETHYL-N-2-PROPYNYLPHENETHYLAMINE HYDROCHLORIDE ◇ (+)-E-250 ◇ (+)-PHENYLISOPROPYLMETHYLPROPYNYLAMINE HYDROCHLORIDE

TOXICITY DATA with REFERENCE
scu-rat LD50:208 mg/kg APACAB 32,377,67
ivn-rat LD50:72500 μg/kg APACAB 32,377,67

SAFETY PROFILE: Poison by subcutaneous and intra-

venous routes. When heated to decomposition it emits toxic fumes of NO_x and HCl. See also AMINES.

DAZ125 CAS:14611-52-0 ***HR: 3***
(−)-DEPRENYL HYDROCHLORIDE
mf: $C_{13}H_{17}N{\bullet}ClH$ mw: 223.77

SYNS: 1-DEPRENIL HYDROCHLORIDE ◇ (−)-N,α-DIMETHYL-N-2-PROPYNYLBENZENEETHANAMINE HYDROCHLORIDE ◇ (−)-E-250 ◇ ELDEPRYL ◇ JUMEX ◇ (−)-PHENYLISOPROPYLMETHYLPROPYNYLAMINE

TOXICITY DATA with REFERENCE
scu-rat LD50:280 mg/kg APACAB 32,377,67
ivn-rat LD50:81 mg/kg APACAB 32,377,67
unr-mus LD50:121 mg/kg APACAB 32,377,67

SAFETY PROFILE: Poison by subcutaneous, intravenous, and possibly other routes. When heated to decomposition it emits toxic fumes of NO_x and HCl.

DAZ135 CAS:17124-74-2 ***HR: 3***
DEPT
mf: $C_{13}H_{21}N_3S{\bullet}2BrH$ mw: 413.27

SYNS: 2-DIETHYLAMINOETHYL-N'-PHENYLISOTHIURONIUM BROMIDE HYDROBROMIDE ◇ 2-(2-(DIETHYLAMINO)ETHYL)-1-PHENYL-2-THIOPSEUDOUREADIHYDROBROMIDE

TOXICITY DATA with REFERENCE
ipr-mus LD50:65900 μg/kg YKKZAJ 88,156,68
scu-mus LD50:100 mg/kg YKKZAJ 88,156,68
ivn-mus LD50:3500 μg/kg YKKZAJ 88,156,68

SAFETY PROFILE: Poison by subcutaneous, intravenous, and intraperitoneal routes. When heated to decomposition it emits toxic fumes of NO_x, SO_x, and HBr.

DAZ140 CAS:10139-98-7 ***HR: 3***
DEPTROPINE METHOBROMIDE
mf: $C_{24}H_{30}NO{\bullet}Br$ mw: 428.46

SYNS: BS 7020a ◇ 3-α-((10,11-DIHYDRO-5H-DIBENZO(a,d)CYCLOHEPTEN-5-YL)OXY)-8-METHYLTROPANIUMBROMIDE

TOXICITY DATA with REFERENCE
orl-rat LD50:800 mg/kg AIPTAK 192,105,71
ipr-rat LD50:7600 μg/kg AIPTAK 192,105,71
ivn-rat LD50:1200 μg/kg AIPTAK 102,105,71
orl-mus LD50:680 mg/kg AIPTAK 192,105,71
ivn-mus LD50:1150 μg/kg AIPTAK 192,105,71
orl-dog LD50:71 mg/kg AIPTAK 192,105,71
orl-rbt LD50:391 mg/kg AIPTAK 192,105,71

SAFETY PROFILE: Poison by ingestion, intravenous, and intraperitoneal routes. When heated to decomposition it emits toxic fumes of Br^- and NO_x.

DBA000 ***HR: 3***
DERRIS ELLIPTICA, root

PROP: The constituents of Derris are rotenone, dequelin, tephrosin and toxicarol.

SYNS: DEGUELIA ROOT ◇ DERRIS RESINS ◇ DERRIS ROOT ◇ TUBA ROOT

TOXICITY DATA with REFERENCE
orl-rat LDLo:100 mg/kg IECHAD 28,815,36
orl-mus LD50:350 mg/kg PSEBAA 54,140,43
orl-dog LDLo:100 mg/kg IECHAD 28,815,36
orl-rbt LDLo:200 mg/kg IECHAD 28,815,36
orl-gpg LDLo:75 mg/kg IECHAD 28,815,36

SAFETY PROFILE: Poison by ingestion. An insecticide. When heated to decomposition it emits acrid smoke and fumes. See also ROTENONE.

DBA175 CAS:49720-72-1 ***HR: 3***
DESACETYLCOLCHICINE-d-TARTRATE
mf: $C_{20}H_{23}NO_5{\bullet}C_4H_6O_6{\bullet}H_2O$ mw: 525.56

SYNS: DEACETYLCOLCHICINE l-TARTRATE ◇ N-DEACETYLCOLCHICINE l-TARTRATE(1:1), HYDRATE ◇ METHYL ETHER-TRIMETHYLCOLCHICINIC ACID-l-TARTRATE ◇ METHYL ETHER TRIMETHYLCOLCHICINIC ACID-d-TARTRATE, HYDRATE ◇ NCI 1136 ◇ NSC-36354 ◇ SKF 250 ◇ TMCA ◇ TMCA METHYL ESTER d-TARTRATE, HYDRATE

TOXICITY DATA with REFERENCE
orl-hmn TDLo:200 μg/kg:GIT,BLD NCISP* JAN86
par-rat LD50:3 mg/kg PMDCAY 9,1,73
ipr-mus LD50:70520 μg/kg NCISP* JAN86
par-mus LD50:8 mg/kg PMDCAY 9,1,73
par-dog LD50:9500 μg/kg PMDCAY 9,1,73

SAFETY PROFILE: Poison by parenteral and intraperitoneal routes. Human systemic effects by ingestion: nausea or vomiting, leukopenia and thrombocytopenia. When heated to decomposition it emits toxic fumes of NO_x. See also CHOLCHICINE.

DBA200 CAS:2731-16-0 ***HR: 3***
N-DESACETYLTHIOCOLCHICINE
mf: $C_{20}H_{23}NO_4S$ mw: 373.50

SYNS: CORPS R. 261 ◇ N-DEACETYLMETHYLTHIOCOLCHICINE ◇ N-DEACETYLTHIOCOLCHICINE ◇ N-DEACETYL-10-THIOCOLCHICINE ◇ R 261

TOXICITY DATA with REFERENCE
ivn-mky TDLo:6 mg/kg (13-14W preg):TER ACEDAB 166,435,72
ivn-hmn TDLo:71 μg/kg/D:GIT,BLD BAFEAG 42,308,55
ipr-rat LD50:175 mg/kg PSEBAA 98,479,58
ims-rat LD50:37 mg/kg PSEBAA 98 479,58
ice-rat LDLo:2 mg/kg AIPTAK 109,386,57
par-rat LDLo:2000 μg/kg FRPSAX 15,533,60
orl-mus LD50:50 mg/kg BAFEAG 42,308,55

ipr-mus LD50:66 mg/kg JMCMAR 24,636,81
ivn-mus LD50:80 mg/kg AIPTAK 107,150,56

SAFETY PROFILE: Poison by ingestion, intraperitoneal, intravenous, intramuscular, parenteral and intracerebral routes. An experimental teratogen. Human systemic effects by intravenous route: hypermotility, diarrhea, nausea, vomiting, and luekopenia (reduced white blood cell count). When heated to decomposition it emits very toxic fumes of NO_x and SO_x. See also COLCHICINE.

DBA250 CAS:57645-49-5 ***HR: 2***
DESBENZYL CLEBOPRIDE
mf: $C_{13}H_{18}ClN_3O_2$ mw: 283.79

SYN: 4-AMINO-5-CHLORO-2-METHOXY-N-(4-PIPERIDYL)BENZAMIDE

TOXICITY DATA with REFERENCE
orl-rat TDLo:3500 mg/kg (35D pre):REP OYYAA2 25,865,83
orl-rat LD50:2550 mg/kg OYYAA2 25,865,83
orl-mus LD50:1750 mg/kg OYYAA2 25,865,83

SAFETY PROFILE: Moderately toxic by ingestion. Experimental reproductive effects. When heated to decomposition it emits toxic fumes of Cl^- and NO_x.

DBA400 CAS:12728-25-5 ***HR: 3***
DESERTOMYCIN

PROP: An antibiotic produced by the strain *Streptomyces flavofungini* (85ERAY 2,1367,78).

TOXICITY DATA with REFERENCE
orl-mus LD50:12000 μg/kg 85ERAY 2,1367,78
ipr-mus LD50:2600 μg/kg 85ERAY 2,1367,78
scu-mus LD50:5300 μg/kg 85ERAY 2,1367,78
ivn-mus LD50:1350 μg/kg 85ERAY 2,1367,78

SAFETY PROFILE: Poison by ingestion, intraperitoneal, subcutaneous, and intravenous routes.

DBA450 ***HR: 3***
DESERT ROSE

PROP: A 6- to 10-foot tall shrub with a swollen trunk and branches. It has thick, dark green leaves and large pink or purple conical flowers. It is cultivated in hot and arid areas of the United States and Jamaica.

SYNS: ADENIUM (VARIOUS SPECIES) ◇ MOCK AZALEA

SAFETY PROFILE: The toxin is a cardiac glycoside similar to digitalis and found throughout the plant. No cases of poisoning have been reported in the United States, however, cardiac glycosides can cause death by their effect on heart function. See also DIGITALIS.

DBA475 ***HR: 3***
DESETHYLAPRINDINE HYDROCHLORIDE
mf: $C_{20}H_{26}N_2•ClH$ mw: 330.94

SYNS: AC 2197 HYDROCHLORIDE ◇ N-ETHYL-N′-2-INDANYL-N′-PHENYL-1,3-PROPANEDIAMINEHYDROCHLORIDE

TOXICITY DATA with REFERENCE
orl-rat LD50:525 mg/kg OYYAA2 27,353,84
ivn-rat LD50:22100 μg/kg OYYAA2 27,353,84
orl-mus LD50:543 mg/kg OYYAA2 27,353,84
ivn-mus LD50:24500 μg/kg OYYAA2 27,353,84

SAFETY PROFILE: Poison by intravenous route. Moderately toxic by ingestion. When heated to decomposition it emits toxic fumes of NO_x and HCl.

DBA500 ***HR: 2***
DESGLUCODIGITONIN

TOXICITY DATA with REFERENCE
ipr-mus LD10: 20 mg/kg:ETA PCJOAU 11,749,77

SAFETY PROFILE: Questionable carcinogen with experimental tumorigenic data. When heated to decomposition it emits acrid smoke and irritating fumes.

DBA600 CAS:5626-16-4 ***HR: 3***
DESMETHYLDOXEPIN
mf: $C_{17}H_{17}NO$ mw: 251.35
/SH,2 SYNS ◇ DIBENZ(b,e)OXEPIN-$\Delta^{11(6H)}$,Γ-PROPYLAMINE ◇ KS 1675

TOXICITY DATA with REFERENCE
ipr-rat LD50:100 mg/kg ARZNAD 15,863,65
orl-mus LD50:118 mg/kg ARZNAD 15,863,65
ipr-mus LD50:92 mg/kg ARZNAD 15,863,65
ivn-mus LD50:29 mg/kg ARZNAD 15,863,65
ivn-rbt LD50:13400 μg/kg ARZNAD 15,863,65

SAFETY PROFILE: Poison by ingestion, intravenous, and intraperitoneal routes. When heated to decomposition it emits toxic fumes of NO_x. See also DOXEPIN.

DBA700 ***HR: 2***
DESMETHYLMISONIDAZOLE
mf: $C_7H_{11}N_3O_4$ mw: 201.21

SYNS: 3-METHOXY-2-(2-NITRO-1-IMIDAZOLYL)-1-PROPANOL ◇ NSC-261036

TOXICITY DATA with REFERENCE
orl-rat LD50:5076 mg/kg NTIS** PB81-121212
ivn-rat LD50:2532 mg/kg NTIS** PB81-121212
ivn-dog LDLo:1800 mg/kg NTIS** PB81-121212

SAFETY PROFILE: Moderately toxic by intravenous route. Mildly toxic by ingestion. When heated to decomposition it emits toxic fumes of NO_x.

DBA750 CAS:54024-22-5 ***HR: D***
DESOGESTREL
mf: $C_{22}H_{30}O$ mw: 310.52

PROP: Crystals. Mp: 109-110°, (α) (20/D) +55° (chloroform). Mixture with ethinyl estradiol.

SYNS: 13-ETHYL-11-METHYLENE-18,19-DINOR-17-α-PREGN-4-EN-20-YN-17-OL ◇ ORG 2969

TOXICITY DATA with REFERENCE
orl-wmn TDLo:6 μg/kg (20D pre):REP CCPTAY 16,51,77

SAFETY PROFILE: Human reproductive effects by ingestion: changes in menstrual cycle and female fertility. Experimental reproductive effects. A steroid. When heated to decomposition it emits acrid smoke and fumes.

DBA800 CAS:300-42-5 ***HR: 3***
DESOXYEPHEDRINE HYDROCHLORIDE
mf: $C_{10}H_{15}N{\cdot}ClH$ mw: 185.72

SYNS: A 884 ◇ AMDRAM ◇ AMEDRINE ◇ AMPHEDROXY ◇ AMPHEDROXYN ◇ APAMINE ◇ BOMBITA ◇ C 6379 ◇ CORVITIN ◇ DAROPERVAMIN ◇ DEA OXO-5 ◇ DEOFED ◇ DEOXYEPHEDRINE ◇ DEPOXIN ◇ DESAMINE ◇ DESFEDRIN ◇ DESOSSIEFEDRINA ◇ DES-OXA-D ◇ DESOXEDRINE ◇ DESOXIN ◇ DESOXO-5 ◇ DESOXYFED ◇ DESOXYN ◇ DESOXYPHED ◇ DESTIM ◇ DETREX ◇ DEXOPHRINE ◇ DEXOVAL ◇ N,α-DIMETHYLPHENETHYLAMINE HYDROCHLORIDE ◇ DOPIDRIN ◇ DOXEPHRIN ◇ DOXYFED ◇ DRINALFA ◇ EFFROXINE ◇ ESTIMULEX ◇ EUPHODRIN ◇ FENYPRIN ◇ GEROBIT ◇ GEROVIT ◇ HEROPON ◇ ISOPHEN ◇ KEMODRIN ◇ LANAZINE ◇ MADRINE ◇ METAMFETAMINA ◇ METAMPHETAMIN ◇ METHAMPHETAMINE HYDROCHLORIDE ◇ METANFETAMINA ◇ METHEDRINE ◇ METHEDRINE HYDROCHLORIDE ◇ METHOXYN ◇ METHYLAMPHETAMINE HYDROCHLORIDE ◇ METHYLBENZEDRIN ◇ METHYLISOMIN ◇ N-METHYL-β-PHENYLISOPROPYLAMINHYDROCHLORID (GERMAN) ◇ METHYLPROPAMINE ◇ NEODRINE ◇ NEOPHARMEDRINE ◇ NORODIN ◇ OXYDRENE ◇ OXYFED ◇ PERVITIN ◇ PHILOPON ◇ PREMODRIN ◇ SEMOXYDRINE ◇ SPEED ◇ STIMULEX ◇ TONEDRIN ◇ VONEDRINE

TOXICITY DATA with REFERENCE
eye-rbt 2% MLD ARZNAD 8,708,58
scu-rat TDLo:210 mg/kg (female 1-21D post):REP DEPBA5 8,397,75
orl-hmn TDLo:17 mg/kg:CNS,CVS KLWOAZ 17,1580,38
orl-rat LD50:29 mg/kg ARZNAD 24,166,74
ipr-rat LD50:19 mg/kg ARZNAD 8,708,58
scu-rat LD50:79700 μg/kg JJPAAZ 35,273,84
ivn-rat LD50:13 mg/kg HEPHD2 55,527,80
orl-mus LD50:15 mg/kg ARZNAD 24,166,74
ipr-mus LD50:4000 μg/kg ARZNAD 11,271,61
scu-mus LD50:9 mg/kg ARZNAD 11,271,61
ivn-mus LD50:15 mg/kg HEPHD2 55,527,80
orl-rbt LD50:10500 μg/kg RCOCB8 3,215,72
scu-rbt LD50:205 mg/kg JPETAB 86,284,46
ivn-rat LD50:3 mg/kg RCOCB8 3,215,72
ims-rbt LD50:220 mg/kg JPETAB 86,284,46

SAFETY PROFILE: Poison by ingestion, intravenous, intraperitoneal, subcutaneous, and intramuscular routes. Human systemic effects by ingestion: altered sleep patterns, anorexia, and change in heart rate. Experimental reproductive effects. An eye irritant. A powerful central nervous system stimulant. When heated to decomposition it emits very toxic fumes of NO_x and HCl. See also BENZEDRINE.

DBA875 CAS:382-67-2 ***HR: D***
DESOXYMETASONE
mf: $C_{22}H_{29}FO_4$ mw: 376.51

PROP: Crystals from ethyl acetate. Mp: 217°. Sol in alc, acetone, chloroform, and hot ethyl acetate; sltly sol in ether, benzene; insol in water, dil aq acids and alkalies.

SYNS: A 41304 ◇ 17-DEOXYMETHANSONE ◇ DEOXYMETHASONE ◇ DESOXIMETASONE ◇ 17-DESOXIMETHASONE ◇ DESOXYMETHASONE ◇ ESPERSON ◇ 9-FLUORO-11-β,21-DIHYDROXY-16-α-METHYLPREGNA-1,4-DIENE-3,20-DIONE ◇ R 2113 ◇ STIEDEX ◇ TOPICORTE ◇ TOPIDERM ◇ TOPISOLON

TOXICITY DATA with REFERENCE
scu-rat TDLo:900 μg/kg (female 8-16D post):REP NYKZAU 71,367,75
scu-mus TDLo:4800 μg/kg (female 7-15D post):TER NYKZAU 71,367,75

SAFETY PROFILE: An experimental teratogen. Other experimental reproductive effects. When heated to decomposition it emits toxic fumes of F^-.

DBB000 CAS:7632-10-2 ***HR: 3***
DESOXYN
mf: $C_{10}H_{15}N$ mw: 149.26

SYNS: ANADREX ◇ DEOXYEPHEDRINE ◇ DESOXYEPHEDRINE ◇ METHAMPHETAMINE ◇ METHEDRINE ◇ METHYLAMPHETAMINE ◇ N-METHYLAMPHETAMINE ◇ N-METHYL-β-PHENYLISOPROPYLAMIN (GERMAN) ◇ N-METHYL-β-PHENYLISOPROPYLAMINE ◇ PERVERTIN ◇ 1-PHENYL-2-METHYLAMINO-PROPAN (GERMAN) ◇ 1-PHENYL-2-METHYLAMINOPROPANE ◇ α-PHENYL-β-METHYLAMINOPROPANE ◇ STIMULEX

TOXICITY DATA with REFERENCE
orl-rat LDLo:70 mg/kg AEPPAE 195,647,40
ipr-rat LDLo:32 mg/kg AEPPAE 195,647,40
scu-rat LD50:30 mg/kg AIPTAK 159,442,66
orl-mus LD50:34 mg/kg JPETAB 131,115,61
ipr-mus LD50:15 mg/kg CPBTAL 22,1459,74
scu-mus LD50:10 mg/kg AEPPAE 241,182,61
ivn-mus LD50:10 mg/kg 27ZIAQ -,-,65
scu-cat LDLo:50 mg/kg 27ZIAQ -,158,73

SAFETY PROFILE: Poison by ingestion, intraperitoneal, subcutaneous, and intravenous routes. A powerful central nervous system stimulant. When heated to decomposition it emits toxic fumes of NO_x. See also BENZEDRINE.

DBB200 CAS:125-33-7 ***HR: 3***
2-DESOXYPHENOBARBITAL
mf: $C_{12}H_{14}N_2O_2$ mw: 218.28

SYNS: 5-AETHYL-5-PHENYL-HEXAHYDROPYRIMIDIN-4,6-DION (GERMAN) ◇ CYRAL ◇ 2-DEOXYPHENOBARBITAL ◇ DESOXY-PHENOBARBITONE ◇ 5-ETHYLDIHYDRO-5-PHENYL-4,6(1H,5H)-PYRIMIDINEDIONE ◇ 5-ETHYLHEXAHYDRO-4,6-DIOXO-5-PHENYLPHRIMIDINE ◇ 5-ETHYLHEXAHYDRO-5-PHENYL-PYRIMIDINE-4,6-DIONE ◇ 5-ETHYL-5-PHENYLHEXAHYDROP-YRIMIDINE-4,6-DIONE ◇ HEXADIONA ◇ HEXAMIDINE ◇ HEXAMID-INE (the antispasmodic) ◇ LEPIMIDIN ◇ LEPSIRAL ◇ MAJSOLIN ◇ MIDONE ◇ MILEPSIN ◇ MISODINE ◇ MISOLYNE ◇ MIZODIN ◇ MIZOLIN ◇ MYLEPSIN ◇ MYLEPSINUM ◇ MYSEDON ◇ MYSOL-INE ◇ NCI-C56360 ◇ 5-PHENYL-5-ETHYL-HEXAHYDROPYRIMIDINE-4,6-DIONE ◇ PRILEPSIN ◇ PRIMACIONE ◇ PRIMACLONE ◇ PRIMACONE ◇ PRIMAKTON ◇ PRIMIDON ◇ PRIMIDONE ◇ PRYSOLINE ◇ PYRIMIDONE MEDI-PETS ◇ ROE 101 ◇ SERTAN

TOXICITY DATA with REFERENCE
mmo-sat 6666 μg/plate ENMUDM 8(Suppl 7),1,86
cyt-hmn:leu 1 mg/L AJOGAH 116,867,73
orl-wmn TDLo:2025 mg/kg (female 1-39W post):REP TJADAB 32,13,85
orl-wmn TDLo:10350 mg/kg (1-39W preg):TER JOPDAB 94,835,79
orl-wmn TDLo:38 g/kg/7Y-I DICPBB 17,551,83
orl-rat LD50:1500 mg/kg NIIRDN 6,691,82
orl-mus LD50:280 mg/kg TXAPA9 34,271,75
ipr-mus LD50:332 mg/kg PCJOAU 15,403,81

SAFETY PROFILE: Poison by ingestion and intraperitoneal routes. Human teratogenic effects include developmental abnormalities of the craniofacial area, skin and skin appendages, and cardiovascular system. Human reproductive effects: effects on newborn including unusual growth statistics, drug dependence, physical and other neonatal changes. Experimental teratogenic and reproductive effects. Human mutation data reported. An addictive drug. When heated to decomposition it emits toxic fumes of NO_x. See also BARBITURATES.

DBB400 CAS:14918-35-5 ***HR: 3***
DESTOMYCIN A
mf: $C_{20}H_{37}N_3O_{13}$ mw: 527.60

SYN: DESTONATE 20

TOXICITY DATA with REFERENCE
orl-mus LD50:50 mg/kg JAJAAA 18,38,65
ivn-mus LD50:5 mg/kg JAJAAA 18,38,65

SAFETY PROFILE: Poison by ingestion and intravenous routes. When heated to decomposition it emits toxic fumes of NO_x.

DBB450 ***HR: D***
DETERGENTS, LIQUID containing AES

PROP: Liquid kitchen detergent containing 8% polyoxyethylene alkyl ether, 7% sodium alkyl ethoxy sulfate, and 4% ethanol as its major constituents SKEZAP 17,158,76

SYN: LIQUID DETERGENTS containing AES

TOXICITY DATA with REFERENCE
skn-mus TDLo:39 g/kg (female 1-14D post):TER SKEZAP 17,158,76

SAFETY PROFILE: An experimental teratogen. When heated to decomposition it emits acrid smoke and irritating fumes.

DBB460 ***HR: D***
DETERGENTS, LIQUID containing LAS

PROP: Liquid kitchen detergent containing 17% linear alkylbenzene sulfonate, 7% ethanol, and 15% urea as its major ingredients (SKEZAP 17,158,76).

TOXICITY DATA with REFERENCE
skn-mus TDLo:39 g/kg (1-13D preg):TER SKEZAP 17,158,76
skn-mus TDLo:39 g/kg (1-13D preg):REP SKEZAP 17,158,76

SAFETY PROFILE: An experimental teratogen. Experimental reproductive effects. When heated to decomposition it emits toxic fumes of SO_x and NO_x. See also UREA, ETHYL ALCOHOL, and ALKYLBENZENE-SULFONATE.

DBB500 CAS:37209-31-7 ***HR: 3***
DETRALFATE

SYNS: APD ◇ DEXTRAN SULFATE SODIUM ALUMINIUM

TOXICITY DATA with REFERENCE
ipr-rat LD50:200 mg/kg IYKEDH 3,24,72
scu-rat LD50:1600 mg/kg IYKEDH 3,24,72
ipr-mus LD50:257 mg/kg IYKEDH 3,24,72
scu-mus LD50:2200 mg/kg IYKEDH 3,24,72

SAFETY PROFILE: Poison by intraperitoneal route. Moderately toxic by subcutaneous route. Used as an antipeptic ulcer agent. When heated to decomposition it emits toxic fumes of SO_x and Na_2O. See also DEXTRAN SULFATE SODIUM.

DBB600 CAS:67293-88-3 ***HR: 3***
DEUTERIOMORPHINE
mf: $C_{17}H_{16}D_3NO_3$ mw: 288.37

TOXICITY DATA with REFERENCE
scu-mus LDLo:400 mg/kg SCIEAS 134,1078,61
ice-mus LD50:11400 μg/kg SCIEAS 134,1078,61

SAFETY PROFILE: Poison by subcutaneous and intracerebral routes. When heated to decomposition it emits toxic fumes of NO_x. See also MORPHINE.

DBB800 CAS:7782-39-0 ***HR: 3***
DEUTERIUM
DOT: UN 1957
mf: D_2 mw: 4.03

PROP: A gas. Chemically the same as hydrogen. Lel: 5%, uel: 75%

SYN: D_2

DOT Classification: Flammable Gas; Label: Flammable Gas

SAFETY PROFILE: Very dangerous fire and explosion hazard when exposed to heat, flame, sparks, and oxidizers. To fight fire, stop flow of gas. See also HYDROGEN.

DBC000 CAS:14333-26-7 ***HR: 3***
DEUTERIUM FLUORIDE
mf: DF mw: 21.01

PROP: Chemically the same as hydrogen fluoride.

TOXICITY DATA with REFERENCE
ihl-rat LC50:1095 ppm/1H AMRL** TR-74-78,74
ihl-mus LC50:324 ppm/1H AMRL** TR-74-78,74

OSHA PEL: TWA 2.5 mg(F)/m^3
ACGIH TLV: TWA 2.5 mg(F)/m^3
NIOSH REL: TWA (Inorganic Fluorides) 2.5 mg(F)/m^3

SAFETY PROFILE: Moderately toxic by inhalation. A dangerously reactive, powerful oxidant. When heated to decomposition it emits toxic fumes of F^-. See also FLUORIDES and HYDROGEN FLUORIDE.

DBC400 CAS:1177-87-3 ***HR: D***
DEXAMETHASONE ACETATE
mf: $C_{24}H_{31}FO_6$ mw: 434.55

SYNS: DECADRON-LA ◇ DECTAN ◇ 9-FLUORO-11-β,17,21-TRIHYDROXY-16-α-METHYLPREGNA-1,4-DIENE-3,20-DIONEACETATE

TOXICITY DATA with REFERENCE
scu-rbt TDLo:6 mg/kg (female 24-26D post):REP BCPCA6 27,1007,78
scu-rat TDLo:1200 μg/kg (12-13D preg):TER TJADAB 23,15,81

CONSENSUS REPORTS: Reported in EPA TSCA Inventory.

SAFETY PROFILE: Experimental teratogenic and reproductive effects. A steroid. When heated to decomposition it emits toxic fumes of F^-.

DBC500 CAS:55541-30-5 ***HR: 3***
DEXAMETHASONE 17,21-DIPROPIONATE
mf: $C_{28}H_{37}FO_7$ mw: 504.65

SYN: DEXAMETHASONE DIPROPIONATE ◇ 9-FLUORO-11-β,17,21-TRIHYDROXY-16-α-METHYLPREGNA-1,4-DIENE-3,20-DIONE-17,21-DIPROPIONATE ◇ THS-101

TOXICITY DATA with REFERENCE
scu-rat TDLo:1100 μg/kg (female 7-17D post):REP IYKEDH 17,310,86
scu-rat TDLo:11 mg/kg (female 7-17D post):TER TJADAB 28,12A,83
ipr-rat LD50:36500 μg/kg OYYAA2 28,687,84
scu-rat LD50:26 mg/kg OYYAA2 28,687,84
ipr-mus LD50:276 mg/kg IYKEDH 18,474,87
scu-mus LD50:183 mg/kg IYKEDH 18,474,87
scu-rbt LD50:7600 μg/kg OYYAA2 28,687,84

SAFETY PROFILE: Poison by subcutaneous and intraperitoneal routes. An experimental teratogen. Other experimental reproductive effects. A steroid. When heated to decomposition it emits toxic fumes of F^-.

DBC510 CAS:2265-64-7 ***HR: 2***
DEXAMETHASONE ISONICOTINATE
mf: $C_{28}H_{32}FNO_6$ mw: 497.61

SYNS: AUXILSON ◇ DEXAMETHASONE-21-ISONICOTINATE ◇ 9-FLUORO-11-β,17,21-TRIHYDROXY-16-α-METHYLPREGNA-1,4-DIENE-3,20-DIONE, 21-ISONICOTINATE ◇ H3 111 ◇ PYRIDIN-4-CARBONSAEURE-(DEXAMETHASON-21')-ESTER (GERMAN) ◇ PYRIDINE-4-CARBOXYLICACID-(DEXAMETHASONE-21') ESTER ◇ VOREN

TOXICITY DATA with REFERENCE
ihl-rat TCLo:4178 mg/m^3/10M (44D pre):REP OYYAA2 8,255,74
orl-rat LD50:3562 mg/kg TOIZAG 20,769,73
ipr-rat LD50:313 mg/kg TOIZAG 20,769,73
scu-rat LD50:3297 mg/kg TOIZAG 20,769,73
orl-mus LD50:3470 mg/kg TOIZAG 20,769,73

SAFETY PROFILE: Moderately toxic by intraperitoneal route. Experimental reproductive effects. A steroid. When heated to decomposition it emits toxic fumes of F^- and NO_x.

DBC525 CAS:14899-36-6 ***HR: D***
DEXAMETHASONE PALMITATE
mf: $C_{38}H_{59}FO_6$ mw: 630.97

SYNS: DEXAMETHASONE-21-PALMITATE ◇ 9-FLUORO-11-β,17,21-TRIHYDROXY-16-α-METHYLPREGNA-1,4-DIENE-3,20-DIONE-21-PALMITATE

TOXICITY DATA with REFERENCE
ivn-rat TDLo:25 μg/kg (multi) :TER KSRNAM 19,4539,85
ivn-rat TDLo:55 μg/kg (female 7-17D post):REP KSRNAM 19,4521,85

SAFETY PROFILE: An experimental teratogen. Other

experimental reproductive effects. A steroid. When heated to decomposition it emits toxic fumes of F^-.

DBC550 CAS:466-11-5 **HR: 2**
DEXAMETHASONE SODIUM SULFATE
mf: $C_{22}H_{29}FO_8S{\cdot}Na$ mw: 495.56

SYNS: DEXAMETHASONE SODIUM HEMISULFATE ◇ DEXA-SCHEROSON (INJECTABLE) ◇ 9-FLUORO-11-β,17,21-TRIHYDROXY-16-α-METHYLPREGNA-1,4-DIENE-3,20-DIONE-21-(HYDROGENSULFATE), MONOSODIUM SALT

TOXICITY DATA with REFERENCE
ipr-rat LD50:2180 mg/kg NIIRDN 6,477,82
scu-rat LD50:3450 mg/kg NIIRDN 6,477,82
ipr-mus LD50:1660 mg/kg NIIRDN 6,477,82
scu-mus LD50:2020 mg/kg NIIRDN 6,477,82

SAFETY PROFILE: Moderately toxic by intraperitoneal and subcutaneous routes. A steroid. When heated to decomposition it emits toxic fumes of F^-, SO_x, and Na_2O.

DBC575 CAS:33755-46-3 **HR: 3**
DEXAMETHASONE VALERATE
mf: $C_{27}H_{37}FO_6$ mw: 476.64

SYNS: DEXAMETHASONE-17-VALERATE ◇ DV-17 ◇ 9-FLUORO-11-β,17,21-TRIHYDROXY-16-α-METHYLPREGNA-1,4-DIENE-3,20-DIONE-17-VALERATE

TOXICITY DATA with REFERENCE
skn-rat TDLo:18 μg/kg (17-22D preg):REP JZKEDZ 8,235,82
scu-rat LD50:117 mg/kg IYKEDH 17,1106,86
skn-mus LD50:1784 mg/kg JZKEDZ 8,23,82
scu-mus LD50:238 mg/kg JZKEDZ 8,23,82

SAFETY PROFILE: Poison by subcutaneous route. Moderately toxic by skin contact. Experimental reproductive effects. A steroid. When heated to decomposition it emits toxic fumes of F^-.

DBC800 CAS:9004-54-0 **HR: 3**
DEXTRAN 1

PROP: A linear water-sol polymer of average molecular weight 200,000 (ARPAAQ 67,589,59).

TOXICITY DATA with REFERENCE
ipr-rat TDLo:2500 mg/kg:NEO,REP AMPLAO 67,589,59
scu-rat TDLo:2500 mg/kg:NEO,REP AMPLAO 67,589,59
ipr-mus TDLo:8000 mg/kg:ETA AMPLAO 67,589,59

CONSENSUS REPORTS: Reported in EPA TSCA Inventory.

SAFETY PROFILE: Questionable carcinogen with experimental neoplastigenic, tumorigenic, and teratogenic data. Other experimental reproductive effects. When heated to decomposition it emits acrid smoke and fumes. See also other dextrans.

DBD000 CAS:9004-54-0 **HR: 3**
DEXTRAN 2

PROP: A linear, water-sol polymer of average molecular weight 100,000 (ARPAAQ 67,589,59).

TOXICITY DATA with REFERENCE
ipr-rat TDLo:2500 mg/kg:NEO,REP AMPLAO 67,589,59
scu-rat TDLo:2500 mg/kg:NEO AMPLAO 67,589,59

CONSENSUS REPORTS: Reported in EPA TSCA Inventory.

SAFETY PROFILE: Questionable carcinogen with experimental neoplastigenic data. When heated to decomposition it emits acrid smoke and fumes. See also other dextrans.

DBD200 CAS:9004-54-0 **HR: 3**
DEXTRAN 5

PROP: A highly branched, water-sol polymer (ARPAAQ 67,589,59).

TOXICITY DATA with REFERENCE
scu-mus TDLo:8000 mg/kg:NEO AMPLAO 67,589,59

CONSENSUS REPORTS: Reported in EPA TSCA Inventory.

SAFETY PROFILE: Questionable carcinogen with experimental neoplastigenic data. When heated to decomposition it emits acrid smoke and fumes. See also other dextrans.

DBD400 CAS:9004-54-0 **HR: 3**
DEXTRAN 10

PROP: A branched, water-sol polymer of average molecular weight 89,400 (ARPAAQ 67,589,59).

TOXICITY DATA with REFERENCE
scu-rat TDLo:2500 mg/kg:CAR AMPLAO 67,589,59
ipr-mus TDLo:8000 mg/kg:CAR AMPLAO 67,589,59

CONSENSUS REPORTS: Reported in EPA TSCA Inventory.

SAFETY PROFILE: Suspected carcinogen with experimental carcinogenic data. When heated to decomposition it emits acrid smoke and fumes. See also other dextrans.

DBD600 CAS:9004-54-0 **HR: 3**
DEXTRAN 11

PROP: A highly branched, water-sol polymer of average molecular weight 71,400 (ARPAAQ 67,589,59).

TOXICITY DATA with REFERENCE
ipr-rat TDLo:2500 mg/kg:NEO AMPLAO 67,589,59
scu-mus TDLo:8000 mg/kg:ETA AMPLAO 67,589,59

CONSENSUS REPORTS: Reported in EPA TSCA Inventory.

SAFETY PROFILE: Questionable carcinogen with experimental neoplastigenic and tumorigenic data. When heated to decomposition it emits acrid smoke and fumes. See also other dextrans.

DBD700 CAS:9004-54-0 ***HR: 2***
DEXTRAN 70

PROP: Chemical and physical properties of the dextrans vary with the methods of production. Native dextrans usually have high molecular weight; lower molecular weight clinical dextrans are usually prepared by depolymerization of native dextrans or by synthesis. All dextrans are composed exclusively of α-d-glucopyranosyl units, differing only in degree of branching and chain length.

SYNS: DEXTRAN ◇ DEXTRAVEN ◇ EXPANDEX ◇ GENTRAN ◇ HEMODEX ◇ INTRADEX ◇ MACROSE ◇ ONKOTIN ◇ PLAVOLEX ◇ POLYGLUCIN ◇ PROMIT

TOXICITY DATA with REFERENCE
ivn-rbt TDLo:675 g/kg (8-16D preg):TER OYYAA2 6,1119,72
ivn-rbt TDLo:3640 g/kg (91D male):REP YACHDS 3,369,75
ivn-wmn TDLo:6 mg/kg:CVS,SKN BMJOAE 2,1502,76
scu-mus LD50:13900 mg/kg OYYAA2 6,1023,72
ivn-mus LD50:12100 mg/kg OYYAA2 6,1023,72
ivn-rbt LD50:17400 mg/kg OYYAA2 6,1023,72

CONSENSUS REPORTS: Reported in EPA TSCA Inventory.

SAFETY PROFILE: Human systemic effects by intravenous route: dermatitis and changes in the vascular and cardiac systems. Experimental reproductive effects. An experimental teratogen. See also various dextrans.

DBD750 CAS:9011-18-1 ***HR: 3***
DEXTRAN SULFATE SODIUM

PROP: White powder from alcohol + ether. Freely sol in water.

SYNS: ASURO ◇ COLYONAL ◇ DEXTRARINE ◇ DEXULATE ◇ DS-M-1 ◇ MDS

TOXICITY DATA with REFERENCE
orl-rat TDLo:330 g/kg/94W-C:CAR CRNGDP 3,353,82
orl-rat TD:335 g/kg/19W-C:CAR JJIND8 66,579,81
orl-rat LD50:20600 mg/kg KSRNAM 13,1318,79
ivn-rat LD50:473 mg/kg KSRNAM 13,1318,79
orl-mus LD50:21 g/kg NIIRDN 6,481,82
ivn-mus LD50:15800 mg/kg NIIRDN 6,481,82
ivn-rbt LD50:19 g/kg NIIRDN 6,481,82

SAFETY PROFILE: Moderately toxic by intravenous route. Questionable carcinogen with experimental carcinogenic data. When heated to decomposition it emits toxic fumes of SO_x and Na_2O. See also other dextran entries.

DBD800 CAS:9004-53-9 ***HR: 1***
DEXTRINS
mf: $(C_6H_{10}O_5)_n \cdot xH_2O$

PROP: An intermediate product formed by the hydrolysis of starches. It describes a class of substances. Yellow or white powder or granules. Sol in water; insol in alc and ether; forms colloids.

SYNS: ARTIFICIAL GUM ◇ DEXTRANS ◇ STARCH GUM ◇ TAPIOCA ◇ VEGETABLE GUM

SAFETY PROFILE: Mildly toxic by intravenous route. When heated to decomposition it emits acrid smoke and irritating fumes.

DBE000 CAS:21888-96-0 ***HR: 3***
DEXTROBENZETIMIDE HYDROCHLORIDE
mf: $C_{23}H_{26}N_2O_2 \cdot ClH$ mw: 398.97

SYNS: (+)-1-BENZYL-4-(2,6-DIOXO-3-PHENYL-3-PIPERIDYL)PIPERIDINE HYDROCHLORIDE ◇ (+)-2-(1-BENZYL-4-PIPERIDYL)-2-PHENYLGLUTARIMIDE HYDROCHLORIDE ◇ (+)-3-(1-BENZYL-4-PIPERIDYL)-3-PHENYLPIPERIDINE-2,6-DIONEHYDROCHLORIDE ◇ DEXBENZETIMIDE HYDROCHLORIDE ◇ DEXETIMIDE HYDROCHLORIDE ◇ (s)-3-PHENYL-1′-(PHENYLMETHYL)-(3,4′-BIPIPERIDINE)-2,6-DIONE HYDROCHLORIDE ◇ R 16470 ◇ TREMBLEX

TOXICITY DATA with REFERENCE
ivn-rat LD50:45 mg/kg 27ZQAG -,219,72
ivn-mus LD50:45 mg/kg ARZNAD 21,1365,71

SAFETY PROFILE: Poison by intravenous route. An anticholinergic agent used to treat Parkinson's disease. When heated to decomposition it emits very toxic fumes of HCl and NO_x.

DBE100 CAS:5653-80-5 ***HR: 3***
DEXTROMETHADONE
mf: $C_{21}H_{27}NO$ mw: 309.49

SYNS: d-6-(DIMETHYLAMINO)-4,4-DIPHENYL-3-HEPTANONE ◇ (s)-6-(DIMETHYLAMINO)-4,4-DIPHENYL-3-HEPTANONE ◇ (+)-METHADONE ◇ d-METHADONE ◇ l-(+)-METHADONE ◇ s-(+)-METHADONE ◇ 6s-METHADONE

TOXICITY DATA with REFERENCE
dni-rat:tes 50 μmol/L BCPCA6 27,123,78
oms-rat:tes 50 μmol/L BCPCA6 27,123,78
scu-rat TDLo:17800 μg/kg (1D male):REP JPETAB 198,340,76

orl-mus LD50:252 mg/kg JPETAB 110,135,54
ipr-mus LD50:74 mg/kg BJPCAL 9,280,54
scu-mus LD50:80 mg/kg BJPCAL 9,280,54

SAFETY PROFILE: Poison by ingestion, subcutaneous, and intraperitoneal routes. Experimental reproductive effects. Mutation data reported. Addictive. When heated to decomposition it emits toxic fumes of NO_x. See also (−)-METHADONE.

DBE150 CAS:125-71-3 ***HR: 3***
DEXTROMETHORPHAN
mf: $C_{18}H_{25}NO$ mw: 271.44

SYNS: BA 2666 ◇ d-METHORPHAN ◇ Δ-METHORPHAN

TOXICITY DATA with REFERENCE
orl-mus LD50:210 mg/kg JPMSAE 60,1523,71
scu-mus LD50:112 mg/kg CPBTAL 7,372,59
ivn-dog LDLo:22 mg/kg CPBTAL 7,372,59

SAFETY PROFILE: Poison by ingestion, subcutaneous, and intravenous routes. When heated to decomposition it emits toxic fumes of NO_x. See also DEXTROMETHORPHAN HYDROBROMIDE.

DBE200 CAS:125-69-9 ***HR: 3***
DEXTROMETHORPHAN HYDROBROMIDE
mf: $C_{18}H_{25}NO•BrH$ mw: 352.36

SYNS: ANTUSSAN ◇ DEMORPHAN ◇ DEXTROMETHORPHAN BROMIDE ◇ DEXTROMETORPHAN HYDROBROMIDE ◇ DORMETHAN ◇ MEDICON ◇ METHORATE HYDROBROMIDE ◇ d-METHORPHAN HYDROBROMIDE ◇ d-3-METHOXY-N-METHYLMORPHINAN HYDROBROMIDE ◇ 3-METHOXY-17-METHYL-9-α,13-α,14-α-MORPHINAN HYDROBROMIDE ◇ METRORAT ◇ RO 1-5470/5 ◇ ROMILAR ◇ ROMILAR HYDROBROMIDE ◇ TUSILAN ◇ TUSSADE

TOXICITY DATA with REFERENCE
orl-chd TDLo:30 mg/kg:CNS PEDIAU 59,117,77
orl-rat LD50:350 mg/kg JPETAB 109,189,53
scu-rat LD50:423 mg/kg OYYAA2 6,1207,72
orl-mus LD50:165 mg/kg JPETAB 109,189,53
scu-mus LD50:153 mg/kg ARZNAD 26,353,76
ivn-mus LD50:34 mg/kg OYYAA2 6,1207,72
ivn-dog LDLo:30 mg/kg ARZNAD 19,1916,69
ivn-cat LD50:19800 μg/kg OYYAA2 6,1207,72
ivn-rbt LD50:15 mg/kg JPETAB 109,189,53
orl-gpg LD50:336 mg/kg OYYAA2 6,1207,72
scu-gpg LD50:150 mg/kg OYYAA2 6,1207,72

SAFETY PROFILE: Poison by ingestion, subcutaneous, and intravenous routes. Human systemic effects by ingestion: ataxia (loss of muscle coordination), excitement and motor activity changes. When heated to decomposition it emits very toxic fumes of NO_x and HBr.

DBE600 CAS:101563-89-7 ***HR: 3***
DEXTROMYCIN HYDROCHLORIDE
mf: $C_{23}H_{46}N_6O_{13}•ClH$ mw: 651.21

TOXICITY DATA with REFERENCE
scu-mus LD50:750 mg/kg 85ERAY 1,743,78
ivn-mus LD50:50 mg/kg 85ERAY 1,743,78

SAFETY PROFILE: Poison by intravenous route. Moderately toxic by subcutaneous route. When heated to decomposition it emits very toxic fumes of HCl and NO_x.

DBE625 CAS:72050-78-3 ***HR: 1***
DEXTROPROPOXYPHENE NAPSYLATE

TOXICITY DATA with REFERENCE
orl-mus TDLo:1920 mg/kg (7-12D preg):REP OYYAA2 4,1031,70
orl-mus TDLo:3600 mg/kg (7-12D preg):TER OYYAA2 4,1031,70
orl-hmn TDLo:28 mg/kg/1W:SKN BMJOAE 2,674,77

SAFETY PROFILE: Human systemic effects by ingestion: dermatitis. An experimental teratogen. Other experimental reproductive effects. See also d-PROPOXYPHENE HYDROCHLORIDE.

DBE800 CAS:125-73-5 ***HR: 3***
DEXTRORPHAN
mf: $C_{17}H_{23}NO$ mw: 257.41

SYNS: (+)-cis-1,3,4,9,10,10A-HEXAHYDRO-11-METHYL-2H-10,4a-IMINOETHANOPHENANTHREN-6-OL ◇ d-3-HYDROXY-N-METHYLMORPHINAN

TOXICITY DATA with REFERENCE
scu-rat TDLo:25750 μg/kg (female 5-12D post):REP JPETAB 200,255,77
scu-rat LD50:800 mg/kg CLDND* 22,1370,57
scu-mus LD50:131 mg/kg JOCEAH 22,1370,57
ivn-mus LD50:65 mg/kg CLDND* 22,1370,57

SAFETY PROFILE: Poison by subcutaneous and intravenous routes. Experimental reproductive effects. When heated to decomposition it emits toxic fumes of NO_x.

DBE825 CAS:143-98-6 ***HR: 3***
DEXTRORPHAN TARTRATE
mf: $C_{17}H_{23}NO•C_4H_6O_6$ mw: 407.51

SYNS: d-3-HYDROXY-N-METHYLMORPHINAN TARTRATE ◇ RO 1-6794

TOXICITY DATA with REFERENCE
orl-rat LD50:1100 mg/kg JPETAB 109,189,53
scu-rat LD50:800 mg/kg JPETAB 109,189,53
orl-mus LD50:385 mg/kg JPETAB 109,189,53
scu-mus LD50:350 mg/kg JPETAB 109,189,53

ivn-mus LD50:42 mg/kg 31ZPAG 2,77,66
ivn-rbt LD50:27500 μg/kg JPETAB 109,189,53

SAFETY PROFILE: Poison by ingestion, subcutaneous, and intravenous routes. When heated to decomposition it emits toxic fumes of NO_x.

DBE835 CAS:70052-12-9 ***HR: 1***
α-DFMO
mf: $C_6H_{12}F_2N_2O_2$ mw: 182.20

SYNS: α-DIFLUOROMETHYLORNITHINE ◇ 2-(DIFLUOROMETHYL) ORNITHINE ◇ RMI 71782

TOXICITY DATA with REFERENCE
dni-rat-ipr 2800 mg/kg BBACAQ 696,179,82
scu-rat TDLo:1 g/kg (16-20D preg):REP NTOTDY 7,57,85
scu-rat TDLo:2500 mg/kg (female 15-17D post):TER TJADAB 33,40C,86
ivn-wmn LDLo:67 mg/kg/40M-C:CVS,BPR AIMEAS 105,141,86

SAFETY PROFILE: Human systemic effects by intravenous route: cardiomyopathy, pulse rate decrease, fall in blood pressure. An experimental teratogen. Other experimental reproductive effects. Mutation data reported. When heated to decomposition it emits toxic fumes of F^- and NO_x.

DBE875 CAS:70711-41-0 ***HR: 3***
DHAQ DIACETATE
mf: $C_{22}H_{28}N_4O_6 \cdot 2C_2H_4O_2$ mw: 564.66

SYNS: 5,8-BIS((2-((2-HYDROXYETHYL)AMINO)ETHYL)AMINO)-1,4-DIHYDROXY-9,10-ANTHRACENEDIONE DIACETATE ◇ 5,8-BIS((2-((2-HYDROXYETHYL)AMINO)ETHYL)AMINO)-1,4-DIHYDROXYANTHRAQUINONE 1,4-DIACETATE ◇ 1,4-DIHYDROXY-5,8-BIS(2-((2-HYDROXYETHYL)AMINO)ETHYLAMINO)-9,10-ANTHRACENEDIONE DIACETATE ◇ NSC 299195

TOXICITY DATA with REFERENCE
mmo-sat 6500 μmol/L CNREA8 41,376,81
mma-sat 6500 μmol/L CNREA8 41,376,81
cyt-ham:ovr 10 nmol/L CNREA8 41,376,81
sce-ham:ovr 10 nmol/L CNREA8 41,376,81
ipr-mus LD50:61300 μg/kg NCISP* JAN86
ivn-dog LDLo:25 mg/kg NTIS** PB297-169

SAFETY PROFILE: Poison by intravenous and intraperitoneal routes. Mutation data reported. When heated to decomposition it emits toxic fumes of NO_x.

DBE885 CAS:24477-37-0 ***HR: 3***
DIABENOR
mf: $C_{20}H_{26}N_4O_5S$ mw: 434.56

SYNS: 1-CYCLOHEXYL-3-((p-(2-(5-METHYL-3-ISOXAZOLECARBOXAMIDO)ETHYL)PHENYL)SULFONYL)UREA ◇ GLISOLAMIDE ◇ N-(4-(β-(5-METILOSSAZOL-3-CARBOSSAMIDO)-ETIL)-BENZENE-SOLFONIL)-N[1]-CICLOESIL-UREA ◇ P.M. 434,526

TOXICITY DATA with REFERENCE
orl-rat TDLo:1500 mg/kg (6-15D preg):TER BCFAAI 117,348,78
ipr-rat LD50:320 mg/kg BCFAAI 117,348,78
scu-rat LD50:900 mg/kg BCFAAI 117,348,78
ipr-mus LD50:230 mg/kg BCFAAI 117,348,78
scu-mus LD50:720 mg/kg BCFAAI 117,348,78

SAFETY PROFILE: Poison by intraperitoneal route. Moderately toxic by subcutaneous route. An experimental teratogen. When heated to decomposition it emits toxic fumes of SO_x and NO_x.

DBF000 CAS:63019-65-8 ***HR: 3***
N-1-DIACETAMIDOFLUORENE
mf: $C_{17}H_{15}NO_2$ mw: 265.33

SYNS: N-FLUOREN-1-YLDIACETAMIDE ◇ N-1-FLUORENYLDIACETAMIDE

TOXICITY DATA with REFERENCE
orl-rat TDLo:5200 mg/kg/52W-C:ETA JNCIAM 24,149,60

SAFETY PROFILE: Questionable carcinogen with experimental tumorigenic data. When heated to decomposition it emits toxic fumes of NO_x.

DBF200 CAS:642-65-9 ***HR: 3***
2-DIACETAMIDOFLUORENE
mf: $C_{17}H_{15}NO_2$ mw: 265.33

SYNS: N-ACETYL-N-9H-FLUOREN-2-YL-ACETAMIDE ◇ N-DIACETYL-2-AMINOFLUORENE ◇ 2-DIACETYLAMINOFLUORENE ◇ N,N-DIACETYL-2-AMINOFLUORENE ◇ N,N-DIACETYL-2-FLUORENAMINE ◇ F-diAA ◇ 2-FLUORENYLDIACETAMIDE ◇ N-FLUOREN-2-YLDIACETAMIDE ◇ N-2-FLUORENYLDIACETAMIDE

TOXICITY DATA with REFERENCE
orl-rat TDLo:970 mg/kg/23W-C:NEO JNCIAM 10,1201,50
skn-rat TDLo:290 mg/kg/77W-I:CAR JNCIAM 10,1201,50
orl-ham TDLo:6174 mg/kg/49W-C:CAR GANNA2 59,239,68
orl-rat TD:475 mg/kg/5W-C:ETA AICCA6 20,1364,64

SAFETY PROFILE: Suspected carcinogen with experimental carcinogenic, neoplastigenic, and tumorigenic data. When heated to decomposition it emits toxic fumes of NO_x.

DBF400 CAS:51325-35-0 ***HR: 3***
2,4-DIACETAMIDO-6-(5-NITRO-2-FURYL)-s-TRIAZINE
mf: $C_{11}H_{10}N_6O_5$ mw: 306.27

SYN: N,N′-(6-(5-NITRO-2-FURYL)-s-TRIAZINE-2,4-DIYL)BISACETAMIDE

TOXICITY DATA with REFERENCE
mma-sat 100 ng/plate MUREAV 48,295,77
orl-rat TDLo:57 g/kg/46W-C:CAR JNCIAM 51,403,73

CONSENSUS REPORTS: EPA Genetic Toxicology Program.

SAFETY PROFILE: Questionable carcinogen with experimental carcinogenic data. Mutation data reported. When heated to decomposition it emits toxic fumes of NO_x.

DBF500 ***HR: D***
2',4'-DIACETATE-DIS-NOGAMYCIN

TOXICITY DATA with REFERENCE
dnd-mam:lym 12 μmol/L CBINA8 36,1,81
dni-mam:lym 12 μmol/L CBINA8 36,1,81
oms-mam:lym 12 μmol/L CBINA8 36,1,81

SAFETY PROFILE: Mutation data reported.

DBF600 CAS:102-62-5 ***HR: 2***
1,3-DIACETIN
mf: $C_7H_{12}O_5$ mw: 176.19

PROP: Crystals. D: 1.178 @ 15°/15°, bp: 280°, mp: 40°.

SYNS: 1,3-DIACETATE GLYCEROL ◇ 1,2-DIACETATE 1,2,3-PROPANETRIOL ◇ DIACETIN ◇ 1,2-DI-ACETIN ◇ 2,3-DIACETIN ◇ DIACETYL GLYCERINE ◇ DIGLYCERIDE ACETIC ACID ◇ GLYCEROL DIACETATE ◇ GLYCERYL-1,3-DIACETATE ◇ (HYDROXYMETHYL) ETHYLENE ACETATE

TOXICITY DATA with REFERENCE
orl-mus LD50:8500 mg/kg JPETAB 65,89,39
scu-mus LD50:3500 mg/kg JPETAB 65,89,39
ivn-mus LD50:2300 mg/kg JPETAB 65,89,39

CONSENSUS REPORTS: Reported in EPA TSCA Inventory.

SAFETY PROFILE: Moderately toxic by subcutaneous and intravenous routes. Mildly toxic by ingestion. When heated to decomposition it emits acrid smoke and irritating fumes.

DBF750 CAS:123-42-2 ***HR: 3***
DIACETONE ALCOHOL
DOT: UN 1148
mf: $C_6H_{12}O_2$ mw: 116.18

PROP: Liquid; faint, pleasant odor. Mp: −47 to −54°, bp: 167.9°, flash p: 148°F, d: 0.9306 @ 25°/4°, autoign temp: 1118°F, vap d: 4.00, vap press: 1.1 mm @ 20°, lel: 1.8%, uel: 6.9%, flash p: (acetone free): 136°F.

SYNS: DIACETONALCOHOL (DUTCH) ◇ DIACETONALCOOL (ITALIAN) ◇ DIACETONALKOHOL (GERMAN) ◇ DIACETONE ◇ DIACETONE-ALCOOL (FRENCH) ◇ DIKETONE ALCOHOL ◇ 4-HYDROXY-2-KETO-4-METHYLPENTANE ◇ 4-HYDROXY-4-METHYL-PENTAN-2-ON (GERMAN, DUTCH) ◇ 4-HYDROXY-4-METHYLPENTANONE-2 ◇ 4-HYDROXY-4-METHYL-2-PENTANONE ◇ 4-HYDROXY-4-METHYL PENTAN-2-ONE ◇ 4-IDROSSI-4-METIL-PENTAN-2-ONE (ITALIAN) ◇ 2-METHYL-2-PENTANOL-4-ONE ◇ PYRANTON ◇ TYRANTON

TOXICITY DATA with REFERENCE
eye-hmn 100 ppm/15M JIHTAB 28,262,46
skn-rbt 10 mg/24H open JIHTAB 30,63,48
skn-rbt 500 mg open MLD UCDS** 6/29/59
eye-rbt 5 mg SEV AJOPAA 29,1363,46
ihl-hmn TCLo:100 ppm:EYE,CNS,GIT JIHTAB 30,63,48
ihl-hmn TCLo:400 ppm:PUL NPIRI* 1,21,74
orl-rat LD50:4000 mg/kg JIHTAB 30,63,48
ipr-mus LD50:933 mg/kg SCCUR* -,3,61
skn-rbt LD50:13500 mg/kg NPIRI* 1,21,74

CONSENSUS REPORTS: Reported in EPA TSCA Inventory.

OSHA PEL: TWA 50 ppm
ACGIH TLV: TWA 50 ppm
DFG MAK: 50 ppm (240 mg/m^3)
NIOSH REL: (Ketones) TWA 240 mg/m^3
DOT Classification: Flammable Liquid; Label: Flammable Liquid; Flammable or Combustible Liquid; Label: Flammable Liquid.

SAFETY PROFILE: Moderately toxic by ingestion and intraperitoneal routes. Mildly toxic by skin contact. Human systemic effects by inhalation: headache, nausea or vomiting, eye and pulmonary changes. A skin, mucous membrane, and severe eye irritant. Can cause anemia and damage to liver and kidneys. Narcotic in high concentration. Flammable when exposed to heat or flame; can react with oxidizing materials. Explosive in the form of vapor when exposed to heat or flame. To fight fire, use alcohol foam, foam, CO_2, dry chemical. When heated to decomposition it emits acrid smoke and irritating fumes. See also KETONES.

DBF800 CAS:1067-33-0 ***HR: 3***
DIACETOXYDIBUTYL STANNANE
mf: $C_{12}H_{24}O_4Sn$ mw: 351.05

PROP: Clear, colorless liquid with a slight acetic acid odor. Bp: decomp, fp: 5°-10°, flash p: 290°F (OC), d: 1.31 @ 25°, vap d: 12.1.

SYNS: BA 2726 ◇ BIS(ACETYLOXY)DIBUTYLSTANNANE ◇ DIACETOXYBUTYLTIN ◇ DIACETOXYDIBUTYLTIN ◇ DIBUTYL TIN DIACETATE ◇ FOMREZ SUL-3 ◇ NCI-C02028 ◇ T 1 (Catalyst)

TOXICITY DATA with REFERENCE
orl-rat LD50:32 mg/kg NCILB* NIH-NCI-E-C-72-3252,73
orl-mus LD50:46 mg/kg NCILB* NIH-NCI-E-C-72-3252,73
ivn-mus LD50:18 mg/kg CSLNX* NX#02348

CONSENSUS REPORTS: NCI Carcinogenesis Bioassay (feed); Inadequate Studies: mouse, rat NCITR*

NCI-CG-TR-183,79. Reported in EPA TSCA Inventory.

OSHA PEL: TWA 0.1 mg(Sn)/m^3 (skin)
ACGIH TLV: TWA 0.1 mg(Sn)/m^3 (skin) (Proposed: TWA 0.1 mg(Sn)/m^3; STEL 0.2 mg(Sn)/m^3 (skin))
NIOSH REL: (Organotin Compounds) TWA 0.1 mg(Sn)/m^3

SAFETY PROFILE: Poison by ingestion and intravenous routes. Combustible when exposed to heat or flame; can react with oxidizing materials. To fight fire, use water, foam, CO_2, dry chemical. When heated to decomposition it emits acrid smoke and irritating fumes. See also TIN COMPOUNDS.

DBF875 CAS:10140-75-7 ***HR: 1***
1,1-DIACETOXY-2,3-DICHLOROPROPANE
mf: $C_7H_{10}Cl_2O_4$ mw: 229.07

SYN: 1,1-PROPANEDIOL, 2,3-DICHLORO-, DIACETATE

TOXICITY DATA with REFERENCE
skn-rbt 10 mg/24H open SEV AIHAAP 23,95,62
orl-rat LD50:320 mg/kg AIHAAP 23,95,62
skn-rbt LD50:1000 mg/kg AIHAAP 23,95,62

SAFETY PROFILE: A poison by ingestion. Mildly toxic by skin contact. A severe skin irritant. When heated to decomposition it emits toxic fumes of Cl^-.

DBG200 CAS:73785-34-9 ***HR: 3***
trans-7,8-DIACETOXY-7,8-DIHYDROBENZO(a) PYRENE
mf: $C_{24}H_{16}O_4$ mw: 368.40

SYN: trans-BP-7,8-DIHYDRODIOL DIACETATE

TOXICITY DATA with REFERENCE
scu-mus TDLo:13 mg/kg:NEO JJIND8 64,617,80

SAFETY PROFILE: Questionable carcinogen with experimental neoplastigenic data. When heated to decomposition it emits acrid smoke and irritating fumes.

DBG300 CAS:2205-78-9 ***HR: D***
3-β,17-β-DIACETOXY-17-α-ETHINYL-19-NOR-$\Delta^{3,5}$-ANDROSTADIENE
mf: $C_{24}H_{30}O_4$ mw: 382.54

TOXICITY DATA with REFERENCE
orl-rat TDLo:75 mg/kg (female 1D post):REP PSEBAA 111,595,62

SAFETY PROFILE: Experimental reproductive effects. A steroid. When heated to decomposition it emits acrid smoke and fumes.

DBG499 ***HR: 3***
2-((2,4-DIACETYL-5-BENZOFURANYL)OXY) TRIETHYLAMINE HYDROCHLORIDE
mf: $C_{18}H_{23}NO_4$•ClH mw: 353.88

TOXICITY DATA with REFERENCE
orl-mus LDLo:150 mg/kg APFRAD 41,603,83

SAFETY PROFILE: A poison by ingestion. When heated to decomposition it emits toxic fumes of NO_x and HCl.

DBG900 CAS:922-89-4 ***HR: 3***
N,N'-DIACETYL-N,N'-DINITRO-1,2-DIAMINOETHANE
mf: $C_6H_{10}N_4O_6$ mw: 234.17

$CH_3CO(NO_2)C_2H_4N(NO_2)OCCH_3$

SAFETY PROFILE: A powerful oxidizer. Decomposes violently at 142°C. Upon decomposition it emits toxic fumes of NO_x. See also NITRO COMPOUNDS of AROMATIC HYDROCARBONS.

DBH000 CAS:95-45-4 ***HR: 3***
DIACETYL DIOXIME
mf: $C_4H_8N_2O_2$ mw: 116.14

SYNS: 2,3-DIISONITROSOBUTANE ◇ DIMETHYLGLYOXIME

TOXICITY DATA with REFERENCE
otr-ham:emb 100 μg/L FCTXAV 18,289,80
orl-rat LDLo:250 mg/kg NCNSA6 5,26,53

CONSENSUS REPORTS: Reported in EPA TSCA Inventory.

SAFETY PROFILE: Poison by ingestion. Mutation data reported. When heated to decomposition it emits toxic fumes of NO_x.

DBH200 CAS:17598-65-1 ***HR: 3***
DIACETYLLANATOSIDE
mf: $C_{47}H_{74}O_{19}$ mw: 943.11

PROP: Crystals from methanol, decomp 265-268°. One part dissolves in 5000 parts water, 200 parts methanol and 2500 parts ethanol; very sltly sol in chloroform; practically insol in ether.

SYNS: DEACETYLLANATOSIDE C ◇ DESACE ◇ DESACETYLLANATOSIDE C ◇ DESACI ◇ DESCETYLDIGILANIDE C ◇ DESLANATOSIDE ◇ DESLANOSIDE ◇ DIACETYLLANATOSID C (GERMAN) ◇ SEDIRANIDO ◇ LANIMERCK

TOXICITY DATA with REFERENCE
ipr-mus LD50:10360 μg/kg NIIRDN 6,489,82
ivn-mus LD50:8100 μg/kg NIIRDN 6,489,82
ivn-cat LD50:230 μg/kg 85ELDJ -,190,63

SAFETY PROFILE: Poison by intravenous and intraperitoneal routes. When heated to decomposition it emits acrid smoke and fumes. See also LANATOSIDE A and LANATOSIDE C.

DBH400 CAS:1502-95-0 ***HR: 3***
DIACETYLMORPHINE HYDROCHLORIDE
mf: $C_{21}H_{23}NO_5{\cdot}ClH$ mw: 405.91

SYNS: DIAMORPHINE HYDROCHLORIDE ◇ HEROIN HYDROCHLORIDE

TOXICITY DATA with REFERENCE
scu-mus TDLo:80 mg/kg (9D preg):TER DGDFA5 22,61,80
ipr-mus LD50:240 mg/kg AIPTAK 122,434,59
ivn-mus LD50:38 mg/kg TXAPA9 6,334,64

SAFETY PROFILE: Poison by intraperitoneal and intravenous routes. An experimental teratogen. A narcotic. Addictive. When heated to decomposition it emits very toxic fumes of NO_x and HCl. See also HEROIN and MORPHINE.

DBH800 CAS:73622-67-0 ***HR: 3***
3,4-DI(ACETYLTHIOMETHYL)-5-HYDROXY-6-METHYLPYRIDINE HYDROBROMIDE
mf: $C_{12}H_{15}NO_3S_2{\cdot}BrH$ mw: 366.32

SYNS: 4,5-DI(MERCAPTOMETHYL)-2-METHYL-3-PYRIDINOL DITHIOACETATE HYDROBROMIDE ◇ 4,5-DIMERCAPTOPYRIDOXINDI-THIOACETAT HYDROBROMID (GERMAN)

TOXICITY DATA with REFERENCE
orl-rat LD50:870 mg/kg ARZNAD 11,922,61
scu-rat LD50:132 mg/kg ARZNAD 11,922,61
ivn-rat LD50:41 mg/kg ARZNAD 11,922,61
orl-mus LD50:409 mg/kg ARZNAD 11,922,61
scu-mus LD50:121 mg/kg ARZNAD 11,922,61
ivn-mus LD50:49 mg/kg ARZNAD 11,922,61

SAFETY PROFILE: Poison by subcutaneous and intravenous routes. Moderately toxic by ingestion. When heated to decomposition it emits very toxic fumes of NO_x, SO_x, and HBr. See also MERCAPTANS and BROMIDES.

DBI099 CAS:10311-84-9 ***HR: 3***
DIALIFOR
mf: $C_{14}H_{17}NO_4PS_2$ mw: 393.86

SYNS: S-(2-CHLORO-1-(1,3-DIHYDRO-1,3-DIOXO-2H-ISOINDOL-2-YL)ETHYL)-O,O-DIETHYL PHOSPHORODITHIOATE ◇ S-(2-CHLORO-1-PHTHALIMIDOETHYL)-O,O-DIETHYLPHOSPHORODITHIOATE ◇ O,O-DIETHYL-S-(2-CHLORO-1-PHTHALIMIDOETHYL)PHOSPHORODITHIOATE ◇ ENT 27,320 ◇ HERCULES 14503 ◇ PHOSPHORODITHIOICACID-S-(2-CHLORO-1-(1,3-DIHYDRO-1,3-DIOXO-2H-ISOINDOL-2-YL)ETHYL-O,O-DIETHYL ESTER ◇ PHOSPHORODITHIOICACID-S-(2-CHLORO-1-PHTHALIMIDOETHYL)-O,)-DIETHYL ESTER ◇ TORAK

TOXICITY DATA with REFERENCE
orl-ham TDLO:100 mg/kg (8D preg):TER,TER TXAPA9 16,24,70
orl-rat LD50:5 mg/kg BESAAT 15,122,69
skn-rat LD50:28 mg/kg FAATDF 7,299,86
orl-mus LD50:39 mg/kg BESAAT 15,122,69
orl-dog LD50:94 mg/kg 85DPAN -,-,71/76
orl-rbt LD50:35 mg/kg BESAAT 15,122,69

SAFETY PROFILE: Poison by ingestion and skin contact. An experimental teratogen. Other experimental reproductive effects. When heated to decomposition it emits toxic fumes of SO_x, PO_x, and NO_x.

DBI159 ***HR: 3***
DIALKYLZINCS
R_2Zn

SAFETY PROFILE: Flammable when exposed to heat or flame. Some may explode on contact with water. Potentially explosive reaction with acyl halides (during the production of ketones); allyl chlorides; and alcohols. When heated to decomposition it emits toxic fumes of ZnO. See also ZINC COMPOUNDS.

DBI200 CAS:2303-16-4 ***HR: 3***
DIALLATE
mf: $C_{10}H_{17}Cl_2NOS$ mw: 270.24

PROP: Brown liquid. Bp: 150° @ 9 mm, mp: 25-30°. Sltly sol in water; sol in organic solvents.

SYNS: AVADEX ◇ CP 15,336 ◇ DATC ◇ 2,3-DCDT ◇ DIALLAAT (DUTCH) ◇ DIALLAT (GERMAN) ◇ 2,3-DICHLORALLYL-N,N-(DIISOPROPYL)-THIOCARBAMAT (GERMAN) ◇ S-(2,3-DICHLORO-ALLIL)-N,N-DIISOPROPIL-MONOTIOCARBAMMATO(ITALIAN) ◇ S-(2,3-DICHLOR-ALLYL)-N,N-DIISOPROPYL-MONOTHIOCARBAMAAT (DUTCH) ◇ DICHLOROALLYL DIISOPROPYLTHIOCARBAMATE ◇ S-2,3-DICHLOROALLYL DIISOPROPYLTHIOCARBAMATE ◇ 2,3-DICHLOROALLYL-N,N-DIISOPROPYLTHIOLCARBAMATE ◇ 2,3-DICHLORO-2-PROPENE-1-THIOL DIISOPROPYLCARBAMATE ◇ S-(2,3-DICHLORO-2-PROPENYL)ESTER, BIS(1-METHYLETHYL) CARBAMOTHIOIC ACID ◇ DI-ISOPROPYLTHIOLOCARBAMATE de S-(2,3-DICHLOROALLYLE) (FRENCH) ◇ PYRADEX ◇ RCRA WASTE NUMBER U062

TOXICITY DATA with REFERENCE
mmo-sat 500 μg/plate MUREAV 85,45,81
cyt-ham:ovr 200 μmol/L MUREAV 85,45,81
orl-rat TDLo:4095 mg/kg/78W-C:ETA JJIND8 67,75,81
orl-mus TDLo:68 g/kg/84W-C:CAR JNCIAM 42,1101,69
orl-rat LD50:395 mg/kg RREVAH 10,97,65
orl-dog LD50:510 mg/kg 28ZEAL 5,71,76
skn-rbt LD50:2000 mg/kg WRPCA2 9,119,70
orl-gpg LD50:420 mg/kg HYSAAV 34(10-12),356,69

CONSENSUS REPORTS: IARC Cancer Review: Group 3 IMEMDT 7,56,87; Animal Limited Evidence IMEMDT 30,235,83; Animal Sufficient Evidence IM-

EMDT 12,69,76. EPA Genetic Toxicology Program. Community Right-To-Know List.

SAFETY PROFILE: Poison by ingestion. Moderately toxic by skin contact. Questionable carcinogen with experimental carcinogenic and tumorigenic data. Mutation data reported. When heated to decomposition it emits very toxic fumes of Cl^-, NO_x, and SO_x. See also CARBAMATES and ALLYL COMPOUNDS.

DBI600 CAS:124-02-7 ***HR: 3***
DIALLYLAMINE
DOT: UN 2359
mf: $C_6H_{11}N$ mw: 97.18

PROP: Liquid, sol in water. D: 0.7889 @ 20°, bp: 112°, fp: −100°, flash p.: 69.8°F.

SYNS: DI-2-PROPENYLAMINE ◇ N-2-PROPENYL-2-PROPEN-1-AMINE

TOXICITY DATA with REFERENCE
skn-rbt 100 μg/24H open AIHAAP 23,95,62
eye-rbt 50 mg/20S rns SEV AEHLAU 1,343,60
ihl-man TCLo:5 ppm/5M:PUL,EYE AEHLAU 1,343,60
orl-rat LD50:650 mg/kg AIHAAP 23,95,62
ihl-rat LCLo:2000 ppm/4H AIHAAP 23,95,62
orl-mus LD50:516 mg/kg AEHLAU 1,343,60
ipr-mus LD50:187 mg/kg AEHLAU 1,343,60
skn-rbt LD50:280 mg/kg AIHAAP 23,95,62
ihl-mam LC50:2100 mg/m^3 TPKVAL 14,80,75

CONSENSUS REPORTS: Reported in EPA TSCA Inventory.

DOT Classification: Flammable Liquid; Label: Flammable Liquid.

SAFETY PROFILE: Poison by skin contact, andintraperitoneal routes. Moderately toxic by ingestion and inhalation. Human systemic effects by inhalation route: eye lacrimation, and changes in the trachea or bronchi. A skin and severe eye irritant. A dangerous fire hazard when exposed to heat or flame. When heated to decomposition it emits toxic fumes of NO_x. See also AMINES and ALLYL COMPOUNDS.

DBI800 CAS:6392-46-7 ***HR: 3***
4-DIALLYLAMINO-3,5-DIMETHYLPHENYL-N-METHYLCARBAMATE
mf: $C_{16}H_{22}N_2O_2$ mw: 274.40

PROP: Powder. Insol water; sol alc, benzene.

SYNS: ALLYLOXYCARB ◇ ALLYXYCARB ◇ APC ◇ BAY 50282 ◇ 4-(DIALLYLAMINO)-3,5-XYLENOL METHYLCARBAMATE (ester) ◇ 4-DIALLYL-AMINO-3,5-XYLYL N-METHYLCARBAMATE ◇ 3,5-DIMETHYL-4-DIALLYLAMINOPHENYL-N-METHYLCARBAMATE ◇ ENZOSE ◇ HYDROL

TOXICITY DATA with REFERENCE
orl-rat LD50:89 mg/kg TXAPA9 21,315,72
orl-mus LDLo:62 mg/kg AECTCV 14,111,85
orl-bwd LD50:13 mg/kg JAFCAU 15,287,67

SAFETY PROFILE: Poison by and ingestion routes. An insecticide. When heated to decomposition it emits toxic fumes of NO_x. See also CARBAMATES, ALLYL COMPOUNDS, and ESTERS.

DBJ100 CAS:13988-24-4 ***HR: 3***
1,1-DIALLYL-3-(1,4-BENZODIOXAN-2-YLMETHYL)-3-METHYLUREA
mf: $C_{17}H_{22}N_2O_3$ mw: 302.41

SYNS: A 2275 ◇ 2-(N,N-DIALLYLCARBAMYLMETHYL)AMINOMETHYL-1,4-BENZODIOXAN ◇ N-((2,3-DIHYDRO-1,4-BENZODIOXIN-2-YL)METHYL)-N-METHYL-N′,N′-DI-2-PROPENYL-UREA (9CI)

TOXICITY DATA with REFERENCE
orl-mus LD50:447 mg/kg JAPMA8 48,409,59
ipr-mus LD50:230 mg/kg JAPMA8 48,409,59
ivn-mus LD50:55 mg/kg JAPMA8 48,409,59

SAFETY PROFILE: Poison by intravenous and intraperitoneal routes. Moderately toxic by ingestion. When heated to decomposition it emits toxic fumes of NO_x. See also ALLYL COMPOUNDS.

DBJ200 CAS:538-08-9 ***HR: 3***
DIALLYLCYANAMIDE
mf: $C_7H_{10}N_2$ mw: 122.19

PROP: Colorless, mobile liquid when pure. Mp: < −70°, bp: 222° (slight decomp), d: 0.9021, vap d: 4.1.

SYN: N-CYANODIALLYLAMINE

TOXICITY DATA with REFERENCE
ipr-mus LDLo:125 mg/kg CBCCT* 4,318,52

CONSENSUS REPORTS: Reported in EPA TSCA Inventory.

SAFETY PROFILE: Poison by intraperitoneal route. Possibly more toxic than cyanamide. When heated to decomposition or on contact with acid or acid fumes it emits highly toxic fumes of CN^-. See CYANAMIDE, ALLYL COMPOUNDS, and AMINES.

DBJ400 CAS:17381-88-3 ***HR: 3***
DIALLYLDIBROMO STANNANE
mf: $C_6H_{10}Br_2Sn$ mw: 360.67

SYN: DIALLYLTIN DIBROMIDE

TOXICITY DATA with REFERENCE
ivn-mus LD50:18 mg/kg CSLNX* NX#03064

OSHA PEL: TWA 0.1 mg(Sn)/m^3 (skin)
ACGIH TLV: TWA 0.1 mg(Sn)/m^3 (skin) (Proposed: TWA 0.1 mg(Sn)/m^3; STEL 0.2 mg(Sn)/m^3 (skin))
NIOSH REL: (Organotin Compounds) TWA 0.1 mg(Sn)/m^3

SAFETY PROFILE: Poison by intravenous route. When heated to decomposition it emits toxic fumes of Br^-. See also ALLYL COMPOUNDS, BROMIDES, and TIN COMPOUNDS.

DBJ600 CAS:37764-25-3 ***HR: 2***
N,N-DIALLYDICHLOROACETAMIDE
mf: $C_8H_{11}Cl_2NO$ mw: 208.10

SYNS: COMPOUND R-25788 ◇ N,N-DIALLYL-2,2-DICHLOROACETAMIDE ◇ 2,2-DICHLORO-N,N-DI-2-PROPENYLACETAMIDE ◇ R-25788 ◇ STAUFFER R-25788

TOXICITY DATA with REFERENCE
orl-rat LD50:2 g/kg 85ARAE 3,108,76/77

CONSENSUS REPORTS: Reported in EPA TSCA Inventory.

SAFETY PROFILE: Moderately toxic by ingestion. When heated to decomposition it emits very toxic fumes of Cl^- and NO_x. See also ALLYL COMPOUNDS.

DBK000 CAS:557-40-4 ***HR: 3***
DIALLYL ETHER
DOT: UN 2360
mf: $C_6H_{10}O$ mw: 98.16

PROP: Liquid, odor of radishes. Bp: 94.3°, d: 0.805, vap d: 3.38, flash p: 20°F (OC).

SYNS: ALLYLETHER ◇ 3,3'-OXYBIS(1-PROPENE) ◇ PROPENYL ETHER

TOXICITY DATA with REFERENCE
skn-rbt 10 mg/24H open MLD JIHTAB 31,60,49
eye-rbt 20 mg open JIHTAB 31,60,49
orl-rat LD50:320 mg/kg JPMSAE 63,1068,74
skn-rbt LD50:600 mg/kg JIHTAB 31,60,49

CONSENSUS REPORTS: Reported in EPA TSCA Inventory.

DOT Classification: Flammable Liquid; Label: Flammable Liquid and Poison.

SAFETY PROFILE: Poison by ingestion. Moderately toxic by skin contact. A skin and eye irritant. A dangerous fire hazard when exposed to heat, flame, or oxidizing materials. To fight fire, use alcohol foam. Reacts with air to form explosive peroxides. Violent explosions have occurred during distillation. When heated to decomposition it emits acrid smoke and fumes. See also ALLYL COMPOUNDS and ETHERS.

DBK100 CAS:5164-11-4 ***HR: 2***
1,1-DIALLYLHYDRAZINE
mf: $C_6H_{12}N_2$ mw: 112.20

SYNS: 1,1-DAH ◇ DIALLYLHYDRAZINE ◇ HYDRAZINE, 1,1-DI-2-PROPENYL-(9CI)

TOXICITY DATA with REFERENCE
orl-mus TDLo:56000 mg/kg/2Y-C:CAR ANTRD4 1,259,81
orl-mus TD:59220 mg/kg/90W-C:CAR ANTRD4 1,259,81

SAFETY PROFILE: Questionable carcinogen with experimental carcinogenic data. When heated to decomposition it emits toxic fumes of NO_x.

DBK120 CAS:26072-78-6 ***HR: 2***
1,2-DIALLYLHYDRAZINE DIHYDROCHLORIDE
mf: $C_6H_{12}N_2 \cdot 2ClH$ mw: 185.12

SYN: 1,2-DAH HYDROCHLORIDE

TOXICITY DATA with REFERENCE
orl-mus TDLo:78176 mg/kg/80W-C:CAR ONCOBS 39,104,82
orl-mus TD:89152 mg/kg/80W-C:CAR ONCOBS 39,104,82

SAFETY PROFILE: Questionable carcinogen with experimental carcinogenic data. When heated to decomposition it emits toxic fumes of NO_x.

DBK200 CAS:999-21-3 ***HR: 3***
DIALLYL MALEATE
mf: $C_{10}H_{12}O_4$ mw: 196.22

PROP: Liquid. Vap d: 6.6.

SYNS: MALEIC ACID, DIALLYL ESTER ◇ SIPOMER DAM

TOXICITY DATA with REFERENCE
skn-rbt 10 mg/24H open MLD JIHTAB 31,60,49
skn-rbt 500 mg MOD SCCUR* -,3,61
eye-rbt 100 mg AJOPAA 29,1363,46
orl-rat LD50:300 mg/kg JIHTAB 31,60,49
orl-mus LD50:493 mg/kg SCCUR* -,3,61
ipr-mus LD50:160 mg/kg SCCUR* -,3,61
skn-rbt LD50:1150 mg/kg JIHTAB 31,60,49

CONSENSUS REPORTS: Reported in EPA TSCA Inventory.

SAFETY PROFILE: Poison by ingestion and intraperitoneal routes. Moderately toxic by skin contact. A skin and eye irritant. When heated to decomposition it emits acrid smoke and irritating fumes. See also ALLYL COMPOUNDS and ESTERS.

DBK400 CAS:15180-03-7 ***HR: 3***
N,N'-DIALLYLNORTOXIFERINIUM DICHLORIDE
mf: $C_{44}H_{50}N_4O_2{\bullet}2Cl$ mw: 737.88

SYNS: ALCUONIUM DICHLORIDE ◇ ALCURONIUM CHLORIDE ◇ ALLOFERIN ◇ DIALFERIN ◇ DIALLYLNORTOXIFERINE DICHLORIDE ◇ 4,4'-DIDEMETHYL-4,4'-DI-2-PROPENYLTOXIFERINE I DICHLORIDE ◇ RO 4-3816

TOXICITY DATA with REFERENCE
orl-rat LD50:27600 µg/kg OYYAA2 3,390,69
ipr-rat LD50:270 µg/kg GNRIDX 1,349,67
scu-rat LD50:280 µg/kg OYYAA2 3,390,69
orl-mus LD50:38500 µg/kg GNRIDX 1,349,67
ipr-mus LD50:610 µg/kg OYYAA2 3,390,69
scu-mus LD50:610 µg/kg GNRIDX 1,349,67
ivn-mus LD50:240 µg/kg OYYAA2 3,390,69

SAFETY PROFILE: Poison by ingestion, intraperitoneal, subcutaneous, and intravenous routes. Used as a skeletal muscle relaxant. When heated to decomposition it emits very toxic fumes of NO_x and Cl^-. See also ALLYL COMPOUNDS.

DBK600 CAS:34037-79-1 ***HR: 3***
DIALLYL PEROXYDICARBONATE
mf: $C_8H_{10}O_6$ mw: 202.16

$(H_2C{=}CHCH_2OCO{\bullet}O\text{-})_2$

SAFETY PROFILE: A powerful oxidizer. A storage hazard. It may explode spontaneously at room temperature. When heated to decomposition it emits acrid smoke and fumes. See also ALLYL COMPOUNDS and PEROXIDES.

DBK800 CAS:3382-99-8 ***HR: 3***
2,6-DIALLYLPHENOL
mf: $C_{12}H_{10}O$ mw: 170.22

TOXICITY DATA with REFERENCE
skn-mus TDLo:4600 mg/kg/12W-I:NEO CNREA8 19,413,59

SAFETY PROFILE: Questionable carcinogen with experimental neoplastigenic data by skin contact. When heated to decomposition it emits acrid smoke and irritating fumes. See also ALLYL COMPOUNDS and PHENOL.

DBL000 CAS:23679-20-1 ***HR: 3***
DIALLYL PHOSPHITE
mf: $C_6H_{11}O_3P$ mw: 162.12

$(H_2C{=}CHCH_2O)_2P(O)H$

SYN: DI-2-PROPENYL PHOSPHONITE

SAFETY PROFILE: A poison. May explode during distillation. When heated to decomposition it emits toxic fumes of PO_x. See also ALLYL COMPOUNDS.

DBL200 CAS:131-17-9 ***HR: 2***
DIALLYL PHTHALATE
mf: $C_{14}H_{14}O_4$ mw: 246.28

PROP: Nearly colorless, oily liquid. Bp: 157°, flash p: 330°F, d: 1.120 @ 20°/20°, vap d: 8.3.

SYNS: DAPON 35 ◇ DAPON R ◇ DI-2-PROPENYL ESTER, 1,2-BENZENEDICARBOXYLIC ACID ◇ NCI-C50657 ◇ PHTHALIC ACID, DIALLYL ESTER ◇ o-PHTHALIC ACID, DIALLYL ESTER

TOXICITY DATA with REFERENCE
eye-rbt 500 mg AJOPAA 29,1363,46
mma-mus:lyms 67200 µg/L SCIEAS 236,933,87
orl-rat LD50:770 mg/kg
cyt-ham:ovr 200 mg/L SCIEAS 236,933,87
orl-rat TDLo:52 g/kg/2Y-I:CAR EVHPAZ 65,271,86
orl-mus TDLo:156 g/kg/2Y-I:CAR EVHPAZ 65,271,86
orl-rat LD50:770 mg/kg FEPRA7 8,299,49
ipr-mus LD50:700 mg/kg 14CYAT 2,1904,63
skn-rbt LDLo:2800 mg/kg FEPRA7 5,191,46
scu-rbt LDLo:1000 mg/kg FEPRA7 5,191,46

CONSENSUS REPORTS: NTP Carcinogenesis Studies (gavage); Equivocal Evidence: rat NTPTR* NTP-TR-284,85. Reported in EPA TSCA Inventory.

SAFETY PROFILE: Suspected carcinogen with experimental carcinogenic data. Moderately toxic by ingestion, skin contact, intraperitoneal, and subcutaneous routes. An eye irritant. Mutation data reported. Combustible when exposed to heat or flame; can react with oxidizing materials. To fight fire use CO_2 or dry chemical. When heated to decomposition it emits acrid smoke and irritating fumes. See also ALLYL COMPOUNDS and ESTERS.

DBL300 CAS:91297-11-9 ***HR: 3***
DIALLYL SELENIDE
mf: $C_6H_{10}Se$ mw: 184.11

SYN: SELENIDE, DIALLYL-

TOXICITY DATA with REFERENCE
ivn-mus LD50:100 mg/kg CSLNX* NX#09268

OSHA PEL: TWA 0.2 mg(Se)/m^3
ACGIH TLV: TWA 0.2 mg(Se)/m^3

SAFETY PROFILE: Poison by intravenous route. When heated to decomposition it emits toxic fumes of Se.

DBL400 ***HR: 3***
DIALLYL SULFATE
mf: $C_6H_{10}O_4S$ mw: 178.20

$(H_2C{=}CHCH_2O)_2SO_2$

SAFETY PROFILE: Explodes when heated. When heated to decomposition it emits toxic fumes of SO_x. See also ALLYL COMPOUNDS and SULFATES.

DBL600 CAS:6601-20-3 ***HR: 2***
DIALLYL THIOUREA
mf: $C_7H_{12}N_2S$ mw: 156.27

TOXICITY DATA with REFERENCE
orl-rat LD50:500 mg/kg JPETAB 90,260,47

CONSENSUS REPORTS: Reported in EPA TSCA Inventory.

SAFETY PROFILE: Moderately toxic by ingestion. When heated to decomposition it emits very toxic fumes of NO_x and SO_x.

DBL649 ***HR: 3***
DIALUMINUM OCTAVANADIUM TRIDECASILICIDE
mf: $Al_2Si_{13}V_8$ mw: 826.61

SAFETY PROFILE: Reacts violently with hydrogen fluoride. When heated to decomposition it emits toxic fumes of VO_x. See also ALUMINUM COMPOUNDS and VANADIUM COMPOUNDS.

DBL700 CAS:21187-98-4 ***HR: 3***
DIAMICRON
mf: $C_{15}H_{21}N_3O_3S$ mw: 323.45

PROP: Crystals from anhydrous ethanol. Mp: 180-182°.

SYNS: 1-(3-AZABICYCLO(3.3.0)OCT-3-YL)-3-(p-TOLYLSULFONYL) UREA ◇ GLICLAZIDE ◇ N-(((HEXAHYDROCYCLOPENTA(c)PYRROL-2(1H)-YL)AMINO)CARBONYL)-4-METHYL-BENZENESULFONAMIDE ◇ 1-(HEXAHYDROCYCLOPENTA(c)PYRROL-2(1H)-YL)-3-(p-TOLYL-SULFONYL)UREA ◇ N-(4-METHYLBENZENESULFONYL)-N′-(3-AZA-BICYCLO(3.3.0)OCT-3-YL)UREA ◇ NORDIALEX ◇ S 852 ◇ S 1702 ◇ SE 1702

TOXICITY DATA with REFERENCE
orl-rat TDLo:4400 mg/kg (7-17D preg):REP YACHDS 9,3551,81
orl-rat TDLo:2200 mg/kg (7-17D preg):TER YACHDS 9,3551,81
orl-rat LD50:5 g/kg YACHDS 8,2661,80
ivn-rat LD50:382 mg/kg YACHDS 8,2661,80
scu-mus LD50:1034 mg/kg YACHDS 8,2661,80
ivn-mus LD50:295 mg/kg YACHDS 8,2661,80

SAFETY PROFILE: Poison by intravenous route. Moderately toxic by subcutaneous route. Mildly toxic by ingestion. Experimental teratogenic and reproductive effects. When heated to decomposition it emits toxic fumes of SO_x and NO_x.

DBL800 CAS:140-64-7 ***HR: 3***
DIAMIDINE
mf: $C_{19}H_{24}N_4O_2{\cdot}C_4H_{12}O_8S_2$ mw: 592.75

SYNS: 4,4′-DIAMIDINODIPHENOXYPENTANE DI(β-HYDROXYETHANESULFONATE ◇ 4,4′-DIAMIDINO-α,omega-DIPHENOXYPENTANE ISETHIONATE ◇ LOMIDIN ◇ LOMIDINE ◇ LOMIDINE ISOETHIONATE ◇ M & B 800 ◇ p,p′-(PENTAMETHYLENEDIOXY) DIBENZAMIDINE BIS(β-HYDROXYETHANESULFONATE) ◇ PENTAMIDINE DIISETHIONATE ◇ PENTAMIDINE ISETHIONATE ◇ 2512 R.P. ◇ R.P. 2512 ◇ USAF XR-10

TOXICITY DATA with REFERENCE
dni-mus:leu 29500 μg/L INNDDK 1,103,83
ims-wmn TDLo:6 mg/kg/D:BLD,SKN CANCAR 34,441,74
ipr-mus LD50:63 mg/kg ANTCAO 2,581,52
scu-mus LD50:120 mg/kg ANTCAO 2,581,52
ivn-mus LD50:15100 μg/kg ANTCAO 2,581,52

SAFETY PROFILE: Poison by intraperitoneal, subcutaneous, and intravenous routes. Human systemic effects by intramuscular route: hemorrhage and dermatitis. Mutation data reported. When heated to decomposition it emits very toxic fumes of NO_x and SO_x.

DBM000 CAS:100-33-4 ***HR: 3***
4,4′-DIAMIDINODIPHENOXYPENTANE
mf: $C_{19}H_{24}N_4O_2$ mw: 340.47

SYNS: 4,4′-DIAMIDINO-α,omega-DIPHENOXYPENTANE ◇ p,p′-(PENTAMETHYLENE-DIOXY)BIS-BENZAMIDINE ◇ p,p′-(PENTAMETHYLENEDIOXY)DIBENZAMIDINE ◇ 4,4′-(PENTAMETHYLENEDIOXY)DIBENZAMIDINE ◇ PENTAMIDINE ◇ 4,4′-(1,5-PENTANEDIYLBIS(OXY))BIS-BENZENECARBOXIMIDAMIDE,(9CI)

TOXICITY DATA with REFERENCE
ims-man TDLo:28 mg/kg/1W-I:KID,MET AIMDAP 145,2247,85
par-man TDLo:4 mg/kg/3D:CNS,MET LANCAO 246,338,44
ipr-mus LD50:50 mg/kg JMCMAR 18,794,75

SAFETY PROFILE: Poison by intraperitoneal route. Human systemic effects by intramuscular and parenteral routes: anesthesia, changes in kidney tubules, fever, and convulsions. When heated to decomposition it emits toxic fumes of NO_x.

DBM400 CAS:6275-69-0 ***HR: 3***
4,4′-DIAMIDINO-1,3-DIPHENOXYPROPANE DIHYDROCHLORIDE
mf: $C_{17}H_{20}N_4O_2{\cdot}2ClH$ mw: 385.33

SYNS: 4,4′-DIAMIDINO-α,Γ-DIPHENOXYPROPANE DIHYDROCHLORIDE ◇ M & B 782 DIHYDROCHLORIDE ◇ PANAMIDIN DIHYDROCHLORIDE ◇ PROPAMIDINE DIHYDROCHLORIDE ◇ 4,4′-(1,3-PROPANEDIYLBIS(OXY))BIS-BENZENECARBOXIMIDAMIDE,DIHYDROCHLORIDE ◇ 4,4′-(TRIMETHYLENEDIOXY)DIBENZAMIDINE DIHYDROCHLORIDE

TOXICITY DATA with REFERENCE
ipr-mus LD50:40 mg/kg ANTCAO 2,581,52

scu-mus LD50:55 mg/kg ATMPA2 37,1,43
ivn-mus LD50:35100 µg/kg ANTCAO 2,581,52

SAFETY PROFILE: Poison by intraperitoneal, subcutaneous, and intravenous routes. When heated to decomposition it emits very toxic fumes of HCl and NO_x.

DBM800 CAS:59-33-6 ***HR: 3***
DIAMINIDE MALEATE
mf: $C_{17}H_{23}N_3O{\cdot}C_4H_4O_4$ mw: 401.51

SYNS: AH ◇ ANISOPYRADAMINE ◇ ANTHISAN MALEATE ◇ ANTIHIST ◇ N-DIMETHYLAMINOETHYL-N-p-METHOXY-α-AMINOPYRIDINE MALEATE ◇ 2-((2-(DIMETHYLAMINO)ETHYL) (p-METHOXYBENZYL)AMINO)PYRIDINE BIMALEATE ◇ 2-((2-(DIMETHYLAMINO)ETHYL)(p-METHOXYBENZYL)AMINO)PYRIDINEMALEATE ◇ N,N-DIMETHYL-N'-(4-METHOXYBENZYL)-N'-(2-PYRIDYL) ETHYLENEDIAMINE MALEATE ◇ HISTATEX ◇ MEPYRAMINE MALEATE ◇ N-p-METHOXYBENZYL-N'-N'-DIMETHYL-N-α-PYRIDYLETHYLENEDIAMINE MALEATE ◇ MINIHIST ◇ NEOANTERGAN MALEATE ◇ PARAMAL ◇ PARAMINYL MALEATE ◇ PYMAFED ◇ PYRA MALEATE ◇ PYRANILAMINE MALEATE ◇ PYRANINYL ◇ PYRANISAMINE MALEATE ◇ PYRILAMINE MALEATE ◇ RENSTAMIN ◇ 2786 R.P. MALEATE ◇ STANGEN MALEATE ◇ STATOMIN MALEATE ◇ THYLOGEN MALEATE

TOXICITY DATA with REFERENCE
dns-rat:lvr 5 µmol/L ENMUDM 3,11,81
orl-mus TDLo:4200 mg/kg (female 1-21D post):REP ARZNAD 18,188,68
orl-rat TDLo:77 g/kg/2Y-C:ETA FCTOD7 22,27,84
orl-chd LDLo:42 mg/kg LANCAO 2,809,52
orl-wmn TDLo:200 mg/kg JAMAAP 257,660,87
orl-rat LD50:365 mg/kg TXAPA9 1,42,59
scu-rat LD50:150 mg/kg CRSBAW 144,887,50
orl-mus LD50:220 mg/kg AIPTAK 155,47,65
ipr-mus LD50:102 mg/kg JPETAB 90,224,47
ivn-mus LD50:23 mg/kg AIPTAK 155,47,65
ivn-gpg LD50:24400 µg/kg AIPTAK 113,313,58

CONSENSUS REPORTS: Reported in EPA TSCA Inventory.

SAFETY PROFILE: A human poison by ingestion. An experimental poison by ingestion, subcutaneous, intravenous, and intraperitoneal routes. Experimental reproductive effects. Questionable carcinogen with experimental tumorigenic data. Mutation data reported. An antihistamine. When heated to decomposition it emits toxic fumes of NO_x.

DBN000 CAS:3407-94-1 ***HR: 3***
2,6-DIAMINOACRIDINE
mf: $C_{13}H_{11}N_3$ mw: 209.27

SYNS: ACRAMINE RED ◇ 2,6-ACRIDINEDIAMINE ◇ 3,7-DIAMINOACRIDINE ◇ DIFLAVINE (ACRIDINE)

TOXICITY DATA with REFERENCE
ipr-mus LD50:300 mg/kg BJEPA5 28,1,47
scu-mus LD50:130 mg/kg BJEPA5 28,1,47

SAFETY PROFILE: Poison by intraperitoneal and subcutaneous routes. When heated to decomposition it emits toxic fumes of NO_x.

DBN200 ***HR: 3***
3,6-DIAMINOACRIDINE HYDROCHLORIDE HEMIHYDRATE
mf: $Cl_3H_{11}N_3{\cdot}ClH{\cdot}1/2H_2O$ mw: 254.74

SYNS: NCI-C04137 ◇ PROFLAVINE MONOHYDROCHLORIDE HEMIHYDRATE

TOXICITY DATA with REFERENCE
orl-rat TDLo:27 g/kg/109W-C:ETA NCITR* NCI-CG-TR-5,77

CONSENSUS REPORTS: IARC Cancer Review: Animal Inadequate Evidence IMEMDT 24,195,80. NCI Carcinogenesis Bioassay (feed); Inadequate Studies: mouse NCITR* NCI-CG-TR-5,77; Clear Evidence: rat NCITR* NCI-CG-TR-5,77.

SAFETY PROFILE: Questionable carcinogen with experimental tumorigenic data. When heated to decomposition it emits very toxic fumes of NO_x and Cl^-.

DBN400 CAS:553-30-0 ***HR: D***
3,6-DIAMINOACRIDINE SULPHATE (1:1)
mf: $C_{13}H_{11}N_3{\cdot}H_2O_4S$ mw: 307.35

SYNS: 3,6-ACRIDINEDIAMINE SULFATE (2:1) ◇ 3,6-ACRIDINEDIAMINE SULPHATE ◇ 3,6-DIAMINOACRIDINE BISULPHATE ◇ 3,6-DIAMINOACRIDINE SULFATE (1:1) ◇ 3,6-DIAMINOACRIDINIUM MONOHYDROGEN SULPHATE ◇ 2,8-DIAMINOACRIDINIUM SULPHATE ◇ FLAVINE ◇ FLAVIN SULPHATE ◇ ISOFLAV ◇ NEUTRAL PROFLAVINE SULPHATE ◇ PANCRIDINE ◇ PROFALVINE SULPHATE ◇ PROFLAVINE (SULFATE) ◇ PROFLAVIN SULFATE ◇ SANOFLAVIN

TOXICITY DATA with REFERENCE
mmo-sat 10 µL/plate ANYAA9 76,475,58
mmo-esc 500 µg/disc APMBAY 6,23,58
dni-omi 5 mg/L MGGEAE 150,301,77
msc-mus:lym 500 µg/L/2H MUREAV 56,61,79

CONSENSUS REPORTS: EPA Genetic Toxicology Program.

SAFETY PROFILE: Mutation data reported. When heated to decomposition it emits very toxic fumes of NO_x and SO_x. See also SULFATES and other diaminoacride entries.

DBN600 CAS:92-62-6 ***HR: 3***
3,6-DIAMINOACRIDINIUM
mf: $C_{13}H_{11}N_3$ mw: 209.27

SYNS: 3,6-ACRIDINEDIAMINE ◇ 3,6-DIAMINOACRIDINE ◇ 2,8-DIAMINOACRIDINE ◇ 2,8-DIAMINOACRIDINIUM ◇ 3,7-DIAMINO-5-AZAANTHRACENE ◇ ISOFLAV BASE ◇ PROFLAVIN ◇ PROFLAVINE ◇ PROFOLIOL ◇ PROFORMIPHEN ◇ PROFUNDOL ◇ PROFURA ◇ PROGARMED ◇ PRO-GEN ◇ PROGESIC

TOXICITY DATA with REFERENCE
dnd-esc 10 mg/L MUREAV 107,1,83
otr-mus:emb 1 μmol/L MOPMA3 21,739,82
dni-mus:emb 1 μmol/L MOPMA3 21,739,82
oms-mus:emb 1 μmol/L MOPMA3 21,739,82
sce-ham:ovr 20 μg/L ENMUDM 4,65,82
scu-mus LD50:140 mg/kg BJEPA5 28,1,47
ivn-cat LDLo:11111 μg/kg LANCAO 196,838,19

CONSENSUS REPORTS: IARC Cancer Review: Animal Inadequate Evidence IMEMDT 24,195,80. Reported in EPA TSCA Inventory. EPA Genetic Toxicology Program.

SAFETY PROFILE: Poison by intravenous and subcutaneous routes. Questionable carcinogen. Mutation data reported. When heated to decomposition it emits toxic fumes of NO_x. See also other diaminoacridine entries.

DBO000 CAS:615-05-4 ***HR: 3***
2,4-DIAMINOANISOLE
mf: $C_7H_{10}N_2O$ mw: 138.19

SYNS: C.I. 76050 ◇ C.I. OXIDATION BASE 12 ◇ 2,4-DAA ◇ 2,4-DI-AMINEANISOLE ◇ 2,4-DIAMINOANISOL ◇ 2,4-DIAMINOANISOLE BASE ◇ m-DIAMINOANISOLE 1,3-DIAMINO-4-METHOXYBENZENE ◇ 2,4-DIAMINO-1-METHOXYBENZENE ◇ FURRO L ◇ 4-METHOXY-1,3-BENZENEDIAMINE ◇ p-METHOXY-m-PHENYLENEDIAMINE ◇ 4-METHOXY-m-PHENYLENEDIAMINE ◇ 4-MMPD ◇ PELAGOL DA ◇ PELAGOL GREY L ◇ PELAGOL L

TOXICITY DATA with REFERENCE
skn-rbt 12500 μg/24H MLD FCTXAV 15,607,77
mma-sat 5 μg/plate MUREAV 79,289,80
dnd-hmn:fbr 50 μmol/L MUREAV 127,107,84
dnd-rat:lvr 500 μmol/L CBINA8 31,35,80
sce-mus-ipr 12 mg/kg MUREAV 108,225,83
scu-uns TDLo:65100 μg/kg (female 10D post):REP JIDOAA 35,387,86
orl-rat LD50:460 mg/kg FCTXAV 15,607,77
ipr-rat LD50:116 mg/kg BCPCA6 30,2715,81

CONSENSUS REPORTS: IARC Cancer Review: Group 2B IMEMDT 7,56,87; Human Limited Evidence IMEMDT 27,103,82. EPA Genetic Toxicology Program. Reported in EPA TSCA Inventory.

DFG MAK: Animal Carcinogen, Suspected Human Carcinogen.
NIOSH REL: (2,4-diaminoanisole) Reduce to lowest feasible level

SAFETY PROFILE: Confirmed carcinogen. Poison by intraperitoneal route. Moderately toxic by ingestion. Experimental reproductive effects. Human mutation data reported. A skin irritant. When heated to decomposition it emits toxic fumes of NO_x. See also other diaminoanisole entries.

DBO400 CAS:39156-41-7 ***HR: 3***
2,4-DIAMINOANISOLE SULPHATE
mf: $C_7H_{10}N_2O{\bullet}xH_2O_4S$ mw: 824.75

SYNS: BASF URSOL SLA ◇ C.I. 76051 ◇ C.I. OXIDATION BASE 12A ◇ 2,4-DAA SULFATE ◇ 2,4-DIAMINOANISOLE SULFATE ◇ 2,4-DI-AMINO-ANISOL SULPHATE ◇ 2,4-DIAMINO-1-METHOXYBENZENE ◇ 1,3-DIAMINO-4-METHOXYBENZENE SULPHATE ◇ 2,4-DIAMINO-1-METHOXYBENZENE SULPHATE ◇ 2,4-DIAMINOSOLE SULPHATE ◇ DURAFUR BROWN MN ◇ FOURAMINE BA ◇ FOURRINE SLA ◇ FURRO SLA ◇ 4-METHOXY-1,3-BENZENEDIAMINE SULFATE ◇ 4-METHOXY-1,3-BENZENEDIAMINE SULPHATE ◇ 4-METHOXY-m-PHENYLENEDIAMINE SULFATE ◇ p-METHOXY-m-PHENYLENEDI-AMINE SULPHATE ◇ 4-METHOXY-m-PHENYLENEDIAMINE SUL-PHATE ◇ 4-MMPD SULPHATE ◇ NAKO TSA ◇ NCI-C01989 ◇ OXIDA-TION BASE 12A ◇ PELAGOL GREY ◇ RENAL SLA ◇ URSOL SLA ◇ ZOBA SLE

TOXICITY DATA with REFERENCE
mma-sat 1 μg/plate ENMUDM 7(Suppl 5),1,85
sln-dmg-orl 15100 μmol/L/3D MUREAV 48,181,77
sln-nsc 150 mg/L MUREAV 167,35,86
mrc-smc 500 mg/L MUREAV 78,243,80
orl-rat TDLo:18 g/kg/10W-C:CAR JJIND8 65,197,80
orl-mus TDLo:157 g/kg/78W-C:CAR NCITR* NCI-CG-TR-84,78
orl-rat TD:33 g/kg/78W-C:ETA NCITR* NCI-CG-TR-84,78
orl-mus TD:1310 g/kg/78W-C:NEO IARC** 27,103,82
ipr-rat LD50:372 mg/kg JTEHD6 2,657,77

CONSENSUS REPORTS: NTP Fifth Annual Report on Carcinogens. IARC Cancer Review: Animal Sufficient Evidence IMEMDT 27,103,82; Animal Inadequate Evidence IMEMDT 16,51,78. NCI Carcinogenesis Bioassay (feed); Clear Evidence: mouse, rat NCITR* NCI-CG-TR-84,78. Reported in EPA TSCA Inventory. EPA Genetic Toxicology Program. Community Right-To-Know List.

SAFETY PROFILE: Confirmed carcinogen with experimental carcinogenic, neoplastigenic, and tumorigenic data. Poison by intraperitoneal route. Mutation data reported. When heated to decomposition it emits very toxic fumes of NO_x and SO_x. See also other diaminoanisole entries.

DBO600 CAS:5327-72-0 ***HR: 2***
1,4-DIAMINOANTHRACENE-9,10-DIOL
mf: $C_{14}H_{12}N_2O_2$ mw: 240.28

SYNS: 1,4-DIAMINO-9,10-ANTHRACENEDIOL ◇ 1,4-DIAMINO-9,10-DIHYDROXYANTHRACEN (CZECH) ◇ LEUKO-1,4-DIAMINOANTHRA-CHINON (CZECH) ◇ LEUCO-1,4-DIAMINOANTHRAQUINONE

TOXICITY DATA with REFERENCE
eye-rbt 100 mg/24H MOD 28ZPAK -,111,72
orl-rat LD50:454 mg/kg 28ZPAK -,111,72

CONSENSUS REPORTS: Reported in EPA TSCA Inventory.

SAFETY PROFILE: Moderately toxic by ingestion. An eye irritant. When heated to decomposition it emits toxic fumes of NO_x.

DBO800 CAS:1758-68-5 ***HR: 3***
1,2-DIAMINOANTHRAQUINONE
mf: $C_{14}H_{10}N_2O_2$ mw: 238.26

SYN: 1,2-DAA (RUSSIAN)

TOXICITY DATA with REFERENCE
mmo-sat 100 μg/plate MUREAV 40,203,76
mma-sat 100 μg/plate MUREAV 40,203,76
ipr-rat LD50:2700 mg/kg GTPZAB 21(12),27,77
ivn-mus LD50:320 mg/kg CSLNX* NX#01102

CONSENSUS REPORTS: Reported in EPA TSCA Inventory.

SAFETY PROFILE: Poison by intravenous route. Moderately toxic by intraperitoneal route. Mutation data reported. When heated to decomposition it emits toxic fumes of NO_x.

DBP000 CAS:128-95-0 ***HR: 2***
1,4-DIAMINOANTHRAQUINONE
mf: $C_{14}H_{10}N_2O_2$ mw: 238.26

SYNS: ACETATE RED VIOLET R ◇ ACETOQUINONE LIGHT HELIOTROPE NL ◇ ACETYLON FAST RED VIOLET R ◇ AMACEL HELIOTROPE R ◇ AMAPLAST RED VIOLET P 2R ◇ 1,4-ANTHRAQUINONYLDIAMINE ◇ ARTISIL VIOLET 2RP ◇ CELANTHRENE RED VIOLET R ◇ CELLITON FAST RED VIOLET ◇ CELLUTATE RED VIOLET RH ◇ C.I. 61100 ◇ CIBACET VIOLET 2R ◇ C.I. DISPERSE VIOLET 1 ◇ CILLA FAST RED VIOLET RN ◇ C.I. SOLVENT VIOLET 11 ◇ DIACELLITON FAST VIOLET 5R ◇ 1,4-DIAMINO-9,10-ANTHRACENEDIONE ◇ 1,4-DIAMINOANTHRACHINON (CZECH) ◇ DISPERSE VIOLET K ◇ DURANOL VIOLET WR ◇ FENACET FAST VIOLET 5R ◇ GRACET VIOLET 2R ◇ GRASOL VIOLET R ◇ INTERCHEM ACETATE VIOLET R ◇ KRISOLAMINE ◇ MICROSETILE VIOLET 3R ◇ MIDETON FAST RED VIOLET R ◇ NACELAN VIOLET 4R ◇ NYLOQUINONE VIOLET R ◇ OIL VIOLET R ◇ ORACET VIOLET 2R ◇ PERLITON VIOLET 3R ◇ RESIREN VIOLET TR ◇ SEACYL VIOLET R ◇ SERISOL BRILLIANT VIOLET 2R ◇ SETACYL VIOLET R ◇ SETILE VIOLET 3R ◇ SUPRACET BRILLIANT VIOLET 3R ◇ TRANSETILE VIOLET P 3R

TOXICITY DATA with REFERENCE
eye-rbt 500 mg/24H MLD 28ZPAK -,121,72
mmo-sat 100 μg/plate MUREAV 40,203,76
mma-sat 100 μg/plate MUREAV 40,203,76
mmo-omi 12800 μg/L HEREAY 99,209,83
orl-rat LDLo:5790 mg/kg 28ZPAK -,121,72
ipr-rat LD50:1300 mg/kg GTPZAB 21(12),27,77

CONSENSUS REPORTS: Reported in EPA TSCA Inventory. EPA Genetic Toxicology Program.

SAFETY PROFILE: Moderately toxic by intraperitoneal route. Mildly toxic by ingestion. Mutation data reported. An eye irritant. When heated to decomposition it emits toxic fumes of NO_x.

DBP200 CAS:129-44-2 ***HR: 2***
1,5-DIAMINOANTHRAQUINONE
mf: $C_{14}H_{10}N_2O_2$ mw: 238.26

SYNS: 1,5-ANTHRAQUINONYLDIAMINE ◇ 1,5-DIAMINOANTHRACHINON (CZECH) ◇ 1,5-DIAMINO-9,10-ANTHRAQUINONE

TOXICITY DATA with REFERENCE
eye-rbt 500 mg/24H MLD 28ZPAK -,122,72
mma-sat 500 μg/plate MUREAV 40,203,76
ipr-rat LD50:1300 mg/kg GTPZAB 21(12),27,77

CONSENSUS REPORTS: Reported in EPA TSCA Inventory.

SAFETY PROFILE: Moderately toxic by intraperitoneal route. Mutation data reported. An eye irritant. When heated to decomposition it emits toxic fumes of NO_x.

DBP400 CAS:129-42-0 ***HR: 1***
1,8-DIAMINOANTHRAQUINONE
mf: $C_{14}H_{10}N_2O_2$ mw: 238.26

SYNS: 1,5-ANTHRAQUINONYLDIAMINE ◇ 1,8-ANTHRAQUINONYLDIAMINE ◇ 1,8-DIAMINOANTHRACHINON (CZECH) ◇ 1,5-DIAMINOANTHRAQUINONE

TOXICITY DATA with REFERENCE
eye-rbt 500 mg/24H MLD 28ZPAK -,122,72

CONSENSUS REPORTS: Reported in EPA TSCA Inventory.

SAFETY PROFILE: An eye irritant. When heated to decomposition it emits toxic fumes of NO_x.

DBP909 CAS:145-49-3 ***HR: 3***
1,5-DIAMINOANTHRARUFIN
mf: $C_{14}H_{10}N_2O_4$ mw: 270.26

SYNS: 4,8-DIAMINOANTHRARUFIN ◇ 1,5-DIAMINO-4,8-DIHYDROXY-9,10-ANTHRACENEDIONE ◇ 4,8-DIAMINO-1,5-DIHYDROXYANTHRAQUINONE ◇ 1,5-DIAMINO-4,8-DIHYDROXYANTHRAQUINONE ◇ leuco-1,5-DIAMINO-4,8-DIHYDROXYANTHRAQUINONE ◇ 1,5-DIHYDROXY-4,8-DIAMINOANTHRACHINON (CZECH) ◇ 1,5-DIHYDROXY-4,8-DIAMINOANTHRAQUINONE

TOXICITY DATA with REFERENCE mg/24H MOD 28ZPAK -,103,72
mmo-sat 50 μg/plate MUREAV 40,203,76
mma-sat 50 μg/plate MUREAV 40,203,76
ivn-mus LD50:56 mg/kg CSLNX* NX#03923

CONSENSUS REPORTS: Reported in EPA TSCA Inventory.

SAFETY PROFILE: Poison by intravenous route. An

eye irritant. Mutation data reported. When heated to decomposition it emits toxic fumes of NO_x.

DBP999 CAS:495-54-5 ***HR: 2***
DIAMINOAZOBENZENE
mf: $C_{12}H_{12}N_4$ mw: 212.28

PROP: White crystals. Mp: 63°, bp: 286°, d: 1.139, vap press: 1 mm @ 99.8°.

SYNS: CHRYSOIDIN A ◇ C.I. 11270 ◇ 2,4-DIAMINOAZOBENZEN (CZECH) ◇ ORANZ ZASADITA 2 (CZECH) ◇ 4-(PHENYLAZO)-m-PHENYLENEDIAMINE

TOXICITY DATA with REFERENCE
eye-rbt 20 mg/24H MOD 28ZPAK -,237,72
orl-rat LD50:1650 mg/kg 28ZPAK -,237,72

CONSENSUS REPORTS: Reported in EPA TSCA Inventory.

SAFETY PROFILE: Moderately toxic by ingestion. An eye irritant. When heated to decomposition it emits toxic fumes of NO_x.

DBQ125 CAS:58338-59-3 ***HR: 3***
2',4-DIAMINOBENZANILIDE
mf: $C_{13}H_{13}N_3O$ mw: 227.26

SYNS: 4-AMINO-N-(2'-AMINOPHENYL)BENZAMIDE ◇ GOE 1734

TOXICITY DATA with REFERENCE
mma-sat 5 mg/plate CTRRDO 69,1415,85
dnd-rat-orl 10 mg/kg CTRRDO 69,1415,85
orl-rat LD50:95 mg/kg CTRRDO 69,1415,85
orl-mus LD50:625 mg/kg CTRRDO 69,1415,85

SAFETY PROFILE: Poison by ingestion. Mutation data reported. When heated to decomposition it emits toxic fumes of NO_x.

DBQ800 CAS:3385-21-5 ***HR: 3***
1,3-DIAMINOCYCLOHEXANE
mf: $C_6H_{14}N_2$ mw: 114.22

SYN: 1,3-CYCLOHEXANEDIAMINE

TOXICITY DATA with REFERENCE
orl-rat LD50:390 mg/kg AMIHAB 17,129,58
orl-mus LD50:543 mg/kg AMIHAB 17,129,58

CONSENSUS REPORTS: Reported in EPA TSCA Inventory.

SAFETY PROFILE: Poison by ingestion. When heated to decomposition it emits toxic fumes of NO_x. See also AMINES.

DBR000 CAS:68772-00-9 ***HR: 3***
2,4-DIAMINO-5-(p-(p-((p-(2,4-DIAMINO-1-ETHYLPYRIMIDINIUM-5-YL)PHENYL)CARBAMOYLCINNAMAMIDO)PHENYL)-1-ETHYLPYRIMIDINIUM, DI-p-TOLUENE SULFONATE
mf: $C_{34}H_{36}N_{10}O_2 \cdot 2C_7H_7O_3S$ mw: 959.20

TOXICITY DATA with REFERENCE
dnd-mus:lym 400 nmol/L JMCMAR 22,134,79
ipr-mus LD10:27 mg/kg JMCMAR 22,134,79

SAFETY PROFILE: Poison by intraperitoneal route. Mutation data reported. When heated to decomposition it emits very toxic fumes of NO_x and SO_x. See also SULFONATES.

DBR200 CAS:68797-80-7 ***HR: 3***
2,4-DIAMINO-5-(p-(p-((-(2,4-DIAMINO-1-METHYLPYRIMIDINIUM-5-YL)PHENYL)CARBAMOYL)CINNAMAMIDO)PHENYL-1-METHYLPYRIMIDINIUM-DI-p-TOLUENE SULFONATE
mf: $C_{32}H_{32}N_{10}O_2 \cdot 2C_7H_7O_3S$ mw: 931.14

TOXICITY DATA with REFERENCE
dnd-mus:lym 400 nmol/L JMCMAR 22,134,79
ipr-mus LD10:50 mg/kg JMCMAR 22,134,79

SAFETY PROFILE: Poison by intraperitoneal route. Mutation data reported. When heated to decomposition it emits very toxic fumes of NO_x and SO_x. See also SULFONATES.

DBR400 CAS:609-20-1 ***HR: 3***
1,4-DIAMINO-2,6-DICHLOROBENZENE
mf: $C_6H_6Cl_2N_2$ mw: 177.04

SYNS: C.I. 37020 ◇ 2,6-DICHLORO-p-PHENYLENEDIAMINE ◇ NCI-C50260

TOXICITY DATA with REFERENCE
mma-sat 1 mg/plate ENMUDM 8(Suppl 7),1,86
orl-mus TDLo:87 g/kg/2Y-C:CAR NTPTR* NTP-TR-219,82
orl-mus TD:260 g/kg/2Y-C:CAR NTPTR* NTP-TR-219,82
orl-rat LD50:700 mg/kg NTPTR* NTP-TR-219,82F

CONSENSUS REPORTS: IARC Cancer Review: Group 3 IMEMDT 7,56,87; Animal Limited Evidence IMEMDT 39,325,86. NTP Carcinogenesis Bioassay (feed); Clear Evidence: mouse NTPTR* NTP-TR-219,82; No Evidence: rat NTPTR* NTP-TR-219,82. Reported in EPA TSCA Inventory.

SAFETY PROFILE: Questionable carcinogen with experimental carcinogenic data. Moderately toxic by ingestion. Mutation data reported. When heated to decomposition it emits very toxic fumes of Cl^- and NO_x.

DBT000 CAS:4702-64-1 ***HR: 1***
1,5-DIAMINO-4,8-DIHYDROXY-3-(p-METHOXYPHENYL)ANTHRAQUINONE
mf: $C_{21}H_{16}N_2O_5$ mw: 376.39

SYN: MODR OSTACETOVA SE-LB (CZECH)

TOXICITY DATA with REFERENCE
eye-rbt 500 mg/24H MLD 28ZPAK -,245,72
orl-rat LD50:6380 mg/kg 28ZPAK -,245,72

CONSENSUS REPORTS: Reported in EPA TSCA Inventory.

SAFETY PROFILE: Mildly toxic by ingestion. An eye irritant. When heated to decomposition it emits toxic fumes of NO_x.

DBT200 CAS:92-26-2 ***HR: 3***
3,6-DIAMINO-2,7-DIMETHYLACRIDINE
mf: $C_{15}H_{15}N_3$ mw: 237.33

SYNS: ACRIDINE YELLOW BASE ◇ 2,8-DIAMINO-3,7-DIMETHYLACRIDINE

TOXICITY DATA with REFERENCE
mma-sat 400 nmol/L ENMUDM 3,11,81
dns-rat:lvr 5 μmol/L ENMUDM 3,11,81
scu-mus LD50:280 mg/kg BJEPA5 28,1,47

CONSENSUS REPORTS: Reported in EPA TSCA Inventory.

SAFETY PROFILE: Poison by subcutaneous route. Mutation data reported. When heated to decomposition it emits toxic fumes of NO_x.

DBT400 CAS:135-49-9 ***HR: D***
3,6-DIAMINO-2,7-DIMETHYLACRIDINE HYDROCHLORIDE
mf: $C_{15}H_{15}N_3•ClH$ mw: 273.79

SYNS: ACRIDINE YELLOW ◇ ACRIDINE YELLOW G ◇ BASIC YELLOW K ◇ 2,7-DIMETHYL-3,6-ACRIDINEDIAMINE MONOHYDROCHLORIDE

TOXICITY DATA with REFERENCE
mmo-omi 1600 μg/L JMOBAK 3,762,61
dnd-mam:lym 10 pph BIPMAA 11,2537,72

CONSENSUS REPORTS: Reported in EPA TSCA Inventory. EPA Genetic Toxicology Program.

SAFETY PROFILE: Mutation data reported. When heated to decomposition it emits very toxic fumes of NO_x and HCl.

DBU600 ***HR: 3***
1,2-DIAMINOETHANEBISTRIMETHYLGOLD
mf: $C_8H_{26}Au_2N_2$ mw: 536.25

$[(CH_3)_3Au:NH_2CH_2\text{-}]_2$

SAFETY PROFILE: Explodes when heated, exposed to light, or on contact with nitric acid. When heated to decomposition it emits toxic fumes of NO_x.

DBU800 CAS:13426-91-0 ***HR: 3***
1,2-DIAMINOETHANE COPPER COMPLEX
DOT: UN 1761
mf: $C_2H_{10}N_2•xCu$ mw: 506.92

SYNS: CUPRIETHYLENE DIAMINE ◇ CUPRIETHYLENEDIAMINE, solution (DOT)

CONSENSUS REPORTS: Copper and its compounds are on the Community Right-To-Know List.

DOT Classification: Corrosive Material; Label: Corrosive, Poison.

SAFETY PROFILE: A corrosive poison. An irritating and corrosive material to the skin, eyes, and mucous membranes. When heated to decomposition it emits toxic fumes of NO_x. See also COPPER COMPOUNDS.

DBV200 CAS:97194-20-2 ***HR: 3***
4,6-DIAMINO-1-(2-ETHYLPHENYL)-2-METHYL-2-PROPYL-s-TRIAZINE HYDROCHLORIDE
mf: $C_{15}H_{23}N_5•ClH$ mw: 309.89

TOXICITY DATA with REFERENCE
orl-mus LD50:700 mg/kg JMCMAR 6,370,63
ipr-mus LD50:78 mg/kg JMCMAR 6,370,63

SAFETY PROFILE: Poison by intraperitoneal route. Moderately toxic by ingestion. When heated to decomposition it emits very toxic fumes of NO_x and HCl.

DBV400 CAS:1239-45-8 ***HR: 3***
2,7-DIAMINO-10-ETHYL-9-PHENYLPHENANTHRIDINIUM BROMIDE
mf: $C_{21}H_{20}N_3•Br$ mw: 394.35

SYNS: 3,8-DIAMINO-5-ETHYL-6-PHENYLPHENANTHRIDINIUM BROMIDE ◇ 2,7-DIAMINO-9-PHENYL-10-ETHYLPHENANTHRIDINIUM BROMIDE ◇ 2,7-DIAMINO-9-PHENYLPHENANTHRIDINE ETHOBROMIDE ◇ DROMILAC ◇ ETHIDIUM BROMIDE ◇ HOMIDIUM BROMIDE ◇ RD 1572

TOXICITY DATA with REFERENCE
mma-sat 200 ng/plate MUREAV 127,31,84
sln-dmg-mul 3 mmol/L MUREAV 138,169,84
dni-hmn:hla 40 μmol/L MUREAV 92,427,82
ipr-mus LD50:20 mg/kg CNREA8 34,2699,74
scu-mus LD50:110 mg/kg ATMPA2 46,285,52

CONSENSUS REPORTS: EPA Genetic Toxicology Program.

SAFETY PROFILE: Poison by intraperitoneal and subcutaneous routes. Human mutation data reported. When heated to decomposition it emits very toxic fumes of NO_x and Br^-. See also BROMIDES.

DBW000 CAS:50309-02-9 ***HR: 3***
4-((4-(((4-(2,4-DIAMINO-1-ETHYLPYRIMIDINIUM-5-YL)PHENYL)AMINO)CARBONYL)PHENYL) AMINO)-1-ETHYLQUINOLINIUM) DIIODIDE
mf: $C_{30}H_{31}N_7O•2I$ mw: 759.48

TOXICITY DATA with REFERENCE
dnd-mus:lym 1240 nmol/L JMCMAR 22,134,79
ipr-mus LD10:32 mg/kg JMCMAR 22,134,79

SAFETY PROFILE: Poison by intraperitoneal route. Mutation data reported. When heated to decomposition it emits very toxic fumes of NO_x and I^-.

DBW100 CAS:10308-83-5 ***HR: 3***
DIAMINOGUANIDINIUM NITRATE
mf: $CH_8N_6O_3$ mw: 152.11

SAFETY PROFILE: Decomposes violently above 260°C. A powerful oxidizer. When heated to decomposition it emits toxic fumes of NO_x. See also NITRATES.

DBW200 CAS:70548-53-7 ***HR: 3***
(l)-2,6-DIAMINO-1-HEXANETHIOL DIHYDROCHLORIDE
mf: $C_6H_{16}N_2S•2ClH$ mw: 221.22

TOXICITY DATA with REFERENCE
orl-mus LD50:450 mg/kg JMCMAR 22,631,79
ipr-mus LD50:88 mg/kg JMCMAR 22,631,79

SAFETY PROFILE: Poison by intraperitoneal route. Moderately toxic by ingestion. When heated to decomposition it emits very toxic fumes of Cl^-, SO_x, and NO_x.

DBW600 CAS:70561-82-9 ***HR: 2***
S-2,6-DIAMINOHEXYL DIHYDROGEN PHOSPHOROTHIOATE DIHYDRATE
mf: $C_6H_{17}N_2O_3PS•2H_2O$ mw: 264.32

SYN: PHOSPHOROTHIOIC ACID-S-2,6-DIAMINOHEXYL ESTER, DIHYDRATE

TOXICITY DATA with REFERENCE
orl-mus LD50:900 mg/kg JMCMAR 22,631,79
ipr-mus LD50:450 mg/kg JMCMAR 22,631,79

SAFETY PROFILE: Moderately toxic by ingestion and intraperitoneal routes. When heated to decomposition it emits very toxic fumes of NO_x, PO_x, and SO_x. See also ESTERS.

DBX000 CAS:2872-48-2 ***HR: 2***
1,4-DIAMINO-2-METHOXYANTHRAQUINONE
mf: $C_{15}H_{12}N_2O_3$ mw: 268.29

SYNS: C.I. 62015 ◇ C.I. DISPERSE RED 11 ◇ C.I. SOLVENT VIOLET 26 ◇ 1,4-DA-2-MOA (RUSSIAN)

TOXICITY DATA with REFERENCE
ipr-rat LD50:700 mg/kg GTPZAB 21(12),27,77

CONSENSUS REPORTS: Reported in EPA TSCA Inventory.

SAFETY PROFILE: Moderately toxic by intraperitoneal route. When heated to decomposition it emits toxic fumes of NO_x.

DBX400 CAS:8048-52-0 ***HR: 3***
3,6-DIAMINO-10-METHYLACRIDINIUM CHLORIDE with 3,6-ACRIDINEDIAMINE
mf: $C_{14}H_{14}N_3•Cl•C_{13}H_{11}N_3$ mw: 469.03

SYNS: ACRIFLAVIN ◇ ACRIFLAVINE mixture with PROFLAVINE ◇ ACRIFLAVINIUM CHLORIDE ◇ ACRIFLAVINIUM CHLORIDUM ◇ ACRIFLAVON ◇ ANGIFLAN ◇ ASSIFLAVINE ◇ AVLON ◇ BIALFLAVINA ◇ BIOACRIDIN ◇ BOVOFLAVIN ◇ BURNOL ◇ BUROFLAVIN ◇ CHOLIFLAVIN ◇ CHROMOFLAVINE ◇ DIACRID ◇ 3,6-DIAMINOACRIDINE mixture with 3,6-DIAMINO-10-METHYLACRIDINIUM CHLORIDE ◇ 2,8-DIAMINO-10-METHYLACRIDINIUM CHLORIDE mixture with 2,8-DIAMINOACRIDINE ◇ EUFLAVINE ◇ FLAVACRIDINUM HYDROCHLORICUM ◇ FLAVINE ◇ FLAVIOFORM ◇ FLAVIPIN ◇ FLAVISEPT ◇ GLYCO-FLAVINE ◇ GONACRINE ◇ ISRAVIN ◇ MEDIFLAVIN ◇ NEUTRAL ACRIFLAVINE ◇ PANFLAVIN ◇ PANTONSILETTEN ◇ TRACHOSEPT ◇ TRIPLA-ETILO ◇ TRYPAFLAVINE ◇ VETAFLAVIN ◇ XANTHACRIDINUM ◇ ZORIFLAVIN

TOXICITY DATA with REFERENCE
mmo-esc 5 μg/plate MUREAV 131,193,84
sln-dmg-mul 5000 ppm MUREAV 138,169,84
dni-hmn:hla 10 μmol/L RAREAE 37,334,69
scu-rat TDLo:160 mg/kg/60W-I:ETA SEMEAS 159,778,52

CONSENSUS REPORTS: IARC Cancer Review: Group 3 IMEMDT 7,56,87; Animal Inadequate Evidence IMEMDT 13,31,77. EPA Genetic Toxicology Program.

SAFETY PROFILE: Questionable carcinogen with experimental tumorigenic data. Poison by subcutaneous route. Human mutation data reported. A topical antiseptic used in the treatment of gonorrhea. When heated to decomposition it emits very toxic fumes of NO_x and Cl^-. See also 3,6-DIAMINO-10-METHYLACRIDINIUM CHLORIDE.

DBX875 CAS:7319-47-3 ***HR: 3***
2,4-DIAMINO-5-METHYL-6-sec-BUTYLPYRIDO(2,3-d)PYRIMIDINE
mf: $C_{12}H_{17}N_5$ mw: 231.34

SYNS: BW 283U ◇ BW 58-283 ◇ BW 58-283b ◇ 2,4-DIAMINO-5-METHYL-6-(BUT-2-YL)PYRIDO(2,3-d)PYRIMIDINE

TOXICITY DATA with REFERENCE
oms-hmn:lym 11 nmol/L BCPCA6 25,1947,76
orl-rat LD50:100 mg/kg 14XBAV -,367,64
orl-mus LD50:57 mg/kg JMCMAR 11,711,68

SAFETY PROFILE: Poison by ingestion. Human mutation data reported. When heated to decomposition it emits toxic fumes of NO_x.

DBY000 CAS:13897-55-7 ***HR: 2***
2,4-DIAMINO-1-METHYLCYCLOHEXANE
mf: $C_7H_{16}N_2$ mw: 128.25

SYNS: 2,4-DIAMINOMETHYLCYCLOHEXANE ◇ 1-METHYL-2,4-CYCLOHEXANEDIAMINE ◇ 4-METHYL-1,3-CYCLOHEXANEDIAMINE

TOXICITY DATA with REFERENCE
orl-rat LD50:1410 mg/kg AIHAAP 30,470,69
skn-rbt LD50:500 mg/kg AIHAAP 30,470,69

SAFETY PROFILE: Moderately toxic by ingestion and skin contact. When heated to decomposition it emits toxic fumes of NO_x. See also AMINES.

DBY100 CAS:18588-50-6 ***HR: D***
2,4-DIAMINO-6-METHYL-5-PHENYLPYRIMIDINE
mf: $C_{11}H_{12}N_4$ mw: 200.27

SYN: 2,4-DIAMINO-5-PHENYL-6-METHYLPYRIMIDINE

TOXICITY DATA with REFERENCE
orl-rat TDLo:50 mg/kg (14D preg):TER SJDBA9 7,45,76
orl-rat TDLo:25 mg/kg (female 9D post):REP AAGEAA 71(7),29,76

SAFETY PROFILE: Experimental teratogenic and reproductive effects. When heated to decomposition it emits toxic fumes of NO_x.

DBY300 CAS:17168-83-1 ***HR: 3***
1,2-DIAMINO-2-METHYLPROPANE AQUADIPEROXO CHROMIUM(IV)
mf: $C_4H_{14}CrN_2O_5$ mw: 222.16

$$[CH_3CH(NH_2)CH_2NH_2Cr(H_{2O})(O_2)_2]\cdot H_2O$$

CONSENSUS REPORTS: Chromium and its compounds are on the Community Right-To-Know List.

SAFETY PROFILE: Explodes at 83°C and is potentially explosive at 20°C. When heated to decomposition it emits toxic fumes of NO_x. See also CHROMIUM COMPOUNDS and PEROXIDES.

DBY500 CAS:18921-70-5 ***HR: 3***
N-(4-(((2,4-DIAMINO-5-METHYL-6-QUINAZOLINYL)METHYL)AMINO)BENZOYL)-l-ASPARTIC ACID
mf: $C_{21}H_{22}N_6O_5$ mw: 438.49

SYNS: N-(p-(((2,4-DIAMINO-5-METHYL-6-QUINAZOLINYL) METHYL)AMINO)BENZOYL)-l-ASPARTIC ACID ◇ METHASQUIN ◇ NSC 122870 ◇ SK 29836

TOXICITY DATA with REFERENCE
oms-hmn:lym 17 nmol/L BCPCA6 25,1947,76
ipr-mus LD50:50110 μg/kg NCISP* JAN86
scu-mus LD50:158 mg/kg NCISP* JAN86

SAFETY PROFILE: Poison by subcutaneous and intraperitoneal routes. Human mutation data reported. When heated to decomposition it emits toxic fumes of NO_x.

DBY600 CAS:3545-88-8 ***HR: 3***
1,7-DIAMINO-8-NAPHTHOL-3,6-DISULPHONIC ACID
mf: $C_{10}H_{10}N_2O_7S_2$ mw: 334.34

SYN: 3,6-NAPHTHALENEDISULFONIC ACID, 1,7-DIAMINO-8-HYDROXY-

TOXICITY DATA with REFERENCE
scu-rat TDLo:25 mg/kg (female 8D post):REP NATUAS 208,1219,65
scu-rat LD50:66 mg/kg NATUAS 208,1219,65

SAFETY PROFILE: Poison by subcutaneous route. Experimental reproductive effects. When heated to decomposition it emits toxic fumes of NO_x and SO_x.

DBY700 CAS:82-33-7 ***HR: 2***
1,4-DIAMINO-5-NITRO ANTHRAQUINONE
mf: $C_{14}H_9N_3O_4$ mw: 283.26

SYN: 9,10-ANTHRACENEDIONE,1,4-DIAMINO-5-NITRO-(9CI)

TOXICITY DATA with REFERENCE
mma-sat 50 μg/plate MUREAV 40,203,76
skn-rat TDLo:22500 mg/kg/65W-I:ETA VINIT* #1684-81
ivn-mus LD50:56 mg/kg CSLNX* NX#01786

CONSENSUS REPORTS: Reported in EPA TSCA Inventory.

SAFETY PROFILE: Poison by intravenous route. Questionable carcinogen with experimental tumorigenic data. Mutation data reported. When heated to decomposition it emits toxic fumes of NO_x.

DBY800 CAS:720-69-4 ***HR: 3***
4,6-DIAMINO-2-(5-NITRO-2-FURYL)-S-TRIAZINE
mf: $C_7H_6N_6O_3$ mw: 222.19

TOXICITY DATA with REFERENCE
mma-sat 100 ng/plate MUREAV 40,9,76
dnr-sat 500 nmol/well CNREA8 34,2266,74
mmo-esc 300 nmol/well CNREA8 34,2266,74
mrc-esc 500 nmol/well CNREA8 34,2266,74
orl-rat TDLo:13 g/kg/46W-C:CAR JNCIAM 51,403,73

CONSENSUS REPORTS: EPA Genetic Toxicology Program.

SAFETY PROFILE: Questionable carcinogen with experimental carcinogenic data. Mutation data reported. When heated to decomposition it emits toxic fumes of NO_x.

DCA200 CAS:95-86-3 ***HR: 3***
2,4-DIAMINOPHENOL
mf: $C_6H_8N_2O$ mw: 124.16

TOXICITY DATA with REFERENCE
mma-sat 10 μg/plate BCPCA6 26,729,77
ipr-mus LDLo:50 mg/kg RBPMAZ 22,1,52

SAFETY PROFILE: Poison by intraperitoneal route. Mutation data reported. When heated to decomposition it emits toxic fumes of NO_x. See also AMINES and PHENOL.

DCA300 CAS:27653-49-2 ***HR: 3***
2,4-DIAMINO-5-PHENYL-6-ETHYLPYRIMIDINE
mf: $C_{12}H_{14}N_4$ mw: 214.30

SYN: 2,4-DIAMINO-6-ETHYL-5-PHENYLPYRIMIDINE

TOXICITY DATA with REFERENCE
orl-rat TDLo:50 mg/kg (14D preg):TER SJDBA9 7,45,76
orl-rat TDLo:1 mg/kg (female 9D post):REP AAGEAA 71(7),29,76
orl-rat LD50:126 mg/kg AAGEAA 71,29,76

SAFETY PROFILE: Poison by ingestion. Experimental teratogenic and reproductive effects. When heated to decomposition it emits toxic fumes of NO_x.

DCA450 CAS:27653-50-5 ***HR: D***
2,4-DIAMINO-5-PHENYL-6-PROPYLPYRIMIDINE
mf: $C_{13}H_{16}N_4$ mw: 228.33

TOXICITY DATA with REFERENCE
orl-rat TDLo:50 mg/kg (14D preg):TER SJDBA9 7,45,76
orl-rat TDLo:5 mg/kg (female 9D post):REP AAGEAA 71(7),29,76

SAFETY PROFILE: Experimental teratogenic and reproductive effects. When heated to decomposition it emits toxic fumes of NO_x.

DCA500 CAS:18588-49-3 ***HR: D***
2,4-DIAMINO-5-PHENYLPYRIMIDINE
mf: $C_{10}H_{10}N_4$ mw: 186.24

TOXICITY DATA with REFERENCE
orl-rat TDLo:50 mg/kg (14D preg):TER SJDBA9 7,45,76
orl-rat TDLo:25 mg/kg (female 9D post):REP AAGEAA 71(7),29,76

SAFETY PROFILE: Experimental teratogenic and reproductive effects. When heated to decomposition it emits toxic fumes of NO_x.

DCA600 CAS:942-31-4 ***HR: 3***
2,4-DIAMINO-5-PHENYLTHIAZOLE HYDROCHLORIDE
mf: $C_9H_9N_3S{\cdot}ClH$ mw: 227.73

SYNS: AMIFENATSOL HYDROCHLORIDE ◇ AMIPHENAZOLE HYDROCHLORIDE ◇ AMPHENAZOLE HYDROCHLORIDE ◇ AMPHISOL HYDROCHLORIDE ◇ DAFTAZOL HYDROCHLORIDE ◇ DAPTAZILE HYDROCHLORIDE ◇ DAPTAZOLE HYDROCHLORIDE ◇ 2,4-DIAMINO 5-PHENYLTHIAZOL CHLORHYDRATE (FRENCH) ◇ 2,4-DIAMINO-5-PHENYLTHIAZOLE MONOHYDROCHLORIDE ◇ DPT ◇ FENAMIZOL HYDROCHLORIDE ◇ PHENAMIZOLE HYDROCHLORIDE ◇ 5-PHENYL-2,4-THIAZOLEDIAMINE MONOHYDROCHLORIDE (9CI)

TOXICITY DATA with REFERENCE
ipr-rat LD50:300 mg/kg 27ZQAG -,201,72
orl-mus LD50:372 mg/kg 27ZQAG -,201,72
ipr-mus LD50:200 mg/kg 27ZQAG -,201,72
scu-mus LD50:310 mg/kg JPPMAB 21,668,69

SAFETY PROFILE: Poison by ingestion, intraperitoneal, and subcutaneous routes. When heated to decomposition it emits very toxic fumes of NO_x, SO_x, and HCl.

DCB000 CAS:38304-91-5 ***HR: 3***
2,4-DIAMINO-6-PIPERIDINOPYRIMIDINE-3-OXIDE
mf: $C_9H_{15}N_5O$ mw: 209.29

SYNS: 6-AMINO-1,2-DIHYDRO-1-HYDROXY-2-IMINO-4-PIPERIDINOPYRIMIDINE ◇ 2,4-DIAMINO-6-PIPERIDINILPIRIMIDINA-3-OSSIDO (ITALIAN) ◇ 2,3-DIHYDRO-3-HYDROXY-2-IMINO-6-(1-PIPERIDINYL)-4-PYRIMIDINAMINE ◇ LONITEN ◇ MINOSSIDILE (ITALIAN) ◇ MINOXIDIL ◇ 6-PIPERIDINO-2,4-DIAMINOPYRIMIDINE-3-OXIDE ◇ 6-(1-PIPERIDINYL)-2,4-PYRIMIDINEDIAMINE-3-OXIDE ◇ U-10,858

TOXICITY DATA with REFERENCE
orl-man TDLo:107 μg/kg/3D-I:BLD AIMEAS 92,874,80
orl-rat LD50:1321 mg/kg TXAPA9 39,1,77
ipr-rat LD50:759 mg/kg TXAPA9 39,1,77
ivn-rat LD50:29 mg/kg MEIEDD 10,888,83
ipr-mus LD50:560 mg/kg BCFAAI 121,16,82
ivn-mus LD50:51 mg/kg TXAPA9 39,1,77

SAFETY PROFILE: Poison by intravenous route. Moderately toxic by ingestion and intraperitoneal routes.

Human systemic effects by ingestion: thrombocytopenia (reduced numbers of blood platelets). An antihypertensive agent. When heated to decomposition it emits toxic fumes of NO_x.

DCC100 ***HR: 3***
m-DI-(2-AMINOPROPYL)BENZENE DIHYDROCHLORIDE
mf: $C_{12}H_{20}N_2•2ClH$ mw: 265.26

SYN: α,α'-DIMETHYL-m-BENZENEBIS(ETHYLAMINE)DIHYDROCHLORIDE

TOXICITY DATA with REFERENCE
ipr-mus LD50:139 mg/kg AITEAT 11,441,63
scu-mus LD50:570 mg/kg AITEAT 11,441,63
ivn-mus LD50:18 mg/kg AITEAT 11,441,63

SAFETY PROFILE: Poison by intravenous and intraperitoneal routes. Moderately toxic by subcutaneous route. When heated to decomposition it emits toxic fumes of NO_x and HCl.

DCC125 CAS:26076-87-9 ***HR: 3***
p-DI-(2-AMINOPROPYL)BENZENE DIHYDROCHLORIDE
mf: $C_{12}H_{20}N_2•2ClH$ mw: 265.26

SYN: α,α'-DIMETHYL-p-BENZENEBIS(ETHYLAMINE)DIHYDROCHLORIDE

TOXICITY DATA with REFERENCE
ipr-mus LD50:177 mg/kg AITEAT 11,441,63
scu-mus LD50:650 mg/kg AITEAT 11,441,63
ivn-mus LD50:23 mg/kg AITEAT 11,441,63

SAFETY PROFILE: Poison by intravenous and intraperitoneal routes. Moderately toxic by subcutaneous route. When heated to decomposition it emits toxic fumes of NO_x and HCl.

DCC200 CAS:50309-03-0 ***HR: 3***
4-((4-(((4-(2,4-DIAMINO-1-PROPYLPYRIMIDINIUM-5-YL)PHENYL)AMINO)CARBONYL)PHENYL)AMINO)-1-PROPYLQUINOLINIUM) DIIODIDE
mf: $C_{32}H_{35}N_7O•2I$ mw: 787.54

TOXICITY DATA with REFERENCE
dnd-mus:lym 1240 nmol/L JMCMAR 22,134,79
ipr-mus LD10:40 mg/kg JMCMAR 22,134,79

SAFETY PROFILE: Poison by intraperitoneal route. Mutation data reported. When heated to decomposition it emits very toxic fumes of NO_x and I^-. See also IODIDES.

DCC400 CAS:71-44-3 ***HR: 3***
DIAMINOPROPYLTETRAMETHYLENEDIAMINE
mf: $C_{10}H_{26}N_4$ mw: 202.40

SYNS: N,N'-BIS(3-AMINOPROPYL)-1,4-BUTANEDIAMINE ◇ 1,4-BIS(AMINOPROPYL) BUTANEDIAMINE ◇ N,N'-BIS(3-AMINOPROPYL)-1,4-DIAMINOBUTANE ◇ GERONTINE ◇ MUSCULAMINE ◇ NEURIDINE ◇ SPERMINE ◇ SPERMINE, PURISS

TOXICITY DATA with REFERENCE
mmo-sat 100 μmol/L AMACCQ 9,77,76
cyt-hmn:hla 2 mmol/L JCLLAX 78,217,71
dni-rat:lvr 100 μmol/L BIJOAK 146,697,75
ivn-rat LD50:65 mg/kg AIPTAK 165,374,67
ipr-mus LDLo:8 mg/kg TXAPA9 23,288,72
ivn-mus LD50:56 mg/kg CSLNX* NX#00641

CONSENSUS REPORTS: Reported in EPA TSCA Inventory.

SAFETY PROFILE: Poison by intraperitoneal and intravenous routes. Human mutation data reported. When heated to decomposition it emits toxic fumes of NO_x. See also AMINES.

DCC600 CAS:7280-83-3 ***HR: 3***
2,6-DIAMINOPURINE SULFATE
mf: $C_5H_6N_6•xH_2O_4S$ mw: 836.73

TOXICITY DATA with REFERENCE
ipr-mus LD50:250 mg/kg NTIS** AD691-490

CONSENSUS REPORTS: Reported in EPA TSCA Inventory.

SAFETY PROFILE: Poison by intraperitoneal route. When heated to decomposition it emits very toxic fumes of SO_x and NO_x. See also SULFATES.

DCC800 CAS:141-86-6 ***HR: 3***
2,6-DIAMINOPYRIDINE
mf: $C_5H_7N_3$ mw: 109.15

PROP: Crystals. Mp: 120.8°, bp: 285°.

TOXICITY DATA with REFERENCE
mma-sat 50 μg/plate ESKHA5 (94),28,76
ipr-mus LD50:100 mg/kg JMCMAR 8,296,65
ivn-mus LD50:56 mg/kg CSLNX* NX#00146

CONSENSUS REPORTS: Reported in EPA TSCA Inventory.

SAFETY PROFILE: Poison by intravenous and intraperitoneal routes. Mutation data reported. When heated to decomposition it emits toxic fumes of NO_x.

DCD000 CAS:54-96-6 ***HR: 3***
3,4-DIAMINOPYRIDINE
mf: $C_5H_7N_3$ mw: 109.15

SYNS: DIAMINO-3,4-PYRIDINE ◇ SC10

TOXICITY DATA with REFERENCE
ipr-mus LD50:20 mg/kg JMCMAR 8,296,65
scu-mus LD50:35 mg/kg AIPTAK 150,413,64
ivn-mus LD50:13 mg/kg APFRAD 26,345,68
orl-bwd LD50:75 mg/kg AECTCV 12,355,83

SAFETY PROFILE: Poison by ingestion, intraperitoneal, intravenous, and subcutaneous routes. When heated to decomposition it emits toxic fumes of NO_x.

DCE000 CAS:636-23-7 ***HR: 3***
2,4-DIAMINOTOLUENE DIHYDROCHLORIDE
mf: $C_7H_{10}N_2 \cdot 2ClH$ mw: 195.11

SYN: METATOLYLENEDIAMINE DIHYDROCHLORIDE

TOXICITY DATA with REFERENCE
orl-rat TDLo:9900 mg/kg/78W-C:NEO JEPTDQ 2,325,78
orl-mam LDLo:3000 mg/kg JIDHAN 13,87,31
ipr-mus LD50:80 mg/kg NCIBR* NIH-NCI-E-68-1311,10,73

CONSENSUS REPORTS: Reported in EPA TSCA Inventory.

SAFETY PROFILE: Poison by intraperitoneal route. Moderately toxic by ingestion. Questionable carcinogen with experimental neoplastigenic data. When heated to decomposition it emits very toxic fumes of NO_x and HCl.

DCE200 CAS:615-45-2 ***HR: 3***
2,5-DIAMINOTOLUENE DIHYDROCHLORIDE
mf: $C_7H_{10}N_2 \cdot 2ClH$ mw: 195.11

SYNS: 2-METHYL-1,4-BENZENEDIAMINEDIHYDROCHLORIDE ◇ p-TOLUENEDIAMINE DIHYDROCHLORIDE

TOXICITY DATA with REFERENCE
ipr-mus TDLo:50 mg/kg (8D preg):TER FCTXAV 15,447,77
orl-rat LDLo:100 mg/kg NCNSA6 5,11,53

CONSENSUS REPORTS: Reported in EPA TSCA Inventory.

SAFETY PROFILE: Poison by ingestion. Experimental teratogenic effects. When heated to decomposition it emits very toxic fumes of NO_x and HCl. See also CHLORIDES.

DCE400 CAS:15481-70-6 ***HR: 3***
2,6-DIAMINOTOLUENE DIHYDROCHLORIDE
mf: $C_7H_{10}N_2 \cdot 2ClH$ mw: 195.11

SYN: NCI-C50317

CONSENSUS REPORTS: Carcinogenesis Bioassay Completed; Results Negative NCITR* NCI-CG-TR-200,80. NCI Carcinogenesis Bioassay (feed); No Evidence: mouse, rat NCITR* NCI-CG-TR-200,80.

SAFETY PROFILE: When heated to decomposition it emits very toxic fumes of NO_x and HCl. See also 2,5,DIAMINOTOLUENE DIHYDROCHLORIDE.

DCE600 CAS:615-50-9 ***HR: 3***
2,5-DIAMINOTOLUENE SULFATE
mf: $C_7H_{10}N_2 \cdot H_2O_4S$ mw: 220.27

SYNS: C.I. 76043 ◇ p-DIAMINOTOLUENE SULFATE ◇ 2,5-DIAMINOTOLUENE SULPHATE ◇ 2-METHYL-1,4-BENZENEDIAMINE SULFATE ◇ 2-METHYL-p-PHENYLENEDIAMINE SULPHATE ◇ NCI-C01832 ◇ p-TOLUENEDIAMINE SULFATE ◇ 2,5-TOLUENEDIAMINE SULFATE ◇ TOLUENE-2,5-DIAMINE, SULFATE (1:1) (8CI) ◇ TOLUENE-2,5-DIAMINE SULPHATE ◇ p-TOLUENEDIAMINE SULPHATE ◇ TOLUYLENE-2,5-DIAMINE SULPHATE ◇ p-TOLUYLENEDIAMINE SULPHATE ◇ p-TOLYLENEDIAMINE SULPHATE

TOXICITY DATA with REFERENCE
orl-mus TDLo:66 g/kg/78W-C:ETA NCITR* NCI-CG-TR-126,78
orl-rat LD50:98 mg/kg JTEHD6 2,657,77
ipr-rat LD50:49 mg/kg JTEHD6 2,657,77

CONSENSUS REPORTS: IARC Cancer Review: Animal Indefinite Evidence IMEMDT 16,97,78. NCI Carcinogenesis Bioassay Completed; Results Indefinite: mouse, rat NCITR* NCI-CG-TR-126,78. Reported in EPA TSCA Inventory.

SAFETY PROFILE: Poison by ingestion and intraperitoneal routes. Questionable carcinogen with experimental tumorigenic data. When heated to decomposition it emits very toxic fumes of NO_x and SO_x. See also SULFATES.

DCE800 CAS:1326-22-3 ***HR: 3***
(4-((4,6-DIAMINO-m-TOLYL)IMINO)-2,5-CYCLOHEXADIEN-1-YLIDENE) DIMETHYLAMMONIUM CHLORIDE
mf: $C_{15}H_{19}N_4 \cdot Cl$ mw: 290.83

SYNS: CHLORIDE of DIAMINOMETHYLPHENYLDIMETHYL-p-BENZOQUINONE-DIIMINE ◇ C.I. 50411 ◇ C.I. 50435 ◇ TOLUYLENE BLUE (BASIC DYE)

TOXICITY DATA with REFERENCE
ivn-mus LD50:56 mg/kg CSLNX* NX#04800

SAFETY PROFILE: Poison by intravenous route. When heated to decomposition it emits very toxic fumes of NO_x and Cl^-, and NH_3.

DCF000 CAS:767-17-9 ***HR: 3***
4,6-DIAMINO-1,3,5-TRIAZINE-2-THIONE
mf: $C_3H_5N_5S$ mw: 143.19

SYNS: 4,6-DIAMINO-s-TRIAZINE-2-THIONE ◇ 4,6-DIAMINO-1,3,5-

TRIAZINE-2(1H)-THIONE ◇ TETRAHYDRO-4,6-DIIMINO-s-TRIAZINE-2(1H)-THIONE ◇ USAF B-45 ◇ USAF CY-14

TOXICITY DATA with REFERENCE
ipr-mus LD50:200 mg/kg NTIS** AD277-689

CONSENSUS REPORTS: Reported in EPA TSCA Inventory.

SAFETY PROFILE: Poison by intraperitoneal route. When heated to decomposition it emits very toxic fumes of NO_x and SO_x.

DCF200 CAS:1455-77-2 ***HR: 3***
3,5-DIAMINO-s-TRIAZOLE
mf: $C_2H_5N_5$ mw: 99.12

SYNS: GUANAZOLE ◇ NCI-C04819 ◇ NSC 1895

TOXICITY DATA with REFERENCE
oms-hmn:oth 500 μg/L CNREA8 32,2661,72
ipr-rat TDLo:2500 mg/kg/7W-I:ETA CANCAR 40(Suppl 4),1935,77
ivn-hmn TDLo:8620 mg/kg/5D:BLD CCROBU 59,1117,75

CONSENSUS REPORTS: NCI Carcinogenesis Studies (ipr); Equivocal Evidence: rat CANCAR 40,1935,77; No Evidence: mouse CANCAR 40,1935,77. Reported in EPA TSCA Inventory.

SAFETY PROFILE: Human systemic effects by intravenous route: leukopenia (reduced white blood cell count) and thrombocytopenia (reduced blood platelet count). Human mutation data reported. Questionable carcinogen with experimental tumorigenic data. When heated to decomposition it emits toxic fumes of NO_x.

DCF600 CAS:6818-18-4 ***HR: 2***
3,10-DIAMINOTRICYCLO(5.2.1.0^{2,6})DECANE
mf: $C_{10}H_{15}N_2O_4$ mw: 227.27

SYN: HEXAHYDRO-1,8-DIAMINO-4,7-METHANOINDAN

TOXICITY DATA with REFERENCE
orl-rat LD50:1070 mg/kg TXAPA9 28,313,74
skn-rbt LD50:530 mg/kg TXAPA9 28,313,74

SAFETY PROFILE: Moderately toxic by ingestion and skin contact. When heated to decomposition it emits toxic fumes of NO_x.

DCF700 CAS:36039-40-4 ***HR: 3***
2,5-DIAMINOTROPONE
mf: $C_7H_8N_2O$ mw: 136.17

SYN: 2,5-DIAMINO-2,4,6-CYCLOPHETATRIEN-1-ONE

TOXICITY DATA with REFERENCE
ipr-mus LD50:174 mg/kg CPBTAL 20,60,72
scu-mus LD50:361 mg/kg CPBTAL 20,60,72
ivn-mus LD50:192 mg/kg CPBTAL 20,60,72

SAFETY PROFILE: Poison by subcutaneous, intravenous, and intraperitoneal routes. When heated to decomposition it emits toxic fumes of NO_x.

DCF710 CAS:34692-97-2 ***HR: 3***
2,5-DIAMINOTROPONE HYDROCHLORIDE
mf: $C_7H_8N_2O\bullet ClH$ mw: 172.63

SYN: 2,5-DIAMINO-2,4,6-CYCLOHEPTATRIEN-1-ONEHDYRO-CHLORIDE

TOXICITY DATA with REFERENCE
ipr-mus LD50:174 mg/kg YKKZAJ 91,1307,71
scu-mus LD50:361 mg/kg YKKZAJ 91,1307,71
ivn-mus LD50:192 mg/kg YKKZAJ 92,19,72

SAFETY PROFILE: Poison by subcutaneous, intravenous, and intraperitoneal routes. When heated to decomposition it emits toxic fumes of NO_x and HCl.

DCF725 CAS:28965-70-0 ***HR: 3***
DIAMMINEBORONIUM HEPTAHYDROTETRABORATE
mf: $B_5H_{15}N_2$ mw: 97.18

SYN: PENTABORANE(9)DIAMMONIATE

SAFETY PROFILE: Decomposes violently at room temperature. When heated to decomposition it emits toxic fumes of NH_3 and NO_x. See also BORON COMPOUNDS and BORANES.

DCF750 CAS:23777-63-1 ***HR: 3***
DIAMMINEBORONIUM TETRAHYDROBORATE
mf: $B_2H_{12}N_2$ mw: 61.73

SAFETY PROFILE: Ignites when heated in air. When heated to decomposition it emits toxic fumes of NO_x. See also BORON COMPOUNDS and BORANES.

DCF800 CAS:41575-87-5 ***HR: 3***
cis-DIAMMINEDINITRATO PLATINUM (II)
mf: $H_6N_4O_4Pt$ mw: 321.17

SYN: PLATINUM(II)DINITRODIAMMINE

TOXICITY DATA with REFERENCE
ipr-mus LDLo:7 mg/kg BICHBX 2,187,73

CONSENSUS REPORTS: Reported in EPA TSCA Inventory.

SAFETY PROFILE: Poison by intraperitoneal route. Explosive decomp @ 200°C. When heated to decomposition it emits toxic fumes of NO_x. See also PLATINUM COMPOUNDS and NITRATES.

DCG000 ***HR: 3***
DIAMMINEMALONATO PLATINUM (II)
mf: $C_3H_8N_2O_4Pt$ mw: 331.22

TOXICITY DATA with REFERENCE
mmo-sat 2 μg/plate MUREAV 77,45,80
mma-sat 2 μg/plate MUREAV 77,45,80
ipr-mus LD50:225 mg/kg CBINA8 5,415,72

CONSENSUS REPORTS: Reported in EPA TSCA Inventory.

SAFETY PROFILE: Poison by intraperitoneal route. Mutation data reported. When heated to decomposition it emits toxic fumes of NO_x. See also PLATINUM COMPOUNDS.

DCG600 CAS:28068-05-5 ***HR: 3***
DIAMMINEPALLADIUM (II) NITRATE
mf: $H_6N_4O_6Pd$ mw: 264.48

$[(H_3N)_2Pd][NO_2]_2$

SAFETY PROFILE: The dry nitrate is a moderately impact-sensitive explosive. When heated to decomposition it emits toxic fumes of NO_x. See also NITRATES.

DCG800 CAS:7784-44-3 ***HR: 3***
DIAMMONIUM HYDROGEN ARSENATE
DOT: UN 1546
mf: $AsH_3O_4•2H_3N$ mw: 176.03

PROP: White powder or crystals. Mp: decomp to yield NH_3.

SYNS: AMMONIUM ACID ARSENATE ◇ AMMONIUM ARSENATE, solid (DOT) ◇ DIAMMONIUM ARSENATE ◇ DIAMMONIUM MONOHYDROGEN ARSENATE ◇ DIBASIC AMMONIUM ARSENATE ◇ SECONDARY AMMONIUM ARSENATE

CONSENSUS REPORTS: Arsenic and its compounds are on the Community Right-To-Know List.

OSHA PEL: Cancer Hazard
ACGIH TLV: TWA 0.2 mg(As)/m^3
NIOSH REL: (Inorganic Arsenic) CL 0.002 mg(As)/m^3/15M
DOT Classification: Poison B; Label: Poison.

SAFETY PROFILE: A poison. See also ARSENIC. When heated to decomposition it emits very toxic fumes of As, NO_x, and NH_3.

DCH000 CAS:3164-29-2 ***HR: 3***
DIAMMONIUM TARTRATE
mf: $C_4H_6O_6•2H_3N$ mw: 184.18

SYNS: AMMONIUM TARTRATE (DOT) ◇ AMMONIUM-d-TARTRATE ◇ 2,3-DIHYDROXYBUTANEDIOIC ACID, DIAMMONIUM SALT ◇ l-TARTARIC ACID, AMMONIUM SALT ◇ TARTARIC ACID, DIAMMONIUM SALT

TOXICITY DATA with REFERENCE
scu-rbt LD50:1130 mg/kg HBAMAK 4,1289,35
ivn-rbt LD50:113 mg/kg HBAMAK 4,1289,35

CONSENSUS REPORTS: Reported in EPA TSCA Inventory.

SAFETY PROFILE: Poison by intravenous route. Moderately toxic by subcutaneous route. When heated to decomposition it emits very toxic fumes of NH_3 and NO_x.

DCH200 CAS:2050-92-2 ***HR: 3***
DIAMYL AMINE
DOT: UN 2841
mf: $C_{10}H_{23}N$ mw: 157.34

PROP: Water-white liquid. Bp: 202°, flash p: 124°F, d: 0.777 @ 20°/20°, vap d: 5.42.

SYNS: DI-n-AMYLAMINE (DOT) ◇ DIPENTYLAMINE ◇ PENTYL PENTYLAMINE

TOXICITY DATA with REFERENCE
skn-rbt 500 mg open SEV UCDS** 8/9/68
orl-rat LD50:270 mg/kg UCDS** 8/9/68
ihl-rat LCLo:63 ppm/4H AIHAAP 23,95,62
skn-rbt LD50:350 mg/kg AIHAAP 23,95,62

CONSENSUS REPORTS: Reported in EPA TSCA Inventory.

DOT Classification: Poison B; Label: St. Andrews Cross, Flammable Liquid.

SAFETY PROFILE: Poison by inhalation, ingestion, and skin contact. A severe skin irritant. See also AMINES. Flammable when exposed to heat or flame; can react with oxidizing materials. To fight fire, use alcohol foam, foam, CO_2, dry chemical. When heated to decomposition it emits toxic fumes of NO_x.

DCH400 CAS:79-74-3 ***HR: 3***
2,5-DI-tert-AMYLHYDROQUINONE
mf: $C_{16}H_{26}O_2$ mw: 250.42

SYNS: 2,5-BIS(1,1-DIMETHYLPROPYL)HYDROQUINONE ◇ 2,5-DI-tert-PENTYLHYDROQUINONE ◇ SANTOUAR A ◇ SANTOVAR A ◇ USAF B-21

TOXICITY DATA with REFERENCE
ipr-mus LD50:200 mg/kg NTIS** AD277-689

CONSENSUS REPORTS: Reported in EPA TSCA Inventory.

SAFETY PROFILE: Poison by intraperitoneal route. When heated to decomposition it emits acrid smoke and irritating fumes.

DCH600 CAS:13256-06-9 ***HR/1***
DI-n-AMYLNITROSAMINE
mf: $C_{10}H_{22}N_2O$ mw: 186.34

SYNS: DIAMYLNITROSAMIN (GERMAN) ◇ DIPENTYLNITROS-

AMINE ◇ DI-n-PENTYLNITROSAMINE ◇ N-NITROSODIPENTYLAMINE ◇ N-NITROSODI-n-PENTYLAMINE

TOXICITY DATA with REFERENCE
mma-sat 465 μg/plate PNASA6 72,5135,75
mma-ham:lng 500 μmol/L IAPUDO 27,179,80
orl-rat TDLo:7733 mg/kg/8W-C:CAR IJCNAW 27,249,81
scu-rat TDLo:11 g/kg/25W-I:ETA NATWAY 49,111,62
orl-rat LD50:1750 mg/kg NATWAY 48,134,61
scu-rat LD50:3000 mg/kg ZEKBAI 69,103,67

CONSENSUS REPORTS: EPA Genetic Toxicology Program.

SAFETY PROFILE: Moderately toxic by ingestion and subcutaneous routes. Questionable carcinogen with experimental carcinogenic and tumorigenic data. Mutation data reported. When heated to decomposition it emits toxic fumes of NO_x. See also NITROSAMINES.

DCH800 CAS:28652-04-2 ***HR: 2***
DIAMYLPHENOL
mf: $C_{16}H_{26}O$ mw: 234.42

PROP: Liquid. Bp: 278°, flash p: 260°F (OC). D: 0.93-0.94, d: 8.1.

SYN: DIPENTYL PHENOL

TOXICITY DATA with REFERENCE
skn-hmn 250 mg/48H MOD AMIHBC 5,311,52
skn-rbt 500 mg SEV AMIHBC 5,311,52
ipr-rat LD50:620 mg/kg AMIHBC 5,311,52

SAFETY PROFILE: Moderately toxic by intraperitoneal route. A severe skin irritant experimentally. A human skin irritant. Combustible when exposed to heat or flame. Can react with oxidizing materials. To fight fire, use CO_2, dry chemical. When heated to decomposition it emits acrid smoke and fumes. See also PHENOL.

DCI000 CAS:25231-47-4 ***HR: 3***
DI-tert-AMYLPHENOL
mf: $C_{16}H_{26}O$ mw: 234.42

SYNS: 2,4-DI-tert-AMYL PHENOL ◇ DI-tert-PENTYLPHENOL ◇ PRODOX 156

TOXICITY DATA with REFERENCE
eye-rbt 100 mg MOD IHFCAY 6,1,67
orl-rat LD50:330 mg/kg IHFCAY 6,1,67

SAFETY PROFILE: Poison by ingestion. An eye irritant. When heated to decomposition it emits acrid smoke and irritating fumes. See also PHENOL.

DCI400 CAS:35865-33-9 ***HR: 3***
DIANEMYCIN
mf: $C_{47}H_{78}O_{14}$ mw: 867.25

TOXICITY DATA with REFERENCE
orl-mus LD50:150 mg/kg 37ASAA 3,47,78
ipr-mus LD50:9 mg/kg 37ASAA 3,47,78
scu-mus LD50:40 mg/kg 85ERAY 1,805,78

SAFETY PROFILE: Poison by ingestion, intraperitoneal, and subcutaneous routes. When heated to decomposition it emits acrid smoke and irritating fumes.

DCI600 CAS:23261-20-3 ***HR: 3***
DIANHYDROGALACTITOL
mf: $C_6H_{10}O_4$ mw: 146.16

SYNS: DAD ◇ DAG ◇ DIANHYDROCULCITOL ◇ 1,2:5,6-DIANHYDRODULCITOL ◇ 1,2:5,6-DIANHYDROGALACTITOL ◇ 1,2:5,6-DIEPOXYDULCITOL ◇ DULCITOLDIEPOXIDE ◇ NSC 132313

TOXICITY DATA with REFERENCE
mmo-sat 100 μg/plate CRNGDP 3,333,82
dni-mus-ipr 5 mg/kg NEOLA4 31,667,84
orl-rat LD50:14 mg/kg CCROBU 56,593,72
ipr-rat LD50:11 mg/kg CCROBU 56,593,72
ivn-rat LD50:16 mg/kg CCROBU 56,593,72
orl-mus LD50:7899 μg/kg NCISP* JAN86
ipr-mus LD50:15 mg/kg CCROBU 56,593,72
scu-mus LD50:16500 μg/kg NCISP* JAN86
ivn-mus LD50:21 mg/kg CCROBU 56,593,72
ivn-dog LDLo:16 mg/kg CTRRDO 60,1585,76

SAFETY PROFILE: Poison by ingestion, intravenous, subcutaneous, and intraperitoneal routes. Mutation data reported. When heated to decomposition it emits acrid smoke and irritating fumes.

DCI800 CAS:19895-66-0 ***HR: 3***
DIANHYDROMANNITOL
mf: $C_6H_{10}O_4$ mw: 146.16

SYNS: 1:2:5:6-DIANHYDRO-d-MANNITOL ◇ NSC-133129

TOXICITY DATA with REFERENCE
dnd-ckn:leu 30 mmol/L TELEAY (29),2477,75
dni-rbt:bmr 274 μmol/L BCPCA6 25,1705,76
ipr-rat LD50:14 mg/kg EJCAAH 4,617,68
ipr-mus LD50:20 mg/kg EJCODS 18,573,82

SAFETY PROFILE: Poison by intraperitoneal route. Mutation data reported. When heated to decomposition it emits acrid smoke and irritating fumes.

DCJ000 CAS:73928-11-7 ***HR: 3***
DIANILINOMERCURY
mf: $C_{12}H_{12}HgN_2$ mw: 384.85

SYN: N,N'-MERCURIDIANILINE

TOXICITY DATA with REFERENCE
ivn-mus LD50:180 mg/kg CSLNX* NX#05148

CONSENSUS REPORTS: Mercury and its compounds are on the Community Right-To-Know List.

OSHA PEL: (Transitional: CL 1 mg/10m^3) CL 0.1 mg(Hg)/m^3 (skin)
ACGIH TLV: TWA 0.1 mg(Hg)/m^3 (skin)
NIOSH REL: (Inorganic Mercury) TWA 0.05 mg(Hg)/m^3

SAFETY PROFILE: Poison by intravenous route. See also MERCURY COMPOUNDS. When heated to decomposition it emits very toxic fumes of NO_x and Hg.

DCJ200 CAS:119-90-4 ***HR: 3***
o-DIANISIDINE
mf: $C_{14}H_{16}N_2O_2$ mw: 244.32

PROP: Colorless crystals. Mp: 137-138°, flash p: 403°F, vap d: 8.5.

SYNS: ACETAMINE DIAZO BLACK RD ◇ AMACEL DEVELOPED NAVY SD ◇ AZOENE FAST BLUE BASE ◇ AZOFIX BLUE B SALT ◇ AZOGNE FAST BLUE B ◇ BLUE BN BALSE ◇ BRENTAMINE FAST BLUE B BASE ◇ CELLITAZOL B ◇ C.I. 24110 ◇ C.I. AZOIC DIAZO COMPONENT 48 ◇ CIBACETE DIAZO NAVY BLUE 2B ◇ C.I. DISPERSE BLACK 6 ◇ DIACELLITON FAST GREY G ◇ DIACEL NAVY DC ◇ o-DIANISIDIN (CZECH, GERMAN) ◇ o-DIANISIDINA (ITALIAN) ◇ 3,3'-DIANISIDINE ◇ O,O'DIANISIDINE ◇ DIATO BLUE BASE B ◇ 3,3'-DIMETHOXYBENZIDIN (CZECH) ◇ 3,3'-DIMETHOXYBENZIDINE ◇ 3,3'-DIMETOSSIBENZODINA (ITALIAN) ◇ FAST BLUE B BASE ◇ HILTONIL FAST BLUE B BASE ◇ HILTOSAL FAST BLUE B SALT ◇ HINDASOL BLUE B SALT ◇ KAKO BLUE B SALT ◇ KAYAKU BLUE B BASE ◇ LAKE BLUE B BASE ◇ MEISEI TERYL DIAZO BLUE HR ◇ MITSUI BLUE B BASE ◇ NAPHTHANIL BLUE B BASE ◇ NEUTROSEL NAVY BN ◇ RCRA WASTE NUMBER U091 ◇ SANYO FAST BLUE SALT B ◇ SETACYL DIAZO NAVY R ◇ SPECTROLENE BLUE B

TOXICITY DATA with REFERENCE
mmo-sat 333 μg/plate ENMUDM 5(Suppl 1),3,83
mma-sat 1 μg/plate IGAYAY 123,18,82
sce-ham:ovr 500 μg/L ENMUDM 7,1,85
oms-dog:oth 100 μmol/L CNREA8 44,1893,84
dnd-dog:oth 100 μmol/L CNREA8 44,1893,84
orl-rat TDLo:12 g/kg/56W-I:ETA GTPZAB 9,18,65
orl-rat LD50:1920 mg/kg 28ZPAK -,119,72
orl-dog LDLo:600 mg/kg AEXPBL 58,167,1907

CONSENSUS REPORTS: NTP Fifth Annual Report on Carcinogens. IARC Cancer Review: Group 2B IMEMDT 7,198,87; Animal Sufficient Evidence IMEMDT 4,41,74. EPA Genetic Toxicology Program. Community Right-To-Know List. Reported in EPA TSCA Inventory.

DFG MAK: Animal Carcinogen, Suspected Human Carcinogen.
NIOSH REL: (Benzidine-based dye) Reduce to lowest feasible level

SAFETY PROFILE: Confirmed carcinogen with experimental tumorigenic data. Moderately toxic by ingestion. Mutation data reported. Combustible when exposed to heat or flame. When heated to decomposition it emits toxic fumes of NO_x.

DCJ400 CAS:91-93-0 ***HR: 3***
DIANISIDINE DIISOCYANATE
mf: $C_{16}H_{12}N_2O_4$ mw: 296.30

SYNS: 4,4'-DIISOCYANATO-3,3'-DIMETHOXY-1,1'-BIPHENYL ◇ 3,3'-DIMETHOXYBENZIDINE-4,4'-DIISOCYANATE ◇ 3,3'-DIMETHOXY-4,4'-BIPHENYLENE DIISOCYANATE ◇ NCI-C02175

TOXICITY DATA with REFERENCE
mmo-sat 3300 ng/plate ENMUDM 7(Suppl 5),1,85
mma-sat 3 μg/plate ENMUDM 7(Suppl 5),1,85
otr-rat:emb 81 μg/plate JJATDK 1,190,81
orl-rat TDLo:565 g/kg/78W-I:CAR NCITR* NCI-CG-TR-128,79
orl-rat TD:1200 g/kg/78W-I:CAR NCITR* NCI-CG-TR-128,79
ivn-mus LD50:180 mg/kg CSLNX* NX#02411

CONSENSUS REPORTS: IARC Cancer Review: Group 3 IMEMDT 7,56,87; Animal Limited Evidence IMEMDT 39,279,86. NCI Carcinogenesis Bioassay (feed); No Evidence: mouse NCITR* NCI-CG-TR-128,79; Clear Evidence: rat NCITR* NCI-CG-TR-128,79.

NIOSH REL: (Diisocyanates) TWA 0.005 ppm; CL 0.02 ppm/10M

SAFETY PROFILE: Poison by intravenous route. Questionable carcinogen with experimental carcinogenic data. When heated to decomposition it emits toxic fumes of NO_x. See also CYANATES.

DCJ450 ***HR: 3***
DIANTHUS SUPERBUS L., extract

PROP: A Southeast Asian carnation belonging to the family *Caryophyllaceae* ZKPAK 2,366,69

TOXICITY DATA with REFERENCE
scu-mus TDLo:20 g/kg (female 2D pre):REP MPHEAE 16,414,67
ipr-mus LD50:100 mg/kg IJEBA6 22,312,84

SAFETY PROFILE: Poison by intraperitoneal route. Experimental reproductive effects. When heated to decomposition it emits acrid smoke and irritating fumes.

DCJ600 CAS:13601-02-0 ***HR: 3***
DIAQUODIAMMINEPLATINUM DINITRATE
mf: $H_{10}N_2O_2Pt \cdot N_2O_6$ mw: 389.23

SYN: cis-DIAQUODIAMMINEPLATINUM(II)DINITRATE

TOXICITY DATA with REFERENCE
mmo-sat 5 μg/plate MUREAV 48,139,77

idr-hmn TDLo:40 mg/kg:SKN CNREA8 35,2766,75
ipr-mus LDLo:5 mg/kg BICHBX 2,187,73

SAFETY PROFILE: Poison by intraperitoneal route. Human systemic skin effects by intradermal route. Mutagenic data reported. When heated to decomposition it emits toxic fumes of NO_x. See also NITRATES and PLATINUM COMPOUNDS.

DCJ800 CAS:61790-53-2 ***HR: 1***
DIATOMACEOUS EARTH

PROP: Composed of skeletons of small aquatic plants related to algae and contains as much as 88% amorphous silica (DTLVS* 4,120,80). White to buff colored solid. Insol in water; sol in hydrofluoric acid.

SYNS: D.E. ◇ DIATOMACEOUS SILICA ◇ DIATOMITE ◇ INFUSORIAL EARTH ◇ KIESELGUHR

CONSENSUS REPORTS: IARC Cancer Review: Group 3 IMEMDT 7,341,87; Animal Inadequate Evidence IMEMDT 42,39,87; Human Inadequate Evidence IMEMDT 42,39,87. Reported in EPA TSCA Inventory.

OSHA PEL: (Transitional: TWA 80 $mg/m^3/\%SiO_2$) TWA 6 mg/m^3
ACGIH TLV: TWA (nuisance particulate) 10 mg/m^3 of total dust (when toxic impurities are not present, e.g., quartz < 1%).

SAFETY PROFILE: A nuisance dust which may cause fibrosis of the lungs. Roasting or calcining at high temperatures produces cristobalite and tridymite, thus increasing the fibrogenicity of the material. A questionable carcinogen.

DCJ850 CAS:7084-07-3 ***HR: 3***
DIATRIN HYDROCHLORIDE
mf: $C_{15}H_{20}N_2S{\cdot}ClH$ mw: 296.89

SYNS: ENSTAMINE HYDROCHLORIDE ◇ METHAPHENILENE HYDROCHLORIDE ◇ NILHISTIN

TOXICITY DATA with REFERENCE
orl-mus LD50:550 mg/kg JPETAB 93,210,48
ipr-mus LD50:117 mg/kg MEIEDD 10,854,83
scu-mus LD50:160 mg/kg JPETAB 93,210,48
ivn-mus LD50:45 mg/kg JPETAB 93,210,48
ivn-rbt LD50:30 mg/kg JPETAB 93,210,48
orl-gpg LD50:900 mg/kg JPETAB 93,210,48
scu-gpg LD50:140 mg/kg JPETAB 93,210,48
ivn-gpg LD50:30 mg/kg JPETAB 93,210,48

SAFETY PROFILE: Poison by subcutaneous, intravenous, and intraperitoneal routes. Moderately toxic by ingestion. When heated to decomposition it emits toxic fumes of SO_x, NO_x, and HCl. See also AMINES.

DCK000 CAS:117-96-4 ***HR: 1***
DIATRIZOIC ACID
mf: $C_{11}H_9I_3N_2O_4$ mw: 613.92

SYNS: AMIDOTRIZOIC ACID ◇ 3,5-BIS(ACETYLAMINO)-2,4,6-TRIIODOBENZOIC ACID ◇ 3,5-DIACETAMIDO-2,4,6-TRIIODOBENZOIC ACID ◇ DIAT (GERMAN) ◇ DIATRIZOESAURE (GERMAN) ◇ ODISTON ◇ UROGRAFIN ACID ◇ UROGRANOIC ACID ◇ UROTRAST

TOXICITY DATA with REFERENCE
ipr-rat LD50:14300 mg/kg ARZNAD 15,222,65
ivn-rat LD50:11300 mg/kg ARZNAD 15,222,65
ivn-mus LD50:8900 mg/kg KSRNAM 19,2411,85
ipr-gpg LD50:13 g/kg ARZNAD 15,222,65

SAFETY PROFILE: Mildly toxic by intraperitoneal and intravenous routes. When heated to decomposition it emits very toxic fumes of NO_x and I^-.

DCK200 CAS:34494-09-2 ***HR: 3***
6,12-DIAZAANTHANTHRENE SULFATE
mf: $C_{20}H_{10}N_2{\cdot}H_2O_4S$ mw: 376.40

SYNS: ACRIDINO(2,1,9,8-klmna)ACRIDINE SULFATE ◇ 6,12-DIAZAANTHANTHRENE SULPHATE

TOXICITY DATA with REFERENCE
imp-rat TDLo:600 mg/kg:ETA NEOLA4 18,591,71

SAFETY PROFILE: Questionable carcinogen with experimental tumorigenic data. When heated to decomposition it emits very toxic fumes of NO_x and SO_x. See also SULFATES.

DCK400 CAS:280-57-9 ***HR: 2***
1,4-DIAZABICYCLO(2,2,2)OCTANE
mf: $C_6H_{12}N_2$ mw: 112.20

$$C_2H_4NC_2H_4NCH_2CH_2$$

PROP: Hygroscopic crystals. Mp: 158°, bp: 174°.

SYNS: BICYCLO(2,2,2)-1,4-DIAZAOCTANE ◇ DABCO ◇ DABCO CRYSTAL ◇ DABCO EG ◇ DABCO 33LV ◇ DABCO R-8020 ◇ DABCO S-25 ◇ D 33LV ◇ 1,4-ETHYLENEPIPERAZINE ◇ TRIETHYLENEDIAMINE

TOXICITY DATA with REFERENCE
skn-rbt 2500 μg open MLD TXAPA9 4,522,62
eye-rbt 25 mg MOD TXAPA9 4,522,62
orl-rat LD50:1700 mg/kg ZHYGAM 20,393,74
orl-rbt LD50:1100 mg/kg GISAAA 45(5),67,80
orl-gpg LD50:2250 mg/kg GISAAA 45(5),67,80

CONSENSUS REPORTS: Reported in EPA TSCA Inventory.

SAFETY PROFILE: Moderately toxic by ingestion. A skin and eye irritant, allergen and skin sensitizer. A powerful base. Forms an explosive complex with hydrogen peroxide. Mixtures with carbon auto-ignite at 230°C. Very exothermic reaction with cellulose nitrate.

When heated to decomposition it emits toxic fumes of NO_x. See also AMINES.

DCK500 ***HR: 3***
1,4-DIAZABICYCLO(2.2.2)OCTANE HYDROGEN PEROXIDATE
mf: $C_6H_{12}N_2•H_2O_2$ mw: 146.19

$C_2H_4NC_2H_4NCH_2CH_2•H_2O_2$

SAFETY PROFILE: Complex explodes when dried at room temperature. Upon decomposition it emits toxic fumes of NO_x. See also PEROXIDES.

DCK700 CAS:283-66-9 ***HR: 3***
1,6-DIAZA-3,4,8,9,12,13-HEXAOXABICYCLO (4.4.4)TETRADECANE
mf: $C_6H_{12}N_2O_6$ mw: 208.17

SYN: HEXAMETHYLENETRIPEROXYDIAMINE

DOT Classification: Forbidden (dry)

SAFETY PROFILE: The dry material is a powerful explosive that is heat- and shock-sensitive. Explodes on contact with bromine or sulfuric acid. When heated to decomposition it emits toxic fumes of NO_x. See also PEROXIDES.

DCK759 CAS:439-14-5 ***HR: 3***
DIAZEPAM
mf: $C_{16}H_{13}ClN_2O$ mw: 284.76

PROP: Plates. Mp: 125-126°.

SYNS: ALBORAL ◇ AMIPROL ◇ ANSIOLISINA ◇ APAURIN ◇ APOZEPAM ◇ ATENSINE ◇ ATILEN ◇ BIALZEPAM ◇ CALMOCITENE ◇ CERCINE ◇ 7-CHLORO-1,3-DIHYDRO-1-METHYL-5-PHENYL-2H-1,4-BENZODIAZEPIN-2-ONE ◇ 7-CHLORO-1-METHYL-5-3H-1,4-BENZODIAZEPIN-2(1H)-ONE ◇ 7-CHLORO-1-METHYL-2-OXO-5-PHENYL-3H-1,4-BENZODIAZEPINE ◇ 7-CHLORO-1-METHYL-5-PHENYL-2H-1,4-BENZODIAZEPIN-2-ONE ◇ 7-CHLORO-1- METHYL-5-PHENYL-1,3-DIHYDRO-2H-1,4-BENZODIAZEPIN-2-ONE ◇ CONDITION ◇ DIACEPAN ◇ DIAPAM ◇ DIAZETARD ◇ DIENPAX ◇ DIPAM ◇ DOMALIUM ◇ DUKSEN ◇ E-PAM ◇ ERIDAN ◇ FAUSTAN ◇ FRUSTAN ◇ GIHITAN ◇ KIATRIUM ◇ LEMBROL ◇ LEVIUM ◇ LIBERETAS ◇ METHYL DIAZEPINONE ◇ 1-METHYL-5-PHENYL-7-CHLORO-1,3-DIHYDRO-2H-1,4-BENZODIAZEPIN-2-ONE ◇ MOROSAN ◇ NSC-77518 ◇ PACITRAN ◇ PAXATE ◇ PLIDAN ◇ QUETINIL ◇ QUIATRIL ◇ RELAMINAL ◇ RELANIUM ◇ RENBORIN ◇ SAROMET ◇ SEDIPAM ◇ SEDUXEN ◇ SERENACK ◇ SERENZIN ◇ SETONIL ◇ SONACON ◇ STESOLID ◇ TENSOPAM ◇ TRANIMUL ◇ TRANQUIRIT ◇ UMBRIUM ◇ UNISEDIL ◇ VALEO ◇ VALITRAN ◇ VALIUM ◇ VATRAN ◇ VIVAL ◇ ZIPAN

TOXICITY DATA with REFERENCE
mma-sat 958 nmol/plate CNREA8 38,4478,78
cyt-wmn-unr 328 mg/kg/78W AJOGAH 107,456,70
cyt-hmn:leu 10 mg/L AJOGAH 103,836,69
orl-wmn TDLo:22800 μg/kg (25-36W preg):REP JOPDAB 90,123,77
ivn-wmn TDLo:400 μg/kg (female 39W post):TER JOGBAS 79,635,72
orl-mus TDLo:42 g/kg/80W-C:ETA TXAPA9 57,39,81
ivn-inf TDLo:150 μg/kg:SKN,BIO BMJOAE 2,298,77
orl-man TDLo:143 μg/kg:EYE BCPHBM 1,335,74
ims-wmn TDLo:181 μg/kg:CNS,CVS BMJOAE 1,144,77
ivn-man TDLo:143 μg/kg:PUL,CNS JAMAAP 238,1052,77
ivn-man TDLo:71 μg/kg/1M-C:CVS DICPBB 17,125,83
orl-rat LD50:352 mg/kg JTCTDW 20,271,83
ipr-rat LD50:300 mg/kg JMCMAR 20,952,77
orl-mus LD50:48 mg/kg PCJOAU 17,30,83
skn-mus LD50:800 mg/kg AREAD8 (4),57,80
ipr-mus LD50:121 mg/kg IJSIDW 44,1,82
scu-mus LD50:300 mg/kg 27ZTAP 3,48,69
ivn-mus LD50:25 mg/kg ARZNAD 31,2180,81
par-mus LD50:150 mg/kg RPTOAN 33,70,70
ivn-rbt LD50:9 mg/kg IJNEAQ 5,305,66

CONSENSUS REPORTS: IARC Cancer Review: Group 3 IMEMDT 7,189,87

CONSENSUS REPORTS: Reported in EPA TSCA Inventory. EPA Genetic Toxicology Program.

SAFETY PROFILE: Poison by ingestion, parenteral, subcutaneous, intravenous, or intraperitoneal routes. Moderately toxic by skin contact. Questionable carcinogen with experimental tumorigenic data. Human systemic effects: dermatitis, effect on inflammation or mediation of inflammation, change in cardiac rate, somnolence, respiratory depression, and other respiratory changes, visual field changes, diplopia (double vision), change in motor activity, muscle contraction or spasticity, ataxia (loss of muscle coordination), an antipsychotic and general anesthetic. A human teratogen by ingestion and intravenous routes which causes developmental abnormalities of the fetal cardiovascular (circulatory) system and postnatal effects. Experimental teratogenic and reproductive effects. Human mutation data reported. An allergen. A drug for the treatment of anxiety. When heated to decomposition it emits very toxic fumes of Cl^- and NO_x.

DCL100 CAS:13556-50-8 ***HR: 3***
1,3-DIAZIDOBENZENE
mf: $C_6H_4N_6$ mw: 160.14

SAFETY PROFILE: Ignites and may explode weakly on contact with concentrated acids. When heated to decomposition it emits toxic fumes of NO_x. See also AZIDES.

DCL125 CAS:2294-47-5 ***HR: 3***
1,4-DIAZIDOBENZENE
mf: $C_6H_4N_6$ mw: 160.14

SYNS: BENZENE, 1,4-DIAZIDO- ◇ p-DIAZIDOBENZENE (DOT) ◇ 1,4-DIAZIDOBENZENE ◇ p-PHENYLENE DIAZIDE

DOT Classification: Forbidden

SAFETY PROFILE: Explodes violently when heated. When heated to decomposition it emits toxic fumes of NO_x. See also AZIDES.

DCL159 ***HR: 3***
2,2-DIAZIDOBUTANE
mf: $C_4H_8N_6$ mw: 140.15

$$CH_3C(N_3)_2C_2H_3$$

SAFETY PROFILE: Potentially explosive. When heated to decomposition it emits toxic fumes of NO_x. See also AZIDES.

DCL200 CAS:67880-17-5 ***HR: 3***
1,2-DIAZIDOCARBONYL HYDRAZINE
mf: $C_2H_2N_8O_2$ mw: 170.09

DOT Classification: Forbidden

SAFETY PROFILE: A heat- and impact-sensitive explosive. When heated to decomposition it emits toxic fumes of NO_x. See also AZIDES.

DCL300 CAS:26157-96-0 ***HR: 3***
2,5-DIAZIDO-3,6-DICHLOROBENZOQUINONE
mf: $C_6Cl_2N_6O_2$ mw: 259.01

$$O{:}CC(N_3){=}CClCO{\cdot}C(N_3){=}CCl$$

SAFETY PROFILE: A moderately impact-sensitive explosive. Upon decomposition it emits toxic fumes of NO_x. See also AZIDES.

DCL350 CAS:4774-73-6 ***HR: 3***
DIAZIDODIMETHYLSILANE
mf: $C_2H_6N_6Si$ mw: 142.20

SAFETY PROFILE: A storage hazard. It may explode spontaneously. When heated to decomposition it emits toxic fumes of NO_x. See also AZIDES and SILANES.

DCL400 CAS:67880-20-0 ***HR: 3***
1,1-DIAZIDOETHANE
mf: $C_2H_4N_6$ mw: 112.09

SAFETY PROFILE: An extremely unstable explosive. Upon decomposition it emits toxic fumes of NO_x. See also AZIDES.

DCL600 CAS:629-13-0 ***HR: 3***
1,2-DIAZIDOETHANE
mf: $C_2H_4N_6$ mw: 112.09

DOT Classification: Forbidden.

SAFETY PROFILE: Explodes on heating or on contact with sulfuric acid. Upon decomposition it emits toxic fumes of NO_x. See also AZIDES.

DCL800 CAS:57512-42-2 ***HR: D***
DIAZIDO ETHIDIUM
mf: $C_{21}H_{16}N_7{\cdot}Br$ mw: 446.35

SYN: 3,8-DIAZIDO-5-ETHYL-6-PHENYLPHENANTHRIDINIUMBROMIDE

TOXICITY DATA with REFERENCE
mma-sat 50 μg/plate MUREAV 121,89,83
mmo-smc 6 μmol/L/10M MUREAV 56,21,77

SAFETY PROFILE: Mutation data reported. When heated to decomposition it emits very toxic fumes of NO_x and Br^-.

DCM000 CAS:67880-21-1 ***HR: 3***
DIAZIDOMALONONITRILE
mf: C_3N_8 mw: 148.09

$$(N_3)_2C(C{\equiv}N)_2$$

SYN: DIAZIDODICYANOMETHANE

CONSENSUS REPORTS: Cyanide and its compounds are on the Community Right-To-Know List.

SAFETY PROFILE: The pure material is an unpredictable explosive. When heated to decomposition it emits toxic fumes of CN^- and NO_x. See also AZIDES and NITRILES.

DCM200 ***HR: 3***
DIAZIDOMETHYLENEAZINE
mf: C_2N_{14} mw: 220.12

SAFETY PROFILE: Very explosive. When heated to decomposition it emits toxic fumes of NO_x. See also AZIDES.

DCM400 CAS:67880-22-2 ***HR: 3***
DIAZIDOMETHYLENECYANAMIDE
mf: C_2N_8 mw: 136.08

SAFETY PROFILE: An explosive solid. Upon decomposition it emits toxic fumes of CN^- and NO_x. See also CYANIDE and AZIDES.

DCM499 ***HR: 3***
1,3-DIAZIDO-2-NITROAZAPROPANE
mf: $C_2H_4N_8O_2$ mw: 172.11

SYN: N,N-BIS(AZIDOMETHYL)NITRICAMIDE

SAFETY PROFILE: An explosive. A high energy component of solid rocket propellants. When heated to decomposition it emits toxic fumes of NO_x. See also AZIDES.

DCM600 CAS:22750-69-2 ***HR: 3***
1,3-DIAZIDOPROPENE
mf: $C_3H_4N_6$ mw: 124.10

$N_3CH{=}CHCH_2N_3$

SAFETY PROFILE: An unpredictable explosive. When heated to decomposition it emits toxic fumes of NO_x and CN^-. See also AZIDES.

DCM700 CAS:74273-75-9 ***HR: 3***
2,6-DIAZIDOPYRAZINE
mf: $C_4H_2N_8$ mw: 162.11

$N_3C{=}NC(N_3){-}CHN{=}CH$

SAFETY PROFILE: A heat- and impact-sensitive explosive. It may be detonated by heating to 200°C or by a hammer blow. Upon decomposition it emits toxic fumes of NO_x. See also AZIDES.

DCM750 CAS:333-41-5 ***HR: 3***
DIAZINON
DOT: NA 2783
mf: $C_{12}H_{21}N_2O_3PS$ mw: 304.38

PROP: Liquid with faint ester-like odor. Bp: 84° @ 0.002 mm, d: 1.116 @ 20°/4°. Miscible in organic solvents.

SYNS: ALFA-TOX ◇ BASUDIN ◇ BASUDIN 10 G ◇ BAZUDEN ◇ DAZZEL ◇ O,O-DIAETHYL-O-(2-ISOPROPYL-4-METHYL-PYRIMIDIN-6-YL)-MONOTHIOPHOSPHAT (GERMAN) ◇ O,O-DI-AETHYL-O-(2-ISOPROPYL-4-METHYL)-6-PYRIMIDYL-THIONO-PHOSPHAT (GERMAN) ◇ DIANON ◇ DIATERR-FOS ◇ DIAZAJET ◇ DIAZATOL ◇ DIAZIDE ◇ DIAZINONE ◇ DIAZITOL ◇ DIAZOL ◇ O,O-DIETHYL-O-(2-ISOPROPYL-4-METHYL-PYRIMIDIN-6-YL)MONOTHIOFOSFAAT (DUTCH) ◇ O,O-DIETHYL-O-(2-ISOPROPYL-4-METHYL-6-PYRIMIDINYL)PHOSPHOROTHIOATE ◇ O,O-DIETHYL-O-(2-ISOPROPYL-6-METHYL-4-PYRIMIDINYL)PHOSPHOROTHIOATE ◇ DIETHYL 4-(2-ISOPROPYL-6-METHYLPYRIMIDINYL)PHOSPHOROTHIONATE ◇ O,O-DIETHYL-O-(2-ISOPROPYL-4-METHYL-6-PYRIMIDYL)PHOSPHOROTHIOATE ◇ O,O-DIETHYL-O-(2-ISOPROPYL-4-METHYL-6-PYRIMIDYL)THIONOPHOSPHATE ◇ O,O-DIETHYL-2-ISOPROPYL-4-METHYLPYRIMIDYL-6-THIOPHOSPHATE ◇ O,O-DIETHYL-O-6-METHYL-2-ISOPROPYL-4-PYRIMIDINYL PHOSPHOROTHIOATE ◇ O,O-DIETIL-O-(2-ISOPROPIL-4-METIL-PIRIMIDIN-6-IL)-MONOTIOFOSFATO (ITALIAN) ◇ DIMPYLATE ◇ DIPOFENE ◇ DIZINON ◇ DYZOL ◇ ENT 19,507 ◇ G 301 ◇ G-24480 ◇ GARDENTOX ◇ GEIGY 24480 ◇ O-2-ISOPROPYL-4-METHYLPYRIMIDYL-O,O-DIETHYL PHOSPHOROTHIOATE ◇ ISOPROPYLMETHYLPYRIMIDYL DIETHYL THIOPHOSPHATE ◇ KAYAZINON ◇ KAYAZOL ◇ NCI-C08673 ◇ NEDCIDOL ◇ NEOCIDOL ◇ NIPSAN ◇ NUCIDOL ◇ SAROLEX ◇ SPECTRACIDE ◇ THIOPHOSPHATE de O,O-DIETHYLE et de o-2-ISOPROPYL-4-METHYL-6-PYRIMIDYLE (FRENCH)

TOXICITY DATA with REFERENCE
skn-rbt 500 mg open MOD CIGET* -,-,77
eye-rbt 100 mg SEV CIGET* -,-,77
cyt-hmn:lym 500 μg/L TSITAQ 18,1490,76
cyt-ham:lng 100 mg/L/27H MUREAV 66,277,79
orl-mus TDLo:3960 μg/kg (female 1-22D post):REP JTEHD6 3,989,77
ipr-rat TDLo:200 mg/kg (female 11D post):TER AEHLAU 16,805,68
orl-hmn TDLo:214 mg/kg:CNS,SKN CTOXAO 12,435,78
orl-rat LD50:66 mg/kg DOEAAH 35,25,79
ihl-rat LC50:3500 mg/m³/4H FMCHA2 -,C75,83
skn-rat LD50:180 mg/kg PMJMAQ -,156,57
ipr-rat LD50:65 mg/kg ARZNAD 5,436,55
orl-mus LD50:17 mg/kg SKEZAP 24,268,83
ihl-mus LC50:1600 mg/m³/4H PSDTAP 15,239,74
skn-mus LD50:2750 mg/kg JTEHD6 9,491,82
ipr-mus LD50:33 mg/kg TXAPA9 2,495,60
scu-mus LD50:58 mg/kg OIZAAV 71,6099,59
ivn-mus LD50:180 mg/kg CSLNX* NX#00023
orl-rbt LD50:143 mg/kg YKYUA6 31,459,80
skn-rbt LD50:180 mg/kg CMEP** -,1,56

CONSENSUS REPORTS: NCI Carcinogenesis Bioassay (feed); No Evidence: mouse, rat NCITR* NCI-CG-TR-137,79. Reported in EPA TSCA Inventory. EPA Genetic Toxicology Program.

OSHA PEL: TWA 0.1 mg/m³ (skin)
ACGIH TLV: TWA 0.1 mg/m³
DFG MAK: 1 mg/m³
DOT Classification: ORM-A; Label: None.

SAFETY PROFILE: Poison by ingestion, skin contact, subcutaneous, intravenous, and intraperitoneal routes. Mildly toxic by inhalation. Human systemic effects by ingestion: changes in motor activity, muscle weakness, and sweating. Experimental teratogenic and reproductive effects. A skin and severe eye irritant. Human mutation data reported. When heated to decomposition it emits very toxic fumes of NO_x, PO_x, and SO_x.

DCM800 CAS:157-22-2 ***HR: 3***
DIAZIRINE
mf: CH_2N_2 mw: 42.04

$CH_2N{=}N$

PROP: Bp: −14°C.

SAFETY PROFILE: The gas explodes when heated. Upon decomposition it emits toxic fumes of NO_x.

DCM875 CAS:76429-98-6 ***HR: 3***
DIAZIRINE-3,3-DICARBOXYLIC ACID
mf: $C_3H_2N_2O_4$ mw: 130.06

SAFETY PROFILE: The potassium salts of this acid are unstable explosives. When heated to decomposition it emits toxic fumes of NO_x.

DCN000 CAS:436-40-8 ***HR: D***
2,5-DIAZIRINO-3,6-DIPROPOXY-p-BENZOQUINONE
mf: $C_{16}H_{22}N_2O_4$ mw: 306.40

SYNS: 2,5-BIS(1-AZIRIDINYL)3,6-DIPROPOXY-2,5-CYCLOHEXADIENE-1,4-DIONE (9CI) ◇ 2,5-BIS(ETHYLENEIMINO)-3,6-DIPROPOXY-1,4-BENZOQUINONE ◇ INPROQUONE

TOXICITY DATA with REFERENCE
dnd-dmg-orl 10 μmol/L CNREA8 30,195,70
sln-dmg-orl 50000 ppm MUREAV 2,29,65

CONSENSUS REPORTS: EPA Genetic Toxicology Program.

SAFETY PROFILE: Mutation data reported. When heated to decomposition it emits toxic fumes of NO_x.

DCN200 CAS:6832-13-9 ***HR: 3***
DIAZOACETALDEHYDE
mf: $C_2H_2N_2O$ mw: 70.05

SAFETY PROFILE: A powerful, heat-sensitive explosive. When heated to decomposition it emits toxic fumes of NO_x. See also AZIDES and ALDEHYDES.

DCN600 CAS:38726-91-9 ***HR: 2***
2-(DIAZOACETAMINO)-N-ETHYLACETAMIDE
mf: $C_6H_{10}N_4O_2$ mw: 170.20

SYNS: 2-((DIAZOACETYL)AMINO)-N-ETHYLACETAMIDE ◇ N-DIAZOACETYLGLYCINEETHYLAMIDE ◇ N-ETHYLDIAZOACETYLGLYCINE AMIDE

TOXICITY DATA with REFERENCE
dns-mus:fbr 2500 mmol/L JCROD7 94,7,79
ipr-mus LD50:3281 mg/kg ARZNAD 23,690,73

CONSENSUS REPORTS: EPA Genetic Toxicology Program.

SAFETY PROFILE: Moderately toxic by intraperitoneal route. Mutation data reported. When heated to decomposition it emits toxic fumes of NO_x.

DCN800 CAS:623-73-4 ***HR: 3***
DIAZOACETIC ESTER
mf: $C_4H_6N_2O_2$ mw: 114.12

$N_2CHCO{\cdot}OC_2H_5$

SYNS: DAAE ◇ DIAZOACETIC ACID, ETHYL ESTER ◇ DIAZOESSIGSAEURE-AETHYLESTER (GERMAN) ◇ EDA ◇ ETHOXYCARBONYLDIAZOMETHANE ◇ ETHYL DIAZOACETATE

TOXICITY DATA with REFERENCE
orl-rat TDLo:2025 mg/kg/81W-I:ETA XENOBH 3,271,73
skn-rat TDLo:4167 mg/kg/48W-I:CAR ARGEAR 55,117,85
ipr-rat TDLo:30 mg/kg/20W-I:CAR ARGEAR 55,117,85
orl-rat LD50:400 mg/kg XENOBH 3,271,73
ivn-rat LD50:280 mg/kg PSEBAA 135,219,70

SAFETY PROFILE: Poison by ingestion and intravenous routes. Questionable carcinogen with experimental carcinogenic and tumorigenic data. Can explode. Explodes on contact with tris(dimethylamino) antimony. When heated to decomposition it emits toxic fumes of NO_x. See also ESTERS.

DCN875 CAS:13138-21-1 ***HR: 3***
DIAZOACETONITRILE
mf: C_2HN_3 mw: 67.05

CONSENSUS REPORTS: Cyanide and its compounds are on the Community Right-To-Know List.

SAFETY PROFILE: The precipitate from concentrated solutions is an extremely friction-sensitive explosive. When heated to decomposition it emits toxic fumes of NO_x and CN^-. See also NITRILES.

DCO200 CAS:38726-90-8 ***HR: 2***
2-((DIAZOACETYL)AMINO)-N-METHYLACETAMIDE
mf: $C_5H_8N_4O_2$ mw: 156.17

SYNS: N-DIAZOACETYLGLYCINE METHYLAMIDE ◇ N-METHYLDIAZOACETYLGLYCINE AMIDE

TOXICITY DATA with REFERENCE
dns-mus:fbr 2500 mmol/L JCROD7 94,7,79
ipr-mus LD50:3672 mg/kg ARZNAD 23,690,73

CONSENSUS REPORTS: EPA Genetic Toxicology Program.

SAFETY PROFILE: Moderately toxic by intraperitoneal route. Mutation data reported. When heated to decomposition it emits toxic fumes of NO_x.

DCO509 CAS:19932-64-0 ***HR: 3***
DIAZOACETYL AZIDE
mf: C_2HN_5O mw: 111.06

PROP: Mp: 7°C

SAFETY PROFILE: The solid and the liquid are powerful explosives sensitive to impact or friction. When heated to decomposition it emits toxic fumes of NO_x. See also AZIDES.

DCO600 CAS:999-29-1 ***HR: 2***
DIAZOACETYLGLYCINE ETHYL ESTER
mf: $C_6H_9N_3O_3$ mw: 132.18

SYN: N-DIAZOACETYLGLYCINE ETHYL ESTER

TOXICITY DATA with REFERENCE
mmo-sat 10 μg/plate AMACCQ 6,655,74

dns-mus:fbr 2500 mmol/L JCROD7 94,7,79
ipr-mus LD50:1149 mg/kg ARZNAD 23,690,73

CONSENSUS REPORTS: EPA Genetic Toxicology Program.

SAFETY PROFILE: Moderately toxic by intraperitoneal route. Mutation data reported. When heated to decomposition it emits toxic fumes of NO_x. See also ESTERS.

DCO800 CAS:820-75-7 ***HR: 3***
N-(DIAZOACETYL)GLYCINE HYDRAZINE
mf: $C_4H_7N_5O_2$ mw: 157.16

SYNS: N-DIAZOACETILGLICINA-IDRAZIDE(ITALIAN) ◇ DIAZOACETYLGLYCINE HYDRAZIDE ◇ N-DIAZOACETYL GLYCYLHYDRAZIDE ◇ NSC-58404

TOXICITY DATA with REFERENCE
mmo-sat 10 μg/plate AMACCQ 6,655,74
mma-sat 10 μg/plate PNASA6 72,5135,75
dns-mus:fbr 2500 μmol/L JCROD7 94,7,79
ipr-mus TDLo:1200 mg/kg/4D-I:CAR BSIBAC 45,227,69
ipr-mus TD:720 mg/kg/4D-I:NEO BSIBAC 45,227,69
ipr-rat LD50:1335 mg/kg CCROBU 53,13,69
ivn-rat LD50:1595 mg/kg CCROBU 53,13,69
orl-mus LD50:3852 mg/kg CCROBU 53,13,69
ipr-mus LD50:2575 mg/kg CCROBU 53,13,69
scu-mus LD50:2985 mg/kg CCROBU 53,13,69
ivn-dog LDLo:39 mg/kg CCROBU 53,13,69
ivn-rbt LDLo:833 mg/kg CCROBU 53,13,69

CONSENSUS REPORTS: EPA Genetic Toxicology Program.

SAFETY PROFILE: Poison by intravenous route. Moderately toxic by ingestion, intraperitoneal, and subcutaneous routes. Questionable carcinogen with experimental carcinogenic and neoplastigenic data. Mutation data reported. When heated to decomposition it emits toxic fumes including NO_x.

DCP200 CAS:1192-27-4 ***HR: 3***
DIAZOCYCLOPENTADIENE
mf: $C_5H_4N_2$ mw: 92.10

CH=CHCH=CHC:N$_2$ (ring joining first CH and final C)

SAFETY PROFILE: Explodes when heated to its boiling point or when cooled to its freezing point. Upon decomposition it emits toxic fumes of NO_x. See also AZIDES.

DCP400 CAS:7008-85-7 ***HR: 3***
5-DIAZOIMIDAZOLE-4-CARBOXAMIDE
mf: $C_4H_3N_5O$ mw: 137.12

SYNS: DIAZO-ICA ◇ DIAZOIMIDAZOLE-4-CARBOXAMIDE ◇ NSC 22420

TOXICITY DATA with REFERENCE
dni-esc 10 μg/L BCPCA6 18,1463,69
oms-esc 500 μg/L BCPCA6 18,1463,69
dni-rat:lvr 100 μmol/L CNREA8 2827,76
orl-rat TDLo:2175 mg/kg/14W-C:ETA JNCIAM 54,951,75
ipr-mus LD50:1002 μg/kg NCISP* JAN86

SAFETY PROFILE: A deadly poison by intraperitoneal route. Questionable carcinogen with experimental tumorigenic data. Mutation data reported. When heated to decomposition it emits toxic fumes of NO_x.

DCP600 CAS:64038-55-7 ***HR: 3***
5-DIAZOIMIDAZOLE-4-CARBOXAMIDE HYDROCHLORIDE
mf: $C_4H_3N_5O•HCl$ mw: 137.12

TOXICITY DATA with REFERENCE
orl-rat TDLo:105 mg/kg/10W-C:ETA JNCIAM 54,951,75

SAFETY PROFILE: Questionable carcinogen with experimental tumorigenic data. When heated to decomposition it emits very toxic fumes of NO_x and HCl.

DCP700 CAS:59348-62-8 ***HR: 3***
DIAZOMALONIC ACID
mf: $C_3H_2N_2O_4$ mw: 130.06

SAFETY PROFILE: The impure acid and the diethyl ester are explosive. When heated to decomposition it emits toxic fumes of NO_x. See also AZIDES.

DCP775 CAS:1618-08-2 ***HR: 3***
DIAZOMALONONITRILE
mf: C_3N_4 mw: 92.06

$N_2C(C{\equiv}N)_2$

SYN: DIAZODICYANOMETHANE

CONSENSUS REPORTS: Cyanide and its compounds are on the Community Right-To-Know List.

SAFETY PROFILE: An explosive sensitive to sparks or heating to 75°C. When heated to decomposition it emits toxic fumes of NO_x and CN^-. See also NITRILES and CYANIDE.

DCP800 CAS:334-88-3 ***HR: 3***
DIAZOMETHANE
mf: CH_2N_2 mw: 42.05

PROP: Yellow gas at ordinary temp. Mp: −145°, bp: −23°, d: 1.45.

SYNS: AZIMETHYLENE ◇ DIAZIRINE

TOXICITY DATA with REFERENCE
mmo-nsc 250 mmol/L HEREAY 35,521,49
ihl-rat TCLo:272 mg/m^3/26W-I:ETA BJCAAI 16,92,62

CONSENSUS REPORTS: IARC Cancer Review: Group 3 IMEMDT 7,56,87; Animal Sufficient Evidence IMEMDT 7,223,74. EPA Genetic Toxicology Program. Community Right-To-Know List.

OSHA PEL: TWA 0.2 ppm
ACGIH TLV: TWA 0.2 ppm
DFG MAK: Animal Carcinogen, Suspected Human Carcinogen.

SAFETY PROFILE: Confirmed carcinogen with experimental tumorigenic data. A poison irritant by inhalation. A powerful allergen. It can cause pulmonary edema and frequently causes hypersensitivity leading to asthmatic symptoms. Mutation data reported. Highly explosive when shocked, exposed to heat or by chemical reaction. Undiluted liquid or gas may explode on contact with alkali metals, rough surfaces, heat (100°C), high intensity light or shock. When heated to decomposition or on contact with acid or acid fumes it emits highly toxic fumes of NO_x. Incompatible with alkali metals; calcium sulfate.

DCP880 CAS:40953-35-3 ***HR: 3***
2-DIAZONIO-4,5-DICYANOIMIDAZOLIDE
mf: C_5N_6 mw: 144.10

NC(CN)=C(CN)N=CN_2

SYN: DIAZODICYANOIMIDAZOLE

CONSENSUS REPORTS: Cyanide and its compounds are on the Community Right-To-Know List.

SAFETY PROFILE: Explodes when heated above 150°C. The dry material is very shock sensitive. When heated to decomposition it emits toxic fumes of NO_x and CN^-. See also CYANIDE.

DCQ200 CAS:13101-58-1 ***HR: 3***
5-DIAZONIOTETRAZOLIDE
mf: CN_6 mw: 96.05

SAFETY PROFILE: Concentrated solutions in water explode at 0°C. When heated to decomposition it emits toxic fumes of NO_x. See also AZIDES.

DCQ400 CAS:157-03-9 ***HR: 3***
6-DIAZO-5-OXONORLEUCINE
mf: $C_6H_9N_3O_3$ mw: 171.18

SYNS: DIAZO-OXO-NORLEUCINE ◇ 6-DIAZO-5-OXO-l-NORLEUCINE ◇ DON ◇ NSC 7365

TOXICITY DATA with REFERENCE
pic-esc 5 μg/plate CNREA8 43,2819,83
ims-mus TDLo:400 μg/kg (female 11D post):TER TJADAB 23,29A,81
ipr-rat TDLo:100 μg/kg (female 7-8D post):REP PSEBAA 94,33,57
ipr-rat LD50:80 mg/kg CPCHAO 18,307,62
orl-mus LD50:197 mg/kg NCISP* JAN86
ipr-mus LDLo:300 mg/kg JOENAK 18,204,59
ivn-mus LD50:74 mg/kg 85ERAY 2,1253,78

CONSENSUS REPORTS: EPA Genetic Toxicology Program.

SAFETY PROFILE: Poison by ingestion, intraperitoneal, and intravenous routes. Experimental teratogenic and reproductive effects. Mutation data reported. When heated to decomposition it emits toxic fumes of NO_x.

DCQ500 CAS:64781-77-7 ***HR: 3***
4-DIAZO-5-PHENYL-1,2,3-TRIAZOLE
mf: $C_8H_5N_5$ mw: 171.16

N_2CN=N-N=CC_6H_5

SAFETY PROFILE: A heat-sensitive explosive. Upon decomposition it emits toxic fumes of NO_x. See also AZIDES.

DCQ550 CAS:2032-04-4 ***HR: 3***
3-DIAZOPROPENE
mf: $C_3H_4N_2$ mw: 68.08

H_2C=$CHCHN_2$

SYN: VINYLDIAZOMETHANE

SAFETY PROFILE: A storage hazard. It is potentially explosive and should be stored in the dark below 0°C. When heated to decomposition it emits toxic fumes of NO_x. See also AZIDES.

DCQ560 ***HR: D***
DIAZO RED RD

TOXICITY DATA with REFERENCE
ipr-mus TDLo:150 mg/kg (female 8D post):REP TCMUD8 4,403,84

SAFETY PROFILE: Experimental reproductive effects. When heated to decomposition it emits acrid smoke and irritating fumes.

DCQ575 ***HR: 2***
3-DIAZOTYRAMINE HYDROCHLORIDE

PROP: A nitrosated product of TYRAMINE
mf: $C_8H_9N_3O$ mw: 163.20

SYNS: 4-(2-AMINOETHYL)-6-DIAZO-2,4-CYCLOHEXADIENONEHYDROCHLORIDE ◇ TYRAMINE, 3-DIAZO-, HYDROCHLORIDE

TOXICITY DATA with REFERENCE
orl-rat TDLo:77700 mg/kg/2Y-C:CAR CRNGDP 8,527,87

SAFETY PROFILE: Questionable carcinogen with experimental carcinogenic data. When heated to decomposition it emits toxic fumes of NO_x.

DCQ600 CAS:2435-76-9 ***HR: 3***
DIAZOURACIL
mf: $C_4H_2N_4O_2$ mw: 138.10

SYNS: 5-DIAZOPYRIMIDINE-2,4(3H)-DIONE ◇ 5-DIAZO-2,4(1H,3H)-PYRIMIDINEDIONE ◇ 5-DIAZOURACIL ◇ 2,4-DIOSSI-5-DIAZOPIRIMIDINA (ITALIAN) ◇ 2,6-DIOXO-5-DIAZOPYRIMIDINE ◇ DU ◇ NSC 23519 ◇ (1,2,3)OXADIAZOLO(5,4-d)PYRIMIDIN-5(4H)-ONE

TOXICITY DATA with REFERENCE
dnd-sat 10 mg/L MILEDM 1,169,76
dnd-esc 10 mg/L MILEDM 1,169,76
dns-rat:lvr 50 μmol/L CALEDQ 13,187,81
ipr-mus LD50:30800 μg/kg NCISP* JAN86
scu-mus LD10:22 mg/kg EJCAAH 10,667,74

CONSENSUS REPORTS: Reported in EPA TSCA Inventory.

SAFETY PROFILE: Poison by subcutaneous and intraperitoneal routes. Mutation data reported. When heated to decomposition it emits toxic fumes of NO_x.

DCQ650 CAS:94362-44-4 ***HR: 2***
DIAZO V

SYN: DIAZO RESIN V

TOXICITY DATA with REFERENCE
scu-rat TDLo:900 mg/kg/26W-I:NEO VINIT* #5689-83

SAFETY PROFILE: Questionable carcinogen with experimental neoplastigenic data. When heated to decomposition it emits acrid smoke and irritating fumes.

DCQ700 CAS:364-98-7 ***HR: 3***
DIAZOXIDE
mf: $C_8H_7ClN_2O_2S$ mw: 230.68

PROP: Crystals from dilute alc. Mp: 330-331°. Sol in alc and alkaline solns; insol in water.

SYNS: 7-CHLORO-3-METHYL-2H-1,2,4-BENZOTHIADIAZINE-1,1-DIOXIDE ◇ 7-CLORO-3-METIL-2H-1,2,4-BENZOTIODIAZINE-1,1-DIOSSIDO (ITALIAN) ◇ DIAZOSSIDO (ITALIAN) ◇ DIZOXIDE ◇ EUDEMINE INJECTION ◇ HYPERSTAT ◇ HYPERTONALUM ◇ MUTABASE ◇ PROGLICEM ◇ SRG 95213

TOXICITY DATA with REFERENCE
ivn-dom TDLo:65 mg/kg (110-114D preg):TER JCNDBK 11,206,71
orl-rat LD50:980 mg/kg AIPTAK 143,446,63
ipr-rat LD50:510 mg/kg AIPTAK 143,446,63
orl-mus LD50:444 mg/kg JPETAB 136,344,62
ipr-mus LD50:326 mg/kg JPETAB 136,344,62
ivn-mus LD50:228 mg/kg JPETAB 136,344,62

SAFETY PROFILE: Poison by intravenous and intraperitoneal routes. Moderately toxic by ingestion. An experimental teratogen. When heated to decomposition it emits toxic fumes of Cl^-, SO_x, and NO_x.

DCQ800 CAS:34493-98-6 ***HR: 3***
DIBEKACIN
mf: $C_{18}H_{37}N_5O_8$ mw: 451.60

SYNS: DEBECACIN ◇ DIDEOXYKANAMYCIN B ◇ 3',4'-DIDEOXYKANAMYCIN B ◇ DKB ◇ ORBICIN

TOXICITY DATA with REFERENCE
ims-rat TDLo:2400 mg/kg (female 10-21D post):REP JJANAX 26,40,73
ims-rat TDLo:900 mg/kg (10-15D preg):TER JJANAX 26,40,73
ipr-rat LD50:16760 μg/kg IYKEDH 6,119,75
scu-rat LD50:23870 μg/kg IYKEDH 6,119,75
ivn-rat LD50:12510 μg/kg IYKEDH 6,119,75
orl-mus LD50:763 mg/kg IYKEDH 6,119,75
ipr-mus LD50:11960 μg/kg IYKEDH 6,119,75
scu-mus LD50:15980 μg/kg IYKEDH 6,119,75
ivn-mus LD50:8950 μg/kg IYKEDH 6,119,75
ims-mus LD50:373 mg/kg JOPHDQ 4,356,81

CONSENSUS REPORTS: EPA Genetic Toxicology Program.

SAFETY PROFILE: Poison by intraperitoneal, subcutaneous, intramuscular, and intravenous routes. Moderately toxic by ingestion. Experimental teratogenic and reproductive effects. An antibacterial agent. When heated to decomposition it emits toxic fumes of NO_x.

DCR000 CAS:51-50-3 ***HR: 3***
DIBENAMINE
mf: $C_{16}H_{18}ClN$ mw: 259.80

SYNS: N-(2-CHLOROETHYL)DIBENZYLAMINE ◇ DIBENZYL CHLORETHYLAMINE ◇ N,N-DIBENZYL-β-CHLOROETHYLAMINE ◇ SYMPATHOLYTIN

TOXICITY DATA with REFERENCE
dni-mus ivg 2 pph JIDEAE 62,378,74
unr-mus TDLo:2 mg/kg (1D pre):REP FESTAS 12,346,61
orl-rat LD50:2400 mg/kg CLDND*
ipr-mus LD50:395 mg/kg AIPTAK 155,69,65
scu-mus LD50:800 mg/kg JPETAB 97,25,49
ivn-mus LD50:50 mg/kg AIPTAK 105,317,56

SAFETY PROFILE: Poison by intravenous and intraperitoneal routes. Moderately toxic by ingestion and subcutaneous routes. Experimental reproductive effects. Can cause leukopenia (reduced white blood cell count).

Mutation data reported. When heated to decomposition it emits very toxic fumes of Cl^- and NO_x. See also AROMATIC AMINES.

DCR200 CAS:55-43-6 ***HR: 3***
DIBENAMINE HYDROCHLORIDE
mf: $C_{16}H_{18}ClN{\cdot}ClH$ mw: 296.26

SYNS: N-(2-CHLOROETHYL)DIBENZYLAMINEHYDROCHLORIDE ◇ DIBENAMINE ◇ N,N-DIBENZYLAMINOETHYL CHLORIDE HYDROCHLORIDE ◇ DIBENZYLCHLORETHAMINE HYDROCHLORIDE ◇ DIBENZYLCHLORETHYLAMINE HYDROCHLORIDE ◇ N,N-DIBENZYL-β-CHLOROETHYLAMINE HYDROCHLORIDE ◇ N,N-DIBENZYL-2-CHLOROETHYLAMINE HYDROCHLORIDE ◇ SYMPATHOLYTIN

TOXICITY DATA with REFERENCE
ipr-mus LD50:75 mg/kg JPETAB 89,167,47
scu-mus LD50:400 mg/kg JPETAB 89,167,47
ivn-mus LD50:50 mg/kg JPETAB 89,167,47
ivn-cat LDLo:35 mg/kg JPETAB 89,167,47

SAFETY PROFILE: Poison by subcutaneous, intravenous, and intraperitoneal routes. An adrenergic blocker and diagnostic aid (pheochromocytoma). When heated to decomposition it emits very toxic fumes of Cl^- and NO_x.

DCR300 CAS:5385-75-1 ***HR: 2***
DIBENZ(a,e)ACEANTHRYLENE
mf: $C_{24}H_{14}$ mw: 302.38

SYNS: DIBENZO(a,e)FLUORANTHENE ◇ 2,3,5,6-DIBENZOFLUORANTHENE

TOXICITY DATA with REFERENCE
mma-sat 500 nmol/L CRNGDP 5,1263,84
dns-mus:emb 1 μmol/L CRNGDP 5,379,84
skn-mus TDLo:2880 μg/kg/15W-I:CAR CRNGDP 8,461,87

CONSENSUS REPORTS: IARC Cancer Review: Group 3 IMEMDT 7,56,87; Animal Limited Evidence IMEMDT 32,321,83

SAFETY PROFILE: Questionable carcinogen with experimental carcinogenic data. Mutation data reported. When heated to decomposition it emits acrid smoke and fumes.

DCR400 CAS:203-20-3 ***HR: 3***
DIBENZ(a,j)ACEANTHRYLENE
mf: $C_{24}H_{14}$ mw: 302.38

SYN: 15,16-BENZDEHYDROCHOLANTHRENE

TOXICITY DATA with REFERENCE
scu-mus TDLo:800 mg/kg/9W-I:ETA AJCAA7 28,334,36
ivn-mus LDLo:10 mg/kg JNCIAM 1,225,40

SAFETY PROFILE: Poison by intravenous route. Questionable carcinogen with experimental tumorigenic data. When heated to decomposition it emits acrid smoke and irritating fumes.

DCR600 CAS:201-42-3 ***HR: 3***
13H-DIBENZ(bc,j)ACEANTHRYLENE
mf: $C_{23}H_{14}$ mw: 290.37

SYNS: 13H-ACENAPHTHO(1,8-ab)PHENANTHRENE ◇ 1′,9-METHYLENE-1,2:5,6-DIBENZANTHRACENE

TOXICITY DATA with REFERENCE
scu-mus TDLo:400 mg/kg:ETA AJCAA7 28,334,36

SAFETY PROFILE: Questionable carcinogen with experimental tumorigenic data. When heated to decomposition it emits acrid smoke and irritating fumes.

DCR800 CAS:517-85-1 ***HR: 3***
4H-DIBENZ(f,g,j)ACEANTHRYLENE, 5,5a,6,7-TETRAHYDRO-
mf: $C_{23}H_{18}$ mw: 294.41

SYNS: ANG.-STERANTHREN (GERMAN) ◇ ANG-STERANTHRENE

TOXICITY DATA with REFERENCE
skn-mus TDLo:1120 mg/kg/35W-I:ETA ZEKBAI 62,217,57

SAFETY PROFILE: Questionable carcinogen with experimental tumorigenic data. When heated to decomposition it emits acrid smoke and irritating fumes.

DCS200 CAS:1977-10-2 ***HR: 3***
DIBENZACEPIN
mf: $C_{18}H_{18}ClN_3O$ mw: 327.84

SYNS: 2-CHLORO-11-(4-METHYL-1-PIPERAZINYL)-DIBENZO(b,f)(1,4)OXAZEPINE ◇ 2-CHLORO-11-(4-METHYL-1-PIPERAZINYL)-DIBENZO(b,f)(1,4)OXOAZEPINE ◇ CL-62362 ◇ CL-71563 ◇ CLOXAZEPINE ◇ DIBENZOAZEPINE ◇ HF3170 ◇ LOXAPINE ◇ LW 3170 ◇ OXILAPINE ◇ S-805 ◇ SUM 3170

TOXICITY DATA with REFERENCE
orl-mus TDLo:6 mg/kg (female 7-12D post):REP OYYAA2 4,305,70
orl-rat TDLo:24 mg/kg (9-14D preg):TER OYYAA2 4,305,70
orl-rat LD50:151 mg/kg OYYAA2 4,293,70
ipr-rat LD50:35 mg/kg OYYAA2 4,293,70
scu-rat LD50:350 mg/kg OYYAA2 4,293,70
ivn-rat LD50:18 mg/kg OYYAA2 4,293,70
orl-mus LD50:40 mg/kg 27ZQAG -,79,72
ipr-mus LD50:27 mg/kg 27ZQAG -,79,72
scu-mus LD50:53 mg/kg OYYAA2 4,293,70
ivn-mus LD50:22 mg/kg OYYAA2 4,293,70

SAFETY PROFILE: Poison by ingestion, intraperitoneal, subcutaneous, and intravenous routes. Experimental teratogenic and reproductive effects. A tranquilizer. Many dibenz-azepine compounds have central nervous system effects. When heated to decomposition it emits very toxic fumes of Cl^- and NO_x.

DCS400 CAS:226-36-8 ***HR: 3***
DIBENZ(a,h)ACRIDINE
mf: $C_{21}H_{13}N$ mw: 279.35

SYNS: 7-AZADIBENZ(a,h)ANTHRACENE ◇ DB(a,h)AC ◇ DIBENZ (a,d)ACRIDINE ◇ 1,2,5,6-DIBENZACRIDINE ◇ 1,2,5,6-DIBENZOACRIDINE ◇ 1,2,5,6-DINAPHTHACRIDINE

TOXICITY DATA with REFERENCE
imp-rat TDLo:5 mg/kg:CAR CALEDQ 20,97,83
orl-mus TDLo:13 g/kg/63W-I:ETA PRLBA4 129,439,40

CONSENSUS REPORTS: NTP Fifth Annual Report on Carcinogens. IARC Cancer Review: Group 2B IMEMDT 7,56,87; Animal Sufficient Evidence IMEMDT 32,277,83; IMEMDT 3,247,73. EPA Genetic Toxicology Program.

SAFETY PROFILE: Confirmed carcinogen with experimental carcinogenic and tumorigenic data. When heated to decomposition it emits toxic fumes of NO_x. See also ANTHRACENE.

DCS600 CAS:224-42-0 ***HR: 3***
DIBENZ(a, j)ACRIDINE
mf: $C_{21}H_{13}N$ mw: 279.35

SYNS: 7-AZADIBENZ(a,j)ANTHRACENE ◇ DB(a,j)AC ◇ DIBENZ(a,f) ACRIDINE ◇ 1,2,7,8-DIBENZACRIDINE ◇ 3,4,5,6-DIBENZACRIDINE ◇ DIBENZO(a,j)ACRIDINE ◇ 3,4,6,7-DINAPHTHACRIDINE

TOXICITY DATA with REFERENCE
mma-sat 5 μg/plate PNASA6 72,5135,75
dnd-esc 10 μmol/L PNCCA2 5,39,65
pic-esc 50 mg/L CNREA8 41,532,81
orl-mus TDLo:2520 mg/kg (21D pre):REP DABSAQ 29,4777,69
skn-mus TDLo:99 mg/kg/99W-I:CAR CALEDQ 37,337,87
scu-mus TDLo:40 mg/kg:ETA JJIND8 1,225,40

CONSENSUS REPORTS: NTP Fifth Annual Report on Carcinogens. IARC Cancer Review: Group 2B IMEMDT 7,56,87; Animal Sufficient Evidence IMEMDT 32,283,83; IMEMDT 3,254,73. EPA Genetic Toxicology Program.

SAFETY PROFILE: Confirmed carcinogen with experimental carcinogenic and tumorigenic data. Experimental reproductive effects. Mutation data reported. When heated to decomposition it emits toxic fumes of NO_x. See also ANTHRACENE.

DCS800 CAS:224-53-3 ***HR: 3***
DIBENZ(c,h)ACRIDINE
mf: $C_{21}H_{13}N$ mw: 279.35

SYNS: 14-AZADIBENZ(a,j)ANTHRACENE ◇ 3,4:5,6-DIBENZACRIDINE ◇ 1,2,7,8-DIBENZACRIDINE (FRENCH)

TOXICITY DATA with REFERENCE
mma-sat 4 μg/plate BJCAAI 37,873,78
skn-mus TDLo:2040 mg/kg/85W-I:ETA PRLBA4 129,439,40

SAFETY PROFILE: Questionable carcinogen with experimental tumorigenic data. Mutation data reported. When heated to decomposition it emits toxic fumes of NO_x. See also ANTHRACENE.

DCT000 CAS:63918-83-2 ***HR: 3***
DIBENZ(a, j)ACRIDINE METHOSULFATE
mf: $C_{21}H_{13}N \cdot C_2H_6O_4S$ mw: 405.49

SYN: 3,4:5,6-DIBENZACRIDINEMETHOSULFATE

TOXICITY DATA with REFERENCE
scu-mus TDLo:3750 mg/kg/56W-I:ETA PRLBA4 129,439,40

SAFETY PROFILE: Questionable carcinogen with experimental tumorigenic data. When heated to decomposition it emits very toxic fumes of SO_x and NO_x. See also SULFONATES.

DCT400 CAS:53-70-3 ***HR: 3***
DIBENZ(a,h)ANTHRACENE
mf: $C_{22}H_{14}$ mw: 278.36

SYNS: 1,2:5,6-BENZANTHRACENE ◇ DBA ◇ DB(a,h)A ◇ 1,2,5,6-DBA ◇ 1,2,5,6-DIBENZANTHRACEEN (DUTCH) ◇ 1,2:5,6-DIBENZANTHRACENE ◇ 1,2:5,6-DIBENZ(a)ANTHRACENE ◇ DIBENZO(a,h)ANTHRACENE ◇ 1,2:5,6-DIBENZOANTHRACENE ◇ RCRA WASTE NUMBER U063

TOXICITY DATA with REFERENCE
dnd-hmn:emb 360 nmol/L CBINA8 22,257,78
dnd-esc 10 μmol/L MUREAV 89,95,81
otr-rat-orl 200 mg/kg CNREA8 40,1157,80
msc-mus:lym 4250 μg/L MUREAV 106,101,82
msc-ham:lng 500 μg/L MUREAV 136,65,84
scu-rat TDLo:2400 μg/kg/50D-I:NEO 85DLAB -,-,75
orl-mus TDLo:4160 mg/kg/26W-I:CAR JPBAA7 49,21,39
skn-mus TDLo:1200 mg/kg/50W-I:CAR 14JTAF -,275,65
mul-mus TDLo:40 mg/kg/12D-I:ETA PHRPA6 52,637,37
ivn-mus LDLo:10 mg/kg JNCIAM 1,225,40

CONSENSUS REPORTS: NTP Fifth Annual Report on Carcinogens. IARC Cancer Review: Group 2A IMEMDT 7,56,87; Animal Sufficient Evidence IMEMDT 32,299,83; IMEMDT 3,178,73. EPA Genetic Toxicology Program. Reported in EPA TSCA Inventory.

SAFETY PROFILE: Confirmed carcinogen with experimental carcinogenic, tumorigenic, and neoplastigenic data. Poison by intravenous route. Human mutation data reported. When heated to decomposition it emits acrid smoke and irritating fumes.

DCT600 CAS:224-41-9 ***HR: 3***
DIBENZ(a,j)ANTHRACENE
mf: $C_{22}H_{14}$ mw: 278.36

SYN: 1,2:7,8-DIBENZANTHRACENE

TOXICITY DATA with REFERENCE
mma-sat 1 μg/plate MUREAV 51,311,78
skn-mus TDLo:252 mg/kg/81W-I:ETA JNCIAM 44,641,70

CONSENSUS REPORTS: IARC Cancer Review: Group 3 IMEMDT 7,56,87; Animal Limited Evidence IMEMDT 32,309,83.

SAFETY PROFILE: Questionable carcinogen with experimental tumorigenic data. Mutation data reported. When heated to decomposition it emits acrid smoke and irritating fumes. See also ANTHRACENE.

DCT800 ***HR: 3***
1,2,5,6-DIBENZANTHRACENECHOLEIC ACID
mf: $C_{96}H_{160}O_{16}{\cdot}C_{22}H_{14}$ mw: 1848.92

SYN: 3-α-12-α-DIHYDROXY-5-β-CHOLAN-24-OIC ACID with DIBENZ (a,h)ANTHRACENE

TOXICITY DATA with REFERENCE
scu-mus TDLo:800 mg/kg/9W-I:ETA JNCIAM 2,99,41

SAFETY PROFILE: Questionable carcinogen with experimental tumorigenic data. When heated to decomposition it emits acrid smoke and irritating fumes. See also ANTHRACENE.

DCU200 CAS:4665-48-9 ***HR: 3***
1,2:5,6-DIBENZANTHRACENE-9,10-endo-α,β-SUCCINIC ACID
mf: $C_{26}H_{18}O_4$ mw: 394.44

SYN: 7,14-DIHYDRO-7,14-ETHANODIBENZ(a,b)ANTHRACENE-15,16-DICARBOXYLIC ACID

TOXICITY DATA with REFERENCE
mmo-esc 2040 mg/L/4H GENTAE 39,141,54
scu-rat TDLo:665 mg/kg/50D-I:CAR,REP 85DLAB -,-,75

SAFETY PROFILE: Questionable carcinogen with experimental carcinogenic data. Experimental reproductive effects. Mutation data reported. When heated to decomposition it emits acrid smoke and irritating fumes. See also ANTHRACENE.

DCU400 CAS:4002-76-0 ***HR: D***
DIBENZ(a,h)ANTHRACEN-5-OL
mf: $C_{22}H_{14}O$ mw: 294.36

SYN: 5-HYDROXY-DIBENZ(a,h)ANTHRACENE

TOXICITY DATA with REFERENCE
mma-sat 3800 ng/plate MUREAV 96,1,82
dnd-mus:oth 1 mg/L CBINA8 4,389,71/72
otr-ham:emb 1 mg/L CNREA8 32,1391,72
dnd-ham:kdy 5 mg/L BCPCA6 20,1297,71
dnd-ham:lng 1 mg/L CBINA8 4,389,71/72

CONSENSUS REPORTS: EPA Genetic Toxicology Program.

SAFETY PROFILE: Mutation data reported. When heated to decomposition it emits acrid smoke and irritating fumes. See also ANTHRACENE.

DCU600 CAS:63041-44-1 ***HR: 3***
DIBENZANTHRANYL GLYCINE COMPLEX
mf: $C_{25}H_{18}N_2O_3$ mw: 394.45

SYNS: 1:2:5:6-DIBENZANTHRACENE-9-CARBAMIDO-ACETICACID ◇ N-(DIBENZ(a,h)ANTHRACEN-7-YLCARBAMOYL)GLYCINE

TOXICITY DATA with REFERENCE
scu-mus TDLo:80 mg/kg/12W-I:ETA AJCAA7 35,203,39

SAFETY PROFILE: Questionable carcinogen with experimental tumorigenic data. When heated to decomposition it emits toxic fumes of NO_x.

DCU800 CAS:116-71-2 ***HR: 2***
DIBENZANTHRONE
mf: $C_{34}H_{16}O_2$ mw: 456.50

SYNS: DINAPHTHO(1,2,3-cd:3′,2′,1′-lm)PERYLENE-5,10-DIONE ◇ VIOLANTHRONE

TOXICITY DATA with REFERENCE
ipr-rat LD50:5000 mg/kg RPTOAN 40,137,77
ipr-mus LD50:2600 mg/kg RPTOAN 40,137,77

CONSENSUS REPORTS: Reported in EPA TSCA Inventory.

SAFETY PROFILE: Moderately toxic by intraperitoneal route. When heated to decomposition it emits acrid smoke and irritating fumes.

DCV000 CAS:116-90-5 ***HR: 2***
4,4′-DIBENZANTHRONIL
mf: $C_{34}H_{18}O_2$ mw: 458.52

SYNS: 4,4′-DI-7H-BENZ(de)ANTHRACEN-7-ONE ◇ 4,4′-DIBENZANTHRONYL

TOXICITY DATA with REFERENCE
eye-rbt 500 mg/24H SEV 28ZPAK -,60,72
ipr-rat LD50:2400 mg/kg RPTOAN 40,137,77
ipr-mus LD50:1100 mg/kg RPTOAN 40,137,77

CONSENSUS REPORTS: Reported in EPA TSCA Inventory.

SAFETY PROFILE: Moderately toxic by intraperitoneal

route. A severe eye irritant. When heated to decomposition it emits acrid smoke and irritating fumes.

DCV200 CAS:298-46-4 ***HR: 3***
5H-DIBENZ(b,f)AZEPINE-5-CARBOXAMIDE
mf: $C_{15}H_{12}N_2O$ mw: 236.29

SYNS: BISTON ◇ CARBAMAZEPEN ◇ CARBAMAZEPINE ◇ CARBAMEZEPINE ◇ 5-CARBAMOYL-5H-DIBENZ(b,f)AZEPINE ◇ 5-CARBAMOYL-5H-DIBENZO(b,f)AZEPINE ◇ 5-CARBAMOYLDIBENZO(b,f) AZEPINE ◇ 5-CARBAMYLDIBENZO(b,f)AZEPINE ◇ 5-CARBAMYL-5H-DIBENZO(b,f)AZEPINE ◇ CARBAZEPINE ◇ FINLEPSIN ◇ G 32883 ◇ GEIGY 32883 ◇ STAZEPIN ◇ TEGRETAL ◇ TEGRETOL ◇ TELESMIN ◇ TIMONIL

TOXICITY DATA with REFERENCE
orl-rat TDLo:3600 mg/kg (14D pre/1-22D preg):REP TJADAB 29(3),33A,84
orl-rat TDLo:765 mg/kg (9-17D preg):TER EAMJAV 60,407,83
orl-man TDLo:160 mg/kg/3W-I:SKN JCPYDR 5,185,85
orl-cld TDLo:1050 mg/kg/6W-I AJPSAO 143,1176,85
orl-man TDLo:253 mg/kg/6W-I:GIT,SYS JCPYDR 6,251,86
orl-cld TDLo:19 mg/kg/4W-I:PNS PEDIAU 73,841,84
orl-wmn TDLo:28 mg/kg/4D-I AJPSAO 143,1328,86
orl-hmn TDLo:43 mg/kg:CNS,GIT BMJOAE 1,754,77
orl-wmn TDLo:100 mg/kg/17D-I:BLD JCLPDE 45,315,84
orl-man LDLo:54 mg/kg/9D-I:CNS,KID,BLD CMAJAX 132,1040,85
orl-wmn LDLo:1920 mg/kg/17W-I:BLD AJPSAO 142,974,85
orl-rat LD50:1957 mg/kg JKXXAF #79-163823
ipr-rat LD50:293 mg/kg ARZNAD 30,477,80
orl-mus LD50:936 mg/kg RPTOAN 32,131,69
ipr-mus LD50:270 mg/kg JKXXAF #79-163823

CONSENSUS REPORTS: EPA Genetic Toxicology Program.

SAFETY PROFILE: A human poison by ingestion. Poison experimentally by intraperitoneal route. Human systemic effects by ingestion: aplastic anemia, sleep, hallucinations, distorted perceptions, nausea or vomiting, somnolence, dermatitis, ataxia (loss of muscle coordination), urine volume increase, and agranulo-cytosis, liver function tests impaired, fasciculations, and thrombocytopenia. Human reproductive effects. Experimental teratogenic and reproductive effects. An analgesic and anticonvulsant. When heated to decomposition it emits toxic fumes of NO_x. See also DIAZEPAM.

DCV400 CAS:28058-62-0 ***HR: 3***
5H-DIBENZ(b,f)AZEPINE, 3-CHLORO-5-(3-(4-CARBAMOYL-4-PIPERIDINOPIPERIDINO) PROPYL)-10,11-DIHYDRO-, DIHYDROCHLORIDE, MONOHYDRATE
mf: $C_{28}H_{37}ClN_4O•2ClH•H_2O$ mw: 571.07

SYN: Y-4153

TOXICITY DATA with REFERENCE
orl-mus TDLo:60 mg/kg (7-12D preg):TER OYYAA2 5,663,71
orl-rat TDLo:7 g/kg (35D male):REP OYYAA2 5,643,71
orl-rat LD50:6800 mg/kg ARZNAD 21,391,71
ipr-rat LD50:125 mg/kg ARZNAD 21,391,71
orl-mus LD50:2550 mg/kg ARZNAD 21,391,71
ipr-mus LD50:160 mg/kg ARZNAD 21,391,71
scu-mus LD50:6500 mg/kg ARZNAD 21,391,71

SAFETY PROFILE: Poison by intraperitoneal route. Moderately toxic by ingestion. Mildly toxic by subcutaneous route. Experimental teratogenic and reproductive effects. A psychotropic drug. When heated to decomposition it emits very toxic fumes of Cl^- and NO_x.

DCV800 CAS:315-72-0 ***HR: 3***
4-(3-(5H-DIBENZ(b,f)AZEPIN-5-YL)PROPYL)-1-PIPERAZINEETHANOL
mf: $C_{23}H_{29}N_3O$ mw: 363.55

SYNS: ENDISON ◇ G 33040 ◇ GR 33040 ◇ 5-(3-(4-(2-HYDROXYETHYL)-1-PIPERAZINYL)PROPYL)-5H-DIBENZ(b,f)AZEPINE ◇ INSIDON ◇ NISIDANA ◇ OPIPRAMOL ◇ OPRAMIDOL

TOXICITY DATA with REFERENCE
orl-rat LD50:1110 mg/kg FRPPAO 25,519,70
ipr-rat LD50:95 mg/kg AIPTAK 148,560,64
scu-rat LD50:497 mg/kg AIPTAK 148,560,64
ivn-rat LD50:32 mg/kg AIPTAK 148,560,64
orl-mus LD50:443 mg/kg FRPPAO 25,519,70
ipr-mus LD50:120 mg/kg AIPTAK 148,560,64
scu-mus LD50:315 mg/kg AIPTAK 148,560,64
ivn-mus LD50:45 mg/kg AIPTAK 148,560,64
ivn-rbt LD50:11 mg/kg AIPTAK 148,560,64

SAFETY PROFILE: Poison by intraperitoneal, intravenous, and subcutaneous routes. Moderately toxic by ingestion. Many dibenz-azepine compounds have central nervous system effects. When heated to decomposition it emits toxic fumes of NO_x.

DCW000 CAS:15727-43-2 ***HR: 3***
DI(BENZENEDIAZONIUM)ZINC TETRACHLORIDE
mf: $C_{12}H_{10}Cl_4N_4Zn$ mw: 417.42

PROP: After drying, it can explode.

SYN: BENZENEDIAZONIUMTETRACHLOROZINCATE

CONSENSUS REPORTS: Zinc and its compounds are on the Community Right-To-Know List.

SAFETY PROFILE: May be a light-, heat-, and shock-sensitive explosive. When heated to decomposition it emits toxic fumes of NO_x, Cl^-, and ZnO. See also ZINC COMPOUNDS.

DCW200 CAS:22755-07-3 ***HR: 3***
DI(BENZENEDIAZO)SULFIDE
mf: $C_{12}H_{10}N_4S$ mw: 242.30

SAFETY PROFILE: The wet solid explodes on impact, heating or contact with air. Upon decomposition it emits toxic fumes of SO_x. See also SULFIDES.

DCW400 CAS:29342-61-8 ***HR: 3***
DIBENZENESULFONYL PEROXIDE
mf: $C_{12}H_{10}O_6S_2$ mw: 314.34

SAFETY PROFILE: Explodes when heated to 53°C and when shocked. Decomposes violently when stored at room temperature. Explodes on contact with boiling water or fuming nitric acid. When heated to decomposition it emits toxic fumes of SO_x. See also PEROXIDES.

DCW600 CAS:4498-32-2 ***HR: 3***
DIBENZEPIN
mf: $C_{18}H_{21}N_3O$ mw: 295.37

SYNS: DIBENZEPINE ◇ 5,10-DIHYDRO-10-(2-(DIMETHYLAMINO) ETHYL)-5-METHYL-11H-DIBENZO(b,e)(1,4)DIAZEPIN-11-ONE ◇ 10-(2-(DIMETHYLAMINO)ETHYL)-5,10-DIHYDRO-5-METHYL-11H-DIBENZO(B,E)(1,4)DIAZEPIN-11-ONE ◇ 10-(2-(DIMETHYLAMINO) ETHYL)-5-METHYL-5H-DIBENZO(b,e)(1,4)DIAZEPIN-11(10H)-ONE ◇ HF 1927

TOXICITY DATA with REFERENCE
orl-rat LD50:220 mg/kg INPHB6 1,214,68
ipr-rat LD50:70 mg/kg INPHB6 1,214,68
scu-rat LD50:542 mg/kg IYKEDH 6,119,75
ivn-rat LD50:22 mg/kg INPHB6 1,214,68
orl-mus LD50:194 mg/kg IYKEDH 6,119,75
ipr-mus LD50:60 mg/kg ARZNAD 21,1727,71
scu-mus LD50:90 mg/kg ARZNAD 19,458,69
ivn-mus LD50:22 mg/kg INPHB6 1,214,68

SAFETY PROFILE: Poison by ingestion, intraperitoneal, intravenous, and subcutaneous routes. Many dibenz-azepine compounds have central nervous system effects. When heated to decomposition it emits toxic fumes of NO_x. See also DIAZEPAM.

DCW800 CAS:315-80-0 ***HR: 3***
DIBENZEPINE HYDROCHLORIDE
mf: $C_{18}H_{21}N_3O{\bullet}ClH$ mw: 331.88

SYNS: DIBENZEPIN HYDROCHLORIDE ◇ HF 1927 ◇ HYDROFLUORIDE-1927 WANDER ◇ 5-METHYL-10-β-DIMETHYLAMINO-AETHYL-10,11-DIHYDRO-11-OXO-5-DIBENZO(b,e)(1,4)DIAZEPIN ◇ NEODALIT ◇ NOVERIL ◇ NOVERYL

TOXICITY DATA with REFERENCE
orl-rat LD50:220 mg/kg 27ZQAG -,70,72
ipr-rat LD50:70 mg/kg 27ZQAG -,70,72
scu-rat LD50:520 mg/kg NIIRDN 6,341,82
ivn-rat LD50:22 mg/kg 27ZQAG -,70,72
orl-mus LD50:174 mg/kg NIIRDN 6,341,82
ipr-mus LD50:64 mg/kg FATOAO 35,274,72
scu-mus LD50:98 mg/kg RFPPAO 25,519,70
ivn-mus LD50:22 mg/kg 27ZQAG -,70,72

SAFETY PROFILE: Poison by ingestion, intravenous, subcutaneous, and intraperitoneal routes. An antidepressant. Many dibenz-azepine compounds have central nervous system effects. When heated to decomposition it emits very toxic fumes of HCl and NO_x. See also DIAZEPAM.

DCX000 CAS:201-65-0 ***HR: 3***
1,2,3,4-DIBENZFLUORENE
mf: $C_{21}H_{14}$ mw: 266.35

SYN: 13H-INDENO(1,2-1)PHENANTHRENE

TOXICITY DATA with REFERENCE
skn-mus TDLo:1040 mg/kg/43W-I:ETA PRLBA4 129,439,40

SAFETY PROFILE: Questionable carcinogen with experimental tumorigenic data. When heated to decomposition it emits acrid smoke and irritating fumes.

DCX400 CAS:193-40-8 ***HR: 3***
DIBENZ(c,f)INDENO(1,2,3-ij)(2,7)NAPHTHYRIDINE
mf: $C_{22}H_{12}N_2$ mw: 304.36

TOXICITY DATA with REFERENCE
skn-mus TDLo:1200 mg/kg/52W-I:NEO BJCAAI 17,266,63

SAFETY PROFILE: Questionable carcinogen with experimental neoplastigenic data. When heated to decomposition it emits toxic fumes of NO_x.

DCX600 CAS:207-84-1 ***HR: 3***
7H-DIBENZO(a,g)CARBAZOLE
mf: $C_{20}H_{13}N$ mw: 267.34

SYN: 1,2,5,6-DIBENZCARBAZOLE

TOXICITY DATA with REFERENCE
skn-mus TDLo:275 mg/kg/23W-I:ETA PRLBA4 122,429,37

SAFETY PROFILE: Questionable carcinogen with experimental tumorigenic data. When heated to decomposition it emits toxic fumes of NO_x.

DCX800 CAS:239-64-5 ***HR: 3***
7H-DIBENZO(a,i)CARBAZOLE
mf: $C_{20}H_{13}N$ mw: 267.34

SYN: 1,2,7,8-DIBENZCARBAZOLE

TOXICITY DATA with REFERENCE
skn-mus TDLo:515 mg/kg/43W-I:ETA PRLBA4 122,429,37

SAFETY PROFILE: Questionable carcinogen with experimental tumorigenic data. When heated to decomposition it emits toxic fumes of NO_x.

DCY000 CAS:194-59-2 ***HR: 3***
7H-DIBENZO(c,g)CARBAZOLE
mf: $C_{20}H_{13}N$ mw: 267.34

PROP: Needles. Mp: 158°.

SYNS: 7-AZA-7H-DIBENZO(c,g)FLUORENE ◇ 7H-DB(c,g)C ◇ 3,4,5,6-DIBENZCARBAZOL ◇ 3,4,5,6-DIBENZCARBAZOLE ◇ 3,4,5,6-DIBENZO-CARBAZOLE ◇ 3,4,5,6-DINAPHTHACARBAZOLE

TOXICITY DATA with REFERENCE
dnd-mus-scu 44 μmol/kg CRNGDP 6,1271,85
scu-rat TDLo:150 mg/kg/17W-I:ETA PRLBA4 122,429,37
skn-mus TDLo:99 mg/kg/99W-I:CAR CALEDQ 37,337,87
itr-ham TDLo:72 mg/kg/18W-I:NEO JTEHD6 3,935,77
ipr-mus LDLo:13 mg/kg BIJOAK 32,1460,38

CONSENSUS REPORTS: NTP Fifth Annual Report on Carcinogens. IARC Cancer Review: Group 2B IMEMDT 7,56,87; Animal Sufficient Evidence IMEMDT 32,315,83; IMEMDT 3,260,73.

SAFETY PROFILE: Confirmed carcinogen with experimental carcinogenic, neoplastigenic, and tumorigenic data. Poison by intraperitoneal route. Mutation data reported. When heated to decomposition it emits toxic fumes of NO_x.

DCY200 CAS:189-64-0 ***HR: 3***
DIBENZO(b,def)CHRYSENE
mf: $C_{24}H_{14}$ mw: 302.38

SYNS: BD(a,h)P ◇ DIBENZO(a,h)PYRENE ◇ 1,2,6,7-DIBENZO-PYRENE ◇ 3,4,8,9-DIBENZOPYRENE ◇ 3,4,8,9-DIBENZPYRENE

TOXICITY DATA with REFERENCE
mma-sat 12500 pmol/plate CNREA8 41,2589,81
msc-ham:lng 30 μg/L CNREA8 42,1646,82
imp-rat TDLo:100 mg/kg:ETA NEOLA4 26,23,79
skn-mus TDLo:287 mg/kg/30W-I:CAR ZKKOBW 89,113,77

CONSENSUS REPORTS: NTP Fifth Annual Report on Carcinogens. IARC Cancer Review: Group 2B IMEMDT 7,56,87; Animal Sufficient Evidence IMEMDT 32,331,83; IMEMDT 3,207,73.

SAFETY PROFILE: Confirmed carcinogen with experimental carcinogenic and tumorigenic data. When heated to decomposition it emits acrid smoke and irritating fumes. Mutation data reported.

DCY400 CAS:191-30-0 ***HR: 3***
DIBENZO(def,p)CHRYSENE
mf: $C_{24}H_{14}$ mw: 302.38

SYNS: BA 51-090462 ◇ DB(a,l)P ◇ DIBENZO(a,d)PYRENE ◇ DIBENZO (a,l)PYRENE ◇ 1,2:3,4-DIBENZOPYRENE ◇ 1,2,9,10-DIBENZOPYRENE ◇ 2,3:4,5-DIBENZOPYRENE ◇ 1,2,3,4-DIBENZPYRENE ◇ 4,5,6,7-DIBENZPYRENE

TOXICITY DATA with REFERENCE
skn-mus TDLo:890 mg/kg/37W-I:ETA PRLBA4 123,343,37

CONSENSUS REPORTS: NTP Fifth Annual Report on Carcinogens. IARC Cancer Review: Group 2B IMEMDT 7,56,87; Animal Sufficient Evidence IMEMDT 32,343,83; Animal Limited Evidence IMEMDT 3,224,73

SAFETY PROFILE: Confirmed carcinogen with experimental tumorigenic data. When heated to decomposition it emits acrid smoke and irritating fumes.

DCY600 CAS:63040-54-0 ***HR: 3***
DIBENZO(b,def)CHRYSENE-7-CARBOXALDEHYDE
mf: $C_{25}H_{14}O$ mw: 330.39

SYN: 5-FORMYL-3,4:8,9-DIBENZOPYRENE

TOXICITY DATA with REFERENCE
scu-mus TDLo:72 mg/kg/9W-I:ETA COREAF 252,1711,61

SAFETY PROFILE: Questionable carcinogen with experimental tumorigenic data. When heated to decomposition it emits acrid smoke and irritating fumes. See also ALDEHYDES.

DCY800 CAS:2869-59-2 ***HR: 2***
DIBENZO(def,p)CHRYSENE-10-CARBOXALDEHYDE
mf: $C_{25}H_{14}O$ mw: 330.39

SYN: 5-FORMYL-1,2:3,4-DIBENZOPYRENE

TOXICITY DATA with REFERENCE
scu-mus TDLo:72 mg/kg/9W-I:ETA COREAF 259,3899,64

SAFETY PROFILE: Questionable carcinogen with experimental tumorigenic data. When heated to decomposition it emits acrid smoke and irritating fumes.

DCZ000 CAS:128-66-5 ***HR: 3***
DIBENZO(b,def)CHRYSENE-7,14-DIONE
mf: $C_{24}H_{12}O_2$ mw: 332.36

PROP: C.I. vat yellow 4 tested in NCITR* NCI-CG-TR-134,79 consists of 18.2% dibenzo(b,def)chrysene-7,14-dione, 30.8% sorbitol, 5.5% lomar twc, 2.7% glycerin and 42.8% water NCITR* NCI-CG-TR-134,79.

SYNS: AHCOVAT PRINTING GOLDEN YELLOW ◇ AMANTHRENE GOLDEN YELLOW ◇ ANTHRAVAT GOLDEN YELLOW ◇ ARLANTHRENE GOLDEN YELLOW ◇ BENZADONE GOLDEN YELLOW ◇ CALCOLOID GOLDEN YELLOW ◇ CALEDON GOLDEN YELLOW ◇ CALEDON PRINTING YELLOW ◇ CARBANTHRENE GOLDEN YELLOW ◇ C.I. 59100 ◇ CIBANONE GOLDEN YELLOW ◇ C.I. VAT YELLOW ◇ DIBENZO(a,b)PYRENE-7,14-DIONE ◇ 2,3,7,8-DIBENZO-

PYRENE-1,6-QUINONE ◇ 1′,2′,6′,7′-DIBENZPYRENE-7,14-QUINONE ◇ FEMANTHREN GOLDEN YELLOW ◇ GOLDEN YELLOW ◇ HELANTHRENE YELLOW ◇ HOSTAVAT GOLDEN YELLOW ◇ INDANTHRENE GOLDEN YELLOW ◇ LEUCOSOL GOLDEN YELLOW ◇ MAYVAT GOLDEN YELLOW ◇ MIKETHRENE GOLD YELLOW ◇ NCI-C03565 ◇ NIHONTHRENE GOLDEN YELLOW ◇ NYANTHRENE GOLDEN YELLOW ◇ PALANTHRENE GOLDEN YELLOW ◇ PARADONE GOLDEN YELLOW ◇ PHARMANTHRENE GOLDEN YELLOW ◇ ROMANTRENE GOLDEN YELLOW ◇ SANDOTHRENE PRINTING YELLOW ◇ SOLANTHRENE BRILLIANT YELLOW ◇ TINON GOLDEN YELLOW ◇ TYRION YELLOW ◇ VAT GOLDEN YELLOW ◇ YELLOW

TOXICITY DATA with REFERENCE
orl-mus TDLo:7420 g/kg/2Y-C:CAR NCITR* NCI-CG-TR-134,79
orl-mus TD:2225 g/kg/106W-C:ETA NCITR* NCI-CG-TR-134,79

CONSENSUS REPORTS: NCI Carcinogenesis Bioassay Completed; Results Positive: Mouse NCITR* NCI-CG-TR-134,79; Negative: Rat NCITR* NCI-CG-TR-134,79. Reported in EPA TSCA Inventory. Community Right-To-Know List.

SAFETY PROFILE: Questionable carcinogen with experimental carcinogenic and tumorigenic data. When heated to decomposition it emits acrid smoke and irritating fumes.

DDA600 CAS:438-60-8 ***HR: 3***
N-3-(5H-DIBENZO(a,d)CYCLOHEPTEN-5-YL) PROPYL-N-METHYLAMINE
mf: $C_{19}H_{21}N$ mw: 263.41

SYNS: 5-(3-METHYLAMINOPROPYL)-5H-DIBENZO(a,d)CYCLOHEPTENE ◇ MK 240 ◇ PROTRIPTYLINE ◇ PROTRYPTYLINE ◇ TRIPTIL ◇ VIVACTIL

TOXICITY DATA with REFERENCE
orl-rat LD50:240 mg/kg FRPPAO 25,519,70
ipr-rat LD50:42 mg/kg FRPPAO 25,519,70
orl-mus LD50:269 mg/kg FRPPAO 25,519,70
ipr-mus LD50:67 mg/kg FRPPAO 25,519,70
scu-mus LD50:192 mg/kg FRPPAO 25,519,70
ivn-mus LD50:30 mg/kg JMCMAR 17,65,74
orl-rbt LD50:310 mg/kg FRPPAO 25,519,70
ivn-rbt LD50:8200 mg/kg FRPPAO 25,519,70

SAFETY PROFILE: Poison by ingestion, intraperitoneal, subcutaneous, and intravenous routes. When heated to decomposition it emits toxic fumes of NO_x.

DDA800 CAS:262-12-4 ***HR: 3***
DIBENZO-p-DIOXIN
mf: $C_{12}H_8O_2$ mw: 184.20

PROP: Crystals. Mp: 123°.

SYNS: DIBENZODIOXIN ◇ DIBENZO(1,4)DIOXIN ◇ DIBENZO(b.e)(1,4)DIOXIN ◇ DIPHENYLENE DIOXIDE ◇ NCI-C03656 ◇ OXANTHRENE ◇ PHENODIOXIN

TOXICITY DATA with REFERENCE
skn-mus TDLo:110 g/kg/58W-I:ETA EVHPAZ 5,163,73

CONSENSUS REPORTS: IARC Cancer Review: Animal Inadequate Evidence IMEMDT 15,41,77. NCI Carcinogenesis Bioassay Completed; Results Negative NCITR* NCI-CG-TR-122,79.

SAFETY PROFILE: Questionable carcinogen with experimental tumorigenic data. When heated to decomposition it emits acrid smoke and irritating fumes.

DDB000 CAS:207-83-0 ***HR: 3***
13H-DIBENZO(a,g)FLUORENE
mf: $C_{21}H_{14}$ mw: 266.35

SYN: 1,2,5,6-DIBENZOFLUORENE

TOXICITY DATA with REFERENCE
skn-mus TDLo:48 mg/kg/15W-I:ETA CNREA8 11,301,51

SAFETY PROFILE: Questionable carcinogen with experimental tumorigenic data. When heated to decomposition it emits acrid smoke and irritating fumes.

DDB200 CAS:239-60-1 ***HR: 3***
13H-DIBENZO(a,i)FLUORENE
mf: $C_{21}H_{14}$ mw: 266.35

SYN: 1,2,7,8-DIBENZFLUORENE

TOXICITY DATA with REFERENCE
skn-mus TDLo:1340 mg/kg/56W-I:ETA PRLBA4 129,439,40

SAFETY PROFILE: Questionable carcinogen with experimental tumorigenic data. When heated to decomposition it emits acrid smoke and irritating fumes.

DDB400 CAS:36115-09-0 ***HR: 2***
1,1′-(2,8-DIBENZOFURADIYL)BIS(2-(DIMETHYLAMINOETHANONE) DIHYDROCHLORIDE HYDRATE (2:5)
mf: $C_{20}H_{22}NO_3 \cdot 2ClH \cdot 5/2H_2O$ mw: 442.35

SYN: RMI 11567 DA

TOXICITY DATA with REFERENCE
orl-mus LD50:2700 mg/kg ALACBI 12,77,79
scu-mus LD50:1000 mg/kg ALACBI 12,77,79

SAFETY PROFILE: Moderately toxic by ingestion and subcutaneous routes. When heated to decomposition it emits very toxic fumes of HCl and NO_x.

DDB600 CAS:3693-22-9 ***HR: 3***
2-DIBENZOFURANAMINE
mf: $C_{12}H_9NO$ mw: 183.22

SYNS: 2-ADO ◇ 3-AMINODIBENZOFURAN ◇ 2-AMINODIPHENYLENE OXIDE

TOXICITY DATA with REFERENCE
dns-mus-orl 80 mg/kg BIJOAK 111,12P,69
orl-rat TDLo:168 mg/kg/90W-I:CAR ZEKBAI 61,45,56
orl-mus TDLo:22 g/kg/52W-C:ETA BECCAN 46,271,68

SAFETY PROFILE: Questionable carcinogen with experimental carcinogenic and tumorigenic data. Mutation data reported. When heated to decomposition it emits toxic fumes of NO_x.

DDB800 CAS:4106-66-5 ***HR: 3***
3-DIBENZOFURANAMINE
mf: $C_{12}H_9NO$ mw: 183.22

SYNS: 2-AMINODIPHENYLENOXYD (GERMAN) ◇ DIBENZOFURANYLAMINE

TOXICITY DATA with REFERENCE
orl-rat TDLo:1400 mg/kg/66W-C:CAR ZEKBAI 61,45,56

SAFETY PROFILE: Questionable carcinogen with experimental carcinogenic data. When heated to decomposition it emits toxic fumes of NO_x.

DDC000 CAS:5834-25-3 ***HR: 3***
N-3-DIBENZOFURANYLACETAMIDE
mf: $C_{14}H_{11}NO_2$ mw: 225.26

SYNS: 3-ACETAMIDODIBENZFURANE ◇ 3-ACETAMIDODIBENZOFURAN ◇ 3-ACETYLAMINODIBENZOFURAN ◇ 3-DIBENZOFURANYLACETAMIDE

TOXICITY DATA with REFERENCE
orl-rat TDLo:4496 mg/kg/35W-C:ETA CNREA8 9,504,49

SAFETY PROFILE: Questionable carcinogen with experimental tumorigenic data. When heated to decomposition it emits toxic fumes of NO_x.

DDC200 CAS:192-47-2 ***HR: 3***
DIBENZO(h,rst)PENTAPHENE
mf: $C_{28}H_{16}$ mw: 352.44

PROP: Pale yellow needles. Mp: 321°.

SYNS: TRIBENZO(a,e,i)PYRENE ◇ (1,2,4,5,7,8)TRIBENZOPYRENE ◇ (1,2,4,5,8,9)TRIBENZOPYRENE ◇ 1,2:4,5:8,9-TRIBENZOPYRENE

TOXICITY DATA with REFERENCE
scu-mus TDLo:72 mg/kg/9W-I:ETA COREAF 259,3899,64

CONSENSUS REPORTS: IARC Cancer Review: Group 3 IMEMDT 7,56,87; Animal Limited Evidence IMEMDT 3,197,73

SAFETY PROFILE: Suspected carcinogen with experimental tumorigenic data. When heated to decomposition it emits acrid smoke and irritating fumes.

DDC400 CAS:188-96-5 ***HR: 3***
DIBENZO(cd,lm)PERYLENE
mf: $C_{26}H_{14}$ mw: 326.40

SYN: PEROPYRENE

TOXICITY DATA with REFERENCE
scu-mus TDLo:72 mg/kg/9W-I:ETA CHDDAT 266,301,68

SAFETY PROFILE: Questionable carcinogen with experimental tumorigenic data. When heated to decomposition it emits acrid smoke and irritating fumes.

DDC600 CAS:215-64-5 ***HR: 3***
DIBENZO(a,c)PHENAZINE
mf: $C_{20}H_{12}N_2$ mw: 280.34

SYN: 1,2,3,4-DIBENZPHENAZINE

TOXICITY DATA with REFERENCE
imp-rat TDLo:7 mg/kg:ETA COREAF 240,1738,55

SAFETY PROFILE: Questionable carcinogen with experimental tumorigenic data. When heated to decomposition it emits toxic fumes of NO_x.

DDC800 CAS:226-47-1 ***HR: 3***
DIBENZO(a,h)PHENAZINE
mf: $C_{20}H_{12}N_2$ mw: 280.34

SYNS: 7,14-DIAZADIBENZ(a,h)ANTHRACENE ◇ DIBENZ(a,h)PHENAZINE ◇ 1,2:5,6-DIBENZPHENAZINE

TOXICITY DATA with REFERENCE
imp-rat TDLo:7 mg/kg:ETA COREAF 240,1738,55

CONSENSUS REPORTS: EPA Genetic Toxicology Program.

SAFETY PROFILE: Questionable carcinogen with experimental tumorigenic data. When heated to decomposition it emits toxic fumes of NO_x.

DDD000 CAS:1785-74-6 ***HR: 3***
DIBENZOSUBERONE OXIME
mf: $C_{15}H_{13}NO$ mw: 223.29

SYN: 10,11-DIHYDRO-5H-DIBENZO(a,d)CYCLOHEPTEN-5-ONE OXIME

TOXICITY DATA with REFERENCE
ipr-mus LD50:350 mg/kg JMCMAR 7,88,64
ivn-mus LD50:100 mg/kg CSLNX* NX#01997

SAFETY PROFILE: Poison by intraperitoneal and intravenous routes. When heated to decomposition it emits toxic fumes of NO_x.

DDD400 CAS:54818-88-1 ***HR: 3***
N-2-DIBENZOTHIENYLACETAMIDE
mf: $C_{14}H_{11}NOS$ mw: 241.32

SYN: 2-ACETYLAMINODIBENZOTHIOPHENE

TOXICITY DATA with REFERENCE
orl-rat TDLo:4680 mg/kg/32W-C:CAR CNREA8 15,188,55

SAFETY PROFILE: Questionable carcinogen with experimental carcinogenic data. When heated to decomposition it emits very toxic fumes of SO_x and NO_x.

DDD600 CAS:64057-52-9 ***HR: 3***
N-3-DIBENZOTHIENYLACETAMIDE
mf: $C_{14}H_{11}NOS$ mw: 241.32

SYNS: 3-ACETAMIDODIBENZTHIOPHENE ◇ 3-ACETAMINODIBENZOTHIOPHENE ◇ 3-ACETYLAMINODIBENZOTHIOPHENE

TOXICITY DATA with REFERENCE
orl-rat TDLo:4739 mg/kg/35W-C:ETA CNREA8 9,504,49
orl-rat LD50:1195 mg/kg JPETAB 99,450,50

SAFETY PROFILE: Moderately toxic by ingestion. Questionable carcinogen with experimental tumorigenic data. When heated to decomposition it emits very toxic fumes of NO_x and SO_x.

DDD800 CAS:63020-21-3 ***HR: 3***
N-3-DIBENZOTHIENYLACETAMIDE-5-OXIDE
mf: $C_{14}H_{11}NO_2S$ mw: 257.32

SYNS: 3-ACETAMIDODIBENZTHIOPHENE OXIDE ◇ 3-ACETYLAMINODIBENZOTHIOPHENE-5-OXIDE

TOXICITY DATA with REFERENCE
orl-rat TDLo:5103 mg/kg/35W-C:ETA CNREA8 9,504,49

SAFETY PROFILE: Questionable carcinogen with experimental tumorigenic data. When heated to decomposition it emits very toxic fumes of SO_x and NO_x.

DDE000 CAS:35556-06-0 ***HR: 2***
1,1'-(2,8-DIBENZOTHIOPHENEDIYL)BIS(2-(DIMETHYLAMINO)ETHANONE) DIHYDROCHLORIDE TRIHYDRATE
mf: $C_{20}H_{22}N_2O_2S{\bullet}2ClH{\bullet}3H_2O$ mw: 481.48

SYN: RMI 11877 DA

TOXICITY DATA with REFERENCE
orl-mus LD50:2930 mg/kg ALACBI 12,77,79
scu-mus LD50:820 mg/kg ALACBI 12,77,79

SAFETY PROFILE: Moderately toxic by ingestion and subcutaneous routes. When heated to decomposition it emits very toxic fumes of HCl, SO_x and NO_x.

DDE200 CAS:257-07-8 ***HR: 3***
DIBENZ(b,f)(1,4)OXAZEPINE
mf: $C_{13}H_9NO$ mw: 195.23

SYNS: CR ◇ EA 3547

TOXICITY DATA with REFERENCE
skn-hmn 500 μg/1H MLD BJDEAZ 90,657,74
eye-man 17 ng APTOA6 35,412,74
eye-rbt 5 mg MLD ARTODN 34,183,75
eye-rbt 1540 ng APTOA6 35,412,74
eye-gpg 682 ng APTOA6 35,412,74
ihl-rat TCLo:2 mg/m³/5M (6-15D preg):TER TXAPA9 29,301,74
ivn-rbt TDLo:47400 μg/kg (14-16D preg):REP TXAPA9 29,301,74
ihl-mus TCLo:236 mg/m³/18W-I:CAR TOLED5 17,13,83
ihl-mus TC:204 mg/m³/18W-I:ETA TOLED5 17,13,83
orl-rat LD50:563 mg/kg IAEC** 17JUN74
ipr-rat LD50:164 mg/kg IAEC** 17JUN74
ivn-rat LD50:26 mg/kg IAEC** 17JUN74
orl-mus LD50:770 mg/kg IAEC** 17JUN74
ihl-mus LCLo:1500 mg/m³/2H TXCYAC 8,347,77
ipr-mus LD50:242 mg/kg IAEC** 17JUN74
ivn-mus LD50:37200 μg/kg IAEC** 17JUN74

CONSENSUS REPORTS: Reported in EPA TSCA Inventory.

SAFETY PROFILE: Poison by intraperitoneal and intravenous routes. Moderately toxic by ingestion and inhalation. Experimental teratogenic and reproductive effects. A human skin and eye irritant. Questionable carcinogen with experimental carcinogenic and tumorigenic data. When heated to decomposition it emits toxic fumes of NO_x.

DDE300 CAS:2743-38-6 ***HR: 1***
DIBENZOYLTARTARIC ACID
mf: $C_{18}H_{14}O_8$ mw: 358.32

PROP: Crystals. Mp: 89-92°.

SYNS: BUTANEDIOIC ACID, 2,3-BIS(BENZOYLOXY)-, (R-(R*,R*))- ◇ TARTARIC ACID, DIBENZOATE

TOXICITY DATA with REFERENCE
eye-rbt 100 mg MOD FCTOD7 20,573,82
eye-rbt 100 mg/4S RNS MLD FCTOD7 20,573,82

CONSENSUS REPORTS: Reported in EPA TSCA Inventory.

SAFETY PROFILE: An eye irritant. When heated to decomposition it emits acrid smoke and irritating fumes.

DDF000 CAS:73926-80-4 ***HR: 3***
DIBENZYLBUTYLSULFONIUM IODIDE MERCURIC IODIDE

SYN: DIBENZYLBUTYLSULFONIUM IODIDE with MERCURY IODIDE (1:1)

TOXICITY DATA with REFERENCE
ivn-mus LD50:56 mg/kg CSLNX* NX#01719

CONSENSUS REPORTS: Mercury and its compounds are on the Community Right-To-Know List.

NIOSH REL: TWA 0.05 mg(Hg)/m³

SAFETY PROFILE: Poison by intravenous route. See

also IODIDES and MERCURY IODIDE. When heated to decomposition it emits very toxic fumes of Hg, I^-, and SO_x.

DDF200 CAS:101833-83-4 ***HR: 3***
1-1,3-DIBENZYLDECAHYDRO-2-OXOIMIDAZO (4,5-c)THIENO(1,2-a)THIOLIUM-2-OXO-10-BORANESULFONATE
mf: $C_{22}H_{25}N_2OS•C_{10}H_{15}O_4S$ mw: 596.86

SYNS: 1-3,4-(1′,3′-DIBENZYL-2′-KETO-IMIDAZOLIDO)-1,2-TRIMETHYLENE THIOPHANIUM CAMPHOR SULFONATE ◇ NU-2221

TOXICITY DATA with REFERENCE
ipr-mus LD50:135 mg/kg JPETAB 97,48,49
ivn-mus LD50:23 mg/kg JPETAB 97,48,49

SAFETY PROFILE: Poison by intraperitoneal and intravenous routes. When heated to decomposition it emits very toxic fumes of SO_x and NO_x.

DDF400 CAS:59766-02-8 ***HR: 3***
7,14-DIBENZYLDIBENZ(a,h)ANTHRACENE
mf: $C_{36}H_{26}$ mw: 458.62

SYN: 9,10-DIBENZYL-1,2,5,6-DIBENZANTHRACENE

TOXICITY DATA with REFERENCE
skn-mus TDLo:1250 mg/kg/52W-I:ETA PRLBA4,111,485,32

SAFETY PROFILE: Questionable carcinogen with experimental tumorigenic data. When heated to decomposition it emits acrid smoke and irritating fumes.

DDF600 CAS:63957-48-2 ***HR: 3***
DIBENZYL(5-DIBENZYLAMINO-2,4-PENTADIENYLIDENE)AMMONIUM CHLORIDE SESQUIHYDRATE
mf: $C_{33}H_{33}N_2•Cl•3/2H_2O$ mw: 520.16

TOXICITY DATA with REFERENCE
orl-mus LD50:500 mg/kg JMCMAR 12,806,69
ipr-mus LD50:200 mg/kg JMCMAR 12,806,69

SAFETY PROFILE: Poison by intraperitoneal route. Moderately toxic by ingestion. When heated to decomposition it emits very toxic fumes of NO_x, NH_3, and Cl^-.

DDG400 CAS:3412-76-8 ***HR: 3***
N,N′-DIBENZYLETHYLENEDIAMINE DIHYDROCHLORIDE
mf: $C_{16}H_{20}N_2•2ClH$ mw: 313.30

SYN: DBED DIHYDROCHLORIDE

TOXICITY DATA with REFERENCE
orl-mus LD50:630 mg/kg ANTCAO 1,504,51
ipr-mus LD50:104 mg/kg ANTCAO 1,504,51
scu-mus LD50:200 mg/kg ARZNAD 9,628,59

SAFETY PROFILE: Poison by intraperitoneal and subcutaneous routes. Moderately toxic by ingestion. When heated to decomposition it emits very toxic fumes of HCl and NO_x.

DDG600 CAS:73926-81-5 ***HR: 3***
DIBENZYLETHYLSULFONIUM IODIDE MERCURIC IODIDE

SYN: DIBENZYLETHYLSULFONIUM IODIDE with MERCURY IODIDE (1:1)

TOXICITY DATA with REFERENCE
ivn-mus LD50:56 mg/kg CSLNX* NX#01718

CONSENSUS REPORTS: Mercury and its compounds are on the Community Right-To-Know List.

NIOSH REL: TWA 0.05 mg(Hg)/m^3

SAFETY PROFILE: Poison by intravenous route. See also IODIDES and MERCURY IODIDE. When heated to decomposition it emits very toxic fumes of Hg, I^- and SO_x.

DDG800 CAS:63-92-3 ***HR: 3***
DIBENZYLINE HYDROCHLORIDE
mf: $C_{18}H_{22}ClNO•ClH$ mw: 340.32

SYNS: 688A ◇ BENSYLYT ◇ 2-(N-BENZYL-2-CHLOROETHYLAMINO)-1-PHENOXYPROPANE HYDROCHLORIDE ◇ BENZYL(2-CHLOROETHYL)(1-METHYL-2-PHENOXYETHYL)AMINEHYDROCHLORIDE ◇ N-BENZYL-N-PHENOXYISOPROPYL-β-CHLORETHYLAMINE HYDROCHLORIDE ◇ BENZYLYT ◇ BLOCADREN ◇ N-(2-CHLOROETHYL)-N-(1-METHYL-2-PHENOXYETHYL) BENZENEMETHANAMINE HYDROCHLORIDE ◇ N-(2-CHLOROETHYL)-N-(1-METHYL-2-PHENOXYETHYL)BENZYLAMINEHYDROCHLORIDE ◇ DIBENZYLENE ◇ DIBENZYLIN ◇ DIBENZYRAN ◇ FENOXYBENZAMIN ◇ NCI-C01661 ◇ PHENOXYBENZAMIDE HYDROCHLORIDE ◇ N-PHENOXYISOPROPYL-N-BENZYL-β-CHLOROETHYLAMINE HYDROCHLORIDE ◇ N-2-PHENOXYISOPROPYL-N-BENZYL-CHLOROETHYLAMINE HYDROCHLORIDE ◇ SKF 688A

TOXICITY DATA with REFERENCE
orl-rat TDLo:12 mg/kg (5-12D preg):TER RCOCB8 7,701,74
par-rat TDLo:24500 μg/kg (male 35D pre):REP CCPTAY 29,189,84
ipr-rat TDLo:780 mg/kg/1Y-I:CAR NCITR* NCI-CG-TR-72,78
ipr-mus TDLo:3900 mg/kg/52W-I:CAR NCITR* NCI-CG-TR-72,78
orl-man TDLo:7143 μg/kg/5D-I:SYS AIMEAS 107,119,87
orl-rat LDLo:800 mg/kg JPETAB 110,463,54
orl-mus LD50:900 mg/kg AIPTAK 108,102,56
ipr-mus LD50:228 mg/kg TXAPA9 28,227,74
scu-mus LD50:105 mg/kg ARZNAD 17,305,67
ivn-mus LD50:63750 μg/kg EJPHAZ 9,289,70

CONSENSUS REPORTS: NTP Fifth Annual Report on Carcinogens. IARC Cancer Review: Group 2B IM-

EMDT 7,56,87; Animal Sufficient Evidence IMEMDT 24,185,80. NCI Carcinogenesis Bioassay Completed; Results Positive: mouse, rat NCITR* NCI-CG-TR-72,78.

SAFETY PROFILE: Confirmed carcinogen with experimental carcinogenic and teratogenic data. Poison by intraperitoneal, intravenous, and subcutaneous routes. Human systemic effects by ingestion: changes in tubules including acute renal failure, acute tubular necrosis. Moderately toxic by ingestion. Other experimental reproductive effects. A long-acting adrenergic blocker. When heated to decomposition it emits very toxic fumes of NO_x and Cl^-.

DDH000 CAS:780-24-5 ***HR: 3***
DIBENZYLMERCURY
mf: $C_{14}H_{14}Hg$ mw: 382.87

PROP: Colorless crystals, sol in organic solvents.

TOXICITY DATA with REFERENCE
ivn-mus LD50:56 mg/kg CSLNX* NX#03272

OSHA PEL: (Transitional: CL 1 mg/10m^3) TWA 0.01 mg(Hg)/m^3; STEL 0.03 mg/m^3 (skin)
ACGIH TLV: TWA 0.01 mg(Hg)/m^3; STEL 0.03 mg(Hg)/m^3
NIOSH REL: (Inorganic Mercury) TWA 0.05 mg(Hg)/m^3

SAFETY PROFILE: Poison by intravenous route. See also MERCURY COMPOUNDS, ORGANIC. When heated to decomposition it emits toxic fumes of Hg.

DDH400 CAS:17176-77-1 ***HR: 3***
DIBENZYL PHOSPHITE
mf: $C_{14}H_{15}O_3P$ mw: 262.25

PROP: Decomposes @ 160°.

SAFETY PROFILE: Potentially explosive decomposition when heated. When heated to decomposition it emits toxic fumes of PO_x.

DDH600 CAS:3666-67-9 ***HR: 3***
2,2-DIBENZYL-4-(2-PIPERIDYL)-1,3-DIOXOLANE HYDROCHLORIDE
mf: $C_{22}H_{27}NO_2 \cdot ClH$ mw: 373.96

TOXICITY DATA with REFERENCE
orl-mus LD50:200 mg/kg JMCMAR 9,127,66
ivn-mus LD50:25 mg/kg JMCMAR 9,127,66

SAFETY PROFILE: Poison by ingestion and intravenous routes. When heated to decomposition it emits very toxic fumes of HCl and NO_x.

DDH800 CAS:621-08-9 ***HR: 2***
DIBENZYLSULFOXIDE
mf: $C_{14}H_{14}OS$ mw: 230.34

SYNS: BENZYL SULFOXIDE ◇ DIBENZYL SULPHOXIDE

TOXICITY DATA with REFERENCE
ipr-mus LD50:600 mg/kg IJRBA3 3,41,61

CONSENSUS REPORTS: Reported in EPA TSCA Inventory.

SAFETY PROFILE: Moderately toxic by intraperitoneal route. When heated to decomposition it emits toxic fumes of SO_x.

DDH900 CAS:2964-06-9 ***HR: 3***
DIBERAL
mf: $C_{12}H_{20}N_2O_3$ mw: 240.34

SYNS: 5-(1,3-DIMETHYLBUTYL)-5-ETHYLBARBITURICACID ◇ DMBEB ◇ 5-(1,3-DIMETHYLBUTYL)-5-ETHYL-2,4,6(1H,3H,5H)-PYRIMIDINETRIONE (9CI)

TOXICITY DATA with REFERENCE
orl-rat LD50:75 mg/kg JMPCAS 1,31,59
orl-mus LD50:70 mg/kg JMPCAS 1,31,59
ipr-mus LD50:17 mg/kg JMPCAS 1,31,59
orl-dog LD50:10 mg/kg JMPCAS 1,31,59
ivn-dog LD50:2500 μg/kg JMPCAS 1,31,59
ivn-rbt LDLo:7 mg/kg JACSAT 58,1354,36

SAFETY PROFILE: Poison by ingestion, intravenous, and intraperitoneal routes. When heated to decomposition it emits toxic fumes of NO_x. See also BARBITURATES.

DDI000 CAS:13084-46-3 ***HR: 3***
DI-1,2-BIS(DIFLUOROAMINO)ETHYL ETHER
mf: $C_4H_6F_8N_4O$ mw: 278.10

$$[F_2NCH_2CH(NF_2)]_2O$$

SAFETY PROFILE: An impact-sensitive explosive. When heated to decomposition it emits toxic fumes of F^- and NO_x. See also ETHERS.

DDI200 CAS:1345-07-9 ***HR: 3***
DIBISMUTH TRISULFIDE
mf: Bi_2S_3 mw: 514.15

SAFETY PROFILE: Possibly explosive during preparation. When heated to decomposition it emits toxic fumes of Bi. See also BISMUTH COMPOUNDS and SULFIDES.

DDI400 ***HR: 3***
DI-(BISTRIFLUOROMETHYLPHOSFIDO)MERCURY
mf: $C_4F_{12}HgP_2$ mw: 537.70

$$[(F_3C)_2P]_2Hg$$

CONSENSUS REPORTS: Mercury and its compounds are on the Community Right-To-Know List.

SAFETY PROFILE: Ignites spontaneously in air. When heated to decomposition it emits toxic fumes of F^-, PO_x, and Hg. See also MERCURY COMPOUNDS and FLUORIDES.

DDI450 CAS:19287-45-7 ***HR: 3***
DIBORANE
DOT: UN 1911
mf: B_2H_6 mw: 27.68

PROP: Colorless gas, sickly sweet odor. Mp: −165.5°, bp: −92.5°, d: 0.447 (liquid @ −112°), 0.577 (solid @ −183°), vap press: 224 mm @ −112°, autoign temp: 38-52°, lel: 0.9%, uel: 98%, flash p: −90°F.

SYNS: BOROETHANE ◇ BORON HYDRIDE ◇ DIBORANE(6) ◇ DIBORON HEXAHYDRIDE

TOXICITY DATA with REFERENCE
ihl-rat LC50:40 ppm/4H 14KTAK -,693-64
ihl-mus LC50:29 ppm/4H TXAPA9 4,215,62
ihl-dog LCLo:125 ppm/2H AMIHAB 13,346,56
ihl-ham LCLo:50 ppm/8H AMIHAB 21,519,60

CONSENSUS REPORTS: Reported in EPA TSCA Inventory. EPA Extremely Hazardous Substances List.

OSHA PEL: TWA 0.1 ppm
ACGIH TLV: TWA 0.1 ppm
DFG MAK: 0.1 ppm (0.1 mg/m^3)
DOT Classification: Flammable Gas; Label: Flammable Gas and Poison; DOT-IMO: Flammable Gas; Label: Poison Gas and Flammable Gas.

SAFETY PROFILE: Poison by inhalation. An irritant to skin, eyes, and mucous membranes comparable to chlorine, fluorine, arsine, and phosgene. The liquid causes local inflammation, blisters, redness, and swelling. Injuries to central nervous system, liver, and kidneys have also been produced in experimental animals. Similar observations have been reported in humans resulting at times in a reaction resembling metal fume fever. Human exposure to pentaborane has produced signs of severe central nervous system irritation such as drowsiness, dizziness, visual disturbances, muscle twitching and in severe cases, painful muscle spasm. Dangerously flammable when exposed to heat or flame or by chemical reaction. On contact with moisture, hydrogen is usually evolved. Highly explosive when exposed to heat or flame. Explosive reaction with air; tetravinyllead; O_2 above 165°C; octanol oxime + sodium hydroxide; benzene vapor; HNO_3; Cl_2. Violent reaction with halocarbon liquids. Other boron hydrides evolve H_2 upon contact with moisture or can propagate a flame rapidly enough to cause an explosion. Heat can cause these materials to decompose violently or at least to evolve H_2. They also react with water or steam to evolve hydrogen. Reaction with Al or Li forms complex hydrides which may ignite spontaneously in air. Powerful oxidizing agents such as chlorine gas, etc., can react violently with boron hydrides. Pentaborane (stable) is spontaneously flammable in air. See also BORANES and HYDRIDES.

DDI500 CAS:12505-77-0 ***HR: 3***
DIBORON OXIDE
mf: B_2O_2 mw: 53.62

SAFETY PROFILE: Violent reaction when heated to 400°C. When heated to decomposition it emits acrid smoke and fumes. See also BORON COMPOUNDS.

DDI600 CAS:13701-67-2 ***HR: 3***
DIBORON TETRACHLORIDE
mf: B_2Cl_4 mw: 163.43

SAFETY PROFILE: May explode on contact with air or during reaction with dimethylmercury. When heated to decomposition it emits fumes of Cl^-. See also BORON COMPOUNDS and CHLORIDES.

DDI800 CAS:13965-73-6 ***HR: 3***
DIBORON TETRAFLUORIDE
mf: B_2F_4 mw: 97.61

SAFETY PROFILE: The gas explodes in the presence of oxygen. It ignites or reacts vigorously with mercury(II) oxide, manganese dioxide, and copper(II) oxide. When heated to decomposition it emits toxic fumes of F^-. See also BORON COMPOUNDS and FLUORIDES.

DDJ000 CAS:10318-26-0 ***HR: 3***
DIBROMDULCITOL
mf: $C_6H_{12}Br_2O_4$ mw: 308.00

SYNS: DBD ◇ 1,6-DIBROMODIDEOXYDULCITOL ◇ 1,6-DIBROMO-1,6-DIDEOXYDULCITOL ◇ 1,6-DIBROMO-1,6-DIDEOXYGALACTITOL ◇ 1,6-DIBROMO-1,6-DIDEOXY-d-GALACTITOL ◇ DIBROMODULCITOL ◇ 1,6-DIBROMODULCITOL ◇ ELOBROMOL ◇ GALACTICOL ◇ MITOLAC ◇ MITOLACTOL ◇ NCI-C04795 ◇ NSC-104800

TOXICITY DATA with REFERENCE
mmo-sat 100 μg/plate CRNGDP 3,333,82
dnd-rat-ipr 110 mg/kg CBINA8 47,133,83
bfa rat/sat 450 mg/kg CRNGDP 3,333,82
sce-ham:oth 5500 ng/L CNREA8 43,4530,83
dnd-mam:lym 150 mmol/L CBINA8 47,133,83
ipr-rat TDLo:5850 mg/kg/26W-I:NEO RRCRBU 52,1,75
orl-hmn TDLo:72 mg/kg/D:ETA ANBCB3 23,50,78
orl-rat LD50:1000 mg/kg CCROBU 56,593,72
ipr-rat LD50:470 mg/kg CCROBU 56,593,72
orl-mus LD50:1238 mg/kg NCISP* JAN86

ipr-mus LD50:550 mg/kg ARZNAD 17,145,67
orl-rbt LD50:300 mg/kg CCROBU 56,593,72

CONSENSUS REPORTS: NCI Carcinogenesis Bioassay Completed; Results Positive: mouse, rat (RRCRBU 52,1,75).

SAFETY PROFILE: Poison by ingestion. Moderately toxic by intraperitoneal route. Questionable carcinogen with experimental carcinogenic, neoplastigenic, and tumorigenic data. Human mutation data reported. An anti-cancer agent taken orally. When heated to decomposition it emits very toxic fumes of Br^-.

DDJ400 CAS:3252-43-5 ***HR: 2***
DIBROMOACETONITRILE
mf: C_2HBr_2N mw: 198.86

TOXICITY DATA with REFERENCE
mma-sat 16 µg/plate ENMUDM 8(Suppl 7),1,86
dnd-hmn:lym 50 µmol/L FAATDF 6,447,86
orl-rat TDLo:750 mg/kg (7-21D post):REP TXCYAC 46,83,87
skn-mus TDLo:2400 mg/kg/2W-I:CAR FAATDF 5,1065,85
orl-rat LD50:245 mg/kg EVHPAZ 69,183,86
orl-mus LD50:289 mg/kg EVHPAZ 69,183,86
ivn-mus LD50:56 mg/kg CSLNX* NX#05210

CONSENSUS REPORTS: Cyanide and its compounds are on the Community Right-To-Know List. Reported in EPA TSCA Inventory.

SAFETY PROFILE: Poison by intravenous route. Questionable carcinogen with experimental carcinogenic data. Experimental reproductive effects. Human mutation data reported. See also NITRILES and BROMIDES. When heated to decomposition it emits very toxic fumes of NO_x, Br^-, and CN^-.

DDJ600 CAS:99-73-0 ***HR: 3***
2,4′-DIBROMOACETOPHENONE
mf: $C_8H_6Br_2O$ mw: 277.96

SYNS: p-BROMPHENACYL-8 ◇ p-BROMOPHENACYL BROMIDE ◇ 4-BROMOPHENACYL BROMIDE ◇ α,p-DIBROMOACETOPHENONE

TOXICITY DATA with REFERENCE
ivn-mus LD50:18 mg/kg CSLNX* NX#02407

CONSENSUS REPORTS: Reported in EPA TSCA Inventory.

SAFETY PROFILE: Poison by intravenous route. See also BROMIDES. When heated to decomposition it emits toxic fumes of Br^-.

DDJ800 CAS:624-61-3 ***HR: 3***
DIBROMOACETYLENE
mf: C_2Br_2 mw: 183.83

PROP: Liquid. Mp: 76° (approx), bp: explodes, d: 2 (approx), vap d: 6.35.

DOT Classification: Forbidden.

SAFETY PROFILE: Ignites spontaneously in air. Explodes when heated. When heated to decomposition it emits toxic fumes of Br^-. See also ACETYLENE COMPOUNDS.

DDK000 CAS:81-98-1 ***HR: 2***
3,9-DIBROMO-7H-BENZ(de)ANTHRACEN-7-ONE
mf: $C_{17}H_8Br_2O$ mw: 388.07

SYNS: 6-Bz-1-DIBROMBENZANTHRON (CZECH) ◇ 3,9-DIBROMBENZANTHRONE ◇ 2,7-DIBROMOMESOBENZANTHRONE

TOXICITY DATA with REFERENCE
eye-rbt 500 mg/24H MLD 28ZPAK -,89,72
ipr-rat LD50:4900 mg/kg RPTOAN 40,137,77
ipr-mus LD50:1410 mg/kg RPTOAN 40,137,77

CONSENSUS REPORTS: Reported in EPA TSCA Inventory.

SAFETY PROFILE: Moderately toxic by intraperitoneal route. An eye irritant. When heated to decomposition it emits toxic fumes of Br^-.

DDK600 CAS:6305-43-7 ***HR: 3***
2,2′-DIBROMOBIACETYL
mf: $C_4H_4Br_2O_2$ mw: 243.90

SYN: α,α′-DIBROMOBIACETYL

TOXICITY DATA with REFERENCE
ipr-mus LD50:9400 µg/kg JNCIAM 31,297,63
ivn-mus LD50:10 mg/kg CSLNX* NX#00598
ivn-dog LD50:21 mg/kg JNCIAM 31,297,63

CONSENSUS REPORTS: Reported in EPA TSCA Inventory.

SAFETY PROFILE: Poison by intravenous and intraperitoneal routes. When heated to decomposition it emits toxic fumes of Br^-. See also BROMIDES.

DDK800 CAS:26637-71-8 ***HR: 3***
DIBROMOBICYCLOHEPTANE (mixed isomers)
mf: $C_7H_{10}Br_2$ mw: 253.99

SYNS: DIBROMOBICYCLOHEPTANE ◇ DIBROMONORBORNANE

TOXICITY DATA with REFERENCE
orl-rat LD50:210 mg/kg AIHAAP 30,470,69
skn-rbt LD50:250 mg/kg AIHAAP 30,470,69

SAFETY PROFILE: Poison by ingestion and skin contact. When heated to decomposition it emits toxic fumes of Br^-. See also BROMIDES.

DDK875 CAS:36333-41-2 ***HR: 3***
1,4-DIBROMO-1,3-BUTADIYNE
mf: C_4Br_2 mw: 207.85

BrC≡CC≡CBr

SAFETY PROFILE: Explodes at room temperature. Upon decomposition it emits toxic fumes of Br^-. See also ACETYLENE COMPOUNDS and BROMIDES.

DDL000 CAS:110-52-1 ***HR: 3***
1,4-DIBROMOBUTANE
mf: $C_4H_8Br_2$ mw: 215.94

SYNS: DBB ◇ 1,4-DIBROMBUTAN (GERMAN)

TOXICITY DATA with REFERENCE
mmo-sat 10 μmol/plate MUREAV 141,11,84
ipr-mus LD50:300 mg/kg ARZNAD 14,668,64

CONSENSUS REPORTS: Reported in EPA TSCA Inventory.

SAFETY PROFILE: Poison by intraperitoneal route. Mutation data reported. When heated to decomposition it emits toxic fumes of Br^-. See also BROMIDES.

DDL400 CAS:6974-12-5 ***HR: 3***
1,4-DIBROMO-2-BUTENE
mf: $C_4H_6Br_2$ mw: 213.92

SYN: TL 80

TOXICITY DATA with REFERENCE
skn-rbt 1 mg/24H AMIHBC 10,61,54
eye-rbt 50 μg open SEV AMIHBC 10,61,54
orl-rat LD50:75 mg/kg AMIHBC 10,61,54
ihl-mus LCLo:1260 mg/m³/10M NDRC** NDCrc-132,Aug,42
ipr-mus LDLo:4 mg/kg CBCCT* 5,338,53

SAFETY PROFILE: Poison by ingestion and intraperitoneal routes. Moderately toxic by inhalation. A skin and severe eye irritant. When heated to decomposition it emits toxic fumes of Br^-. See also BROMIDES.

DDL600 CAS:821-06-7 ***HR: 3***
trans-1,4-DIBROMOBUT-2-ENE
mf: $C_4H_6Br_2$ mw: 213.92

SYNS: DIBROMOBUTENE ◇ 1,4-trans-DIBROMOBUTENE-2

TOXICITY DATA with REFERENCE
skn-rbt 500 mg SEV SCCUR* -,3,61
orl-rat LD50:62 mg/kg SCCUR* -,9,61
orl-mus LD50:29 mg/kg SCCUR* -,9,61

CONSENSUS REPORTS: Reported in EPA TSCA Inventory.

SAFETY PROFILE: Poison by ingestion. A severe skin irritant. When heated to decomposition it emits toxic fumes of Br^-. See also BROMIDES.

DDL800 CAS:96-12-8 ***HR: 3***
1,2-DIBROMO-3-CHLOROPROPANE
DOT: UN 2872
mf: $C_3H_5Br_2Cl$ mw: 236.35

PROP: Bp: 196°, flash p: 170°F (TOC).

SYNS: BBC 12 ◇ 1-CHLORO-2,3-DIBROMOPROPANE ◇ 3-CHLORO-1,2-DIBROMOPROPANE ◇ DBCP ◇ DIBROMCHLORPROPAN (GERMAN) ◇ 1,2-DIBROM-3-CHLOR-PROPAN (GERMAN) ◇ DIBROMOCHLOROPROPANE ◇ 1,2-DIBROMO-3-CLORO-PROPANO (ITALIAN) ◇ 1,2-DIBROOM-3-CHLOORPROPAAN (DUTCH) ◇ FUMAGON ◇ FUMAZONE ◇ NCI-C00500 ◇ NEMABROM ◇ NEMAFUME ◇ NEMAGON ◇ NEMAGONE ◇ NEMAGON SOIL FUMIGANT ◇ NEMANAX ◇ NEMAPAZ ◇ NEMASET ◇ NEMATOCIDE ◇ NEMATOX ◇ NEMAZON ◇ OS 1897 ◇ OXY DBCP ◇ RCRA WASTE NUMBER 7066 ◇ SD 1897

TOXICITY DATA with REFERENCE
skn-rbt 10 g SEV TXAPA9 3,545,61
eye-rbt 1% MLD TXAPA9 3,545,61
dni-hmn:hla 10 mmol/L MUREAV 92,427,82
mma-sat 500 ng/plate ENMUDM 7(Suppl 3),15,85
spm-rbt-orl 375 mg/kg/10W-I FAATDF 6,628,86
orl-rat TDLo:50 mg/kg (male 5D pre):TER MUREAV 77,71,80
orl-rat TDLo:250 mg/kg (male 5D pre):REP MUREAV 101,321,82
orl-rat TDLo:5475 mg/kg/73W-I:CAR NCITR* NCI-CG-TR-28,78
ihl-rat TCLo:600 ppb/6H/2Y-I:CAR BJCAAI 42,772,80
orl-rat LD50:170 mg/kg FMCHA2 -,C76,83
ihl-rat LC50:103 ppm/8H FEPRA7 15,448,56
scu-rat LD50:100 mg/kg TXCYAC 27,287,83
orl-mus LD50:257 mg/kg GUCHAZ 6,172,73
orl-rbt LD50:180 mg/kg TXAPA9 3,545,61
skn-rbt LD50:1400 mg/kg TXAPA9 3,545,61
orl-ckn LD50:60 mg/kg TXAPA9 3,545,61

CONSENSUS REPORTS: NTP Fifth Annual Report on Carcinogens. IARC Cancer Review: Group 2B IMEMDT 7,191,87; Animal Sufficient Evidence IMEMDT 15,139,77; Human Limited Evidence IMEMDT 20,83,79; Animal Sufficient Evidence IMEMDT 20,83,79. NCI Carcinogenesis Bioassay Completed; Results Positive: mouse, rat NCITR* NCI-CG-TR-28,78. EPA Genetic Toxicology Program. Community Right-To-Know List. Reported in EPA TSCA Inventory.

OSHA PEL: TWA 0.001 ppm; Cancer Hazard.
DFG MAK: Animal Carcinogen, Suspected Human Carcinogen.
NIOSH REL: (Dibromochloropropane) CL 0.01 ppm/30M
DOT Classification: Poison B; Label: St. Andrews Cross

SAFETY PROFILE: Confirmed human carcinogen with experimental carcinogenic and teratogenic data. Poison by ingestion, inhalation, and subcutaneous routes. Moderately toxic by skin contact. An eye and severe skin irritant. A suspected human carcinogenic. Narcotic in high concentrations. Has been implicated in causing human male sterility in factory workers. Human mutation data reported. A soil fumigant. Flammable when exposed to heat or flame. When heated to decomposition it emits toxic fumes of Cl^- and Br^-. See also CHLORIDES and BROMIDES.

DDM000 CAS:10222-01-2 ***HR: 3***
α,α-DIBROMO-α-CYANOACETAMIDE
mf: $C_3H_2Br_2N_2O$ mw: 241.89

SYNS: DBNPA ◇ DIBROMOCYANOACETAMIDE ◇ 2,2-DIBROMO-3-NITRILOPROPIONAMIDE

TOXICITY DATA with REFERENCE
skn-rbt 500 mg SEV PHMCAA 15,226,73
eye-rbt 100 mg SEV PHMCAA 15,226,73
ivn-mus LD50:10 mg/kg CSLNX* NX#07898
orl-mam LD50:118 mg/kg PHMCAA 15,226,73

CONSENSUS REPORTS: Cyanide and its compounds are on the Community Right-To-Know List. Reported in EPA TSCA Inventory.

SAFETY PROFILE: Poison by ingestion and intravenous routes. A severe skin and eye irritant. When heated to decomposition it emits very toxic fumes of Br^- and NO_x. See also NITRILES.

DDM200 CAS:1689-99-2 ***HR: 3***
2,6-DIBROMO-4-CYANOPHENYL OCTANOATE
mf: $C_{15}H_{17}Br_2NO_2$ mw: 403.15

SYNS: BROMOXYNIL OCTANOATE ◇ 3,5-DIBROMO-4-OCTANOYLOXYBENZONITRILE ◇ NCR CE EE DOV7

TOXICITY DATA with REFERENCE
orl-rat LD50:250 mg/kg 28ZEAL 5,30,76
orl-mus LD50:245 mg/kg 28ZEAL 5,30,76
orl-rbt LD50:2 g/kg GUCHAZ 6,56,73

CONSENSUS REPORTS: Cyanide and its compounds are on the Community Right-To-Know List. Reported in EPA TSCA Inventory.

SAFETY PROFILE: Poison by ingestion. When heated to decomposition it emits very toxic fumes of NO_x and Br^-. See also NITILES.

DDM400 CAS:996-08-7 ***HR: 3***
DIBROMODIBUTYLSTANNANE
mf: $C_8H_{18}Br_2Sn$ mw: 392.77

PROP: Mp: 20°.

SYNS: DIBROMODIBUTYLTIN ◇ DIBUTYL TIN DIBROMIDE

TOXICITY DATA with REFERENCE
orl-rbt LDLo:150 mg/kg SAIGBL 15,3,73
skn-rbt LDLo:1000 mg/kg SAIGBL 15,3,73

OSHA PEL: TWA 0.1 mg(Sn)/m^3 (skin)
ACGIH TLV: TWA 0.1 mg(Sn)/m^3 (skin) (Proposed: TWA 0.1 mg(Sn)/m^3; STEL 0.2 mg(Sn)/m^3 (skin))
NIOSH REL: (Organotin Compounds) TWA 0.1 mg(Sn)/m^3

SAFETY PROFILE: Poison by ingestion. Moderately toxic by skin contact. See also TIN COMPOUNDS. When heated to decomposition it emits toxic fumes of Br^-.

DDM500 CAS:35691-65-7 ***HR: D***
1,2-DIBROMO-2,4-DICYANOBUTANE
mf: $C_6H_6Br_2N_2$ mw: 265.96

SYNS: 2-BROMO-2-(BROMOMETHYL)GLUTARONITRILE ◇ GLUTARONITRILE, 2-BROMO-2-(BROMOMETHYL)- ◇ METHYLDIBROMOGLUTARONITRILE ◇ PENTANEDINITRILE, 2-BROMO-2-(BROMOMETHYL)-

TOXICITY DATA with REFERENCE
orl-rat TDLo:1750 mg/kg (female 6-15D post):REP TOLED5 18(Suppl 1),158,83

CONSENSUS REPORTS: Reported in EPA TSCA Inventory.

SAFETY PROFILE: Experimental reproductive effects. When heated to decomposition it emits toxic fumes of NO_x and Br^-.

DDM600 CAS:77966-70-2 ***HR: 3***
2',6'-DIBROMO-2-(DIETHYLAMINO)-p-ACETOTOLUIDIDE HYDROCHLORIDE
mf: $C_{13}H_{18}Br_2N_2O{\cdot}ClH$ mw: 414.61

SYNS: C 3039 ◇ 2',6'-DIBROMO-2-(DIETHYLAMINO)-4'-METHYLACETANILIDE HYDROCHLORIDE

TOXICITY DATA with REFERENCE
eye-rbt 2% MLD ARZNAD 8,270,58
ipr-rat LD50:234 mg/kg ARZNAD 8,270,58
scu-mus LD50:505 mg/kg ARZNAD 8,270,58

SAFETY PROFILE: Poison by intraperitoneal route. Moderately toxic by subcutaneous route. An eye irritant. When heated to decomposition it emits very toxic fumes of Br^-, NO_x, and HCl.

DDM800 CAS:52400-80-3 ***HR: 3***
5,6-DIBROMO-2-(2-(2-(DIETHYLAMINO)ETHYLAMINO)ETHYL)-2-METHYL-1,3-BENZODIOXOLE DIHYDROCHLORIDE
mf: $C_{16}H_{24}Br_2N_2O_2{\cdot}2ClH$ mw: 509.16

TOXICITY DATA with REFERENCE
ivn-rat LD50:40 mg/kg EJMCA5 12,413,77
ipr-mus LD50:150 mg/kg EJMCA5 12,413,77

SAFETY PROFILE: Poison by intravenous and intraperitoneal routes. When heated to decomposition it emits very toxic fumes of HCl, NO_x, and Br^-.

DDN100 CAS:51877-12-4 ***HR: 3***
1,2-DIBROMO-1,2-DIISOCYANATOETHANE POLYMERS
mf: $(C_4H_2Br_2N_2O_2)_{2 \text{ OR } 3}$

$(O{=}N{=}CCHBrCHBrC{=}N{=}O)_{2 \text{ OR } 3}$

CONSENSUS REPORTS: Cyanide and its compounds are on the Community Right-To-Know List.

SAFETY PROFILE: Vigorous or explosive reaction on heating with 2-phenyl-2-propyl hydroperoxide. When heated to decomposition it emits toxic fumes of CN^-, Br^-, and NO_x. See also CYANIDE.

DDN150 CAS:72957-64-3 ***HR: 2***
2,2-DIBROMO-1,3-DIMETHYLCYCLOPROPANOIC ACID
mf: $C_6H_8Br_2O_2$ mw: 271.94

$CH_3CHCBr_2C(CH_3)CO{\bullet}OH$ (ring bond from CH to C(CH₃))

SAFETY PROFILE: Vigorous exothermic reaction on contact with tert-butylamine. When heated to decomposition it emits toxic fumes of Br^-.

DDN200 CAS:4713-59-1 ***HR: 3***
DIBROMODIPHENYLSTANNANE
mf: $C_{12}H_{10}Br_2Sn$ mw: 432.73

PROP: Colorless crystals. Sol in alc and ether. Mp: 38°; bp: 230° @ 42 mm.

SYNS: DIPHENYLDIBROMOTIN ◇ DIPHENYLTIN DIBROMIDE

TOXICITY DATA with REFERENCE
ivn-mus LD50:71 mg/kg CSLNX* NX#05803

CONSENSUS REPORTS: Polybrominated biphenyl compounds are on the Community Right-To-Know List.

OSHA PEL: TWA 0.1 mg(Sn)/m^3 (skin)
ACGIH TLV: TWA 0.1 mg(Sn)/m^3 (skin) (Proposed: TWA 0.1 mg(Sn)/m^3; STEL 0.2 mg(Sn)/m^3 (skin))
NIOSH REL: (Organotin Compounds) TWA 0.1 mg(Sn)/m^3

SAFETY PROFILE: Poison by intravenous route. See also TIN COMPOUNDS and BROMIDES. When heated to decomposition it emits toxic fumes of Br^-.

DDN700 CAS:56411-66-6 ***HR: 3***
2,3-DIBROMO-5,6-EPOXY-7,8-DIOXABICYCLO (2.2.2)OCTANE
mf: $C_6H_6Br_2O_3$ mw: 285.92

SAFETY PROFILE: Explodes on heating. When heated to decomposition it emits toxic fumes of Br^-. See also PEROXIDES.

DDN800 CAS:557-91-5 ***HR: 2***
1,1-DIBROMOETHANE
mf: $C_2H_4Br_2$ mw: 187.88

PROP: Liquid. Insol in water; sol in organic solvents. D: 1.089 @ 20.5°/4°, bp: 110°.

SYNS: ETHYLIDENE BROMIDE ◇ ETHYLIDENE DIBROMIDE

TOXICITY DATA with REFERENCE
mma-sat 10 μmol/plate EVHPAZ 21,79,77
dnr-esc 10 μL/plate EVHPAZ 21,79,77
rec-rbt LDLo:1250 mg/kg JPETAB 34,223,28

CONSENSUS REPORTS: Reported in EPA TSCA Inventory.

SAFETY PROFILE: Moderately toxic by rectal route. Mutation data reported. Violent reaction with magnesium. When heated to decomposition it emits toxic fumes of Br^-. See also BROMIDES.

DDO200 CAS:596-03-2 ***HR: D***
DIBROMOFLUORESCEIN
mf: $C_{20}H_{10}Br_2O_5$ mw: 490.12

SYNS: ACID ORANGE 11 ◇ C.I. 45370:1 ◇ C.I. SOLVENT RED 72 ◇ D&C ORANGE NO. 5 ◇ 4′,5′-DIBROMOFLUORORESCEIN ◇ FLUORESCEIN, 4′,5′-DIBROMO- ◇ SOLVENT RED 72 ◇ SPIRO(ISOBENZOFURAN-1(3H),9′-(9H)XANTHEN)-3-ONE, 4′,5′-DIBROMO-3′,6′-DIHYDROXY- (9CI)

TOXICITY DATA with REFERENCE
mmo-sat 100 ng/plate ENMUDM 1,116,79
orl-rat TDLo:10500 mg/kg (female 21D pre):REP TRENAF 27,145,76

CONSENSUS REPORTS: Reported in EPA TSCA Inventory.

SAFETY PROFILE: Experimental reproductive effects. Mutation data reported. When heated to decomposition it emits toxic fumes of Br^-.

DDO400 CAS:20404-94-8 ***HR: 3***
1,2-DIBROMOHEPTAFLUOROISOBUTYL METHYL ETHER
mf: $C_5H_3Br_2F_7O$ mw: 371.90

TOXICITY DATA with REFERENCE
orl-mus LD50:1150 mg/kg TXAPA9 14,114,69
ipr-mus LD50:140 mg/kg TXAPA9 14,114,69

SAFETY PROFILE: Poison by intraperitoneal route. Moderately toxic by ingestion. See also ETHERS, BROMIDES, and FLUORIDES. When heated to decomposition it emits very toxic fumes of Br^- and F^-.

DDO800 CAS:629-03-8 ***HR: 3***
1,6-DIBROMOHEXANE
mf: $C_6H_{12}Br_2$ mw: 244.00

SYNS: DBH ◇ 1,6-DIBROMOHEXAN (GERMAN)

TOXICITY DATA with REFERENCE
skn-rbt 500 mg MLD FCTOD7 20,563,82
eye-rbt 100 mg MLD FCTOD7 20,573,82
eye-rbt 100 mg/30S rns MLD FCTOD7 20,573,82
mmo-sat 10 μmol/plate MUREAV 141,11,84
ipr-mus LD50:270 mg/kg ARZNAD 14,668,64

CONSENSUS REPORTS: Reported in EPA TSCA Inventory.

SAFETY PROFILE: Poison by intraperitoneal route. A skin and eye irritant. Mutation data reported. When heated to decomposition it emits very toxic fumes of Br^-. See also BROMIDES.

DDP000 CAS:1689-84-5 ***HR: 3***
3,5-DIBROMO-4-HYDROXYBENZONITRILE
mf: $C_7H_3Br_2NO$ mw: 276.93

SYNS: BRITTOX ◇ BROMINAL ◇ BROMINEX ◇ BROMINIL ◇ BROMOXYNIL ◇ BROXYNIL ◇ BUCTRIL ◇ BUCTRIL INDUSTRIAL ◇ BUTILCHLOROFOS ◇ CHIPCO BUCTRIL ◇ CHIPCO CRAB-KLEEN ◇ 2,6-DIBROMO-4-CYANOPHENOL ◇ 3,5-DIBROMO-4-HYDROXYPHENYLCYANIDE ◇ ENT 20,852 ◇ 4-HYDROXY-3,5-DIBROMOBENZONITRILE ◇ MB 10064 ◇ ME4 BROMINAL ◇ NU-LAWN WEEDER ◇ OXYTRIL M

TOXICITY DATA with REFERENCE
orl-rat LD50:190 mg/kg WRPCA2 9,119,70
orl-mus LD50:110 mg/kg GUCHAZ 6,55,73
ivn-mus LD50:56 mg/kg CSLNX* NX#02212
orl-rbt LD50:260 mg/kg 85DPAN -,-,71/76
orl-gpg LD50:63 mg/kg GUCHAZ 6,55,73
orl-dck LD50:200 mg/kg DOEAAH 35,25,79

CONSENSUS REPORTS: Cyanide and its compounds are on the Community Right-To-Know List.

SAFETY PROFILE: Poison by ingestion and intravenous routes. An herbicide. When heated to decomposition it emits highly toxic fumes of NO_x, CN^-, and Br^-. See also NITRILES.

DDP200 CAS:3562-84-3 ***HR: 3***
3,5-DIBROMO-4-HYDROXYPHENYL-2-ETHYL-3-BENZOFURANYL KETONE
mf: $C_{17}H_{12}Br_2O_3$ mw: 424.11

SYNS: BENZBROMARON ◇ BENZBROMARONE ◇ DESURIC ◇ 3-(3,5-DIBROMO-4-HYDROXYBENZOYL-2-ETHYLBENZOFURAN ◇ (3,5-DIBROMO-4-HYDROXYPHENYL)(2-ETHYL-3-BENZOFURANYL) METHANONE ◇ EXURATE ◇ L2214 ◇ MINURIC ◇ MJ 10061 ◇ URICOVAC

TOXICITY DATA with REFERENCE
orl-rat TDLo:360 mg/kg (9-14D preg):TER SHNSAS 16,1521,79
orl-rat TDLo:480 mg/kg (9-14D preg):REP SHNSAS 16,1521,79
orl-rat LD50:248 mg/kg IYKEDH 10,232,79
ipr-rat LD50:239 mg/kg IYKEDH 10,232,79
scu-rat LD50:1230 mg/kg IYKEDH 10,232,79
orl-mus LD50:618 mg/kg IYKEDH 10,232,79
ipr-mus LD50:146 mg/kg OYYAA2 6,341,72
scu-mus LD50:4120 mg/kg IYKEDH 10,232,79
ivn-mus LD50:77 mg/kg OYYAA2 6,341,72

CONSENSUS REPORTS: Reported in EPA TSCA Inventory.

SAFETY PROFILE: Poison by ingestion, intravenous and intraperitoneal routes. Moderately toxic by subcutaneous route. Experimental teratogenic and reproductive effects. A uricosuric agent which promotes the excretion of uric acid in the urine. When heated to decomposition it emits toxic fumes of Br^-. See also KETONES.

DDP400 CAS:1122-10-7 ***HR: 3***
DIBROMOMALEINIMIDE
mf: $C_4HBr_2NO_2$ mw: 254.88

TOXICITY DATA with REFERENCE
ipr-mus TDLo:3100 μg/kg (9D preg):REP ARTODN 37,15,76
ipr-mus TDLo:3100 μg/kg (9D preg):TER ARTODN 37,15,76
ipr-mus LD50:11 mg/kg ARTODN 37,15,76

SAFETY PROFILE: Poison by intraperitoneal route. An experimental teratogen. Other experimental reproductive effects. When heated to decomposition it emits very toxic fumes of Br^- and NO_x.

DDP600 CAS:488-41-5 ***HR: 3***
1,6-DIBROMOMANNITOL
mf: $C_6H_{12}Br_2O_4$ mw: 308.00

SYNS: DBM ◇ DIBROMANNIT ◇ DIBROMANNITOL ◇ d-DIBROMANNITOL ◇ 1,6-DIBROMO-1,6-DIDEOXY-d-MANNITOL ◇ 1,6-DIBROMO-1,6-d-DIDESOXYMANNITOL ◇ MIEOBROMOL ◇ MITOBRONITOL ◇ MYEBROL ◇ MYELOBROMOL ◇ NCI-C04762 ◇ NSC-94100 ◇ R 54

TOXICITY DATA with REFERENCE
mmo-sat 1 mg/plate CNREA8 43,4530,83
mma-sat 667 μg/plate ENMUDM 8(Suppl 7),1,86
sce-hmn:lym 10 nmol/L NGCJAK 15,1085,80
cyt-mus-ivn 90 mg/kg MUREAV 60,329,79
sce-ham:oth 1300 ng/L CNREA8 43,4530,83

orl-mus TDLo:150 mg/kg (female 11D post):TER KSRNAM 6,30,72
orl-rat TDLo:300 mg/kg (female 9D post):REP KSRNAM 6,30,72
ipr-rat TDLo:9750 mg/kg/26W-I:NEO RRCRBU 52,1,75
orl-rat LD50:1500 mg/kg NIIRDN 6,810,82
ipr-rat LD50:900 mg/kg EJCAAH 4,617,68
scu-rat LD50:1240 mg/kg NIIRDN 6,810,82
ivn-rat LD50:1370 mg/kg IYKEDH 8,680,77
orl-mus LD50:1380 mg/kg NIIRDN 6,810,82
ipr-mus LD50:900 mg/kg NIIRDN 6,810,82
scu-mus LD50:2200 mg/kg NIIRDN 6,810,82
ivn-mus LD50:2200 mg/kg IYKEDH 8,680,77
orl-rbt LD50:1080 mg/kg OYYAA2 6,831,72

CONSENSUS REPORTS: NCI Carcinogenesis Bioassay Completed; Results Positive: mouse, rat (RRCRBU 52,1,75).

SAFETY PROFILE: Moderately toxic by ingestion, intravenous, intraperitoneal, and subcutaneous routes. Questionable carcinogen with experimental carcinogenic and neoplastigenic data. Experimental teratogenic and reproductive effects. Human mutation data reported. When heated to decomposition it emits toxic fumes of Br^-.

DDP800 CAS:74-95-3 ***HR: 3***
DIBROMOMETHANE
DOT: UN 2664
mf: CH_2Br_2 mw: 173.85

PROP: Colorless, heavy liquid. Bp: 95.6°-97.4°, fp: <50°, d: 2.485 @ 25°/25°, vap d: 6.05.

SYNS: METHYLENE BROMIDE ◇ METHYLENE DIBROMIDE ◇ RCRA WASTE NUMBER U068

TOXICITY DATA with REFERENCE
mmo-sat 100 ng/plate BECTA6 24,590,80
ihl-rat LC50:40 g/m^3/2H 85GMAT -,82,82
scu-mus LD50:3738 mg/kg TXAPA9 4,354,62
rec-rbt LDLo:5000 mg/kg JPETAB 34,223,28

CONSENSUS REPORTS: Community Right-To-Know List. Reported in EPA TSCA Inventory.

DOT Classification: Poison B; Label: St. Andrews Cross.

SAFETY PROFILE: A poison. Moderately toxic by subcutaneous route. Mildy toxic by inhalation. Mutation data reported. Mixtures with potassium explode on light impact. When heated to decomposition it emits toxic fumes of Br^-. See also BROMIDES.

DDQ100 CAS:10218-83-4 ***HR: 3***
N,N-DIBROMOMETHYLAMINE
mf: CH_3Br_2N mw: 188.85

SAFETY PROFILE: An explosive very sensitive to impact or shock. Upon decomposition it emits toxic fumes of Br^- and NO_x. See also AMINES.

DDQ125 ***HR: 3***
DIBROMOMETHYLBORANE
mf: CH_3BBr_2 mw: 185.65

SAFETY PROFILE: Ignites spontaneously in air. Explodes on mixing with sodium-potassium alloys. When heated to decomposition it emits toxic fumes of Br^-. See also BORANES.

DDQ400 CAS:3296-90-0 ***HR: 2***
DIBROMONEOPENTYL GLYCOL
mf: $C_5H_{10}Br_2O_2$ mw: 261.97

SYN: 2,2-BIS(BROMOMETHYL)-1,3-PROPANEDIOL ◇ DIBROMONEOPENTYL GLYCOL ◇ DIBROMOPENTAERYTHRITOL ◇ FR 1138 ◇ NCI-C55516 ◇ PENTAERYTHRITOL DIBROMIDE ◇ PENTAERYTHRITOL DIBROMOHYDRIN

TOXICITY DATA with REFERENCE
cyt-ham:ovr 800 mg/L EMMUEG 10(Suppl 10),1,87
orl-rat LD50:3458 mg/kg JCTODH 7,77,80

CONSENSUS REPORTS: Reported in EPA TSCA Inventory.

SAFETY PROFILE: Moderately toxic by ingestion. Mutation data reported. When heated to decomposition it emits toxic fumes of Br^-.

DDQ800 CAS:57541-73-8 ***HR: 3***
3,4-DIBROMONITROSOPIPERIDINE
mf: $C_5H_8Br_2N_2O$ mw: 271.97

SYN: N-NITROSO-3,4-DIBROMOPIPERIDINE

TOXICITY DATA with REFERENCE
mmo-sat 200 μg/plate MUREAV 56,131,77
mma-sat 1 μmol/plate MUREAV 56,131,77
pic-esc 10 mg/L TCMUE9 1,91,84
sln-dmg-orl 1 mmol/L/24H MUREAV 67,27,79
orl-rat TDLo:1090 mg/kg/27W-C:ETA CNREA8 35,3209,75

CONSENSUS REPORTS: EPA Genetic Toxicology Program.

SAFETY PROFILE: Mutation data reported. Questionable carcinogen with experimental tumorigenic data. Many N-nitroso compounds are carcinogens. When heated to decomposition it emits very toxic fumes of Br^- and NO_x. See also N-NITROSO COMPOUNDS.

DDR000 CAS:111-24-0 ***HR: D***
1,5-DIBROMOPENTANE
mf: $C_5H_{10}Br_2$ mw: 229.97

SYNS: PENTAMETHYLENE BROMIDE ◇ PENTAMETHYLENE DIBROMIDE

TOXICITY DATA with REFERENCE
mmo-sat 10 μmol/plate MUREAV 141,11,84
mma-sat 10 μmol/plate EVHPAZ 21,79,77
dnr-esc 10 μL/disc CNREA8 34,2576,74

CONSENSUS REPORTS: Reported in EPA TSCA Inventory. EPA Genetic Toxicology Program.

SAFETY PROFILE: Mutation data reported. When heated to decomposition it emits toxic fumes of Br^-. See also PENTANE and BROMIDES.

DDR100 ***HR: 3***
2-(3,5-DIBROMO-2-PENTYLOXYBENZYLOXY) TRIETHYLAMINE
mf: $C_{18}H_{29}Br_2NO_2$ mw: 451.30

TOXICITY DATA with REFERENCE
orl-rat LD50:350 mg/kg JPETAB 121,210,57
ipr-rat LD50:90 mg/kg JPETAB 121,210,57
orl-mus LD50:160 mg/kg JPETAB 121,210,57
ipr-mus LD50:90 mg/kg JPETAB 121,210,57

SAFETY PROFILE: Poison by ingestion and intraperitoneal routes. When heated to decomposition it emits toxic fumes of Br^- and NO_x. See also AMINES and BROMIDES.

DDR200 CAS:696-24-2 ***HR: 3***
DIBROMOPHENYLARSINE
mf: $C_6H_5AsBr_2$ mw: 311.85

SYNS: PHENYLARSONOUS DIBROMIDE ◇ PHENYLDIBROMOARSINE

TOXICITY DATA with REFERENCE
skn-rat LD50:15 mg/kg JPBAA7 58,411,46
skn-rbt LD50:4 mg/kg JPBAA7 58,411,46
ivn-rbt LD50:500 μg/kg JPBAA7 58,411,46
skn-gpg LD50:6 mg/kg JPBAA7 58,411,46

CONSENSUS REPORTS: Arsenic and its compounds are on the Community Right-To-Know List.

OSHA PEL: TWA 0.5 mg(As)/m^3

SAFETY PROFILE: Poison by skin contact and intravenous routes. See also ARSENIC COMPOUNDS. When heated to decomposition it emits very toxic fumes of As and Br^-.

DDR400 CAS:78-75-1 ***HR: 3***
1,2-DIBROMOPROPANE
mf: $C_3H_6BR_2$ mw: 201.91

PROP: Colorless liquid. Mp: −55°, bp: 140-142°, n (20/D) 1.5203, d: 1.933. Sltly soluble in water; miscible with organic solvents.

SYN: PROPYLENE DIBROMIDE

TOXICITY DATA with REFERENCE
mmo-sat 10 μmol/plate ENMUDM 2,59,80
mma-sat 10 μmol/plate ENMUDM 2,59,80
sln-dmg-orl 5 mmol/L EXPEAM 30,621,74
orl-rat LD50:1373 mg/kg GTPZAB 19(9),36,75
ihl-rat LC50:15344 mg/m^3 GTPZAB 19(9),36,75
ipr-mus LD50:75 mg/kg NTIS** AD691-490

CONSENSUS REPORTS: EPA Genetic Toxicology Program. Reported in EPA TSCA Inventory.

SAFETY PROFILE: Poison by intraperitoneal route. Moderately toxic by ingestion. Mildly toxic by inhalation. Mutation data reported. When heated to decomposition it emits toxic fumes of Br^-. See also PROPANE and BROMIDES.

DDR600 CAS:78-75-1 ***HR: 3***
2,3-DIBROMOPROPANE
mf: $C_3H_6Br_2$ mw: 201.91

PROP: Colorless liquid. Mp: −55°, bp: 139.6°-142.6°, fp: < −75°, d: 1.940 @ 25°/25°, vap d: 7.0.

SYN: PROPYLENE DIBROMIDE

TOXICITY DATA with REFERENCE
sln-dmg-orl 5 mmol/L EXPEAM 30,621,74
ipr-mus LD50:75 mg/kg NTIS** AD691-490

CONSENSUS REPORTS: Reported in EPA TSCA Inventory.

SAFETY PROFILE: Poison by intraperitoneal route. Mutation data reported. When heated to decomposition it emits toxic fumes of Br^-. See also PROPANE and BROMIDES.

DDR800 CAS:96-21-9 ***HR: 3***
1,3-DIBROMO-2-PROPANOL
mf: $C_3H_6Br_2O$ mw: 217.91

SYN: GLYCEROL-α,Γ-DIBROMOHYDRINE

TOXICITY DATA with REFERENCE
mmo-sat 100 nmol/plate ENMUDM 2,59,80
ipr-mus LD50:150 mg/kg ARZNAD 17,145,67

CONSENSUS REPORTS: Reported in EPA TSCA Inventory.

SAFETY PROFILE: Poison by intraperitoneal route. Mutation data reported. When heated to decomposition it emits toxic fumes of Br^-.

DDS000 CAS:96-13-9 ***HR: 3***
2,3-DIBROMOPROPANOL
mf: $C_3H_6Br_2O$ mw: 217.91

SYNS: 2,3-DIBROMO-1-PROPANOL ◇ NCI-C55436 ◇ USAF DO-42

TOXICITY DATA with REFERENCE
sln-dmg-orl 500 ppm ENMUDM 7,349,85
trn-dmg-orl 500 ppm ENMUDM 7,349,85
dnd-rat:oth 1 μmol/L CRNGDP 6,705,85
otr-ham:emb 500 nmol/L CRNGDP 6,705,85
msc-ham:lng 20 μmol/L MUREAV 124,213,83
ipr-mus LDLo:125 mg/kg NTIS** AD277-689

CONSENSUS REPORTS: Reported in EPA TSCA Inventory. EPA Genetic Toxicology Program.

SAFETY PROFILE: Poison by intraperitoneal route. Mutation data reported. When heated to decomposition it emits toxic fumes of Br^-. See also BROMIDES.

DDS100 CAS:18791-02-1 ***HR: 2***
2,3-DIBROMOPROPANOYL CHLORIDE
mf: $C_3H_3Br_2ClO$ mw: 250.33

SYNS: α,β-DIBROMOPROPIONYL CHLORIDE ◇ 2,3-DIBROMOPROPIONYL CHLORIDE

TOXICITY DATA with REFERENCE
orl-rat LD50:1200 mg/kg GISAAA 49(4),90,84
ihl-rat LCLo:1360 mg/m^3 GISAAA 49(4),90,84
ihl-mus LC50:19200 mg/m^3 GISAAA 49(4),90,84

SAFETY PROFILE: Moderately toxic by inhalation and ingestion. When heated to decomposition it emits toxic fumes of Cl^- and Br^-.

DDS200 CAS:513-31-5 ***HR: 3***
2,3-DIBROMOPROPENE
mf: $C_3H_4Br_2$ mw: 199.89

TOXICITY DATA with REFERENCE
mmo-sat 100 nmol/plate ENMUDM 2,59,80
mma-sat 100 nmol/plate ENMUDM 2,59,80
ivn-mus LD50:100 mg/kg CSLNX* NX#03619

CONSENSUS REPORTS: Reported in EPA TSCA Inventory.

SAFETY PROFILE: Poison by intravenous route. Mutation data reported. When heated to decomposition it emits toxic fumes of Br^-.

DDS400 CAS:5221-17-0 ***HR: 3***
2,3-DIBROMOPROPIONALDEHYDE
mf: $C_3H_4Br_2O$ mw: 215.89

SYN: DIBROMOPROPANAL

TOXICITY DATA with REFERENCE
mmo-sat 1 nmol/plate MUREAV 78,113,80
mma-sat 1 nmol/plate MUREAV 78,113,80
ipr-mus LD50:5 mg/kg JAFCAU 30,627,82
ivn-mus LD50:56 mg/kg CSLNX* NX#02408

SAFETY PROFILE: Poison by intravenous route. Mutagenic data reported. When heated to decomposition it emits toxic fumes of Br^-. See also ALDEHYDES and BROMIDES.

DDS600 CAS:521-74-4 ***HR: 3***
5,7-DIBROMO-8-QUINOLINOL
mf: $C_9H_5Br_2NO$ mw: 302.97

SYNS: BRODIAR ◇ BROXYKINOLIN ◇ BROXYQUINOLINE ◇ COLEPUR ◇ COLIPAR ◇ 5,7-DIBROMO-8-HYDROXYQUINOLINE ◇ DIBROMOXYQUINOLINE ◇ FENILOR ◇ PARAMIBE

TOXICITY DATA with REFERENCE
orl-chd TDLo:1000 mg/kg/27D:CNS,PUL LANCAO 1,922,68
orl-rat LDLo:10 g/kg KSRNAM 4,27,70
ipr-rat LD50:1140 mg/kg KSRNAM 4,27,70
orl-mus LD50:7420 mg/kg KSRNAM 4,27,70
ipr-mus LD50:325 mg/kg KSRNAM 4,27,70

CONSENSUS REPORTS: Reported in EPA TSCA Inventory. EPA Genetic Toxicology Program.

SAFETY PROFILE: Poison by intraperitoneal route. Mildly toxic by ingestion. Human systemic effects by ingestion: muscle weakness, ataxia (loss of muscle coordination), and gastritis. When heated to decomposition it emits very toxic fumes of Br^- and NO_x.

DDT000 CAS:15091-30-2 ***HR: 3***
3,4-DIBROMOSULFOLANE
mf: $C_4H_6Br_2O_2S$ mw: 277.98

SYN: 3,4-DIBROMOTETRAHYDROTHIOPHENE-1,1-DIOXIDE

TOXICITY DATA with REFERENCE
ipr-mus LD50:9500 μg/kg RPTOAN 41,257,78
ivn-mus LD50:56 mg/kg CSLNX* NX#03181

CONSENSUS REPORTS: Reported in EPA TSCA Inventory.

SAFETY PROFILE: Poison by intraperitoneal and intravenous routes. See also BROMIDES. When heated to decomposition it emits very toxic fumes of Br^- and SO_x.

DDT200 CAS:85-79-0 ***HR: 3***
DIBUCAINE
mf: $C_{20}H_{29}N_3O_2$ mw: 343.52

SYNS: 2-BUTOXY-N-(β-DIETHYLAMINOETHYL)CINCHONINAMIDE ◇ 2-BUTOXY-N-(2-(DIETHYLAMINO)ETHYL)CINCHONINAMIDE ◇ 2-BUTOXYQUINOLINE-4-CARBOXYLIC ACID DIETHYLAMINOETHYLAMIDE ◇ α-BUTYLOXYCINCHONINIC ACID DIETHYLETHYLENEDIAMIDE ◇ 2-BUTYOXY-N-(2-(DIETHYLAMINO)ETHYL)-4-QUINOLINECARBOXAMIDE ◇ CINCHOCAINE ◇ DERMACAINE ◇ N-(2-(DIETHYLAMINO)ETHYL)-2-BUTOXYCINCHONINAMIDE ◇ NUPERCAINAL ◇ NUPERCAINE ◇ SOVCAINE

TOXICITY DATA with REFERENCE
orl-chd LDLo:50 mg/kg 34ZIAG -,209,69
ipr-rat LDLo:7 mg/kg TXAPA9 1,156,59
ipr-mus LD50:24500 μg/kg ARZNAD 10,925,60

scu-mus LD50:28500 μg/kg ARZNAD 10,925,60
ivn-mus LDLo:6 mg/kg JAPMA8 39,4,50
scu-rbt LD50:8500 μg/kg PSEBAA 29,368,32
ivn-rbt LD50:2500 μg/kg PSEBAA 29,368,32
scu-gpg LDLo:112 mg/kg PHREA7 12,190,32

SAFETY PROFILE: A human poison by ingestion. Poison experimentally by subcutaneous, intravenous, and intraperitoneal routes. When heated to decomposition it emits toxic fumes of NO_x.

DDT250 CAS:37235-82-8 ***HR: 3***
DIBUSMUTH DICHROMIUM NONAOXIDE
mf: $Bi_2Cr_2O_9$ mw: 665.95

SYN: BISMUTH CHROMATE

CONSENSUS REPORTS: Chromium compounds are on the Community Right-To-Know List.

SAFETY PROFILE: May ignite on contact with H_2S. When heated to decomposition it emits acrid smoke and fumes. See also CHROMIUM COMPOUNDS and BISMUTH COMPOUNDS.

DDT300 CAS:519-88-0 ***HR: 3***
DIBUTAMIDE
mf: $C_{17}H_{28}N_2O_2$ mw: 292.47

PROP: Rods from ethanol + 10% ether. Mp: 134°. Practically insol in water; sol in ethanol, isopropanol, glacial acetic acid.

SYNS: AMBUCETAMID ◇ AMBUCETAMIDE ◇ BERSEN ◇ DIBUTAMID (GERMAN) ◇ α-DIBUTYL-AMINO-4-METHOXYBENZENEACETAMIDE (9CI) ◇ α-DIBUTYL-AMINO-p-METHOXYPHENYLACETAMIDE ◇ α-DIBUTYLAMINO-α-(p-METHOXYPHENYL)ACETAMIDE ◇ 2-DIBUTYLAMINO-2-(p-METHOXYPHENYL)ACETAMIDE ◇ α-p-METHOXYPHENYL-α-DI-n-BUTYLAMINOACETAMIDE ◇ MERITIN ◇ R 5

TOXICITY DATA with REFERENCE
ivn-rat LD50:61 mg/kg ARZNAD 11,929,61
orl-mus LD50:813 mg/kg JAPMA8 46,564,57
ipr-mus LD50:92 mg/kg JAPMA8 46,564,57
ivn-mus LD50:62200 μg/kg JAPMA8 46,564,57

SAFETY PROFILE: Poison by intravenous and intraperitoneal routes. Moderately toxic by ingestion. When heated to decomposition it emits toxic fumes of NO_x.

DDT400 CAS:871-22-7 ***HR: 2***
1,1-DIBUTOXYETHANE
mf: $C_{10}H_{22}O_2$ mw: 174.32

SYNS: ACETALDEHYDE DIBUTYL ACETAL ◇ DIBUTYL ACETAL ◇ 1,1′-(ETHYLIDENEBIS(OXY)BISBUTANE

TOXICITY DATA with REFERENCE
skn-rbt 10 mg/24H open SEV AMIHBC 10,61,54
eye-rbt 500 mg open AMIHBC 10,61,54
orl-rat LD50:8790 mg/kg AMIHBC 10,61,54

SAFETY PROFILE: Mildly toxic by ingestion. An eye and severe skin irritant. Combustible. To fight fire, use water, foam, CO_2, dry chemical. When heated to decomposition it emits acrid smoke and irritating fumes. See also ALDEHYDES.

DDT600 ***HR: 3***
DI-sec-BUTYLAMINE
mf: $C_8H_{19}N$ mw: 129.25

PROP: Liquid. Bp: 134°, flash p: 75.2°F (OC), d: 0.75, vap d: 4.5.

SAFETY PROFILE: See also n-DIBUTYLAMINE. Dangerous fire hazard when exposed to heat or flame; can react with oxidizing materials. To fight fire, use alcohol foam, foam, CO_2, dry chemical.

DDT800 CAS:111-92-2 ***HR: 3***
n-DIBUTYLAMINE
DOT: UN 2248
mf: $C_8H_{19}N$ mw: 129.28

PROP: Liquid. Mp: −59°; bp: 159°, flash p: 125°F (OC), d: 0.76, vap d: 4.46, vap press: 2 mm @ 20°.

SYNS: N-BUTYL-1-BUTANAMINE ◇ DI-n-BUTYLAMINE ◇ DI(n-BUTYL)AMINE (DOT)

TOXICITY DATA with REFERENCE
skn-rbt 10 mg/24H open SEV AMIHBC 10,61,54
skn-rbt 500 mg open MOD UCDS** 3/25/70
eye-rbt 250 μg open SEV AMIHBC 10,61,54
cyt-ham:fbr 200 mg/L/48H MUREAV 48,337,77
orl-rat LD50:220 mg/kg ZHYGAM 20,393,74
ihl-rat LCLo:500 ppm/4H AEHLAU 1,343,60
scu-rat LDLo:330 mg/kg JPETAB 20,435,23
orl-mus LD50:290 mg/kg GISAAA 40(11),21,75
skn-rbt LD50:1010 mg/kg AMIHBC 10,61,54
orl-gpg LD50:230 mg/kg GISAAA 40(11),21,75

CONSENSUS REPORTS: Reported in EPA TSCA Inventory. EPA Genetic Toxicology Program.

DOT Classification: IMO: Corrosive Material; Label: Corrosive, Flammable Liquid.

SAFETY PROFILE: Poison by ingestion and subcutaneous routes. Moderately toxic by skin contact and inhalation. Corrosive. A severe skin and eye irritant. Mutation data reported. Flammable when exposed to heat or flame; can react with oxidizing materials. To fight fire, use alcohol foam, foam, CO_2, dry chemical. Exothermic reaction with cellulose nitrate does not proceed to ignition. When heated to decomposition it emits toxic fumes of NO_x.

DDU000 CAS:12107-76-5 ***HR: 3***
DIBUTYLAMINE TETRAFLUOROBORATE
mf: $C_8H_{20}N \cdot BF_4$ mw: 217.10

SYN: DI-n-BUTYLAMMONIUMTETRAFLUOROBORATE

TOXICITY DATA with REFERENCE
ivn-mus LD50:56 mg/kg CSLNX* NX#02191

CONSENSUS REPORTS: Reported in EPA TSCA Inventory.

SAFETY PROFILE: Poison by intravenous route. See also BORON COMPOUNDS. When heated to decomposition it emits very toxic fumes of NO_x, NH_3, and F^-.

DDU200 CAS:77966-79-1 ***HR: 3***
2-(DIBUTYLAMINO)-2',6'-ACETOXYLIDIDE HYDROCHLORIDE
mf: $C_{18}H_{30}N_2O \cdot ClH$ mw: 326.96

SYN: C 3103

TOXICITY DATA with REFERENCE
eye-rbt 2% SEV ARZNAD 8,407,58
ipr-rat LD50:221 mg/kg ARZNAD 8,407,58
scu-mus LD50:805 mg/kg ARZNAD 8,407,58

SAFETY PROFILE: Poison by intraperitoneal route. Moderately toxic by subcutaneous route. A severe eye irritant. When heated to decomposition it emits very toxic fumes of NO_x and HCl. See also AMINES.

DDU600 CAS:102-81-8 ***HR: 3***
2-N-DIBUTYLAMINOETHANOL
DOT: UN 2873
mf: $C_{10}H_{23}NO$ mw: 173.34

PROP: Liquid. Bp: 222°, flash p: 220°F (OC), d: 0.85, vap d: 6.0.

SYNS: BU2AE ◇ DIBUTYLAMINOETHANOL ◇ 2-DIBUTYLAMINOETHANOL ◇ 2-DI-n-BUTYLAMINOETHANOL ◇ N,N-DI-n-BUTYLAMINOETHANOL (DOT) ◇ β-N-DIBUTYLAMINOETHYL ALCOHOL ◇ N,N-DIBUTYLETHANOLAMINE ◇ N,N-DIBUTYL-N-(2-HYDROXYETHYL)AMINE

TOXICITY DATA with REFERENCE
skn-rbt 10 mg/24H open AMIHBC 10,61,54
skn-rbt 500 mg open SEV UCDS** 11/29/63
eye-rbt 20 mg/24H open SEV AMIHBC 10,61,54
orl-rat LD50:1070 mg/kg AMIHBC 10,61,54
ipr-rat LD50:144 mg/kg TXAPA9 12,486,68
ipr-mus LD50:52 mg/kg RCRVAB 38,975,69
skn-rbt LD50:1680 mg/kg AMIHBC 10,61,54
ipr-mam LD50:120 mg/kg TXAPA9 8,344,66

CONSENSUS REPORTS: Reported in EPA TSCA Inventory.

OSHA PEL: TWA 2 ppm
ACGIH TLV: TWA 2 ppm (skin)
DOT Classification: Poison B; Label; St. Andrews Cross.

SAFETY PROFILE: Poison by intraperitoneal route. Moderately toxic by ingestion and skin contact. A severe eye and skin irritant. Flammable when exposed to heat or flame; can react with oxidizing materials. To fight fire, use CO_2, dry chemical. When heated to decomposition it emits toxic fumes of NO_x. See also AMINES and ALCOHOLS.

DDV200 CAS:102-83-0 ***HR: 3***
3-(DIBUTYLAMINO)PROPYLAMINE
mf: $C_{11}H_{26}N_2$ mw: 186.39

TOXICITY DATA with REFERENCE
skn-rbt 100 μg/24H open AIHAAR 23,95,62
orl-rat LD50:820 mg/kg AIHAAP 23,95,62
skn-rbt LD50:270 mg/kg AIHAAP 23,95,62

CONSENSUS REPORTS: Reported in EPA TSCA Inventory.

SAFETY PROFILE: Poison by skin contact. Moderately toxic by ingestion. A skin irritant. Ignites on contact with cellulose nitrate. When heated to decomposition it emits toxic fumes of NO_x.

DDV225 CAS:7128-68-9 ***HR: 3***
DI-N-BUTYLAMMONIUM HEXAFLUOROARSENATE
mf: $C_8H_{19}N \cdot AsF_6H$ mw: 319.21

SYN: DIBUTYLAMINE, HEXAFLUOROARSENATE(1-)

TOXICITY DATA with REFERENCE
ivn-mus LD50:180 mg/kg CSLNX* NX#04252

OSHA PEL: TWA 0.5 mg(As)/m^3

SAFETY PROFILE: Poison by intravenous route. When heated to decomposition it emits toxic fumes of NO_x, As, and F^-.

DDV250 CAS:2850-61-5 ***HR: 3***
DIBUTYLARSINIC ACID
mf: $C_8H_{19}AsO_2$ mw: 222.19

SYNS: ARSINE OXIDE, DIBUTYLHYDROXY- ◇ ARSINIC ACID, DIBUTYL-(9CI)

TOXICITY DATA with REFERENCE
ivn-mus LD50:18 mg/kg CSLNX* NX#01190

OSHA PEL: TWA 0.5 mg(As)/m^3

SAFETY PROFILE: Poison by intravenous route. When heated to decomposition it emits toxic fumes of As.

DDV400 CAS:17013-41-1 ***HR: 3***
5,5-DIBUTYLBARBITURIC ACID
mf: $C_{12}H_{20}N_2O_3$ mw: 240.34

SYNS ◇ DIBUTYLBARBITURIC ACID ◇ 5,5-DIBUTYL-2,4,6(1H,3H,5H)-PYRIMIDINETRIONE

TOXICITY DATA with REFERENCE
ipr-mus LD50:232 mg/kg JPETAB 89,356,47
orl-cat LDLo:350 mg/kg JPETAB 26,371,25
scu-rbt LDLo:500 mg/kg JACSAT 45,243,23

SAFETY PROFILE: Poison by ingestion and intraperitoneal routes. Moderately toxic by subcutaneous route. An hypnotic agent. When heated to decomposition it emits toxic fumes of NO_x. See also BARBITURATES.

DDV600 CAS:77-58-7 ***HR: 3***
DIBUTYLBIS(LAUROYLOXY)STANNANE
mf: $C_{32}H_{64}O_4Sn$ mw: 631.65

PROP: Pale yellow liquid to colorless solid (when pure). Mp: 23°, bp: non-distillable @ 10 mm, flash p: 455°F (OC), d: 1.066 @ 20°/20°, vap d: 21.8.

SYNS: BIS(DODECANOYLOXY)DI-n-BUTYLSTANNANE ◇ BIS(LAUROYLOXY)DIBUTYLSTANNANE ◇ BIS(LAUROYLOXY) DI(n-BUTYL)STANNANE ◇ BUTYNORATE ◇ DBTL ◇ DIBUTYLBIS (LAUROYLOXY)TIN ◇ DI-n-BUTYLTIN DI(DODECANOATE) ◇ DIBUTYLTIN DILAURATE (USDA) ◇ DIBUTYLTIN LAURATE ◇ DIBUTYL-ZINN-DILAURAT (GERMAN) ◇ FOMREZ SUL-4 ◇ LAUDRAN DI-n-BUTYLCINICITY (CZECH) ◇ LAURIC ACID, DIBUTYLSTANNYLENE derivative ◇ LAURIC ACID, DIBUTYL-STANNYLENE SALT ◇ STABILIZER D-22 ◇ THERM CHEK 820 ◇ TIN DIBUTYL DILAURATE ◇ TINOSTAT

TOXICITY DATA with REFERENCE
skn-rbt 500 mg/24H MLD 28ZPAK -,230,72
eye-rbt 100 mg/24H MOD 28ZPAK -,230,72
orl-rat LD50:175 mg/kg ARZNAD 10,44,60
ipr-rat LDLo:85 mg/kg BJPCAL 10,16,55
orl-mus LDLo:710 mg/kg AECTCV 14,111,85

CONSENSUS REPORTS: Reported in EPA TSCA Inventory.

OSHA PEL: TWA 0.1 mg(Sn)/m^3 (skin)
ACGIH TLV: TWA 0.1 mg(Sn)/m^3 (skin) (Proposed: TWA 0.1 mg(Sn)/m^3; STEL 0.2 mg(Sn)/m^3 (skin))
NIOSH REL: (Organotin Compounds) TWA 0.1 mg(Sn)/m^3

SAFETY PROFILE: Poison by ingestion and intraperitoneal routes. A skin and eye irritant. Avoid the vapor produced by heating. Combustible when exposed to heat or flame; reacts with oxidizers. When heated to decomposition it emits acrid smoke and fumes. See also TIN COMPOUNDS.

DDV800 CAS:78-46-6 ***HR: 3***
DIBUTYL BUTANEPHOSPHONATE
mf: $C_{12}H_{27}O_3P$ mw: 250.36

PROP: Colorless liquid, mild odor. Bp: 128° @ 2.5 mm, flash p: 311° (COC), d: 8.62.

SYN: DIBUTYL BUTYLPHOSPHONATE

TOXICITY DATA with REFERENCE
ipr-mus LDLo:125 mg/kg CBCCT* 7,789,55
ivn-mus LD50:56 mg/kg CSLNX* NX#03463

CONSENSUS REPORTS: Reported in EPA TSCA Inventory.

SAFETY PROFILE: Poison by intraperitoneal and intravenous routes. Combustible when exposed to heat or flame. It can react vigorously with oxidizing materials. To fight fire, use foam, CO_2, or dry chemical. When heated to decomposition it emits toxic fumes of PO_x.

DDW000 CAS:532-49-0 ***HR: 3***
DI-n-BUTYL-CARBAMYLCHOLINE SULPHATE
mf: $C_{30}H_{66}N_4O_4 \cdot O_4S$ mw: 643.06

SYNS: DIBULINESULFAT ◇ DIBULINE SULFATE ◇ DIBUTOLINE ◇ DIBUTOLINE SULFATE ◇ 1-(((DIBUTYLAMINO)CARBONYL)OXY)-N-ETHYL-N,N-DIMETHYLETHANAMINIUM SULFATE (2:1) ◇ (2-DIBUTYLCARBAMYLOXYETHYL)-DIMETHYLETHYLAMMONIUM SULFATE ◇ DIMETHYL-ETHYL-β-HYDROXYETHYL-AMMONIUM-SULFATE-DI-n-BUTYLCARBAMATE ◇ DIMETHYLETHYL-β-HYDROXYETHYLAMMONIUM SULFATE DIBUTYLURETHAN ◇ ETHYL(2-HYDROXYETHYL)DIMETHYL-AMMONIUM SULFATE (SALT), BIS(DIBUTYLCARBAMATE)

TOXICITY DATA with REFERENCE
ipr-rat LD50:22 mg/kg JPETAB 84,105,45
scu-mus LD50:49 mg/kg CLDND*

SAFETY PROFILE: Poison by intraperitoneal and subcutaneous routes. See also CARBAMATES and SULFATES. When heated to decomposition it emits very toxic fumes of NO_x, NH_3, and SO_x.

DDW200 CAS:112-73-2 ***HR: 2***
DIBUTYL CARBITOL
mf: $C_{12}H_{26}O_3$ mw: 218.38

PROP: Practically colorless liquid, characteristic odor, sltly sol in water. D: 0.8853 @ 20°/20°, bp: 256°, fp: −60.2°, flash p: 245°F (OC).

SYNS: BIS(BUTOXYETHYL) ETHER ◇ BIS(2-BUTOXYETHYL) ETHER ◇ BUTYL DIGLYME ◇ 2,2′-DIBUTOXYETHYL ETHER ◇ DIETHYLENEGLYCOL DIBUTYL ETHER ◇ DIETHYLENEGLYCOL DI-n-BUTYL ETHER ◇ 1,1′-(OXYBIS(2,1-ETHANEDIYLOXY))BISBUTANE ◇ 5,8,11-TRIOXAPENTADECANE

TOXICITY DATA with REFERENCE
skn-rbt 500 mg open MLD UCDS** 4/21/67
eye-rbt 500 mg open AMIHBC 10,61,54

orl-mus TDLo:16 g/kg (7-14D preg):REP EVHPAZ 57,141,84
orl-rat LD50:3900 mg/kg AMIHBC 10,61,54
skn-rbt LD50:4040 mg/kg AMIHBC 10,61,54

CONSENSUS REPORTS: Reported in EPA TSCA Inventory. Glycol ether compounds are on the Community Right-To-Know List.

SAFETY PROFILE: Moderately toxic by ingestion. Mildly toxic by skin contact. Experimental reproductive effects. A skin and eye irritant. See also GLYCOL ETHERS. Combustible when exposed to heat or flame. To fight fire, use foam or alcohol foam. When heated to decomposition it emits acrid smoke and irritating fumes.

DDW400 CAS:112-48-1 ***HR: 2***
DIBUTYL CELLOSOLVE
mf: $C_{10}H_{22}O_2$ mw: 174.32

SYNS: 1,2-DIBUTOXYETHANE ◇ 1,1′-(1,2-ETHANEDIYLBIS(OXY)) BIS-BUTANE ◇ ETHYLENE GLYCOL DIBUTYL

TOXICITY DATA with REFERENCE
skn-rbt 500 mg open MLD UCDS** 11/7/57
eye-rbt 500 mg open AMIHBC 10,61,54
orl-rat LD50:3250 mg/kg UCDS** 11/7/57
skn-rbt LD50:3560 mg/kg AMIHBC 10,61,54

SAFETY PROFILE: Moderately toxic by ingestion and skin contact. A skin and eye irritant. When heated to decomposition it emits acrid smoke and irritating fumes.

DDX000 CAS:497-39-2 ***HR: 2***
4,6-DI-tert-BUTYL-m-CRESOL
mf: $C_{15}H_{24}O$ mw: 220.39

PROP: Yellow, crystalline solid. Mp: 62.1°, bp: 282°, flash p: 262°F (OC), d: 0.912 @ 80°/4°.

SYN: 2,4-DI-tert-BUTYL-5-METHYLPHENOL

TOXICITY DATA with REFERENCE
orl-mus LD50:1420 mg/kg JAPMA8 38,366,49

CONSENSUS REPORTS: Reported in EPA TSCA Inventory.

SAFETY PROFILE: Moderately toxic by ingestion. Combustible when exposed to heat or flame. Can react with oxidizing materials. To fight fire, use foam, CO_2, dry chemical. When heated to decomposition it emits acrid smoke and fumes. See also CRESOL.

DDX200 CAS:63041-48-5 ***HR: 2***
9,10-DI-n-BUTYL-1,2,5,6-DIBENZANTHRACENE
mf: $C_{30}H_{30}$ mw: 390.60

TOXICITY DATA with REFERENCE
skn-mus TDLo:1250 mg/kg/52W-I:ETA PRLBA4 117,318,35

SAFETY PROFILE: Questionable carcinogen with experimental tumorigenic data. When heated to decomposition it emits acrid smoke and irritating fumes.

DDX600 CAS:28660-63-1 ***HR: 3***
DI-n-BUTYL(DIBUTYRYLOXY)STANNANE
mf: $C_{16}H_{32}O_4Sn$ mw: 407.17

SYNS: DI-n-BUTYLTIN DIBUTYRATE ◇ MASELNAN DI-n-BUTYLCINICITY (CZECH)

TOXICITY DATA with REFERENCE
skn-rbt 500 mg/24H SEV 28ZPAK -,228,72
eye-rbt 20 mg/24H MOD 28ZPAK -,228,72
orl-rat LD50:90700 μg/kg 28ZPAK -,228,72

OSHA PEL: TWA 0.1 mg(Sn)/m^3 (skin)
ACGIH TLV: TWA 0.1 mg(Sn)/m^3 (skin) (Proposed: TWA 0.1 mg(Sn)/m^3; STEL 0.2 mg(Sn)/m^3 (skin))
NIOSH REL: (Organotin Compounds) TWA 0.1 mg(Sn)/m^3

SAFETY PROFILE: Poison by ingestion. An eye and severe skin irritant. See also TIN COMPOUNDS. When heated to decomposition it emits acrid smoke and irritating fumes.

DDY000 CAS:4593-81-1 ***HR: 3***
DIBUTYLDICHLOROGERMANE
mf: $C_8H_{18}Cl_2Ge$ mw: 257.75

SYNS: DI-n-BUTYLGERMANEDICHLORIDE ◇ DICHLORODIBUTYLGERMANE

TOXICITY DATA with REFERENCE
msc-ham:ovr 100 mg/L TXAPA9 64,482,82
ipr-rat LDLo:100 mg/kg CHDDAT 262,1302,66
ipr-mus LDLo:96 mg/kg CHDDAT 262,1302,66

CONSENSUS REPORTS: Reported in EPA TSCA Inventory.

SAFETY PROFILE: Poison by intraperitoneal route. Mutation data reported. See also GERMANIUM COMPOUNDS. When heated to decomposition it emits very toxic fumes of Cl^-.

DDY200 CAS:683-18-1 ***HR: 3***
DIBUTYLDICHLOROSTANNANE
mf: $C_8H_{18}Cl_2Sn$ mw: 303.85

PROP: White, crystalline solid. Mp: 43°, bp: 135° @ 10 mm, flash p: 335°F (OC), d: 1.36 @ 50°, vap press: 2 mm @ 100°, vap d: 10.5.

SYNS: CHLORID DI-n-BUTYLCINICITY (CZECH) ◇ D.B.T.C. ◇ DIBUTYLDICHLOROTIN ◇ DIBUTYLTIN CHLORIDE ◇ DIBUTYLTIN DICHLORIDE ◇ DI-n-BUTYLTIN DICHLORIDE ◇ DI-n-BUTYLZINN-DICHLORID (GERMAN) ◇ DICHLORODIBUTYLSTANNANE ◇ DICHLORODIBUTYLTIN

TOXICITY DATA with REFERENCE
skn-rbt 500 mg/24H SEV 28ZPAK -,226,72
eye-rbt 50 μg/24H SEV 28ZPAK -,226,72
msc-ham:ovr 100 μg/L TXAPA9 64,482,82
dns-rbt:oth 10 μg/L JTEHD6 16,229,85
dni-rbt:oth 100 μg/L JTEHD6 16,229,85
orl-rat LD50:100 mg/kg ARZNAD 10,44,60
ipr-rat LDLo:7500 μg/kg JOCMA7 2,183,60
ivn-rat LDLo:10 mg/kg BJIMAG 15,15,58
orl-mus LD50:70 mg/kg PHARAT 39,572,84
ivn-mus LD50:180 mg/kg CSLNX* NX#00182
orl-rbt LDLo:680 mg/kg SAIGBL 15,3,73
skn-rbt LDLo:1360 mg/kg SAIGBL 15,3,73
ivn-rbt LDLo:5 mg/kg BJIMAG 15,15,58
ivn-gpg LDLo:5 mg/kg BJIMAG 15,15,58

CONSENSUS REPORTS: Reported in EPA TSCA Inventory.

OSHA PEL: TWA 0.1 mg(Sn)/m^3 (skin)
ACGIH TLV: TWA 0.1 mg(Sn)/m^3 (skin) (Proposed: TWA 0.1 mg(Sn)/m^3; STEL 0.2 mg(Sn)/m^3 (skin))
NIOSH REL: (Organotin Compounds) TWA 0.1 mg(Sn)/m^3

SAFETY PROFILE: Poison by ingestion, intravenous, and intraperitoneal routes. Moderately toxic by skin contact. A severe skin and eye irritant. Mutation data reported. See also TIN COMPOUNDS. Combustible when exposed to heat or flame. A dangerous material; emits highly toxic fumes of HCl; will react with water or steam to produce heat and toxic fumes; can react vigorously with oxidizing materials. To fight fire, use water, foam, CO_2, dry chemical.

DDY400 CAS:7483-25-2 ***HR: 1***
DIBUTYL (DIETHYLENE GLYCOL BISPHTHALATE)

SYNS: DIETHYLENE GLYCOL, DIESTER with BUTYLPHTHALATE ◇ HOWFLEX GBP

TOXICITY DATA with REFERENCE
orl-rat LD50:11 g/kg EVHPAZ 4,3,73
ipr-rat LD50:11200 mg/kg FCTXAV 4,383,66
orl-mus LD50:10300 mg/kg FCTXAV 4,383,66
ipr-mus LD50:8400 mg/kg FCTXAV 4,383,66

SAFETY PROFILE: Mildly toxic by ingestion and intraperitoneal routes. When heated to decomposition it emits acrid smoke and irritating fumes.

DDY600 CAS:10584-98-2 ***HR: 2***
DIBUTYLDI(2-ETHYLHEXYLOXYCARBONYLMETHYLTHIO)STANNANE
mf: $C_{28}H_{56}O_4S_2Sn$ mw: 639.65

SYNS: BIS(2-ETHYLHEXYLOXYCARBONYLMETHYLTHIO) DIBUTYLSTANNANE ◇ BIS(2-ETHYLHEXYLTHIOGLYCOLATE)DIBUTYLTIN ◇ DI-n-BUTYLTIN DI-2-ETHYLHEXYLTHIOGLYCOLATE ◇ DI-n-BUTYL-ZINN DI-2-AETHYLHEXYL THIOGLYKOLAT (GERMAN)

TOXICITY DATA with REFERENCE
orl-rat LD50:510 mg/kg ARZNAD 19,934,69

CONSENSUS REPORTS: Reported in EPA TSCA Inventory.

OSHA PEL: TWA 0.1 mg(Sn)/m^3 (skin)
ACGIH TLV: TWA 0.1 mg(Sn)/m^3 (skin) (Proposed: TWA 0.1 mg(Sn)/m^3; STEL 0.2 mg(Sn)/m^3 (skin))
NIOSH REL: (Organotin Compounds) TWA 0.1 mg(Sn)/m^3

SAFETY PROFILE: Moderately toxic by ingestion. See also TIN COMPOUNDS. When heated to decomposition it emits toxic fumes of SO_x.

DDY800 CAS:563-25-7 ***HR: 3***
DIBUTYLDIFLUOROSTANNANE
mf: $C_8H_{18}F_2Sn$ mw: 270.95

SYN: DIBUTYLTIN DIFLUORIDE

TOXICITY DATA with REFERENCE
orl-rbt LDLo:200 mg/kg SAIGBL 15,3,73

CONSENSUS REPORTS: Reported in EPA TSCA Inventory.

OSHA PEL: TWA 0.1 mg(Sn)/m^3 (skin)
ACGIH TLV: TWA 0.1 mg(Sn)/m^3 (skin) (Proposed: TWA 0.1 mg(Sn)/m^3; STEL 0.2 mg(Sn)/m^3 (skin))
NIOSH REL: (Organotin Compounds) TWA 0.1 mg(Sn)/m^3

SAFETY PROFILE: Poison by ingestion. See also TIN COMPOUNDS and FLUORIDES. When heated to decomposition it emits toxic fumes of F^-.

DDZ000 CAS:7392-96-3 ***HR: 3***
DIBUTYL(DIFORMYLOXY)STANNANE
mf: $C_{10}H_{20}O_4Sn$ mw: 322.99

SYNS: DI-n-BUTYLTIN DIFORMATE ◇ MRAVENCAN DI-n-BUTYLCINICITY (CZECH)

TOXICITY DATA with REFERENCE
skn-rbt 500 mg/24H SEV 28ZPAK -,227,72
eye-rbt 5 mg/24H SEV 28ZPAK -,227,72
orl-rat LD50:60900 μg/kg 28ZPAK -,227,72

OSHA PEL: TWA 0.1 mg(Sn)/m^3 (skin)
ACGIH TLV: TWA 0.1 mg(Sn)/m^3 (skin) (Proposed: TWA 0.1 mg(Sn)/m^3; STEL 0.2 mg(Sn)/m^3 (skin))
NIOSH REL: (Organotin Compounds) TWA 0.1 mg(Sn)/m^3

SAFETY PROFILE: Poison by ingestion. A severe skin

and eye irritant. When heated to decomposition it emits acrid and irritating fumes. See also TIN COMPOUNDS.

DEA000 CAS:2865-19-2 ***HR: 3***
DIBUTYLDIIODOSTANNANE
mf: $C_8H_{18}I_2Sn$ mw: 486.75

SYN: DIBUTYLTIN DIIODIDE

TOXICITY DATA with REFERENCE
orl-rbt LDLo:150 mg/kg SAIGBL 15,3,73
skn-rbt LDLo:1000 mg/kg SAIGBL 15,3,73

OSHA PEL: TWA 0.1 $mg(Sn)/m^3$ (skin)
ACGIH TLV: TWA 0.1 $mg(Sn)/m^3$ (skin) (Proposed: TWA 0.1 $mg(Sn)/m^3$; STEL 0.2 $mg(Sn)/m^3$ (skin))
NIOSH REL: (Organotin Compounds) TWA 0.1 $mg(Sn)/m^3$

SAFETY PROFILE: Poison by ingestion. Moderately toxic by skin contact. See also TIN COMPOUNDS and IODIDES. When heated to decomposition it emits toxic fumes of I^-.

DEA200 CAS:3231-93-4 ***HR: 3***
2,2-DIBUTYL-1,3-DIOXA-2-STANNA-7,9-DITHIACYCLODODECAN-4,12-DIONE
mf: $C_{15}H_{28}O_4S_2Sn$ mw: 455.24

SYN: 2,2-DIBUTYL-1,3-DIOXA-7,9-DITHIA-2-STANNACYCLODODECAN

TOXICITY DATA with REFERENCE
ivn-mus LD50:320 mg/kg CSLNX* NX#02853

OSHA PEL: TWA 0.1 $mg(Sn)/m^3$ (skin)
ACGIH TLV: TWA 0.1 $mg(Sn)/m^3$ (skin) (Proposed: TWA 0.1 $mg(Sn)/m^3$; STEL 0.2 $mg(Sn)/m^3$ (skin))
NIOSH REL: (Organotin Compounds) TWA 0.1 $mg(Sn)/m^3$

SAFETY PROFILE: Poison by intravenous route. See also TIN COMPOUNDS. When heated to decomposition it emits toxic fumes of SO_x.

DEA400 CAS:4981-24-2 ***HR: 3***
2,2-DIBUTYL-1,3-DIOXA-2-STANNA-7-THIACYCLODECAN-4,10-DIONE
mf: $C_{14}H_{26}O_4SSn$ mw: 409.15

SYNS: DIBUTYLDIHYDROXYSTANNANE-3,3'-THIODIPROPIONATE ◇ 2,2-DIBUTYL-1,3,7,2-DIOXATHIASTANNECANE-4,10-DIONE ◇ DIBUTYLTIN 3,3'-THIODIPROPIONATE

TOXICITY DATA with REFERENCE
ivn-mus LD50:180 mg/kg CSLNX* NX#02851

OSHA PEL: TWA 0.1 $mg(Sn)/m^3$ (skin)
ACGIH TLV: TWA 0.1 $mg(Sn)/m^3$ (skin) (Proposed: TWA 0.1 $mg(Sn)/m^3$; STEL 0.2 $mg(Sn)/m^3$ (skin))
NIOSH REL: (Organotin Compounds) TWA 0.1 $mg(Sn)/m^3$

SAFETY PROFILE: Poison by intravenous route. See also TIN COMPOUNDS. When heated to decomposition it emits toxic fumes of SO_x.

DEA600 CAS:3465-74-5 ***HR: 3***
DIBUTYLDIPENTANOYLOXYSTANNANE
mf: $C_{18}H_{36}O_4Sn$ mw: 435.23

SYNS: DI-n-BUTYLTIN DIPENTANOATE ◇ DI(PENTANOYLOXY)DIBUTYLSTANNANE ◇ VALERAN DI-n-BUTYLCINICITY (CZECH)

TOXICITY DATA with REFERENCE
skn-rbt 500 mg/24H SEV 28ZPAK -,228,72
eye-rbt 5 mg/24H SEV 28ZPAK -,228,72
orl-rat LD50:134 mg/kg 28ZPAK -,228,72

OSHA PEL: TWA 0.1 $mg(Sn)/m^3$ (skin)
ACGIH TLV: TWA 0.1 $mg(Sn)/m^3$ (skin) (Proposed: TWA 0.1 $mg(Sn)/m^3$; STEL 0.2 $mg(Sn)/m^3$ (skin))
NIOSH REL: (Organotin Compounds) TWA 0.1 $mg(Sn)/m^3$

SAFETY PROFILE: Poison by ingestion. A severe skin and eye irritant. See also TIN COMPOUNDS. When heated to decomposition it emits acrid and irritating fumes.

DEA800 CAS:3236-56-4 ***HR: 3***
DI-tert-BUTYL DIPEROXYCARBONATE
mf: $C_9H_{18}O_5$ mw: 206.23

$$[(CH_3)_3COO]_2CO$$

SAFETY PROFILE: Potentially explosive when heated to 135°C. When heated to decomposition it emits acrid smoke and fumes. See also PEROXIDES.

DEB000 CAS:14666-77-4 ***HR: 3***
DI-tert-BUTYL DIPEROXYOXALATE
mf: $C_{10}H_{18}O_6$ mw: 234.25

$$[(CH_3)_3COOCO\bullet\text{-}]_2$$

SAFETY PROFILE: When removed from freezing mixture it exploded. When heated to decomposition it emits acrid smoke and fumes. See also PEROXIDES.

DEB200 CAS:2155-71-7 ***HR: 3***
DI-tert-BUTYL DIPEROXYPHTHALATE
mf: $C_{16}H_{22}O_6$ mw: 310.35

$$C_6H_4[CO\bullet OOC(CH_3)_3]_2$$

SAFETY PROFILE: A shock-sensitive explosive. Upon decomposition it emits acrid smoke and fumes.

DEB400 CAS:3465-73-4 ***HR: 3***
DIBUTYLDIPROPIONYLOXYSTANNANE
mf: $C_{14}H_{28}O_4Sn$ mw: 379.11

SYNS: DI-n-BUTYLTIN DIPROPIONATE ◇ PROPINAN DI-n-BUTYLCINICITY (CZECH)

TOXICITY DATA with REFERENCE
skn-rbt 500 mg/24H SEV 28ZPAK -,228,72
eye-rbt 5 mg/24H SEV 28ZPAK -,228,72
orl-rat LD50:70900 μg/kg 28ZPAK -,228,72

OSHA PEL: TWA 0.1 mg(Sn)/m^3 (skin)
ACGIH TLV: TWA 0.1 mg(Sn)/m^3 (skin) (Proposed: TWA 0.1 mg(Sn)/m^3; STEL 0.2 mg(Sn)/m^3 (skin))
NIOSH REL: (Organotin Compounds) TWA 0.1 mg(Sn)/m^3

SAFETY PROFILE: Poison by ingestion. A severe skin and eye irritant. See also TIN COMPOUNDS. When heated to decomposition it emits acrid smoke and irritating fumes.

DEB600 CAS:67057-34-5 ***HR: 3***
DIBUTYLDITHIOCARBAMIC ACID-S-TRIBUTYLSTANNYL ESTER
mf: $C_{21}H_{45}NS_2Sn$ mw: 494.48

SYNS: ((DIBUTYLDITHIOCARBAMOYL)OXY)TRIBUTYLSTANNANE ◇ TRIBUTYLTIN-S,S'-DIBUTYLDITHIOCARBAMATE

TOXICITY DATA with REFERENCE
ivn-mus LD50:56 mg/kg CSLNX* NX#04817

OSHA PEL: TWA 0.1 mg(Sn)/m^3 (skin)
ACGIH TLV: TWA 0.1 mg(Sn)/m^3 (skin) (Proposed: TWA 0.1 mg(Sn)/m^3; STEL 0.2 mg(Sn)/m^3 (skin))
NIOSH REL: (Organotin Compounds) TWA 0.1 mg(Sn)/m^3

SAFETY PROFILE: Poison by intravenous route. See also TIN COMPOUNDS, CARBAMATES, and ESTERS. When heated to decomposition it emits very toxic fumes of NO_x and SO_x.

DEB800 CAS:26818-53-1 ***HR: 3***
N,N-DI-sec-BUTYL DITHIOOXAMIDE
mf: $C_{10}H_{20}N_2S_2$ mw: 232.44

TOXICITY DATA with REFERENCE
ivn-rat LDLo:5 mg/kg JPETAB 121,32,57
ivn-dog LDLo:5 mg/kg JPETAB 121,32,57
ivn-cat LDLo:5 mg/kg JPETAB 121,32,57
ivn-rbt LD50:2.3 mg/kg JPETAB 121,32,57
ivn-gpg LDLo:5 mg/kg JPETAB 121,32,57

SAFETY PROFILE: Poison by intravenous route. When heated to decomposition it emits very toxic fumes of NO_x and SO_x.

DEC000 CAS:625-22-9 ***HR: 3***
DIBUTYL ESTER SULFURIC ACID
mf: $C_8H_{18}O_4S$ mw: 210.32

SYNS: DIBUTYL SULFATE ◇ DI-n-BUTYLSULFAT (GERMAN)

TOXICITY DATA with REFERENCE
orl-rat TDLo:12 g/kg/24W-I:ETA ZEKBAI 74,241,70
scu-rat LD50:5000 mg/kg ZEKBAI 74,241,70
orl-rbt LDLo:192 mg/kg AEXPBL 47,113,02

SAFETY PROFILE: Poison by ingestion. Mildly toxic by subcutaneous route. Questionable carcinogen with experimental tumorigenic data. See also ESTERS and SULFATES. When heated to decomposition it emits toxic fumes of SO_x.

DEC200 CAS:625-17-2 ***HR: 3***
DI-sec-BUTYL FLUOROPHOSPHONATE
mf: $C_8H_{18}FO_3P$ mw: 212.23

SYNS: DI-sec-BUTYL ESTER PHOSPHOROFLUORIDIC ACID ◇ DI-sec-BUTYLFLUOROPHOSPHATE ◇ T-1835 ◇ TL 1266

TOXICITY DATA with REFERENCE
ihl-man TCLo:1 ppm/5M:EYE,CNS,PUL JCSOA9 -,635,49
ihl-rat LC50:4 g/kg/m^3/10M NTIS** PB158-508
ihl-mus LC50:540 mg/m^3/10M JCSOA9 -,635,49
ihl-dog LC50:4 g/m^3/10M NTIS** PB-158-508
ihl-mky LC50:100 mg/m^3//2M NTIS** PB158-508
ihl-cat LC50:6 g/m^3/10M NTIS** PB158-508
ihl-rbt LC50:5 g/m^3/10M NTIS** PB158-508

SAFETY PROFILE: Poison by inhalation. Human systemic effects by inhalation including: miosis (pupillary constriction), somnolence, and respiratory changes. When heated to decomposition it emits very toxic fumes of F^- and PO_x.

DEC400 CAS:761-65-9 ***HR: 3***
N,N-DI-n-BUTYLFORMAMIDE
mf: $C_9H_{19}NO$ mw: 157.29

SYN: DBF

TOXICITY DATA with REFERENCE
skn-rat TDLo:1200 mg/kg (10D preg):TER TXAPA9 41,35,77
ipr-rat LD50:390 mg/kg TXAPA9 26,596,73
ipr-mus LD50:300 mg/kg TXAPA9 26,596,73

CONSENSUS REPORTS: Reported in EPA TSCA Inventory.

SAFETY PROFILE: Poison by intraperitoneal route.

An experimental teratogen. When heated to decomposition it emits toxic fumes of NO_x.

DEC600 CAS:105-75-9 ***HR: 3***
DIBUTYL FUMARATE
mf: $C_{12}H_{20}O_4$ mw: 228.32

PROP: Colorless, clear, mobile liquid; typical odor. Bp: 285.1°, fp: −19°, flash p: 300°F (OC), d: 0.986 @ 20°/20°, vap d: 7.88.

SYN: FUMARIC ACID, DIBUTYL ESTER

TOXICITY DATA with REFERENCE
skn-rbt 10 mg/24H open MLD AMIHBC 4,119,51
eye-rbt 500 mg open AMIHBC 4,119,51
orl-rat LD50:8530 mg/kg AMIHBC 4,119,51
ipr-mus LD50:250 mg/kg NTIS** AD691-490
skn-rbt LD50:16 g/kg AMIHBC 4,119,51

CONSENSUS REPORTS: Reported in EPA TSCA Inventory.

SAFETY PROFILE: Poison by intraperitoneal route. Mildly toxic by ingestion and skin contact. An eye, skin, and mucous membrane irritant. Combustible when exposed to heat or flame; can react with oxidizing materials. To fight fire, use foam, CO_2, dry chemical. When heated to decomposition it emits acrid smoke and fumes.

DEC699 CAS:4835-11-4 ***HR: 3***
N,N'-DIBUTYLHEXAMETHYLENEDIAMINE
mf: $C_{14}H_{32}N_2$ mw: 228.48

SYNS: DBHMD ◇ DIBUTYLHEXAMETHYLENEDIAMINE ◇ N,N'-DIBUTYL-1,6-HEXANEDIAMINE ◇ 1,6-N,N'-DIBUTYLHEXANEDIAMINE

TOXICITY DATA with REFERENCE
ihl-rat LD50:220 mg/m³/4H FCTOD7 22,425,84

CONSENSUS REPORTS: EPA Extremely Hazardous Substances List. Reported in EPA TSCA Inventory.

SAFETY PROFILE: Poison by inhalation. A corrosive alkali. A severe eye, skin and mucous membrane irritant. Strong alkalies are markedly corrosive and penetrating to the skin, and mucous membranes. Human systemic effects by ingestion: acute circulatory shock; burns in the mouth, throat, and esophagus; suffocation due to glottal or laryngeal swelling; perforation and inflammation of the esophagus and the tracheobronchial tree; aspiration pneumonia. Scar formation can cause delayed problems with swallowing, and stomach filling and emptying. The immediate symptoms of ingestion are: visible burns in mouth, drooling, gagging, vomiting, chest and upper abdominal pain, difficulty in breathing or apnea (respiratory arrest), collapse and cardiac arrest may occur. Flammable or poisonous gases may accumulate in tanks or hopper cars. This material may react violently with water. To fight small fires, use dry chemical, carbon dioxide, water spray, or foam. To fight large fires, use water spray, fog, or foam. When heated to decomposition it emits toxic fumes of NO_x. See also AMINES.

DEC725 CAS:7422-80-2 ***HR: 2***
1,1-DIBUTYLHYDRAZINE
mf: $C_8H_{20}N_2$ mw: 144.30

SYNS: 1,1-DBH ◇ N,N-DIBUTYLHYDRAZINE ◇ 1,1-DI-n-BUTYLHYDRAZINE

TOXICITY DATA with REFERENCE
orl-mus TDLo:49280 mg/kg/2Y-C:CAR CRNGDP 2,651,81

SAFETY PROFILE: Questionable carcinogen with experimental carcinogenic data. When heated to decomposition it emits toxic fumes of NO_x.

DEC775 CAS:78776-28-0 ***HR: 2***
1,2-DI-n-BUTYLHYDRAZINE DIHYDROCHLORIDE
mf: $C_8H_{20}N_2 \cdot 2ClH$ mw: 217.22

TOXICITY DATA with REFERENCE
orl-mus TDLo:92 g/kg/90W-C:CAR EXPEAM 37,773,81
orl-mus TD:142 g/kg/90W-C:CAR EXPEAM 37,773,81

SAFETY PROFILE: Questionable carcinogen with experimental carcinogenic data. When heated to decomposition it emits toxic fumes of NO_x and HCl.

DED000 CAS:10537-47-0 ***HR: 3***
(3,5-DI-tert-BUTYL-4-HYDROXYBENZYLIDENE) MALONONITRILE
mf: $C_{18}H_{22}N_2O$ mw: 282.42

SYNS: ((3,5-BIS(1,1-DIMETHYLETHYL)-4-HYDROXYPHENYL) METHYLENE)PROPANEDINITRILE ◇ 2-((3,5-BIS(1,1-DIMETHYL)-4-HYDROXYPHENYL)METHYLENE)PROPANEDINITRILE ◇ ENT 27,910 ◇ GCP 5126 ◇ GULF S-15126 ◇ MALONOBEN ◇ S-15126

TOXICITY DATA with REFERENCE
orl-rat LD50:87 mg/kg SPEADM 78-1,21,78
skn-rbt LD50:226 mg/kg SPEADM 74-1,-,74

CONSENSUS REPORTS: Cyanide and its compounds are on the Community Right-To-Know List.

SAFETY PROFILE: Poison by ingestion and skin contact. See also NITRILES. When heated to decomposition it emits toxic fumes of NO_x and CN^-.

DED200 CAS:2109-64-0 ***HR: 2***
N,N-DIBUTYL(2-HYDROXYPROPYL)AMINE
mf: $C_{11}H_{25}NO$ mw: 187.37

TOXICITY DATA with REFERENCE
skn-rbt 10 mg/24H open SEV AIHAAP 23,95,62
orl-rat LD50:1990 mg/kg AIHAAP 23,95,62

SAFETY PROFILE: Moderately toxic by ingestion. A severe skin irritant. When heated to decomposition it emits toxic fumes of NO_x.

DED400 CAS:2587-84-0 ***HR: 3***
DIBUTYL LEAD DIACETATE
mf: $C_{12}H_{24}O_4Pb$ mw: 439.55

SYN: DIACETOXYDIBUTYLPLUMBANE

TOXICITY DATA with REFERENCE
orl-rat LD50:34 mg/kg JJATDK 1,247,81
ipr-rat LDLo:10 mg/kg CRSBAW 164,209,70
orl-mus LD50:115 mg/kg CRSBAW 162,1456,68
ipr-mus LD50:6 mg/kg CRSBAW 164,209,70
ivn-mus LD50:6 mg/kg CRSBAW 164,209,70
orl-dom LDLo:30 mg/kg REMVAY 22,85,69

CONSENSUS REPORTS: Lead and its compounds are on the Community Right-To-Know List.

SAFETY PROFILE: Poison by ingestion, intraperitoneal, and intravenous routes. See also LEAD COMPOUNDS. When heated to decomposition it emits toxic fumes of Pb.

DED600 CAS:105-76-0 ***HR: 3***
DIBUTYL MALEATE
mf: $C_{12}H_{20}O_4$ mw: 228.32

PROP: Liquid. Mp: −85° (sets to a glass), bp: 281°, flash p: 285°F (OC), d: 0.9964 @ 20°/20°, vap d: 7.9.

SYNS: 2-BUTENEDIOIC ACID, DIBUTYL ESTER ◇ DBM ◇ MALEIC ACID, DIBUTLY ESTER ◇ RC COMONOMER DBM ◇ STAFLEX DBM

TOXICITY DATA with REFERENCE
skn-rbt 500 mg open MLD UCDS** 11/27/63
eye-rbt 500 mg open AMIHBC 10,61,54
orl-rat LD50:3730 mg/kg UCDS** 11/27/63
orl-mus LD50:2400 mg/kg ARZNAD 14,670,64
ipr-mus LD50:150 mg/kg NTIS** AD691-490
skn-rbt LD50:10 g/kg NPIRI* 2,19,75

CONSENSUS REPORTS: Reported in EPA TSCA Inventory.

SAFETY PROFILE: Poison by intraperitoneal route. Moderately toxic by ingestion. Mildly toxic by skin contact. An eye and skin irritant. See also ESTERS and BUTYL ALCOHOL. Combustible when exposed to heat or flame; can react with oxidizing materials. To fight fire, use foam, CO_2, dry chemical, alcohol foam. When heated to decomposition it emits acrid smoke and irritating fumes.

DED800 CAS:15535-69-0 ***HR: 3***
DIBUTYLMALOYLOXYSTANNANE
mf: $C_{12}H_{22}O_5Sn$ mw: 365.03

SYN: DIBUTYLTIN MALATE

TOXICITY DATA with REFERENCE
ivn-mus LD50:56 mg/kg CSLNX* NX#03637

OSHA PEL: TWA 0.1 mg(Sn)/m^3 (skin)
ACGIH TLV: TWA 0.1 mg(Sn)/m^3 (skin) (Proposed: TWA 0.1 mg(Sn)/m^3; STEL 0.2 mg(Sn)/m^3 (skin))
NIOSH REL: (Organotin Compounds) TWA 0.1 mg(Sn)/m^3

SAFETY PROFILE: Poison by intravenous route. See also TIN COMPOUNDS. When heated to decomposition it emits acrid smoke and irritating fumes.

DEE000 CAS:629-35-6 ***HR: 3***
DIBUTYLMERCURY
mf: $C_8H_{18}Hg$ mw: 314.85

PROP: Liquid. Bp: 105° @ 10 mm, d: 1.779, vap d: 10.8.

TOXICITY DATA with REFERENCE
ipr-mus LDLo:8 mg/kg CBCCT* 4,230,52

CONSENSUS REPORTS: Mercury and its compounds are on the Community Right-To-Know List. Reported in EPA TSCA Inventory.

OSHA PEL: (Transitional: CL 1 mg/10m^3) TWA 0.01 mg(Hg)/m^3; STEL 0.03 mg/m^3 (skin)
ACGIH TLV: TWA 0.01 mg(Hg)/m^3; STEL 0.03 mg(Hg)/m^3
NIOSH REL: (Inorganic Mercury) TWA 0.05 mg(Hg)/m^3

SAFETY PROFILE: Poison by intraperitoneal route. See also MERCURY COMPOUNDS, ORGANIC. Flammable when exposed to heat or flame. Can react vigorously with oxidizing materials. When heated to decomposition or on contact with acid or acid fumes it emits highly toxic fumes of mercury.

DEE200 CAS:691-88-3 ***HR: 3***
DI-sec-BUTYLMERCURY
mf: $C_8H_{18}Hg$ mw: 314.85

TOXICITY DATA with REFERENCE
ipr-mus LDLo:31 mg/kg CBCCT* 4,230,52

CONSENSUS REPORTS: Mercury and its compounds are on the Community Right-To-Know List.

OSHA PEL: (Transitional: CL 1 mg/10m^3) TWA 0.01 mg(Hg)/m^3; STEL 0.03 mg/m^3 (skin)
ACGIH TLV: TWA 0.01 mg(Hg)/m^3; STEL 0.03 mg(Hg)/m^3
NIOSH REL: (Inorganic Mercury) TWA 0.05 mg(Hg)/m^3

SAFETY PROFILE: Poison by intraperitoneal route.

See also MERCURY COMPOUNDS, ORGANIC. When heated to decomposition it emits toxic fumes of Hg.

DEE400 CAS:3405-45-6 ***HR: 2***
N,N-DIBUTYLMETHYLAMINE
mf: $C_9H_{21}N$ mw: 143.31

PROP: Colorless liquid, amine odor. Insol in water; sol in alcohol and ether, miscible with hydrocarbons. D: 0.7613 @ 20°/20°, bp: 159.6°, fp: −62°, flash p: 125°F (OC).

TOXICITY DATA with REFERENCE
skn-rbt 10 mg/24H open MLD AIHAAP 23,95,62
orl-rat LD50:540 mg/kg AIHAAP 23,95,62
ihl-rbt LCLo:250 ppm/4H AIHAAP 23,95,62
skn-rbt LDLo:880 mg/kg AIHAAP 23,95,62

CONSENSUS REPORTS: Reported in EPA TSCA Inventory.

SAFETY PROFILE: Moderately toxic by ingestion, inhalation, and skin contact. A skin irritant. Combustible when exposed to heat or flame. To fight fire, use dry chemical, fog, mist, CO_2. When heated to decomposition it emits toxic fumes of NO_x. See also AMINES.

DEE600 CAS:1301-14-0 ***HR: 3***
2,6-DI-tert-BUTYLNAPHTHALENESULFONIC ACID SODIUM SALT
mf: $C_{18}H_{23}O_3S{\cdot}Na$ mw: 342.46

SYNS: BECANTAL ◇ BECANTEX ◇ BECANTYL ◇ 2,6-DI-tert-BUTYL NAPHTALENE SULFONATE SODIQUE (FRENCH) ◇ KEUTEN ◇ L. 1633 ◇ LINCTUSSAL ◇ SODIUM-2,6-DI-tert-BUTYLNAPHTHALENESULFONATE

TOXICITY DATA with REFERENCE
orl-rat LDLo:5000 mg/kg CLDND*
scu-gpg LDLo:250 mg/kg AIPTAK 97,34,54

SAFETY PROFILE: Poison by subcutaneous route. Mildly toxic by ingestion. When heated to decomposition it emits toxic fumes of SO_x and Na_2O. See also SULFONATES.

DEE800 CAS:728-40-5 ***HR: 3***
2,6-DI-tert-BUTYL-4-NITROPHENOL
mf: $C_{14}H_{21}NO_3$ mw: 251.33

$[(CH_3)_3C]_2O_2NC_6H_2OH$

SAFETY PROFILE: Explodes when heated to 100°C. May explode spontaneously. See also NITRO COMPOUNDS of AROMATIC HYDROCARBONS and PHENOLS.

DEF000 CAS:56654-52-5 ***HR: D***
N,N′-DIBUTYL-N-NITROSOUREA
mf: $C_9H_{19}N_3O_2$ mw: 201.31

SYN: 1,3-DIBUTYL-3-NITROSOUREA

TOXICITY DATA with REFERENCE
cyt-ham:fbr 125 mg/L/22H MUREAV 48,337,77
cyt-ham:lng 79 mg/L GMCRDC 27,95,81

SAFETY PROFILE: Mutation data reported. Many N-nitroso compounds are carcinogens. When heated to decomposition it emits toxic fumes of NO_x. See also N-NITROSO COMPOUNDS.

DEF150 CAS:27371-95-5 ***HR: 3***
2,2-DIBUTYL-1,3,2-OXATHIASTANNOLANE
mf: $C_{10}H_{22}OSSn$ mw: 309.07

SYN: 1,3,2-OXATHIASTANNOLANE,2,2-DIBUTYL-

TOXICITY DATA with REFERENCE
ivn-mus LD50:180 mg/kg CSLNX* NX#02078

OSHA PEL: TWA 0.1 mg(Sn)/m^3 (skin)
ACGIH TLV: TWA 0.1 mg(Sn)/m^3; STEL 0.2 mg/m^3 (skin)
NIOSH REL: (Organotin Compounds): 10H TWA 0.1 mg(Sn)/m^3

SAFETY PROFILE: Poison by intravenous route. When heated to decomposition it emits toxic fumes of SO_x, and Sn.

DEF200 CAS:78-20-6 ***HR: 2***
2,2-DIBUTYL-1,3,2-OXATHIASTANNOLANE-5-OXIDE
mf: $C_{10}H_{20}O_2SSn$ mw: 323.05

SYNS: DIBUTYL(THIOACETOXY)STANNANE ◇ DI-n-BUTYLZINN THIOGLYKOLAT (GERMAN)

TOXICITY DATA with REFERENCE
orl-rat LD50:510 mg/kg TRIPA7 -,1,73

CONSENSUS REPORTS: Reported in EPA TSCA Inventory.

OSHA PEL: TWA 0.1 mg(Sn)/m^3 (skin)
ACGIH TLV: TWA 0.1 mg(Sn)/m^3 (skin) (Proposed: TWA 0.1 mg(Sn)/m^3; STEL 0.2 mg(Sn)/m^3 (skin))
NIOSH REL: (Organotin Compounds) TWA 0.1 mg(Sn)/m^3

SAFETY PROFILE: Moderately toxic by ingestion. See also TIN COMPOUNDS. When heated to decomposition it emits toxic fumes of SO_x.

DEF400 CAS:818-08-6 ***HR: 3***
DIBUTYLOXOSTANNANE
mf: $C_8H_{18}OSn$ mw: 248.95

PROP: White, amorphous powder. Mp: decomp without melting, bulk density: 0.5, vap d: 8.6.

SYNS: DBOT ◇ DIBUTYLOXIDE of TIN ◇ DIBUTYLOXOTIN

◇ DIBUTYLSTANNANE OXIDE ◇ DIBUTYLTIN OXIDE ◇ DI-n-BUTYLTIN OXIDE ◇ DI-n-BUTYL-ZINN-OXYD (GERMAN) ◇ KYSLICNIK DI-n-BUTYLCINICITY (CZECH)

TOXICITY DATA with REFERENCE
skn-rbt 500 mg/24H MLD 28ZPAK -,226,72
eye-rbt 100 mg/24H MOD 28ZPAK -,226,72
orl-rat LD50:44900 μg/kg 28ZPAK -,226,72
ipr-rat LD50:40 mg/kg FCTXAV 7,47,69
orl-rbt LDLo:1500 mg/kg SAIGBL 15,3,73

CONSENSUS REPORTS: Reported in EPA TSCA Inventory.

OSHA PEL: TWA 0.1 mg(Sn)/m^3 (skin)
ACGIH TLV: TWA 0.1 mg(Sn)/m^3 (skin) (Proposed: TWA 0.1 mg(Sn)/m^3; STEL 0.2 mg(Sn)/m^3 (skin))
NIOSH REL: (Organotin Compounds) TWA 0.1 mg(Sn)/m^3

SAFETY PROFILE: Poison by ingestion and intraperitoneal routes. A skin and eye irritant. Flammable when exposed to flame; can react with oxidizing materials. To fight fire use dry chemical, fog, CO_2. When heated to decomposition it emits acrid smoke and irritating fumes. See also TIN COMPOUNDS.

DEF600 CAS:2167-23-9 ***HR: 3***
2,2-DI(tert-BUTYLPEROXY)BUTANE
mf: $C_{12}H_{26}O_4$ mw: 234.34

$[(CH_3)_3COO]_2C(CH_3)CH_2CH_3$

SAFETY PROFILE: Pure material explodes on heating to 130°C, sparking or on impact. When heated to decomposition it emits acrid smoke and fumes. See also PEROXIDES.

DEF800 CAS:5510-99-6 ***HR: 3***
2,6-DI-sec-BUTYLPHENOL
mf: $C_{14}H_{22}O$ mw: 206.36

PROP: Amber liquid. Bp: 152°-165° @ 25 mm, fp: −50°, flash p: 280°F, d: 0.936 @ 25°/4°.

SYN: 2,6-DI-sec-BUTYLFENOL (CZECH)

TOXICITY DATA with REFERENCE
skn-rbt 500 mg/24H SEV 28ZPAK -,56,72
eye-rbt 50 μg/24H SEV 28ZPAK -,56,72
orl-rat LD50:1320 mg/kg 28ZPAK -,56,72
ivn-mus LD50:60 mg/kg JMCMAR 23,1350,80
ivn-rbt LDLo:10 mg/kg JMCMAR 23,1350,80

SAFETY PROFILE: Poison by intravenous route. Moderately toxic by ingestion. A severe skin and eye irritant. Combustible when exposed to heat or flame; can react with oxidizing materials. To fight fire, use foam, CO_2, dry chemical. When heated to decomposition it emits acrid and irritating fumes. See also PHENOL.

DEG000 CAS:96-76-4 ***HR: 3***
2,4-DI-tert-BUTYLPHENOL
mf: $C_{14}H_{22}O$ mw: 206.36

PROP: Tan crystals. Mp: 51°, bp: 260.8°, flash p: 265°F, d: 0.907 @ 60°/4°, vap press: 1 mm @ 84.5°.

SYNS: ANTIOXIDANT No. 33 ◇ PRODOX 146 ◇ PRODOX 146A-85X

TOXICITY DATA with REFERENCE
ipr-mus LD50:25 mg/kg NTIS** AD691-490
ivn-mus LD50:100 mg/kg JMCMAR 23,1350,80

CONSENSUS REPORTS: Reported in EPA TSCA Inventory.

SAFETY PROFILE: Poison by intraperitoneal and intravenous routes. Combustible when exposed to heat or flame. Can react with oxidizing materials. Violent reaction with HNO_3. To fight fire, use foam, CO_2, dry chemical. When heated to decomposition it emits acrid smoke and fumes. See also PHENOL.

DEG200 CAS:101-96-2 ***HR: 3***
N,N'-DI-sec-BUTYL-p-PHENYLENEDIAMINE
mf: $C_{14}H_{24}N_2$ mw: 220.40

PROP: Liquid. Mp: 17.8°, flash p: 285°F (OC), d: 0.94-0.95 @ 24°/24°.

SYN: TENAMENE 2

TOXICITY DATA with REFERENCE
orl-rat LDLo:200 mg/kg KODAK* -,-,71
ihl-rat LCLo:600 mg/m^3/6H KODAK* -,-,71
skn-gpg LD50:5000 mg/kg RCTEA4 45(3),627,72

CONSENSUS REPORTS: Reported in EPA TSCA Inventory.

SAFETY PROFILE: Poison by ingestion. Moderately toxic by inhalation and skin contact. Corrosive to skin. A mild allergen. Symptoms of exposure are sweating, flushing, shortness of breath and slow pulse. Combustible when exposed to heat or flame; can react with oxidizing materials. To fight fire, use foam, CO_2, dry chemical. When heated to decomposition it emits toxic fumes of NO_x. See also AMINES.

DEG400 CAS:2655-19-8 ***HR: 2***
3,5-DI-tert-BUTYLPHENYLMETHYLCARBAMATE
mf: $C_{16}H_{25}NO_2$ mw: 263.42

SYNS: BUTACARB ◇ BUTACARBE (FRENCH)

TOXICITY DATA with REFERENCE
orl-rat LD50:1800 mg/kg SPEADM 78-1,57,78
orl-mus LD50:3200 mg/kg SPEADM 78-1,57,78
orl-dog LD50:1000 mg/kg SPEADM 78-1,57,78

SAFETY PROFILE: Moderately toxic by ingestion.

When heated to decomposition it emits toxic fumes of NO_x. See also CARBAMATES.

DEG600 CAS:2528-36-1 **HR: 2**
DIBUTYL PHENYL PHOSPHATE
mf: $C_{14}H_{23}O_4P$ mw: 286.34

SYN: PHOSPHORIC ACID, DIBUTYL PHENYL ESTER

TOXICITY DATA with REFERENCE
orl-rat LD50:2140 mg/kg GTPZAB 25(4),46,81
orl-mus LD50:1790 mg/kg GTPZAB 25(4),46,81

CONSENSUS REPORTS: Reported in EPA TSCA Inventory.

ACGIH TLV: TWA 0.3 ppm (skin)

SAFETY PROFILE: Moderately toxic by ingestion. When heated to decomposition it emits toxic fumes of PO_x.

DEG700 CAS:107-66-4 **HR: 2**
DIBUTYL PHOSPHATE
mf: $C_8H_{19}PO_4$ mw: 210.2

PROP: Pale amber liquid. Bp: decomp > 100°.

SYNS: DIBUTYL ACID PHOSPHATE ◇ DIBUTYL HYDROGEN PHOSPHATE ◇ DIBUTYL PHOSPHATE ◇ DI-n-BUTYL PHOSPHATE

TOXICITY DATA with REFERENCE
orl-rat LD50:3200 mg/kg 14CYAT -,1918

CONSENSUS REPORTS: Reported in EPA TSCA Inventory.

OSHA PEL: TWA 1 ppm; STEL 2 ppm
ACGIH TLV: TWA 1 ppm; STEL 2 ppm

SAFETY PROFILE: Moderately toxic by ingestion. When heated to decomposition it emits toxic fumes of PO_x. See also PHOSPHATES.

DEG800 CAS:1809-19-4 **HR: 2**
DIBUTYL PHOSPHITE
mf: $C_8H_{19}O_3P$ mw: 194.24

PROP: Liquid. Bp: 115° @ 10 mm, flash p: 120°F, d: 0.971 @ 35°/4°, vap press: <1 mm @ 20°, vap d: 6.7.

SYNS: BUTYL ALCOHOL HYDROGEN PHOSPHITE ◇ DIBUTYL HYDROGEN PHOSPHITE ◇ MOBIL DBHP

TOXICITY DATA with REFERENCE
skn-rbt 10 mg/24H open MLD JIHTAB 31,60,49
eye-rbt 250 μg open SEV JIHTAB 31,60,49
orl-rat LD50:3200 mg/kg ALBRW* #OPB-3,84
skn-rbt LD50:1990 mg/kg JIHTAB 31,60,49

CONSENSUS REPORTS: Reported in EPA TSCA Inventory.

SAFETY PROFILE: Moderately toxic by ingestion and skin contact. A skin and severe eye irritant. Combustible when exposed to heat or flame or by chemical reaction. Many phosphites decompose to evolve phosphine when heated. Explosion Hazard: See PHOSPHINE. Can react vigorously with oxidizing materials. To fight fire, use foam, CO_2, dry chemical. Dangerous; when heated to decomposition or on contact with acid or acid fumes it emits highly toxic fumes of PO_x.

DEH200 CAS:84-74-2 **HR: 3**
DIBUTYL PHTHALATE
mf: $C_{16}H_{22}O_4$ mw: 278.38

PROP: Oily liquid, mild odor. Bp: 340°, fp: −35°, flash p: 315°F (CC), d: 1.047-1.049 @ 20°/20°, autoign temp: 757°F, vap d: 9.58.

SYNS: o-BENZENEDICARBOXYLIC ACID, DIBUTYL ESTER ◇ BENZENE-o-DICARBOXYLIC ACID DI-n-BUTYL ESTER ◇ n-BUTYL PHTHALATE (DOT) ◇ CELLUFLEX DPB ◇ DBP ◇ DIBUTYL-1,2-BENZENEDICARBOXYLATE ◇ DI-n-BUTYL PHTHALATE ◇ ELAOL ◇ HEXAPLAS M/B ◇ PALATINOL C ◇ POLYCIZER DBP ◇ PX 104 ◇ RCRA WASTE NUMBER U069 ◇ STAFLEX DBP ◇ WITCIZER 300

TOXICITY DATA with REFERENCE
mmo-sat 100 μg/plate JTEHD6 16,61,85
cyt-ham:fbr 30 mg/L/24H MUREAV 48,337,77
ipr-rat TDLo:6 g/kg (female 3-9D post):REP EVHPAZ 3,91,73
orl-rat TDLo:2520 mg/kg (1-21D preg):TER TXAPA9 26,253,73
orl-hmn TDLo:140 mg/kg:CNS,GIT,KID SMWOAS 84,1243,54
orl-rat LD50:8000 mg/kg FMCHA2 -,C76,83
skn-rat LDLo:6 g/kg 85GMAT -,44,82
ipr-rat LD50:3050 mg/kg JPMSAE 61,51,72
orl-mus LD50:5289 mg/kg GTPZAB 17(11),51,73
ihl-mus LC50:25 g/m³/2H 85GMAT -,44,82
ivn-mus LD50:720 mg/kg KEKHB8 (3),19,73

CONSENSUS REPORTS: On EPA Extremely Hazardous Substances List by error. On the Community Right-To-Know List. EPA Genetic Toxicology Program. Reported in EPA TSCA Inventory.

OSHA PEL: TWA 5 mg/m^3
ACGIH TLV: TWA 5 mg/m^3

SAFETY PROFILE: Moderately toxic by intraperitoneal and intravenous routes. Mildly toxic by ingestion. Human systemic eye effects by ingestion: hallucinations, distorted perceptions, nausea or vomiting and kidney, ureter or bladder changes. Experimental teratogenic and reproductive effects. Mutation data reported. Combustible when exposed to heat or flame; can react with oxidizing materials. Violent reaction with Cl_2. Incompatible with chlorine. To fight fire, use CO_2, dry chemical. When heated to decomposition it emits acrid smoke and fumes. See also ESTERS, PHTHALIC ACID, and BUTYL ALCOHOL.

DEH600 CAS:109-43-3 ***HR: 1***
DIBUTYL SEBACATE
mf: $C_{18}H_{34}O_4$ mw: 314.52

PROP: Clear liquid. Bp: 180° @ 3 mm, fp: −11°, flash p: 353°F (COC), d: 0.936 @ 20°/20°, vap d: 10.8.

SYNS: BIS(n-BUTYL)SEBACATE ◇ DECANEDIOIC ACID, DIBUTYL ESTER ◇ DI-n-BUTYL SEBACATE ◇ KODAFLEX DBS ◇ MONOPLEX DBS ◇ POLYCIZER DBS ◇ PX 404 ◇ SEBACIC ACID, DIBUTYL ESTER ◇ STAFLEX DBS

TOXICITY DATA with REFERENCE
orl-rat TDLo:418 g/kg (10W male/10D pre):REP AMIHBC 7,310,53
orl-rat LD50:16 g/kg NPIRI* 2,22,75

CONSENSUS REPORTS: Reported in EPA TSCA Inventory.

SAFETY PROFILE: Mildly toxic by ingestion. Experimental reproductive effects. Combustible liquid when exposed to heat or flame; can react with oxidizing materials. To fight fire, use CO_2, dry chemical. When heated to decomposition it emits acrid smoke and fumes. See also ESTERS and BUTYL ALCOHOL.

DEH650 CAS:7399-02-2 ***HR: 3***
2,2'-((DIBUTYLSTANNYLENE)BIS(THIO)) BISACETIC ACID DINONYL ESTER
mf: $C_{30}H_{60}O_4S_2Sn$ mw: 667.71

SYNS: ACETIC ACID, 2,2'-((DIBUTYLSTANNYLENE)BIS(THIO))BIS-, DINONYL ESTER ◇ ACETIC ACID, ((DIBUTYLSTANNYLENE)DITHIO) DI-, DINONYL ESTER (8CI) ◇ MELLITE 131 ◇ 8-OXA-3,5-DITHIA-4-STANNAHEPTADECANOIC ACID, 4,4-DIBUTYL-7-OXO-, NONYLESTER (9CI)

TOXICITY DATA with REFERENCE
orl-mus LD50:150 mg/kg ERNFA7 11,424,66

OSHA PEL: 8H TWA 0.1 mg(Sn)/m^3 (skin)
ACGIH TLV: TWA 0.1 mg(Sn)/m^3; STEL 0.2 mg/m^3 (skin)
NIOSH REL: (Organotin Compounds): 10H TWA 0.1 mg(Sn)/m^3

SAFETY PROFILE: Poison by ingestion. When heated to decomposition it emits toxic fumes of SO_x and Sn.

DEH800 CAS:23535-89-9 ***HR: 3***
DIBUTYL(TETRACHLOROPHTHALATO)STANNANE
mf: $C_{16}H_{18}Cl_4O_4Sn$ mw: 534.83

SYNS: 3,3-DIBUTYL-6,7,8,9-TETRACHLORO-2,4,3-BENZODIOXASTANNEPIN-1,5-DIONE ◇ DIBUTYLTIN TETRACHLOROPHTHALATE

TOXICITY DATA with REFERENCE
ivn-mus LD50:180 mg/kg CSLNX* NX#02077

OSHA PEL: TWA 0.1 mg(Sn)/m^3 (skin)
ACGIH TLV: TWA 0.1 mg(Sn)/m^3 (skin) (Proposed: TWA 0.1 mg(Sn)/m^3; STEL 0.2 mg(Sn)/m^3 (skin))
NIOSH REL: (Organotin Compounds) TWA 0.1 mg(Sn)/m^3

SAFETY PROFILE: Poison by intravenous route. See also TIN COMPOUNDS. When heated to decomposition it emits toxic fumes of Cl^-.

DEI000 CAS:109-46-6 ***HR: 3***
1,3-DIBUTYLTHIOUREA
mf: $C_9H_{20}N_2S$ mw: 188.37

PROP: White to light tan powder. Mp: 60°, vap d: 6.5.

SYNS: N,N'-DIBUTYLTHIOUREA ◇ 1,3-DI-n-BUTYL-2-THIOUREA ◇ 1,3-DIBUTYL-2-THIOUREA ◇ PENNZONE B ◇ THIATE U ◇ USAF EK-2138

TOXICITY DATA with REFERENCE
orl-rat LD50:350 mg/kg JPETAB 90,260,47
ipr-mus LD50:800 mg/kg NTIS** AD277-689

CONSENSUS REPORTS: Reported in EPA TSCA Inventory.

SAFETY PROFILE: Poison by ingestion. Moderately toxic by intraperitoneal route. When heated to decomposition it emits very toxic fumes of NO_x and SO_x.

DEI200 CAS:4253-22-9 ***HR: 3***
DIBUTYLTHIOXOSTANNANE
mf: $C_8H_{18}SSn$ mw: 265.01

SYNS: DIBUTYLTIN SULFIDE ◇ TIN DIBUTYL MERCAPTIDE

TOXICITY DATA with REFERENCE
orl-rat LD50:145 mg/kg UBZHD4 50,695,78
orl-mus LD50:145 mg/kg UBZHD4 50,695,78

CONSENSUS REPORTS: Reported in EPA TSCA Inventory.

OSHA PEL: TWA 0.1 mg(Sn)/m^3 (skin)
ACGIH TLV: TWA 0.1 mg(Sn)/m^3 (skin) (Proposed: TWA 0.1 mg(Sn)/m^3; STEL 0.2 mg(Sn)/m^3 (skin))
NIOSH REL: (Organotin Compounds) TWA 0.1 mg(Sn)/m^3

SAFETY PROFILE: Poison by ingestion. See also TIN COMPOUNDS and SULFIDES. When heated to decomposition it emits toxic fumes of SO_x.

DEI400 CAS:73927-86-3 ***HR: 3***
DI-n-BUTYLTIN BISMETHANESULFONATE
mf: $C_{10}H_{24}O_6S_2Sn$ mw: 423.15

SYN: BIS(METHYLSULFONYL)OXY)DIBUTYLSTANNANE

TOXICITY DATA with REFERENCE
ivn-mus LD50:10 mg/kg CSLNX* NX#02276

OSHA PEL: TWA 0.1 mg(Sn)/m^3 (skin)
ACGIH TLV: TWA 0.1 mg(Sn)/m^3 (skin) (Proposed: TWA 0.1 mg(Sn)/m^3; STEL 0.2 mg(Sn)/m^3 (skin))
NIOSH REL: (Organotin Compounds) TWA 0.1 mg(Sn)/m^3

SAFETY PROFILE: Poison by intravenous route. See also TIN COMPOUNDS and SULFONATES. When heated to decomposition it emits toxic SO_x.

DEI600 CAS:19706-58-2 ***HR: 3***
DI-n-BUTYL TIN DI(HEXADECYLMALEATE)
mf: $C_{48}H_{88}O_8Sn$ mw: 912.05

SYN: HEXADECYLMALEINAN DI-n-BUTYLCINICITY (CZECH)

TOXICITY DATA with REFERENCE
eye-rbt 100 mg/24H MOD 28ZPAK -,231,72
orl-rat LD50:386 mg/kg 28ZPAK -,231,72

OSHA PEL: TWA 0.1 mg(Sn)/m^3 (skin)
ACGIH TLV: TWA 0.1 mg(Sn)/m^3 (skin) (Proposed: TWA 0.1 mg(Sn)/m^3; STEL 0.2 mg(Sn)/m^3 (skin))
NIOSH REL: (Organotin Compounds) TWA 0.1 mg(Sn)/m^3

SAFETY PROFILE: Poison by ingestion. An eye irritant. See also TIN COMPOUNDS. When heated to decomposition it emits acrid smoke and irritating fumes.

DEI800 CAS:69239-37-8 ***HR: 3***
DI-n-BUTYLTIN DI(MONONONYL)MALEATE
mf: $C_{42}H_{76}O_8Sn$ mw: 827.87

SYNS: BIS(NONYLOXYMALEOYLOXY)DIOCTYLSTANNANE ◇ DI-n-BUTYL-ZINN-DI(MONONONYL)MALEINAT(GERMAN) ◇ DIOCTYLBIS(NONYLOXYMALEOYLOXY)STANNANE

TOXICITY DATA with REFERENCE
orl-rat LD50:170 mg/kg ARZNAD 19,934,69

OSHA PEL: TWA 0.1 mg(Sn)/m^3 (skin)
ACGIH TLV: TWA 0.1 mg(Sn)/m^3 (skin) (Proposed: TWA 0.1 mg(Sn)/m^3; STEL 0.2 mg(Sn)/m^3 (skin))
NIOSH REL: (Organotin Compounds) TWA 0.1 mg(Sn)/m^3

SAFETY PROFILE: Poison by ingestion. See also TIN COMPOUNDS. When heated to decomposition it emits acrid smoke and irritating fumes.

DEJ000 CAS:13323-62-1 ***HR: 3***
DIBUTYLTIN DIOLEATE
mf: $C_{44}H_{54}O_4Sn$ mw: 765.67

SYNS: BIS(OLEOYLOXY)DIBUTYLSTANNANE ◇ CN 447 ◇ DIBUTYLBIS(OLEOYLOXY)STANNANE ◇ DIBUTYLBIS((1-OXO-9-OCTADECENYL)OXY)STANNANE(Z,Z)

TOXICITY DATA with REFERENCE
ivn-mus LD50:32 mg/kg CSLNX* NX#03563

CONSENSUS REPORTS: Reported in EPA TSCA Inventory.

OSHA PEL: TWA 0.1 mg(Sn)/m^3 (skin)
ACGIH TLV: TWA 0.1 mg(Sn)/m^3 (skin) (Proposed: TWA 0.1 mg(Sn)/m^3; STEL 0.2 mg(Sn)/m^3 (skin))
NIOSH REL: (Organotin Compounds) TWA 0.1 mg(Sn)/m^3

SAFETY PROFILE: Poison by intravenous route. See also TIN COMPOUNDS. When heated to decomposition it emits acrid smoke and irritating fumes.

DEJ200 CAS:78-06-8 ***HR: 3***
DIBUTYLTIN MERCAPTOPROPIONATE
mf: $C_{11}H_{22}O_2SSn$ mw: 337.08

SYNS: 2,2-DIBUTYLDIHYDRO-6H-1,3,2-OXATHIASTANNIN-6-ONE ◇ 2,2-DIBUTYL-1-OXA-2-STANNA-3-THIACYCLOHEXAN-6-ONE ◇ DIBUTYLTIN-S,O-3-MERCAPTOPROPIONATE ◇ DIBUTYLTIN-S,O-β-MERCAPTOPROPIONATE ◇ DIBUTYLTIN-O,S-MERCAPTOPROPIONATE ◇ DIBUTYL(3-MERCAPTOPROPIONATO(2-))TIN ◇ MERCAPTOPROPIONIC ACID, DIBUTYLTIN SALT

TOXICITY DATA with REFERENCE
ivn-mus LD50:100 mg/kg CSLNX* NX#02852

CONSENSUS REPORTS: Reported in EPA TSCA Inventory.

OSHA PEL: TWA 0.1 mg(Sn)/m^3 (skin)
ACGIH TLV: TWA 0.1 mg(Sn)/m^3 (skin) (Proposed: TWA 0.1 mg(Sn)/m^3; STEL 0.2 mg(Sn)/m^3 (skin))
NIOSH REL: (Organotin Compounds) TWA 0.1 mg(Sn)/m^3

SAFETY PROFILE: Poison by intravenous route. See also TIN COMPOUNDS and MERCAPTANS. When heated to decomposition it emits toxic fumes of SO_x.

DEJ300 ***HR: D***
DIBUTYRYL CYCLIC AMP
mf: $C_{18}H_{23}N_5O_8P \cdot Na$ mw: 491.42

SYNS: BUCLADESINE ◇ SODIUM-N^6,2′-o-DIBYTYRYLADENOSINE 3′,5′-CYCLIC PHOSPHATE

TOXICITY DATA with REFERENCE
ivn-rat TDLo:138 mg/kg (female 7-17D post):REP OYYAA2 27,585,84
ivn-rat TDLo:630 mg/kg (male 9W pre):TER OYYAA2 27,571,84

SAFETY PROFILE: Experimental teratogenic and reproductive effects. When heated to decomposition it emits toxic fumes of PO_x, NO_x, and Na_2O. See also AMIDES.

DEJ400 CAS:31052-46-7 ***HR: 3***
DICAESIUM SELENIDE
mf: Cs_2Se mw: 344.78
SYNS: CESIUM SELENIDE ◇ DICESIUM SELENIDE

CONSENSUS REPORTS: Selenium and its compounds are on the Community Right-To-Know List.

OSHA PEL: TWA 0.2 mg(Se)/m^3
ACGIH TLV: TWA 0.2 mg(Se)/m^3
DFG MAK: 0.1 mg(Se)/m^3

SAFETY PROFILE: Ignites in air when warmed. When heated to decomposition it emits toxic fumes of Se. See also SELENIUM COMPOUNDS.

DEJ600 ***HR: 3***
DICARBADODECABORANYLMETHYLETHYL SULFIDE
mf: $C_5H_{18}B_{10}S$ mw: 218.39
SYN: CARBORANYLMETHYLETHYL SULFIDE

TOXICITY DATA with REFERENCE
skn-rbt 1% SEV NTIS** AD-A041-973
orl-rat TDLo:660 mg/kg (6-16D preg):TER AEHA** 51-044-74/76
orl-rat LD50:2085 mg/kg AEHA** 51-044-74/76
skn-rbt LD50:3890 mg/kg AEHA** 51-044-74/76
ivn-rbt LDLo:320 mg/kg AEHA** 51-044-74/76

SAFETY PROFILE: Poison by intravenous route. Moderately toxic by ingestion and skin contact. An experimental teratogen. A severe skin irritant. See also BORON COMPOUNDS. When heated to decomposition it emits toxic fumes of SO_x.

DEJ800 ***HR: 3***
DICARBADODECABORANYLMETHYLPROPYL SULFIDE
mf: $C_6H_{20}B_{10}S$ mw: 232.42
SYN: CARBORANYLMETHYLPROPYL SULFIDE

TOXICITY DATA with REFERENCE
skn-rbt 1% SEV NTIS** AD-A041-973
orl-rat TDLo:2090 mg/kg (6-16D preg):TER AEHA** 51-044-74/76
orl-rat TDLo:2090 mg/kg (6-16D preg):REP AEHA** 51-044-74/76
orl-rat LD50:3440 mg/kg AEHA** 51-044-74/76
skn-rbt LD50:3160 mg/kg AEHA** 51-044-74/76
ivn-rbt LDLo:320 mg/kg AEHA** 51-044-74/76

SAFETY PROFILE: Poison by intravenous route. Moderately toxic by ingestion and skin contact. An experimental teratogen. Other experimental reproductive effects. A severe skin irritant. See also BORON COMPOUNDS. When heated to decomposition it emits toxic fumes of SO_x.

DEJ849 CAS:68348-85-6 ***HR: 3***
DICARBONYL MOLYBDENUM DIAZIDE
mf: $C_2MoN_6O_2$ mw: 236.00

SAFETY PROFILE: An extremely sensitive explosive. It may be initiated by touch or on contact with traces of water. When heated to decomposition it emits toxic fumes of NO_x. See also MOLYBDENUM COMPOUNDS, CARBONYLS, and AZIDES.

DEJ859 ***HR: 3***
DICARBONYLPYRAZINE RHODIUM(I) PERCHLORATE
mf: $C_6H_4ClN_2O_6Rh$

$$[(OC)_2RhC_4H_4N_2]_n\ [ClO4]_n$$

SAFETY PROFILE: The complex explodes violently when heated. When heated to decomposition it emits toxic fumes of Cl^- and NO_x. See also PERCHLORATES, CARBONYLS, and RHODIUM.

DEJ880 CAS:68379-32-8 ***HR: 3***
DICARBONYLTUNGSTEN DIAZIDE
mf: $C_2N_6O_2W$ mw: 323.91

SAFETY PROFILE: An extremely sensitive explosive. It may be initiated by touch or on contact with traces of water. When heated to decomposition it emits toxic fumes of NO_x. See also TUNGSTEN COMPOUNDS, CARBONYLS, and AZIDES.

DEK000 CAS:56455-90-4 ***HR: 3***
DICARBOXIDINE HYDROCHLORIDE
mf: $C_{20}H_{24}N_2O_6 \cdot 2ClH$ mw: 461.38
SYNS: 4,4'-((4,4'-DIAMINO-(1,1'-BIPHENYL)-3,3'-DIYL)BIS(OXY) BISBUTANOIC ACID, DIHYDROCHLORIDE ◇ HYDROCHLORIC ACID DICARBOXIDE

TOXICITY DATA with REFERENCE
scu-rat TDLo:21250 mg/kg/2Y-I:ETA JJIND8 62,301,79

SAFETY PROFILE: Questionable carcinogen with experimental tumorigenic data. When heated to decomposition it emits very toxic fumes of NO_x and HCl.

DEK200 CAS:6362-79-4 ***HR: 2***
3,5-DICARBOXYBENZENESULFONIC ACID, SODIUM SALT
mf: $C_8H_5O_7S \cdot Na$ mw: 268.18

SYN: 3,5-DIKARBOXYBENZENSULFONAN SODNY (CZECH)

TOXICITY DATA with REFERENCE
eye-rbt 20 mg/24H SEV 28ZPAK -,185,72
orl-rat LD50:6450 mg/kg 28ZPAK -,185,72

CONSENSUS REPORTS: Reported in EPA TSCA Inventory.

SAFETY PROFILE: Mildly toxic by ingestion. A severe eye irritant. When heated to decomposition it emits toxic fumes of SO_x and Na_2O.

DEK400 CAS:73758-56-2 ***HR: 3***
DICARBOXYDINE
mf: $C_{20}H_{24}N_2O_6$ mw: 388.46

SYNS: Γ,Γ'-,3,3'-BENZIDINE DIOXYDIBUTYRIC ACID ◇ 3,3'-BENZIDINE-Γ,Γ'-DIOXYDIBUTYRIC ACID ◇ 4,4'-(3,3'-DIAMINO-p,p'-BIPHENYLENEDIOXY)DIBUTYRIC ACID

TOXICITY DATA with REFERENCE
scu-rat TDLo:19 g/kg/2Y-I:ETA JJIND8 62,301,79

SAFETY PROFILE: Questionable carcinogen with experimental tumorigenic data. When heated to decomposition it emits toxic fumes of NO_x.

DEK600 CAS:12014-93-6 ***HR: 3***
DICERIUM TRISULFIDE
mf: Ce_2S_2 mw: 344.37

SYN: CERIUM TRISULFIDE

SAFETY PROFILE: The powder explodes spontaneously in air. When heated to decomposition it emits toxic fumes of SO_x. See also SULFIDES and CERIUM COMPOUNDS.

DEL000 CAS:79-43-6 ***HR: 2***
DICHLORACETIC ACID
DOT: UN 1764
mf: $C_2H_2Cl_2O_2$ mw: 128.94

PROP: Colorless, corrosive liquid; pungent odor. Mp (a): 10°, (b): −4°, bp: 194°, d: 1.5634 @ 20°/4°, vap press: 1 mm @ 44.0°, vap d: 4.45.

SYNS: BICHLORACETIC ACID ◇ DCA ◇ DICHLORETHANOIC ACID ◇ 2,2-DICHLOROACETIC ACID ◇ DICHLOROETHANOIC ACID ◇ URNER'S LIQUID

TOXICITY DATA with REFERENCE
skn-rbt 10 mg/24H open MOD AMIHBC 4,119,51
skn-rbt 2 mg/24H SEV 85JCAE-,570,86
eye-rbt 50 μg open SEV AMIHBC 4,119,51
orl-mus TDLo:427 g/kg/61W-C:CAR TXAPA9 90,183,87
orl-rat LD50:2820 mg/kg AMIHBC 4,119,51
skn-rbt LD50:510 mg/kg AMIHBC 4,119,51

CONSENSUS REPORTS: Reported in EPA TSCA Inventory.

DOT Classification: Corrosive Material; Label: Corrosive.

SAFETY PROFILE: Moderately toxic by skin contact and ingestion. It is corrosive to the skin, eyes, and mucous membranes. Questionable carcinogen with experimental tumorigenic data. Will react with water or steam to produce toxic and corrosive fumes. When heated to decomposition it emits toxic fumes of Cl^-. See also CHLORIDES.

DEL200 CAS:50264-69-2 ***HR: 2***
1-(2,4-DICHLORBENZYL)INDAZOLE-3-CARBOXYLIC ACID
mf: $C_{15}H_{10}Cl_2N_2O_2$ mw: 321.17

SYNS: AF 1890 ◇ DICA ◇ 1-(2,4-DICHLOROBENZYL)-1H-INDAZOLE-3-CARBOXYLIC ACID ◇ 1-((2,4-DICHLOROPHENYL)METHYL)-1H-INDAZOLE-3-CARBOXYLIC ACID ◇ DICLONDAZOLIC ACID ◇ LONIDAMINE

TOXICITY DATA with REFERENCE
spm-rat-orl 250 mg/kg/5D-C EXMPA6 23,357,75
spm-mky-orl 250 mg/kg/5D JRPFA4 52,275,78
orl-rat TDLo:90 mg/kg (female 6-15D post):TER ARTODN 5,197,82
orl-mky TDLo:250 mg/kg (male 5D pre):REP JRPFA4 52,275,78
orl-rat LD50:1700 mg/kg CHTHBK 27,91,81
ipr-rat LD50:525 mg/kg CHTHBK 27,91,81
orl-mus LD50:900 mg/kg CHTHBK 27,91,81
ipr-mus LD50:435 mg/kg CHTHBK 27,91,81

SAFETY PROFILE: Moderately toxic by ingestion and intraperitoneal routes. Experimental teratogenic and reproductive effects. Mutation data reported. When heated to decomposition it emits very toxic fumes of Cl^- and NO_x.

DEL600 CAS:7791-21-1 ***HR: 3***
DICHLORINE OXIDE
mf: Cl_2O mw: 86.906

SAFETY PROFILE: The liquid at 2°C is an unstable spark- and touch-sensitive explosive. The gas may explode when heated above 42°C. A powerful oxidizing agent. Explodes on contact with alcohols, ammonia, antimony, antimony sulfide, arsenic, barium sulfide, calcium phosphide, carbon, carbon disulfide vapor, charcoal, cork, dicyanogen, ethers, hydrogen sulfide, mercury sulfide, nitrogen oxide, paper, phosphine, phosphorus, potassium, rubber, sulfur, tin sulfide, turpentine, and other oxidizable materials. Self-explodes. Incompatible with carbon, dicyanogen, diphenylmercury, nitrogen oxide, oxidizable materials, and potassium. Explosive reaction when heated above 50°C with many hydrocarbons, (e.g., butadiene, ethane,; ethylene, methane, propane).

DEL800 CAS:17496-59-2 ***HR: 3***
DICHLORINE TRIOXIDE
mf: Cl_2O_3 mw: 118.91

SAFETY PROFILE: An unstable explosive gas. When heated to decomposition it emits toxic fumes of Cl^-.

DEM000 CAS:5571-97-1 ***HR: 3***
DICHLORMETHAZANONE
mf: $C_{11}H_{11}Cl_2NO_3S$ mw: 308.19

SYNS: DICHLORMEZANONE ◇ 2-(3,4-DICHLOROPHENYL)-3-METHYL-4-METATHIAZANONE-1,1-DIOXIDE ◇ 2-(3,4-DICHLOROPHENYL)TETRAHYDRO-3-METHYL-4H-1,3-THIAZIN-4-ONE-1,1-DIOXIDE ◇ WIN 12267

TOXICITY DATA with REFERENCE
orl-rat LD50:1050 mg/kg TXAPA9 1,168,59
orl-mus LD50:840 mg/kg JPETAB 122,517,57
ipr-mus LD50:570 mg/kg TXAPA9 1,168,59
orl-cat LD50:300 mg/kg TXAPA9 1,168,59
ipr-cat LD50:400 mg/kg TXAPA9 1,168,59

SAFETY PROFILE: Poison by ingestion and intraperitoneal routes. When heated to decomposition it emits very toxic fumes of Cl^-, SO_x, and NO_x.

DEM200 CAS:79-02-7 ***HR: D***
2,2-DICHLOROACETALDEHYDE
mf: $C_2H_2Cl_2O$ mw: 112.94

PROP: Colorless liquid, polymerizes slowly to white solid. Bp: 88°, fp: −50°, flash p: 140°F (CC), d: 1.436 @ 25°/4°. Vap press: 50 mm @ 20°, vap d: 3.9.

SYNS: CHLORALDEHYDE ◇ DICHLOROACETALDEHYDE ◇ α,α-DICHLOROACETALDEHYDE

TOXICITY DATA with REFERENCE
mmo-sat 10 mg/plate CBINA8 30,9,80
mmo-omi 10 μL/plate CBINA8 30,9,80
sln-asn 10 mmol/L MUREAV 138,33,84

CONSENSUS REPORTS: Reported in EPA TSCA Inventory.

SAFETY PROFILE: Mutation data reported. Flammable when exposed to heat or flame. To fight fire, use water, foam, CO_2, dry chemical. When heated to decomposition it emits toxic fumes of Cl^-. See also ACETALDEHYDE and CHLORIDES.

DEM800 CAS:116-54-1 ***HR: 3***
DICHLOROACETIC ACID METHYL ESTER
DOT: UN 2299
mf: $C_3H_4Cl_2O_2$ mw: 142.97

PROP: Colorless liquid, ethereal odor. Bp: 143.0°, d: 1.3809 @ 19.2°/19.2°, vap d: 4.93.

SYNS: METHYL DICHLOROACETATE (DOT) ◇ METHYL DICHLOROETHANOATE

TOXICITY DATA with REFERENCE
ihl-cat LCLo:2000 ppm/30M TXAPA9 19,1,71

CONSENSUS REPORTS: Reported in EPA TSCA Inventory.

DOT Classification: Corrosive Material; Label: Corrosive; IMO: Poison B; Label: St. Andrews Cross.

SAFETY PROFILE: Poisonous irritant to the skin, eyes, and mucous membranes. Hydrolyzes upon contact with moisture to form a product corrosive to tissue. See also DICHLOROACETIC ACID and ESTERS. Dangerous; when heated to decomposition it emits highly toxic fumes of phosgene and Cl^-.

DEM825 CAS:4124-30-5 ***HR: 2***
DICHLOROACETIC ANHYDRIDE
mf: $C_4H_2Cl_4O_3$ mw: 239.86

TOXICITY DATA with REFERENCE
skn-rbt 10 mg/24H open MOD AMIHBC 4,119,51
eye-rbt 50 μg open SEV AMIHBC 4,119,51
orl-rat LD50:2820 mg/kg AMIHBC 4,119,51
skn-rbt LD50:470 mg/kg AMIHBC 4,119,51

CONSENSUS REPORTS: Reported in EPA TSCA Inventory.

SAFETY PROFILE: Moderately toxic by ingestion and skin contact. A skin and severe eye irritant. When heated to decomposition it emits toxic fumes of Cl^-. See also ANHYDRIDES and CHLORIDES.

DEN000 CAS:3018-12-0 ***HR: 3***
DICHLOROACETONITRILE
mf: C_2HCl_2N mw: 109.94

SYN: DICHLOROMETHYL CYANIDE

TOXICITY DATA with REFERENCE
mmo-sat 1 nmol/plate ENMUDM 5,447,83
dnd-hmn:lym 50 μmol/L FAATDF 6,447,86
orl-rat TDLo:825 mg/kg (female 7-21D post):REP TXCYAC 46,83,87
orl-rat TDLo:325 mg/kg (female 6-18D post):TER TJADAB 35,58A,87
orl-rat LD50:330 mg/kg EVHPAZ 69,183,86
orl-mus LD50:270 mg/kg EVHPAZ 69,183,86

CONSENSUS REPORTS: Cyanide and its compounds are on the Community Right-To-Know List. EPA Genetic Toxicology Program.

SAFETY PROFILE: Poison by ingestion. An experimental teratogen. Other experimental reproductive effects. Human mutation data reported. When heated to decomposition it emits toxic fumes of Cl^-, CN^-, and NO_x. See also NITRILES and CHLORIDES.

DEN200 CAS:2648-61-5 ***HR: 3***
2,2-DICHLOROACETOPHENONE
mf: $C_8H_6Cl_2O$ mw: 189.04

PROP: Crystals. Mp: 21°, bp: 247° (decomp), d: 1.34 @ 15°, vap d: 6.5.

SYNS: α,α-DICHLOROACETOPHENONE ◇ ω,ω-DICHLOROACETOPHENONE ◇ PHENACYLIDENE CHLORIDE

TOXICITY DATA with REFERENCE
ihl-mus LCLo:940 mg/m^3/10M NDRC** NDCrc-132,Aug,42
ivn-mus LD50:100 mg/kg CSLNX* NX#03021

CONSENSUS REPORTS: Reported in EPA TSCA Inventory.

SAFETY PROFILE: Poison by intravenous route. Moderately toxic by inhalation. When heated to decomposition it emits toxic fumes of Cl^-.

DEN300 CAS:75084-25-2 ***HR: D***
8-DICHLOROACETOXY-9-HYDROXY-8,9-DIHYDRO-AFLATOXIN B1
mf: $C_{19}H_{14}Cl_2O_8$ mw: 441.23

TOXICITY DATA with REFERENCE
mmo-sat 1 mg/L CRNGDP 1,79,80
dns-hmn:hla 50 μmol/L CRNGDP 1,79,80
dnd-mam:lym 3 g/L CRNGDP 1,79,80

SAFETY PROFILE: Human mutation data reported. See also AFLATOXIN B1.

DEN400 CAS:79-36-7 ***HR: 3***
DICHLOROACETYL CHLORIDE
DOT: UN 1765
mf: C_2HCl_3O mw: 147.38

PROP: Fuming liquid, acrid odor, misc in ether. D: 1.5315 @ 16°/4°, bp: 108°, flash p: 151°F, vap d: 5.8.

SYNS: CHLORURE de DICHLORACETYLE (FRENCH) ◇ DICHLORACETYL CHLORIDE ◇ α,α-DICHLOROACETYL CHLORIDE ◇ 2,2-DICHLOROACETYL CHLORIDE ◇ DICHLOROACETYL CHLORIDE (DOT) ◇ DICHLOROETHANOYL CHLORIDE

TOXICITY DATA with REFERENCE
skn-rbt 100 μg/24H open MOD AMIHBC 4,119,51
eye-rbt 50 μg open SEV AMIHBC 4,119,51
scu-mus TDLo:2 mg/kg/80W-I:ETA CNREA8 43,159,83
orl-rat LD50:2460 mg/kg AMIHBC 4,119,51
ihl-rat LCLo:2000 ppm/4H AMIHBC 4,119,51
skn-rbt LD50:650 mg/kg AMIHBC 4,119,51

CONSENSUS REPORTS: Reported in EPA TSCA Inventory.

DOT Classification: Corrosive Material; Label: Corrosive.

SAFETY PROFILE: Questionable carcinogen with experimental tumorigenic data. Moderately toxic by ingestion, inhalation, and skin contact. Corrosive to the skin, eyes, and mucous membranes. Flammable when exposed to heat or flame. When heated to decomposition it emits toxic fumes of Cl^-. See also CHLORIDES.

DEN600 CAS:7572-29-4 ***HR: 3***
DICHLOROACETYLENE
mf: C_2Cl_2 mw: 94.92

SYN: DICHLOROETHYNE

TOXICITY DATA with REFERENCE
mmo-sat 4000 ppm MUREAV 117,21,83
mma-sat 4000 ppm MUREAV 117,21,83
ihl-rat TCLo:14 ppm/6H/77W-I:CAR CRNGDP 5,1411,84
ihl-mus TCLo:2 ppm/24H/77W-I:CAR CRNGDP 5,1411,84
ihl-mus LC50:19 ppm/6H FCTXAV 13,511,75
ihl-rbt LCLo:307 ppm/1H FCTXAV 16,227,78

CONSENSUS REPORTS: IARC Cancer Review: Group 3 IMEMDT 7,56,87; Animal Limited Evidence IMEMDT 39,369,86

OSHA PEL: CL 0.1 ppm
ACGIH TLV: CL 0.1 ppm
DFG MAK: Animal Carcinogen, Suspected Human Carcinogen.
DOT Classification: Forbidden.

SAFETY PROFILE: Confirmed carcinogen with experimental carcinogenic data. Poison by inhalation. Central nervous system effects. Can be formed by thermal decomposition (>70°) from trichloroethylene. Symptoms include a disabling nausea and intense jaw pain. Strong explosive when shocked or exposed to heat or air. Can react vigorously with oxidizing materials. When heated to decomposition or on contact with acid or acid fumes it emits highly toxic fumes of Cl^-. See also ACETYLENE COMPOUNDS and CHLORINATED HYDROCARBONS, ALIPHATIC.

DEN800 ***HR: 3***
DICHLOROACETYLENE mixed with ETHER (1:9)

SYN: DCA-ETHER (1:9)

TOXICITY DATA with REFERENCE
ihl-rat LC50:219 ppm/4H TXAPA9 18,168,71
ihl-gpg LC50:52 ppm/4H TXAPA9 18,168,71

SAFETY PROFILE: Poison by inhalation. A very dangerous fire and explosion hazard. See also DICHLOROACETYLENE and ETHERS. When heated to decomposition it emits toxic fumes of Cl^-.

DEO200 CAS:68594-17-2 ***HR: 3***
1-DICHLOROAMINOTETRAZOLE
mf: $CHCl_2N_5$ mw: 153.96

SAFETY PROFILE: A sensitive explosive. Upon decomposition it emits toxic fumes of Cl^- and NO_x.

DEO300 CAS:95-76-1 ***HR: 3***
3,4-DICHLOROANILINE
mf: $C_6H_5Cl_2N$ mw: 162.02

PROP: Crystals. Mp: 71-72°, bp: 272°. Practically insol in water; very sol in alcohol, ether; sltly sol in benzene.

SYNS: 1-AMINO-3,4-DICHLOROBENZENE ◇ DCA ◇ 3,4-DCA ◇ 3,4-DICHLORANILIN ◇ 3,4-DICHLORANILINE ◇ 4,5-DICHLOROANILINE ◇ 3,4-DICHLOROBENZENAMINE (9CI)

TOXICITY DATA with REFERENCE
skn-rbt 500 mg/24H SEV 28ZPAK -,96,72
eye-rbt 250 μg/24H SEV 28ZPAK -,96,72
mmo-asn 200 mg/L CJMIAZ 16,369,70
orl-rat LD50:648 mg/kg 28ZPAK -,96,72
ipr-rat LD50:280 mg/kg LPPTAK 27,306,79
orl-mus LD50:740 mg/kg GTPZAB 13(5),29,69
ipr-mus LD50:310 mg/kg LPPTAK 27,306,79
skn-cat LD50:700 mg/kg GTPZAB 13(5),29,69
orl-bwd LD50:237 mg/kg AECTCV 12,355,83

CONSENSUS REPORTS: EPA Genetic Toxicology Program. Reported in EPA TSCA Inventory.

SAFETY PROFILE: Poison by ingestion and intraperitoneal routes. Moderately toxic by skin contact. A severe eye and skin irritant. Mutation data reported. When heated to decomposition it emits toxic fumes of Cl^- and NO_x. See also ANILINE DYES and CHLORIDES.

DEO400 CAS:95-82-9 ***HR: 3***
2,5-DICHLOROANILINE
mf: $C_6H_5Cl_2N$ mw: 162.02

SYNS: AMARTHOL FAST SCARLETT GG BASE ◇ AZOBASE DCA ◇ AZOEN FAST SCARLET 2G BASE ◇ AZOFIX SCARLET GG SALT ◇ C.I. 37010 ◇ C.I. AZOIC DIAZO COMPONENT 3 ◇ DEVOL SCARLET A (FREE BASE) ◇ 2,5-DICHLOROANILIN (CZECH) ◇ 2,5-DICHLOROBENZENEAMINE ◇ DURGASOL SCARLET GG SALT ◇ FAST RED SGG BASE ◇ HILTONIL FAST SCARLET 2G BASE ◇ HILTOSAL FAST SCARLET 2G SALT ◇ HINDAMINE SCARLET GG ◇ KAKO SCARLET GG SALT ◇ KAMBAMINE SCARLET GG BASE ◇ KAYAKU SCARLET GG BASE ◇ LAKE SCARLET GG BASE ◇ MEISEI SCARLET GG SALT ◇ MITUSI SCARLET GG BASE ◇ NAPHTHANIL SCARLET 2G BASE ◇ NAPHTOELAN MITSUI SCARLET GG SALT ◇ SANYO FAST SCARLET GG BASE ◇ SCARLET BASE CIBA I ◇ SPECTROLENE SCARLET 2G ◇ SYMULON SCARLET 2G SALT

TOXICITY DATA with REFERENCE
orl-rat LD50:2900 mg/kg MarJV# 29MAR77
ivn-mus LD50:56 mg/kg CSLNX* NX#00202

CONSENSUS REPORTS: Reported in EPA TSCA Inventory.

SAFETY PROFILE: Poison by intravenous route. Moderately toxic by ingestion. Explodes spontaneously. When heated to decomposition it emits highly toxic fumes of Cl^- and NO_x. See also ANILINE and CHLORIDES.

DEO500 CAS:70278-00-1 ***HR: 3***
N,N-DICHLOROANILINE
mf: $C_6H_5CL_2N$ mw: 162.02

SAFETY PROFILE: A poison. An oil which explodes spontaneously at room temperature. When heated to decomposition it emits toxic fumes of Cl^- and NO_x. See also ANILINE and CHLORIDES.

DEO600 CAS:15307-79-6 ***HR: 3***
(o-((2,6-DICHLOROANILINO)PHENYL)ACETIC ACID SODIUM SALT
mf: $C_{14}H_{10}Cl_2NO_2 \cdot Na$ mw: 318.14

SYNS: (o-(2,6-DICHLOROANILINO)PHENYL)ACETIC ACID MONOSODIUM SALT ◇ 2-((2,6-DICHLOROPHENYL)AMINO)BENZENEACETIC ACID MONOSODIUM SALT ◇ DICHRONIC ◇ DICLOFENAC SODIUM ◇ DICLOPHENAC SODIUM ◇ GP 45840 ◇ KRIPLEX ◇ NERIODIN ◇ PROPHENATIN ◇ SODIUM (o-(2,6-DICHLOROANILINO)PHENYL)ACETATE ◇ SODIUM (o-((2,6-DICHLOROPHENYL)AMINO)PHENYL)ACETATE ◇ TSUDOHMIN ◇ VALETAN ◇ VOLTAREN ◇ VOLTAROL

TOXICITY DATA with REFERENCE
orl-rat TDLo:24 mg/kg (female 9-14D post):REP KSRNAM 6,1673,72
orl-rat TDLo:6 mg/kg (9-14D preg):TER KSRNAM 6,1673,72
orl-wmn TDLo:180 mg/kg/13W-I BMJOAE 295,182,87
orl-rat LD50:53 mg/kg TOIZAG 28,99,81
ipr-rat LD50:25 mg/kg NIIRDN 6,311,82
scu-rat LD50:83 mg/kg IYKEDH 5,106,74
ivn-rat LD50:117 mg/kg IYKEDH 5,106,74
orl-mus LD50:125 mg/kg ARZNAD 34,280,84
ipr-mus LD50:130 mg/kg IYKEDH 5,106,74
scu-mus LD50:390 mg/kg NIIRDN 6,311,82
ivn-mus LD50:116 mg/kg IYKEDH 5,106,74
orl-dog LD50:59 mg/kg KSRNAM 6,1521,72

SAFETY PROFILE: Poison by ingestion, intravenous, intraperitoneal, and subcutaneous routes. Experimental teratogenic and reproductive effects. An anti-inflammatory agent. When heated to decomposition it emits very toxic fumes of Cl^-, Na_2O, and NO_x.

DEO700 CAS:82-46-2 ***HR: 1***
1,5-DICHLORO-9,10-ANTHRAQUINONE
mf: $C_{14}H_6Cl_2O_2$ mw: 277.10

PROP: Mp: 245-247°.

SYNS: 9,10-ANTHRACENEDIONE, 1,5-DICHLORO- ◇ 1,5-DICHLORANTHRACHINON ◇ 1,5-DICHLOROANTHRAQUINONE

TOXICITY DATA with REFERENCE
eye-rbt 500 mg/24H MLD 85JCAE-,566,86

CONSENSUS REPORTS: Reported in EPA TSCA Inventory.

SAFETY PROFILE: An eye irritant. When heated to decomposition it emits toxic fumes of Cl^-.

DEO750 CAS:82-43-9 ***HR: 1***
1,8-DICHLORO-9,10-ANTHRAQUINONE
mf: $C_{14}H_6Cl_2O_2$ mw: 277.10

PROP: Mp: 201.5-203°.

SYNS: 9,10-ANTHRACENEDIONE, 1,8-DICHLORO- ◇ 1,8-DICHLORANTHRACHINON ◇ 1,8-DICHLOROANTHRAQUINONE

TOXICITY DATA with REFERENCE
eye-rbt 500 mg/24H MLD 85JCAE-,566,86

CONSENSUS REPORTS: Reported in EPA TSCA Inventory.

SAFETY PROFILE: An eye irritant. When heated to decomposition it emits toxic fumes of Cl^-.

DEP400 CAS:63834-20-8 ***HR: 3***
2-DICHLOROARSINOPHENOXATHIIN
mf: $C_{12}H_7AsCl_2OS$ mw: 345.07

SYN: TL 472

TOXICITY DATA with REFERENCE
orl-rat LDLo:250 mg/kg NCNSA6 5,13,53
ihl-mus LCLo:400 mg/m³/10M NDRC** NDCrc-132,Dec,42

CONSENSUS REPORTS: Arsenic and its compounds are on the Community Right-To-Know List.

OSHA PEL: TWA 0.5 mg(As)/m³

SAFETY PROFILE: Poison by ingestion and inhalation. See also ARSENIC COMPOUNDS. When heated to decomposition it emits very toxic fumes of As, Cl^-, and SO_x.

DEP600 CAS:95-50-1 ***HR: 3***
o-DICHLOROBENZENE
DOT: UN 1591
mf: $C_6H_4Cl_2$ mw: 147.00

PROP: Clear liquid. Mp: −17.5°, bp: 180-183°, fp: −22°, flash p: 151°F, d: 1.307 @ 20°/20°, vap d: 5.05, autoign temp: 1198°F, lel: 2.2%, uel: 9.2%.

SYNS: CHLOROBEN ◇ CHLORODEN ◇ CLOROBEN ◇ DCB ◇ o-DICHLORBENZENE ◇ o-DICHLOR BENZOL ◇ 1,2-DICHLOROBENZENE (MAK) ◇ DICHLOROBENZENE, ORTHO, liquid (DOT) ◇ DILANTIN DB ◇ DILATIN DB ◇ DIZENE ◇ DOWTHERM E ◇ NCI-C54944 ◇ ODB ◇ ODCB ◇ ORTHODICHLOROBENZENE ◇ ORTHODICHLOROBENZOL ◇ RCRA WASTE NUMBER U070 ◇ SPECIAL TERMITE FLUID ◇ TERMITKIL

TOXICITY DATA with REFERENCE
eye-rbt 100 mg/30S rns MLD AMIHAB 17,180,58
spm-rat-ipr 250 mg/kg JACTDZ 4(2),224,85
ipr-rat TDLo:50 mg/kg (1D male):REP JACTDZ 4(1),224,85
ihl-rat TCLo:200 ppm/6H (6-15D preg):TER FAATDF 5,190,85
orl-rat LD50:500 mg/kg WRPCA2 7,135,68
ihl-rat LCLo:821 ppm/7H AMIHAB 17,180,58
ipr-rat LD50:840 mg/kg MEPAAX 20,519,69
orl-mus LD50:4386 g/kg YKYUA6 32,471,81
ivn-mus LDLo:400 mg/kg JPBAA7 44,281,37
orl-rbt LD50:500 mg/kg 85ARAE 3,32,76/77
ivn-rbt LDLo:250 mg/kg JPBAA7 44,281,37
orl-gpg LDLo:2000 mg/kg 14CYAT 2,1336,63
ihl-gpg LCLo:800 ppm/24H JPBAA7 44,281,37

CONSENSUS REPORTS: IARC Cancer Review: Group 3 IMEMDT 7,192,87; Animal Inadequate Evidence IMEMDT 7,231,74, IMEMDT 29,213,82; Human Inadequate Evidence IMEMDT 7,231,74, IMEMDT 29,213,82. Reported in EPA TSCA Inventory. Community Right-To-Know List.

OSHA PEL: CL 50 ppm
ACGIH TLV: CL 50 ppm; (Proposed: TWA 25 ppm, STEL 50 ppm)
DFG MAK: 50 ppm (300 mg/m³)
DOT Classification: ORM-A; Label: None; IMO: Poison B; Label: St. Andrews Cross.

SAFETY PROFILE: Poison by ingestion and intravenous routes. Moderately toxic by inhalation and intraperitoneal routes. An experimental teratogen. Other experimental reproductive effects. An eye, skin, and mucous membrane irritant. Causes liver and kidney injury. Questionable carcinogen. Mutation data reported. A pesticide. Flammable when exposed to heat or flame. Can react vigorously with oxidizing materials. To fight fire, use water, foam, CO_2, or dry chemical. Slow reaction with aluminum may lead to explosion during storage in a sealed aluminum container. When heated to decomposition it emits toxic fumes of Cl^-. See also BENZENE CHLORIDE and CHLORINATED HYDROCARBONS, AROMATIC.

DEP699 CAS:541-73-1 ***HR: 3***
m-DICHLOROBENZENE
mf: $C_6H_4Cl_2$ mw: 147.00

SYN: 1,3-DICHLOROBENZENE

TOXICITY DATA with REFERENCE
mrc-smc 5 ppm NTIS** PB84-138973

CONSENSUS REPORTS: Reported in EPA TSCA Inventory. Community Right-To-Know List.

SAFETY PROFILE: A poison. Mutation data reported. When heated to decomposition it emits toxic fumes of Cl^-. See also o-DICHLOROBENZENE and p-DICHLOROBENZENE.

DEP800 CAS:106-46-7 ***HR: 3***
p-DICHLOROBENZENE
DOT: UN 1592
mf: $C_6H_4Cl_2$ mw: 147.00

PROP: White crystals, penetrating odor. Mp: 53°, bp: 173.4°, flash p: 150°F (CC), d: 1.4581 @ 20.5°/4°, vap press: 10 mm @ 54.8°, vap d: 5.08.

SYNS: p-CHLOROPHENYL CHLORIDE ◇ p-DICHLOORBENZEEN (DUTCH) ◇ 1,4-DICHLOORBENZEEN (DUTCH) ◇ p-DICHLORBENZOL (GERMAN) ◇ 1,4-DICHLOR-BENZOL (GERMAN) ◇ DI-CHLORICIDE ◇ 1,4-DICHLOROBENZENE (MAK) ◇ p-DICLOROBENZENE (ITALIAN) ◇ DICHLOROBENZENE, PARA, solid (DOT) ◇ 1,4-DICLOROBENZENE (ITALIAN) ◇ p-DICHLOROBENZOL ◇ EVOLA ◇ NCI-C54955 ◇ PARACIDE ◇ PARA CRYSTALS ◇ PARADI ◇ PARADICHLORBENZOL (GERMAN) ◇ PARADICHLOROBENZENE ◇ PARADICHLOROBENZOL ◇ PARADOW ◇ PARAMOTH ◇ PARANUGGETS ◇ PARAZENE ◇ PDB ◇ PDCB ◇ PERSIA-PERAZOL ◇ RCRA WASTE NUMBER U070 ◇ RCRA WASTE NUMBER U071 ◇ RCRA WASTE NUMBER U072 ◇ SANTOCHLOR

TOXICITY DATA with REFERENCE
eye-hmn 80 ppm AMIHAB 14,138,56
mmo-asn 200 mg/L CJMIAZ 16,369,70
orl-rat TDLo:10 g/kg (female 6-15D post):TER BECTA6 37,164,86
orl-rat TDLo:155 g/kg/2Y-I:CAR NTPTR* NTP-TR-319,87
orl-mus TDLo:155 g/kg/2Y-I:CAR NTPTR* NTP-TR-319,87
orl-hmn TDLo:300 mg/kg:EYE,PUL,GIT PCOC** -,851,66
orl-hmn LDLo:857 mg/kg 34ZIAG-,210,69
unr-hmn LDLo:357 mg/kg YKYUA6 31,1499,80
unr-man LDLo:221 mg/kg 85DCAI 2,73,70
orl-rat LD50:500 mg/kg WRPCA2 9,119,70
ipr-rat LD50:2562 mg/kg JAPMA8 38,124,49
orl-mus LD50:2950 mg/kg GUCHAZ 6,183,73
ipr-mus LD50:2 g/kg MUTAEX 2,111,87
scu-mus LD50:5145 mg/kg TOIZAG 20,772,73
orl-rbt LD50:2830 mg/kg YKYUA6 29,453,78
orl-gpg LDLo:2800 mg/kg 14CYAT 2,1338,63

CONSENSUS REPORTS: NTP Fifth Annual Report on Carcinogens. IARC Cancer Review: Group 2B IMEMDT 7,192,87; Animal Inadequate Evidence IMEMDT 7,231,74; IMEMDT 29,213,82. Human Inadequate Evidence IMEMDT 7,231,74; Reported in EPA TSCA Inventory. EPA Genetic Toxicology Program. Community Right-To-Know List.

OSHA PEL: (Transitional: TWA 75 ppm) TWA 75 ppm; STEL 110 ppm
ACGIH TLV: TWA 75 ppm; STEL 110 ppm; (Proposed: 10 ppm; Suspected Human Carcinogen)
DFG MAK: 75 ppm (450 mg/m^3)
DOT Classification: ORM-A; Label: None; IMO: Poison B; Label: St. Andrews Cross.

SAFETY PROFILE: Confirmed carcinogen with experimental carcinogenic data. An experimental teratogen. A human poison by an unspecified route. Moderately toxic to humans by ingestion. Moderately toxic experimentally by ingestion, subcutaneous, and intraperitoneal routes. Mildly toxic by subcutaneous route. Other experimental reproductive effects. Human systemic effects by ingestion: unspecified changes in the eyes, lungs, thorax and respiration, and decreased motility or constipation. Can cause liver injury in humans. A human eye irritant. Mutation data reported. A fumigant. Flammable when exposed to heat, flame, or oxidizers. Dangerous; can react vigorously with oxidizing materials. To fight fire, use water, foam, CO_2, dry chemical. When heated to decomposition it emits toxic fumes of Cl^-. See also CHLORINATED HYDROCARBONS, AROMATIC.

DEQ000 CAS:5836-73-7 ***HR: 3***
3,4-DICHLOROBENZENE DIAZOTHIOUREA
mf: $C_7H_6Cl_2N_4S$ mw: 249.13

SYNS: CHLOROPROMURITE ◇ (3,4-DICHLOOR-FENYL-AZO)-THIOUREUM (DUTCH) ◇ 1-(3',4'-DICHLOROBENZENEDIAZOL)-2-THIOUREA ◇ 3,4-DICHLOROBENZENE DIAZOTHIOCARBAMID ◇ 3,4-DICHLOROPHENYLAZOTHIOUREA ◇ 3,4-DICHLOROPHENYLAZOTHIOUREE (FRENCH) ◇ (3,4-DICHLOR-PHENYL-AZO)-THIOHARNSTOFF (GERMAN) ◇ (3,4-DICLORO-FENIL-AZO)-TIOUREA (ITALIAN) ◇ MURITAN ◇ PROMURIT ◇ PROMURITE

TOXICITY DATA with REFERENCE
orl-rat LD50:280 μg/kg FEPRA7 8,282,49
ipr-rat LD50:200 μg/kg FEPRA7 8,282,49
orl-mus LD50:1 mg/kg 28ZEAL 5,188,76
ipr-mus LD50:1350 μg/kg FEPRA7 8,282,49
orl-dog LD50:1 mg/kg 28ZEAL 5,188,76
ipr-rbt LD50:1750 μg/kg FEPRA7 8,282,49
ipr-gpg LD50:1900 μg/kg FEPRA7 8,282,49

SAFETY PROFILE: A deadly poison by ingestion, and intraperitoneal routes. When heated to decomposition it emits very toxic fumes of Cl^-, NO_x, and SO_x.

DEQ200 CAS:120-97-8 ***HR: 3***
4,5-DICHLORO-m-BENZENEDISULFONAMIDE
mf: $C_6H_6Cl_2N_2O_4S_2$ mw: 305.16

SYNS: CB 8000 ◇ DARANIDE ◇ DASANIDE ◇ DICHLOFENAMIDE ◇ 4,5-DICHLORO-1,3-BENZENEDISULFONAMIDE ◇ 4,5-DICHLORO-1,3-DISULFAMOYLBENZENE ◇ DICHLOROPHENAMIDE ◇ 3,4-DICHLORO-5-SULFAMYLBENZENESULFONAMIDE ◇ DICHLORPHENAMIDE ◇ 1,3-DISULFAMYL-4,5-DICHLOROBENZENE ◇ ORATROL

TOXICITY DATA with REFERENCE
scu-mus TDLo:96 mg/kg (female 15D post):TER DEBIAO 27,395,72
orl-rat LD50:2600 mg/kg 29ZVAB -,41,69
orl-mus LD50:1710 mg/kg 29ZVAB -,41,69
ipr-mus LD50:304 mg/kg THERAP 19,1423,64
ivn-mus LD50:643 mg/kg 29ZVAB -,41,69
ivn-dog LD50:200 mg/kg 29ZVAB -,41,69

SAFETY PROFILE: Poison by intravenous and intraperitoneal routes. Moderately toxic by ingestion. An experimental teratogen. When heated to decomposition it emits very toxic fumes of Cl^-, NO_x, and SO_x.

DEQ400 CAS:84-68-4 ***HR: D***
2,2'-DICHLOROBENZIDINE
mf: $C_{12}H_{10}Cl_2N_2$ mw: 253.14

PROP: Needle-like crystals. Mp: 165°, vap d: 8.73. Insol in water; sol in alc and ether.

SYN: 2,2'-DICHLORO-(1,1'-BIPHENYL)-4,4'-DIAMINE

TOXICITY DATA with REFERENCE
dns-hmn:hla 100 nmol/L CNREA8 38,2621,78

CONSENSUS REPORTS: EPA Genetic Toxicology Program.

SAFETY PROFILE: Human mutation data reported. An allergen. When heated to decomposition it emits very toxic fumes of Cl^- and NO_x.

DEQ600 CAS:91-94-1 ***HR: 3***
3',3'-DICHLOROBENZIDINE
mf: $C_{12}H_{10}Cl_2N_2$ mw: 253.14

PROP: Crystals. Mp: 133°. Insol in water; sol in alc, benzene, and glacial acetic acid.

SYNS: C.I. 23060 ◇ CURITHANE C126 ◇ DCB ◇ 4,4'-DIAMINO-3,3'-DICHLOROBIPHENYL ◇ 4,4'-DIAMINO-3,3'-DICHLORODIPHENYL ◇ 3,3'-DICHLORBENZIDIN (CZECH) ◇ 3,3'-DICHLOROBENZIDINA (SPANISH) ◇ DICHLOROBENZIDINE ◇ 3,3'-DICHLOROBENZIDENE ◇ o,o'-DICHLOROBENZIDINE ◇ DICHLOROBENZIDINE BASE ◇ 3,3'-DICHLORO-4,4'-BIPHENYLDIAMINE ◇ 3,3'-DICHLOROBIPHENYL-4,4'-DIAMINE ◇ 3,3'-DICHLORO-4,4'-DIAMINOBIPHENYL ◇ 3,3'-DICHLORO-4,4'-DIAMINO(1,1-BIPHENYL) ◇ RCRA WASTE NUMBER U073

TOXICITY DATA with REFERENCE
mma-sat 5 μg/plate ENMUDM 6,145,84
dns-hmn:hla 100 nmol/L CNREA8 38,2621,78
bfa-rat/sat 40 mg/kg SAIGBL 23,426,81
otr-ham:kdy 80 μg/L BJCAAI 37,873,78
dnd-mam:lym 25500 nmol/L CBINA8 38,369,82
orl-rat TDLo:17 g/kg/50W-C:CAR TXAPA9 31,159,75
scu-rat TDLo:7 g/kg/43W-I:ETA VOONAW 5(5),524,59
scu-mus TDLo:320 mg/kg (15-21D preg):NEO,TER BEXBAN 78,1402,75
orl-rat TD:21 g/kg/50W-C:CAR TXAPA9 31,159,75
orl-rat LD50:5250 mg/kg 28ZPAK -,98,72

CONSENSUS REPORTS: NTP Fifth Annual Report on Carcinogens. IARC Cancer Review: Group 2B IMEMDT 7,193,87; Human Inadequate Evidence IMEMDT 29, 239,82; Animal Sufficient Evidence IMEMDT 29,239, 82; IMEMDT 4,49,74. Reported in EPA TSCA Inventory. Community Right-To-Know List. EPA Genetic Toxicology Program.

OSHA PEL: Cancer Suspect Agent
ACGIH TLV: Suspected Human Carcinogen.
DFG TRK: 0.1 mg/m^3, Animal Carcinogen, Suspected Human Carcinogen.
NIOSH REL: (Benzidine-based Dye) Reduce to lowest feasible level.

SAFETY PROFILE: Confirmed carcinogen with experimental carcinogenic and tumorigenic data. Mildly toxic by ingestion. An experimental teratogen. Human mutation data reported. When heated to decomposition it emits very toxic fumes of Cl^- and NO_x.

DEQ800 CAS:612-83-9 ***HR: 3***
3,3'-DICHLOROBENZIDINE DIHYDROCHLORIDE
mf: $C_{12}H_{10}Cl_2N_2 \cdot 2ClH$ mw: 326.06

SYN: 3,3'-DICHLORO-(1,1'-BIPHENYL)-4,4'-DIAMINE DIHYDROCHLORIDE

TOXICITY DATA with REFERENCE
mmo-sat 10 μg/plate ENMUDM 5(Suppl 1),3,83
mma-sat 1 μg/plate ENMUDM 5(Suppl 1),3,83
orl-rat LD50:3820 mg/kg 34ZIAG -,211,69

CONSENSUS REPORTS: NTP Fifth Annual Report on Carcinogens. Reported in EPA TSCA Inventory.

OSHA PEL: Cancer Suspect Agent

SAFETY PROFILE: Confirmed carcinogen. Moderately toxic by ingestion. Mutation data reported. When heated to decomposition it emits very toxic fumes of Cl^- and NO_x.

DER000 CAS:510-15-6 ***HR: 3***
4,4'-DICHLOROBENZILIC ACID ETHYL ESTER
mf: $C_{16}H_{14}Cl_2O_3$ mw: 325.20

PROP: Viscous liquid, sometimes yellow, sltly sol in water. Bp: 156-158°, vap press: 2.2×10^{-6} mm @ 20°.

SYNS: ACAR ◇ ACARABEN 4E ◇ AKAR ◇ BENZILAN ◇ BENZ-o-CHLOR ◇ CHLORBENZILATE ◇ CHLOROBENZYLATE ◇ COMPOUND 338 ◇ 4,4'-DICHLORBENZILSAEUREAETHYLESTER (GERMAN) ◇ 4,4'-DICHLOROBENZILATE ◇ ENT 18,596 ◇ ETHYL 4-CHLORO-α-(4-CHLOROPHENYL)-α-HYDROXYBENZENEACETATE ◇ ETHYL-p,p'-DICHLOROBENZILATE ◇ ETHYL-4,4'-DICHLOROBENZILATE ◇ ETHYL-4,4'-DICHLORODIPHENYL GLYCOLLATE

◇ ETHYL-4,4'-DICHLOROPHENYL GLYCOLLATE ◇ ETHYL ESTER of 4,4'-DICHLOROBENZILIC ACID ◇ ETHYL-2-HYDROXY-2,2-BIS(4-CHLOROPHENYL)ACETATE ◇ FOLBEX ◇ FOLBEX SMOKE-STRIPS ◇ G 338 ◇ G 23992 ◇ GEIGY 338 ◇ KOP MITE ◇ NCI-C00408 ◇ NCI-C60413 ◇ RCRA WASTE NUMBER U038

TOXICITY DATA with REFERENCE
skn-rbt 125 mg open MLD CIGET* -,-,77
eye-rbt 25 mg MOD CIGET* -,-,77
orl-rat TDLo:5475 mg/kg/2Y-C:CAR CTOXAO 16,67,80
orl-mus TDLo:71 g/kg/82W-C:CAR JNCIAM 42,1101,69
orl-rat TD:72 g/kg/78W-C:ETA NCITR* NCI-CG-TR-75,78
orl-rat TD:1752 mg/kg/2Y-C:NEO CTOXAO 16,67,80
orl-rat LD50:700 mg/kg WRPCA2 9,119,70
orl-mus LD50:729 mg/kg GUCHAZ 6,106,73
orl-ham LD50:700 mg/kg TXAPA9 48,A192,79

CONSENSUS REPORTS: IARC Cancer Review: Group 3 IMEMDT 7,56,87; Animal Limited Evidence IMEMDT 30,73,83; Animal Sufficient Evidence IMEMDT 5,75,74. NCI Carcinogenesis Bioassay Completed; Results Positive: mouse NCITR* NCI-CG-TR-75,78. NCI Carcinogenesis Bioassay Completed; Results Indefinite: rat NCITR* NCI-CG-TR-75,78. Community Right-To-Know List. Reported in EPA TSCA Inventory.

SAFETY PROFILE: Suspected carcinogen with experimental carcinogenic, neoplastigenic, and tumorigenic data. Moderately toxic by ingestion. A skin and eye irritant. A pesticide. When heated to decomposition it emits toxic fumes of Cl^-.

DER400 CAS:50-79-3 ***HR: 2***
2,5-DICHLOROBENZOIC ACID
mf: $C_7H_4Cl_2O_2$ mw: 191.01

TOXICITY DATA with REFERENCE
scu-mus LD50:1200 mg/kg BCPCA6 13,1538,64

CONSENSUS REPORTS: Reported in EPA TSCA Inventory.

SAFETY PROFILE: Moderately toxic by subcutaneous route. When heated to decomposition it emits toxic fumes of Cl^-.

DER600 CAS:51-44-5 ***HR: 3***
3,4-DICHLOROBENZOIC ACID
mf: $C_7H_4Cl_2O_2$ mw: 191.01

SYNS: SYNSTIGMINE ◇ SYNTOSTIGMIN ◇ VAGOSTIGMIN

TOXICITY DATA with REFERENCE
scu-mus LD50:400 mg/kg BCPCA6 13,1538,64

CONSENSUS REPORTS: Reported in EPA TSCA Inventory.

SAFETY PROFILE: Poison by subcutaneous route. When heated to decomposition it emits toxic fumes of Cl^-.

DER800 CAS:1194-65-6 ***HR: 3***
2,6-DICHLOROBENZONITRILE
mf: $C_7H_3Cl_2N$ mw: 172.01

PROP: White solid. Mp: 144°. Almost insol in water; sol in organic solvents.

SYNS: CARSORON ◇ CASORON 133 ◇ CODE H 133 ◇ 2,6-DBN ◇ DBN (the herbicide) ◇ DCB ◇ DECABANE ◇ DICHLOBENIL (DOT) ◇ 2,6-DICHLORBENZONITRIL (GERMAN) ◇ DU-SPREX ◇ H 133 ◇ H 1313 ◇ NIA 5996 ◇ NIAGARA 5006 ◇ NIAGARA 5,996

TOXICITY DATA with REFERENCE
ipr-mus TDLo:260 μg/kg/39D-I:ETA AAATAP 122,107,79
orl-rat LD50:2710 mg/kg RREVAH 10,97,65
orl-mus LD50:2056 mg/kg 28ZEAL 5,73,76
skn-rbt LD50:1350 mg/kg GUCHAZ 6,177,73
orl-gpg LD50:681 mg/kg PCOC** -,337,66

CONSENSUS REPORTS: Cyanide and its compounds are on the Community Right-To-Know List. EPA Genetic Toxicology Program.

SAFETY PROFILE: Moderately toxic by ingestion and skin contact. Questionable carcinogen with experimental tumorigenic data. Does not hydrolyze to HCN in body. Less toxic than most aliphatic nitriles. When heated to decomposition it emits toxic fumes of Cl^-, CN^-, and NO_x. See also BENZONITRILE, CHLORIDES, and NITRILES.

DES000 CAS:90-98-2 ***HR: 3***
p,p'-DICHLOROBENZOPHENONE
mf: $C_{13}H_8Cl_2O$ mw: 251.11

SYNS: DBP ◇ DCB ◇ 4,4'-DICHLOROBENZOPHENONE ◇ USAF DO-4

TOXICITY DATA with REFERENCE
ipr-mus LD50:200 mg/kg NTIS** AD277-689

CONSENSUS REPORTS: Reported in EPA TSCA Inventory.

SAFETY PROFILE: Poison by intraperitoneal route. When heated to decomposition it emits toxic fumes of Cl^-.

DES400 CAS:697-91-6 ***HR: 2***
2,6-DICHLORO-p-BENZOQUINONE
mf: $C_6H_2Cl_2O_2$ mw: 176.98

SYNS: 2,6-DICHLOQUINONE ◇ 2,6-DICHLORO-2,5-CYCLOHEXADIENE-1,4-DIONE ◇ 2,6-DICHLORO-p-QUINONE

TOXICITY DATA with REFERENCE
orl-rat LDLo:500 mg/kg NCNSA6 5,20,53

CONSENSUS REPORTS: Reported in EPA TSCA Inventory.

SAFETY PROFILE: Moderately toxic by ingestion. Mutation data reported. When heated to decomposition it emits very toxic fumes of Cl^-.

DET000 CAS:12041-76-8 ***HR: 3***
DICHLOROBENZYL ALCOHOL
mf: $C_7H_6Cl_2O$ mw: 177.03

PROP: Crystals. Vap d: 6.1.

SYNS: BAYER 4245 ◇ RAPIDOSEPT

TOXICITY DATA with REFERENCE
orl-rat LD50:810 mg/kg AIHAAP 30,470,69
skn-rbt LD50:400 mg/kg AIHAAP 30,470,69

SAFETY PROFILE: Poison by skin contact. Moderately toxic by ingestion. An insecticide. When heated to decomposition it emits toxic fumes of Cl^-. See also ALCOHOLS and CHLORIDES.

DET125 CAS:67230-61-9 ***HR: D***
7-((3,4-DICHLOROBENZYL)AMINOACTINOMYCIN D
mf: $C_{69}H_{91}Cl_2N_{13}O_{16}$ mw: 1429.63

SYN: 7-(((3,4-DICHLOROPHENYL)METHYL)AMINO)-ACTINOMYCIN D

TOXICITY DATA with REFERENCE
dni-mus:lym 418 nmol/L JMCMAR 24,1052,81
oms-mus:lym 96 nmol/L JMCMAR 24,1052,81
dnd-mam:lym 3 μmol/L JMCMAR 26,448,83

SAFETY PROFILE: Mutation data reported. When heated to decomposition it emits toxic fumes of Cl^- and NO_x.

DET400 CAS:1966-58-1 ***HR: 2***
3,4-DICHLOROBENZYL METHYLCARBAMATE
mf: $C_9H_9Cl_2NO_2$ mw: 234.09

SYNS: 3,4-DICHLOROBENZENEMETHANOLMETHYLCARBAMATE ◇ ROWMATE ◇ SIRMATE ◇ UC 22,463

TOXICITY DATA with REFERENCE
orl-rat LD50:1870 mg/kg WRPCA2 9,119,70
orl-mus LD50:1620 mg/kg 31ZOAD 1,141,68

SAFETY PROFILE: Moderately toxic by ingestion. When heated to decomposition it emits very toxic fumes of Cl^- and NO_x. See also CARBAMATES.

DET600 CAS:62046-37-1 ***HR: 2***
3,4-DICHLOROBENZYL METHYLCARBAMATE with 2,3-DICHLOROBENZYL METHYLCARBAMATE (80:20)
mf: $C_9H_9Cl_2NO_2$ mw: 234.09

SYNS: CHLORXYLAM ◇ 2,3(or 3,4)-DICHLOROBENZENEMETHANOL METHYL CARBAMATE ◇ ENT 25,736 ◇ ROWMATE ◇ SIRMATE ◇ U-17004 ◇ UC 22,463

TOXICITY DATA with REFERENCE
orl-rat LD50:1870 mg/kg 28ZEAL 4,144,69

SAFETY PROFILE: Moderately toxic by ingestion. When heated to decomposition it emits very toxic fumes of Cl^- and NO_x. See also individual components and CARBAMATES.

DET800 CAS:13029-08-8 ***HR: D***
2,2'-DICHLOROBIPHENYL
mf: $C_{12}H_8Cl_2$ mw: 223.10

SYNS: 2,2'-DICHLORBIPHENYL (GERMAN) ◇ 2,2'-DICHLORO-1,1'-BIPHENYL

TOXICITY DATA with REFERENCE
orl-mus TDLo:1125 mg/kg (1-3D preg):TER ARTODN 40,249,78
orl-mus TDLo:1125 mg/kg (1-3D preg):REP ARTODN 40,249,78

SAFETY PROFILE: Experimental teratogenic and reproductive effects. When heated to decomposition it emits toxic fumes of Cl^-. See also POLYCHLORINATED BIPHENYLS.

DEU000 CAS:6358-85-6 ***HR: 3***
2,2'-((3,3'-DICHLORO(1,1'-BIPHENYL)-4,4'-DIYL)BIS(AZO))BIS(3-OXO-N-PHENYL)BUTANAMIDE
mf: $C_{32}H_{26}Cl_2N_6O_4$ mw: 629.54

SYNS: AMAZON YELLOW X2485 ◇ BENZIDINE LACQUER YELLOW G ◇ BENZIDINE YELLOW ◇ BENZIDINE YELLOW TONER YT-378 ◇ BIS(ACETYL-N-PHENYLCARBAMYLMETHYL)-4,4'-DISAZO-3,3'-DICHLOROBIPHENYL ◇ BRILLIANT YELLOW SLURRY ◇ CARNELIO YELLOW GX ◇ C.I. 21090 ◇ C.I. PIGMENT YELLOW 12 ◇ DAINICHI BENZIDINE YELLOW GRT ◇ DAIRYLIDE YELLOW AAA ◇ DALTOLITE FAST YELLOW GT ◇ DIARYLANILIDE YELLOW ◇ 2,2'-((3,3'-DICHLORO(1,1'-DIPHENYL)-4,4'-DIYL)BIS(AZO)BIS(3-OXO-N-PHENYLBUTANAMIDE ◇ ELJON YELLOW BG ◇ GRAPHTOL YELLOW A-HG ◇ HANCOCK YELLOW 10010 ◇ HELIC YELLOW GW ◇ IRGALITE YELLOW BO ◇ ISOL BENZIDINE YELLOW G ◇ KROMON YELLOW MTB ◇ LIGHT YELLOW JB ◇ LODESTONE YELLOW YB-57 ◇ MONOLITE YELLOW GT ◇ NCI-C03269 ◇ No. 49 CONCENTRATED BENZIDINE YELLOW ◇ PERMANENT YELLOW GHG ◇ PIGMENT YELLOW GT ◇ RANGOON YELLOW ◇ RECOLITE YELLOW GB ◇ SANYO BENZIDINE YELLOW-B ◇ SEGNALE LIGHT YELLOW 2GR ◇ SILOTON YELLOW GTX ◇ SYMULER FAST YELLOW GF ◇ VERONA YELLOW X-1791 ◇ VULCAFOR FAST YELLOW GTA

CONSENSUS REPORTS: NCI Carcinogenesis Bioassay Completed; Results Negative NCITR* NCI-CG-Tr-30,78. Reported in EPA TSCA Inventory.

SAFETY PROFILE: No toxicity information. When heated to decomposition it emits very toxic fumes of Cl^- and NO_x. See also CHLORIDES and AMIDES.

DEU100 CAS:70134-26-8 ***HR: 3***
DICHLOROBIS(2-CHLOROCYCLOHEXYL)SELENIUM
mf: $C_{12}H_{20}Cl_4Se$ mw: 385.08

SYN: SELENIUM,DICHLOROBIS(2-CHLOROCYCLOHEXYL)-

TOXICITY DATA with REFERENCE
ivn-mus LD50:56 mg/kg CSLNX* NX#04576

OSHA PEL: TWA 0.2 mg(Se)/m^3
ACGIH TLV: TWA 0.2 mg(Se)/m^3

SAFETY PROFILE: Poison by intravenous route. When heated to decomposition it emits toxic fumes of Se and Cl^-.

DEU115 CAS:18252-65-8 ***HR: 3***
cis-DICHLOROBIS(DIMETHYLSELENIDE)PLATINUM(II)
mf: $C_4H_{12}Cl_2PtSe_2$ mw: 484.07

SYNS: NSC 271675 ◇ PLATINUM (II), BIS(METHYL SELENIDE) DICHLORO-, cis- ◇ PLATINUM, DICHLOROBIS(METHYL SELENIDE)-, cis- ◇ PLATINUM, DICHLOROBIS(SELENOBIS(METHANE))-(SP-4-2)

TOXICITY DATA with REFERENCE
ivn-mus LD50:50 mg/kg CTRRDO 61,1519,77 85INA8 5,492,86

OSHA PEL: TWA 0.2 mg(Se)/m^3
ACGIH TLV: TWA 0.2 mg(Se)/m^3; TWA 0.002 mg(Pt)/m^3

SAFETY PROFILE: Poison by intravenous route. When heated to decomposition it emits toxic fumes of Se, Pt, and Cl^-.

DEU125 CAS:74037-18-6 ***HR: 3***
DICHLOROBIS(2-ETHOXYCYCLOHEXYL)SELENIUM
mf: $C_{16}H_{30}Cl_2O_2Se$ mw: 404.32

SYN: SELENIUM,DICHLOROBIS(2-ETHOXYCYCLOHEXYL)-

TOXICITY DATA with REFERENCE
ivn-mus LD50:180 mg/kg CSLNX* NX#04578

OSHA PEL: TWA 0.2 mg(Se)/m^3
ACGIH TLV: TWA 0.2 mg(Se)/m^3

SAFETY PROFILE: Poison by intravenous route. When heated to decomposition it emits toxic fumes of Se and Cl^-.

DEU200 CAS:38780-42-6 ***HR: 3***
cis-DICHLOROBIS(PYRROLIDINE)PLATINUM(II)
mf: $C_8H_{18}Cl_2N_2Pt$ mw: 408.27

SYN: cis-DIPYRROLIDINEDICHLOROPLATINUM(II)

TOXICITY DATA with REFERENCE
mmo-sat 100 nmol/plate CNREA8 39,913,79
scu-rat LD:89 mg/kg/6W-I:ETA CNREA8 39,913,79
ipr-mus LD50:240 mg/kg CBINA8 5,415,72

SAFETY PROFILE: Poison by intraperitoneal route. Questionable carcinogen with experimental tumorigenic data. Mutation data reported. See also PLATINUM COMPOUNDS. When heated to decomposition it emits very toxic fumes of Cl^- and NO_x.

DEU259 CAS:2899-02-7 ***HR: 3***
N,N'-DICHLOROBIS(2,4,6-TRICHLOROPHENYL)UREA
mf: $C_{13}H_4Cl_8N_2O$ mw: 487.81

$O{:}C(NClC_6H_2Cl_3)_2$

SAFETY PROFILE: Violent or explosive reaction on mixing with dimethyl sulfoxide. Ignites on contact with ammonia, ammonium carbonate, or organic amines. A fabric treatment mixture of the urea with 1-(4-nitrophenylazo)-2-naphthol + zinc oxide may ignite spontaneously in storage, especially if heated. When heated to decomposition it emits toxic fumes of Cl^- and NO_x.

DEU300 CAS:10325-39-0 ***HR: 3***
DICHLOROBORANE
mf: BCl_2H mw: 82.72

SAFETY PROFILE: A poison. Ignites spontaneously in air. When heated to decomposition it emits toxic fumes of Cl^-. See also BORON COMPOUNDS, BORANES, and CHLORIDES.

DEU375 CAS:28577-62-0 ***HR: D***
DICHLORO-1,3-BUTADIENE
mf: $C_4H_4Cl_2$ mw: 122.98

SYNS: DCDB ◇ DICHLOROBUTADIENE

TOXICITY DATA with REFERENCE
ihl-rat TCLo:51 mg/m^3/24H (female 1-4D post):TER ZKMAAX 17,16,77
ihl-rat TCLo:14 mg/m^3/4H (male 16W pre):REP GISAAA 41(8),96,76

SAFETY PROFILE: An experimental teratogen. Experimental reproductive effects by inhalation. When heated to decomposition it emits toxic fumes of Cl^-. See also CHLORINATED HYDROCARBONS, ALIPHATIC.

DEU400 CAS:1653-19-6 ***HR: 2***
2,3-DICHLORO-1,3-BUTADIENE
mf: $C_4H_4Cl_2$ mw: 122.98

SYN: 2,3-DICHLOR-1,3-BUTADIEN(CZECH)

TOXICITY DATA with REFERENCE
skn-rbt 500 mg/24H SEV 28ZPAK -,29,72

eye-rbt 500 mg/24H MLD 28ZPAK -,29,72
orl-rat LD50:422 mg/kg 28ZPAK -,29,72

CONSENSUS REPORTS: Reported in EPA TSCA Inventory.

SAFETY PROFILE: Moderately toxic by ingestion. An eye and severe skin irritant. When heated to decomposition it emits toxic fumes of Cl^-. See also CHLORINATED HYDROCARBONS, ALIPHATIC.

DEU509 CAS:51104-87-1 ***HR: 3***
1,4-DICHLORO-1,3-BUTADIYNE
mf: C_4Cl_2 mw: 118.95

ClC≡CC≡CCl

SAFETY PROFILE: Explodes when heated above 70°C. When heated to decomposition it emits toxic fumes of Cl^-. See also CHLORINATED HYDROCARBONS, ALIPHATIC; and ACETYLENE COMPOUNDS.

DEU600 CAS:26761-81-9 ***HR: 2***
mixo-DICHLOROBUTANE
mf: $C_4H_8Cl_2$ mw: 122.98

PROP: Flash p: 69.8°F (1,2 and 1,3 isomers).

SAFETY PROFILE: A dangerous fire hazard when exposed to heat or flame. When heated to decomposition it emits toxic fumes of Cl^-. See also CHLORINATED HYDROCARBONS, ALIPHATIC.

DEV000 CAS:764-41-0 ***HR: 3***
1,4-DICHLORO-2-BUTENE
mf: $C_4H_6Cl_2$ mw: 125.00

PROP: Colorless liquid. Mp: 1-3°; bp: 156°; d: 1.183 @ 25°/4°.

SYNS: DCB ◇ 1,4-DCB ◇ 1,4-DICHLOROBUTENE-2 (MAK) ◇ RCRA WASTE NUMBER U074

TOXICITY DATA with REFERENCE
skn-rbt 10 mg/24H open SEV AMIHBC 4,119,51
eye-rbt 20 mg open SEV AMIHBC 4,119,51
mmo-sat 1 mmol/L ARTODN 41,249,79
mma-sat 1 mmol/L ARTODN 41,249,79
sln-dmg-orl 2 mmol/L/3D-I 35WYAM -,63,76
cyt-rat-ihl 1700 μg/m³/30D-I ZKMAAX 25,335,85
orl-rat TDLo:750 μg/ (75D male pre):REP GISAAA 51(7),77,86
ihl-rat TCLo:5 ppm/6H (6-15D preg):TER TXAPA9 64,125,82
ihl-rat TCLo:1 ppm/6H/82W-I:CAR EPASR* 8EHQ-0985-0567
ihl-rat TC:100 ppb/6H/82W-I:NEO EPASR* 8EHQ-0985-0567
orl-rat LD50:89 mg/kg AMIHBC 4,119,51
ihl-rat LCLo:62 ppm/4H AMIHBC 4,119,51
orl-mus LD50:190 mg/kg GTPZAB 29(4),49,85
ihl-mus LC50:920 mg/m³ GTPZAB 29(4),49,85
ivn-mus LD50:56 mg/kg CSLNX* NX#01103
skn-rbt LD50:620 mg/kg AMIHBC 4,119,51

CONSENSUS REPORTS: Reported in EPA TSCA Inventory. EPA Genetic Toxicology Program.

DFG MAK: Animal Carcinogen, Suspected Human Carcinogen.

SAFETY PROFILE: Confirmed carcinogen with experimental carcinogenic and neoplastigenic data. Poison by ingestion, inhalation, and intravenous routes. Moderately toxic by skin contact. An experimental teratogen. Other experimental reproductive effects. Mutation data reported. A severe skin and eye irritant. When heated to decomposition it emits toxic fumes of Cl^-. See also CHLORINATED HYDROCARBONS, ALIPHATIC.

DEV100 CAS:760-23-6 ***HR: 2***
3,4-DICHLORO-1-BUTENE
mf: $C_4H_6Cl_2$ mw: 125.00

SYN: 1-BUTENE, 3,4-DICHLORO-

TOXICITY DATA with REFERENCE
cyt-rat-ihl 13700 μg/m³/30D-I ZKMAAX 25,335,85
orl-rat TDLo:75 μg/kg (male 75D pre):REP GISAAA 51(7),77,86
orl-mus LD50:724 mg/kg GISAAA 51(7),77,86

CONSENSUS REPORTS: Reported in EPA TSCA Inventory.

SAFETY PROFILE: Moderately toxic by ingestion. Experimental reproductive effects. Mutation data reported. When heated to decomposition it emits toxic fumes of Cl^-.

DEV300 CAS:42520-97-8 ***HR: 3***
2,2'-DICHLORO-N-BUTYLDIETHYLAMINE
mf: $C_8H_{17}Cl_2N$ mw: 198.16

SYNS: N-N-BIS(2-CHLOROETHYL)BUTYLAMINE ◇ N-BUTYL-2,2'-DICHLORODIETHYLAMINE ◇ TL 513

TOXICITY DATA with REFERENCE
dns-rat-ipr 10 mg/kg CRNGDP 1,621,80
dns-rat-orl 10 mg/kg CRNGDP 1,621,80
unr-rat LD50:1 mg/kg PHBUA9 1,297,53
ihl-mus LCLo:350 mg/m³/10M NDRC** NDCrc-132,Dec,42

SAFETY PROFILE: Poison by inhalation and possibly other routes. Mutation data reported. When heated to decomposition it emits toxic fumes of Cl^- and NO_x.

DEV400 CAS:821-10-3 ***HR: 3***
1,4-DICHLORO-2-BUTYNE
mf: $C_4H_4Cl_2$ mw: 122.98

SYN: 1,4-DICHLOROBUTYNE

TOXICITY DATA with REFERENCE
ivn-mus LD50:56 mg/kg CSLNX* NX#02969

CONSENSUS REPORTS: Reported in EPA TSCA Inventory.

SAFETY PROFILE: Poison by intravenous route. When heated to decomposition it emits toxic fumes of Cl^-. Probably a dangerous fire and explosion hazard. See also ACETYLENE COMPOUNDS; and CHLORINATED HYDROCARBONS, ALIPHATIC.

DEV600 CAS:1918-18-9 ***HR: 2***
3,4-DICHLOROCARBANILIC ACID METHYL ESTER
mf: $C_8H_7Cl_2NO_2$ mw: 220.06

SYNS: (3,4-DICHLOROPHENYL)CARBAMIC ACID METHYL ESTER ◇ MCC ◇ METHYL-3,4-DICHLOROCARBANILATE ◇ METHYL-N-(3,4-DICHLOROPHENYL) CARBAMATE ◇ NIA 2,995 ◇ NIA 2995J ◇ SWEP

TOXICITY DATA with REFERENCE
orl-rat LD50:522 mg/kg GUCHAZ 6,477,73
skn-rbt LD50:2480 mg/kg WRPCA2 9,119,70

SAFETY PROFILE: Moderately toxic by ingestion and skin contact. A pesticide. See also CARBAMATES and ESTERS. When heated to decomposition it emits very toxic fumes of Cl^- and NO_x.

DEV800 CAS:101-05-3 ***HR: 3***
2,4-DICHLORO-6-o-CHLORANILINO-s-TRIAZINE
mf: $C_9H_5Cl_3N_4$ mw: 275.53

PROP: White to tan crystals, insol in water. Mp: 160°.

SYNS: ANILAZIN ◇ ANILAZINE ◇ B-622 ◇ BORTRYSAN ◇ 2-(2-CHLORANILIN)-4,6-DICHLOR-1,3,5-TRIAZIN (GERMAN) ◇ (o-CHLOROANILINO)DICHLOROTRIAZINE ◇ 2,4-DICHLORO-6-(o-CHLOROANILINO)-s-TRIAZINE ◇ 2,4-DICHLORO-6-(2-CHLOROANILINO)-1,3,5-TRIAZINE ◇ 4,6-DICHLORO-N-(2-CHLOROPHENYL)-1,3,5-TRIAZIN-2-AMINE ◇ DIREZ ◇ DYRENE ◇ DYRENE 50W ◇ ENT 26,058 ◇ KEMATE ◇ NCI-C08684 ◇ TRIASYN ◇ TRIAZIN ◇ TRIAZINE (PESTICIDE) ◇ ZINOCHLOR

TOXICITY DATA with REFERENCE
skn-man 0.1% MOD LANCAO 2,1252,80
skn-rbt 500 mg SEV 34ZIAG -,235,69
otr-rat:emb 990 ng/plate JJATDK 1,190,81
orl-rat LD50:2700 mg/kg ARSIM* 20,9,66
ipr-rat LD50:25 mg/kg JAFCAU 21,140,73
ipr-mus LD50:50 mg/kg JAFCAU 21,140,73
orl-rbt LD50:400 mg/kg 34ZIAG -,235,69

CONSENSUS REPORTS: NCI Carcinogenesis Bioassay Completed; Results Negative NCITR* NCI-CG-TR-104,78. EPA Genetic Toxicology Program.

SAFETY PROFILE: Poison by ingestion and intraperitoneal routes. A human skin irritant. A severe skin irritant experimentally. Mutation data reported. A fungicide. When heated to decomposition it emits very toxic fumes of Cl^- and NO_x.

DEW000 CAS:333-25-5 ***HR: 3***
DICHLORO(2-CHLOROVINYL)ARSINE OXIDE

SYN: LEWISITE I OXIDE

TOXICITY DATA with REFERENCE
orl-rat LD50:5 mg/kg JPBAA7 58,411,46
orl-rbt LD50:3 mg/kg JPBAA7 58,411,46
ivn-rbt LD50:1 mg/kg JPBAA7 58,411,46
orl-gpg LD50:2 mg/kg JPBAA7 58,411,46
scu-gpg LD50:200 μg/kg JPBAA7 58,411,46

CONSENSUS REPORTS: Arsenic and its compounds are on the Community Right-To-Know List.

OSHA PEL: TWA 0.5 mg(As)/m^3

SAFETY PROFILE: Poison by ingestion, intravenous, and subcutaneous routes. See also ARSENIC COMPOUNDS. When heated to decomposition it emits very toxic fumes of Cl^- and As. See also CHLOROVINYLARSINE DICHLORIDE (Lewisite).

DEW200 CAS:26270-58-6 ***HR: 3***
5-(3,4-DICHLOROCINNAMOYL)-4,7-DIMETHOXY-6-(2-DIMETHYLAMINOETHOXY)BENZOFURAN MALEATE
mf: $C_{24}H_{23}Cl_2NO_5 \cdot C_4H_4O_4$ mw: 592.46

TOXICITY DATA with REFERENCE
orl-mus LD50:320 mg/kg CHTPBA 8,479,73
ivn-mus LD50:24 mg/kg CHTPBA 8,479,73

SAFETY PROFILE: Poison by ingestion and intravenous routes. When heated to decomposition it emits very toxic fumes of Cl^- and NO_x.

DEW400 CAS:20373-56-2 ***HR: 3***
2,6-DICHLORO-N-CYCLOPROPYL-N-ETHYL ISONICOTINAMIDE
mf: $C_{11}H_{12}Cl_2N_2O$ mw: 259.15

SYN: ABBOTT-28440

TOXICITY DATA with REFERENCE
orl-rat LD50:78 mg/kg 27ZQAG -,196,72
ipr-rat LD50:54 mg/kg 27ZQAG -,196,72
orl-mus LD50:123 mg/kg 27ZQAG -,196,72
ipr-mus LD50:129 mg/kg 27ZQAG -,196,72

SAFETY PROFILE: Poison by ingestion and intraperi-

toneal routes. When heated to decomposition it emits very toxic fumes of Cl^- and NO_x.

DEX000 CAS:14913-33-8 ***HR: 3***
trans-DICHLORODIAMMINEPLATINUM(II)
mf: $C_{12}H_6N_2Pt$ mw: 300.07

SYNS: trans-DIAMMINEDICHLOROPLATINUM(II) ◇ trans-PLATINUM(II)DIAMMINEDICHLORIDE

TOXICITY DATA with REFERENCE
mma-sat 2 μg/plate MUREAV 77,45,80
dnd-hmn:fbr 50 μmol/L/4H CNREA8 42,145,82
dnd-hmn:lng 100 μmol/L CBINA8 36,345,81
dnd-hmn:oth 20 mg/L CNREA8 45,6232,85
msc-ham:lng 100 mg/L CNREA8 44,3270,84
ipr-mus TDLo:32408 μg/kg/10W-I:ETA CNREA8 41,4368,81
ipr-mus LD50:27 mg/kg CBINA8 5,415,72

CONSENSUS REPORTS: EPA Genetic Toxicology Program.

SAFETY PROFILE: Poison by intraperitoneal route. Questionable carcinogen with experimental tumorigenic data. Human mutation data reported. See also PLATINUM COMPOUNDS. When heated to decomposition it emits toxic fumes of NO_x and Cl^-.

DEX400 CAS:84-58-2 ***HR: 3***
2,3-DICHLORO-5,6-DICYANOBENZOQUINONE
mf: $C_8Cl_2N_2O_2$ mw: 227.00

SYN: 4,5-DICHLORO-3,6-DIOXO-1,4-CYCLOHEXADIENE-1,2-DICARBONITRILE

TOXICITY DATA with REFERENCE
ivn-mus LD50:13 mg/kg CSLNX* NX#07894

CONSENSUS REPORTS: Reported in EPA TSCA Inventory. Cyanide and its compounds are on the Community Right-To-Know List.

SAFETY PROFILE: Poison by intravenous route. When heated to decomposition it emits very toxic fumes of Cl^-, CN^-, and NO_x. See also NITRILES.

DEX600 CAS:17751-20-1 ***HR: 3***
2',6'-DICHLORO-2-(DIETHYLAMINO)ACETANILIDE HYDROCHLORIDE
mf: $C_{12}H_{16}Cl_2N_2O$•ClH mw: 311.66

SYN: C 3053

TOXICITY DATA with REFERENCE
eye-rbt 2% MLD ARZNAD 8,270,58
ipr-rat LD50:420 mg/kg ARZNAD 8,270,58
ipr-mus LD50:365 mg/kg ARZNAD 8,270,58
scu-mus LD50:775 mg/kg ARZNAD 8,270,58

SAFETY PROFILE: Poison by intraperitoneal route. Moderately toxic by subcutaneous route. An eye irritant. When heated to decomposition it emits very toxic fumes of Cl^- and NO_x.

DEX800 CAS:41572-59-2 ***HR: 3***
7,8-DICHLORO-10-(2-(DIETHYLAMINO)ETHYL) ISOALLOXAZINE HYDROCHLORIDE
mf: $C_{16}H_{17}Cl_2N_5O_2$•ClH mw: 418.74

TOXICITY DATA with REFERENCE
scu-mus LD50:15 mg/kg CMTRAG 2,96,61
ivn-mus LD50:22 mg/kg CMTRAG 2,96,61

SAFETY PROFILE: Poison by subcutaneous and intravenous routes. When heated to decomposition it emits very toxic fumes of NO_x and Cl^-.

DEY000 CAS:77791-63-0 ***HR: 3***
2',6'-DICHLORO-2-(2-(DIETHYLAMINO)ETHYL) METHYLAMINOACETANILIDE DIHYDROCHLORIDE
mf: $C_{15}H_{23}Cl_2N_3O$•2ClH mw: 405.23

SYN: C 5365

TOXICITY DATA with REFERENCE
ipr-rat LD50:140 mg/kg ARZNAD 9,262,59
scu-mus LD50:585 mg/kg ARZNAD 9,262,59

SAFETY PROFILE: Poison by intraperitoneal route. Moderately toxic by subcutaneous route. When heated to decomposition it emits very toxic fumes of Cl^- and NO_x.

DEY200 CAS:101651-69-8 ***HR: 3***
2',6'-DICHLORO-2-(2-(DIETHYLAMINO)ETHYL) THIOACETANILIDE HYDROCHLORIDE
mf: $C_{14}H_{20}Cl_2N_2OS$•ClH mw: 371.78

SYN: C 4910

TOXICITY DATA with REFERENCE
ipr-rat LD50:172 mg/kg ARZNAD 9,683,59
scu-mus LD50:340 mg/kg ARZNAD 9,683,59

SAFETY PROFILE: Poison by intraperitoneal and subcutaneous routes. When heated to decomposition it emits very toxic fumes of Cl^-, NO_x, and SO_x.

DEY400 CAS:93405-68-6 ***HR: 3***
7,8-DICHLORO-10-(3-(DIETHYLAMINO)-2-HYDROXYPROPYL)ISOALLOXAZINE SULFATE
mf: $C_{17}H_{19}Cl_2N_5O_3$•H_2O_4S mw: 510.39

TOXICITY DATA with REFERENCE
ipr-rat LD50:18 mg/kg CMTRAG 2,96,61
scu-mus LD50:74 mg/kg CMTRAG 2,96,61
ivn-mus LD50:75 mg/kg CMTRAG 2,96,61

SAFETY PROFILE: Poison by intraperitoneal, intravenous, and subcutaneous routes. When heated to decomposition it emits very toxic fumes of SO_x, Cl^-, and NO_x.

DEY600 CAS:101652-01-1 ***HR: 3***
7,8-DICHLORO-10-(4-(DIETHYLAMINO)-1-METHYLBUTYL)ISOALLOXAZINE HYDROCHLORIDE
mf: $C_{19}H_{23}Cl_2N_5O_2 \cdot ClH$ mw: 460.83

TOXICITY DATA with REFERENCE
scu-mus LD50:18 mg/kg CMTRAG 2,96,61
ivn-mus LD50:48 mg/kg CMTRAG 2,96,61

SAFETY PROFILE: Poison by subcutaneous and intravenous routes. When heated to decomposition it emits very toxic fumes of HCl and NO_x.

DEY800 CAS:1719-53-5 ***HR: 3***
DICHLORODIETHYLSILANE
DOT: UN 1767
mf: $C_4H_{10}Cl_2Si$ mw: 157.13

PROP: Liquid. Mp: −96°; bp: 131.0°, d: 1.05, vap d: 5.41; flash p: 75.2°F.

SYN: DIETHYLDICHLOROSILANE (DOT)

TOXICITY DATA with REFERENCE
orl-rat LDLo:1000 mg/kg JIHTAB 30,332,48
ipr-rat LDLo:100 mg/kg JIHTAB 30,332,48

CONSENSUS REPORTS: Reported in EPA TSCA Inventory.

DOT Classification: Flammable Liquid; Label: Flammable Liquid; IMO: Corrosive Material; Label: Corrosive, Flammable Liquid.

SAFETY PROFILE: Poison by intraperitoneal route. Moderately toxic by ingestion. Corrosive to tissue. Dangerous fire hazard when exposed to heat, flame, or oxidizers. Can react vigorously with oxidizing materials. To fight fire, use foam, CO_2, dry chemical. When heated to decomposition or in reaction with water or steam it emits toxic and corrosive fumes of Cl^-. See also CHLOROSILANES.

DEZ000 CAS:866-55-7 ***HR: 3***
DICHLORODIETHYLSTANNANE
mf: $C_4H_{10}Cl_2Sn$ mw: 247.73

PROP: Water-white crystals. Mp: 85°, bp: 220°.

SYNS: DIAETHYLZINNDICHLORID (GERMAN) ◇ DICHLORODIETHYLTIN ◇ DIETHYLDICHLOROSTANNANE ◇ DIETHYLSTANNYL DICHLORIDE ◇ DIETHYLTIN CHLORIDE ◇ DIETHYLTIN DICHLORIDE

TOXICITY DATA with REFERENCE
orl-rat LDLo:160 mg/kg BJIMAG 15,15,58
ivn-rat LD50:20600 μg/kg AEPPAE 242,370,61

OSHA PEL: TWA 0.1 mg(Sn)/m³ (skin)
ACGIH TLV: TWA 0.1 mg(Sn)/m³ (skin) (Proposed: TWA 0.1 mg(Sn)/m³; STEL 0.2 mg(Sn)/m³ (skin))
NIOSH REL: (Organotin Compounds) TWA 0.1 mg(Sn)/m³

SAFETY PROFILE: Poison by ingestion and intravenous routes. See also TIN COMPOUNDS and CHLORIDES. When heated to decomposition it emits toxic fumes of Cl^-.

DFA000 CAS:1649-08-7 ***HR: 1***
1,2-DICHLORO-1,1-DIFLUOROETHANE
mf: $C_2H_2Cl_2F_2$ mw: 134.94

TOXICITY DATA with REFERENCE
ihl-rat TCLo:2000 ppm/6H (male 14W pre):REP EPASR* 8EHQ-0587-0676
ihl-rat LCLo:20000 ppm/4H TXAPA9 19,1,71

CONSENSUS REPORTS: Reported in EPA TSCA Inventory.

SAFETY PROFILE: Experimental reproductive effects. Mildly toxic by inhalation. When heated to decomposition it emits very toxic fumes of Cl^- and F^-.

DFA200 CAS:27156-03-2 ***HR: 2***
DICHLORODIFLUOROETHYLENE
DOT: NA 9018
mf: $C_2Cl_2F_2$ mw: 132.92

PROP: Liquid. Vap d: 4.6.

DOT Classification: ORM-A; Label: None.

SAFETY PROFILE: Moderately toxic by inhalation. A skin, eye, and mucous membrane irritant. Will react with water or steam to produce toxic and corrosive fumes. When heated to decomposition it emits toxic fumes of F^- and Cl^-.

DFA300 CAS:79-35-6 ***HR: 2***
1,1-DICHLORO-2,2-DIFLUOROETHYLENE
mf: $C_2Cl_2F_2$ mw: 132.92

SYNS: 1,1-DIFLUORO-2,2-DICHLOROETHYLENE ◇ GENETRON 1112A ◇ GENETRONE 1112A

TOXICITY DATA with REFERENCE
ihl-rat LC50:505 mg/m³/4H GTPZAB 21(5),36,77
ihl-mus LC50:610 mg/m³/4H GTPZAB 21(5),36,77
ihl-gpg LC50:700 mg/m³/4H GTPZAB 21(5),36,77

SAFETY PROFILE: Moderately toxic by inhalation. When heated to decomposition it emits toxic fumes of F^- and Cl^-.

DFA400 CAS:76-38-0 ***HR: 2***
2,2-DICHLORO-1,1-DIFLUOROETHYL METHYL ETHER
mf: $C_3H_4Cl_2F_2O$ mw: 164.97

SYNS: ANALGIZER ◇ ANECOTAN ◇ 2,2-DICHLORO-1,1-DIFLUORO-1-METHOXYETHANE ◇ INGALAN ◇ INGALAN (RUSSIAN) ◇ INHALAN ◇ METHOFLURANE ◇ METHOXANE ◇ METHOXYFLUORAN ◇ METHOXYFLUORANE ◇ METHOXYFLURANE ◇ METOFANE ◇ METOXFLURAN ◇ METOXIFLURAN ◇ MOF ◇ NSC-110432 ◇ PENTHRANE ◇ PENTRAN ◇ PENTRANE

TOXICITY DATA with REFERENCE
eye-rbt 100 mg MOD FEPRA7 35,729,76
cyt-hmn:lym 200 ppm/24H ENVRAL 12,366,76
oms-ham:fbr 1 pph ANESAV 43,21,75
ihl-rat TCLo:100 ppm/8H (1-21D preg):TER ANESAV 48,11,78
ihl-hmn TCLo:3500 ppm/1H:KID CANJAE 21,294,74
orl-rat LD50:3600 mg/kg 85GMAT -,53,82
ihl-rat LC50:33500 mg/m³/4H 85GMAT -,53,82
ihl-mus LC50:118 g/m³/2H FATOAO 35,271,72
orl-mam LD50:3600 mg/kg GTPZAB 24(4),55,80

CONSENSUS REPORTS: IARC Cancer Review: Animal Inadequate Evidence IMEMDT 7,93,87. EPA Genetic Toxicology Program.

NIOSH REL: (Waste Anesthetic Gases and Vapors) CL 2 ppm/1H

SAFETY PROFILE: Moderately toxic by ingestion. Mildly toxic by inhalation. Human systemic effects by inhalation: depressed renal function. An experimental teratogen. Human mutation data reported. An eye irritant. See also ETHERS. When heated to decomposition it emits very toxic fumes of Cl^- and F^-.

DFA600 CAS:75-71-8 ***HR: 1***
DICHLORODIFLUOROMETHANE
DOT: UN 1028
mf: CCl_2F_2 mw: 120.91

PROP: Colorless, almost odorless gas. Mp: −158°, bp: −29°, vap press: 5 atm @ 16.1°.

SYNS: ALGOFRENE TYPE 2 ◇ ARCTON 6 ◇ DIFLUORODICHLOROMETHANE ◇ DWUCHLORODWUFLUOROMETAN (POLISH) ◇ ELECTRO-CF 12 ◇ ESKIMON 12 ◇ F 12 ◇ FC 12 ◇ FLUOROCARBON-12 ◇ FREON F-12 ◇ FRIGEN 12 ◇ GENETRON 12 ◇ HALON ◇ ISCEON 122 ◇ ISOTRON 12 ◇ KAISER CHEMICALS 12 ◇ LEDON 12 ◇ PROPELLANT 12 ◇ RCRA WASTE NUMBER U075 ◇ R 12 (DOT) ◇ REFRIGERANT 12 ◇ UCON 12 ◇ UCON 12/HALOCARBON 12

TOXICITY DATA with REFERENCE
ihl-hmn TCLo:200000 ppm/30M:EYE,PUL,LIV EJTXAZ 9,385,76
ihl-rat LC50:80 pph/30M EJTXAZ 9,385,76
ihl-mus LC50:76 pph/30M EJTXAZ 9,385,76
ihl-rbt LC50:80 pph/30M EJTXAZ 9,385,76
ihl-gpg LC50:80 pph/30M EJTXAZ 9,385,76

CONSENSUS REPORTS: Reported in EPA TSCA Inventory. EPA Genetic Toxicology Program.

OSHA PEL: TWA 1000 ppm
ACGIH TLV: TWA 1000 ppm
DFG MAK: 1000 ppm (5000 mg/m³)
DOT Classification: Nonflammable Gas; Label: Nonflammable gas.

SAFETY PROFILE: Human systemic effects by inhalation: conjunctiva irritation, fibrosing alveolitis, and liver changes. Narcotic in high concentrations. Nonflammable Gas. Can react violently with Al. When heated to decomposition it emits highly toxic fumes of phosgene, Cl^-, and F^-.

DFB000 CAS:70281-30-0 ***HR: 1***
DICHLORODIFLUOROMETHANE mixed with CHLORODIFLUOROMETHANE
DOT: UN 1078/NA 1954

SYN: DICHLORODIFLUOROMETHANE-CHLORODIFLUOROMETHANE MIXTURE (DOT)

DOT Classification: Nonflammable Gas; Label: Nonflammable Gas (UN 1078); Flammable Gas; Label; Flammable Gas (NA 1954).

SAFETY PROFILE: A simple asphyxiant. See individual components as listed. When heated to decomposition it emits very toxic fumes of Cl^- and F^-.

DFB400 CAS:56275-41-3 ***HR: 1***
DICHLORODIFLUOROMETHANE with 1,1-DIFLUOROETHANE
DOT: UN 1078/NA 1954
mf: $C_2H_4F_2 \cdot CCl_2F_2$ mw: 186.97

SYNS: DICHLORODIFLUOROMETHANE and DIFLUOROETHANE mixture (constant boiling mixture) (DOT) ◇ FREON 500 ◇ UCON 500/HALOCARBON 500

DOT Classification: Nonflammable Gas; Label: Nonflammable Gas (UN 1078); Flammable Gas; Label: Flammable Gas (NA 1954).

SAFETY PROFILE: A simple asphyxiant. See also components as listed. When heated to decomposition it emits very toxic fumes of Cl^- and F^-.

DFB800 ***HR: 1***
DICHLORODIFLUOROMETHANE mixed with TRICHLOROFLUOROMETHANE (1:1)
mf: $CCl_3F \cdot CCl_2F_2$ mw: 258.27

SYN: DICHLOROFLUOROMETHANE-TRICHLOROFLUOROMETHANE (DOT)

TOXICITY DATA with REFERENCE
ihl-rat LC50:30 pph/30M JETOAS 9,385,76
ihl-mus LC50:22 pph/30M JETOAS 9,385,76
ihl-gpg LC50:50 pph/30M JETOAS 9,385,76

SAFETY PROFILE: Mildly toxic by inhalation. When heated to decomposition it emits very toxic fumes of Cl^- and F^-.

DFC000 ***HR: 3***
DICHLORODIFLUOROMETHANE with TRICHLOROTRIFLUOROETHANE
DOT: UN 1078/ NA 1954
mf: $CCl_3F{\bullet}CCl_2F_2$ mw: 258.27

SYN: DICHLORODIFLUOROMETHANE-TRICHLOROTRIFLUOROETHANE MIXTURE (DOT)

TOXICITY DATA with REFERENCE
ihl-rat LC50:30 pph/30M EJTXAZ 9,385,76
ihl-mus LC50:22 pph/30M EJTXAZ 9,385,76
ihl-gpg LC50:50 pph/30M EJTXAZ 9,385,76

CONSENSUS REPORTS: NTP Fifth Annual Report On Carcinogens.

DOT Classification: Nonflammable Gas; Label: Nonflammable Gas (UN1078); Flammable Gas; Label: Flammable Gas (NA1954)

SAFETY PROFILE: Confirmed carcinogen. Very mildly toxic by inhalation. See components as listed. When heated to decomposition it emits very toxic fumes of Cl^- and F^-.

DFC200 CAS:2767-41-1 ***HR: 3***
DICHLORODIHEXYLSTANNANE
mf: $C_{12}H_{26}Cl_2Sn$ mw: 359.97

SYN: DIHEXYLTIN DICHLORIDE

TOXICITY DATA with REFERENCE
orl-rat LDLo:160 mg/kg BJIMAG 15,15,58
ivn-rat LDLo:10 mg/kg BJIMAG 15,15,58

OSHA PEL: TWA 0.1 mg(Sn)/m^3 (skin)
ACGIH TLV: TWA 0.1 mg(Sn)/m^3 (skin) (Proposed: TWA 0.1 mg(Sn)/m^3; STEL 0.2 mg(Sn)/m^3 (skin))
NIOSH REL: (Organotin Compounds) TWA 0.1 mg(Sn)/m^3

SAFETY PROFILE: Poison by ingestion and intravenous routes. See also TIN COMPOUNDS. When heated to decomposition it emits toxic fumes of Cl^-.

DFC600 CAS:6837-97-4 ***HR: D***
4,8-DICHLORO-1,5-DIHYDROXYANTHRAQUINONE
mf: $C_{14}H_6Cl_2O_4$ mw: 309.10

SYN: DICHLOROANTHRARUFIN

TOXICITY DATA with REFERENCE
mmo-sat 100 μg/plate BCSTB5 5,1489,77
mma-sat 100 μg/plate BCSTB5 5,1489,77

CONSENSUS REPORTS: EPA Genetic Toxicology Program. Reported in EPA TSCA Inventory.

SAFETY PROFILE: Mutation data reported. When heated to decomposition it emits very toxic fumes of Cl–.

DFC800 CAS:33770-60-4 ***HR: 3***
(2,5-DICHLORO-3,6-DIHYDROXY-p-BENZOQUINOLATO)MERCURY
mf: $C_6Cl_2HgO_4$ mw: 407.55

SYNS: 2,5-DICHLORO-3,6-DIHYDROXY-p-BENZOQUINONE,MERCURY SALT ◇ (2,5-DICHLORO-3,6-DIHYDROXY-p-BENZOQUINONE), MERCURY SALT

TOXICITY DATA with REFERENCE
ivn-mus LD50:10 mg/kg CSLNX* NX#04223

CONSENSUS REPORTS: Mercury and its compounds are on the Community Right-To-Know List.

OSHA PEL: (Transitional: CL 1 mg/10m^3) CL 0.1 mg(Hg)/m^3 (skin)
ACGIH TLV: TWA 0.1 mg(Hg)/m^3 (skin)
NIOSH REL: (Inorganic Mercury) TWA 0.05 mg (Hg)/m^3

SAFETY PROFILE: Poison by intravenous route. See also MERCURY COMPOUNDS. When heated to decomposition it emits very toxic fumes of Cl^- and Hg.

DFD000 CAS:10331-57-4 ***HR: 3***
5,5'-DICHLORO-2,2'-DIHYDROXY-3,3'-DINITROBIPHENYL
mf: $C_{12}H_6Cl_2N_2O_6$ mw: 345.10

SYNS: BAY 9015 ◇ BAYER 9015 ◇ BILEVON M ◇ 4,4'-DICHLORO-6,6'-DINITRO-O,O'-BIPHENOL ◇ 3,3'-DICHLORO-5,5'-DINITRO-O,O'-BIPHENOL (FRENCH) ◇ 5,5'-DICHLORO-3,3'-DINITRO(1,1'-BIPHENYL)-2,2'-DIOL ◇ ME 3625 ◇ MENICHLOPHOLAN ◇ NICLOFOLAN

TOXICITY DATA with REFERENCE
orl-ham TDLo:10 mg/kg/(8D preg):TER JETOAS 4,525,71
orl-rat LD50:10 mg/kg TXAPA9 21,315,72
orl-ham LD50:50 mg/kg JETOAS 4,525,71
orl-dom LDLo:15 mg/kg FAZMAE 17,108,73
orl-bwd LD50:13 mg/kg TXAPA9 21,315,72

SAFETY PROFILE: A poison by ingestion. An experimental teratogen. When heated to decomposition it emits very toxic fumes of Cl^- and NO_x. See also NITRO COMPOUNDS of AROMATIC HYDROCARBONS and CHLORIDES.

DFD200 CAS:29202-04-8 ***HR: 3***
3,4-DICHLORO-2,5-DILITHIOTHIOPHENE
mf: $C_4Cl_2Li_2S$ mw: 164.89

ClC=C(Li)SC(Li)=CCl

SAFETY PROFILE: The dry material is a slightly shock-sensitive explosive. When heated to decomposition it emits toxic fumes of Cl^- and SO_x. See also LITHIUM COMPOUNDS.

DFD400 CAS:17010-61-6 ***HR: 2***
3',4'-DICHLORO-4-DIMETHYLAMINOAZOBENZENE
mf: $C_{14}H_{13}Cl_2N_3$ mw: 294.20

SYNS: BENZENAMINE,4-((3,4-DICHLOROPHENYL)AZO)-N,N-DIMETHYL-(9CI) ◇ 3',4'-Cl2-DAB ◇ p-((3,4-DICHLOROPHENYL)AZO)-N,N-DIMETHYLANILINE

TOXICITY DATA with REFERENCE
orl-rat TDLo:15120 mg/kg/36W-C:CAR CBINA8 53,107,85
orl-rat TD:11 g/kg/17W-I:ETA CNREA8 30,1520,70

SAFETY PROFILE: Questionable carcinogen with experimental carcinogenic and tumorigenic data. When heated to decomposition it emits toxic fumes of Cl^- and NO_x.

DFD600 CAS:101652-02-2 ***HR: 3***
7,8-DICHLORO-10-(2-(DIMETHYLAMINO)ETHYL)ISOALLOXAZINE SULFATE
mf: $C_{14}H_{13}Cl_2N_5O_2 \cdot H_2O_4S$ mw: 452.30

TOXICITY DATA with REFERENCE
scu-mus LD50:30 mg/kg CMTRAG 2,96,61
ivn-mus LD50:18 mg/kg CMTRAG 2,96,61

SAFETY PROFILE: Poison by subcutaneous and intravenous routes. See also SULFATES and CHLORIDES. When heated to decomposition it emits very toxic fumes of SO_x, NO_x, and Cl^-.

DFE000 CAS:97864-38-5 ***HR: 3***
7,8-DICHLORO-10-(3-(DIMETHYLAMINO)PROPYL)ISOALLOXAZINE HYDROCHLORIDE
mf: $C_{15}H_{15}Cl_2N_5O_2 \cdot ClH$ mw: 404.71

TOXICITY DATA with REFERENCE
ipr-rat LD50:22 mg/kg CMTRAG 2,96,61
scu-mus LD50:23 mg/kg CMTRAG 2,96,61
ivn-mus LD50:34 mg/kg CMTRAG 2,96,61

SAFETY PROFILE: Poison by intraperitoneal, subcutaneous, and intravenous routes. When heated to decomposition it emits very toxic fumes of NO_x and Cl^-.

DFE100 CAS:594-84-3 ***HR: 3***
2,2-DICHLORO-3,3-DIMETHYLBUTANE
mf: $C_6H_{12}Cl_2$ mw: 155.07

$CH_3CCl_2C(CH_3)_3$

SAFETY PROFILE: Violent reaction with sodium hydroxide. When heated to decomposition it emits toxic fumes of Cl^-. See also CHLORINATED HYDROCARBONS, ALIPHATIC.

DFE200 CAS:118-52-5 ***HR: 2***
1,3-DICHLORO-5,5-DIMETHYL HYDANTOIN
mf: $C_5H_6Cl_2N_2O_2$ mw: 197.03

$ClNCO \cdot NClCO \cdot C(CH_3)_2$ (ring)

PROP: Crystals, liberates chlorine on contact with hot water. Mp: 132°; subl @ 100°, conflagrates @ 212°; d: 1.5 @ 20°, vap d: 6.8.

SYNS: DACTIN ◇ DAKTIN ◇ DANTOIN ◇ DCA ◇ DICHLORANTIN ◇ DICHLORODIMETHYLHYDANTOIN ◇ 1,3-DICHLORO-5,5-DIMETHYL-2,4-IMIDAZOLIDINEDIONE ◇ 1,3-DICHLORO-5,5'-METHYLHYDANTOIN ◇ HALANE ◇ HYDAN ◇ HYDAN (antiseptic) ◇ NCI-C03054 ◇ OMCHLOR

TOXICITY DATA with REFERENCE
skn-rbt 500 mg/24H SEV EPASR* 8EHQ-0281-0382
skn-rbt 100 mg/24H SEV EPASR* 8EHQ-0281-0382
sln-dmg-par 250 ppm ENMUDM 7,677,85
otr-rat:emb 6300 ng/plate JJATDK 1,190,81
orl-rat LD50:542 mg/kg DTLVS* 4,129,80
ihl-rat LCLo:20 g/m³/1H EPASR* 8EHQ-0281-0382
orl-rbt LD50:1520 mg/kg GISAAA 47(6),76,82
orl-gpg LD50:1350 mg/kg GISAAA 47(6),76,82

CONSENSUS REPORTS: Reported in EPA TSCA Inventory.

OSHA PEL: TWA 0.2 mg/m³; STEL 0.4 mg/m³
ACGIH TLV: TWA 0.2 mg/m³; STEL 0.4 mg/m³

SAFETY PROFILE: Moderately toxic by ingestion. Mildly toxic by inhalation. A severe skin irritant. Mutation data reported. Avoid excessive contact because of effects of active chlorine on skin. Some of the hydantoins are central nervous system depressants. Mixtures with xylene may explode. Will react with water or steam to produce toxic and corrosive fumes. When heated to decomposition it emits toxic fumes of Cl^- and NO_x. See also CHLORIDES.

DFE229 CAS:40580-75-4 ***HR: 3***
DICHLORO(4,5-DIMETHYL-o-PHENYLENEDIAMMINE)PLATINUM(II)
mf: $C_8H_{12}Cl_2N_2Pt$ mw: 402.21

SYN: cis-DICHLORO(4,5-DIMETHYL-O-PHENYLENEDIAMMINE) PLATINUM(II)

TOXICITY DATA with REFERENCE
mmo-sat 10200 nmol/L JMCMAR 23,459,80
ipr-mus LD50:283 mg/kg CBINA8 11,145,75

CONSENSUS REPORTS: Reported in EPA TSCA Inventory.

SAFETY PROFILE: Poison by intraperitoneal route.

Mutation data reported. See also TIN COMPOUNDS. When heated to decomposition it emits toxic fumes of Cl^-.

DFE259 CAS:75-78-5 ***HR: 3***
DICHLORODIMETHYLSILANE
DOT: UN 1162
mf: $C_2H_6Cl_2Si$ mw: 129.06

CONSENSUS REPORTS: Reported in EPA TSCA Inventory.

DOT Classification: Flammable Liquid; Label:Flammable Liquid

SAFETY PROFILE: Probably a skin, eye, and mucous membrane irritant. Violent reaction on contact with water. When heated to decomposition it emits toxic fumes of Cl^-. See also CHLOROSILANES.

DFE300 CAS:59183-17-4 ***HR: 3***
3,6-DICHLORO-3,6-DIMETHYLTETRAOXANE
mf: $C_4H_6Cl_2O_4$ mw: 189.00

$Cl(CH_3)COOCCl(CH_3)OO$ (ring)

SAFETY PROFILE: An extremely shock- and heat-sensitive explosive. When heated to decomposition it emits toxic fumes of Cl^-. See also PEROXIDES.

DFE469 CAS:58270-08-9 ***HR: 3***
(trans-4)-DICHLORO(4,4-DIMETHYLZINC 5((((METHYLAMINO)CARBONYL)OXY)IMINO) PENTANENITRILE)
mf: $C_9H_{15}Cl_2N_3O_2Zn$ mw: 333.54

SYNS: AC 85258 ◇ ETHIENOCARB

TOXICITY DATA with REFERENCE
orl-rat LD50:9 mg/kg KSKZAN 16(2),65,78
skn-rat LD50:857 mg/kg KSKZAN 16(2),65,78

CONSENSUS REPORTS: EPA Extremely Hazardous Substances List. Reported in EPA TSCA Inventory. Zinc and its compounds, as well as cyanide and its compounds, are on the Community Right-To-Know List.

SAFETY PROFILE: Poison by ingestion. Moderately toxic by skin contact. When heated to decomposition it emits toxic fumes of Cl^-, NO_x, CN^-, and ZnO. See also ZINC COMPOUNDS and NITRILES.

DFE550 CAS:1587-41-3 ***HR: 3***
DICHLORODINITROMETHANE
mf: $CCl_2N_2O_4$ mw: 174.93

SAFETY PROFILE: Explodes when heated. Upon decomposition it emits toxic fumes of Cl^- and NO_x. See also NITROMETHANE.

DFE600 CAS:3883-43-0 ***HR: 3***
trans-2,3-DICHLORO-1,4-DIOXANE
mf: $C_4H_6Cl_2O_2$ mw: 157.00

SYN: trans-2,3-DICHLORO-p-DIOXANE

TOXICITY DATA with REFERENCE
skn-mus TDLo:4080 mg/kg/68W-I:ETA JNCIAM 53,695,74
scu-mus TDLo:1260 mg/kg/63W-I:NEO JNCIAM 53,695,74
orl-rat LD50:1410 mg/kg AIHAAP 30,470,69
skn-rbt LD50:440 mg/kg AIHAAP 30,470,69

SAFETY PROFILE: Moderately toxic by ingestion and skin contact. Questionable carcinogen with experimental neoplastigenic and tumorigenic data. When heated to decomposition it emits toxic fumes of Cl^-.

DFE700 CAS:3646-61-5 ***HR: D***
2,3-DICHLORO-6,12-DIPHENYL-DIBENZO(b,f) (1,5)DIAZOCINE
mf: $C_{26}H_{16}Cl_2N_2$ mw: 427.34

SYNS: 2,8-DICHLORO-6,12-DIPHENYL-DIBENZO(b,f)(1,5)DIAZOCINE ◇ U-10293

TOXICITY DATA with REFERENCE
orl-rat TDLo:17500 μg/kg (female 7D pre):REP PSEBAA 120,725,65

SAFETY PROFILE: Experimental reproductive effects. When heated to decomposition it emits toxic fumes of Cl^- and NO_x.

DFE800 CAS:28675-08-3 ***HR: 2***
DICHLORO DIPHENYL OXIDE
mf: $C_{12}H_8Cl_2O$ mw: 239.10

PROP: Liquid. Vap d: 8.2.

SYNS: DICHLOROPHENYL ETHER ◇ PHENYL ETHER DICHLORO

TOXICITY DATA with REFERENCE
orl-gpg LDLo:1000 mg/kg 14CYAT 2,1707,63

OSHA PEL: TWA 0.5 mg/m^3

SAFETY PROFILE: Moderately toxic by ingestion. When heated to decomposition it emits toxic fumes of Cl^-. See also ETHERS and CHLORIDES.

DFF000 CAS:80-10-4 ***HR: 3***
DICHLORO DIPHENYLSILANE
DOT: UN 1769
mf: $C_{12}H_{10}Cl_2Si$ mw: 253.21

PROP: Colorless liquid. Mp: −22°, bp: 303°, d: 1.19 @ 20°, vap d: 8.45.

SYN: DIPHENYL DICHLOROSILANE (DOT)

TOXICITY DATA with REFERENCE
skn-rbt 500 mg/24H MOD 28ZPAK -,221,72
eye-rbt 5 mg/24H SEV 28ZPAK -,221,72

CONSENSUS REPORTS: Reported in EPA TSCA Inventory.

DOT Classification: Corrosive Material; Label: Corrosive.

SAFETY PROFILE: A poison irritant to skin, eyes, and mucous membranes. See also CHLOROSILANES. Can react vigorously with oxidizing materials. When heated to decomposition or on contact with acid or acid fumes it emits it emits toxic fumes of Cl^-.

DFF200 CAS:77791-64-1 ***HR: 2***
2',6'-DICHLORO-2-(DIPROPYLAMINO)ACETANILIDE HYDROCHLORIDE
mf: $C_{14}H_{20}Cl_2N_2O \cdot ClH$ mw: 339.72

SYN: C 3057

TOXICITY DATA with REFERENCE
eye-rbt 2% MOD ARZNAD 8,407,58
ipr-rat LD50:450 mg/kg ARZNAD 8,407,58
ipr-mus LD50:550 mg/kg ARZNAD 8,407,58
scu-mus LD50:2250 mg/kg ARZNAD 8,407,58

SAFETY PROFILE: Moderately toxic by intraperitoneal and subcutaneous routes. An eye irritant. When heated to decomposition it emits very toxic fumes of Cl^- and NO_x.

DFF400 CAS:867-36-7 ***HR: 3***
DICHLORODIPROPYLSTANNANE
mf: $C_6H_{14}Cl_2Sn$ mw: 275.79

PROP: Colorless crystals. Sol in organic solvents. Mp: 81°.

SYNS: DICHLORODIPROPYLTIN ◇ DIPROPYLTIN CHLORIDE ◇ DIPROPYLTIN DICHLORIDE ◇ DI-n-PROPYLTIN DICHLORIDE

TOXICITY DATA with REFERENCE
orl-rat LDLo:160 mg/kg BJIMAG 15,15,58

OSHA PEL: TWA 0.1 mg(Sn)/m^3 (skin)
ACGIH TLV: TWA 0.1 mg(Sn)/m^3 (skin) (Proposed: TWA 0.1 mg(Sn)/m^3; STEL 0.2 mg(Sn)/m^3 (skin))
NIOSH REL: (Organotin Compounds) TWA 0.1 mg(Sn)/m^3

SAFETY PROFILE: Poison by ingestion. See also TIN COMPOUNDS and CHLORIDES. When heated to decomposition it emits toxic fumes of Cl^-.

DFF500 CAS:15227-42-6 ***HR: 3***
cis-DICHLORO(DIPYRIDINE)PLATINUM(II)
mf: $C_{10}H_{10}Cl_2N_2Pt$ mw: 424.21

SYN: DICHLORODIPYRIDINEPLATINUM(II)(Z)

TOXICITY DATA with REFERENCE
mmo-sat 100 μg/plate MUREAV 95,79,82
mma-sat 100 μg/plate MUREAV 95,79,82
dnd-mam:lym 30 μmol/L CBINA8 7,305,73
ipr-mus LDLo:131 mg/kg JPMSAE 65,315,76

SAFETY PROFILE: Poison by intraperitoneal route. Mutation data reported. When heated to decomposition it emits toxic fumes of Cl^- and NO_x. See also PLATINUM COMPOUNDS.

DFF600 CAS:3583-47-9 ***HR: 2***
1,4-DICHLORO-2,3-EPOXYBUTANE
mf: $C_4H_6Cl_2O$ mw: 141.00

TOXICITY DATA with REFERENCE
skn-rbt 10 mg/24H open MLD AIHAAP 23,95,62
mmo-klp 5 mmol/L MUREAV 89,269,81
mma-sat 1 mmol/L ARTODN 41,249,79
orl-rat LDLo:710 mg/kg AIHAAP 23,95,62
skn-rbt LDLo:2830 mg/kg AIHAAP 23,95,62

SAFETY PROFILE: Moderately toxic by by ingestion and skin contact. A skin irritant. Mutation data reported. When heated to decomposition it emits toxic fumes of Cl^-.

DFF800 CAS:1300-21-6 ***HR: 2***
DICHLOROETHANE
mf: $C_2H_4Cl_2$ mw: 98.96

PROP: Lel: 5.6%, uel: 11.4%.

TOXICITY DATA with REFERENCE
ihl-rat TCLo:57 mg/m^3/4H (female 26W pre):REP GTPZAB 19(7),20,75
ihl-rat TCLo:15 mg/m^3/4H (female 96D pre):TER AKGIAO 53(2),57,77
orl-rat LD50:1120 mg/kg HYSAAV 32,349,67
orl-mus LD50:625 mg/kg HYSAAV 32,349,67
ihl-mus LCLo:10 g/m^3 GISAAA 20(8),19,55
skn-rbt LD50:3890 mg/kg UCDS** 3/23/70

SAFETY PROFILE: Moderately toxic by ingestion and skin contact. Mildly toxic by inhalation. An experimental teratogen. Other experimental reproductive effects by inhalation. When heated to decomposition it emits very toxic fumes of Cl^-. See also 1,2-DICHLOROETHANE; and CHLORINATED HYDROCARBONS, ALIPHATIC.

DFF809 CAS:75-34-3 ***HR: 3***
1,1-DICHLOROETHANE
DOT: UN 2362
mf: $C_2H_4Cl_2$ mw: 98.96

PROP: Colorless liquid; aromatic, ethereal odor; hot,

saccharine taste. Mp: −97.7°, lel: 5.6%, bp: 57.3°, flash p: 22°F (TOC), d: 1.174 @ 20°/4°, vap press: 230 mm @ 25°, vap d: 3.44, autoign temp: 856°F.

SYNS: AETHYLIDENCHLORID (GERMAN) ◇ CHLORINATED HYDROCHLORIC ETHER ◇ CHLORURE d'ETHYLIDENE (FRENCH) ◇ CLORURO di ETILIDENE (ITALIAN) ◇ 1,1-DICHLOORETHAAN (DUTCH) ◇ 1,1-DICHLORAETHAN (GERMAN) ◇ 1,1-DICLOROETANO (ITALIAN) ◇ ETHYLIDENE CHLORIDE ◇ ETHYLIDENE DICHLORIDE ◇ NCI-C04535 ◇ RCRA WASTE NUMBER U076

TOXICITY DATA with REFERENCE
ihl-rat TCLo:6000 ppm/7H (6-15D preg):TER TXAPA9 28,452,74
orl-mus TDLo:185 g/kg/78W-I:ETA,TER NCITR* NCI-CG-TR-66,78
orl-mus TD:1300 g/kg/78W-I:ETA,TER NCITR* NCI-CG-TR-66,78
orl-rat LD50:725 mg/kg HYSAAV 32,349,67
ihl-rat LCLo:16000 ppm/4H JIDHAN 31,343,49

CONSENSUS REPORTS: NCI Carcinogenesis Bioassay (gavage); Inadequate Studies: mouse, rat NCITR* NCI-CG-TR-66,78. Reported in EPA TSCA Inventory.

OSHA PEL: TWA 100 ppm
ACGIH TLV: TWA 200 ppm; STEL 250 ppm; (Proposed: 100 ppm)
DFG MAK: 100 ppm (400 mg/m^3)
NIOSH REL: (1,1-Dichloroethane): handle with caution
DOT Classification: Flammable Liquid; Label: Flammable Liquid.

SAFETY PROFILE: Moderately toxic by ingestion. Experimental teratogenic effects. Questionable carcinogen with experimental tumorigenic data. Liver damage reported in experimental animals. A very dangerous fire hazard and moderate explosion hazard when exposed to heat or flame; can react vigorously with oxidizing materials. To fight fire, use alcohol foam, water, foam, CO_2, dry chemical. When heated to decomposition it emits highly toxic fumes of phosgene and Cl^-.

DFG000 CAS:598-38-9 ***HR: D***
2,2-DICHLOROETHANOL
mf: $C_2H_4Cl_2O$ mw: 114.96

TOXICITY DATA with REFERENCE
mmo-omi 80 μL/plate CBINA8 30,9,80
mmo-asn 20 μL/plate/2H CBINA8 30,9,80
sln-asn 37 mmol/L MUREAV 138,33,84

CONSENSUS REPORTS: Reported in EPA TSCA Inventory.

SAFETY PROFILE: Mutation data reported. When heated to decomposition it emits toxic fumes of Cl^-.

DFG159 CAS:10140-87-1 ***HR: 3***
1,2-DICHLOROETHANOL ACETATE
mf: $C_4H_6Cl_2O_2$ mw: 157.00

PROP: Water-white liquid. Bp: 58-65°C/13 mm Hg; d: 1.296/20°C. Flash p: 307°F. Insoluble in water.

SYN: 1,2-DICHLOROETHYL ACETATE

TOXICITY DATA with REFERENCE
ihl-rat LCLo:16 ppm/4H JIHTAB 31,343,49

CONSENSUS REPORTS: EPA Extremely Hazardous Substances List.

SAFETY PROFILE: Poison by inhalation. Combustible when exposed to heat or flame. The vapor is potentially explosive. To fight small fires, use dry chemical, carbon dioxide, water spray, or foam. To fight large fires, use water spray, fog, or foam. May explode on heating with nitrates. When heated to decomposition it emits toxic fumes of Cl^- and phosgene.

DFG200 CAS:72-00-4 ***HR: 3***
2,2-DICHLOROETHENYL DIETHYL PHOSPHATE
mf: $C_6H_{11}Cl_2O_4P$ mw: 249.04

SYNS: 2,2-DICHLOROVINYL DIETHYL PHOSPHATE ◇ O-(2,2-DICHLORVINYL)-O,O-DIETHYLPHOSPHAT (GERMAN) ◇ DICHLORVOS-ETHYL

TOXICITY DATA with REFERENCE
mmo-sat 5 μL/plate MUREAV 28,405,75
orl-mus LDLo:42 mg/kg AECTCV 14,111,85
ipr-mus LD50:12 mg/kg ARZNAD 5,746,55

SAFETY PROFILE: Poison by ingestion and intraperitoneal routes. Mutation data reported. See also PHOSPHATES and ESTERS. When heated to decomposition it emits very toxic fumes of Cl^- and PO_x.

DFG400 CAS:72595-96-1 ***HR: D***
DICHLORO(4-ETHOXY-O-PHENYLENEDIAMMINE)PLATINUM(II)
mf: $C_8H_{12}Cl_2N_2OPt$ mw: 418.21

TOXICITY DATA with REFERENCE
mmo-sat 2500 nmol/L JMCMAR 23,459,80
mma-sat 2500 nmol/L JMCMAR 23,459,80

SAFETY PROFILE: Mutation data reported. See also PLATINUM COMPOUNDS. When heated to decomposition it emits very toxic fumes of Cl^- and NO_x.

DFG600 CAS:14689-97-5 ***HR: 3***
DI(2-CHLOROETHYL) ACETAL
mf: $C_6H_{12}Cl_2O_2$ mw: 187.08

SYN: 1,1'-(ETHYLIDENE)BIS(OXY)BIS(2-CHLOROETHANE)

TOXICITY DATA with REFERENCE
orl-rat LD50:310 mg/kg AIHAAP 30,470,69
skn-rbt LD50:200 mg/kg AIHAAP 30,470,69

SAFETY PROFILE: Poison by ingestion and skin contact. When heated to decomposition it emits toxic fumes of Cl^-.

DFG700 CAS:5960-88-3 ***HR: 3***
2,2-DICHLOROETHYLAMINE
mf: $C_2H_5Cl_2N$ mw: 113.97

SAFETY PROFILE: A poison. Solutions in ether are violently explosive at 80°C and 260 mbar. When heated to decomposition it emits toxic fumes of Cl^- and NO_x.

DFH000 CAS:10072-25-0 ***HR: 3***
9-(2-(DI(2-CHLOROETHYL)AMINO)ETHYLAMINO)-6-CHLORO-2-METHOXYACRIDINE
mf: $C_{20}H_{22}Cl_3N_3O \cdot 2ClH \cdot H_2O$ mw: 517.74

SYNS: 9-(2-(BIS(2-CHLOROETHYL)AMINO)ETHYLAMINO)-6-CHLORO-2-METHOXYACRIDINE DIHYDROCHLORIDE ◇ ICR-48b ◇ NSC-34372 ◇ QUINACRINE ETHYL MUSTARD

TOXICITY DATA with REFERENCE
ipr-mus TDLo:16 mg/kg/4W:CAR JNCIAM 36,915,66

SAFETY PROFILE: Questionable carcinogen with experimental carcinogenic data. When heated to decomposition it emits very toxic fumes of Cl^- and NO_x.

DFH100 ***HR: 3***
o-(p-DI-(2-CHLOROETHYL)AMINOPHENYL)-dl-TYROSINE DIHYDROCHLORIDE
mf: $C_{19}H_{22}Cl_2N_2O_3 \cdot 2ClH$ mw: 470.25

SYN: o-(p-DI(2-CHLORAETHYL)-AMINOPHENYL)-dl-TYROSIN-DIHYDROCHLORID (GERMAN)

TOXICITY DATA with REFERENCE
orl-rat LD50:620 mg/kg GWXXBX #2644941
ipr-rat LD50:62 mg/kg GWXXBX #2644941
ivn-rat LD50:62 mg/kg GWXXBX #2644941
orl-mus LD50:360 mg/kg GWXXBX #2644941
ipr-mus LD50:110 mg/kg GWXXBX #2644941
ivn-mus LD50:30 mg/kg GWXXBX #2644941

SAFETY PROFILE: Poison by ingestion, intravenous, and intraperitoneal routes. When heated to decomposition it emits toxic fumes of NO_x and Cl^-.

DFH200 CAS:598-14-1 ***HR: 3***
DICHLOROETHYLARSINE
DOT: UN 1892
mf: $C_2H_5AsCl_2$ mw: 174.89

PROP: Colorless liquid; fruity, biting, irritating odor. Mp: −65°, bp: 156° decomp, d: 1.742 @ 14°, vap press: 2.29 mm @ 21.5°, vap d: 6.03.

SYNS: ARSENIC DICHLOROETHANE ◇ DICK (GERMAN) ◇ ED ◇ ETHYLARSONOUS DICHLORIDE ◇ ETHYLIDICHLORARSINE ◇ ETHYLIDICHLOROARSINE (DOT) ◇ TL 214

TOXICITY DATA with REFERENCE
ihl-hmn LCLo:14 ppm/30M NTIS** PB214-270
ihl-mus LC50:1555 mg/m^3/10M NTIS** PB158-508
skn-mus LDLo:20 mg/kg NTIS** PB158-508
ihl-cat LCLo:12 ppm/40M ZGEMAZ 13,523,21
scu-cat LDLo:1 mg/kg ZGEMAZ 13,523,21

CONSENSUS REPORTS: Arsenic and its compounds are on the Community Right-To-Know List.

OSHA PEL: TWA 0.5 mg(As)/m^3
DOT Classification: Poison B; Label: Poison.

SAFETY PROFILE: A human poison by inhalation. Experimentally, a deadly poison by inhalation and subcutaneous routes, and probably by ingestion. A severe irritant. A military poison gas. Can react with oxidizing materials. Will react with water or steam to produce toxic and corrosive fumes. Dangerous; on contact with acid or acid fumes it emits highly toxic fumes of Cl^-, As and phosgene. See also ARSENIC COMPOUNDS.

DFH300 CAS:1739-53-3 ***HR: 3***
DICHLOROETHYLBORANE
mf: $C_2H_5BCl_2$ mw: 110.78

SAFETY PROFILE: Ignites spontaneously in air. When heated to decomposition it emits toxic fumes of Cl^-. See also BORANES and BORON COMPOUNDS.

DFH600 CAS:321-55-1 ***HR: 2***
O,O-DI(2-CHLOROETHYL)-O-(3-CHLORO-4-METHYLCOUMARIN-7-YL) PHOSPHATE
mf: $C_{14}H_{14}Cl_3O_6P$ mw: 415.60

SYNS: O,O-BIS(2-CHLOROETHYL)-O-(3-CHLORO-4-METHYL-7-COUMARINYL) PHOSPHATE ◇ 2-CHLOROETHANOL HYDROGEN PHOSPHATE ESTER with 3-CHLORO-7-HYDROXY-4-METHYLCOUMARIN ◇ 2-CHLOROETHANOL PHOSPHATE DIESTER ESTER with 3-CHLORO-7-HYDROXY-4-METHYLCOUMARIN ◇ 3-CHLORO-7-HYDROXY-4-METHYLCOUMARINBIS(2-CHLOROETHYL)PHOSPHATE ◇ 3-CHLORO-4-METHYL-UMBELLIFERONE BIS(2-CHLOROETHYL)PHOSPHATE ◇ DI-(2-CHLOROETHYL)-3-CHLORO-4-METHYL-7-COUMARINYL PHOSPHATE ◇ DI-(2-CHLOROETHYL)-3-CHLORO-4-METHYLCOUMARIN-7-YL PHOSPHATE ◇ EUSTIDIL ◇ GALLOXON ◇ GALOXANE ◇ 96H60 ◇ HALOXON ◇ HELMIRANE ◇ HELMIRON ◇ HELMIRONE ◇ LOXON ◇ LUXON ◇ LXON

TOXICITY DATA with REFERENCE
dni-hmn:oth 10 mg/L JTEHD6 10,143,82
orl-rat LD50:900 mg/kg FAZMAE 17,108,73
ipr-ckn LD50:800 mg/kg BCPCA6 16,1183,67
orl-dom LD50:763 mg/kg AJVRAH 41,1857,80

SAFETY PROFILE: Moderately toxic by ingestion and intraperitoneal routes. Human mutation data reported.

When heated to decomposition it emits very toxic fumes of PO_x and Cl^-. See also other coumarin entries.

DFH800 CAS:25323-30-2 ***HR: 3***
DICHLOROETHYLENE
DOT: UN 1150
mf: $C_2H_2Cl_2$ mw: 96.94

TOXICITY DATA with REFERENCE
ihl-mus LCLo:76 g/m^3/2H AEXPBL 83,235,18
ihl-gpg LCLo:155 g/m^3/1H AEXPBL 83,235,18
orl-mam LDLo:2500 mg/kg UGLAAD 121,375,59

DOT Classification: Flammable Liquid; Label: Flammable Liquid.

SAFETY PROFILE: Moderately toxic by ingestion. Mildly toxic by inhalation. Flammable when exposed to heat or flame. When heated to decomposition it emits toxic fumes of Cl^-. See also 1,1-DICHLOROETHYLENE.

DFI100 CAS:540-59-0 ***HR: 3***
1,2-DICHLOROETHYLENE
mf: $C_2H_2Cl_2$ mw: 96.94

SYNS: ACETYLENE DICHLORIDE ◇ DIOFORM ◇ 1,2-DICHLORAETHEN (GERMAN) ◇ DICHLORO-1,2-ETHYLENE (FRENCH) ◇ sym-DICHLOROETHYLENE ◇ NCI-C56031

TOXICITY DATA with REFERENCE
ipr-mus LD50:2 g/kg EJTXAZ 7,247,74
ihl-frg LCLo:117 mg/m^3/1H AISFAR 15,1,37
orl-rat LD50:770 mg/kg ARSIM* 20,10,66

CONSENSUS REPORTS: Reported in EPA TSCA Inventory. Community Right-To-Know List.

OSHA PEL: TWA 200 ppm
ACGIH TLV: TWA 200 ppm
DFG MAK: 200 ppm (790 mg/m^3)

SAFETY PROFILE: Poison by inhalation. Moderately toxic by ingestion and other routes. When heated to decomposition it emits highly toxic fumes of Cl^-. See also ACETYLENE COMPOUNDS; and CHLORINATED HYDROCARBONS, ALIPHATIC.

DFI200 CAS:156-59-2 ***HR: 1***
cis-DICHLOROETHYLENE
mf: $C_2H_2Cl_2$ mw: 96.94

HCCl=CHCl

PROP: Colorless liquid, pleasant odor. Mp: −80.5°, bp: 59°, lel: 9.7%, uel: 12.8%, flash p: 39°F, d: 1.2743 @ 25°/4°, vap press: 400 mm @ 41.0°, vap d: 3.34.

SYN: 1,2-DICHLOROETHYLENE

TOXICITY DATA with REFERENCE
mmo-smc 100 mmol/L TCMUD8 4,365,84
mma-smc 40 mmol/L TCMUD8 4,365,84
mrc-smc 100 mmol/L TCMUD8 4,365,84
dns-rat:lvr 4300 μmol/L CRNGDP 5,1629,84
ihl-mus LCLo:65000 mg/m^3/2H AHBAAM 116,131,36
ihl-cat LCLo:20000 mg/m^3/6H AHBAAM 116,131,36

CONSENSUS REPORTS: Reported in EPA TSCA Inventory.

DFG MAK: 200 ppm (790 mg/m^3)

SAFETY PROFILE: Mildly toxic by ingestion and inhalation. In high concentration it is irritating and narcotic. Has produced liver and kidney injury in experimental animals. Mutation data reported. Sometimes thought to be nonflammable, however, it is a dangerous fire hazard when exposed to heat or flame. Reaction with solid caustic alkalies or their concentrated solutions produces chloracetylene gas which ignites spontaneously in air. Reacts violently with N_2O_4, KOH, Na, NaOH. Moderate explosion hazard in the form of vapor when exposed to flame. Can react vigorously with oxidizing materials. To fight fire, use water spray, foam, CO_2, dry chemical. When heated to decomposition it emits toxic fumes of Cl^-. See also 1,1-DICHLOROETHYLENE and CHLORINATED HYDROCARBONS, ALIPHATIC.

DFI800 CAS:3967-55-3 ***HR: 3***
1,2-DICHLOROETHYLENE CARBONATE
mf: $C_3H_2Cl_2O_3$ mw: 156.95

SYN: 4,5-DICHLORO-2-OXO-1,3-DIOXOLANE

TOXICITY DATA with REFERENCE
scu-mus TDLo:648 mg/kg/54W-I:ETA JNCIAM 48,1431,72

SAFETY PROFILE: Questionable carcinogen with experimental tumorigenic data. When heated to decomposition it emits toxic fumes of Cl^-.

DFJ000 CAS:14096-51-6 ***HR: 3***
DICHLORO(ETHYLENEDIAMMINE)PLATINUM(II)
mf: $C_2H_8Cl_2N_2Pt$ mw: 326.11

SYNS: ETHYLENEDIAMINEDICHLORIDE PLATINUM (II) ◇ PLATINUM ETHYLENEDIAMMINE DICHLORIDE

TOXICITY DATA with REFERENCE
mmo-sat 2 μg/plate MUREAV 77,45,80
dni-hmn:oth 25 μmol/L IJCNAW 6,207,70
ipr-mus LDLo:14 mg/kg BCPCA6 2,187,73

SAFETY PROFILE: Poison by intraperitoneal route. Human mutation data reported. See also PLATINUM COMPOUNDS. When heated to decomposition it emits very toxic fumes of Cl^- and NO_x.

DFJ050 CAS:111-44-4 ***HR: 3***
DICHLOROETHYL ETHER
DOT: UN 1916
mf: $C_4H_8Cl_2O$ mw: 143.02

PROP: Colorless, stable liquid. Bp: 178.5°, fp: −51.9°, flash p: 131°F (CC), d: 1.2220 @ 20°/20°, autoign temp: 696°F, vap press: 0.7 mm @ 20°, vap d: 4.93.

SYNS: BIS(2-CHLOROETHYL) ETHER ◇ BIS(β-CHLOROETHYL) ETHER ◇ CHLOREX ◇ 1-CHLORO-2-(β-CHLOROETHOXY)ETHANE ◇ CHLOROETHYL ETHER ◇ CLOREX ◇ DCEE ◇ 2,2′-DICHLOROETHYLETHER (DUTCH) ◇ 2,2′-DICHLOR-DIAETHYLAETHER (GERMAN) ◇ 2,2′-DICHLORETHYL ETHER ◇ β,β-DICHLORODIETHYL ETHER ◇ DICHLOROETHER ◇ DI(β-CHLOROETHYL)ETHER ◇ β,β′-DICHLOROETHYL ETHER ◇ sym-DICHLOROETHYL ETHER ◇ 2,2′-DICHLOROETHYL ETHER (MAK) ◇ DICHLOROETHYL OXIDE ◇ 2,2′-DICLOROETILETERE (ITALIAN) ◇ DWUCHLORODWUETYLOWY ETER (POLISH) ◇ ENT 4,504 ◇ ETHER DICHLORE (FRENCH) ◇ 1,1′-OXYBIS(2-CHLORO)ETHANE ◇ OXYDE de CHLORETHYLE (FRENCH) ◇ RCRA WASTE NUMBER U025

TOXICITY DATA with REFERENCE
skn-rbt 10 mg/24H open JIHTAB 30,63,48
skn-rbt 500 mg open MLD UCDS** 12/29/71
eye-rbt 20 mg AJOPAA 29,1363,46
mmo-sat 1 mL/plate/2H DHEFDK FDA-78-1046,78
mma-sat 1 mg/plate ENMUDM 8 (Suppl 7),1,86
orl-mus TDLo:33 g/kg/79W-C:CAR JNCIAM 42,1101,69
scu-mus TDLo:2400 mg/kg/60W-I:ETA JNCIAM 48,1431,72
orl-rat LD50:75 mg/kg JIHTAB 30,63,48
ihl-rat LC50:330 mg/m³/4H 85GMAT -,45,82
orl-mus LD50:112 mg/kg 85GMAT -,45,82
ihl-mus LC50:650 mg/m³/2H 85GMAT -,45,82
skn-rbt LD50:720 mg/kg UCDS** 12/29/71
skn-gpg LD50:300 mg/kg JIHTAB 30,63,48

CONSENSUS REPORTS: IARC Cancer Review: Group 3 IMEMDT 7,56,87; Animal Sufficient Evidence IMEMDT 9,117,75. Reported in EPA TSCA Inventory. On Community Right-To-Know List. On EPA Extremely Hazardous Substances List.

OSHA PEL: (Transitional: CL 15 ppm (skin)) TWA 5 ppm; STEL 10 ppm (skin)
ACGIH TLV: TWA 5 ppm; STEL 10 ppm (skin)
DFG MAK: 10 ppm (60 mg/m³)
DOT Classification: IMO: Poison B; Label: Poison.

SAFETY PROFILE: A poison by ingestion, skin contact, and inhalation. A skin, eye, and mucous membrane irritant. Questionable carcinogen with experimental carcinogenic and tumorigenic data. Mutation data reported. Exposure to 1000 ppm for 30 to 60 minutes may result in death within days. The odor is easily detectable at 35 ppm which causes only slight irritation. Flammable when exposed to heat, flame, or oxidants. Dangerous explosion hazard; reacts vigorously with oleum, chlorosulfonic acid. Reacts with water or steam to evolve toxic and corrosive fumes. Can react vigorously with oxidizing materials. To fight fire, use water, foam, mist, fog, spray, dry chemical. When heated to decomposition it emits toxic fumes of Cl^-. See also ETHERS.

DFJ100 CAS:90584-32-0 ***HR: 3***
1,2-DICHLOROETHYL HYDROPEROXIDE
mf: $C_2H_4Cl_2O_2$ mw: 130.96

SAFETY PROFILE: Undergoes rapid exothermic decomposition at room temperature. When heated to decomposition it emits toxic fumes of Cl^-. See also PEROXIDES.

DFJ200 CAS:63917-06-6 ***HR: 3***
DI-2-CHLOROETHYL MALEATE
mf: $C_8H_{10}Cl_2O_4$ mw: 241.08

SYN: DI(2-CHLOROETHYL) ESTER, MALEIC ACID

TOXICITY DATA with REFERENCE
orl-rat LD50:71 mg/kg TXAPA9 28,313,74
skn-rbt LD50:140 mg/kg TXAPA9 28,313,74

SAFETY PROFILE: Poison by ingestion and skin contact. When heated to decomposition it emits toxic fumes of Cl^-.

DFJ400 CAS:20198-77-0 ***HR: 3***
2,3-DICHLORO-N-ETHYLMALEINIMIDE
mf: $C_6H_5Cl_2NO_2$ mw: 194.02

SYN: N-ETHYL-DICHLOROMALEINIMIDE

TOXICITY DATA with REFERENCE
ipr-mus TDLo:6200 μg/kg/(9D preg):REP ARTODN 37,15,76
ipr-mus TDLo:6200 μg/kg/(9D preg):TER ARTODN 37,15,76
ipr-mus LD50:15 mg/kg ARTODN 37,15,76
ivn-mus LD50:5600 μg/kg CSLNX* NX#03694

SAFETY PROFILE: Poison by intraperitoneal and intravenous routes. An experimental teratogen. Experimental reproductive effects. When heated to decomposition it emits very toxic fumes of Cl^- and NO_x.

DFJ500 CAS:10232-90-3 ***HR: 2***
2-(1,2-DICHLOROETHYL)-4-METHYL-1,3-DIOXOLANE
mf: $C_6H_{10}Cl_2O_2$ mw: 185.06

SYN: 1,3-DIOXOLANE,2-(1,2-DICHLOROETHYL)-4-METHYL-

TOXICITY DATA with REFERENCE
skn-rbt 10 mg/24H open MLD AIHAAP 23,95,62
orl-rat LD50:620 mg/kg AIHAAP 23,95,62
skn-rbt LD50:1010 mg/kg AIHAAP 23,95,62

SAFETY PROFILE: Moderately toxic by ingestion and

skin contact. A skin irritant. When heated to decomposition it emits toxic fumes of Cl^-.

DFJ800 CAS:1125-27-5 ***HR: 3***
DICHLOROETHYLPHENYLSILANE
DOT: UN 2435
mf: $C_8H_{10}Cl_2Si$ mw: 205.17

PROP: Liquid.

SYN: ETHYL PHENYL DICHLOROSILANE (DOT)

TOXICITY DATA with REFERENCE

CONSENSUS REPORTS: Reported in EPA TSCA Inventory.

DOT Classification: Corrosive Material; Label: Corrosive.

SAFETY PROFILE: Poison by ingestion and inhalation. A poison irritant to skin, eyes, and mucous membranes. Corrosive. Will react with water or steam to produce toxic and corrosive fumes. Can react with oxidizing materials. When heated to decomposition it emits toxic fumes of Cl^- and phenol. See also CHLOROSILANES.

DFK000 CAS:1789-58-8 ***HR: 3***
DICHLOROETHYLSILANE
DOT: UN 1183
mf: $C_2H_6Cl_2Si$ mw: 129.07

PROP: Liquid. Vap d: 4.45, flash p: < 73.4°F.

SYN: ETHYL DICHLOROSILANE (DOT)

CONSENSUS REPORTS: Reported in EPA TSCA Inventory.

DOT Classification: Flammable Liquid; Label: Flammable Liquid; IMO: Flammable Liquid; Label: Flammable Liquid, Corrosive.

SAFETY PROFILE: Poison by ingestion and inhalation. A severe irritant to skin, eyes, and mucous membranes. Corrosive. Dangerous fire hazard if exposed to heat, open flames or powerful oxidizers. Will react with water or steam to produce heat and toxic and corrosive fumes. To fight fire, use foam, dry chemical, mist, spray. When heated to decomposition it emits toxic fumes of Cl^- and phosgene. See also CHLOROSILANES.

DFK200 CAS:63918-89-8 ***HR: 3***
2-2'-DI(3-CHLOROETHYLTHIO)DIETHYL ETHER
mf: $C_8H_{16}Cl_2OS_2$ mw: 263.26

SYNS: BIS(β-CHLOROETHYLTHIOETHYL) ETHER ◇ BIS(2-CHLOROETHYLTHIOETHYL) ETHER ◇ 1,1'-OXYBIS(2-(2-CHLOROETHYL)THIOETHANE

TOXICITY DATA with REFERENCE
sln-dmg-ihl 100 pph/5M PREBA3 62B,284,46/47
ihl-hmn LCLo:400 mg/m³ SCJUAD 4,33,67
ihl-mus LC50:1650 mg/m³/10M NTIS** PB158-508

CONSENSUS REPORTS: EPA Genetic Toxicology Program.

SAFETY PROFILE: A human poison by inhalation. Mutation data reported. When heated to decomposition it emits very toxic fumes of SO_x and Cl^-. See also ETHERS and CHLORIDES.

DFK400 CAS:10138-21-3 ***HR: 2***
DICHLOROETHYLVINYLSILANE
mf: $C_4H_8Cl_2Si$ mw: 155.11

SYN: ETHYLVINYLDICHLOROSILANE

TOXICITY DATA with REFERENCE
skn-rbt 100 μg/24H open AIHAAP 23,95,62
orl-rat LDLo:2830 mg/kg AIHAAP 23,95,62
ihl-rat LCLo:8000 ppm/4H AIHAAP 23,95,62
skn-rbt LD50:750 mg/kg AIHAAP 23,95,62

CONSENSUS REPORTS: Reported in EPA TSCA Inventory.

SAFETY PROFILE: Moderately toxic by ingestion and skin contact. Mildly toxic by inhalation. A skin irritant. When heated to decomposition it emits toxic fumes of Cl^-. See also CHLOROSILANES.

DFK600 CAS:97-17-6 ***HR: 3***
DICHLOROFENTHION
mf: $C_{10}H_{13}Cl_2O_3PS$ mw: 315.16

PROP: A nonvolatile, residual organic phosphate nematocide and insecticide. Bp: 166° @ 0.1 mm, d: 1.3. Insol in water; sol in most organic solvents.

SYNS: BROMEX ◇ O,O-DIAETHYL-O-2,4-DICHLOR-PHENYL-MONOTHIOPHOSPHAT (GERMAN) ◇ O,O-DIAETHYL-O-2,4-DICHLORPHENYL-THIONOPHOSPHAT (GERMAN) ◇ DICHLOFENTHION ◇ DICHLOFENTION ◇ 2,4-DICHLORO-PHENOL-O-ESTER with O,O-DIETHYL PHOSPHOROTHIOATE ◇ O-2,4-DICHLOROPHENYL-O,O-DIETHYL PHOSPHOROTHIOATE ◇ 2,4-DICHLOROPHENYL DIETHYL PHOSPHOROTHIONATE ◇ O,O-DIETHYL-O-(2,4-DICHLOOR-FENYL)-MONOTHIOFOSFAAT ◇ O,O-DIETHYL-O-(2,4-DICHLOROPHENYL) PHOSPHOROTHIOATE ◇ DIETHYL 2,4-DICHLOROPHENYL PHOSPHOROTHIONATE ◇ O,O-DIETHYL-O-2,4-DICHLOROPHENYL THIOPHOSPHATE ◇ O,O-DIETIL-O-(2,4-DICLORO-FENIL)-MONOTIOFOSFATO (ITALIAN) ◇ ECP ◇ ENT 17,470 ◇ HEXA-NEMA ◇ MOBILAWN ◇ NEMACIDE ◇ THIOPHOSPHATE de O-2,4-DICHLOROPHENYLE et de O,O-DIETHYLE (FRENCH) ◇ TRI-VC 13 ◇ VC13 NEMACIDE

TOXICITY DATA with REFERENCE
orl-rat LD50:250 mg/kg WRPCA2 9,119,70
skn-rbt LD50:6000 mg/kg 31ZOAD 1,136,68
orl-pgn LD50:75 mg/kg ASTTA8 (680),157,79
orl-ckn LD50:148 mg/kg TXAPA9 11,49,67
orl-qal LD50:316 mg/kg ASTTA8 (680),157,79

orl-mam LD50:270 mg/kg FMCHA2 -,C160,83
orl-bwd LD50:14 mg/kg TXAPA9 21,315,72

SAFETY PROFILE: Poison by ingestion. Mildly toxic by skin contact. A very toxic insecticide. See also ESTERS and PARATHION. When heated to decomposition it emits very toxic fumes of PO_x, SO_x, and Cl^-.

DFL000 CAS:75-43-4 **HR: 1**
DICHLOROFLUOROMETHANE
DOT: UN 1029
mf: $CHCl_2F$ mw: 102.92

PROP: Heavy, colorless gas. Mp: −135°, bp: 8.9°, d: 1.48, vap press: 2 atm @ 28.4°, vap d: 3.82.

SYNS: ALGOFRENE TYPE 5 ◇ ARCTON 7 ◇ DICHLOROMONOFLUOROMETHANE (OSHA, DOT) ◇ DWUCHLOROFLUOROMETAN (POLISH) ◇ FLUORODICHLOROMETHANE ◇ FREON 21 ◇ GENETRON 21

TOXICITY DATA with REFERENCE
ihl-rat TCLo:1 pph/6H (6-15D preg):REP TXAPA9 45,293,78
ihl-rat LC50:49900 ppm/4H DTLVS* 4,132,80
ihl-gpg LCLo:10 pph/1H FLCRAP 1,197,67

CONSENSUS REPORTS: Reported in EPA TSCA Inventory.

OSHA PEL: (Transitional: TWA 1000 ppm) TWA 10 ppm
ACGIH TLV: TWA 10 ppm
DFG MAK: 10 ppm (45 mg/m^3)
DOT Classification: Nonflammable Gas; Label: Nonflammable Gas.

SAFETY PROFILE: Mildly toxic by inhalation. Experimental reproductive effects. When heated to decomposition it emits very toxic fumes of Cl^- and F^-.

DFL200 CAS:1085-98-9 **HR: 2**
N-(DICHLOROFLUOROMETHYLTHIO)-N',N'-DIMETHYL-N-PHENYLSULFAMIDE
mf: $C_9H_{11}Cl_2FN_2O_2S_2$ mw: 333.24

SYNS: BAY 47531 ◇ BAYER 47531 ◇ DICHLOFLUANID ◇ DICHLOFLUANIDE ◇ N-DICHLORFLUORMETHYLTHIO-N',N'-DIMETHYLAMINOSULFONSAEUREANILID (GERMAN) ◇ N-(DICHLOR-FLUORMETHYL-THIO)-N',N'-DIMETHYL-N-PHENYL-SCHWEFEL-SAEUREDIAMID (GERMAN) ◇ 1,1-DICHLORO-N-((DIMETHYLAMINO)SULFONYL)-1-FLUORO-N-PHENYLMETHANE SULFENAMIDE ◇ N-((DICHLOROFLUOROMETHYL)THIO)-N-((DIMETHYLAMINO)SULFONYL)ANILINE ◇ N-(DICHLOROFLUOROMETHYLTHIO)-N-(DIMETHYLSULFAMOYL)ANILINE ◇ N,N-DIMETHYL-N'-PHENYL-N'-FLUORODICHLOROMETHYLTHIOSULFAMIDE ◇ ELVARON ◇ EPAREN ◇ EUPAREN ◇ EUPARENE ◇ KU 13-032-C ◇ KUE 13032c

TOXICITY DATA with REFERENCE
mmo-esc 10 µg/plate MUREAV 116,185,83
mma-esc 100 µg/plate MUREAV 116,185,83
orl-rat LD50:500 mg/kg WRPCA2 9,119,70
skn-rat LD50:1000 mg/kg GUCHAZ 6,179,73
orl-mus LD50:1250 mg/kg MEIEDD 10,442,83
orl-cat LD50:1 g/kg 85GYAZ -,97,71

CONSENSUS REPORTS: EPA Genetic Toxicology Program.

SAFETY PROFILE: Moderately toxic by ingestion and skin contact. Mutation data reported. A pesticide. When heated to decomposition it emits very toxic fumes of Cl^-, F^-, NO_x, and SO_x.

DFL400 CAS:731-27-1 **HR: 3**
N'-DICHLOROFLUOROMETHYLTHIO-N,N-DIMETHYL-N'-(4-TOLYL)SULFAMIDE
mf: $C_{10}H_{13}Cl_2FN_2O_2S_2$ mw: 347.27

SYNS: N,N-DIMETHYL-N'-(4-TOLYL)-N'-(DICHLORFLUORMETHYLTHIO)SULFAMID (GERMAN) ◇ N,N-DIMETHYL-N-(4-TOLYL)-N-(DICHLOROFLUOR-METHYLTHIO)SULFAMIDE

TOXICITY DATA with REFERENCE
orl-rat LD50:1000 mg/kg GUCHAZ 6,505,73
skn-rat LD50:500 mg/kg GUCHAZ 6,505,73
orl-rbt LD50:500 mg/kg 28ZEAL 5,225,76
orl-gpg LD50:250 mg/kg 85DPAN -,-,71/76
orl-brd LD50:1000 mg/kg 28ZEAL 5,225,76

SAFETY PROFILE: Poison by ingestion. Moderately toxic by skin contact. When heated to decomposition it emits very toxic fumes of Cl^-, F^-, NO_x, and SO_x.

DFL600 CAS:15230-48-5 **HR: 3**
DICHLOROGERMANE
mf: Cl_2GeH_2 mw: 145.51

PROP: Colorless liquid. Mp: −68.0°, bp: 69.5°, d: 1.90 @ 68°, vap d: 5.0.

SAFETY PROFILE: Reaction with ammonia forms heat-sensitive explosive product. See also HYDROCHLORIC ACID and GERMANIUM COMPOUNDS. When heated to decomposition it emits toxic fumes of Cl^-.

DFL709 CAS:58941-14-3 **HR: 3**
N,N-DICHLOROGLYCINE
mf: $C_2H_3Cl_2NO_2$ mw: 143.96

SAFETY PROFILE: Explodes when heated to 65°C. When heated to decomposition it emits toxic fumes of Cl^- and NO_x.

DFL800 CAS:16260-59-6 **HR: 3**
1,6-DICHLORO-2,4-HEXADIYNE
mf: $C_6H_4Cl_2$ mw: 147.00

$ClCH_2(C{\equiv}C)_2CH_2Cl$

SAFETY PROFILE: An extremely shock-sensitive explosive. Upon decomposition it emits toxic fumes of Cl^-. See also CHLORINATED HYDROCARBONS, ALIPHATIC.

DFM000 CAS:303-04-8 ***HR: 3***
2,3-DICHLOROHEXAFLUOROBUTENE-2
mf: $C_4Cl_2F_6$ mw: 232.94

SYNS: DCHFB ◇ 2,3-DICHLOROHEXAFLUORO-2-BUTENE ◇ 2,3-DICHLORO-1,1,1,4,4,4-HEXAFLUOROBUTENE-2

TOXICITY DATA with REFERENCE
orl-rat LDLo:1000 mg/kg DOWCC* -,-,63
ihl-rat LC50:16 ppm/4H BJANAD 37,716,65
ihl-mus LC50:26 ppm/4H BJANAD 37,716,65
ihl-dog LC50:182 ppm/4H JETOAS 4,517,71
ihl-mky LC50:54 ppm/3H ANESAV 26,140,65

SAFETY PROFILE: Poison by inhalation. Moderately toxic by ingestion. When heated to decomposition it emits very toxic fumes of Cl^- and F^-.

DFM099 ***HR: 3***
4,5-DICHLORO-3,3,4,5,6,6-HEXAFLUORO-1,2-DIOXANE
mf: $C_4Cl_2F_6O_2$ mw: 264.94

$F_2C(CFCl)_2CF_2OO$ (ring)

SAFETY PROFILE: Explodes violently when heated. When heated to decomposition it emits toxic fumes of F^- and Cl^-. See also CHLORIDES and FLUORIDES.

DFM200 CAS:13442-13-2 ***HR: 3***
6,7-DICHLORO-4-(HYDROXYAMINO)QUINOLINE-1-OXIDE
mf: $C_9H_6Cl_2N_2O_2$ mw: 245.07

TOXICITY DATA with REFERENCE
scu-mus TDLo:120 mg/kg/50D-I:ETA BCPCA6 16,631,67

SAFETY PROFILE: Questionable carcinogen with experimental tumorigenic data. When heated to decomposition it emits very toxic fumes of Cl^- and NO_x.

DFM600 CAS:101652-05-5 ***HR: 3***
6,7-DICHLORO-10-(3-(N-(2-HYDROXYETHYL)ETHYLAMINO)ISOALLOXAZINE SULFATE
mf: $C_{17}H_{19}Cl_2N_5O_3 \cdot H_2O_4S$ mw: 510.39

TOXICITY DATA with REFERENCE
ipr-rat LD50:30 mg/kg CMTRAG 2,96,61
scu-mus LD50:38 mg/kg CMTRAG 2,96,61
ivn-mus LD50:90 mg/kg CMTRAG 2,96,61

SAFETY PROFILE: Poison by intraperitoneal, subcutaneous, and intravenous routes. See also SULFATES and CHLORIDES. When heated to decomposition it emits very toxic fumes of SO_x, Cl^-, and NO_x.

DFM800 CAS:101652-07-7 ***HR: 3***
6,7-DICHLORO-10-(3-(N-(2-HYDROXYETHYL)METHYLAMINO)PROPYL) ISOALLOXAZINE SULFATE
mf: $C_{16}H_{17}Cl_2N_5O_3 \cdot H_2O_4S$ mw: 496.36

TOXICITY DATA with REFERENCE
ipr-rat LD50:45 mg/kg CMTRAG 2,96,61
ipr-mus LD50:75 mg/kg CMTRAG 2,96,61
scu-mus LD50:97 mg/kg CMTRAG 2,96,61
ivn-mus LD50:75 mg/kg CMTRAG 2,96,61
ims-mus LD50:40 mg/kg CMTRAG 2,96,61

SAFETY PROFILE: Poison by intraperitoneal, intravenous, intramuscular, and subcutaneous routes. When heated to decomposition it emits very toxic fumes of SO_x, NO_x, and Cl^-.

DFM875 CAS:90742-91-9 ***HR: 3***
1-(2,5-DICHLORO-6-(1-(1H-IMIDAZOL-1-YL)VINYL)PHENOXY)-3-(ISOPROPYLAMINO)-2-PROPANOL HYDROCHLORIDE
mf: $C_{17}H_{21}Cl_2N_3O_2 \cdot ClH$ mw: 406.77

SYNS: 711389-S ◇ 1-(1-(2-(3-ISOPROPYLAMINO-2-HYDROXYPROPOXY)-3,6-DICHLOROPHENYL)VINYL)-1H-IMIDAZOLEHCl

TOXICITY DATA with REFERENCE
unr-rat LD50:245 mg/kg DRFUD4 10,472,85
ivn-mus LD50:19800 μg/kg JMCMAR 27,1142,84
unr-mus LD50:171 mg/kg DRFUD4 10,472,85

SAFETY PROFILE: Poison by intravenous and possibly other routes. When heated to decomposition it emits toxic fumes of Cl^- and NO_x.

DFN400 CAS:59-61-0 ***HR: 3***
3,4-DICHLORO-α-(ISOPROPYLAMINO)METHYL) BENZYL ALCOHOL
mf: $C_{11}H_{15}Cl_2NO$ mw: 248.17

SYNS: DCI ◇ DICHLORISOPRENALINE (GERMAN) ◇ DICHLORISOPROTERENOL ◇ 3,4-DICHLOR-ISOPROTERENOL (GERMAN) ◇ 3,4-DICHLORO-α-(((1-METHYLETHYL)AMINO)METHYL) BENZENEMETHANOL ◇ N-(β-(3,4-DICHLOROPHENYL)-β-HYDROXYETHYL)ISOPROPYLAMINE ◇ 1-(3,4-DICHLOROPHENYL)-2-ISOPROPYLAMINOETHANOL ◇ β-HYDROXY-N-ISOPROPYL-3,4-DICHLOROPHENETHYLAMINE

TOXICITY DATA with REFERENCE
orl-mus LD50:165 mg/kg ARZNAD 18,48,68
ivn-mus LD50:39 mg/kg ARZNAD 18,48,68

SAFETY PROFILE: Poison by ingestion and intravenous routes. When heated to decomposition it emits very toxic fumes of Cl^- and NO_x.

DFN500 CAS:36417-16-0 ***HR: 3***
DICHLOROLAWSONE
mf: $C_{13}H_8Cl_2O_3$ mw: 283.11

SYNS: DCL ◇ DICHLOROALLYL LAWSONE ◇ 2-HYDROXY-3-(3,3-DICHLOROALLYL)-1,4-NAPHTHOQUINONE ◇ NSC 126771

TOXICITY DATA with REFERENCE
oms-mus:leu 6300 nmol/L NCREA8 39,4868,79
orl-rat LD50:281 mg/kg NCISP* JAN86
orl-mus LD50:192 mg/kg NCISP* JAN86
ipr-mus LD50:37780 μg/kg NCISP* JAN86

SAFETY PROFILE: Poison by ingestion and intraperitoneal routes. Mutation data reported. When heated to decomposition it emits toxic fumes of Cl^-. See also ALLYL COMPOUNDS.

DFN700 CAS:1122-17-4 ***HR: 3***
DICHLOROMALEIC ANHYDRIDE
mf: $C_4Cl_2O_3$ mw: 166.95

ClC=CClCO•OCO (ring)

SAFETY PROFILE: Mixtures with sodium chloride + urea undergo vigorous exothermic reaction above 118°C. When heated to decomposition it emits toxic fumes of Cl^-. See also ANHYDRIDES.

DFN800 CAS:1193-54-0 ***HR: 3***
DICHLOROMALEIMIDE
mf: $C_4HCl_2NO_2$ mw: 165.96

CO•CCl=CClCO•NH (ring)

SYNS: DICHLOROMALEINIMIDE ◇ 3,4-DICHLORO-2,5-PYRROLIDINEDIONE

TOXICITY DATA with REFERENCE
ipr-mus TDLo:25 mg/kg (9D preg):REP ARTODN 37,15,76
ipr-mus TDLo:25 mg/kg (9D preg):TER ARTODN 37,15,76
ipr-mus LD50:31 mg/kg ARTODN 37,15,76

SAFETY PROFILE: Poison by intraperitoneal route. Experimental teratogenic and reproductive effects. When heated to decomposition it emits very toxic fumes of Cl^- and NO_x.

DFO000 CAS:528-74-5 ***HR: 2***
3'5'-DICHLOROMETHOTREXATE
mf: $C_{20}H_{20}Cl_2N_8O_5$ mw: 523.38

SYNS: DCM ◇ DICHLOROAMETHOPTERIN ◇ 3',5'-DICHLOROAMETHOPTERIN ◇ 3',5'-DICHLORO-4-AMINO-4-DEOXY-N_{10}-METHYLPTEROGLUTAMIC ACID ◇ N-(3,5-DICHLORO-4-((2,4-DIAMINO-6-PTERIDINYL METHYL)METHYLAMINO)BENZOYL)GLUTAMICACID ◇ DICHLOROMETHOTREXATE ◇ NCI-C04875 ◇ NSC-29630

TOXICITY DATA with REFERENCE
mma-sat 1 mg/plate ENMUDM 5(Suppl 1),3,83
ipr-mus TDLo:500 mg/kg (female 9D post):REP TCMUD8 7,7,87
ipr-rat TDLo:75 mg/kg/7W-I:ETA CANCAR 40(Suppl 4),1935,77
ipr-mus LD50:655 mg/kg NCISP* JAN86
ivn-mus LD50:1021 mg/kg NTIS** PB82-172644

CONSENSUS REPORTS: NCI Carcinogenesis Studies (ipr): Equivocal Evidence: rat; No Evidence: mouse CANCAR 40,1935,77

SAFETY PROFILE: Moderately toxic by intraperitoneal and intravenous routes. Questionable carcinogen with experimental tumorigenic data. Experimental reproductive effects. Mutation data reported. When heated to decomposition it emits very toxic fumes of NO_x and Cl^-.

DFO200 CAS:72595-99-4 ***HR: D***
DICHLORO(4-METHOXYCARBONYL-O-PHENYLENEDIAMMINE)PLATINUM(II)
mf: $C_8H_{10}Cl_2N_2O_2Pt$ mw: 432.19

TOXICITY DATA with REFERENCE
mmo-sat 100 nmol/L JMCMAR 23,459,80
mma-sat 100 nmol/L JMCMAR 23,459,80

SAFETY PROFILE: Mutation data reported. See also PLATINUM COMPOUNDS. When heated to decomposition it emits very toxic fumes of Cl^- and NO_x.

DFO400 CAS:72595-97-2 ***HR: D***
DICHLORO(4-METHOXY-O-PHENYLENEDIAMMINE)PLATINUM(II)
mf: $C_7H_{10}Cl_2N_2OPt$ mw: 404.18

TOXICITY DATA with REFERENCE
mmo-sat 2200 nmol/L JMCMAR 23,459,80
mma-sat 2200 nmol/L JMCMAR 23,459,80

SAFETY PROFILE: Mutation data reported. See also PLATINUM COMPOUNDS. When heated to decomposition it emits very toxic fumes of Cl^- and NO_x.

DFO600 CAS:56776-25-1 ***HR: 3***
(2,3-DICHLORO-4-METHOXYPHENYL)-2-FURANYLMETHANONE)-O-(2-(DIETHYLAMINO)ETHYL) OXIME,MONOMETHANE SULFONATE
mf: $C_{18}H_{22}Cl_2N_2O_3•CH_4O_3S$ mw: 481.43

SYNS: ANP 4364 ◇ (DICHLORO-2,3-METHOXY-4) PHENYL FURYL-2-O-(DIETHYLAMINOETHYL)-CETONE-OXIME(FRENCH)

TOXICITY DATA with REFERENCE
ipr-mus LD50:110 mg/kg EJTXAZ 8,122,75
ivn-mus LD50:6 mg/kg EJTXAZ 8,188,75

SAFETY PROFILE: Poison by intraperitoneal and in-

travenous routes. When heated to decomposition it emits very toxic fumes of SO_x, NO_x, and Cl^-.

DFO800 CAS:2164-09-2 ***HR: 2***
3',4'-DICHLORO-2-METHYLACRYLANILIDE
mf: $C_{10}H_9Cl_2NO$ mw: 230.10

PROP: Solid. Insol in water but sol in acetone, alcohol, isophorone, DMSO. Mp: 128°.

SYNS: CHLORANOCRYL ◇ DCM ◇ DCMA ◇ 3,4-DICHLOROANILIDE-α-METHYLACRYLIC ACID ◇ 3',4'-DICHLORO-2-METHACRYLANILIDE ◇ N-(3,4-DICHLOROPHENYL)METHACRYLAMIDE ◇ N-(3,4-DICHLOROPHENYL)-2-METHYL-2-PROPENAMIDE ◇ DICRYL ◇ METHACRYLIC ACID-3,4-DICHLOROANILIDE ◇ NIA 4556 ◇ NIAGARA 4556

TOXICITY DATA with REFERENCE
orl-rat LD50:1800 mg/kg WRPCA2 9,119,70
skn-rat LD50:1780 mg/kg 31ZOAD 1,155,68
orl-mus LD50:410 mg/kg GTPZAB 21(12),30,77
ipr-mus LD50:3000 mg/kg ARZNAD 14,668,64
skn-rbt LD50:10 g/kg PCOC** -,375,66

SAFETY PROFILE: Moderately toxic by ingestion, skin contact, and intraperitoneal routes. An herbicide. When heated to decomposition it emits very toxic fumes of Cl^- and NO_x.

DFO900 CAS:7651-91-4 ***HR: 3***
N,N-DICHLOROMETHYLAMINE
mf: CH_3Cl_2N mw: 99.95

SAFETY PROFILE: Explodes on contact with water, sodium sulfide, or calcium hypochlorite. When heated to decomposition it emits toxic fumes of Cl^- and NO_x.

DFP200 CAS:593-89-5 ***HR: 3***
DICHLOROMETHYLARSINE
DOT: NA 1556
mf: CH_3AsCl_2 mw: 160.86

PROP: Colorless liquid. Bp: 134.5°, fp: −59°, flash p: > 221°F, d: 1.838 @ 20°/4°, vap press: 10 mm @ 24.3°, vap d: 5.40.

SYNS: METHYLARSINE DICHLORIDE ◇ METHYLARSONOUS DICHLORIDE ◇ METHYLDICHLORARSINE ◇ METHYLDICHLOROARSINE (DOT) ◇ TL 294

TOXICITY DATA with REFERENCE
ihl-mus LC50:2700 mg/m³/10M NTIS** PB158-508

CONSENSUS REPORTS: Arsenic and its compounds are on the Community Right-To-Know List.

OSHA PEL: TWA 0.5 mg/(As)/m³
DOT Classification: Poison A; Label: Poison Gas.

SAFETY PROFILE: Poison irritant to skin, eyes, and mucous membranes and poison by ingestion and inhalation. A blistering type of military poison. It is rapidly detoxified in the body. A moderately persistent gas. Combustible when exposed to heat or flame. To fight fire, use water, foam, CO_2, dry chemical. Explosive reaction with chlorine. Can react vigorously with oxidizing materials. Dangerous; when heated to decomposition or on contact with acid or acid fumes it emits highly toxic fumes of Cl^- and As. See also CHLOROVINYL ARSINE DICHLORIDE and ARSENIC COMPOUNDS.

DFP500 ***HR: 3***
1-(2,4-DICHLORO-β-(p-METHYLBENZYLOXY) PHENETHYL)IMIDAZOLE NITRATE
mf: $C_{19}H_{18}Cl_2N_2O{\bullet}HNO_3$ mw: 424.31

TOXICITY DATA with REFERENCE
orl-rat LD50:915 mg/kg IYKEDH 12,933,81
ipr-rat LD50:240 mg/kg IYKEDH 12,933,81
scu-rat LD50:1420 mg/kg IYKEDH 12,933,81
ivn-rat LD50:50 mg/kg IYKEDH 12,933,81
orl-mus LD50:720 mg/kg IYKEDH 12,933,81
ipr-mus LD50:180 mg/kg IYKEDH 12,933,81
scu-mus LD50:840 mg/kg IYKEDH 12,933,81
ivn-mus LD50:42 mg/kg IYKEDH 12,933,81

SAFETY PROFILE: Poison by intravenous and intraperitoneal routes. Moderately toxic by ingestion and subcutaneous routes. When heated to decomposition it emits toxic fumes of Cl^- and NO_x. See also NITRATES.

DFP600 CAS:58-54-8 ***HR: 3***
2,3-DICHLORO-4-(2-METHYLENEBUTYRL) PHENOXY ACETIC ACID
mf: $C_{13}H_{12}Cl_2O_4$ mw: 303.15

SYNS: CRINURYL ◇ (2,3-DICHLORO-4-(2-METHYLENEBUTYRYL) PHENOXY)ACETIC ACID ◇ (2,3-DICHLORO-4-(2-METHYLENE-1-OXOBUTYL)PHENOXY)ACETIC ACID ◇ EDECRIL ◇ EDECRIN ◇ EDECRINA ◇ ENDECRIL ◇ ETACRINIC ACID ◇ ETAKRINIC ACID ◇ ETHACRYNIC ACID ◇ HIDROMEDIN ◇ HYDROMEDIN ◇ (4-(2-METHYLENEBUTYRYL)-2,3-DICHLOROPHENOXY)ACETIC ACID ◇ METHYLENEBUTYRYL PHENOXYACETIC ACID ◇ MINGIT ◇ MK-595 ◇ OTACRIL ◇ REOMAX ◇ TALADREN ◇ UREGIT

TOXICITY DATA with REFERENCE
orl-wmn TDLo:4 mg/kg:EAR AIMDAP 117,715,66
orl-man TDLo:3 mg/kg:EAR,KID AIMDAP 117,715,66
ivn-wmn TDLo:3 mg/kg:EAR AIMDAP 117,715,66
orl-mus LD50:627 mg/kg MEIEDD 10,539,83
ivn-mus LD50:176 mg/kg MEIEDD 10,539,83

SAFETY PROFILE: Poison by intravenous route. Moderately toxic by ingestion. Human systemic effects by ingestion and intravenous routes: urine volume increase, impaired hearing, and tinnitus (ringing in the ears). A diuretic. When heated to decomposition it emits toxic fumes of Cl^-.

DFP800 CAS:1123-61-1 ***HR: 3***
DICHLORO-N-METHYLMALEIMIDE
mf: $C_5H_3Cl_2NO_2$ mw: 179.99

SYNS: 2,3-DICHLORO-N-METHYLMALEIMIDE ◇ N-METHYLDICHLOROMALEINIMIDE

TOXICITY DATA with REFERENCE
ipr-mus TDLo:3100 μg/kg (9D preg):REP ARTODN 37,15,76
ipr-mus TDLo:3100 μg/kg (9D preg):TER ARTODN 37,15,76
ipr-mus LD50:4 mg/kg ARTODN 37,15,76
ivn-mus LD50:10 mg/kg CSLNX* NX#03682

SAFETY PROFILE: Poison by intraperitoneal and intravenous routes. An experimental teratogen. Other experimental reproductive effects. When heated to decomposition it emits very toxic fumes of Cl^- and NO_x.

DFQ000 CAS:4885-02-3 ***HR: 3***
α,α-DICHLOROMETHYL METHYL ETHER
mf: $C_2H_4Cl_2O$ mw: 114.96

SYNS: BIS(CHLOROPHENYL) ETHER ◇ α,α-DICHLOROMETHYL ETHER

TOXICITY DATA with REFERENCE
skn-mus TDLo:40 mg/kg:ETA ANYAA9 163,633,69

CONSENSUS REPORTS: Reported in EPA TSCA Inventory.

SAFETY PROFILE: Questionable carcinogen with experimental tumorigenic data. See also ETHERS and CHLORIDES. When heated to decomposition it emits toxic fumes of Cl^-.

DFQ100 CAS:76738-28-8 ***HR: 2***
d-threo-2-(DICHLOROMETHYL)-α-(p-NITROPHENYL)-2-OXAZOLINE-4-METHANOL
mf: $C_{11}H_{10}Cl_2N_2O_4$ mw: 305.13

SYN: d-threo-2-DICLOROMETIL-4-((4'-NITROFENIL)-OSSIMETIL)-2-OSSAZOLINA (ITALIAN)

TOXICITY DATA with REFERENCE
orl-rat LD50:6 g/kg FRPSAX 10,3,55
orl-mus LD50:5700 mg/kg FRPSAX 10,3,55
ipr-mus LD50:4 g/kg FRPSAX 10,3,55
orl-gpg LD50:1000 mg/kg FRPSAX 10,3,55

SAFETY PROFILE: Moderately toxic by ingestion and intraperitoneal routes. When heated to decomposition it emits toxic fumes of Cl^- and NO_x.

DFQ200 CAS:84-57-1 ***HR: 2***
2,5-DICHLORO-4-(3-METHYL-5-OXO-2-PYRAZOLIN-1-YL) BENZENESULFONIC ACID
mf: $C_{10}H_8Cl_2N_2O_4S$ mw: 323.16

SYNS: 2,5-DICHLORO-4-(4,5-DIHYDRO-3-METHYL-5-OXO-1H-PYRAZOL-1-YL)BENZENESULFONIC ACID ◇ DICHLORSULFOFENYLMETHYLPYRAZOLON (CZECH) ◇ KYSELINA 2,5-DICHLOR-4-(3'-METHYL-5'-PYRAZOLON-1'-YL)BENZENSULFONOVA (CZECH)

TOXICITY DATA with REFERENCE
eye-rbt 500 mg/24H SEV 28ZPAK -,186,72

CONSENSUS REPORTS: Reported in EPA TSCA Inventory.

SAFETY PROFILE: A severe eye irritant. See also SULFONATES. When heated to decomposition it emits very toxic fumes of SO_x, NO_x, and Cl^-.

DFQ400 CAS:57948-13-7 ***HR: 3***
DICHLORO(4-METHYL-O-PHENYLENEDIAMMINE)PLATINUM(II)
mf: $C_7H_{10}Cl_2N_2Pt$ mw: 388.18

TOXICITY DATA with REFERENCE
mmo-sat 2500 nmol/L JMCMAR 23,459,80
mma-sat 2500 nmol/L JMCMAR 23,459,80
ipr-mus LD50:23 mg/kg RCRVAB 50,353,81

SAFETY PROFILE: Poison by intraperitoneal route. Mutation data reported. See also PLATINUM COMPOUNDS. When heated to decomposition it emits very toxic fumes of Cl^- and NO_x.

DFQ800 CAS:149-74-6 ***HR: 3***
DICHLOROMETHYLPHENYLSILANE
DOT: UN 2437
mf: $C_7H_8Cl_2Si$ mw: 191.14

SYNS: METHYLPHENYLDICHLOROSILANE (DOT) ◇ PHENYLMETHYLDICHLOROSILANE

TOXICITY DATA with REFERENCE
ipr-rat LDLo:100 mg/kg 85GMAT -,99,82
ihl-mus LCLo:200 mg/m^3/2H 85GMAT -,99,82
ipr-mus LDLo:100 mg/kg 85GMAT -,99,82
scu-mus LDLo:100 mg/kg 85GMAT -,99,82

CONSENSUS REPORTS: Reported in EPA TSCA Inventory. EPA Extremely Hazardous Substances List.

DOT Classification: IMO: Flammable or Combustible Liquid; Label: Flammable Liquid, Corrosive.

SAFETY PROFILE: Poison by inhalation, subcutaneous, and intraperitoneal routes. Corrosive to eyes, skin, and mucous membranes. When heated to decomposition it emits toxic fumes of Cl^-. See also CHLOROSILANES.

DFR400 CAS:10141-22-7 ***HR: 3***
2,3-DICHLORO-2-METHYLPROPIONALDEHYDE
mf: $C_4H_6Cl_2O$ mw: 141.00

TOXICITY DATA with REFERENCE
skn-rbt 10 mg/24H open MLD AIHAAP 23,95,62

orl-rat LD50:1620 mg/kg AIHAAP 23,95,62
ihl-rat LCLo:250 ppm/4H AIHAAP 23,95,62
skn-rbt LD50:360 mg/kg AIHAAP 23,95,62

SAFETY PROFILE: Poison by skin contact. Moderately toxic by ingestion and inhalation. A skin irritant. When heated to decomposition it emits toxic fumes of Cl^-. See also ALDEHYDES.

DFS000 CAS:75-54-7 ***HR: 3***
DICHLOROMETHYLSILANE
DOT: UN 1242
mf: CH_4Cl_2Si mw: 115.04

PROP: Colorless liquid, sol in benzene, ether, and heptane. Bp: 41°, d: 1.10 @ 27°, flash p: −26°F.

SYNS: METHYL DICHLOROSILANE (DOT) ◇ METHYL-DICHLOR-SILAN (CZECH)

TOXICITY DATA with REFERENCE
skn-rbt 500 mg/24H SEV 28ZPAK -,216,72
eye-rbt 20 mg/24H MOD 28ZPAK -,216,72
ihl-rat LCLo:300 ppm/4H 28ZPAK -,216,72

CONSENSUS REPORTS: Reported in EPA TSCA Inventory.

DOT Classification: Flammable Liquid; Label: Flammable Liquid

SAFETY PROFILE: Moderately toxic by inhalation. Corrosive. A severe irritant to skin, eyes, and mucous membranes. Ignites spontaneously in air. A very dangerous fire hazard when exposed to heat or flame. Forms impact-sensitive explosive mixtures with potassium permanganate, lead(II) oxide, lead(IV) oxide, copper oxide, silver oxide. To fight fire, use water, foam, CO_2, mist. When heated to decomposition it emits toxic fumes of Cl^-. See also CHLOROSILANE.

DFS200 CAS:2700-89-2 ***HR: 3***
1,2-DICHLORO-1-(METHYLSULFONYL)ETHYLENE
mf: $C_3H_4Cl_2O_2S$ mw: 175.03

SYN: CHEMAGRO D-113

TOXICITY DATA with REFERENCE
skn-rat 500 mg SEV 34ZIAG -,161,69
orl-rat LD50:61 mg/kg 34ZIAG -,161,69
skn-rat LD50:500 mg/kg 34ZIAG -,161,69
ipr-rat LD50:12500 μg/kg 34ZIAG -,161,69
ipr-mus LD50:12500 μg/kg 34ZIAG -,161,69
orl-gpg LD50:40 mg/kg 34ZIAG -,161,69
ipr-gpg LD50:12500 μg/kg 34ZIAG -,161,69

SAFETY PROFILE: Poison by ingestion and intraperitoneal routes. Moderately toxic by skin contact. A severe skin irritant. When heated to decomposition it emits very toxic fumes of SO_x and Cl^-.

DFS600 CAS:31335-41-8 ***HR: 2***
DICHLOROMETHYL TRICHLOROMETHYLTHIOSULFONE
mf: $C_2HCl_5O_2S_2$ mw: 298.40

SYNS: DICHLOROMETHANETHIOSULFONICACID-S-TRICHLOROMETHYL ESTER ◇ TRICHLORMETHYLESTER KYSELINY DICHLORMETHANTHIOSULFONOVE (CZECH)

TOXICITY DATA with REFERENCE
skn-rbt 500 mg/24H MLD 28ZPAK -,198,72
eye-rbt 20 mg/24H MOD 28ZPAK -,198,72
orl-rat LD50:3620 mg/kg 28ZPAK -,198,72

SAFETY PROFILE: Moderately toxic by ingestion. A skin and eye irritant. See also SULFONATES. When heated to decomposition it emits very toxic fumes of Cl^- and SO_x.

DFS800 CAS:124-70-9 ***HR: 3***
DICHLOROMETHYLVINYLSILANE
mf: $C_3H_6Cl_2Si$ mw: 141.08

PROP: Flash p: −1°C.

TOXICITY DATA with REFERENCE
ivn-mus LD50:56 mg/kg CSLNX* NX#03620

CONSENSUS REPORTS: Reported in EPA TSCA Inventory.

SAFETY PROFILE: Poison by intravenous route. A very dangerous fire hazard when exposed to heat, flame or oxidizers. When heated to decomposition it emits toxic fumes of Cl^-. See also CHLOROSILANES.

DFT000 CAS:117-80-6 ***HR: 3***
2,3-DICHLORO-1,4-NAPHTHOQUINONE
mf: $C_{10}H_4Cl_2O_2$ mw: 227.04

PROP: Golden-yellow crystals. Mp: 193°, vap d: 7.8. Insol in water; moderately sol in organic solvents.

SYNS: ALGISTAT ◇ COMPOUND 604 ◇ DICHLONE (DOT) ◇ 2,3-DICHLOR-1,4-NAPHTHOCHINON (GERMAN) ◇ 2,3-DICHLORO-1,4-NAPHTHALENEDIONE ◇ 2,3-DICHLORO-1,4-NAPHTHAQUINONE ◇ DICHLORONAPHTHOQUINONE ◇ 2,3-DICHLORONAPHTHOQUINONE ◇ 2,3-DICHLORO-α-NAPHTHOQUINONE ◇ 2,3-DICHLORONAPHTHOQUINONE-1,4 ◇ ENT 3,776 ◇ PHYGON ◇ PHYGON PASTE ◇ PHYGON SEED PROTECTANT ◇ PHYGON XL ◇ QUINTAR ◇ QUINTAR 540F ◇ SANQUINON ◇ UNIROYAL ◇ USR 604 ◇ U.S. RUBBER 604

TOXICITY DATA with REFERENCE
orl-mus TDLo:3300 mg/kg/78W-I:NEO NTIS** PB223-159
scu-mus TDLo:22 mg/kg:CAR NTIS** PB223-159
orl-rat LD50:160 mg/kg GTPZAB 16(5),52,72

ipr-mus LD50:30 mg/kg JMCMAR 26,570,83
skn-rbt LD50:5000 mg/kg FMCHA2 -,C77,83

CONSENSUS REPORTS: Reported in EPA TSCA Inventory.

SAFETY PROFILE: Poison by ingestion and intraperitoneal routes. Mildly toxic by skin contact. A skin, eye, and mucous membrane irritant. Large doses can cause central nervous system depression. Questionable carcinogen with experimental carcinogenic and neoplastigenic data. A fungicide and algaecide. When heated to decomposition it emits toxic fumes of Cl^-. See also CHLORIDES.

DFT400 CAS:89-61-2 ***HR: 2***
1,4-DICHLORO-2-NITROBENZENE
mf: $C_6H_3Cl_2NO_2$ mw: 192.00

SYNS: 2,5-DICHLORNITROBENZEN (CZECH) ◇ 2,5-DICHLORONITROBENZENE ◇ NITRO-p-DICHLOROBENZENE

TOXICITY DATA with REFERENCE
skn-rbt 500 mg/24H MLD 28ZPAK -,94,72
eye-rbt 100 mg/24H MOD 28ZPAK -,94,72
mmo-sat 205 μg/plate MUREAV 116,217,83
orl-rat LD50:1210 mg/kg 28ZPAK -,94,72
orl-mus LD50:2850 mg/kg GTPZAB 25(8),50,81

CONSENSUS REPORTS: Reported in EPA TSCA Inventory.

SAFETY PROFILE: Moderately toxic by ingestion. A skin and eye irritant. Mutation data reported. See also CHLORINATED HYDROCARBONS, AROMATIC; and NITRO COMPOUNDS of AROMATIC HYDROCARBONS. When heated to decomposition it emits very toxic fumes of Cl^- and NO_x.

DFT600 CAS:99-54-7 ***HR: 2***
1,2-DICHLORO-4-NITROBENZENE
mf: $C_6H_3Cl_2NO_2$ mw: 192.00

PROP: Liquid. Vap d: 6.6.

SYNS: DCNB ◇ 3,4-DICHLORNITROBENZEN (CZECH) ◇ 3,4-DICHLORONITROBENZENE

TOXICITY DATA with REFERENCE
skn-rbt 500 mg/24H MLD 28ZPAK -,94,72
eye-rbt 100 mg/24H MOD 28ZPAK -,94,72
mmo-sat 500 μg/plate AECTCV 9,533,80
mma-sat 500 μg/plate AECTCV 9,533,80
sln-dmg-par 200 ppm ENMUDM 7,677,85
orl-rat LD50:643 mg/kg 28ZPAK -,94,72

CONSENSUS REPORTS: Reported in EPA TSCA Inventory.

SAFETY PROFILE: Moderately toxic by ingestion. A skin and eye irritant. Mutation data reported. Potentially explosive reactions when heated with hydrogen + a catalyst. When heated to decomposition it emits very toxic fumes of NO_x and Cl^-. See also CHLORINATED HYDROCARBONS, AROMATIC; and NITRO COMPOUNDS of AROMATIC HYDROCARBONS.

DFT800 CAS:1836-75-5 ***HR: 3***
2,4-DICHLORO-4'-NITRODIPHENYL ETHER
mf: $C_{12}H_7Cl_2NO_3$ mw: 284.10

SYNS: 2,4-DECHLOROPHENYL-p-NITROPHENYL ETHER ◇ 2',4'-DICHLORO-4-NITROBIPHENYL ETHER ◇ 2,4-DICHLORO-1-(4-NITROPHENOXY)BENZENE ◇ 4-(2,4-DICHLOROPHENOXY)NITROBENZENE ◇ 2,4-DICHLOROPHENYL-p-NITROPHENYL ETHER ◇ 2,4-DICHLOROPHENYL-4-NITROPHENYL ETHER ◇ 2,4,-DICHLORPHENYL-4-NITROPHENYLAETHER (GERMAN) ◇ FW 925 ◇ MEZOTOX ◇ NCI-C00420 ◇ NICLOFEN ◇ NIP ◇ NITOFEN ◇ NITRAFEN ◇ NITRAPHEN ◇ NITROCHLOR ◇ 4'-NITRO-2,4-DICHLORODIPHENYL ETHER ◇ NITROFEN ◇ NITROFENE (FRENCH) ◇ NITROPHEN ◇ NITROPHENE ◇ PREPARATION 125 ◇ TOK ◇ TOK-2 ◇ TOK E ◇ TOK E-25 ◇ TOK E 40 ◇ TOKKORN ◇ TOK WP-50 ◇ TRIZILIN

TOXICITY DATA with REFERENCE
skn-rbt 500 mg/24H MOD 28ZPAK -,84,72
eye-rbt 100 mg/24H SEV 28ZPAK -,84,72
mmo-sat 33300 ng/plate ENMUDM 7(Suppl 5),1,85
mma-sat 10 μg/plate ENMUDM 7(Suppl 5),1,85
otr-rat:emb 1500 ng/plate JJATDK 1,190,81
orl-mus TDLo:1100 mg/kg (female 7-17D post):REP SCIEAS 215,293,82
orl-rat TDLo:150 mg/kg (female 7-12D post):TER TCMUD8 6,339,86
orl-rat TDLo:42 g/kg/94W-C:CAR NCITR* NCI-CG-TR-26,78
orl-mus TDLo:24 g/kg/12W-C:CAR JJIND8 65,937,80
orl-rat LD50:740 mg/kg HYSAAV 32,20,67
skn-rat LD50:5000 mg/kg AEHLAU 28,316,74
unk-rat LD50:3000 mg/kg 30ZDA9 -,109,71
orl-mus LD50:450 mg/kg HYSAAV 32,20,67
orl-cat LDLo:300 mg/kg HYSAAV 32,20,67
ihl-cat LCLo:620 mg/m³/4H HYSAAV 32,20,67
orl-rbt LD50:1620 mg/kg 28ZEAL 5,166,76

CONSENSUS REPORTS: NTP Fifth Annual Report on Carcinogens. IARC Cancer Review: Group 2B IMEMDT 7,56,87; Animal Sufficient Evidence IMEMDT 30,271,83. NCI Carcinogenesis Bioassay (feed); No Evidence: rat NCITR* NCI-CG-TR-184,79; Clear Evidence: mouse, rat NCITR* NCI-CG-TR-26,78; Clear Evidence: mouse NCITR* NCI-CG-TR-184,79. EPA Genetic Toxicology Program. Community Right-To-Know List. Reported in EPA TSCA Inventory.

SAFETY PROFILE: Confirmed carcinogen with experimental carcinogenic data. Poison by ingestion. Moderately toxic by inhalation and possibly other routes. Mildly toxic by skin contact. Experimental teratogenic and reproductive effects. A skin and severe eye irritant.

Mutation data reported. A broad spectrum herbicide. See also NITRO COMPOUNDS of AROMATIC HYDROCARBONS and ETHERS. When heated to decomposition it emits very toxic fumes of Cl^- and NO_x.

DFU000 CAS:594-72-9 ***HR: 3***
1,1-DICHLORO-1-NITROETHANE
DOT: UN 2650
mf: $C_2H_3Cl_2NO_2$ mw: 143.96

PROP: Liquid. Bp: 124°, flash p: 168°F(OC), d: 1.4153 @ 20°/20°, vap d: 4.97.

SYNS: 1,1-DICHLOOR-1-NITROETHAAN (DUTCH) ◇ 1,1-DICHLOR-1-NITROAETHAN (GERMAN) ◇ DICHLORONITROETHANE ◇ 1,1-DICLORO-1-NITROETANO (ITALIAN) ◇ ETHIDE

TOXICITY DATA with REFERENCE
orl-rat LD50:410 mg/kg BESAAT 12,161,66
ipr-mus LD50:240 mg/kg KHFZAN 10(6),53,76
orl-rbt LDLo:150 mg/kg JIHTAB 27,95,45
ihl-rbt LCLo:580 mg/m^3/6H JIHTAB 27,95,45
ihl-gpg LCLo:580 mg/m^3/6H JIHTAB 27,95,45

OSHA PEL: (Transitional: CL 10 ppm) TWA 2 ppm
ACGIH TLV: TWA 2 ppm
DFG MAK: 10 ppm (60 mg/m^3)
DOT Classification: Poison B; Label: Poison.

SAFETY PROFILE: Poison by ingestion and intraperitoneal routes. Moderately toxic by inhalation. A strong irritant. Inhalation causes pulmonary edema. A fumigant for produce. Flammable when exposed to heat, flame, or oxidizers. Can react vigorously with oxidizing materials. To fight fire, use water, CO_2, dry chemical. When heated to decomposition it emits highly toxic fumes of Cl^- and NO_x.

DFU400 CAS:6240-55-7 ***HR: 3***
1,2-DICHLORO-3-NITRONAPHTHALENE
mf: $C_{10}H_5Cl_2NO_2$ mw: 242.06

TOXICITY DATA with REFERENCE
orl-rat TDLo:13 g/kg/52W-I:ETA,REP JNCIAM 41,985,68

SAFETY PROFILE: Questionable carcinogen with experimental tumorigenic data. Experimental reproductive effects. When heated to decomposition it emits very toxic fumes of Cl^- and NO_x. See also NITRO COMPOUNDS of AROMATIC HYDROCARBONS.

DFU600 CAS:609-89-2 ***HR: 3***
2,4-DICHLORO-6-NITROPHENOL
mf: $C_6H_3Cl_2NO_3$ mw: 208.00
SYN: 2,4-DICHLOR-6-NITROFENOL (CZECH)

TOXICITY DATA with REFERENCE
eye-rbt 100 mg/24H SEV 28ZPAK -,80,72
orl-rat LD50:129 mg/kg 28ZPAK -,80,72

CONSENSUS REPORTS: Chlorophenol compounds are on the Community Right-To-Know List.

SAFETY PROFILE: Poison by ingestion. A severe eye irritant. When heated to decomposition it emits very toxic fumes of Cl^- and NO_x. See also CHLOROPHENOLS and NITRO COMPOUNDS of AROMATIC HYDROCARBONS.

DFU800 CAS:37169-10-1 ***HR: 3***
2,4-DICHLORO-6-NITROPHENOL ACETATE
mf: $C_8H_5Cl_2NO_4$ mw: 250.04

SYN: 2,4-DICHLOR-6-NITROFENYLESTER KYSELINY OCTIVE (CZECH)

TOXICITY DATA with REFERENCE
skn-rbt 500 mg/24H MLD 28ZPAK -,93,72
eye-rbt 20 mg/24H MOD 28ZPAK -,93,72
orl-rat LD50:96 mg/kg 28ZPAK -,93,72

CONSENSUS REPORTS: Chlorophenol compounds are on the Community Right-To-Know List.

SAFETY PROFILE: Poison by ingestion. A skin and eye irritant. See also CHLOROPHENOLS and NITRO COMPOUNDS of AROMATIC HYDROCARBONS. When heated to decomposition it emits very toxic fumes of Cl^- and NO_x.

DFV000 CAS:72596-02-2 ***HR: D***
DICHLORO(4-NITRO-o-PHENYLENEDIAMINE)PLATINUM(II)
mf: $C_6H_7Cl_2N_3O_2Pt$ mw: 419.15

TOXICITY DATA with REFERENCE
mmo-sat 2 nmol/L JMCMAR 23,459,80
mma-sat 2 nmol/L JMCMAR 23,459,80

SAFETY PROFILE: Mutation data reported. See also PLATINUM COMPOUNDS and NITRO COMPOUNDS of AROMATIC HYDROCARBONS. When heated to decomposition it emits very toxic fumes of Cl^- and NO_x.

DFV200 CAS:14094-48-5 ***HR: 3***
6,7-DICHLORO-4-NITROQUINOLINE-1-OXIDE
mf: $C_9H_4Cl_2N_2O_3$ mw: 259.05

TOXICITY DATA with REFERENCE
scu-mus TDLo:120 mg/kg/50D-I:ETA BCPCA6 16,631,67

SAFETY PROFILE: Questionable carcinogen with experimental tumorigenic data. When heated to decomposition it emits very toxic fumes of Cl^- and NO_x. See also

NITRO COMPOUNDS of AROMATIC HYDROCARBONS.

DFV400 CAS:50-65-7 ***HR: 3***
2',5-DICHLORO-4'-NITROSALICYLANILIDE
mf: $C_{13}H_8Cl_2N_2O_4$ mw: 327.13

SYNS: BAY 2353 ◇ BAYER 73 ◇ BAYER 2353 ◇ BAYLUSCID ◇ CHEMAGRO 2353 ◇ 5-CHLORO-N-(2-CHLORO-4-NITROPHENYL)-2-HYDROXYBENZAMIDE ◇ 5-CHLORO-2'-CHLORO-4'-NITROSALICYLANILIDE ◇ 2-CHLORO-4-NITROPHENYLAMIDE-6-CHLOROSALICYLIC ACID ◇ N-(2-CHLORO-4-NITROPHENYL)-5-CHLOROSALICYLAMIDE ◇ CLONITRALID ◇ 2',5-DICHLOR-4'-NITRO-SALIZYLSAEUREANILID (GERMAN) ◇ DICHLOSALE ◇ ENT 25,823 ◇ FENASAL ◇ HL 2447 ◇ 2-HYDROXY-5-CHLORO-N-(2-CHLORO-4-NITROPHENYL)BENZAMIDE ◇ IOMESAN ◇ IOMEZAN ◇ NICLOSAMIDE ◇ PHENASAL ◇ VERMITIN ◇ YOMESAN

TOXICITY DATA with REFERENCE
cyt-hmn:lym 6 mL/L MUREAV 173,81,86
sce-hmn:lym 8 mg/L MUREAV 173,81,86
orl-mus TDLo:300 mg/kg (male 5D pre):REP MUREAV 204,269,88
orl-rat LDLo:10 g/kg ZTMPA5 13,1,62
ipr-rat LD50:250 mg/kg ZTMPA5 13,1,62
orl-mus LD50:1000 mg/kg 85DPAN -,-,71/76
ivn-mus LD50:7500 μg/kg ARZNAD 10,884,60

SAFETY PROFILE: Poison by intraperitoneal and intravenous routes. Moderately toxic by ingestion. Experimental reproductive effects. Human mutation data reported. When heated to decomposition it emits very toxic fumes of Cl^- and NO_x.

DFV600 CAS:1420-04-8 ***HR: 3***
2',5-DICHLORO-4'-NITROSALICYLANILIDE-2-AMINOETHANOL SALT
mf: $C_{13}H_8Cl_2N_2O_4 \cdot C_2H_7NO$ mw: 388.23

SYNS: BAYER 73 ◇ BAYER 25648 ◇ BAYLUSCID ◇ BAYLUSCIDE ◇ 5-CHLORO-N-(2-CHLORO-4-NITROPHENYL)-2-HYDROXYBENZAMIDE with 2-AMINOETHANOL (1:1) ◇ CLONITARLID ◇ 5,2'-DICHLORO-4'-NITROSALICYLANILIDE ETHANOLAMINE SALT ◇ 5,2-DICHLORO-4-NITROSALICYLIC ANILIDE-2-AMINOETHANOL SALT ◇ 2',5-DICHLORO-4'-NITROSALICYLOYLANILIDE ETHANOLAMINE SALT ◇ ETHANOLAMINE SALT of 5,2'-DICHLORO-4'-NITROSALICYCLICANILIDE ◇ M 73 ◇ MOLLUSCICIDE BAYER 73 ◇ NCI-C00431 ◇ NICLOSAMIDE ◇ SR 73

TOXICITY DATA with REFERENCE
orl-rat LD50:500 mg/kg 85ARAE 3,103,76/77
ipr-rat LD50:250 mg/kg GUCHAZ 6,126,73

CONSENSUS REPORTS: NCI Carcinogenesis Bioassay (feed); Inadequate Studies: mouse, rat NCITR* NCI-CG-TR-91,78.

SAFETY PROFILE: Poison by intraperitoneal route. Moderately toxic by ingestion. Many N-nitroso compounds are carcinogens. A pesticide. When heated to decomposition it emits very toxic fumes of NO_x and Cl^-. See also N-NITROSO COMPOUNDS.

DFV800 CAS:100836-84-8 ***HR: D***
3,4-DICHLORO-N-NITROSOCARBANILIC ACID METHYL ESTER
mf: $C_8H_6Cl_2N_2O_3$ mw: 249.06

SYN: NITROSOSWEP

TOXICITY DATA with REFERENCE
mmo-sat 1 μL/plate MUREAV 48,225,77

SAFETY PROFILE: Mutation data reported. Many N-nitroso compounds are carcinogens. When heated to decomposition it emits very toxic fumes of Cl^- and NO_x. See also N-NITROSO COMPOUNDS.

DFW000 CAS:69112-96-5 ***HR: 3***
2,2'-DICHLORO-N-NITROSODIPROPYLAMINE
mf: $C_6H_{12}Cl_2N_2O$ mw: 199.10

SYN: NITROSOBIS(2-CHLOROPROPYL)AMINE

TOXICITY DATA with REFERENCE
mmo-sat 10 μg/plate MUREAV 66,1,79
mma-sat 10 μg/plate MUREAV 66,1,79
orl-rat TDLo:1360 mg/kg/20W-I:ETA EESADV 2,421,78

SAFETY PROFILE: Questionable carcinogen with experimental tumorigenic data. Many N-nitroso compounds are carcinogens. Mutation data reported. When heated to decomposition it emits very toxic fumes of Cl^- and NO_x. See also N-NITROSO COMPOUNDS.

DFW200 CAS:57541-72-7 ***HR: 3***
3,4-DICHLORONITROSOPIPERIDINE
mf: $C_5H_8Cl_2N_2O$ mw: 183.05

SYN: N-NITROSO-3,4-DICHLOROPIPERIDINE

TOXICITY DATA with REFERENCE
mmo-sat 200 μg/plate MUREAV 56,131,77
mma-sat 10 nmol/plate MUREAV 57,85,78
sln-dmg-orl 200 μmol/L/24H MUREAV 67,27,79
sce-hmn:lym 100 μmol/L TCMUE9 1,129,84
orl-rat TDLo:169 mg/kg/30W-I:ETA ZKKOBW 92,221,78

CONSENSUS REPORTS: EPA Genetic Toxicology Program.

SAFETY PROFILE: Questionable carcinogen with experimental tumorigenic data. Human mutation data reported. Many N-nitroso compounds are carcinogens. See also N-NITROSO COMPOUNDS. When heated to decomposition it emits very toxic fumes of Cl^- and NO_x.

DFW600 CAS:59863-59-1 ***HR: 3***
3,4-DICHLORO-N-NITROSOPYRROLIDINE
mf: $C_4H_6Cl_2N_2O$ mw: 169.02

TOXICITY DATA with REFERENCE
mma-sat 250 μg/plate MUREAV 89,35,81
orl-rat TDLo:1550 mg/kg/31W-I:ETA CNREA8 36,1988,76

SAFETY PROFILE: Questionable carcinogen with experimental tumorigenic data. Mutation data reported. Many N-nitroso compounds are carcinogens. When heated to decomposition it emits very toxic fumes of Cl^- and NO_x. See also N-NITROSO COMPOUNDS.

DFW800 CAS:10213-09-9 ***HR: D***
DICHLOROOXOVANADIUM
mf: Cl_2OV mw: 137.84

PROP: Dark green, syrupy mass. D: 2.88 @ 13°.

SYNS: VANADIUM CHLORIDE OXIDE ◇ VANADIUM DICHLORIDE OXIDE ◇ VANADIUM OXYCHLORIDE ◇ VANADIUM OXYDICHLORIDE

TOXICITY DATA with REFERENCE
mrc-bcs 400 mmol/L MUREAV 77,109,80

CONSENSUS REPORTS: EPA Genetic Toxicology Program.

ACGIH TLV: TWA 0.05 mg(V_2O_5)/m^3
NIOSH REL: (Vanadium Compounds) CL 0.05 mg(V)/m^3/15M

SAFETY PROFILE: Mutation data reported. See also VANADIUM COMPOUNDS and CHLORIDES. Reacts violently with K. When heated to decomposition it emits toxic fumes of Cl^- and VO_x.

DFX000 CAS:30586-10-8 ***HR: 2***
DICHLOROPENTANE
DOT: UN 1152
mf: $C_5H_{10}Cl_2$ mw: 141.05

PROP: Clear, light yellow liquid. Bp: 130°, flash p: 106°F (OC), vap d: 4.86, d: 1.06-1.08 @ 20°.

SYN: DICHLOROPENTANES (DOT)

DOT Classification: Flammable or Combustible Liquid; Label: Flammable Liquid.

SAFETY PROFILE: Flammable when exposed to heat or flame. Can react vigorously with oxidizing materials. To fight fire, use water, foam, CO_2, dry chemical. When heated to decomposition it emits highly toxic fumes of Cl^- and phosgene. See also 1,5-DICHLOROPENTANE; and CHLORINATED HYDROCARBONS, ALIPHATIC.

DFX200 CAS:628-76-2 ***HR: 3***
1,5-DICHLOROPENTANE
mf: $C_5H_{10}Cl_2$ mw: 141.05

PROP: Insol in water. D: 1.1, vap d: 4.9, bp: 180°, flash p: > 80°F (OC).

TOXICITY DATA with REFERENCE
ipr-mus LDLo:64 mg/kg CBCCT* 2,189,50

CONSENSUS REPORTS: Reported in EPA TSCA Inventory.

SAFETY PROFILE: Poison by by intraperitoneal route. Dangerous fire hazard when exposed to heat or flame. To fight fire, use alcohol foam or spray. Use of water is ineffective except as a blanket. When heated to decomposition it emits toxic fumes of Cl^-. See also CHLORINATED HYDROCARBONS, ALIPHATIC.

DFX400 CAS:536-29-8 ***HR: 3***
DICHLOROPHENARSINE HYDROCHLORIDE
mf: $C_6H_6AsCl_2NO \cdot ClH$ mw: 290.41

SYNS: 2-AMINO-4-DICHLOROARSINOPHENOLHYDROCHLORIDE ◇ (3-AMINO-4-HYDROXYPHENYL)ARSONOUS DICHLORIDE MONOHYDROCHLORIDE ◇ 3-AMINO-4-HYDROXYPHENYL DICHLORARSINE HYDROCHLORIDE ◇ (3-AMINO-4-HYDROXYPHENYL)DICHLOROARSINE HYDROCHLORIDE ◇ ARSECLOR ◇ CHLORARSOL ◇ CHLORASEN ◇ CLORARSEN ◇ DICHLOROMAPHARSEN ◇ FILARSEN ◇ FONTARSOL ◇ HALARSOL ◇ R.P. 2591

TOXICITY DATA with REFERENCE
par-hmn TDLo:957 μg/kg:GIT JPETAB 73,412,41
orl-rat LDLo:500 mg/kg NCNSA6 5,12,53
ivn-rat LD50:24500 μg/kg AMIUAG 8,196,54
ipr-mus LD50:41 mg/kg PSEBAA 78,392,51
unr-mus LD50:44 mg/kg CNREA8 9,626,49
ivn-rbt LDLo:15 mg/kg JPETAB 73,412,41

CONSENSUS REPORTS: Arsenic and its compounds, as well as chlorophenol compounds, are on the Community Right-To-Know List.

OSHA PEL: TWA 0.5 mg(As)/m^3
ACGIH TLV: TWA 0.2 mg(As)/m^3

SAFETY PROFILE: Poison by intravenous, intraperitoneal, and possibly other routes. Moderately toxic by ingestion. Human systemic effects by parenteral route: hypermotility, diarrhea, nausea, vomiting. See also ARSENIC COMPOUNDS and CHLOROPHENOLS. When heated to decomposition it emits very toxic fumes of As, NO_x, and Cl^-.

DFX800 CAS:120-83-2 ***HR: 3***
2,4-DICHLOROPHENOL
mf: $C_6H_4Cl_2O$ mw: 163.00

PROP: Colorless crystals. Mp: 45°, bp: 210°, flash p: 237°F, d: 1.383 @ 60°/25°, vap d: 5.62, vap press: 1 mm @ 53.0°.

SYNS: DCP ◇ 2,4-DCP ◇ NCI-C55345 ◇ RCRA WASTE NUMBER U081

TOXICITY DATA with REFERENCE
sln-ham:lng 500 umol/L MUREAV 182,135,87
orl-rat TDLo:7500 mg/kg (female 6-15D post):TER TOXID9 4,167,84
skn-mus TDLo:16 g/kg/39W-I:CAR CNREA8 19,413,59
orl-rat LD50:580 mg/kg FEPRA7 2,76,43
ipr-rat LD50:430 mg/kg BJPCAL 13,20,58
scu-rat LD50:1730 mg/kg FEPRA7 2,76,43
orl-mus LD50:1276 mg/kg FAATDF 5,478,85
ipr-mus LD50:153 mg/kg JMCMAR 18,868,75

CONSENSUS REPORTS: IARC Cancer Review: Human Limited Evidence IMEMDT 41,319,86. Reported in EPA TSCA Inventory. EPA Genetic Toxicology Program. Community Right-To-Know List.

SAFETY PROFILE: Suspected carcinogen with experimental carcinogenic and teratogenic data. Poison by intraperitoneal route. Moderately toxic by ingestion and subcutaneous routes. An experimental teratogen. Mutation data reported. Combustible when exposed to heat or flame. Can react vigorously with oxidizing materials. To fight fire, use alcohol foam, foam, CO_2, dry chemical. When heated to decomposition, or on contact with acid or acid fumes it emits highly toxic fumes of Cl^-. See also CHLOROPHENOLS.

DFY000 CAS:87-65-0 ***HR: 3***
2,6-DICHLOROPHENOL
mf: $C_6H_4Cl_2O$ mw: 163.00

SYNS: 2,6-DICHLORFENOL (CZECH) ◇ RCRA WASTE NUMBER U082

TOXICITY DATA with REFERENCE
skn-rbt 500 mg/24H SEV 28ZPAK -,79,72
eye-rbt 250 μg/24H SEV 28ZPAK -,79,72
orl-rat LD50:2940 mg/kg 28ZPAK -,79,72
ipr-rat LD50:390 mg/kg BJPCAL 13,20,58
scu-rat LD50:1730 mg/kg FEPRA7 2,76,43
orl-mus LD50:2120 mg/kg TOLED5 29,39,85

CONSENSUS REPORTS: Reported in EPA TSCA Inventory. EPA Genetic Toxicology Program. Chlorophenol compounds are on the Community Right-To-Know List.

SAFETY PROFILE: Poison by intraperitoneal route. Moderately toxic by ingestion and subcutaneous routes. A severe skin and eye irritant. When heated to decomposition it emits toxic fumes of Cl^-. See also CHLOROPHENOLS.

DFY400 CAS:97-16-5 ***HR: 3***
2,4-DICHLOROPHENOL BENZENESULFONATE
mf: $C_{12}H_8Cl_2O_3S$ mw: 303.16

SYNS: COMPOUND 923 ◇ 2,4-DICHLOROPHENYL BENZENESULFONATE ◇ 2,4-DICHLOROPHENYL BENZENESULPHONATE ◇ 2,4-DICHLOROPHENYL ESTER of BENZENESULFONIC ACID ◇ 2,4-DICHLOROPHENYL ESTER BENZENESULPHONIC ACID ◇ DPBS ◇ EM 923 ◇ GENITE ◇ GENITOL

TOXICITY DATA with REFERENCE
orl-mus TDLo:260 g/kg/78W-I:ETA NTIS** PB223-159
scu-mus TDLo:1000 mg/kg:CAR NTIS** PB223-159
orl-rat LDLo:1000 mg/kg BESAAT 12,117,66
unk-rat LD50:1400 mg/kg 30ZDA9 -,274,71
orl-dog LDLo:620 mg/kg AIPTAK 121,306,59
orl-rbt LD50:700 mg/kg PCOC** -,556,66
ivn-rbt LD50:115 mg/kg AIPTAK 121,306,59

CONSENSUS REPORTS: Chlorophenol compounds are on the Community Right-To-Know List.

SAFETY PROFILE: Poison by intravenous route. Moderately toxic by ingestion and possibly other routes. Questionable carcinogen with experimental carcinogenic and tumorigenic data. An irritant. A pesticide. See also CHLOROPHENOLS. When heated to decomposition it emits very toxic fumes of Cl^- and SO_x.

DFY500 CAS:588-22-7 ***HR: D***
3,4-DICHLOROPHENOXYACETIC ACID
mf: $C_8H_6Cl_2O_3$ mw: 221.04

SYNS: ACETIC ACID, (3,4-DICHLOROPHENOXY)- ◇ 3,4-D ◇ 3,4-DA

TOXICITY DATA with REFERENCE
orl-mus TDLo:1989 mg/kg (female 7-15D post):TER AECTCV 6,33,77

SAFETY PROFILE: Experimental teratogenic effects. When heated to decomposition it emits toxic fumes of Cl^-.

DFY709 CAS:1929-73-3 ***HR: 2***
(2,4-DICHLOROPHENOXY)ACETIC ACID BUTOXYETHYL ESTER
mf: $C_{14}H_{18}Cl_2O_4$ mw: 321.22

SYNS: BUTOXYETHYL-2,4-DICHLOROPHENOXYACETATE ◇ 2,4-D BUTOXYETHANOL ESTER ◇ 2,4-D BUTOXYETHYL ESTER ◇ 2,4-D 2-BUTOXYETHYL ESTER

TOXICITY DATA with REFERENCE
orl-rat TDLo:1500 mg/kg (6-15D preg):TER TXAPA9 22,14,72
orl-rat LD50:831 mg/kg FAATDF 9,423,87

SAFETY PROFILE: Moderately toxic by ingestion. An experimental teratogen. When heated to decomposition it emits toxic fumes of Cl^-. See also ESTERS.

DFY800 CAS:2008-39-1 ***HR: 3***
(2,4-DICHLOROPHENOXY)ACETIC ACID DIMETHYLAMINE
mf: $C_{10}H_{11}Cl_2NO_3$ mw: 264.12

SYNS: 2,4-D ACETATE ◇ 2,4-D AMINE SALT ◇ BLADEX G ◇ 2,4-D

DIMETHYLAMINE SALT ◇ DEFY ◇ DEMISE ◇ (2,4-DICHLOROPHENOXY)ACETATE DIMETHYLAMINE ◇ DIMETHYLAMINE SALT of 2,4-D ◇ DIMETHYLAMMONIUM 2,4-DICHLOROPHENOXYACETATE ◇ FORMULA 40 ◇ HORMIN ◇ PHORDENE ◇ REED AMINE 400

TOXICITY DATA with REFERENCE
cyt-hmn:lyms 500 umol/L MUTAEX 1,241,86
orl-rat TDLo:3 g/kg (6-15D preg):TER TXAPA9 22,14,72
unr-rat LD50:1000 mg/kg HYSAAV 31(9),383,66
unr-mus LD50:300 mg/kg HYSAAV 31(9),383,66
skn-rbt LD50:2115 mg/kg FMCHA2 -,C73,83

SAFETY PROFILE: Poison by unreported route. Moderately toxic by skin contact. An experimental teratogen. Human mutation data reported. A weed killer. When heated to decomposition it emits very toxic fumes of Cl^-, NH_3, and NO_x.

DFZ000 CAS:1928-45-6 ***HR: 2***
2,4-DICHLOROPHENOXYACETIC ACID PROPYLENE GLYCOL BUTYL ETHER ESTER
mf: $C_{15}H_{20}Cl_2O_4$ mw: 335.25

SYNS: 2,4-D PGBE ◇ 2,4-D PROPYLENE GLYCOL BUTYL ETHER ESTER

TOXICITY DATA with REFERENCE
orl-rat TDLo:114 mg/kg (6-15D preg):REP FCTXAV 9,801,71
orl-rat TDLo:114 mg/kg (6-15D preg):TER FCTXAV 9,801,71
orl-rat LD50:500 mg/kg NTIS** PB85-143766

CONSENSUS REPORTS: Glycol ether compounds are on the Community Right-To-Know List.

SAFETY PROFILE: Moderately toxic by ingestion. An experimental teratogen. Other experimental reproductive effects. A pesticide. When heated to decomposition it emits toxic fumes of Cl^-. See also GLYCOL ETHERS.

DGA000 CAS:94-82-6 ***HR: 2***
4-(2,4-DICHLOROPHENOXY)BUTYRIC ACID
mf: $C_{10}H_{10}Cl_2O_3$ mw: 249.10

SYNS: BUTOXON ◇ BUTOXONE ◇ BUTOXONE AMINE ◇ BUTOXONE ESTER ◇ BUTYRAC ◇ BUTYRAC ESTER ◇ 2,4-DB ◇ 2,4-D BUTYRIC ◇ Γ-(2,4-DICHLOROPHENOXY)BUTYRIC ACID ◇ EMBUTOX ◇ EMBUTOX KLEAN-UP ◇ LEGUMEX D

TOXICITY DATA with REFERENCE
dnr-esc 5 mg/disc NTIS** PB80-133226
dnr-bcs 5 mg/disc NTIS** PB80-133226
orl-rat TDLo:17 mg/kg (1-7D preg):REP GISAAA 41(2),20,76
orl-rat TDLo:416 mg/kg (female 9D post):TER GISAAA 41(2),20,76
orl-rat LD50:700 mg/kg RREVAH 10,97,65
skn-rat LD50:800 mg/kg WRPCA2 9,119,70

CONSENSUS REPORTS: EPA Genetic Toxicology Program.

SAFETY PROFILE: Moderately toxic by ingestion and skin contact. An experimental teratogen. Other experimental reproductive effects. Mutation data reported. An herbicide. When heated to decomposition it emits toxic fumes of Cl^-.

DGA200 CAS:14255-88-0 ***HR: 3***
5,6-DICHLORO-1-PHENOXYCARBONYL-2-TRIFLUOROMETHYLBENZIMIDAZOLE
mf: $C_{15}H_7Cl_2F_2N_2O_2$ mw: 375.14

SYNS: 5,6-DICHLORO-2-TRIFLUOROMETHYLBENZIMIDAZOLE-1-CARBOXYLATE ◇ 5,6-DICHLORO-2-(TRIFLUOROMETHYL)-1H-BENZIMIDAZOLE-1-CARBOXYLIC ACID PHENYL ESTER ◇ ENT 27,438 ◇ FENAZAFLOR ◇ FENOFLURAZOLE ◇ FENOZAFLOR ◇ FENZAFLOR ◇ FISONS NC 5016 ◇ LOVOZAL ◇ NC 5016 ◇ NSC 191025 ◇ PHENYL-5,6-DICHLORO-2-TRIFLUOROMETHYL-BENZIMIDAZOLE-1-CARBOXYLATE ◇ TARZOL

TOXICITY DATA with REFERENCE
orl-rat LD50:283 mg/kg FMCHA2 -,C103,83
skn-rat LD50:700 mg/kg WRPCA2 9,119,70
ipr-rat LD50:168 mg/kg GUCHAZ 6,275,73
orl-mus LD50:1600 mg/kg MRLAB3 33,839,68
ipr-mus LD50:42 mg/kg BCPCA6 18,1389,69
orl-dog LD50:50 mg/kg 85DPAN -,-,71/76
orl-rbt LD50:28 mg/kg GUCHAZ 6,275,73
orl-gpg LD50:59 mg/kg 31ZOAD 1,222,68
orl-ckn LD50:50 mg/kg MRLAB3 33,839,68
orl-mam LD50:3717 mg/kg NTIS** PB288-416

SAFETY PROFILE: Poison by ingestion, and intraperitoneal routes. Moderately toxic by skin contact. When heated to decomposition it emits very toxic fumes of F^-, Cl^-, and NO_x.

DGA400 CAS:73986-95-5 ***HR: 2***
2,4-DICHLOROPHENOXY ETHANEDIOL
mf: $C_8H_8Cl_2O_3$ mw: 223.06

SYN: 2,4-DICHLOROPHENOXY-1,2-ETHANEDIOL

TOXICITY DATA with REFERENCE
skn-rbt 10 mg/24H open MLD AMIHBC 4,119,51
eye-rbt 50 μg open SEV AMIHBC 4,119,51
orl-rat LD50:1070 mg/kg AMIHBC 4,119,51
skn-rbt LD50:420 mg/kg AMIHBC 4,119,51

SAFETY PROFILE: Moderately toxic by ingestion and skin contact. A skin and severe eye irritant. When heated to decomposition it emits very toxic fumes of Cl^-.

DGA425 CAS:73791-41-0 ***HR: 3***
3-(2,4-DICHLOROPHENOXY)-2-HYDROXYPROPYL-o-CHLOROPHENYL ARSINIC ACID
mf: $C_{15}H_{14}AsCl_3O_4$ mw: 439.56

SYNS: ARSINE OXIDE, (o-CHLOROPHENYL)(3-(2,4-DICHLOROPHENOXY)-2-HYDROXYPROPYL)HYDROXY-◇ (o-CHLOROPHENYL)(3-(2,4-DICHLOROPHENOXY)-2-HYDROXYPROPYL)HYDROXYARSINEOXIDE

TOXICITY DATA with REFERENCE
ivn-mus LD50:56 mg/kg CSLNX* NX#06928

OSHA PEL: TWA 0.5 mg(As)/m^3

SAFETY PROFILE: Poison by intravenous route. When heated to decomposition it emits toxic fumes of As and Cl^-.

DGA800 CAS:23712-05-2 ***HR: 3***
2-((3,4-DICHLOROPHENOXY)METHYL)-2-IMIDAZOLINE HYDROCHLORIDE
mf: $C_{10}H_{10}Cl_2N_2O$ mw: 245.12

SYNS: DH-524 ◇ 2-((3,4-DICHLOROPHENOXY)METHYL-2-IMIDAZOLINE MONOHYDROCHLORIDE

TOXICITY DATA with REFERENCE
orl-rat LD50:100 mg/kg 27ZQAG -,220,72
orl-mus LD50:111 mg/kg 27ZQAG -,220,72
ipr-mus LD50:41 mg/kg USXXAM #4020167

SAFETY PROFILE: Poison by ingestion and intraperitoneal routes. When heated to decomposition it emits very toxic fumes of Cl^- and NO_x.

DGB000 CAS:120-36-5 ***HR: 3***
2-(2,4-DICHLOROPHENOXY) PROPIONIC ACID
mf: $C_9H_8Cl_2O_3$ mw: 235.07

SYNS: ACIDE-2-(2,4-DICHLORO-PHENOXY)PROPIONIQUE (FRENCH) ◇ ACIDO-2-(2,4-DICLORO-FENOSSI)-PROPIONICO (ITALIAN) ◇ CORNOX RD ◇ CORNOX RK ◇ DESORMONE ◇ 2-(2,4-DICHLOR-FENOXY)-PROPIONZUUR (DUTCH) ◇ α-(2,4-DICHLOROPHENOXY) PROPIONIC ACID ◇ DICHLOROPROP ◇ 2-(2,4-DICHLORPHENOXY)-PROPIONSAEURE (GERMAN) ◇ DICHLORPROP ◇ 2,4-DP ◇ 2-(2,4-DP) ◇ HEDONAL ◇ HEDONAL DP ◇ HORMATOX ◇ KILDIP ◇ POLYCLENE ◇ POLYMONE ◇ POLYTOX ◇ RD 406 ◇ SERITOX 50 ◇ U46 ◇ U46 DP-FLUID ◇ VISKO-RHAP ◇ WEEDONE DP ◇ WEEDONE 170

TOXICITY DATA with REFERENCE
mmo-smc 700 mg/L ZAPOAK 9,483,69
orl-rat TDLo:20 mg/kg (female 4-18D post):REP TJADAB 33,11A,86
orl-mus TDLo:3 g/kg (female 6-15D post):TER ARZNAD 33,1479,83
orl-rat LD50:800 mg/kg WRPCA2 9,119,70
skn-rat LD50:1400 mg/kg WRPCA2 9,119,70
orl-mus LD50:309 mg/kg RPZHAW 31,373,80
skn-mus LD50:1400 mg/kg 28ZEAL 5,76,76

CONSENSUS REPORTS: IARC Cancer Review: Group 2B IMEMDT 7,156,87; Human Limited Evidence IMEMDT 41,357,86. Reported in EPA TSCA Inventory.

SAFETY PROFILE: Suspected carcinogen. Poison by ingestion. Moderately toxic by skin contact. An experimental teratogen. Other experimental reproductive effects. Mutation data reported. A fumigant. When heated to decomposition it emits toxic fumes of Cl^-.

DGB100 ***HR: 1***
(+)-2-(2,4-DICHLOROPHENOXY)PROPIONIC ACID
mf: $C_9H_8Cl_2O_3$ mw: 235.07

SYN: (+)-2-(2,4-DICHLORPHENOXY)PROPIONSAFEURE(GERMAN)

TOXICITY DATA with REFERENCE
orl-mus TDLo:3 g/kg (6-15D preg):TER ARZNAD 33,1479,83
orl-mus TDLo:4 g/kg (6-15D preg):REP ARZNAD 33,1479,83
orl-rat LD50:1 g/kg FMCHA2 -,C100,89

SAFETY PROFILE: Slightly toxic by ingestion. Experimental teratogenic and reproductive effects. When heated to decomposition it emits toxic fumes of Cl^-. See also CHLORINATED HYDROCARBONS, AROMATIC.

DGB200 CAS:6965-71-5 ***HR: 3***
2-(2,5-DICHLOROPHENOXY)PROPIONIC ACID
mf: $C_9H_8Cl_2O_3$ mw: 235.07

SYN: α-(2,5-DICHLOROPHENOXY)PROPIONIC ACID

TOXICITY DATA with REFERENCE
scu-mus TDLo:100 mg/kg:ETA NTIS** PB223-159

SAFETY PROFILE: Questionable carcinogen with experimental tumorigenic data. When heated to decomposition it emits toxic fumes of Cl^-.

DGB400 CAS:39637-16-6 ***HR: 3***
(2,4-DICHLOROPHENOXY)TRIBUTYLSTANNANE
mf: $C_{18}H_{30}Cl_2OSn$ mw: 452.07

SYN: TRI-n-BUTYL-2,4-DICHLOROPHENOXYTIN

TOXICITY DATA with REFERENCE
ivn-mus LD50:56 mg/kg CSLNX* NX#01883

OSHA PEL: TWA 0.1 mg(Sn)/m^3 (skin)
ACGIH TLV: TWA 0.1 mg(Sn)/m^3 (skin) (Proposed: TWA 0.1 mg(Sn)/m^3; STEL 0.2 mg(Sn)/m^3 (skin))
NIOSH REL: (Organotin Compounds) TWA 0.1 mg(Sn)/m^3

SAFETY PROFILE: Poison by intravenous route. Tributyl tin compounds are extremely toxic to marine life. See also TIN COMPOUNDS and CHLORIDES. When heated to decomposition it emits toxic fumes of Cl^-.

DGB500 CAS:4205-90-7 ***HR: 3***
2-(2,6-DICHLOROPHENYLAMINO)-2-IMIDAZOLINE
mf: $C_9H_9Cl_2N_3$ mw: 230.11

SYNS: 734571A ◇ BENZENAMINE, 2,6-DICHLORO-N-2-IMIDAZOLIDINYLIDENE- (9CI) ◇ CLONIDIN ◇ CLONIDINE ◇ 2-IMIDAZOLINE, 2-(2,6-DICHLOROANILINO)-

TOXICITY DATA with REFERENCE
scu-rat TDLo:2080 μg/kg (female 8-20D post):REP NRTXDN 9,559,88
orl-mus TDLo:20 mg/kg (female 11D post):TER TJADAB 32,19A,85
orl-man TDLo:2857 ng/kg:BPR AIMDAP 143,2195,83
orl-rat LD50:157 mg/kg IYKEDH 18,366,87
scu-rat LD50:108 mg/kg IYKEDH 18,366,87
orl-mus LD50:108 mg/kg IYKEDH 18,366,87
scu-mus LD50:364 mg/kg IYKEDH 18,366,87

SAFETY PROFILE: Poison by ingestion and subcutaneous routes. Human systemic effects by ingestion: blood pressure lowering. An experimental teratogen. Other experimental reproductive effects. When heated to decomposition it emits toxic fumes of NO_x and Cl^-.

DGB600 CAS:696-28-6 ***HR: 3***
DICHLOROPHENYLARSINE
DOT: NA 1556
mf: $C_6H_5AsCl_2$ mw: 222.93

PROP: Colorless gas or liquid, changes to yellow. Bp: 255-275°, fp: −15.6°, d: 1.654 @ 20°, vap press: 0.021 mm @ 20°, vap d: 7.7.

SYNS: FDA ◇ FENILDICLOROARSINA (ITALIAN) ◇ PHENYLARSINEDICHLORIDE ◇ PHENYLARSONOUS DICHLORIDE ◇ PHENYL DICHLOROARSINE (DOT) ◇ RCRA WASTE NUMBER P036 ◇ TL 69

TOXICITY DATA with REFERENCE
skn-rat LD50:16 mg/kg JPBAA7 58,411,46
ihl-mus LC50:3300 mg/m³/10M NTIS** PB158-508
skn-mus LD50:4 mg/kg NTIS** PB158-508
ivn-mus LD50:500 μg/kg JPBAA7 58,411,46
skn-rbt LD50:5 mg/kg JPBAA7 58,411,46
ivn-rbt LD50:500 mg/kg JPBAA7 58,411,46
skn-gpg LD50:4 mg/kg JPBAA7 58,411,46

CONSENSUS REPORTS: Arsenic and its compounds are on the Community Right-To-Know List. Reported in EPA TSCA Inventory. EPA Extremely Hazardous Substances List.

OSHA PEL: TWA 0.5 mg(As)/m³
DOT Classification: Poison B; Label: Poison.

SAFETY PROFILE: Poison by inhalation, ingestion, skin contact, and intravenous routes. See also ARSENIC. A lachrymator type of military poison gas. When exposed to heat, water, or steam it reacts to produce corrosive fumes of Cl^-. When heated to decomposition it emits highly toxic fumes of arsenic.

DGB800 CAS:15460-48-7 ***HR: 3***
N-(3,4-DICHLOROPHENYL)-1-AZIRIDINECARBOXAMIDE
mf: $C_9H_8Cl_2N_2O$ mw: 231.09

SYN: 3,4-DICHLOROPHENYL-N-CARBAMOYLAZIRIDINE

TOXICITY DATA with REFERENCE
ipr-mus TDLo:20 mg/kg/4W-I:NEO CNREA8 29,2184,69

SAFETY PROFILE: Questionable carcinogen with experimental neoplastigenic data. When heated to decomposition it emits very toxic fumes of Cl^- and NO_x.

DGB875 CAS:873-51-8 ***HR: 3***
DICHLOROPHENYLBORANE
mf: $C_6H_5BCl_2$ mw: 158.82

SAFETY PROFILE: The hot borane ignites in air. When heated to decomposition it emits toxic fumes of Cl^-. See also BORANES.

DGC000 CAS:10140-84-8 ***HR: 2***
2,4-DICHLOROPHENYL "CELLOSOLVE"
mf: $C_{10}H_{12}Cl_2O_2$ mw: 235.12

TOXICITY DATA with REFERENCE
skn-rbt 500 mg open MLD UCDS** 12/29/71
eye-rbt 1 mg SEV UCDS** 12/29/71
orl-rat LD50:1410 mg/kg UCDS** 12/29/71
skn-rbt LD50:1250 mg/kg UCDS** 12/29/71

SAFETY PROFILE: Moderately toxic by ingestion and skin contact. A skin and severe eye irritant. When heated to decomposition it emits very toxic fumes of Cl^-.

DGC100 CAS:76714-88-0 ***HR: 2***
(E)-1-(2,4-DICHLOROPHENYL)-4,4-DIMETHYL-2-(1,2,4-TRIAZOL-1-YL)PENTEN-3-OL
mf: $C_{15}H_{17}Cl_2N_3O$ mw: 326.25

SYNS: S 3308 ◇ 1H-1,2,4-TRIAZOLE-1-ETHANOL, β-((2,4-DICHLOROPHENYL)METHYLENE)-α-(1,1-DIMETHYLETHYL)-, (E)-

TOXICITY DATA with REFERENCE
eye-rbt 100 mg MLD EPASR* 8EHQ-0485-0548
orl-rat LD50:474 mg/kg EPASR* 8EHQ-0485-0548

SAFETY PROFILE: Moderately toxic by ingestion. An eye irritant. When heated to decomposition it emits toxic fumes of NO_x and Cl^-.

DGC600 CAS:38780-39-1 ***HR: 3***
cis-DICHLORO(o-PHENYLENEDIAMINE)PLATINUM(II)
mf: $C_6H_8Cl_2N_2Pt$ mw: 374.15

SYN: DICHLORO(1,2-PHENYLENEDIAMMINE)PLATINUM(II)

TOXICITY DATA with REFERENCE
mmo-sat 2300 nmol/L JMCMAR 23,459,80
mma-sat 2300 nmol/L JMCMAR 23,459,80
ipr-mus LD50:48 mg/kg CBINA8 5,415,72

SAFETY PROFILE: Poison by intraperitoneal route. Mutation data reported. See PLATINUM COMPOUNDS. When heated to decomposition it emits very toxic fumes of Cl^- and NO_x.

DGC800 CAS:34643-46-4 ***HR: 2***
O-(2,4-DICHLOROPHENYL)-O-ETHYL-S-PROPYLPHOSPHORODITHIOATE
mf: $C_{11}H_{15}Cl_2O_2PS_2$ mw: 345.25

SYNS: BAY NTN 8629 ◇ BIDERON ◇ O-ETHYL-O-(2,4-DICHLOROPHENYL)-S-n-PROPYL-DITHIOPHOSPHATE ◇ NTN-8629 ◇ PROTHIOPHOS ◇ TOKUTHION

TOXICITY DATA with REFERENCE
orl-rat LD50:925 mg/kg FMCHA2 -,C2370,83
skn-rat LD50:3900 mg/kg KONODE 20,94,76
orl-mus LD50:940 mg/kg NEZAAQ 33,221,78
skn-mus LD50:1600 mg/kg KONODE 20,94,76

SAFETY PROFILE: Moderately toxic by ingestion and skin contact. When heated to decomposition it emits very toxic fumes of Cl^-, PO_x, and SO_x.

DGD075 CAS:33175-34-7 ***HR: 3***
3,4-DICHLOROPHENYL HYDROXYLAMINE
mf: $C_6H_5Cl_2NO$ mw: 178.02

$Cl_2C_6H_3NHOH$

SAFETY PROFILE: Decomposes exothermically at 80°C. It is a chemical intermediate in the production of aniline from 3,4-dichloronitrobenzene and its thermal decomposition has caused violent explosions in plant-scale reactors. When heated to decomposition it emits toxic fumes of Cl^- and NO_x. See also AROMATIC AMINES.

DGD100 ***HR: 3***
1-(3,4-DICHLOROPHENYL)-5-ISOPROPYLBIGUANIDE HYDROCHLORIDE
mf: $C_{11}H_{15}Cl_2N_5$•ClH mw: 324.67

SYNS: N^1-3,4-DICHLOROPHENYL-N^5-ISOPROPYLDIGUANIDE HYDROCHLORIDE ◇ M5943

TOXICITY DATA with REFERENCE
orl-mus LD50:100 mg/kg BJPCAL 5,438,50
ipr-mus LD50:25 mg/kg BJPCAL 5,438,50
ivn-mus LD50:25 mg/kg BJPCAL 5,438,50

SAFETY PROFILE: Poison by ingestion, intravenous, and intraperitoneal routes. When heated to decomposition it emits toxic fumes of NO_x and HCl.

DGD400 CAS:3687-13-6 ***HR: 2***
2,4-DICHLOROPHENYLMETHANESULFONATE
mf: $C_7H_6Cl_2O_3S$ mw: 241.09

SYNS: SD 7727 ◇ SHELL SD 7,727

TOXICITY DATA with REFERENCE
orl-rat LD50:2793 mg/kg 28ZEAL 4,153,69
orl-mus LDLo:1070 mg/kg AECTCV 14,111,85
skn-rbt LD50:2500 mg/kg 28ZEAL 5,204,76

SAFETY PROFILE: Moderately toxic by ingestion and skin contact. When heated to decomposition it emits very toxic fumes of Cl^- and SO_x. See also SULFONATES.

DGD600 CAS:330-55-2 ***HR: 3***
3-(3,4-DICHLOROPHENYL)-1-METHOXYMETHYLUREA
mf: $C_9H_{10}Cl_2N_2O_2$ mw: 249.11

PROP: Solid. Mp: 93-94°. Sltly sol in water; partially sol in acetone and alc.

SYNS: 3-(3,4-DICHLOOR-FENYL)-1-METHOXY-1-METHYLUREUM (DUTCH) ◇ 3-(3,4-DICHLORO-FENIL)-1-METOSSI-1-METIL-UREA (ITALIAN) ◇ 3-(3,4-DICHLOROPHENYL)-1-METHOXY-1-METHYLUREA ◇ N'-(3,4-DICHLOROPHENYL)-N-METHOXY-N-METHYLUREA ◇ 1-(3,4-DICHLOROPHENYL)3-METHOXY-3-METHYLUREE (FRENCH) ◇ N-(3,4-DICHLOROPHENYL)-N'-METHYL-N'-METHOXYUREA ◇ 3-(3,4-DICHLOR-PHENYL)-1-METHOXY-1-METHYL-HARNSTOFF (GERMAN) ◇ 3-(4,5-DICHLORPHENYL)-1-METHOXY-1-METHYLHARNSTOFF (GERMAN) ◇ DU PONT 326 ◇ DUPONT HERBICIDE 326 ◇ GARNITAN ◇ HERBICIDE 326 ◇ HOE 2810 ◇ LINEX 4L ◇ LINOROX ◇ LINUREX ◇ LINURON ◇ LINURON (herbicide) ◇ LOREX ◇ LOROX ◇ LOROX LINURON WEED KILLER ◇ METHOXYDIURON ◇ 1-METHOXY-1-METHYL-3-(3,4-DICHLOROPHENYL)UREA ◇ PREMALIN ◇ SARCLEX ◇ SCARCLEX ◇ SINURON

TOXICITY DATA with REFERENCE
dni-mus-orl 500 mg/kg MUREAV 58,353,78
orl-rat LD50:1500 mg/kg 85ARAE 2,150,77
ihl-rat LD50:48 mg/m³/4H 85GMAT -,48,82
orl-mus LD50:2400 mg/kg 85GMAT -,48,82
orl-dog LD50:500 mg/kg GUCHAZ 6,317,73
orl-ckn LD50:3765 mg/kg VETNAL 58(7),63,82

CONSENSUS REPORTS: Reported in EPA TSCA Inventory. EPA Genetic Toxicology Program.

SAFETY PROFILE: Poison by inhalation. Moderately toxic by ingestion. Mutation data reported. A selective herbicide used in farming. When heated to decomposition it emits very toxic fumes of Cl^- and NO_x. See also 3-(p-CHLOROPHENYL)-1,1-DIMETHYLUREA.

DGD800 CAS:299-85-4 ***HR: 3***
O-(2,4-DICHLOROPHENYL)-O-METHYLISOPROPYLPHOSPHORAMIDOTHIOATE
mf: $C_{10}H_{14}Cl_2NO_2PS$ mw: 314.18

SYNS: O-(2,4-DICHLOROPHENYL)-O-METHYL-N-ISOPROPYL-PHOSPHORAMIDOTHIOATE ◇ DMPA ◇ DOW 1329 ◇ DOWCO 118 ◇ ENT 25,647 ◇ ISOPROPYLPHOSPHORAMIDOTHIOIC ACID-O-2,4-DICHLOROPHENYL-O-METHYL ESTER ◇ K 22023 ◇ (1-METHYLETHYL)PHOSPHORAMIDOTHIOIC ACID O-(2,4-DICHLOROPHENYL)-O-METHYL ESTER ◇ OMS 115 ◇ ZYTRON

TOXICITY DATA with REFERENCE
orl-rat LDLo:270 mg/kg FMCHA2 -,C260,83
unr-dog LD50:1000 mg/kg 30ZDA9 -,353,71
skn-rbt LD50:1680 mg/kg GUCHAZ 6,242,73
orl-gpg LD50:210 mg/kg 31ZOAD 1,191,68
orl-ckn LD50:1357 mg/kg TXAPA9 6,147,64
orl-bwd LD50:100 mg/kg TXAPA9 21,315,72

SAFETY PROFILE: Poison by ingestion. Moderately toxic by skin contact. An herbicide and plant growth regulator. When heated to decomposition it emits very toxic fumes of Cl^-, NO_x, PO_x, and SO_x.

DGE200 CAS:13412-64-1 ***HR: 2***
3-(2,6-DICHLOROPHENYL)-5-METHYL-4-ISOXAZOLYL PENICILLIN SODIUM MONOHYDRATE
mf: $C_{19}H_{16}Cl_2N_3O_5S{\cdot}Na{\cdot}H_2O$ mw: 510.35

SYNS: BLP-1011 ◇ BRISPEN ◇ BRL-1702 ◇ CONSTAPHYL ◇ DICHLOR STAPENOR ◇ DICLOCIL ◇ DICLOXACILLIN SODIUM MONOHYDRATE ◇ DICLOXACILLIN SODIUM SALT ◇ DYCILL ◇ DYNAPEN ◇ MDI-PC ◇ NOXABEN ◇ P 1011 ◇ PATHOCIL ◇ PENSINT ◇ SODIUM DICLOXACILLIN ◇ SODIUM DICLOXACILLIN MONOHYDRATE ◇ STAMPEN ◇ STPHCILLIN A BANYU ◇ SYNTARPEN ◇ VERACILLIN

TOXICITY DATA with REFERENCE
orl-rat LD50:3580 mg/kg TXAPA9 18,185,71
ipr-rat LD50:630 mg/kg ARZNAD 15,322,65
ivn-rat LD50:520 mg/kg JJANAX 21,274,68
orl-mus LD50:4560 mg/kg JJANAX 21,274,68
ipr-mus LD50:1000 mg/kg JJANAX 21,274,68
scu-mus LD50:1100 mg/kg JJANAX 21,274,68
ivn-mus LD50:875 mg/kg JJANAX 21,274,68
ivn-rbt LD50:600 mg/kg JJANAX 21,274,68

SAFETY PROFILE: Moderately toxic by ingestion, intraperitoneal, intravenous, and subcutaneous routes. An antibacterial agent. When heated to decomposition it emits very toxic fumes of Na_2O, NO_x, SO_x, and Cl^-. See also other penicillin entries.

DGE400 CAS:644-97-3 ***HR: 3***
DICHLOROPHENYLPHOSPHINE
DOT: UN 2798
mf: $C_6H_5Cl_2P$ mw: 178.98

PROP: Fuming liquid. Bp: 225°, d: 1.319, vap d: 6.17.

SYNS: BENZENE PHOSPHORUS DICHLORIDE (DOT) ◇ PHOSPHENYL CHLORIDE

CONSENSUS REPORTS: Reported in EPA TSCA Inventory.

DOT Classification: Corrosive Material; Label: Corrosive.

SAFETY PROFILE: A poison irritant to skin, eyes, and mucous membranes and poison by ingestion and inhalation. When heated to decomposition it emits very toxic fumes of Cl^- and PO_x. See also PHOSPHINE.

DGE800 CAS:1698-53-9 ***HR: 2***
4,5-DICHLORO-2-PHENYL-3(2H)-PYRIDAZINONE
mf: $C_{10}H_6Cl_2N_2O$ mw: 241.08

SYN: 1-FENYL-4,5-DICHLOR-6-PYRIDAZINON(CZECH)

TOXICITY DATA with REFERENCE
eye-rbt 20 mg/24H SEV 28ZPAK -,151,72
orl-rat LD50:2520 mg/kg 28ZPAK -,151,72

SAFETY PROFILE: Moderately toxic by ingestion. A severe eye irritant. When heated to decomposition it emits very toxic fumes of Cl^- and NO_x.

DGF000 CAS:24096-53-5 ***HR: 3***
N-(3,5-DICHLOROPHENYL)SUCCINIMIDE
mf: $C_{10}H_7Cl_2NO_2$ mw: 244.08

SYNS: 1-(3,5-DICHLOROPHENYL)-2,5-PYRROLIDINEDIONE ◇ DIMETHACHLON ◇ OHRIC

TOXICITY DATA with REFERENCE
orl-rat TDLo:17 g/kg/8W-C:ETA GANNA2 67,147,76
orl-mus LD50:1250 mg/kg FMCHA2 -,C172,83

SAFETY PROFILE: Moderately toxic by ingestion. Questionable carcinogen with experimental tumorigenic data. When heated to decomposition it emits very toxic fumes of Cl^- and NO_x.

DGF200 CAS:27137-85-5 ***HR: 3***
(DICHLOROPHENYL)TRICHLOROSILANE
DOT: UN 1766
mf: $C_6H_3Cl_5Si$ mw: 280.43

PROP: Straw-colored liquid, sol in benzene and perchloroethylene. (mixture of isomers). D: 1.562, bp: 260°, flash p: 286°F.

SYNS: DICHLOROPHENYLTRICHLOROSILANE(DOT) ◇ TRICHLORO(DICHLOROPHENYL)SILANE

TOXICITY DATA with REFERENCE
ipr-rat LDLo:100 mg/kg 85GMAT -,48,82
orl-mus LDLo:100 mg/kg 85GMAT -,48,82
ihl-mus LCLo:80 mg/m^3/2H 85GMAT -,48,82
ipr-mus LDLo:100 mg/kg 85GMAT -,48,82
scu-mus LDLo:100 mg/kg 85GMAT -,48,82
ihl-mam LCLo:80 mg/m^3 CHABA8 57,12828a,62

CONSENSUS REPORTS: Reported in EPA TSCA Inventory. EPA Extremely Hazardous Substances List.

DOT Classification: Corrosive Material; Label: Corrosive.

SAFETY PROFILE: Poison by ingestion, inhalation, subcutaneous, and intraperitoneal routes. Corrosive to the eyes, skin, and mucous membranes. On contact with moisture it releases corrosive HCl. Combustible when exposed to heat or flame. When heated to decomposition it emits toxic fumes of Cl^-. See also CHLOROSILANES.

DGF400 CAS:78-99-9 ***HR: 1***
1,1-DICHLOROPROPANE
mf: $C_3H_6Cl_2$ mw: 112.99

PROP: Flash p: 69.8°F, lel: 3.1%.

SYN: PROPYLIDENE CHLORIDE

TOXICITY DATA with REFERENCE
eye-rbt 500 mg AMIHBC 10,61,54
orl-rat LD50:6500 mg/kg AMIHBC 10,61,54
ihl-rat LCLo:4000 ppm/4H AMIHBC 10,61,54
skn-rbt LD50:14 g/kg AMIHBC 10,61,54

SAFETY PROFILE: Mildly toxic by ingestion, inhalation, and skin contact. An eye irritant. A very dangerous fire hazard when exposed to heat, flame, or oxidizers. When heated to decomposition it emits toxic fumes of Cl^-. See also 1,2-DICHLOROPROPANE; and CHLORINATED HYDROCARBONS, ALIPHATIC.

DGF600 ***HR: 3***
1,2-DICHLOROPROPANE mixed with DICHLOROPROPENE
DOT: NA 2047

SYN: DICHLOROPROPENE and PROPYLENE DICHLORIDE MIXTURE (DOT)

DOT Classification: Flammable Liquid; Label: Flammable Liquid

SAFETY PROFILE: Probably a poison. Flammable liquid. See 1,2-DICHLOROPROPANE and 2,3-DICHLOROPROPENE. When heated to decomposition it emits toxic fumes of Cl^-.

DGF800 CAS:142-28-9 ***HR: 2***
1,3-DICHLOROPROPANE
mf: $C_3H_6Cl_2$ mw: 112.99

PROP: Colorless liquid. Bp: 125°, d: 1.201 @ 15°, vap d: 3.90, flash p: 69.8°F.

SYN: TRIMETHYLENE DICHLORIDE

TOXICITY DATA with REFERENCE
mmo-sat 10 μmol/plate ENMUDM 2,59,80
mma-sat 10 μmol/plate ENMUDM 2,59,80
orl-dog LDLo:3000 mg/kg AJHYA2 16,325,32

CONSENSUS REPORTS: Reported in EPA TSCA Inventory.

SAFETY PROFILE: Moderately toxic by ingestion. Mutation data reported. A very dangerous fire hazard when exposed to heat or flame. When heated to decomposition it emits highly toxic fumes of Cl^- and phosgene. See also CHLORINATED HYDROCARBONS, ALIPHATIC; and 1,2-DICHLOROPROPANE.

DGF900 CAS:594-20-7 ***HR: 3***
2,2-DICHLOROPROPANE
mf: $C_3H_6Cl_2$ mw: 112.99

$H_3CCCl_2CH_3$

SAFETY PROFILE: Reacts explosively with dimethylzinc. When heated to decomposition it emits toxic fumes of Cl^-. See also 1,2-DICHLOROPROPANE; and CHLORINATED HYDROCARBONS, ALIPHATIC.

DGG000 CAS:8003-19-8 ***HR: 3***
DICHLOROPROPANE-DICHLOROPROPENE MIXTURE
mf: $C_3H_6Cl_2 \cdot C_3H_4Cl_2$ mw: 223.96

PROP: D-D Soil fumigant consists of chlorinated C_3 hydrocarbons (100%), 1,3-dichloropropene, 3,3-dichloropropene, 1,2-dichloropropane, 2,3-dichloropropene, and related C_3 chlorinated hydrocarbons (SHELL*).

SYNS: D-D ◇ DD MIXTURE ◇ DD SOIL FUMIGANT ◇ 1,3-DICHLOROPROPENE and 1,2-DICHLOROPROPANE MIXTURE ◇ DICHLORPROPAN-DICHLORPROPENGEMISCH (GERMAN) ◇ DOWFUME N ◇ ENT 8,420 ◇ NEMAFENE ◇ TELONE ◇ VIDDEN D

TOXICITY DATA with REFERENCE
skn-rbt 500 mg/24H SEV AMIHBC 7,118,53
eye-rbt 5 mg SEV AMIHBC 7,118,53
mma-sat 500 μg/plate CNREA8 37,1915,77
mmo-sat 500 μg/plate CNREA8 37,1915,77
orl-rat LD50:140 mg/kg ARSIM* 20,8,66
ihl-rat LC50:1000 ppm/4H AMIHBC 7,118,53
skn-rat LD50:2100 mg/kg BESAAT 15,99,69
orl-mus LD50:3 mg/kg ARSIM* 20,8,66
skn-rbt LD50:2100 mg/kg PCOC** -,371,66

SAFETY PROFILE: Poison by ingestion and inhalation. Moderately toxic by skin contact. Severe skin and eye irritant. Mutation data reported. A fumigant. When heated to decomposition it emits toxic fumes of Cl^-. See also 1,2-DICHLOROPROPANE; and CHLORINATED HYDROCARBONS, ALIPHATIC.

DGG400 CAS:96-23-1 ***HR: 3***
1,3-DICHLORO-2-PROPANOL
DOT: UN 2750
mf: $C_3H_6Cl_2O$ mw: 128.99

PROP: Colorless liquid, ether-like odor. Bp: 174°, d: 1.367 @ 20°/4°, vap press: 1 mm @ 28.0°, vap d: 4.45, flash p: 165°F (OC), mp: −4°.

SYNS: DICHLOROHYDRIN ◇ α-DICHLOROHYDRIN ◇ sym-DICHLOROISOPROPYL ALCOHOL ◇ 1,3-DICHLOROPROPANOL-2 (DOT) ◇ GLYCEROL α,Γ-DICHLOROHYDRIN ◇ sym-GLYCEROL DICHLOROHYDRIN ◇ U 25,354

TOXICITY DATA with REFERENCE
skn-rbt 10 mg/24H open MLD AIHAAP 23,95,62
mmo-sat 1 μmol/plate ENMUDM 2,59,80
mma-sat 100 μg/plate SCIEAS 200,785,78
dni-hmn:hla 2500 μmol/L MUREAV 92,427,82
orl-rat LD50:110 mg/kg AIHAAP 23,95,62
ihl-rat LCLo:125 ppm/4H AIHAAP 23,95,62
orl-mus LD50:100 mg/kg 85GMAT -,46,82
skn-rbt LD50:800 mg/kg AIHAAP 23,95,62

CONSENSUS REPORTS: Reported in EPA TSCA Inventory. EPA Genetic Toxicology Program.

DOT Classification: Poison B; Label: Poison.

SAFETY PROFILE: Poison by ingestion and inhalation. Moderately toxic by skin contact. Human mutation data reported. A skin irritant. Action may be similar to carbon tetrachloride, but more irritating to mucous membranes. Flammable when exposed to heat, flame, or oxidizers. To fight fire, use alcohol foam, dry chemical, fog, mist, or spray. Dangerous; when heated to decomposition it emits highly toxic fumes of Cl^- and phosgene.

DGG600 CAS:616-23-9 ***HR: 3***
2,3-DICHLOROPROPANOL
mf: $C_3H_6Cl_2O$ mw: 128.99

SYNS: 1,2-DICHLORO-3-PROPANOL ◇ 1,2-DICHLOROPROPANOL-3 ◇ 2,3-DICHLORO-1-PROPANOL ◇ GLYCEROL-α,β-DICHLOROHYDRIN

TOXICITY DATA with REFERENCE
skn-rbt 10 mg/24H JIHTAB 30,63,48
eye-rbt 6800 μg SEV AJOPAA 29,1363,46
mma-sat 262 μg/plate MUREAV 57,381,78
mma-esc 131 μg/plate MUREAV 57,381,78
orl-rat LD50:90 mg/kg JIHTAB 30,63,48
ihl-rat LCLo:500 ppm/4H JIHTAB 31,343,49
skn-rbt LD50:200 mg/kg JIHTAB 30,63,48

CONSENSUS REPORTS: Reported in EPA TSCA Inventory.

SAFETY PROFILE: Poison by ingestion and skin contact. Moderately toxic by inhalation. A skin and severe eye irritant. Mutation data reported. When heated to decomposition it emits toxic fumes of Cl^-. See also CHLORINATED HYDROCARBONS, AROMATIC.

DGG800 CAS:563-54-2 ***HR: 2***
1,2-DICHLOROPROPENE
mf: $C_3H_4Cl_2$ mw: 110.97

PROP: Liquid. Bp: 75°, vap d: 3.83.

SYNS: DICHLOR ◇ 1,2-DICHLOROPROPYLENE ◇ DICHLORPROPEN-GEMISCH (GERMAN) ◇ PDC ◇ PROPYLENE DICHLORIDE ◇ RCRA WASTE NUMBER U083

TOXICITY DATA with REFERENCE
mmo-sat 10 μL/plate JSFAAE 32,826,81
mma-sat 10 μL/plate JSFAAE 32,826,81
mmo-asn 10 μL/plate PMRSDJ 2,87,81
orl-rat LD50:2 g/kg 85ARAE 3,20,76/77
skn-rbt LD50:8750 mg/kg 34ZIAG -,744,69

SAFETY PROFILE: Moderately toxic by ingestion. Mildly toxic by skin contact. Mutation data reported. When heated to decomposition it emits toxic fumes of Cl^-. See also CHLORINATED HYDROCARBONS, ALIPHATIC.

DGG950 CAS:542-75-6 ***HR: 3***
1,3-DICHLOROPROPENE
mf: $C_3H_4Cl_2$ mw: 110.97

PROP: Liquid. Bp: 103-110°, flash p: 95°F, d: 1.22, vap d: 3.8.

SYNS: α-CHLOROALLYL CHLORIDE ◇ Γ-CHLOROALLYL CHLORIDE ◇ 1,3-DICHLOROPROPENE-1 ◇ α,Γ-DICHLOROPROPYLENE ◇ 1,3-DICHLOROPROPYLENE ◇ NCI-C03985 ◇ RCRA WASTE NUMBER U084 ◇ TELONE ◇ TELONE II SOIL FUMIGANT ◇ VIDDEN D

TOXICITY DATA with REFERENCE
mmo-sat 33 μg/plate ENMUDM 5(Suppl 1),3,83
mma-sat 1 μmol/plate ENMUDM 2,59,80
sln-dmg-orl 5750 ppm ENMUDM 7,325,85
sce-ham:ovr 900 nmol/L CNJGA8 22,681,80
orl-rat TDLo:15600 mg/kg/2Y-I:CAR NTPTR* NTP-TR-269,85
orl-mus TDLo:31200 mg/kg/2Y-I:CAR NTPTR* NTP-TR-269,85
orl-rat LD50:250 mg/kg GUCHAZ 6,192,73
ihl-mus LC50:4650 mg/m³/2H 85GMAT -,48,82
skn-rbt LD50:504 mg/kg DTLVS* 4,133,80

CONSENSUS REPORTS: NTP Fifth Annual Report on Carcinogens. IARC Cancer Review: Group 2B IMEMDT 7,195,87; Human Inadequate Evidence IMEMDT 41,113,86; Animal Sufficient Evidence IMEMDT 41,113, 86. NTP Carcinogenesis Studies (gavage); Clear Evidence: mouse, rat NTPTR* NTP-TR-269,86. Reported in EPA TSCA Inventory. EPA Genetic Toxicology Program. Community Right-To-Know List.

OSHA PEL: TWA 1 ppm (skin)
ACGIH TLV: TWA 1 ppm (skin)
DFG MAK: Animal Carcinogen, Suspected Human Carcinogen.

SAFETY PROFILE: Confirmed carcinogen with experimental carcinogenic data. Poison by ingestion. Moderately toxic by skin contact. Mildly toxic by inhalation. A strong irritant. Mutation data reported. A pesticide. Dangerous fire hazard when exposed to heat, flame, or oxidizers. Reacts vigorously with oxidizing materials. To fight fire, use water, foam, CO_2, dry chemical. When heated to decomposition it emits toxic fumes of Cl^-. See also ALLYL COMPOUNDS and CHLORIDES.

DGH000 CAS:10061-02-6 ***HR: 3***
trans-1,3-DICHLOROPROPENE
mf: $C_3H_4Cl_2$ mw: 110.97

PROP: Flash point 21°C.

SYNS: (E)-1,3-DICHLOROPROPENE ◇ trans-1,3-DICHLOROPROPYLENE

TOXICITY DATA with REFERENCE
mmo-sat 20 μg/plate CNREA8 37,1915,77
mma-sat 20 μg/plate CNREA8 37,1915,77
dns-hmn:hla 100 μmol/L CALEDQ 20,263,83

CONSENSUS REPORTS: EPA Genetic Toxicology Program.

DFG MAK: Animal Carcinogen, Suspected Human Carcinogen.

SAFETY PROFILE: Human mutation data reported. A dangerous fire hazard when exposed to heat, flame, or oxidizers. When heated to decomposition it emits toxic fumes of Cl^-. See also CHLORINATED HYDROCARBONS, ALIPHATIC.

DGH200 CAS:10061-01-5 ***HR: 3***
cis-1,3-DICHLOROPROPENE
mf: $C_3H_4Cl_2$ mw: 110.97

PROP: Flash point 21°C.

SYNS: (Z)-1,3-DICHLOROPROPENE ◇ cis-1,3-DICHLOROPROPYLENE

TOXICITY DATA with REFERENCE
mmo-sat 20 μg/plate CNREA8 37,1915,77
mma-sat 20 μg/plate CNREA8 37,1915,77
dns-hmn:hla 100 μmol/L CALEDQ 20,263,83
scu-mus TDLo:9240 mg/kg/77W-I:NEO JJIND8 63,1433,79

CONSENSUS REPORTS: EPA Genetic Toxicology Program.

DFG MAK: Animal Carcinogen, Suspected Human Carcinogen.

SAFETY PROFILE: Confirmed carcinogen with experimental neoplastigenic data. Human mutation data reported. A dangerous fire hazard when exposed to heat, flame, or oxidizers. When heated to decomposition it emits toxic fumes of Cl^-. See also CHLORINATED HYDROCARBONS, ALIPHATIC.

DGH400 CAS:78-88-6 ***HR: 3***
2,3-DICHLOROPROPENE
mf: $C_3H_4Cl_2$ mw: 110.97

PROP: Flash p: 50°F.

SYNS: 2,3-DICHLORO-1-PROPENE ◇ 2,3-DICHLOROPROPYLENE

TOXICITY DATA with REFERENCE
skn-rbt 10 mg/24H open SEV AIHAAP 23,95,62
mmo-sat 20 μg/plate CNREA8 37,1915,77
mma-sat 20 μg/plate CNREA8 37,1915,77
dns-hmn:hla 100 μmol/L CALEDQ 20,263,83
orl-rat LD50:320 mg/kg AIHAAP 23,95,62
ihl-rat LCLo:500 ppm/4H AIHAAP 23,95,62
ihl-mus LC50:3100 mg/m³/2H 85GMAT -,48,82
skn-rbt LD50:1580 mg/kg AIHAAP 23,95,62

CONSENSUS REPORTS: Reported in EPA TSCA Inventory. EPA Genetic Toxicology Program.

SAFETY PROFILE: Poison by ingestion. Moderately toxic by inhalation and skin contact. Human mutation data reported. A severe skin irritant. A very dangerous fire hazard when exposed to heat, flame, or oxidizers. When heated to decomposition it emits toxic fumes of Cl^-. See also CHLORINATED HYDROCARBONS, ALIPHATIC.

DGH500 CAS:66826-73-1 ***HR: 3***
trans-1,3-DICHLOROPROPENE OXIDE
mf: $C_3H_4Cl_2O$ mw: 126.97

SYNS: trans-2-CHLORO-3-(CHLOROMETHYL)OXIRANE ◇ trans-DCPO ◇ trans-1,3-DICHLORO-1,2-EPOXYPROPANE

TOXICITY DATA with REFERENCE
otr-ham:emb 10 μmol/L JJIND8 69,531,82
skn-mus TDLo:400 mg/kg/73W-I:CAR CNREA8 43,159,83
scu-mus TDLo:20 mg/kg/71W-I:CAR CNREA8 43,159,83

SAFETY PROFILE: Questionable carcinogen with experimental carcinogenic data. Mutation data reported. When heated to decomposition it emits toxic fumes of Cl^-.

DGH600 CAS:2736-73-4 ***HR: D***
2,3-DICHLORO-2-PROPEN-1-OL
mf: $C_3H_4Cl_2O$ mw: 126.97

TOXICITY DATA with REFERENCE
mmo-sat 1 nmol/plate MUREAV 78,113,80
mma-sat 1 nmol/plate MUREAV 78,113,80

SAFETY PROFILE: Mutation data reported. When heated to decomposition it emits toxic fumes of Cl^-.

DGH800 CAS:10140-89-3 ***HR: 3***
2,3-DICHLORO PROPIONALDEHYDE
mf: $C_3H_4Cl_2O$ mw: 126.97

PROP: Liquid. Vap d: 4.4.

SYNS: 1,2-DICHLORO-3-PROPIONAL ◇ α,β-DICHLOROPROPIONALDEHYDE

TOXICITY DATA with REFERENCE
skn-rbt 10 mg/24H open SEV AMIHBC 4,119,51
skn-rbt 500 mg MOD SCCUR* -,3,61
eye-rbt 50 μg open SEV AMIHBC 4,119,51
mmo-sat 1 nmol/plate MUREAV 78,113,80
mma-sat 1 nmol/plate MUREAV 78,113,80
orl-rat LD50:160 mg/kg AMIHBC 4,119,51
ihl-rat LCLo:2500 ppb/4H SCCUR* -,3,61
orl-mus LD50:250 mg/kg SCCUR* -,3,61
ihl-mus LCLo:9300 ppm/15M SCCUR* -,3,61
skn-rbt LD50:78 mg/kg AMIHBC 4,119,51

SAFETY PROFILE: Poison by ingestion and skin contact. Mildly toxic by inhalation. A severe skin and eye irritant. Mutation data reported. When heated to decomposition it emits toxic fumes of Cl^-. See also ALDEHYDES and CHLORIDES.

DGI000 CAS:709-98-8 ***HR: 3***
DICHLOROPROPIONANILIDE
mf: $C_9H_9Cl_2NO$ mw: 218.09

PROP: Light brown solid (pure); liquid (technical grade). Mp (pure): 85-89°, bp (technical grade): 91-95°.

SYNS: BAY 30130 ◇ CHEM RICE ◇ CRYSTAL PROPANIL-4 ◇ DCPA ◇ N-(3,4-DICHLOROPHENYL)PROPANAMIDE ◇ N-(3,4-DICHLOROPHENYL)PROPIONAMIDE ◇ 3,4-DICHLOROPROPIONANILIDE ◇ 3',4'-DICHLOROPROPIONANILIDE ◇ DIPRAM ◇ DPA ◇ FARMCO PROPANIL ◇ FW 734 ◇ GRASCIDE ◇ HERBAX TECHNICAL ◇ MONTROSE PROPANIL ◇ PROPANEX ◇ PROPANID ◇ PROPANIDE ◇ PROPANIL ◇ PROPIONIC ACID-3,4-DICHLOROANILIDE ◇ PROP-JOB ◇ RISELECT ◇ ROGUE ◇ ROSANIL ◇ S 10165 ◇ STAM ◇ STAM F 34 ◇ STAM LV 10 ◇ STAM M-4 ◇ STAMPEDE ◇ STAMPEDE 3E ◇ STAM SUPERNOX ◇ STREL ◇ SUPERNOX ◇ SURCOPUR ◇ SURPUR ◇ VERTAC

TOXICITY DATA with REFERENCE
dnr-bcs 100 μg/disc NTIS** PB80-133226
cyt-mus-unr 100 mg/kg TGANAK 14(6),41,80
cyt-mus-orl 100 mg/kg CYGEDX 14(6),38,80
orl-rat LD50:560 mg/kg WRPCA2 9,119,70
orl-mus LD50:360 mg/kg GTPZAB 21(12),30,77
orl-dog LD50:1217 mg/kg TXAPA9 23,650,72
skn-rbt LD50:4830 mg/kg FMCHA2 -,C197,83
orl-mam LD50:2527 mg/kg NTIS** PB288-416

CONSENSUS REPORTS: EPA Genetic Toxicology Program.

SAFETY PROFILE: Poison by ingestion. Moderately toxic by an unspecified route. Mildly toxic by skin contact. Mutation data reported. When heated to decomposition it emits very toxic fumes of Cl^- and NO_x.

DGI400 CAS:75-99-0 ***HR: 2***
2,2-DICHLOROPROPIONIC ACID
DOT: NA 1760
mf: $C_3H_4Cl_2O_2$ mw: 142.97

PROP: White to tan powder.

SYNS: BASFAPON ◇ BASFAPON B ◇ BASFAPON/BASFAPON N ◇ BASINEX ◇ BH DALAPON ◇ CRISAPON ◇ DALAPON (USDA) ◇ DALAPON 85 ◇ DED-WEED ◇ DEVIPON ◇ α-DICHLOROPROPIONIC ACID ◇ α,α-DICHLOROPROPIONIC ACID ◇ DOWPON ◇ DOWPON M ◇ GRAMEVIN ◇ KENAPON ◇ LIROPON ◇ PROPROP ◇ RADAPON ◇ REVENGE ◇ UNIPON

TOXICITY DATA with REFERENCE
skn-rbt 100 μg/24H open AIHAAP 23,95,62
mmo-omi 500 ppm IJEBA6 11,114,73
orl-rat LD50:970 mg/kg FMCHA2 -,C69,83

CONSENSUS REPORTS: EPA Genetic Toxicology Program. Reported in EPA TSCA Inventory.

OSHA PEL: TWA 1 ppm
ACGIH TLV: TWA 1 ppm
DFG MAK: 1 ppm (6 mg/m^3)
DOT Classification: Corrosive Material; Label: Corrosive.

SAFETY PROFILE: Moderately toxic by ingestion. Corrosive. A skin irritant. Mutation data reported. When heated to decomposition it emits toxic fumes of Cl^-.

DGI 600 CAS: 127-20-8 ***HR: 2***
α,α-DICHLOROPROPIONIC ACID SODIUM SALT
mf: $C_3H_3Cl_2O_2 \cdot Na$ mw: 164.95

SYNS: BASFAPON B ◇ DALAPON ◇ DALAPON SODIUM ◇ DALAPON SODIUM SALT ◇ 2,2-DICHLOROPROPIONIC ACID, SODIUM SALT ◇ DOWPON ◇ 2,2-DPA ◇ GRAMEVIN ◇ NATRIUMSALZ DER 2,2-DICHLORPROPIONSAURE ◇ RADAPON ◇ SODIUM DALAPON ◇ SODIUM-α,α-DICHLOROPROPIONATE ◇ SODIUM-2,2-DICHLOROPROPIONATE ◇ UNIPON

TOXICITY DATA with REFERENCE
cyt-mus-unr 200 mg/kg TGANAK 16(1),45,82
orl-rat LD50:3860 mg/kg WRPCA2 9,119,70
unr-mus LD50:3650 mg/kg TGANAK 16(1),45,82
orl-rbt LD50:3400 mg/kg 85DPAN -,-,71/76
orl-gpg LD50:3400 mg/kg 85DPAN -,-,71/76
orl-ckn LD50:5600 mg/kg DOEAAH 35,25,79
orl-mam LD50:4 g/kg 85GYAZ -,86,71

CONSENSUS REPORTS: Reported in EPA TSCA Inventory. EPA Genetic Toxicology Program.

SAFETY PROFILE: Moderately toxic by ingestion. Mutation data reported. When heated to decomposition it emits toxic fumes of Na_2O and Cl^-.

DGJ100 CAS:1702-17-6 ***HR: 2***
3,6-DICHLORO-2-PYRIDINECARBOXYLIC ACID
mf: $C_6H_3Cl_2NO_2$ mw: 192.00

SYNS: CLOPYRALID ◇ 3,6-DICHLOROPICOLINIC ACID ◇ DOWCO 290 ◇ KYSELINA 3,6-DICHLORPIKOLINOVA ◇ LONTREL ◇ LONTREL 3 ◇ MATRIGON ◇ PICOLINIC ACID, 3,6-DICHLORO- ◇ 2-PYRIDINECARBOXYLIC ACID, 3,6-DICHLORO-(9CI) ◇ XRM 3972

TOXICITY DATA with REFERENCE
orl-rat TDLo:150 mg/kg (female 6-15D post):TER FAATDF 4,91,84
orl-rat LD50:4300 mg/kg 85JFAN A433,85
orl-dck LD50:1465 mg/kg PEMNDP 8,189,87

SAFETY PROFILE: Moderately toxic by ingestion. An experimental teratogen. When heated to decomposition it emits toxic fumes of NO_x and Cl^-.

DGJ200 CAS:3428-24-8 ***HR: 3***
4,5-DICHLOROPYROCATECHOL
mf: $C_6H_4Cl_2O_2$ mw: 179.00

SYNS: 4,5-DICHLORO-1,2-BENZENEDIOL ◇ 4,5-DICHLOROCATECHOL

TOXICITY DATA with REFERENCE
mmo-smc 75 mg/L MUREAV 119,273,83
ivn-mus LD50:42 mg/kg CSLNX* NX#07864

SAFETY PROFILE: Poison by intravenous route. Mutation data reported. When heated to decomposition it emits toxic fumes of Cl^-.

DGK000 CAS:1919-43-3 ***HR: 2***
2,3-DICHLOROQUINOXALINE-6-CARBONYLCHLORIDE
mf: $C_9H_3Cl_3N_2O$ mw: 261.49

SYN: 2,3-DICHLOROCHINOXALIN-6-KARBONYLCHLORID(CZECH)

TOXICITY DATA with REFERENCE
skn-rbt 500 mg/24H MOD 28ZPAK -,150,72
eye-rbt 100 mg/24H SEV 28ZPAK -,150,72
orl-rat LD50:2240 mg/kg 28ZPAK -,150,72

CONSENSUS REPORTS: Reported in EPA TSCA Inventory.

SAFETY PROFILE: Moderately toxic by ingestion. A skin and severe eye irritant. When heated to decomposition it emits very toxic fumes of Cl^- and NO_x.

DGK100 CAS:53-85-0 ***HR: D***
5,6-DICHLORO-1-β-d-RIBOFURANOSYLBENZIMIDAZOLE
mf: $C_{12}H_{12}Cl_2N_2O_4$ mw: 319.16

SYN: DRB

TOXICITY DATA with REFERENCE
oms-oin:oth 65 μmol/L BBACAQ 697,213,82
oms-hmn:hla 40 μmol/L PNASA6 79,2569,82
oms-hmn:hla 40 μmol/L PNASA6 79,2569,82

SAFETY PROFILE: Human mutation data reported. When heated to decomposition it emits toxic fumes of NO_x.

DGK200 CAS:320-72-9 ***HR: 3***
3,5-DICHLOROSALICYLIC ACID
mf: $C_7H_4Cl_2O_3$ mw: 207.01

SYN: USAF DO-68

TOXICITY DATA with REFERENCE
ipr-mus LD50:50 mg/kg NTIS** AD277-689

CONSENSUS REPORTS: Reported in EPA TSCA Inventory.

SAFETY PROFILE: Poison by intraperitoneal route. When heated to decomposition it emits toxic fumes of Cl^-.

DGK300 CAS:4109-96-0 ***HR: 3***
DICHLOROSILANE
DOT: UN 2189
mf: Cl_2H_2Si mw: 101.01

CONSENSUS REPORTS: Reported in EPA TSCA Inventory.

DOT Classification: Poison A; Label: Poison Gas and Flammable Gas

SAFETY PROFILE: Ignites spontaneously in air. Confined mixtures with air are spontaneously explosive. When heated to decomposition it emits toxic fumes of Cl^-. See also CHLOROSILANES.

DGK400 CAS:73926-91-7 ***HR: 3***
2,2'-DICHLORO-4,4'-STILBENEDIAMINE
mf: $C_{14}H_{12}Cl_2N_2$ mw: 279.18

SYNS: 4:4'-DIAMINO-2:2'-DICHLOROSTILBENE ◇ 2,2'-DICHLORO-4,4'-STILBENAMINE

TOXICITY DATA with REFERENCE
scu-rat TDLo:1400 mg/kg/W-I:ETA BMBUAQ 14,141,58

SAFETY PROFILE: Questionable carcinogen with experimental tumorigenic data. When heated to decomposition it emits very toxic fumes of Cl^- and NO_x.

DGK600 CAS:73926-92-8 ***HR: 3***
3,3'-DICHLORO-4,4'-STILBENEDIAMINE
mf: $C_{14}H_{12}Cl_2N_2$ mw: 279.18

SYN: 4:4'-DIAMINO-3:3'-DICHLOROSTILBENE

TOXICITY DATA with REFERENCE
scu-rat TDLo:200 mg/kg/W-I:ETA BMBUAQ 14,141,58

SAFETY PROFILE: Questionable carcinogen with experimental tumorigenic data. When heated to decomposition it emits very toxic fumes of Cl^- and NO_x.

DGK800 CAS:6607-45-0 ***HR: 2***
DICHLOROSTYRENE
mf: $C_8H_6Cl_2$ mw: 173.04

PROP: Liquid. Flash p: 225°F (OC), vap d: 6.0.

SYNS: α,β-DICHLOROSTYRENE ◇ DWUCHLOROSTYREN (POLISH)

TOXICITY DATA with REFERENCE
skn-rbt 10 mg/24H JIHTAB 30,63,48
eye-rbt 20 mg SEV JIHTAB 30,63,48
orl-rat LD50:4000 mg/kg JIHTAB 30,63,48
skn-rbt LD50:9 g/kg JIHTAB 30,63,48

SAFETY PROFILE: Moderately toxic by ingestion. Mildly toxic by skin contact. A skin and severe eye irritant. Combustible when exposed to heat, flame or oxidizers. Dangerous; see CHLORIDES. Can react vigorously with oxidizing materials. When heated to decomposition it emits toxic fumes of Cl^-. To fight fire, use CO_2, dry chemical.

DGK900 CAS:2736-23-4 ***HR: 2***
2,4-DICHLORO-5-SULFAMOYLBENZOIC ACID
mf: $C_7H_5Cl_2NO_4S$ mw: 270.09

SYNS: BENZOIC ACID, 5-(AMINOSULFONYL)-2,4-DICHLORO- ◇ BENZOIC ACID, 2,4-DICHLORO-5-SULFAMOYL- ◇ 2,4-DICHLORO-5-SULPHAMOYLBENZOIC ACID

TOXICITY DATA with REFERENCE
eye-rbt 100 mg SEV FCTOD7 20,573,82
eye-rbt 100 mg/4S RNS MLD FCTOD7 20,573,82
ipr-mus LD50:15 g/kg PCJOAU 19,697,85

SAFETY PROFILE: Slightly toxic by intraperitoneal route. A severe eye irritant. When heated to decomposition it emits toxic fumes of NO_x, SO_x, and Cl^-.

DGL200 CAS:3001-57-8 ***HR: 3***
3,4-DICHLOROSULFOLANE
mf: $C_4H_6Cl_2O_2S$ mw: 189.06

SYNS: DAC PRO ◇ 3,4-DICHLOROTETRAHYDROTHIOPHENE-1,1-DIOXIDE ◇ DICHLOROTHIOLANE DIOXIDE ◇ PRD EXPERIMENTAL NEMATOCIDE

TOXICITY DATA with REFERENCE
orl-rat LD50:482 mg/kg 28ZEAL 5,76,76
ipr-mus LD50:23 mg/kg RPTOAN 41,257,78
ivn-mus LD50:56 mg/kg CSLNX* NX#03183
skn-rbt LD50:1130 mg/kg TXAPA9 28,313,74

SAFETY PROFILE: Poison by intravenous and intraperitoneal routes. Moderately toxic by ingestion and skin contact. When heated to decomposition it emits very toxic fumes of Cl^- and SO_x.

DGL400 CAS:127-21-9 ***HR: 3***
DICHLOROTETRAFLUOROACETONE
mf: $C_3Cl_2F_4O$ mw: 198.93

PROP: A colorless liquid. Miscible with water and many organic solvents. Bp: 45.2°, fp: < −100°.

SYNS: ACETONE-1,3-DICHLORO-1,1,3,3-TETRAFLUOROACETONE ◇ sym-DICHLOROTETRAFLUOROACETONE ◇ 1,3-DICHLORO-1,1,3,3-TETRAFLUORO-2-PROPANONE

TOXICITY DATA with REFERENCE
orl-rat LD50:61 mg/kg TXAPA9 7,592,65
ihl-rat LCLo:50 ppm/6H TXAPA9 7,592,65
skn-rat LD50:91 mg/kg TXAPA9 7,592,65
ivn-mus LD50:180 mg/kg CSLNX* NX#01749
skn-rbt LD50:146 mg/kg 34ZIAG -,213,69

SAFETY PROFILE: Poison by ingestion, skin contact, inhalation, and intravenous routes. When heated to decomposition it emits very toxic fumes of Cl^- and F^-.

DGL600 CAS:1320-37-2 ***HR: 1***
DICHLOROTETRAFLUOROETHANE
DOT: UN 1958
mf: $C_2Cl_2F_4$ mw: 170.92

PROP: Colorless gas. Bp: 3.5°.

SYNS: DWUCHLOROCZTEROFLUOROETAN (POLISH) ◇ TETRAFLUORODICHLOROETHANE

TOXICITY DATA with REFERENCE
ihl-mus LCLo:700000 ppm/30M AMPMAR 30,447,69

CONSENSUS REPORTS: Reported in EPA TSCA Inventory.

OSHA PEL: TWA 1000 ppm
ACGIH TLV: TWA 1000 ppm
DOT Classification: Nonflammable Gas; Label: Nonflammable Gas.

SAFETY PROFILE: A mildly toxic irritant; narcotic in high concentrations. An asphyxiant. Reacts violently with alcohol. When heated to decomposition it emits toxic fumes of F^- and Cl^-.

DGL800 CAS:3511-19-1 ***HR: 3***
2,3-DICHLOROTETRAHYDROFURAN
mf: $C_4H_6Cl_2O$ mw: 141.00

TOXICITY DATA with REFERENCE
scu-mus TDLo:888 mg/kg/74W-I:ETA JNCIAM 48,1431,72

SAFETY PROFILE: Questionable carcinogen with experimental tumorigenic data. When heated to decomposition it emits toxic fumes of Cl^-. See also TETRAHYDROFURAN and CHLORIDES.

DGL875 CAS:6522-40-3 ***HR: 3***
endo-2,5-DICHLORO-7-THIABICYCLO(2.2.1) HEPTANE
mf: $C_6H_8Cl_2S$ mw: 183.10

SAFETY PROFILE: Mixtures with dimethylformamide + sodium tetrahydroborate explode when heated. When heated to decomposition it emits toxic fumes of Cl^- and SO_x.

DGM600 CAS:1918-13-4 ***HR: 3***
2,6-DICHLOROTHIOBENZAMIDE
mf: $C_7H_5Cl_2NS$ mw: 206.09

SYNS: CHLOROTHIAMIDE ◇ DCBN ◇ 2,6-DICHLOROBENZENECARBOTHIOAMIDE ◇ SD 7961 ◇ WL-5792

TOXICITY DATA with REFERENCE
cyt-mus-unr 500 mg/kg TGANAK 14(6),41,80
cyt-mus-orl 500 mg/kg CYGEDX 14(6),38,80
orl-rat LD50:757 mg/kg WRPCA2 9,119,70
skn-rat LD50:1000 mg/kg WRPCA2 9,119,70
ipr-rat LD50:242 mg/kg ATXKA8 23,42,67
orl-mus LD50:500 mg/kg ATXKA8 23,42,67
orl-rbt LD50:300 mg/kg ATXKA8 23,42,67
orl-ckn LD50:500 mg/kg 28ZEAL 5,50,76
orl-dom LDLo:125 mg/kg ATXKA8 23,42,67

CONSENSUS REPORTS: EPA Genetic Toxicology Program.

SAFETY PROFILE: Poison by ingestion and intraperitoneal route. Moderately toxic by skin contact. Mutation data reported. An herbicide. When heated to decomposition it emits very toxic fumes of Cl^-, NO_x, and SO_x.

DGM875 CAS:644-62-2 ***HR: 3***
N-(2,6-DICHLORO-m-TOLYL)ANTHRANILIC ACID
mf: $C_{14}H_{11}Cl_2NO_2$ mw: 296.16

PROP: White crystals from acetone lose water. Mp: 257-259°. Solubility (mg/mL): water 0.03; 0.1N NaOH 28, pH of saturated aq soln: approx. 6.9.

SYNS: ARQUEL ◇ 2-((2,6-DICHLORO-3-METHYLPHENYL)AMINO)-BENZOIC ACID (9CI) ◇ INF 4668 ◇ MECLOFENAMIC ACID ◇ MECLOPHENAMIC ACID

TOXICITY DATA with REFERENCE
ims-rbt TDLo:2750 μg/kg (female 1-3D post):REP BIREBV 14,451,76
orl-rat LD50:100 mg/kg AGACBH 7,481,77
ipr-rat LD50:109 mg/kg JPETAB 148,422,65

SAFETY PROFILE: Poison by ingestion. Experimental reproductive effects. When heated to decomposition it emits toxic fumes of Cl^- and NO_x.

DGN000 CAS:29098-15-5 ***HR: 3***
N-(2,6-DICHLORO-m-TOLYL)ANTHRANILIC ACID ETHOXYMETHYL ESTER
mf: $C_{17}H_{17}Cl_2NO_3$ mw: 354.25

SYNS: 2-((2,6-DICHLORO-3-METHYLPHENYL)AMINO)BENZOIC ACID ETHOXYMETHYL ESTER ◇ ESTERE ETOSSIMETILICO dell' ACIDO N-(2,6-DICLORO-m-TOLIL)ANTRANILICO (ITALIAN) ◇ ETHOXYMETHYL-N-(2,6-DICHLORO-m-TOLYL)ANTHRANILATE ◇ ETOCLOFENE ◇ ETOFEN ◇ TEROFENAMATE

TOXICITY DATA with REFERENCE
orl-mus LD50:918 mg/kg DRFUD4 1,421,76
ipr-mus LD50:300 mg/kg DRFUD4 1,421,76

SAFETY PROFILE: Poison by intraperitoneal route. Moderately toxic by ingestion. An analgesic and anti-inflammatory agent. See also ESTERS. When heated to decomposition it emits very toxic fumes of Cl^- and NO_x.

DGN200 CAS:2782-57-2 ***HR: 2***
1,3-DICHLORO-s-TRIAZINE-2,4,6(1H,3H,5H)-TRIONE
DOT: UN 2465
mf: $C_3H_2Cl_2N_3O_3$ mw: 198.98

PROP: White crystals, chlorine odor, moderately sol in water. Mp: 225°.

SYNS: ACL 70 ◇ DICHLOROISOCYANURIC ACID ◇ DICHLOROISOCYANURIC ACID, dry (DOT) ◇ KYSELINA DICHLORISOKYANUROVA (CZECH) ◇ TROCLOSENE

TOXICITY DATA with REFERENCE
skn-rbt 500 mg SEV 34ZIAG -,167,69
eye-rbt 100 MG SEV 34ZIAG -,167,69
orl-hmn LDLo:3570 mg/kg:GIT 34ZIAG -,167,69
orl-rat LD50:1173 mg/kg MarJV# 29MAR77

CONSENSUS REPORTS: Reported in EPA TSCA Inventory.

DOT Classification: Oxidizer; Label: Oxidizer.

SAFETY PROFILE: Moderately toxic by ingestion. Human systemic effects by ingestion: ulceration or bleeding from stomach. Autopsy findings include gastrointestinal tract irritation, tissue edema, liver and kidney congestion. A severe eye and skin irritant. When heated to decomposition it emits chlorides and carbon monoxide.

DGN400 CAS:4499-01-8 ***HR: 2***
2-(4,6-DICHLORO-s-TRIAZIN-2-YLAMINO)-4-(4-AMINO-3-SULFO-1-ANTHRAQUINONYLAMINO)BENZENESULFONIC ACID, DISODIUM SALT
mf: $C_{23}H_{12}Cl_2N_6O_8S_2•2Na$ mw: 681.41

SYN: MODR BRILANTNI OSTAZINOVA S-R (CZECH)

TOXICITY DATA with REFERENCE
eye-rbt 100 mg/24H SEV 28ZPAK -,242,72
orl-rat LD50:8980 mg/kg 28ZPAK -,242,72

CONSENSUS REPORTS: Reported in EPA TSCA Inventory.

SAFETY PROFILE: Mildly toxic by ingestion. A severe eye irritant. When heated to decomposition it emits very toxic fumes of Cl^-, NO_x, Na_2O, and SO_x.

DGN600 CAS:73826-58-1 ***HR: 1***
4-(4,6-DICHLORO-s-TRIAZIN-2-YLAMINO)-5-HYDROXY-6-(2-HYDROXY-5-NITROPHENYLAZO)-2,7-NAPHTHALENEDISULFONIC ACID
mf: $C_{19}H_{11}Cl_2N_7O_{10}S_2$ mw: 632.39

SYN: CERN OSTAZINOVA H-N (CZECH)

TOXICITY DATA with REFERENCE
skn-rbt 500 mg/24H MLD 28ZPAK -,234,72
eye-rbt 500 mg/24H MLD 28ZPAK -,234,72
orl-rat LD50:9120 mg/kg 28ZPAK -,234,72

SAFETY PROFILE: Mildly toxic by ingestion. A skin and eye irritant. When heated to decomposition it emits very toxic fumes of SO_x, NO_x, and Cl^-.

DGN800 CAS:6522-86-7 ***HR: 1***
5-(3,5-DICHLORO-s-TRIAZINYLAMINO)-4-HYDROXY-3-PHENYLAZO-2,7-NAPHTHALENEDISULFONIC ACID
mf: $C_{19}H_{12}Cl_2N_6O_7S_2$ mw: 571.39

SYN: CERVEN BRILANTNI OSTAZINOVA S-5B (CZECH)

TOXICITY DATA with REFERENCE
eye-rbt 100 mg/24H MOD 28ZPAK -,235,72
orl-rat LD50:7460 mg/kg 28ZPAK -,235,72

SAFETY PROFILE: Mildly toxic by ingestion. An eye irritant. When heated to decomposition it emits very toxic fumes of Cl^-, NO_x, and SO_x.

DGO000 CAS:73816-75-8 ***HR: 1***
2-(6-(4,6-DICHLORO-s-TRIAZINYL)METHYLAMINO-1-HYDROXY-3-SULFONAPHTHYLAZO)-1,5-NAPHTHALENEDISULFONIC ACID
mf: $C_{24}H_{16}Cl_2N_6O_{10}S_3$ mw: 715.54

SYN: ORANZ BRILANTNI OSTAZINOVA S-2R (CZECH)

TOXICITY DATA with REFERENCE
skn-rbt 500 mg/24H MLD 28ZPAK -,237,72
eye-rbt 500 mg/24H MLD 28ZPAK -,237,72
orl-rat LD50:8500 mg/kg 28ZPAK -,237,72

SAFETY PROFILE: Mildly toxic by ingestion. A skin and eye irritant. When heated to decomposition it emits very toxic fumes of Cl^-, NO_x and SO_x.

DGO200 CAS:72596-01-1 ***HR: D***
DICHLORO(4,5,6-TRICHLORO-o-PHENYLENEDIAMMINE)PLATINUM(II)
mf: $C_6H_5Cl_5N_2Pt$ mw: 477.47

TOXICITY DATA with REFERENCE
mmo-sat 19 nmol/L JMCMAR 23,459,80
mma-sat 2 nmol/L JMCMAR 23,459,80

SAFETY PROFILE: Mutation data reported. See also PLATINUM COMPOUNDS. When heated to decomposition it emits very toxic fumes of Cl^- and NO_x.

DGO400 CAS:3615-21-2 ***HR: 3***
4,5-DICHLORO-2-TRIFLUOROMETHYLBENZIMIDAZOLE
mf: $C_8H_3Cl_2F_3N_2$ mw: 255.03

SYNS: CHLORFLURAZOLE ◇ CHLOROFLURAZOLE ◇ NC 3363

TOXICITY DATA with REFERENCE
orl-rat LD50:13080 μg/kg PSSCBG 15,31,84
ipr-mus LD50:14 mg/kg BCPCA6 18,1389,69
orl-ckn LD50:34 mg/kg GUCHAZ 6,101,73

CONSENSUS REPORTS: EPA Extremely Hazardous Substances List.

SAFETY PROFILE: Poison by ingestion and intraperitoneal routes. A pesticide. When heated to decomposition it emits very toxic fumes of Cl^-, NO_x, and F^-.

DGO600 CAS:64048-90-4 ***HR: 3***
DICHLORO(m-TRIFLUOROMETHYLPHENYL)ARSINE
mf: $C_7H_5AsCl_2F_3$ mw: 291.94

TOXICITY DATA with REFERENCE
ihl-mus LCLo:380 mg/m^3 NDRC** -,12,43
ihl-hmn LCLo:28 ppm/10M NTIS** PB214-270

CONSENSUS REPORTS: Arsenic and its compounds are on the Community Right-To-Know List.

OSHA PEL: TWA 0.5 mg(As)/m^3

SAFETY PROFILE: A human poison by inhalation. See also ARSENIC COMPOUNDS. When heated to decomposition it emits very toxic fumes of As, Cl^-, and F^-.

DGO800 CAS:594-31-0 **HR: 3**
DICHLOROTRIPHENYLANTIMONY
mf: $C_{18}H_{15}Cl_2Sb$ mw: 423.98

SYNS: ANTIMONY TRIPHENYLDICHLORIDE ◇ DICHLOROTRIPHENYLSTIBINE ◇ TRIPHENYLANTIMONY DICHLORIDE

TOXICITY DATA with REFERENCE
orl-rat LD50:195 mg/kg MarJV# 29MAR77

CONSENSUS REPORTS: Reported in EPA TSCA Inventory. Antimony and its compounds are on the Community Right-To-Know List.

OSHA PEL: TWA 0.5 mg(Sb)/m^3
ACGIH TLV: TWA 0.5 mg(Sb)/m^3
NIOSH REL: TWA 0.5 mg/m^3

SAFETY PROFILE: Poison by ingestion. See also ANTIMONY COMPOUNDS. When heated to decomposition it emits very toxic fumes of Cl^- and Sb.

DGP000 CAS:627-72-5 **HR: 3**
S-DICHLOROVINYL-l-CYSTEINE
mf: $C_5H_7Cl_2NO_2S$ mw: 216.09

SYNS: S-(1,2-DICHLOROETHYLENEYL)-l-CYSTEINE ◇ l-3-((1,2-DICHLOROVINYL)THIO)ALANINE

TOXICITY DATA with REFERENCE
mma-sat 1 μg/plate CBINA8 54,15,85
ipr-rat LDLo:50 mg/kg FCTXAV 3,67,65
ipr-mus LD50:45 mg/kg FCTXAV 3,67,65
ivn-rbt LDLo:10 mg/kg FCTXAV 3,67,65
ipr-gpg LDLo:20 mg/kg FCTXAV 3,67,65

SAFETY PROFILE: Poison by intravenous and intraperitoneal routes. Mutation data reported. When heated to decomposition it emits very toxic fumes of Cl^-, NO_x, and SO_x.

DGP125 CAS:13419-46-0 **HR: D**
S-(trans-1,2-DICHLOROVINYL)-l-CYSTEINE
mf: $C_5H_7Cl_2NO_2S$ mw: 216.09

SYN: (E)-S-(1,2-DICHLOROETHENYL)-l-CYSTEINE(9CI)

TOXICITY DATA with REFERENCE
dnd-rbt-ivn 20 mg/kg TXCYAC 35,25,85
dnd-rbt-ipr 100 mg/kg TXCYAC 35,25,85
dnd-rbt:kdy 100 μmol/L TXCYAC 35,25,85

SAFETY PROFILE: Mutation data reported. When heated to decomposition it emits toxic fumes of Cl^-, NO_x, and SO_x.

DGP200 CAS:626-16-4 **HR: 3**
α,α'-DICHLORO-m-XYLENE
mf: $C_8H_8Cl_2$ mw: 175.06

SYNS: 1,3-BIS(CHLOROMETHYL)BENZENE ◇ m-XYLYLENE DICHLORIDE

TOXICITY DATA with REFERENCE
ivn-mus LD50:100 mg/kg CSLNX* NX#03828

CONSENSUS REPORTS: Reported in EPA TSCA Inventory.

SAFETY PROFILE: Poison by intravenous route. See also CHLORINATED AROMATIC HYDROCARBONS. When heated to decomposition it emits toxic fumes of Cl^-.

DGP400 CAS:612-12-4 **HR: 3**
α,α'-DICHLORO-o-XYLENE
mf: $C_8H_8Cl_2$ mw: 175.06

SYNS: 1,2-BIS(CHLOROMETHYL)BENZENE ◇ o-XYLYLENE DICHLORIDE

TOXICITY DATA with REFERENCE
ivn-mus LD50:320 mg/kg CSLNX* NX#03225

CONSENSUS REPORTS: Reported in EPA TSCA Inventory.

SAFETY PROFILE: Poison by intravenous route. See also CHLORINATED AROMATIC HYDROCARBONS. When heated to decomposition it emits toxic Cl^-.

DGP600 CAS:623-25-6 **HR: 2**
α,α'-DICHLORO-p-XYLENE
mf: $C_8H_8Cl_2$ mw: 175.06

TOXICITY DATA with REFERENCE
skn-rbt 500 mg MOD 34ZIAG -,213,69
eye-rbt 100 mg MOD 34ZIAG -,213,69
orl-rat LD50:1780 mg/kg 34ZIAG -,213,69

CONSENSUS REPORTS: Reported in EPA TSCA Inventory.

SAFETY PROFILE: Moderately toxic by ingestion. A skin and eye irritant. See also CHLORINATED HYDROCARBONS, AROMATIC. When heated to decomposition it emits toxic fumes of Cl^-.

DGP800 CAS:120-67-2 **HR: 2**
2,4-DICHLORPHENYL "CELLOSOLVE"
mf: $C_8H_8Cl_2O_2$ mw: 207.06

SYN: 2-(2,4-DICHLOROPHENOXY)ETHANOL

TOXICITY DATA with REFERENCE
orl-rat LD50:1410 mg/kg UCDS** 12/29/71
skn-rbt LD50:1250 mg/kg UCDS** 12/29/71

SAFETY PROFILE: Moderately toxic by skin contact and ingestion. When heated to decomposition it emits toxic fumes of Cl^-.

DGP900 CAS:62-73-7 ***HR: 3***
DICHLORVOS
DOT: NA 2783
mf: $C_4H_7Cl_2O_4P$ mw: 220.98

PROP: Liquid. Bp: 120° @ 14 mm, bp: 77° @ 1 mm. Sltly sol in water and glycerin; misc with aromatic and chlorinated hydrocarbon solvents and alc.

SYNS: APAVAP ◇ ASTROBOT ◇ ATGARD ◇ BAY 19149 ◇ BENFOS ◇ BIBESOL ◇ BREVINYL ◇ CANOGARD ◇ CEKUSAN ◇ CHLORVINPHOS ◇ CYANOPHOS ◇ CYPONA ◇ DDVF ◇ DDVP ◇ DEDEVAP ◇ DERIBAN ◇ DERRIBANTE ◇ DEVIKOL ◇ (2,2-DICHLOOR-VINYL)-DIMETHYL-FOSFAAT (DUTCH) ◇ DICHLOORVO (DUTCH) ◇ DICHLORFOS (POLISH) ◇ 2,2-DICHLOROETHENOL DIMETHYL PHOSPHATE ◇ 2,2-DICHLOROETHENYL DIMETHYL PHOSPHATE ◇ 2,2-DICHLOROETHENYL PHOSPHORIC ACID DIMETHYL ESTER ◇ DICHLOROPHOS ◇ DICHLOROVAS ◇ (2,2-DICHLORO-VINIL)DIMETILFOSFATO (ITALIAN) ◇ 2,2-DICHLOROVINYL ALCOHOL, DIMETHYL PHOSPHATE ◇ 2,2-DICHLOROVINYL DIMETHYL PHOSPHATE ◇ 2,2-DICHLOROVINYL DIMETHYL PHOSPHORIC ACID ESTER ◇ DICHLOROVOS ◇ DICHLORPHOS ◇ (2,2-DICHLOR-VINYL)-DIMETHYL-PHOSPHAT (GERMAN) ◇ O-(2,2-DICHLOR-VINYL)-O,O-DIMETHYLPHOSPHAT (GERMAN) ◇ DIMETHYL-2,2-DICHLOROETHENYL PHOSPHATE ◇ DIMETHYL DICHLOROVINYL PHOSPHATE ◇ DIMETHYL-2,2-DICHLOROVINYL PHOSPHATE ◇ O,O-DIMETHYL DICHLOROVINYL PHOSPHATE ◇ O,O-DIMETHYL-O-2,2-DICHLOROVINYL PHOSPHATE ◇ O,O-DIMETHYL-O-(2,2-DICHLOR-VINYL)-PHOSPHAT (GERMAN) ◇ DIVIPAN ◇ DQUIGARD ◇ DUO-KILL ◇ DURAVOS ◇ ENT 20,738 ◇ EQUIGEL ◇ ESTROSEL ◇ ESTROSOL ◇ FECAMA ◇ FLY-DIE ◇ FLY FIGHTER ◇ HERKAL ◇ KRECALVIN ◇ LINDAN ◇ MAFU ◇ MARVEX ◇ MOPARI ◇ NCI-C00113 ◇ NERKOL ◇ NOGOS ◇ NO-PEST ◇ NO-PEST STRIP ◇ NSC-6738 ◇ NUVA ◇ OKO ◇ OMS 14 ◇ PHOSPHATE de DIMETHYLE et de 2,2-DICHLOROVINYLE (FRENCH) ◇ PHOSPHORIC ACID-2,2-DICHLOROETHENYL DIMETHYL ESTER ◇ PHOSVIT ◇ SD-1750 ◇ SZKLARNIAK ◇ TAP 9VP ◇ TASK ◇ TASK TABS ◇ TENAC ◇ TETRAVOS ◇ VAPONA ◇ VAPONITE ◇ VERDICAN ◇ VERDIPOR ◇ VINYLOFOS ◇ VINYLOPHOS

TOXICITY DATA with REFERENCE
dns-hmn:oth 65 mmol/L PSSCBG 15,439,84
dni-hmn:lym 62 mg/L TUMOAB 66,425,80
cyt-ham-ipr 3 mg/kg ARTODN 58,152,85
orl-rat TDLo:39200 μg/kg (14-21D preg):REP NUPOBT 15,255,77
ihl-rbt TCLo:4 mg/m³/23H (female 1-28D post):TER ATXKA8 30,29,72
orl-rat LD50:25 mg/kg DOEAAH 35,25,79
ihl-rat LC50:15 mg/m³/4H GISAAA 33(12),35,68
skn-rat LD50:70400 μg/kg APYPAY 32,507,81
ipr-rat LD50:23300 μg/kg IJPPAZ 31,19,87
scu-rat LD50:10800 μg/kg APYPAY 32,507,81
orl-mus LD50:61 mg/kg OSDIAF 15,553,66
ihl-mus LC50:13 mg/m³/4H GISAAA 33(12),35,68
skn-mus LD50:206 mg/kg ABCHA6 27,684,63
ipr-mus LD50:22 mg/kg JAFCAU 11,91,63
scu-mus LD50:24 mg/kg IMSUAI 31,170,62
ivn-mus LD50:18 mg/kg CSLNX* NX#00004
orl-dog LD50:100 mg/kg 85JFAN A141,84
orl-rbt LD50:10 mg/kg BEXBAN 83,32,77
skn-rbt LD50:107 mg/kg BESAAT 12,117,66

CONSENSUS REPORTS: IARC Cancer Review: Group 3 IMEMDT 7,56,87; Animal Inadequate Evidence IMEMDT 20,97,79. NCI Carcinogenesis Bioassay (feed); No Evidence: mouse, rat NCITR* NCI-CG-TR-10,77. EPA Genetic Toxicology Program. Community Right-To-Know List. EPA Extremely Hazardous Substances List.

OSHA PEL: TWA 1 mg/m³ (skin)
ACGIH TLV: TWA 0.1 ppm (skin)
DFG MAK: 0.1 ppm (1 mg/m³)
DOT Classification: Poison B; Label: Poison; Poison B; Label: Poison.

SAFETY PROFILE: Poison by ingestion, inhalation, skin contact, subcutaneous, intravenous, and intraperitoneal routes. Experimental teratogenic and reproductive effects. Human mutation data reported. A cholinesterase inhibitor, it is used in flea (pest) collars for pets. No neurotoxicity has been observed. It is very rapidly metabolized and excreted. When heated to decomposition it emits very toxic fumes of Cl^- and PO_x. See also PARATHION.

DGQ200 CAS:116-52-9 ***HR: 1***
DICLORALUREA
mf: $C_5H_6Cl_6N_2O_3$ mw: 354.83

SYNS: 1,3-BIS(1-HYDROXY-2,2,2-TRICHLOROETHYL)UREA ◇ 1,3-BIS(2,2,2-TRICHLORO-1-HYDROXYETHYL)UREA ◇ CRAG DCU-73w ◇ CRAG EXPERIMENTAL HERBICIDE 2 ◇ CRAG HERBICIDE 2 ◇ DCM ◇ DCU ◇ DICHLORAL UREA ◇ DKhM ◇ EH2 ◇ EXPERIMENTAL HERBICIDE 2

TOXICITY DATA with REFERENCE
skn-rbt 50 mg open MLD UCDS** 4/25/68
cyt-mus-orl 5 g/kg CYGEDX 11(4),62,77
orl-rat LD50:6680 mg/kg 28ZEAL 5,74,76

SAFETY PROFILE: Mildly toxic by ingestion. A skin irritant. Mutation data reported. A pesticide. When heated to decomposition it emits very toxic fumes of Cl^- and NO_x.

DGQ300 CAS:12045-01-1 ***HR: 3***
DICOBALT BORIDE
mf: BCo_2 mw: 128.68

CONSENSUS REPORTS: Cobalt compounds are on the Community Right-To-Know List.

SAFETY PROFILE: Ignites spontaneously in air when dry. See also COBALT COMPOUNDS and BORON COMPOUNDS.

DGQ400 CAS:36499-65-7 ***HR: 3***
DICOBALT EDETATE
mf: $C_{10}H_{12}CoN_2O_8{\cdot}Co$ mw: 406.10

SYNS: Ba 2724 ◇ COBALT(2)-EDATHAMIL ◇ DICOBALT EDTA ◇ ((ETHYLENEDINITRILO)TETRAACETATO(2−))-COBALTATE(2−) COBALT(2+) SALT ◇ KOBALT-EDTA (GERMAN)

TOXICITY DATA with REFERENCE
ipr-rat LD50:100 mg/kg AEPPAE 243,254,62
ivn-rat LD50:43 mg/kg AIPTAK 143,219,63

CONSENSUS REPORTS: Cobalt and its compounds are on the Community Right-To-Know List.

NIOSH REL: (Cobalt) TWA: Insufficient evidence for recommending limit

SAFETY PROFILE: Poison by intravenous and intraperitoneal routes. See also COBALT COMPOUNDS. When heated to decomposition it emits toxic fumes of NO_x.

DGQ500 CAS:965-52-6 ***HR: 3***
DICOFERIN
mf: $C_{12}H_9N_3O_5$ mw: 275.24

PROP: Crystals from pyridine. Mp: 298°. Practically insol in water.

SYNS: BACIFURANE ◇ DIARLIDAN ◇ ERCEFUROL ◇ ERCEFURYL ◇ 4-HYDROXY-BENZOIC ACID ((5-NITRO-2-FURANYL)METHYLENE)HYDRAZIDE ◇ p-HYDROXYBENZOIC ACID (5-NITROFURFURYLIDENE)HYDRAZIDE ◇ NIFUROXAZID ◇ NIFUROXAZIDE ◇ (NITRO-5' FURFURYLIDENE-2') HYDROXY-4 BENZHYDRAZIDE (FRENCH) ◇ PENTOFURYL ◇ R.C. 27-109 ◇ RC 30-109

TOXICITY DATA with REFERENCE
mmo-sat 500 ng/plate MUREAV 157,1,85
mma-sat 500 ng/plate MUREAV 157,1,85
orl-mus LDLo:6000 mg/kg APFRAD 21,287,63
ipr-mus LD50:100 mg/kg JPPMAB 16,663,64

SAFETY PROFILE: Poison by intraperitoneal route. Mildly toxic by ingestion. Mutation data reported. When heated to decomposition it emits toxic fumes of NO_x.

DGQ600 CAS:1117-94-8 ***HR: 3***
DICOPPER(I) ACETYLIDE
mf: C_2Cu_2 mw: 151.10

CuC≡CCu

CONSENSUS REPORTS: Copper and its compounds are on the Community Right-To-Know List.

SAFETY PROFILE: An unstable material. It explodes on impact or heating to 100°C. If warmed in air or oxygen it explodes on subsequent contact with acetylene. The sensitivity of the acetylide when precipitated from solution increases with acidity of the solution. It ignites on contact with chlorine; bromine vapor; or finely divided iodine. Reaction with silver nitrate solutions produces a sensitive, explosive mixture of silver acetylide and silver. When heated to decomposition it emits acrid smoke and fumes. See also COPPER COMPOUNDS and ACETYLIDES.

DGQ625 CAS:86425-12-9 ***HR: 3***
DICOPPER(I)-1,5-HEXADIYNIDE
mf: $C_6H_4Cu_2$ mw: 203.19

$(\text{-}CH_2C{\equiv}CCu)_2$

CONSENSUS REPORTS: Copper and its compounds are on the Community Right-To-Know List.

SAFETY PROFILE: The dry material explodes at room temperature. When heated to decomposition it emits acrid smoke and fumes. See also COPPER COMPOUNDS and ACETYLIDES.

DGQ650 CAS:41084-90-6 ***HR: 3***
DICOPPER(I) KETENIDE
mf: C_2Cu_2O mw: 167.11

CONSENSUS REPORTS: Copper and its compounds are on the Community Right-To-Know List.

SAFETY PROFILE: Mildly explosive when dry. See also COPPER COMPOUNDS.

DGQ700 CAS:27134-24-3 ***HR: 2***
DICRESOL
mf: $C_{14}H_{14}O_2$ mw: 214.28

SYN: ar,ar'-DIMETHYL-(1,1'-BIPHENYL)-ar,ar'-DIOL (9CI)

TOXICITY DATA with REFERENCE
orl-rat LD50:1625 mg/kg GTPZAB 20(9),53,76
skn-rat LD50:825 mg/kg GTPZAB 20(9),53,76
orl-mus LD50:651 mg/kg GTPZAB 20(9),53,76

SAFETY PROFILE: Moderately toxic by ingestion and skin contact. When heated to decomposition it emits acrid smoke and fumes.

DGQ859 ***HR: 3***
DICROTONYL PEROXIDE
mf: $C_8H_{10}O_4$ mw: 170.16

$CH_3CH{=}CHCO{\cdot}O\text{-})_2$

SAFETY PROFILE: A very shock-sensitive explosive. When heated to decomposition it emits acrid smoke and fumes. See also PEROXIDES.

DGQ875 CAS:141-66-2 ***HR: 3***
DICROTOPHOS
mf: $C_8H_{16}NO_5P$ mw: 237.22

SYNS: BIDIRL ◇ BIDRIN ◇ C 709 ◇ CARBICRON ◇ CIBA 709 ◇ DIAPADRIN ◇ DICROTOFOS (DUTCH) ◇ 3-(DIMETHOXYPHOSPHINYLOXY)-N,N-DIMETHYL-cis-CROTONAMIDE ◇ 3-(DIMETHOXYPHOSPHINYLOXY)-N,N-DIMETHYLISOCROTONAMIDE ◇ 3-(DIMETHYLAMINO)-1-METHYL-3-OXO-1-PROPENYL DIMETHYL PHOSPHATE ◇ cis-2-DIMETHYLCARBAMOYL-1-METHYLVINYL DIMETHYLPHOSPHATE ◇ O,O-DIMETHYL-O-(2-DIMETHYL-CARBAMOYL-1-METHYL-VINYL)PHOSPHAT (GERMAN) ◇ O,O-DIMETHYL-O-(N,N-DIMETHYLCARBAMOYL-1-METHYLVINYL) PHOSPHATE ◇ O,O-DIMETHYL-O-(1,4-DIMETHYL-3-OXO-4-AZA-PENT-1-ENYL)FOSFAAT (DUTCH) ◇ O,O-DIMETHYL-O-(1,4-DIMETHYL-3-OXO-4-AZA-PENT-1-ENYL)PHOSPHATE ◇ DIMETHYLPHOSPHATE ESTER with 3-HYDROXY-N,N-DIMETHYL-cis-CROTONAMIDE ◇ DIMETHYL PHOSPHATE of 3-HYDROXY-N,N-DIMETHYL-cis-CROTONAMIDE ◇ O,O-DIMETIL-O-(1,4-DIMETIL-3-OXO-4-AZA-PENT-1-ENIL)-FOSFATO (ITALIAN) ◇ EKTAFOS ◇ ENT 24,482 ◇ 3-HYDROXYDIMETHYL CROTONAMIDE DIMETHYL PHOSPHATE ◇ 3-HYDROXY-N,N-DIMETHYL-cis-CROTONAMIDE DIMETHYL PHOSPHATE ◇ PHOSPHATE de DIMETHYLE et de 2-DIMETHYLCARBAMOYL-1-METHYL VINYLE (FRENCH) ◇ SD 3562 ◇ SHELL SD-3562

TOXICITY DATA with REFERENCE
mma-sat 500 μg/plate JTEHD6 16,403,85
mrc-smc 30 mmol/L/5H MUREAV 32,133,75
orl-rat LD50:13 mg/kg ARSIM* 20,6,66
ihl-rat LC50:90 mg/m^3/4H PSDTAP 15,239,74
skn-rat LD50:42 mg/kg WRPCA2 9,119,70
scu-rat LD50:8137 μg/kg BJPCBM 40,124,70
orl-mus LD50:11 mg/kg GUCHAZ 6,196,73
ipr-mus LD50:9500 μg/kg TXAPA9 16,446,70
scu-mus LD50:11500 μg/kg JPPMAB 19,612,67
ivn-mus LD50:9900 μg/kg JPPMAB 19,612,67
skn-rbt LD50:168 mg/kg GUCHAZ 6,196,73

CONSENSUS REPORTS: EPA Farm Worker Reentry (39 FR 16888,74). EPA Genetic Toxicology Program. EPA Extremely Hazardous Substances List.

OSHA PEL: TWA 0.25 mg/m^3 (skin)
ACGIH TLV: TWA 0.25 mg/m^3 (skin)

SAFETY PROFILE: Poison by ingestion, inhalation, skin contact, subcutaneous, intravenous, and intraperitoneal routes. Mutation data reported. Used to control the coffee borer and certain economically important pests of cotton. When heated to decomposition it emits very toxic fumes of NO_x and PO_x. See also ESTERS.

DGR200 CAS:12001-89-7 ***HR: 3***
DICUMENE CHROMIUM
mf: $C_{18}H_{24}\cdot Cr$ mw: 292.42

SYNS: BIS(CUMENE)CHROMIUM ◇ BIS(pi-CUMENE)CHROMIUM ◇ BIS(ISOPROPYLBENZENE)CHROMIUM ◇ DICUMENYLCHROMIUM

TOXICITY DATA with REFERENCE
skn-rbt 500 mg open MOD UCDS** 4/21/67
eye-rbt 15 mg MOD UCDS** 4/21/67
orl-rat LD50:810 mg/kg AIHAAP 30,470,69
ivn-mus LD50:2200 μg/kg CSLNX* NX#06778
skn-rbt LD50:22 mg/kg UCDS** 4/21/67

CONSENSUS REPORTS: Chromium and its compounds are on the Community Right-To-Know List.

OSHA PEL: TWA 1 mg(Cr)/m^3
ACGIH TLV: TWA 0.05 mg(Cr)/m^3
NIOSH REL: (Chromium(VI)) CL 1 μg(Cr(VI))/m^3

SAFETY PROFILE: Poison by skin contact and intravenous routes. Moderately toxic by ingestion. A skin and eye irritant. See also CHROMIUM COMPOUNDS. When heated to decomposition it emits acrid and irritating fumes.

DGR400 CAS:25566-92-1 ***HR: 3***
DICUMYLMETHANE
mf: $C_{19}H_{24}$ mw: 252.43

SYN: DKM (RUSSIAN)

TOXICITY DATA with REFERENCE
orl-mus LDLo:15 g/kg TPKVAL 6,73,64
ihl-mus LCLo:122 mg/m^3 TPKVAL 6,73,64
scu-mus LDLo:19 g/kg TPKVAL 6,73,64

SAFETY PROFILE: Poison by inhalation. Mildly toxic by ingestion and subcutaneous routes. When heated to decomposition it emits acrid smoke and irritating fumes.

DGR600 CAS:80-43-3 ***HR: 1***
DI-α-CUMYL PEROXIDE
DOT: UN 2121
mf: $C_{18}H_{22}O_2$ mw: 270.40

SYNS: ACTIVE DICUMYL PEROXIDE ◇ BIS(α,α-DIMETHYLBENZYL)PEROXIDE ◇ CUMENE PEROXIDE ◇ CUMYL PEROXIDE ◇ DICUMYL PEROXIDE (DOT) ◇ DI-CUP ◇ DI-CUP 40 KF ◇ DI-CUPR ◇ DIISOPROPYLBENZENE PEROXIDE ◇ ISOPROPYLBENZENE PEROXIDE ◇ LUPERCO ◇ LUPEROX ◇ LUPEROX 500R ◇ LUPEROX 500T ◇ VAROX DCP-R ◇ VAROX DCP-T

TOXICITY DATA with REFERENCE
orl-rat LD50:4100 mg/kg BSPII* 1/75-19B

CONSENSUS REPORTS: Reported in EPA TSCA Inventory.

DOT Classification: Organic Peroxide, Label: Organic Peroxide.

SAFETY PROFILE: Mildly toxic by ingestion. See also PEROXIDES. When heated to decomposition it emits acrid smoke and irritating fumes.

DGS000 CAS:1071-98-3 ***HR: 3***
DICYANOACETYLENE
mf: C_4N_2 mw: 76.09

$$N\equiv CC\equiv C\equiv N$$

SYN: 2-BUTYNEDINITRILE

CONSENSUS REPORTS: Cyanide and its compounds are on the Community Right-To-Know List.

SAFETY PROFILE: The pure material and concentrated solutions are potenially explosive. Ignites in air at 130°C. The flame temperature in oxygen can reach 4700°C. When heated to decomposition it emits toxic fumes of CN^- and NO_x. See also NITRILES and ACETYLENE COMPOUNDS.

DGS200 CAS:1119-69-3 ***HR: 3***
1,4-DICYANO-2-BUTENE
mf: $C_6H_6N_2$ mw: 106.13

$$N{\equiv}CCH_2CHCHCH_2C{\equiv}N$$

PROP: Accelerated polymerization decomp of dicyanobutene.

CONSENSUS REPORTS: Cyanide and its compounds are on the Community Right-To-Know List.

SAFETY PROFILE: Decomposes violently when heated. Upon decomposition it emits toxic fumes of NO_x and CN^-. See also CYANIDE.

DGS300 CAS:1557-57-9 ***HR: 3***
DICYANODIAZENE
mf: C_2N_4 mw: 80.05

SYN: AZOCARBONITRILE

CONSENSUS REPORTS: Cyanide and its compounds are on the Community Right-To-Know List.

SAFETY PROFILE: An explosive sensitive to shock or heating in a closed container. Upon decomposition it emits toxic fumes of CN^- and NO_x. See also CYANIDE.

DGS600 CAS:111-97-7 ***HR: 3***
DI(2-CYANOETHYL)SULFIDE
mf: $C_6H_8N_2S$ mw: 140.22

PROP: White crystals. Mp: (α) 28.65°,(β): 22.10°, d: 1.1095 @ 30°.

SYNS: β,β'-DICYANODIETHYL SULFIDE ◇ NITRIL KYSELINY β,β'-THIODIPROPIONOVE (CZECH) ◇ β,β'-THIODIPROPIONITRILE ◇ USAF HA-5

TOXICITY DATA with REFERENCE
skn-rbt 500 mg/24H MLD 28ZPAK -,172,72
eye-rbt 500 mg AJOPAA 29,1363,46
eye-rbt 500 mg/24H MOD 28ZPAK -,172,72
orl-cat LD50:4210 mg/kg JIHTAB 31,60,49
ipr-mus LD50:300 mg/kg NTIS** AD277-689
orl-rat LD50:4500 mg/kg JIHTAB 31,60,49

CONSENSUS REPORTS: Cyanide and its compounds are on the Community Right-To-Know List. Reported in EPA TSCA Inventory.

SAFETY PROFILE: Poison by intraperitoneal route. Moderately toxic by ingestion. A skin and eye irritant. When heated to decomposition it emits very toxic fumes of NO_x, SO_x, and CN^-. See also NITRILES.

DGS700 CAS:55644-07-0 ***HR: 3***
DICYANOFURAZAN
mf: C_4N_4O mw: 120.07

$$N{\equiv}CC{=}NON{=}CC{\equiv}N$$

CONSENSUS REPORTS: Cyanide and its compounds are on the Community Right-To-Know List.

SAFETY PROFILE: A powerful but insensitive explosive. Explodes on contact with nitrogenous bases (e.g., hydrazine, mono- or di-methylhydrazine, piperidine, piperazine, diethylamine). Upon decomposition it emits toxic fumes of NO_x and CN^-. See also CYANIDE.

DGS800 CAS:55644-07-0 ***HR: 3***
DICYANOFURAZAN-N-OXIDE
mf: $C_4N_4O_2$ mw: 136.07

$$N{\equiv}CC{=}NON(O){=}CC{\equiv}N$$

SYN: DICYANOFUROXAN

CONSENSUS REPORTS: Cyanide and its compounds are on the Community Right-To-Know List.

SAFETY PROFILE: Explodes on contact with nitrogenous bases (e.g., hydrazine; mono- or di- methylhydrazine; piperidine; piperazine; diethylamine). Upon decomposition it emits toxic fumes of NO_x and CN^-. See also CYANIDE.

DGT000 CAS:4331-98-0 ***HR: 3***
DICYANOGEN-N,N-DIOXIDE
mf: $C_2N_2O_2$ mw: 84.04

$$(-C{\equiv}NO)_2$$

CONSENSUS REPORTS: Cyanide and its compounds are on the Community Right-To-Know List.

SAFETY PROFILE: Decomposes explosively at −45°C under vacuum. Upon decomposition it emits toxic fumes of NO_x and CN^-. See also CYANIDE.

DGT200 CAS:38780-37-9 ***HR: 3***
cis-DICYCLOBUTYLAMMINEDICHLORO-PLATINUM(II)
mf: $C_8H_{18}Cl_2N_2Pt$ mw: 408.27

SYNS: cis-BIS(CYCLOBUTYLAMMINE)DICHLOROPLATINUM(II) ◇ cis-DICHLOROBIS(CYCLOBUTYLAMMINE)PLATINUM(II)

TOXICITY DATA with REFERENCE
mmo-sat 1 μmol/plate CBINA8 26,179,79
ipr-mus LD50:90 mg/kg CBINA8 5,415,72

SAFETY PROFILE: Poison by intraperitoneal route. Mutagenic data reported. See also PLATINUM COMPOUNDS. When heated to decomposition it emits very toxic fumes of Cl^- and NO_x.

DGT300 CAS:17455-23-1 ***HR: 3***
DICYCLOHEXANO-24-CROWN-8
mf: $C_{24}H_{44}O_8$ mw: 380.61

SYN: TETRACOSAHYDRODIBENZ(b,n)(1,4,7,10,13,16,19,22)OCTAOXACYCLOTETRACOSIN

TOXICITY DATA with REFERENCE
skn-rbt 100 mg/24H MLD DCTODJ 8,451,85
eye-rbt 50 mg MOD DCTODJ 8,451,85
orl-rat LD50:75 mg/kg DCTODJ 8,451,85
ipr-rat LD50:10 mg/kg DCTODJ 8,451,85
ipr-mus LD50:12 mg/kg DCTODJ 8,451,85

SAFETY PROFILE: Poison by ingestion and intraperitoneal routes. An eye and skin irritant. When heated to decomposition it emits acrid smoke and fumes.

DGT400 CAS:6600-31-3 ***HR: 3***
3,9,DI-(3-CYCLOHEXENYL)-2,4,8,10-TETRAOXASPIRO(5,5)UNDECANE
mf: $C_{19}H_{28}O_4$ mw: 320.47

TOXICITY DATA with REFERENCE
ivn-mus LD50:320 mg/kg CSLNX* NX#01914

CONSENSUS REPORTS: Reported in EPA TSCA Inventory.

SAFETY PROFILE: Poison by intravenous route. When heated to decomposition it emits acrid smoke and irritating fumes.

DGT500 CAS:849-99-0 ***HR: 1***
DICYCLOHEXYL ADIPATE
mf: $C_{18}H_{30}O_4$ mw: 310.48

SYN: ERGOPLAST ADC

TOXICITY DATA with REFERENCE
ipr-rat TDLo:1020 mg/kg (5-15D preg):TER JPMSAE 62,1596,73
ipr-rat TDLo:170 mg/kg (female 5-15D post):REP JPMSAE 62,1596,73
orl-rat LD50:16 g/kg RPZHAW 18,283,67
ipr-rat LD50:5101 mg/kg JPMSAE 62,1596,73

CONSENSUS REPORTS: Reported in EPA TSCA Inventory.

SAFETY PROFILE: Mildly toxic by ingestion and intraperitoneal routes. Experimental teratogenic and reproductive effects. When heated to decomposition it emits acrid smoke and fumes.

DGT600 CAS:101-83-7 ***HR: 3***
N,N-DICYCLOHEXYLAMINE
DOT: UN 2565
mf: $C_{12}H_{23}N$ mw: 181.36

PROP: Liquid, fishy odor. Mp: −1°, bp: 256°, flash p: >210°F (OC), d: 0.910, vap d: 6.27.

SYNS: CDHA ◇ N-CYCLOHEXYLCYCLOHEXANAMINE ◇ DICYCLOHEXYLAMINE (DOT) ◇ DICYKLOHEXYLAMIN (CZECH) ◇ DODECAHYDRODIPHENYLAMINE

TOXICITY DATA with REFERENCE
cyt-hmn:leu 200 μg/L INHEAO 9,188,71
skn-rbt 500 mg/24H SEV 28ZPAK -,68,72
eye-rbt 750 μg/24H SEV 28ZPAK -,68,72
orl-rat TDLo:40 g/kg/52W-I:ETA VOONAW 4,659,58
orl-rat LD50:373 mg/kg 28ZPAK -,68,72
scu-mus LD50:135 mg/kg VOONAW 4,659,58
scu-rbt LDLo:500 mg/kg IECHAD 29,1247,37

CONSENSUS REPORTS: IARC Cancer Review: Group 3 IMEMDT 7,178,87; Animal Inadequate Evidence IMEMDT 22,55,80. Reported in EPA TSCA Inventory.

DOT Classification: Corrosive Material; Label: Corrosive.

SAFETY PROFILE: Poison by ingestion and subcutaneous routes. Corrosive. A severe skin and eye irritant. Questionable carcinogen with experimental tumorigenic data. Human mutation data reported. Combustible when exposed to heat or flame; can react with oxidizing materials. To fight fire, use alcohol foam, CO_2, dry chemical. When heated to decomposition it emits toxic fumes of NO_x. See also CYCLOHEXYLAMINE.

DGU200 CAS:3129-91-7 ***HR: 3***
DICYCLOHEXYLAMINE NITRITE
mf: $C_{12}H_{23}N{\bullet}HNO_2$ mw: 228.38

SYNS: DECHAN ◇ DICHAN (CZECH) ◇ DICYCLOHEXYLAMINONITRITE ◇ DICYCLOHEXYLAMMONIUM NITRITE ◇ DICYKLOHEXYLAMIN NITRIT (CZECH) ◇ DICYNIT (CZECH) ◇ DODECAHYDROPHENYLAMINE NITRITE ◇ DUSITAN DICYKLOHEXYLAMINU (CZECH)

TOXICITY DATA with REFERENCE
scu-rat TDLo:2400 mg/kg/48W-I:ETA VOONAW 4,659,58
orl-rat LD50:284 mg/kg 28ZPAK -,68,72
orl-mus LD50:80 mg/kg GISAAA 30(8),35,65
scu-mus LD50:155 mg/kg VOONAW 4,659,58
orl-gpg LD50:350 mg/kg UCPHAQ 2,231,49

CONSENSUS REPORTS: Reported in EPA TSCA Inventory.

SAFETY PROFILE: Poison by ingestion and subcutaneous routes. Questionable carcinogen with experimental tumorigenic data. When heated to decomposition it emits very toxic fumes of HNO_2 and NO_x. See also NITRITES.

DGU400 CAS:63915-52-6 ***HR: 2***
DICYCLOHEXYLAMINE PENTANOATE
mf: $C_{12}H_{23}N \cdot C_5H_{10}O_2$ mw: 283.51

SYNS: DICYKLOHEXYLAMINKAPRONAT (CZECH) ◇ KAPRONAN DICYKLOHEXYLAMINU (CZECH)

TOXICITY DATA with REFERENCE
skn-rbt 500 mg/24H MOD 28ZPAK -,68,72
eye-rbt 250 μg/24H SEV 28ZPAK -,68,72
orl-rat LD50:3290 mg/kg 28ZPAK -,68,72

SAFETY PROFILE: Moderately toxic by ingestion. A skin and severe eye irritant. When heated to decomposition it emits toxic fumes of NO_x.

DGU709 CAS:38780-35-7 ***HR: 3***
cis-DICYCLOHEXYLAMMINEDICHLOROPLATINUM(II)
mf: $C_{12}H_{26}Cl_2N_2Pt$ mw: 464.39

TOXICITY DATA with REFERENCE
mma-sat 30 μg/plate MUREAV 95,79,82
dni-ham:ovr 40 mg/L CBINA8 35,189,81
cyt-ham:ovr 40 mg/L CBINA8 35,189,81
ipr-mus LD50:12 mg/kg BJCAAI 36,420,77

SAFETY PROFILE: Poison by intraperitoneal route. Mutation data reported. See also PLATINUM COMPOUNDS. When heated to decomposition it emits very toxic fumes of Cl^- and NO_x.

DGU800 CAS:4979-32-2 ***HR: 1***
N,N-DICYCLOHEXYL-2-BENZOTHIAZOLESULFENAMIDE
mf: $C_{19}H_{26}N_2S_2$ mw: 346.59

SYN: N,N-DICYKLOHEXYLBENZTHIAZOLSULFENAMID(CZECH)

TOXICITY DATA with REFERENCE
skn-rbt 500 mg/24H MOD 28ZPAK -,203,72
eye-rbt 500 mg/24H MLD 28ZPAK -,203,72
orl-rat LDLo:6420 mg/kg 28ZPAK -,203,72

CONSENSUS REPORTS: Reported in EPA TSCA Inventory.

SAFETY PROFILE: Mildly toxic by ingestion. An eye and skin irritant. When heated to decomposition it emits very toxic fumes of SO_x and NO_x.

DGV000 ***HR: 3***
DICYCLOHEXYLCARBONYL PEROXIDE
mf: $C_{14}H_{22}O_4$ mw: 254.33

$(C_6H_{11}CO \cdot O\text{-})_2$

SAFETY PROFILE: Large quantities may spontaneously explode. When heated to decomposition it emits acrid smoke and fumes. See also PEROXIDES, ORGANIC.

DGV100 CAS:16069-36-6 ***HR: 3***
DICYCLOHEXYL-18-CROWN-6
mf: $C_{20}H_{36}O_6$ mw: 372.56

SYNS: DICYCLOHEXANO-18-CROWN-6 ◇ EICOSAHYDRO DIBENZO(b,k)(1,4,7,10,13,16)HEXAOXACYCLOOCTADECIN

TOXICITY DATA with REFERENCE
skn-rbt 100 mg/24H MLD DCTODJ 8,451,85
eye-rbt 50 mg MOD DCTODJ 8,451,85
orl-rat TDLo:105 mg/kg (female 1-21D post):TER GISAAA 52(11),72,87
orl-rat LD50:176 mg/kg DCTODJ 8,451,85
skn-rat LDLo:130 mg/kg MEIEDD 10,373,83
ipr-rat LD50:55 mg/kg DCTODJ 8,451,85
orl-mus LD50:192 mg/kg GISAAA 52(11),72,87
ipr-mus LD50:53 mg/kg DCTODJ 8,451,85

CONSENSUS REPORTS: Reported in EPA TSCA Inventory.

SAFETY PROFILE: Poison by ingestion, skin contact and intraperitoneal routes. An experimental teratogen. An eye and skin irritant. When heated to decomposition it emits acrid smoke and fumes.

DGV200 CAS:587-15-5 ***HR: 3***
DICYCLOHEXYL FLUOROPHOSPHONATE
mf: $C_{12}H_{22}FO_3P$ mw: 264.31

SYN: DICYCLOHEXYLFLUOROPHOSPHATE

TOXICITY DATA with REFERENCE
ihl-rat LD50:1200 mg/m^3/10M NTIS** PB158-508
ihl-mus LD50:800 mg/m^3/10M NTIS** PB158-508
ihl-dog LD50:1 g/m^3/10M NTIS** PB158-508
ihl-rbt LD50:1200 mg/m^3/10M NTIS** PB158-508
ihl-gpg LD50:6 g/m^3/10M NTIS** PB158-508

SAFETY PROFILE: Poison by inhalation. See also FLUORIDES. When heated to decomposition it emits very toxic fumes of F^- and PO_x.

DGV600 CAS:119-60-8 ***HR: 3***
DICYCLOHEXYL KETONE
mf: $C_{13}H_{22}O$ mw: 194.35

TOXICITY DATA with REFERENCE
ivn-mus LD50:56 mg/kg CSLNX* NX#06751

CONSENSUS REPORTS: Reported in EPA TSCA Inventory.

SAFETY PROFILE: Poison by intravenous route. See also KETONES. When heated to decomposition it emits acrid and irritating fumes.

DGV800 CAS:1212-29-9 ***HR: 1***
DICYCLOHEXYL THIOUREA
mf: $C_{13}H_{24}N_2S$ mw: 240.45

SYNS: N,N'-DICYCLOHEXYLTHIOUREA ◇ NCI-C04524

TOXICITY DATA with REFERENCE
eye-rbt 100 mg MLD 34ZIAG -,214,69
otr-rat:emb 250 ng/plate JJATDK 1,190,81

CONSENSUS REPORTS: NCI Carcinogenesis Bioassay (feed); No Evidence: mouse, rat NCITR* NCI-CG-TR-56,78. Reported in EPA TSCA Inventory.

SAFETY PROFILE: Mutation data reported. An eye irritant. When heated to decomposition it emits very toxic fumes of NO_x and SO_x.

DGV900 CAS:22771-17-1 ***HR: 3***
DICYCLOHEXYLTIN OXIDE
mf: $C_{12}H_{22}OSn$ mw: 301.03

SYN: STANNANE, DICYCLOHEXYLOXO-

TOXICITY DATA with REFERENCE
orl-rat LD50:355 mg/kg GISAAA 48(3),55,83

OSHA PEL FINAL: TWA 0.1 mg(Sn)/m^3 (skin)
ACGIH TLV: TWA 0.1 mg(Sn)/m^3 (skin)

SAFETY PROFILE: Poison by ingestion. When heated to decomposition it emits toxic fumes of Sn.

DGW000 CAS:77-73-6 ***HR: 3***
DICYCLOPENTADIENE
DOT: UN 2048
mf: $C_{10}H_{12}$ mw: 132.22

PROP: Colorless crystals. Mp: 32.9°, bp: 166.6°, d: 0.976 @ 35°, vap press: 10 mm @ 47.6°, vap d: 4.55, flash p: 90°F (OC).

SYNS: BICYCLOPENTADIENE ◇ BISCYCLOPENTADIENE ◇ 1,3-CYCLOPENTADIENE, DIMER ◇ DICYKLOPENTADIEN (CZECH) ◇ DIMER CYKLOPENTADIENU (CZECH) ◇ 3a,4,7,7a-TETRAHYDRO-4,7-METHANOINDENE

TOXICITY DATA with REFERENCE
skn-rbt 10 mg/24H open SEV AMIHBC 10,61,54
skn-rbt 9300 μg/24H open SEV AIHAAP 23,95,62
skn-rbt 500 mg/24H MOD 28ZPAK -,27,72
eye-rbt 500 mg open AMIHBC 10,61,54
eye-rbt 500 mg/24H MLD 28ZPAK -,27,72
orl-rat LD50:353 mg/kg TXAPA9 20,552,71
ihl-rat LC50:372 ppm/4H 28ZPAK -,27,72
ipr-rat LD50:200 mg/kg NCIUS* PH 43-64-886,JAN,65
orl-mus LD50:190 mg/kg 40QBA3 -,448,78
ipr-mus LD50:200 mg/kg NCIUS* PH 43-64-886,JAN,65
skn-rbt LD50:5080 mg/kg TXAPA9 20,552,71
orl-brd LD50:1010 mg/kg NTIS** AD-A087-257

CONSENSUS REPORTS: Reported in EPA TSCA Inventory. EPA Genetic Toxicology Program.

OSHA PEL: TWA 5 ppm
ACGIH TLV: TWA 5 ppm
DOT Classification: IMO: Flammable or Combustible Liquid; Label: Flammable Liquid.

SAFETY PROFILE: Poison by ingestion and intraperitoneal routes. Moderately toxic by inhalation. Mildly toxic by skin contact. A severe skin and moderate eye irritant. Dangerous fire hazard when exposed to heat or flame; can react with oxidizing materials. To fight fire, use alcohol foam. When heated to decomposition it emits acrid smoke and fumes.

DGW200 CAS:1271-19-8 ***HR: 3***
DICYCLOPENTADIENYLDICHLOROTITANIUM
mf: $C_{10}H_{10}Cl_2Ti$ mw: 249.00

SYNS: DICHLOROBIS(ETA^5-2,4-CYCLOPENTADIEN-1-YL-TITANIUM (9CI) ◇ DICHLORODICYCLOPENTADIENYLTITANIUM ◇ DICHLORODI-pi-CYCLOPENTADIENYLTITANIUM ◇ DICHLOROTITANOCENE ◇ DICYCLOPENTADIENYLTITANIUMDICHLORIDE ◇ NCI-C04502 ◇ TITANIUM FERROCENE ◇ TITANOCENE ◇ TITANOCENE, DICHLORIDE

TOXICITY DATA with REFERENCE
mmo-sat 100 μg/plate ENMUDM 5(Suppl 1),3,83
mma-sat 1 mg/plate ENMUDM 5(Suppl 1),3,83
otr-rat:emb 2960 μg/L JJIND8 67,1303,81
otr-mus:fbr 800 μg/L JJIND8 67,1303,81
otr-ham:emb 100 μg/L JJIND8 67,1303,81
ipr-mus TDLo:30 mg/kg (10D preg):TER TXCYAC 33,171,84
ims-rat TDLo:720 mg/kg/2Y-I:NEO NCIUS* PH 43-64-886,JUL,68
ims-rat TD:430 mg/kg/81W-I:ETA NCIUS* PH 43-64-886,AUG,69
ipr-rat LD50:25 mg/kg NCIUS* PH 43-64-886,JAN,65
ipr-mus LD50:60 mg/kg NCIUS* PH 43-64-886,AUG,64
ivn-mus LD50:180 mg/kg CSLNX* NX#00774

CONSENSUS REPORTS: Reported in EPA TSCA Inventory. EPA Genetic Toxicology Program.

SAFETY PROFILE: Poison by intravenous and intraperitoneal routes. Questionable carcinogen with experimental neoplastigenic, tumorigenic, and teratogenic data. Mutation data reported. See also TITANIUM COMPOUNDS. When heated to decomposition it emits toxic fumes of Cl^-.

DGW300 ***HR: 2***
DICYCLOPENTA(c,lmn)PHENANTHREN-1(9H)-ONE, 2,3-DIHYDRO-
mf: $C_{18}H_{12}O$ mw: 244.30

SYNS: 2,3-DIHYDRODICYCLOPENTA(c,lmn)PHENANTHREN-1(9H)-ONE ◇ 15,16-DIHYDRO-1,11-METHANOCYCLOPENTA(a)PHENANTHREN-17-ONE

TOXICITY DATA with REFERENCE
skn-mus TDLo:16 mg/kg:CAR CRNGDP 5,1485,84

SAFETY PROFILE: Questionable carcinogen with experimental carcinogenic data. When heated to decomposition it emits acrid smoke and irritating fumes.

DGW400 CAS:50976-02-8 ***HR: 1***
DICYCLOPENTENYL ACRYLATE
mf: $C_{13}H_{14}O_2$ mw: 202.27

SYN: ACRYLIC ACID ((3a,4,7,7a-TETRAHYDRO)-4,7-METHANOINDENYL) ESTER

TOXICITY DATA with REFERENCE
skn-rbt 500 mg open MOD UCDS** 3/28/72
orl-rat LD50:11 g/kg UCDS** 3/28/72
skn-rbt LD50:45200 mg/kg TXAPA9 28,313,74

CONSENSUS REPORTS: EPA Extremely Hazardous Substances List. Reported in EPA TSCA Inventory.

SAFETY PROFILE: Mildly toxic by ingestion and skin contact. A skin irritant. See also ESTERS. When heated to decomposition it emits acrid smoke and irritating fumes.

DGW450 CAS:75662-22-5 ***HR: 1***
DICYCLOPENTENYLOXYETHYL METHACRYLATE
mf: $C_{16}H_{22}O_3$ mw: 262.38

SYNS: 2-PROPENOIC ACID, 2-METHYL-, 2-((3a,4,5,6,7,7a-HEXAHYDRO-4,7-METHANO-1H-INDEN-5-YL)OXY) ETHYL ESTER ◇ QM 657

TOXICITY DATA with REFERENCE
skn-rbt 500 mg/24H MLD JTEHD6 16,39,85

SAFETY PROFILE: A skin irritant. When heated to decomposition it emits acrid smoke and irritating fumes.

DGW600 CAS:2001-81-2 ***HR: 3***
(2-(DICYCLOPENTYLACETOXY)ETHYL) TRIETHYLAMMONIUM BROMIDE
mf: $C_{20}H_{38}NO_2 \cdot Br$ mw: 404.50

SYNS: AETHOBROMID DES α,α-DICYCLOPENTYLESSIGSAEURE-β'-DIAETHYLAMINO AETHYLESTER (GERMAN) ◇ DICYCLOPENTYLACETIC ACID-β-DIETHYLAMINOETHYL ESTER ETHOBROMIDE ◇ 2-((DICYCLOPENTYLACETYL)OXY)-N,N,N-TRIETHYL-ETHANAMINIUM BROMIDE ◇ DIETHYLAMINOETHYL-α,α-DICYCLOPENTYLACETATE ETHOBROMIDE ◇ DIPENINBROMID (GERMAN) ◇ HL 267 ◇ Sa 267 ◇ TRIETHYL(2-HYDROXYETHYL)-AMMONIUM BROMIDE DICYCLOPENTYLACETATE ◇ UNOSPASTON

TOXICITY DATA with REFERENCE
orl-rat LD50:780 mg/kg ARZNAD 15,878,65
ipr-rat LD50:84400 μg/kg OYYAA2 2,70,68
scu-rat LD50:299 mg/kg OYYAA2 2,70,68
ivn-rat LD50:6600 μg/kg ARZNAD 15,878,65
orl-mus LD50:570 mg/kg NIIRDN 6,350,82
ipr-mus LD50:88 mg/kg ARZNAD 10,911,60
scu-mus LD50:80 mg/kg NIIRDN 6,350,82
ivn-mus LD50:6200 μg/kg NIIRDN 6,350,82

SAFETY PROFILE: Poison by subcutaneous, intravenous, and intraperitoneal routes. Moderately toxic by ingestion. See also BROMIDES. An antispasmodic agent. When heated to decomposition it emits very toxic fumes of Br^- and NO_x.

DGW875 CAS:16102-24-2 ***HR: 3***
DICYCLOPROPYLDIAZOMETHANE
mf: $C_7H_{10}N_2$ mw: 122.17

$(C_3H_5)_2CN_2$

SAFETY PROFILE: Decomposes violently above −15°C. When heated to decomposition it emits toxic fumes of NO_x.

DGX000 CAS:5232-99-5 ***HR: 3***
3,3-DICYCLOPROPYL-2-(ETHOXYCARBONYL) ACRYLONITRILE
mf: $C_{18}H_{15}NO_2$ mw: 277.34

SYNS: α-CARBETHOXY-β,β-BISCYCLOPROPYL ACRYLONITRILE ◇ 2-CYANO-3,3-DIPHENYLACRYLIC ACID, ETHYL ESTER ◇ 2-CYANO-3,3-DIPHENYL-2-PROPENOIC ACID, ETHYL ESTER ◇ USAF A -15972 ◇ UV ABSORBER-2

TOXICITY DATA with REFERENCE
ipr-mus LD50:100 mg/kg NTIS** AD277-689

CONSENSUS REPORTS: Cyanide and its compounds are on the Community Right-To-Know List. Reported in EPA TSCA Inventory.

SAFETY PROFILE: Poison by intraperitoneal route. See also ESTERS and NITRILES. When heated to decomposition it emits toxic fumes of NO_x.

DGX200 CAS:7173-51-5 ***HR: 3***
DIDECYL DIMETHYL AMMONIUM CHLORIDE
mf: $C_{22}H_{48}N \cdot Cl$ mw: 362.16

SYNS: ALIQUAT 203 ◇ BARDAC 22 ◇ BIO-DAC 50-22 ◇ BTC 1010 ◇ N-DECYL-N,N-DIMETHYL-1-DECANAMINIUM CHLORIDE (CI) ◇ DIMETHYLDIDECYLAMMONIUM CHLORIDE ◇ QUATERNIUM-12

TOXICITY DATA with REFERENCE
skn-rbt 500 mg SEV NTIS** AD867-663
orl-rat LD50:84 mg/kg NTIS** AD867-663

ipr-rat LD50:45 mg/kg NTIS** AD867-663
orl-mus LD50:268 mg/kg NTIS** AD867-663
ipr-mus LD50:11 mg/kg NTIS** AD867-663
ipr-gpg LDLo:7 mg/kg NTIS** AD867-663

CONSENSUS REPORTS: Reported in EPA TSCA Inventory.

SAFETY PROFILE: Poison by ingestion and intraperitoneal routes. A severe skin irritant. A fungicide. When heated to decomposition it emits very toxic fumes of NO_x, NH_3 and Cl^-.

DGX600 CAS:84-77-5 ***HR: 2***
DIDECYL PHTHALATE
mf: $C_{28}H_{46}O_4$ mw: 446.74

PROP: A clear liquid. Bp: 252° @ 4 mm, fp: −53°, flash p: 450°F (COC), d: 0.964-0.968 @ 20°/20°.

SYN: DI-N-DECYL PHTHALATE

TOXICITY DATA with REFERENCE
skn-rbt 10 mg/24H open MLD AIHAAP 23,95,62
ipr-mus LDLo:2233 mg/kg JPMSAE 56,1446,67
skn-rbt LD50:17 g/kg AIHAAP 23,95,62

CONSENSUS REPORTS: Reported in EPA TSCA Inventory.

SAFETY PROFILE: Moderately toxic by intraperitoneal route. Mildly toxic by skin contact. See also ESTERS and PHTHALIC ACID. A skin irritant. Combustible when exposed to heat or flame; can react with oxidizing materials. To fight fire, use foam, CO_2, dry chemical.

DGY000 CAS:19763-77-0 ***HR: 3***
7,8-DIDEHYDRO-4,5-α-EPOXY-14-HYDROXY-3-METHOXY-17-METHYLMORPHINAN-6-ONE-N-OXIDE
mf: $C_{18}H_{21}NO_5$ mw: 331.40

SYN: DIHYDROOXYCODEINON-N-OXYD (GERMAN)

TOXICITY DATA with REFERENCE
ivn-mus LD50:800 mg/kg ARZNAD 7,594,57
ivn-rbt LD50:75 mg/kg ARZNAD 7,594,57

SAFETY PROFILE: Poison by intravenous route. When heated to decomposition it emits toxic fumes of NO_x.

DHA200 CAS:74193-14-9 ***HR: D***
7,8-DIDEHYDRORETINOIC ACID
mf: $C_{20}H_{26}O_2$ mw: 298.46

SYNS: trans-3,7-DIMETHYL-9-(2,6,6-TRIMETHYL-1-CYCLOHEXEN-1-YL)-7-YNE-2,4,6-NON ATRIENOIC ACID ◇ RETINOIC ACID, 7,8-DIDEHYDRO-

TOXICITY DATA with REFERENCE
orl-ham TDLo:74 mg/kg (female 8D post):TER TXAPA9 83,563,86
orl-ham TDLo:74 mg/kg (female 8D post):REP TXAPA9 83,563,86

SAFETY PROFILE: An experimental teratogen. Experimental reproductive effects. When heated to decomposition it emits acrid smoke and irritating fumes.

DHA300 CAS:77327-05-0 ***HR: 3***
DIDEMNIN B
mf: $C_{57}H_{89}N_7O_{15}$ mw: 1112.53

SYN: NSC-325319

TOXICITY DATA with REFERENCE
dni-mus:leu 2500 ng/L CALEDQ 23,279,84
dni-mus:oth 400 μg/L CNREA8 44,1796,84
oms-mus:oth 400 μg/L CNREA8 44,1796,84
oms-mus:leu 2500 ng/L CALEDQ 23,279,84
ivn-rat LD50:860 μg/kg NTIS** PB84-192251
ivn-mus LD50:1530 μg/kg NTIS** PB84-192251
ivn-dog LDLo:418 μg/kg NTIS** PB84-192251

SAFETY PROFILE: A deadly poison by intravenous route. Mutation data reported. When heated to decomposition it emits toxic fumes of NO_x.

DHA325 CAS:4097-22-7 ***HR: 2***
2',3'-DIDEOXYADENOSINE
mf: $C_{10}H_{13}N_5O_2$ mw: 235.28

TOXICITY DATA with REFERENCE
dni-esc 37 mg/L AMACCQ 19,424,81
dni-mus 144 mg/L AMACCQ 19,424,81
dni-omi 19 mg/L AMACCQ 19,424,81
orl-mus LD50:5 g/kg AMACCQ 19,424,81
scu-mus LD50:1320 mg/kg AMACCQ 19,424,81

SAFETY PROFILE: Moderately toxic by subcutaneous route. Mildly toxic by ingestion. Mutation data reported. When heated to decomposition it emits toxic fumes of NO_x.

DHA400 CAS:64070-13-9 ***HR: 3***
3',4'-DIDEOXYKANAMYCIN B SULFATE
mf: $C_{18}H_{37}N_5O_8 \cdot O_4S$ mw: 547.66

SYN: DKB SULFATE

TOXICITY DATA with REFERENCE
ipr-rat LD50:799 mg/kg JJANAX 26,221,73
scu-rat LD50:1376 mg/kg JJANAX 26,221,73
ivn-rat LD50:140 mg/kg JJANAX 26,221,73
ims-rat LD50:560 mg/kg JJANAX 26,221,73
ipr-mus LD50:431 mg/kg JJANAX 26,221,73
scu-mus LD50:521 mg/kg JJANAX 26,221,73

ivn-mus LD50:63 mg/kg JJANAX 26,221,73
ims-mus LD50:396 mg/kg JJANAX 26,221,73

SAFETY PROFILE: Poison by intramuscular and intravenous routes. Moderately toxic by intraperitoneal and subcutaneous routes. When heated to decomposition it emits very toxic fumes of NO_x and SO_x.

DHA425 CAS:53866-33-4 ***HR: 2***
2,4-DIDEUTERIOESTRADIOL
mf: $C_{18}H_{22}D_2O_2$ mw: 274.42

SYN: ESTRA-1,3,5(10)-TRIENE-2,4-D2-3,17-DIOL,(17-β)-

TOXICITY DATA with REFERENCE
imp-ham TDLo:360 mg/kg/15W-I:CAR MOPMA3 23,278,83

SAFETY PROFILE: Questionable carcinogen with experimental carcinogenic data. When heated to decomposition it emits acrid smoke and irritating fumes.

DHA450 CAS:14621-84-2 ***HR: 3***
DIDEUTERODIAZOMETHANE
mf: CD_2N_2 mw: 44.06

SAFETY PROFILE: An explosive. When heated to decomposition it emits toxic fumes of NO_x. See also DIAZOMETHANE.

DHB309 ***HR: 1***
DIEFFENBACHIA (VARIOUS SPECIES)

PROP: Tall, unbranched ornamentals with long, ivory-spotted leaves. They are extremely popular as indoor plants, and are grown outdoors in southern Florida and Hawaii.

SYNS: CAMILICHIGUI (MEXICO) ◇ CANNE-A-GRATTER (HAITI) ◇ CANNE-MADERE (HAITI) ◇ DICHA (CUBA) ◇ DIEFFENBACHIA MACUALTA ◇ DIEFFENBACHIA SEQUINE ◇ DUMBCANE ◇ DUMB PLANT ◇ MOTHER-IN-LAW'S TONGUE PLANT ◇ PELA PUERCO (DOMINICAN REPUBLIC) ◇ RABANO (PUERTO RICO) ◇ TUFT ROOT

SAFETY PROFILE: The leaf contains calcium oxalate crystals and chewing them results in burning pain in the lips, mouth and throat, possibly followed by inflammation and blistering. The cut or crushed leaves may also cause contact dermatitis or conjunctivitis. Systemic effects are usually not seen because of the insolubility of calcium oxalate. See also OXALATES.

DHB400 CAS:60-57-1 ***HR: 3***
DIELDRIN
DOT: UN 2761
mf: $C_{12}H_8Cl_6O$ mw: 380.90

PROP: White crystals; odorless. Mp: 150°, vap d: 13.2. Insol in water; sol in common organic solvents.

SYNS: ALVIT ◇ COMPOUND 497 ◇ DIELDREX ◇ DIELDRINE (FRENCH) ◇ DIELDRITE ◇ ENT 16,225 ◇ HEOD ◇ HEXACHLORO-EPOXYOCTAHYDRO-endo,exo-DIMETHANONAPHTHALENE ◇ 3,4,5,6,9,9-HEXACHLORO-1a,2,2a,3,6,6a,7,7a-OCTAHYDRO-2,7:3,6-DIMETHANONAPHTH(2,3-b)OXIRENE ◇ ILLOXOL ◇ INSECTICIDE No. 497 ◇ NCI-C00124 ◇ OCTALOX ◇ PANORAM D-31 ◇ QUINTOX ◇ RCRA WASTE NUMBER P037

TOXICITY DATA with REFERENCE
mmo-sat 1 mg/L JOHEA8 68,184,77
mma-hmn:fbr 1 μmol/L MUREAV 42,161,77
dns-hmn:fbr 1 μmol/L MUREAV 42,161,77
dni-hmn:hla 400 μmol/L MUREAV 92,427,82
otr-rat-orl 5 mg/kg CNREA8 40,1157,80
orl-rat TDLo:14 μg/kg (multi) :REP PCBPBS 13,20,80
orl-ham TDLo:30 mg/kg (female 8D post):TER TJADAB 9,11,74
orl-rat TDLo:200 mg/kg/2Y-C:ETA FCTXAV 2,551,64
orl-mus TDLo:546 mg/kg/65W-C:CAR ARTODN Suppl.2,197,79
orl-mus TD:11 g/kg/3Y-C:NEO FCTXAV 11,415,73
orl-mus TD:714 mg/kg/85W-C:CAR LAINAW 44,392,81
orl-man LDLo:65 mg/kg 34ZIAG -,215,69
unk-hmn LDLo:28 mg/kg ATXKA8 22,115,66
orl-rat LD50:38300 μg/kg JAFCAU 3,402,55
ihl-rat LC50:13 mg/m^3/4H 85GMAT -,73,82
skn-rat LD50:56 mg/kg RPZHAW 18,161,67
ipr-rat LD50:35 mg/kg CBPCBB 85,437,86
ivn-rat LD50:9 mg/kg BJIMAG 21,269,64
orl-mus LD50:38 mg/kg SPEADM 78-1,13,78
ipr-mus LDLo:26 mg/kg TXAPA9 23,288,72
ivn-mus LD50:10.5 mg/kg TXAPA9 23,408,72
orl-dog LD50:65 mg/kg GUCHAZ 6,198,73
orl-mky LD50:3 mg/kg 32ZDAL -,79,70
ihl-cat LC50:80 mg/m^3/4H GTPZAB 8(4),30,64

CONSENSUS REPORTS: IARC Cancer Review: Group 3 IMEMDT 7,196,87; Human Inadequate Evidence IMEMDT 5,125,74; Animal Sufficient Evidence IMEMDT 5,125,74. NCI Carcinogenesis Bioassay (feed); Clear Evidence: mouse NCITR* NCI-CG-TR-21,78; No Evidence: rat NCITR* NCI-CG-TR-22,78; Inadequate Studies: rat NCITR* NCI-CG-TR-21,78.

OSHA PEL: TWA 0.25 mg/m^3 (skin)
ACGIH TLV: TWA 0.25 mg/m^3 (skin)
DFG MAK: 0.25 mg/m^3
NIOSH REL: (Dieldrin) Lowest reliable detectable level.
DOT Classification: ORM-A; Label: None.

SAFETY PROFILE: A human poison by ingestion and possibly other routes. Poison experimentally by inhalation, ingestion, skin contact, intravenous, and intraperitoneal routes. Experimental teratogenic and reproductive data. Absorbed readily through the skin and by other routes. It is a central nervous system stimulant. Questionable carcinogen with experimental carcinogenic, neoplastigenic, and tumorigenic data. Human mutation data reported. An insecticide. Dieldrin is considerably more toxic than DDT by ingestion and skin contact. Dieldrin or its derivatives may accumulate in the body from chronic low dosages. When heated to decomposition it emits toxic fumes of Cl^-. See also ALDRIN.

DHB500 CAS:84-19-5 ***HR: D***
DIENESTROL DIACETATE
mf: $C_{22}H_{22}O_4$ mw: 350.44

SYNS: DIENOESTROL DIACETATE ◇ 4,4'-(1,2-DIETHYLIDENE-1,2-ETHANEDIYL)BIS(PHENOL) DIACETATE ◇ 4,4'-(DIETHYLIDENE-ETHYLENE)DIPHENOL DIACETATE ◇ FARMACYROL ◇ LIPAMONE ◇ RETALON-ORAL

TOXICITY DATA with REFERENCE
orl-uns TDLo:663 μg/kg (female 1-53D post):REP AJVRAH 32,1599,71

SAFETY PROFILE: Experimental reproductive effects. When heated to decomposition it emits acrid smoke and fumes.

DHB550 CAS:13029-44-2 ***HR: 2***
(E,E)-DIENESTROL
mf: $C_{18}H_{18}O_2$ mw: 266.36

SYNS: α-DIENESTROLPHENOL, 4,4'-(DIETHYLIDENEETHYLENE) DI-, trans-, (E,E)- ◇ PHENOL, 4,4'-(1,2-DIETHYLIDENE-1,2-ETHANEDIYL)BIS-, (E,E)-(9CI)

TOXICITY DATA with REFERENCE
mnt-ham:emb 10 mg/L TOLED5 31(Suppl),204,86
imp-ham TDLo:640 mg/kg/38W-I:ETA CNREA8 43,5200,83

SAFETY PROFILE: Questionable carcinogen with experimental tumorigenic data. Mutation data reported. When heated to decomposition it emits acrid smoke and irritating fumes.

DHB600 CAS:298-18-0 ***HR: 3***
dl-DIEPOXYBUTANE
mf: $C_4H_6O_2$ mw: 86.10

PROP: Colorless liquid. Bp: 138°, mp: 4°, d: 1.112 @ 18°/4°.

SYNS: dl-BUTADIENE DIOXIDE ◇ 1,2:3,4-DIANHYDRO-dl-THREITOL ◇ (±)-1,2:3,4-DIEPOXYBUTANE ◇ dl-1,2:3,4-DIEPOXYBUTANE

TOXICITY DATA with REFERENCE
mmo-ssp 20 mmol/L ADWMAX -,193,62
dns-ham:lvr 1 μmol/L ENMUDM 6,1,84
scu-rat TDLo:335 mg/kg/67W-I:ETA JNCIAM 37,825,66
skn-mus TDLo:132 mg/kg/66W-I:CAR CNREA8 43,159,83
scu-mus TDLo:13200 μg/kg/68W-I:CAR CNREA8 43,159,83
scu-mus TD:260 mg/kg/65W-I:NEO JNCIAM 37,825,66
orl-rat LD50:210 mg/kg IHFCAY 6,1,67
ihl-rat LC50:56 ppm/4H IHFCAY 6,1,67
skn-mus LDLo:400 mg/kg JNCIAM 31,41,63
skn-rbt LD50:800 mg/kg IHFCAY 6,1,67

CONSENSUS REPORTS: IARC Cancer Review: Group 2B IMEMDT 7,56,87; Animal Sufficient Evidence IMEMDT 11,115,76. EPA Genetic Toxicology Program.

SAFETY PROFILE: Suspected carcinogen with experimental carcinogenic, neoplastigenic, and tumorigenic data. Poison by ingestion, inhalation, and skin contact. Mutation data reported. When heated to decomposition it emits acrid smoke and irritating fumes.

DHB800 CAS:564-00-1 ***HR: 3***
meso-1,2,3,4-DIEPOXYBUTANE
mf: $C_4H_6O_2$ mw: 86.10

SYNS: (R*,S*)-2,2'-BIOXIRANE ◇ 1,2:3,4-DIANHYDROERYTHRITOL ◇ meso-DIEPOXYBUTANE ◇ (R*,S*)-DIEPOXYBUTANE ◇ ERYTHRITOL ANHYDRIDE

TOXICITY DATA with REFERENCE
mmo-sat 100 μg/plate ENMUDM 6(Suppl 2),1,84
mmo-ssp 51 mmol/L ADWMAX -,193,62
skn-mus TDLo:26 g/kg/22W-I:NEO 14JTAF -,275,64
skn-mus TD:26 g/kg/22W-I:ETA JNCIAM 31,41,63
skn-mus LDLo:400 mg/kg JNCIAM 31,41,63

CONSENSUS REPORTS: IARC Cancer Review: Group 2B IMEMDT 7,56,87; Animal Sufficient Evidence IMEMDT 11,115,76.

SAFETY PROFILE: Suspected carcinogen with experimental carcinogenic, neoplastigenic, and tumorigenic data. Poison by skin contact. Mutation data reported. When heated to decomposition it emits acrid smoke and irritating fumes.

DHB875 ***HR: D***
trans-1,2,3,4-DIEPOXYCYCLOHEXANE
mf: $C_5H_8O_2$ mw: 100.11

SYN: (E)-3,8-DIOXATRICYCLO(5.1.0.0^{2,4})OCTANE

TOXICITY DATA with REFERENCE
mmo-smc 32 mmol/L MUREAV 117,213,83
mma-smc 8 mmol/L MUREAV 117,213,83
mrc-smc 8 mmol/L MUREAV 117,213,83

SAFETY PROFILE: Mutation data reported. When heated to decomposition it emits acrid smoke and fumes.

DHC000 CAS:24854-67-9 ***HR: 3***
1,2,9,10-DIEPOXYDECANE
mf: $C_{10}H_{18}O_2$ mw: 170.28

TOXICITY DATA with REFERENCE
unk-mus TDLo:510 mg/kg:ETA RARSAM 3,193,63

SAFETY PROFILE: Questionable carcinogen with experimental tumorigenic data. When heated to decomposition it emits acrid smoke and irritating fumes.

DHC200 CAS:63869-17-0 ***HR: 3***
DIEPOXYDIHYDROMYRCENE
mf: $C_{10}H_{18}O_2$ mw: 170.28

SYN: DIEPOXYDIHYDRO-7-METHYL-3-METHYLENE-1,6-OCTADIENE

TOXICITY DATA with REFERENCE
unk-mus TDLo:5100 mg/kg:ETA RARSAM 3,193,63

SAFETY PROFILE: Questionable carcinogen with experimental tumorigenic data. When heated to decomposition it emits acrid smoke and irritating fumes.

DHC309 CAS:56411-67-7 ***HR: 3***
2,3:5,6-DIEPOXY-7,8-DIOXABICYCLO[2.2.2] OCTANE
mf: $C_6H_6O_4$ mw: 142.11

SAFETY PROFILE: Explodes when heated. When heated to decomposition it emits acrid smoke and fumes. See also PEROXIDES.

DHC400 CAS:39079-58-8 ***HR: 3***
2,5-DI(1,2-EPOXYETHYL)TETRAHYDRO-2H-PYRAN
mf: $C_9H_{14}O_3$ mw: 170.23

TOXICITY DATA with REFERENCE
orl-rat LD50:1410 mg/kg TXAPA9 28,313,74
skn-rbt LD50:320 mg/kg TXAPA9 28,313,74

SAFETY PROFILE: Poison by skin contact. Moderately toxic by ingestion. When heated to decomposition it emits acrid smoke and irritating fumes.

DHC600 CAS:4247-19-2 ***HR: 3***
1,2,6,7-DIEPOXYHEPTANE
mf: $C_7H_{12}O_2$ mw: 128.19

TOXICITY DATA with REFERENCE
skn-mus TDLo:2400 mg/kg/20W-I:NEO JNCIAM 35,707,65

SAFETY PROFILE: Questionable carcinogen with experimental neoplastigenic data. When heated to decomposition it emits acrid smoke.

DHC800 CAS:1888-89-7 ***HR: 3***
1,2:5,6-DIEPOXYHEXANE
mf: $C_6H_{10}O_2$ mw: 114.16

TOXICITY DATA with REFERENCE
mmo-ssp 116 mmol/L ADWMAX -,193,62
skn-mus TDLo:6960 mg/kg/29W-I:ETA JNCIAM 39,1217,67
scu-mus TDLo:2068 mg/kg/47W-I:NEO JNCIAM 37,825,66

CONSENSUS REPORTS: EPA Genetic Toxicology Program.

SAFETY PROFILE: Questionable carcinogen with experimental neoplastigenic and tumorigenic data. Mutation data reported. When heated to decomposition it emits acrid smoke and irritating fumes.

DHD200 CAS:6341-85-1 ***HR: 3***
1,2,3,4-DIEPOXY-2-METHYLBUTANE
mf: $C_5H_8O_2$ mw: 100.13

SYN: 2-METHYL-2,2'-BIOXIRANE (9CI)

TOXICITY DATA with REFERENCE
mmo-sat 7500 μmol/L MUREAV 156,77,85
unk-mus TDLo:800 mg/kg:ETA RARSAM 3,193,63

SAFETY PROFILE: Questionable carcinogen with experimental tumorigenic data. Mutation data reported. When heated to decomposition it emits acrid smoke and irritating fumes.

DHD400 CAS:24829-11-6 ***HR: 3***
1,2:8,9-DIEPOXYNONANE
mf: $C_9H_{16}O_2$ mw: 156.25

TOXICITY DATA with REFERENCE
unr-mus TDLo:3750 mg/kg:ETA RARSAM 3,193,63

SAFETY PROFILE: Questionable carcinogen with experimental tumorigenic data. When heated to decomposition it emits acrid smoke and irritating fumes.

DHD600 CAS:3012-69-9 ***HR: 3***
9,10:12,13-DIEPOXYOCTADECANOIC ACID
mf: $C_{18}H_{32}O_4$ mw: 312.50

SYN: 9,10:12,13-DIEPOXYSTEARIC ACID

TOXICITY DATA with REFERENCE
skn-mus TDLo:3360 mg/kg/28W-I:ETA JNCIAM 31,41,63

SAFETY PROFILE: Questionable carcinogen with experimental tumorigenic data. When heated to decomposition it emits acrid smoke and irritating fumes.

DHD800 CAS:2426-07-5 ***HR: 3***
1,2,7,8-DIEPOXYOCTANE
mf: $C_8H_{14}O_2$ mw: 142.22

SYN: 1,2-EPOXY-7,8-EPOXYOCTANE

TOXICITY DATA with REFERENCE
mmo-klp 500 μmol/L MUREAV 89,269,81

mma-sat 990 μg/plate PNASA6 72,5135,75
mmo-nsc 75 mmol/L CNREA8 32,1890,72
cyt-ham:lng 6 μmol/L JEPTDQ 2(2),587,79
msc-ham:lng 6 μmol/L JEPTDQ 2(2),587,79
skn-rbt 100 μg/24H open AIHAAP 23,95,62
skn-mus TDLo:6600 mg/kg/55W-I:ETA JNCIAM 39,1217,67
orl-rat LD50:1070 mg/kg AIHAAP 23,95,62
skn-rbt LD50:320 mg/kg AIHAAP 23,95,62

CONSENSUS REPORTS: EPA Genetic Toxicology Program.

SAFETY PROFILE: Poison by skin contact. Moderately toxic by ingestion. A skin irritant. Questionable carcinogen with experimental tumorigenic data. Mutation data reported. When heated to decomposition it emits acrid and irritating fumes.

DHE000 CAS:4051-27-8 ***HR: 3***
1,2,4,5-DIEPOXYPENTANE
mf: $C_5H_8O_2$ mw: 100.13

SYNS: 2,2′-METHYLENEBIS OXIRANE (9CI) ◇ 1:4-PENTADIENE DIOXIDE

TOXICITY DATA with REFERENCE
mmo-nsc 75 mmol/L CNREA8 32,1890,72
cyt-rat-ipr 2500 mg/kg BJPCAL 6,235,51
skn-mus TDLo:70 g/kg/58W-I:CAR JNCIAM 35,707,65
skn-mus TD:76 g/kg/63W-I:NEO 14JTAF -,275,64

SAFETY PROFILE: Questionable carcinogen with experimental carcinogenic and neoplastigenic data. Mutation data reported. When heated to decomposition it emits acrid smoke and irritating fumes.

DHE100 ***HR: 3***
DIEPOXYPIPERAZINE
mf: $C_{10}H_{18}N_2O_2$ mw: 198.30

SYNS: N,N′-BIS(2,3-EPOXYPROPYL)PIPERAZINE ◇ 1,4-BIS-(2,3-ETHOXYPROPYL)PIPERAZINE ◇ DIGLYCIDYL PIPERAZINE ◇ EPOXYPIPERAZINE ◇ NSC 74437

TOXICITY DATA with REFERENCE
orl-mus LD50:91380 μg/kg NCISP* JAN86
ipr-mus LD50:40770 μg/kg NCISP* JAN86
scu-mus LD50:199 mg/kg NCISP* JAN86

SAFETY PROFILE: Poison by ingestion, subcutaneous, and intraperitoneal routes. When heated to decomposition it emits toxic fumes of NO_x.

DHE485 ***HR: 2***
DIESEL EXHAUST

TOXICITY DATA with REFERENCE
ihl-rat TCLo:4900 μg/m³/8H/2Y-C:CAR DTESD7 13,349,86
ihl-rat TC:7 mg/m³/7H/2Y-I:CAR FAATDF 9,208,87

SAFETY PROFILE: Questionable carcinogen with experimental carcinogenic data. When heated to decomposition it emits acrid smoke and irritating fumes.

DHE500 ***HR: D***
DIESEL EXHAUST EXTRACT

PROP: Dichloromethane extract of diesel particles absorbed onto carbon black (DTESD7 10,265,82).

TOXICITY DATA with REFERENCE
sce-mus-unr 800 mg/kg DTESD7 10,265,82
sce-ham-ihl 12 mg/m³/14W DTESD7 10,265,82
sce-ham-itr 13300 μg/kg DTESD7 10,265,82

SAFETY PROFILE: Mutation data reported.

DHE700 ***HR: 2***
DIESEL EXHAUST PARTICLES

PROP: Particulate samples collected from the exhaust of a 1979 2.3 L diesel powered automobile running on No. 2 diesel fuel TXAPA9 56,110,80

TOXICITY DATA with REFERENCE
mmo-sat 200 μg/plate TXAPA9 56,110,80
mma-sat 200 μg/plate TXAPA9 56,110,80
msc-hmn:lyms 100 mg/L DTESD7 10,277,82
sce-mus-uns 300 mg/kg DTESD7 10,265,82
ihl-rat TCLo:2200 μg/m³/16H/2Y-I:NEO DTESD7 13,471,86
ihl-rat TC:8300 μg/kg/6H/86W-I:ETA AIHAAP 42,382,81
NIOSH REL: (Diesel Exhaust) TWA reduce to lowest feasible level

SAFETY PROFILE: Questionable carcinogen with experimental neoplastigenic and tumorigenic data. Human mutation data reported. When heated to decomposition it emits acrid smoke and irritating fumes.

DHE800 CAS:77650-28-3 ***HR: 2***
DIESEL FUEL MARINE

PROP: Brown, sltly viscous liquid. Flash p: 100°F; d: <1; autoign temp: 494°F.

SYNS: DFM ◇ FUEL OIL #2 ◇ NCI-C54795

SAFETY PROFILE: Dangerous fire hazard when exposed to heat, flame, or oxidizers. To fight fire, use foam, CO_2, dry chemical. When heated to decomposition it emits acrid smoke and irritating fumes. See also MINERAL OILS and KEROSENE.

DHF000 CAS:111-42-2 ***HR: 2***
DIETHANOLAMINE
mf: $C_4H_{11}NO_2$ mw: 105.16

PROP: A faintly colored, viscous liquid. Mp: 28°, bp: 269.1° (decomp), flash p: 305°F (OC), d: 1.0919 @

30°/20°, autoign temp: 1224°F, vap press: 5 mm @ 138°, vap d: 3.65.

SYNS: BIS(2-HYDROXY ETHYL)AMINE ◇ DEA ◇ DIAETHANOLAMIN (GERMAN) ◇ DIETHANOLAMIN (CZECH) ◇ DIETHYLOLAMINE ◇ 2,2'-DIHYDROXYDIETHYLAMINE ◇ DI(2-HYDROXYETHYL)AMINE ◇ DIOLAMINE ◇ 2,2'-IMINOBISETHANOL ◇ 2,2'-IMINODIETHANOL ◇ NCI-C55174

TOXICITY DATA with REFERENCE
skn-rbt 50 mg open MLD UCDS** 12/29/71
skn-rbt 500 mg/24H MLD 28ZPAK -,109,72
eye-rbt 5500 mg SEV AJOPAA 29,1363,46
eye-rbt 750 μg/24H SEV 28ZPAK -,109,72
orl-rat LD50:710 mg/kg TXAPA9 17,498,70
ipr-rat LD50:2300 mg/kg NPIRI* 1,24,74
orl-mus LD50:3300 mg/kg GISAAA 29(11),25,64
ipr-mus LD50:2300 mg/kg TXAPA9 22,175,72
scu-mus LD50:3553 mg/kg ARZNAD 4,649,54
skn-rbt LD50:12200 mg/kg NPIRI* 1,24,74
orl-gpg LD50:2 g/kg DTLVS* 4,140,80

CONSENSUS REPORTS: Community Right-To-Know List. Reported in EPA TSCA Inventory.

OSHA PEL: TWA 3 ppm
ACGIH TLV: TWA 3 ppm

SAFETY PROFILE: Moderately toxic by ingestion, intraperitoneal, and subcutaneous routes. Mildly toxic by skin contact. A severe eye and mild skin irritant. Combustible when exposed to heat or flame; can react with oxidizing materials. To fight fire, use alcohol foam, water, CO_2, dry chemical. When heated to decomposition it emits toxic fumes such as NO_x. See also AMINES.

DHF200 CAS:5716-15-4 ***HR: 3***
DIETHANOLAMMONIUM MALEIC HYDRAZIDE
mf: $C_4H_{11}NO_2{\cdot}C_4H_4N_2O_2$ mw: 217.26

SYNS: 6-HYDROXY-3-(2H)-PYRIDAZINONEDIETHANOLAMINE ◇ 2,2'-IMINODI-ETHANOL with 1,2-DIHYDRO-3,6-PYRIDAZINEDIONE (1:1) ◇ MALEIC HYDRAZIDE DIETHANOLAMINE SALT ◇ MH-30 ◇ NCI-C54660

TOXICITY DATA with REFERENCE
mma-sat 50 μL/plate MUREAV 66,247,79
cyt-ham:lng 20 g/L MUREAV 67,249,79
scu-rat TDLo:1300 mg/kg/65W-I:ETA TXCYAC 1,301,73
orl-rat LD50:2340 mg/kg PCOC** -,673,66

CONSENSUS REPORTS: Reported in EPA TSCA Inventory.

SAFETY PROFILE: Moderately toxic by ingestion. Questionable carcinogen with experimental tumorigenic data. Mutation data reported. When heated to decomposition it emits toxic fumes of NO_x and NH_3.

DHF400 CAS:91-99-6 ***HR: 2***
DIETHANOL-m-TOLUIDINE
mf: $C_{11}H_{17}NO_2$ mw: 195.29

SYNS: N,N-BIS(β-HYDROXYETHYL)-3-METHYLANILINE ◇ N,N-BIS(2-HYDROXYETHYL)-3-METHYLANILINE ◇ N,N-BIS(2-HYDROXYETHYL)-m-TOLUIDINE ◇ N,N-DIHYDROXYETHYL-m-TOLUIDINE ◇ EMERY 5709 ◇ 2,2'-((3-METHYLPHENYL)IMINO)BISETHANOL ◇ m-TOLYLDIETHANOLAMINE ◇ 2,2'-(m-TOLYLIMINO)DIETHANOL

TOXICITY DATA with REFERENCE
orl-rat LD50:3100 mg/kg LONZA# 02JUN80

CONSENSUS REPORTS: Reported in EPA TSCA Inventory.

SAFETY PROFILE: Moderately toxic by ingestion. When heated to decomposition it emits toxic fumes of NO_x.

DHF600 CAS:341-70-8 ***HR: 3***
DIETHAZINE HYDROCHLORIDE
mf: $C_{18}H_{22}N_2S{\cdot}ClH$ mw: 334.94

SYNS: ANTIPAR ◇ APARKAZIN ◇ CASANTIN ◇ DEPARKIN ◇ 10-(2-DIETHYLAMINO)ETHYLPHENOTHIAZINEHYDROCHLORIDE ◇ DIPARCOL ◇ LABITON ◇ THIANTAN ◇ THIONTAN

TOXICITY DATA with REFERENCE
dnd-esc 20 μmol/L MUREAV 89,95,81
orl-mus LD50:450 mg/kg 27ZQAG -,20,72
ipr-mus LD50:225 mg/kg 27ZQAG -,20,72
scu-mus LD50:450 mg/kg THERAP 2,115,47
ivn-mus LD50:5 mg/kg THERAP 2,115,47
scu-rbt LD50:150 mg/kg 27ZQAG -,20,72
ivn-rbt LD50:2500 ug/kg THERAP 2,115,47
scu-gpg LD50:450 mg/kg AIPTAK 137,375,62

SAFETY PROFILE: Poison by intraperitoneal, subcutaneous, and intravenous routes. Moderately toxic by ingestion. Mutation data reported. When heated to decomposition it emits very toxic fumes of NO_x, SO_x, and HCl.

DHF800 CAS:6485-91-2 ***HR: 2***
DIETHOXYCHLOROSILANE
mf: $C_4H_{11}ClO_2Si$ mw: 154.69

PROP: Liquid. Vap d: 5.33.

SYN: CHLORODIETHOXYSILANE

TOXICITY DATA with REFERENCE
skn-rbt 10 mg/24H open MLD JIHTAB 31,60,49
eye-rbt 250 μg open SEV JIHTAB 31,60,49
orl-rat LD50:6300 mg/kg JIHTAB 31,60,49
ihl-rat LCLo:1000 ppm/4H JIHTAB 31,343,49

SAFETY PROFILE: A severe eye and mild skin irritant. Mildly toxic by inhalation and ingestion. Reacts with water or steam to produce heat and toxic and corrosive fumes of HCl. When heated to decomposition it emits toxic fumes of Cl^-. See also CHLOROSILANES.

DHG000 CAS:78-62-6 ***HR: 1***
DIETHOXYDIMETHYLSILANE
DOT: UN 2380
mf: $C_6H_{16}O_2Si$ mw: 148.31

$(CH_3CH_2O)_2Si(CH_3)_2$

PROP: Liquid. Bp: 113.5°, d: 0.834, vap press: 10 mm @ 13.3°, vap d: 5.1. flash p.: <73.4°F.

SYNS: DIMETHYL-DIETHOXYSILAN (CZECH) ◇ DIMETHYLDIETHOXYSILANE (DOT)

TOXICITY DATA with REFERENCE
skn-rbt 500 mg/24H MLD 28ZPAK -,218,72
eye-rbt 83 mg JIHTAB 30,332,48
eye-rbt 500 mg/24H MLD 28ZPAK -,218,72
orl-rat LDLo:9280 mg/kg 28ZPAK -,218,72
ihl-rat LCLo:8000 ppm/4H 28ZPAK -,218,72

CONSENSUS REPORTS: Reported in EPA TSCA Inventory.

DOT Classification: Flammable Liquid; Label: Flammable Liquid.

SAFETY PROFILE: Mildly toxic by inhalation and ingestion. A skin and eye irritant. A dangerous fire hazard when exposed to heat, flame, or oxidizers. When heated to decomposition it emits acrid smoke and irritating fumes. See also SILANES.

DHH200 CAS:21548-32-3 ***HR: 3***
(DIETHOXYPHOSPHINYLIMINO)-1,3-DITHIETANE
mf: $C_6H_{12}NO_3PS_2$ mw: 241.28

SYNS: AC 64475 ◇ ACCONEM ◇ CL 64475 ◇ DIETHOXYPHOSPHINYLIMINO-2-DITHIETANNE-1,3 (FRENCH) ◇ 1,3-DITHIETAN-2-YLIDENE PHOSPHORAMIDIC ACID DIETHYL ESTER ◇ FOSTHIETAN ◇ GEOFOS ◇ NEM-A-TAK

TOXICITY DATA with REFERENCE
orl-rat LD50:4700 μg/kg FMCHA2 -,C167,83
orl-mus LD50:18 mg/kg MEIEDD 10,607,83
skn-rbt LD50:27400 μg/kg FMCHA2 -,C167,83

CONSENSUS REPORTS: Reported in EPA TSCA Inventory. EPA Extremely Hazardous Substances List.

SAFETY PROFILE: Poison by ingestion and skin contact. When heated to decomposition it emits very toxic fumes of NO_x, PO_x, and SO_x.

DHH400 CAS:950-10-7 ***HR: 3***
2-(DIETHOXYPHOSPHINYLIMINO)-4-METHYL-1,3-DITHIOLANE
mf: $C_8H_{16}NO_3PS_2$ mw: 269.34

SYNS: AC 47470 ◇ AMERICAN CYANAMID CL-47470 ◇ CL-47,470 ◇ CYCLIC PROPYLENE (DIETHOXYPHOSPHINYL)DITHIOIMIDOCARBONATE ◇ CYTROLANE ◇ p,p-DIETHYL CYCLIC PROPYLENE ESTER of PHOSPHONODITHIOIMIDOCARBONIC ACID ◇ DIETHYL (4-METHYL-1,3-DITHIOLAN-2-YLIDENE)PHOSPHOROAMIDATE ◇ EI-47470 ◇ ENT 25,991 ◇ MEPHOSFOLAN ◇ (4-METHYL-1,3-DITHIOLAN-2-YLIDENE)PHOSPHORAMIDIC ACID, DIETHYL ESTER

TOXICITY DATA with REFERENCE
orl-rat LD50:9 mg/kg BESAAT 15,122,69
orl-mus LD50:11 mg/kg BESAAT 15,122,69
skn-rbt LD50:28700 ug/kg FMCHA2 -,C68,83
orl-ckn LD50:2800 ug/kg EXPEAM 30,63,74

CONSENSUS REPORTS: EPA Extremely Hazardous Substances List. Reported in EPA TSCA Inventory. EPA Genetic Toxicology Program.

SAFETY PROFILE: Poison by ingestion and skin contact. When heated to decomposition it emits very toxic fumes of NO_x, PO_x, and SO_x.

DHH600 CAS:74038-45-2 ***HR: 3***
(α-(DIETHOXYPHOSPHINYL)-p-METHOXYBENZYL)BIS(2-CHLOROPROPYL)ANTIMONITE
mf: $C_{18}H_{30}Cl_2O_7PSb$ mw: 582.10

SYN: (α-HYDROXY-p-METHOXYBENZYL)-PHOSPHONIC ACID DIETHYL ESTER, ESTER with BIS(2-CHLOROPROPYL) ANTIMONATE(III)

TOXICITY DATA with REFERENCE
ivn-mus LD50:180 mg/kg CSLNX* NX#01812

CONSENSUS REPORTS: Antimony and its compounds are on the Community Right-To-Know List.

OSHA PEL: TWA 0.5 mg(Sb)/m^3
ACGIH TLV: TWA 0.5 mg(Sb)/m^3
NIOSH REL: (Antimony) TWA 0.5 mg/m^3

SAFETY PROFILE: Poison by intravenous route. See also ANTIMONY COMPOUNDS. When heated to decomposition it emits fumes of Sb, PO_x, and Cl^-.

DHH800 CAS:3054-95-3 ***HR: 3***
3,3-DIETHOXYPROPENE
mf: $C_7H_{14}O_2$ mw: 124.19

$H_2C{=}CHCH(OCH_2CH_3)_2$

PROP: Flash p: < 73.4°.

SAFETY PROFILE: A dangerous fire hazard when exposed to heat, flame or oxidizers. When heated to decomposition it emits acrid smoke and fumes.

DHI000 CAS:97-96-1 ***HR: 3***
DIETHYL ACETALDEHYDE
DOT: UN 1178
mf: $C_6H_{12}O$ mw: 100.18

PROP: Colorless liquid; ungent odor. Bp: 116.8°, flash p: 70°F (OC), fp: −89°, d: 0.808-0.814, vap press: 13.7 mm @ 20°, vap d: 3.45, lel: 1.2%, uel: 7.7%. Misc in alc, ether; sltly sol in water.

SYNS: ALDEHYDE-2-ETHYLBUTYRIQUE (FRENCH) ◇ 2-ETHYLBUTANAL ◇ 2-ETHYLBUTRIC ALDEHYDE ◇ ETHYL BUTYRALDEHYDE ◇ α-ETHYLBUTYRALDEHYDE ◇ ETHYL BUTYRALDEHYDE (DOT) ◇ 2-ETHYLBUTYRALDEHYDE (DOT,FCC) ◇ FEMA No. 2426

TOXICITY DATA with REFERENCE
skn-rbt 500 mg open MLD UCDS** 12/14/71
orl-rat LD50:3980 mg/kg AMIHBC 4,119,51
ihl-rat LCLo:8000 ppm/4H AMIHBC 4,119,51

CONSENSUS REPORTS: Reported in EPA TSCA Inventory.

DOT Classification: Flammable Liquid; Label: Flammable Liquid.

SAFETY PROFILE: Moderately toxic by ingestion. Mildly toxic by inhalation. A skin irritant. Flammable liquid. Can react vigorously with oxidizing materials. To fight fire, use alcohol foam, CO_2, dry chemical. When heated to decomposition it emits acrid smoke and fumes. See also ALDEHYDES.

DHI200 CAS:685-91-6 ***HR: 3***
N,N-DIETHYLACETAMIDE
mf: $C_6H_{13}NO$ mw: 115.20

PROP: Liquid. Mp: <65°, bp: 180°, flash p: 170°F, d: 0.92, vap d: 4.0.

TOXICITY DATA with REFERENCE
orl-rat TDLo:910 mg/kg/73W-I:ETA JNCIAM 35,949,65
orl-rat LD50:1500 mg/kg ARZNAD 19,1073,69
ipr-rat LD50:1840 mg/kg ARZNAD 20,1242,70
ivn-rat LD50:1 g/kg ARZNAD 28,1571,78
ipr-mus LD50:1690 mg/kg ARZNAD 20,1242,70
ivn-dog LD50:1 g/kg ARZNAD 28,1571,78
ivn-rbt LDLo:1920 mg/kg ARZNAD 20,1242,70
ivn-ckn LDLo:3900 mg/kg ARZNAD 20,1242,70

CONSENSUS REPORTS: Reported in EPA TSCA Inventory.

SAFETY PROFILE: Moderately toxic by ingestion, intravenous, and intraperitoneal routes. Questionable carcinogen with experimental tumorigenic data. Flammable when exposed to heat or flame. To fight fire, use foam, mist, CO_2, dry chemical. When heated to decomposition it emits toxic fumes of NO_x.

DHI400 CAS:88-09-5 ***HR: 3***
DIETHYLACETIC ACID
mf: $C_6H_{12}O_2$ mw: 116.18

PROP: Colorless, volatile liquid; rancid odor. Mp: −93°, bp: 121.0°, flash p: 78°F (CC), d: 0.917, vap press: 10 mm @ 15.3°, vap d: 4.0, autoign temp: 865°F. Misc in alc, ether, water.

SYNS: 2-ETHYL BUTANOIC ACID ◇ α-ETHYLBUTYRIC ACID ◇ 2-ETHYLBUTYRIC ACID (FCC) ◇ FEMA No. 2429 ◇ 3-PENTANECARBOXYLIC ACID

TOXICITY DATA with REFERENCE
skn-rbt 10 mg/24H open MLD AMIHBC 10,61,54
eye-rbt 250 μg open SEV AMIHBC 10,61,54
orl-rat LD50:2200 mg/kg AMIHBC 10,61,54
skn-rbt LD50:520 mg/kg AMIHBC 10,61,54

CONSENSUS REPORTS: Reported in EPA TSCA Inventory.

SAFETY PROFILE: Moderately toxic by ingestion and skin contact. An irritant to skin and mucous membranes. A severe eye irritant. See also ESTERS. Narcotic in high concentrations. Flammable liquid. To fight fire, use CO_2, dry chemical, alcohol foam. When heated to decomposition it emits acrid smoke and fumes.

DHI600 CAS:2235-46-3 ***HR: 1***
DIETHYLACETOACETAMIDE
mf: $C_8H_{15}NO_2$ mw: 157.24

PROP: Misc liquid. D: 0.995 @ 20°/20°, bp: decomp, fp: −70°, flash p: 250°F (COC).

SYNS: N,N-DIETHYLACETOACETAMIDE ◇ N,N-DIETHYL-3-OXOBUTANAMIDE (9CI)

TOXICITY DATA with REFERENCE
orl-rat LD50:4760 mg/kg AIHAAP 23,95,62

CONSENSUS REPORTS: Reported in EPA TSCA Inventory.

SAFETY PROFILE: Mildly toxic by ingestion. Combustible when exposed to heat or flame. When heated to decomposition it emits toxic fumes of NO_x.

DHI800 CAS:63019-57-8 ***HR: 3***
1-DIETHYLACETYLAZIRIDINE
mf: $C_8H_{15}NO$ mw: 141.24

SYN: DIETHYLACETYLETHYLENEIMINE

TOXICITY DATA with REFERENCE
cyt-rat-ipr 50 mg/kg BJPCAL 9,306,54
scu-rat TDLo:400 mg/kg/35W-I:NEO BJPCAL 9,306,54

SAFETY PROFILE: Questionable carcinogen with experimental neoplastigenic data. Mutation data reported. When heated to decomposition it emits toxic fumes of NO_x.

DHI850 CAS:762-21-0 ***HR: 3***
DIETHYL ACETYLENE DICARBOXYLATE
mf: $C_8H_{10}O_4$ mw: 170.16

$$CH_3CH_2OCO \cdot OC \equiv CCO \cdot OCH_2CH_3$$

SAFETY PROFILE: Mixture with 1,3,5-cyclooctatriene explodes when heated to 60°C. When heated to decomposition it emits acrid smoke and fumes. See also ACETYLENE COMPOUNDS.

DHI875 CAS:2014-30-4 ***HR: 2***
l-N,N-DIETHYLALANINE-6-CHLORO-o-TOLYL ESTER HYDROCHLORIDE
mf: $C_{14}H_{20}ClNO_2 \cdot ClH$ mw: 306.26

SYN: FC 676

TOXICITY DATA with REFERENCE
skn-rbt 200 mg MOD BCFAAI 107,310,68
eye-rbt 1 g MOD BCFAAI 107,310,68
scu-mus LD50:2000 mg/kg BCFAAI 107,310,68

SAFETY PROFILE: Moderately toxic by subcutaneous route. An eye and skin irritant. When heated to decomposition it emits toxic fumes of NO_x and HCl. See also ESTERS.

DHI880 CAS:760-19-0 ***HR: 3***
DIETHYLALUMINUM BROMIDE
mf: $C_4H_{10}AlBr$ mw: 165.01

$(CH_3CH_2)_2AlBr$

SAFETY PROFILE: Ignites on contact with nitromethane. When heated to decomposition it emits toxic fumes of Br^-. See also ALUMINUM COMPOUNDS and BROMIDES.

DHI885 CAS:96-10-6 ***HR: 3***
DIETHYLALUMINUM CHLORIDE
mf: $C_4H_{10}AlCl$ mw: 110.56

$(CH_3CH_2)_2AlCl$

SAFETY PROFILE: Reaction with chlorine azide may form an explosive product. When heated to decomposition it emits toxic fumes of Cl^-. See also ALUMINUM COMPOUNDS and CHLORIDES.

DHJ200 CAS:109-89-7 ***HR: 3***
DIETHYLAMINE
DOT: UN 1154
mf: $C_4H_{11}N$ mw: 73.16

PROP: Colorless liquid, ammoniacal odor. Mp: −38.9°, bp: 55.5°, flash p: −0.4°F, d: 0.7108 @ 20°/20°, autoign temp: 594°F, vap press: 400 mm @ 38.0°, vap d: 2.53, lel: 1.8%, uel: 10.1%.

SYNS: 2-AMINOPENTANE ◇ DIAETHYLAMIN (GERMAN) ◇ N,N-DIETHYLAMINE ◇ DIETILAMINA (ITALIAN) ◇ DWUETYLOAMINA (POLISH) ◇ N-ETHYL-ETHANAMINE

TOXICITY DATA with REFERENCE
skn-rbt 10 mg/24H MLD AMIHBC 4,119,51
skn-rbt 500 mg open MLD UCDS** 5/21/71
eye-rbt 50 μg open SEV AMIHBC 4,119,51
ihl-mam LC50:5000 mg/m³ TPKVAL 14,80,75
orl-rat LD50:540 mg/kg AEHLAU 1,343,60
ihl-rat LC50:4000 ppm/4H AEHLAU 1,343,60
skn-rbt LD50:820 mg/kg UCDS** 5/21/71

CONSENSUS REPORTS: Reported in EPA TSCA Inventory.

OSHA PEL: (Transitional: TWA 25 ppm) TWA 10 ppm; STEL 25 ppm
ACGIH TLV: TWA 10 ppm; STEL 25 ppm
DFG MAK: 10 ppm (30 mg/m³)
DOT Classification: Flammable Liquid; Label: Flammable Liquid.

SAFETY PROFILE: Moderately toxic by ingestion, inhalation, and skin contact. A skin and severe eye irritant. Exposure to strong vapor can cause severe cough and chest pains. Contact with liquid can damage eyes, possibly permanently; contact with skin causes necrosis and vesiculation. See also AMINES. A very dangerous fire hazard when exposed to heat, flame, or oxidizers. To fight fire, use alcohol foam, CO_2, dry chemical. Explodes on contact with dicyanofurazan. Violent reaction with sulfuric acid. Ignites on contact with cellulose nitrate of sufficiently high surface area. When heated to decomposition it emits toxic fumes of NO_x.

DHJ400 CAS:3213-15-8 ***HR: 3***
2-(DIETHYLAMINO)ACETANILIDE
mf: $C_{12}H_{18}N_2O$ mw: 206.32

SYNS: (DIETHYLAMINO)ACYLANILIDE ◇ 2-(DIETHYLAMINO)-N-PHENYLACETAMIDE ◇ V 343

TOXICITY DATA with REFERENCE
eye-rbt 2% MLD ARZNAD 8,270,58
ipr-rat LD50:465 mg/kg ARZNAD 8,270,58
ipr-mus LD50:235 mg/kg ARZNAD 8,270,58
scu-mus LD50:800 mg/kg ARZNAD 8,270,58

SAFETY PROFILE: Poison by intraperitoneal route. Moderately toxic by subcutaneous route. An eye irritant. When heated to decomposition it emits toxic fumes of NO_x.

DHJ600 CAS:3010-02-4 ***HR: 3***
N,N-DIETHYLAMINOACETONITRILE
mf: $C_6H_{12}N_2$ mw: 112.20

SYNS: (DIETHYLAMINO)ACETONITRILE ◇ N,N-DIETHYLGLYCINONITRILE ◇ NITRIL KISELINY DIETHYLAMINOOCTOVE (CZECH)

TOXICITY DATA with REFERENCE
skn-rbt 10 mg/24H open MLD AMIHBC 4,119,51
skn-rbt 10 mg/24H open MLD AIHAAP 23,95,62
eye-rbt 750 μg open SEV AMIHBC 4,119,51
orl-rat LD50:92300 ug/kg 28ZPAK -,162,72
ihl-rat LC50:125 ppm/4H AMIHBC 4,119,51
skn-rbt LD50:360 mg/kg AMIHBC 4,119,51

CONSENSUS REPORTS: Cyanide and its compounds are on the Community Right-To-Know List. Reported in EPA TSCA Inventory.

SAFETY PROFILE: Poison by ingestion and skin contact. Moderately toxic by inhalation. A skin and severe eye irritant. When heated to decomposition it emits toxic fumes of NO_x. See also NITRILES.

DHJ800 CAS:77966-26-8 ***HR: 2***
2-(DIETHYLAMINO)-o-ACETOPHENETIDIDE, HYDROCHLORIDE
mf: $C_{14}H_{22}N_2O_2 \cdot ClH$ mw: 286.84

SYNS: C 3095 ◇ 2-(DIETHYLAMINO)-2'-ETHOXYACETANILIDE, HYDROCHLORIDE

TOXICITY DATA with REFERENCE
eye-rbt 2% MLD ARZNAD 8,270,58
ipr-rat LD50:443 mg/kg ARZNAD 8,270,58
scu-mus LD50:1295 mg/kg ARZNAD 8,270,58

SAFETY PROFILE: Moderately toxic by intraperitoneal and subcutaneous routes. An eye irritant. When heated to decomposition it emits very toxic fumes of NO_x and HCl.

DHK000 CAS:77966-27-9 ***HR: 3***
2-(DIETHYLAMINO)-p-ACETOPHENETIDIDE, HYDROCHLORIDE
mf: $C_{14}H_{22}N_2O_2 \cdot ClH$ mw: 286.84

SYN: C 3094

TOXICITY DATA with REFERENCE
eye-rbt 2% MLD ARZNAD 8,270,58
ipr-rat LD50:349 mg/kg ARZNAD 8,270,58
scu-mus LD50:720 mg/kg ARZNAD 8,270,58

SAFETY PROFILE: Poison by intraperitoneal route. Moderately toxic by subcutaneous route. An eye irritant. When heated to decomposition it emits very toxic fumes of NO_x and HCl.

DHK200 CAS:6304-07-0 ***HR: 3***
2-(DIETHYLAMINO)-o-ACETOTOLUIDIDE HYDROCHLORIDE
mf: $C_{13}H_{20}N_2O \cdot ClH$ mw: 256.81

SYNS: C 3080 ◇ 2-(DIMETHYLAMINO)-2'-METHYLACETANILIDE HYDROCHLORIDE

TOXICITY DATA with REFERENCE
ipr-rat LD50:430 mg/kg ARZNAD 8,270,58
ipr-mus LD50:280 mg/kg ARZNAD 8,270,58
scu-mus LD50:825 mg/kg ARZNAD 8,270,58

SAFETY PROFILE: Poison by intraperitoneal route. Moderately toxic by subcutaneous route. When heated to decomposition it emits very toxic fumes of NO_x and HCl.

DHK400 CAS:137-58-6 ***HR: 3***
2-(DIETHYLAMINO)-2',6'-ACETOXYLIDIDE
mf: $C_{14}H_{22}N_2O$ mw: 234.38

SYNS: ANESTACON ◇ DIETHYLAMINOACETO-2,6-XYLIDIDE ◇ α-DIETHYLAMINOACETO-2,6-XYLIDIDE ◇ α-DIETHYLAMINO-2,6-ACETOXYLIDIDE ◇ DIETHYLAMINOACET-2,6-XYLIDIDE ◇ α-DIETHYLAMINO-2,6-DIMETHYLACETANILIDE ◇ omega-DIETHYLAMINO-2,6-DIMETHYLACETANILIDE ◇ α-DIETILAMINO-2,6-DIMETILACETANILIDE (ITALIAN) ◇ DUNCAINE ◇ GRAVOCAIN ◇ ISICAINA ◇ LEOSTESIN ◇ LIDA-MANTLE ◇ LIDOCAINE ◇ LIGNOCAINE ◇ MARICAINE ◇ RUCAINA ◇ SOLCAIN ◇ XILOCAINA (ITALIAN) ◇ XYCIANE ◇ XYLESTESIN ◇ XYLOCAIN ◇ XYLOCITIN ◇ XYLOTOX

TOXICITY DATA with REFERENCE
mma-sat 50 umol/plate JAFCAU 34,157,86
ims-rat TDLo:6 mg/kg (female 11D post):REP NETOD7 8,61,86
imp-rat TDLo:7500 mg/kg (female 3-17D post):TER TJADAB 33,73C,86
orl-wmn TDLo:39 mg/kg:CNS,CVS NEJMAG 306,381,82
ivn-wmn TDLo:16 mg/kg:CVS,PUL EJCPAS 22,129,82
ivn-man TDLo:8643 μg/kg/4H-C AIMEAS 97,149,82
ivn-hmn TDLo:23 mg/kg ATXKA8 28,72,71
orl-rat LD50:317 mg/kg BCFAAI 110,330,71
scu-rat LD50:335 mg/kg EJMCA5 9,188,74
ivn-rat LDLo:25 mg/kg BJANAD 23,153,51
orl-mus LD50:220 mg/kg ARZNAD 16,1275,66
ipr-mus LD50:102 mg/kg JMCMAR 24,1059,81
scu-mus LD50:238 mg/kg JMCMAR 28,714,85

CONSENSUS REPORTS: Reported in EPA TSCA Inventory.

SAFETY PROFILE: Poison by ingestion, intravenous, intraperitoneal, and subcutaneous routes. Human systemic effects by ingestion: excitement, hallucinations, distorted perceptions, and changes in heart rate, dyspnea. An experimental teratogen. Other experimental reproductive effects. A local anesthetic. Mutation data reported. When heated to decomposition it emits toxic fumes of NO_x.

DHK600 CAS:73-78-9 ***HR: 3***
2-DIETHYLAMINO-2',6'-ACETOXYLIDIDE HYDROCHLORIDE
mf: $C_{14}H_{22}N_2O \cdot ClH$ mw: 270.84

SYNS: ANESTACON HYDROCHLORIDE ◇ 2-(DIETHYLAMINO)-2',6'-ACETOXYLIDIDE MONOHYDROCHLORIDE ◇ α-DIETHYLAMINO-2,5-ACETOXYLIDINE HYDROCHLORIDE ◇ omega-DIETHYLAMINO-2,6-DIMETHYLACETANILIDE HYDROCHLORIDE ◇ 2-(DIETHYLAMINO)-N-(2,6-DIMETHYLPHENYL)ACETAMIDEMONOHYDROCHLORIDE ◇ DUNCAINE HYDROCHLORIDE ◇ GRAVOCAIN HYDROCHLORIDE ◇ ISICAINE HYDROCHLORIDE ◇ LEOSTESIN HYDROCHLORIDE ◇ LIDOCAINE HYDROCHLORIDE ◇ LIDOTHESIN HYDROCHLORIDE

◇ LIGNOCAINE HYDROCHLORIDE ◇ RUCAINA HYDROCHLORIDE ◇ S 202 ◇ XYCAINE HYDROCHLORIDE ◇ XYLESTESIN HYDROCHLORIDE ◇ XYLOCAINE HYDROCHLORIDE ◇ XYLOCARD ◇ XYLOCITIN HYDROCHLORIDE ◇ XYLONEURAL ◇ XYLOTOX HYDROCHLORIDE

TOXICITY DATA with REFERENCE
skn-rbt 3% MLD AIPTAK 137,410,62
eye-rbt 3% MLD AIPTAK 137,410,62
imp-rat TDLo:7500 mg/kg (female 3-17D post):TER ANESAV 65,626,86
imp-rat TDLo:7500 mg/kg (female 3-17D post):REP ANESAV 65,626,86
ivn-cld TDLo:60 mg/kg/1H JTCTDW 24,51,86
ivn-man TDLo:9 mg/kg/4H-C:CVS DICPBB 19,669,85
ivn-man TDLo:7143 μg/kg:BPR CHETBF 61,682,72
imp-man TDLo:5714 μg/kg CMAJAX 137,219,87
ipr-rat LD50:122 mg/kg JPETAB 111,224,54
scu-rat LD50:570 mg/kg RPOBAR 2,299,70
ivn-rat LD50:21 mg/kg RPOBAR 2,299,70
orl-mus LD50:220 mg/kg RPOBAR 2,298,70
ipr-mus LD50:63 mg/kg AIPTAK 274,253,85
scu-mus LD50:163 mg/kg PSEBAA 103,353,60
ivn-mus LD50:15 mg/kg JPPMAB 14(Suppl),48T,62
ims-mus LD50:260 mg/kg RPOBAR 2,298,70
itr-rbt LD50:28 mg/kg AIPTAK 200,359,72

CONSENSUS REPORTS: Reported in EPA TSCA Inventory. EPA Genetic Toxicology Program.

SAFETY PROFILE: Poison by ingestion, intraperitoneal, intravenous, subcutaneous, intramuscular, and intratracheal routes. Human systemic effects: somnolence, respiratory depression, low blood pressure, cardiomyopathy including infarction, pulse rate increase. An experimental teratogen. Other experimental reproductive effects. A skin and eye irritant. An anesthetic. When heated to decomposition it emits very toxic fumes of NO_x and HCl.

DHK800 CAS:77966-82-6 ***HR: 3***
2-(DIETHYLAMINO)-3',5'-ACETOXYLIDIDE HYDROCHLORIDE
mf: $C_{14}H_{22}N_2O \cdot ClH$ mw: 270.84

SYNS: C 3065 ◇ 2-(DIETHYLAMINO)-3',5'-DIMETHYLACETANILIDE HYDROCHLORIDE

TOXICITY DATA with REFERENCE
eye-rbt 2% MOD ARZNAD 8,270,58
ipr-rat LD50:260 mg/kg ARZNAD 8,270,58
scu-mus LD50:550 mg/kg ARZNAD 8,270,58

SAFETY PROFILE: Poison by intraperitoneal route. Moderately toxic by subcutaneous route. An eye irritant. When heated to decomposition it emits very toxic fumes of NO_x and HCl.

DHL200 CAS:77967-24-9 ***HR: 2***
N-(2-DIETHYLAMINO)ACETYLANTHRANILIC ACID, ETHYL ESTER HYDROCHLORIDE
mf: $C_{15}H_{22}N_2O_3 \cdot ClH$ mw: 314.85

SYNS: C 3102 ◇ N-((DIETHYLAMINO)ACETYL)ANTHRANILIC ACID, ETHYL ESTER, HYDROCHLORIDE

TOXICITY DATA with REFERENCE
eye-rbt 2% MOD ARZNAD 8,270,58
ipr-rat LD50:447 mg/kg ARZNAD 8,270,58
scu-mus LD50:2150 mg/kg ARZNAD 8,270,58

SAFETY PROFILE: Moderately toxic by intraperitoneal and subcutaneous routes. An eye irritant. See also ESTERS. When heated to decomposition it emits very toxic fumes of NO_x and HCl.

DHL400 CAS:77967-25-0 ***HR: 2***
N-(2-DIETHYLAMINO)ACETYLANTHRANILIC ACID, METHYL ESTER, HYDROCHLORIDE
mf: $C_{14}H_{20}N_2O_3 \cdot ClH$ mw: 300.82

SYNS: C 3089 ◇ N-((DIETHYLAMINO)ACETYL)ANTHRANILIC ACID, METHYL ESTER, HYDROCHLORIDE

TOXICITY DATA with REFERENCE
eye-rbt 2% MLD ARZNAD 8,270,58
ipr-rat LD50:615 mg/kg ARZNAD 8,270,58
scu-mus LD50:2400 mg/kg ARZNAD 8,270,58

SAFETY PROFILE: Moderately toxic by intraperitoneal and subcutaneous routes. An eye irritant. See also ESTERS. When heated to decomposition it emits very toxic fumes of HCl and NO_x.

DHL600 ***HR: 3***
β-4-(1-DIETHYLAMINOACETYL-2-PIPERIDYL)-2,2-DIPHENYL-1,3-DIOXOLANE HYDROCHLORIDE
mf: $C_{26}H_{34}N_2O_3 \cdot ClH$ mw: 459.08

TOXICITY DATA with REFERENCE
orl-mus LD50:150 mg/kg JMCMAR 9,127,66
ivn-mus LD50:12 mg/kg JMCMAR 9,127,66

SAFETY PROFILE: Poison by ingestion and intravenous routes. When heated to decomposition it emits very toxic fumes of NO_x and HCl.

DHL800 CAS:1027-14-1 ***HR: 3***
DIETHYLAMINOACETYL-2,4,6-TRIMETHYLANILINE HYDROCHLORIDE
mf: $C_{15}H_{24}N_2O \cdot ClH$ mw: 284.87

SYNS: 2-DIETHYLAMINO-2',4',6'-TRIMETHYLACETANILIDE HYDROCHLORIDE ◇ 2-(DIETHYLAMINO)-2',4',6'-TRIMETHYLACETANILIDE MONOHYDROCHLORIDE ◇ 2-(DIETHYLAMINO)-N-(2,4,6-TRIMETHYLPHENYL)ACETAMIDEMONOHYDROCHLORIDE ◇ MESIDICAINE HYDROCHLORIDE ◇ MESOCAINE HYDROCHLORIDE ◇ MESOKAIN HYDROCHLORIDE ◇ TRIMECAINE ◇ TRIMECA-

INE HYDROCHLORIDE ◇ TRIMEKAIN HYDROCHLORIDE ◇ N-sym-TRIMETHYLPHENYLDIETHYLAMINOACETAMIDEHYDROCHLORIDE

TOXICITY DATA with REFERENCE
ipr-mus LD50:172 mg/kg AFPCAG 29,81,76
scu-mus LD50:295 mg/kg MEIEDD 10,1386,83

SAFETY PROFILE: Poison by intraperitoneal and subcutaneous routes. When heated to decomposition it emits very toxic fumes of NO_x and HCl.

DHM000 CAS:5123-63-7 ***HR: 1***
3-(DIETHYLAMINO)BENZENESULFONIC ACID, SODIUM SALT
mf: $C_{10}H_{14}NO_3S{\bullet}Na$ mw: 251.30

SYN: N,N-DIETHYLMETANILAN SODNY (CZECH)

TOXICITY DATA with REFERENCE
skn-rbt 500 mg/24H MLD 28ZPAK -,185,72
eye-rbt 20 mg/24H MOD 28ZPAK -,185,72
orl-rat LD50:7470 mg/kg 28ZPAK -,185,72

CONSENSUS REPORTS: Reported in EPA TSCA Inventory.

SAFETY PROFILE: Mildly toxic by ingestion. A skin and eye irritant. When heated to decomposition it emits very toxic fumes of NO_x, Na_2O, and SO_x. See also SULFONATES.

DHM200 CAS:1809-53-6 ***HR: 3***
1-DIETHYLAMINO-1-BUTEN-3-YNE
mf: $C_8H_{13}N$ mw: 123.20

$(CH_3CH_2)_2NCH{=}CHC{\equiv}CH$

SAFETY PROFILE: The residue from distillation is explosive. When heated to decomposition it emits toxic fumes of NO_x. See also ACETYLENE COMPOUNDS.

DHM309 CAS:1877-24-3 ***HR: 3***
2-(DIETHYLAMINO)BUTYRIC ACID-2,6-XYLYL ESTER HYDROCHLORIDE
mf: $C_{16}H_{25}NO_2{\bullet}ClH$ mw: 299.88

SYN: FC 455

TOXICITY DATA with REFERENCE
skn-rbt 200 mg MOD BCFAAI 107,310,68
eye-rbt 1 g MOD BCFAAI 107,310,68
scu-mus LD50:230 mg/kg BCFAAI 107,310,68
ivn-mus LD50:11 mg/kg BCFAAI 107,310,68

SAFETY PROFILE: Poison by subcutaneous and intravenous routes. An eye and skin irritant. When heated to decomposition it emits toxic fumes of NO_x and HCl. See also ESTERS.

DHM400 CAS:77985-21-8 ***HR: 3***
3-(DIETHYLAMINO)-2′,6′-BUTYROXYLIDIDE HYDROCHLORIDE
mf: $C_{16}H_{26}N_2O{\bullet}ClH$ mw: 298.90

SYN: C 5125

TOXICITY DATA with REFERENCE
ipr-rat LD50:40 mg/kg ARZNAD 8,544,58
ipr-mus LD50:30 mg/kg ARZNAD 8,544,58
scu-mus LD50:86 mg/kg ARZNAD 8,544,58

SAFETY PROFILE: Poison by intraperitoneal and subcutaneous routes. When heated to decomposition it emits very toxic fumes of HCl and NO_x.

DHM500 CAS:2869-83-2 ***HR: 2***
3-(DIETHYLAMINO)-7-((p-(DIMETHYLAMINO) PHENYL)AZO)-5-PHENYLPHENAZINIUM CHLORIDE
mf: $C_{30}H_{31}N_6{\bullet}Cl$ mw: 511.12

SYNS: C.I. 11050 ◇ JANUS GREEN B ◇ JANUS GREEN V

TOXICITY DATA with REFERENCE
cyt-ham:ovr 20 μmol/L/5H-C ENMUDM 1,27,79
scu-rat TDLo:960 mg/kg/13W-I:ETA GANNA2 44,293,53

CONSENSUS REPORTS: Reported in EPA TSCA Inventory.

SAFETY PROFILE: Questionable carcinogen with experimental tumorigenic data. Mutation data reported. When heated to decomposition it emits toxic fumes of NO_x.

DHN800 CAS:77791-20-9 ***HR: 3***
2-(DIETHYLAMINO)-N,N-DIPHENYLACETAMIDE HYDROCHLORIDE
mf: $C_{18}H_{22}N_2O{\bullet}ClH$ mw: 318.88

SYN: C 3135

TOXICITY DATA with REFERENCE
ipr-rat LD50:210 mg/kg ARZNAD 8,609,58
scu-mus LD50:222 mg/kg ARZNAD 8,609,58

SAFETY PROFILE: Poison by intraperitoneal and subcutaneous routes. When heated to decomposition it emits very toxic fumes of NO_x and HCl.

DHO000 CAS:59960-90-6 ***HR: 3***
2-(DIETHYLAMINO)-N-(DIPHENYLMETHYL) ACETAMIDE HYDROCHLORIDE
mf: $C_{19}H_{24}N_2O{\bullet}ClH$ mw: 332.91

SYN: C 3209

TOXICITY DATA with REFERENCE
eye-rbt 2% MOD ARZNAD 8,609,58
ipr-rat LD50:165 mg/kg ARZNAD 8,609,58

scu-mus LD50:399 mg/kg APSXAS 5,429,68
ivn-mus LD50:26 mg/kg APSXAS 5,429,68

SAFETY PROFILE: Poison by intraperitoneal, intravenous, and subcutaneous routes. An eye irritant. When heated to decomposition it emits very toxic fumes of NO_x and HCl.

DHO400 CAS:1942-52-5 ***HR: 3***
DIETHYLAMINOETHANETHIOL HYDROCHLORIDE
mf: $C_6H_{15}NS•ClH$ mw: 169.74

SYN: USAF E-4

TOXICITY DATA with REFERENCE
ipr-mus LD50:100 mg/kg NTIS** AD277-689

CONSENSUS REPORTS: Reported in EPA TSCA Inventory.

SAFETY PROFILE: Poison by intraperitoneal route. When heated to decomposition it emits very toxic fumes of Cl^-, SO_x, and NO_x.

DHO500 CAS:100-37-8 ***HR: 3***
2-DIETHYLAMINOETHANOL
DOT: UN 2686
mf: $C_6H_{15}NO$ mw: 117.22

PROP: Colorless, hygroscopic liquid. Bp: 162°, flash p: 140°F (OC), d: 0.8851 @ 20°/20°, vap press: 1.4 mm @ 20°, vap d: 4.03.

SYNS: DEAE ◇ DIAETHYLAMINOAETHANOL (GERMAN) ◇ DIETHYLAMINOETHANOL ◇ β-DIETHYLAMINOETHANOL ◇ N-DIETHYLAMINOETHANOL ◇ 2-(DIETHYLAMINO)ETHANOL ◇ 2-N-DIETHYLAMINOETHANOL ◇ DIETHYLAMINOETHANOL (DOT) ◇ β-DIETHYLAMINOETHYL ALCOHOL ◇ DIETHYLETHANOLAMINE ◇ N,N-DIETHYLETHANOLAMINE ◇ N,N-DIETHYL-N-(β-HYDROXYETHYL) AMINE ◇ 2-HYDROXYTRIETHYLAMINE

TOXICITY DATA with REFERENCE
skn-rbt 10 mg/24H open JIHTAB 26,269,44
skn-rbt 500 mg open MLD UCDS** 6/11/63
eye-rbt 5 mg SEV UCDS** 6/11/63
ihl-hmn TCLo:200 ppm:GIT 34ZIAG -,216,69
orl-rat LD50:1300 mg/kg JIHTAB 26,269,44
ihl-rat LCLo:4500 mg/m^3/4H GTPZAB 14(11),52,70
ipr-rat LD50:1220 mg/kg TXAPA9 12,486,68
ihl-mus LC50:5000 mg/m^3 GTPZAB 14(11),52,70
ipr-mus LD50:192 mg/kg JPETAB 94,249,48
scu-mus LD50:1561 mg/kg ARZNAD 4,649,54
ivn-mus LD50:188 mg/kg ARZNAD 9,31,59
ims-mus LD50:416 mg/kg ARZNAD 9,31,59
skn-gpg LD50:884 mg/kg JIHTAB 26,269,44

CONSENSUS REPORTS: Reported in EPA TSCA Inventory.

OSHA PEL: TWA 10 ppm (skin)
ACGIH TLV: TWA 10 ppm (skin)
DFG MAK: 10 ppm (50 mg/m^3)
DOT Classification: Flammable or Combustible Liquid; Label: Flammable Liquid.

SAFETY PROFILE: Poison by intraperitoneal and intravenous routes. Moderately toxic by ingestion, skin contact, subcutaneous, and intramuscular routes. Human systemic effects by inhalation: nausea or vomiting. A skin and severe eye skin irritant. Combustible liquid. Flammable when exposed to heat or flame; can react with oxidizing materials. To fight fire, use alcohol foam, CO_2, dry chemical. When heated to decomposition it emits toxic fumes of NO_x. See also AMINES.

DHO600 CAS:487-53-6 ***HR: 3***
DIETHYLAMINOETHANOL-p-AMINOSALICYLATE
mf: $C_{13}H_{20}N_2O_3$ mw: 252.35

SYNS: DIETHYLAMINOETHYL-p-AMINOSALICYLATE ◇ 2-DIETHYLnAMINOETHYL-p-AMINOSALICYLATE ◇ 2-DIETHYLAMINO- ETHYL-4-AMINOSALICYLATE ◇ DIETHYLAMINOETHYL-3-HYDROXY-4-AMINOBENZOATE ◇ HYDROXYPROCAINE ◇ m-HYDROXY- PROCAINE ◇ METAHYDROXYPROCAINE ◇ OXYCAINE ◇ OXYPROCAIN ◇ OXYPROCAINE

TOXICITY DATA with REFERENCE
ipr-rat LD50:190 mg/kg KLWOAZ 31,97,53
ivn-rat LD50:36 mg/kg KLWOAZ 31,97,53
ipr-mus LD50:125 mg/kg ARZNAD 2,112,52
scu-mus LD50:280 mg/kg KLWOAZ 31,97,53
ivn-mus LD50:47 mg/kg KLWOAZ 31,97,53

SAFETY PROFILE: Poison by intravenous and intraperitoneal routes. When heated to decomposition it emits toxic fumes of NO_x.

DHO700 CAS:14426-20-1 ***HR: 2***
2-DIETHYLAMINOETHANOL HYDROCHLORIDE
mf: $C_6H_{15}NO•ClH$ mw: 153.68

TOXICITY DATA with REFERENCE
ipr-mus LD50:1162 mg/kg APBDAJ 290,131,57
scu-mus LD50:1260 mg/kg AIPTAK 112,36,57
ivn-mus LD50:458 mg/kg APBDAJ 290,131,57

CONSENSUS REPORTS: Reported in EPA TSCA Inventory.

SAFETY PROFILE: Moderately toxic by intraperitoneal, intravenous, and subcutaneous routes. When heated to decomposition it emits toxic fumes of NO_x and HCl.

DHO800 CAS:102207-84-1 ***HR: 3***
2-(2-(DIETHYLAMINO)ETHOXY)-2',6'-ACETOXYLIDIDE HYDROCHLORIDE
mf: $C_{16}H_{26}N_2O_2{\cdot}ClH$ mw: 314.90

SYN: C 3054

TOXICITY DATA with REFERENCE
eye-rbt 2% MLD ARZNAD 9,113,59
ipr-rat LD50:245 mg/kg ARZNAD 9,113,59
scu-mus LD50:500 mg/kg ARZNAD 9,113,59

SAFETY PROFILE: Poison by intraperitoneal route. Moderately toxic by subcutaneous route. An eye irritant. When heated to decomposition it emits very toxic fumes of NO_x and HCl.

DHP000 CAS:54099-23-9 ***HR: 3***
2-(2-(DIETHYLAMINO)ETHOXY)ADAMANTANE ETHYL IODIDE
mf: $C_{18}H_{34}NO{\cdot}I$ mw: 407.43

SYN: (2-(2-ADAMANTYLOXY)ETHYL)TRIETHYL-AMMONIUMIODIDE

TOXICITY DATA with REFERENCE
orl-mus LD50:400 mg/kg FRPSAX 32,129,77
ipr-mus LD50:30 mg/kg FRPSAX 32,129,77

SAFETY PROFILE: Poison by ingestion and intraperitoneal routes. When heated to decomposition it emits very toxic fumes of NO_x, NH_3, and I^-. See also IODIDES.

DHP200 CAS:6376-26-7 ***HR: 3***
o-(DIETHYLAMINOETHOXY)BENZANILIDE
mf: $C_{19}H_{24}N_2O_2$ mw: 312.45

SYNS: o-DIAETHYLAMINOAETHOXY-BENZANILID (GERMAN) ◇ 2-(2-(DIETHYLAMINO)ETHOXY)BENZANILIDE

TOXICITY DATA with REFERENCE
scu-mus TDLo:1425 mg/kg (female 1-19D post):REP ARZNAD 18,658,68
scu-mus TDLo:950 mg/kg (female 1-19D post):TER ARZNAD 18,658,68
orl-mus LD50:335 mg/kg ARZNAD 16,1127,66
scu-mus LD50:225 mg/kg ARZNAD 18,658,68
ivn-mus LD50:29 mg/kg ARZNAD 16,1127,66

SAFETY PROFILE: Poison by ingestion, subcutaneous, and intravenous routes. An experimental teratogen. Other experimental reproductive effects. When heated to decomposition it emits toxic fumes of NO_x.

DHP400 CAS:5014-35-7 ***HR: 3***
2-(2-(DIETHYLAMINO)ETHOXY)-5-BROMO-BENZANILIDE
mf: $C_{19}H_{23}BrN_2O_2$ mw: 391.35

SYN: 5-BROMO-2-(2-(DIETHYLAMINO)ETHOXY)BENZANILIDE

TOXICITY DATA with REFERENCE
scu-mus TDLo:950 mg/kg (female 1-19D post):TER ARZNAD 18,658,68
scu-mus TDLo:285 mg/kg (1-19D preg):REP ARZNAD 18,658,68
ipr-mus LD50:250 mg/kg BCFAAI 101,785,62
scu-mus LD50:830 mg/kg ARZNAD 18,658,68

SAFETY PROFILE: Poison by intraperitoneal route. Moderately toxic by subcutaneous route. Experimental teratogenic and reproductive effects. When heated to decomposition it emits toxic fumes of Br^- and NO_x.

DHP450 CAS:17822-72-9 ***HR: 3***
2-(2-(DIETHYLAMINO)ETHOXY)-2'-CHLORO-BENZANILIDE
mf: $C_{19}H_{23}ClN_2O_2$ mw: 346.89

SYN: BENZANILIDE, 2'-CHLORO-2-(2-(DIETHYLAMINO)EHOXY)-

TOXICITY DATA with REFERENCE
scu-mus TDLo:950 mg/kg (female 1-19D post):TER ARZNAD 18,658,68
scu-mus LD50:215 mg/kg ARZNAD 18,658,68

SAFETY PROFILE: Poison by subcutaneous route. An experimental teratogen. When heated to decomposition it emits toxic fumes of NO_x and Cl^-.

DHP500 CAS:17822-73-0 ***HR: 3***
2-(2-(DIETHYLAMINO)ETHOXY)-3'-CHLORO-BENZANILIDE
mf: $C_{19}H_{23}ClN_2O_2$ mw: 346.89

TOXICITY DATA with REFERENCE
scu-mus TDLo:570 mg/kg (1-19D preg):REP ARZNAD 18,658,68
scu-mus TDLo:TDLo:950 mg/kg (1-19D preg):TER ARZNAD 18,658,68
scu-mus LD50:380 mg/kg ARZNAD 18,658,68

SAFETY PROFILE: Poison by subcutaneous route. Experimental teratogenic and reproductive effects. When heated to decomposition it emits toxic fumes of Cl^- and NO_x.

DHP550 CAS:17822-71-8 ***HR: 3***
2-(2-(DIETHYLAMINO)ETHOXY)-4'-CHLORO-BENZANILIDE
mf: $C_{19}H_{23}ClN_2O_2$ mw: 346.89

SYN: BENZANILIDE, 4'-CHLORO-2-(2-(DIETHYLAMINO)ETHOXY)-

TOXICITY DATA with REFERENCE
scu-mus TDLo:570 mg/kg (female 1-19D post):REP ARZNAD 18,658,68
scu-mus LD50:260 mg/kg ARZNAD 18,658,68

SAFETY PROFILE: Poison by subcutaneous route. Experimental reproductive effects. When heated to decomposition it emits toxic fumes of NO_x and Cl^-.

DHP600 CAS:26270-62-2 ***HR: 3***
6-(2-(DIETHYLAMINOETHOXY)-4,7-DIMETHOXY-5-CINNAMOYLBENZOFURANMALEATE
mf: $C_{25}H_{29}NO_5 \cdot C_4H_4O_4$ mw: 539.63

TOXICITY DATA with REFERENCE
orl-mus LD50:150 mg/kg CHTPBA 8,479,73
ivn-mus LD50:11 mg/kg CHTPBA 8,479,73

SAFETY PROFILE: Poison by ingestion and intravenous routes. When heated to decomposition it emits toxic fumes of NO_x.

DHQ000 CAS:41226-18-0 ***HR: 3***
6-(2-DIETHYLAMINOETHOXY)-4,7-DIMETHOXY-5-(p-METHOXYCINNAMOYL)BENZOFURAN OXALATE
mf: $C_{26}H_{31}NO_6 \cdot C_2H_2O_4$ mw: 543.62

TOXICITY DATA with REFERENCE
orl-mus LD50:240 mg/kg CHTPBA 8,479,73
ivn-mus LD50:15 mg/kg CHTPBA 8,479,73

SAFETY PROFILE: Poison by ingestion and intravenous routes. When heated to decomposition it emits toxic fumes of NO_x. See also OXALATES.

DHQ200 CAS:468-61-1 ***HR: 3***
2-(2-DIETHYLAMINOETHOXY)ETHYL-2-ETHYL-2-PHENYLBUTYRATE
mf: $C_{20}H_{33}NO_3$ mw: 335.54

SYNS: 2-(2-(DIETHYLAMINO)ETHOXY)ETHANOL-2-ETHYL-2-PHENYLBUTYRATE ◇ 2-(2-DIETHYLAMINOETHOXY)ETHYL α,α-DIETHYLPHENYLACETATE ◇ α,α-DIETHYLBENZENEACETIC ACID 2-(2-(DIETHYLAMINO)ETHOXY)ETHYL ESTER ◇ 2-ETHYL-2-PHENYLBUTYRIC ACID 2-(2-DIETHYLAMINOETHOXY)ETHYL ESTER ◇ OXELADIN ◇ PECTAMOL

TOXICITY DATA with REFERENCE
orl-rat LD50:183 mg/kg JPPMAB 9,446,57
orl-mus LD50:130 mg/kg JPPMAB 9,446,57
ivn-mus LD50:13 mg/kg JPPMAB 9,446,57

SAFETY PROFILE: Poison by ingestion and intravenous routes. When heated to decomposition it emits toxic fumes of NO_x.

DHQ500 CAS:1045-21-2 ***HR: 3***
DIETHYLAMINOETHOXYETHYL-1-PHENYL-1-CYCLOPENTANE CARBOXYLATE HYDROCHLORIDE
mf: $C_{20}H_{31}NO_3 \cdot ClH$ mw: 369.98

SYN: 1-PHENYLCYCLOPENTANECARBOXYLIC ACID (2-(2-(DIETHYLAMINO)ETHOXY)ETHYL) ESTER HYDROCHLORIDE

TOXICITY DATA with REFERENCE
orl-rat LD50:830 mg/kg AIPTAK 103,200,55
ivn-rat LD50:25 mg/kg AIPTAK 103,200,55
orl-mus LD50:230 mg/kg AIPTAK 103,200,55
ivn-mus LD50:26500 μg/kg AIPTAK 103,200,55

SAFETY PROFILE: Poison by ingestion and intravenous routes. When heated to decomposition it emits toxic fumes of NO_x and HCl. See also ESTERS.

DHQ600 CAS:75348-40-2 ***HR: 3***
6-(2-DIETHYLAMINOETHOXY)-N-(o-METHOXYPHENYL)NICOTINAMIDE HYDROCHLORIDE
mf: $C_{19}H_{25}N_3O_3 \cdot ClH$ mw: 379.93

TOXICITY DATA with REFERENCE
orl-mus LD50:500 mg/kg CHTPBA 8,226,73
ivn-mus LDLo:100 mg/kg CHTPBA 8,226,73

SAFETY PROFILE: Poison by intravenous route. Moderately toxic by ingestion. When heated to decomposition it emits very toxic fumes of HCl and NO_x.

DHQ800 CAS:17822-74-1 ***HR: 3***
2-(2-(DIETHYLAMINO)ETHOXY)-3-METHYLBENZANILIDE
mf: $C_{20}H_{25}N_2O_2$ mw: 325.47

TOXICITY DATA with REFERENCE
scu-mus TDLo:570 mg/kg (1-19D preg):REP ARZNAD 18,658,68
scu-mus TDLo:570 mg/kg (1-19D preg):TER ARZNAD 18,658,68
scu-mus LD50:285 mg/kg ARZNAD 18,658,68

SAFETY PROFILE: Poison by subcutaneous route. An experimental teratogen. When heated to decomposition it emits toxic fumes of NO_x.

DHR000 CAS:5372-13-4 ***HR: 3***
5-(2-(DIETHYLAMINO)ETHOXY)-3-METHYL-1-PHENYLPYRAZOLE
mf: $C_{16}H_{23}N_3O$ mw: 273.42

SYN: P-314

TOXICITY DATA with REFERENCE
orl-mus LD50:416 mg/kg ARZNAD 17,214,67
scu-mus LD50:306 mg/kg ARZNAD 17,214,67

SAFETY PROFILE: Poison by ingestion and subcutaneous routes. When heated to decomposition it emits toxic fumes of NO_x.

DHR800 CAS:101692-44-8 ***HR: 3***
1-DIETHYLAMINO-3-(p-ETHOXYPHENYL) INDAN CITRATE
mf: $C_{21}H_{27}NO \cdot C_6H_8O_7$ mw: 501.63

SYNS: 1-p-AETHOXYPHENYL-3-DIAETHYLAMINO-INDANCITRAT (GERMAN) ◇ LABOR-NR 2683

TOXICITY DATA with REFERENCE
orl-rat LD50:860 mg/kg ARZNAD 11,915,61
ivn-rat LD50:97 mg/kg ARZNAD 11,915,61
orl-mus LD50:347 mg/kg ARZNAD 11,915,61
ivn-mus LD50:137 mg/kg ARZNAD 11,915,61

SAFETY PROFILE: Poison by ingestion and intravenous routes. When heated to decomposition it emits toxic fumes of NO_x.

DHS000 CAS:67-98-1 ***HR: 3***
(p-2-DIETHYLAMINOETHOXYPHENYL)-1-PHENYL-2-p-ANISYLETHANOL
mf: $C_{27}H_{33}NO_3$ mw: 419.61

SYNS: 1-(p-2-DIETHYLAMINOETHOXYPHENYL)-1-PHENYL-2-p-ANISYLETHANOL ◇ 1-(4-(2-DIETHYLAMINOETHOXY)PHENYL)-1-PHENYL-2-(p-ANISYL)ETHANOL ◇ 1-(p-(2-(DIETHYLAMINO)ETHOXY)PHENYL)-1-PHENYL-2-(p-METHOXYPHENYL)ETHANOL ◇ ETHAMOXYTRIPHETOL ◇ ETHANOXYTRIPHETOL ◇ MER 25

TOXICITY DATA with REFERENCE
unr-rbt TDLo:150 mg/kg (female 21-26D post):REP JRPMAP 4,137,70
orl-rbt TDLo:25 mg/kg (female 1D post):TER ENDOAO 65,339,59
scu-mus TDLo:120 mg/kg (female 12-15D post):ETA,TER IRLCDZ 4,379,76
scu-mus TDLo:30 mg/kg/5D-I:ETA IRLCDZ 4,379,76
orl-mus LD50:1700 mg/kg ENDOAO 63,295,58

SAFETY PROFILE: Moderately toxic by ingestion. Questionable carcinogen with experimental tumorigenic data. Experimental teratogenic and reproductive effects. When heated to decomposition it emits toxic fumes of NO_x.

DHS200 CAS:2192-21-4 ***HR: 3***
1-(2-(2-(DIETHYLAMINO)ETHOXY)PHENYL)-3-PHENYL-1-PROPANONE HYDROCHLORIDE
mf: $C_{21}H_{27}NO_2 \cdot ClH$ mw: 361.95

SYNS: ASAMEDOL ◇ BAXACOR ◇ CORODILAN ◇ DIALICOR ◇ (o-β-DIETHYLAMINOETHOXY)-PHENYL PROPIOPHENONE HYDROCHLORIDE ◇ ETAFENONE HYDROCHLORIDE ◇ HETAPHENONE ◇ L.G. 11,457 HYDROCHLORIDE ◇ PAGANO-COR ◇ β-PHENYL-o-(DIETHYLAMINOETHOXY)PROPIOPHENONE HYDROCHLORIDE ◇ RELICOR ◇ RELICOR HYDROCHLORIDE

TOXICITY DATA with REFERENCE
orl-rat LD50:716 mg/kg ARZNAD 19,1664,69
ivn-rat LD50:20800 μg/kg ARZNAD 19,1664,69
orl-mus LD50:352 mg/kg NIIRDN 6,109,82
orl-dog LD50:50 mg/kg ARZNAD 19,1664,69

SAFETY PROFILE: Poison by ingestion and intravenous routes. When heated to decomposition it emits very toxic fumes of NO_x and HCl.

DHS400 CAS:17066-89-6 ***HR: 3***
2-(2-(DIETHYLAMINO)ETHOXY)-N-(2,6-XYLYL) CINCHONINAMIDE, HYDROCHLORIDE
mf: $C_{24}H_{29}N_3O_2 \cdot ClH$ mw: 428.02

SYNS: C 3223 ◇ 2-(2-(DIETHYLAMINO)ETHOXY)-N-(2,6-XYLYL)-4-QUINOLINECARBOXAMIDE, HYDROCHLORIDE

TOXICITY DATA with REFERENCE
eye-rbt 2% SEV ARZNAD 8,708,58
ipr-rat LD50:175 mg/kg ARZNAD 8,708,58
scu-mus LD50:465 mg/kg ARZNAD 8,708,58

SAFETY PROFILE: Poison by intraperitoneal route. Moderately toxic by subcutaneous route. A severe eye irritant. When heated to decomposition it emits very toxic fumes of NO_x and HCl.

DHS600 CAS:102207-85-2 ***HR: 3***
2-(N'-(2-(DIETHYLAMINO)ETHYL)ACETAMIDO)-2',6'-ACETOXYLIDIDE HYDROCHLORIDE
mf: $C_{18}H_{29}N_3O_2 \cdot ClH$ mw: 355.96

SYN: C 3150

TOXICITY DATA with REFERENCE
ipr-rat LD50:92 mg/kg ARZNAD 9,167,59
scu-mus LD50:212 mg/kg ARZNAD 9,167,59

SAFETY PROFILE: Poison by intraperitoneal and subcutaneous routes. When heated to decomposition it emits very toxic fumes of NO_x and HCl.

DHS800 CAS:77966-71-3 ***HR: 3***
2-(DIETHYLAMINO)-N-ETHYL-o-ACETOTOLUIDIDE HYDROCHLORIDE
mf: $C_{15}H_{24}N_2O \cdot ClH$ mw: 284.87

SYN: C 5123

TOXICITY DATA with REFERENCE
ipr-rat LD50:140 mg/kg ARZNAD 8,708,58
scu-mus LD50:170 mg/kg ARZNAD 8,708,58

SAFETY PROFILE: Poison by intraperitoneal and subcutaneous routes. When heated to decomposition it emits very toxic fumes of NO_x and HCl.

DHT000 CAS:77966-80-4 ***HR: 3***
N-(2-(DIETHYLAMINO)ETHYL)-2',6'-ACETOXYLIDIDE HYDROCHLORIDE
mf: $C_{16}H_{26}N_2O \cdot ClH$ mw: 298.90

SYN: C 3137

TOXICITY DATA with REFERENCE
ipr-rat LD50:125 mg/kg ARZNAD 8,708,58
scu-mus LD50:190 mg/kg ARZNAD 8,708,58

SAFETY PROFILE: Poison by intraperitoneal and subcutaneous routes. When heated to decomposition it emits very toxic fumes of NO_x and HCl.

DHT125 CAS:2426-54-2 ***HR: 3***
DIETHYLAMINOETHYL ACRYLATE
mf: $C_9H_{17}NO_2$ mw: 171.27

SYNS: ACRYLIC ACID-N,N-DIETHYLAMINOETHYL ESTER ◇ AGEFLEX FA-2 ◇ N,N-DIETHYLAMINOETHYL ACRYLATE ◇ β-DIETHYLAMINOETHYL ACRYLATE ◇ 2-(DIETHYLAMINO) ETHYL ACRYLATE

TOXICITY DATA with REFERENCE
skn-rbt 100 μg/24H open AIHAAP 23,95,62
eye-rbt 1 mg MLD UCDS** 12/30/71
orl-rat LD50:770 mg/kg UCDS** 9/15/64
ipr-mus LDLo:31 mg/kg CBCCT* 4,226,52
skn-rbt LD50:200 mg/kg UCDS** 12/30/71

CONSENSUS REPORTS: Reported in EPA TSCA Inventory.

SAFETY PROFILE: Poison by skin contact and intraperitoneal routes. Moderately toxic by ingestion. A skin and eye irritant. When heated to decomposition it emits toxic fumes of NO_x.

DHT200 CAS:54099-14-8 ***HR: 3***
N-(2-(DIETHYLAMINO)ETHYL)-1-ADAMANTANEACETAMIDE ETHYL IODIDE
mf: $C_{18}H_{32}N_2O•C_2H_5I$ mw: 448.49

TOXICITY DATA with REFERENCE
orl-mus LD50:600 mg/kg FRPSAX 32,129,77
ipr-mus LD50:35 mg/kg FRPSAX 32,129,77

SAFETY PROFILE: Poison by intraperitoneal route. Moderately toxic by ingestion. See also IODIDES. When heated to decomposition it emits very toxic fumes of NO_x and I^-.

DHT300 CAS:133-16-4 ***HR: 3***
2-(DIETHYLAMINO)ETHYL-4-AMINO-2-CHLOROBENZOATE
mf: $C_{13}H_{19}ClN_2O_2$ mw: 270.79

SYN: CHLOROPROCAINE

TOXICITY DATA with REFERENCE
ipr-mus LD50:266 mg/kg TXAPA9 54,501,80
scu-mus LD50:1069 mg/kg ANESAV 54,177,81
ivn-gpg LDLo:64 mg/kg JMPCAS 3,525,61

SAFETY PROFILE: Poison by intravenous and intraperitoneal routes. Moderately toxic by subcutaneous route. When heated to decomposition it emits toxic fumes of NO_x and Cl^-.

DHT400 CAS:52401-04-4 ***HR: 3***
2-(2-(2-DIETHYLAMINO)ETHYLAMINO)ETHYL-2,5-DIMETHYL-1,3-BENZODIOXOLE DIMALEATE
mf: $C_{17}H_{28}N_2O_2•2C_4H_4O_4$ mw: 524.63

TOXICITY DATA with REFERENCE
ivn-rat LD50:40 mg/kg EJMCA5 12,413,77
ipr-mus LD50:150 mg/kg EJMCA5 12,413,77

SAFETY PROFILE: Poison by intravenous and intraperitoneal routes. When heated to decomposition it emits toxic fumes of NO_x.

DHT600 CAS:52400-99-4 ***HR: 3***
2-(2-(2-(DIETHYLAMINO)ETHYLAMINO)ETHYL)-2-METHYL-1,3-BENZODIOXOLE, DIMALEATE
mf: $C_{16}H_{26}N_2O_2•2C_4H_4O_4$ mw: 510.60

TOXICITY DATA with REFERENCE
ivn-rat LD50:30 mg/kg EJMCA5 12,413,77
ipr-mus LD50:98 mg/kg EJMCA5 12,413,77

SAFETY PROFILE: Poison by intravenous and intraperitoneal routes. When heated to decomposition it emits toxic fumes of NO_x.

DHT800 CAS:52400-82-5 ***HR: 3***
2-(2-(DIETHYLAMINO)ETHYLAMINOMETHYL)-2-METHYL-1,3-BENZODIOXOLE DIMALEATE
mf: $C_{15}H_{24}N_2O_2•2C_4H_4O_4$ mw: 496.57

TOXICITY DATA with REFERENCE
ivn-rat LD50:40 mg/kg EJMCA5 12,413,77
ipr-mus LD50:150 mg/kg EJMCA5 12,413,77

SAFETY PROFILE: Poison by intravenous and intraperitoneal routes. When heated to decomposition it emits toxic fumes of NO_x.

DHU000 CAS:479-50-5 ***HR: 3***
1-(2'-DIETHYLAMINO)ETHYLAMINO-4-METHYLTHIOXANTHENONE
mf: $C_{20}H_{24}N_2OS$ mw: 340.52

SYNS: 1-((2-(DIETHYLAMINO)ETHYL)AMINO)-4-METHYL-9H-THIOXANTHEN-9-ONE ◇ LUCANTHON ◇ LUCANTHONE ◇ MIRACIL D ◇ NILODIN

TOXICITY DATA with REFERENCE
mmo-sat 500 μmol/plate SCIEAS 186,647,74
mma-sat 50 μg/plate TXAPA9 52,237,80
dnd-hmn:hla 3 mg/L/15M ECREAL 103,175,76
cyt-hmn:lym 4 mg/L/24H MUREAV 55,43,78
cyt-hmn:leu 340 μg/L/24H MUREAV 55,43,78
ivn-mus LD50:56 mg/kg CSLNX* NX#01110
ims-mus LD50:400 mg/kg CSHCAL 4,445,77

CONSENSUS REPORTS: EPA Genetic Toxicology Program.

SAFETY PROFILE: Poison by intravenous route. Human mutation data reported. When heated to decomposition it emits very toxic fumes of NO_x and SO_x.

DHU400 CAS:65268-91-9 ***HR: 3***
2-(3-(2-(DIETHYLAMINO)ETHYLAMINO)PROPYL)-2-METHYL-1,3-BENZODIOXOLE DIMALEATE
mf: $C_{17}H_{28}N_2O_2 \cdot 2C_4H_4O_4$ mw: 524.63

TOXICITY DATA with REFERENCE
ivn-rat LD50:30 mg/kg EJMCA5 12,413,77
ipr-mus LD50:88 mg/kg EJMCA5 12,413,77

SAFETY PROFILE: Poison by intravenous and intraperitoneal routes. When heated to decomposition it emits toxic fumes of NO_x.

DHU600 CAS:52479-15-9 ***HR: 3***
2-(3-(2-(DIETHYLAMINO)ETHYLAMINO)PROPYL)-2-METHYL-1,3-NAPHTHOL(2,3-d) DIOXOLE DIHYDROCHLORIDE
mf: $C_{21}H_{30}N_2O_2 \cdot 2ClH$ mw: 415.45

TOXICITY DATA with REFERENCE
ivn-rat LD50:30 mg/kg EJMCA5 12,413,77
ipr-mus LD50:150 mg/kg EJMCA5 12,413,77

SAFETY PROFILE: Poison by intravenous and intraperitoneal routes. When heated to decomposition it emits very toxic fumes of NO_x and HCl.

DHU900 CAS:302-40-9 ***HR: 3***
DIETHYLAMINOETHYL BENZILATE
mf: $C_{20}H_{25}NO_3$ mw: 327.46

PROP: Crystals. Mp: 51°. Crystals from acetone, mp: 177-178°. Solubility in water (25°): 14.9/100 mL. Practically insol in ether.

SYNS: BENACTIZINA (ITALIAN) ◇ BENACTYZIN ◇ BENACTYZINE ◇ BENZILIC ACID-β-DIETHYLAMINOETHYL ESTER ◇ DIAZIL ◇ β-DIETHYLAMINOETHYL BENZILATE ◇ 2-(DIETHYLAMINO) ETHYL BENZILATE ◇ 2-(DIETHYLAMINO)ETHYL DIPHENYLGLYCOLATE ◇ DIPHENYLGLYCOLIC ACID 2-(DIETHYLAMINO)ETHYL ESTER ◇ α-HYDROXY-α-PHENYLBENZENEACETIC ACID-2-(DIETHYLAMINO)ETHYL ESTER

TOXICITY DATA with REFERENCE
cyt-ham-ipr 10 mg/kg ACNSAX 17,253,75
ims-rat LD50:135 mg/kg NTIS** AD-A099-745
ipr-mus LD50:100 mg/kg BCFAAI 111,293,72
scu-mus LD50:159 mg/kg RPTOAN 32,311,69

SAFETY PROFILE: Poison by subcutaneous, intramuscular, and intraperitoneal routes. Mutation data reported. When heated to decomposition it emits toxic fumes of NO_x.

DHV200 CAS:77985-23-0 ***HR: 3***
N-(2-DIETHYLAMINO)ETHYL)CARBAMIC ACID-6-CHLORO-o-TOLYL ESTER HYDROCHLORIDE
mf: $C_{14}H_{21}ClN_2O_2 \cdot ClH$ mw: 321.28

SYN: C 5307

TOXICITY DATA with REFERENCE
eye-rbt 2% MLD ARZNAD 8,708,58
ipr-rat LD50:52 mg/kg ARZNAD 8,708,58
scu-mus LD50:84 mg/kg ARZNAD 8,708,58

SAFETY PROFILE: Poison by intraperitoneal and subcutaneous routes. An eye irritant. When heated to decomposition it emits very toxic fumes of Cl^- and NO_x. See also CARBAMATES.

DHV400 CAS:77985-24-1 ***HR: 3***
N-(2-(DIETHYLAMINO)ETHYLCARBAMIC ACID MESITYL ESTER HYDROCHLORIDE
mf: $C_{16}H_{26}N_2O_2 \cdot ClH$ mw: 314.90

SYN: C 5308

TOXICITY DATA with REFERENCE
ipr-rat LD50:62 mg/kg ARZNAD 8,708,58
scu-mus LD50:178 mg/kg ARZNAD 8,708,58

SAFETY PROFILE: Poison by intraperitoneal and subcutaneous routes. See also CARBAMATES. When heated to decomposition it emits very toxic fumes of NO_x and HCl.

DHV600 CAS:77985-25-2 ***HR: 3***
N-(2-(DIETHYLAMINO)ETHYL)CARBAMIC ACID-2,6-XYLYL ESTER HYDROCHLORIDE
mf: $C_{15}H_{24}N_2O_2 \cdot ClH$ mw: 300.87

SYN: C 3221

TOXICITY DATA with REFERENCE
ipr-rat LD50:40 mg/kg ARZNAD 8,708,58
scu-mus LD50:136 mg/kg ARZNAD 8,708,58

SAFETY PROFILE: Poison by intraperitoneal and subcutaneous routes. See also CARBAMATES. When heated to decomposition it emits very toxic fumes of NO_x and HCl.

DHW200 CAS:902-83-0 ***HR: 3***
2-(DIETHYLAMINO)ETHYLCHLORODIPHENYLACETATE HYDROCHLORIDE
mf: $C_{20}H_{24}ClNO_2 \cdot ClH$ mw: 382.36

SYNS: 2-CHLORO-2,2-DIPHENYLACETICACID-2-(DIETHYLAMINO)ETHYL ESTER HYDROCHLORIDE ◇ DIAMINOPHEN ◇ DIAPHEN (NEUROPLEGIC)

TOXICITY DATA with REFERENCE
ipr-mus LD50:76 mg/kg JPETAB 74,274,42

SAFETY PROFILE: Poison by intraperitoneal route. When heated to decomposition it emits very toxic fumes of Cl^- and NO_x.

DHW400 CAS:3737-35-7 ***HR: 3***
2-DIETHYLAMINOETHYL CYCLOPENTYL(2-THIENYL)GLYCOLATE HYDROCHLORIDE
mf: $C_{17}H_{27}NO_3S•ClH$ mw: 361.97

SYNS: α-CYCLOPENTYL-2-THIOPHENEGLYCOLIC ACID-2-(DIETHYLAMINO)ETHYL ESTER HYDROCHLORIDE ◇ WIN-2299 HYDROCHLORIDE

TOXICITY DATA with REFERENCE
ivn-hmn TDLo:29 μg/kg:EYE,CNS AJPSAO 113,887,57
orl-mus LD50:770 mg/kg JPETAB 110,282,54
ivn-mus LD50:80 mg/kg JPETAB 110,282,54

SAFETY PROFILE: Poison by intravenous route. Moderately toxic by ingestion. Human systemic effects by intravenous route: dilation of the pupils, somnolence, and hallucinations or distorted perceptions. When heated to decomposition it emits very toxic fumes of HCl, SO_x, and NO_x.

DHW600 CAS:9015-73-0 ***HR: 3***
DIETHYLAMINOETHYL-DEXTRAN

SYNS: DEAE-D ◇ DIETHYLAMINOETHYLDEXTRAN POLYMER

TOXICITY DATA with REFERENCE
scu-mus TDLo:1200 mg/kg/30W-I:NEO JNCIAM 50,387,73

SAFETY PROFILE: Questionable carcinogen with experimental neoplastigenic data. When heated to decomposition it emits toxic fumes of NO_x. See also various dextrans.

DHX600 CAS:77945-09-6 ***HR: 3***
N-(2-(DIETHYLAMINO)ETHYL)-2',6'-DIMETHYLCROTONANILIDE HYDROCHLORIDE
mf: $C_{18}H_{28}N_2O•ClH$ mw: 324.94

SYNS: C 3141 ◇ N-(2-(DIETHYLAMINO)ETHYL)-2',6'-CROTONXYLILIDE HYDROCHLORIDE

TOXICITY DATA with REFERENCE
eye-rbt 2% MLD ARZNAD 8,708,58
ipr-rat LD50:58 mg/kg ARZNAD 8,708,58
scu-mus LD50:70 mg/kg ARZNAD 8,708,58

SAFETY PROFILE: Poison by intraperitoneal and subcutaneous routes. An eye irritant. When heated to decomposition it emits very toxic fumes of HCl and NO_x.

DHX800 CAS:64-95-9 ***HR: 3***
2-DIETHYLAMINOETHYL DIPHENYLACETATE
mf: $C_{20}H_{25}NO_2$ mw: 311.46

SYNS: ADIPHENIN ◇ ADIPHENINE ◇ BENZENEACETIC ACID, α-PHENYL-, 2-(DIETHYLAMINO)ETHYL ESTER, (9CI) ◇ 2-DIETHYLAMINOETHYLESTER KYSELINY DIFENYLOCTOVE ◇ DIFACIL ◇ DIPHACIL ◇ DIPHACYL ◇ DIPHENYLACETIC ACID DIETHYLAMINOETHYL ESTER ◇ DIPHENYLACETIC ACID, 2-(DIETHYLAMINO)ETHYL ESTER ◇ DIPHENYLACETYLDIETHYLAMINOETHANOL ◇ ESTER DWUETYLOAMINOETYLOWY KWASU DWUFENYLOOCTOWEGO ◇ PATROVINE ◇ SPASMOLYTIN ◇ TRANSENTINE ◇ TRANZETIL ◇ TRASENTIN ◇ TRASENTINE ◇ TRAZENTYNA ◇ VEGANTINE ◇ WEGANTYNA

TOXICITY DATA with REFERENCE
orl-rat LDLo:1600 mg/kg APPNAH 1,4,50
scu-rat LDLo:1600 mg/kg APPNAH 1,4,50
ivn-rat LD50:27 mg/kg AJDDAL 18,241,51
orl-mus LD50:600 mg/kg CLDND* 2,201,68
scu-mus LD50:400 mg/kg BJPCAL 14,559,59
ivn-mus LD50:21500 μg/kg JPETAB 104,269,52
ivn-dog LD50:35 mg/kg JPETAB 89,131,47
ivn-rbt LD50:30 mg/kg CLDND*

SAFETY PROFILE: Poison by intravenous and intraperitoneal routes. Moderately toxic by ingestion and subcutaneous routes. See also ESTERS. When heated to decomposition it emits toxic fumes of NO_x.

DHY200 CAS:668-37-1 ***HR: 3***
2-DIETHYLAMINOETHYLDIPHENYLCARBAMATE HYDROCHLORIDE
mf: $C_{19}H_{24}N_2O_2•ClH$ mw: 348.91

SYNS: DIAETHYLAMINOAETHYL DIPHENYLCARBAMAT HYDROCHLORID (GERMAN) ◇ DIAMFEN HYDROCHLORIDE ◇ DIETHYLAMINOETHYLDIPHENYL CARBAMATE HYDROCHLORIDE ◇ DIETHYL(2-HYDROXYETHYL)AMMONIUM CHLORIDE DIPHENYLCARBAMATE ◇ DIPHENYLCARBAMIC ACID-2-(DIETHYLAMINO)ETHYL ESTER HYDROCHLORIDE ◇ 1352 HC ◇ SD 25

TOXICITY DATA with REFERENCE
orl-rat LD50:710 mg/kg ARZNAD 15,534,65
ivn-rat LD50:27 mg/kg ARZNAD 15,534,65
orl-mus LD50:270 mg/kg ARZNAD 15,534,65
ipr-mus LD50:76 mg/kg ARZNAD 15,534,65
scu-mus LD50:180 mg/kg AIPTAK 125,311,60
ivn-mus LD50:41 mg/kg ARZNAD 15,534,65

SAFETY PROFILE: Poison by ingestion, subcutaneous, intravenous, and intraperitoneal routes. When heated to decomposition it emits very toxic fumes of NO_x, NH_3, and HCl. See also CARBAMATES.

DHY300 ***HR: 3***
3-(2-(DIETHYLAMINO)ETHYL)-5,5-DIPHENYLHYDANTOIN
mf: $C_{21}H_{25}N_2O_2$ mw: 337.44

SYN: DIETHYLAMINOETHYLDIPHENYLHYDANTOINE(FRENCH)

TOXICITY DATA with REFERENCE
orl-rat LD50:1300 mg/kg THERAP 11,1159,56
scu-mus LD50:1300 mg/kg THERAP 11,1159,56
ivn-mus LD50:37 mg/kg THERAP 11,1159,56

SAFETY PROFILE: Poison by intravenous route. Moderately toxic by ingestion and subcutaneous routes. When heated to decomposition it emits toxic fumes of NO_x.

DHY400 CAS:548-68-5 ***HR: 3***
S-(2-(DIETHYLAMINO)ETHYL)DIPHENYLTHIOACETIC ACID HYDROCHLORIDE
mf: $C_{20}H_{25}NOS{\cdot}ClH$ mw: 363.98

SYNS: 23B ◇ S-(2-(DIETHYLAMINO)ETHYL)DIPHENYLTHIOACETIC ACID HYDROCHLORIDE ◇ β-DIETHYLAMINOETHYL DIPHENYLTHIOACETATE HYDROCHLORIDE ◇ 2-(DIETHYLAMINO)ETHYL DIPHENYLTHIOACETATE HYDROCHLORIDE ◇ 2-DIETHYLAMINOETHYL DIPHENYLTHIOLACETATE HYDROCHLORIDE ◇ DIPHENYLTHIOLACETIC ACID-2-DIETHYLAMINOETHYL ESTER HYDROCHLORIDE ◇ THIPHEN ◇ THIPHENAMIL HYDROCHLORIDE ◇ TIFEN ◇ TIPHEN ◇ TROCINAT ◇ TROCINATE ◇ TROCINATE HYDROCHLORIDE

TOXICITY DATA with REFERENCE
orl-rat LD50:2720 mg/kg 29ZVAB -,117,69
ipr-rat LD50:1500 mg/kg JPETAB 89,131,47
imp-rat LD50:1500 mg/kg 29ZVAB -,117,69
orl-mus LD50:443 mg/kg JPETAB 89,131,47
ipr-mus LD50:187 mg/kg JPETAB 89,131,47
ivn-mus LD50:30 mg/kg JPETAB 89,131,47
orl-dog LD50:1500 mg/kg 29ZVAB -,117,69
ivn-dog LD50:30 mg/kg 29ZVAB -,117,69
ivn-rbt LD50:19 mg/kg 29ZVAB -,117,69

SAFETY PROFILE: Poison by intravenous and intraperitoneal routes. Moderately toxic by ingestion and implant. See also ESTERS. When heated to decomposition it emits very toxic fumes of NO_x, SO_x, and HCl.

DHZ000 CAS:41542-50-1 ***HR: 2***
S-DIETHYLAMINOETHYL ESTER-p-BROMOTHIOBENZO HYDROXIMIC ACID HYDROCHLORIDE
mf: $C_{13}H_{19}BrNOS{\cdot}ClH$ mw: 353.76

SYNS: p-BROMOBENZOYLTHIOHYDROXIMIC ACID-5-DIETHYLAMINOETHYL ESTER HYDROCHLORIDE ◇ p-BROMOTHIOBENZOHYDROXIMIC ACID-S-DIETHYLAMINOETHYL ESTER HYDROCHLORIDE ◇ DIETHYXIME

TOXICITY DATA with REFERENCE
ims-rat LD50:920 mg/kg RPTOAN 38,105,75
ims-mus LD50:810 mg/kg RPTOAN 38,105,75
ims-cat LD50:500 mg/kg RPTOAN 38,105,75
ims-rbt LD50:1 g/kg BEXBAN 83,32,77

SAFETY PROFILE: Moderately toxic by intramuscular route. When heated to decomposition it emits toxic fumes of Br^-, SO_x, NO_x, and HCl. See also ESTERS.

DHZ100 CAS:4929-77-5 ***HR: 3***
1-(2-(DIETHYLAMINO)ETHYL)-2-((p-ETHOXYPHENYL)THIO)BENZIMIDAZOLE HYDROCHLORIDE
mf: $C_{21}H_{27}N_3OS{\cdot}ClH$ mw: 406.03

TOXICITY DATA with REFERENCE
orl-mus LD50:230 mg/kg YKKZAJ 87,296,67
ipr-mus LD50:84 mg/kg YKKZAJ 85,962,65
scu-mus LD50:190 mg/kg YKKZAJ 87,296,67
ivn-mus LD50:24 mg/kg YKKZAJ 87,296,67

SAFETY PROFILE: Poison by ingestion, subcutaneous, intravenous, and intraperitoneal routes. When heated to decomposition it emits toxic fumes of SO_x, NO_x, and HCl.

DIA000 CAS:77945-03-0 ***HR: 3***
N-(2-(DIETHYLAMINO)ETHYL)-2-ETHOXY-N-(2,6-XYLYL)CINCHONINAMIDE HYDROCHLORIDE
mf: $C_{26}H_{33}N_3O_2{\cdot}ClH$ mw: 456.08

SYNS: C 3222 ◇ N-(2-(DIETHYLAMINO)ETHYL)-2-ETHOXY-N-(2,6-XYLYL)-4-QUINOLINECARBOXAMIDEHYDROCHLORIDE

TOXICITY DATA with REFERENCE
eye-rbt 2% SEV ARZNAD 8,708,58
ipr-rat LD50:150 mg/kg ARZNAD 8,708,58
scu-mus LD50:250 mg/kg ARZNAD 8,708,58

SAFETY PROFILE: Poison by intraperitoneal and subcutaneous routes. A severe eye irritant. When heated to decomposition it emits very toxic fumes of NO_x and HCl.

DIA400 CAS:102129-21-5 ***HR: 3***
3-(2-(DIETHYLAMINO)ETHYL)-5-(2-FURYL)-1-PHENYL-1H-PYRAZOLINE HYDROCHLORIDE
mf: $C_{19}H_{25}N_3O{\cdot}ClH$ mw: 347.93

SYN: 1-PHENYL-3-(β-DIAETHYLAMINO-AETHYL)-5-FURYL-PYRAZOLIN-HYDROCHLORID (GERMAN)

TOXICITY DATA with REFERENCE
ipr-mus LD50:95 mg/kg ARZNAD 10,925,60
scu-mus LD50:180 mg/kg ARZNAD 10,925,60

SAFETY PROFILE: Poison by intraperitoneal and subcutaneous routes. When heated to decomposition it emits very toxic fumes of NO_x and HCl.

DIA800 CAS:39367-89-0 ***HR: 3***
N-(2-(DIETHYLAMINO)ETHYL)-N-MESITYL-β-ISODURYLAMIDE HYDROCHLORIDE
mf: $C_{25}H_{36}N_2O{\cdot}ClH$ mw: 417.09

SYNS: C 3230 ◇ N-(2-(DIETHYLAMINO)ETHYL-2,2′,4,4′,6,6′-HEXAMETHYLBENZANILIDE HYDROCHLORIDE

TOXICITY DATA with REFERENCE
eye-rbt 2% MLD ARZNAD 8,708,58
ipr-rat LD50:210 mg/kg ARZNAD 8,708,58
scu-mus LD50:700 mg/kg ARZNAD 8,708,58

SAFETY PROFILE: Poison by intraperitoneal route. Moderately toxic by subcutaneous route. An eye irritant. When heated to decomposition it emits very toxic fumes of HCl and NO_x.

DIB000 CAS:78372-03-9 ***HR: 3***
1-(2-(DIETHYLAMINO)ETHYL)-3-MESITYL-1-METHYLUREA HYDROCHLORIDE
mf: $C_{17}H_{29}N_3O{\cdot}ClH$ mw: 327.95

SYN: C 3234

TOXICITY DATA with REFERENCE
eye-rbt 2% MLD ARZNAD 8,664,58
ipr-rat LD50:160 mg/kg ARZNAD 8,664,58
scu-mus LD50:287 mg/kg ARZNAD 8,664,58

SAFETY PROFILE: Poison by intraperitoneal and subcutaneous routes. An eye irritant. When heated to decomposition it emits very toxic fumes of NO_x and HCl.

DIB200 CAS:97702-94-8 ***HR: 3***
2-(DIETHYLAMINO)-N-ETHYL-N-(1-MESITYLOXY-2-PROPYL)ACETAMIDE HYDROCHLORIDE
mf: $C_{20}H_{34}N_2O_2{\cdot}ClH$ mw: 371.02

SYN: C 2046

TOXICITY DATA with REFERENCE
eye-rbt 2% MLD ARZNAD 9,70,59
scu-mus LD50:140 mg/kg ARZNAD 9,70,59

SAFETY PROFILE: Poison by subcutaneous route. An eye irritant. When heated to decomposition it emits very toxic fumes of NO_x and HCl.

DIB600 CAS:1227-61-8 ***HR: 3***
N-(2-(DIETHYLAMINO)ETHYL)-2-(p-METHOXYPHENOXY)ACETAMIDE
mf: $C_{15}H_{24}N_2O_3$ mw: 280.41

SYNS: MEPHEXAMIDE ◇ 2-(p-METHOXYPHENOXY)-N-(2-(DIETHYLAMINO)ETHYL)ACETAMIDE ◇ MEXEPHENAMIDE

TOXICITY DATA with REFERENCE
orl-mus LD50:1500 μg/kg CHTPBA 1,444,66
ivn-mus LD50:168 mg/kg 27ZQAG -,396,72
ivn-rbt LD50:135 mg/kg AMPYAT 123,141,65

SAFETY PROFILE: Poison by ingestion and intravenous routes. When heated to decomposition it emits toxic fumes of NO_x.

DIB800 CAS:97702-95-9 ***HR: 3***
2-(DIETHYLAMINO)-N-ETHYL-N-(1-(p-METHOXYPHENOXY)-2-PROPYL)ACETAMIDE HYDROCHLORIDE
mf: $C_{18}H_{30}N_2O_3{\cdot}ClH$ mw: 358.96

SYN: C 2098

TOXICITY DATA with REFERENCE
eye-rbt 2% MLD ARZNAD 9,70,59
scu-mus LD50:355 mg/kg ARZNAD 9,70,59

SAFETY PROFILE: Poison by subcutaneous route. An eye irritant. When heated to decomposition it emits very toxic fumes of NO_x and HCl.

DIC000 CAS:77791-67-4 ***HR: 3***
2-(2-(DIETHYLAMINO)ETHYL)METHYLAMINO-o-ACETANISIDIDE DIHYDROCHLORIDE
mf: $C_{16}H_{27}N_3O_2{\cdot}2ClH$ mw: 366.38

SYN: C 5416

TOXICITY DATA with REFERENCE
ipr-rat LD50:182 mg/kg ARZNAD 9,262,59
scu-mus LD50:850 mg/kg ARZNAD 9,262,59

SAFETY PROFILE: Poison by intraperitoneal route. Moderately toxic by subcutaneous route. When heated to decomposition it emits very toxic fumes of NO_x and HCl.

DIC200 CAS:77966-72-4 ***HR: 3***
2-(2-(DIETHYLAMINO)ETHYL)METHYLAMINO-o-ACETOTOLUIDIDE DIHYDROCHLORIDE
mf: $C_{16}H_{27}N_3O{\cdot}2ClH$ mw: 350.38

SYN: C 5346

TOXICITY DATA with REFERENCE
ipr-rat LD50:150 mg/kg ARZNAD 9,262,59
scu-mus LD50:800 mg/kg ARZNAD 9,262,59

SAFETY PROFILE: Poison by intraperitoneal route. Moderately toxic by subcutaneous route. When heated to decomposition it emits very toxic fumes of NO_x and HCl.

DIC400 CAS:77966-81-5 ***HR: 3***
2-(2-(DIETHYLAMINO)ETHYL)METHYLAMINO-2',6'-ACETOXYLIDIDE DIHYDROCHLORIDE
mf: $C_{17}H_{29}N_3O{\cdot}2ClH$ mw: 364.41

SYN: C 5342

TOXICITY DATA with REFERENCE
ipr-rat LD50:125 mg/kg ARZNAD 9,262,59
scu-mus LD50:500 mg/kg ARZNAD 9,262,59

SAFETY PROFILE: Poison by intraperitoneal route. Moderately toxic by subcutaneous route. When heated to decomposition it emits very toxic fumes of NO_x and HCl.

DIC600 CAS:52479-14-8 ***HR: 3***
2-((2-(DIETHYLAMINO)ETHYL)METHYLAMINO)ETHYL-2-METHYL-1,3-BENZODIOXOLE DIMALEATE
mf: $C_{17}H_{28}N_2O_2 \cdot 2C_4H_4O_4$ mw: 524.63

TOXICITY DATA with REFERENCE
ivn-rat LD50:8500 μg/kg EJMCA5 12,413,77
ipr-mus LD50:75 mg/kg EJMCA5 12,413,77

SAFETY PROFILE: Poison by intravenous and intraperitoneal routes. When heated to decomposition it emits toxic fumes of NO_x.

DIC800 CAS:65210-31-3 ***HR: 3***
2-(2-(DIETHYLAMINO)ETHYL)-2-METHYL-1,3-BENZODIOXOLE HYDROCHLORIDE
mf: $C_{14}H_{21}NO_2 \cdot ClH$ mw: 271.82

TOXICITY DATA with REFERENCE
ivn-rat LD50:45 mg/kg EJMCA5 12,413,77
ipr-mus LD50:100 mg/kg EJMCA5 12,413,77

SAFETY PROFILE: Poison by intravenous and intraperitoneal routes. When heated to decomposition it emits very toxic fumes of HCl and NO_x.

DID800 CAS:17617-13-9 ***HR: 3***
1-DIETHYLAMINOETHYL-2-METHYL-3-PHENYL-1,2,3,4-TETRAHYDRO-4-QUINAZOLINONE
mf: $C_{21}H_{27}N_3O \cdot C_2H_2O_4$ mw: 427.55

SYN: 2,3-DIHYDRO-1-DIETHYLAMINOETHYL-2-METHYL-3-PHENYL-4(3H)-QUINAZOLINONEOXALATE

TOXICITY DATA with REFERENCE
orl-mus LD50:678 mg/kg JMCMAR 11,788,68
ipr-mus LD50:180 mg/kg JMCMAR 11,788,68

SAFETY PROFILE: Poison by intraperitoneal route. Moderately toxic by ingestion. When heated to decomposition it emits toxic fumes of NO_x.

DIE200 CAS:34963-48-9 ***HR: 3***
1-DIETHYLAMINOETHYL-2-METHYL-1,2,3,4-TETRAHYDRO-3-(o-TOLYL)-4-QUINAZOLINONE OXALATE
mf: $C_{22}H_{29}N_3O \cdot C_2H_2O_4$ mw: 441.58

SYN: 2,3-DIHYDRO-1-DIETHYLAMINOETHYL-2-METHYL-3-(o-TOLYL)-4(3H)-QUINAZOLINONEOXALATE

TOXICITY DATA with REFERENCE
orl-mus LD50:620 mg/kg JMCMAR 11,788,68
ipr-mus LD50:156 mg/kg JMCMAR 11,788,68

SAFETY PROFILE: Poison by intraperitoneal route. Moderately toxic by ingestion. When heated to decomposition it emits toxic fumes of NO_x. See also OXALATES.

DIE300 ***HR: 3***
DIETHYLAMINOETHYLMORPHINE
mf: $C_{23}H_{32}N_2O_3$ mw: 384.57

SYN: 7,8-DIDEHYDRO-3-(2-(DIETHYLAMINO)ETHOXY)-4,5-α-EPOXY-17-METHYLMORPHINAN-6-α-OL

TOXICITY DATA with REFERENCE
ipr-mus LD50:40 mg/kg APFRAD 8,261,50
scu-mus LD50:110 mg/kg THERAP 7,21,52
ivn-mus LD50:30 mg/kg THERAP 7,21,52

SAFETY PROFILE: Poison by subcutaneous, intravenous, and intraperitoneal routes. When heated to decomposition it emits toxic fumes of NO_x. See also MORPHINE.

DIE400 CAS:97702-97-1 ***HR: 3***
2-(DIETHYLAMINO)-N-ETHYL-N-(1-PHENOXY-2-PROPYL)ACETAMIDE HYDROCHLORIDE
mf: $C_{17}H_{28}N_2O_2 \cdot ClH$ mw: 328.93

SYN: C 2053

TOXICITY DATA with REFERENCE
eye-rbt 2% MLD ARZNAD 9,70,59
scu-mus LD50:315 mg/kg ARZNAD 9,70,59

SAFETY PROFILE: Poison by subcutaneous route. An eye irritant. When heated to decomposition it emits very toxic fumes of NO_x and HCl.

DIE600 CAS:102128-91-6 ***HR: 3***
3-(DIETHYLAMINO)-N-ETHYL-N-(1-PHENOXY-2-PROPYL)PROPIONAMIDE HYDROCHLORIDE
mf: $C_{18}H_{30}N_2O_2 \cdot ClH$ mw: 342.96

SYN: C 2052

TOXICITY DATA with REFERENCE
eye-rbt 2% HLD ARZNAD 9,70,59
scu-mus LD50:387 mg/kg ARZNAD 9,70,59

SAFETY PROFILE: Poison by subcutaneous route. An eye irritant. When heated to decomposition it emits very toxic fumes of HCl and NO_x.

DIF200 CAS:6009-67-2 ***HR: 3***
3-(β-DIETHYLAMINOETHYL)-3-PHENYL-2-BENZOFURANONE HYDROCHLORIDE
mf: $C_{20}H_{23}NO_2 \cdot ClH$ mw: 345.90

SYNS: ABBOTT'S A.P. 43 ◇ AMETHONE HYDROCHLORIDE ◇ AMOLANONE HYDROCHLORIDE ◇ AP 43 ◇ 3-PHENYL-3-DIETHYLAMINOETHYLBENZOFURANONE-2-HYDROCHLORIDE

TOXICITY DATA with REFERENCE
orl-rat LD50:600 mg/kg JPETAB 84,387,45
orl-mus LD50:660 mg/kg JPETAB 84,387,45
ipr-mus LD50:205 mg/kg JPETAB 84,387,45
scu-mus LD50:395 mg/kg JPETAB 84,387,45
scu-rbt LD50:250 mg/kg JPETAB 84,387,45
ivn-rbt LD50:23 mg/kg JPETAB 84,387,45

SAFETY PROFILE: Poison by intraperitoneal, subcutaneous, and intravenous routes. Moderately toxic by ingestion. An anesthetic. When heated to decomposition it emits very toxic fumes of NO_x and HCl.

DIF600 CAS:14557-50-7 ***HR: 3***
2-(2-(DIETHYLAMINO)ETHYL)-2-PHENYL-4-PENTENOIC ACID ETHYL ESTER
mf: $C_{19}H_{29}NO_2$ mw: 303.49

SYN: UCB 6249

TOXICITY DATA with REFERENCE
orl-rat LD50:200 mg/kg 27ZQAG -,367,72
ipr-rat LD50:90 mg/kg 27ZQAG -,367,72
ivn-rat LD50:41 mg/kg 27ZQAG -,367,72
orl-mus LD50:200 mg/kg 27ZQAG -,367,72
ipr-mus LD50:60 mg/kg 27ZQAG -,367,72
ivn-mus LD50:28 mg/kg 27ZQAG -,367,72

SAFETY PROFILE: Poison by ingestion, intraperitoneal, and intravenous routes. See also ESTERS. When heated to decomposition it emits toxic fumes of NO_x.

DIG400 CAS:302-33-0 ***HR: D***
2-DIETHYLAMINOETHYLPROPYLDIPHENYL ACETATE
mf: $C_{23}H_{31}NO_2$ mw: 353.55

SYNS: BCTB ◇ 2-DIETHYLAMINOETHYL-2,2-DIPHENYLVALERATE ◇ 2,2-DIPHENYL-VALERIC ACID-2-(DIETHYLAMINO)ETHYL) ESTER ◇ HL 8727 ◇ NSC-39690 ◇ α-PHENYL-α-PROPYL-BENZENE-ACETIC ACID-S-(DIETHYLAMINO)ETHYL ESTER ◇ PROADIFEN ◇ RP5171 ◇ SKF-525-A

TOXICITY DATA with REFERENCE
ipr-mus TDLo:240 mg/kg (8-13D preg):REP JPETAB 202,455,77

SAFETY PROFILE: Experimental reproductive effects. When heated to decomposition it emits toxic fumes of NO_x.

DIG800 CAS:53-18-9 ***HR: D***
1-(2-(DIETHYLAMINO)ETHYL)RESERPINE
mf: $C_{39}H_{53}N_3O_9$ mw: 707.95

SYNS: BIETASERPINE ◇ N-DIAETHYLAMINOAETHYL-RESERPIN (GERMAN) ◇ DL 152

TOXICITY DATA with REFERENCE
orl-mus TDLo: mg/kg (7-12D preg):TER KSRNAM 4,2937,70

SAFETY PROFILE: An experimental teratogen. When heated to decomposition it emits toxic fumes of NO_x.

DIH000 CAS:1111-44-0 ***HR: 3***
1-(2-(DIETHYLAMINO)ETHYL)RESERPINE BITARTRATE
mf: $C_{39}H_{53}N_3O_9 \cdot 2C_4H_6O_6$ mw: 1008.15

SYNS: ◇ BIETASERPINE BITARTRATE ◇ D. L. 152 ◇ TENSIBAR

TOXICITY DATA with REFERENCE
orl-rat TDLo:24 g/kg (30D pre):REP KSRNAM 5,81,71
orl-rat LD50:4885 mg/kg KSRNAM 5,81,71
ipr-rat LD50:496 mg/kg KSRNAM 5,81,71
scu-rat LD50:888 mg/kg KSRNAM 5,81,71
orl-mus LD50:620 mg/kg ARZNAD 14,1040,64
ipr-mus LD50:430 mg/kg ARZNAD 14,1040,64
scu-mus LD50:1061 mg/kg KSRNAM 5,81,71
ivn-mus LD50:215 mg/kg ARZNAD 14,1040,64

SAFETY PROFILE: Poison by intravenous route. Moderately toxic by ingestion, subcutaneous, and intraperitoneal routes. An experimental teratogen. Experimental reproductive effects. An antihypertensive agent. When heated to decomposition it emits toxic fumes of NO_x.

DIH600 CAS:17140-68-0 ***HR: 3***
7-(2-(DIETHYLAMINO)ETHYL)THEOPHYLLINE HYDROCHLORIDE
mf: $C_{13}H_{21}N_5O_2 \cdot ClH$ mw: 315.85

SYNS: CHLORHYDRATE de DIETHYLAMINOETHYLTHEOPHYLLINE (FRENCH) ◇ 7-(β-DIAETHYLAMINO-AETHYL)-THEOPHYLLIN-HYDROCHLORID (GERMAN) ◇ 7-(2-(DIETHYLAMINO)ETHYL)-3,7-DIHYDRO-1,3-DIMETHYL-1H-PURINE-2,6-DIONEHYDROCHLORIDE ◇ 7-(2-(N,N-DIETHYLAMINO)ETHYL)THEOPHYLLINE HYDROCHLORIDE ◇ ETAMIPHYLLINE HYDROCHLORIDE ◇ ETAMIPHYLLIN HYDROCHLORIDE ◇ SOLUFILINA

TOXICITY DATA with REFERENCE
ipr-mus LD50:182 mg/kg ARZNAD 8,190,58
scu-mus LD50:182 mg/kg ARZNAD 4,649,54
ivn-mus LD50:127 mg/kg AIPTAK 103,146,55

SAFETY PROFILE: Poison by intravenous, intraperitoneal, and subcutaneous routes. When heated to decomposition it emits very toxic fumes of HCl and NO_x. See also THEOPHYLLIN.

DIH800 CAS:101670-51-3 ***HR: 3***
2-(2-DIETHYLAMINO)ETHYL)THIOACETANILIDE HYDROCHLORIDE
mf: $C_{14}H_{22}N_2OS \cdot ClH$ mw: 302.90

SYN: C 4928

TOXICITY DATA with REFERENCE
ipr-rat LD50:220 mg/kg ARZNAD 9,683,59
scu-mus LD50:1500 mg/kg ARZNAD 9,683,59

SAFETY PROFILE: Poison by intraperitoneal route. Moderately toxic by subcutaneous route. When heated to decomposition it emits very toxic fumes of NO_x, SO_x, and HCl.

DII000 CAS:102489-61-2 ***HR: 3***
2-(2-(DIETHYLAMINO)ETHYL)THIO-o-ACETO-TOLUIDIDE HYDROCHLORIDE
mf: $C_{15}H_{24}N_2OS \cdot ClH$ mw: 316.93

SYN: C 4924

TOXICITY DATA with REFERENCE
ipr-rat LD50:310 mg/kg ARZNAD 9,683,59
scu-mus LD50:1350 mg/kg ARZNAD 9,683,59

SAFETY PROFILE: Poison by intraperitoneal route. Moderately toxic by subcutaneous route. When heated to decomposition it emits very toxic fumes of NO_x, SO_x and HCl.

DII200 CAS:60-91-3 ***HR: 3***
N-(DIETHYLAMINOETHYL)THIODIPHENYL-AMINE
mf: $C_{18}H_{22}N_2S$ mw: 298.48

SYNS: ANTIPAR ◇ CASANTIN ◇ DIETHAZIN ◇ DIETHAZINE ◇ N-(2'-DIETHYLAMINOETHYL)DIBENZOPARATHIAZINE ◇ 10-(2-DIETHYLAMINOETHYL)PHENOTHIAZINE ◇ DINEZIN ◇ DIPARCOL ◇ DOLISINA ◇ DOLSIMA ◇ EAZAMINE ◇ FOURNEAU 2987 ◇ LATIBON ◇ LODIBON ◇ 2987 R.P. ◇ THIANTAN

TOXICITY DATA with REFERENCE
ivn-rat LD50:28500 μg/kg AIPTAK 103,371,55
orl-mus LD50:450 mg/kg AIPTAK 134,255,61
ipr-mus LD50:225 mg/kg ARZNAD 4,171,54
scu-mus LD50:450 mg/kg PSCBAY 2,17,63
ivn-mus LD50:65 mg/kg AIPTAK 120,450,59
ivn-rbt LD50:20 mg/kg AIPTAK 120,450,59

SAFETY PROFILE: Poison by intravenous and intraperitoneal routes. Moderately toxic by ingestion and subcutaneous routes. When heated to decomposition it emits very toxic fumes of NO_x and SO_x.

DII400 CAS:78109-87-2 ***HR: 3***
N-(2-(DIETHYLAMINO)ETHYL)-2,4,6-TRIMETHYLBENZAMIDE HYDROCHLORIDE
mf: $C_{16}H_{26}N_2O \cdot ClH$ mw: 298.90

SYNS: C 3235 ◇ N-(2-(DIETHYLAMINO)ETHYL)-β-ISODURYLAMIDE HYDROCHLORIDE

TOXICITY DATA with REFERENCE
eye-rbt 2% MLD ARZNAD 8,708,58
ipr-rat LD50:171 mg/kg ARZNAD 8,708,58
scu-mus LD50:390 mg/kg ARZNAD 8,708,58

SAFETY PROFILE: Poison by intraperitoneal and subcutaneous routes. An eye irritant. When heated to decomposition it emits very toxic fumes of HCl and NO_x.

DII600 CAS:78280-29-2 ***HR: 3***
N-(2-(DIETHYLAMINO)ETHYL)-2,6-XYLIDINE DIHYDROCHLORIDE
mf: $C_{14}H_{24}N_2 \cdot 2ClH$ mw: 293.32

SYN: C 3144

TOXICITY DATA with REFERENCE
eye-rbt 2% MLD ARZNAD 8,708,58
ipr-rat LD50:120 mg/kg ARZNAD 8,708,58
scu-mus LD50:400 mg/kg ARZNAD 8,708,58

SAFETY PROFILE: Poison by intraperitoneal and subcutaneous route. An eye irritant. When heated to decomposition it emits very toxic fumes of NO_x and HCl.

DII800 CAS:97702-98-2 ***HR: 3***
2-(DIETHYLAMINO)-N-ETHYL-N-(1-(2,4-XYLYL-OXY)-2-PROPYL)ACETAMIDE HYDROCHLORIDE
mf: $C_{19}H_{32}N_2O_2 \cdot ClH$ mw: 356.99

SYN: C 2039

TOXICITY DATA with REFERENCE
eye-rbt 2% MLD ARZNAD 9,70,59
scu-mus LD50:185 mg/kg ARZNAD 9,70,59

SAFETY PROFILE: Poison by subcutaneous route. An eye irritant. When heated to decomposition it emits very toxic fumes of NO_x and HCl.

DIJ000 CAS:97702-99-3 ***HR: 3***
2-(DIETHYLAMINO)-N-ETHYL-N-(1-(3,5-XYLYL-OXY)-2-PROPYL) ACETAMIDE HYDROCHLORIDE
mf: $C_{19}H_{32}N_2O_2 \cdot ClH$ mw: 356.99

SYN: C 2057

TOXICITY DATA with REFERENCE
eye-rbt 2% MLD ARZNAD 9,70,59
scu-mus LD50:112 mg/kg ARZNAD 9,70,59

SAFETY PROFILE: Poison by subcutaneous route. An eye irritant. When heated to decomposition it emits very toxic fumes of NO_x and HCl.

DIJ200 CAS:78372-04-0 ***HR: 3***
1-(2-(DIETHYLAMINO)ETHYL)-3-(2,6-XYLYL) UREA HYDROCHLORIDE
mf: $C_{15}H_{25}N_3O \cdot ClH$ mw: 299.89

SYN: C 3145

TOXICITY DATA with REFERENCE
ipr-rat LD50:260 mg/kg ARZNAD 8,664,58
scu-mus LD50:445 mg/kg ARZNAD 8,664,58

SAFETY PROFILE: Poison by intraperitoneal route. Moderately toxic by subcutaneous route. When heated to decomposition it emits very toxic fumes of NO_x and HCl.

DIK000 CAS:77-51-0 ***HR: 3***
4-(DIETHYLAMINO)-2-ISOPROPYL-2-PHENYLVALERONITRILE
mf: $C_{16}H_{24}N_2$ mw: 244.42

SYN: TAT-1

TOXICITY DATA with REFERENCE
ipr-mus LD50:77 mg/kg OYYAA2 2,323,68
scu-mus LD50:127 mg/kg OYYAA2 2,323,68

CONSENSUS REPORTS: Cyanide and its compounds are on the Community Right-To-Know List.

SAFETY PROFILE: Poison by intraperitoneal and subcutaneous routes. See also NITRILES. When heated to decomposition it emits very toxic fumes of NO_x and CN^-.

DIK200 CAS:97703-02-1 ***HR: 3***
2-(DIETHYLAMINO)-N-(1-MESITYLOXY-2-PROPYL)-N-METHYLACETAMIDE HYDROCHLORIDE
mf: $C_{19}H_{32}N_2O_2 \cdot ClH$ mw: 356.99

SYN: C 2047

TOXICITY DATA with REFERENCE
eye-rbt 2% MLD ARZNAD 9,70,59
scu-mus LD50:180 mg/kg ARZNAD 9,70,59

SAFETY PROFILE: Poison by subcutaneous route. An eye irritant. When heated to decomposition it emits very toxic fumes of NO_x and HCl.

DIK400 CAS:97703-04-3 ***HR: 2***
2-(DIETHYLAMINO)-N-(1-(p-METHOXYPHENOXY)-2-PROPYL)-N-METHYLACETAMIDE HYDROCHLORIDE
mf: $C_{17}H_{28}N_2O_3 \cdot ClH$ mw: 344.93

SYN: C 2097

TOXICITY DATA with REFERENCE
eye-rbt 2% MLD ARZNAD 9,70,59
scu-mus LD50:590 mg/kg ARZNAD 9,70,59

SAFETY PROFILE: Moderately toxic by subcutaneous route. An eye irritant. When heated to decomposition it emits very toxic fumes of NO_x and HCl.

DIK600 CAS:77966-73-5 ***HR: 3***
2-(DIETHYLAMINO)-N-METHYL-o-ACETOTOLUIDIDE HYDROCHLORIDE
mf: $C_{14}H_{22}N_2O \cdot ClH$ mw: 270.84

SYN: V 346

TOXICITY DATA with REFERENCE
ipr-rat LD50:260 mg/kg ARZNAD 8,609,58
ipr-mus LD50:200 mg/kg ARZNAD 8,609,58
scu-mus LD50:670 mg/kg ARZNAD 8,609,58

SAFETY PROFILE: Poison by intraperitoneal route. Moderately toxic by subcutaneous route. When heated to decomposition it emits very toxic fumes of NO_x and HCl.

DIK800 CAS:77966-83-7 ***HR: 3***
2-(DIETHYLAMINO)-N-METHYL-2',6'-ACETOXYLIDIDE HYDROCHLORIDE
mf: $C_{15}H_{24}N_2O \cdot ClH$ mw: 284.87

SYN: V 317

TOXICITY DATA with REFERENCE
eye-rbt 2% MLD ARZNAD 8,609,58
ipr-rat LD50:138 mg/kg ARZNAD 8,609,58
scu-mus LD50:280 mg/kg ARZNAD 8,609,58

SAFETY PROFILE: Poison by intraperitoneal and subcutaneous routes. An eye irritant. When heated to decomposition it emits very toxic fumes of NO_x and HCl.

DIL000 CAS:102259-67-6 ***HR: 3***
6-((4-(DIETHYLAMINO)-1-METHYLBUTYL)AMINO)-2-METHYL-4,5,8-TRIMETHOXYQUINOLINE-1,5-NAPHTHALENE DISULFONATE
mf: $C_{22}H_{35}N_3O_3 \cdot C_{10}H_8O_6S_2$ mw: 677.90

TOXICITY DATA with REFERENCE
ipr-mus LD50:60 mg/kg ARZNAD 20,1775,70
ivn-brd LD50:51 mg/kg ARZNAD 20,1775,70

SAFETY PROFILE: Poison by intraperitoneal and intravenous routes. When heated to decomposition it emits very toxic fumes of NO_x and SO_x.

DIL200 CAS:72820-33-8 ***HR: 3***
6((-4-(DIETHYLAMINO)-1-METHYLBUTYL)-5,8-DIMETHOXY-2,4-DIMETHYLQUINOLINE-1,5-NAPHTHALENE DISULFONATE
mf: $C_{22}H_{35}N_3O_2 \cdot C_{10}H_8O_6S_2$ mw: 661.90

TOXICITY DATA with REFERENCE
ipr-mus LD50:45 mg/kg ARZNAD 20,1775,70
ivn-brd LD50:47 mg/kg ARZNAD 20,1775,70

SAFETY PROFILE: Poison by intraperitoneal and in-

travenous routes. When heated to decomposition it emits very toxic fumes of NO_x and SO_x.

DIL400 CAS:91-44-1 ***HR: 3***
7-DIETHYLAMINO-4-METHYLCOUMARIN
mf: $C_{14}H_{17}NO_2$ mw: 231.32

SYNS: COUMARIN 1 ◇ 7-(DIETHYLAMINO)-4-METHYL-2H-1-BENZOPYRAN-2-ONE

TOXICITY DATA with REFERENCE
orl-rat LD50:5 g/kg MVCRB3 2,193,73
orl-mus LD50:2400 mg/kg MPHEAE 17,497,67
ivn-mus LD50:180 mg/kg CSLNX* NX#03230

CONSENSUS REPORTS: Reported in EPA TSCA Inventory.

SAFETY PROFILE: Poison by intravenous route. Moderately toxic by ingestion. When heated to decomposition it emits toxic fumes of NO_x.

DIL600 CAS:67210-66-6 ***HR: 2***
7-DIETHYLAMINO-4-METHYLCOUMARIN, HYDROGEN SULFATE
mf: $C_{14}H_{17}NO_2{\cdot}H_2O_4S$ mw: 329.40

TOXICITY DATA with REFERENCE
orl-rat LD50:1600 mg/kg MarJV# 29MAR77

CONSENSUS REPORTS: Reported in EPA TSCA Inventory.

SAFETY PROFILE: Moderately toxic by ingestion. When heated to decomposition it emits very toxic fumes of SO_x and NO_x.

DIM000 ***HR: 3***
2-(DIETHYLAMINO)-N-METHYL-N-(2-MESITYLOXYETHYL)ACETAMIDE HYDROCHLORIDE
mf: $C_{18}H_{30}N_2O_2{\cdot}ClH$ mw: 342.96

SYN: C 2060

TOXICITY DATA with REFERENCE
eye-rbt 2% MLD ARZNAD 8,761,58
scu-mus LD50:280 mg/kg ARZNAD 8,761,58

SAFETY PROFILE: Poison by subcutaneous route. An eye irritant. When heated to decomposition it emits very toxic fumes of NO_x and HCl.

DIM200 CAS:30185-90-1 ***HR: 2***
5-(DIETHYLAMINO)METHYL-3-(1-METHYL-5-NITROIMIDAZOL-2-YLMETHYLENE AMINO)-2-OXAZOLIDINONE HYDROCHLORIDE
mf: $C_{13}H_{20}N_6O_4{\cdot}ClH$ mw: 360.85

TOXICITY DATA with REFERENCE
orl-mus LD50:3400 mg/kg JMCMAR 14,94,71
ipr-mus LD50:970 mg/kg JMCMAR 14,94,71

SAFETY PROFILE: Moderately toxic by ingestion and intraperitoneal routes. When heated to decomposition it emits very toxic fumes of NO_x and HCl.

DIM400 CAS:16417-75-7 ***HR: 3***
2-(DIETHYLAMINO)-N-(2-METHYL-1-NAPHTHYL)ACETAMIDE HYDROCHLORIDE
mf: $C_{17}H_{22}N_2O{\cdot}ClH$ mw: 306.87

SYN: V331

TOXICITY DATA with REFERENCE
eye-rbt 2% MLD ARZNAD 8,609,58
ipr-rat LD50:167 mg/kg ARZNAD 8,609,58
scu-mus LD50:300 mg/kg ARZNAD 8,609,58

SAFETY PROFILE: Poison by intraperitoneal and subcutaneous routes. An eye irritant. When heated to decomposition it emits very toxic fumes of NO_x and HCl.

DIM600 CAS:30096-82-3 ***HR: 3***
4-DIETHYLAMINOMETHYL-2-(5-NITRO-2-THIENYL)THIAZOLE HYDROCHLORIDE
mf: $C_{12}H_{15}N_3O_2S_2{\cdot}ClH$ mw: 333.88

TOXICITY DATA with REFERENCE
orl-mus LD50:300 mg/kg JMCMAR 18,794,75
ipr-mus LD50:100 mg/kg JMCMAR 18,794,75

SAFETY PROFILE: Poison by ingestion and intraperitoneal routes. When heated to decomposition it emits very toxic fumes of NO_x, SO_x, and HCl.

DIM800 CAS:97703-08-7 ***HR: 3***
2-(DIETHYLAMINO)-N-METHYL-N-(2-PHENETHYL)ACETAMIDE HYDROCHLORIDE
mf: $C_{15}H_{24}N_2O{\cdot}CClH$ mw: 284.87

SYN: C 6608

TOXICITY DATA with REFERENCE
ipr-rat LD50:232 mg/kg ARZNAD 9,113,59
scu-mus LD50:630 mg/kg ARZNAD 9,113,59

SAFETY PROFILE: Poison by intraperitoneal route. Moderately toxic by subcutaneous route. When heated to decomposition it emits very toxic fumes of NO_x and HCl.

DIN000 CAS:102128-92-7 ***HR: 2***
3-(DIETHYLAMINO)-N-METHYL-N-(1-PHENOXY-2-PROPYL)PROPIONAMIDE HYDROCHLORIDE
mf: $C_{17}H_{28}N_2O_2{\cdot}ClH$ mw: 328.93

SYN: C 1863

TOXICITY DATA with REFERENCE
eye-rbt 2% MLD ARZNAD 9,70,59
scu-mus LD50:480 mg/kg ARZNAD 9,70,59

SAFETY PROFILE: Moderately toxic by subcutaneous route. An eye irritant. When heated to decomposition it emits very toxic fumes of HCl and NO_x.

DIN200 CAS:97703-11-2 ***HR: 3***
2-(DIETHYLAMINO)-N-METHYL-N-(3-PHENYLPROPYL)ACETAMIDE HYDROCHLORIDE
mf: $C_{16}H_{26}N_2O•ClH$ mw: 298.90

SYN: C 6610

TOXICITY DATA with REFERENCE
eye-rbt 2% MLD ARZNAD 9,113,59
ipr-rat LD50:100 mg/kg ARZNAD 9,113,59
scu-mus LD50:350 mg/kg ARZNAD 9,113,59

SAFETY PROFILE: Poison by intraperitoneal and subcutaneous routes. An eye irritant. When heated to decomposition it emits very toxic fumes of NO_x and HCl.

DIN400 CAS:77791-27-6 ***HR: 3***
2-(DIETHYLAMINO)-N-(4-METHYL-2-PYRIDYL) ACETAMIDE DIHYDROCHLORIDE
mf: $C_{12}H_{19}N_3O•2ClH$ mw: 294.26

SYN: C 5334

TOXICITY DATA with REFERENCE
ipr-rat LD50:183 mg/kg ARZNAD 8,609,58
scu-mus LD50:370 mg/kg ARZNAD 8,609,58

SAFETY PROFILE: Poison by intraperitoneal and subcutaneous routes. When heated to decomposition it emits very toxic fumes of NO_x and HCl.

DIN600 CAS:23505-41-1 ***HR: 3***
2-DIETHYLAMINO-6-METHYLPYRIMIDIN-4-YL DIETHYLPHOSPHOROTHIONATE
mf: $C_{13}H_{24}N_3O_3PS$ mw: 333.43

SYNS: O-(2-(DIETHYLAMINO)-6-METHYL-4-PYRIMIDINYL)-O,O-DIETHYL PHOSPHOROTHIOATE ◇ O,O-DIETHYL O-(2-DIETHYLAMINO-6-METHYL-4-PYRIMIDINYL)PHOSPHOROTHIOATE ◇ FERNEX ◇ PIRIMIFOSETHYL ◇ PIRIMIPHOS-ETHYL ◇ PP211 ◇ PRIMICID ◇ PRIMOTEC ◇ PRINICID

TOXICITY DATA with REFERENCE
mmo-sat 5 μL/plate MUREAV 28,405,75
mmo-esc 5 μL/plate MUREAV 28,405,75
orl-rat LD50:140 mg/kg GUCHAZ 6,421,73
skn-rat LD50:1000 mg/kg 28ZEAL 5,184,76
orl-mus LD50:105 mg/kg 28ZEAL 5,184,76
orl-cat LD50:25 mg/kg 28ZEAL 5,184,76
orl-gpg LD50:50 mg/kg 28ZEAL 5,184,76
orl-bwd LD50:7500 μg/kg AECTCV 12,355,83

CONSENSUS REPORTS: EPA Extremely Hazardous Substances List.

SAFETY PROFILE: Poison by ingestion. Moderately toxic by skin contact. Mutation data reported. When heated to decomposition it emits very toxic fumes of NO_x, PO_x, and SO_x.

DIN800 CAS:29232-93-7 ***HR: 2***
2-DIETHYLAMINO-6-METHYLPYRIMIDIN-4-YL DIMETHYL PHOSPHOROTHIONATE
mf: $C_{11}H_{20}N_3O_3PS$ mw: 305.37

SYNS: ACTELIC ◇ ACTELLIC ◇ ACTELLIFOG ◇ BLEX ◇ O-(2-(DIETHYLAMINO)-6-METHYL-4-PYRIMIDINYL)-O,O-DIMETHYLPHOSPHOROTHIOATE ◇ O-(2-DIETHYLAMINO-6-METHYLPYRIMIDIN-4-YL)-O,O-DIMETHYL PHOSPHOROTHIOATE ◇ ENT 27,699GC ◇ METHYL PIRIMIPHOS ◇ PIRIMIFOS-METHYL ◇ PLANT PROTECTION PP511 ◇ PP511 ◇ PYRIMIDINE PHOSPHATE ◇ PYRIMIPHOS METHYL ◇ SILOSAN

TOXICITY DATA with REFERENCE
mmo-sat 5 μL/plate MUREAV 28,405,75
cyt-mus-unr 500 mg/kg TGANAK 16(1),45,82
orl-rat LD50:1250 mg/kg SSCMBX 20,33,83
orl-mus LD50:1180 mg/kg 28ZEAL 5,184,76
orl-rbt LD50:1150 mg/kg 28ZEAL 5,184,76
orl-gpg LD50:1000 mg/kg 28ZEAL 5,184,76

SAFETY PROFILE: Moderately toxic by ingestion. Mutation data reported. When heated to decomposition it emits very toxic fumes of NO_x, PO_x, and SO_x.

DIO200 CAS:15421-84-8 ***HR: 3***
7-DIETHYLAMINO-5-METHYL-s-TRIAZOLO (1,5-a)PYRIMIDINE
mf: $N_5C_{10}H_{15}$ mw: 205.30

SYNS: AR 12008 ◇ N,N-DIETHYL-5-METHYL-(1,2,4)TRIAZOLO (1,5-a)PYRIMIDINE-7-AMINE ◇ 5-METHYL-7-DIETHYLAMINO-s-TRIAZOLO-(1,5-a)PYRIMIDINE ◇ ROCORNAL ◇ TRAPIDIL ◇ TRAPYMIN

TOXICITY DATA with REFERENCE
orl-mus TDLo:240 mg/kg (female 7-13D post):REP IYKEDH 6,418,75
orl-rat TDLo:420 mg/kg (8-14D preg):TER IYKEDH 6,418,75
orl-rat LD50:235 mg/kg MEIEDD 10,1369,83
ipr-rat LD50:100 mg/kg MEIEDD 10,1369,83
scu-rat LD50:100 mg/kg MEIEDD 10,1369,83
ivn-rat LD50:76 mg/kg MEIEDD 10,1369,83
orl-mus LD50:380 mg/kg MEIEDD 10,1369,83
ipr-mus LD50:155 mg/kg MEIEDD 10,1369,83
scu-mus LD50:132 mg/kg MEIEDD 10,1369,83
ivn-mus LD50:101 mg/kg IYKEDH 10,232,79

SAFETY PROFILE: Poison by ingestion, intraperitoneal, subcutaneous, and intravenous routes. An experimental teratogen. Other experimental reproductive effects. A coronary vasodilator. When heated to decomposition it emits toxic fumes of NO_x. See also AMINES.

DIO300 CAS:25953-06-4 ***HR: 3***
5-(DIETHYLAMINO)-2-NITROSOPHENOL HYDROCHLORIDE
mf: $C_{10}H_{14}N_2O_2{\cdot}ClH$ mw: 230.72

TOXICITY DATA with REFERENCE
orl-rat LD50:315 mg/kg JMCMAR 13,370,70
scu-rat LD50:84 mg/kg JMCMAR 13,370,70
ivn-rat LD50:52 mg/kg JMCMAR 13,370,70
orl-mus LD50:122 mg/kg JMCMAR 13,370,70
ipr-mus LD50:36800 μg/kg JMCMAR 13,370,70
ivn-mus LD50:34500 μg/kg JMCMAR 13,370,70

SAFETY PROFILE: Poison by ingestion, intraperitoneal, subcutaneous, and intravenous routes. When heated to decomposition it emits toxic fumes of NO_x and HCl.

DIP000 CAS:7347-49-1 ***HR: 3***
4-((4-(DIETHYLAMINO)PHENYL)AZO)PYRIDINE-1-OXIDE
mf: $C_{15}H_{18}N_4O$ mw: 270.37

SYN: N,N-DIETHYL-4-(4′-(PYRIDYL-1′-OXIDE)AZO)ANILINE

TOXICITY DATA with REFERENCE
orl-rat TDLo:6426 mg/kg/52W-C:NEO JNCIAM 37,365,66

SAFETY PROFILE: Questionable carcinogen with experimental neoplastigenic data. When heated to decomposition it emits toxic fumes of NO_x.

DIP100 CAS:5185-78-4 ***HR: 3***
2-(p-(DIETHYLAMINOPHENYL)-1,3,2-DITHIARSENOLANE
mf: $C_{12}H_{18}AsNS_2$ mw: 315.35

SYN: 1,3,2-DITHIARSENOLANE,2-(p-(DIETHYLAMINO)PHENYL)-

TOXICITY DATA with REFERENCE
ipr-mus LD50:16 mg/kg JMCMAR 9,221,66

OSHA PEL: TWA 0.5 mg(As)/m^3

SAFETY PROFILE: Poison by intraperitoneal route. When heated to decomposition it emits toxic fumes of NO_x, SO_x, and As.

DIP400 CAS:90-84-6 ***HR: 3***
2-(DIETHYLAMINO)PROPIOPHENONE
mf: $C_{13}H_{19}NO$ mw: 205.33

SYNS: α-DIETHYLAMINOPROPIOPHENONE ◇ 1-PHENYL-2-DIETHYLAMINO-1-PROPANONE

TOXICITY DATA with REFERENCE
orl-man TDLo:11 mg/kg:ANS,CNS,CVS THERAP 34,205,79
ipr-mus LD50:120 mg/kg TXAPA9 41,329,77
orl-mus LD50:160 mg/kg THERAP 20,297,65

SAFETY PROFILE: Poison by ingestion and intraperitoneal routes. Human systemic effects by ingestion: autonomic nervous systemic effects, somnolence, increase in blood pressure. When heated to decomposition it emits toxic fumes of NO_x.

DIP600 CAS:134-80-5 ***HR: 3***
2-DIETHYLAMINOPROPIOPHENONE HYDROCHLORIDE
mf: $C_{13}H_{19}NO{\cdot}ClH$ mw: 241.79

SYNS: AMPHEPRAMONUM HYDROCHLORIDE ◇ α-BENZOYLTRIETHYLAMINE HYDROCHLORIDE ◇ α-BENZOYLTRIETHYLAMMONIUM CHLORIDE ◇ 2-(DIETHYLAMINO)-1-PHENYL-1-PROPANONE HYDROCHLORIDE ◇ DIETHYLPROPIONE HYDROCHLORIDE ◇ DIETHYLPROPION HYDROCHLORIDE ◇ 1-PHENYL-2-DIETHYLAMINOPROPANONE-1-HYDROCHLORIDE ◇ 1-PHENYL-2-DIETHYLAMINO-1-PROPANONE HYDROCHLORIDE ◇ REGENON ◇ REGONON HYDROCHLORIDE ◇ TENUATE ◇ TENUATE HYDROCHLORIDE ◇ TEPANIL

TOXICITY DATA with REFERENCE
orl-rat LD50:500 mg/kg JPETAB 137,365,62
ipr-rat LD50:139 mg/kg APTOA6 17,121,60
orl-mus LD50:450 mg/kg JPETAB 137,365,62
ipr-mus LD50:190 mg/kg AEPPAE 237,171,59
scu-mus LD50:450 mg/kg APSXAS 4,37,67
ivn-mus LD50:50 mg/kg APTOA6 17,182,60

SAFETY PROFILE: Poison by intraperitoneal and intravenous routes. Moderately toxic by ingestion and subcutaneous routes. When heated to decomposition it emits very toxic fumes of HCl, NH_3, and NO_x.

DIP800 CAS:41240-93-1 ***HR: 3***
2-(DIETHYLAMINO)-2′-PROPOXYACETANILIDE HYDROCHLORIDE
mf: $C_{15}H_{24}N_2O_2{\cdot}ClH$ mw: 300.87

SYN: C 3127

TOXICITY DATA with REFERENCE
eye-rbt 2% MOD ARZNAD 8,270,58
ipr-rat LD50:350 mg/kg ARZNAD 8,270,58
scu-mus LD50:1950 mg/kg ARZNAD 8,270,58

SAFETY PROFILE: Poison by intraperitoneal route. Moderately toxic by subcutaneous route. An eye irritant. When heated to decomposition it emits very toxic fumes of NO_x and HCl.

DIQ100 CAS:14642-66-1 ***HR: 3***
3-DIETHYLAMINOPROPYLAMINE
mf: $C_7H_{18}N_2$ mw: 130.23

$$(CH_3CH_2)_2NN(CH_2)_3NH_2$$

SAFETY PROFILE: Ignites spontaneously on cellulose nitrate of high surface area. When heated to decomposition it emits toxic fumes of NO_x. See also AMINES.

DIR000 CAS:522-00-9 ***HR: 3***
10-(2-DIETHYLAMINOPROPYL)PHENOTHIAZINE
mf: $C_{19}H_{24}N_2S$ mw: 312.51

SYNS: AETHOPROPROPAZIN ◇ ATHAPROPAZINE ◇ ATHOPROPAZIN ◇ DIBUTIL ◇ 10-(2-DIETHYLAMINO-2-METHYLETHYL)PHENOTHIAZINE ◇ 2-DIETHYLAMINO-1-PROPYL-N-DIBENZOPARATHIAZINE ◇ N,N-DIETHYL-α-METHYL-10H-PHENOTHIAZINE-10-ETHANAMINE ◇ ETHOPROMAZINE ◇ ETOPROPEZINA ◇ FEMPROPAZINE ◇ FENPROPAZINA ◇ ISOTAZIN ◇ ISOTHAZINE ◇ ISOTHIAZINE ◇ LYSIVANE ◇ PARCIDOL ◇ PARDIDOL ◇ PARDISOL ◇ PARFEZINE ◇ PARKIN ◇ PARKISOL ◇ PARPHEZEIN ◇ PARSIDOL ◇ PARSITAN ◇ PHENOPROPAZINE ◇ PHENOPROZINE ◇ PRODICTAZIN ◇ PRODIERAZINE ◇ PROFENAMINA (ITALIAN) ◇ PROFENAMINUM ◇ ROCHIPEL ◇ ROCIPEL ◇ RODIPAL ◇ RP 3356 ◇ SC 2538 ◇ SKF 2538 ◇ TOMIL ◇ W 483

TOXICITY DATA with REFERENCE
orl-rat LD50:1700 mg/kg NIIRDN 6,733,82
ipr-rat LD50:390 mg/kg NIIRDN 6,733,82
scu-rat LD50:200 mg/kg PSCBAY 2,17,63
ivn-rat LD50:15 mg/kg 27ZQAG -,41,72
orl-mus LD50:300 mg/kg BCFAAI 111,293,72
ipr-mus LD50:160 mg/kg NIIRDN 6,733,82
scu-mus LD50:500 mg/kg 27ZQAG -,41,72
ivn-mus LD50:50 mg/kg 27ZQAG -,41,72
scu-rbt LD50:200 mg/kg 27ZQAG -,41,72
ivn-rbt LD50:15 mg/kg 27ZQAG -,41,72

SAFETY PROFILE: Poison by ingestion, subcutaneous, intraperitoneal, and intravenous routes. An anticholinergic agent used to treat Parkinsons disease. When heated to decomposition it emits very toxic fumes of NO_x and SO_x.

DIR800 CAS:102206-93-9 ***HR: D***
3-DIETHYLAMINO-5H-PYRIDO(4,3-b)INDOLE
mf: $C_{15}H_{17}N_3$ mw: 239.2

TOXICITY DATA with REFERENCE
mma-sat 2500 nmol/L PNASA6 77,1427,80
dnd-mam:lym 100 μmol/L PNASA6 77,1427,80

SAFETY PROFILE: Mutation data reported. When heated to decomposition it emits toxic fumes of NO_x.

DIR875 CAS:38078-09-0 ***HR: 3***
DIETHYLAMINOSULFUR TRIFLUORIDE
mf: $C_4H_{10}F_3NS$ mw: 161.18

$(CH_3CH_2)_2NSF_3$

SAFETY PROFILE: Decomposes violently above 90°C. Decomposes explosively on contact with water. When heated to decomposition it emits toxic fumes of F^-, SO_x, and NO_x. See also FLUORIDES.

DIS000 CAS:77791-66-3 ***HR: 2***
2-(DIETHYLAMINO)-2',4',6'-TRICHLOROACETANILIDE HYDROCHLORIDE
mf: $C_{12}H_{15}Cl_3N_2O•ClH$ mw: 346.10

SYN: V 340

TOXICITY DATA with REFERENCE
eye-rbt 2% MLD ARZNAD 8,270,58
ipr-rat LD50:487 mg/kg ARZNAD 8,270,58
ipr-mus LD50:490 mg/kg ARZNAD 8,270,58
scu-mus LD50:1165 mg/kg ARZNAD 8,270,58

SAFETY PROFILE: Moderately toxic by intraperitoneal and subcutaneous routes. An eye irritant. When heated to decomposition it emits very toxic fumes of Cl^- and NO_x.

DIS200 CAS:2150-48-3 ***HR: 3***
(6-(DIETHYLAMINO)-3H-XANTEN-3-YLIDENE)DIETHYLAMMONIUM CHLORIDE
mf: $C_{21}H_{27}N2O•Cl$ mw: 358.95

SYNS: C.I. 45010 ◇ N-(6-(DIETHYLAMINO)-3H-XANTHEN-3-YLIDINE)-N-ETHYLETHANAMINIUM CHLORIDE ◇ ETHANIMINIUM-N-(6-DIETHYLAMINO)-3H-XANTHEN-3-YLIDENE)-N-ETHYLCHLORIDE ◇ NSC 44690 ◇ PYRONIN B ◇ E TETRAETHYLPYRONIN

TOXICITY DATA with REFERENCE
sln-dmg-orl 1000 ppm AMNTA4 87,295,53
dnd-esc 20 μmol/L MUREAV 89,95,81
sln-dmg-orl:1000 ppm AMNTA4 87,295,53
ipr-mus LD50:15100 μg/kg NCISP* Jan 86

CONSENSUS REPORTS: Reported in EPA TSCA Inventory.

SAFETY PROFILE: Poison by intraperitoneal route. Mutation data reported. When heated to decomposition it emits toxic fumes of Cl^-, NO_x, and NH_3.

DIS400 ***HR: 3***
2-DIETHYLAMMONIOETHYL NITRATE
mf: $C_6H_{15}N_3O_6$ mw: 225.21

SAFETY PROFILE: Explodes during vacuum distillation with nitric acid. When heated to decomposition it emits toxic fumes of NO_x and NH_3. See also NITRATES.

DIS500 CAS:660-68-4 ***HR: 3***
DIETHYLAMMONIUM CHLORIDE
mf: $C_4H_{11}N•ClH$ mw: 109.62

SYNS: N-ETHYL-ETHANAMINE HYDROCHLORIDE (9CI) ◇ HYDROCHLORIDE DIETHYLAMINE

TOXICITY DATA with REFERENCE
ipr-mus LD50:960 mg/kg JJPAAZ 17,475,67
scu-mus LD50:1130 mg/kg AIPTAK 112,36,57
ivn-mus LD50:320 mg/kg AIPTAK 112,36,57

CONSENSUS REPORTS: EPA Genetic Toxicology Program. Reported in EPA TSCA Inventory.

SAFETY PROFILE: Poison by intravenous route. Moderately toxic by subcutaneous and intraperitoneal routes. When heated to decomposition it emits toxic fumes of NO_x, NH_3, and HCl.

DIS600 CAS:2624-44-4 ***HR: 2***
DIETHYLAMMONIUM-2,5-DIHYDROXYBENZENE SULFONATE
mf: $C_{10}H_{17}NO_5S$ mw: 263.34

SYNS: AGLUMIN ◇ ALTODOR ◇ CYCLOHEXADIENOL-4-ONE-1-SULFONATE de DIETHYLAMINE (FRENCH) ◇ CYCLONAMINE ◇ DICYNENE ◇ DICYNONE ◇ DIETHYLAMMONIUM CYCLOHEXADIEN-4-OL-1-ONE-4-SULFONATE ◇ 2,5-DIHYDROXYBENZENESULFONIC ACID with N-ETHYLETHANAMINE ◇ DIIDROXI-1,4-BENZENESULFONATO-3-DI-ETILAMMONIUM (ITALIAN) ◇ E 141 ◇ ESELIN ◇ ETAMSYLATE ◇ ETHAMSYLATE ◇ 1-HYDROXY-4-OXO-2,5-CYCLOHEXADIENE-1-SULFONIC ACID compound with DIETHYLAMINE ◇ MD 141

TOXICITY DATA with REFERENCE
sln-asn 1 mg/L MUREAV 26,159,74
ivn-rat LD50:1350 mg/kg THERAP 15,110,60
ivn-mus LD50:785 mg/kg YHTPAD 15(12),17,80
ivn-cat LDLo:1 g/kg THERAP 15,110,60

CONSENSUS REPORTS: Reported in EPA TSCA Inventory.

SAFETY PROFILE: Moderately toxic by intravenous route. Mutation data reported. When heated to decomposition it emits very toxic fumes of SO_x, NH_3, and NO_x. See also SULFONATES and AMINES.

DIS700 CAS:91-66-7 ***HR: 2***
N,N-DIETHYLANILINE
DOT: UN 2432
mf: $C_{10}H_{15}N$ mw: 149.26

PROP: Colorless to yellow liquid. D: (25/4) 0.9302, bp: 215-216°, mp: −38°, n (24/D) 1.5394. Volatile with steam. Sltly sol in alc, chloroform, ether. One gram dissolves in 70 mL water at 12°.

SYNS: DIAETHYLANILIN (GERMAN) ◇ N,N-DIETHYLAMINOBENZENE ◇ N,N-DIETHYLANILIN (CZECH) ◇ DIETHYLANILINE ◇ DIETHYLPHENYLAMINE

TOXICITY DATA with REFERENCE
orl-rat LD50:782 mg/kg MarJV# 29MAR77
ipr-rat LD50:420 mg/kg AGGHAR 15,447,57

CONSENSUS REPORTS: Reported in EPA TSCA Inventory.

DOT Classification: Poison B; Label: St. Andrews Cross.

SAFETY PROFILE: Moderately toxic by ingestion, and intraperitoneal routes. When heated to decomposition it emits toxic fumes of NO_x. See also ANILINE DYES.

DIS775 CAS:5185-76-2 ***HR: 3***
N,N-DIETHYL-p-ARSANILIC ACID
mf: $C_{10}H_{16}AsNO_3$ mw: 273.19

SYN: p-ARSANILIC ACID, N,N-DIETHYL-

TOXICITY DATA with REFERENCE
ipr-mus LD50:7102 μg/kg JMCMAR 9,221,66

OSHA PEL: TWA 0.5 mg(As)/m^3

SAFETY PROFILE: Poison by intraperitoneal route. When heated to decomposition it emits toxic fumes of NO_x and As.

DIS800 CAS:692-42-2 ***HR: 3***
DIETHYL ARSINE
mf: $C_4H_{11}As$ mw: 134.05

$$(CH_3CH_2)_2AsH$$

CONSENSUS REPORTS: Arsenic and its compounds are on the Community Right-To-Know List.

SAFETY PROFILE: A poison. A dangerous fire hazard. It spontaneously ignites in air above 0°C. Upon decomposition it emits toxic fumes of As. See also ARSENIC COMPOUNDS.

DIS850 CAS:4964-27-6 ***HR: 3***
DIETHYL ARSINIC ACID
mf: $C_4H_{11}AsO_2$ mw: 166.07

SYNS: ARSINE OXIDE, DIETHYLHYDROXY- ◇ DIETHYLHYDROXY ARSINE OXIDE

TOXICITY DATA with REFERENCE
ivn-mus LD50:180 mg/kg CSLNX* NX#03818

OSHA PEL: TWA 0.5 mg(As)/m^3

SAFETY PROFILE: Poison by intravenous route. When heated to decomposition it emits toxic fumes of As.

DIT300 CAS:1972-28-7 ***HR: 3***
DIETHYL AZOFORMATE
mf: $C_6H_{10}N_2O_4$ mw: 174.16

$$CH_3CH_2OCO{\cdot}N{=}NCO{\cdot}OCH_2CH_3$$

SYN: 1,2-DIETHOXYCARBONYLDIAZENE

SAFETY PROFILE: A shock-sensitive explosive which also burns explosively. When heated to decomposition it emits toxic fumes of NO_x. See also AZIDES.

DIT350 CAS:5256-74-6 ***HR: 3***
DIETHYL AZOMALONATE
mf: $C_7H_{10}N_2O_4$ mw: 186.17

$N_2C(CO{\bullet}OCH_2CH_3)_2$

SAFETY PROFILE: A poison. Potentially explosive when heated. When heated to decomposition it emits toxic fumes of NO_x.

DIT400 CAS:36911-94-1 ***HR: 3***
6,8-DIETHYLBENZ(a)ANTHRACENE
mf: $C_{22}H_{20}$ mw: 284.42

TOXICITY DATA with REFERENCE
ims-rat TDLo:50 mg/kg:ETA JMCMAR 15,905,72

SAFETY PROFILE: Questionable carcinogen with experimental tumorigenic data. When heated to decomposition it emits acrid smoke and irritating fumes.

DIT600 CAS:36911-95-2 ***HR: 3***
8,12-DIETHYLBENZ(a)ANTHRACENE
mf: $C_{22}H_{20}$ mw: 284.42

TOXICITY DATA with REFERENCE
ims-rat TDLo:50 mg/kg:ETA JMCMAR 15,905,72

SAFETY PROFILE: Questionable carcinogen with experimental tumorigenic data. When heated to decomposition it emits acrid smoke and irritating fumes.

DIT800 CAS:16354-52-2 ***HR: 3***
9,10-DIETHYL-1,2-BENZANTHRACENE
mf: $C_{22}H_{20}$ mw: 284.42

SYN: 7,12-DIETHYLBENZ(a)ANTHRACENE

TOXICITY DATA with REFERENCE
skn-mus TDLo:380 mg/kg/16W-I:ETA PRLBA4 129,439,40

SAFETY PROFILE: Questionable carcinogen with experimental tumorigenic data. When heated to decomposition it emits acrid smoke and irritating fumes.

DIU000 CAS:25340-17-4 ***HR: 2***
DIETHYL BENZENE
DOT: UN 2049
mf: $C_{10}H_{14}$ mw: 134.24

PROP: Colorless, mobile liquid. Bp: 183.8°, flash p: 134°F, d: 0.868 @ 25°/25°, autoign temp: 743-842°F, vap press: 1 mm @ 20.7°, vap d: 4.62.

TOXICITY DATA with REFERENCE
skn-rbt 100% MOD AMIHAB 14,387,56
eye-rbt 88 mg MLD AMIHAB 14,387,56
orl-rat LDLo:5000 mg/kg 28ZRAQ -,57,60

CONSENSUS REPORTS: Reported in EPA TSCA Inventory.

DOT Classification: Flammable or Combustible Liquid; Label: Flammable Liquid.

SAFETY PROFILE: Mildly toxic by ingestion. A skin and eye irritant. Flammable when exposed to heat or flame; can react with oxidizing materials. To fight fire, use CO_2, dry chemical. When heated to decomposition it emits acrid smoke and fumes. See also ETHYL BENZENE.

DIU200 CAS:141-93-5 ***HR: 2***
m-DIETHYLBENZENE
mf: $C_{10}H_{14}$ mw: 134.24

TOXICITY DATA with REFERENCE
orl-rat LDLo:5000 mg/kg 28ZRAQ -,57,60

CONSENSUS REPORTS: Reported in EPA TSCA Inventory.

SAFETY PROFILE: Moderately toxic by ingestion. When heated to decomposition it emits acrid and irritating fumes. See also DIETHYLBENZENE.

DIU400 CAS:1709-50-8 ***HR: 2***
N,N-DIETHYLBENZENESULFONAMIDE
mf: $C_{10}H_{15}NO_2S$ mw: 213.32

TOXICITY DATA with REFERENCE
orl-rat TDLo:300 mg/kg (female 9D post):REP TXAPA9 23,376,72
orl-rat TDLo:300 mg/kg (female 9D post):TER TXAPA9 23,376,72
orl-rat LD50:890 mg/kg TXAPA9 23,376,72

SAFETY PROFILE: Moderately toxic by ingestion. An experimental teratogen. Other experimental reproductive effects. When heated to decomposition it emits very toxic fumes of SO_x and NO_x.

DIU500 CAS:75889-62-2 ***HR: 3***
DIETHYL-4-(BENZOTHIAZOL-2-YL)BENZYLPHOSPHONATE
mf: $C_{18}H_{20}NO_3PS$ mw: 361.42

SYNS: (p-(2-BENZOTHIAZOLYL)BENZYL)PHOSPHONIC ACID DIETHYL ESTER ◇ KB-944

TOXICITY DATA with REFERENCE
orl-rat TDLo:3 g/kg (30D pre):REP ARZNAD 32,1071,82
orl-rat LD50:1555 mg/kg ARZNAD 32,1068,82
ipr-rat LD50:479 mg/kg ARZNAD 32,1068,82
scu-rat LD50:2216 mg/kg ARZNAD 32,1068,82
orl-mus LD50:2807 mg/kg ARZNAD 32,1068,82
ipr-mus LD50:777 mg/kg ARZNAD 32,1068,82
ivn-dog LD50:76 mg/kg ARZNAD 32,1068,82

SAFETY PROFILE: Poison by intravenous route. Moderately toxic by ingestion, subcutaneous, and intraperitoneal routes. Experimental reproductive effects. A calcium antagonist. When heated to decomposition it emits toxic fumes of PO_x, SO_x, and NO_x. See also ESTERS.

DIU600 CAS:1080-32-6 ***HR: 3***
DIETHYL BENZYLPHOSPHONATE
mf: $C_{11}H_{17}O_3P$ mw: 228.25

TOXICITY DATA with REFERENCE
ivn-mus LD50:180 mg/kg CSLNX* NX#01630

CONSENSUS REPORTS: Reported in EPA TSCA Inventory.

SAFETY PROFILE: Poison by intravenous route. When heated to decomposition it emits toxic fumes of PO_x.

DIU800 CAS:13286-32-3 ***HR: 3***
O,O-DIETHYL-S-BENZYL THIOPHOSPHATE
mf: $C_{11}H_{17}O_3PS$ mw: 260.31

SYNS: IBP ◇ KITAZIN ◇ PHOSPHOROTHIOIC ACID-S-BENZYL-O,O-DIETHYL ESTER ◇ RICID

TOXICITY DATA with REFERENCE
orl-rat LD50:660 mg/kg 85ARAE 4,102,76/77
orl-mus LD50:230 mg/kg CHYCDW 14,197,80
ipr-mus LD50:96 mg/kg CHYCDW 14,197,80
ivn-mus LD50:570 mg/kg CHYCDW 14,197,80

SAFETY PROFILE: Poison by ingestion and intraperitoneal routes. Moderately toxic by intravenous route. When heated to decomposition it emits very toxic fumes of PO_x and SO_x.

DIV000 CAS:542-63-2 ***HR: 3***
DIETHYLBERYLLIUM
mf: $C_4H_{10}Be$ mw: 67.13

PROP: Colorless liquid. Mp: 12°; bp: 110° @ 15 mm; vap d: 2.3.

CONSENSUS REPORTS: Beryllium and its compounds are on the Community Right-To-Know List.

OSHA PEL: (Transitional: TWA 0.002 mg(Be)/m^3; CL 0.005; Pk 0.025/30M/8H) TWA 0.002 mg(Be)/m^3; STEL 0.005 mg(Be)/m^3/30M; CL 0.025 mg(Be)/m^3
ACGIH TLV: TWA 0.002 mg/m^3, Suspected Human Carcinogen.

SAFETY PROFILE: Confirmed human carcinogen. Very poisonous. Dangerous fire hazard when exposed to heat or flame. Spontaneously flammable in air. Can react vigorously with oxidizing materials. To fight fire, use special extinguishing agents, dry chemical. Explodes on contact with water. Upon decomposition it emits poisonous fumes of BeO. See also BERYLLIUM COMPOUNDS.

DIV200 CAS:28616-48-0 ***HR: 3***
DIETHYL BIS-DIMETHYLPYROPHOSPHORADIAMIDE (symmetrical)
mf: $C_8H_{22}N_2O_5P_2$ mw: 288.26

SYNS: sym-DIETHYLBIS(DIMETHYLAMIDO)PYROPHOSPHATE ◇ DIETHYL DI(DIMETHYLAMIDO)PYROPHOSPHATE (symmetrical) ◇ N,N,N',N'-TETRAMETHYL-p,p'-DIAMIDODIPHOSPHORIC ACID DIETHYL ESTER

TOXICITY DATA with REFERENCE
orl-rat LD50:12400 μg/kg JPETAB 107,464,53
skn-rat LD50:10 mg/kg JPETAB 107,464,53
ipr-rat LD50:10 mg/kg JPETAB 107,464,53
ipr-mus LD50:16400 μg/kg JPETAB 107,464,53
ivn-dog LD50:20 mg/kg JPETAB 107,464,53
ipr-gpg LD50:13 mg/kg JPETAB 107,464,53

SAFETY PROFILE: Poison by ingestion, skin contact, intravenous, and intraperitoneal routes. When heated to decomposition it emits very toxic fumes of NO_x and PO_x.

DIV400 CAS:65313-34-0 ***HR: 3***
DIETHYLBISMUTH CHLORIDE
mf: $C_4H_{10}BiCl$ mw: 302.45

$(CH_3CH_2)_2BiCl$

SAFETY PROFILE: A dangerous fire hazard; ignites spontaneously in air. When heated to decomposition it emits toxic fumes of Bi. See also BISMUTH COMPOUNDS and CHLORIDES.

DIV600 CAS:2641-56-7 ***HR: 3***
DIETHYLBIS(OCTANOYLOXY)STANNANE
mf: $C_{20}H_{40}O_4Sn$ mw: 463.29

SYNS: DIETHYLBIS(1-OXOOCTYL)OXY)STANNANE ◇ DIETHYLTIN DICAPRYLATE ◇ DIETHYLTIN DIOCTANOATE

TOXICITY DATA with REFERENCE
orl-rat LD50:330 mg/kg UBZHAZ 50,695,78
orl-mus LD50:330 mg/kg UBZHAZ 50,695,78

OSHA PEL: TWA 0.1 mg(Sn)/m^3 (skin)
ACGIH TLV: TWA 0.1 mg(Sn)/m^3 (skin) (Proposed: TWA 0.1 mg(Sn)/m^3; STEL 0.2 mg(Sn)/m^3 (skin))
NIOSH REL: (Organotin Compounds) TWA 0.1 mg(Sn)/m^3

SAFETY PROFILE: Poison by ingestion. See also TIN COMPOUNDS. When heated to decomposition it emits acrid and irritating fumes.

DIV800 ***HR: 3***
DIETHYLCADMIUM
mf: $C_4H_{10}Cd$ mw: 170.5

PROP: An oil; decomp by moisture. D: 1.6562, mp: −21°, bp: 64°.

CONSENSUS REPORTS: Cadmium and its compounds are on the Community Right-To-Know List.

OSHA PEL: TWA 0.1 mg(Cd)/m^3; CL 0.6 mg(Cd)/m^3 (fume)
ACGIH TLV: TWA 0.05 mg(Cd)/m^3 (Proposed: TWA 0.01 mg(Cd)/m^3 (dust), Suspected Human Carcinogen; 0.002 mg(Cd)/m^3 (respirable dust), Suspected Human Carcinogen); BEI: 10 μg/g creatinine in urine; 10 μg/L in blood.
DFG BAT: Blood 1.5 μg/dL; Urine 15 μg/dL, Suspected Carcinogen.
NIOSH REL: (Cadmium) Reduce to lowest feasible level

SAFETY PROFILE: Confirmed human carcinogen. A poison. See also CADMIUM COMPOUNDS. A dangerous fire and explosion hazard. Explodes when heated rapidly to 130°C. On exposure to air it forms white fumes which turn brown and explode. The vapor explodes when heated to 180°C. When heated to decomposition it emits highly toxic fumes of cadmium.

DIW000 CAS:90-89-1 ***HR: 3***
DIETHYLCARBAMAZINE
mf: $C_{10}H_{21}N_3O$ mw: 199.34

PROP: Crystals. Mp: 47-49°, bp: 108.5-111°.

SYNS: BITIRAZINE ◇ CARBAMAZINE ◇ CARBILAZINE ◇ CARICIDE ◇ CATACIDE ◇ CYPIP ◇ 1-DIETHYLCARBAMOYL-4-METHYLPIPERAZINE ◇ 1-DIETHYLCARBAMYL-4-METHYLPIPERZINE ◇ N,N-DIETHYL-4-METHYL-1-PIPERAZINECARBOXAMIDE ◇ DITRAZINE BASE ◇ ETHODRYL ◇ 84L ◇ 1-METHYL-4-DIETHYLCARBAMYLPIPERAZINE ◇ NOTEZINE ◇ RP 3799 ◇ SPATONIN

TOXICITY DATA with REFERENCE
oms-nml 10 mmol/L CUSCAM 52,303,83
orl-rat TDLo:630 mg/kg (7-11D preg):TER CYLPDN 4,201,83
orl-wmn TDLo:7300 mg/kg/1Y-I:SKN BMJOAE 291,632,85
ipr-mus LD50:240 mg/kg JPPMAB 22,306,70

CONSENSUS REPORTS: Reported in EPA TSCA Inventory.

SAFETY PROFILE: Poison by intraperitoneal route. Human systemic effects by ingestion: allergic dermatitis. An experimental teratogen. Mutation data reported. When heated to decomposition it emits toxic fumes of NO_x and HCl. An additive permitted in the food and drinking water of animals and/or for the treatment of food-producing animals.

DIW200 CAS:1642-54-2 ***HR: 3***
DIETHYLCARBAMAZINE ACID CITRATE
mf: $C_{10}H_{21}N_3O{\cdot}C_6H_8O_7$ mw: 391.48

SYNS: BANOCIDE ◇ CARICIDE ◇ CARITROL ◇ DICAROCIDE ◇ DIETHYLCARBAMAZANE CITRATE ◇ DIETHYLCARBAMAZINE CITRATE ◇ DIETHYLCARBAMAZINE HYDROGEN CITRATE ◇ 1-DIETHYLCARBAMOYL-4-METHYLPIPERAZINE DIHYDROGEN CITRATE ◇ N,N-DIETHYL-4-METHYL-1-PIPERAZINE CARBOXAMIDE CITRATE ◇ N,N-DIETHYL-4-METHYL-1-PIPERAZINECARBOXAMIDE DIHYDROGEN CITRATE ◇ N,N-DIETHYL-4-METHYL-1-PIPERAZINECARBOXAMIDE-2-HYDROXY-1,2,3-PROPANETIRCARBOXYLATE ◇ DITRAZIN ◇ DITRAZIN CITRATE ◇ DITRAZINE ◇ DITRAZINE CITRATE ◇ ETHODRYL CITRATE ◇ FRANOCIDE ◇ FRANOZAN ◇ HETRAZAN ◇ LOXURAN ◇ 1-METHYL-4-DIETHYLCARBAMOYLPIPERAZINECITRATE

TOXICITY DATA with REFERENCE
orl-rat TDLo:1800 mg/kg (8-16D preg):TER IJMRAQ 60,1529,72
orl-rat LD50:1400 mg/kg FAZMAE 17,108,73
ihl-rat LC50:309 mg/m^3/4H 85GMAT -,63,82
scu-rat LD50:1136 mg/kg NIIRDN 6,306,82
orl-mus LD50:660 mg/kg 29ZVAB -,41,69
ipr-mus LD50:425 mg/kg IJOCAP 26,748,87
scu-mus LD50:608 mg/kg NIIRDN 6,306,82
ivn-mus LD50:180 mg/kg CSLNX* NX#00466

CONSENSUS REPORTS: Reported in EPA TSCA Inventory.

SAFETY PROFILE: Poison by inhalation and intravenous routes. Moderately toxic by ingestion, subcutaneous, and intraperitoneal routes. An experimental teratogen. When heated to decomposition it emits toxic fumes of NO_x.

DIW300 CAS:5348-97-0 ***HR: 3***
DIETHYLCARBAMAZINE HYDROCHLORIDE
mf: $C_{10}H_{21}N_{3O}{\cdot}ClH$ mw: 235.80

SYN: 1-DIETHYLCARBAMYL-4-METHYLPIPERAZINEHYDROCHLORIDE

TOXICITY DATA with REFERENCE
orl-rat LD50:1380 mg/kg ANYAA9 50,141,48
ipr-rat LD50:285 mg/kg CLDND*
ivn-rat LD50:150 mg/kg ANYAA9 50,141,48
orl-mus LD50:660 mg/kg ANYAA9 50,141,48
ipr-mus LD50:248 mg/kg ANYAA9 50,141,48
ivn-mus LD50:82 mg/kg ANYAA9 50,141,48

CONSENSUS REPORTS: Reported in EPA TSCA Inventory.

SAFETY PROFILE: A poison by intraperitoneal and intravenous routes. Moderately toxic by ingestion. When heated to decomposition it emits very toxic fumes of NO_2 and HCl.

DIW400 CAS:88-10-8 ***HR: 2***
DIETHYLCARBAMOYL CHLORIDE
mf: $C_5H_{10}ClNO$ mw: 135.61

PROP: Liquid. Mp: −44°, bp: 190-195°, vap d: 4.1.

SYNS: DIETHYLCARBAMIC CHLORIDE ◇ DIETHYLCARBAMIDOYL CHLORIDE ◇ N,N-DIETHYLCARBAMOYL CHLORIDE ◇ DIETHYLCARBAMYL CHLORIDE

TOXICITY DATA with REFERENCE
skn-mus TDLo:43200 mg/kg/72W-I:CAR JACTDZ 6(4),479,87
ipr-mus LD50:750 mg/kg NTIS** AD691-490

CONSENSUS REPORTS: Reported in EPA TSCA Inventory.

DFG MAK: Suspected Carcinogen.

SAFETY PROFILE: Suspected carcinogen with experimental carcinogenic data. Moderately toxic by intraperitoneal route. Mutation data reported. Reacts with water or steam to produce toxic and corrosive fumes. When heated to decomposition it emits highly toxic fumes of Cl^- and NO_x. See also CARBAMATES and CHLORIDES.

DIW600 CAS:2425-25-4 ***HR: 3***
O,O-DIETHYL-S-(CARBETHOXY)METHYL PHOSPHOROTHIOLATE
mf: $C_8H_{17}O_5PS$ mw: 256.28

SYNS: ACETOPHOS ◇ ACETOXON ◇ O,O-DIETHYL-S-CARBOETHOXYMETHYL PHOSPHOROTHIOATE ◇ O,O-DIETHYL-S-CARBOETHOXYMETHYL THIOPHOSPHATE ◇ PHOSPHOROTHIOIC ACID-O,O-DIETHYL ESTER-S-ESTER with ETHYL MERCAPTOACETATE

TOXICITY DATA with REFERENCE
orl-rat LD50:45 mg/kg HYSAAV 31,18,66
ipr-mus LD50:214 mg/kg JEENAI 51,714,58

SAFETY PROFILE: Poison by ingestion, and intraperitoneal routes. When heated to decomposition it emits very toxic fumes of PO_x and SO_x.

DIW800 CAS:112-36-7 ***HR: 3***
DIETHYL CARBITOL
mf: $C_8H_{18}O_3$ mw: 162.26

PROP: Colorless liquid; sol in water and hydrocarbons. D: 0.9082 @ 20°/20°, bp: 189°, fp: −44°, flash p: 180°F (OC), vap d: 5.6.

SYNS: BIS(2-ETHOXYETHYL)ETHER ◇ DIETHYLENE GLYCOL DIETHYL ETHER ◇ 1-ETHOXY-2-(β-ETHOXYETHOXY)ETHANE ◇ ETHYL DIGLYME ◇ 3,6,9-TRIOXAUNDECANE

TOXICITY DATA with REFERENCE
eye-rbt 50 mg MOD UCDS** 4/25/58
orl-mus TDLo:24 g/kg (7-14D preg):REP EVHPAZ 57,141,84
orl-mus TDLo:3 g/kg (female 6-15D post):TER NTIS** PB88-168257
orl-rat LD50:4970 mg/kg UCDS** 4/25/58
orl-gpg LD50:1850 mg/kg JIHTAB 23,259,41

CONSENSUS REPORTS: Reported in EPA TSCA Inventory. Glycol ether compounds are on the Community Right-To-Know List.

SAFETY PROFILE: Moderately toxic by ingestion. An experimental teratogen. Other experimental reproductive effects. An eye irritant. Flammable when exposed to heat or flame. When heated to decomposition it emits acrid smoke and irritating fumes. See also GLYCOL ETHERS.

DIX000 CAS:919-54-0 ***HR: 2***
O,O-DIETHYL-S-CARBOETHOXYMETHYL DITHIOPHOSPHATE
mf: $C_8H_{17}O_4PS_2$ mw: 272.34

SYNS: ACETHION ◇ ACETHIONE ◇ ((DIETHOXYPHOSPHINOTHIOYL)THIO)ACETIC ACID, ETHYL ESTER ◇ O,O-DIETHYL-S-CARBOETHOXYMETHYL PHOSPHORODITHIOATE ◇ ENT 25,650 ◇ ETHOXYPHAS ◇ ETHOXYPHOS ◇ ETHYL ((DIETHOXYPHOSPHINOTHIOYL)THIO)ACETATE ◇ HERCULES 4580 ◇ PHOSPHORODITHIOIC ACID-O,O-DIETHYL ESTER-S-ESTER with ETHYL MERCAPTOACETATE

TOXICITY DATA with REFERENCE
orl-mus LD50:1200 mg/kg ARSIM* 20,12,66
ipr-mus LD50:1280 mg/kg JEENAI 51,714,58

SAFETY PROFILE: Moderately toxic by ingestion and intraperitoneal routes. When heated to decomposition it emits very toxic fumes of PO_x and SO_x. See also MERCAPTANS and ESTERS.

DIX200 CAS:105-58-8 ***HR: 3***
DIETHYL CARBONATE
DOT: UN 2366
mf: $C_5H_{10}O_3$ mw: 118.15

PROP: Colorless liquid, mild odor. Mp: 43°, bp: 125.8°, flash p: 77°F (OC), d: 0.975 @ 20°/4°, vap press: 10 mm @ 23.8°, vap d: 4.07.

SYNS: DEC ◇ DIAETHYLCARBONAT (GERMAN) ◇ DIETHYL CARBONATE (DOT) ◇ ETHOXYFORMIC ANHYDRIDE ◇ ETHYL CARBONATE ◇ EUFIN ◇ NCI-C60899

TOXICITY DATA with REFERENCE
ipr-ham TDLo:496 mg/kg (female 8D post):TER CNREA8 27,1696,67
orl-mus TDLo:500 mg/kg:ETA BCPCA6 2,168,59

orl-rat LDLo:15 g/kg FAONAU 53A,52,74
scu-rat LD50:8500 mg/kg CLDND* 49,172.102,88

CONSENSUS REPORTS: Reported in EPA TSCA Inventory.

DOT Classification: IMO: Flammable or Combustible Liquid; Label: Flammable Liquid.

SAFETY PROFILE: Mildly toxic by subcutaneous route. Questionable carcinogen with experimental tumorigenic and teratogenic data. A dangerous fire hazard when exposed to heat or flame; can react with oxidizing materials. To fight fire, use foam, CO_2, dry chemical. When heated to decomposition it emits acrid smoke and fumes. See also ANHYDRIDES.

DIX400 CAS:2315-36-8 ***HR: 2***
N,N-DIETHYLCHLORACETAMIDE
mf: $C_6H_{12}ClNO$ mw: 149.64

SYN: N-CHLOROACETYLDIETHYLAMINE

TOXICITY DATA with REFERENCE
orl-rat LD50:500 mg/kg RREVAH 10,97,65
ihl-mus LCLo:660 mg/m^3/10M NDRC** NDCrc-132,Mar,42

CONSENSUS REPORTS: Reported in EPA TSCA Inventory.

SAFETY PROFILE: Moderately toxic by ingestion and inhalation. When heated to decomposition it emits very toxic fumes of Cl^- and NO_x.

DIX600 CAS:1757-18-2 ***HR: 3***
O,O-DIETHYL-O-(2-CHLORO-1,2,5-DICHLOROPHENYLVINYL) PHOSPHOROTHIOATE
mf: $C_{12}H_{14}Cl_3O_3PS$ mw: 375.64

SYNS: AKTON ◇ AXIOM ◇ O-(2-CHLORO-1-(2,5-DICHLOROPHENYL)-O,O-DIETHYL ESTER PHOSPHOROTHIOIC ACID ◇ O-(2-CHLORO-1-(2,5-DICHLOROPHENYL)VINYL)-O,O-DIETHYLPHOSPHOROTHIOATE ◇ ENT 27,102 ◇ SD 9098 ◇ SHELL SD-9098

TOXICITY DATA with REFERENCE
orl-rat LD50:146 mg/kg WRPCA2 9,119,70
orl-mus LD50:89 mg/kg SPEADM 78-1,41,78
skn-rbt LD50:177 mg/kg SPEADM 78-1,41,78
orl-bwd LD50:75 mg/kg AECTCV 12,355,83

SAFETY PROFILE: Poison by ingestion and skin contact. An insecticide. When heated to decomposition it emits very toxic fumes of Cl^-, PO_x, and SO_x.

DIX800 CAS:7173-84-4 ***HR: 3***
O,O-DIETHYL-S-p-CHLOROPHENYL THIOMETHYLPHOSPHOROTHIOATE
mf: $C_{11}H_{16}ClO_3PS_2$ mw: 326.81

SYNS: S-((p-CHLOROPHENYLTHIO)METHYL)-O,O-DIETHYLPHOSPHORODITHIOATE ◇ DANIFOS

TOXICITY DATA with REFERENCE
orl-mus LD50:165 mg/kg BESAAT 15,118,69
skn-mus LD50:220 mg/kg BESAAT 15,118,69

SAFETY PROFILE: Poison by ingestion and skin contact. When heated to decomposition it emits very toxic fumes of Cl^-, PO_x, and SO_x.

DIY000 CAS:814-49-3 ***HR: 3***
DIETHYL CHLOROPHOSPHATE
mf: $C_4H_{10}ClO_3P$ mw: 172.56

PROP: Water white liquid. Bp: 60° @ 2 mm, d: 1.1915 @ 25°/25°, vap d: 5.94.

SYNS: CHLOROPHOSPHORIC ACID DIETHYL ESTER ◇ DIETHOXYPHOSPHORUS OXYCHLORIDE

TOXICITY DATA with REFERENCE
skn-rbt 10 mg/24H open MLD AIHAAP 23,95,62
orl-rat LD50:110 mg/kg AIHAAP 23,95,62
skn-rbt LD50:790 μg/kg AIHAAP 23,95,62

CONSENSUS REPORTS: Reported in EPA TSCA Inventory.

SAFETY PROFILE: Deadly poison by skin contact. Poison by ingestion. A cholinesterase inhibitor. See also PARATHION. Trace HCl catalyzes a hazardous reaction during the preparation of diethyl phosphate from diethyl chlorophosphate. When heated to decomposition it emits very toxic fumes of Cl^- and PO_x.

DIY200 CAS:1331-43-7 ***HR: 2***
DIETHYLCYCLOHEXANE (mixed isomers)
mf: $C_{10}H_{20}$ mw: 140.30

PROP: Liquid, insol in water. D: 0.8037 @ 20°/20°, bp: 174°, fp: −100°, flash p: 120°F (OC), autoign temp: 465°F, lel: 0.8% @ 140°F, uel: 6.0% @ 230°F.

TOXICITY DATA with REFERENCE
skn-rbt 10 mg/24H open MLD AIHAAP 23,95,62
ihl-rat LCLo:2000 ppm/4H AIHAAP 23,95,62
orl-rat LD50:64 g/kg AIHAAP 23,95,62
skn-rbt LDLo:2500 mg/kg AIHAAP 23,95,62

SAFETY PROFILE: Moderately toxic by inhalation and skin contact. Mildly toxic by ingestion. A skin irritant. Combustible when exposed to heat, flame, or oxidizers. To fight fire, use foam, mist, dry chemical. When heated to decomposition it emits acrid smoke and fumes.

DIY600 CAS:100-38-9 ***HR: 3***
N,N-DIETHYL CYSTEAMINE
mf: $C_6H_{15}NS$ mw: 133.28

SYNS: N-DIAETHYL CYSTEAMIN (GERMAN) ◇ DIETHYLAMINOETHANETHIOL ◇ 2-(DIETHYLAMINO)ETHANETHIOL ◇ 2-(DIETHY-

LAMINO)ETHYLMERCAPTAN ◇ β-DIETHYLAMINOETHYL MERCAPTAN ◇ DIETHYLCYSTEAMIN ◇ DIETHYLCYSTEAMINE ◇ N-DIETHYL CYSTEAMINE ◇ DIETHYL(2-MERCAPTOETHYL)AMINE

TOXICITY DATA with REFERENCE
orl-mus LD50:231 mg/kg GTPZAB 13(7),26,69
ihl-mus LD50:42500 mg/m^3 GTPZAB 13(7),26,69
ipr-mus LD50:96 mg/kg BCPCA6 14,289,65
scu-mus LD50:120 mg/kg AIPTAK 109,108,57

SAFETY PROFILE: Poison by ingestion, intraperitoneal, and subcutaneous routes. Mildly toxic by inhalation. When heated to decomposition it emits very toxic fumes of NO_x and SO_x. See also MERCAPTANS and AMINES.

DIY800 CAS:104-78-9 ***HR: 3***
N,N-DIETHYL-1,3-DIAMINOPROPANE
DOT: UN 2684
mf: $C_7H_{18}N_2$ mw: 130.27

PROP: Liquid. Bp: 165°-170°, flash p: 138°F (OC), d: 0.82, vap d: 4.48.

SYNS: 1-AMINO-3-(DIETHYLAMINO)PROPANE ◇ N-(3-DIETHYLAMINOPROPYL)AMINE ◇ N,N-DIETHYLAMINOPROPYLAMINE ◇ 3-(DIETHYLAMINO)PROPYLAMINE (DOT) ◇ DIETHYLAMINOTRIMETHYLENAMINE

TOXICITY DATA with REFERENCE
skn-rbt 100 μg/24H open AIHAAP 23,95,62
orl-rat LD50:1410 mg/kg UCDS** 2/27/67
skn-rbt LD50:750 mg/kg AIHAAP 23,95,62

CONSENSUS REPORTS: Reported in EPA TSCA Inventory.

DOT Classification: Corrosive Material; Label: Corrosive, Flammable Liquid.

SAFETY PROFILE: Moderately toxic by ingestion and skin contact. Corrosive to the eyes, skin, and mucous membranes. A sensitizer. See also AMINES. Flammable when exposed to heat or flame; can react with oxidizing materials. To fight fire, use foam, CO_2, dry chemical. When heated to decomposition it emits toxic fumes of NO_x.

DIZ100 CAS:1609-47-8 ***HR: 3***
DIETHYL DICARBONATE
mf: $C_6H_{10}O_5$ mw: 162.16

PROP: Viscous liquid; fruity odor. D: 1.12, visc (20°): 1.97 cp. Soluble in alc, esters, ketones, and hydrocarbons.

SYNS: BAYCOVIN ◇ DEPC ◇ DICARBONIC ACID DIETHYL ESTER ◇ DIETHYL ESTER of PYROCARBONIC ACID ◇ DIETHYL OXYDIFORMATE ◇ DIETHYL PYROCARBONATE ◇ DIETHYL PYROCARBONIC ACID ◇ DKD ◇ ETHYL PYROCARBONATE ◇ OXYDIFORMIC ACID DIETHYL ESTER ◇ PIREF ◇ PYROCARBONATE d'ETHYLE (FRENCH) ◇ PYROCARBONIC ACID, DIETHYL ESTER ◇ PYROKOHLENSAEURE DIAETHYL ESTER (GERMAN)

TOXICITY DATA with REFERENCE
orl-rat LD50:850 mg/kg FAONAU 51A,69,72
ipr-rat LD50:100 mg/kg ZLUFAR 114,292,61
orl-mus LD50:2027 mg/kg ZLUFAR 139,287,69
orl-dog LD50:500 mg/kg FAONAU 53A,52,74
orl-cat LDLo:100 mg/kg FAONAU 51A,69,72
orl-rbt LDLo:500 mg/kg FAONAU 51A,69,72

CONSENSUS REPORTS: Reported in EPA TSCA Inventory.

SAFETY PROFILE: Poison by ingestion and intraperitoneal routes. Concentrated DEPC is irritating to eyes, mucous membranes, and skin. When heated to decomposition it emits acrid smoke and fumes. See also ESTERS.

DJA200 CAS:3152-41-8 ***HR: 3***
O,O-DIETHYL-S-(3,4-DICHLOROPHENYL-THIO) METHYL PHOSPHOROTHIOATE
mf: $C_{11}H_{15}Cl_2O_2PS_3$ mw: 377.31

SYNS: S-((3,4-DICHLOROPHENYLTHIO)METHYL)-O,O-DIETHYL PHOSPHORODITHIOATE ◇ ENT 25,555-X ◇ G 27365 ◇ GEIGY G-27365

TOXICITY DATA with REFERENCE
orl-rat LD50:89 mg/kg ARSIM* 20,10,66
orl-mus LD50:175 mg/kg ARSIM* 20,10,66
orl-ckn LD50:143 mg/kg TXAPA9 11,49,67

SAFETY PROFILE: Poison by ingestion. When heated to decomposition it emits very toxic fumes of Cl^-, PO_x, and SO_x.

DJA300 CAS:1474-80-2 ***HR: 3***
DIETHYL DI(DIMETHYLAMIDO)PYROPHOSPHATE (unsymmetrical)
mf: $C_8H_{22}N_2O_5P_2$ mw: 288.26

SYNS: COMPOUND 6515 ◇ DIETHYL BIS-DIMETHYL PYROPHOSPHORDIAMIDE asym ◇ unsym-DIETHYL BIS(DIMETHYLAMIDO)PYROPHOSPHATE

TOXICITY DATA with REFERENCE
orl-rat LD50:3800 μg/kg JPETAB 107,464,53
skn-rat LD50:5 mg/kg JPETAB 107,464,53
ipr-rat LD50:2400 μg/kg JPETAB 107,464,53
ipr-mus LD50:4700 μg/kg JPETAB 107,464,53
ivn-dog LD50:10 mg/kg JPETAB 107,464,53
ipr-gpg LD50:5 mg/kg JPETAB 107,464,53

SAFETY PROFILE: Poison by ingestion, skin contact, intravenous, and intraperitoneal routes. When heated to decomposition it emits toxic fumes of PO_x and NO_x. See also ESTERS.

DJA325 CAS:10161-85-0 ***HR: 3***
O,O-DIETHYL Se-(2-DIETHYLAMINOETHYL) PHOSPHOROSELENOATE
mf: $C_{10}H_{24}NO_3PSe$ mw: 316.28

SYN: PHOSPHOROSELENOIC ACID, Se-(2-(DIETHYLAMINO) ETHYL) O,O-DIETHYL ESTER

TOXICITY DATA with REFERENCE
scu-mus LD50:60 μg/kg JMCMAR 10,115,67

OSHA PEL: TWA 0.2 mg(Se)/m^3
ACGIH TLV: TWA 0.2 mg(Se)/m^3

SAFETY PROFILE: Poison by subcutaneous route. When heated to decomposition it emits toxic fumes of NO_x, PO_x, and Se.

DJA400 CAS:78-53-5 ***HR: 3***
O,O-DIETHYL-S-(2-DIETHYLAMINOETHYL) THIOPHOSPHATE
mf: $C_{10}H_{24}NO_3PS$ mw: 269.38

PROP: Liquid. Bp: 110° @ 0.2 mm, mp: 98°.

SYNS: AMITON ◇ CHIPMAN 6200 ◇ CITRAM ◇ S-(DIETHYLAMINOETHYL)-O,O-DIETHYL PHOSPHOROTHIOATE ◇ S-(2-(DIETHYLAMINO)ETHYL)PHOSPHOROTHIOICACID-O,O-DIETHYL ESTER ◇ DIETHYL-S-2-DIETHYLAMINOETHYL PHOSPHOROTHIOATE ◇ (2-DIETHYLAMINO)ETHYLPHOSPHOROTHIOIC ACID-O,O-DIETHYL ESTER ◇ O,O-DIETHYL-S-(β-DIETHYLAMINO)ETHYL PHOSPHOROTHIOLATE ◇ O,O-DIETHYL-S-DIETHYLAMINOETHYL PHOSPHOROTHIOLATE ◇ O,O-DIETHYL-S-2-DIETHYLAMINOETHYL PHOSPHOROTHIOATE ◇ O,O-DIETHYL-S-2-DIETHYLAMINOETHYL PHOSPHOROTHIOLATE ◇ DSDP ◇ ENT 24,980-X ◇ INFERNO ◇ METRAMAC ◇ METRAMAK ◇ R-5,158 ◇ RHODIA-6200 ◇ TETRAM

TOXICITY DATA with REFERENCE
orl-rat LD50:3300 μg/kg ARSIM* 20,7,66
scu-rat LD50:150 μg/kg CJPPA3 44,745,66
scu-mus LD50:190 μg/kg CJPPA3 46,109,68
scu-rbt LD50:125 μg/kg CJJPA3 46,109,68
scu-gpg LD50:80 μg/kg CJPPA3 46,109,68
scu-ham LD50:210 μg/kg CJPPA3 46,109,68

CONSENSUS REPORTS: EPA Extremely Hazardous Substances List.

SAFETY PROFILE: A deadly poison by ingestion and subcutaneous routes. A cholinesterase inhibitor. An insecticide. See also PARATHION. When heated to decomposition it emits very toxic fumes of NO_x, PO_x, and SO_x.

DJA600 CAS:10140-08-6 ***HR: 3***
1-(2-(2-(2,6-DIETHYL-α-(2,6-DIETHYLPHENYL) BENZYLOXY)ETHOXY)ETHYL)-4-METHYLPIPERAZINE
mf: $C_{30}H_{46}N_2O_2$ mw: 466.78

TOXICITY DATA with REFERENCE
ipr-mus LD50:325 mg/kg ARZNAD 16,1557,66
scu-mus LD50:500 mg/kg ARZNAD 16,1557,66

SAFETY PROFILE: Poison by intraperitoneal route. Moderately toxic by subcutaneous route. When heated to decomposition it emits toxic fumes of NO_x.

DJA800 CAS:106-20-7 ***HR: 2***
2,2′-DIETHYLDIHEXYLAMINE
mf: $C_{16}H_{35}N$ mw: 241.52

PROP: Water-white liquid with sltly ammoniacal odor. Flash p: 270°F (OC), d: 0.8062 @ 20°/20°, vap d: 8.35, bp: 281.1°.

SYNS: BIS-2-ETHYLHEXYLAMIN ◇ DI(2-ETHYLHEXYL)AMINE ◇ 2-ETHYL-N-(2-ETHYLHEXYL)-1-HEXANAMINE

TOXICITY DATA with REFERENCE
skn-rbt 10 mg/24H open SEV JIHTAB 31,60,49
skn-rbt 500 mg open MLD UCDS** 8/9/68
skn-rbt 500 mg/24H SEV 28ZPAK -,62,72
eye-rbt 750 μg SEV AJOPAA 29,1363,46
eye-rbt 50 μg/24H SEV 28ZPAK -,62,72
orl-rat LD50:1640 mg/kg JIHTAB 31,60,49
ipr-mus LD50:800 mg/kg CBCCT* 2,184,50
skn-rbt LD50:1190 mg/kg UCDS** 8/9/68

CONSENSUS REPORTS: Reported in EPA TSCA Inventory.

SAFETY PROFILE: Moderately toxic by ingestion, skin contact, and intraperitoneal routes. A severe skin and eye irritant. Combustible when exposed to heat or flame; can react with oxidizing materials. To fight fire, use alcohol foam, foam, CO_2, dry chemical. When heated to decomposition it emits toxic fumes of NO_x. See also AMINES.

DJB000 CAS:2767-55-7 ***HR: 3***
DIETHYLDIIODOSTANNANE
mf: $C_4H_{10}I_2Sn$ mw: 430.63

PROP: Very sltly sol white crystals. Mp: 45°, bp: 240-245° (decomp).

SYN: DIETHYLTIN DIIODIDE

TOXICITY DATA with REFERENCE
orl-rat LDLo:100 mg/kg BJPCAL 10,16,55
ipr-rat LDLo:26 mg/kg BJPCAL 10,16,55

OSHA PEL: TWA 0.1 mg(Sn)/m^3 (skin)
ACGIH TLV: TWA 0.1 mg(Sn)/m^3 (skin) (Proposed: TWA 0.1 mg(Sn)/m^3; STEL 0.2 mg(Sn)/m^3 (skin))
NIOSH REL: (Organotin Compounds) TWA 0.1 mg(Sn)/m^3

SAFETY PROFILE: Poison by ingestion and intraperitoneal routes. See also TIN COMPOUNDS and IO-

DIDES. When heated to decomposition it emits toxic fumes of I^-.

DJB200 CAS:7773-34-4 ***HR: 3***
α,α'-DIETHYL-4,4'-DIMETHOXYSTILBENE
mf: $C_{20}H_{24}O_2$ mw: 296.44

SYNS: 3,4-BIS(p-METHOXYPHENYL)-3-HEXENE ◇ DEPOT-OESTROMENINE ◇ DEPOT-OESTROMON ◇ 3,4-DIANISYL-3-HEXENE ◇ trans-α,α'-DIETHYL-4,4'-DIMETHOXYSTILBENE ◇ (E)-1,1'-(1,2-DIETHYL-1,2-ETHENE-DIYL)BIS(4-METHOXYBENZENE) ◇ DIETHYLSTILBESTROL DIMETHYL ETHER ◇ DIMESTROL ◇ 4,4'-DIMETHOXY-α,β-DIETHYLSTILBENE ◇ STILBESTROL DIMETHYL ETHER ◇ SYNTHILA

TOXICITY DATA with REFERENCE
scu-ham TDLo:560 g/kg/35W-I:ETA CNREA8 31,1251,71

SAFETY PROFILE: Questionable carcinogen with experimental tumorigenic data. When heated to decomposition it emits acrid and irritating fumes. See also DIETHYLSTILBESTEROL.

DJB400 CAS:17010-64-9 ***HR: 2***
3',4'-DIETHYL-4-DIMETHYLAMINOAZOBENZENE
mf: $C_{18}H_{23}N_3$ mw: 281.44

SYNS: BENZENAMINE,4-((3,4-DIETHYLPHENYL)AZO)-N,N-DIMETHYL-(9CI) ◇ p-((3,4-DIETHYLPHENYL)AZO)-N,N-DIMETHYLANILINE ◇ N,N-DIMETHYL-p-((3,4-DIETHYLPHENYL)AZO)ANILINE ◇ 3',4'-Et2-DAB

TOXICITY DATA with REFERENCE
orl-rat TDLo:8467 mg/kg/36W-C:CAR CBINA8 53,107,85
orl-rat TD:2511 mg/kg/17W-I:ETA CNREA8 30,1520,70

SAFETY PROFILE: Questionable carcinogen with experimental carcinogenic and tumorigenic data. When heated to decomposition it emits toxic fumes of NO_x.

DJB600 CAS:64048-13-1 ***HR: 3***
p-DIETHYL-p'-DIMETHYLTHIOPYROPHOSPHATE
mf: $C_6H_{16}O_5P_2S_2$ mw: 294.28

SYN: p'-DIETHYL-p-DIMETHYLTHIOPYROPHOSPHATE

TOXICITY DATA with REFERENCE
ims-rat LD50:1 mg/kg CJCHAG 34,1819,56
ipr-mus LD50:1800 μg/kg CJCHAG 34,1819,56
ims-mus LD50:1 mg/kg CJCHAG 34,1819,56

SAFETY PROFILE: A deadly poison by intramuscular and intraperitoneal routes. When heated to decomposition it emits very toxic fumes of SO_x and PO_x.

DJB800 CAS:7346-14-7 ***HR: 3***
N,N'-DIETHYL-N,N'-DINITROSOETHYLENEDIAMINE
mf: $C_6H_{14}N_4O_2$ mw: 174.24

SYNS: N,N'-DINITROSO-N,N'-DIETHYLETHYLENEDIAMINE ◇ NSC 62579

TOXICITY DATA with REFERENCE
orl-rat TDLo:12 mg/kg/1Y-I:CAR JNCIAM 41,985,68
orl-rat TD:34 mg/kg/50W-I:CAR JNCIAM 41,985,68

SAFETY PROFILE: Questionable carcinogen with experimental carcinogenic data. When heated to decomposition it emits toxic fumes of NO_x. See also N-NITROSO COMPOUNDS.

DJC000 CAS:72-56-0 ***HR: 3***
DIETHYLDIPHENYL DICHLOROETHANE
mf: $C_{18}H_{20}Cl_2$ mw: 307.28

SYNS: 1,1-BIS(p-ETHYLPHENYL)-2,2-DICHLOROETHANE ◇ 2,2-BIS (p-ETHYLPHENYL)-1,1-DICHLOROETHANE ◇ 1,1-DICHLORO-2,2-BIS (p-ETHYLPHENYL)ETHANE ◇ 1,1-DICHLORO-2,2-BIS(4-ETHYLPHENYL)ETHANE ◇ 2,2-DICHLORO-1,1-BIS(p-ETHYLPHENYL)ETHANE ◇ α,α-DICHLORO-2,2-BIS(p-ETHYLPHENYL)ETHANE ◇ DI(p-ETHYLPHENYL)DICHLOROETHANE ◇ ETHYLAN ◇ p,p-ETHYL DDD ◇ p,p'-ETHYL-DDD ◇ NCI-C02868 ◇ PERTHANE ◇ Q-137

TOXICITY DATA with REFERENCE
mma-sat 333 μg/plate NTPTB* APR 82
scu-mus TDLo:900 mg/kg (6-14D preg):TER NTIS** PB223-160
scu-mus TDLo:900 mg/kg (6-14D preg):REP NTIS** PB223-160
orl-mus TDLo:210 g/kg/2Y-C:CAR TUMOAB 66,277,80
orl-mus TD:547 g/kg/2Y-C:ETA NCITR* NCI-CG-TR-156,79
orl-rat LD50:6600 mg/kg SPEADM 78-1,16,78
ivn-rat LD50:73 mg/kg AIPTAK 103,404,55
orl-mus LD50:6600 mg/kg 31ZOAD 1,142,68
ivn-mus LD50:173 mg/kg AIPTAK 103,404,55
orl-bwd LD50:9000 mg/kg DOEAAH 35,25,79

CONSENSUS REPORTS: NCI Carcinogenesis Bioassay (feed); Clear Evidence: mouse NCITR* NCI-CG-TR-156,79; No Evidence: rat NCITR* NCI-CG-TR-156,79.

SAFETY PROFILE: Poison by intravenous route. Mildly toxic by ingestion. Questionable carcinogen with experimental carcinogenic and tumorigenic data. Experimental teratogenic and reproductive effects. Mutation data reported. A pesticide. When heated to decomposition it emits toxic fumes of Cl^-. See also CHLORINATED HYDROCARBONS, ALIPHATIC.

DJC200 CAS:41365-24-6 ***HR: D***
N,N-DIETHYL-N,N-DIPHENYLTHIURAMDISULFIDE
mf: $C_{18}H_{20}N_2S_4$ mw: 392.64

SYNS: DIETHYLDIPHENYLTHIURAM DISULFIDE ◇ EKAGON TE ◇ THIOPEROXYDICARBONIC DIAMIDE, N,N'-DIETHYL-N,N'-DIPHENYL- ◇ THIURAM EF

TOXICITY DATA with REFERENCE
cyt-rat-orl 3500 mg/kg/10W GTPZAB 24(3),20,80
dlt-rat-orl 3500 mg/kg/10W GTPZAB 24(3),20,80
orl-rat TDLo:3800 mg/kg (male 76D pre):REP GTPZAB 24(3),20,80

SAFETY PROFILE: Experimental reproductive effects. Mutation data reported. When heated to decomposition it emits toxic fumes of NO_x and SNO_x.

DJC400 CAS:85-98-3 ***HR: 3***
1,3-DIETHYL-1,3-DIPHENYLUREA
mf: $C_{17}H_{20}N_2O$ mw: 268.39

PROP: Colorless crystals. Mp: 73°, d: 1.12, bp: 326°, flash p: 302°F (CC), vap d: 9.3.

SYNS: BIS(N-ETHYL-N-PHENYL)UREA ◇ N,N-DIETHYLCARBANILIDE ◇ N,N'-DIETHYL-N,N'-DIPHENYLUREA ◇ sym-DIETHYLDIPHENYLUREA ◇ USAF EK-1047

TOXICITY DATA with REFERENCE
ipr-mus LD50:200 mg/kg NTIS** AD277-689

CONSENSUS REPORTS: Reported in EPA TSCA Inventory.

SAFETY PROFILE: Poison by intraperitoneal route. Combustible when exposed to heat or flame. Probably a slight explosion hazard, although it is a component of smokeless explosive mixtures. When heated to decomposition it burns and emits very toxic fumes of NO_x. To fight fire, use dry chemical, CO_2, spray or mist.

DJC600 CAS:110-81-6 ***HR: 3***
DIETHYLDISULFIDE
mf: $C_4H_{10}S_2$ mw: 122.26

PROP: Liquid. Bp: 154°, fp: −101.5°, d: 0.99267 @ 20°/4°, vap d: 4.22, vap press: 4.28 mm @ 25°.

SYN: DIETHYLDISULFID (CZECH)

TOXICITY DATA with REFERENCE
skn-rbt 500 mg/24H MLD 28ZPAK -,171,72
eye-rbt 100 mg/24H MOD 28ZPAK -,171,72
orl-rat LD50:2030 mg/kg 28ZPAK -,171,72

CONSENSUS REPORTS: Reported in EPA TSCA Inventory.

SAFETY PROFILE: Moderately toxic by ingestion. A skin and eye irritant. Flammable when exposed to heat or flame; can react vigorously with oxidizing materials. To fight fire, use foam or dry chemical. See also SULFIDES and SULFATES.

DJC800 CAS:147-84-2 ***HR: 2***
DIETHYLDITHIOCARBAMIC ACID
mf: $C_5H_{11}NS_2$ mw: 149.29

SYNS: DIETHYLDITHIOCARBAMINIC ACID ◇ DIETHYLDITHIONE

TOXICITY DATA with REFERENCE
ivn-rat LD50:1250 mg/kg PHMCAA 3,62,61
ivn-mus LD50:1750 mg/kg PHMCAA 3,62,61
ivn-dog LD50:1000 mg/kg PHMCAA 3,62,61

CONSENSUS REPORTS: Reported in EPA TSCA Inventory.

SAFETY PROFILE: Moderately toxic by intravenous route. See also CARBAMATES. When heated to decomposition it emits very toxic fumes of NO_x and SO_x.

DJC875 CAS:69654-93-9 ***HR: 3***
DIETHYLDITHIOCARBAMIC ACID ANHYDROSULFIDE with DIMETHYLTHIOCARBAMIC ACID
mf: $C_8H_{16}N_2OS_2$ mw: 220.38

SYNS: DDS ◇ DIMETHYLCARBAMYL DIETHYLTHIOCARBAMYL SULFIDE

TOXICITY DATA with REFERENCE
ipr-rat LD50:900 μg/kg AIPTAK 108,27,56
ipr-mus LD50:800 μg/kg AIPTAK 108,27,56
ivn-dog LD50:200 μg/kg AIPTAK 108,27,56

SAFETY PROFILE: A deadly poison by intravenous and intraperitoneal routes. When heated to decomposition it emits toxic fumes of SO_x and NO_x. See also CARBAMATES and SULFIDES.

DJD000 CAS:1518-58-7 ***HR: 2***
DIETHYLDITHIOCARBAMIC ACID DIETHYLAMINE SALT
mf: $C_9H_{21}N_2S_2$ mw: 221.44

SYN: USAF EK-2635

TOXICITY DATA with REFERENCE
ipr-mus LD50:500 mg/kg NTIS** AD277-689

CONSENSUS REPORTS: Reported in EPA TSCA Inventory.

SAFETY PROFILE: Moderately toxic by intraperitoneal route. When heated to decomposition it emits very toxic fumes of NO_x and SO_x. See also CARBAMATES.

DJD200 CAS:17549-30-3 ***HR: 2***
DIETHYLDITHIOCARBAMIC ACID LEAD(II) SALT
mf: $C_{10}H_{20}N_2S_4 \cdot Pb$ mw: 503.75

TOXICITY DATA with REFERENCE
ipr-rat LDLo:500 mg/kg NCNSA6 5,30,53

NIOSH REL: (Inorganic Lead) TWA 0.10 mg(Pb)/m^3

CONSENSUS REPORTS: Lead and its compounds are on the Community Right-To-Know List.

SAFETY PROFILE: Moderately toxic by intraperitoneal route. See also LEAD COMPOUNDS and CARBAMATES. When heated to decomposition it emits very toxic fumes of Pb, SO_x, and NO_x.

DJD400 CAS:136-92-5 ***HR: 3***
DIETHYLDITHIOCARBAMIC ACID SELENIUM(II) SALT
mf: $C_{10}H_{20}N_2S_4 \cdot Se$ mw: 375.52

SYNS: ETHYL SELENAC ◇ SELENIUM DIETHYLDITHIOCARBAMATE

TOXICITY DATA with REFERENCE
orl-rat LDLo:250 mg/kg NCNSA6 5,40,53
ipr-rat LDLo:50 mg/kg NCNSA6 5,40,53

CONSENSUS REPORTS: Selenium and its compounds are on the Community Right-To-Know List.

OSHA PEL: TWA 0.2 mg(Se)/m^3
ACGIH TLV: TWA 0.2 mg(Se)/m^3
DFG MAK: 0.1 mg(Se)/m^3

SAFETY PROFILE: Poison by ingestion and intraperitoneal routes. See also SELENIUM COMPOUNDS and CARBAMATES. When heated to decomposition it emits very toxic fumes of NO_x, SO_x, and Se.

DJD600 CAS:111-46-6 ***HR: 3***
DIETHYLENE GLYCOL
mf: $C_4H_{10}O_3$ mw: 106.14

$(HOC_2H_4)_2O$

PROP: Clear, colorless, practically odorless, syrupy liquid. Bp: 245.8°, fp: −8°, flash p: 255°F, d: 1.1184 @ 20°/20°, autoign temp: 444°F, vap press: 1 mm @ 91.8°, vap d: 3.66.

SYNS: BIS(2-HYDROXYETHYL) ETHER ◇ BRECOLANE NDG ◇ CARBITOL ◇ DEACTIVATOR E ◇ DEACTIVATOR H ◇ DEG ◇ DICOL ◇ DIGLYCOL ◇ DIHYDROXYDIETHYL ETHER ◇ β,β′-DIHYDROXYDIETHYL ETHER ◇ 2,2′-DIHYDROXYETHYL ETHER ◇ DISSOLVANT APV ◇ ETHYLENE DIGLYCOL ◇ GLYCOL ETHER ◇ GLYCOL ETHYL ETHER ◇ 3-OXAPENTANE-1,5-DIOL ◇ 3-OXA-1,5-PENTANEDIOL ◇ 2,2′-OXYBISETHANOL ◇ 2,2′-OXYDIETHANOL ◇ TL4N

TOXICITY DATA with REFERENCE
skn-hmn 112 mg/3D-I MLD 85DKA8 -,127,77
skn-rbt 500 mg MLD 34ZIAG -,731,69
eye-rbt 50 mg MLD JPETAB 42,355,31
orl-rat TDLo:50 g/kg (1-20D preg):TER OYYAA2 27,801,84
orl-rat TDLo:890 g/kg/53W-C:CAR JIHTAB 28,40,46
orl-rat TD:1752 g/kg/2Y-C:ETA IMSUAI 36,55,67
scu-rat TDLo:2500 mg/kg/82W-I:NEO VINIT* #6801-83
orl-hmn LD50:1000 mg/kg JIHTAB 21,173,39
orl-cld TDLo:2400 mg/kg JOPDAB 109,731,86
orl-rat LD50:12565 mg/kg NPIRI* 1,25,74
ipr-rat LD50:7700 mg/kg 38MKAJ 2C,3836,82
scu-rat LD50:18800 mg/kg 38MKAJ 2C,3836,82
orl-mus LD50:23700 mg/kg FEPRA7 4,142,45
ihl-mus LCLo:130 mg/m^3/2H GTPZAB 10(12),30,66
ipr-mus LD50:9719 mg/kg FEPRA7 6,342,47
scu-mus LDLo:5 g/kg JPETAB 42,355,31
orl-dog LD50:9000 mg/kg JPETAB 67,101,39
orl-cat LD50:3300 mg/kg JIHTAB 21,173,39
skn-rbt LD50:11890 mg/kg NPIRI* 1,25,74
ivn-rbt LD50:2000 mg/kg JPETAB 59,93,37

CONSENSUS REPORTS: Reported in EPA TSCA Inventory. Glycol ether compounds are on the Community Right-To-Know List.

SAFETY PROFILE: Moderately toxic to humans by ingestion. Poison by experimentally inhalation. Moderately toxic by ingestion and intravenous routes. Questionable carcinogen with experimental carcinogenic, tumorigenic, and teratogenic data. An eye and human skin irritant. Combustible when exposed to heat or flame; can react with oxidizing materials. To fight fire, use alcohol foam, water, CO_2, dry chemical. Mixtures with sodium hydroxide decompose exothermically when heated to 230°C and release explosive hydrogen gas. When heated to decomposition it emits acrid smoke and irritating fumes. See also GLYCOL ETHERS.

DJD700 CAS:13988-26-6 ***HR: 2***
DIETHYLENE GLYCOL BISPHTHALATE
mf: $C_{12}H_{12}O_5$ mw: 236.24

SYNS: 2,5,8-BENZOTRIOXACYCLOUNDECIN-1,9-DIONE,3,4,6,7-TETRAHYDRO-(9CI) ◇ HOWFLEX GBP

TOXICITY DATA with REFERENCE
orl-rat TDLo:45 g/kg/13W-C:ETA EJCAAH 5,415,69

CONSENSUS REPORTS: Reported in EPA TSCA Inventory.

SAFETY PROFILE: Questionable carcinogen with experimental tumorigenic data. When heated to decomposition it emits acrid smoke and irritating fumes.

DJD800 CAS:4246-51-9 ***HR: 2***
DIETHYLENE GLYCOL DI(3-AMINOPROPYL) ETHER
mf: $C_{10}H_{24}N_2O_3$ mw: 220.36

SYNS: DI(3-AMINOPROPYL) ETHER of DIETHYLENE GLYCOL ◇ 3,3′-(OXYBIS(2,1-ETHANEDIYLOXY))BIS-1-PROPANAMINE

TOXICITY DATA with REFERENCE
orl-rat LD50:4290 mg/kg AIHAAP 30,470,69
skn-rbt LD50:2500 mg/kg AIHAAP 30,470,69

CONSENSUS REPORTS: Reported in EPA TSCA Inventory. Glycol ether compounds are on the Community Right-To-Know List.

SAFETY PROFILE: Moderately toxic by skin contact. Mildly toxic by ingestion. When heated to decomposition it emits toxic fumes of NO_x. See also GLYCOL ETHERS.

DJE000 CAS:120-55-8 ***HR: 2***
DIETHYLENE GLYCOL DIBENZOATE
mf: $C_{18}H_{18}O_5$ mw: 314.36

PROP: Crystals. Mp: 70°, bp: 210°, flash p: 365°F, vap d: 9.38.

SYNS: BENZOIC ACID, DIESTER with DIETHYLENE GLYCOL ◇ DIBENZOYLDIETHYLENEGLYCOL ESTER

TOXICITY DATA with REFERENCE
orl-rat LD50:2830 mg/kg AIHAAP 23,95,62

CONSENSUS REPORTS: Reported in EPA TSCA Inventory. Glycol ether compounds are on the Community Right-To-Know List.

SAFETY PROFILE: Moderately toxic by ingestion. Combustible when exposed to heat or flame; can react with oxidizing materials. To fight fire use water, foam, CO_2, or dry chemical. When heated to decomposition it emits acrid smoke and fumes. See also GLYCOL ETHERS.

DJE200 CAS:4206-61-5 ***HR: D***
DIETHYLENE GLYCOL DIGLYCIDYL ETHER
mf: $C_{10}H_{18}O_5$ mw: 218.28

SYN: BIS(2-(2,3-EPOXYPROPOXY)ETHYL)ETHER

TOXICITY DATA with REFERENCE
cyt-rat-ipr 100 mg/kg BJPCAL 6,235,51

CONSENSUS REPORTS: Glycol ether compounds are on the Community Right-To-Know List.

SAFETY PROFILE: Mutation data reported. When heated to decomposition it emits acrid smoke and irritating fumes. See also GLYCOL ETHERS.

DJE400 CAS:693-21-0 ***HR: 3***
DIETHYLENE GLYCOL DINITRATE
DOT: UN 0075
mf: $C_4H_8N_2O_7$ mw: 196.14

$$O(C_2H_4ONO_2)_2$$

PROP: Liquid. Vap d: 6.76.

SYNS: BIS(HYDROXYAETHYL)-AETHER-DINITRAT(GERMAN) ◇ DIETHYLENEGLYCOL DINITRATE, containing at least 25% phlegmatizer (DOT) ◇ DIETHYLENGLYKOLDINITRATE (CZECH) ◇ DIGLYCOLDINITRAAT (DUTCH) ◇ DIGLYCOL (DINITRATE de) (FRENCH) ◇ DIGLYKOLDINITRAT (GERMAN) ◇ DI(HYDROXYETHYL) ETHER DINITRATE ◇ DINITRATE de DIETHYLENE-GLYCOL (FRENCH) ◇ DINITRODIGLICOL (ITALIAN) ◇ DINITRODIGLYKOL (CZECH)

TOXICITY DATA with REFERENCE
orl-rat LD50:777 mg/kg 28ZPAK -,116,72

CONSENSUS REPORTS: Reported in EPA TSCA Inventory. Glycol ether compounds are on the Community Right-To-Know List.

DOT Classification: Forbidden; Class A Explosive; Label: Explosive A (UN0075)

SAFETY PROFILE: Moderately toxic by ingestion. Ingestion of this compound can cause a drop in blood pressure and cardiac disturbances. A dangerous fire hazard when exposed to heat or flame; can react vigorously with oxidizing or reducing materials. A dangerous explosive sensitive to heat, shock, and vibration. Used in low freezing dynamites and some permissible explosives. Upon decomposition it emits toxic fumes of NO_x. See also GLYCOL ETHERS; NITRATES and EXPLOSIVES, HIGH.

DJE600 CAS:764-99-8 ***HR: 2***
DIETHYLENE GLYCOL DIVINYL ETHER
mf: $C_8H_{14}O_3$ mw: 158.22

SYNS: BIS(2-VINYLOXYETHYL)ETHER ◇ DVEDEG (RUSSIAN)

TOXICITY DATA with REFERENCE
skn-rbt 10 mg/24H open MLD AIHAAP 23,95,62
orl-rat LD50:3730 mg/kg AIHAAP 23,95,62
orl-mus LD50:2570 mg/kg GISAAA 42(3),12,77
skn-rbt LD50:14100 mg/kg AIHAAP 23,95,62

CONSENSUS REPORTS: Reported in EPA TSCA Inventory. Glycol ether compounds are on the Community Right-To-Know List.

SAFETY PROFILE: Moderately toxic by ingestion. Mildly toxic by skin contact. A skin irritant. When heated to decomposition it emits acrid smoke and fumes. See also GLYCOL ETHERS.

DJE800 CAS:1002-67-1 ***HR: 1***
DIETHYLENE GLYCOL ETHYL METHYL ETHER
mf: $C_7H_{16}O_3$ mw: 148.23

SYN: 2-ETHOXYETHYL-2-METHOXYETHYLETHER

TOXICITY DATA with REFERENCE
skn-rbt 10 mg/24H open MLD AIHAAP 23,95,62
orl-rat LD50:6500 mg/kg AIHAAP 23,95,62
skn-rbt LD50:7070 mg/kg AIHAAP 23,95,62

CONSENSUS REPORTS: Glycol ether compounds are on the Community Right-To-Know List.

SAFETY PROFILE: Mildly toxic by ingestion and skin contact. A skin irritant. When heated to decomposition it emits acrid smoke and irritating fumes. See also GLYCOL ETHERS.

DJF000 CAS:10143-53-0 ***HR: 1***
DIETHYLENE GLYCOL ETHYLVINYL ETHER
mf: $C_8H_{16}O_3$ mw: 160.24

SYN: 2-ETHOXYETHYL-2-(VINYLOXY)ETHYLETHER

TOXICITY DATA with REFERENCE
skn-rbt 10 mg/24H open MLD AIHAAP 23,95,62
orl-rat LD50:11 g/kg AIHAAP 23,95,62
skn-rbt LD50:8410 mg/kg AIHAAP 23,95,62

CONSENSUS REPORTS: Reported in EPA TSCA Inventory. Glycol ether compounds are on the Community Right-To-Know List.

SAFETY PROFILE: Mildly toxic by ingestion and skin contact. A skin irritant. When heated to decomposition it emits acrid smoke and fumes. See also GLYCOL ETHERS.

DJF200 CAS:112-34-5 ***HR: 2***
DIETHYLENE GLYCOL MONOBUTYL ETHER
mf: $C_8H_{18}O_3$ mw: 162.26

PROP: Colorless liquid. Mp: $-68.1°$, bp: 230.6°, flash p: 172°F, d: 0.9553 @ 20°/4°, autoign temp: 442°F, vap press: 0.02 mm @ 20°, vap d: 5.58.

SYNS: BUCB ◇ BUTOXYDIETHYLENE GLYCOL ◇ BUTOXYDIGLYCOL ◇ 2-(2-BUTOXYETHOXY)ETHANOL ◇ BUTYL CARBITOL ◇ o-BUTYL DIETHYLENE GLYCOL ◇ BUTYL DIOXITOL ◇ DIETHYLENE GLYCOL-n-BUTYL ETHER ◇ DIGLYCOL MONOBUTYL ETHER ◇ DOWANOL DB ◇ EKTASOLVE DB ◇ GLYCOL ETHER DB ◇ JEFFERSOL DB ◇ POLY-SOLV DB

TOXICITY DATA with REFERENCE
eye-rbt 5 mg SEV AJOPAA 29,1363,46
orl-rat LD50:6560 mg/kg UCDS** 1/31/66
unr-rat LD50:4500 mg/kg GISAAA 46(2),14,81
ipr-mus LD50:850 mg/kg FEPRA7 6,342,47
unr-mus LD50:6050 mg/kg GISAAA 46(2),14,81
skn-rbt LD50:4120 mg/kg UCDS** 1/31/66
orl-gpg LD50:2000 mg/kg JIHTAB 23,259,41

CONSENSUS REPORTS: Reported in EPA TSCA Inventory. Glycol ether compounds are on the Community Right-To-Know List.

SAFETY PROFILE: Moderately toxic by ingestion and intraperitoneal routes. Mildly toxic by skin contact. A severe eye irritant. Combustible when exposed to heat or flame; can react with oxidizing materials. To fight fire, use alcohol foam, CO_2, or dry chemical. When heated to decomposition it emits acrid smoke and irritating fumes. See also GLYCOL ETHERS.

DJF400 CAS:16672-39-2 ***HR: 1***
DI(ETHYLENE GLYCOL MONOBUTYL ETHER) PHTHALATE
mf: $C_{24}H_{38}O_8$ mw: 454.62

TOXICITY DATA with REFERENCE
orl-rat LD50:9700 mg/kg JIHTAB 23,259,41

CONSENSUS REPORTS: Reported in EPA TSCA Inventory. Glycol ether compounds are on the Community Right-To-Know List.

SAFETY PROFILE: Mildly toxic by ingestion. When heated to decomposition it emits acrid smoke and irritating fumes. See also GLYCOL ETHERS.

DJF600 CAS:10143-54-1 ***HR: 2***
DIETHYLENE GLYCOL MONO-2-CYANOETHYL ETHER
mf: $C_7H_{13}NO_3$ mw: 159.21

TOXICITY DATA with REFERENCE
skn-rbt 10 mg/24H open MLD AIHAAP 23,95,62
orl-rat LD50:13400 mg/kg AIHAAP 23,95,62

CONSENSUS REPORTS: Cyanide and its compounds as well as glycol ether compounds are on the Community Right-To-Know List.

SAFETY PROFILE: Mildly toxic by ingestion. A skin irritant. When heated to decomposition it emits toxic fumes of NO_x and CN^-. See also GLYCOL ETHERS and NITRILES.

DJF800 CAS:18912-80-6 ***HR: 2***
DIETHYLENE GLYCOL MONOISOBUTYL ETHER
mf: $C_8H_{18}O_3$ mw: 162.26

SYNS: EKTASOLVE DIB ◇ ISOBUTOXY-2-ETHOXY-2-ETHANOL ◇ 2-(2-ISOBUTOXYETHOXY)ETHANOL

TOXICITY DATA with REFERENCE
orl-rat LD50:4920 mg/kg TXAPA9 28,313,74
ihl-rat LClo:6700 ppm/6H KODAK* 21MAY71
skn-rbt LD50:3560 mg/kg TXAPA9 28,313,74

CONSENSUS REPORTS: Reported in EPA TSCA Inventory. Glycol ether compounds are on the Community Right-To-Know List.

SAFETY PROFILE: Moderately toxic by skin contact. Mildly toxic by ingestion and inhalation. When heated to decomposition it emits acrid smoke and irritating fumes. See also GLYCOL ETHERS.

DJG000 CAS:111-77-3 ***HR: 2***
DIETHYLENE GLYCOL MONOMETHYL ETHER
mf: $C_5H_{12}O_3$ mw: 120.17

PROP: Hygroscopic, water-white liquid. Bp: 194.2°, flash p: 200°F (OC), d: 1.0354 @ 20°/4°, vap press: 0.2 mm @ 20°, vap d: 4.14, mp: < −84°.

SYNS: DIETHYLENE GLYCOL METHYL ETHER ◇ DIGLYCOL MONOMETHYL ETHER ◇ DOWANOL DM ◇ ETHYLENE DIGLYCOL MONOMETHYL ETHER ◇ MECB ◇ METHOXYDIGLYCOL ◇ 2-(2-METHOXYETHOXY)ETHANOL ◇ β-METHOXY-β'-HYDROXYDIETHYL ETHER ◇ METHYL CARBITOL ◇ POLY-SOLV DM

TOXICITY DATA with REFERENCE
eye-rbt 500 mg MOD UCDS** 4/21/67
eye-rbt 500 mg/24H MLD 85JCAE -,628,86
orl-mus TDLo:32 g/kg (female 7-14D post):REP EVHPAZ 57,141,84
orl-rat TDLo:21650 mg/kg (female 7-16D post):TER FAATDF 6,430,86
orl-rat LD50:5500 mg/kg 38MKAJ 2C,3957,82
ipr-rat LDLo:3000 mg/kg JPPMAB 11,150,59
skn-rbt LD50:650 mg/kg UCDS** 4/21/67
orl-gpg LD50:4160 mg/kg JIHTAB 23,259,41

CONSENSUS REPORTS: Reported in EPA TSCA Inventory. Glycol ether compounds are on the Community Right-To-Know List.

SAFETY PROFILE: Moderately toxic by skin contact and intraperitoneal routes. Mildly toxic by ingestion. An experimental teratogen. Other experimental reproductive effects. An eye irritant. Combustible when exposed to heat or flame; can react with oxidizing materials. Reacts violently with $Ca(OCl)_2$; chlorosulfonic acid; oleum. To fight fire, use dry chemical, alcohol foam, water spray or mist, CO_2. When heated to decomposition it emits acrid smoke and irritating fumes. See also GLYCOL ETHERS.

DJG200 CAS:10143-56-3 ***HR: 2***
DIETHYLENEGLYCOL-MONO-2-METHYLPENTYL ETHER
mf: $C_{10}H_{22}O_3$ mw: 190.32

SYNS: DIETHYLENE GLYCOL MONOMETHYLPENTYL ETHER ◇ 2-METHYLPENTYL CARBITOL ◇ 2-(2-((2-METHYLPENTYL)OXY)ETHOXY)ETHANOL

TOXICITY DATA with REFERENCE
skn-rbt 10 mg/24H open MLD AIHAAP 23,95,62
eye-rbt 100 mg SEV 34ZIAG -,730,69
orl-rat LD50:5660 mg/kg AIHAAP 23,95,62
skn-rbt LD50:1580 mg/kg AIHAAP 23,95,62

CONSENSUS REPORTS: Glycol ether compounds are on the Community Right-To-Know List.

SAFETY PROFILE: Moderately toxic by skin contact. Mildly toxic by ingestion. A skin and severe eye irritant. When heated to decomposition it emits acrid smoke and irritating fumes. See also GLYCOL ETHERS.

DJG400 CAS:929-37-3 ***HR: 1***
DIETHYLENE GLYCOL MONOVINYL ETHER
mf: $C_6H_{12}O_3$ mw: 132.18

SYNS: DEGMVE (RUSSIAN) ◇ DIETHYLENEGLYCOL VINYL ETHER ◇ DIETHYLEN-GLYCOL MONOVINYL ESTER ◇ 2-(2-(ETHENYLOXY)ETHOXY)ETHANOL

TOXICITY DATA with REFERENCE
orl-rat LD50:4930 mg/kg GISAAA 42(3),12,77
orl-mus LD50:4450 mg/kg GISAAA 39(11),94,74

CONSENSUS REPORTS: Glycol ether compounds are on the Community Right-To-Know List.

SAFETY PROFILE: Mildly toxic by ingestion. When heated to decomposition it emits acrid smoke and irritating fumes. See also GLYCOL ETHERS.

DJG600 CAS:111-40-0 ***HR: 3***
DIETHYLENETRIAMINE
DOT: UN 2079
mf: $C_4H_{13}N_3$ mw: 103.20

$$HN(C_2H_4NH_2)_2$$

PROP: Yellow, viscous liquid; mild ammoniacal odor. Mp: −39°, bp: 207°, flash p: 215°F (OC), d: 0.9586 @ 20°/20°, autoign temp: 750°F, vap press: 0.22 mm @ 20°, vap d: 3.48.

SYNS: AMINOETHYLETHANDIAMINE ◇ N-(2-AMINOETHYL) ETHYLENEDIAMINE ◇ 3-AZAPENTANE-1,5-DIAMINE ◇ BIS(2-AMINOETHYL)AMINE ◇ BIS(β-AMINOETHYL)AMINE ◇ D.E.H. 20 ◇ DETA ◇ 2,2'-DIAMINODIETHYLAMINE ◇ 2,2'-IMINOBISETHYLAMINE

TOXICITY DATA with REFERENCE
skn-rbt 10 mg/24H open SEV JIHTAB 31,60,49
skn-rbt 500 mg open MOD UCDS** 12/30/71
skn-rbt 500 mg IYKEDH 6,170,75
eye-rbt 750 μg open SEV JIHTAB 31,60,49
orl-rat LD50:1080 mg/kg AMIHAB 17,129,58
ipr-rat LD50:74 mg/kg AMIHAB 17,129,58
ipr-mus LD50:71 mg/kg AMIHAB 17,129,58
skn-rbt LD50:1090 mg/kg JIHTAB 31,60,49
skn-gpg LD50:162 mg/kg JIHTAB 26,269,44

CONSENSUS REPORTS: Reported in EPA TSCA Inventory.

OSHA PEL: TWA 1 ppm
ACGIH TLV: TWA 1 ppm (skin)
DOT Classification: Corrosive Material; Label: Corrosive.

SAFETY PROFILE: Poison by skin contact and intraperitoneal routes. Moderately toxic by ingestion. Corrosive. A severe skin and eye irritant. High concentration of vapors causes irritation of respiratory tract, nausea, and vomiting. Repeated exposures can cause asthma and sensitization of skin. Combustible when exposed to heat or flame; can react with oxidizing materials. Mixture with nitromethane is a shock-sensitive explosive. Ignites on contact with cellulose nitrate of high surface area. To fight fire, use alcohol foam. When heated to decomposition it emits toxic fumes of NO_x. See also AMINES.

DJG800 CAS:67-43-6 ***HR: 2***
(DIETHYLENETRINITRILO)PENTAACETIC ACID
mf: $C_{14}H_{23}N_3O_{10}$ mw: 393.40

SYNS: ((CARBOXYMETHYLIMINO)BIS(ETHYLENENITRILO)) TETRAACETIC ACID ◇ CHEL 330 ◇ CHEL 330 ACID ◇ CHEL DTPA ◇ DIETHYLENETRIAMINEPENTAACETIC ACID ◇ 1,1,4,7,7-DIETHYLENETRIAMINEPENTAACETIC ACID ◇ DTPA ◇ HAMP-EX ACID ◇ MONAQUEST ◇ PENTHAMIL ◇ PERMA KLEER ◇ 3,6,9-TRIS(CARBOXYMETHYL)-3,6,9-TRIAZAUNDECANEDIOICACID

TOXICITY DATA with REFERENCE
ipr-rat LDLo:665 mg/kg AHRTAN 13,295,62
orl-mus LD50:4840 mg/kg ARZNAD 17,748,67
ipr-mus LD50:543 mg/kg ARTODN 57,212,85

CONSENSUS REPORTS: Reported in EPA TSCA Inventory.

SAFETY PROFILE: Moderately toxic by intraperitoneal route. Mildly toxic by ingestion. When heated to decomposition it emits toxic fumes of NO_x.

DJH200 CAS:7316-37-2 ***HR: 3***
DIETHYL-β,γ-EPOXYPROPYLPHOSPHONATE
mf: $C_7H_{15}O_4P$ mw: 194.19

TOXICITY DATA with REFERENCE
ipr-mus TDLo:13 g/kg/64W-I:ETA JNCIAM 53,695,74
scu-mus TDLo:13 g/kg/63W-I:NEO JNCIAM 53,695,74

SAFETY PROFILE: Questionable carcinogen with experimental neoplastigenic and tumorigenic data. When heated to decomposition it emits toxic fumes of PO_x.

DJH500 CAS:2651-85-6 ***HR: 3***
DIETHYL ETHANE PHOSPHONITE
mf: $C_6H_{15}O_2P$ mw: 150.16

$(CH_3CH_2O)_2PCH_2CH_3$

SAFETY PROFILE: Ignites spontaneously in air when a large surface area is exposed (e.g., on filter paper). When heated to decomposition it emits toxic fumes of PO_x.

DJH800 ***HR: 3***
DIETHYLETHEROXODIPEROXOCHROMIUM(VI)
mf: $C_2H_6CrO_6$ mw: 178.07

PROP: Blue solid explodes powerfully @ −30°.

CONSENSUS REPORTS: Chromium and its compounds are on the Community Right-To-Know List.

SAFETY PROFILE: A very unstable explosive. See also CHROMIUM COMPOUNDS and PEROXIDES.

DJI000 CAS:2595-54-2 ***HR: 3***
O,O-DIETHYL-S-(N-ETHOXYCARBONYL-N-METHYLCARBAMOYLMETHYL) PHOSPHORODITHIOATE
mf: $C_{10}H_{20}NO_5PS_2$ mw: 329.40

SYNS: AFOS ◇ O,O-DIAETHYL-S-(3-METHYL-2,4-DIOXO-5-OXA-3-AZA-HEPTYL)-DITHIOPHOSPHAT (GERMAN) ◇ O,O-DIETHYL S-(N-ETHOXYCARBONYL-N-METHYLCARBAMOYLMETHYL)PHOSPHOROTHIOLOTHIONATE ◇ O,O-DIETHYL S-(N-METHYL-N-CARBOETHOXYCARBAMOYLMETHYL)DITHIOPHOSPHATE ◇ O,O-DIETHYL-S-(3-METHYL-2,4-DIOXO-5-OXA-3-AZA-HEPTYL)-DITHIOFOSFAAT (DUTCH) ◇ O,O-DIETIL-S-(N-ETOSSI-CARBONIL-N-METIL-CARBAMOIL-METIL)-DITIOFOSFATO (ITALIAN) ◇ DITHIOPHOSPHATE de O,O-DIETHYLE et de S-N-METHYL-N-CARBOETHOXY CARBAMOYLMETHYLE (FRENCH) ◇ N-ETHOXYCARBONYL-N-METHYLCARBAMOYLMETHYL-O,O-DIETHYLPHOSPHORODITHIOATE ◇ S-((ETHOXYCARBONYL)METHYLCARBAMOYL)METHYL-O,O-DIETHYL PHOSPHORODITHIOATE ◇ S-(N-ETHOXYCARBONYL-N-METHYLCARBAMOYLMETHYL)-DIETHYL PHOSPHORODITHIOATE ◇ MARFOTOKS ◇ MC 474 ◇ MECARBAM ◇ MS 1053 ◇ MS 1143 ◇ MURATOX ◇ MURFOTOX ◇ MUROTOX ◇ MURPHOTOX ◇ MURUTOX ◇ PENNSALT TD-72 ◇ PESTAN

TOXICITY DATA with REFERENCE
orl-rat LD50:36 mg/kg 28ZEAL 5,144,76
skn-rat LD50:380 mg/kg WRPCA2 9,119,70
orl-mus LD50:106 mg/kg GUCHAZ 6,323,73
orl-gpg LDLo:25 mg/kg JEENAI 62,934,69
scu-gpg LDLo:50 mg/kg JEENAI 62,934,69

CONSENSUS REPORTS: EPA Genetic Toxicology Program.

SAFETY PROFILE: Poison by ingestion, skin contact, and subcutaneous routes. An insecticide. When heated to decomposition it emits very toxic fumes of SO_x, PO_x, and NO_x.

DJI200 CAS:52400-60-9 ***HR: 3***
N,N-DIETHYL-N'-(2-(2-ETHYL-1,3-BENZODIOXOL-2-YL)ETHYL) ETHYLENEDIAMINE DIMALEATE
mf: $C_{17}H_{28}N_2O_2 \cdot 2C_4H_4O_4$ mw: 524.63

SYN: 2-(2-(2-(DIETHYLAMINO)ETHYLAMINO)ETHYL)-2-ETHYL-1,3-BENZODIOXOLE DIMALEATE

TOXICITY DATA with REFERENCE
ivn-rat LD50:55 mg/kg EJMCA5 12,413,77
ipr-mus LD50:175 mg/kg EJMCA5 12,413,77

SAFETY PROFILE: Poison by intravenous and intraperitoneal routes. When heated to decomposition it emits toxic fumes of NO_x.

DJI250 CAS:66877-41-6 ***HR: D***
4,4'-(1,2-DIETHYLETHYLENE)BIS(2-AMINOPHENOL)
mf: $C_{18}H_{24}N_2O_2$ mw: 300.44

SYN: PHENOL, 4,4'-(1,2-DIETHYLETHYLENE)BIS(2-AMINO-)

TOXICITY DATA with REFERENCE
scu-mus TDLo:12 mg/kg (female 3D pre):REP JMCMAR 26,1137,83

SAFETY PROFILE: Experimental reproductive effects. When heated to decomposition it emits toxic fumes of NO_x.

DJI300 CAS:85720-57-6 ***HR: D***
4,4'-(1,2-DIETHYLETHYLENE)DI-m-CRESOL
mf: $C_{20}H_{26}O_2$ mw: 298.46

SYN: m-CRESOL, 4,4'-(1,2-DIETHYLETHYLENE)DI-

TOXICITY DATA with REFERENCE
scu-mus TDLo:120 μg/kg (female 3D pre):REP JMCMAR 26,1137,83

SAFETY PROFILE: Experimental reproductive effects. When heated to decomposition it emits acrid smoke and irritating fumes.

DJI350 CAS:10465-10-8 ***HR: D***
4,4'-(1,2-DIETHYLETHYLENE)DI-o-CRESOL
mf: $C_{20}H_{26}O_2$ mw: 298.46

SYNS: o-CRESOL, 4,4'-(1,2-DIETHYLETHYLENE)DI- ◇ meso-3,4-BIS(4-HYDROXY-3-METHYLPHENYL)HEXANE

TOXICITY DATA with REFERENCE
scu-mus TDLo:120 μg/kg (female 3D pre):REP JMCMAR 26,1137,83

SAFETY PROFILE: Experimental reproductive effects. When heated to decomposition it emits acrid smoke and irritating fumes.

DJI375 CAS:85720-47-4 ***HR: D***
4,4'-(1,2-DIETHYLETHYLENE)DIRESORCINOL
mf: $C_{18}H_{22}O_4$ mw: 302.40

TOXICITY DATA with REFERENCE
scu-mus TDLo:120 μg/kg (female 3D pre):REP JMCMAR 26,1137,83

SAFETY PROFILE: Experimental reproductive effects. When heated to decomposition it emits acrid smoke and irritating fumes.

DJI400 CAS:100-36-7 ***HR: 3***
N,N-DIETHYLETHYLENEDIAMINE
DOT: UN 2685
mf: $C_6H_{16}N_2$ mw: 116.24

PROP: Liquid. Bp: 60° @ 40 mm, flash p: 115°F (OC), d: 0.82 @ 20°/20°, vap d: 4.00.

SYNS: N,N-DIETHYL-1,2-ETHANEDIAMINE ◇ USAF AM-1

TOXICITY DATA with REFERENCE
skn-rbt 10 mg/24H open AMIHBC 10,61,54
eye-rbt 50 μg open SEV AMIHBC 10,61,54
orl-rat LD50:2830 mg/kg AMIHBC 10,61,54
ipr-mus LD50:300 mg/kg NTIS** AD277-689
skn-rbt LD50:820 mg/kg AMIHBC 10,61,54

CONSENSUS REPORTS: Reported in EPA TSCA Inventory.

DOT Classification: Corrosive Material; Label: Corrosive, Flammable Liquid

SAFETY PROFILE: Poison by intraperitoneal route. Moderately toxic by by ingestion and skin contact. A skin and severe eye irritant. Flammable when exposed to heat or flame; can react with oxidizing materials. To fight fire, use alcohol foam, CO_2, dry chemical. When heated to decomposition it emits toxic fumes of NO_x. See also AMINES.

DJJ400 CAS:358-74-7 ***HR: 3***
DIETHYL FLUOROPHOSPHATE
mf: $C_4H_{10}FO_3P$ mw: 156.11

PROP: A liquid with a sweet or fruity odor. Mp: low, bp: 170°, d: 1.15 (approx), vap d: 5.38.

SYNS: FLUOPHOSPHORIC ACID, DIETHYL ESTER ◇ PHOSPHOROFLUORIDIC ACID, DIETHYL ESTER ◇ T-1036 ◇ TL 345

TOXICITY DATA with REFERENCE
ihl-rat LD50:7 g/m^3/10M NTIS** PB158-508
ihl-mus LC50:500 mg/m^3/10M JIHTAB 30,307,48
skn-mus LD50:35 mg/kg NTIS** PB158-508
ihl-gpg LC50:7 g/m^3/10M NTIS** PB158-508

SAFETY PROFILE: Poison by inhalation and skin contact. See also FLUORIDES. When heated to decomposition or on contact with acid or acid fumes it emits highly toxic fumes of F^- and PO_x.

DJJ600 CAS:617-84-5 ***HR: 2***
DIETHYL FORMAMIDE
mf: $C_5H_{11}NO$ mw: 101.17

TOXICITY DATA with REFERENCE
ipr-rat LD50:1740 mg/kg BIJOAK 85,72,62
ipr-mus LDLo:3200 mg/kg THERAP 26,409,71

CONSENSUS REPORTS: Reported in EPA TSCA Inventory.

SAFETY PROFILE: Moderately toxic by intraperitoneal route. When heated to decomposition it emits toxic fumes of NO_x.

DJJ800 CAS:623-91-6 ***HR: 2***
DIETHYL FUMARATE
mf: $C_8H_{12}O_4$ mw: 172.20

PROP: White crystals or liquid. Mp: 0.6°, bp: 218.5°, flash p: 220°F, d: 1.0529 @ 20°/20°, vap press: 1 mm @ 53.2°, vap d: 5.93.

SYN: ETHYL FUMARATE

TOXICITY DATA with REFERENCE
orl-rat LD50:1780 mg/kg AIHAAP 23,95,62

CONSENSUS REPORTS: Reported in EPA TSCA Inventory.

SAFETY PROFILE: Moderately toxic by ingestion. Combustible when exposed to heat or flame; can react with oxidizers. To fight fire, use alcohol foam, foam, CO_2, dry chemical. When heated to decomposition it emits acrid smoke and fumes. See also ESTERS, FUMARIC ACID, and ETHYL ALCOHOL.

DJJ829 ***HR: 3***
DIETHYL GALLIUM HYDRIDE
mf: $C_4H_{11}Ga$ mw: 128.85

SAFETY PROFILE: Ignites spontaneously in air. Reacts violently with water. When heated to decomposition it emits acrid smoke and fumes. See also GALLIUM and HYDRIDES.

DJJ850 CAS:26645-10-3 ***HR: 3***
DIETHYL GOLD BROMIDE
mf: $C_4H_{10}AuBr$ mw: 334.994

DOT Classification: Forbidden

SAFETY PROFILE: Explodes at 70°C. When heated to decomposition it emits toxic fumes of Br^-. See also GOLD COMPOUNDS and BROMIDES.

DJJ875 ***HR: 2***
DIETHYLGUANIDINE HYDROCHLORIDE DIHYDRATE
mf: $C_5H_{13}N_3 \cdot ClH \cdot 2H_2O$ mw: 187.71

TOXICITY DATA with REFERENCE
orl-rat LD50:1580 mg/kg GISAAA 49(1),71,84
orl-mus LD50:1600 mg/kg GISAAA 49(1),71,84
orl-gpg LD50:890 mg/kg GISAAA 49(1),71,84

SAFETY PROFILE: Moderately toxic by ingestion. When heated to decomposition it emits toxic fumes of NO_x and HCl.

DJK000 CAS:3071-70-3 ***HR: 3***
3,3′-DIETHYLHEPTAMETHINETHIACYANINE IODIDE
mf: $C_{25}H_{25}N_2S_2 \cdot I$ mw: 544.54

TOXICITY DATA with REFERENCE
ipr-mus LD50:100 μg/kg JMCMAR 10,897,67

CONSENSUS REPORTS: Reported in EPA TSCA Inventory.

SAFETY PROFILE: Poison by intraperitoneal route. When heated to decomposition it emits very toxic fumes of I^-, NO_x, and SO_x. See also IODIDES.

DJK100 CAS:424-40-8 ***HR: 2***
DIETHYL HEXAFLUOROGLUTARATE
mf: $C_9H_{10}F_6O_4$ mw: 296.19

SYNS: DIETHYL PERFLUOROGLUTARATE ◇ HEXAFLUOROGLUTARIC ACID DIETHYL ESTER ◇ HEXAFLUOROPENTANEDIOIC ACID DIETHYL ESTER

TOXICITY DATA with REFERENCE
orl-rat LD50:5 g/kg 85GMAT -,53,82
ihl-rat LC50:1300 mg/m^3/4H 85GMAT -,53,82
orl-mus LD50:4200 mg/kg 85GMAT -,53,82
ihl-mus LCLo:10 g/m^3/2H 85GMAT -,53,82

SAFETY PROFILE: Moderately toxic by inhalation. Mildly toxic by ingestion. When heated to decomposition it emits toxic fumes of F^-. See also ESTERS.

DJK200 CAS:101-07-5 ***HR: 2***
2-DI-(2-ETHYLHEXYL)AMINOETHANOL
mf: $C_{18}H_{39}NO$ mw: 285.58

SYN: 2-(BIS(2-ETHYLHEXYL)AMINO)ETHANOL

TOXICITY DATA with REFERENCE
skn-rbt 10 mg/24H open MLD AMIHBC 10,61,54
eye-rbt 500 mg open AMIHBC 10,61,54
orl-rat LD50:4920 mg/kg AMIHBC 10,61,54
skn-rbt LD50:2520 mg/kg AMIHBC 10,61,54

SAFETY PROFILE: Moderately toxic by skin contact. Mildly toxic by ingestion. A skin and eye irritant. When heated to decomposition it emits toxic fumes of NO_x. See also PARATHION.

DJK400 CAS:5810-88-8 ***HR: 2***
O,O′-DI(2-ETHYLHEXYL) DITHIOPHOSPHORIC ACID
mf: $C_{16}H_{35}O_2PS_2$ mw: 354.60

TOXICITY DATA with REFERENCE
skn-rbt 100 μg/24H open AIHAAP 23,95,62
orl-rat LD50:4920 mg/kg AIHAAP 23,95,62
skn-rbt LD50:1250 mg/kg AIHAAP 23,95,62

CONSENSUS REPORTS: Reported in EPA TSCA Inventory.

SAFETY PROFILE: Moderately toxic by skin contact. Mildly toxic by ingestion. A skin irritant. When heated to decomposition it emits very toxic fumes of PO_x and SO_x.

DJK600 CAS:10143-60-9 ***HR: 1***
DI-(2-ETHYLHEXYL) ETHER
mf: $C_{16}H_{34}O$ mw: 242.50

SYNS: BIS(2-ETHYLHEXYL)ETHER ◇ 1,1'-OXYBIS(2-ETHYLHEXANE)

TOXICITY DATA with REFERENCE
skn-rbt 10 mg/24H open MLD AMIHBC 10,61,54
eye-rbt 500 mg open AMIHBC 10,61,54
orl-rat LD50:34 g/kg AMIHBC 10,61,54

CONSENSUS REPORTS: Reported in EPA TSCA Inventory.

SAFETY PROFILE: Mildly toxic by ingestion. A skin and eye irritant. When heated to decomposition it emits acrid smoke and irritating fumes. See also ETHERS.

DJK800 CAS:16111-62-9 ***HR: 2***
DI(2-ETHYLHEXYL) PEROXYDICARBONATE
DOT: UN 2122/UN 2123/UN 2960
mf: $C_{18}H_{34}O_6$ mw: 346.52

SYNS: BIS(2-ETHYLHEXYL) ESTER, PEROXYDICARBONIC ACID ◇ DI(2-ETHYLHEXYL) ESTER, PEROXYDICARBONIC ACID ◇ DI-(2-ETHYLHEXYL)PEROXYDICARBONATE, technical pure (DOT)

TOXICITY DATA with REFERENCE
orl-rat LD50:1020 mg/kg BSPII* 1/75-19B

CONSENSUS REPORTS: Reported in EPA TSCA Inventory.

DOT Classification: Organic Peroxide; Label: Organic Peroxide.

SAFETY PROFILE: Moderately toxic by ingestion. When heated to decomposition it emits acrid smoke and irritating fumes. See also PEROXIDES, ORGANIC.

DJL000 CAS:577-11-7 ***HR: 3***
DI-(2-ETHYLHEXYL) SODIUM SULFOSUCCINATE
mf: $C_{20}H_{38}O_7S{\cdot}Na$ mw: 445.63

PROP: White, waxlike, plastic solid; octyl alcohol odor. Sol in hexane, glycerin, alc; sltly sol in water.

SYNS: AEROSOL GPG ◇ ALCOPOL O ◇ ALPHASOL OT ◇ BEROL 478 ◇ BIS(ETHYLHEXYL) ESTER of SODIUM SULFOSUCCINIC ACID ◇ BIS(2-ETHYLHEXYL)SODIUM SULFOSUCCINATE ◇ BIS(2-ETHYLHEXYL)-S-SODIUM SULFOSUCCINATE ◇ 1,4-BIS(2-ETHYLHEXYL) SODIUM SULFOSUCCINATE ◇ 1,4-BIS(2-ETHYLHEXYL)SULFOBUTANEDIOIC ACID ESTER, SODIUM SALT ◇ CELANOL DOS 75 ◇ CLESTOL ◇ COLACE ◇ COMPLEMIX ◇ CONSTONATE ◇ COPROL ◇ DEFILIN ◇ DIOCTLYN ◇ DIOCTYLAL ◇ DIOCTYL ESTER of SODIUM SULFOSUCCINATE ◇ DIOCTYL ESTER of SODIUM SULFOSUCCINIC ACID ◇ DIOCTYL-MEDO FORTE ◇ DIOCTYL SODIUM SULFOSUCCINATE (FCC) ◇ DIOCTYL SULFOSUCCINATE SODIUM SALT ◇ DIOMEDICONE ◇ DIOSUCCIN ◇ DIOTILAN ◇ DIOVAC ◇ DOCUSATE SODIUM ◇ DOXINATE ◇ DOXOL ◇ DSS ◇ DULSIVAC ◇ DUOSOL ◇ 2-ETHYLHEXYL SULFOSUCCINATE SODIUM ◇ HUMIFEN WT 27G ◇ KONLAX ◇ KOSATE ◇ LAXINATE ◇ MANOXAL OT ◇ MERVAMINE ◇ MODANE SOFT ◇ MOLATOC ◇ MOLCER ◇ MOLOFAC ◇ MONAWET MD 70E ◇ NEKAL WT-27 ◇ NEVAX ◇ NIKKOL OTP 70 ◇ NORVAL ◇ OBSTON ◇ RAPISOL ◇ REGUTOL ◇ REQUTOL ◇ REVAC ◇ SANMORIN OT 70 ◇ SBO ◇ SOBITAL ◇ SODIUM BIS(2-ETHYLHEXYL) SULFOSUCCINATE ◇ SODIUM DI-(2-ETHYLHEXYL) SULFOSUCCINATE ◇ SODIUM DIOCTYL SULFOSUCCINATE ◇ SODIUM DIOCTYL SULPHOSUCCINATE ◇ SODIUM-2-ETHYLHEXYLSULFOSUCCINATE ◇ SODIUM SULFODI-(2-ETHYLHEXYL)SULFOSUCCINATE ◇ SOFTIL ◇ SOLIWAX ◇ SOLUSOL-75% ◇ SOLUSOL-100% ◇ SULFIMEL DOS ◇ TEX WET 1001 ◇ TRITON GR-5 ◇ VATSOL OT ◇ VELMOL ◇ WAXSOL ◇ WETAID SR

TOXICITY DATA with REFERENCE
skn-rbt 10 mg/24H MOD JPETAB 82,377,44
eye-rbt 250 μg MLD AROPAW 34,99,45
eye-rbt 1% SEV JAPMA8 38,428,49
orl-rat LD50:1900 mg/kg JSCCA5 13,469,62
ipr-rat LD50:590 mg/kg BCTKAG 7,161,74
orl-mus LD50:2640 mg/kg DCTODJ 1,89,77
ivn-mus LD50:60 mg/kg JAPMA8 38,428,49

CONSENSUS REPORTS: Reported in EPA TSCA Inventory.

SAFETY PROFILE: Poison by intravenous route. Moderately toxic by ingestion and intraperitoneal routes. A skin and severe eye irritant. See also ESTERS. When heated to decomposition it emits toxic fumes of SO_x and Na_2O.

DJL200 CAS:25430-97-1 ***HR: 3***
DI-2-ETHYLHEXYLTIN DICHLORIDE
mf: $C_{16}H_{34}Cl_2Sn$ mw: 416.09

PROP: Crystals.

SYNS: DICHLORODI(2-ETHYLHEXYL)STANNANE ◇ DI(2-ETHYLHEXYL)TIN DICHLORIDE

TOXICITY DATA with REFERENCE
ivn-rat LD50:5 mg/kg JOCMA7 2,183,60

OSHA PEL: TWA 0.1 mg(Sn)/m^3 (skin)
ACGIH TLV: TWA 0.1 mg(Sn)/m^3 (skin) (Proposed: TWA 0.1 mg(Sn)/m^3; STEL 0.2 mg(Sn)/m^3 (skin))
NIOSH REL: (Organotin Compounds) TWA 0.1 mg(Sn)/m^3

SAFETY PROFILE: Poison by intravenous route. See also TIN COMPOUNDS. When heated to decomposition it emits highly toxic fumes of Cl^-.

DJL400 CAS:1615-80-1 ***HR: 3***
1,2-DIETHYLHYDRAZINE
mf: $C_4H_{12}N_2$ mw: 88.18

PROP: Bp: 86°, d: 0.797 @ 26°. Sol in alc and ether.

SYNS: 1,2-DIAETHYLHYDRAZINE (GERMAN) ◇ N-N'-

DIETHYLHYDRAZINE ◇ sym-DIETHYLHYDRAZINE ◇ HYDRAZOETHANE ◇ HYDROAZOETHANE ◇ RCRA WASTE NUMBER U086 ◇ SDEH

TOXICITY DATA with REFERENCE
ivn-rat TDLo:500 mg/kg (15D preg):TER IARCCD 4,45,73
scu-rat TDLo:700 mg/kg/28W-I:ETA NATWAY 53,557,66
ivn-rat TDLo:50 mg/kg (15D preg):CAR,TER IARCCD 4,45,73
ivn-rat TD:50 mg/kg (15D preg):ETA,TER FCTXAV 6,584,68

CONSENSUS REPORTS: IARC Cancer Review: Group 2B IMEMDT 7,56,87; Animal Sufficient Evidence IMEMDT 4,153,74.

SAFETY PROFILE: Suspected carcinogen with experimental carcinogenic, tumorigenic, and teratogenic data. It is also a transplacental carcinogen. When heated to decomposition it emits toxic fumes of NO_x. See also HYDRAZINE.

DJL600 CAS:7699-31-2 ***HR: 3***
1,2-DIETHYLHYDRAZINE DIHYDROCHLORIDE
mf: $C_4H_{12}N_2 \cdot 2ClH$ mw: 161.10

TOXICITY DATA with REFERENCE
ivn-rat TDLo:50 mg/kg (female 15D post):ETA,TER EXPEAM 24,561,68

SAFETY PROFILE: Questionable carcinogen with experimental tumorigenic and teratogenic data. When heated to decomposition it emits very toxic fumes of HCl and NO_x. See also 1,2-DIETHYLHYDRAZINE.

DJM800 CAS:53-46-3 ***HR: 3***
DIETHYL(2-HYDROXYETHYL)METHYLAMMONIUM BROMIDE XANTHENE-9-CARBOXYLATE
mf: $C_{21}H_{26}NO_3 \cdot Br$ mw: 420.39

SYNS: ASABAINE ◇ AVAGAL ◇ BANTHIN ◇ BANTHINE BROMIDE ◇ β-DIETHYLAMINOETHYL XANTHENE-9-CARBOXYLATE METHOBROMIDE ◇ β-DIETHYLAMINOETHYL-9-XANTHENECARBOXYLATE METHOBROMIDE ◇ N,N-DIETHYL-N-METHYL-2-((9H-XANTHEN-9-YLCARBONYL)OXY)ETHANAMINIUMBROMIDE ◇ DOLADENE ◇ FRENOGASTRICO ◇ GASTRON ◇ GASTROSEDAN ◇ MANTHELINE ◇ METANTYL ◇ METAXAN ◇ METHANIDE ◇ METHANTHELINE BROMIDE ◇ METHELINA ◇ MTB 51 ◇ RESOBANTIN ◇ SC 2910 ◇ ULCINE ◇ ULCUDEXTER ◇ VAGAMIN ◇ VAGANTIN ◇ XANTELINE ◇ XANTHENE-9-CARBOXYLIC ACID ESTER with DIETHYL(2-HYDROXYETHYL) METHYLAMMONIUM BROMIDE

TOXICITY DATA with REFERENCE
orl-rat LD50:1660 mg/kg NIIRDN 6,357,82
orl-mus LD50:460 mg/kg NIIRDN 6,357,82
ipr-mus LD50:46 mg/kg JPETAB 106,141,52
scu-mus LDLo:600 mg/kg ARZNAD 8,107,58
ivn-mus LD50:4300 μg/kg AIPTAK 105,221,56
ivn-dog LDLo:23 mg/kg PSEBAA 78,576,51

SAFETY PROFILE: Poison by intraperitoneal and intravenous routes. Moderately toxic by ingestion and subcutaneous routes. Unspecified human reproductive effects. When heated to decomposition it emits very toxic fumes of NO_x, NH_3 and Br^-.

DJN000 CAS:3710-84-7 ***HR: 3***
DIETHYLHYDROXYLAMINE
mf: $C_4H_{11}NO$ mw: 89.16

SYNS: DEHA ◇ N,N-DIETHYLHYDROXYLAMINE

TOXICITY DATA with REFERENCE
sln-dmg-orl 200 ppm CAES** 492-78,78
dns-hmn:leu 4000 ppm ENVRAL 20,99,79
dlt-rat-par 180 mg/kg CAES** 492-78,78
par-rat TDLo:18 mg/kg (1D male):REP ENVRAL 20,99,79
orl-rat LDLo:1600 mg/kg KODAK* 21MAY71
skn-rat LDLo:100 mg/kg KODAK* 21MAY71
orl-mus LD80:2150 mg/kg 34ZIAG -,217,69
ipr-mus LDLo:1750 mg/kg 34ZIAG -,217,69
skn-rbt LDLo:2000 mg/kg 34ZIAG -,217,69

CONSENSUS REPORTS: Reported in EPA TSCA Inventory. EPA Genetic Toxicology Program.

SAFETY PROFILE: Poison by skin contact. Moderately toxic by ingestion and intraperitoneal routes. Experimental reproductive effects. Human mutation data reported. When heated to decomposition it emits toxic fumes of NO_x. See also AMINES.

DJN400 CAS:64048-99-3 ***HR: 3***
DIETHYL(m-HYDROXYPHENYL)ARSINE METHIODIDE

SYNS: DIETHYL(m-HYDROXYPHENYL)METHYLARSONIUMIODIDE ◇ TL 1503

TOXICITY DATA with REFERENCE
scu-mus LDLo:20 mg/kg NDRC** 30101,9,45

CONSENSUS REPORTS: Arsenic and its compounds are on the Community Right-To-Know List.

OSHA PEL: TWA 0.5 mg(As)/m^3

SAFETY PROFILE: Poison by subcutaneous route. See also ARSENIC COMPOUNDS and IODIDES. When heated to decomposition it emits very toxic fumes of As and I^-.

DJN489 ***HR: 3***
DIETHYL HYDROXYTIN HYDROPEROXIDE
mf: $C_4H_{12}O_3Sn$ mw: 226.85

$(CH_3CH_2)Sn(OH)OOH$

SAFETY PROFILE: An explosive. When heated to decomposition it emits acrid smoke and fumes. See also TIN COMPOUNDS and PEROXIDES.

DJN600 CAS:78-52-4 **HR: 3**
O,O-DIETHYL-S-2-ISOPROPYLMERCAPTOMETHYLDITHIOPHOSPHATE
mf: $C_8H_{19}O_2PS_3$ mw: 274.42

SYNS: AMERICAN CYANAMID 12,008 ◇ O,O-DIETHYL-S-(ISOPROPYLMERCAPTOMETHYL) PHOSPHORODITHIOATE ◇ O,O-DIETHYL-S-(ISOPROPYLTHIOMETHYL) PHOSPHORODITHIOATE ◇ ENT 22,865 ◇ EXPERIMENTAL INSECTICIDE 12008 ◇ (ISOPROPYLTHIO)-METHANETHIOL-S-ESTER with O,O-DIETHYL PHOSPHORODITHIOATE ◇ TM 12008

TOXICITY DATA with REFERENCE
orl-rat LD50:1100 μg/kg ARSIM* 20,1,66
scu-rat LD50:2 mg/kg JEENAI 50,356,57

SAFETY PROFILE: Poison by ingestion and subcutaneous routes. When heated to decomposition it emits very toxic fumes of PO_x and SO_x. See also MERCAPTANS.

DJN700 CAS:24264-08-2 **HR: 3**
DIETHYLKETENE
mf: $C_6H_{10}O$ mw: 98.14

$(CH_3CH_2)_2C{=}C{=}O$

SYN: 2-ETHYL-1-BUTENE-1-ONE

SAFETY PROFILE: Reacts with air to form explosive peroxides. When heated to decomposition it emits acrid smoke and fumes. See also PEROXIDES.

DJN750 CAS:96-22-0 **HR: 3**
DIETHYL KETONE
DOT: UN 1156
mf: $C_5H_{10}O$ mw: 86.15

$(CH_3CH_2)_2C{=}O$

PROP: Colorless, mobile liquid; acetone-like odor. Mp: −42°, bp: 101°, flash p: 55°F, d: 0.8159 @ 19°/4°, vap d: 2.96, autoign temp: 842°F, lel: 1.6%. Sol in water; misc in alc and ether.

SYNS: DEK ◇ DIETHYLCETONE (FRENCH) ◇ DIMETHYLACETONE ◇ METACETONE ◇ METHACETONE ◇ PENTANONE-3 ◇ 3-PENTANONE ◇ PROPIONE

TOXICITY DATA with REFERENCE
skn-rbt 410 mg open MLD UCDS** 4/25/58
eye-rbt 50 mg MOD UCDS** 4/25/58
mrc-smc 14800 ppm MUREAV 149,339,85
sln-smc 14800 ppm MUREAV 149,339,85
orl-rat LD50:2140 mg/kg UCDS** 4/25/58
ihl-rat LCLo:8000 ppm/4H AMIHBC 10,61,54
ipr-rat LDLo:1250 mg/kg JIHTAB 27,1,45
ivn-mus LD50:513 mg/kg JPMSAE 67,566,78
skn-rbt LD50:20 g/kg AMIHBC 10,61,54

CONSENSUS REPORTS: Reported in EPA TSCA Inventory.

OSHA PEL: TWA 200 ppm
ACGIH TLV: TWA 200 ppm
DOT Classification: Label: Flammable Liquid.

SAFETY PROFILE: Moderately toxic by ingestion, intraperitoneal, and intravenous routes. A skin and eye irritant. Mutation data reported. Dangerous fire hazard when exposed to heat or flame; can react vigorously with oxidizing materials. To fight fire, use alcohol foam, foam, CO_2, dry chemical. Reacts with hydrogen peroxide + nitric acid to form a shock- and heat-sensitive explosive peroxide. When heated to decomposition it emits acrid smoke and irritating fumes. See also KETONES.

DJN800 CAS:15773-47-4 **HR: 3**
DIETHYL LEAD DIACETATE
mf: $C_8H_{16}O_4Pb$ mw: 383.43

TOXICITY DATA with REFERENCE
orl-mus LD50:130 mg/kg CRSBAW 162,1456,68

CONSENSUS REPORTS: Lead and its compounds are on the Community Right-To-Know List.

SAFETY PROFILE: Poison by ingestion. See also LEAD COMPOUNDS. When heated to decomposition it emits toxic fumes of Pb.

DJN875 CAS:17498-10-1 **HR: 3**
DIETHYL LEAD DINITRATE
mf: $C_4H_{10}N_2O_6Pb$ mw: 389.33

CONSENSUS REPORTS: Lead and its compounds are on the Community Right-To-Know List.

SAFETY PROFILE: A poison. Unstable above 0°C and explodes when heated. When heated to decomposition it emits toxic fumes of NO_x. See also LEAD COMPOUNDS and NITRATES.

DJO000 CAS:50-37-3 **HR: 3**
N,N-DIETHYLLYSERGAMIDE
mf: $C_{20}H_{25}N_3O$ mw: 323.48

SYNS: ACID ◇ CUBES ◇ DELYSID ◇ 9,10-DIDEHYDRO-N,N-DIETHYL-6-METHYL-ERGOLINE-8-β-CARBOXAMIDE ◇ HEAVENLY BLUE ◇ LSD ◇ d-LSD ◇ LSD-25 ◇ LYSERGAMID ◇ LYSERGAURE DIETHYLAMID ◇ d-LYSERGIC ACID DIETHYLAMIDE ◇ LYSERGIC ACID DIETHYLAMIDE-25 ◇ LYSERGIDE ◇ LYSERGSAUEREDIAETHYLAMID ◇ PEARLY GATES ◇ ROYAL BLUE ◇ WEDDING BELLS

TOXICITY DATA with REFERENCE
mmo-sat 100 mg/L MUREAV 10,269,70
dlt-dmg-ipr 1 mg/kg ACNSAX 13,212,73
cyt-mus-orl 7500 ng/kg/5W-I NULSAK 19,153,76

cyt-ham-orl 375 μg/kg/5W NULSAK 21,206,78
cyt-ham:lng 1180 mg/L GMCRDC 27,95,81
orl-wmn TDLo:2500 mg/kg (1D pre):REP JAMAAP 212,1483,70
scu-ham TDLo:84 ng/kg (female 8D post):TER SCIEAS 158,265,67
orl-hmn TDLo:700 ng/kg:CNS JPETAB 120,340,57
orl-hmn TDLo:2857 ng/kg:CNS 34ZIAG -,356,69
orl-hmn TDLo:857 ng/kg:CNS,GIT ARZNAD 16,220,66
ims-hmn TDLo:750 ng/kg:CNS PSYPAG 3,219,62
ivn-rat LD50:16 mg/kg IRMEA9 172,702,59
ipr-mus LD50:50 mg/kg NTIS** AD277-689
ivn-mus LD50:46 mg/kg IRMEA9 172,702,59
ivn-rbt LD50:300 μg/kg ANYAA9 66,668,57
scu-gpg LD50:16 mg/kg AIPTAK 137,375,62
orl-bwd LD50:1800 μg/kg TXAPA9 21,315,72

CONSENSUS REPORTS: EPA Genetic Toxicology Program.

SAFETY PROFILE: Poison by ingestion, subcutaneous, intraperitoneal, and intravenous routes. Mutation data reported. Human systemic effects by ingestion and intramuscular routes: euphoria, hallucinations, distorted perceptions, excitement, anorexia, nausea and vomiting. An experimental teratogen. Other experimental reproductive effects. Mutation data reported. A much abused hallucinogen. A federally regulated substance. When heated to decomposition it emits toxic fumes of NO_x.

DJO100 CAS:557-18-6 ***HR: 3***
DIETHYL MAGNESIUM
DOT: UN 1367
mf: $C_4H_{10}Mg$ mw: 82.43

CONSENSUS REPORTS: Reported in EPA TSCA Inventory.

DOT Classification: Flammable Solid; Label:Spontaneously Combustible

SAFETY PROFILE: Ignites on contact with moist air, water, or carbon dioxide. See also MAGNESIUM COMPOUNDS.

DJO200 CAS:141-05-9 ***HR: 2***
DIETHYL MALEATE
mf: $C_8H_{12}O_4$ mw: 172.20

PROP: Water-white liquid. Mp: −11.5°, bp: 225.0°, flash p: 250°F (OC), d: 1.0687 @ 20°, vap press: 1 mm @ 57.3°, vap d: 5.93.

SYNS: (Z)-2-BUTENEDIOIC ACID DIETHYL ESTER ◇ ETHYL MALEATE ◇ MALEIC ACID, DIETHYL ESTER

TOXICITY DATA with REFERENCE
skn-rbt 10 mg/24H open MLD JIHTAB 31,60,49
skn-rbt 530 mg open MLD UCDS** 10/29/57
eye-rbt 500 mg open JIHTAB 31,60,49
orl-rat LD50:3200 mg/kg JIHTAB 31,60,49
ipr-rat LD50:3070 mg/kg TXAPA9 52,422,80
skn-rbt LD50:4000 mg/kg UCDS** 10/29/57

CONSENSUS REPORTS: Reported in EPA TSCA Inventory.

SAFETY PROFILE: Moderately toxic by ingestion, skin contact, and intraperitoneal routes. A skin and eye irritant. Combustible when exposed to heat or flame; can react with oxidizing materials. To fight fire, use CO_2, dry chemical. When heated to decomposition it emits acrid smoke and irritating fumes.

DJO400 CAS:627-44-1 ***HR: 3***
DIETHYL MERCURY
mf: $C_4H_{10}Hg$ mw: 258.73

PROP: Colorless liquid, hazel-like odor. Bp: 159°, d: 2.4660 @ 20°.

TOXICITY DATA with REFERENCE
dlt-rat-ihl 6 μg/m³/24H GISAAA 47(5),8,82
ihl-rat TCLo:6 μg/m³/24H (16W pre):TER GISAAA 47(5),8,82
ihl-hmn LCLo:1040 μg/m³/14W CJPEA4 34,158,43
orl-rat LD50:51 mg/kg 85GMAT -,52,82
ihl-rat LC50:258 mg/m³ GISAAA 38(1),100,73
orl-mus LD50:44 mg/kg 85GMAT -,52,82
ihl-mus LC50:91 mg/m³ GISAAA 38(1),100,73
ipr-mus LD50:45 mg/kg YKKZAJ 79,579,59

CONSENSUS REPORTS: Reported in EPA TSCA Inventory. Mercury and its compounds are on the Community Right-To-Know List.

OSHA PEL: (Transitional: CL 1 mg/10m³) TWA 0.01 mg(Hg)/m³; STEL 0.03 mg/m³ (skin)
ACGIH TLV: TWA 0.01 mg(Hg)/m³; STEL 0.03 mg(Hg)/m³

SAFETY PROFILE: A deadly human poison by inhalation. Poison by ingestion and intraperitoneal route. An experimental teratogen. See also MERCURY COMPOUNDS, ORGANIC. Flammable when exposed to heat or flame; can react with oxidizing materials. When heated to decomposition or on contact with acid or acid fumes it emits highly toxic fumes of Hg.

DJO800 CAS:50-11-3 ***HR: 2***
5,5-DIETHYL-1-METHYLBARBITURIC ACID
mf: $C_9H_{14}N_2O_3$ mw: 198.25

SYNS: AN 23 ◇ 5,5-DIETHYL-1-METHYL-2,4,6(1H,3H,5H)-PYRIMIDINETRIONE ◇ ENDIEMALUM ◇ GEMONIL ◇ GEMONIT ◇ METABARBITAL ◇ METHARBITAL ◇ METHARBITONE ◇ METHARBUTAL ◇ METHYLBARBITAL ◇ N-METHYLBARBITAL ◇ 1-METHYLBARBITAL ◇ SCH 412

TOXICITY DATA with REFERENCE
orl-hmn TDLo:30 mg/kg:CNS 27ZXA3 -,315,80
orl-mus LD50:500 mg/kg 27ZIAQ -,-,65
ipr-mus LD50:500 mg/kg 27ZIAQ -,159,73

SAFETY PROFILE: Moderately toxic by ingestion and intraperitoneal routes. Human systemic effects by ingestion: sleep, hallucinations, distorted perceptions, and changes in motor activity. When heated to decomposition it emits toxic fumes of NO_x. See also BARBITURATES.

DJP000 CAS:2050-24-0 ***HR: 1***
1,3-DIETHYL-5-METHYLBENZENE
mf: $C_{11}H_{16}$ mw: 148.27

TOXICITY DATA with REFERENCE
orl-rat LDLo:5000 mg/kg AMIHAB 19,403,59

CONSENSUS REPORTS: Reported in EPA TSCA Inventory.

SAFETY PROFILE: Mildly toxic by ingestion. When heated to decomposition it emits acrid smoke and irritating fumes.

DJP500 CAS:132-19-4 ***HR: 3***
N,N-DIETHYL-1-METHYL-3,3-DI-2-THIENYLALLYLAMINE HYDROCHLORIDE
mf: $C_{16}H_{21}NS_2 \cdot ClH$ mw: 327.96

SYNS: 191C49 HYDROCHLORIDE ◇ DIETHIBUTIN HYDROCHLORIDE ◇ 3-DIETHYLAMINO-1,1-DI(2'-THIENYL)BUT-1-ENE HYDROCHLORIDE ◇ N,N-DIETHYL-4,4-DI-2-THIENYL-3-BUTEN-2-AMINE HYDROCHLORIDE ◇ N,N-DIETHYL-3,3-DI-2-THIENYL-1-METHYLALLYLAMINE HYDROCHLORIDE ◇ DIETHYLTHIAMBUTENE HYDROCHLORIDE ◇ NIH-4185 HYDROCHLORIDE ◇ THEMALON HYDROCHLORIDE ◇ THIAMBUIENE HYDROCHLORIDE

TOXICITY DATA with REFERENCE
scu-rat LD50:45 mg/kg BJPCAL 8,2,53
orl-mus LD50:204 mg/kg MEIEDD 10,1330,83
scu-mus LD50:81 mg/kg BJPCAL 8,2,53
ivn-mus LD50:16 mg/kg BJPCAL 8,2,53

SAFETY PROFILE: Poison by ingestion, subcutaneous, and intravenous routes. An analgesic and narcotic used in veterinary medicine. When heated to decomposition it emits very toxic fumes of NO_x, SO_x, and HCl. See also ALLYL COMPOUNDS.

DJP600 CAS:50285-72-8 ***HR: 3***
1,1-DIETHYL-3-METHYL-3-NITROSOUREA
mf: $C_6H_{13}N_3O_2$ mw: 159.22

SYNS: NITROSO-1,1-DIETHYL-3-METHYLUREA ◇ NITROSOMETHYLDIAETHYLHARNSTOFF ◇ NITROSOMETHYLDIETHYLUREA ◇ 1-NITROSO-1-METHYL-3,3-DIETHYLUREA

TOXICITY DATA with REFERENCE
mma-sat 250 μg/plate JJIND8 67,1117,81
sce-ham:lng 500 μmol/L MUREAV 126,259,84
orl-rat TDLo:1400 mg/kg/50W-I:ETA ZKKOBW 83,315,75
scu-ham TD:623 mg/kg/22W-I:CAR CALEDQ 23,177,84
scu-ham LD50:283 mg/kg CALEDQ 23,177,84

SAFETY PROFILE: Poison by subcutaneous route. Questionable carcinogen with experimental carcinogenic and tumorigenic data. Mutation data reported. When heated to decomposition it emits toxic fumes of NO_x. See also N-NITROSO COMPOUNDS.

DJP700 CAS:73671-86-0 ***HR: D***
N,N-DIETHYL-4-METHYL-3-OXO-5-α-4-AZAANDROSTANE-17-β-CARBOXAMIDE
mf: $C_{24}H_{40}N_2O_2$ mw: 388.66

SYNS: 17-β-N,N-DIETHYLCARBAMOYL-4-METHYL-4-AZA-5α-ANDROSTAN-3-ONE ◇ 4-MA

TOXICITY DATA with REFERENCE
scu-rat TDLo:700 mg/kg (14D male):REP PSEBAA 169,67,82
scu-rat TDLo:135 mg/kg (13-21D preg):TER PSEBAA 169,67,82

SAFETY PROFILE: Experimental teratogenic and reproductive effects. A steroid. When heated to decomposition it emits toxic fumes of NO_x.

DJQ200 CAS:1605-58-9 ***HR: 3***
DIETHYLMETHYLPHOSPHINE
mf: $C_5H_{13}P$ mw: 104.14

SAFETY PROFILE: May ignite spontaneously after long exposure to air. When heated to decomposition it emits toxic fumes of PO_x. See also PHOSPHINE.

DJQ300 CAS:7310-87-4 ***HR: 3***
3,3'-DIETHYL-9-METHYLSELENOCARBOCYANINE IODIDE
mf: $C_{22}H_{23}N_2Se_2 \cdot I$ mw: 600.29

SYN: BENZOSELENAZOLIUM,3-ETHYL-2-(3-(3-ETHYL-2-BENZOSELENAZOLINYLIDENE)-2-METHYLPROPENYL)-,IODIDE

TOXICITY DATA with REFERENCE
ivn-mus LD50:18 mg/kg CSLNX* NX#02907

OSHA PEL: TWA 0.2 mg(Se)/m^3
ACGIH TLV: TWA 0.2 mg(Se)/m^3

SAFETY PROFILE: Poison by intravenous route. When heated to decomposition it emits toxic fumes of NO_x, Se, and I^-.

DJQ800 CAS:19481-39-1 **HR: 3**
DIETHYLMETHYLSULFONIUM IODIDEMERCURIC IODIDE (ADDITION COMPOUND)

SYN: DIETHYLMETHYL SULFONIUM IODINE with MERCURY IODIDE (1:1)

TOXICITY DATA with REFERENCE
ivn-mus LD50:18 mg/kg CSLNX* NX#01853

CONSENSUS REPORTS: Mercury and its compounds are on the Community Right-To-Know List.

NIOSH REL: TWA 0.05 mg(Hg)/m^3

SAFETY PROFILE: Poison by intravenous route. See also MERCURY COMPOUNDS, IODIDES, and SULFONATES. When heated to decomposition it emits very toxic fumes of SO_x, I^-, and Hg.

DJR200 CAS:3065-79-0 **HR: 3**
3,3'-DIETHYL-9-METHYLTHIACARBOCYANINE IODIDE
mf: $C_{22}H_{23}N_2S_2{\cdot}I$ mw: 506.49

TOXICITY DATA with REFERENCE
ivn-mus LD50:3600 μg/kg CSLNX* NX#03432

CONSENSUS REPORTS: Reported in EPA TSCA Inventory.

SAFETY PROFILE: Poison by intravenous route. When heated to decomposition it emits very toxic fumes of I^-, SO_x, and NO_x.

DJR700 CAS:2511-10-6 **HR: 3**
O,S-DIETHYL METHYLTHIOPHOSPHONATE
mf: $C_5H_{13}O_2PS$ mw: 168.21

SYNS: O,S-DIETHYL METHYLPHOSPHONOTHIOATE ◇ LG 61 ◇ METHYLPHOSPHONOTHIOIC ACID-O,S-DIETHYL ESTER ◇ OSDMP

TOXICITY DATA with REFERENCE
orl-rat LD50:6 mg/kg JAFCAU 32,774,84
ivn-mus LD50:1 mg/kg IAEC** 17JUN74
ivn-dog LD50:5620 μg/kg IAEC** 17JUN74
ivn-rbt LD50:2480 μg/kg IAEC** 17JUN74

SAFETY PROFILE: Poison by ingestion and intravenous routes. When heated to decomposition it emits toxic fumes of PO_x and SO_x.

DJR800 CAS:52-60-8 **HR: 3**
O,O-DIETHYL-O-(4-(METHYLTHIO)-3,5-XYLYL) PHOSPHOROTHIOATE
mf: $C_{13}H_{21}O_3PS_2$ mw: 320.43

SYNS: BAY 9017 ◇ BAY 37341 ◇ BAYER 37341 ◇ O-(3,5-DIMETHYL-4-(METHYLTHIO)PHENYL O,O-DIETHYL ESTER PHOSPHOROTHIOIC ACID ◇ O-(3,5-DIMETHYL-4-(METHYLTHIO)PHENYL)-O,O-DIETHYL PHOSPHOROTHIOATE ◇ ENT 25,673 ◇ 4-(METHYLTHIO)-3,5-XYLENOL-O-ESTER with O,O-DIETHYL PHOSPHOROTHIOATE

TOXICITY DATA with REFERENCE
orl-rat LD50:375 mg/kg TXAPA9 21,315,72
ipr-mus LDLo:500 mg/kg CBCCT* 5,337,53
orl-ckn LD50:18 mg/kg TXAPA9 11,49,67
orl-bwd LD50:4200 μg/kg TXAPA9 21,315,72

SAFETY PROFILE: Poison by ingestion. Moderately toxic by intraperitoneal route. When heated to decomposition it emits very toxic fumes of PO_x and SO_x.

DJS100 CAS:3060-37-5 **HR: 3**
α,α-DIETHYL-1-NAPHTHALENEACETIC ACID SODIUM SALT
mf: $C_{16}H_{17}O_2{\cdot}Na$ mw: 264.32

TOXICITY DATA with REFERENCE
ipr-rat LD50:466 mg/kg AIPTAK 154,297,65
scu-rat LD50:809 mg/kg AIPTAK 154,297,65
ivn-rat LD50:433 mg/kg AIPTAK 154,297,65
ipr-mus LD50:470 mg/kg AIPTAK 154,297,65
scu-mus LD50:964 mg/kg AIPTAK 154,297,65
ivn-mus LD50:324 mg/kg AIPTAK 154,297,65

SAFETY PROFILE: Poison by intravenous route. Moderately toxic by subcutaneous and intraperitoneal other routes. When heated to decomposition it emits toxic fumes of Na_2O.

DJS200 CAS:59-26-7 **HR: 3**
N,N-DIETHYLNICOTINAMIDE
mf: $C_{10}H_{14}N_2O$ mw: 178.26

SYNS: ANACARDONE ◇ ANACORDONE ◇ ASTROCAR ◇ BETAPYRIMIDUM ◇ CAMPHOZONE ◇ CARBAMIDAL ◇ CARDAMINE ◇ CARDIAGEN ◇ CARDIAMID ◇ CARDIAMINA ◇ CARDIAMINE ◇ CARDIMON ◇ CITOCOR ◇ CORACON ◇ CORAETHAMIDE ◇ CORAETHAMIDUM ◇ CORALEPT ◇ CORAMINE ◇ CORAVITA ◇ CORAZONE ◇ CORDIAMID ◇ CORDIAMIN ◇ CORDIAMINE ◇ CORDITON ◇ CORDYNIL ◇ COREDIOL ◇ CORESPIN ◇ CORETHAMIDE ◇ CORETONE ◇ CORMED ◇ CORMID ◇ CORMOTYL ◇ CORNOTONE ◇ COROTONIN ◇ COROVIT ◇ CORVITAN ◇ CORVITOL ◇ CORVITONE ◇ CORYWAS ◇ DANAMINE ◇ DIAETHYL-NICOTINAMID (GERMAN) ◇ DIETHYL-NICOTAMIDE ◇ N,N-DIETHYL-3-PYRIDINECARBOXAMIDE ◇ DIETILAMIDE-CARBOPIRIDINA ◇ DINACORYL ◇ DYNACORYL ◇ DYNAMICARDE ◇ ELITONE ◇ EUCORAN ◇ HANSACOR ◇ INICARDIO ◇ KARDIAMID ◇ KARDONYL ◇ KORDIAMIN ◇ LEPTAMIN ◇ MEDIAMID ◇ NIAMINE ◇ NICAMIDE ◇ NICETAMIDE ◇ NICETHAMIDE ◇ NICOR ◇ NICORDAMIN ◇ NICORINE ◇ NICORYL ◇ NICOTINIC ACID DIETHYLAMIDE ◇ NIKARDIN ◇ NIKETAMID ◇ NIKETHAROL ◇ NIKETHYL ◇ NIKETILAMID ◇ NIKORIN ◇ NIQUETAMIDA ◇ NISETAMIDE ◇ PERCORAL ◇ PROCARDINE ◇ PROCORMAN ◇ PYRICAROYL ◇ PYRIDINE-3-CARBOXYDIETHYLAMIDE ◇ PYRIDINE-3-CARBOXYLIC ACID DIETHYLAMIDE ◇ REFORMIN ◇ REHORMIN ◇ SALVACARD ◇ SALVACORIN ◇ SOLYACORD ◇ STELLAMINE ◇ STIMINOL ◇ STIMULIN ◇ TONOCARD ◇ TONOCOR ◇ VASAZOL ◇ VENTRAMINE

TOXICITY DATA with REFERENCE
ipr-rat LD50:272 mg/kg PSDTAP 4,132,64
scu-rat LD50:240 mg/kg PSEBAA 62,19,46
ivn-rat LD50:191 mg/kg 27ZIAQ -,173,73
orl-mus LD50:188 mg/kg RPOBAR 1,423,64
ipr-mus LD50:174 mg/kg JPETAB 128,176,60
scu-mus LD50:200 mg/kg BCFAAI 111,293,72
ivn-mus LD50:180 mg/kg SMWOAS 85,305,55
scu-dog LDLo:175 mg/kg 27ZWAY -,-,37
ivn-dog LDLo:175 mg/kg 27ZWAY -,-,37
ipr-rbt LD50:225 mg/kg 27ZWAY -,-,37
scu-rbt LDLo:300 mg/kg KLWOAZ 12,1860,33
ivn-rbt LDLo:150 mg/kg JPETAB 66,260,39
ipr-gpg LDLo:250 mg/kg 27ZWAY -,-,37
scu-gpg LDLo:300 mg/kg 27ZWAY -,-,37

SAFETY PROFILE: Poison by ingestion, intravenous, intraperitoneal, and subcutaneous routes. When heated to decomposition it emits toxic fumes of NO_x.

DJS500 CAS:7119-92-8 ***HR: 3***
DIETHYLNITRAMINE
mf: $C_4H_{10}N_2O_2$ mw: 118.16

SYNS: N-ETHYL-N-NITROETHANAMINE (9CI) ◇ N-NITRO-DIETHYLAMINE

TOXICITY DATA with REFERENCE
mmo-esc 20 μmol/plate IAPUDO 57,485,84
hma-rat/sat 200 mg/kg CNREA8 41,3205,81
orl-rat TDLo:18200 mg/kg/2Y-C:ETA ARGEAR 52,629,82
ipr-mus LD50:730 mg/kg PCJOAU 10,1504,76

SAFETY PROFILE: Moderately toxic by intraperitoneal route. Questionable carcinogen with experimental tumorigenic data. Mutation data reported. When heated to decomposition it emits toxic fumes of NO_x.

DJS800 CAS:3270-86-8 ***HR: 3***
O,O-DIETHYL-S-(4-NITROPHENYL)THIOPHOSPHATE
mf: $C_{10}H_{14}NO_5PS$ mw: 291.28

SYNS: O,O-DIETHYL-S-p-NITROFENYLESTER KYSELINY THIOFOSFORECNE (CZECH) ◇ O,O-DIETHYL-S-(4-NITROPHENYL) PHOSPHOROTHIOATE ◇ O,O-DIETHYL-S-(4-NITROPHENYL)PHOSPHOROTHIOIC ACID ESTER ◇ PARATHION S ◇ S-PHENYL PARATHION

TOXICITY DATA with REFERENCE
orl-rat LD50:4410 μg/kg 28ZPAK -,208,72
orl-mus LD50:11900 μg/kg 28ZPAK -,208,72
ipr-mus LD50:107 μg/kg PHARAT 35,806,80
scu-mus LD50:1250 μg/kg AMIHAB 11,487,55

SAFETY PROFILE: A deadly poison by ingestion, subcutaneous, and intraperitoneal routes. When heated to decomposition it emits very toxic fumes of NO_x, PO_x, and SO_x.

DJT000 CAS:597-88-6 ***HR: 3***
O,S-DIETHYL-O-(4-NITROPHENYL)THIOPHOSPHATE
mf: $C_{10}H_{14}NO_5PS$ mw: 291.28

SYNS: O,S-DIETHYL-O-(p-NITROPHENYL)PHOSPHOROTHIOATE ◇ O,S-DIETHYL-O-(4-NITROPHENYL)PHOSPHOROTHIOATE ◇ O,S-DIETHYL-O-(4-NITROPHENYL)PHOSPHOROTHIOIC ACID ESTER ◇ O,S-DIETHYL-O-(p-NITROPHENYL)PHOSPHOROTHIOIC ACID ESTER ◇ S-ETHYL PARATHION ◇ ISOPARATHION

TOXICITY DATA with REFERENCE
orl-rat LD50:17900 μg/kg 28ZPAK -,209,72
skn-rat LD50:75 mg/kg TXAPA9 2,523,60
ipr-rat LD50:5500 μg/kg AMIHBC 6,9,52
ims-rat LD50:10 mg/kg AIHAAP 19,190,58
orl-mus LD50:25 mg/kg JPETAB 105,156,52
ipr-mus LD50:50 μg/kg PHARAT 35,806,80
scu-mus LD50:20 mg/kg PAREAQ 11,636,59
orl-gpg LD50:32 mg/kg JPETAB 105,156,52

SAFETY PROFILE: Poison by ingestion, skin contact, intraperitoneal, subcutaneous, and intramuscular routes. When heated to decomposition it emits very toxic fumes of NO_x. See also PARATHION.

DJT200 CAS:95-92-1 ***HR: 3***
DIETHYL OXALATE
DOT: UN 2525
mf: $C_6H_{10}O_4$ mw: 146.16

PROP: Colorless, oily, aromatic liquid; decomp in water. Mp: −40.6°, bp: 185.4°, flash p: 168°F (OC), d: 1.08426 @ 15°, 1.0785 @ 20°/4°, vap d: 5.04.

SYNS: DIETHYL ETHANEDIOATE ◇ ETHYL OXALATE ◇ ETHYL OXALATE (DOT) ◇ OXALIC ACID, DIETHYL ESTER

TOXICITY DATA with REFERENCE
orl-rat LD50:400 mg/kg 14CYAT 2,1882,63
orl-mus LD50:2000 mg/kg GISAAA 46(5),87,81

CONSENSUS REPORTS: Reported in EPA TSCA Inventory.

DOT Classification: Poison B; Label: St. Andrews Cross.

SAFETY PROFILE: Poison by ingestion. Flammable when exposed to heat or flame; can react with oxidizing materials. To fight fire, use foam, CO_2, dry chemical. When heated to decomposition it emits acrid smoke and fumes. See also OXALATES and ESTERS.

DJT400 CAS:702-54-5 ***HR: 3***
5,5-DIETHYL-1,3-OXAZIN-2,4-DIONE
mf: $C_8H_{13}NO_3$ mw: 171.22

SYNS: DIETADIONE (ITALIAN) ◇ DIETHADION ◇ DIETHADIONE ◇ 5,5-DIETHYLDIHYDRO-2H-1,3-OXAZINE-2,4(3H)-DIONE ◇ 5,5-DIETHYL-1,3-OXAZINE-2,4-DIONE ◇ 5,5-DIETHYLTETRAHYDRO-2H-1,3-OXAZINE-2,4(3H)-DIONE ◇ 5,5-DIETILDIIDRO-1,3-OSSAZIN-2,4-DIONE (ITALIAN) ◇ DIETROXINE ◇ DIHYDRO-5,5-DIETHYL-2H-

1,3-OXAZINE-2,4(3H)-DIONE ◇ DIIDRO-5,5-DIETIL-2H-1,3-OSSAZIN-2,4(3H)-DIONE (ITALIAN) ◇ DIOXONE ◇ L 1811 ◇ LEDOSTEN ◇ LEPTON ◇ PERSISTEN ◇ TOCE ◇ TOCEN

TOXICITY DATA with REFERENCE
orl-rat LD50:71 mg/kg JPPMAB 13,244,61
ipr-rat LD50:32 mg/kg JPPMAB 13,244,61
scu-rat LD50:39 mg/kg JPPMAB 13,244,61
orl-mus LD50:81 mg/kg RPOBAR 2,282,70
ipr-mus LD50:49 mg/kg RPOBAR 2,281,70
scu-mus LD50:61 mg/kg RPOBAR 2,281,70
ivn-mus LD50:32 mg/kg JPPMAB 13,244,61
ims-mus LD50:45 mg/kg RPOBAR 2,281,70

CONSENSUS REPORTS: Reported in EPA TSCA Inventory.

SAFETY PROFILE: Poison by ingestion, intravenous, intraperitoneal, subcutaneous, and intramuscular routes. An analeptic (central nervous system stimulant). When heated to decomposition it emits toxic fumes of NO_x.

DJT800 CAS:514-73-8 ***HR: 3***
3,3'-DIETHYLPENTAMETHINETHIACYANINE IODIDE
mf: $C_{23}H_{24}N_2S_2 \cdot I$ mw: 519.51

SYNS: ABMINTHIC ◇ ANELMID ◇ ANGUIFUGAN ◇ COMPOUND 01748 ◇ DEJO ◇ DELVEX ◇ DIETHYLTHIADICARBOCYANINE IODIDE ◇ 3,3'-DIETHYLTHIADICARBOCYANINE IODIDE ◇ DILOMBRIN ◇ DITHIAZANINE IODIDE ◇ DITHIAZANIN IODIDE ◇ DITHIAZININE ◇ EASTMAN 7663 ◇ 3-ETHYL-2-(5-(3-ETHYL-2-BENZOTHIAZOLINYLIDENE)-1,3-PENTADIENYL)BENZOTHIAZOLIUM IODIDE ◇ L-01748 ◇ NETOCYD ◇ NK 136 ◇ OMNI-PASSIN ◇ PARTEL ◇ TELMICID ◇ TELMID ◇ TELMIDE ◇ VERCIDON

TOXICITY DATA with REFERENCE
orl-mus LD50:20 mg/kg BSIBAC 44,1032,68
ipr-mus LD50:3 mg/kg JMCMAR 10,897,67
ivn-mus LD50:1 mg/kg CSLNX* NX#02015

CONSENSUS REPORTS: Reported in EPA TSCA Inventory.

SAFETY PROFILE: Poison by ingestion, intraperitoneal, and intravenous routes. When heated to decomposition it emits very toxic fumes of I^-, SO_x, and NO_x. See also IODIDES.

DJU000 CAS:1067-20-5 ***HR: 3***
3,3-DIETHYLPENTANE
mf: C_9H_{20} mw: 128.26

PROP: Flash p: <69.8°F, lel: 0.7%, uel: 7.7%.

SAFETY PROFILE: A very dangerous fire and explosion hazard when exposed to heat or flame; can react vigorously with oxidizing materials. When heated to decomposition it emits acrid smoke and fumes.

DJU200 CAS:512-48-1 ***HR: 3***
2,2-DIETHYL-4-PENTENAMIDE
mf: $C_9H_{17}NO$ mw: 155.27

SYNS: DIAETHYLALLYLACETAMIDE (GERMAN) ◇ EPINOVAL ◇ NOVONAL

TOXICITY DATA with REFERENCE
orl-hmn LDLo:300 mg/kg DMWOAX 102,1591,77
orl-hmn TDLo:100 mg/kg:CVS,CVS,PUL DMWOAX 102,1591,77
orl-rat LD50:400 mg/kg DMWOAX 102,1591,77
ipr-rat LD50:217 mg/kg ITMZBJ 17,305,80
orl-mus LDLo:300 mg/kg LDTU** -,-,31
orl-dog LD50:300 mg/kg DMWOAX 102,1591,77
rec-gpg LDLo:150 mg/kg LDTU** -,-,31

SAFETY PROFILE: Poison to humans by ingestion. An experimental poison by ingestion, rectal, and intraperitoneal routes. Human systemic effects by ingestion: muscle spasms, cardiac arrythmias, and respiratory depression. When heated to decomposition it emits toxic fumes of NO_x.

DJU400 CAS:628-37-5 ***HR: 3***
DIETHYL PEROXIDE
mf: $C_4H_{10}O_2$ mw: 90.12

$CH_3CH_2OOCH_2CH_3$

PROP: Lel: 2.3%, d: 0.8, vap d: 7.7, bp: 65°.

SAFETY PROFILE: A shock- and heat-sensitive explosive. Reacts violently with O_2. When heated to decomposition it emits acrid smoke and fumes. See also PEROXIDES, ORGANIC.

DJU600 CAS:14666-78-5 ***HR: 3***
DIETHYL PEROXYDICARBONATE
DOT: UN 2175
mf: $C_6H_{10}O_6$ mw: 178.14

$CH_3CH_2OCO \cdot OOCO \cdot OCH_2CH_3$

SYNS: DIETHYL PEROXYDIFORMATE ◇ ETHYL PEROXYCARBONATE

DOT Classification: Organic Peroxide; Label: Organic Peroxide: Forbidden (> 27% in solution).

SAFETY PROFILE: The impure material is a powerful explosive extremely sensitive to heat or impact. When heated to decomposition it emits acrid smoke and fumes. See also PEROXIDES.

DJU800 CAS:52400-58-5 ***HR: 3***
N,N-DIETHYL-N'-(2-(2-PHENYL-1,3-BENZODIOXOL-2-YL)ETHYL)ETHYLENEDIAMINE DIMALEATE
mf: $C_{21}H_{28}N_2O_2 \cdot 2C_4H_4O_4$ mw: 572.67

SYN: 2-(2-(2-(DIETHYLAMINO)ETHYLAMINO)ETHYL)-1,3-BENZODIOXOLE, DIMALEATE

TOXICITY DATA with REFERENCE
ivn-rat LD50:10 mg/kg EJMCA5 12,413,77
ipr-mus LD50:90 mg/kg EJMCA5 12,413,77

SAFETY PROFILE: Poison by intravenous and intraperitoneal routes. When heated to decomposition it emits toxic fumes of NO_x.

DJV000 CAS:23564-06-9 ***HR: 2***
DIETHYL-4,4′-o-PHENYLENEBIS(3-THIOALLOPHANATE)
mf: $C_{14}H_{18}N_4O_4S_2$ mw: 370.48

SYNS: BAS 3220 ◇ 1,2-BIS-(3-ETHOXYCARBONYLTHIOUREIDO) BENZENE ◇ 1,2-BIS(3-ETHOXYCARBONYL-2-THIOUREIDO) BENZENE ◇ CERCOBIN ◇ CLEARY 3336 ◇ ENOVIT ◇ ETHYL THIOPHANATE ◇ NF 35 (fungicide) ◇ PELT SOL ◇ (1,2-PHENYLENEBIS(IMINOCARBONOTHIOYL))BISCARBAMIC ACID DIETHYL ESTER ◇ THIOFANATE ◇ THIOPHANAT (GERMAN) ◇ THIOPHANATE ETHYL ◇ THIOPHENITE ◇ TIOFANATE ETILE (ITALIAN) ◇ TOPSIN ◇ 3336 TURF FUNGICIDE

TOXICITY DATA with REFERENCE
mmo-smc 5 ppm RSTUDV 6,161,76
sln-asn 100 mg/L EVHPAZ 31,81,79
orl-mus TDLo:336 g/kg (MGN):REP OYYAA2 4,23,70
ipr-rat LD50:2400 mg/kg OYYAA2 4,5,70
ipr-mus LD50:3750 mg/kg OYYAA2 4,5,70

CONSENSUS REPORTS: EPA Genetic Toxicology Program.
SAFETY PROFILE: Moderately toxic by intraperitoneal route. Experimental reproductive effects. Mutation data reported. A fungicide. When heated to decomposition it emits very toxic fumes of NO_x and SO_x. See also CARBAMATES.

DJV200 CAS:93-05-0 ***HR: 3***
DIETHYL-p-PHENYLENEDIAMINE
mf: $C_{10}H_{16}N_2$ mw: 164.28

SYN: N,N-DIETHYL-p-PHENYLENEDIAMINE

TOXICITY DATA with REFERENCE
skn-hmn TDLo:73 μg/kg:SKN,BLD JIDHAN 4,386,23
scu-rat LDLo:100 mg/kg JIDHAN 4,386,23
ivn-dog LDLo:70 mg/kg JIDHAN 4,386,23
orl-cat LDLo:300 mg/kg JIDHAN 4,386,23
orl-rbt LDLo:450 mg/kg JIDHAN 4,386,23
skn-rbt LDLo:125 mg/kg JIDHAN 4,386,23
scu-rbt LDLo:250 mg/kg JIDHAN 4,386,23

CONSENSUS REPORTS: Reported in EPA TSCA Inventory.

SAFETY PROFILE: Poison by ingestion, skin contact, subcutaneous, and intravenous routes. Human systemic skin effects by skin contact: hemorrhage, allergic dermatitis, and primary irritantion. When heated to decomposition it emits toxic fumes of NO_x. See also AMINES.

DJV250 CAS:6283-63-2 ***HR: 3***
N,N′-DIETHYL-p-PHENYLENEDIAMINE SULFATE
mf: $C_{10}H_{16}N_2 \cdot H_2O_4S$ mw: 262.36

SYNS: p-PHENYLENEDIAMINE, N,N-DIETHYL-, SULFATE (1:1) ◇ SIRAN N,N-DIETHYL-p-FENYLENDIAMINU

TOXICITY DATA with REFERENCE
eye-rbt 500 mg/24H MLD 85JCAE -,479,86
orl-uns TDLo:36 mg/kg (male 60D pre):REP GISAAA 51(9),71,86
unr-rat LD50:450 mg/kg GISAAA 51(9),71,86
unr-mus LD50:318 mg/kg GISAAA 51(9),71,86
unr-gpg LD50:560 mg/kg GISAAA 51(9),71,86

SAFETY PROFILE: Poison by an unspecified route. Experimental reproductive effects. Eye irritant. When heated to decomposition it emits toxic fumes of SO_x.

DJV600 CAS:18854-01-8 ***HR: 3***
O,O-DIETHYL-O-(5-PHENYL-3-ISOXAZOLYL) PHOSPHOROTHIOATE
mf: $C_{13}H_{16}NO_4PS$ mw: 313.33

SYNS: O,O-DIETHYL-O-(3-(5-PHENYL)-1,2-ISOXAZOLYL)PHOSPHOROTHIOATE ◇ O,O-DIETHYL-O-(5-PHENYL-3-ISOXAZOLYL) PHOSPHOROTHIOIC ACID ESTER ◇ E-48 ◇ ISOXATHION ◇ KARPHOS ◇ SI-6711

TOXICITY DATA with REFERENCE
orl-rat LD50:112 mg/kg BESAAT 15,121,69
skn-rat LD50:450 mg/kg BESAAT 15,121,69
orl-mus LD50:79100 μg/kg SKKNAJ 29,1,77
skn-mus LD50:193 mg/kg BESAAT 15,121,69
ipr-mus LD50:105 mg/kg SKKNAJ 29,1,77
scu-mus LD50:720 mg/kg SKKNAJ 29,1,77
orl-ckn LD50:21600 μg/kg SKKNAJ 29,1,77

SAFETY PROFILE: Poison by ingestion, skin contact, and intraperitoneal routes. Moderately toxic by subcutaneous route. When heated to decomposition it emits very toxic fumes of NO_x, PO_x and SO_x.

DJV800 CAS:64036-46-0 ***HR: 3***
DIETHYL PHENYLTIN ACETATE
mf: $C_{12}H_{18}O_2Sn$ mw: 312.99

SYN: ACETOXYDIETHYLPHENYLSTANNANE

TOXICITY DATA with REFERENCE
orl-rat LDLo:50 mg/kg BJPCAL 10,16,55

OSHA PEL: TWA 0.1 mg(Sn)/m^3 (skin)
ACGIH TLV: TWA 0.1 mg(Sn)/m^3 (skin) (Proposed: TWA 0.1 mg(Sn)/m^3; STEL 0.2 mg(Sn)/m^3 (skin))
NIOSH REL: (Organotin Compounds) TWA 0.1 mg(Sn)/m^3

SAFETY PROFILE: Poison by ingestion. See also TIN COMPOUNDS. When heated to decomposition it emits acrid and irritating fumes.

DJW000 CAS:627-49-6 ***HR: 3***
DIETHYL PHOSPHINE
mf: $C_4H_{11}P$ mw: 90.11

$(CH_3CH_2)_2PH$

PROP: Bp: 85°, d: 1, vap d: 3.11.

SAFETY PROFILE: Poison by ingestion and inhalation. Flammable when exposed to heat or flame; spontaneously flammable in air. Can react vigorously with oxidizing materials. To fight fire, use foam, CO_2, dry chemical. When heated to decomposition it emits toxic fumes of PO_x. See also PHOSPHINE.

DJW200 CAS:7531-39-7 ***HR: 3***
DIETHYLPHOSPHINIC ACID-p-NITROPHENYL ESTER
mf: $C_{10}H_{14}NO_4P$ mw: 243.22

SYN: p-NITROPHENYL ESTER of DIETHYLPHOSPHINIC ACID

TOXICITY DATA with REFERENCE
scu-rat LD50:3400 μg/kg FATOAO 42(3),299,79
ivn-rat LD50:3350 μg/kg FATOAO 42(3),299,79
scu-mus LD50:3400 μg/kg RPTOAN 42,106,79
ivn-mus LD50:3350 μg/kg RPTOAN 42,106,79

SAFETY PROFILE: Poison by subcutaneous and intravenous routes. See also ESTERS. When heated to decomposition it emits very toxic fumes of NO_x and PO_x.

DJW400 CAS:762-04-9 ***HR: 2***
DIETHYL PHOSPHITE
mf: $C_4H_{11}O_3P$ mw: 138.12

$(CH_3CH_2O)_2P(:O)H$

SYNS: DIETHYL HYDROGEN PHOSPHITE ◇ PHOSPHOROUS ACID, DIETHYL ESTER

TOXICITY DATA with REFERENCE
orl-rat LD50:3900 mg/kg ALBRW* #OPB-3,84
skn-rbt LD50:2165 mg/kg AIHAAP 30,470,69

CONSENSUS REPORTS: Reported in EPA TSCA Inventory.

SAFETY PROFILE: Moderately toxic by ingestion and skin contact. Mixtures with 4-nitrophenol may explode if heated. When heated to decomposition it emits toxic fumes of PO_x.

DJW600 CAS:2524-04-1 ***HR: 3***
O,O-DIETHYLPHOSPHOROCHLORIDOTHIOATE
DOT: UN 2751
mf: $C_4H_{10}ClO_2PS$ mw: 188.62

SYNS: CHLORO-PHOSPHONOTHIOIC ACID-O,O-DIETHYL ESTER ◇ DIETHYLCHLOROTHIOPHOSPHATE ◇ DIETHYLCHLORTHIOFOSFAT (CZECH) ◇ DIETHYLTHIOPHOSPHORYL CHLORIDE (DOT)

TOXICITY DATA with REFERENCE
orl-rat LDLo:1000 mg/kg 34ZIAG -,393,69
ihl-rat LCLo:20 ppm/4H 28ZPAK -,214,72
orl-mus LD50:910 mg/kg HYSAAV 33(12),334,68
ihl-mus LC50:725 mg/m^3/2H 85GMAT -,51,82
orl-rbt LD50:900 mg/kg HYSAAV 33(12),334,68
skn-rbt LDLo:250 mg/kg 34ZIAG -,393,69
orl-gpg LD50:810 mg/kg HYSAAV 33(12),334,68

CONSENSUS REPORTS: Reported in EPA TSCA Inventory.

DOT Classification: Corrosive Material; Label: Corrosive, Flammable Liquid.

SAFETY PROFILE: Poison by inhalation and skin contact. Moderately toxic by ingestion. Corrosive. Probably a severe eye and skin irritant. See also ESTERS. When heated to decomposition it emits very toxic fumes of Cl^-, PO_x, and SO_x.

DJW800 CAS:2942-58-7 ***HR: 3***
DIETHYL PHOSPHOROCYANIDATE
mf: $C_5H_{10}NO_3P$ mw: 163.13

SYNS: DIETHOXYPHOSPHORYL CYANIDE ◇ DIETHYLCYANOPHOSPHATE ◇ DIETHYL CYANOPHOSPHONATE

TOXICITY DATA with REFERENCE
ipr-mus LD50:1400 μg/kg PAREAQ 11,636,59
scu-mus LD50:25 mg/kg JCSOA9 -,699,48
ivn-rbt LD50:4 mg/kg JCSOA9 -,699,48

SAFETY PROFILE: Poison by intravenous, intraperitoneal, and subcutaneous routes. When heated to decomposition it emits very toxic fumes of NO_x and PO_x.

DJW875 CAS:1068-22-0 ***HR: 1***
O,O-DIETHYL PHOSPHORODITHIOATE AMMONIUM
mf: $H_4N \cdot C_4H_{10}O_2PS_2$ mw: 203.28

TOXICITY DATA with REFERENCE
skn-rbt 500 mg MLD 34ZIAG -,97,69
eye-rbt 100 mg MLD 34ZIAG -,97,69
orl-rat LD50:7900 mg/kg 34ZIAG -,97,69

SAFETY PROFILE: Mildly toxic by ingestion. An eye

and skin irritant. When heated to decomposition it emits toxic fumes of PO_x, SO_x, NH_3, and NO_x.

DJX000 CAS:84-66-2 ***HR: 3***
DIETHYL PHTHALATE
mf: $C_{12}H_{14}O_4$ mw: 222.26

PROP: Clear, colorless liquid. Mp: −40.5°, bp: 302°, flash p: 325°F (OC), d: 1.110, vap d: 7.66.

SYNS: ANOZOL ◇ 1,2-BENZENEDICARBOXYLIC ACID, DIETHYL ESTER ◇ DIETHYL-o-PHTHALATE ◇ ESTOL 1550 ◇ ETHYL PHTHALATE ◇ NCI-C60048 ◇ NEANTINE ◇ PALATINOL A ◇ PHTHALIC ACID, DIETHYL ESTER ◇ PHTHALOL ◇ PHTHALSAEUREDIAETHYLESTER (GERMAN) ◇ PLACIDOL E ◇ RCRA WASTE NUMBER U088 ◇ SOLVANOL

TOXICITY DATA with REFERENCE
eye-rbt 112 mg JPETAB 82,377,44
mmo-sat 200 μg/plate JTEHD6 16,61,85
orl-mus TDLo:171 g/kg (male 7D pre):REP TXAPA9 88,255,87
ipr-rat TDLo:506 mg/kg (5-15D preg):TER JPMSAE 61,51,72
ihl-hmn TCLo:1000 mg/m³:EYE,PUL AGGHAR 5,1,33
orl-rat LD50:8600 mg/kg GTPZAB 24(3),25,80
ipr-rat LD50:5058 mg/kg JPMSAE 61,51,72
orl-mus LD50:6172 mg/kg GTPZAB 17(11),51,73
ipr-mus LD50:2749 mg/kg FEPRA7 6,342,47
orl-rbt LDLo:1000 mg/kg 14CYAT 2,1904,63
ivn-rbt LDLo:100 mg/kg AGGHAR 5,1,33
orl-gpg LD50:8600 mg/kg GTPZAB 24(3),25,80
scu-gpg LDLo:3000 mg/kg AGGHAR 5,1,33

CONSENSUS REPORTS: Reported in EPA TSCA Inventory.

OSHA PEL: TWA 5 mg/m³
ACGIH TLV: TWA 5 mg/m³

SAFETY PROFILE: Poison by intravenous route. Moderately toxic by ingestion, subcutaneous, and intraperitoneal routes. Human systemic effects by inhalation: lachrimation, respiratory obstruction, and other unspecified respiratory system effects. An eye irritant and systemic irritant by inhalation. An experimental teratogen. Other experimental reproductive effects. Narcotic in high concentrations. Combustible when exposed to heat or flame. To fight fire, use water spray, mist, foam. When heated to decomposition it emits acrid smoke and irritating fumes.

DJX200 CAS:5131-24-8 ***HR: 2***
O,O-DIETHYLPHTHALIMIDOPHOSPHONOTHIOATE
mf: $C_{12}H_{14}NO_4PS$ mw: 299.30

SYNS: O,O-DIAETHYL-N-PHTALIMIDOTHIOPHOSPHAT (GERMAN) ◇ O,O-DIETHYL-(1,2-DIHYDRO-1,3-DIOXO-2H-ISOINDOL-2-YL)PHOSPHONOTHIOATE ◇ O,O-DIETHYL PHTHALIMIDOTHIOPHOSPHATE

TOXICITY DATA with REFERENCE
orl-rat LD50:5660 mg/kg 85ARAE 4,104,76
unk-rat LD50:5000 mg/kg 30ZDA9 -,353,71
orl-rbt LD50:1000 mg/kg 85DPAN -,-,71/76
orl-gpg LD50:5660 mg/kg 28ZEAL 5,86,76
orl-ckn LD50:4500 mg/kg 31ZOAD 1,160,68

SAFETY PROFILE: Moderately toxic by ingestion. When heated to decomposition it emits very toxic fumes of NO_x, PO_x and SO_x.

DJX300 CAS:77650-95-4 ***HR: 3***
N,N-DIETHYL-N'-((8-α)-6-PROPYLERGOLIN-8-YL)UREA
mf: $C_{22}H_{32}N_4O$ mw: 368.58

SYNS: 1-((5R,8S,10R)-6-PROPYL-8-ERGOLINYL)-3,3-DIETHYLUREA ◇ PROTERGURIDE ◇ UREA, N,N-DIETHYL-N'-((8-α)-6-PROPYLERGOLIN-8-YL)-

TOXICITY DATA with REFERENCE
orl-rat TDLo:10 μg/kg (female 5D post):REP CCCCAK 52,2983,87
ivn-mus LD50:37 mg/kg CCCCAK 52,2983,87

SAFETY PROFILE: Poison by intravenous route. Experimental reproductive effects. When heated to decomposition it emits toxic fumes of NO_x.

DJX350 CAS:96860-89-8 ***HR: 3***
N,N-DIETHYL-N'-((8-α)-6-PROPYLERGOLIN-8-YL)UREA (Z)-2-BUTENEDIOATE
mf: $C_{22}H_{32}N_4O•C_4H_4O_4$ mw: 484.66

SYN: UREA, N,N-DIETHYL-N'-((8-α)-6-PROPYLERGOLIN-8-YL)-,(Z)-2-BUTENEDIOATE (1:1)

TOXICITY DATA with REFERENCE
orl-rat TDLo:52600 ng/kg (lactating female 4D post):REP CCCCAK 49,2828,84
ivn-mus LD50:48650 μg/kg CCCCAK 49,2828,84

SAFETY PROFILE: Poison by intravenous route. Experimental reproductive effects. When heated to decomposition it emits toxic fumes of NO_x.

DJX400 CAS:5826-91-5 ***HR: 3***
DIETHYL PROPYLMETHYLPYRIMIDYL THIOPHOSPHATE
mf: $C_{12}H_{21}N_2O_3PS$ mw: 304.38

SYNS: O,O-DIETHYL-O-(2-PROPYL-4-METHYLPYRIMIDINYL-6) PHOSPHOROTHIOATE ◇ O,O-DIETHYL-O-(2-N-PROPYL-4-METHYLPYRIMIDYL-6)PHOSPHOROTHIOATE ◇ O,O-DIETHYL-O-(2-PROPYL-4-METHYL-6-PYRIMIDYL)PHOSPHOROTHIOIC ACID ESTER ◇ G-24622 ◇ PIRAZINON ◇ RCRA WASTE NUMBER P040

TOXICITY DATA with REFERENCE
orl-rat LD50:261 mg/kg PCOC** -,913,66
orl-mus LD50:50 mg/kg PCOC** -,913,66
skn-rbt LD50:3500 mg/kg 27ZTAP 3,53,69

SAFETY PROFILE: Poison by ingestion. Moderately toxic by skin contact. When heated to decomposition it emits very toxic fumes of NO_x, PO_x, and SO_x.

DJX600 CAS:4231-35-0 ***HR: 2***
N,N-DIETHYL-1-PROPYNYLAMINE
mf: $C_7H_{13}N$ mw: 111.21

TOXICITY DATA with REFERENCE
orl-rat LD50:840 mg/kg TXAPA9 28,313,74
skn-rbt LD50:570 mg/kg TXAPA9 28,313,74

SAFETY PROFILE: Moderately toxic by ingestion and skin contact. When heated to decomposition it emits toxic fumes of NO_x.

DJX800 CAS:4079-68-9 ***HR: 2***
N,N-DIETHYL-2-PROPYNYLAMINE
mf: $C_7H_{13}N$ mw: 111.21

TOXICITY DATA with REFERENCE
orl-rat LD50:1540 mg/kg TXAPA9 28,313,74
ihl-rat LCLo:1000 ppm/4H TXAPA9 28,313,74
skn-rbt LD50:570 mg/kg TXAPA9 28,313,74

CONSENSUS REPORTS: Reported in EPA TSCA Inventory.

SAFETY PROFILE: Moderately toxic by ingestion and skin contact. Mildly toxic by inhalation. When heated to decomposition it emits toxic fumes of NO_x.

DJY000 CAS:21600-43-1 ***HR: 3***
3,3-DIETHYL-1-(m-PYRIDYL)TRIAZENE
mf: $C_9H_{14}N_4$ mw: 178.27

SYNS: PYDT ◇ 1-(PYRIDYL-3-)-3,3-DIAETHYL-TRIAZEN (GERMAN) ◇ 1-PYRIDYL-3,3-DIETHYLTRIAZENE ◇ 1-(PYRIDYL-3)-3,3-DIETHYL-TRIAZENE ◇ m-PYRIDYL-DIETHYL-TRIAZENE ◇ 1-(3-PYRIDYL)-3,3-DIETHYLTRIAZENE

TOXICITY DATA with REFERENCE
sln-dmg-orl 2 mmol/L/3D-I ARTODN 43,201,80
mrc-smc 21 mmol/L MUREAV 21,123,73
cyt-hmn:leu 25 µmol/L MUREAV 21,123,73
hma-mus/smc 1600 µmol/kg MUREAV 21,123,73
cyt-ham:lng 10 mg/L MUREAV 88,197,81
ivn-rat TDLo:167 mg/kg (15D preg):TER IARCCD 4,45,73
orl-rat TDLo:660 mg/kg/73W-I:ETA ZKKOBW 77,217,72
scu-rat TDLo:500 mg/kg/50W-I:NEO ZKKOBW 81,285,74
ivn-rat TDLo:55 mg/kg (15D preg):CAR,TER IARCCD 4,45,73
orl-rat LD50:210 mg/kg ZKKOBW 77,217,72
scu-rat LD50:210 mg/kg ZKKOBW 81,285,74

SAFETY PROFILE: Poison by ingestion and subcutaneous routes. Questionable carcinogen with experimental carcinogenic, neoplastigenic, tumorigenic, and teratogenic data. Human mutation data reported. A transplacental carcinogen. When heated to decomposition it emits toxic fumes of NO_x.

DJY050 ***HR: 2***
DIETHYL PYROCARBONATE mixed with AMMONIA
mf: $C_6H_{10}O_5 \cdot 2H_3N$ mw: 196.24

SYN: DEPC and AMMONIA

TOXICITY DATA with REFERENCE
orl-mus TDLo:1936 mg/kg/4W-I:ETA JCREA8 97,205,80

SAFETY PROFILE: Questionable carcinogen with experimental tumorigenic data. When heated to decomposition it emits toxic fumes of ammonia.

DJY100 CAS:60842-44-6 ***HR: 3***
3-(2-(DIETHYLPYRROLIDINO)ETHOXY)-6-METHOXY-2-PHENYLBENZOFURAN HYDROCHLORIDE
mf: $C_{25}H_{31}NO_2 \cdot ClH$ mw: 414.03

SYNS: DBF ◇ 2-PHENYL-3-DIETHYLPYRROLIDINOETHOXY-6-METHOXYBENZOFURANHYDROCHLORIDE

TOXICITY DATA with REFERENCE
orl-rat TDLo:20 mg/kg (8D preg):REP IJEBA6 5,80,67
ipr-rat LD50:388 mg/kg IJEBA6 5,80,67

SAFETY PROFILE: Poison by intraperitoneal route. Experimental reproductive effects. When heated to decomposition it emits toxic fumes of NO_x and HCl.

DJY200 CAS:13593-03-8 ***HR: 3***
O,O-DIETHYL-O-2-QUINOXALYLTHIOPHOSPHATE
mf: $C_{12}H_{15}N_2O_3PS$ mw: 298.32

SYNS: BAY 5821 ◇ BAY 77049 ◇ BAYRUSIL ◇ CHINALPHOS ◇ O,O-DIAETHYL-O-(CHINOXALYL-(2))-MONOTHIOPHOSPHAT(GERMAN) ◇ DIETHQUINALPHION ◇ DIETHQUINALPHIONE ◇ O,O-DIETHYL-O-(2-CHINOXALYL)PHOSPHOROTHIOATE ◇ O,O-DIETHYL-O-QUINOXALIN-2-YL PHOSPHOROTHIOATE ◇ O,O-DIETHYL-O-(2-QUIN-OXALINYL) PHOSPHOROTHIOATE ◇ O,O-DIETHYL-O-(2-QUINOXALYL) PHOSPHOROTHIOATE ◇ EKALUX ◇ ENT 27,394 ◇ NSC 190986 ◇ QUINALPHOS ◇ SAN 6538 I ◇ SANDOZ 6538 ◇ SPENCER S-6538 ◇ SRA 7312 ◇ WIE OBEN

TOXICITY DATA with REFERENCE
orl-gpg TDLo:30 mg/kg (female 15D post):REP BECTA6 24,739,80
orl-rat LD50:26 mg/kg ARSIM* 20,21,66

ihl-rat LC50:175 mg/m^3 85DPAN -,-,71/76
skn-rat LD50:300 mg/kg BESAAT 15,124,69
orl-mus LD50:107 mg/kg 85JCAE -,1166,86
orl-dog LD50:100 mg/kg 85JCAE -,1166,86
par-ckn LD50:10250 μg/kg IVEJAC 62,86,85

SAFETY PROFILE: Poison by ingestion, inhalation, skin contact, parenteral, and intraperitoneal routes. Experimental reproductive effects. An insecticide. When heated to decomposition it emits very toxic fumes of NO_x, PO_x, and SO_x.

DJY400 CAS:19311-91-2 ***HR: 3***
N,N-DIETHYLSALICYLAMIDE
mf: $C_{11}H_{15}NO_2$ mw: 193.27

SYNS: N,N-DIETHYL-2-HYDROXYBENZAMIDE ◇ o-HYDROXY-N,N-DIETHYLBENZAMIDE ◇ SALICYLDIETHYLAMIDE

TOXICITY DATA with REFERENCE
orl-rat LD50:580 mg/kg JPETAB 108,450,53
ipr-rat LD50:350 mg/kg JPETAB 108,450,53
orl-mus LD50:850 mg/kg THERAP 8,237,53
ipr-mus LD50:200 mg/kg NTIS** AD691-490

SAFETY PROFILE: Poison by intraperitoneal route. Moderately toxic by ingestion. When heated to decomposition it emits toxic fumes of NO_x.

DJY600 CAS:110-40-7 ***HR: 1***
DIETHYL SEBACATE
mf: $C_{14}H_{26}O_4$ mw: 258.40

PROP: Colorless to sltly yellow liquid; faint fruity odor. D: 0.960-0.965, refr index: 1.435. Misc with alc, ether, other organic solvents, fixed oils; insol in water @ 302°.

SYNS: DIETHYL DECANEDIOATE ◇ DIETHYL-1,10-DECANEDIOATE ◇ ETHYL SEBACATE ◇ FEMA No. 2376 ◇ SEBACIC ACID, DIETHYL ESTER

TOXICITY DATA with REFERENCE
skn-rbt 500 mg/24H MLD FCTXAV 16,637,78
orl-rat LD50:14470 mg/kg FCTXAV 2,327,64
orl-gpg LD50:7280 mg/kg FCTXAV 2,327,64

CONSENSUS REPORTS: Reported in EPA TSCA Inventory.

SAFETY PROFILE: Mildly toxic by ingestion. A skin irritant. See also ESTERS. When heated to decomposition it emits acrid smoke and irritating fumes.

DJY800 CAS:5117-17-9 ***HR: 3***
N,N-DIETHYLSELENOUREA
mf: $C_5H_{12}N_2Se$ mw: 179.15

SYNS: 1,1-DIETHYL-2-SELENOUREA ◇ USAF B-100

TOXICITY DATA with REFERENCE
ipr-mus LD50:10 mg/kg NTIS** AD277-689

CONSENSUS REPORTS: Reported in EPA TSCA Inventory. Community Right-To-Know List.

OSHA PEL: TWA 0.2 mg(Se)/m^3
ACGIH TLV: TWA 0.2 mg(Se)/m^3
DFG MAK: 0.1 mg(Se)/m^3

SAFETY PROFILE: Poison by intraperitoneal route. See also SELENIUM COMPOUNDS. When heated to decomposition it emits very toxic fumes of NO_x and Se.

DKA000 CAS:40193-47-3 ***HR: 3***
N,N-DIETHYL-4-STILBENAMINE
mf: $C_{18}H_{21}N$ mw: 251.40

SYNS: DIETHYLAMINO STILBENE ◇ 4-STILBENYL-N,N-DIETHYLAMINE

TOXICITY DATA with REFERENCE
scu-rat TDLo:160 mg/kg/9W-I:NEO XPHPAW 149,328,57
scu-rat TD:180 mg/kg/8W-I:ETA PTRMAD 241,147,48

SAFETY PROFILE: Questionable carcinogen with experimental neoplastigenic and tumorigenic data. When heated to decomposition it emits toxic fumes of NO_x.

DKA200 CAS:522-40-7 ***HR: 2***
α,α'-DIETHYL-(E)-4,4'-STILBENEDIOL BIS(DIHYDROGEN PHOSPHATE)
mf: $C_{18}H_{22}O_8P_2$ mw: 428.34

SYNS: DESdp ◇ 4,4'-(1,2-DIETHYL-1,2-ETHENEDIYL)BISPHENOL-(E)-BIS(DIHYDROGEN PHOSPHATE) ◇ DIETHYLSTILBESTEROL DIPHOSPHATE ◇ DIETHYLSTILBESTROL DIPHOSPHATE ◇ DIETHYLSTILBESTROL PHOSPHATE ◇ DIETHYLSTILBESTRYL DIPHOSPHATE ◇ FOSFESTROL ◇ HONVAN ◇ PHOSPHESTROL ◇ ST52-ASTA ◇ STILBESTROL DIPHOSPHATE ◇ STILPHOSTROL

TOXICITY DATA with REFERENCE
cyt-mus-ivn 10 mg/kg ENMUDM 1,184,79
dnd-ham:ovr 400 nmol/L ENMUDM 1,163,79
ipr-mus TDLo:400 mg/kg (6D preg):TER TJADAB 9,229,74
orl-rat TDLo:48 mg/kg (6D male):REP OYYAA2 20,1141,80
ivn-man TDLo:11 mg/kg/3D-I:BPR,SKN SMJOAV 75,248,82
ivn-mus LD50:630 mg/kg ARZNAD 18,666,68

CONSENSUS REPORTS: EPA Genetic Toxicology Program.

SAFETY PROFILE: Moderately toxic by intravenous route. An experimental teratogen. Other experimental reproductive effects. Human systemic effects by intravenous routes: blood pressure lowering, blood clotting factor change, sweating. Mutation data reported. When heated to decomposition it emits toxic fumes of PO_x. See also DIETHYLSTILBESTROL.

DKA400 CAS:63528-82-5 **HR: 3**
α,α'-DIETHYL-4,4'-STILBENEDIOL DISODIUM SALT
mf: $C_{18}H_{18}O_2 \cdot 2Na$ mw: 312.34

SYNS: DES DISODIUM SALT ◇ DIETHYLSTILBESTROL DISODIUM SALT

TOXICITY DATA with REFERENCE
scu-mus TDLo:10 mg/kg (17D preg):REP CNREA8 37,1099,77
scu-mus TDLo:10 mg/kg/(15D preg):NEO,REP CNREA8 37,1099,77
scu-mus TDLo:10 mg/kg (female 15D post):NEO,TER CNREA8 37,1099,77

SAFETY PROFILE: Experimental teratogenic and reproductive effects. Questionable carcinogen with experimental neoplastigenic data. When heated to decomposition it emits toxic fumes of Na2O. See also DIETHYLSTILBESTEROL.

DKA600 CAS:56-53-1 **HR: 3**
DIETHYLSTILBESTEROL
mf: $C_{18}H_{20}O_2$ mw: 268.38

PROP: Small crystals. Mp: 171°.

SYNS: ACNESTROL ◇ AGOSTILBEN ◇ ANTIGESTIL ◇ BIO-DES ◇ 3,4-BIS(p-HYDROXYPHENYL)-3-HEXENE ◇ BUFON ◇ CLIMATERINE ◇ COMESTROL ◇ COMESTROL ESTROBENE ◇ CYREN ◇ DAWE'S DESTROL ◇ DEB ◇ DES (synthetic estrogen) ◇ DESMA ◇ DESTROL ◇ DIASTYL ◇ DIBESTROL ◇ DICORVIN ◇ DI-ESTRYL ◇ trans-4,4'-(1,2-DIETHYL-1,2-ETHENEDIYL)BISPHENOL ◇ 4,4'-(1,2-DIETHYL-1,2-ETHENEDIYL)BIS-PHENOL ◇ α,α'-DIETHYLSTILBENEDIOL ◇ α,α'-DIETHYL-(E)-4,4'-STILBENEDIOL ◇ α,α'-DIETHYL-4,4'-STILBENEDIOL ◇ trans-α,α'-DIETHYL-4,4'-STILBENEDIOL ◇ 2,2'-DIETHYL-4, 4'-STILBENEDIOL ◇ trans-DIETHYLSTILBESTEROL ◇ DIETHYLSTILBESTROL ◇ trans-DIETHYLSTILBESTROL ◇ DIETHYLSTILBOESTEROL ◇ trans-DIETHYLSTILBOESTEROL ◇ DIETILESTILBESTROL (SPANISH) ◇ 4,4'-DIHYDROXYDIETHYLSTILBENE ◇ 4,4'-DIHYDROXY-α,β-DIETHYLSTILBENE ◇ 3,4'(4,4'-DIHYDROXYPHENYL)HEX-3-ENE ◇ DISTILBENE ◇ DOMESTROL ◇ DYESTROL ◇ ESTILBEN ◇ ESTRIL ◇ ESTROBENE ◇ ESTROGEN ◇ ESTROMENIN ◇ ESTROSYN ◇ FOLLIDIENE ◇ FONATOL ◇ GRAFESTROL ◇ GYNOPHARM ◇ HIBESTROL ◇ IDROESTRIL ◇ ISCOVESCO ◇ MAKAROL ◇ MENOSTILBEEN ◇ MICREST ◇ MICROEST ◇ MILESTROL ◇ NEO-OESTRANOL 1 ◇ NSC-3070 ◇ OEKOLP ◇ OESTROGENINE ◇ OESTROL VETAG ◇ OESTROMENIN ◇ OESTROMENSIL ◇ OESTROMENSYL ◇ OESTROMIENIN ◇ OESTROMON ◇ PABESTROL ◇ PALESTROL ◇ PERCUTATRINE OESTROGENIQUE ISCOVESCO ◇ PROTECTONA ◇ RCRA WASTE NUMBER U089 ◇ RUMESTROL 1 ◇ RUMESTROL 2 ◇ SEDESTRAN ◇ SERRAL ◇ SEXOCRETIN ◇ SIBOL ◇ SINTESTROL ◇ STIBILIUM ◇ STIL ◇ STILBESTROL ◇ STILBESTRONE ◇ STILBETIN ◇ STILBOEFRAL ◇ STILBOESTROFORM ◇ STILBOESTROL ◇ STILBOFOLLIN ◇ STILBOL ◇ STILKAP ◇ STIL-ROL ◇ SYNESTRIN ◇ SYNTHOESTRIN ◇ SYNTHOFOLIN ◇ SYNTOFOLIN ◇ TAMPOVAGAN STILBOESTROL ◇ TYLOSTERONE ◇ VAGESTROL

TOXICITY DATA with REFERENCE
mma-esc 50 mg/L MUREAV 130,97,84
dnd-hmn:fbr 300 μmol/L ENMUDM 7,267,85
cyt-hmn:lym 100 μg/L PMRSDJ 5,457,85
orl-wmn TDLo:1730 mg/kg (1-39W preg):REP NEJMAG 292,334,75
unr-wmn TDLo:150 mg/kg (female 7-34W post):TER JOURAA 117,477,77
orl-wmn TDLo:21 mg/kg (female 13-15W post):CAR,TER AJOGAH 137,220,80
orl-wmn TDLo:7655 μg/kg/4Y-C:CAR,REP BJOGAS 82,417,75
unr-man TDLo:184 mg/kg/12Y-C:CAR,PUL,LIV JAMAAP 240,1510,78
mul-man TDLo:25 mg/kg/2Y-C:CAR,SKN UROTAQ 19,180,52
orl-rat TDLo:103 g/kg/2Y-C:CAR AEHLAU 19,489,69
scu-rat TDLo:6 μg/kg (15-18D preg):NEO JTEHD6 5,1059,79
imp-rat TDLo:11500 μg/kg:CAR JJIND8 67,455,81
ipr-mus TDLo:1 mg/kg (17D preg):CAR,TER JJIND8 67,455,81
scu-mus TDLo:800 μg/kg (9-16D preg):CAR,TER CNREA8 45,5145,85
scu-mus TDLo:800 μg/kg (9-16D preg):CAR,TER CNREA8 40,3988,80
par-mus TDLo:340 mg/kg/34W-I:ETA CNREA8 2,759,42
par-dog TDLo:12 mg/kg/23W-I:ETA,TER AJEBAK 40,139,62
orl-gpg TDLo:144 mg/kg/12W-I:ETA,TER RSABAC 25,215,49
imp-gpg TDLo:1080 μg/kg:ETA,TER BSBSAS 8,142,51
orl-ham TDLo:40 mg/kg/(15D preg):ETA,TER CALEDQ 1,139,76
scu-rat TD:1 mg/kg (19D preg):NEO,TER CALEDQ 6,107,79
orl-man TD:45990 μg/kg/3Y-C:CAR,KID BJURAN 51,6,79
skn-hmn TDLo:60 μg/kg/14D:SKN AIHAAP 20,469,59
ipr-rat LD50:34 mg/kg TXAPA9 24,142,73
orl-mus LDLo:2500 mg/kg KLWOAZ 18,156,39
ipr-mus LD50:67 mg/kg TXAPA9 24,37,73
scu-mus LDLo:50 mg/kg KLWOAZ 18,156,39
ivn-mus LD50:630 mg/kg ARZNAD 18,666,68

CONSENSUS REPORTS: NTP Fifth Annual Report on Carcinogens. IARC Cancer Review: Group 1 IMEMDT 7,273,87; Human Limited Evidence IMEMDT 6,55,74; IMEMDT 21,173,79; Animal Sufficient Evidence IMEMDT 21,173,79; IMEMDT 6,55,74. EPA Genetic Toxicology Program. Reported in EPA TSCA Inventory.

SAFETY PROFILE: Confirmed carcinogen producing skin, liver, and lung tumors in exposed humans as well as uterine and other reproductive system tumors in the female offspring of exposed women. Experimental carcinogenic, neoplastigenic, tumorigenic, and teratogenic data. A transplacental carcinogen. A human teratogen by many routes. Poison by intraperitoneal and subcutaneous routes. Moderately toxic by ingestion and other routes. It causes glandular system effects by skin contact. Human reproductive effects by ingestion: abnormal spermatogenesis; changes in testes, epididymis and sperm duct; menstrual cycle changes or disorders; changes in female fertility; unspecified maternal effects; developmental abnormalities of the fetal urogenital system; germ cell effects in offspring; and delayed effects in newborn. Implicated in male impotence and enlargement of male breasts. Other experimental reproductive effects. Mutation data reported. When heated to decomposition it emits acrid smoke and fumes. See also 19-NOR-17-a-PREGNA-1,3,5(10)-TRIEN-2-YNE-3,17-DIOL (ESTROGEN).

DKA800 CAS:63019-08-9 ***HR: 3***
DIETHYLSTILBESTROL DIPALMITATE
mf: $C_{50}H_{80}O_4$ mw: 745.30

SYNS: α,α'-DIETHYL-4,4'-STILBENEDIOL DIPALMITATE ◇ 4,4'-DIHYDROXY-α,β-DIETHYLSTILBENE PALMITATE

TOXICITY DATA with REFERENCE
scu-rbt TDLo:535 μg/kg (5D male):REP ENDOAO 60,519,57
ims-rbt TDLo:38 mg/kg/45W-I:ETA,TER CANCAR 10,500,57

SAFETY PROFILE: Experimental teratogenic and reproductive effects. Questionable carcinogen with experimental tumorigenic data. When heated to decomposition it emits acrid smoke and irritating fumes. See also DIETHYLSTILBESTEROL.

DKB000 CAS:130-80-3 ***HR: 3***
DIETHYLSTILBESTROL DIPROPIONATE
mf: $C_{24}H_{28}O_4$ mw: 380.52

PROP: Crystals. Mp: 104°.

SYNS: CLINESTROL ◇ CYREN B ◇ DESD ◇ DIBESTIL ◇ trans-4,4'-(1,2-DIETHYL-1,2-ETHENEDIYL)BISPHENOL DIPROPIONATE ◇ α,α'-DIETHYL-4,4'-STILBENEDIOL, DIPROPIONATE ◇ α,α'-DIETHYL-4,4'-STILBENEDIOL trans-DIPROPIONATE ◇ trans-α,α'-DIETHYL-4,4'-STILBENEDIOL DIPROPIONATE ◇ DIETHYLSTILBENE DIPROPIONATE ◇ α,α'-DIETHYL-4,4'-STILBENEDIOL DIPROPIONYL ESTER ◇ DIETHYLSTILBESTEROL DIPROPIONATE ◇ DIETHYLSTILBESTROL PROPIONATE ◇ DIHYDROXYDIETHYLSTILBENE DIPROPIONATE ◇ 4,4'-DIHYDROXY-α,β-DIETHYLSTILBENE DIPROPIONATE ◇ DIPROPIONATO de ESTILBENE (SPANISH) ◇ p,p'-DIPROPIONOXY-trans-α,β-DIETHYLSTILBENE ◇ DISTILBENE ◇ ESTILBEN ◇ ESTILBIN ◇ ESTROBEN ◇ ESTROBENE ◇ ESTROGENIN ◇ ESTROSTILBEN ◇ EUVESTIN ◇ GYNOLETT ◇ HORFEMINE ◇ NEO-OESTRANOL II ◇ OESTROGYNAEDRON ◇ ORESTOL ◇ PABESTROL ◇ SINCICLAN ◇ STILBESTROL DIETHYL DIPROPIONATE ◇ STILBESTROL DIPROPIONATE ◇ STILBESTROL PROPIONATE ◇ STILBESTRONATE ◇ STILBOESTROL DIPROPIONATE ◇ STILBOFAX ◇ STILRONATE ◇ SYNESTRIN ◇ SYNOESTRON ◇ SYNTESTRIN ◇ SYNTESTRINE ◇ WILLESTROL

TOXICITY DATA with REFERENCE
cyt-hmn:lym 1300 nmol/L TGANAK 16(2),24,82
sln-mus-ipr 50 mg/kg MUREAV 144,27,85
cyt-ham:fbr 5 mg/L CRNGDP 3,499,82
sce-ham:fbr 5 mg/L CRNGDP 3,499,82
orl-rat TDLo:60 μg/kg (3-4D preg):REP APTSAI 26(Suppl 1),1,68
unr-mus TDLo:200 μg/kg (female 13D post):TER CRSBAW 164,2372,70
scu-rat TDLo:80 mg/kg:ETA HMMRA2 7,228,75
imp-gpg TDLo:5 mg/kg:ETA,REP RSABAC 25,215,49
scu-frg TDLo:133 mg/kg/32W-I:NEO BEXBAN 81,898,76

CONSENSUS REPORTS: IARC Cancer Review: Animal Sufficient Evidence IMEMDT 21,173,79. EPA Genetic Toxicology Program.

SAFETY PROFILE: Confirmed carcinogen with experimental tumorigenic data. An experimental teratogen. Other experimental reproductive effects. Human mutagenic data. When heated to decomposition it emits acrid smoke and irritating fumes. See also DIETHYLSTILBESTEROL.

DKB100 CAS:6052-82-0 ***HR: 2***
DIETHYLSTILBOESTROL-3,4-OXIDE
mf: $C_{18}H_{20}O_3$ mw: 284.38

SYNS: DES-α,β-OXIDE ◇ DES-3,4-OXIDE ◇ α,α'-DIETHYL-α,α'-EPOXYBIBENZYL-4,4'-DIOL ◇ DIETHYLSTILBOESTROL-α,β-OXIDE

TOXICITY DATA with REFERENCE
dnd-hmn:leu 300 μmol/L BCPCA6 34,3251,85
sce-hmn:fbr 40 nmol/L NATUAS 281,392,79
imp-ham TDLo:640 mg/kg/38W-I:ETA CNREA8 43,5200,83

SAFETY PROFILE: Questionable carcinogen with experimental tumorigenic data. Human mutation data reported. When heated to decomposition it emits acrid smoke and fumes. See also DIETHYLSTILBESTEROL.

DKB110 CAS:64-67-5 ***HR: 3***
DIETHYL SULFATE
DOT: UN 1594
mf: $C_4H_{10}O_4S$ mw: 154.20

PROP: Colorless, oily liquid; faint ethereal odor. Mp: −25°, Bp: 209.5° (decomp to ethyl ether), flash p: 220°F (CC), d: 1.172 @ 25°/4°, autoign temp: 817°F, vap press: 1 mm @ 47.0°, vap d: 5.31. Insol in water; decomp by hot water; misc with alc and ether.

SYNS: DIAETHYLSULFAT (GERMAN) ◇ DIETHYL ESTER SULFURIC ACID ◇ ETHYL SULFATE

TOXICITY DATA with REFERENCE
skn-rbt 10 mg/24H SEV JIHTAB 31,60,49
skn-rbt 500 mg open MLD UCDS** 12/30/71
eye-rbt 2 mg SEV JIHTAB 31,60,49
mmo-sat 5 mg/plate MUREAV 57,141,78
mma-sat 5 mg/plate MUREAV 57,141,78
slt-mus-ipr 100 mg/kg/10W MUREAV 75,63,80
cyt-mus-ipr 150 mg/kg MUREAV 75,63,80
msc-ham:ovr 1 mmol/L MUREAV 57,217,78
ivn-rat TDLo:340 mg/kg (female 15D post):TER IARCCD 4,45,73
orl-rat TDLo:3700 mg/kg/81W-I:ETA ZEKBAI 74,241,70
ivn-rat TDLo:85 mg/kg (15D preg):CAR IARCCD 4,45,73
orl-rat LD50:880 mg/kg JIDHAN 31,60,49
ihl-rat LCLo:250 ppm/4H JIHTAB 31,343,49
scu-rat LD50:350 mg/kg ZEKBAI 74,241,70
orl-mus LD50:647 mg/kg NTIS** PB214-270
skn-rbt LD50:600 mg/kg UCDS** 12/30/71

CONSENSUS REPORTS: NTP Fifth Annual Report on Carcinogens. IARC Cancer Review: Group 2A IMEMDT 7,198,87; Animal Sufficient Evidence IMEMDT 4,277,74. EPA Genetic Toxicology Program. Community Right-To-Know List. Reported in EPA TSCA Inventory.

DFG TRK: 0.03 ppm; Animal Carcinogen, Human Suspected Carcinogen.
DOT Classification: Poison B; Label: Poison

SAFETY PROFILE: Confirmed with experimental carcinogenic and tumorigenic data. Poison by inhalation and subcutaneous route. Moderately toxic by ingestion and skin contact. A severe skin irritant. An experimental teratogen. Mutation data reported. Combustible when exposed to heat or flame; can react with oxidizing materials. Moisture causes liberation of H_2SO_4. Violent reaction with potassium tert-butoxide. Reacts violently with 3,8-dinitro-6-phenylphenanthridine + water. Reaction with iron + water forms the explosive hydrogen gas. To fight fire, use alcohol foam, H_2O foam, CO_2, dry chemicals. When heated to decomposition it emits toxic fumes of SO_x. See also SULFATES.

DKB119 CAS:623-81-4 ***HR: 3***
DIETHYL SULFITE
mf: $C_4H_{10}O_3S$ mw: 138.18

SAFETY PROFILE: Reaction with chlorine fluoride gives the violently explosive ethyl fluorosulfate. When heated to decomposition it emits toxic fumes of SO_x. See also SULFITES.

DKB139 CAS:1912-30-7 ***HR: D***
DIETHYLSULFONATE
mf: $C_4H_{10}O_3S$ mw: 138.20

SYNS: EES ◇ ETHYL ETHANE SULFONATE

TOXICITY DATA with REFERENCE
mmo-sat 80 nmol/L CPBTAL 32,3626,84
mmo-omi 110 nmol/L/30M-C PNASA6 47,540,61
ivn-rat TDLo:10 mg/kg (8D preg):REP IARCCD 4,100,73

CONSENSUS REPORTS: EPA Genetic Toxicology Program.

SAFETY PROFILE: Experimental reproductive effects. Mutation data reported. When heated to decomposition it emits toxic fumes of SO_x. See also SULFONATES and ESTERS.

DKB150 CAS:627-54-3 ***HR: 3***
DIETHYL TELLURIDE
mf: $C_4H_{10}Te$ mw: 185.72

SAFETY PROFILE: Ignites spontaneously in air. See also TELLURIUM COMPOUNDS.

DKB175 CAS:22392-07-0 ***HR: 3***
DIETHYL THALLIUM PERCHLORATE
mf: $C_4H_{10}ClO_4Tl$ mw: 361.96

CONSENSUS REPORTS: Thallium and its compounds are on the Community Right-To-Know List.

SAFETY PROFILE: A poison. Explodes at its mp 250°C. When heated to decomposition it emits toxic fumes of Cl^-. See also THALLIUM COMPOUNDS and PERCHLORATES.

DKB600 CAS:5827-03-2 ***HR: 3***
N,N-DIETHYLTHIOCARBAMYL-O,O-DIISOPROPYLDITHIOPHOSPHATE
mf: $C_{11}H_{24}NO_2PS_3$ mw: 329.51

SYNS: DIETHYLDITHIOCARBAMIC ANHYDRIDE of O,O-DIISOPROPYL THIONOPHOSPHORIC ACID ◇ DIETHYLDITHIOCARBAMIC ANHYDROSULFIDE ◇ O,O-DIISOPROPYL-S-DIETHYLDITHIOCARBAMOYLPHOSPHORODITHIOATE ◇ O,O-DIIOSPROPYL DITHIOPHOSPHORIC ACID ESTER of-N,N-S-DIETHYLTHIOCARBAMOYL-O,O-DIISOPROPYLPHOSPHOROTHIOATE ◇ DIISOPROPYL ESTER of DITHIOCARBAMYL PHOSPHOROTHIOIC ACID ◇ ENT 24,725

TOXICITY DATA with REFERENCE
orl-rat LD60:320 mg/kg 28ZEAL 4,175,69
orl-mus LD50:290 mg/kg ARSIM* 20,13,66

SAFETY PROFILE: Poison by ingestion. See also CARBAMATES and ANHYDRIDES. When heated to decomposition it emits very toxic fumes of PO_x, NO_x, and SO_x.

DKC200 CAS:69226-06-8 ***HR: 3***
2,2-DIETHYL-3-THIOMORPHOLINONE
mf: $C_8H_{15}NOS$ mw: 173.30

TOXICITY DATA with REFERENCE
orl-mus LD50:2250 mg/kg JMCMAR 6,136,63
ipr-mus LD50:652 mg/kg JMCMAR 6,136,63
ivn-mus LD50:137 mg/kg JMCMAR 6,136,63
orl-rbt LD50:1500 mg/kg JMCMAR 6,136,63
ivn-rbt LD50:185 mg/kg JMCMAR 6,136,63

SAFETY PROFILE: Poison by intravenous route. Moderately toxic by ingestion and intraperitoneal route. When heated to decomposition it emits very toxic fumes of NO_x and SO_x.

DKC400 CAS:105-55-5 ***HR: 3***
1,3-DIETHYLTHIOUREA
mf: $C_5H_{12}N_2S$ mw: 132.25

SYNS: N,N'-DIETHYLTHIOCARBAMIDE ◇ N,N'-DIETHYLTHIOUREA ◇ 1,3-DIETHYL-2-THIOUREA ◇ NCI-C03816 ◇ PENNZONE E ◇ THIATE H ◇ U 15030 ◇ USAF EK-1803

TOXICITY DATA with REFERENCE
orl-rat TDLo:11 g/kg/2Y-C:CAR NCITR* NCI-CG-TR-149,79
orl-rat LD50:316 mg/kg NCILB* NCI-E-C-72-3252,73
orl-mus LDLo:62 mg/kg AECTCV 14,111,85
ipr-mus LD50:500 mg/kg NTIS** AD277-689

CONSENSUS REPORTS: NCI Carcinogenesis Bioassay (feed); Clear Evidence: rat NCITR* NCI-CG-TR-149,79; No Evidence: mouse NCITR* NCI-CG-TR-149,79. Reported in EPA TSCA Inventory. EPA Genetic Toxicology Program.

SAFETY PROFILE: Poison by ingestion. Moderately toxic by intraperitoneal route. Questionable carcinogen with experimental carcinogenic data. When heated to decomposition it emits very toxic fumes of NO_x and SO_x.

DKC600 CAS:73940-85-9 ***HR: 3***
DIETHYLTIN DI(10-CAMPHORSULFONATE)
mf: $C_{24}H_{40}O_8S_2Sn$ mw: 639.45

SYNS: BIS(10-CAMPHORSULFONATO)DIETHYLSTANNANE ◇ BIS(2-OXO-9-BORNANESULFONIC ACID) DIETHYLSTANNYL ESTER

TOXICITY DATA with REFERENCE
ivn-mus LD50:89 mg/kg CSLNX* NX#03150

OSHA PEL: TWA 0.1 mg(Sn)/m^3 (skin)
ACGIH TLV: TWA 0.1 mg(Sn)/m^3 (skin) (Proposed: TWA 0.1 mg(Sn)/m^3; STEL 0.2 mg(Sn)/m^3 (skin))
NIOSH REL: (Organotin Compounds) TWA 0.1 mg(Sn)/m^3

SAFETY PROFILE: Poison by intravenous route. See also TIN COMPOUNDS; SULFONATES; and CAMPHOR. When heated to decomposition it emits toxic fumes of SO_x.

DKC800 CAS:134-62-3 ***HR: 3***
DIETHYL-m-TOLUAMIDE
mf: $C_{12}H_{17}NO$ mw: 191.30

PROP: A liquid, sol in water, alc, and ether. Bp: 160° @ 19 mm, d: 0.996 @ 20°/4°.

SYNS: AI 3-22542 ◇ AUTAN ◇ BAKER'S ANTIFOL ◇ CHEMFORM ◇ DEET ◇ DELPHENE ◇ m-DELPHENE ◇ DET ◇ m-DET ◇ m-DETA ◇ DETAMIDE ◇ DIELTAMID ◇ N,N-DIETHYL-3-METHYLBENZAMIDE ◇ DIETHYLTOLUAMIDE ◇ N,N-DIETHYL-m-TOLUAMIDE ◇ ENT 20,218 ◇ ENT 22,542 ◇ FLYPEL ◇ METADELPHENE ◇ 3-METHYL-N,N-DIETHYLBENZAMIDE ◇ MGK DIETHYLTOLUAMIDE ◇ NAUGATUCK DET ◇ OFF ◇ REPEL ◇ REPPER-DET ◇ REPUDIN-SPECIAL ◇ m-TOLUIC ACID DIETHYLAMIDE

TOXICITY DATA with REFERENCE
skn-rbt 500 mg MOD NTIS** AD-A106-944
eye-rbt 100 mg NTIS** AD-A107-736
spm-rat-ihl 1500 mg/m^3 AEHA** 75-51-0034-80
skn-rat TDLo:19 g/kg (1-19D preg):REP FATOAO 38,202,75
orl-wmn LDLo:950 mg/kg JAMAAP 258,1509,87
orl-man LDLo:679 mg/kg JAMAAP 258,1509,87
orl-cld TDLo:4750 mg/kg JAMAAP 258,1509,87
orl-wmn TDLo:950 mg/kg:EYE,PUL JAMAAP 258,1509,87
skn-hmn TDLo:35 mg/kg/5D:SKN TXAPA9 1,97,59
orl-rat LD50:1950 mg/kg SPEADM 78-1,53,78
ihl-rat LC50:5950 mg/m^3 AEHA** 75-51-0034-80
skn-rat LD50:5000 mg/kg MPPBAB 45,65,76
skn-mus LD50:3170 mg/kg YKYUA6 31,309,80
orl-rbt LD50:1584 mg/kg NTIS** AD-A082-131
skn-rbt LD50:3180 mg/kg TXAPA9 28,313,74
ivn-rbt LDLo:75 mg/kg TXAPA9 1,97,59

CONSENSUS REPORTS: Reported in EPA TSCA Inventory.

SAFETY PROFILE: Poison by intravenous route. Moderately toxic by ingestion and skin contact. Human systemic effects by skin contact: dermatitis, pupillary dilation, coma pulminary changes. An eye and skin irritant. Experimental reproductive effects by skin contact. Mutation data reported. Can cause central nervous system disturbances. A pesticide. DEET is the active ingredient in most commercial insect repellents. When heated to decomposition it emits toxic fumes of NO_x.

DKD000 CAS:2728-04-3 ***HR: 2***
N,N-DIETHYL-o-TOLUAMIDE
mf: $C_{12}H_{17}NO$ mw: 191.30

SYN: 2-METHYL-N,N-DIETHYLBENZAMIDE

TOXICITY DATA with REFERENCE
skn-hmn 2500 μg/5D MLD TXAPA9 1,97,59
skn-rbt 2 mg/kg TXAPA9 1,97,59
orl-rat LD50:1210 mg/kg TXAPA9 7,772,65

SAFETY PROFILE: Moderately toxic by ingestion. A human skin irritant. When heated to decomposition it emits toxic fumes of NO_x.

DKD200 CAS:63980-20-1 ***HR: 1***
DIETHYL TRIAZENE
mf: $C_4H_{11}N_3$ mw: 101.18

TOXICITY DATA with REFERENCE
orl-rat TDLo:400 mg/kg/20W-I:CAR CALEDQ 35,129,87
scu-rat TDLo:110 mg/kg (15D post):NEO,TER XENOBH 3,271,73
scu-rat LD50:220 mg/kg 85JCAE-,498,86

SAFETY PROFILE: Poison by subcutaneous route. Questionable carcinogen with experimental carcinogenic and tumorigenic data. An experimental teratogen. When heated to decomposition it emits toxic fumes of NO_x.

DKD400 CAS:3282-85-7 ***HR: 2***
3,9-DIETHYLTRIDECYL-6-SULFATE
mf: $C_{17}H_{35}O_4S{\cdot}Na$ mw: 358.57

SYNS: SODIUM-4-ETHYL-1-(3-ETHYLPENTYL)-1-OCTYLSULFATE ◇ TERGITOL 7

TOXICITY DATA with REFERENCE
skn-rbt 500 mg open MLD UCDS** 1/20/72
eye-rbt 250 μg MLD AROPAW 34,99,45
orl-rat LD50:1430 mg/kg JIHTAB 23,478,41
skn-rbt LD50:3560 mg/kg UCDS** 1/20/72
orl-gpg LD50:425 mg/kg JIHTAB 23,478,41
skn-gpg LDLo:425 mg/kg JIHTAB 23,478,41

CONSENSUS REPORTS: Reported in EPA TSCA Inventory.

SAFETY PROFILE: Moderately toxic by ingestion and skin contact. A skin and eye irritant. See also SULFATES. When heated to decomposition it emits toxic fumes of SO_x and Na_2O.

DKD600 CAS:3218-37-9 ***HR: 3***
N,N-DIETHYL-4-(α-(α,α,α-TRIFLUORO-o-TOLYL) BENZYLOXY)PENTYLAMINE CITRATE
mf: $C_{23}H_{30}F_3NO{\cdot}C_6H_8O_7$ mw: 585.68

TOXICITY DATA with REFERENCE
orl-mus LDLo:300 mg/kg ARZNAD 14,964,64
ivn-mus LDLo:20 mg/kg ARZNAD 14,964,64

SAFETY PROFILE: Poison by ingestion and intravenous routes. See also FLUORIDES. When heated to decomposition it emits very toxic fumes of F^- and NO_x.

DKE200 CAS:304-84-7 ***HR: 3***
N,N-DIETHYLVANILLAMIDE
mf: $C_{12}H_{17}NO_3$ mw: 223.30

SYNS: DIETHYLAMIDE de VANILLIQUE ◇ 3-METHOXY-4-HYDROXYBENZOIC ACID DIETHYLAMIDE ◇ VANILLIC ACID DIETHYLAMIDE ◇ VANILLIC ACID-N,N-DIETHYLAMIDE ◇ VANILLINSAEURE-DIAETHYLAMID (GERMAN)

TOXICITY DATA with REFERENCE
ipr-rat LD50:28 mg/kg COREAF 243,609,56
orl-mus LD50:67 mg/kg RPOBAR 1,423,64
ipr-mus LD50:38 mg/kg RPOBAR 1,423,64
scu-mus LD50:57 mg/kg RPOBAR 1,423,64
ivn-mus LD50:15 mg/kg 27ZQAG -,377,72
orl-dog LD50:300 mg/kg 27ZQAG -,377,72
ivn-dog LD50:30 mg/kg 27ZQAG -,377,72
orl-rbt LDLo:50 mg/kg ARZNAD 20,367,70

SAFETY PROFILE: Poison by ingestion, intravenous, subcutaneous, and intraperitoneal routes. When heated to decomposition it emits toxic fumes of NO_x.

DKE400 CAS:1851-77-0 ***HR: 3***
DI(ETHYLXANTHOGEN)TRISULFIDE
mf: $C_6H_{10}O_2S_5$ mw: 274.46

SYNS: BEXT ◇ BIS(ETHOXYTHIOCARBONYL)TRISULFIDE ◇ BIS-(ETHYLXANTHOGEN) TRISULFIDE ◇ DEFOLIANT 713 ◇ DI-ETHOXY-THIOKARBONYL-TRISULFID ◇ TRISULFIDE, BIS(ETHOXYTHIO-CARBONYL)-

TOXICITY DATA with REFERENCE
orl-rat LD50:235 mg/kg 28ZEAL 5,69,76
orl-mus LD50:140 mg/kg 28ZEAL 5,69,76
ihl-mus LC50:300 mg/m³/3H 85JCAE-,1018,86

SAFETY PROFILE: Poison by ingestion. Moderately toxic by inhalation. When heated to decomposition it emits toxic fumes of SO_x.

DKE600 CAS:557-20-0 ***HR: 3***
DIETHYLZINC
DOT: UN 1366/UN 2845
mf: $C_4H_{10}Zn$ mw: 123.51

PROP: Liquid. Mp: −28°, bp: 118°, d: 1.2065 @ 20°/4°.

SYNS: ZINC ETHIDE ◇ ZINC ETHYL (DOT)

CONSENSUS REPORTS: Reported in EPA TSCA Inventory. Zinc and its compounds are on the Community Right-To-Know List.

DOT Classification: Flammable Liquid; Label: Flammable Liquid; IMO: Flammable Liquid; Label: Spontaneously Combustible.

SAFETY PROFILE: Presumed to be a poison. Ignites spontaneously in air. Dangerously flammable by spontaneous chemical reaction in air, or with oxidizing materials. A dangerous explosion hazard. Explosive reaction with alkenes + diiodomethane; sulfur dioxide. Reacts violently with bromine; water; nitro compounds. Ignites on contact with air; ozone; methanol; or hydrazine. Reacts violently with non-metal halides (e.g., arsenic trichloride or phosphorus trichloride to produce pyrophoric triethyl arsine or triethyl phosphine. To fight fire, do not use water, foam, or halogenated extinguishing agents. Use dry materials, such as graphite, sand, etc. When heated to decomposition it emits toxic fumes of ZnO. See also ZINC COMPOUNDS.

DKE800 CAS:63868-62-2 ***HR: 3***
DIETROL
mf: $C_{12}H_{17}NO{\cdot}2C_4H_6O_6$ mw: 491.50

SYNS: ADPHEN ◇ BACARATE ◇ 3,4-DIMETHYL-2-PHENYLMORPHOLINE BITARTRATE ◇ HOURBESE ◇ LIMIT ◇ MINUS ◇ NEONILOREX ◇ OBEPAR ◇ PHENAZINE ◇ PHENDIMETRAZINE BITARTRATE ◇ PLEGINE ◇ REDUCTO ◇ STATOBEX ◇ SYMETRA ◇ TRIMSTAT ◇ TRIMTABS

TOXICITY DATA with REFERENCE
orl-mus LD50:285 mg/kg 27ZQAG -,285,72
ipr-mus LD50:410 mg/kg 27ZQAG -,285,72
ivn-mus LD50:170 mg/kg 27ZQAG -,285,72

SAFETY PROFILE: Poison by ingestion, intraperitoneal, and intravenous routes. When heated to decomposition it emits toxic fumes of NO_x.

DKF125 CAS:33564-31-7 ***HR: D***
DIFLORASONE DIACETATE
mf: $C_{26}H_{32}F_2O_7$ mw: 494.58

SYNS: 17,21-BIS(ACETYLOXY)-6,9-DIFLUORO-11-HYDROXY-16-METHYLPREGNA-1,4-DIENE-3,20-DIONE(6-α,11-β,16-β)- ◇ FLORONE

TOXICITY DATA with REFERENCE
scu-rat TDLo:495 μg/kg (female 7-17D post):REP OYYAA2 28,207,84
scu-rat TDLo:495 μg/kg (female 7-17D post):TER OYYAA2 28,207,84

SAFETY PROFILE: An experimental teratogen. Experimental reproductive effects at very low doses. A steroid. When heated to decomposition it emits toxic fumes of F^-.

DKF130 CAS:59198-70-8 ***HR: 3***
DIFLUCORTOLONE VALERATE
mf: $C_{27}H_{36}F_2O_5$ mw: 478.63

SYNS: DFV ◇ DIFLUCORTOLONE 21-VALERATE ◇ DIFLUCORTOLONVALERIANAT (GERMAN) ◇ 6-α,9-DIFLUORO-11-β-HYDROXY-16-α-METHYL-21-VALERYLOXY-1,4-PREGNADIENE-3,20-DIONE ◇ 6-α,2-DIFLUORO-11-β-HYDROXY-21-VALERYLOXY-16-α-METHYL-1,4-PREGNADIENE-3,20-DIONE ◇ NERISONA ◇ NERISONE ◇ TEMETEX

TOXICITY DATA with REFERENCE
scu-rat TDLo:700 μg/kg (female 7-13D post):REP YKRYAH 10,1357,77
scu-rat TDLo:700 μg/kg (female 7-13D post):TER YKRYAH 10,1357,77
orl-rat LD50:3100 mg/kg ARZNAD 26,1476,76
ipr-rat LD50:19500 μg/kg IYKEDH 8,165,77
scu-rat LD50:13 mg/kg ARZNAD 26,634,76
orl-mus LD50:4750 mg/kg IYKEDH 11,811,80
ipr-mus LD50:432 mg/kg ARZNAD 26,634,76
scu-mus LD50:140 mg/kg IYKEDH 11,811,80
orl-dog LD50:1 g/kg ARZNAD 26,1476,76
scu-rbt TDLo:26300 μg/kg NIIRDN 6,339,82

SAFETY PROFILE: Poison by subcutaneous route. Moderately toxic by ingestion and intraperitoneal routes. An experimental teratogen. Experimental reproductive effects. When heated to decomposition it emits toxic fumes of F^-.

DKF170 CAS:404-42-2 ***HR: 3***
2,4′-DIFLUOROACETANILIDE
mf: $C_8H_7F_2NO$ mw: 171.16

TOXICITY DATA with REFERENCE
orl-rat LD50:2 mg/kg JSFAAE 8,400,57
ipr-rat LD50:5 mg/kg JSFAAE 8,400,57
orl-mus LD50:25 mg/kg YKKZAJ 88,1620,68

SAFETY PROFILE: Poison by ingestion and intraperitoneal routes. When heated to decomposition it emits toxic fumes of F^- and NO_x.

DKF200 CAS:381-73-7 ***HR: 3***
DIFLUOROACETIC ACID
mf: $C_2H_2F_2O_2$ mw: 96.04

TOXICITY DATA with REFERENCE
ivn-mus LD50:180 mg/kg CSLNX* NX#03015

CONSENSUS REPORTS: Reported in EPA TSCA Inventory.

SAFETY PROFILE: Poison by intravenous route. Explosive reaction with fluorine + cesium fluoride. When heated to decomposition it emits toxic fumes of F^-. See also FLUORIDES.

DKF400 CAS:10405-27-3 ***HR: 3***
DIFLUOROAMINE
mf: F_2HN mw: 53.01

SYN: FLUORIMIDE

SAFETY PROFILE: A shock-sensitive explosive. Tends to explode in the solid state or upon freezing. Explosive reaction with cesium fluoride + Benzenediazonium tetrafluoroborate. Violent reaction with ClO_2. When heated to decomposition it emits toxic fumes of F^- and NO_x. See also FLUORIDES and AMINES.

DKF600 CAS:17224-08-7 ***HR: 3***
3-DIFLUOROAMINO-1,2,3-TRIFLUORODIAZIRIDINE
mf: CF_5N_3 mw: 149.02

SAFETY PROFILE: A strong oxidant. An explosive sensitive to shock and phase change. When heated to decomposition it emits toxic fumes of F^- and NO_x. See also FLUORIDES.

DKF620 CAS:56533-30-3 ***HR: 3***
DIFLUOROAMMONIUM HEXAFLUOROARSENATE
mf: AsF_8H_2N mw: 242.93

CONSENSUS REPORTS: Arsenic and its compounds are on the Community Right-To-Know List.

SAFETY PROFILE: Arsenic compounds are generally poisons. A storage hazard; decomposes at room temperature after a delay period. When heated to decomposition it emits toxic fumes of F^-, NO_x, NH_3, and As. See also ARSENIC COMPOUNDS.

DKF800 CAS:372-18-9 ***HR: 3***
m-DIFLUOROBENZENE
mf: $C_6H_4F_2$ mw: 114.09

PROP: Flash p: < 32°C.

SAFETY PROFILE: A very dangerous fire hazard when exposed to heat or flame; can react vigorously with oxidizing materials. When heated to decomposition it emits toxic fumes of F^-.

DKG000 CAS:540-36-3 ***HR: 3***
p-DIFLUOROBENZENE
mf: $C_6H_4F_2$ mw: 114.09

PROP: Flash p: 23°F.

SAFETY PROFILE: A very dangerous fire hazard when exposed to heat or flame; can react vigorously with oxidizing materials. When heated to decomposition it emits toxic fumes of F^-.

DKG100 CAS:368-68-3 ***HR: 3***
3,4-DIFLUOROBENZENEARSONIC ACID
mf: $C_6H_5AsF_2O_3$ mw: 238.03

SYN: BENZENEARSONIC ACID, 3,4-DIFLUORO-

TOXICITY DATA with REFERENCE
ipr-mus LDLo:31300 μg/kg CBCCT* 6,53,54

OSHA PEL: TWA 0.5 mg(As)/m^3

SAFETY PROFILE: Poison by intraperitoneal route. When heated to decomposition it emits toxic fumes of As and F^-.

DKG400 CAS:61735-78-2 ***HR: 3***
2,10-DIFLUOROBENZO(rst)PENTAPHENE
mf: $C_{24}H_{12}F_2$ mw: 338.36

SYN: 2,10-DIFLUORODIBENZO(a,i)PYRENE

TOXICITY DATA with REFERENCE
mma-sat 12500 pmol/plate CNREA8 41,2589,81
scu-mus TDLo:20 mg/kg:ETA PAACA3 13,37,72

SAFETY PROFILE: Questionable carcinogen with experimental tumorigenic data. Mutation data reported. When heated to decomposition it emits toxic fumes of F^-.

DKG600 CAS:10578-16-2 ***HR: 3***
DIFLUORODIAZENE
mf: F_2N_2 mw: 66.01

FN=NF

SAFETY PROFILE: Very unstable explosive. Explodes when frozen or condensed from gas. Explosive reaction with hydrogen above 90°C. When heated to decomposition it emits toxic fumes of F^- and NO_x.

DKG700 CAS:693-85-6 ***HR: 3***
DIFLUORODIAZIRINE
mf: CF_2N_2 mw: 78.02

SAFETY PROFILE: An explosive sensitive to heat or exposure to light. When heated to decomposition it emits toxic fumes of F^- and NO_x.

DKG800 CAS:75-82-1 ***HR: 1***
DIFLUORODIBROMOETHANE
mf: $C_2H_2Br_2F_2$ mw: 223.86

SYN: 1,1-DIFLUORO-1,2-DIBROMOETHANE

TOXICITY DATA with REFERENCE
ihl-rat LCLo:5000 ppm/18H AIHOAX 2,335,50

SAFETY PROFILE: Mildly toxic by inhalation. When heated to decomposition it emits very toxic fumes of F^- and Br^-.

DKG850 CAS:75-61-6 ***HR: 1***
DIFLUORODIBROMOMETHANE
DOT: UN 1941
mf: CBr_2F_2 mw: 209.83

PROP: Colorless, heavy liquid. Bp: 23.2°, fp: −141°, d: 2.288 @ 15°/4°.

SYNS: DIBROMODIFLUOROMETHANE ◇ FREON 12-B2 ◇ HALON 1202

TOXICITY DATA with REFERENCE
ihl-rat LCLo:54630 ppm/15M MRLR** No.23,50
ihl-mus LCLo:67 g/m^3/15M MRLR** #107,52

CONSENSUS REPORTS: Reported in EPA TSCA Inventory.

OSHA PEL: TWA 100 ppm
ACGIH TLV: TWA 100 ppm
DFG MAK: 100 ppm (860 mg/m^3)
DOT Classification: ORM-A; Label: None.

SAFETY PROFILE: Mildly toxic by inhalation. When heated to decomposition it emits very toxic fumes of Br^- and F^-.

DKG980 CAS:351-63-3 ***HR: 2***
2′,4′-DIFLUORO-4-DIMETHYLAMINOAZOBENZENE
mf: $C_{14}H_{13}F_2N_3$ mw: 261.30

TOXICITY DATA with REFERENCE
orl-rat TDLo:2727 mg/kg/14W-I:ETA CNREA8 18,469,58

SAFETY PROFILE: Questionable carcinogen with experimental tumorigenic data. When heated to decomposition it emits toxic fumes of F^- and NO_x.

DKH000 CAS:349-37-1 ***HR: 3***
2′,5′-DIFLUORO-4-DIMETHYLAMINOAZOBENZENE
mf: $C_{14}H_{13}F_2N_3$ mw: 261.30

SYN: N,N-DIMETHYL-p-(2,5-DIFLUOROPHENYLAZO)ANILINE

TOXICITY DATA with REFERENCE
orl-rat TDLo:3400 mg/kg/12W-C:ETA CNREA8 13,93,53

SAFETY PROFILE: Questionable carcinogen with experimental tumorigenic data. When heated to decomposition it emits very toxic fumes of F^- and NO_x.

DKH100 CAS:350-87-8 ***HR: 2***
3′,5′-DIFLUORO-4-DIMETHYLAMINOAZOBENZENE
mf: $C_{14}H_{13}F_2N_3$ mw: 261.30

SYNS: ANILINE,N,N-DIMETHYL-p-(3,5-DIFLUOROPHENYLAZO)- ◇ p-((3,5-DIFLUOROPHENYL)AZO)-N,N-DIMETHYLANILINE ◇ N,N-DIMETHYL-p-(3,5-DIFLUOROPHENYLAZO)ANILINE

TOXICITY DATA with REFERENCE
orl-rat TDLo:3400 mg/kg/12W-C:ETA CNREA8 13,93,53

SAFETY PROFILE: Questionable carcinogen with experimental tumorigenic data. When heated to decomposition it emits toxic fumes of F^- and NO_x.

DKH200 CAS:3582-17-0 ***HR: 3***
DIFLUORODIMETHYLSTANNANE
mf: $C_2H_6F_2Sn$ mw: 186.77

PROP: White crystals. Water-sol. Bp: decomp < 360°.

SYNS: DIMETHYLTIN DIFLUORIDE ◇ DIMETHYLTIN FLUORIDE

TOXICITY DATA with REFERENCE
ivn-mus LD50:56 mg/kg CSLNX* NX#00014

CONSENSUS REPORTS: Reported in EPA TSCA Inventory.

OSHA PEL: TWA 0.1 mg(Sn)/m^3 (skin)
ACGIH TLV: TWA 0.1 mg(Sn)/m^3 (skin) (Proposed: TWA 0.1 mg(Sn)/m^3; STEL 0.2 mg(Sn)/m^3 (skin))
NIOSH REL: (Organotin Compounds) TWA 0.1 mg(Sn)/m^3

SAFETY PROFILE: Poison by intravenous route. See also TIN COMPOUNDS and FLUORIDES. When heated to decomposition it emits toxic fumes of F^-.

DKH600 CAS:24937-79-9 ***HR: 1***
1,1-DIFLUOROETHYLENE POLYMERS (PYROLYSIS)
mf: $(C_2H_2F_2)_x$

SYN: POLYVINYLIDENE FLUORIDE (PYROLYSIS)

TOXICITY DATA with REFERENCE
ihl-mus LC50:99 g/m^3/30M PWPSA8 21,167,78

CONSENSUS REPORTS: Reported in EPA TSCA Inventory.

SAFETY PROFILE: Very mildly toxic by inhalation. When heated to decomposition it emits very toxic fumes of F^-.

DKH825 CAS:338-66-9 ***HR: 3***
DIFLUORO-N-FLUOROMETHANIMINE
mf: CF_3N mw: 83.01

SAFETY PROFILE: Explodes when exposed to flame. When heated to decomposition it emits toxic fumes of F^- and NO_x.

DKH830 CAS:16282-67-0 ***HR: 3***
DIFLUOROMETHYLENE DIHYPOFLUORITE
mf: CF_4O_2 mw: 120.00

SAFETY PROFILE: A strong oxidant. Potentially hazardous reactions with organic or easily oxidized materials. Potentially explosive reactions with trans-dichloroethylene; tetrafluoroethylene and possibly other haloalkanes. When heated to decomposition it emits toxic fumes of F^-.

DKH875 CAS:67037-37-0 ***HR: 3***
dl-α-DIFLUOROMETHYLORNITHINE
mf: $C_6H_{12}F_2N_2O_2$ mw: 182.20

SYNS: α-DFMO ◇ EFLORNITHINE ◇ RMI 71782

TOXICITY DATA with REFERENCE
msc-hmn:lym 1250 μmol/L CNREA8 44,4272,84

cyt-ham:ovr 500 μmol/L CNREA8 44,4535,84
scu-ham TDLo:150 mg/kg (female 7D post):TER CCPTAY 28,159,83
orl-mus TDLo:3952 mg/kg (female 5-8D post):REP SCIEAS 208,505,80
ice-rat LD50:1364 μg/kg APTOA6 56,250,85

SAFETY PROFILE: Poison by intracerebral route. An experimental teratogen. Experimental reproductive effects. Human mutation data reported. When heated to decomposition it emits toxic fumes of F^- and NO_x. An ornithine-decarboxylase inhibitor.

DKI200 ***HR: 3***
3,4-DIFLUORO-2-NITROBENZENEDIAZONIUM-6-OXIDE
mf: $C_6HF_2N_3O_3$ mw: 201.09

SAFETY PROFILE: An explosive sensitive to heat, ignition, or impact. When heated to decomposition it emits toxic fumes of F^- and NO_x. See also NITRO COMPOUNDS of AROMATIC HYDROCARBONS.

DKI289 ***HR: 3***
3,6-DIFLUORO-2-NITROBENZENEDIAZONIUM-4-OXIDE
mf: $C_6HF_2N_3O_3$ mw: 201.09

SAFETY PROFILE: An explosive sensitive to heat, ignition, or impact. When heated to decomposition it emits toxic fumes of F^- and NO_x. See also NITRO COMPOUNDS of AROMATIC HYDROCARBONS.

DKI400 CAS:368-97-8 ***HR: 3***
DIFLUOROPHENYLARSINE
mf: $C_6H_5AsF_2$ mw: 190.03

SYN: PHENYLDIFLUOROARSINE

TOXICITY DATA with REFERENCE
skn-rat LD50:15 mg/kg JPBAA7 58,411,46
skn-rbt LD50:4 mg/kg JPBAA7 58,411,46
ivn-rbt LD50:500 μg/kg JPBAA7 58,411,46
skn-gpg LD50:10 mg/kg JPBAA7 58,411,46

CONSENSUS REPORTS: Arsenic and its compounds are on the Community Right-To-Know List.

OSHA PEL: TWA 0.5 mg(As)/m^3

SAFETY PROFILE: Poison by skin contact and intravenous routes. See also FLUORIDES and ARSENIC COMPOUNDS. When heated to decomposition it emits very toxic fumes of As and F^-.

DKI600 CAS:22494-42-4 ***HR: 3***
5-(2,4-DIFLUOROPHENYL)SALICYLIC ACID
mf: $C_{13}H_8F_2O_3$ mw: 250.21

SYNS: DIFLUNISAL ◇ 2',4'-DIFLUORO-4-HYDROXY-(1,1'-BIPHENYL)-3-CARBOXYLIC ACID ◇ 2',4'-DIFLUORO-4-HYDROXY-3-BIPHENYLCARBOXYLIC ACID ◇ 2',4'-DIFLUORO-4-HYDROXY-(1',1-DIPHENYL)-3-CARBOXYLIC ACID ◇ DOLOBID ◇ DOLOBIL ◇ DOLOBIS ◇ FLOVACIL ◇ FLUNIGET ◇ 2-(HYDROXY)-5-(2,4-DIFLUOROPHENYL) BENZOIC ACID ◇ MK 647

TOXICITY DATA with REFERENCE
orl-rbt TDLo:520 mg/kg (female 6-18D post):TER TJADAB 30,319,84
orl-rbt TDLo:780 mg/kg (female 6-18D post):REP TJADAB 30,319,84
orl-hmn TDLo:71 mg/kg/5D-I:CNS,LIV BMJOAE 2,736,78
orl-man TDLo:429 mg/kg/17W-I:BLD ANZJB8 16,811,86
orl-rat LD50:392 mg/kg IYKEDH 15,688,84
ipr-rat LD50:159 mg/kg IYKEDH 15,688,84
scu-rat LD50:185 mg/kg IYKEDH 15,688,84
orl-mus LD50:439 mg/kg BCPHBM 4,19S,77
ipr-mus LD50:124 mg/kg IYKEDH 15,688,84
scu-mus LD50:220 mg/kg IYKEDH 15,688,84
orl-rbt LD50:603 mg/kg BCPHBM 4,19S,77

SAFETY PROFILE: Poison by ingestion, subcutaneous, and intraperitoneal routes. Human systemic effects by ingestion: tolerance, and cholestatic jaundice (due to the stoppage of the flow of bile), agranulocytosis, increased body temperature. An experimental teratogen. Other experimental reproductive effects. An analgesic and anti-inflammatory agent. When heated to decomposition it emits toxic fumes of F^-. See also FLUORIDES.

DKI800 CAS:453-13-4 ***HR: 3***
1,3-DIFLUORO-2-PROPANOL
mf: $C_3H_6F_2O$ mw: 96.09

TOXICITY DATA with REFERENCE
ivn-mus LD50:178 mg/kg CSLNX* NX#00467

CONSENSUS REPORTS: Reported in EPA TSCA Inventory.

SAFETY PROFILE: Poison by intravenous route. When heated to decomposition it emits toxic fumes of F^-.

DKJ200 CAS:314-04-5 ***HR: 3***
3,8-DIFLUOROTRICYCLOQUINAZOLINE
mf: $C_{21}H_{10}F_2N_4$ mw: 356.35

TOXICITY DATA with REFERENCE
skn-mus TDLo:1200 mg/kg/50W-I:NEO BCPCA6 14,323,65

SAFETY PROFILE: Questionable carcinogen with experimental neoplastigenic data. See also FLUORIDES. When heated to decomposition it emits very toxic fumes of F^- and NO_x.

DKJ225 CAS:1510-31-2 ***HR: 3***
1,1-DIFLUOROUREA
mf: $CH_2F_2N_2O$ mw: 96.04

SAFETY PROFILE: Concentrated aqueous solutions decompose above −20°C and evolve the explosive gases tetrafluorohydrazine and difluoramine. When heated to decomposition it emits toxic fumes of F^- and NO_x.

DKJ300 CAS:23674-86-4 ***HR: 3***
DIFLUPREDNATE
mf: $C_{27}H_{34}F_2O_7$ mw: 508.61

PROP: Crystals from methylene chloride/ether/petr ether. Mp: 191-194°.

SYNS: 6-α,9-α-DIFLUOROPREDNISOLONE 17-BUTYRATE 21-ACETATE ◇ 6-α,9-DIFLUORO-11-β,17,21-TRIHYDROXYPREGNA-1,4-DIENE-3,20-DIONE-21-ACETATE-17-BUTYRATE ◇ W 6309 ◇ EPITOPIC

TOXICITY DATA with REFERENCE
scu-rat TDLo:1100 ng/kg (female 7-17D post):REP IYKEDH 15,1035,84
scu-rbt TDLo:1300 ng/kg (female 6-18D post):TER IYKEDH 15,1055,84
ipr-rat LD50:1 g/kg IYKEDH 15,1066,84
scu-rat LD50:2315 mg/kg IYKEDH 15,1066,84
ipr-mus LD50:146 mg/kg IYKEDH 15,1066,84
scu-mus LD50:105 mg/kg IYKEDH 15,1066,84

SAFETY PROFILE: Poison by subcutaneous and intraperitoneal routes. An experimental teratogen. Other experimental reproductive effects. When heated to decomposition it emits toxic fumes of F^-.

DKJ600 CAS:628-36-4 ***HR: 3***
1,2-DIFORMYLHYDRAZINE
mf: $C_2H_4N_2O_2$ mw: 88.08

SYNS: 1,2-DIFORMYLHYDRAZIN (GERMAN) ◇ HYDRAZODIFORMIC ACID

TOXICITY DATA with REFERENCE
orl-mus TDLo:5800 mg/kg/66W-C:CAR ZKKOBW 92,11,78

CONSENSUS REPORTS: Reported in EPA TSCA Inventory.

SAFETY PROFILE: Questionable carcinogen with experimental carcinogenic data. When heated to decomposition it emits toxic fumes of NO_x. See also HYDRAZINE.

DKK100 CAS:2315-20-0 ***HR: D***
DIFURAN
mf: $C_{14}H_{12}N_6O_6 \cdot ClH$ mw: 396.78

SYNS: AC 1692-40 ◇ 1,5-BIS(5-NITRO-2-FURYL)-1,4-PENTADIENE-3-AMINOHYDRAZONE HYDROCHLORIDE ◇ NITROVIN ◇ PANAZON ◇ PAYZONE

TOXICITY DATA with REFERENCE
mma-sat 500 ng/plate MUREAV 77,21,80
mrc-bcs 100 ng/disc MUREAV 77,21,80

CONSENSUS REPORTS: EPA Genetic Toxicology Program.

SAFETY PROFILE: Mutation data reported. When heated to decomposition it emits very toxic fumes of HCl and NO_x.

DKK200 CAS:19247-68-8 ***HR: 3***
N,N-DIFURFURAL-n-PHENYLENEDIAMINE
mf: $C_{14}H_{12}N_2O_2$ mw: 240.28

TOXICITY DATA with REFERENCE
orl-rat LD50:1220 mg/kg 85GMAT -,54,82
ihl-rat LCLo:70 mg/m³/4H 85GMAT -,54,82
orl-mus LD50:400 mg/kg 85GMAT -,54,82

SAFETY PROFILE: Poison by inhalation and ingestion. When heated to decomposition it emits toxic fumes of NO_x.

DKK400 CAS:25639-45-6 ***HR: 3***
DI-2-FUROYL PEROXIDE
mf: $C_{10}H_6O_6$ mw: 122.16

PROP: Violently explodes on friction and heating.

SAFETY PROFILE: A powerful explosive sensitive to friction or heat. When heated to decomposition it emits acrid smoke and fumes. See also PEROXIDES.

DKK800 CAS:63906-88-7 ***HR: 3***
DIGAMMACAINE
mf: $C_{21}H_{26}N_2O \cdot ClH$ mw: 358.95

SYNS: 1-BENZAMIDO-1-PHENYL-3-PIPERIDINOPROPANEHYDROCHLORIDE ◇ N-(3-BENZAMIDO-3-PHENYL)PROPYL PIPERIDINE HYDROCHLORIDE

TOXICITY DATA with REFERENCE
scu-rat LDLo:112 mg/kg CLDND*
ivn-rat LDLo:16 mg/kg CLDND*
orl-mus LDLo:99 mg/kg CLDND*
scu-mus LDLo:105 mg/kg CLDND*
ivn-mus LD50:17 mg/kg CLDND*
ims-mus LDLo:96 mg/kg CLDND*

SAFETY PROFILE: Poison by ingestion, intramuscular, subcutaneous, and intravenous routes. When heated to decomposition it emits very toxic fumes of HCl and NO_x.

DKL000 ***HR: 3***
DIGERMANE
mf: Ge_2H_6 mw: 151.25

H_3GeGeH_3

PROP: Liquid or gas. Bp: 29°, mp: −109°, d (gas): 6.74 g/L @ 20°, d (liquid): 1.98 @ −109°.

SAFETY PROFILE: May ignite spontaneously in air. Autoignition temperature is about 50°C in air. Combustion may be explosive. See also HYDRIDES and GERMANIUM COMPOUNDS.

DKL200 CAS:8031-42-3 ***HR: 3***
DIGITALIS

PROP: Dried whole leaf of *Digitalis purpurea* (27ZTAP 3,54,69). Composition: digitoxin (0.2-0.4%), etc.

SYNS: DIGITANNOID ◇ DIGITALIS PURPUREA, LEAF ◇ FOXGLOVE

TOXICITY DATA with REFERENCE
ivn-man TDLo:12 μg/kg JAMAAP 241,164,79
unr-man LDLo:29 mg/kg 85DCAI 2,73,70
ipr-rat LDLo:50 mg/kg TXAPA9 1,156,59
ivn-mus LDLo:200 mg/kg AEXPBL 82,1,17
orl-cat LD50:244 mg/kg JAPMA8 44,607,55
ivn-cat LDLo:62 mg/kg JAPMA8 44,607,55
ipr-pgn LD50:220 mg/kg JPPMAB 14,96,62
ivn-pgn LDLo:670 mg/kg JPPMAB 14,96,62

SAFETY PROFILE: A deadly human poison by intravenous route. An experimental poison by ingestion, intravenous, and intraperitoneal routes. 2.5 grams or 30 cc of the tincture is a toxic dose. An overdose can be fatal. it has been implicated in aplastic anemia. It contains digitalin, digitalein, digitonin, and digitoxin (the most toxic component). See also individual components.

DKL300 CAS:90028-00-5 ***HR: 3***
DIGITALIS LANATA STANDARD

TOXICITY DATA with REFERENCE
orl-cat LD50:103 mg/kg JAPMA8 44,607,55
ivn-cat LDLo:29900 μg/kg JAPMA8 44,607,55
ivn-gpg LDLo:81200 μg/kg ARZNAD 17,1237,67
ivn-pgn LDLo:300 μg/kg JPHAA3 25,611,36

SAFETY PROFILE: Poison by ingestion and intravenous routes. See also DIGITALIS.

DKL400 CAS:11024-24-1 ***HR: 3***
DIGITONIN
mf: $C_{56}H_{92}O_{29}$ mw: 1229.48

SYN: DIGITIN

TOXICITY DATA with REFERENCE
sln-smc 250 ppb ANYAA9 407,186,83
ivn-rat LD50:4 mg/kg ARZNAD 12,815,62
orl-mus LDLo:90 mg/kg HBAMAK 4,1289,35
ipr-mus LD10:20 mg/kg PCJOAU 11,749,77
scu-mus LDLo:200 mg/kg HBAMAK 4,1289,35
ivn-mus LDLo:10 mg/kg HBAMAK 4,1289,35

CONSENSUS REPORTS: Reported in EPA TSCA Inventory.

SAFETY PROFILE: Poison by ingestion, intravenous, intraperitoneal, and subcutaneous routes. Mutation data reported. See also DIGITALIS. When heated to decomposition it emits acrid smoke and irritating fumes.

DKL800 CAS:71-63-6 ***HR: 3***
DIGITOXIN
mf: $C_{41}H_{64}O_{13}$ mw: 765.05

SYNS: ACEDOXIN ◇ ASTHENTHILO ◇ CARDIDIGIN ◇ CARDIGIN ◇ CARDITOXIN ◇ CRISTAPURAT ◇ CRYSTALLINE DIGITALIN ◇ CRYSTODIGIN ◇ DIGILONG ◇ DIGIMED ◇ DIGIMERCK ◇ DIGISIDIN ◇ DIGITALIN ◇ DIGITALINE (FRENCH) ◇ DIGITALINE CRISTALLISEE ◇ DIGITALINE NATIVELLE ◇ DIGITALINUM VERUM ◇ DIGITOPHYLLIN ◇ DIGITOXIGENIN-TRIDIGITOXOSID (GERMAN) ◇ DIGITOXIGENIN TRIDIGITOXOSIDE ◇ DITAVEN ◇ GLUCODIGIN ◇ LANATOXIN ◇ MONO-GLYCOCOARD ◇ MYODIGIN ◇ PURODIGIN ◇ PURPURID ◇ TARDIGAL ◇ TRI-DIGITOXOSIDE (GERMAN) ◇ UNIDIGIN

TOXICITY DATA with REFERENCE
eye-dog 1%/1D AJOPAA 56,889,63
orl-wmn TDLo:200 μg/kg (27-30W preg):REP AJCDAG 6,834,60
orl-inf TDLo:150 μg/kg:CVS,GIT AJDCAI 130,425,76
unk-man LDLo:44 μg/kg 85DCAI 2,73,70
orl-rat LD50:56 mg/kg TXAPA9 18,185,71
ivn-rat LD50:3900 μg/kg ARZNAD 11,848,61
orl-mus LD50:4950 μg/kg ARZNAD 8,557,58
ipr-mus LD50:3900 μg/kg AIPTAK 155,165,65
scu-mus LD50:22180 μg/kg AIPTAK 153,436,65
ivn-mus LD50:4100 μg/kg ARZNAD 8,557,58
ivn-dog LDLo:500 μg/kg HBAMAK 4,1343,35
orl-cat LD50:180 μg/kg AIPTAK 159,1,66
ipr-cat LD50:170 μg/kg AIPTAK 155,165,65
ivn-cat LDLo:180 μg/kg AEPPEA 184,181,37
idu-cat LDLo:373 μg/kg ARZNAD 19,687,69
iat-cat LDLo:440 μg/kg AEPPAE 141,329,29
ivn-rbt LDLo:1 mg/kg HBAMAK 4,1343,35
ivn-pig LDLo:400 μg/kg ARZNAD 20,229,70
idu-pig LDLo:550 μg/kg ARZNAD 20,229,70

CONSENSUS REPORTS: Reported in EPA TSCA Inventory. EPA Extremely Hazardous Substances List.

SAFETY PROFILE: A deadly poison by most routes. Human systemic effects by ingestion and possibly other routes: cardiac arrythmias, nausea and vomiting. Human reproductive effects by ingestion: reduced viability of newborn. An eye irritant. When heated to decomposition it emits acrid smoke and irritating fumes. See also DIGITALIS.

DKL875 CAS:1339-93-1 ***HR: 3***
DIGITOXOSIDE
mf: $C_{41}H_{64}O_{17}$ mw: 765.05

SYN: MONO-DIGITOXID (GERMAN)

TOXICITY DATA with REFERENCE
ivn-cat LDLo:270 μg/kg AEPPAE 237,222,59
ivn-gpg LDLo:310 μg/kg AEPPAE 237,222,59
ivn-pgn LDLo:400 μg/kg APFRAD 9,730,51

SAFETY PROFILE: A deadly poison by intravenous route. When heated to decomposition it emits acrid smoke and fumes. See also DIGITALIS.

DKM100 CAS:32144-31-3 ***HR: D***
DIGLYCIDYLANILINE
mf: $C_{12}H_{15}NO_2$ mw: 205.28

SYN: BIS(OXIRANYLMETHYL)BENZENAMINE (9CI)

TOXICITY DATA with REFERENCE
mmo-sat 100 μg/plate MUREAV 135,159,84
cyt-ham:ovr 8 mg/L MUREAV 135,159,84

SAFETY PROFILE: Mutation data reported. When heated to decomposition it emits toxic fumes of NO_x. See also ANILINE DYES.

DKM120 CAS:2095-06-9 ***HR: 2***
N-N-DIGLYCIDYLANILINE
mf: $C_{12}H_{15}NO_2$ mw: 205.28

SYNS: BIS(2,3-EPOXYPROPYL)ANILINE ◇ N,N-BIS(2,3-EPOXYPROPYL)ANILINE ◇ BIS(EPOXYPROPYL)PHENYLAMINE ◇ N,N-DIGLYCIDYLANILIN (CZECH) ◇ N-N-DIGLYCIDYLPHENYLAMINE ◇ N-(OXIRANYLMETHYL)-N-PHENYL-OXIRANEMETHANAMINE (9CI)

TOXICITY DATA with REFERENCE
skn-rbt 500 mg/24H SEV 28ZPAK -,136,72
eye-rbt 500 mg/24H MLD 28ZPAK -,136,72
orl-rat LD50:1620 mg/kg AIHAAP 30,470,69
skn-rbt LD50:3560 mg/kg AIHAAP 30,470,69

CONSENSUS REPORTS: Reported in EPA TSCA Inventory.

SAFETY PROFILE: Moderately toxic by ingestion and skin contact. An eye and severe skin irritant. When heated to decomposition it emits toxic fumes of NO_x. See also ANILINE DYES.

DKM130 CAS:15336-81-9 ***HR: 3***
N,N'-DIGLYCIDYL-5,5-DIMETHYLHYDANTOIN
mf: $C_{11}H_{16}N_2O_4$ mw: 240.29

SYNS: 1,3-BIS(2,3-EPOXYPROPYL)-5,5-DIMETHYLHYDANTOIN ◇ 5,5-DIMETHYL-1,3-BIS(OXIRANYLMETHYL)-2,4-IMIDAZOLIDINEDIONE ◇ 5,5-DIMETHYL-1,3-BIS(OXIRANYLMETHYL)-2,4-IMIDAZOLIDINEDIONE (9CI) ◇ XB 2793

TOXICITY DATA with REFERENCE
skn-mus TDLo:3513 mg/kg/93W-I:ETA NTIS** ORNL-5762

CONSENSUS REPORTS: Reported in EPA TSCA Inventory.

SAFETY PROFILE: Questionable carcinogen with experimental tumorigenic data by skin contact. When heated to decomposition it emits toxic fumes of NO_x.

DKM200 CAS:2238-07-5 ***HR: 3***
DIGLYCIDYL ETHER
mf: $C_6H_{10}O_3$ mw: 130.16

PROP: Liquid.

SYNS: BIS(2,3-EPOXYPROPYL)ETHER ◇ DGE ◇ DI(2,3-EPOXYPROPYL) ETHER

TOXICITY DATA with REFERENCE
skn-rbt 563 mg/3D SEV AMIHAB 14,250,56
eye-rbt 113 mg SEV AMIHAB 14,250,56
mmo-sat 50 μg/plate MUREAV 66,367,79
mma-sat 50 μg/plate MUREAV 66,367,79
unr-mus TDLo:1300 mg/kg:ETA RARSAM 3,193,63
orl-rat LD50:450 mg/kg AMIHAB 14,250,56
orl-mus LD50:170 mg/kg AMIHAB 14,250,56
ihl-mus LC50:30 ppm/4H AMIHAB 14,250,56
ivn-mus LD50:100 mg/kg CSLNX* NX#00108
skn-rbt LD50:1500 mg/kg AMIHAB 14,250,56
ivn-rbt LDLo:200 mg/kg AEHLAU 2,31,61

CONSENSUS REPORTS: EPA Extremely Hazardous Substances List. Reported in EPA TSCA Inventory. EPA Genetic Toxicology Program.

OSHA PEL: (Transitional: CL 0.5 ppm) TWA 0.1 ppm
ACGIH TLV: TWA 0.1 ppm
DFG MAK: 0.1 ppm (0.6 mg/m^3); Suspected Carcinogen.
NIOSH REL: (Glycidyl Ethers) CL 1 mg/m^3/15M

SAFETY PROFILE: Suspected carcinogen with experimental tumorigenic data. Poison by ingestion, inhalation, and intravenous routes. Moderately toxic by skin contact. A severe eye and skin irritant. Mutation data reported. Chronic exposure can cause bone marrow depression. When heated to decomposition it emits acrid smoke and fumes. See also ETHERS.

DKM400 CAS:63041-01-0 ***HR: 3***
DIGLYCIDYL ETHER of N,N-BIS(2-HYDROXYPROPYL)-tert-BUTYLAMINE
mf: $C_{16}H_{31}NO_4$ mw: 301.48

SYN: 2,2'-BIS(2,3-EPOXYPROPOXY)-N-tert-BUTYLDIPROPYLAMINE

TOXICITY DATA with REFERENCE
scu-mus TDLo:3600 mg/kg/9W-I:ETA FCTXAV 4,365,66

SAFETY PROFILE: Questionable carcinogen with experimental tumorigenic data. See also ETHERS. When heated to decomposition it emits toxic fumes of NO_x.

DKM600 CAS:7195-45-1 ***HR: 2***
DIGLYCIDYL PHTHALATE
mf: $C_{14}H_{14}O_6$ mw: 278.28

SYN: PHTHALIC ACID, DIGLYCIDYL ESTER

TOXICITY DATA with REFERENCE
orl-rat LD50:1245 mg/kg MarJV# 29MAR77

CONSENSUS REPORTS: Reported in EPA TSCA Inventory.

SAFETY PROFILE: Moderately toxic by ingestion. See also ESTERS. When heated to decomposition it emits acrid smoke and irritating fumes.

DKM800 CAS:63040-98-2 ***HR: 3***
N,N-DIGLYCIDYL-p-TOLUENESULPHONAMIDE
mf: $C_{13}H_{17}NO_4S$ mw: 283.37

SYNS: N,N-BIS(2,3-EPOXYPROPYL)-p-TOLUENESULFONAMIDE ◇ N,N-DIGLYCIDYL-p-TOLUENESULFONAMIDE

TOXICITY DATA with REFERENCE
scu-mus TDLo:17 g/kg/43W-I:ETA FCTXAV 4,365,66

SAFETY PROFILE: Questionable carcinogen with experimental tumorigenic data. When heated to decomposition it emits very toxic fumes of NO_x and SO_x.

DKN000 CAS:628-89-7 ***HR: 2***
DIGLYCOL CHLORHYDRIN
mf: $C_4H_9ClO_2$ mw: 124.58

PROP: Colorless liquid. Bp: 196.8°, flash p: 225°F (OC), d: 1.1698, vap press: 0.17 mm @ 20°.

SYN: 2-(2-CHLOROETHOXY)ETHANOL

TOXICITY DATA with REFERENCE
skn-rbt 10 mg/24H open JIHTAB 26,269,44
eye-rbt 5 mg SEV AJOPAA 29,1363,46
orl-rat LD50:6300 mg/kg JIHTAB 26,269,44
skn-gpg LD50:3000 mg/kg JIHTAB 26,269,44

CONSENSUS REPORTS: Reported in EPA TSCA Inventory.

SAFETY PROFILE: Moderately toxic by skin contact. Mildly toxic by ingestion. A skin and severe eye irritant. Combustible when exposed to heat or flame; can react with oxidizing materials. To fight fire, use alcohol foam, CO_2, dry chemical. Will react with water or steam to produce toxic and corrosive fumes. When heated to decomposition it emits toxic fumes of Cl^-.

DKN250 CAS:54086-41-8 ***HR: 3***
DIGOLD(I) KETENIDE
mf: C_2Au_2O mw: 433.96

SAFETY PROFILE: The dry material is a shock-sensitive explosive. Forms heat-sensitive explosive complexes with tertiary heterocyclic bases (e.g., pyridine, methyl pyridines, 2,6-dimethylpyridine, quinoline). See other gold compounds.

DKN300 CAS:1672-46-4 ***HR: 3***
DIGOXIGENINE
mf: $C_{23}H_{34}O_5$ mw: 390.57

PROP: Dihydrate: Prismatic rods from dilute alc. Anhydrous: As stout prisms from ethyl acetate. Mp: 222°.

SYNS: CARDOGENEN-(20:22)-TRIOL-(3-β,12,14) (GERMAN) ◇ DIGOXIGENIN ◇ LANADIGENIN ◇ $\Delta^{-20:22}$-3-β,12-β,14,21-TETRAHYDROXYNORCHOLENIC ACID LACTONE ◇ 3-β,12-β,14-TRIHYDROXYCARD-20(22)-ENOLIDE ◇ 3-β,12,14-TRIOXY-CARDEN-(20:22)-OLID (GERMAN) ◇ 3-β,12,14-TRIOXY-DIGEN-(20:22)-OLID (GERMAN)

TOXICITY DATA with REFERENCE
ivn-gpg LDLo:3545 μg/kg AIPTAK 153,536,65
par-pgn LDLo:1620 μg/kg CPBTAL 8,18,60

SAFETY PROFILE: A deadly poison by intravenous and parenteral routes. When heated to decomposition it emits acrid smoke and fumes. See also DIGITALIS.

DKN400 CAS:20830-75-5 ***HR: 3***
DIGOXIN
mf: $C_{41}H_{64}O_{14}$ mw: 781.05

PROP: White, crystalline powder. Mp: 265°. Glycoside isolated from *Digitalis lanata* (JPETAB 52,1,34).

SYNS: CHLOROFORMIC DIGITALIN ◇ DIGACIN ◇ DIGITALIS GLYCOSIDE ◇ DIGOXIGENIN-TRIDIGITOXOSID (GERMAN) ◇ DIGOXINE ◇ HOMOLLE'S DIGITALIN ◇ LANICOR ◇ LANOXIN ◇ ROUGOXIN ◇ SK-DIGOXIN

TOXICITY DATA with REFERENCE
ims-dom TDLo:810 μg/kg (1-18W preg):TER AJOGAH 121,1100,75
orl-wmn TDLo:100 μg/kg:CNS,CVS,GIT NZMJAX 84,443,76
orl-man TDLo:333 μg/kg:EYE,CVS BMJOAE 287,392,83
orl-cld TDLo:127 μg/kg:BPR AFPYAE 34(1),137,86
ivn-inf LDLo:200 μg/kg:BPR PEDIAU 77,848,86
ivn-inf TDLo:50 μg/kg:BPR AFPYAE 34(1),137,86
scu-rat LD50:30 mg/kg PPHAD4 1,97,80
scu-rat LD50:30 mg/kg PPHAD4 1,97,80
ivn-rat LD50:25 mg/kg AIPTAK 155,165,65
orl-mus LD50:17780 μg/kg AIPTAK 153,436,65
ipr-mus LD50:3964 μg/kg CHTHBK 16,371,71
scu-mus LD50:12880 μg/kg AIPTAK 153,436,65
ivn-mus LD50:7670 μg/kg PLRCAT 6,417,74
par-mus LDLo:2947 μg/kg SIHSD8 4(2),32,81
orl-dog LDLo:300 μg/kg CRSBAW 109,279,32

CONSENSUS REPORTS: EPA Extremely Hazardous Substances List. Reported in EPA TSCA Inventory. EPA Genetic Toxicology Program.

SAFETY PROFILE: A deadly poison by most routes.

Human systemic effects by ingestion: anorexia, cardiac arrythmias, nausea and vomiting, visual field changes, pulse rate decrease, fall in blood pressure. An experimental teratogen. When heated to decomposition it emits acrid and irritating fumes. See also DIGITALIS.

DKN600 ***HR: 3***
ε-DIGOXIN ACETATE
mf: $C_{43}H_{66}O_{15}$ mw: 823.09

SYN: ε-ACETYLDIGOXIN (GERMAN)

TOXICITY DATA with REFERENCE
orl-gpg LD50:2500 μg/kg ARZNAD 15,481,65
ivn-gpg LD50:2900 μg/kg ARZNAD 15,481,65

SAFETY PROFILE: Poison by ingestion and intravenous routes. See also DIGOXIN. When heated to decomposition it emits acrid and irritating fumes.

DKN875 ***HR: 3***
DIGOXIN PENTAFORMATE
mf: $C_{46}H_{64}O_{19}$ mw: 921.10

SYN: PENTAFORMYL DIGOXIN

TOXICITY DATA with REFERENCE
orl-mus LD50:9660 μg/kg AIPTAK 153,436,65
scu-mus LD50:9220 μg/kg AIPTAK 153,436,65
orl-gpg LDLo:8788 μg/kg AIPTAK 153,436,65
ivn-gpg LDLo:3515 μg/kg AIPTAK 153,436,65

SAFETY PROFILE: Poison by ingestion, subcutaneous, and intravenous routes. When heated to decomposition it emits acrid smoke and fumes. See also DIGOXIN.

DKO000 CAS:51622-02-7 ***HR: 3***
DIHEPTYLMERCURY
mf: $C_{14}H_{30}Hg$ mw: 399.03

TOXICITY DATA with REFERENCE
ipr-mus LDLo:16 mg/kg CBCCT* 4,230,52

OSHA PEL: (Transitional: CL 1 mg/10m^3) TWA 0.01 mg(Hg)/m^3; STEL 0.03 mg/m^3 (skin)

CONSENSUS REPORTS: Mercury and its compounds are on the Community Right-To-Know List.

ACGIH TLV: TWA 0.01 mg(Hg)/m^3; STEL 0.03 mg(Hg)/m^3
NIOSH REL: (Inorganic Mercury) TWA 0.05 mg(Hg)/m^3

SAFETY PROFILE: Poison by intraperitoneal route. See also MERCURY COMPOUNDS, ORGANIC. When heated to decomposition it emits toxic fumes of Hg.

DKO400 CAS:2400-59-1 ***HR: 3***
DIHEXANOYL PEROXIDE
mf: $C_{12}H_{22}O_4$ mw: 230.31

SAFETY PROFILE: Explodes when heated to 85°C. When heated to decomposition it emits acrid smoke and fumes. See also PEROXIDES.

DKO600 CAS:143-16-8 ***HR: 3***
DIHEXYLAMINE
mf: $C_{12}H_{27}N$ mw: 185.40

PROP: Liquid. Bp: 233-243°, flash p: 220°F (OC), d: 0.78, vap d: 6.38.

SYN: DI-N-HEXYLAMINE

TOXICITY DATA with REFERENCE
skn-rbt 500 mg open MLD UCDS** 11/29/63
orl-rat LD50:380 mg/kg UCDS** 11/29/63
ivn-mus LD50:10 mg/kg CSLNX* NX#05177
skn-rbt LD50:170 mg/kg AEHLAU 1,343,60

CONSENSUS REPORTS: Reported in EPA TSCA Inventory.

SAFETY PROFILE: Poison by ingestion, skin contact, and intravenous routes. A skin irritant. Flammable when exposed to heat or flame; can react with oxidizing materials. To fight fire, use CO_2, dry chemical. When heated to decomposition it emits toxic fumes of NO_x. See also AMINES.

DKO800 CAS:112-58-3 ***HR: 2***
DIHEXYL ETHER
mf: $C_{12}H_{26}O$ mw: 186.38

PROP: Mp: −43.0°, bp: 227°, flash p: 170°F (OC), d: 0.794, autoign temp: 365°F, vap d: 6.4.

SYNS: HEXYL ETHER ◇ N-HEXYL ETHER ◇ 1,1′-OXYBISHEXANE

TOXICITY DATA with REFERENCE
skn-rbt 10 mg/24H open SEV AMIHBC 10,61,54
skn-rbt 500 mg open MLD UCDS** 7/13/71
eye-rbt 500 mg open AMIHBC 10,61,54
orl-rat LD50:30900 mg/kg AMIHBC 10,61,54
skn-rbt LD50:6900 mg/kg AMIHBC 10,61,54

CONSENSUS REPORTS: Reported in EPA TSCA Inventory.

SAFETY PROFILE: Mildly toxic by ingestion and skin contact. An eye and severe skin irritant. Combustible when exposed to heat or flame; can react with oxidizing materials. To fight fire, use foam, CO_2, dry chemical. See also ETHERS for explosion hazard.

DKP000 CAS:19139-31-2 ***HR: 1***
DIHEXYL FUMARATE
mf: $C_{16}H_{28}O_4$ mw: 284.44

SYNS: DIHEXYL trans-BUTENEDIOATE ◇ FUMARIC ACID DIHEXYL ESTER ◇ HEXYL FUMARATE

TOXICITY DATA with REFERENCE
skn-rbt 500 mg/24H MLD FCTXAV 16,709,78

CONSENSUS REPORTS: Reported in EPA TSCA Inventory.

SAFETY PROFILE: A skin irritant. When heated to decomposition it emits acrid smoke and irritating fumes.

DKP200 CAS:18279-21-5 ***HR: 3***
DIHEXYL LEAD DIACETATE
mf: $C_{16}H_{32}O_4Pb$ mw: 495.67

TOXICITY DATA with REFERENCE
orl-mus LD50:215 mg/kg CRSBAW 162,1456,68

CONSENSUS REPORTS: Lead and its compounds are on the Community Right-To-Know List.

SAFETY PROFILE: Poison by ingestion. See also LEAD COMPOUNDS. When heated to decomposition it emits toxic fumes of Pb.

DKP400 CAS:105-52-2 ***HR: 1***
DIHEXYL MALEATE
mf: $C_{16}H_{28}O_4$ mw: 284.44

PROP: Liquid, sol in water less than 0.01% by weight @ 20°. D: 0.9602 @ 20°/20°, bp: 179° @ 10 mm, vap press: < 0.01 mm @ 20°, fp: −70°, flash p: 290°F (OC).

SYNS: 2-BUTENEDIOIC ACID BIS(1,3-DIMETHYLBUTYL) ESTER ◇ DI(4-METHYL-2-AMYL) MALEATE ◇ DI(4-METHYL-2-PENTYL) MALEATE ◇ DMAM ◇ MALEIC ACID DI(1,3-DIMETHYLBUTYL) ESTER ◇ MALEIC ACID DIHEXYL ESTER

TOXICITY DATA with REFERENCE
skn-rbt 10 mg/24H open MLD AIHAAP 23,95,62
eye-rbt 500 mg open AMIHBC 10,61,54
orl-rat LD50:7340 mg/kg AIHAAP 23,95,62
skn-rbt LD50:12 g/kg AMIHBC 10,61,54

CONSENSUS REPORTS: Reported in EPA TSCA Inventory.

SAFETY PROFILE: Mildly toxic by ingestion and skin contact. A skin and eye irritant. Combustible material when exposed to heat of flame. See also ESTERS and MALEIC ACID. To fight fire, use dry chemical, CO_2, water, fog, mist. When heated to decomposition it emits acrid smoke and irritating fumes.

DKP600 CAS:84-75-3 ***HR: 1***
DI-n-HEXYL PHTHALATE
mf: $C_{20}H_{30}O_4$ mw: 334.50

PROP: Liquid. Mp: −58°, bp: 210° @ 5 mm, flash p: 350°F, d: 0.995 @ 20°/20°, vap d: 11.5.

SYNS: 1,2-BENZENEDICARBOXYLIC ACID DIHEXYL ESTER ◇ DIHEXYL PHTHALATE ◇ PHTHALIC ACID DIHEXYL ESTER

TOXICITY DATA with REFERENCE
eye-rbt 500 mg open AMIHBC 10,61,54
orl-mus TDLo:13 g/kg (male 7D pre):REP TXAPA9 88,255,87
orl-rat LD50:29600 mg/kg EVHPAZ 3,131,73
skn-rbt LD50:20 g/kg EVHPAZ 4,3,73

CONSENSUS REPORTS: Reported in EPA TSCA Inventory.

SAFETY PROFILE: Very mildly toxic by ingestion and skin contact. Experimental reproductive effects. An eye irritant. Combustible when exposed to heat or flame; can react with oxidizing materials. To fight fire, use foam, CO_2, dry chemical. See also PHTHALIC ACID and ESTERS.

DKP800 CAS:3006-15-3 ***HR: 1***
DIHEXYL SODIUM SULFOSUCCINATE
mf: $C_{16}H_{30}O_7S{\cdot}Na$ mw: 389.51

PROP: Clear, viscous liquid.

SYN: SULFOSUCCINIC ACID DIHEXYL ESTER, SODIUM SALT

CONSENSUS REPORTS: Reported in EPA TSCA Inventory.

SAFETY PROFILE: Flammable when exposed to heat, flame, oxidizers. To fight fire, use water, spray, foam, mist, dry chemical. When heated to decomposition it emits toxic fumes of SO_x and Na_2O.

DKQ000 CAS:57-41-0 ***HR: 3***
DIHYDANTOIN
mf: $C_{15}H_{12}N_2O_2$ mw: 252.29

SYNS: ALEVIATIN ◇ ANTISACER ◇ AURANILE ◇ CAUSOIN ◇ CITRULLAMON ◇ COMITAL ◇ CONVUL ◇ DANTEN ◇ DANTINAL ◇ DANTOINAL KLINOS ◇ DANTOINE ◇ DENYL ◇ DIDAN-TDC-250 ◇ DIFENILHIDANTOINA (SPANISH) ◇ DIFENIN ◇ DIFHYDAN ◇ DIHYCON ◇ DI-HYDAN ◇ DILANTIN ◇ DILANTINE ◇ DINTOIN ◇ DIPHANTOIN ◇ DIPHEDAL ◇ DIPHENINE ◇ DIPHENTOIN ◇ DIPHENYLAN ◇ DIPHENYLHYDANTOIN ◇ 5,5-DIPHENYLHYDANTOIN ◇ DIPHENYLHYDANTOINE (FRENCH) ◇ 5,5-DIPHENYLIMIDAZOLIDIN-2,4-DIONE ◇ 5,5-DIPHENYL-2,4-IMIDAZOLIDINEDIONE ◇ DI-PHETINE ◇ DITOINATE ◇ DPH ◇ EKKO CAPSULES ◇ ELEPSINDON ◇ ENKELFEL ◇ EPAMIN ◇ EPANUTIN ◇ EPASMIR "5" ◇ EPDANTOINE SIMPLE ◇ EPELIN ◇ EPIFENYL ◇ EPIHYDAN ◇ EPILAN ◇ EPILANTIN ◇ EPINAT ◇ EPISED ◇ EPTAL ◇ EPTOIN ◇ FENANTOIN ◇ FENIDANTOIN "S" ◇ FENYLEPSIN ◇ FENYTOINE ◇ GEROT-EPILAN-D ◇ HIDAN ◇ HIDANTILO ◇ HIDANTINA SENOSIAN ◇ HIDANTINA VITORIA ◇ HIDANTOMIN ◇ HYDANTAL

◇ HYDANTOIN ◇ ICTALIS SIMPLE ◇ IDANTOIN ◇ KESSODANTEN ◇ LABOPAL ◇ LEHYDAN ◇ LEPITOIN ◇ LEPSIN ◇ MINETOIN ◇ NCI-C55765 ◇ NEOS-HIDANTOINA ◇ NOVANTOINA ◇ OM-HYDANTOINE ◇ OXYLAN ◇ PHANANTIN ◇ PHENATOINE ◇ RITMENAL ◇ SACERIL ◇ SANEPIL ◇ SILANTIN ◇ SODANTON ◇ SOLANTIN ◇ SYLANTOIC ◇ TACOSAL ◇ THILOPHENYL ◇ TOIN UNICELLES ◇ ZENTRONAL ◇ ZENTROPIL

TOXICITY DATA with REFERENCE
dnd-esc 50 μmol/L MUREAV 89,95,81
dni-hmn:lym 360 μmol/L TXAPA9 33,38,75
mnt-mus-ivn 500 μg/kg MUREAV 141,183,84
orl-wmn TDLo:1365 mg/kg (female 1-39W post):REP JOPDAB 89,662,76
orl-wmn TDLo:1620 mg/kg (female 1-39W post):TER LANCAO 2,481,80
orl-wmn TDLo:1620 mg/kg (female 1-39W post):CAR,TER JAMAAP 244,1464,80
orl-wmn TDLo:730 mg/kg/1Y-C:CAR,BLD NKGZAE 46,1,83
unr-wmn TDLo:1620 mg/kg (female 1-39W post):CAR,TER JAMAAP 244,1464,80
unr-wmn TDLo:1620 mg/kg (female 1-39W post):CAR,TER LANCAO 2,481,80
orl-rat TDLo:1500 mg/kg:ETA CNREA8 26,619,66
orl-wmn TD:730 mg/kg/1Y-C:CAR,BLD NKGZAE 43,711,80
orl-chd TD:6023 mg/kg/1Y-C:NEO,BLD,SKN AJDEBP 22,28,81
orl-chd TDLo:3 mg/kg/3W:SKN CLPTAT 20,48,76
orl-chd LDLo:67 mg/kg NEURAI 11,138,61
orl-chd TDLo:11 mg/kg/D:CNS AJDCAI 130,75,76
orl-man TDLo:31 mg/kg/4D-I:CNS,PSY JCGADC 9,337,87
orl-wmn TDLo:106 mg/kg/2W-I:CNS,PSY JCGADC 9,337,87
orl-man TDLo:1300 mg/kg:BLD AIMEAS 69,557,68
orl-chd TDLo:140 mg/kg/:LIV,BLD,SKN PEDIAU 28,943,61
ivn-cld TDLo:15 mg/kg CPEDAM 24,467,85
ivn-cld TDLo:15 mg/kg PEDIAU 72,831,83
unr-cld TDLo:18 mg/kg/3D-I PEDIAU 72,831,83
orl-rat LD50:1635 mg/kg ARZNAD 33,1168,83
ipr-rat LD50:200 mg/kg JPMSAE 67,600,78
ivn-rat LD50:101 mg/kg JKXXAF #79-163823
orl-mus LD50:150 mg/kg JMCMAR 18,383,75
ipr-mus LD50:100 mg/kg JMCMAR 24,465,81
scu-mus LD50:138 mg/kg ARZNAD 4,723,54

CONSENSUS REPORTS: NTP Fifth Annual Report on Carcinogens. IARC Cancer Review: Group 2B IMEMDT 7,319,87; Human Limited Evidence IMEMDT 13,201,77; Animal Sufficient Evidence IMEMDT 13,201,77. EPA Genetic Toxicology Program.

SAFETY PROFILE: Confirmed carcinogen producing lymphoma, Hodgkin's disease, tumors of the skin and appendages. Experimental carcinogenic and tumorigenic data. A human poison by ingestion. Poison experimentally by ingestion, subcutaneous, intravenous, and intraperitoneal routes. Moderately toxic by an unspecified route. Experimental teratogenic and reproductive effects. Human systemic effects by ingestion: dermatitis, change in motor activity (specific assay), ataxia (loss of muscle coordination), degenerative brain changes, encephalitis, hallucinations, distorted perceptions, irritability, and jaundice. Human teratogenic effects by ingestion: developmental abnormalities of the central nervous system, cardiovascular (circulatory) system, musculoskeletal system, craniofacial area, skin and skin appendages, eye, ear, other developmental abnormalities. Effects on newborn include abnormal growth statistics (e.g., reduced weight gain), physical abnormalities, other postnatal measures or effects, and delayed effects. Human mutation data reported. A drug for the treatment of grand mal and psychomotor seizures. When heated to decomposition it emits toxic fumes of NO_x.

DKQ200 CAS:63868-75-7 ***HR: 3***
DIHYDRALAZINE HYDROCHLORIDE
mf: $C_8H_{10}N_6 \cdot ClH$ mw: 226.70

SYNS: C-7441 ◇ 1,4-DIHYDRAZINOPHTHALAZINE HYDROCHLORIDE

TOXICITY DATA with REFERENCE
orl-rat LD50:350 mg/kg RPOBAR 2,283,70
ipr-rat LD50:270 mg/kg RPOBAR 2,283,70
ivn-rat LD50:167 mg/kg RPOBAR 2,283,70
ipr-mus LD50:290 mg/kg JPETAB 101,368,51

SAFETY PROFILE: Poison by ingestion, intraperitoneal, and intravenous routes. When heated to decomposition it emits very toxic fumes of NO_x and HCl.

DKQ400 ***HR: 3***
DIHYDRAZINECOBALT(II) CHLORATE
mf: $Cl_2CoH_8N_4O_6$ mw: 298.0

CONSENSUS REPORTS: Cobalt and its compounds are on the Community Right-To-Know List.

SAFETY PROFILE: A powerful explosive extremely sensitive to light impact, friction or heating to 90°C. When heated to decomposition it emits toxic fumes of Cl^- and NO_x. See also COBALT COMPOUNDS and CHLORATES.

DKQ600 CAS:7327-87-9 ***HR: 3***
1,4-DIHYDRAZINOPHTHALAZINE SULFATE
mf: $C_8H_{10}N_6 \cdot H_2O_4S$ mw: 288.32

SYN: DIHYDRALAZINE SULFATE

TOXICITY DATA with REFERENCE
mmo-sat 100 μg/plate TCMUD8 5,339,85

mma-sat 100 μg/plate TCMUD8 5,339,85
dnr-rat:lvr 500 μmol/L SCIEAS 210,329,80
dnd-mus-ipr 159 mg/kg ENMUDM 4,605,82
sce-mus-ipr 79500 μg/kg ENMUDM 4,605,82
orl-rat LD50:820 mg/kg RPOBAR 2,284,70
orl-mus LD50:400 mg/kg RPOBAR 2,284,70
ipr-mus LD50:159 mg/kg RPOBAR 2,283,70
scu-mus LD50:223 mg/kg RPOBAR 2,283,70

SAFETY PROFILE: Poison by ingestion, intraperitoneal, and subcutaneous routes. Mutation data reported. When heated to decomposition it emits very toxic fumes of NO_x and SO_x.

DKQ650 CAS:75034-93-4 ***HR: 2***
DIHYDREL

SYN: DIMETHYLHYDRAZINIUM derivative of 2-CHLOROETHYLPHOSPHONIC ACID

TOXICITY DATA with REFERENCE
eye-rbt 40% GISAAA 51(10),85,86
orl-rat LD50:3500 mg/kg AIPTAK 123,48,59
orl-mus LD50:2333 mg/kg GISAAA 51(10),85,86

SAFETY PROFILE: Moderately toxic by ingestion. An eye irritant. When heated to decomposition it emits toxic fumes of NO_x, PO_x, and Cl^-.

DKR000 CAS:6258-06-6 ***HR: 1***
9,10-DIHYDRO-1-AMINO-4-BROMO-9,10-DIOXO-2-ANTHRACENE SULFONIC ACID SODIUM SALT
mf: $C_{14}H_7BrNO_5S$•Na mw: 404.18

SYNS: 1-AMINO-4-BROMANTHRACHINON-2-SULFONANSODNY (CZECH) ◇ BROMANMINAN SODNY (CZECH)

TOXICITY DATA with REFERENCE
eye-rbt 500 mg/24H MLD 28ZPAK -,193,72

CONSENSUS REPORTS: Reported in EPA TSCA Inventory.

SAFETY PROFILE: An eye irritant. When heated to decomposition it emits very toxic fumes of Br^-, Na_2O, NO_x, and SO_x.

DKR200 CAS:3567-76-8 ***HR: 2***
2,3-DIHYDRO-6-AMINO-2-(2-CHLOROETHYL)-4H-1,3-BENZOXAZIN-4-ONE
mf: $C_{10}H_{11}ClN_2O_2$ mw: 226.68

SYNS: A 350 ◇ 6-AMINO-2-(2-CHLOROETHYL)-2,3-DIHYDRO-4H-1,3-BENZOXAZIN-4-ONE ◇ AMINOCHLORTHENOXAZINE ◇ 2-(β-CHLOROETHYL)-2,3-DIHYDRO-4-OXO-6-AMINO-1,3-BENZOXAZINE

TOXICITY DATA with REFERENCE
orl-rat LD50:1958 mg/kg JPPMAB 16,502,64
ipr-rat LD50:1500 mg/kg ARZNAD 13,884,63
orl-mus LD50:10 g/kg ARZNAD 13,884,63
ipr-mus LD50:1941 mg/kg ARZNAD 13,884,63

SAFETY PROFILE: Moderately toxic by ingestion and intraperitoneal routes. An antipyretic and analgesic. When heated to decomposition it emits very toxic fumes of Cl^- and NO_x.

DKR400 CAS:40847-64-1 ***HR: 1***
9,10-DIHYDRO-1-AMINO-4-(3-(2-HYDROXYETHYL)AMINOSULFONYL-4-METHYLPHENYLAMINO)-9,10-DIOXO-2-ANTHRACENE SULFONIC ACID SODIUM SALT
mf: $C_{23}H_{20}N_3O_8S_2$•Na mw: 553.57

SYN: MODR BRILANTNI ALIZARINOVA BRL (CZECH)

TOXICITY DATA with REFERENCE
eye-rbt 500 mg/24H MLD 28ZPAK -,241,72

CONSENSUS REPORTS: Reported in EPA TSCA Inventory.

SAFETY PROFILE: An eye irritant. When heated to decomposition it emits very toxic fumes of NO_x, Na_2O, and SO_x.

DKS400 CAS:60968-08-3 ***HR: 3***
1,2-DIHYDROBENZO(a)ANTHRACENE
mf: $C_{18}H_{14}$ mw: 230.32

SYN: 1,2-DIHYDROBENZ(a)ANTHRACENE

TOXICITY DATA with REFERENCE
skn-mus TDLo:18 mg/kg:NEO CNREA8 38,1705,78

SAFETY PROFILE: Questionable carcinogen with experimental neoplastigenic data by skin contact. When heated to decomposition it emits acrid smoke and irritating fumes.

DKS600 CAS:60968-01-6 ***HR: 3***
3,4-DIHYDROBENZO(a)ANTHRACENE
mf: $C_{18}H_{14}$ mw: 230.32

SYN: 3,4-DIHYDROBENZ(a)ANTHRACENE

TOXICITY DATA with REFERENCE
skn-mus TDLo:3700 μg/kg:NEO CNREA8 38,1705,78

SAFETY PROFILE: Questionable carcinogen with experimental neoplastigenic data by skin contact. When heated to decomposition it emits acrid smoke and irritating fumes.

DKS800 CAS:10023-25-3 ***HR: 3***
6,13-DIHYDROBENZO(e)(1)BENZOTHIOPYRANO(4,3-b)INDOLE
mf: $C_{19}H_{13}NS$ mw: 287.39

TOXICITY DATA with REFERENCE
mma-sat 100 μg/plate MUREAV 66,307,79
scu-mus TDLo:72 mg/kg/9W-I:NEO MUREAV 66,307,79

SAFETY PROFILE: Questionable carcinogen with experimental neoplastigenic data. Mutation data reported. When heated to decomposition it emits very toxic fumes of SO_x and NO_x.

DKS909 CAS:64709-48-4 ***HR: D***
5,6-DIHYDRO-7H-BENZO(c)CARBAZOLE-8-CARBOXYLIC ACID-2-(DIETHYLAMINO)ETHYL ESTER HYDROCHLORIDE
mf: $C_{23}H_{26}N_2O_2$•ClH mw: 398.97

TOXICITY DATA with REFERENCE
mmo-esc 10 mmol/L/80M GNKAA5 13,1621,77
mmo-omi 50 mmol/L GNKAA5 13,1621,77

SAFETY PROFILE: Mutation data reported. When heated to decomposition it emits very toxic fumes of HCl and NO_x.

DKT000 CAS:64709-49-5 ***HR: D***
5,6-DIHYDRO-7H-BENZO(c)CARBAZOLE-9-CARBOXYLIC ACID-2-(DIETHYLAMINO) ETHYL ESTER HYDROCHLORIDE
mf: $C_{23}H_{26}N_2O_2$•ClH mw: 398.97

TOXICITY DATA with REFERENCE
mmo-esc 500 μmol/L/30M GNKAA5 13,1621,77
mmo-omi 25 mmol/L GNKAA5 13,1621,77

SAFETY PROFILE: Mutation data reported. When heated to decomposition it emits very toxic fumes of Cl^- and NO_x.

DKT200 CAS:64709-50-8 ***HR: D***
5,6-DIHYDRO-7H-BENZO(c)CARBAZOLE-10-CARBOXYLIC ACID-2-(DIETHYLAMINO)ETHYL ESTER HYDROCHLORIDE
mf: $C_{23}H_{26}N_2O_2$•ClH mw: 398.97

TOXICITY DATA with REFERENCE
mmo-esc 3 mmol/L/40M GNKAA5 13,1621,77
mmo-omi 25 mmol/L GNKAA5 13,1621,77

SAFETY PROFILE: Mutation data reported. When heated to decomposition it emits very toxic fumes of HCl and NO_x.

DKT400 CAS:100466-04-4 ***HR: 3***
2,3-DIHYDRO-1H-BENZO(h,i)CHRYSENE
mf: $C_{21}H_{16}$ mw: 268.37

SYNS: 5:10-TRIMETHYLENE-1:2-BENZANTHRACENE ◇ 1:12-TRIMETHYLENECHRYSENE

TOXICITY DATA with REFERENCE
skn-mus TDLo:3000 mg/kg/43W-I:ETA AKBNAE 62(2),30,41

SAFETY PROFILE: Questionable carcinogen with experimental tumorigenic data. When heated to decomposition it emits acrid smoke and fumes. See also CHRYSENE.

DKU000 CAS:17573-23-8 ***HR: 3***
7,8-DIHYDROBENZO(a)PYRENE
mf: $C_{20}H_{14}$ mw: 254.34

TOXICITY DATA with REFERENCE
mma-sat 10 μg/plate PNASA6 72,5135,75
scu-mus TDLo:9 mg/kg:NEO JJIND8 64,617,80

CONSENSUS REPORTS: EPA Genetic Toxicology Program.

SAFETY PROFILE: Questionable carcinogen with experimental neoplastigenic data. Mutation data reported. When heated to decomposition it emits acrid smoke and irritating fumes.

DKU400 CAS:66788-01-0 ***HR: 3***
9,10-DIHYDROBENZO(e)PYRENE
mf: $C_{20}H_{14}$ mw: 254.34

SYN: 9,10-H2 B(e)P

TOXICITY DATA with REFERENCE
mmo-sat nmol/plate CNREA8 40,1985,80
mma-sat 10 nmol/plate JBCHA3 254,4408,79
skn-mus TDLo:25 mg/kg:NEO CNREA8 40,203,80
scu-mus TDLo:9 mg/kg:ETA JJIND8 64,617,80

SAFETY PROFILE: Questionable carcinogen with experimental neoplastigenic and tumorigenic data. Mutation data reported. When heated to decomposition it emits acrid smoke and irritating fumes.

DKU875 ***HR: 3***
3,4-DIHYDRO-2H-1,4-BENZOTHIAZINE HYDROCHLORIDE
mf: C_8H_9NS•ClH mw: 187.70

SYNS: BTZ ◇ DIIDROBENZO(1-4)TIAZINA CLORIDRATO (ITALIAN)

TOXICITY DATA with REFERENCE
orl-mus LD50:800 mg/kg AIPTAK 89,55,52
ipr-mus LD50:300 mg/kg AIPTAK 89,55,52
ivn-rbt LDLo:200 mg/kg AIPTAK 89,55,52

SAFETY PROFILE: Poison by intraperitoneal and intravenous routes. Moderately toxic by ingestion. When heated to decomposition it emits toxic fumes of SO_x, NO_x, and HCl.

DKV125 CAS:102366-79-0 ***HR: 2***
7,8-DIHYDRO-N-BENZYLADENINE
mf: $C_{12}H_{13}N_5$ mw: 227.30

TOXICITY DATA with REFERENCE
orl-rat LD50:2100 mg/kg SHBOAO 32,530,78
scu-rat LD50:480 mg/kg SHBOAO 32,530,78
orl-mus LD50:2000 mg/kg SHBOAO 32,530,78
scu-mus LD50:1100 mg/kg SHBOAO 32,530,78

SAFETY PROFILE: Moderately toxic by ingestion and subcutaneous routes. When heated to decomposition it emits toxic fumes of NO_x.

DKV150 CAS:619-01-2 ***HR: 2***
DIHYDROCARVEOL
mf: $C_{10}H_{18}O$ mw: 154.28

PROP: Colorless, oily liquid; spearmint odor. D: 0.921-0.926, refr index: 1.477-1.481, flash p: +153°F. Sol in alc, fixed oils; insol in water.

SYNS: 1,6-DIHYDROCARVEOL ◇ FEMA No. 2379 ◇ 8-p-MENTHEN-2-OL ◇ 6-METHYL-3-ISOPROPYLCYCLOHEXANOL

TOXICITY DATA with REFERENCE
skn-rbt 500 mg/24H MOD FCTXAV 17,771,79

CONSENSUS REPORTS: Reported in EPA TSCA Inventory.

SAFETY PROFILE: A moderate skin and eye irritant. A combustible liquid. When heated to decomposition it emits acrid smoke and irritating fumes.

DKV160 CAS:20777-49-5 ***HR: 1***
DIHYDROCARVEYL ACETATE
mf: $C_{12}H_{20}O_2$ mw: 196.32

SYNS: CYCLOHEXANOL, 2-METHYL-5-(1-METHYLETHENYL)-, ACETATE,(1-α-2-β,5α—(9CI) ◇ DIHYDROCARVEOL ACETATE ◇ DIHYDROCARVYL ACETATE ◇ p-MENTH-8-EN-2-OL, ACETATE ◇ p-MENTH-8-EN-2-YL ACETATE ◇ 2-METHYL-5-(1-METHYLETHENYL)CYCLOHEXYL ACETATE

TOXICITY DATA with REFERENCE
skn-rbt 500 mg/24H MLD FCTOD7 21,843,83

CONSENSUS REPORTS: Reported in EPA TSCA Inventory.

SAFETY PROFILE: A skin irritant. When heated to decomposition it emits acrid smoke and irritating fumes.

DKV175 CAS:7764-50-3 ***HR: 2***
d-DIHYDROCARVONE
mf: $C_{10}H_{16}O$ mw: 152.26

PROP: Colorless liquid; spearmint-like odor. D: 0.923-0.928, refr index: 1.470-1.474. Sol in alc, fixed oils; insol in water.

SYNS: FEMA No. 3565 ◇ 8-p-MENTHEN-2-ONE ◇ p-MENTH-8-EN-2-ONE ◇ d-2-METHYL-5-(1-METHYLENENYL)-CYCLOHEXANONE

TOXICITY DATA with REFERENCE
scu-mus LD50:2900 mg/kg FCTXAV 18,665,80

CONSENSUS REPORTS: Reported in EPA TSCA Inventory.

SAFETY PROFILE: Moderately toxic by subcutaneous route. When heated to decomposition it emits acrid smoke and irritating fumes.

DKV200 CAS:1016-75-7 ***HR: 2***
2,3-DIHYDRO-6-CHLORO-2-(2-CHLOROETHYL)-4H-1,3-BENZOXAZIN-4-ONE
mf: $C_{10}H_9Cl_2NO_2$ mw: 246.10

TOXICITY DATA with REFERENCE
ipr-rat LD50:2000 mg/kg ARZNAD 13,884,63
orl-mus LD50:11765 mg/kg ARZNAD 13,884,63
ipr-mus LD50:3143 mg/kg ARZNAD 13,884,63

SAFETY PROFILE: Moderately toxic by intraperitoneal route. Mildly toxic by ingestion. When heated to decomposition it emits very toxic fumes of Cl^- and NO_x.

DKV309 CAS:28058-62-0 ***HR: 3***
1'-(3-(10,11-DIHYDRO-3-CHLORO-5H-DIBENZ(b,f)AZEPIN-5-YL)PROPYL)-(1,4'-BIPIPERIDINE)-4'-CARBOXAMIDE DIHYDROCHLORIDE
mf: $C_{28}H_{37}ClN_4O \cdot 2ClH$ mw: 552.99

SYNS: 3-CHLOROCARPIPRAMINE DIHYDROCHLORIDE ◇ CLOCAPRAMINE DIHYDROCHLORIDE ◇ CLOCARPRAMINE DIHYDROCHLORIDE

TOXICITY DATA with REFERENCE
orl-rat LD50:6200 mg/kg NIIRDN 6,231,82
ipr-rat LD50:105 mg/kg NIIRDN 6,231,82
orl-mus LD50:2550 mg/kg NIIRDN 6,231,82
ipr-mus LD50:160 mg/kg NIIRDN 6,231,82
scu-mus LD50:6300 mg/kg NIIRDN 6,231,82

SAFETY PROFILE: Poison by intraperitoneal route. Moderately toxic by ingestion. When heated to decomposition it emits toxic fumes of NO_x and HCl.

DKV400 CAS:40762-15-0 ***HR: 2***
1,3-DIHYDRO-7-CHLORO-5-(o-FLUOROPHENYL)-3-HYDROXY-1-(2-HYDROXYETHYL)-2H-1,4-BENZODIAZEPIN-2-ONE
mf: $C_{17}H_{14}ClFN_2O_3$ mw: 348.78

SYN: SAS 643

TOXICITY DATA with REFERENCE
unr-hmn TDLo:71 μg/kg:CNS DRFUD4 3,145,78
orl-rat LD50:2550 mg/kg ARZNAD 25,1294,75
ipr-rat LD50:586 mg/kg ARZNAD 25,1294,75
ipr-mus LD50:760 mg/kg JPPMAB 26,566,74

SAFETY PROFILE: Moderately toxic by ingestion and intraperitoneal routes. Human systemic effects by an unspecified route: somnolence. When heated to decomposition it emits very toxic fumes of Cl^-, F^-, and NO_x.

DKV700 CAS:5991-71-9 ***HR: 3***
2,3-DIHYDRO-7-CHLORO-2-OXO-5-PHENYL-1H-1,4-BENZODIAZEPINE-3-CARBOXYLIC ACID MONOPOTASSIUM SALT
mf: $C_{16}H_{10}ClN_2O_3 \cdot K$ mw: 352.80

SYN: 4311 CB

TOXICITY DATA with REFERENCE
orl-mam LDLo:870 mg/kg CHTPBA 4,239,69
ipr-mam LDLo:300 mg/kg CHTPBA 4,239,69
scu-mam LDLo:450 mg/kg CHTPBA 4,239,69
ivn-mam LDLo:220 mg/kg CHTPBA 4,239,69

SAFETY PROFILE: Poison by intravenous and intraperitoneal routes. Moderately toxic by ingestion and subcutaneous routes. When heated to decomposition it emits toxic fumes of Cl^-, NO_x, and K_2O.

DKV800 CAS:63041-49-6 ***HR: 3***
meso-DIHYDROCHOLANTHRENE
mf: $C_{20}H_{16}$ mw: 256.36

SYN: 6,12,b-DIHYDROCHOLANTHRENE

TOXICITY DATA with REFERENCE
scu-mus TDLo:40 mg/kg:ETA CNREA8 1,695,41

SAFETY PROFILE: Questionable carcinogen with experimental tumorigenic data. A cholinergic agent. When heated to decomposition it emits acrid smoke and irritating fumes.

DKW000 CAS:360-68-9 ***HR: 3***
DIHYDROCHOLESTEROL
mf: $C_{27}H_{48}O$ mw: 388.75

SYNS: (3-β,5-β)-CHOLESTAN-3-OL ◇ 3-β-CHOLESTANOL ◇ COPROSTANOL ◇ COPROSTAN-3-β-OL ◇ COPROSTEROL ◇ 3-β-HYDROXYCHOLESTANE ◇ KOPROSTERIN (GERMAN) ◇ STERCORIN ◇ XYMOSTANOL

TOXICITY DATA with REFERENCE
scu-mus TDLo:800 mg/kg/4W-I:NEO NATWAY 60,525,73

SAFETY PROFILE: Questionable carcinogen with experimental neoplastigenic data. When heated to decomposition it emits acrid smoke and irritating fumes.

DKW200 CAS:41593-31-1 ***HR: 3***
1,2-DIHYDROCHRYSENE
mf: $C_{18}H_{14}$ mw: 230.32

SYN: DIHYDROCHRYSENE

TOXICITY DATA with REFERENCE
ipr-mus TDLo:59 mg/kg/15D-I:CAR CNREA8 39,5063,79

SAFETY PROFILE: Questionable carcinogen with experimental carcinogenic data. When heated to decomposition it emits acrid smoke and irritating fumes. See also CHRYSENE.

DKW400 CAS:71435-43-3 ***HR: 3***
3,4-DIHYDROCHRYSENE
mf: $C_{18}H_{14}$ mw: 230.32

TOXICITY DATA with REFERENCE
ipr-mus TDLo:59 mg/kg/15D-I:ETA CNREA8 39,5063,79

SAFETY PROFILE: Questionable carcinogen with experimental tumorigenic data. When heated to decomposition it emits acrid smoke and irritating fumes. See also CHRYSENE.

DKW800 CAS:125-28-0 ***HR: 3***
DIHYDROCODEINE
mf: $C_{18}H_{23}NO_3$ mw: 301.42

SYNS: CODHYDRINE ◇ COHYDRIN ◇ DF 118 ◇ DEHACODIN ◇ DIDRATE ◇ DIHYDRIN ◇ 7,8-DIHYDROCODEINE ◇ DIHYDRONEOPINE ◇ DROCODE ◇ HYDROCODIN ◇ 6-HYDROXY-3-METHOXY-N-METHYL-4,5-EPOXYMORPHINAN ◇ NADEINE ◇ NOVICODIN ◇ PARACODIN ◇ PARACODINE ◇ PARZONE ◇ RAPACODIN

TOXICITY DATA with REFERENCE
orl-hmn TDLo:6500 mg/kg/26W:EYE,CNS BMJOAE 1,1594,78
scu-wmn TDLo:6 mg/kg/36H-I:PUL BMJOAE 1,211,59
orl-rat LD50:240 mg/kg TXAPA9 1,42,59
scu-mus LD50:135 mg/kg THERAP 6,146,51
ivn-mus LD50:80 mg/kg THERAP 6,146,51
orl-rbt LDLo:400 mg/kg HBAMAK 4,1289,35
scu-gpg LDLo:80 mg/kg HBAMAK 4,1289,35

SAFETY PROFILE: Poison by ingestion, intravenous, and subcutaneous routes. Human systemic effects by ingestion and subcutaneous routes: somnolence, miosis (pupillary constriction), and respiratory depression. An analgesic. Can cause drug dependency with repeated doses. When heated to decomposition it emits toxic fumes of NO_x. See also CODEINE.

DKX000 CAS:5965-13-9 ***HR: 3***
DIHYDROCODEINE BITARTRATE
mf: $C_{18}H_{23}NO_3 \cdot C_4H_6O_6$ mw: 451.52

SYNS: DF 118 ◇ DIHYDROCODEINE ACID TARTRATE ◇ DIHYDROCODEINE TARTRATE ◇ DIHYDROCODEINE TARTRATE (1:1)

TOXICITY DATA with REFERENCE
ivn-hmn TDLo:357 μg/kg/5M-C:CNS LANCAO 1,1425,82
orl-rat LD50:240 mg/kg TXAPA9 1,42,59
ipr-mus LD50:252 mg/kg AIPTAK 136,333,62
scu-mus LDLo:350 mg/kg JPETAB 51,35,34

scu-rbt LDLo:39 mg/kg JPETAB 66,182,39
par-rbt LDLo:116 mg/kg JPETAB 66,182,39

SAFETY PROFILE: Poison by ingestion, intraperitoneal, subcutaneous, and parenteral routes. Human systemic effects by intravenous route: irritability. When heated to decomposition it emits toxic fumes of NO_x. See also CODEINE.

DKX050 CAS:34195-34-1 ***HR: 3***
DIHYDROCODEINONE BITARTRATE
mf: $C_{18}H_{21}NO_3 \cdot C_4H_6O_6$ mw: 449.50

SYNS: CODEINONE, DIHYDRO-, TARTRATE ◇ HYDROCODONE BITARTRATE ◇ MORPHINAN-6-ONE, 4,5-α-EPOXY-3-METHOXY-17-METHYL-, TARTRATE (1:1)

TOXICITY DATA with REFERENCE
scu-ham TDLo:153 mg/kg (female 8D post):TER AJOGAH 123,705,75
orl-rat LD50:250 mg/kg TXAPA9 1,42,59

SAFETY PROFILE: Poison by ingestion. An experimental teratogen. When heated to decomposition it emits toxic fumes of NO_x.

DKX100 CAS:2111-75-3 ***HR: 2***
DIHYDROCUMINYL ALDEHYDE
mf: $C_{10}H_{14}O$ mw: 150.24

PROP: d-Form: Liquid. Bp: (745) 237°, d: (20/4) 0.953, n (20/D) 1.5058. l-Form: Liquid. Bp: (10) 104-105°, d: (20/4) 0.9645, n (20/D) 1.5069.

SYNS: 4-ISOPROPENYL-1-CYCLOHEXENE-1-CARBOXALDEHYDE ◇ p-MENTHA-1,8-DIEN-7-AL ◇ 4-(1-METHYLETHENYL)-1-CYCLOHEXENE-1-CARBOXALDEHYDE (9CI) ◇ PERILLA ALDEHYDE ◇ PERILLAL ◇ PERILLALDEHYDE ◇ PERILLYL ALDEHYDE

TOXICITY DATA with REFERENCE
skn-gpg 100%/24H MOD FCTOD7 20(Suppl),799,82
cyt-ham:fbr 50 mg/L FCTOD7 22,623,84
orl-mus LD50:1720 mg/kg FCTOD7 20(Suppl),799,82

CONSENSUS REPORTS: Reported in EPA TSCA Inventory.

SAFETY PROFILE: Moderately toxic by ingestion. A skin irritant. Mutation data reported. When heated to decomposition it emits acrid smoke and fumes. See also ALDEHYDES.

DKX600 CAS:427-00-9 ***HR: 3***
DIHYDRODEOXYMORPHINE
mf: $C_{17}H_{21}NO_2$ mw: 271.39

SYNS: 6-DEOXY-7,8-DIHYDROMORPHINE ◇ DESOMORPHINE ◇ DIHYDRODESOXYMORPHINE-D ◇ 4,5-EPOXY-3-HYDROXY-N-METHYLMORPHINAN ◇ 4,5-α-EPOXY-17-METHYLMORPHINAN-3-OL ◇ PERMONID

TOXICITY DATA with REFERENCE
scu-mus LDLo:104 mg/kg JPETAB 55,257,35
ivn-mus LD50:27 mg/kg YKKZAJ 84,268,64

SAFETY PROFILE: Poison by subcutaneous and intravenous routes. See also MORPHINE. When heated to decomposition it emits toxic fumes of NO_x.

DKX800 CAS:73651-49-7 ***HR: 1***
9,10-DIHYDRO-4,5-DIAMINO-1-HYDROXY-2,7-ANTHRACENE DISULFONIC ACID DISODIUM SALT
mf: $C_{14}H_8N_2O_9S_2 \cdot 2Na$ mw: 458.34

SYN: AZUROL ALIZARINOVY SW (CZECH)

TOXICITY DATA with REFERENCE
skn-rbt 500 mg/24H MLD 28ZPAK -,239,72
eye-rbt 500 mg/24H MLD 28ZPAK -,239,72

SAFETY PROFILE: A skin and eye irritant. When heated to decomposition it emits very toxic fumes of NO_x, Na_2O, and SO_x.

DKX875 CAS:74339-98-3 ***HR: 2***
trans-1,2-DIHYDRODIBENZ(a,e)ACEANTHRYLENE-1,2-DIOL
mf: $C_{24}H_{16}O_2$ mw: 336.40

SYN: trans-12,13-DIHYDRO-12,13-DIHYDROXYDIBENZO(a,e) FLUORANTHENE

TOXICITY DATA with REFERENCE
skn-mus TDLo:1200 μg/kg:ETA CRNGDP 8,461,87

SAFETY PROFILE: Questionable carcinogen with experimental tumorigenic data. When heated to decomposition it emits acrid smoke and irritating fumes.

DKX900 CAS:74340-04-8 ***HR: 2***
trans-10,11-DIHYDRODIBENZ(a,e)ACEANTHRYLENE-10,11-DIOL
mf: $C_{24}H_{16}O_2$ mw: 336.40

SYN: trans-3,4-DIHYDRO-3,4-DIHYDROXYDIBENZO(a,e) FLUORANTHENE

TOXICITY DATA with REFERENCE
mma-sat 1200 nmol/L CRNGDP 5,1263,84
dnd-man:lyms 208 nmol CRNGDP 4,27,83
skn-mus TDLo:1200 μg/kg:ETA CRNGDP 8,461,87

SAFETY PROFILE: Questionable carcinogen with experimental tumorigenic data. Mutation data reported. When heated to decomposition it emits acrid smoke and irritating fumes.

DKY000 CAS:153-34-4 ***HR: 3***
5,6-DIHYDRODIBENZ(a,h)ANTHRACENE
mf: $C_{22}H_{16}$ mw: 280.38

TOXICITY DATA with REFERENCE
mma-sat 1 μg/plate MUREAV 51,311,78
skn-mus TDLo:130 mg/kg/44W-I:NEO JNCIAM 34,1,65
scu-mus TDLo:16 mg/kg:ETA JNCIAM 44,641,70

SAFETY PROFILE: Questionable carcinogen with experimental neoplastigenic and tumorigenic data. Mutation data reported. When heated to decomposition it emits acrid smoke and irritating fumes.

DKY200 CAS:16361-01-6 ***HR: 3***
5,6-DIHYDRODIBENZ(a, j)ANTHRACENE
mf: $C_{22}H_{16}$ mw: 280.38

TOXICITY DATA with REFERENCE
mma-sat 1 μg/plate MUREAV 51,311,78
skn-mus TDLo:268 mg/kg/85W-I:ETA JNCIAM 44,641,70

SAFETY PROFILE: Questionable carcinogen with experimental tumorigenic data. Mutation data reported. When heated to decomposition it emits acrid smoke and irritating fumes.

DKY400 CAS:57816-08-7 ***HR: 3***
7,14-DIHYDRODIBENZ(a,h)ANTHRACENE
mf: $C_{22}H_{16}$ mw: 280.38

SYN: 9,10-DIHYDRO-1,2,5,6-DIBENZANTHRACENE

TOXICITY DATA with REFERENCE
mma-sat 1 μg/plate MUREAV 51,311,78
skn-mus TDLo:1150 mg/kg/48W-I:ETA PRLBA4 129,439,40

CONSENSUS REPORTS: EPA Genetic Toxicology Program.

SAFETY PROFILE: Questionable carcinogen with experimental tumorigenic data. Mutation data reported. When heated to decomposition it emits acrid smoke and irritating fumes.

DKY800 CAS:494-19-9 ***HR: 3***
10,11-DIHYDRO-5-DIBENZ(b, f)AZEPINE
mf: $C_{14}H_{13}N$ mw: 195.28

SYN: IMINODIBENZYL

TOXICITY DATA with REFERENCE
eye-rbt 100 mg MLD FCTOD7 20,573,82
ivn-mus LD50:320 mg/kg CSLNX* NX#01352

CONSENSUS REPORTS: Reported in EPA TSCA Inventory.

SAFETY PROFILE: Poison by intravenous route. An eye irritant. When heated to decomposition it emits toxic fumes of NO_x.

DLA000 CAS:63077-00-9 ***HR: 3***
3,4-DIHYDRO-1,2,5,6-DIBENZCARBAZOLE
mf: $C_{20}H_{15}N$ mw: 269.36

SYN: 12,13-DIHYDRO-7H-DIBENZO(a,g)CARBAZOLE

TOXICITY DATA with REFERENCE
scu-mus TDLo:120 mg/kg/9W-I:ETA BAFEAG 42,3,55

SAFETY PROFILE: Questionable carcinogen with experimental tumorigenic data. When heated to decomposition it emits toxic fumes of NO_x.

DLA100 ***HR: 2***
5,8-DIHYDRODIBENZO(a,def)CHRYSENE
mf: $C_{24}H_{16}$ mw: 304.40

SYN: 5,8-DIHYDRO-3,4:9,10-DIBENZOPYRENE

TOXICITY DATA with REFERENCE
scu-mus TDLo:72 mg/kg/9W-I:ETA COREAF 251,1322,60

SAFETY PROFILE: Questionable carcinogen with experimental tumorigenic data. When heated to decomposition it emits acrid smoke and irritating fumes.

DLA120 CAS:7350-86-9 ***HR: 2***
7,14-DIHYDRODIBENZO(b,def)CHRYSENE
mf: $C_{24}H_{16}$ mw: 304.40

SYN: 5,10-DIHYDRO-3,4:8,9-DIBENZOPYRENE

TOXICITY DATA with REFERENCE
scu-mus TDLo:72 mg/kg/9W-I:ETA COREAF 251,1322,60

SAFETY PROFILE: Questionable carcinogen with experimental tumorigenic data. When heated to decomposition it emits acrid smoke and irritating fumes.

DLB400 CAS:84-16-2 ***HR: 3***
DIHYDRODIETHYLSTILBESTROL
mf: $C_{18}H_{22}O_2$ mw: 270.40

PROP: Needles from benzene, thin plates from dilute alc. Mp: 185-188°. Freely sol in ether; sol in acetone, alc, methanol; sltly sol in benzene, chloroform. Sol in dilute solns of alkali hydroxides. Practically insol in water and in dil mineral acids.

SYNS: meso-3,4-BIS(p-HYDROXYPHENYL)-n-HEXANE ◇ 3,4-BIS(p-HYDROXYPHENYL)HEXANE ◇ CYCLOESTROL ◇ 4,4'-(1,2-DIETHYLETHYLENE)DIPHENOL ◇ DIHYDROSTILBESTROL ◇ 4,4'-DIHYDROXY-α,β-DIETHYLDIPHENYLETHANE ◇ 4,4'-DIHYDROXY-Γ,Δ-DIPHENYLHEXANE ◇ Γ,Δ-DI(p-HYDROXYPHENYL)-HEXANE ◇ meso-3,4-DI(p-HYDROXYPHENYL)-n-HEXANE ◇ EXTRA-PLEX ◇ HEXANOESTROL ◇ HEXESTROL ◇ meso-HEXESTROL ◇ HEXOESTROL ◇ HORMOESTROL ◇ SINESTROL ◇ SYNESTROL ◇ SYNTHOVO ◇ SYNTROGENE ◇ VITESTROL

TOXICITY DATA with REFERENCE
dns-ham:emb 1 mg/L CNREA8 44,184,84
scu-rat TDLo:3 mg/kg (female 19-21D post):REP BEXBAN 90,1597,80

scu-rat TDLo:3 mg/kg (female 19-21D post):TER BEXBAN 90,1597,80
scu-mus TDLo:74 mg/kg/56W-I:CAR VRDEA5 (6),46,62
ivg-mus TDLo:18 mg/kg/17W-I:NEO VOONAW 22(3),68,76
imp-gpg TDLo:540 μg/kg:ETA BSBSAS 8,142,51
ipr-rat LD50:200 mg/kg NIIRDN 6,743,82
scu-rat LD50:1000 mg/kg NIIRDN 6,743,82
orl-mus LD50:1000 mg/kg NIIRDN 6,743,82
ipr-mus LD50:200 mg/kg NIIRDN 6,743,82

SAFETY PROFILE: Poison by intraperitoneal route. Moderately toxic by ingestion and subcutaneous routes. Questionable carcinogen with experimental carcinogenic and neoplastigenic data. Experimental teratogenic and reproductive effects. Mutation data reported. See also diethylstilbestrols.

DLB800 CAS:28622-84-6 ***HR: 2***
4,5-DIHYDRO-4,5-DIHYDROXYBENZO(a)PYRENE
mf: $C_{20}H_{14}O_2$ mw: 286.34

SYNS: BENZO(a)PYRENE, 4,5-DIHYDROXY-4,5-DIHYDRO- ◇ BP-4,5-DIHYDRODIOL

TOXICITY DATA with REFERENCE
dnd-hmn:fbr 30 μmol/L CBINA8 41,155,82
skn-mus TDLo:4580 μg/kg:NEO CRNGDP 3,371,78

SAFETY PROFILE: Questionable carcinogen with experimental neoplastigenic data. Mutation data reported. When heated to decomposition it emits acrid smoke and irritating fumes.

DLC000 CAS:24909-09-9 ***HR: 3***
9,10-DIHYDRO-9,10-DIHYDROXYBENZO(a)PYRENE
mf: $C_{20}H_{14}O_2$ mw: 286.34

SYN: 9,10-DIHYDROBENZO(a)PYRENE-9,10-DIOL

TOXICITY DATA with REFERENCE
dnd-hmn:fbr 30 μmol/L CBINA8 41,155,82
skn-mus TDLo:1 mg/kg:NEO BJCAAI 34,523,76
skn-mus TD:4580 μg/kg:ETA CCSUDL 3,371,78

CONSENSUS REPORTS: EPA Genetic Toxicology Program.

SAFETY PROFILE: Questionable carcinogen with experimental neoplastigenic and tumorigenic data by skin contact. Human mutation data reported. When heated to decomposition it emits acrid smoke and irritating fumes.

DLC200 CAS:58886-98-9 ***HR: D***
trans-9,10-DIHYDRO-9,10-DIHYDROXYBENZO(a) PYRENE
mf: $C_{20}H_{12}O_2$ mw: 284.32

SYN: trans-9,10-DIHYDROXY-9,10-DIHYDROBENZO(a)PYRENE

TOXICITY DATA with REFERENCE
mma-ham:lng 3500 nmol/L PNASA6 73,607,76
otr-ham:emb 10 mg/L IJCNAW 19,814,77
sce-ham:ovr 8 mg/L MUREAV 50,367,78

SAFETY PROFILE: Mutation data reported. When heated to decomposition it emits acrid smoke and irritating fumes.

DLC400 CAS:58030-91-4 ***HR: 3***
(±)-trans-9,10-DIHYDRO-9,10-DIHYDROXY-BENZO(a)PYRENE
mf: $C_{20}H_{14}O_2$ mw: 286.34

SYN: BP-9,10-DIHYDRODIOL

TOXICITY DATA with REFERENCE
mma-sat 50 μg/plate CNREA8 41,270,81
skn-mus TDLo:4600 μg/kg:ETA CALEDQ 3,23,77

SAFETY PROFILE: Questionable carcinogen with experimental tumorigenic data. Mutation data reported. When heated to decomposition it emits acrid smoke and irritating fumes.

DLC600 CAS:37571-88-3 ***HR: 3***
trans-4,5-DIHYDRO-4,5-DIHYDROXYBENZO(a) PYRENE
mf: $C_{20}H_{14}O_2$ mw: 286.34

SYNS: (E)-BENZO(a)PYRENE-4,5-DIHYDRODIOL ◇ trans-4,5-DIHYDROBENZO(a)PYRENE-4,5-DIOL ◇ trans-4,5-DIHYDROXY-4,5-DIHYDROBENZO(a)PYRENE

TOXICITY DATA with REFERENCE
mma-sat 30 mg/L ENMUDM 7,839,85
dni-omi 2 mg/L PNASA6 74,1378,77
otr-ham:emb 1 mg/L IJCNAW 19,814,77
sce-ham:ovr 8 mg/L MUREAV 50,367,78
msc-ham:lng 25 mg/L CNREA8 36,3350,76
skn-mus TDLo:1 mg/kg:NEO BJCAAI 34,523,76

CONSENSUS REPORTS: EPA Genetic Toxicology Program.

SAFETY PROFILE: Questionable carcinogen with experimental neoplastigenic data by skin contact. Mutation data reported. When heated to decomposition it emits acrid smoke and irritating fumes.

DLC800 CAS:57404-88-3 ***HR: D***
trans-7,8-DIHYDRO-7,8-DIHYDROXYBENZO(a) PYRENE
mf: $C_{20}H_{14}O_2$ mw: 286.34

SYNS: (E)-BENZO(a)PYRENE-7,8-DIHYDRODIOL ◇ (E)-7,8-DIHYDROBENZO(a)PYRENE-7,8-DIOL ◇ trans-7,8-DIHYDROXY-7,8-DIHYDRO-BENZO(a)PYRENE

TOXICITY DATA with REFERENCE
mma-sat 1 μg/plate CRNGDP 6,1235,85
hma-mus/sat 100 mg/kg MUREAV 156,163,85
sce-ham:ovr 500 μg/L MUREAV 50,367,78

SAFETY PROFILE: Mutation data reported. When heated to decomposition it emits acrid smoke and irritating fumes.

DLD000 CAS:72485-26-8 ***HR: D***
7,8-DIHYDRO-7,8-DIHYDROXYBENZO(a)PYRENE-9,10-OXIDE
mf: $C_{20}H_{12}O_3$ mw: 300.32

SYN: 7,8-DIHYDROXY-9,10-EPOXY-7,8-DIHYDROBENZO(a)PYRENE

TOXICITY DATA with REFERENCE
mma-sat 4500 nmol/L BBRCA9 66,693,75
dnd-mam:lym 9700 pmol IJBBBQ 17,96,80

SAFETY PROFILE: Mutation data reported. When heated to decomposition it emits acrid smoke and irritating fumes.

DLD200 CAS:64920-31-6 ***HR: D***
trans-1,2-DIHYDRO-1,2-DIHYDROXYCHRYSENE
mf: $C_{18}H_{14}O_2$ mw: 262.32

SYNS: (E)-1,2-DIHYDRO-1,2-CHRYSENEDIOL ◇ trans-1,2-DIHYDROCHRYSENE-1,2-DIOL ◇ trans-1,2-DIHYDROXY-1,2-DIHYDROCHRYSENE

TOXICITY DATA with REFERENCE
mmo-sat 5 μg/plate CNREA8 44,3408,84
mma-sat 37500 pmol/plate BBRCA9 78,847,77
ipr-mus TDLo:67 mg/kg/15D-I:NEO CNREA8 39,5063,79

SAFETY PROFILE: Questionable carcinogen with neoplastigenic data. Mutation data reported. When heated to decomposition it emits acrid smoke and irritating fumes.

DLD400 CAS:66267-19-4 ***HR: 2***
trans-3,4-DIHYDRO-3,4-DIHYDROXYDIBENZ(a,h) ANTHRACENE
mf: $C_{22}H_{16}O_2$ mw: 312.38

SYNS: trans-DBA-3,4-DIHYDRODIOL ◇ trans-3,4-DIHYDRO-3,4-DIHYDROXYDIBENZO(a,h)ANTHRACENE

TOXICITY DATA with REFERENCE
skn-mus TDLo:500 μg/kg:NEO CNREA8 39,1310,79

SAFETY PROFILE: Questionable carcinogen with experimental neoplastigenic data by skin contact. When heated to decomposition it emits acrid smoke and irritating fumes.

DLD600 CAS:68162-13-0 ***HR: 3***
trans-3,4-DIHYDRO-3,4-DIHYDROXY-7,12-DIMETHYLBENZ(a)ANTHRACENE
mf: $C_{20}H_{18}O_2$ mw: 290.38

SYN: trans-3,4-DIHYDRO-3,4-DIHYDROXYDMBA

TOXICITY DATA with REFERENCE
mma-sat 2500 nmol/L CBINA8 32,257,80
mma-sat 5 μmol/L BBRCA9 83,1468,78
otr-mus:fbr 120 μg/L BBRCA9 85,357,78
sce-ham:ovr 2 mg/L CALEDQ 7,45,79
msc-ham:lng 120 μg/L/3H BJCAAI 39,540,79
skn-mus TDLo:105 μg/kg:NEO CNREA8 39,1934,79
skn-mus TD:34846 ng/kg:ETA CNREA8 40,3661,80

SAFETY PROFILE: Questionable carcinogen with experimental neoplastigenic and tumorigenic data by skin contact. Mutation data reported. When heated to decomposition it emits acrid smoke and irritating fumes.

DLD800 CAS:65763-32-8 ***HR: 3***
trans-8,9-DIHYDRO-8,9-DIHYDROXY-7,12-DIMETHYLBENZ(a)ANTHRACENE
mf: $C_{20}H_{18}O_2$ mw: 290.38

SYNS: (E)-8,9-DIHYDRO-8,9-DIHYDROXY-7,12-DIMETHYLBENZ(a)ANTHRACENE ◇ trans-8,9-DIHYDRO-8,9-DIHYDROXY DMBA

TOXICITY DATA with REFERENCE
mma-sat 5 μmol/L BBRCA9 83,1468,78
otr-mus:fbr 250 mg/L BBRCA9 85,357,78
sce-ham:ovr 8 mg/L CALEDQ 7,45,79
msc-ham:lng 250 μg/L/3H BJCAAI 39,540,79
skn-mus TDLo:1050 μg/kg:ETA CNREA8 39,1934,79

SAFETY PROFILE: Questionable carcinogen with experimental tumorigenic data by skin contact. Mutation data reported. When heated to decomposition it emits acrid smoke and irritating fumes.

DLD875 CAS:68162-14-1 ***HR: D***
trans-10,11-DIHYDRO-10,11-DIHYDROXY-7,12-DIMETHYLBENZ(a)ANTHRACENE
mf: $C_{20}H_{18}O_2$ mw: 290.38

SYN: trans-10,11-DIHYDROXY-7,12-DIMETHYL-10,11-DIHYDROBENZ(a)ANTHRACENE

TOXICITY DATA with REFERENCE
mma-sat 25 μmol/L BBRCA9 83,1468,78
sce-ham:ovr 8 mg/L CALEDQ 7,45,79
msc-ham:lng 250 mg/L BBRCA9 85,357,78

SAFETY PROFILE: Mutation data reported. When heated to decomposition it emits acrid smoke and fumes.

DLE000 CAS:64598-80-7 ***HR: 3***
(±)-(1R,2S,3R,4R)-3,4-DIHYDRO-3,4-DIHYDROXY-1,2-EPOXYBENZ(a)ANTHRACENE
mf: $C_{18}H_{14}O_3$ mw: 278.32

SYNS: BA-3,4-DIOL-1,2-EPOXIDE-1 ◇ BA-3,4-DIOL-1,2-EPOXIDE-2 ◇ BENZ(a)ANTHRACENE 3,4-DIOL-1,2-EPOXIDE-2 ◇ (±)-3-α,4-β-DIHYDROXY-1-α,2-α-EPOXY-1,2,3,4-TETRAHYDROBENZ(a)ANTHRACENE ◇ (E)-1,2,3,4-TETRAHYDRO-3-α,4-β-DIHYDROXY-1-α,2-α-EPOXYBENZ(a)ANTHRACENE

TOXICITY DATA with REFERENCE
mmo-sat 100 pmol/plate CNREA8 43,1656,83
mma-sat 150 pmol/plate CRNGDP 4,1631,83
msc-ham:lng 20 μmol/L CNREA8 43,1656,83
skn-mus TDLo:22 mg/kg:ETA CNREA8 38,1699,78

CONSENSUS REPORTS: EPA Genetic Toxicology Program.

SAFETY PROFILE: Questionable carcinogen with experimental tumorigenic data by skin contact. Mutation data reported. When heated to decomposition it emits acrid smoke and irritating fumes.

DLE200 CAS:64598-81-8 ***HR: 3***
(±)-(1S,2R,3R,4R)-3,4-DIHYDRO-3,4-DIHYDROXY-1,2-EPOXYBENZ(a)ANTHRACENE
mf: $C_{18}H_{14}O_3$ mw: 278.32

SYN: BA-3,4-DIOL-1,2-EPOXIDE-1

TOXICITY DATA with REFERENCE
ipr-mus TDLo:3100 μg/kg/15D-I:NEO JJIND8 63,201,79

SAFETY PROFILE: Questionable carcinogen with experimental neoplastigenic data. When heated to decomposition it emits acrid smoke and irritating fumes.

DLE400 CAS:102420-56-4 ***HR: 2***
trans-1,2-DIHYDRO-1,2-DIHYDROXYINDENO(1,2,3-cd)PYRENE
mf: $C_{22}H_{14}O_2$ mw: 310.36

SYN: IP-1,2-DIOL

TOXICITY DATA with REFERENCE
skn-mus TDLo:40 mg/kg/20D-I:ETA CRNGDP 7,1761,86

SAFETY PROFILE: Questionable carcinogen with experimental tumorigenic data. When heated to decomposition it emits acrid smoke and irritating fumes.

DLE500 CAS:83876-50-0 ***HR: 2***
cis-5,6-DIHYDRO-5,6-DIHYDROXY-12-METHYLBENZ(a)ACRIDINE
mf: $C_{18}H_{15}NO_2$ mw: 277.34

SYN: BENZ(a)ACRIDINE-5,6-DIOL, 5,6-DIHYDRO-12-METHYL-, (Z)-

TOXICITY DATA with REFERENCE
scu-mus TDLo:72 mg/kg/12W-I:ETA JMCMAR 26,303,83

SAFETY PROFILE: Questionable carcinogen with experimental tumorigenic data. When heated to decomposition it emits toxic fumes of NO_x.

DLE600 CAS:64521-13-7 ***HR: D***
trans-1,2-DIHYDRO-1,2-DIHYDROXY-7-METHYLBENZ(a)ANTHRACENE
mf: $C_{19}H_{17}O_2$ mw: 277.36

SYN: (E)-1,2-DIHYDRO-1,2-DIHYDROXY-7-METHYLBENZ(a)ANTHRACENE

TOXICITY DATA with REFERENCE
mma-sat 30 μg/L BBRCA9 75,427,77
sce-ham:ovr 8 mg/L MUREAV 50,367,78

SAFETY PROFILE: Mutation data reported. When heated to decomposition it emits acrid smoke and irritating fumes.

DLF000 CAS:16053-71-7 ***HR: D***
trans-5,6-DIHYDRO-5,6-DIHYDROXY-7-METHYLBENZ(a)ANTHRACENE
mf: $C_{19}H_{17}O_2$ mw: 277.36

SYN: (E)-5,6-DIHYDRO-5,6-DIHYDROXY-7-METHYL-BENZ(a)ANTHRACENE

TOXICITY DATA with REFERENCE
mma-sat 50 μg/L BBRCA9 75,427,77
sce-ham:ovr 8 mg/L MUREAV 50,367,78

SAFETY PROFILE: Mutation data reported. When heated to decomposition it emits acrid smoke and irritating fumes.

DLF200 CAS:64521-15-9 ***HR: 3***
trans-8,9-DIHYDRO-8,9-DIHYDROXY-7-METHYLBENZ(a)ANTHRACENE
mf: $C_{18}H_{16}O_2$ mw: 264.34

TOXICITY DATA with REFERENCE
mma-sat 30 μg/L BBRCA9 75,427,77
otr-mus:fbr 1 mg/L IJCNAW 19,828,77
sce-ham:ovr 8 mg/L MUREAV 50,367,78
msc-ham:lng 1 mg/L IJCNAW 19,828,77
skn-mus TDLo:1000 μg/kg:ETA CALEDQ 3,247,77

CONSENSUS REPORTS: EPA Genetic Toxicology Program.

SAFETY PROFILE: Questionable carcinogen with experimental tumorigenic data by skin contact. Mutation data reported. When heated to decomposition it emits acrid smoke and irritating fumes.

DLF300 CAS:68688-84-6 ***HR: D***
cis-2-α,3-DIHYDRODIHYDROXY-3-METHYLCHOLANTHRENE

SYN: (Z)-1,2-DIHYDRO-3-METHYL-BENZ(j)ACEANTHRYLENE-2a,3(3H)-DIOL

TOXICITY DATA with REFERENCE
mma-sat 20 μmol/L BBRCA9 85,1568,78
otr-mus:fbr 1 mg/L BBRCA9 85,1568,78
msc-ham:lng 1 mg/L BBRCA9 85,1568,78

CONSENSUS REPORTS: EPA Genetic Toxicology Program.

SAFETY PROFILE: Mutation data reported. When heated to decomposition it emits acrid smoke and fumes.

DLF400 CAS:67411-81-8 ***HR: 3***
1,2-DIHYDRO-1,2-DIHYDROXY-5-METHYLCHRYSENE
mf: $C_{19}H_{16}O_2$ mw: 276.35

SYN: 1,2-DIHYDRO-5-METHYL-1,2-CHRYSENEDIOL

TOXICITY DATA with REFERENCE
mma-sat 7200 pmol/plate CNREA8 38,2191,78
skn-mus TDLo:36 μg/kg:CAR CNREA8 45,6406,79
skn-mus TDLo:1200 μg/kg/18D-I:NEO CNREA8 40,1396,80

SAFETY PROFILE: Questionable carcinogen with experimental carcinogenic and neoplastigenic data. Mutation data reported. When heated to decomposition it emits acrid smoke and irritating fumes.

DLF600 CAS:67523-22-2 ***HR: 3***
7,8-DIHYDRO-7,8-DIHYDROXY-5-METHYLCHRYSENE
mf: $C_{19}H_{16}O_2$ mw: 276.35

SYN: 7,8-DIHYDRO-5-METHYL-7,8-CHRYSENEDIOL

TOXICITY DATA with REFERENCE
mma-sat 2700 pmol/plate CNREA8 38,2191,78
skn-mus TDLo:1200 μg/kg/18D-I:NEO CNREA8 40,1396,80

SAFETY PROFILE: Questionable carcinogen with experimental neoplastigenic data by skin contact. Mutation data reported. When heated to decomposition it emits acrid and irritating fumes.

DLF700 CAS:81840-15-5 ***HR: 3***
3,4-DIHYDRO-6-(4-(3,4-DIMETHOXYBENZOYL)-1-PIPERAZINYL)-2(1H)-QUINOLINONE
mf: $C_{22}H_{25}N_3O_4$ mw: 395.50

SYNS: 1-(3,4-DIMETHOXYBENZOYL)-4-(1,2,3,4-TETRAHYDRO-2-OXO-6-QUINOLINYL)PIPERA ZINE ◇ OPC-8212 ◇ PIPERAZINE, 1-(3,4-DIMETHOXYBENZOYL)-4-(1,2,3,4-TETRAHYDRO-2-OXO-6-QUINOLINYL)-

TOXICITY DATA with REFERENCE
orl-rat TDLo:1100 mg/kg (female 7-17D post):REP IYKEDH 18,875,87
orl-rbt TDLo:3900 mg/kg (female 6-18D post):TER IYKEDH 18,898,87
ivn-rat LD50:79300 μg/kg IYKEDH 18,922,87
ivn-mus LD50:56300 μg/kg IYKEDH 18,922,87
ivn-dog LD50:63300 μg/kg IYKEDH 18,922,87

SAFETY PROFILE: Poison by intravenous route. An experimental teratogen. Experimental reproductive effects. When heated to decomposition it emits toxic fumes of NO_x.

DLG000 CAS:22797-20-2 ***HR: 3***
5,10-DIHYDRO-10-(2-(DIMETHYLAMINO)ETHYL)-8-ETHYLSULFONYL-5-METHYL-11H-DIBENZO(b,e) (1,4)DIAZEPIN-11-ONE
mf: $C_{20}H_{25}N_3O_3S$ mw: 387.54

SYN: SM-307

TOXICITY DATA with REFERENCE
orl-mus LD50:857 mg/kg JJPAAZ 21,47,71
ipr-mus LD50:354 mg/kg JJPAAZ 21,47,71
ivn-mus LD50:74 mg/kg JJPAAZ 21,47,71

SAFETY PROFILE: Poison by intraperitoneal and intravenous routes. Moderately toxic by ingestion. When heated to decomposition it emits very toxic fumes of NO_x and SO_x.

DLH200 CAS:739-71-9 ***HR: 3***
10,11-DIHYDRO-5-(3-DIMETHYLAMINO-2-METHYLPROPYL)-5H-DIBENZ (b,f)AZEPINE
mf: $C_{20}H_{26}N_2$ mw: 294.48

SYNS: 10,11-DIHYDRO-N,N,β-TRIMETHYL-5H-DIBENZ(b,f)AZEPINE-5-PROPANAMINE ◇ 1-(3-DIMETHYLAMINO-2-METHYLPROPYL)-4,5-DIHYDRO-2,3:6,7-DIBENZAZEPINE ◇ 5-(3-(DIMETHYLAMINO)-2-METHYLPROPYL)-10,11-DIHYDRO-5H-DIBENZ(b,f)AZEPINE ◇ 5-(Γ-DIMETHYLAMINO-β-METHYLPROPYL)-10,11-DIHYDRO-5H-DIBENZO (b,f)AZEPINE ◇ FI 6120 ◇ IL 6001 ◇ 2'-METIL-3'-DIMETILAMINO-PROPIL-5-IMINODIBENZILE (ITALIAN) ◇ 7162 RP ◇ SAPILENT ◇ SURMONTIL ◇ TRIMEPRIMINA (ITALIAN) ◇ TRIME- PROPIMINE ◇ TRIMIPRAMINE

TOXICITY DATA with REFERENCE
orl-hmn TDLo:18 mg/kg:CVS PSDTAP 6,171,65
orl-mus LD50:250 mg/kg BCFAAI 102,753,63
ipr-mus LD50:145 mg/kg CRSBAW 155,307,61
scu-mus LD50:200 mg/kg BCFAAI 102,753,63
ivn-mus LD50:42 mg/kg CRSBAW 155,307,61

CONSENSUS REPORTS: EPA Genetic Toxicology Program.

SAFETY PROFILE: Poison by ingestion, intraperi-

toneal, subcutaneous, and intravenous routes. Human systemic effects by ingestion: unspecified heart effects. When heated to decomposition it emits toxic fumes of NO_x.

DLH600 CAS:50-49-7 ***HR: 3***
5,6-DIHYDRO-N-(3-(DIMETHYLAMINO)PROPYL)-11H-DIBENZ(b,e)AZEPINE
mf: $C_{19}H_{24}N_2$ mw: 280.45

SYNS: ANTIDEPRIN ◇ BERKOMINE ◇ CENSTIN ◇ 10,11-DIHYDRO-5-(3-(DIMETHYLAMINO)PROPYL)-5H-DIBENZ(b,f)AZEPINE◇ 2,2′-(3-DIMETHYLAMINOPROPYLAMINO)BIBENZYL ◇ 1-(3-DIMETHYLAMINOPROPYL)-4,5-DIHYDRO-2,3,6,7-DIBENZAZEPINE◇ 5-(3-DIMETHYLAMINOPROPYL)-10,11-DIHYDRO-5H-DIBENZO(b,f)AZEPINE ◇ 2,2′-(3-DIMETHYLAMINOPROPYLIMINO)DIBENZYL ◇ N-(Γ-DIMETHYLAMINOPROPYL)IMINODIBENZYL◇ DIMIPRESSIN ◇ DPID ◇ DYNAPRIN ◇ DYNA-ZINA ◇ EUPRAMIN ◇ G 22355 ◇ IM ◇ IMIDOBENZYLE ◇ IMIPRAMINA (ITALIAN) ◇ IMIPRAMINE ◇ IMIPRIN ◇ IMIZIN ◇ IMIZINUM ◇ IMPRAMINE ◇ INTALPRAM ◇ IRAMIL ◇ IRMIN ◇ MELIPRAMIN ◇ MELIPRAMINE ◇ NELIPRAMIN ◇ PRAZEPINE ◇ PROMIBEN ◇ SURPLIX ◇ TIMOLET ◇ TOFRANIL

TOXICITY DATA with REFERENCE
dni-oin-unr 10 g/L JCLBA3 47,182a,70
cyt-oin-unr 10 g/L JCLBA3 47,182a,70
orl-rat TDLo:378 mg/kg (lactating female 21D post):REP PSCHDL 56,93,78
scu-rbt TDLo:55 mg/kg (female 6-16D post):TER ARZNAD 15,1218,65
orl-wmn LDLo:2 mg/kg/1D:PUL JPPMAB 16,265,64
orl-man LD50:30 mg/kg HEPHD2 55,527,80
orl-cld LD50:40 mg/kg HEPHD2 55,527,80
orl-wmn TDLo:3 mg/kg/32H-I:SKN JAMAAP 254,357,85
orl-chd TDLo:30 mg/kg:CNS AJDCAI 130,507,76
orl-cld LDLo:35 mg/kg SMWOAS 99,1157,69
orl-hmn LDLo:450 mg/kg:SKN BMJOAE 1,722,79
orl-hmn LD50:40 mg/kg PSDTAP 6,171,65
orl-man TDLo:8 mg/kg/3D-I:CNS LANCAO 2,568,59
orl-rat LD50:250 mg/kg PCJOAU 14,773,80
ipr-rat LD50:79 mg/kg AIPTAK 148,560,64
scu-rat LD50:250 mg/kg AIPTAK 148,560,64
ivn-rat LD50:15900 μg/kg ARZNAD 29,193,79
orl-mus LD50:188 mg/kg PCJOAU 14,773,80
ipr-mus LD50:51600 μg/kg BRXXAA #1460700
scu-mus LD50:195 μg/kg PCJOAU 15,412,81
ivn-mus LD50:21 mg/kg AIPTAK 245,283,80
orl-dog LDLo:100 mg/kg 27ZQAG -,78,72

CONSENSUS REPORTS: EPA Genetic Toxicology Program.

SAFETY PROFILE: A human poison by ingestion. An experimental poison by ingestion, subcutaneous, intravenous, and intraperitoneal routes. Human systemic effects by ingestion: somnolence, hallucinations, distorted perceptions, changes in motor activity, ataxia (loss of muscle coordination), coma, nausea, and vomiting, irritative dermatitis. An experimental teratogen by ingestion. Other experimental reproductive effects. Mutation data reported. When heated to decomposition it emits toxic fumes of NO_x.

DLH630 CAS:113-52-0 ***HR: 3***
10,11-DIHYDRO-5-(3-(DIMETHYLAMINO)PROPYL)-5H-DIBENZ(b,f)AZEPINE HYDROCHLORIDE
mf: $C_{19}H_{24}N_2 \cdot ClH$ mw: 316.91

SYNS: ANTIDEPRIN HYDROCHLORIDE ◇ BERKOMINE ◇ CENSTIM ◇ CENSTIN ◇ CHIMOREPTIN ◇ CHRYTEMIN ◇ CO CAP IMIPRAMINE 25 ◇ DEPRINOL ◇ 10,11-DIHYDRO-N,N-DIMETHYL-5H-DIBENZ(b,f)AZEPINE-5-PROPANAMINEMONOHYDROCHLORIDE ◇ 5-(3-DIMETHYLAMINOPROPYL)-10,11-DIHYDRO-5H-DIBENZ(b,f) AZEPINE HYDROCHLORIDE ◇ N-(3-DIMETHYLAMINO- PROPYL)IMINODIBENZYL HYDROCHLORIDE ◇ N-(Γ-DIMETILAMINOPROPIL)-IMINODIBENZILE CLORIDRATO (ITALIAN) ◇ DIMIPRESSIN ◇ DYNAZINA ◇ EFURANOL ◇ EUPRAMIN ◇ FEINALMIN ◇ G 22150 ◇ G 22355 ◇ IA-PRAM ◇ IMAVATE ◇ IMIDOBENZYLE ◇ IMIDOL ◇ IMILANYLE ◇ IMIPRAMINA (ITALIAN) ◇ IMIPRAMINE ◇ IMIPRAMINE HYDROCHLORIDE ◇ IMIPRAMINE MONOHYDROCHLORIDE ◇ IMIPRIN ◇ IMP HYDROCHLORIDE ◇ INTALPRAM ◇ IPROGEN ◇ IRAMIL ◇ JANIMINE ◇ LOFEPRAMINE ◇ MELIPRAMIN ◇ MELIPRAMINE ◇ MELIPRAMINE HYDROCHLORIDE ◇ MELIPRAMIN HYDROCHLORIDE ◇ NSC 114900 ◇ PERSAMINE ◇ PERTOFRAM ◇ PRESAMINE ◇ PROMIBEN ◇ PYRLEUGAN ◇ SK-PRAMINE ◇ SK-PRAMINE HYDROCHLORIDE ◇ SURPLIX ◇ TEPERINE ◇ TIMOLET ◇ TOFRANIL ◇ TOFRANILE

TOXICITY DATA with REFERENCE
dni-oin-unr 10 g/L JCLBA3 47,182a,70
cyt-oin-unr 10 g/L JCLBA3 47,182a,70
orl-rat TDLo:285 mg/kg (female 14D pre-21D post):REP PSYPAG 41,237,75
scu-mus TDLo:125 mg/kg (female 9D post):TER DGDFA5 22,61,80
orl-man TDLo:2143 μg/kg/2D-I JCLPDE 44,225,83
orl-chd TDLo:25 mg/kg:CNS,PUL JAMAAP 179,456,62
orl-chd LDLo:15 mg/kg BMJOAE 1,261,74
orl-chd TDLo:27 mg/kg:CNS,KID JAMAAP 230,1405,74
orl-wmn TDLo:107 mg/kg:CNS,CVS NEJMAG 268,33,63
orl-wmn TDLo:30 mg/kg:PNS,CNS BMJOAE 2,1458,59
orl-rat LD50:305 mg/kg TXAPA9 18,185,71
ipr-rat LD50:72 mg/kg ARZNAD 21,391,71
scu-rat LD50:217 mg/kg ARZNAD 21,391,71
ivn-rat LD50:18 mg/kg ATXKA8 21,30,65
orl-mus LD50:275 mg/kg THERAP 20,67,65
ipr-mus LD50:104 mg/kg ARZNAD 24,166,74
scu-mus LD50:189 mg/kg FRPPAO 25,519,70
ivn-mus LD50:27 mg/kg ARZNAD 31,75,81
ivn-mky LDLo:25 mg/kg IJEBA6 22,539,84
ipr-gpg LD50:85 mg/kg PHARAT 38,749,83
scu-gpg LD50:190 mg/kg AIPTAK 137,375,62

CONSENSUS REPORTS: Reported in EPA TSCA Inventory.

SAFETY PROFILE: Human poison by ingestion. An experimental poison by ingestion, intravenous, subcutaneous, and intraperitoneal routes. An experimental teratogen. Human systemic effects by ingestion: sleep, somnolence, convulsions, muscle contraction or spasticity, coma, blood pressure decrease, dyspnea (difficulty in breathing), paresthesia (abnormal sensations) and kidney changes. Experimental reproductive effects. Mutation data reported. Used in the treatment of depression. When heated to decomposition it emits very toxic fumes of NO_x and HCl. See also DIAZEPAM.

DLH800 CAS:35281-29-9 ***HR: 3***
5,6-DIHYDRO-7,12-DIMETHYLBENZ(a)ANTHRACENE
mf: $C_{20}H_{18}$ mw: 258.38

TOXICITY DATA with REFERENCE
skn-mus TDLo:128 mg/kg/50W-I:CAR ZKKOBW 77,226,72

SAFETY PROFILE: Questionable carcinogen with experimental carcinogenic data by skin contact. When heated to decomposition it emits acrid smoke and irritating fumes.

DLI000 CAS:52171-93-4 ***HR: 3***
3,4-DIHYDRO-1,11-DIMETHYLCHRYSENE
mf: $C_{20}H_{18}$ mw: 258.38

TOXICITY DATA with REFERENCE
mma-sat 20 μg/plate CNREA8 36,4525,76
skn-mus TDLo:120 mg/kg/50W-I:NEO CNREA8 34,1315,74

CONSENSUS REPORTS: EPA Genetic Toxicology Program.

SAFETY PROFILE: Questionable carcinogen with experimental neoplastigenic data by skin contact. Mutation data reported. When heated to decomposition it emits acrid smoke and irritating fumes. See also CHRYSENE.

DLI200 CAS:5831-16-3 ***HR: 3***
16,17-DIHYDRO-11,17-DIMETHYLCYCLOPENTA(a)PHENANTHRENE
mf: $C_{19}H_{17}$ mw: 245.36

SYN: 11,17-DIMETHYL-16,17-DIHYDRO-15H-CYCLOPENTA(a)PHENANTHRENE

TOXICITY DATA with REFERENCE
skn-mus TDLo:108 mg/kg/1Y-I:ETA PEXTAR 11,69,69

CONSENSUS REPORTS: EPA Genetic Toxicology Program.

SAFETY PROFILE: Questionable carcinogen with experimental tumorigenic data by skin contact. When heated to decomposition it emits acrid smoke and irritating fumes.

DLI300 CAS:85616-56-4 ***HR: 2***
15,16-DIHYDRO-7,11-DIMETHYL-17H-CYCLOPENTA(a)PHENANTHREN-17-ONE
mf: $C_{19}H_{16}O$ mw: 260.35

SYN: 7,11-DIMETHYL-15,16-DIHYDROCYCLOPENTA(a)PHENANTHREN-17-ONE

TOXICITY DATA with REFERENCE
skn-mus TDLo:40 mg/kg/10W-I:CAR CNREA8 46,1817,86

SAFETY PROFILE: Questionable carcinogen with experimental carcinogenic data. When heated to decomposition it emits acrid smoke and irritating fumes.

DLI400 CAS:894-52-0 ***HR: 3***
15,16-DIHYDRO-11,12-DIMETHYLCYCLOPENTA(a)PHENANTHREN-17-ONE
mf: $C_{19}H_{16}O$ mw: 260.35

TOXICITY DATA with REFERENCE
mma-sat 50 μg/plate CNREA8 36,4525,76
skn-mus TDLo:108 mg/kg/1Y-I:CAR PEXTAR 11,69,69
scu-mus TDLo:360 mg/kg:CAR PEXTAR 11,69,69

CONSENSUS REPORTS: EPA Genetic Toxicology Program.

SAFETY PROFILE: Questionable carcinogen with experimental carcinogenic data. Mutation data reported. When heated to decomposition it emits acrid smoke and irritating fumes.

DLI600 CAS:1920-21-4 ***HR: 2***
2,3-DIHYDRO-2,5-DIMETHYL-2-FORMYL-1,4-PYRAN
mf: $C_8H_{12}O_2$ mw: 140.20

SYNS: 3,4-DIHYDRO-2,5-DIMETHYL-2H-PYRAN-2-CARBOXALDEHYDE ◇ METHACROLEIN DIMER ◇ METHACRYLALDEHYDE DIMER

TOXICITY DATA with REFERENCE
skn-rbt 10 mg/24H open MLD AMIHBC 10,61,54
eye-rbt 20 mg open AMIHBC 10,61,54
orl-rat LD50:2460 mg/kg AMIHBC 10,61,54

SAFETY PROFILE: Moderately toxic by ingestion. A skin and eye irritant. When heated to decomposition it emits acrid smoke and irritating fumes.

DLI650 CAS:78499-27-1 ***HR: 3***
10,11-DIHYDRO-α-8-DIMETHYL-11-OXODIBENZ(b,f)OXEPIN-2-ACETIC ACID
mf: $C_{18}H_{16}O_4$ mw: 296.34

SYNS: AD-1590 ◇ DIBENZ(b,f)OXEPIN-2-ACETIC ACID, 10,11-

DIHYDRO-α-8-DIMETHYL-11-OXO- ◇ 2-(8-METHYL-10,11-DIHYDRO-11-OXODIBENZ(b,f)OXEPIN-2-YL)PROPIONIC ◇ PROPIONIC ACID, 2-(8-METHYL-10,11-DIHYDRO-11-OXODIBENZ(b,f)OXEPIN-2-YL)-

TOXICITY DATA with REFERENCE
orl-rat TDLo:33 mg/kg (female 7-17D post):REP YACHDS 16,2797,88
orl-rat LD50:147 mg/kg JMCMAR 25,1065,82
ipr-rat LD50:391 mg/kg YACHDS 16,2701,88
scu-rat LD50:483 mg/kg YACHDS 16,2701,88
ivn-rat LDLo:160 mg/kg YKKZAJ 108,788,88
orl-mus LD50:212 mg/kg YACHDS 16,2701,88
ipr-mus LD50:191 mg/kg YACHDS 16,2701,88
scu-mus LD50:387 mg/kg YACHDS 16,2701,88

SAFETY PROFILE: Poison by ingestion, intraperitoneal, and subcutaneous routes. Experimental reproductive effects. When heated to decomposition it emits acrid smoke and irritating fumes.

DLJ000 CAS:34375-78-5 ***HR: 3***
4,5-DIHYDRO-1,3-DIMETHYL-1H-(PYRAZOLO (3,4-b)((1,4)-BENZOXAZEPINE) HYDROCHLORIDE

TOXICITY DATA with REFERENCE
orl-rat LD50:1360 mg/kg AIPTAK 202,119,73
orl-mus LD50:1276 mg/kg AIPTAK 202,119,73
ipr-mus LD50:465 mg/kg AIPTAK 202,119,73
ivn-mus LD50:220 mg/kg AIPTAK 202,119,73

CONSENSUS REPORTS: EPA Extremely Hazardous Substances List.

SAFETY PROFILE: Poison by intravenous route. Moderately toxic by ingestion and intraperitoneal routes. When heated to decomposition it emits very toxic fumes of NO_x and HCl.

DLJ500 CAS:66289-74-5 ***HR: 3***
endo,endo-DIHYDRODI(NORBORNADIENE)
mf: $C_{14}H_{18}$ mw: 186.32

SYNS: 4,7-METHANO-2,3,8-METHENOCYCLOPENT(a)INDENE, DODECAHYDRO-, stereoisomer ◇ RJ 5 ◇ SHELLOYNE H

TOXICITY DATA with REFERENCE
ihl-rat TCLo:150 mg/m^3/6H/1Y-I:ETA NTIS** AD-A134-150
orl-mus LDLo:250 mg/kg AMRL** TR-74-78,74

SAFETY PROFILE: Poison by ingestion. Questionable carcinogen with experimental tumorigenic data. When heated to decomposition it emits acrid smoke and irritating fumes.

DLJ600 CAS:14938-42-2 ***HR: 2***
9,10-DIHYDRO-9,10-DIOXO-1,8-ANTHRACENE-DISULFONIC ACID DIPOTASSIUM SALT
mf: $C_{14}H_6O_8S_2$•2K mw: 444.52

SYNS: ANTHRACHINON-1,8-DISULFONAN DRASELNY (CZECH) ◇ ANTHRAQUINONEDISULFONIC ACID, DIPOTASSIUM SALT

TOXICITY DATA with REFERENCE
skn-rbt 500 mg/24H MLD 28ZPAK -,193,72
eye-rbt 100 mg/24H MOD 28ZPAK -,193,72
orl-rat LD50:15 g/kg 28ZPAK -,193,72

SAFETY PROFILE: Mildly toxic by ingestion. A skin and eye irritant. When heated to decomposition it emits toxic fumes of SO_x and K_2O.

DLJ700 CAS:853-35-0 ***HR: 1***
9,10-DIHYDRO-9,10-DIOXO-1,5-ANTHRACENE DISULFONIC ACID DISODIUM SALT
mf: $C_{14}H_6O_8S_2$•2Na mw: 412.30

SYNS: ANTHRACHINON-1,5-DISULFONAN SODNY (CZECH) ◇ DISODIUM ANTHRAQUINONE-1,5-DISULFONATE ◇ SODIUM ANTHRAQUINONE-1,5-DISULFONATE

TOXICITY DATA with REFERENCE
eye-rbt 500 mg/24H MLD 28ZPAK -,193,72

CONSENSUS REPORTS: Reported in EPA TSCA Inventory.

SAFETY PROFILE: An eye irritant. When heated to decomposition it emits toxic fumes of SO_x and Na_2O. See also SULFONATES.

DLJ800 CAS:128-56-3 ***HR: 1***
9,10-DIHYDRO-9,10-DIOXO-1-ANTHRACENE SULFONIC ACID SODIUM SALT
mf: $C_{14}H_7O_5S$•Na mw: 310.26

SYNS: ANTHRACHINON-1-SULFONAN SODNY (CZECH) ◇ SODIUM ANTHRAQUINONE-1-SULFONATE

TOXICITY DATA with REFERENCE
eye-rbt 500 mg/24H MLD 28ZPAK -,192,72
orl-rat LD50:20 g/kg GISAAA 45(3),73,80
orl-mus LD50:32 g/kg GISAAA 45(3),73,80
orl-rbt LD50:14 g/kg GISAAA 45(3),73,80
orl-gpg LD50:32 g/kg GISAAA 45(3),73,80

CONSENSUS REPORTS: Reported in EPA TSCA Inventory.

SAFETY PROFILE: Very mildly toxic by ingestion. When heated to decomposition it emits toxic fumes of SO_x and Na_2O.

DLK000 CAS:521-24-4 ***HR: 2***
3,4-DIHYDRO-3,4-DIOXO-1-NAPHTHALENE SULFONIC ACID SODIUM SALT
mf: $C_{10}H_6O_5S{\cdot}Na$ mw: 261.21

SYNS: β-NAPHTHOQUINONE-4-SULFONATE SODIUM SALT ◇ SODIUM-β-NAPHTHOQUINONE-4-SULFONATE ◇ SODIUM-1,2-NAPHTHOQUINONE-4-SULFONATE

TOXICITY DATA with REFERENCE
ipr-mus LD50:625 mg/kg NTIS** AD691-490

CONSENSUS REPORTS: Reported in EPA TSCA Inventory.

SAFETY PROFILE: Moderately toxic by intraperitoneal route. When heated to decomposition it emits toxic fumes of SO_x and Na_2O.

DLK200 CAS:3347-22-6 ***HR: 3***
5,10-DIHYDRO-5,10-DIOXONAPHTHO(2,3-b)-p-DITHIIN-2,3-DICARBONITRILE
mf: $C_{14}H_4N_2O_2S_2$ mw: 296.32

SYNS: DELAN ◇ DELAN-COL ◇ 2,3-DICARBONITRILO-1,4-DIATHIAANTHRACHINON (GERMAN) ◇ 2,3-DICYANO-1,4-DITHIA-ANTHRAQUINONE ◇ 2,3-DINITRILO-1,4-DITHIA-ANTHRAQUINONE ◇ 2,3-DINITRILO-1,4-DITHIOANTHRACHINON (GERMAN) ◇ 1,4-DITHIAANTHRAQUINONE-2,3-DICARBONITRILE ◇ 1,4-DITHIA-ANTHRAQUINONE-2,3-DINITRILE ◇ DITHIANON ◇ DITHIANONE ◇ DTA ◇ IT 931 ◇ MV 119A ◇ STAUFFER MV-119A ◇ THYNON

TOXICITY DATA with REFERENCE
orl-rat LD50:638 mg/kg FMCHA2 -,C86,83
unk-rat LD50:1015 mg/kg 30ZDA9 -,425,71
orl-mus LD50:1140 mg/kg 31ZOAD 1,189,68
orl-gpg LD50:110 mg/kg 28ZEAL 5,86,76

CONSENSUS REPORTS: Cyanide and its compounds are on the Community Right-To-Know List.

SAFETY PROFILE: Poison by ingestion. Moderately toxic by an unspecified route. A fungicide. See also NITRILES. When heated to decomposition it emits very toxic fumes of NO_x, SO_x, and CN^-.

DLK600 CAS:63041-56-5 ***HR: 3***
7,14-DIHYDRO-7,14-DIPROPYLDIBENZ(a,h)ANTHRACENE-7,14-DIOL
mf: $C_{28}H_{28}O_2$ mw: 396.56

SYNS: 9,10-DIHYDRO-9,10-DIHYDROXY-9,10-DI-n-PROPYL-1,25,6-DIBENZANTHRACENE ◇ 9,10-DIHYDROXY-9,10-DI-n-PROPYL-9,10-DIHYDRO-1,2:5,6-DIBENZANTHRACENE ◇ 9,10-DI-n-PROPYL-9-10-DIHYDROXY-9,10-DIHYDRO-1,2,5,6-DIBENZANTHRACENE

TOXICITY DATA with REFERENCE
scu-rat TDLo:8 mg/kg:ETA JOCEAH 2,175,37

SAFETY PROFILE: Questionable carcinogen with experimental tumorigenic data. When heated to decomposition it emits acrid smoke and irritating fumes.

DLK700 CAS:24519-85-5 ***HR: 3***
5,6-DIHYDRO-p-DITHIIN-2,3-DICARBOXIMIDE
mf: $C_6H_5NO_2S_2$ mw: 187.24

SYN: 3,6-DITHIA-3,4,5,6-TETRAHYDROPHTHALIMIDE

TOXICITY DATA with REFERENCE
ipr-mus TDLo:25 mg/kg (9D preg):REP MOPMA3 13,133,77
ipr-mus TDLo:50 mg/kg (9D preg):TER MOPMA3 13,133,77
ipr-mus LD50:64 mg/kg DPHFAK 23,113,71

SAFETY PROFILE: Poison by intraperitoneal route. An experimental teratogen. Experimental reproductive effects. When heated to decomposition it emits toxic fumes of SO_x and NO_x.

DLK750 ***HR: 2***
1,2-DIHYDRO-1,2-EPOXYINDENO(1,2,3-cd) PYRENE
mf: $C_{22}H_{12}O$ mw: 292.34

SYN: INDENO(1,2,3-cd)PYRENE-1,2-OXIDE

TOXICITY DATA with REFERENCE
mmo-sat 1 μg/plate CNREA8 45,5421,85
skn-mus TDLo:40 mg/kg/20D-I:ETA CRNGDP 7,1761,86

SAFETY PROFILE: Questionable carcinogen with experimental tumorigenic data. Mutation data reported. When heated to decomposition it emits acrid smoke and irritating fumes.

DLK800 CAS:511-12-6 ***HR: 3***
DIHYDROERGOTAMINE
mf: $C_{33}H_{37}N_5O_5$ mw: 583.65

SYN: DEHYDROERGOTAMINE

TOXICITY DATA with REFERENCE
ivn-rat LD50:110 mg/kg HPPAAL 2,48,44
scu-mus LD50:80 mg/kg RPTOAN 31,53,68

SAFETY PROFILE: Poison by intravenous and subcutaneous routes. When heated to decomposition it emits toxic fumes of NO_x.

DLL000 CAS:5989-77-5 ***HR: 3***
DIHYDROERGOTAMINE TARTRATE (2:1)
mf: $C_{66}H_{74}N_{10}O_{10}{\cdot}C_4H_6O_6$ mw: 1317.60

TOXICITY DATA with REFERENCE
ivn-rat LD50:110 mg/kg BSAMA5 2,1,46
ipr-mus LD50:210 mg/kg RPOBAR 2,284,70
ivn-mus LD50:118 mg/kg BSAMA5 2,1,46
scu-cat LD50:68 mg/kg BSAMA5 2,1,46
ivn-rbt LD50:25 mg/kg BSAMA5 2,1,46

SAFETY PROFILE: Poison by intravenous, intraperitoneal, and subcutaneous routes. When heated to decomposition it emits toxic fumes of NO_x.

DLL400 CAS:8067-24-1 ***HR: 3***
DIHYDROERGOTOXINE METHANE SULFONATE

SYNS: CCK 179 ◇ CIRCANOL ◇ CO-DERGOCRINE MESYLATE ◇ DIHYDROERGOTOXINE MESYLATE ◇ DIHYDROERGOTOXINE METHANESULPHONATE ◇ DIHYDROERGOTOXINE MONOMETHANESULFONATE (SALT) ◇ DIHYDROERGOTOXIN MESYLATE ◇ DIHYDROERGOTOXIN METHANESULFONATE ◇ HYDERGIN ◇ HYDERGINE ◇ ISCHELIUM ◇ REDERGIN

TOXICITY DATA with REFERENCE
dlt-mus-ipr 100 mg/kg MUREAV 50,317,78
ipr-rat TDLo:10 mg/kg (female 12-21D post):REP AJPHAP 180,296,55
ivn-mus LD50:180 mg/kg NIIRDN 6,329,82

CONSENSUS REPORTS: EPA Genetic Toxicology Program.

SAFETY PROFILE: Poison by intravenous route. Experimental reproductive effects. Mutation data reported. When heated to decomposition it emits very toxic fumes of SO_x and NO_x.

DLL600 CAS:29734-68-7 ***HR: 3***
DIHYDRO-β-ERYTHROIDINE HYDROBROMIDE
mf: $C_{16}H_{21}NO_3•BrH$ mw: 356.30

TOXICITY DATA with REFERENCE
ipr-rat LDLo:30 mg/kg JPETAB 75,270,42
ivn-rat LD50:8900 μg/kg JPETAB 82,266,44
ivn-dog LD50:1100 μg/kg JPETAB 82,266,44
ivn-rbt LD50:2100 μg/kg JPETAB 82,266,44

SAFETY PROFILE: A deadly poison by intravenous and intraperitoneal routes. When heated to decomposition it emits very toxic fumes of Br^- and NO_x.

DLM000 CAS:42028-27-3 ***HR: 3***
15,16-DIHYDRO-11-ETHYLCYCLOPENTA(a)PHENANTHREN-17-ONE
mf: $C_{19}H_{16}O$ mw: 260.35

TOXICITY DATA with REFERENCE
mma-sat 50 μg/plate CNREA8 36,4525,76
skn-mus TDLo:120 mg/kg/50W-I:NEO CNREA8 33,832,73

CONSENSUS REPORTS: EPA Genetic Toxicology Program.

SAFETY PROFILE: Questionable carcinogen with experimental neoplastigenic data. Mutation data reported. When heated to decomposition it emits acrid smoke and irritating fumes.

DLM600 CAS:5096-24-2 ***HR: 3***
2,3-DIHYDRO-3-ETHYL-6-METHYL-1H-CYCLOPENTA(a)ANTHRACENE
mf: $C_{20}H_{20}$ mw: 260.40

SYN: 3-ETHYL-2,3-DIHYDRO-6-METHYL-1H-CYCLOPENT(a)ANTHRACENE

TOXICITY DATA with REFERENCE
orl-rat TDLo:1000 mg/kg:ETA CNREA8 26,619,66
orl-rat LDLo:1000 mg/kg CNREA8 26,619,66

SAFETY PROFILE: Moderately toxic by ingestion. Questionable carcinogen with experimental tumorigenic data. When heated to decomposition it emits acrid smoke and irritating fumes.

DLN000 CAS:52831-41-1 ***HR: 3***
6,13-DIHYDRO-2-FLUOROBENZO(g)(1)BENZOTHIOPYRANO(4,3-b)INDOLE
mf: $C_{19}H_{12}FNS$ mw: 305.38

TOXICITY DATA with REFERENCE
mma-sat 30 μg/plate MUREAV 66,307,79
scu-mus TDLo:78 mg/kg/9W-I:NEO MUREAV 66,307,79

SAFETY PROFILE: Questionable carcinogen with experimental neoplastigenic data. Mutation data reported. When heated to decomposition it emits very toxic fumes of F^-, NO_x, and SO_x.

DLN200 CAS:52831-55-7 ***HR: 3***
6,13-DIHYDRO-3-FLUOROBENZO(e)(1)BENZOTHIOPYRANO(4,3-b)INDOLE
mf: $C_{19}H_{12}FNS$ mw: 305.38

TOXICITY DATA with REFERENCE
mma-sat 30 μg/plate MUREAV 66,307,79
scu-mus TDLo:72 mg/kg/9W-I:NEO MUREAV 66,307,79

SAFETY PROFILE: Questionable carcinogen with experimental neoplastigenic data. Mutation data reported. When heated to decomposition it emits very toxic fumes of F^-, NO_x, and SO_x.

DLN400 CAS:52831-67-1 ***HR: 3***
6,13-DIHYDRO-4-FLUOROBENZO(e)(1)BENZOTHIOPYRANO(4,3-b)INDOLE
mf: $C_{19}H_{12}FNS$ mw: 305.38

TOXICITY DATA with REFERENCE
mma-sat 30 μg/plate MUREAV 66,307,79
scu-mus TDLo:72 mg/kg/9W-I:NEO MUREAV 66,307,79

SAFETY PROFILE: Questionable carcinogen with experimental neoplastigenic data. Mutation data reported. When heated to decomposition it emits very toxic fumes of F^-, NO_x, and SO_x.

DLO000 CAS:22298-04-0 ***HR: 3***
6,11-DIHYDRO-2-FLUORO(1)BENZOTHIOPYRANO(4,3-b)INDOLE
mf: $C_{15}H_{10}FNS$ mw: 255.32

SYN: 6,11-DIHYDRO-2-FLUORO-THIOPYRANO(4,3-b)BENZ(e)INDOLE

TOXICITY DATA with REFERENCE
mma-sat 100 μg/plate MUREAV 66,307,79
scu-mus TDLo:72 mg/kg/9W-I:NEO MUREAV 66,307,79

SAFETY PROFILE: Questionable carcinogen with experimental neoplastigenic data. Mutation data reported. When heated to decomposition it emits very toxic fumes of F^-, SO_x, and NO_x.

DLO200 CAS:21243-26-5 ***HR: 3***
6,11-DIHYDRO-4-FLUORO(1)BENZOTHIOPYRANO(4,3-b)INDOLE
mf: $C_{15}H_{10}FNS$ mw: 255.32

TOXICITY DATA with REFERENCE
mma-sat 90 μg/plate MUREAV 66,307,79
scu-mus TDLo:78 mg/kg/9W-I:NEO MUREAV 66,307,79

SAFETY PROFILE: Questionable carcinogen with experimental neoplastigenic data. Mutation data reported. When heated to decomposition it emits very toxic fumes of F^-, SO_x, and NO_x.

DLO400 CAS:18497-13-7 ***HR: 3***
DIHYDROGEN HEXACHLOROPLATINATE HEXAHYDRATE
mf: $Cl_6Pt{\bullet}2H{\bullet}6H_2O$ mw: 521.97

SYN: PLATINATE(2-), HEXACHLORO-, DIHYDROGEN, HEXAHYDRATE

TOXICITY DATA with REFERENCE
ipr-mus LD50:82 mg/kg TXAPA9 49,41,79

OSHA PEL: TWA 0.002 mg(Pt)/m^3
ACGIH TLV: TWA 0.002 mg(Pt)/m^3

SAFETY PROFILE: Poison by intraperitoneal route. When heated to decomposition it emits toxic fumes of Pt and Cl^-.

DLO800 CAS:812-00-0 ***HR: D***
DIHYDROGEN METHYL PHOSPHATE
mf: CH_5O_4P mw: 112.0

SYNS: METHYL PHOSPHATE ◇ o-METHYLPHOSPHATE ◇ MONOMETHYL DIHYDROGEN PHOSPHATE

TOXICITY DATA with REFERENCE
mmo-klp 1000 ppm MUREAV 16,413,72

CONSENSUS REPORTS: Reported in EPA TSCA Inventory.

SAFETY PROFILE: Mutation data reported. When heated to decomposition it emits toxic fumes of PO_x.

DLO875 CAS:34257-95-9 ***HR: 3***
DIHYDROHELENALIN
mf: $C_{15}H_{20}O_4$ mw: 264.35

SYNS: 11,13-DIHYDROHELENALIN ◇ PLENOLIN

TOXICITY DATA with REFERENCE
dni-mus:ast 2143 μmol/L JPMSAE 67,1235,78
orl-mus LD50:123 mg/kg PLMEAA 45,131,82
ipr-mus LD50:31 mg/kg PLMEAA 45,131,82

SAFETY PROFILE: Poison by ingestion and intraperitoneal routes. Mutation data reported.

DLO880 CAS:14168-01-5 ***HR: 2***
β-DIHYDROHEPTACHLOR
mf: $C_{10}H_7Cl_7$ mw: 375.32

SYNS: β-DHC ◇ DILOR ◇ GL 2487

TOXICITY DATA with REFERENCE
orl-rat LD50:2000 mg/kg VETNAL 52(7),100,76
orl-mus LD50:1890 mg/kg VETNAL 52(7),100,76
orl-ckn LD50:2000 mg/kg VETNAL 52(7),100,76

SAFETY PROFILE: Moderately toxic by ingestion. When heated to decomposition it emits toxic fumes of Cl^-. See also HEPTACHLOR.

DLO950 CAS:83053-63-8 ***HR: 2***
15,16-DIHYDRO-11-HYDROXYCYCLOPENTA(a)PHENANTHREN-17-ONE
mf: $C_{17}H_{12}O_2$ mw: 248.29

TOXICITY DATA with REFERENCE
skn-mus TDLo:400 mg/kg:ETA CRNGDP 3,677,82

SAFETY PROFILE: Questionable carcinogen with experimental tumorigenic data. When heated to decomposition it emits acrid smoke and irritating fumes.

DLP000 CAS:63918-74-1 ***HR: 3***
6,7-DIHYDRO-6-(2-HYDROXYETHYL)-5H-DIBENZ(c,e)AZEPINE
mf: $C_{16}H_{17}NO$ mw: 239.34

SYN: RO 2-3599

TOXICITY DATA with REFERENCE
ipr-mus LD50:135 mg/kg JPETAB 103,10,51
ivn-mus LD50:32 mg/kg JPETAB 103,10,51

SAFETY PROFILE: Poison by intravenous and intraperitoneal routes. When heated to decomposition it emits toxic fumes of NO_x.

DLP200 CAS:55651-36-0 ***HR: 3***
15,16-DIHYDRO-11-HYDROXYMETHYL-17H-CYCLOPENTA(a)PHENANTHREN-17-ONE

TOXICITY DATA with REFERENCE
mma-sat 1 μg/plate CNREA8 40,882,80
skn-mus TDLo:16 mg/kg:ETA CNREA8 40,882,80

SAFETY PROFILE: Questionable carcinogen with experimental tumorigenic data. Mutation data reported. When heated to decomposition it emits acrid smoke and fumes.

DLP400 CAS:55651-31-5 ***HR: 3***
15,16-DIHYDRO-15-HYDROXY-11-METHYL-17H-CYCLOPENTA(a)PHENANTHREN-17-ONE

TOXICITY DATA with REFERENCE
mma-sat 1 μg/plate CNREA8 40,882,80
skn-mus TDLo:16 mg/kg:ETA CNREA8 40,882,80

SAFETY PROFILE: Questionable carcinogen with experimental tumorigenic data. Mutation data reported. When heated to decomposition it emits acrid smoke and fumes.

DLP600 CAS:24684-56-8 ***HR: 3***
15,16-DIHYDRO-16-HYDROXY-11-METHYL-CYCLOPENTA(a)PHENANTHREN-17-ONE
mf: $C_{18}H_{14}O_2$ mw: 262.32

SYN: 15,16-DIHYDRO-16-HYDROXY-11-METHYL-17H-CYCLOPENTA(a)PHENANTHREN-17-ONE

TOXICITY DATA with REFERENCE
mma-sat 1 μg/plate CNREA8 40,882,80
skn-mus TDLo:115 mg/kg/48W-I:NEO CNREA8 33,832,73

CONSENSUS REPORTS: EPA Genetic Toxicology Program.

SAFETY PROFILE: Questionable carcinogen with experimental neoplastigenic data. Mutation data reported. When heated to decomposition it emits acrid smoke and irritating fumes.

DLP800 CAS:31499-72-6 ***HR: 1***
DIHYDRO-α-IONONE
mf: $C_{13}H_{22}O$ mw: 194.35

SYN: 4-(2,6,6-TRIMETHYL-2-CYCLOHEXEN-1-YL)-2-BUTANONE

TOXICITY DATA with REFERENCE
skn-rbt 500 mg/24H MOD FCTXAV 16,711,78

CONSENSUS REPORTS: Reported in EPA TSCA Inventory.

SAFETY PROFILE: A skin irritant. When heated to decomposition it emits acrid smoke and irritating fumes. See also KETONES.

DLQ000 CAS:6414-38-6 ***HR: 3***
DIHYDROISOCODEINE ACID TARTRATE
mf: $C_{18}H_{23}NO_3 \cdot C_4H_6O_6$ mw: 451.52

SYNS: DIHYDROISOCODEINE TARTRATE ◇ DIHYDRO ISOCODEINE TARTRATE (1:1)

TOXICITY DATA with REFERENCE
scu-mus LDLo:1500 mg/kg JPETAB 51,35,34
scu-rbt LDLo:196 mg/kg JPETAB 66,182,39

SAFETY PROFILE: Poison by subcutaneous route. See also CODEINE. When heated to decomposition it emits toxic fumes of NO_x.

DLQ400 CAS:37795-69-0 ***HR: 3***
2,3-DIHYDRO-9H-ISOXAZOLO(3,2-b)QUINAZOLIN-9-ONE
mf: $C_{10}H_8N_2O_2$ mw: 188.20

SYN: W-2429

TOXICITY DATA with REFERENCE
orl-rat LD50:395 mg/kg ARZNAD 27,770,77
ivn-rat LD50:278 mg/kg ARZNAD 27,770,77
orl-mus LD50:420 mg/kg ARZNAD 27,770,77
ivn-mus LD50:212 mg/kg ARZNAD 27,770,77
orl-dog LD50:700 mg/kg ARZNAD 27,793,77

SAFETY PROFILE: Poison by ingestion and intravenous routes. When heated to decomposition it emits toxic fumes of NO_x. See also KETONES.

DLQ600 CAS:1128-08-1 ***HR: 1***
DIHYDROJASMONE
mf: $C_{11}H_{18}O$ mw: 166.29

SYNS: 3-METHYL-2-n-PENTANYL-2-CYCLOPENTEN-1-ONE ◇ 3-METHYL-2-PENTYL-2-CYCLOPENTEN-1-ONE ◇ 2-PENTYL-3-METHYL-2-CYCLOPENTEN-1-ONE

TOXICITY DATA with REFERENCE
skn-rbt 500 mg/24H FCTXAV 12,517,74

CONSENSUS REPORTS: Reported in EPA TSCA Inventory.

SAFETY PROFILE: A skin irritant. When heated to decomposition it emits acrid smoke and irritating fumes. See also KETONES.

DLQ800 CAS:21842-58-0 ***HR: 3***
12,β,13,α-DIHYDROJERVINE
mf: $C_{26}H_{39}NO_3$ mw: 413.66

TOXICITY DATA with REFERENCE
orl-ham TDLo:113 mg/kg (7D preg):REP 41CIAR-,409,78

orl-ham TDLo:75 mg/kg (7D preg):TER JAFCAU 26,561,78
orl-ham LDLo:75 mg/kg JAFCAU 26,561,78

SAFETY PROFILE: Poison by ingestion. Experimental teratogenic and reproductive effects. When heated to decomposition it emits toxic fumes of NO_x.

DLR000 CAS:587-63-3 ***HR: 3***
DIHYDROKAVAIN
mf: $C_{14}H_{16}O_3$ mw: 232.30

PROP: Extracted from the roots of *Piper methysticum forst* AIPTAK 138,505,62).

SYNS: 7,8-DIHYDROKAWAIN ◇ 5,6-DIHYDRO-4-METHOXY-6-PHENETHYL-2H-PYRAN-2-ONE ◇ (S)-5,6-DIHYDRO-4-METHOXY-6-(2-PHENYLETHYL)-2H-PYRAN-2-ONE (9CI) ◇ DHK ◇ MARINDININ

TOXICITY DATA with REFERENCE
orl-mus LD50:920 mg/kg AIPTAK 138,505,62
ipr-mus LD50:325 mg/kg AIPTAK 138,505,62
ivn-mus LD50:53 mg/kg AIPTAK 177,261,69
ipr-rbt LD50:350 mg/kg AIPTAK 138,505,62

SAFETY PROFILE: Poison by intravenous and intraperitoneal routes. Moderately toxic by ingestion. When heated to decomposition it emits acrid smoke and irritating fumes.

DLR100 CAS:37686-84-3 ***HR: 3***
9,10-α-DIHYDROLISURIDE
mf: $C_{20}H_{28}N_4O$ mw: 340.52

SYNS: trans-DIHYDROLISURIDE ◇ 1-((5R,8S,10R)-6-METHYL-8-ERGOLINYL)-3,3-DIETHYLUREA ◇ TERGURID ◇ TERGURIDE ◇ UREA, N,N-DIETHYL-N'-((8-α)-6-METHYLERGOLIN-8-YL)-

TOXICITY DATA with REFERENCE
orl-wmn TDLo:30 μg/kg (lactating female 3D post):REP GOBIDS 26,33,88
ivn-mus LD50:75 mg/kg CCCCAK 52,2983,87

SAFETY PROFILE: Poison by intravenous route. Experimental reproductive effects. When heated to decomposition it emits toxic fumes of NO_x.

DLR150 CAS:37686-85-4 ***HR: 3***
trans-9,10-DIHYDROLISURIDE HYDROGEN MALEATE
mf: $C_{20}H_{28}N_4O{\bullet}C_4H_4O_4$ mw: 456.60

SYNS: N,N-DIETHYL-N'-((8-α)-6-METHYLERGOLIN-8-YL)UREA (Z)-2-BUTENEDIOATE ◇ DIRONYL ◇ 1-((5R,8S,10R)-6-METHYL-8-ERGOLINYL)-3,3-DIETHYLUREA HYDROGEN MALEATE ◇ TERGURIDE HYDROGEN MALEATE ◇ VUFB 6638 ◇ UREA, N,N-DIETHYL-N'-((8-α)-6-METHYLERGOLIN-8-YL)-, (Z)-2-BUTENEDIOATE (1:1)

TOXICITY DATA with REFERENCE
orl-rat TDLo:1180 μg/kg (lactating female 4D post):REP CCCCAK 49,2828,84
ivn-mus LD50:100 mg/kg CCCCAK 49,2828,84

SAFETY PROFILE: Poison by intravenous route. Experimental reproductive effects. When heated to decomposition it emits toxic fumes of NO_x.

DLR200 CAS:5836-85-1 ***HR: 3***
15,16-DIHYDRO-11-METHOXYCYCLOPENTA(a) PHENANTHREN-17-ONE
mf: $C_{18}H_{14}O_2$ mw: 262.32

SYN: 11-METHOXY-15,16-DIHDYROCYCLOPENTA(a)PHENANTHREN-17-ONE

TOXICITY DATA with REFERENCE
skn-mus TDLo:108 mg/kg/1Y-I:CAR PEXTAR 11,69,69
skn-mus TD:1600 μg/kg:ETA CRNGDP 3,677,82

SAFETY PROFILE: Questionable carcinogen with experimental carcinogenic and tumorigenic data. When heated to decomposition it emits acrid smoke and irritating fumes. See also KETONES.

DLR600 CAS:30835-61-1 ***HR: 3***
15,16-DIHYDRO-11-METHOXY-7-METHYL-CYCLOPENTA(a)PHENANTHREN-17-ONE
mf: $C_{19}H_{16}O_2$ mw: 276.35

SYN: 15,16-DIHYDRO-11-METHOXY-7-METHYL-17H-CYCLOPENTA(a)PHENANTHREN-17-ONE

TOXICITY DATA with REFERENCE
mma-sat 20 μg/plate CNREA8 36,4525,76
skn-mus TDLo:96 mg/kg/40W-I:NEO CNREA8 33,832,73

CONSENSUS REPORTS: EPA Genetic Toxicology Program.

SAFETY PROFILE: Questionable carcinogen with experimental neoplastigenic data. Mutation data reported. When heated to decomposition it emits acrid smoke and irritating fumes. See also KETONES.

DLS000 CAS:70301-64-3 ***HR: 3***
10,11-DIHYDRO-11-(p-METHOXYPHENYL)-2-(4-METHYL-1-PIPERAZINYL)PYRIDAZINO(3,4-b)(1,4)BENZOXAZEPINE
mf: $C_{23}H_{25}N_5O_2$ mw: 403.53

TOXICITY DATA with REFERENCE
scu-rat LD50:255 mg/kg PCJOAU 13,256,79
scu-mus LD50:255 mg/kg PCJOAU 13,256,79

SAFETY PROFILE: Poison by subcutaneous route. When heated to decomposition it emits toxic fumes of NO_x.

DLS275 CAS:16879-01-9 ***HR: D***
3,4-DIHYDRO-7-METHOXY-3-PHENYL-4-(4-(2-(1-PYRROLIDINYL)ETHOXY)PHENYL)-2H-1-BENZOTHIOPYRAN-4-OL HYDROCHLORIDE
mf: $C_{28}H_{31}NO_3S{\bullet}ClH$ mw: 498.12

TOXICITY DATA with REFERENCE
orl-mus TDLo:90 mg/kg (female 4D pre):REP JMCMAR 14,1185,71

SAFETY PROFILE: Experimental reproductive effects. When heated to decomposition it emits toxic fumes of SO_x, NO_x, and HCl.

DLS600 CAS:58-28-6 ***HR: 3***
10,11-DIHYDRO-5-(3-(METHYLAMINO)PROPYL)-5H-DIBENZ(b,f)AZEPINE HYDROCHLORIDE
mf: $C_{18}H_{22}N_2{\cdot}ClH$ mw: 302.88

SYNS: DESIPRAMINE HYDROCHLORIDE ◇ DESMETHYLIMIPRAMINE HYDROCHLORIDE ◇ DIMETHYLIMIPRAMINE HYDROCHLORIDE ◇ DMI HYDROCHLORIDE ◇ EX 4355 ◇ G 35020 ◇ GMI ◇ IMIPRAMINEDEMETHYL HYDROCHLORIDE ◇ IRENE ◇ JB 8181 ◇ N-(Γ-METHYLAMINOPROPYL)IMINODIBENZYL HYDROCHLORIDE ◇ NORPRAMIN ◇ NORTIMIL ◇ NSC-114901 ◇ PERTOFRAN ◇ PERTOFRANE ◇ RMI9,384A

TOXICITY DATA with REFERENCE
cyt-oin-unr 10 g/L JCLBA3 47,182a,70
scu-rat TDLo:130 mg/kg (female 10-22D post):REP NETOD7 7,493,85
scu-mus TDLo:90 mg/kg (female 9D post):TER DGDFA5 22,61,80
orl-man TDLo:5 mg/kg/5D-I AJPSAO 142,386,85
orl-wmn TDLo:70 mg/kg/4W-I AJPSAO 142,386,85
orl-man TDLo:490 μg/kg JCLPDE 47,210,86
orl-wmn TDLo:14 mg/kg/1W-I:KID,SYS JCLPDE 44,153,83
orl-rat LD50:55 mg/kg WMWOA4 112,558,62
ipr-rat LD50:55 mg/kg 27ZQAG -,70,72
scu-rat LD50:250 mg/kg WMWOA4 112,558,62
ivn-rat LD50:19 mg/kg 27ZQAG -,70,72
orl-mus LD50:315 mg/kg 27ZQAG -,70,72
ipr-mus LD50:88 mg/kg 27ZQAG -,70,72
scu-mus LD50:180 mg/kg WMWOA4 112,558,62

SAFETY PROFILE: Poison by ingestion, intraperitoneal, subcutaneous, and intravenous routes. Human systemic effects by ingestion: decreased urine volume, sodium level changes, chlorine level changes. An experimental teratogen. Other experimental reproductive effects. Mutation data reported. When heated to decomposition it emits very toxic fumes of NO_x and HCl. See also 10,11-DIHYDRO-5-(3-METHYLAMINOPROPYL)-5H-DIBENZ(b,f)AZEPINE HYDROCHLORIDE and DIAZEPAM.

DLS800 CAS:1563-67-3 ***HR: 3***
2,3-DIHYDRO-2-METHYLBENZOPYRANYL-7,N-METHYLCARBAMATE
mf: $C_{11}H_{13}NO_3$ mw: 207.25

SYNS: A 468 ◇ BAY 48130 ◇ BAY 62863 ◇ BAYER 62863 ◇ C 1120 ◇ DECARBOFURAN ◇ ENT 27,324

TOXICITY DATA with REFERENCE
orl-rat LD50:43 mg/kg JAFCAU 20,923,72
orl-gpg LDLo:25 mg/kg JEENAI 61,1261,68
scu-gpg LDLo:25 mg/kg JEENAI 61,1261,68

SAFETY PROFILE: Poison by ingestion and subcutaneous routes. When heated to decomposition it emits toxic fumes of NO_x. See also CARBAMATES.

DLT000 CAS:7499-32-3 ***HR: 3***
9,10-DIHYDRO-7-METHYLBENZO(a)PYRENE
mf: $C_{21}H_{16}$ mw: 268.37

SYN: 1′:2′-DIHYDRO-4′-METHYL-3:4-BENZPYRENE

TOXICITY DATA with REFERENCE
imp-mus TDLo:520 mg/kg/10W-I:ETA AJCAA7 36,211,39

SAFETY PROFILE: Questionable carcinogen with experimental tumorigenic data. When heated to decomposition it emits acrid smoke and irritating fumes.

DLT200 CAS:63041-50-9 ***HR: 3***
meso-DIHYDRO-3-METHYLCHOLANTHRENE
mf: $C_{21}H_{18}$ mw: 270.39

SYN: 6,12b-DIHYDRO-3-METHYLCHOLANTHRENE

TOXICITY DATA with REFERENCE
scu-mus TDLo:40 mg/kg:ETA CNREA8 1,695,41
scu-mus TDLo:200 mg/kg CNREA8 1,695,41

SAFETY PROFILE: Questionable carcinogen with experimental tumorigenic data. When heated to decomposition it emits acrid smoke and irritating fumes.

DLT400 CAS:25486-92-4 ***HR: 3***
11,12-DIHYDRO-3-METHYLCHOLANTHRENE
mf: $C_{21}H_{18}$ mw: 270.39

TOXICITY DATA with REFERENCE
mma-sat 1 μg/plate MUREAV 51,311,78
skn-mus TDLo:168 mg/kg/57W-I:ETA JNCIAM 44,641,70

SAFETY PROFILE: Questionable carcinogen with experimental tumorigenic data. Mutation data reported. When heated to decomposition it emits acrid smoke and irritating fumes.

DLT600 ***HR: 3***
9,10-DIHYDRO-3-METHYLCHOLANTHRENE-1,9,10-TRIOL
mf: $C_{21}H_{18}O_3$ mw: 318.39

SYNS: (E)-1-HYDROXY-MC-9,10-DIHYDRODIOL ◇ (E)-1-HYDROXY-3-METHYLCHOLANTHRENE9,10-DIHYDRODIOL

TOXICITY DATA with REFERENCE
skn-mus TDLo:127 μg/kg:NEO CNREA8 39,3549,79

SAFETY PROFILE: Questionable carcinogen with experimental neoplastigenic data. When heated to decomposition it emits acrid smoke and irritating fumes.

DLT800 CAS:40951-13-1 ***HR: 3***
15,16-DIHYDRO-11-METHYL-17H-CYCLOPENTA(a)PHENANTHREN-17-OL
mf: $C_{18}H_{16}O$ mw: 248.34

TOXICITY DATA with REFERENCE
skn-mus TDLo:91 mg/kg/38W-I:NEO CNREA8 33,832,73

SAFETY PROFILE: Questionable carcinogen with experimental neoplastigenic data. When heated to decomposition it emits acrid smoke and irritating fumes.

DLU200 CAS:24684-42-2 ***HR: 3***
16,17-DIHYDRO-11-METHYLCYCLOPENTA(a)PHENANTHREN-15-ONE
mf: $C_{18}H_{14}O$ mw: 246.32

TOXICITY DATA with REFERENCE
mma-sat 50 μg/plate CNREA8 36,4525,76
skn-mus TDLo:120 mg/kg/50W-I:ETA CNREA8 33,832,73

CONSENSUS REPORTS: EPA Genetic Toxicology Program.

SAFETY PROFILE: Questionable carcinogen with experimental tumorigenic data. Mutation data reported. When heated to decomposition it emits acrid smoke and irritating fumes. See also KETONES.

DLU400 CAS:30835-65-5 ***HR: 3***
15,16-DIHYDRO-7-METHYLCYCLOPENTA(a)PHENANTHREN-17-ONE
mf: $C_{18}H_{14}O$ mw: 246.32

SYN: 15,16-DIHYDRO-7-METHYL-17H-CYCLOPENTA(a)PHENANTHREN-17-ONE

TOXICITY DATA with REFERENCE
skn-mus TDLo:72 mg/kg/30W-I:ETA CNREA8 33,832,73

CONSENSUS REPORTS: EPA Genetic Toxicology Program.

SAFETY PROFILE: Questionable carcinogen with experimental tumorigenic data. When heated to decomposition it emits acrid smoke and irritating fumes. See also KETONES.

DLU600 CAS:5837-17-2 ***HR: 3***
16,17-DIHYDRO-17-METHYLENE-15H-CYCLOPENTA(a)PHENANTHRENE
mf: $C_{18}H_{14}$ mw: 230.32

TOXICITY DATA with REFERENCE
skn-mus TDLo:125 mg/kg/52W-I:ETA NATUAS 210,1281,66

SAFETY PROFILE: Questionable carcinogen with experimental tumorigenic data. When heated to decomposition it emits acrid smoke and irritating fumes.

DLU700 CAS:83053-62-7 ***HR: 2***
15,16-DIHYDRO-11-METHYL-15-METHOXYCYCLOPENTA(a)PHENANTHREN-17-ONE
mf: $C_{19}H_{16}O_2$ mw: 276.35

TOXICITY DATA with REFERENCE
skn-mus TDLo:1600 μg/kg:ETA CRNGDP 3,677,82

SAFETY PROFILE: Questionable carcinogen with experimental tumorigenic data. When heated to decomposition it emits acrid smoke and irritating fumes.

DLU800 CAS:29676-95-7 ***HR: 3***
1,4-DIHYDRO-1-METHYL-7-(2-(5-NITRO-2-FURYL)VINYL)-4-OXO-1,8-NAPHTHYRIDINE-3-CARBOXYLIC ACID, POTASSIUM SALT
mf: $C_{16}H_{10}N_3O_6 \cdot K$ mw: 379.39

SYN: NFN

TOXICITY DATA with REFERENCE
orl-mus TDLo:1411 mg/kg/14W-C:CAR JJIND8 69,1317,82
orl-mus TDLo:3150 mg/kg/25W-C:CAR CIZAAZ 50,249,74

SAFETY PROFILE: Questionable carcinogen with experimental carcinogenic data. When heated to decomposition it emits toxic fumes of NO_x and K_2O.

DLU900 CAS:27016-91-7 ***HR: 3***
2,3-DIHYDRO-N-METHYL-7-NITRO-2-OXO-5-PHENYL-1H-1,4-BENZODIAZEPINE-1-CARBOXAMIDE
mf: $C_{17}H_{14}N_4O_4$ mw: 338.35

SYN: D 58SI

TOXICITY DATA with REFERENCE
orl-rat TDLo:32500 mg/kg (25D pre):REP TAKHAA 30,85,71
orl-rat LD50:2280 mg/kg TAKHAA 29,153,70
ipr-rat LD50:345 mg/kg TAKHAA 29,153,70
orl-mus LD50:1230 mg/kg TAKHAA 29,134,70
ipr-mus LD50:370 mg/kg TAKHAA 29,134,70

SAFETY PROFILE: Poison by intraperitoneal route. Moderately toxic by ingestion. Experimental reproductive effects. When heated to decomposition it emits toxic fumes of NO_x.

DLV000 CAS:2011-67-8 ***HR: 2***
1,3-DIHYDRO-1-METHYL-7-NITRO-5-PHENYL-2H-1,4-BENZODIAZEPIN-2-ONE
mf: $C_{16}H_{13}N_3O_3$ mw: 295.32

SYNS: ELIMIN ◇ HYPNON ◇ 1-METHYLNITRAZEPAM ◇ 1-METHYL-7-NITRO-5-PHENYL-1,3-DIHYDRO-2H-1,4-BENZODIAZEPIN-2-ONE ◇ 1-METHYL-5-PHENYL-7-NITRO-1,3-DIHYDRO-2H-1,4-BENZODIAZEPIN-2-ONE ◇ NIMETAZEPAM ◇ S 1530

TOXICITY DATA with REFERENCE
orl-rat TDLo:700 mg/kg (8-14D preg):REP RCOCB8 46,437,84
orl-rat LD50:970 mg/kg ARZNAD 22,534,72
ipr-rat LD50:970 mg/kg ARZNAD 22,534,72
orl-mus LD50:750 mg/kg ARZNAD 22,534,72
ipr-mus LD50:510 mg/kg OYYAA2 7,705,73

SAFETY PROFILE: Moderately toxic by ingestion and intraperitoneal routes. Experimental reproductive effects. An anticonvulsant and muscle relaxant. Related to diazepam. When heated to decomposition it emits toxic fumes of NO_x.

DLV200 CAS:5259-88-1 ***HR: 2***
5,6-DIHYDRO-2-METHYL-1,4-OXATHIIN-3-CARBOXANILIDE-4,4-DIOXIDE
mf: $C_{12}H_{13}NO_4S$ mw: 267.32

SYNS: DCMOD ◇ 2,3-DIHYDRO-5-CARBOXANILIDO-6-METHYL-1,4-OXATHIIN-4,4-DIOXIDE ◇ 5,6-DIHYDRO-2-METHYL-3-CARBOXANILIDO-1,4-OXATHIIN-4,4-DIOXID (GERMAN) ◇ 5,6-DIHYDRO-2-METHYL-N-PHENYL-1,4-OXATHIIN-3-CARBOXAMIDE-4,4-DIOXIDE ◇ DIOXIDE of VITAVAX ◇ F461 ◇ OXYCARBOXIN ◇ OXYCARBOXINE ◇ PLANTVAX ◇ PLANT WAX ◇ VITAVEX

TOXICITY DATA with REFERENCE
mrc-asn 8000 ppm ENMUDM 2,359,80
orl-rat LD50:2000 mg/kg WRPCA2 9,119,70

SAFETY PROFILE: Moderately toxic by ingestion. Mutation data reported. A pesticide. When heated to decomposition it emits very toxic fumes of NO_x and SO_x.

DLV400 CAS:70301-54-1 ***HR: 3***
10,11-DIHYDRO-2-(4-METHYL-1-PIPERAZINYL)-11-(2-ATHIAZOLYL)-PYRIDAZINO(3,4-b)(1,4)BENZOXAZEPINE
mf: $C_{19}H_{20}N_6OS$ mw: 380.51

TOXICITY DATA with REFERENCE
scu-rat LD50:217 mg/kg PCJOAU 13,256,79
scu-mus LD50:217 mg/kg PCJOAU 13,256,79

SAFETY PROFILE: Poison by subcutaneous route. When heated to decomposition it emits very toxic fumes of SO_x and NO_x.

DLV600 CAS:70301-68-7 ***HR: 3***
10,11-DIHYDRO-2-(4-METHYL-1-PIPERAZINYL)-11-(3,4-XYLYL)PYRIDAZINO(3,4-b)(1,4)BENZOXAZEPINE MALEATE
mf: $C_{24}H_{27}N_5O \cdot C_4H_4O_4$ mw: 517.64

TOXICITY DATA with REFERENCE
scu-rat LD50:155 mg/kg PCJOAU 13,256,79
scu-mus LD50:155 mg/kg PCJOAU 13,256,79

SAFETY PROFILE: Poison by subcutaneous route. When heated to decomposition it emits toxic fumes of NO_x.

DLV800 CAS:3978-86-7 ***HR: 3***
6,11-DIHYDRO-11-(1-METHYL-4-PIPERIDYLIDENE)-5H-BENZO(5,6)CYCLOHEPTA (1,2-b) PYRIDINE DIMALEATE
mf: $C_{20}H_{22}N_2 \cdot 2C_4H_4O_4$ mw: 522.60

SYNS: AZATADINE DIMALEATE ◇ AZATADINE MELEATE ◇ IDULIAN ◇ OPTIMINE ◇ SCH 10649 ◇ ZADINE

TOXICITY DATA with REFERENCE
orl-rat LD50:440 mg/kg TXAPA9 18,185,71
ipr-rat LD50:166 mg/kg JZKEDZ 1,173,75
scu-rat LD50:1000 mg/kg JZKEDZ 1,173,75
orl-mus LD50:184 mg/kg JZKEDZ 1,173,75
ipr-mus LD50:105 mg/kg JZKEDZ 1,173,75
scu-mus LD50:178 mg/kg JZKEDZ 1,173,75
scu-mus LDLo:50 mg/kg JZKEDZ 1,79,75

SAFETY PROFILE: Poison by ingestion, subcutaneous, and intraperitoneal routes. An antihistamine. When heated to decomposition it emits toxic fumes of NO_x.

DLW600 CAS:466-99-9 ***HR: 3***
DIHYDROMORPHINONE
mf: $C_{17}H_{19}NO_3$ mw: 285.37

SYNS: DIMO ◇ HYDROMORPHONE ◇ HYMORPHAN ◇ LAUDICON ◇ PARAMORPHAN

TOXICITY DATA with REFERENCE
orl-hmn LDLo:1428 μg/kg 34ZIAG -,223,69
scu-mus LD50:84 mg/kg JPETAB 52,468,34
ivn-mus LD50:104 mg/kg YKKZAJ 84,268,64

SAFETY PROFILE: A deadly human poison by ingestion. An experimental poison by ingestion and subcutaneous route. When heated to decomposition it emits toxic fumes of NO_x. See also (−)MORPHINE.

DLX000 CAS:18479-58-8 ***HR: 2***
DIHYDROMYRCENOL
mf: $C_{10}H_{20}O$ mw: 156.30

SYN: 2,6-DIMETHYL-7-OCTEN-2-OL

TOXICITY DATA with REFERENCE
skn-rbt 500 mg/24H MLD FCTXAV 12,525,74
orl-rat LD50:3600 mg/kg FCTXAV 12,525,74

CONSENSUS REPORTS: Reported in EPA TSCA Inventory.

SAFETY PROFILE: Moderately toxic by ingestion. A skin irritant. When heated to decomposition it emits acrid smoke and irritating fumes.

DLX100 CAS:88969-41-9 ***HR: 2***
DIHYDROMYRCENYL ACETATE
mf: $C_{12}H_{22}O_2$ mw: 198.34

SYNS: 3-METHYLENE-7-METHYLOCTAN-7-YLACETATE ◇ 2-METHYL-6-METHYLENE-2-OCTANOL ACETATE (ESTER)

TOXICITY DATA with REFERENCE
skn-rbt 500 mg/24H MLD FCTOD7 21,847,83
eye-rbt 10% MLD FCTOD7 21,847,83
skn-rbt LD50:2800 mg/kg FCTOD7 21,847,83

SAFETY PROFILE: Moderately toxic by skin contact. A skin and eye irritant. When heated to decomposition it emits acrid smoke and fumes.

DLX200 CAS:529-34-0 ***HR: 2***
3,4-DIHYDRO-1(2H)-NAPHTHALENONE
mf: $C_{10}H_{10}O$ mw: 146.20

SYNS: α-TETRALONE ◇ 1-TETRALONE

TOXICITY DATA with REFERENCE
orl-rat LD50:810 mg/kg AIHAAP 30,470,69

CONSENSUS REPORTS: Reported in EPA TSCA Inventory.

SAFETY PROFILE: Moderately toxic by ingestion. When heated to decomposition it emits acrid smoke and irritating fumes.

DLX300 CAS:31785-60-1 ***HR: 2***
2,3-DIHYDRO-2-(1-NAPHTHYL)-4(1H)-QUINAZOLINONE
mf: $C_{18}H_{14}N_2O$ mw: 274.34

SYNS: 2,3-DIHYDRO-2-(1-NAPHTHALENYL)-4(1H)-QUINAZOLINONE ◇ NSC-145669 ◇ U-29,409

TOXICITY DATA with REFERENCE
orl-rat TDLo:480 mg/kg (8D male):REP PSEBAA 137,532,71
ipr-mus LD50:998 mg/kg NCISP* JAN86

SAFETY PROFILE: Moderately toxic by intraperitoneal route. Experimental reproductive effects. When heated to decomposition it emits toxic fumes of NO_x.

DLX400 CAS:124-90-3 ***HR: 3***
DIHYDRONE HYDROCHLORIDE
mf: $C_{18}H_{21}NO_4{\bullet}ClH$ mw: 351.86

SYNS: DIHYDROOXYCODEINONE HYDROCHLORIDE ◇ DIHYDROXYCODEINONE HYDROCHLORIDE ◇ DINARKON ◇ EUBINE ◇ EUCODAL ◇ EUKODAL ◇ EUTAGEN ◇ 14-HYDROXYDIHYDROCODEINONE HYDROCHLORIDE ◇ OXIKON ◇ OXYCODONE HYDROCHLORIDE ◇ OXYCODON HYDROCHLORIDE ◇ OXYCON ◇ OXYKODAL ◇ OXYKON ◇ PANCODINE ◇ PERCODAN HYDROCHLORIDE ◇ STUPENONE ◇ TECODIN ◇ TECODINE ◇ TEKODIN ◇ THECODIN ◇ THECODINE ◇ THEKODIN

TOXICITY DATA with REFERENCE
scu-mus LDLo:350 μg/kg AEPPAE 194,296,40
ivn-cat LDLo:2500 μg/kg AEPPAE 194,296,40
scu-rbt LDLo:80 mg/kg HBAMAK 4,1289,35
ivn-rbt LDLo:45 mg/kg HBAMAK 4,1289,35
scu-frg LDLo:500 mg/kg HBAMAK 4,1289,35

SAFETY PROFILE: Poison by intravenous and subcutaneous routes. When heated to decomposition it emits very toxic fumes of NO_x and HCl. See also (-)MORPHINE.

DLX800 CAS:17247-77-7 ***HR: 3***
1,2-DIHYDRO-2-(5'-NITROFURYL)-4-HYDROXYQUINAZOLINE-3-OXIDE
mf: $C_{12}H_9N_3O_5$ mw: 275.24

SYN: 1,2-DIHYDRO-2-(5'-NITROFURYL)-4-HYDROXY-CHINAZOLIN-3-OXID (GERMAN)

TOXICITY DATA with REFERENCE
orl-rat TDLo:40 g/kg/26W-C:ETA ZKKOBW 79,165,73

SAFETY PROFILE: Questionable carcinogen with experimental tumorigenic data. When heated to decomposition it emits toxic fumes of NO_x.

DLY000 CAS:146-22-5 ***HR: 3***
1,3-DIHYDRO-7-NITRO-5-PHENYL-2H-1,4-BENZODIAZEPIN-2-ONE
mf: $C_{15}H_{11}N_3O_3$ mw: 281.29

SYNS: BENZALIN ◇ CALSMIN ◇ EATAN ◇ EPIBENZALIN ◇ EPINELBON ◇ EUNOCTIN ◇ HIPNAX ◇ HIPSAL ◇ LA 1 ◇ MOGADAN ◇ NELBON ◇ NEOZEPAM ◇ NEUCHLONIC ◇ NITRADOS ◇ NITRAZEPAM ◇ NITRENPAX ◇ 7-NITRO-5-PHENYL-2,3-DIHYDRO-1H-1,4-BENZODIAZEPIN-2-ONE◇ NSC-58775 ◇ PAXISYN ◇ PELSON ◇ RADEDORM ◇ RELACT ◇ RO 4-5360 ◇ RO 5-3059 ◇ SOMNASED ◇ SOMNIBEL ◇ SOMNITE ◇ SONEBON ◇ SONNOLIN ◇ SUREM ◇ UNISOMNIA

TOXICITY DATA with REFERENCE
sln-dmg-orl 2 mg/9D SOGEBZ 11,718,75
spm-mus-orl 300 mg/kg/15D-C CYTBAI 36,45,83
orl-mus TDLo:126 mg/kg (female 1-21D post):REP PLRCAT 9,325,77
orl-rat LD50:825 mg/kg TXAPA9 18,185,71
ipr-rat LD50:733 mg/kg JMCMAR 20,952,77
orl-mus LD50:550 mg/kg VINIT* #3206-79

ipr-mus LD50:275 mg/kg 27ZQAG -,165,72
ivn-mus LD50:130 mg/kg CSLNX* NX#01434
ivn-rbt LD50:520 mg/kg 27ZQAG -,165,72

CONSENSUS REPORTS: EPA Genetic Toxicology Program.

SAFETY PROFILE: Poison by intraperitoneal and intravenous routes. Moderately toxic by ingestion. Experimental reproductive effects. Mutation data reported. An anticonvulsant and hypnotic agent. When heated to decomposition it emits toxic fumes of NO_x. See also DIAZEPAM.

DLY200 CAS:33389-33-2 ***HR: 3***
1,2-DIHYDRO-2-(5-NITRO-2-THIENYL) QUINAZOLIN-4(3H)-ONE
mf: $C_{12}H_9N_3O_3S$ mw: 275.30

SYN: 1,2-DIHYDRO-2-(5-NITRO-2-THIENYL)-4(3H)-QUINAZOLINONE

TOXICITY DATA with REFERENCE
orl-rat TDLo:13 g/kg/49W-C:CAR JNCIAM 57,277,76

CONSENSUS REPORTS: EPA Genetic Toxicology Program.

SAFETY PROFILE: Questionable carcinogen with experimental carcinogenic data. When heated to decomposition it emits very toxic fumes of SO_x and NO_x.

DLY400 CAS:5413-60-5 ***HR: 1***
DIHYDRONORDICYCLOPENTADIENYL ACETATE
mf: $C_{12}H_{16}O_2$ mw: 192.28

SYNS: 3a,4,5,6,7,7a-HEXAHYDRO-4,7-METHANO-1H-INDEN-6-OL ACETATE ◇ TRICYCLODECEN-4-YL-8-ACETATE ◇ VERDYL ACETATE

TOXICITY DATA with REFERENCE
skn-rbt 500 mg/24H MOD FCTXAV 14,889,76

CONSENSUS REPORTS: Reported in EPA TSCA Inventory.

SAFETY PROFILE: A skin irritant. When heated to decomposition it emits acrid smoke and irritating fumes.

DLY700 CAS:3686-43-9 ***HR: 3***
3,6-DIHYDRO-1,2,2H-OXAZINE
mf: C_4H_7NO mw: 85.11

$$\overline{ONHCH_2CH{=}CHCH_2}$$

SAFETY PROFILE: Reaction with nitric acid forms an explosive product. When heated to decomposition it emits toxic fumes of NO_x.

DLY800 CAS:7374-66-5 ***HR: 3***
5,13-DIHYDRO-5-OXOBENZO(e)(2)BENZOPYRANO(4,3-b)INDOLE
mf: $C_{19}H_{11}NO_2$ mw: 285.31

SYNS: 5-OXO-5H-BENZO(E)ISOCHROMENO(4,3-b)INDOLE ◇ 5-OXO-5,13-DIHYDROBENZO(E)(2)BENZOPYRANO(4,3-b)INDOLE

TOXICITY DATA with REFERENCE
scu-mus TDLo:72 mg/kg/9W-I:ETA SCIEAS 158,387,67

SAFETY PROFILE: Questionable carcinogen with experimental tumorigenic data. When heated to decomposition it emits toxic fumes of NO_x.

DLZ000 CAS:56179-83-0 ***HR: 3***
1,2-DIHYDROPHENANTHRENE
mf: $C_{14}H_{12}$ mw: 180.26

TOXICITY DATA with REFERENCE
skn-mus TDLo:72 mg/kg:ETA CNREA8 39,4069,79

SAFETY PROFILE: Questionable carcinogen with experimental tumorigenic data. When heated to decomposition it emits acrid smoke and irritating fumes.

DMA000 CAS:28622-66-4 ***HR: 3***
1,2-DIHYDRO-1,2-PHENANTHRENEDIOL
mf: $C_{14}H_{12}O_2$ mw: 212.26

SYN: PHENANTHRENE-1,2-DIHYDRODIOL

TOXICITY DATA with REFERENCE
skn-mus TDLo:85 mg/kg:ETA CNREA8 39,4069,79

SAFETY PROFILE: Questionable carcinogen with experimental tumorigenic data. When heated to decomposition it emits acrid smoke and irritating fumes.

DMA400 CAS:18264-88-5 ***HR: 3***
N-(9,10-DIHYDRO-2-PHENANTHRYL)ACETAMIDE
mf: $C_{16}H_{15}NO$ mw: 237.32

SYN: 2-ACETYLAMINO-9,10-DIHYDROPHENANTHRENE

TOXICITY DATA with REFERENCE
orl-rat TDLo:4608 mg/kg/32W-C:CAR CNREA8 15,188,55

SAFETY PROFILE: Questionable carcinogen with experimental carcinogenic data. When heated to decomposition it emits toxic fumes of NO_x.

DMB000 CAS:21820-82-6 ***HR: 3***
5-(2-(3,6-DIHYDRO-4-PHENYL-1(2H)-PYRIDYL) ETHYL)-3-METHYL-2-OXAZOL IDINONE
mf: $C_{17}H_{22}N_2O_2$ mw: 286.41

SYN: AHR-1680

TOXICITY DATA with REFERENCE
orl-rat LD50:340 mg/kg 27ZQAG -,200,72
ipr-rat LD50:140 mg/kg 27ZQAG -,200,72
ivn-rat LD50:71 mg/kg 27ZQAG -,200,72
orl-mus LD50:349 mg/kg 27ZQAG -,200,72
ipr-mus LD50:180 mg/kg 27ZQAG -,200,72
ivn-mus LD50:91 mg/kg 27ZQAG -,200,72
orl-dog LD50:300 mg/kg 27ZQAG -,200,72
ipr-gpg LD50:189 mg/kg 27ZQAG -,200,72

SAFETY PROFILE: Poison by ingestion, intraperitoneal, and intravenous routes. When heated to decomposition it emits toxic fumes of NO_x.

DMB200 CAS:66731-42-8 ***HR: 3***
2,3-DIHYDROPHORBOL MYRISTATE ACETATE
mf: $C_{36}H_{58}O_8$ mw: 618.94

SYNS: 2,3-DIHYDROPHORBOL ACETATE MYRISTATE ◇ DPMA

TOXICITY DATA with REFERENCE
skn-mus TDLo:37 mg/kg/31W-I:NEO CNREA8 38,921,78

SAFETY PROFILE: Questionable carcinogen with experimental neoplastigenic data. When heated to decomposition it emits acrid smoke and irritating fumes.

DMC000 CAS:68-94-0 ***HR: 2***
1,7-DIHYDRO-6H-PURIN-6-ONE
mf: $C_5H_4N_4O$ mw: 136.13

SYNS: HYPOXANTHINE ◇ 9H-PURIN-6-OL ◇ PURIN-6(3H)-ONE ◇ 6(1H)-PURINONE

TOXICITY DATA with REFERENCE
ipr-mus TDLo:600 mg/kg (female 13D post):TER JJPAAZ 22,201,72
ipr-mus LD50:750 mg/kg NTIS** AD691-490

CONSENSUS REPORTS: Reported in EPA TSCA Inventory.

SAFETY PROFILE: Moderately toxic by intraperitoneal route. An experimental teratogen. When heated to decomposition it emits toxic fumes of NO_x.

DMC200 CAS:110-87-2 ***HR: 3***
DIHYDROPYRAN
DOT: UN 2376
mf: C_5H_8O mw: 84.13

$O(CH_2)_3CH{=}CH$ (ring: O bonded to terminal CH)

PROP: Colorless, mobile liquid; ethereal odor. Bp: 85.6°, flash p: 0°F, d: 0.923 @ 20°/4°, vap d: 2.90.

SYNS: Δ^2-DIHYDROPYRAN ◇ 3,4-DIHYDROPYRAN ◇ 2H-3,4-DIHYDROPYRAN

TOXICITY DATA with REFERENCE
ipr-mus LDLo:256 mg/kg CBCCT* 2,135,50

CONSENSUS REPORTS: Reported in EPA TSCA Inventory.

DOT Classification: Flammable Liquid; Label: Flammable Liquid.

SAFETY PROFILE: Poison by intraperitoneal route. Very dangerous fire hazard when exposed to heat or flame; can react vigorously with oxidizing materials. Keep away from heat and open flame. To fight fire, use alcohol foam, CO_2, or dry chemical. When heated to decomposition it emits acrid smoke and irritating fumes.

DMC600 CAS:123-33-1 ***HR: 3***
1,2-DIHYDROPYRIDAZINE-3,6-DIONE
mf: $C_4H_4N_2O_2$ mw: 112.10

PROP: Crystals. Mp: > 300°. Sol in water and alc.

SYNS: 1,2-DIHYDRO-3,6-PYRIDAZINEDIONE ◇ ENT 18,870 ◇ 6-HYDROXY-3(2H)-PYRIDAZINONE ◇ MALEIC ACID HYDRAZIDE ◇ MALEIC HYDRAZIDE ◇ N,N-MALEOYLHYDRAZINE ◇ 1,2,3,6-TETRAHYDRO-3,6-DIOXOPYRIDAZINE

TOXICITY DATA with REFERENCE
cyt-grh-orl 5 mg CYTOAN 37,345,72
mma-sat 50 μL/plate MUREAV 66,247,79
dns-esc 30 μmol/L ZKKOBW 92,177,78
sln-dmg-orl 4000 ppm MUREAV 55,15,78
sln-dmg-par 4000 ppm NATUAS 207,439,65
cyt-mus-ipr 5000 ppm CISCB7 20,28,76
scu-rat TDLo:2600 mg/kg/65W-I:ETA BJCAAI 19,392,65
orl-rat LD50:3800 mg/kg WRPCA2 9,119,70

CONSENSUS REPORTS: IARC Cancer Review: Group 3 IMEMDT 7,56,87; Animal Inadequate Evidence IMEMDT 4,173,74. Reported in EPA TSCA Inventory.

SAFETY PROFILE: Moderately toxic by ingestion. Questionable carcinogen with experimental tumorigenic data. Mutation data reported. Can cause chronic liver damage and acute central nervous system effects. When heated to decomposition emits highly toxic fumes of NO_x. See also HYDRAZINE.

DMC800 ***HR: 3***
1,2-DIHYDROPYRIDO(2,1,e)TETRAZOLE
mf: $C_5H_4N_4$ mw: 120.12

CH=CHCH=CHNC=NN=N (ring structure)

PROP: Explodes on touching with a hot rod.

SAFETY PROFILE: A friction- and heat-sensitive explosive. When heated to decomposition it emits toxic fumes of NO_x.

DMD000 CAS:480-18-2 ***HR: 2***
2,3-DIHYDROQUERCETIN
mf: $C_{15}H_{12}O_7$ mw: 304.27

SYNS: CATECHIN HYDRATE ◇ DIHYDROQUERCETIN ◇ (+)-DIHYDROQUERCETIN ◇ (2R,3R)-DIHYDROQUERCETIN ◇ 2,3-DIHYDRO-3,3',4',5,7-PENTAHYDROXYFLAVONE ◇ 2-(3,4-DIHYDROXYPHENYL)-2,3-DIHYDRO-3,5,7-TRIHYDROXY-4H-1-BENZOPYRAN-4-ONE ◇ (2R-trans)-2-(3,4-DIHYDROXYPHENYL)-2,3-DIHYDRO-3,5,7-TRIHYDROXY-4H-1-BENZOPYRAN-4-ONE ◇ DISTYLIN ◇ 3,3',4',5,7-PENTAHYDROXYFLAVANONE ◇ TAXIFOLIN ◇ TAXIFOLIOL

TOXICITY DATA with REFERENCE
mmo-sat 100 μg/plate ENMUDM 3,401,81
mma-sat 1660 nmol/plate MUREAV 54,297,78
cyt-ham:fbr 1 g/L/48H MUREAV 48,337,77
ipr-rat LD50:1200 mg/kg JJPAAZ 21,377,71
ipr-mus LD50:985 mg/kg RPTOAN 38,213,75

CONSENSUS REPORTS: Reported in EPA TSCA Inventory.

SAFETY PROFILE: Moderately toxic by intraperitoneal route. Mutation data reported. When heated to decomposition it emits acrid smoke and irritating fumes.

DMD100 CAS:51077-50-0 ***HR: D***
7,8-DIHYDRORETINOIC ACID
mf: $C_{20}H_{30}O_2$ mw: 302.50

SYNS: trans-3,7-DIMETHYL-9-(2,6,6-TRIMETHYL-1-CYCLOHEXEN-1-YL)-2,4,6-NONATRIEN OIC ACID ◇ RETINOIC ACID, 7,8-DIHYDRO-

TOXICITY DATA with REFERENCE
orl-ham TDLo:38 mg/kg (female 8D post):TER TXAPA9 83,563,86

SAFETY PROFILE: An experimental teratogen. When heated to decomposition it emits acrid smoke and irritating fumes.

DMD200 CAS:102338-88-5 ***HR: 3***
4a,5-DIHYDRO-RIBOFLAVIN-5'-PHOSPHATE SODIUM SALT
mf: $C_{17}H_{22}N_4O_6P \cdot xNa$ mw: 570.33

TOXICITY DATA with REFERENCE
scu-mus LD50:375 mg/kg CMTRAG 2,96,61
ivn-mus LD50:420 mg/kg CMTRAG 2,96,61

SAFETY PROFILE: Poison by subcutaneous and intravenous routes. When heated to decomposition it emits very toxic fumes of NO_x, PO_x, and Na_2O.

DMD600 CAS:94-58-6 ***HR: 3***
DIHYDROSAFROLE
mf: $C_{10}H_{12}O_2$ mw: 164.22

PROP: An oily liquid. Bp: 228°, d: 1.0695 @ 20°.

SYNS: 1,2-(METHYLENEDIOXY)-4-PROPYLBENZENE ◇ 5-PROPYL-1,3-BENZODIOXOLE ◇ 4-PROPYL-1,2-METHYLENEDIOXYBENZENE ◇ RCRA WASTE NUMBER U090

TOXICITY DATA with REFERENCE
skn-rbt 500 mg/24H MLD FCTXAV 12,527,74
orl-mus TDLo:101 g/kg/81W-C:CAR FCTXAV 19,130,81
orl-mus TDLo:163 g/kg/81W-C:CAR JNCIAM 42,1101,69
orl-rat LD50:2260 mg/kg TXAPA9 7,18,65
orl-mus LD50:3700 mg/kg TXAPA9 7,18,65
ipr-mus LD50:2830 mg/kg COREAF 250,1148,60

CONSENSUS REPORTS: IARC Cancer Review: Group 2B IMEMDT 7,56,87; Animal Sufficient Evidence IMEMDT 10,231,76; Animal Limited Evidence IMEMDT 1,169,72. Reported in EPA TSCA Inventory. EPA Genetic Toxicology Program.

SAFETY PROFILE: Suspected carcinogen with experimental carcinogenic data. Moderately toxic by ingestion and intraperitoneal routes. A skin irritant. When heated to decomposition it emits acrid smoke and irritating fumes.

DME000 CAS:128-46-1 ***HR: 3***
DIHYDROSTREPTOMYCIN
mf: $C_{21}H_{41}N_7O_{12}$ mw: 583.69

SYNS: DHMS ◇ DST

TOXICITY DATA with REFERENCE
cyt-mus-par 100 mg/kg NULSAK 2,161,71
unr-wmn TDLo:260 mg/kg (19-22W preg):TER SJRDAH 50,61,69
ipr-mus TDLo:12 g/kg (female 1-20D post):REP KJMEA9 10,31,61
ivn-rat LD50:200 mg/kg JOBAAY 53,205,47
ipr-mus LD50:533 mg/kg UPJOH* 2(6),-,71
scu-rat LD50:1100 mg/kg ARZNAD 12,597,62
ivn-mus LD50:200 mg/kg 85GDA2 1,96,80
ipr-mus LD50:533 mg/kg UPJOH* 2(6),-,71
scu-mus LD50:1180 mg/kg ACHTA6 11,2,63
ivn-mus TDLo:200 mg/kg 85GDA2 1,96,80
ims-mus LD50:350 mg/kg AIMDAP 119,493,67

CONSENSUS REPORTS: EPA Genetic Toxicology Program.

SAFETY PROFILE: Poison by intravenous and intramuscular routes. Moderately toxic by subcutaneous and intraperitoneal routes. Human teratogenic effects by unspecified route: developmental abnormalities of the eye and ear. An experimental teratogen. Other experimental reproductive effects. Mutation data reported. A derivative of streptomycin; has anesthetic properties. When heated to decomposition it emits toxic fumes of NO_x.

DME200 CAS:1425-61-2 ***HR: 3***
DIHYDROSTREPTOMYCIN SULFATE
mf: $C_{21}H_{41}N_7O_{12} \cdot 7H_2O_4S$ mw: 1270.25

PROP: White or practically white powder, odorless or slt odor, freely sol in water, very sltly sol in alcohol, practically insol in chloroform.

TOXICITY DATA with REFERENCE
ipr-mus LD50:1380 mg/kg RPOBAR 2,285,70
scu-mus LD50:761 mg/kg ANTBAL 18,444,73
ivn-mus LD50:137 mg/kg ANTBAL 18,444,73

SAFETY PROFILE: Poison by intravenous route. Moderately toxic by intraperitoneal and subcutaneous routes. An additive permitted in the feed and drinking water of animals, and/or for the treatment of food-producing animals. When heated to decomposition it emits toxic fumes of NO_x. See also DIHYDROSTREPTOMYCIN.

DME300 CAS:67-96-9 ***HR: 3***
DIHYDROTACHYSTEROL
mf: $C_{28}H_{46}O$ mw: 398.74

PROP: Needles from 90% methanol. Mp: 125-127°. Insol in water. Easily sol in organic solvents.

SYNS: ANTITANIL ◇ ANTI-TETANY SUBSTANCE 10 ◇ A.T. 10 ◇ CALCAMINE ◇ DHT_2 ◇ DICHYSTROLUM ◇ DIHYDROTACHYSTEROL$_2$ ◇ DYGRATYL ◇ HYTAKEROL ◇ PARTEROL ◇ (E-β,5E,7E,10-α,22E)-9,10-SECOERGOSTA-5,7,22-TRIEN-3-OL(9CI)

TOXICITY DATA with REFERENCE
orl-rat TDLo:150 mg/kg (1-5D preg):REP CUSCAM 41,181,72
orl-mus LD50:288 mg/kg NIIRDN 6,330,82
ipr-mus LD50:104 mg/kg NIIRDN 6,330,82

SAFETY PROFILE: Poison by ingestion and intraperitoneal routes. Experimental reproductive effects. When heated to decomposition it emits acrid smoke and irritating fumes.

DME400 CAS:80-25-1 ***HR: 1***
DIHYDROTERPINYL ACETATE
mf: $C_{12}H_{22}O_2$ mw: 198.34

SYNS: ACETIC ACID DIHYDROTERPINYL ESTER ◇ ACETIC ACID-p-MENTHAN-8-OL ESTER ◇ p-MENTHAN-8-OL ACETATE

TOXICITY DATA with REFERENCE
skn-rbt 500 mg/24H MLD FCTXAV 12,807,74

CONSENSUS REPORTS: Reported in EPA TSCA Inventory.

SAFETY PROFILE: A skin irritant. When heated to decomposition it emits acrid smoke and irritating fumes. See also ESTERS.

DME500 CAS:521-18-6 ***HR: D***
4-DIHYDROTESTOSTERONE
mf: $C_{19}H_{30}O_2$ mw: 290.49

PROP: Crystals from ethyl acetate + hexane. Sublimes @ 0.01 135°, mp: 181°. Sol in acetone, ether, alc, ethyl acetate. Practically insol in water.

SYNS: ANABOLEEN ◇ ANABOLEX ◇ ANAPROTIN ◇ ANDRACTIM ◇ ANDROLONE ◇ ANDROSTANOLONE ◇ 5-α-ANDROSTAN-17-β-OL-3-ONE ◇ CRISTERONA MB ◇ DHT ◇ DIHYDROTESTOSTERONE ◇ 5-α-DIHYDROTESTOSTERONE ◇ 4,5-α-DIHYDROTESTOSTERONE ◇ 17-β-HYDROXY-5-α-ANDROSTAN-3-ONE ◇ (5-α,17-β)-17-HYDROXY-ANDROSTAN-3-ONE (9CI) ◇ NEODROL ◇ PROTEINA ◇ PROTONA ◇ STANAPROL ◇ STANOLONE

TOXICITY DATA with REFERENCE
spm-nml-par 12 mg/8W-I JEZOAO 205,403,78
scu-rat TDLo:17500 μg/kg (35D pre):REP RPHRA6 20,395,64
ims-rat TDLo:80 mg/kg (female 13-20D post):TER ENDOAO 99,1490,76

SAFETY PROFILE: Experimental teratogenic and reproductive effects. Mutation data reported. A steroid. When heated to decomposition it emits acrid smoke and irritating fumes. See also TESTOSTERONE.

DME525 CAS:855-22-1 ***HR: D***
5-α-DIHYDROTESTOSTERONE PROPIONATE
mf: $C_{22}H_{34}O_3$ mw: 346.56

SYNS: ANDROSTANOLONE PROPIONATE ◇ DHTP ◇ DIHYDROTESTOSTERONE PROPIONATE ◇ DIHYDROTESTOSTERONE-17-β-PROPIONATE

TOXICITY DATA with REFERENCE
scu-gpg TDLo:150 mg/kg (female 28-58D post):REP HOBEAO 6,139,75

SAFETY PROFILE: Experimental reproductive effects. A steroid. When heated to decomposition it emits acrid smoke and irritating fumes. See also TESTOSTERONE.

DME600 CAS:63681-01-6 ***HR: 3***
1,2-DIHYDRO-2,2,4,6-TETRAMETHYLPYRIDINE
mf: $C_9H_{15}N$ mw: 137.25

TOXICITY DATA with REFERENCE
skn-rbt 500 mg SEV SCCUR* -,4,61
orl-rat LDLo:600 mg/kg SCCUR* -,4,61
orl-mus LD50:640 mg/kg SCCUR* -,4,61
skn-rbt LDLo:140 mg/kg SCCUR* -,4,61

SAFETY PROFILE: Poison by skin contact. Moderately toxic by ingestion. A severe skin irritant. When heated to decomposition it emits toxic fumes of NO_x.

DME700 CAS:21457-22-7 ***HR: 3***
6,7-DIHYDRO-3,5,5,7-TETRAMETHYL-5H-THIAZOLO(3,2-a)PYRIMIDIN-7-OL HYDROCHLORIDE
mf: $C_{10}H_{16}N_2OS•ClH$ mw: 248.80

TOXICITY DATA with REFERENCE
orl-mus LD50:820 mg/kg PHARAT 24,572,69
ipr-mus LD50:360 mg/kg PHARAT 24,572,69
scu-mus LD50:795 mg/kg PHARAT 24,572,69

SAFETY PROFILE: Poison by intraperitoneal route. Moderately toxic by ingestion and subcutaneous routes. When heated to decomposition it emits toxic fumes of SO_x, NO_x, and HCl.

DMF000 CAS:77-79-2 ***HR: 2***
2,5-DIHYDROTHIOPHENE DIOXIDE
mf: $C_4H_6O_2S$ mw: 118.16

SYNS: BUTADIENE SULFONE ◇ 2,5-DIHYDROTHIOPHENE-1,1-DIOXIDE ◇ 2,5-DIHYDROTHIOPHENE SULFONE ◇ NCI-C04557 ◇ SULFOL-3-ENE ◇ β-SULFOLENE ◇ 3-SULFOLENE

TOXICITY DATA with REFERENCE
orl-rat LD50:2830 mg/kg TXAPA9 28,313,74
ipr-mus LD50:1700 mg/kg PCJOAU 12,1568,78

CONSENSUS REPORTS: NCI Carcinogenesis Bioassay (gavage); No Evidence: mouse, rat NCITR* NCI-CG-TR-102,78. Reported in EPA TSCA Inventory.

SAFETY PROFILE: Moderately toxic by ingestion and intraperitoneal routes. When heated to decomposition it emits toxic fumes of SO_x.

DMF400 CAS:63979-37-3 ***HR: 2***
1,2-DIHYDRO-s-TRIAZINE-4,6-DIAMINO-2,2-DIMETHYL-1-PHENYL-2,4,5-TRICHLOROPHENOXYACETATE
mf: $C_{11}H_{15}N_5 \cdot C_8H_5Cl_3O_3$ mw: 472.79

SYN: 2,3,5-TRICHLORFENOXYOCTAN1-FENYL-2,2-DIMETHYL-4,6-DIAMINO-1,2-DIHYDRO-1,3,5-TRIAZINU(CZECH)

TOXICITY DATA with REFERENCE
eye-rbt 500 mg/24H SEV 28ZPAK -,156,72
orl-rat LD50:447 mg/kg 28ZPAK -,156,72

SAFETY PROFILE: Moderately toxic by ingestion. A severe eye irritant. When heated to decomposition it emits very toxic fumes of Cl^- and NO_x.

DMF600 CAS:5831-17-4 ***HR: 3***
16,17-DIHYDRO-11,12,17-TRIMETHYLCYCLOPENTA(a)PHENANTHRENE
mf: $C_{20}H_{19}$ mw: 259.39

SYN: 11,12,17-TRIMETHYL-16,17-DIHYDRO-15H-CYCLOPENTA(a) PHENANTHRENE

TOXICITY DATA with REFERENCE
skn-mus TDLo:108 mg/kg/1Y-I:CAR PEXTAR 11,69,69

SAFETY PROFILE: Questionable carcinogen with experimental carcinogenic data. When heated to decomposition it emits acrid smoke and irritating fumes.

DMF800 CAS:35764-73-9 ***HR: 3***
cis-(±)-9,10-DIHYDRO-N,N,10-TRIMETHYL-2-(TRIFLUOROMETHYL)-9-ANTHRACENE PROPANAMINE
mf: $C_{21}H_{24}F_3N$ mw: 347.46

SYNS: (+ −)-9,10-DIHYDRO-N,N,10-TRIMETHYL-2-(TRIFLUORMETHYL)-9-ANTHRACENPROPANAMIN (GERMAN) ◇ FLUOTRACEN ◇ SKF 28175

TOXICITY DATA with REFERENCE
orl-rat LD50:487 mg/kg ARZNAD 27,1589,77
orl-mus LD50:353 mg/kg ARZNAD 27,1589,77

SAFETY PROFILE: Poison by ingestion. An antipsychotic agent. When heated to decomposition it emits very toxic fumes of F^- and NO_x.

DMG000 CAS:68151-18-8 ***HR: D***
1,2-DIHYDROTRIPHENYLENE

TOXICITY DATA with REFERENCE
mmo-sat 1 nmol/plate CNREA8 40,1985,80
mma-sat 1 nmol/plate CNREA8 40,1985,80

SAFETY PROFILE: Mutation data reported. When heated to decomposition it emits acrid smoke and irritating fumes.

DMG400 CAS:89-84-9 ***HR: 2***
2',4'-DIHYDROXYACETOPHENONE
mf: $C_8H_8O_3$ mw: 152.16

SYNS: 4-ACETYLRESORCINOL ◇ 2,4-DIHYDROXYACETOPHENONE ◇ 1-(2,4-DIHYDROXYPHENYL)ETHANONE ◇ RESACETOPHENONE ◇ β-RESACETOPHENONE ◇ RESOACETOPHENONE

TOXICITY DATA with REFERENCE
eye-rbt 500 mg SEV IHFCAY 6,1,67
scu-rat TDLo:20 mg/kg (4D pre):REP JSICAZ 19,264,60
orl-rat LD50:2830 mg/kg IHFCAY 6,1,67

CONSENSUS REPORTS: Reported in EPA TSCA Inventory.

SAFETY PROFILE: Moderately toxic by ingestion. Experimental reproductive effects. A severe eye irritant. When heated to decomposition it emits acrid smoke and irritating fumes.

DMG600 CAS:490-78-8 ***HR: 2***
2',5'-DIHYDROXYACETOPHENONE
mf: $C_8H_8O_3$ mw: 152.16

SYNS: ACETYLHYDROQUINONE ◇ 2-ACETYLHYDROQUINONE ◇ 2,5-DIHYDROXYACETOPHENONE ◇ QUINACETOPHENONE

TOXICITY DATA with REFERENCE
ipr-mus LDLo:500 mg/kg CBCCT* 5,140,53

CONSENSUS REPORTS: Reported in EPA TSCA Inventory.

SAFETY PROFILE: Moderately toxic by intraperitoneal route. When heated to decomposition it emits acrid smoke and irritating fumes.

DMG700 ***HR: D***
3-α,17-β-DIHYDROXY-5-α-ANDROSTANE
mf: $C_{19}H_{32}O_2$ mw: 292.51

TOXICITY DATA with REFERENCE
scu-rat TDLo:48 mg/kg (6D male):REP JSTBBK 18,397,83
ims-rat TDLo:40 mg/kg (13-20D preg):TER ENDOAO 99,1490,76

SAFETY PROFILE: Experimental teratogenic and reproductive effects. A steroid. When heated to decomposition it emits acrid smoke and irritating fumes.

DMG800 CAS:72-48-0 ***HR: 3***
1,2-DIHYDROXY-9,10-ANTHRAQUINONE
mf: $C_{14}H_8O_4$ mw: 240.22

PROP: Orange-red crystals. Bp: 430° (sublimes); mp: 289°.

SYNS: ALIZARIN ◇ ALIZARINA ◇ ALIZARINE ◇ ALIZARINE LAKE RED 2P ◇ ALIZARINE L PASTE ◇ ALIZARINE NAC ◇ ALIZARINE RED ◇ 1,2-ANTHRAQUINONEDIOL ◇ CERTIQUAL ALIZARINE ◇ C.I. 58000 ◇ D&C ORANGE NUMBER 15 ◇ DEEP CRIMSON MADDER 10821 ◇ 1,2-DIHYDROXYANTHRACHINON (CZECH) ◇ 1,2-DIHYDROXYANTHRAQUINONE ◇ ELJON MADDER ◇ MITSUI ALIZARINE B ◇ SANYO CARMINE L2B ◇ TURKEY RED

TOXICITY DATA with REFERENCE
eye-rbt 500 mg/24H MLD 28ZPAK -,101,72
mmo-sat 100 μg/plate MUREAV 40,203,76
mma-sat 100 μg/plate MUREAV 40,203,76
dnr-bcs 2 mg/disc TRENAF 27,153,76
orl-bwd LD50:316 mg/kg AECTCV 12,355,83

CONSENSUS REPORTS: Reported in EPA TSCA Inventory. EPA Genetic Toxicology Program.

SAFETY PROFILE: Poison by ingestion. Mutation data reported. An eye irritant. Flammable when exposed to oxidizers and heat. When heated to decomposition it emits acrid smoke and irritating fumes.

DMH000 CAS:81-64-1 ***HR: 3***
1,4-DIHYDROXYANTHRAQUINONE
mf: $C_{14}H_8O_4$ mw: 240.22

PROP: Crystals. Mp: 194°, bp: 450.0°, vap press: 1 mm @ 196.7°, vap d: 8.3.

SYNS: 1,4-DIHYDROXYANTHRACHINON (CZECH) ◇ 1,4-DIHYDROXY-9,10-ANTHRAQUINONE ◇ 1,4-DIOXYANTHRAQUINONE (RUSSIAN) ◇ QUINIZARIN

TOXICITY DATA with REFERENCE
eye-rbt 500 mg/24H MLD 28ZPAK -,102,72
mmo-sat 100 μg/plate MUREAV 40,203,76
mma-sat 100 μg/plate MUREAV 40,203,76
ipr-rat LD50:2100 mg/kg GTPZAB 21(12),27,77
ivn-mus LD50:320 mg/kg CSLNX* NX#03274

CONSENSUS REPORTS: Reported in EPA TSCA Inventory.

SAFETY PROFILE: Poison by intravenous route. Moderately toxic by intraperitoneal route. Mutation data reported. An eye irritant. A weak allergen. When heated to decomposition it emits acrid smoke and irritating fumes.

DMH200 CAS:117-12-4 ***HR: 1***
1,5-DIHYDROXYANTHRAQUINONE
mf: $C_{14}H_8O_4$ mw: 240.22

PROP: Green to yellow crystals. Mp: 280°, bp: subl, vap d: 8.3.

SYNS: ANTHRARUFIN ◇ 1,5-DIHYDROXYANTHRACHINON (CZECH) ◇ 1,5-DIHYDROXY-9,10-ANTHRAQUINONE

TOXICITY DATA with REFERENCE
eye-rbt 500 mg/24H MLD 28ZPAK -,102,72
mmo-sat 50 μg/plate MUREAV 40,203,76
mma-sat 50 μg/plate MUREAV 40,203,76

CONSENSUS REPORTS: Reported in EPA TSCA Inventory.

SAFETY PROFILE: An eye irritant. Mutation data reported. When heated to decomposition it emits acrid smoke and irritating fumes.

DMH400 CAS:117-10-2 ***HR: 3***
1,8-DIHYDROXYANTHRAQUINONE
mf: $C_{14}H_8O_4$ mw: 240.22

PROP: Crystals. Mp: 193°, vap d: 8.3.

SYNS: ALTAN ◇ ANTRAPUROL ◇ CHRYSAZIN ◇ DANTHRON ◇ DANTRON ◇ DIAQUONE ◇ 1,8-DIHYDROXY-9,10-ANTHRACENEDIONE ◇ 1,8-DIHYDROXYANTHRACHINON (CZECH) ◇ DIONONE ◇ DORBANE ◇ DORBANEX ◇ DUOLAX ◇ ISTIN ◇ LAXANORM ◇ LAXANTHREEN ◇ LAXIPUR ◇ LAXIPURIN ◇ LTAN ◇ MODANE ◇ USAF ND-59 ◇ ZWITSALAX

TOXICITY DATA with REFERENCE
eye-rbt 500 mg/24H MLD 28ZPAK -,102,72
mmo-sat 100 μg/plate MUREAV 40,203,76
dns-mus:lvr 20 μmol/L CNREA8 44,2918,84
orl-rat TDLo:292 g/kg/70W-C:CAR BJCAAI 52,781,85
orl-mus TDLo:129 g/kg/77W-C:CAR JJCREP 77,871,86
orl-mus TD:130 g/kg/77W-C:NEO TOLED5 31(Suppl),206,86
ipr-mus LD50:500 mg/kg NTIS** AD277-689

CONSENSUS REPORTS: Reported in EPA TSCA Inventory.

SAFETY PROFILE: Moderately toxic by intraperi-

toneal route. An eye irritant. Questionable carcinogen with experimental carcinogenic and neoplastigenic data. Human mutation data reported. A laxative. When heated to decomposition it emits acrid smoke and irritating fumes.

DMH600 CAS:84-60-6 ***HR: 3***
2,6-DIHYDROXYANTHRAQUINONE
mf: $C_{14}H_8O_4$ mw: 240.22

SYNS: ANTHRAFLAVIC ACID ◇ ANTHRAFLAVIN ◇ NSC-33531

TOXICITY DATA with REFERENCE
mma-sat 100 μg/plate MUREAV 40,203,76
ivn-mus LD50:180 mg/kg CSLNX* NX#06773

CONSENSUS REPORTS: Reported in EPA TSCA Inventory.

SAFETY PROFILE: Poison by intravenous route. Mutation data reported. When heated to decomposition it emits acrid smoke and irritating fumes. See other dihydroxyanthraquinone entries.

DMI300 CAS:20123-80-2 ***HR: 2***
2,5-DIHYDROXYBENZENESULFONIC ACID CALCIUM SALT
mf: $C_{12}H_{10}O_{10}S_2{\cdot}Ca$ mw: 418.42

PROP: White, powdery crystals from water. Mp: >300° (decomp). Color deepens to pink upon exposure to air. Very soluble in water and alc; practically insol in ether, benzene, chloroform.

SYNS: CALCIUM DOBESILATE ◇ DEXIUM ◇ DOBESILATE CALCIUM ◇ DOXIUM ◇ HYDROQUINONE CALCIUM SULFONATE

TOXICITY DATA with REFERENCE
orl-rat LD50:9400 mg/kg APFRAD 30,415,72
orl-mus LD50:7700 mg/kg APFRAD 20,415,72
ivn-mus LD50:775 mg/kg APFRAD 30,415,72

SAFETY PROFILE: Moderately toxic by intravenous route. Mildly toxic by ingestion. When heated to decomposition it emits toxic fumes of SO_x. See also SULFONATES and CALCIUM COMPOUNDS.

DMI400 CAS:2373-98-0 ***HR: 3***
3,3′-DIHYDROXYBENZIDINE
mf: $C_{12}H_{12}N_2O_2$ mw: 216.26

SYNS: 6,6′-DIAMINO-m,m′-BIPHENOL ◇ 4,4′-DIAMINO-3,3′-BIPHENYLDIOL ◇ 3,3′-DIOXYBENZIDINE ◇ 3,3′-DWUOKSYBENZYDYNA (POLISH)

TOXICITY DATA with REFERENCE
pic-esc 100 mmol/L MDMIAZ 31,11,79
orl-rat TDLo:9950 mg/kg/52W-I:NEO VOONAW 7(2),33,61
scu-rat TDLo:5900 mg/kg/43W-I:CAR VOONAW 7(2),33,61
mul-rat TDLo:6900 mg/kg/43W-I:CAR VOONAW 7(2),33,61
orl-mus TDLo:11 g/kg/47W-I:ETA VOONAW 7(2),33,61

SAFETY PROFILE: Suspected carcinogen with experimental carcinogenic, neoplastigenic, and tumorigenic data. Mutation data reported. When heated to decomposition it emits toxic fumes of NO_x.

DMI600 CAS:131-56-6 ***HR: 3***
2,4-DIHYDROXYBENZOPHENONE
mf: $C_{13}H_{10}O_3$ mw: 214.23

SYNS: 2,4-DIHYDROXYBENZOFENON (CZECH) ◇ EASTMAN INHIBITOR DHPB ◇ QUINSORB 010 ◇ SYNTASE 100 ◇ UF 1 ◇ USAF DO-28 ◇ USAF ND-54 ◇ UVINUL 400

TOXICITY DATA with REFERENCE
eye-rbt 100 mg/24H MOD 28ZPAK -,101,72
orl-rat LD50:7220 mg/kg 28ZPAK -,101,72
ipr-mus LD50:100 mg/kg NTIS** AD277-689
ivn-mus LD50:85 mg/kg BJPCAL 22,221,64

CONSENSUS REPORTS: Reported in EPA TSCA Inventory.

SAFETY PROFILE: Poison by intravenous and intraperitoneal routes. Mildly toxic by ingestion. An eye irritant. When heated to decomposition it emits acrid smoke and irritating fumes.

DMJ000 CAS:143-62-4 ***HR: 3***
3,β,14-DIHYDROXY-5,β-CARD-20(22)ENOLIDE
mf: $C_{23}H_{34}O_4$ mw: 374.57

SYNS: CARDOGENEN-(20:22)-DIOL-(3-β,14) (GERMAN) ◇ CERBERIGENIN ◇ DIGITOXIGENIN ◇ DIGITOXIGENINE ◇ (3-β,5-β)-3,14-DIHYDROXY-CARD-20(22)-ENOLIDE ◇ 3-β,14-DIOXY-CARDEN-(20:22)-OLID (GERMAN) ◇ 3-β,14-DIOXY-DIGEN-(20:22)-OLID (GERMAN) ◇ ECHUJETIN ◇ EVONOGENIN ◇ THEVETIGENIN ◇ $\Delta^{-20:22}$-3,14,21-TRIHYDROXYNORCHOLENIC ACID LACTONE

TOXICITY DATA with REFERENCE
ivn-rat LD50:1600 μg/kg ARZNAD 11,848,61
orl-mus LD50:26170 μg/kg AIPTAK 153,436,65
scu-mus LD50:11820 μg/kg AIPTAK 153,436,65
ivn-mus LD50:1131 μg/kg LIFSAK 37,775,85
ivn-cat LDLo:420 mg/kg AEPPAE 184,181,37
ivn-gpg LDLo:1419 μg/kg AIPTAK 153,436,65
par-pgn LDLo:600 μg/kg CPBTAL 8,18,60

SAFETY PROFILE: Poison by ingestion, subcutaneous, intravenous, and parenteral routes. When heated to decomposition it emits acrid smoke and irritating fumes.

DMJ200 CAS:128-13-2 ***HR: 3***
3-α,7-β-DIHYDROXY-6-β-CHOLAN-24-OIC ACID
mf: $C_{24}H_{40}O_4$ mw: 392.64

SYNS: CHOLIT-URSAN ◇ DELURSAN ◇ DESTOLIT ◇ DEURSIL ◇ 3-α,7-β-DIHYDROXYCHOLANIC ACID ◇ 3,7-DIHYDROXYCHOLAN-

24-OIC ACID ◇ 3-α,7-β-DIHYDROXY-5-β-CHOLANOIC ACID ◇ (3-α,5-β,7-β)-3,7-DIHYDROXYCHOLAN-24-OIC ACID ◇ 3-α,7-α-DIHYDROXYCHOLANSAEURE (GERMAN) ◇ 3-α,7-β-DIOXYCHOLANIC ACID ◇ 17-β-(1-METHYL-3-CARBOXYPROPYL)ETIOCHOLANE-3-α,7-β-DIOL ◇ UDCA ◇ URSACOL ◇ URSO ◇ URSOCHOL ◇ URSODEOXYCHOL ◇ URSODEOXYCHOLIC ACID ◇ URSODESOXYCHOLIC ACID ◇ URSOFALK ◇ URSOLVAN

TOXICITY DATA with REFERENCE
mmo-sat 40 mg/L MUREAV 158,45,85
orl-rat TDLo:11 g/kg (female 7-17D post):REP OYYAA2 15,931,78
orl-rat TDLo:11900 mg/kg (4-20D preg):TER AIPTAK 246,149,80
ipr-rat LD50:890 mg/kg NIIRDN 6,95,82
ivn-rat LD50:310 mg/kg NIIRDN 6,95,82
ivn-mus LD50:240 mg/kg NIIRDN 6,95,82

SAFETY PROFILE: Poison by intraperitoneal and intravenous routes. Experimental teratogenic and reproductive effects. Mutation data reported. Stimulates the flow of bile to the duodenum (a cholagogic). When heated to decomposition it emits acrid smoke and irritating fumes.

DMJ400 CAS:32222-06-3 ***HR: 3***
1a,25-DIHYDROXYCHOLECALCIFEROL
mf: $C_{27}H_{44}O_3$ mw: 416.71

SYNS: CALCITRIOL ◇ 1-α,25-DIHYDROXYCHOLECALCIFEROL ◇ 1,25-DIHYDROXYCHOLECALCIFEROL ◇ DIHYDROXYVITAMIN D3 ◇ 1-α,25-DIHYDROXYVITAMIN D3 ◇ Ro 215535 ◇ ROCALTROL ◇ (5Z,7E)-9,10-SECOCHESTA-5.7.10(19)-TRIENE-1-α,3-β,25-TRIOL ◇ (1-α,3-β,5Z,7E)-9,10-SECOCHOLESTA-5,7,10(19)-TRIENE-1,3,25-TRIOL

TOXICITY DATA with REFERENCE
dns-mus:fbr 2 μg/L CNREA8 46,604,86
scu-rat TDLo:15 μg/kg (female 16-21D post):REP ACATA5 111,343,81
orl-rat TDLo:55 μg/kg (7-17D preg):TER YACHDS 11,4221,83
scu-rat LD50:66 μg/kg YACHDS 11,4175,83
orl-mus LD50:1350 μg/kg YACHDS 11,4175,83
ipr-mus LD50:1900 μg/kg YACHDS 11,4175,83
scu-mus LD50:145 μg/kg YACHDS 11,4175,83

SAFETY PROFILE: A deadly poison by ingestion, intraperitoneal, and subcutaneous routes. Experimental teratogenic and reproductive effects. Mutation data reported. Enhances intestinal calcium transport and bone mineral mobilization. When heated to decomposition it emits acrid smoke and irritating fumes.

DMJ600 CAS:2892-51-5 ***HR: 3***
3,4-DIHYDROXY-3-CYCLOBUTENE-1,2-DIONE
mf: $C_4H_2O_4$ mw: 114.06

SYNS: DIHYDROXYCYCLOBUTENEDIONE ◇ 3,4-DIHYDROXYCYCLOBUTENE-1,2-DIONE ◇ QUADRATIC ACID ◇ SQUARIC ACID

TOXICITY DATA with REFERENCE
scu-mus TDLo:368 mg/kg/92W-I:ETA JNCIAM 46,143,71

CONSENSUS REPORTS: Reported in EPA TSCA Inventory.

SAFETY PROFILE: Questionable carcinogen with experimental tumorigenic data. When heated to decomposition it emits acrid smoke and fumes.

DMJ800 CAS:128-59-6 ***HR: 1***
DIHYDROXYDIBENZANTHRONE
mf: $C_{34}H_{16}O_4$ mw: 488.50

SYNS: 16,17-DIHYDROXYDIBENZANTHRONE ◇ DIHYDROXYVIOLANTHRON (CZECH) ◇ 16,17-DIHYDROXYVIOLANTHRONE

TOXICITY DATA with REFERENCE
eye-rbt 500 mg/24H MLD 28ZPAK -,104,72

CONSENSUS REPORTS: Reported in EPA TSCA Inventory.

SAFETY PROFILE: An eye irritant. When heated to decomposition it emits acrid smoke and fumes.

DMK200 CAS:66267-18-3 ***HR: 3***
trans-1,2-DIHYDROXY-1,2-DIHYDROBENZO(a,h) ANTHRACENE
mf: $C_{22}H_{16}O_2$ mw: 312.38

SYNS: DBA-1,2-DIHYDRODIOL ◇ (E)-1,2-DIHYDRO-1,2-DIHYDROXYDIBENZ(a,h)ANTHRACENE ◇ trans-1,2-DIHYDROXY-1,2-DIHYDROBENZ(a,h)ANTHRACENE

TOXICITY DATA with REFERENCE
skn-mus TDLo:2000 μg/kg:ETA CNREA8 39,1310,79

SAFETY PROFILE: Questionable carcinogen with experimental tumorigenic data. When heated to decomposition it emits acrid smoke and irritating fumes.

DMK400 CAS:24961-49-7 ***HR: 3***
trans-4,5-DIHYDROXY-4,5-DIHYDROBENZO(e) PYRENE
mf: $C_{20}H_{14}O_2$ mw: 286.34

SYNS: BENZO(e)PYRENE-4,5-DIHDYRODIOL ◇ 4,5-DIHYDRO-4,5-DIHYDROXYBENZO(e)PYRENE ◇ B(e)P-4,5-DIHYDRODIOL

TOXICITY DATA with REFERENCE
mma-sat 10 nmol/plate JBCHA3 254,4408,79
ipr-mus TDLo:32 mg/kg/15D-I:ETA CNREA8 40,203,80

SAFETY PROFILE: Questionable carcinogen with experimental tumorigenic data. Mutation data reported. When heated to decomposition it emits acrid smoke and irritating fumes.

DMK600 CAS:66788-06-5 ***HR: 3***
trans-9,10-DIHYDROXY-9,10-DIHYDROBENZO(e)PYRENE
mf: $C_{20}H_{14}O_2$ mw: 286.34

SYN: B(E)P 9,10-DIHYDRODIOL

TOXICITY DATA with REFERENCE
mma-sat 10 nmol/plate JBCHA3 254,4408,79
ipr-mus TDLo:32 mg/kg/15D-I:NEO CNREA8 40,203,80

SAFETY PROFILE: Questionable carcinogen with experimental neoplastigenic data. Mutation data reported. When heated to decomposition it emits acrid smoke and fumes.

DML000 CAS:61443-57-0 ***HR: 3***
(+,−)-trans-7,8-DIHYDROXY-7,8-DIHYDROBENZO(a)PYRENE
mf: $C_{20}H_{14}O_2$ mw: 286.34

SYN: BP-7,8-DIHYDRODIOL

TOXICITY DATA with REFERENCE
dnd-rat:lvr 20 μmol/L CRNGDP 3,861,82
msc-mus:fbr 200 nmol/L CNREA8 42,1866,82
orl-mus TDLo:206 mg/kg/6W-I:CAR JJIND8 62,1103,79
skn-mus TDLo:34 mg/kg/60W-I:NEO CNREA8 37,3356,77
skn-mus TD:22 mg/kg/25W-I:ETA PNASA6 73,3867,76

CONSENSUS REPORTS: EPA Genetic Toxicology Program.

SAFETY PROFILE: Questionable carcinogen with experimental carcinogenic, neoplastigenic, and tumorigenic data by skin contact. Mutation data reported. When heated to decomposition it emits toxic fumes of NO_x.

DML200 CAS:60864-95-1 ***HR: 3***
(−)-trans-7,8-DIHYDROXY-7,8-DIHYDROBENZO(a)PYRENE
mf: $C_{20}H_{14}O_2$ mw: 286.34

SYN: BP-7,8-DIHYDRODIOL

TOXICITY DATA with REFERENCE
mmo-sat 8 μg/plate MUREAV 58,361,78
otr-rat:lvr 10 mg/L CNREA8 40,1281,80
msc-ham:lng 40 nmol/L/2D CALEDQ 4,35,77
skn-mus TDLo:573 μg/kg:NEO CNREA8 37,2721,77

CONSENSUS REPORTS: EPA Genetic Toxicology Program.

SAFETY PROFILE: Questionable carcinogen with experimental neoplastigenic data. Mutation data reported. When heated to decomposition it emits acrid smoke and fumes.

DML400 CAS:62314-67-4 ***HR: 3***
(+)-trans-7,8-DIHYDROXY-7,8-DIHYDROBENZO(a)PYRENE
mf: $C_{20}H_{14}O_2$ mw: 286.34

TOXICITY DATA with REFERENCE
mma-sat 4 μg/plate MUREAV 58,361,78
otr-rat:lvr 10 mg/L CNREA8 40,1281,80
msc-ham:lng 1200 nmol/L/2D CALEDQ 4,35,77
skn-mus TDLo:573 μg/kg:NEO CNREA8 37,2721,77

CONSENSUS REPORTS: EPA Genetic Toxicology Program.

SAFETY PROFILE: Questionable carcinogen with experimental neoplastigenic data. Mutation data reported. When heated to decomposition it emits acrid smoke and fumes.

DML775 CAS:69260-85-1 ***HR: 2***
trans-3,4-DIHYDROXY-3,4-DIHYDRO-7,12-DIHYDROXYMETHYLBENZ(a)ANTHRACENE
mf: $C_{20}H_{18}O_4$ mw: 322.38

SYN: (E)-3,4-DIHYDROXY-3,4-DIHYDROBENZ(a)ANTHRACENE-7,12-DIMETHANOL

TOXICITY DATA with REFERENCE
mma-sat 50 nmol/plate CNREA8 40,3661,80
skn-mus TDLo:116 μg/kg:ETA CNREA8 40,3661,80

SAFETY PROFILE: Questionable carcinogen with experimental tumorigenic data. Mutation data reported. When heated to decomposition it emits acrid smoke and irritating fumes.

DML800 CAS:3343-12-2 ***HR: 3***
11,12-DIHYDROXY-11,12-DIHYDRO-3-METHYLCHOLANTHRENE (E)
mf: $C_{21}H_{18}O_2$ mw: 302.39

SYNS: trans-11,12-DIHYDRO-11,12-DIHYDROXY-3-METHYLCHOLANTHRENE ◇ (E)-MC 11,12-DIHYDRODIOL ◇ (E)-11,12-DIHYDRO-3-METHYLCHOLANTHRENE-11,12-DIOL ◇ (E)-3-METHYLCHOLANTHRENE-11,12-DIHYDRODIOL ◇ trans-3-METHYL-11,12-DIHYDROCHOLANTHRENE-11,12-DIOL

TOXICITY DATA with REFERENCE
mma-sat 20 μmol/L BBRCA9 85,1568,78
sce-ham:ovr 1 mg/L CALEDQ 7,45,79
msc-ham:lng 1 mg/L BBRCA9 85,1568,78
skn-mus TDLo:121 μg/kg:ETA CNREA8 39,3549,79

CONSENSUS REPORTS: EPA Genetic Toxicology Program.

SAFETY PROFILE: Questionable carcinogen with experimental tumorigenic data. Mutation data reported. When heated to decomposition it emits acrid smoke and fumes.

DMM200 CAS:68151-04-2 ***HR: D***
trans-1,2,-DIHYDROXY-1,2-DIHYDROTRIPHENYLENE
mf: $C_{18}H_{14}O_2$ mw: 262.2

SYN: (E)-1,2-DIHYDRO-1,2-DIHYDROXYTRIPHENYLENE

TOXICITY DATA with REFERENCE
mmo-sat 1 nmol/plate CNREA8 40,1985,80
mma-sat 1 nmol/plate CNREA8 40,1985,80

SAFETY PROFILE: Mutation data reported. When heated to decomposition it emits acrid smoke and fumes.

DMM400 CAS:3179-90-6 ***HR: 2***
5,8-DIHYDROXY-1,4-DIHYDROXYETHYL-AMINOANTHRAQUINONE
mf: $C_{18}H_{18}N_2O_6$ mw: 358.38

SYNS: 1,4-DIOXYETHYLAMINO-5,8-DIOXYANTHRAQUINONE (RUSSIAN) ◇ 1,4-DOEA-5,8-DAPFA (RUSSIAN)

TOXICITY DATA with REFERENCE
ipr-rat LD50:700 mg/kg GTPZAB 21(12),27,77

CONSENSUS REPORTS: Reported in EPA TSCA Inventory.

SAFETY PROFILE: Moderately toxic by intraperitoneal route. When heated to decomposition it emits toxic fumes of NO_x.

DMM600 CAS:10232-92-5 ***HR: 3***
2,4-DIHYDROXY-3,3-DIMETHYLBUTYRONITRILE
mf: $C_6H_{11}NO_2$ mw: 129.18

TOXICITY DATA with REFERENCE
skn-rbt 500 mg/24H MLD 85JCAE -,917,86
eye-rbt 2 mg/24H SEV 85JCAE -,917,86
ihl-man TCLo:13 mg/m³ (5Y male):REP GTPZAB 24(5),28,80
orl-rat LD50:310 mg/kg AIHAAP 23,95,62
skn-rbt LD50:130 mg/kg AIHAAP 23,95,62

CONSENSUS REPORTS: Cyanide and its compounds are on the Community Right-To-Know List.

SAFETY PROFILE: Poison by ingestion and skin contact. Human reproductive effects by inhalation: impaired spermatogenesis. When heated to decomposition it emits toxic fumes of NO_x and CN^-. See also NITRILES.

DMN000 CAS:4418-66-0 ***HR: 3***
2,2'-DIHYDROXY-3,3'-DIMETHYL-5,5'-DICHLORODIPHENYL SULFIDE
mf: $C_{14}H_{12}Cl_2O_2S$ mw: 315.22

SYNS: CHLORBISAN ◇ 2,2'-THIOBIS(4-CHLORO-6-METHYLPHENOL)

TOXICITY DATA with REFERENCE
orl-rat LD50:1300 μg/kg PCOC** -,224,66
ipr-rat LD50:850 μg/kg PCOC** -,224,66

CONSENSUS REPORTS: EPA Extremely Hazardous Substances List. Chlorophenol compounds are on the Community Right-To-Know List.

SAFETY PROFILE: Poison by ingestion and intraperitoneal routes. When heated to decomposition it emits very toxic fumes of SO_x and Cl^-. See also CHLOROPHENOLS and SULFIDES.

DMN200 CAS:17088-73-2 ***HR: D***
DIHYDROXYDIMETHYL PEROXIDE
mf: $C_2H_6O_4$ mw: 94.08

SYNS: BIS(HYDROXYMETHYL)PEROXIDE ◇ BIS-(1-HYDROXYMETHYL)PEROXIDE ◇ DIHYDROXYMETHYL PEROXIDE ◇ DIOXYBIS METHANOL ◇ DIOXYDIMETHANOL

TOXICITY DATA with REFERENCE
sln-dmg-par 20 mmol/L NATUAS 177,979,56
oms-mus/ast 20 mg/kg BSBGAQ 81,180,72

CONSENSUS REPORTS: EPA Genetic Toxicology Program.

SAFETY PROFILE: Mutation data reported. When heated to decomposition it emits acrid smoke and fumes. See also PEROXIDES.

DMN400 CAS:81-55-0 ***HR: 3***
1,8-DIHYDROXY-4,5-DINITROANTHRAQUINONE
mf: $C_{14}H_6N_2O_8$ mw: 330.22
SYN: NCI-C60742

TOXICITY DATA with REFERENCE
mmo-sat 100 μg/plate MUREAV 40,203,76
mma-sat 100 μg/plate MUREAV 40,203,76
ivn-mus LD50:180 mg/kg CSLNX* NX#01788

CONSENSUS REPORTS: Reported in EPA TSCA Inventory.

SAFETY PROFILE: Poison by intravenous route. Mutation data reported. When heated to decomposition it emits toxic fumes of NO_x.

DMO500 CAS:64551-89-9 ***HR: 2***
(±)-cis-3,4-DIHYDROXY-1,2-EPOXY-1,2,3,4-TETRAHYDROBENZ(a)ANTHRACENE
mf: $C_{18}H_{10}O_3$ mw: 274.28

SYNS: BENZ(a)ANTHRACENE,3,4-DIHYDROXY-1,2-EPOXY-1,2,3,4-TETRAHYDRO-, (Z), (+)- ◇ (±)-cis-3,4-DIHYDROXY-1,2-EPOXY-1,2,3,4-TETRAHYDROBENZO(a)ANTHRACENE ◇ DIOL-EPOXIDE-1

TOXICITY DATA with REFERENCE
skn-mus TDLo:4400 μg/kg:NEO CNREA8 38,1705,78

SAFETY PROFILE: Questionable carcinogen with experimental neoplastigenic data. When heated to decomposition it emits acrid smoke and irritating fumes.

DMO600 CAS:64598-83-0 ***HR: 3***
(±)trans-8-β,9-α-DIHYDROXY-10-α,11-α-EPOXY-8,9,10,11-TETRAHYDROBENZ(a)ANTHRACENE
mf: $C_{18}H_{14}O_3$ mw: 278.32

SYNS: BA-8,9-DIOL-10,11-EPOXIDE-1 ◇ (E)-8,9,10,11-TETRAHYDRO-8-β,9-α-DIHYDROXY-10-α,11-α-BENZ(a)ANTHRACENE

TOXICITY DATA with REFERENCE
skn-mus TDLo:22 mg/kg:ETA CNREA8 38,1699,78

CONSENSUS REPORTS: EPA Genetic Toxicology Program.

SAFETY PROFILE: Questionable carcinogen with experimental tumorigenic data. When heated to decomposition it emits acrid smoke and fumes.

DMO800 CAS:63438-26-6 ***HR: 3***
(+)-trans-3,4-DIHYDROXY-1,2-EPOXY-1,2,3,4-TETRAHYDROBENZ(a) ANTHRACENE
mf: $C_{18}H_{10}O_3$ mw: 274.28

SYNS: (E)-(+)-3,4-DIHYDROXY-1,2-EPOXY-1,2,3,4-TETRAHYDROBENZ(a)ANTHRACENE ◇ (+)-trans-3,4-DIHYDROXY-1,2-EPOXY-1,2,3,4-TETRAHYDROBENZO(a)ANTHRACENE ◇ DIOL-EPOXIDE 2

TOXICITY DATA with REFERENCE
skn-mus TDLo:4400 μg/kg:NEO CNREA8 38,1705,78

SAFETY PROFILE: Questionable carcinogen with experimental neoplastigenic data. When heated to decomposition it emits acrid smoke and fumes.

DMP000 CAS:64838-75-1 ***HR: 3***
(±)-trans-1,β,2,α-DIHYDROXY-3,α,4,α-EPOXY-1,2,3,4-TETRAHYDROBENZ(a)ANTHRACENE
mf: $C_{18}H_{14}O_3$ mw: 278.32

SYN: BA-1,2-DIOL-3,4-EPOXIDE-1

TOXICITY DATA with REFERENCE
skn-mus TDLo:22 mg/kg:ETA CNREA8 38,1699,78

SAFETY PROFILE: Questionable carcinogen with experimental tumorigenic data. When heated to decomposition it emits acrid smoke and fumes.

DMP200 CAS:64598-82-9 ***HR: 3***
(±)-trans-8-β,9-α-DIHYDROXY-10-β,11-β-EPOXY-8,9,10,11-TETRAHYDROBENZ(a)ANTHRACENE
mf: $C_{18}H_{14}O_3$ mw: 278.32

SYN: (E)-(±)-8,9,10,11-TETRAHYDRO-8-β,9-α-DIHYDROXY-10-β,11-β-EPOXYBENZ(a)ANTHRACENE

TOXICITY DATA with REFERENCE
skn-mus TDLo:22 mg/kg:ETA CNREA8 38,1699,78

CONSENSUS REPORTS: EPA Genetic Toxicology Program.

SAFETY PROFILE: Questionable carcinogen with experimental tumorigenic data. When heated to decomposition it emits acrid smoke and fumes.

DMP600 CAS:63323-29-5 ***HR: 3***
(+)cis-7,α,8,β-DIHYDROXY-9,α,10,α-EPOXY-7,8,9,10-TETRAHYDROBENZO(a)PYRENE
mf: $C_{20}H_{14}O_3$ mw: 302.34

SYNS: (+)-BP-7,α,8-β-DIOL-9,α,10,α-EPOXIDE 1 ◇ (+)-Z-7,8,9,10-TETRAHYDRO-7-α,8-β-DIHDYROXY-9-α,10-α-EPOXYBENZO(a)PYRENE ◇ (+)-cis-7,8,9,10-TETRAHYDRO-7-β,8-α-DIHYDROXY-9-β,10-β-EPOXYBENZO(a)PYRENE

TOXICITY DATA with REFERENCE
mmo-sat 100 pmol/plate BBRCA9 77,1389,77
skn-mus TDLo:1200 μg/kg:ETA CNREA8 39,67,79

SAFETY PROFILE: Questionable carcinogen with experimental tumorigenic data. Mutation data reported. When heated to decomposition it emits acrid smoke and fumes.

DMP800 CAS:63357-09-5 ***HR: 3***
(−)-cis-7,β,8,α-DIHYDROXY-9,β,10,β-EPOXY-7,8,9,10-TETRAHYDROBENZO(a)PYRENE
mf: $C_{20}H_{14}O_3$ mw: 302.34

SYNS: (−)BP-7,β,8,α-DIOL-9,β,10,β-EPOXIDE 1 ◇ (−)-Z-7,8,9,10-TETRAHYDRO-7-α,8-β-DIHDYROXY-9-α,10-α-EPOXYBENZO(a)PYRENE ◇ (−)-Z-7,8,9,10-TETRAHYDRO-7-β,8-α-DIHYDROXY-9-β,10-β-EPOXYBENZO(a)PYRENE

TOXICITY DATA with REFERENCE
mmo-sat 100 pmol/plate BBRCA9 77,1389,77
dnd-mus-skn 8 μmol/kg CNREA8 44,1081,84
msc-ham:lng 1 μmol/L MUREAV 44,313,77
skn-mus TDLo:1200 μg/kg:ETA CNREA8 39,67,79

SAFETY PROFILE: Questionable carcinogen with experimental tumorigenic data. Mutation data reported. When heated to decomposition it emits acrid smoke and irritating fumes.

DMP900 CAS:58917-67-2 ***HR: 2***
(±)-(E)-7,8-DIHYDROXY-9,10-EPOXY-7,8,9,10-TETRAHYDROBENZO(a)PYRENE
mf: $C_{20}H_{14}O_3$ mw: 302.34

SYN: BP 7,8-DIOL-9,10-EPOXIDE 2

TOXICITY DATA with REFERENCE
skn-mus TDLo:2400 μg/kg:ETA CNREA8 39,67,79

SAFETY PROFILE: Questionable carcinogen with experimental tumorigenic data. When heated to decomposition it emits acrid smoke and irritating fumes.

DMQ000 CAS:58917-67-2 ***HR: 3***
(±)-trans-7,8-DIHYDROXY-9,10-EPOXY-7,8,9,10-TETRAHYDRO-BENZO(a)PYRENE
mf: $C_{20}H_{14}O_3$ mw: 302.34

SYNS: anti-BENZO(a)PYRENE-DIOLEPOXIDE ◇ anti-BP-DIOLEPOXIDE

TOXICITY DATA with REFERENCE
dni-omi 200 μg/L PNASA6 74,1378,77
dni-omi 100 μg/L PNASA6 74,1378,77
msc-ham:ovr 100 nmol/L MUREAV 112,329,83

SAFETY PROFILE: Mutation data reported. When heated to decomposition it emits acrid smoke and fumes.

DMQ600 CAS:58917-67-2 ***HR: D***
(±)-7,α,8,β-DIHYDROXY-9,β,10,β-EPOXY-7,8,9,10-TETRAHYDROBENZO(a)PYRENE
mf: $C_{20}H_{14}O_3$ mw: 302.34

SYNS: anti-BPDE ◇ (±)-7,8,9,10-TETRAHYDRO-7-α,8-β-DIHYDROXY-9-β,10-β-EPOXYBENZO(a)PYRENE ◇ (±)-7,8,9,10-TETRAHYDRO-9-β,10-β-EPOXY-7-α,8-β-DIHYDROXYBENZO(a)PYRENE

TOXICITY DATA with REFERENCE
dnd-mus-skn 20 μmol/kg CRNGDP 3,1135,82
sce-ham:lng 600 μg/L IJCNAW 24,485,79

SAFETY PROFILE: Mutation data reported. When heated to decomposition it emits acrid smoke and fumes.

DMQ800 CAS:60268-85-1 ***HR: D***
7-β,8-α-DIHYDROXY-9-α,10-α-EPOXY-7,8,9,10-TETRAHYDROBENZO(a)PYRENE
mf: $C_{20}H_{14}O_3$ mw: 302.34

SYN: R-7,t-8-DIHYDROXY-t-9,10-OXY-7,8,9,10-TETRAHYDROBENZO(a)PYRENE

TOXICITY DATA with REFERENCE
otr-hmn:fbr 330 nmol/L PNASA6 80,7219,83
dnd-hmn:fbr 200 nmol/L PNASA6 77,5933,80
sce-hmn:fbr 100 nmol/L CNREA8 45,2600,85
msc-hmn:fbr 100 nmol/L PNASA6 77,5933,80

SAFETY PROFILE: Human mutation data reported. When heated to decomposition it emits acrid smoke and fumes.

DMR000 CAS:58917-91-2 ***HR: 3***
(±)-7,β,8,α-DIHYDROXY-9,β,10,β-EPOXY-7,8,9,10-TETRAHYDROBENZO(a)PYRENE
mf: $C_{20}H_{14}O_3$ mw: 302.34

SYNS: BPDE-syn ◇ B(a)P EPOXIDE I ◇ (±)-7-α,8-β-DIHYDROXY-9-α,10-α-EPOXY-7,8,9,10-TETRAHYDROBENZO(a)PYRENE ◇ (±)-7,8,8a,9a-TETRAHYDROBENZO(10,11)CHYRSENO(3,4-b)OXIRENE-7,8-DIOL ◇ (±)-7,8,9,10-TETRAHYDRO-7-α,8-β-DIHYDROXY-9-α,10-α-EPOXYBENZO(a)PYRENE

TOXICITY DATA with REFERENCE
mmo-sat 300 pmol/plate CNREA8 36,3358,76
mma-sat 300 pmol/plate CNREA8 36,3358,76
dnr-hmn:fbr 1 μmol/L CBINA8 20,279,78
dnd-mus-skn 20 μmol/kg CRNGDP 3,1135,82
oms-mus-skn 20 μmol/kg CRNGDP 3,1135,82
dnd-mam:lym 600 nmol CRNGDP 3,267,82
skn-mus TDLo:2420 μg/kg:ETA CCSUDL 3,371,78

SAFETY PROFILE: Questionable carcinogen with experimental tumorigenic data. Human mutation data reported. When heated to decomposition it emits acrid smoke and fumes.

DMR150 ***HR: 2***
(±)-9-α-10-β-DIHYDROXY-11-β,12-β-EPOXY-9,10,11,12-TETRAHYD ROBENZO(e)PYRENE
mf: $C_{20}H_{14}O_3$ mw: 302.34

SYNS: BENZO(e)PYRENE, 9,10-DIOL-11,12-EPOXIDE 1 (cis) ◇ B(e)P DIOL EPOXIDE-1 ◇ B(e)P 9,10-DIOL-11,12-EPOXIDE-1

TOXICITY DATA with REFERENCE
ipr-mus TDLo:476 mg/kg:CAR CNREA8 41,915,81

SAFETY PROFILE: Questionable carcinogen with experimental carcinogenic data. When heated to decomposition it emits acrid smoke and irritating fumes.

DMR200 ***HR: 3***
(±)-9,β,10,α-DIHYDROXY-11,α,12,α-EPOXY-9,10,11,12-TETRAHYDROBENZO(e)PYRENE
mf: $C_{20}H_{14}O_3$ mw: 302.3

SYN: B(E)P DIOL EPOXIDE-2

TOXICITY DATA with REFERENCE
mmo-sat 1 nmol/plate CNREA8 40,1985,80
mma-sat 1 nmol/plate CNREA8 40,1985,80
msc-ham:lng 1 nmol/L CNREA8 40,1985,80
ipr-mus TDLo:476 mg/kg:CAR CNREA8 41,915,81

SAFETY PROFILE: Questionable carcinogen with experimental carcinogenic data. Mutation data reported. When heated to decomposition it emits acrid smoke and fumes.

DMR400 CAS:74465-36-4 ***HR: D***
(±)-9,β,10,α-DIHYDROXY-11,β,12,β-EPOXY-9,10,11,12-TETRAHYDROBENZO(e)PYRENE
mf: $C_{20}H_{14}O_3$ mw: 302.34

SYN: B(e)P DIOL EPOXIDE-1

TOXICITY DATA with REFERENCE
mmo-sat 1 nmol/plate CNREA8 40,1985,80
mma-sat 1 nmol/plate CNREA8 40,1985,80
msc-ham:lng 1 nmol/L CNREA8 40,1985,80

SAFETY PROFILE: Mutation data reported. When heated to decomposition it emits acrid smoke and fumes.

DMS000 ***HR: 3***
trans-1,2-DIHYDROXY-anti-3,4-EPOXY-1,2,3,4-TETRAHYDROCHRYSENE
mf: $C_{18}H_{14}O$ mw: 278.32

SYN: (+)-(E)-3,4-EPOXY-1,2,3,4-TETRAHYDRO-CHRYSENEDIOL

TOXICITY DATA with REFERENCE
skn-mus TDLo:20 mg/kg:NEO CNREA8 40,1981,80

SAFETY PROFILE: Questionable carcinogen with experimental neoplastigenic data. When heated to decomposition it emits acrid smoke and fumes.

DMS200 CAS:72074-67-0 ***HR: 3***
(±)-1,β,2,α-DIHYDROXY-3,α,4,α-EPOXY-1,2,3,4-TETRAHYDROCHRYSENE
mf: $C_{18}H_{14}O_3$ mw: 278.32

SYN: (±)-1,2,3,4-TETRAHYDRO-3,α,4,α-EPOXY-1,β,2,α-CHRYSENE-DIOL

TOXICITY DATA with REFERENCE
mmo-sat 1 nmol/L CRNGDP 6,237,85
mma-sat 1 nmol/plate CNREA8 39,4069,79
ipr-mus TDLo:72 mg/kg/15D-I:CAR CNREA8 39,5063,79

SAFETY PROFILE: Questionable carcinogen with experimental carcinogenic data. Mutation data reported. When heated to decomposition it emits acrid smoke and fumes.

DMS400 CAS:72074-66-9 ***HR: 3***
(±)-1,β,2,α-DIHYDROXY-3,β,4,β-EPOXY-1,2,3,4-TETRAHYDROCHRYSENE
mf: $C_{18}H_{14}O_3$ mw: 278.32

SYN: (±)-1,2,3,4-TETRAHYDRO-3,β,4,β-EPOXY-1-β,2-α-CHRYSENE-DIOL

TOXICITY DATA with REFERENCE
mma-sat 1 nmol/plate CNREA8 39,4069,79
msc-ham:lng 1 nmol/plate CNREA8 39,4069,79
ipr-mus TDLo:72 mg/kg/15D-I:ETA CNREA8 39,5063,79

SAFETY PROFILE: Questionable carcinogen with experimental tumorigenic data. Mutation data reported. When heated to decomposition it emits acrid smoke and fumes.

DMS500 CAS:78919-11-6 ***HR: D***
(±)-1-β,2-α-DIHYDROXY-3-α,4-α-EPOXY-1,2,3,4-TETRAHYDRODIBENZO(a,h)PYRENE
mf: $C_{24}H_{16}O_3$ mw: 352.40

SYN: (±)-3-α,4-α-EPOXY-1,2,3,4-TETRAHYDROBENZO(b,def) CHRYSENE-1-β,2-α-DIOL

TOXICITY DATA with REFERENCE
mmo-sat 150 pmol/L CRNGDP 6,237,85
mma-sat 100 pmol/plate CNREA8 41,2589,81
msc-ham:lng 500 nmol/L CNREA8 41,2589,81

SAFETY PROFILE: Mutation data reported. When heated to decomposition it emits acrid smoke and fumes.

DMS600 CAS:72074-68-1 ***HR: D***
(±)-1,β,2,β-DIHYDROXY-3,α,4,α-EPOXY-1,2,3,4-TETRAHYDROPHENANTHRENE
mf: $C_{14}H_{12}O_3$ mw: 228.26

TOXICITY DATA with REFERENCE
mma-sat 1 nmol/plate CNREA8 39,4069,79
msc-ham:lng 1 nmol/plate CNREA8 39,4069,79

SAFETY PROFILE: Mutation data reported. When heated to decomposition it emits acrid smoke and fumes.

DMS800 CAS:74465-39-7 ***HR: D***
(+ −)-1,β,2,α-DIHYDROXY-3,α,4,α-EPOXY-1,2,3,4-TETRAHYDROTRIPHENYLENE
mf: $C_{18}H_{14}O_3$ mw: 278.2

SYN: TP DIOL EPOXIDE-2

TOXICITY DATA with REFERENCE
mmo-sat 1 nmol/plate CNREA8 40,1985,80
mma-sat 1 nmol/plate CNREA8 40,1985,80

SAFETY PROFILE: Mutation data reported. When heated to decomposition it emits acrid smoke and fumes.

DMT000 CAS:74465-38-6 ***HR: D***
(+ −)-1,β,2,α-DIHYDROXY-3,β,4,β-EPOXY-1,2,3,4-TETRAHYDROTRIPHENYLENE
mf: $C_{18}H_{14}O_3$ mw: 278.2

SYN: TP DIOL-EPOXIDE-1

TOXICITY DATA with REFERENCE
mmo-sat 1 nmol/plate CNREA8 40,1985,80
mma-sat 1 nmol/plate CNREA8 40,1985,80

SAFETY PROFILE: Mutation data reported. When heated to decomposition it emits acrid smoke and fumes.

DMT100 CAS:1038-19-3 ***HR: D***
2-β,17-β-DIHYDROXY-2-α-ETHINYL-A-NOR(5-α) ANDROSTANE
mf: $C_{20}H_{30}O_2$ mw: 302.50

TOXICITY DATA with REFERENCE
orl-rat TDLo:75 mg/kg (female 1D post):REP PSEBAA 111,595,62

SAFETY PROFILE: Experimental reproductive effects. A steroid. When heated to decomposition it emits acrid smoke and fumes.

DMT200 CAS:17526-17-9 ***HR: D***
1,3-DIHYDROXY-2-ETHOXYMETHYLANTHRAQUINONE
mf: $C_{17}H_{14}O_5$ mw: 298.31

SYNS: 2-(ETHOXYMETHYL)-1,3-DIHYDROXY-9,10-ANTHRACENEDIONE ◇ LUCIDIN ETHYL ETHER

TOXICITY DATA with REFERENCE
mmo-sat 10 μg/plate BCSTB5 5,1489,77
mma-sat 10 μg/plate BCSTB5 5,1489,77

CONSENSUS REPORTS: EPA Genetic Toxicology Program.

SAFETY PROFILE: Mutation data reported. When heated to decomposition it emits acrid smoke and fumes. See also ETHERS.

DMT400 CAS:4500-29-2 ***HR: 2***
N,N-DI(2-HYDROXYETHYL)CYCLOHEXYLAMINE
mf: $C_{10}H_{21}NO_2$ mw: 187.32

SYNS: ABBOMEEN E-2 ◇ 2,2′-CYCLOHEXYLIMINODIETHANOL

TOXICITY DATA with REFERENCE
eye-rbt 100 mg SEV 34ZIAG -,61,69
orl-rat LD50:2600 mg/kg 34ZIAG -,61,69

CONSENSUS REPORTS: Reported in EPA TSCA Inventory.

SAFETY PROFILE: Moderately toxic by ingestion. A severe eye irritant. When heated to decomposition it emits toxic fumes of NO_x.

DMT800 CAS:28005-74-5 ***HR: 2***
DI-(HYDROXYETHYL)-o-TOLYLAMINE
mf: $C_{11}H_{17}NO_2$ mw: 195.29

SYNS: EMERY 5712 ◇ 2,2′-((2-METHYLPHENYL)IMINO)BISETHANOL ◇ 2,2′-(o-TOYLYIMINO)DIETHANOL ◇ o-TOLYLDIETHANOLAMINE

TOXICITY DATA with REFERENCE
orl-rat LDLo:2200 mg/kg AIHAAP 23,95,62
skn-rbt LDLo:1000 mg/kg AIHAAP 23,95,62

CONSENSUS REPORTS: Reported in EPA TSCA Inventory.

SAFETY PROFILE: Moderately toxic by ingestion and skin contact. When heated to decomposition it emits toxic fumes of NO_x. See also AMINES.

DMU000 CAS:1069-23-4 ***HR: 3***
3,4-DIHYDROXY-1,5-HEXADIENE
mf: $C_6H_{10}O_2$ mw: 114.16

TOXICITY DATA with REFERENCE
orl-rat LD50:1620 mg/kg AIHAAP 30,470,69
skn-rbt LD50:400 mg/kg AIHAAP 30,470,69

SAFETY PROFILE: Poison by skin contact. Moderately toxic by ingestion. When heated to decomposition it emits acrid smoke and irritating fumes.

DMU600 CAS:481-72-1 ***HR: D***
1,8-DIHYDROXY-3-HYDROXYMETHYLANTHRAQUINONE
mf: $C_{15}H_{10}O_5$ mw: 270.25

SYNS: ALOE-EMODIN ◇ 1,8-DIHYDROXY-3-(HYDROXYMETHYL)-9,10-ANTHRACENEDIONE ◇ HYDROXYMETHYLCHRYSAZIN ◇ RHABARBERONE

TOXICITY DATA with REFERENCE
mmo-sat 100 μg/plate BCSTB5 5,1489,77
mma-sat 100 μg/plate BCSTB5 5,1489,77

CONSENSUS REPORTS: EPA Genetic Toxicology Program.

SAFETY PROFILE: Mutation data reported. A cathartic. When heated to decomposition it emits acrid smoke and fumes.

DMV200 CAS:485-47-2 ***HR: 3***
2,2-DIHYDROXY-1,3-INDANDIONE
mf: $C_9H_6O_4$ mw: 178.15

PROP: Crystals. Turns reddish @ 125°, swells @ 139°, decomp @ 240°.

SYNS: 2,2-DIHYDROXY-1H-INDENE-1,3(2H)-DIONE ◇ 1,2,3-INDANTRIONE-2-HYDRATE ◇ 1,2,3-INDANTRIONE MONOHYDRATE ◇ NINHYDRIN ◇ NINHYDRIN HYDRATE ◇ TRIKETOHYDRINDENE HYDRATE

TOXICITY DATA with REFERENCE
orl-rat LDLo:250 mg/kg NCNSA6 5,28,53
ipr-mus LD50:78 mg/kg CRSBAW 151,719,57

CONSENSUS REPORTS: Reported in EPA TSCA Inventory.

SAFETY PROFILE: Poison by ingestion and intraperitoneal routes. When heated to decomposition it emits acrid smoke and fumes.

DMV400 CAS:99-11-6 ***HR: 2***
2,6-DIHYDROXYISONICOTINIC ACID
mf: $C_6H_5NO_4$ mw: 155.12

PROP: Buff to gray powder.

SYNS: CITRAZINIC ACID ◇ 2,6-DIHYDROXY-4-CARBOXYPYRIDINE

TOXICITY DATA with REFERENCE
ipr-rat LDLo:800 mg/kg KODAK* -,-,71

CONSENSUS REPORTS: Reported in EPA TSCA Inventory.

SAFETY PROFILE: Moderately toxic by intraperitoneal route. A moderately irritating organic acid with some allergenic properties. When heated to decomposition it emits toxic fumes of NO_x.

DMV600 CAS:7683-59-2 ***HR: 3***
3,4-DIHYDROXY-α-((ISOPROPYLAMINO) METHYL)BENZYL ALCOHOL
mf: $C_{11}H_{17}NO_3$ mw: 211.29

SYNS: A 21 ◇ ALEUDRIN ◇ ALUDRINE ◇ ASIPRENOL ◇ ASMALAR ◇ ASSIPRENOL ◇ BELLASTHMAN ◇ BRONKEPHRINE ◇ DIHYDROXYPHENYLETHANOLISOPROPYLAMINE ◇ 1-(3,4-DIHYDROXYPHENYL)-2-ISOPROPYLAMINOETHANOL ◇ EPINEPHRINE ISOPROPYL HOMOLOG ◇ 4-(1-HYDROXY-2-((1-METHYLETHYL) AMINO)ETHYL)-1,2-BENZENEDIOL ◇ IPA ◇ ISONORENE ◇ ISOPRENALINE ◇ ISOPROPYDRIN ◇ ISOPROPYLADRENALINE ◇ ISOPROPYLAMINOMETHYL-3,4-DIHYDROXYPHENYL CARBINOL ◇ α-(ISOPROPYLAMINOMETHYL)PROTOCATECHUYL ALCOHOL ◇ ISOPROPYLARTERENOL ◇ N-ISOPROPYL-β-DIHYDROXYPHENYL-β-HYDROXYETHYLAMINE ◇ ISOPROPYL NORADRENALINE ◇ 1-ISOPROPYLNORADRENALINE ◇ N-ISOPROPYLNORADRENALINE ◇ ISOPROTERENOL ◇ l-ISOPROTERENOL ◇ ISORENIN ◇ ISUPREL ◇ ISUPREN ◇ LOMUPREN ◇ NEODRENAL ◇ NEO-EPININE ◇ NORISODRINE ◇ NOVODRIN ◇ PROTERNOL ◇ RESPIFRAL ◇ SAVENTRINE ◇ VAPO-N-ISO ◇ WIN 5162

TOXICITY DATA with REFERENCE
oms-rat-par 100 mg/kg BEXBAN 94,1458,82
dns-mus-ipr 341 mg/kg JCLBA3 56,605,73
scu-ham TDLo:3 μg/kg (female 8D post):TER PSEBAA 130,1168,69
scu-ham TDLo:3 μg/kg (female 8D post):REP PSEBAA 130,1168,69
ims-hmn TDLo:14 μg/kg:CVS KLWOAZ 19,1303,40
orl-rat LD50:355 mg/kg USXXAM #4026897
ipr-rat LDLo:100 mg/kg FCTXAV 3,597,65
scu-rat LD50:600 μg/kg TOXID9 4,77,84
ivn-rat LD50:57 mg/kg TXAPA9 16,303,70
orl-mus LD50:450 mg/kg 27ZIAQ -,139,73
ipr-mus LD50:440 mg/kg JPETAB 164,290,68
scu-mus LD50:400 mg/kg ARZNAD 26,1404,76
ivn-mus LD50:83 mg/kg JPETAB 97,14,49
orl-dog LD50:600 mg/kg TXAPA9 8,353,66
ivn-dog LD50:50 mg/kg JPETAB 164,290,68
orl-rbt LD50:3070 mg/kg TXAPA9 8,353,66
ivn-rbt LD50:27 mg/kg TXAPA9 8,353,66
orl-gpg LD50:270 μg/kg JPETAB 164,290,68
scu-gpg LD50:320 μg/kg JPETAB 164,290,68

SAFETY PROFILE: Poison by ingestion, subcutaneous, intravenous, and intraperitoneal routes. An experimental teratogen. Other experimental reproductive effects. Human systemic effects by intramuscular route: increased pulse and cardiac rate. A bronchodilator. Mutation data reported. When heated to decomposition it emits toxic fumes of NO_x.

DMV800 CAS:586-06-1 ***HR: 3***
3,5-DIHYDROXY-α-((ISOPROPYLAMINO) METHYL)BENZYL ALCOHOL
mf: $C_{11}H_{17}NO_3$ mw: 211.29

SYNS: METAPROTERENOL ◇ ORCIPRENALINE

TOXICITY DATA with REFERENCE
orl-wmn TDLo:800 μg/kg (1D pre):REP RDCNBM 5,31,81
orl-man TDLo:286 μg/kg AJMSA9 291,168,86
orl-rat LD50:3370 mg/kg TXAPA9 8,353,66
ivn-rat LD50:67200 μg/kg TXAPA9 8,353,66
ipr-mus LDLo:240 mg/kg APTOA6 31,33,72
scu-mus LD50:406 mg/kg ARZNAD 26,1404,76
ivn-mus LD50:86 mg/kg APTOA6 38,474,76
orl-dog LD50:125 mg/kg TXAPA9 8,353,66
ivn-dog LD50:30 mg/kg TXAPA9 8,353,66
orl-rbt LD50:3110 mg/kg TXAPA9 8,353,66
ivn-rbt LD50:81300 μg/kg TXAPA9 8,353,66

SAFETY PROFILE: Poison by ingestion, intraperitoneal and intravenous routes. Moderately toxic by subcutaneous route. Human reproductive effects by ingestion: changes in the uterus, cervix and vagina. When heated to decomposition it emits toxic fumes of NO_x.

DMW000 CAS:7361-61-7 ***HR: 3***
5,6-DIHYDRO-2-(2,6-XYLIDINO)-4H-1,3-THIAZINE
mf: $C_{12}H_{16}N_2S$ mw: 220.36

SYNS: BAY 1470 ◇ BAY VA 1470 ◇ N-(5,6-DIHYDRO-4H-1,3-THIAZINYL)-2,6-XYLIDINE ◇ 2-(2,6-DIMETHYLANILINO)-5,6-DIHYDRO-4H-1,3-THIAZINE ◇ 2-(2,6-DIMETHYLPHENYLAMINO)-4H-5,6-DIHYDRO-1,3-THIAZINE ◇ N-(2,6-DIMETHYLPHENYL)-5,6-DIHYDRO-4H-1,3-THIAZIN-2-AMINE ◇ N-(2,6-DIMETHYLPHENYL)-5,6-DIHYDRO-4H-1,3-THIAZINE-2-AMINE (9CI) ◇ ROMPUN ◇ WH 7286 ◇ XYLZIN ◇ XYLAZINE (USDA)

TOXICITY DATA with REFERENCE
orl-rat LD50:130 mg/kg DTTIAF 75,565,68
orl-mus LD50:240 mg/kg DTTIAF 75,565,68
scu-mus LD50:121 mg/kg DTTIAF 75,565,68
ivn-mus LD50:18 mg/kg CSLNX* NX#10054

SAFETY PROFILE: Poison by ingestion, subcutaneous, and intravenous routes. When heated to decomposition it emits very toxic fumes of NO_x and SO_x.

DMW200 CAS:526-84-1 ***HR: 3***
DIHYDROXYMALEIC ACID
mf: $C_4H_4O_6$ mw: 148.07

$(:C(OH)CO•OH)_2$

SYN: DIHYDROXYBUTENEDIOIC ACID

SAFETY PROFILE: A storage hazard. It may explode in a sealed container. Slowly decomposes to release carbon dioxide. When heated to decomposition it emits acrid smoke and fumes.

DMX000 CAS:69260-83-9 ***HR: 2***
(E)-3,4-DIHYDROXY-7-METHYL-3,4-DIHYDROBENZ(a)ANTHRACENE-12-METHANOL
mf: $C_{20}H_{18}O_3$ mw: 306.38

SYNS: trans-3,4-DIHYDRO-12-(HYDROXYMETHYL)-7-METHYLBENZ(a)ANTHRACENE-3,4-DIOL ◇ trans-3,4-DIHYDROXY-3,4-DIHYDRO-7-METHYL-12-HYDROXYMETHYLBENZ(a)ANTHRACENE

TOXICITY DATA with REFERENCE
mma-sat 35 nmol/plate CNREA8 40,3661,80
mma-ham:lng 400 nmol/L PNASA6 76,862,79
skn-mus TDLo:110 μg/kg:ETA CNREA8 40,3661,80

SAFETY PROFILE: Questionable carcinogen with experimental tumorigenic data. Mutation data reported. When heated to decomposition it emits acrid smoke and fumes.

DMX200 CAS:2318-18-5 ***HR: 3***
2,12-DIHYDROXY-4-METHYL-11,16-DIOXOSENECIONANIUM
mf: $C_{19}H_{28}NO_6$ mw: 366.48

SYNS: trans-15-ETHYLIDENE-12-β-HYDROXY-4,12-α,13-β-TRIMETHYL 8-OXO-4,8 SECOSENEC-1-ENINE ◇ 12-HYDROXY-4-METHYL-4,8-SECOSENECIONAN-8,11,16-TRIONE ◇ NSC-89945 ◇ RENARDIN ◇ RENARDINE ◇ SENKIRKIN ◇ SENKIRKINE

TOXICITY DATA with REFERENCE
mma-sat 1 mg/plate MUREAV 68,211,79
sln-dmg-orl 10 μmol/L/3D-I FCTOD7 22,223,84
dns-rat:lvr 2 μmol/L CNREA8 45,3125,85
dns-mus:lvr 20 μmol/L CNREA8 45,3125,85
dns-ham:lvr 2 μmol/L CNREA8 45,3125,85
sce-ham:lng 60 μg/L MUREAV 142,209,85
ipr-rat TDLo:1320 mg/kg/56W-I:NEO JJIND8 63,469,79
orl-rat LDLo:200 mg/kg NATUAS 227,401,70
ipr-rat LD50:220 mg/kg JJIND8 63,469,79

CONSENSUS REPORTS: IARC Cancer Review: Group 3 IMEMDT 7,56,87; Animal Limited Evidence IMEMDT 31,231,83; Animal Inadequate Evidence IMEMDT 10,327,76.

SAFETY PROFILE: Poison by ingestion and intraperitoneal routes. Questionable carcinogen with experimental neoplastigenic data. Mutation data reported. When heated to decomposition it emits toxic fumes of NO_x.

DMX800 CAS:2033-94-5 ***HR: 3***
3,4-DIHYDROXY-3-METHYL-4-PHENYL-1-BUTYNE
mf: $C_{11}H_{12}O_2$ mw: 176.23

SYNS: 3-METHYL-3,4-DIHYDROXY-4-PHENYL-BUTIN-1(GERMAN) ◇ 3-METHYL-3,4-DIHYDROXY-4-PHENYL-1-BUTYNE ◇ 2-METHYL-1-PHENYL-3-BUTYNE-1,2-DIOL

TOXICITY DATA with REFERENCE
orl-rat LD50:730 mg/kg NYKZAU 64,351,68
scu-rat LD50:610 mg/kg NYKZAU 64,351,68
orl-mus LD50:710 mg/kg ARZNAD 13,728,63
ivn-mus LD50:240 mg/kg ARZNAD 13,728,63

SAFETY PROFILE: Poison by intravenous route. Moderately toxic by subcutaneous and ingestion routes. When heated to decomposition it emits acrid smoke and fumes. See also ACETYLENE COMPOUNDS.

DMZ000 CAS:2277-92-1 ***HR: 3***
2,2′-DIHYDROXY-3,3′,5,5′,6-PENTACHLOROBENZANILIDE
mf: $C_{13}H_6Cl_5NO_3$ mw: 401.45

SYNS: DIPLIN ◇ ICI 46638 ◇ OXYCLOZANID ◇ OXYCLOZANIDE ◇ 3,5,6,3′,5′-PENTACHLORO-2,2′-DIHYDROXYBENZANILIDE ◇ 3,3′,5,5′,6-PENTACHLORO-2,2′-DIHYDROXYBENZANILIDE ◇ 3,3′,5,5′,6-PENTACHLORO-2′-HYDROXYSALICYLANILIDE ◇ 2,3,5-TRICHLORO-N-(3,5-DICHLORO-2-HYDROXYPHENYL)-6-HYDROXYBENZAMIDE ◇ ZANIL ◇ ZANILOX

TOXICITY DATA with REFERENCE
orl-rat LD50:1000 mg/kg NATUAS 210,744,66
orl-dom LDLo:60 mg/kg VETRAX 78,267,66
orl-ctl LDLo:60 mg/kg VETRAX 78,267,66
ivn-dom LDLo:10 mg/kg VETRAX 78,267,66

SAFETY PROFILE: Poison by ingestion and intravenous routes. When heated to decomposition it emits very toxic fumes of Cl^- and NO_x.

DNA200 CAS:59-92-7 ***HR: 3***
l-DIHYDROXYPHENYL-l-ALANINE
mf: $C_9H_{11}NO_4$ mw: 197.21

SYNS: 2-AMINO-3-(3,4-DIHYDROXYPHENYL)PROPANOICACID ◇ BENDOPA ◇ BIODOPA ◇ BROCADOPA ◇ CEREPAP ◇ CIDANDOPA ◇ DA ◇ DEADOPA ◇ DIHYDROXY-l-PHENYLALANINE ◇ (−)-3-(3,4-DIHYDROXYPHENYL)-l-ALANINE ◇ β-(3,4-DIHYDROXYPHENYL)-α-ALANINE ◇ l-α-DIHYDROXYPHENYLALANINE ◇ l-β-(3,4-DIHYDROXYPHENYL)ALANINE ◇ l-3,4-DIHYDROXYPHENYL-α-ALANINE ◇ β-(3,4-DIHYDROXYPHENYL)-l-ALANINE ◇ 3-(3,4-DIHYDROXYPHENYL)-l-ALANINE ◇ 3,4-DIHYDROXYPHENYLALANINE ◇ (−)-3,4-DIHYDROXYPHENYLALANINE ◇ 3,4-DIHYDROXYPHENYL-l-ALANINE ◇ 3,4-DIHYDROXY-l-PHENYLALANINE ◇ l-3,4-DIHYDROXYPHENYLALANINE ◇ (−)-DOPA ◇ l-DOPA ◇ DOPAFLEX ◇ DOPAL ◇ DOPARKINE ◇ DOPASOL ◇ DOPRIN ◇ ELDOPAL ◇ EURODOPA ◇ HELFO DOPA ◇ l-o-HYDROXYTYROSINE ◇ 3-HYDROXY-l-TYROSINE ◇ INSULAMINA ◇ LARODOPA ◇ MAIPEDOPA ◇ PARDA ◇ RO 4-6316 ◇ SOBIODOPA ◇ VELDOPA

TOXICITY DATA with REFERENCE
dnr-bcs 500 μg/disc MUREAV 137,17,84
dni-hmn:fbr 3 mmol/L CNREA8 42,3783,82

orl-rat TDLo:3600 mg/kg (female 9-14D post):REP KSRNAM 4,2877,70
orl-mus TDLo:4800 mg/kg (female 7-12D post):TER IYKEDH 3,53,72
orl-man TDLo:87520 mg/kg/1.5Y-C:CAR,SKN NEURAI 24,340,74
orl-wmn TDLo:320 mg/kg/4D-I AHJOA2 110,488,85
orl-hmn TDLo:156 g/kg/10Y:CNS JNNPAU 34,502,71
orl-hmn TDLo:13 g/kg/1Y:CNS,PUL JNNPAU 34,668,71
orl-rat LD50:1780 mg/kg TXAPA9 28,1,74
ipr-rat LD50:624 mg/kg TXAPA9 28,1,74
orl-mus LD50:2363 mg/kg TXAPA9 28,1,74
ipr-mus LD50:588 mg/kg TXAPA9 28,1,74
scu-mus LD50:4449 mg/kg IYKEDH 3,186,72
ivn-mus LD50:450 mg/kg TXAPA9 28,1,74
orl-rbt LD50:609 mg/kg TXAPA9 28,1,74
orl-bwd LD50:100 mg/kg AECTCV 12,355,83

CONSENSUS REPORTS: Reported in EPA TSCA Inventory.

SAFETY PROFILE: Poison by ingestion. Moderately toxic by intravenous and intraperitoneal routes. Human systemic effects by ingestion: somnolence, hallucinations and distorted perceptions, toxic psychosis, motor activity changes, ataxia, dyspnea. Experimental teratogenic and reproductive effects. Questionable human carcinogen producing skin tumors. Human mutation data reported. An anticholinergic agent used as an anti-Parkinsonian drug. When heated to decomposition it emits toxic fumes of NO_x.

DNA300 CAS:65561-73-1 ***HR: D***
3,4-DIHYDROXYPHENYLGLYOXIME
mf: $C_8H_8N_2O_4$ mw: 196.18

SYN: DPG

TOXICITY DATA with REFERENCE
dnd-omi 250 μmol/L ABCHA6 42,1019,78
dnd-hmn:hla 250 μmol/L ABCHA6 42,1019,78
dnd-rat:lng 250 μmol/L ABCHA6 42,1019,78

SAFETY PROFILE: Human mutation data reported. When heated to decomposition it emits toxic fumes of NO_x.

DNA600 CAS:13055-82-8 ***HR: 3***
7-(3-(2-(3,5-DIHYDROXYPHENYL-2-HYDROXY-ETHYLAMINO)PROPYL)THEOPHYLLINE HYDROCHLORIDE
mf: $C_{18}H_{23}N_5O_5$•ClH mw: 425.92

SYNS: BRONCHODIL ◇ BRONCHOSPASMIN ◇ REPROTEROL HYDROCHLORIDE ◇ 7-(3-((β,3,5-TRIHYDROXYPHENETHYL)AMINO)PROPYL)THEOPHYLLINE MONOHYDROCHLORIDE ◇ W-2946M

TOXICITY DATA with REFERENCE
orl-rat TDLo:35 g/kg (16-22D preg/28D post):REP ARZNAD 27,45,77
ivn-rat LD50:142 mg/kg ARZNAD 27,45,77
ivn-mus LD50:148 mg/kg ARZNAD 27,45,77
ivn-dog LD50:160 mg/kg ARZNAD 27,45,77

SAFETY PROFILE: Poison by intravenous route. Experimental reproductive effects. When heated to decomposition it emits very toxic fumes of HCl and NO_x. See also THEOPHYLLINE.

DNA800 CAS:555-30-6 ***HR: 3***
l-(−)-3-(3,4-DIHYDROXYPHENYL)-2-METHYLALANINE
mf: $C_{10}H_{13}NO_4$ mw: 211.24

SYNS: ALDOMET ◇ ALDOMETIL ◇ ALDOMIN ◇ ALPHA MEDOPA ◇ AMD ◇ BAYER 1440 L ◇ BAYPRESOL ◇ l(−)-β-(3,4-DIHYDROXYPHENYL)-α-METHYLALANINE ◇ DOPAMET ◇ DOPEGYT ◇ DOPTAEC ◇ 3-HYDROXY-α-METHYL-l-TYROSINE ◇ HYPERPAX ◇ l-(α-MD) ◇ MEDOMET ◇ MEDOPREN ◇ METHOPLAIN ◇ α-METHYL-l-3,4-DIHYDROXYPHENYLALANINE ◇ l-α-METHYL-3,4-DIHYDROXYPHENYLALANINE ◇ α-METHYL-β-(3,4-DIHYDROXYPHENYL)-l-ALANINE ◇ l-(−)-α-METHYL-β-(3,4-DIHYDROXYPHENYL)ALANINE ◇ METHYLDOPA ◇ l-α-METHYLDOPA ◇ α-METHYL-l-DOPA ◇ MK. B51 ◇ MK 351 ◇ NCI-C55721 ◇ NR.C 2294 ◇ PRESINOL ◇ PRESOLISIN ◇ SEDOMETIL ◇ SEMBRINA

TOXICITY DATA with REFERENCE
dlt-mus-orl 960 mg/kg CYTBAI 41,151,84
cyt-ham:lng 37 mg/L CMCRDC 27,95,81
unr-wmn TDLo:1680 mg/kg (28-36W preg):REP LANCAO 2,498,82
orl-rat TDLo:7500 mg/kg (female 6-20D post):TER NTIS** PB86-245321
orl-wmn TDLo:900 mg/kg/13W-I NEURAI 35,1668,85
orl-wmn TDLo:1830 mg/kg/17W-I:PNS SAMJAF 65,194,84
orl-man TDLo:1071 mg/kg/22W-I:SKN CUTIBC 38,187,86
orl-wmn TDLo:44 g/kg/3Y-I:GIT AHJOA2 105,1037,83
orl-rat LD50:5000 mg/kg 27ZQAG -,348,72
ipr-rat LD50:300 mg/kg 27ZQAG -,348,72
ipr-mus LD50:150 mg/kg JMCMAR 20,1378,77
ivn-mus LD50:1700 mg/kg NYKZAU 56,1103,60
orl-rbt LD50:713 mg/kg 27ZIAQ -,162,73
ivn-rbt LD50:713 mg/kg 27ZQAG -,348,72

CONSENSUS REPORTS: Reported in EPA TSCA Inventory. EPA Genetic Toxicology Program.

SAFETY PROFILE: Poison by intraperitoneal route. Moderately toxic by ingestion and intravenous routes. Human systemic effects by ingestion: fasciculations, hallucinations, distorted perceptions, tremors, allergic dermatitis, necrotic gastrointestinal changes. An experimental teratogen. Human reproductive effects: menstrual cycle changes or disorders, effects on newborn including abnormal neonatal measures and growth statistics, biochemical and metabolic changes. Experimental reproductive effects. Mutation data reported. When heated to decomposition it emits toxic fumes of NO_x.

DNB000 CAS:2589-47-1 ***HR: 3***
17R,21-α-DIHYDROXY-4-PROPYLAJMALANIUM HYDROGEN TARTRATE
mf: $C_{23}H_{32}N_2O_2 \cdot C_4H_6O_6$ mw: 518.67

SYNS: GT-1012 ◇ NEO-GILURYTMAL ◇ NPA ◇ PRAJMALINE BITARTRATE ◇ PRAJMALINE HYDROGEN TARTRATE ◇ N-PROPYLAJMALINE BITARTRATE ◇ N-PROPYLAJMALINE HYDROGEN TARTRATE ◇ N-PROPYLAJMALINIUM BITARTRATE ◇ N-PROPYLAJMALINIUMHYDROGENTARTRAT (GERMAN) ◇ N^4-PROPYLAJMALINIUM HYDROGEN TARTRATE

TOXICITY DATA with REFERENCE
orl-rbt TDLo:65 mg/kg (6-18D preg):TER ARZNAD 22,2085,72
orl-wmn LDLo:22 mg/kg ARTODN 37,135,77
orl-man TDLo:1400 μg/kg:CNS BMJOAE 2,675,77
orl-rat LD50:54 mg/kg ARZNAD 22,2085,72
ivn-rat LD50:3400 μg/kg ARZNAD 22,2085,72
orl-mus LD50:43 mg/kg MEIEDD 10,1107,83
ivn-mus LD50:1700 μg/kg MEIEDD 10,1107,83

SAFETY PROFILE: Poison by ingestion and intravenous routes. An experimental teratogen. Human systemic effects by ingestion: hallucinations and distorted perceptions. Experimental reproductive effects. An anti-arrhythmic agent. When heated to decomposition it emits toxic fumes of NO_x.

DNB200 CAS:53609-64-6 ***HR: 3***
DI(2-HYDROXY-n-PROPYL)AMINE
mf: $C_6H_{14}N_2O_3$ mw: 162.22

SYNS: BHP ◇ N-BIS(2-HYDROXYPROPYL)NITROSAMINE ◇ 2,2'-BISHYDROXYPROPYLNITROSAMINE ◇ DHPN ◇ 2,2'-DIHYDROXY-DI-n-PROPYLNITROSOAMINE ◇ N,N-DI-(2-HYDROXYPROPYL)NITROSAMINE ◇ DIISOPROPANOLNITROSAMINE ◇ DIPN ◇ N-NITROSOBIS(2-HYDROXYPROPYL)AMINE ◇ N-NITROSO-N,N-DI(2-HYDROXYPROPYL)AMINE ◇ N-NITROSO-1,1'-IMINODI-2-PROPANOL ◇ 1,1'-NITROSOIMINODI-2-PROPANOL

TOXICITY DATA with REFERENCE
mma-sat 250 μg/plate MUREAV 111,135,83
otr-hmn:oth 5 mg/L BANRDU 12,15,82
dns-rat:lvr 5 mmol/L MUREAV 144,197,85
msc-ham:lng 700 μmol/L CNREA8 40,3463,80
orl-rat TDLo:4600 mg/kg/42W-C:CAR CRNGDP 5,167,84
ipr-rat TDLo:3000 mg/kg:CAR CALEDQ 6,115,79
orl-mus TDLo:28 g/kg/16W-I:NEO CALEDQ 3,255,77
ims-mus TDLo:125 mg/kg:ETA IGSBAL 109,99,84
scu-ham TDLo:100 mg/kg (14D post):NEO,TER ZEKBAI 90,119,77
scu-rat LD50:5000 mg/kg JJIND8 63,181,79
scu-mus LD50:5160 mg/kg CALEDQ 9,257,80
scu-gpg LD50:4900 mg/kg JNCIAM 58,387,77

SAFETY PROFILE: Suspected carcinogen with experimental carcinogenic, neoplastigenic, tumorigenic, and teratogenic data. Moderately toxic by subcutaneous route. Human mutation data reported. When heated to decomposition it emits toxic fumes of NO_x. See also NITROSAMINES.

DNB600 CAS:33372-40-6 ***HR: 3***
4-(2,3-DIHYDROXYPROPYLAMINO)-2-(5-NITRO-2-THIENYL)QUINAZOLINE
mf: $C_{15}H_{14}N_4O_4S$ mw: 346.39

TOXICITY DATA with REFERENCE
mma-sat 1250 μg/plate CNREA8 35,3611,75
orl-rat TDLo:8313 mg/kg/47W-C:CAR JNCIAM 57,277,76

SAFETY PROFILE: Questionable carcinogen with experimental carcinogenic data. Mutation data reported. When heated to decomposition it emits very toxic fumes of NO_x and SO_x.

DNC000 CAS:479-18-5 ***HR: 3***
7-(2,3-DIHYDROXYPROPYL)THEOPHYLLINE
mf: $C_{10}H_{14}N_4O_4$ mw: 254.28

SYNS: AFI-PHYLLIN ◇ ARISTOPHYLLIN ◇ ASTMAMASIT ◇ ASTROPHYLLIN ◇ CIRCAIN ◇ CIRCAIR ◇ CORONAL ◇ CORONARIN ◇ CORPHYLLIN ◇ COR-THEOPHYLLINE ◇ 7-(2,3-DIHYDROXYPROPYL)-3,7-DIHYDRO-1,3-DIMETHYL-1H-PURINE-2,5-DIONE ◇ DIHYDROXYPROPYL THEOPYLIN (GERMAN) ◇ DIHYDROXYPROPYL THEOPHYLLINE ◇ (1,2-DIHYDROXY-3-PROPYL)THIOPHYLLIN ◇ DILOR ◇ 1,3-DIMETHYL-7-(2,3-DIHYDROXYPROPYL)XANTHINE ◇ 7-(2,3-DIOXYPROPYL)THEOPHYLLINE ◇ DIPHYLLIN ◇ DIPROFILLIN ◇ DIPROFILLINE ◇ DIPROPHYLLIN ◇ DIPROPHYLLINE ◇ DT ◇ DYPHYLLINE ◇ GLYFYLLIN ◇ GLYPHYLLIN ◇ GLYPHYLLINE ◇ HIDROXITEOFILLINA ◇ HIPHYLLIN ◇ HYPHYLLINE ◇ LUFYLLIN ◇ NEOPHYLLIN ◇ NEOPHYLLINE ◇ NEOPHYLLIN M ◇ NEOSTENOVASAN ◇ NEOTHYLLINE ◇ NEOTILINA ◇ NEO-VASOPHYLINE ◇ NEUFIL ◇ NEUTRAFIL ◇ NEUTRAFILLINA ◇ NEUTRAPHYLLIN ◇ NEUTRAPHYLLINE ◇ NEUTROXANTINA ◇ PROPYLPHYLLIN ◇ PROTHEOPHYLLINE ◇ PURIFILIN ◇ SIBEPHYLLIN ◇ SIBEPHYLLINE ◇ SOLUFILIN ◇ SOLUFYLLIN ◇ SYNTHOPHYLLINE ◇ TEFILAN ◇ THEAL ◇ THEAL AMPULES ◇ THEFYLAN

TOXICITY DATA with REFERENCE
scu-rat LD50:1253 mg/kg AEPPAE 230,194,57
ivn-rat LD50:860 mg/kg AEPPAE 230,194,57
orl-mus LD50:1954 mg/kg JPETAB 116,343,56
ipr-mus LD50:1052 mg/kg ARZNAD 8,190,58
scu-mus LD50:120 mg/kg ARZNAD 6,601,56
ivn-mus LD50:1080 mg/kg RPOBAR 2,288,70

CONSENSUS REPORTS: Reported in EPA TSCA Inventory.

SAFETY PROFILE: Poison by subcutaneous route. Moderately toxic by ingestion, intraperitoneal, and intravenous routes. A smooth muscle relaxant. When heated to decomposition it emits toxic fumes of NO_x. See also THEOPHYLLINE.

DNC200 CAS:59-00-7 ***HR: 3***
4,8-DIHYDROXYQUINALDIC ACID
mf: $C_{10}H_7NO_4$ mw: 205.18

PROP: Sulfur-yellow crystals. Mp: 286°. Insol in water; sol in aqueous alkali, hydroxides, and hot dil HCl.

SYNS: 4,8-DIHYDROXYQUINALDINIC ACID ◇ 4,8-DIHYDROXY-QUINOLINE-2-CARBOXYLIC ACID ◇ XANTHURENIC ACID

TOXICITY DATA with REFERENCE
imp-mus TDLo:160 mg/kg:NEO ANYAA9 108,924,63

SAFETY PROFILE: Questionable carcinogen with experimental neoplastigenic data. When heated to decomposition it emits toxic fumes of NO_x.

DNC400 CAS:66788-03-2 ***HR: 3***
trans-9,10-DIHYDROXY-9,10,11,12-TETRA-HYDROBENZO(e)PYRENE
mf: $C_{20}H_{16}O_2$ mw: 288.36

SYN: B(E)P H4-9,10-DIOL

TOXICITY DATA with REFERENCE
mma-sat 10 nmol/plate JBCHA3 254,4408,79
skn-mus TDLo:69 mg/kg:ETA CNREA8 40,203,80

SAFETY PROFILE: Questionable carcinogen with experimental tumorigenic data. Mutation data reported. When heated to decomposition it emits acrid smoke and fumes.

DNC600 CAS:73771-79-6 ***HR: 3***
trans-1,2-DIHYDROXY-1,2,3,4-TETRAHYDRO-CHRYSENE
mf: $C_{18}H_{16}O_2$ mw: 264.34

SYN: trans-1,2,3,4-TETRAHYDROCHRYSENE-1,2,-DIOL

TOXICITY DATA with REFERENCE
skn-mus TDLo:42 mg/kg:NEO CNREA8 38,1831,78

SAFETY PROFILE: Questionable carcinogen with experimental neoplastigenic data. When heated to decomposition it emits acrid smoke and irritating fumes.

DNC800 CAS:70443-38-8 ***HR: 3***
trans-3,4-DIHYDROXY-1,2,3,4-TETRAHYDRO-DIBENZ(a,h)ANTHRACENE
mf: $C_{22}H_{18}O_2$ mw: 314.40

SYN: trans-3,4-DIHYDROXY-1,2,3,4-TETRAHYDRODIBENZO(a,h) ANTHRACENE

TOXICITY DATA with REFERENCE
mma-sat 60 μg/plate MUREAV 96,1,82
skn-mus TDLo:2010 mg/kg:ETA CNREA8 39,1310,79

SAFETY PROFILE: Questionable carcinogen with experimental tumorigenic data. Mutation data reported. When heated to decomposition it emits acrid smoke and fumes.

DND000 CAS:74444-58-9 ***HR: D***
trans-1,2-DIHYDROXY-1,2,3,4-TETRAHYDRO-TRIPHENYLENE
mf: $C_{18}H_{16}O_2$ mw: 264.2

SYN: TP H4-1,2-DIOL

TOXICITY DATA with REFERENCE
mmo-sat 1 nmol/plate CNREA8 40,1985,80
mma-sat 1 nmol/plate CNREA8 40,1985,80

SAFETY PROFILE: Mutation data reported. When heated to decomposition it emits acrid smoke and fumes.

DND400 CAS:1246-87-3 ***HR: 3***
1,1'-(2,3-DIHYDROXYTETRAMETHYLENE)BIS(4-FORMYLPYRIDINIUM) DIPERCHLORATE, DIOXIME
mf: $C_{16}H_{20}N_4O_4{\cdot}2ClO_4$ mw: 531.30

SYN: 1,4-BIS(4-HYDROXYIMINOMETHYL-PYRIDINIUM-(1))BUTANEDIOL(2,3)-DIPERCHLORAT(GERMAN)

TOXICITY DATA with REFERENCE
ipr-mus LD50:190 mg/kg ARZNAD 14,870,64
ivn-mus LD50:65 mg/kg ARZNAD 14,870,64

SAFETY PROFILE: Poison by intraperitoneal and intravenous routes. Can explode when heated. When heated to decomposition it emits very toxic fumes of NO_x and Cl^-.

DND900 ***HR: 3***
5,7-DIHYDROXYTETRAZOLO(1,5-a)PYRIDINE-6-CARBONITRILE
mf: $C_6H_3N_5O_2$ mw: 177.14

SYNS: 5,7-DIHYDROXY-PYRIDOTETRAZOLE-6-CARBONITRILE ◇ 1,3-DIOXY-2-NICOTINSAEURENITRIL-TETRAZOL (GERMAN)

TOXICITY DATA with REFERENCE
scu-mus LDLo:500 mg/kg BDVU** -,-,37
scu-gpg LDLo:346 mg/kg BDVU** -,-,37
scu-frg LDLo:3800 mg/kg BDVU** -,-,37

SAFETY PROFILE: Poison by subcutaneous route. When heated to decomposition it emits toxic fumes of NO_x.

DNE000 CAS:488-17-5 ***HR: 3***
2,3-DIHYDROXYTOLUENE
mf: $C_7H_8O_2$ mw: 124.15

SYNS: 3-METHYL-1,2-BENZENEDIOL ◇ 3-METHYLCATECHOL ◇ 3-METHYLPYROCATECHOL ◇ 2,3-TOLUENEDIOL

TOXICITY DATA with REFERENCE
ivn-mus LD50:56 mg/kg CSLNX* NX#07878

CONSENSUS REPORTS: Reported in EPA TSCA Inventory.

SAFETY PROFILE: Poison by intravenous route. When heated to decomposition it emits acrid smoke and fumes.

DNE200 CAS:452-86-8 ***HR: D***
3,4-DIHYDROXYTOLUENE
mf: $C_7H_8O_2$ mw: 124.15

SYNS: HOMOCATECHOL ◇ HOMOPYROCATECHOL ◇ 4-METHYL-1,2-BENZENEDIOL ◇ 4-METHYLCATECHOL ◇ p-METHYLPYROCATECHOL ◇ 4-METHYLPYROCATECHOL ◇ TOLUENE-3,4-DIOL

TOXICITY DATA with REFERENCE
mrc-smc 300 mg/L MUREAV 135,109,84
cyt-ham:ovr 10 mg/L CALEDQ 14,251,81

SAFETY PROFILE: Mutation data reported. When heated to decomposition it emits acrid smoke and fumes.

DNE400 CAS:3468-11-9 ***HR: 3***
1,3-DIIMINOISOINDOLINE
mf: $C_8H_7N_3$ mw: 145.18

SYNS: AFASTOGEN BLUE 5040 ◇ 1,3-DIIMINOISOINDOLIN (CZECH) ◇ FASTOGEN BLUE FP-3100 ◇ FASTOGEN BLUE SH-100 ◇ MODR FRALOSTANOVA 3G (CZECH) ◇ PHTHALIMIDIMIDE ◇ PHTHALOCYANINE BLUE 01206 ◇ PHTHALOGEN

TOXICITY DATA with REFERENCE
skn-rbt 500 mg/24H SEV 28ZPAK -,143,72
eye-rbt 250 μg/24H SEV 28ZPAK -,143,72
scu-rat TDLo:990 mg/kg/44W-I:ETA VOONAW 21(11),75,75
scu-mus TDLo:140 mg/kg/51W-I:CAR VOONAW 21(11),75,75
orl-rat LD50:273 mg/kg 28ZPAK -,143,72

CONSENSUS REPORTS: Reported in EPA TSCA Inventory.

SAFETY PROFILE: Poison by ingestion. A severe eye and skin irritant. Questionable carcinogen with experimental carcinogenic and tumorigenic data. When heated to decomposition it emits toxic fumes of NO_x.

DNE500 CAS:624-74-8 ***HR: 3***
DIIODOACETYLENE
mf: C_2I_2 mw: 277.83

SYN: DIIODOETHYNE

DOT Classification: Forbidden

SAFETY PROFILE: An explosive sensitive to impact, crushing, or heating to 84°C. When heated to decomposition it emits toxic fumes of I^-. See also IODIDES and ACETYLENE COMPOUNDS.

DNE600 ***HR: 3***
DIIODOAMINE
mf: HI_2N mw: 268.82

SYNS: DIIODAMINE ◇ IODIMIDE

SAFETY PROFILE: An explosive formed by reaction of nitrogen triiodide with water. When heated to decomposition it emits toxic fumes of I^- and NO_x. See also IODIDES and AMINES.

DNE700 CAS:615-42-9 ***HR: 3***
1,2-DIIODOBENZENE
mf: $C_6H_4I_2$ mw: 329.91

SAFETY PROFILE: Explodes violently when heated to 181°C in a sealed container. When heated to decomposition it emits toxic fumes of I^-. See also IODIDES.

DNE800 CAS:4460-32-6 ***HR: 2***
N-2,5-DIIODOBENZOYL-N′,N′,N″,N″-DIETHYLENEPHOSPHORTRIAMIDE
mf: $C_{11}H_{12}I_2N_3O_2P$ mw: 503.03

SYNS: p,p-BIS(1-AZIRIDINYL)-2,5-DIIODOBENZOYLPHOSPINIC AMIDE ◇ N-(BIS(1-AZIRIDINYL)PHOSPHINYL)-2,5-DIIODOBENZAMIDE ◇ DIIODBENZOTEPH ◇ DIIODOBENZOTEF

TOXICITY DATA with REFERENCE
orl-rat LD50:500 mg/kg PCJOAU 12,689,78
ipr-rat LD50:500 mg/kg RPTOAN 41,135,78

SAFETY PROFILE: Moderately toxic by ingestion and intraperitoneal routes. When heated to decomposition it emits very toxic fumes of PO_x, NO_x, and I^-.

DNE875 CAS:53214-97-4 ***HR: 3***
1,4-DIIODO-1,3-BUTADIYNE
mf: C_4I_2 mw: 174.95

$$IC\equiv CC\equiv CI$$

SAFETY PROFILE: Explodes at 100°C. When heated to decomposition it emits toxic fumes of I^-. See also IODIDES.

DNF000 CAS:15978-93-5 ***HR: 2***
cis-DIIODODIAMMINEPLATIUM (II)
mf: $H_6I_2N_2Pt$ mw: 482.97

TOXICITY DATA with REFERENCE
idr-hmn TDLo:40 mg/kg:SKN CNREA8 35,2766,75
pic-esc 1 μg/plate BBRCA9, 90,209,79

CONSENSUS REPORTS: Reported in EPA TSCA Inventory.

SAFETY PROFILE: Human systemic effects by intradermal route: unspecified effects on the skin. Mutation data reported. When heated to decomposition it emits very toxic fumes of I^- and NO_x. See also IODIDES and PLATINUM COMPOUNDS.

DNF200 CAS:1955-21-1 ***HR: 3***
2,6-DIIODOHYDROQUINONE
mf: $C_6H_4I_2O_2$ mw: 361.90

SYNS: 2,6-DIIODO-1,4-BENZENEDIOL ◇ 2,6-DIIODOQUINOL

TOXICITY DATA with REFERENCE
ipr-mus LD50:237 mg/kg BCPCA6 12,885,63

CONSENSUS REPORTS: Reported in EPA TSCA Inventory.

SAFETY PROFILE: Poison by intraperitoneal route. When heated to decomposition it emits toxic fumes of I^-.

DNF400 CAS:2961-61-7 ***HR: 3***
3,5-DIIODO-4-HYDROXYBENZONITRILE, LITHIUM SALT
mf: $C_7H_2I_2NO•Li$ mw: 376.84

SYNS: BENTROL ◇ CERTOL ◇ 4-CYANO-2,6-DIJODPHENOL LITHIUMSALZ (GERMAN) ◇ 3,5-DIJOD-4-HYDROXY-BENZONITRILE LITHIUMSALZ (GERMAN) ◇ IOXYNIL, LITHIUM SALT

TOXICITY DATA with REFERENCE
orl-rat LD50:71 mg/kg RREVAH 10,97,65
skn-rat LD50:87 mg/kg 85GYAZ -,93,71
orl-mus LD50:190 mg/kg 85GYAZ -,93,71
orl-dog LD50:140 mg/kg PCOC** -,616,66
orl-ckn LD50:120 mg/kg 85GYAZ -,93,71

CONSENSUS REPORTS: Cyanide and its compounds are on the Community Right-To-Know List.

SAFETY PROFILE: Poison by ingestion and skin contact. When heated to decomposition it emits very toxic fumes of I^-, NO_x, CN^-, and Li_2O. See also NITRILES and LITHIUM COMPOUNDS.

DNF600 CAS:83-73-8 ***HR: 3***
DIIODOHYDROXYQUIN
mf: $C_9H_5I_2NO$ mw: 396.95

SYNS: DIIODOHYDROXYQUINOLINE ◇ 5,7-DIIODO-8-HYDROXYQUINOLINE ◇ 5,7-DIIODO-OXINE ◇ 5,7-DIIODO-8-QUINOLINOL ◇ DIODOHYDROXYQUIN ◇ 8-HYDROXY-5,7-DIIODOQUINOLINE

TOXICITY DATA with REFERENCE
orl-chd TDLo:120 g/kg/2Y-I:EYE LANCAO 1,261,66
ipr-mus LDLo:128 mg/kg CBCCT* 2,242,50
ivn-mus LD50:56 mg/kg CSLNX* NX#03304
orl-cat LDLo:300 mg/kg AJTMAQ 24,29,44
orl-gpg LDLo:50 mg/kg AJTMAQ 24,29,44

CONSENSUS REPORTS: Reported in EPA TSCA Inventory.

SAFETY PROFILE: Poison by ingestion, intravenous, and intraperitoneal routes. Human systemic effects by ingestion: eye effects. When heated to decomposition it emits very toxic fumes of I^- and NO_x.

DNF800 CAS:75-11-6 ***HR: 3***
DIIODOMETHANE
mf: CH_2I_2 mw: 267.83

PROP: Light straw-colored to clear, heavy, refractive liquid. Mp: 5-6°, bp: 181°, d: 3.325 @ 20°/4°, vap d: 9.25.

SYNS: METHYLENE DIIODIDE ◇ METHYLENE IODIDE ◇ MI-GEE

TOXICITY DATA with REFERENCE
ipr-rat LD50:403 mg/kg 34ZIAG -,756,69
ipr-mus LD50:467 mg/kg 34ZIAG -,756,69
scu-mus LD50:830 mg/kg TXAPA9 4,354,62

CONSENSUS REPORTS: Reported in EPA TSCA Inventory.

SAFETY PROFILE: Moderately toxic by intraperitoneal and subcutaneous routes. Probably an irritant and narcotic in high concentration. Potentially explosive reaction with diethyl zinc + alkenes. Violent reaction with copper-zinc alloys + ether. Forms very shock-sensitive explosive mixtures with potassium, potassium-sodium alloys, and lithium. When heated to decomposition it emits toxic fumes of I^-. See also IODIDES.

DNG000 CAS:305-85-1 ***HR: 3***
2,6-DIIODO-4-NITROPHENOL
mf: $C_6H_3I_2NO_3$ mw: 390.90

SYNS: ANCYLOL ◇ DIISOPHENOL ◇ DISOFEN ◇ DNP ◇ DISOPHENOL

TOXICITY DATA with REFERENCE
orl-rat LD50:170 mg/kg TXAPA9 6,232,64
ipr-rat LD50:105 mg/kg TXAPA9 6,232,64
scu-rat LD50:122 mg/kg TXAPA9 6,232,64
ivn-rat LD50:105 mg/kg TXAPA9 6,232,64
orl-mus LD50:212 mg/kg TXAPA9 6,232,64
ipr-mus LD50:107 mg/kg TXAPA9 6,232,64
scu-mus LD50:110 mg/kg TXAPA9 6,232,64
ivn-mus LD50:88 mg/kg TXAPA9 6,232,64
par-uns LDLo:36 mg/kg FAZMAE 17,108,73

CONSENSUS REPORTS: Reported in EPA TSCA Inventory.

SAFETY PROFILE: Poison by ingestion, intraperitoneal, subcutaneous, intravenous, and parenteral routes. An anthelmintic. When heated to decomposition it emits very toxic fumes of I^- and NO_x. See also NITRO COMPOUNDS of AROMATIC HYDROCARBONS.

DNG200 CAS:3861-47-0 ***HR: 3***
3,5-DIIODO-4-OCTANOYLOXYBENZONITRILE
mf: $C_{15}H_{17}I_2NO_2$ mw: 497.13

SYNS: 4-CYANO-2,6-DIJODPHENOL CAPRYSAEUREESTER (GERMAN) ◇ 3,5-DIIODO-4-HYDROXYBENZONITRILE OCTANOATE ◇ 3,5-DIJOD-4-HYDROXY-BENZONITRIL CAPRYSAEUREESTER (GERMAN) ◇ IOXYNIL OCTANOATE ◇ M&B 11,461 ◇ RIP-15830 ◇ TOTRIL

TOXICITY DATA with REFERENCE
mrc-bcs 20 μg/disc/24H MUREAV 40,19,76
orl-rat LD50:190 mg/kg GUCHAZ 6,305,73
orl-mus LD50:240 mg/kg GUCHAZ 6,305,73

CONSENSUS REPORTS: EPA Genetic Toxicology Program. Cyanide and its compounds are on the Community Right-To-Know List.

SAFETY PROFILE: Poison by ingestion. Mutation data reported. An herbicide. When heated to decomposition it emits very toxic fumes of I^-, NO_x, and CN^-. See also NITRILES.

DNG400 CAS:300-37-8 ***HR: 2***
3,5-DIIODO-4-OXO-1(4H)PYRIDINEACETIC ACID-2,2'-IMINODIETHANOL SALT
mf: $C_7H_5I_2O_3 \cdot C_4H_{11}NO_2$ mw: 510.09

SYNS: CARDIOTRAST ◇ DIAETHANOLAMIN-3,5-DIJODPYRIDON-(4)-ESSIGSAEURE (GERMAN) ◇ DIATRAST ◇ DIETHANOLAMINE-3,5-DIIODO-4-PYRIDONE-N-ACETATE ◇ 3,5-DIIODO-4-PYRIDONE-N-ACETATE BIS(HYDROXYETHYL)AMMONIUM ◇ 3,5-DIIODO-4-PYRIDONE-N-ACETIC ACID, DIETHANOLAMINE SALT ◇ DIODON ◇ DIODONE ◇ DIODRAST ◇ ETHANOL,2,2'-IMINODI-,3,5-DIIODO-4-OXO-1(4H)-PYRIDINEACETATE (salt) ◇ ETHANOL,2,2'-IMINODI- with 3,5-DIIODO-4-OXO-1(4H)-PYRIDINEACETIC ACID (1:1) ◇ IODOPYRACET ◇ IODURON B ◇ IOPYRACIL ◇ METHYLGLUCAMINE-3,5-DIIODO-4-PYRIDONE-N-ACETATE ◇ MOSYLAN ◇ NEOMETHIODAL ◇ NEO-SKIODAN ◇ NEO-TENEBRYL ◇ NOSYDRAST ◇ OPARENOL ◇ PELVIRAN ◇ PER-ABRODIL ◇ PER-RADIOGRAPHOL ◇ PYELOSIL ◇ PYLUMBRIN ◇ PYRACETON ◇ RP 3203 ◇ SAVAC ◇ UMBRADIL ◇ URIODONE ◇ VASIODONE ◇ XUMBRADIL

TOXICITY DATA with REFERENCE
ivn-rat LD50:5400 mg/kg AEPPAE 222,584,54
ivn-mus LD50:6400 mg/kg JPETAB 116,394,56
ivn-dog LD50:2 g/kg BJRAAP 6,304,33
ivn-cat LD50:2800 mg/kg MECHAN 6,344,63
ivn-rbt LD50:4700 mg/kg MECHAN 6,344,63

SAFETY PROFILE: Moderately toxic by intravenous route. When heated to decomposition it emits very toxic fumes of NO_x and I^-.

DNG800 CAS:20389-01-9 ***HR: 3***
DIIODOQUINONE
mf: $C_6H_2I_2O_2$ mw: 359.88

SYN: 2,6-DIIODO-p-BENZOQUINONE

TOXICITY DATA with REFERENCE
ipr-mus LD50:84 mg/kg BCPCA6 12,885,63

CONSENSUS REPORTS: Reported in EPA TSCA Inventory.

SAFETY PROFILE: Poison by intraperitoneal route. When heated to decomposition it emits toxic fumes of I^-.

DNH000 CAS:133-91-5 ***HR: 2***
3,5-DIIODOSALICYLIC ACID
mf: $C_7H_4I_2O_3$ mw: 389.91

PROP: White to pale pink, crystalline powder; sltly sol in water.

TOXICITY DATA with REFERENCE
orl-rat LDLo:500 mg/kg NCNSA6 5,8,53
orl-mus LD50:450 mg/kg QJPPAL 19,483,46

CONSENSUS REPORTS: Reported in EPA TSCA Inventory.

SAFETY PROFILE: Moderately toxic by ingestion. A trace mineral added to animal feeds. When heated to decomposition it emits toxic fumes of I^-.

DNH125 CAS:141-04-8 ***HR: 2***
DIISOBUTYL ADIPATE
mf: $C_{14}H_{26}O_4$ mw: 258.40

SYNS: DIBA ◇ FTAFLEX DIBA ◇ ISOBUTYL ADIPATE

TOXICITY DATA with REFERENCE
ipr-rat TDLo:1190 mg/kg (5-15D preg):TER JPMSAE 62,1596,73
ipr-rat LD50:5950 mg/kg JPMSAE 62,1596,73
orl-gpg LD50:12300 mg/kg GWXXBX #2703360

CONSENSUS REPORTS: Reported in EPA TSCA Inventory.

SAFETY PROFILE: Moderately toxic by intraperitoneal route. Mildly toxic by ingestion. Experimental teratogenic effects. When heated to decomposition it emits acrid smoke and fumes.

DNH400 CAS:110-96-3 ***HR: 3***
DIISOBUTYLAMINE
DOT: UN 2361
mf: $C_8H_{19}N$ mw: 129.28

PROP: Water-white liquid, amine odor. Mp: −70°, bp: 139°, flash p: 69.8°F, d: 0.745 @ 20°/4°, vap press: 10 mm @ 30.6°, vap d: 4.46.

SYN: 2-METHYL-N-(2-METHYLPROPYL)-1-PROPANAMINE

TOXICITY DATA with REFERENCE
orl-rat LD50:258 mg/kg HYSAAV 34(7-9),426,69

orl-mus LD50:629 mg/kg HYSAAV 34(7-9),426,69
orl-gpg LD50:620 mg/kg HYSAAV 34(7-9),426,69

CONSENSUS REPORTS: Reported in EPA TSCA Inventory.

DOT Classification: Flammable or Combustible Liquid; Label: Flammable Liquid.

SAFETY PROFILE: Poison by ingestion. A dangerous fire hazard when exposed to heat or flame; can react vigorously with oxidizing materials. To fight fire, use alcohol foam, CO_2, dry chemical. When heated to decomposition it emits toxic fumes of NO_x.

DNH500 CAS:102367-57-7 ***HR: 3***
DIISOBUTYLAMINOBENZOYLOXYPROPYL THEOPHYLLINE
mf: $C_{25}H_{35}N_5O_4$ mw: 469.65

SYN: α-((DIISOBUTYLAMINO)METHYL)THEOPHYLLINE-8-ETHANOL BENZOATE (ester)

TOXICITY DATA with REFERENCE
ipr-mus LD50:2567 mg/kg NIIRDN 6,306,82
ipr-mus LD50:1835 mg/kg NIIRDN 6,306,82
ivn-mus LD50:273 mg/kg NIIRDN 6,306,82

SAFETY PROFILE: Poison by intravenous route. Moderately toxic by ingestion and intraperitoneal routes. When heated to decomposition it emits toxic fumes of NO_x.

DNH800 CAS:108-82-7 ***HR: 2***
DIISOBUTYL CARBINOL
mf: $C_9H_{20}O$ mw: 144.29

PROP: Colorless liquid. Bp: 173.3°, fp: −65°, flash p: 165°F, d: 0.8121 @ 20°/20°, vap press: 0.3 mm @ 20°, vap d: 4.98, lel: 0.8% @ 212°F, uel: 6.1% @ 212°F.

SYNS: 2,6-DIMETHYL-4-HEPTANOL ◇ 2,6-DIMETHYL HEPTANOL-4 ◇ sec-NONYL ALCOHOL

TOXICITY DATA with REFERENCE
skn-rbt 10 mg/24H open MLD JIHTAB 31,60,49
skn-rbt 500 mg open MLD UCDS** 12/30/71
eye-rbt 500 mg open JIHTAB 31,60,49
orl-rat LD50:3160 mg/kg JIHTAB 31,60,49
ipr-rat LD50:800 mg/kg NPIRI* 1,22,74
orl-mus LD50:3530 mg/kg SCCUR* -,4,61
skn-rbt LD50:4600 mg/kg NPIRI* 1,22,74

CONSENSUS REPORTS: Reported in EPA TSCA Inventory.

SAFETY PROFILE: Moderately toxic by ingestion and intraperitoneal routes. Mildly toxic by skin contact. A powerful systemic irritant by inhalation. A skin and eye irritant. Can cause central nervous system and liver damage when ingested. Combustible when exposed to heat or flame; can react with oxidizing materials. To fight fire, use alcohol foam, foam, CO_2, dry chemical. When heated to decomposition it emits acrid smoke and fumes.

DNI200 CAS:63919-00-6 ***HR: 2***
DIISOBUTYLENE OXIDE
mf: $C_8H_{16}O$ mw: 128.24

SYN: EP-185

TOXICITY DATA with REFERENCE
skn-rbt 500 mg open MLD UCDS** 4/10/68
eye-rbt 500 mg open AMIHBC 10,61,54
orl-rat LD50:4920 mg/kg AMIHBC 10,61,54
ihl-rat LCLo:4000 ppm/4H AMIHBC 10,61,54
skn-rbt LD50:14 g/kg AMIHAB 14,250,56

SAFETY PROFILE: Moderately toxic by inhalation. Mildly toxic by ingestion and skin contact. A skin and eye irritant. When heated to decomposition it emits acrid smoke and fumes.

DNI400 CAS:7283-69-4 ***HR: 1***
DIISOBUTYL FUMARATE
mf: $C_{12}H_{20}O_4$ mw: 228.32

TOXICITY DATA with REFERENCE
skn-rbt 10 mg/24H open MLD AIHAAP 23,95,62
orl-rat LD50:8120 mg/kg AIHAAP 23,95,62
skn-rbt LD50:7490 mg/kg AIHAAP 23,95,62

SAFETY PROFILE: Mildly toxic by ingestion and skin contact. A skin irritant. When heated to decomposition it emits acrid smoke and fumes. See also ESTERS.

DNI600 CAS:1191-15-7 ***HR: 3***
DIISOBUTYLHYDROALUMINUM
mf: $C_8H_{19}Al$ mw: 142.25

$[(CH_3)_2CHCH_2]_2AlH$

PROP: Colorless pyroforic liquid, misc in hydrocarbon solvents, bp: 105° @ 2 mm, fp: −80°, d: 0.798.

SYNS: AL-ALCHILI (ITALIAN) ◇ AL-DIISOBUTYL ◇ BIS(ISOBUTYL)HYDROALUMINUM ◇ DIISOBUTYLALUMINIUM HYDRIDE ◇ DIISOBUTYLALUMINUM HYDRIDE ◇ HYDROBIS(2-METHYLPROPYL)ALUMINUM ◇ HYDRODIISOBUTYLALUMINUM

TOXICITY DATA with REFERENCE
ihl-gpg LCLo:70 g/m³/1H MELAAD 57,188,66

CONSENSUS REPORTS: Reported in EPA TSCA Inventory.

ACGIH TLV: TWA 2 mg(Al)/m³

SAFETY PROFILE: Mildly toxic by inhalation. Dangerous fire hazard; ignites spontaneously in air. To fight fire, do not use water, foam, or halogenated extinguishing agents. See also HYDRIDES and ALUMINUM COMPOUNDS.

DNI800 CAS:108-83-8 *HR: 2*
DIISOBUTYL KETONE
DOT: UN 1157
mf: $C_9H_{18}O$ mw: 142.27

PROP: Liquid. Bp: 166°, flash p: 140°F, d: 0.81, vap d: 4.9, lel: 0.8% @ 212°F, uel: 6.2% @ 212°F.

SYNS: DIISOBUTILCHETONE (ITALIAN) ◇ DI-ISOBUTYLCETONE (FRENCH) ◇ DIISOBUTYLKETON (DUTCH, GERMAN) ◇ s-DIISOPROPYLACETONE ◇ 2,6-DIMETHYL-HEPTAN-4-ON (DUTCH, GERMAN) ◇ 2,6-DIMETHYLHEPTAN-4-ONE ◇ 2,6-DIMETHYL-4-HEPTANONE ◇ 2,6-DIMETIL-EPTAN-4-ONE (ITALIAN) ◇ ISOBUTYL KETONE ◇ ISOVALERONE ◇ VALERONE

TOXICITY DATA with REFERENCE
eye-hmn 25 ppm/15M MLD JIHTAB 28,262,46
skn-rbt 10 mg/24H open MLD JIHTAB 31,60,49
skn-rbt 500 mg open MLD UCDS** 12/15/71
eye-rbt 500 mg AJOPAA 29,1363,46
ihl-hmn TCLo:50 ppm:EYE,CNS,GIT JIHTAB 30,63,48
orl-rat LD50:5750 mg/kg NPIRI* 1,23,74
ihl-rat LCLo:2000 ppm/4H JIHTAB 31,343,49
orl-mus LD50:1416 mg/kg SCCUR* -,4,61
skn-rbt LD50:16 g/kg NPIRI* 1,23,74

CONSENSUS REPORTS: Reported in EPA TSCA Inventory.

OSHA PEL: (Transitional: TWA 50 ppm) TWA 25 ppm
ACGIH TLV: TWA 25 ppm
DFG MAK: 50 ppm (290 mg/m^3)
NIOSH REL: (Ketones) TWA 140 mg/m^3
DOT Classification: Flammable or Combustible Liquid; Label: Flammable Liquid; Combustible Liquid; Label: None.

SAFETY PROFILE: Moderately toxic by ingestion and inhalation. Mildly toxic by skin contact. Human systemic effects by inhalation: headache, nausea or vomiting, and unspecified eye effects. An eye and skin irritant. Narcotic in high concentrations. Flammable when exposed to heat or flame; can react with oxidizing materials. To fight fire, use CO_2, dry chemical, water spray, mist or fog. When heated to decomposition it emits acrid smoke and fumes. See also KETONES.

DNJ000 CAS:61947-30-6 *HR: 3*
DIISOBUTYLOXOSTANNANE
mf: $C_8H_{18}OSn$ mw: 248.95

SYNS: DIISOBUTYLTIN OXIDE ◇ KYSLICNIK DIISOBUTYLCINICITY (CZECH)

TOXICITY DATA with REFERENCE
skn-rbt 500 mg/24H SEV 28ZPAK -,226,72
eye-rbt 100 mg/24H MOD 28ZPAK -,226,72
orl-rat LD50:53200 μg/kg 28ZPAK -,226,72

OSHA PEL: TWA 0.1 mg(Sn)/m^3 (skin)
ACGIH TLV: TWA 0.1 mg(Sn)/m^3 (skin) (Proposed: TWA 0.1 mg(Sn)/m^3; STEL 0.2 mg(Sn)/m^3 (skin))
NIOSH REL: (Organotin Compounds) TWA 0.1 mg(Sn)/m^3

SAFETY PROFILE: Poison by ingestion. An eye and severe skin irritant. When heated to decomposition it emits acrid smoke and fumes. See also TIN COMPOUNDS.

DNJ400 CAS:84-69-5 *HR: 2*
DIISOBUTYL PHTHALATE
mf: $C_{16}H_{22}O_4$ mw: 278.38

PROP: Liquid. Mp: −64°, flash p: 385°F, d: 1.039-1.043, vap d: 9.59.

SYNS: DIBP ◇ HEXAPLAS M/1B ◇ PALATINOL IC

TOXICITY DATA with REFERENCE
ipr-rat TDLo:375 mg/kg (female 5-15D post):TER JPMSAE 61,51,72
orl-mus TDLo:32 g/kg (female 6-13D post):REP TCMUD8 7,29,87
orl-rat LD50:15 g/kg EVHPAZ 3,131,73
ipr-rat LD50:3749 mg/kg JPMSAE 61,51,72
orl-mus LD50:13 g/kg EVHPAZ 4,3,73
ipr-mus LD50:3990 mg/kg JSCCA5 28,667,77
skn-gpg LD50:10 g/kg EVHPAZ 4,3,73

CONSENSUS REPORTS: Reported in EPA TSCA Inventory.

SAFETY PROFILE: Moderately toxic by intraperitoneal route. Mildly toxic by ingestion and skin contact. Experimental teratogenic and reproductive effects. Combustible when exposed to heat or flame. To fight fire, use foam, CO_2, dry chemical. When heated to decomposition it emits acrid smoke and fumes.

DNJ600 CAS:3437-84-1 *HR: 3*
DIISOBUTYRYL PEROXIDE
DOT: UN 2182
mf: $C_8H_{14}O_4$ mw: 174.20

$$[(CH_3)_2CHCO\bullet O\text{-}]_2$$

CONSENSUS REPORTS: Reported in EPA TSCA Inventory.

DOT Classification: Organic Peroxide; Label:Organic Peroxide

SAFETY PROFILE: May explode when dried at room temperature. When heated to decomposition it emits acrid smoke and fumes. See also PEROXIDES.

DNJ800 CAS:822-06-0 **HR: 3**
1,6-DIISOCYANATOHEXANE
DOT: UN 2281
mf: $C_8H_{12}N_2O_2$ mw: 168.22

O:N:C(CH_2)$_6$C:N:O

SYNS: HEXAMETHYLENE DIISOCYANATE ◇ HEXAMETHYLENE-1,6-DIISOCYANATE ◇ 1,6-HEXAMETHYLENE DIISOCYANATE (MAK) ◇ HEXAMETHYLENEDIISOCYANATE (DOT) ◇ 1,6-HEXANEDIOL DIISOCYANATE ◇ HMDI ◇ ISOCYANIC ACID, DIESTER with 1,6-HEXANEDIOL ◇ ISOCYANIC ACID, HEXAMETHYLENE ESTER ◇ METHYLENO-BIS-FENYLOIZOCYJANIAN (POLISH) ◇ SZESCIO-METYLENODWUIZOCYJANIAN (POLISH) ◇ TL 78

TOXICITY DATA with REFERENCE
orl-rat LD50:738 mg/kg AIHAAP 30,470,69
ihl-rat LCLo:60 mg/m^3/4H GIPZAB 12(10),40,68
ihl-mus LCLo:1570 mg/m^3/10M NDRC** NDCrc-132,Aug,42
ivn-mus LD50:5600 μg/kg CSLNX* NX#07805
skn-rbt LD50:593 mg/kg AIHAAP 30,470,69

CONSENSUS REPORTS: Reported in EPA TSCA Inventory.

ACGIH TLV: TWA 0.005 ppm
DFG MAK: 0.01 ppm (0.07 mg/m^3)
NIOSH REL: (Diisocyanates) TWA 0.005 ppm; CL 0.02 ppm/10M
DOT Classification: Poison B; Label: Poison.

SAFETY PROFILE: Poison by inhalation and intravenous routes. Moderately toxic by ingestion and skin contact. Potentially explosive reaction with alcohols + base. When heated to decomposition it emits toxic fumes of NO_x. See also CYANATES.

DNK100 CAS:4747-90-4 **HR: 3**
DIISOCYANATOMETHANE
mf: $C_3H_2N_2O_2$ mw: 98.06

SYN: METHYLENE DIISOCYANATE

SAFETY PROFILE: Polymerizes violently on contact with dimethyl formamide (DMF). When heated to decomposition it emits toxic fumes of NO_x.

DNK200 CAS:1321-38-6 **HR: 2**
DIISOCYANATOMETHYLBENZENE
mf: $C_9H_6N_2O_2$ mw: 174.17

SYN: NIAX ISOCYANATE TDI

TOXICITY DATA with REFERENCE
skn-rbt 500 mg open SEV UCDS** 7/11/67
orl-rat LD50:6170 mg/kg UCDS** 7/11/67
ihl-rat LCLO:600 ppm/6H UCDS** 7/11/67

SAFETY PROFILE: Mildly toxic by ingestion and inhalation. A severe skin irritant. When heated to decomposition it emits toxic fumes of NO_x.

DNK800 CAS:27215-10-7 **HR: 2**
DIISOOCTYL ACID PHOSPHATE
DOT: UN 1902
mf: $C_{16}H_{35}O_4P$ mw: 322.48

PROP: A corrosive liquid.

SYN: DIISOOCTYL PHOSPHATE (DOT)

CONSENSUS REPORTS: Reported in EPA TSCA Inventory.

DOT Classification: Corrosive Material; Label: Corrosive.

SAFETY PROFILE: Moderately toxic by irritation to skin, eyes, and mucous membranes. A corrosive compound. When heated to decomposition it emits toxic fumes of PO_x. See also PHOSPHATES.

DNL200 CAS:24423-68-5 **HR: 3**
DIISOPENTYLMERCURY
mf: $C_{10}H_{22}Hg$ mw: 342.91

SYN: DIISOAMYLMERCURY

TOXICITY DATA with REFERENCE
ipr-mus LDLo:16 mg/kg CBCCT* 4,230,52

CONSENSUS REPORTS: Mercury and its compounds are on the Community Right-To-Know List.

OSHA PEL: (Transitional: CL 1 mg/10m^3) TWA 0.01 mg(Hg)/m^3; STEL 0.03 mg/m^3 (skin)
ACGIH TLV: TWA 0.01 mg(Hg)/m^3; STEL 0.03 mg(Hg)/m^3
NIOSH REL: (Inorganic Mercury) TWA 0.05 mg (Hg)/m^3

SAFETY PROFILE: Poison by intraperitoneal route. Violent reaction on contact with iodine. When heated to decomposition it emits toxic fumes of Hg. See also MERCURY COMPOUNDS.

DNL400 CAS:63979-62-4 **HR: 3**
DIISOPENTYLOXOSTANNANE
mf: $C_{10}H_{22}OSn$ mw: 277.01

SYNS: DIISOPENTYLTIN OXIDE ◇ KYSLICNIK DIISOAMYLCINICITY (CZECH)

TOXICITY DATA with REFERENCE
skn-rbt 500 mg/24H SEV 28ZPAK -,227,72
eye-rbt 20 mg/24H MOD 28ZPAK -,227,72
orl-rat LD50:64500 μg/kg 28ZPAK -,227,72

OSHA PEL: TWA 0.1 mg(Sn)/m^3 (skin)
ACGIH TLV: TWA 0.1 mg(Sn)/m^3 (skin) (Proposed: TWA 0.1 mg(Sn)/m^3; STEL 0.2 mg(Sn)/m^3 (skin))
NIOSH REL: (Organotin Compounds) TWA 0.1 mg (Sn)/m^3

SAFETY PROFILE: Poison by ingestion. An eye and severe skin irritant. When heated to decomposition it emits acrid smoke and fumes. See also TIN COMPOUNDS.

DNL600 CAS:110-97-4 ***HR: 3***
DIISOPROPANOLAMINE
mf: $C_6H_{15}NO_2$ mw: 133.22

PROP: Mp: 42°, bp: 249°, flash p: 260°F (OC), d: 0.9890 @ 45°/20°, vap d: 4.59.

SYNS: BIS(2-HYDROXYPROPYL)AMINE ◇ BIS(2-PROPANOL) AMINE ◇ DIPA ◇ DIPROPYL-2,2'-DIHYDROXYAMINE ◇ 1,1'-IMINODI-2-PROPANOL

TOXICITY DATA with REFERENCE
skn-rbt 500 mg open MLD UCDS** 5/21/71
eye-rbt 50 mg SEV UCDS** 5/21/71
mma-sat 250 μg/plate MUREAV 111,135,83
orl-rat LD50:6720 mg/kg UCDS** 5/21/71
ipr-mus LD50:96 mg/kg AIMJA9 30,23,79

CONSENSUS REPORTS: Reported in EPA TSCA Inventory.

SAFETY PROFILE: Poison by intraperitoneal route. Mildly toxic by ingestion. Mutation data reported. A skin and severe eye irritant. Combustible when exposed to heat or flame; can react with oxidizing materials. To fight fire, use alcohol foam, CO_2, dry chemical. When heated to decomposition it emits toxic fumes of NO_x.

DNL800 CAS:6938-94-9 ***HR: 2***
DIISOPROPYL ADIPATE
mf: $C_{12}H_{22}O_4$ mw: 230.34

SYNS: ADIPIC ACID DIISOPROPYL ESTER ◇ CERAPHYL 230 ◇ HEXANEDIOIC ACID, BIS(1-METHYLETHYL) ESTER ◇ ISOPROPYL ADIPATE ◇ STANDAMUL DIPA ◇ WICKENOL 116

TOXICITY DATA with REFERENCE
ivn-rat LD50:640 mg/kg MRLR** No.256,54

CONSENSUS REPORTS: Reported in EPA TSCA Inventory.

SAFETY PROFILE: Moderately toxic by intravenous route. When heated to decomposition it emits acrid smoke and fumes. See also ESTERS.

DNM200 CAS:108-18-9 ***HR: 3***
DIISOPROPYLAMINE
DOT: UN 1158
mf: $C_6H_{15}N$ mw: 101.22

PROP: Colorless liquid. Bp: 83-84°, flash p: 19.4°F. D: 0.722 @ 220.0°, vap d: 3.5.

SYNS: DIPA ◇ N-(1-METHYLETHYL)-2-PROPANAMINE

TOXICITY DATA with REFERENCE
eye-rbt 750 μg open SEV AMIHBC 10,61,54
mma-sat 1 μg/plate NUCADQ 3,129,82
orl-rat LD50:770 mg/kg AIHLAU 1,343,60
ihl-rat LC50:4800 mg/m³/2H 85GMAT -,54,82
orl-mus LD50:2120 mg/kg GISAAA 45(3),79,80
ihl-mus LC50:4200 mg/m³/2H 85GMAT -,54,82
ihl-cat LCLo:2207 ppm/72M JIHTAB 31,142,49
orl-rbt LD50:4700 mg/kg GISAAA 45(3),79,80
ihl-rbt LCLo:2207 ppm/150M JIHTAB 31,142,49
orl-gpg LD50:2800 mg/kg GISAAA 45(3),79,80
ihl-gpg LCLo:2207 ppm/82M JIHTAB 31,142,49
scu-gpg LDLo:1400 mg/kg JIHTAB 31,142,49
ihl-mam LC50:4200 mg/m³ TPKVAL 14,80,75

CONSENSUS REPORTS: Reported in EPA TSCA Inventory.

OSHA PEL: TWA 5 ppm (skin)
ACGIH TLV: TWA 5 ppm (skin)
DOT Classification: Flammable Liquid; Label: Flammable Liquid.

SAFETY PROFILE: Moderately toxic by ingestion, and subcutaneous routes. Mildly toxic by inhalation. Mutation data reported. A severe eye irritant. Inhalation of fumes can cause pulmonary edema. A very dangerous fire hazard when exposed to heat or flame; can react vigorously with oxidizing materials. To fight fire, use alcohol foam, foam, CO_2, dry chemical. When heated to decomposition it emits toxic fumes of NO_x. See also AMINES.

DNM400 CAS:660-27-5 ***HR: 2***
DIISOPROPYLAMINE DICHLORACETATE
mf: $C_6H_{15}N \cdot C_2H_2Cl_2O_2$ mw: 230.16

SYNS: DICHLOROACETIC ACID, DIISOPROPYLAMINE SALT ◇ DIISOPROPYLAMINE with DICHLOROACETIC ACID (1:1) ◇ DIISOPROPYLAMINE DICHLOROETHANOATE ◇ DIISOPROPYLAMMONIUM DICHLOROACETATE ◇ DIISOPROPYLAMMONIUM DICHLOROETHANOATE

TOXICITY DATA with REFERENCE
mma-sat 10 μg/plate NUCADQ 3,129,82
ipr-rat LD50:840 mg/kg THERAP 16,136,61
orl-mus LD50:1700 mg/kg ARZNAD 13,109,63
ipr-mus LD50:750 mg/kg BCFAAI 97,608,58
scu-mus LD50:1330 mg/kg ARZNAD 13,109,63

CONSENSUS REPORTS: Reported in EPA TSCA Inventory.

SAFETY PROFILE: Moderately toxic by intraperitoneal, subcutaneous, and ingestion routes. Mutation data reported. When heated to decomposition it emits very toxic fumes of Cl^-, NH_3, and NO_x.

DNM600 CAS:77966-84-8 ***HR: 3***
2-(DIISOPROPYLAMINO)-2',6'-ACETOXYLIDIDE HYDROCHLORIDE
mf: $C_{16}H_{26}N_2O{\bullet}ClH$ mw: 298.90

SYN: V 377

TOXICITY DATA with REFERENCE
eye-rbt 2% MLD ARZNAD 8,407,58
ipr-rat LD50:90 mg/kg ARZNAD 8,407,58
scu-mus LD50:197 mg/kg ARZNAD 8,407,58

SAFETY PROFILE: Poison by subcutaneous and intraperitoneal routes. An eye irritant. When heated to decomposition it emits very toxic fumes of NO_x and HCl.

DNN000 CAS:14549-32-7 ***HR: 3***
2-(2-(DIISOPROPYLAMINO)ETHOXY)BUTYROPHENONE HYDROCHLORIDE
mf: $C_{18}H_{29}NO_2{\bullet}ClH$ mw: 327.94

SYNS: 2-BUTYRYL-β-(N,N-DIISOPROPYL)PHENOXYETHYLAMINE HYDROCHLORIDE ◇ KETOCAINE HYDROCHLORIDE ◇ REC 7-0518

TOXICITY DATA with REFERENCE
orl-rat LD50:446 mg/kg ARZNAD 16,1275,66
scu-rat LD50:935 mg/kg ARZNAD 16,1275,66
orl-mus LD50:147 mg/kg ARZNAD 16,1275,66
ipr-mus LD50:102 mg/kg ARZNAD 16,1275,66
scu-mus LD50:217 mg/kg ARZNAD 16,1275,66
ivn-mus LD50:14 mg/kg ARZNAD 16,1275,66
ivn-cat LD50:6 mg/kg ARZNAD 16,1275,66

SAFETY PROFILE: Poison by ingestion, intravenous, subcutaneous, and intraperitoneal routes. When heated to decomposition it emits very toxic fumes of NO_x and HCl.

DNN600 CAS:3737-09-5 ***HR: 3***
α-(2-(DIISOPROPYLAMINO)ETHYL)-α-PHENYL-2-PYRIDINEACETAMIDE
mf: $C_{21}H_{29}N_3O$ mw: 339.53

SYNS: DICORANTIL ◇ Γ-DIISOPROPYLAMINO-α-PHENYL-α-(2-PYRIDYL)BUTYRAMIDE ◇ DISOPYRAMIDE ◇ H 3292 ◇ RITMODAN ◇ SC 7031 ◇ SEARLE 703

TOXICITY DATA with REFERENCE
orl-rat TDLo:420 mg/kg (female 9-15D post):REP KSRNAM 5,1641,71
orl-rat TDLo:1260 mg/kg (female 9-15D post):TER KSRNAM 5,1641,71
orl-hmn TDLo:4286 μg/kg:CVS,PUL NEJMAG 302,614,80
orl-rat LD50:333 mg/kg ARZNAD 38,1398,88
ipr-rat LD50:170 mg/kg NIIRDN 6,319,82
scu-rat LD50:800 mg/kg IYKEDH 9,829,78
orl-mus LD50:409 mg/kg NIIRDN 6,319,82
ipr-mus LD50:114 mg/kg KSRNAM 5,1628,71
scu-mus LD50:305 mg/kg NIIRDN 6,319,82
ivn-mus LD50:30 mg/kg JMCMAR 27,1142,84
ivn-dog LDLo:36 mg/kg JPETAB 136,114,62

SAFETY PROFILE: Poison by ingestion, intraperitoneal, intravenous, and subcutaneous routes. An experimental teratogen. Other experimental reproductive effects. Human systemic effects by ingestion: dyspnea, cardiac and pulmonary changes. Experimental reproductive effects. When heated to decomposition it emits toxic fumes of NO_x.

DNN709 CAS:25321-09-9 ***HR: 1***
DIISOPROPYLBENZENE
mf: $C_{12}H_{18}$ mw: 162.30

TOXICITY DATA with REFERENCE
orl-rat LD50:6500 mg/kg TXAPA9 28,313,74
ihl-rat LCLo:5300 mg/m^3/4H 85GMAT -,55,82
ihl-mus LCLo:5300 mg/m^3/2H 85GMAT -,55,82
skn-rbt LD50:16 g/kg TXAPA9 28,313,74

CONSENSUS REPORTS: Reported in EPA TSCA Inventory.

SAFETY PROFILE: Mildly toxic by ingestion and skin contact. When heated to decomposition it emits acrid smoke and fumes.

DNN800 CAS:577-55-9 ***HR: 2***
o-DIISOPROPYLBENZENE
mf: $C_{12}H_{18}$ mw: 162.30

PROP: Clear, colorless liquid. Mp: $< -55°$, bp: 205°, flash p: 170°F (OC), d: 0.863-0.867 @ 25°/25°, autoign temp: 840°F, vap d: 5.6.

TOXICITY DATA with REFERENCE
orl-rat LDLo:5000 mg/kg 28ZRAQ -,57,60

CONSENSUS REPORTS: Reported in EPA TSCA Inventory.

SAFETY PROFILE: Mildly toxic by ingestion. Combustible when exposed to heat or flame; can react with oxidizing materials. To fight fire, use foam, CO_2, dry chemical, water spray or mist. When heated to decomposition it emits acrid smoke and irritating fumes. See also CUMENE.

DNN829 CAS:99-62-7 ***HR: 2***
1,3-DIISOPROPYLBENZENE
mf: $C_{12}H_{18}$ mw: 162.30

SYN: m-DIISOPROPYLBENZENE

TOXICITY DATA with REFERENCE
orl-rat LD50:7400 mg/kg 85GMAT -,54,82
orl-mus LD50:3100 mg/kg 85GMAT -,54,82
ipr-mus LD50:1650 mg/kg 85GMAT -,54,82

CONSENSUS REPORTS: Reported in EPA TSCA Inventory.

SAFETY PROFILE: Moderately toxic by ingestion and intraperitoneal routes. When heated to decomposition it emits acrid smoke and fumes.

DNN840 ***HR: 3***
1,3-DIISOPROPYLBENZENE SODIUM SALT, DIHYDROPEROXIDE
mf: $C_{12}H_{18}•Na•2H_2O_2$ mw: 253.33

SYN: SODIUM-m-DIISOPROPYLBENZOL (Na-m) DIHYDROPEROXIDE (RUSSIAN)

TOXICITY DATA with REFERENCE
unr-rat LD50:1050 mg/kg GISAAA 42(4),11,77
unr-mus LD50:415 mg/kg GISAAA 42(4),11,77
unr-rbt LD50:320 mg/kg GISAAA 42(4),11,77

SAFETY PROFILE: Poison by an unspecified route. When heated to decomposition it emits toxic fumes of Na_2O. See also PEROXIDES, ORGANIC.

DNN850 ***HR: 2***
1,4-DIISOPROPYLBENZENE SODIUM SALT, DIISOPEROXIDE
mf: $C_{12}H_{18}•Na•2H_2O_2$ mw: 253.33

SYN: SODIUM-p-DIISOPROPYLBENZOL (Na-p) DIHYDROPEROXIDE (RUSSIAN)

TOXICITY DATA with REFERENCE
unr-rat LD50:1250 mg/kg GISAAA 42(4),11,77
unr-mus LD50:660 mg/kg GISAAA 42(4),11,77
unr-rbt LD50:450 mg/kg GISAAA 42(4),11,77

SAFETY PROFILE: Moderately toxic by unspecified route. When heated to decomposition it emits toxic fumes of Na_2O. See also PEROXIDES, ORGANIC.

DNO200 CAS:15721-33-2 ***HR: 3***
DIISOPROPYLBERYLLIUM
mf: $C_6H_{14}Be$ mw: 95.19

$((CH_3)_2CH)_2Be$

CONSENSUS REPORTS: Beryllium and its compounds are on the Community Right-To-Know List.

OSHA PEL: (Transitional: TWA 0.002 mg(Be)/m^3; CL 0.005; Pk 0.025/30M/8H) TWA 0.002 mg(Be)/m^3; STEL 0.005 mg(Be)/m^3/30M; CL 0.025 mg(Be)/m^3
ACGIH TLV: TWA 0.002 mg/m^3, Suspected Human Carcinogen.
DFG TRK: 0.002 mg(Be)/m^3. Animal Carcinogen, Suspected Human Carcinogen.

SAFETY PROFILE: Confirmed human carcinogen. Explosive reaction on contact with water. When heated to decomposition it emits toxic fumes of BeO. See also BERYLLIUM COMPOUNDS.

DNO400 CAS:693-13-0 ***HR: 3***
DIISOPROPYLCARBODIIMIDE
mf: $C_7H_{14}N_2$ mw: 126.23

TOXICITY DATA with REFERENCE
ivn-mus LD50:36 mg/kg CSLNX* NX#05886

CONSENSUS REPORTS: Reported in EPA TSCA Inventory.

SAFETY PROFILE: Poison by intravenous route. When heated to decomposition it emits toxic fumes of NO_x.

DNO800 CAS:741-58-2 ***HR: 3***
N-(2-(O,O-DIISOPROPYLDITHIOPHOSPHORYL) ETHYL)BENZENESULFONAMIDE
mf: $C_{14}H_{24}NO_4PS_3$ mw: 397.54

SYNS: BENSULIDE ◇ BENZULFIDE ◇ BETAMEC ◇ BETASAN ◇ O,O-BIS(1-METHYLETHYL)-S-(2-((PHENYLSULFONYL) AMINO) ETHYL)PHEOSPHORODITHIOATE ◇ N-(β-O,O-DIISOPROPYLDITHIOPHOSPHORYLETHYL)BEZENESULFONAMIDE ◇ S-(O,O-DIISOPROPYL PHOSPHORODITHIOATE) ESTER of N-(2-MERCAPTOETHYL) BENZENESULFONAMIDE ◇ DISAN ◇ EXPORSAN ◇ N-(2-MERCAPTOETHYLBENZENESULFONAMIDE-S-(O,O-DIISOPROPYLPHOSPHORODITHIOATE) ◇ PHOSPHORODITHIOIC ACID-O,O-BIS(1-METHYLETHYL)-S-(2-((PHENYLSULFONYL)AMINO)ETHYLESTER ◇ PREFAR ◇ PRE-SAN ◇ R-4461

TOXICITY DATA with REFERENCE
orl-rat LD50:271 mg/kg FMCHA2 -,C31,83
skn-rat LD50:3950 mg/kg 31ZOAD 1,34,68
skn-rbt LD50:2000 mg/kg WRPCA2 9,119,70

SAFETY PROFILE: Poison by ingestion. Moderately toxic by skin and contact. An herbicide. When heated to decomposition it emits very toxic fumes of NO_x, SO_x, and PO_x. See also ESTERS.

DNO900 CAS:2973-10-6 ***HR: 3***
DIISOPROPYL ESTER SULFURIC ACID
mf: $C_6H_{14}O_4S$ mw: 182.26

SYNS: DI-ISOPROPYLSULFAT (GERMAN) ◇ DI-ISOPROPYLSULFATE ◇ ISOPROPYL SULFATE

TOXICITY DATA with REFERENCE
scu-rat TDLo:300 mg/kg:ETA ZKKOBW 79,135,73
orl-rat LD50:1090 mg/kg AIHAAP 30,470,69
skn-rbt LD50:1410 mg/kg AIHAAP 30,470,69

SAFETY PROFILE: Moderately toxic by ingestion and skin contact. Questionable carcinogen with experimental tumorigenic data. When heated to decomposition it emits toxic fumes of SO_x. See also ESTERS and SULFATES.

DNP000 CAS:96-80-0 ***HR: 2***
N,N-DIISOPROPYL ETHANOLAMINE
DOT: UN 2825
mf: $C_8H_{19}NO$ mw: 145.28

SYNS: 2-DIISOPROPYLAMINOETHANOL ◇ DIISOPROPYL ETHANOLAMINE

TOXICITY DATA with REFERENCE
skn-rbt 500 mg open MLD UCDS** 6/6/69
eye-rbt 750 μg open SEV AMIHBC 10,61,54
orl-rat LD50:1070 mg/kg UCDS** 6/6/69
skn-rbt LD50:450 mg/kg AMIHBC 10,61,54

CONSENSUS REPORTS: Reported in EPA TSCA Inventory.

DOT Classification: Corrosive Material; Label: Corrosive.

SAFETY PROFILE: Moderately toxic by skin contact and ingestion. A skin and severe eye irritant.

DNP600 CAS:20652-39-5 ***HR: 3***
N,N-DIISOPROPYL ETHYL CARBAMATE
mf: $C_9H_{19}NO_2$ mw: 173.29

SYNS: DIISOPROPYLCARBAMIC ACID, ETHYL ESTER ◇ DIISOPROPYL ETHYL CARBAMATE

TOXICITY DATA with REFERENCE
ipr-mus TDLo:6500 mg/kg/13W-I:ETA JNCIAM 9,35,48

SAFETY PROFILE: Questionable carcinogen with experimental tumorigenic data. When heated to decomposition it emits toxic fumes of NO_x. See also CARBAMATES.

DNQ200 CAS:7283-70-7 ***HR: 2***
DIISOPROPYL FUMARATE
mf: $C_{10}H_{16}O_4$ mw: 200.26

SYN: FUMARIC ACID, DIISOPROPYL ESTER

TOXICITY DATA with REFERENCE
skn-rbt 10 mg/24H open MLD AMIHBC 10,61,54
eye-rbt 500 mg open AMIHBC 10,61,54
orl-rat LD50:3250 mg/kg AMIHBC 10,61,54
skn-rbt LD50:10 g/kg AMIHBC 10,61,54

SAFETY PROFILE: Moderately toxic by ingestion. Mildly toxic by skin contact. A skin and eye irritant. When heated to decomposition it emits acrid smoke and fumes.

DNQ600 CAS:1809-20-7 ***HR: 2***
DIISOPROPYL HYDROGEN PHOSPHITE
mf: $C_6H_{15}O_3P$ mw: 166.18

SYNS: DIISOPROPYL PHOSPHITE ◇ DIISOPROPYLPHOSPHONATE ◇ O,O-DIISOPROPYL PHOSPHONATE ◇ ISOPROPYL PHOSPHONATE ◇ PHOSPHONIC ACID, BIS(1-METHYLETHYL) ESTER

TOXICITY DATA with REFERENCE
mmo-sat 5 μL/plate MUREAV 28,405,75
orl-rat LD50:3100 mg/kg ALBRW* #OPB-3,84
skn-rbt LD50:5700 mg/kg ALBRW* #OPB-3,84

CONSENSUS REPORTS: Reported in EPA TSCA Inventory.

SAFETY PROFILE: Moderately toxic by ingestion. Mildly toxic by skin contact. Mutation data reported. When heated to decomposition it emits toxic fumes of PO_x. See also ESTERS.

DNQ700 CAS:86886-16-0 ***HR: 3***
DIISOPROPYL HYPONITRITE
mf: $C_6H_{14}N_2O_2$ mw: 146.19

$(CH_3)_2CHON{=}NOCH(CH_3)_2$

SYN: BIS(2-PROPYLOXY)DIAZENE

SAFETY PROFILE: An impact-sensitive explosive. When heated to decomposition it emits toxic fumes of NO_x. See also NITRITES.

DNQ800 CAS:1071-39-2 ***HR: 3***
DIISOPROPYLMERCURY
mf: $C_6H_{14}Hg$ mw: 286.79

PROP: Liquid. Bp: 63° @ 10 mm, d: 2.0024, vap d: 9.9.

CONSENSUS REPORTS: Mercury and its compounds are on the Community Right-To-Know List.

OSHA PEL: (Transitional: CL 1 mg/10m³) TWA 0.01 mg(Hg)/m³; STEL 0.03 mg/m³ (skin)
ACGIH TLV: TWA 0.01 mg(Hg)/m³; STEL 0.03 mg(Hg)/m³
NIOSH REL: (Inorganic Mercury) TWA 0.05 mg(Hg)/m³

SAFETY PROFILE: Mercury compounds are poisons. When heated to decomposition it emits toxic fumes of Hg. See also MERCURY COMPOUNDS, ORGANIC.

DNQ875 CAS:1445-75-6 ***HR: 2***
DIISOPROPYL METHYLPHOSPHONATE
mf: $C_7H_{17}O_3P$ mw: 180.21

SYNS: DIISOPROPYL METHANEPHOSPHONATE ◇ DIMP

TOXICITY DATA with REFERENCE
orl-rat LD50:826 mg/kg 40QBA3 -,450,78
orl-mus LD50:1041 mg/kg 40QBA3 -,450,78
orl-dck LD50:1490 mg/kg NTIS** AD-A087-257
orl-ctl LD50:750 mg/kg NTIS** AD-A093-673
orl-mam LD50:503 mg/kg NTIS** AD-A087-257
orl-brd LD50:1000 mg/kg NTIS** AD-A087-257

CONSENSUS REPORTS: Reported in EPA TSCA Inventory.

SAFETY PROFILE: Moderately toxic by ingestion. When heated to decomposition it emits toxic fumes of PO_x. See also ESTERS and PHOSPHONIC ACID.

DNR200 CAS:23668-76-0 ***HR: 3***
DIISOPROPYLOXOSTANNANE
mf: $C_6H_{14}OSn$ mw: 220.89

PROP: Solid. Insol in water.

SYNS: DIISOPROPYLTIN OXIDE ◇ KYSLICNIK DIISOPROPYLCINICITY (CZECH)

TOXICITY DATA with REFERENCE
skn-rbt 500 mg/24H SEV 28ZPAK -,225,72
eye-rbt 20 mg/24H MOD 28ZPAK -,225,72
orl-rat LD50:57700 μg/kg 28ZPAK -,225,72

OSHA PEL: TWA 0.1 mg(Sn)/m^3 (skin)
ACGIH TLV: TWA 0.1 mg(Sn)/m^3 (skin) (Proposed: TWA 0.1 mg(Sn)/m^3; STEL 0.2 mg(Sn)/m^3 (skin))
NIOSH REL: (Organotin Compounds) TWA 0.1 mg (Sn)/m^3

SAFETY PROFILE: Poison by ingestion. An eye and severe skin irritant. When heated to decomposition it emits acrid smoke and irritating fumes. See also TIN COMPOUNDS.

DNR309 CAS:3254-66-8 ***HR: 3***
DIISOPROPYL PARAOXON
mf: $C_{12}H_{18}NO_6P$ mw: 303.28

SYNS: DIISOPROPYL-p-NITROPHENYL PHOSPHATE ◇ O,O-DIISOPROPYL-o,p-NITROPHENYL PHOSPHATE ◇ MIOTICOL ◇ PROPICOL

TOXICITY DATA with REFERENCE
orl-mus LD50:143 mg/kg JAFCAU 17,243,69
ipr-mus LD50:33 mg/kg JAFCAU 12,318,64

SAFETY PROFILE: Poison by ingestion and intraperitoneal routes. When heated to decomposition it emits very toxic fumes of NO_x and PO_x.

DNR400 CAS:105-64-6 ***HR: 3***
DIISOPROPYL PERDICARBONATE
DOT: UN 2133/UN 2134
mf: $C_8H_{14}O_6$ mw: 206.22

$[(CH_3)_2CHOCO•O-)_2$

PROP: Colorless, crystalline solid. Rapid decomp @ 63°F, mp: 8°-10°, d: 1.080 @ 15.5°/4°. Almost insol in water; miscible with aliphatic and aromatic hydrocarbons, esters, ethers, and chlorinated hydrocarbons.

SYNS: DIISOPROPYL PEROXYDICARBONATE ◇ ISOPROPYL PERCARBONATE ◇ ISOPROPYL PERCARBONATE, stabilized (DOT) ◇ ISOPROPYL PERCARBONATE, unstabilized (DOT) ◇ ISOPROPYL PEROXYDICARBONATE ◇ ISOPROPYL PEROXYDICARBONATE, technically pure (DOT) ◇ PEROXYDICARBONATE D'ISOPROPYLE (FRENCH) ◇ PEROXYDICARBONIC ACID, BIS(1-METHYLETHYL) ESTER

TOXICITY DATA with REFERENCE
eye-rbt 500 mg SEV IHFCAY 6,1,67
orl-rat LD50:2140 mg/kg IHFCAY 6,1,67
skn-rbt LD50:2025 mg/kg BSPII* 1/75-19B

CONSENSUS REPORTS: Reported in EPA TSCA Inventory.

DOT Classification: Organic Peroxide; Label: Organic Peroxide.

SAFETY PROFILE: Moderately toxic by ingestion and skin contact. A severe eye irritant. Very dangerous fire hazard. Dangerously unstable above 10°C. An impact- and heat-sensitive explosive. Solutions may spontaneously explode (the hazard increases with concentration). Storage in sealed containers may be dangerous. Explodes on contact with amines or potassium iodide. May explode on contact with organic matter. When heated to decomposition it emits acrid smoke and fumes. See also PEROXIDES, ORGANIC.

DNR800 CAS:2078-54-8 ***HR: 3***
2,6-DIISOPROPYLPHENOL
mf: $C_{12}H_{18}O$ mw: 178.30

PROP: A colorless liquid or solid. Bp: 242.4°, fp: 17.9°, flash p: 235°F (CC), d: 0.955 @ 20°/4°.

TOXICITY DATA with REFERENCE
unr-wmn TDLo:2800 μg/kg (female 39W post):REP BJANAD 62,649,89
unr-man TDLo:2857 μg/kg/1D-I ANASAB 43,170,88
ipr-mus LD50:170 mg/kg JMPCAS 2,201,60
ivn-mus LD50:50 mg/kg JMCMAR 23,1350,80
ivn-rbt LDLo:20 mg/kg JMCMAR 23,1350,80

CONSENSUS REPORTS: Reported in EPA TSCA Inventory.

SAFETY PROFILE: Poison by intravenous and intraperitoneal routes. Experimental reproductive effects. Combustible when exposed to heat or flame; can react with oxidizing materials. To fight fire, use foam, CO_2, dry chemical. When heated to decomposition it emits acrid smoke and fumes. See also PHENOL.

DNS000 CAS:26762-93-6 ***HR: 3***
DIISOPROPYLPHENYLHYDROPEROXIDE (solution)
DOT: UN 2171
mf: $C_{12}H_{19}O_2$ mw: 195.31

PROP: Colorless to pale yellow liquid.

SYN: DIISOPROPYLBENZENE HYDROPEROXIDE, not more than 72% in solution (DOT)

TOXICITY DATA with REFERENCE
unr-mus TDLo:391 mg/kg:ETA RARSAM 3,193,63

CONSENSUS REPORTS: Reported in EPA TSCA Inventory.

DOT Classification: Organic Peroxide; Label: Organic Peroxide; Forbidden (> 72% in solution).

SAFETY PROFILE: Questionable carcinogen with experimental tumorigenic data. A powerful oxidizer. When heated to decomposition it emits acrid smoke and fumes. See also PEROXIDES, ORGANIC.

DNS200 CAS:330-64-3 ***HR: 3***
3,5-DIISOPROPYLPHENYL-N-METHYLCARBAMATE
mf: $C_{14}H_{21}NO_2$ mw: 235.36

SYNS: 3,5-BIS(1-METHYLETHYL)PHENOLMETHYLCARBAMATE ◇ 3,5-BIS(1-METHYLETHYL)PHENYL ESTER METHYL CARBAMIC ACID ◇ 3,5-BIS(1-METHYLETHYL)PHENYL METHYLCARBAMATE ◇ 3,5-DIISOPROPYLPHENOL METHYLCARBAMATE ◇ 3,5-DIISOPROPYLPHENYL METHYLCARBAMATE ◇ DIP ◇ ENT 25,780 ◇ HOOKER HRS-1422

TOXICITY DATA with REFERENCE
orl-rat LD50:200 mg/kg TXAPA9 14,515,69
ipr-rat LD50:267 mg/kg BWHOA6 44(1-3),241,71
ivn-rat LD50:29700 μg/kg BWHOA6 44(1-3),241,71
ipr-mus LD50:31 mg/kg BECTA6 2,163,67
orl-bwd LD50:10 mg/kg TXAPA9 21,315,72

SAFETY PROFILE: Poison by ingestion, intraperitoneal, and intravenous routes. When heated to decomposition it emits toxic fumes of NO_x. See also CARBAMATES.

DNS600 CAS:2303-17-5 ***HR: 3***
N-DIISOPROPYLTHIOCARBAMIC ACID S-2,3,3-TRICHLORO-2-PROPENYL ESTER
mf: $C_{10}H_{16}Cl_3NOS$ mw: 304.68

SYNS: AVADEX BW ◇ CP 23426 ◇ N-DIISOPROPYLTHIOCARBAMIC ACID-S-2,3,3-TRICHLOROALLYL ESTER ◇ N,N-DIISOPROPYL-2,3,3-TRICHLORALLYL-THIOLCARBAMAT (GERMAN) ◇ DIISOPROPYLTRICHLOROALLYLTHIOCARBAMATE ◇ DIPTHAL ◇ FAR-GO ◇ 2,3,3-THICHLORO-2-PROPENE-1-THIOL, DIISOPROPYLCARBAMATE ◇ TRIALLAT (GERMAN) ◇ TRIALLATE ◇ 2,3,3-TRICHLORALLYL-N,N-(DIISOPROPYL)-THIOCARBAMAT (GERMAN) ◇ 2,3,3-TRICHLOROALLYL DIISOPROPYLTHIOCARBAMATE ◇ S-2,3,3-TRICHLOROALLYL-N,N-DIISOPROPYLTHIOCARBAMATE

TOXICITY DATA with REFERENCE
mma-sat 100 nmol/L BCPCA6 32,3739,83
mma-bcs 50 μg/plate JAFCAU 29,268,81
cyt-ham:ovr 100 μmol/L MUREAV 85,45,81
sce-ham:ovr 20 μmol/L MUREAV 85,45,81
orl-rat LD50:1471 mg/kg HYSAAV 33,41,68
skn-rat LDLo:3500 mg/kg GISAAA 33(7)37,68
orl-mus LD50:930 mg/kg GISAAA 33(7),37,68
ihl-cat LCLo:400 mg/m³/4H GISAAA 33(7),37,68
skn-rbt LD50:2225 mg/kg 28ZEAL 5,226,76

CONSENSUS REPORTS: EPA Genetic Toxicology Program.

SAFETY PROFILE: Poison by inhalation. Moderately toxic by ingestion and skin contact. Mutation data reported. An herbicide. When heated to decomposition it emits very toxic fumes of Cl^-, NO_x, and SO_x. See also CARBAMATES and ESTERS.

DNS800 CAS:2986-17-6 ***HR: 2***
DIISOPROPYL THIOUREA
mf: $C_7H_{16}N_2S$ mw: 160.31

SYN: 1,3-DIISOPROPYL-2-THIOUREA

TOXICITY DATA with REFERENCE
orl-rat LD50:450 mg/kg JPETAB 90,260,47
orl-mus LDLo:1070 mg/kg AECTCV 14,111,85

CONSENSUS REPORTS: Reported in EPA TSCA Inventory.

SAFETY PROFILE: Moderately toxic by ingestion. When heated to decomposition it emits very toxic fumes of NO_x and SO_x.

DNT000 CAS:38802-82-3 ***HR: 3***
DIISOPROPYLTIN DICHLORIDE
mf: $C_6H_{14}Cl_2Sn$ mw: 275.79

PROP: Colorless crystals. Sol in water. Mp: 84°

SYN: DICHLORODIISOPROPYLSTANNANE

TOXICITY DATA with REFERENCE
ivn-rat LD50:15 mg/kg JOCMA7 2,183,60

OSHA PEL: TWA 0.1 mg(Sn)/m³ (skin)
ACGIH TLV: TWA 0.1 mg(Sn)/m³ (skin) (Proposed: TWA 0.1 mg(Sn)/m³; STEL 0.2 mg(Sn)/m³ (skin))
NIOSH REL: (Organotin Compounds) TWA 0.1 mg (Sn)/m³

SAFETY PROFILE: Poison by intravenous route. When heated to decomposition it emits toxic fumes of Cl^-. See also TIN COMPOUNDS and CHLORIDES.

DNT200 CAS:49538-98-9 ***HR: 2***
O,O-DIISOPROPYL-S-TRICYCLOHEXYLTIN PHOSPHORODITHIOATE
mf: $C_{24}H_{47}O_2PS_2Sn$ mw: 581.49

SYNS: ((DIISOPROPROXYPHOSPHINOTHIOYL)THIO)TRICYCLOHEXYL STANNANE ◇ R-28627

TOXICITY DATA with REFERENCE
orl-rat LD50:860 mg/kg SPEADM 74-1,-,74

OSHA PEL: TWA 0.1 mg(Sn)/m^3 (skin)
ACGIH TLV: TWA 0.1 mg(Sn)/m^3 (skin) (Proposed: TWA 0.1 mg(Sn)/m^3; STEL 0.2 mg(Sn)/m^3 (skin))
NIOSH REL: (Organotin Compounds) TWA 0.1 mg(Sn)/m^3

SAFETY PROFILE: Moderately toxic by ingestion. When heated to decomposition it emits very toxic fumes of SO_x and PO_x. See also TIN COMPOUNDS.

DNU000 CAS:630-93-3 ***HR: 3***
DILANTIN
mf: $C_{15}H_{11}N_2O_2 \cdot Na$ mw: 274.27

SYNS: ALEPSIN ◇ ANTILEPSIN ◇ ANTISACER ◇ AURANILE ◇ CITRULLAMON ◇ DANTEN ◇ DANTOIN ◇ DENYL ◇ DENYLSODIUM ◇ DERIZENE ◇ DIFENIN ◇ DIFETOIN ◇ DIFHYDAN ◇ DIHYDAN ◇ DIHYDANTOIN ◇ DILANTIN SODIUM ◇ DI-LEN ◇ DINTOINA ◇ DIPHANTOINE SODIUM ◇ DIPHEDAN ◇ DIPHENATE ◇ DIPHENIN ◇ DIPHENINE SODIUM ◇ DIPHENTOIN ◇ DIPHENYLAN SODIUM ◇ DIPHENYLHYDANTOIN SODIUM ◇ 5,5-DIPHENYLHYDANTOIN SODIUM ◇ 5,5-DIPHENYL-2,4-IMIDAZOLIDINE-DIONE, MONOSODIUM SALT ◇ DI-PHETINE ◇ DITOIN ◇ DIVULSAN ◇ DPH ◇ ENKEFAL ◇ EPAMIN ◇ EPANUTIN ◇ EPELIN ◇ EPIFENYL ◇ EPIHYDAN ◇ EPILAN-D ◇ EPILANTIN ◇ EPINAT ◇ EPTOIN ◇ FENANTOIN ◇ FENITOIN ◇ FENYTOINE ◇ HYDANTIN SODIUM ◇ HYDANTOIN SODIUM ◇ IDANTOIL ◇ IDANTOINAL ◇ LEPITOIN ◇ LEPITOIN SODIUM ◇ MINETOIN ◇ NOVANTOINA ◇ NOVODIPHENYL ◇ OM-HYDANTOINE SODIUM ◇ PHENYTOIN SODIUM ◇ SACERIL ◇ SDPH ◇ SODANTON ◇ SODIUM DIPHENYLHYDANTOIN ◇ SODIUM DIPHENYL HYDANTOINATE ◇ SODIUM-5,5-DIPHENYLHYDANTOINATE ◇ SODIUM-5,5-DIPHENYL-2,4-IMIDAZOLIDINEDIONE ◇ SOLANTOIN ◇ SOLANTYL ◇ SOLUBLE PHENYTOIN ◇ SYLANTOIC ◇ TACOSAL ◇ THILOPHENYT ◇ ZENTROPIL

TOXICITY DATA with REFERENCE
dnd-esc 50 μmol/L MUREAV 89,95,81
pic-esc 100 mg/L VIRLAX 99,257,79
orl-rat TDLo:3 g/kg (9-13D preg):REP TJADAB 27,149,83
scu-mus TDLo:50 mg/kg (female 12D post):TER PSEBAA 168,175,81
orl-man TDLo:70 mg/kg/17D-I:CNS,GIT,SKN NEJMAG 242,897,50
orl-wmn TDLo:4 mg/kg/D:END,SKN JAMAAP 176(6),491,61
orl-wmn LDLo:78 mg/kg:SKN,BLD,PUL ADSYAF 46,856,42
orl-man LDLo:647 mg/kg/21W-I:SKN AIMDAP 81,605,48
unr-man LDLo:29 mg/kg 85DCAI 2,73,70
orl-rat LD50:1530 mg/kg JPETAB 138,224,62
ipr-rat LD50:138 mg/kg IJPPAZ 10,5,66
ivn-rat LD50:104 mg/kg ARZNAD 33,1155,83
scu-rat LD50:230 mg/kg NYKZAU 56,377,60
orl-mus LD50:165 mg/kg IJEBA6 19,1047,81
ipr-mus LD50:103 mg/kg ARZNAD 30,12,80
scu-mus LD50:400 mg/kg BCPCA6 17,369,68
ivn-mus LD50:110 mg/kg ARZNAD 30,477,80
ivn-dog LDLo:90 mg/kg ARPAAQ 28,761,39

CONSENSUS REPORTS: IARC Cancer Review: Animal Sufficient Evidence IMEMDT 13,201,77. Reported in EPA TSCA Inventory.

SAFETY PROFILE: Confirmed carcinogen. Experimental teratogen. Other experimental reproductive effects. Poison by ingestion, subcutaneous, intravenous, and intraperitoneal routes. Human systemic effects by ingestion: anorexia, respiratory depression, nausea or vomiting, hemorrhage, dermatitis, and endocrine effects. Mutation data reported. An anticonvulsant and cardiac depressant used for the treatment of grand mal and psychomotor seizures. When heated to decomposition it emits very toxic fumes of NO_x and Na_2O.

DNU100 CAS:456-59-7 ***HR: 2***
DILATIN
mf: $C_{17}H_{24}O_3$ mw: 276.41

PROP: Crystals. Mp: 50-53°, bp: 192-194°. Practically insol in water; sol in lipoids and their solvents.

SYNS: ARTO-ESPASMOL ◇ BS 572 ◇ CAPILAN ◇ CICLOSPASMOL ◇ CLANDILON ◇ CYCLANDELATE ◇ CYCLERGINE ◇ CYCLOBRAL ◇ CYCLOLYT ◇ CYCLOMANDOL ◇ CYCLOSPASMOL ◇ α-HYDROXYBENZENEACETIC ACID 3,3,5-TRIMETHYLCYCLOHEXYL ESTER (9CI) ◇ NATIL ◇ NOVODIL ◇ PEREBRAL ◇ SAICLATE ◇ SANCYCLAN ◇ SEPYRON ◇ SPASMIONE ◇ SPASMOCYCLON ◇ SPASMOCYCLONE ◇ 3,3,5-TRIMETHYLCYCLOHEXANOL-α-PHENYL-α-HYDROXYACETATE ◇ 3,5,5-TRIMETHYLCYCLOHEXYL AMYGDALATE ◇ 3,3,5-TRIMETHYLCYCLOHEXYL MANDELATE

TOXICITY DATA with REFERENCE
orl-rat LD50:5 g/kg NIIRND 6,310,82
ipr-rat LD50:2570 mg/kg AIPTAK 105,145,56
ipr-mus LD50:3780 mg/kg AIPTAK 105,145,56
ipr-dog LD50:2000 mg/kg AIPTAK 105,145,56
orl-gpg LD50:3950 mg/kg AIPTAK 105,145,56
ipr-gpg LD50:2480 mg/kg AIPTAK 105,145,56

CONSENSUS REPORTS: Reported in EPA TSCA Inventory.

SAFETY PROFILE: Moderately toxic by ingestion and intraperitoneal routes. When heated to decomposition it emits acrid smoke and fumes. See also ESTERS.

DNU200 CAS:849-55-8 ***HR: 3***
DILATOL HYDROCHLORIDE
mf: $C_{19}H_{25}NO_2 \cdot ClH$ mw: 335.91

SYNS: ARLIDIN HYDROCHLORIDE ◇ BUPHENINE HYDROCHLORIDE ◇ DILATYL ◇ p-HYDROXY-α-(1-((1-METHYL-3-PHENYLPROPYL)AMINO)ETHYL)BENZYL ALCOHOL HYDROCHLORIDE ◇ 1-p-HYDROXYPHENYL-2-(1′-METHYL-3′-PHENYLPROPYLAMINO)-1-PROPANOL HYDROCHLORIDE ◇ NYLIDRIN HYDROCHLORIDE ◇ SUPRIFEN PSB HYDROCHLORIDE ◇ VERINA

TOXICITY DATA with REFERENCE
ipr-rat LD50:380 mg/kg 27ZQAG -,351,72
orl-mus LD50:250 mg/kg 27ZQAG -,351,72

ipr-mus LD50:136 mg/kg YKYUA6 24,431,73
ivn-mus LD50:40 mg/kg 27ZQAG -,351,72

SAFETY PROFILE: Poison by ingestion, intraperitoneal, and intravenous routes. When heated to decomposition it emits very toxic fumes of Cl^- and NO_x.

DNU300 CAS:71-68-1 ***HR: 3***
DILAUDID
mf: $C_{17}H_{19}NO_3{\bullet}ClH$ mw: 321.83

SYNS: DIHYDROMORPHINONE HYDROCHLORIDE ◇ DILAUDID HYDROCHLORIDE ◇ 4,5-α-EPOXY-3-HYDROXY-17-METHYLMORPHINAN-6-ONE HYDROCHLORIDE ◇ HYDROMORPHONE HYDROCHLORIDE ◇ HYMORPHAN

TOXICITY DATA with REFERENCE
scu-ham TDLo:21500 μg/kg (8D preg):TER AJOGAH 123,705,75
scu-rat LD50:51 mg/kg ARZNAD 3,238,53
scu-mus LD50:120 mg/kg ARZNAD 3,238,53
ivn-mus LD50:55 mg/kg TXAPA9 6,334,64
ivn-cat LDLo:3 mg/kg AEPPAE 194,296,40
ivn-rbt LDLo:2500 μg/kg AEPPAE 194,296,40

SAFETY PROFILE: Poison by subcutaneous and intravenous routes. Experimental teratogenic effects. A powerful analgesic. When heated to decomposition it emits very toxic fumes of NO_x and HCl. See also MORPHINE.

DNU310 CAS:1421-28-9 ***HR: 3***
DILAUDID HYDROCHLORIDE
mf: $C_{17}H_{21}NO_3{\bullet}ClH$ mw: 323.85

SYNS: DIHYDROMORPHINE HYDROCHLORIDE ◇ PARAMORFAN

TOXICITY DATA with REFERENCE
scu-mus LDLo:149 mg/kg JPETAB 52,468,34
ivn-mus LD50:55 mg/kg TXAPA9 6,334,64
orl-rbt LDLo:800 mg/kg HBAMAK 4,1289,35
scu-rbt LD50:50 mg/kg JPETAB 66,182,39
par-rbt LDLo:142 mg/kg JPETAB 66,182,39
scu-gpg LDLo:500 mg/kg HBAMAK 4,1289,35

SAFETY PROFILE: Poison by subcutaneous, intravenous, and parenteral routes. Moderately toxic by ingestion. When heated to decomposition it emits very toxic fumes of NO_x and HCl. See also MORPHINE.

DNU325 CAS:2592-85-0 ***HR: 3***
1,3-DILITHIOBENZENE
mf: $C_6H_4Li_2$ mw: 89.98

SAFETY PROFILE: An unstable explosive. See also LITHIUM COMPOUNDS and EXPLOSIVES.

DNU350 CAS:15114-92-8 ***HR: 3***
DILITHIUM-1,1-BIS(TRIMETHYLSILYL) HYDRAZIDE
mf: $C_6H_{18}Li_2N_2Si_2$ mw: 188.27

SAFETY PROFILE: Ignites spontaneously in air. Ignites or explodes on contact with nitric acid; fluorine gas; or liquid ozone + oxygen. When heated to decomposition it emits toxic fumes of NO_x. See also LITHIUM COMPOUNDS.

DNU400 CAS:8006-75-5 ***HR: 1***
DILL SEED OIL, EUROPEAN TYPE

PROP: From steam distillation of the dried ripe fruit of *Anethum graveolens* L. (Fam. *Umbelliferae.* Yellowish liquid; caraway odor and taste. D: 0.890-0.915, refr index: 1.4836 @ 20°. Sol in fixed oils, mineral oil, and propylene glycol; insol in glycerin.

SYNS: DILL FRUIT OIL ◇ DILL HERB OIL ◇ DILL OIL ◇ DILL SEED OIL ◇ DILL WEED OIL

TOXICITY DATA with REFERENCE
skn-rbt 500 mg/24H MOD FCTOD7 20 (Suppl),673,82
mma-sat 1 mg/plate JOPHDQ 3,236,80
orl-rat LD50:4040 mg/kg FCTXAV 14,659,76

CONSENSUS REPORTS: Reported in EPA TSCA Inventory.

SAFETY PROFILE: Mildly toxic by ingestion. A skin irritant. Mutation data reported. When heated to decomposition it emits acrid smoke and fumes.

DNU600 CAS:33286-22-5 ***HR: 3***
DILTIAZEM HYDROCHLORIDE
mf: $C_{22}H_{26}N_2O_4S{\bullet}ClH$ mw: 451.02

SYNS: ANGINYL ◇ CADIZEM ◇ CARDIEM ◇ CRD-401 ◇ CRP-401 ◇ DILZEM ◇ HERBESSER ◇ TILDIEM

TOXICITY DATA with REFERENCE
orl-rat TDLo:700 mg/kg (female 15-21D post):REP KSRNAM 8,3401,74
orl-mus TDLo:25 mg/kg (female 10D post):TER KSRNAM 8,3401,74
orl-man TDLo:1286 μg/kg/1D AIMEAS 99,794,83
orl-wmn TDLo:19 mg/kg:SKN PGMJAO 64,467,88
orl-man TDLo:36 mg/kg/13D-I:LVR GASTAB 88,1260,85
orl-rat LD50:560 mg/kg JJPAAZ 22,467,72
scu-rat LD50:520 mg/kg JJPAAZ 22,467,72
ivn-rat LD50:38 mg/kg JMGZAI 11(1),12,74
orl-mus LD50:640 mg/kg JJPAAZ 22,467,72
ipr-mus LD50:177 mg/kg JMCMAR 29,820,86
scu-mus LD50:260 mg/kg JJPAAZ 22,467,72
ivn-mus LD50:58 mg/kg JJPAAZ 22,467,72

SAFETY PROFILE: Poison by subcutaneous, intravenous, and intraperitoneal routes. Moderately toxic by ingestion. Human systemic effects by ingestion: fall in blood pressure, pulse rate decrease, gastrointestinal effects, dermatitis, fibrous hepatitis. Experimental reproductive effects. An experimental teratogen. When heated to decomposition it emits toxic fumes of SO_x, NO_x and HCl.

DNU850 CAS:14465-96-4 ***HR: 3***
DIMATIF
mf: $C_4H_{10}N_3PS$ mw: 163.20

SYNS: BIS(1-AZIRIDINYL)AMINOPHOSPINE SULFIDE ◇ p,p-BIS-(1-AZIRIDINYL)PHOSPHINOTHIOIC AMIDE ◇ DIETHYLENEIMINE-AMIDOTHIOPHOSPHORIC ACID ◇ ENT 61,969

TOXICITY DATA with REFERENCE
mmo-smc 1 mg/L TGANAK 18,455,84
cyt-hmn:lym 400 mg/L TGANAK 18,455,84
sce-hmn:lym 2 mg/L TGANAK 16(2),34,82
cyt-mus-unr 1 mg/kg TGANAK 18,455,84
orl-rat TDLo:50 mg/kg (6-15D preg):TER GISAAA 48(5),75,83
orl-rat TDLo:10 mg/kg (1-20D preg):REP GISAAA 48(5),75,83
orl-rat LD50:66 mg/kg GISAAA 48(5),75,83
orl-qal LD50:100 mg/kg JRPFA4 48,371,76

SAFETY PROFILE: Poison by ingestion. An experimental teratogen. Other experimental reproductive effects. Human mutation data reported. When heated to decomposition it emits toxic fumes of PO_x, NO_x, and SO_x.

DNU860 CAS:4076-02-2 ***HR: 2***
DIMAVAL
mf: $C_3H_7O_3S_3 \cdot Na$ mw: 210.27

SYNS: DIMAYAL ◇ 2,3-DIMERCAPTOPROPANE SODIUM SULPHONATE ◇ 2,3-DIMERCAPTOPROPANESULFONIC ACID SODIUM SALT ◇ 2,3-DIMERCAPTO-1-PROPANESULFONIC ACID SODIUM SALT ◇ DMPS ◇ meso-DIMERCAPTOSUCCINIC ACID SODIUM SALT ◇ SODIUM-2,3-DIMERCAPTOPROPANE-1-SULFONATE ◇ SODIUM-2,3-DITHIOLPROPANESULFONATE ◇ UNITHIOL ◇ UNITIOL ◇ UNITOL

TOXICITY DATA with REFERENCE
cyt-hmn:lym 1 mmol/L CYGEDX 8(4),31,74
orl-rat TDLo:28 g/kg (26W pre-21D post):REP ARZNAD 30,1291,80
ipr-rat LD50:1055 mg/kg ARZNAD 30,1291,80
ipr-mus LD50:1098 mg/kg TXAPA9 61,385,81

SAFETY PROFILE: Moderately toxic by intraperitoneal route. Experimental reproductive effects. Human mutation data reported. When heated to decomposition it emits toxic fumes of SO_x and Na_2O.

DNU875 ***HR: 3***
DIMEBON DIHYDROCHLORIDE
mf: $C_{22}H_{22}N_3 \cdot 2ClH$ mw: 401.39

SYN: 9-(2-(2-METHYLPYRIDYL-5)ETHYL)-3,6-DIMETHYL-1,2,3,4-TETRAHYDRO-Γ-CARBOLINE 2HCl

TOXICITY DATA with REFERENCE
orl-rat LD50:1132 mg/kg RPTOAN 48,103,85
ipr-rat LD50:160 mg/kg RPTOAN 48,103,85
ivn-rat LD50:59 mg/kg RPTOAN 48,103,85
orl-mus LD50:486 mg/kg FATOAO 47(3),75,84
ipr-mus LD50:145 mg/kg RPTOAN 48,103,85
scu-mus LD50:465 mg/kg FATOAO 47(3),75,84
ivn-mus LD50:90500 μg/kg FATOAO 47(3),75,84

SAFETY PROFILE: Poison by intravenous and intraperitoneal routes. Moderately toxic by ingestion and subcutaneous routes. When heated to decomposition it emits toxic fumes of NO_x and HCl.

DNV000 CAS:1165-48-6 ***HR: 3***
DIMEFLINE
mf: $C_{20}H_{21}NO_3$ mw: 323.42

SYNS: 8-(DIMETHYLAMINOMETHYL)-7-METHOXY-3-METHYLFLAVONE ◇ 8-((DIMETHYLAMINO)METHYL)-7-METHOXY-3-METHYL-2-PHENYLFLAVONE ◇ DW 62 ◇ MALIVAN ◇ N-(7-METHOXY-3-METHYL-4-OXO-2-PHENYL-4H-CHROMEN-8-YL)METHYL-N,N-DIMETHYLAMINE ◇ REANIMIL ◇ REC 7/0267 ◇ REMEFLIN

TOXICITY DATA with REFERENCE
orl-rat LD50:40 mg/kg RPOBAR 1,423,64
ipr-rat LD50:6 mg/kg TXAPA9 18,185,71
ivn-rat LD50:1800 μg/kg RPOBAR 1,423,64
rec-rat LD50:10 mg/kg RPOBAR 1,423,64
orl-mus LD50:12 mg/kg RPOBAR 1,423,64
ipr-mus LD50:4800 μg/kg JMPCAS 3,471,61
scu-mus LD50:4 mg/kg RPOBAR 1,423,64
ivn-dog LDLo:1 mg/kg RPOBAR 1,423,64

SAFETY PROFILE: Poison by ingestion, intraperitoneal, intravenous, rectal, and subcutaneous routes. When heated to decomposition it emits very toxic fumes of NO_x.

DNV200 CAS:2740-04-7 ***HR: 3***
DIMEFLINE HYDROCHLORIDE
mf: $C_{20}H_{21}NO_3 \cdot ClH$ mw: 359.88

SYNS: DEMEFLINE ◇ 8-((DIMETHYLAMINO)METHYL)-7-METHOXY-3-METHYLFLAVONE HYDROCHLORIDE ◇ 8-((DIMETHYLAMINO)METHYL)-7-METHOXY-3-METHYL-2-PHENYL-4H-1-BENZOPYRAN-4-ONE HYDROCHLORIDE ◇ DW 62 ◇ 3-METHYL-7- METHOXY-8-(DIMETHYLAMINO-METHYL)-FLAVONEHYDROCHLORIDE ◇ NSC-114650 ◇ REC 7/0267 ◇ REMEFLIN

TOXICITY DATA with REFERENCE
orl-rat LD50:14 mg/kg NIIRDN 6,345,82
orl-mus LD50:12 mg/kg JPETAB 128,176,60
ipr-mus LD50:5 mg/kg JPETAB 128,176,60
scu-mus LD50:4 mg/kg JPETAB 128,176,60

SAFETY PROFILE: Poison by ingestion, intraperitoneal, and subcutaneous routes. When heated to decomposition it emits very toxic fumes of HCl and NO_x.

DNV600 CAS:27292-46-2 ***HR: 2***
2,3-DIMERCAPTOPROPYL-p-TOLYSULFIDE
mf: $C_{10}H_{14}S_3$ mw: 230.42

SYN: ANTARSIN

TOXICITY DATA with REFERENCE
orl-rat LDLo:2000 mg/kg FATOAO 30(2),226,67
ipr-rat LDLo:1000 mg/kg FATOAO 30(2),226,67
scu-rat LDLo:7000 mg/kg FATOAO 30(2),226,67

SAFETY PROFILE: Moderately toxic by ingestion and intraperitoneal routes. Mildly toxic by subcutaneous route. When heated to decomposition it emits toxic fumes of SO_x. See also MERCAPTANS.

DNV800 CAS:304-55-2 ***HR: 2***
meso-2,3-DIMERCAPTOSUCCINIC ACID
mf: $C_4H_6O_4S_2$ mw: 182.22

SYNS: (R*,S*)-2,3-DIMERCAPTOBUTANEDIOIC ACID ◇ meso-DIMERCAPTOSUCCINIC ACID ◇ DIM-SA ◇ DMS ◇ DMSA ◇ DTS ◇ Ro 1-7977 ◇ SUCCIMER

TOXICITY DATA with REFERENCE
ipr-mus LD50:500 mg/kg NTIS** AD691-490
scu-mus LD50:1725 mg/kg AIPTAK 131,283,61
ivn-rbt LDLo:2700 mg/kg AIPTAK 131,283,61

CONSENSUS REPORTS: Reported in EPA TSCA Inventory.

SAFETY PROFILE: Moderately toxic by intraperitoneal, intravenous, and subcutaneous routes. When heated to decomposition it emits toxic fumes of SO_x. See also MERCAPTANS.

DNW000 CAS:63869-15-8 ***HR: 3***
DIMERCUROUS METHANE ARSONATE
mf: $CH_3AsO_3 \cdot 2Hg$ mw: 539.14

SYN: METHANEARSONIC ACID DIMERCURY SALT

TOXICITY DATA with REFERENCE
ipr-mus LD50:75 mg/kg NTIS** AD691-490

CONSENSUS REPORTS: Arsenic and its compounds, as well as mercury and its compounds, are on the Community Right-To-Know List.

OSHA PEL: TWA 0.5 mg(As)/m^3; (Transitional: CL 1 mg/10m^3) CL 0.1 mg(Hg)/m^3 (skin)
ACGIH TLV: TWA 0.2 mg(As)/m^3; 0.1 mg(Hg)/m^3 (skin)
NIOSH REL: (Inorganic Mercury) TWA 0.05 mg(Hg)/m^3

SAFETY PROFILE: Poison by intraperitoneal route. When heated to decomposition it emits very toxic fumes of As and Hg. See also MERCURY COMPOUNDS and ARSENIC COMPOUNDS.

DNW200 CAS:12529-66-7 ***HR: 3***
DIMERCURY IMIDE OXIDE
mf: $(HHg_2NO)_n$

$$(Hg:N^-:HgOH^-)_n$$

SYNS: MILLON'S BASE ANHYDRIDE ◇ POLY(DIMERCURYIMMONIUM HYDOXIDE)

CONSENSUS REPORTS: Mercury and its compounds are on the Community Right-To-Know List.

SAFETY PROFILE: A severe explosion hazard if touched or heated. When heated to decomposition it emits toxic fumes of Hg and NO_x. See also MERCURY COMPOUNDS.

DNW400 CAS:125-64-4 ***HR: 3***
DIMERIN
mf: $C_{10}H_{17}NO_2$ mw: 183.28

SYNS: 3,3-DIETHYL-2,4-DIOXO-5-METHYLPIPERIDINE ◇ 3,3-DIETHYL-5-METHYL-2,4-PIPERIDINEDIONE ◇ 3,3-DIETHYL-5-METHYLPIPERIDINE-2,4-DIONE ◇ 2,4-DIOXY-3,3-DIETHYL-5-METHYLPIPERIDINE ◇ METHYPROLON ◇ METHYPRYLON ◇ METIPRILONE ◇ NOCTAN ◇ NOLUDAR ◇ RO 1-6463

TOXICITY DATA with REFERENCE
orl-wmn TDLo:600 mg/kg:CNS,PUL JAMAAP 198,1213,66
orl-hmn TDLo:26 mg/kg:CNS CTOXAO 6,563,73
orl-rat LD50:860 mg/kg 27ZQAG -,264,72
scu-rat LD50:400 mg/kg 27ZQAG -,264,72
ivn-rat LD50:380 mg/kg 27ZQAG -,264,72
orl-mus LD50:890 mg/kg JPETAB 118,139,56
ipr-mus LD50:1000 mg/kg 27ZQAG -,264,72
ivn-mus LD50:275 mg/kg 27ZQAG -,264,72
orl-dog LD50:300 mg/kg CLDND* -,264,72
scu-rbt LD50:500 mg/kg 27ZQAG -,264,72
ivn-rbt LD50:315 mg/kg 27ZQAG -,264,72

SAFETY PROFILE: Poison by ingestion, subcutaneous, and intravenous routes. Moderately toxic by intraperitoneal routes. Human systemic effects by ingestion: general anesthesia, sleep disorder, motor activity and pulmonary changes. When heated to decomposition it emits toxic fumes of NO_x.

DNW700 CAS:4757-55-5 ***HR: 3***
DIMETACRINE
mf: $C_{20}H_{26}N_2$ mw: 294.48

PROP: Free base. Bp: (1) 200°.

SYNS: DIMETHACIN ◇ DIMETHACINE ◇ 9,9-DIMETHYL-10-(3-(DIMETHYLAMINO)PROPYL)ACRIDAN

TOXICITY DATA with REFERENCE
ipr-rat TDLo:91 mg/kg (91D male):REP OYYAA2 8,949,74
orl-rat LD50:1850 mg/kg IYKEDH 6,530,75

ipr-rat LD50:181 mg/kg IYKEDH 6,530,75
scu-rat LD50:1076 mg/kg IYKEDH 6,530,75
orl-mus LD50:1293 mg/kg IYKEDH 6,530,75
ipr-mus LD50:206 mg/kg IYKEDH 6,530,75
scu-mus LD50:676 mg/kg IYKEDH 6,530,75
ivn-mus LD50:39600 μg/kg IYKEDH 6,530,75

SAFETY PROFILE: Poison by intravenous and intraperitoneal routes. Moderately toxic by ingestion and subcutaneous routes. Experimental reproductive effects. When heated to decomposition it emits toxic fumes of NO_x.

DNW759 CAS:32865-01-3 ***HR: 3***
dl-DIMETANE MALEATE
mf: $C_{16}H_{19}BrN_2 \cdot C_4H_4O_4$ mw: 435.36

SYNS: (±)-2-(p-BROMO-α-(2-(DIMETHYLAMINO)ETHYL)BENZYL)PYRIDINE MALEATE ◇ dl-BROMOPHENIRAMINE MALEATE ◇ (±)-BROMPHENIRAMINE MALEATE ◇ dl-BROMPHENIRAMINE MALEATE ◇ (±)-(Z)-Γ-(4-BROMOPHENYL)-N,N-DIMETHYL-2-PYRIDINEPROPANAMINE 2-BUTENEDIOATE (1:1)

TOXICITY DATA with REFERENCE
orl-rat LD50:161 mg/kg CMTRAG 3,120,61
ipr-rat LD50:113 mg/kg CMTRAG 3,120,61
orl-mus LD50:147 mg/kg CMTRAG 3,120,61
ipr-mus LD50:109 mg/kg CMTRAG 3,120,61
ivn-mus LD50:26 mg/kg CMTRAG 3,120,61
orl-gpg LD50:245 mg/kg CMTRAG 3,120,61

SAFETY PROFILE: Poison by ingestion, intravenous, and intraperitoneal routes. When heated to decomposition it emits toxic fumes of Br^- and NO_x.

DNW800 CAS:2303-47-1 ***HR: 3***
cis-1,4-DIMETHANE SULFONOXY-2-BUTENE
mf: $C_6H_{12}O_6S_2$ mw: 244.30

TOXICITY DATA with REFERENCE
sln-dmg-par 10 mmol/L JOGNAU 54,146,56
skn-mus TDLo:480 mg/kg/10W-I:NEO CNREA8 17,64,57

SAFETY PROFILE: Questionable carcinogen with experimental neoplastigenic data. Mutation data reported. When heated to decomposition it emits toxic fumes of SO_x.

DNX000 CAS:1953-56-6 ***HR: 2***
trans-1,4-DIMETHANE SULFONOXY-2-BUTENE
mf: $C_6H_{12}O_6S_2$ mw: 244.30

SYNS: CB 2095 ◇ 2-BUTENE-1,4-DIOL, DIMETHANESULFONATE, (E)-

TOXICITY DATA with REFERENCE
sln-dmg-par 10 mmol/L JOGNAU 54,146,56
skn-mus TDLo:480 mg/kg/10W-I:NEO CNREA8 17,64,57

SAFETY PROFILE: Questionable carcinogen with experimental neoplastigenic data. Mutation data reported. When heated to decomposition it emits toxic fumes of SO_x.

DNX200 CAS:2917-96-6 ***HR: 3***
1,4-DIMETHANESULFONOXY-2-BUTYNE
mf: $C_6H_{10}O_6S_2$ mw: 242.28

SYN: CB2058

TOXICITY DATA with REFERENCE
sln-dmg-par 10 mmol/L JOGNAU 54,146,56
sln-dmg-unk 10 mmol/L ANYAA9 160,228,69
skn-mus TDLo:320 mg/kg/5W-I:NEO CNREA8 17,64,57

SAFETY PROFILE: Questionable carcinogen with experimental neoplastigenic data by skin contact. Mutation data reported. When heated to decomposition it emits toxic fumes of SO_x. See also ACETYLENE COMPOUNDS.

DNX300 CAS:1001-62-3 ***HR: 3***
DIMETHANESULFONYL PEROXIDE
mf: $C_2H_6O_6S_2$ mw: 190.19

SAFETY PROFILE: Decomposes explosively after melting at 79°C. When heated to decomposition it emits toxic fumes of SO_x. See also PEROXIDES.

DNX400 CAS:2773-92-4 ***HR: 3***
DIMETHISOQUIN HYDROCHLORIDE
mf: $C_{17}H_{24}N_2O \cdot ClH$ mw: 308.89

SYNS: 3-BUTYL-1-(2-(DIMETHYLAMINO)ETHOXY)ISOQUINOLINE HYDROCHLORIDE ◇ 2-((3-BUTYL-1-ISOQUINOLINYL)OXY)-N,N-DIMETHYLETHANAMINE MONOHYDROCHLORIDE ◇ 1-(β-DIMETHYLAMINOETHOXY)-3-N-BUTYLISOQUINOLINEHYDROCHLORIDE ◇ 1-(β-DIMETHYLAMINOETHOXY)-3-N-BUTYLISOQUINOLINE MONOHYDROCHLORIDE ◇ ISOCHINOL ◇ PRURALGAN ◇ PRURALGIN ◇ QUOTANE ◇ QUOTANE HYDROCHLORIDE

TOXICITY DATA with REFERENCE
ipr-rat LD50:45 mg/kg JPETAB 103,306,51
ivn-mus LD50:8 mg/kg ARZNAD 18,729,68
ivn-rbt LD50:5 mg/kg JPETAB 103,306,51

SAFETY PROFILE: Poison by intraperitoneal and intravenous routes. A topical anesthetic. When heated to decomposition it emits very toxic fumes of HCl and NO_x.

DNX500 CAS:8015-19-8 ***HR: 3***
DIMETHISTERONE and ETHINYL ESTRADIOL
mf: $C_{23}H_{32}O_2 \cdot C_{20}H_{24}O_2$ mw: 636.99

SYNS: ETHINYL ESTRADIOL and DIMETHISTERONE ◇ ORACON ◇ OVIN ◇ SECROVIN

TOXICITY DATA with REFERENCE
orl-wmn TDLo:273 mg/kg (77W pre):REP AJOGAH 107,717,70

orl-wmn TDLo:244 mg/kg/8Y-I:CAR AJOGAH 123,299,75
orl-wmn TDLo:92 mg/kg/3Y-I:CAR OBGNAS 47,639,76

SAFETY PROFILE: Suspected human carcinogen producing uterine tumors. Human reproductive effects by ingestion: abnormalities of the uterus, cervix, and vagina. A steroid. When heated to decomposition it emits acrid smoke and irritating fumes.

DNX600 CAS:116-01-8 ***HR: 3***
DIMETHOATE-ETHYL
mf: $C_6H_{14}NO_3PS_2$ mw: 243.30

SYNS: AMERICAN CYANAMID 18706 ◇ B/77 ◇ O,O-DIMETHYL-S-(N-ETHYLCARBAMOYLMETHYL) DITHIOPHOSPHATE ◇ O,O-DIMETHYL-S-(N-ETHYLCARBAMOYLMETHYL)PHOSPHORODITHIOATE ◇ EI-18706 ◇ ENT 25,506 ◇ ETHOATE METHYL ◇ S-(2-(ETHYLAMINO-2-OXOETHYL)-O,O-DIMETHYL PHOSPHORODITHIOATE ◇ S-(N-ETHYLCARBAMOYLMETHYL) DIMETHYL PHOSPHORODITHIOATE ◇ FITIOS ◇ FITIOS B/77 ◇ N-MONOETHYLAMIDE of O,O-DIMETHYLDITHIOPHOSPHORYLACETIC ACID ◇ PHOSHOROTHIOIC ACID-S-(2-(ETHYLAMINO)-2-OXOETHYL)-O,O-DIMETHYLESTER

TOXICITY DATA with REFERENCE
orl-rat LD50:125 mg/kg WRPCA2 9,119,70
skn-rat LD50:2000 mg/kg WRPCA2 9,119,70
ims-rat LD50:250 mg/kg FRPSAX 21,443,66
orl-mus LD50:350 mg/kg SPEADM 74-1,-,74

SAFETY PROFILE: Poison by ingestion and intramuscular routes. Moderately toxic by skin contact. A pesticide. When heated to decomposition it emits very toxic fumes of NO_x, PO_x, and SO_x.

DNX800 CAS:1113-02-6 ***HR: 3***
DIMETHOATE OXYGEN ANALOG
mf: $C_5H_{12}NO_4PS$ mw: 213.21

SYNS: O-ANALOG of DIMETHOATE ◇ BAY 45432 ◇ BAYER 45,432 ◇ DIMETHOATE O-ANALOG ◇ DIMETHOATE PO ISOLOGUE ◇ DIMETHOXON ◇ O,O-DIMETHYL-S-((N-METHYL-CARBAMOYL)-METHYL)MONOTHIOFOSFAAT (DUTCH) ◇ O,O-DIMETHYL-S-(N-METHYL-CARBAMOYL)-METHYL-MONOTHIOPHOSPHAT(GERMAN) ◇ O,O-DIMETHYL-S-((METHYLCARBAMOYL)METHYL) PHOSPHOROTHIOATE ◇ O,O-DIMETHYL-S-(N-METHYLCARBAMOYLMETHYL)PHOSPHOROTHIOATE ◇ O,O-DIMETHYL-S-(N-METHYLCARBAMOYLMETHYL)PHOSPHOROTHIOLATE ◇ DIMETHYL-S-(N-METHYL-CARBAMOYL-METHYL)PHOSPHOROTHIOLATE ◇ O,O-DIMETHYL-S-(N-METHYLCARBAMOYLMETHYL) THIOPHOSPHATE ◇ O,O-DIMETHYL-S-(2-OXO-3-AZABUTYL)-MONOTHIOPHOSPHATE ◇ O,O-DIMETIL-S-(N-METIL-CARBAMOIL)-METIL-MONOTIOFOSFATO (ITALIAN) ◇ ENT 25,776 ◇ FOLIMAT ◇ OMETHOAT ◇ OMETHOATE ◇ PHOSPHOROTHIOIC ACID, O,O-DIMETHYL S-(2-(METHYLAMINO)-2-OXOETHYL) ESTER ◇ PO-DIMETHOATE ◇ THIOPHOSPHATE de O,O-DIETHYLE et de S-(N-METHYLCARBAMOYL) METHYLE (FRENCH)

TOXICITY DATA with REFERENCE
mmo-esc 5 μL/plate MUREAV 28,405,75
orl-rat LDLo:50 mg/kg ARSIM* 20,5,66
skn-rat LD50:700 mg/kg WRPCA2 9,119,70
ipr-mus LD50:180 mg/kg ACPMAP 16,7,63
orl-cat LD50:50 mg/kg 85DPAN -,-,71/76
orl-rbt LD50:50 mg/kg 85DPAN -,-,71/76

SAFETY PROFILE: Poison by ingestion and intraperitoneal routes. Moderately toxic by skin contact. Mutation data reported. An insecticide. When heated to decomposition it emits very toxic fumes of NO_x, PO_x, and SO_x.

DNY000 CAS:94-15-5 ***HR: 3***
DIMETHOCAINE
mf: $C_{16}H_{26}N_2O_2$ mw: 278.44

SYNS: 3-(DIETHYLAMINO)-2,2-DIMETHYL-1-PROPANOL-p-AMINOBENZOATE ◇ LAROCAINE

TOXICITY DATA with REFERENCE
scu-mus LDLo:380 mg/kg AEPPAE 168,447,32
ivn-mus LDLo:40 mg/kg PHREA7 12,190,32
scu-rbt LDLo:150 mg/kg PHREA7 12,190,32
ivn-rbt LDLo:150 mg/kg PHREA7 12,190,32
scu-gpg LDLo:200 mg/kg PHREA7 12,190,32
par-frg LDLo:200 mg/kg AEPPAE 168,447,32

SAFETY PROFILE: Poison by subcutaneous, intravenous, and parenteral routes. When heated to decomposition it emits toxic fumes of NO_x.

DNY400 CAS:17210-48-9 ***HR: 3***
3,4'-DIMETHOXY-4-AMINOAZOBENZENE
mf: $C_{14}H_{15}N_3O_2$ mw: 257.32

SYN: 4-((p-METHOXYPHENYL)AZO)-o-ANISIDINE

TOXICITY DATA with REFERENCE
orl-rat TDLo:10 g/kg/24W-C:ETA GANNA2 59,131,68
orl-man TDLo:642 μg/kg:BLD,BIO IARC** 27,39,82

CONSENSUS REPORTS: IARC Cancer Review: Human Inadequate Evidence IMEMDT 27,39,82; Animal Inadequate Evidence IMEMDT 27,39,82.

SAFETY PROFILE: Questionable carcinogen with experimental tumorigenic data. Human systemic effects by ingestion: methemoglobinemia-carboxhemoglobinemia, and changes in porphyrin metabolism. When heated to decomposition it emits toxic fumes of NO_x.

DNY500 CAS:2735-04-8 ***HR: 2***
2,4-DIMETHOXYANILINE
mf: $C_8H_{11}NO_2$ mw: 153.20

TOXICITY DATA with REFERENCE
mma-sat 10 μg/plate ENMUDM 7(Suppl 5),1,85
otr-rat:emb 55 μg/plate JJATDK 1,190,81
orl-rat LD50:464 mg/kg NCILB* NCI-E-C-72-3252,73
orl-mus LD50:1 g/kg NCILB* NCI-E-C-72-3252,73

CONSENSUS REPORTS: Reported in EPA TSCA Inventory.

SAFETY PROFILE: Moderately toxic by ingestion. Mutation data reported. When heated to decomposition it emits toxic fumes of NO_x. See also ANILINE DYES.

DNY800 CAS:6448-90-4 ***HR: 1***
1,5-DIMETHOXYANTHRAQUINONE
mf: $C_{16}H_{12}O_4$ mw: 268.28

SYNS: 1,5-DIMETHOXY-9,10-ANTHRACENEDIONE ◇ 1,5-DIMETHOXYANTHRACHINON (CZECH)

TOXICITY DATA with REFERENCE
eye-rbt 500 mg/24H MLD 28ZPAK -,113,72
orl-rat LD50:16600 mg/kg 28ZPAK -,113,72

CONSENSUS REPORTS: Reported in EPA TSCA Inventory.

SAFETY PROFILE: Mildly toxic by ingestion. An eye irritant. When heated to decomposition it emits acrid smoke and irritating fumes.

DNZ100 CAS:476-70-0 ***HR: 3***
1,10-DIMETHOXY-6a-α-APORPHINE-2,9-DIOL
mf: $C_{19}H_{21}NO_4$ mw: 327.41

PROP: d-Form: Crystals from ether. Mp: 162-164°. Very sltly sol in water or ether; sol in alcohol, chloroform, dilute acids. dl-Form: Mp: 159-162°.

SYNS: BOLDIN ◇ BOLDINE ◇ (+)-BOLDINE ◇ (S)-BOLDINE ◇ (+)-(S)-BOLDINE ◇ (S)-5,6,6a,7-TETRAHYDRO-1,10-DIMETHOXY-6-METHYL-4H-DIBENZO(de,g)QUINOLINE-2,9-DIOL ◇ UNIBOLDINA

TOXICITY DATA with REFERENCE
orl-mus LD50:450 mg/kg APFRAD 38,537,80
ipr-mus LD50:170 mg/kg APFRAD 38,537,80
ivn-mus LD50:90 mg/kg APFRAD 38,537,80

CONSENSUS REPORTS: Reported in EPA TSCA Inventory.

SAFETY PROFILE: Poison by intravenous and intraperitoneal routes. Moderately toxic by ingestion. When heated to decomposition it emits toxic fumes of NO_x.

DOA000 CAS:16354-53-3 ***HR: 3***
7,12-DIMETHOXYBENZ(a)ANTHRACENE
mf: $C_{20}H_{16}O_2$ mw: 288.36

TOXICITY DATA with REFERENCE
ims-rat TDLO:50 mg/kg:NEO CNREA8 29,506,69

SAFETY PROFILE: Questionable carcinogen with experimental neoplastigenic data. When heated to decomposition it emits acrid smoke and irritating fumes.

DOA200 CAS:91-16-7 ***HR: 2***
o-DIMETHOXYBENZENE
mf: $C_8H_{10}O_2$ mw: 138.18

SYNS: 1,2-DIMETHOXYBENZENE ◇ PYROCATECHOL DIMETHYL ETHER ◇ VERATROL ◇ VERATROLE

TOXICITY DATA with REFERENCE
orl-rat LD50:890 mg/kg GTPZAB 26(2),54,82
orl-mus LD50:700 mg/kg GTPZAB 26(2),54,82

CONSENSUS REPORTS: Reported in EPA TSCA Inventory.

SAFETY PROFILE: Moderately toxic by ingestion. When heated to decomposition it emits acrid smoke and irritating fumes.

DOA400 CAS:150-78-7 ***HR: 3***
p-DIMETHOXYBENZENE
mf: $C_8H_{10}O_2$ mw: 138.18

PROP: Colorless leaflets, odor of sweet clover. Bp: 212.6°, d: 1.053 @ 55°/55°.

SYNS: DIMETHYL ETHER HYDROQUINONE ◇ DIMETHYLHYDROQUINONE ◇ DIMETHYLHYDROQUINONE ETHER ◇ DMB ◇ QUINOL DIMETHYL ETHER ◇ USAF AN-9 ◇ USAF UCTL-1791

TOXICITY DATA with REFERENCE
skn-rbt 6 g/12D-I MLD JIHTAB 31,79,49
skn-rbt 500 mg/24H MOD FCTXAV 16,715,78
skn-gpg 40%/24H MOD FCTXAV 16,715,78
ipr-rat LD50:1100 mg/kg JIHTAB 31,79,49
orl-rat LD50:3600 mg/kg FCTXAV 16,715,78
ipr-mus LD50:100 mg/kg NTIS** AD277-689

CONSENSUS REPORTS: Reported in EPA TSCA Inventory.

SAFETY PROFILE: Poison by intraperitoneal route. Moderately toxic by ingestion. A skin irritant. Flammable when exposed to heat or flame; can react with oxidizing materials. See also ETHERS.

DOA800 CAS:20325-40-0 ***HR: 3***
3,3'-DIMETHOXYBENZIDINE DIHYDROCHLORIDE
mf: $C_{14}H_{16}N_2O_2{\cdot}2ClH$ mw: 317.24

SYNS: C.I. DISPERSE BLACK-6-DIHYDROCHLORIDE ◇ o-DIANISIDINE DIHYDROCHLORIDE ◇ 3,3-DIMETHOXY-(1,1'-BIPHENYL)-4,4'-DIAMINE DIHYDROCHLORIDE

TOXICITY DATA with REFERENCE
mmo-sat 100 nmol/plate MUREAV 136,33,84
mma-sat 100 nmol/plate MUREAV 136,33,84
orl-rat TDLo:1040 mg/kg/1Y-I:CAR,REP JNCIAM 41,985,68
orl-mus TDLo:5760 mg/kg/2Y-I:ETA VOONAW 25(7),43,79
scu-mus TDLo:1152 mg/kg/2Y-I:ETA,REP VOONAW 25(7),43,79
orl-rat TD:11 g/kg/51W-I:CAR JNCIAM 41,985,68

CONSENSUS REPORTS: NTP Carcinogenesis Studies (Gavage); Clear Evidence: Rat NCITR* NTP-TR-372,90. Reported in EPA TSCA Inventory.

NIOSH REL: (Benzidine-Based Dye) Reduce to lowest feasible level

SAFETY PROFILE: Questionable carcinogen with experimental carcinogenic and tumorigenic data. Experimental reproductive data. Mutation data reported. When heated to decomposition it emits very toxic fumes of NO_x and HCl.

DOA875 ***HR: 3***
β-(2,4-DIMETHOXY-5-BENZYLBENZOYL)PROPIONIC ACID SODIUM SALT
mf: $C_{19}H_{19}O_5$•Na mw: 350.34

SYNS: 3-(4,6-DIMETHOXY-α-PHENYL-m-TOLUOYL)-PROPIONIC ACID SODIUM SALT ◇ SC-2657

TOXICITY DATA with REFERENCE
orl-mus LD50:1140 mg/kg JPETAB 100,421,50
ipr-mus LD50:225 mg/kg JPETAB 100,421,50
ivn-dog LDLo:113 mg/kg JPETAB 100,421,50

SAFETY PROFILE: Poison by intravenous and intraperitoneal routes. Moderately toxic by ingestion. When heated to decomposition it emits toxic fumes of Na_2O.

DOB200 CAS:10143-66-5 ***HR: 2***
1,3-DIMETHOXYBUTANE
mf: $C_6H_{14}O_2$ mw: 118.20

TOXICITY DATA with REFERENCE
skn-rbt 10 mg/24H open MLD AIHAAP 23,95,62
orl-rat LD50:3730 mg/kg AIHAAP 23,95,62
ihl-rat LCLo:8000 ppm AIHAAP 23,95,62
skn-rbt LD50:10 g/kg AIHAAP 23,95,62

SAFETY PROFILE: Moderately toxic by ingestion. Mildly toxic by skin contact. A skin irritant. When heated to decomposition it emits acrid smoke and irritating fumes.

DOB275 ***HR: 3***
β-(2,4-DIMETHOXY-5-CYCLOHEXYLBENZOYL) PROPIONIC ACID
mf: $C_{18}H_{24}O_5$ mw: 320.42

TOXICITY DATA with REFERENCE
ivn-rat LD50:300 mg/kg AIPTAK 116,154,58
orl-mus LD50:400 mg/kg AIPTAK 116,154,58
scu-mus LD50:300 mg/kg AIPTAK 116,154,58

SAFETY PROFILE: Poison by ingestion, subcutaneous, and intravenous routes. When heated to decomposition it emits acrid smoke and fumes.

DOB300 ***HR: 3***
β-(2,4-DIMETHOXY-5-CYCLOHEXYLBENZOYL) PROPIONIC ACID SODIUM SALT
mf: $C_{18}H_{23}O_5$•Na mw: 342.36

SYN: SC-2644

TOXICITY DATA with REFERENCE
orl-mus LD50:1020 mg/kg JPETAB 100,421,50
ipr-mus LD50:214 mg/kg JPETAB 100,421,50
ivn-dog LDLo:107 mg/kg JPETAB 100,421,50

SAFETY PROFILE: Poison by intravenous and intraperitoneal routes. Moderately toxic by ingestion. When heated to decomposition it emits toxic fumes of Na_2O.

DOB325 ***HR: 3***
β-(2,4-DIMETHOXY-5-CYCLOPENTYLMETHYLBENZOYL)PROPIONIC ACID SODIUM SALT
mf: $C_{18}H_{23}O_5$•Na mw: 342.36

SYNS: 3-(α-CYCLOPENTYL-4,6-DIMETHOXY-m-TOLUOYL)-PROPIONIC ACID SODIUM SALT ◇ SC-2798

TOXICITY DATA with REFERENCE
orl-mus LD50:1060 mg/kg JPETAB 100,421,50
ipr-mus LD50:148 mg/kg JPETAB 100,421,50
ivn-dog LDLo:74 mg/kg JPETAB 100,421,50

SAFETY PROFILE: Poison by intravenous and intraperitoneal routes. Moderately toxic by ingestion. When heated to decomposition it emits toxic fumes of Na_2O.

DOB600 CAS:63040-49-3 ***HR: 3***
5,6-DIMETHOXYDIBENZ(a,h)ANTHRACENE
mf: $C_{24}H_{18}O_2$ mw: 338.42

SYNS: 3,4-DIMETHOXY-DBA ◇ 3,4-DIMETHOXY-1,2:5,6-DIBENZANTHRACENE

TOXICITY DATA with REFERENCE
skn-mus TDLo:600 mg/kg/60W-I:NEO CNREA8 22,78,62
scu-mus TDLo:80 mg/kg/4W-I:ETA CNREA8 22,78,62

SAFETY PROFILE: Questionable carcinogen with experimental neoplastigenic and tumorigenic data. When heated to decomposition it emits acrid smoke and irritating fumes.

DOC000 CAS:41226-20-4 ***HR: 3***
4,7-DIMETHOXY-6-(2-DIISOPROPYLAMINOETHOXY)-5-(p-METHOXYCINNAMOYL) BENZOFURAN OXALATE
mf: $C_{28}H_{35}NO_6$•$C_2H_2O_4$ mw: 571.68

TOXICITY DATA with REFERENCE
orl-mus LD50:185 mg/kg CHTPBA 8,479,73
ivn-mus LD50:13 mg/kg CHTPBA 8,479,73

SAFETY PROFILE: Poison by ingestion and intrave-

nous routes. When heated to decomposition it emits toxic fumes of NO_x.

DOC800 CAS:26270-60-0 ***HR: 3***
4,7-DIMETHOXY-6-(2-DIMETHYLAMINOETHOXY)-5-(p-FLUOROCINNAMOYL)BENZOFURAN MALEATE
mf: $C_{23}H_{25}FNO_5 \cdot C_4H_4O_4$ mw: 530.57

TOXICITY DATA with REFERENCE
orl-mus LD50:430 mg/kg CHTPBA 8,479,73
ivn-mus LD50:14 mg/kg CHTPBA 8,479,73

SAFETY PROFILE: Poison by intravenous route. Moderately toxic by ingestion. When heated to decomposition it emits very toxic fumes of NO_x and F^-.

DOD200 CAS:52171-41-2 ***HR: 3***
4-,7-DIMETHOXY-6-(2-DIMETHYLAMINOETHOXY)-5-(p-HYDROXYCINNAMOYL)BENZOFURAN OXALATE
mf: $C_{23}H_{25}NO_6 \cdot C_2H_2O_4$ mw: 501.53

TOXICITY DATA with REFERENCE
orl-mus LD50:640 mg/kg CHTPBA 8,479,73
ivn-mus LD50:35 mg/kg CHTPBA 8,479,73

SAFETY PROFILE: Poison by intravenous route. Moderately toxic by ingestion. When heated to decomposition it emits toxic fumes of NO_x.

DOD400 CAS:52171-37-6 ***HR: 3***
4,7-DIMETHOXY-6-(2-DIMETHYLAMINOETHOXY)-5-(p-ISOPROPOXYCINNAMOYL)BENZOFURAN MALEATE
mf: $C_{26}H_{31}NO_6 \cdot C_4H_4O_4$ mw: 569.66

TOXICITY DATA with REFERENCE
orl-mus LD50:300 mg/kg CHTPBA 8,479,73
ivn-mus LD50:27 mg/kg CHTPBA 8,479,73

SAFETY PROFILE: Poison by ingestion and intravenous routes. When heated to decomposition it emits toxic fumes of NO_x.

DOD600 CAS:26270-59-7 ***HR: 3***
4,7-DIMETHOXY-6-(2-DIMETHYLAMINOETHOXY)-5-(p-METHOXYCINNAMOYL)BENZOFURAN MALEATE
mf: $C_{24}H_{27}NO_6 \cdot C_4H_4O_4$ mw: 541.60

TOXICITY DATA with REFERENCE
orl-mus LD50:255 mg/kg CHTPBA 8,479,73
ivn-mus LD50:25 mg/kg CHTPBA 8,479,73

SAFETY PROFILE: Poison by ingestion and intravenous routes. When heated to decomposition it emits toxic fumes of NO_x.

DOE000 CAS:7549-37-3 ***HR: 1***
1,1-DIMETHOXY-3,7-DIMETHYL-2,6-OCTADIENE (cis and trans)
mf: $C_{12}H_{22}O_2$ mw: 198.34

SYNS: CITRAL DIMETHYL ACETAL ◇ 1,1-DIMETHOXY-3,7-DIMETHYL-2,6-OCTADIENE

TOXICITY DATA with REFERENCE
skn-rbt 500 mg/24H FCTXAV 11,1065,73

CONSENSUS REPORTS: Reported in EPA TSCA Inventory.

SAFETY PROFILE: A skin irritant. When heated to decomposition it emits acrid smoke and irritating fumes.

DOE200 CAS:120-20-7 ***HR: 3***
3,4-DIMETHOXYDOPAMINE
mf: $C_{10}H_{15}NO_2$ mw: 181.26

PROP: Colorless to pale yellow liquid. Mp: 15°, bp: 156° @ 10 mm, d: 1.08 @ 28°/4°, vap d: 6.25.

SYNS: DIMETHYOXYDOPAMINE ◇ 3,4-DIMETHOXYPHENETHYLAMINE ◇ 3,4-DIMETHOXY-β-PHENETHYLAMINE ◇ DIMETHOXYPHENYLETHYLAMINE ◇ 3,4-DIMETHOXYPHENYLETHYLAMINE ◇ 3,4-DIMETHOXY-β-PHENYLETHYLAMINE ◇ β-(3,4-DIMETHOXYPHENYL)ETHYLAMINE ◇ 2-(3,4-DIMETHOXYPHENYL)ETHYLAMINE ◇ 3,4-DIMETHOXYPHENYLETHYLAMINE (base) ◇ DIMETHYLMESCALINE ◇ DIMPEA ◇ DMPE ◇ DMPEA ◇ HOMOVERATRYLAMINE

TOXICITY DATA with REFERENCE
ipr-mus LDLo:420 mg/kg JPHYA7 76,224,32
ivn-mus LD50:56 mg/kg CSLNX* NX#04483

CONSENSUS REPORTS: Reported in EPA TSCA Inventory.

SAFETY PROFILE: Poison by intravenous route. Moderately toxic by intraperitoneal route. When heated to decomposition it emits toxic fumes of NO_x.

DOE600 CAS:110-71-4 ***HR: 3***
1,2-DIMETHOXYETHANE
DOT: UN 2252
mf: $C_4H_{10}O_2$ mw: 90.14

$CH_3OC_2H_4OCH_3$

PROP: Liquid; sharp, ethereal odor. D: 0.86877, mp: −58°, bp: 82-83°, n (24/D) 1.3739, flash p: 4.5°C (40°F). Miscible with water and alc; sol in hydrocarbon solvents.

SYNS: DIMETHOXYETHANE ◇ α,β-DIMETHOXYETHANE ◇ 1,2-DIMETHOXYETHANE (DOT) ◇ DIMETHYLCELLOSOLVE ◇ 2,5-DIOXAHEXANE ◇ EGDME ◇ ETHYLENE DIMETHYL ETHER ◇ ETHYLENE GLYCOL DIMETHYL ETHER ◇ GLYCOL DIMETHYL ETHER ◇ GLYME ◇ MONOETHYLENE GLYCOL DIMETHYL ETHER ◇ MONOGLYME

TOXICITY DATA with REFERENCE
orl-mus TDLo:1960 mg/kg (7-10D preg):TER NISFAY 32,113,80
orl-mus TDLo:16 g/kg (female 7-14D post):REP EVHPAZ 57,141,84

CONSENSUS REPORTS: Reported in EPA TSCA Inventory. Glycol ether compounds are on the Community Right-To-Know List.

DOT Classification: Flammable Liquid; Label: Flammable Liquid.

SAFETY PROFILE: An experimental teratogen. Other experimental reproductive effects. Readily forms an explosive peroxide. A very dangerous fire hazard when exposed to heat, flame, or oxidizers. Mixture with lithium tetrahydroaluminate may ignite or explode if heated. When heated to decomposition it emits acrid smoke and fumes. See also GLYCOL ETHERS.

DOF000 CAS:10232-93-6 ***HR: 2***
DI(2-METHOXYETHYL) MALEATE
mf: $C_{10}H_{16}O_6$ mw: 232.26

SYN: BIS(2-METHOXYETHYL)ESTER MALEIC ACID

TOXICITY DATA with REFERENCE
skn-rbt 10 mg/24H open MLD AMIHBC 4,119,51
eye-rbt 100 mg open AMIHBC 4,119,51
orl-rat LD50:3340 mg/kg AMIHBC 4,119,51
skn-rbt LD50:1940 mg/kg AMIHBC 4,119,51

SAFETY PROFILE: Moderately toxic by ingestion and skin contact. A skin and eye irritant. When heated to decomposition it emits acrid smoke and irritating fumes.

DOF200 CAS:22575-95-7 ***HR: 3***
DI(2-METHOXYETHYL)PEROXYDICARBONATE
mf: $C_8H_{14}O_8$ mw: 238.20

$(CH_3OC_2H_4OCO{\cdot}O\text{-})_2$

SAFETY PROFILE: Explodes when heated to 34°C. When heated to decomposition it emits acrid smoke and fumes. See also PEROXIDES.

DOF400 CAS:117-82-8 ***HR: 3***
DIMETHOXY ETHYL PHTHALATE
mf: $C_{14}H_{18}O_6$ mw: 282.32

PROP: Light-colored, clear liquid; mild aromatic odor. Mp: −40° (forms gel), bp: 190°-210° @ 4 mm, flash p: 360°F, d: 1.171 @ 20°/20°, vap press: 0.3 mm @ 150°, vap d: 9.75.

SYNS: 1,2-BENZENEDICARBOXYLIC ACID BI(2-METHOXYETHYL) ESTER (9CI) ◇ BIS(METHOXYETHYL) PHTHALATE ◇ BIS(2-METHOXYETHYL) PHTHALATE ◇ DI(2-METHOXYETHYL)PHTHALATE ◇ DMEP ◇ KESSCOFLEX MCP ◇ 2-METHOXYETHYL PHTHALATE ◇ PHTHALIC ACID BIS(2-METHOXYETHYL) ESTER

TOXICITY DATA with REFERENCE
dlt-mus-ipr 1190 mg/kg TXAPA9 29,35,74
ipr-mus TDLo:2380 mg/kg (male 1D pre):TER EVHPAZ 3,81,73
ipr-mus TDLo:2380 mg/kg (male 1D pre):REP EVHPAZ 3,81,73
orl-rat LDLo:2750 mg/kg 29ZWAE -,356,68
ihl-rat LCLo:1595 ppm/16H 14CYAT 2,1904,63
ipr-rat LD50:3735 mg/kg JPMSAE 61,51,72
ipr-mus LD50:2510 mg/kg 34ZIAG -,691,69
orl-gpg LD50:1600 mg/kg 14CYAT 2,1904,63

CONSENSUS REPORTS: EPA Genetic Toxicology Program. Reported in EPA TSCA Inventory.

SAFETY PROFILE: Moderately toxic by ingestion and intraperitoneal routes. Mildly toxic by inhalation. Experimental teratogenic and reproductive effects. Mutation data reported. A pesticide. Combustible when exposed to heat or flame; can react with oxidizing materials. To fight fire, use water, foam, CO_2, dry chemical. When heated to decomposition it emits acrid smoke and irritating fumes.

DOF600 CAS:134-96-3 ***HR: 2***
3,5-DIMETHOXY-4-HYDROXYBENZALDEHYDE
mf: $C_9H_{10}O_4$ mw: 182.19

SYNS: GALLALDEHYDE-3,5-DIMETHYL ETHER ◇ SYRINGALDEHYDE ◇ SYRINGEALDEHYDE ◇ SYRINGIC ALDEHYDE ◇ SYRINGYLALDEHYDE

TOXICITY DATA with REFERENCE
ipr-mus LD50:1000 mg/kg JMCMAR 7,178,64

CONSENSUS REPORTS: Reported in EPA TSCA Inventory.

SAFETY PROFILE: Moderately toxic by intraperitoneal route. When heated to decomposition it emits acrid smoke and irritating fumes. See also ALDEHYDES.

DOF800 CAS:52171-42-3 ***HR: 3***
4,7-DIMETHOXY-5-(p-HYDROXYCINNAMOYL)-6-(2-PYRROLIDINYLETHOXY)BENZOFURAN MALEATE
mf: $C_{25}H_{27}NO_6{\cdot}C_4H_4O_4$ mw: 553.61

TOXICITY DATA with REFERENCE
orl-mus LD50:1700 mg/kg CHTPBA 8,479,73
ivn-mus LD50:40 mg/kg CHTPBA 8,479,73

SAFETY PROFILE: Poison by intravenous route. Moderately toxic by ingestion. When heated to decomposition it emits toxic fumes of NO_x.

DOG600 CAS:15589-00-1 ***HR: 3***
2,5-DIMETHOXY-4-METHYLAMPHETAMINE HYDROCHLORIDE
mf: $C_{12}H_{19}NO_2{\cdot}ClH$ mw: 245.78

SYNS: 2,5-DIMETHOXY-α,4-DIMETHYLPHENETHYLAMINE HYDROCHLORIDE ◇ 1-(2,5-DIMETHOXY-4-METHYLPHENYL)-2-AMINOPROPANE

TOXICITY DATA with REFERENCE
ipr-rat LD50:32500 μg/kg TXAPA9 45(1),49,78
orl-mus LD50:330 mg/kg TXAPA9 45(1),49,78
ipr-mus LD50:89 mg/kg JMCMAR 13,26,70
ivn-mus LD50:36 mg/kg TXAPA9 45(1),49,78
ivn-dog LD50:7200 μg/kg TXAPA9 45(1),49,78

SAFETY PROFILE: Poison by ingestion, intraperitoneal, and intravenous routes. A central nervous system stimulant. When heated to decomposition it emits very toxic fumes of NO_x and HCl. See also BENZEDRINE.

DOG700 CAS:1125-88-8 ***HR: 2***
DIMETHOXYMETHYLBENZENE
mf: $C_9H_{12}O_2$ mw: 152.21

SYNS: BENZALDEHYDE, DIMETHYL ACETAL ◇ DIMETHOXYPHENYLMETHANE

TOXICITY DATA with REFERENCE
skn-rbt 500 mg/24H MLD FCTXAV 17,711,79
orl-rat LD50:1220 mg/kg FCTXAV 17,711,79

CONSENSUS REPORTS: Reported in EPA TSCA Inventory.

SAFETY PROFILE: Moderately toxic by ingestion. A skin irritant. When heated to decomposition it emits acrid smoke and irritating fumes.

DOH400 CAS:3027-21-2 ***HR: 2***
DIMETHOXYMETHYLPHENYLSILANE
mf: $C_9H_{14}O_2Si$ mw: 182.3

SYNS: DIMETHOXYPHENYLMETHYLSILANE ◇ METHYLPHENYLDIMETHOXYSILANE ◇ PHENYLMETHYLDIMETHOXYSILANE

TOXICITY DATA with REFERENCE
orl-rat LD50:892 mg/kg GTPZAB 22(2),50,78

CONSENSUS REPORTS: Reported in EPA TSCA Inventory.

SAFETY PROFILE: Moderately toxic by ingestion. When heated to decomposition it emits acrid smoke and irritating fumes.

DOI400 CAS:635-85-8 ***HR: 3***
3,4-DIMETHOXYPHENETHYLAMINE HYDROCHLORIDE
mf: $C_{10}H_{15}NO_2{\cdot}ClH$ mw: 217.72

SYN: 3,4-DIMETHOXY-β-PHENYLETHYLAMINE HYDROCHLORIDE

TOXICITY DATA with REFERENCE
ipr-rat LD50:146 mg/kg TXAPA9 25,299,73
ipr-mus LD50:363 mg/kg TXAPA9 25,299,73
ivn-dog LD50:122 mg/kg TXAPA9 25,299,73
ivn-mky LD50:220 mg/kg TXAPA9 25,299,73
ipr-gpg LD50:375 mg/kg TXAPA9 25,299,73

SAFETY PROFILE: Poison by intraperitoneal and intravenous routes. When heated to decomposition it emits very toxic fumes of NO_x and HCl.

DOJ200 CAS:91-10-1 ***HR: 3***
2,6-DIMETHOXYPHENOL
mf: $C_8H_{10}O_3$ mw: 154.18

SYNS: ALDRICH ◇ 1,3-DIMETHYL PYROGALLATE ◇ PYROGALLOL DIMETHYLETHER ◇ PYROGALLOL-1,3-DIMETHYL ETHER ◇ SYRINGOL

TOXICITY DATA with REFERENCE
orl-mus LD50:2500 mg/kg BCTKAG 14,301,84
ivn-cat LDLo:100 mg/kg BJPCBM 53,93,75

CONSENSUS REPORTS: Reported in EPA TSCA Inventory.

SAFETY PROFILE: Poison by intravenous route. Moderately toxic by ingestion. When heated to decomposition it emits acrid smoke and irritating fumes. See also ETHERS.

DOJ400 CAS:69782-26-9 ***HR: 3***
(2-(2,5-DIMETHOXYPHENOXY)ETHYL)HYDRAZINE HYDROCHLORIDE
mf: $C_{10}H_{16}N_2O_3{\cdot}ClH$ mw: 248.74

TOXICITY DATA with REFERENCE
orl-mus LD50:125 mg/kg JMCMAR 6,63,63
ipr-mus LD50:125 mg/kg JMCMAR 6,63,63

SAFETY PROFILE: Poison by ingestion and intraperitoneal routes. When heated to decomposition it emits very toxic fumes of Cl^-, and NO_x. See also HYDRAZINE.

DOJ600 CAS:69782-18-9 ***HR: 3***
(2-(3,4-DIMETHOXYPHENOXY)ETHYL)HYDRAZINE HYDROCHLORIDE
mf: $C_{10}H_{16}N_2O_3{\cdot}ClH$ mw: 248.74

TOXICITY DATA with REFERENCE
orl-mus LD50:90 mg/kg JMCMAR 6,63,63
ipr-mus LD50:90 mg/kg JMCMAR 6,63,63

SAFETY PROFILE: Poison by ingestion and intraperitoneal routes. When heated to decomposition it emits very toxic fumes of Cl^- and NO_x. See also HYDRAZINE.

DOJ700 CAS:27318-87-2 ***HR: 2***
3-(3,5-DIMETHOXYPHENOXY)-1,2-PROPANEDIOL
mf: $C_{11}H_{16}O_5$ mw: 228.27

SYN: 3-(3',5'-DIMETHOXYPHENOXY)PROPANEDIOL-(1,2)

TOXICITY DATA with REFERENCE
orl-rat LD50:2100 mg/kg ARZNAD 24,111,74
ipr-rat LD50:955 mg/kg ARZNAD 24,111,74
orl-mus LD50:2070 mg/kg ARZNAD 24,111,74
ipr-mus LD50:780 mg/kg ARZNAD 24,111,74

SAFETY PROFILE: Moderately toxic by ingestion and intraperitoneal routes. When heated to decomposition it emits acrid smoke and fumes.

DOJ800 CAS:24973-25-9 ***HR: 3***
1-(2,5-DIMETHOXYPHENYL)-2-AMINOPROPANE
mf: $C_{11}H_{17}NO_2{\cdot}ClH$ mw: 231.75

SYNS: 2,5-DIMETHOXYAMPHETAMINE HYDROCHLORIDE ◇ 2,5-DIMETHOXY-α-METHYLBENZENEETHANAMINEHYDROCHLORIDE ◇ 2,5-DIMETHOXY-α-METHYLPHENETHYLAMINE HYDROCHLORIDE ◇ 2,5-DIMETHOXY-α-METHYL-β-PHENYLETHYLAMINE HYDROCHLORIDE ◇ β-(2,5-DIMETHOXYPHENYL)ISOPROPYLAMINE HYDROCHLORIDE

TOXICITY DATA with REFERENCE
mmo-sat 10 mg/plate MUREAV 56,199,77
ipr-rat LD50:63 mg/kg TXAPA9 45,49,78
ipr-mus LD50:135 mg/kg JMCMAR 13,26,70
ivn-mus LD50:39 mg/kg TXAPA9 45,49,78
ivn-dog LD50:26 mg/kg TXAPA9 45,49,78

SAFETY PROFILE: Poison by intraperitoneal and intravenous routes. Mutation data reported. When heated to decomposition it emits very toxic fumes of NO_x and HCl.

DOK000 CAS:13078-75-6 ***HR: 3***
1-(3,4-DIMETHOXYPHENYL)-2-AMINOPROPANE
mf: $C_{11}H_{17}NO_2{\cdot}ClH$ mw: 231.75

SYNS: 3,4-DIMETHOXYAMPHETAMINE HYDROCHLORIDE ◇ 3,4-DIMETHOXY-α-METHYL-β-PHENYLETHYLAMINEHYDROCHLORIDE

TOXICITY DATA with REFERENCE
ipr-rat LD50:48 mg/kg TXAPA9 25,299,73
ipr-mus LD50:168 mg/kg TXAPA9 25,299,73
ivn-dog LD50:59 mg/kg TXAPA9 25,299,73
ivn-mky LD50:53 mg/kg TXAPA9 25,299,73
ipr-gpg LD50:195 mg/kg TXAPA9 25,299,73

SAFETY PROFILE: Poison by intraperitoneal and intravenous routes. When heated to decomposition it emits very toxic fumes of NO_x and HCl.

DOK200 CAS:6358-53-8 ***HR: 3***
1-((2,5-DIMETHOXYPHENYL)AZO)-2-NAPHTHOL
mf: $C_{18}H_{16}N_2O_3$ mw: 308.36

PROP: Mp: 156°. Sltly water-sol; mod sol in alc.

SYNS: C.I. 12156 ◇ C.I. SOLVENT RED 80 ◇ CITRUS RED No. 2 ◇ 2,5-DIMETHOXYBENZENEAZO-β-NAPHTHOL ◇ 1-((2,5-DIMETHOXYPHENYL)AZO)-2-NAPHTHALENOL ◇ 2,5-DIMETHOXY-1-(PHENYLAZO)-2-NAPHTHOL ◇ 1-(1-(2,5-DIMETHOXYPHENYL)AZO)-2-NAPHTHOL ◇ 1-(2,5-DIMETHYLOXYPHENYLAZO)-2-NAPHTHOL

TOXICITY DATA with REFERENCE
mmo-sat 500 μg/plate MUREAV 56,249,78
scu-mus TDLo:20 g/kg/80W-C:CAR FCTXAV 4,493,66
imp-mus TDLo:80 mg/kg:CAR BJCAAI 22,825,68

CONSENSUS REPORTS: IARC Cancer Review: Group 2B IMEMDT 7,56,87; Animal Sufficient Evidence IMEMDT 8,101,75. EPA Genetic Toxicology Program.

SAFETY PROFILE: Suspected carcinogen with experimental carcinogenic data. Mutation data reported. When heated to decomposition it emits toxic fumes of NO_x.

DOK400 CAS:26011-83-6 ***HR: 3***
3-(2,4-DIMETHOXYPHENYL)CROTONIC ACID MAGNESIUM SALT
mf: $C_{24}H_{26}O_8{\cdot}Mg$ mw: 466.81

SYNS: DIMECROTIC ACID MAGNESIUM SALT ◇ 2,4-DIMETHOXY-β-METHYLCINNAMIC ACID MAGNESIUM SALT ◇ HEPADIAL

TOXICITY DATA with REFERENCE
ipr-rat LD50:1000 μg/kg MEIEDD 10,466,83
ipr-mus LD50:1300 μg/kg MEIEDD 10,466,83

SAFETY PROFILE: A deadly poison by intraperitoneal route. When heated to decomposition it emits acrid smoke and irritating fumes. Stimulates the production of bile by the liver. See also MAGNESIUM COMPOUNDS.

DOK600 CAS:2801-68-5 ***HR: 3***
2-(2,5-DIMETHOXYPHENYL)ISOPROPYLAMINE
mf: $C_{11}H_{17}NO_3$ mw: 211.29

SYN: C 1739

TOXICITY DATA with REFERENCE
ipr-rat LD50:170 mg/kg ARZNAD 8,708,58
scu-mus LD50:375 mg/kg ARZNAD 8,708,58

SAFETY PROFILE: Poison by intraperitoneal and subcutaneous routes. When heated to decomposition it emits toxic fumes of NO_x.

DOL400 CAS:64050-54-0 ***HR: 3***
2-((DIMETHOXYPHOSPHINYL)OXY)-1H-BENZ(d,e)ISOQUINOLINE-1,3(2H)-DIONE
mf: $C_{14}H_{12}NO_6P$ mw: 321.24

SYN: PHOSPHORIC ACID, DIMETHYL ESTER, ester with N-HYDROXYNAPHTHALIMIDE

TOXICITY DATA with REFERENCE
orl-rat LD50:70 mg/kg TXAPA9 21,315,72
orl-bwd LD50:2400 μg/kg TXAPA9 21,315,72

SAFETY PROFILE: Poison by ingestion. When heated to decomposition it emits very toxic fumes of PO_x and NO_x. See also ESTERS.

DOL800 CAS:25601-84-7 ***HR: 3***
3-(DIMETHOXYPHOSPHINYLOXY)-N-METHYL-N-METHOXY-cis-CROTONAMIDE
mf: $C_8H_{16}NO_6P$ mw: 253.22

SYNS: CIBA C-2307 ◇ ENT 27,625 ◇ 3-HYDROXY-N-METHOXY-N-METHYL-cis-CROTONAMIDE, DIMETHYL PHOSPHATE ◇ METHOCROTOPHOS ◇ (E)-(3-(METHOXYMETHYLAMINO)-1-METHYL-3-OXO-1-PROPENYL)DIMETHYL PHOSPHATE ◇ NSC 195154

TOXICITY DATA with REFERENCE
orl-rat LD50:2 mg/kg BESAAT 15,107,69
orl-mus LD50:2 mg/kg BESAAT 15,107,69
orl-dog LD50:7 mg/kg BESAAT 15,107,69
orl-rbt LD50:11 mg/kg BESAAT 15,107,69
skn-rbt LD50:107 mg/kg BESAAT 15,107,69

SAFETY PROFILE: A poison by ingestion and skin contact. When heated to decomposition it emits very toxic fumes of NO_x and PO_x.

DOM100 CAS:24991-55-7 ***HR: D***
DIMETHOXY POLYETHYLENE GLYCOL
mf: $(C_2H_4O)_n{\cdot}C_2H_6O$

SYNS: GLYCOLS, POLYETHYLENE, DIMETHYL ETHER ◇ GLYME-23 ◇ α-omega-METHOXYPOLY(ETHYLENE OXIDE) ◇ POLYETHYLENE GLYCOL DIMETHYL ETHER ◇ POLY(OXY-1,2-ETHANEDIYL), α-METHYL-omega-METHOXY-(9CI) ◇ POLYOXYETHYLENE DIMETHYL ETHER ◇ SELEXOL

TOXICITY DATA with REFERENCE
orl-rbt TDLo:2500 mg/kg (male 1D pre):REP TOXID9 4,139,84

CONSENSUS REPORTS: Reported in EPA TSCA Inventory.

SAFETY PROFILE: Experimental reproductive effects. When heated to decomposition it emits acrid smoke and irritating fumes.

DOM200 CAS:4744-10-9 ***HR: 2***
1,1-DIMETHOXYPROPANE
mf: $C_5H_{12}O_2$ mw: 104.15

$(CH_3O)_2CHCH_2CH_3$

PROP: Flash p: 50°F.

SAFETY PROFILE: A very dangerous fire hazard when exposed to heat, flame or oxidizers. When heated to decomposition it emits acrid smoke and fumes.

DOM400 CAS:77-76-9 ***HR: 3***
2,2-DIMETHOXYPROPANE
mf: $C_5H_{12}O_2$ mw: 104.15

$(CH_3O)_2C(CH_3)_2$

PROP: Flash p: 19.4°F.

SAFETY PROFILE: A very dangerous fire hazard when exposed to heat, flame or oxidizers. When heated to 210°C it burns with a cool flame and then explodes. Explosive reaction with metal perchlorates (e.g., manganese(II) perchlorate; nickel(II) perchlorate above 65°C. When heated to decomposition it emits acrid smoke and fumes.

DOM600 CAS:6044-68-4 ***HR: 2***
3,3-DIMETHOXYPROPENE
mf: $C_5H_{10}O_2$ mw: 102.14

$H_2C{=}CHCH(OCH_3)_2$

PROP: Flash p: 66.2°F.

SAFETY PROFILE: A very dangerous fire hazard when exposed to heat, flame oxidizers. May form dangerous peroxides upon exposure to air. When heated to decomposition it emits acrid smoke and fumes. See also ALLYL COMPOUNDS.

DON000 CAS:26270-61-1 ***HR: 3***
4,7-DIMETHOXY-6-(2-PYRROLIDINYLETHOXY)-5-CINNAMOYLBENZOFURAN MALEATE
mf: $C_{25}H_{27}NO_5{\cdot}C_4H_4O_4$ mw: 537.61

TOXICITY DATA with REFERENCE
orl-mus LD50:250 mg/kg CHTPBA 8,479,73
ivn-mus LD50:10 mg/kg CHTPBA 8,479,73

SAFETY PROFILE: Poison by ingestion and intravenous routes. When heated to decomposition it emits toxic fumes of NO_x.

DON200 CAS:15233-65-5 ***HR: 3***
2,6-DIMETHOXYQUINOL
mf: $C_8H_{10}O_4$ mw: 170.18

SYNS: 2,6-DIMETHOXY-1,4-BENZENEDIOL ◇ 2,6-DIMETHOXYHYDROQUINONE ◇ 3,5-DIMETHOXYHYDROQUINONE

TOXICITY DATA with REFERENCE
ivn-mus LD50:35 mg/kg BJPCBM 53,93,75
ivn-cat LDLo:30 mg/kg BJPCBM 53,93,75

SAFETY PROFILE: Poison by intravenous route.

When heated to decomposition it emits acrid smoke and irritating fumes.

DON400 CAS:23435-31-6 ***HR: 3***
2',5'-DIMETHOXYSTILBENAMINE
mf: $C_{16}H_{17}NO_2$ mw: 255.34

SYNS: (trans)-2,5-DIMETHOXY-4'-AMINOSTILBENE ◇ 4-(2,5-DIMETHOXYPHENETHYL)ANILINE ◇ 4-(2-(2,5-DIMETHOXYPHENYL) ETHYL)BENZENAMINE ◇ 4-(2,5-DIMETHOXY)STILBENAMINE ◇ 2,5-DIMETHOXY-4'-STILBENAMINE

TOXICITY DATA with REFERENCE
orl-rat TDLo:2360 mg/kg/78W-C:CAR JEPTDQ 2,325,78
orl-mus TDLo:130 g/kg/78W-C:CAR JEPTDQ 2,325,78

SAFETY PROFILE: Suspected carcinogen with experimental carcinogenic data. When heated to decomposition it emits toxic fumes of NO_x.

DON700 CAS:1230-33-7 ***HR: 2***
3,6-DIMETHOXY-4-SULFANILAMIDOPYRIDAZINE
mf: $C_{12}H_{14}N_4O_4S$ mw: 310.36

SYNS: CS-61 ◇ N-(3,6-DIMETHOXY-4-PYRIDAZINYL)SULFANILAMIDE ◇ 4-SULFANILAMIDO-3,6-DIMETHOXYPYRIDAZINE

TOXICITY DATA with REFERENCE
orl-rat TDLo:3 g/kg (9-14D preg):TER TXAPA9 27,20,74
orl-rat TDLo:4500 mg/kg (9-14D preg):REP TXAPA9 27,20,74
orl-mus LD50:2050 mg/kg ARZNAD 15,1441,65

SAFETY PROFILE: Moderately toxic by ingestion. Experimental teratogenic and reproductive effects. When heated to decomposition it emits toxic fumes of SO_x and NO_x.

DON800 CAS:696-59-3 ***HR: 3***
2,5-DIMETHOXYTETRAHYDROFURAN
mf: $C_6H_{12}O_3$ mw: 131.16

$OCH(OCH_3)C_2H_4CHOCH_3$ (ring)

PROP: Colorless liquid. Bp: 35° @ 10 mm, vap d: 4.56, flash p: <50°F.

SAFETY PROFILE: A very dangerous fire hazard when exposed to heat, flame, or oxidizers. When heated to decomposition it emits acrid smoke and fumes. See also TETRAHYDROFURAN.

DOO400 CAS:6483-64-3 ***HR: 3***
3,3'-DIMETHOXYTRIPHENYLMETHANE-4,4'-BIS(1''-AZO-2''-NAPHTHOL)
mf: $C_{41}H_{32}N_4O_4$ mw: 644.77

SYN: 1,1'-(BENZYLIDENEBIS((2-METHOXY-p-PHENYLENE)(AZO)) DI-2-NAPHTHOL

TOXICITY DATA with REFERENCE
orl-rat TDLo:40 g/kg/83W-C:ETA ZEKBAI 57,530,51

CONSENSUS REPORTS: Reported in EPA TSCA Inventory.

SAFETY PROFILE: Questionable carcinogen with experimental tumorigenic data. When heated to decomposition it emits toxic fumes of NO_x.

DOO600 CAS:534-15-6 ***HR: 3***
DIMETHYLACETAL
DOT: UN 2377
mf: $C_4H_{10}O_2$ mw: 90.14

PROP: Colorless liquid; strong aromatic odor. Bp: 61.8°, flash p: 34°F, d: 0.848 @ 25°, vap d: 3.1.

SYNS: ACETALDEHYDE DIMETHYL ACETAL ◇ 1,1-DIMETHOXYETHANE (DOT) ◇ DIMETHYL ALDEHYDE ◇ ETHYLIDENE DIMETHYL ETHER ◇ METHYL FORMYL

TOXICITY DATA with REFERENCE
skn-rbt 10 mg/24H open MLD JIHTAB 31,60,49
eye-rbt 20 mg open JIHTAB 31,60,49
orl-rat LD50:6500 mg/kg JIHTAB 31,60,49
ihl-rat LC50:3000 ppm/4H AMIHAB 12,623,55
orl-rbt LD50:4507 mg/kg PSEBAA 29,730,32
skn-rbt LD50:20 g/kg JIHTAB 31,60,49

CONSENSUS REPORTS: Glycol ether compounds are on the Community Right-To-Know List. Reported in EPA TSCA Inventory.

DOT Classification: Flammable Liquid; Label: Flammable Liquid.

SAFETY PROFILE: Mildly toxic by inhalation, ingestion, and skin contact. A skin and eye irritant. A very dangerous fire hazard when exposed to heat, flame, or oxidizers. When exposed to heat or flame it can react vigorously with oxidizing materials. To fight fire, use foam, CO_2, dry chemical. When heated to decomposition it emits acrid smoke and irritating fumes. See also GLYCOL ETHERS.

DOO800 CAS:127-19-5 ***HR: 2***
N,N-DIMETHYLACETAMIDE
mf: C_4H_9NO mw: 87.14

PROP: Liquid. Mp: −20°, bp: 165°, d: 0.9448 @ 15.5°, vap d: 3.01, vap press: 1.3 mm @ 25°, flash p: 171°F (TOC), lel: 2.0%, uel: 11.5% @ 740 mm and 160°.

SYNS: ACETDIMETHYLAMIDE ◇ ACETIC ACID DIMETHYLAMIDE ◇ DIMETHYLACETAMIDE ◇ DIMETHYLACETONE AMIDE ◇ DIMETHYLAMIDE ACETATE ◇ DMA ◇ DMAC ◇ NSC 3138 ◇ U-5954

TOXICITY DATA with REFERENCE
skn-rbt 10 mg/24H open MLD AIHAAP 23,95,62
eye-rbt 100 mg MLD DCTODJ 9,147,86
dni-mus-unr 4400 mg/kg APHGAO 56,97,86

ihl-rat TCLo:300 ppm/6H (female 10W pre-3W post):REP FAATDF 7,132,86
skn-rat TDLo:2400 mg/kg (female 10-11D post):TER TXAPA9 41,35,77
orl-rat LD50:4930 mg/kg DCTODJ 9,147,86
ihl-rat LC50:2475 ppm/1H DCTODJ 9,147,86
ipr-rat LD50:2750 mg/kg JRPFA4 4,219,62
ivn-rat LD50:2640 mg/kg ARZNAD 26,1581,76
orl-mus LD50:4620 mg/kg ARZNAD 26,1581,76
ihl-mus LC50:7200 mg/m^3 CHYCDW 13,29,79
skn-mus LD50:9600 mg/kg CHYCDW 13,29,79
ipr-mus LD50:2800 mg/kg YKIGAK 31,327,80
ivn-mus LD50:3020 mg/kg ARZNAD 26,1581,76
skn-rbt LD50:2240 mg/kg AIHAAP 23,95,62

CONSENSUS REPORTS: Reported in EPA TSCA Inventory.

OSHA PEL: TWA 10 ppm (skin)
ACGIH TLV: TWA 10 ppm (skin)
DFG MAK: 10 ppm (35 mg/m^3)

SAFETY PROFILE: Moderately toxic by skin contact, inhalation, intravenous, and intraperitoneal routes. Mildly toxic by ingestion. Experimental teratogenic and reproductive effects. A skin and eye irritant. Less toxic than dimethylformamide. Mutation data reported. Combustible when exposed to heat and flame. A moderate explosion hazard. Violent reaction with halogenated compounds (e.g., carbon tetrachloride, hexachlorocyclohexane) when heated above 90°C. Iron powder catalyzes the reaction so that it initiates at 71°C. When heated to decomposition it emits toxic fumes of NO_x.

DOP000 CAS:2044-64-6 ***HR: 1***
N,N-DIMETHYLACETOACETAMIDE
mf: $C_6H_{11}NO_2$ mw: 129.18

PROP: Liquid, misc in water and organic solvents. Bp: 220°, d: 1.049 – 1.052 @ 20°/20°, flash p: 252°F (COC).

SYN: N,N-DIMETHYL-3-OXOBUTANAMIDE

TOXICITY DATA with REFERENCE
orl-rat LD50:23 g/kg AIHAAP 23,95,62
skn-rbt LD50:14 g/kg AIHAAP 23,95,62

CONSENSUS REPORTS: Reported in EPA TSCA Inventory.

SAFETY PROFILE: Mildly toxic by ingestion and skin contact. Combustible when exposed to heat or flame. To fight fire use water, fog, mist, CO_2, foam. When heated to decomposition it emits toxic fumes of NO_x.

DOP200 CAS:13265-60-6 ***HR: 3***
O,O-DIMETHYL-S-(2-(ACETYLAMINO)ETHYL) DITHIOPHOSPHATE
mf: $C_6H_{14}NO_3PS_2$ mw: 243.30

SYNS: S-(2-(ACETYLAMINO)ETHYL)-O,O-DIMETHYLPHOSPHORODITHIOATE ◇ AMIPHOS ◇ CP 49674 ◇ DAEP ◇ O,O-DIMETHYL-S-(2-ACETAMIDOETHYL) ESTER PHOSPHORODITHIOIC ACID ◇ O,O-DIMETHYL-S-(2-ACETYLAMINOETHYL)PHOSPHORODITHIOATE ◇ N-((O,O-DIMETHYLPHOSPHORODITHIOYL)ETHYL)ACETAMIDE ◇ ENT 27,346 ◇ MONSANTO CP-49674 ◇ NSC 190945 ◇ PHOSPHORODITHIOIC ACID, O,O-DIMETHYL ESTER, S-ESTER with N-(2-MERCAPTOETHYL)ACETAMIDE

TOXICITY DATA with REFERENCE
orl-mus TDLo:600 mg/kg (1-15D preg):TER OYYAA2 6,621,72
orl-mus TDLo:600 mg/kg (1-15D preg):REP OYYAA2 6,621,72
orl-rat LD50:220 mg/kg ARSIM* 20,16,66
ihl-rat LCLo:40 mg/m^3/4H 85GMAT -,55,82
skn-rat LD50:375 mg/kg 85GMAT -,55,82
orl-mus LD50:146 mg/kg 85GMAT -,55,82
skn-mus LD50:472 mg/kg OYYAA2 1,57,67
ipr-mus LD50:117 mg/kg OYYAA2 1,57,67
scu-mus LD50:245 mg/kg OYYAA2 1,57,67
skn-mky LD50:400 mg/kg OYYAA2 1,57,67

SAFETY PROFILE: Poison by ingestion, skin contact, intraperitoneal, and subcutaneous routes. Experimental teratogenic and reproductive effects. When heated to decomposition it emits very toxic NO_x, PO_x, and SO_x. See also MERCAPTANS.

DOP400 CAS:762-42-5 ***HR: 3***
DIMETHYL ACETYLENEDICARBOXYLIC ACID
mf: $C_6H_6O_4$ mw: 142.12

$$CH_3OCO{\cdot}C{\equiv}CCO{\cdot}OCH_3$$

SYNS: ACETYLENEDICARBOXYLIC ACID DIMETHYL ESTER ◇ DIMETHYL-2-BUTYNEDIOATE

TOXICITY DATA with REFERENCE
orl-rat LDLo:50 mg/kg NCNSA6 5,13,53

CONSENSUS REPORTS: Reported in EPA TSCA Inventory.

SAFETY PROFILE: Poison by ingestion. A dienophile. Explosive reaction with 1-methylsilacyclopenta-2,4-diene at 150°C. Octakis(trifluorophosphine)dirhodium catalyzes explosive polymerization of the acid above 20°C. When heated to decomposition it emits acrid smoke and irritating fumes. See also ESTERS.

DOP600 CAS:30560-19-1 ***HR: 3***
O,S-DIMETHYLACETYLPHOSPHOROAMIDOTHIOATE
mf: $C_4H_{10}NO_3PS$ mw: 183.18

SYNS: ACEPHAT (GERMAN) ◇ ACEPHATE ◇ ACETYLPHOSPHORAMIDOTHIOIC ACID-O,S-DIMETHYL ESTER ◇ CHEVRON RE 12,420 ◇ ENT 27,822 ◇ ORTHENE ◇ ORTHENE-755 ◇ ORTHO 12420 ◇ ORTRAN ◇ ORTRIL ◇ RE 12420 ◇ 75 SP

TOXICITY DATA with REFERENCE
mmo-sat 3 mg/plate NTIS** PB80-133226
mrc-smc 50000 ppm NTIS** PB80-133226
mmo-esc 5 μL/plate MUREAV 28,405,75
dns-hmn:fbr 1 g/L NTIS** PB80-133226
msc-mus:lym 1 g/L NTIS** PB84-138973
orl-rat LD50:700 mg/kg MEIEDD 10,5,83
orl-mus LD50:233 mg/kg CHYCDW 14,226,80
ihl-mus LCLo:2200 mg/m³/5H TXAPA9 45,232,78
orl-dog LDLo:681 mg/kg GUCHAZ 6,1,73
skn-rbt LD50:2000 mg/kg 85DPAN -,-,71/76
orl-ckn LD50:852 mg/kg 28ZEAL 5,4,76
orl-dck LD50:350 mg/kg 28ZEAL 5,4,76
orl-mam LD50:321 mg/kg AECTCV 13,483,84
orl-brd LD50:106 mg/kg AECTCV 10,185,81

CONSENSUS REPORTS: EPA Genetic Toxicology Program.

SAFETY PROFILE: Poison by ingestion. Moderately toxic by skin contact and inhalation. Human mutation data reported. When heated to decomposition it emits very toxic fumes of NO_x, PO_x, and SO_x. See also ESTERS.

DOP800 CAS:2680-03-7 ***HR: 2***
N,N-DIMETHYLACRYLAMIDE
mf: C_5H_9NO mw: 99.15

SYN: N,N-DIMETHYL-2-PROPENAMIDE

TOXICITY DATA with REFERENCE
orl-mus LD50:677 mg/kg ARTODN 47,179,81
orl-rat LD50:700 mg/kg 14CYAT 2,1833,63

CONSENSUS REPORTS: Reported in EPA TSCA Inventory.

SAFETY PROFILE: Moderately toxic by ingestion. When heated to decomposition it emits toxic fumes of NO_x.

DOQ300 CAS:627-93-0 ***HR: 2***
DIMETHYL ADIPATE
mf: $C_8H_{14}O_4$ mw: 174.22
SYNS: DIMETHYL HEXANEDIOATE ◇ METHYL ADIPATE

TOXICITY DATA with REFERENCE
ipr-rat TDLo:181 mg/kg (5-15D preg):REP JPMSAE 62,1596,73
ipr-rat TDLo:362 mg/kg (5-15D preg):TER JPMSAE 62,1596,73
ipr-rat LD50:1809 mg/kg JPMSAE 62,1596,73

CONSENSUS REPORTS: Reported in EPA TSCA Inventory.

SAFETY PROFILE: Moderately toxic by intraperitoneal route. Experimental teratogenic and reproductive effects. When heated to decomposition it emits acrid smoke and irritating fumes.

DOQ350 CAS:1191-16-8 ***HR: 1***
3,3-DIMETHYLALLYL ACETATE
mf: $C_7H_{12}O_2$ mw: 128.19

SYN: 2-BUTEN-1-OL, 3-METHYL-, ACETATE ◇ DIMETHYLALLYL ACETATE ◇ Γ,Γ-DIMETHYLALLYL ACETATE ◇ ISOPENT-2-ENYL ACETATE ◇ 3-METHYL-2-BUTENYL ACETATE ◇ PRENYL ACETATE

TOXICITY DATA with REFERENCE
skn-rbt 500 mg/24H MLD FCTOD7 20,817,82
orl-rat LD50:3 g/kg FCTOD7 20,817,82

CONSENSUS REPORTS: Reported in EPA TSCA Inventory.

SAFETY PROFILE: Mildly toxic by ingestion. A skin irritant. When heated to decomposition it emits acrid smoke and irritating fumes.

DOQ400 CAS:359-83-1 ***HR: 3***
2-(3,3-DIMETHYLALLYL)CYCLAZOCINE
mf: $C_{19}H_{27}NO$ mw: 285.47

SYNS: 2-DIMETHYLALLYL-5,9-DIMETHYL-2'-HYDORXYBENZOMORPHAN ◇ 2-(3,3-DIMETHYLALLYL)-2',2'-HYDROXY-5,9-DIMETHYL-6,7-BENZOMORPHAN ◇ FORTALGESIC ◇ FORTALIN ◇ FORTRAL ◇ 1,2,3,4,5,6-HEXAHYDRO-6,11-DIMETHYL-3-(3-METHYL-2-BUTENYL)-2,6-METHANO-3-BENZAZOCINE ◇ 2'-HYDROXY-5,9-DIMETHYL-2-(3,3-DIMETHYLALLYL)-6,7-BENZOMORPHAN ◇ dl-2'-HYDROXY-5,9-DIMETHYL-2-(3,3-DIMETHYLALLYL)-6,7-BENZOMORPHAN ◇ II-C-2 ◇ KF-1820 ◇ LITICON ◇ 3-(3-METHYL-2-BUTENYL)-1,2,3,4,5,6-HEXAHYDRO-6,11-DIMETHYL-2,6-METHANO-3-BENZAZOCIN-8-OL ◇ NIH 7958 ◇ NSC-107430 ◇ PENTAGIN ◇ PENTAZOCINE ◇ SOSIGON ◇ TALWAN ◇ TALWIN ◇ WIN 20228

TOXICITY DATA with REFERENCE
ims-rat TDLo:1500 mg/kg (female 30D pre):REP IYKEDH 13,198,82
orl-hmn LDLo:18 mg/kg CTOXAO 10,327,77
ims-wmn TDLo:1400 mg/kg/3Y:CNS,SKN JAMAAP 231,271,75
ivn-hmn TDLo:300 mg/kg/D:CNS BMJOAE 2,21,78
ivn-man TDLo:3 mg/kg/2D-I:CNS JPETAB 143,149,64
ims-man TDLo:83 mg/kg/4Y-I:MUS AIMDAP 143,2203,83
ims-hmn TDLo:571 μg/kg:CNS,GIT JPETAB 143,149,64
orl-rat LD50:1110 mg/kg KSRNAM 4,2145,70
scu-rat LD50:61 mg/kg AMOKAG 35,179,81
ivn-rat LD50:21 mg/kg 31ZPAG 2,174,66
ims-rat LD50:175 mg/kg AIPTAK 190,124,71
orl-mus LD50:205 mg/kg AMOKAG 35,179,81
ipr-mus LD50:85 mg/kg CPBTAL 24,2912,76
scu-mus LD50:80 mg/kg AMOKAG 35,179,81

ivn-mus LD50:19800 μg/kg NIIRDN 6,777,82
ims-mus LD50:98 mg/kg AIPTAK 190,124,71

SAFETY PROFILE: Poison by ingestion, subcutaneous, intramuscular, intraperitoneal, and intravenous routes. Experimental reproductive effects. Human systemic effects by intramuscular and intravenous routes: wakefulness, euphoria, hallucinations or distorted perceptions, tremors, convulsions, excitement, motor activity changes, muscle weakness, analgesia, withdrawal, parasympathomimetic effects, nausea or vomiting, and dermititis. Can cause drug dependency and other central nervous system effects. An analgesic. When heated to decomposition it emits toxic fumes of NO_x. See also ALLYL COMPOUNDS.

DOQ600 CAS:3639-66-5 ***HR: 3***
2-(3,3-DIMETHYLALLYL)-5-ETHYL-2'-HYDROXY-9-METHYL-6,7-BENZOMORPHAN
mf: $C_{20}H_{29}NO$ mw: 299.50

SYN: 5-ETHYL-2'-HYDROXY-2(N)-(3-METHYL-2-BUTENYL)-9-METHYL-6,7-BENZOMORPHAN

TOXICITY DATA with REFERENCE
scu-rat LD50:128 mg/kg JPETAB 143,141,64
ivn-rat LD50:16 mg/kg JPETAB 143,141,64
scu-mus LD50:116 mg/kg JPETAB 143,141,64
ivn-mus LD50:16 mg/kg 31ZPAG 2,175,66

SAFETY PROFILE: Poison by subcutaneous and intravenous routes. When heated to decomposition it emits toxic fumes of NO_x. See also ALLYL COMPOUNDS.

DOQ700 CAS:1184-58-3 ***HR: 3***
DIMETHYLALUMINUM CHLORIDE
mf: C_2H_6AlCl mw: 92.50

SAFETY PROFILE: Ignites spontaneously in air. Violent reaction on contact with water. When heated to decomposition it emits toxic fumes of Cl^-. See also ALUMINUM COMPOUNDS and CHLORIDES.

DOQ750 CAS:865-37-2 ***HR: 3***
DIMETHYLALUMINUM HYDRIDE
mf: C_2H_7Al mw: 58.06

SAFETY PROFILE: Ignites on contact with traces of air or moisture. See also ALUMINUM COMPOUNDS and HYDRIDES.

DOQ800 CAS:124-40-3 ***HR: 3***
DIMETHYLAMINE
DOT: UN 1032/UN 1160
mf: C_2H_7N mw: 45.10

SYNS: DIMETHYLAMINE, anhydrous (DOT) ◇ DIMETHYLAMINE, aqueous solution (DOT) ◇ DIMETHYLAMINE, solution (DOT) ◇ DMA ◇ N-METHYLMETHANAMINE ◇ RCRA WASTE NUMBER U092

TOXICITY DATA with REFERENCE
eye-rbt 50 mg/5M BJIMAG 23,153,66
cyt-rat-ihl 50 μg/m³ GISAAA 36(11),9,71
orl-rat LD50:698 mg/kg HYSAAV 32,329,67
ihl-rat LC50:4540 ppm/6H AIHAAP 43,411,82
orl-mus LD50:316 mg/kg HYSAAV 32,329,67
ihl-mus LC50:7650 ppm/2H AIHAAP 43,411 82
orl-rbt LD50:240 mg/kg HYSAAV 32,329,67
ivn-rbt LD50:4 g/kg MEIEDD 10,470,83
orl-gpg LD50:240 mg/kg HYSAAV 32,329,67
ihl-mam LC50:3700 mg/m³ TPKVAL 14,80,75

CONSENSUS REPORTS: EPA Genetic Toxicology Program. Reported in EPA TSCA Inventory.

OSHA PEL: TWA 10 ppm
ACGIH TLV: TWA 10 ppm; (Proposed: TWA 5 ppm, STEL 15 ppm)
DFG MAK: 10 ppm (18 mg/m³)
DOT Classification: Flammable Gas; Label: Flammable Gas (UN1032); Flammable Liquid; Label: Flammable Liquid (UN1160).

SAFETY PROFILE: Poison by ingestion. Moderately toxic by inhalation and intravenous routes. Mutation data reported. An eye irritant. Corrosive to the eyes, skin, and mucous membranes. A flammable gas. When heated to decomposition it emits toxic fumes of NO_x. Incompatible with acrylaldehyde, fluorine, and maleic anhydride

DOR000 CAS:124-40-3 ***HR: 3***
DIMETHYLAMINE (anhydrous)
mf: C_2H_7N mw: 45.10

PROP: Colorless gas. Bp: 6.88°, flash p: 0°F, fp: −92.19°, d: 0.6804 @ 0°/4°, autoign temp: 752°F, vap d: 1.55, lel: 2.8%, uel: 14.4%. Sol in water, ether, alc.

CONSENSUS REPORTS: Reported in EPA TSCA Inventory.

DOT Classification: Flammable Gas; Label: Flammable Gas.

SAFETY PROFILE: Poison by ingestion. An irritant. See also DIMETHYLAMINE. A very dangerous fire hazard when exposed to heat or flame; can react vigorously with oxidizing materials. Moderately explosive when exposed to flame. To fight fire, stop flow of gas, foam, CO_2, dry chemical. When heated to decomposition it emits toxic fumes of NO_x.

DOR200 CAS:74-94-2 ***HR: 3***
DIMETHYLAMINE BORANE
mf: $C_2H_7N{\bullet}BH_3$ mw: 58.94

SYNS: BORANE with DIMETHYLAMINE (1:1) ◇ DMAB ◇ N-METHYLMETHANAMINE with BORANE (1:1)

TOXICITY DATA with REFERENCE
skn-rbt 50 mg MLD JOCMA7 1,46,59
eye-rbt 10 mg JOCMA7 1,46,59
orl-rat LD50:59 mg/kg AIHQA5 16,280,55
ipr-rat LD50:39 mg/kg 14KTAK -,693,64
ipr-mus LD50:200 mg/kg JPMSAE 69,1025,80
ivn-mus LD50:56 mg/kg CSLNX* NX#05150
ipr-rbt LD50:35100 μg/kg 14KTAK -,693,64
ipr-gpg LD50:55900 μg/kg 14KTAK -,693,64

CONSENSUS REPORTS: Reported in EPA TSCA Inventory.

SAFETY PROFILE: Poison by ingestion, intraperitoneal, and intravenous routes. A skin and eye irritant. When heated to decomposition it emits toxic fumes of NO_x. See also DIMETHYLAMINE and BORANE.

DOR400 CAS:2032-59-9 ***HR: 3***
4-DIMETHYLAMINE m-CRESYL METHYLCARBAMATE
mf: $C_{11}H_{16}N_2O_2$ mw: 208.29

SYNS: A 363 ◇ AMINOCARB ◇ AMINOCARBE (FRENCH) ◇ BAY 44646 ◇ BAYER 5080 ◇ BAYER 44646 ◇ 4-DIMETHYLAMINO-3-CRESYL METHYLCARBAMATE ◇ 4-(DIMETHYLAMINO)-3-METHYLPHENOL METHYL CARBAMATE (ester) ◇ (4-DIMETHYLAMINO-3-METHYL-PHENYL)N-METHYL-CARBAMAAT (DUTCH) ◇ (4-DIMETHYLAMINO-3-METHYL-PHENYL)N-METHYL-CARBAMAT (GERMAN) ◇ (4-DIMETHYLAMINO-3-METHYL-PHENYL)N-METHYL-CARBAMATE ◇ 4-(DIMETHYLAMINO)-m-TOLYL METHYLCARBAMATE ◇ (4-DIMETILAMINO-3-METIL-FENIL)-N-METIL-CARBAMMATO(ITALIAN) ◇ ENT 25,784 ◇ MATACIL ◇ N-METHYLCARBAMATE de 4-DIMETHYLAMINO-3-METHYL PHENYLE (FRENCH) ◇ MITACIL

TOXICITY DATA with REFERENCE
mma-sat 5 mmol/L ENMUDM 5,384,83
cyt-ham:ovr 5 mmol/L ENMUDM 5,384,83
orl-rat LD50:30 mg/kg TXAPA9 21,315,72
skn-rat LD50:275 mg/kg WRPCA2 9,119,70
orl-mus LDLo:94 mg/kg AECTCV 14,111,85
ipr-mus LD50:7 mg/kg TXAPA9 6,402,64
orl-gpg LDLo:50 mg/kg JEENAI 60(3),733,67
scu-gpg LDLo:50 mg/kg JEENAI 60(3),733,67

SAFETY PROFILE: Poison by ingestion, skin contact, intraperitoneal, and subcutaneous routes. Mutation data reported. An insecticide used for forest insect control. When heated to decomposition it emits toxic fumes of NO_x. See also CARBAMATES.

DOR600 CAS:506-59-2 ***HR: 3***
DIMETHYLAMINE HYDROCHLORIDE
mf: $C_2H_7N{\cdot}ClH$ mw: 81.56

SYNS: DIMETHYLAMMONIUM CHLORIDE ◇ HYDROCHLORIC ACID DIMETHYLAMINE ◇ N-METHYLMETHANAMINE HYDROCHLORIDE

TOXICITY DATA with REFERENCE
orl-mus TDLo:12 g/kg/Y-C:NEO GISAAA 44(8),15,79
orl-rat LD50:1070 mg/kg GISAAA 32(6),12,67
orl-mus LD50:8100 mg/kg GISAAA 32(6),12,67
ipr-mus LD50:1570 mg/kg JJPAAZ 17,475,67
scu-mus LD50:2000 mg/kg AIPTAK 112,36,57
ivn-mus LD50:1210 mg/kg AIPTAK 112,36,57
orl-rbt LD50:1600 mg/kg GISAAA 32(6),12,67
orl-gpg LD50:1600 mg/kg GISAAA 32(6),12,67

CONSENSUS REPORTS: Reported in EPA TSCA Inventory. EPA Genetic Toxicology Program.

SAFETY PROFILE: Moderately toxic by ingestion, intravenous, subcutaneous, and intraperitoneal routes. Questionable carcinogen with experimental neoplastigenic data. When heated to decomposition it emits very toxic fumes of NO_x, NH_3, and HCl.

DOR800 CAS:3426-62-8 ***HR: 2***
DIMETHYLAMINE-2,3,6-TRICHLOROBENZOATE
mf: $C_7H_3Cl_3O_2{\cdot}C_2H_7N$ mw: 270.55

SYNS: BENZAC 1281 ◇ 2,3,6-TRICHLOROBENZOIC ACID DIMETHYLAMINE SALT

TOXICITY DATA with REFERENCE
skn-gpg 500 mg open MOD DUPON*
orl-rat LD50:1644 mg/kg DUPON*

SAFETY PROFILE: Moderately toxic by ingestion. A skin irritant. When heated to decomposition it emits very toxic fumes of Cl^- and NO_x.

DOS000 CAS:315-18-4 ***HR: 3***
4-(DIMETHYLAMINE)-3,5-XYLYL-N-METHYLCARBAMATE
DOT: NA 2757
mf: $C_{12}H_{18}N_2O_2$ mw: 222.32

PROP: Crystals. Mp: 85°, vap press: <0.1 mm @ 139°.

SYNS: 4-(DIMETHYLAMINO)-3,5-DIMETHYLPHENOLMETHYLCARBAMATE (ESTER) ◇ 4-(DIMETHYLAMINO)-3,5-DIMETHYLPHENYL ESTER, METHYLCARBAMIC ACID ◇ 4-(DIMETHYLAMINO)-3,5-DIMETHYLPHENYL-N-METHYLCARBAMATE ◇ 4-(DIMETHYLAMINO)-3,5-XY-LENOL METHYLCARBAMATE (ESTER) ◇ 4-(DIMETHYLAMINO)-3,5-XYLYL ESTER METHYLCARBAMIC ACID ◇ 4-DIMETHYLAMINO-3,5-XYLYL METHYLCARBAMATE ◇ 4-DIMETHYLAMINO-3,5-XYLYL-N-METHYLCARBAMATE ◇ 4-(N,N-DIMETHYLAMINO)-3,5-XYLYL N-METHYLCARBAMATE ◇ DOWCO 139 ◇ ENT 25,766 ◇ METHYL-4-DIMETHYLAMINO-3,5-XYLYL CARBAMATE ◇ METHYL-4-DIMETHYLAMINO-3,5-XYLYL ESTER of CARBAMIC ACID ◇ MEXACARBATE (DOT) ◇ NCI-C00544 ◇ OMS-47 ◇ ZACTRAN ◇ ZECTANE ◇ ZECTRAN ◇ ZEXTRAN

TOXICITY DATA with REFERENCE
scu-mus TDLo:90 mg/kg (6-14D preg):TER NTIS** PB223-160
orl-mus TDLo:1200 mg/kg/78W-I:NEO NTIS** PB223-159

orl-rat LD50:14 mg/kg JEENAI 62,1307,69
orl-mus LD50:12 mg/kg PSSCBG 2,10,71
skn-mus LD50:107 mg/kg JAFCAU 15,479,67
ipr-mus LD50:7800 μg/kg JAFCAU 16,561,68
orl-dog LD50:22 mg/kg FMCHA2 -,C196,89
orl-rbt LD50:37 mg/kg SPEADM 78-1,59,78
orl-gpg LD50:15 mg/kg PCOC** -,1232,66
orl-pgn LD50:5620 μg/kg ASTTA8 (680),157,79 JEENAI 62,1307,69

CONSENSUS REPORTS: IARC Cancer Review: Group 3 IMEMDT 7,56,87; Animal Inadequate Evidence IMEMDT 12,237,76. NCI Carcinogenesis Bioassay (feed); No Evidence: mouse, rat NCITR* NCI-CG-TR-147,78. EPA Extremely Hazardous Substances List.

DOT Classification: Poison B; Label: Poison.

SAFETY PROFILE: Poison by ingestion, skin contact, and intraperitoneal routes. Experimental teratogenic effects. Questionable carcinogen with experimental neoplastigenic data. When heated to decomposition it emits toxic fumes of NO_x. See also ESTERS and CARBAMATES.

DOS200 CAS:926-64-7 ***HR: 3***
DIMETHYLAMINOACETONITRILE
DOT: UN 2378
mf: $C_4H_8N_2$ mw: 84.14

PROP: Flash p: <73.4°F.

SYNS: N-(CYANOMETHYL)DIMETHYLAMINE ◇ N,N-DIMETHYLGLYCINONITRILE

TOXICITY DATA with REFERENCE
orl-rat LD50:50 mg/kg AIHAAP 23,95,62
ihl-rat LCLo:250 ppm/4H AIHAAP 23,95,62
skn-rbt LD50:170 mg/kg EPASR* FYI-OTS-0483,0238
ocu-rbt LDLo:100 mg/kg EPASR* FYI-OTS-0482,0238

CONSENSUS REPORTS: Cyanide and its compounds are on the Community Right-To-Know List.

DOT Classification: Flammable or Combustible Liquid; Label: Flammable and Poison.

SAFETY PROFILE: Poison by ingestion, skin contact, and ocular routes. Moderately toxic by inhalation. A dangerous fire hazard when exposed to heat or flame. When heated to decomposition it emits toxic fumes of NO_x and CN^-. See also NITRILES.

DOS300 CAS:24869-88-3 ***HR: 3***
(DIMETHYLAMINO)ACETYLENE
mf: C_4H_7N mw: 69.11

$$(CH_3)_2NC\equiv CH$$

SAFETY PROFILE: Reacts vigorously with water. When heated to decomposition it emits toxic fumes of NO_x. See also ACETYLENE COMPOUNDS.

DOS800 CAS:13365-38-3 ***HR: 3***
9-(p-DIMETHYLAMINOANILINO)ACRIDINE
mf: $C_{21}H_{19}N_3$ mw: 313.43

SYNS: N'-9-ACRIDINYL-N,N-DIMETHYL-1,4-BENZENEDIAMINE ◇ 9-((p-DIMETHYLAMINO)PHENYL)AMINO)ACRIDINE ◇ NSC 13002 ◇ WIN 1701

TOXICITY DATA with REFERENCE
dnd-mam:lym 2 μmol/L JMCMAR 24,170,81
ivn-rat LD50:15 mg/kg NCIAL* -,326,67
ipr-mus LD50:80970 μg/kg NCISP* JAN86
ivn-dog LDLo:5 mg/kg NCIAL* -,326,67
ivn-mky LDLo:5 mg/kg NCIAL* -,326,67

SAFETY PROFILE: Poison by intravenous and intraperitoneal routes. Mutation data reported. When heated to decomposition it emits toxic fumes of NO_x.

DOT000 CAS:58-15-1 ***HR: 3***
DIMETHYLAMINOANTIPYRINE
mf: $C_{13}H_{17}N_3O$ mw: 231.33

PROP: Colorless leaflets, somewhat water-sol. Mp: 107-109°.

SYNS: AMIDAZOPHEN ◇ AMIDOFEBRIN ◇ AMIDOPHEN ◇ AMIDOPHENAZONE ◇ AMIDOPYRAZOLINE ◇ AMIDOPYRIN ◇ AMINOFENAZONE (ITALIAN) ◇ AMINOPHENAZONE ◇ AMINOPYRINE ◇ ANAFEBRINA ◇ BRUFANEUXOL ◇ DAP ◇ DEREUMA ◇ DIMAPYRIN ◇ DIMETHYLAMINO-ANALGESINE ◇ 4-(DIMETHYLAMINO)ANTIPYRINE ◇ DIMETHYLAMINOAZOPHENE ◇ 4-(DIMETHYLAMINO)-1,2-DIHYDRO-1,5-DIMETHYL-2-PHENYL-3H-PYRAZOL-3-ONE ◇ 4-DIMETHYLAMINO-2,3-DIMETHYL-1-PHENYL-3-PYRAZOLIN-5-ONE ◇ 4-DIMETHYLAMINO-2,3-DIMETHYL-1-PHENYL-5-PYRAZOLONE ◇ DIMETHYLAMINOPHENAZON (GERMAN) ◇ DIMETHYLAMINOPHENAZONE ◇ 4-DIMETHYLAMINOPHENAZONE ◇ DIMETHYLAMINOPHENYLDIMETHYLPYRAZOLIN ◇ 4-DIMETHYLAMINO-1-PHENYL-2,3-DIMETHYLPYRAZOLONE ◇ 3-keto-1,5-DIMETHYL-4-DIMETHYLAMINO-2-PHENYL-2,3-DIHYDROPYRAZOLE ◇ 1,5-DIMETHYL-4-DIMETHYLAMINO-2-PHENYL-3-PYRAZOLONE ◇ 2,3-DIMETHYL-4-DIMETHYLAMINO-1-PHENYL-5-PYRAZOLONE ◇ DIPIRIN ◇ DIPYRIN ◇ FEBRININA ◇ FEBRON ◇ ITAMIDONE ◇ MAMALLET-A ◇ NETSUSARIN ◇ NOVAMIDON ◇ 1-PHENYL-2,3-DIMETHYL-4-DIMETHYLAMINOPYRAZOL-5-ONE ◇ 1-PHENYL-2,3-DIMETHYL-4-DIMETHYLAMINOPYRAZOLONE-5 ◇ PIRAMIDON ◇ PIRIDOL ◇ PIROMIDINA ◇ POLINALIN ◇ PYRADONE ◇ PYRAMIDON ◇ PYRAMIDONE

TOXICITY DATA with REFERENCE
mma-sat 31 μmol/plate MUREAV 66,33,79
dni-mus:oth 100 mg/L ONCODU 19,183,80
otr-ham-orl 100 mg/kg IAPUDO 41,585,82
cyt-ham:fbr 3 mmol/L HDSKEK 10,63,85
msc-ham-orl 100 mg/kg IAPUDO 41,585,82
orl-rat TDLo:60 mg/kg (9-14D preg):REP SKNEA7 22,109,72
scu-mus TDLo:630 mg/kg (female 9-11D post):TER TJADAB 16,118,77

unr-man LDLo:220 mg/kg 85DCAI 2,73,70
orl-rat LD50:600 mg/kg ARZNAD 21,719,71
ipr-rat LD50:190 mg/kg NYKZAU 68,442,72
scu-rat LD50:295 mg/kg ARZNAD 8,229,58
ivn-rat LD50:98 mg/kg OYYAA2 16,1011,78
ims-rat LD50:340 mg/kg ARZNAD 10,665,60
orl-mus LD50:358 mg/kg CHTPBA 5,211,70
ipr-mus LD50:169 mg/kg RPTOAN 36,293,72
scu-mus LD50:248 mg/kg IYKEDH 8,494,77
ivn-mus LD50:78 mg/kg OYYAA2 16,1011,78
ims-mus LD50:306 mg/kg OYYAA2 13,109,77

CONSENSUS REPORTS: Reported in EPA TSCA Inventory. EPA Genetic Toxicology Program.

SAFETY PROFILE: Human poison by unspecified route. Experimental poison by ingestion, subcutaneous, intramuscular, intravenous, and intraperitoneal routes. Moderately toxic by parenteral route. Experimental teratogenic and reproductive effects. Questionable carcinogen when mixed with $NaNO_2$ (1:1). Mutation data reported. Can cause bone marrow depression resulting in leucopenia. Has been implicated in development of aplastic anemia. A tranquilizer. When heated to decomposition it emits toxic fumes of NO_x.

DOT200 ***HR: 3***
4-(DIMETHYLAMINO)ANTIPYRINE mixed with SODIUM NITRITE (1:1)
mf: $C_{13}H_{17}N_3O{\bullet}NNaO_2$ mw: 300.33

SYNS: AMINOPHENAZONE mixed with SODIUM NITRITE (1:1) ◇ AMINOPYRINE mixed with SODIUM NITRITE (1:1) ◇ SODIUM NITRITE mixed with AMINOPYRINE (1:1) ◇ SODIUM NITRITE mixed with 4-(DIMETHYLAMINO)ANTIPYRINE (1:1)

TOXICITY DATA with REFERENCE
mmo-sat 1 mg/plate TOLED5 12,281,82
mma-sat 1 mg/plate TOLED5 12,281,82
cyt-rat-orl 600 mg/kg MFEPDX 1,225,79
hma-mus/sat 2 mmol/L/kg ATSUDG (4),49,80
hma-mus/esc 10 mg/kg CBINA8 35,199,81
orl-rat TDLo:3438 mg/kg/50W-I:CAR NATUAS 244,176,73
orl-rat TD:10 g/kg/40W-I:CAR IARCCD 14,461,76

SAFETY PROFILE: Questionable carcinogen with experimental carcinogenic data. Mutation data reported. When heated to decomposition it emits toxic fumes of NO_x and Na_2O. See also SODIUM NITRITE.

DOT300 CAS:60-11-7 ***HR: 3***
4-DIMETHYLAMINOAZOBENZENE
mf: $C_{14}H_{15}N_3$ mw: 225.32

PROP: Yellow, crystalline tablets; insol in water; sol in strong mineral acids and oils.

SYNS: ATUL FAST YELLOW R ◇ BENZENEAZODIMETHYLANILINE ◇ BRILLIANT FAST YELLOW ◇ BUTTER YELLOW ◇ CERASINE YELLOW GG ◇ C.I. 11020 ◇ C.I. SOLVENT YELLOW 2 ◇ DAB ◇ p-DIMETHYLAMINOAZOBENZEN (CZECH) ◇ DIMETHYLAMINOAZOBENZENE ◇ N,N-DIMETHYL-4-AMINOAZOBENZENE ◇ N,N-DIMETHYL-p-AMINOAZOBENZENE ◇ p-DIMETHYLAMINOAZOBENZENE ◇ 4-(N,N-DIMETHYLAMINO)AZOBENZENE ◇ DIMETHYLAMINOAZOBENZOL ◇ p-DIMETHYLAMINO-AZOBENZOL (GERMAN) ◇ 4-DIMETHYLAMINOAZOBENZOL ◇ 4-DIMETHYLAMINOPHENYLAZOBENZENE ◇ N,N-DIMETHYL-p-AZOANILINE ◇ N,N-DIMETHYL-p-PHENYLAZOANILINE ◇ N,N-DIMETHYL-4-(PHENYLAZO)BENZAMINE ◇ N,N-DIMETHYL-4-(PHENYLAZO) BENZENAMINE ◇ DIMETHYL YELLOW ◇ DIMETHYL YELLOW-N,N-DIMETHYLANILINE ◇ DMAB ◇ ENIAL YELLOW 2G ◇ FAST OIL YELLOW B ◇ FAT YELLOW ◇ GRASAL BRILLIANT YELLOW ◇ JAUNE de BEURRE (FRENCH) ◇ METHYL YELLOW ◇ OIL YELLOW ◇ OLEAL YELLOW 2G ◇ ORGANOL YELLOW ADM ◇ ORIENT OIL YELLOW GG ◇ P.D.A.B. ◇ PETROL YELLOW WT ◇ RCRA WASTE NUMBER U093 ◇ RESINOL YELLOW GR ◇ RESOFORM YELLOW GGA ◇ SILOTRAS YELLOW T2G ◇ SOMALIA YELLOW A ◇ STEAR YELLOW JB ◇ SUDAN YELLOW ◇ TOYO OIL YELLOW G ◇ USAF EK-338 ◇ WAXOLINE YELLOW AD ◇ YELLOW G SOLUBLE in GREASE ◇ ZLUT MASELNA (CZECH)

TOXICITY DATA with REFERENCE
dnr-esc 80 mg/L MUREAV 119,135,83
dni-hmn:hla 100 μmol/L MUREAV 92,427,82
ipr-mus TDLo:3 g/kg (5D male):REP PMRSDJ 1,712,81
scu-mus TDLo:200 mg/kg/(10D preg):TER OFAJAE 36,195,60
orl-rat TDLo:5426 mg/kg/17W-C:CAR CBINA8 53,107,85
skn-rat TDLo:1440 mg/kg/90W-I:NEO CNREA8 26,2406,66
scu-mus TDLo:4000 mg/kg (15-21D preg):CAR,TER BEXBAN 78,1402,75
mul-mus TDLo:400 mg/kg/I:ETA CNREA8 1,397,41
orl-rat LD50:200 mg/kg ZEKBAI 69,103,67
ipr-rat LD50:230 mg/kg CNREA8 34,2274,74
orl-mus LD50:300 mg/kg GANNA2 54,455,63
ipr-mus LD50:230 mg/kg CNREA8 34,2274,74

CONSENSUS REPORTS: NTP Fifth Annual Report on Carcinogens. IARC Cancer Review: Group 2B IMEMDT 7,56,87; Animal Sufficient Evidence IMEMDT 8,125,75. EPA Genetic Toxicology Program. Community Right-To-Know List. Reported in EPA TSCA Inventory.

OSHA PEL: Cancer Suspect Agent
NIOSH REL: (4-Dimethylaminoazobenzene): TWA use 29 CFR 1910.1015

SAFETY PROFILE: Confirmed carcinogen with experimental carcinogenic, neoplastigenic, and tumorigenic data. Poison by ingestion and intraperitoneal routes. Experimental teratogenic and reproductive effects. Human mutation data reported. When heated to decomposition it emits toxic fumes of NO_x.

DOT400 CAS:100-10-7 ***HR: 2***
p-(DIMETHYLAMINO)BENZALDEHYDE
mf: $C_9H_{11}NO$ mw: 149.21

PROP: Small, granular, lemon-colored crystals (may turn pink upon exposure to light); sltly water-sol; sol in alc, ether, chloroform, acetic acid and many other organic solvents. Mp: 73°, bp: 176-177° @ 17 mm.

SYNS: 4-(DIMETHYLAMINO) BENZALDEHYDE ◇ 4-DIMETHYLAMINOBENZENECARBONAL ◇ EHRLICH'S REAGENT ◇ p-FORMYLDIMETHYLANILINE

TOXICITY DATA with REFERENCE
orl-rat LDLo:500 mg/kg JPETAB 90,260,47
ipr-rat LD50:620 mg/kg HINEL* AF33(657)-11756,64

CONSENSUS REPORTS: Reported in EPA TSCA Inventory.

SAFETY PROFILE: Moderately toxic by ingestion and intraperitoneal routes. When heated to decomposition it emits toxic fumes of NO_x. See also ALDEHYDES.

DOT600 CAS:443-30-1 ***HR: 3***
1-(4-DIMETHYLAMINOBENZAL)INDENE
mf: $C_{18}H_{17}N$ mw: 247.36

SYNS: DABI ◇ (4-DIMETHYLAMINOBENZYLIDENE)INDENE ◇ N,N-DIMETHYL-α-INDOLYLIDENE-p-TOLUIDINE ◇ 4-(1H-INDEN-1-YLIDENEMETHYL)-N,N-DIMETHYLBENZENAMINE ◇ NSC-80087

TOXICITY DATA with REFERENCE
orl-rat TDLo:270 mg/kg/15D-I:ETA,REP NATUAS 222,383,69
orl-mus TDLo:4000 mg/kg/26W-C:ETA PTEUA6 7,229,72
ipr-rat LDLo:2000 mg/kg JMCMAR 13,770,70

SAFETY PROFILE: Moderately toxic by intraperitoneal route. Experimental reproductive effects. Questionable carcinogen with experimental tumorigenic data. When heated to decomposition it emits toxic fumes of NO_x.

DOT800 CAS:536-17-4 ***HR: 3***
p-DIMETHYLAMINOBENZALRHODANINE
mf: $C_{12}H_{12}N_2OS_2$ mw: 264.38

SYNS: p-(DIMETHYLAMINO)BENZAL-5-RHODANINE ◇ 5-(p-DIMETHYLAMINOBENZAL)RHODANINE ◇ 5-(p-DIMETHYLAMINOBENZOYLIDENE)RHODANINE ◇ p-DIMETHYLAMINOBENZYLIDENE RHODAMINE ◇ USAF PD-20

TOXICITY DATA with REFERENCE
ipr-mus LD50:150 mg/kg NTIS** AD603-561

CONSENSUS REPORTS: Reported in EPA TSCA Inventory.

SAFETY PROFILE: Poison by intraperitoneal route. When heated to decomposition it emits very toxic fumes of NO_x and SO_x.

DOU000 CAS:53004-03-8 ***HR: 3***
5(4-DIMETHYLAMINOBENZENEAZO) TETRAZOLE
mf: $C_9H_{11}N_7$ mw: 217.24

$$HNN{=}NN{=}CN{=}NC_6H_4N(CH_3)_2$$

SAFETY PROFILE: Explodes when heated to 155°C. When heated to decomposition it emits toxic fumes of NO_x.

DOU600 CAS:140-56-7 ***HR: 3***
p-DIMETHYLAMINOBENZENEDIAZOSODIUM SULPHONATE
mf: $C_8H_{10}N_3O_3S{\cdot}Na$ mw: 251.26

PROP: Yellow-brown crystals.

SYNS: BAYER 5072 ◇ DAPA ◇ DAS ◇ DEKSONAL ◇ DEXON ◇ p-DIMETHYLAMINOBENZENE DIAZO SODIUM SULFONATE ◇ p-(DIMETHYLAMINO)BENZENEDIAZOSULFONATE ◇ p-DIMETHYLAMINOBENZENEDIAZOSULFONIC ACID, SODIUM SALT ◇ 4-DIMETHYLAMINOBENZENEDIAZOSULFONIC ACID, SODIUM SALT ◇ p-(DIMETHYLAMINO)BENZENEDIAZOSULPHONATE ◇ p-(DIMETHYLAMINO)BENZENEDIAZOSULPHONIC ACID, SODIUM SALT ◇ 4-DIMETHYLAMINOBENZENEDIAZOSULPHONIC ACID, SODIUM SALT ◇ p-DIMETHYLAMINOBENZOLDIAZOSULFONAT (NATRIUMSALZ) (GERMAN) ◇ (4-(DIMETHYLAMINO)PHENYL)DIAZENESULFONIC ACID, SODIUM SALT ◇ 4-((DIMETHYLAMINO)PHENYL)DIAZENESULFONIC ACID, SODIUM SALT ◇ p-(DIMETHYLAMINO)-PHENYLDIAZO-NATRIUMSULFONAT(GERMAN) ◇ N,N-DIMETHYL-p-ANILINEDIAZOSULFONIC ACID SODIUM SALT ◇ FENAMINOSULF ◇ GOLD ORANGE MP ◇ LESAN ◇ NCI-C03010 ◇ SODIUM-p-(DIMETHYLAMINO)BENZENEDIAZOSULFONATE ◇ SODIUM-4-(DIMETHYLAMINO)BENZENEDIAZOSULFONATE ◇ SODIUM-p-(DIMETHYLAMINO)BENZENEDIAZOSULPHONATE ◇ SODIUM-4-(DIMETHYLAMINO)BENZENEDIAZOSULPHONATE ◇ SODIUM-(4-(DIMETHYLAMINO)PHENYL)DIAZENESULFONATE ◇ TROPAEO- LIN D

TOXICITY DATA with REFERENCE
mmo-sat 25 μg/plate YACHDS 13,4923,85
mmo-esc 25 μg/plate YACHDS 13,4923,85
orl-rat TDLo:25 mg/kg (7-11D preg):TER CYLPDN 4,201,83
orl-rat LD50:60 mg/kg WRPCA2 9,119,70
ipr-rat LD50:10300 μg/kg 34ZIAG -,202,69
orl-mus LDLo:140 mg/kg AECTCV 14,111,85
ipr-mus LD50:60 mg/kg 34ZIAG -,202,69
ivn-mus LD50:56 mg/kg CSLNX* NX#00143
ipr-dog LDLo:5 mg/kg JPETAB 95,262,49
orl-rbt LD50:150 mg/kg 85DPAN -,-,71/76
ipr-rbt LDLo:10 mg/kg JPETAB 95,262,49

CONSENSUS REPORTS: IARC Cancer Review: Group 3 IMEMDT 7,56,87; Animal Inadequate Evidence IMEMDT 8,147,75. NCI Carcinogenesis Bioassay (feed); No Evidence: mouse, rat NCITR* NCI-CG-TR-101,78. EPA Genetic Toxicology Program.

SAFETY PROFILE: Poison by ingestion, intravenous, and intraperitoneal routes. Experimental teratogenic effects. Human mutation data reported. Questionable carcinogen. A fungicide. When heated to decomposition it emits very toxic fumes of NO_x, Na_2O, and SO_x.

DOV000 CAS:63918-82-1 ***HR: 3***
p-DIMETHYLAMINOBENZYLIDENE-3,4,5,6-DIBENZ-9-METHYLACRIDINE
mf: $C_{31}H_{24}N_2$ mw: 424.57

SYN: 14-(p-(DIMETHYLAMINO)STYRYL)DIBENZ(a,j)ACRIDINE

TOXICITY DATA with REFERENCE
scu-mus TDLo:200 mg/kg:ETA VOONAW 1,52,55

SAFETY PROFILE: Questionable carcinogen with experimental tumorigenic data. When heated to decomposition it emits toxic fumes of NO_x.

DOV200 CAS:13629-82-8 ***HR: 3***
3,3'-DIMETHYL-4-AMINOBIPHENYL
mf: $C_{14}H_{15}N$ mw: 197.30

SYNS: 3,3'-DIMETHYL-4-AMINODIPHENYL ◇ 3,3'-DIMETHYL-4-BIPHENYLAMINE

TOXICITY DATA with REFERENCE
scu-rat TDLo:960 mg/kg/21W-I:CAR ANZJA7 29,38,59
scu-rat TD:2400 mg/kg/W-I:ETA BMBUAQ 14,141,58

SAFETY PROFILE: Questionable carcinogen with experimental carcinogenic and tumorigenic data. When heated to decomposition it emits toxic fumes of NO_x.

DOV400 CAS:63019-93-2 ***HR: 3***
4-(DIMETHYLAMINO)-3-BIPHENYLOL
mf: $C_{14}H_{15}NO$ mw: 213.30

SYN: 4-DIMETHYLAMINO-3-HYDROXYDIPHENYL

TOXICITY DATA with REFERENCE
imp-mus TDLo:80 mg/kg:ETA BJCAAI 11,212,57

SAFETY PROFILE: Questionable carcinogen with experimental tumorigenic data. When heated to decomposition it emits toxic fumes of NO_x.

DOV600 CAS:1195-69-3 ***HR: D***
DIMETHYLAMINO-BIS(1-AZIRIDINYL)PHOSPHINE OXIDE
mf: $C_6H_{14}N_3OP$ mw: 175.20

SYNS: p,p-BIS(1-AZIRIDINYL)-N,N-DIMETHYLAMINOPHOSPHINE OXIDE ◇ p,p-BIS(1-AZIRIDINYL)-N,N-DIMETHYLPHOSPHINIC AMIDE ◇ DIMETHYLAMIDE DIETHYLENEIMIDE PHOSPHORIC ACID ◇ ENT 50,990

TOXICITY DATA with REFERENCE
sln-dmg-par 500 pmol/L IMSUAI 38,442,69
cyt-oin-par 24 mmol/L CNJGA8 11,648,69
sce-hmn:lym 2 mg/L TGANAK 16(2),34,82

CONSENSUS REPORTS: EPA Genetic Toxicology Program.

SAFETY PROFILE: Human mutation data reported. When heated to decomposition it emits very toxic fumes of NO_x and PO_x.

DOV800 CAS:64246-07-7 ***HR: 3***
4-DIMETHYLAMINO-1,1-BIS((3,4-(METHYLENEDIOXY)PHENOXY)METHYL)-1-BUTANOL, METHYLCARBAMATE (ester), CITRATE
mf: $C_{24}H_{30}N_2O_8 \cdot C_6H_8O_7$ mw: 666.70

TOXICITY DATA with REFERENCE
orl-mus LD50:525 mg/kg FRPSAX 32,502,77
ivn-mus LD50:72 mg/kg FRPSAX 32,502,77

SAFETY PROFILE: Poison by intravenous route. Moderately toxic by ingestion. When heated to decomposition it emits toxic fumes of NO_x. See also CARBAMATES and ESTERS.

DOW875 ***HR: 3***
p-DIMETHYLAMINO-CARVACROLDIMETHYLURETHANE METHIODIDE
mf: $C_{16}H_{27}N_2O_2 \cdot I$ mw: 406.35

SYN: (CARBOXYMETHYL)TRIMETHYLAMMONIUMIODIDE-5-(DIMETHYLAMINO)-4-ISOPROPYL-o-TOLYLESTER

TOXICITY DATA with REFERENCE
orl-mus LDLo:95 mg/kg FEPRA7 5,184,46
scu-mus LDLo:4 μg/kg FEPRA7 5,184,46
scu-dog LDLo:345 μg/kg FEPRA7 5,184,46

SAFETY PROFILE: Poison by ingestion and subcutaneous routes. When heated to decomposition it emits toxic fumes of I^-, NH_3, and NO_x. See also IODIDES.

DOX000 CAS:5913-82-6 ***HR: 3***
3-β-(DIMETHYLAMINO)CON-5-ENINE-DIHYDROBROMIDE
mf: $C_{24}H_{40}N_2 \cdot 2BrH$ mw: 356.66

SYNS: CONESSINE DIHYDROBROMIDE ◇ KONESSIN DIHYDROBROMIDE ◇ NERIINE DIHYDRBROMIDE ◇ ROQUESSINE DIHYDROBROMIDE ◇ WRIGHTINE DIHYDROBROMIDE

TOXICITY DATA with REFERENCE
orl-mus LD50:390 mg/kg CHTPBA 5,129,70
ipr-mus LD50:85 mg/kg CHTPBA 5,129,70
ivn-mus LD50:27 mg/kg CHTPBA 5,129,70

SAFETY PROFILE: Poison by ingestion, intraperi-

toneal, and intravenous routes. When heated to decomposition it emits toxic fumes of NO_x and HBr.

DOX100 ***HR: 3***
1-DIMETHYLAMINO-3-CYANO-3-PHENYL-4-METHYLHEXANE HYDROCHLORIDE
mf: $C_{16}H_{24}N_2 \cdot ClH$ mw: 280.88

SYNS: 2-(2-(DIMETHYLAMINO)ETHYL)-3-METHYL-2-PHENYLVALERONITRILE HYDROCHLORIDE ◇ Z-4

TOXICITY DATA with REFERENCE
orl-rat LD50:407 mg/kg JPETAB 117,451,56
orl-mus LD50:382 mg/kg JPETAB 117,451,56
ipr-mus LD50:186 mg/kg JPETAB 117,451,56

CONSENSUS REPORTS: Cyanide and its compounds are on the Community Right-To-Know List.

SAFETY PROFILE: Poison by ingestion and intraperitoneal routes. When heated to decomposition it emits toxic fumes of NO_x, CN^-, and HCl. See also NITRILES.

DOX200 CAS:23273-02-1 ***HR: 3***
DIMETHYLAMINODIBORANE
mf: $C_2H_7N \cdot B_2H_6$ mw: 72.78

SYN: DIMETHYLAMINE with DIBORANE (1:1)

TOXICITY DATA with REFERENCE
ihl-rat LC50:248 mg/m^3/4H 14KTAK -,693,64
ihl-mus LC50:182 mg/m^3/4H 14KTAK -,693,64

SAFETY PROFILE: Poison by inhalation. Ignites spontaneously in air. When heated to decomposition it emits toxic fumes of NO_x. See also BORANES and AMINES.

DOX400 CAS:17268-47-2 ***HR: 2***
3-DIMETHYLAMINO-N,N-DIMETHYLPROPIONAMIDE
mf: $C_7H_{16}N_2O$ mw: 144.25

TOXICITY DATA with REFERENCE
orl-rat LD50:3080 mg/kg TXAPA9 28,313,74
skn-rbt LD50:790 mg/kg TXAPA9 28,313,74

CONSENSUS REPORTS: Reported in EPA TSCA Inventory.

SAFETY PROFILE: Moderately toxic by ingestion and skin contact. When heated to decomposition it emits toxic fumes of NO_x.

DOX600 CAS:23103-98-2 ***HR: 3***
2-(DIMETHYLAMINO)-5,6-DIMETHYL-4-PYRIMIDINYLDIMETHYLCARBAMATE
mf: $C_{11}H_{18}N_4O_2$ mw: 238.33

SYNS: ABOL ◇ AFICIDA ◇ APHOX ◇ DIMETHYLCARBAMIC ACID 2-(DIMETHYLAMINO)-5,6-DIMETHYL-4-PYRIMIDINYL ESTER ◇ 5,6-DIMETHYL-2-DIMETHYLAMINO-4-PYRIMIDINYLDIMETHYLCARBAMATE ◇ ENT 27,766 ◇ FERNOS ◇ PIRIMICARB ◇ PIRIMOR ◇ PP 062 ◇ PYRIMOR ◇ RAPID

TOXICITY DATA with REFERENCE
cyt-hmn:lym 10 mg/L CYGEDX 15(2),74,81
cyt-mus-unr 2 mg/kg TGANAK 14(6),41,80
orl-rat LD50:147 mg/kg CHINAG 30,1018,69
orl-mus LD50:107 mg/kg 28ZEAL 5,184,76
orl-dog LD50:100 mg/kg 28ZEAL 5,184,76

CONSENSUS REPORTS: EPA Genetic Toxicology Program.

SAFETY PROFILE: Poison by ingestion. Human mutation data reported. An insecticide. When heated to decomposition it emits toxic fumes of NO_x. See also CARBAMATES.

DOY400 CAS:60-46-8 ***HR: 3***
4-(DIMETHYLAMINO)-2,2-DIPHENYLVALERAMIDE
mf: $C_{19}H_{24}N_2O$ mw: 296.45

SYNS: AMINOPENTAMIDE ◇ BL 139 ◇ CENTRINE ◇ α-(2-(DIMETHYLAMINO)PROPYL)-α-PHENYLBENZENEACETAMIDE ◇ DIMEVAMIDE ◇ α,α-DIPHENYL-Γ-DIMETHYLAMINOVALERAMIDE ◇ 3-METHYL-4-DIMETHYLAMINO-2,2-DIPHENYLBUTYRAMIDE ◇ VALERAMIDE-OM

TOXICITY DATA with REFERENCE
orl-mus LDLo:441 mg/kg CLDND*
ipr-mus LDLo:121 mg/kg CLDND*
ivn-mus LDLo:46 mg/kg CLDND*

SAFETY PROFILE: Poison by intraperitoneal, and intravenous routes. Moderately toxic by ingestion. An anticholinergic. Used in veterinary medicine as an anticonvulsant and anti-emetic. When heated to decomposition it emits toxic fumes of NO_x.

DOY600 CAS:13242-44-9 ***HR: 3***
2-DIMETHYLAMINO ETHANETHIOL HYDROCHLORIDE
mf: $C_4H_{11}NS \cdot ClH$ mw: 141.68

SYNS: CAPTAMINE HYDROCHLORIDE ◇ N-DIMETHYLCYSTEAMINE HYDROCHLORIDE ◇ N-(2-MERCAPTOETHYL)DIMETHYLAMINE HYDROCHLORIDE ◇ NSC-45463

TOXICITY DATA with REFERENCE
ipr-mus LD50:280 mg/kg YKKZAJ 93,25,73

CONSENSUS REPORTS: Reported in EPA TSCA Inventory.

SAFETY PROFILE: Poison by intraperitoneal route. When heated to decomposition it emits very toxic fumes of Cl^-, SO_x, and NO_x.

DOY800 CAS:108-01-0 ***HR: 3***
N-DIMETHYLAMINOETHANOL
DOT: UN 2051
mf: $C_4H_{11}NO$ mw: 89.16

$HOC_2H_4N(CH_3)_2$

PROP: Liquid. Bp: 131°, flash p: 105°F (OC), d: 0.8866 @ 20°/4°, vap d: 3.03.

SYNS: DEANOL ◇ DIMETHYLAETHANOLAMIN (GERMAN) ◇ DIMETHYLAMINOAETHANOL (GERMAN) ◇ DIMETHYLAMINOETHANOL ◇ β-DIMETHYLAMINOETHANOL ◇ N,N-DIMETHYLAMINOETHANOL ◇ 2-(DIMETHYLAMINO)ETHANOL ◇ β-DIMETHYLAMINOETHYL ALCOHOL ◇ DIMETHYLETHANOLAMINE ◇ N,N-DIMETHYLETHANOLAMINE ◇ DIMETHYLETHANOLAMINE (DOT) ◇ N,N-DIMETHYL-2-HYDROXYETHYLAMINE ◇ N,N-DIMETHYL-N-(2-HYDROXYETHYL)AMINE ◇ DMAE ◇ β-HYDROXYETHYLDIMETHYLAMINE

TOXICITY DATA with REFERENCE
skn-rbt 445 mg open MLD UCDS** 12/15/71
eye-rbt 750 μg open SEV AMIHBC 4,119,51
orl-rat LD50:2 g/kg ZHYGAM 20,393,74
ihl-rat LCLo:4500 mg/m³/4H GTPZAB 14(11),52,70
ipr-rat LD50:1080 mg/kg TXAPA9 12,486,68
ihl-mus LC50:3250 mg/m³ GTPZAB 14(11),52,70
ipr-mus LD50:234 mg/kg JPETAB 94,249,48
scu-mus LD50:961 mg/kg AEPPAE 225,428,55
skn-rbt LD50:1370 mg/kg AMIHBC 4,119,51

CONSENSUS REPORTS: Reported in EPA TSCA Inventory.

DOT Classification: Flammable or Combustible Liquid; Label: Flammable Liquid.

SAFETY PROFILE: Moderately toxic by ingestion, inhalation, skin contact, intraperitoneal, and subcutaneous routes. A skin and severe eye irritant. Used medically as a central nervous system stimulant. Flammable when exposed to heat or flame; can react vigorously with oxidizing materials. Ignites spontaneously in contact with cellulose nitrate of high surface area. To fight fire, use alcohol foam, foam, CO_2, dry chemical. When heated to decomposition it emits toxic fumes of NO_x.

DOZ000 CAS:3635-74-3 ***HR: 2***
2-DIMETHYLAMINOETHANOL-p-ACETAMIDOBENZOATE
mf: $C_{13}H_{18}N_2O_2$ mw: 234.33

SYNS: 4-(ACETYLAMINO)BENZOIC ACID with 2-(DIMETHYLAMINO)ETHANOL (1:1) ◇ CERVOXAN ◇ DAYFEN ◇ DEANER ◇ DEANOL ACETAMIDOBENZOATE ◇ DEANOL-p-ACETAMIDOBENZOATE ◇ DIFORENE ◇ DMAE p-ACETAMIDOBENZOATE ◇ ELEVAN ◇ NERVOTON

TOXICITY DATA with REFERENCE
ipr-rat LD50:800 mg/kg 27ZQAG -,419,72
orl-mus LD50:3918 mg/kg 27ZQAG -,419,72
ipr-mus LD50:1020 mg/kg 27ZQAG -,419,72

SAFETY PROFILE: Moderately toxic by ingestion and intraperitoneal routes. An antidepressant. When heated to decomposition it emits toxic fumes of NO_x.

DPA000 CAS:63980-59-6 ***HR: 2***
2-(DIMETHYLAMINO)ETHANOL BITARTRATE
mf: $C_4H_{11}NO•2C_4H_4O_6$ mw: 385.32

SYNS: ATROL ◇ DIMETHAEN ◇ LIPARON ◇ RECREIN

TOXICITY DATA with REFERENCE
orl-rat LD50:2590 mg/kg 27ZQAG -,419,72
ipr-rat LD50:459 mg/kg 27ZQAG -,419,72
scu-rat LD50:1098 mg/kg 27ZQAG -,419,72
orl-mus LD50:3100 mg/kg SCIEAS 126,610,57

SAFETY PROFILE: Moderately toxic by ingestion, intraperitoneal and subcutaneous routes. When heated to decomposition it emits toxic fumes of NO_x.

DPA200 CAS:5988-51-2 ***HR: 2***
2-(DIMETHYLAMINO)ETHANOL TARTRATE
mf: $C_4H_{11}NO•C_4H_6O_6$ mw: 239.26

SYN: DMAE TARTRATE

TOXICITY DATA with REFERENCE
orl-rat LD50:2600 mg/kg 27ZQAG -,420,72
orl-mus LD50:3100 mg/kg 27ZQAG -,420,72

CONSENSUS REPORTS: Reported in EPA TSCA Inventory.

SAFETY PROFILE: Moderately toxic by ingestion. When heated to decomposition it emits toxic fumes of NO_x.

DPA500 ***HR: 3***
2-(2-DIMETHYLAMINOETHOXY)CHALCONE CITRATE
mf: $C_{19}H_{21}NO_2•C_6H_8O_7$ mw: 487.55

TOXICITY DATA with REFERENCE
orl-mus LD50:603 mg/kg JAPMA8 47,640,58
ipr-mus LD50:158 mg/kg JAPMA8 47,640,58
ivn-mus LD50:40800 μg/kg JAPMA8 47,640,58
ivn-dog LDLo:41 mg/kg JAPMA8 47,640,58

SAFETY PROFILE: Poison by intravenous and intraperitoneal routes. Moderately toxic by ingestion. When heated to decomposition it emits toxic fumes of NO_x.

DPA600 CAS:1704-62-7 ***HR: 3***
2-(2-DIMETHYLAMINOETHOXY)ETHANOL
mf: $C_6H_{15}NO_2$ mw: 133.22

TOXICITY DATA with REFERENCE
orl-rat LD50:2460 mg/kg TXAPA9 28,313,74
par-mus LDLo:240 mg/kg CBCCT* 7,687,55
skn-rbt LD50:1410 mg/kg TXAPA9 28,313,74

CONSENSUS REPORTS: Reported in EPA TSCA Inventory.

SAFETY PROFILE: Poison by parenteral route. Moderately toxic by ingestion and skin contact. When heated to decomposition it emits toxic fumes of NO_x.

DPA800 CAS:13877-99-1 ***HR: 3***
2-(2-(DIMETHYLAMINO)ETHOXY)ETHYL-1-PHENYLCYCLOPENTANECARBOXYLATE
mf: $C_{18}H_{27}NO_3$ mw: 305.46

SYNS: 2-(2-DIMETHYLAMINOETHOXY)ETHANOL-1-PHENYLCYCLOPENTYLCARBOXYLATE ◇ MINEPENTATE ◇ UCB 1549

TOXICITY DATA with REFERENCE
orl-hmn TDLo:360 μg/kg:CNS BMJOAE 2,1112,66
orl-rat LD50:1270 mg/kg BMJOAE 2,1112,66
ivn-rat LD50:42 mg/kg BMJOAE 2,1112,66

SAFETY PROFILE: Poison by intravenous route. Moderately toxic by ingestion. Human systemic effects by ingestion: central nervous system disorders. When heated to decomposition it emits toxic fumes of NO_x.

DPB200 CAS:55118-19-9 ***HR: D***
6-(2-DIMETHYLAMINOETHOXY)-2-(5-NITRO-1-METHYL-2-IMIDAZOLYL)-METHYLENE)-1-TETRALON SULFATE
mf: $C_{19}H_{22}N_4O_4 \cdot H_2O_4S$ mw: 468.53

SYNS: ZK 26173 ◇ ZK-Nr.26173

TOXICITY DATA with REFERENCE
mmo-esc 28 μmol/L JEPTDQ 2(3),657,79
sln-dmg-orl 10 mmol/L JEPTDQ 2(3),657,79
mmo-nsc 400 μg/plate MUREAV 53,297,78
mmo-smc 10 μmol/L JEPTDQ 2(3),657,79

CONSENSUS REPORTS: EPA Genetic Toxicology Program.

SAFETY PROFILE: Mutation data reported. When heated to decomposition it emits very toxic fumes of NO_x and SO_x.

DPB400 CAS:54099-13-7 ***HR: 3***
N-(2-(DIMETHYLAMINO)ETHYL)-1-ADAMANTANEACETAMIDE ETHYL IODIDE
mf: $C_{16}H_{28}N_2O \cdot C_2H_5I$ mw: 420.43

TOXICITY DATA with REFERENCE
orl-mus LD50:600 mg/kg FRPSAX 32,129,77
ipr-mus LD50:75 mg/kg FRPSAX 32,129,77

SAFETY PROFILE: Poison by intraperitoneal route. Moderately toxic by ingestion. When heated to decomposition it emits very toxic fumes of NO_x and I^-. See also IODIDES.

DPC000 CAS:108-00-9 ***HR: 3***
2-DIMETHYLAMINOETHYLAMINE
mf: $C_4H_{12}N_2$ mw: 88.15

PROP: Flash p: 51.8°F.

SAFETY PROFILE: A very dangerous fire hazard when exposed to heat, flame or oxidizers. When heated to decomposition it emits toxic fumes of NO_x.

DPC200 CAS:17599-02-9 ***HR: 3***
2-(DIMETHYLAMINO)ETHYL-p-AMINOBENZOATE HYDROCHLORIDE
mf: $C_{11}H_{16}N_2O_2 \cdot ClH$ mw: 244.75

SYNS: 2-(DIMETHYLAMINO)ETHYLESTER-p-AMINOBENZOIC ACID HYDROCHLORIDE ◇ HCL SALZ des p-AMINO-BENZOESAEURE-DIMETHYLAMINO-AETHYL-ESTER(GERMAN)

TOXICITY DATA with REFERENCE
ipr-rat LDLo:370 mg/kg ARZNAD 1,154,51
ivn-rat LDLo:72 mg/kg ARZNAD 1,154,51

SAFETY PROFILE: Poison by intraperitoneal and intravenous routes. When heated to decomposition it emits very toxic fumes of HCl and NO_x.

DPC400 CAS:52400-61-0 ***HR: 3***
2-(2-(2-(DIMETHYLAMINO)ETHYLAMINO)ETHYL)-2-METHYL-1,3-BENZODIOXOLEDI HYDROCHLORIDE
mf: $C_{14}H_{22}N_2O_2 \cdot 2ClH$ mw: 323.30

TOXICITY DATA with REFERENCE
ivn-rat LD50:51 mg/kg EJMCA5 12,413,77
ipr-mus LD50:175 mg/kg EJMCA5 12,413,77

SAFETY PROFILE: Poison by intravenous and intraperitoneal routes. When heated to decomposition it emits very toxic fumes of HCl and NO_x.

DPD200 CAS:942-46-1 ***HR: 3***
α-(1-(DIMETHYLAMINO)ETHYL)BENZYL ALCOHOL HYDROCHLORIDE
mf: $C_{11}H_{17}NO \cdot ClH$ mw: 215.75

SYNS: METHYLEPHEDRINE HYDROCHLORIDE ◇ l-METHYLEPHEDRINE HYDROCHLORIDE ◇ N-METHYLEPHEDRINE HYDROCHLORIDE

TOXICITY DATA with REFERENCE
ipr-mus LD50:185 mg/kg AIPTAK 138,209,62
scu-mus LD50:699 mg/kg NIIRDN 6,827,82
ivn-rbt LDLo:55 mg/kg JPETAB 36,363,29
ivn-gpg LDLo:139 mg/kg AIPTAK 125,236,60

SAFETY PROFILE: Poison by intravenous and intraperitoneal routes. Moderately toxic by subcutaneous route. When heated to decomposition it emits very toxic fumes of NO_x and Cl^-.

DPD400 CAS:6152-43-8 ***HR: 3***
(2-(DIMETHYLAMINO)ETHYL)(o-BENZYLPHENOXY)ETHERHYDROCHLORIDE
mf: $C_{17}H_{21}NO \cdot ClH$ mw: 291.85

SYNS: N-(2'-DIMETHYLAMINOAETHYL)-(o-BENZYLPHENOL)-AETHER HYDROCHLORID (GERMAN) ◇ HL 2153

TOXICITY DATA with REFERENCE
orl-mus LD50:305 mg/kg ARZNAD 8,219,58
ipr-mus LD50:164 mg/kg ARZNAD 8,219,58
ivn-mus LD50:60 mg/kg ARZNAD 8,219,58

SAFETY PROFILE: Poison by ingestion, intraperitoneal and intravenous routes. When heated to decomposition it emits very toxic fumes of HCl and NO_x.

DPE000 CAS:51-68-3 ***HR: 3***
DIMETHYLAMINOETHYL-4-CHLOROPHENOXYACETIC ACID
mf: $C_{12}H_{16}ClNO_3$ mw: 257.74

SYNS: ACEPHENE ◇ ANALUX ◇ ANP 235 ◇ AT SEFEN ◇ CENTROFENOXINA ◇ CEREBON ◇ CLOFENOXIN ◇ CLOPHENOXATE ◇ p-CHLOROPHENOXYACETIC ACID-β-DIMETHYLAMINOETHYL ESTER ◇ DEANOL-p-CHLOROPHENOXYACETATE ◇ DEANOLESTERE ◇ DIMETHYLAMINOETHYL-p-CHLOROPHENOXYACETATE ◇ EN 1627 ◇ HELFERGIN ◇ LICIDRIL ◇ LUCIDRYL ◇ MECLOFENOXANE ◇ MECLOPHENOXATE ◇ MUCIDRIL ◇ PROSERYL

TOXICITY DATA with REFERENCE
orl-rat LD50:2600 mg/kg 27ZQAG -,386,72
orl-mus LD50:1750 mg/kg 27ZQAG -,386,72
ipr-mus LD50:800 mg/kg BCFAAI 111,293,72
ivn-rbt LD50:150 mg/kg 27ZQAG -,386,72

SAFETY PROFILE: Poison by intravenous route. Moderately toxic by ingestion and intraperitoneal routes. When heated to decomposition it emits very toxic fumes of Cl^- and NO_x.

DPE100 CAS:33232-39-2 ***HR: 3***
5-(2-(DIMETHYLAMINO)ETHYL)-2,3-DIHYDRO-3-HYDROXY-2-(p-METHOXYPHENYL)1,5-BENZOTHIAZEPIN-4(5H)-ONE-ACETATE (ESTER) HYDROCHLORIDE
mf: $C_{22}H_{26}N_2O_4S \cdot ClH$ mw: 451.02

TOXICITY DATA with REFERENCE
orl-rat LD50:560 mg/kg IYKEDH 5,106,74
scu-rat LD50:520 mg/kg IYKEDH 5,106,74
ivn-rat LD50:38 mg/kg IYKEDH 5,106,74
orl-mus LD50:640 mg/kg IYKEDH 5,106,74
scu-mus LD50:280 mg/kg IYKEDH 5,106,74
ivn-mus LD50:58 mg/kg IYKEDH 5,106,74

SAFETY PROFILE: Poison by subcutaneous and intravenous routes. Moderately toxic by ingestion. When heated to decomposition it emits toxic fumes of SO_x, NO_x, and HCl. See also ESTERS.

DPE200 CAS:2424-75-1 ***HR: 3***
DIMETHYLAMINOETHYLDIPHENYLETHOXY ACETATE HYDROCHLORIDE
mf: $C_{20}H_{25}NO_3 \cdot ClH$ mw: 363.92

SYNS: AESTOCIN ◇ DIMENOXADOL HYDROCHLORIDE ◇ β'-DIMETHYLAMINOETHYL-α,α-DIPHENYL-α-ETHOXY-ACETATEHYDROCHLORIDE ◇ 2-(DIMETHYLAMINO) ETHYL ESTER HYDROCHLORIDE ETHOXYDIPHENYLACETIC ACID ◇ 2,2-DIPHENYL-2-ETHOXYACETIC ACID (2-(DIMETHYLAMINO) ETHYL) ESTER HYDROCHLORIDE ◇ ESTOCINE ◇ ESTOTSIN ◇ LOKARIN ◇ PROPALGYL

TOXICITY DATA with REFERENCE
ivn-rat LD50:66 mg/kg PCJOAU 8,189,74
orl-mus LD50:700 mg/kg MEIEDD 10,467,84
ipr-mus LD50:175 mg/kg PCJOAU 4,7,70
scu-mus LD50:179 mg/kg RPTOAN 32,317,69
ivn-mus LDLo:40 mg/kg JMCMAR 8,571,65

SAFETY PROFILE: Poison by intravenous, intraperitoneal, and subcutaneous routes. Moderately toxic by ingestion. An analgesic and anticonvulsant. When heated to decomposition it emits very toxic fumes of NO_x and HCl.

DPE800 CAS:78372-05-1 ***HR: 3***
1-(2-(DIMETHYLAMINO)ETHYL)-1-ETHYL-3-MESITYLUREA HYDROCHLORIDE
mf: $C_{16}H_{27}N_3O \cdot ClH$ mw: 313.92

TOXICITY DATA with REFERENCE
ipr-rat LD50:123 mg/kg ARZNAD 8,664,58
scu-mus LD50:230 mg/kg ARZNAD 8,664,58

SAFETY PROFILE: Poison by intraperitoneal and subcutaneous routes. When heated to decomposition it emits very toxic fumes of NO_x and HCl.

DPF200 CAS:69884-15-7 ***HR: 2***
N-(4-(1-(DIMETHYLAMINO)ETHYLIDENE) AMINO)PHENYL)-2-METHOXYACETAMIDE HYDROCHLORIDE
mf: $C_{13}H_{19}N_3O_2 \cdot ClH$ mw: 285.81

SYNS: AMIDANTEL ◇ BAY d8815 ◇ N-(4-((1-(DIMETHYLAMINO)-AETHYLIDEN)AMINO)PHENYL)-2-METHOXYACETAMID-HYDROCHLORID (GERMAN)

TOXICITY DATA with REFERENCE
orl-rat LD50:4693 mg/kg ARZNAD 29,31,79
orl-mus LD50:1207 mg/kg ARZNAD 29,31,79
scu-mus LD50:569 mg/kg ARZNAD 29,31,79
orl-dog LD50:500 mg/kg ARZNAD 29,31,79
orl-cat LD50:750 mg/kg ARZNAD 29,31,79
orl-rbt LD50:500 mg/kg ARZNAD 29,31,79

SAFETY PROFILE: Moderately toxic by ingestion and subcutaneous routes. When heated to decomposition it emits very toxic fumes of NO_x and HCl.

DPF600 CAS:61-50-7 ***HR: 3***
3-(2-(DIMETHYLAMINO)ETHYL)INDOLE
mf: $C_{12}H_{16}N_2$ mw: 188.30

SYNS: N,N-DIMETHYLTRYPTAMINE ◇ DMT

TOXICITY DATA with REFERENCE
ims-man TDLo:1 mg/kg:EYE,CNS,CVS PSYPAG 4,39,63
ipr-mus LD50:47 mg/kg YKKZAJ 94,1620,74
ivn-mus LD50:32 mg/kg CSLNX* NX#00740

SAFETY PROFILE: Poison by intravenous and intraperitoneal routes. Human systemic effects by intramuscular route: pupiliary dilation, hallucinations and distorted perceptions, blood pressure increase. When heated to decomposition it emits toxic fumes of NO_x.

DPG000 CAS:101831-88-3 ***HR: 3***
3-(2-(DIMETHYLAMINO)ETHYL)INDOLESULFOSALICYLATE

SYN: N,N-DIMETHYL-β-3-INDOLYLETHYLAMINESULFOSALICYLATE

TOXICITY DATA with REFERENCE
ipr-mus LD50:153 mg/kg RPTOAN 33,180,70
ivn-mus LD50:69 mg/kg RPTOAN 33,180,70

SAFETY PROFILE: Poison by intraperitoneal and intravenous routes. When heated to decomposition it emits very toxic fumes of NO_x and SO_x.

DPG109 CAS:487-93-4 ***HR: 3***
3-(2-DIMETHYLAMINOETHYL)-5-INDOLOL
mf: $C_{12}H_{16}N_2O$ mw: 204.30

SYNS: BUFOTENIN ◇ 3-(β-DIMETHYLAMINOETHYL)-5-HYDROXYINDOLE ◇ N,N-DIMETHYL-5-HYDROXYTRYPTAMINE ◇ N,N-DIMETHYLSEROTONIN ◇ 5-HYDROXY-N,N-DIMETHYLTRYPTAMINE

TOXICITY DATA with REFERENCE
ivn-hmn TDLo:57 μg/kg:PSY SCIEAS 123,886,56
ipr-mus LD50:290 mg/kg PSYPAG 16,385,70

SAFETY PROFILE: Poison by intraperitoneal route. Human systemic effects with very small amounts taken by intravenous route: psychotropic effects. A modified natural neurotransmitter. When heated to decomposition it emits toxic fumes of NO_x.

DPG200 CAS:101652-11-3 ***HR: 3***
10-(2-(DIMETHYLAMINO)ETHYL)ISOALLOXAZINE SULFATE
mf: $C_{14}H_{15}N_5O_2 \cdot H_2O_4S$ mw: 383.42

TOXICITY DATA with REFERENCE
scu-mus LD50:90 mg/kg CMTRAG 2,96,61
ivn-mus LD50:75 mg/kg CMTRAG 2,96,61

SAFETY PROFILE: Poison by subcutaneous and intravenous routes. When heated to decomposition it emits very toxic fumes of SO_x and NO_x.

DPG400 CAS:78372-06-2 ***HR: 3***
1-(2-(DIMETHYLAMINO)ETHYL)-1-ISOPROPYL-3-(2,6-XYLYL)UREA HYDROCHLORIDE
mf: $C_{16}H_{27}N_3O \cdot ClH$ mw: 313.92

SYN: C 3215

TOXICITY DATA with REFERENCE
eye-rbt 2% MLD ARZNAD 8,664,58
ipr-rat LD50:41 mg/kg ARZNAD 8,664,58
scu-mus LD50:57 mg/kg ARZNAD 8,664,58

SAFETY PROFILE: Poison by intraperitoneal and subcutaneous routes. An eye irritant. When heated to decomposition it emits very toxic fumes of NO_x and HCl.

DPG600 CAS:2867-47-2 ***HR: 3***
DIMETHYLAMINOETHYL METHACRYLATE
DOT: UN 2522
mf: $C_8H_{15}NO_2$ mw: 157.24

PROP: Liquid, sol in water and organic solvents. D: 0.933 @ 25°, bp: 182-190°, flash p: 165°F (TOC), vap d: 5.4.

SYNS: AGEFLEX FM-1 ◇ 2-(DIMETHYLAMINO)ETHANOL METHACRYLATE ◇ 2-(DIMETHYLAMINO)ETHYL ESTER METHACRYLIC ACID ◇ N,N-DIMETHYLAMINOETHYL METHACRYLATE ◇ β-DIMETHYLAMINOETHYL METHACRYLATE ◇ 2-(DIMETHYLAMINO) ETHYL METHACRYLATE ◇ USAF RH-3

TOXICITY DATA with REFERENCE
orl-rat LD50:1751 mg/kg 85GMAT -,55,82
ihl-rat LC50:620 mg/m³/4H 85GMAT -,55,82
ihl-mus LC50:1800 mg/m³/2H 85GMAT -,55,82
ipr-mus LD50:25 mg/kg NTIS** AD277-689

CONSENSUS REPORTS: Reported in EPA TSCA Inventory.

DOT Classification: Poison B; Label: Poison.

SAFETY PROFILE: Poison by intraperitoneal route. Moderately toxic by ingestion and inhalation. A skin, eye, and mucous membrane irritant. A powerful lachrymator. Flammable when exposed to sparks, heat, open flame, or oxidizers. To fight fire, use alcohol foam, dry chemical, spray. When heated to decomposition it emits toxic fumes of NO_x. See also ESTERS.

DPH000 CAS:4724-58-7 ***HR: 3***
2-DIMETHYLAMINOETHYL-2-METHYL-BENZHYDRYL ETHER CITRATE
mf: $C_{18}H_{23}NO•C_6H_8O_7$ mw: 461.56

SYNS: N,N-DIMETHYL-2-((o-METHYL-α-PHENYL-BENZYL)OXY)-ETHYLAMINE CITRATE ◇ ORPHENADRINE CITRATE

TOXICITY DATA with REFERENCE
dnd-esc 50 μmol/L MUREAV 89,95,81
ivn-rat LD50:26 mg/kg 27ZQAG -,373,72
ims-rat LD50:208 mg/kg 27ZQAG -,373,72
orl-mus LD50:150 mg/kg 29ZVAB -,83,69
ivn-mus LD50:37 mg/kg 27ZQAG -,373,72
ivn-rbt LD50:22 mg/kg IJNEAQ 5,305,66

SAFETY PROFILE: Poison by ingestion, intramuscular, and intravenous routes. Mutation data reported. When heated to decomposition it emits toxic fumes of NO_x. See also ETHERS.

DPH200 CAS:65210-30-2 ***HR: 3***
2-(2-(DIMETHYLAMINO)ETHYL)-2-METHYL-1,3-BENZODIOXOLE HYDROCHLORIDE
mf: $C_{12}H_{17}NO_2•ClH$ mw: 243.76

TOXICITY DATA with REFERENCE
ivn-rat LD50:60 mg/kg EJMCA5 12,413,77
ipr-mus LD50:200 mg/kg EJMCA5 12,413,77

SAFETY PROFILE: Poison by intravenous and intraperitoneal routes. When heated to decomposition it emits very toxic fumes of HCl and NO_x.

DPH400 CAS:104-19-8 ***HR: 3***
1-(2-(DIMETHYLAMINO)ETHYL)-4-METHYLPIPERAZINE
mf: $C_9H_{21}N_3$ mw: 171.33

TOXICITY DATA with REFERENCE
skn-rbt 100 μg/24H open AIHAAP 23,95,62
orl-rat LD50:1420 mg/kg AIHAAP 23,95,62
skn-rbt LD50:390 mg/kg AIHAAP 23,95,62

CONSENSUS REPORTS: Reported in EPA TSCA Inventory.

SAFETY PROFILE: Poison by skin contact. Moderately toxic by ingestion. A skin irritant. When heated to decomposition it emits toxic fumes of NO_x.

DPH600 CAS:4985-15-3 ***HR: 3***
5-DIMETHYLAMINOETHYLOXYIMINO-5H-DIBENZO(a,d)CYCLOHEPTA-1,4-DIENE HYDROCHLORIDE
mf: $C_{19}H_{22}N_2O•ClH$ mw: 330.89

SYNS: AGEDAL ◇ BAY 1521 ◇ 5-(DIMETHYLAMINOAETHYL-OXYIMINO)-5H-DIBENZO(a,d)CYCLOHEPTA-1,4-DIENHYDROCHLORID (GERMAN) ◇ 5-(DIMETHYLAMINOOXYIMINO)-5H-DIBENZO(a,b) CYCLOHEPTA-1,4-DIENE HYDROCHLORIDE ◇ 5-(DIMETILAMINO-ETILOXIMINO-5H-DIBENZO(a,d)CICLOEPTA-1,4-DIENE)CLORIDRATO (ITALIAN) ◇ NOGEDAL ◇ NOXIPTILINE HYDROCHLORIDE ◇ NOXIPTILIN HYDROCHLORID (GERMAN) ◇ NOXIPTYLINE HYDROCHLORIDE

TOXICITY DATA with REFERENCE
orl-rat TDLo:16800 mg/kg (24W male):REP KSRNAM 6,1897,72
orl-rat LD50:607 mg/kg FRPPAO 25,519,70
ipr-rat LD50:149 mg/kg FRPPAO 25,519,70
scu-rat LD50:985 mg/kg 27ZQAG -,84,72
ivn-rat LD50:12 mg/kg KSRNAM 6,1897,72
ims-rat LD50:209 mg/kg FRPPAO 25,519,70
orl-mus LD50:275 mg/kg KSRNAM 6,1897,72
ipr-mus LD50:93 mg/kg FRPPAO 25,519,70
scu-mus LD50:212 mg/kg KSRNAM 6,1897,72
ivn-mus LD50:21300 μg/kg KSRNAM 6,1897,72
ims-mus LD50:144 mg/kg FRPPAO 25,519,70
orl-dog LD50:800 mg/kg ARZNAD 19,846,69
scu-dog LD50:100 mg/kg ARZNAD 19,846,69

SAFETY PROFILE: Poison by ingestion, subcutaneous, intravenous, intramuscular, and intraperitoneal routes. Experimental reproductive effects. When heated to decomposition it emits very toxic fumes of NO_x and HCl.

DPI000 CAS:5934-20-3 ***HR: 3***
N-DIMETHYLAMINOETHYLPHENOTHIAZINE HYDROCHLORIDE
mf: $C_{16}H_{18}N_2S•ClH$ mw: 306.88

SYNS: N-(β-DIMETHYLAMINOETHYL)-PHENOTHIAZINEHYDROCHLORIDE ◇ FENETHAZINE HYDROCHLORIDE ◇ LISERGAN HYDROCHLORIDE ◇ RUTERGAN HYDROCHLORIDE

TOXICITY DATA with REFERENCE
ipr-mus LD50:115 mg/kg MEIEDD 10,572,83
scu-mus LD50:210 mg/kg 27ZQAG -,38,72

SAFETY PROFILE: Poison by intraperitoneal and subcutaneous routes. An antihistamine. When heated to decomposition it emits very toxic fumes of NO_x, and SO_x, and HCl.

DPI400 CAS:52401-02-2 ***HR: 3***
2-(2-(DIMETHYLAMINO)ETHYL)-2-PHENYL-1,3-BENZODIOXOLE HYDROCHLORIDE
mf: $C_{17}H_{19}NO_2•ClH$ mw: 305.83

TOXICITY DATA with REFERENCE
ivn-rat LD50:23 mg/kg EJMCA5 12,413,77
ipr-mus LD50:83 mg/kg EJMCA5 12,413,77

SAFETY PROFILE: Poison by intravenous and intraperitoneal routes. When heated to decomposition it emits very toxic fumes of HCl and NO_x.

DPI600 CAS:74758-13-7 ***HR: 2***
N-(2-(DIMETHYLAMINO)ETHYL)-N-(3-PHENYL-1-INDOLYL)ACETAMIDE HYDROCHLORIDE
mf: $C_{20}H_{23}N_3O{\bullet}ClH$ mw: 357.92

TOXICITY DATA with REFERENCE
orl-rat LD50:1100 mg/kg ARZNAD 30,919,80
orl-mus LD50:1000 mg/kg ARZNAD 30,919,80

SAFETY PROFILE: Moderately toxic by ingestion. When heated to decomposition it emits very toxic fumes of NO_x and HCl.

DPI700 ***HR: 3***
β-DIMETHYLAMINOETHYL-2-PHENYLTETRAHYDROBENZOATE HYDROCHLORIDE
mf: $C_{17}H_{23}NO_2{\bullet}ClH$ mw: 309.87

SYNS: 2-PHENYL-3-CYCLOHEXENE-1-CARBOXYLIC ACID 2-DIMETHYLAMINOETHYL ESTER HYDROCHLORIDE ◇ S 187

TOXICITY DATA with REFERENCE
scu-rat LDLo:800 mg/kg APPNAH 1,4,50
ivn-rat LDLo:80 mg/kg APPNAH 1,4,50
ivn-rbt LDLo:40 mg/kg APPNAH 1,4,50
scu-gpg LDLo:267 mg/kg APPNAH 1,4,50
ivn-gpg LDLo:27 mg/kg APPNAH 1,4,50

SAFETY PROFILE: Poison by intravenous and subcutaneous routes. When heated to decomposition it emits toxic fumes of NO_x and HCl.

DPI750 CAS:96811-96-0 ***HR: 3***
2-((2-(DIMETHYLAMINO)ETHYL)(SELENOPHENE-2-YLMETHYL)AMINO)PYRIDINE
mf: $C_{14}H_{19}N_3Se$ mw: 308.32

SYN: PYRIDINE,2-((2-(DIMETHYLAMINO)ETHYL)(SELENOPHENE-2-YLMETHYL)AMINO)-

TOXICITY DATA with REFERENCE
par-mus LD50:90 mg/kg 43FLAV 4(3),559,80

OSHA PEL: TWA 0.2 mg(Se)/m^3
ACGIH TLV: TWA 0.2 mg(Se)/m^3

SAFETY PROFILE: Poison by parenteral route. When heated to decomposition it emits toxic fumes of NO_x and Se.

DPJ200 CAS:91-79-2 ***HR: 3***
2-((2-DIMETHYLAMINOETHYL)-3-THENYLAMINO)PYRIDINE
mf: $C_{14}H_{19}N_3S$ mw: 261.42

SYNS: DIETHYLENDIAMINE ◇ N-(2-DIMETHYLAMINOETHYL)-N-2-PYRIDYL-3-THENYLAMINE ◇ METHAPYRILENE ◇ NCI-C60640 ◇ N-(α-PYRIDYL)-N-(β-THENYL)-N',N'-DIMETHYLETHYLENEDIAMINE ◇ TENFIDIL ◇ THEFANIL ◇ THENFADIL ◇ THENYLDIAMINE ◇ WIN-2848

TOXICITY DATA with REFERENCE
dnd-esc 30 μmol/L MUREAV 89,95,81
orl-hmn TDLo:50 mg/kg:CNS,GIT CTOXAO 11,287,77
ipr-mus LD50:77 mg/kg CLDND*

SAFETY PROFILE: Poison by intraperitoneal route. Human systemic effects by ingestion: hallucinations and distorted perceptions, gastrointestinal changes. Mutation data reported. An antihistamine. When heated to decomposition it emits very toxic fumes of NO_x and SO_x.

DPJ400 CAS:135-23-9 ***HR: 3***
2-((2-(DIMETHYLAMINO)ETHYL)-2-THENYLAMINO)PYRIDINE HYDROCHLORIDE
mf: $C_{14}H_{19}N_3S{\bullet}ClH$ mw: 297.88

SYNS: BARHIST ◇ CAPATHYN ◇ CORYZOL ◇ N,N-DIMETHYL-N'-2-PYRIDINYL-N'-(2-THIENYLMETHYL)-1,2-ETHANEDIAMINEMONOHYDROCHLORIDE ◇ N,N-DIMETHYL-N'-(2-PYRIDYL)-N'-THENYLETHYLENEDIAMINE HYDROCHLORIDE ◇ N,N-DIMETHYL-N'-(2-THENYL)-N'-(2-PYRIDYL-ETHYLENE-DIAMINE HYDROCHLORIDE) ◇ DOZAR ◇ HISTADYL HYDROCHLORIDE ◇ HISTAFED ◇ HISTIDYL ◇ LULLAMIN ◇ METHACON ◇ METHAPYRILENE HYDROCHLORIDE ◇ METHAPYRILENE HYDROCHLORIDE (L.A.) ◇ METHAPYRILENE HYDROCHLORIDE (S.A.) ◇ METHOXYLENE ◇ PYRATHYN ◇ N-(2-PYRIDYL)-N-(2-THIENYL)-N,N'-DIMETHYLETHYLENEDIAMINE HYDROCHLORIDE ◇ SEMIKON ◇ SEMIKON HYDROCHLORIDE ◇ SOMNICAPS ◇ TEM-HISTINE ◇ TERALIN ◇ THENYLENE ◇ THENYLENE HYDROCHLORIDE ◇ THENYLPYRAMINE HYDROCHLORIDE ◇ W-53 HYDROCHLORIDE ◇ WIN 2848 HYDROCHLORIDE SALT

TOXICITY DATA with REFERENCE
mmo-smc 500 mg/L IAPUDO 57,721,84
mrc-smc 1250 mg/L IAPUDO 57,721,84
dns-rat :lvr 1 μmol/L CNREA8 42,3010,82
oms-rat:lvr 100 μmol/L MUREAV 135,131,84
orl-rat TDLo:9100 mg/kg/26/W-C:CAR TXAPA9 66,252,82
orl-rat TD:18200 mg/kg/26W-C:CAR TXAPA9 66,252,82
orl-rat TD:4813 mg/kg/2Y-C:ETA FCTOD7 22,27,84
orl-man TDLo:429 μg/kg:GIT JPETAB 90,83,47
orl-rat LD50:521 mg/kg 29ZVAB -,73,69
scu-rat LD50:150 mg/kg 29ZVAB -,73,69
orl-mus LD50:182 mg/kg JPETAB 90,83,47
scu-mus LD50:75 mg/kg JPETAB 93,210,48
ivn-mus LD50:17500 μg/kg JPETAB 93,210,48
ivn-gpg LD50:14600 μg/kg AIPTAK 113,313,58

CONSENSUS REPORTS: NCI Carcinogenesis Studies

(feed): Clear Evidence: rat FCTOD7 22,27,84; Clear Evidence: rat SCIEAS 209,817,80; (gavage); No Evidence: guinea pig,ham JTEHD6 12,653,83. Reported in EPA TSCA Inventory.

SAFETY PROFILE: Poison by ingestion, intravenous, and subcutaneous routes. Human systemic effects by ingestion: gastritis. Questionable carcinogen with experimental carcinogenic and tumorigenic data. Mutation data reported. An antihistamine. When heated to decomposition it emits very toxic fumes of Cl^-, SO_x, and NO_x.

DPJ600 CAS:13261-62-6 ***HR: 3***
2-DIMETHYLAMINOFLUORENE
mf: $C_{15}H_{15}N$ mw: 209.31

SYNS: 2-DIMETHYLAMINO-FLUOREN (GERMAN) ◇ N,N-DIMETHYL-2-AMINOFLUORENE ◇ 2-FLUORENYLDIMETHYLAMINE

TOXICITY DATA with REFERENCE
orl-rat TDLo:1370 mg/kg/47W-I:ETA BJCAAI 6,89,52

SAFETY PROFILE: Questionable carcinogen with experimental tumorigenic data. When heated to decomposition it emits toxic fumes of NO_x.

DPJ800 CAS:87-01-4 ***HR: 2***
7-DIMETHYLAMINO-4-METHYLCOUMARIN
mf: $C_{12}H_{13}NO_2$ mw: 203.26

TOXICITY DATA with REFERENCE
orl-rat LDLo:1500 mg/kg CNREA8 26,619,66

CONSENSUS REPORTS: Reported in EPA TSCA Inventory.

SAFETY PROFILE: Moderately toxic by ingestion. When heated to decomposition it emits toxic fumes of NO_x.

DPK000 ***HR: 3***
anti-8-(N,N-DIMETHYLAMINOMETHYL)DIBENZOBICYCLO(3.2.1)OCTADIENE HYDROCHLORIDE

SYN: 10,11-DIHYDRO-N,N-DIMETHYL-5,10-METHANO-5H-DIBENZO (a,d)CYCLOHEPTENE-12-METHANAMINE HCl

TOXICITY DATA with REFERENCE
orl-mus LD50:151 mg/kg DRFUD4 3,142,78
ivn-mus LD50:29 mg/kg DRFUD4 3;142,78

SAFETY PROFILE: Poison by ingestion and intravenous routes. When heated to decomposition it emits very toxic fumes of NO_x and HCl.

DPK400 CAS:142-25-6 ***HR: 3***
2-DIMETHYLAMINO-N-METHYLETHYLAMINE
mf: $C_5H_{14}N_2$ mw: 102.18

$(CH_3)_2NC_2H_4NHCH_3$

PROP: Flash p: 57.2°F.

SAFETY PROFILE: A very dangerous fire hazard when exposed to heat, flame or oxidizers. When heated to decomposition it emits toxic fumes of NO_x. See also AMINES.

DPL000 CAS:55738-54-0 ***HR: 3***
trans-2-((DIMETHYLAMINO)METHYLIMINO)-5-(2-(5-NITRO-2-FURYL)VINYL)-1,3,4-OXADIAZOLE
mf: $C_{11}H_{12}N_5O_4$ mw: 277.27

TOXICITY DATA with REFERENCE
mma-sat 100 ng/plate MUREAV 40,9,76
orl-rat TDLo:42 g/kg/46W-I:CAR JNCIAM 51,403,73
orl-rat TD:42 g/kg/46W-C:CAR JNCIAM 54,841,75

CONSENSUS REPORTS: IARC Cancer Review: Group 2B IMEMDT 7,56,87; Animal Limited Evidence IMEMDT 7,147,74. EPA Genetic Toxicology Program.

SAFETY PROFILE: Suspected carcinogen with experimental carcinogenic data. Mutation data reported. When heated to decomposition it emits toxic fumes of NO_x.

DPL200 CAS:2914-77-4 ***HR: 3***
2-DIMETHYLAMINOMETHYL-1-(m-METHOXYPHENYL)CYCLOHEXANOL
mf: $C_{16}H_{25}NO_2$ mw: 263.42

TOXICITY DATA with REFERENCE
orl-mus LD50:395 mg/kg ARZNAD 28,107,78

SAFETY PROFILE: Poison by ingestion route. When heated to decomposition it emits toxic fumes of NO_x.

DPL900 CAS:63982-47-8 ***HR: 3***
3-DIMETHYLAMINO-4-METHYLPHENYL ESTER-N-METHYLCARBAMIC ACID HYDROCHLORIDE
mf: $C_{11}H_{16}N_2O_2{\cdot}ClH$ mw: 244.75

SYNS: N-METHYLURETHANE of HYDROCHLORIDE of 2-DIMETHYLAMINO-p-CRESOL ◇ T-1768

TOXICITY DATA with REFERENCE
orl-mus LD50:60 mg/kg JCSOA9 -,182,47
scu-mus LD50:10 mg/kg JCSOA9 -,182,47

SAFETY PROFILE: Poison by ingestion and subcutaneous routes. When heated to decomposition it emits toxic fumes of NO_x and HCl. See also CARBAMATES and ESTERS.

DPM200 CAS:1477-79-8 ***HR: 3***
3-DIMETHYLAMINO-2-METHYL-1-PHENYL-o-TOLYPROPANOL HYDROCHLORIDE
mf: $C_{19}H_{25}NO{\cdot}ClH$ mw: 319.91

SYNS: α-(2-(DIMETHYLAMINO)-1-METHYLETHYL)-2-METHYL-α-PHENYLBENZENEETHANOL, HCl ◇ SKF 70643-A

TOXICITY DATA with REFERENCE
orl-rat LD50:185 mg/kg JPPMAB 17,509,65
orl-mus LD50:245 mg/kg JPPMAB 17,509,65
orl-mam LD50:250 mg/kg JMCMAR 8,836,65

SAFETY PROFILE: Poison by ingestion. When heated to decomposition it emits very toxic fumes of HCl and NO_x.

DPM400 CAS:7005-47-2 ***HR: 3***
2-DIMETHYLAMINO-2-METHYL-1-PROPANOL
mf: $C_6H_{15}NO$ mw: 117.22

SYNS: DMAMP ◇ USAF CS-1

TOXICITY DATA with REFERENCE
ipr-mus LDLo:25 mg/kg NTIS** AD277-689

CONSENSUS REPORTS: Reported in EPA TSCA Inventory.

SAFETY PROFILE: Poison by intraperitoneal route. When heated to decomposition it emits toxic fumes of NO_x.

DPN200 CAS:605-65-2 ***HR: 3***
5-(DIMETHYLAMINO)-1-NAPHTHALENESULFONYL CHLORIDE
mf: $C_{12}H_{12}ClNO_2$ mw: 237.70

SYNS: 1-CHLOROSULFONYL-5-DIMETHYLAMINONAPHTHALENE ◇ DANSYL ◇ DANSYL CHLORIDE ◇ DIMETHYLAMINONAPHTHALENESULFONYL CHLORIDE ◇ 1-DIMETHYLAMINONAPHTHALENE-5-SULFONYL CHLORIDE ◇ 1-(DIMETHYLAMINO)-5-NAPHTHALENESULFONYLCHLORIDE ◇ 5-DIMETHYLAMINONA- PHTHYL-5-SULFONYL CHLORIDE

TOXICITY DATA with REFERENCE
ivn-mus LD50:56 mg/kg CSLNX* NX#00262

CONSENSUS REPORTS: Reported in EPA TSCA Inventory.

SAFETY PROFILE: Poison by intravenous route. When heated to decomposition it emits very toxic fumes of Cl^- and NO_x.

DPN400 CAS:6632-68-4 ***HR: 3***
1,3-DIMETHYL-4-AMINO-5-NITROSOURACIL
mf: $C_6H_8N_4O_3$ mw: 184.18

SYN: DANU

TOXICITY DATA with REFERENCE
scu-rat TDLo:16 g/kg/34W-I:ETA ZKKOBW 80,297,73
ipr-mus LD50:2000 mg/kg ZKKOBW 80,297,73

CONSENSUS REPORTS: Reported in EPA TSCA Inventory.

SAFETY PROFILE: Moderately toxic by intraperitoneal route. Questionable carcinogen with experimental tumorigenic data. When heated to decomposition it emits toxic fumes of NO_x.

DPN800 CAS:5882-48-4 ***HR: 3***
4-DIMETHYLAMINOPHENOL HYDROCHLORIDE
mf: $C_8H_{11}NO{\bullet}ClH$ mw: 173.66

SYN: p-DIMETHYLAMINOPHENOL HYDROCHLORIDE

TOXICITY DATA with REFERENCE
orl-rat LD50:689 mg/kg TXCYAC 31,165,84
ipr-rat LD50:90 mg/kg ARTODN 34,333,75
ivn-rat LD50:57 mg/kg ARTODN 34,337,75
orl-mus LD50:946 mg/kg TXCYAC 31,165,84
ipr-mus LD50:83 mg/kg ARTODN 49,191,82
ivn-mus LD50:70 mg/kg TXCYAC 31,165,84
orl-gpg LD50:1032 mg/kg TXCYAC 31,165,84

SAFETY PROFILE: Poison by intraperitoneal and intravenous routes. Moderately toxic by ingestion. When heated to decomposition it emits very toxic fumes of NO_x and HCl.

DPO100 CAS:93407-11-5 ***HR: 3***
3-DIMETHYLAMINO-2-PHENOXYPROPIOPHENONE HYDROCHLORIDE
mf: $C_{17}H_{19}NO_2{\bullet}ClH$ mw: 305.83

SYN: U-0172

TOXICITY DATA with REFERENCE
skn-rbt 2500 ppm MLD AIPTAK 137,410,62
eye-rbt 5000 ppm MLD AIPTAK 137,410,62
ipr-mus LD50:178 mg/kg AIPTAK 137,410,62

SAFETY PROFILE: Poison by intraperitoneal route. A skin and eye irritant. When heated to decomposition it emits toxic fumes of NO_x and HCl.

DPO200 CAS:539-17-3 ***HR: 3***
4-(p-DIMETHYLAMINOPHENYLAZO)ANILINE
mf: $C_{14}H_{16}N_4$ mw: 240.34

SYNS: ACETILE DIAZO BLACK N ◇ ADAB ◇ p-AMINOBENZENEAZODIMETHYLANILINE ◇ 4-AMINO-DAB ◇ 4-AMINO-4'-DIMETHYLAMINOAZOBENZENE ◇ 4'-AMINO-N,N-DIMETHYL-4-AMINOAZOBENZENE ◇ 4-((4-AMINOPHENYL)AZO)-N,N-DIME- THYLBENZENAMINE ◇ C.I. 11025 ◇ C.I. DISPERSE BLACK 3 ◇ DIAZO NERO MICROSETILE G ◇ INTERCHEM ACETATE DEVELOPED BLACK ◇ MEISEI TERYL DIAZO BLACK CR ◇ MICROSETILE DIAZO BLACK G ◇ SUPRACET DIAZO BLACK A

TOXICITY DATA with REFERENCE
mma-sat 20 μg/plate CALEDQ 17,263,83
otr-rat:lvr 50 μmol/L CNREA8 43,5087,83
otr-ham:kdy 2500 μg/L BJCAAI 38,34,78

orl-rat TDLo:1127 mg/kg (male 10D pre):REP ESKHA5 (99),156,81
ipr-rat LD50:350 mg/kg CNREA8 34,2274,74
ipr-mus LD50:350 mg/kg CNREA8 34,2274,74

CONSENSUS REPORTS: Reported in EPA TSCA Inventory.

SAFETY PROFILE: Poison by intraperitoneal route. Experimental reproductive effects. Mutation data reported. When heated to decomposition it emits toxic fumes of NO_x.

DPO275 CAS:73688-85-4 ***HR: 3***
4-(p-DIMETHYLAMINOPHENYLAZO)-BENZENEARSONIC ACID HYDROCHLORIDE
mf: $C_{14}H_{16}AsN_3O_3$•ClH mw: 385.71

PROP: Mp: 203° decomposes.

SYN: BENZENEARSONIC ACID, 4-(p-DIMETHYLAMINOPHENYLAZO)-, HYDROCHLORIDE

TOXICITY DATA with REFERENCE
ivn-mus LD50:25 mg/kg CSLNX* NX#05710

OSHA PEL: TWA 0.5 mg(As)/m^3

SAFETY PROFILE: Poison by intravenous route. When heated to decomposition it emits toxic fumes of NO_x, As, and HCl.

DPO400 CAS:18463-85-9 ***HR: 3***
6-((p-(DIMETHYLAMINO)PHENYL)AZO) BENZOTHIAZOLE
mf: $C_{15}H_{14}N_4S$ mw: 282.39

SYNS: 6-DIMETHYLAMINOPHENYLAZOBENZOTHIAZOLE ◇ 6-DIMETHYLAMINOPHENYLAZOBENZTHIAZOLE ◇ N,N-DIMETHYL-p-(6-BENZTHIAZOLYLAZO)ANILINE ◇ N,N-DIMETHYL-4-(6′-BENZTHIAZOLYLAZO)ANILINE

TOXICITY DATA with REFERENCE
dns-rat-unr 10 mg/kg CRNGDP 6,611,85
dns-rat-orl 10 mg/kg MUREAV 156,1,85
orl-rat TDLo:540 mg/kg/4W-C:ETA JMCMAR 11,1074,68

SAFETY PROFILE: Questionable carcinogen with experimental tumorigenic data. Mutation data reported. When heated to decomposition it emits very toxic fumes of SO_x and NO_x.

DPO600 CAS:18559-92-7 ***HR: 3***
7-((p-(DIMETHYLAMINO)PHENYL)AZO)BENZOTHIAZOLE
mf: $C_{15}H_{14}N_4S$ mw: 282.39

SYNS: N,N-DIMETHYL-p-(7-BENZTHIAZOLYLAZO)ANILINE ◇ N,N-DIMETHYL-4-(7′-BENZTHIAZOLYLAZO)ANILINE

TOXICITY DATA with REFERENCE
mma-sat 4 μg/plate MUREAV 93,67,82
orl-rat TDLo:1620 mg/kg/13W-C:ETA JMCMAR 11,1074,68

SAFETY PROFILE: Questionable carcinogen with experimental tumorigenic data. Mutation data reported. When heated to decomposition it emits very toxic fumes of NO_x and SO_x.

DPO800 CAS:63040-63-1 ***HR: 3***
4-((p-(DIMETHYLAMINO)PHENYL)AZO)ISOQUINOLINE
mf: $C_{17}H_{16}N_4$ mw: 276.37

SYN: N,N-DIMETHYL-4-(4′-ISOQUINOLINYLAZO)ANILINE

TOXICITY DATA with REFERENCE
orl-rat TDLo:3276 mg/kg/26W-C:ETA AICCA6 19,531,63

SAFETY PROFILE: Questionable carcinogen with experimental tumorigenic data. When heated to decomposition it emits toxic fumes of NO_x.

DPP000 CAS:63040-64-2 ***HR: 3***
5-((p-(DIMETHYLAMINO)PHENYL)AZO)ISOQUINOLINE
mf: $C_{17}H_{16}N_4$ mw: 276.37

SYN: N,N′-DIMETHYL-4-(5′-ISOQUINOLINYLAZO)ANILINE

TOXICITY DATA with REFERENCE
orl-rat TDLo:1092 mg/kg/26W-C:ETA AICCA6 19,531,63

SAFETY PROFILE: Questionable carcinogen with experimental tumorigenic data. When heated to decomposition it emits toxic fumes of NO_x.

DPP200 CAS:63040-65-3 ***HR: 3***
7-(p-(DIMETHYLAMINO)PHENYL)AZO)ISOQUINOLINE
mf: $C_{17}H_{16}N_4$ mw: 276.37

SYN: N,N-DIMETHYL-4-(7′-ISOQUINOLINYLAZO)ANILINE

TOXICITY DATA with REFERENCE
orl-rat TDLo:3276 mg/kg/26W-C:ETA AICCA6 19,531,63

SAFETY PROFILE: Questionable carcinogen with experimental tumorigenic data. When heated to decomposition it emits toxic fumes of NO_x.

DPP400 CAS:10318-23-7 ***HR: 3***
5-((p-(DIMETHYLAMINO)PHENYL)AZO)ISOQUINOLINE-2-OXIDE
mf: $C_{17}H_{16}N_4O$ mw: 292.37

SYN: N,N-DIMETHYL-4-(5′-ISOQUINOLYL-2′-OXIDE)AZOANILINE

TOXICITY DATA with REFERENCE
orl-rat TDLo:720 mg/kg/17W-C:ETA AICCA6 19,531,66

SAFETY PROFILE: Questionable carcinogen with experimental tumorigenic data. When heated to decomposition it emits toxic fumes of NO_x.

DPP600 CAS:19471-27-3 ***HR: 3***
4-((p-(DIMETHYLAMINO)PHENYL)AZO)-2,5-LUTIDINE 1-OXIDE
mf: $C_{15}H_{18}N_4O$ mw: 270.37

SYN: N,N-DIMETHYL-4-(4'-(2',5'-DIMETHYLPYRIDYL-1'-OXIDE)AZO)ANILINE

TOXICITY DATA with REFERENCE
orl-rat TDLo:2646 mg/kg/21W-C:ETA JNCIAM 41,855,68

SAFETY PROFILE: Questionable carcinogen with experimental tumorigenic data. When heated to decomposition it emits toxic fumes of NO_x.

DPP709 CAS:19456-77-0 ***HR: 3***
4-((p-(DIMETHYLAMINO)PHENYL)AZO)-3,5-LUTIDINE-1-OXIDE
mf: $C_{15}H_{18}N_4O$ mw: 270.37

SYNS: N,N-DIMETHYL-4-(4'-(3',5'-DIMETHYLPYRIDYL-1'-OXIDE)AZO)ANILINE ◇ N,N-DIMETHYL-4-(3',5'-LUTIDYL-1'-OXIDE)AZO)ANILINE

TOXICITY DATA with REFERENCE
orl-rat TDLo:5292 mg/kg/21W-C:ETA JNCIAM 41,855,68

SAFETY PROFILE: Questionable carcinogen with experimental tumorigenic data. When heated to decomposition it emits toxic fumes of NO_x.

DPP800 CAS:7349-99-7 ***HR: 3***
4-((4-(DIMETHYLAMINO)PHENYL)AZO)-2,6-LUTIDINE-1-OXIDE
mf: $C_{15}H_{18}N_4O$ mw: 270.37

SYNS: N,N-DIMETHYL-4-(4'-(2',6'-DIMETHYLPYRIDYL-1'-OXIDE)AZO)ANILINE ◇ 2,6-DIMETHYLPYRIDINE-1-OXIDE-4-AZO-p-DIMETHYLANILINE

TOXICITY DATA with REFERENCE
orl-rat TDLo:714 mg/kg/17W-C:NEO JNCIAM 37,365,66
orl-rat TD:4300 mg/kg/17W-C:ETA CNREA8 14,715,54

SAFETY PROFILE: Questionable carcinogen with experimental neoplastigenic and tumorigenic data. When heated to decomposition it emits toxic fumes of NO_x.

DPQ200 CAS:33804-48-7 ***HR: 3***
4-((p-(DIMETHYLAMINO)PHENYL)AZO)-N-METHYLACETANILIDE
mf: $C_{17}H_{20}N_4O$ mw: 296.41

SYNS: N'-ACETYL-N'-METHYL-4'-AMINO-N,N-DIMETHYL-4-AMINOAZOBENZENE ◇ 4-(N-ACETYL-N-METHYL)AMINO-4'-(N',N'-DIMETHYLAMINO)AZOBENZENE ◇ N',N'-DIMETHYL-4'-AMINO-N-ACETYL-N-MONOMETHYL-4-AMINOAZOBENZENE ◇ N-(4-((4-(DIMETHYLAMINO)PHENYL)AZO)PHENYL)-N-METHYLACETAMIDE

TOXICITY DATA with REFERENCE
mma-sat 250 nmol/plate CNREA8 46,1654,86
dns-rat:lvr 1 μmol/L CNREA8 46,1654,86
orl-rat TDLo:1630 mg/kg/21W-C:ETA CNREA8 34,2274,74
ipr-rat LD50:370 mg/kg CNREA8 34,2274,74

SAFETY PROFILE: Poison by intraperitoneal route. Questionable carcinogen with experimental tumorigenic data. Mutation data reported. When heated to decomposition it emits toxic fumes of NO_x.

DPQ400 CAS:17400-65-6 ***HR: 3***
5-((p-(DIMETHYLAMINO)PHENYL)AZO)-7-METHYLQUINOLINE
mf: $C_{18}H_{18}N_4$ mw: 290.40

SYNS: N,N-DIMETHYL-4-(5'-(7'-METHYLQUINOLYL)AZO)ANILINE ◇ 7'-METHYL-5'-(p-DIMETHYLAMINOPHENYLAZO)QUINOLINE

TOXICITY DATA with REFERENCE
orl-rat TDLo:540 mg/kg/30D-C:CAR JNCIAM 40,891,68

SAFETY PROFILE: Questionable carcinogen with experimental carcinogenic data. When heated to decomposition it emits toxic fumes of NO_x.

DPQ600 CAS:17416-18-1 ***HR: 3***
5-((p-(DIMETHYLAMINO)PHENYL)AZO)QUINALDINE
mf: $C_{18}H_{18}N_4$ mw: 290.40

SYN: 2'-METHYL-5'-(p-DIMETHYLAMINOPHENYLAZO)QUINOLINE

TOXICITY DATA with REFERENCE
orl-rat TDLo:540 mg/kg/30D-C:CAR JNCIAM 40,891,68

SAFETY PROFILE: Questionable carcinogen with experimental carcinogenic data. When heated to decomposition it emits toxic fumes of NO_x.

DPQ800 CAS:17416-17-0 ***HR: 3***
5-((p-DIMETHYLAMINO)PHENYL)AZO)QUINOLINE
mf: $C_{17}H_{16}N_4$ mw: 276.37

SYNS: N,N-DIMETHYL-p-(5'-QUINOLYLAZO)ANILINE ◇ N,N-DIMETHYL-4-(5'-QUINOLYLAZO)ANILINE

TOXICITY DATA with REFERENCE
orl-rat TDLo:714 mg/kg/17W-C:CAR JNCIAM 40,891,68
orl-rat TD:720 mg/kg/17W-C:ETA AICCA6 19,531,63

SAFETY PROFILE: Questionable carcinogen with experimental carcinogenic and tumorigenic data. When heated to decomposition it emits toxic fumes of NO_x.

DPR000 CAS:30041-69-1 ***HR: 3***
6-((p-(DIMETHYLAMINO)PHENYL)AZO)QUINOLINE
mf: $C_{17}H_{16}N_4$ mw: 276.37

SYNS: N,N-DIMETHYL-4-(6'-QUINOLYLAZO)ANILINE ◇ QUINOLINE-6-AZO-p-DIMETHYLANILINE

TOXICITY DATA with REFERENCE
dns-rat-orl 40 mg/kg CALEDQ 27,115,85
orl-rat TDLo:714 mg/kg/17W-C:ETA AICCA6 19,531,63

CONSENSUS REPORTS: EPA Genetic Toxicology Program.

SAFETY PROFILE: Questionable carcinogen with experimental tumorigenic data. Mutation data reported. When heated to decomposition it emits toxic fumes of NO_x.

DPR200 CAS:22750-85-2 ***HR: 3***
5-((p-(DIMETHYLAMINO)PHENYL)AZO)QUINOLINE-1-OXIDE
mf: $C_{17}H_{16}N_4O$ mw: 292.37

SYN: N,N-DIMETHYL-4-((5'-QUINOLYL-1'-OXIDE)AZO)ANILINE

TOXICITY DATA with REFERENCE
orl-rat TDLo:714 mg/kg/17W-C:ETA AICCA6 19,531,63

SAFETY PROFILE: Questionable carcinogen with experimental tumorigenic data. When heated to decomposition it emits toxic fumes of NO_x.

DPR400 CAS:22750-86-3 ***HR: 3***
6-((p-(DIMETHYLAMINO)PHENYL)AZO)QUINOLINE-1-OXIDE
mf: $C_{17}H_{16}N_4O$ mw: 292.37

SYN: N,N'-DIMETHYL-4-((6'-QUINOLYL-1'-OXIDE)AZO)ANILINE

TOXICITY DATA with REFERENCE
orl-rat TDLo:714 mg/kg/17W-C:ETA AICCA6 19,531,63

SAFETY PROFILE: Questionable carcinogen with experimental tumorigenic data. When heated to decomposition it emits toxic fumes of NO_x.

DPS200 CAS:24220-18-6 ***HR: 3***
2-(p-DIMETHYLAMINOPHENYL)-1,6-DIMETHYLQUINOLINIUM CHLORIDE
mf: $C_{19}H_{21}N_2 \cdot Cl$ mw: 312.87

TOXICITY DATA with REFERENCE
orl-mus LD50:50 mg/kg JMCMAR 13,122,70
ipr-mus LD50:10 mg/kg JMCMAR 13,122,70

SAFETY PROFILE: Poison by ingestion and intraperitoneal routes. When heated to decomposition it emits very toxic fumes of NO_x and Cl^-.

DPS600 CAS:2150-58-5 ***HR: 3***
4-(p-DIMETHYLAMINOPHENYL)IMINO-2,5-CYCLOHEXADIENE-1-ONE
mf: $C_{14}H_{14}N_2O$ mw: 226.30

TOXICITY DATA with REFERENCE
ipr-mus LD50:80 mg/kg JMCMAR 21,11,78

CONSENSUS REPORTS: Reported in EPA TSCA Inventory.

SAFETY PROFILE: Poison by intraperitoneal route. When heated to decomposition it emits toxic fumes of NO_x.

DPT200 CAS:16032-41-0 ***HR: 2***
2-(4-DIMETHYLAMINOPHENYL)QUINOLINE
mf: $C_{17}H_{16}N_2$ mw: 248.35

SYN: 2-(p-(DIMETHYLAMINO)PHENYL)QUINOLINE

TOXICITY DATA with REFERENCE
skn-mam 20 µg SEV JACSAT 72,2181,50

SAFETY PROFILE: A severe skin irritant. When heated to decomposition it emits toxic fumes of NO_x.

DPT800 CAS:108-16-7 ***HR: 2***
1,1-DIMETHYLAMINOPROPANOL-2
mf: $C_5H_{14}N_2O$ mw: 118.21

PROP: Bp: 122.5-126.2°, flash p: 90°F, fp: < −20°, d: 0.850 @ 25°/25°, vap d: 3.52.

SYNS: 1,1-DIMETHYLAMINOPROPAN-2-OL ◇ DIMETHYL(2-HYDROXYPROPYL)AMINE ◇ DIMETHYLISOPROPANOLAMINE

TOXICITY DATA with REFERENCE
skn-rbt 10 mg/24H MLD AMIHBC 10,61,54
skn-rbt 500 mg MOD FCTOD7 20 563,82
eye-rbt 250 µg SEV AMIHBC 10,61,54
eye-rbt 100 mg/4S rns SEV FCTOD7 20,573,82
orl-rat LD50:1890 mg/kg AMIHBC 10,61,54

CONSENSUS REPORTS: Reported in EPA TSCA Inventory.

SAFETY PROFILE: Moderately toxic by ingestion. A skin and severe eye irritant. Used in boiler water condensate corrosion control. Dangerous fire hazard when exposed to heat or flame. Can react vigorously with oxidizers. To fight fire, use foam, CO_2, dry chemical. When heated to decomposition it emits toxic fumes of NO_x. See also AMINES.

DPU000 CAS:1738-25-6 ***HR: 3***
3-(DIMETHYLAMINO)PROPIONITRILE
mf: $C_5H_{10}N_2$ mw: 98.17

PROP: Liquid. Mp: −43°, bp: 170°, d: 0.8617, vap d: 3.35. flash p: 145°F.

SYN: β-DIMETHYLAMINOPROPIONITRILE

TOXICITY DATA with REFERENCE
orl-rat LD50:2600 mg/kg DCTODJ 2,223,79
ivn-mus LD50:180 mg/kg CSLNX* NX#00201
skn-rbt LD50:1410 mg/kg AIHAAP 23,95,62

CONSENSUS REPORTS: Reported in EPA TSCA Inventory. Cyanide and its compounds are on the Community Right-To-Know List.

SAFETY PROFILE: Poison by intravenous route. Moderately toxic by ingestion and skin contact. Combustible liquid when exposed to heat, flame, or oxidizers; can react with oxidizing materials. To fight fire, use foam, CO_2, dry chemical. When heated to decomposition it emits highly toxic fumes of NO_x and CN^-. See also NITRILES.

DPU400 CAS:879-72-1 ***HR: 3***
3-(DIMETHYLAMINO)PROPIOPHENONE HYDROCHLORIDE
mf: $C_{11}H_{15}NO•ClH$ mw: 213.73

SYN: β-DIMETHYLAMINOPROPIOPHENONEHYDROCHLORIDE

TOXICITY DATA with REFERENCE
orl-mus LD50:100 mg/kg AITEAT 15,249,67
ipr-mus LD50:70 mg/kg AITEAT 15,249,67

CONSENSUS REPORTS: Reported in EPA TSCA Inventory.

SAFETY PROFILE: Poison by ingestion and intraperitoneal routes. When heated to decomposition it emits very toxic fumes of HCl and NO_x.

DPU600 CAS:102571-36-8 ***HR: 3***
1-(3-(DIMETHYLAMINO)PROPOXY)ADAMANTANE ETHYL IODIDE
mf: $C_{17}H_{32}NO•I$ mw: 393.40

SYN: (2-(1-ADAMANTYLOXY)PROPYL)DIMETHYLETHYLAMMONIUM IODIDE

TOXICITY DATA with REFERENCE
orl-mus LD50:400 mg/kg FRPSAX 32,129,77
ipr-mus LD50:100 mg/kg FRPSAX 32,129,77

SAFETY PROFILE: Poison by ingestion and intraperitoneal routes. When heated to decomposition it emits very toxic fumes of I^-, NH_3, and NO_x. See also IODIDES.

DPU800 CAS:15083-53-1 ***HR: 3***
5-(3-(DIMETHYLAMINO)PROPOXY)-3-METHYL-1-PHENYLPYRAZOLE
mf: $C_{15}H_{21}N_3O$ mw: 259.39

SYN: P-329

TOXICITY DATA with REFERENCE
orl-mus LD50:636 mg/kg ARZNAD 17,214,67
scu-mus LD50:350 mg/kg ARZNAD 17,214,67

SAFETY PROFILE: Poison by subcutaneous route. Moderately toxic by ingestion. When heated to decomposition it emits toxic fumes of NO_x.

DPW600 CAS:303-54-8 ***HR: 3***
5-(3-(DIMETHYLAMINO)PROPYL)-5H-DIBENZ(b,f)AZEPINE
mf: $C_{19}H_{22}N_2$ mw: 278.43

TOXICITY DATA with REFERENCE
ivn-rat LD50:23 mg/kg AIPTAK 120,450,59
ivn-mus LD50:41 mg/kg AIPTAK 120,450,59
ivn-rbt LD50:6 mg/kg AIPTAK 120,450,59

SAFETY PROFILE: Poison by intravenous route. When heated to decomposition it emits toxic fumes of NO_x.

DPX200 CAS:5560-72-5 ***HR: 3***
5-(3-(DIMETHYLAMINO)PROPYL)-6,7,8,9,10,11-HEXAHYDRO-5H-CYCLOOCT(b)INDOLE
mf: $C_{19}H_{28}N_2$ mw: 284.49

SYNS: GALATUR ◇ IPRINDOLE ◇ PRAMINDOLE ◇ PRONDOL ◇ WY-3263

TOXICITY DATA with REFERENCE
orl-rat LD50:484 mg/kg TXAPA9 18,185,71
ipr-rat LD50:187 mg/kg 27ZQAG -,130,72
orl-mus LD50:759 mg/kg FRPPAO 25,519,70
ipr-mus LD50:195 mg/kg 27ZQAG -,130,72
ivn-mus LD50:32 mg/kg CSLNX* NX#01206

SAFETY PROFILE: Poison by intravenous and intraperitoneal routes. Moderately toxic by ingestion. When heated to decomposition it emits toxic fumes of NO_x.

DPX400 CAS:303-70-8 ***HR: 3***
5-(3-(DIMETHYLAMINO)PROPYL)-2-HYDROXY-10,11-DIHYDRO-5H-DIBENZ(b,f)AZEPINE
mf: $C_{19}H_{24}N_2O$ mw: 296.45

SYNS: GP 33679 ◇ 2-HYDROXYINIPRAMINE

TOXICITY DATA with REFERENCE
scu-rat TDLo:300 mg/kg (female 8-12D post):TER ARZNAD 19,1617,69
orl-rat LD50:2980 mg/kg ARZNAD 19,1617,69
ivn-rat LD50:19 mg/kg ARZNAD 19,1617,69
orl-mus LD50:733 mg/kg ARZNAD 19,1617,69
ivn-mus LD50:29 mg/kg ARZNAD 19,1617,69
ivn-rbt LD50:12500 μg/kg ARZNAD 19,1617,69

SAFETY PROFILE: Poison by intravenous route. Moderately toxic by ingestion. Experimental teratogenic effects. When heated to decomposition it emits toxic fumes of NO_x.

DPX800 CAS:6202-23-9 ***HR: 3***
5-(3-DIMETHYLAMINOPROPYLIDENE)-5H-DIBENZO-(a,d)CYCLOHEPTENE HYDROCHLORIDE
mf: $C_{20}H_{21}N \cdot ClH$ mw: 311.88

SYNS: CYCLOBENZAPRINE HYDROCHLORIDE ◇ 3-(5H-DIBENZO(a,d)CYCLOHEPTEN-5-YLIDENE)-N,N-DIMETHYL-1-PROPANAMINE HYDROCHLORIDE ◇ N,N-DIMETHYL-5H-DIBENZO(a,d) CYCLOHEPTENE-$\Delta^{5,\Gamma}$-PROPYLAMINE HYDROCHLORIDE ◇ FLEXERIL ◇ FLEXIBAN ◇ PROHEPTATRIENE HYDROCHLORIDE ◇ PROHEPTATRIEN MONOHYDROCHLORIDE

TOXICITY DATA with REFERENCE
orl-mus LD50:250 mg/kg 27ZQAG -,87,72
ivn-mus LD50:36 mg/kg AIPTAK 144,481,63

SAFETY PROFILE: Poison by ingestion and intravenous routes. A muscle relaxant. When heated to decomposition it emits very toxic fumes of NO_x and HCl.

DPY200 CAS:897-15-4 ***HR: 3***
11-(3-DIMETHYLAMINOPROPYLIDENE-6,11-DIHYDRODIBENZO(b,e)THIEPINE HYDROCHLORIDE
mf: $C_{19}H_{21}NS \cdot ClH$ mw: 331.93

SYNS ◇ 3-DIBENZO(b,e)THIEPIN-11(6H)-YLIDENE-N,N-DIMETHYL-1-PROPANAMINE, HYDROCHLORIDE ◇ N,N-DIMETHYLDIBENZO(b,e)THIEPIN-$\Delta^{11(6H),\Gamma}$-PROPYLAMINE HYDROCHLORIDE ◇ DOSULEPIN CHLORIDE ◇ DOSULEPIN HYDROCHLORIDE ◇ DOTHEIPIN HYDROCHLORIDE ◇ PROTHIADENE HYDROCHLORIDE ◇ PROTHIADEN HYDROCHLORIDE

TOXICITY DATA with REFERENCE
orl-rat TDLo:440 mg/kg (MGN):REP OYYAA2 27,1103,84
orl-rat TDLo:440 mg/kg (MGN):TER OYYAA2 27,1103,84
orl-wmn LDLo:40 mg/kg PGMJAO 60,442,84
orl-cld TDLo:90 mg/kg:CVS PGMJAO 60,442,84
orl-rat LD50:260 mg/kg 27ZQAG -,88,72
ipr-rat LD50:105 mg/kg IYKEDH 14,192,83
scu-rat LD50:760 mg/kg 27ZQAG -,88,72
ivn-rat LD50:24 mg/kg 27ZQAG -,88,72
orl-mus LD50:209 mg/kg 27ZQAG -,88,72
ipr-mus LD50:116 mg/kg IYKEDH 14,192,83
scu-mus LD50:620 mg/kg IYKEDH 14,192,83
ivn-mus LD50:29200 μg/kg IYKEDH 14,192,83

SAFETY PROFILE: Poison by ingestion, intraperitoneal, and intravenous routes. Moderately toxic by subcutaneous route. Human systemic effects by ingestion: cardiomyopathy. Experimental teratogenic and reproductive effects. An antidepressant. When heated to decomposition it emits very toxic fumes of NO_x, SO_x and HCl.

DPY600 CAS:51003-81-7 ***HR: 3***
5-DIMETHYLAMINO-6-PROPYL-5H-INDENO(5,6-d)-1,3-DIOXOLE HYDROCHLORIDE
mf: $C_{15}H_{19}NO_2 \cdot ClH$ mw: 281.81

SYNS: pr-MDI ◇ 2-PROPYL-3-DIMETHYLAMINO-5,6-METHYLENEDIOXYINDENE HYDROCHLORIDE ◇ 2-N-PROPYL-3-DIMETHYLAMINO-5,6-METHYLENEDIOXYINDENEHYDROCHLORIDE

TOXICITY DATA with REFERENCE
ipr-rat LD50:175 mg/kg RCOCB8 26,85,79
ipr-mus LD50:185 mg/kg RCOCB8 26,85,79
ivn-mus LD50:40 mg/kg RCOCB8 26,85,79

SAFETY PROFILE: Poison by intraperitoneal and intravenous routes. When heated to decomposition it emits very toxic fumes of NO_x and HCl.

DQA400 CAS:60-87-7 ***HR: 3***
10-(2-(DIMETHYLAMINO)PROPYL)PHENOTHIAZINE
mf: $C_{17}H_{20}N_2S$ mw: 284.45

SYNS: A-91033 ◇ APROBIT ◇ ATOSIL ◇ AVOMINE ◇ DIMAPP ◇ DIMETHYLAMINO-ISOPROPYL-PHENTHIAZIN (GERMAN) ◇ (2-DIMETHYLAMINO-2-METHYL)ETHYL-N-DIBENZOPARATHIAZINE ◇ N-(2'-DIMETHYLAMINO-2'-METHYL)ETHYLPHENOTHIAZINE ◇ 10-(2-(DIMETHYLAMINO)-2-METHYLETHYL)PHENOTHIAZINE ◇ N-DIMETHYLAMINO-2-METHYLETHYL THIODIPHENYLAMINE ◇ (DIMETHYLAMINO-2-PROPYL-10-PHENOTHIAZINE HYDROCHLORIDE (FRENCH) ◇ DIPRAZINE ◇ DIPROZIN ◇ FARGAN ◇ FENAZIL ◇ FENERGAN ◇ FENETAZINA ◇ HIBERNA ◇ HISTARGAN ◇ IERGIGAN ◇ ISOPHENERGAN ◇ ISOPROMETHAZINE ◇ LERCIGAN ◇ LERGIGAN ◇ LILLY 1516 ◇ LILLY 01516 ◇ NCI-C60673 ◇ PHARGAN ◇ PHENERGAN ◇ PHENSEDYL ◇ PILPOPHEN ◇ PIPOLPHEN ◇ PROAZAIMINE ◇ PROAZAMINE ◇ PROCIT ◇ PROMAZINAMIDE ◇ PROMETASIN ◇ PROMETAZIN ◇ PROMETHIAZINE ◇ PROMEZATHINE ◇ PROREX ◇ PROTAZINE ◇ PROTHAZIN ◇ PROVIGAN ◇ PYRETHIA ◇ PYRETHIAZINE ◇ ROMERGAN ◇ 3277 RP ◇ 3389 R.P. ◇ 4182 R.P. ◇ SKF 1498 ◇ SYNALGOS ◇ TANIDIL ◇ THIERGAN ◇ VALLERGINE ◇ WY 509

TOXICITY DATA with REFERENCE
eye-rbt 100 mg SEV FCTOD7 20 573,82
eye-rbt 100 mg/4S rns MLD FCTOD7 20573,82
dni-hmn:fbr 80 nmol/L DNSYAG 29,829,68
otr-ham:emb 10 mg/L ENMUDM 8(Suppl 6),4,86
msc-ham:lng 10 mg/L SHIGAZ 70,943,83
ipr-rat TDLo:350 mg/kg (10-16D preg):TER JPHYA7 164,138,62
par-rat TDLo:350 mg/kg (female 4D post):REP FATOAO 48(6),89,85
skn-cld TDLo:13 mg/kg:EYE,PSY CMAJAX 130,1460,84
ipr-rat LDLo:140 mg/kg TXAPA9 1,156,59
scu-rat LD50:700 mg/kg CRSBAW 144,887,50
ivn-rat LD50:45 mg/kg AIPTAK 120,450,59
ims-rat LD50:169 mg/kg TXAPA9 18,185,71
orl-mus LD50:326 mg/kg ARZNAD 7,237,57
ipr-mus LD50:124 mg/kg JPETAB 108,201,53
scu-mus LD50:225 mg/kg PRPHA8 2,53,47
ivn-mus LD50:40 mg/kg TXAPA9 18,185,71

ims-mus LD50:175 mg/kg TXAPA9 18,185,71
orl-rbt LD50:580 mg/kg PHARAT 25,91,70
ivn-rbt LD50:19 mg/kg AIPTAK 120,450,59

SAFETY PROFILE: Poison by ingestion, intravenous, intramuscular, intraperitoneal, and subcutaneous routes. Human systemic effects by ingestion: pupillary dilation, wakefulness, hallucinations, and distorted perceptions. An experimental teratogen. Other experimental reproductive effects. Human mutation data reported. A severe eye irritant. When heated to decomposition it emits very toxic fumes of NO_x and SO_x.

DQA600 CAS:58-40-2 ***HR: 3***
10-(3-(DIMETHYLAMINO)PROPYL)PHENOTHIAZINE
mf: $C_{17}H_{20}N_2S$ mw: 284.45

SYNS: AMPAZINE ◇ N,N-DIMETHYL-10H-PHENOTHIAZINE-10-PROPANAMINE ◇ ESPARIN ◇ LIRANOL ◇ NEO-HIBERNEX ◇ PROMAZINE ◇ PROTACTYL ◇ SPARINE ◇ VEROPHEN ◇ WY 1094

TOXICITY DATA with REFERENCE
scu-rat TDLo:80 mg/kg (1-18D preg):REP JNEUAY 3,295,62
orl-rat LD50:350 mg/kg ARZNAD 8,507,58
ipr-rat LDLo:210 mg/kg TXAPA9 1,156,59
scu-rat LD50:192 mg/kg ARZNAD 24,1798,74
ivn-rat LD50:14500 μg/kg AIPTAK 119,311,59
orl-mus LD50:401 mg/kg ARZNAD 8,489,58
ipr-mus LD50:140 mg/kg AIPTAK 155,69,65
scu-mus LD50:110 mg/kg ARZNAD 24,1798,74
ivn-mus LD50:45 mg/kg APTOA6 19,87,62
ivn-rbt LD50:21 mg/kg AIPTAK 120,450,59

CONSENSUS REPORTS: EPA Genetic Toxicology Program.

SAFETY PROFILE: Human poison by unspecified route. An experimental poison by ingestion, subcutaneous, intravenous, and intraperitoneal routes. Experimental reproductive effects. When heated to decomposition it emits very toxic fumes of NO_x and SO_x.

DQB309 CAS:13713-13-8 ***HR: 3***
2-(3-DIMETHYLAMINOPROPYL)-3a,4,7,7a-TETRAHYDRO-4,7-ETHANOISOINDOLINE DIMETHIODIDE
mf: $C_{19}H_{36}N_2 \cdot 2I$ mw: 546.37

TOXICITY DATA with REFERENCE
orl-rat LD50:1730 mg/kg SKNEA7 10,15,60
ivn-rat LD50:137 mg/kg SKNEA7 10,15,60
orl-mus LD50:2334 mg/kg SKNEA7 10,15,60
ivn-mus LD50:156 mg/kg SKNEA7 10,15,60
ivn-rbt LD50:82200 μg/kg SKNEA7 10,15,60

SAFETY PROFILE: Poison by intravenous route. Moderately toxic by ingestion. When heated to decomposition it emits toxic fumes of I^- and NO_x. See also IODIDES.

DQB600 CAS:1122-58-3 ***HR: 3***
4-DIMETHYLAMINOPYRIDINE
mf: $C_7H_{10}N_2$ mw: 122.19

SYNS: 4-DIMETHYLAMINEPYRIDINE ◇ Γ-(DIMETHYLAMINO)PYRIDINE ◇ p-DIMETHYLAMINOPYRIDINE

TOXICITY DATA with REFERENCE
orl-mus LDLo:470 mg/kg AECTCV 14,111,85
ivn-mus LD50:56 mg/kg CSLNX* NX#04228

CONSENSUS REPORTS: Reported in EPA TSCA Inventory.

SAFETY PROFILE: Poison by intravenous route. Moderately toxic by ingestion. An acylation catalyst. When heated to decomposition it emits toxic fumes of NO_x.

DQB800 CAS:5585-67-1 ***HR: 3***
2-(DIMETHYLAMINO) RESERPILINATE
mf: $C_{26}H_{35}N_3O_5$ mw: 469.64

SYNS: ANTIPRESSINE DIHYDROCHLORIDE ◇ 2-(DIMETHYLAMINO) RESERPILIN-24-OIC ACID ETHYL ESTER

TOXICITY DATA with REFERENCE
orl-rat LD50:2350 mg/kg OYYAA2 3,390,69
ipr-rat LD50:330 mg/kg OYYAA2 3,390,69
scu-rat LD50:1000 mg/kg OYYAA2 3,390,69
orl-mus LD50:2100 mg/kg OYYAA2 3,390,69
ipr-mus LD50:410 mg/kg OYYAA2 3,390,69
scu-mus LD50:980 mg/kg OYYAA2 3,390,69

SAFETY PROFILE: Poison by intraperitoneal route. Moderately toxic by ingestion and subcutaneous routes. When heated to decomposition it emits toxic fumes of NO_x.

DQC000 CAS:63019-60-3 ***HR: 3***
7-(p-(DIMETHYLAMINO)STYRYL)BENZ(c)ACRIDINE
mf: $C_{27}H_{22}N_2$ mw: 374.51

SYN: p-DIMETHYLAMINOBENZYLIDEN-3,4-BENZ-9-METHYLACRIDINE

TOXICITY DATA with REFERENCE
scu-mus TDLo:200 mg/kg:ETA VOONAW 1,52,55

SAFETY PROFILE: Questionable carcinogen with experimental tumorigenic data. When heated to decomposition it emits toxic fumes of NO_x.

DQC200 CAS:63019-59-0 ***HR: 3***
12-(p-DIMETHYLAMINO)STYRYLBENZ(a)ACRIDINE
mf: $C_{27}H_{22}N_2$ mw: 374.51

SYN: p-DIMETHYLAMINOBENZYLIDEN-1,2-BENZ-9-METHYL-ACRIDINE

TOXICITY DATA with REFERENCE
scu-mus TDLo:200 mg/kg:ETA VOONAW 1,52,55

SAFETY PROFILE: Questionable carcinogen with experimental tumorigenic data. When heated to decomposition it emits toxic fumes of NO_x.

DQC400 CAS:1628-58-6 ***HR: 3***
2-(p-(DIMETHYLAMINO)STYRYL)BENZOTHIAZOLE
mf: $C_{17}H_{16}N_2S$ mw: 280.41

SYN: 2-(4-DIMETHYLAMINOSTYRYL)BENZOTHIAZOLE

TOXICITY DATA with REFERENCE
orl-rat TDLo:35 g/kg/1Y-I:NEO,REP JNCIAM 41,985,68

CONSENSUS REPORTS: Reported in EPA TSCA Inventory.

SAFETY PROFILE: Experimental reproductive effects. Questionable carcinogen with experimental neoplastigenic data. When heated to decomposition it emits very toxic fumes of NO_x and SO_x.

DQC600 CAS:19716-21-3 ***HR: 3***
4-(p-(DIMETHYLAMINO)STYRYL)-6,8-DIMETHYLQUINOLINE
mf: $C_{21}H_{22}N_2$ mw: 302.45

SYN: 6,8-DIMETHYL-(4-p-(DIMETHYLAMINO)STYRYL)QUINOLINE

TOXICITY DATA with REFERENCE
orl-rat TDLo:38 mg/kg/51W-I:ETA,REP JNCIAM 41,985,68

SAFETY PROFILE: Experimental reproductive effects. Questionable carcinogen with experimental tumorigenic data. When heated to decomposition it emits toxic fumes of NO_x.

DQD000 CAS:897-55-2 ***HR: 3***
4-(4-DIMETHYLAMINOSTYRYL)QUINOLINE
mf: $C_{19}H_{18}N_2$ mw: 274.39

SYNS: 2-(4-N,N-DIMETHYLAMINOSTYRYL)QUINOLINE ◇ 4-(p-(DIMETHYLAMINO)STYRYL)QUINOLINE

TOXICITY DATA with REFERENCE
ivn-mus TDLo:100 mg/kg:NEO,REP CNREA8 25,938,65
ivn-mus LDLo:160 mg/kg CNREA8 25,938,65

CONSENSUS REPORTS: Reported in EPA TSCA Inventory.

SAFETY PROFILE: Poison by intravenous route. Experimental reproductive effects. Questionable carcinogen with experimental neoplastigenic data. When heated to decomposition it emits toxic fumes of NO_x.

DQD200 CAS:21970-53-6 ***HR: 3***
4-(p-(DIMETHYLAMINO)STYRYL)QUINOLINE MONOHYDROCHLORIDE
mf: $C_{19}H_{18}N_2 \cdot ClH$ mw: 310.85

SYN: NSC 63346

TOXICITY DATA with REFERENCE
orl-rat TDLo:115 mg/kg/1Y-I:NEO,REP JNCIAM 41,985,68

SAFETY PROFILE: Experimental reproductive effects. Questionable carcinogen with experimental neoplastigenic data. When heated to decomposition it emits very toxic fumes of NO_x and HCl.

DQD400 CAS:1596-84-5 ***HR: 3***
DIMETHYLAMINOSUCCINAMIC ACID
mf: $C_6H_{12}N_2O_3$ mw: 160.20

SYNS: ALAR ◇ ALAR-85 ◇ AMINOZIDE ◇ B 995 ◇ BERNSTEINSAEURE-2,2-DIMETHYLHYDRAZID (GERMAN) ◇ B-NINE ◇ BUTANEDIOIC ACID MONO(2,2-DIMETHYLHYDRAZIDE) ◇ DAMINOZIDE (USDA) ◇ DIMAS ◇ N-DIMETHYL AMINO-β-CARBAMYL PROPIONIC ACID ◇ N-(DIMETHYLAMINO)SUCCINAMIC ACID ◇ N-DIMETHYLAMINO-SUCCINAMIDSAEURE (GERMAN) ◇ DMASA ◇ DMSA ◇ KYLAR ◇ NCI-C03827 ◇ SADH ◇ SUCCINIC ACID-2,2-DIMETHYLHYDRAZIDE ◇ SUCCINIC-1,1-DIMETHYL HYDRAZIDE

TOXICITY DATA with REFERENCE
mma-mus:lyms 1650 mg/L EMMUEG 12,85,88
msc-mus:lyms 156 mg/L EMMUEG 12,85,88
orl-rat TDLo:182 g/kg/2Y-C:CAR,REP NCITR* NCI-CG-TR-83,78
orl-mus TDLo:2600 g/kg/62W-C:CAR CNREA8 37,3497,77
orl-mus TD:873 g/kg/2Y-C:ETA NCITR* NCI-CG-TR-83,78
orl-rat LD50:8400 mg/kg FMCHA2 -,D10,80
orl-mus LD50:6300 mg/kg CHABA8 84,100483n,76
ipr-mus LD50:1325 mg/kg CHABA8 22,126,73

CONSENSUS REPORTS: EPA Genetic Toxicology Program. NCI Carcinogenesis Bioassay (feed); Clear Evidence: mouse, rat NCITR* NCI-CG-TR-83,78.

SAFETY PROFILE: Suspected carcinogen with experimental carcinogenic and tumorigenic data. Moderately toxic by ingestion and intraperitoneal routes. Experimental reproductive effects. When heated to decomposition it emits toxic fumes of NO_x.

DQD500 ***HR: 3***
p-DIMETHYLAMINOTHYMOLDIMETHYLURETHANE METHIODIDE
mf: $C_{16}H_{27}N_2O_2 \cdot I$ mw: 406.35

SYN: (CARBOXYMETHYL)TRIMETHYLAMMONIUMIODIDE-6-(DIMETHYLAMINO)-4-ISOPROPYL-m-TOLYLESTER

TOXICITY DATA with REFERENCE
orl-mus LDLo:77500 μg/kg FEPRA7 5,184,46

scu-mus LDLo:15 μg/kg FEPRA7 5,184,46
scu-dog LDLo:4600 μg/kg FEPRA7 5,184,46

SAFETY PROFILE: Poison by ingestion and subcutaneous routes. When heated to decomposition it emits toxic fumes of I^-, NH_3, and NO_x. See also IODIDES.

DQD600 CAS:7347-47-9 ***HR: 3***
4-((4-(DIMETHYLAMINO)-m-TOLYL)AZO)-2-PICOLINE-1-OXIDE
mf: $C_{14}H_{16}N_4O$ mw: 256.34

SYN: N,N-DIMETHYL-2-METHYL-4-(4'-(2'-METHYLPYRIDYL-1'-OXIDE)AZO)ANILINE

TOXICITY DATA with REFERENCE
orl-rat TDLo:2142 mg/kg/17W-C:NEO JNCIAM 37,365,66

SAFETY PROFILE: Questionable carcinogen with experimental neoplastigenic data. When heated to decomposition it emits toxic fumes of NO_x.

DQD800 CAS:7347-48-0 ***HR: 3***
4-((4-(DIMETHYLAMINO)-o-TOLYL)AZO)-2-PICOLINE-1-OXIDE
mf: $C_{15}H_{18}N_4O$ mw: 270.37

SYNS: N,N'DIMETHYL-3-METHYL-4-(4'-(2'-METHYLPYRIDYL-1'OXIDE)AZO)ANILINE ◇ N,N'-DIMETHYL-4-(4'-(2'-METHYLPYRIDYL-1-OXIDE)AZO)-o-TOLUIDINE

TOXICITY DATA with REFERENCE
orl-rat TDLo:2142 mg/kg/17W-C:NEO JNCIAM 37,365,66

SAFETY PROFILE: Questionable carcinogen with experimental neoplastigenic data. When heated to decomposition it emits toxic fumes of NO_x.

DQE000 CAS:19456-74-7 ***HR: 3***
4-((4-(DIMETHYLAMINO)-m-TOLYL)AZO)-3-PICOLINE-1-OXIDE
mf: $C_{15}H_{18}N_4O$ mw: 270.37

SYN: N,N,2-TRIMETHYL-4-(4'-(3'-METHYLPYRIDYL-1'-OXIDE) AZO)ANILINE

TOXICITY DATA with REFERENCE
orl-rat TDLo:4284 mg/kg/17W-C:ETA JNCIAM 41,85,68

SAFETY PROFILE: Questionable carcinogen with experimental tumorigenic data. When heated to decomposition it emits toxic fumes of NO_x.

DQE200 CAS:19471-28-4 ***HR: 3***
4-((4-(DIMETHYLAMINO)-o-TOLYL)AZO)-3-PICOLINE-1-OXIDE
mf: $C_{15}H_{18}N_4O$ mw: 270.37

SYN: N,N,3-TRIMETHYL-4-(4'-(3'-METHYLPYRIDYL-1'-OXIDE)AZO) ANILINE

TOXICITY DATA with REFERENCE
orl-rat TDLo:4284 mg/kg/17W-C:ETA JNCIAM 41,855,68

SAFETY PROFILE: Questionable carcinogen with experimental tumorigenic data. When heated to decomposition it emits toxic fumes of NO_x.

DQE400 CAS:17400-68-9 ***HR: 3***
5-((4-(DIMETHYLAMINO)-m-TOLYL)AZO)QUINOLINE
mf: $C_{18}H_{18}N_4$ mw: 290.40

SYNS: N,N-DIMETHYL-4-(5'-QUINOLYLAZO)-m-TOLUIDINE ◇ 3-METHYL-5'-(p-DIMETHYLAMINOPHENYLAZO)QUINOLINE

TOXICITY DATA with REFERENCE
orl-rat TDLo:2142 mg/kg/17W-C:CAR JNCIAM 40,891,68

SAFETY PROFILE: Questionable carcinogen with experimental carcinogenic data. When heated to decomposition it emits toxic fumes of NO_x.

DQE600 CAS:17416-21-6 ***HR: 3***
5-((4-(DIMETHYLAMINO)-o-TOLYL)AZO)QUINOLINE
mf: $C_{18}H_{18}N_4$ mw: 290.40

SYN: 2-METHYL-5'-(p-DIMETHYLAMINOPHENYLAZO)QUINOLINE

TOXICITY DATA with REFERENCE
orl-rat TDLo:2142 mg/kg/17W-C:CAR JNCIAM 40,891,68

SAFETY PROFILE: Questionable carcinogen with experimental carcinogenic data. When heated to decomposition it emits toxic fumes of NO_x.

DQE800 CAS:14144-91-3 ***HR: 3***
5-DIMETHYLAMINO-4-TOLYL METHYLCARBAMATE
mf: $C_{11}H_{16}N_2O_2$ mw: 208.29

SYNS: BAY 42696 ◇ 4-METHYL-3-DIMETHYLAMINOPHENYL ESTER-N-METHYLCARBAMIC ACID

TOXICITY DATA with REFERENCE
orl-rat LD50:46 mg/kg ATXKA8 27,311,71
scu-mus LDLo:20 mg/kg NTIS** PB158-508

SAFETY PROFILE: Poison by ingestion and subcutaneous routes. A strong oxidizing agent. When heated to decomposition it emits toxic fumes of NO_x. See also CARBAMATES and ESTERS.

DQE900 CAS:2083-91-2 ***HR: 3***
DIMETHYLAMINOTRIMETHYLSILANE
mf: $C_5H_{15}NSi$ mw: 117.27

SAFETY PROFILE: Explosive reaction with xenon difluoride below 0°C. When heated to decomposition it emits toxic fumes of NO_x. See also SILANE.

DQF000 CAS:6120-10-1 ***HR: 3***
4-DIMETHYLAMINO-3,5-XYLENOL
mf: $C_{10}H_{15}NO$ mw: 165.26

TOXICITY DATA with REFERENCE
orl-mus TDLo:33 g/kg/78W-I:ETA NTIS** PB223-159

SAFETY PROFILE: Questionable carcinogen with experimental tumorigenic data. When heated to decomposition it emits toxic fumes of NO_x.

DQF200 CAS:19456-73-6 ***HR: 3***
4-((4-(DIMETHYLAMINO)-2,3-XYLYL)AZO)PYRIDINE-1-OXIDE
mf: $C_{15}H_{18}N_4O$ mw: 270.37

SYN: N,N,2,3-TETRAMETHYL-4-(4'-(PYRIDYL-1'-OXIDE)AZO)ANILINE

TOXICITY DATA with REFERENCE
orl-rat TDLo:6552 mg/kg/26W-C:ETA JNCIAM 41,855,68

SAFETY PROFILE: Questionable carcinogen with experimental tumorigenic data. When heated to decomposition it emits toxic fumes of NO_x.

DQF400 CAS:19456-75-8 ***HR: 3***
4-((4-(DIMETHYLAMINO)-2,5-XYLYL)AZO)PYRIDINE-1-OXIDE
mf: $C_{15}H_{18}N_4O$ mw: 270.37

SYN: N,N,2,5-TETRAMETHYL-4-(4'-(PYRIDYL-1'-OXIDE)AZO)ANILINE

TOXICITY DATA with REFERENCE
orl-rat TDLo:3276 mg/kg/26W-C:ETA JNCIAM 41,855,68

SAFETY PROFILE: Questionable carcinogen with experimental tumorigenic data. When heated to decomposition it emits toxic fumes of NO_x.

DQF600 CAS:19595-66-5 ***HR: 3***
4-((4-(DIMETHYLAMINO)-3,5-XYLYL)AZO)PYRIDINE-1-OXIDE
mf: $C_{15}H_{18}N_4O$ mw: 270.37

SYN: N,N,2,6-TETRAMETHYL-4-(4'-(PYRIDYL-1'-OXIDE)AZO)ANILINE

TOXICITY DATA with REFERENCE
orl-rat TDLo:6552 mg/kg/26W-C:ETA JNCIAM 41,855,68

SAFETY PROFILE: Questionable carcinogen with experimental tumorigenic data. When heated to decomposition it emits toxic fumes of NO_x.

DQF650 CAS:14488-49-4 ***HR: 3***
DIMETHYLAMMONIUM PERCHLORATE
mf: $C_2H_8ClNO_4$ mw: 145.54

SAFETY PROFILE: An explosive salt. When heated to decomposition it emits toxic fumes of Cl^-, NH_3, and NO_x. See also PERCHLORATES.

DQF700 CAS:4063-41-6 ***HR: D***
4,5'-DIMETHYL ANGELICIN
mf: $C_{13}H_{10}O_3$ mw: 214.23

SYN: 4,8-DIMETHYL-2H-FURO(2,3-h)-1-BENZOPYRAN-2-ONE

TOXICITY DATA with REFERENCE
dnd-esc 20 μmol/L CBINA8 21,103,78
dnd-mam:lym 20 μmol/L CBINA8 21,103,78

CONSENSUS REPORTS: IARC Cancer Review: Animal Inadequate Evidence IMEMDT 40,291,86.

SAFETY PROFILE: Questionable carcinogen. Mutation data reported. When heated to decomposition it emits acrid smoke and fumes.

DQF800 CAS:121-69-7 ***HR: 3***
N,N-DIMETHYLANILINE
DOT: UN 2253
mf: $C_8H_{11}N$ mw: 121.20

PROP: Liquid. Mp: 2.5°, bp: 193.1°, flash p: 145°F (CC), d: 0.9557 @ 20°/4°, ULC: 20-25, autoign temp: 700°F, vap press: 1 mm @ 29.5°, vap d: 4.17.

SYNS: (DIMETHYLAMINO)BENZENE ◇ N,N-DIMETHYLBENZENEAMINE ◇ DIMETHYLPHENYLAMINE ◇ N,N-DIMETHYLPHENYLAMINE ◇ DWUMETYLOANILINA (POLISH) ◇ NCI-C56428 ◇ VERSNELLER NL 63/10

TOXICITY DATA with REFERENCE
skn-rbt 10 mg/24H open MLD AIHAAP 23,95,62
orl-hmn LDLo:50 mg/kg NCPBBY Jan/Feb,69
orl-rat LD50:1410 mg/kg AIHAAP 23,95,62
ihl-rat LCLo:250 mg/m^3/4H GISAAA 37(4),35,72
skn-rbt LD50:1770 mg/kg AIHAAP 23,95,62

CONSENSUS REPORTS: Reported in EPA TSCA Inventory. Community Right-To-Know List.

OSHA PEL: (Transitional: TWA 5 ppm (skin)) TWA 5 ppm; STEL 10 ppm (skin)
ACGIH TLV: TWA 5 ppm; STEL 10 ppm (skin)
DFG MAK: 5 ppm (25 mg/m^3)
DOT Classification: Poison B; Label: Poison.

SAFETY PROFILE: Human poison by ingestion. Moderately toxic by inhalation and skin contact. A skin irritant. Physiological action is similar to, but less toxic than aniline. A central nervous system depressant. Flammable when exposed to heat, flame, or oxidizers. Explodes on contact with benzoyl peroxide or diisopropyl peroxydicarbonate. To fight fire, use foam, CO_2, dry chemical. When heated to decomposition it emits highly toxic fumes of aniline and NO_x. See also ANILINE.

DQG000 CAS:41217-05-4 ***HR: 3***
6,12-DIMETHYLANTHANTHRENE
mf: $C_{24}H_{16}$ mw: 304.40

SYN: 6,12-DIMETHYL-BIBENZO(def,mno)CHRYSENE

TOXICITY DATA with REFERENCE
dnd-mam:lym 30 μmol/L CBINA8 47,87,83
scu-mus TDLo:72 mg/kg/9W-I:ETA COREAF 246,1477,58

SAFETY PROFILE: Questionable carcinogen with experimental tumorigenic data. Mutation data reported. When heated to decomposition it emits acrid smoke and irritating fumes.

DQG200 CAS:781-43-1 ***HR: 3***
9,10-DIMETHYLANTHRACENE
mf: $C_{16}H_{14}$ mw: 206.30

TOXICITY DATA with REFERENCE
mmo-sat 20 μg/plate CRNGDP 6,1483,85
mma-esc 10 μg/plate PMRSDJ 1,387,81
dnr-esc 500 mg/L PMRSDJ 1,195,81
sln-dmg-par 5 mmol/L MUREAV 125,243,84
mrc-smc 200 ppm PMRSDJ 1,481,81
skn-mus TDLo:40 mg/kg/20D-I:CAR CRNGDP 6,1483,85
skn-mus TDLo:1100 mg/kg/46W-I:ETA CNREA8 2,157,42

CONSENSUS REPORTS: EPA Genetic Toxicology Program.

SAFETY PROFILE: Questionable carcinogen with experimental carcinogenic and tumorigenic data. Mutation data reported. When heated to decomposition it emits acrid smoke and irritating fumes.

DQG400 CAS:18380-68-2 ***HR: 3***
DIMETHYLANTIMONY CHLORIDE
mf: C_2H_6ClSb mw: 187.27

CONSENSUS REPORTS: Antimony and its compounds are on the Community Right-To-Know List.

SAFETY PROFILE: A poison. Ignites at 40°C in air. When heated to decomposition it emits toxic fumes of Cl^-. See also ANTIMONY COMPOUNDS and CHLORIDES.

DQG600 CAS:593-57-7 ***HR: 3***
DIMETHYLARSINE
mf: C_2H_7As mw: 106.07

PROP: Colorless liquid. Bp: 36°, d: 1.213 @ 29°/4°, vap d: 3.65.

SYN: CACODYL HYDRIDE

CONSENSUS REPORTS: Arsenic and its compounds are on the Community Right-To-Know List.

SAFETY PROFILE: Arsenic compounds are generally poisons. Ignites spontaneously in air. It is more toxic than its oxidation products; reacts vigorously with oxidizing agents. To fight fire, exclude O_2, allow fire to burn, or apply water, foam, dry chemical, water spray, or CO_2. When heated to decomposition it emits toxic fumes of As. See also ARSINE and ARSENIC COMPOUNDS.

DQG700 CAS:13367-92-5 ***HR: 3***
DIMETHYL ARSINIC SULFIDE
mf: $C_2H_6As_2S_2$ mw: 244.04

SYNS: ARSINE SULFIDE, DIMETHYLDI- ◇ DIMETHYLDIARSINE SULFIDE

TOXICITY DATA with REFERENCE
ivn-mus LD50:10 mg/kg CSLNX* NX#03919

OSHA PEL: TWA 0.5 mg(As)/m^3

SAFETY PROFILE: Poison by intravenous route. When heated to decomposition it emits toxic fumes of SO_x and As.

DQH000 CAS:28842-05-9 ***HR: 3***
3,6′-DIMETHYLAZOBENZENE
mf: $C_{14}H_{14}N_2$ mw: 210.30

SYNS: 2:3′-AZOTOLUENE ◇ 2,3′-DIMETHYLAZOBENZENE

TOXICITY DATA with REFERENCE
orl-rat TDLo:26 g/kg/36W-C:NEO JPBAA7 58,275,46
mul-mus TDLo:400 mg/kg/I:ETA CNREA8 1,397,41

SAFETY PROFILE: Questionable carcinogen with experimental neoplastigenic and tumorigenic data. When heated to decomposition it emits toxic fumes of NO_x.

DQH200 CAS:35077-51-1 ***HR: 3***
N,N′-DIMETHYL-4,4′-AZODIACETANILIDE
mf: $C_{18}H_{20}N_4O_2$ mw: 324.42

SYNS: N′-ACETYL-N′-MONOMETHYL-4′-AMINO-N-ACETYL-N-MONOMETHYL-4-AMINOAZOBENZENE ◇ N,N′-(AZODI-4,1-PHENYLENE)BIS(N-METHYLACETAMIDE) ◇ 4,4′-BIS(N-ACETYL-N-METHYLAMINO)AZOBENZENE

TOXICITY DATA with REFERENCE
dns-rat:lvr 1 μmol/L CNREA8 46,1654,86
orl-rat TDLo:2200 mg/kg/13W-C:ETA CNREA8 34,2274,74
ipr-rat LD50:480 mg/kg CNREA8 34,2274,74

SAFETY PROFILE: Moderately toxic by intraperitoneal route. Questionable carcinogen with experimental tumorigenic data. Mutation data reported. When heated to decomposition it emits toxic fumes of NO_x.

DQH509 CAS:2446-84-6 ***HR: 3***
DIMETHYL AZODIFORMATE
mf: $C_4H_6N_2O_4$ mw: 146.11

$CH_3OCO{\cdot}N{=}NCO{\cdot}OCH_3$

PROP: Shock-sensitive, burns explosively.

SYNS: DIMETHYL AZOFORMATE ◇ DIMETHYL DIZENEDICARBOXYLATE

SAFETY PROFILE: A shock-sensitive explosive. It burns explosively when ignited. When heated to decomposition it emits toxic fumes of NO_x.

DQH550 ***HR: 2***
1,3-DIMETHYLBENZ(e)ACEPHENANTHRYLENE
mf: $C_{22}H_{18}$ mw: 282.40

SYN: 1,3-DIMETHYLBENZO(b)FLUORANTHENE

TOXICITY DATA with REFERENCE
mma-sat 63 nmol/plate CRNGDP 6,1023,85
skn-mus TDLo:452 μg/kg/20D-I:ETA CRNGDP 6,1023,85

SAFETY PROFILE: Questionable carcinogen with experimental tumorigenic data. Mutation data reported. When heated to decomposition it emits acrid smoke and irritating fumes.

DQH600 CAS:3518-05-6 ***HR: 3***
1,10-DIMETHYL-5,6-BENZACRIDINE
mf: $C_{19}H_{15}N$ mw: 257.35

SYN: 8,12-DIMETHYLBENZ(a)ACRIDINE

TOXICITY DATA with REFERENCE
scu-mus TDLo:200 mg/kg/4W-I:ETA ACRSAJ 4,315,56

SAFETY PROFILE: Questionable carcinogen with experimental tumorigenic data. When heated to decomposition it emits toxic fumes of NO_x.

DQH800 CAS:17401-48-8 ***HR: 3***
2,10-DIMETHYL-5,6-BENZACRIDINE
mf: $C_{19}H_{15}N$ mw: 257.35

SYN: 9,12-DIMETHYLBENZ(a)ACRIDINE

TOXICITY DATA with REFERENCE
skn-mus TDLo:540 mg/kg/45W-I:ETA ACRSAJ 4,315,56

SAFETY PROFILE: Questionable carcinogen with experimental tumorigenic data. When heated to decomposition it emits toxic fumes of NO_x.

DQI200 CAS:963-89-3 ***HR: 3***
7,9-DIMETHYLBENZ(c)ACRIDINE
mf: $C_{19}H_{15}N$ mw: 257.35

SYN: 3,10-DIMETHYL-7,8-BENZACRIDINE(FRENCH)

TOXICITY DATA with REFERENCE
mma-sat 10 μg/plate ENMUDM 6(Suppl 2),1,84
otr-rat:emb 40800 μg/L JJIND8 67,1303,81
otr-ham:emb 100 μg/L JJIND8 67,1303,81
skn-mus TDLo:180 mg/kg/15W-I:ETA ACRSAJ 4,315,56

CONSENSUS REPORTS: EPA Genetic Toxicology Program.

SAFETY PROFILE: Questionable carcinogen with experimental tumorigenic data. Mutation data reported. When heated to decompositions it emits toxic fumes of NO_x.

DQI400 CAS:32740-01-5 ***HR: 3***
7,11-DIMETHYLBENZ(c)ACRIDINE
mf: $C_{19}H_{15}N$ mw: 257.35

SYN: 1,10-DIMETHYL-7,8-BENZACRIDINE(FRENCH)

TOXICITY DATA with REFERENCE
mma-sat 1 nmol/plate GANNA2 70,749,79
skn-mus TDLo:360 mg/kg/30W-I:ETA ACRSAJ 4,315,56

SAFETY PROFILE: Questionable carcinogen with experimental tumorigenic data. Mutation data reported. When heated to decomposition it emits toxic fumes of NO_x.

DQI600 CAS:53-69-0 ***HR: 3***
5,7-DIMETHYL-1,2-BENZACRIDINE
mf: $C_{19}H_{15}N$ mw: 257.35

SYN: 8,10-DIMETHYL-BENZ(a)ACRIDINE

TOXICITY DATA with REFERENCE
mma-sat 500 nmol/L ENMUDM 3,11,81
dns-rat:lvr 50 μmol/L ENMUDM 3,11,81
irn-frg LDLo:11 mg/kg CNREA8 24,1969,64

SAFETY PROFILE: Poison by intrarenal route. Mutation data reported. When heated to decomposition it emits toxic fumes of NO_x.

DQI800 CAS:2381-40-0 ***HR: 3***
6,9-DIMETHYL-1,2-BENZACRIDINE
mf: $C_{19}H_{15}N$ mw: 257.35

SYNS: 7,10-DIMETHYLBENZ(c)ACRIDINE ◇ 2,10-DIMETHYL-7,8-BENZACRIDINE (FRENCH)

TOXICITY DATA with REFERENCE
mma-sat 50 μg/plate PNASA6 72,5135,75
otr-ham:emb 2 mg/L EJCAAH 17,179,81
otr-ham:kdy 80 μg/L BJCAAI 37,873,78
skn-mus TDLo:190 mg/kg/16W-I:ETA ACRSAJ 4,315,56

CONSENSUS REPORTS: EPA Genetic Toxicology Program.

SAFETY PROFILE: Questionable carcinogen with experimental tumorigenic data. Mutation data reported. When heated to decomposition it emits toxic fumes of NO_x.

DQJ000 CAS:611-74-5 ***HR: 2***
N,N-DIMETHYLBENZAMIDE
mf: $C_9H_{11}NO$ mw: 149.21

SYN: DIMETHYL BENZMIDE

TOXICITY DATA with REFERENCE
orl-mus LD50:960 mg/kg TXAPA9 19,20,71

CONSENSUS REPORTS: Reported in EPA TSCA Inventory.

SAFETY PROFILE: Moderately toxic by ingestion. When heated to decomposition it emits toxic fumes of NO_x.

DQJ200 CAS:57-97-6 ***HR: 3***
DIMETHYLBENZANTHRACENE
mf: $C_{20}H_{16}$ mw: 256.36

SYNS: DBA ◇ DIMETHYLBENZ(a)ANTHRACENE ◇ 7,12-DIMETHYLBENZANTHRACENE ◇ 7,12-DIMETHYLBENZ(a)ANTHRACENE ◇ 9,10-DIMETHYL-BENZANTHRACENE ◇ 9,10-DIMETHYLBENZ(a)ANTHRACENE ◇ 9,10-DIMETHYL-1,2-BENZANTHRACENE ◇ 9,10-DIMETHYL-1,2-BENZANTHRAZEN (GERMAN) ◇ DIMETHYLBENZANTHRENE ◇ 7,12-DIMETHYLBENZO(a)ANTHRACENE ◇ 1,4-DIMETHYL-2,3-BENZPHENANTHRENE ◇ DMBA ◇ 7,12-DMBA ◇ NCI-C03918 ◇ RCRA WASTE NUMBER U094

TOXICITY DATA with REFERENCE
skn-mus 64 μg MLD CALEDQ 4,333,78
dnd-hmn:emb 220 nmol/L MUREAV 89,95,81
dni-hmn:lvr 1 mmol/L VOONAW 28(11),53,82
otr-mus:emb 300 μg/L PMRSDJ 5,659,85
ipr-rat TDLo:6060 μg/kg (female 8-12D post):REP IRLCDZ 7,358,79
orl-rat TDLo:672 mg/kg (15D pre-5D post):TER DOESD6 54,410,81
orl-rat TDLo:37500 μg/kg (female 14-20D post):ETA,TER CRNGDP 3,413,78
orl-rat TDLo:15 mg/kg:CAR LIFSAK 7,259,68
skn-rat TDLo:60 mg/kg/30W-I:ETA JCUPBN 7,277,80
ipr-rat TDLo:24 mg/kg (20D preg):ETA,TER CCSUDL 3,413,78
scu-rat TDLo:500 μg/kg:CAR CRNGDP 6,769,85
ivn-rat TDLo:15 mg/kg (21D preg):ETA,TER JNCIAM 52,1365,74
ivn-rat TDLo:30 mg/kg:NEO KIDZAK 23(Suppl 1),34,71
ivn-rat TDLo:15 mg/kg (21D preg):ETA,TER NEOLA4 23,285,76
imp-rat TDLo:11 μg/kg:ETA,TER NISFAY 34,1853,82
ipc-rat TDLo:1250 μg/kg:ETA,TER GANNA2 62,55,71
orl-mus TDLo:640 mg/kg/75D-I:NEO,REP CNREA8 48,425,88
ipr-mus TDLo:112 mg/kg (female 14-21D post):NEO,TER IJCNAW 4,219,69
scu-mus TDLo:60 mg/kg (13-17D preg):NEO,TER LIFSAK 26,1955,80
ivn-mus TDLo:120 mg/kg (18-20D preg):NEO,REP VOONAW 20(8),65,74
ivn-mus TDLo:120 mg/kg (female 18-20D post):NEO,TER VOONAW 20(8),65,74
ivg-mus TDLo:744 mg/kg/31W-I:CAR,REP BJCAAI 20,184,66
ivg-mus TDLo:40 mg/kg (19D preg):CAR,TER VOONAW 22(6),44,76
ivn-rbt TDLO:20 mg/kg (25D preg):NEO,TER BEXBAN 85,369,78
orl-ham TDLo:25 mg/kg/(15D preg):NEO,TER PAACA3 18,1,77
ivn-ham TDLo:24 mg/kg/13W-I:ETA,REP INURAQ 15,42,77
orl-rat LD50:327 mg/kg GANNA2 68,237,77
ivn-rat LD50:54 mg/kg SCIEAS 147,1153,65
orl-mus LD50:340 mg/kg SCIEAS 147,1153,65
ipr-mus LD50:54 mg/kg PWPSA8 24,177,81
itr-mus LD50:22500 μg/kg PWPSA8 24,177,81
scu-gpg LDLo:20 mg/kg COREAF 252,1236,61

CONSENSUS REPORTS: Reported in EPA TSCA Inventory. EPA Genetic Toxicology Program.

SAFETY PROFILE: Suspected carcinogen with experimental carcinogenic, neoplastigenic, tumorigenic, and teratogenic data. A transplacental carcinogen. Poison by ingestion, intravenous, subcutaneous, intraperitoneal, and intratracheal routes. Other experimental reproductive effects. Human mutation data reported. A skin irritant. When heated to decomposition it emits acrid smoke and irritating fumes.

DQJ400 CAS:313-74-6 ***HR: 3***
1,12-DIMETHYLBENZ(a)ANTHRACENE
mf: $C_{20}H_{16}$ mw: 256.36

SYN: 1′,9-DIMETHYL-1,2-BENZANTHRACENE

TOXICITY DATA with REFERENCE
scu-mus TDLo:72 mg/kg/9W-I:ETA BAFEAG 49,312,62

SAFETY PROFILE: Questionable carcinogen with experimental tumorigenic data. When heated to decomposition it emits acrid smoke and irritating fumes. See also DIMETHYLBENZANTHRACENE.

DQJ600 CAS:18429-70-4 ***HR: 3***
4,5-DIMETHYLBENZ(a)ANTHRACENE
mf: $C_{20}H_{16}$ mw: 256.36

SYN: 3,4′-DIMETHYL-1,2-BENZANTHRACENE

TOXICITY DATA with REFERENCE
scu-rat TDLo:18 mg/kg:ETA PSEBAA 128,720,68

SAFETY PROFILE: Questionable carcinogen with experimental tumorigenic data. When heated to decomposition it emits acrid smoke and irritating fumes. See also DIMETHYLBENZANTHRACENE.

DQJ800 CAS:20627-28-5 ***HR: 3***
6,7-DIMETHYLBENZ(a)ANTHRACENE
mf: $C_{20}H_{16}$ mw: 256.36

SYN: 4,10-DIMETHYL-1,2-BENZANTHRACENE

TOXICITY DATA with REFERENCE
scu-mus TDLo:40 mg/kg:ETA JNCIAM 1,303,40

SAFETY PROFILE: Questionable carcinogen with experimental tumorigenic data. When heated to decomposition it emits acrid smoke and fumes. See also DIMETHYLBENZANTHRACENE.

DQK000 CAS:317-64-6 ***HR: 3***
6,8-DIMETHYLBENZ(a)ANTHRACENE
mf: $C_{20}H_{16}$ mw: 256.36

SYN: 6,8-DIMETHYL-1,2-BENZANTHRACENE

TOXICITY DATA with REFERENCE
ims-rat TDLo:50 mg/kg:NEO CNREA8 29,506,69
skn-mus TDLo:240 mg/kg/37W-I:ETA CNREA8 11,892,51

SAFETY PROFILE: Questionable carcinogen with experimental neoplastigenic and tumorigenic data. When heated to decomposition it emits acrid smoke and irritating fumes. See also DIMETHYLBENZANTHRACENE.

DQK200 CAS:568-81-0 ***HR: 3***
6,12-DIMETHYLBENZ(a)ANTHRACENE
mf: $C_{20}H_{16}$ mw: 256.36

SYN: 4,9-DIMETHYL-1,2-BENZANTHRACENE

TOXICITY DATA with REFERENCE
imp-mus TDLo:80 mg/kg:ETA JNCIAM 2,241,41

SAFETY PROFILE: Questionable carcinogen with experimental tumorigenic data. When heated to decomposition it emits acrid smoke and irritating fumes. See also DIMETHYLBENZANTHRACENE.

DQK400 CAS:35187-28-1 ***HR: 3***
7,11-DIMETHYLBENZ(a)ANTHRACENE
mf: $C_{20}H_{16}$ mw: 256.36

SYN: 8,10-DIMETHYL-1,2-BENZANTHRACENE

TOXICITY DATA with REFERENCE
scu-mus TDLo:40 mg/kg:ETA CNREA8 6,454,46

SAFETY PROFILE: Questionable carcinogen with experimental tumorigenic data. When heated to decomposition it emits acrid smoke and irritating fumes. See also DIMETHYLBENZANTHRACENE.

DQK600 CAS:58430-00-5 ***HR: 3***
5,6-DIMETHYL-1,2-BENZANTHRACENE
mf: $C_{20}H_{16}$ mw: 256.36

SYN: 8,9-DIMETHYLBENZ(a)ANTHRACENE

TOXICITY DATA with REFERENCE
skn-mus TDLo:500 mg/kg/21W-I:ETA PRLBA4 117,318,35

SAFETY PROFILE: Questionable carcinogen with experimental tumorigenic data. When heated to decomposition it emits acrid smoke and irritating fumes. See also DIMETHYLBENZANTHRACENE.

DQK800 CAS:20627-31-0 ***HR: 3***
5,9-DIMETHYL-1,2-BENZANTHRACENE
mf: $C_{20}H_{16}$ mw: 256.36

SYNS: 5:9-DIMETHYL-1:2-BENZANTHRACENE ◇ 8,12-DIMETHYLBENZ(a)ANTHRACENE

TOXICITY DATA with REFERENCE
scu-mus TDLo:40 mg/kg:ETA AJCAA7 33,499,38

SAFETY PROFILE: Questionable carcinogen with experimental tumorigenic data. When heated to decomposition it emits acrid smoke and irritating fumes. See also DIMETHYLBENZANTHRACENE.

DQK900 CAS:71964-72-2 ***HR: 2***
7,12-DIMETHYLBENZ(a)ANTHRACENE-3,4-DIOL
mf: $C_{20}H_{16}O_2$ mw: 288.36

TOXICITY DATA with REFERENCE
mma-sat 10 nmol/plate 46OJAN-,675,81
skn-mus TDLo:35 μg/kg:ETA CNREA8 40,3661,80

SAFETY PROFILE: Questionable carcinogen with experimental tumorigenic data. Mutation data reported. When heated to decomposition it emits acrid smoke and irritating fumes.

DQL000 CAS:604-81-9 ***HR: 3***
5,10-DIMETHYL-1,2-BENZANTHRACENE
mf: $C_{20}H_{16}$ mw: 256.36

SYN: 7,8-DIMETHYLBENZ(a)ANTHRACENE

TOXICITY DATA with REFERENCE
scu-mus TDLo:40 mg/kg:ETA AJCAA7 33,499,38

SAFETY PROFILE: Questionable carcinogen with experimental tumorigenic data. When heated to decomposition it emits acrid smoke and irritating fumes. See also DIMETHYLBENZANTHRACENE.

DQL200 CAS:58429-99-5 ***HR: 3***
6,7-DIMETHYL-1,2-BENZANTHRACENE
mf: $C_{20}H_{16}$ mw: 256.36

SYN: 9,10-DIMETHYLBENZ(a)ANTHRACENE

TOXICITY DATA with REFERENCE
mma-sat 2500 μg/plate BJCAAI 37,873,78
dns-rat:lvr 50 μmol/L ENMUDM 3,11,81
dnd-mus-skn 110 mg/L CNREA8 32,643,72
dni-mus-ipr 100 mg/kg MUREAV 46,305,77
otr-ham:kdy 80 μg/L BJCAAI 37,873,78
skn-mus TDLo:1250 mg/kg/52W-I:ETA PRLBA4 117,318,35

SAFETY PROFILE: Questionable carcinogen with experimental tumorigenic data. Mutation data reported. When heated to decomposition it emits acrid smoke and irritating fumes. See also DIMETHYLBENZANTHRACENE.

DQL400 CAS:32976-87-7 ***HR: 3***
7,12-DIMETHYLBENZ(a)ANTHRACENE, DEUTERATED
mf: $C_{20}D_{16}$ mw: 272.36

SYN: 7,12-DIMETHYLBENZ(a)ANTHRACENE-D16

TOXICITY DATA with REFERENCE
scu-mus TDLo:60 mg/kg/9W-I:ETA NATWAY 58,371,71

SAFETY PROFILE: Questionable carcinogen with experimental tumorigenic data. When heated to decomposition it emits acrid smoke and irritating fumes. See also DIMETHYLBENZANTHRACENE.

DQL600 CAS:39834-38-3 ***HR: D***
7,12-DIMETHYLBENZ(a)ANTHRACENE-5,6-OXIDE
mf: $C_{20}H_{16}O$ mw: 272.36

SYNS: 5,6-DIHYDRO-7,12-DIMETHYL-5,6-EPOXYBENZ(a)ANTHRACENE ◇ 7,12-DIMETHYL-5,6-EPOXY-5,6-DIHYDROBENZ(a)ANTHRACENE ◇ EDMBA ◇ 5,6-EPOXY-5,6-DIHYDRO-7,12-DIMETHYLBENZ(a) ANTHRACENE

TOXICITY DATA with REFERENCE
dnr-esc 100 μmol/L ZKKOBW 92,157,78
dnd-hmn:fbr 20 mmol/L MUREAV 43,117,77
otr-mus:oth 50 μg/L IJCNAW 13,304,74

CONSENSUS REPORTS: EPA Genetic Toxicology Program.

SAFETY PROFILE: Human mutation data reported. When heated to decomposition it emits acrid smoke and irritating fumes.

DQL800 CAS:63019-25-0 ***HR: 3***
9:10-DIMETHYL-1:2-BENZANTHRACENE-9:10-OXIDE
mf: $C_{20}H_{16}O$ mw: 272.36

SYN: 9:10-DIMETHYL-9-10-DIHYDRO-1,2-BENZANTHRACENE-9,10-OXIDE

TOXICITY DATA with REFERENCE
skn-mus TDLo:860 mg/kg/36W-I:ETA PRLBA4, 129,439,40

SAFETY PROFILE: Questionable carcinogen with experimental tumorigenic data. When heated to decomposition it emits acrid smoke and irritating fumes.

DQL899 CAS:68596-88-3 ***HR: 3***
3,5-DIMETHYLBENZENEDIAZONIUM-2-CARBOXYLATE
mf: $C_9H_8N_2O_2$ mw: 176.18

PROP: Explodes on melting.

SAFETY PROFILE: A powerful, heat-sensitive explosive. When heated to decomposition it emits toxic fumes of NO_x.

DQL959 ***HR: 3***
4,6-DIMETHYLBENZENEDIAZONIUM-2-CARBOXYLATE
mf: $C_9H_8N_2O_2$ mw: 176.18

SAFETY PROFILE: A sensitive, high explosive. When heated to decomposition it emits toxic fumes of NO_x. See also EXPLOSIVES, HIGH.

DQM000 CAS:612-82-8 ***HR: D***
3,3'-DIMETHYLBENZI DINEDIHYDROCHLORIDE
mf: $C_{14}H_{16}N_2$•2ClH mw: 285.24

TOXICITY DATA with REFERENCE
mma-sat 50 μg/plate CALEDQ 4,21,77
sln-dmg-orl 14 pph ENMUDM 7,325,85
sln-dmg-par 2750 ppm ENMUDM 7,325,85

CONSENSUS REPORTS: Reported in EPA TSCA Inventory.

SAFETY PROFILE: Mutation data reported. When heated to decomposition it emits very toxic fumes of HCl and NO_x.

DQM200 CAS:18463-86-0 ***HR: 3***
N,N-DIMETHYL-p-(4-BENZIMIDAZOLYAZO)ANILINE
mf: $C_{15}H_{15}N_5$ mw: 265.35

SYNS: 4-((p-(DIMETHYLAMINO)PHENYL)AZO)BENZIMIDAZOLE ◇ N,N-DIMETHYL-4(4'-BENZIMIDAZOLYLAZO)ANILINE

TOXICITY DATA with REFERENCE
orl-rat TDLo:1080 mg/kg/9W-C:ETA JMCMAR 11,1074,68

SAFETY PROFILE: Questionable carcinogen with experimental tumorigenic data. When heated to decomposition it emits toxic fumes of NO_x.

DQM400 CAS:4699-26-7 ***HR: 3***
6,12-DIMETHYLBENZO(1,2-b:5,4-b')BIS(1) BENZOTHIOPHENE
mf: $C_{20}H_{14}S_2$ mw: 318.46

TOXICITY DATA with REFERENCE
scu-mus TDLo:80 mg/kg:ETA JNCIAM 18,555,57

SAFETY PROFILE: Questionable carcinogen with experimental tumorigenic data. When heated to decomposition it emits toxic fumes of SO_x.

DQM600 CAS:22781-23-3 ***HR: 3***
2,2-DIMETHYL-1,3-BENZODIOXOL-4-OL METHYLCARBAMATE
mf: $C_{11}H_{13}NO_4$ mw: 223.25

SYNS: BENCARBATE ◇ BENDIOCARB ◇ BICAM ULV ◇ 2,2-DIMETHYL-1,3-BENZDIOXOL-4-YL-N-METHYLCARBAMATE ◇ 2,2-DIMETHYLBENZO-1,3-DIOXOL-4-YL METHYLCARBAMATE ◇ 2,2-DIMETHYL-4-(N-METHYLAMINOCARBOXYLATO)-1,3-BENZODIOXOLE ◇ 2,2-DIMETHYL-4-(N-METHYLCARBAMATO)-1,3-BENZODIOXOLE ◇ DYCARB ◇ FICAM ◇ GARVOX ◇ 2,3-ISOPROPYLIDENEDIOXYPHENYL METHYLCARBAMATE ◇ MC6897 ◇ METHYLCARBAMIC ACID-2,3-(ISOPROPYLIDENEDIOXY)PHENYL ESTER ◇ MULTAMAT ◇ NIOMIL ◇ ROTATE ◇ TATTOO ◇ TURCAM

TOXICITY DATA with REFERENCE
orl-rat LD50:40 mg/kg FMCHA2 -,C29,83
skn-rat LD50:1000 mg/kg GUCHAZ 6,31,73
orl-mam LD50:35 mg/kg MEIEDD 10,147,83

CONSENSUS REPORTS: EPA Genetic Toxicology Program.

SAFETY PROFILE: Poison by ingestion. Moderately toxic by skin contact. When heated to decomposition it emits toxic fumes of NO_x. See also CARBAMATES.

DQM800 CAS:37750-86-0 ***HR: 2***
6,12-DIMETHYLBENZO(1,2-b:4,5-b')DITHIONAPHTHENE
mf: $C_{20}H_{14}S_2$ mw: 318.46

TOXICITY DATA with REFERENCE
mnt-mus:fbr 1500 μg/L NULSAK 6,17,63
cyt-mus:fbr 1500 μg/L NULSAK 6,17,63
scu-mus TDLo:80 mg/kg:ETA JNCIAM 18,555,57

SAFETY PROFILE: Questionable carcinogen with experimental tumorigenic data. Mutation data reported. When heated to decomposition it emits toxic fumes of SO_x.

DQN000 CAS:16757-85-0 ***HR: 2***
1,2-DIMETHYLBENZO(a)PYRENE
mf: $C_{22}H_{16}$ mw: 280.38

TOXICITY DATA with REFERENCE
cyt-ckn:leu 1 pph/30M BBRCA9 93,954,80
scu-mus TDLo:72 mg/kg/13W-I:ETA IJCNAW 3,238,68

SAFETY PROFILE: Questionable carcinogen with experimental tumorigenic data. Mutation data reported. When heated to decomposition it emits acrid smoke and irritating fumes.

DQN200 CAS:16757-86-1 ***HR: 2***
1,3-DIMETHYLBENZO(a)PYRENE
mf: $C_{22}H_{16}$ mw: 280.38

TOXICITY DATA with REFERENCE
scu-mus TDLo:72 mg/kg/13W-I:ETA IJCNAW 3,238,68

SAFETY PROFILE: Questionable carcinogen with experimental tumorigenic data. When heated to decomposition it emits acrid smoke and irritating fumes.

DQN400 CAS:16757-88-3 ***HR: 2***
1,4-DIMETHYLBENZO(a)PYRENE
mf: $C_{22}H_{16}$ mw: 280.38

TOXICITY DATA with REFERENCE
scu-mus TDLo:72 mg/kg/13W-I:ETA IJCNAW 3,238,68

SAFETY PROFILE: Questionable carcinogen with experimental tumorigenic data. When heated to decomposition it emits acrid smoke and irritating fumes.

DQN600 CAS:16757-90-7 ***HR: 2***
1,6-DIMETHYLBENZO(a)PYRENE
mf: $C_{22}H_{16}$ mw: 280.38

TOXICITY DATA with REFERENCE
scu-mus TDLo:72 mg/kg/13W-I:ETA IJCNAW 3,238,68

SAFETY PROFILE: Questionable carcinogen with experimental tumorigenic data. When heated to decomposition it emits acrid smoke and irritating fumes.

DQN800 CAS:16757-87-2 ***HR: 2***
2,3-DIMETHYLBENZO(a)PYRENE
mf: $C_{22}H_{16}$ mw: 280.38

TOXICITY DATA with REFERENCE
scu-mus TDLo:72 mg/kg/13W-I:ETA IJCNAW 3,238,68

SAFETY PROFILE: Questionable carcinogen with experimental tumorigenic data. When heated to decomposition it emits acrid smoke and irritating fumes.

DQO000 CAS:16757-91-8 ***HR: 2***
3,6-DIMETHYLBENZO(a)PYRENE
mf: $C_{22}H_{16}$ mw: 280.38

TOXICITY DATA with REFERENCE
scu-mus TDLo:72 mg/kg/13W-I:ETA IJCNAW 3,238,68

SAFETY PROFILE: Questionable carcinogen with experimental tumorigenic data. When heated to decomposition it emits acrid smoke and irritating fumes.

DQO200 CAS:16757-84-9 ***HR: 2***
3,12-DIMETHYLBENZO(a)PYRENE
mf: $C_{22}H_{16}$ mw: 280.38

TOXICITY DATA with REFERENCE
scu-mus TDLo:72 mg/kg/13W-I:ETA IJCNAW 3,238,68

SAFETY PROFILE: Questionable carcinogen with experimental tumorigenic data. When heated to decomposition it emits acrid smoke and irritating fumes.

DQO400 CAS:16757-89-4 ***HR: 2***
4,5-DIMETHYLBENZO(a)PYRENE
mf: $C_{22}H_{16}$ mw: 280.38

TOXICITY DATA with REFERENCE
scu-mus TDLo:72 mg/kg/13W-I:ETA IJCNAW 3,238,68

SAFETY PROFILE: Questionable carcinogen with experimental tumorigenic data. When heated to decomposition it emits acrid smoke and irritating fumes.

DQO600 CAS:2818-89-5 ***HR: 2***
2,5-DIMETHYLBENZOSELENAZOLE
mf: C_9H_9NSe mw: 210.15

SYN: 2,5-DIMETHYLBENZSELENAZOL (CZECH)

TOXICITY DATA with REFERENCE
skn-rbt 500 mg/24H MLD 28ZPAK -,222,72
eye-rbt 500 mg/24H MLD 28ZPAK -,222,72
orl-rat LD50:1060 mg/kg 28ZPAK -,222,72

CONSENSUS REPORTS: Community Right-To-Know List.

OSHA PEL: TWA 0.2 mg(Se)/m^3
ACGIH TLV: TWA 0.2 mg(Se)/m^3
DFG MAK: 0.1 mg(Se)/m^3

SAFETY PROFILE: Moderately toxic by ingestion. A skin and eye irritant. When heated to decomposition it emits very toxic fumes of NO_x and Se. See also SELENIUM COMPOUNDS.

DQO650 CAS:2626-34-8 ***HR: 3***
5,6-DIMETHYL-2,1,3-BENZOSELENODIAZOLE
mf: $C_8H_8N_2Se$ mw: 211.14

SYN: 2,1,3-BENZOSELENADIAZOLE, 5,6-DIMETHYL-

TOXICITY DATA with REFERENCE
ivn-mus LD50:56 mg/kg CSLNX* NX#02249

OSHA PEL: TWA 0.2 mg(Se)/m^3
ACGIH TLV: TWA 0.2 mg(Se)/m^3

SAFETY PROFILE: Poison by intravenous route. When heated to decomposition it emits toxic fumes of NO_x and Se.

DQO800 CAS:95-26-1 ***HR: 2***
2,5-DIMETHYLBENZOTHIAZOLE
mf: C_9H_9NS mw: 163.25

SYN: 2,5-DIMETHYLBENZTHIAZOL (CZECH)

TOXICITY DATA with REFERENCE
skn-rbt 500 mg/24H MOD 28ZPAK -,202,72
eye-rbt 20 mg/24H MOD 28ZPAK -,202,72
orl-rat LD50:957 mg/kg 28ZPAK -,202,72

CONSENSUS REPORTS: Reported in EPA TSCA Inventory.

SAFETY PROFILE: Moderately toxic by ingestion. A skin and eye irritant. When heated to decomposition it emits very toxic fumes of SO_x and NO_x.

DQP000 CAS:18648-22-1 ***HR: 3***
N,N-DIMETHYLBENZO(b)THIOPHENE-3-ETHYLAMINE HYDROCHLORIDE
mf: $C_{12}H_{15}NS•ClH$ mw: 241.80

SYN: N,N-DIMETHYL-β-3-AMINOETHYLBENZOTHIOPHENE HYDROCHLORIDE

TOXICITY DATA with REFERENCE
ipr-mus LDLo:150 mg/kg JSOOAX 12,1612,69
ivn-mus LD50:100 mg/kg CSLNX* NX#01828

SAFETY PROFILE: Poison by intraperitoneal and intravenous routes. When heated to decomposition it emits very toxic fumes of Cl^-, SO_x, and NO_x.

DQP100 ***HR: 3***
1,2-DIMETHYL-1H-BENZOTRIAZOLIUM IODIDE
mf: $C_8H_{10}N_3•I$ mw: 275.11

SYN: 1,2-DIMETHYLBENZOTRIAZOLIUM JODID (GERMAN)

TOXICITY DATA with REFERENCE
scu-mus LDLo:436 mg/kg SDMU** -,-,36
ivn-mus LDLo:105 mg/kg SDMU** -,-,36
scu-frg LDLo:380 mg/kg SDMU** -,-,36

SAFETY PROFILE: Poison by intravenous route. Moderately toxic by subcutaneous route. When heated to decomposition it emits toxic fumes of I^- and NO_x. See also IODIDES.

DQP125 CAS:22713-35-5 ***HR: 3***
1,3-DIMETHYL-3H-BENZOTRIAZOLIUM IODIDE

SYN: 1,3-DIMETHYLBENZOTRIAZOLIUM JODID (GERMAN)

TOXICITY DATA with REFERENCE
scu-mus LDLo:316 mg/kg SDMU** -,-,36
ivn-mus LDLo:127 mg/kg SDMU** -,-,36
scu-frg LDLo:250 mg/kg SDMU** -,-,36

SAFETY PROFILE: Poison by intravenous and subcutaneous routes. When heated to decomposition it emits toxic fumes of I^- and NO_x. See also IODIDES.

DQP400 CAS:32362-68-8 ***HR: 3***
4,9-DIMETHYL-2,3-BENZTHIOPHANTHRENE
mf: $C_{18}H_{14}S$ mw: 262.38

TOXICITY DATA with REFERENCE
skn-mus TDLo:440 mg/kg/11W-I:ETA XPHPAW 149,477,51

SAFETY PROFILE: Questionable carcinogen with experimental tumorigenic data. When heated to decomposition it emits toxic fumes of SO_x.

DQP800 CAS:103-83-3 ***HR: 3***
N,N-DIMETHYLBENZYLAMINE
DOT: UN 2619
mf: $C_9H_{13}N$ mw: 135.23

SYNS: ARALDITE ACCELERATOR 062 ◇ BDMA ◇ BENZYLDIMETHYLAMINE ◇ BENZYL-N,N-DIMETHYLAMINE ◇ N-BENZYLDIMETHYLAMINE ◇ N,N-DIMETHYLBENZENEMETHANAMINE ◇ N-(PHENYLMETHYL)DIMETHYLAMINE ◇ SUMINE 2015

TOXICITY DATA with REFERENCE
skn-rbt 500 mg/4H SEV DCTODJ 8,43,85
eye-rbt 5 mg SEV DCTODJ 8,43,85
orl-rat LD50:265 mg/kg KorCJ# 22AUG74
ihl-rat LC50:2062 mg/m^3/4H DCTODJ 8,43,85
ihl-mus LC50:1800 mg/m^3/2H 85GMAT -,56,82
skn-rbt LD50:1660 mg/kg DCTODJ 8,43,85

CONSENSUS REPORTS: Reported in EPA TSCA Inventory.

DOT Classification: Corrosive Material; Label: Corrosive, Flammable Liquid.

SAFETY PROFILE: Poison by ingestion. Moderately toxic by inhalation and skin contact. Corrosive. A severe eye and skin irritant. Flammable when exposed to heat of flame. When heated to decomposition it emits toxic fumes of NO_x.

DQQ000 CAS:1875-92-9 ***HR: 3***
DIMETHYLBENZYLAMINE HYDROCHLORIDE
mf: $C_9H_{13}N{\bullet}ClH$ mw: 171.69

SYNS: DIMETHYLBENZYLAMMONIUM CHLORIDE ◇ USAF EL-78

TOXICITY DATA with REFERENCE
ipr-mus LD50:200 mg/kg NTIS** AD277-689

SAFETY PROFILE: Poison by intraperitoneal route. When heated to decomposition it emits toxic fumes of NO_x, NH_3, and HCl.

DQQ200 CAS:100-86-7 ***HR: 2***
DIMETHYL BENZYL CARBINOL
mf: $C_{10}H_{14}O$ mw: 150.24

PROP: White crystalline solid; floral odor. D: 0.972-0.977, flash p: 198°F. Sol in fixed oils, mineral oil, propylene glycol; insol in glycerin.

SYNS: BENZYL DIMETHYL CARBINOL ◇ α,α-DIMETHYLPHENETHYL ALCOHOL ◇ 1,1-DIMETHYL-2-PHENYLETHANOL ◇ DMBC ◇ FEMA No. 2393

TOXICITY DATA with REFERENCE
orl-rat LD50:1280 mg/kg FCTXAV 2,327,64
orl-gpg LD50:988 mg/kg FCTXAV 2,327,64

CONSENSUS REPORTS: Reported in EPA TSCA Inventory.

SAFETY PROFILE: Moderately toxic by ingestion. Combustible liquid. When heated to decomposition it emits acrid smoke and irritating fumes.

DQQ375 ***HR: 1***
DIMETHYL BENZYL CARBINYL ACETATE
mf: $C_{12}H_{16}O_2$ mw: 192.26

PROP: Colorless liquid to solid at room temp; floral, fruity odor. D: 0.995-1.002, refr index: 1.490-1.495, flash p: +212°F. Sol in fixed oils; sltly sol in propylene glycol; insol in water.

SYNS: α,α-DIMETHYLPHENETHYL ACETATE ◇ FEMA No. 2392

SAFETY PROFILE: Combustible liquid. When heated to decomposition it emits acrid smoke and irritating fumes.

DQQ380 ***HR: 2***
DIMETHYL BENZYL CARBINYL BUTYRATE
mf: $C_{14}H_{20}O_2$ mw: 220.31

PROP: Colorless liquid; prunelike odor. D: 0.960-0.981, refr index: 1.473-1.493 @ 25°, flash p: +151°F. Sol in alc, fixed oils; insol in water, propylene glycol.

SYNS: α,α-DIMETHYLPHENRTHYL BUTYRATE ◇ FEMA No. 2394

SAFETY PROFILE: Combustible liquid. Use in accordance with good manufacturing practice.

DQQ400 CAS:67785-77-7 ***HR: 1***
DIMETHYL BENZYL CARBINYL PROPIONATE
mf: $C_{13}H_{18}O_2$ mw: 206.31

SYNS: BENZYLISOPROPYL PROPIONATE ◇ α,α-DIMETHYLPHENETHYL ALCOHOL PROPIONATE

TOXICITY DATA with REFERENCE
skn-rbt 500 mg/24H MOD FCTXAV 18,669,80

CONSENSUS REPORTS: Reported in EPA TSCA Inventory.

SAFETY PROFILE: A skin irritant. When heated to decomposition it emits acrid smoke and irritating fumes.

DQQ600 CAS:6280-75-7 ***HR: 3***
N,N'-DI(α-METHYLBENZYL)ETHYLENEDIAMINE
mf: $C_{18}H_{22}N_2$ mw: 266.42

SYN: N,N'-BIS(α-METHYLBENZYL)ETHYLENEDIAMINE

TOXICITY DATA with REFERENCE
skn-rbt 10 mg/24H open MLD AMIHBC 10,61,54
eye-rbt 20 mg open SEV AMIHBC 10,61,54
orl-rat LD50:1290 mg/kg AMIHBC 10,61,54
scu-mus LD50:200 mg/kg ARZNAD 9,628,59
skn-rbt LD50:530 mg/kg AMIHBC 10,61,54

SAFETY PROFILE: Poison by subcutaneous route. Moderately toxic by ingestion and skin contact. A skin and severe eye irritant. When heated to decomposition it emits toxic fumes of NO_x.

DQQ700 CAS:42609-52-9 ***HR: 1***
1-(α,α-DIMETHYLBENZYL)-3-METHYL-3-PHENYLUREA
mf: $C_{17}H_{20}N_2O$ mw: 268.39

TOXICITY DATA with REFERENCE
orl-rat LD50:6130 mg/kg SHBOAO 32,488,78
scu-rat LD50:7810 mg/kg SHBOAO 32,488,78
orl-mus LD50:6830 mg/kg SHBOAO 32,488,78
scu-mus LD50:7600 mg/kg SHBOAO 32,488,78

SAFETY PROFILE: Mildly toxic by ingestion and subcutaneous routes. When heated to decomposition it emits toxic fumes of NO_x.

DQR200 CAS:506-63-8 ***HR: 3***
DIMETHYL BERYLLIUM
mf: C_2H_6Be mw: 39.09

PROP: White needles. Bp: sublimes @ 200°.

CONSENSUS REPORTS: Beryllium and its compounds are on the Community Right-To-Know List.

OSHA PEL: (Transitional: TWA 0.002 mg(Be)/m^3; CL 0.005; Pk 0.025/30M/8H) TWA 0.002 mg(Be)/m^3; STEL 0.005 mg(Be)/m^3/30M; CL 0.025 mg(Be)/m^3
ACGIH TLV: TWA 0.002 mg/m^3, Suspected Human Carcinogen.

SAFETY PROFILE: Confirmed human carcinogen. A poison. Flammable when exposed to heat or flame; can react with oxidizing materials. Explosive reaction on contact with water. Ignites on contact with moist air or carbon dioxide. Upon decomposition it emits highly toxic fumes of BeO. See also BERYLLIUM COMPOUNDS.

DQR289 ***HR: 3***
DIMETHYLBERYLLIUM-1,2-DIMETHOXYETHANE
mf: $C_2H_6Be \cdot C_4H_{10}O_2$ mw: 129.21

CONSENSUS REPORTS: Beryllium and its compounds are on the Community Right-To-Know List.

SAFETY PROFILE: Confirmed human carcinogen. Ignites spontaneously in air. Upon decomposition it emits highly toxic fumes of BeO. See also BERYLLIUM COMPOUNDS.

DQR600 CAS:657-24-9 ***HR: 3***
1,1-DIMETHYLBIGUANIDE
mf: $C_4H_{11}N_5$ mw: 129.20

SYNS: N,N-DIMETHYLBIGUANIDE ◇ N,N-DIMETHYLDIGUANIDE ◇ FLUMAMINE ◇ GLUCOPHAGE ◇ GLUCOPHAGE LA 6023 ◇ GLUEOPHOGE ◇ LA 6023 ◇ MELBIN ◇ METFORMIN ◇ NNDG

TOXICITY DATA with REFERENCE
cyt-ham:lng 2 g/L/48H GMCRDC 27,95,81
orl-rat TDLo:6 g/kg (1-12D preg):TER COREAF 253,321,61
scu-mus LD50:230 mg/kg AITDAQ 2,1,54
ipr-mus LD50:247 mg/kg JMCMAR 10,521,67
scu-gpg LD50:146 mg/kg MEXPAG 8,237,63
par-frg LD50:5000 mg/kg AITDAQ 2,1,54

SAFETY PROFILE: Poison by subcutaneous and intraperitoneal routes. Mildly toxic by parenteral route. Experimental teratogenic effects. Mutation data reported. When heated to decomposition it emits toxic fumes of NO_x.

DQR800 CAS:1115-70-4 ***HR: 3***
1,1-DIMETHYLBIGUANIDE HYDROCHLORIDE
mf: $C_4H_{11}N_5 \cdot ClH$ mw: 165.66

SYNS: DIABEFAGOS ◇ DIMETHYLBIGUANIDE HYDROCHLORIDE ◇ N,N-DIMETHYLIMIDODICARBONIMIDIC DIAMIDE MONOHYDROCHLORIDE ◇ GLUCOPHAGE ◇ HAURYMELLIN ◇ MEGUAN ◇ METFORMIN HYDROCHLORIDE ◇ METIGUANIDE

TOXICITY DATA with REFERENCE
skn-rbt 500 mg MLD FCTOD7 20,563,82
eye-rbt 100 mg MLD FCTOD7 20 573,82
eye-rbt 100 mg/4S rns MLD FCTOD7 20,573,82
orl-rat LD50:1 g/kg MEIEDD 10,849,83
scu-rat LD50:300 mg/kg MEIEDD 10,849,83
orl-mus LD50:1450 mg/kg NIIRDN 6,841,82
ipr-mus LD50:420 mg/kg NIIRDN 6,841,82

scu-mus LD50:620 mg/kg NIIRDN 6,841,82
ivn-mus LD50:180 mg/kg CSLNX* NX#04012

SAFETY PROFILE: Poison by intravenous and subcutaneous routes. Moderately toxic by ingestion. An eye and skin irritant. When heated to decomposition it emits toxic fumes of NO_x and Cl^-.

DQS000 CAS:91-97-4 ***HR: 3***
3,3'-DIMETHYL-4,4'-BIPHENYLENE DIISOCYANATE
mf: $C_{16}H_{12}N_2O_2$ mw: 264.30

SYNS: 4,4'-DIISOCYANATO-3,3'-DIMETHYL-1,1'-BIPHENYL ◇ ISOCYANIC ACID, 3,3'-DIMETHYL-4,4'-BIPHENYLENE ESTER

TOXICITY DATA with REFERENCE
ivn-mus LD50:56 mg/kg CSLNX* NX#02412

CONSENSUS REPORTS: Reported in EPA TSCA Inventory.

NIOSH REL: TWA (Diisocyanates) 0.005 ppm; CL 0.02 ppm/10M

SAFETY PROFILE: Poison by intravenous route. When heated to decomposition it emits toxic fumes of NO_x. See also CYANATES and ESTERS.

DQS600 CAS:63977-49-1 ***HR: 3***
DIMETHYL-BIS(β-CHLOROETHYL)AMMONIUM CHLORIDE
mf: $C_6H_{14}Cl_2N\cdot Cl$ mw: 206.56

SYNS: 2-CHLORO-N-(2-CHLOROETHYL)-N,N-DIMETHYLETHANAMINIUM CHLORIDE ◇ TL 379

TOXICITY DATA with REFERENCE
ihl-mus LCLo:1700 mg/m³/10M NDRC** NDCrc-132,Sept,42
ipr-mus LD50:67 mg/kg CANCAR 2,1055,49
scu-mus LDLo:100 mg/kg JPETAB 91,224,47

SAFETY PROFILE: Poison by intraperitoneal and subcutaneous routes. Moderately toxic by inhalation. When heated to decomposition it emits very toxic fumes of Cl^-, NO_x, and NH_3.

DQT000 ***HR: 2***
DIMETHYLBISMUTH CHLORIDE
mf: C_2H_6BiCl mw: 274.50

SAFETY PROFILE: Ignites spontaneously in air when warmed. When heated to decomposition it emits toxic fumes of Bi and Cl^-. See also BISMUTH COMPOUNDS and CHLORIDES.

DQT100 ***HR: 3***
N,N-DIMETHYL-(2-BROMOETHYL)HYDRAZINIUM BROMIDE
mf: $C_4H_{12}BrN_2\cdot Br$ mw: 248.00

SYN: N,N-DIMETHYL-(2-BROMAETHYL)-HYDRAZINIUMBROMID (GERMAN)

TOXICITY DATA with REFERENCE
orl-mus LD50:230 mg/kg ABMGAJ 27,663,71
ipr-mus LD50:315 mg/kg ABMGAJ 27,663,71
ivn-mus LD50:72 mg/kg ABMGAJ 27,663,71

SAFETY PROFILE: Poison by ingestion, intravenous, and intraperitoneal routes. When heated to decomposition it emits toxic fumes of Br^- and NO_x.

DQT150 CAS:513-81-5 ***HR: 3***
2,3-DIMETHYL-1,3-BUTADIENE
mf: C_6H_{10} mw: 82.14

$H_2C{=}C(CH_3)C(CH_3){=}CH_2$

SAFETY PROFILE: Forms explosive polymeric peroxides on exposure to air. Explodes on contact with thiazyl fluoride. Ignites on contact with oxygen + ozone above −78°C. When heated to decomposition it emits acrid smoke and fumes.

DQT200 CAS:75-83-2 ***HR: 3***
2,2-DIMETHYLBUTANE
DOT: UN 1208
mf: C_6H_{14} mw: 86.20

PROP: Liquid. Bp: 49.7°, mp: −98.2°, flash p: −54°F, fp: −101.9°, d: 0.649, autoign temp: 797°F, vap press: 400 mm @ 31.0°, vap d: 3.00, lel: 1.2%, uel: 7.0%.

SYN: NEOHEXANE (DOT)

CONSENSUS REPORTS: Reported in EPA TSCA Inventory.

OSHA PEL: TWA 500 ppm; STEL 1000 ppm
ACGIH TLV: TWA 500 ppm; STEL 1000 ppm
NIOSH REL: (Alkanes) TWA 350 mg/m³
DOT Classification: Flammable Liquid; Label: Flammable Liquid.

SAFETY PROFILE: Probably an irritant and narcotic in high concentration. A very dangerous fire and explosion hazard when exposed to heat or flame; can react vigorously with oxidizing materials. Keep away from heat or open flame. To fight fire, use foam, CO_2, dry chemical. When heated to decomposition it emits acrid smoke and irritating fumes.

DQT400 CAS:79-29-8 ***HR: 3***
2,3-DIMETHYLBUTANE
DOT: UN 2457
mf: C_6H_{14} mw: 86.20

PROP: Liquid. Mp: −135°, bp: 58.0°, flash p: −20°F, d: 0.662 @ 20°/4°, autoign temp: 788°F, vap press: 400 mm @ 39.0°, vap d: 3.0, lel: 1.2%, uel: 7.0%.

SYN: ISOHEXANE

CONSENSUS REPORTS: Reported in EPA TSCA Inventory.

OSHA PEL: TWA 500 ppm; STEL 1000 ppm
ACGIH TLV: TWA 500 ppm; STEL 1000 ppm
NIOSH REL: TWA (Alkanes) 350 mg/m^3
DOT Classification: Flammable Liquid; Label: Flammable Liquid.

SAFETY PROFILE: Probably an irritant and narcotic in high concentration. A very dangerous fire and explosion hazard when exposed to heat or flame; can react vigorously with oxidizing materials. Keep away from heat and open flame. To fight fire, use foam, CO_2, dry chemical. When heated to decomposition it emits acrid smoke and irritating fumes.

DQT800 ***HR: 2***
2,2-DIMETHYL-3-BUTANONE
mf: $C_6H_{12}O$ mw: 100.16

PROP: Flash p: 53.6°F.

SAFETY PROFILE: A very dangerous fire hazard when exposed to heat, flame or oxidizers. When heated to decomposition it emits acrid smoke and fumes. See also KETONES.

DQU000 CAS:75-97-8 ***HR: 2***
3,3-DIMETHYL-2-BUTANONE
mf: $C_6H_{12}O$ mw: 100.18

$CH_3CO{\bullet}C(CH_3)_3$

PROP: Flash p: 53.6°F.

SYNS: tert-BUTYL METHYL KETONE ◇ METHYL-tert-BUTYL KETONE ◇ PINACOLIN ◇ PINACOLINE ◇ PINACOLONE ◇ PINAKOLIN (GERMAN)

TOXICITY DATA with REFERENCE
scu-gpg LDLo:700 mg/kg MEIEDD 10,1072,83

CONSENSUS REPORTS: Reported in EPA TSCA Inventory.

SAFETY PROFILE: Moderately toxic by subcutaneous route. A dangerous fire hazard when exposed to heat, flame or oxidizers. When heated to decomposition it emits acrid smoke and irritating fumes. See also KETONES.

DQU200 CAS:3625-18-1 ***HR: 3***
5-(1,3-DIMETHYL-2-BUTENYL)-5-ETHYL BARBITURIC ACID
mf: $C_{12}H_{18}N_2O_3$ mw: 238.32

SYNS: 5-(1,3-DIMETHYL-2-BUTENYL)-5-ETHYL-2,4,6(1H,3H,5H) PYRIMIDINETRIONE ◇ MCNEIL 481

TOXICITY DATA with REFERENCE
orl-mus LD50:18 mg/kg 27ZQAG -,177,72
ipr-mus LD50:3500 μg/kg PMDCAY 8,61,71
orl-dog LD50:2 mg/kg 27ZQAG -,177,72
ivn-dog LDLo:250 μg/kg 27ZQAG -,177,72
ivn-rbt LDLo:250 μg/kg 27ZQAG -,177,72

SAFETY PROFILE: Poison by ingestion, intravenous, and intraperitoneal routes. When heated to decomposition it emits toxic fumes of NO_x. See also BARBITURATES.

DQU400 CAS:36798-79-5 ***HR: 2***
1-(2-(1,3-DIMETHYL-2-BUTENYLIDENE)HYDRAZINO)PHTHALAZINE
mf: $C_{14}H_{16}N_4$ mw: 240.34

SYNS: BUDRALAZINE ◇ BUTERAZINE ◇ DJ-1461 ◇ MESITYL OXIDE (1-PHTHALAZINYL)HYDRAZONE ◇ 4-METHYL-3-PENTEN-2-ONE (1-PHTHALAZINYL)HYDRAZONE ◇ 1(2H)-PHTHALAZINONE (1,3-DIMETHYL-2-BUTENYLIDENE)HYDRAZONE

TOXICITY DATA with REFERENCE
orl-rat TDLo:55 mg/kg (7-17D preg):REP OYYAA2 21,321,81
orl-rbt TDLo:65 mg/kg (female 6-18D post):TER OYYAA2 21,343,81
orl-rat LD50:620 mg/kg TXAPA9 44,431,78
ipr-rat LD50:3570 mg/kg TXAPA9 44,431,78
orl-mus LD50:1820 mg/kg TXAPA9 44,431,78
ipr-mus LD50:4020 mg/kg TXAPA9 44,431,78

SAFETY PROFILE: Moderately toxic by ingestion and intraperitoneal routes. Experimental teratogenic and reproductive effects. An antihypertensive agent. When heated to decomposition it emits toxic fumes of NO_x.

DQU600 CAS:108-09-8 ***HR: 3***
1,3-DIMETHYL BUTYLAMINE
DOT: UN 2379
mf: $C_6H_{15}N$ mw: 101.22

PROP: A liquid. Bp: 106-109°, flash p: 55°F (OC), d: 0.750 @ 20°/20°.

TOXICITY DATA with REFERENCE
orl-rat LDLo:600 mg/kg SCCUR* -,4,61
orl-mus LD50:470 mg/kg SCCUR* -,4,61
ihl-mus LCLo:1278 ppm/15M SCCUR* -,4,61
ivn-mus LD50:80 mg/kg CSLNX* NX#03558
skn-rbt LDLo:600 mg/kg SCCUR* -,4,61

CONSENSUS REPORTS: Reported in EPA TSCA Inventory.

DOT Classification: Flammable Liquid; Label: Flammable Liquid

SAFETY PROFILE: Poison by intravenous route. Moderately toxic by ingestion and skin contact. Mildly toxic by inhalation. A dangerous fire and explosion hazard when exposed to heat or flame; can react vigorously with oxidizing materials. To fight fire, use foam, CO_2, dry chemical. When heated to decomposition it emits toxic fumes of NO_x. See also AMINES.

DQU800 CAS:98-19-1 ***HR: 1***
1,3-DIMETHYL-5-tert-BUTYLBENZENE
mf: $C_{12}H_{18}$ mw: 162.30

SYN: 5-tert-BUTYL-m-XYLENE

TOXICITY DATA with REFERENCE
orl-rat LDLo:5000 mg/kg 28ZRAQ -,58,60

CONSENSUS REPORTS: Reported in EPA TSCA Inventory.

SAFETY PROFILE: Mildly toxic by ingestion. When heated to decomposition it emits acrid smoke and irritating fumes.

DQV000 CAS:17874-34-9 ***HR: 1***
4,6-DIMETHYL-8-tert-BUTYLCOUMARIN

SYN: 8-tert-BUTYL-4,6-DIMETHYLCOUMARIN

TOXICITY DATA with REFERENCE
skn-gpg 100% MLD FCTXAV 18,671,80

CONSENSUS REPORTS: Reported in EPA TSCA Inventory.

SAFETY PROFILE: A skin irritant. When heated to decomposition it emits toxic fumes of NO_x.

DQV200 CAS:6592-90-1 ***HR: 3***
5-(1,3-DIMETHYLBUTYL)-5-ETHYL BARBITURIC ACID, SODIUM SALT
mf: $C_{12}H_{19}N_2O_3•Na$ mw: 262.32

TOXICITY DATA with REFERENCE
ipr-rat LDLo:20 mg/kg JAPMA8 29,509,40
scu-rat LD50:22 mg/kg QJPPAL 12,657,39
par-rat LDLo:10 mg/kg JACSAT 58,585,36
ipr-mus LD50:24 mg/kg QJPPAL 12,657,39
scu-mus LD50:27 mg/kg QJPPAL 12,657,39
ivn-rbt LD50:20 mg/kg QJPPAL 12,657,39
ivn-gpg LD50:22 mg/kg QJPPAL 12,657,39

SAFETY PROFILE: Poison by intraperitoneal, subcutaneous, intravenous and parenteral routes. When heated to decomposition it emits toxic fumes of NO_x and Na_2O. See also BARBITURATES.

DQW600 CAS:1607-30-3 ***HR: 3***
DI-2-METHYLBUTYRYL PEROXIDE
mf: $C_{10}H_{18}O_4$ mw: 202.25

$(CH_3CH_2CH(CH_3)CO•O-)_2$

SAFETY PROFILE: The pure material is unstable and explodes at room temperature. When heated to decomposition it emits acrid smoke and fumes. See also PEROXIDES.

DQW800 CAS:506-82-1 ***HR: 3***
DIMETHYLCADMIUM
mf: C_2H_6Cd mw: 142.47

PROP: Oil, decomp by water, foul odor. D: 1.984; mp: −4.5°; bp: 106°.

CONSENSUS REPORTS: Cadmium and its compounds are on Community Right-To-Know List.

OSHA PEL: TWA 0.1 mg(Cd)/m^3; CL 0.6 mg(Cd)/m^3 (fume)
ACGIH TLV: TWA 0.05 mg(Cd)/m^3 (Proposed: TWA 0.01 mg(Cd)/m^3 (dust), Suspected Human Carcinogen; 0.002 mg(Cd)/m^3 (respirable dust), Suspected Human Carcinogen); BEI: 10 μg/g creatinine in urine; 10 μg/L in blood.
DFG BAT: Blood 1.5 μg/dL; Urine 15 μg/dL, Suspected Carcinogen.
NIOSH REL: (Cadmium) Reduce to lowest feasible level

SAFETY PROFILE: Confirmed human carcinogen. Contact with air produces the friction-sensitive explosive dimethyl cadmium peroxide. Explodes when heated above 150°C. Ignition may occur on contact with air if the surface area is large. See also CADMIUM COMPOUNDS.

DQX000 CAS:14433-76-2 ***HR: 3***
N,N-DIMETHYLCAPRAMIDE
mf: $C_{12}H_{25}NO$ mw: 199.38

SYN: N,N-DIMETHYLDECANAMIDE

TOXICITY DATA with REFERENCE
ipr-mus LD50:800 mg/kg AIHAAP 32,539,71
ivn-mus LD50:40 mg/kg AIHAAP 32,539,71
ipr-rbt LD50:2000 mg/kg AIHAAP 32,539,71
ivn-rbt LD50:29 mg/kg AIHAAP 32,539,71

CONSENSUS REPORTS: Reported in EPA TSCA Inventory.

SAFETY PROFILE: Poison by intravenous route. Moderately toxic by intraperitoneal route. When heated to decomposition it emits toxic fumes of NO_x.

DQX200 CAS:5830-30-8 ***HR: 3***
N,N-DIMETHYLCAPROAMIDE
mf: $C_8H_{17}NO$ mw: 143.26

SYN: N,N-DIMETHYLHEXANAMIDE

TOXICITY DATA with REFERENCE
ipr-mus LD50:425 mg/kg AIHAAP 32,539,71
ivn-mus LD50:90 mg/kg AIHAAP 32,539,71

orl-rbt LD50:1100 mg/kg AIHAAP 32,539,71
ipr-rbt LD50:520 mg/kg AIHAAP 32,539,71
ivn-rbt LD50:64 mg/kg AIHAAP 32,539,71

CONSENSUS REPORTS: Reported in EPA TSCA Inventory.

SAFETY PROFILE: Poison by intravenous route. Moderately toxic by ingestion and intraperitoneal routes. When heated to decomposition it emits toxic fumes of NO_x.

DQX300 CAS:3938-45-2 ***HR: 3***
N,N-DIMETHYLCARBAMIC ACID, m-ISOPROPYL PHENYL ESTER
mf: $C_{12}H_{17}NO_2$ mw: 207.30

SYN: CARBAMIC ACID, N,N-DIMETHYL-, m-ISOPROPYLPHENYL ESTER

TOXICITY DATA with REFERENCE
skn-rbt 10 mg/24H open MLD AIHAAP 23,95,62
orl-rat LD50:160 mg/kg AIHAAP 23,95,62
skn-rbt LD50:280 mg/kg AIHAAP 23,95,62

SAFETY PROFILE: Poison by ingestion and skin contact. When heated to decomposition it emits toxic fumes of NO_x.

DQX800 CAS:63884-71-9 ***HR: 3***
DIMETHYLCARBAMIC ESTER of HORDENINE HYDROCHLORIDE
mf: $C_{13}H_{20}N_2O_2$•ClH mw: 272.81

SYNS: AR-41 ◇ N,N-DIMETHYLCARBAMIC ACID-4-(β-DIMETHYLAMINOETHYL)PHENYL ESTER, HYDROCHLORIDE ◇ N,N-DIMETHYLCARBAMIC ACID-p-(β-DIMETHYLAMINOETHYL) PHENYL ESTER, HYDROCHLORIDE ◇ N,N-DIMETHYL-p- (N',N'-DIMETHYLCARBAMOYLOXY)PHENETHYLAMINE, HYDROCHLORIDE

TOXICITY DATA with REFERENCE
orl-mus LDLo:75 mg/kg JPETAB 43,413,31
ivn-mus LD80:15 mg/kg NTIS** PB158-508

SAFETY PROFILE: Poison by ingestion and intravenous routes. When heated to decomposition it emits very toxic fumes of NO_x and HCl. See also ESTERS and CARBAMATES.

DQY000 CAS:63884-67-3 ***HR: 3***
DIMETHYLCARBAMIC ESTER of 2-OXYBENZYLDIETHYLAMINE HYDROCHLORIDE
mf: $C_{14}H_{22}N_2O_2$•ClH mw: 286.84

SYN: DIMETHYLCARBAMIC ACID-(α-(DIETHYLAMINO))-o-TOLYL ESTER, HYDROCHLORIDE

TOXICITY DATA with REFERENCE
orl-mus LDLo:5 mg/kg JPETAB 43,413,31
ivn-mus LDLo:1500 μg/kg NTIS** PB158-508

SAFETY PROFILE: Poison by ingestion and intravenous routes. When heated to decomposition it emits very toxic fumes of NO_x and HCl. See also ESTERS and CARBAMATES.

DQY400 CAS:63680-76-2 ***HR: 3***
DIMETHYLCARBAMIC ESTER of 8-OXYMETHYLQUINOLINIUM METHYLSULFATE
mf: $C_{13}H_{15}N_2O_2$•CH_3O_4S mw: 342.40

SYNS: N,N-DIMETHYLCARBAMIC ACID-8-QUINOLINYL ESTER METHOSULFATE ◇ 8-HYDROXY-1-METHYLQUINOLINIUM METHYLSULFATE DIMETHYLCARBAMATE

TOXICITY DATA with REFERENCE
orl-mus LDLo:200 mg/kg JPETAB 43,413,31
ivn-mus LDLo:500 μg/kg JPETAB 43,413,31

SAFETY PROFILE: Poison by ingestion and intravenous routes. When heated to decomposition it emits very toxic fumes of NO_x and SO_x. See also ESTERS and CARBAMATES.

DQY909 CAS:51-60-5 ***HR: 3***
3-(DIMETHYLCARBAMOXY)PHENYL TRIMETHYLAMMONIUM METHYL SULFATE
mf: $C_{12}H_{19}N_2O_2$•CH_3O_4S mw: 334.43

SYNS: AR-32 ◇ N,N-DIMETHYLCARBAMIC ACID-3-DIMETHYLAMINOPHENYL ESTER METHOSULFATE ◇ DIMETHYLCARBAMIC ACID ESTER with (m-HYDROXYPHENYL)TRIMETHYLAMMONIUM METHYL SULFATE ◇ N,N-DIMETHYLCARBAMIC ACID-3-(TRIMETHYLAMMONIO)PHENYL ESTER METHYLSULFATE ◇ DIMETHYLCARBAMIC ESTER of 3-OXYPHENYLTRIMETHYLAMMONIUM METHYLSULFATE ◇ (3-(DIMETHYLCARBAMOYLOXY)PHENYL)TRIMETHYLAMMONIUM METHYLSULFATE ◇ EUSTIGMIN METHYLSULFATE ◇ HODOSTIN ◇ (m-HYDROXYPHENYL) TRIMETHYLAMMONIUM METHYL SULFATE DIMETHYLCARBAMATE ◇ (3-HYDROXYPHENYL)TRIMETHYLAMMONIUM METHYL SULFATE DIMETHYLCARBAMIC ESTER ◇ KIRKSTIGMINE METHYL SULFATE ◇ LEOSTIGMINE METHYL SULFATE ◇ NEOESERINE METHYL SULFATE ◇ NEOSTIGMETH ◇ NEOSTIGMINE METHOSULFATE ◇ NEOSTIGMINE METHYL SULFATE ◇ NEOSTIGMINE MONOMETHYLSULFATE ◇ NORMASTIGMIN ◇ PHILOSTIGMIN METHYL SULFATE ◇ POLSTIGMINE ◇ PROSERIN ◇ PROSERINE METHYL SULFATE ◇ PROSTIGMINE METHYLSULFATE ◇ SB-23 ◇ STIGMANOL METHYL SULFATE ◇ STIGMOSAN METHYL SULFATE ◇ SYNTHOSTIGMINE METHYL SULFATE ◇ TL-1394 ◇ VAGOSTIGMINE METHYL SULFATE

TOXICITY DATA with REFERENCE
scu-mus TDLo:1400 μg/kg (7-13D preg):REP KSRNAM 8,1986,74
scu-rat LD50:334 μg/kg TXAPA9 25,569,73
orl-mus LD50:7500 μg/kg JPETAB 99,16,50
ipr-mus LD50:230 μg/kg ATXKA8 29,39,72
scu-mus LD50:420 μg/kg JPETAB 99,16,50
ivn-mus LD50:160 μg/kg JPETAB 99,16,50
scu-dog LD50:50 μg/kg NTIS** PB158-508
scu-cat LDLo:50 μg/kg NTIS** PB158-508
ims-rbt LD50:310 μg/kg AIPTAK 81,276,50

SAFETY PROFILE: A deadly poison by ingestion, intravenous, subcutaneous, intraperitoneal, and intramuscular routes. Experimental reproductive effects. When heated to decomposition it emits very toxic fumes of SO_x, NH_3, and NO_x. See also CARBAMATES.

DQY950 CAS:79-44-7 ***HR: 3***
DIMETHYLCARBAMOYL CHLORIDE
DOT: UN 2262
mf: C_3H_6ClNO mw: 107.55

PROP: Liquid. Mp: −33°, bp: 165-167°, d: 1.678 @ 20°/4°, vap d: 3.73.

SYNS: CHLOROFORMIC ACID DIMETHYLAMIDE ◇ DDC ◇ (DIMETHYLAMINO)CARBONYL CHLORIDE ◇ DIMETHYLCARBAMIC ACID CHLORIDE ◇ DIMETHYLCARBAMIC CHLORIDE ◇ DIMETHYLCARBAMIDOYL CHLORIDE ◇ N,N-DIMETHYLCARBAMOYL CHLORIDE (DOT) ◇ DIMETHYLCARBAMYL CHLORIDE ◇ N,N-DIMETHYLCARBAMYL CHLORIDE ◇ DMCC ◇ RCRA WASTE NUMBER U097 ◇ TL 389

TOXICITY DATA with REFERENCE
mma-sat 300 ng/plate ENMUDM 6(Suppl 2),1,84
mmo-esc 100 μg/plate ENMUDM 6(Suppl 2),1,84
ihl-rat TCLo:1 ppm/6H/6W-I:CAR CALEDQ 33,175,86
skn-mus TDLo:17280 mg/kg/72W-I:CAR JACTDZ 6(4),479,87
ipr-mus TDLo:2560 mg/kg/64W-I:NEO JNCIAM 53,695,74
ihl-rat TC:1 ppm/6H/6W-I:ETA PAACA3 21,106,80
orl-rat LD50:1000 mg/kg ZAARAM 24(3),71,74
ihl-rat LC50:180 ppm/6H JEPTDQ 4(1),107,80
ihl-mus LCLo:1000 mg/m^3/10M NDRC** NDCrc-132,Oct,42
ipr-mus LD50:300 mg/kg ZAARAM 24,71,74

CONSENSUS REPORTS: NTP Fifth Annual Report on Carcinogens. IARC Cancer Review: Group 2A IMEMDT 7,199,87; Animal Sufficient Evidence IMEMDT 12,77,76; Human Inadequate Evidence IMEMDT 12,77,76. EPA Genetic Toxicology Program. Community Right-To-Know List. Reported in EPA TSCA Inventory.

ACGIH TLV: Suspected Human Carcinogen
DFG MAK: Animal Carcinogen, Suspected Human Carcinogen.
DOT Classification: Corrosive Material; Label: Corrosive.

SAFETY PROFILE: Confirmed carcinogen with experimental carcinogenic, neoplastigenic, and tumorigenic data. Poison by intraperitoneal route. Moderately toxic by inhalation and ingestion. Human mutation data reported. Can cause skin and papillary tumors by skin contact, and squamous cell carcinoma by inhalation. Will react with water or steam to produce toxic and corrosive fumes. A powerful lachrymator. When heated to decomposition it emits very toxic fumes of Cl^- and NO_x. See also CHLORIDES.

DQZ000 CAS:644-64-4 ***HR: 3***
1-DIMETHYLCARBAMOYL-5-METHYL-3-PYRAZOLYL DIMETHYLCARBAMATE
mf: $C_{10}H_{16}N_4O_3$ mw: 240.30

SYNS: DIMETHYLCARBAMICACID-1-((DIMETHYLAMINO)CARBONYL)-5-METHYL-1H-PYRAZOL-3-YL ESTER ◇ DIMETHYLCARBAMIC ACID ESTER with 3-HYDROXY-N,N,5-TRIMETHYLPYRAZOLE-1-CARBOXAMIDE ◇ DIMETHYLCARBAMIC ACID-5-METHYL-1H-PYRAZOL-3-YL ESTER ◇ 2-DIMETHYLCARBAMOYL-3-METHYL-PYRAZOLYL-(5)-N,N-DIMETHYLCARBAMAT ◇ 2-DIMETHYLCARBAMOYL-3-METHYL-5-PYRAZOLYLDIMETHYLCARBAMATE ◇ 2-(N,N-DIMETHYLCARBAMYL)-3-METHYLPYRAZOLYL-5 N,N-DIMETHYLCARBAMATE ◇ DIMETHYL 2-CARBAMYL-3-METHYLPYRAZOLYLDIMETHYLCARBAMATE (GERMAN) ◇ DIMETILAN ◇ DIMETILANE ◇ ENT 25595-X ◇ ENT 25,922 ◇ GEIGY 22870 ◇ 3-HYDROXY-N,N,5-TRIMETHYLPYRAZOLE-1-CARBOXAMIDE DIMETHYLCARBAMATE (ESTER) ◇ 5-METHYL-1H-PYRAZOL-3-YL DIMETHYLCARBAMATE ◇ SNIP ◇ SNIP FLY ◇ SNIP FLY BANDS

TOXICITY DATA with REFERENCE
orl-rat LD50:25 mg/kg PHJOAV 185,361,60
skn-rat LD50:600 mg/kg PHJOAV 185,361,60
orl-mus LD50:60 mg/kg GUCHAZ 6,222,73
ipr-mus LD50:12 mg/kg BECTA6 2,163,67
skn-rbt LD50:2000 mg/kg PCOC** -,393,66
orl-gpg LD50:63 mg/kg 85DPAN -,-,71/76

CONSENSUS REPORTS: EPA Extremely Hazardous Substances List.

SAFETY PROFILE: Poison by ingestion and intraperitoneal routes. Moderately toxic by skin contact. An insecticide. When heated to decomposition it emits toxic fumes of NO_x. See also CARBAMATES.

DRB200 CAS:611-92-7 ***HR: 2***
N,N'-DIMETHYL CARBANILIDE
mf: $C_{15}H_{16}N_2O$ mw: 240.33

TOXICITY DATA with REFERENCE
orl-rat LDLo:500 mg/kg JPETAB 90,260,47

CONSENSUS REPORTS: Reported in EPA TSCA Inventory.

SAFETY PROFILE: Moderately toxic by ingestion. When heated to decomposition it emits toxic fumes of NO_x.

DRB400 CAS:5826-73-3 ***HR: 2***
DIMETHYL CARBATE
mf: $C_{11}H_{14}O_4$ mw: 210.25

SYNS: cis-BICYCLO(2.2.1)HEPT-5-ENE-2,3-DICARBOXYLICACID, DIMETHYL ESTER ◇ (endo,endo)-BICYCLO(2.2.1)HEPT-5-ENE-2,3-DICARBOXYLIC ACID DIMETHYL ESTER ◇ cis-BICYCLO(2,2,1-HEPTENE-2,3-DICARBOXYLIC ACID) METHYL ESTER ◇ COMPOUND-3916 ◇ DIMALONE ◇ DIMELONE ◇ DIMETHYL cis-BICYCLO(2,2,1)-5-HEPTENE-2,3-DICARBOXYLATE ◇ cis-3,6-ENDOMETHYLENE-Δ^4-TETRAHYDROPHTHALIC ACID DIMETHYL ESTER ◇ NISY ◇ cis-5-NORBORNENE-2,3-DICARBOXYLIC ACID DIMETHYL ESTER

TOXICITY DATA with REFERENCE
eye-rbt 500 mg AJOPAA 29,1363,46
orl-rat LD50:1000 mg/kg GUCHAZ 6,211,73
orl-mus LD50:1400 mg/kg JPETAB 93,26,48

SAFETY PROFILE: Moderately toxic by ingestion. An eye irritant. An insect repellent. Combustible. When heated to decomposition it emits acrid smoke and irritating fumes.

DRB600 CAS:2088-72-4 ***HR: 3***
O,O-DIMETHYL-S-CARBOETHOXYMETHYL THIOPHOSPHATE
mf: $C_6H_{13}O_5PS$ mw: 228.22

SYNS: ((DIMETHOXYPHOSPHINYL)THIO)ACETIC ACID ETHYL ESTER ◇ O,O-DIMETHYL-S-(CARBETHOXY)METHYL PHOSPHOROTHIOLATE ◇ O,O-DIMETHYL ESTER PHOSPHOROTHIOIC ACID-S-ESTER with ETHYL MERCAPTOACETATE ◇ METHYLACETAPHOS ◇ METHYL ACETOPHOS ◇ METHYL ACETOXON

TOXICITY DATA with REFERENCE
orl-rat LD50:385 mg/kg HYSAAV 31,18,66
skn-rat LD50:220 mg/kg GISAAA 33(8),107,68
unr-rat LD50:1000 mg/kg 30ZDA9 -,351,71
orl-mus LD50:314 mg/kg GISAAA 33(8),107,68

SAFETY PROFILE: Poison by ingestion and skin contact. Moderately toxic by unspecified route. When heated to decomposition it emits very toxic fumes of PO_x and SO_x. See also ESTERS.

DRB800 CAS:64038-38-6 ***HR: 3***
7,11-DIMETHYL-10-CHLOROBENZ(c)ACRIDINE
mf: $C_{19}H_{14}ClN$ mw: 291.79

SYNS: 2-CHLORO-1,10-DIMETHYL-7,8-BENZACRIDINE(FRENCH) ◇ 1,10-DIMETHYL-2-CHLORO-7,8-BENZACRIDINE (FRENCH)

TOXICITY DATA with REFERENCE
skn-mus TDLo:250 mg/kg/21W-I:ETA ACRSAJ 4,315,56

SAFETY PROFILE: Questionable carcinogen with experimental tumorigenic data. When heated to decomposition it emits very toxic fumes of NO_x and Cl^-.

DRC000 CAS:4584-46-7 ***HR: 3***
DIMETHYL(2-CHLOROETHYL)AMINE HYDROCHLORIDE
mf: $C_4H_{10}ClN{\cdot}ClH$ mw: 144.06

SYNS: 2-CHLORO-N,N-DIMETHYLETHYLAMINEHYDROCHLORIDE ◇ DIMETHYL-β-CHLOROETHYLAMINE HYDROCHLORIDE

TOXICITY DATA with REFERENCE
mmo-sat 1 mg/L ENMUDM 3,33,81
mmo-esc 1 μmol/L JPPMAB 31,67P,79
sln-dmg-orl 1700 mmol/L MUREAV 95,237,82
dns-rat:lvr 5 μmol/L ENMUDM 3,33,81
ipr-mus TDLo:720 mg/kg/8W-I:NEO CNREA8 39,391,79
ipr-mus LD50:280 mg/kg CANCAR 2,1055,49
scu-mus LD50:250 mg/kg JPETAB 97,25,49

CONSENSUS REPORTS: Reported in EPA TSCA Inventory.

SAFETY PROFILE: Poison by intraperitoneal and subcutaneous routes. Questionable carcinogen with experimental neoplastigenic data. Mutation data reported. When heated to decomposition it emits very toxic fumes of Cl^- and NO_x.

DRC400 CAS:13508-53-7 ***HR: 2***
DIMETHYLCHLOROMETHYLETHOXYSILANE
mf: $C_5H_{13}ClOSi$ mw: 152.72

SYN: DIMETHYL-CHLORMETHYL-ETHOXYSILAN(CZECH)

TOXICITY DATA with REFERENCE
skn-rbt 500 mg/24H MOD 28ZPAK -,217,72
eye-rbt 500 mg/24H MOD 28ZPAK -,217,72
orl-rat LD50:1550 mg/kg 28ZPAK -,217,72
ihl-rat LCLo:2560 ppm/4H 28ZPAK -,217,72

CONSENSUS REPORTS: Reported in EPA TSCA Inventory.

SAFETY PROFILE: Moderately toxic by ingestion and inhalation. A skin and eye irritant. When heated to decomposition it emits toxic fumes of Cl^-. See also SILANE.

DRC500 ***HR: 3***
DIMETHYL-2-CHLORO-4-NITROPHENYLTHIONOPHOSPHATE
mf: $C_8H_9ClNO_5PS$ mw: 297.65

$(CH_3O)_2P(S)OC_6H_3(Cl)NO_2$

SAFETY PROFILE: Ignites during thermal decomposition when heated to 270°C. When heated to decomposition it emits toxic fumes of Cl^-, NO_x, PO_x and SO_x. See also NITRO COMPOUNDS of AROMATIC HYDROCARBONS.

DRC600 CAS:10389-72-7 ***HR: 3***
α,α-DIMETHYL-o-CHLOROPHENETHYLAMINE HYDROCHLORIDE
mf: $C_{10}H_{14}ClN{\cdot}ClH$ mw: 220.16

SYNS: 2-CHLORO-α,α-DIMETHYLBENZENEETHANIAMINE HYDROCHLORIDE ◇ O-CHLORO-α,α-DIMETHYLPHENETHYLAMINE HYDROCHLORIDE ◇ CLORTERMINE HYDROCHLORIDE ◇ S 77 ◇ SU-10568 ◇ VORANIL

TOXICITY DATA with REFERENCE
orl-rat LD50:332 mg/kg TXAPA9 18,185,71
ipr-rat LD50:92 mg/kg APTOA6 17,121,60

SAFETY PROFILE: Poison by ingestion and intraperi-

toneal routes. When heated to decomposition it emits very toxic fumes of Cl^- and NO_x.

DRC800 CAS:3789-77-3 ***HR: 3***
N,N-DIMETHYL-p-((m-CHLOROPHENYL)AZO)ANILINE
mf: $C_{14}H_{14}ClN_3$ mw: 259.76

SYN: 3′-CHLORO-4-DIMETHYLAMINOAZOBENZENE

TOXICITY DATA with REFERENCE
orl-rat TDLo:4900 mg/kg/17W-C:NEO JEMEAV 87,139,48

SAFETY PROFILE: Questionable carcinogen with experimental neoplastigenic data. When heated to decomposition it emits very toxic fumes of Cl^- and NO_x.

DRD000 CAS:3010-47-7 ***HR: 3***
N,N-DIMETHYL-p-((o-CHLOROPHENYL)AZO)ANILINE
mf: $C_{14}H_{14}ClN_3$ mw: 259.76

SYN: 2′-CHLORO-4-DIMETHYLAMINOAZOBENZENE

TOXICITY DATA with REFERENCE
orl-rat TDLo:4900 mg/kg/17W-C:NEO JEMEAV 87,139,48

SAFETY PROFILE: Questionable carcinogen with experimental neoplastigenic data. When heated to decomposition it emits very toxic fumes of Cl^- and NO_x.

DRD800 CAS:63041-62-3 ***HR: 3***
2,3-DIMETHYLCHOLANTHRENE
mf: $C_{22}H_{18}$ mw: 282.40

SYN: 16:20-DIMETHYLCHOLANTHRENE

TOXICITY DATA with REFERENCE
scu-mus TDLo:400 mg/kg:ETA AJCAA7 28,334,36

SAFETY PROFILE: Questionable carcinogen with experimental tumorigenic data. When heated to decomposition it emits acrid smoke and irritating fumes.

DRD850 CAS:85923-37-1 ***HR: 2***
3,6-DIMETHYLCHOLANTHRENE
mf: $C_{22}H_{18}$ mw: 282.40

SYN: BENZ(j)ACEANTHRYLENE,1,2-DIHYDRO-3,6-DIMETHYL-(9CI)

TOXICITY DATA with REFERENCE
mma-ham:lng 50 μg/L CALEDQ 28,223,85
msc-ham:lng 100 μg/L PAACA3 24,94,83
skn-mus TDLo:706 ng/kg:ETA CALEDQ 28,223,85

SAFETY PROFILE: Questionable carcinogen with experimental tumorigenic data. Mutation data reported. When heated to decomposition it emits acrid smoke and irritating fumes.

DRE000 CAS:63041-61-2 ***HR: 3***
15,20-DIMETHYLCHOLANTHRENE
mf: $C_{22}H_{18}$ mw: 282.40

SYN: 1,3-DIMETHYLCHOLANTHRENE

TOXICITY DATA with REFERENCE
scu-mus TDLo:200 μg/kg:ETA JNCIAM 2,99,41

SAFETY PROFILE: Questionable carcinogen with experimental tumorigenic data. When heated to decomposition it emits acrid smoke and irritating fumes.

DRE200 CAS:15914-23-5 ***HR: 3***
1,2-DIMETHYLCHRYSENE
mf: $C_{20}H_{16}$ mw: 256.36

TOXICITY DATA with REFERENCE
skn-mus TDLo:800 mg/kg/33W-I:ETA PRLBA4 129,439,40

SAFETY PROFILE: Questionable carcinogen with experimental tumorigenic. When heated to decomposition it emits acrid smoke and irritating fumes. See also CHRYSENE.

DRE400 CAS:52171-92-3 ***HR: 3***
1,11-DIMETHYLCHRYSENE
mf: $C_{20}H_{16}$ mw: 256.36

SYN: 5,7-DIMETHYLCHRYSENE

TOXICITY DATA with REFERENCE
mma-sat 20 μg/plate CNREA8 36,4525,76
skn-mus TDLo:120 mg/kg/50W-I:NEO CNREA8 34,1315,74
ipr-mus LD50:300 mg/kg AACRAT 38,265,59
scu-mus LD50:900 mg/kg AACRAT 38,265,59

CONSENSUS REPORTS: EPA Genetic Toxicology Program.

SAFETY PROFILE: Poison by intraperitoneal and intravenous routes. Mutation data reported. Questionable carcinogen with experimental neoplastigenic data. When heated to decomposition it emits acrid smoke and irritating fumes. See also CHRYSENE.

DRE600 CAS:63019-23-8 ***HR: 3***
4,5-DIMETHYLCHRYSENE
mf: $C_{20}H_{16}$ mw: 256.36

TOXICITY DATA with REFERENCE
scu-mus TDLo:80 mg/kg:ETA CNREA8 3,606,43

SAFETY PROFILE: Questionable carcinogen with experimental tumorigenic data. When heated to decomposition it emits acrid smoke and irritating fumes. See also CHRYSENE.

DRE800 CAS:3697-27-6 ***HR: 3***
5,6-DIMETHYLCHRYSENE
mf: $C_{20}H_{16}$ mw: 256.36

TOXICITY DATA with REFERENCE
scu-mus TDLo:80 mg/kg:ETA CNREA8 3,606,43

SAFETY PROFILE: Questionable carcinogen with experimental tumorigenic data. When heated to decomposition it emits acrid smoke and irritating fumes. See also CHRYSENE.

DRF000 CAS:14207-78-4 ***HR: 3***
5,11-DIMETHYLCHRYSENE
mf: $C_{20}H_{16}$ mw: 256.36

TOXICITY DATA with REFERENCE
skn-mus TDLo:400 μg/kg/20D-I:ETA CALEDQ 8,65,79

SAFETY PROFILE: Questionable carcinogen with experimental tumorigenic data. When heated to decomposition it emits acrid smoke and irritating fumes. See also CHRYSENE.

DRF200 CAS:617-54-9 ***HR: 1***
DIMETHYL CITRACONATE
mf: $C_7H_{10}O_4$ mw: 158.17

SYNS: DIMETHYL METHYL MALEATE ◇ cis-2-METHYL-2-BUTENEDIOIC ACID, DIMETHYL ESTER ◇ METHYLMALEIC ACID, DIMETHYL ESTER

TOXICITY DATA with REFERENCE
skn-rbt 500 mg/24H MLD FCTXAV 14,659,76

CONSENSUS REPORTS: Reported in EPA TSCA Inventory.

SAFETY PROFILE: A skin irritant. When heated to decomposition it emits acrid smoke and irritating fumes. See also ESTERS.

DRF400 CAS:675-09-2 ***HR: 2***
4,6-DIMETHYLCOUMARIN
mf: $C_8H_{12}O_2$ mw: 140.20

TOXICITY DATA with REFERENCE
ipr-mus LD50:750 mg/kg APTOA6 2,109,46

CONSENSUS REPORTS: Reported in EPA TSCA Inventory.

SAFETY PROFILE: Moderately toxic by intraperitoneal route. When heated to decomposition it emits acrid smoke and irritating fumes. See also COUMARIN.

DRF600 CAS:1467-79-4 ***HR: 3***
DIMETHYLCYANAMIDE
mf: $C_3H_6N_2$ mw: 70.11

PROP: Colorless, mobile liquid. Mp: −41.0°, bp: 160°, flash p: 160°F (TCC), d: 0.8767 @ 30°, vap press: 40 mm @ 80°, vap d: 2.55.

TOXICITY DATA with REFERENCE
orl-rat LD50:146 mg/kg GTPZAB 19(11),23,75
ihl-rat LC50:2500 mg/m^3 GTPZAB 19(11),23,75
orl-mus LD50:73 mg/kg GTPZAB 19(11),23,75
ihl-mus LC50:2800 mg/m^3 GTPZAB 19(11),23,75
skn-mus LD50:125 mg/kg GTPZAB 19(11),23,75
ipr-mus LD50:40 mg/kg NTIS** AD691-490
orl-gpg LD50:146 mg/kg GTPZAB 19,23,75

CONSENSUS REPORTS: Reported in EPA TSCA Inventory.

SAFETY PROFILE: Poison by ingestion, skin contact, and intraperitoneal routes. Moderately toxic by inhalation. Flammable when exposed to heat, flame, or oxidizers. Can react with oxidizing materials. To fight fire, use foam, CO_2, or dry chemical. When heated to decomposition or in reaction with water or steam it produces toxic fumes of NO_x and CN^- and flammable vapors. See also CYANIDES.

DRF709 CAS:98-94-2 ***HR: 3***
N,N-DIMETHYLCYCLOHEXANAMINE
DOT: UN 2264
mf: $C_8H_{17}N$ mw: 127.26

SYNS: CYCLOHEXYLDIMETHYLAMINE ◇ N-CYCLOHEXYLDIMETHYLAMINE ◇ (DIMETHYLAMINO)CYCLOHEXANE ◇ N,N-DIMETHYLAMINOCYCLOHEXANE ◇ DIMETHYLCYCLOHEXYLAMINE ◇ N,N-DIMETHYLCYCLOHEXYLAMINE (DOT) ◇ POLYCAT 8

TOXICITY DATA with REFERENCE
orl-rat LD50:348 mg/kg ZHYGAM 20,393,74
ihl-rat LC50:1889 mg/m^3/2H GTPZAB 28(5),54,84
orl-mus LD50:320 mg/kg GTPZAB 28(5),54,84
ihl-mus LC50:1100 mg/m^3/2H GTPZAB 28(5),54,84
orl-rbt LD50:620 mg/kg ZHYGAM 20,393,74
orl-gpg LD50:520 mg/kg ZHYGAM 20,393,74

CONSENSUS REPORTS: Reported in EPA TSCA Inventory.

DOT Classification: Corrosive Material; Label: Corrosive, Flammable Liquid.

SAFETY PROFILE: Poison by ingestion. Moderately toxic by inhalation. When heated to decomposition it emits toxic fumes of NO_x.

DRF800 CAS:583-57-3 ***HR: 3***
cis-1,2-DIMETHYLCYCLOHEXANE
DOT: UN 2263
mf: C_8H_{16} mw: 112.22

PROP: Flash p: 61.8°F.

SYNS: o-DIMETHYLCYCLOHEXANE ◇ 1,2-DIMETHYLCYCLOHEXANE (DOT)

DOT Classification: Flammable Liquid; Label: Flammable Liquid.

SAFETY PROFILE: A very dangerous fire hazard when exposed to heat, flame, or oxidizers. When heated to decomposition it emits acrid smoke and fumes.

DRG000 CAS:591-21-9 ***HR: 3***
1,3-DIMETHYLCYCLOHEXANE
DOT: UN 2263
mf: C_8H_{16} mw: 112.22

PROP: Flash p: 42.8°F.

SYN: m-DIMETHYLCYCLOHEXANE

DOT Classification: Flammable Liquid; Label: Flammable Liquid.

SAFETY PROFILE: A very dangerous fire hazard when exposed to heat, flame or oxidizers. When heated to decomposition it emits acrid smoke and fumes.

DRG200 CAS:589-90-2 ***HR: 3***
1,4-DIMETHYLCYCLOHEXANE
DOT: UN 2263
mf: C_8H_{16} mw: 112.24

PROP: Liquid. Mp: 86°, bp: 119.5°, flash p: 50°F (CC), d: 0.77, vap press: 10 mm @ 10.2°, vap d: 3.86.

DOT Classification: Flammable Liquid; Label: Flammable Liquid.

SAFETY PROFILE: Dangerous fire hazard when exposed to heat or flame; can react vigorously with oxidizing materials. Keep away from heat and open flame. To fight fire, use foam, CO_2, dry chemical.

DRG400 ***HR: 3***
trans-1,2-DIMETHYLCYCLOHEXANE
mf: C_8H_{16} mw: 112.22

PROP: Flash p: 41.6°F.

SAFETY PROFILE: A very dangerous fire hazard when exposed to heat, flame or oxidizers. When heated to decomposition it emits acrid smoke and fumes.

DRG600 ***HR: 2***
trans-1,4-DIMETHYLCYCLOHEXANE
mf: C_8H_{16} mw: 112.22

PROP: Flash p: 60.8°F.

SAFETY PROFILE: A very dangerous fire hazard when exposed to heat, flame or oxidizers. When heated to decomposition it emits acrid smoke and fumes.

DRH200 CAS:5831-10-7 ***HR: 3***
11,17-DIMETHYL-15H-CYCLOPENTA(a)PHENANTHRENE
mf: $C_{19}H_{16}$ mw: 244.35

TOXICITY DATA with REFERENCE
mma-sat 50 µg/plate CNREA8 36,4525,76
skn-mus TDLo:108 mg/kg/1Y-I:CAR PEXTAR 11,69,69

CONSENSUS REPORTS: EPA Genetic Toxicology Program.

SAFETY PROFILE: Questionable carcinogen with experimental carcinogenic data. Mutation data reported. When heated to decomposition it emits acrid smoke and irritating fumes.

DRH400 CAS:5831-09-4 ***HR: 3***
12,17-DIMETHYL-15H-CYCLOPENTA(a)PHENANTHRENE
mf: $C_{19}H_{16}$ mw: 244.35

TOXICITY DATA with REFERENCE
mma-sat 50 µg/plate CNREA8 36,4525,76
skn-mus TDLo:108 mg/kg/1Y-I:ETA PEXTAR 11,69,69

CONSENSUS REPORTS: EPA Genetic Toxicology Program.

SAFETY PROFILE: Questionable carcinogen with experimental tumorigenic data. Mutation data reported. When heated to decomposition it emits acrid smoke and irritating fumes.

DRH600 ***HR: 3***
1,2-DIMETHYLCYCLOPENTENE OZONIDE
mf: $C_7H_{12}O_3$ mw: 144.17

SAFETY PROFILE: Distillation residue explodes violently when heated to 130°C. When heated to decomposition it emits acrid smoke and fumes.

DRH800 CAS:63020-69-9 ***HR: 3***
3,4-DIMETHYL-1,2-CYCLOPENTENOPHENANTHRENE
mf: $C_{19}H_{18}$ mw: 246.37

SYN: 16,17-DIHYDRO-11,12-DIMETHYL-15H-CYCLOPENTA(a)PHENANTHRENE

TOXICITY DATA with REFERENCE
skn-mus TDLo:1260 mg/kg/39W-I:ETA ARGEAR 6,1,53

SAFETY PROFILE: Questionable carcinogen with experimental tumorigenic data. When heated to decomposition it emits acrid smoke and irritating fumes.

DRI400 CAS:3546-11-0 ***HR: 3***
3,3'-DIMETHYL-N,N'-DIACETYLBENZIDINE
mf: $C_{18}H_{20}N_2O_2$ mw: 296.40

SYNS: N,N'-DIACETYL-3,3'-DIMETHYLBENZIDINE ◇ 3',3'''-DIMETHYL-4',4'''-BIACETANILIDE

TOXICITY DATA with REFERENCE
mma-sat 10 μg/plate SAIGBL 23,168,81
orl-rat TDLo:7900 mg/kg/43W-C:CAR CNREA8 16,525,56

SAFETY PROFILE: Questionable carcinogen with experimental carcinogenic data. Mutation data reported. When heated to decomposition it emits toxic fumes of NO_x.

DRI700 CAS:53534-20-6 ***HR: 3***
1,1-DIMETHYLDIAZENIUM PERCHLORATE
mf: $C_2H_8ClN_2O_4$ mw: 159.55

SAFETY PROFILE: An impact-sensitive salt. When heated to decomposition it emits toxic fumes of Cl^- and NO_x. See also PERCHLORATES.

DRI800 CAS:35335-07-0 ***HR: 3***
9,10-DIMETHYL-1,2,5,6-DIBENZANTHRACENE
mf: $C_{24}H_{18}$ mw: 306.42

SYNS: 9,10-DIMETHYL-DBA ◇ 7,14-DIMETHYLDIBENZ(a,h)ANTHRACENE

TOXICITY DATA with REFERENCE
msc-ham:lng 25 μg/L MUREAV 136,65,84
skn-mus TDLo:200 mg/kg/20W-I:NEO CNREA8 22,78,62
scu-mus TDLo:20 mg/kg:ETA CNREA8 22,78,62

SAFETY PROFILE: Questionable carcinogen with experimental neoplastigenic and tumorigenic data. Mutation data reported. When heated to decomposition it emits acrid smoke and irritating fumes.

DRJ000 CAS:63042-50-2 ***HR: 3***
4,9-DIMETHYL-2,3,5,6-DIBENZOTHIOPHENTHRENE
mf: $C_{22}H_{16}S$ mw: 312.44

SYN: 7,13-DIMETHYLBENZO(b)PHENANTHRO(3,2-d)THIOPHENE

TOXICITY DATA with REFERENCE
scu-mus TDLo:80 mg/kg:ETA JNCIAM 18,555,57

SAFETY PROFILE: Questionable carcinogen with experimental tumorigenic data. When heated to decomposition it emits toxic fumes of SO_x.

DRJ200 CAS:16924-32-6 ***HR: 3***
1,1-DIMETHYLDIBORANE
mf: $C_2H_{10}B_2$ mw: 55.724

$(CH_3)_2B{:}H_2{:}BH_2$

PROP: Colorless gas, decomp by water. Mp: −150°; bp: −2.6° flash p: <14°F.

SAFETY PROFILE: A very dangerous fire and explosion hazard when exposed to heat, flame, or oxidizers. See also BORON COMPOUNDS and BORANES.

DRJ400 CAS:17156-88-6 ***HR: 3***
1,2-DIMETHYLDIBORANE
mf: $C_2H_{10}B_2$ mw: 55.724

$CH_3HB{:}H_2{:}BHCH_3$

PROP: Colorless gas, decomp by water. Mp: −125°; bp: −49° flash p: < −67°.

SAFETY PROFILE: A very dangerous fire and explosion hazard when exposed to heat, flame or oxidizers. See also BORON COMPOUNDS and BORANES.

DRJ800 CAS:78-63-7 ***HR: 2***
2,5-DIMETHYL-2,5-DI-(tert-BUTYLPEROXY)HEXANE
DOT: UN 2155/UN 2156
mf: $C_{16}H_{34}O_4$ mw: 290.50

PROP: Colorless to light yellow liquid. D: 0.85, fp: 8°, flash p: >180°F (MOC), bp: 250°. Insol in water; sol in many organic solvents.

SYNS: 2,5-DIMETHYL-2,5-DI-(tert-BUTYLPEROXY)HEXANE,technically pure (DOT) ◇ TRIGONOX 101-101/45 ◇ VAROX

TOXICITY DATA with REFERENCE
ipr-mus LDLo:1700 mg/kg BSPII* 1/75-19B

CONSENSUS REPORTS: Reported in EPA TSCA Inventory.

DOT Classification: Organic Peroxide; Label: Organic Peroxide.

SAFETY PROFILE: Moderately toxic by intraperitoneal route. Combustible when exposed to heat, flames, or reducing agents. To fight fire, use water spray, foam, dry chemical. When heated to decomposition it emits acrid smoke and irritating fumes. Used in the polymerization of styrene and in cross-linking of various grades of polyethylene. See also PEROXIDES, ORGANIC.

DRK200 CAS:1812-53-9 ***HR: D***
DIMETHYLDICETYLAMMONIUM CHLORIDE
mf: $C_{34}H_{72}N{\bullet}Cl$ mw: 530.52

SYNS: ALIQUAT 206 ◇ AMMONIUM DIHEXADECYLDIMETHYL-, CHLORIDE ◇ DICETYLDIMETHYLAMMONIUM CHLORIDE ◇ 1-HEXADECANAMINIUM, N-HEXADECYL-N,N-DIMETHYL-, CHLORIDE (9CI)

TOXICITY DATA with REFERENCE
scu-mus TDLo:50 mg/kg (female 7D post):TER FCTXAV 18,189,80

SAFETY PROFILE: Experimental reproductive effects.

When heated to decomposition it emits toxic fumes of NO_x and Cl^-.

DRK400 CAS:42149-31-5 ***HR: 3***
2,5-DIMETHYL-1,2,5,6-DIEPOXYHEX-3-YNE
mf: $C_8H_{10}O_2$ mw: 138.18

TOXICITY DATA with REFERENCE
scu-mus TDLo:1040 mg/kg/26W-I:NEO JNCIAM 53,695,74

SAFETY PROFILE: Questionable carcinogen with experimental neoplastigenic data. When heated to decomposition it emits acrid smoke and irritating fumes. See also ACETYLENE COMPOUNDS.

DRK500 CAS:34983-45-4 ***HR: 2***
trans-4,4′-DIMETHYL-α-α′-DIETHYLSTILBENE
mf: $C_{20}H_{24}$ mw: 264.44

SYNS: STILBENE, α-α′-DIETHYL-4,4′-DIMETHYL-, (E)- ◇ DMES

TOXICITY DATA with REFERENCE
orl-dog TDLo:6 g/kg/30D-C:ETA TXAPA9 21,582,72

SAFETY PROFILE: Questionable carcinogen with experimental tumorigenic data. When heated to decomposition it emits acrid smoke and irritating fumes.

DRK600 CAS:19072-57-2 ***HR: 3***
2,6-DIMETHYL-1,1-DIETHYLPIPERIDINIUM BROMIDE
mf: $C_{11}H_{24}BrN$ mw: 250.27

SYNS: AGILENE ◇ SC-1950

TOXICITY DATA with REFERENCE
orl-rat LD50:2000 mg/kg JPETAB 99,435,50
orl-mus LD50:365 mg/kg JPETAB 99,435,50
ipr-mus LD50:40 mg/kg JPETAB 99,435,50
ivn-dog LDLo:25 mg/kg JPETAB 99,435,50
ivn-rbt LDLo:25 mg/kg JPETAB 99,435,50

SAFETY PROFILE: Poison by ingestion, intraperitoneal, and intravenous routes. When heated to decomposition it emits very toxic fumes of Br^- and NO_x.

DRK800 CAS:578-32-5 ***HR: 3***
N,N-DIMETHYL-2,5-DIFLUORO-p-(2,5-DIFLUOROPHENYLAZO)ANILINE
mf: $C_{14}H_{11}F_4N_3$ mw: 297.28

SYN: 2,5,2′,5′-TETRAFLUORO-4-DIMETHYLAMINOAZOBENZENE

TOXICITY DATA with REFERENCE
orl-rat TDLo:6400 mg/kg/21W-C:NEO CNREA8 17,387,57

SAFETY PROFILE: Questionable carcinogen with experimental neoplastigenic data. When heated to decomposition it emits very toxic fumes of F^- and NO_x.

DRL000 CAS:351-65-5 ***HR: 3***
N,N-DIMETHYL-p-(3,4-DIFLUOROPHENYLAZO) ANILINE
mf: $C_{14}H_{13}F_2N_3$ mw: 261.30

SYNS: 3′,4′-DIFLUORO-4-DIMETHYLAMINOAZOBENZENE ◇ N,N-DIMETHYL-3′,4′-DIFLUORO-4-(PHENYLAZO)BENZENEAMINE

TOXICITY DATA with REFERENCE
orl-rat TDLo:2356 mg/kg/17W-C:CAR CBINA8 53,107,85
orl-rat TD:3400 mg/kg/13W-C:NEO CNREA8 17,387,57

SAFETY PROFILE: Questionable carcinogen with experimental carcinogenic and neoplastigenic data. When heated to decomposition it emits very toxic fumes of F^- and NO_x.

DRL200 CAS:122-15-6 ***HR: 3***
5,5-DIMETHYLDIHYDRORESORCINOL DIMETHYLCARBAMATE
mf: $C_{11}H_{17}NO_3$ mw: 211.29

SYNS: DIMETAN ◇ DIMETHYLCARBAMATE de 5,5-DIMETHYL DIHYDRORESORCINOL (FRENCH) ◇ DIMETHYLCARBAMIC ACID ester with 3-HYDROXY-5,5-DIMETHYL-2-CYCLOHEXEN-1-ONE ◇ 5,5-DIMETHYL-DIHYDRORESORCINOL-N,N-DIMETHYLCARBAMAT (GERMAN) ◇ 5,5-DIMETHYL-4,5-DIHYDRO-3-RESORCYL-DIMETHYL-CARBAMAT (GERMAN) ◇ (5,5-DIMETHYL-3-OXO-CYCLOHEX-1-EN-YL)-N,N-DIMETHYL-CARBAMAAT (DUTCH) ◇ 5,5-DIMETHYL-3-OXO-1-CYCLOHEXEN-1-YL DIMETHYLCARBAMATE ◇ 5,5-DIMETHYL-3-OXOCYCLOHEX-1-ENYL DIMETHYLCARBAMATE ◇ (5,5-DIMETHYL-3-OXO-CYCLOHEX-1-EN-YL)-N,N-DIMETHYL-CARBAMAT(GERMAN) ◇ (5,5-DIMETIL-3-OXO-CICLOES-1-EN-IL)-N,N-DIMETIL-CARBAMMATO (ITALIAN) ◇ ENT 24,738 ◇ GEIGY 19258 ◇ 3-HYDROXY-5,5-DIMETHYL-2-CYCLOHEXEN-1-ONEDIMETHYLCARBAMATE

TOXICITY DATA with REFERENCE
orl-rat LD50:120 mg/kg WRPCA2 9,119,70
orl-mus LD50:90 mg/kg 85DPAN -,-,71/76
orl-dog LD50:50 mg/kg 85GYAZ -,66,71

SAFETY PROFILE: Poison by ingestion. When heated to decomposition it emits toxic fumes of NO_x. See also CARBAMATES.

DRL400 CAS:35653-70-4 ***HR: 3***
2,4′-DIMETHYL-4-DIMETHYLAMINOAZOBENZENE
mf: $C_{16}H_{19}N_3$ mw: 253.38

TOXICITY DATA with REFERENCE
orl-rat TDLo:8940 mg/kg/35W-C:ETA ARZNAD 12,270,62

SAFETY PROFILE: Questionable carcinogen with experimental tumorigenic data. When heated to decomposition it emits toxic fumes of NO_x.

DRL600 CAS:38035-28-8 ***HR: 3***
2,3-DIMETHYL-8-(DIMETHYLAMINOMETHYL)-7-METHOXYCHROMONE HYDROCHLORIDE
mf: $C_{15}H_{19}NO_3 \cdot ClH$ mw: 297.81

TOXICITY DATA with REFERENCE
orl-rat LD50:7800 μg/kg 27ZQAG -,156,72
ipr-rat LD50:3300 μg/kg 27ZQAG -,157,72
scu-rat LD50:2200 μg/kg 27ZQAG -,157,72
ipr-mus LD50:3300 μg/kg JMPCAS 3,471,61

SAFETY PROFILE: Poison by ingestion, intraperitoneal, and subcutaneous routes. When heated to decomposition it emits very toxic fumes of NO_x and HCl.

DRM000 CAS:3759-07-7 ***HR: 3***
9,9-DIMETHYL-10-DIMETHYLAMINOPROPYLACRIDAN HYDROGEN TARTRATE
mf: $C_{20}H_{26}N_2 \cdot C_4H_4O_6$ mw: 442.56

SYNS: DIMETACRINE BITARTRATE ◇ DIMETACRIN HYDROGENTARTRATE ◇ DIMETHACRINE TARTRATE ◇ 10-(3-(DIMETHYLAMINO)PROPYL)-9,9-DIMETHYLACRIDAN TARTRATE (1:1) ◇ 9,9-DIMETHYL-10-(3-DIMETHYLAMINO)PROPYLACRIDINETARTRATE ◇ ISOTONIL ◇ ISTONYL ◇ MIROISTONIL ◇ MO 709 ◇ SD 709 ◇ (R-R*,R*))-N,N,9,9-TETRAMETHYL-10(9H)-ACRIDINEPROPANAMINE-2,3-DIHDYROXYBUTANEDIOATE (1:1)

TOXICITY DATA with REFERENCE
orl-mus TDLo:175 mg/kg (female 7-13D post):REP OYYAA2 4,855,70
ipr-rat TDLo:315 mg/kg (female 6-12D post):TER OYYAA2 5,129,71
orl-rat LD50:1671 mg/kg ARZNAD 24,1098,74
ipr-rat LD50:203 mg/kg OYYAA2 4,855,70
scu-rat LD50:1214 mg/kg OYYAA2 4,855,70
ivn-rat LD50:38 mg/kg WKWOAO 78,21,66
orl-mus LD50:860 mg/kg WKWOAO 78,21,66
ipr-mus LD50:175 mg/kg WKWOAO 78,21,66
scu-mus LD50:798 mg/kg OYYAA2 4,855,70
ivn-mus LD50:40900 μg/kg OYYAA2 4,855,70
orl-cat LD50:150 mg/kg WKWOAO 78,21,66
ivn-cat LD50:40 mg/kg WKWOAO 78,21,66

SAFETY PROFILE: Poison by ingestion, intravenous, and intraperitoneal routes. Moderately toxic by subcutaneous route. Experimental teratogenic and reproductive effects. When heated to decomposition it emits toxic fumes of NO_x.

DRM600 CAS:4100-38-3 ***HR: 3***
3,4-DIMETHYL-4-(3,4-DIMETHYL-5-ISOXAZOLYAZO)-ISOXAZOLIN-5-ONE
mf: $C_{10}H_{12}N_4O_3$ mw: 236.26

O:CON=CCH₃C(CH₃)N=NC=CCH₃C(CH₃)=NO

SAFETY PROFILE: Explodes if heated rapidly to 100°C but is stable to impact or friction. When heated to decomposition it emits toxic fumes of NO_x.

DRM800 CAS:63886-45-3 ***HR: 3***
2,5-DIMETHYL-1-(5-(2,5-DIMETHYLPYRROLIDINO)-2,4-PENTADIENYLIDENE) PYRROLIDINIUM CHLORIDE SESQUIHYDRATE
mf: $C_{17}H_{29}N_2 \cdot Cl \cdot 3/2H_2O$ mw: 323.96

TOXICITY DATA with REFERENCE
orl-mus LD50:50 mg/kg JMCMAR 12,806,69
ipr-mus LD50:10 mg/kg JMCMAR 12,806,69

SAFETY PROFILE: Poison by ingestion and intraperitoneal routes. When heated to decomposition it emits very toxic fumes of NO_x and Cl^-.

DRN200 CAS:40487-42-1 ***HR: 2***
3,4-DIMETHYL-2,6-DINITRO-N-(1-ETHYLPROPYL)ANILINE
mf: $C_{13}H_{19}N_3O_4$ mw: 281.35

SYNS: AC 92553 ◇ N-(1-AETHYLPROPYL)-3,4-DIMETHYL-2,6-DINITROANILIN (GERMAN) ◇ N-(1-AETHYLPROPYL)-2,6-DINITRO-3,4-XYLIDIN (GERMAN) ◇ 2,5-DINITRO-N-(1-ETHYLPROPYL)-3,4-XYLIDINE ◇ N-(1-ETHYLPROPYL)-3,4-DIMETHYL-2,6-DINITRO-BENZENAMINE ◇ HERBADOX ◇ HORBADOX ◇ PAY-OFF ◇ PENDIMETHALIN ◇ PENOXALINE ◇ PHENOXALIN ◇ PROWL ◇ STOMP ◇ TENDIMETHALIN

TOXICITY DATA with REFERENCE
orl-rat LD50:1250 mg/kg 85ARAE 2,53,77

SAFETY PROFILE: Moderately toxic by ingestion. An herbicide. When heated to decomposition it emits toxic fumes of NO_x.

DRN300 CAS:14760-99-7 ***HR: 3***
N,N'-DIMETHYL-N,N'-DINITROOXAMIDE
mf: $C_4H_6N_4O_6$ mw: 206.11

SAFETY PROFILE: An explosive. When heated to decomposition it emits toxic fumes of NO_x. See also EXPLOSIVES.

DRN400 CAS:3844-60-8 ***HR: 3***
1,6-DIMETHYL-1,6-DINITROSOBIUREA
mf: $C_4H_8N_6O_4$ mw: 204.18

SYNS: N,N'-DIMETHYL-N,N'-DINITROSO-1,2-HYDRAZINEDICARBOXAMIDE ◇ HYDRAZODICARBONSAEUREABIS(METHYLNITROSAMID) (GERMAN) ◇ HYDRAZODICARBOXYLIC ACID BIS(METHYLNITROSAMIDE) ◇ HYDROAZODICARBOXYBIS(METHYLNITRO- SAMIDE) ◇ NSC 409425 ◇ SRI 1666

TOXICITY DATA with REFERENCE
scu-rat TDLo:420 mg/kg/28W-I:ETA ZEKBAI 69,103,67
scu-rat LD50:200 mg/kg ZEKBAI 69,103,67
ipr-mus LD50:56570 μg/kg NCISP* JAN86

SAFETY PROFILE: Poison by subcutaneous and intraperitoneal routes. Questionable carcinogen with experimental tumorigenic data. Many N-nitroso compounds are carcinogens. When heated to decomposition it emits toxic fumes of NO_x. See also N-NITROSO COMPOUNDS.

DRN600 CAS:7601-87-8 **HR: 3**
N,N'-DIMETHYL-N,N'-DINITROSOOXAMIDE
mf: $C_4H_6N_4O_4$ mw: 174.14

$[CH_3N(N{:}O)CO\bullet\text{-}]_2$

SYNS: DIMETHYLDINITROSOOXAMID (GERMAN) ◇ N,N'-DINITROSO-N,N'-DIMETHYLOXAMID (GERMAN)

TOXICITY DATA with REFERENCE
mmo-smc 1 mmol/L/10M ZEVBA5 95,82,64
orl-rat LD50:96 mg/kg ZEKBAI 69,103,676

SAFETY PROFILE: Poison by ingestion. Mutation data reported. Many N-nitroso compounds are carcinogens. See also NITRATES for fire hazard. A heat- and shock-sensitive explosive. Can react vigorously with reducing materials. When heated to decomposition it emits highly toxic fumes of NO_x. See also N-NITROSO COMPOUNDS and EXPLOSIVES, HIGH.

DRN800 CAS:55556-88-2 **HR: 3**
2,5-DIMETHYLDINITROSOPIPERAZINE
mf: $C_6H_{14}N_4O_2$ mw: 174.24

PROP: Mixture approximately 25% cis and 75% trans conformers (CNREA8 35,1270,75).

SYNS: 2,5-DIMETHYL-1,4-DINITROSOPIPERAZINE ◇ 2,5-DIMETHYL-DNPZ ◇ DINITROSO-2,5-DIMETHYLPIPERAZINE

TOXICITY DATA with REFERENCE
mma-sat 25 μg/plate TCMUE9 1,13,84
mma-smc 50 μmol/plate MUREAV 77,143,80
orl-rat TDLo:2740 mg/kg/50W-I:ETA CNREA8 35,1270,75

CONSENSUS REPORTS: EPA Genetic Toxicology Program.

SAFETY PROFILE: Many N-nitroso compounds are carcinogens. Questionable carcinogen with experimental tumorigenic data. Mutation data reported. When heated to decomposition it emits toxic fumes of NO_x. See also N-NITROSO COMPOUNDS.

DRO000 CAS:55380-34-2 **HR: 3**
2,6-DIMETHYLDINITROSOPIPERAZINE
mf: $C_6H_{14}N_4O_2$ mw: 174.24

SYNS: 2,6-DIMETHYL-DNPZ ◇ DINITROSO-2,6-DIMETHYLPIPERAZINE ◇ N,N'-DINITROSO-2,6-DIMETHYLPIPERAZINE ◇ 1,4-DINITROSO-2,6-DIMETHYLPIPERAZINE ◇ DNDMP

TOXICITY DATA with REFERENCE
mma-smc 50 μmol/plate TCMUE9 1,13,84
mma-sat 50 μg/plate MUREAV 77,143,80
orl-rat TDLo:240 mg/kg/20W-I:CAR CRNGDP 4,1165,83
orl-gpg TDLo:4800 mg/kg/50W-I:ETA CNREA8 40,1879,80
orl-ham TDLo:1960 mg/kg35W-I:NEO CRNGDP 4,1165,83
orl-rat TD:1200 mg/kg/20W-I:CAR CRNGDP 4,1165,83

CONSENSUS REPORTS: EPA Genetic Toxicology Program.

SAFETY PROFILE: Questionable carcinogen with experimental carcinogenic, neoplastigenic, and tumorigenic data. Mutation data reported. A model carcinogen and carcinogenic metabolite. When heated to decomposition it emits toxic fumes of NO_x. See also N-NITROSO COMPOUNDS.

DRO200 CAS:6972-76-5 **HR: 3**
N,N'-DIMETHYL-N,N'-DINITROSO-1,3-PROPANEDIAMINE
mf: $C_5H_{12}N_4O_2$ mw: 160.21

SYNS: DINITROSODIMETHYLPROPANEDIAMINE ◇ N,N'-DINITROSO-N,N'-DIMETHYL-1,3-PROPANEDIAMINE ◇ NSC 62580

TOXICITY DATA with REFERENCE
orl-rat TDLo:360 mg/kg/48W-I:NEO JNCIAM 41,985,68

SAFETY PROFILE: Questionable carcinogen with experimental neoplastigenic data. Many N-nitroso compounds are carcinogens. When heated to decomposition it emits toxic fumes of NO_x. See also N-NITROSO COMPOUNDS.

DRO400 CAS:133-55-1 **HR: D**
N,N'-DIMETHYL-N,N'-DINITROSOTEREPHTHALAMIDE
mf: $C_{10}H_{10}N_4O_4$ mw: 250.24

SYNS: N,N'-DIMETHYL-N,N'-DINITROSO-1,4-BENZENEDICARBOXAMIDE ◇ N,N'-DINITROSO-N,N'-DIMETHYLTEREPHTALSAUREAMID (GERMAN)

TOXICITY DATA with REFERENCE
mmo-smc 30 μmol/L ZEVBA5 97,55,65
mrc-smc 800 nmol/L ZEVBA5 98,230,66

CONSENSUS REPORTS: EPA Genetic Toxicology Program. Reported in EPA TSCA Inventory.

SAFETY PROFILE: Mutation data reported. Many N-nitroso compounds are carcinogens. When heated to decomposition it emits toxic fumes of NO_x. See also N-NITROSO COMPOUNDS.

DRO800 CAS:25136-55-4 ***HR: 3***
DIMETHYL DIOXANE
DOT: UN 2707
mf: $C_6H_{12}O_2$ mw: 116.18

PROP: Water-white liquid. Bp: 117.5°, flash p: 75°F, d: 0.9268, vap press: 15.4 mm @ 20°, vap d: 4.0.

SYN: DIMETHYL-p-DIOXANE (DOT)

TOXICITY DATA with REFERENCE
skn-rbt 10 mg/24H open JIHTAB 30,63,48
eye-rbt 20 mg open SEV JIHTAB 30,63,48
ihl-rat TCLo:500 mg/m^3/24H (30D pre):REP TPKVAL 12,64,71
orl-rat LD50:3000 mg/kg JIHTAB 30,63,48
ihl-rat LCLo:8000 ppm/4H JIHTAB 30,63,48

DOT Classification: Flammable Liquid; Label: Flammable Liquid

SAFETY PROFILE: Moderately toxic by ingestion. Mildly toxic by inhalation. Experimental reproductive effects. A skin and severe eye irritant. A very dangerous fire hazard when exposed to heat or flame; can react vigorously with oxidizing materials. To fight fire, use foam, CO_2, dry chemical. When heated to decomposition it emits acrid smoke and irritating fumes.

DRP000 CAS:10138-17-7 ***HR: 2***
2,6-DIMETHYL-1,4-DIOXANE
mf: $C_6H_{12}O_2$ mw: 116.18

PROP: Flash p: 75.2°F

TOXICITY DATA with REFERENCE
eye-rbt 5 mg SEV AJOPAA 29,1363,46
ihl-rat LCLo:8000 ppm/4H JIHTAB 31,343,49

SAFETY PROFILE: Moderately toxic by inhalation. A severe eye irritant. A very dangerous fire hazard when exposed to heat, flame or oxidizers. When heated to decomposition it emits acrid smoke and irritating fumes.

DRP200 CAS:2033-24-1 ***HR: 3***
2,2-DIMETHYL-m-DIOXANE-4,6-DIONE
mf: $C_6H_8O_4$ mw: 144.14

PROP: Pale yellow, crystalline solid. Mp: 96-104°.

SYNS: 2,2-DIMETHYL-1,3-DIOXANE-4,6-DIONE ◇ 2,2-DIMETHYL-4,6-DIOXO-m-DIOXANE

TOXICITY DATA with REFERENCE
ivn-mus LD50:180 mg/kg CSLNX* NX#04102

CONSENSUS REPORTS: Reported in EPA TSCA Inventory.

SAFETY PROFILE: Poison by intravenous route. When heated to decomposition it emits acrid smoke and irritating fumes.

DRP400 CAS:2916-31-6 ***HR: 3***
2,2-DIMETHYL-1,3-DIOXOLAN
mf: $C_5H_{10}O_2$ mw: 102.14

$OC(CH_3)_2OCH_2CH_2$ (ring)

PROP: Flash p: 30.2°F.

SAFETY PROFILE: A very dangerous fire hazard when exposed to heat, flame or oxidizers. When heated to decomposition it emits acrid smoke and fumes.

DRP600 CAS:7122-04-5 ***HR: 3***
2-(4,5-DIMETHYL-1,3-DIOXOLAN-2-YL)PHENYL-N-METHYLCARBAMATE
mf: $C_{13}H_{17}NO_4$ mw: 251.31

SYNS: C-10015 ◇ CIBA-GEIGY C-10015 ◇ ENT 27,410 ◇ FONDAREN ◇ NSC 191000 ◇ SAPRECON C

TOXICITY DATA with REFERENCE
orl-rat LD50:110 mg/kg FMCHA2 -,C210,83
orl-dog LD50:300 mg/kg 28ZEAL 5,118,76

SAFETY PROFILE: Poison by ingestion. When heated to decomposition it emits toxic fumes of NO_x. See also CARBAMATES.

DRP800 CAS:957-51-7 ***HR: 2***
N,N-DIMETHYL-2,2-DIPHENYLACETAMIDE
mf: $C_{16}H_{17}NO$ mw: 239.34

PROP: White solid; very sltly sol in water; mod sol in acetone, dimethyl formamide, and phenyl cellosolve. Mp: 134.5-135.5°.

SYNS: DIAMIDE ◇ DIF 4 ◇ N,N-DIMETHYLDIPHENYLACETAMIDE ◇ N,N-DIMETHYL-α,α-DIPHENYLACETAMIDE ◇ N,N-DIMETHYL-α-PHENYLBENZENEACETAMIDE ◇ DIMID ◇ DIPHENAMID ◇ DIPHENAMIDE ◇ DIPHENYLAMIDE ◇ 2,2-DIPHENYL-N,N-DIMETHYLACETAMIDE ◇ DYMID ◇ ENIDE ◇ FDN ◇ FENAM ◇ LILLY 34,314 ◇ U 4513

TOXICITY DATA with REFERENCE
cyt-mus-unr 10 mg/kg TGANAK 16(1),45,82
orl-rat LD50:685 mg/kg JDGRAX 12(1-2),155,80
orl-mus LD50:600 mg/kg PCOC** -,431,66
ipr-mus LD50:500 mg/kg GUCHAZ 6,233,73
scu-mus LD50:800 mg/kg GUCHAZ 6,233,73
orl-dog LD50:1000 mg/kg 28ZEAL 5,84,76
orl-mky LD50:1000 mg/kg 28ZEAL 5,84,76
orl-rbt LD50:1500 mg/kg 28ZEAL 5,84,76

SAFETY PROFILE: Moderately toxic by ingestion, intraperitoneal, and subcutaneous routes. Mutation data reported. A pesticide. When heated to decomposition it emits toxic fumes of NO_x.

DRP875 ***HR: 3***
N,N-DIMETHYL-2-(p-(1,2-DIPHENYL-1-BUTENYL)PHENOXY)ETHYLAMINE CITRATE
mf: $C_{26}H_{29}NO{\bullet}C_6H_8O_7$ mw: 563.70

TOXICITY DATA with REFERENCE
orl-rat LD50:1550 mg/kg IYKEDH 12,933,81
ipr-rat LD50:660 mg/kg IYKEDH 12,933,81
ivn-rat LD50:76 mg/kg IYKEDH 12,933,81
orl-mus LD50:6500 mg/kg IYKEDH 12,933,81
ipr-mus LD50:218 mg/kg IYKEDH 12,933,81
ivn-mus LD50:95 mg/kg IYKEDH 12,933,81

SAFETY PROFILE: Poison by intravenous and intraperitoneal routes. Moderately toxic by ingestion. When heated to decomposition it emits toxic fumes of NO_x. See also AMINES.

DRQ000 CAS:13865-57-1 ***HR: 3***
N,N-DIMETHYL-4-(DIPHENYLMETHYL)ANILINE
mf: $C_{21}H_{21}N$ mw: 287.43

SYNS: 4-DIMETHYLAMINOTRIPHENYLMETHANE ◇ 4-DIMETHYLAMINOTRIPHENYLMETHAN (GERMAN)

TOXICITY DATA with REFERENCE
scu-rat TDLo:1620 mg/kg/12W-I:ETA NATWAY 42,215,55

SAFETY PROFILE: Questionable carcinogen with experimental tumorigenic data. When heated to decomposition it emits toxic fumes of NO_x.

DRQ200 CAS:997-95-5 ***HR: 3***
2,2'-DIMETHYLDIPROPYLINITROSOAMINE
mf: $C_8H_{18}N_2O$ mw: 158.28

SYNS: DI-ISO-BUTYLNITROSAMINE ◇ DMDPN ◇ NITROSODIISOBUTYLAMINE ◇ N-NITROSODIISOBUTYLAMINE ◇ N-NITROSODI-ISO-BUTYLAMINE ◇ N-NITROSO-2,2'-DIMETHYLDI-n-PROPYLAMINE

TOXICITY DATA with REFERENCE
mma-sat 25 μg/plate TCMUE9 1,13,84
orl-rat TDLo:1750 mg/kg/30W-I:ETA JJIND8 62,407,79
scu-ham LD50:5600 mg/kg JNCIAM 55,1209,75

SAFETY PROFILE: Mildly toxic by subcutaneous route. Questionable carcinogen with experimental neoplastigenic and tumorigenic data. Mutation data reported. Many nitrosamines compounds are carcinogens. When heated to decomposition it emits toxic fumes of NO_x. See also NITROSAMINES.

DRQ400 CAS:624-92-0 ***HR: 3***
DIMETHYLDISULFIDE
DOT: UN 2381
mf: $C_2H_6S_2$ mw: 94.20

PROP: Liquid. flash p: 44.6°F. Bp: 109.7°, d: 1.0569 @ 25°, vap press: 28.6 mm @ 25°, vap d: 3.24.

TOXICITY DATA with REFERENCE
ihl-rat LC50:15850 $\mu g/m^3$/2H GTPZAB 16(6),46,72
ihl-mus LC50:12300 $\mu g/m^3$/2H GTPZAB 16(6),46,72

CONSENSUS REPORTS: Reported in EPA TSCA Inventory. EPA Extremely Hazardous Substances List.

DOT Classification: Flammable Liquid; Label: Flammable Liquid.

SAFETY PROFILE: Poison by inhalation. A very dangerous fire hazard when exposed to heat, flame, or oxidizers. Can react vigorously with oxidizing materials. See also SULFIDES.

DRQ600 CAS:598-64-1 ***HR: 3***
DIMETHYLDITHIOCARBAMIC ACID with DIMETHYLAMINE (1:1)
mf: $C_5H_{12}N_2S_2$ mw: 164.31

SYNS: DIMETHYLDITHIOCARBAMIC ACID DIMETHYL AMINE SALT ◇ DIMETHYLDITHIOCARBAMIC ACID DIMETHYLAMMONIUM SALT

TOXICITY DATA with REFERENCE
orl-mus TDLo:29 g/kg/78W-I:ETA NTIS** PB223-159

CONSENSUS REPORTS: Reported in EPA TSCA Inventory.

SAFETY PROFILE: Questionable carcinogen with experimental tumorigenic data. When heated to decomposition it emits very toxic fumes of NO_x, NH_3 and SO_x. See also CARBAMATES.

DRR000 CAS:26419-73-8 ***HR: 3***
2,4-DIMETHYL-1,3-DITHIOLANE-2-CARBOXALDEHYDE O-(METHYLCARBAMOYL)OXIME
mf: $C_8H_{14}N_2O_2S_2$ mw: 234.36

SYNS: 2,4-DIMETHYL-1,3-DITHIOLANE-2-CARBOXALDEHYDEO-((METHYLAMINO)CARBONYL)OXIME ◇ 2,4-DIMETHYL-2-FORMYL-1,3-DITHIOLANE OXIME METHYLCARBAMATE ◇ ENT 27,696 ◇ MBR 6168 ◇ 3M MBR 6168 ◇ TIRPATE

TOXICITY DATA with REFERENCE
orl-rat LD50:1 mg/kg WRPCA2 9,119,70
skn-rat LD50:300 mg/kg GUCHAZ 6,213,73

CONSENSUS REPORTS: EPA Extremely Hazardous Substances List.

SAFETY PROFILE: Poison by ingestion and skin contact. A pesticide. When heated to decomposition it emits very toxic fumes of NO_x and SO_x. See also CARBAMATES and ALDEHYDES.

DRR200 CAS:2540-82-1 ***HR: 3***
O,O-DIMETHYL DITHIOPHOSPHORYLACETIC ACID-N-METHYL-N-FORMYLAMIDE
mf: $C_6H_{12}NO_4PS_2$ mw: 257.28

SYNS: AFLIX ◇ ANTHIO ◇ ANTIO ◇ CP 53926 ◇ O,O-DIMETHYL-S-(N-FORMYL-N-METHYLCARBAMOYLMETHYL)PHOSPHORODITHIOATE ◇ O,O-DIMETHYL-S-(3-METHYL-2,4-DIOXO-3-AZA-BUTYL)-DITHIOFOSFAAT (DUTCH) ◇ O,O-DIMETHYL-S-(3-METHYL-2,4-DIOXO-3-AZA-BUTYL)-DITHIOPHOSPHAT (GERMAN) ◇ O,O-DIMETHYL-S-(N-METHYL-N-FORMYL-CARBAMOYLMETHYL)-DITHIOPHOSPHAT ◇ O,O-DIMETHYL-S-(N-METHYL-N-FORMYLCARBAMOYLMETHYL)PHOSPHORODITHIOATE ◇ O,O-DIMETHYL PHOSPHORODITHIOATE N-FORMYL-2-MERCAPTO-N-METHYLACETAMIDE-S-ESTER ◇ O,O-DIMETIL-S-(N-FORMIL-N-METIL-CARBAMOIL-METIL)-DITIOFOSFATO(ITALIAN) ◇ ENT 27,257 ◇ FORMOTHION ◇ S-(2-(FORMYLMETHYLAMINO)-2-OXOETHYL)-O,O-DIMETHYLPHOSPHORODITHIOATE ◇ N-FORMYL-N-METHYLCARBAMOYLMETHYL-O,O-DIMETHYLPHOSPHORODITHIOATE ◇ S-(N-FORMYL-N-METHYLCARBAMOYLMETHYL)-O,O-DIMETHYL PHOSPHORODITHIOATE ◇ S-(N-FORMYL-N-METHYLCARBAMOYLMETHYL) DIMETHYL PHOSPHOROTHIOLOTHIONATE ◇ S 6900 ◇ SAN 244 I ◇ SAN 6913 I ◇ SAN 7107 I ◇ SPENCER S-6900 ◇ VEL 4284

TOXICITY DATA with REFERENCE
mmo-sat 5 mg/plate MUREAV 116,185,83
mma-sat 5 mg/plate MUREAV 116,185,83
orl-rat LD50:250 mg/kg IRGGAJ 21,92,64
skn-rat LD50:353 mg/kg BJIMAG 26,59,69
ivn-rat LD50:35 mg/kg IRGGAJ 22,246,66
orl-mus LD50:190 mg/kg SPEADM 78-1,31,78
ihl-mus LC50:27 mg/m^3 GISAAA 40(4),110,75
orl-cat LD50:210 mg/kg 85DPAN -,-,71/76
orl-rbt LD50:420 mg/kg SPEADM 78-1,31,78

CONSENSUS REPORTS: EPA Extremely Hazardous Substances List.

SAFETY PROFILE: Poison by ingestion, inhalation, skin contact, and intravenous routes. Mutation data reported. When heated to decomposition it emits very toxic fumes of NO_x, PO_x, and SO_x. See also ESTERS.

DRR400 CAS:2597-03-7 ***HR: 3***
(O,O-DIMETHYLDITHIOPHOSPHORYLPHENYL)ACETIC ACID ETHYL ESTER
mf: $C_{12}H_{17}O_4PS_2$ mw: 320.38

SYNS: AIMSAN ◇ BAY 33051 ◇ BAYER 18510 ◇ CIDEMUL ◇ CIDIAL ◇ DIMEPHENTHIOATE ◇ DIMEPHENTHOATE ◇ O,O-DIMETHYL-S-(1-CARBOETHOXYBENZYL) DITHIOPHOSPHATE ◇ O,O-DIMETHYL-S-α-ETHOXY-CARBONYLBENZYL PHOSPHORODITHIOATE ◇ O,O-DIMETHYL-S-(PHENYLACETIC ACID ETHYL ESTER) PHOSPHORODITHIOATE ◇ O,O-DIMETHYL-S-(PHENYL)(CARBOETHOXY) METHYL PHOSPHORODITHIOATE ◇ (DIMETHYL-S-(PHENYLETHOXYCARBONYLMETHYL)PHOSPHOROTHIOLOTHIONATE) ◇ ELSAN ◇ ENT 23,438 ◇ ENT 27,386GC ◇ S-α-ETHOXYCARBONYLBENZYL-O,O-DIMETHYL PHOSPHORODITHIOATE ◇ S-α-ETHOXYCARBONYLBENZYL DIMETHYL PHOSPHOROTHIOLOTHIONATE ◇ ETHYL-α-((DIMETHOXYPHOSPHENOTHIOYL)THIO)BENZENEACETATE ◇ ETHYL-O,O-DIMETHYL PHOSPHORODITHIOYLPHENYL ACETATE ◇ ETHYL ESTER of O,O-DIMETHYLDITHIOPHOSPHORYL α-PHENYL ACETATE ACID ◇ ETHYL MERCAPTOPHENYLACETATE-O,O-DIMETHYLPHOSPHOROCITHIOATE ◇ FENTHOATE ◇ L-561 ◇ MONTECATINI L-561 ◇ NSC 190978 ◇ OMS 1075 ◇ PAP ◇ PAPTHION ◇ PHENDAL ◇ PHENTHOATE ◇ ROGODIAL ◇ S 2940 ◇ TANONE ◇ TH 346-1 ◇ TSIDIAL

TOXICITY DATA with REFERENCE
orl-rat LD50:200 mg/kg WRPCA2 9,119,70
skn-rat LD50:700 mg/kg WRPCA2 9,119,70
orl-mus LD50:150 mg/kg GUCHAZ 6,207,73
skn-mus LD50:2620 mg/kg GUCHAZ 6,217,73
orl-dog LD50:500 mg/kg SPEADM 78-1,40,78
orl-rbt LD50:72 mg/kg GUCHAZ 6,207,73

SAFETY PROFILE: Poison by ingestion. Moderately toxic by skin contact. An insecticide used for control of crop pests and mosquitoes. When heated to decomposition it emits very toxic fumes of PO_x and SO_x. See also ESTERS.

DRR500 CAS:18539-34-9 ***HR: 3***
N,N-DIMETHYL-2-(DI-2,6-XYLYLMETHOXY) ETHYLAMINE HYDROCHLORIDE
mf: $C_{21}H_{29}NO{\cdot}ClH$ mw: 347.97

SYNS: BS 5933 ◇ β-DIMETHYLAMINOETHYL-2,6,2′,6′-TETRAMETHYLBENZHYDRYL ETHER HYDROCHLORIDE

TOXICITY DATA with REFERENCE
orl-mus LD50:250 mg/kg AIPTAK 135,442,62
ipr-mus LD50:80 mg/kg AIPTAK 135,442,62
scu-mus LD50:140 mg/kg AIPTAK 135,442,62
ivn-mus LD50:35 mg/kg AIPTAK 135,442,62
ivn-cat LD50:15 mg/kg AIPTAK 135,442,62
orl-gpg LD50:100 mg/kg AIPTAK 135,442,62

SAFETY PROFILE: Poison by ingestion, subcutaneous, intravenous, and intraperitoneal routes. When heated to decomposition it emits toxic fumes of NO_x and HCl. See also AMINES.

DRR600 CAS:3007-53-2 ***HR: 3***
N,N-DIMETHYLDODECANAMIDE
mf: $C_{14}H_{29}NO$ mw: 227.44

SYN: N,N-DIMETHYLLAURAMIDE

TOXICITY DATA with REFERENCE
ipr-mus LD50:1000 mg/kg AIHAAP 32,539,71
ivn-mus LD50:75 mg/kg AIHAAP 32,539,71
ipr-rbt LD50:1000 mg/kg AIHAAP 32,539,71
ivn-rbt LD50:50 mg/kg AIHAAP 32,539,71

CONSENSUS REPORTS: Reported in EPA TSCA Inventory.

SAFETY PROFILE: Poison by intravenous route. Moderately toxic by intraperitoneal route. When heated to decomposition it emits toxic fumes of NO_x.

DRR800 CAS:112-18-5 ***HR: 2***
N,N-DIMETHYLDODECYLAMINE
mf: $C_{14}H_{31}N$ mw: 213.46

SYNS: ADMA 2 ◇ ARMEEN DM-12D ◇ BARLENE 125 ◇ DDA ◇ N,N-DIMETHYL-1-DODECANAMINE ◇ N,N-DIMETHYLLAURYLAMINE ◇ DODECYLDIMETHYLAMINE ◇ N-DODECYLDIMETHYLAMINE ◇ LAURYLDIMETHYLAMINE ◇ N-LAURYLDIMETHYLAMINE ◇ MONOLAURYL DIMETHYLAMINE ◇ RC 5629

TOXICITY DATA with REFERENCE
skn-rbt 500 mg/24H SEV 28ZPAK -,63,72
eye-rbt 50 μg/24H SEV 28ZPAK -,63,72
orl-rat LD50:740 mg/kg CMEP** -,1,56

CONSENSUS REPORTS: Reported in EPA TSCA Inventory.

SAFETY PROFILE: Moderately toxic by ingestion. A severe skin and eye irritant. When heated to decomposition it emits toxic fumes of NO_x.

DRS000 CAS:1920-05-4 ***HR: 2***
DIMETHYLDODECYLAMINE ACETATE
mf: $C_{14}H_{31}N \cdot C_2H_4O_2$ mw: 273.52

SYNS: N,N-DIMETHYLDODECYLAMINE ACETATE ◇ PENAR ◇ TRI-PENAR

TOXICITY DATA with REFERENCE
orl-rat LD50:800 mg/kg 28ZEAL 4,186,69

CONSENSUS REPORTS: Reported in EPA TSCA Inventory.

SAFETY PROFILE: Moderately toxic by ingestion. When heated to decomposition it emits toxic fumes of NO_x.

DRS200 CAS:1643-20-5 ***HR: 2***
DIMETHYLDODECYLAMINE-N-OXIDE
mf: $C_{14}H_{31}NO$ mw: 229.46

SYNS: AMMONYX LO ◇ AMONYX AO ◇ AROMOX DMMC-W ◇ CONCO XAL ◇ DDNO ◇ N,N-DIMETHYLDODECYLAMINE OXIDE ◇ N,N-DIMETHYL-DODECYLAMINOXID (CZECH) ◇ DODECYLDIMETHYLAMINE OXIDE ◇ N-DODECYLDIMETHYLAMINE OXIDE ◇ LAURYLDIMETHYLAMINE OXIDE ◇ NCI-C55129

TOXICITY DATA with REFERENCE
skn-rbt 500 mg/24H SEV 28ZPAK -,76,72
eye-rbt 50 μg/24H SEV 28ZPAK -,76,72

CONSENSUS REPORTS: Reported in EPA TSCA Inventory.

SAFETY PROFILE: A severe skin and eye irritant. When heated to decomposition it emits toxic fumes of NO_x.

DRS400 CAS:41892-01-7 ***HR: 3***
N,N-DIMETHYL-n-DODECYL(2-HYDROXY-3-CHLOROPROPYL)AMMONIUM CHLORIDE
mf: $C_{17}H_{37}ClNO \cdot Cl$ mw: 342.45

TOXICITY DATA with REFERENCE
orl-rat LD50:1070 mg/kg TXAPA9 28,313,74
skn-rbt LD50:200 mg/kg TXAPA9 28,313,74

SAFETY PROFILE: Poison by skin contact. Moderately toxic by ingestion. When heated to decomposition it emits very toxic fumes of Cl^-, NH_3, and NO_x.

DRS600 CAS:38094-02-9 ***HR: 3***
N,N-DIMETHYL-n-DODECYL(3-HYDROXYPROPENYL) AMMONIUM CHLORIDE
mf: $C_{17}H_{35}NO \cdot Cl$ mw: 304.98

TOXICITY DATA with REFERENCE
orl-rat LD50:1070 mg/kg TXAPA9 28,313,74
skn-rbt LD50:89 mg/kg TXAPA9 28,313,74

SAFETY PROFILE: Poison by skin contact. Moderately toxic by ingestion. When heated to decomposition it emits very toxic fumes of NO_x, NH_3, and Cl^-.

DRS800 CAS:120-08-1 ***HR: 3***
6,7-DIMETHYLESCULETIN
mf: $C_{11}H_{10}O_4$ mw: 206.21

SYNS: AESCULETIN DIMETHYL ETHER ◇ 6,7-DIMETHOXYBENZOPYRAN-2-ONE ◇ 6,7-DIMETHOXYCOUMARIN ◇ ESCOPARONE ◇ ESCULETIN DIMETHYL ETHER ◇ SCOPARON ◇ SCOPARONE

TOXICITY DATA with REFERENCE
orl-rat TDLo:1180 mg/kg (8W male/2W pre-3W post):REP IJEBA6 17,740,79
orl-rat LD50:292 mg/kg DRFUD4 3,550,78
ipr-rat LD50:190 mg/kg DRFUD4 3,550,78
orl-mus LD50:280 mg/kg DRFUD4 3,550,78
ipr-mus LD50:180 mg/kg IJMRAQ 60,763,72

SAFETY PROFILE: Poison by ingestion and intraperitoneal routes. Experimental reproductive effects. An antihypertensive agent. When heated to decomposition it emits acrid smoke and irritating fumes.

DRT000 ***HR: 3***
7,14-DIMETHYL-7,14-ETHANODIBENZ(a,b)ANTHRACENE-15,16-DICARBOXYLIC ACID
mf: $C_{24}H_{20}O_4$ mw: 372.44

SYN: 7,12-DIMETHYLBENZANTHRACENE-7,12-endo-α,β-SUCCINIC ACID

TOXICITY DATA with REFERENCE
scu-rat TDLo:600 mg/kg/50D-I:CAR,REP 85DLAB -,-,75
ipr-rat LDLo:297 mg/kg 85DLAB -,-,75
ipr-mus LDLo:247 mg/kg 85DLAB -,-,75

SAFETY PROFILE: Poison by intraperitoneal route. Experimental reproductive effects. Questionable carcinogen with experimental carcinogenic data. When heated to decomposition it emits acrid smoke and irritating fumes.

DRT089 ***HR: 3***
(DIMETHYL ETHER)OXODIPEROXO CHROMIUM(VI)
mf: $C_2H_6CrO_6$ mw: 178.06

CONSENSUS REPORTS: Chromium and its compounds are on the Community Right-To-Know List.

SAFETY PROFILE: The solid material explodes violently above −30°C. See also CHROMIUM COMPOUNDS; PEROXIDES and ETHERS.

DRT200 CAS:79-64-1 ***HR: 1***
6-α,21-DIMETHYLETHISTERONE
mf: $C_{23}H_{32}O_2$ mw: 340.55

SYNS: DIMETHESTERONE ◇ DIMETHISTERON ◇ DIMETHISTERONE ◇ 6-α,21-DIMETHYL-17-β-HYDROXY-17-α-PREG-4-EN-20-YN-3-ONE ◇ 6-α,21-DIMETHYL-17-β-HYDROXY-17-α-PREGN-4-EN-20-YN-3-ONE ◇ 17-α-ETHYNYL-6-α,21-DIMETHYLTESTOSTERONE ◇ 17-α-ETHYNYL-17-HYDROXY-6-α,21-DIMETHYLANDROST-4-EN-3-ONE ◇ (6-α,17-β)-17-HYDROXY-6-METHYL-17-(1-PROPYNYL)-ANDROST-4-EN-3-ONE ◇ 17-β-HYDROXY-6-α-METHYL-17-(1-PROPYNYL)ANDROST-4-EN-3-ONE ◇ LUTOGAN ◇ LUTOSAN ◇ 6-α-METHYL-17-α-PROPYNYLTESTOSTERONE ◇ 6-α-METHYL-17-(1-PROPYNYL)TESTOSTERONE ◇ P-5048 ◇ SECROSTERON

TOXICITY DATA with REFERENCE
scu-rat TDLo:70 mg/kg (female 14D pre):REP CCPTAY 5,57,72
orl-rat TDLo:100 mg/kg (17-20D preg):TER ECJPAE 24,77,77
orl-mus LD50:7650 mg/kg MEIEDD 10,469,83

CONSENSUS REPORTS: IARC Cancer Review: Animal Inadequate Evidence IMEMDT 21,377,79.

SAFETY PROFILE: Mildly toxic by ingestion. Questionable carcinogen. Experimental teratogenic and reproductive effects. A steroid used as a progestin and in the treatment of menstrual disorders. When heated to decomposition it emits acrid smoke and irritating fumes.

DRT400 CAS:67262-78-6 ***HR: 3***
2′,6′-DIMETHYL-2-(2-ETHOXYETHYLAMINO)ACETANILIDE
mf: $C_{14}H_{22}N_2O_2$ mw: 250.38

SYN: 2-(2-ETHOXYETHYLAMINO)-2′,6′-ACETOXYLIDIDE

TOXICITY DATA with REFERENCE
ipr-mus LD50:130 mg/kg JPMSAE 67,595,78
ivn-mus LD50:35 mg/kg JPMSAE 67,595,78

SAFETY PROFILE: Poison by intraperitoneal and intravenous routes. When heated to decomposition it emits toxic fumes of NO_x.

DRT600 CAS:102207-86-3 ***HR: 3***
2′,6′-DIMETHYL-2-(2-ETHOXYETHYLAMINO)ACETANILIDE HYDROCHLORIDE
mf: $C_{14}H_{22}N_2O_2 \cdot ClH$ mw: 286.84

SYN: 2-(2-ETHOXYETHYLAMINO)-2′,6′-ACETOXYLIDIDE HYDROCHLORIDE

TOXICITY DATA with REFERENCE
ipr-mus LD50:150 mg/kg JPMSAE 67,595,78
ivn-mus LD50:35 mg/kg JPMSAE 67,595,78

SAFETY PROFILE: Poison by intraperitoneal and intravenous routes. When heated to decomposition it emits very toxic fumes of NO_x and HCl.

DRU000 CAS:3837-54-5 ***HR: 3***
N,N-DIMETHYL-p-((3-ETHOXYPHENYL)AZO)ANILINE
mf: $C_{16}H_{19}N_3O$ mw: 269.38

SYN: 3′-ETHOXY-4-DIMETHYLAMINOAZOBENZENE

TOXICITY DATA with REFERENCE
orl-rat TDLo:9202 mg/kg/30W-C:NEO JEMEAV 87,139,48

SAFETY PROFILE: Questionable carcinogen with experimental neoplastigenic data. When heated to decomposition it emits toxic fumes of NO_x.

DRU200 CAS:1825-58-7 ***HR: 2***
DIMETHYLETHOXYPHENYLSILANE
mf: $C_{10}H_{16}OSi$ mw: 180.35

SYN: DIMETHYL-FENYL-ETHOXYSILAN(CZECH)

TOXICITY DATA with REFERENCE
skn-rbt 500 mg/24H MOD 28ZPAK -,221,72
eye-rbt 500 mg/24H MLD 28ZPAK -,221,72
orl-rat LD50:2460 mg/kg 28ZPAK -,221,72

CONSENSUS REPORTS: Reported in EPA TSCA Inventory.

SAFETY PROFILE: Moderately toxic by ingestion. A skin and eye irritant. When heated to decomposition it emits acrid smoke and irritating fumes. See also SILANE.

DRU400 CAS:2669-32-1 ***HR: 3***
O,O-DIMETHYL-S-(5-ETHOXY-1,3,4-THIADIAZOLINYL-3-METHYL)DITHIOPHOSPHATE
mf: $C_7H_{13}N_2O_4PS_3$ mw: 316.37

SYNS: O,O-DIMETHYL-S-(5-ETHOXY-1,3,4-THIADIAZOL-2(3H)-ONYL-(3)-METHYL)DITHIOPHOSPHATE ◇ O,O-DIMETHYL-S-(5-ETHOXY-1,3,4-THIADIAZOL-2(3H)-ONYL-(3)-METHYL)PHOSPHORODITHIOATE ◇ ENT 27,238 ◇ GEIGY 12968 ◇ LYTHIDATHION ◇ NC-2962

TOXICITY DATA with REFERENCE
-rat LD50:268 mg/kg 28ZEAL 4,186,69

SAFETY PROFILE: Poison by ingestion. A pesticide. When heated to decomposition it emits very toxic fumes of NO_x, PO_x, and SO_x.

DRU600 CAS:63021-00-1 ***HR: 3***
DIMETHYL ETHYL ALLENOLIC ACID METHYL ETHER
mf: $C_{16}H_{18}O_3$ mw: 258.34

SYNS: ACIDE DIMETHYL-ETHYL-ALLENOLIQUE ETHER METHYLIQUE (FRENCH) ◇ α,α-DIMETHYL-2-(6-METHOXYNAPHTHYL)PROPIONIC ACID

TOXICITY DATA with REFERENCE
orl-mus TDLo:139 mg/kg/24W-I:ETA CRSBAW 146,916,52

SAFETY PROFILE: Questionable carcinogen with experimental tumorigenic data. When heated to decomposition it emits acrid smoke and irritating fumes. See also ETHERS.

DRU875 ***HR: D***
N-(5-(((1,1-DIMETHYLETHYL)AMINO)SULFONYL)-1,3,4-THIADIAZOL-2-YL)ACETAMIDE MONOSODIUM SALT
mf: $C_8H_{14}N_4O_3S_2 \cdot Na$ mw: 301.37

SYN: CL 13,850 SODIUM

TOXICITY DATA with REFERENCE
scu-rat TDLo:200 mg/kg (10-11D preg):REP JHMJAX 130,95,72
scu-rat TDLo:400 mg/kg (10-11D preg):TER JHMJAX 130,95,72

SAFETY PROFILE: Experimental reproductive effects. When heated to decomposition it emits toxic fumes of SO_x, NO_x, and Na_2O.

DRV000 CAS:529-05-5 ***HR: 2***
1,4-DIMETHYL-7-ETHYLAZULENE
mf: $C_{14}H_{16}$ mw: 184.30

SYNS: BA 2784 ◇ CAMUZULENE ◇ CHAMAZULEN ◇ CHAMAZULENE ◇ DIMETHULENE ◇ DIMETHWLEN ◇ 7-ETHYL-1,4-DIMETHYLAZULENE ◇ KAMILLENOEL (GERMAN)

TOXICITY DATA with REFERENCE
orl-rat LD50:10 g/kg ARZNAD 19,615,69
ims-mus LD50:3 g/kg MEIEDD 10,283,83

SAFETY PROFILE: Moderately toxic by intramuscular route. Mildly toxic by ingestion. An anti-inflammatory and antipyretic agent. When heated to decomposition it emits acrid smoke and irritating fumes.

DRV200 CAS:1420-07-1 ***HR: 3***
2-(1,1-DIMETHYLETHYL)-4,6-DINITROPHENOL
mf: $C_{10}H_{12}N_2O_5$ mw: 240.24

SYNS: o-tert-BUTYL-4,6-DINITROPHENOL ◇ 2,4-DINITRO-6-tert-BUTYLPHENOL ◇ DINOTERB ◇ DNTBP ◇ HERBOGIL

TOXICITY DATA with REFERENCE
orl-rat LD50:62 mg/kg FMCHA2 -,C84,83
orl-mus LD50:25 mg/kg 28ZEAL 5,82,76
skn-gpg LD50:150 mg/kg 28ZEAL 5,82,76

CONSENSUS REPORTS: EPA Extremely Hazardous Substances List.

SAFETY PROFILE: Poison by ingestion and skin contact. A pesticide. When heated to decomposition it emits toxic fumes of NO_x. See also NITRO COMPOUNDS of AROMATIC HYDROCARBONS.

DRV500 ***HR: 3***
N,N-DIMETHYL-N'-ETHYL-N'-1-NAPHTHYLETHYLENEDIAMINE
mf: $C_{16}H_{22}N_2$ mw: 242.40

TOXICITY DATA with REFERENCE
ipr-rat LDLo:80 mg/kg BJPCAL 11,1,56
ipr-mus LD50:121 mg/kg BJPCAL 11,1,56
scu-mus LD50:443 mg/kg BJPCAL 11,1,56

SAFETY PROFILE: Poison by intraperitoneal route. Moderately toxic by subcutaneous route. When heated to decomposition it emits toxic fumes of NO_x. See also AMINES.

DRV550 ***HR: 3***
N,N-DIMETHYL-N'-ETHYL-N'-2-NAPHTHYLETHYLENEDIAMINE
mf: $C_{16}H_{22}N_2$ mw: 242.40

TOXICITY DATA with REFERENCE
ipr-rat LDLo:50 mg/kg BJPCAL 11,1,56
ipr-mus LD50:53 mg/kg BJPCAL 11,1,56
scu-mus LD50:287 mg/kg BJPCAL 11,1,56

SAFETY PROFILE: Poison by subcutaneous and intraperitoneal routes. When heated to decomposition it emits toxic fumes of NO_x. See also AMINES.

DRV600 CAS:50285-71-7 ***HR: 3***
1,1-DIMETHYL-3-ETHYL-3-NITROSOUREA
mf: $C_5H_{11}N_3O_2$ mw: 145.19

SYNS: NITROSOAETHYLDIMETHYLHARNSTOFF ◇ NITROSO-1,1-DIMETHYL-3-ETHYLUREA ◇ NITROSOETHYLDIMETHYLUREA ◇ 1-NITROSO-1-ETHYL-3,3-DIMETHYLUREA

TOXICITY DATA with REFERENCE
mma-sat 250 μg/plate JJIND8 67,1117,81
orl-rat TDLo:1230 mg/kg/50W-I:ETA ZKKOBW 83,315,75

SAFETY PROFILE: Questionable carcinogen with experimental tumorigenic data. Mutation data reported. Many N-nitroso compounds are carcinogens. When heated to decomposition it emits toxic fumes of NO_x. See also N-NITROSO COMPOUNDS.

DRV850 CAS:27692-91-7 ***HR: 3***
N,N-DIMETHYL-N'-ETHYL-N'-PHENYLETHYLENEDIAMINE
mf: $C_{19}H_{20}N_2$ mw: 192.34

SYN: 2325 RP

TOXICITY DATA with REFERENCE
ipr-rat LDLo:350 mg/kg BJPCAL 11,1,56
ipr-mus LD50:500 mg/kg BJPCAL 11,1,56
scu-mus LD50:1150 mg/kg BJPCAL 11,1,56

SAFETY PROFILE: Poison by intraperitoneal route. Moderately toxic by subcutaneous route. When heated to decomposition it emits toxic fumes of NO_x. See also AMINES.

DRW000 CAS:66967-65-5 ***HR: 3***
DIMETHYLETHYL(3-(10H-PYRIDO(3,2-b)(1,4)BENZOTHIAZIN-10-YL)PROPYLAMMONIUM ETHYL SULFATE
mf: $C_{18}H_{24}N_3S{\cdot}C_2H_5O_4S$ mw: 439.64

SYN: D 268

TOXICITY DATA with REFERENCE
orl-mus LD50:494 mg/kg ARZNAD 8,489,58
ipr-mus LD50:58 mg/kg ARZNAD 8,489,58

SAFETY PROFILE: Poison by intraperitoneal route. Moderately toxic by ingestion. When heated to decomposition it emits very toxic fumes of NO_x, NH_3, and SO_x.

DRX400 CAS:5581-40-8 ***HR: 3***
DIMETHYL FANDANE
mf: $C_{17}H_{19}N$ mw: 237.37

SYNS: 2,3-DIHYDRO-N,N-DIMETHYL-3-PHENYL-1H-INDEN-1-AMINE ◇ DIMEFADANE ◇ N,N-DIMETHYL-3-PHENYL-1-INDANAMINE ◇ SK+F 1340

TOXICITY DATA with REFERENCE
orl-rat LD50:176 mg/kg TXAPA9 21,315,72
orl-bwd LD50:75 mg/kg TXAPA9 21,315,72

SAFETY PROFILE: Poison by ingestion. When heated to decomposition it emits toxic fumes of NO_x.

DRX600 CAS:23339-04-0 ***HR: 3***
2,3-DIMETHYLFLUORANTHENE
mf: $C_{18}H_{14}$ mw: 230.32

TOXICITY DATA with REFERENCE
skn-mus TDLo:40 mg/kg/20D:NEO JNCIAM 49,1165,72

CONSENSUS REPORTS: IARC Cancer Review: Animal No Evidence IMEMDT 32,355,83.

SAFETY PROFILE: Questionable carcinogen with experimental neoplastigenic data. When heated to decomposition it emits acrid smoke and irritating fumes.

DRX800 CAS:38048-87-2 ***HR: 3***
7,8-DIMETHYLFLUORANTHENE
mf: $C_{18}H_{14}$ mw: 230.32

TOXICITY DATA with REFERENCE
skn-mus TDLo:40 mg/kg/20D:NEO JNCIAM 49,1165,72

SAFETY PROFILE: Questionable carcinogen with experimental neoplastigenic data. An initiator. When heated to decomposition it emits acrid smoke and irritating fumes.

DRY000 CAS:25889-63-8 ***HR: 3***
8,9-DIMETHYLFLUORANTHENE
mf: $C_{18}H_{14}$ mw: 230.32

TOXICITY DATA with REFERENCE
skn-mus TDLo:40 mg/kg/20D-I:ETA JNCIAM 49,1165,72

SAFETY PROFILE: Questionable carcinogen with experimental tumorigenic data. An initiator. When heated to decomposition it emits acrid smoke and irritating fumes.

DRY100 CAS:17057-98-6 ***HR: 2***
1,9-DIMETHYLFLUORENE
mf: $C_{15}H_{14}$ mw: 194.29

SYN: 9H-FLUORENE, 1,9-DIMETHYL-

TOXICITY DATA with REFERENCE
mma-sat 10 μg/plate MUREAV 91,167,81
ipr-mus TDLo:42 mg/kg/3D-I:NEO JTEHD6 21,525,87

SAFETY PROFILE: Questionable carcinogen with experimental neoplastigenic data. Mutation data reported. When heated to decomposition it emits acrid smoke and irritating fumes.

DRY289 CAS:420-23-5 ***HR: 3***
DIMETHYLFLUOROARSINE
mf: C_2H_6AsF mw: 123.99

CONSENSUS REPORTS: Arsenic and its compounds are on the Community Right-To-Know List.

SAFETY PROFILE: Arsenic compounds are poisons by

many routes. Ignites spontaneously in air. When heated to decomposition it emits toxic fumes of F^- and As. See also ARSENIC COMPOUNDS.

DRY400 CAS:737-22-4 ***HR: 3***
7,12-DIMETHYL-4-FLUOROBENZ(a)ANTHRACENE
mf: $C_{20}H_{15}F$ mw: 274.35

SYN: 4-FLUORO-7,12-DIMETHYLBENZ(a)ANTHRACENE

TOXICITY DATA with REFERENCE
ims-rat TDLo:10 mg/kg:NEO NATUAS 273,566,78

SAFETY PROFILE: Questionable carcinogen with experimental neoplastigenic data. When heated to decomposition it emits toxic fumes of F^-.

DRY600 CAS:794-00-3 ***HR: 3***
7,12-DIMETHYL-5-FLUOROBENZ(a)ANTHRACENE
mf: $C_{20}H_{15}F$ mw: 274.35

SYN: 5-FLUORO-7,12-DIMETHYLBENZ(a)ANTHRACENE

TOXICITY DATA with REFERENCE
dni-hmn:hla 70 μmol/L MUREAV 92,427,82
scu-rat TDLo:823 mg/kg/10W-I:ETA JMCMAR 21,1076,78

CONSENSUS REPORTS: EPA Genetic Toxicology Program.

SAFETY PROFILE: Questionable carcinogen with experimental tumorigenic data. Human mutation data reported. An initiator. When heated to decomposition it emits toxic fumes of F^-.

DRY800 CAS:2023-60-1 ***HR: 3***
7,12-DIMETHYL-8-FLUOROBENZ(a)ANTHRACENE
mf: $C_{20}H_{15}F$ mw: 274.35

SYN: 8-FLUORO-7,12-DIMETHYLBENZ(a)ANTHRACENE

TOXICITY DATA with REFERENCE
ims-rat TDLo:10 mg/kg:NEO NATUAS 273,566,78

SAFETY PROFILE: Questionable carcinogen with experimental neoplastigenic data. When heated to decomposition it emits toxic fumes of F^-.

DRZ000 CAS:2023-61-2 ***HR: 3***
7,12-DIMETHYL-11-FLUOROBENZ(a)ANTHRACENE
mf: $C_{20}H_{15}F$ mw: 274.35

SYN: 11-FLUORO-7,12-DIMETHYLBENZ(a)ANTHRACENE

TOXICITY DATA with REFERENCE
ims-rat TDLo:10 mg/kg:NEO NATUAS 273,566,78
skn-mus TDLo:110 μg/kg:ETA CNREA8 39,411,79

CONSENSUS REPORTS: EPA Genetic Toxicology Program.

SAFETY PROFILE: Questionable carcinogen with experimental neoplastigenic and tumorigenic data. An initiator. When heated to decomposition it emits toxic fumes of F^-.

DSA000 CAS:150-74-3 ***HR: 3***
N,N-DIMETHYL-p-((p-FLUOROPHENYL)AZO)ANILINE
mf: $C_{14}H_{14}FN_3$ mw: 243.31

SYNS: 4-(DIMETHYLAMINO)-4'-FLUOROAZOBENZENE ◇ 4'-FLUORO-N,N-DIMETHYL-4-AMINOAZOBENZENE ◇ 4'-FLUORO-p-DIMETHYLAMINOAZOBENZENE ◇ 4'-FLUORO-4-DIMETHYLAMINOAZOBENZENE ◇ 4'-FLUORO-N,N-DIMETHYL-p-PHENYLAZOANILINE ◇ p-((p-FLUOROPHENYL)AZO)-N,N-DIMETHYLANILINE ◇ 4-((4-FLUOROPHENYL)AZO)-N,N-DIMETHYLBENZENAMINE

TOXICITY DATA with REFERENCE
dns-rat-orl 2520 mg/kg/12W-I CNREA8 29,2039,69
ipr-mus TDLo:400 mg/kg (8-9D preg):TER KAIZAN 37,179,62
ipr-mus TDLo:400 mg/kg (8-9D preg):REP KAIZAN 37,179,62
orl-rat TDLo:2720 mg/kg/14W-I:ETA CNREA8 18,469,58

SAFETY PROFILE: Experimental teratogenic and reproductive effects. Questionable carcinogen with experimental tumorigenic data. Mutation data reported. When heated to decomposition it emits very toxic fumes of F^- and NO_x.

DSA600 CAS:23456-94-2 ***HR: 3***
3,3-DIMETHYL-1-(p-FLUOROPHENYL)TRIAZENE
mf: $C_8H_{10}FN_3$ mw: 167.21

SYN: 1-p-FLUORFENYL-3,3-DIMETHYLTRIAZEN(CZECH)

TOXICITY DATA with REFERENCE
orl-rat LD50:326 mg/kg 28ZPAK -,97,72
ipr-mus LD50:445 mg/kg JMCMAR 19,1299,76

SAFETY PROFILE: Poison by ingestion. Moderately toxic by intraperitoneal route. When heated to decomposition it emits very toxic fumes of F^- and NO_x.

DSA800 CAS:5954-50-7 ***HR: 3***
DIMETHYL FLUOROPHOSPHATE
mf: $C_2H_6FO_3P$ mw: 128.05

PROP: Liquid. Mp: low, bp: 149°, d: 1.28, vap d: 4.42.

SYNS: FLUOPHOSPHORIC ACID, DIMETHYL ESTER ◇ PHOSPHOROFLUORIDIC ACID, DIMETHYL ESTER ◇ PF-1 ◇ T-1035 ◇ TL 311

TOXICITY DATA with REFERENCE
ihl-rat LC50:1800 mg/m^3/1M NTIS** PB158-508

ihl-mus LC50:290 mg/m³/10M JIHTAB 30,307,48
skn-mus LD50:36 mg/kg NTIS** PB158-508
ipr-mus LD50:3 mg/kg NTIS** PB158-508
ivn-mus LD50:450 μg/kg NTIS** PB158-508
ihl-dog LC50:6 g/m³/1M NTIS** PB158-508
ivn-dog LD50:1 mg/kg NTIS** PB158-508
ihl-cat LC50:6 g/m³/1M NTIS** PB158-508

SAFETY PROFILE: Poison by inhalation, skin contact, and intravenous routes. When heated to decomposition it emits toxic fumes of F^- and PO_x. See also ESTERS, FLUORIDES, and PHOSPHATES.

DSB000 CAS:68-12-2 ***HR: 3***
DIMETHYLFORMAMIDE
DOT: UN 2265
mf: C_3H_7NO mw: 73.11

$(CH_3)_2NCO{\bullet}H$

PROP: Colorless, mobile liquid. Bp: 152.8°, lel: 2.2% @ 100°, uel: 15.2% @ 100°, flash p: 136°, fp: −61°, d: 0.9445 @ 25°/4°, autoign temp: 833°F, vap press: 3.7 mm @ 25°, vap d: 2.51.

SYNS: DIMETHYLFORMAMID (GERMAN) ◇ N,N-DIMETHYL FORMAMIDE ◇ N,N-DIMETHYLFORMAMIDE (DOT) ◇ DIMETILFORMAMIDE (ITALIAN) ◇ DIMETYLFORMAMIDU (CZECH) ◇ DMF ◇ DMFA ◇ DWUMETHYLOFORMAMID (POLISH) ◇ N-FORMYLDIMETHYLAMINE ◇ NCI-C60913 ◇ NSC 5356 ◇ U-4224

TOXICITY DATA with REFERENCE
skn-hmn 100%/24H MLD BJIMAG 13,51,56
skn-rbt 10 mg/24H open JIHTAB 30,63,48
eye-rbt 20 mg open JIHTAB 30,63,48
mma-sat 600 ug/plate PMRSDJ 1,343,81
cyt-hmn:lym 100 nmol/L CHPUA4 31,548,81
ihl-rat TDLo:600 mg/m³/24H (1-19D preg):REP TPKVAL 13,75,73
ihl-rat TCLo:4 mg/m³/4H (1-19D preg):TER TPKVAL 14,32,75
orl-rat LD50:2800 mg/kg ZEKBAI 69,103,67
ipr-rat LD50:1400 mg/kg BJIMAG 13,51,56
scu-rat LD50:3800 mg/kg ARZNAD 15,618,65
ivn-rat LD50:2000 mg/kg ZEKBAI 69,103,67
orl-mus LD50:3750 mg/kg TPKVAL 1,54,61
ihl-mus LC50:9400 mg/m³/2H TPKVAL 1,54,61
ipr-mus LD50:650 mg/kg CNCRA6 30,9,63
scu-mus LD50:4500 mg/kg ARZNAD 15,618,65
ivn-mus LD50:2500 mg/kg ARZNAD 15,618,65
ims-mus LD50:3800 mg/kg ARZNAD 15,618,65
ivn-dog LD50:470 mg/kg ARZNAD 15,618,65
ipr-cat LD50:500 mg/kg BJIMAG 13,51,56
skn-rbt LD50:4720 mg/kg AIHAAP 30,470,69

CONSENSUS REPORTS: IARC Cancer Review: Group 2B IMEMDT 47,171,89; Human Limited Evidence IMEMDT 47,171,89; Animal Inadequate Evidence IMEMDT 47,171,89. EPA Genetic Toxicology Program. Reported in EPA TSCA Inventory.

OSHA PEL: TWA 10 ppm (skin)
ACGIH TLV: TWA 10 ppm (skin); BEI: 40 mg(N-methylformamide)/g creatinine at end of shift.
DFG MAK: 20 ppm (60 mg/m³)
DOT Classification: Flammable or Combustible Liquid; Label: Flammable Liquid.

SAFETY PROFILE: Suspected carcinogen. Moderately toxic by ingestion, intravenous, subcutaneous, intramuscular, and intraperitoneal routes. Mildly toxic by skin contact and inhalation. Experimental teratogenic and reproductive effects. A skin and eye irritant. Human mutation data reported. Flammable when exposed to heat or flame; can react with oxidizing materials. Explosion hazard when exposed to flame. Explosive reaction with bromine; potassium permanganate; triethylaluminum + heat. Forms explosive mixtures with lithium azide (shock sensitive above 200°C); uranium perchlorate. Ignition on contact with chromium trioxide. Violent reaction with chlorine; sodium hydroborate + heat; diisocyanatomethane; carbon tetrachloride + iron; 1,2,3,4,5,6-hexachlorocyclohexane + iron. Vigorous exothermic reaction with magnesium nitrate; sodium + heat; sodium hydride + heat; sulfinyl chloride + traces of iron or zinc; 2,4,6-trichloro-1,3,5-triazine (with gas evolution); and many other materials. Avoid contact with halogenated hydrocarbons; inorganic and organic nitrates; (2,5-dimethyl pyrrole + $P(OCl)_3$); C_6Cl_6; methylene diisocyanates; P_2O_3. To fight fire, use foam, CO_2, dry chemical. When heated to decomposition it emits toxic fumes of NO_x.

DSB200 CAS:533-74-4 ***HR: 3***
DIMETHYLFORMOCARBOTHIALDINE
mf: $C_5H_{10}N_2S_2$ mw: 162.29

PROP: Crystals, sol in alc. Mp: 107°.

SYNS: BASAMID ◇ BASAMID G ◇ BASAMID-GRANULAR ◇ BASAMID P ◇ BASAMID-PUDER ◇ CARBOTHIALDIN ◇ CARBOTHIALDINE ◇ CRAG 974 ◇ CRAG FUNGICIDE 974 ◇ CRAG NEMACIDE ◇ CRAG 85W ◇ DAZOMET ◇ 3,5-DIMETHYLPERHYDRO-1,3,5-THIADIAZIN-2-THION (CZECH, GERMAN) ◇ 3,5-DIMETHYL-TETRAHYDRO-1,3,5-THIADIAZINE-2-THIONE ◇ 3,5-DIMETHYL-1,2,3,5-TETRAHYDRO-1,3,5-THIADIAZINETHIONE-2 ◇ 3,5-DIMETHYLTETRAHYDRO-1,3,5-2H-THIADIAZINE-2-THIONE ◇ 3,5-DIMETHYL-1,3,5-2H-TETRAHYDROTHIADIAZINE-2-THIONE ◇ 3,5-DIMETHYLTETRAHYDRO-2H-1,3,5-THIADIAZINE-2-THIONE ◇ 3,5-DIMETHYL-2-THIONOTETRAHYDRO-1,3,5-THIADIAZINE ◇ 3,5-DIMETIL-PERIDRO-1,3,5-TIHADIAZIN-2-TIONE (ITALIAN) ◇ DMTT ◇ FENNOSAN B 100 ◇ MICOFUME ◇ MYLON (CZECH) ◇ MYLONE ◇ MYLONE 85 ◇ N 521 ◇ NALCON 243 ◇ NEFUSAN ◇ PREZERVIT ◇ STAUFFER N 521 ◇ TETRAHYDRO-2H-3,5-DIMETHYL-1,3,5-THIADIAZINE-2-THIONE ◇ TETRAHYDRO-3,5-DIMETHYL-2H-1,3,5-THIADIAZINE-2-THIONE ◇ THIAZON ◇ THIAZONE ◇ 2-THIO-3,5-DIMETHYLTETRAHYDRO-1,3,5-THIADIAZINE ◇ TIAZON ◇ TROYSAN 142 ◇ UCC 974

TOXICITY DATA with REFERENCE
skn-rbt 500 mg/24H MLD 28ZPAK -,204,72
eye-rbt 500 mg/24H SEV 28ZPAK -,204,72
orl-rat LD50:363 mg/kg 28ZPAK -,204,72
ipr-rat LD50:87 mg/kg TXAPA9 9,521,66
orl-mus LD50:180 mg/kg TXAPA9 9,521,66
ipr-mus LDLo:50 mg/kg ARZNAD 21,121,71
scu-mus LDLo:500 mg/kg AIPTAK 12,447,04
ipr-dog LD50:47 mg/kg TXAPA9 9,521,66
orl-rbt LD50:120 mg/kg TXAPA9 9,521,66
ipr-rbt LD50:127 mg/kg TXAPA9 9,521,66
orl-gpg LD50:160 mg/kg TXAPA9 9,521,66

CONSENSUS REPORTS: Reported in EPA TSCA Inventory.

SAFETY PROFILE: Poison by ingestion and intraperitoneal routes. Moderately toxic by subcutaneous route. A skin and severe eye irritant. A mild primary skin irritant and sensitizer. When heated to decomposition it emits very toxic fumes of NO_x and SO_x.

DSB400 CAS:2175-91-9 ***HR: 3***
6,6-DIMETHYLFULVENE
mf: C_8H_{10} mw: 106.17

SAFETY PROFILE: Peroxidizes in air to form a heat-sensitive explosive, insoluble peroxide. The peroxide ignites on contact with ether. When heated to decomposition it emits acrid smoke and fumes.

DSB600 CAS:624-49-7 ***HR: 2***
DIMETHYL FUMARATE
mf: $C_6H_8O_4$ mw: 144.14

SYNS: ALLOMALEIC ACID DIMETHYL ESTER ◇ BOLETIC ACID DIMETHYL ESTER ◇ trans-BUTENEDIOIC ACID DIMETHYL ESTER ◇ trans-1,2-ETHYLENEDICARBOXYLIC ACID DIMETHYL ESTER ◇ FUMARIC ACID, DIMETHYL ESTER

TOXICITY DATA with REFERENCE
orl-rat LD50:2240 mg/kg AIHAAP 30,470,69
skn-rbt LD50:1250 mg/kg AIHAAP 30,470,69

CONSENSUS REPORTS: Reported in EPA TSCA Inventory.

SAFETY PROFILE: Moderately toxic by ingestion and skin contact. When heated to decomposition it emits acrid smoke and irritating fumes. See also ESTERS.

DSB800 CAS:28802-49-5 ***HR: 2***
DIMETHYL FURANE
mf: C_6H_8O mw: 96.14

SYN: DIMETHYL FURAN

TOXICITY DATA with REFERENCE
skn-rbt 10 mg/24H open JIHTAB 26,269,44
eye-rbt 5 mg SEV AJOPAA 29,1363,46
orl-rat LD50:3000 mg/kg JIHTAB 26,269,44
skn-gpg LD50:1000 mg/kg JIHTAB 26,269,44

SAFETY PROFILE: Moderately toxic by ingestion and skin contact. A skin and severe eye irritant. When heated to decomposition it emits acrid smoke and irritating fumes.

DSC000 CAS:625-86-5 ***HR: 2***
2,5-DIMETHYL FURANE
mf: C_6H_8O mw: 96.14

PROP: Colorless liquid. Bp: 94°, flash p: 60.8°F. D: 0.9026 @ 17.7°/4°, vap d: 3.31.

TOXICITY DATA with REFERENCE
ihl-rat LCLo:500 ppm/4H JIHTAB 31,343,49

CONSENSUS REPORTS: Reported in EPA TSCA Inventory.

SAFETY PROFILE: Moderately toxic by inhalation. A very dangerous fire hazard when exposed to heat or flame; can react vigorously with oxidizing materials. Keep away from heat and open flame. To fight fire, use alcohol foam, foam, CO_2, dry chemical. When heated to decomposition it emits acrid smoke and irritating fumes.

DSC200 ***HR: 3***
DIMETHYLGLYCINE HYDROCHLORIDE mixed with SODIUM NITRITE (3 : 1)
mf: $C_4H_8NO_2 \cdot ClH \cdot 1/3(NNaO_2)$ mw: 161.57

TOXICITY DATA with REFERENCE
mmo-sat 30 μmol/plate PSEBAA 164,9,80
mma-sat 30 μmol/plate PSEBAA 164,9,80

SAFETY PROFILE: Sodium nitrite is a poison. Mutation data reported. When heated to decomposition it emits very toxic fumes of Cl^-, NO_x, and Na_2O. See also SODIUM NITRITE.

DSC400 ***HR: 3***
DIMETHYLGOLD SELENOCYANATE
mf: C_3H_6AuNSe mw: 332.02

$$(CH_3)_2AuSeC{\equiv}N$$

CONSENSUS REPORTS: Cyanide and its compounds as well as selenium and its compounds are on the Community Right-To-Know List.

OSHA PEL: TWA 0.2 mg(Se)/m^3
ACGIH TLV: TWA 0.2 mg(Se)/m^3
DFG MAK: 0.1 mg(Se)/m^3

SAFETY PROFILE: A very shock-sensitive explosive. It explodes when precipitated from aqueous solutions. When heated to decomposition it emits toxic fumes of

Se, CN^- and NO_x. See also SELENIUM COMPOUNDS and CYANIDE.

DSC800 CAS:22583-29-5 ***HR: 3***
as-DIMETHYLGUANIDINE HYDROCHLORIDE
mf: $C_3H_9N_3•ClH$ mw: 123.61

TOXICITY DATA with REFERENCE
scu-rat LDLo:284 mg/kg JPETAB 28,251,26

CONSENSUS REPORTS: Reported in EPA TSCA Inventory.

SAFETY PROFILE: Poison by subcutaneous route. When heated to decomposition it emits toxic fumes of HCl and NO_x.

DSD000 ***HR: 3***
2,6-DIMETHYL-2,5-HEPTADIEN-4-ONE DIOZONIDE
mf: $C_9H_{14}O_7$ mw: 234.21

SAFETY PROFILE: "Phorone" diozonide ignites spontaneously at room temperature. When heated to decomposition it emits acrid smoke and fumes.

DSD200 ***HR: 2***
2,5-DIMETHYLHEPTANE
mf: C_9H_{20} mw: 128.26

PROP: Liquid. Bp: 136°, flash p: 75.2°F, d: 0.715 @ 20°, vap d: 4.42.

SAFETY PROFILE: Probably narcotic in high concentration. A mild irritant. A very dangerous fire hazard when exposed to heat or flame; can react vigorously with oxidizing materials. To fight fire, use foam, CO_2, dry chemical. When heated to decomposition it emits acrid smoke and irritating fumes.

DSD400 CAS:926-82-9 ***HR: 2***
3,5-DIMETHYLHEPTANE
mf: C_9H_{20} mw: 128.26

$CH_3CH_2CH_3CHCH_2CHCH_3CH_2CH_3$

PROP: Liquid. Bp: 136°, flash p: 73.5°F, d: 0.723 @ 20°, vap press: 9.5 mm @ 25°, vap d: 4.42.

SAFETY PROFILE: No toxicity information. A probable irritant and narcotic in high concentration. A very dangerous fire hazard when exposed to heat or flame; can react vigorously with oxidizing materials. To fight fire, use CO_2, dry chemical. When heated to decomposition it emits acrid smoke and irritating fumes.

DSD600 ***HR: 2***
4,4-DIMETHYLHEPTANE
mf: C_9H_{20} mw: 128.26

$(CH_3)_2C(CH_2CH_2CH_3)_2$

PROP: Liquid. Bp: 135.2°, flash p: 69.8°F, d: 0.72 @ 25°/4°, vap press: 10.4 mm @ 25°, vap d: 4.42.

SAFETY PROFILE: No toxicity information. A probable irritant and narcotic in high concentration. A very dangerous fire hazard when exposed to heat or flame; can react vigorously with oxidizing materials. To fight fire, use CO_2, dry chemical. When heated to decomposition it emits acrid smoke and irritating fumes.

DSD775 CAS:106-72-9 ***HR: 1***
2,6-DIMETHYL-5-HEPTENAL
mf: $C_9H_{16}O$ mw: 140.23

PROP: Pale yellow liquid; melon odor. D: 0.852-0.858, refr index: 1.443-1.448

SYN: FEMA No. 2497

SAFETY PROFILE: Skin and eye irritant. When heated to decomposition it emits acrid smoke and irritating fumes.

DSD800 CAS:2738-18-3 ***HR: 2***
2,6-DIMETHYL-3-HEPTENE
mf: C_9H_{18} mw: 126.23

$(CH_3)_2CHCH{=}CHCH_2CH(CH_3)_2$

PROP: Clear liquid. Bp: 128.5-129°, flash p: 59.8°F, d: 0.722 @ 15.5°/15.5°, vap press: 28.4 mm @ 38°, vap d: 4.38.

SAFETY PROFILE: A probable irritant and narcotic in high concentration. A very dangerous fire hazard when exposed to heat or flame; can react vigorously with oxidizing materials. To fight fire, use foam, CO_2, dry chemical. When heated to decomposition it emits acrid smoke and irritating fumes.

DSE489 CAS:7226-23-5 ***HR: 3***
1,3-DIMETHYLHEXAHYDROPYRIMIDONE
mf: $C_6H_{12}N_2O$ mw: 128.17

$CH_3NCO•N(CH_3)C_2H_4CH_2$

SYN: DIMETHYLPROPYLENEUREA

SAFETY PROFILE: Explodes on contact with chromium trioxide. When heated to decomposition it emits toxic fumes of NO_x.

DSE509 CAS:584-94-1 ***HR: 2***
2,3-DIMETHYLHEXANE
mf: C_8H_{18} mw: 115.67

$(CH_3)_2CHCHCH_3CH_2CH_2CH_3$

PROP: A clear liquid. Bp: 116°, flash p: 41.6°F, d: 0.716 @ 15.5°/15.5°, vap d: 4.1, autoign temp: 820°F.

SAFETY PROFILE: A probable irritant and narcotic in

high concentration. A very dangerous fire hazard when exposed to heat or flame; can react vigorously with oxidizing materials. To fight fire, use foam, CO_2, dry chemical. When heated to decomposition it emits acrid smoke and irritating fumes.

DSE600 CAS:589-43-5 ***HR: 2***
2,4-DIMETHYLHEXANE
mf: C_8H_{18} mw: 115.67

$(CH_3)_2CHCH_2CHCH_3CH_2CH_3$

PROP: A liquid. Bp: 109°, flash p: 50°F(OC), d: 0.705 @ 15.5°/15.5°, vap d: 3.9.

SAFETY PROFILE: A probable irritant and narcotic in high concentration. A very dangerous fire hazard when exposed to heat or flame; can react vigorously with oxidizing materials. To fight fire, use foam, CO_2, dry chemical. When heated to decomposition it emits acrid smoke and irritating fumes.

DSE800 CAS:3025-88-5 ***HR: 3***
DIMETHYLHEXANE DIHYDROPEROXIDE (dry)
DOT: UN 2174

PROP: Fine, white crystals; insol in hydrocarbons; sltly sol in water, esters and glycerin; sol in other organic solvents. Mp: 104°.

SYN: DIMETHYLHEXANE DIHYDROPEROXIDE (with 18% or more water) (DOT)

CONSENSUS REPORTS: Reported in EPA TSCA Inventory.

DOT Classification: Forbidden, Dry; Organic Peroxide; Label: Organic Peroxide.

SAFETY PROFILE: A reactive peroxide. When heated to decomposition it emits acrid smoke and fumes. See also PEROXIDES, ORGANIC.

DSF100 CAS:25234-79-1 ***HR: D***
3,4-DIMETHYL-2,5-HEXANEDIONE
mf: $C_8H_{14}O_2$ mw: 142.22

SYN: 2,5-HEXANEDIONE, 3,4-DIMETHYL-

TOXICITY DATA with REFERENCE
orl-rat TDLo:980 mg/kg (male 28D pre):REP TXAPA9 88,370,87

SAFETY PROFILE: Experimental reproductive effects. When heated to decomposition it emits acrid smoke and irritating fumes.

DSF200 CAS:53306-53-9 ***HR: 1***
DI(3-METHYLHEXYL)PHTHALATE
mf: $C_{22}H_{34}O_4$ mw: 362.56

SYN: BIS(3-METHYLHEXYL)PHTHALIC ACID ESTER

TOXICITY DATA with REFERENCE
orl-rat LD50:35500 mg/kg GTPZAB 24(3),25,80
orl-mus LD50:35500 mg/kg GTPZAB 24(3),25,80
orl-gpg LD50:35500 mg/kg GTPZAB 24(3),25,80

SAFETY PROFILE: Mildly toxic by ingestion. When heated to decomposition it emits acrid smoke and irritating fumes.

DSF300 CAS:77-71-4 ***HR: 2***
5,5-DIMETHYLHYDANTOIN
mf: $C_5H_8N_2O_2$ mw: 128.15

SYNS: 5,5-DIMETHYL-2,4-IMIDAZOLIDINEDIONE ◇ DMH

TOXICITY DATA with REFERENCE
orl-rat LD50:7800 mg/kg GISAAA 47(6),76,82
scu-mus LD50:2800 mg/kg ARZNAD 4,723,54
unr-mus LD50:10 g/kg GISAAA 46(5),69,81
orl-rbt LD50:12660 mg/kg GISAAA 47(6),76,82
orl-gpg LD50:8430 mg/kg GISAAA 47(6),76,82
unr-gpg LD50:5600 mg/kg GISAAA 46(5),69,81
unr-mam LD50:15950 mg/kg GISAAA 46(5),69,81

CONSENSUS REPORTS: Reported in EPA TSCA Inventory.

SAFETY PROFILE: Moderately toxic by subcutaneous route. Mildly toxic by ingestion. When heated to decomposition it emits toxic fumes of NO_x.

DSF400 CAS:57-14-7 ***HR: 3***
1,1-DIMETHYLHYDRAZINE
DOT: UN 1163
mf: $C_2H_8N_2$ mw: 60.12

PROP: Colorless liquid, ammonia-like odor. Hygroscopic, water-misc. Bp: 63.3°, fp: −58°, flash p: 5°F, d: 0.782 @ 25°/4°, vap press: 157 mm @ 25°, vap d: 1.94, autoign temp: 480°F, lel: 2%, uel: 95%.

SYNS: DIMAZINE ◇ DIMETHYLHYDRAZINE ◇ asym-DIMETHYLHYDRAZINE ◇ N,N-DIMETHYLHYDRAZINE ◇ uns-DIMETHYLHYDRAZINE ◇ unsym-DIMETHYLHYDRAZINE ◇ 1,1-DIMETHYLHYDRAZINE (GERMAN) ◇ DIMETHYLHYDRAZINE, unsymmetrical (DOT) ◇ DMH ◇ NIESYMETRYCZNA DWU METYLOHYDRAZYNA (POLISH) ◇ RCRA WASTE NUMBER U098 ◇ UDMH (DOT)

TOXICITY DATA with REFERENCE
otr-hmn:fbr 167 μmol/L PNASA6 80,7219,83
dnd-hmn:fbr 300 μmol/L ENMUDM 7,267,85
ipr-rat TDLo:600 mg/kg (6-15D preg):TER JTEHD6 13,125,84
ipr-rat TDLo:600 mg/kg (6-15D preg):REP JTEHD6 13,125,84
orl-rat TDLo:150 mg/kg/7W-I:ETA NATUAS 246,491,73
orl-mus TDLo:5880 mg/kg/42W-C:CAR JNCIAM 50,181,73

scu-mus TDLo:420 mg/kg/21W-I:NEO CALEDQ 39,69,88
orl-ham TDLo:228 g/kg/48W-I:CAR CANCAR 40,2427,77
orl-rat LD50:122 mg/kg MEPAAX 24,71,73
ihl-rat LC50:252 ppm/4H AMIHAB 12,609,55
ipr-rat LD50:102 mg/kg TXAPA9 6,371,64
ivn-rat LD50:119 mg/kg MEPAAX 24,71,73
ice-rat LDLo:27 mg/kg BCPCA6 14,1901,65
orl-mus LD50:265 mg/kg MEPAAX 24,71,73
ihl-mus LC50:172 ppm/4H AMIHAB 12,609,55
ipr-mus LD50:113 mg/kg PSEBAA 124,172,67
ivn-mus LD50:250 mg/kg MEPAAX 24,71,73
ihl-dog LC50:3580 ppm/15M AIHAAP 24,137,63

CONSENSUS REPORTS: NTP Fifth Annual Report on Carcinogens. IARC Cancer Review: Group 2B IMEMDT 7,56,87; Animal Sufficient Evidence IMEMDT 4,137,74. EPA Genetic Toxicology Program. Community Right-To-Know List. EPA Extremely Hazardous Substances List. Reported in EPA TSCA Inventory.

OSHA PEL: TWA 0.5 ppm (skin)
ACGIH TLV: TWA 0.5 ppm (skin); Suspected Human Carcinogen; (Proposed: TWA 0.01 ppm (skin); Suspected Human Carcinogen)
DFG MAK: Animal Carcinogen, Suspected Human Carcinogen.
NIOSH REL: (Hydrazines) CL 0.15 $mg/m^3/2H$
DOT Classification: Flammable Liquid; Label: Flammable Liquid and Poison; Flammable Liquid; Label: Flammable Liquid, Corrosive.

SAFETY PROFILE: Confirmed carcinogen with experimental carcinogenic, tumorigenic, and teratogenic data. Other experimental reproductive effects. Poison by ingestion, intraperitoneal, intravenous, and intracerebral routes. Moderately toxic by inhalation and skin contact. Human mutation data reported. A plant growth control agent. Corrosive. A powerful reducing agent. A dangerous fire hazard. It is hypergolic with many oxidants (e.g., dinitrogen tetroxide; hydrogen peroxide; and nitric acid). Dangerous when exposed to heat, flame or oxidizers; can react vigorously with oxidizing materials such as air; fuming HNO_3; ($HNO_3 + N_2O_4$); No. A high energy propellant for liquid fueled rockets. To fight fire, use alcohol foam, CO_2, dry chemical. When heated to decomposition it emits highly toxic fumes of NO_x. See also HYDRAZINE.

DSF600 CAS:540-73-8 ***HR: 3***
1,2-DIMETHYLHYDRAZINE
DOT: UN 2382
mf: $C_2H_8N_2$ mw: 60.12

PROP: Clear, colorless, flammable, hygro liquid; fishy ammonia odor. Flash p: < 73.4°F, bp: 81°, mp: −9°, d: 0.8274 @ 20°/4°.

SYNS: 1,2-DIMETHYLHYDRAZIN (GERMAN) ◇ N,N'-DIMETHYLHYDRAZINE ◇ sym-DIMETHYLHYDRAZINE ◇ DIMETHYLHYDRAZINE, symmetrical (DOT) ◇ DMH ◇ symetryczna DWUMETYLOHYDRAZYNA (POLISH) ◇ HYDRAZOMETHANE ◇ RCRA WASTE NUMBER U099 ◇ SDMH

TOXICITY DATA with REFERENCE
otr-hmn:fbr 230 μmol/L CALEDQ 29,265,85
dns-hmn:lng 100 μL/L NTIS** AD-A041-973
sce-mus-rec 20 mg/kg ENMUDM 8(Suppl 6),41,86
hma-mus/esc 50 μmol/kg MUREAV 148,1,85
dnd-ham:lng 2 mmol/L MUREAV 173,157,86
ipr-rat TDLo:100 mg/kg (6-15D preg):TER JTEHD6 13,125,84
orl-rat TDLo:120 mg/kg/4W-I:ETA AJCNAC 30,176,77
scu-rat TDLo:90 mg/kg/28W-I:CAR BIMDB3 32,41,80
orl-mus TDLo:12500 μg/kg/24W-I:NEO TXAPA9 72,313,84
scu-mus TDLo:200 mg/kg/10W-I:CAR JJIND8 63,1081,79
orl-rat LD50:100 mg/kg NATWAY 54,285,67
ihl-rat LCLo:280 ppm/4H AMIHAB 12,609,55
ipr-rat LD50:163 mg/kg MEPAAX 24,71,73
scu-rat LD50:220 mg/kg XENOBH 3,271,73
ivn-rat LD50:176 mg/kg MEPAAX 24,71,73
orl-mus LD50:36 mg/kg MEPAAX 24,71,73
ipr-mus LD50:35 mg/kg MEPAAX 24,71,73
scu-mus LD50:24 mg/kg TOLED5 8,87,81
ivn-mus LD50:29 mg/kg MEPAAX 24,71,73
ivn-dog LD50:100 mg/kg MEPAAX 24,71,73
ims-ham LD50:95 mg/kg ARZNAD 19,1891,69

CONSENSUS REPORTS: IARC Cancer Review: Group 2B IMEMDT 7,56,87; Animal Sufficient Evidence IMEMDT 4,145,74. EPA Genetic Toxicology Program.

DFG MAK: Animal Carcinogen, Suspected Human Carcinogen.
DOT Classification: Flammable Liquid; Label: Flammable Liquid, Poison.

SAFETY PROFILE: Confirmed carcinogen with experimental carcinogenic, neoplastigenic, tumorigenic, and teratogenic data. Poison by ingestion, intraperitoneal, intravenous, subcutaneous, and intramuscular routes. Moderately toxic by inhalation. Human mutation data reported. A very dangerous fire hazard when exposed to heat, flame, or oxidizers. A high energy propellant for liquid fueled rockets. When heated to decomposition it emits toxic fumes of NO_x. See also 1,1-DIMETHYL HYDRAZINE.

DSF800 CAS:306-37-6 ***HR: 3***
1,2-DIMETHYLHYDRAZINE DIHYDROCHLORIDE
mf: $C_2H_8N_2•2ClH$ mw: 133.04

SYNS: N,N'-DIMETHYLHYDRAZINE DIHYDROCHLORIDE ◇ sym-DIMETHYLHYDRAZINE DIHYDROCHLORIDE ◇ DMH

TOXICITY DATA with REFERENCE
otr-rat-ipr 100 mg/kg CALEDQ 26,191,82
dnd-rat-orl 1700 μg/kg/2D-C CRNGDP 4,529,83
dns-rat-orl 1700 μg/kg/4D-C CRNGDP 4,529,83
dni-rat-orl 1700 μg/kg/2D-C CRNGDP 4,529,83
dni-mus:oth 500 μmol/L JJIND8 68,1015,82
ims-ham TDLo:200 mg/kg (12D preg):REP APTOD9 19,A71,80
orl-rat TDLo:35 mg/kg:CAR CALEDQ 14,47,81
ipr-rat TDLo:160 mg/kg/16W-I:CAR NUCADQ 4,146,82
ipr-mus TDLo:212 mg/kg/8W-I:ETA JNCIAM 42,337,69
olr-ham TDLo:446 mg/kg/51W-C:NEO CNREA8 32,804,72
orl-rat LD50:100 mg/kg 23HZAR -,267,70
scu-rat LD50:122 mg/kg NTIS** AD-A062-138
scu-mus LD50:25400 μg/kg CRNGDP 3,603,82
scu-ham LD50:50 mg/kg NTIS** AD-A062-138

CONSENSUS REPORTS: Reported in EPA TSCA Inventory. EPA Genetic Toxicology Program.

SAFETY PROFILE: Suspected carcinogen with experimental carcinogenic, neoplastigenic, and tumorigenic data. Poison by ingestion and subcutaneous routes. Experimental reproductive effects. Mutation data reported. A rocket fuel. When heated to decomposition it emits very toxic fumes of HCl and NO_x. See also 1,1-DIMETHYL HYDRAZINE.

DSG000 CAS:593-82-8 ***HR: 3***
1,1-DIMETHYLHYDRAZINE HYDROCHLORIDE
mf: $C_2H_8N_2$•ClH mw: 96.58

TOXICITY DATA with REFERENCE
orl-rat TDLo:35 g/kg/73W-C:ETA ZEKBAI 69,103,67
orl-rat LD50:196 mg/kg AMIHAB 13,34,56
ipr-rat LD50:210 mg/kg AMIHAB 13,34,56
ivn-rat LD50:191 mg/kg AMIHAB 13,34,56
orl-mus LD50:426 mg/kg AMIHAB 13,34,56
ipr-mus LD50:466 mg/kg AMIHAB 13,34,56
ivn-mus LD50:402 mg/kg AMIHAB 13,34,56
ivn-dog LD50:96 mg/kg AMIHAB 13,34,56

NIOSH REL: (Hydrazines) CL 0.15 mg/m^3/2H

SAFETY PROFILE: Poison by ingestion, intraperitoneal, and intravenous routes. Questionable carcinogen with experimental tumorigenic data. When heated to decomposition it emits very toxic fumes of HCl and NO_x. See also 1,1-DIMETHYL HYDRAZINE.

DSG200 CAS:56400-60-3 ***HR: 3***
1,2-DIMETHYLHYDRAZINE HYDROCHLORIDE
mf: $C_2H_8N_2$•ClH mw: 96.58

SYNS: sym-DIMETHYLHYDRAZINE HYDROCHLORIDE ◇ DMH

TOXICITY DATA with REFERENCE
dns-mus-orl 20 mg/kg FEPRA7 33,596,74
orl-rat TDLo:150 mg/kg/5W-I:NEO JJIND8 63,1089,79
scu-rat TDLo:300 mg/kg/20W-I:CAR CANCAR 40,2502,77
scu-rat TDLo:120 mg/kg/6W-I:ETA CNREA8 33,940,73
orl-rat LD50:257 mg/kg AMIHAB 13,34,56
ipr-rat LD50:262 mg/kg AMIHAB 13,34,56
ivn-rat LD50:281 mg/kg AMIHAB 13,34,56
orl-mus LD50:58 mg/kg AMIHAB 13,34,56
ipr-mus LD50:56 mg/kg AMIHAB 13,34,56
scu-mus LD50:12 mg/kg BIJOAK 122,121,71
ivn-mus LD50:47 mg/kg AMIHAB 13,34,56
ivn-dog LD50:161 mg/kg AMIHAB 13,34,56

SAFETY PROFILE: Poison by ingestion, intraperitoneal, subcutaneous, and intravenous routes. Questionable carcinogen with experimental carcinogenic, neoplastigenic, and tumorigenic data. Mutation data reported. When heated to decomposition it emits very toxic fumes of HCl and NO_x. See also 1,1-DIMETHYL HYDRAZINE.

DSG400 CAS:26049-69-4 ***HR: 3***
2-(2,2-DIMETHYLHYDRAZINO)-4-(5-NITRO-2-FURYL)THIAZOLE
mf: $C_9H_{10}N_4O_3S$ mw: 254.29

SYN: DMNT

TOXICITY DATA with REFERENCE
mma-sat 100 ng/plate MUREAV 40,9,76
dnr-sat 500 nmol/well CNREA8 34,2266,74
mmo-esc 300 nmol/well CNREA8 34,2266,74
mrc-esc 500 nmol/well CNREA8 34,2266,74
pic-esc 1 mg/L MUREAV 26,3,74
orl-rat TDLo:4800 mg/kg/46W-I:CAR CNREA8 30,897,70
orl-mus TDLo:15 g/kg/17W-C:CAR CNREA8 33,1593,73

CONSENSUS REPORTS: EPA Genetic Toxicology Program.

SAFETY PROFILE: Suspected carcinogen with experimental carcinogenic data. Mutation data reported. When heated to decomposition it emits very toxic fumes of NO_x and SO_x.

DSG600 CAS:868-85-9 ***HR: 3***
DIMETHYLHYDROGENPHOSPHITE
mf: $C_2H_7O_3P$ mw: 110.06

SYNS: DIMETHYLESTER KYSELINY FOSFORITE (CZECH) ◇ NCI-C54773 ◇ PHOSPHONIC ACID, DIMETHYL ESTER

TOXICITY DATA with REFERENCE
skn-rbt 500 mg/24H MLD 28ZPAK -,215,72
eye-rbt 20 mg/24H MOD 28ZPAK -,215,72
mma-sat 7500 μg/plate ENMUDM 8(Suppl 7),1,86
orl-rat TDLo:103 g/kg/2Y-I:CAR NTPTR* NTP-TR-287,85

orl-rat LD50:3050 mg/kg ALBRW* #OPB-3,84
skn-rbt LD50:2400 mg/kg ALBRW* #OPB-3,84

CONSENSUS REPORTS: Reported in EPA TSCA Inventory. NTP Carcinogenesis Studies (gavage); No Evidence: mouse NTPTR* NTP-TR-287,85; Clear Evidence: rat NTPTR* NTP-TR-287,85.

SAFETY PROFILE: Moderately toxic by ingestion and skin contact. A skin and eye irritant. Questionable carcinogen with experimental carcinogenic data. Mutation data reported. When heated to decomposition it emits toxic fumes of PO_x.

DSG700 CAS:654-42-2 ***HR: 3***
2,6-DIMETHYLHYDROQUINONE
mf: $C_8H_{10}O_2$ mw: 138.18

SYNS: 2,6-DIMETHYL-1,4-BENZENEDIOL (9CI) ◇ DMHQ ◇ m-XHQ ◇ m-XYLOHYDROQUINONE ◇ 2,6-XYLOHYDROQUINONE ◇ 2,6-XYLOQUINOL

TOXICITY DATA with REFERENCE
orl-wmn TDLo:84 mg/kg (30W pre):REP ACEDAB 28,83,56
orl-mus TDLo:40 mg/kg (female 1D pre):TER JOENAK 14,228,56
orl-mus LD50:186 mg/kg IJEBA6 2,23,64
ipr-mus LD50:117 mg/kg IJEBA6 2,23,64
ipr-cat LDLo:20 mg/kg IJEBA6 2,23,64

SAFETY PROFILE: Poison by ingestion and intraperitoneal routes. An experimental teratogen. Human reproductive effects by ingestion: impaired spermatogenesis in men and changes in fertility in women. Experimental reproductive effects. When heated to decomposition it emits acrid smoke and irritating fumes.

DSH000 CAS:2019-14-9 ***HR: 3***
DIMETHYL(2-HYDROXYETHYL)OCTYLAMMONIUM BROMIDE BENZILATE
mf: $C_{26}H_{38}NO_3{\cdot}Br$ mw: 492.56

SYNS: AD-205 ◇ (2-BENZILOXYETHYL)DIMETHYLOCTYLAMMONIUM BROMIDE ◇ BENZILSAEURE-DIMETHYL-OCTYL-AMMONIUM-AETHYLESTER BROMIDE (GERMAN)

TOXICITY DATA with REFERENCE
orl-mus LD50:1150 mg/kg TXAPA9 5,225,63
ipr-mus LD50:104 mg/kg TXAPA9 5,225,63
scu-mus LD50:850 mg/kg TXAPA9 5,225,63
ivn-mus LD50:17500 μg/kg TXAPA9 5,225,63

SAFETY PROFILE: Poison by intraperitoneal and intravenous routes. Moderately toxic by ingestion and subcutaneous routes. When heated to decomposition it emits very toxic fumes of NO_x, NH_3, and Br^-.

DSH200 CAS:56927-39-0 ***HR: 3***
DIMETHYL(2-HYDROXYETHYL)PENTYLAMMONIUM BROMIDE BENZILATE
mf: $C_{23}H_{32}NO_3{\cdot}Br$ mw: 450.47

SYN: BENZILSAEURE-DIMETHYL-PENTYL-AMMONIUM-AETHYLESTER BROMIDE (GERMAN)

TOXICITY DATA with REFERENCE
scu-mus LD50:230 mg/kg ARZNAD 10,763,60
ivn-mus LD50:12 mg/kg ARZNAD 10,763,60

SAFETY PROFILE: Poison by subcutaneous and intravenous routes. When heated to decomposition it emits very toxic fumes of NO_x, NH_3 and Br^-.

DSH400 CAS:69928-30-9 ***HR: 2***
3,5-DIMETHYL-3-HYDROXYHEXANE-4-CARBOXYLIC ACID-β-LACTONE
mf: $C_9H_{16}O_2$ mw: 156.25

SYN: 4-ETHYL-3-ISOPROPYL-4-METHYL-1-OXACYCLOBUTAN-2-ONE

TOXICITY DATA with REFERENCE
eye-rbt 20 mg open SEV AMIHBC 10,61,54
orl-rat LD50:2700 mg/kg AMIHBC 10,61,54

SAFETY PROFILE: Moderately toxic by ingestion. A severe eye irritant. When heated to decomposition it emits acrid smoke and irritating fumes.

DSH600 CAS:58344-42-6 ***HR: 2***
4-(4,4-DIMETHYL-3-HYDROXY-1-PENTENYL)-2-METHOXYPHENOL
mf: $C_{14}H_{20}O_3$ mw: 236.34

TOXICITY DATA with REFERENCE
skn-rbt 10 mg/24H MLD AMIHBC 4,119,51
eye-rbt 750 μg SEV AMIHBC 4,119,51
ipr-mus LD50:1800 mg/kg EJMCA5 13,41,78

SAFETY PROFILE: Moderately toxic by intraperitoneal route. A skin and severe eye irritant. When heated to decomposition it emits acrid smoke and irritating fumes.

DSH700 CAS:66634-53-5 ***HR: 3***
N,N-DIMETHYL-β-HYDROXYPHENETHYLAMINE
mf: $C_{10}H_{15}NO$ mw: 165.26

SYNS: α-(DIMETHYLAMINOMETHYL)BENZYL ALCOHOL ◇ β-HYDROXY-β-PHENYLETHYL DIMETHYLAMINE

TOXICITY DATA with REFERENCE
scu-mus LDLo:826 mg/kg AIPTAK 47,96,34
ivn-mus LD50:56 mg/kg CSLNX* NX#04125
ivn-rbt LDLo:132 mg/kg AIPTAK 47,96,34

SAFETY PROFILE: Poison by intravenous route.

Moderately toxic by subcutaneous route. When heated to decomposition it emits toxic fumes of NO_x. See also AMINES.

DSH800 CAS:27945-43-3 ***HR: 3***
3-(3,5-DIMETHYL-4-HYDROXYPHENYL)-2-METHYL-4(3H)-QUINAZOLINONE
mf: $C_{17}H_{16}N_2O_2$ mw: 280.35

SYNS: 3-(4-HYDROXY-3,5-XYLYL)-2-METHYL-4(3H)-QUINAZOLINONE ◇ 2-METHYL-3-(3,5-DIMETHYL-4-HYDROXYPHENYL)-3,4-DIHYDROQUINAZOLIN-4-ONE ◇ SRC-226

TOXICITY DATA with REFERENCE
orl-rat LD50:230 mg/kg IJPAAO 37,109,75
orl-mus LD50:775 mg/kg IJPAAO 37,109,75

SAFETY PROFILE: Poison by ingestion. When heated to decomposition it emits toxic fumes of NO_x.

DSI000 CAS:1505-26-6 ***HR: 3***
1,2-DIMETHYL-3-(m-HYDROXYPHENYL)-3-PROPYLPYRROLIDINE
mf: $C_{15}H_{23}N$ mw: 217.39

SYN: m-(1,2-DIMETHYL-3-PROPYL-3-PYRROLIDINYL)PHENOL

TOXICITY DATA with REFERENCE
ipr-rat LDLo:97 mg/kg JMCMAR 8,316,65
orl-mus LDLo:300 mg/kg CHTPBA 7,450,72
ipr-mus LDLo:300 mg/kg CHTPBA 7,450,72

SAFETY PROFILE: Poison by ingestion and intraperitoneal routes. When heated to decomposition it emits toxic fumes of NO_x.

DSI200 CAS:66941-43-3 ***HR: 3***
(2,2-DIMETHYL-3-HYDROXYPROPYL)TRIETHYLAMMONIUM BROMIDE TROPATE (ESTER)
mf: $C_{20}H_{34}NO_3 \cdot Br$ mw: 416.46

SYN: TROPASAEUREESTER DES 3-TRIAETHYLAMMONIUM-2,2-DIMETHYL-1-PROPANOLBROMID(GERMAN)

TOXICITY DATA with REFERENCE
ivn-rat LDLo:80 mg/kg AEPPAE 173,86,33
ivn-mus LDLo:70 mg/kg AEPPAE 173,86,33
unr-frg LDLo:3000 mg/kg AEPPAE 173,86,33

SAFETY PROFILE: Poison by intravenous route. Moderately toxic by an unspecified route. When heated to decomposition it emits very toxic fumes of NO_x, NH_3, and Br^-. See also ESTERS.

DSI489 CAS:29128-41-4 ***HR: 3***
DIMETHYL HYPONITRILE
mf: $C_2H_6N_2O_2$ mw: 90.08

SYN: DIMETHOXYDIAZENE

CONSENSUS REPORTS: Cyanide and its compounds are on the Community Right-To-Know List.

SAFETY PROFILE: A dangerously unpredictable explosive. When heated to decomposition it emits toxic fumes of CN^- and NO_x. See also NITRILES.

DSI709 CAS:50-47-5 ***HR: 3***
DIMETHYLIMIPRAMINE
mf: $C_{18}H_{22}N_2$ mw: 266.42

SYNS: DEMETHYLIMIPRAMINE ◇ DESIMIPRAMINE ◇ DESIPRAMIN ◇ DESIPRAMINE (D4) ◇ DESMETHYLIMIPRAMINE ◇ DMI ◇ DMI 50475 ◇ METHYLAMINOPROPYLIMINODIBENZYL ◇ MONODEMETHYLIMIPRAMINE ◇ NORIMIPRAMINE ◇ PENTOFRAN ◇ PERTOFRAN ◇ PERTOFRANE ◇ SERTOFRAN

TOXICITY DATA with REFERENCE
dnd-esc 20 μmol/L MUREAV 89,95,81
cyt-oin-unr 10 g/L JCLBA3 47,182a,70
scu-rat TDLo:16250 μg/kg (female 8-20D post):REP ARTODN 7,504,84
orl-rat TDLo:500 mg/kg (female 7-16D post):TER ARZNAD 19,1617,69
orl-wmn LDLo:30 mg/kg:BRN,CNS 34ZIAG -,201,69
orl-chd LDLo:125 mg/kg:CNS,PUL PSYPAG 10,431,67
orl-hmn LDLo:30 mg/kg:CNS,BRN,PUL DMWOAX 93,117,68
orl-rat LD50:375 mg/kg ARZNAD 19,1617,69
ipr-rat LD50:48 mg/kg ARZNAD 20,1561,70
scu-rat LD50:183 mg/kg ARZNAD 20,1561,70
ivn-rat LD50:29 mg/kg AIPTAK 148,560,64
orl-mus LD50:448 mg/kg JJPAAZ 21,47,71
ipr-mus LD50:85 mg/kg ARZNAD 21,1727,71
scu-mus LD50:214 mg/kg FRPPAO 25,519,70
ivn-mus LD50:22 mg/kg APSXAS 12,173,75

SAFETY PROFILE: Human poison by ingestion. Experimental poison by ingestion, intraperitoneal, subcutaneous, and intravenous routes. Human systemic effects by ingestion: degenerative brain changes, tremors, coma, and cyanosis. An experimental teratogen. Other experimental reproductive effects. Mutation data reported. An antidepressant. Related to diazepam. When heated to decomposition it emits toxic fumes of NO_x.

DSI800 CAS:17309-87-4 ***HR: 3***
N,N-DIMETHYL-p-(6-INDAZYLAZO)ANILINE
mf: $C_{15}H_{15}N_5$ mw: 265.35

SYNS: 6-((p-(DIMETHYLAMINO)PHENYL)AZO)-1H-INDAZOLE ◇ N,N-DIMETHYL-4-(6′-1H-INDAZYLAZO)ANILINE

TOXICITY DATA with REFERENCE
mma-sat 20 μg/plate MUREAV 93,67,82
orl-rat TDLo:2700 mg/kg/21W-C:CAR JMCMAR 12,1113,69

SAFETY PROFILE: Questionable carcinogen with ex-

perimental carcinogenic data. Mutation data reported. When heated to decomposition it emits toxic fumes of NO_x.

DSI889 ***HR: 3***
DIMETHYLIODOARSINE
mf: C_2H_6AsI mw: 231.90

CONSENSUS REPORTS: Arsenic and its compounds are on the Community Right-To-Know List.

SAFETY PROFILE: Ignites when heated in air. When heated to decomposition it emits toxic fumes of I^- and As. See also ARSENIC COMPOUNDS and IODIDES.

DSJ200 CAS:10143-20-1 ***HR: 1***
2,8-DIMETHYL-6-ISOBUTYLNONANOL-4
mf: $C_{15}H_{32}O$ mw: 228.47

TOXICITY DATA with REFERENCE
skn-rbt 10 mg/24H open MLD AMIHBC 10,61,54
eye-rbt 500 mg open AMIHBC 10,61,54
orl-rat LD50:16 g/kg AMIHBC 10,61,54

SAFETY PROFILE: Mildly toxic by ingestion. A skin and eye irritant. When heated to decomposition it emits acrid smoke and irritating fumes.

DSJ800 CAS:489-84-9 ***HR: 2***
1,4-DIMETHYL-7-ISOPROPYLAZULENE
mf: $C_{15}H_{18}$ mw: 198.33

SYNS: AZULON ◇ s-GUAIAZULENE

TOXICITY DATA with REFERENCE
orl-rat LD50:1550 mg/kg ARZNAD 19,615,69
orl-mus LD50:1300 mg/kg ARZNAD 19,615,69

CONSENSUS REPORTS: Reported in EPA TSCA Inventory.

SAFETY PROFILE: Moderately toxic by ingestion. When heated to decomposition it emits acrid smoke and irritating fumes.

DSK200 CAS:119-38-0 ***HR: 3***
DIMETHYL-5-(1-ISOPROPYL-3-METHYL-PYRAZOLYL)CARBAMATE
mf: $C_{10}H_{17}N_3O_2$ mw: 211.30

SYNS: DIMETHYLCARBAMATE-d'1-ISOPROPYL-3-METHYL-5-PYRAZOLYLE (FRENCH) ◇ DIMETHYLCARBAMIC ACID 3-METHYL-1-(1-METHYLETHYL)-1H-PYRAZOL-5-YL ESTER ◇ ENT 19,060 ◇ GEIGY G-23611 ◇ ISOLAN ◇ ISOLANE (FRENCH) ◇ (1-ISOPROPIL-3-METIL-1H-PIRAZOL-5-IL)-N,N-DIMETIL-CARBAMMATO(ITALIAN) ◇ (1-ISOPROPYL-3-METHYL-1H-PYRAZOL-5-YL)-N,N-DIMETHYL-CARBAMAAT (DUTCH) ◇ (1-ISOPROPYL-3-METHYL-1H-PYRAZOL-5-YL)-N,N-DIMETHYL-CARBAMAT (GERMAN) ◇ ISOPROPYLMETHYLPYRAZOLYL DIMETHYLCARBAMATE ◇ 1-ISOPROPYL-3-METHYL-5-PYRAZOLYL DIMETHYLCARBAMATE ◇ 1-ISOPROPYL-3-METHYLPYRAZOLYL-(5)-DIMETHYLCARBAMATE ◇ 5-METHYL-2-ISOPROPYL-3-PYRAZOLYL DIMETHYLCARBAMATE ◇ PRIMIN ◇ SAOLAN

TOXICITY DATA with REFERENCE
mmo-smc 5 ppm RSTUDV 6,161,76
orl-mus TDLo:6600 mg/kg/78W-I:ETA NTIS** PB223-159
orl-rat LD50:10800 μg/kg PESTD5 17,351,76
skn-rat LD50:5600 μg/kg 85DPAN -,-,71/76
ipr-rat LD50:2150 μg/kg PESTD5 17,351,76
orl-mus LD50:9800 μg/kg PESTD5 17,351,76
ipr-mus LD50:1 mg/kg TXAPA9 6,402,64
orl-bwd LD50:8600 μg/kg TXAPA9 21,315,72

CONSENSUS REPORTS: EPA Extremely Hazardous Substances List.

SAFETY PROFILE: Poison by ingestion, skin contact, and intraperitoneal routes. Questionable carcinogen with experimental tumorigenic data. Mutation data reported. An insecticide. When heated to decomposition it emits toxic fumes of NO_x. See also CARBAMATES.

DSK600 CAS:2674-91-1 ***HR: 3***
O,O-DIMETHYL-S-ISOPROPYL-2-SULFINYLETHYLPHOSPHOROTHIOATE
mf: $C_7H_{17}O_4PS_2$ mw: 260.33

SYNS: S-2-AETHYLSULFINYL-1-METHYLAETHYL-O,O DIMETHYL-MONOTHIOPHOSPHAT ◇ BAY 23655 ◇ BAYER 23655 ◇ ENT 25,674 ◇ ESP ◇ ESTON ◇ ESTOX ◇ S-2-ETHYL-SULFINYL-1-METHYL-ETHYL-O,O-DIMETHYL-MONOTHIOFOSFAAT ◇ S-2-ETHYL-SULPHINYL-1-METHYL-ETHYL-O,O-DIMETHYLPHOSPHOROTHIOLATE ◇ S-2-ETIL-SULFINIL-1-METIL-ETIL-O,O-DIMETIL-MONOTIOFOSFATO ◇ METASYSTOX-S ◇ OXYDEPROFOS ◇ OXYPHIONFOS ◇ PHOSPHOROTHIOIC ACID, O,O-DIMETHYL S-(ETHYLSULFINYL-(2-ISOPROPYL)) ESTER ◇ S410 ◇ THIOMETAN ◇ THIOPHOSPHATE de O,O-DIMETHYLE ET DE S-2-(ISOPROPYLSULFINYL)-ETHYLE

TOXICITY DATA with REFERENCE
mmo-sat 50 mg/plate MUREAV 116,185,83
mma-sat 50 mg/plate MUREAV 116,185,83
orl-rat LD50:103 mg/kg ARSIM* 20,3,66
skn-rat LD50:1000 mg/kg GUCHAZ 6,215,73
ipr-rat LD50:50 mg/kg GUCHAZ 6,215,73
unr-rat LD50:105 mg/kg 30ZDA9-,349,71
orl-mus LD50:58700 μg/kg YKYUA6 30,623,79
ipr-mus LD50:30 mg/kg TXAPA9 4,621,62

SAFETY PROFILE: Poison by ingestion and intraperitoneal routes. Mutation data reported. When heated to decomposition it emits very toxic fumes of PO_x and SO_x.

DSK800 CAS:36614-38-7 ***HR: 3***
O,O-DIMETHYL-S-2-(ISOPROPYLTHIO)ETHYL-PHOSPHORODITHIOATE
mf: $C_7H_{17}O_2PS_3$ mw: 260.39

SYNS: HODSON ◇ HOSALON ◇ HOSDON GRANULE ◇ S-2-ISOPROPYLTHIOETHYL-O,O-DIMETHYLPHOSPHORODITHIOATE

◇ ISOTHIOATE ◇ PHOSPHORODITHIOIC ACID-O,O-DIMETHYL-S-(2-((1-METHYLETHYL)THIO)ETHYL)ESTER

TOXICITY DATA with REFERENCE
orl-rat LD50:150 mg/kg 28ZEAL 5,134,76
orl-mus LD50:50 mg/kg FMCHA2 -,C128,83
skn-mus LD50:240 mg/kg FMCHA2 -,C128,83

SAFETY PROFILE: Poison by ingestion and skin contact. When heated to decomposition it emits very toxic fumes of PO_x and SO_x.

DSK900 CAS:534-13-4 **HR: 2**
1,3-DIMETHYLISOTHIOUREA
mf: $C_3H_8N_2S$ mw: 104.19

PROP: Colorless, exceedingly deliquescent crystals. Mp: 60-62°. Very sol in water, alc, acetone; sparingly sol in benzene, ether, carbon disulfide; very sltly in petr ether.

SYNS: DIMETHYLTHIOCARBAMIDE ◇ N,N'-DIMETHYLTHIOCARBAMIDE ◇ sym-DIMETHYLTHIOUREA ◇ 1,3-DIMETHYLTHIOUREA

TOXICITY DATA with REFERENCE
orl-rat TDLo:2 g/kg (14D preg):TER TJADAB 23,335,81
orl-rat TDLo:2 g/kg (14D preg):REP TJADAB 23,335,81
orl-mus LDLo:500 mg/kg TJADAB 23,335,81

CONSENSUS REPORTS: Reported in EPA TSCA Inventory.

SAFETY PROFILE: Moderately toxic by ingestion. Experimental teratogenic and reproductive effects. When heated to decomposition it emits toxic fumes of SO_x and NO_x.

DSL000 CAS:6155-81-3 **HR: 2**
N(1)-(3,4-DIMETHYL-5-ISOXAZOLYL)SULFANILAMIDE LITHIUM SALT
mf: $C_{11}H_{13}N_3O_3S{\cdot}Li$ mw: 274.27

TOXICITY DATA with REFERENCE
orl-mus LD50:1 g/kg JPETAB 88,47,46
scu-mus LD50:5000 mg/kg JPETAB 88,47,46
ivn-mus LD50:2500 mg/kg JPETAB 88,47,46

SAFETY PROFILE: Moderately toxic by ingestion and intravenous routes. Mildly toxic by subcutaneous route. When heated to decomposition it emits very toxic fumes of NO_x and SO_x. See also LITHIUM COMPOUNDS.

DSL200 CAS:2200-44-4 **HR: 2**
N(1)-(3,4-DIMETHYL-5-ISOXAZOLYL)SULFANILAMIDE SODIUM SALT
mf: $C_{11}H_{13}N_3O_3S{\cdot}Na$ mw: 290.32

TOXICITY DATA with REFERENCE
orl-rat LD50:1 g/kg JPETAB 88,47,46
ipr-rat LD50:3200 mg/kg JPETAB 88,47,46
orl-mus LD50:1 g/kg JPETAB 88,47,46
ivn-mus LD50:2300 mg/kg JPETAB 88,47,46

SAFETY PROFILE: Moderately toxic by ingestion, intraperitoneal and intravenous routes. When heated to decomposition it emits very toxic fumes of NO_x, Na_2O, and SO_x.

DSL289 CAS:598-26-5 **HR: 3**
DIMETHYLKETENE
mf: C_4H_6O mw: 70.09

$(CH_3)_2C{=}C{=}O$

SYN: 2-METHYL-1-PROPENE-1-ONE

SAFETY PROFILE: Upon exposure to air it forms the very unstable explosive peroxide poly(peroxyisobutyrolactone). The peroxide is heat- and friction-sensitive and will also explode upon evaporation. When heated to decomposition it emits acrid smoke and fumes. See also PEROXIDES.

DSL400 CAS:20917-34-4 **HR: 3**
DIMETHYL LEAD DIACETATE
mf: $C_6H_{12}O_4Pb$ mw: 355.37

SYN: DIACETOXYDIMETHYLPLUMBANE

TOXICITY DATA with REFERENCE
orl-mus LD50:120 mg/kg CRSBAW 162,1456,68

CONSENSUS REPORTS: Lead and its compounds are on the Community Right-To-Know List.

SAFETY PROFILE: Poison by ingestion. When heated to decomposition it emits toxic fumes of Pb. See also LEAD COMPOUNDS.

DSL600 CAS:2999-74-8 **HR: 3**
DIMETHYLMAGNESIUM
DOT: UN 1368
mf: C_2H_6Mg mw: 54.38

CONSENSUS REPORTS: Reported in EPA TSCA Inventory.

DOT Classification: Flammable Solid; Label:Spontaneously Combustible

SAFETY PROFILE: The solid and its solution in ether ignite on contact with water. The powder ignites on contact with moist air. When heated to decomposition it emits irritating fumes of MgO. See also MAGNESIUM COMPOUNDS.

DSL800 CAS:624-48-6 **HR: 2**
DIMETHYL MALEATE
mf: $C_6H_8O_4$ mw: 144.14

PROP: Liquid. Mp: −17.5°, bp: 205.0°, flash p: 235°F (OC), d: 1.153, vap press: 1 mm @ 45.7°, vap d: 4.97.

SYN: METHYL MALEATE

TOXICITY DATA with REFERENCE
eye-rbt 100 mg AJOPAA 29,1363,46
orl-rat LDLo:1410 mg/kg AIHAAP 23,95,62
skn-rbt LD50:530 mg/kg AIHAAP 23,95,62

CONSENSUS REPORTS: Reported in EPA TSCA Inventory.

SAFETY PROFILE: Moderately toxic by ingestion and skin contact. An eye irritant. Combustible when exposed to heat or flame; can react with oxidizing materials. To fight fire use CO_2, dry chemical. See also ESTERS and MALEIC ACID.

DSM000 CAS:766-39-2 ***HR: 3***
α,β-DIMETHYLMALEIC ANHYDRIDE
mf: $C_6H_6O_3$ mw: 126.12

SYN: DIMETHYLMALEIC ANHYDRIDE

TOXICITY DATA with REFERENCE
scu-rat TDLo:2600 mg/kg/65W-I:ETA BJCAAI 19,392,65

CONSENSUS REPORTS: Reported in EPA TSCA Inventory.

SAFETY PROFILE: Questionable carcinogen with experimental tumorigenic data. When heated to decomposition it emits acrid smoke and irritating fumes. See also ANHYDRIDES.

DSM200 CAS:108-59-8 ***HR: 1***
DIMETHYL MALONATE
mf: $C_5H_8O_4$ mw: 132.13

SYNS: DIMETHYL PROPANEDIOATE ◇ MALONIC ACID DIMETHYL ESTER ◇ METHYL MALONATE ◇ PROPANEDIOIC ACID DIMETHYL ESTER (9CI)

TOXICITY DATA with REFERENCE
skn-rbt 500 mg/24H FCTXAV 17,363,79
orl-rat LD50:5331 mg/kg FCTXAV 17,363,79

CONSENSUS REPORTS: Reported in EPA TSCA Inventory.

SAFETY PROFILE: Mildly toxic by ingestion. A skin irritant. Violent reaction with CH_3N_3 occurred with $NaOCH_3$ present. When heated to decomposition it emits acrid smoke and irritating fumes. See also ESTERS.

DSM289 CAS:33212-68-9 ***HR: 3***
DIMETHYL MANGANESE
mf: C_2H_6Mn mw: 85.01

CONSENSUS REPORTS: Manganese and its compounds are on the Community Right-To-Know List.

SAFETY PROFILE: An unstable explosive. Ignites spontaneously in air. See also MANGANESE COMPOUNDS.

DSM450 CAS:593-74-8 ***HR: 3***
DIMETHYL MERCURY
mf: C_2H_6Hg mw: 230.67

PROP: Volatile, colorless liquid. Bp: 92°, d: 3.1874. Insoluble in water; very sol in alc and ether.

SYN: MERCURY, DIMETHYL

TOXICITY DATA with REFERENCE
dnd-mmo-omi 600 mg/L NATUAS 257,422,75
oth-mus:oth 25 mg/L MUREAV 17,93,73

CONSENSUS REPORTS: Reported in EPA TSCA Inventory.

OSHA PEL: (Transitional: 0.01 mg/10m^3) 0.01 mg(Hg)/m^3; CL 0.03 mg(Hg)/m^3 (skin)
ACGIH TLV: TWA 0.01 mg(Hg)/m^3; STEL 0.03 mg(Hg)/m^3

SAFETY PROFILE: Highly toxic. Mutation data reported. Easily flammable. When heated to decomposition it emits toxic fumes of Hg.

DSM600 CAS:63938-21-6 ***HR: 3***
2,3-DIMETHYL-7-METHOXY-8-(MORPHOLINOMETHYL)CHROMONE HYDROCHLORIDE
mf: $C_{17}H_{21}NO_4{\cdot}ClH$ mw: 339.85

TOXICITY DATA with REFERENCE
orl-rat LD50:44 mg/kg 27ZQAG -,156,72
ipr-rat LD50:17 mg/kg 27ZQAG -,156,72
scu-rat LD50:19 mg/kg 27ZQAG -,156,72
ipr-mus LD50:17400 μg/kg JMPCAS 3,471,61

SAFETY PROFILE: Poison by ingestion, intraperitoneal, and subcutaneous routes. When heated to decomposition it emits very toxic fumes of NO_x and HCl.

DSM800 CAS:3613-30-7 ***HR: 1***
3,7-DIMETHYL-7-METHOXY-1-OCTANAL
mf: $C_{11}H_{22}O_2$ mw: 186.33

SYNS: HYDROXYCITRONELLA METHYL ETHER ◇ METHOXYCITRONELLAL METHYL ETHER

TOXICITY DATA with REFERENCE
skn-rbt 500 mg/24H MOD FCTXAV 14,807,76

CONSENSUS REPORTS: Reported in EPA TSCA Inventory.

SAFETY PROFILE: A skin irritant. When heated to decomposition it emits acrid smoke and irritating fumes. See also ETHERS.

DSN000 CAS:3009-55-0 ***HR: 3***
N,N-DIMETHYL-p-(2-METHOXYPHENYLAZO) ANILINE
mf: $C_{15}H_{17}N_3O$ mw: 255.35

SYN: 2′-METHOXY-4-DIMETHYLAMINOAZOBENZENE

TOXICITY DATA with REFERENCE
orl-rat TDLo:6600 mg/kg/26W-C:NEO CNREA8 17,387,57

SAFETY PROFILE: Questionable carcinogen with experimental neoplastigenic data. When heated to decomposition it emits toxic fumes of NO_x.

DSN200 CAS:20691-83-2 ***HR: 3***
N,N-DIMETHYL-p-(3-METHOXYPHENYLAZO) ANILINE
mf: $C_{15}H_{17}N_3O$ mw: 255.35

SYN: 3′-METHOXY-4-DIMETHYLAMINOAZOBENZENE

TOXICITY DATA with REFERENCE
orl-rat TDLo:2800 mg/kg/13W-C:NEO CNREA8 17,387,57

SAFETY PROFILE: Questionable carcinogen with experimental neoplastigenic data. When heated to decomposition it emits toxic fumes of NO_x.

DSN400 CAS:3009-50-5 ***HR: 3***
N,N-DIMETHYL-p-(4-METHOXYPHENYLAZO) ANILINE
mf: $C_{15}H_{17}N_3O$ mw: 255.35

SYN: 4′-METHOXY-4-DIMETHYLAMINOAZOBENZENE

TOXICITY DATA with REFERENCE
orl-rat TDLo:6600 mg/kg/26W-C:NEO CNREA8 17,387,57
orl-rat TD:10204 mg/kg/48W-C:ETA ARZNAD 12,270,62

SAFETY PROFILE: Questionable carcinogen with experimental neoplastigenic and tumorigenic data. When heated to decomposition it emits toxic fumes of NO_x.

DSN600 CAS:7203-92-1 ***HR: 3***
3,3-DIMETHYL-1-p-METHOXYPHENYL-TRIAZENE
mf: $C_9H_{13}N_3O$ mw: 179.25

SYNS: 1-p-METHOXYFENYL-3,3-DIMETHYLTRIAZEN (CZECH) ◇ 1-(p-METHOXYPHENYL)-3,3-DIMETHYLTRIAZENE ◇ 1-(4-METHYLOXYPHENYL)-3,3-DIMETHYLTRIAZINE

TOXICITY DATA with REFERENCE
sln-dmg-orl 1 mmol/L CBINA8 9,365,74
mrc-smc 1 mmol/L/1H CBINA8 9,365,74
hma-mus/smc 10 mmol/L CBINA8 9,365,74
scu-rat TDLo:1700 mg/kg/45W-I:CAR ZKKOBW 81,285,74
orl-rat LD50:347 mg/kg 28ZPAK -,119,72
scu-rat LD50:450 mg/kg ZKKOBW 81,285,74

CONSENSUS REPORTS: EPA Genetic Toxicology Program.

SAFETY PROFILE: Poison by ingestion. Moderately toxic by subcutaneous route. Questionable carcinogen with experimental carcinogenic data. Mutation data reported. When heated to decomposition it emits toxic fumes of NO_x.

DSN800 CAS:67262-79-7 ***HR: 3***
2′,6′-DIMETHYL-2-(2-METHOXYPROPYLAMINO)ACETANILIDE HYDROCHLORIDE
mf: $C_{14}H_{22}N_2O_2$ mw: 250.38

SYN: 2-(2-METHOXYPROPYLAMINO)-2,6′-ACETOXYLIDIDE HYDROCHLORIDE

TOXICITY DATA with REFERENCE
ipr-mus LD50:120 mg/kg JPMSAE 67,595,78
ivn-mus LD50:40 mg/kg JPMSAE 67,595,78

SAFETY PROFILE: Poison by intraperitoneal and intravenous routes. When heated to decomposition it emits toxic fumes of NO_x and HCl.

DSO000 CAS:950-37-8 ***HR: 3***
O,O-DIMETHYL-S-(5-METHOXY-1,3,4-THIADIAZOLINYL-3-METHYL) DITHIOPHOSPHATE
mf: $C_6H_{11}N_2O_4PS_3$ mw: 302.34

SYNS: CIBA-GEIGY GS 13005 ◇ S-(2,3-DIHYDRO-5-METHOXY-2-OXO-1,3,4-THIADIAZOL-3-METHYL) ◇ (O,O-DIMETHYL)-S-(-2-METHOXY-Δ^2-1,3,4-THIADIAZOLIN-5-ON-4-YLMETHYL)DITHIOPHOSPHATE DIMETHYL PHOSPHOROTHIOLOTHIONATE ◇ O,O-DIMETHYL-S-(2-METHOXY-1,3,4-THIADIAZOL-5-(4H)-ONYL-(4)-METHYL)-DITHIOPHOSPHAT (GERMAN) ◇ O,O-DIMETHYL-S-(2-METHOXY-1,3,4-THIADIAZOL-5(4H)-ONYL-(4)-METHYL)PHOSPHORODITHIOATE ◇ O,O-DIMETHYL-S-((2-METHOXY-1,3,4 (4H)-THIODIAZOL-5-ON-4-YL)-METHYL)DITHIOFOSFAAT (DUTCH) ◇ O,O-DIMETIL-S-((2-METOSSI-1,3,4-(4H)-TIADIZAOL-5-ON-4-IL)-METIL)-DITIFOSFATO (ITALIAN) ◇ DMTP (JAPAN) ◇ ENT 27,193 ◇ FISONS NC 2964 ◇ GEIGY 13005 ◇ METHIDATHION ◇ S-((5-METHOXY-2-OXO-1,3,4-THIADIAZOL-3(2H)-YL)METHYL)-O,O-DIMETHYLPHOSPHORODITHIOATE ◇ SOMONIL ◇ SURPRACIDE ◇ ULTRACIDE

TOXICITY DATA with REFERENCE
eye-rbt 34 mg SEV CIGET* 8/1/73
sce-hmn:lym 40 mg/L MUREAV 88,307,81
sce-ham:lng 40 mg/L MUREAV 88,307,81
orl-rat LD50:20 mg/kg WRPCA2 9,119,70
ihl-rat LC50:3600 mg/m^3/4H FMCHA2 -,C224,83
skn-rat LD50:25 mg/kg WRPCA2 9,119,70
orl-mus LD50:25 mg/kg BESAAT 15,122,69
orl-rbt LD50:63 mg/kg 31ZOAD 1,293,68
skn-rbt LD50:200 mg/kg FMCHA2 -,C224,83

orl-gpg LD50:25 mg/kg 31ZOAD 1,293,68
orl-ham LD50:30 mg/kg 31ZOAD 1,293,68
orl-ckn LD50:80 mg/kg 31ZOAD 1,293,68

CONSENSUS REPORTS: EPA Extremely Hazardous Substances List.

SAFETY PROFILE: Poison by ingestion and skin contact. Moderately toxic by inhalation. Human mutation data reported. A severe eye irritant. An insecticide. When heated to decomposition it emits very toxic fumes of NO_x, PO_x, and SO_x.

DSO200 CAS:23422-53-9 ***HR: 3***
N,N-DIMETHYL-N'-(((METHYLAMINO)CARBONYL)OXY)PHENYLMETHANIMIDAMIDE MONOHYDROCHLORIDE
mf: $C_{11}H_{15}N_3O_2$•ClH mw: 257.75

SYNS: CARZOL SP ◇ DICARZOL ◇ m-(((DIMETHYLAMINO)METHYLENE)AMINO)PHENYLMETHYLCARBAMATE,HYDROCHLORIDE ◇ 3-DIMETHYLAMINOMETHYLENEIMINOPHENYL-N-METHYLCARBAMATE, HYDROCHLORIDE ◇ ENT 27,566 ◇ EP-332 ◇ FORMETANATE HYDROCHLORIDE ◇ MORTON EP332 ◇ NOR-AM EP 332 ◇ SCHERING 36056 ◇ SN 36056

TOXICITY DATA with REFERENCE
orl-mus LD50:18 mg/kg 28ZEAL 5,118,76
orl-dog LD50:19 mg/kg 28ZEAL 5,118,76
skn-rbt LD50:10200 mg/kg 28ZEAL 5,118,76
orl-ckn LD50:21500 μg/kg 28ZEAL 5,118,76

CONSENSUS REPORTS: Reported in EPA TSCA Inventory. EPA Extremely Hazardous Substances List.

SAFETY PROFILE: Poison by ingestion. Mildly toxic by skin contact. When heated to decomposition it emits very toxic fumes of NO_x and HCl.

DSO400 CAS:64050-20-0 ***HR: 3***
5,7-DIMETHYL-1-(2-METHYLAMINOPROPYL)-2-PHENYLADAMANTANE HYDROCHLORIDE
mf: $C_{23}H_{35}N$•ClH mw: 362.05

SYN: 1-(5,7-DIMETHYL-2-PHENYL-1-ADAMANTYL)-N-METHYL-2-PROPYLAMINE HYDROCHLORIDE

TOXICITY DATA with REFERENCE
orl-mus LD50:1200 mg/kg JMCMAR 17,602,74
ipr-mus LD50:50 mg/kg JMCMAR 17,602,74

SAFETY PROFILE: Poison by intraperitoneal route. Moderately toxic by ingestion. When heated to decomposition it emits very toxic fumes of NO_x and HCl.

DSO800 CAS:2449-49-2 ***HR: 2***
N,N-DIMETHYL-α-METHYLBENZYLAMINE
mf: $C_{10}H_{15}N$ mw: 149.26

SYN: N,N,α-TRIMETHYLBENZYLAMINE

TOXICITY DATA with REFERENCE
skn-rbt 10 mg/24H open AMIHBC 10,61,54
eye-rbt 20 mg open AMIHBC 10,61,54
orl-rat LD50:420 mg/kg AMIHBC 10,61,54
ihl-rat LCLo:125 ppm/4H AMIHBC 10,61,54
skn-rbt LD50:890 mg/kg AMIHBC 10,61,54

SAFETY PROFILE: Moderately toxic by ingestion, skin contact and inhalation. A skin and eye irritant. When heated to decomposition it emits toxic fumes of NO_x.

DSP200 CAS:66941-08-0 ***HR: 3***
1,5-DIMETHYL-5-(1-METHYLBUTYL)BARBITURIC ACID
mf: $C_{11}H_{18}N_2O_3$ mw: 226.31

TOXICITY DATA with REFERENCE
ipr-mus LDLo:350 mg/kg JACSAT 58,1358,36
ivn-rbt LDLo:90 mg/kg JACSAT 58,1354,36

SAFETY PROFILE: Poison by intraperitoneal and intravenous routes. When heated to decomposition it emits toxic fumes of NO_x. See also BARBITURATES.

DSP400 CAS:60-51-5 ***HR: 3***
O,O-DIMETHYL METHYLCARBAMOYLMETHYL PHOSPHORODITHIOATE
mf: $C_5H_{12}NO_3PS_2$ mw: 229.27

SYNS: AC-12682 ◇ AMERICAN CYANAMID 12880 ◇ BI-58 ◇ CEKUTHOATE ◇ CL 12880 ◇ CYGON ◇ CYGON INSECTICIDE ◇ DAPHENE ◇ DE-FEND ◇ DEMOS-L40 ◇ DEVIGON ◇ DIMATE 267 ◇ DIMETATE ◇ DIMETHOAAT (DUTCH) ◇ DIMETHOAT (GERMAN) ◇ DIMETHOATE (USDA) ◇ DIMETHOAT TECHNISCH 95% ◇ DIMETHOGEN ◇ O,O-DIMETHYLDITHIOPHOSPHORYLACETIC ACID-N-MONOMETHYLAMIDE SALT ◇ O,O-DIMETHYL-DITHIOPHOSPHORYLESSIGSAEURE MONOMETHYLAMID (GERMAN) ◇ O,O-DIMETHYL-S-(2-(METHYLAMINO)-2-OXOETHYL)PHOSPHORODITHIOATE ◇ O,O-DIMETHYL-S-(N-METHYL-CARBAMOYL)-METHYL-DITHIOFOSFAAT (DUTCH) ◇ (O,O-DIMETHYL-S-(N-METHYL-CARBAMOYL-METHYL)-DITHIOPHOSPHAT) (GERMAN) ◇ O,O-DIMETHYL-S-(N-METHYLCARBAMOYLMETHYL)DITHIOPHOSPHATE ◇ O,O-DIMETHYL-S-(N-METHYLCARBAMOYLMETHYL) PHOSPHORODITHIOATE ◇ O,O-DIMETHYL-S-(N-METHYLCARBAMYLMETHYL) THIOTHIONOPHOSPHATE ◇ O,O-DIMETHYL-S-(N-MONOMETHYL)-CARBAMYLMETHYLDITHIOPHOSPHATE ◇ O,O-DIMETHYL-S-(2-OXO-3-AZA-BUTYL)-DITHIOPHOSPHAT (GERMAN) ◇ O,O-DIMETIL-S-(N-METIL-CARBAMOIL-METIL)-DITIOFOSFATO (ITALIAN) ◇ DIMETON ◇ DIMEVUR ◇ DITHIOPHOSPHATE de O,O-DIMETHYLE et de S(-N-METHYLCARBAMOYL-METHYLE) (FRENCH) ◇ EI-12880 ◇ ENT 24,650 ◇ EXPERIMENTAL INSECTICIDE 12,880 ◇ FERKETHION ◇ FORTION NM ◇ FOSFAMID ◇ FOSFOTOX ◇ FOSTION MM ◇ L-395 ◇ LURGO ◇ S-METHYLCARBAMOYLMETHYL-O,O-DIMETHYL PHOSPHORODITHIOATE ◇ N-MONOMETHYLAMIDE of O,O-DIMETHYLDITHIOPHOSPHORYLACETIC ACID ◇ NC-262 ◇ NCI-C00135 ◇ PERFECTHION ◇ PHOSPHAMID ◇ PHOSPHORODITHIOIC ACID-O,O-DIMETHYL-S-(2-(METHYLAMINO)-2-OXOETHYL) ESTER ◇ RACUSAN ◇ RCRA WASTE NUMBER P044 ◇ REBELATE ◇ ROGODIAL ◇ ROGOR ◇ ROXION U.A. ◇ SINORATOX ◇ TRIMETION

TOXICITY DATA with REFERENCE
mma-sat 500 μg/plate JTEHD6 16,403,85
mma-hmn:fbr 100 μmol/L MUREAV 42,161,77
sce-hmn:lym 20 mg/L MUREAV 88,307,81
orl-mus TDLo:1050 mg/kg (MGN):REP TXAPA9 26,29,73
ipr-mus TDLo:40 mg/kg (female 1D post):TER BIRUAA 13,238,75
orl-rat TDLo:256 mg/kg/4W-I:CAR ARGEAR 41,311,73
ims-rat TDLo:176 mg/kg/6W-I:CAR ARGEAR 41,311,73 GUCHAZ 6,209,73
orl-man TDLo:300 mg/kg JTCTDW 24,69,86
orl-rat LD50:60 mg/kg YKYUA6 30,623,79
skn-rat LD50:353 mg/kg BJIMAG 26,59,69
ipr-rat LD50:100 mg/kg BJIMAG 21,52,64
scu-rat LD50:350 mg/kg BJIMAG 21,52,64
ivn-rat LD50:450 mg/kg BJIMAG 21,52,64
orl-mus LD50:60 mg/kg BJIMAG 21,52,64
orl-mus LD50:60 mg/kg BJIMAG 21,52,64
ipr-mus LD50:45 mg/kg BJIMAG 21,52,64
orl-dog LD50:400 mg/kg SPEADM 78-1,31,78

CONSENSUS REPORTS: NCI Carcinogenesis Bioassay (feed); No Evidence: mouse, rat NCITR* NCI-CG-TR-4,77. Reported in EPA TSCA Inventory. EPA Genetic Toxicology Program. EPA Extremely Hazardous Substances List.

SAFETY PROFILE: A deadly human poison. Poison by ingestion, skin contact, intraperitoneal, and subcutaneous routes. Moderately toxic by intravenous route. Questionable carcinogen with experimental carcinogenic data. Experimental teratogenic and reproductive effects. Human mutation data reported. When heated to decomposition it emits very toxic fumes of NO_x, PO_x, and SO_x. See also ESTERS.

DSP600 CAS:23135-22-0 ***HR: 3***
N′,N′-DIMETHYL-N-((METHYLCARBAMOYL)OXY)-1-METHYLTHIOOXAMIMIDIC ACID
mf: $C_7H_{13}N_3O_3S$ mw: 219.29

SYNS: D-1410 ◇ 2-(DIMETHYLAMINO)-N-(((METHYLAMINO)CARBONYL)OXY)-2-OXOETHANIMIDOTHIOIC ACID METHYL ESTER ◇ 2-DIMETHYLAMINO-1-(METHYLTHIO)GLYOXAL-o-METHYLCARBAMOYLMONOXIME ◇ N,N-DIMETHYL-α-METHYLCARBAMOYLOXYIMINO-α-(METHYLTHIO)ACETAMIDE ◇ N′,N′-DIMETHYL-N-((METHYLCARBAMOYL)OXY)-1-THIOOXAMIMIDIC ACID METHYL ESTER ◇ DPX 1410 ◇ INSECTICIDE-NEMATICIDE 1410 ◇ METHYL-2-(DIMETHYLAMINO)-N-(((METHYLAMINO)CARBONYL)OXY)-2-OXOETHANIMIDOTHIOATE ◇ METHYL-1-(DIMETHYLCARBAMOYL)-N-(METHYLCARBAMOYLOXY)THIOFORMIMIDATE ◇ S-METHYL-1-(DIMETHYLCARBAMOYL)-N-((METHYLCARBAMOYL)OXY)THIOFORMIMIDATE ◇ METHYL-N′,N′-DIMETHYL-N-((METHYLCARBAMOYL)OXY)-1-THIOOXAMIMIDATE ◇ OXAMYL ◇ THIOXAMYL ◇ VYDATE ◇ VYDATE L INSECTICIDE/NEMATICIDE ◇ VYDATE L OXAMYL INSECTICIDE/NEMATOCIDE

TOXICITY DATA with REFERENCE
orl-rat TDLo:945 mg/kg (male 12W pre):REP FAATDF 7,106,86
orl-rat LD50:2500 μg/kg FAATDF 6,423,86
ihl-rat LC50:170 mg/m^3/1H 85DPAN -,-,71/76
skn-rat LDLo:300 mg/kg FAATDF 6,423,86
orl-mus LD50:2300 μg/kg FAATDF 6,423,86
skn-rbt LD50:740 mg/kg SPEADM 78-1,61,78
orl-qal LD50:4180 μg/kg 85DPAN -,-,71/76

CONSENSUS REPORTS: EPA Extremely Hazardous Substances List.

SAFETY PROFILE: Poison by ingestion, skin contact, and inhalation. Experimental reproductive effects. Moderately toxic by skin contact. When heated to decomposition it emits very toxic fumes of NO_x and SO_x.

DSP650 CAS:69462-47-1 ***HR: 1***
2,6-DIMETHYL-1-((2-METHYLCYCLOHEXYL)CARBONYL)PIPERIDINE
mf: $C_{15}H_{27}NO$ mw: 237.43

SYNS: AI3-36561 ◇ PIPERIDINE, 2,6-DIMETHYL-1-((2-METHYLCYCLOHEXYL)CARBONYL)-

TOXICITY DATA with REFERENCE
skn-rbt 500 mg/24H MLD AEHA** 51-029-76
eye-rbt 100 mg/24H MOD AEHA** 51-029-76

SAFETY PROFILE: A skin and eye irritant. When heated to decomposition it emits toxic fumes of NO_x.

DSQ000 CAS:122-14-5 ***HR: 3***
DIMETHYL-3-METHYL-4-NITROPHENYLPHOSPHOROTHIONATE
mf: $C_9H_{12}NO_5PS$ mw: 277.25

SYNS: ACCOTHION ◇ ACEOTHION ◇ AGRIA 1050 ◇ AGRIYA 1050 ◇ AGROTHION ◇ AMERICAN CYANAMID CL-47,300 ◇ ARBOGAL ◇ BAY 41831 ◇ BAYER 41831 ◇ BAYER S 5660 ◇ CEKUTROTHION ◇ CL 47300 ◇ CP 47114 ◇ CYFEN ◇ CYTEL ◇ CYTEN ◇ O,O-DIMETHYL-O-(3-METHYL-4-NITROFENYL)-MONOTHIOFOSFAAT(DUTCH) ◇ O,O-DIMETHYL-O-(3-METHYL-4-NITRO-PHENYL)-MONOTHIOPHOSPHAT (GERMAN) ◇ O,O-DIMETHYL-O-(3-METHYL-4-NITROPHENYL) PHOSPHOROTHIOATE ◇ O,O-DIMETHYL-O-(3-METHYL-4-NITROPHENYL) THIOPHOSPHATE ◇ O,O-DIMETHYL-O-(3-METHYL) PHOSPHOROTHIOATE ◇ O,O-DIMETHYL-O-(4-NITRO-3-METHYLPHENYL)THIOPHOSPHATE ◇ O,O-DIMETHYL-O-4-NITRO-m-TOLYL PHOSPHOROTHIOATE ◇ O,O-DIMETIL-O-(3-METIL-4-NITRO-FENIL)-MONOTIOFOSFATO (ITALIAN) ◇ EI 47300 ◇ ENT 25,715 ◇ FALITHION ◇ FENITOX ◇ FENITROTHION ◇ FENITROTION (HUNGARIAN) ◇ FOLETHION ◇ H-35-F 87 (BVM) ◇ 8057HC ◇ KOTION ◇ MEP (Pesticide) ◇ METATHIONE ◇ METATION ◇ METHYLNITROPHOS ◇ MONSANTO CP 47114 ◇ NITROPHOS ◇ NOVATHION ◇ NUVANOL ◇ OLEOSUMIFENE ◇ OMS 43 ◇ OVADOFOS ◇ PENNWALT C-4852 ◇ PHENITROTHION ◇ S 112A ◇ S 5660 ◇ SUMITHIAN ◇ THIOPHOSPHATE de O,O-DIMETHYLE et de O-(3-METHYL-4-NITROPHENYLE) (FRENCH) ◇ VERTHION

TOXICITY DATA with REFERENCE
mmo-sat 500 μg/plate MUREAV 116,185,83

dni-omi 100 ppm NNGADV 9,325,84
orl-wmn TDLo:800 mg/kg:GIT,PUL ARTODN 56,136,84
orl-rat LD50:250 mg/kg TXAPA9 21,315,72
ihl-rat LC50:378 mg/m^3/4H EGESAQ 24,173,80
skn-rat LD50:750 mg/kg SRTCDF -,101,77
ipr-rat LD50:300 mg/kg TXAPA9 63,91,82
ivn-rat LD50:33 mg/kg ABCHA6 27,669,63
itr-rat LD50:950 mg/kg TXAPA9 63,91,82
orl-mus LD50:715 mg/kg HYSAAV 31,13,66
skn-mus LD50:2500 mg/kg ABCHA6 25,605,61
scu-mus LD50:1000 mg/kg ABCHA6 25,605,61

CONSENSUS REPORTS: EPA Genetic Toxicology Program. EPA Extremely Hazardous Substances List.

SAFETY PROFILE: Poison by ingestion, inhalation, intravenous, and intraperitoneal routes. Moderately toxic by skin contact, intratracheal, and subcutaneous routes. Human systemic effects by ingestion: hypermotility, diarrhea, nausea or vomiting, and dyspnea. Mutation data reported. When heated to decomposition it emits very toxic fumes of NO_x, PO_x, and SO_x.

DSQ600 CAS:3572-74-5 ***HR: 3***
N,N-DIMETHYL-2-(α-METHYL-α-PHENYLBENZYLOXY)ETHYLAMINE
mf: $C_{18}H_{23}NO$ mw: 269.42

SYNS: N,N-DIMETHYL-2-((α-METHYL-α-PHENYLBENZYL)OXY)ETHYLAMINE ◇ SUBSTANZ NR. 1934 (GERMAN)

TOXICITY DATA with REFERENCE
ivn-mus LD50:43 mg/kg ARZNAD 4,189,54
scu-gpg LD50:54 mg/kg ARZNAD 4,189,54

SAFETY PROFILE: Poison by intravenous and subcutaneous routes. When heated to decomposition it emits toxic fumes of NO_x.

DSQ800 CAS:66-79-5 ***HR: 3***
3,3-DIMETHYL-6-(((5-METHYL-3-PHENYL-4-ISOXAZOLECARBOXAMIDE-7-OXO-4-THIA-1-AZABICYCLO(3.2.0)HEPTANE-2-CARBOXYLIC ACID

SYNS: BRL 1400 ◇ 5-METHYL-3-PHENYL-4-ISOXAZOLYL-PENICILLIN ◇ MPI-PC ◇ MPI-PENICILLIN ◇ OXACILLIN ◇ OXAZOCILLIN ◇ PENICILLIN P-12 ◇ PROSTAPHLYN ◇ STAPENOR

TOXICITY DATA with REFERENCE
ivn-wmn TDLo:5560 mg/kg/20D-I:BLD SMJOAV 70,1245,77
orl-mus LD50:6500 mg/kg 85GMAT -,95,82
ivn-mus LD50:1500 mg/kg ARZNAD 15,322,65
orl-cat LDLo:750 mg/kg ARZNAD 15,322,65

SAFETY PROFILE: Poison by an unspecified route. Moderately toxic by ingestion and intravenous routes. Human systemic effects by intravenous route: angranulocytosis. When heated to decomposition it emits very toxic fumes of NO_x and SO_x. See also other penicillin entries.

DSR200 CAS:20241-03-6 ***HR: 3***
3,3-DIMETHYL-1-(m-METHYLPHENYL) TRIAZENE
mf: $C_9H_{13}N_3$ mw: 163.25

SYNS: 3,3-DIMETHYL-1-(m-TOLYL)TRIAZENE ◇ 1-(m-METHYLPHENYL)-3,3-DIMETHYLTRIAZENE ◇ 1-(3-METHYLPHENYL)-3,3-DIMETHYLTRIAZENE

TOXICITY DATA with REFERENCE
mma-sat 400 nmol/L JMCMAR 22,473,79
orl-rat TDLo:250 mg/kg:CAR ZKKOBW 81,285,74
scu-rat TDLo:500 mg/kg:CAR ZKKOBW 81,285,74
orl-rat LD50:300 mg/kg ZKKOBW 81,285,74
scu-rat LD50:500 mg/kg ZKKOBW 81,285,74
ipr-mus LD50:201 mg/kg JMCMAR 19,1299,76

SAFETY PROFILE: Poison by ingestion and intraperitoneal routes. Moderately toxic by subcutaneous route. Questionable carcinogen with experimental carcinogenic data. Mutation data reported. When heated to decomposition it emits toxic fumes of NO_x.

DSR400 CAS:756-79-6 ***HR: 2***
DIMETHYL METHYLPHOSPHONATE
mf: $C_3H_9O_3P$ mw: 124.09

SYNS: DMMP ◇ METHYLPHOSPHONIC ACID DIMETHYL ESTER ◇ NCI-C56762

TOXICITY DATA with REFERENCE
dlt-mus-orl 65 g/kg/13W-C MUREAV 138,213,84
cyt-ham:ovr 250 mg/L NTIS** AD-A124-785
orl-mus TDLo:33 g/kg (female 7-14D post):REP NTIS** PB85-220143
orl-rat TDLo:515 g/kg/2Y-C:CAR FAATDF 11,91,88

CONSENSUS REPORTS: Reported in EPA TSCA Inventory.

SAFETY PROFILE: Experimental reproductive effects. Questionable carcinogen with experimental carcinogenic data. Mutation data reported. An experimental nerve gas stimulant. A flame retardant. When heated to decomposition it emits toxic fumes of PO_x.

DSR600 CAS:50308-86-6 ***HR: 3***
1,3-DIMETHYL-4-(p-((p-((1-METHYLPYRIDINIUM-4-YL)AMINO)PHENYL)CARBAMOYL)ANILINOQUINOLINIUM), DIBROMIDE
mf: $C_{30}H_{29}N_5O \cdot 2Br$ mw: 635.46

TOXICITY DATA with REFERENCE
dnd-mus:lym 4900 nmol/L JMCMAR 22,134,79
ipr-mus LD10:5 mg/kg JMCMAR 22,134,79

SAFETY PROFILE: Poison by intraperitoneal route. Mutation data reported. When heated to decomposition it emits very toxic fumes of NO_x and Br^-.

DSR800 CAS:50308-87-7 ***HR: 3***
1,6-DIMETHYL-4-(p-((p-((1-METHYLPYRIDINIUM-4-YL)AMINO)PHENYL) CARBAMOYL) ANILINO)QUINOLINIUM) DI-p-TOLUENESULFONATE
mf: $C_{30}H_{29}N_5O{\bullet}2C_7H_7O_3S$ mw: 818.04

TOXICITY DATA with REFERENCE
dnd-mus:lym 1050 nmol/L JMCMAR 22,134,79
ipr-mus LD10:15 mg/kg JMCMAR 22,134,79

SAFETY PROFILE: Poison by intraperitoneal route. Mutation data reported. When heated to decomposition it emits very toxic fumes of NO_x and SO_x.

DSS000 CAS:50425-34-8 ***HR: 3***
1,8-DIMETHYL-4-(p-((p-((1-METHYLPYRIDINIUM-4-YL)AMINO)PHENYL)CARBAMOYL) ANILINO)QUINOLINIUM)DI-p-TOLUENESULFONATE
mf: $C_{30}H_{29}N_5O{\bullet}2C_7H_7O_3S$ mw: 818.04

TOXICITY DATA with REFERENCE
dnd-mus:lym 870 nmol/L JMCMAR 22,134,79
ipr-mus LD10:60 mg/kg JMCMAR 22,134,79

SAFETY PROFILE: Poison by intraperitoneal route. Mutation data reported. When heated to decomposition it emits very toxic fumes of NO_x and SO_x.

DSS200 CAS:7347-46-8 ***HR: 3***
N,N-DIMETHYL-4-(2-METHYL-4-PYRIDYLAZO)ANILINE-N-OXIDE
mf: $C_{14}H_{16}N_4O$ mw: 256.34

SYNS: 4-((4-(DIMETHYLAMINO)PHENYL)AZO)-2-PICOLINE-1-OXIDE ◇ N,N-DIMETHYL-4-((2-METHYL-4-PYRIDINYL)AZO)BENZENAMINE-N-OXIDE ◇ N,N-DIMETHYL-4-(4′-(2′-METHYLPYRIDYL-1′-OXIDE)AZO)ANILINE ◇ 2′-MePO4′ ◇ 2-METHYLPYRIDINE-1-OXIDE-4-AZO-p-DIMETHYLANILINE

TOXICITY DATA with REFERENCE
orl-rat TDLo:714 mg/kg/17W-C:NEO JNCIAM 37,365,66
orl-rat TD:2142 mg/kg/17W-C:ETA JNCIAM 41,855,68

SAFETY PROFILE: Questionable carcinogen with experimental neoplastigenic and tumorigenic data. When heated to decomposition it emits toxic fumes of NO_x.

DSS400 CAS:3761-41-9 ***HR: 3***
O,O-DIMETHYL-O-(4-(METHYLSULFINYL)-m-TOLYL) PHOSPHOROTHIOATE
mf: $C_{10}H_{15}O_4PS_2$ mw: 294.34

SYN: O,O-DIMETHYL-O-((4-METHYLTHIO)-m-TOLYL)PHOSPHOROTHIOATE SULFOXIDE

TOXICITY DATA with REFERENCE
orl-rat LD50:125 mg/kg GUCHAZ 6,279,73
ipr-rat LDLo:250 mg/kg TXAPA9 6,86,64

SAFETY PROFILE: Poison by ingestion and intraperitoneal routes. When heated to decomposition it emits very toxic fumes of PO_x and SO_x.

DSS600 CAS:3773-37-3 ***HR: 3***
N,N-DIMETHYL-10-(3-(4-(METHYLSULFONYL)-1-PIPERAZINYL)PROPYL) PHENOTHIAZINE-2-SULFONAMIDE
mf: $C_{22}H_{30}N_4O_4S_3$ mw: 510.74
SYN: 9260 RP

TOXICITY DATA with REFERENCE
orl-hmn TDLo:2 mg/kg:PSY,GIT PSYPAG 2,209,61
orl-mus LD50:1300 mg/kg PSYPAG 2,209,61
ipr-mus LD50:480 mg/kg PSYPAG 2,209,61
ivn-mus LD50:310 mg/kg PSYPAG 2,209,61

SAFETY PROFILE: Poison by intravenous route. Moderately toxic by ingestion and intraperitoneal routes. Human effects by ingestion: sleep, excitement, nausea or vomiting. When heated to decomposition it emits very toxic fumes of NO_x and SO_x.

DSS800 CAS:3761-42-0 ***HR: 3***
O,O-DIMETHYL-o-(4-(METHYLSULFONYL)-m-TOLYL) PHOSPHOROTHIOATE
mf: $C_{10}H_{15}O_5PS_2$ mw: 310.34

SYNS: O,O-DIMETHYL-o-((4-METHYLTHIO)-m-TOLYL)PHOSPHOROTHIOATE SULFONE ◇ FENTHIONE SULFONE

TOXICITY DATA with REFERENCE
orl-rat LD50:125 mg/kg GUCHAZ 6,279,73
ipr-rat LDLo:250 mg/kg TXAPA9 6,86,64
orl-mus LD50:210 mg/kg JOPHDQ 9,697,86

SAFETY PROFILE: Poison by ingestion and intraperitoneal routes. When heated to decomposition it emits very toxic fumes of PO_x and SO_x.

DST000 CAS:2032-65-7 ***HR: 3***
3,5-DIMETHYL-4-METHYLTHIOPHENYL-N-METHYLCARBAMATE
DOT: NA 2757
mf: $C_{11}H_{15}NO_2S$ mw: 225.33

SYNS: BAY 9026 ◇ BAYER 37344 ◇ 3,5-DIMETHYL-4-(METHYLTHIO)PHENOL METHYLCARBAMATE ◇ 3,5-DIMETHYL-4-METHYLTHIOPHENYL-N-CARBAMAT (GERMAN) ◇ DRAZA ◇ ENT 25,726 ◇ H 321 ◇ MERCAPTODIMETHUR (DOT) ◇ MESUROL ◇ METHIOCARB ◇ METHYL CARBAMIC ACID-4-(METHYLTHIO)-3,5-XYLYL ESTER ◇ 4-METHYLMERCAPTO-3,5-DIMETHYLPHENYL N-METHYLCARBAMATE ◇ 4-METHYLMERCAPTO-3,5-XYLYL METHYLCARBAMATE ◇ 4-METHYLTHIO-3,5-DIMETHYLPHENYL METHYLCARBAMATE ◇ 4-(METHYLTHIO)-3,5-XYLENOL METHYLCARBAMATE

◇ 4-(METHYLTHIO)-3,5-XYLYL METHYLCARBAMATE ◇ METMERCAPTURON ◇ OMS-93

TOXICITY DATA with REFERENCE
orl-rat LD50:15 mg/kg FMCHA2 -,C150,83
skn-rat LD50:350 mg/kg PCOC** -,105,66
orl-mus LD50:34 mg/kg FATOAO 35,356,72
ipr-mus LD50:16 mg/kg TXAPA9 6,402,64
orl-gpg LD50:40 mg/kg 85DPAN -,-,71/76

CONSENSUS REPORTS: EPA Extemely Hazardous Substances List.

SAFETY PROFILE: Poison by ingestion, skin contact, and intraperitoneal routes. Used as an insecticide, molluscicide, and bird repellant. When heated to decomposition it emits very toxic fumes of NO_x and SO_x. See also ESTERS and CARBAMATES.

DST200 CAS:55-37-8 ***HR: 3***
O,O-DIMETHYL-O-4-(METHYLTHIO)-3,5-XYLYL PHOSPHOROTHIOATE
mf: $C_{11}H_{17}O_3PS_2$ mw: 292.37

SYNS: BAY 37342 ◇ BAYER 9013 ◇ BAYER 37342 ◇ O,O-DIMETHYL-O-(3,5-DIMETHYL-4-METHYLTHIOPHENYL)PHOSPHOROTHIOATE ◇ O-(3,5-DIMETHYL-4-(METHYLTHIO)PHENYL)-O,O-DIMETHYL PHOSPHOROTHIOATE ◇ ENT 25,684 ◇ G 347

TOXICITY DATA with REFERENCE
orl-rat LD50:1000 mg/kg TXAPA9 21,315,72
orl-mus LDLo:1070 mg/kg AECTCV 14,111,85
orl-ckn LD50:103 mg/kg TXAPA9 11,49,67
orl-bwd LD50:10 mg/kg TXAPA9 21,315,72

SAFETY PROFILE: Poison by ingestion. When heated to decomposition it emits very toxic fumes of PO_x and SO_x.

DST400 CAS:22120-39-4 ***HR: 3***
N,N-DIMETHYL-5-METHYLTRYPTAMINE
mf: $C_{13}H_{18}N_2$ mw: 202.33

SYN: 3-(2-DIMETHYLAMINOETHYL)-5-METHYLINDOLE

TOXICITY DATA with REFERENCE
ipr-mus LD50:110 mg/kg PSYPAG 16,385,70
ivn-mus LD50:71 mg/kg CSLNX* NX#01632

SAFETY PROFILE: Poison by intraperitoneal and intravenous routes. When heated to decomposition it emits toxic fumes of NO_x.

DST600 CAS:141-91-3 ***HR: 2***
2,6-DIMETHYLMORPHOLINE
mf: $C_6H_{13}NO$ mw: 115.20

PROP: Liquid. D: 0.9346, bp: 146.6, fp: −85°, flash p: 112°F (OC), vap d: 4.0.

SYN: 2,6-DIMETHYL-2,3,5,6-TETRAHYDRO-4H-1,4-OXAZINE

TOXICITY DATA with REFERENCE
skn-rbt 10 mg/24H open MLD AIHAAP 23,95,62
orl-rat LD50:2830 mg/kg UCDS** 11/13/61
skn-rbt LD50:710 mg/kg AIHAAP 23,95,62

CONSENSUS REPORTS: Reported in EPA TSCA Inventory.

SAFETY PROFILE: Moderately toxic by ingestion and skin contact. A skin irritant. Flammable when exposed to heat, flame, or oxidizers. To fight fire, use alcohol foam. When heated to decomposition it emits toxic fumes of NO_x.

DST800 CAS:597-25-1 ***HR: 3***
DIMETHYLMORPHOLINOPHOSPHONATE
mf: $C_6H_{14}NO_4P$ mw: 195.18

SYNS: DIMETHYL MORPHOLINOPHOSPHORAMIDATE ◇ DMMPA ◇ MORPHOLINOPHOSPHONIC ACID DIMETHYL ESTER ◇ 4-MORPHOLINYLPHOSPHONIC ACID DIMETHYL ESTER ◇ NCI-C54740

TOXICITY DATA with REFERENCE
msc-mus:lym 2200 mg/L NTPTR* NTP-TR-298,86
cyt-ham:ovr 3 g/L NTPTR* NTP-TR-298,86
sce-ham:ovr 3 g/L NTPTR* NTP-TR-298,86
orl-rat TDLo:309 g/kg/2Y-I:CAR NTPTR* NTP-TR-298,86
orl-rat LD50:6 g/kg NTPTR* NTP-TR-298,86
ipr-rat LD50:2400 mg/kg NTPTR* NTP-TR-298,86
ims-rat LD50:5200 mg/kg NTPTR* NTP-TR-298,86
orl-mus LD50:3300 mg/kg NTPTR* NTP-TR-298,86
ipr-mus LD50:5 g/kg NTPTR* NTP-TR-298,86
ivn-mus LD50:400 mg/kg NTPTR* NTP-TR-298,86
ims-mus LD50:4800 mg/kg NTPTR* NTP-TR-298,86
ivn-rbt LD50:350 mg/kg NTPTR* NTP-TR-298,86

CONSENSUS REPORTS: NTP Carcinogenesis Studies (gavage); Some Evidence: rat NTPTR* NTP-TR-298,86; No Evidence: mouse NTPTR* NTP-TR-298,86.

SAFETY PROFILE: Poison by intravenous route. Moderately toxic by ingestion and intraperitoneal routes. Questionable carcinogen with experimental carcinogenic data. Mutation data reported. When heated to decomposition it emits very toxic fumes of NO_x and PO_x. See also ESTERS.

DSU000 CAS:55-93-6 ***HR: 3***
DIMETHYLMYLERAN
mf: $C_8H_{18}O_6S_2$ mw: 272.36

SYNS: DDM ◇ 2,5-DIMETHANESULFOMYLOXYHEXANE ◇ 1,4-DIMETHANESULFONOXY-1,4-DIMETHYLBUTANE◇ 2,5-HEXANEDIOL DIMETHYLSULFONATE ◇ NSC-23890

TOXICITY DATA with REFERENCE
sln-dmg-orl 1 pph ZEVBA5 90,457,59
dlt-ofs-ipr 4 mg/kg MUREAV 58,263,78
spm-mus-ipr 4 mg/kg EXPEAM 30,178,74
dlt-mus-ipr 8 mg/kg IRLCDZ 5,341,77

ipr-mus LD50:16 mg/kg JNCIAM 56,609,76
ivn-dog LDLo:1 mg/kg CCSUBJ 2,203,65
ivn-mky LDLo:1 mg/kg CCSUBJ 2,203,65

CONSENSUS REPORTS: EPA Genetic Toxicology Program.

SAFETY PROFILE: Poison by intravenous and intraperitoneal routes. Mutation data reported. Used for treatment of chronic granulocytic leukemia. When heated to decomposition it emits very toxic fumes of SO_x.

DSU100 CAS:33447-90-4 ***HR: D***
meso-DIMETHYLMYLERAN
mf: $C_8H_{18}O_6S_2$ mw: 274.38

SYN: (R*,S*)-DIMETHANESULFONATE-meso-2,5-HEXANEDIOL(9CI)

TOXICITY DATA with REFERENCE
dlt-ofs-ipr 4 mg/kg MUREAV 58,263,78
spm-mus-ipr 4 mg/kg EXPEAM 30,178,74
ipr-rat TDLo:4 mg/kg (1D male):REP CBINA8 36,331,81

SAFETY PROFILE: Experimental reproductive effects. Mutation data reported. When heated to decomposition it emits toxic fumes of SO_x.

DSU200 CAS:3015-65-4 ***HR: 3***
N,N-DIMETHYLMYRISTAMIDE
mf: $C_{16}H_{33}NO$ mw: 255.50

SYN: N,N-DIMETHYLTETRADECANAMIDE

TOXICITY DATA with REFERENCE
ipr-mus LD50:1430 mg/kg AIHAAP 32,539,71
ivn-mus LD50:150 mg/kg AIHAAP 32,539,71
ipr-rbt LD50:1500 mg/kg AIHAAP 32,539,71
ivn-rbt LD50:84 mg/kg AIHAAP 32,539,71

CONSENSUS REPORTS: Reported in EPA TSCA Inventory.

SAFETY PROFILE: Poison by intravenous route. Moderately toxic by intraperitoneal route. When heated to decomposition it emits toxic fumes of NO_x.

DSU400 CAS:86-56-6 ***HR: 3***
N,N-DIMETHYL-1-NAPHTHYLAMINE
mf: $C_{12}H_{13}N$ mw: 171.26

SYNS: 1-DIMETHYLAMINONAPHTHALENE ◇ DIMETHYL-α-NAPHTHYLAMINE ◇ α-DIMETHYLNAPHTHYLAMINE ◇ N,N-DIMETHYL-α-NAPHTHYLAMINE

TOXICITY DATA with REFERENCE
orl-rat LDLo:500 mg/kg JPETAB 90,260,47
ipr-mus LD50:75 mg/kg NTIS** AD691-490

CONSENSUS REPORTS: Reported in EPA TSCA Inventory.

SAFETY PROFILE: Poison by intraperitoneal route. Moderately toxic by ingestion. When heated to decomposition it emits toxic fumes of NO_x.

DSU600 CAS:607-59-0 ***HR: 3***
N,N-DIMETHYL-p-(1-NAPHTHYLAZO)ANILINE
mf: $C_{18}H_{17}N_3$ mw: 275.38

SYNS: DAN ◇ p-DIMETHYLAMINOBENZENEAZO-1-NAPHTHALENE ◇ p-DIMETHYLAMINOBENZENE-1-AZO-1-NAPHTHALENE

TOXICITY DATA with REFERENCE
dns-rat:lvr 100 μmol/L MUREAV 136,255,84
dns-rat-orl 100 mg/kg ENMUDM 7,101,85
dns-ham:lvr 10 μmol/L MUREAV 136,255,84
orl-rat TDLo:25 g/kg/79W-C:ETA JNCIAM 13,57,52
scu-rat TDLo:90 mg/kg/2W-I:CAR JNCIAM 18,843,57

SAFETY PROFILE: Questionable carcinogen with experimental carcinogenic and tumorigenic data. Mutation data reported. When heated to decomposition it emits toxic fumes of NO_x.

DSU800 CAS:613-65-0 ***HR: 3***
N,N-DIMETHYL-4(2'-NAPHTHYLAZO)ANILINE
mf: $C_{18}H_{17}N_2$ mw: 261.37

SYNS: DA-2-N ◇ p-DIMETHYLAMINOBENZENE-1-AZO-2-NAPHTHALENE ◇ 2-(4-DIMETHYLAMINOPHENYLAZO)NAPHTHALENE

TOXICITY DATA with REFERENCE
dns-rat:lvr 10 μmol/L MUREAV 136,255,84
dns-rat-orl 100 mg/kg ENMUDM 7,101,85
dns-ham:lvr 2 μmol/L MUREAV 136,255,84
orl-rat TDLo:8630 mg/kg/230D-C:CAR JNCIAM 14,571,53
skn-mus TDLo:4800 mg/kg:ETA JNCIAM 13,1259,53

SAFETY PROFILE: Questionable carcinogen with experimental carcinogenic and tumorigenic data. Mutation data reported. When heated to decomposition it emits toxic fumes of NO_x.

DSV000 CAS:63019-14-7 ***HR: 3***
N,N-DIMETHYL-p-(2-(1-NAPHTHYL)VINYL)ANILINE
mf: $C_{20}H_{19}N$ mw: 273.40

SYN: 1-(4'-DIMETHYLAMINOPHENYL)-2-(1'-NAPHTHYL)ETHYLENE

TOXICITY DATA with REFERENCE
scu-rat TDLo:215 mg/kg/W-I:ETA PTRMAD 241,147,48
scu-mus TDLo:320 mg/kg/W-I:ETA,REP PTRMAD 241,147,48

SAFETY PROFILE: Experimental reproductive effects. Questionable carcinogen with experimental tumorigenic data. When heated to decomposition it emits toxic fumes of NO_x.

DSV200 CAS:4164-28-7 ***HR: 3***
DIMETHYLNITRAMINE
mf: $C_2H_6N_2O_2$ mw: 90.10

SYNS: DIMETHYLNITRAMIN (GERMAN) ◇ DIMETHYLNITROAMINE ◇ DMNM ◇ DMNO ◇ N-NITRODIMETHYLAMINE ◇ N-NITRO-DMA

TOXICITY DATA with REFERENCE
mma-sat 250 μmol/plate CRNGDP 5,809,84
hma-rat/sat 200 mg/kg CNREA8 41,3205,81
orl-rat TDLo:20 g/kg/1Y-C:ETA JJIND8 64,1435,80
orl-rat LD50:1095 mg/kg TXAPA9 33,185,75
ipr-rat LD50:897 mg/kg TXAPA9 33,185,75
ipr-mus LD50:399 mg/kg TODJ 1,363,78

SAFETY PROFILE: Poison by intraperitoneal route. Moderately toxic by ingestion. Questionable carcinogen with experimental tumorigenic data. Mutation data reported. When heated to decomposition it emits toxic fumes of NO_x.

DSV289 CAS:22691-91-4 ***HR: 3***
3,3-DIMETHYL-1-NITRO-1-BUTYNE
mf: $C_6H_9NO_2$ mw: 127.14

$O_2NC{\equiv}CC(CH_3)_3$

SYN: tert-BUTYLNITROACETATE

SAFETY PROFILE: Ignites and then explodes on contact with primary, secondary, and tertiary amines. When heated to decomposition it emits toxic fumes of NO_x. See also ACETYLENE COMPOUNDS and NITRO COMPOUNDS.

DSV400 CAS:59-35-8 ***HR: 3***
4,6-DIMETHYL-2-(5-NITRO-2-FURYL)PYRIMIDINE
mf: $C_{10}H_9N_3O_3$ mw: 219.22

TOXICITY DATA with REFERENCE
orl-rat TDLo:8988 mg/kg/49W-C:CAR JNCIAM 57,277,76

SAFETY PROFILE: Questionable carcinogen with experimental carcinogenic data. When heated to decomposition it emits toxic fumes of NO_x.

DSV600 CAS:13230-04-1 ***HR: D***
1,2-DIMETHYL-4-NITRO-1H-IMIDAZOLE
mf: $C_5H_7N_3O_2$ mw: 141.15

TOXICITY DATA with REFERENCE
mmo-sat 1 mmol/L MUREAV 66,207,79
mmo-klp 5 mmol/L/20H MUREAV 66,207,79

SAFETY PROFILE: Mutation data reported. When heated to decomposition it emits toxic fumes of NO_x. See also other dimethyl nitroimidazole entries.

DSV800 CAS:551-92-8 ***HR: 3***
1,2-DIMETHYL-5-NITROIMIDAZOLE
mf: $C_5H_7N_3O_2$ mw: 141.15

SYNS: 1,2-DIMETHYL-5-NITRO-1H-IMIDAZOLE ◇ DIMETRIDAZOLE ◇ EMTRYL ◇ EMTRYLVET ◇ EMTRYMIX ◇ 8595 R.P.

TOXICITY DATA with REFERENCE
mmo-sat 25 μg/plate MUREAV 38,203,76
bfa-rat/sat 800 mg/kg MUREAV 97,171,82
orl-rat TDLo:50 g/kg/46W-C:NEO JNCIAM 51,403,73

CONSENSUS REPORTS: EPA Genetic Toxicology Program.

SAFETY PROFILE: Questionable carcinogen with experimental neoplastigenic data. Mutation data reported. When heated to decomposition it emits toxic fumes of NO_x.

DSW500 CAS:5213-47-8 ***HR: 3***
4,5-DIMETHYL-2-NITROIMIDAZOLE
mf: $C_5H_7N_3O_2$ mw: 141.15

TOXICITY DATA with REFERENCE
orl-mus LD50:330 mg/kg AACHAX -,478,65
ipr-mus LD50:158 mg/kg AACHAX -,478,65
scu-mus LD50:297 mg/kg AACHAX -,478,65

SAFETY PROFILE: Poison by ingestion, subcutaneous, and intraperitoneal routes. When heated to decomposition it emits toxic fumes of NO_x.

DSW600 CAS:3837-55-6 ***HR: 3***
N,N-DIMETHYL-p-((m-NITROPHENYL)AZO)ANILINE
mf: $C_{14}H_{14}N_4O_2$ mw: 270.32

SYN: 3′-NITRO-4-DIMETHYLAMINOAZOBENZENE

TOXICITY DATA with REFERENCE
orl-rat TDLo:5184 mg/kg/17W-C:NEO JEMEAV 87,139,48

SAFETY PROFILE: Questionable carcinogen with experimental neoplastigenic data. When heated to decomposition it emits toxic fumes of NO_x.

DSW800 CAS:3010-38-6 ***HR: 3***
N,N-DIMETHYL-p-((o-NITROPHENYL)AZO)ANILINE
mf: $C_{14}H_{14}N_4O_2$ mw: 270.32

SYN: 2′-NITRO-4-DIMETHYLAMINOAZOBENZENE

TOXICITY DATA with REFERENCE
orl-rat TDLo:5184 mg/kg/17W-C:NEO JEMEAV 87,139,48

SAFETY PROFILE: Questionable carcinogen with ex-

perimental neoplastigenic data. When heated to decomposition it emits toxic fumes of NO_x.

DSX400 CAS:7227-92-1 ***HR: 3***
3,3-DIMETHYL-1-(p-NITROPHENYL)TRIAZENE
mf: $C_8H_{10}N_4O_2$ mw: 194.22

SYNS: 1-p-NITROFENYL-3,3-DIMETHYLTRIAZEN (CZECH) ◇ 1-(p-NITROPHENYL-3,3-DIMETHYL-TRIAZEN (GERMAN) ◇ 1-(p-NITROPHENYL)-3,3-DIMETHYL-TRIAZENE ◇ 1-(4-NITROPHENYL)-3,3-DIMETHYLTRIAZENE

TOXICITY DATA with REFERENCE
scu-rat TDLo:3250 mg/kg/72W-I:NEO ZKKOBW 81,285,74
scu-rat TD:330 mg/kg:ETA ZKKOBW 81,285,74
orl-rat LD50:1660 mg/kg 28ZPAK -,133,72
scu-rat LD50:350 mg/kg ZKKOBW 81,285,74

SAFETY PROFILE: Poison by subcutaneous route. Moderately toxic by ingestion. Questionable carcinogen with experimental neoplastigenic and tumorigenic data. When heated to decomposition it emits toxic fumes of NO_x.

DSX800 CAS:37699-43-7 ***HR: 3***
2,3-DIMETHYL-4-NITROPYRIDINE-1-OXIDE
mf: $C_7H_8N_2O_3$ mw: 168.17

TOXICITY DATA with REFERENCE
mmo-sat 100 nmol/plate GANNA2 70,799,79
dnr-esc 500 μg/well CNREA8 32,2369,72
scu-mus TDLo:1760 mg/kg/15W-I:ETA GANNA2 70,799,79

SAFETY PROFILE: Questionable carcinogen with experimental tumorigenic data. Mutation data reported. When heated to decomposition it emits toxic fumes of NO_x.

DSY000 CAS:21816-42-2 ***HR: 3***
2,5-DIMETHYL-4-NITROPYRIDINE-1-OXIDE
mf: $C_7H_8N_2O_3$ mw: 168.17

TOXICITY DATA with REFERENCE
mmo-sat 100 nmol/plate GANNA2 70,799,79
mmo-esc 500 μmol/L GANNA2 70,799,79
dnd-mus:fbr 500 μmol/L CNREA8 35,521,75
scu-mus TDLo:1800 mg/kg/15W-I:ETA GANNA2 70,799,79

SAFETY PROFILE: Questionable carcinogen with experimental tumorigenic data. Mutation data reported. When heated to decomposition it emits toxic fumes of NO_x.

DSY200 CAS:14248-66-9 ***HR: D***
3,5-DIMETHYL-4-NITROPYRIDINE 1-OXIDE
mf: $C_7H_8N_2O_3$ mw: 168.17

TOXICITY DATA with REFERENCE
mmo-sat 2500 nmol/plate GANNA2 70,799,79
dnd-mus:fbr 500 μmol/L CNREA8 35,521,75

SAFETY PROFILE: Mutation data reported. When heated to decomposition it emits acrid smoke and irritating fumes.

DSY600 CAS:138-89-6 ***HR: 3***
N,N-DIMETHYL-p-NITROSOANILINE
DOT: UN 1369
mf: $C_8H_{10}N_2O$ mw: 150.20

$(CH_3)_2NC_6H_4N{:}O$

SYNS: ACCELERINE ◇ p-(DIMETHYLAMINO)NITROSOBENZENE ◇ 4-(DIMETHYLAMINO)NITROSOBENZENE ◇ DIMETHYL-p-NITROSOANILINE (DOT) ◇ N,N-DIMETHYL-4-NITROSOBENZENAMINE ◇ DIMETHYL(p-NITROSOPHENYL)AMINE ◇ NCI-C01821 ◇ NDMA ◇ p-NITROSO-N,N-DIMETHYLANILINE ◇ 4-NITROSODIMETHYLANILINE ◇ p-NITROSODIMETHYLANILINE (DOT) ◇ PARANITROSODIMETHYLANILIDE ◇ ULTRA BRILLIANT BLUE P

TOXICITY DATA with REFERENCE
mmo-sat 10 μg/plate ENMUDM 8(Suppl 7),1,86
mma-sat 33 μg/plate ENMUDM 8(Suppl 7),1,86
orl-rat TDLo:7300 mg/kg/1Y-C:ETA PUOMA5 46,68,68
orl-rat LD50:65 mg/kg NCIMR* NIH-71-E-2144
orl-gpg LDLo:650 mg/kg JIDHAN 13,87,31
orl-mam LDLo:650 mg/kg JIDHAN 13,87,31

CONSENSUS REPORTS: Reported in EPA TSCA Inventory.

DOT Classification: Flammable Solid; Label: Spontaneously Combustible.

SAFETY PROFILE: Poison by ingestion. Mutation data reported. Questionable carcinogen with experimental tumorigenic data. Flammable when exposed to heat, flame, or oxidizers. Violent reaction with acetic anhydride + acetic acid. When heated to decomposition it emits toxic fumes of NO_x.

DSY800 CAS:70786-64-0 ***HR: 3***
3,2′-DIMETHYL-4-NITROSOBIPHENYL
mf: $C_{14}H_{13}NO$ mw: 211.28

TOXICITY DATA with REFERENCE
mmo-sat 120 μmol/plate JMCMAR 22,981,79
mma-sat 80 nmol/plate JMCMAR 22,981,79
scu-ham TDLo:1173 mg/kg/37W-I:CAR CALEDQ 22,981,79

SAFETY PROFILE: Questionable carcinogen with experimental carcinogenic data. Mutation data reported. When heated to decomposition it emits toxic fumes of NO_x.

DSY889 ***HR: 3***
1,2-DIMETHYLNITROSOHYDRAZINE
mf: $C_2H_7N_3$ mw: 73.10

SAFETY PROFILE: The liquid deflagrates on heating. When heated to decomposition it emits toxic fumes of NO_x. See also HYDRAZINE.

DSZ000 CAS:16339-12-1 ***HR: 3***
N,O-DIMETHYL-N-NITROSOHYDROXYLAMINE
mf: $C_2H_6N_2O_2$ mw: 90.10

SYNS: N-METHOXY-N-NITROSOMETHYLAMINE ◇ N-NITROSOMETHOXYMETHYLAMINE ◇ N-NITROSOMETHYLMETHOXYAMINE ◇ N-NITROSO-N-METHYL-o-METHYLHYDROXYLAMIN (GERMAN) ◇ N-NITROSO-N-METHYL-o-METHYL-HYDROXYLAMINE

TOXICITY DATA with REFERENCE
mmo-sat 1 μg/plate MUREAV 51,319,78
mma-sat 1 μg/plate MUREAV 51,319,78
mmo-omi 1 pph/72H-C SOGEBZ 10,522,74
orl-rat TDLo:6000 mg/kg/50W-I:ETA ZKKOBW 89,31,77
ivn-rat LD50:130 mg/kg ZEKBAI 69,103,67

SAFETY PROFILE: Poison by intravenous route. Questionable carcinogen with experimental tumorigenic data. Mutation data reported. Many N-nitroso compounds are carcinogens. When heated to decomposition it emits toxic fumes of NO_x. See also N-NITROSO COMPOUNDS.

DTA000 CAS:1456-28-6 ***HR: 3***
2,6-DIMETHYLNITROSOMORPHOLINE
mf: $C_6H_{12}N_2O_2$ mw: 144.20

SYNS: DIMETHYLNITROSOMORPHOLINE ◇ 2,6-DIMETHYL-N-NITROSOMORPHOLINE ◇ DMNM ◇ Me_2NMOR ◇ NITROSO-2,6-DIMETHYLMORPHOLINE ◇ N-NITROSO-2,6-DIMETHYLMORPHOLINE

TOXICITY DATA with REFERENCE
mmo-sat 1 mg/plate TCMUD8 1,295,80
mma-sat 50 nmol/plate MUREAV 57,1,78
dns-rat:lvr 1 mmol/L MUREAV 144,197,85
orl-rat TDLo:135 mg/kg/30W-I:ETA ZKKOBW 92,221,78
scu-rat TDLo:1684 mg/kg/18W-I:CAR CALEDQ 13,159,81
orl-gpg TDLo:960 mg/kg/12W-I:CAR CNREA8 40,1879,80
scu-rat TD:1198 mg/kg/20W-I:NEO CALEDQ 13,159,81
scu-rat LD50:387 mg/kg CALEDQ 13,159,81
orl-gpg LD50:280 mg/kg JJIND8 64,529,80
orl-ham LD50:367 mg/kg JNCIAM 58,429,77
scu-ham LD50:320 mg/kg JNCIAM 60,197,78

CONSENSUS REPORTS: EPA Genetic Toxicology Program.

SAFETY PROFILE: Suspected carcinogen with experimental carcinogenic, tumorigenic, and neoplastigenic data. Poison by ingestion and subcutaneous routes. Mutation data reported. Used as a model carcinogenic and carcinogenic metabolite. When heated to decomposition it emits toxic fumes of NO_x. See also N-NITROSO COMPOUNDS.

DTA050 ***HR: 2***
2,6-DIMETHYL-4-NITROSOMORPHOLINE cis and trans mixture (2:1)
mf: $C_6H_{12}N_2O_2$ mw: 144.20

TOXICITY DATA with REFERENCE
orl-rat TDLo:75 g/kg/50W-I:CAR CRNGDP 3,911,82
orl-rat TD:138 g/kg/50W-I:CAR CRNGDP 3,911,82

SAFETY PROFILE: Questionable carcinogen with experimental carcinogenic data. When heated to decomposition it emits toxic fumes of NO_x.

DTA400 CAS:17721-95-8 ***HR: 3***
2,6-DIMETHYLNITROSOPIPERIDINE
mf: $C_7H_{14}N_2O$ mw: 142.23

SYN: N-NITROSO-2,6-DIMETHYLPIPERIDINE

TOXICITY DATA with REFERENCE
orl-rat TDLo:2813 mg/kg/50W-I:ETA IJCNAW 16,318,75

CONSENSUS REPORTS: EPA Genetic Toxicology Program.

SAFETY PROFILE: Questionable carcinogen with experimental tumorigenic data. Many N-nitroso compounds are carcinogens. When heated to decomposition it emits toxic fumes of NO_x. See also N-NITROSO COMPOUNDS.

DTA600 CAS:65445-59-2 ***HR: 3***
3,5-DIMETHYLNITROSOPIPERIDINE
mf: $C_7H_{14}N_2O$ mw: 142.23

SYNS: 3,5-DIMETHYL-1-NITROSOPIPERIDINE ◇ N-NITROSO-3,5-DIMETHYLPIPERIDINE

TOXICITY DATA with REFERENCE
mma-sat 1 μg/plate MUREAV 56,131,77
sln-dmg-orl 5 mmol/L/24H MUREAV 67,27,79
mma-smc 50 mmol/L/24H MUREAV 57,155,78
orl-rat TDLo:3100 mg/kg/50W-I:ETA JJIND8 68,989,82

CONSENSUS REPORTS: EPA Genetic Toxicology Program.

SAFETY PROFILE: Questionable carcinogen with experimental tumorigenic data. Mutation data reported. Many N-nitroso compounds are carcinogens. When heated to decomposition it emits toxic fumes of NO_x. See also N-NITROSO COMPOUNDS.

DTA690 CAS:78338-31-5 ***HR: 2***
cis-3,5-DIMETHYL-1-NITROSOPIPERIDINE
mf: $C_7H_{14}N_2O$ mw: 142.23

SYNS: NITROSO-3,5-DIMETHYLPIPERIDINE cis-isomer ◇ PIPERIDINE, 3,5-DIMETHYL-1-NITROSO-, (Z)-

TOXICITY DATA with REFERENCE
mma-sat 100 μg/plate TCMUD8 1,295,80
orl-rat TDLo:2550 mg/kg/50W-I:ETA JJIND8 68,989,82

SAFETY PROFILE: Questionable carcinogen with experimental tumorigenic data. Mutation data reported. When heated to decomposition it emits toxic fumes of NO_x.

DTA700 CAS:78338-32-6 ***HR: 2***
trans-3,5-DIMETHYL-1-NITROSOPIPERIDINE
mf: $C_7H_{14}N_2O$ mw: 142.23

SYNS: NITROSO-3,5-DIMETHYLPIPERIDINE trans-isomer ◇ PIPERIDINE, 3,5-DIMETHYL-1-NITROSO-, (E)-

TOXICITY DATA with REFERENCE
mma-sat 50 μg/plate TCMUE9 1,129,84
orl-rat TDLo:500 mg/kg/50W-I:ETA JJIND8 68,989,82

SAFETY PROFILE: Questionable carcinogen with experimental tumorigenic data. Mutation data reported. When heated to decomposition it emits toxic fumes of NO_x.

DTA800 CAS:55556-86-0 ***HR: 3***
2,5-DIMETHYL-N-NITROSOPYRROLIDINE
mf: $C_6H_{12}N_2O$ mw: 128.20

TOXICITY DATA with REFERENCE
orl-rat TDLo:5625 mg/kg/50W-I:ETA CNREA8 36,1988,76

SAFETY PROFILE: Questionable carcinogen with experimental tumorigenic data. Many N-nitroso compounds are carcinogens. See also N-NITROSO COMPOUNDS. When heated to decomposition it emits toxic fumes of NO_x.

DTB200 CAS:13256-32-1 ***HR: 3***
1,3-DIMETHYLNITROSOUREA
mf: $C_3H_7N_3O_2$ mw: 117.13

SYNS: DIMETHYLNITROSOHARNSTOFF (GERMAN) ◇ N,N'-DIMETHYLNITROSOUREA ◇ 1,3-DIMETHYL-N-NITROSOUREA ◇ NITROSODIMETHYLUREA ◇ N-NITROSODIMETHYLUREA

TOXICITY DATA with REFERENCE
mmo-omi 1 pph ANTBAL 27,738,82
dni-mus-ipr 80 mg/kg INSSDM 19,85,81
par-rat TDLo:50 mg/kg (female 9D post):TER IARCCD 4,112,73
orl-rat TDLo:1300 mg/kg/65W-C:ETA ZEKBAI 69,103,67
scu-rat TDLo:836 mg/kg/20W-I:NEO AMOKAG 32,119,78
scu-mus TDLo:720 mg/kg/9W-I:CAR GANNA2 62,135,71
scu-ham TDLo:680 mg/kg/17W-I:CAR AMOKAG 28,333,74
orl-rat LD50:280 mg/kg ZEKBAI 69,103,67
ivn-rat LD50:280 mg/kg ZEKBAI 69,103,67

CONSENSUS REPORTS: EPA Genetic Toxicology Program.

SAFETY PROFILE: Suspected carcinogen with experimental carcinogenic, neoplastigenic, tumorigenic, and teratogenic data. Poison by ingestion and intravenous routes. Mutation data reported. When heated to decomposition it emits toxic fumes of NO_x. See also N-NITROSO COMPOUNDS.

DTB800 CAS:128-50-7 ***HR: 2***
6,6-DIMETHYL-2-NORPINENE-2-ETHANOL
mf: $C_{11}H_{18}O$ mw: 166.29

SYNS: 6,6-DIMETHYLBICYCLO-(3.1.1)-2-HEPTENE-2-ETHANOL ◇ HOMOMYRETENOL ◇ NOPOL ◇ NOPOL (TERPENE)

TOXICITY DATA with REFERENCE
skn-rbt 500 mg/24H MOD FCTXAV 17,879,79
orl-rat LD50:890 mg/kg FCTXAV 17,879,79
ims-mus LD50:500 mg/kg JSICAZ 21,342,62

CONSENSUS REPORTS: Reported in EPA TSCA Inventory.

SAFETY PROFILE: Moderately toxic by ingestion and intramuscular routes. A skin irritant. When heated to decomposition it emits acrid smoke and irritating fumes.

DTC000 CAS:128-51-8 ***HR: 2***
6,6-DIMETHYL-2-NORPINENE-2-ETHANOL ACETATE
mf: $C_{13}H_{20}O_2$ mw: 208.33

SYNS: CITROVIOL ◇ 6,6-DIMETHYLBICYCLO(3.1.1)-2-HEPTENE-2-ETHYL ACETATE ◇ LIGNYL ACETATE ◇ NOPOL ACETATE ◇ NOPYL ACETATE ◇ 2-PINENE-10-METHYL ACETATE

TOXICITY DATA with REFERENCE
skn-rbt 500 mg/24H MLD FCTXAV 12,943,74
orl-rat LD50:3000 mg/kg FCTXAV 12,943,74

CONSENSUS REPORTS: Reported in EPA TSCA Inventory.

SAFETY PROFILE: Moderately toxic by ingestion. A skin irritant. When heated to decomposition it emits acrid smoke and irritating fumes.

DTC200 CAS:3886-90-6 ***HR: 3***
N,N-DIMETHYLOCTADECANAMIDE
mf: $C_{20}H_{41}NO$ mw: 311.62

SYN: N,N-DIMETHYLSTEARAMIDE

TOXICITY DATA with REFERENCE
ipr-mus LD50:2000 mg/kg AIHAAP 32,539,71
ivn-mus LD50:153 mg/kg AIHAAP 32,539,71
ivn-rbt LD50:71 mg/kg AIHAAP 32,539,71

CONSENSUS REPORTS: Reported in EPA TSCA Inventory.

SAFETY PROFILE: Poison by intravenous route. Moderately toxic by intraperitoneal route. When heated to decomposition it emits toxic fumes of NO_x.

DTC400 CAS:124-28-7 ***HR: 3***
N,N-DIMETHYLOCTADECYLAMINE
mf: $C_{20}H_{43}N$ mw: 297.64

SYN: N,N-DIMETHYLOKTADECYLAMIN(CZECH)

TOXICITY DATA with REFERENCE
skn-rbt 500 mg/24H MOD 28ZPAK -,63,72
eye-rbt 20 mg/24H MOD 28ZPAK -,63,72
orl-rat LDLo:500 mg/kg NCNSA6 5,11,53
ipr-rat LDLo:100 mg/kg NCNSA6 5,11,53

CONSENSUS REPORTS: Reported in EPA TSCA Inventory.

SAFETY PROFILE: Poison by intraperitoneal route. Moderately toxic by ingestion. A skin and eye irritant. When heated to decomposition it emits toxic fumes of NO_x.

DTC600 CAS:122-19-0 ***HR: 3***
DIMETHYLOCTADECYLBENZYLAMMONIUM CHLORIDE
mf: $C_{27}H_{50}N \cdot Cl$ mw: 424.23

SYNS: AMMONYX 4 ◇ AMMONYX CA SPECIAL ◇ ARQUAD DM18B-90 ◇ BARQUAT SB-25 ◇ BENZYLDIMETHYLSTEARYLAMMONIUM CHLORIDE ◇ BENZYLSTEARYLDIMETHYLAMMONIUM CHLORIDE ◇ CARSOQUAT SDQ-25 ◇ DEHYQUART STC-25 ◇ DIMETHYLBENZYLOCTADECYLAMMONIUM CHLORIDE ◇ INTEXAN SB-85 ◇ J SOFT C 4 ◇ KATAMINE AB ◇ NISSAN CATION S2-100 ◇ N-OCTADECYL-N-BENZYL-N,N-DIMETHYLAMMONIUMCHLORIDE◇ OCTADECYLDIMETHYLBENZYLAMMONIUM CHLORIDE ◇ ORTHOSAN MB ◇ QUATERNOL 1 ◇ STEARALKONIUM CHLORIDE ◇ STEARYLDIMETHYLBENZYLAMMONIUM CHLORIDE ◇ STEBAC ◇ TALLOW BENZYL DIMETHYLAMMONIUM CHLORIDE ◇ TRITON X-40 ◇ VARISOFT SDC

TOXICITY DATA with REFERENCE
skn-hmn 3 mg/3D-I MLD 85DKA8 -,127,77
skn-man 125 mg/2D MLD PSTGAW 20,16,53
skn-rbt 1 mg/24H OYYAA2 6,329,72
eye-rbt 200 μg SEV PSTGAW 20,16,53
orl-rat LD50:1250 mg/kg JACTDZ 1(2),57,82
ipr-rat LD50:280 mg/kg KHFZAN 12(12),61,78
orl-mus LD50:760 mg/kg JACTDZ 1(2),57,82
ipr-mus LD50:175 mg/kg KHFZAN 12(12),61,78
orl-gpg LD50:500 mg/kg GISAAA 49(8),90,84

CONSENSUS REPORTS: Reported in EPA TSCA Inventory.

SAFETY PROFILE: Poison by intraperitoneal route. Moderately toxic by ingestion. A human skin irritant and severe experimental eye irritant. When heated to decomposition it emits very toxic fumes of NO_x, NH_3, and Cl^-.

DTC800 CAS:5392-40-5 ***HR: 2***
3,7-DIMETHYL-2,6-OCTADIENAL
mf: $C_{10}H_{16}O$ mw: 152.26

PROP: Mobile, pale yellow liquid; strong lemon odor. D: 0.891-0.897 @ 15°, refr index: 1.486-1.490, flash p: 198°F. Sol in 5 volumes of 60% alc; sol in all proportions of benzyl benzoate, diethyl phthalate, glycerin, propylene glycol, mineral oil, fixed oils and 95% alc; insol in water.

SYNS: BUTOBEN ◇ BUTYL p-HYDROXYBENZOATE ◇ CITRAL (FCC) ◇ FEMA No. 2203 ◇ NCI-C56348 ◇ NERAL

TOXICITY DATA with REFERENCE
skn-hmn 40 mg/24H MLD FCTXAV 17,259,79
skn-rbt 500 mg/24H MOD FCTXAV 17,259,79
skn-gpg 1%/48H MOD JSCCA5 28,357,77
skn-rat TDLo:27600 mg/kg (female 60D pre):REP JRPFA4 55,347,79
orl-rat LD50:4960 mg/kg FCTXAV 2,327,64
ipr-rat LD50:460 mg/kg JRPFA4 55,347,79
orl-mus LD50:6000 mg/kg BIJOAK 34,1196,40

CONSENSUS REPORTS: Reported in EPA TSCA Inventory.

SAFETY PROFILE: Moderately toxic by intraperitoneal route. Mildly toxic by ingestion. Experimental reproductive effects. A human and experimental skin irritant. Combustible liquid. When heated to decomposition it emits acrid smoke and irritating fumes.

DTD000 CAS:106-24-1 ***HR: 3***
3,7-DIMETHYL-(E)-2,6-OCTADIEN-1-OL
mf: $C_{10}H_{18}O$ mw: 154.28

PROP: Colorless to pale yellow, oily liquid; pleasant geranium odor. D: 0.870-0.890 @ 15°, refr index: 1.469-1.478, mp: 15°, bp: 230°, flash p: 214°F. Sol in fixed oils, propylene glycol; sltly sol in water; insol in glycerin @ 230°.

SYNS: 2,6-DIMETHYL-trans-2,6-OCTADIEN-8-OL◇ 3,7-DIMETHYL-trans-2,6-OCTADIEN-1-OL ◇ FEMA No. 2507 ◇ GERANIOL (FCC) ◇ GERANIOL ALCOHOL ◇ GERANIOL EXTRA ◇ GERANYL ALCOHOL ◇ GUANIOL ◇ LEMONOL

TOXICITY DATA with REFERENCE
orl-rat LD50:3600 mg/kg FCTXAV 2,327,64

ims-mus LD50:4000 mg/kg JSICAZ 21,342,62
ivn-rbt LDLo:50 mg/kg NYKZAU 58,394,62

CONSENSUS REPORTS: Reported in EPA TSCA Inventory.

SAFETY PROFILE: Poison by intravenous route. Moderately toxic by ingestion and intramuscular routes. Combustible liquid. When heated to decomposition it emits acrid smoke and irritating fumes.

DTD200 CAS:106-25-2 ***HR: 2***
2-cis-3,7-DIMETHYL-2,6-OCTADIEN-1-OL
mf: $C_{10}H_{18}O$ mw: 154.28

PROP: Colorless liquid; sweet, rose odor. D: 0.875-0.880, refr index: 1.467-1.478. Sol in alc, chloroform, ether, water @ 227°.

SYNS: 3,7-DIMETHYL-(Z)-2,6-OCTADIEN-1-OL ◇ FEMA No. 2770 ◇ NEROL (FCC)

TOXICITY DATA with REFERENCE
skn-rbt 500 mg/24H MOD FCTXAV 14,623,76
orl-rat LD50:4500 mg/kg FCTXAV 14,623,76
ims-mus LD50:3000 mg/kg JSICAZ 21,342,62

CONSENSUS REPORTS: Reported in EPA TSCA Inventory.

SAFETY PROFILE: Moderately toxic by intramuscular route. Mildly toxic by ingestion. A skin irritant. When heated to decomposition it emits acrid smoke and irritating fumes.

DTD400 CAS:5986-38-9 ***HR: 2***
2,6-DIMETHYL-5,7-OCTADIEN-2-OL
mf: $C_{10}H_{18}O$ mw: 154.28

SYN: OCIMENOL

TOXICITY DATA with REFERENCE
orl-rat LD50:1700 mg/kg FCTXAV 14,817,76

CONSENSUS REPORTS: Reported in EPA TSCA Inventory.

SAFETY PROFILE: Moderately toxic by ingestion. When heated to decomposition it emits acrid smoke and irritating fumes.

DTD800 CAS:105-87-3 ***HR: 1***
trans-3,7-DIMETHYL-2,6-OCTADIEN-1-OL ACETATE
mf: $C_{12}H_{20}O_2$ mw: 196.32

PROP: Colorless, sweet, clear liquid; odor of lavender. D: 0.907-0.918 @ 15°, refr index: 1.458-1.464, bp: 128-129° @ 16 mm, flash p: 219°F. Sol in alc, fixed oils, ether; sltly sol in propylene glycol; insol in water and glycerol.

SYNS: ACETIC ACID GERANIOL ESTER ◇ 3,7-DIMETHYL-2-trans-6-OCTADIENYL ACETATE ◇ trans-3,7-DIMETHYL-2,6-OCTADIEN-1-YL ACETATE ◇ trans-2,6-DIMETHYL-2,6-OCTADIEN-8-YL ETHANOATE ◇ FEMA No. 2509 ◇ GERANIOL ACETATE ◇ GERANYL ACETATE (FCC) ◇ NCI-C54728

TOXICITY DATA with REFERENCE
orl-rat LD50:6330 mg/kg FCTXAV 2,327,64

CONSENSUS REPORTS: Reported in EPA TSCA Inventory.

SAFETY PROFILE: Mildly toxic by ingestion. Combustible liquid. When heated to decomposition it emits acrid smoke and irritating fumes. See also ESTERS.

DTE000 CAS:56172-46-4 ***HR: 1***
3,7-DIMETHYL-2-trans-6-OCTADIENYL CROTONATE
mf: $C_{14}H_{22}O_2$ mw: 222.36

SYNS: CROTONIC ACID GERNAIOL ESTER ◇ trans-3,7-DIMETHYL-2,6-OCTADIEN-1-OL-2-BUTENOATE ◇ 3-7-DIMETHYL-2,6-OCTADIENYL ESTER-2-BUTENOIC ACID ◇ GERANIOL CROTONATE ◇ GERANYL-2-BUTENOATE ◇ GERANYL CROTONATE

TOXICITY DATA with REFERENCE
skn-rbt 500 mg/24H MOD FCTXAV 12,891,74

CONSENSUS REPORTS: Reported in EPA TSCA Inventory.

SAFETY PROFILE: A skin irritant. When heated to decomposition it emits acrid smoke and irritating fumes. See also ESTERS.

DTE200 CAS:1118-92-9 ***HR: 3***
N,N-DIMETHYLOCTANAMIDE
mf: $C_{10}H_{21}NO$ mw: 171.32

SYN: N,N-DIMETHYLCAPRYLAMIDE

TOXICITY DATA with REFERENCE
ipr-mus LD50:620 mg/kg AIHAAP 32,539,71
ivn-mus LD50:36 mg/kg AIHAAP 32,539,71

CONSENSUS REPORTS: Reported in EPA TSCA Inventory.

SAFETY PROFILE: Poison by intravenous route. Moderately toxic by intraperitoneal route. When heated to decomposition it emits toxic fumes of NO_x.

DTE400 CAS:107-74-4 ***HR: 1***
3,7-DIMETHYL-1,2-OCTANEDIOL
mf: $C_{10}H_{22}O_2$ mw: 174.32

SYNS: 3,7-DIMETHYL-7-HYDROXY-1-OCTANOL ◇ HYDROXYCITRONELLOL ◇ 7-HYDROXY-3,7-DIMETHYLOCTAN-1-OL

TOXICITY DATA with REFERENCE
skn-rbt 500 mg/24H MLD FCTXAV 12,923,74

CONSENSUS REPORTS: Reported in EPA TSCA Inventory.

SAFETY PROFILE: A skin irritant. When heated to decomposition it emits acrid smoke and irritating fumes.

DTE600 CAS:106-21-8 ***HR: 2***
DIMETHYLOCTANOL
mf: $C_{10}H_{22}O$ mw: 158.32

PROP: Colorless liquid; sweet, rose odor. D: 0.26-0.842, refr index: 1.435. Sol in fixed oils, propylene glycol; insol in glycerin.

SYNS: DIHYDROCITRONELLOL ◇ 2,6-DIMETHYL-8-OCTANOL ◇ 3,7-DIMETHYL-1-OCTANOL (FCC) ◇ FEMA No. 2391 ◇ GERANIOL TETRAHYDRIDE ◇ PELARGOL ◇ PERHYDROGERANIOL ◇ TETRAHYDROGERANIOL

TOXICITY DATA with REFERENCE
skn-rbt 500 mg/24H FCTXAV 13,517,75
skn-rbt LD50:2400 mg/kg FCTXAV 12,535,74

CONSENSUS REPORTS: Reported in EPA TSCA Inventory.

SAFETY PROFILE: Moderately toxic by skin contact. A skin irritant. When heated to decomposition it emits acrid smoke and irritating fumes.

DTE800 CAS:20780-49-8 ***HR: 1***
3,7-DIMETHYLOCTANYL ACETATE
mf: $C_{12}H_{24}O_2$ mw: 200.36

SYNS: DIHYDROCITRONELLYL ACETATE ◇ 3,7-DIMETHYLOCTYL ACETATE ◇ TETRAHYDROGERANYL ACETATE

TOXICITY DATA with REFERENCE
skn-rbt 500 mg/24H MOD FCTXAV 18,673,80
skn-rbt LD50:5000 mg/kg FCTXAV 18,673,80

CONSENSUS REPORTS: Reported in EPA TSCA Inventory.

SAFETY PROFILE: Mildly toxic by skin contact. A skin irritant. When heated to decomposition it emits acrid smoke and irritating fumes.

DTF000 CAS:67874-80-0 ***HR: 1***
3,7-DIMETHYLOCTANYL BUTYRATE
mf: $C_{14}H_{28}O_2$ mw: 228.42

SYNS: 3,7-DIMETHYLOCTYL ESTER BUTANOIC ACID ◇ TETRAHYDROGERANYL BUTYRATE

TOXICITY DATA with REFERENCE
skn-rbt 500 mg/24H MOD FCTXAV 18,649,80

CONSENSUS REPORTS: Reported in EPA TSCA Inventory.

SAFETY PROFILE: A skin irritant. When heated to decomposition it emits acrid smoke and irritating fumes.

DTF200 CAS:29714-87-2 ***HR: 1***
DIMETHYLOCTATRIENE
mf: $C_{10}H_{16}$ mw: 136.26

SYNS: DIMETHYLOCTATRIENE (mixed isomer) ◇ OCIMENE

TOXICITY DATA with REFERENCE
skn-rbt 500 mg/24H MOD FCTXAV 16,829,78

SAFETY PROFILE: A skin irritant. When heated to decomposition it emits acrid smoke and irritating fumes.

DTF400 CAS:141-25-3 ***HR: 2***
2,6-DIMETHYL-1-OCTEN-8-OL
mf: $C_{10}H_{20}O$ mw: 156.30

PROP: Flash p: +212°F.

SYNS: α-CITRONELLOL ◇ 3,7-DIMETHYL-7-OCTEN-1-OL ◇ FEMA No. 2981 ◇ RHODINOL (FCC)

TOXICITY DATA with REFERENCE
ims-mus LD50:4000 mg/kg JSICAZ 21,342,62

CONSENSUS REPORTS: Reported in EPA TSCA Inventory.

SAFETY PROFILE: Moderately toxic by intramuscular route. Combustible liquid. When heated to decomposition it emits acrid smoke and irritating fumes.

DTF800 CAS:141-16-2 ***HR: 1***
2,6-DIMETHYL-2-OCTEN-8-YL BUTYRATE
mf: $C_{14}H_{26}O_2$ mw: 226.40

SYNS: BUTYRIC ACID-3,7-DIMETHYL-6-OCTENYL ESTER ◇ 3,7-DIMETHYL-6-OCTEN-1-OL BUTYRATE ◇ 2,6-DIMETHYL-2-OCTEN-8-OL-BUTYRATE ◇ RHODINYL BUTYRATE

TOXICITY DATA with REFERENCE
skn-rbt 500 mg MLD FCTXAV 14,849,76

CONSENSUS REPORTS: Reported in EPA TSCA Inventory.

SAFETY PROFILE: A skin irritant. When heated to decomposition it emits acrid smoke and irritating fumes.

DTG000 CAS:1854-26-8 ***HR: 2***
DIMETHYLOL DIHYDROXYETHYLENE UREA
mf: $C_4H_{10}N_2O_5$ mw: 178.17

SYNS: ARKOFIX NG ◇ CASSURIT LR ◇ DEPREMOL G ◇ (4,5-DIHYDROXY-1,3-BIS(HYDROXYMETHYL)-2-IMIDAZOLIDINONE◇ DIMETHYLOLGLYOXALUREA ◇ DMDHEU ◇ FIRMATEX RK ◇ FIXAPRET CP ◇ HYLITE LF ◇ KNITTEX LE ◇ NCI-C60322 ◇ NEUPERM GFN ◇ NS 11 ◇ PERMAFRESH 183 ◇ PROTOCOL C ◇ PROX DW ◇ READPRET KPN ◇ SARCOSET GM ◇ SUMITEX FSK ◇ SUMITEX NS ◇ VERAPRET DH ◇ WNM

TOXICITY DATA with REFERENCE
skn-rbt 500 mg/24H SEV 28ZPAK -,269,72
eye-rbt 500 mg/24H MLD 28ZPAK -,269,72

CONSENSUS REPORTS: Reported in EPA TSCA Inventory.

SAFETY PROFILE: An eye and severe skin irritant. When heated to decomposition it emits toxic fumes of NO_x.

DTG200 CAS:10143-22-3 ***HR: 2***
N,N-DIMETHYLOL-2-METHOXYETHYL CARBAMATE
mf: $C_6H_{13}NO_5$ mw: 179.20

TOXICITY DATA with REFERENCE
skn-rbt 500 mg open MLD UCDS** 5/19/66
eye-rbt 15 mg SEV UCDS** 3/7/66
orl-rat LD50:11 g/kg UCDS** 5/19/66

CONSENSUS REPORTS: Reported in EPA TSCA Inventory.

SAFETY PROFILE: Mildly toxic by ingestion. A skin and severe eye irritant. When heated to decomposition it emits toxic fumes of NO_x. See also CARBAMATES.

DTG400 CAS:126-30-7 ***HR: 2***
DIMETHYLOLPROPANE
mf: $C_5H_{12}O_2$ mw: 104.17

PROP: White, crystalline solid. Mp: 121-126°, bp: 207-212°.

SYNS: 2,2-DIMETHYL-1,3-PROPANEDIOL ◇ DIMETHYLTRIMETHYLENE GLYCOL ◇ NEOL ◇ NEOPENTYLENE GLYCOL ◇ NEOPENTYL GLYCOL ◇ NPG

TOXICITY DATA with REFERENCE
orl-rat LDLo:3200 mg/kg KODAK* -,-,71

CONSENSUS REPORTS: Reported in EPA TSCA Inventory.

SAFETY PROFILE: Moderately toxic by ingestion. Used in polyester manufacture. An insect repellent. Combustible when exposed to heat or flame; can react with oxidizing materials. When heated to decomposition it emits acrid smoke and irritating fumes. See also GLYCOLS.

DTG600 CAS:3084-25-1 ***HR: 2***
DIMETHYLOL THIOUREA
mf: $C_3H_8N_2O_2S$ mw: 136.19

SYN: USAF B-74

TOXICITY DATA with REFERENCE
ipr-mus LD50:1000 mg/kg NTIS** AD277-689

CONSENSUS REPORTS: Reported in EPA TSCA Inventory.

SAFETY PROFILE: Moderately toxic by intraperitoneal route. When heated to decomposition it emits very toxic fumes of NO_x and SO_x.

DTG700 CAS:140-95-4 ***HR: 2***
1,3-DIMETHYLOLUREA
mf: $C_3H_8N_2O_3$ mw: 120.13

PROP: Crystals from alc. Mp: 137-139°. Very sol in cold water, hot ethanol, methanol.

SYNS: N,N'-BIS(HYDROXYMETHYL)UREA ◇ 1,3-BIS(HYDROXYMETHYL)UREA ◇ CAURITE ◇ CSI PASTE ◇ N,N'-DIHYDROXYMETHYLUREA ◇ DMU ◇ FINISH EN ◇ KAURIT S ◇ KNITTEX ASL ◇ METHURAL ◇ METHURIN (RUSSIAN) ◇ METURAL ◇ OXYMETHUREA ◇ PERMAFRESH 477 ◇ PROTESINE DMU ◇ UREOL P

TOXICITY DATA with REFERENCE
orl-rat LD50:3400 mg/kg GISAAA 44(3),68,79
orl-mus LD50:1795 mg/kg GISAAA 44(3),68,79
orl-rbt LD50:3200 mg/kg GISAAA 44(3),68,79

CONSENSUS REPORTS: Reported in EPA TSCA Inventory.

SAFETY PROFILE: Moderately toxic by ingestion. When heated to decomposition it emits toxic fumes of NO_x.

DTG750 CAS:51200-87-4 ***HR: 2***
DIMETHYL OXAZOLIDINE
mf: $C_5H_{11}NO$ mw: 101.17

SYNS: 4,4-DIMETHYLOXAZOLIDINE ◇ OXAZOLIDINE A

TOXICITY DATA with REFERENCE
orl-rat LD50:950 mg/kg CTOIDG 96(3),79,81
ihl-rat LC50:11700 mg/m^3 CTOIDG 96(3),79,81
skn-rbt LD50:1400 mg/kg CTOIDG 96(3),79,81

CONSENSUS REPORTS: Reported in EPA TSCA Inventory.

SAFETY PROFILE: Moderately toxic by ingestion and skin contact. Mildly toxic by inhalation. When heated to decomposition it emits toxic fumes of NO_x.

DTH000 CAS:1955-45-9 ***HR: 3***
3,3-DIMETHYL-2-OXETHANONE
mf: $C_5H_8O_2$ mw: 100.13

SYNS: 3,3-DIMETHYL-2-OXETANONE ◇ DIMETHYL PROPIOLACTONE ◇ 3,3-DIMETHYL-β-PROPIOLACTONE ◇ NCI-C04126 ◇ PIVALIC ACID LACTONE ◇ PIVALOLACTONE

TOXICITY DATA with REFERENCE
mma-sat 333 μg/plate ENMUDM 7(Suppl 5),1,85
mma-esc 333 μg/plate ENMUDM 7(Suppl 5),1,85
orl-rat TDLo:216 g/kg/2Y-I:CAR NCITR* NCI-CG-TR-140,78
orl-rat TD:72 g/kg/69W:ETA NCITR* NCI-CG-TR-140,78
orl-rat LD50:1470 mg/kg NCILB* NIH-NCI-E-C-72-3252,73
orl-mus LD50:316 mg/kg NCILB* NIH-NCI-E-C-72-3252,73

CONSENSUS REPORTS: NCI Carcinogenesis Bioassay (gavage); No Evidence: mouse NCITR* NCI-CG-TR-140,78; Clear Evidence: rat NCITR* NCI-CG-TR-140,78. Reported in EPA TSCA Inventory.

SAFETY PROFILE: Poison by ingestion. Questionable carcinogen with experimental carcinogenic and tumorigenic data. Mutation data reported. When heated to decomposition it emits acrid smoke and irritating fumes.

DTH200 CAS:2873-97-4 ***HR: 2***
N-(1,1-DIMETHYL-3-OXOBUTYL)ACRYLAMIDE
mf: $C_9H_{15}NO_2$ mw: 169.25

SYNS: DIACETONE ACRYLAMIDE ◇ N-(1,1-DIMETHYL-3-OXOBUTYL)-2-PROPENAMIDE ◇ N-(2-(2-METHYL-4-OXOPENTYL)) ACRYLAMIDE

TOXICITY DATA with REFERENCE
orl-mus LD50:1303 mg/kg ARTODN 47,179,81

CONSENSUS REPORTS: Reported in EPA TSCA Inventory.

SAFETY PROFILE: Moderately toxic by ingestion. When heated to decomposition it emits toxic fumes of NO_x.

DTH400 CAS:2273-45-2 ***HR: 3***
DIMETHYLOXOSTANNANE
mf: C_2H_6OSn mw: 164.77

PROP: White powder. Insol in water.

SYN: DIMETHYLTIN OXIDE

TOXICITY DATA with REFERENCE
ivn-mus LD50:100 mg/kg CSLNX* NX#03809

CONSENSUS REPORTS: Reported in EPA TSCA Inventory.

OSHA PEL: TWA 0.1 mg(Sn)/m^3 (skin)
ACGIH TLV: TWA 0.1 mg(Sn)/m^3 (skin) (Proposed: TWA 0.1 mg(Sn)/m^3; STEL 0.2 mg(Sn)/m^3 (skin))
NIOSH REL: (Organotin Compounds) TWA 0.1 mg(Sn)/m^3

SAFETY PROFILE: Poison by intravenous route. When heated to decomposition it emits acrid smoke and irritating fumes. See also TIN COMPOUNDS.

DTH600 CAS:63951-48-4 ***HR: 2***
α,γ-DIMETHYL-α-OXYMETHYL GLUTARALDEHYDE
mf: $C_8H_{14}O_3$ mw: 158.22

SYN: 2-(HYDROXYMETHYL)-2,4-DIMETHYLPENTANEDIAL

TOXICITY DATA with REFERENCE
skn-rbt 500 mg SEV SCCUR* -,4,61
orl-rat LD50:2040 mg/kg SCCUR* -,4,61
orl-mus LD50:570 mg/kg SCCUR* -,4,61

SAFETY PROFILE: Moderately toxic by ingestion. A severe skin irritant. When heated to decomposition it emits acrid smoke and irritating fumes. See also ALDEHYDES.

DTH800 CAS:3820-53-9 ***HR: 3***
DIMETHYL PARANITROPHENYL THIONOPHOSPHATE
mf: $C_8H_{10}NO_5PS$ mw: 263.22

PROP: Crystals. Vap d: 9.1, mp: 38°, d: 1.235 @ 20°/4°.

SYNS: O,O-DIMETHYL-S-p-NITROFENYL ESTER KYSELINY THIOFOSFORECEN (CZECH) ◇ O,O-DIMETHYL-S-(p-NITROPHENYL) PHOSPHOROTHIOATE ◇ O,O-DIMETHYL-S-(4-NITROPHENYL)THIOPHOSPHATE

TOXICITY DATA with REFERENCE
orl-rat LD50:43 mg/kg 28ZPAK -,208,72
scu-mus LD50:8 mg/kg AMIHAB 11,487,55

SAFETY PROFILE: Poison by ingestion and subcutaneous routes. When heated to decomposition it emits very toxic fumes of NO_x, PO_x, and SO_x.

DTI000 ***HR: 2***
2,3-DIMETHYLPENTANE
mf: C_7H_{16} mw: 94.21

PROP: Liquid. Mp: −135°, bp: 89.8°, d: 0.69 @ 15.5°/15.5°, autoign temp: 635°F, flash p: 21.2°F, lel: 1.1%, uel: 6.7%, vap press: 40 mm @ 13.9°, vap d: 3.45.

SYN: DIETHYLDIMETHYLMETHANE

SAFETY PROFILE: Probable irritant and narcotic in high concentration. A very dangerous fire hazard and explosion hazard when exposed to heat, flame, or oxidizers. Keep away from heat and open flame; can react vigorously with oxidizing materials. To fight fire, use foam, CO_2, dry chemical. When heated to decomposition it emits acrid smoke and irritating fumes.

DTI200 CAS:565-59-3 ***HR: 2***
2,4-DIMETHYLPENTANE
mf: C_7H_{16} mw: 94.21

$(CH_3)_2CHCHCH_3CH_2CH_3$

PROP: Flash p: 10.4°F.

SAFETY PROFILE: A very dangerous fire hazard when exposed to heat, flame, or oxidizers. When heated to decomposition it emits acrid smoke and fumes.

DTI400 CAS:10143-23-4 ***HR: 2***
2,3-DIMETHYL-1-PENTANOL
mf: $C_7H_{16}O$ mw: 116.23

SYN: 2,3-DIMETHYLPENTANOL

TOXICITY DATA with REFERENCE
skn-rbt 10 mg/24H open MLD AIHAAP 23,95,62
orl-rat LD50:2380 mg/kg AIHAAP 23,95,62
skn-rbt LD50:2500 mg/kg AIHAAP 23,95,62

SAFETY PROFILE: Moderately toxic by ingestion and skin contact. A skin irritant. When heated to decomposition it emits acrid smoke and irritating fumes.

DTI600 CAS:565-80-0 ***HR: 2***
2,4-DIMETHYL-3-PENTANONE
mf: $C_7H_{14}O$ mw: 108.19

PROP: Flash p: 59°C.

SAFETY PROFILE: A very dangerous fire hazard when exposed to heat, flame, or oxidizers. When heated to decomposition it emits acrid smoke and fumes.

DTI709 CAS:71901-54-7 ***HR: 3***
S,S-DIMETHYLPENTASULFUR HEXANITRIDE
mf: $C_2H_6N_6S_5$ mw: 274.41

SAFETY PROFILE: A powerful explosive. Upon decomposition it emits toxic fumes of SO_x and NO_x. See also NITRIDES.

DTJ000 CAS:690-02-8 ***HR: 3***
DIMETHYL PEROXIDE
mf: $C_2H_6O_2$ mw: 62.07

SAFETY PROFILE: Both the liquid and the vapor are powerful explosives extremely sensitive to heat or shock. Rough handling may cause ignition. When heated to decomposition it emits acrid smoke and fumes. See also PEROXIDES.

DTJ159 CAS:15411-45-7 ***HR: 3***
DIMETHYLPEROXYCARBONATE
mf: $C_4H_6O_6$ mw: 150.09

$CH_3OCO{\bullet}OOCO{\bullet}OCH_3$

SAFETY PROFILE: Explodes when heated above 55°C or on impact. When heated to decomposition it emits acrid smoke and fumes. See also PEROXIDES.

DTJ200 CAS:22349-59-3 ***HR: 2***
1,4-DIMETHYLPHENANTHRENE
mf: $C_{16}H_{14}$ mw: 206.30

TOXICITY DATA with REFERENCE
mma-sat 50 µg/plate MUREAV 116,91,83
skn-mus TDLo:40 mg/kg/20D-I:CAR CNREA8 41,3441,81

CONSENSUS REPORTS: IARC Cancer Review: Animal Inadequate Evidence IMEMDT 32,349,83

SAFETY PROFILE: Questionable carcinogen with experimental carcinogenic data. Mutation data reported. When heated to decomposition it emits acrid smoke and irritating fumes.

DTJ400 CAS:122-09-8 ***HR: 3***
α,α-DIMETHYLPHENETHYLAMINE
mf: $C_{10}H_{15}N$ mw: 149.26

SYNS: α,α-DIMETHYLBENZEETHANAMINE ◇ 1,1-DIMETHYL-2-PHENYLETHANAMINE ◇ α,α-DIMETHYL-β-PHENYLETHYLAMINE ◇ DUROMINE ◇ LIPOPILL ◇ LONAMIN ◇ MG 18370 ◇ MG 18570 ◇ MIRAPRONT ◇ PHENTERMINE ◇ 2-PHENYL-tert-BUTYLAMINE ◇ RCRA WASTE NUMBER P046 ◇ WILPO

TOXICITY DATA with REFERENCE
sln-asn 1 mg/L MUREAV 26,159,74
orl-man TDLo:1429 µg/kg:ANS THERAP 34,205,79
orl-mus LD50:105 mg/kg AIPTAK 178,62,69
ipr-mus LD50:71 mg/kg RCOCB8 14,677,76
ivn-mus LD50:14 mg/kg CSLNX* NX#03232

CONSENSUS REPORTS: Reported in EPA TSCA Inventory.

SAFETY PROFILE: Poison by ingestion, intravenous, and intraperitoneal routes. Human systemic effects by ingestion: sympathomimetic. Mutation data reported. When heated to decomposition it emits toxic fumes of NO_x.

DTK200 CAS:14543-76-1 ***HR: 3***
3,4-DIMETHYLPHENISOPROPYLAMINE SULFATE
mf: $C_{11}H_{17}N{\bullet}1/2H_2O_4S$ mw: 212.33

TOXICITY DATA with REFERENCE
orl-man TDLo:1500 µg/kg:CNS,GIT,SKN JPETAB 100,298,50
ipr-mus LD50:83 mg/kg JPETAB 100,298,50

SAFETY PROFILE: Poison by intraperitoneal route. Human systemic effects by ingestion of very small amounts: anorexia, nausea or vomiting, and sweating. When heated to decomposition it emits very toxic fumes of NO_x and SO_x.

DTK300 CAS:27691-62-9 ***HR: 1***
N,N-DIMETHYL-3-PHENOTHIAZINESULFONAMIDE
mf: $C_{14}H_{14}N_2O_2S_2$ mw: 306.42

SYNS: 3-DIMETHYLSULPHAMIDOPHENOTHIAZINE ◇ 3-PHENOTHIAZINESULFONAMIDE, N,N-DIMETHYL- ◇ 10H-PHENOTHIAZINE-3-SULFONAMIDE, N,N-DIMETHYL-

TOXICITY DATA with REFERENCE
eye-rbt 100 mg MLD FCTOD7 20,573,82

SAFETY PROFILE: An eye irritant. When heated to decomposition it emits toxic fumes of NO_x and SO_x.

DTK600 CAS:2747-31-1 ***HR: 3***
N,N-DIMETHYL-p-PHENYLAZOANILINE-N-OXIDE
mf: $C_{14}H_{15}N_3O$ mw: 241.32

SYNS: DAB-N-OXIDE ◇ 4-DIMETHYLAMINOAZOBENZENE AMINE-N-OXIDE ◇ N,N-DIMETHYLAMINOAZOBENZENE-N-OXIDE

TOXICITY DATA with REFERENCE
orl-rat TDLo:6300 mg/kg/30W-C:ETA GANNA2 54,455,63
orl-rat LD50:2200 mg/kg GANNA2 54,455,63
ipr-rat LD50:155 mg/kg GANNA2 54,455,63
orl-mus LD50:760 mg/kg GANNA2 54,455,63
ipr-mus LD50:175 mg/kg GANNA2 54,455,63

SAFETY PROFILE: Poison by intraperitoneal route. Moderately toxic by ingestion. Questionable carcinogen with experimental tumorigenic data. When heated to decomposition it emits toxic fumes of NO_x.

DTK800 CAS:2438-49-5 ***HR: 3***
N,N-DIMETHYL-4-PHENYLAZO-o-ANISIDINE
mf: $C_{15}H_{17}N_3O$ mw: 255.35

SYN: 3-METHOXY-4-DIMETHYLAMINOAZOBENZENE

TOXICITY DATA with REFERENCE
orl-rat TDLo:9800 mg/kg/34W-C:ETA CNREA8 21,1068,61

SAFETY PROFILE: Questionable carcinogen with experimental tumorigenic data. When heated to decomposition it emits toxic fumes of NO_x.

DTL000 CAS:36576-23-5 ***HR: 3***
2,3-DIMETHYL-4-(PHENYLAZO)BENZENAMINE
mf: $C_{14}H_{15}N_3$ mw: 225.32

SYN: 2,3-DIMETHYL-4-PHENYLAZOANILINE

TOXICITY DATA with REFERENCE
orl-mus TDLo:6000 mg/kg/26W-C:NEO FCTXAV 11,415,73

SAFETY PROFILE: Questionable carcinogen with experimental neoplastigenic data. When heated to decomposition it emits toxic fumes of NO_x.

DTL200 CAS:126-27-2 ***HR: 3***
2-DI(N-METHYL-N-PHENYL-tert-BUTYL-CARBAMOYLMETHYL)AMINOETHANOL
mf: $C_{28}H_{41}N_3O_3$ mw: 467.72

SYNS: BETALGIL ◇ N,N-BIS(N-METHYL-N-PHENYL-tert-BUTYLACETAMIDO)-β-HYDROXYETHYLAMINE ◇ EMOREN ◇ FH 099 ◇ H4 099 ◇ 2,2′-((2-HYDROXYETHYL)IMINO BIS(N-(α,α-DIMETHYLPHENETHYL)-N-METHYL-ACETAMIDE ◇ 2,2′-((2-HYDROXYETHYL)IMINO)BIS(N-(1,1-DIMETHYL-2-PHENYLETHYL)-N-METHYLACETAMIDE) ◇ MUCAINE ◇ MUCOXIN ◇ MUTHESA ◇ OXAINE ◇ OXETACAINE ◇ OXETHACAINA (ITALIAN) ◇ OXETHAZINE ◇ STOMACAIN ◇ TEPILTA ◇ TOPICAIN ◇ WY 806

TOXICITY DATA with REFERENCE
sln-asn 1 mg/L MUREAV 26,159,74
orl-rat LD50:625 mg/kg 29ZVAB -,85,69
ipr-rat LD50:30 mg/kg GMITAB 134,642,75
ivn-rat LD50:1300 μg/kg 29ZVAB -,85,69
imp-rat LD50:502 mg/kg 29ZVAB -,85,69
orl-mus LD50:400 μg/kg 29ZVAB -,85,69
ipr-mus LD50:27 mg/kg GMITAB 134,642,75
scu-mus LD50:58 mg/kg GMITAB 134,642,75
ivn-mus LD50:3600 μg/kg 29ZVAB -,85,69
ims-mus LD50:247 mg/kg TXAPA9 5,184,63
imp-mus LD50:2470 μg/kg 29ZVAB -,85,69
ivn-rbt LD50:540 μg/kg 29ZVAB -,85,69
itr-rbt LD50:420 μg/kg TXAPA9 5,184,63

SAFETY PROFILE: Poison by ingestion, intratracheal, intravenous, intraperitoneal, intramuscular, subcutaneous, and implant routes. Mutation data reported. When heated to decomposition it emits toxic fumes of NO_x.

DTL600 CAS:99-98-9 ***HR: 3***
N,N-DIMETHYL-p-PHENYLENEDIAMINE
mf: $C_8H_{12}N_2$ mw: 136.22

PROP: Reddish-violet crystals. Mp: 53°, bp: 262°. Sol in water, alc, chloroform, ether.

SYN: DIMETHYL-p-PHENYLENEDIAMINE

TOXICITY DATA with REFERENCE
skn-hmn TDLo:14 μg/kg:SKN JIDHAN 4,386,23
orl-rat LDLo:50 mg/kg NCNSA6 5,11,53
scu-rat LDLo:50 mg/kg JIDHAN 4,386,23
ipr-mus LDLo:50 mg/kg RBPMAZ 22,1,52
skn-dog LDLo:84 mg/kg JIDHAN 4,386,23
ivn-dog LDLo:51 mg/kg JIDHAN 4,386,23
orl-cat LDLo:20 mg/kg JIDHAN 4,386,23
orl-rbt LDLo:150 mg/kg JIDHAN 4,386,23
ihl-rbt LCLo:500 ppb JIDHAN 4,386,23
skn-rbt LDLo:60 mg/kg JIDHAN 4,386,23
scu-rbt LDLo:60 mg/kg JIDHAN 4,386,23
ihl-gpg LCLo:240 ppb JIDHAN 4,386,23
scu-gpg LDLo:100 mg/kg JIDHAN 4,386,23

CONSENSUS REPORTS: Reported in EPA TSCA Inventory. EPA Extremely Hazardous Substances List.

SAFETY PROFILE: Poison by ingestion, inhalation, skin contact, subcutaneous, intraperitoneal and intravenous routes. Human systemic effects by skin contact: primary skin irritation, allergic dermatitis and hemorrhage. When heated to decomposition it emits toxic fumes of NO_x. See also AROMATIC AMINES.

DTL800 CAS:105-10-2 ***HR: 3***
N,N-DIMETHYL-p-PHENYLENEDIAMINE
mf: $C_8H_{12}N_2$ mw: 136.22

SYNS: p-AMINODIMETHYLANILINE ◇ C.I. 76075 ◇ p-DIMETHYLAMINOPHENYLAMINE ◇ N,N-DIMETHYL-1,4-BENZENEDIAMINE ◇ DIMETHYL-p-PHENYLENEDIAMINE ◇ DMPD

TOXICITY DATA with REFERENCE
mma-sat 5 μg/plate AEMIDF 42,641,81
dns-rat:lvr 100 μmol/L MUREAV 135,255,84
dns-ham:lvr 100 μmol/L MUREAV 135,255,84
ipr-rat LD50:21 mg/kg JPETAB 95,262,49
ipr-mus LD50:25 mg/kg JPETAB 95,262,49
ipr-dog LDLo:10 mg/kg JPETAB 95,262,49
ipr-rbt LD50:100 mg/kg JPETAB 95,262,49
ipr-gpg LD50:45 mg/kg JPETAB 95,262,49

SAFETY PROFILE: Poison by intraperitoneal route. Mutation data reported. When heated to decomposition it emits toxic fumes of NO_x.

DTM000 CAS:536-46-9 ***HR: 3***
N,N-DIMETHYL-p-PHENYLENEDIAMINE DIHYDROCHLORIDE
mf: $C_8H_{12}N_2$•2ClH mw: 209.14

SYNS: p-DIMETHYLAMINOANILINE DIHYDROCHLORIDE ◇ DIMETHYL-p-PHENYLENEDIAMINE HYDROCHLORIDE ◇ USAF EK-7423

TOXICITY DATA with REFERENCE
ipr-mus LDLo:25 mg/kg NTIS** AD404-218

CONSENSUS REPORTS: Reported in EPA TSCA Inventory.

SAFETY PROFILE: Poison by intraperitoneal route. When heated to decomposition it emits very toxic fumes of NO_x and HCl.

DTM200 CAS:60160-75-0 ***HR: 3***
N,N-DIMETHYL-p-PHENYLENEDIAMINE HEMISULFATE

TOXICITY DATA with REFERENCE
orl-rat LDLo:75 mg/kg NCNSA6 5,11,53

CONSENSUS REPORTS: Reported in EPA TSCA Inventory.

SAFETY PROFILE: Poison by ingestion. When heated to decomposition it emits very toxic fumes of NO_x and SO_x.

DTM400 CAS:2052-46-2 ***HR: 3***
N,N-DIMETHYL-p-PHENYLENEDIAMINE MONOHYDROCHLORIDE
mf: $C_8H_{12}N_2$•ClH mw: 172.68

TOXICITY DATA with REFERENCE
orl-rat LDLo:100 mg/kg NCNSA6 5,11,53

CONSENSUS REPORTS: Reported in EPA TSCA Inventory.

SAFETY PROFILE: Poison by ingestion. When heated to decomposition it emits very toxic fumes of NO_x and HCl.

DTM600 CAS:154-99-4 ***HR: 3***
o,p-DIMETHYL-β-PHENYLETHYLHYDRAZINE DIHYDROGEN SULFATE

SYNS: β-(2,4-DIMETHYLPHENYL)ETHYLHYDRAZINEDIHYDROGEN SULPHATE ◇ LON 41

TOXICITY DATA with REFERENCE
scu-mus TDLo:160 mg/kg (female 7-10D post):REP JOENAK 49,635,71
orl-mus LD50:250 mg/kg JOENAK 30,205,64
scu-mus LD50:250 mg/kg JOENAK 30,205,64

SAFETY PROFILE: Poison by ingestion and subcutaneous routes. Experimental reproductive effects. When heated to decomposition it emits very toxic fumes of SO_x and NO_x.

DTM800 CAS:10158-43-7 ***HR: 3***
DIMETHYLPHENYLETHYNYLTHALLIUM
mf: $C_{10}H_{11}Tl$ mw: 335.57

$(CH_3)_2TlC\equiv CPh$

CONSENSUS REPORTS: Thallium and its compounds are on the Community Right-To-Know List.

SAFETY PROFILE: An explosive sensitive to heating, stirring or impact. See also THALLIUM COMPOUNDS and ACETYLENE COMPOUNDS.

DTN000 ***HR: 2***
2,4-DIMETHYLPHENYLMALEIMIDE
mf: $C_{12}H_{11}NO_2$ mw: 201.24

SYN: 2,4-DIMETHYL-N-PHENYLMALEIMIDE

TOXICITY DATA with REFERENCE
skn-rbt 100 μg/24H open AIHAAP 23,95,62
orl-rat LD50:710 mg/kg AIHAAP 23,95,62

SAFETY PROFILE: Moderately toxic by ingestion. A skin irritant. When heated to decomposition it emits toxic fumes of NO_x.

DTN100 CAS:617-94-7 ***HR: 2***
DIMETHYLPHENYLMETHANOL
mf: $C_9H_{12}O$ mw: 136.21

SYNS: α-CUMYL ALCOHOL ◇ α,α-DIMETHYLBENZENEMETHANOL ◇ α,α-DIMETHYLBENZYL ALCOHOL ◇ DIMETHYLPHENYLCARBINOL ◇ 1-HYDROXYCUMENE ◇ PHENYLDIMETHYLCARBINOL ◇ 2-PHENYLISOPROPANOL

TOXICITY DATA with REFERENCE
skn-rbt 500 mg/24H SEV FCTOD7 20(Suppl), 675,82
orl-rat LD50:1300 mg/kg FCTOD7 20(Suppl),675,82
orl-mus LD50:1950 mg/kg FCTOD7 20(Suppl),675,82
skn-rbt LD50:4300 mg/kg FCTOD7 20(Suppl),675,82

CONSENSUS REPORTS: Reported in EPA TSCA Inventory.

SAFETY PROFILE: Moderately toxic by ingestion. Mildly toxic by skin contact. A severe skin irritant. When heated to decomposition it emits acrid smoke and irritating fumes.

DTN200 CAS:2655-14-3 ***HR: 3***
3,5-DIMETHYLPHENYL-N-METHYLCARBAMATE
mf: $C_{10}H_{13}NO_2$ mw: 179.24

SYNS: DRC 3340 ◇ H-69 ◇ MACBAL ◇ MAQBARL ◇ 3,5-XMC ◇ 3,5-XYLENOL METHYLCARBAMATE ◇ 3,5-XYLENYL-N-METHYLCARBAMATE ◇ 3,5-XYLYL-N-METHYLCARBAMATE

TOXICITY DATA with REFERENCE
orl-rat LD50:542 mg/kg 85ARAE 1,44,77
orl-mus LD50:280 mg/kg OYYAA2 3,74,69
orl-rbt LD50:445 mg/kg SPEADM 78-1,56,78

CONSENSUS REPORTS: EPA Genetic Toxicology Program.

SAFETY PROFILE: Poison by ingestion. When heated to decomposition it emits toxic fumes of NO_x. See also CARBAMATES.

DTN400 CAS:58139-33-6 ***HR: D***
3,4-DIMETHYLPHENYL-N-METHYL-N-NITROSOCARBAMATE
mf: $C_{10}H_{12}N_2O_3$ mw: 208.24

SYNS: METHYLNITROSOCARBAMIC ACID-3,4-DIMETHYLPHENYL ESTER ◇ NITROSO-MPMC

TOXICITY DATA with REFERENCE
mmo-esc 10 μg/plate BECTA6 14,389,75
dnr-esc 1500 pmol/plate/30M MUREAV 54,283,78
cyt-ham:fbr 15 mg/L/24H MUREAV 48,337,77

CONSENSUS REPORTS: EPA Genetic Toxicology Program.

SAFETY PROFILE: Mutation data reported. Many N-nitroso compounds are carcinogens. When heated to decomposition it emits toxic fumes of NO_x. See also CARBAMATES and N-NITROSO COMPOUNDS.

DTN775 CAS:42013-48-9 ***HR: 3***
5,5-DIMETHYL-2-PHENYLMORPHOLINE
mf: $C_{12}H_{17}NO$ mw: 191.30

SYNS: G 130 ◇ GP 130 ◇ 2-PHENYL-5-DIMETHYLTETRAHYDRO-1,4-OXAZINE

TOXICITY DATA with REFERENCE
orl-rat LD50:480 mg/kg ARZNAD 23,810,73
orl-mus LD50:380 mg/kg ARZNAD 23,810,73
ipr-mus LD50:100 mg/kg ARZNAD 23,810,73

SAFETY PROFILE: Poison by ingestion and intraperitoneal routes. When heated to decomposition it emits toxic fumes of NO_x.

DTN800 CAS:7635-51-0 ***HR: 3***
3,4-DIMETHYL-2-PHENYLMORPHOLINEHYDROCHLORIDE
mf: $C_{12}H_{17}NO \cdot ClH$ mw: 227.76

SYNS: PHENDIMETRAZINE HYDROCHLORIDE ◇ d-2-PHENYL-3,4-DIMETHYLMORPHOLINE HYDROCHLORIDE

TOXICITY DATA with REFERENCE
orl-rat LD50:455 mg/kg 27ZQAG -,285,72
ipr-rat LD50:245 mg/kg TXAPA9 2,589,60
scu-rat LD50:435 mg/kg TXAPA9 2,589,60
orl-mus LD50:340 mg/kg TXAPA9 2,589,60
ipr-mus LD50:195 mg/kg TXAPA9 2,589,60
scu-mus LD50:270 mg/kg 27ZQAG -,285,72
ivn-mus LD50:92 mg/kg TXAPA9 2,589,60

SAFETY PROFILE: Poison by ingestion, intraperitoneal, subcutaneous, and intravenous routes. When heated to decomposition it emits very toxic fumes of NO_x and HCl.

DTN875 CAS:72586-68-6 ***HR: 2***
1,3-DIMETHYL-3-PHENYL-1-NITROSOUREA
mf: $C_9H_{11}N_3O_2$ mw: 193.23

SYN: N,N'-DIMETHYL-N-NITROSO-N'-PHENYLUREA

TOXICITY DATA with REFERENCE
mmo-sat 1 μmol/plate CRNGDP 4,409,83
sce-ham:lng 100 nmol/L MUREAV 126,259,84
orl-rat TDLo:58 mg/kg/20W-I:ETA CRNGDP 8,237,87

SAFETY PROFILE: Questionable carcinogen with experimental tumorigenic data. Mutation data reported. Many N-nitroso compounds are carcinogens. When heated to decomposition it emits toxic fumes of NO_x. See also N-NITROSO COMPOUNDS.

DTN896 CAS:672-66-2 ***HR: 3***
DIMETHYLPHENYLPHOSPHINE
mf: $C_8H_{11}P$ mw: 150.09

$CH_3OCO•OOCO•OCH_3$

SAFETY PROFILE: Explodes when heated above 55°C or on impact. When heated to decomposition it emits acrid smoke and fumes. See also PEROXIDES.

DTO000 CAS:54-77-3 ***HR: 3***
1,1-DIMETHYL-4-PHENYLPIPERAZINE IODIDE
mf: $C_{12}H_{19}N_2•I$ mw: 318.23

SYNS: 1,1-DIMETHYL-4-PHENYLPIPERAZINIUM IODIDE ◇ DMPP ◇ DMPP IODIDE

TOXICITY DATA with REFERENCE
ipr-mus LD50:18500 μg/kg AIPTAK 97,186,54
ivn-mus LD50:1600 μg/kg EJPHAZ 11,75,70
ims-mus LD50:28 mg/kg JPETAB 103,330,51
ivn-rbt LD50:1 mg/kg JPETAB 103,330,51

SAFETY PROFILE: Poison by intravenous, intraperitoneal, and intramuscular routes. When heated to decomposition it emits very toxic fumes of NO_x and I^-.

DTO100 CAS:10125-85-6 ***HR: 3***
1,1-DIMETHYL-4-PHENYLPIPERIDINIUM IODIDE
mf: $C_{13}H_{20}N•I$ mw: 317.24

TOXICITY DATA with REFERENCE
ipr-mus LD50:17 mg/kg AIPTAK 97,186,54
ivn-mus LD50:1333 μg/kg JMPCAS 2,449,60
ims-mus LD50:28 mg/kg AIPTAK 97,186,54

SAFETY PROFILE: Poison by intramuscular, intravenous, and intraperitoneal routes. When heated to decomposition it emits toxic fumes of NO_x and I^-.

DTO200 CAS:3734-17-6 ***HR: 3***
1,2-DIMETHYL-3-PHENYL-3-PYRROLIDYL PROPIONATE
mf: $C_{15}H_{21}NO_2$ mw: 247.37

SYNS: A-1981 ◇ COGESIC ◇ 1,2-DIMETHYL-3-PHENYL-3-PYRROLIDINOL PROPIONATE (ester) ◇ PRODILIDINE

TOXICITY DATA with REFERENCE
orl-rat LD50:253 mg/kg JPETAB 134,332,61
ipr-rat LDLo:133 mg/kg JMPCAS 5,441,62
scu-rat LD50:188 mg/kg JPETAB 134,332,61
ivn-rat LD50:74 mg/kg JPETAB 134,332,61
orl-mus LD50:318 mg/kg JPETAB 134,332,61
scu-mus LD50:194 mg/kg JPETAB 134,332,61
ivn-mus LD50:91 mg/kg JPETAB 134,332,61

SAFETY PROFILE: Poison by ingestion, intravenous, intraperitoneal, and subcutaneous routes. When heated to decomposition it emits toxic fumes of NO_x.

DTO600 CAS:1176-08-5 ***HR: 3***
N,N-DIMETHYL-2-(α-PHENYL-o-TOLOXY) ETHYLAMINE DIHYDROGEN CITRATE
mf: $C_{17}H_{21}NO•C_6H_8O_7$ mw: 447.53

SYN: PHENYLTOLOXAMINE DIHYDROGEN CITRATE

TOXICITY DATA with REFERENCE
orl-rat LD50:840 mg/kg TXAPA9 1,42,59
ipr-mus LD50:246 mg/kg JAPMA8 42,587,53

CONSENSUS REPORTS: Reported in EPA TSCA Inventory.

SAFETY PROFILE: Poison by intraperitoneal route. Moderately toxic by ingestion. When heated to decomposition it emits toxic fumes of NO_x.

DTO800 CAS:6152-43-8 ***HR: 3***
N,N-DIMETHYL-2-(α-PHENYL-o-TOLOXY) ETHYLAMINE HYDROCHLORIDE
mf: $C_{17}H_{21}NO•ClH$ mw: 291.85

SYNS: BRISTAMIN HYDROCHLORIDE ◇ PHENYLTOLOXAMINE HYDROCHLORIDE

TOXICITY DATA with REFERENCE
orl-mus LD50:305 mg/kg ARZNAD 8,219,58
ipr-mus LD50:163 mg/kg JAPMA8 42,587,53
ivn-mus LD50:33 mg/kg JAPMA8 42,587,53

SAFETY PROFILE: Poison by ingestion, intravenous, and intraperitoneal routes. Moderately toxic by ingestion. When heated to decomposition it emits very toxic fumes of HCl and NO_x.

DTP000 CAS:7227-91-0 ***HR: 3***
3,3-DIMETHYL-1-PHENYLTRIAZENE
mf: $C_8H_{11}N_3$ mw: 149.22

$C_6H_5N{=}NN(CH_3)_2$

SYNS: 3,3-DIMETHYL-1-PHENYL-1-TRIAZENE ◇ DMPT ◇ 1-FENYL-3,3-DIMETHYLTRIAZIN ◇ NSC 3094 ◇ PDMT ◇ PDT ◇ 1-PHENYL-3,3-DIMETHYLTRIAZENE ◇ PHENYLDIMETHYLTRIAZINE ◇ X 119

TOXICITY DATA with REFERENCE
mmo-sat 1 μg/plate JNCIAM 62,873,79
mrc-smc 900 ppm JNCIAM 62,901,79
cyt-hmn:leu 25 μmol/L MUREAV 77,123,73
otr-ham:emb 100 μg/L NCIMAV 58,243,81
cyt-ham:lng 10 mg/L MUREAV 88,197,81
scu-rat TDLo:125 mg/kg (female 15D post):TER IARCCD 4,45,73
orl-rat TDLo:310 mg/kg:CAR ZKKOBW 81,285,74
scu-rat TDLo:1250 mg/kg/59W-I:CAR ZKKOBW 81,285,74
ivn-rat TDLo:30500 μg/kg/16W-I:ETA ZAPPAN 115,8,72

orl-rat LD50:310 mg/kg ZKKOBW 81,285,74
ipr-rat LD50:180 mg/kg CPCHAO 18,307,62
orl-mus LD50:200 mg/kg NCISP* JAN86
ipr-mus LD50:190 mg/kg JMCMAR 19,1299,76

CONSENSUS REPORTS: EPA Genetic Toxicology Program.

SAFETY PROFILE: Poison by ingestion and intraperitoneal routes. Questionable carcinogen with experimental carcinogenic and tumorigenic data. Experimental teratogenic effects. Human mutation data reported. Decomposes explosively on attempted distillation at atmospheric pressure. When heated to decomposition it emits toxic fumes of NO_x.

DTP400 CAS:101-42-8 ***HR: 2***
1,1-DIMETHYL-3-PHENYLUREA
mf: $C_9H_{12}N_2O$ mw: 164.23

PROP: White crystals. Mp: 127-129°. Insol in water; sltly sol in hydrocarbons.

SYNS: BEET-KLEEN ◇ DIBAR ◇ N,N-DIMETHYL-N'-PHENYLUREA ◇ DYBAR ◇ FENIDIN ◇ FENULON ◇ FENURON ◇ N-PHENYL-N',N'-DIMETHYLUREA ◇ 1-PHENYL-3,3-DIMETHYLUREA ◇ 3-PHENYL-1,1-DIMETHYLUREA ◇ PDU ◇ PUD (HERBICIDE)

TOXICITY DATA with REFERENCE
dni-mus-orl 500 mg/kg MUREAV 58,353,78
orl-rat LD50:6400 mg/kg FMCHA2 -,D137,80
orl-mus LD50:4700 mg/kg GISAAA 47(3),82,82
orl-rbt LD50:4700 mg/kg GISAAA 47(3),82,82
orl-gpg LD50:3200 mg/kg GISAAA 40(10),22,75

CONSENSUS REPORTS: Reported in EPA TSCA Inventory.

SAFETY PROFILE: Moderately toxic by ingestion. Mutation data reported. When heated to decomposition it emits toxic fumes of NO_x.

DTP600 CAS:13171-22-7 ***HR: 3***
DIMETHYL PHOSPHATE ESTER with 2-CHLORO-N-ETHYL-3-HYDROXYCROTONAMIDE
mf: $C_8H_{15}ClNO_5P$ mw: 271.66

SYNS: C-776 ◇ 2-CHLORO-3-(ETHYLAMINO)-1-METHYL-3-OXO-1-PROPENYL DIMETHYL ESTER PHOSPHORIC ACID ◇ 2-CHLORO-3-(ETHYLAMINO)-1-METHYL-3-OXO-1-PROPENYL DIMETHYL PHOSPHATE ◇ CIBA C-776 ◇ ENT 27,358 ◇ NSC 190956

TOXICITY DATA with REFERENCE
orl-rat LD50:37 mg/kg ARSIM* 20,7,66
ipr-mus LD50:7800 μg/kg TXAPA9 13,37,68
scu-gpg LDLo:100 mg/kg JEENAI 62(4),934,69

SAFETY PROFILE: Poison by ingestion, intraperitoneal, and subcutaneous routes. When heated to decomposition it emits very toxic fumes of Cl^-, NO_x, and PO_x.

DTP800 CAS:34491-04-8 ***HR: 3***
DIMETHYL PHOSPHATE ESTER with 2-CHLORO-N-METHYL-3-HYDROXYCROTONAMIDE
mf: $C_7H_{13}ClNO_5P$ mw: 257.63

SYNS: CIBA C-768 ◇ ENT 27,357 ◇ NSC 190955

TOXICITY DATA with REFERENCE
orl-rat LD50:33 mg/kg ARSIM* 20,7,66
orl-gpg LDLo:100 mg/kg JEENAI 62(4),934,69
scu-gpg LDLo:50 mg/kg JEENAI 62(4),934,69

SAFETY PROFILE: Poison by ingestion and subcutaneous routes. When heated to decomposition it emits very toxic Cl^-, NO_x, and PO_x. See also ESTERS.

DTQ089 CAS:676-59-5 ***HR: 3***
DIMETHYL PHOSPHINE
mf: C_2H_7P mw: 64.05

SAFETY PROFILE: Ignites spontaneously in air. When heated to decomposition it emits toxic fumes of PO_x and phosphine. See also PHOSPHINE.

DTQ400 CAS:10265-92-6 ***HR: 3***
O,S-DIMETHYL PHOSPHORAMIDOTHIOATE
mf: $C_2H_8NO_2PS$ mw: 141.14

PROP: Crystals. Mp: 40°. Sltly water-sol; sol in alc.

SYNS: ACEPHATE-MET ◇ BAY 71628 ◇ BAYER 71628 ◇ CHEVRON 9006 ◇ CHEVRON ORTHO 9006 ◇ O,S-DIMETHYL ESTER AMIDE of AMIDOTHIOATE ◇ ENT 27,396 ◇ HAMIDOP ◇ METAMIDOFOS ESTRELLA ◇ METHAMIDOPHOS ◇ MONITOR ◇ MTD ◇ NSC 190987 ◇ ORTHO 9006 ◇ PILLARON ◇ SRA 5172 ◇ TAHMABON ◇ TAMARON ◇ THIOPHOSPHORSAEURE-O,S-DIMETHYLESTERAMID (GERMAN)

TOXICITY DATA with REFERENCE
orl-rat TDLo:10 mg/kg (female 6-15D post):TER VEMJA8 34,357,86
orl-man TDLo:257 mg/kg:PNS,EYE,SKN NEJMAG 306,125,82
orl-wmn TDLo:360 mg/kg:PNS,EYE,SKN NEJMAG 306,125,82
orl-rat LD50:7500 μg/kg ARSIM* 20,7,66
ihl-rat LD50:9 mg/kg TXAPA9 45,232,78
skn-rat LD50:50 mg/kg 28ZEAL 5,149,76
ipr-rat LD50:15 mg/kg PCBPBS 13,267,80
orl-mus LD50:14 mg/kg PCBPBS 7,83,77
ihl-mus LD50:19 mg/kg TXAPA9 45,232,78
orl-rbt LD50:10 mg/kg 28ZEAL 5,149,76
skn-rbt LD50:118 mg/kg GUCHAZ 6,333,73

CONSENSUS REPORTS: EPA Extremely Hazardous Substances List.

SAFETY PROFILE: Poison by ingestion, inhalation, skin contact, subcutaneous, and intraperitoneal routes. Human systemic effects by ingestion: fasciculations, pupillary constriction and sweating. A cholinesterase inhibitor type of insecticide. When heated to decomposition it emits very toxic fumes of NO_x, PO_x, and SO_x. See also PARATHION.

DTQ600 CAS:2524-03-0 ***HR: 3***
O,O-DIMETHYLPHOSPHOROCHLORIDOTHIOATE
DOT: NA 2267
mf: $C_2H_6ClO_2PS$ mw: 160.56

SYNS: CHLOROPHOSPHONOTHIOIC ACID-O,O-DIMETHYL ESTER ◇ DIMETHYL CHLOROTHIOPHOSPHATE (DOT) ◇ DIMETHYL-CHLORTHIOFOSAT (CZECH) ◇ O,O-DIMETHYLESTER KYSELINY CHLORTHIOFOSFORECNE (CZECH) ◇ DIMETHYL PHOSPHOROCHLORIDOTHIOATE (DOT) ◇ METHYL PCT ◇ PHOSPHOROCHLORIDOTHIOIC ACID-O,O-DIMETHYL ESTER

TOXICITY DATA with REFERENCE
orl-rat LDLo:1000 mg/kg 34ZIAG -,393,69
ihl-rat LC50:340 mg/m^3/4H 85GMAT -,56,82
orl-mus LD50:1800 mg/kg 85GMAT -,56,82
ihl-mus LC50:320 mg/m^3/2 85GMAT -,56,82
skn-rbt LDLo:750 mg/kg 34ZIAG -,393,69

CONSENSUS REPORTS: Reported in EPA TSCA Inventory. EPA Extremely Hazardous Substances List.

DOT Classification: Corrosive Material; Label: Corrosive.

SAFETY PROFILE: Poison by inhalation. Moderately toxic by ingestion and skin contact. Corrosive. When heated to decomposition it emits very toxic fumes of Cl^-, PO_x, and SO_x.

DTQ800 CAS:3581-11-1 ***HR: 3***
O,O-DIMETHYL PHOSPHOROTHIOATE-O-ESTER with 4-HYDROXY-m-ANISONITRILE
mf: $C_{10}H_{12}NO_4PS$ mw: 273.26

SYNS: B 11163 ◇ O-(4-CYANO-2-METHOXYPHENYL)-O,O-DIMETHYL PHOSPHOROTHIOATE ◇ ENT 27,230 ◇ PHOSPHOROTHIOICACID-O,O-DIMETHYL-O-(4-CYANO-2-METHOXYPHENYL) ESTER ◇ PHOSPHOROTHIOIC ACID-O,O-DIMETHYL ESTER-O-ESTER with VANNILLONITRILE ◇ STAUFFER B-11163 ◇ TP540

TOXICITY DATA with REFERENCE
orl-rat LD50:2710 mg/kg ARSIM* 20,21,66
orl-mus LD50:4200 mg/kg TDKNAF 24,221,65
scu-gpg LDLo:100 mg/kg JEENAI 61,1261,68

CONSENSUS REPORTS: Cyanide and its compounds are on the Community Right-To-Know List.

SAFETY PROFILE: Poison by subcutaneous route. Moderately toxic by ingestion. When heated to decomposition it emits very toxic fumes of NO_x, PO_x, CN^-, and SO_x. See also ESTERS and NITRILES.

DTR200 CAS:131-11-3 ***HR: 2***
DIMETHYL PHTHALATE
mf: $C_{10}H_{10}O_4$ mw: 194.20

PROP: Colorless, odorless liquid. Bp: 283.7°, flash p: 295°F (CC), d: 1.189 @ 25°/25°, autoign temp: 1032°F, vap d: 6.69, vap press: 1 mm @ 100.3°.

SYNS: AVOLIN ◇ 1,2-BENZENEDICARBOXYLIC ACID DIMETHYL ESTER ◇ DIMETHYL-1,2-BENZENEDICARBOXYLATE ◇ DIMETHYL BENZENEORTHODICARBOXYLATE ◇ DMP ◇ ENT 262 ◇ FERMINE ◇ METHYL PHTHALATE ◇ MIPAX ◇ NTM ◇ PALATINOL M ◇ PHTHALIC ACID METHYL ESTER ◇ PHTHALSAEUREDIMETHYLESTER (GERMAN) ◇ RCRA WASTE NUMBER U102 ◇ SOLVANOM ◇ SOLVARONE

TOXICITY DATA with REFERENCE
eye-rbt 119 mg JPETAB 82,377,44
mmo-sat 200 μg/plate JTEHD6 16,61,85
cyt-rat-skn 25 g/kg/4W-I FATOAO 40,454,77
ipr-rat TDLo:1125 mg/kg (5-15D preg):TER JPMSAE 61,51,72
ipr-rat TDLo:338 mg/kg (5-15D preg):REP JPMSAE 61,51,72
orl-rat LD50:6800 mg/kg GTPZAB 24(3),25,80
ipr-rat LD50:3375 mg/kg JPMSAE 61,51,72
orl-mus LD50:6800 mg/kg GTPZAB 24(3),25,80
ipr-mus LD50:1380 mg/kg IPSTB3 3,93,76
scu-mus LDLo:6500 mg/kg EDWU** -,-,37
ihl-cat LCLo:9630 mg/m^3/6H EDWU** -,-,37
orl-rbt LD50:4400 mg/kg JPETAB 93,26,48
orl-gpg LD50:2400 mg/kg JPETAB 93,26,48
orl-ckn LD50:8500 mg/kg JPETAB 93,26,48

CONSENSUS REPORTS: On EPA Extremely Hazardous Substances List by error. Reported in EPA TSCA Inventory. Community Right-To-Know List.

OSHA PEL: TWA 5 mg/m^3
ACGIH TLV: TWA 5 mg/m^3

SAFETY PROFILE: Moderately toxic by ingestion and intraperitoneal routes. Mildly toxic by inhalation. Experimental teratogenic and reproductive effects. Mutation data reported. An eye irritant. A pesticide and insect repellent. Combustible when exposed to heat or flame; can react with oxidizing materials. To fight fire, use CO_2, dry chemical. When heated to decomposition it emits acrid smoke and irritating fumes. See also ESTERS.

DTR400 CAS:106-55-8 ***HR: 2***
2,5-DIMETHYLPIPERAZINE
mf: $C_6H_{14}N_2$ mw: 114.22

TOXICITY DATA with REFERENCE
eye-rbt 750 μg SEV AMIHBC 4,119,51
orl-rat LD50:3160 mg/kg AMIHBC 4,119,51
skn-rbt LD50:800 mg/kg AMIHBC 4,119,51

CONSENSUS REPORTS: Reported in EPA TSCA Inventory.

SAFETY PROFILE: Moderately toxic by ingestion and skin contact. A severe eye irritant. When heated to decomposition it emits toxic fumes of NO_x.

DTR800 CAS:77966-85-9 ***HR: 3***
2-(2,6-DIMETHYLPIPERIDINO)-2',6'-ACETOXYLIDIDE HYDROCHLORIDE
mf: $C_{17}H_{26}N_2O{\cdot}ClH$ mw: 310.91

SYN: V 374

TOXICITY DATA with REFERENCE
eye-rbt 2% MLD ARZNAD 8,407,58
ipr-rat LD50:53 mg/kg ARZNAD 8,407,58
scu-mus LD50:125 mg/kg ARZNAD 8,407,58

SAFETY PROFILE: Poison by intraperitoneal and subcutaneous routes. An eye irritant. When heated to decomposition it emits very toxic fumes of NO_x and HCl.

DTR850 ***HR: 1***
DIMETHYLPOLYSILOXANE
mf: $[(CH_3)_2SiO-]$

PROP: Clear, colorless viscous liquid. D: 0.96, refr index: 1.400. Sol in hydrocarbon solvents; insol in water.

SYNS: DIMETHYL SILICONE ◇ POLYDIMETHYLSILOXANE

SAFETY PROFILE: Combustible liquid. When heated to decomposition it emits acrid smoke and irritating fumes.

DTS400 CAS:3282-30-2 ***HR: 3***
2,2-DIMETHYLPROPANOYL CHLORIDE
DOT: UN 2438
mf: C_5H_9ClO mw: 120.59

SYNS: 2,2-DIMETHYLPROPIONYL CHLORIDE ◇ NEOPANTANOYL CHLORIDE ◇ PIVALIC ACID CHLORIDE ◇ PIVALOLYL CHLORIDE ◇ PIVALOYL CHLORIDE ◇ PIVALYL CHLORIDE ◇ TRIMETHYL ACETYL CHLORIDE (DOT)

CONSENSUS REPORTS: Reported in EPA TSCA Inventory.

DOT Classification: Corrosive Material; Label: Corrosive; Flammable Liquid.

SAFETY PROFILE: A corrosive irritant to skin, eyes, and mucous membranes. The liquid is flammable when exposed to heat, flame, or oxidizers. When heated to decomposition it emits toxic fumes of Cl^-.

DTS600 CAS:758-96-3 ***HR: 2***
N,N-DIMETHYLPROPIONAMIDE
mf: $C_5H_{11}NO$ mw: 101.17

TOXICITY DATA with REFERENCE
ipr-mus LD50:875 mg/kg AIHAAP 32,539,71
ivn-mus LD50:820 mg/kg AIHAAP 32,539,71

CONSENSUS REPORTS: Reported in EPA TSCA Inventory.

SAFETY PROFILE: Moderately toxic by intraperitoneal and intravenous routes. When heated to decomposition it emits toxic fumes of NO_x.

DTS625 CAS:95619-40-2 ***HR: 1***
2,6-DIMETHYL-4-PROPOXY-BENZOIC ACID 2-METHYL-2-(1-PYRROLIDINYL)PROPYLESTER
mf: $C_{20}H_{31}NO_3{\cdot}ClH$ mw: 369.98

SYNS: BENZOIC ACID, 2,6-DIMETHYL-4-PROPOXY-, 2-METHYL-2-(1-PYRROLIDINYL)PROPYL ESTER, HYDROCHLORIDE ◇ U-2363

TOXICITY DATA with REFERENCE
skn-rbt 5 pph MLD AIPTAK 137,410,62
eye-rbt 5000 ppm MLD AIPTAK 137,410,62
ipr-mus LD50:55700 mg/kg AIPTAK 137,410,62

SAFETY PROFILE: Slightly toxic by intraperitoneal route. A skin and eye irritant. When heated to decomposition it emits toxic fumes of NO_x and HCl.

DTT400 CAS:24690-46-8 ***HR: 3***
N,N-DIMETHYL-p-((p-PROPYLPHENYL)AZO)ANILINE
mf: $C_{17}H_{21}N_3$ mw: 267.41

SYN: 4'-N-PROPYL-4-DIMETHYLAMINOAZOBENZENE

TOXICITY DATA with REFERENCE
orl-rat TDLo:4284 mg/kg/17W-C:ETA JNCIAM 27,663,61

SAFETY PROFILE: Questionable carcinogen with experimental tumorigenic data. When heated to decomposition it emits toxic fumes of NO_x.

DTT600 CAS:23950-58-5 ***HR: 3***
N-(1,1-DIMETHYLPROPYNYL)-3,5-DICHLOROBENZAMIDE
mf: $C_{12}H_{11}Cl_2NO$ mw: 256.14

SYNS: 3,5-DICHLORO-N-(1,1-DIMETHYL-2-PROPYNYL)BENZAMIDE ◇ KERB ◇ PROMAMIDE ◇ PRONAMIDE ◇ PROPYZAMIDE ◇ RCRA WASTE NUMBER U192 ◇ RH 315

TOXICITY DATA with REFERENCE
orl-rat TDLo:1092 mg/kg/2Y-C:ETA ENVRAL 23,1,80
orl-mus TDLo:65520 mg/kg/78W-C:CAR ENVRAL 23,1,80
orl-rat LD50:5620 mg/kg 85ARAE 2,217,77

SAFETY PROFILE: Mildly toxic by ingestion. Questionable carcinogen with experimental carcinogenic and tumorigenic data. An herbicide. When heated to decomposition it emits very toxic fumes of Cl^- and NO_x.

DTT800 ***HR: 3***
DIMETHYL-1-PROPYNYLTHALLIUM
mf: C_5H_9Tl mw: 273.50

$$(CH_3)_2TlC \equiv CCH_3$$

CONSENSUS REPORTS: Thallium and its compounds are on the Community Right-To-Know List.

SAFETY PROFILE: Explodes on heating, stirring or impact. See also THALLIUM COMPOUNDS.

DTU200 CAS:2825-00-5 ***HR: 3***
3,5-DIMETHYL-4H-PYRAN-4-ONE-2-METHOXY-6-(TETRAHYDRO-4-(β-METHYL-p-NITROCINNAMYLIDENE)-2-FURYL)
mf: $C_{22}H_{23}NO_6$ mw: 397.46

SYNS: AUREOTHIN ◇ MYCOLUTEIN

TOXICITY DATA with REFERENCE
orl-mus LD50:3 mg/kg 85GDA2 5,390,81
ipr-mus LD50:1 mg/kg 85GDA2 5,388,81
scu-mus LD50:2 mg/kg 85GDA2 5,388,81
ivn-mus LD50:1260 μg/kg CSLNX* NX#02084

SAFETY PROFILE: Poison by ingestion, intravenous, intraperitoneal, and subcutaneous routes. When heated to decomposition it emits toxic fumes of NO_x.

DTU400 CAS:5910-89-4 ***HR: 2***
2,3-DIMETHYLPYRAZINE
mf: $C_6H_8N_2$ mw: 108.16

PROP: Colorless liquid; nutty cocoa odor. D: 1.000-1.022 @ 20°, refr index: 1.506-1.509, flash p: 147°F (OC), d: 0.99, vap d: 3.72, bp: 182.2°. Misc with water, organic solvents.

SYNS: 2,3-DIMETHYL-1,4-DIAZINE ◇ FEMA No. 3271

TOXICITY DATA with REFERENCE
orl-rat LD50:613 mg/kg DCTODJ 3,249,80
ipr-mus LD50:1390 mg/kg TXAPA9 17,244,70

CONSENSUS REPORTS: Reported in EPA TSCA Inventory.

SAFETY PROFILE: Moderately toxic by ingestion and intraperitoneal routes. Combustible liquid. When heated to decomposition it emits toxic fumes of NO_x.

DTU600 CAS:123-32-0 ***HR: 2***
2,5-DIMETHYLPYRAZINE
mf: $C_6H_8N_2$ mw: 108.16

PROP: Colorless liquid; potato taste. D: 0.980-1.000, refr index: 1.497-1.501, flash p: 147°F (OC), d: 0.99, vap d: 3.72, bp: 182.2°. Misc with water, organic solvents.

SYNS: 2,5-DIMETHYL-1,4-DIAZINE ◇ FEMA No. 3272

TOXICITY DATA with REFERENCE
mmo-smc 3300 μg/L FCTXAV 18,581,80
cyt-ham:ovr 2500 μg/L FCTXAV 18,581,80
orl-rat LD50:1020 mg/kg DCTODJ 3,249,80
ipr-mus LD50:1350 mg/kg TXAPA9 17,244,70

CONSENSUS REPORTS: Reported in EPA TSCA Inventory.

SAFETY PROFILE: Moderately toxic by ingestion and intraperitoneal routes. Mutation data reported. Combustible liquid when exposed to heat, open flame, spark, oxidizers. To fight fire, use water spray, mist, dry chemical, CO_2, foam. When heated to decomposition it emits toxic fumes of NO_x.

DTU800 CAS:108-50-9 ***HR: 2***
2,6-DIMETHYLPYRAZINE
mf: $C_6H_8N_2$ mw: 108.16

PROP: White to yellow crystals; nutty, coffee odor. Mp: 48°, d:.965 @ 50°. Sol in water, organic solvents @ 155°.

SYN: FEMA No. 3273

TOXICITY DATA with REFERENCE
ipr-mus LD50:1080 mg/kg TXAPA9 17,244,70

CONSENSUS REPORTS: Reported in EPA TSCA Inventory.

SAFETY PROFILE: Moderately toxic by intraperitoneal route. When heated to decomposition it emits toxic fumes of NO_x.

DTV089 CAS:1073-23-0 ***HR: 3***
2,6-DIMETHYLPYRIDINE-N-OXIDE
mf: C_7H_9NO mw: 123.15

$$CH_3CH{=}CHCH{=}CHCCH_3{=}N{:}O$$

SAFETY PROFILE: Explosive reaction with phosphoryl chloride. When heated to decomposition it emits toxic fumes of NO_x.

DTV200 CAS:21600-42-0 ***HR: 3***
(3,3-DIMETHYL-1-(m-PYRIDYL-N-OXIDE)) TRIAZENE
mf: $C_7H_{10}N_4O$ mw: 166.21

SYNS: 3-(3′,3′-DIMETHYLTRIAZENO)-PYRIDIN-N-OXID(GERMAN) ◇ 3-(3′,3′-DIMETHYLTRIAZENO)PYRIDINE-N-OXIDE ◇ PYNDT

◇ 1-(PYRIDYL-3-N-OXID)-3,3-DIMETHYL-TRIAZEN (GERMAN)
◇ 1-(PYRIDYL-3-N-OXIDE)-3,3-DIMETHYLTRIAZENE

TOXICITY DATA with REFERENCE
sln-dmg-orl 700 μmol/L CBINA8 9,365,74
cyt-hmn:leu 25 μmol/L MUREAV 77,123,73
hma-mus/smc 400 μmol/L/Kg AGACBH 3,99,73
ivn-rat TDLo:490 mg/kg/38W-I:CAR ZKKOBW 81,285,74
ivn-rat TD:540 mg/kg/36W-I:ETA XENOBH 3,271,73
scu-rat LD50:200 mg/kg ZKKOBW 81,285,74
ivn-rat LD50:230 mg/kg ZKKOBW 81,285,74

CONSENSUS REPORTS: EPA Genetic Toxicology Program.

SAFETY PROFILE: Poison by intravenous and subcutaneous routes. Questionable carcinogen with experimental carcinogenic and tumorigenic data. Human mutation data reported. When heated to decomposition it emits toxic fumes of NO_x.

DTV400 CAS:333-40-4 ***HR: 3***
S-(4,6-DIMETHYL-2-PYRIMIDINYL)-O,O-DI-ETHYL PHOSPHORODITHIOATE
mf: $C_{10}H_{17}N_2O_2PS_2$ mw: 292.38

SYNS: ENT 25,737 ◇ STAUFFER R-3413

TOXICITY DATA with REFERENCE
orl-rat LD50:59 mg/kg ARSIM* 20,23,66
orl-ckn LD50:41 mg/kg TXAPA9 7,606,65

SAFETY PROFILE: Poison by ingestion. When heated to decomposition it emits very toxic fumes of NO_x, PO_x, and SO_x.

DTY200 CAS:17025-30-8 ***HR: 3***
N,N-DIMETHYL-4-(4′-QUINOLYLAZO)ANILINE
mf: $C_{17}H_{16}N_4$ mw: 276.37

SYN: 4-((p-(DIMETHYLAMINO)PHENYL)AZO)QUINOLINE

TOXICITY DATA with REFERENCE
orl-rat TDLo:2142 mg/kg/17W-C:ETA JNCIAM 26,1461,61

SAFETY PROFILE: Questionable carcinogen with experimental tumorigenic data. When heated to decomposition it emits toxic fumes of NO_x.

DTY400 CAS:63042-68-2 ***HR: 3***
N,N-DIMETHYL-4-((4′-QUINOLYL-1′-OXIDE)AZO)ANILINE
mf: $C_{17}H_{16}N_4O$ mw: 292.37

SYN: 4-((p-(DIMETHYLAMINO)PHENYL)AZO)QUINOLINE-1-OXIDE

TOXICITY DATA with REFERENCE
orl-rat TDLo:2142 mg/kg/17W-C:ETA JNCIAM 26,1461,61

SAFETY PROFILE: Questionable carcinogen with experimental tumorigenic data. When heated to decomposition it emits toxic fumes of NO_x.

DTY600 CAS:70324-23-1 ***HR: 3***
3,3-DIMETHYL-1(3-QUINOLYL)TRIAZENE
mf: $C_{11}H_{12}N_4$ mw: 144.22

$(CH_3)_2NN{=}NC_9H_6N$

SAFETY PROFILE: Crude material decomposes violently when dried. The pure material explodes at 131°C. When heated to decomposition it emits toxic fumes of NO_x.

DUA200 CAS:23521-13-3 ***HR: 3***
N,N-DIMETHYL-p-(5-QUINOXALYLAZO)ANILINE
mf: $C_{16}H_{15}N_5$ mw: 277.36

SYN: 5-((p-(DIMETHYLAMINO)PHENYL)AZO)QUINOXALINE

TOXICITY DATA with REFERENCE
orl-rat TDLo:2200 mg/kg/17W-C:CAR JMCMAR 12,1113,69

SAFETY PROFILE: Questionable carcinogen with experimental carcinogenic data. When heated to decomposition it emits toxic fumes of NO_x.

DUA400 CAS:23521-14-4 ***HR: 3***
N,N-DIMETHYL-p-(6-QUINOXALYAZO)ANILINE
mf: $C_{16}H_{15}N_5$ mw: 277.36

SYNS: 6-((p-(DIMETHYLAMINO)PHENYL)AZO)QUINOXALINE ◇ N,N-DIMETHYL-p-(6-QUINOXALINYLAZO)ANILINE

TOXICITY DATA with REFERENCE
orl-rat TDLo:1100 mg/kg/60D-C:CAR JMCMAR 12,1113,69

SAFETY PROFILE: Questionable carcinogen with experimental carcinogenic data. When heated to decomposition it emits toxic fumes of NO_x.

DUA600 CAS:101652-10-2 ***HR: 2***
7,8-DIMETHYL-10-(d-RIBO-2,3,4,5-TETRAHYDROXYPENTYL)-4a,5-DIHYDROISOALLOXAZINE
mf: $C_{17}H_{22}N_4O_6$ mw: 378.43

TOXICITY DATA with REFERENCE
ipr-rat LD50:965 mg/kg CMTRAG 2,96,61
ipr-mus LD50:800 mg/kg CMTRAG 2,96,61

SAFETY PROFILE: Moderately toxic by intraperitoneal route. When heated to decomposition it emits toxic fumes of NO_x.

DUA800 CAS:1778-08-1 ***HR: 2***
N,N-DIMETHYLSALICYLAMIDE
mf: $C_9H_{11}NO_2$ mw: 165.21

SYNS: SALICYLDIMETHYLAMIDE ◇ SAM

TOXICITY DATA with REFERENCE
orl-rat LD50:2300 mg/kg JPETAB 108,450,53
ipr-rat LD50:2000 mg/kg JPETAB 108,450,53
ipr-mus LD50:1100 mg/kg YKKZAJ 86,120,66

SAFETY PROFILE: Moderately toxic by ingestion and intraperitoneal routes. When heated to decomposition it emits toxic fumes of NO_x.

DUB000 CAS:6918-51-0 ***HR: 3***
DIMETHYL SELENATE
mf: $C_2H_6O_4Se$ mw: 173.03

CONSENSUS REPORTS: Selenium and its compounds are on the Community Right-To-Know List.

OSHA PEL: TWA 0.2 mg(Se)/m^3
ACGIH TLV: TWA 0.2 mg(Se)/m^3
DFG MAK: 0.1 mg(Se)/m^3

SAFETY PROFILE: Explodes when heated to 150°C. When heated to decomposition it emits toxic fumes of Se. See also SELENIUM COMPOUNDS.

DUB200 CAS:593-79-3 ***HR: 2***
DIMETHYL SELENIDE
mf: C_2H_6Se mw: 109.04

PROP: Liquid. Bp: 58°, d: 1.4077 @ 14.6°/4°, vap d: 3.75.

SYNS: DIMETHYLSELENIUM ◇ METHYL SELENIDE ◇ METHYL SELENIUM

TOXICITY DATA with REFERENCE
ipr-rat LD50:2200 mg/kg PSEBAA 79,230,52
scu-rat LDLo:2180 mg/kg ARTODN 45,207,80
ipr-mus LD50:1800 mg/kg PSEBAA 79,230,52

CONSENSUS REPORTS: Selenium and its compounds are on the Community Right-To-Know List.

OSHA PEL: TWA 0.2 mg(Se)/m^3
ACGIH TLV: TWA 0.2 mg(Se)/m^3
DFG MAK: 0.1 mg(Se)/m^3

SAFETY PROFILE: Moderately toxic by intraperitoneal and subcutaneous routes. When heated to decomposition it emits toxic fumes of Se. See also SELENIUM COMPOUNDS.

DUB600 CAS:63148-62-9 ***HR: 3***
DIMETHYL SILOXANE

PROP: Viscosity 100 at 25 degrees (ISMJAV 22,15,63).

SYN: DOW-CORNING 200 FLUID-LOT No. AA-4163

TOXICITY DATA with REFERENCE
scu-mus TDLo:120 g/kg:ETA ISMJAV 22,15,63

CONSENSUS REPORTS: Reported in EPA TSCA Inventory.

SAFETY PROFILE: Questionable carcinogen with experimental tumorigenic data.

DUB689 ***HR: 3***
(DIMETHYL SILYLMETHYL)TRIMETHYL LEAD
mf: $C_6H_{18}PbSi$ mw: 325.49

$(CH_3)_2SiHCH_2Pb(CH_3)_3$

CONSENSUS REPORTS: Lead and its compounds are on the Community Right-To-Know List.

SAFETY PROFILE: Lead compounds are generally poisons. Decomposes violently when heated above 100°C in the presence of oxygen. When heated to decomposition it emits toxic fumes of Pb. See also LEAD COMPOUNDS.

DUB800 CAS:1145-73-9 ***HR: 3***
N,N-DIMETHYL-4-STILBENAMINE
mf: $C_{16}H_{17}N$ mw: 223.34

SYNS: 4-DIMETHYLAMINOSTILBEN (GERMAN) ◇ N,N-DIMETHYL-4-AMINOSTILBENE ◇ N,N-DIMETHYL-p-STYRYLANILINE ◇ STILBENYL-N,N-DIMETHYLAMINE

TOXICITY DATA with REFERENCE
mma-sat 10 μg/plate JJIND8 71,293,83
mmo-bcs 5 g/L MUREAV 42,19,77
dnr-bcs 5 g/L MUREAV 42,19,77
orl-rat TDLo:50 mg/kg:CAR CNREA8 26,619,66
ipr-rat TDLo:180 mg/kg/9W-I:ETA BJCAAI 6,392,52
orl-rat TD:4200 mg/kg/20W-C:CAR GANNA2 61,367,70
orl-rat LDLo:50 mg/kg CNREA8 26,619,66
ipr-rat LD50:70 mg/kg ZEKBAI 65,272,63

SAFETY PROFILE: Poison by ingestion and intraperitoneal routes. Questionable carcinogen with experimental carcinogenic and tumorigenic data. Mutation data reported. When heated to decomposition it emits toxic fumes of NO_x.

DUC000 CAS:838-95-9 ***HR: 3***
(E)-N,N-DIMETHYL-4-STILBENAMINE
mf: $C_{16}H_{17}N$ mw: 223.34

SYNS: trans-p-(DIMETHYLAMINO)STILBENE ◇ trans-4-DIMETHYLAMINOSTILBENE ◇ (E)-N,N,-DIMETHYL-4-(2-PHENYLETHENYL) BENZENAMINE ◇ 4-DIMETHYLAMINO-trans-STILBENE ◇ trans-N,N-DIMETHYL-4-STILBENAMINE

TOXICITY DATA with REFERENCE
mma-sat 10 μg/plate PNASA6 72,5135,75

sln-dmg-orl 1500 μmol/L BIZNAT 102,271,83
oms-rat-orl 25 μmol/kg CBINA8 24,355,79
dns-rat:lvr 1 μmol/L ENMUDM 7,101,85
dns-rat-orl 40 mg/kg ENMUDM 7,101,85
orl-rat TDLo:240 mg/kg/20W-I:CAR ZEKBAI 74,200,70
scu-rat TDLo:135 mg/kg/W-I:ETA PTRMAD 241,147,48
orl-rat LD50:50 mg/kg ARTODN 56,151,85

CONSENSUS REPORTS: EPA Genetic Toxicology Program.

SAFETY PROFILE: Poison by ingestion and subcutaneous routes. Questionable carcinogen with experimental carcinogenic and tumorigenic data. Mutation data reported. When heated to decomposition it emits toxic fumes of NO_x.

DUC200 CAS:14301-11-2 ***HR: 3***
(Z)-N,N-DIMETHYL-4-STILBENAMINE
mf: $C_{16}H_{17}N$ mw: 223.34

SYNS: cis-4-DIMETHYLAMINOSTILBENE ◇ cis-N,N-DIMETHYL-4-STILBENAMINE

TOXICITY DATA with REFERENCE
orl-rat TDLo:310 mg/kg/20W-I:NEO ZEKBAI 74,200,70

SAFETY PROFILE: Questionable carcinogen with experimental neoplastigenic data. When heated to decomposition it emits toxic fumes of NO_x.

DUC300 CAS:552-80-7 ***HR: D***
DIMETHYLSTILBESTROL
mf: $C_{16}H_{16}O_2$ mw: 240.32

SYNS: (E)-4,4'-(1,2-DIMETHYL-1,2-ETHENEDIYL)BIS-PHENOL(9CI) ◇ DIMETHYLSILBOESTROL ◇ (E)-α,α'-DIMETHYL-4,4'-STILBENEDIOL ◇ DMS

TOXICITY DATA with REFERENCE
otr-ham:emb 10 mg/L CNREA8 42,3040,82
orl-rat TDLo:2250 μg/kg (female 4-6D post):REP ACENA7 49,83,65

SAFETY PROFILE: Experimental reproductive effects. Mutation data reported. When heated to decomposition it emits acrid smoke and irritating fumes. See also DIETHYL STILBESTEROL.

DUC400 CAS:7456-24-8 ***HR: 3***
DIMETHYLSULFAMIDO-3-(DIMETHYLAMINO-2-PROPYL)-10-PHENOTHIAZINE
mf: $C_{19}H_{25}N_3O_2S_2$ mw: 391.59

SYNS: DIMETHOTHIAZINE ◇ 10-(2-(DIMETHYLAMINO)PROPYL)-N,N-DIMETHYLPHENOTHIAZINE-2-SULFONAMIDE ◇ 3-DIMETHYLSULFONAMIDO-10-(2-DIMETHYLAMINOPROPYL)PHENOTHIAZINE

TOXICITY DATA with REFERENCE
orl-rat TDLo:600 mg/kg (female 9-14D post):REP OYYAA2 4,381,70
orl-mus LD50:740 mg/kg AIPTAK 159,70,66
ipr-mus LD50:190 mg/kg AIPTAK 159,70,66
scu-mus LD50:475 mg/kg AIPTAK 159,70,66
ivn-mus LD50:100 mg/kg AIPTAK 159,70,66

SAFETY PROFILE: Poison by intraperitoneal and intravenous routes. Moderately toxic by ingestion and subcutaneous routes. Experimental reproductive effects. When heated to decomposition it emits very toxic fumes of NO_x and SO_x.

DUC600 CAS:15020-57-2 ***HR: 2***
p-(N,N-DIMETHYLSULFAMOYL)PHENOL
mf: $C_8H_{11}NO_3S$ mw: 201.26

SYN: N,N-DIMETHYL-HYDROXYBENZENESULFONAMIDE

TOXICITY DATA with REFERENCE
orl-mus LD50:2290 mg/kg JAFCAU 15,845,67

CONSENSUS REPORTS: Reported in EPA TSCA Inventory.

SAFETY PROFILE: Moderately toxic by ingestion. When heated to decomposition it emits very toxic fumes of SO_x and NO_x.

DUD000 CAS:121-58-4 ***HR: 2***
N,N-DIMETHYLSULFANILIC ACID
mf: $C_8H_{11}NO_3S$ mw: 201.26

TOXICITY DATA with REFERENCE
par-mus LDLo:4000 mg/kg CBCCT* 7,695,55

CONSENSUS REPORTS: Reported in EPA TSCA Inventory.

SAFETY PROFILE: Moderately toxic by parenteral route. When heated to decomposition it emits very toxic fumes of NO_x and SO_x.

DUD100 CAS:77-78-1 ***HR: 3***
DIMETHYL SULFATE
DOT: UN 1595
mf: $C_2H_6O_4S$ mw: 126.14

PROP: Colorless, odorless liquid. Mp: −31.8°, bp: 188°, flash p: 182°F (OC), d: 1.3322 @ 20°/4°, vap d: 4.35, autoign temp: 370°F.

SYNS: DIMETHYLESTER KYSELINY SIROVE (CZECH) ◇ DIMETHYL MONOSULFATE ◇ DIMETHYLSULFAAT (DUTCH) ◇ DIMETHYLSULFAT (CZECH) ◇ DIMETILSOLFATO (ITALIAN) ◇ DMS ◇ DMS(METHYL SULFATE) ◇ DWUMETYLOWY SIARCZAN (POLISH) ◇ METHYLE (SULFATE de) (FRENCH) ◇ METHYL SULFATE (DOT) ◇ RCRA WASTE NUMBER U103 ◇ SULFATE de METHYLE (FRENCH) ◇ SULFATE DIMETHYLIQUE (FRENCH) ◇ SULFURIC ACID, DIMETHYL ESTER

TOXICITY DATA with REFERENCE
skn-rbt 10 mg/24H open SEV AMIHBC 4,119,51
eye-rbt 100 mg/4S rns SEV FCTOD7 20,573,82

eye-rbt 50 μg/24H SEV 28ZPAK -,177,72
mma-sat 4300 nmol/L/1H PNASA6 75,4465,78
dnr-omi 640 μg/plate BIZNAT 95,463,76
dnd-hmn:lym 1 mmol/L JACTDZ 1(3),125,82
ivn-rat TDLo:100 mg/kg (15D preg):TER IARCCD 4,45,73
ihl-rat TCLo:17 mg/m^3/19W-I:ETA ZEKBAI 74,241,70
ivn-rat TDLo:20 mg/kg (15D preg):CAR,REP IARCCD 4,45,73
ihl-hmn LCLo:97 ppm/10M 34ZIAG -,226,69
orl-rat LD50:205 mg/kg GTPZAB 23(3),28,79
ihl-rat LC50:45 mg/m^3/4H GTPZAB 24(11),55,80
scu-rat LD50:100 mg/kg ZEKBAI 74,241,70
orl-mus LD50:140 mg/kg GTPZAB 23(3),28,79
ihl-mus LC50:280 mg/m^3 GTPZAB 23(3),28,79
orl-rbt LDLo:45 mg/kg AEXPBL 47,113,02
scu-rbt LDLo:53 mg/kg AEXPBL 47,113,02
ivn-rbt LDLo:50 mg/kg AEXPBL 47,113,02

CONSENSUS REPORTS: NTP Fifth Annual Report on Carcinogens. IARC Cancer Review: Group 2A IMEMDT 7,200,87; Animal Sufficient Evidence IMEMDT 4,271, 74; Human Inadequate Evidence IMEMDT 4,271,74. EPA Genetic Toxicology Program. Community Right To Know List. EPA Extremely Hazardous Substances List. Reported in EPA TSCA Inventory.

OSHA PEL: (Transitional: TWA 1 ppm (skin)) TWA 0.1 ppm (skin)
ACGIH TLV: TWA 0.1 ppm (skin); Suspected Human Carcinogen.
DFG TRK: Production: 0.02 ppm; Use: 0.04 ppm; Animal Carcinogen, Suspected Human Carcinogen.
DOT Classification: Corrosive Material; Label: Corrosive; Poison B; Label: Poison

SAFETY PROFILE: Confirmed carcinogen with experimental carcinogenic, tumorigenic, and teratogenic data. Human poison by inhalation. Experimental poison by ingestion, inhalation, intravenous, and subcutaneous routes. Other experimental reproductive effects. Human mutation data reported. A corrosive irritant to skin, eyes, and mucous membranes. There is no odor or initial irritation to give warning of exposure. On brief, mild exposures, conjunctivitis, catarrhal inflammation of the mucous membranes of the nose, throat, larynx, and trachea and possibly some reddening of the skin develop after the latent period. With longer, heavier exposures, the cornea shows clouding, the irritation changes to the nasopharynx are more marked and after 6 to 8 hours pulmonary edema may develop. Death may occur in 3 or 4 days. The liver and kidneys are frequently damaged. Spilling of the liquid on the skin can cause ulceration and local necrosis. In patients surviving severe exposure, there may be serious injury of the liver and kidneys, with suppression of urine, jaundice, albuminuria and hematuria appearing. Death, resulting from the kidney or liver damage, may be delayed for several weeks. Flammable when exposed to heat, flame, or oxidizers. Can react with oxidizing materials. Violent reaction with NH_4OH and NaN_3. To fight fire, use water, foam, CO_2, dry chemical. When heated to decomposition it emits toxic fumes of SO_x. See also SULFATES.

DUD400 CAS:1003-78-7 ***HR: 3***
2,4-DIMETHYL SULFOLANE
mf: $C_6H_{12}O_2S$ mw: 148.24

PROP: Solid. Bp: 280°, flash p: 290°F (OC), d: 1.1362 @ 20°/4°, vap press: 0.006 mm @ 20°.

SYNS: DMS ◇ TETRAHYDRO-2,4-DIMETHYLTHIOPHENE-1,1-DIOXIDE

TOXICITY DATA with REFERENCE
orl-rat LDLo:100 mg/kg SCCUR* -,4,61
orl-mus LD50:140 mg/kg SCCUR* -,4-61
ipr-mus LD50:72 mg/kg AIHAAP 32,539,71
ivn-mus LD50:61 mg/kg AIHAAP 32,539,71
orl-rbt LD50:115 mg/kg SCCUR* -,4,61
skn-rbt LDLo:3600 mg/kg SCCUR* -,4,61
ivn-rbt LD50:36 mg/kg AIHAAP 32,539,71

SAFETY PROFILE: Poison by ingestion, intraperitoneal, and intravenous routes. Moderately toxic by skin contact. Combustible when exposed to heat or flame; can react with oxidizing materials. To fight fire, use water, foam, CO_2, dry chemical. When heated to decomposition it emits toxic fumes of SO_x. See also SULFATES.

DUD800 CAS:67-68-5 ***HR: 3***
DIMETHYL SULFOXIDE
mf: C_2H_6OS mw: 78.14

PROP: Clear, water-white, hygroscopic liquid. Mp: 18.5°, bp: 189°, flash p: 203°F (OC), d: 1.100 @ 20°, vap press: 0.37 mm @ 20°, lel: 2.6%, uel: 28.5%, autoign temp: 419°F.

SYNS: A 10846 ◇ DELTAN ◇ DEMASORB ◇ DEMAVET ◇ DEMESO ◇ DEMSODROX ◇ DERMASORB ◇ DIMETHYL SULPHOXIDE ◇ DIMEXIDE ◇ DIPIRARTRIL-TROPICO ◇ DMS-70 ◇ DMS-90 ◇ DMSO ◇ DOLICUR ◇ DOLIGUR ◇ DOMOSO ◇ DROMISOL ◇ DURASORB ◇ GAMASOL 90 ◇ HYADUR ◇ INFILTRINA ◇ M 176 ◇ METHYLSULFINYLMETHANE ◇ METHYL SULFOXIDE ◇ NSC-763 ◇ RIMSO-50 ◇ SOMIPRONT ◇ SQ 9453 ◇ SULFINYLBIS(METHANE) ◇ SYNTEXAN ◇ TOPSYM

TOXICITY DATA with REFERENCE
skn-rbt 10 mg/24H open MLD AIHAAP 23,95,62
skn-rbt 500 mg/24H MLD 28ZPAK -,177,72
eye-rbt 500 mg/24H MLD 28ZPAK -,177,72
mmo-esc 551 g/L MUREAV 130,97,84
oms-hmn:lym 140 mmol/L PNASA6 79,1171,82

ipr-ham TDLo:4400 mg/kg (female 8D post):TER LANCAO 1,208,66
ipr-rat TDLo:56 g/kg (6-12D preg):REP ANYAA9 141,110,67
orl-rat TDLo:59 g/kg/81W-I:ETA GTPZAB 28(5),39,84
ivn-man TDLo:606 mg/kg:GIT,LIV LANCAO 2,1004,80
orl-rat LD50:14500 mg/kg TXAPA9 15,74,69
ipr-rat LD50:8200 mg/kg FCTOD7 22,665,84
scu-rat LD50:12 g/kg ARZNAD 14,1050,64
ivn-rat LD50:5360 mg/kg TXAPA9 7,104,65
orl-mus LD50:7920 mg/kg CHTPBA 3,10,68
ipr-mus LD50:2500 mg/kg RPTOAN 35,300,72
ivn-mus LD50:3800 mg/kg 34ZIAG -,656,69
ivn-dog LD50:2500 mg/kg CNCRA6 31,7,63

CONSENSUS REPORTS: Reported in EPA TSCA Inventory. EPA Genetic Toxicology Program.

SAFETY PROFILE: Poison by ingestion. Moderately toxic by intravenous and intraperitoneal routes. Mildly toxic by subcutaneous route. Human systemic effects by intravenous route: nausea or vomiting and jaundice. Experimental teratogenic and reproductive effects. A skin and eye irritant. Questionable carcinogen with experimental tumorigenic data. Human mutation data reported. Can cause an anaphylactic reaction, and corneal opacity. It freely penetrates the skin and may carry dissolved chemicals with it into the body. Combustible when exposed to heat or flame; can react with oxidizing materials. To fight fire, use foam, alcohol foam, CO_2, dry chemical. Violent or explosive reaction with many acyl, aryl and non-metal halides (e.g., acetyl chloride, benzenesulfonyl chloride, bromobenzoyl acetanilide, cyanuric chloride, iodine pentafluoride, $Mg(ClO_4)_2$, CH_3Br, NIO_4, oxalyl chloride, P_2O_3, phosphorous trichloride, phosphoryl chloride, silver fluoride, silver difluoride, sodium hydride, sulfur dichloride, disulfur dichloride, sulfuryl chloride, tetrachlorosilane, thionyl chloride. Violent or explosive reaction with boron compounds [e.g., borane, nonahydrononaborate(2-) ion], 4(4'-bromobenzoyl)acetanilide, carbonyl diisothiocyanate, dinitrogen tetraoxide, hexachlorocyclotriphosphazine, copper + trichloroacetic acid, metal alkoxides (e.g., potassium tert-butoxide, sodium isopropoxide), trifluoroacetic acid anhydride. Incompatible with magnesium perchlorate, metal oxosalts, perchloric acid, periodic acid, sulfur trioxide. Forms powerfully explosive mixtures with metal salts of oxoacids (e.g., aluminum perchlorate, sodium perchlorate, iron(III) nitrate). When heated to decomposition it emits toxic fumes of SO_x.

DUE000 CAS:120-61-6 ***HR: 3***
DIMETHYL TEREPHTHALATE
mf: $C_{10}H_{10}O_4$ mw: 194.20

SYNS: 1,4-BENZENE DICARBOXYLIC ACID DIMETHYL ESTER (9CI) ◇ DIMETHYL-1,4-BENZENE DICARBOXYLATE ◇ METHYL-4-CARBOMETHOXY BENZOATE ◇ NCI-C50055 ◇ TEREPHTHALIC ACID METHYL ESTER

TOXICITY DATA with REFERENCE
eye-rbt 500 mg/24H MOD 28ZPAK -,47,72
orl-mus TDLo:216 g/kg/103W-C:CAR NCITR* NCI-CG-TR-121,79
orl-mus TD:433 g/kg/103W-C:CAR NCITR* NCI-CG-TR-121,79
orl-rat LD50:4390 mg/kg 28ZPAK -,47,72
ipr-rat LD50:3900 mg/kg AIHAAP 34,455,73

CONSENSUS REPORTS: NCI Carcinogenesis Bioassay (feed): Clear Evidence: mouse NCITR* NCI-CG-TR-121,79; No Evidence: rat NCITR* NCI-CG-TR-121,79. Reported in EPA TSCA Inventory.

SAFETY PROFILE: Moderately toxic by intraperitoneal route. Mildly toxic by ingestion. An eye irritant. Questionable carcinogen with experimental carcinogenic data. When heated to decomposition it emits acrid smoke and irritating fumes.

DUE600 CAS:3862-21-3 ***HR: 3***
DIMETHYL-1,2,2,2-TETRACHLOROETHYL PHOSPHATE
mf: $C_4H_7Cl_4O_4P$ mw: 291.88

TOXICITY DATA with REFERENCE
orl-rat LD50:14 mg/kg JAFCAU 8,196,60
ipr-mus LD50:5 mg/kg PAREAQ 11,636,59

SAFETY PROFILE: Poison by ingestion and intraperitoneal routes. When heated to decomposition it emits very toxic fumes of Cl^- and PO_x.

DUF000 CAS:25486-91-3 ***HR: 3***
7,12-DIMETHYL-8,9,10,11-TETRAHYDROBENZ(a)ANTHRACENE
mf: $C_{20}H_{20}$ mw: 260.40

SYN: 8,9,10,11-TETRAHYDRO-7,12-DIMETHYLBENZ(a)ANTHRACENE

TOXICITY DATA with REFERENCE
skn-mus TDLo:212 mg/kg/65W-I:ETA JNCIAM 44,641,70

SAFETY PROFILE: Questionable carcinogen with experimental tumorigenic data. When heated to decomposition it emits acrid smoke and irritating fumes.

DUF200 CAS:52171-94-5 ***HR: 3***
1,11-DIMETHYL-1,2,3,4-TETRAHYDROCHRYSENE
mf: $C_{20}H_{20}$ mw: 260.40

TOXICITY DATA with REFERENCE
mma-sat 20 μg/plate CNREA8 36,4525,76
skn-mus TDLo:120 mg/kg/50W-I:ETA CNREA8 34,1315,74

CONSENSUS REPORTS: EPA Genetic Toxicology Program.

SAFETY PROFILE: Questionable carcinogen with experimental tumorigenic data. Mutation data reported. When heated to decomposition it emits acrid smoke and irritating fumes.

DUF400 CAS:4336-19-0 ***HR: 2***
DIMETHYL TETRAHYDROPHTHALATE
mf: $C_{10}H_{14}O_4$ mw: 198.24

PROP: Crystals. Vap d: 6.83.

SYN: 1-CYCLOHEXENE-1,2-DICARBOXYLIC ACID DIMETHYL ESTER

TOXICITY DATA with REFERENCE
eye-rbt 500 mg AJOPAA 29,1363,46
orl-rat LD50:700 mg/kg JIHTAB 31,60,49

SAFETY PROFILE: Moderately toxic by ingestion. An eye irritant. Combustible when exposed to heat or flame; can react with oxidizing materials. When heated to decomposition it emits acrid smoke and irritating fumes. See also ESTERS.

DUF800 ***HR: 3***
3,6-DIMETHYL-1,2,4,5-TETRAOXANE
mf: $C_4H_8O_4$ mw: 120.11

$CH_3CHOOCHCH_3OO$ (ring structure)

SAFETY PROFILE: An extremely shock-sensitive explosive. May explode if touched. When heated to decomposition it emits acrid smoke and fumes.

DUG000 ***HR: 3***
DIMETHYLTHALLIUM FULMINATE
mf: C_3H_6NOTl mw: 276.46

$(CH_3)_2TlC{\equiv}N{\cdot}O$

CONSENSUS REPORTS: Thallium and its compounds are on the Community Right-To-Know List.

SAFETY PROFILE: Highly explosive. See also THALLIUM COMPOUNDS and FULMINATES.

DUG089 ***HR: 3***
DIMETHYLTHALLIUM-N-METHYLACETOHYDROXAMATE
mf: $C_5H_{12}NO_2Tl$ mw: 276.46

$(CH_3)_2TlON(CH_3)CO{\cdot}CH_3$

CONSENSUS REPORTS: Thallium and its compounds are on the Community Right-To-Know List.

SAFETY PROFILE: Highly explosive. It may explode below 160°C. See also THALLIUM COMPOUNDS and FULMINATES.

DUG200 CAS:541-58-2 ***HR: 3***
2,4-DIMETHYLTHIAZOLE
mf: C_5H_7NS mw: 113.19

TOXICITY DATA with REFERENCE
ipr-mus LD50:250 mg/kg NTIS** AD691-490

CONSENSUS REPORTS: Reported in EPA TSCA Inventory.

SAFETY PROFILE: Poison by intraperitoneal route. When heated to decomposition it emits very toxic fumes of NO_x and SO_x.

DUG400 CAS:298-93-1 ***HR: D***
3-(4,5-DIMETHYLTHIAZOLYL-2)-2,5-DIPHENYLTETRAZOLIUM BROMIDE
mf: $C_{18}H_{16}N_5S$ mw: 334.45

TOXICITY DATA with REFERENCE
mmo-sat 5 μg/plate MUREAV 68,107,79
mmo-esc 5 μg/plate MUREAV 68,107,79

SAFETY PROFILE: Mutation data reported. When heated to decomposition it emits very toxic fumes of NO_x and SO_x. See also BROMIDES.

DUG500 CAS:152-20-5 ***HR: 3***
DIMETHYLTHIOMETHYLPHOSPHATE
mf: $C_3H_9O_3PS$ mw: 156.15

$((CH_3O)_2(CH_3S)PO)$

SYNS: HC7901 ◇ METHYLPHOSPHOROTHIOATE ◇ O,O,S-TRIMETHYL PHOSPHOROTHIOATE

TOXICITY DATA with REFERENCE
orl-rat TDLo:40 mg/kg (female 20D post):REP ARTODN 61,378,88
orl-rat TDLo:40 mg/kg (female 8-10D post):TER TOLED5 32,185,86
orl-rat LD50:15 mg/kg JAFCAU 27,463,79
ipr-rat LD50:51 mg/kg ARTODN 51,221,82
ivn-rat LD50:45 mg/kg DTESD7 8,631,80
orl-mus LD50:38 mg/kg PCBPBS 24,251,85
ipr-mus LD50:5 mg/kg ACPMAP 16,7,63
ivn-mus LD50:123 mg/kg DTESD7 8,631,80

SAFETY PROFILE: Poison by ingestion, intravenous, and intraperitoneal routes. An experimental teratogen. Other experimental reproductive effects. When heated to decomposition it emits toxic fumes of PO_x and SO_x.

DUG600 CAS:50847-92-2 ***HR: 3***
2,2-DIMETHYL-3-THIOMORPHOLINONE
mf: $C_6H_{11}NOS$ mw: 145.24

SYN: 2,2-DIMETHYL-3-THIOMORPHOLONE

TOXICITY DATA with REFERENCE
orl-mus LD50:3423 mg/kg JMCMAR 6,136,63
ipr-mus LD50:400 mg/kg NTIS** AD691-490

SAFETY PROFILE: Poison by intraperitoneal route. Moderately toxic by ingestion. When heated to decomposition it emits very toxic fumes of NO_x and SO_x.

DUG800 CAS:2767-47-7 ***HR: 3***
DIMETHYLTIN DIBROMIDE
mf: $C_2H_6Br_2Sn$ mw: 308.59

PROP: Colorless crystals. Sol in water and organic solvents. Mp: 76°; bp: 208-213°.

SYN: DIBROMODIMETHYLSTANNANE

TOXICITY DATA with REFERENCE
ivn-mus LD50:56200 μg/kg CSLNX* NX#02289

OSHA PEL: TWA 0.1 mg(Sn)/m^3 (skin)
ACGIH TLV: TWA 0.1 mg(Sn)/m^3 (skin) (Proposed: TWA 0.1 mg(Sn)/m^3; STEL 0.2 mg(Sn)/m^3 (skin))
NIOSH REL: (Organotin Compounds) TWA 0.1 mg(Sn)/m^3

SAFETY PROFILE: Poison by intravenous route. When heated to decomposition it emits toxic fumes of Br^-. See also TIN COMPOUNDS and BROMIDES.

DUG889 CAS:40237-34-1 ***HR: 3***
DIMETHYLTIN DINITRATE
mf: $C_2H_6N_6O_6Sn$ mw: 272.79

SAFETY PROFILE: Explodes when heated. When heated to decomposition it emits toxic fumes of NO_x. See also TIN COMPOUNDS and NITRATES.

DUH000 CAS:599-69-9 ***HR: 2***
N,N-DIMETHYL-p-TOLUENESULFONAMIDE
mf: $C_9H_{13}NO_2S$ mw: 199.29

TOXICITY DATA with REFERENCE
orl-rat LDLo:500 mg/kg JPETAB 90,260,47

CONSENSUS REPORTS: Reported in EPA TSCA Inventory.

SAFETY PROFILE: Moderately toxic by ingestion. When heated to decomposition it emits very toxic fumes of NO_x and SO_x.

DUH200 CAS:609-72-3 ***HR: 3***
N,N-DIMETHYL-o-TOLUIDINE
mf: $C_9H_{13}N$ mw: 135.23

SYNS: DIMETHYL-o-TOLUIDINE ◇ o-METHYLDIMETHYLANILINE ◇ N,N,2-TRIMETHYLANILINE

TOXICITY DATA with REFERENCE
ipr-mus LD50:338 mg/kg AISFAR 1,284,51
orl-rat LDLo:500 mg/kg JPETAB 90,260,47

CONSENSUS REPORTS: Reported in EPA TSCA Inventory.

SAFETY PROFILE: Poison by intraperitoneal route. Moderately toxic by ingestion. When heated to decomposition it emits toxic fumes of NO_x.

DUH400 CAS:3010-57-9 ***HR: 3***
N,N-DIMETHYL-4-(p-TOLYLAZO)ANILINE
mf: $C_{15}H_{17}N_3$ mw: 239.35

SYNS: N,N-DIMETHYL-4-((4-METHYLPHENYL)AZO)BENZENAMINE ◇ p'-METHYL-p-DIMETHYLAMINOAZOBENZENE ◇ 4'-METHYL-4-DIMETHYLAMINOAZOBENZENE

TOXICITY DATA with REFERENCE
mma-sat 500 nmol/plate MUREAV 121,95,83
dns-rat:lvr 10 μmol/L CNREA8 46,1654,86
ipr-mus TDLo:1 g/kg (8-9D preg):TER KAIZAN 37,179,62
ipr-mus TDLo:1 g/kg (8-9D preg):REP KAIZAN 37,179,62
orl-rat TDLo:7776 mg/kg/35W-C:ETA CNREA8 5,227,45

CONSENSUS REPORTS: EPA Genetic Toxicology Program.

SAFETY PROFILE: Questionable carcinogen with experimental tumorigenic and teratogenic data. Experimental reproductive effects. Mutation data reported. When heated to decomposition it emits toxic fumes of NO_x.

DUH600 CAS:55-80-1 ***HR: 3***
N,N-DIMETHYL-p-(m-TOLYLAZO)ANILINE
mf: $C_{15}H_{17}N_3$ mw: 239.35

SYNS: 4-(N,N-DIMETHYLAMINO)-3'-METHYLAZOBENZENE ◇ N,N-DIMETHYL-p-(3'-METHYLPHENYLAZO)ANILINE ◇ N,N-DIMETHYL-4-((3-METHYLPHENYL)AZO)BENZENAMINE ◇ MDAB ◇ 3'-MDAB ◇ 3'-METHYLBUTTERGELB (GERMAN) ◇ 3'-METHYL-DAB ◇ 3'-METHYL-4-DIMETHYLAMINOAZOBENZEN (CZECH) ◇ M'-METHYL-p-DIMETHYLAMINOAZOBENZENE ◇ 3'-METHYL-4-DIMETHYLAMINOAZOBENZENE ◇ 3'-METHYL-N,N-DIMETHYL-4-AMINOAZOBENZENE ◇ 3'-METHYLDIMETHYLAMINOAZOBENZOL (GERMAN)

TOXICITY DATA with REFERENCE
otr-rat:lvr 240 μmol/L AMOKAG 39,231,85
dns-rat:lvr 1 μmol/L CNREA8 46,1654,86
orl-rat TDLo:1800 mg/kg/7W-C:CAR JJIND8 71,855,83
scu-rat TDLo:2750 mg/kg/1Y-I:ETA CNREA8 35,3798,75
scu-mus TDLo:7179 mg/kg (15-19D preg):NEO,TER CALEDQ 17,321,83
orl-rat TD:2419 mg/kg/12W-C:CAR CBINA8 53,107,85
orl-rat LDLo:1500 mg/kg 28ZPAK -,236,72

orl-mus LD50:17700 mg/kg JKXXAF #80-157517
ipr-mus LD50:1530 mg/kg JJIND8 62,911,79

CONSENSUS REPORTS: Reported in EPA TSCA Inventory. EPA Genetic Toxicology Program.

SAFETY PROFILE: Moderately toxic by ingestion. An experimental teratogen. Questionable carcinogen with experimental carcinogenic, neoplastigenic, and tumorigenic data. Mutation data reported. When heated to decomposition it emits toxic fumes of NO_x.

DUH800 CAS:3731-39-3 ***HR: 3***
N,N-DIMETHYL-p-((o-TOLYL)AZO)ANILINE
mf: $C_{15}H_{17}N_3$ mw: 239.35

SYNS: N,N-DIMETHYL-p-(2'-METHYLPHENYLAZO)ANILINE ◇ N,N-DIMETHYL-4-((2-METHYLPHENYL)AZO)BENZENAMINE ◇ o'-METHYL-p-DIMETHYLAMINOAZOBENZENE ◇ 2'-METHYL-4-DIMETHYLAMINOAZOBENZENE ◇ 2-METHYL-N,N-DI-METHYL-4-AMINOAZOBENZENE

TOXICITY DATA with REFERENCE
mma-sat 100 nmol/plate CALEDQ 1,91,75
dns-rat:lvr 10 μmol/L CNREA8 46,1654,86
orl-rat TDLo:14414 mg/kg/52W-C:CAR CBINA8 53,107,85
orl-rat TD:5856 mg/kg/26W-C:ETA CNREA8 5,227,45

SAFETY PROFILE: Questionable carcinogen with experimental carcinogenic and tumorigenic data. Mutation data reported. When heated to decomposition it emits toxic fumes of NO_x.

DUI000 CAS:1933-50-2 ***HR: 3***
4'-(3,3-DIMETHYL-1-TRIAZENO)ACETANILIDE
mf: $C_{10}H_{14}N_4O$ mw: 206.28

SYNS: AC 24055 ◇ 1-(p-ACETAMIDOPHENYL)-3,3-DIMETHYLTRIAZENE ◇ 1-(4-ACETAMINOPHENYL)-3,3-DIMETHYLTRIAZENE ◇ AMERICAN CYANIMID 24,055 ◇ AMERICAN CYANAMID CL-24055 ◇ ANTIFEEDANT 24005 ◇ ANTIFEEDING COMPOUND 24,055 ◇ CL 24055 ◇ CYANAMID 24055 ◇ 1,1-DIMETHYL-3-(p-ACETAMIDO-PHENYL)TRIAZENE ◇ 4'-DIMETHYLTRIAZENOACETANILIDE ◇ N-(4-(3,3-DIMETHYL-1-TRIAZENYL)PHENYL)ACETAMIDE ◇ ENT 25,651

TOXICITY DATA with REFERENCE
mma-sat 1 μmol/L JMCMAR 22,473,79
sln-dmg-orl 1 μmol/L CBINA8 9,365,74
mrc-smc 10 mmol/L CBINA8 9,365,74
hma-mus/smc 1 mmol/L CBINA8 9,365,74
orl-rat LD50:510 mg/kg TXAPA9 21,315,72
skn-rbt LD50:1400 mg/kg 28ZEAL 4,195,69
orl-bwd LD50:56 mg/kg TXAPA9 21,315,72

CONSENSUS REPORTS: EPA Genetic Toxicology Program.

SAFETY PROFILE: Poison by ingestion. Moderately toxic by skin contact. Mutation data reported. When heated to decomposition it emits toxic fumes of NO_x.

DUI200 CAS:80266-48-4 ***HR: 3***
4'-(3-(3,3-DIMETHYL-1-TRIAZENO)-9-ACRIDINYLAMINO)METHANESULFONANILIDE
mf: $C_{22}H_{22}N_6O_2S$ mw: 434.56

TOXICITY DATA with REFERENCE
mmo-sat 32 μmol/L JMCMAR 23,269,80
ipr-mus LD10:11 mg/kg JMCMAR 23,269,80

SAFETY PROFILE: Poison by intraperitoneal route. Mutation data reported. When heated to decomposition it emits very toxic fumes of NO_x and SO_x.

DUI400 CAS:64038-56-8 ***HR: D***
5-(3,3-DIMETHYL-1-TRIAZENO)IMIDAZOLE-4-CARBOXAMIDE CITRATE
mf: $C_6H_{10}N_6O{\bullet}C_6H_8O_7$ mw: 374.36

SYN: DTIC CITRATE

TOXICITY DATA with REFERENCE
ipr-rbt TDLo:130 mg/kg (female 6-18D post):REP TXAPA9 33,281,75
ipr-rat TDLo:225 mg/kg (male):TER TXAPA9 33,281,75

SAFETY PROFILE: An experimental teratogen. Other experimental reproductive effects. When heated to decomposition it emits toxic fumes of NO_x.

DUI600 CAS:7227-93-2 ***HR: D***
p-(3,3-DIMETHYLTRIAZENO)PHENOL
mf: $C_8H_{11}N_3O$ mw: 165.22

SYNS: 4-(3,3-DIMETHYL-1-TRIAZENYL)PHENOL ◇ 1-(4-HYDROXYPHENYL)-3,3-DIMETHYLTRIAZINE

TOXICITY DATA with REFERENCE
sln-dmg-orl 600 μmol/L CBINA8 9,365,74
mrc-smc 2500 μmol/L CBINA8 9,365,74

CONSENSUS REPORTS: EPA Genetic Toxicology Program.

SAFETY PROFILE: Mutation data reported. When heated to decomposition it emits toxic fumes of NO_x.

DUI709 CAS:3585-32-8 ***HR: 3***
1,3-DIMETHYLTRIAZINE
mf: $C_2H_7N_3$ mw: 73.10

SYN: DIAZOAMINOMETHANE

SAFETY PROFILE: Explodes violently on contact with flame. Upon decomposition it emits toxic fumes of NO_x.

DUI800 CAS:50355-75-4 ***HR: D***
3,3-DIMETHYL-1-(2,4,6-TRIBROMOPHENYL) TRIAZENE
mf: $C_8H_8Br_3N_3$ mw: 385.92

SYN: 1-(2,4,6-TRIBROMOPHENYL)-3,3-DIMETHYLTRIAZENE

TOXICITY DATA with REFERENCE
mma-sat 5 mmol/L MUREAV 36,1,76
hma-mus/smc 10 mmol/L CBINA8 9,365,74

CONSENSUS REPORTS: EPA Genetic Toxicology Program.

SAFETY PROFILE: Mutation data reported. When heated to decomposition it emits very toxic fumes of Br^- and NO_x. See also BROMIDES.

DUJ000 CAS:25724-50-9 ***HR: 3***
3,5-DIMETHYL-1-(TRICHLOROMETHYLMERCAPTO)PYRAZOLE
mf: $C_6H_7Cl_3N_2S$ mw: 245.56

TOXICITY DATA with REFERENCE
orl-rat LD50:570 mg/kg AIHAAP 30,470,69
skn-rbt LD50:200 mg/kg AIHAAP 30,470,69

SAFETY PROFILE: Poison by skin contact. Moderately toxic by ingestion. When heated to decomposition it emits very toxic fumes of Cl^-, NO_x, SO_x. See also MERCAPTANS.

DUJ400 CAS:24602-86-6 ***HR: 2***
2,6-DIMETHYL-4-TRIDECYLMORPHOLINE
mf: $C_{19}H_{39}NO$ mw: 297.59

SYNS: BAS 2205-F ◇ E-236 ◇ N-TRIDECYL-2,6-DIMETHYLMORPHOLIN (GERMAN) ◇ N-TRIDECYL-2,6-DIMETHYLMORPHOLINE ◇ 4-TRIDECYL-2,6-DIMETHYLMORPHOLINE

TOXICITY DATA with REFERENCE
orl-rat TDLo:5400 μg/kg (female 7-15D post):TER VPITAR 39(6),55,81
orl-rat LD50:650 mg/kg GUCHAZ 6,522,73
orl-mus LD50:1560 mg/kg VPITAR 39(6),55,81
orl-cat LD50:540 mg/kg VPITAR 39(6),55,81
orl-rbt LD50:750 mg/kg 28ZEAL 5,229,76
orl-gpg LD50:1 g/kg VPITAR 39(6),55,81

SAFETY PROFILE: Moderately toxic by ingestion. Experimental teratogenic effects. When heated to decomposition it emits toxic fumes of NO_x.

DUJ800 CAS:61471-62-3 ***HR: 3***
α,N-DIMETHYL-m-TRIFLUOROMETHYLPHENETHYLAMINE
mf: $C_{11}H_{14}F_3N$ mw: 217.26

SYN: α,N-DIMETHYL-3-TRIFLUOROMETHYLPHENETHYLAMINE

TOXICITY DATA with REFERENCE
orl-mus LD50:250 mg/kg ARZNAD 27,116,77
ipr-mus LD50:130 mg/kg ISYAM* -,21,70

SAFETY PROFILE: Poison by ingestion and intraperitoneal routes. When heated to decomposition it emits very toxic fumes of F^- and NO_x.

DUK000 CAS:53780-34-0 ***HR: 2***
2',4'-DIMETHYL-5-((TRIFLUOROMETHYL)SULFONAMIDO)ACETANILIDE
mf: $C_{11}H_{13}F_3N_2O_3S$ mw: 310.32

SYNS: N-(2,4-DIMETHYL-5-(((TRIFLUOROMETHYL)SULFONYL)AMINO)PHENYL)ACETAMIDE ◇ EMBARK ◇ EMBARK PLANT GROWTH REGULATOR ◇ MBR 12325 ◇ MEFLUIDIDE ◇ VEL 3973 ◇ VISTAR ◇ VISTAR HERBICIDE

TOXICITY DATA with REFERENCE
orl-rat LD50:4 g/kg 85ARAE 3,66,76/77
orl-mus LD50:1920 mg/kg FMCHA2 -,C148,83

SAFETY PROFILE: Moderately toxic by ingestion. An herbicide and plant growth regulator. When heated to decomposition it emits very toxic fumes of F^-, NO_x, and SO_x.

DUK200 CAS:343-75-9 ***HR: 3***
N,N-DIMETHYL-p-(2,4,6-TRIFLUOROPHENYLAZO)ANILINE
mf: $C_{14}H_{12}F_3N_3$ mw: 279.29

SYN: 2',4',6'-TRIFLUORO-4-DIMETHYLAMINOAZOBENZENE

TOXICITY DATA with REFERENCE
orl-rat TDLo:2700 mg/kg/13W-C:ETA CNREA8 13,93,53

SAFETY PROFILE: Questionable carcinogen with experimental tumorigenic data. When heated to decomposition it emits very toxic fumes of F^- and NO_x.

DUK800 CAS:2164-17-2 ***HR: 3***
1,1-DIMETHYL-3-(α,α,α-TRIFLUORO-m-TOLYL) UREA
mf: $C_{10}H_{11}F_3N_2O$ mw: 232.23

SYNS: C 2059 ◇ CIBA 2059 ◇ COTORAN ◇ COTORAN MULTI 50WP ◇ COTTONEX ◇ 1,1-DIMETHYL-3-(3-TRIFLUOROMETHYLPHENYL) UREA ◇ N,N-DIMETHYL-N'-(3-TRIFLUOROMETHYLPHENYL)UREA ◇ FLUOMETURON ◇ HERBICIDE C-2059 ◇ LANEX ◇ NCI-C08695 ◇ PAKHTARAN ◇ 3-(5-TRIFLUORMETHYLPHENYL)-,1-DIMETHYLHARNSTOFF (GERMAN) ◇ N-(m-TRIFLUOROMETHYLPHENYL)-N',N'-DIMETHYLUREA ◇ N-(3-TRIFLUOROMETHYLPHENYL)-N'-N'-DIMETHYLUREA ◇ 3-(m-TRIFLUOROMETHYLPHENYL)-1,1-DIMETHYLUREA

TOXICITY DATA with REFERENCE
mma-sat 1 μg/plate MUREAV 58,353,78
otr-rat:emb 56 μg/plate JJATDK 1,190,81
dni-mus-orl 1 g/kg MUREAV 58,353,78

orl-mus TDLo:87 g/kg/2Y-C:CAR NCITR* NCI-CG-TR-195,80
orl-rat LD50:6416 mg/kg PESTD5 17,351,76
ipr-rat LD50:685 mg/kg PESTD5 17,351,76
orl-mus LD50:900 mg/kg CIGET* -,-,77
ipr-mus LD50:552 mg/kg PESTD5 17,351,76
orl-rbt LD50:2500 mg/kg 85GMAT -,116,82
orl-gpg LD50:810 mg/kg 85GMAT -,116,82

CONSENSUS REPORTS: EPA Genetic Toxicology Program. IARC Cancer Review: Group 3 IMEMDT 7,56,87; Animal Inadequate Evidence IMEMDT 30,245, 83. NCI Carcinogenesis Bioassay (feed); No Evidence: rat NCITR* NCI-CG-TR-195,80; Equivocal Evidence: mouse NCITR* NCI-CG-TR-195,80. Reported in EPA TSCA Inventory.

SAFETY PROFILE: Moderately toxic by ingestion and intraperitoneal routes. Questionable carcinogen with experimental carcinogenic data. Mutation data reported. When heated to decomposition it emits very toxic fumes of F^- and NO_x.

DUL200 CAS:2223-82-7 ***HR: 3***
2,2-DIMETHYLTRIMETHYLENE ACRYLATE
mf: $C_{11}H_{16}O_4$ mw: 212.27

SYNS: DIMETHYLOLPROPANE DIACRYLATE ◇ 2,2-DIMETHYL-1,3-PROPANEDIOL DIACRYLATE ◇ 2,2-DIMETHYLTRIMETHYLENE ESTER ACRYLIC ACID ◇ NEOPENTYL GLYCOL DIACRYLATE ◇ 2-PROPENOIC ACID-2,2-DIMETHYL-1,3-PROPANEDIYL ESTER ◇ SR 247

TOXICITY DATA with REFERENCE
skn-rbt 500 mg open SEV UCDS** 11/30/71
skn-mus TDLo:46800 mg/kg/28W-I:CAR JTEHD6 16,55,85
orl-rat LD50:6730 mg/kg UCDS** 11/30/71
skn-rbt LD50:400 mg/kg TXAPA9 28,313,74

CONSENSUS REPORTS: Reported in EPA TSCA Inventory.

SAFETY PROFILE: Poison by skin contact. Mildly toxic by ingestion. A severe skin irritant. Questionable carcinogen with experimental carcinogenic data. When heated to decomposition it emits acrid smoke and irritating fumes. See also ESTERS.

DUL400 CAS:34522-40-2 ***HR: 3***
N,N-DIMETHYL-4-(3,4,5-TRIMETHYLPHENYL) AZOANILINE
mf: $C_{17}H_{21}N_3$ mw: 267.41

SYNS: N,N-DIMETHY[illegible] 4-((3,4,5-TRIMETHYLPHENYL)AZO) BENZENAMINE ◇ N,N-3',4',5'-PENTAMETHYLAMINOAZOBENZENE

TOXICITY DATA with REFERENCE
orl-rat TDLo:4320 mg/kg/17W-C:ETA JMCMAR 15,212,72

SAFETY PROFILE: Questionable carcinogen with experimental tumorigenic data. When heated to decomposition it emits toxic fumes of NO_x.

DUL500 CAS:13271-93-7 ***HR: 3***
1,2-DIMETHYL-2-TRIMETHYLSILYLHYDRAZINE
mf: $C_5H_{16}N_2Si$ mw: 132.28

$CH_3NHN(CH_3)SI(CH_3)_3$

SAFETY PROFILE: Explosive or violent reaction on contact with: 50/50 mixture of nitric + sulfuric acids; fuming nitric acid; fluorine; ozone + oxygen. When heated to decomposition it emits toxic fumes of NO_x.

DUL550 CAS:26464-99-3 ***HR: 3***
DIMETHYLTRIMETHYLSILYLPHOSPHINE
mf: $C_5H_{15}PSi$ mw: 134.23

SAFETY PROFILE: Ignites spontaneously in air. Reaction with water forms the spontaneously flammable dimethylphosphine. When heated to decomposition it emits toxic fumes of PO_x. See also PHOSPHINE and SILANE.

DUL589 ***HR: 3***
1,4-DIMETHYL-2,3,7-TRIOXABICYCLO[2.2.1] HEPT-5-ENE
mf: $C_6H_8O_3$ mw: 128.13

SYN: 2,5-DIMETHYL-2,5-DIHYDROFURAN-2,5-ENDOPEROXIDE

SAFETY PROFILE: A very unstable explosive. When heated to decomposition it emits acrid smoke and fumes. See also PEROXIDES.

DUL800 CAS:5152-30-7 ***HR: 3***
o,o'-DIMETHYLTUBOCURARINE
mf: $C_{40}H_{48}N_2O_6$ mw: 652.90

SYNS: DIMETHYL TUBOCURARINE ◇ o,o-DIMETHYLTUBOCURARINE ◇ N,N',o,o-TETRAMETHYL-(+)-TUBOCURINE

TOXICITY DATA with REFERENCE
ivn-dog LD50:120 μg/kg RISSAF 13,339,50
ivn-rbt LD50:35 μg/kg RISSAF 13,339,50

SAFETY PROFILE: Poison by intravenous route. When heated to decomposition it emits toxic fumes of NO_x.

DUM000 CAS:7601-55-0 ***HR: 3***
DIMETHYL TUBOCURARINE IODIDE
mf: $C_{40}H_{48}N_2O_6 \cdot 2I$ mw: 906.70

SYNS: (+)-o,o'-DIMETHYLCHONDROCURARINE DIIODIDE ◇ DIMETHYLETHER of d-TUBOCURARINE IODIDE ◇ 6,6',7',12'-TETRAMETHOXY-2,2,2',2'-TETRAMETHYLTUBOCURARANIUM

DIIODIDE ◇ TUBOCURARINE DIMETHYL ETHER IODIDE ◇ d-TUBOCURARINE IODIDE DIMETHYL ETHER

TOXICITY DATA with REFERENCE
ivn-rat LD50:35 μg/kg JLCMAK 34,516,49
ivn-mus LD50:238 μg/kg JLCMAK 34,516,49
ivn-rbt LD50:32 μg/kg JLCMAK 34,516,49
ivn-gpg LD50:50 mg/kg JLCMAK 34,516,49

SAFETY PROFILE: Poison by intravenous route. When heated to decomposition it emits very toxic fumes of I^- and NO_x. See also IODIDES and ETHER.

DUM100 CAS:13265-01-5 ***HR: 3***
α,3-DIMETHYLTYROSINE METHYL ESTER HYDROCHLORIDE
mf: $C_{12}H_{17}NO_3 \cdot ClH$ mw: 259.76

SYN: H 59/64

TOXICITY DATA with REFERENCE
orl-rat LD50:1400 mg/kg PSDTAP 10,206,69
orl-mus LD50:700 mg/kg PSDTAP 10,206,69
ivn-mus LD50:140 mg/kg PSDTAP 10,206,69

SAFETY PROFILE: Poison by intravenous route. Moderately toxic by ingestion. When heated to decomposition it emits toxic fumes of NO_x and HCl.

DUM200 CAS:96-31-1 ***HR: 2***
1,3-DIMETHYLUREA
mf: $C_3H_8N_2O$ mw: 88.13

PROP: Colorless crystals, water- and alc-sol. D: 1.14, mp: 106°, bp: 270°.

SYNS: N,N'-DIMETHYLHARNSTOFF (GERMAN) ◇ N,N'-DIMETHYLUREA ◇ sym-DIMETHYLUREA ◇ SYMMETRIC DIMETHYLUREA

TOXICITY DATA with REFERENCE
mmo-clr 400 mmol/L FOMIAZ 20,452,75
dni-hmn:lym 40 mmol/L PNASA6 79,1171,82
orl-rat TDLo:2 g/kg (12D preg):TER TJADAB 23,335,81
orl-rat TDLo:2 g/kg (12D preg):REP TJADAB 23,335,81
ipr-mus LDLo:4962 mg/kg JPETAB 54,188,35

CONSENSUS REPORTS: Reported in EPA TSCA Inventory.

SAFETY PROFILE: Moderately toxic by intraperitoneal route. Experimental teratogenic and reproductive effects. Human mutation data reported. When heated to decomposition it emits toxic fumes of NO_x.

DUM400 ***HR: 3***
DIMETHYLUREA and SODIUM NITRITE

SYNS: DIMETHYLHARNSTOFF and NATRIUMNITRIT (GERMAN) ◇ SODIUM NITRITE and DIMETHYLUREA

TOXICITY DATA with REFERENCE
orl-rat TDLo:14 g/kg/56D-C:ETA ARZNAD 21,1707,71

SAFETY PROFILE: Questionable carcinogen with experimental tumorigenic data. When heated to decomposition it emits toxic fumes of NO_x and Na_2O. See also SODIUM NITRITE.

DUM600 CAS:63019-76-1 ***HR: 3***
p-N,N-DIMETHYLUREIDOAZOBENZENE
mf: $C_{15}H_{16}N_4O$ mw: 268.35

TOXICITY DATA with REFERENCE
scu-rat TDLo:2600 mg/kg/83D-I:NEO BJPCAL 9,306,54

SAFETY PROFILE: Questionable carcinogen with experimental neoplastigenic data. When heated to decomposition it emits toxic fumes of NO_x.

DUM800 CAS:4849-32-5 ***HR: 3***
m-(3,3-DIMETHYLUREIDO)PHENYL-tert-BUTYL CARBAMATE
mf: $C_{14}H_{21}N_3O_3$ mw: 279.38

SYNS: tert-BUTYLCARBAMIC ACID ESTER with 3-(m-HYDROXYPHENYL-1,1-DIMETHYLUREA ◇ 3-(((DIMETHYLAMINO)CARBONYL) AMINO)PHENYL-1,1-DIMETHYLETHYL)CARBAMATE ◇ 1,1-DIMETHYL-3-(3-N-tert-BUTYLCARBAMYLOXY)-PHENYL)UREA ◇ m-(3,3-DIMETHYLHARNSTOFF)-PHENYL-tert-BUTYLCARBAMAT(GERMAN) ◇ FMC 11092 ◇ KARBUTILATE ◇ NIA 11092 ◇ TANDEX

TOXICITY DATA with REFERENCE
orl-rat LD50:3000 mg/kg GUCHAZ 6,310,72
ivn-mus LD50:320 mg/kg CSLNX* NX#03896

SAFETY PROFILE: Poison by intravenous route. Moderately toxic by ingestion. An herbicide. When heated to decomposition it emits toxic fumes of NO_x. See also CARBAMATES.

DUN200 CAS:6225-06-5 ***HR: 3***
N,N-DIMETHYLVALERAMIDE
mf: $C_7H_{15}NO$ mw: 129.23

SYN: N,N-DIMETHYLPENTANAMIDE

TOXICITY DATA with REFERENCE
ipr-mus LD50:500 mg/kg AIHAAP 32,539,71
ivn-mus LD50:230 mg/kg AIHAAP 32,539,71

CONSENSUS REPORTS: Reported in EPA TSCA Inventory.

SAFETY PROFILE: Poison by intravenous route. Moderately toxic by intraperitoneal route. When heated to decomposition it emits toxic fumes of NO_x.

DUN300 CAS:33499-84-2 ***HR: D***
6,8-o-DIMETHYLVERSICOLORIN A
mf: $C_{20}H_{14}O_7$ mw: 366.34

TOXICITY DATA with REFERENCE
mmo-sat 500 nmol/plate MUREAV 143,121,85

mma-sat 500 nmol/plate MUREAV 143,121,85
dns-rat:lvr 500 nmol/L MUREAV 143,121,85

SAFETY PROFILE: Mutation data reported. When heated to decomposition it emits acrid smoke and irritating fumes.

DUN310 ***HR: D***
6,8-o-DIMETHYLVERSICOLORIN B
mf: $C_{20}H_{14}O_7$ mw: 366.34

TOXICITY DATA with REFERENCE
mmo-sat 500 nmol/plate MUREAV 143,121,85
mma-sat 500 nmol/plate MUREAV 143,121,85
dns-rat:lvr 5 μmol/L MUREAV 143,121,85

SAFETY PROFILE: Mutation data reported. When heated to decomposition it emits acrid smoke and irritating fumes.

DUN400 CAS:63141-79-7 ***HR: 3***
α-(2,2-DIMETHYLVINYL)-α-ETHYNYL-p-CRESOL
mf: $C_{13}H_{13}O$ mw: 185.26

SYN: DIMETHYLVINYLETHINYL-p-HYDROXYPHENYLMETHANE

TOXICITY DATA with REFERENCE
ihl-rat TCLo:8900 $\mu g/m^3$/24H (1-18D preg):REP GISAAA 41(1),95,76
ihl-mus LC50:220 mg/m^3 GISAAA 41(1),95,76

SAFETY PROFILE: Poison by inhalation. Experimental reproductive effects. When heated to decomposition it emits acrid smoke and irritating fumes.

DUN600 CAS:1468-37-7 ***HR: 3***
DIMETHYLXANTHOGEN DISULFIDE
mf: $C_4H_6O_2S_4$ mw: 214.34

SYNS: BIS(METHYLXANTHOGEN) DISULFIDE ◇ DI(METHOXYTHIOCARBONYL) DISULFIDE ◇ o,o-DIMETHYL DITHIOBIS (THIOFORMATE) ◇ DIMETHYL DIXANTHOGEN ◇ DIMETHYL XANTHIC DISULFIDE ◇ DIMEXAN ◇ DIMEXANO ◇ DI(THIONOCARBOMETHOXY) DISULFIDE ◇ THIOPEROXYDICARBONIC ACID DIMETHYL ESTER ◇ TRIDEX ◇ TRI-PE

TOXICITY DATA with REFERENCE
orl-rat LD50:240 mg/kg WRPCA2 9,119,70

SAFETY PROFILE: Poison by ingestion. A pesticide. When heated to decomposition it emits toxic fumes of SO_x.

DUN800 CAS:18997-62-1 ***HR: 3***
N,N-DIMETHYL-p-(2,3,XYLYLAZO)ANILINE
mf: $C_{16}H_{19}N_3$ mw: 253.38

SYNS: 2′,3′-DIMETHYL-4-DIMETHYLAMINOAZOBENZENE ◇ N,N-DIMETHYL-p-(2′,3′-DIMETHYLPHENYLAZO)ANILINE

TOXICITY DATA with REFERENCE
mma-sat 4 μg/plate MUREAV 93,67,82
orl-rat TDLo:1075 mg/kg/8W-C:CAR CBINA8 53,107,85
orl-rat TDLo:1080 mg/kg/30D-I:ETA JMCMAR 11,1234,68

SAFETY PROFILE: Questionable carcinogen with experimental carcinogenic and tumorigenic data. Mutation data reported. When heated to decomposition it emits toxic fumes of NO_x.

DUO000 CAS:3025-73-8 ***HR: 3***
N,N-DIMETHYL-p-(3,4-XYLYLAZO)ANILINE
mf: $C_{16}H_{19}N_3$ mw: 253.38

SYNS: 3′,4′-DIMETHYL-4-DIMETHYLAMINOZOBENZENE ◇ N,N-DIMETHYL-p-(3′,4′-DIMETHYLPHENYLAZO)ANILINE

TOXICITY DATA with REFERENCE
orl-rat TDLo:9030 mg/kg/25W-C:CAR CBINA8 53,107,85
orl-rat TD:6600 mg/kg/26W-C:ETA CNREA8 17,387,57
orl-rat TD:6480 mg/kg/26W-C:NEO JMCMAR 15,212,72

SAFETY PROFILE: Questionable carcinogen with experimental carcinogenic, neoplastigenic, and tumorigenic data. When heated to decomposition it emits toxic fumes of NO_x.

DUO200 CAS:544-97-8 ***HR: 3***
DIMETHYLZINC
DOT: UN 1370
mf: C_2H_6Zn mw: 95.44

CONSENSUS REPORTS: Zinc and its compounds are on the Community Right-To-Know List.

DOT Classification: Flammable Solid; Label: Spontaneously Combustible.

SAFETY PROFILE: A poison. Ignites spontaneously in air. Explodes in an oxygen atmosphere. Explosive reaction on contact with water; 2,2-dichloropropane. When heated to decomposition it emits toxic fumes of ZnO. See also ZINC COMPOUNDS.

DUO300 CAS:89591-51-5 ***HR: 3***
DIMETPRAMIDE
mf: $C_{16}H_{26}N_4O_4$•ClH mw: 374.87

SYNS: N-(2-(DIETHYLAMINO)ETHYL)-4-(DIMETHYLAMINO)-2-METHOXY-5-NITROBENZAMIDE MONOHYDROCHLORIDE ◇ DIMETHPRAMIDE ◇ 4-DIMETHYLAMINO-5-NITRO-2-METHOXY-N-(2-DIETHYLAMINOETHYL)BENZAMIDE HYDROCHLORIDE

TOXICITY DATA with REFERENCE
eye-rbt 500 mg MLD RPTOAN 48,201,85
orl-rat LD50:856 mg/kg RPTOAN 48,201,85
orl-mus LD50:230 mg/kg RPTOAN 48,201,85
ipr-mus LD50:159 mg/kg RPTOAN 48,201,85
orl-gpg LD50:636 mg/kg RPTOAN 48,201,85

SAFETY PROFILE: Poison by ingestion and intraperitoneal routes. An eye irritant. When heated to decomposition it emits toxic fumes of NO_x and HCl.

DUO350 CAS:642-15-9 ***HR: 3***
DIMIDIN
mf: $C_{28}H_{40}N_2O_9$ mw: 548.62

PROP: Crystals from ethyl acetate + Skellysolve B. Mp: 149-150°. Freely sol in alc, ether, acetone, chloroform. Very sltly sol in petr ether, benzene, carbon tetrachloride. Practically insol in water and in 5% aq solns of hydrochloric acid, sodium carbonate and sodium bicarbonate.

SYNS: ANTIMYCIN A1 ◇ DIHYDROSAMIDIN ◇ ISOVALERIC ACID-8-ESTER with 3-FORMAMIDO-N-(7-HEXYL-8-HYDROXY-4,9-DIMETHYL-2,6-DIOXO-1,5-DIOXONAN-3-YL)SALICYLAMIDE

TOXICITY DATA with REFERENCE
orl-rat LD50:1469 mg/kg JDGRAX 7(2),1,75
ipr-mus LD50:7600 μg/kg 85FZAT -,144,67
scu-mus LD50:25 mg/kg 85FZAT -,144,67
ivn-mus LD50:900 μg/kg 85FZAT -,144,67
ims-mus LD50:1000 mg/kg JDGRAX 7(2),1,75

SAFETY PROFILE: Poison by subcutaneous, intravenous, and intraperitoneal routes. Moderately toxic by ingestion and intramuscular routes. When heated to decomposition it emits toxic fumes of NO_x. See also ESTERS.

DUO400 CAS:119-48-2 ***HR: 3***
DIMORPHOLAMINE
mf: $C_{20}H_{38}N_4O_4$ mw: 398.62

SYNS: AMIPAN T ◇ N,N'-DIBUTYL-N,N'-DICARBOXYETHYLENE DIAMINEMORPHOLIDE ◇ N,N'-DIBUTYL-N,N'-DICARBOXY-MORPHOLIDE-ETHYLENEDIAMINE ◇ N,N'-DI-n-BUTYLETHYLENEDIAMINE-N,N'-DICARBOXYBISMORPHOLIDE ◇ N,N'-1,2-ETHANEDIYLBIS(N-BUTYL-4-MORPHOLINECARBOXAMIDE) ◇ N,N'-ETHYLENEBIS(N-BUTYL-4-MORPHOLINECARBOXAMIDE) ◇ PRONTODIN ◇ 1064 TH ◇ THERALEPTIQUE ◇ THERAPTIQUE

TOXICITY DATA with REFERENCE
orl-rat LD50:270 mg/kg NIIRDN 6,347,82
scu-rat LD50:190 mg/kg NIIRDN 6,347,82
ivn-rat LD50:24 mg/kg NIIRDN 6,347,82
ims-rat LD50:122 mg/kg NIIRDN 6,347,82
orl-mus LD50:150 mg/kg MEIEDD 10,476,83
ipr-mus LD50:80 mg/kg AIPTAK 163,133,66
scu-mus LD50:104 mg/kg AIPTAK 163,133,66
ivn-mus LD50:42200 μg/kg NIIRDN 6,347,82

SAFETY PROFILE: Poison by ingestion, intraperitoneal, intravenous, intramuscular, and subcutaneous routes. An analeptic agent (stimulant). When heated to decomposition it emits toxic fumes of NO_x. See also AMINES.

DUO500 CAS:69853-15-2 ***HR: 3***
DIMORPHOLINIUM HEXACHLOROSTANNATE
mf: $C_8H_{10}Cl_6N_2O_2Sn$ mw: 497.59

SYN: MORPHOLINIUM, HEXACHLOROSTANNATE(2-) (2:1)

TOXICITY DATA with REFERENCE
ivn-mus LD50:18 mg/kg CSLNX* NX#02489

OSHA PEL: TWA 2 mg(Sn)/m^3
ACGIH TLV: TWA 2 mg(Sn)/m^3

SAFETY PROFILE: Poison by intravenous route. When heated to decomposition it emits toxic fumes of NO_x, Sn, and Cl^-.

DUO600 CAS:13071-27-7 ***HR: 3***
1,5-DIMORPHOLINO-3-(1-NAPHTHYL)-PENTANE
mf: $C_{23}H_{32}N_2O_2$ mw: 368.57

SYNS: 4,4'-(3-(1-NAPHTHALENYL)-1,5-PENTANEDIYL)BISMORPHOLINE ◇ 4,4'-(3-(1-NAPHTHYL)-1,5-PENTAMETHYLENEDIMORPHOLINE

TOXICITY DATA with REFERENCE
orl-rat LD50:708 mg/kg ARZNAD 18,1127,68
scu-rat LD50:830 mg/kg ARZNAD 18,1127,68
ivn-rat LD50:26 mg/kg ARZNAD 18,1127,68
orl-mus LD50:1700 mg/kg ARZNAD 18,1127,68
ipr-mus LD50:452 mg/kg JMCMAR 13,418,70
ivn-mus LD50:98 mg/kg ARZNAD 18,1127,68

SAFETY PROFILE: Poison by intravenous route. Moderately toxic by ingestion, subcutaneous, and intraperitoneal routes. When heated to decomposition it emits toxic fumes of NO_x.

DUO800 CAS:74749-73-8 ***HR: 1***
DIMYRCETOL

PROP: A mixture of dihydromyrcenol and dihydromyrcenyl formate (FCTXAV 18,649,80).

TOXICITY DATA with REFERENCE
skn-rbt 500 mg/24H MOD FCTXAV 18,679,80
orl-rat LD50:4100 mg/kg FCTXAV 18,649,80

SAFETY PROFILE: Mildly toxic by ingestion. A skin irritant.

DUP000 CAS:258-76-4 ***HR: 3***
DINAPHTHAZINE
mf: $C_{20}H_{12}N_2$ mw: 280.34

SYN: DINAPHTAZIN (GERMAN)

TOXICITY DATA with REFERENCE
scu-mus TDLo:200 mg/kg:ETA ZEKBAI 58,56,51

SAFETY PROFILE: Questionable carcinogen with ex-

perimental tumorigenic data. When heated to decomposition it emits toxic fumes of NO_x.

DUP200 CAS:29903-04-6 ***HR: 3***
DI-(1-NAPHTHOYL)PEROXIDE (DOT)
mf: $C_{22}H_{14}O_4$ mw: 342.35

DOT Classification: Forbidden.

SAFETY PROFILE: A friction-sensitive explosive. When heated to decomposition it emits acrid smoke and fumes. See also PEROXIDES.

DUP300 CAS:148-01-6 ***HR: 3***
DINITOLMIDE
mf: $C_8H_7N_3O_5$ mw: 225.18

PROP: Yellowish solid. Mp: 177°. Very sltly sol in water; sol in acetone, acetonitrile, and dimethyl formamide.

SYNS: COCCIDINE A ◇ COCCIDOT ◇ DINITOLMID ◇ 3,5-DINITRO-o-TOLUAMIDE ◇ D.O.T. ◇ 2-METHYL-3,5-DINITROBENZAMIDE ◇ ZOALENE ◇ ZOAMIX

TOXICITY DATA with REFERENCE
mmo-esc 500 μg/plate MUREAV 77,21,80
mrc-bcs 1 mg/disc MUREAV 77,21,80
orl-rat LD50:600 mg/kg 29ZVAB -,537,69
ivn-dog LD50:75 mg/kg PCOC** -,1252,66

OSHA PEL: TWA 5 mg/m^3
ACGIH TLV: TWA 5 mg/m^3

SAFETY PROFILE: Poison by intravenous route. Moderately toxic by ingestion. Mutation data reported. A strong exothermic reaction above 248°C has caused industrial explosions. When heated to decomposition it emits toxic fumes of NO_x. See also NITRO COMPOUNDS of AROMATIC HYDROCARBONS.

DUP400 CAS:96-91-3 ***HR: 3***
4,6-DINITRO-2-AMINOPHENOL
mf: $C_6H_5N_3O_5$ mw: 199.14

PROP: Red crystals. Mp: 168°; flash p: 410°F. Soluble in alc, benzene; glacial acetic acid, aniline, and ether; sparingly sol in water.

SYNS: ACIDE PICRAMIQUE (FRENCH) ◇ 2-AMINO-4,6-DINITROPHENOL ◇ C.I. OXIDATION BASE 21 ◇ FOURRINE 93 ◇ FOURRINE 4R ◇ FURRO 4R ◇ PICRAMIC ACID ◇ ZOBA 4R

TOXICITY DATA with REFERENCE
mmo-sat 5 μmol/plate NEZAAQ 38,533,83
mma-sat 5 μmol/plate NEZAAQ 38,533,83
ivn-dog LDLo:150 mg/kg AIPTAK 50,20,35
ipr-pgn LDLo:140 mg/kg AIPTAK 50,20,35

SAFETY PROFILE: Poison by intravenous and intraperitoneal routes. Mutation data reported. Combustible when exposed to heat, flame or oxidizers. A powerful explosive when dry. May explode when shocked or heated. When heated to decomposition it emits toxic fumes of NO_x. See also EXPLOSIVES, HIGH.

DUP600 CAS:97-02-9 ***HR: 3***
2,4-DINITROANILINE
mf: $C_6H_5N_3O_4$ mw: 183.14

PROP: Yellow, needle-like crystals; insol in water. Mp: 188°, flash p: 435°F (CC), d: 1.615, vap d: 6.31.

SYNS: 2,4-DINITRANILINE ◇ 2,4-DINITROANILIN (GERMAN) ◇ 2,4-DINITROANILINA (ITALIAN) ◇ 2,4-DINITROBENZENAMIME ◇ DNA ◇ NCI-C60753

TOXICITY DATA with REFERENCE
eye-rbt 500 mg/24H MLD 28ZPAK -,132,72
mmo-sat 10 μg/plate ENMUDM 5(Suppl 1),3,83
mma-sat 2 μg/plate MUREAV 67,1,79
ihl-rat TCLo:17 mg/m^3/4H (1-22D preg):TER GTPZAB 26(4),47,82
ihl-rat TCLo:17 mg/m^3/4H (1-7D preg):REP GTPZAB 26(4),47,82
ipr-rat LDLo:250 mg/kg NCNSA6 5,32,53
orl-mus LD50:370 mg/kg GTPZAB 25(8),50,81
ipr-mus LDLo:400 mg/kg JAPMA8 48,419,59
orl-gpg LD50:1050 mg/kg GISAAA 47(10),15,82

CONSENSUS REPORTS: Reported in EPA TSCA Inventory.

SAFETY PROFILE: Poison by ingestion and intraperitoneal routes. Experimental teratogenic and reproductive effects. Mutation data reported. An eye irritant. Combustible and explosive when exposed to heat or flame; can react with oxidizing materials. To fight fire, use CO_2, dry chemical. Mixtures with charcoal ignite at 350°C. Vigorous reaction with chlorine + hydrochloric acid evolves gases. When heated to decomposition it emits highly toxic fumes of NO_x. See also NITROANILINES.

DUP800 CAS:119-27-7 ***HR: 3***
2,4-DINITROANISOL
mf: $C_7H_6N_2O_5$ mw: 198.15

PROP: Colorless to yellow crystals. Mp: 89°, bp: sublimes, d: 1.341 @ 20°/4°, vap d: 6.83.

SYNS: 2,4-DINITROANISOLE ◇ α-DINITROANISOLE ◇ 2,4-DINITROPHENYLMETHYL ETHER ◇ 1-METHOXY-2,4-DINITROBENZENE

TOXICITY DATA with REFERENCE
mmo-sat 10 μg/plate BCPCA6 26,729,77
mma-sat 20 μmol/plate MUREAV 58,11,78
orl-rat LDLo:100 mg/kg NCNSA6 5,16,53

CONSENSUS REPORTS: Reported in EPA TSCA Inventory.

SAFETY PROFILE: Poison by ingestion. Mutation data reported. When heated to decomposition it emits toxic fumes of NO_x. See also NITRO COMPOUNDS of AROMATIC HYDROCARBONS and NITRATES.

DUQ000 CAS:82-35-9 ***HR: 2***
1,5-DINITROANTHRAQUINONE
mf: $C_{14}H_6N_2O_6$ mw: 298.22

SYNS: 1,5-DINITRO-9,10-ANTHRACENEDIONE ◇ 1,5-DINITRO-ANTHRACHINON (CZECH)

TOXICITY DATA with REFERENCE
eye-rbt 500 mg/24H MLD 28ZPAK -,121,72
ipr-rat LD50:3130 mg/kg GISAAA 49(4),90,84
orl-mus LD50:4750 mg/kg GISAAA 49(4),90,84

CONSENSUS REPORTS: Reported in EPA TSCA Inventory.

SAFETY PROFILE: Moderately toxic by intraperitoneal route. Mildly toxic by ingestion. An eye irritant. When heated to decomposition it emits toxic fumes of NO_x.

DUQ180 CAS:25154-54-5 ***HR: 3***
DINITROBENZENE
DOT: UN 1597
mf: $C_6H_4N_2O_4$ mw: 168.12

SYNS: DINITROBENZENE, solution (DOT) ◇ DINITROBENZOL, solid (DOT)

OSHA PEL: TWA 1 mg/m^3 (skin)
ACGIH TLV: TWA 0.15 ppm (skin)
DFG MAK: 0.15 ppm (1 mg/m^3); Suspected Carcinogen
DOT Classification: Poison B; Label: Poison, Solid and Solution.

SAFETY PROFILE: Suspected carcinogen. A poison. When heated to decomposition it emits toxic fumes of NO_x. See also o-DINITROBENZENE.

DUQ200 CAS:99-65-0 ***HR: 3***
m-DINITROBENZENE
DOT: UN 1597
mf: $C_6H_4N_2O_4$ mw: 168.12

PROP: Yellowish crystals. Mp: 89°, bp: 301°.

SYNS: BINITROBENZENE ◇ 1,3-DINITROBENZENE ◇ 2,4-DINITROBENZENE ◇ 1,3-DINITROBENZOL ◇ DWUNITROBENZEN (POLISH)

TOXICITY DATA with REFERENCE
mmo-sat 3300 ng/plate ENMUDM 2,531,80
mma-sat 100 nmol/plate MUREAV 58,11,78
orl-rat TDLo:150 mg/kg (male 10W pre):REP JTEHD6 19,477,86
orl-hmn LDLo:28 mg/kg 34ZIAG -226,69
skn-man TDLo:4 mg/kg/2D-I:CNS,PUL LANCAO 2,582,01
orl-rat LD50:83 mg/kg NTIS** AD-A066-307
ipr-rat LD50:28 mg/kg AEPPAE 207,446,49
orl-dog LDLo:600 mg/kg NTIS** AD-A066-307
ivn-dog LD50:10 mg/kg NTIS** AD-A066-307
orl-cat LDLo:27 mg/kg LANCAO 2,582,01
orl-rbt LDLo:400 mg/kg NTIS** AD-A066-307
orl-bwd LD50:42 mg/kg TXAPA9 21,315,72

CONSENSUS REPORTS: Reported in EPA TSCA Inventory. EPA Genetic Toxicology Program.

OSHA PEL: TWA 1 mg/m^3 (skin)
ACGIH TLV: TWA 0.15 ppm (skin)
DFG MAK: 0.15 ppm (1 mg/m^3); Suspected Carcinogen
DOT Classification: Poison B; Label: Poison.

SAFETY PROFILE: Suspected carcinogen. Human poison by ingestion. Experimental poison by ingestion, intraperitoneal, and intravenous routes. Human systemic effects by skin contact: cyanosis and motor activity changes. Experimental reproductive effects. Mutation data reported. Mixture with nitric acid is a high explosive. Mixture with tetranitromethane is a high explosive very sensitive to sparks. When heated to decomposition it emits toxic fumes of NO_x. See also o- and p-DINITROBENZENE.

DUQ400 CAS:528-29-0 ***HR: 3***
o-DINITROBENZENE
DOT: UN 1597
mf: $C_6H_4N_2O_4$ mw: 168.12

PROP: Colorless needles or plates. Mp: 118°, bp: 319°, flash p: 302°F (CC), d: 1.571 @ 0°/4°, vap d: 5.79.

SYN: 1,2-DINITROBENZENE

OSHA PEL: TWA 1 mg/m^3 (skin)
ACGIH TLV: TWA 0.15 ppm (skin)
DFG MAK: 0.15 ppm (1 mg/m^3); Suspected Carcinogen
DOT Classification: Poison B; Label: Poison.

SAFETY PROFILE: Suspected carcinogen. Poison by inhalation and ingestion. Moderately toxic by skin contact. Can cause liver, kidney, and central nervous system injury. Combustible when exposed to heat or flame; can react vigorously with oxidizing materials. A severe explosion hazard when shocked or exposed to heat or flame. It is used in bursting charges and to fill artillery shells. Mixtures with nitric acid are highly explosive. To fight fire, use water, CO_2, dry chemical. Dangerous; when heated to decomposition it emits highly toxic fumes of NO_x and explodes. See also m- and p-DINITROBENZENE and NITRO COMPOUNDS of AROMATIC HYDROCARBONS.

DUQ600 CAS:100-25-4 ***HR: 3***
p-DINITROBENZENE
DOT: UN 1597
mf: $C_6H_4N_2O_4$ mw: 168.12

PROP: White crystals. Mp: 173°, bp: 299°. Volatile with steam.

SYN: DITHANE A-4

TOXICITY DATA with REFERENCE
mmo-sat 5 μg/plate CRNGDP 6,727,85
mma-sat 25 μg/plate CRNGDP 6,727,85
orl-cat LDLo:29 mg/kg MEIEDD 10,477,83

CONSENSUS REPORTS: Reported in EPA TSCA Inventory.

OSHA PEL: TWA 1 mg/m^3 (skin)
ACGIH TLV: TWA 0.15 ppm (skin)
DFG MAK: 0.15 ppm (1 mg/m^3); Suspected Carcinogen
DOT Classification: Poison B; Label: Poison.

SAFETY PROFILE: Suspected carcinogen. Poison by ingestion. Mutation data reported. Mixture with nitric acid is a high explosive. When heated to decomposition it emits toxic fumes of NO_x. See also o- and m-DINITROBENZENE.

DUQ800 ***HR: 3***
4,6-DINITROBENZENEDIAZONIUM-2-OXIDE
mf: $C_6H_2N_4O_5$ mw: 210.11

SAFETY PROFILE: A very powerful priming explosive as sensitive as mercury fulminate. When heated to decomposition it emits toxic fumes of NO_x.

DUR200 CAS:528-76-7 ***HR: 3***
2,4-DINITROBENZENESULFENYL CHLORIDE
mf: $C_6H_3ClN_2O_4S$ mw: 234.62

SAFETY PROFILE: A heat-sensitive explosive. When heated to decomposition it emits toxic fumes of Cl^-, SO_x and NO_x. See also NITRO COMPOUNDS of AROMATIC HYDROCARBONS.

DUR400 CAS:89-02-1 ***HR: 3***
2,4-DINITROBENZENESULFONIC ACID
mf: $C_6H_4N_2O_7S$ mw: 248.18

SYN: KYSELINA-2,4-DINITROBENZENSULFONOVA(CZECH)

TOXICITY DATA with REFERENCE
skn-rbt 500 mg/24H MOD 28ZPAK -,180,72
eye-rbt 500 mg/24H MLD 28ZPAK -,180,72
orl-rat LD50:5800 mg/kg 28ZPAK -,180,72
ivn-mus LD50:320 mg/kg CSLNX* NX#01550

CONSENSUS REPORTS: Reported in EPA TSCA Inventory.

SAFETY PROFILE: Poison by intravenous route. Mildly toxic by ingestion. A skin and eye irritant. When heated to decomposition it emits very toxic fumes of NO_x and SO_x.

DUR500 CAS:5128-28-9 ***HR: 3***
4,6-DINITROBENZOFURAZAN-N-OXIDE
mf: $C_6H_2N_4O_6$ mw: 226.10

SAFETY PROFILE: A powerful, high explosive as sensitive as picric acid. Forms impact-, friction- or electric shock-sensitive explosive complexes with cysteine and nucleophiles (e.g., potassium hydrogen carbonate in water or methanol; potassium hydroxide in methanol; ammonia; hydroxylamine; hydrazine hydrate). Forms explosive adducts with furan; N-methylindole; and N-methylpyrrole. Forms adducts with ketones (e.g., acetone; cyclopentanone; cyclopentanedione; 2,4-pentanedione; 3-methyl-2,4-pentanedione) which can produce shock-sensitive, high explosive potassium salts. Upon decomposition it emits toxic fumes of NO_x. See also EXPLOSIVES, HIGH.

DUR800 CAS:87-31-0 ***HR: 3***
5,7-DINITRO-1,2,3-BENZOXADIAZOLE
DOT: UN 0074
mf: $C_6H_2N_4O_5$ mw: 210.12

SYNS: DDNP ◇ DIAZO ◇ 2-DIAZO-4,6-DINITROBENZENE-1-OXIDE ◇ DIAZODINITROPHENOL (DOT) ◇ DIAZODINITROPHENOL, containing, by weight, at least 40% water (DOT) ◇ DIAZODINITROPHENOL, dry (DOT) ◇ INITIATING EXPLOSIVE DIAZODINITROPHENOL (DOT)

DOT Classification: Class A Explosive; Label: Explosive A; Forbidden, Dry.

SAFETY PROFILE: An explosive. When heated to decomposition it emits toxic fumes of NO_x. See also NITRO COMPOUNDS of AROMATIC HYDROCARBONS; and EXPLOSIVES, HIGH.

DUS000 CAS:1528-74-1 ***HR: 3***
4,4'-DINITROBIPHENYL
mf: $C_{12}H_8N_2O_4$ mw: 244.22

SYN: 4,4'-DINITROBIFENYL (CZECH)

TOXICITY DATA with REFERENCE
eye-rbt 500 mg/24H MLD 28ZPAK -,61,72
mmo-sat 2500 μg/plate MUREAV 91,321,81
mma-sat 5 μg/plate MUREAV 91,321,81
orl-rat TDLo:950 mg/kg/W-I:ETA TXAPA9 6,352,64
orl-rat LD50:13900 mg/kg 28ZPAK -,61,72

CONSENSUS REPORTS: Reported in EPA TSCA Inventory.

SAFETY PROFILE: Mildly toxic by ingestion. An eye irritant. Questionable carcinogen with experimental tumorigenic data. Mutation data reported. When heated to decomposition it emits toxic fumes of NO_x.

DUS200 CAS:1817-73-8 ***HR: 2***
2,4-DINITRO-6-BROMOANILINE
mf: $C_6H_4BrN_3O_4$ mw: 262.04

SYN: 2,4-DINITRO-6-BROMANILIN(CZECH)

TOXICITY DATA with REFERENCE
eye-rbt 500 mg/24H SEV 28ZPAK -,94,72
orl-rat LD50:4490 mg/kg 28ZPAK -,94,72

CONSENSUS REPORTS: Reported in EPA TSCA Inventory.

SAFETY PROFILE: Mildly toxic by ingestion. A severe eye irritant. When heated to decomposition it emits very toxic fumes of Br^- and NO_x. See also 2,4-DINITROANILINE.

DUS400 CAS:28103-68-6 ***HR: 3***
2,3-DINITRO-2-BUTENE
mf: $C_4H_6N_2O_4$ mw: 146.11

$$CH_3C(NO_2)=C(NO_2)CH_3$$

SAFETY PROFILE: Potentially explosive at 135°C/14 mbar. When heated to decomposition it emits toxic fumes of NO_x.

DUS500 CAS:29110-68-7 ***HR: 3***
2,4-DINITRO-6-tert-BUTYLPHENYL METHANESULFONATE
mf: $C_{11}H_{14}N_2O_7S$ mw: 318.33

SYNS: HE 166 ◇ PREPARATION HE 166

TOXICITY DATA with REFERENCE
orl-rat TDLo:1820 mg/kg (52W pre):REP JTSCDR 9,161,84
orl-rat TDLo:280 mg/kg/32W-C:ETA JTSCDR 9,161,84

SAFETY PROFILE: Questionable carcinogen with experimental tumorigenic data. Experimental reproductive effects. An herbicide. When heated to decomposition it emits toxic fumes of SO_x and NO_x. See also SULFONATES.

DUS600 CAS:2401-85-6 ***HR: 3***
2,4-DINITRO-1-CHLORO-NAPHTHALENE
mf: $C_{10}H_5ClN_2O_4$ mw: 252.62

SYN: 1-CHLORO-2,4-DINITRONAPHTHALENE

TOXICITY DATA with REFERENCE
orl-rat TDLo:5000 mg/kg:CAR CNREA8 26,619,66
orl-rat TD:5000 mg/kg:NEO CNREA8 28,924,68
unr-mam LD50:250 mg/kg 30ZDA9 -,81,71

CONSENSUS REPORTS: Reported in EPA TSCA Inventory.

SAFETY PROFILE: Poison by unspecified route. Questionable carcinogen with experimental carcinogenic and neoplastigenic data. When heated to decomposition it emits very toxic fumes of Cl^- and NO_x. See also 2,4-DINITROANILINE.

DUS700 CAS:534-52-1 ***HR: 3***
DINITRO-o-CRESOL
mf: $C_7H_6N_2O_5$ mw: 198.15

PROP: Yellow, prismatic crystals. Mp: 85.8°, vap d: 6.82.

SYNS: ANTINONIN ◇ ARBOROL ◇ CAPSINE ◇ CHEMSECT DNOC ◇ DEGRASSAN ◇ DEKRYSIL ◇ DETAL ◇ DINITROCRESOL ◇ 2,4-DINITRO-o-CRESOL ◇ 4,6-DINITRO-o-CRESOL ◇ 4,6-DINITRO-o-CRESOLO (ITALIAN) ◇ DINITRODENDTROXAL ◇ 3,5-DINITRO-2-HYDROXYTOLUENE ◇ 4,6-DINITRO-o-KRESOL (CZECH) ◇ 4,6-DINITROKRESOL (DUTCH) ◇ DINITROL ◇ DINITROMETHYL CYCLOHEXYLTRIENOL ◇ 2,4-DINITRO-6-METHYLPHENOL ◇ DINOC ◇ DINURANIA ◇ DITROSOL ◇ DN-DRY MIX No.2 ◇ DNOK (CZECH) ◇ DWUNITRO-o-KREZOL (POLISH) ◇ EFFUSAN ◇ ELGETOL ◇ ELIPOL ◇ ENT 154 ◇ EXTRAR ◇ HEDOLIT ◇ K III ◇ KRENITE (OBS.) ◇ KRESAMONE ◇ KREZOTOL 50 ◇ LE DINITROCRESOL-4,6 (FRENCH) ◇ LIPAN ◇ 2-METHYL-4,6-DINITROPHENOL ◇ NITRADOR ◇ NITROFAN ◇ PROKARBOL ◇ RAFEX ◇ RAPHATOX ◇ RCRA WASTE NUMBER P047 ◇ SANDOLIN ◇ SELINON ◇ SINOX ◇ TRIFOCIDE ◇ TRIFRINA ◇ WINTERWASH ◇ ZAHLREICHE BEZEICHNUNGEN (GERMAN)

TOXICITY DATA with REFERENCE
skn-rbt 105 mg/9D-I MLD JIHTAB 30,10,48
eye-rbt 20 mg/24H MOD 28ZPAK -,107,72
mma-sat 1 μmol/plate AIDZAC 10,305,82
sln-dmg-orl 250 μmol/L ARTODN Suppl. 4,59,80
orl-man TDLo:7500 μg/kg/7D:CNS CMEP** -,1,56
ihl-hmn TCLo:1 mg/m³:BRN,CVS,GIT HYSAAV 30,197,65
unr-man LDLo:29 mg/kg 85DCAI 2,73,70
orl-rat LD50:10 mg/kg 85ARAE 3,54,76
skn-rat LD50:200 mg/kg WRPCA2 9,119,70
ipr-rat LDLo:28 mg/kg TXAPA9 1,156,59
scu-rat LD50:25600 μg/kg JPPMAB 4,1062,52
orl-mus LD50:47 mg/kg HYSAAV 30,197,65
ipr-mus LD50:19 mg/kg BCPCA6 18,1389,69
ivn-dog LDLo:15 mg/kg AIPTAK 50,20,35
ihl-cat LCLo:40 mg/m³ HYSAAV 30,197,65

CONSENSUS REPORTS: Reported in EPA TSCA Inventory. EPA Genetic Toxicology Program. Community Right-To-Know List. EPA Extremely Hazardous Substances List.

OSHA PEL: TWA 0.2 mg/m³ (skin)
ACGIH TLV: TWA 0.2 mg/m³ (skin)
DFG MAK: 0.2 mg/m³)
NIOSH REL: (Dinitro-Ortho-Cresol) TWA 0.2 mg/m³

SAFETY PROFILE: Human poison by unspecified route. Experimental poison by ingestion, inhalation, skin contact, intraperitoneal, and intravenous routes. Human systemic effects by ingestion and inhalation: somnolence, headache, brain recordings from specific areas of the central nervous system, cardiac and gastrointestinal changes. Mutation data reported. An eye and skin irritant. Less toxic than the para form, but is still highly toxic. A pesticide. See also NITRO COMPOUNDS of AROMATIC HYDROCARBONS and other dinitro cresol entries.

DUT000 CAS:497-56-3 ***HR: 3***
3,5-DINITRO-o-CRESOL
mf: $C_7H_6N_2O_5$ mw: 198.15

TOXICITY DATA with REFERENCE
scu-rat LDLo:40 mg/kg XPHBAO 271,146,41

NIOSH REL: (Dinitro-Ortho-Cresol) TWA 0.2 mg/m^3

SAFETY PROFILE: Poison by subcutaneous route. When heated to decomposition it emits toxic fumes of NO_x. See also NITRO COMPOUNDS of AROMATIC HYDROCARBONS and other dinitrocresol entries.

DUT200 CAS:63989-82-2 ***HR: 3***
3,5-DINITRO-p-CRESOL
mf: $C_7H_6N_2O_5$ mw: 198.15

PROP: Crystals.

TOXICITY DATA with REFERENCE
ipr-pgn LDLo:20 mg/kg AIPTAK 50,20,35

SAFETY PROFILE: Poison by intraperitoneal route. Strong irritant to eyes, skin, and mucous membranes. Can cause brain, liver, and kidney damage by various routes. When heated to decomposition it emits toxic fumes of NO_x. See also 4,6-DINITRO-o-CRESOL.

DUT600 CAS:609-93-8 ***HR: 3***
2,6-DINITRO-p-CRESOL
mf: $C_7H_6N_2O_5$ mw: 198.15

SYNS: DINITRO-p-CRESOL ◇ DNPC ◇ VICTORIA ORANGE ◇ VICTORIA YELLOW

TOXICITY DATA with REFERENCE
mmo-sat 1 nmol/plate MUREAV 58,1,78
mma-esc 100 mg/L BJCAAI 37(Suppl 3),124,78
ipr-mus LD50:24.8 mg/kg JPPMAB 5,497,53

CONSENSUS REPORTS: Reported in EPA TSCA Inventory.

SAFETY PROFILE: Poison by intraperitoneal route. Mutation data reported. When heated to decomposition it emits toxic fumes of NO_x. See also other dinitrocresol entries.

DUT800 CAS:2980-64-5 ***HR: 3***
4,6-DINITRO-o-CRESOL AMMONIUM SALT
mf: $C_7H_6N_2O_5 \cdot H_3N$ mw: 215.19

SYNS: AMMONIUM DNOC ◇ DINOZOL ◇ DINOZOL 50 ◇ DNOC AMMONIUM SALT ◇ ERBITOX ◇ KRESONIT E ◇ KREZAMON ◇ KREZONIT E ◇ 2-METHYL-4,6-DINITROPHENOL, AMMONIUM SALT ◇ SUPERELGETOL

TOXICITY DATA with REFERENCE
cyt-hmn:leu 20 μg/L EESADV 2,243,78
cyt-mus-orl 20 mg/kg EESADV 5,38,81
cyt-mus-ipr 10 mg/kg PHABDI 18,77,78
dlt-mus-ipr 10 mg/kg EESADV 2,401,78
spm-mus-ipr 10 mg/kg EESADV 2,243,78
orl-mus TDLo:20 mg/kg (9-12D preg):TER EESADV 5,38,81
orl-rat LDLo:50 mg/kg JPPMAB 4,1062,52
scu-rat LD50:27500 μg/kg JPPMAB 4,1062,52

SAFETY PROFILE: Poison by ingestion and subcutaneous routes. An experimental teratogen. Human mutation data reported. An herbicide. When heated to decomposition it emits toxic fumes of NO_x and NH_3. See also other dinitrocresol entries.

DUU000 ***HR: 3***
4,6-DINITRO-o-CRESOL DIETHYLAMINE SALT
mf: $C_7H_6N_2O_4 \cdot C_4H_{11}N$ mw: 255.31

TOXICITY DATA with REFERENCE
orl-rat LDLo:50 mg/kg JPPMAB 4,1062,52
scu-rat LD50:36500 μg/kg JPPMAB 4,1062,52

SAFETY PROFILE: Poison by ingestion and subcutaneous routes. When heated to decomposition it emits toxic fumes of NO_x. See also other dinitrocresol entries.

DUU200 CAS:63989-84-4 ***HR: 3***
4,6-DINITRO-o-CRESOL METHYLAMINE (1:1)

TOXICITY DATA with REFERENCE
ipr-mus LDLo:31 mg/kg CBCCT* 6,146,54

NIOSH REL: (Dinitro-Ortho-Cresol) TWA 0.2 mg/m^3

SAFETY PROFILE: Poison by intraperitoneal route. When heated to decomposition it emits toxic fumes of NO_x. See also NITRO COMPOUNDS of AROMATIC HYDROCARBONS and other dinitrocresol entries.

DUU400 CAS:63989-85-5 ***HR: 3***
4,6-DINITRO-o-CRESOL MORPHOLINE (1:1)
mf: $C_7H_6N_2O_5 \cdot C_4H_9NO$ mw: 285.29

TOXICITY DATA with REFERENCE
ipr-mus LDLo:25 mg/kg CBCCT* 6,146,54

NIOSH REL: (Dinitro-Ortho-Cresol) TWA 0.2 mg/m^3

SAFETY PROFILE: Poison by intraperitoneal route.

When heated to decomposition it emits toxic fumes of NO_x. See also NITRO COMPOUNDS of AROMATIC HYDROCARBONS and other dinitrocresol entries.

DUU600 CAS:2312-76-7 ***HR: 3***
4,6-DINITRO-o-CRESOL SODIUM SALT
mf: $C_7H_5N_2O_5$•Na mw: 220.13

PROP: Brilliant, orange-yellow dye.

SYNS: CORODINOC ◇ CRESOTOL ◇ DINITRO-o-CRESOL SODIUM SALT ◇ 3,5-DINITRO-o-CRESOL SODIUM SALT ◇ 2,4-DINITRO-6-METHYLPHENOL SODIUM SALT ◇ DINOC ◇ DNOC SOLDIUM SALT ◇ DYNOSOL ◇ EK 54 ◇ ELGETOL ◇ KRENITE (OBS.) ◇ KREZONITE ◇ 2-METHYL-4,6-DINITROPHENOL SODIUM SALT ◇ SINOX ◇ SODIUM-4,6-DINITRO-o-CRESOXIDE ◇ SODIUM SALT of 4,6-DINITRO-o-CRESOL

TOXICITY DATA with REFERENCE
orl-rat LD50:26 mg/kg SPEADM 74-1,-,74
skn-rat LD50:200 mg/kg SPEADM 74-1,-,74
scu-rat LDLo:20 mg/kg JPETAB 76,245,42
orl-dom LD50:200 mg/kg 85GYAZ -,75,71
orl-mam LD50:200 mg/kg GUCHAZ 6,243,73

CONSENSUS REPORTS: Reported in EPA TSCA Inventory.

NIOSH REL: (Dinitro-Ortho-Cresol) TWA 0.2 mg/m^3

SAFETY PROFILE: Poison by ingestion, skin contact, and subcutaneous routes. Flammable. A pesticide. When heated to decomposition it emits toxic fumes of Na_2O. See also other dinitrocresol entries.

DUU800 CAS:505-71-5 ***HR: 3***
N,N'-DINITRO-1,2-DIAMINOETHANE
mf: $C_2H_6N_4O_4$ mw: 150.10

$(O_2NNHCH_2\text{-})_2$

SAFETY PROFILE: A relatively insensitive explosive. Decomposes violently at 202°C. Forms very impact-sensitive lead and silver salts. When heated to decomposition it emits toxic fumes of NO_x. See also NITRO COMPOUNDS.

DUV000 CAS:25240-93-1 ***HR: 3***
DINITRODIAZOMETHANE
mf: CN_4O_4 mw: 132.03

SAFETY PROFILE: Explodes on impact, rapid heating, or on contact with sulfuric acid. Upon decomposition it emits toxic fumes of NO_x. See also AZIDES.

DUV089 ***HR: 3***
5,6-DINITRO-2-DIMETHYLAMINOPYRIMIDINONE
mf: C6H7N_5O_5 mw: 229.15

PROP: Melting point 190°C.

SAFETY PROFILE: Decomposes violently when heated to 190°C. When heated to decomposition it emits toxic fumes of NO_x. See also NITRO COMPOUNDS of AROMATIC HYDROCARBONS.

DUV400 CAS:52129-71-2 ***HR: 2***
3',5'-DINITRO-4'-(DI-n-PROPYLAMINO)ACETOPHENONE
mf: $C_{14}H_{19}N_3O_5$ mw: 309.36

SYN: BUBAN 37

TOXICITY DATA with REFERENCE
orl-rat LD50:750 mg/kg FMCHA2 -,C38,83

CONSENSUS REPORTS: Reported in EPA TSCA Inventory.

SAFETY PROFILE: Moderately toxic by ingestion. When heated to decomposition it emits toxic fumes of NO_x.

DUV600 CAS:1582-09-8 ***HR: 3***
2,6-DINITRO-N,N-DIPROPYL-4-(TRIFLUOROMETHYL)BENZENAMINE
mf: $C_{13}H_{16}F_3N_3O_4$ mw: 335.32

PROP: Technical product contains 84-88 ppm dipropylnitrosoamine NCITR* NCI-CG-TR-34,78.

SYNS: AGREFLAN ◇ AGRIFLAN 24 ◇ CRISALIN ◇ DIGERMIN ◇ 2,6-DINITRO-N,N-DI-N-PROPYL-α,α,α-TRIFLURO-p-TOLUIDINE ◇ 2,6-DINITRO-4-TRIFLUORMETHYL-N,N-DIPROPYLANILIN (GERMAN) ◇ 4-(DI-N-PROPYLAMINO)-3,5-DINITRO-1-TRIFLUOROMETHYLBENZENE ◇ N,N-DI-N-PROPYL-2,6-DINITRO-4-TRIFLUOROMETHYLANILINE ◇ N,N-DIPROPYL-4-TRIFLUOROMETHYL-2,6-DINITROANILINE ◇ ELANCOLAN ◇ L-36352 ◇ LILLY 36,352 ◇ NCI-C00442 ◇ NITRAN ◇ OLITREF ◇ SU SEGURO CARPIDOR ◇ TREFANOCIDE ◇ TREFICON ◇ TREFLAM ◇ TREFLAN ◇ TREFLANOCIDE ELANCOLAN ◇ TRIFLUORALIN (USDA) ◇ TRIFLURALIN ◇ TRIFLURALINE ◇ α,α,α-TRIFLUORO-2,6-DINITRO-N,N-DIPROPYL-p-TOLUIDINE ◇ TRIFUREX ◇ TRIKEPIN ◇ TRIM

TOXICITY DATA with REFERENCE
mma-sat 1 mg/plate ENMUDM 8(Suppl 7),1,86
mrc-asn 100 μg/plate AISSAW 18,123,82
cyt-hmn:lym 2 ppm PATHAB 73,707,81
sce-hmn:lym 1 mg/L BSIBAC 60,2149,84
cyt-mus-ipr 200 mg/kg EESADV 4,263,80
orl-mus TDLo:10 mg/kg (6-15D preg):REP TJADAB 15,15A,77
ipr-mus TDLo:200 mg/kg (1D male):TER EESADV 4,263,80
orl-mus TDLo:180 g/kg/78W-C:CAR NCITR* NCI-CG-TR-34,78
ipr-mus TDLo:2600 μg/kg/39D-I:ETA PATHAB 73,707,81
orl-mus TD:340 g/kg/78W-C:CAR NCITR* NCI-CG-TR-34,78
orl-mus LD50:5000 mg/kg GUCHAZ 6,524,73

ipr-mus LDLo:1500 mg/kg BECTA6 20,554,78
orl-mam LD50:3700 mg/kg PCOC** -,1188,66

CONSENSUS REPORTS: NCI Carcinogenesis Bioassay (feed); Clear Evidence: mouse NCITR* NCI-CG-TR-34,78; No Evidence: rat NCITR* NCI-CG-TR-34,78. EPA Genetic Toxicology Program. Community Right-To-Know List.

SAFETY PROFILE: Moderately toxic by ingestion and intraperitoneal routes. Experimental teratogenic and reproductive effects. Questionable carcinogen with experimental carcinogenic and tumorigenic data. Human mutation data reported. When heated to decomposition it emits very toxic fumes of F^- and NO_x. See also FLUORIDES and DIPROPYLNITROSO AMINE.

DUV800 CAS:505-71-5 ***HR: 2***
N,N'-DINITROETHYLENEDIAMINE

SYNS: ETHYLENEDINITRAMINE ◇ ETHYLENEDINITROAMINE

TOXICITY DATA with REFERENCE
ipr-mus LD50:540 mg/kg PCJOAU 10,1504,76

CONSENSUS REPORTS: Reported in EPA-TSCA Inventory.

SAFETY PROFILE: Moderately toxic by intraperitoneal route. When heated to decomposition it emits toxic fumes of NO_x.

DUW100 CAS:105735-71-5 ***HR: 2***
3,7-DINITROFLUORANTHENE
mf: $C_{16}H_8N_2O_4$ mw: 292.26

TOXICITY DATA with REFERENCE
mmo-sat 250 pg/plate MUREAV 191,85,87
mma-sat 1 μg/plate MUREAV 191,85,87
dnr-bcs 10 ng/disc MUREAV 191,85,87
scu-rat TDLo:5 mg/kg/10W-I:CAR CRNGDP 8,1919,87

SAFETY PROFILE: Questionable carcinogen with experimental carcinogenic data. Mutation data reported. When heated to decomposition it emits toxic fumes of NO_x.

DUW120 CAS:22506-53-2 ***HR: 2***
3,9-DINITROFLUORANTHENE
mf: $C_{16}H_8N_2O_4$ mw: 292.26

SYN: 4,12-DINITROFLUORANTHENE

TOXICITY DATA with REFERENCE
mmo-sat 250 pg/plate MUREAV 191,85,87
mma-sat 1 μg/plate MUREAV 191,85,87
dnr-bcs 10 ng/disc MUREAV 191,85,87
scu-rat TDLo:5 mg/kg/10W-I:CAR CRNGDP 8,1919,87

SAFETY PROFILE: Questionable carcinogen with experimental carcinogenic data. Mutation data reported. When heated to decomposition it emits toxic fumes of NO_x.

DUW200 CAS:5405-53-8 ***HR: 3***
2,7-DINITROFLUORENE
mf: $C_{13}H_8N_2O_4$ mw: 256.23

SAFETY PROFILE: A very shock-sensitive explosive which also explodes above its melting point of 152°C. Upon decomposition it emits toxic fumes of NO_x. See also NITRO COMPOUNDS of AROMATIC HYDROCARBONS.

DUW400 CAS:70-34-8 ***HR: 3***
2,4-DINITRO-1-FLUOROBENZENE
mf: $C_6H_3FN_2O_4$ mw: 186.11

PROP: Crystals; sol in ether, benzene, propylene glycol. Mp: 26°, bp: 137° @ 20 mm.

SYNS: 2,4-DINITROFLUOROBENZENE ◇ 2,4-DNFB ◇ 1,2,4-FLUORODINITROBENZENE ◇ 1-FLUORO-2,4-DINITROBENZENE

TOXICITY DATA with REFERENCE
mmo-sat 500 nmol/L ENMUDM 3,11,81
mma-sat 33 μg/plate ENMUDM 5(Suppl 1),3,83
mmo-esc 5 μmol/L CRNGDP 3,139,82
mrc-smc 320 μmol/L MGGEAE 168,125,79
otr-ham:kdy 80 μg/L BJCAAI 37,873,78
orl-rat LDLo:50 mg/kg NCNSA6 5,17,53
skn-mus LDLo:100 mg/kg CNREA8 29,179,69
scu-mus LDLo:100 mg/kg BIJOAK 41,558,47

CONSENSUS REPORTS: Reported in EPA TSCA Inventory. EPA Genetic Toxicology Program.

SAFETY PROFILE: Poison by ingestion, skin contact, and subcutaneous routes. A powerful irritant and vesicant. Mutation data reported. Solutions in ether may explode when evaporated. When heated to decomposition it emits highly toxic fumes of NO_x and F^-. See also NITRO COMPOUNDS of AROMATIC HYDROCARBONS and FLUORIDES.

DUX509 ***HR: 3***
3,5-DINITRO-2-METHYLBENZENEDIZAONIUM-4-OXIDE
mf: $C_7H_4N_4O_5$ mw: 224.13

SAFETY PROFILE: A very shock-sensitive explosive. When heated to decomposition it emits toxic fumes of NO_x. See also EXPLOSIVES, HIGH.

DUX560 CAS:70343-15-6 ***HR: 3***
2,5-DINITRO-3-METHYLBENZOIC ACID
mf: $C_8H_6N_2O_6$ mw: 226.15

SAFETY PROFILE: Mixtures with oleum + sodium

azide are potentially explosive. When heated to decomposition it emits toxic fumes of NO_x. See also NITRO COMPOUNDS of AROMATIC HYDROCARBONS.

DUX600 CAS:10308-90-4 ***HR: 3***
N,N'-DINITRO-N-METHYL-1,2-DIAMINOETHANE
mf: $C_3H_8N_4O_4$ mw: 164.14

$CH_3N(NO_2)C_2H_4NHNO_2$

SAFETY PROFILE: Decomposes violently at 210°C. When heated to decomposition it emits toxic fumes of NO_x.

DUX700 CAS:605-71-0 ***HR: 3***
1,5-DINITRONAPHTHALENE
mf: $C_{10}H_6N_2O_4$ mw: 218.17

DFG MAK: Suspected Carcinogen. (all isomers)

SAFETY PROFILE: Mixtures with sulfur or sulfuric acid (used in commercial reactions) may explode if heated to 120°C. Initiation temperature depends on the quality of the dinitronaphthalene. When heated to decomposition it emits toxic fumes of NO_x. See also NITRO COMPOUNDS of AROMATIC HYDROCARBONS.

DUX800 CAS:605-69-6 ***HR: 3***
2,4-DINITRO-1-NAPHTHOL
mf: $C_{10}H_6N_2O_5$ mw: 234.18

PROP: Yellow needles or leaflets. Mp: 138°, vap d: 8.08.

SYNS: 2-4 DINITRO-α-NAPHTOL (FRENCH) ◇ GOLDEN YELLOW ◇ MANCHESTER YELLOW ◇ MARITUS YELLOW ◇ NAPHTHOL YELLOW ◇ NAPHTHYLENE YELLOW ◇ SAFFRON YELLOW

TOXICITY DATA with REFERENCE
skn-hmn TDLo:50 mg/kg:SKN XPHBAO 271,187,41
ivn-mus LD50:180 mg/kg CSLNX* NX#03278
ivn-dog LDLo:13300 μg/kg AIPTAK 35,63,28
scu-gpg LDLo:80 mg/kg HBTXAC 1,118,56
ipr-pgn LDLo:15 mg/kg HBTXAC 1,118,56
ivn-pgn LDLo:15 mg/kg AIPTAK 50,20,35
ims-pgn LD50:1850 μg/kg HBTXAC 1,118,56
scu-frg LDLo:60 mg/kg HBTXAC 1,118,56

CONSENSUS REPORTS: Reported in EPA TSCA Inventory.

SAFETY PROFILE: Poison by subcutaneous, intramuscular, intravenous, and intraperitoneal routes. Human reproductive effects by skin contact: toxic to the skin. For fire, disaster, and explosion hazards, see NITRATES.

DUY200 ***HR: 3***
2,6-DINITRO-4-PERCHLORYLPHENOL
mf: $C_6H_3ClN_2O_8$ mw: 266.56

SAFETY PROFILE: A very shock-sensitive explosive. An analog of picric acid. When heated to decomposition it emits toxic fumes of Cl^- and NO_x. See also PICRIC ACID and PERCHLORATES.

DUY400 CAS:610-54-8 ***HR: 3***
2,4-DINITROPHENETOLE
mf: $C_8H_8N_2O_5$ mw: 212.18

PROP: Crystals. Vap d: 7.32.

TOXICITY DATA with REFERENCE
mmo-sat 500 nmol/L ENMUDM 3,11,81
orl-rat LDLo:250 mg/kg NCNSA6 5,16,53

CONSENSUS REPORTS: Reported in EPA TSCA Inventory.

SAFETY PROFILE: Poison by ingestion. Mutation data reported. When heated to decomposition it emits toxic fumes of NO_x.

DUY600 CAS:25550-58-7 ***HR: 3***
DINITROPHENOL
DOT: UN 0076/UN 1320/UN 1599
mf: $C_6H_4N_2O_5$ mw: 184.12

TOXICITY DATA with REFERENCE
orl-rat LDLo:30 mg/kg 28ZEAL 4,198,69
orl-dog LDLo:30 mg/kg JPETAB 49,187,33
scu-rbt LDLo:30 mg/kg JPETAB 49,187,33

DOT Classification: Poison B; Label: Poison (UN1599); Class A Explosive; Label: EXPLOSIVE A and Poison (UN0076); Poison B; Label: Flammable Liquid and Poison (UN1599); Flammable Solid; Label: Flammable Solid and Poison (UN1320).

SAFETY PROFILE: Poison by ingestion and subcutaneous routes. An explosive. When heated to decomposition it emits toxic fumes of NO_x. See also NITRO COMPOUNDS of AROMATIC HYDROCARBONS.

DUY810 CAS:25550-58-7 ***HR: 3***
DINITROPHENOL (solution)

SYN: DINITROPHENOL, solution (DOT)

DOT Classification: Poison B; Label: Poison.

SAFETY PROFILE: A poison. When heated to decomposition it emits toxic fumes of NO_x. See also 2,4-DINITROPHENOL.

DUY800 CAS:66-56-8 ***HR: 3***
2,3-DINITROPHENOL
mf: $C_6H_4N_2O_5$ mw: 184.12

PROP: Yellow needles. Mp: 144°, d: 1.681 @ 20°, vap d: 6.35.

TOXICITY DATA with REFERENCE
unr-rat LD50:190 mg/kg JPPMAB 11,462,59
unr-mus LD50:200 mg/kg JPPMAB 11,462,59
unr-dog LDLo:1000 mg/kg JPPMAB 11,462,59

SAFETY PROFILE: Poison by unspecified route. Inhalation of dust can be fatal. A skin irritant and an allergen. A powerful stimulant of the metabolism by excessive oxidation. For fire hazard, see NITRATES. Highly explosive when exposed to heat. It is used as a component of some shell and bomb charges. See also NITRO COMPOUNDS of AROMATIC HYDROCARBONS; and EXPLOSIVES, HIGH.

DUZ000 CAS:51-28-5 ***HR: 3***
2,4-DINITROPHENOL
mf: $C_6H_4N_2O_5$ mw: 184.12

PROP: Yellow crystals. Mp: 112°, d: 1.683 @ 24°, vap d: 6.35.

SYNS: ALDIFEN ◇ CHEMOX PE ◇ 2,4-DINITROFENOL (DUTCH) ◇ DINITROFENOLO (ITALIAN) ◇ α-DINITROPHENOL ◇ 2,4-DNP ◇ FENOXYL CARBON N ◇ 1-HYDROXY-2,4-DINITROBENZENE ◇ MAROXOL-50 ◇ NITRO KLEENUP ◇ NSC 1532 ◇ RCRA WASTE NUMBER P048 ◇ SOLFO BLACK B ◇ SOLFO BLACK BB ◇ SOLFO BLACK 2B SUPRA ◇ SOLFO BLACK G ◇ SOLFO BLACK SB ◇ TERTROSULPHUR BLACK PB ◇ TERTROSULPHUR PBR

TOXICITY DATA with REFERENCE
skn-rbt 300 mg/4W-I MLD JIHTAB 30,10,48
oms-ofs:oth 100 μmol/L AEEXAH (3),279,72
cyt-mus-ipr 10 g/kg IJMRAQ 59,1442,71
orl-rat TDLo:2040 mg/kg (8D pre-21D post):REP PSEBAA 32,678,35
ipr-mus TDLo:40800 μg/kg (10-12D preg):TER FCTXAV 11,31,73
orl-hmn LDLo:4300 μg/kg JAMAAP 101,1333,33
orl-rat LD50:30 mg/kg TXAPA9 21,315,72
ipr-rat LD50:20 mg/kg JPPMAB 17,814,65
scu-rat LD50:25 mg/kg JPETAB 49,187,33
orl-mus LD50:45 mg/kg FATOAO 28,493,65
ipr-mus LD50:26 mg/kg BCPCA6 18,1389,69
orl-dog LDLo:30 mg/kg JPETAB 49,187,33
ihl-dog LCLo:300 mg/m^3/30M 85GMAT -,62,82
scu-gpg LDLo:25 mg/kg AEPPAE 192,331,39

CONSENSUS REPORTS: Reported in EPA TSCA Inventory. EPA Genetic Toxicology Program.

SAFETY PROFILE: A deadly human poison by ingestion. An experimental poison by ingestion, inhalation, intravenous, intraperitoneal, subcutaneous, and intramuscular routes. Moderately toxic by skin contact. Experimental teratogenic and reproductive effects. A skin irritant. Mutation data reported. Phytotoxic. A pesticide. An explosive. Forms explosive salts with alkalies and ammonia. When heated to decomposition it emits toxic fumes of NO_x. See also NITRO COMPOUNDS of AROMATIC HYDROCARBONS.

DVA000 CAS:329-71-5 ***HR: 3***
2,5-DINITROPHENOL
mf: $C_6H_4N_2O_5$ mw: 184.12

PROP: Yellow crystals, mp: 108°. Sltly sol in cold water; sltly sol in alc but sol in hot alc, ether, alkali hydroxides.

SYNS: Γ-DINITROPHENOL ◇ 2,5-DNP

TOXICITY DATA with REFERENCE
unk-rat LD50:150 mg/kg JPPMAB 11,462,59
unk-mus LD50:273 mg/kg JPPMAB 11,462,59
unk-dog LDLo:100 mg/kg JPPMAB 11,462,59

SAFETY PROFILE: A poison by an unspecified route. When heated to decomposition it emits toxic fumes of NO_x. See also 2,4-DINITROPHENOL.

DVA200 CAS:573-56-8 ***HR: 3***
2,6-DINITROPHENOL
mf: $C_6H_4N_2O_5$ mw: 184.12

PROP: Yellow crystals. Mp: 63°, vap d: 6.35. Sltly sol in cold water, alc. Very sol in chloroform, ether or boiling alc; also sol in fixed alkali solns.

SYN: β-DINITROPHENOL

TOXICITY DATA with REFERENCE
ims-pgn LDLo:40 mg/kg JPETAB 49,187,33

SAFETY PROFILE: Poison by intramuscular route. Moderately explosive when exposed to heat. See also 2,4-DINITROPHENOL.

DVA400 CAS:577-71-9 ***HR: 3***
3,4-DINITROPHENOL
mf: $C_6H_4N_2O_5$ mw: 184.12

TOXICITY DATA with REFERENCE
unk-rat LD50:98 mg/kg JPPMAB 11,462,59
unk-mus LD50:112 mg/kg JPPMAB 11,462,59
unk-dog LDLo:500 mg/kg JPPMAB 11,462,59

SAFETY PROFILE: A poison by unspecified routes. When heated to decomposition it emits toxic fumes of NO_x. See also 2,4-DINITROPHENOL.

DVA600 CAS:586-11-8 ***HR: 3***
3,5-DINITROPHENOL
mf: $C_6H_4N_2O_5$ mw: 184.12

TOXICITY DATA with REFERENCE
unk-rat LD50:45 mg/kg JPPMAB 11,462,59
unk-mus LD50:50 mg/kg JPPMAB 11,462,59
unk-dog LDLo:500 mg/kg JPPMAB 11,462,59

SAFETY PROFILE: A poison by unspecified routes. When heated to decomposition it emits toxic fumes of NO_x. See also 2,4-DINITROPHENOL.

DVA800 CAS:1011-73-0 ***HR: 3***
2,4-DINITROPHENOL SODIUM SALT
mf: $C_6H_3N_2O_5$•Na mw: 206.10

SYN: SODIUM-2,4-DINITROPHENOL ◇ SODIUM-2,4-DINITROPHENOLATE ◇ SODIUM DNP

TOXICITY DATA with REFERENCE
scu-rat TDLo:8 mg/kg (female 9D post):TER AEHLAU 8,648,64
scu-rat TDLo:30 mg/kg (female 10D post):REP AEHLAU 8,648,64
scu-rat LDLo:10 mg/kg JPETAB 48,410,33
scu-mus LD50:50 mg/kg NYKZAU 56,23,60
scu-dog LDLo:25 mg/kg JPETAB 48,410,33
ivn-dog LDLo:20 mg/kg AIPTAK 50,20,35
ivn-pgn LDLo:15 mg/kg AIPTAK 50,20,35

SAFETY PROFILE: Poison by subcutaneous and intravenous route. An experimental teratogen. Other experimental reproductive effects. When heated to decomposition it emits toxic fumes of NO_x and Na_2O. See also 2,4-DINITROPHENOL.

DVB200 ***HR: 2***
2,4-DINITROPHENYLACETYL CHLORIDE
mf: $C_8H_5ClN_2O_5$ mw: 244.60

SAFETY PROFILE: Potentially explosive when heated. When heated to decomposition it emits toxic fumes of Cl^- and NO_x. See also NITRO COMPOUNDS of AROMATIC HYDROCARBONS.

DVB800 CAS:3468-63-1 ***HR: D***
1-((2,4-DINITROPHENYL)AZO)-2-NAPHTHOL
mf: $C_{16}H_{10}N_4O_5$ mw: 338.30

SYNS: BRILLIANT TANGERINE 13030 ◇ CALCOTONE ORANGE 2R ◇ CARNELIO RED 2G ◇ CHROMATEX ORANGE R ◇ C.I. 12075 ◇ C.I. PIGMENT ORANGE 5 ◇ DAINICHI PERMANENT RED GG ◇ D&C ORANGE No. 17 ◇ DINITRANILINE ORANGE ◇ DINITROANILINE ORANGE ND-204 ◇ DINITROANILINE RED ◇ FASTOAN RED 2G ◇ GRAPHTOL RED 2GL ◇ HANSA ORANGE RN ◇ HELIO FAST ORANGE RN ◇ IRGALITE FAST RED 2GL ◇ ISOL FAST RED 2G ◇ LAKE RED 2GL ◇ LIGHT ORANGE R ◇ LUTETIA FAST ORANGE R ◇ MONOLITE FAST ORANGE R ◇ NIPPON ORANGE X-881 ◇ ORALITH RED 2GL ◇ ORANGE No. 203 ◇ ORANGE PIGMENT X ◇ PERMANENT ORANGE ◇ PERMATONE ORANGE ◇ PIGMENT FAST ORANGE ◇ SEGNALE LIGHT ORANGE RNG ◇ SIGNAL ORANGE ORANGE Y-17 ◇ SILOPOL ORANGE R ◇ SYTON FAST RED 2G ◇ TERTROPIGMENT ORANGE LRN ◇ VERSAL ORANGE RNL

TOXICITY DATA with REFERENCE
mmo-sat 5 μg/plate ESKGA2 29,212,83
mma-sat 50 μg/plate MUREAV 66,181,79

CONSENSUS REPORTS: Reported in EPA TSCA Inventory. EPA Genetic Toxicology Program.

SAFETY PROFILE: Mutation data reported. When heated to decomposition it emits toxic fumes of NO_x.

DVC200 CAS:2600-55-7 ***HR: 3***
2,4-DINITROPHENYL-2,4-DINITRO-6-sec-BUTYLPHENYL CARBONATE
mf: $C_{17}H_{14}N_4O_{11}$ mw: 450.35

SYNS: B 377 ◇ CARBONIC ACID-2-sec-BUTYL-4,6-DINITROPHENYL-2,4-DINITROPHENYL ESTER (8CI) ◇ TRIBONATE

TOXICITY DATA with REFERENCE
orl-rat LD50:108 mg/kg WRPCA2 7,135,68

SAFETY PROFILE: Poison by ingestion. When heated to decomposition it emits toxic fumes of NO_x. See also NITRO COMPOUNDS of AROMATIC HYDROCARBONS.

DVC400 CAS:119-26-6 ***HR: 3***
2,4-DINITROPHENYLHYDRAZINE
mf: $C_6H_6N_4O_4$ mw: 198.16

PROP: Red, crystalline powder; sltly sol in water and alcohol. Mp: 200° (approx).

SYNS: 2,4-DINITROFENYLHYDRAZIN (CZECH) ◇ 2,4-DNPH

TOXICITY DATA with REFERENCE
eye-rbt 500 mg/24H MOD 28ZPAK -,132,72
mmo-sat 5 μmol/L ENMUDM 3,11,81
mma-sat 1 μmol/plate MUREAV 58,11,78
mmo-omi 6 mg/L MUREAV 173,233,86
dnd-mus-ipr 1900 μmol/kg CNREA8 41,1469,81
orl-rat LD50:654 mg/kg 28ZPAK -,132,72

CONSENSUS REPORTS: Reported in EPA TSCA Inventory. EPA Genetic Toxicology Program.

SAFETY PROFILE: Moderately toxic by ingestion. An eye irritant. Mutation data reported. When heated to decomposition it emits toxic fumes of NO_x. See also HYDRAZINE. A dangerous explosive.

DVC600 ***HR: 3***
2,4-DINITROPHENYLHYDRAZINIUMPERCHLORATE
mf: $C_6H_7ClN_4O_8$ mw: 219.48

SAFETY PROFILE: Explosive decomposition may occur during concentration by evaporation. When heated to decomposition it emits toxic fumes of Cl^- and NO_x. See also NITRO COMPOUNDS of AROMATIC HYDROCARBONS.

DVC700 CAS:17508-17-7 ***HR: 3***
o-(2,4-DINITROPHENYL)HYDROXYLAMINE
mf: $C_6H_5N_3O_5$ mw: 199.12

SAFETY PROFILE: Potentially explosive reaction with potassium hydride in THF solution. When heated to decomposition it emits toxic fumes of NO_x. See also NITRO COMPOUNDS of AROMATIC HYDROCARBONS.

DVC800 CAS:63732-56-9 ***HR: 3***
2,4-DINITROPHENYLMORPHINE HYDROCHLORIDE
mf: $C_{23}H_{21}N_3O_7 \cdot ClH$ mw: 487.93

SYN: 2,4-DINITROPHENYL ETHER of MORPHINE

TOXICITY DATA with REFERENCE
ipr-rat LDLo:40 mg/kg UCPHAQ 1,59,38
ipr-mus LDLo:300 mg/kg UCPHAQ 1,59,38
scu-mus LD50:700 mg/kg JPETAB 67,127,39
scu-rbt LDLo:100 mg/kg UCPHAQ 1,59,38
ivn-rbt LDLo:3 mg/kg UCPHAQ 1,59,38
par-frg LDLo:2000 mg/kg UCPHAQ 1,59,38

SAFETY PROFILE: Poison by intravenous, intraperitoneal, and subcutaneous routes. Moderately toxic by parenteral route. When heated to decomposition it emits very toxic fumes of HCl and NO_x. See also MORPHINE and ETHERS.

DVD000 CAS:2736-80-3 ***HR: 3***
2,2-DINITRO-1,3-PROPANEDIOL
mf: $C_3H_6N_2O_6$ mw: 166.11

TOXICITY DATA with REFERENCE
ipr-mus LD50:76 mg/kg KHFZAN 11(1),73,77

CONSENSUS REPORTS: Reported in EPA TSCA Inventory.

SAFETY PROFILE: Poison by intraperitoneal route. When heated to decomposition it emits toxic fumes of NO_x.

DVD200 CAS:918-52-5 ***HR: 3***
2,2-DINITROPROPANOL
mf: $C_3H_6N_2O_5$ mw: 150.11

SYNS: 2,2-DINITRO-1-PROPANOL ◇ DNPOH ◇ NPOH

TOXICITY DATA with REFERENCE
ipr-mus LD50:280 mg/kg KHFZAN 11(1),73,77

CONSENSUS REPORTS: Reported in EPA TSCA Inventory.

SAFETY PROFILE: Poison by intraperitoneal route. When heated to decomposition it emits toxic fumes of NO_x.

DVD400 CAS:75321-20-9 ***HR: 3***
1,3-DINITROPYRENE
mf: $C_{16}H_8N_2O_4$ mw: 292.26

SYN: DINITROPYRENE

TOXICITY DATA with REFERENCE
mmo-esc 80 ng/plate MUREAV 142,163,85
msc-ham:lng 2500 μg/L CRNGDP 3,917,82
msc-ham:ovr 500 μg/L MUREAV 119,387,83
ipr-rat TDLo:23 mg/kg/4W-I:CAR DTESD7 13,279,86
scu-rat TDLo:16 mg/kg/10W-I:CAR CRNGDP 5,583,84

CONSENSUS REPORTS: IARC Cancer Review: Group 3 IMEMDT 46,201,89; Animal Limited Evidence IMEMDT 46,201,89; Human No Adequate Data IMEMDT 46,201,89.

DFG MAK: Suspected Carcinogen.

SAFETY PROFILE: Suspected carcinogen with experimental carcinogenic data. Mutation data reported. When heated to decomposition it emits toxic fumes of NO_x.

DVD600 CAS:42397-64-8 ***HR: 3***
1,6-DINITROPYRENE
mf: $C_{16}H_8N_2O_4$ mw: 292.26

SYN: DINITROPYRENE

TOXICITY DATA with REFERENCE
mmo-sat 600 pg/plate JJIND8 73,1359,84
dns-hmn:oth 500 nmol/L TXAPA9 79,28,85
dns-hmn:lvr 80 nmol/L ENMUDM 5,488,83
ipr-rat TDLo:23 mg/kg/4W-I:CAR DTESD7 13,279,86
scu-rat TDLo:9206 μg/kg/8W-I:CAR DTESD7 13,279,86

CONSENSUS REPORTS: IARC Cancer Review: Group 2B IMEMDT 46,215,89; Animal Sufficient Evidence IMEMDT 46,215,89; Human No Adequate Data IMEMDT 46,215,89.

DFG MAK: Suspected Carcinogen.

SAFETY PROFILE: Suspected carcinogen with experimental carcinogenic data. Human mutation data reported. When heated to decomposition it emits toxic fumes of NO_x.

DVD800 CAS:42397-65-9 ***HR: 3***
1,8-DINITROPYRENE
mf: $C_{16}H_8N_2O_4$ mw: 292.26

SYN: DINITROPYRENE

TOXICITY DATA with REFERENCE
mmo-sat 1 nmol/plate SCIEAS 209,1039,80
mma-sat 1 μg/plate MUREAV 91,321,81
msc-hmn:lym 100 μg/L ENMUDM 5,457,83
msc-mus:lym 500 μg/L EVSRBT 25,397,82

ipr-rat TDLo:23 mg/kg/4W-I:CAR DTESD7 13,279,86
scu-rat TDLo:16 mg/kg/10W-I:CAR CRNGDP 5,583,84

CONSENSUS REPORTS: IARC Cancer Review: Group 2B IMEMDT 46,231,89; Animal Sufficient Evidence IMEMDT 46,231,89; Human No Adequate Data IMEMDT 46,231,89.

DFG MAK: Suspected Carcinogen.

SAFETY PROFILE: Suspected carcinogen with experimental carcinogenic data. Human mutation data reported. When heated to decomposition it emits toxic fumes of NO_x.

DVE000 CAS:1596-52-7 ***HR: 3***
4,6-DINITROQUINOLINE-1-OXIDE
mf: $C_9H_5N_3O_5$ mw: 235.17

TOXICITY DATA with REFERENCE
cyt-omi 170 μmol/L GANNA2 60,155,69
mmo-smc 100 mg/L IGSBAL 85,127,72
cyt-hmn:lvr 5260 nmol/L JNCIAM 47,367,71
scu-mus TDLo:560 mg/kg/I:ETA CPBTAL 17,544,69

SAFETY PROFILE: Questionable carcinogen with experimental carcinogenic data. Human mutation data reported. When heated to decomposition it emits toxic fumes of NO_x.

DVE200 CAS:13442-17-6 ***HR: 3***
4,7-DINITROQUINOLINE-1-OXIDE
mf: $C_9H_5N_3O_5$ mw: 235.17

TOXICITY DATA with REFERENCE
skn-mus TDLo:300 mg/kg/25W-I:NEO GANNA2 60,523,69

SAFETY PROFILE: Questionable carcinogen with experimental neoplastigenic data. When heated to decomposition it emits toxic fumes of NO_x.

DVE300 CAS:82038-92-4 ***HR: D***
DINITROSOCIMETIDINE
mf: $C_{10}H_{16}N_8O_2S$ mw: 312.40

TOXICITY DATA with REFERENCE
mmo-sat 200 μg/plate CRNGDP 2,261,81
cyt-ham:ovr 2600 nmol/L/2H CRNGDP 2,261,81

SAFETY PROFILE: Mutation data reported. Many N-nitroso compounds are carcinogens. When heated to decomposition it emits toxic fumes of SO_x and NO_x. See also N-NITROSO COMPOUNDS.

DVE400 CAS:13256-12-7 ***HR: 3***
N,N′-DINITROSO-N,N′-DIMETHYLETHYLENEDIAMINE
mf: $C_4H_{10}N_4O_2$ mw: 146.18

SYNS: DIMETHYL-DI-NITROSO-AETHYLENDIAMIN(GERMAN) ◇ DIMETHYLDINITROSOETHYLENEDIAMINE ◇ N,N′-DIMETHYL-N,N′-DINITROSOETHYLENEDIAMINE

TOXICITY DATA with REFERENCE
mmo-omi 5000 ppm/24H-C SOGEBZ 10,522,74
orl-rat TDLo:570 mg/kg/43W-C:ETA ARZNAD 19,1077,69
orl-rat LD50:150 mg/kg NATWAY 50,100,63

SAFETY PROFILE: Poison by ingestion. Questionable carcinogen with experimental tumorigenic data. Mutation data reported. Many N-nitroso compounds are carcinogens. When heated to decomposition it emits toxic fumes of NO_x. See also N-NITROSO COMPOUNDS.

DVE600 CAS:55557-00-1 ***HR: 3***
DINITROSOHOMOPIPERAZINE
mf: $C_5H_{10}N_4O_2$ mw: 158.19

SYN: HEXAHYDRO-1,4-DINITROSO-1H-1,4-DIAZEPINE

TOXICITY DATA with REFERENCE
mmo-sat 1 μg/plate MUREAV 51,319,78
mma-sat 1 μg/plate MUREAV 51,319,78
mma-esc 16700 μmol/L CNREA8 36,4099,76
msc-ham:ovr 5 μmol/L TCMUE9 1,129,84
orl-rat TDLo:65 mg/kg/2Y-I:CAR EESADV 6,513,82
orl-rat TD:105 mg/kg/30W-I:CAR EESADV 6,513,82
orl-rat TD:1530 mg/kg/31W-I:ETA CNREA8 35,1270,75

CONSENSUS REPORTS: EPA Genetic Toxicology Program.

SAFETY PROFILE: Questionable carcinogen with experimental carcinogenic and tumorigenic data. Mutation data reported. When heated to decomposition it emits toxic fumes of NO_x. See also N-NITROSO COMPOUNDS.

DVF000 CAS:15973-99-6 ***HR: 3***
DI(N-NITROSO)-PERHYDROPYRIMIDINE
mf: $C_4H_8N_4O_2$ mw: 144.16

TOXICITY DATA with REFERENCE
ipr-rat TDLo:13 mg/kg/66W-I:CAR JCROD7 105,191,83
ipr-rat TDLo:525 mg/kg/35W-I:NEO CALEDQ 6,57,79
ipr-rat LD50:300 mg/kg CALEDQ 6,57,79

SAFETY PROFILE: Poison by intraperitoneal route. Questionable carcinogen with experimental carcinogenic and neoplastigenic data. When heated to decomposition it emits toxic fumes of NO_x. See also N-NITROSO COMPOUNDS.

DVF200 CAS:140-79-4 ***HR: 3***
DINITROSOPIPERAZINE
mf: $C_4H_8N_4O_2$ mw: 144.16

PROP: White crystals. Mp: 158°, vap d: 4.97.

SYNS: DINITROSOPIPERAZIN (GERMAN) ◇ N,N'-DINITROSOPIPERAZINE ◇ 1,4-DINITROSOPIPERAZINE ◇ DNPZ ◇ NSC 339 ◇ USAF DO-36

TOXICITY DATA with REFERENCE
mma-smc 50 μmol/plate MUREAV 77,143,80
sce-hmn:lym 10 mmol/L TCMUE9 1,129,84
orl-mus TDLo:140 mg/kg (15-21D preg):REP CNREA8 40,2925,80
orl-rat TDLo:1040 mg/kg/1Y-I:CAR JNCIAM 41,985,68
scu-rat TDLo:1070 mg/kg/53W-I:ETA ZEKBAI 69,103,67
orl-mus TDLo:140 mg/kg (15-21D preg):CAR,TER CNREA8 40,2925,80
orl-mus TDLo:1568 mg/kg/28W-C:NEO JNCIAM 46,1029,71
orl-rat LD50:160 mg/kg ZEKBAI 69,103,67
scu-rat LD50:160 mg/kg ZEKBAI 69,103,67
ipr-mus LD50:100 mg/kg NTIS** AD277-689

CONSENSUS REPORTS: EPA Genetic Toxicology Program. Reported in EPA TSCA Inventory.

SAFETY PROFILE: Suspected carcinogen with experimental carcinogenic, neoplastigenic, tumorigenic, and teratogenic data. Poison by ingestion, subcutaneous, and intraperitoneal routes. Experimental reproductive effects. Human mutation data reported. When heated to decomposition it emits toxic fumes of NO_x. See also N-NITROSO COMPOUNDS.

DVF400 CAS:101-25-7 ***HR: 3***
3,7-DINITROSO-1,3,5,7-TETRAAZABICYCLO [3.3.1]NONANE
mf: $C_5H_{10}N_6O_2$ mw: 186.18

SYNS: ACETO DNPT 40 ◇ ACETO DNPT 80 ◇ ACETO DNPT 100 ◇ CHKHZ 18 ◇ DINITROSOPENTAMETHYLENETETRAMINE ◇ N,N-DINITROSOPENTAMETHYLENETETRAMINE ◇ N[1],N[3]-DINITROSOPENTAMETHYLENETETRAMINE ◇ 3,4-DI-N-NITROSOPENTAMETHYLENETETRAMINE ◇ 3,7-DI-N-NITROSOPENTAMETHYLENETETRAMINE ◇ DNPMT ◇ DNPT ◇ 1,5-METHYLENE-3,7-DINITROSO-1,3,5,7-TETRAAZACYCLOOCTAINE ◇ 1,5-METHYLENE-3,7-DINITROSO-1,3,5,7-TETRAAZACYCLOOCTANE ◇ POROFOR CHKHC-18 ◇ POROPHOR B ◇ UNICEL-ND ◇ UNICEL NDX ◇ VULCACEL B-40 ◇ VULCACEL BN

TOXICITY DATA with REFERENCE
mmo-sat 500 μg/plate PMRSDJ 1,302,81
pic-esc 100 mg/L PMRSDJ 1,224,81
dnd-bcs 2 mg/disc PMRSDJ 1,175,81
sce-ham:ovr 80 mg/L PMRSDJ 1,538,81
otr-ham:kdy 73500 μg/L PMRSDJ 1,626,81
orl-rat LD50:940 mg/kg MELAAD 58,22,67
ipr-rat LD50:220 mg/kg MELAAD 58,22,67
scu-rat LD50:220 mg/kg APACAB 28,209,65
ipr-mus LD50:130 mg/kg APACAB 28,209,65
scu-mus LD50:140 mg/kg APACAB 28,209,65
ivn-mus LD50:120 mg/kg APACAB 28,209,65
ivn-rbt LD50:130 mg/kg APACAB 28,209,65

CONSENSUS REPORTS: IARC Cancer Review: Group 3 IMEMDT 7,56,87; Animal No Evidence IMEMDT 11,241,76. Reported in EPA TSCA Inventory. EPA Genetic Toxicology Program.

SAFETY PROFILE: Poison by intravenous, intraperitoneal, and subcutaneous routes. Moderately toxic by ingestion. Questionable carcinogen. Mutation data reported. Can ignite when handled and burns very rapidly. Many N-nitroso compounds are carcinogens. A blowing agent. When heated to decomposition it emits toxic fumes of NO_x. See also N-NITROSO COMPOUNDS.

DVF600 CAS:128-42-7 ***HR: 2***
4,4'-DINITRO-2,2'-STILBENEDISULFONIC ACID
mf: $C_{14}H_{10}N_2O_{10}S_2$ mw: 430.38

PROP: Yellow paste or brownish crystals.

SYNS: DINITROSTILBENEDISULFONIC ACID ◇ KYSELINA-4,4'-DINITROSTILBEN-2,2'-DISULFONOVA (CZECH)

TOXICITY DATA with REFERENCE
skn-rbt 500 mg/24H MOD 28ZPAK -,194,72
eye-rbt 500 mg/24H SEV 28ZPAK -,194,72
orl-mus LD50:47 g/kg GISAAA 45(3),73,80
orl-rbt LD50:30 g/kg GISAAA 45(3),73,80
orl-gpg LD50:71 g/kg GISAAA 45(3),73,80
orl-rat LD50 12,600 mg/kg 28ZPAK -,194,72

CONSENSUS REPORTS: Reported in EPA TSCA Inventory.

SAFETY PROFILE: Mildly toxic by ingestion. A skin and severe eye irritant. Can react vigorously with reducing materials. When heated to decomposition it emits very toxic fumes of NO_x and SO_x.

DVF800 CAS:1594-56-5 ***HR: 3***
2,4-DINITRO-1-THIOCYANOBENZENE
mf: $C_7H_3N_3O_4S$ mw: 225.19

SYNS: 2,4-DINITROPHENYL THIOCYANATE ◇ 2,4-DINITRO-RHODANBENZOL (GERMAN) ◇ 2,4-DINITROTHIOCYANATOBENZENE ◇ 2,4-DINITROTHIOCYANOBENZENE ◇ DNRB ◇ DNTB ◇ DRB ◇ GRYZBOL ◇ GRZYBOL ◇ NBT ◇ NIRIT ◇ NITRITE ◇ RHODANDINITROBENZOL ◇ RODATOX 60 ◇ TRIRODAZEEN ◇ TRIRODAZENE ◇ 2317-W

TOXICITY DATA with REFERENCE
mma-sat 50 μg/plate MUREAV 40,19,76
hma-ofs/sat 450 μg/L CALEDQ 19,147,83
ipr-rat LDLo:30 mg/kg ARZNAD 16,870,66
orl-mus LD50:2750 mg/kg FMCHA2 -,D219,80
ipr-mus LDLo:30 mg/kg ARZNAD 21,121,71
orl-gpg LD50:1650 mg/kg 85GMAT -,62,82

CONSENSUS REPORTS: EPA Genetic Toxicology Program. Community Right-To-Know List.

SAFETY PROFILE: Poison by intraperitoneal route.

Moderately toxic by ingestion. Mutation data reported. When heated to decomposition it emits very toxic fumes of NO_x, CN^-, and SO_x. See also CYANIDE.

DVG000 CAS:5347-12-6 ***HR: 3***
2,4-DINITROTHIOPHENE
mf: $C_4H_2N_2O_4S$ mw: 174.14

TOXICITY DATA with REFERENCE
ipr-mus LDLo:1 mg/kg HBTXAC 5,171,59
ivn-mus LD50:32 mg/kg CBCCT* 6,143,54
par-mus LDLo:40 mg/kg CBCCT* 7,695,55

SAFETY PROFILE: Poison by intraperitoneal, intravenous, and parenteral routes. When heated to decomposition it emits very toxic fumes of NO_x and SO_x. See also NITRO COMPOUNDS.

DVG200 CAS:303-21-9 ***HR: 3***
2,6-DINITROTHYMOL
mf: $C_{10}H_{12}N_2O_5$ mw: 240.24

SYNS: 2,4-DINITRO-6-ISOBROPYL-m-CRESOL ◇ DINITROTHYMOL 1-2-4 (FRENCH)

TOXICITY DATA with REFERENCE
orl-rat LDLo:100 mg/kg NCNSA6 5,36,53
orl-mus TDLo:140 mg/kg AECTCV 14,111,85
ivn-dog LD50:15 mg/kg AIPTAK 50,20,35
ipl-pgn LD50:10 mg/kg AIPTAK 50,20,35

CONSENSUS REPORTS: Reported in EPA TSCA Inventory.

SAFETY PROFILE: Poison by ingestion, intravenous, and implant routes. When heated to decomposition it emits toxic fumes of NO_x. See also NITRO COMPOUNDS of AROMATIC HYDROCARBONS.

DVG600 CAS:25321-14-6 ***HR: 3***
DINITROTOLUENE
DOT: UN 1600/UN 2038
mf: $C_7H_6N_2O_4$ mw: 182.15

SYNS: DINITROPHENYLMETHANE ◇ ar,ar-DINITROTOLUENE ◇ DINITROTOLUENE, liquid (DOT) ◇ DINITROTOLUENE, molten (DOT) ◇ DINITROTOLUENE, solid (DOT) ◇ METHYLDINITROBENZENE

TOXICITY DATA with REFERENCE
dns-rat-orl 100 mg/kg CRNGDP 3,241,82
orl-rat TDLo:196 mg/kg (7-20D preg):REP CIIT** DOCKET #10992,82
orl-rat TDLo:1050 mg/kg (7-20D preg):TER CIIT** DOCKET #1099/82
orl-rat TDLo:12775 mg/kg/Y-C:ETA PAACA3 24,91,83

CONSENSUS REPORTS: Reported in EPA TSCA Inventory. EPA Genetic Toxicology Program.

OSHA PEL: TWA 1.5 mg/m³ (skin)
ACGIH TLV: TWA 1.5 mg/m³ (skin); (Proposed: 0.15 mg/m³ (skin); Suspected Human Carcinogen)
DFG MAK: Animal Carcinogen, Suspected Human Carcinogen.
NIOSH REL: (Dinitrotoluene): Reduce to lowest level
DOT Classification: Poison B; Label: Poison

SAFETY PROFILE: Confirmed carcinogen with experimental tumorigenic and teratogenic data. A poison. Experimental reproductive effects. Mutation data reported. Flammable. When heated to decomposition it emits toxic fumes of NO_x. See also 2,4-DINITROTOLUENE.

DVG800 CAS:602-01-7 ***HR: 2***
2,3-DINITROTOLUENE
mf: $C_7H_6N_2O_4$ mw: 182.15

SYNS: 1-METHYL-2,3-DINITRO-BENZENE (9CI) ◇ 2,3-DNT

TOXICITY DATA with REFERENCE
skn-rbt 500 mg/24H MLD NTIS** AD-B011-150
mmo-sat 50 μg/plate ENMUDM 4,163,82
mma-sat 1 mg/plate NTIS** AD-A080-146
dnd-rat:lvr 300 μmol/L SinJF# 26OCT82
orl-rat LD50:911 mg/kg NTIS** AD-A080-146
orl-mus LD50:1072 mg/kg NTIS** PB214-270

CONSENSUS REPORTS: Reported in EPA TSCA Inventory.

OSHA PEL: TWA 1.5 mg/m³ (skin)
NIOSH REL: (Dinitrotoluene): Reduce to lowest level

SAFETY PROFILE: Moderately toxic by ingestion. Mutation data reported. A skin irritant. When heated to decomposition it emits toxic fumes of NO_x. See also 2,4-DINITROTOLUENE.

DVH000 CAS:121-14-2 ***HR: 3***
2,4-DINITROTOLUENE
mf: $C_7H_6N_2O_4$ mw: 182.15

PROP: Yellow needles. Mp: 69.5°, bp: 300°, d: 1.521 @ 15°, vap d: 6.27, flash p: 404°F.

SYNS: 2,4-DINITROTOLUOL ◇ DNT ◇ 2,4-DNT ◇ 1-METHYL-2,4-DINITROBENZENE ◇ NCI-C01865 ◇ RCRA WASTE NUMBER U105

TOXICITY DATA with REFERENCE
mma-sat 125 μg/plate ENMUDM 7(Suppl 5),1,85
dnd-rat:lvr 3 mmol/L SinJF# 26OCT82
oms-rat-orl 10 mg/kg JTEHD6 11,555,83
mmo-sat 10 μg/plate NTIS** AD-A080-146
cyt-mus-orl 840 μg/kg MUREAV 38,387,76
dlt-mus-orl 2 mg/kg MUREAV 38,387,76
orl-rat TDLo:3094 mg/kg (13W male):REP NTIS** AD-A077-692

orl-rat TDLo:2620 mg/kg/78W-C:NEO NCITR* NCI-CG-TR-54,78
orl-mus TDLo:10080 mg/kg/2Y-C:CAR JACTDZ 4(4),257,85
orl-rat TD:28 g/kg/2Y-C:CAR NTIS** AD-A080-146
orl-rat TD:12775 mg/kg/2Y-C:ETA NCITR* NCI-CG-TR-54,78
orl-rat LD50:268 mg/kg NTIS** PB214-270
orl-mus LD50:790 mg/kg GTPZAB 25(8),50,81
scu-cat LDLo:25 mg/kg XPHBAO 271,110,41
orl-gpg LD50:1300 mg/kg GISAAA 42(10),12,77

CONSENSUS REPORTS: NCI Carcinogenesis Bioassay (feed); No Evidence: mouse NCITR* NCI-CG-TR-54,78; Some Evidence: rat NCITR* NCI-CG-TR-54,78. Reported in EPA TSCA Inventory.

OSHA PEL: TWA 1.5 mg/m^3 (skin)
NIOSH REL: (Dinitrotoluene): Reduce to lowest level

SAFETY PROFILE: Suspected carcinogen with experimental carcinogenic and neoplastigenic data. Poison by ingestion and subcutaneous routes. Experimental reproductive effects. Mutation data reported. An irritant and an allergen. Can cause anemia, methemoglobinemia, cyanosis, and liver damage. Combustible when exposed to heat or flame; can react with oxidizing materials. To fight fire, use water spray or mist, dry chemical. Decomposes when heated to 250°C. There are instances of explosion during manufacture or storage. Mixture with nitric acid is a high explosive. Mixture with sodium carbonate can decompose with significant pressure increase at 210°C. Mixtures with other alkalies may have the same effect. Ignites on contact with sodium oxide. When heated to decomposition it emits toxic fumes of NO_x. See also TRINITROTOLUENE.

DVH200 CAS:619-15-8 ***HR: 2***
2,5-DINITROTOLUENE
mf: $C_7H_6N_2O_4$ mw: 182.15

SYNS: 2,5-DNT ◇ 2-METHYL-1,4-DINITROBENZENE

TOXICITY DATA with REFERENCE
skn-rbt 500 mg/24H MOD NTIS** AD-B011-150
mmo-sat 10 μg/plate NTIS** AD-A080-146
mma-sat 10 μg/plate NTIS** AD-A080-146
orl-rat LD50:517 mg/kg NTIS** AD-A080-146
orl-mus LD50:652 mg/kg NTIS** AD-A080-146

CONSENSUS REPORTS: Reported in EPA TSCA Inventory.

OSHA PEL: TWA 1.5 mg/m^3 (skin)
NIOSH REL: (Dinitrotoluene): Reduce to lowest level

SAFETY PROFILE: Moderately toxic by ingestion. Mutation data reported. A skin irritant. When heated to decomposition it emits toxic fumes of NO_x. See also 2,4-DINITROTOLUENE.

DVH400 CAS:606-20-2 ***HR: 3***
2,6-DINITROTOLUENE
mf: $C_7H_6N_2O_4$ mw: 182.15

SYNS: 2,6-DNT ◇ 2-METHYL-1,3-DINITROBENZENE ◇ RCRA WASTE NUMBER U106

TOXICITY DATA with REFERENCE
skn-rbt 500 mg/24H MLD NTIS** AD-B011-150
dnd-rat-orl 10 mg/kg JTEHD6 11,555,83
dns-rat-orl 5 mg/kg CRNGDP 3,241,82
orl-rat TDLo:5110 mg/kg/Y-C:ETA PAACA3 24,91,83
orl-rat LD50:177 mg/kg NTIS** PB214-270
orl-mus LD50:621 mg/kg NTIS** AD-A080-146

CONSENSUS REPORTS: Reported in EPA TSCA Inventory.

OSHA PEL: TWA 1.5 mg/m^3 (skin)
NIOSH REL: (Dinitrotoluene): Reduce to lowest level

SAFETY PROFILE: Poison by ingestion. A skin irritant. Questionable carcinogen with experimental tumorigenic data. Mutation data reported. When heated to decomposition it emits toxic fumes of NO_x. See also 2,4-DINITROTOLUENE.

DVH600 CAS:610-39-9 ***HR: 3***
3,4-DINITROTOLUENE
mf: $C_7H_6N_2O_4$ mw: 182.15

SYNS: 3,4-DNT ◇ 4-METHYL-1,2-DINITROBENZENE

TOXICITY DATA with REFERENCE
skn-rbt 500 mg/24H MLD NTIS** AD-B011-150
mmo-sat 1 mg/plate NTIS** AD-A080-146
mma-sat 10 μg/plate ENMUDM 4,163,82
dnd-rat:lvr 300 μmol/L SinJF# 26OCT82
orl-rat LD50:177 mg/kg NTIS** PB214-270
orl-mus LD50:747 mg/kg NTIS** AD-A080-146

CONSENSUS REPORTS: Reported in EPA TSCA Inventory.

OSHA PEL: TWA 1.5 mg/m^3 (skin)
NIOSH REL: (Dinitrotoluene): Reduce to lowest level

SAFETY PROFILE: Poison by ingestion. Mutation data reported. A skin irritant. When heated to decomposition it emits toxic fumes of NO_x. See also 2,4-DINITROTOLUENE.

DVH800 CAS:618-85-9 ***HR: 3***
3,5-DINITROTOLUENE
mf: $C_7H_6N_2O_4$ mw: 182.15

SYNS: 3,5-DNT ◇ 1-METHYL-3,5-DINITRO-BENZENE

TOXICITY DATA with REFERENCE
mmo-sat 100 μg/plate NTIS** AD-A080-146
mma-sat 500 μg/plate NTIS** AD-A080-146
orl-rat LD50:216 mg/kg NTIS** AD-A080-146
orl-mus LD50:607 mg/kg NTIS** AD-A080-146

OSHA PEL: TWA 1.5 mg/m^3 (skin)
NIOSH REL: (Dinitrotoluene): Reduce to lowest level

SAFETY PROFILE: Poison by ingestion. Mutation data reported. Flammable when exposed to heat or flame; can react with oxidizing materials. A moderate explosion hazard when exposed to heat. To fight fire use water, CO_2, dry chemical. When heated to decomposition it emits toxic fumes of NO_x. See also 2,4-DINITROTOLUENE; EXPLOSIVES, HIGH; and NITRATES.

DVI100 CAS:6393-42-6 ***HR: 3***
2,6-DINITRO-p-TOLUIDINE
mf: $C_7H_7N_3O_4$ mw: 197.17

SYNS: 4-AMINO-3,5-DINITROTOLUENE ◇ 2,6-DINITRO-4-METHYLANILINE ◇ 4-METHYL-2,6-DINITROANILINE ◇ 4-METHYL-2,6-DINITROBENZENAMINE

TOXICITY DATA with REFERENCE
mmo-sat 1 μg/plate ENMUDM 4,163,82
mma-sat 1 μg/plate ENMUDM 4,163,82
ivn-mus LD50:320 mg/kg CSLNX* NX#03362

SAFETY PROFILE: Poison by intravenous route. Mutation data reported. When heated to decomposition it emits toxic fumes of NO_x. See also NITRO COMPOUNDS of AROMATIC HYDROCARBONS.

DVI600 CAS:6379-46-0 ***HR: 3***
4,6-DINITRO-1,2,3-TRICHLOROBENZENE
mf: $C_6HCl_3N_2O_4$ mw: 271.44

SYNS: 1,2,3-TRICHLORO-4,6-DINITROBENZENE ◇ VANCIDE PB

TOXICITY DATA with REFERENCE
orl-mus TDLo:13 g/kg/78W-I:CAR NTIS** PB223-159
scu-mus TDLo:10 mg/kg:CAR NTIS** PB223-159

SAFETY PROFILE: Questionable carcinogen with experimental carcinogenic data. When heated to decomposition it emits very toxic fumes of NO_x and Cl^-. See also NITRTO COMPOUNDS of AROMATIC HYDROCARBONS.

DVI800 CAS:8069-76-9 ***HR: 2***
DINOCTON-O
mf: $C_{16}H_{22}N_2O_7$ mw: 354.40

PROP: A mixture of methyl-2,4-dinitro-6-(1-ethylhexyl) phenyl carbonate and methyl-2,4-dinitro-6-(1-propylpentyl)phenyl carbonate (30ZDA9 -,100,71).

SYNS: DINOCTON-6 ◇ MC 1945

TOXICITY DATA with REFERENCE
orl-rat LD50:1250 mg/kg 28ZEAL 5,82,76
skn-rat LD50:3000 mg/kg 31ZOAD 1,176,68

SAFETY PROFILE: Moderately toxic by ingestion and skin contact. When heated to decomposition it emits toxic fumes of NO_x.

DVJ000 CAS:84-76-4 ***HR: 2***
DI-n-NONYL PHTHALATE
mf: $C_{26}H_{42}O_4$ mw: 418.68

SYNS: BISOFLEX 91 ◇ DINONYL-1,2-BENZENEDICARBOXYLATE ◇ DIONONYL PHTHALATE

TOXICITY DATA with REFERENCE
orl-rat LD50:2000 mg/kg 14CYAT 2,1904,63
unr-rat LD50:21500 mg/kg GTPZAB 24(3),25,80
unr-mus LD50:21500 mg/kg GTPZAB 24(3),25,80

CONSENSUS REPORTS: Reported in EPA TSCA Inventory.

SAFETY PROFILE: Moderately toxic by ingestion. Used as a plasticizer. When heated to decomposition it emits acrid smoke and irritating fumes. See also ESTERS.

DVJ100 CAS:33854-16-9 ***HR: 3***
DINOPROST METHYL ESTER
mf: $C_{21}H_{36}O_5$ mw: 368.57

SYNS: PGF2 METHYL ESTER ◇ PGF2-α METHYL ESTER ◇ PROSTAGLANDIN F2-α METHYL ESTER ◇ (5Z,9-α,11-α,13E,15S)-9,11,15-TRIHYDROXY-PROSTA-5,13-DIEN-1-OIC ACID, METHYL ESTER

TOXICITY DATA with REFERENCE
ivg-wmn TDLo:80 μg/kg (7W preg):TER JOPDAB 102,620,83
scu-ham TDLo:480 μg/kg (4D preg):REP JPETAB 186,67,73
ivg-wmn TDLo:80 μg/kg:GIT JOPDAB 102,620,83

SAFETY PROFILE: A human teratogen by intravaginal route with developmental abnormalities of the central nervous system and musculoskeletal system. Human systemic effects by intravaginal route: nausea or vomiting. Experimental reproductive effects. See also ESTERS.

DVJ200 CAS:363-24-6 ***HR: 3***
DINOPROSTONE
mf: $C_{20}H_{32}O_5$ mw: 352.52

SYNS: (5Z,11-α,13E,15S)-11,15-DIHYDROXY-9-OXOPROSTA-5,13-DIEN-1-OIC ACID ◇ 7-(3-HYDROXY-2-(3-HYDROXY-1-OCTENYL)-5-OXOCYCLOPENTYL)-5-HEPTENOIC ACID ◇ PGE2 ◇ PROSTAGLANDIN E2 ◇ (−)-PROSTAGLANDIN E2 ◇ (15S)-PROSTAGLANDIN E2 ◇ PROSTIN E2 ◇ U-12062

TOXICITY DATA with REFERENCE
oms-mus:oth 100 nmol/L JIDEAE 66,313,76
spm-mus-ipr 3 mg/kg INJFA3 21,82,76
dns-gpg:lng 1 mg/L PSEBAA 171,109,82
ivg-wmn TDLo:2400 μg/kg (female 14W post):REP JOGBAS 78,294,71
orl-mus TDLo:36 mg/kg (female 7-12D post):TER OYYAA2 8,787,74
orl-rat LD50:500 mg/kg OYYAA2 8,787,74
scu-rat LD50:31600 μg/kg OYYAA2 8,787,74
ivn-rat LD50:59500 μg/kg OYYAA2 8,787,74
orl-mus LD50:750 mg/kg OYYAA2 8,787,74
scu-mus LD50:19700 μg/kg OYYAA2 8,787,74
ivn-mus LD50:23200 μg/kg OYYAA2 8,787,74
ipr-ham LD50:1 mg/kg PLMEDD 17,309,85

SAFETY PROFILE: Poison by subcutaneous and intravenous routes. Moderately toxic by ingestion and intraperitoneal routes. An experimental teratogen. Human reproductive effects by intravenous, intraplacental, and intravaginal routes: changes in the uterus, cervix and vagina; termination of pregnancy; and changes in fertility. Experimental reproductive effects. Mutation data reported. When heated to decomposition it emits acrid smoke and irritating fumes.

DVJ400 CAS:3204-27-1 ***HR: 3***
DINOTERB ACETATE
mf: $C_{12}H_{14}N_2O_6$ mw: 282.28

SYNS: 2-tert-BUTYL-4,6-DINITROPHENYL ACETATE ◇ 2-(1,1-DIMETHYLETHYL)-4,6-DINITROPHENOL ACETATE ◇ MC 1108 ◇ P-1108

TOXICITY DATA with REFERENCE
orl-rat LD50:62 mg/kg WRPCA2 9,119,70
orl-rbt LD50:100 mg/kg 28ZEAL 5,83,76

SAFETY PROFILE: Poison by ingestion. When heated to decomposition it emits toxic fumes of NO_x. See also 2-(1,1-DIMETHYLETHYL)-4,6-DINITROPHENOL (dinoterb).

DVJ500 CAS:8015-43-8 ***HR: 3***
DIOCIDE
mf: $C_{21}H_{38}N \cdot C_2H_5ClHgO \cdot Br$ mw: 665.62

SYNS: DIOCID ◇ 1-HEXADECYL-PYRIDINIUM BROMIDE mixture with CHLORO(2-HYDROXYETHYL)MERCURY

TOXICITY DATA with REFERENCE
orl-rat LD50:172 mg/kg PCJOAU 11,918,77
ipr-rat LD50:21700 mg/kg PCJOAU 11,918,77
orl-mus LD50:54 mg/kg PCJOAU 11,918,77

OSHA PEL: (Transitional: CL 1 mg/10m^3) TWA 0.01 mg(Hg)/m^3; STEL 0.03 mg/m^3 (skin)
ACGIH TLV: TWA 0.01 mg(Hg)/m^3; STEL 0.03 mg(Hg)/m^3

CONSENSUS REPORTS: Mercury and its compounds are on the Community Right-To-Know List.

SAFETY PROFILE: Poison by ingestion. Mildly toxic by intraperitoneal route. When heated to decomposition it emits toxic fumes of Br^-, Cl^-, NO_x, and Hg. See also MERCURY COMPOUNDS.

DVJ600 CAS:1120-48-5 ***HR: 3***
DIOCTYLAMINE
mf: $C_{16}H_{35}N$ mw: 241.52

TOXICITY DATA with REFERENCE
ipr-mus LDLo:4 mg/kg CBCCT* 2,133,50

CONSENSUS REPORTS: Reported in EPA TSCA Inventory.

SAFETY PROFILE: Poison by intraperitoneal route. When heated to decomposition it emits toxic fumes of NO_x. See also AMINES.

DVJ800 CAS:3648-18-8 ***HR: 3***
DIOCTYLDI(LAUROYLOXY)STANNANE
mf: $C_{40}H_{80}O_4Sn$ mw: 743.89

SYNS: BIS(DODECANOLOXY)DIOCTYLSTANNANE ◇ BIS(LAUROYLOXY)DIOCTYLSTANNANE ◇ DIDODECANOYLOXYDIOCTYLSTANNANE ◇ DIOCTYLDIDODECANOYLOXYSTANNANE ◇ DIOCTYLTIN DILAURATE ◇ DI-n-OCTYLTIN DILAURATE ◇ DI-n-OCTYL-ZINN DILAURAT (GERMAN) ◇ DIOCYTLBIS(LAUROYLOXY)STANNANE

TOXICITY DATA with REFERENCE
orl-rat LD50:6450 mg/kg ARZNAD 19,934,69
ipr-rat LD50:95 mg/kg ARZNAD 19,934,69

CONSENSUS REPORTS: Reported in EPA TSCA Inventory.

OSHA PEL: TWA 0.1 mg(Sn)/m^3 (skin)
ACGIH TLV: TWA 0.1 mg(Sn)/m^3 (skin) (Proposed: TWA 0.1 mg(Sn)/m^3; STEL 0.2 mg(Sn)/m^3 (skin))
NIOSH REL: (Organotin Compounds) TWA 0.1 mg(Sn)/m^3

SAFETY PROFILE: Poison by intraperitoneal route. Mildly toxic by ingestion. When heated to decomposition it emits acrid smoke and irritating fumes. See also TIN COMPOUNDS.

DVK200 CAS:16091-18-2 ***HR: 2***
2,2-DIOCTYL-1,3,2-DIOXASTANNEPIN-4,7-DIONE
mf: $C_{20}H_{36}O_4Sn$ mw: 459.25

SYN: ESTABEX U 18

TOXICITY DATA with REFERENCE
orl-rat LD50:4500 mg/kg ARZNAD 19,934,69
orl-mus LD50:775 mg/kg ERNFA7 11,424,66

CONSENSUS REPORTS: Reported in EPA TSCA Inventory.

OSHA PEL: TWA 0.1 mg(Sn)/m^3 (skin)
ACGIH TLV: TWA 0.1 mg(Sn)/m^3 (skin) (Proposed: TWA 0.1 mg(Sn)/m^3; STEL 0.2 mg(Sn)/m^3 (skin))
NIOSH REL: (Organotin Compounds) TWA 0.1 mg(Sn)/m^3

SAFETY PROFILE: Moderately toxic by ingestion. When heated to decomposition it emits acrid smoke and irritating fumes. See also TIN COMPOUNDS.

DVK400 CAS:101-67-7 ***HR: 1***
4,4'-DIOCTYLDIPHENYLAMINE
mf: $C_{28}H_{43}N$ mw: 393.72

TOXICITY DATA with REFERENCE
orl-rat LD50:8000 mg/kg TXAPA9 42,417,77

CONSENSUS REPORTS: Reported in EPA TSCA Inventory.

SAFETY PROFILE: Mildly toxic by ingestion. When heated to decomposition it emits toxic fumes of NO_x. See also AMINES.

DVK600 CAS:141-02-6 ***HR: 3***
DIOCTYL FUMARATE
mf: $C_{20}H_{36}O_4$ mw: 340.56

PROP: Clear, mobile liquid; mild odor. Bp: 211-220°, flash p: 365°F (COC), d: 0.942 @ 20°/20°.

SYNS: BIS(2-ETHYLHEXYL) FUMARATE ◇ 2-BUTENEDIOIC ACID BIS(2-ETHYLHEXYL) ESTER ◇ DI(2-ETHYLHEXYL) FUMARATE ◇ DOF ◇ 2-ETHYLHEXYL FUMARATE ◇ RC COMONOMER DOF

TOXICITY DATA with REFERENCE
skn-rbt 10 mg/24H open SEV AMIHBC 10,61,54
eye-rbt 500 mg open AMIHBC 10,61,54
ipr-mus LD50:250 mg/kg NTIS** AD691-490

CONSENSUS REPORTS: Reported in EPA TSCA Inventory.

SAFETY PROFILE: Poison by intraperitoneal route. An eye and severe skin irritant. Combustible when exposed to heat or flame; can react with oxidizing materials. To fight fire, use foam, CO_2, dry chemical. See also ESTERS and FUMARIC ACID.

DVK709 ***HR: 1***
DIOCTYLISOPENTYLPHOSPHINE OXIDE
mf: $C_{21}H_{45}OP$ mw: 344.63

TOXICITY DATA with REFERENCE
orl-rat LD50:8369 mg/kg GISAAA 47(8),27,82
ihl-rat LC50:3311 g/m^3 GISAAA 47(8),27,82
orl-mus LD50:9500 mg/kg GISAAA 47(8),27,82
ihl-mus LC50:1288 g/m^3 GISAAA 47(8),27,82

SAFETY PROFILE: Mildly toxic by inhalation and ingestion. When heated to decomposition it emits toxic fumes of PO_x and phosphine. See also PHOSPHINE.

DVK800 CAS:2915-53-9 ***HR: 1***
DIOCTYL MALEATE
mf: $C_{20}H_{36}O_4$ mw: 340.56

SYN: DI-N-OCTYL MALEATE

TOXICITY DATA with REFERENCE
orl-rat LD50:14200 mg/kg NPIRI* 2,43,75

CONSENSUS REPORTS: Reported in EPA TSCA Inventory.

SAFETY PROFILE: Mildly toxic by ingestion. When heated to decomposition it emits acrid smoke and fumes. See also ESTERS.

DVL000 CAS:1116-76-3 ***HR: 3***
N,N-DIOCTYL-1-OCTANAMINE
mf: $C_{24}H_{51}N$ mw: 353.76

SYNS: ALAMINE 308 ◇ ALAMINE 336 ◇ TRICAPRYLYLAMINE ◇ TRI-n-OCTYLAMINE

TOXICITY DATA with REFERENCE
ipr-rat LD50:1 g/kg HYDRDA 3,201,78
ipr-mus LDLo:63 mg/kg CBCCT* 4,323,52

CONSENSUS REPORTS: Reported in EPA TSCA Inventory.

SAFETY PROFILE: Poison by intraperitoneal route. When heated to decomposition it emits toxic fumes of NO_x. See also AMINES.

DVL200 CAS:15535-79-2 ***HR: 2***
2,2-DIOCTYL-1,3,2-OXATHIASTANNOLANE-5-OXIDE
mf: $C_{18}H_{36}O_2SSn$ mw: 435.29

SYNS: DIOCTYLTHIOACETOXYSTANNANE ◇ DIOCTYLTIN THIOGLYCOLATE ◇ DI-n-OCTYLTIN THIOGLYCOLATE ◇ DI-n-OCTYL-ZINN THIOGLYKOLAT (GERMAN)

TOXICITY DATA with REFERENCE
orl-rat LD50:945 mg/kg ARZNAD 19,943,69

CONSENSUS REPORTS: Reported in EPA TSCA Inventory.

OSHA PEL: TWA 0.1 mg(Sn)/m^3 (skin)
ACGIH TLV: TWA 0.1 mg(Sn)/m^3 (skin) (Proposed: TWA 0.1 mg(Sn)/m^3; STEL 0.2 mg(Sn)/m^3 (skin))
NIOSH REL: (Organotin Compounds) TWA 0.1 mg(Sn)/m^3

SAFETY PROFILE: Moderately toxic by ingestion. When heated to decomposition it emits toxic fumes of SO_x. See also TIN COMPOUNDS.

DVL400 CAS:870-08-6 ***HR: 2***
DIOCTYLOXOSTANNANE
mf: $C_{16}H_{34}OSn$ mw: 361.19

SYNS: DIOCTYLTIN OXIDE ◇ DI-n-OCTYLTIN OXIDE ◇ DI-n-OCTYL-ZINN OXYD (GERMAN) ◇ OXODIOCTYLSTANNANE

TOXICITY DATA with REFERENCE
orl-rat LD50:2500 mg/kg ARZNAD 19,934,69

CONSENSUS REPORTS: Reported in EPA TSCA Inventory.

OSHA PEL: TWA 0.1 mg(Sn)/m^3 (skin)
ACGIH TLV: TWA 0.1 mg(Sn)/m^3 (skin) (Proposed: TWA 0.1 mg(Sn)/m^3; STEL 0.2 mg(Sn)/m^3 (skin))
NIOSH REL: (Organotin Compounds) TWA 0.1 mg(Sn)/m^3

SAFETY PROFILE: Moderately toxic by ingestion. When heated to decomposition it emits acrid smoke and irritating fumes. See also TIN COMPOUNDS.

DVL600 CAS:117-84-0 ***HR: 2***
n-DIOCTYL PHTHALATE
mf: $C_{24}H_{38}O_4$ mw: 390.62

SYNS: o-BENZENEDICARBOXYLIC ACID DIOCTYL ESTER ◇ 1,2-BENZENEDICARBOXYLIC ACID DIOCYTL ESTER ◇ CELLUFLEX DOP ◇ DINOPOL NOP ◇ DIOCTYL-o-BENZENEDICARBOXYLATE ◇ DIOCTYL PHTHALATE ◇ DNOP ◇ OCTYL PHTHALATE ◇ n-OCTYL PHTHALATE ◇ PX-138 ◇ RCRA WASTE NUMBER U107 ◇ VINICIZER 85

TOXICITY DATA with REFERENCE
skn-rbt 500 mg/24H MLD 28ZPAK -,48,72
eye-rbt 5 mg SEV AJOPAA 29,1363,46
eye-rbt 500 mg/24H MLD 28ZPAK -,48,72
orl-mus TDLo:78 g/kg (7-14D preg):REP NTIS** PB85-220143
ipr-rat TDLo:5 g/kg (5-15D preg):TER JPMSAE 61,51,72
orl-mus LD50:6513 mg/kg GTPZAB 17(10),51,73
ipr-mus LD50:65 g/kg JSCCA5 28,667,77

CONSENSUS REPORTS: On EPA Extremely Hazardous Substances List by error. Reported in EPA TSCA Inventory.

SAFETY PROFILE: Mildly toxic by ingestion. Experimental teratogenic and reproductive effects. A skin and severe eye irritant. Used as a plasticizer. When heated to decomposition it emits acrid smoke and irritating fumes. See also ESTERS.

DVL700 CAS:117-81-7 ***HR: 3***
DI-sec-OCTYL PHTHALATE
mf: $C_{24}H_{38}O_4$ mw: 390.62

SYNS: BEHP ◇ BIS(2-ETHYLHEXYL)-1,2-BENZENEDICARBOXYLATE ◇ BIS(2-ETHYLHEXYL)PHTHALATE ◇ BISOFLEX 81 ◇ BISOFLEX DOP ◇ COMPOUND 889 ◇ DAF 68 ◇ DEHP ◇ DI(2-ETHYLHEXYL)ORTHOPHTHALATE ◇ DI(2-ETHYLHEXYL)PHTHALATE ◇ DIOCTYL PHTHALATE ◇ DOP ◇ ETHYLHEXYL PHTHALATE ◇ ERGOPLAST FDO ◇ 2-ETHYLHEXYL PHTHALATE ◇ EVIPLAST 80 ◇ EVIPLAST 81 ◇ FLEXIMEL ◇ FLEXOL DOP ◇ FLEXOL PLASTICIZER DOP ◇ GOOD-RITE GP 264 ◇ HATCOL DOP ◇ HERCOFLEX 260 ◇ KODAFLEX DOP ◇ MOLLAN O ◇ NCI-C52733 ◇ NUOPLAZ DOP ◇ OCTOIL ◇ OCTYL PHTHALATE ◇ PALATINOL AH ◇ PHTHALIC ACID DIOCTYL ESTER ◇ PITTSBURGH PX-138 ◇ PLATINOL AH ◇ PLATINOL DOP ◇ RC PLASTICIZER DOP ◇ RCRA WASTE NUMBER U028 ◇ REOMOL DOP ◇ REOMOL D 79P ◇ SICOL 150 ◇ STAFLEX DOP ◇ TRUFLEX DOP ◇ VESTINOL AH ◇ VINICIZER 80 ◇ WITCIZER 312

TOXICITY DATA with REFERENCE
skn-rbt 500 mg/24H MLD 28ZPAK -,48,72
eye-rbt 500 mg AJOPAA 29,1363,46
eye-rbt 500 mg/24H MLD 28ZPAK -,48,72
dns-rat:lvr 500 μmol/L PMRSDJ 5,371,85
sln-ham:lvr 50 mg/L PMRSDJ 5,397,85
orl-mus TDLo:1 g/kg (female 7D post):TER EVHPAZ 45,71,82
ipr-mus TDLo:24 g/kg (female 7-9D post):REP ARTODN 56,263,85
orl-rat TDLo:216 g/kg/2Y-C:CAR NTPTR* NTP-TR-217,82
orl-mus TDLo:260 g/kg/2Y-C:CAR NTPTR* NTP-TR-217,82
orl-mus TD:120 g/kg/24W-C:ETA CRNGDP 4,1021,83
orl-man TDLo:143 mg/kg:GIT JIHTAB 27,130,45
orl-rat LD50:30600 mg/kg EVHPAZ 3,131,73
ipr-rat LD50:30700 mg/kg JIHTAB 27,130,45
ivn-rat LD50:250 mg/kg TXAPA9 45,230,78
orl-mus LD50:30 g/kg TJADAB 14,259,76
ipr-mus LD50:14 g/kg JPMSAE 55,158,66
ivn-mus LD50:1060 mg/kg NTIS** PB250-102
orl-rbt LD50:34 g/kg EVHPAZ 4,3,73
skn-rbt LD50:25 g/kg JIHTAB 27,130,45
skn-gpg LD50:10 g/kg EVHPAZ 4,3,73
skn-gpg LD50:10 g/kg EVHPAZ 4,3,73

CONSENSUS REPORTS: NTP Fifth Annual Report on Carcinogens. IARC Cancer Review: Group 2B IMEMDT 7,56,87; Human Inadequate Evidence IMEMDT 29,269,82; Animal Sufficient Evidence IMEMDT 29,269,82. NTP Carcinogenesis Bioassay (feed); Clear Evidence: mouse, rat NTPTR* NTP-TR-217,82. EPA Genetic Toxicology Program. Reported in EPA TSCA Inventory. Community Right-To-Know List.

OSHA PEL: (Transitional: TWA 5 mg/m^3) TWA 5 mg/m^3; STEL 10 mg/m^3
ACGIH TLV: TWA 5 mg/m^3; STEL 10 mg/m^3
DFG MAK: 10 mg/m^3
NIOSH REL: (DEHP) Reduce to lowest feasible level

SAFETY PROFILE: Confirmed carcinogen with experimental carcinogenic and tumorigenic data. Experimental teratogen data. Other experimental reproductive effects. Poison by intravenous route. Human systemic effects by ingestion: gastrointestinal tract effects. A mild skin and eye irritant. When heated to decomposition it emits acrid smoke.

DVL800 CAS:69226-45-5 ***HR: 3***
DIOCTYL(1,2-PROPYLENEDIOXYBIS(MALE-OYLDIOXY))STANNANE
mf: $C_{27}H_{42}O_8Sn$ mw: 613.38

SYNS: DI-n-OCTYLTINDI(1,2-PROPYLENEGLYCOLMALEATE) ◇ DI-n-OCTYL-ZINN-DI-(1,2-PROPYLENGLYKOLMALEINAT)(GERMAN)

TOXICITY DATA with REFERENCE
orl-rat LD50:4775 mg/kg ARZNAD 19,934,69
ipr-rat LD50:50 mg/kg ARZNAD 19,934,69

OSHA PEL: TWA 0.1 mg(Sn)/m^3 (skin)
ACGIH TLV: TWA 0.1 mg(Sn)/m^3 (skin) (Proposed: TWA 0.1 mg(Sn)/m^3; STEL 0.2 mg(Sn)/m^3 (skin))
NIOSH REL: (Organotin Compounds) TWA 0.1 mg(Sn)/m^3

SAFETY PROFILE: Poison by intraperitoneal route. Mildly toxic by ingestion. When heated to decomposition it emits acrid smoke and irritating fumes. See also TIN COMPOUNDS.

DVM000 CAS:3572-47-2 ***HR: 3***
DIOCTYLTHIOXOSTANNANE
mf: $C_{16}H_{34}SSn$ mw: 377.25

SYN: DI-n-OCTYLTIN SULFIDE

TOXICITY DATA with REFERENCE
ivn-mus LD50:180 mg/kg CSLNX* NX#01771

CONSENSUS REPORTS: Reported in EPA TSCA Inventory.

OSHA PEL: TWA 0.1 mg(Sn)/m^3 (skin)
ACGIH TLV: TWA 0.1 mg(Sn)/m^3 (skin) (Proposed: TWA 0.1 mg(Sn)/m^3; STEL 0.2 mg(Sn)/m^3 (skin))
NIOSH REL: (Organotin Compounds) TWA 0.1 mg(Sn)/m^3

SAFETY PROFILE: Poison by intravenous route. When heated to decomposition it emits toxic fumes of SO_x. See also SULFIDES and TIN COMPOUNDS.

DVM200 CAS:27107-88-6 ***HR: 2***
DI-n-OCTYLTIN BIS(BUTYL MERCAPTOACETATE)
mf: $C_{28}H_{56}O_4S_2Sn$ mw: 639.65

SYN: BIS(MERCAPTOACETATE)DIOCTYLTIN BIS(BUTYL) ESTER

TOXICITY DATA with REFERENCE
orl-mus LD50:1140 mg/kg ATXKA8 26,196,70

OSHA PEL: TWA 0.1 mg(Sn)/m^3 (skin)
ACGIH TLV: TWA 0.1 mg(Sn)/m^3 (skin) (Proposed: TWA 0.1 mg(Sn)/m^3; STEL 0.2 mg(Sn)/m^3 (skin))
NIOSH REL: (Organotin Compounds) TWA 0.1 mg(Sn)/m^3

SAFETY PROFILE: Moderately toxic by ingestion. When heated to decomposition it emits toxic fumes of SO_x. See also TIN COMPOUNDS.

DVM400 CAS:22205-30-7 ***HR: 2***
DI-n-OCTYLTIN BIS(DODECYL MERCAPTIDE)
mf: $C_{44}H_{88}O_4S_2Sn$ mw: 864.13

SYN: BIS(MERCAPTO)DIOCTYLTIN BIS(DODECYL) ESTER

TOXICITY DATA with REFERENCE
orl-mus LD50:4000 mg/kg ATXKA8 26,196,70

CONSENSUS REPORTS: Reported in EPA TSCA Inventory.

OSHA PEL: TWA 0.1 mg(Sn)/m^3 (skin)
ACGIH TLV: TWA 0.1 mg(Sn)/m^3 (skin) (Proposed: TWA 0.1 mg(Sn)/m^3; STEL 0.2 mg(Sn)/m^3 (skin))
NIOSH REL: (Organotin Compounds) TWA 0.1 mg(Sn)/m^3

SAFETY PROFILE: Moderately toxic by ingestion. When heated to decomposition it emits toxic fumes of SO_x. See also TIN COMPOUNDS and SULFIDES.

DVM600 CAS:10039-33-5 ***HR: 2***
DI-n-OCTYLTIN BIS(2-ETHYLHEXYL MALEATE)
mf: $C_{40}H_{72}O_8Sn$ mw: 799.81

SYNS: BIS(HYDROGEN MALEATO)DIOCTYLTIN BIS(2-ETHYLHEXYL) ESTER ◇ DI-n-OCTYL-ZINN-BIS(2-AETHYLHEXYLMALEINAT) (GERMAN)

TOXICITY DATA with REFERENCE
orl-rat LD50:2760 mg/kg ARZNAD 19,934,69
orl-mus LD50:2700 mg/kg FCTXAV 8,655,70

CONSENSUS REPORTS: Reported in EPA TSCA Inventory.

OSHA PEL: TWA 0.1 mg(Sn)/m^3 (skin)
ACGIH TLV: TWA 0.1 mg(Sn)/m^3 (skin) (Proposed: TWA 0.1 mg(Sn)/m^3; STEL 0.2 mg(Sn)/m^3 (skin))
NIOSH REL: (Organotin Compounds) TWA 0.1 mg(Sn)/m^3

SAFETY PROFILE: Moderately toxic by ingestion. When heated to decomposition it emits acrid smoke and irritating fumes. See also TIN COMPOUNDS.

DVM800 CAS:15571-58-1 ***HR: 2***
DI-n-OCTYLTIN BIS(2-ETHYLHEXYL) MERCAPTOACETATE
mf: $C_{36}H_{72}O_4S_2Sn$ mw: 751.89

SYNS: BIS(2-ETHYLHEXYLTHIOGLYCOLATE)DIOCTYLTIN ◇ BIS-(MERCAPTOACETATE)DIOCTYLTIN BIS(2-ETHYLHEXYL) ESTER ◇ 10-ETHYL-4,4-DIOCTYL-7-OXO-8-OXA-3,5-DITHIA-4-STANNATETRADECANOIC ACID-2-ETHYLHEXYL ESTER ◇ DI-N-OCTYLTIN-ITHIOGLYCOLIC ACID 2-ETHYLHEXYL ESTER ◇ DI-N-OCTYLTIN-2-ETHYLHEXYLDIMERCAPTOETHANOATE ◇ OTS 11

TOXICITY DATA with REFERENCE
orl-rat LD50:2100 mg/kg NAHRAR 13,343,69
orl-mus LD50:2010 mg/kg ATXKA8 26,196,70

CONSENSUS REPORTS: Reported in EPA TSCA Inventory.

OSHA PEL: TWA 0.1 mg(Sn)/m^3 (skin)
ACGIH TLV: TWA 0.1 mg(Sn)/m^3 (skin) (Proposed: TWA 0.1 mg(Sn)/m^3; STEL 0.2 mg(Sn)/m^3 (skin))
NIOSH REL: (Organotin Compounds) TWA 0.1 mg(Sn)/m^3

SAFETY PROFILE: Moderately toxic by ingestion. When heated to decomposition it emits toxic fumes of SO_x. See also TIN COMPOUNDS and ESTERS.

DVN000 CAS:69226-43-3 ***HR: 2***
DI-n-OCTYLTIN BIS(LAURYLTHIOGLYCOLATE)
mf: $C_{44}H_{88}O_4S_2Sn$ mw: 864.13

SYNS: BIS(LAUROYLOXYCARBONYLMETHYLTHIO)DIOCTYLSTANNANE ◇ DI-n-OCTYL-ZINN-BIS(LAURYL-THIOGLYKOLAT) (GERMAN)

TOXICITY DATA with REFERENCE
orl-rat LD50:3700 mg/kg ARZNAD 19,934,69

OSHA PEL: TWA 0.1 mg(Sn)/m^3 (skin)
ACGIH TLV: TWA 0.1 mg(Sn)/m^3 (skin) (Proposed: TWA 0.1 mg(Sn)/m^3; STEL 0.2 mg(Sn)/m^3 (skin))
NIOSH REL: (Organotin Compounds) TWA 0.1 mg(Sn)/m^3

SAFETY PROFILE: Moderately toxic by ingestion. When heated to decomposition it emits toxic fumes of SO_x. See also TIN COMPOUNDS.

DVN200 CAS:69226-46-6 ***HR: 2***
DI-n-OCTYLTIN-1,4-BUTANEDIOL-BIS-MERCAPTOACETATE
mf: $C_{24}H_{46}O_4S_2Sn$ mw: 581.51

SYN: DI-n-OCTYL-ZINN-1,4-BUTANDIOL-BIS-MERCAPTOACETAT (GERMAN)

TOXICITY DATA with REFERENCE
orl-rat LD50:2950 mg/kg ARZNAD 19,934,69

OSHA PEL: TWA 0.1 mg(Sn)/m^3 (skin)
ACGIH TLV: TWA 0.1 mg(Sn)/m^3 (skin) (Proposed: TWA 0.1 mg(Sn)/m^3; STEL 0.2 mg(Sn)/m^3 (skin))
NIOSH REL: (Organotin Compounds) TWA 0.1 mg(Sn)/m^3

SAFETY PROFILE: Moderately toxic by ingestion. When heated to decomposition it emits toxic fumes of SO_x. See also TIN COMPOUNDS and MERCAPTANS.

DVN400 CAS:69226-44-4 ***HR: 2***
DI-n-OCTYLTIN ETHYLENEGLYCOL DITHIOGLYCOLATE
mf: $C_{22}H_{42}O_4S_2Sn$ mw: 553.45

SYNS: DIOCTYL(ETHYLENEDIOXYBIS(CARBONYLMETHYLTHIO))STANNANE ◇ DI-n-OCTYL-ZINN AETHYLENGLYKOL-DITHIOGLYKOLAT (GERMAN)

TOXICITY DATA with REFERENCE
orl-rat LD50:2950 mg/kg ARZNAD 19,934,69

OSHA PEL: TWA 0.1 mg(Sn)/m^3 (skin)
ACGIH TLV: TWA 0.1 mg(Sn)m^3 (skin) (Proposed: TWA 0.1 mg(Sn)/m^3; STEL 0.2 mg(Sn)/m^3 (skin))
NIOSH REL: (Organotin Compounds) TWA 0.1 mg(Sn)/m^3

SAFETY PROFILE: Moderately toxic by ingestion. When heated to decomposition it emits toxic fumes of SO_x. See also TIN COMPOUNDS.

DVN600 CAS:58229-88-2 ***HR: 2***
DI-n-OCTYLTIN MERCAPTIDE

SYNS: DIOCTYLTIN MERCAPTIDE ◇ ERGOTERM OTGO

TOXICITY DATA with REFERENCE
orl-rat LDLo:750 mg/kg RPZHAW 19,329,68

OSHA PEL: TWA 0.1 mg(Sn)/m^3 (skin)
ACGIH TLV: TWA 0.1 mg(Sn)/m^3 (skin) (Proposed: TWA 0.1 mg(Sn)/m^3; STEL 0.2 mg(Sn)/m^3 (skin))
NIOSH REL: (Organotin Compounds) TWA 0.1 mg(Sn)/m^3

SAFETY PROFILE: Moderately toxic by ingestion. When heated to decomposition it emits toxic fumes of SO_x. See also TIN COMPOUNDS and MERCAPTANS.

DVN800 CAS:3033-29-2 ***HR: 3***
DI-n-OCTYLTIN β-MERCAPTOPROPIONATE
mf: $C_{19}H_{38}O_2SSn$ mw: 449.32

SYNS: DIHYDRO-2,2-DIOCTYL-6H-1,3,2-OXATHIASTANNIN-6-ONE

◇ DIOCTYLTIN-β-MERCAPTOPROPIONATE ◇ DI-n-OCTYL-ZINN β-MERCAPTOPROPIONAT (GERMAN)

TOXICITY DATA with REFERENCE
orl-rat LD50:1850 mg/kg ARZNAD 19,934,69
ipr-rat LD50:6600 μg/kg TXAPA9 35,63,76

CONSENSUS REPORTS: Reported in EPA TSCA Inventory.

OSHA PEL: TWA 0.1 mg(Sn)/m^3 (skin)
ACGIH TLV: TWA 0.1 mg(Sn)/m^3 (skin) (Proposed: TWA 0.1 mg(Sn)/m^3; STEL 0.2 mg(Sn)/m^3 (skin))
NIOSH REL: (Organotin Compounds) TWA 0.1 mg(Sn)/m^3

SAFETY PROFILE: Poison by intraperitoneal route. Moderately toxic by ingestion. When heated to decomposition it emits toxic fumes of SO_x. See also TIN COMPOUNDS and MERCAPTANS.

DVN909 CAS:3594-15-8 ***HR: 3***
DIOCTYLTIN-3,3'-THIODIPROPIONATE
mf: $C_{22}H_{42}O_4SSn$ mw: 521.39

SYN: 2,2-DIOCTYL-1,3-DIOXA-2-STANNA-7-THIADECAN-4,10-DIONE

TOXICITY DATA with REFERENCE
ivn-mus LD50:320 mg/kg CSLNX* NX#02854

OSHA PEL: TWA 0.1 mg(Sn)/m^3 (skin)
ACGIH TLV: TWA 0.1 mg(Sn)/m^3 (skin) (Proposed: TWA 0.1 mg(Sn)/m^3; STEL 0.2 mg(Sn)/m^3 (skin))
NIOSH REL: (Organotin Compounds) TWA 0.1 mg(Sn)/m^3

SAFETY PROFILE: Poison by intravenous route. When heated to decomposition it emits toxic fumes of SO_x. See also TIN COMPOUNDS.

DVO000 ***HR: 3***
DIOCYDE

TOXICITY DATA with REFERENCE
orl-rat LD50:172 mg/kg KHFZAN 11(7),44,77
orl-mus LD50:54 mg/kg KHFZAN 11(7),44,77
ipr-mus LD50:21700 μg/kg KHFZAN 11(7),44,77

SAFETY PROFILE: Poison by ingestion and intraperitoneal routes.

DVO175 CAS:63323-30-8 ***HR: 3***
anti-DIOLEPOXIDE
mf: $C_{20}H_{12}O_3$ mw: 300.32

SYNS: (−)-7-α,8-β-DIHYDROXY-9-β,10-β-EPOXY-7,8,9,10-TETRAHYDROBENZO(a)PYRENE ◇ (−)-BP 7-α,8-β-DIOL-9-β,10-β-EPOXIDE 2 ◇ (−)-7,8,9,10-TETRAHYDRO-7-β,8-α-DIHYDROXY-9-α,10-α-EPOXY-BENZO(a)PYRENE

TOXICITY DATA with REFERENCE
mmo-sat 100 pmol/plate BBRCA9 77,1389,77
cyt-ham:lng 300 μg/L IJCNAW 24,485,79
sce-ham:lng 600 μg/L IJCNAW 24,485,79
msc-ham:lng 10 μmol CRNGDP 3,1223,82
skn-mus TDLo:2400 μg/kg:ETA CNREA8 39,67,79

SAFETY PROFILE: Questionable carcinogen with experimental tumorigenic data by skin contact. Mutation data reported. When heated to decomposition it emits toxic fumes of NO_x.

DVO600 CAS:702-62-5 ***HR: 2***
2-4-DIONE-1,3-DIAZASPIRO(4.5)DECANE
mf: $C_8H_{12}N_2O_2$ mw: 168.22

SYNS: CYCLOHEXANESPIRO-5'-HYDANTOIN ◇ SPIRO(CYCLOHEXANE-1,5'-HYDANTOIN)

TOXICITY DATA with REFERENCE
orl-mus LD50:420 mg/kg JMCMAR 8,239,65

CONSENSUS REPORTS: Reported in EPA TSCA Inventory.

SAFETY PROFILE: Moderately toxic by ingestion. When heated to decomposition it emits toxic fumes of NO_x.

DVO700 CAS:125-30-4 ***HR: 3***
DIONIN HYDROCHLORIDE
mf: $C_{19}H_{23}NO_3$•ClH mw: 349.89

SYNS: CODETHYLINE HYDROCHLORIDE ◇ DIONINE HYDROCHLORIDE ◇ ETHYLMORPHINE HYDROCHLORIDE ◇ o-ETHYLMORPHINE HYDROCHLORIDE

TOXICITY DATA with REFERENCE
orl-rat LD50:950 mg/kg ARZNAD 21,719,71
scu-rat LD50:200 mg/kg ARZNAD 21,727,71
orl-mus LD50:520 mg/kg ARZNAD 21,719,71
scu-mus LD50:265 mg/kg APFRAD 15,640,57

SAFETY PROFILE: Poison by subcutaneous route. Moderately toxic by ingestion. When heated to decomposition it emits toxic fumes of NO_x and HCl.

DVO809 CAS:17667-23-1 ***HR: 3***
DIOSPYROL
mf: $C_{22}H_{18}O_4$ mw: 346.40

SYN: 6,6'-DIMETHYL-(2,2'-BINAPHTHALENE)-1,1',8,8'-TETROL

TOXICITY DATA with REFERENCE
orl-rat LD50:3000 mg/kg DRFUD4 5,438,80
orl-mus LD50:3000 mg/kg DRFUD4 5,438,80
orl-dog LD50:2000 mg/kg DRFUD4 5,438,80
ipr-mky LD50:100 mg/kg DRFUD4 5,438,80
orl-ham LD50:3000 mg/kg DRFUD4 5,438,80

SAFETY PROFILE: Poison by intraperitoneal route. Moderately toxic by ingestion. When heated to decomposition it emits acrid smoke and fumes.

DVO819 CAS:101-08-6 ***HR: 3***
DIOTHANE
mf: $C_{22}H_{27}N_3O_4$ mw: 397.52

SYNS: 3-PIPERIDINO-1,2-PROPANEDIOLDICARBANILATE ◇ 3-(1-PIPERIDYL)-1,2-PROPANE DICARBANILATE

TOXICITY DATA with REFERENCE
scu-mus LDLo:1200 mg/kg JPETAB 47,255,33
scu-rbt LDLo:300 mg/kg JPETAB 47,255,33
ivn-rbt LDLo:15 mg/kg JPETAB 47,255,33
scu-gpg LDLo:400 mg/kg JPETAB 47,255,33

SAFETY PROFILE: Poison by subcutaneous and intravenous routes. When heated to decomposition it emits toxic fumes of NO_x.

DVP400 CAS:631-06-1 ***HR: 3***
d-DIOXADROL HYDROCHLORIDE
mf: $C_{20}H_{23}NO_2 \cdot ClH$ mw: 345.90

SYNS: CL-911C ◇ DEXOXADROL HYDROCHLORIDE ◇ d-2-(2,2-DIPHENYL-1,3-DIOXOLAN-4-YL)PIPERIDINEHYDROCHLORIDE ◇ d-2,2-DIPHENYL-4-(2-PIPERIDYL)-1,3-DIOXOLANE HYDROCHLORIDE ◇ NSC-526062 ◇ RELANE ◇ U-22,559A

TOXICITY DATA with REFERENCE
orl-hmn TDLo:400 µg/kg:CNS PSEBAA 118,352,65
orl-rat LD50:280 mg/kg AIPTAK 153,105,65
orl-mus LD50:380 mg/kg AIPTAK 153,105,65
ivn-rbt LD50:33 mg/kg AIPTAK 153,105,65

SAFETY PROFILE: Poison by ingestion and intravenous routes. Human systemic effects by ingestion: somnolence, hallucinations or distorted perceptions, and analgesia. When heated to decomposition it emits very toxic fumes of HCl and NO_x.

DVP600 CAS:505-22-6 ***HR: 3***
m-DIOXAN
mf: $C_4H_8O_2$ mw: 88.12

$O(CH_2)_3OCH_2$

PROP: Flash p: 33.8°F. lel: 2%, uel: 22%.

SYN: 1,3-DIOXANE

SAFETY PROFILE: A very dangerous fire and explosion hazard when exposed to heat or flame; can react with oxidizing materials. Can form dangerous peroxides when exposed to air. When heated to decomposition it emits acrid smoke and fumes.

DVQ000 CAS:123-91-1 ***HR: 3***
DIOXANE
DOT: UN 1165
mf: $C_4H_8O_2$ mw: 88.11

$OC_2H_4OCH_2CH_2$

PROP: Colorless liquid, pleasant odor. Mp: 12°, bp: 101.1°, lel: 2.0%, uel: 22.2%, flash p: 54°F (CC), d: 1.0353 @ 20°/4°, autoign temp: 356°F, vap press: 40 mm @ 25.2°, vap d: 3.03.

SYNS: DIETHYLENE DIOXIDE ◇ 1,4-DIETHYLENE DIOXIDE ◇ DIETHYLENE ETHER ◇ DI(ETHYLENE OXIDE) ◇ DIOKAN ◇ DIOKSAN (POLISH) ◇ DIOSSANO-1,4 (ITALIAN) ◇ DIOXAAN-1,4 (DUTCH) ◇ 1,4-DIOXACYCLOHEXANE ◇ DIOXAN-1,4 (GERMAN) ◇ p-DIOXAN (CZECH) ◇ 1,4-DIOXANE (MAK) ◇ p-DIOXANE ◇ DIOXANNE (FRENCH) ◇ DIOXYETHYLENE ETHER ◇ GLYCOL ETHYLENE ETHER ◇ NCI-C03689 ◇ RCRA WASTE NUMBER U108 ◇ TETRAHYDRO-p-DIOXIN ◇ TETRAHYDRO-1,4-DIOXIN

TOXICITY DATA with REFERENCE
eye-hmn 300 ppm/15M JIHTAB 28,262,46
skn-rbt 515 mg open MLD UCDS** 12/17/71
eye-rbt 21 mg AJOPAA 29,1363,46
eye-gpg 10 µg MOD JPPMAB 11,150,59
dnd-rat:lvr 300 µmol/L SinJF# 26OCT82
oms-rat-ivn 50 mg/kg ARTODN 49,29,81
orl-rat TDLo:10 g/kg (6-15D preg):TER TOLED5 26,85,85
orl-rat TDLo:185 g/kg/2Y-C:CAR NCITR* NCI-CG-TR-80,78
ihl-rat TCLo:111 ppm/7H/2Y-C:ETA TXAPA9 30,287,74
orl-mus TDLo:239 g/kg/90W-C:CAR NCITR* NCI-CG-TR-80,78
ipr-mus TDLo:12 g/kg/8W-I:NEO TXAPA9 82,19,86
ihl-hmn TCLo:470 ppm:CNS,CVS,GIT AMIHAB 20,445,59
ihl-hmn TCLo:5500 ppm/1M:EYE,PUL PHRPA6 45,2023,30
ihl-hmn LCLo:470 ppm/3D PLENBW 7,22,75
ihl-rat LC50:46 g/m³/2H KBAMAJ 11(6),53,77
ipr-rat LD50:799 mg/kg ENVRAL 40,411,86
orl-mus LD50:5700 mg/kg JIHTAB 21,173,39
ihl-mus LC50:37 g/m³/2H 85GMAT -,63,82
ipr-mus LD50:790 mg/kg FEPRA7 6,342,47
orl-cat LD50:2000 mg/kg JIHTAB 21,173,39
ihl-cat LCLo:44 g/m³/7H KDPU** -,-,37
orl-rbt LD50:2000 mg/kg JIHTAB 21,173,39
skn-rbt LD50:7600 mg/kg UCDS** 12/17/71
ivn-rbt LDLo:1500 mg/kg JOHYAY 35,540,35
orl-gpg LD50:3150 mg/kg JIHTAB 23,259,41

CONSENSUS REPORTS: NTP Fifth Annual Report on Carcinogens. IARC Cancer Review: Group 2B IMEMDT 7,201,87; Animal Sufficient Evidence IMEMDT 11,247,76. NCI Carcinogenesis Bioassay (oral); Clear Evidence: mouse, rat NCITR* NCI-CG-TR-80,78. EPA

Genetic Toxicology Program. Glycol ether compounds are on the Community Right-To-Know List. Reported in EPA TSCA Inventory.

OSHA PEL: (Transitional: TWA 100 ppm (skin)) TWA 25 ppm (skin)
ACGIH TLV: TWA 25 ppm (skin)
DFG MAK: 50 ppm (180 mg/m^3); Suspected Carcinogen.
NIOSH REL: CL (Dioxane) 1 ppm/30M
DOT Classification: Flammable Liquid; Label: Flammable Liquid.

SAFETY PROFILE: Confirmed carcinogen with experimental carcinogenic, neoplastigenic, tumorigenic, and teratogenic data. Poison by intraperitoneal route. Moderately toxic by ingestion and inhalation. Mildly toxic by skin contact. Human systemic effects by inhalation: lachrimation, conjunctiva irritation, convulsions, high blood pressure, unspecified respiratory and gastrointestinal system effects. Mutation data reported. An eye and skin irritant. The irritant effects probably provide sufficient warning, in acute exposures, to enable a worker to leave exposure before being seriously affected. Repeated exposure to low concentrations has resulted in human fatalities, the organs chiefly affected being the liver and kidneys.

A very dangerous fire and explosion hazard when exposed to heat or flame; can react vigorously with oxidizing materials. Violent reaction with (H_2 + Raney Ni), $AgClO_4$. Can form dangerous peroxides when exposed to air. Potentially explosive reaction with nitric acid + perchloric acid; Raney nickel catalyst (above 210°C). Forms explosive mixtures with decaborane (impact-sensitive); triethynylaluminum (sensitive to heating or drying). Violent reaction with sulfur trioxide. Incompatible with sulfur trioxide. To fight fire, use alcohol foam, CO_2, dry chemical. When heated to decomposition it emits acrid smoke and irritating fumes. See also GLYCOL ETHERS.

DVQ400 CAS:766-15-4 ***HR: 2***
m-DIOXANE-4,4-DIMETHYL
mf: $C_6H_{12}O_2$ mw: 116.18

SYN: 4,4-DIMETHYLDIOXANE-1,3

TOXICITY DATA with REFERENCE
skn-rbt 10 mg/24H open MLD AIHAAP 23,95,62
ihl-rat TCLo:10 μg/m^3/24H (16W pre):TER GISAAA 43(9),16,78
ihl-rat TCLo:10 μg/m^3/24H (16W pre):REP GISAAA 43(9),16,78
orl-rat LD50:3730 mg/kg AIHAAP 23,95,62
ihl-rat LCLo:8000 ppm/4H AIHAAP 23,95,62
orl-mus LDLo:1 g/kg GISAAA 25(6),85,60
skn-rbt LD50:3540 mg/kg AIHAAP 23,95,62

SAFETY PROFILE: Moderately toxic by ingestion and skin contact. An experimental teratogen. A skin irritant. When heated to decomposition it emits acrid smoke and fumes.

DVQ600 CAS:16088-56-5 ***HR: 3***
cis-2,3-p-DIOXANEDITHIOL-S,S-BIS(O,O-DIETHYLPHOSPHORODITHIOATE)
mf: $C_{12}H_{26}O_6P_2S_4$ mw: 456.56

SYN: (E)-PHOSPHORODITHIOIC ACID-O,O-DIETHYL ESTER-S,S-DIESTER with p-DIOXANE-2,3-DIETHIOL

TOXICITY DATA with REFERENCE
scu-rat LD50:66 mg/kg TXAPA9 5,605,63

SAFETY PROFILE: Poison by subcutaneous route. When heated to decomposition it emits very toxic fumes of PO_x and SO_x. See also ESTERS.

DVQ709 CAS:78-34-2 ***HR: 3***
DIOXATHION
mf: $C_{12}H_{26}O_6P_2S_4$ mw: 456.56

PROP: Nonvolatile, stable solid. Nonflammable. Insol in water.

SYNS: BIS(DITHIOPHOSPHATE de O,O-DIETHYLE) de S,S'-(1,4-DIOXANNE-2,3-DIYLE) (FRENCH) ◇ DELNAV ◇ 1,4-DIOSSAN-2,3-DIYL-BIS(O,O-DIETIL-DITIOFOSFATO) (ITALIAN) ◇ 1,4-DIOXAAN-2,3-DIYL-BIS(O,O-DIETHYL-DITHIOFOSFAAT) (DUTCH) ◇ 2,3-p-DIOXANDITHIOL S,S-BIS(O,O-DIETHYL PHOSPHORODITHIOATE) ◇ 1,4-DIOXAN-2,3-DIYL-BIS(O,O-DIAETHYL-DITHIOPHOSPHAT) (GERMAN) ◇ 1,4-DIOXAN-2,3-DIYL-BIS(O,O-DIETHYLPHOSPHOROTHIOLOTHIONATE) ◇ 1,4-DIOXAN-2,3-DIYL-O,O,O',O'-TETRAETHYL DI(PHOSPHOROMITHIOATE) ◇ 2,3-p-DIOXANE-S,S-BIS(O,O-DIETHYLPHOSPHOROITHIOATE) ◇ p-DIOXANE-2,3-DITHIOL-S,S-DIESTER with O,O-DIETHYL PHOSPHORODITHIOATE ◇ p-DIOXANE-2,3-DIYL ETHYL PHOSPHORODITHIOATE ◇ ENT 22,897 ◇ NCI-C00395 ◇ PHOSPHORODITHIOIC ACID-S,S'-1,4-DIOXANE-2,3-DIYL O,O,O',O'-TETRAETHYL ESTER

TOXICITY DATA with REFERENCE
skn-rat LD50:63 mg/kg TXAPA9 5,605,63
orl-rat LD50:20 mg/kg WRPCA2 9,119,70
ihl-rat LC50:1398 mg/m^3/1H TXAPA9 5,605,63
ipr-rat LD50:30 mg/kg TXAPA9 5,605,63
orl-mus LD50:176 mg/kg TXAPA9 5,605,63
ihl-mus LC50:340 mg/m^3/1H TXAPA9 5,605,63
ipr-mus LD50:33 mg/kg PSEBAA 129,699,68
orl-dog LD50:10 mg/kg PCOC** -,427,66
skn-rbt LD50:85 mg/kg PCOC** -,427,66
orl-ckn LD50:170 mg/kg 32ZXAD 37,A10,75

CONSENSUS REPORTS: NCI Carcinogenesis Bioassay (feed); No Evidence: mouse, rat NCITR* NCI-CG-TR-125,78.

OSHA PEL: TWA 0.2 mg/m^3 (skin)
ACGIH TLV: TWA 0.2 mg/m^3 (skin)

SAFETY PROFILE: Poison by ingestion, inhalation,

skin contact, and intraperitoneal routes. A cholinesterase inhibitor. When heated to decomposition it emits very toxic fumes of PO_x and SO_x. See also PARATHION.

DVQ759 CAS:59261-17-5 **HR: 3**
cis-1,4-DIOXENEDIOXETANE
mf: $C_4H_6O_4$ mw: 118.09

SYN: 2,5,7,8-TETRAOXA[4.2.0]BICYCLOOCTANE

SAFETY PROFILE: Explodes at room temperature. When heated to decomposition it emits acrid smoke and fumes. See also PEROXIDES.

DVQ800 CAS:17311-31-8 **HR: 2**
1,4-DI-N-OXIDE of DIHYDROXYMETHYLQUINOXALINE
mf: $C_{10}H_{12}N_2O_4$ mw: 222.22

SYNS: 2,3-BIS(HYDROXYMETHYL)QUINOXALINEDI-N-OXIDE ◇ 1,4-DI-N-OXIDE 2,3-BIS(OXYMETHYL)QUINOXLINE ◇ 1,4-DIOXIDE-2,3-QUINOXALINEDIMETHANOL ◇ DIOXIDIN ◇ DIOXIDINE ◇ DIOXYDINE

TOXICITY DATA with REFERENCE
mmo-sat 750 μg/L CYGEDX 14(1),57,80
mmo-esc 4 mg/L CYGEDX 14(1),57,80
dnr-esc 100 μg/L PCJOAU 15,721,82
dnd-esc 100 μg/L KHFZAN 16(10),11,82
pic-esc 30 mg/L TGANAK 16(6),38,82
scu-rat TDLo:300 mg/kg (11D preg):TER FATOAO 45(6),85,82
scu-rat TDLo:300 mg/kg (11D preg):REP FATOAO 45(6),85,82
ipr-mus LD50:750 mg/kg PCJOAU 14,440,80

CONSENSUS REPORTS: EPA Genetic Toxicology Program.

SAFETY PROFILE: Moderately toxic by intraperitoneal route. Experimental teratogenic and reproductive effects. Mutation data reported. When heated to decomposition it emits toxic fumes of NO_x.

DVR000 CAS:107-61-9 **HR: 3**
4,4-DIOXIDE-1,4-OXATHIANE
mf: $C_4H_8O_3S$ mw: 136.18

SYNS: p-OXATHIANE-4,4-DIOXIDE ◇ USAF DO-38

TOXICITY DATA with REFERENCE
ipr-mus LD50:200 mg/kg NTIS** AD277-689

CONSENSUS REPORTS: Reported in EPA TSCA Inventory.

SAFETY PROFILE: Poison by intraperitoneal route. When heated to decomposition it emits toxic fumes of SO_x.

DVR200 CAS:105-11-3 **HR: 3**
DIOXIME-p-BENZOQUINONE
mf: $C_6H_6N_2O_2$ mw: 138.14

SYNS: ACTOR Q ◇ 1,4-BENZOQUINONE DIOXINE ◇ 2,5-CYCLOHEXADIENE-1,4-DIONE DIOXIME ◇ DIBENZO PQD ◇ DIOXIME-1,4-CYCLOHEXADIENEDIONE ◇ DIOXIME-2,5-CYCLOHEXADIENE-1,4-DIONE ◇ G-M-F ◇ NCI-C03850 ◇ PQD ◇ QDO ◇ QUINONE DIOXIME ◇ p-QUINONE DIOXIME ◇ p-QUINONE OXIME

TOXICITY DATA with REFERENCE
mmo-sat 3300 ng/plate ENMUDM 7(Suppl 5),1,85
mma-sat 10 μg/plate ENMUDM 7(Suppl 5),1,85
dnr-bcs 1 mg/disc SAIGBL 26,147,84
orl-rat TDLo:14 g/kg/2Y-C:NEO NCITR* NCI-CG-TR-179,79
orl-mus TDLo:131 g/kg/104W-C:ETA NCITR* NCI-CG-TR-179,79
orl-rat LD50:464 mg/kg NCILB* NIH-NCI-E-C-72-3252,73
orl-mus LD50:1542 mg/kg GTPZAB 8(7),39,64

CONSENSUS REPORTS: IARC Cancer Review: Group 3 IMEMDT 7,56,87; Animal Limited Evidence IMEMDT 29,185,82. NCI Carcinogenesis Bioassay (feed); Clear Evidence: rat NCITR* NCI-CG-TR-179,79; No Evidence: mouse NCITR* NCI-CG-TR-179,79. Reported in EPA TSCA Inventory.

SAFETY PROFILE: Moderately toxic by ingestion. Questionable carcinogen with experimental neoplastigenic and tumorigenic data. Mutation data reported. When heated to decomposition it emits toxic fumes of NO_x.

DVR400 CAS:6483-86-9 **HR: D**
9,10-DIOXO-9,10-DIHYDRO-1-NITRO-6-ANTHRACENESULFONIC ACID
mf: $C_{14}H_7NO_7S$ mw: 333.28

SYN: 1-NITRO-6(7)SULFONATOANTHRAQUINONE

TOXICITY DATA with REFERENCE
mmo-sat 50 μg/plate MUREAV 40,203,76
mma-sat 50 μg/plate MUREAV 40,203,76

SAFETY PROFILE: Mutation data reported. When heated to decomposition it emits very toxic fumes of NO_x and SO_x.

DVR600 CAS:100-79-8 **HR: 1**
DIOXOLAN
DOT: UN 1166
mf: $C_6H_{12}O_3$ mw: 132.18

PROP: Water-white liquid. Mp: −26.4°, bp: 75°, flash p: 35°F (OC), d: 1.065, vap press: 70 mm @ 20°, vap d: 2.6.

SYNS: CYCLIC (HYDROXYMETHYL)ETHYLENE ACETAL ACETONE ◇ 2,2-DIMETHYL-1,3-DIOXOLANE-4-METHANOL ◇ 2,2-DIMETHYL-5-HYDROXYMETHYL-1,3-DIOXOLANE ◇ 2,2-DIMETHYL-4-

OXYMETHYL-1,3-DIOXOLANE ◇ DIOXOLANE (DOT) ◇ GIE ◇ GLYCEROLACETONE ◇ GLYCEROL DIMETHYLKETAL ◇ 4-HYDROXYMETHYL-2,2-DIMETHYL-1,3-DIOXOLANE ◇ ISOPROPYLIDENE GLYCEROL ◇ 1,2-o-ISOPROPYLIDENE GLYCEROL ◇ SOLKETAL

TOXICITY DATA with REFERENCE
eye-rbt 100 mg TXAPA9 39,129,77
mnt-mus-ipr 1500 mg/kg TOLED5 21,349,84

CONSENSUS REPORTS: Reported in EPA TSCA Inventory.

DOT Classification: Flammable Liquid; Label: Flammable Liquid.

SAFETY PROFILE: An eye irritant. Mutation data reported. A very dangerous fire hazard when exposed to heat or flame; can react vigorously with oxidizing materials. To fight fire, use alcohol foam, CO_2, dry chemical. When heated to decomposition it emits acrid smoke and fumes.

DVR800 CAS:646-06-0 ***HR: 2***
1,3-DIOXOLANE
mf: $C_3H_6O_2$ mw: 74.09

PROP: Flash p: 35.6°F.

SYNS: 1,3-DIOXACYCLOPENTANE ◇ 1,3-DIOXOLAN ◇ ETHYLAENE GLYCOL FORMAL ◇ FORMAL GLYCOL ◇ GLYCOL FORMAL

TOXICITY DATA with REFERENCE
skn-rbt 530 mg open MLD UCDS** 12/17/71
eye-rbt 750 μg open SEV JIHTAB 31,60,49
orl-rat LD50:3000 mg/kg HIHTAB 31,60,49
ihl-rat LD50:20650 mg/m³/4H 85GMAT -,70,82
ipr-rat LDLo:500 mg/kg JPPMAB 11,150,59
orl-mus LD50:3200 mg/kg 85GMAT -,70,82
ihl-mus LC50:104 g/m³ GTPZAB 19(8),45,75
ihl-rbt LCLo:32000 ppm/4H UCDS** 12/17/71
skn-rbt LD50:8480 mg/kg UCDS** 12/17/71

CONSENSUS REPORTS: Reported in EPA TSCA Inventory.

SAFETY PROFILE: Moderately toxic by ingestion and intraperitoneal routes. Mildly toxic by skin contact and inhalation. A skin and severe eye irritant. A very dangerous fire hazard when exposed to heat or flame; can react with oxidizers. Used in lithium batteries. Potentially explosive reaction with lithium perchlorate. When heated to decomposition it emits acrid smoke and irritating fumes.

DVR909 CAS:5464-28-8 ***HR: 1***
1,3-DIOXOLANE-4-METHANOL
mf: $C_4H_8O_3$ mw: 104.12

SYNS: GF ◇ GLYCERINFORMALE ◇ GLYCEROL FORMAL ◇ SERICOSOL-N

TOXICITY DATA with REFERENCE
orl-rat TDLo:6600 mg/kg (female 7-17D post):REP TCMUD8 7,73,87
orl-rat TDLo:6 g/kg (6-15D preg):TER TXAPA9 56,93,80
ipr-rat LD50:9500 mg/kg APFRAD 44,293,86
orl-mus LD50:8 g/kg APFRAD 44,293,86
ipr-mus LD50:7500 mg/kg APFRAD 44,293,86

SAFETY PROFILE: Mildly toxic by intraperitoneal routes. An experimental teratogen. Other experimental reproductive effects. When heated to decomposition it emits acrid smoke and fumes.

DVS000 CAS:6988-21-2 ***HR: 3***
o-(1,3-DIOXOLAN-2-YL)PHENYL METHYLCARBAMATE
mf: $C_{11}H_{13}NO_4$ mw: 223.25

SYNS: CIBA 8353 ◇ DIOXACARB ◇ 2-(1,3-DIOXOLANE-2-YL)PHENYL N-METHYLCARBAMATE ◇ 2-(1,3-DIOXOLAN-2-YL)PHENYL-N-METHYLCARBAMAT ◇ DU PONT INSECTICIDE 1519 ◇ ELOCRON ◇ ENT 27,389 ◇ FAMID ◇ NSC 190981

TOXICITY DATA with REFERENCE
orl-rat LD50:60 mg/kg FMCHA2 -,D123,80
skn-rat LD50:3000 mg/kg FMCHA2 -,C94,83
ipr-rat LD50:8300 μg/kg PESTD5 17,351,76
orl-mus LD50:48 mg/kg PESTD5 17,351,76
skn-mus LD50:1660 mg/kg BESAAT 15,133,69
ipr-mus LD50:20 mg/kg PESTD5 17,351,76
skn-rbt LD50:1950 mg/kg BESAAT 15,133,69

SAFETY PROFILE: Poison by ingestion and intraperitoneal routes. Moderately toxic by skin contact. Mutation data reported. A toxic contact and systemic insecticide. When heated to decomposition it emits toxic fumes of NO_x.

DVS100 CAS:76059-11-5 ***HR: 3***
3-(2-(1,3-DIOXO-2-METHYLINDANYL)) GLUTARIMIDE
mf: $C_{15}H_{13}NO_4$ mw: 271.29

SYN: GLUTARIMIDE,3-(1,3-DIOXO-2-METHYLINDAN-2-YL)-

TOXICITY DATA with REFERENCE
ipr-mus TDLo:10 mg/kg (female 9D post):REP ARPMAS 313,481,80
ipr-mus TDLo:10 mg/kg (female 9D post):TER ARPMAS 313,481,80
ipr-mus LD50:28 mg/kg ARPMAS 313,481,80

SAFETY PROFILE: Poison by intraperitoneal route. An experimental teratogen. Other experimental reproductive effects. When heated to decomposition it emits toxic fumes of NO_x.

DVS200 CAS:82-50-8 ***HR: D***
9,10-DIOXO-1-NITRO-9,10-DIHYDRO-5-ANTHRACENESULFONIC ACID
mf: $C_{14}H_7NO_7S$ mw: 333.28

SYN: 1-NITRO-5-SULFONATO-ANTHRAQUINONE

TOXICITY DATA with REFERENCE
mmo-sat 500 μg/plate MUREAV 40,203,76
mma-sat 500 μg/plate MUREAV 40,203,76

SAFETY PROFILE: Mutation data reported. When heated to decomposition it emits very toxic fumes of NO_x and SO_x.

DVS300 CAS:76059-13-7 ***HR: 3***
3-(2-(1,3-DIOXO-2-PHENYLINDANYL)) GLUTARIMIDE
mf: $C_{20}H_{15}NO_4$ mw: 333.36

SYN: GLUTARIMIDE,3-(1,3-DIOXO-2-PHENYLINDAN-2-YL)-

TOXICITY DATA with REFERENCE
ipr-mus TDLo:2500 μg/kg (female 9D post):REP ARPMAS 313,481,80
ipr-mus TDLo:2500 μg/kg (female 9D post):TER ARPMAS 313,481,80
ipr-mus LD50:14 mg/kg ARPMAS 313,481,80

SAFETY PROFILE: Poison by ingestion. An experimental teratogen. Other experimental reproductive effects. When heated to decomposition it emits toxic fumes of NO_x.

DVS400 CAS:76059-14-8 ***HR: 3***
3-(2-(1,3-DIOXO-2-PHENYL-4,5,6,7-TETRAHYDRO-4,7-DITHIAINDANYL))GLUTARIMIDE
mf: $C_{18}H_{15}NO_4S_2$ mw: 373.46

SYN: GLUTARIMIDE,3-(5,7-DIOXO-6-PHENYL-2,3,6,7-TETRAHYDRO-5H-CYCLOPENTA-p-DITHIIN-6-YL)-

TOXICITY DATA with REFERENCE
ipr-mus TDLo:1200 μg/kg (female 9D post):REP ARPMAS 313,481,80
ipr-mus TDLo:2500 μg/kg (female 9D post):TER ARPMAS 313,481,80
ipr-mus LD50:7 mg/kg ARPMAS 313,481,80

SAFETY PROFILE: Poison by intraperitoneal route. An experimental teratogen. Other experimental reproductive effects. When heated to decomposition it emits toxic fumes of NO_x and SO_x.

DVS600 CAS:26581-81-7 ***HR: 2***
2-(2,6-DIOXOPIPERIDEN-3-YL) PHTHALIMIDINE
mf: $C_{13}H_9N_2O_3$ mw: 241.24

SYNS: 3-(1,3-DIHYDRO-1-OXO-2H-ISOINDOL-2-YL)-2,6-DIOXO PIPERIDINE ◇ EM 12

TOXICITY DATA with REFERENCE
orl-rat TDLo:1 g/kg (9-12D preg):TER TJADAB 5,233,72
orl-rat TDLo:1 g/kg (9-12D preg):REP TJADAB 5,233,72
orl-mus LD50:6000 mg/kg ARZNAD 31,941,81
ipr-mus LD50:1830 mg/kg ARZNAD 31,941,81

SAFETY PROFILE: Moderately toxic by intraperitoneal route. Mildly toxic by ingestion. Experimental teratogenic and reproductive effects. When heated to decomposition it emits toxic fumes of NO_x.

DVT200 CAS:731-40-8 ***HR: D***
(±)-N-(2,6-DIOXO-3-PIPERIDYL)PHTHALIMIDE
mf: $C_{13}H_{10}N_2O_4$ mw: 258.25

SYN: (+,−)-THALIDOMIDE

TOXICITY DATA with REFERENCE
orl-rbt TDLo:900 mg/kg (7-12D preg):REP NATUAS 215,296,67
orl-rbt TDLo:900 mg/kg (7-12D preg):TER NATUAS 215,296,67

SAFETY PROFILE: An experimental teratogen. Other experimental reproductive effects. When heated to decomposition it emits toxic fumes of NO_x.

DVT400 CAS:13754-56-8 ***HR: 3***
DIOXOPROMETHAZINE HYDROCHLORIDE
mf: $C_{17}H_{20}N_2O_2S•ClH$ mw: 352.91

SYN: 5,5-DIOXO-10-(2-(DIMETHYLAMINO)PROPYL)PHENOTHIAZINE HYDROCHLORIDE

TOXICITY DATA with REFERENCE
orl-mus LD50:70 mg/kg PHARAT 25,91,70
ipr-mus LD50:49 mg/kg PHARAT 25,91,70
scu-mus LD50:96 mg/kg PHARAT 25,91,70
ivn-mus LD50:19 mg/kg PHARAT 25,91,70
orl-rbt LD50:165 mg/kg PHARAT 25,91,70
orl-gpg LD50:225 mg/kg PHARAT 25,91,70
scu-gpg LD50:33 mg/kg PHARAT 25,91,70
ivn-gpg LD50:8 mg/kg PHARAT 25,91,70

SAFETY PROFILE: Poison by ingestion, intraperitoneal, intravenous, and subcutaneous routes. When heated to decomposition it emits very toxic fumes of NO_x, SO_x, and HCl.

DVT459 CAS:31083-55-3 ***HR: 3***
1,3-DIOXO-2-(3-PYRIDYLMETHYLENE)INDAN
mf: $C_{15}H_9NO_2$ mw: 235.25

TOXICITY DATA with REFERENCE
ipr-mus TDLo:80 mg/kg (9D preg):REP ARPMAS 313,481,80
ipr-mus TDLo:40 mg/kg (9D preg):TER ARPMAS 313,481,80
ipr-mus LD50:200 mg/kg ARPMAS 313,481,80

SAFETY PROFILE: Poison by intraperitoneal route. An experimental teratogen. Other experimental reproductive effects. When heated to decomposition it emits toxic fumes of NO_x.

DVT500 **HR: 1**
DIOXYBIS(2,2'-DI-tert-BUTYLBUTANE
mf: $C_{24}H_{50}O_2$ mw: 370.74

SYN: 2,2-BIS-DI-tert-BUTYLPEROXYBUTANE

TOXICITY DATA with REFERENCE
skn-rbt 500 mg MLD SCCUR* -,2,61
orl-rat LD50:12500 mg/kg FEPRA7 7,252,48
orl-mus LD50:17500 mg/kg FEPRA7 7,252,48

SAFETY PROFILE: Mildly toxic by ingestion. A skin irritant. When heated to decomposition it emits acrid smoke and fumes. See also PEROXIDES.

DVT800 CAS:12228-13-6 **HR: 3**
DIOXYGENYL TETRAFLUOROBORATE
mf: BFN_4O_2 mw: 118.81

SAFETY PROFILE: A very powerful oxidant. Explodes in methane or ethane. Ignites in benzene or 2-propanol. When heated to decomposition it emits toxic fumes of F^-. See also BORON COMPOUNDS and FLUORIDES.

DVU000 CAS:580-74-5 **HR: 3**
1,4-DI-p-OXYPHENYL-2,3-DI-ISONITRILO-1,3-BUTADIENE
mf: $C_{18}H_{12}N_2O_2$ mw: 288.32

SYNS: OPHTHOCILLIN ◇ XANTHOCILLIN ◇ XANTYRID

TOXICITY DATA with REFERENCE
orl-mus LD50:45 mg/kg ARZNAD 7,98,57
ipr-mus LD50:20 mg/kg ARZNAD 7,98,57

CONSENSUS REPORTS: Cyanide and its compounds are on the Community Right-To-Know List.

SAFETY PROFILE: Poison by ingestion and intraperitoneal routes. When heated to decomposition it emits toxic fumes of NO_x and CN^-. See also NITRILES.

DVU100 CAS:519-65-3 **HR: 2**
DIOXYPYRAMIDON
mf: $C_{13}H_{17}N_3O_3$ mw: 263.33

PROP: Orthorhombic, translucent prisms from water. Somewhat bitter taste. Mp: 105.5° (softens at 96°), bp: (2) 194-201°. Solubility in water at 20° = 7.69 g/100 mL; at 37° = 48.2 g/100 mL. Also sol in alc.

SYNS: 1-ACETYL-2-PHENYL-1,5,5-TRIMETHYL-SEMIOXAMAZIDE ◇ (DIMETHYLAMINO)OXO-ACETIC ACID 2-ACETYL-2-METHYL-1-PHENYLHYDRAZIDE (9CI) ◇ DIOXOAMINOPYRINE ◇ DIOXYAMINOPYRINE

TOXICITY DATA with REFERENCE
orl-mus LD50:1631 mg/kg AEPPAE 213,501,51
scu-mus LD50:1066 mg/kg AEPPAE 213,501,51
ivn-mus LD50:698 mg/kg AEPPAE 213,501,51

SAFETY PROFILE: Moderately toxic by ingestion, subcutaneous, and intravenous routes. When heated to decomposition it emits toxic fumes of NO_x.

DVU200 **HR: 3**
DIPALLADIUM TRIOXIDE
mf: Pd_2O_3 mw: 260.80

SAFETY PROFILE: Hydrated oxide explodes when heated. See also PALLADIUM.

DVU300 **HR: 2**
DIPENICILLIN-G-ALUMINIUM-SULPHAMETHOXYPYRIDAZINE
mf: $C_{43}H_{45}AlN_8O_{11}S_3$ mw: 973.12

SYNS: AB 109 ◇ DIPENICILLINA-G-ALLUMINIO-SULFAMETOSSIPIRIDAZINA (ITALIAN)

TOXICITY DATA with REFERENCE
ipr-mus LD50:3580 mg/kg BCFAAI 98,453,59
scu-mus LD50:5720 mg/kg BCFAAI 98,453,59
ipr-gpg LD50:3000 mg/kg BCFAAI 98,453,59
scu-gpg LD50:4500 mg/kg BCFAAI 98,453,59

ACGIH TLV: TWA 2 mg(Al)/m^3

SAFETY PROFILE: Moderately toxic by intraperitoneal route. Mildly toxic by subcutaneous route. When heated to decomposition it emits toxic fumes of SO_x and NO_x. See also PENICILLIN and ORGANOMETALS.

DVU600 CAS:18279-20-4 **HR: 3**
DIPENTYL LEAD DIACETATE
mf: $C_{14}H_{28}O_4Pb$ mw: 467.61

TOXICITY DATA with REFERENCE
orl-mus LD50:90 mg/kg CRSBAW 162,1456,68

CONSENSUS REPORTS: Lead and its compounds are on the Community Right-To-Know List.

SAFETY PROFILE: Poison by ingestion. When heated to decomposition it emits toxic fumes of Pb. See also LEAD COMPOUNDS.

DVV000 CAS:2273-46-3 **HR: 3**
DIPENTYLOXOSTANNANE
mf: $C_{10}H_{22}OSn$ mw: 277.01

SYNS: DIPENTYLTIN OXIDE ◇ KYSLICNIK DI-n-AMYLCINICITY (CZECH)

TOXICITY DATA with REFERENCE
skn-rbt 500 mg/24H SEV 28ZPAK -,227,72

eye-rbt 100 mg/24H MOD 28ZPAK -,227,72
orl-rat LD50:55200 μg/kg 28ZPAK -,227,72

OSHA PEL: TWA 0.1 mg(Sn)/m^3 (skin)
ACGIH TLV: TWA 0.1 mg(Sn)/m^3 (skin) (Proposed: TWA 0.1 mg(Sn)/m^3; STEL 0.2 mg(Sn)/m^3 (skin))
NIOSH REL: (Organotin Compounds) TWA 0.1 mg(Sn)/m^3

SAFETY PROFILE: Poison by ingestion. An eye and severe skin irritant. When heated to decomposition it emits acrid smoke and fumes. See also TIN COMPOUNDS.

DVV109 CAS:13403-01-5 ***HR: 3***
2-(2,4-DI-tert-PENTYLPHENOXY)BUTYRIC ACID

TOXICITY DATA with REFERENCE
ipr-rat LD50:400 mg/kg KODAK* -,-,71

CONSENSUS REPORTS: Reported in EPA TSCA Inventory.

SAFETY PROFILE: Poison by intraperitoneal route. When heated to decomposition it emits acrid smoke and irritating fumes.

DVV200 CAS:1118-42-9 ***HR: 3***
DIPENTYLTIN DICHLORIDE
mf: $C_{10}H_{22}Cl_2Sn$ mw: 331.91

SYN: DICHLORODIPENTYLSTANNANE

TOXICITY DATA with REFERENCE
ivn-rat LDLo:10 mg/kg BJIMAG 15,15,58

OSHA PEL: TWA 0.1 mg(Sn)/m^3 (skin)
ACGIH TLV: TWA 0.1 mg(Sn)/m^3 (skin) (Proposed: TWA 0.1 mg(Sn)/m^3; STEL 0.2 mg(Sn)/m^3 (skin))
NIOSH REL: (Organotin Compounds) TWA 0.1 mg(Sn)/m^3

SAFETY PROFILE: Poison by intravenous route. When heated to decomposition it emits toxic fumes of Cl^-. See also TIN COMPOUNDS and CHLORIDES.

DVV400 ***HR: 3***
2,6-DIPERCHLORYL-4,4'-DIPHENOQUINONE
mf: $C_{12}H_6Cl_2O_8$ mw: 369.09

SAFETY PROFILE: A shock-sensitive explosive. When heated to decomposition it emits toxic fumes of Cl^-.

DVV500 CAS:537-12-2 ***HR: 3***
DIPERODON HYDROCHLORIDE
mf: $C_{22}H_{27}N_3O_4 \cdot ClH$ mw: 433.98

PROP: Crystals; bitter taste followed by a sense of numbness. Decomp @ 195-200°. Sol in alc; sltly sol in water, acetone and ethyl acetate; insol in benzene or ether.

SYNS: DIOTHANE HYDROCHLORIDE ◇ DIPERDON HYDROCHLORIDE ◇ 3-PIPERIDINO-1,2-PROPANEDIOL DICARBANILATE HYDROCHLORIDE ◇ 3-(1-PIPERIDYL)-1,2-PROPANEDIOL DICARBANILATE HYDROCHLORIDE ◇ PROCTODON

TOXICITY DATA with REFERENCE
scu-mus LD50:890 mg/kg JAPMA8 48,398,59
scu-rbt LDLo:300 mg/kg JPETAB 47,255,33
ivn-rbt LDLo:15 mg/kg JPETAB 47,255,33
scu-gpg LDLo:400 mg/kg JPETAB 47,255,33

SAFETY PROFILE: Poison by subcutaneous and intravenous routes. When heated to decomposition it emits toxic fumes of NO_x and HCl.

DVV550 CAS:1711-42-8 ***HR: 3***
DIPEROXYTEREPHTHALIC ACID
mf: $C_8H_6O_6$ mw: 198.13

SAFETY PROFILE: An impact- and heat-sensitive explosive. When heated to decomposition it emits acrid smoke and fumes. See also PEROXIDES.

DVV600 CAS:82-66-6 ***HR: 3***
DIPHENADIONE
mf: $C_{23}H_{16}O_3$ mw: 340.39

PROP: Pale yellow crystals. Mp: 147°. Sol in acetone and acetic acid.

SYNS: DIDANDIN ◇ DIPAXIN ◇ DIPHACIN ◇ DIPHACINONE ◇ DIPHENACIN ◇ 2-DIPHENYLACETYL-1,3-DIKETOHYDRINDENE ◇ 2-DIPHENYLACETYL-1,3-INDANDIONE ◇ 2-(DIPHENYLACETYL) INDAN-1,3-DIONE ◇ 2-(DIPHENYLACETYL)-1H-INDENE-1,3(2H)-DIONE ◇ PID ◇ PROMAR ◇ RAMIK ◇ RATINDAN 1 ◇ U 1363

TOXICITY DATA with REFERENCE
orl-rat LD50:1500 μg/kg 85DPAN -,-,71/76
orl-mus LD50:340 mg/kg 85DPAN -,-,71/76
orl-dog LD50:3 mg/kg 28ZEAL 5,84,76
orl-cat LD50:15 mg/kg PCOC** -,429,66
orl-rbt LD50:35 mg/kg 85DPAN -,-,71/76
orl-pig LD50:150 mg/kg 28ZEAL 5,84,76
orl-mam LD50:910 μg/kg SCIEAS 177,806,72

CONSENSUS REPORTS: EPA Extremely Hazardous Substances List.

SAFETY PROFILE: Poison by ingestion. Inhibits blood clotting, leading to hemorrhages. Action similar to coumadin (warfarin). A pesticide used in rodent control. When heated to decomposition it emits acrid smoke and irritating fumes.

DVV800 CAS:1210-05-5 ***HR: 2***
DIPHENALDEHYDE
mf: $C_{14}H_{10}O_2$ mw: 210.24

SYNS: 2,2′-BIPHENYLDICARBOXALDEHYDE ◇ 2,2′-DIFORMYL-BIPHENYL

TOXICITY DATA with REFERENCE
skn-rbt 500 mg MOD IHFCAY 6,1,67
eye-rbt 500 mg SEV IHFCAY 6,1,67
orl-rat LD50:2830 mg/kg IHFCAY 6,1,67

SAFETY PROFILE: Moderately toxic by ingestion. A skin and severe eye irritant. When heated to decomposition it emits acrid smoke and irritating fumes. See also ALDEHYDES.

DVW000 CAS:1798-49-8 ***HR: 3***
DIPHENCHLOXAZINE HYDROCHLORIDE
mf: $C_{19}H_{22}ClNO_2•ClH$ mw: 368.33

SYN: PHENYL-p-CHLOROPHENYLTETRAHYDRO-1,4-OXAZINYLETHOXYMETHANE HYDROCHLORIDE

TOXICITY DATA with REFERENCE
orl-rat LD50:710 mg/kg 27ZQAG -,223,72
scu-mus LD50:368 mg/kg YKKZAJ 87,1109,67
ivn-mus LD50:140 mg/kg 27ZQAG -,223,72

SAFETY PROFILE: Poison by subcutaneous and intravenous routes. Moderately toxic by ingestion. When heated to decomposition it emits very toxic fumes of NO_x and Cl^-.

DVW100 CAS:82-21-3 ***HR: 1***
1,5-DIPHENOXYANTHRAQUINONE
mf: $C_{26}H_{16}O_4$ mw: 392.42

SYNS: 9,10-ANTHRACENEDIONE, 1,5-DIPHENOXY- ◇ 1,5-DIFENOXYANTHRACHINON

TOXICITY DATA with REFERENCE
eye-rbt 500 mg/24H MLD 85JCAE-,705,86

CONSENSUS REPORTS: Reported in EPA TSCA Inventory.

SAFETY PROFILE: An eye irritant. When heated to decomposition it emits acrid smoke and irritating fumes.

DVW600 CAS:622-04-8 ***HR: 3***
1,3-DIPHENOXY-2-PROPANOL
mf: $C_{15}H_{16}O_3$ mw: 244.31

SYN: GLYCERYL-α,Γ-DIPHENYL ETHER

TOXICITY DATA with REFERENCE
skn-rbt 10 mg/24H open MLD AMIHBC 4,119,51
orl-rat LD50:1450 mg/kg AMIHBC 4,119,51
ivn-mus LD50:180 mg/kg CSLNX* NX#04474
skn-rbt LD50:15800 mg/kg AMIHBC 4,119,51

CONSENSUS REPORTS: Reported in EPA TSCA Inventory.

SAFETY PROFILE: Poison by intravenous route. Moderately toxic by ingestion. Mildly toxic by skin contact. A skin irritant. When heated to decomposition it emits acrid smoke and irritating fumes.

DVW700 CAS:132-18-3 ***HR: 3***
DIPHENPYRALINE HYDROCHLORIDE
mf: $C_{19}H_{23}NO•ClH$ mw: 317.89

SYNS: DIAFEN ◇ DIAFEN (antihistamine) ◇ DIAFEN HYDROCHLORIDE ◇ DIAPHEN ◇ DIPHENYLPYRALINE HYDROCHLORIDE ◇ HISPRIL HYDROCHLORIDE ◇ SUMADIL

TOXICITY DATA with REFERENCE
orl-rat LD50:698 mg/kg NIIRDN 6,333,82
ivn-rat LD50:28800 μg/kg NIIRDN 6,333,82
scu-rat LD50:278 mg/kg NIIRDN 6,333,82
ims-rat LD50:180 mg/kg NIIRDN 6,333,82
orl-mus LD50:202 mg/kg NIIRDN 6,333,82
scu-mus LD50:112 mg/kg NIIRDN 6,333,82
ivn-mus LD50:31800 μg/kg NIIRDN 6,333,82
ims-mus LD50:64300 μg/kg NIIRDN 6,333,82

CONSENSUS REPORTS: Reported in EPA TSCA Inventory.

SAFETY PROFILE: Poison by ingestion, subcutaneous, intramuscular, and intravenous routes. When heated to decomposition it emits toxic fumes of NO_x and HCl.

DVW800 CAS:117-34-0 ***HR: 3***
DIPHENYLACETIC ACID
mf: $C_{14}H_{12}O_2$ mw: 212.26

PROP: White crystals. Mp: 147-148.2°, bp: sublimes, vap d: 7.3.

SYNS: α,α-DIPHENYLACETIC ACID ◇ α-PHENYLBENZENEACETIC ACID

TOXICITY DATA with REFERENCE
orl-rat LD50:5540 mg/kg GNAMAP 17,48,78
orl-mus LD50:3200 mg/kg GNAMAP 17,48,78
ipr-mus LD50:500 mg/kg FRPSAX 13,286,58
scu-mus LD50:400 mg/kg AIPTAK 116,154,58

CONSENSUS REPORTS: Reported in EPA TSCA Inventory.

SAFETY PROFILE: Poison by subcutaneous route. Moderately toxic by ingestion and intraperitoneal routes. When heated to decomposition it emits acrid smoke and irritating fumes.

DVW900 CAS:1984-87-8 ***HR: 2***
DIPHENYLACETIC ACID 2-(BIS(2-HYDROXYETHYL)AMINO)ETHYL ESTER HYDROCHLORIDE
mf: $C_{20}H_{25}NO_4•ClH$ mw: 379.92

SYN: ACETIC ACID, DIPHENYL-, 2-(BIS(2-HYDROXYETHYL) AMINO)ETHYL ESTER, HYDROCHLORIDE

TOXICITY DATA with REFERENCE
eye-cat 5% APPHAX 21,557,64
scu-mus LD50:1380 mg/kg APPHAX 21,557,64

SAFETY PROFILE: Moderately toxic by subcutaneous route. An eye irritant. When heated to decomposition it emits toxic fumes of NO_x and HCl.

DVX200 CAS:86-29-3 ***HR: 3***
DIPHENYLACETONITRILE
mf: $C_{14}H_{11}N$ mw: 193.26

SYNS: BENZYHYDRYLCYANIDE ◇ α-CYANODIPHENYLMETHANE ◇ DIPAN ◇ DIPHENATRILE ◇ DIPHENYL-α-CYANOMETHANE ◇ DIPHENYLMETHYLCYANIDE ◇ α-PHENYLBENZYLCYANIDE ◇ α-PHENYLPHENYLACETONITRILE ◇ USAF KF-13

TOXICITY DATA with REFERENCE
orl-mus TDLo:61 g/kg/78W-I:ETA NTIS** PB223-159
scu-mus TDLo:464 mg/kg:CAR NTIS** PB223-159
orl-rat LD50:3500 mg/kg RREVAH 10,97,65
ipr-mus LD50:200 mg/kg NTIS** AD691-490
ivn-mus LD50:100 mg/kg CSLNX* NX#04134

CONSENSUS REPORTS: Reported in EPA TSCA Inventory. Cyanide and its compounds are on the Community Right-To-Know List.

SAFETY PROFILE: Poison by ingestion, intraperitoneal, and intravenous routes. Moderately toxic by subcutaneous route. Questionable carcinogen with experimental carcinogenic and tumorigenic data. When heated to decomposition it emits toxic fumes of NO_x and CN^-. See also NITRILES.

DVX600 CAS:2510-95-4 ***HR: 3***
2,3-DIPHENYLACRYLONITRILE
mf: $C_{15}H_{11}N$ mw: 205.27

SYNS: BENZAL-(BENZYL-CYANID) (GERMAN) ◇ BENZYLIDENEPHENYLACETONITRILE ◇ α-CYANOSTILBENE ◇ α,β-DIPHENYLACRYLONITRILE ◇ F 2387 ◇ α-PHENYLCINNAMONITRILE ◇ α-(PHENYLMETHYLENE)BENZENEACETONITRILE ◇ α-STILBENECARBONITRILE ◇ USAF A-9789

TOXICITY DATA with REFERENCE
ipr-mus LD50:100 mg/kg NTIS** AD277-689

CONSENSUS REPORTS: Reported in EPA TSCA Inventory. Cyanide and its compounds are on the Community Right-To-Know List.

SAFETY PROFILE: Poison by intraperitoneal route. When heated to decomposition it emits toxic fumes of NO_x and CN^-. See also NITRILES.

DVX800 CAS:122-39-4 ***HR: 3***
DIPHENYLAMINE
mf: $C_{12}H_{11}N$ mw: 169.24

PROP: Crystals; floral odor. Mp: 52.9°, bp: 302.0°, flash p: 307°F (CC), d: 1.16, autoign temp: 1173°F, vap press: 1 mm @ 108.3°, vap d: 5.82. Sol in benzene, ether, and carbon disulfide.

SYNS: ANILINOBENZENE ◇ BIG DIPPER ◇ C.I. 10355 ◇ DFA ◇ N,N-DIPHENYLAMINE ◇ DPA ◇ NO SCALD ◇ N-PHENYLANILINE ◇ N-PHENYLBENEZENAMINE ◇ SCALDIP

TOXICITY DATA with REFERENCE
orl-rat TDLo:7500 mg/kg (17-22D preg):TER PEREBL 4,448,70
orl-rat LDLo:3000 mg/kg CNREA8 26,619,66
orl-gpg LD50:300 mg/kg FMCHA2 -,C85,83

CONSENSUS REPORTS: Reported in EPA TSCA Inventory. EPA Genetic Toxicology Program.

OSHA PEL: TWA 10 mg/m^3
ACGIH TLV: TWA 10 mg/m^3

SAFETY PROFILE: Poison by ingestion. Experimental teratogenic effects. Action similar to aniline but less severe. Combustible when exposed to heat or flame. Can react violently with hexachloromelamine or trichloromelamine. Can react with oxidizing materials. To fight fire, use CO_2, dry chemical. When heated to decomposition it emits highly toxic fumes of NO_x. See also ANILINE, AMINES, and AROMATIC AMINES.

DVY000 CAS:587-84-8 ***HR: 3***
DIPHENYLAMINE HYDROGEN SULFATE
mf: $C_{12}H_{11}N \cdot H_2O_4S$ mw: 267.32

PROP: White to yellowish powder. Mp: 123-125°. Insol in water; sol in alcohol and sulfuric acid.

SYNS: DIPHENYLAMINE SULFATE ◇ USAF EK-743

TOXICITY DATA with REFERENCE
ipr-mus LD50:200 mg/kg NTIS** AD277-689

CONSENSUS REPORTS: Reported in EPA TSCA Inventory.

SAFETY PROFILE: Poison by intraperitoneal route. When heated to decomposition it emits toxic fumes of NO_x. See also SULFATES.

DVY100 CAS:6217-24-9 ***HR: 3***
DIPHENYLARSINOUS ACID
mf: $C_{12}H_{11}AsO$ mw: 246.15

SYNS: ARSINE, DIPHENYLHYDROXY- ◇ ARSINE, HYDROXYDIPHENYL- ◇ DIPHENYLHYDROXYARSINE

TOXICITY DATA with REFERENCE
ivn-mus LD50:10 mg/kg PHBUA9 2,19,54

OSHA PEL: TWA 0.5 mg(As)/m^3

SAFETY PROFILE: Poison by intravenous route. When heated to decomposition it emits toxic fumes of As.

DVY800 CAS:1103-05-5 ***HR: 3***
DIPHENYLBIS(PHENYLTHIO)TIN
mf: $C_{24}H_{20}S_2Sn$ mw: 491.25

SYN: DIPHENYLBIS(PHENYLTHIO)STANNANE

TOXICITY DATA with REFERENCE
ivn-mus LD50:18 mg/kg CSLNX* NX#01644

OSHA PEL: TWA 0.1 mg(Sn)/m^3 (skin)
ACGIH TLV: TWA 0.1 mg(Sn)/m^3 (skin) (Proposed: TWA 0.1 mg(Sn)/m^3; STEL 0.2 mg(Sn)/m^3 (skin))
NIOSH REL: (Organotin Compounds) TWA 0.1 mg(Sn)/m^3

SAFETY PROFILE: Poison by intravenous route. When heated to decomposition it emits toxic fumes of SO_x. See also TIN COMPOUNDS.

DVY875 CAS:76487-65-5 ***HR: D***
trans-2-(p-(1,2-DIPHENYL-1-BUTENYL)PHENOXY)-N,N-DIMETHYLETHYLAMINE CITRATE
mf: $C_{26}H_{29}NO \cdot C_6H_8O_7$ mw: 563.70

SYN: ICI 47,699 CITRATE

TOXICITY DATA with REFERENCE
orl-rat TDLo:70 mg/kg (7D pre):REP JRPFA4 13,101,67

SAFETY PROFILE: Experimental reproductive effects. When heated to decomposition it emits toxic fumes of NO_x.

DVY900 CAS:20930-10-3 ***HR: 2***
1,1-DIPHENYL-2-BUTYNYL-N-CYCLOHEXYLCARBAMATE
mf: $C_{23}H_{25}NO_2$ mw: 347.49

SYN: CYCLOHEXANECARBAMIC ACID, 1,1-DIPHENYL-2-BUTYNYL ESTER

TOXICITY DATA with REFERENCE
orl-rat TDLo:14800 mg/kg/73W-C:ETA JJIND8 71,211,83

SAFETY PROFILE: Questionable carcinogen with experimental tumorigenic data. When heated to decomposition it emits toxic fumes of NO_x.

DVZ000 CAS:102-09-0 ***HR: 3***
DIPHENYL CARBONATE
mf: $C_{13}H_{10}O_3$ mw: 214.23

SYNS: CARBONIC ACID, DIPHENYL ESTER ◇ PHENYL CARBONATE

TOXICITY DATA with REFERENCE
orl-mus TDLo:28 g/kg/78W-I:ETA NTIS** PB223-159
scu-mus TDLo:1000 mg/kg:NEO NTIS** PB223-159

CONSENSUS REPORTS: Reported in EPA TSCA Inventory.

SAFETY PROFILE: Questionable carcinogen with experimental neoplastigenic and tumorigenic data. When heated to decomposition it emits acrid smoke and irritating fumes.

DWA400 CAS:25868-47-7 ***HR: 3***
DIPHENYLDICHLORO TIN DIPYRIDINE complex
mf: $C_{22}H_{20}Cl_2N_2Sn$ mw: 502.03

SYN: DICHLORODIPHENYLSTANNANE complex with PYRIDINE (1:2)

TOXICITY DATA with REFERENCE
ivn-mus LD50:14 mg/kg CSLNX* NX#02208

OSHA PEL: TWA 0.1 mg(Sn)/m^3 (skin)
ACGIH TLV: TWA 0.1 mg(Sn)/m^3 (skin) (Proposed: TWA 0.1 mg(Sn)/m^3; STEL 0.2 mg(Sn)/m^3 (skin))
NIOSH REL: (Organotin Compounds) TWA 0.1 mg(Sn)/m^3

SAFETY PROFILE: Poison by intravenous route. When heated to decomposition it emits very toxic fumes of Cl^- and NO_x. See also TIN COMPOUNDS.

DWA500 CAS:2652-77-9 ***HR: 3***
DIPHENYLDIKETOPYRAZOLIDINE
mf: $C_{15}H_{15}N_2O_2$ mw: 252.29

SYNS: DA 339 ◇ DIFENILDICHETOPIRAZOLIDINA (ITALIAN) ◇ 1,2-DIPHENYL-3,5-DIOXOPYRAZOLIDIN (GERMAN) ◇ 1,2-DIPHENYL-3,5-PYRAZOLIDINEDIONE ◇ G 14744

TOXICITY DATA with REFERENCE
orl-rat LD50:1 g/kg MPHEAE 16,536,67
ipr-rat LD50:338 mg/kg AIPTAK 132,16,61
orl-mus LD50:574 mg/kg MPHEAE 16,536,67

SAFETY PROFILE: Poison by intraperitoneal route. Moderately toxic by ingestion. When heated to decomposition it emits toxic fumes of NO_x.

DWA600 CAS:14148-99-3 ***HR: 3***
1,2-DIPHENYL-1-(DIMETHYLAMINO)ETHANE
mf: $C_{16}H_{19}N \cdot ClH$ mw: 261.82

SYNS: (R) (−)-N,N-DIMETHYL-1,2-DIPHENYLETHYLAMINE HYDROCHLORIDE ◇ (R)-N,N-DIMETHYL-α-PHENYLBENZENEETHANAMINE, HYDROCHLORIDE ◇ SPA

TOXICITY DATA with REFERENCE
orl-rat LD50:300 mg/kg DRFUD4 2,39,77
scu-rat LD50:148 mg/kg AIPTAK 221,105,76
orl-mus LD50:176 mg/kg DRFUD4 2,39,77

scu-mus LD50:104 mg/kg DRFUD4 2,39,77
ivn-mus LD50:32600 μg/kg DRFUD4 2,39,77

SAFETY PROFILE: Poison by ingestion, intravenous, and subcutaneous routes. When heated to decomposition it emits very toxic fumes of HCl and NO_x.

DWA700 CAS:56767-15-8 ***HR: 2***
3,3-DIPHENYL-3-DIMETHYLCARBAMOYL-1-PROPYNE
mf: $C_{18}H_{17}NO$ mw: 263.36

SYN: N,N-DIMETHYL-α-ETHYNYL-α-PHENYLBENZENEACETAMIDE

TOXICITY DATA with REFERENCE
orl-rat TDLo:780 mg/kg/21W-I:CAR CNREA8 35,2469,75

SAFETY PROFILE: Questionable carcinogen with experimental carcinogenic data. When heated to decomposition it emits toxic fumes of NO_x.

DWA800 ***HR: 3***
1,3-DIPHENYL-1,3-EPIDIOXY-1,3-DIHYDROISOBENZOFURAN
mf: $C_{20}H_{14}O_3$ mw: 302.33

SAFETY PROFILE: Explosive. When heated to decomposition it emits acrid smoke and fumes.

DWB000 CAS:5959-42-2 ***HR: 3***
DIPHENYLETHANOLAMINE HYDROCHLORIDE
mf: $C_{14}H_{15}NO{\bullet}ClH$ mw: 249.76

SYNS: α-(α-AMINOBENZYL)BENZYL ALCOHOL HYDROCHLORIDE ◇ 2-AMINO-1,2-DIPHENYLETHANOL HYDROCHLORIDE ◇ β-AMINO-α-PHENYLBENZENEETHANOL HDYROCHLORIDE

TOXICITY DATA with REFERENCE
scu-mus LDLo:336 mg/kg JAPMA8 39,354,50
ivn-rbt LDLo:60 mg/kg JACSAT 52,3317,30
scu-gpg LDLo:450 mg/kg JACSAT 52,3317,30

SAFETY PROFILE: Poison by intravenous and subcutaneous routes. When heated to decomposition it emits very toxic fumes of Cl^- and NO_x.

DWB300 CAS:24301-89-1 ***HR: 3***
1,2-DIPHENYLETHYLAMINE HYDROCHLORIDE
mf: $C_{14}H_{15}N{\bullet}ClH$ mw: 233.76

TOXICITY DATA with REFERENCE
ivn-rat LD50:45 mg/kg JPETAB 77,317,43
ipr-mus LD50:175 mg/kg JPHYA7 98,424,40
scu-mus LDLo:222 mg/kg JAPMA8 39,354,50

SAFETY PROFILE: Poison by subcutaneous, intravenous, and intraperitoneal routes. When heated to decomposition it emits toxic fumes of NO_x and HCl. See also AMINES.

DWB800 CAS:1241-94-7 ***HR: 3***
DIPHENYL-2-ETHYLHEXYL PHOSPHATE
mf: $C_{20}H_{27}O_4P$ mw: 362.44

SYNS: 2-ETHYL-1-HEXANOL ESTER with DIPHENYL PHOSPHATE ◇ 2-ETHYLHEXYL DIPHENYL ESTER PHOSPHORIC ACID ◇ 2-ETHYLHEXYL DIPHENYLPHOSPHATE ◇ SANTICIZER 141 (MONSANTO)

TOXICITY DATA with REFERENCE
ivn-rbt LDLo:272 mg/kg AMIHBC 8,170,53

CONSENSUS REPORTS: Reported in EPA TSCA Inventory.

SAFETY PROFILE: Poison by intravenous route. When heated to decomposition it emits toxic fumes of PO_x.

DWB875 ***HR: 3***
4,4-DIPHENYL-1-ETHYLPIPERIDINE MALEATE
mf: $C_{19}H_{23}N{\bullet}C_4H_4O_4$ mw: 381.51

TOXICITY DATA with REFERENCE
orl-mus LD50:150 mg/kg ARZNAD 34,233,84
ivn-mus LD50:45 mg/kg ARZNAD 34,233,84
scu-mus LD50:85 mg/kg ARZNAD 34,233,84

SAFETY PROFILE: Poison by ingestion, subcutaneous, and intravenous routes. When heated to decomposition it emits toxic fumes of NO_x.

DWC000 ***HR: 3***
N-(1,2-DIPHENYLETHYL)-2-(PYRROLIDINYL) ACETAMIDE HYDROCHLORIDE
mf: $C_{20}H_{24}N_2O{\bullet}ClH$ mw: 344.92

SYN: C 3155

TOXICITY DATA with REFERENCE
eye-rbt 2% SEV ARZNAD 8,609,58
ipr-rat LD50:84 mg/kg ARZNAD 8,609,58
scu-mus LD50:247 mg/kg ARZNAD 8,609,58

SAFETY PROFILE: Poison by intraperitoneal and subcutaneous routes. A severe eye irritant. When heated to decomposition it emits very toxic fumes of NO_x and HCl.

DWC050 CAS:53067-74-6 ***HR: 3***
3,3-DIPHENYL-2-ETHYL-1-PYRROLINE
mf: $C_{18}H_{19}N$ mw: 249.38

TOXICITY DATA with REFERENCE
ipr-rat LD50:335 mg/kg AIPTAK 115,332,58
ivn-rat LD50:42 mg/kg AIPTAK 115,332,58
orl-mus LD50:1350 mg/kg AIPTAK 115,332,58
ipr-mus LD50:118 mg/kg AIPTAK 115,332,58
ivn-mus LD50:57 mg/kg AIPTAK 115,332,58

SAFETY PROFILE: Poison by intravenous and intra-

peritoneal routes. Moderately toxic by ingestion. When heated to decomposition it emits toxic fumes of NO_x.

DWC100 CAS:60662-79-5 ***HR: 3***
3,3-DIPHENYL-3-(ETHYLSULFONYL)-N,N,1-TRIMETHYLPROPYLAMINE HYDROCHLORIDE
mf: $C_{20}H_{27}NO_2S{\cdot}ClH$ mw: 382.00

SYNS: Γ-(ETHYLSULFONYL)-N,N,α-TRIMETHYL-Γ-PHENYLBENZENEPROPANAMINE HYDROCHLORIDE ◇ I-C 26 ◇ WIN-1161-3

TOXICITY DATA with REFERENCE
ipr-mus LD50:36 mg/kg CPBTAL 6,109,58
scu-mus LD50:102 mg/kg CPBTAL 6,109,58
ivn-dog LD50:26 mg/kg CPBTAL 6,109,58

SAFETY PROFILE: Poison by subcutaneous, intravenous, and intraperitoneal routes. When heated to decomposition it emits toxic fumes of SO_x, NO_x, and HCl.

DWC600 CAS:102-06-7 ***HR: 3***
DIPHENYLGUANIDINE
mf: $C_{13}H_{13}N_3$ mw: 211.29

PROP: White powder. Mp: 145°, d: 1.115 @ 25°.

SYNS: N,N'-DIPHENYLGUANIDINE ◇ 1,3-DIPHENYLGUANIDINE ◇ DPG ◇ DPG ACCELERATOR ◇ DWUFENYLOGUANIDYNA (POLISH) ◇ MELANILINE ◇ NCI-C60924 ◇ USAF B-19 ◇ USAF EK-1270 ◇ VULCACID D ◇ VULKACIT D/C ◇ VULKAZIT

TOXICITY DATA with REFERENCE
mmo-sat 360 ng/plate JEPOEC 6,293,85
mma-sat 33 μg/plate ENMUDM 8(Suppl 7),1,86
bfa-mus/sat 36 ng/kg JEPOEC 6,293,85
orl-mus TDLo:56 mg/kg (male 7D pre):REP JTEHD6 11,869,83
orl-mus TDLo:76 mg/kg (1-19D preg):TER JEPTDQ 4(1),451,80
orl-rat LD50:507 mg/kg MEPAAX 16,35,65
ipr-rat LD50:75 mg/kg MEPAAX 16,35,65
orl-mus LD50:290 mg/kg HYSAAV 29,37,64
ipr-mus LD50:25 mg/kg NTIS** AD277-689
orl-mam LDLo:250 mg/kg JIDHAN 13,87,31

CONSENSUS REPORTS: Reported in EPA TSCA Inventory.

SAFETY PROFILE: Poison by ingestion and intraperitoneal routes. Experimental teratogenic and reproductive effects. Mutation data reported. When heated to decomposition it emits toxic fumes of NO_x.

DWC650 CAS:4657-20-9 ***HR: D***
2,6-DIPHENYL-2,4,6,6,8,8-HEXAMETHYLCYCLOTETRASILOXANE
mf: $C_{18}H_{28}O_4Si_4$ mw: 420.82

SYN: CYCLOTETRASILOXANE,2,6-DIPHENYL-2,4,4,6,8,8-HEXAMETHYL-

TOXICITY DATA with REFERENCE
orl-rat TDLo:5 mg/kg (female 1D pre):REP TXAPA9 21,68,72

CONSENSUS REPORTS: Reported in EPA TSCA Inventory.

SAFETY PROFILE: Experimental reproductive effects. When heated to decomposition it emits acrid smoke and irritating fumes.

DWD000 ***HR: 3***
DIPHENYLHYDANTOIN and PHENOBARBITAL

SYNS: PHENOBARBITOL and DIPHENYLHDANTOIN ◇ PHENOBARBITONE and PHENOBARBITONE ◇ PHENYTOIN and PHENOBARBITONE

TOXICITY DATA with REFERENCE
orl-wmn TDLo:3203 mg/kg (1-44W preg):REP AJDCAI 127,758,74
unr-wmn TDLo:1884 mg/kg (1-39W preg):TER JOPDAB 84,254,74

SAFETY PROFILE: A human teratogen. Human reproductive effects: developmental abnormalities of the eye, ear, central nervous system, craniofacial, musculoskeletal, urogenital and cardiovascular system. Effects on newborn including growth statistics and physical effects. See also BARBITURATES.

DWD200 CAS:53421-38-8 ***HR: 3***
3,3-DIPHENYL-3-HYDROXYPROPIONIC ACID DIETHYLAMINOETHYL ESTER HYDROCHLORIDE
mf: $C_{20}H_{27}NO_3{\cdot}ClH$ mw: 365.94

SYN: DPE-HCl

TOXICITY DATA with REFERENCE
orl-rat LD50:2310 mg/kg NIIRDN 6,332,82
scu-rat LD50:330 mg/kg NIIRDN 6,332,82
ivn-rat LD50:27500 μg/kg NIIRDN 6,332,82
orl-mus LD50:418 mg/kg NIIRDN 6,332,82
ipr-mus LD50:140 mg/kg JPETAB 74,274,42
scu-mus LD50:280 mg/kg NIIRDN 6,332,82
ivn-mus LD50:27500 μg/kg NIIRDN 6,332,82

SAFETY PROFILE: Poison by subcutaneous, intraperitoneal, and intravenous routes. Moderately toxic by ingestion. When heated to decomposition it emits very toxic fumes of HCl and NO_x.

DWD400 ***HR: 3***
3,4-DIPHENYL-1-ISOBUTYLPYRAZOLE-5-ACETIC ACID SODIUM SALT
mf: $C_{21}H_{21}N_2O_2{\cdot}Na$ mw: 356.43

SYNS: 1-ISOBUTYL-3,4-DIPHENYLPYRAZOLE-5-ACETIC ACID SODIUM SALT ◇ LM 22070

TOXICITY DATA with REFERENCE
orl-rat LD50:46 mg/kg AIPTAK 238,305,79
orl-mus LD50:320 mg/kg AIPTAK 238,305,79

SAFETY PROFILE: Poison by ingestion. When heated to decomposition it emits toxic fumes of NO_x and Na_2O.

DWD800 CAS:587-85-9 ***HR: 3***
DIPHENYLMERCURY
mf: $C_{12}H_{10}Hg$ mw: 354.81

PROP: White crystals, insol in water. D: 2.318, mp: 122° (sublimes), bp: 204° @ 10.5 mm.

TOXICITY DATA with REFERENCE
orl-rat LDLo:500 mg/kg NCNSA6 5,30,53
ipr-rat LDLo:50 mg/kg NCNSA6 5,30,53
ipr-mus LDLo:250 mg/kg CBCCT* 4,231,52

CONSENSUS REPORTS: Reported in EPA TSCA Inventory. Mercury and its compounds are on the Community Right-To-Know List.

OSHA PEL: (Transitional: CL 1 mg/10m^3) CL 0.1 mg(Hg)/m^3 (skin)
ACGIH TLV: TWA 0.1 mg(Hg)/m^3 (skin)
NIOSH REL: (Inorganic Mercury) TWA 0.05 mg(Hg)/m^3

SAFETY PROFILE: Poison by intraperitoneal route. Moderately toxic by ingestion. Incompatible with nonmetal oxides. When heated to decomposition it emits toxic fumes of Hg. See also MERCURY COMPOUNDS.

DWE200 CAS:18656-25-2 ***HR: 3***
2-(DIPHENYLMETHOXY)ACETAMIDOXIME HYDROGEN MALEATE
mf: $C_{15}H_{16}N_2O_2 \cdot C_4H_4O_4$ mw: 372.41

TOXICITY DATA with REFERENCE
orl-mus LD50:450 mg/kg ARZNAD 17,1446,67
ivn-mus LD50:33 mg/kg ARZNAD 17,1446,67
ivn-cat LDLo:8 mg/kg ARZNAD 17,1446,67

SAFETY PROFILE: Poison by intravenous route. Moderately toxic by ingestion. When heated to decomposition it emits toxic fumes of NO_x.

DWE800 CAS:524-83-4 ***HR: 3***
3-(DIPHENYLMETHOXY)-8-ETHYLNORTROPANE
mf: $C_{22}H_{27}NO$ mw: 321.50

SYNS: 3-α-(DIPHENYLMETHOXY)-8-ETHYLNORTROPANE ◇ endo-3-(DIPHENYLMETHOXY)-8-ETHYL-8-AZABICYCLO(3.2.1)OCTANE ◇ ETHYBENZTROPINE ◇ ETHYLBENATROPINE ◇ ETHYLBENZTROPINE ◇ N-ETHYLNORTROPINE BENZHYDRYL ETHER ◇ ETYBENZATROPINE ◇ PANOLID ◇ PONALID ◇ PONALIDE ◇ UK-738 ◇ VK-738

TOXICITY DATA with REFERENCE
orl-rat LDLo:560 mg/kg 27ZQAG -,228,72
orl-mus LDLo:66 mg/kg 27ZQAG -,228,72
ivn-mus LD50:12 mg/kg EJPHAZ 9,304,70
orl-rbt LDLo:215 mg/kg 27ZQAG -,228,72
ivn-rbt LDLo:6 mg/kg 27ZQAG -,228,72

SAFETY PROFILE: Poison by ingestion and intravenous routes. When heated to decomposition it emits toxic fumes of NO_x. See also ETHERS.

DWF000 CAS:606-90-6 ***HR: 3***
4-(DIPHENYLMETHOXY)-1-METHYLPIPERIDINE CHLOROTHEOPHYLLINE
mf: $C_{19}H_{23}NO \cdot C_7H_7ClN_4O_2$ mw: 496.06

SYN: P 284

TOXICITY DATA with REFERENCE
orl-mus LD50:275 mg/kg ARZNAD 5,185,55
ivn-mus LD50:75 mg/kg ARZNAD 5,185,55

SAFETY PROFILE: Poison by ingestion and intravenous routes. When heated to decomposition it emits very toxic fumes of NO_x and Cl^-. See also THEOPHYLLINE.

DWF200 CAS:1146-95-8 ***HR: 3***
2-DIPHENYLMETHYLENEBUTYLAMINE HYDROCHLORIDE
mf: $C_{17}H_{19}N \cdot ClH$ mw: 273.83

SYNS: EDPA HYDROCHLORIDE ◇ 2-ETHYL-3,3-DIPHENYL-2-PROPENYLAMINE HYDROCHLORIDE ◇ ETIFELMIN HYDROCHLORIDE ◇ GILUTENSIN ◇ Na III HYDROCHLORIDE ◇ TENSINASE D

TOXICITY DATA with REFERENCE
orl-rat LD50:139 mg/kg NIIRDN 6,112,82
ipr-rat LD50:23400 μg/kg NIIRDN 6,112,82
scu-rat LD50:91300 μg/kg NIIRDN 6,112,82
orl-mus LD50:115 mg/kg NIIRDN 6,112,82
ipr-mus LD50:45500 μg/kg NIIRDN 6,112,82
scu-mus LD50:73400 μg/kg NIIRDN 6,112,82
ivn-mus LD50:33900 μg/kg NIIRDN 6,112,82

SAFETY PROFILE: Poison by ingestion, intravenous, intraperitoneal, and subcutaneous routes. When heated to decomposition it emits very toxic fumes of NO_x and HCl.

DWF300 CAS:52236-34-7 ***HR: D***
4-(DIPHENYLMETHYLENE)-2-ETHYL-3-METHYLCYCLOHEXANOL ACETATE
mf: $C_{24}H_{28}O_2$ mw: 348.52

SYNS: 1-DIPHENYLMETHYLENYL-2-METHYL-3-ETHYL-4-ACETOXYCYCLOHEXANE ◇ ORF 8511

TOXICITY DATA with REFERENCE
orl-rat TDLo:750 μg/kg (female 1-3D post):REP
CCPTAY 9,393,74

SAFETY PROFILE: Experimental reproductive effects. When heated to decomposition it emits acrid smoke and irritating fumes.

DWF700 CAS:57726-65-5 ***HR: 3***
2-(3,3-DIPHENYL-3-(5-METHYL-1,3,4-OXADIAZOL-2-YL)PROPYL)-2-AZABICYCLO(2.2.2)OCTANE
mf: $C_{25}H_{29}N_3O$ mw: 387.57

SYNS: 2-(3-(5-METHYL-1,3,4-OXADIAZOL-2-YL)-3,3-DIPHENYLPROPYL)-2-AZABICYCLO(2.2.2)OCTANE ◇ NUFENOXOLE ◇ SC 27166 ◇ SEARLE 27166

TOXICITY DATA with REFERENCE
ivn-rat LD50:21 mg/kg JPETAB 210,327,79
orl-mus LD50:399 mg/kg NICHAS 28,1621,79
ipr-mus LD50:95 mg/kg NICHAS 38,1621,79
scu-mus LD50:258 mg/kg NICHAS 38,1621,79
ivn-mus LD50:12 mg/kg NICHAS 38,1621,79

SAFETY PROFILE: Poison by ingestion, intravenous, subcutaneous, and intraperitoneal routes. When heated to decomposition it emits toxic fumes of NO_x.

DWF869 ***HR: 3***
4,4-DIPHENYL-1-METHYLPIPERIDINE MALEATE
mf: $C_{18}H_{21}N \cdot C_4H_4O_4$ mw: 367.48

TOXICITY DATA with REFERENCE
orl-mus LD50:140 mg/kg ARZNAD 34,233,84
scu-mus LD50:80 mg/kg ARZNAD 34,233,84
ivn-mus LD50:45 mg/kg ARZNAD 34,233,84

SAFETY PROFILE: Poison by ingestion, subcutaneous, and intravenous routes. When heated to decomposition it emits toxic fumes of NO_x.

DWF865 CAS:19841-73-7 ***HR: 3***
4-DIPHENYLMETHYLPIPERIDINE
mf: $C_{18}H_{21}N$ mw: 251.40

TOXICITY DATA with REFERENCE
ipr-mus LD50:30 mg/kg JMCMAR 15,690,72

CONSENSUS REPORTS: Reported in EPA TSCA Inventory.

SAFETY PROFILE: Poison by intraperitoneal route. When heated to decomposition it emits toxic fumes of NO_x.

DWF875 CAS:101564-67-4 ***HR: 3***
5,5-DIPHENYL-3-(3-(2-METHYLPIPERIDINO)PROPYL)-2-THIOHYDANTOIN HYDROCHLORIDE
mf: $C_{24}H_{29}N_3OS \cdot ClH$ mw: 444.08

SYNS: 5,5-DIPHENYL-3-(3-(2-METHYLPIPERIDINO)PROPYL)-2-THIOHYDANTOIN MONOHYDROCHLORIDE ◇ 3-Γ-(α-METIL-1-PIPERIDINO)PROPIL-5,5-DIFENILTIOIDANTOINACLORIDRATO (ITALIAN)

TOXICITY DATA with REFERENCE
ipr-mus LD50:25 mg/kg FRPSAX 15,809,60
scu-mus LD50:25 mg/kg FRPSAX 15,809,60
ivn-rbt LD50:3630 μg/kg FRPSAX 15,809,60

SAFETY PROFILE: Poison by subcutaneous, intravenous, and intraperitoneal routes. When heated to decomposition it emits toxic fumes of SO_x, NO_x and HCl.

DWG600 ***HR: 3***
α-2,2-DIPHENYL-4-(1-METHYL-2-PIPERIDYL)-1,3-DIOXOLANE HYDROCHLORIDE
mf: $C_{21}H_{25}NO_2 \cdot ClH$ mw: 359.93

TOXICITY DATA with REFERENCE
orl-mus LD50:400 mg/kg JMCMAR 9,127,66
ivn-mus LD50:25 mg/kg JMCMAR 9,127,66

SAFETY PROFILE: Poison by ingestion and intravenous routes. When heated to decomposition it emits very toxic fumes of NO_x and HCl.

DWH000 ***HR: 3***
β-2,2-DIPHENYL-4-(1-METHYL-2-PIPERIDYL)-1,3-DIOXOLANE HYDROCHLORIDE
mf: $C_{21}H_{25}NO_2 \cdot ClH$ mw: 359.93

TOXICITY DATA with REFERENCE
orl-mus LD50:300 mg/kg JMCMAR 9,127,66
ivn-mus LD50:75 mg/kg JMCMAR 9,127,66

SAFETY PROFILE: Poison by ingestion and intravenous routes. When heated to decomposition it emits very toxic fumes of NO_x and HCl.

DWH200 ***HR: 3***
α-2,2-DIPHENYL-4-(1-METHYL-2-PIPERIDYL)-1,3-DIOXOLANE METHYLIODIDE
mf: $C_{21}H_{25}NO_2 \cdot CH_3I$ mw: 465.41

TOXICITY DATA with REFERENCE
orl-mus LD50:800 mg/kg JMCMAR 9,127,66
ipr-mus LD50:75 mg/kg JMCMAR 9,127,66

SAFETY PROFILE: Poison by intraperitoneal route. Moderately toxic by ingestion. When heated to decomposition it emits very toxic fumes of NO_x and I^-. See also IODIDES.

DWH500 CAS:102280-81-9 *HR: 3*
3,3-DIPHENYL-2-METHYL-1-PYRROLINE
mf: $C_{17}H_{17}N$ mw: 235.35

TOXICITY DATA with REFERENCE
ipr-rat LD50:310 mg/kg AIPTAK 115,332,58
orl-mus LD50:560 mg/kg AIPTAK 115,332,58
ivn-mus LD50:47 mg/kg AIPTAK 115,332,58

SAFETY PROFILE: Poison by intravenous and intraperitoneal routes. Moderately toxic by ingestion. When heated to decomposition it emits toxic fumes of NO_x.

DWH550 CAS:778-25-6 *HR: D*
DIPHENYLMETHYLSILANOL
mf: $C_{13}H_{14}OSi$ mw: 214.36

SYN: SILANOL, DIPHENYLMETHYL-

TOXICITY DATA with REFERENCE
orl-rat TDLo:700 mg/kg (male 7D pre):REP TXAPA9 21,55,72

CONSENSUS REPORTS: Reported in EPA TSCA Inventory.

SAFETY PROFILE: Experimental reproductive effects. When heated to decomposition it emits acrid smoke and irritating fumes.

DWH600 CAS:545-91-5 *HR: 3*
4,4-DIPHENYL-6-MORPHOLINO-3-HEPTANONE HYDROCHLORIDE
mf: $C_{23}H_{29}NO_2$•ClH mw: 387.99

TOXICITY DATA with REFERENCE
scu-rat LD50:132 mg/kg JPETAB 98,305,50
orl-mus LD50:208 mg/kg FSTEAI 5,185,50
ipr-mus LD50:131 mg/kg JACSAT 71,57,49
scu-mus LD50:110 mg/kg JPETAB 98,121,50
ivn-mus LD50:48 mg/kg JPETAB 98,305,50

SAFETY PROFILE: Poison by ingestion, intravenous, subcutaneous, and intraperitoneal routes. When heated to decomposition it emits very toxic fumes of Cl^- and NO_x.

DWH800 CAS:63765-88-8 *HR: 3*
4,4-DIPHENYL-6-MORPHOLINO-3-HEXANONE HYDROCHLORIDE
mf: $C_{22}H_{27}NO_2$•ClH mw: 373.96

TOXICITY DATA with REFERENCE
orl-mus LD50:208 mg/kg BJPCAL 5,125,50
ipr-mus LD50:114 mg/kg JACSAT 71,57,49
scu-mus LD50:220 mg/kg JPPMAB 2,418,50
ivn-mus LD50:43 mg/kg BJPCAL 5,125,50

SAFETY PROFILE: Poison by ingestion, intraperitoneal, subcutaneous and intravenous routes. When heated to decomposition it emits very toxic fumes of HCl and NO_x.

DWH875 *HR: 3*
2,3-DIPHENYL-3H-NAPHTHO(1,2-d)TRIAZOLIUM CHLORIDE
mf: $C_{22}H_{16}N_3$•Cl mw: 357.86

SYN: α-β-NAPHTHO-2,3-DIPHENYL-TRIAZOLIUM CHLORID (GERMAN)

TOXICITY DATA with REFERENCE
scu-mus LDLo:2800 μg/kg SDMU** -,-,36
ivn-mus LDLo:2400 μg/kg SDMU** -,-,36
ivn-rbt LDLo:40 mg/kg SDMU** -,-,36
scu-gpg LDLo:12 mg/kg SDMU** -,-,36
scu-frg LDLo:45 mg/kg SDMU** -,-,36

SAFETY PROFILE: Poison by subcutaneous and intravenous routes. When heated to decomposition it emits toxic fumes of Cl^- and NO_x.

DWI000 CAS:86-30-6 *HR: 3*
DIPHENYLNITROSAMINE
mf: $C_{12}H_{10}N_2O$ mw: 198.24

PROP: Green crystals. Mp: 144°.

SYNS: CURETARD A ◇ DELAC J ◇ DIPHENYLNITROSAMIN (GERMAN) ◇ DIPHENYL N-NITROSOAMINE ◇ N,N-DIPHENYLNITROSAMINE ◇ NAUGARD TJB ◇ NCI-C02880 ◇ NDPA ◇ NDPhA ◇ N-NITROSODIFENYLAMIN (CZECH) ◇ NITROSODIPHENYLAMINE ◇ N-NITROSODIPHENYLAMINE ◇ N-NITROSO-N-PHENYLANILINE ◇ NITROUS DIPHENYLAMIDE ◇ REDAX ◇ RETARDER J ◇ TJB ◇ VULCALENT A ◇ VULCATARD ◇ VULKALENT A (CZECH) ◇ VULTROL

TOXICITY DATA with REFERENCE
eye-rbt 500 mg/24H MLD 28ZPAK -,134,72
mma-sat 50 μg/plate CANCAR 49,1970,82
dnd-hmn:fbr 3 mmol/L ENMUDM 7,267,85
dns-rat:lvr 500 nmol/L CNREA8 42,3010,82
otr-ham:emb 6300 μg/L NCIMAV 58,243,81
orl-rat TDLo:140 g/kg/2Y-C:CAR NCITR* NCI-CG-TR-164,79
skn-mus TDLo:800 mg/kg/20W-I:ETA EJCAAH 16,695,80
orl-rat TD:170 g/kg/2Y-C:CAR NCITR* NCI-CG-TR-164,79
orl-rat LD50:1650 mg/kg 28ZPAK -,134,72
orl-mus LD50:3850 mg/kg GTPZAB 10(4),60,66
ipr-mus LD50:1000 mg/kg PMRSDJ 1,682,81

CONSENSUS REPORTS: IARC Cancer Review: Group 3 IMEMDT 7,56,87; Animal Limited Evidence IMEMDT 27,213,82. NCI Carcinogenesis Bioassay (feed); Clear Evidence: rat NCITR* NCI-CG-TR-164,79; No Evidence: mouse NCITR* NCI-CG-TR-164,79. Reported in EPA TSCA Inventory. EPA Genetic Toxicology Program. Community Right-To-Know List.

SAFETY PROFILE: Moderately toxic by ingestion. An eye irritant. Questionable carcinogen with experimental carcinogenic and tumorigenic data. Human mutation data reported. Dangerous fire hazard when exposed to heat, flame, or oxidizing materials. Can react vigorously with oxidizing materials. When heated to decomposition it emits highly toxic fumes of NO_x. See also NITROSAMINES.

DWI200 CAS:92-71-7 ***HR: 2***
2,5-DIPHENYLOXAZOLE
mf: $C_{15}H_{11}NO$ mw: 221.27

SYN: USAF EK-6775

TOXICITY DATA with REFERENCE
ipr-mus LD50:750 mg/kg NTIS** AD277-689

CONSENSUS REPORTS: Reported in EPA TSCA Inventory.

SAFETY PROFILE: Moderately toxic by intraperitoneal route. When heated to decomposition it emits toxic fumes of NO_x.

DWI400 CAS:16230-71-0 ***HR: 3***
3,3-DIPHENYL-2-OXETANONE
mf: $C_{15}H_{12}O_2$ mw: 224.27

SYNS: 2,2-DIPHENYL-3-HYDROXYPROPIONIC ACID LACTONE ◇ α,α-DIPHENYL-β-PROPIOLACTONE

TOXICITY DATA with REFERENCE
scu-rat TDLo:380 mg/kg/19W-I:ETA BJCAAI 15,85,61

SAFETY PROFILE: Questionable carcinogen with experimental tumorigenic data. When heated to decomposition it emits acrid smoke and irritating fumes.

DWI800 ***HR: 3***
1,5-DIPHENYL-1,4-PENTAZDIENE
mf: $C_{12}H_{11}N_5$ mw: 225.26

SAFETY PROFILE: Explodes violently on warming, impact, or friction. When heated to decomposition it emits toxic fumes of NO_x.

DWJ300 CAS:21413-28-5 ***HR: 2***
5,5-DIPHENYL-1-PHENYLSULFONYLHYDANTOIN
mf: $C_{21}H_{16}N_2O_4S$ mw: 392.45

SYN: 5,5-DIPHENYL-1-(PHENYLSULFONYL)-2,4-IMIDAZOLIDINEDIONE (9CI)

TOXICITY DATA with REFERENCE
orl-rat LD50:1700 mg/kg ARZNAD 20,1579,70
ipr-rat LD50:695 mg/kg ARZNAD 20,1579,70
orl-mus LD50:1700 mg/kg ARZNAD 20,1579,70
ipr-mus LD50:1037 mg/kg ARZNAD 20,1579,70

SAFETY PROFILE: Moderately toxic by ingestion and intraperitoneal routes. When heated to decomposition it emits toxic fumes of SO_x and NO_x.

DWJ400 CAS:3736-92-3 ***HR: 3***
1,2-DIPHENYL-4-PHENYLTHIOETHYL-3,5-PYRAZOLIDINEDIONE
mf: $C_{23}H_{20}N_2O_2S$ mw: 388.51

SYN: 4-(PHENYLTHIOETHYL)-1,2-DIPHENYL-3,5-PYRAZOLIDINEDIONE

TOXICITY DATA with REFERENCE
orl-rat LD50:450 mg/kg ANYAA9 86,263,60
ivn-rat LD50:190 mg/kg ANYAA9 86,263,60
orl-mus LD50:560 mg/kg ANYAA9 86,263,60
ivn-mus LD50:178 mg/kg ANYAA9 86,263,60
ivn-rbt LD50:100 mg/kg MDCHAG 5,391,65

SAFETY PROFILE: Poison by intravenous route. Moderately toxic by ingestion. When heated to decomposition it emits very toxic fumes of NO_x and SO_x.

DWK200 CAS:972-02-1 ***HR: 3***
α,α-DIPHENYL-1-PIPERIDINEBUTANOL
mf: $C_{21}H_{27}NO$ mw: 309.49

SYNS: DIFENIDOL ◇ DIPHENIDOL ◇ NOMETIC ◇ α-(3-PIPERIDINOPROPYL)BENZHYDROL ◇ SKF 478 ◇ SK&F No. 478-A ◇ VONTROL

TOXICITY DATA with REFERENCE
orl-rat LD50:815 mg/kg PHMCAA 5,265,63
scu-rat LD50:50 mg/kg TXAPA9 18,185,71
orl-mus LD50:450 mg/kg PHMCAA 5,265,63
ivn-mus LD50:32 mg/kg MPHEAE 13,325,65

SAFETY PROFILE: Poison by intravenous and subcutaneous routes. Moderately toxic by ingestion. When heated to decomposition it emits toxic fumes of NO_x.

DWK400 CAS:467-60-7 ***HR: 3***
α,α-DIPHENYL-2-PIPERIDINEMETHANOL
mf: $C_{18}H_{21}NO$ mw: 267.40

SYNS: DETARIL ◇ GERODYL ◇ MERATRAN ◇ MRD 108 ◇ α-(2-PIPERIDYL)BENZHYDROL ◇ PIPRADOL ◇ α-PIPRADOL ◇ PIRIDROL ◇ PYRIDROL ◇ PYRIDROLE

TOXICITY DATA with REFERENCE
orl-rat LD50:180 mg/kg 27ZIAQ -,-,65
orl-mus LD50:74 mg/kg FRPSAX 12,853,57
ipr-mus LD50:74 mg/kg FRPSAX 12,853,57
ivn-cat LD50:15 mg/kg DAZEA2 122,283,82

SAFETY PROFILE: Poison by ingestion, intravenous, and intraperitoneal routes. When heated to decomposition it emits toxic fumes of NO_x.

DWK500 CAS:101564-69-6 ***HR: 3***
5,5-DIPHENYL-3-(3-PIPERIDINOPROPYL)-2-THIOHYDANTOIN HYDROCHLORIDE
mf: $C_{23}H_{27}N_3OS{\cdot}ClH$ mw: 430.05

SYN: 3-Γ-(1-PIPERIDINO)PROPIL-5,5-DIFENILTIOIDANTOINA CLORIDRATO (ITALIAN)

TOXICITY DATA with REFERENCE
ipr-mus LD50:35 mg/kg FRPSAX 15,809,60
scu-mus LD50:30 mg/kg FRPSAX 15,809,60
ivn-rbt LD50:3350 μg/kg FRPSAX 15,809,60

SAFETY PROFILE: Poison by subcutaneous, intravenous, and intraperitoneal routes. When heated to decomposition it emits toxic fumes of SO_x, NO_x, and HCl.

DWK700 CAS:24050-58-6 ***HR: 3***
3-(3,3-DIPHENYLPROPYLAMINO)PROPYL-3',4',5'-TRIMETHOXYBENZOATE HYDROCHLORIDE
mf: $C_{28}H_{33}NO_5{\cdot}ClH$ mw: 500.08

SYNS: PF-26 ◇ 3,4,5-TRIMETHOXYBENZOIC ACID 3-((3,3-DIPHENYLPROPYL)AMINO)PROPYL ESTER HYDROCHLORIDE

TOXICITY DATA with REFERENCE
orl-rat TDLo:9 g/kg (30D male):REP GNRIDX 5,256,71
orl-rat TDLo:300 mg/kg (9-14D preg):TER GNRIDX 5,271,71
ipr-rat LD50:138 mg/kg ARZNAD 21,1628,71
ivn-rat LD50:31500 μg/kg ARZNAD 21,1628,71
ivn-mus LD50:35 mg/kg ARZNAD 21,1628,71

SAFETY PROFILE: Poison by intravenous and intraperitoneal routes. Experimental teratogenic and reproductive effects. When heated to decomposition it emits toxic fumes of NO_x and HCl. See also ESTERS.

DWK900 CAS:23903-11-9 ***HR: 3***
3-(N-d,d-DIPHENYLPROPYL-N-METHYL) AMINOPROPAN-1-OL HYDROCHLORIDE
mf: $C_{19}H_{25}NO{\cdot}ClH$ mw: 319.91

SYN: PF-82

TOXICITY DATA with REFERENCE
orl-rat LD50:1400 mg/kg ARZNAD 24,166,74
ipr-rat LD50:160 mg/kg ARZNAD 24,166,74
orl-mus LD50:428 mg/kg ARZNAD 24,166,74
ipr-mus LD50:100 mg/kg ARZNAD 24,166,74
scu-mus LD50:364 mg/kg ARZNAD 24,166,74
ivn-mus LD50:51 mg/kg ARZNAD 24,166,74

SAFETY PROFILE: Poison by subcutaneous, intravenous, and intraperitoneal routes. Moderately toxic by ingestion. When heated to decomposition it emits toxic fumes of NO_x and HCl.

DWL200 CAS:69-43-2 ***HR: 3***
N-(3,3-DIPHENYLPROPYL)-α-METHYL-PHENETHYLAMINE LACTATE
mf: $C_{24}H_{27}N{\cdot}C_3H_6O_3$ mw: 419.61

SYN: PRENYLAMINE LACTATE

TOXICITY DATA with REFERENCE
orl-rat LD50:1000 mg/kg ARZNAD 26,2127,76
ipr-rat LD50:40 mg/kg ARZNAD 26,2127,76
orl-mus LD50:580 mg/kg ARZNAD 26,2127,76
ipr-mus LD50:40 mg/kg ARZNAD 26,2127,76

SAFETY PROFILE: Poison by intraperitoneal route. Moderately toxic by ingestion. When heated to decomposition it emits toxic fumes of NO_x.

DWL400 CAS:10087-89-5 ***HR: 3***
1,1-DIPHENYL-2-PROPYNYL-N-CYCLOHEXYL-CARBAMATE
mf: $C_{22}H_{23}NO_2$ mw: 333.46

SYNS: 1,1-DIPHENYL-2-PROPYN-1-OLCYCLOHEXANECARBAMATE ◇ 1,1-DIPHENYL-2-PROPYNYL ESTER CYCLOHEXANECARBAMIC ACID ◇ ENPROMATE

TOXICITY DATA with REFERENCE
mma-sat 100 μg/plate PNASA6 72,5135,75
dns-rat:lvr 10 μmol/L ENMUDM 3,11,81
orl-rat TDLo:1690 mg/kg/27W-C:ETA PAACA3 10,35,69
orl-mus LD50:1 g/kg GANMAX 2,261,66
ipr-mus LD50:374 mg/kg JMCMAR 11,1155,68

CONSENSUS REPORTS: EPA Genetic Toxicology Program.

SAFETY PROFILE: Poison by intraperitoneal route. Moderately toxic by ingestion. Questionable carcinogen with experimental tumorigenic data. Mutation data reported. When heated to decomposition it emits toxic fumes of NO_x. See also CARBAMATES.

DWL500 CAS:10473-64-0 ***HR: 2***
1,1-DIPHENYL-2-PROPYNYL-N-ETHYLCARBAMATE
mf: $C_{18}H_{17}NO_2$ mw: 279.36

SYN: ETHYLCARBAMIC ACID 1,1-DIPHENYL-2-PROPYNYL ESTER

TOXICITY DATA with REFERENCE
orl-rat TDLo:8800 mg/kg/28W-C:ETA JJIND8 71,211,83

SAFETY PROFILE: Questionable carcinogen with experimental tumorigenic data. When heated to decomposition it emits toxic fumes of NO_x.

DWL525 CAS:10473-98-0 ***HR: 2***
1,1-DIPHENYL-2-PROPYNYL 1-PYRROLIDINECARBOXYLATE
mf: $C_{20}H_{19}NO_2$ mw: 305.40

SYN: 3,3-DIPHENYL-3-(PYRROLIDINE-CARBONYLOXY)-1-PROPYNE

TOXICITY DATA with REFERENCE
orl-rat TDLo:2 g/kg/35W-C:ETA JJIND8 71,211,83

SAFETY PROFILE: Questionable carcinogen with experimental tumorigenic data. When heated to decomposition it emits toxic fumes of NO_x.

DWL600 CAS:3426-01-5 ***HR: 2***
1,4-DIPHENYL-3,5-PYRAZOLIDINEDIONE
mf: $C_{15}H_{11}N_2O_2$ mw: 251.28

SYN: 1,4-DIPHENYL-3,5-DIOXO-PYRAZOLIDIN(GERMAN)

TOXICITY DATA with REFERENCE
scu-mus LD50:1003 mg/kg ARZNAD 4,249,54
ivn-mus LD50:740 mg/kg AEPPAE 233,365,58

SAFETY PROFILE: Moderately toxic by subcutaneous and intravenous routes. When heated to decomposition it emits toxic fumes of NO_x.

DWL800 CAS:4845-49-2 ***HR: 2***
1,3-DIPHENYL-5-PYRAZOLONE
mf: $C_{15}H_{12}N_2O$ mw: 236.29

TOXICITY DATA with REFERENCE
ipr-mus LDLo:512 mg/kg CBCCT* 2,135,50

CONSENSUS REPORTS: Reported in EPA TSCA Inventory.

SAFETY PROFILE: Moderately toxic by intraperitoneal route. When heated to decomposition it emits toxic fumes of NO_x.

DWM000 CAS:57-96-5 ***HR: 3***
DIPHENYLPYRAZONE
mf: $C_{23}H_{20}N_2O_3S$ mw: 404.51

SYNS: 1,2-DIPHENYL-4-(2'-PHENYLSULFINETHYL)-3,5-PYRAZOLIDINEDIONE ◇ 4-(PHENYLSULFOXYETHYL)-1,2-DIPHENYL-3,5-PYRAZOLIDINEDIONE ◇ SULFINPYRAZINE ◇ SULFOXYPHENYLPYRAZOLIDINE ◇ USAF GE-13

TOXICITY DATA with REFERENCE
orl-hmn TDLo:29 mg/kg NEJMAG 303,702,80
orl-rat LD50:375 mg/kg ANYAA9 86,263,60
ivn-rat LD50:154 mg/kg ANYAA9 86,263,60
orl-mus LD50:298 mg/kg ANYAA9 86,263,60
ipr-mus LD50:100 mg/kg NTIS** AD414-344
ivn-mus LD50:240 mg/kg ANYAA9 86,263,60
ivn-rbt LD50:195 mg/kg MDCHAG 5,391,65

SAFETY PROFILE: Poison by ingestion, intravenous, and intraperitoneal routes. When heated to decomposition it emits very toxic fumes of NO_x and SO_x.

DWM400 CAS:10447-38-8 ***HR: 3***
α,α-DIPHENYL-3-QUINUCLIDINEMETHANOL HYDROCHLORIDE
mf: $C_{20}H_{23}NO{\cdot}ClH$ mw: 329.90

SYNS: FENCAROL ◇ FENKAROL ◇ PHENCAROL ◇ QUINUCLIDYL-3-DIPHENYLCARBINOL HYDROCHLORIDE

TOXICITY DATA with REFERENCE
orl-rat LD50:440 mg/kg RPTOAN 40,42,77
orl-mus LD50:370 mg/kg RPTOAN 40,42,77
ivn-mus LD50:62 mg/kg RPTOAN 40,42,77
orl-gpg LD50:860 mg/kg FATOAO 43,148,80

SAFETY PROFILE: Poison by ingestion and intravenous routes. When heated to decomposition it emits very toxic fumes of NO_x and HCl.

DWM600 CAS:10504-99-1 ***HR: 1***
DIPHENYLSELENONE
mf: $C_{12}H_{10}O_2Se$ mw: 265.17

CONSENSUS REPORTS: Selenium and its compounds are on the Community Right-To-Know List.

OSHA PEL: TWA 0.2 mg(Se)/m^3
ACGIH TLV: TWA 0.2 mg(Se)/m^3
DFG MAK: 0.1 mg(Se)/m^3

SAFETY PROFILE: Explodes weakly when heated. When heated to decomposition it emits toxic fumes of Se. See also SELENIUM COMPOUNDS.

DWM800 CAS:1011-95-6 ***HR: 3***
DIPHENYLSTANNANE
mf: $C_{12}H_{12}Sn$ mw: 274.93

PROP: Yellow powder, insol in water. Mp: 226°.

SYNS: DIPHENYLTIN ◇ DIPHENYLTIN DIHYDRIDE

TOXICITY DATA with REFERENCE
ipr-rat LDLo:15 mg/kg BJIMAG 23,222,66

OSHA PEL: TWA 0.1 mg(Sn)/m^3 (skin)
ACGIH TLV: TWA 0.1 mg(Sn)/m^3 (skin) (Proposed: TWA 0.1 mg(Sn)/m^3; STEL 0.2 mg(Sn)/m^3 (skin))
NIOSH REL: (Organotin Compounds) TWA 0.1 mg(Sn)/m^3

SAFETY PROFILE: Poison by intraperitoneal route. Ignites on contact with fuming nitric acid. When heated to decomposition it emits acrid smoke and irritating fumes. See also TIN COMPOUNDS.

DWN000 CAS:945-51-7 ***HR: 2***
DIPHENYL SULFOXIDE
mf: $C_{12}H_{10}OS$ mw: 202.28

PROP: Crystals. Vap d: 7.0.

TOXICITY DATA with REFERENCE
ipr-mus LD50:750 mg/kg IJRBA3 3,41,61

CONSENSUS REPORTS: Reported in EPA TSCA Inventory.

SAFETY PROFILE: Moderately toxic by intraperitoneal route. A fungicide. When heated to decomposition it emits toxic fumes of SO_x.

DWN100 CAS:31751-59-4 ***HR: D***
trans-2,4-DIPHENYL-2,4,6,6-TETRAMETHYL-CYCLOTRISILOXANE
mf: $C_{16}H_{22}O_3Si_3$ mw: 346.65

SYN: CYCLOTRISILOXANE,2,4-DIPHENYL-2,4,6,6-TETRAMETHYL-, (E)-

TOXICITY DATA with REFERENCE
orl-rat TDLo:30 mg/kg (female 3D pre):REP TXAPA9 21,68,72

SAFETY PROFILE: Experimental reproductive effects. When heated to decomposition it emits acrid smoke and irritating fumes.

DWN150 CAS:56-33-7 ***HR: D***
1,3-DIPHENYL-1,1,3,3-TETRAMETHYLDISILOXANE
mf: $C_{16}H_{22}OSi_2$ mw: 286.56

SYN: DISILOXANE,1,3-DIPHENYL-1,1,3,3-TETRAMETHYL-

TOXICITY DATA with REFERENCE
orl-rat TDLo:700 mg/kg (male 7D pre):REP TXAPA9 21,55,72

CONSENSUS REPORTS: Reported in EPA TSCA Inventory.

SAFETY PROFILE: Experimental reproductive effects. When heated to decomposition it emits acrid smoke and irritating fumes.

DWN200 CAS:60-10-6 ***HR: 3***
DIPHENYLTHIOCARBAZONE
mf: $C_{13}H_{12}N_4S$ mw: 256.35

PROP: Bluish-black, crystalline powder. Insol in water; sparingly sol in alcohol; freely sol in carbon tetrachloride and chloroform.

SYNS: DITHIZON ◇ DITHIZONE ◇ USAF EK-3092

TOXICITY DATA with REFERENCE
ipr-mus LD50:200 mg/kg NTIS** AD277-689

CONSENSUS REPORTS: Reported in EPA TSCA Inventory.

SAFETY PROFILE: Poison by intraperitoneal route. Can cause eye injury and glycosuria. When heated to decomposition it emits highly toxic fumes of NO_x and SO_x.

DWN400 CAS:622-03-7 ***HR: 3***
1,5-DIPHENYL-3-THIOCARBOHYDRAZIDE
mf: $C_{13}H_{14}N_4S$ mw: 258.37

SYNS: DIPHENYL THIOCARBAZIDE ◇ USAF EK-3110

TOXICITY DATA with REFERENCE
orl-rat LD50:1500 mg/kg JPETAB 90,260,47
ipr-mus LD50:200 mg/kg NTIS** AD277-689

CONSENSUS REPORTS: Reported in EPA TSCA Inventory.

SAFETY PROFILE: Poison by intraperitoneal route. Moderately toxic by ingestion. When heated to decomposition it emits very toxic fumes of NO_x and SO_x.

DWN600 CAS:21083-47-6 ***HR: 3***
5,5-DIPHENYL-2-THIOHYDANTOIN
mf: $C_{15}H_{12}N_2OS$ mw: 268.35

TOXICITY DATA with REFERENCE
orl-rat TDLo:3500 mg/kg:ETA CNREA8 28,924,68

CONSENSUS REPORTS: Reported in EPA TSCA Inventory.

SAFETY PROFILE: Questionable carcinogen with experimental tumorigenic data. When heated to decomposition it emits very toxic fumes of NO_x and SO_x.

DWN800 CAS:102-08-9 ***HR: 2***
DIPHENYLTHIOUREA
mf: $C_{13}H_{12}N_2S$ mw: 228.33

PROP: White to faint gray powder. Mp: 154°, bp: decomp, d: 1.32 @ 25°.

SYNS: DFT ◇ N,N'-DIPHENYLTHIOCARBAMIDE ◇ sym-DIPHENYLTHIOCARBAMIDE ◇ N,N'-DIPHENYLTHIOUREA ◇ sym-DIPHENYLTHIOUREA ◇ 1,3-DIPHENYLTHIOUREA ◇ 1,3-DIPHENYL-2-THIOUREA ◇ 2-FENYLOTIOMOCZNIK (POLISH) ◇ RHENOCURE CA ◇ STABILISATOR C ◇ SULFOCARBANILIDE ◇ THIOCARBANILIDE ◇ USAF EK-245 ◇ VALKACIT CA

TOXICITY DATA with REFERENCE
ipr-rat LD50:1000 mg/kg MEPAAX 16,35,65
ipr-mus LD50:500 mg/kg NTIS** AD277-689
orl-cat LDLo:720 mg/kg JPETAB 17,349,21
orl-rbt LDLo:1500 mg/kg MEIEDD 10,487,83

CONSENSUS REPORTS: Reported in EPA TSCA Inventory.

SAFETY PROFILE: Moderately toxic by ingestion and intraperitoneal routes. When heated to decomposition it emits highly toxic fumes of SO_x and NO_x.

DWO000 CAS:3898-08-6 ***HR: 3***
1,1-DIPHENYL-2-THIOUREA
mf: $C_{13}H_{12}N_2S$ mw: 228.33

SYN: USAF EK-7087

TOXICITY DATA with REFERENCE
ipr-mus LD50:200 mg/kg NTIS** AD277-689

CONSENSUS REPORTS: Reported in EPA TSCA Inventory.

SAFETY PROFILE: Poison by intraperitoneal route. When heated to decomposition it emits very toxic fumes of NO_x and SO_x.

DWO400 CAS:1135-99-5 ***HR: 2***
DIPHENYLTIN DICHLORIDE
mf: $C_{12}H_{10}Cl_2Sn$ mw: 343.81

PROP: Colorless crystals. Decomp by water. Mp: 42°, bp: 333-337° (decomp).

SYN: DICHLORODIPHENYLSTANNANE

TOXICITY DATA with REFERENCE
orl-rat LDLo:410 mg/kg BJIMAG 23,222,66
orl-mus LDLo:470 mg/kg AECTCV 14,111,85

OSHA PEL: TWA 0.1 mg(Sn)/m^3 (skin)
ACGIH TLV: TWA 0.1 mg(Sn)/m^3 (skin) (Proposed: TWA 0.1 mg(Sn)/m^3; STEL 0.2 mg(Sn)/m^3 (skin))
NIOSH REL: (Organotin Compounds) TWA 0.1 mg(Sn)/m^3

SAFETY PROFILE: Moderately toxic by ingestion. When heated to decomposition it emits toxic fumes of Cl^-. See also TIN COMPOUNDS and CHLORIDES.

DWO600 CAS:31671-16-6 ***HR: 3***
DIPHENYLTIN OXIDE POLYMER
mf: $(C_{12}H_{10}OSn)_x$

SYNS: DIPHENYLOXOSTANNANE, POLYMER ◇ OXODIPHENYLSTANNANE, POLYMER

TOXICITY DATA with REFERENCE
ivn-mus LD50:180 mg/kg CSLNX* NX#03497

OSHA PEL: TWA 0.1 mg(Sn)/m^3 (skin)
ACGIH TLV: TWA 0.1 mg(Sn)/m^3 (skin) (Proposed: TWA 0.1 mg(Sn)/m^3; STEL 0.2 mg(Sn)/m^3 (skin))
NIOSH REL: (Organotin Compounds) TWA 0.1 mg(Sn)/m^3

SAFETY PROFILE: Poison by intravenous route. When heated to decomposition it emits acrid smoke and irritating fumes. See also TIN COMPOUNDS.

DWO800 CAS:136-35-6 ***HR: 3***
1,3-DIPHENYLTRIAZENE
mf: $C_{12}H_{11}N_3$ mw: 197.26

PROP: Golden yellow crystals. Mp: 98-99°, bp: explodes, vap d: 6.8.

SYNS: CELLOFOR (CZECH) ◇ DAAB ◇ DIAZOAMINOBENZEN (CZECH) ◇ DIAZOAMINOBENZENE ◇ p-DIAZOAMINOBENZENE ◇ DIAZOAMINOBENZOL (GERMAN) ◇ N-(PHENYLAZO)ANILINE

TOXICITY DATA with REFERENCE
orl-mus TDLo:1480 mg/kg/59D-C:ETA GANNA2 29,209,35
orl-rat LD50:224 mg/kg 28ZPAK -,77,72

CONSENSUS REPORTS: EPA Genetic Toxicology Program. Reported in EPA TSCA Inventory.

SAFETY PROFILE: Poison by ingestion. Questionable carcinogen with experimental tumorigenic data. Strongly explosive when shocked or heated to 98°C. Mixture with acetic anhydride explodes when warmed. When heated to decomposition it emits toxic fumes of NO_x.

DWO875 CAS:34177-12-3 ***HR: 3***
5,6-DIPHENYL-as-TRIAZIN-3-OL
mf: $C_{15}H_{11}N_3O$ mw: 249.29

TOXICITY DATA with REFERENCE
ipr-rat LD50:120 mg/kg AIPTAK 95,123,53
ivn-mus LD50:100 mg/kg CSLNX* NX#05146
ivn-rbt LDLo:175 mg/kg AIPTAK 95,123,53

SAFETY PROFILE: Poison by intravenous and intraperitoneal routes. When heated to decomposition it emits toxic fumes of NO_x.

DWO950 CAS:2039-06-7 ***HR: 3***
3,5-DIPHENYL-s-TRIAZOLE
mf: $C_{14}H_{11}N_3$ mw: 221.28

SYNS: 3,5-DIPHENYL-1H-1,2,4-TRIAZOLE ◇ 3,5-DIPHENYL-1,2,4-TRIAZOLE ◇ 3,5-DPT ◇ s-TRIAZOLE, 3,5-DIPHENYL- ◇ 1H-1,2,4-TRIAZOLE, 3,5-DIPHENYL-

TOXICITY DATA with REFERENCE
scu-ham TDLo:30 mg/kg (female 4-8D post):REP JMCMAR 26,1187,83
ipr-mus LD50:300 mg/kg JPMSAE 50,597,61
ivn-mus LD50:90 mg/kg JPMSAE 50,597,61

SAFETY PROFILE: Poison by intraperitoneal and intravenous routes. Experimental reproductive effects. When heated to decomposition it emits acrid smoke and irritating fumes.

DWP000 CAS:2971-22-4 ***HR: 2***
2,2-DIPHENYL-1,1,1-TRICHLOROETHANE
mf: $C_{14}H_{11}Cl_3$ mw: 285.60

SYN: DT

TOXICITY DATA with REFERENCE
orl-rat LDLo:1000 mg/kg JPETAB 88,359,46
orl-mus LD50:2000 mg/kg HCACAV 29,1317,46

CONSENSUS REPORTS: Reported in EPA TSCA Inventory.

SAFETY PROFILE: Moderately toxic by ingestion. When heated to decomposition it emits very toxic fumes of Cl^-.

DWP229 CAS:13445-50-6 ***HR: 3***
DIPHOSPHANE
mf: H_4P_2 mw: 65.98

PROP: Ignites in air; causes other flammable gases to ignite in air.

SAFETY PROFILE: Ignites spontaneously in air. A concentration of 0.2% will cause flammable gases to ignite. When heated to decomposition it emits toxic fumes of PO_x and phosphine. See also PHOSPHINE.

DWP250 CAS:5518-62-7 ***HR: 3***
1,2-DIPHOSPHINOETHANE
mf: $C_2H_8P_2$ mw: 94.03

SAFETY PROFILE: Ignites spontaneously in air. When heated to decomposition it emits toxic fumes of PO_x. See also PHOSPHINE.

DWP300 ***HR: 3***
DIPHTHERIA TOXIN

SYN: TOXIN, BACTERIUM CORYNE-BACTERIUM DIPHTHERIAE, DIPTHERIA

TOXICITY DATA with REFERENCE
ipr-ham TDLo:11 μg/kg (1D pre):REP PSEBAA 162,170,79
par-cld LDLo:488 ng/kg/1W-I:PUL,KID,SKN ANYAA9 88,1093,60
ipr-mus LD50:300 ng/kg SCIEAS 144,1100,64
ipr-ham LD50:6500 ng/kg PSEBAA 162,170,79

SAFETY PROFILE: Poison by intraperitoneal route. Human systemic effects by parenteral routes: lungs consolidation, changes in kidney tubules, acute tubular necrosis, corrosive to skin. Experimental reproductive effects. A corrosive.

DWP500 CAS:63665-41-8 ***HR: 3***
2-β,16-β-DIPIPERIDINO-5-α-ANDROSTAN-3-α,17-β-DIOL DIPIVALATE HYDROCHLORIDE
mf: $C_{35}H_{62}N_2O_4$•2ClH mw: 647.91

TOXICITY DATA with REFERENCE
ipr-rat LD50:373 mg/kg CALEDQ 2,267,77
ipr-mus LD50:424 mg/kg CALEDQ 2,267,77
ipr-ham LD50:308 mg/kg CALEDQ 2,267,77

SAFETY PROFILE: Poison by intraperitoneal route. When heated to decomposition it emits toxic fumes of NO_x and HCl.

DWP559 CAS:64019-93-8 ***HR: 3***
DIPIVEFRIN HYDROCHLORIDE
mf: $C_{19}H_{29}NO_5$•ClH mw: 387.95

SYNS: (±)-2,2-DIMETHYL-PROPANOICACID-4-(1-HYDROXY-2-(METHYLAMINO)ETHYL)-1,2-PHENYLENE ESTER, HYDROCHLORIDE ◇ DIPIVEFRINE HYDROCHLORIDE

TOXICITY DATA with REFERENCE
scu-rat TDLo:13500 μg/kg (female 17-22D post):REP OYYAA2 31,1083,86
orl-rat LD50:183 mg/kg KSRNAM 20,25,86
ipr-rat LD50:8500 μg/kg KSRNAM 20,25,86
scu-rat LD50:21200 μg/kg KSRNAM 20,25,86
orl-mus LD50:224 mg/kg KSRNAM 20,25,86
ipr-mus LD50:32700 μg/kg KSRNAM 20,25,86
scu-mus LD50:35 mg/kg KSRNAM 20,25,86
ivn-mus LD50:4 mg/kg KSRNAM 20,25,86

SAFETY PROFILE: Poison by ingestion, subcutaneous, intravenous and intraperitoneal routes. Experimental reproductive effects. When heated to decomposition it emits toxic fumes of NO_x and HCl.

DWP900 CAS:78831-88-6 ***HR: 3***
DIPOTASSIUM CYCLOOCTATETRAENE
mf: $C_8H_8K_2$ mw: 182.35

CH=CH(CH=CH)2CH=CH•K2

SAFETY PROFILE: The dry solid explodes on contact with air or oxygen. Reacts with oxygen in THF solution to form a shock-sensitive explosive product. When heated to decomposition it emits toxic fumes of K_2O.

DWP950 CAS:76429-97-5 ***HR: 3***
DIPOTASSIUM DIAZIRINE-3,3-DICARBOXYLATE
mf: $C_3K_2N_2O_4$ mw: 206.24

N=NC(CO•OK)$_2$

SAFETY PROFILE: Explodes when triturated. Upon decomposition it emits toxic fumes of NO_x and K_2O.

DWQ000 CAS:7727-21-1 ***HR: 3***
DIPOTASSIUM PERSULFATE
DOT: UN 1492
mf: $H_2O_8S_2$•2K mw: 272.34

PROP: White, odorless crystals. Mp: decomp @ 100°d: 2.477.

SYNS: ANTHION ◇ PEROXYDISULFURIC ACID DIPOTASSIUM SALT ◇ POTASSIUM PEROXYDISULFATE ◇ POTASSIUM PEROXYDISULPHATE ◇ POTASSIUM PERSULFATE (DOT)

CONSENSUS REPORTS: Reported in EPA TSCA Inventory.
ACGIH TLV: TWA 5 mg(S_2O_8)/m^3
DOT Classification: Oxidizer; Label: Oxidizer.

SAFETY PROFILE: Moderately toxic. An irritant and allergen. A powerful oxidizer. Flammable when exposed to heat or by chemical reaction. Can react with reducing materials. It liberates oxygen above 100° when dry or @ about 50° when in solution. When heated to decomposition it emits highly toxic fumes of SO_x, S_2O_8, and K_2O.

DWQ800 CAS:3248-28-0 ***HR: 3***
DIPROPIONYL PEROXIDE
DOT: UN 2132
mf: $C_6H_{10}O_4$ mw: 146.15

$CH_3CH_2CO{\bullet}OOCO{\bullet}CH_2CH_3$

SYNS: BIS(1-OXOPROPYL)PEROXIDE ◇ PROPIONYL PEROXIDE (DOT) ◇ PROPIONYL PEROXIDE, not more than 28% in solution

DOT Classification: Forbidden (> 28% in solution); Organic Peroxide; Label: Organic Peroxide (UN2132).

SAFETY PROFILE: The pure material explodes at room temperature. When heated to decomposition it emits acrid smoke and fumes. See also PEROXIDES.

DWQ850 CAS:90729-15-0 ***HR: 2***
4,5-DIPROPOXY-2-IMIDAZOLIDINONE
mf: $C_9H_{18}N_2O_3$ mw: 202.29

SYN: SRC-7

TOXICITY DATA with REFERENCE
orl-mus LD50:1430 mg/kg CPBTAL 12,843,64
ipr-mus LD50:890 mg/kg CPBTAL 12,843,64
scu-mus LD50:980 mg/kg CPBTAL 12,843,64

SAFETY PROFILE: Moderately toxic by ingestion, subcutaneous, and intraperitoneal routes. When heated to decomposition it emits toxic fumes of NO_x.

DWQ875 CAS:106-19-4 ***HR: 2***
DIPROPYL ADIPATE
mf: $C_{12}H_{22}O_4$ mw: 230.34

SYN: DI-n-PROPYL ADIPATE

TOXICITY DATA with REFERENCE
ipr-rat TDLo:757 mg/kg (5-15D preg):TER JPMSAE 62,1596,73
ipr-rat TDLo:1262 mg/kg (5-15D preg):REP JPMSAE 62,1596,73
ipr-rat LD50:3786 mg/kg JPMSAE 62,1596,73

CONSENSUS REPORTS: Reported in EPA TSCA Inventory.

SAFETY PROFILE: Moderately toxic by some routes. Experimental reproductive effects. An experimental teratogen. When heated to decomposition it emits acrid smoke and fumes.

DWR000 CAS:142-84-7 ***HR: 3***
DIPROPYLAMINE
DOT: UN 2383
mf: $C_6H_{15}N$ mw: 101.22

PROP: Water-white liquid, amine odor. Mp: −40°, bp: 105°, flash p: 63°F (OC), d: 0.741 @ 20°, vap d: 3.5.

SYNS: DI-n-PROPYLAMINE ◇ n-DIPROPYLAMINE ◇ N-PROPYL-1-PROPANAMINE ◇ RCRA WASTE NUMBER U110

TOXICITY DATA with REFERENCE
skn-rbt 100 μg/24H open AIHAAP 23,95,62
orl-rat LD50:460 mg/kg 85GMAT-,63,82
ihl-rat LC50:4400 mg/m^3/4H 85GMAT -,63,82
ipr-rat LDLo:75 mg/kg FATOAO 31,238,68
ihl-mus LC50:3070 mg/m^3/2H 85GMAT -,63,82
skn-rbt LD50:1250 mg/kg AIHAAP 23,95,62
ihl-mam LC50:4400 mg/m^3 TPKVAL 14,80,75

CONSENSUS REPORTS: Reported in EPA TSCA Inventory.

DOT Classification: Flammable Liquid; Label: Flammable Liquid.

SAFETY PROFILE: Poision by ingestion. Moderately toxic by skin contact and inhalation. A skin irritant. A very dangerous fire hazard, when exposed to heat or flame. Can react with oxidizers. Explosion Hazard is unknown. Keep away from heat and open flame. To fight fire, use foam, CO_2, dry chemical. When heated to decomposition it emits toxic fumes of NO_x. See also AMINES.

DWR200 CAS:64140-51-8 ***HR: 3***
1-DIPROPYLAMINOACETYLINDOLINE
mf: $C_{16}H_{24}N_2O$ mw: 260.42

SYN: 1-(N,N-DIPROPYLGLYCYL)INDOLINE

TOXICITY DATA with REFERENCE
eye-rbt 100 mg PCJOAU 11,785,77
ipr-mus LD50:240 mg/kg PCJOAU 11,785,77

SAFETY PROFILE: Poison by intraperitoneal route. An eye irritant. When heated to decomposition it emits toxic fumes of NO_x.

DWS200 CAS:29091-21-2 ***HR: 1***
N^3,N^3-DIPROPYL-2,4-DINITRO-6-TRIFLUORO-METHYL-m-PHENYLENEDIAMINE
mf: $C_{13}H_{17}F_3N_4O_4$ mw: 350.34

SYN: USB-3153

TOXICITY DATA with REFERENCE
orl-rat LD50:15380 mg/kg 85ARAE 2,48,77

CONSENSUS REPORTS: Reported in EPA TSCA Inventory.

SAFETY PROFILE: Mildly toxic by ingestion. When heated to decomposition it emits very toxic fumes of F^- and NO_x.

DWS400 CAS:94-91-7 ***HR: 1***
α,α'-DIPROPYLENEDINITRILODI-o-CRESOL
mf: $C_{17}H_{18}N_2O_2$ mw: 282.37

SYNS: DISALICYLALPROPYLENEDIIMINE ◇ N,N'-DISALICYLIDENE-1,2-DIAMINOPROPANE ◇ N,N'-DISALICYCLIDENE-1,2-PROPANEDIAMINE

TOXICITY DATA with REFERENCE
orl-rat LD50:4560 mg/kg AEHLAU 6,324,63

CONSENSUS REPORTS: Reported in EPA TSCA Inventory.

SAFETY PROFILE: Mildly toxic by ingestion. When heated to decomposition it emits toxic fumes of NO_x.

DWS600 CAS:29911-28-2 ***HR: 2***
DIPROPYLENE GLYCOL BUTYL ETHER
mf: $C_{10}H_{22}O_3$ mw: 190.32

TOXICITY DATA with REFERENCE
orl-rat LDLo:2000 mg/kg 14CYAT 2,1576,63

CONSENSUS REPORTS: Reported in EPA TSCA Inventory. Glycol ethers are on the Community Right-To-Know List.

SAFETY PROFILE: Moderately toxic by ingestion. When heated to decomposition it emits acrid smoke and irritating fumes. See also GLYCOL ETHERS.

DWS650 CAS:57472-68-1 ***HR: 2***
DIPROPYLENE GLYCOL DIACRYLATE
mf: $C_{12}H_{18}O_5$ mw: 242.30

SYN: 2-PROPENOIC ACID, OXYBIS(METHYL-2,1-ETHANEDIYL) ESTER

TOXICITY DATA with REFERENCE
skn-rbt 500 mg SEV EPASR* 8EHQ-0184-0459
eye-rbt 100 mg SEV EPASR* 8EHQ-0184-0459
orl-rat LD50:4600 mg/kg EPASR* 8EHQ-0184-0459

SAFETY PROFILE: Mildly toxic by ingestion. A severe skin and eye irritant. When heated to decomposition it emits acrid smoke and irritating fumes.

DWS800 CAS:94-51-9 ***HR: 1***
DIPROPYLENE GLYCOL DIBENZOATE
mf: $C_{20}H_{22}O_5$ mw: 342.42

SYNS: BENZOFLEX 9-88 ◇ BENZOFLEX 9-98 ◇ BENZOFLEX 9-88 SG ◇ BENZOIC ACID DIESTER with DIPROPYLENE GLYCOL ◇ BENZOIC ACID-n-DIPROPYLENE GLYCOL DIESTER ◇ DIBENZOYL DIPROPYLENE GLYCOL ESTER ◇ DIPROPANEDIOL DIBENZOATE ◇ K-FLEX DP ◇ 3,3'-OXYDI-1-PROPANOL DIBENZOATE

TOXICITY DATA with REFERENCE
orl-rat LD50:9800 mg/kg AIHAAP 23,95,62

SAFETY PROFILE: Mildly toxic by ingestion. When heated to decomposition it emits acrid smoke and fumes. See also ESTERS.

DWT000 CAS:63716-17-6 ***HR: 2***
DIPROPYLENE GLYCOL DIPELARGONATE

SYNS: EMERY X-88-R ◇ NONANOIC ACID OXYDI-3,1-PROPANEDIYL ESTER (9CI) ◇ NONANOIC ACID OXYDIPROPYLENE ESTER

TOXICITY DATA with REFERENCE
ivn-rat LD50:1060 mg/kg MRLR** No. 256,54
ivn-rbt LD50:1060 mg/kg MRLR** No.256,54

SAFETY PROFILE: Moderately toxic by intravenous route. When heated to decomposition it emits acrid smoke and irritating fumes.

DWT200 CAS:34590-94-8 ***HR: 2***
DIPROPYLENE GLYCOL METHYL ETHER
mf: $C_7H_{16}O_3$ mw: 148.23

PROP: Liquid. Bp: 190°, d: 0.951, vap d: 5.11, flash p: 185°F.

SYNS: ARCOSOLV ◇ DIPROPYLENE GLYCOL MONOMETHYL ETHER ◇ DOWANOL DPM ◇ DOWANOL-50B ◇ UCAR SOLVENT 2LM

TOXICITY DATA with REFERENCE
skn-rbt 500 mg open MLD UCDS** 11/15/71
eye-rbt 238 mg MLD AMIHBC 9,509,54
orl-rat LD50:5660 mg/kg AIHAAP 23,95,62
orl-dog LD50:7500 mg/kg JPETAB 102,79,51
skn-rat LD50:9500 mg/kg DTLVS* 4,157,80

CONSENSUS REPORTS: Reported in EPA TSCA Inventory. Glycol ether compounds are on the Community Right-To-Know List.

OSHA PEL: (Transitional: TWA 100 ppm (skin)) TWA 100 ppm; STEL 150 ppm (skin)
ACGIH TLV: TWA 100 ppm; STEL 150 ppm (skin)
DFG MAK: 50 ppm (300 mg/m^3)

SAFETY PROFILE: Mildly toxic by ingestion and skin

contact. A skin and eye irritant. A mild allergen. Combustible when exposed to heat or flame; can react with oxidizing materials. To fight fire, use dry chemical, CO_2, mist, foam. When heated to decomposition it emits acrid smoke and irritating fumes. See also GLYCOL ETHERS.

DWT400 CAS:6976-50-7 ***HR: 3***
N,N-DI-n-PROPYL ETHYL CARBAMATE
mf: $C_9H_{19}NO_2$ mw: 173.29

SYN: DIPROPYLCARBAMIC ACID ETHYL ESTER

TOXICITY DATA with REFERENCE
ipr-mus TDLo:6500 mg/kg/13W-I:ETA JNCIAM 9,35,48

SAFETY PROFILE: Questionable carcinogen with experimental tumorigenic data. When heated to decomposition it emits toxic fumes of NO_x. See also CARBAMATES.

DWT600 CAS:123-19-3 ***HR: 2***
DIPROPYL KETONE
DOT: UN 2710
mf: $C_7H_{14}O$ mw: 114.21

PROP: Colorless, refractive liquid. Bp: 144°, mp: −32.6°, vap press: 5.2 mm @ 20°, flash p: 120°F (CC), d: 0.815, vap d: 3.93.

SYNS: BUTYRONE (DOT) ◇ GBL ◇ 4-HEPTANONE ◇ HEPTAN-4-ONE ◇ PROPYL KETONE

TOXICITY DATA with REFERENCE
orl-rat LD50:3730 mg/kg TXAPA9 28,313,74
ihl-rat LCLo:4000 ppm/4H TXAPA9 28,313,74
skn-rbt LD50:5660 mg/kg TXAPA9 28,313,74

CONSENSUS REPORTS: Reported in EPA TSCA Inventory.

OSHA PEL: TWA 50 ppm
ACGIH TLV: TWA 50 ppm
DOT Classification: Flammable or Combustible Liquid; Label: Flammable Liquid.

SAFETY PROFILE: Moderately toxic by ingestion, inhalation, and skin contact. Flammable when exposed to heat or flame; can react with oxidizing materials. To fight fire, use CO_2, dry chemical, alcohol foam, fog, and mist. When heated to decomposition it emits acrid smoke and fumes. See also KETONES.

DWU000 CAS:628-85-3 ***HR: 3***
DIPROPYL MERCURY
mf: $C_6H_{14}Hg$ mw: 286.79

PROP: Colorless liquid, immisc in water. D: 2.0208, bp: 190°.

TOXICITY DATA with REFERENCE
ipr-mus LDLo:2 mg/kg CBCCT* 4,320,52

CONSENSUS REPORTS: Mercury and its compounds are on the Community Right-To-Know List.

OSHA PEL: (Transitional: CL 1 mg/10m^3) TWA 0.01 mg(Hg)/m^3; STEL 0.03 mg/m^3 (skin)
ACGIH TLV: TWA 0.01 mg(Hg)/m^3; STEL 0.03 mg(Hg)/m^3
NIOSH REL: (Inorganic Mercury) TWA 0.05 mg (Hg)/m^3

SAFETY PROFILE: Poison by intraperitoneal route. Violent reaction with iodine. When heated to decomposition it emits toxic fumes of Hg. See also MERCURY COMPOUNDS, ORGANIC.

DWU200 CAS:996-05-4 ***HR: 3***
S,S-DIPROPYL METHYLPHOSPHONOTRITHIOATE
mf: $C_7H_{17}PS_3$ mw: 228.39

SYNS: ENT 25,979 ◇ V-C 3-670 ◇ VIRGINIA-CAROLINA 3-670

TOXICITY DATA with REFERENCE
orl-rat LD50:18 mg/kg ARSIM* 20,26,66
orl-ckn LD50:12 mg/kg TXAPA9 11,49,67

SAFETY PROFILE: Poison by ingestion. When heated to decomposition it emits very toxic fumes of PO_x and SO_x.

DWU400 CAS:60580-30-5 ***HR: 3***
O,O-DI-n-PROPYL-O-(4-METHYLTHIOPHENYL)PHOSPHATE
mf: $C_{13}H_{21}O_4PS$ mw: 304.37

SYNS: NK-1158 ◇ PHOSPHORIC ACID DIPROPYL-4-METHYLTHIOPHENYL ESTER

TOXICITY DATA with REFERENCE
orl-rat LD50:70 mg/kg HDIZAB 26,91,78
ipr-rat LD50:35 mg/kg HDIZAB 26,91,78

SAFETY PROFILE: Poison by ingestion and intraperitoneal routes. When heated to decomposition it emits very toxic fumes of PO_x and SO_x. See also ESTERS.

DWU800 CAS:53230-00-5 ***HR: 3***
α-DIPROPYLNITROSAMINE METHYL ETHER

SYNS: 1-METHOXY-N-NITROSO-N-PROPYLPROPYLAMINE ◇ 1-METHOXYPROPYLPROPYLNITROSAMIN (GERMAN) ◇ 1-METHOXYPROPYLPROPYLNITROSAMINE ◇ 1-MPPN

TOXICITY DATA with REFERENCE
scu-ham TDLo:555 mg/kg/37W-I:NEO ZKKOBW 90,215,77
scu-ham LD50:458 mg/kg ZKKOBW 90,215,77

SAFETY PROFILE: Questionable carcinogen with ex-

perimental neoplastigenic data. Moderately toxic by subcutaneous route. Many nitrosamines are carcinogens. When heated to decomposition it emits toxic fumes of NO_x. See also NITROSAMINES and ETHERS.

DWV000 CAS:7664-98-4 ***HR: 3***
DIPROPYLOXOSTANNANE
mf: $C_6H_{14}OSn$ mw: 220.89

SYNS: DIPROPYLTIN OXIDE ◇ KYSLICNIK DI-N-PROPYLCINICITY (CZECH)

TOXICITY DATA with REFERENCE
skn-rbt 500 mg/24H SEV 28ZPAK -,225,72
eye-rbt 100 mg/24H MOD 28ZPAK -,225,72
orl-rat LD50:36800 μg/kg 28ZPAK -,225,72

OSHA PEL: TWA 0.1 mg(Sn)/m^3 (skin)
ACGIH TLV: TWA 0.1 mg(Sn)/m^3 (skin) (Proposed: TWA 0.1 mg(Sn)/m^3; STEL 0.2 mg(Sn)/m^3 (skin))
NIOSH REL: (Organotin Compounds) TWA 0.1 mg(Sn)/m^3

SAFETY PROFILE: Poison by ingestion. An eye and severe skin irritant. When heated to decomposition it emits acrid smoke and irritating fumes. See also TIN COMPOUNDS.

DWV200 CAS:29914-92-9 ***HR: 3***
DIPROPYL PEROXIDE
mf: $C_6H_{14}O_2$ mw: 118.18

$CH_3CH_2CH_2OOCH_2CH_2CH_3$

SAFETY PROFILE: Potentially explosive. When heated to decomposition it emits acrid smoke and fumes. See also PEROXIDES.

DWV400 CAS:16066-38-9 ***HR: 2***
DI-n-PROPYL PEROXYDICARBONATE
DOT: UN 2176
mf: $C_8H_{14}O_6$ mw: 206.22

SYNS: PEROXYDICARBONIC ACID DIPROPYL ESTER ◇ n-PROPYL PERCARBONATE

TOXICITY DATA with REFERENCE
orl-rat LD50:3400 mg/kg BSPII* 1/75-19B
skn-rbt LD50:3500 mg/kg BSPII* 1/75-19B

CONSENSUS REPORTS: Reported in EPA TSCA Inventory.

DOT Classification: Organic Peroxide; Label: Organic Peroxide.

SAFETY PROFILE: Moderately toxic by ingestion and skin contact. When heated to decomposition it emits acrid smoke and irritating fumes.

DWV500 CAS:131-16-8 ***HR: 2***
DIPROPYL PHTHALATE
mf: $C_{14}H_{18}O_4$ mw: 250.32

PROP: Bp: 317.5°, d: 1.078, Flash p: >230° F.

SYNS: 1,2-BENZENEDICARBOXYLIC ACID, DIPROPYL ESTER ◇ DI-n-PROPYL PHTHALATE ◇ PHTHALIC ACID, DIPROPYL ESTER

TOXICITY DATA with REFERENCE
orl-mus TDLo:630 g/kg (male 15W pre): REP FAATDF 12,508,89
orl-mus TDLo:1260 g/kg (male 15W pre): REP FAATDF 12,508,89
orl-mus TDLo:1260 g/kg (male 15W pre): REP FAATDF 12,508,89
ipr-mus LDLo:1251 mg/kg JPMSAE 56,1446,67

CONSENSUS REPORTS: Reported in EPA TSCA Inventory.

SAFETY PROFILE: Moderately toxic by intraperitoneal route. Experimental reproductive effects. An irritant. Combustible when exposed to heat and flame. When heated to decomposition it emits acrid smoke and irritating fumes.

DWV800 CAS:925-15-5 ***HR: 3***
DIPROPYL SUCCINATE
mf: $C_{10}H_{18}O_4$ mw: 202.28

SYNS: BUTANEDIOIC ACID DIPROPYL ESTER ◇ DI-N-PROPYL SUCCINATE ◇ SUCCINIC ACID DIPROPYL ESTER

TOXICITY DATA with REFERENCE
eye-rbt 500 mg AMIHBC 10,61,54
orl-rat LD50:6490 mg/kg AMIHBC 10,61,54
ipr-rat LD50:290 mg/kg NEPSBV 1,286,59

SAFETY PROFILE: Poison by intraperitoneal route. Mildly toxic by ingestion. An eye irritant. When heated to decomposition it emits acrid smoke and irritating fumes. See also ESTERS.

DWW000 CAS:57-66-9 ***HR: 3***
p-(DIPROPYLSULFAMOYL)BENZOIC ACID
mf: $C_{13}H_{19}NO_4S$ mw: 285.39

SYNS: APURINA ◇ BENECID ◇ BENEMID ◇ BENURYL ◇ 4-((DIPROPYLAMINO)SULFONYL)BENZOIC ACID ◇ 4-(DIPROPYLSULFAMOYL) BENZOIC ACID ◇ p-(DIPROPYLSULFAMYL)BENZOIC ACID ◇ ETHAMIDE ◇ NCI-C56097 ◇ PROBECID ◇ PROBEN ◇ PROBENECID ACID ◇ PROBENEMID ◇ PROLONGINE ◇ SYNERGID R ◇ TUBOPHAN ◇ URICOSID

TOXICITY DATA with REFERENCE
hma-rat/smc 60 mg/kg MUREAV 28,57,75
orl-rat LD50:1600 mg/kg MEIEDD 10,1116,83
ipr-rat LDLo:394 mg/kg CLDND*
scu-rat LDLo:611 mg/kg CLDND*
orl-mus LDLo:1666 mg/kg NIIRDN 6,735,82

ipr-mus LD50:1000 mg/kg CPBTAL 16,1655,68
scu-mus LDLo:1156 mg/kg CLDND*
ivn-mus LDLo:458 mg/kg CLDND*
ivn-dog LDLo:230 mg/kg CLDND*

CONSENSUS REPORTS: Reported in EPA TSCA Inventory. EPA Genetic Toxicology Program.

SAFETY PROFILE: Poison by intraperitoneal and intravenous routes. Moderately toxic by ingestion and subcutaneous route. Mutation data reported. An uricosuric which promotes the secretion of uric acid in the urine. When heated to decomposition it emits very toxic fumes of SO_x and NO_x.

DWW200 CAS:23795-03-1 ***HR: 3***
p-(DIPROPYLSULFAMOYL)BENZOIC ACID SODIUM SALT
mf: $C_{13}H_{18}NO_4S \cdot Na$ mw: 307.37

SYNS: p-(DI-N-PROPYLSULFAMYL)BENZOIC ACID SODIUM SALT ◇ PROBENECID SODIUM SALT

TOXICITY DATA with REFERENCE
orl-man TDLo:630 mg/kg/6W:KID ARPAAQ 94,241,72
orl-rat LD50:1604 mg/kg JPETAB 102,208,51
ipr-rat LD50:394 mg/kg JPETAB 102,208,51
scu-rat LD50:611 mg/kg JPETAB 102,208,51
orl-mus LD50:1666 mg/kg JPETAB 102,208,51
ipr-mus LD50:500 mg/kg TXAPA9 24,37,73
scu-mus LD50:1156 mg/kg JPETAB 102,208,51
ivn-mus LD50:458 mg/kg JPETAB 102,208,51
ivn-dog LD50:270 mg/kg JPETAB 102,208,51
ivn-rbt LD50:304 mg/kg JPETAB 102,208,51

SAFETY PROFILE: Poison by intraperitoneal and intravenous routes. Moderately toxic by ingestion and subcutaneous routes. Human systemic effects by ingestion: proteinurea and damage to the kidney (glomeruli), ureter and bladder. When heated to decomposition it emits very toxic fumes of NO_x, Na_2O and SO_x.

DWW400 CAS:73927-87-4 ***HR: 3***
DI-n-PROPYLTIN BISMETHANESULFONATE
mf: $C_8H_{20}O_6S_2Sn$ mw: 395.09

SYN: BIS((METHYLSULFONYL)OXY)DIPROPYLSTANNANE

TOXICITY DATA with REFERENCE
ivn-mus LD50:17800 μg/kg CSLNX* NX#02277

OSHA PEL: TWA 0.1 mg(Sn)/m^3 (skin)
ACGIH TLV: TWA 0.1 mg(Sn)/m^3 (skin) (Proposed: TWA 0.1 mg(Sn)/m^3; STEL 0.2 mg(Sn)/m^3 (skin))
NIOSH REL: (Organotin Compounds) TWA 0.1 mg(Sn)/m^3

SAFETY PROFILE: Poison by intravenous route. When heated to decomposition it emits toxic fumes of SO_x. See also TIN COMPOUNDS and SULFONATES.

DWW500 CAS:628-91-1 ***HR: 3***
DIPROPYL ZINC
mf: $C_6H_{14}Zn$ mw: 151.57

CONSENSUS REPORTS: Zinc and its compounds are on the Community Right-To-Know List.

SAFETY PROFILE: Ignites in air if a large surface area is exposed. When heated to decomposition it emits toxic fumes of ZnO. See also ZINC COMPOUNDS.

DWW600 ***HR: 3***
DIPYRIDINESODIUM
mf: $C_{10}H_{10}N_2Na$ mw: 181.19

SAFETY PROFILE: Ignites spontaneously in air. When heated to decomposition it emits toxic fumes of Na_2O.

DWW700 CAS:67730-10-3 ***HR: 3***
DIPYRIDO(1,2-a:3',2'-d)IMIDAZOL-2-AMINE
mf: $C_{10}H_8N_4$ mw: 184.22

SYNS: 2-AMINODIPYRIDO(1,2-a:3',2'-d)-IMIDAZOLE ◇ GLU-P-2

TOXICITY DATA with REFERENCE
mma-sat 100 ng/plate MUREAV 136,23,84
pic-sat 10 μg/plate MUREAV 110,243,83
sce-hmn:lym 10 μmol/L MUREAV 116,137,83
dns-rat:lvr 1 μmol/L CALEDQ 20,283,83
dnd-mus:lvr 200 μmol/L JJCREP 76,835,85
orl-rat TDLo:12500 mg/kg/2Y-C:CAR,REP EVHPAZ 67,129,86
orl-mus TDLo:23100 mg/kg/83W-C:CAR EVHPAZ 67,129,86
orl-mus TD:24024 mg/kg/66W-C:CAR CRNGDP 5,815,84

CONSENSUS REPORTS: IARC Cancer Review: Group 2B IMEMDT 7,56,87; Animal Sufficient Evidence IMEMDT 40,235,86

SAFETY PROFILE: Suspected carcinogen with experimental carcinogenic data. Experimental reproductive effects. Human mutation data reported. When heated to decomposition it emits toxic fumes of NO_x.

DWX000 CAS:21000-42-0 ***HR: 3***
DIPYRIDYL HYDROGEN PHOSPHATE
mf: $C_{12}H_{14}N_2 \cdot 2C_2H_6O_4P$ mw: 436.2

SYN: DIPYRIDYL PHOSPHATE

TOXICITY DATA with REFERENCE
orl-rat LD50:280 mg/kg GTPZAB 24(9),48,80
skn-rat LD50:460 mg/kg GTPZAB 24(9),48,80
orl-mus LD50:240 mg/kg GTPZAB 24(9),48,80
orl-rbt LD50:295 mg/kg GTPZAB 24(9),48,80
skn-rbt LD50:404 mg/kg GTPZAB 24(9),48,80

SAFETY PROFILE: Poison by ingestion. Moderately

toxic by skin contact. When heated to decomposition it emits very toxic fumes of PO_x and NO_x.

DWX200 CAS:20738-78-7 ***HR: 3***
DI-3-PYRIDYLMERCURY
mf: $C_{10}H_8HgN_2$ mw: 356.79

TOXICITY DATA with REFERENCE
ivn-mus LD50:180 mg/kg CSLNX* NX#05152

CONSENSUS REPORTS: Mercury and its compounds are on the Community Right-To-Know List.

OSHA PEL: (Transitional: CL 1 mg/10m³) CL 0.1 mg(Hg)/m³ (skin)
ACGIH TLV: TWA 0.1 mg(Hg)/m³ (skin)
NIOSH REL: (Inorganic Mercury) TWA 0.05 mg(Hg)/m³

SAFETY PROFILE: Poison by intravenous route. When heated to decomposition it emits very toxic fumes of NO_x and Hg. See also MERCURY COMPOUNDS, ORGANIC.

DWX600 CAS:51-73-0 ***HR: 3***
1,4-DIPYRROLIDINYL-2-BUTYNE
mf: $C_{12}H_{20}N_2$ mw: 192.34

SYNS: BIOFERMIN ◇ 1,1′-(2-BUTYNYLENE)DIPYRROLIDINE ◇ 1,4-DIPYRROLIDINYL-2-BUTYNE ◇ TREMORINE

TOXICITY DATA with REFERENCE
ipr-mus LD50:25 mg/kg ARZNAD 21,1727,71
ivn-mus LD50:112 mg/kg CSLNX* NX12070
scu-cat LDLo:5 mg/kg AIPTAK 135,447,62
orl-bwd LD50:100 mg/kg TXAPA9 21,315,72

SAFETY PROFILE: Poison by ingestion, intraperitoneal, intravenous, and subcutaneous routes. When heated to decomposition it emits toxic fumes of NO_x. See also ACETYLENE COMPOUNDS.

DWX800 CAS:85-00-7 ***HR: 3***
DIQUAT
mf: $C_{12}H_{12}N_2$•2Br mw: 344.08

PROP: Yellow crystals. Mp: 355°. Sol in water.

SYNS: AQUACIDE ◇ DEIQUAT ◇ DEXTRONE ◇ 9,10-DIHYDRO-8a,10,-DIAZONIAPHENANTHRENE DIBROMIDE ◇ 9,10-DIHYDRO-8a,10a-DIAZONIAPHENANTHRENE(1,1′-ETHYLENE-2,2′-BIPYRIDYLIUM)DIBROMIDE ◇ 5,6-DIHYDRO-DIPYRIDO(1,2a;2,1c)PYRAZINIUM DIBROMIDE ◇ 6,7-DIHYDROPYRIDO(1,2a;2′,1′-C)PYRAZINEDIUM DIBROMIDE ◇ DIQUAT DIBROMIDE ◇ 1,1′-ETHYLENE-2,2′-BIPYRIDYLIUM DIBROMIDE ◇ ETHYLENE DIPYRIDYLIUM DIBROMIDE ◇ 1,1-ETHYLENE 2,2-DIPYRIDYLIUM DIBROMIDE ◇ 1,1′-ETHYLENE-2,2′-DIPYRIDYLIUM DIBROMIDE ◇ FB/2 ◇ FEGLOX ◇ PREEGLONE ◇ REGLON ◇ REGLONE ◇ WEEDTRINE-D

TOXICITY DATA with REFERENCE
skn-rbt 400 mg/kg/20D MLD BJIMAG 27,51,70
eye-rbt 10 mg MLD BJIMAG 27,51,70
mmo-sat 100 nmol/plate TOLED5 3,169,79
dns-hmn:fbr 1 μmol/L MUREAV 42,161,77
ipr-rat TDLo:7 mg/kg (7D preg):TER 26UZAB 6,257,68
ivn-rat TDLo:15 mg/kg (female 17D post):REP TXAPA9 33,450,75
orl-rat LD50:120 mg/kg PRKHDK 1,31,75
skn-rat LD50:433 mg/kg FAATDF 7,299,86
ipr-rat LDLo:500 mg/kg PAREAQ 14,225,62
scu-rat LD50:20 mg/kg PAREAQ 14,225,62
orl-mus LD50:233 mg/kg BJIMAG 27,51,70
orl-dog LDLo:187 mg/kg BJIMAG 27,51,70
orl-rbt LD50:188 mg/kg BJIMAG 27,51,70

CONSENSUS REPORTS: EPA Genetic Toxicology Program.

OSHA PEL: TWA 0.5 mg/m³
ACGIH TLV: TWA 0.5 mg/m³; (Proposed: Total Dust: TWA 0.5 mg/m³; Respirable Dust: 0.1 mg/m³)

SAFETY PROFILE: Poison by ingestion, subcutaneous, intravenous, and intraperitoneal routes. Experimental teratogenic and reproductive effects. A skin and eye irritant. Human mutation data reported. When heated to decomposition it emits very toxic fumes of NO_x and Br^-. See also PARAQUAT.

DWY000 CAS:4032-26-2 ***HR: 3***
DIQUAT DICHLORIDE
mf: $C_{12}H_{12}N_2$•2Cl mw: 255.16

SYN: 1,1′-ETHYLENE-2,2′-DIPYRIDINIUM DICHLORIDE

TOXICITY DATA with REFERENCE
scu-rat LD50:19 mg/kg BJIMAG 27,51,70
ivn-mus LD50:180 mg/kg CSLNX* NX#00223
orl-mky LDLo:100 mg/kg TXAPA9 51,277,79

SAFETY PROFILE: Poison by ingestion, intravenous, and subcutaneous routes. When heated to decomposition it emits very toxic fumes of NO_x and Cl^-.

DWY200 CAS:94-93-9 ***HR: 3***
N,N′-DISALICYLIDENE ETHYLENEDIAMINE
mf: $C_{16}H_{16}N_2O_2$ mw: 268.34

SYNS: N,N′-ETHYLENE DIIMINO DI(o-CRESOL) ◇ USAF DO-63

TOXICITY DATA with REFERENCE
orl-rat LDLo:500 mg/kg JPETAB 90,260,47
ipr-mus LD50:100 mg/kg NTIS** AD277-689

CONSENSUS REPORTS: Reported in EPA TSCA Inventory.

SAFETY PROFILE: Poison by intraperitoneal route. Moderately toxic by ingestion. When heated to decomposition it emits toxic fumes of NO_x.

DWY400 CAS:63990-56-7 ***HR: 3***
α,α'-DISELENOBIS-o-ACETOTOLUIDIDE

TOXICITY DATA with REFERENCE
orl-rat LDLo:25 mg/kg NCNSA6 5,10,53
ipr-rat LDLo:25 mg/kg NCNSA6 5,10,53

CONSENSUS REPORTS: Selenium and its compounds are on the Community Right-To-Know List.

OSHA PEL: TWA 0.2 mg(Se)/m^3
ACGIH TLV: TWA 0.2 mg(Se)/m^3
DFG MAK: 0.1 mg(Se)/m^3

SAFETY PROFILE: Poison by ingestion and intraperitoneal routes. When heated to decomposition it emits toxic fumes of Se. See also SELENIUM COMPOUNDS.

DWY600 CAS:64046-56-6 ***HR: 3***
2,2'-DISELENOBIS(N-PHENYLACETAMIDE)
mf: $C_{16}H_{16}N_2O_2Se_2$ mw: 426.26

TOXICITY DATA with REFERENCE
orl-rat LDLo:50 mg/kg NCNSA6 5,10,53
ipr-rat LDLo:25 mg/kg NCNSA6 5,10,53

CONSENSUS REPORTS: Selenium and its compounds are on the Community Right-To-Know List.

OSHA PEL: TWA 0.2 mg(Se)/m^3
ACGIH TLV: TWA 0.2 mg(Se)/m^3
DFG MAK: 0.1 mg(Se)/m^3

SAFETY PROFILE: Poison by ingestion and intraperitoneal routes. When heated to decomposition it emits very toxic fumes of NO_x and Se. See also SELENIUM COMPOUNDS.

DWY800 CAS:1464-43-3 ***HR: 3***
3,3'-DISELENODIALANINE
mf: $C_6H_{12}N_2O_4Se_2$ mw: 334.12

SYNS: SELENIUM CYSTINE ◇ SELENOCYSTINE

TOXICITY DATA with REFERENCE
slt-dmg-orl 10 μmol/L CNJGA8 17,55,75
sln-dmg-orl 2 μmol/L CNJGA8 11,677,69
ipr-rat LDLo:4 mg/kg CTOXAO 17,171,80

CONSENSUS REPORTS: Selenium and its compounds are on the Community Right-To-Know List.

OSHA PEL: TWA 0.2 mg(Se)/m^3
ACGIH TLV: TWA 0.2 mg(Se)/m^3
DFG MAK: 0.1 mg(Se)/m^3

SAFETY PROFILE: Poison by intraperitoneal route. Mutation data reported. When heated to decomposition it emits very toxic fumes of NO_x and Se. See also SELENIUM COMPOUNDS.

DWZ000 CAS:35507-35-8 ***HR: 3***
p,p'-DISELENODIANILINE
mf: $C_{12}H_{12}N_2Se_2$ mw: 342.18

TOXICITY DATA with REFERENCE
orl-rat LDLo:100 mg/kg NCNSA6 5,11,53
ipr-rat LDLo:25 mg/kg NCNSA6 5,11,53

CONSENSUS REPORTS: Selenium and its compounds are on the Community Right-To-Know List.

OSHA PEL: TWA 0.2 mg(Se)/m^3
ACGIH TLV: TWA 0.2 mg(Se)/m^3
DFG MAK: 0.1 mg(Se)/m^3

SAFETY PROFILE: Poison by ingestion and intraperitoneal routes. When heated to decomposition it emits very toxic fumes of NO_x and Se. See also SELENIUM COMPOUNDS.

DWZ100 CAS:70145-55-0 ***HR: 3***
β,β'-DISELENODIPROPIONIC ACID, SODIUM SALT
mf: $C_6H_9O_4Se_2$•Na mw: 326.06

SYN: PROPIONIC ACID, 3,3'-DISELENODI-, SODIUM SALT

TOXICITY DATA with REFERENCE
ipr-rat LDLo:25 mg(Se)/kg JPETAB 63,357,38

OSHA PEL: TWA 0.2 mg(Se)/m^3
ACGIH TLV: TWA 0.2 mg(Se)/m^3

SAFETY PROFILE: Poison by intraperitoneal route. When heated to decomposition it emits toxic fumes of Se.

DXA000 ***HR: 3***
DISILANE
mf: H_6Si_2 mw: 62.22

PROP: Gas, repulsive odor. Mp: −132.5°, bp: −14.5°, d: 0.686 @ −25°/4°.

SYN: SILICOETHANE

SAFETY PROFILE: Poison by inhalation. Dangerous when exposed to heat or flame or by chemical reaction; can react with oxidizing materials. Ignites spontaneously in air. Reacts violently with CCl_4, $CHCl_3$, O_2, and SF_6. See also HYDRIDES and SILANES.

DXA500 ***HR: 3***
DISILVER CYANAMIDE
mf: CAg_2N_2 mw: 255.76

CONSENSUS REPORTS: Silver and its compounds are on the Community Right-To-Know List.

SAFETY PROFILE: A heat- and light-sensitive explosive. When heated to decomposition it emits toxic fumes of NO_x. See also SILVER COMPOUNDS.

DXA600 ***HR: 3***
DISILVER KETENIDE
mf: C_2Ag_2O mw: 255.76

$$Ag_2C=C=O$$

CONSENSUS REPORTS: Silver and its compounds are on the Community Right-To-Know List.

SAFETY PROFILE: The ketenide and its pyridine complex are heat- and impact-sensitive explosives. When heated to decomposition it emits acrid smoke and fumes. See also SILVER COMPOUNDS.

DXA800 ***HR: 3***
DISILVER PENTATIN UNDECAOXIDE
mf: $Ag_2Sn_5O_{11}$ mw: 985.19

SYN: SILVER BETA-STANNATE

CONSENSUS REPORTS: Silver and its compounds are on the Community Right-To-Know List.

SAFETY PROFILE: Can explode on heating. See also SILVER COMPOUNDS and TIN COMPOUNDS.

DXA900 CAS:64267-45-4 ***HR: D***
DISNOGALAMYCINIC ACID
mf: $C_{29}H_{47}N0_{16}$ mw: 773.86

TOXICITY DATA with REFERENCE
dnd-mam:lym 12 μmol/L CBINA8 36,1,81
oms-mam:lym 12 μmol/L CBINA8 36,1,81

SAFETY PROFILE: Mutation data reported. When heated to decomposition it emits toxic fumes of NO_x.

DXB400 CAS:25295-51-6 ***HR: 3***
DISODIUM-4,4'-BIS((4-AMINO-6-(2-HYDROXYETHYL)AMINO-s-TRIAZIN-2-YL)AMINO)-2,2'-STILBENDISULFONIC ACID
mf: $C_{24}H_{26}N_{12}O_8S_2{\cdot}2Na$ mw: 720.72

TOXICITY DATA with REFERENCE
ipr-rat LD50:1000 mg/kg CTOXAO 13,171,78
scu-mus LD50:500 mg/kg CTOXAO 13,171,78
ivn-mus LD50:100 mg/kg CTOXAO 13,171,78

SAFETY PROFILE: Poison by intravenous route. Moderately toxic by intraperitoneal and subcutaneous routes. When heated to decomposition it emits very toxic fumes of SO_x, Na_2O, and NO_x. See also SULFONATES.

DXC200 CAS:7775-11-3 ***HR: 3***
DISODIUM CHROMATE
mf: $CrO_4{\cdot}2Na$ mw: 161.98

SYNS: CHROMATE of SODA ◇ CHROMIUM DISODIUM OXIDE ◇ CHROMIUM SODIUM OXIDE ◇ NEUTRAL SODIUM CHROMATE ◇ SODIUM CHROMATE (DOT) ◇ SODIUM CHROMATE (VI)

TOXICITY DATA with REFERENCE
mmo-sat 33 μg/plate ENMUDM 7,185,85
dnr-sat 50 mmol/L TOLED5 7,439,81
sce-ham:lng 32 μg/L CRNGDP 4,605,83
ipr-rat LD50:57 mg/kg AIPTAK 154,243,65
ipr-mus LD50:32 mg/kg COREAF 257,791,63
ivn-dog LDLo:235 mg/kg EQSSDX 1,1,75
ivn-cat LD50:164 mg/kg AGSOA6 8,51,67
scu-rbt LDLo:243 mg/kg EQSFAP 1,1,75
ivn-rbt LDLo:32 mg/kg EQSSDX 1,1,75
idr-rbt LDLo:250 mg/kg JAPHAR 11,285,1877
skn-gpg LDLo:206 mg/kg AEHLAU 11,201,65
ipr-gpg LDLo:206 mg/kg AEHLAU 11,201,65
scu-gpg LDLo:30 mg/kg EQSSDX 1,1,75
idr-gpg LDLo:382 mg/kg JAPHAR 11,285,1877

CONSENSUS REPORTS: NTP Fifth Annual Report on Carcinogens. IARC Cancer Review: Animal Inadequate Evidence IMEMDT 23,205,80; Human Inadequate Evidence IMEMDT 23,205,80. Reported in EPA TSCA Inventory. EPA Genetic Toxicology Program. Chromium and its compounds are on the Community Right-To-Know List.

OSHA PEL: Cl 0.1 $mg(CrO_3)/m^3$
ACGIH TLV: TWA 0.05 $mg(CrO_2)/m^3$
NIOSH REL: (Chromium(VI)) TWA 25 $\mu g(Cr(VI))/m^3$; CL 50 $\mu g/m^3/15M$

SAFETY PROFILE: Confirmed carcinogen. Poison by skin contact, intraperitoneal, intravenous, subcutaneous, and intradermal routes. A suspected human carcinogenic. Mutation data reported. A powerful oxidizer. When heated to decomposition it emits toxic fumes of Na_2O. See also CHROMIUM COMPOUNDS.

DXC400 CAS:144-33-2 ***HR: 3***
DISODIUM CITRATE
mf: $C_6H_6O_7{\cdot}2Na$ mw: 236.10

PROP: White crystals or granular powder; odorless. Mp: loses water @ 150°, bp: decomp @ red heat. Sol in water; insol in alc.

SYNS: DISODIUM HYDROGEN CITRATE ◇ NATRIUM CITRICUM (GERMAN) ◇ SODIUM CITRATE (FCC)

TOXICITY DATA with REFERENCE
ipr-rat LD50:1724 mg/kg JPETAB 94,65,48
ipr-mus LD50:1771 mg/kg JPETAB 94,65,48
scu-mus LD50:2580 mg/kg ARZNAD 15,852,65
ivn-mus LD50:71 mg/kg JPETAB 94,65,48
ivn-rbt LD50:418 mg/kg JPETAB 94,65,48

CONSENSUS REPORTS: Reported in EPA TSCA Inventory.

SAFETY PROFILE: Poison by intravenous route. Moderately toxic by intraperitoneal and subcutaneous routes. When heated to decomposition it emits toxic fumes of Na_2O.

DXC600 ***HR: 3***
DISODIUM-1,3-DIHYDROXY-1,3-BIS-(aci-NITROMETHYL)-2,2,4,4-TETRAMETHYL-CYCLO BUTANE
mf: $C_{10}H_{16}N_2Na_2O_6$ mw: 306.23

SAFETY PROFILE: Explodes on contact with water. When heated to decomposition it emits toxic fumes of Na_2O.

DXC800 ***HR: 3***
N,N'-DISODIUM N,N'-DIMETHOXYSULFONYL-DIAMIDE
mf: $C_2H_6N_2Na_2O_2S$ mw: 168.13

SAFETY PROFILE: An unstable explosive. When heated to decomposition it emits toxic fumes of Na_2O.

DXD000 CAS:129-67-9 ***HR: 3***
DISODIUM-3,6-ENDOXOHEXAHYDROPHTHALATE
mf: $C_8H_8O_5•2Na$ mw: 230.14

PROP: Water-sol solid. Mp: 144°.

SYNS: ACCELERATE ◇ AGUATHOL ◇ DES-I-CATE ◇ DINATRIUM-(3,6-EPOXY-CYCLOHEXAAN-1,2-DICARBOXYLAAT)(DUTCH) ◇ DINATRIUM-(3,6-EPOXY-CYCLOHEXAN-1,2-DICARBOXYLAT) (GERMAN) ◇ DISODIUM-3,6-EPOXYCYCLOHEXANE-1,2-DICARBOXYLATE ◇ DISODIUM-7-OXABICYCLO(2.2.1)HEPTANE-2,3-DICARBOXYLATE ◇ DISODIUM SALT of ENDOTHALL ◇ DISODIUM SALT of 7-OXABICYCLO(2.2.1)HEPTANE-2,3-DICARBOXYLIC ACID ◇ ENDOTAL ◇ ENDOTHAL ◇ ENDOTHAL-NATRIUM (DUTCH) ◇ ENDOTHAL-SODIUM ◇ ENDOTHAL WEED KILLER ◇ 3,6-ENDOXOHEXAHYDROPHTHALIC ACID DISODIUM SALT ◇ (3,6-EPOSSI-CICLOESAN-1,2-DICARBOSSILATO) DISODICO (ITALIAN) ◇ 3,6-EPOXY-CYCLOHEXANE 1,2-CARBOXYLATE DISODIQUE (FRENCH) ◇ HERBICIDE 273 ◇ HYDOUT ◇ HYDROTHOL ◇ NIAGARATHAL ◇ RCRA WASTE NUMBER P088 ◇ RIPENTHOL ◇ TRI-ENDOTHAL

TOXICITY DATA with REFERENCE
orl-rat LD50:51 mg/kg GUCHAZ 6,248,73
skn-rat LD50:750 mg/kg PHJOAV 185,361,60
ivn-dog LDLo:5 mg/kg FEPRA7 11,349,52
skn-rbt LD50:100 mg/kg AFDOAQ 16,3,52
ivn-rbt LDLo:5 mg/kg FEPRA7 11,349,52
orl-gpg LDLo:250 mg/kg HYSAAV 31,225,66

SAFETY PROFILE: Poison by ingestion, skin contact, and intravenous routes. Very irritating to eyes, skin, and mucous membranes. A defoliant and an herbicide. When heated to decomposition it emits toxic fumes of Na_2O.

DXD200 CAS:142-59-6 ***HR: 3***
DISODIUM ETHYLENE-1,2-BISDITHIOCARBAMATE
mf: $C_4H_6N_2S_4•2Na$ mw: 256.34

PROP: Crystals. Sol in water.

SYNS: CARBON D ◇ CHEM BAM ◇ DINATRIUM-AETHYLEN-BISDITHIOCARBAMAT (GERMAN) ◇ DINATRIUM-(N,N'-AETHYLEN-BIS(DITHIOCARBAMAT)) (GERMAN) ◇ DINATRIUM-(N,N'-ETHYLE-EN-BIS(DITHIOCARBAMAAT)) (DUTCH) ◇ DISODIUM ETHYLENEBIS (DITHIOCARBAMATE) ◇ DITHANE A-40 ◇ DITHANE D-14 ◇ DSE ◇ 1,2-ETHANEDIYLBISCARBAMODITHIOIC ACID DISODIUM SALT ◇ N,N'-ETHYLENE BIS(DITHIOCARBAMATE de SODIUM) (FRENCH) ◇ ETHYLENEBIS(DITHIOCARBAMATE) DISODIUM SALT ◇ ETHYLENEBIS(DITHIOCARBAMIC ACID) DISODIUM SALT ◇ N,N'-ETILEN-BIS(DITIOCARBAMMATO) di SODIO (ITALIAN) ◇ NABAM ◇ NABAME (FRENCH) ◇ PARZATE ◇ SPRING-BAK

TOXICITY DATA with REFERENCE
mmo-omi 1000 ppm MMAPAP 50,233,73
scu-mus TDLo:194 mg/kg (6-14D preg):TER NTIS** Pb 223,-,160
scu-mus TDLo:418 mg/kg (6-14D preg):REP NTIS** PB223-160
orl-rat LD50:395 mg/kg FEPRA7 11,391,52
ipr-rat LD50:500 mg/kg 85DPAN -,-,71/76
orl-mus LD50:580 mg/kg PCOC** -,777,66

CONSENSUS REPORTS: Reported in EPA TSCA Inventory. EPA Genetic Toxicology Program.

SAFETY PROFILE: Poison by ingestion. Moderately toxic by intraperitoneal route. Experimental teratogenic and reproductive effects. Mutation data reported. When heated to decomposition it emits very toxic fumes of NO_x, Na_2O, and SO_x. See also CARBAMATES.

DXD400 CAS:7414-83-7 ***HR: 3***
DISODIUM ETIDRONATE
mf: $C_2H_6O_7P_2•2Na$ mw: 249.99

SYNS: DIDRONEL R ◇ DISODIUM DIHYDROGEN-(1-HYDROXYETHYLIDENE)DIPHOSPHONATE ◇ DISODIUM ETHANOL-1,1-DIPHOSPHONATE ◇ DISODIUM ETHYDRONATE ◇ EITDRONATE DISODIUM ◇ ETHANE-1-HYDROXY-1,1-DIPHOSPHONIC ACID DISODIUM SALT ◇ (1-HYDROXYETHYLIDENE)DIPHOSPHONIC ACID DISODIUM SALT ◇ SODIUM ETHIDRONATE ◇ SODIUM ETIDRONATE ◇ SODIUM ETHYDRONATE

TOXICITY DATA with REFERENCE
orl-rat TDLo:2500 mg/kg (6-15D preg):REP TXAPA9 18,548,71
orl-rbt TDLo:1500 μg/kg (female 2-16D post):TER TXAPA9 18,548,71
orl-rat LD50:1340 mg/kg TXAPA9 22,661,72
orl-mus LD50:2050 mg/kg SIIPD4 10,447,83

ivn-mus LD50:200 mg/kg PCJOAU 12,574,78
ivn-dog LDLo:32 mg/kg JMPSAE 73,1097,84

CONSENSUS REPORTS: Reported in EPA TSCA Inventory.

SAFETY PROFILE: Poison by intravenous route. Moderately toxic by ingestion. An experimental teratogen. Other experimental reproductive effects. When heated to decomposition it emits toxic fumes of PO_x and Na_2O.

DXD600 CAS:10163-15-2 ***HR: 3***
DISODIUM FLUOROPHOSPHATE
mf: $FO_3P \cdot 2Na$ mw: 143.95

SYNS: DISODIUM MONOFLUOROPHOSPHATE ◇ DISODIUM PHOSPHOROFLUORIDATE ◇ SODIUM FLUOROPHOSPHATE (Na_2PO_3F) ◇ SODIUM PHOSPHOROFLUORIDATE ◇ SODIUM PHOSPHOROFLURIDATE

TOXICITY DATA with REFERENCE
dlt-dmg-orl 20 μmol/L GENTAE 87,67,77
orl-rat LD50:570 mg/kg JDREAF 29,529,50
ipr-rat LD50:220 mg/kg JDREAF 29,529,50
orl-mus LD50:710 mg/kg CAREBK 12,177,78

CONSENSUS REPORTS: IARC Cancer Review: Group 3 IMEMDT 7,208,87; Animal Inadequate Evidence IMEMDT 27,237,82; Human Inadequate Evidence IMEMDT 27,237,82.

OSHA PEL: TWA 2.5 mg(F)/m^3
NIOSH REL: (Inorganic Fluorides) TWA 2.5 mg(F)/m^3

SAFETY PROFILE: Poison by intraperitoneal route. Moderately toxic by ingestion. Mutation data reported. Questionable carcinogen. An anticaries ingredient in dentifrices for children's teeth. When heated to decomposition it emits very toxic fumes of F^-, PO_x, and Na_2O.

DXD800 CAS:17013-01-3 ***HR: 2***
DISODIUM FUMARATE
mf: $C_4H_2O_4 \cdot 2Na$ mw: 160.64

SYN: SODIUM FUMARATE

TOXICITY DATA with REFERENCE
orl-hmn TDLo:215 mg/kg:GIT JAPMA8 31,1,42
ipr-rat LDLo:2420 mg/kg JAPMA8 35,298,46
orl-mus LDLo:3680 mg/kg JAPMA8 31,12,42
ivn-rbt LDLo:500 mg/kg JAPMA8 31,1,42

SAFETY PROFILE: Moderately toxic by ingestion, intravenous, and intraperitoneal routes. Human systemic effects by ingestion: hypermotility, diarrhea, nausea or vomiting, and other gastrointestinal changes. When heated to decomposition it emits toxic fumes of Na_2O.

DXD875 CAS:71277-79-7 ***HR: 3***
DISODIUM GLYCYRRHIZIN
mf: $C_{42}H_{62}O_{16} \cdot 2Na$ mw: 869.02

SYNS: DISODIUM GLYCYRRHIZINATE ◇ GLYCYRRHIZINIC ACID DISODIUM SALT

TOXICITY DATA with REFERENCE
cyt-ham:fbr 4 g/L FCTOD7 22,623,84
cyt-ham:lng 1700 mg/L GMCRDC 27,95,81
ipr-mus LD50:144 mg/kg YKYUA6 32,1367,81

SAFETY PROFILE: Poison by intraperitoneal route. Mutation data reported. When heated to decomposition it emits toxic fumes of Na_2O.

DXE000 CAS:16893-85-9 ***HR: 3***
DISODIUM HEXAFLUOROSILICATE
DOT: UN 2674
mf: $F_6Si \cdot 2Na$ mw: 188.07

SYNS: DESTRUXOL APPLEX ◇ (2-)-DISODIUM HEXAFLUOROSILICATE ◇ DISODIUM SILICOFLUORIDE ◇ ENS-ZEM WEEVIL BAIT ◇ ENT 1,501 ◇ FLUOSILICATE de SODIUM ◇ NATRIUMSILICOFLUORID (GERMAN) ◇ ORTHO EARWIG BAIT ◇ ORTHO WEEVIL BAIT ◇ PRODAN ◇ PSC CO-OP WEEVIL BAIT ◇ SAFSAN ◇ SALUFER ◇ SILICON SODIUM FLUORIDE ◇ SODIUM FLUOROSILICATE ◇ SODIUM FLUOSILICATE ◇ SODIUM HEXAFLUOROSILICATE ◇ SODIUM HEXAFLUOSILICATE ◇ SODIUM SILICOFLUORIDE (DOT) ◇ SUPER PRODAN

TOXICITY DATA with REFERENCE
skn-rbt 500 mg MLD FCTOD7 20,563,82
eye-rbt 100 mg SEV FCTOD7 20,573,82
eye-rbt 100 mg/4S rns SEV FCTOD7 20,573,82
orl-rat LD50:125 mg/kg ARSIM* 20,21,66
scu-rat LDLo:70 mg/kg JPETAB 39,246,30
orl-rbt LDLo:125 mg/kg JPETAB 39,246,30
scu-frg LDLo:448 mg/kg CRSBAW 124,133,37

CONSENSUS REPORTS: Reported in EPA TSCA Inventory.

OSHA PEL: TWA 2.5 mg(F)/m^3
NIOSH REL: (Inorganic Fluorides) TWA 2.5 mg(F)/m^3
DOT Classification: Poison B; Label: St. Andrews Cross.

SAFETY PROFILE: Poison by ingestion and subcutaneous route. A skin and severe eye irritant. An insecticide. When heated to decomposition it emits very toxic fumes of F^- and Na_2O.

DXE200 CAS:928-72-3 ***HR: 1***
DISODIUM IMINODIACETATE
mf: $C_4H_5NO_4 \cdot 2Na$ mw: 177.08

SYN: IMINODIOCTAN SODNY (CZECH)

TOXICITY DATA with REFERENCE
eye-rbt 500 mg/24H MLD 28ZPAK -,128,72
orl-rat LD50:8070 mg/kg 28ZPAK -,128,72

CONSENSUS REPORTS: Reported in EPA TSCA Inventory.

SAFETY PROFILE: Mildly toxic by ingestion. An eye irritant. When heated to decomposition it emits toxic fumes of NO_x and Na_2O.

DXE500 CAS:4691-65-0 ***HR: 2***
DISODIUM INOSINATE
mf: $C_{10}H_{11}N_4O_8P{\cdot}2Na$ mw: 392.20

PROP: Colorless to white crystals; characteristic taste. Sol in water; sltly sol in alc; insol in ether.

SYNS: DISODIUM IMP ◇ DISODIUM-5′-INOSINATE ◇ DISODIUM INOSINE-5′-MONOPHOSPHATE ◇ DISODIUM INOSINE-5′-PHOSPHATE ◇ IMP DISODIUM SALT ◇ 5′-IMP DISODIUM SALT ◇ IMP SODIUM SALT ◇ INOSINE-5′-MONOPHOSPHATE DISODIUM ◇ INOSIN-5′-MONOPHOSPHATE DISODIUM ◇ SODIUM INOSINATE ◇ SODIUM-5′-INOSINATE

TOXICITY DATA with REFERENCE
cyt-ham:fbr 1 g/L FCTOD7 22,623,84
ipr-mus TDLo:500 mg/kg (13D preg):TER JJPAAZ 22,201,72
orl-rat LD50:15900 mg/kg AJINO* -,-,73
ipr-rat LD50:4850 mg/kg AJINO* -,-,73
scu-rat LD50:3900 mg/kg AJINO* -,-,73
ivn-rat LD50:2730 mg/kg AJINO* -,-,73
orl-mus LD50:12 g/kg TIDZAH 24,553,66
ipr-mus LD50:5400 mg/kg TIDZAH 24,553,66
scu-mus LD50:5480 mg/kg AJINO* -,-,73
ivn-mus LD50:3300 mg/kg TIDZAH 24,553,66

CONSENSUS REPORTS: Reported in EPA TSCA Inventory.

SAFETY PROFILE: Moderately toxic by several routes. An experimental teratogen. Mutation data reported. When heated to decomposition it emits toxic fumes of PO_x, NO_x, and Na_2O.

DXE600 CAS:144-21-8 ***HR: 3***
DISODIUM METHANEARSENATE
mf: $CH_3AsO_3{\cdot}2Na$ mw: 183.94

PROP: Crystals, water-sol hydrate. Mp: 132-139°, bp: 165°, fp: −6°, d: 1.15.

SYNS: ANSAR 184 ◇ ANSAR DSMA LIQUID ◇ ARRHENAL ◇ ARSINYL ◇ ARSYNAL ◇ CACODYL NEW ◇ CHIPCO CRAB KLEEN ◇ CLOUT ◇ CRAB-E-RAD ◇ CRALO-E-RAD ◇ DAL-E-RAD 100 ◇ DIARSEN ◇ DIMET ◇ DINATE ◇ DISODIUM METHANEARSONATE ◇ DISODIUM METHYLARSENATE ◇ DISODIUM METHYLARSONATE ◇ DISODIUM MONOMETHYLARSONATE ◇ DISOMAR ◇ DI-TAC ◇ DMA ◇ DREXEL DSMA LIQUID ◇ DSMA LIQUID ◇ JON-TROL ◇ MAA SODIUM SALT ◇ METHAR ◇ METHARSINAT ◇ NAMATE ◇ NEOASYCODILE ◇ SODAR ◇ SODIUM METHANEARSONATE ◇ SODIUM METHARSONATE ◇ SODIUM METHYLARSONATE ◇ SOMAR ◇ STENOSINE ◇ TONARSEN ◇ VERSAR DSMA LQ ◇ WEED BROOM ◇ WEED-E-RAD ◇ WEED-HOE

TOXICITY DATA with REFERENCE
ipr-ham TDLo:500 mg/kg (12D preg):TER BECTA6 29,679,82
orl-rat LD50:821 mg/kg FAATDF 7,299,86
skn-rbt LD50:10 g/kg FMCHA2 -,C114,89

CONSENSUS REPORTS: EPA Genetic Toxicology Program. Arsenic and its compounds are on the Community Right-To-Know List.

OSHA PEL: TWA 0.5 mg(As)/m^3
ACGIH TLV: TWA 0.2 mg(As)/m^3

SAFETY PROFILE: Confirmed human carcinogen. Moderately toxic by ingestion. Experimental teratogen. Dangerous fire hazard by spontaneous chemical reaction. Ignites spontaneously in dry air. Can react vigorously with oxidizing materials, i.e., air, Cl_2. An herbicide. When heated to decomposition it emits toxic fumes of As and Na_2O. See also ARSENIC COMPOUNDS.

DXE800 CAS:7631-95-0 ***HR: 3***
DISODIUM MOLYBDATE
mf: $MoO_4{\cdot}2Na$ mw: 205.92

SYNS: MOLYBDIC ACID, DISODIUM SALT ◇ NATRIUMMOLYBDAT (GERMAN) ◇ SODIUM MOLYBDATE ◇ SODIUM MOLYBDATE(VI)

TOXICITY DATA with REFERENCE
pic-esc 16 mmol/L ENMUDM 6,59,84
sln-smc 80 mmol/L MUTAEX 1,21,86
itt-mus TDLo:16474 μg/kg (1D male):REP JRPFA4 7,21,64
ipr-rat LD50:576 mg/kg EQSSDX 1,1,75
ipr-mus LD50:303 mg/kg EQSSDX 1,1,75
scu-mus LD50:570 mg/kg AEPPAE 244,17,62
ivn-cat LD50:917 mg/kg AGSOA6 8,51,67

CONSENSUS REPORTS: Reported in EPA TSCA Inventory.

OSHA PEL: TWA 5 mg(Mo)/m^3
ACGIH TLV: TWA 5 mg(Mo)/m^3

SAFETY PROFILE: Poison by intraperitoneal route. Moderately toxic by subcutaneous and intravenous routes. Experimental reproductive effects. Mutation data reported. When heated to decomposition it emits toxic fumes of Na_2O. See also MOLYBDENUM COMPOUNDS.

DXE875 CAS:10102-40-6 ***HR: 3***
DISODIUM MOLYBDATE DIHYDRATE
mf: $MoO_4{\cdot}_2Na{\cdot}_2H_2O$ mw: 241.96

SYN: SODIUM MOLYBDATE DIHYDRATE

TOXICITY DATA with REFERENCE
ivn-mus TDLo:968 g/kg (8D preg):TER ENVRAL 33,47,84

ipr-rat LD50:520 mg/kg AIPTAK 154,243,65
ipr-mus LD50:257 mg/kg AIPTAK 154,243,65

OSHA PEL: (Transitional: TWA Total Dust: 15 mg/m^3; Respirable Fraction: 5 mg/m^3) TWA Total Dust: 10 mg/m^3; Respirable Fraction: 5 mg/m^3
ACGIH TLV: TWA 10 mg(Mo)/m^3

SAFETY PROFILE: Poison by intraperitoneal route. An experimental teratogen. When heated to decomposition it emits toxic fumes of Na_2O. See also MOLYBDENUM COMPOUNDS.

DXF000 CAS:15467-20-6 ***HR: 2***
DISODIUM NITRILOTRIACETATE
mf: $C_6H_7NO_6 \cdot 2Na$ mw: 235.12

SYNS: KIRESUTO NTB ◇ NITRILOTRIACETIC ACID, DISODIUM SALT

TOXICITY DATA with REFERENCE
orl-rat LD50:1460 mg/kg TXAPA9 18,398,71

CONSENSUS REPORTS: Reported in EPA TSCA Inventory.

SAFETY PROFILE: Moderately toxic by ingestion. When heated to decomposition it emits toxic fumes of NO_x and Na_2O.

DXF200 CAS:12008-41-2 ***HR: 2***
DISODIUM OCTABORATE, TETRAHYDRATE
mf: $B_8Na_2O_{13} \cdot 4H_2O$ mw: 412.54

SYNS: POLYBOR ◇ POLYBOR 3

TOXICITY DATA with REFERENCE
orl-rat LD50:2000 mg/kg FMCHA2 -,C191,83
orl-gpg LD50:5300 mg/kg 28ZEAL 5,85,76

CONSENSUS REPORTS: Reported in EPA TSCA Inventory.

SAFETY PROFILE: Moderately toxic by ingestion. A insecticide. When heated to decomposition it emits toxic fumes of Na_2O. See also BORON COMPOUNDS.

DXF400 CAS:53778-51-1 ***HR: 2***
DISODIUM-2-(p-(γ-PHENYLPROPYLAMINO) BENZENESULFONAMIDO) PYRIDINE
mf: $C_{20}H_{19}N_3O_8S_3 \cdot 2Na$ mw: 571.58

SYNS: DISODIUM CINNAMYLIDENE BISULFITE derivative of SULFAPYRIDINE ◇ 1-PHENYL-3-(p-2-PYRIDYLSULFAMOYLANILINO)-1,3-PROPANEDISULFONIC ACID DISODIUM SALT ◇ SOLUPYRIDINE ◇ SULFAPYRIDINE NEUTRAL SOLUBLE

TOXICITY DATA with REFERENCE
orl-mus LD50:7500 mg/kg JPETAB 84,203,45
scu-mus LD50:2680 mg/kg JPETAB 84,203,45
ivn-mus LD50:1280 mg/kg JPETAB 84,203,45

SAFETY PROFILE: Moderately toxic by ingestion, intravenous, and subcutaneous routes. When heated to decomposition it emits very toxic fumes of NO_x, Na_2O, and SO_x.

DXF600 CAS:26016-99-9 ***HR: 3***
DISODIUM PHOSPHONOMYCIN
mf: $C_3H_7O_4P \cdot 2Na$ mw: 184.05

SYNS: DISODIUM FOSFOMYCIN ◇ (1R,2S)(−)-(1,2-EPOXYPROPYL) PHOSPHONIC ACID DISODIUM SALT ◇ FOM-Na ◇ FOSFOMYCIN DISODIUM ◇ FOSFOMYCIN DISODIUM SALT ◇ FOSFOMYCIN SODIUM SALT ◇ (2R-cis)-(3-METHYLOXIRANYL)PHOSPHONIC ACID DISODIUM SALT ◇ PHOSPHONOMYCIN DISODIUM SALT ◇ PHOSPHONOMYCIN SODIUM ◇ SODIUM FOSFOMYCIN

TOXICITY DATA with REFERENCE
mmo-klp 20 μmol/L MUREAV 89,269,81
ipr-rat TDLo:1375 mg/kg (female 7-17D post):REP JJANAX 32,155,79
ipr-rat TDLo:1375 mg/kg (female 7-17D post):TER JJANAX 32,155,79
orl-rat TD50:4550 mg/kg JJANAX 32,61,79
ipr-rat LD50:2000 mg/kg JJANAX 32,61,79
scu-rat LD50:4320 mg/kg JJANAX 32,61,79
ivn-rat LD50:1560 mg/kg JJANAX 32,61,79
ims-rat LD50:2460 mg/kg JJANAX 32,61,79
orl-mus LD50:7300 mg/kg JJANAX 32,61,79
ipr-mus LD50:2175 mg/kg IYKEDH 12,668,81
scu-mus LD50:5100 mg/kg IYKEDH 12,668,81
ivn-mus LD50:1225 mg/kg JJANAX 32,61,79
ims-mus LD50:2625 mg/kg JJANAX 32,61,79

SAFETY PROFILE: Poison by intraperitoneal and subcutaneous routes. Moderately toxic by intravenous and intramuscular routes. Mildly toxic by ingestion. Experimental teratogenic and reproductive effects. Mutation data reported. When heated to decomposition it emits toxic fumes of PO_x and Na_2O.

DXF700 CAS:50865-01-5 ***HR: 3***
DISODIUM PROTOPORPHYRIN
mf: $C_{34}H_{32}H_4O_4 \cdot 2Na$ mw: 606.68

SYNS: 7,12-DIETHYENYL-3,8,13,17-TETRAMETHYL-21H,23H-PORPHINE-2,18-DIPROPANOIC ACID DISODIUM SALT ◇ PROTOPORPHYRIN DISODIUM ◇ PROTOPORPHYRIN SODIUM ◇ PROTOPORPHYRIN SODIUM SALT

TOXICITY DATA with REFERENCE
ivn-rat LD50:240 mg/kg NIIRDN 6,729,82
ipr-mus LD50:1029 mg/kg NIIRDN 6,729,82
scu-mus LD50:1147 mg/kg NIIRDN 6,729,82
ivn-mus LD50:484 mg/kg NIIRDN 6,729,82

SAFETY PROFILE: Poison by intravenous route. Moderately toxic by intraperitoneal and subcutaneous routes. When heated to decomposition it emits toxic fumes of NO_x and Na_2O.

DXF800 CAS:7758-16-9 ***HR: 3***
DISODIUM PYROPHOSPHATE
mf: $H_2O_7P_2 \cdot Na_2$ mw: 221.94

PROP: White, crystalline powder. D: 1.862, mp: 220° (decomp). Sol in water.

SYNS: DINATRIUMPYROPHOSPHAT (GERMAN) ◇ DIPHOSPHORIC ACID, DISODIUM SALT ◇ DISODIUM DIHYDROGEN PYROPHOSPHATE ◇ DISODIUM DIPHOSPHATE ◇ SODIUM ACID PYROPHOSPHATE (FCC) ◇ SODIUM PYROPHOSPHATE

TOXICITY DATA with REFERENCE
orl-mus LD50:2650 mg/kg ARZNAD 7,445,57
scu-mus LD50:480 mg/kg ARZNAD 7,445,57
ivn-mus LD50:59 mg/kg ARZNAD 7,445,57
ivn-rbt LDLo:50 mg/kg AEPPAE 169,238,33

CONSENSUS REPORTS: Reported in EPA TSCA Inventory.

SAFETY PROFILE: Poison by intravenous route. Moderately toxic by ingestion and subcutaneous routes. An irritant to skin, eyes, and mucous membranes. When heated to decomposition it emits toxic fumes of PO_x and Na_2O.

DXG000 CAS:13410-01-0 ***HR: 3***
DISODIUM SELENATE
mf: $O_4Se \cdot 2Na$ mw: 188.94

PROP: Colorless, rhombic crystals. D: 3.098.

SYNS: NATRIUMSELENIAT (GERMAN) ◇ P-40 ◇ SEL-TOX SSO2 and SS-20 ◇ SODIUM SELENATE

TOXICITY DATA with REFERENCE
mmo-sat 40 µmol/L ENVRAL 36,379,85
mma-sat 2 µmol/plate MUREAV 66,175,79
dnr-sat 10 µg/plate CALEDQ 10,75,80
mrc-bcs 50 µmol/plate MUREAV 66,175,79
dns-rat:lvr 100 µmol/L CALEDQ 10,75,80
orl-mus TDLo:60 mg/kg (female 8-12D post):REP TCMUD8 6,361,86
orl-mus TDLo:14 mg/kg (14D post):TER 32XPAD -,83,75
orl-rat TDLo:128 mg/kg/2Y-C:CAR JONUAI 101,1531,71
orl-wmn TDLo:53 mg/kg:CVS,GIT,LIV NZMJAX 87,354,78
orl-rat LD50:1600 µg/kg GISAAA 49(9),66,84
ipr-rat LDLo:8973 µg/kg JPETAB 58,454,36
scu-rat LDLo:11336 µg/kg ARTODN 45,207,80
ivn-rat LDLo:4786 µg/kg JPETAB 60,449,37
scu-cat LDLo:20 mg/kg HBAMAK 4,1289,35
orl-rbt LD50:2250 µg/kg GISAAA 49(9),66,84
ivn-rbt LDLo:3600 µg/kg JPETAB 60,449,37

CONSENSUS REPORTS: IARC Cancer Review: Group 3 IMEMDT 7,56,87. Selenium and its compounds are on the Community Right-To-Know List. EPA Genetic Toxicology Program. Reported in EPA TSCA Inventory.

OSHA PEL: TWA 0.2 mg(Se)/m^3
ACGIH TLV: TWA 0.2 mg(Se)/m^3
DFG MAK: 0.1 mg(Se)/m^3

SAFETY PROFILE: Poison by ingestion, intravenous, subcutaneous, and intraperitoneal routes. Questionable carcinogen with experimental carcinogenic and teratogenic data. Human systemic effects by ingestion: EKG changes, hypermotility, diarrhea and liver impairment. Experimental reproductive effects. Effects similar to arsenic. Mutation data reported. A pesticide. When heated to decomposition it emits toxic fumes of Se and Na_2O. See also SELENIUM COMPOUNDS and ARSENIC COMPOUNDS.

DXG025 CAS:2583-80-4 ***HR: 1***
DISODIUM 2-(4-STYRYL-3-SULFOPHENYL)-7-SULFO-2H-NAPHTHO(1,2-d)TRIAZOLE
mf: $C_{24}H_{15}N_3O_6S_2 \cdot 2Na$ mw: 551.52

SYNS: 2H-NAPHTHO(1,2-d)TRIAZOLE,2-(4-STYRYL-3-SULFOPHENYL)-7-SULFO-, DISODIUMSALT ◇ NAPHTHO(1,2-d)TRIAZOLE-7-SULFONICACID,2-(4-(2-PHENYLETHENYL)-3-SULFOPHENYL)-, DISODIUM ◇ 2-STILBENESULFONIC ACID, 4-(7-SULFO-2H-NAPHTHO(1,2-d)TRIAZOL-2-YL)-, DISODIUMSALT

TOXICITY DATA with REFERENCE
skn-rbt 500 mg/24H MLD MVCRB3 2,193,73
eye-rbt 100 mg MLD MVCRB3 2,193,73

CONSENSUS REPORTS: Reported in EPA TSCA Inventory.

SAFETY PROFILE: A skin and eye irritant. When heated to decomposition it emits toxic fumes of NO_x and SO_x

DXG050 CAS:68594-24-1 ***HR: 3***
DISODIUM-5-TETRAZOLAZOCARBOXYLATE
mf: $C_2N_6Na_2O_2$ mw: 186.04

SAFETY PROFILE: An explosive. When heated to decomposition it emits toxic fumes of NO_x and Na_2O. See also EXPLOSIVES, HIGH.

DXG100 CAS:2391-03-9 ***HR: 3***
DISOMER MALEATE
mf: $C_{16}H_{19}BrN_2 \cdot C_4H_4O_4$ mw: 435.36

SYNS: (+)-2-(p-BROMO-α-(2-(DIMETHYLAMINO)ETHYL)BENZYL) PYRIDINE MALEATE ◇ (S)-Γ-(4-BROMOPHENYL)-N,N-DIMETHYL-2-PYRIDINEPROPANAMINE (Z)-2-BUTENEDIOATE (1:1) ◇ d-BROMPHENIRAMINE MALEATE ◇ DEXBROMPHENIRAMINE MALEATE

TOXICITY DATA with REFERENCE
orl-rat LD50:191 mg/kg CMTRAG 3,120,61
ipr-rat LD50:104 mg/kg CMTRAG 3,120,61
orl-mus LD50:176 mg/kg CMTRAG 3,120,61

ipr-mus LD50:106 mg/kg CMTRAG 3,120,61
ivn-mus LD50:25 mg/kg CMTRAG 3,120,61
orl-gpg LD50:259 mg/kg CMTRAG 3,120,61

SAFETY PROFILE: Poison by ingestion, intramuscular and intraperitoneal routes. When heated to decomposition it emits toxic fumes of Br^- and NO_x.

DXG150 ***HR: 3***
DISPHOLIDUS TYPHUS VENOM

SYN: VENOM, SNAKE, DISPHOLIDUS TYPHUS

TOXICITY DATA with REFERENCE
scu-mus LDLo:10 mg/kg SAMJAF 14,236,40
ivn-mus LD50:67 μg/kg 23EIAT 1,437,68
ims-mus LDLo:15 mg/kg SAMJAF 14,236,40
ivn-rbt LDLo:5 μg/kg SAMJAF 14,236,40
ims-rbt LDLo:500 μg/kg SAMJAF 14,236,40
ipr-pgn LDLo:667 ng/kg SAMJAF 14,236,40
scu-pgn LDLo:33 μg/kg SAMJAF 14,236,40
ivn-pgn LDLo:667 ng/kg SAMJAF 14,236,40
ims-pgn LDLo:33 μg/kg SAMJAF 14,236,40

SAFETY PROFILE: Poison by subcutaneous, intramuscular, intravenous, and intraperitoneal routes.

DXG200 ***HR: 3***
3,6-DI(SPIROCYCLOHEXANE)TETRAOXANE
mf: $C_{12}H_{20}O_4$ mw: 228.29

SAFETY PROFILE: Explodes on impact. When heated to decomposition it emits acrid smoke and irritating fumes.

DXG400 CAS:39389-47-4 ***HR: D***
DISTAMYCIN

TOXICITY DATA with REFERENCE
mmo-bcs 90 mg/L CMMUAO 9,165,84
dnd-mam:lym 680 nmol/L JMCMAR 22,134,79

SAFETY PROFILE: Mutation data reported.

DXG500 CAS:35967-49-8 ***HR: D***
DISTAMYCIN A/5
mf: $C_{34}H_{39}N_{13}O_6{\cdot}ClH$ mw: 762.32

TOXICITY DATA with REFERENCE
dnd-esc 200 μmol EJBCAI 26,81,72
dnd-omi 200 μmol EJBCAI 26,81,72
dni-mus:lym 600 nmol/L CBINA8 8,183,74

SAFETY PROFILE: Mutation data reported. When heated to decomposition it emits toxic fumes of NO_x and HCl.

DXG600 CAS:6576-51-8 ***HR: 3***
DISTAMYCIN A HYDROCHLORIDE
mf: $C_{22}H_{27}N_9O_4{\cdot}ClH$ mw: 518.04

SYNS: N''-(2-AMIDINOETHYL)-4-FORMAMIDO-1,1',1''-TRIMETHYL-(N,4':N',4''-TERPYRROLE)-2-CARBOXAMIDE HYDROCHLORIDE ◇ HEPERAL ◇ STALLIMYCIN HYDROCHLORIDE

TOXICITY DATA with REFERENCE
dni-mus:lym 1900 nmol/L CBINA8 8,183,74
oms-mus:lym 3200 nmol/L CBINA8 8,183,74
dnd-mam:lym 200 μmol/ EJBCAI 26,81,72L
ipr-rat LD50:169 mg/kg MDACAP 13,319,77
ipr-mus LD50:160 mg/kg MDACAP 13,319,77
ivn-mus LD50:75 mg/kg MEIEDD 11,1383,89

SAFETY PROFILE: Poison by intraperitoneal and intravenous routes. Mutation data reported. When heated to decomposition it emits very toxic fumes of NO_x and HCl.

DXG800 CAS:15876-67-2 ***HR: 3***
DISTIGMINE BROMIDE
mf: $C_{22}H_{32}N_4O_4{\cdot}2Br$ mw: 576.40

SYNS: HEXAMARIUM ◇ 3,3'-(1,6-HEXANEDIYLBIS-((METHYLIMINO)CARBONYL)OXY)BIS(1-METHYLPYRIDINIUMDIBROMIDE) ◇ 3-HYDROXY-1-METHYLPYRIDINIUM BROMIDE HEXAMETHYLENEBIS(METHYLCARBAMATE) ◇ UBRETID ◇ UBRITIL

TOXICITY DATA with REFERENCE
orl-rat LD50:10 mg/kg OYYAA2 3,68,69
ipr-rat LD50:740 μg/kg DRUGAY 6,349,82
scu-rat LD50:1080 μg/kg DRUGAY 6,349,82
ivn-rat LD50:740 μg/kg OYYAA2 3,68,69
orl-mus LD50:10500 μg/kg OYYAA2 3,68,69
ipr-mus LD50:310 μg/kg DRUGAY 6,349,82
ivn-mus LD50:300 μg/kg OYYAA2 3,68,69

SAFETY PROFILE: Poison by ingestion, intravenous, and intraperitoneal routes. When heated to decomposition it emits very toxic fumes of NO_x and Br^-.

DXH200 CAS:150-60-7 ***HR: 2***
DISULFIDE DIBENZYL
mf: $C_{14}H_{14}S_2$ mw: 246.40

SYN: DIBENZYLDISULFID (CZECH)

TOXICITY DATA with REFERENCE
skn-rbt 500 mg/24H MLD 28ZPAK -,173,72
eye-rbt 500 mg/24H MLD 28ZPAK -,173,72
orl-rat LD50:3780 mg/kg 28ZPAK -,173,72

CONSENSUS REPORTS: Reported in EPA TSCA Inventory.

SAFETY PROFILE: Moderately toxic by ingestion. A skin and eye irritant. When heated to decomposition it emits toxic fumes of SO_x.

DXH250 CAS:97-77-8 ***HR: 3***
DISULFIRAM
mf: $C_{10}H_{20}N_2S_4$ mw: 296.56

PROP: Yellow-white crystals; mp: 72°.

SYNS: ABSTENSIL ◇ ABSTINYL ◇ ALCOPHOBIN ◇ ALK-AUBS ◇ ANTABUS ◇ ANTABUSE ◇ ANTADIX ◇ ANTAENYL ◇ ANTAETHAN ◇ ANTAETHYL ◇ ANTAETIL ◇ ANTALCOL ◇ ANTETAN ◇ ANTETHYL ◇ ANTETIL ◇ ANTEYL ◇ ANTIAETHAN ◇ ANTIETANOL ◇ ANTI-ETHYL ◇ ANTIETIL ◇ ANTIKOL ◇ ANTIVITIUM ◇ AVERSAN ◇ AVERZAN ◇ BIS(DIETHYLAMINO)THIOXOMETHYL) DISULPHIDE ◇ BIS(N,N-DIETHYLTHIOCARBAMOYL) DISULFIDE ◇ BIS(DIETHYLTHIOCARBAMOYL) DISULFIDE ◇ BIS(N,N-DIETHYLTHIOCARBAMOYL)DISULPHIDE ◇ BONIBAL ◇ CONTRALIN ◇ CONTRAPOT ◇ CRONETAL ◇ DICUPRAL ◇ DISETIL ◇ DISULFAN ◇ DISULFURAM ◇ DISULPHURAM ◇ 1,1'-DITHIOBIS(N,N-DIETHYLTHIOFORMAMIDE) ◇ EKAGOM TEDS ◇ EPHORRAN ◇ ESPENAL ◇ ESPERAL ◇ ETABUS ◇ ETHYLDITHIOURAME ◇ ETHYLDITHIURAME ◇ ETHYL THIRAM ◇ ETHYL THIUDAD ◇ ETHYL THIURAD ◇ ETHYL TUADS ◇ ETHYL TUEX ◇ EXHORAN ◇ EXHORRAN ◇ HOCA ◇ KROTENAL ◇ NCI-C02959 ◇ NOCBIN ◇ NOXAL ◇ REFUSAL ◇ RO-SULFIRAM ◇ STOPAETHYL ◇ STOPETHYL ◇ STOPETYL ◇ TATD ◇ TENURID ◇ TENUTEX ◇ TETD ◇ TETIDIS ◇ TETRADIN ◇ TETRADINE ◇ TETRAETHYLTHIOPEROXYDICARBONIC DIAMIDE ◇ TETRAETHYLTHIRAM DISULPHIDE ◇ TETRAETHYLTHIURAM ◇ TETRAETHYLTHIURAM DISULFIDE ◇ TETRAETHYLTHIURAM DISULPHIDE ◇ N,N,N',N'-TETRAETHYLTHIURAM DISULPHIDE ◇ TETRAETIL ◇ TETURAM ◇ TETURAMIN ◇ THIOSAN ◇ THIOSCABIN ◇ THIRERANIDE ◇ THIURAM E ◇ THIURANIDE ◇ TILLRAM ◇ TIURAM ◇ TTD ◇ TTS ◇ USAF B-33

TOXICITY DATA with REFERENCE
eye-rbt 100 mg MLD FCTOD7 20,573,82
mmo-sat 25 μg/plate CBINA8 49,329,84
dni-ckn:emb 120 nmol/L BBACAQ 519,65,78
orl-mus TDLo:39200 mg/kg (female 7-14D post):REP NTIS** PB86-197605
par-rat TDLo:400 mg/kg (female 4-11D post):TER BEXBAN 93,107,82
orl-mus TDLo:35 g/kg/78W-I:NEO NTIS** PB223-159
orl-hmn LDLo:160 mg/kg BMJOAE 2,94,77
orl-wmn TDLo:90 mg/kg/18D-I:LVR JCLPDE 46,67,85
orl-man TDLo:150 mg/kg/6W-I:MUS ARHEAW 25,1494,82
orl-chd TDLo:150 mg/kg 34ZIAG -,230,69
orl-hmn LDLo:160 mg/kg BMJOAE 2,94,77
orl-cld TDLo:150 mg/kg 34ZIAG -,230,69
orl-rat LD50:500 mg/kg ATXKA8 22,12,66
orl-mus LD50:1980 mg/kg AIPTAK 112,36,57
ipr-mus LD50:75 mg/kg NTIS** AD691-490

CONSENSUS REPORTS: IARC Cancer Review: Group 3 IMEMDT 7,56,87; Animal Inadequate Evidence IMEMDT 12,85,76. NCI Carcinogenesis Bioassay (feed); No Evidence: mouse, rat NCITR* NCI-CG-TR-16,79. Reported in EPA TSCA Inventory.

OSHA PEL: TWA 2 mg/m^3
ACGIH TLV: TWA 2 mg/m^3
DFG MAK: 2 mg/m^3

SAFETY PROFILE: A human poison by ingestion. An experimental poison by intraperitoneal route. Toxic symptoms when accompanied by ingestion of alcohol. Human systemic effects by ingestion: jaundice, joint changes. An experimental teratogen. Other experimental reproductive effects. Questionable carcinogen with experimental neoplastigenic data. See also BIS(DIMETHYLTHIOCARBAMYL)DISULFIDE.

DXH300 CAS:149-45-1 ***HR: 1***
3,5-DISULFOCATECHOL DISODIUM SALT
mf: $C_6H_6O_8S_2 \cdot 2Na$ mw: 316.22

PROP: Crystals, nonhygroscopic. Very freely sol in water. Produces water-sol, colored compounds with metal salts.

SYNS: 4,5-DIHYDROXY-1,3-BENZENEDISULFONIC ACID DISODIUM SALT ◇ SDD ◇ TIFERRON ◇ TIRON

TOXICITY DATA with REFERENCE
mmo-sat 100 μg/plate ABCHA6 45,327,81
mma-sat 100 μg/plate ABCHA6 45,327,81
ipr-mus LD50:6103 mg/kg TOLED5 26,95,85

CONSENSUS REPORTS: Reported in EPA TSCA Inventory.

SAFETY PROFILE: Mildly toxic by intraperitoneal route. Mutation data reported. Colorimetric reagent for iron, manganese, titanium, molybdenum. When heated to decomposition it emits toxic fumes of SO_x and Na_2O. See also SULFONATES.

DXH325 CAS:298-04-4 ***HR: 3***
DISULFOTON
DOT: NA 2783
mf: $C_8H_{19}O_2PS_3$ mw: 274.42

SYNS: BAYER 19639 ◇ O,O-DIAETHYL-S-(2-AETHYLTHIO-AETHYL)-DITHIOPHOSPHAT (GERMAN) ◇ O,O-DIAETHYL-S-(3-THIA-PENTYL)-DITHIOPHOSPHAT (GERMAN) ◇ O,O-DIETHYL-S-(2-ETHTHIOETHYL) PHOSPHORODITHIOATE ◇ O,O-DIETHYL-S-(2-ETHTHIOETHYL) THIOTHIONOPHOSPHATE ◇ O,O-DIETHYL-S-(2-ETHYLMERCAPTOETHYL) DITHIOPHOSPHATE ◇ O,O-DIETHYL-S-(2-ETHYLTHIO-ETHYL)-DITHIOFOSFAAT (DUTCH) ◇ O,O-DIETHYL-2-ETHYLTHIOETHYL PHOSPHORODITHIOATE ◇ O,O-DIETHYL-S-2-(ETHYLTHIO)ETHYL PHOSPHORODITHIOATE ◇ O,O-DIETIL-S-(2-ETILTIO-ETIL)-DITIOFOSFATO (ITALIAN) ◇ DIMAZ ◇ DISULFATON ◇ DI-SYSTON ◇ DISYSTOX ◇ DITHIODEMETON ◇ DITHIOPHOSPHATE de O,O-DIETHYLE et de S-(2-ETHYLTHIO-ETHYLE) (FRENCH) ◇ DITHIOSYSTOX ◇ ENT 23,437 ◇ O,O-ETHYL-S-2(ETHYLTHIO)ETHYL PHOSPHORODITHIOATE ◇ S-2-(ETHYLTHIO)ETHYL O,O-DIETHYL ESTER of PHOSPHORODITHIOIC ACID ◇ FRUMIN AL ◇ M-74 ◇ RCRA WASTE NUMBER P039 ◇ S 276 ◇ SOLVIREX ◇ THIODEMETON ◇ THIODEMETRON

TOXICITY DATA with REFERENCE
mma-sat 5 mg/plate MUREAV 116,185,83
mmo-esc 5 μL/plate MUREAV 28,405,75
orl-rat LD50:2 mg/kg FMCHA2 -,C85,83
ihl-rat LC50:200 mg/m^3 85GYAZ -,25,71

skn-rat LD50:6 mg/kg TXAPA9 14,515,69
ipr-rat LD50:2 mg/kg AMIHAB 17,192,58
ivn-rat LD50:5500 mg/kg 13ZGAF -,206,62
orl-mus LD50:5500 μg/kg 85DPAN -,-,71/76
ipr-mus LD50:5500 μg/kg AMIHAB 17,192,58
orl-gpg LD50:10800 μg/kg AMIHAB 17,192,58
ipr-gpg LD50:7 mg/kg AMIHAB 17,192,58
orl-qal LD50:12 mg/kg EESADV 8,551,84
orl-dck LD50:6500 μg/kg DOEAAH 35,25,79

CONSENSUS REPORTS: EPA Extremely Hazardous Substances List. EPA Genetic Toxicology Program.

OSHA PEL: TWA 0.1 mg/m^3 (skin)
ACGIH TLV: TWA 0.1 mg/m^3
DOT Classification: Poison B; Label: Poison; Poison B; Label: Poison, dry or liquid.

SAFETY PROFILE: Poison by ingestion, inhalation, skin contact, intraperitoneal, and intravenous routes. Human mutation data reported. When heated to decomposition it emits very toxic SO_x and PO_x. See also various demeton entries and ESTERS.

DXH350 CAS:13172-31-1 ***HR: 3***
DISULFUR DIBROMIDE
mf: Br_2S_2 mw: 223.93

SAFETY PROFILE: Violent reaction with potassium, sodium, aluminum, and antimony. Violent reaction with iron at 650°C. The moist bromide reacts violently with oxidants. When heated to decomposition it emits toxic fumes of Br^- and SO_x. See also BROMIDES.

DXH400 CAS:25474-92-4 ***HR: 3***
DISULFUR DINITRIDE
mf: N_2S_2 mw: 92.14

SN=S=N (ring)

SAFETY PROFILE: An explosive sensitive to shock, friction, pressure, or heating above 30°C. Potentially explosive reaction with complexes with beryllium chloride or titanium tetrachloride. When heated to decomposition it emits toxic fumes of SO_x and NO_x. See also NITRIDES.

DXH600 ***HR: 1***
DISULFUR HEPTAOXIDE
mf: O_7S_2 mw: 208.19

SYN: PEROXYDISULFIRIC ANHYDRIDE

SAFETY PROFILE: Explodes when exposed to moist air. When heated to decomposition it emits toxic fumes of SO_x. See also PEROXIDES.

DXH800 ***HR: 3***
DISULFURYL DIAZIDE
mf: $N_6O_5S_2$ mw: 228.17

$N_3SO_2OSO_2N_3$

SAFETY PROFILE: An unstable explosive. Incompatible with dilute alkali. When heated to decomposition it emits toxic fumes of SO_x and NO_x. See also AZIDES.

DXI000 ***HR: 3***
1,2-DI(5-TETRAZOLYL)HYDRAZINE
mf: $C_2H_4N_{10}$ mw: 168.12

(HNN=NN=CNH-)$_2$ (ring)

SAFETY PROFILE: Explodes on heating. When heated to decomposition it emits toxic fumes of NO_x. See also HYDRAZINE.

DXI200 CAS:56929-36-3 ***HR: 3***
1,3-DI(5-TETRAZOYL)TRIAZENE
mf: $C_2H_3N_{11}$ mw: 181.12

HNN=NN=CN=NNHC=NN=NNH (rings)

SAFETY PROFILE: The metal salts are heat- and friction-sensitive explosives (e.g., barium, copper, silver). The barium salt is a weak explosive. When heated to decomposition it emits toxic fumes of NO_x.

DXI400 CAS:8018-01-7 ***HR: 2***
DITHANE M-45
mf: $C_4H_6MnN_2S_4 \cdot C_4H_6N_2S_4Zn$ mw: 541.03

PROP: Contains 16% manganese, 2% zinc, and 62% ethylenebisdithiocarbamate ion/manganese ethylenebisdithiocarbamate plus zinc ion (85ARAE 4,52,76).

SYNS: CARMAZINE ◇ DITHANE S60 ◇ DITHANE SPC ◇ DITHANE ULTRA ◇ ETHYLENEBIS(DITHIOCARBAMIC ACID MANGANESE ZINC COMPLEX (8CI) ◇ F 2966 ◇ FORE ◇ GREEN-DAISEN M ◇ KARAMATE ◇ MANCOFOL ◇ MANCOZEB ◇ MANEB-ZINC ◇ MANEB-ZINEB-KOMPLEX (GERMAN) ◇ MANGAN-ZINK-AETHYLENDIAMIN-BIS-DITHIO-CARBAMAT (GERMAN) ◇ MANOSEB ◇ MANZATE 200 ◇ MANZEB ◇ MANZIN 80 ◇ MARZIN ◇ NEMISPOR ◇ POLICAR MZ ◇ POLICAR S ◇ TRIZIMAN ◇ TRIZIMAN D ◇ ZIMANAT ◇ ZIMANEB ◇ ZIMMAN-DITHANE

TOXICITY DATA with REFERENCE
cyt-hmn:lym 4 mg/L MUREAV 116,341,83
cyt-rat-ipr 2500 μg/kg MUREAV 116,341,83
ihl-rat TCLo:500 mg/m^3/6H (female 6-10D post):REP TXAPA9 84,355,86
orl-rat TDLo:1320 mg/kg (11D preg):TER TJADAB 14,171,76
orl-rat LD50:5 g/kg 85JFAN A251,86

CONSENSUS REPORTS: EPA Genetic Toxicology Program. Zinc and its compounds, as well as manganese and its compounds, are on the Community Right-To-Know List.

OSHA PEL: (Transitional: CL 5 mg(Mn)/m^3) CL 5 mg(Mn)/m^3
ACGIH TLV: TWA 5 mg(Mn)/m^3

SAFETY PROFILE: An experimental teratogen. Other experimental reproductive effects. Mutation data reported. When heated to decomposition it emits very toxic fumes of SO_x, ZnO and NO_x. See also CARBAMATES, MANGANESE COMPOUNDS, and ZINC COMPOUNDS.

DXI500 CAS:503-41-3 ***HR: 3***
2,4-DITHIA-1,3-DIOXANE-2,2,4,4-TETRAOXIDE
mf: $C_2H_4O_6S_2$ mw: 188.17

SYN: CARBYL SULFATE

SAFETY PROFILE: Potentially violent reaction with N-methyl-4-nitroaniline. When heated to decomposition it emits toxic fumes of SO_x.

DXI600 CAS:7187-55-5 ***HR: 3***
DITHIAZANINE
mf: $C_{23}H_{23}N_2S_2$ mw: 391.60

TOXICITY DATA with REFERENCE
orl-mus LD50:2550 μg/kg RPOBAR 2,288,70
ipr-mus LD50:780 μg/kg RPOBAR 2,288,70
ivn-mus LD50:1800 μg/kg CSLNX* NX#01838

SAFETY PROFILE: Poison by ingestion, intravenous, and intraperitoneal routes. When heated to decomposition it emits very toxic fumes of NO_x and SO_x.

DXI800 CAS:66788-41-8 ***HR: 3***
(R,S)-α-(1-((3,3-DI-3-THIENYLALLYL)AMINO)ETHYL)BENZYL ALCOHOL (+)-(α)-HYDROCHLORIDE
mf: $C_{20}H_{21}NOS_2$•ClH mw: 392.00

SYNS: 1-3,3-BIS(3'-THIENYL)-2-PROPENYL-(3-HYDROXY-3-PHENYLPROPYL-2)AMINE ◇ (+)-α-(1-(93,3-DI-3-THIENYLALLYL)AMINO) ETHYL)BENZYL ALCOHOL HYDROCHLORIDE ◇ d-8955 HYDROCHLORIDE ◇ NOVOCEBRIN HYDROCHLORIDE ◇ TINOFEDRINE HYDROCHLORIDE

TOXICITY DATA with REFERENCE
orl-rat LD50:6600 mg/kg DRFUD4 4,286,79
orl-mus LD50:1890 mg/kg DRFUD4 4,286,79
ipr-mus LD50:14 mg/kg DRFUD4 4,286,79
ivn-mus LD50:20150 μg/kg DRFUD4 4,286,79
ivn-dog LD50:20 mg/kg DRFUD4 4,286,79

SAFETY PROFILE: Poison by intraperitoneal and intravenous routes. Moderately toxic by ingestion. A cerebral vasodilator. When heated to decomposition it emits very toxic fumes of SO_x and NO_x.

DXJ100 ***HR: 3***
1-((3,3-DI-2-THIENYL-1-METHYL)ALLYL)PYRROLIDINE HYDROCHLORIDE
mf: $C_{16}H_{19}NS_2$•ClH mw: 325.94

TOXICITY DATA with REFERENCE
scu-rat LD50:92 mg/kg BJPCAL 8,2,53
scu-mus LD50:120 mg/kg BJPCAL 8,2,53
ivn-mus LD50:25 mg/kg BJPCAL 8,2,53

SAFETY PROFILE: Poison by subcutaneous and intravenous routes. When heated to decomposition it emits toxic fumes of NO_x, SO_x, and HCl.

DXJ400 CAS:64059-02-5 ***HR: 3***
2,2'-DITHIOBIS(N-(1-ADAMANTYLMETHYL) ACETAMIDINE) DIHYDROCHLORIDE HEMIHYDRATE
mf: $C_{26}H_{42}N_4S_2$•2ClH•1/2H_2O mw: 556.77

TOXICITY DATA with REFERENCE
orl-mus LD50:325 mg/kg JMCMAR 15,1313,72
ipr-mus LD50:100 mg/kg JMCMAR 15,1313,72

SAFETY PROFILE: Poison by ingestion and intraperitoneal routes. When heated to decomposition it emits very toxic fumes of NO_x, SO_x, and HCl.

DXJ800 CAS:1141-88-4 ***HR: 3***
2,2'-DITHIOBISANILINE
mf: $C_{12}H_{12}N_2S_2$ mw: 248.38

SYNS: BIS(2-AMINOPHENYL)DISULFIDE ◇ BIS(o-AMINOPHENYL) DISULFIDE ◇ 1,1'-BIS(2-AMINOPHENYL)DISULFIDE ◇ O,O'-DIAMINO DIPHENYL DISULFIDE ◇ O,O-DITHIO-BIS-ANILINE ◇ 2,2'-DITHIODIANILINE ◇ USAF AB-315

TOXICITY DATA with REFERENCE
eye-rbt 50 μg/24H SEV 28ZPAK -,172,72
orl-rat LD50:558 mg/kg 28ZPAK -,172,72
ipr-mus LD50:50 mg/kg NTIS** AD277-689
ivn-mus LD50:178 mg/kg CSLNX* NX#00263

CONSENSUS REPORTS: Reported in EPA TSCA Inventory.

SAFETY PROFILE: Poison by intravenous and intraperitoneal routes. Moderately toxic by ingestion. A severe eye irritant. When heated to decomposition it emits very toxic fumes of NO_x and SO_x.

DXL200 CAS:2127-10-8 ***HR: D***
2,2'-DITHIOBIS(5-NITROPYRIDINE)
mf: $C_{10}H_6N_4O_4S_2$ mw: 312.34

TOXICITY DATA with REFERENCE
mmo-sat 50 μg/plate MUREAV 67,123,79
mma-sat 50 μg/plate MUREAV 67,123,79

SAFETY PROFILE: Mutation data reported. When

heated to decomposition it emits very toxic fumes of SO_x and NO_x.

DXL400 CAS:4540-66-3 ***HR: D***
2,2'-DITHIOBIS(PYRIDINE-1-OXIDE)MAGNESIUM SULFATE TRIHYDRATE
mf: $C_{10}H_8N_2O_2S_2 \cdot MgOS \cdot 3H_2O$ mw: 378.75

SYN: MDS

TOXICITY DATA with REFERENCE
skn-pig TDLo:750 mg/kg (female 8-32D post):TER TXAPA9 42,561,77
orl-rat TDLo:1 g/kg (female 6-15D post):REP TXAPA9 42,561,77

SAFETY PROFILE: Experimental teratogenic and reproductive effects. When heated to decomposition it emits very toxic fumes of SO_x and NO_x.

DXL800 CAS:541-53-7 ***HR: 3***
DITHIOBIURET
mf: $C_2H_5N_3S_2$ mw: 135.22

PROP: Crystals. Mp: 181°, bp: decomp, d: 1.522 @ 30°.

SYNS: DTB ◇ RCRA WASTE NUMBER P049 ◇ 2-THIO-1-(THIOCARBAMOYL)UREA ◇ USAF B-44 ◇ USAF EK-P-6281

TOXICITY DATA with REFERENCE
orl-rat LD50:5 mg/kg JPETAB 90,260,47
ipr-rat LD50:29 mg/kg TXAPA9 57,63,81
ipr-mus LDLo:50 mg/kg NTIS** AD277-689

CONSENSUS REPORTS: Reported in EPA TSCA Inventory. EPA Extremely Hazardous Substances List.

SAFETY PROFILE: Poison by ingestion and intraperitoneal routes. When heated to decomposition it emits highly toxic fumes of SO_x and NO_x.

DXM000 CAS:36551-21-0 ***HR: 3***
DITHIOCARBONIC ACID-o-sec-BUTYL ESTER SODIUM SALT
mf: $C_5H_9OS_2 \cdot Na$ mw: 172.25

SYN: sec-BUTYLXANTHIC ACID SODIUM SALT

TOXICITY DATA with REFERENCE
ipr-mus LDLo:500 mg/kg CBCCT* 6,230,54
par-mus LDLo:400 mg/kg CBCCT* 7,696,55

CONSENSUS REPORTS: Reported in EPA TSCA Inventory.

SAFETY PROFILE: Poison by parenteral route. Moderately toxic by intraperitoneal route. When heated to decomposition it emits toxic fumes of SO_x and Na_2O. See also ESTERS.

DXM100 ***HR: 3***
DITHIOCARBOXYMETHYL-p-CARBAMIDOPHENYLARSENOUS OXIDE
mf: $C_9H_{11}AsN_2O_4S_2$ mw: 350.26

SYNS: ((p-ARSONOPHENYL)CARBAMOYL)DITHIOCARBAMIC ACID ◇ N-((p-ARSONPHENYL)CARBAMOYL)DITHIOGLYCINE

TOXICITY DATA with REFERENCE
orl-rat LD50:1 g/kg FEPRA7 6,306,47
ipr-rat LD50:75 mg/kg FEPRA7 6,306,47
ipr-mus LD50:100 mg/kg FEPRA7 6,306,47

CONSENSUS REPORTS: Arsenic and its compounds are on the Community Right-To-Know List.

SAFETY PROFILE: Poison by intraperitoneal route. Moderately toxic by ingestion. When heated to decomposition it emits toxic fumes of SO_x, NO_x, and As. See also ARSENIC COMPOUNDS.

DXM600 CAS:1892-29-1 ***HR: 3***
DITHIODIGLYCOL
mf: $C_4H_{10}O_2S_2$ mw: 154.26

SYNS: 2,2-DITHIODIETHANOL ◇ USAF TH-9

TOXICITY DATA with REFERENCE
ipr-mus LD50:100 mg/kg NTIS** AD277-689

CONSENSUS REPORTS: Reported in EPA TSCA Inventory.

SAFETY PROFILE: Poison by intraperitoneal route. When heated to decomposition it emits toxic fumes of SO_x.

DXN300 CAS:3696-28-4 ***HR: D***
2,2'-DITHIODIPYRIDINE-1,1'-DIOXIDE
mf: $C_{10}H_8N_2O_2S_2$ mw: 252.32

SYNS: DS ◇ OLIN

TOXICITY DATA with REFERENCE
scu-mus TDLo:1600 μg/kg (6-15D preg):REP OYYAA2 15,1169,78
scu-mus TDLo:1600 μg/kg (6-15D preg):TER OYYAA2 15,1169,78

SAFETY PROFILE: Experimental reproductive effects. An experimental teratogen. When heated to decomposition it emits toxic fumes of NO_x and SO_x.

DXN400 CAS:14807-75-1 ***HR: 3***
1,1'-DITHIOFORMAMIDINE DIHYDROCHLORIDE
mf: $C_2H_6N_4S_2 \cdot 2ClH$ mw: 223.16

SYNS: GUANYL DISULFIDE DIHYDROCHLORIDE ◇ USAF A-11074

TOXICITY DATA with REFERENCE
ipr-mus LD50:100 mg/kg NTIS** AD277-689

CONSENSUS REPORTS: Reported in EPA TSCA Inventory.

SAFETY PROFILE: Poison by intraperitoneal route. When heated to decomposition it emits very toxic fumes of NO_x and Cl^-.

DXN600 CAS:333-29-9 **_HR: 3_**
DITHIOLANE IMINOPHOSPHATE
mf: $C_7H_{14}NO_2PS_3$ mw: 271.37

SYNS: AC-43064 ◇ AMERICAN CYANAMID AC 43,064 ◇ CL-43,064 ◇ CYALANE ◇ CYCLIC ETHYLENE (DIETHOXYPHOSPHINOTHIOYL)DITHIOIMIDOCARBONATE ◇ CYCLIC ETHYLENE ESTER of (DIETHOXYPHOSPHINOTHIOYL)DITHIOIMIDOCARBONICACID ◇ CYLAN ◇ CYOLAN ◇ CYOLANE INSECTICIDE ◇ 2-(DIETHOXYPHOSPHINYLIMINO)-1,3-DITHIOLANE ◇ DIETHYL-N-1,3-DITHIOLANYL-2-IMINO PHOSPHATE ◇ O,O-DIETHYL 1,3-DITHIOLAN-2-YLIDENEPHOSPHORAMIDOTHIOATE ◇ DITHIOLANE ◇ 1,3-DITHIOLAN-2-YLIDENE-PHOSPHORAMIDOTHIOIC ACID DIETHYL ESTER ◇ 1,3-DITHIOLAN-2-YLIDENE-PHOSPHORAMIDOTHIOIC ACID-O,O-DIETHYL ESTER ◇ ENT 25,809 ◇ IMINOPHOSPHATE ◇ PHOSFOLAN

TOXICITY DATA with REFERENCE
orl-rat LD50:14 mg/kg IPCLBZ 22,40,80
orl-mus LD50:62 mg/kg AECTCV 14,111,85
skn-rbt LD50:23 mg/kg GUCHAZ 6,200,73
orl-ckn LD50:5200 μg/kg EXPEAM 30,63,74
orl-bwd LD50:1800 μg/kg TXAPA9 26,154,73
skn-bwd LD50:10 mg/kg TXAPA9 26,154,73

SAFETY PROFILE: Poison by ingestion and skin contact. An insecticide. When heated to decomposition it emits very toxic fumes of NO_x and SO_x.

DXN709 CAS:940-69-2 **_HR: 2_**
1,2-DITHIOLANE-3-VALERAMIDE
mf: $C_8H_{15}NOS_2$ mw: 205.36

SYNS: 1,2-DITHIOLANE-3-PENTANAMIDE (9CI) ◇ LIPAMIDE ◇ α-LIPAMIDE ◇ LIPOACIN ◇ LIPOAMID ◇ LIPOAMIDE ◇ α-LIPOAMIDE ◇ LIPOCTON ◇ α-LIPOIC ACID AMIDE ◇ LIPOICIN ◇ LIPOZYME ◇ LYPOARAN ◇ PATHOCLON ◇ THIOAMI ◇ THIOCTAMID ◇ THIOCTAMIDE ◇ THIOCTIC ACID AMIDE ◇ THIOTOMIN ◇ TICOLIN ◇ TIOCTAN

TOXICITY DATA with REFERENCE
orl-rat LD50:1980 mg/kg NIIRDN 6,454,82
ipr-rat LD50:250 mg/kg NIIRDN 6,454,82
scu-rat LD50:1700 mg/kg NIIRDN 6,454,82
orl-mus LD50:620 mg/kg NIIRDN 6,454,82
ipr-mus LD50:310 mg/kg NIIRDN 6,454,82
scu-mus LD50:3500 mg/kg NIIRDN 6,454,82

SAFETY PROFILE: Poison by intraperitoneal route. Moderately toxic by ingestion and subcutaneous routes. When heated to decomposition it emits toxic fumes of SO_x and NO_x.

DXN800 CAS:62-46-4 **_HR: 3_**
1,2-DITHIOLANE-3-VALERIC ACID
mf: $C_8H_{14}O_2S_2$ mw: 206.34

SYNS: ACETATE-REPLACING FACTOR ◇ BILETAN ◇ 5-(1,2-DITHIOLAN-3-YL)VALERIC ACID ◇ 6,8-DITHIOOCTANOIC ACID ◇ HEPARLIPON ◇ LIPOIC ACID ◇ α-LIPOIC ACID ◇ α-LIPONIC ACID ◇ α-LIPONSAEURE (GERMAN) ◇ LIPOSAN ◇ LIPOTHION ◇ THIOCTACID ◇ THIOCTIC ACID ◇ 6-THIOCTIC ACID ◇ 6,8-THIOCTIC ACID ◇ THIOCTIDASE ◇ THIOCTSAN ◇ THIOKTSAFURE (GERMAN) ◇ THIOOCTANOIC ACID ◇ 6-THIOTIC ACID ◇ 6,8-THIOTIC ACID ◇ TIOCTACID ◇ TIOCTIDASI ◇ TIOCTIDASI ACETATE REPLACING FACTOR

TOXICITY DATA with REFERENCE
scu-rat TDLo:480 mg/kg (30D male):REP NYKZAU 68,265,72
orl-rat LD50:1130 mg/kg NYKZAU 68,265,72
ipr-rat LD50:200 mg/kg NYKZAU 68,265,72
scu-rat LD50:230 mg/kg NYKZAU 68,265,72
ivn-rat LD50:180 mg/kg NYKZAU 68,265,72
orl-mus LD50:502 mg/kg ARTODN 41,79,78
ipr-mus LD50:160 mg/kg NIIRDN 6,454,82
scu-mus LD50:200 mg/kg NIIRDN 6,454,82
ivn-mus LD50:197 mg/kg NYKZAU 60,278,64

SAFETY PROFILE: Poison by intravenous, subcutaneous, and intraperitoneal routes. Moderately toxic by ingestion. Experimental reproductive effects. When heated to decomposition it emits toxic fumes of SO_x.

DXN850 CAS:3706-77-2 **_HR: 3_**
1,3-DITHIOLIUM PERCHLORATE
mf: $C_3H_3ClO_4S_2$ mw: 250.67

SAFETY PROFILE: A friction- and heat-sensitive explosive. Explodes at 250°C. When heated to decomposition it emits toxic fumes of Cl^- and SO_x. See also PERCHLORATES.

DXO000 CAS:572-48-5 **_HR: 3_**
DITHION
mf: $C_{17}H_{21}O_5PS$ mw: 368.41

PROP: Crystals nearly insol in water. Mp: 88°.

SYNS: O,O-DIETHYL-7-HYDROXY-3,4-TETRAMETHYLENE COUMARINYL PHOSPHOROTHIOATE ◇ O,O-DIETHYL-O-(7,8,9,10-TETRAHYDRO-6-OXOBENZO(C)CHROMAN-3-YL)PHOSPHOROTHIOATE ◇ O,O-DIETHYL-O-(7,8,9,10-TETRAHYDRO-6-OXO-6H-DIBENZO (b,d)PYRAN-3-YL)PHOSPHOROTHIOATE ◇ O,O-DIETHYL-O-(3,4-TETRAMETHYLENECOUMARINYL-7) THIOPHOSPHATE ◇ DITHIONE ◇ ENT 24,986 ◇ 7-HYDROXY-3,4-TETRAMETHYLENE-COUMARIN-O,O-DIETHYL THIOPHOSPHATE

TOXICITY DATA with REFERENCE
orl-rat LD50:67 mg/kg PHJOAV 185,361,60
orl-mus LD50:3800 mg/kg 28ZEAL 5,56,76
orl-dog LD50:400 mg/kg 28ZEAL 5,56,76
orl-rbt LD50:500 mg/kg 28ZEAL 5,56,76
orl-gpg LD50:200 mg/kg 28ZEAL 5,56,76

SAFETY PROFILE: Poison by ingestion. When heated to decomposition it emits very toxic fumes of PO_x and SO_x. See also PARATHION.

DXO200 CAS:79-40-3 ***HR: 3***
DITHIOOXAMIDE
mf: $C_2H_4N_2S_2$ mw: 120.20

SYNS: DITHIOOXALDIIMIDIC ACID ◇ DITHIOXAMIDE ◇ ETHANEDITHIOAMIDE ◇ HYDRORUBEANIC ACID ◇ RUBEANE ◇ RUBEANIC ACID ◇ RVK ◇ USAF EK-4394 ◇ USAF MK-6

TOXICITY DATA with REFERENCE
orl-rat LDLo:500 mg/kg NCNSA6 5,43,53
orl-mus LD50:350 mg/kg FATOAO 28,230,65
ipr-mus LD50:100 mg/kg NTIS** AD277-689
ivn-mus LD50:56 mg/kg CSLNX* NX#04475

CONSENSUS REPORTS: Reported in EPA TSCA Inventory.

SAFETY PROFILE: Poison by ingestion, intraperitoneal, and intravenous routes. When heated to decomposition it emits very toxic fumes of NO_x and SO_x.

DXO300 CAS:59-58-5 ***HR: 3***
DITHIOPROPYLTHIAMINE
mf: $C_{15}H_{24}N_4O_2S_2$ mw: 356.55

PROP: Prisms from benzene. Decomp @ 128-129°. Sparingly sol in water. Soluble in organic solvents and lipids.

SYNS: ALINAMIN ◇ ALITON ◇ ANEURIMEC ◇ ARINAMINE ◇ AUSOVIT ◇ BETATRON ◇ BINOVA ◇ DITIOVIT ◇ LIPONEURINA ◇ MARINEURINA ◇ NERVILON ◇ NEURVITA ◇ NEVRITON ◇ NUVELBI V.C.A. ◇ OROBETINA ◇ PRONEURIN ◇ PROSULTHIAMINE ◇ PROSULTIAMINE ◇ SINTOTIAMINA ◇ THIAMINE PROPYL DISULFIDE ◇ THIAMIN PROPYL DISULFIDE ◇ TIOTIAMINE ◇ TIOVIT-B_1 ◇ TIPIDI ◇ TPD ◇ VITAMIN B_1 PROPYL DISULFIDE

TOXICITY DATA with REFERENCE
orl-mus LD50:2750 mg/kg NIIRDN 6,724,82
ipr-mus LD50:650 mg/kg NIIRDN 6,724,82
ivn-mus LD50:320 mg/kg NIIRDN 6,724,82

SAFETY PROFILE: Poison by intravenous route. Moderately toxic by ingestion and intraperitoneal routes. When heated to decomposition it emits toxic fumes of SO_x and NO_x.

DXO400 CAS:973-99-9 ***HR: 3***
DITHIOPROPYLTHIAMINE HYDROCHLORIDE
mf: $C_{15}H_{24}N_4O_2S_2•ClH$ mw: 393.01

SYNS: DTP HYDROCHLORIDE ◇ THIAMINE PROPYL DISULFIDE HYDROCHLORIDE

TOXICITY DATA with REFERENCE
orl-mus LD50:2512 mg/kg IZVIAK 37,82,67
ivn-mus LD50:302 mg/kg IZVIAK 37,82,67

SAFETY PROFILE: Poison by intravenous route. Moderately toxic by ingestion. When heated to decomposition it emits very toxic fumes of HCl, SO_x, and NO_x.

DXO600 CAS:1076-98-8 ***HR: 3***
DITHIOTEREPHTHALIC ACID
mf: $C_8H_6O_2S_2$ mw: 198.26

TOXICITY DATA with REFERENCE
ivn-mus LD50:320 mg/kg CSLNX* NX#01918

CONSENSUS REPORTS: Reported in EPA TSCA Inventory.

SAFETY PROFILE: Poison by intravenous route. When heated to decomposition it emits toxic fumes of SO_x.

DXO775 CAS:27565-41-9 ***HR: 3***
DITHIOTHREITOL
mf: $C_4H_{10}O_2S_2$ mw: 154.26

SYNS: CLELAND'S REAGENT ◇ (R*,R*)-(±)-1,4-DIMERCAPTO-2,3-BUTANEDIOL (9CI) ◇ dl-threo-DIMERCAPTO-2,3-BUTANEDIOL ◇ dl-DITHIOTHREITOL ◇ dl-1,4-DITHIOTHREITOL ◇ rac-DITHIOTHREITOL ◇ DTT

TOXICITY DATA with REFERENCE
mmo-sat 100 μg/plate ABCHA6 45,327,81
mma-sat 100 μg/plate ABCHA6 45,327,81
dnd-omi 1 mmol/L BBRCA9 77,1150,77
mnt-nml:emb 2 mmol/L JCLBA3 60,497,74
cyt-nml:emb 2 mmol/L JCLBA3 60,497,74
dni-rat:lvr 1 mmol/L ABBIA4 166,400,75
ipr-mus LD50:169 mg/kg JMCMAR 15,600,72
scu-mus LD50:333 mg/kg ARZNAD 22,1434,72
ivn-mus LD50:94 mg/kg ARZNAD 22,1434,72

SAFETY PROFILE: Poison by subcutaneous, intravenous, and intraperitoneal routes. Mutation data reported. When heated to decomposition it emits toxic fumes of SO_x.

DXO800 CAS:3483-12-3 ***HR: 3***
d-1,4-DITHIOTHREITOL
mf: $C_4H_{10}O_2S_2$ mw: 154.26

SYNS: CLELAND'S REAGENT ◇ D-DTT ◇ d-threo-1,4-DIMERCAPTO-2,3-BUTANEDIOL ◇ 1,4-DITHIOTHREITOL ◇ DTT ◇ SPUTOLYSIN

TOXICITY DATA with REFERENCE
ipr-mus LD50:154 mg/kg YKKZAJ 94,1419,74
ims-mus LD50:108 mg/kg JPPMAB 1,576,49

CONSENSUS REPORTS: Reported in EPA TSCA Inventory.

SAFETY PROFILE: Poison by intraperitoneal and intramuscular routes. When heated to decomposition it emits toxic fumes of SO_x.

DXP000 CAS:27755-15-3 ***HR: 3***
DITOLYLETHANE
mf: $C_{16}H_{18}$ mw: 210.34

TOXICITY DATA with REFERENCE
orl-mus TDLo:9200 mg/kg/73W-C:ETA GTPZAB 15(5),49,71
orl-mus LD50:725 mg/kg GTPZAB 15(5),49,71
scu-mus LD50:10 g/kg GTPZAB 15(5),49,71

SAFETY PROFILE: Moderately toxic by ingestion. Questionable carcinogen with experimental tumorigenic data. When heated to decomposition it emits acrid smoke and irritating fumes.

DXP200 CAS:97-39-2 ***HR: 3***
DI-o-TOLYLGUANIDINE
mf: $C_{15}H_{17}N_3$ mw: 239.35

PROP: White crystals. Mp: 179°, d: 1.10 @ 20°/4°, vap d: 8.24.

SYNS: DIORTHOTOLYLGUANIDINE ◇ 1,3-DI-o-TOLYLGUANIDINE ◇ DOTG ACCELERATOR ◇ USAF A-6598 ◇ VULKACIT DOTG/C

TOXICITY DATA with REFERENCE
orl-rat LD50:500 mg/kg JPETAB 90,260,47
ipr-mus LD50:25 mg/kg NTIS** AD277-689
orl-rbt LDLo:120 mg/kg JIDHAN 13,87,31
orl-gpg LDLo:120 mg/kg JIDHAN 13,87,31
orl-mam LDLo:120 mg/kg JIDHAN 13,87,31

CONSENSUS REPORTS: Reported in EPA TSCA Inventory.

SAFETY PROFILE: Poison by ingestion and intraperitoneal routes. When heated to decomposition it emits toxic fumes of NO_x.

DXP400 CAS:15017-02-4 ***HR: 3***
N,N'-DI-o-TOLYL-p-PHENYLENE DIAMINE
mf: $C_{20}H_{20}N_2$ mw: 288.42

SYN: USAF GY-1

TOXICITY DATA with REFERENCE
ipr-mus LD50:300 mg/kg NTIS** AD277-689

CONSENSUS REPORTS: Reported in EPA TSCA Inventory.

SAFETY PROFILE: Poison by intraperitoneal route. When heated to decomposition it emits toxic fumes of NO_x.

DXP600 CAS:137-97-3 ***HR: 3***
DI-o-TOLYLTHIOUREA
mf: $C_{15}H_{16}N_2S$ mw: 256.39

PROP: Crystals. Mp: 178°, vap d: 8.85.

SYNS: N,N'-BIS(2-METHYLPHENYLTHIOUREA ◇ 1,3-BIS(o-TOLYL)-2-THIOUREA ◇ 2,2'-DIMETHYLTHIOCARBANILIDE ◇ DI-o-TOLUYLTHIOUREA ◇ USAF EK-1651

TOXICITY DATA with REFERENCE
ipr-mus LD50:100 mg/kg NTIS** AD277-689
orl-rbt LDLo:3000 mg/kg JPETAB 17,349,21

CONSENSUS REPORTS: Reported in EPA TSCA Inventory.

SAFETY PROFILE: Poison by intraperitoneal route. Moderately toxic by ingestion. When heated to decomposition it emits very toxic fumes of NO_x and SO_x.

DXP800 CAS:8015-54-1 ***HR: 3***
DITRAN
mf: $C_{20}H_{29}NO_3 \cdot C_{20}H_{29}NO_3 \cdot 2ClH$ mw: 735.92

SYN: JB 329

TOXICITY DATA with REFERENCE
orl-hmn TDLo:40 μg/kg:EYE,CNS,GIT JNMDAN 131,428,60
ims-man TDLo:150 μg/kg:CNS FEPRA7 32,250,73
ipr-rat LD50:25 mg/kg 27ZQAG -,224,72
ivn-rat LD50:19 mg/kg 27ZQAG -,224,72
ims-rat LD50:1193 μg/kg BJPCBM 39,822,70
orl-mus LD50:300 mg/kg ARZNAD 21,1727,71
ipr-mus LD50:60 mg/kg 27ZQAG -,224,72
ivn-mus LD50:10 mg/kg 27ZQAG -,224,72
ims-mus LD50:1634 μg/kg BJPCBM 39,822,70
ivn-rbt LD50:12 mg/kg 27ZQAG -,224,72
ivn-gpg LD50:45 mg/kg 27ZQAG -,224,72
ims-gpg LD50:327 μg/kg BJPCBM 39,822,70

SAFETY PROFILE: Poison by intravenous, intraperitoneal, and intramuscular routes. Human systemic effects by ingestion and intramuscular routes: visual field changes, hallucinations, distorted perceptions, nausea and vomiting. When heated to decomposition it emits toxic fumes of NO_x and Cl^-.

DXQ000 CAS:5910-75-8 ***HR: 2***
DITRIDECYLAMINE
mf: $C_{26}H_{55}N$ mw: 381.82

TOXICITY DATA with REFERENCE
skn-rbt 100 μg/24H open AIHAAP 23,95,62
orl-rat LD50:9850 mg/kg AIHAAP 23,95,62
skn-rbt LD50:3540 mg/kg AIHAAP 23,95,62

CONSENSUS REPORTS: Reported in EPA TSCA Inventory.

SAFETY PROFILE: Moderately toxic by skin contact. Mildly toxic by ingestion. A skin irritant. When heated to decomposition it emits toxic fumes of NO_x. See also AMINES.

DXQ200 CAS:119-06-2 ***HR: 1***
DITRIDECYL PHTHALATE
mf: $C_{34}H_{58}O_4$ mw: 530.92

PROP: D: 0.951 @ 20°/20°, bp: >285° @ 5 mm, flash p: 470°F (OC).

SYNS: 1,2-BENZENEDICARBOXYLIC ACID, DITRIDECYL ESTER ◇ DTDP ◇ JAYFLEX DTDP ◇ NUOPLAZ ◇ PHTHALIC ACID, DITRIDECYL ESTER ◇ POLYCIZER 962-BPA ◇ STALFEX DTDP ◇ 1-TRIDECANOL PHTHALATE ◇ TRUFLEX DTDP

TOXICITY DATA with REFERENCE
skn-rbt 10 mg/24H open MLD AIHAAP 23,95,62

CONSENSUS REPORTS: Reported in EPA TSCA Inventory.

SAFETY PROFILE: A skin irritant. Combustible when exposed to heat or flame. To fight fire, use dry chemical, CO_2. When heated to decomposition it emits acrid smoke and irritating fumes.

DXQ339 ***HR: 3***
DI[TRIS-1,2-DIAMINOETHANECHROMIUM(III)]TRIPEROXODISULFATE
mf: $C_{12}H_{24}Cr_2N_{12}O_{24}S_6$ mw: 1016.78

CONSENSUS REPORTS: Chromium and its compounds are on the Community Right-To-Know List.

SAFETY PROFILE: Explodes. When heated to decomposition it emits toxic fumes of SO_x and NO_x. See also CHROMIUM COMPOUNDS, PEROXIDES, and SULFATES.

DXQ369 ***HR: 3***
DI[TRIS-1,2-DIAMINOETHANECOBALT(III)]TRIPEROXODISULFATE
mf: $C_{12}H_{24}Co_2N_{12}O_{24}S_6$ mw: 1030.66

CONSENSUS REPORTS: Cobalt and its compounds are on the Community Right-To-Know List.

SAFETY PROFILE: Explodes. When heated to decomposition it emits toxic fumes of SO_x and NO_x. See also COBALT COMPOUNDS, PEROXIDES, and SULFATES.

DXQ400 CAS:3648-20-2 ***HR: 1***
DIUNDECYL PHTHALATE
mf: $C_{30}H_{50}O_2$ mw: 442.80

SYN: SANTICIZER 711

TOXICITY DATA with REFERENCE
eye-rbt 100 mg MLD MONS** 11/4/75

CONSENSUS REPORTS: Reported in EPA TSCA Inventory.

SAFETY PROFILE: An eye irritant. When heated to decomposition it emits acrid smoke and irritating fumes. See also ESTERS.

DXQ500 CAS:330-54-1 ***HR: 3***
DIURON
mf: $C_9H_{10}Cl_2N_2O$ mw: 233.11

PROP: Crystals. Mp: 159°. Sltly sol in water and hydrocarbon solvents.

SYNS: AF 101 ◇ CEKIURON ◇ CRISURON ◇ DAILON ◇ DCMU ◇ DIATER ◇ 3-(3,4-DICHLOOR-FENYL)-1,1-DIMETHYLUREUM (DUTCH) ◇ DICHLORFENIDIM ◇ 3-(3,4-DICHLOROPHENOL)-1,1-DIMETHYLUREA ◇ N'-(3,4-DICHLOROPHENYL)-N,N-DIMETHYLUREA ◇ 1-(3,4-DICHLOROPHENYL)-3,3-DIMETHYLUREE (FRENCH) ◇ 3-(3,4-DICHLOR-PHENYL)-1,1-DIMETHYL-HARNSTOFF (GERMAN) ◇ 3-(3,4-DICLORO-FENYL)-1,1-DIMETIL-UREA (ITALIAN) ◇ 1,1-DIMETHYL-3-(3,4-DICHLOROPHENYL)UREA ◇ DI-ON ◇ DIREX 4L ◇ DIUREX ◇ DIUROL ◇ DIURON 4L ◇ DMU ◇ DREXEL ◇ DREXEL DIURON 4L ◇ DURAN ◇ DYNEX ◇ FARMCO DIURON ◇ HERBATOX ◇ HW 920 ◇ KARMEX ◇ KARMEX DIURON HERBICIDE ◇ KARMEX DW ◇ MARMER ◇ SUP'R FLO ◇ TELVAR ◇ TELVAR DIURON WEED KILLER ◇ UNIDRON ◇ USAF P-7 ◇ USAF XR-42 ◇ VONDURON

TOXICITY DATA with REFERENCE
mma-sat 3 μg/plate MUREAV 58,353,78
dni-mus-orl 1 g/kg MUREAV 58,353,78
scu-mus TDLo:1935 mg/kg (female 6-14D post):TER NTIS** PB223-160
orl-mus TDLo:153 g/kg/78W-I:ETA NTIS** PB223-159
orl-rat LD50:1017 mg/kg JAFCAU 18,1104,70
ipr-mus LDLo:500 mg/kg NTIS** AD277-689

CONSENSUS REPORTS: Reported in EPA TSCA Inventory. EPA Genetic Toxicology Program. Chlorophenol compounds are on The Community Right-To-Know List.

OSHA PEL: TWA 10 mg/m³
ACGIH TLV: TWA 10 mg/m³

SAFETY PROFILE: Moderately toxic by ingestion and intraperitoneal routes. Questionable carcinogen with experimental tumorigenic and teratogenic data. Mutation data reported. When heated to decomposition it emits highly toxic fumes of Cl^- and NO_x. See also CHLOROPHENOLS.

DXQ745 CAS:108-57-6 ***HR: 2***
DIVINYLBENZENE
mf: $C_{10}H_{10}$ mw: 130.20

PROP: Pale straw-colored liquid. Bp: 195-200°, mp: −87°, d: 0.918, flash p: 165F°. Not misc in water; sol in ether and methanol.

SYN: m-DIVINYLBENZEN ◇ m-DIVINYLBENZENE ◇ m-VINYLSTYRENE

TOXICITY DATA with REFERENCE
eye-rbt 500 mg/24H MLD 85JCAE-,38,86

CONSENSUS REPORTS: Reported in EPA TSCA Inventory.

OSHA PEL: 10 ppm
ACGIH TLV: 10 ppm

SAFETY PROFILE: An mild eye irritant. Combustible. When heated to decomposition it emits acrid smoke and irritating fumes.

DXQ850 CAS:6928-74-1 ***HR: 3***
DIVINYL MAGNESIUM
mf: C_4H_6Mg mw: 78.40

$(H_2C{=}CH)_2Mg$

SAFETY PROFILE: May ignite spontaneously in air. See also MAGNESIUM COMPOUNDS.

DXR000 CAS:78-19-3 ***HR: 3***
3,9-DIVINYLSPIROBI(m-DIOXANE)
mf: $C_{11}H_{16}O_4$ mw: 212.27

SYN: 3,9-DIVINYL-2,4,8,10-TETRAOXASPIRO(5.5)UNDECANE

TOXICITY DATA with REFERENCE
orl-rat LD50:4066 mg/kg AIHAAP 30,470,69
ivn-mus LD50:320 mg/kg CSLNX* NX#01186
skn-rbt LD50:9908 mg/kg AIHAAP 30,470,69

SAFETY PROFILE: Poison by intravenous route. Mildly toxic by ingestion and skin contact. When heated to decomposition it emits acrid smoke and irritating fumes.

DXR200 CAS:77-77-0 ***HR: 3***
DIVINYL SULFONE
mf: $C_4H_6O_2S$ mw: 118.16

SYNS: TL 797 ◇ VINYL SULFONE

TOXICITY DATA with REFERENCE
skn-rbt 50 mg open MOD UCDS** 3/18/65
eye-rbt 50 mg MOD UCDS** 3/18/65
orl-rat LD50:32 mg/kg AIHAAP 23,95,62
ipr-rat LD50:3 mg/kg TXAPA9 31,222,75
scu-rat LD50:14 mg/kg JPETAB 93,1,48
ivn-rat LD50:12 mg/kg JPETAB 93,1,48
scu-mus LD50:16 mg/kg JPETAB 93,1,48
ihl-mus LCLo:990 mg/m^3/10M NDRC** No.9-4-1-9,43
ivn-mus LD50:11 mg/kg JPETAB 93,1,48
skn-rbt LD50:22 mg/kg AIHAAP 23,95,62

SAFETY PROFILE: Poison by ingestion, skin contact, intravenous, subcutaneous, and intraperitoneal routes. A skin and eye irritant. See also SULFONATES.

DXR400 CAS:25724-33-8 ***HR: 2***
2,5-DIVINYLTETRAHYDROPYRAN
mf: $C_9H_{14}O$ mw: 138.23

SYNS: 2,5-DIVINYLTETRAHYDRO-2H-PYRAN ◇ TETRAHYDRO-2,5-DIVINYL-2H-PYRAN ◇ TETRAHYDRO-2,5-DIVINYLPYRAN

TOXICITY DATA with REFERENCE
orl-rat LD50:2460 mg/kg AIHAAP 30,470,69
skn-rbt LD50:1410 mg/kg AIHAAP 30,470,69

SAFETY PROFILE: Moderately toxic by ingestion and skin contact. When heated to decomposition it emits acrid smoke and irritating fumes.

DXR500 CAS:1119-22-8 ***HR: 3***
DIVINYL ZINC
mf: C_4H_6Zn mw: 119.48

$(H_2C{=}CH)_2Zn$

CONSENSUS REPORTS: Zinc and its compounds are on the Community Right-To-Know List.

SAFETY PROFILE: Ignites spontaneously in air. When heated to decomposition it emits toxic fumes of ZnO. See also ZINC COMPOUNDS.

DXR800 CAS:60539-20-0 ***HR: 3***
DIXYRAZINE DIHYDROCHLORIDE
mf: $C_{24}H_{33}N_3O_2S{\cdot}2ClH$ mw: 500.58

SYN: 2-(2-(4-(2-METHYL-3-PHENOTHIAZIN-10-YLPROPYL)-1-PIPERAZINYL)ETHOXY)ETHANOLDIHYDROCHLORIDE

TOXICITY DATA with REFERENCE
orl-mus LD50:420 mg/kg 27ZQAG -,22,72
ipr-mus LD50:102 mg/kg 27ZQAG -,22,72
ivn-mus LD50:48 mg/kg 27ZQAG -,22,72

SAFETY PROFILE: Poison by intravenous and intraperitoneal routes. Moderately toxic by ingestion. When heated to decomposition it emits very toxic fumes of Cl^-, SO_x, and NO_x.

DXS000 CAS:38001-34-2 ***HR: 2***
(α-DIYLENE)POLY(p-AMINOBENZALDEHYDE-N)
mf: $C_{14}H_{12}N_2O{\cdot}(C_7H_5N)_n$

SYN: POLY-p-AMINOBENZALDEHYD (CZECH)

TOXICITY DATA with REFERENCE
eye-rbt 100 mg/24H SEV 28ZPAK -,254,72

SAFETY PROFILE: A severe eye irritant. When heated to decomposition it emits toxic fumes of NO_x. See also ALDEHYDES.

DXS200 CAS:38222-35-4 ***HR: 3***
DMA
mf: $C_8H_{13}NO_3$ mw: 171.22

SYNS: 2,5-DIHYDROXY-3-DIMETHYLAMINO-5-METHYL-2-CYCLOPENTEN-1-ONE ◇ DIMETHYLAMINO HEXOSE REDUCTIONE

TOXICITY DATA with REFERENCE
orl-mus LD50:400 mg/kg PSEBAA 106,656,61
ipr-mus LD50:300 mg/kg PSEBAA 106,656,61

SAFETY PROFILE: Poison by ingestion and intraperitoneal routes. When heated to decomposition it emits toxic fumes of NO_x.

DXS300 CAS:70244-12-1 ***HR: D***
trans-DMS-DIACETATE
mf: $C_{20}H_{20}O_4$ mw: 324.40

SYNS: (E)-α,α'-DIMETHYL-4,4'-STILBENEDIOL DIACETATE (ester) ◇ trans-DIMETHYLSTILBESTROL DIACETATE ◇ trans-DIMETHYLSTILBOESTROL DIACETATE

TOXICITY DATA with REFERENCE
scu-mus TDLo:12240 μg/kg (4-6D preg):REP JRPFA4 5,239,63

SAFETY PROFILE: Experimental reproductive effects. When heated to decomposition it emits acrid smoke and fumes.

DXS375 CAS:52663-81-7 ***HR: 3***
DOBUTAMINE HYDROCHLORIDE
mf: $C_{18}H_{23}NO_3{\cdot}ClH$ mw: 337.88

SYNS: dl-3,4-DIHYDROXY-N-3-(4-HYDROXYPHENYL)-1-METHYL-n-PROPYL PHENETHYLAMINE HYDROCHLORIDE ◇ DOBUTREX ◇ (±)-4-(2-((3-(4-HYDROXYPHENYL)-1-METHYLPROPYL)AMINO) ETHYL)-1,2-BENZENEDIOL HYDROCHLORIDE ◇ 4-(2-((3-(p-HYDROXYPHENYL)-1-METHYLPROPYL)AMINO)ETHYL)PYROCATECHOL HYDROCHLORIDE ◇ INOTREX ◇ S-1000

TOXICITY DATA with REFERENCE
ivn-rat TDLo:300 mg/kg (female 17-22D post):REP YACHDS 7,1742,79
ivn-rbt TDLo:390 mg/kg (female 6-18D post):TER YACHDS 7,1731,79
orl-rat LD50:2296 mg/kg IYKEDH 13,349,82
ipr-rat LD50:260 mg/kg IYKEDH 13,349,82
scu-rat LD50:368 mg/kg IYKEDH 13,349,82
ivn-rat LD50:59600 μg/kg YACHDS 7,627,79
orl-mus LD50:1324 mg/kg IYKEDH 13,349,82
ipr-mus LD50:243 mg/kg YACHDS 7,627,79
scu-mus LD50:341 mg/kg YACHDS 7,627,79
ivn-mus LD50:34300 μg/kg YACHDS 7,627,79

SAFETY PROFILE: Poison by subcutaneous, intravenous, and intraperitoneal routes. Moderately toxic by ingestion. Experimental reproductive effects. An experimental teratogen. When heated to decomposition it emits toxic fumes of NO_x and HCl.

DXS400 ***HR: 3***
DODECACARBONYLDIVANADIUM
mf: $C_{12}O_{12}V_2$ mw: 438.02

SAFETY PROFILE: Ignites spontaneously in air. See also VANADIUM COMPOUNDS.

DXS600 ***HR: 3***
DODECACARBONYLTRIIRON
mf: $C_{12}Fe_3O_{12}$ mw: 503.67

SAFETY PROFILE: On prolonged storage it will ignite spontaneously in air. When heated to decomposition it emits acrid smoke and fumes.

DXS700 ***HR: 2***
Δ-DODECALACTONE
mf: $C_{12}H_{22}O_2$ mw: 198.31

PROP: Colorless to yellow liquid; coconut-fruity odor. Refr index: 1.458-1.461, flash p: +151°F. Very sol in alc, propylene glycol, veg. oil; insol in water.

SYN: FEMA No. 2401

SAFETY PROFILE: Combustible liquid. When heated to decomposition it emits acrid smoke and irritating fumes.

DXT000 CAS:112-54-9 ***HR: 1***
1-DODECANAL
mf: $C_{12}H_{24}O$ mw: 184.36

PROP: Reported in pine-needle, lime, sweet-orange, and a dozen other essential oils (FCTXAV 11,477,73). Colorless to light yellow liquid; fatty odor. D: 0.826-0.836, refr index: 1.433-1.439, flash p: 180°F. Sol in alc, fixed oils, propylene glycol; insol in glycerin, water.

SYNS: C-12 ALDEHYDE, LAURIC ◇ 1-DODECYL ALDEHYDE ◇ DUODECYLIC ALDEHYDE ◇ FEMA No. 2615 ◇ LAURYL ALDEHYDE (FCC)

TOXICITY DATA with REFERENCE
skn-hmn 5 mg/48H MLD FCTXAV 11,1079,73
skn-rbt 500 mg/24H MOD FCTXAV 11,1079,73
orl-rat LD50:23 g/kg FCTXAV 11,483,73

CONSENSUS REPORTS: Reported in EPA TSCA Inventory.

SAFETY PROFILE: Mildly toxic by ingestion. A human and experimental skin irritant. Combustible liquid. When heated to decomposition it emits acrid smoke and irritating fumes. See also ALDEHYDES.

DXT200 CAS:112-40-3 ***HR: 3***
DODECANE
mf: $C_{12}H_{26}$ mw: 170.38

SYNS: ADAKANE 12 ◇ BIHEXYL ◇ DIHEXYL ◇ n-DODECAN (GERMAN) ◇ DUODECANE

TOXICITY DATA with REFERENCE
skn-mus TDLo:11 g/kg/22W-I:ETA TXAPA9 9,70,66

CONSENSUS REPORTS: Reported In EPA TSCA Inventory.

SAFETY PROFILE: Questionable carcinogen with experimental tumorigenic data. When heated to decomposition it emits acrid smoke and irritating fumes.

DXT400 CAS:2437-25-4 ***HR: 2***
DODECANENITRILE
mf: $C_{12}H_{23}N$ mw: 181.36

SYN: LAURONITRILE

TOXICITY DATA with REFERENCE
orl-rat LDLo:500 mg/kg JPETAB 90,260,47

CONSENSUS REPORTS: Reported in EPA TSCA Inventory. Cyanide and its compounds are on the Community Right-To-Know List.

SAFETY PROFILE: Moderately toxic by ingestion. When heated to decomposition it emits toxic fumes of NO_x and CN-. See also NITRILES.

DXT800 CAS:25103-58-6 ***HR: 3***
tert-DODECANETHIOL
mf: $C_{12}H_{26}S$ mw: 202.44

PROP: White to light yellow liquid. Bp: 200-235°, flash p: 205°F (OC), d: 0.85 @ 25°/25°, vap d: 6.98.

SYNS: tert-DODECYLMERCAPTAN ◇ terc.DODECYLMERKAPTAN (CZECH) ◇ tert-DODECYLTHIOL ◇ 2,3,3,4,4,5-HEXAMETHYL-2-HEXANETHIOL

TOXICITY DATA with REFERENCE
skn-rbt 500 mg/24H MOD 28ZPAK -,169,72
eye-rbt 500 mg/24H MLD 28ZPAK -,169,72
orl-rat LD50:309 mg/kg 28ZPAK -,169,72
ipr-rat LD50:1833 mg/kg 85GMAT -,64,82

CONSENSUS REPORTS: Reported in EPA TSCA Inventory.

SAFETY PROFILE: Poison by ingestion. Moderately toxic by intraperitoneal route. A skin and eye irritant. Combustible when exposed to heat or flame; can react vigorously with oxidizing materials. To fight fire, use foam, CO_2, dry chemical. When heated to decomposition it emits toxic fumes of SO_x. See also MERCAPTANS and SULFATES.

DXU200 CAS:18186-71-5 ***HR: 3***
DODECATRIETHYLAMMONIUM BROMIDE
mf: $C_{18}H_{40}N{\bullet}Br$ mw: 350.50

TOXICITY DATA with REFERENCE
ipr-rat LD50:30 mg/kg JPMSAE 57,1431,68
orl-mus LD50:7 mg/kg JPMSAE 57,1431,68
ipr-mus LD50:8380 μg/kg JPMSAE 57,1431,68

SAFETY PROFILE: Poison by ingestion and intraperitoneal routes. When heated to decomposition it emits very toxic fumes of NO_x, NH_3, and Br^-. See also BROMIDES.

DXU250 CAS:76379-66-3 ***HR: 3***
5,7,11-DODECATRIYN-1-OL
mf: $C_{12}H_{16}O$ mw: 176.28

TOXICITY DATA with REFERENCE
skn-rbt 500 mg/24H MLD JJATDK 8,35,88
orl-rat TDLo:510 mg/kg (female 2D pre)-15D post):REP JJATDK 8,35,88
orl-rat LD50:250 mg/kg JJATDK 8,35,88

SAFETY PROFILE: Poison by ingestion. Experimental reproductive effects. A skin irritant. When heated to decomposition it emits acrid smoke and irritating fumes.

DXU280 CAS:4826-62-4 ***HR: 2***
2-DODECENAL
mf: $C_{12}H_{22}O$ mw: 182.34

SYN: β-OCTYL ACROLEIN

TOXICITY DATA with REFERENCE
skn-rbt 500 mg/24H SEV FCTOD7 21,849,83
skn-rbt LDLo:5 g/kg FCTOD7 21,849,83

CONSENSUS REPORTS: Reported in EPA TSCA Inventory.

SAFETY PROFILE: A severe skin irritant. When heated to decomposition it emits acrid smoke and irritating fumes.

DXU400 CAS:2855-19-8 ***HR: 3***
DODECENE EPOXIDE
mf: $C_{12}H_{24}O$ mw: 184.36

SYN: 1,2-EPOXYDODECANE

TOXICITY DATA with REFERENCE
unr-mus TDLo:147 mg/kg:ETA RARSAM 3,193,63

CONSENSUS REPORTS: Reported in EPA TSCA Inventory.

SAFETY PROFILE: Questionable carcinogen with experimental tumorigenic data. When heated to decomposition it emits acrid smoke and irritating fumes.

DXU600 CAS:56530-48-4 ***HR: 2***
9a-DODECENOATE
mf: $C_{32}H_{49}O_6$ mw: 528.80

SYN: 12-DEOXYPHORBOL-13-DODECENOATE

TOXICITY DATA with REFERENCE
skn-mus 42 ng open ARTODN 44,279,80
skn-mus 30 ng/4H APTOA6 37,250,75
skn-mus 100 ng/24H APTOA6 37,250,75

SAFETY PROFILE: A skin irritant requiring only very small amounts for effect. When heated to decomposition it emits acrid smoke and irritating fumes.

DXU800 CAS:20056-92-2 ***HR: 1***
(Z)-7-DODECEN-1-OL
mf: $C_{12}H_{24}O$ mw: 184.36

SYN: LOOPLURE INHIBITOR

TOXICITY DATA with REFERENCE
skn-rbt 500 mg/24H TXAPA9 31,421,75

SAFETY PROFILE: A skin irritant. When heated to decomposition it emits acrid smoke and irritating fumes.

DXV000 CAS:25377-73-5 ***HR: 3***
DODECENYLSUCCINIC ANHYDRIDE
mf: $C_{16}H_{27}O_3$ mw: 266.38

PROP: Light yellow, clear, visc oil. Bp: 180-182° @ 5 mm, flash p: 352°F (COC), d: 1.002 @ 25°/4°.

TOXICITY DATA with REFERENCE
ipr-mus LD50:320 mg/kg NTIS** AD441-640

CONSENSUS REPORTS: Reported in EPA TSCA Inventory.

SAFETY PROFILE: Poison by intraperitoneal route. An irritant and sensitizer. Combustible when exposed to heat or flame; can react with oxidizing materials. To fight fire, use foam, CO_2, dry chemical. When heated to decomposition it emits acrid smoke and irritating fumes.

DXV400 CAS:112-66-3 ***HR: 1***
n-DODECYL ACETATE
mf: $C_{14}H_{28}O_2$ mw: 228.42

SYNS: ACETATE C-12 ◇ ACETIC ACID, DODECYL ESTER ◇ DODECANOL ACETATE ◇ 1-DODECANOL ACETATE ◇ DODECAN-1-YL ACETATE ◇ DODECYL ACETATE ◇ DODECYL ALCOHOL ACETATE ◇ LAURYL ACETATE

TOXICITY DATA with REFERENCE
skn-rbt 500 mg/24H MLD FCTXAV 14,659,76

CONSENSUS REPORTS: Reported in EPA TSCA Inventory.

SAFETY PROFILE: A skin irritant. When heated to decomposition it emits acrid smoke and irritating fumes.

DXV600 CAS:112-53-8 ***HR: 3***
DODECYL ALCOHOL
mf: $C_{12}H_{26}O$ mw: 186.38

PROP: Colorless solid, liquid above 21°; floral odor. D: 0.830-0.836, refr index: 1.440-1.444, mp: 24°, bp: 259°, flash p: 260°F, autoign temp: 527°F. Sol in 2 parts of 70% alc, fixed oils, propylene glycol; insol in water, glycerin.

SYNS: ALCOHOL C-12 ◇ ALFOL 12 ◇ CACHALOT L-50 ◇ CO 12 ◇ CO-1214 ◇ n-DODECANOL ◇ 1-DODECANOL ◇ n-DODECYL ALCOHOL ◇ DUODECYL ALCOHOL ◇ DYTOL J-68 ◇ EPAL 12 ◇ FEMA No. 2617 ◇ LAURIC ALCOHOL ◇ LAURINIC ALCOHOL ◇ LAURYL 24 ◇ LAURYL ALCOHOL (FCC) ◇ n-LAURYL ALCOHOL, PRIMARY ◇ LOROL ◇ MA-1214 ◇ SIPOL L12

TOXICITY DATA with REFERENCE
skn-hmn 75 mg/3D-I SEV 85DKA8 -,127,77
skn-mus TDLo:19 g/kg/39W-I:ETA TXAPA9 9,70,66
orl-rat LD50:12800 mg/kg FCTXAV 11,95,73
ipr-rat LD50:800 mg/kg 14CYAT 2,1469,63

CONSENSUS REPORTS: Reported in EPA TSCA Inventory.

SAFETY PROFILE: Moderately toxic by intraperitoneal route. Mildly toxic by ingestion. A severe human skin irritant. Questionable carcinogen with experimental tumorigenic data. Combustible when exposed to heat or flame; can react with oxidizing materials. To fight fire, use dry chemical, CO_2. When heated to decomposition it emits acrid smoke and irritating fumes.

DXW000 CAS:124-22-1 ***HR: 3***
DODECYLAMINE
mf: $C_{12}H_{27}N$ mw: 185.40

PROP: Oil, amine odor. Fp: 28.3°, vap press: 64 mm @ 170°.

SYN: LAURYLAMINE

TOXICITY DATA with REFERENCE
skn-rbt 500 mg/24H SEV 28ZPAK -,62,72
eye-rbt 50 μg/24H SEV 28ZPAK -,52,72
eye-rbt 2 mg AEPPAE 219,119,53
orl-rat LD50:1020 mg/kg AEPPAE 219,119,53
orl-mus LD50:1160 mg/kg AEPPAE 219,119,53
ipr-mus LD50:50 mg/kg NTIS** AD691-490

CONSENSUS REPORTS: Reported in EPA TSCA Inventory.

SAFETY PROFILE: Poison by intraperitoneal route. Moderately toxic by ingestion. A severe skin and eye irritant. When heated to decomposition it emits toxic fumes of NO_x. See also AMINES.

DXW200 CAS:25155-30-0 ***HR: 3***
DODECYL BENZENE SODIUM SULFONATE
mf: $C_{18}H_{29}O_3S{\cdot}Na$ mw: 348.52

PROP: White to light yellow flakes, granules, or powder.

SYNS: AA-9 ◇ ABESON NAM ◇ BIO-SOFT D-40 ◇ CALSOFT F-90 ◇ CONCO AAS-35 ◇ CONOCO C-50 ◇ DETERGENT HD-90 ◇ DODECYLBENZENESULFONIC ACID SODIUM SALT

◇ DODECYLBENZENESULPHONATE, SODIUM SALT ◇ DODECYLBENZENSULFONAN SODNY (CZECH) ◇ MERCOL 25 ◇ NACCANOL NR ◇ NECCANOL SW ◇ PILOT HD-90 ◇ PILOT SF-40 ◇ RICHONATE 1850 ◇ SANTOMERSE 3 ◇ SODIUM DODECYLBENZENESULFONATE (DOT) ◇ SODIUM DODECYLBENZENESULFONATE, dry ◇ SODIUM LAURYLBENZENESULFONATE ◇ SOLAR 40 ◇ SOL SODOWA KWASU LAURYLOBENZENOSULFONOWEGO (POLISH) ◇ SULFAPOL ◇ SULFAPOLU (POLISH) ◇ SULFRAMIN 85 ◇ SULFRAMIN 40 FLAKES ◇ SULFRAMIN 40 GRANULAR ◇ SULFRAMIN 1238 SLURRY ◇ p-1',1',4',4'-TETRAMETHYLOKTYLBENZENSULFONAN SODNY (CZECH) ◇ ULTRAWET K

TOXICITY DATA with REFERENCE
skn-rbt 500 mg/24H MOD 28ZPAK -,195,72
eye-rbt 250 μg/24H SEV 28ZPAK -,195,72
eye-rbt 1% SEV JAPMA8 38,428,49
orl-rat LD50:1260 mg/kg FSASAX 63,938,61
orl-mus LD50:2000 mg/kg JAPMA8 38,428,49
ivn-mus LD50:105 mg/kg JAPMA8 38,428,49

CONSENSUS REPORTS: Reported in EPA TSCA Inventory.

SAFETY PROFILE: Poison by intravenous route. Moderately toxic by ingestion. A skin and severe eye irritant. When heated to decomposition it emits toxic fumes of Na_2O. See also SULFONATES.

DXW400 CAS:1886-81-3 ***HR: 2***
DODECYL BENZENESULFONATE
mf: $C_{18}H_{30}O_3S$ mw: 326.54

SYN: BENZENESULFONIC ACID, DODECYL ESTER

TOXICITY DATA with REFERENCE
orl-rat LD50:650 mg/kg ARTODN 32,245,74

CONSENSUS REPORTS: Reported in EPA TSCA Inventory.

SAFETY PROFILE: Moderately toxic by ingestion. When heated to decomposition it emits toxic fumes of SO_x.

DXW600 CAS:28061-21-4 ***HR: 1***
DODECYLBENZYL CHLORIDE
mf: $C_{19}H_{31}Cl$ mw: 294.95

SYN: CONOCO DBCL

TOXICITY DATA with REFERENCE
skn-rbt 500 mg MLD 34ZIAG -,233,69
eye-rbt 100 mg MLD 34ZIAG -,233,69
skn-rbt LDLo:12500 mg/kg 34ZIAG -,233,69

CONSENSUS REPORTS: Reported in EPA TSCA Inventory.

SAFETY PROFILE: Mildly toxic by skin contact. A skin and eye irritant. When heated to decomposition it emits toxic fumes of Cl^-.

DXW800 CAS:2783-17-7 ***HR: 3***
DODECYLDIAMINE
mf: $C_{12}H_{28}N_2$ mw: 200.42

SYNS: 1,12'-DIAMINODECANE ◇ 1,12'-DODECAMETHYLENEDIAMINE ◇ 1,12-DODECANEDIAMINE ◇ 1,12'-DODECANEDIAMINE ◇ 1,12'-DODECYLENEDIAMINE

TOXICITY DATA with REFERENCE
orl-rat LD50:1105 mg/kg GISAAA 43(5),18,78
orl-mus LD50:1088 mg/kg GISAAA 43(5),18,78
orl-rbt LD50:330 mg/kg GISAAA 43(5),18,78
orl-gpg LD50:440 mg/kg GISAAA 43(5),18,78

CONSENSUS REPORTS: Reported in EPA TSCA Inventory.

SAFETY PROFILE: Poison by ingestion. When heated to decomposition it emits toxic fumes of NO_x. See also AMINES.

DXX000 CAS:538-71-6 ***HR: 3***
DODECYLDIMETHYL(2-PHENOXYETHYL)AMMONIUM BROMIDE
mf: $C_{22}H_{40}NO{\cdot}Br$ mw: 414.54

SYNS: PHENODODECINIUM BROMIDE ◇ β-PHENOXYETHYLDIMETHYLDODECYLAMMONIUM BROMIDE

TOXICITY DATA with REFERENCE
ipr-rat LD50:40 mg/kg FEPRA7 6,307,47
ivn-rat LD50:18 mg/kg FEPRA7 6,307,47
ivn-mus LD50:31 mg/kg FEPRA7 6,307,47
ivn-rbt LDLo:11 mg/kg FEPRA7 6,307,47
ipr-gpg LDLo:10 mg/kg HBTXAC 1,222,56

CONSENSUS REPORTS: Reported in EPA TSCA Inventory.

SAFETY PROFILE: Poison by intraperitoneal and intravenous routes. When heated to decomposition it emits very toxic fumes of NO_x, NH_3, and Br^-. See also BROMIDES.

DXX100 CAS:56501-30-5 ***HR: 3***
1-DODECYL-1-ETHYLPIPERIDINIUM BROMIDE
mf: $C_{19}H_{40}N{\cdot}Br$ mw: 362.51

TOXICITY DATA with REFERENCE
orl-mus LD50:198 mg/kg PSDTAP 15,331,74
ipr-mus LD50:19980 μg/kg PSDTAP 15,331,74
ivn-mus LD50:4595 μg/kg PSDTAP 15,331,74

SAFETY PROFILE: Poison by ingestion, intravenous, and intraperitoneal routes. When heated to decomposition it emits toxic fumes of NO_x and Br^-.

DXX200 CAS:1166-52-5 ***HR: 2***
DODECYL GALLATE
mf: $C_{19}H_{30}O_5$ mw: 338.49

SYNS: LAURYL GALLATE ◇ NIPAGALLIN LA ◇ PROGALLIN LA

TOXICITY DATA with REFERENCE
orl-mus LD50:1600 mg/kg 14CYAT 2,1897,63
orl-rat LD50:6500 mg/kg FOMAAB 26,99,51

CONSENSUS REPORTS: Reported in EPA TSCA Inventory.

SAFETY PROFILE: Moderately toxic by ingestion. When heated to decomposition it emits acrid smoke and irritating fumes.

DXX400 CAS:2439-10-3 ***HR: 3***
N-DODECYLGUANIDINE ACETATE
mf: $C_{13}H_{29}N_3 \cdot C_2H_4O_2$ mw: 287.51

PROP: Crystals, sol in hot water and alc. Mp: 136°.

SYNS: AC 5223 ◇ AMERICAN CYANAMID 5223 ◇ APADODINE ◇ CARPENE ◇ CURITAN ◇ CYPREX ◇ CYPREX 65W ◇ N-DODECYLGUANIDINACETAT (GERMAN) ◇ DODECYLGUANIDINE ACETATE ◇ DODGUADINE ◇ DODINE ◇ DODINE ACETATE ◇ DODINE, mixture with GLYODIN ◇ DOGQUADINE ◇ ENT 16,436 ◇ EXPERIMENTAL FUNGICIDE 5223 ◇ LAURYLGUANIDINE ACETATE ◇ MELPREX ◇ MILPREX ◇ SYLLIT ◇ TSITREX ◇ VENTUROL ◇ VONDODINE

TOXICITY DATA with REFERENCE
eye-rbt 100 mg SEV 34ZIAG -,234,69
scu-mus TDLo:1000 mg/kg:ETA NTIS** PB223-159
orl-rat LD50:566 mg/kg WRPCA2 7,135,68
orl-mus LD50:1200 mg/kg CALEDQ 5,107,78
orl-rbt LD50:535 mg/kg 85GMAT -,64,82
skn-rbt LD50:1500 mg/kg 28ZEAL 5,88,76
orl-gpg LD50:176 mg/kg 85GMAT -,64,82
skn-gpg LDLo:2 g/kg 85GMAT -,64,82

CONSENSUS REPORTS: Reported in EPA TSCA Inventory. EPA Genetic Toxicology Program.

SAFETY PROFILE: Poison by ingestion. Moderately toxic by skin contact. A severe eye irritant. Questionable carcinogen with experimental tumorigenic data. A pesticide. When heated to decomposition it emits very toxic fumes of NO_x.

DXX600 ***HR: 3***
DODECYLGUANIDINE ACETATE with SODIUM NITRITE (3:5)

SYNS: DODINE with SODIUM NITRITE (3:5) ◇ SODIUM NITRITE with DODECYL GUANIDINE ACETATE (5:3)

TOXICITY DATA with REFERENCE
orl-mus TDLo:112 mg/kg/6D-I:CAR CALEDQ 5,107,78
orl-mus TDLo:112 mg/kg (15-21D preg):CAR,TER CALEDQ 5,107,78

SAFETY PROFILE: Questionable carcinogen with experimental carcinogenic data. An experimental teratogen. When heated to decomposition it emits very toxic fumes of NO_x and Na_2O. See also SODIUM NITRITE.

DXX800 CAS:13590-97-1 ***HR: 3***
DODECYLGUANIDINE HYDROCHLORIDE
mf: $C_{13}H_{29}N_3 \cdot ClH$ mw: 263.91

TOXICITY DATA with REFERENCE
unr-mus LDLo:13 mg/kg ATMPA2 32,177,38

CONSENSUS REPORTS: Reported in EPA TSCA Inventory.

SAFETY PROFILE: Poison by unspecified route. When heated to decomposition it emits very toxic fumes of HCl and NO_x.

DXX875 CAS:56501-36-1 ***HR: 3***
1-DODECYLHEXAHYDRO-1H-AZEPINE-1-OXIDE
mf: $C_{18}H_{37}NO$ mw: 283.56

SYNS: 1-DODECYLHEXAMETHYLENIMINE-N-OXIDE ◇ HEXAHYDRO-1-DODECYL-1H-AZEPINE-1-OXIDE

TOXICITY DATA with REFERENCE
orl-mus LD50:94 mg/kg PESTD5 16,236,75
ipr-mus LD50:76 mg/kg PESTD5 16,236,75
ivn-mus LD50:33 mg/kg PESTD5 16,236,75

SAFETY PROFILE: Poison by ingestion, intravenous, and intraperitoneal routes. When heated to decomposition it emits toxic fumes of NO_x.

DXY000 CAS:9002-92-0 ***HR: 3***
α-DODECYL-ω-HYDROXY-POLYOXYETHYLENE
mf: $(C_2H_4O)_n \cdot C_{12}H_{26}O$

SYNS: ALDOSPERSE L 9 ◇ ATLAS G-2133 ◇ BASE LP 12 ◇ BL 9 ◇ BRIJ 30 ◇ CHIMIPAL AE 3 ◇ CIMAGEL ◇ DEHYDOL LS 4 ◇ DO 9 ◇ DODECANOL, ETHOXYLATE ◇ DODECANOL-ETHYLENE OXIDE (9.5 moles) CONDENSATE ◇ DODECANOL, POLYETHOXYLATED ◇ DODECYL ALCOHOL, ETHOXYLATED ◇ DODECYL-POLYAETHYLENOXYD-AETHER (GERMAN) ◇ DODECYL POLY(OXYETHYLENE)ETHER ◇ DU PONT WK ◇ EMULGEN 100 ◇ ETHAL LA-X ◇ ETHOSPERSE LA-4 ◇ ETHOXYLATED LAURYL ALCOHOL ◇ G 3707 ◇ G-2130A ◇ HYDROXYPOLYETHOXYDODECANE ◇ LA ◇ LAURETH ◇ LAUROMACROGOL 400 ◇ LAURYL ALCOHOL, ETHOXYLATED ◇ LAURYL POLYETHYLENE GLYCOL ETHER ◇ LIPAL 4LA ◇ LIPOCOL L-4 ◇ LUBROL 12A9 ◇ MARLIPAL 1217 ◇ MCI-C54875 ◇ NIKKOL BL ◇ NOIGEN 160 ◇ OXYETHYLENATED DODECYL ALCOHOL ◇ PLURAFAC RA 43 ◇ POLIDOCANOL ◇ POLYETHYLENE GLYCOL DODECYL ETHER ◇ POLY(ETHYLENE OXIDE) DODECYL ETHER ◇ POLYOXYETHYLENE LAURIC ALCOHOL ◇ POLYOXYETHYLENE LAURYL ETHER ◇ ROKANOL L ◇ SIMULOL 330 M ◇ SIPONIC L ◇ SURFACTANT WK ◇ TRYCOL LAL SERIES

TOXICITY DATA with REFERENCE
skn-hmn 6 mg/3D-I MOD 85DKA8 -,127,77
skn-rbt 500 mg/24H MLD 28ZPAK -,301,72
skn-rbt 75 mg/24H MLD TXAPA9 7,206,65
skn-rbt 500 mg/24H MOD TXAPA9 19,276,71
eye-rbt 100 mg TXAPA9 19,276,71
eye-rbt 750 μg/24H SEV 28ZPAK -,301,72
dnd-esc 50 mg/L MUREAV 89,95,81
orl-rat LD50:1 g/kg FCTXAV 8,125,70

ipr-rat LD50:125 mg/kg FCTXAV 8,125,70
orl-mus LD50:1170 mg/kg ARZNAD 7,162,57
ivn-mus LD50:100 mg/kg TXAPA9 7,206,65
ivn-rbt LD50:36 mg/kg ARZNAD 7,162,57
orl-gpg LD50:384 mg/kg ARZNAD 7,162,57
ivn-gpg LD50:38 mg/kg ARZNAD 7,162,57

CONSENSUS REPORTS: Reported in EPA TSCA Inventory. Glycol ethers are on the Community Right-To-Know List.

SAFETY PROFILE: Poison by ingestion, intraperitoneal, and intravenous routes. A human and experimental skin irritant. A severe eye irritant. Mutation data reported. When heated to decomposition it emits acrid smoke and irritating fumes. See also GLYCOL ETHERS.

DXY200 CAS:142-90-5 ***HR: 1***
DODECYL METHACRYLATE
mf: $C_{16}H_{30}O_2$ mw: 254.46

SYNS: AGEFLEX FM 246 ◇ DODECYL-2-METHYL-2-PROPENOATE ◇ LAURYLESTER KYSELINYMETHAKRYLOVE (CZECH) ◇ LAURYL METHACRYLATE ◇ METHACRYLIC ACID DODECYL ESTER ◇ METHACRYLIC ACID LAURYL ESTER ◇ 2-METHYLACRYLIC ACID DODECYL ESTER

TOXICITY DATA with REFERENCE
skn-rbt 500 mg/24H MLD 28ZPAK -,44,72
eye-rbt 500 mg/24H MLD 28ZPAK -,44,72
ipr-rat LD50:12 g/kg AMPMAR 36,58,75
ipr-mus LD50:25 g/kg JPMSAE 62,778,73

CONSENSUS REPORTS: Reported in EPA TSCA Inventory.

SAFETY PROFILE: A skin and eye irritant. When heated to decomposition it emits acrid smoke and irritating fumes.

DXY300 CAS:2530-46-3 ***HR: 3***
4-DODECYLMORPHOLINE-4-OXIDE
mf: $C_{16}H_{33}NO_2$ mw: 271.50

SYN: 4-DODECYLMORPHOLINE-N-OXIDE

TOXICITY DATA with REFERENCE
orl-mus LD50:1880 mg/kg PESTD5 16,236,75
ipr-mus LD50:110 mg/kg PESTD5 16,236,75
ivn-mus LD50:68 mg/kg PESTD5 16,236,75

SAFETY PROFILE: Poison by intravenous and intraperitoneal routes. When heated to decomposition it emits toxic fumes of NO_x.

DXY400 ***HR: 3***
N-(2-DODECYLOXYETHYL)-N-METHYL-2-(PYRROLIDINYL)ACETAMIDE HYDROCHLORIDE
mf: $C_{21}H_{42}N_2O_2 \cdot ClH$ mw: 391.11

SYN: C 6606

TOXICITY DATA with REFERENCE
eye-rbt 2% SEV ARZNAD 9,113,59
scu-mus LD50:400 mg/kg ARZNAD 9,113,59

SAFETY PROFILE: Poison by subcutaneous route. A severe eye irritant. When heated to decomposition it emits very toxic fumes of NO_x and HCl.

DXY600 CAS:27193-86-8 ***HR: 2***
DODECYLPHENOL
mf: $C_{18}H_{30}O$ mw: 262.48

PROP: Straw-colored liquid, phenolic odor. Bp: 154-168°, flash p: 325°F (OC), d: 0.93 @ 20°/20°, vap d: 9.04.

TOXICITY DATA with REFERENCE
orl-rat LD50:2140 mg/kg AIHAAP 23,95,62
skn-rbt LD50:5000 mg/kg AIHAAP 23,95,62

CONSENSUS REPORTS: Reported in EPA TSCA Inventory.

SAFETY PROFILE: Moderately toxic by ingestion. Mildly toxic by skin contact. Combustible when exposed to heat or flame; can react with oxidizing materials. To fight fire, use CO_2, dry chemical. When heated to decomposition it emits acrid smoke and irritating fumes.

DXY700 CAS:56501-35-0 ***HR: 3***
1-DODECYLPIPERIDINE-N-OXIDE
mf: $C_{17}H_{35}NO$ mw: 269.53

SYN: 1-DODECYLPIPERIDINE-1-OXIDE

TOXICITY DATA with REFERENCE
orl-mus LD50:619 mg/kg PSDTAP 15,331,74
ipr-mus LD50:67920 μg/kg PSDTAP 15,331,74
ivn-mus LD50:23 mg/kg PESTD5 16,236,75

SAFETY PROFILE: Poison by intravenous and intraperitoneal routes. Moderately toxic by ingestion. When heated to decomposition it emits toxic fumes of NO_x.

DXY750 CAS:55257-88-0 ***HR: 2***
1-DODECYL-2-PYRROLIDINONE
mf: $C_{16}H_{31}NO$ mw: 253.48

PROP: Bp: 202-205°/11 mm, d: 0.890, Flash p: >230° F.

SYNS: N-DODECYLPYRROLIDINONE ◇ 2-PYRROLIDINONE, 1-DODECYL-

TOXICITY DATA with REFERENCE
skn-rbt 500 mg SEV FCTOD7 26,475,88
eye-rbt 100 mg MOD FCTOD7 26,475,88
orl-rat LDLo:5 g/kg FCTOD7 26,475,88

SAFETY PROFILE: Mildly toxic by ingestion. A severe skin and moderate eye irritant. Combustible when exposed to heat and flame. When heated to decomposition it emits toxic fumes of NO_x.

DXZ000 CAS:7631-98-3 ***HR: 3***
N-DODECYLSARCOSINE SODIUM SALT
mf: $C_{15}H_{30}NO_2{\cdot}Na$ mw: 279.45

SYN: SODIUM-N-LAURYL SARCOSINE

TOXICITY DATA with REFERENCE
ivn-mus LD50:180 mg/kg CSLNX* NX#00171

CONSENSUS REPORTS: Reported in EPA TSCA Inventory.

SAFETY PROFILE: Poison by intravenous route. When heated to decomposition it emits toxic fumes of NO_x and Na_2O.

DYA000 CAS:15826-16-1 ***HR: 2***
DODECYL SODIUM ETHOXYSULFATE
mf: $C_{14}H_{30}O_5S{\cdot}Na$ mw: 333.49

SYNS: 2-(DODECYLOXY)ETHANOL HYDROGEN SULFATE SODIUM SALT ◇ SODIUM LAURYL ETHOXYSULPHATE

TOXICITY DATA with REFERENCE
orl-rat LD50:1995 mg/kg FCTXAV 5,763,67

CONSENSUS REPORTS: Reported in EPA TSCA Inventory.

SAFETY PROFILE: Moderately toxic by ingestion. When heated to decomposition it emits very toxic fumes of SO_x and Na_2O.

DYA200 CAS:765-15-1 ***HR: 2***
n-DODECYL THIOCYANATE
mf: $C_{13}H_{25}NS$ mw: 227.45

SYNS: ENT 114 ◇ LAURYL RHODANATE ◇ LAURYL THIOCYANATE ◇ LORO ◇ LOROL THIOCYANATE ◇ 1-THIOCYANATODODECANE ◇ THIOCYANIC ACID, DODECYL ESTER

TOXICITY DATA with REFERENCE
skn-rat 350 mg/7D-I open MOD JPTLAS 49,363,39
orl-rat LD50:1250 mg/kg 28ZEAL 4,205,69
scu-mus LDLo:20 g/kg JIHTAB 18,310,36
orl-cat LDLo:2000 mg/kg JIHTAB 18,310,36

SAFETY PROFILE: Moderately toxic by ingestion. A skin irritant. When heated to decomposition it emits very toxic fumes of NO_x and SO_x. See also THIOCYANATES.

DYA400 CAS:5416-74-0 ***HR: 3***
(DODECYLTHIO)PHENYLMERCURY
mf: $C_{18}H_{30}HgS$ mw: 479.13

SYN: DODECYL PHENYLMERCURI SULFIDE

TOXICITY DATA with REFERENCE
ipr-mus LDLo:16 mg/kg CBCCT* 1,46,49

CONSENSUS REPORTS: Mercury and its compounds are on the Community Right-To-Know List.

OSHA PEL: (Transitional: CL 1 mg/10m^3) CL 0.1 mg (Hg)/m^3 (skin)
ACGIH TLV: TWA 0.1 mg(Hg)/m^3 (skin)
NIOSH REL: (Inorganic Mercury) TWA 0.05 mg(Hg)/m^3

SAFETY PROFILE: Poison by intraperitoneal route. When heated to decomposition it emits very toxic fumes of Hg and SO_x. See also MERCURY COMPOUNDS.

DYA600 CAS:1399-80-0 ***HR: 3***
DODECYL-p-TOLYL TRIMETHYL AMMONIUM CHLORIDE

PROP: Aqueous preparation containing approximately 40% methyl dodecylbenzyl trimethyl ammonium chloride, and 10% methyl dodecylxylene bis(trimethyl ammonium chloride).

SYNS: ALKYL(C9-15)TOLYL METHYLTRIMETHYL AMMONIUM CHLORIDE ◇ HYAMINE 2389 ◇ METHYL DODECYL BENZYL AMMONIUM CHLORIDE

TOXICITY DATA with REFERENCE
skn-man 50 mg/2D MLD PSTGAW 20,16,53
eye-rbt 100 μg SEV PSTGAW 20,16,53
orl-rat LD50:389 mg/kg PCOC** -,592,66
ipr-rat LD50:10 mg/kg AEHA** 5176T6-66/67
ivn-rat LD50:3060 μg/kg AEHA** 5176T6-66/67

SAFETY PROFILE: Poison by ingestion, intraperitoneal, and intravenous routes. A skin and severe eye irritant. When heated to decomposition it emits very toxic fumes of NO_x, NH_3, and Cl^-.

DYA800 CAS:4484-72-4 ***HR: 3***
DODECYLTRICHLOROSILANE
DOT: UN 1771
mf: $C_{12}H_{25}Cl_3Si$ mw: 303.81

PROP: Colorless to yellow liquid, readily hydrolyzed by moisture with the production of hydrochloric acid. Bp: 288°, d: 1.026 @ 25°/25°.

SYN: TRICHLORODODECYLSILANE

CONSENSUS REPORTS: Reported in EPA TSCA Inventory.

DOT Classification: Corrosive Material; Label: Corrosive.

SAFETY PROFILE: A poison. A corrosive irritant to the eyes, skin and mucous membranes. When heated to decomposition it emits toxic fumes of Cl^-. See also CHLOROSILANES.

DYA850 ***HR: 2***
DODICIN HYDROCHLORIDE
mf: $C_{18}H_{39}N_3O_2{\cdot}ClH$ mw: 366.06

SYNS: DODECYLBIS(AMINOETHYL)GLYCINEHYDROCHLORIDE ◇ LEBON 15 HYDROCHLORIDE ◇ TEGO 51

TOXICITY DATA with REFERENCE
ipr-rat LD50:534 mg/kg KSRNAM 12,1821,78
scu-rat LD50:5262 mg/kg KSRNAM 12,1821,78
ipr-mus LD50:590 mg/kg KSRNAM 12,1821,78
scu-mus LD50:2145 mg/kg KSRNAM 12,1821,78

SAFETY PROFILE: Moderately toxic by intraperitoneal and subcutaneous routes. When heated to decomposition it emits toxic fumes of NO_x and HCl.

DYA875 ***HR: 3***
DOG LAUREL

PROP: Deciduous or evergreen shrubs with alternating leaves and clusters of white or pink flowers. They grow wild in the coastal states from Virginia to Florida, Louisiana, Tennessee, and California.

SYNS: DOG HOBBLE ◇ FETTER BUSH ◇ LEUCOTHOE (VARIOUS SPECIES) ◇ PEPPER BUSH ◇ SWEET BELLS ◇ SWITCH IVY ◇ WHITE OSTER

SAFETY PROFILE: The leaves and nectar contain poisonous grayanotoxins (andromedotoxins). Ingestion of these plant parts results in immediate pain in the mouth and may be followed several hours later by vomiting, diarrhea, headache, impaired vision, irregular heartbeat, severe low blood pressure, coma, convulsions, and death.

DYB000 CAS:482-49-5 ***HR: 3***
DOISYNOLIC ACID
mf: $C_{18}H_{24}O_3$ mw: 288.42

SYNS: ACIDO DOISYNOLICO (SPANISH) ◇ 1-ETHYL-7-HYDROXY-2-METHYL-1,2,3,4,4a,9,10,10a-OCTAHYDROPHENANTHRENE-2-CARBOXYLIC ACID ◇ 1-ETHYL-1,2,3,4,4a,9,10,10a-OCTAHYDRO-7-HYDROXY-2-METHYL-2-PHENANTHRENECARBOXYLIC ACID ◇ 3-HYDROXY-16,7-SECOESTRA-1,3,5(10)-TRIEN-17-OICACID

TOXICITY DATA with REFERENCE
orl-gpg TDLo:72 mg/kg/12W-I:ETA,REP RSABAC 25,215,49

SAFETY PROFILE: Experimental reproductive effects. Questionable carcinogen with experimental tumorigenic data. When heated to decomposition it emits acrid smoke and irritating fumes.

DYB250 CAS:4956-15-4 ***HR: 3***
DOLANTIN-N-OXIDE HYDROCHLORIDE
mf: $C_{15}H_{21}NO_3{\cdot}ClH$ mw: 299.83

SYN: 1-METHYL-4-PHENYLISONIPECOTIC ACID ETHYL ESTER-1-OXIDE HYDROCHLORIDE

TOXICITY DATA with REFERENCE
orl-mus LD50:2 g/kg AITEAT 15,290,67
ipr-mus LD50:575 mg/kg AITEAT 15,290,67
scu-mus LD50:975 mg/kg AITEAT 15,290,67
ivn-mus LD50:300 mg/kg AITEAT 15,290,67

SAFETY PROFILE: Poison by intravenous. Moderately toxic by ingestion, subcutaneous, and intraperitoneal routes. When heated to decomposition it emits toxic fumes of NO_x and HCl. See also ESTERS.

DYB300 CAS:96081-07-1 ***HR: 2***
DOLOCONEURASE

TOXICITY DATA with REFERENCE
orl-rat LD50:3681 mg/kg ACTTDZ 10,365,84
ivn-rat LD50:1301 mg/kg ACTTDZ 10,365,84
ims-rat LD50:3088 mg/kg ACTTDZ 10,365,84
orl-mus LD50:3273 mg/kg ACTTDZ 10,365,84
ivn-mus LD50:1213 mg/kg ACTTDZ 10,365,84
ims-mus LD50:2926 mg/kg ACTTDZ 10,365,84

SAFETY PROFILE: Moderately toxic by ingestion, intravenous, and intramuscular routes.

DYB400 CAS:17140-78-2 ***HR: 3***
DOLOXENE
mf: $C_{10}H_8O_3S{\cdot}C_{22}H_{29}NO_2$ mw: 547.76

SYNS: DARVON-N ◇ 4-(DIMETHYLAMINO)-3-METHYL-1,2-DIPHENYL-2-BUTANOLPROPIONATE-2-NAPHTHALENESULFONATE ◇ PROPOXYPHENE N ◇ PROPOXYPHENE-2-NAPHTHALENE-SULFONATE ◇ PROPOXYPHENE NAPSYLATE

TOXICITY DATA with REFERENCE
orl-rat TDLo:3 g/kg (female 14-22D post):REP TXAPA9 19,471,71
orl-rat LD50:647 mg/kg TXAPA9 19,445,71
orl-mus LD50:915 mg/kg TXAPA9 19,445,71
orl-rbt LDLo:183 mg/kg TXAPA9 19,445,71

SAFETY PROFILE: Poison by ingestion. Experimental reproductive effects. When heated to decomposition it emits very toxic fumes of NO_x and SO_x. See also SULFONATES.

DYB600 CAS:303-69-5 ***HR: 3***
DOMINAL
mf: $C_{16}H_{19}N_3S$ mw: 285.44

SYNS: D 206 ◇ DIMETHYLAMINO-N-PROPYL-THIOPHENYL-PYRIDYLAMINE ◇ PROTHIPENDYL

TOXICITY DATA with REFERENCE
orl-mus LD50:415 mg/kg ARZNAD 8,489,58
ipr-mus LD50:155 mg/kg ARZNAD 8,489,58
scu-mus LD50:102 mg/kg AIPTAK 149,374,64

SAFETY PROFILE: Poison by subcutaneous and intraperitoneal routes. Moderately toxic by ingestion.

When heated to decomposition it emits very toxic fumes of SO_x and NO_x.

DYB800 CAS:1225-65-6 ***HR: 3***
DOMINAL HYDROCHLORIDE
mf: $C_{16}H_{19}N_3S \cdot ClH$ mw: 321.90

SYNS: 10-(3-DIMETHYLAMINOPROPYL)-1-AZAPHENOTHIAZINE HYDROCHLORIDE ◇ ((4-DIMETHYLAMINOPROPYLPYRIDO (3,2b) BENZOTHIAZINE)) HYDROCHLORIDE ◇ PROTHIPENDYL HYDROCHLORIDE

TOXICITY DATA with REFERENCE
orl-rat LD50:610 mg/kg 27ZQAG -,292,72
ipr-rat LD50:115 mg/kg 27ZQAG -,292,72
scu-rat LD50:1323 mg/kg 27ZQAG -,292,72
ivn-rat LD50:110 mg/kg 27ZQAG -,292,72

SAFETY PROFILE: Poison by intravenous and intraperitoneal routes. Moderately toxic by ingestion and subcutaneous routes. When heated to decomposition it emits very toxic fumes of NO_x, SO_x, and HCl.

DYB875 CAS:57808-66-9 ***HR: 3***
DOMPERIDONE
mf: $C_{22}H_{24}ClN_5O_2$ mw: 425.96

PROP: Crystals from DMF/water. Mp: 242.5°.

SYNS: 5-CHLORO-1-(1-(3-(2-OXO-1-BENZIMIDAZOLINYL)PROPYL)-4-PIPERIDYL)-2-BENZIMIDAZOLINONE ◇ KW-5338 ◇ MOTILIUM

TOXICITY DATA with REFERENCE
orl-wmn TDLo:4800 μg/kg/3D-I:REP BMJOAE 286,1395,83
orl-mus TDLo:700 mg/kg (female 6-15D post):TER YACHDS 8,4045,80
orl-inf TDLo:1170 μg/kg/1D-I:EYE JOPDAB 108,630,86
ivn-hmn TDLo:714 μg/kg:CNS BMJOAE 288,1728,84
ivn-hmn LDLo:429 μg/kg:CNS BMJOAE 288,1728,84
ivn-wmn TDLo:1 mg/kg:CVS BMJOAE 289,1579,84
ivn-wmn TDLo:600 μg/kg/1D:CVS LANCAO 2,385,85
ims-man TDLo:200 μg/kg:CNS,EYE LANCAO 2,1259,80
rec-wmn TDLo:7200 μg/kg/30H-I:EYE JOPDAB 105,852,84
orl-rat LD50:5243 mg/kg YACHDS 8,3991,80
ipr-rat LD50:61200 μg/kg IYKEDH 13,1128,82
ivn-rat LD50:41700 μg/kg YACHDS 8,3991,80
ivn-mus LD50:46500 μg/kg YACHDS 8,3991,80
ivn-dog LD50:42700 μg/kg AIPTAK 244,130,80
ivn-gpg LD50:42900 μg/kg AIPTAK 244,130,80

SAFETY PROFILE: Poison by intravenous and intraperitoneal routes. Human systemic effects by intravenous and intramuscular routes: eye effects, convulsions, muscle spasms and cardiac arrhythmias. Experimental teratogenic and reproductive effects. When heated to decomposition it emits toxic fumes of Cl^- and NO_x.

DYC000 CAS:87-52-5 ***HR: 3***
DONAXINE
mf: $C_{11}H_{14}N_2$ mw: 174.27

SYNS: β-DIMETHYLAMINOMETHYLINDOLE ◇ 3-(DIMETHYLAMINOMETHYL)INDOLE ◇ GRAMIN ◇ GRAMINE

TOXICITY DATA with REFERENCE
ipr-rat LDLo:250 mg/kg NCNSA6 5,11,53
ipr-mus LD50:122 mg/kg PSYPAG 16,385,70
ivn-mus LD50:46 mg/kg JPETAB 105,130,52

CONSENSUS REPORTS: Reported in EPA TSCA Inventory.

SAFETY PROFILE: Poison by intraperitoneal and intravenous routes. When heated to decomposition it emits toxic fumes of NO_x.

DYC200 CAS:5796-14-5 ***HR: 3***
l-DOPA HYDROCHLORIDE
mf: $C_9H_{11}NO_4 \cdot ClH$ mw: 233.67

SYNS: l-3-(3,4-DIHYDROXYPHENYL)ALANINE ◇ 3-HYDROXY-l-TYROSINE HYDROCHLORIDE

TOXICITY DATA with REFERENCE
orl-mus LD50:1460 mg/kg TXAPA9 28,1,74
ivn-mus LD50:527 mg/kg TXAPA9 28,1,74
orl-rbt LDLo:950 mg/kg TXAPA9 28,1,74
ivn-rbt LD50:144 mg/kg TXAPA9 28,1,74

SAFETY PROFILE: Poison by intravenous route. Moderately toxic by ingestion. Used to treat Parkinson's disease. When heated to decomposition it emits very toxic fumes of NO_x and HCl.

DYC300 CAS:7101-51-1 ***HR: 2***
l-DOPA METHYL ESTER
mf: $C_{10}H_{13}NO_4$ mw: 211.24

SYN: l-3-(3,4-DIHYDROXYPHENYL)ALANINE METHYL ESTER

TOXICITY DATA with REFERENCE
dni-hmn:oth 3300 nmol/L CNREA8 40,1414,80
dni-mus:oth 3300 nmol/L CNREA8 40,1414,80
ipr-mus LD50:1000 mg/kg CTRRDO 63,991,79

SAFETY PROFILE: Moderately toxic by intraperitoneal route. Human mutation data reported. When heated to decomposition it emits toxic fumes of NO_x. See also l-3,4-DIHYDROXYPHENYLALANINE and ESTERS.

DYC400 CAS:51-61-6 ***HR: 3***
DOPAMINE
mf: $C_8H_{11}NO_2$ mw: 153.20

SYN: 4-(2-AMINOETHYL)PYROCATECHOL

TOXICITY DATA with REFERENCE
dnd-esc 10 mg/L MUREAV 124,9,83

dnd-hmn:fbr 50 mg/L MUREAV 137,17,84
scu-rat TDLo:100 mg/kg (female 10-14D post):REP AJOGAH 104,578,69
scu-rat TDLo:100 mg/kg (female 10-14D post):TER AJOGAH 104,578,69
ipr-rat LD50:163 mg/kg TXAPA9 88,433,87
ipr-mus LD50:950 mg/kg OYYAA2 8,835,74
ivn-mus LD50:59 mg/kg OYYAA2 8,835,74
icv-mus LD50:74 mg/kg TYKNAQ 27,131,80
ivn-dog LD50:79 mg/kg ARZNAD 28,2208,78

SAFETY PROFILE: Poison by intravenous, intracervical, and intraperitoneal routes. Experimental teratogenic and reproductive effects. Human mutation data reported. A neurotransmitter. An adrenergic agent. When heated to decomposition it emits toxic fumes of NO_x.

DYC600 CAS:62-31-7 ***HR: 3***
DOPAMINE HYDROCHLORIDE
mf: $C_8H_{11}NO_2 \cdot ClH$ mw: 189.66

SYNS: 4-(2-AMINOETHYL)-1,2-BENZENEDIOLHYDROCHLORIDE ◇ 4-(2-AMINOETHYL)PYROCATECHOL HYDROCHLORIDE ◇ DOPAMINE CHLORIDE ◇ m-HYDROXYTYRAMINE HYDROCHLORIDE ◇ 3-HYDROXYTYRAMINE HYDROCHLORIDE ◇ INTROPIN

TOXICITY DATA with REFERENCE
dnd-mus:lvr 6 μmol/L JNCIAM 62,947,79
msc-mus:lym 93800 μg/L MUREAV 124,9,83
ivn-rat TDLo:300 mg/kg (female 6-15D post):REP ARZNAD 28,2208,78
ivn-dom TDLo:361 mg/kg (female 15W post):TER AJOGAH 130,211,78
orl-rat LD50:2859 mg/kg KSRNAM 8,2311,74
ipr-rat LD50:597 mg/kg KSRNAM 8,2311,74
scu-rat LD50:647 mg/kg KSRNAM 8,2311,74
ivn-rat LD50:4800 μg/kg KSRNAM 3,2311,74
orl-mus LD50:4361 mg/kg KSRNAM 8,2311,74
ipr-mus LD50:914 mg/kg KSRNAM 8,2311,74
scu-mus LD50:1366 mg/kg KSRNAM 8,2311,74
ivn-mus LD50:156 mg/kg KSRNAM 8,2311,74
ivn-rbt LD50:90 mg/kg KSRNAM 3,2311,74

SAFETY PROFILE: Poison by intravenous route. Moderately toxic by ingestion, intraperitoneal, and subcutaneous routes. Experimental teratogenic and reproductive effects. Mutation data reported. An antihypotensive. When heated to decomposition it emits very toxic fumes of NO_x and HCl. See also DOPAMINE.

DYC700 CAS:520-09-2 ***HR: 3***
DOPAN
mf: $C_9H_{13}Cl_2N_3O_2$ mw: 266.15

PROP: Snow-white crystals. Decomp @ 178-179°. Practically insol in cold water, acetone, and benzene; sltly sol in alc.

SYNS: 5-(BIS(2-CHLOROETHYL)AMINO)-6-METHYLURACIL ◇ 5-(BIS(2-CLOROETIL)-AMINO)-6-METILURACILE (ITALIAN) ◇ CHLOROETHYLAMINOURACIL ◇ 2,6-DIHYDROXY-4-METHYL-5-BIS(2-CHLOROETHYL)AMINOPYRIMIDINE ◇ DOPANE ◇ ELDERFIELD PYRIMIDINE MUSTARD ◇ ENT 50,698 ◇ 4-METHYL-5-(BIS(β-CHLOROETHYL)AMINO)URACIL ◇ 6-METHYL-5-(BIS(2-CHLOROETHYL)AMINO)URACIL

TOXICITY DATA with REFERENCE
cyt-hmn:leu 2500 μg/L TSITAQ 15,1505,73
dni-mus:ast 2 mg/L NEOLA4 22,105,75
oms-mus:ast 2 mg/L NEOLA4 22,105,75
cyt-mky:kdy 100 mg/L TSITAQ 15,1505,73
orl-rat LD50:2400 μg/kg BCFAAI 101,630,62
ipr-mus LDLo:3 mg/kg FATOAO 34,83,71

SAFETY PROFILE: Poison by ingestion and intraperitoneal routes. Human mutation data reported. When heated to decomposition it emits toxic fumes of Cl^- and NO_x.

DYC800 CAS:77-21-4 ***HR: 3***
DORIDEN
mf: $C_{13}H_{15}NO_2$ mw: 217.29

PROP: dl-Form: Crystals from ether or from ethyl acetate + petr ether. Mp: 84°. Freely sol in ethyl acetate, acetone, ether, chloroform; sol in ethanol, methanol. Practically insol in water. d-Form: Crystals. Mp: 102.5-103°, refractive index: (α) (20/D) +176° (methanol). l-Form: Crystals. Mp: 102-103°, refr index: (α) (20/D) -181° (methanol).

SYNS: ALFIMID ◇ CC 11511 ◇ DORIDEN-SED ◇ ELRODORM ◇ 3-ETHYL-3-PHENYL-2,6-DIKETOPIPERIDINE ◇ 3-ETHYL-3-PHENYL-2,6-DIOXOPIPERIDINE ◇ α-ETHYL-α-PHENYLGLUTARIMIDE ◇ 2-ETHYL-2-PHENYLGLUTARIMIDE ◇ 3-ETHYL-3-PHENYL-2,6-PIPERIDINEDIONE ◇ GIMID ◇ GLIMID ◇ GLUTATHIMID ◇ GLUTETHIMID ◇ GLUTETHIMIDE ◇ GLUTETIMIDE ◇ NOXYRON ◇ 3-PHENYL-3-ETHYL-2,6-DIKETOPIPERIDINE ◇ 3-PHENYL-3-ETHYL-2,6-DIOXOPIPERIDINE ◇ α-PHENYL-α-ETHYLGLUTARIC ACID IMIDE ◇ 2-PHENYL-2-ETHYLGLUTARIC ACID IMIDE ◇ α-PHENYL-α-ETHYLGLUTARIMIDE ◇ SARODORMIN

TOXICITY DATA with REFERENCE
orl-rat TDLo:2250 mg/kg (6-14D preg):REP COREAF 256,1841,63
orl-rbt TDLo:1350 mg/kg (female 8-16D post):TER TXAPA9 10,244,67
orl-chd TDLo:25 mg/kg:CNS AJDCAI 130,507,76
orl-hmn TDLo:171 mg/kg JAMAAP 214,1704,70
unr-man LDLo:147 mg/kg 85DCAI 2,73,70
orl-rat LD50:600 mg/kg 27ZQAG -,233,72
orl-mus LD50:360 mg/kg 27ZQAG -,233,72
ipr-mus LD50:208 mg/kg APPHAX 32,243,75
orl-dog LD50:500 mg/kg SMWOAS 85,305,55
orl-mky LDLo:300 mg/kg TXAPA9 29,75,74

CONSENSUS REPORTS: EPA Genetic Toxicology Program.

SAFETY PROFILE: Poison by ingestion, and intraperitoneal routes. Human systemic effects by ingestion: pupillary dilation, ataxia, somnolence, coma, and blood pressure depression. An experimental teratogen. Other experimental reproductive effects. When heated to decomposition it emits toxic fumes of NO_x. Caution: May be habit forming. This is a controlled substance (depressant) listed in the U.S. Code of Federal Regulations, Title 21 Part 1308.13 (1985).

DYC875 CAS:113-53-1 ***HR: 3***
DOSULEPIN
mf: $C_{19}H_{21}NS$ mw: 295.47

PROP: Bp: (0.05) 171-172°, mp: 55-57°.

SYNS: 3-DIBENZO(b,e)THIEPIN-11(6H)-YLIDENE-N,N-DIMETHYL-1-PROPAMINE ◇ 11-(3-DIMETHYLAMINOPROPYLIDENE)-6,11-DIHYDRODIBENZO(b,e)THIEPIN ◇ N,N-DIMETHYLDIBENZO(b,e)THIEPIN-$\Delta^{11(6H),\gamma}$PROPYLAMINE ◇ DOTHIEPIN ◇ IZ 914 ◇ PROTHIADEN ◇ PROTHIADEN SPOFA

TOXICITY DATA with REFERENCE
cyt-mus-ipr 100 mg/kg ACNSAX 15,114,73
dlt-mus-ipr 150 mg/kg ACNSAX 15,114,73
ipr-mus TDLo:150 mg/kg (6D male):REP ACNSAX 15,114,73
orl-inf TDLo:100 mg/kg:CNS,CVS BMJOAE 288,1800,84
orl-wmn TDLo:4500 μg/kg:CNS BMJOAE 2,97,77

CONSENSUS REPORTS: EPA Genetic Toxicology Program.

SAFETY PROFILE: Human systemic effects by ingestion: central nervous system including coma, and cardiac arrhythmia. Mutation data reported. Experimental reproductive effects. When heated to decomposition it emits toxic fumes of NO_x and SO_x.

DYD200 CAS:35944-83-3 ***HR: 3***
DOWCO 133
mf: $C_{10}H_{11}Cl_3NO_2PS$ mw: 346.60

SYN: N-ISOPROPYLPHOSPHORAMIDOTHIOICACID-O-(2,4,5-TRICHLOROPHENYL)ESTER

TOXICITY DATA with REFERENCE
orl-rat LDLo:31 mg/kg TXAPA9 21,315,72
orl-bwd LD50:7 mg/kg TXAPA9 21,315,72

SAFETY PROFILE: Poison by ingestion. A pesticide. When heated to decomposition it emits very toxic fumes of Cl^-, NO_x, PO_x, and SO_x.

DYD400 CAS:2213-84-5 ***HR: 3***
DOWCO 159
mf: $C_9H_{11}Cl_3NO_3P$ mw: 318.53

SYN: ETHYLPHOSPHORAMIDIC ACID, METHYL-(2,4,5-TRICHLOROPHENYL) ESTER

TOXICITY DATA with REFERENCE
orl-bwd LD50:56 mg/kg TXAPA9 21,315,72

SAFETY PROFILE: Poison by ingestion. When heated to decomposition it emits very toxic fumes of Cl^-, NO_x, and PO_x. See also ESTERS.

DYD600 CAS:35944-82-2 ***HR: 3***
DOWCO 160
mf: $C_8H_9Cl_3NO_3P$ mw: 304.50

SYN: ETHYLPHOSPHORAMIDICACID-(2,4,5-TRICHLOROPHENYL) ESTER

TOXICITY DATA with REFERENCE
orl-bwd LD50:5600 μg/kg TXAPA9 21,315,72

SAFETY PROFILE: Poison by ingestion. When heated to decomposition it emits very toxic fumes of Cl^-, NO_x, and PO_x. See also ESTERS.

DYD800 CAS:35944-84-4 ***HR: 3***
DOWCO 177
mf: $C_4H_{12}NO_2PS$ mw: 169.20

SYN: N-METHYL-PHOSPHORAMIDOTHIOICACID-O-ISOPROPYL ESTER

TOXICITY DATA with REFERENCE
orl-rat LD50:800 mg/kg TXAPA9 21,315,72
orl-bwd LD50:100 mg/kg TXAPA9 21,315,72

SAFETY PROFILE: Poison by ingestion. When heated to decomposition it emits very toxic fumes of NO_x, PO_x, and SO_x. See also ESTERS.

DYE200 CAS:4492-96-0 ***HR: 3***
DOWCO-183
mf: $C_{13}H_{20}ClNO_3P$ mw: 305.77

SYNS: 2-CHLORO-4-(1,1-DIMETHYLPROPYL)PHENYLMETHYL METHYLPHOSPHORAMIDATE ◇ ENT 27,192 ◇ METHYLPHOSPHORAMIDIC-2-CHLORO-4-(1,1-DIMETHYLPROPYL)PHENYL METHYL ESTER

TOXICITY DATA with REFERENCE
orl-rat LDLo:500 mg/kg ARSIM* 20,9,66
scu-gpg LDLo:100 mg/kg JEENAI 61,1261,68

SAFETY PROFILE: Poison by subcutaneous route. Moderately toxic by ingestion. When heated to decomposition it emits very toxic fumes of Cl^- and NO_x.

DYE400 CAS:53908-27-3 ***HR: 3***
DOWFUME EB-5

PROP: Contains ethylene dichloride (30%), carbon tetrachloride (63%), and ethylene dibromide (7%) (AMIHAB 13,1,56).

TOXICITY DATA with REFERENCE
orl-rat LD50:660 mg/kg AMIHAB 13,1,56

ihl-rat LCLo:920 ppm/4H JAFCAU 2,1318,54
orl-rbt LD50:290 mg/kg JAFCAU 2,1318,54
skn-rbt LDLo:500 mg/kg JAFCAU 2,1318,54
orl-gpg LD50:280 mg/kg JAFCAU 2,1318,54
orl-ckn LD50:780 mg/kg JAFCAU 2,1318,54

SAFETY PROFILE: Poison by ingestion. Mildly toxic by skin contact and inhalation. When heated to decomposition it emits very toxic fumes of Cl^- and Br^-. See also components as listed.

DYE409 CAS:1668-19-5 ***HR: 3***
DOXEPIN
mf: $C_{19}H_{21}NO$ mw: 279.41

SYNS: 11-(3-DIMETHYLAMINOPROPYLIDENE)-6,11-DIHYDRODIBENZ(b,e)OXIPIN ◇ N,N-DIMETHYLDIBENZ(b,e)OXEPIN-$\Delta^{11(6H)}$-Γ-PROPYLAMINE

TOXICITY DATA with REFERENCE
orl-hmn LDLo:60 mg/kg CTOXAO 10,327,77
orl-rat LD50:147 mg/kg ARZNAD 15,863,65
ipr-rat LD50:182 mg/kg ARZNAD 15,863,65
ivn-rat LD50:16 mg/kg ARZNAD 15,863,65
orl-mus LD50:135 mg/kg ARZNAD 15,863,65
ipr-mus LD50:79 mg/kg ARZNAD 15,863,65
ivn-mus LD50:26 mg/kg ARZNAD 15,863,65
ivn-rbt LD50:11 mg/kg ARZNAD 15,863,65

SAFETY PROFILE: Human poison by ingestion. Experimental poison by ingestion, intravenous, and intraperitoneal routes. A sedative and hypnotic used as an antianxiety agent. When heated to decomposition it emits toxic fumes of NO_x.

DYE415 CAS:3094-09-5 ***HR: 2***
DOXIFLURIDINE
mf: $C_9H_{11}FN_2O_5$ mw: 246.22

SYNS: 5′-DEOXY-5-FLUOROURIDINE ◇ 5′-DFUR ◇ RO 21-9738

TOXICITY DATA with REFERENCE
orl-rat TDLo:1100 mg/kg (7-17D preg):REP YACHDS 13(Suppl 2),481,85
orl-rat TDLo:550 mg/kg (7-17D preg):TER YACHDS 13(Suppl 2),481,85
orl-rat LD50:3390 mg/kg YACHDS 13(Suppl 2),209,85
orl-mus LDLo:5 g/kg YACHDS 13(Suppl 2),209,85

SAFETY PROFILE: Moderately toxic by ingestion. An experimental teratogen. Other experimental reproductive effects. When heated to decomposition it emits toxic fumes of F^- and NO_x.

DYE425 CAS:564-25-0 ***HR: 2***
DOXYCYCLINE
mf: $C_{22}H_{24}N_2O_8$ mw: 444.48

SYNS: α-6-DEOXY-5-HYDROXYTETRACYCLINE ◇ α-6-DEOXYOXYTETRACYCLINE ◇ 6-α-DEOXY-5-OXYTETRACYCLINE ◇ DOXICICLINA (ITALIAN) ◇ GS-3065 ◇ 5-HYDROXY-α-6-DEOXYTETRACYCLINE ◇ LIVIATIN ◇ VIBRAMYCIN

TOXICITY DATA with REFERENCE
dni-hmn:lym 3750 μg/L BCPHBM 16,127,83
orl-mus TDLo:750 mg/kg (female 2D post):TER ARZNAD 36,219,86
ipr-rat LD50:378 mg/kg THERAP 23,575,68
ivn-rat LD50:228 mg/kg PBPSDY 2,305,79
orl-mus LD50:1870 mg/kg GICTAL 17,276,70
ipr-mus LD50:410 mg/kg GICTAL 17,276,70

CONSENSUS REPORTS: EPA Genetic Toxicology Program.

SAFETY PROFILE: Moderately toxic by ingestion and intraperitoneal routes. An experimental teratogen. Human mutation data reported. When heated to decomposition it emits toxic fumes of NO_x. See also TETRACYCLINE.

DYE500 CAS:469-21-6 ***HR: 3***
DOXYLAMINE
mf: $C_{17}H_{22}N_2O$ mw: 270.41

PROP: Liquid. Bp: (0.5 mm Hg) 137-141°. Sol in acids.

SYNS: 2-(α-(2-(DIMETHYLAMINO)ETHOXY)-α-METHYLBENZYL) PYRIDINE ◇ 2-DIMETHYLAMINOETHOXYPHENYLMETHYL-2-PICOLINE ◇ N,N-DIMETHYL-2-(1-PHENYL-1-(2-PYRIDINYL)ETHOXY) ETHANAMINE (9CI) ◇ NCI C60684 ◇ PHENYL-2-PYRIDYLMETHYL-β-N,N-DIMETHYLAMINOETHYL ETHER

TOXICITY DATA with REFERENCE
scu-rat LD50:440 mg/kg DPIRDU 2(5),17,82
orl-mus LD50:470 mg/kg DPIRDU 2(5),17,82
scu-mus LD50:460 mg/kg DPIRDU 2(5),17,82
ivn-mus LD50:62 mg/kg DPIRDU 2(5),17,82
orl-rbt LD50:250 mg/kg DPIRDU 2(5),17,82
ivn-rbt LD50:49 mg/kg DPIRDU 2(5),17,82

SAFETY PROFILE: Poison by ingestion and intravenous routes. Moderately toxic by subcutaneous route. When heated to decomposition it emits toxic fumes of NO_x.

DYE550 CAS:54622-43-4 ***HR: 2***
DPF
mf: $C_7H_{22}N_2O_{13}P_4$ mw: 466.19

SYNS: ((2-HYDROXY-1,3-PROPANEDIYL)BIS(NITRILOBIS(METHYLENE)))TETRAKISPHOSPHONIC ACID ◇ (2-HYDROXY-1,3-PROPYLENEDIAMINE-N,N,N^1,N$'^1$-TETRAMETHYLENEPHOSPHORIC ACID ◇ 2-OXY-1,3-PROPYLENEDIAMINE-N,N,N′,N′-TETRAMETHYLENEPHOSPHONIC ACID

TOXICITY DATA with REFERENCE
orl-rat LD50:4300 mg/kg GTPZAB 29(4),45,85
orl-mus LD50:2800 mg/kg GISAAA 49(5),81,84
orl-gpg LD50:3800 mg/kg GTPZAB 29(4),45,85

CONSENSUS REPORTS: Cyanide and its compounds are on the Community Right-To-Know List.

SAFETY PROFILE: Moderately toxic by ingestion. When heated to decomposition it emits toxic fumes of PO_x and NO_x. See also NITRILES.

DYE600 CAS:523-87-5 ***HR: 3***
DRAMAMINE
mf: $C_{17}H_{21}NO \cdot C_7H_7ClN_4O_2$ mw: 470.02

SYNS: AMOSYT ◇ ANAUTINE ◇ ANDRAMINE ◇ AVIOMARIN ◇ o-BENZHYDRYLDIMETHYLAMINOETHANOL-8-CHLOROTHEOPHYLLINATE ◇ 2-(BENZHYDRYLOXY)-N,N-DIMETHYLETHYLAMINE with 8-CHLOROTHEOPHYLLINE ◇ CHLOR- ANAUTINE ◇ DIAMARIN ◇ DIMENHYDRINATE ◇ DIPHENHYDRINATE ◇ DRAMAMIN ◇ DRAMARIN ◇ DRAMYL ◇ DROMYL ◇ ELDODRAM ◇ ETHYLAMINE-2-(DIPHENYLMETHOXY)-N,N-DIMETHYL, compound with 8-CHLOROTHEOPHYLLINE (1:1) ◇ GRAVINOL ◇ GRAVOL ◇ MENHYDRINATE ◇ NCI-C60639 ◇ NEO-NAVIGAN ◇ NOVAMIN ◇ NOVAMINE ◇ PERMITAL ◇ REISE-ENGLETTEN ◇ SUPREMAL ◇ TEODRAMIN ◇ TRAVELIN ◇ TRAVELMIN ◇ VOMEX A ◇ XAMAMINA

TOXICITY DATA with REFERENCE
mmo-sat 333 μg/plate ENMUDM 9(Suppl 9),1,87
orl-man TDLo:11400 μg/kg:CNS AJPSAO 128,1012,72
orl-rat LD50:1320 mg/kg NIIRDN 6,346,82
ivn-rat LD50:200 mg/kg CLDND* 6,346,82
ipr-mus LD50:149 mg/kg NIIRDN 6,346,82

CONSENSUS REPORTS: Reported in EPA TSCA Inventory.

SAFETY PROFILE: Poison by intraperitoneal and intravenous route. Moderately toxic by ingestion. A drug much used for motion sickness. Human systemic effects by ingestion: hallucinations and distorted perceptions. Mutation data reported. When heated to decomposition it emits very toxic fumes of NO_x and Cl^-. See also AMINES.

DYE700 CAS:526-18-1 ***HR: 3***
DRIOL
mf: $C_{13}H_{11}NO_3$ mw: 229.25

PROP: Crystals. Mp: 179°. Practically insol in cold water and acetic acid; sltly sol in warm water, benzene, and toluene; freely sol in methanol, ethanol, ether, and acetone.

SYNS: AUXOBIL ◇ BICHOL ◇ BILENE ◇ BILOCOL ◇ DRIBAZIL ◇ DRIOL-LABAZ ◇ ENIDRAN ◇ HYDROXYPHENYL SALICYLAMIDE ◇ N-(p-HYDROXYPHENYL)SALICYLAMIDE ◇ N-(4-HYDROXYPHENYL)SALICYLAMIDE ◇ p-HYDROXYPHENYLSALICYLAMIDE ◇ p'-HYDROXYSALICYLANILIDE ◇ 4'-HYDROXYSALICYLANILIDE ◇ KANOCHOL ◇ L 1718 ◇ OSALMID ◇ OSALMIDE ◇ OXAPHENAMID ◇ OXAPHENAMIDE ◇ PHPS ◇ SALMIDOCHOL ◇ SARYUURIN

TOXICITY DATA with REFERENCE
orl-rat LD50:6702 mg/kg NIIRDN 6,159,82
ipr-rat LD50:1621 mg/kg NIIRDN 6,159,82
orl-mus LD50:1050 mg/kg JJPAAZ 22,235,72
ipr-mus LD50:517 mg/kg NIIRDN 6,159,82
scu-mus LD50:1900 mg/kg AIPTAK 116,154,58
ivn-mus LD50:180 mg/kg CSLNX* NX#03234

SAFETY PROFILE: Poison by intravenous route. Moderately toxic by ingestion and intraperitoneal, and subcutaneous routes. When heated to decomposition it emits toxic fumes of NO_x.

DYF000 CAS:125-72-4 ***HR: 3***
l-DROMORAN TARTRATE
mf: $C_{17}H_{23}NO \cdot C_4H_6O_6$ mw: 407.51

SYNS: l-3-HYDROXY-N-METHYLMORPHINAN BITARTRATE ◇ l-3-HYDROXY-N-METHYLMORPHINAN TARTRATE ◇ LEMORAN ◇ LEVORPHAN TARTRATE ◇ LEVORPHANOL TARTRATE ◇ RO 1-5431/7

TOXICITY DATA with REFERENCE
oms-rat-ipr 2500 μg/kg RCOCB8 3,105,72
scu-mus TDLo:25 mg/kg (9D preg):TER DGD5A5 22,61,80
ivn-dog LD50:46 mg/kg AIPTAK 149,571,64
orl-rat LD50:150 mg/kg JPETAB 109,189,53
ipr-rat LD50:48 mg/kg TXCVAC 14,217,79
scu-rat LD50:110 mg/kg JPETAB 109,189,53
ivn-rat LD50:27 mg/kg AIPTAK 108,171,57
orl-mus LD50:285 mg/kg JPETAB 109,189,53
ipr-mus LD50:92 mg/kg JJPAAZ 11,101,62
scu-mus LD50:187 mg/kg JPETAB 109,189,53
ivn-mus LD50:32 mg/kg AIPTAK 105,221,56
ivn-rbt LD50:20 mg/kg AIPTAK 109,171,57

SAFETY PROFILE: Poison by ingestion, intravenous, intraperitoneal and subcutaneous routes. Experimental teratogenic effects. Mutation data reported. When heated to decomposition it emits toxic fumes of NO_x.

DYF200 CAS:548-73-2 ***HR: 3***
DROPERIDOL
mf: $C_{22}H_{22}FN_3O_2$ mw: 379.47

SYNS: DEHIDROBENZPERIDOL ◇ DEHYDROBENZPERIDOL ◇ DEIDROBENZPERIDOLO ◇ DHBP ◇ DIHIDROBENZPERIDOL ◇ DRIDOL ◇ DROLEPTAN ◇ 1-(1-(3-(p-FLUOROBENZOYL)PROPYL)-1,2,3,6-TETRAHYDRO-4-PYRIDYL)-2-BENZIMIDAZOLINONE ◇ 1-(1-(4-(p-FLUOROPHENYL-4-OXOBUTYL)-1,2,3,6-TETRAHYDRO-4-PYRIDYL)-2-BENZIMIDAZOLINONE ◇ HALKAN ◇ INAPPIN ◇ INAPSIN ◇ INNOVAN ◇ INNOVAR ◇ INOPSIN ◇ INOVAL ◇ LEPTANAL ◇ LEPTOFEN ◇ MCN-JR-4749 ◇ PROPERIDOL ◇ R 4749 ◇ SINTOSIAN ◇ THALAMONAL ◇ VETKALM

TOXICITY DATA with REFERENCE
orl-hmn TDLo:223 μg/kg/12D-I:CNS ARZNAD 22,93,72
orl-rat LDLo:700 mg/kg TXAPA9 18,185,71
scu-rat LD50:20520 mg/kg NIIRDN 6,440,82
ivn-rat LD50:30 mg/kg 27ZQAG -,187,72
ipr-mus LD50:80 mg/kg EJMCA5 13,387,78
scu-mus LD50:125 mg/kg TXAPA9 6,37,64
ivn-mus LD50:20 mg/kg 27ZQAG -,187,72

SAFETY PROFILE: Poison by intravenous, subcutaneous, and intraperitoneal routes. Moderately toxic by ingestion. Human systemic effects by ingestion: wakefulness, tremors, and muscle weakness. An antipsychotic agent. When heated to decomposition it emits very toxic fumes of F^- and NO_x.

DYF400 CAS:1403-47-0 ***HR: D***
DUAZOMYCIN

TOXICITY DATA with REFERENCE
unr-rat TDLo:1600 μg/kg (11-18D preg):TER 85DJA5 -,95,71
unr-rat TDLo:1600 μg/kg (11-18D preg):REP 85DJA5 -,95,71

SAFETY PROFILE: An experimental teratogen. Other experimental reproductive effects.

DYF450 CAS:62-90-8 ***HR: 2***
DURABOLIN
mf: $C_{27}H_{34}O_3$ mw: 406.61

PROP: Crystals. Mp: 95-96°.

SYNS: ACTIVIN ◇ DURABOL ◇ FENOBOLIN ◇ 17-β-HYDROXY-ESTR-4-ENE-3-ONE-3-PHENYLPROPIONATE ◇ 17-β-HYDROXY-ESTRA-4-EN-3-ONE-17-PHENYLPROPIONATE ◇ NANDROLIN ◇ NANDROLONE PHENPROPIONATE ◇ NANDROLONE PHENYLPROPIONATE ◇ NEROBIL ◇ NEROBIOLIL ◇ NORANDROLONE PHENYLPROPIONATE ◇ NORANDROSTENOLONE PHENYLPROPIONATE ◇ 19-NORTESTOSTERONE PHENYLPROPIONATE ◇ NPP ◇ NTPP ◇ 19NTPP ◇ (17-β)-17-(1-OXO-3-PHENYLPROPOXY)-ESTR-4-EN-3-ONE (9CI) ◇ PHENOBOLIN ◇ 17-β-PHENYLPROPIONYLOXY-4-ESTREN-3-ONE ◇ STRABOLENE ◇ SUPERANABOLON

TOXICITY DATA with REFERENCE
ims-rat TDLo:3750 μg/kg (female 9-14D post):REP OYYAA2 7,723,73
ims-rat TDLo:15 mg/kg (female 9-14D post):TER OYYAA2 7,723,73
ipr-rat LD50:595 mg/kg PCJOAU 20,143,86

SAFETY PROFILE: Moderately toxic by intraperitoneal routes. An experimental teratogen. Other experimental reproductive effects. When heated to decomposition it emits acrid smoke and fumes.

DYF700 CAS:50370-12-2 ***HR: D***
DURICEF
mf: $C_{16}H_{17}N_3O_5S$ mw: 363.42

SYNS: BIDOCEF ◇ BL-S 578 ◇ S-578

TOXICITY DATA with REFERENCE
orl-rat TDLo:9200 mg/kg (17-20D preg):REP JJANAX 33,503,80
orl-rbt TDLo:1300 mg/kg (female 6-18D post):TER JJANAX 33,497,80

SAFETY PROFILE: Experimental teratogenic and reproductive effects. When heated to decomposition it emits toxic fumes of SO_x and NO_x.

DYF759 CAS:152-62-5 ***HR: D***
DYDROGESTERONE
mf: $C_{21}H_{28}O_2$ mw: 312.49

PROP: Crystals from acetone and hexane. Mp: 169-170 °.

SYNS: 6-DEHYDRO-RETRO-PROGESTERONE ◇ DIPHASTON ◇ DUFASTON ◇ DUPHASTON ◇ DUVARON ◇ GESTATRON ◇ GYNOREST ◇ HYDROGESTRONE ◇ ISOPREGNENONE ◇ 9-β,10-α-PREGNA-4,6-DIENE-3,20-DIONE ◇ (9-β,10-α)-PREGNA-4,6-DIENE-3,20-DIONE (9CI) ◇ PRODEL ◇ RETRO-6-DEHYDROPROGESTERONE ◇ RETRONE ◇ Δ^6-RETROPROGESTERONE ◇ TEROLUT

TOXICITY DATA with REFERENCE
scu-rbt TDLo:40 mg/kg (female 10D pre):REP FESTAS 23,196,72
orl-rbt TDLo:1600 mg/kg (female 13-28D post):TER ACENA7 43,539,63

SAFETY PROFILE: An experimental teratogen. Other experimental reproductive effects.

DYF800 ***HR: D***
DYE C

PROP: A red impurity contained in commercial trypan blue (TJADAB 3,371,70).

TOXICITY DATA with REFERENCE
ipr-mus TDLo:180 μg/kg (female 8D post):TER TJADAB 3,371,70

SAFETY PROFILE: An experimental teratogen.

DYG000 ***HR: 3***
DYNAMITE

PROP: Major constituent is nitroglycerin (85ESA3 8,739,68).

SYN: GELATINE DYNAMITE

DOT Classification: Class A Explosive; Label: Explosive A.

SAFETY PROFILE: A high explosive used industrially in construction and mining. The name generally refers to a mixture containing as its principal explosive ingredient either glyceryl trinitrate (nitroglycerin) or ammonium nitrate, suitably sensitized. It does not apply to black blasting powders, chlorate powders, and other deflagrating mixtures. While this material is a powerful explosive when detonated by shock or heat, it is only moderately hazardous. It can react vigorously with oxidizing materials. Dangerous; shock and heat will explode it; when heated to decomposition it emits highly toxic fumes of NO_x and CO, etc. See also NITROGLYCERIN; EXPLOSIVES, HIGH; and NITRATES.

An ordinary blasting cap or an electric blasting cap is used for detonating a charge of dynamite. The various classes and grades of dynamite are made from mixtures composed of an explosive compound or a mixture of explosive compounds, a "dope", and an antiacid. If any of the explosive ingredients are in a liquid state they are referred to as the "explosive oil," which is usually composed of glyceryl trinitrate (nitroglycerin) and about 25-30% of ethylene glycol dinitrate. The latter compound depresses the freezing point of the nitroglycerin and renders the dynamite low-freezing. Other compounds may also be used as freezing point depressants. The explosive oil is absorbed by carbonaceous materials that have entirely replaced kieselguhr (diatomaceous earth), formerly used exclusively as the absorbent or "dope" in dynamites. This type of "dope" does not enter into the explosive reaction. Wood pulp is now most commonly used as the absorber, either alone or mixed in suitable proportions with flour, starch, etc.

The absorbents may be mixed with an oxidizer such as sodium nitrate, in which case an active "dope" is formed. For neutralizing any acid that may be present, about 1% of an antiacid (calcium carbonate or zinc oxide) is added to the mixture. The explosive oil is mixed into the "dope". The strength of a kieselguhr dynamite, when detonated, is derived only from the explosive oil, since kieselguhr is inert. A mixture of this kind is known as a straight dynamite. On the other hand, an active "dope", (an admixture of carbonaceous absorbents with an oxidizer), furnishes explosive strength in addition to that derived from the explosive ingredients.

By replacing a part of the explosive oil of a straight dynamite with ammonium nitrate, so that the latter becomes the principal explosive ingredient, a mixture known as an ammonia dynamite is obtained.

When the explosive oil is gelatinized the explosive is known as a gelatin or an ammonia gelatin dynamite.

Blasting gelatin is a gelatinized mass of an elastic nature obtained by incorporating nitrocotton with an explosive oil into which is mixed about 1% of antiacid.

Dynamites may be in bulk form (bag powder) or in cartridge form, the most common size being 11/4 inch in diameter and 8 inches long, although for holes of small diameter, cartridges as small as 7/8 inch in diameter are also used. In large diameter well-drill holes for quarry blasting, cartridge diameters up to 10 inches and lengths up to 30 inches may be used. These upper limits or 50 pounds in weight of each cartridge are imposed by the DOT Regulations, and the maximum length of 30 inches applies to all cartridge diameters between 4 and 10 inches.

An integral part of a stick of dynamite is the paraffined paper wrapper that not only holds the ingredients together but enters into the explosive reaction.

The wrapper also affords some measure of protection from moderate exposure to dampness. For blasting in wet operations, a gelatinized dynamite which resists the absorption of water is used.

The strength of straight dynamite is graded by its explosive oil content (% by weight), while for any other class of dynamite, the strength is determined experimentally in comparison with the various grades of the straight dynamites. For example, a 40% straight dynamite is one which contains 40% of explosive oil; a 40% strength ammonia dynamite, as determined by tests, equals a 40% straight dynamite in strength. In other words a 40% strength ammonia dynamite will release the same energy as an equivalent weight of a 40% straight dynamite.

DYG400 ***HR: 3***
DYSPROSIUM
aw: Dy aw: 162.5

PROP: Bright, lustrous, silvery metal. Mp: 1409°, bp: 2335°, d: 8.540 @ 25°.

SAFETY PROFILE: It may exhibit an anticoagulant effect. Flammable; an active reducing agent. Reacts violently in air and to halogens. See also RARE EARTHS.

DYG600 CAS:10025-74-8 ***HR: 3***
DYSPROSIUM CHLORIDE
mf: Cl_3Dy mw: 268.85

PROP: Shiny, yellow crystals. D: 3.67 @ 0°/4°, mp: 718°, bp: 1500°. A sol salt.

TOXICITY DATA with REFERENCE
orl-mus LD50:5443 mg/kg EQSSDX 1,1,75
ipr-mus LD50:343 mg/kg AEHLAU 5,437,62
ipr-gpg LD50:196 mg/kg AEHLAU 5,437,62

CONSENSUS REPORTS: Reported in EPA TSCA Inventory.

SAFETY PROFILE: Poison by intraperitoneal route. Mildly toxic by ingestion. When heated to decomposition it emits toxic fumes of Cl^-. See also CHLORIDES and RARE EARTHS.

DYG800 CAS:13074-91-4 ***HR: 3***
DYSPROSIUM CITRATE
mf: $C_6H_5O_7{\cdot}Dy$ mw: 351.61

SYN: CITRIC ACID, DYSPROSIUM(3+) salt (1:1)

TOXICITY DATA with REFERENCE
ipr-mus LD50:113 mg/kg AEHLAU 5,437,62
ipr-gpg LD50:54 mg/kg AEHLAU 5,437,62

SAFETY PROFILE: Poison by intraperitoneal route. When heated to decomposition it emits acrid smoke and irritating fumes. See also RARE EARTHS.

DYH000 CAS:35725-30-5 ***HR: 3***
DYSPROSIUM(III) NITRATE HEXAHYDRATE(1:3:6)
mf: $N_3O_9 \cdot Dy \cdot 6H_2O$ mw: 456.65

SYN: NITRIC ACID DYSPROSIUM(3+) SALT HEXAHYDRATE

TOXICITY DATA with REFERENCE
orl-rat LD50:3100 mg/kg TXAPA9 5,750,63
ipr-rat LD50:295 mg/kg TXAPA9 5,750,63
ipr-mus LD50:310 mg/kg TXAPA9 5,750,63

SAFETY PROFILE: Poison by intraperitoneal route. Moderately toxic by ingestion. When heated to decomposition it emits very toxic fumes of NO_x. See also NITRATES and RARE EARTHS.

DYH100 CAS:10143-38-1 ***HR: 3***
DYSPROSIUM TRINITRATE
mf: $Dy(NO_3)_3$ mw: 348.60

SYN: DYSPROSIUM NITRATE

TOXICITY DATA with REFERENCE
orl-rat LD50:2386 mg/kg EQSSDX 1,1,75
ipr-rat LD50:227 mg/kg EQSSDX 1,1,75
ipr-mus LD50:238 mg/kg EQSSDX 1,1,75

SAFETY PROFILE: Poison by intraperitoneal route. Moderately toxic by ingestion. When heated to decomposition it emits toxic fumes of NO_x. See also NITRATES.

E

EAB100 CAS:55620-97-8 ***HR: D***
E 785
mf: $C_{19}H_{26}O_3$ mw: 302.45

SYN: α,α-DIMETHYL-6-METHOXY-β-PROPYL-3,4-DIHYDRO-2-NAPHTHALENEPROPIONIC ACID

TOXICITY DATA with REFERENCE
orl-rat TDLo:70 μg/kg (female 1-7D post):REP IJEBA6 13,79,75

SAFETY PROFILE: Experimental reproductive effects. When heated to decomposition it emits acrid smoke and fumes.

EAB200 ***HR: 3***
EASTERN COTTONMOUTH VENOM

SYNS: AGKISTRODON PISCIVORUS PISCIVORUS VENOM ◇ VENOM, SNAKE, AGKISTRODON PISCIVORUS PISCIVORUS

TOXICITY DATA with REFERENCE
ipr-mus LD50:4844 μg/kg TOXIA6 9,131,71
scu-mus LD50:11 mg/kg TOXIA6 17,237,79
ivn-mus LD50:2044 μg/kg TOXIA6 9,131,71

SAFETY PROFILE: Poison by subcutaneous, intravenous, and intraperitoneal routes.

EAB225 ***HR: 3***
EASTERN DIAMOND-BACK RATTLESNAKE VENOM

SYNS: C. ADAMANTEUS VENOM ◇ CROTALUS ADAMANTEUS VENOM ◇ VENOM, SNAKE, CROTALUS ADAMANTEUS

TOXICITY DATA with REFERENCE
ims-rat LDLo:25 mg/kg AJMSA9 236,204,58
scu-mus LD50:7700 μg/kg TOXIA6 17,661,79
ivn-mus LD50:1333 μg/kg TOXIA6 9,131,71
ims-mus LD50:28 mg/kg AJMSA9 239,1,60
scu-dog LDLo:5 mg/kg SURGAZ 47,975,60
ivn-dog LDLo:500 μg/kg 19DDA6 1,269,67
ims-dog LDLo:640 μg/kg AJMSA9 236,204,58
ims-mky LDLo:640 μg/kg AJMSA9 236,204,58
ivn-cat LDLo:80 μg/kg TOXIA6 1,99,63
ivn-rbt LDLo:125 μg/kg SCIEAS 117,47,53
ims-rbt LDLo:940 μg/kg AJMSA9 236,204,58
ims-pgn LDLo:1400 μg/kg AJMSA9 236,204,58
ims-ckn LDLo:1400 μg/kg AJMSA9 236,204,58
ipr-frg LDLo:190 mg/kg ANREAK 139,305,61

SAFETY PROFILE: A deadly poison by subcutaneous, intramuscular, intravenous, and intraperitoneal routes.

EAB500 ***HR: 1***
EAU de BROUTS ABSOLUTE

PROP: Extracted from mixture of orange-flower water and petigrain-bigarde water (FCTXAV 14,659,76).

TOXICITY DATA with REFERENCE
skn-rbt 500 mg/24H MOD FCTXAV 14,659,76

SAFETY PROFILE: A skin irritant. When heated to decomposition it emits acrid smoke and fumes.

EAC500 CAS:520-68-3 ***HR: 3***
ECHIMIDINE
mf: $C_{20}H_{31}NO_7$ mw: 397.52

TOXICITY DATA with REFERENCE
sln-dmg-par 20 μmol/L ZEVBA5 91,74,60
ipr-rat LDLo:200 mg/kg CBINA8 12,299,76

SAFETY PROFILE: Poison by intraperitoneal route. Mutation data reported. When heated to decomposition it emits toxic fumes of NO_x.

EAD500 CAS:512-64-1 ***HR: 3***
ECHINOMYCIN
mf: $C_{51}H_{64}N_{12}O_{12}S_2$ mw: 1101.39

SYNS: ECHINOMYCIN A ◇ LEVOMYCIN ◇ NSC 526417 ◇ QUINOMYCIN A ◇ S-426-S (LEPETIT)

TOXICITY DATA with REFERENCE
pic-esc 600 ng/plate CNREA8 43,2819,83
dnd-mam:lym 400 nmol/L NATUAS 252,653,74
dni-mam:lym 27 μmol/L HXPHAU 38(Pt 2),623,75
ipr-mus LD50:400 μg/kg 85ERAY 1,311,78
scu-mus LD50:3800 μg/kg 85ERAY 1,311,78
ivn-mus LD50:629 μg/kg NTIS** PB83-114298
ivn-dog LDLo:89 μg/kg NTIS** PB82-114298

SAFETY PROFILE: A deadly poison by intraperitoneal, intravenous, and subcutaneous routes. Mutation data reported. When heated to decomposition it emits very toxic fumes of SO_x and NO_x.

EAD600 ***HR: 3***
ECHIS CARINATUS VENOM

SYNS: E. CARINATUS VENOM ◇ VENOM, SNAKE, ECHIS CARINATUS

TOXICITY DATA with REFERENCE
scu-rat LD50:5130 μg/kg AITPAK 137,299,62
ipr-mus LD50:196 μg/kg TOXIA6 18,384,80
ivn-mus LD50:120 μg/kg TOXIA6 22,373,84
ivn-dog LDLo:100 μg/kg TOXIA6 6,51,68
ivn-rbt LDLo:100 μg/kg TOXIA6 2,5,64
ivn-mam LD50:2300 μg/kg CLPTAT 8,849,67

SAFETY PROFILE: A deadly poison by subcutaneous, intravenous, and intraperitoneal routes.

EAD650 ***HR: 3***
ECHIS COLORATA VENOM

SYNS: ECHIS COLORATUS VENOM ◇ VENOM, SNAKE, ECHIS COLORATUS

TOXICITY DATA with REFERENCE
ipr-mus LD50:1550 μg/kg AJTHAB 9,391,60
scu-mus LD50:5167 μg/kg AJTHAB 9,391,60
ivn-mus LD50:425 μg/kg TOXIA6 14,146,76

SAFETY PROFILE: A deadly poison by subcutaneous, intravenous, and intraperitoneal routes.

EAE000 CAS:28558-28-3 ***HR: 3***
ECONAZOLE NITRATE
mf: $C_{18}H_{15}Cl_3N_2O{\bullet}HNO_3$ mw: 444.72

SYNS: 1-(2-((4-CHLOROPHENYL)METHOXY)-2-(2,4-DICHLOROPHENYL)ETHYL)-1H-IMIDAZOLE NITRATE ◇ 1-(2,4-DICHLORO-β-((p-CHLOROBENZYL)OXY)-PHENETHYL)IMIDAZOLE NITRATE ◇ ECOSTATIN ◇ EPI-PEVARYL ◇ GYNO-PEVARYL ◇ IFENEC ◇ PALAVALE ◇ PEVARYL ◇ R 14,827 ◇ SPECTAZOLE

TOXICITY DATA with REFERENCE
eye-rbt 1% MLD IYKEDH 10,16,79
scu-rat TDLo:22 mg/kg (female 7-17D post):REP IYKEDH 9,928,78
scu-mus TDLo:160 mg/kg (female 7-14D post):TER IYKEDH 9,955,78
orl-rat LD50:668 mg/kg ARZNAD 25,224,75
ipr-rat LD50:240 mg/kg IYKEDH 9,643,78
scu-rat LD50:1360 mg/kg IYKEDH 9,643,78
ivn-rat LD50:50 mg/kg IYKEDH 9,643,78
orl-mus LD50:463 mg/kg ARZNAD 25,224,75
ipr-mus LD50:180 mg/kg IYKEDH 9,643,78
scu-mus LD50:750 mg/kg IYKEDH 9,643,78
ivn-mus LD50:38 mg/kg IYKEDH 9,643,78
orl-rbt LD50:431 mg/kg YAKUD5 23,1253,81
scu-rbt LD50:650 mg/kg IYKEDH 9,643,78
ivn-rbt LD50:85 mg/kg IYKEDH 9,643,78
orl-gpg LD50:272 mg/kg ARZNAD 25,224,75

SAFETY PROFILE: Poison by ingestion, intravenous, and intraperitoneal routes. Moderately toxic by subcutaneous route. Experimental teratogenic and reproductive effects. An eye irritant. An antimycotic drug. When heated to decomposition it emits very toxic fumes of Cl^- and NO_x. See also NITRATES.

EAE400 CAS:11006-90-9 ***HR: D***
EDEINE

TOXICITY DATA with REFERENCE
mmo-esc 400 μg/L AMIGB9 3,29,71
mrc-nsc 150 mg/L MGGEAE 136,309,75
mmo-eug 4 mg/L NEOLA4 19,579,72

SAFETY PROFILE: Mutation data reported.

EAE500 CAS:5036-03-3 ***HR: 3***
EDROFURADENE
mf: $C_{10}H_{12}N_4O_5$ mw: 268.26

SYNS: 1-(2-HYDROXYETHYL)-3-((5-NITROFURFURYLIDENE)AMINO)-2-IMIDAZOLIDINONE ◇ NF 1010 ◇ NIFURDAZIL

TOXICITY DATA with REFERENCE
mma-sat 100 ng/plate MUREAV 48,295,77
orl-rat TDLo:27 g/kg/46W-C:CAR JNCIAM 51,403,73

CONSENSUS REPORTS: EPA Genetic Toxicology Program.

SAFETY PROFILE: Questionable carcinogen with experimental carcinogenic data. Mutation data reported. When heated to decomposition it emits toxic fumes of NO_x.

EAE600 CAS:116-38-1 ***HR: 3***
EDROPHONIUM CHLORIDE
mf: $C_{10}H_{16}NO{\bullet}Cl$ mw: 201.72

SYNS: ANTIREX ◇ DIMETHYLETHYL(m-HYDROXYPHENYL)AMMONIUM CHLORIDE ◇ N-ETHYL-3-HYDROXY-N,N-DIMETHYLBENZENAMINIUM CHLORIDE (9CI) ◇ ETHYL (m-HYDROXYPHENYL)DIMETHYLAMMONIUM CHLORIDE ◇ TENSILON ◇ TENSILON CHLORIDE

TOXICITY DATA with REFERENCE
ipr-mus LD50:30 mg/kg PCJOAU 10,327,76
scu-mus LD50:52 mg/kg JPETAB 123,121,58
ivn-mus LD50:15 mg/kg NIIRDN 6,133,82

SAFETY PROFILE: Poison by subcutaneous, intravenous, and intraperitoneal routes. When heated to decomposition it emits toxic fumes of Cl^-, NH_3, and NO_x. See also CHLORIDES.

EAE675 CAS:55294-15-0 ***HR: 3***
EDRUL
mf: $C_{11}H_{11}Cl_2N_3O$ mw: 272.15

PROP: Crystals from methanol. Mp: 127-129°.

SYNS: 3-AMINO-1-(3,4-DICHLORO-α-METHYLBENZYL)-2-PYRAZOLIN-5-ONE ◇ 5-AMINO-2-(1-(3,4-DICHLOROPHENYL)ETHYL)-2,4-DIHYDRO-3H-PYRAZOL-3-ONE ◇ BAY G 2821 ◇ 2,4-

DIHYDRO-5-AMINO-2-(1-(3,4-DICHLOROPHENYL)ETHYL)-3H-PYRAZOL-3-ONE ◇ LUTROL ◇ MUZOLIMINE

TOXICITY DATA with REFERENCE
orl-rat LD50:1559 mg/kg CMROCX 4,716,77
ivn-rat LD50:221 mg/kg CLNHBI 19(Suppl 1),20,83
orl-mus LD50:1794 mg/kg CMROCX 4,716,77
ivn-mus LD50:310 mg/kg CLNHBI 19(Suppl 1),20,83
orl-dog LD50:2000 mg/kg CMROCX 4,716,77
orl-rbt LD50:1250 mg/kg CMROCX 4,716,77

SAFETY PROFILE: Poison by intravenous route. Moderately toxic by ingestion. When heated to decomposition it emits toxic fumes of Cl^- and NO_x.

EAE775 CAS:68278-23-9 ***HR: D***
EFLORNITHINE HYDROCHLORIDE
mf: $C_6H_{12}F_2N_2O_2 \cdot ClH$ mw: 218.66

SYNS: α-DIFLUOROMETHYLORNITHINE HYDROCHLORIDE ◇ 2-(DIFLUOROMETHYL)-dl-ORNITHINE HYDROCHLORIDE ◇ dl-ORNITHINE, 2-(DIFLUOROMETHYL)-, MONOHYDROCHLORIDE

TOXICITY DATA with REFERENCE
orl-rat TDLo:11250 mg/kg (female 10-18D post):TER TJADAB 35,55A,87
orl-rbt TDLo:900 mg/kg (female 11-19D post):REP TJADAB 35,55A,87

SAFETY PROFILE: An experimental teratogen. Other experimental reproductive effects. When heated to decomposition it emits toxic fumes of NO_x and F^-.

EAE875 CAS:1765-48-6 ***HR: 3***
EICOSAFLUOROUNDECANOIC ACID
mf: $C_{11}H_2F_{20}O_2$ mw: 546.13

SYNS: 2,2,3,3,4,4,5,5,6,6,7,7,8,8,9,9,10,10,11,11-EICOSAFLUOROUNDECANOIC ACID ◇ 11-H-EICOSAFLUORUNDEKANSAEURE (GERMAN) ◇ omega-H-EICOSAFLUORUNDEKANSAEURE (GERMAN)

TOXICITY DATA with REFERENCE
orl-rat LD50:518 mg/kg ZHYGAM 26,9,80
ipr-mus LD50:181 mg/kg ZHYGAM 26,9,80
orl-rbt LD50:470 mg/kg ZHYGAM 26,9,80
orl-gpg LD50:470 mg/kg ZHYGAM 26,9,80

SAFETY PROFILE: Poison by intraperitoneal route. Moderately toxic by ingestion. When heated to decomposition it emits toxic fumes of F^-.

EAF000 CAS:506-30-9 ***HR: 3***
EICOSANOIC ACID
mf: $C_{20}H_{40}O_2$ mw: 312.60

PROP: White leaflets. Mp: 77°, bp: 328°. Slt decomp in water; very sol in hot abs alc and ether.

SYNS: ARACHIC ACID ◇ ARACHIDIC ACID

TOXICITY DATA with REFERENCE
imp-mus TDLo:1000 mg/kg:NEO CNREA8 26,105,66

CONSENSUS REPORTS: Reported in EPA TSCA Inventory.

SAFETY PROFILE: Questionable carcinogen with experimental neoplastigenic data by implant route. When heated to decomposition it emits acrid smoke and fumes.

EAF100 CAS:21708-94-1 ***HR: D***
EIPW 111 CITRATE
mf: $C_{26}H_{28}N_2O_3 \cdot C_6H_8O_7$ mw: 608.70

SYNS: 1-(p-(β-DIETHYLAMINOETHOXY)PHENYL)-2-NITRO-1,2-DIPHENYLETHYLENE CITRATE ◇ N,N-DIETHYL-2-(4-(2-NITRO-1,2-DIPHENYLETHENYL)PHENOXY)ETHANAMINE CITRATE ◇ 2-(p-(2-NITRO-1,2-DIPHENYLVINYL)PHENOXY)TRIETHYLAMINECITRATE

TOXICITY DATA with REFERENCE
orl-mus TDLo:3 mg/kg (female 1D post):REP IJEBA6 7,117,69

SAFETY PROFILE: Experimental reproductive effects. When heated to decomposition it emits toxic fumes of NO_x.

EAF500 CAS:112-79-8 ***HR: 3***
ELAIDIC ACID
mf: $C_{18}H_{34}O_2$ mw: 282.52

PROP: (A) Leaflets from ethyl alcohol. Mp: 48-9°, bp: 235 @ 12 mm. (B) Needles from ethyl alcohol. Mp: 72°, bp: slt decomp CS_2. Sol in hot acetic acid.

SYNS: trans-Δ^9-OCTADECENOIC ACID ◇ trans-OCTADEC-9-ENOIC ACID ◇ trans-9-OCTADECENOIC ACID

TOXICITY DATA with REFERENCE
ipr-mus LDLo:512 mg/kg CBCCT* 2,188,50
ivn-mus LD50:100 mg/kg CSLNX* NX#00371

SAFETY PROFILE: Poison by intravenous route. Moderately toxic by intraperitoneal route. When heated to decomposition it emits acrid smoke and fumes.

EAG000 CAS:23315-05-1 ***HR: 3***
ELAIOMYCIN
mf: $C_{13}H_{26}N_2O_3$ mw: 258.41

PROP: Metabolite of *Streptomyces hepaticus* (NATUAS 221,765,69).

SYN: d-threo-METHOXY-3-(1-OCTENYL-ONN-AZOXY)-2-BUTANOL

TOXICITY DATA with REFERENCE
orl-rat TDLo:35 mg/kg:ETA NATUAS 221,765,69
scu-mus LD50:63 mg/kg ANTCAO 4,338,54
ivn-mus LD50:44 mg/kg ANTCAO 4,338,54

SAFETY PROFILE: Poison by intravenous and subcutaneous routes. Questionable carcinogen with experimental tumorigenic data. Causes tumors of the brain. When heated to decomposition it emits toxic fumes of NO_x.

EAG500 CAS:50814-62-5 ***HR: 3***
ELASIOMYCIN

TOXICITY DATA with REFERENCE
ivn-rat TDLo:50 mg/kg (18D preg):ETA,TER IARCCD 4,100,73

SAFETY PROFILE: Questionable carcinogen with experimental tumorigenic and teratogenic data. When heated to decomposition it emits acrid smoke and fumes.

EAG875 CAS:9004-06-2 ***HR: 3***
ELASTASE

PROP: White, lyophilized powder with a sltly yellowish shade.

SYNS: E.C. 3.4.4.7 ◇ E.C. 3.4.2.1.11 ◇ ELASZYM ◇ PANCREATOPEPTIDASE E

TOXICITY DATA with REFERENCE
ipr-rat LD50:78 mg/kg IYKEDH 7,108,76
ivn-rat LD50:85 mg/kg IYKEDH 7,108,76
ipr-mus LD50:37 mg/kg IYKEDH 7,108,76
ivn-mus LD50:57 mg/kg IYKEDH 7,108,76

CONSENSUS REPORTS: Reported in EPA TSCA Inventory.

SAFETY PROFILE: Poison by intravenous and intraperitoneal routes. When heated to decomposition it emits toxic fumes of NO_x.

EAH100 CAS:3877-86-9 ***HR: 3***
ELATERICIN A
mf: $C_{30}H_{44}O_7$ mw: 516.74

SYN: CUCURBITACINE (D)

TOXICITY DATA with REFERENCE
orl-rat LD50:8200 μg/kg AIPTAK 130,315,61
ipr-rat LD50:1300 μg/kg AIPTAK 130,315,61
scu-rat LD50:3400 μg/kg AIPTAK 130,315,61
orl-mus LD50:5 mg/kg CHTPBA 4,459,69
ipr-mus LD50:1750 μg/kg AIPTAK 130,315,61
scu-mus LD50:4600 μg/kg AIPTAK 130,315,61
ivn-mus LD50:960 μg/kg AIPTAK 130,315,61
ivn-dog LD50:1000 μg/kg AIPTAK 130,315,61
ivn-cat LD50:900 μg/kg AIPTAK 130,315,61

SAFETY PROFILE: Poison by ingestion, subcutaneous, intravenous, and intraperitoneal routes. When heated to decomposition it emits acrid smoke and fumes.

EAH500 CAS:50-48-6 ***HR: 3***
ELAVIL
mf: $C_{20}H_{23}N$ mw: 277.44

SYNS: AMITRIPTILINE ◇ AMITRIPTYLIN (GERMAN) ◇ AMITRIPTYLINE ◇ DAMILAN ◇ 3,10-DIHYDRO-5H-DIBENZO(a,d)CYCLOHEPTEN-5-YLIDENE-N,N-DIMETHYL-1-PROPANAMINE◇ 10,11-DIHYDRO-5-(Γ-DIMETHYLAMINOPROPYLIDENE)-5H-DIBENZO(a,d)CYCLOHEPTENE ◇ 10,11-DIHYDRO-N,N-DIMETHYL-5H-DIBENZO(a,d)HEPTALENE-Δ^5-Γ-PROPYLAMINE ◇ 5-(3′-DIMETHYLAMINOPROPYLIDENE)-DIBENZO-(a,d)(1,4)-CYCLOHEPTADIENE◇ 5-(Γ-DIMETHYLAMINOPROPYLIDENE)-5H-DIBENZO(a,d)-10,11-DIHYDROCYCLOHEPTENE ◇ 5-(3-DIMETHYLAMINOPROPYLIDENE)-10,11-DIHYDRO-5H-DIBENZO(a,d)CYCLOHEPTENE◇ 5-(Γ-DIMETHYLAMINOPROPYLIDENE)-10,11-DIHYDRO-5H-DIBENZO(A,D)CYCLOHEPTENE ◇ 5-(3-DIMETHYLPROPYLIDENE)DIBENZO(a,d)(1,4)CYCLOHEPTADIENE ◇ ELANIL ◇ LAROXIL ◇ LAROXYL ◇ PROHEPTADIENE ◇ TRYPTIZOL

TOXICITY DATA with REFERENCE
orl-rbt TDLo:540 mg/kg (female 8-16D post):REP PSDTAP 10,235,69
orl-rat TDLo:75 mg/kg (4-18D preg):TER JNMDBO 2,271,71
orl-chd TDLo:4500 μg/kg:CNS BMJOAE 3,663,67
orl-wmn TDLo:16800 μg/kg/2W-I:PNS,CNS LANCAO 1,426,68
orl-hmn TDLo:14 mg/kg:CVS PSDTAP 6,171,65
orl-inf TDLo:50 mg/kg:CNS,CVS AJDCAI 130,507,76
orl-wmn TDLo:13500 μg/kg/3D-C:CNS LANCAO 1,390,68
ims-wmn TDLo:240 μg/kg/36H-C:CNS LANCAO 1,390,68
unr-man TDLo:45 mg/kg/21D:CVS,CNS LANCAO 2,1202,72
orl-rat LD50:320 mg/kg ARZNAD 15,863,65
ipr-rat LD50:72 mg/kg PJPPAA 27,503,75
orl-mus LD50:140 mg/kg THERAP 26,459,71
ipr-mus LD50:56 mg/kg ARZNAD 15,863,65
scu-mus LD50:140 mg/kg ARZNAD 19,458,69
ivn-mus LD50:16 mg/kg JMCMAR 17,65,74
scu-dog LDLo:50 mg/kg PHMCAA 11,283,69
ivn-rbt LD50:8600 μg/kg ARZNAD 15,863,65

SAFETY PROFILE: Human poison by an unspecified route. Poison experimentally by ingestion, intraperitoneal, subcutaneous, intramuscular, and intravenous routes. Human systemic effects by ingestion, and intramuscular routes: changes in sleep, headache, paresthesia, convulsions, excitement, somnolence, muscle contractions, change in heart rate, and other cardiac changes. An experimental teratogen. Other experimental reproductive effects. An antidepressant. When heated to decomposition it emits toxic fumes of NO_x.

EAI000 CAS:549-18-8 ***HR: 3***
ELAVIL HYDROCHLORIDE
mf: $C_{20}H_{23}N \cdot ClH$ mw: 313.90

SYNS: AMITID ◇ AMITRIL ◇ AMITRIPTYLINE CHLORIDE ◇ AMITRYPTYLINE HYDROCHLORIDE ◇ DAMILEN HYDROCHLORIDE ◇ DEPREX ◇ 10,11-DIHYDRO-N,N-DIMETHYL-5H-DIBENZO(a,d)-CYCLOHEPTENE-$\Delta^{5,\Gamma}$-PROPYLAMINE HCL ◇ 3-(3-DIMETHYLAMINOPROPYLIDENE)-1:2-4:5-DIBENZOCYCLOHEPTA-1:4-DIENE ◇ 5-(3-DIMETHYLAMINOPROPYLIDENE)DIBENZO(a,d)(1,4)CYCLOHEPTADIENE HYDROCHLORIDE ◇ DOMICAL ◇ ELAVIL ◇ ENDEP

◇ LENTIZOL ◇ MIKETORIN ◇ PROHEPTADIEN MONOHYDROCHLORIDE ◇ SAROTEN ◇ SAROTENE ◇ SK-AMITRIPTYLINE ◇ TRYPTIZOL ◇ TRYPTIZOL HYDROCHLORIDE

TOXICITY DATA with REFERENCE
sln-dmg-orl 2 mg/2D SOGEBZ 11,718,75
scu-mus TDLo:60 mg/kg (female 9D post):TER DGDFA5 22,61,80
orl-rat TDLo:3300 mg/kg (female 33D pre):REP OYYAA2 6,899,72
orl-wmn TDLo:10 mg/kg CMAJAX 129,1203,83
orl-chd LDLo:62500 μg/kg:CNS,CVS,PUL NEJMAG 267,1031,62
orl-chd TDLo:4167 μg/kg:CNS NEJMAG 267,1031,62
orl-hmn TDLo:200 mg/kg:CNS JAMAAP 230,1405,74
orl-chd LDLo:62500 μg/kg:CNS,PUL AJDCAI 106,501,63
orl-wmn LDLo:19 mg/kg:CNS,PUL LANCAO 2,543,63
orl-man TDLo:14286 μg/kg:CNS,PUL LANCAO 2,543,63
orl-rat LD50:240 mg/kg OYYAA2 6,889,72
ipr-rat LD50:72 mg/kg 27ZQAG -,61,72
scu-rat LD50:1290 mg/kg 27ZQAG -,61,72
ivn-rat LD50:14 mg/kg 27ZQAG -,61,72
orl-mus LD50:140 mg/kg WMWOA4 111,256,61
ipr-mus LD50:65 mg/kg JDGRAX 16,7,85
scu-mus LD50:80 mg/kg OYYAA2 6,889,72
ivn-mus LD50:21 mg/kg ARZNAD 21,808,71
ivn-mky LDLo:20300 μg/kg IJEBA6 22,539,84
ivn-rbt LD50:9900 μg/kg 27ZQAG -,61,72

CONSENSUS REPORTS: Reported in EPA TSCA Inventory.

SAFETY PROFILE: Poison by ingestion, intraperitoneal, intravenous, and subcutaneous routes. Human systemic effects by ingestion: convulsions, respiratory depression, changes in sleep, hallucinations, muscle contractions, somnolence, blood pressure decrease, coma, cyanosis, dyspnea, and ataxia. An experimental teratogen. Other experimental reproductive effects. Mutation data reported. Used in the treatment of depression. When heated to decomposition it emits very toxic fumes of HCl and NO_x. See also ELAVIL.

EAI100 ***HR: 2***
ELDERBERRY

PROP: A shrub which may grow to 12 feet. The compound leaves have about 7 serrated, oval leaflets. It bears clusters of small, white flowers and purple-black berries. Both flowers and berries may be on the plant simultaneously. Various species are found in every state in the United States, most of Canada, and the Greater Antilles.

SYNS: AMERICAN ELDER ◇ FLEUR SUREAU (CANADA, HAITI) ◇ SAMBUCUS CAERULEA ◇ SAMBUCUS CANADENSIS ◇ SAMBUCUS MELANOCARPA ◇ SAMBUCUS MEXICANA ◇ SAMBUCUS RACEMOSA ◇ SAUCO (MEXICO, PUERTO RICO) ◇ SAUCO BLANCO (CUBA) ◇ SUREAU (CANADA, HAITI)

SAFETY PROFILE: The whole plant contains poisonous cyanogenetic glycosides. Ingestion of small amounts of the raw or cooked fruit is harmless. Ingestion of the leaves, roots, bark or unripe berries may cause prolonged, severe diarrhea. Juice from the pressed berries may cause nausea, vomiting, abdominal cramps, dizziness, numbness and stupor. See also CYANIDE.

EAI500 CAS:8023-89-0 ***HR: 2***
ELEMI OIL

PROP: Distilled from the crude resin of *Canarium commune* or *Canarium luzonicum* (FCTXAV 14,659,76).

SYN: ELEMI

TOXICITY DATA with REFERENCE
skn-rbt 500 mg/24H FCTXAV 14,755,76
orl-rat LD50:3370 mg/kg FCTXAV 14,755,76
skn-rbt LD50:5 g/kg FCTXAV 14,755,76

CONSENSUS REPORTS: Reported in EPA TSCA Inventory.

SAFETY PROFILE: Moderately toxic by ingestion. Mildly toxic by skin contact. A skin irritant. When heated to decomposition it emits acrid smoke and fumes.

EAI600 ***HR: 1***
ELEPHANT'S EAR

PROP: Popular ornamental houseplant which has large, heart-shaped leaves and edible roots. They are grown indoors and outdoors in southern Florida, southern California, Hawaii, Guam, and the West Indies. One species, *A. macrorrhiza*, is grown for food.

SYNS: AHE POI (HAWAII) ◇ ALOCASIA (VARIOUS SPECIES) ◇ 'APE (HAWAII) ◇ CABEZA de BURRO ◇ CARAIBE (HAITI) ◇ CHINE APE ◇ COLOCASIA ESCULENTA ◇ COLOCASIA GIGANTEA ◇ DASHEEN ◇ EDDO ◇ KALO ◇ MALANGA CARA de CHIVO (CUBA) ◇ MALANGA de JARDIN (CUBA) ◇ MALANGA DEUX PALLES (HAITI) ◇ MALANGA ISLENA (CUBA) ◇ PAPAO-APAKA (GUAM) ◇ PAPAO-ATOLONG (GUAM) ◇ TARO ◇ TAYO BAMBOU (HAITI) ◇ YAUTIA MALANGA (PUERTO RICO)

SAFETY PROFILE: The leaves and stems contain calcium oxalate raphides. Chewing of these plant parts results in immediate burning pain in the lips, mouth and throat followed by inflammation and blistering. Systemic effects are usually not seen because the immediate pain prevents ingestion and the low solubility of calcium oxalate reduces its absorption. See also OXALATES and OXALIC ACID.

EAI800 CAS:79818-59-0 *HR: 1*
ELEX 334

TOXICITY DATA with REFERENCE
orl-rat TDLo:1638 mg/kg (91D male):REP OYYAA2 11,901,76
orl-rat LD50:9150 mg/kg OYYAA2 11,901,76
orl-mus LD50:8910 mg/kg OYYAA2 11,901,76

SAFETY PROFILE: Mildly toxic by ingestion. Experimental reproductive effects.

EAI850 CAS:519-23-3 *HR: 3*
ELLIPTISINE
mf: $C_{17}H_{14}N_2$ mw: 246.33

PROP: Bright yellow needles from ethyl acetate. Mp: 311-315° (decomp).

SYNS: 5,11-DIMETHYL-6H-PYRIDO(4,3-b)CARBAZOLE ◇ ELLIPTICINE ◇ ICIG 770 ◇ NSC-71795

TOXICITY DATA with REFERENCE
mmo-sat 1 μg/plate BICMBE 60,1011,78
pic-esc 500 ng/plate CNREA8 43,2819,83
sce-ham:ovr 100 μg/L CNREA8 43,577,83
orl-mus LD50:178 mg/kg MEIEDD 10,512,83
ipr-mus LD50:150 mg/kg BIMDB3 21,101,74
ivn-mus LD50:19500 μg/kg MEIEDD 10,512,83

SAFETY PROFILE: Poison by ingestion, intravenous, and intraperitoneal routes. Mutation data reported. When heated to decomposition it emits toxic fumes of NO_x.

EAI875 CAS:3818-88-0 *HR: 3*
ELORINE CHLORIDE
mf: $C_{20}H_{32}NO \cdot Cl$ mw: 337.98

SYNS: COMPOUND 14045 METHOCHLORIDE ◇ (±)-N-((3-CYCLOHEXYL-3-HYDROXY-3-PHENYL)PROPYL)-N-METHYLPYRROLIDINIUM CHLORIDE ◇ 1-(3-CYCLOHEXYL-3-HYDROXY-3-PHENYLPROPYL)-1-METHYL-PYRROLIDINIUM CHLORIDE ◇ 1-CYCLOHEXYL-1-PHENYL-3-PYRROLIDINO-1-PROPANOLMETHYL CHLORIDE ◇ LERGINE CHLORIDE ◇ TRICOLOID CHLORIDE ◇ TRICYCLAMOL CHLORIDE ◇ TRICYCLAMOL METHOCHLORIDE

TOXICITY DATA with REFERENCE
orl-rat LD50:984 mg/kg 29ZVAB -,119,69
orl-mus LD50:395 mg/kg JAPMA8 43,408,54
ivn-mus LD50:11700 μg/kg JAPMA8 43,408,54

SAFETY PROFILE: Poison by ingestion and intravenous routes. When heated to decomposition it emits toxic fumes of Cl^- and NO_x.

EAJ000 CAS:548-43-6 *HR: 3*
ELYMOCLAVINE
mf: $C_{16}H_{18}N_2O$ mw: 254.36

PROP: A close chemical relative of LSD found in ergot fungi and bindweeds of the genus *Ipomoea, fm. convolvulaceae* (JANSAG 41,1700,75).

SYNS: ELIMOCLAVIN ◇ ELYMOCLAVIN

TOXICITY DATA with REFERENCE
ipr-mus TDLo:60 mg/kg (10D preg):REP JANSAG 41,1700,75
ipr-mus TDLo:30 mg/kg (10D preg):TER JANSAG 41,1700,75
ipr-mus LD50:67 mg/kg JANSAG 41,1700,75
ivn-mus LD50:45 mg/kg NYKZAU 58,386,62

SAFETY PROFILE: A poison by intraperitoneal and intravenous routes. Experimental teratogenic and reproductive effects. When heated to decomposition it emits acrid smoke and fumes. See also N,N-DIETHYLLYSERGAMIDE (LSD).

EAJ100 CAS:79458-80-3 *HR: 2*
EM 255
mf: $C_{13}H_{14}N_2O_2$ mw: 230.29

SYN: 3-(1,3-DIHYDRO-1-OXO-2H-ISOINDOL-2-YL)-2-OXOPIPERIDINE

TOXICITY DATA with REFERENCE
orl-rbt TDLo:400 mg/kg (8-15D preg):TER ARZNAD 31,941,81
orl-mus LD50:1200 mg/kg ARZNAD 31,941,81
ipr-mus LD50:494 mg/kg ARZNAD 31,941,81

SAFETY PROFILE: Moderately toxic by ingestion and other routes. An experimental teratogen. When heated to decomposition it emits toxic fumes of NO_x.

EAJ500 CAS:19526-81-9 *HR: 3*
EMAZOL RED B
$C_{18}H_{16}N_2O_{10}S_3 \cdot 2Na$ mw: 562.52

TOXICITY DATA with REFERENCE
orl-rat TDLo:78 g/kg/86W-I:ETA TKORAS 3,53,67

CONSENSUS REPORTS: Reported in EPA TSCA Inventory.

SAFETY PROFILE: Questionable carcinogen with experimental tumorigenic data. When heated to decomposition it emits very toxic fumes of NO_x, Na_2O, and SO_x.

EAJ600 CAS:550-24-3 *HR: 3*
EMBELIC ACID
mf: $C_{17}H_{26}O_4$ mw: 294.43

PROP: Glistening orange plates from alc, benzene, or acetic acid. Mp: 142-143°. Sol in the usual hot organic solvents or in alkali hydroxide solns; very sltly sol in petr ether; practically insol in water.

SYNS: 2,5-DIHYDROXY-3-UNDECYL-1,4-BENZOQUINONE ◇ 2,5-DI-

HYDROXY-3-UNDECYL-2,5-CYCLOHEXADIENE-1,4-DIONE(9CI) ◇ EMBELIN ◇ EMBERLINE

TOXICITY DATA with REFERENCE
orl-rbt TDLo:50 mg/kg (female 5D pre):REP JRIHDC 11(4),84,76
ipr-uns LD50:44 mg/kg IJPPAZ 21,31,77

SAFETY PROFILE: Poison by intraperitoneal route. Experimental reproductive effects. When heated to decomposition it emits acrid smoke and fumes.

EAK000 CAS:10429-82-0 ***HR: 3***
EMBITOL
mf: $C_{11}H_{15}Cl_2N{\cdot}ClH$ mw: 268.63

SYNS: BENZYL-BIS(2-CHLOROETHYL)AMINEHYDROCHLORIDE ◇ N,N-BIS(2-CHLOROETHYL)BENZENEMETHANAMINE HYDROCHLORIDE ◇ N,N-BIS(2-CHLOROETHYL)BENZYLAMINE HYDROCHLORIDE

TOXICITY DATA with REFERENCE
mmo-asn 2500 μmol/L SOGEBZ 6,220,70
ipr-mus LD50:28 mg/kg CANCAR 2,1055,49

SAFETY PROFILE: Poison by intraperitoneal route. Mutation data reported. When heated to decomposition it emits very toxic fumes of Cl^- and NO_x.

EAK500 CAS:10433-59-7 ***HR: 3***
EMBUTOX
mf: $C_{10}H_9Cl_2O_3{\cdot}Na$ mw: 271.08

SYNS: BUTOXONE SB ◇ BUTYRAC ◇ 2,4-DB SODIUM SALT ◇ Γ-(2,4-DICHLOROPHENOXY)BUTYRIC ACID, SODIUM SALT ◇ 2,4-DICHLOROPHENOXYBUTYRIC ACID, SODIUM SALT ◇ MB 2878

TOXICITY DATA with REFERENCE
orl-rat LD50:700 mg/kg GUCHAZ 6,153,73
orl-mus LD50:400 mg/kg PCOC** -,367,66

CONSENSUS REPORTS: EPA Genetic Toxicology Program.

SAFETY PROFILE: Poison by ingestion. An herbicide. When heated to decomposition it emits toxic fumes of Cl^- and Na_2O.

EAL000 CAS:8034-17-1 ***HR: 3***
EMCOL 888

PROP: Polyalkyl naphthalene pyridinium chloride, average molecular weight 382 (JAPMA8 38,428,49).

TOXICITY DATA with REFERENCE
eye-rbt 1% SEV JAPMA8 38,428,49
orl-mus LD50:470 mg/kg JAPMA8 38,428,49
ivn-mus LD50:8 mg/kg JAPMA8 38,428,49

SAFETY PROFILE: Poison by intravenous route. Moderately toxic by ingestion. A severe eye irritant. When heated to decomposition it emits very toxic fumes of Cl^- and NO_x.

EAL100 CAS:1302-74-5 ***HR: 3***
EMERY
mf: Al_2O_3 mw: 101.96

PROP: A varicolored mineral. D: 3.95-4.10.

SYNS: ALUMINUM OXIDE ◇ CORUNDUM ◇ ELECTROCORUNDUM ◇ EN 237 ◇ KER 710 ◇ KO 7 ◇ KORUND ◇ KU 5-3 ◇ MP 1 (refractory)

TOXICITY DATA with REFERENCE
ipr-rat TDLo:225 mg/kg/1W-I:CAR ZHPMAT 162,467,76

OSHA PEL: (Transitional: TWA Total Dust: 15 mg/m^3; Respirable Fraction: 5 mg/m^3) TWA Total Dust: 10 mg/m^3; Respirable Fraction: 5 mg/m^3
ACGIH TLV: TWA (nuisance particulate) 10 mg/m^3 of total dust (when toxic impurities are not present, e.g., quartz $< 1\%$).

SAFETY PROFILE: Questionable carcinogen with experimental carcinogenic data. May cause a pneumoconiosis. It is mainly a nuisance dust.

EAL500 CAS:483-18-1 ***HR: 3***
EMETINE
mf: $C_{29}H_{40}N_2O_4$ mw: 480.71

PROP: White powder or lumps, bitter taste, darkens on exposure. Mp: 74°.

SYNS: CEPHAELINE METHYL ETHER ◇ (−)-EMETINE ◇ NSC-33669 ◇ 6′,7′,10,11-TETRAMETHOXYEMETAN

TOXICITY DATA with REFERENCE
skn-rbt 5 mg/24H rns TXCYAC 14,117,79
skn-rbt 1%/24H MOD NTIS** PB-274-082
eye-rbt 2000 ppm SEV AJOPAA 31,837,48
dnd-mam:lym 6600 μmol/L FOMIAZ 15,76,70
scu-man TDLo:10 mg/kg/10D:GIT,CNS,CVS CCROBU 57,423,73
unr-man LDLo:2941 μg/kg 85DCAI 2,73,70
orl-rat LD50:68 mg/kg ARZNAD 13,474,63
ipr-rat LD50:12 mg/kg JPETAB 104,421,52
scu-rat LD50:95 mg/kg ARZNAD 13,474,63
ipr-mus LD50:12 mg/kg JPETAB 104,421,52
scu-mus LD10:25 mg/kg EJCAAH 10,667,74
scu-cat LDLo:8 mg/kg HBAMAK 4,1289,35
ivn-cat LDLo:10 mg/kg HBAMAK 4,1289,35
scu-rbt LDLo:30 mg/kg HBAMAK 4,1289,35
ivn-rbt LDLo:2 mg/kg HBAMAK 4,1289,35
scu-gpg LDLo:70 mg/kg HBAMAK 4,1289,35
ivn-gpg LDLo:7 mg/kg HBAMAK 4,1289,35

CONSENSUS REPORTS: NCI Carcinogenesis Studies (ipr); Inadequate Studies: mouse, rat NCITR* NCI-CG-TR-43,77

SAFETY PROFILE: A human poison by an unspecified route. A experimental poison by ingestion, intraperitoneal, intravenous, and subcutaneous routes. Human systemic effects by subcutaneous route: muscle weakness, cardiac arrythmias and gastrointestinal effects. Mutation data reported. A severe eye and moderate skin irritant. It is one of the two potent alkaloids obtained from the Brazilian plant ipecac. The therapeutic use of various ipecac preparations has caused many cases of poisoning, in some instances with fatal results. The toxic effects are particularly prominent if the drug is given intravenously. The symptoms of intoxication are gastrointestinal irritation and salivation, as well as general edema which follows renal insufficiency, hemoptysis (blood stained sputum), flaccid paralysis, peripheral neuritis (inflammation of the nerve endings), aphonia (loss of voice), difficulties in swallowing, delirium, coma and failure of the heart. The fatal dose is considered to be approximately 2 grams, whether administered over a short or relatively long period. The drug seems to have a cumulative effect. A severe eye irritant. When heated to decomposition it emits highly toxic fumes of NO_x.

EAM000 ***HR: 3***
EMETINE ANTIMONY IODIDE

PROP: Percentage composition = 34% emetine and 14% antimony (AJTMAQ 10,249,30).

SYN: ANTIMONY EMETINE IODIDE

TOXICITY DATA with REFERENCE
orl-hmn TDLo:471 μg/kg:GIT AJTMAQ 10,249,30
orl-cat LDLo:15 mg/kg AJTMAQ 10,249,30
orl-rbt LDLo:10 mg/kg AJTMAQ 10,249,30

CONSENSUS REPORTS: Antimony and its compounds are on the Community Right-To-Know List.

OSHA PEL: TWA 0.5 mg(Sb)/m^3
ACGIH TLV: TWA 0.5 mg(Sb)/m^3
NIOSH REL: (Antimony) TWA 0.5 mg/m^3

SAFETY PROFILE: Poison by ingestion. Human systemic effects by ingestion: nausea, vomiting, and other gastrointestinal effects. When heated to decomposition it emits very toxic fumes of I^-, NO_x, and Sb. See also ANTIMONY COMPOUNDS, EMETINE and IODIDES.

EAM500 CAS:8001-15-8 ***HR: 3***
EMETINE BISMUTH IODIDE
mf: $C_{29}H_{40}N_2O_4 \cdot BiI_3$ mw: 1070.39

PROP: Composition = 25% emetine and 17% bismuth (AJTMAQ 10,249,30).

SYNS: BISMUTH EMETINE IODIDE ◇ EMETINE with BISMUTH(III) TRIIODIDE ◇ EMETINE TRIIODOBISMUTH(III) ◇ NSC 44185 ◇ TRIIODO(6′,7′,10,11-TETRAMETHOXYEMETAN)BISMUTH

TOXICITY DATA with REFERENCE
orl-mus LD50:74 mg/kg JPPMAB 16,65,64
ipr-mus LD50:126 mg/kg NCISP* JAN86
orl-cat LDLo:20 mg/kg AJTMAQ 10,249,30
orl-rbt LDLo:40 mg/kg AJTMAQ 10,249,30

SAFETY PROFILE: A poison by ingestion and intraperitoneal routes. When heated to decomposition it emits very toxic fumes of I^- and NO_x. See also BISMUTH COMPOUNDS, EMETINE, and IODIDES.

EAN000 CAS:316-42-7 ***HR: 3***
1-EMETINE DIHYDROCHLORIDE
mf: $C_{29}H_{40}N_2O_4 \cdot 2ClH$ mw: 553.63

SYNS: AMEBICIDE ◇ EMETINE, DIHYDROCHLORIDE ◇ (−)-EMETINE DIHYDROCHLORIDE ◇ EMETINE HYDROCHLORIDE ◇ NSC-33669

TOXICITY DATA with REFERENCE
eye-hmn 3250 μg BJIMAG 4,111,47
orl-rat LD50:12 μg/kg ANTCAO 8,297,58
ipr-rat LD50:17 mg/kg JPETAB 94,431,48
orl-mus LD50:15 μg/kg ANTCAO 8,297,58
ipr-mus LD50:62 mg/kg JPETAB 94,431,48
scu-mus LD50:37 mg/kg JPPMAB 16,65,64
scu-mky LDLo:13886 μg/kg IMGAAY 49,85,14
orl-cat LDLo:15 mg/kg AJTMAQ 10,249,30
orl-rbt LDLo:15 mg/kg AJTMAQ 10,249,30
scu-rbt LDLo:13886 μg/kg IMGAAY 49,85,14
ivn-rbt LDLo:2500 μg/kg JAMAAP 62,501,14
orl-bwd LD50:56200 μg/kg AECTCV 12,355,83

CONSENSUS REPORTS: EPA Extremely Hazardous Substances List. Reported in EPA TSCA Inventory.

SAFETY PROFILE: A poison by ingestion, intraperitoneal, subcutaneous, and intravenous routes. A human eye irritant. When heated to decomposition it emits very toxic fumes of Cl^- and NO_x.

EAN500 CAS:73713-75-4 ***HR: 3***
EMETINE DIHYDROCHLORIDE TETRAHYDRATE
mf: $C_{29}H_{40}N_2O_4 \cdot 2ClH \cdot 4H_2O$ mw: 625.71

SYNS: NCI-C01605 ◇ 6′,7′,10,11-TETRAMETHOXYEMETAN DIHYDROCHLORIDE TETRAHYDRATE

CONSENSUS REPORTS: NCI Carcinogenesis Bioassay (ipr); Inadequate Studies: mouse, rat NCITR* NCI-CG-TR-43,77.

SAFETY PROFILE: Probably a poison. When heated to decomposition it emits very toxic fumes of NO_x and HCl. See also EMETINE DIHYDROCHLORIDE.

EAN600 CAS:82-92-8 ***HR: 3***
EMOQUIL
mf: $C_{18}H_{22}N_2$ mw: 266.42

PROP: Light-sensitive crystals from petr ether. Mp: 105.5-107.5°. Solubility in g/mL at 25°: chloroform 1.1; ether 0.17; ethanol 0.17; water <0.001.

SYNS: (N-BENZHYDRYL)(N'-METHYL)DIETHYLENEDIAMINE ◇ N-BENZHYDRYL-N-METHYL PIPERAZINE ◇ BW 47-83 ◇ CICLIZINA ◇ COMPOUND 47-83 ◇ CYCLIZINE ◇ 1-DIPHENYLMETHYL-4-METHYLPIPERAZINE ◇ MARAZINE ◇ MAREZINE ◇ MARZINE ◇ N-METHYL-N'-BENZYHYDRYLPIPERAZINE ◇ NAUTAZINE ◇ NEO-DEVOMIT ◇ VALOID ◇ WELLCOME PREPARATION 47-83

TOXICITY DATA with REFERENCE
orl-rat TDLo:450 mg/kg (10-15D preg):REP BCFAAI 109,323,70
orl-rat TDLo:450 mg/kg (10-15D preg):TER BCFAAI 109,323,70
orl-wmn TDLo:126 mg/kg/6W-I:BLD BMJOAE 292,174,86
orl-mus LD50:147 mg/kg 27ZQAG -,218,72

SAFETY PROFILE: Poison by ingestion. Human systemic effects: unspecified blood effects. An experimental teratogen. Other experimental reproductive effects. When heated to decomposition it emits toxic fumes of NO_x.

EAN700 CAS:38957-41-4 ***HR: 3***
EMORFAZONE
mf: $C_{11}H_{17}N_3O_3$ mw: 239.31

PROP: Crystals from methanol/isopropyl ether. Mp: 89-91°.

SYNS: 4-ETHOXY-2-METHYL-5-MORPHOLINO-3(2H)-PYRIDAZINONE ◇ 4-ETHOXY-2-METHYL-5-(4-MORPHOLINYL)-3(2H)-PYRIDAZINONE ◇ M73101 ◇ NANDRON ◇ PENTOYL

TOXICITY DATA with REFERENCE
dlt-mus-orl 500 mg/kg IYKEDH 11,559,80
orl-rat TDLo:700 mg/kg (female 9-15D post):REP JTSCDR 3,69,78
orl-rat TDLo:4200 mg/kg (9-15D preg):TER JTSCDR 3,69,78
orl-rat LD50:1900 mg/kg IYKEDH 14,838,83
ipr-rat LD50:600 mg/kg IYKEDH 14,838,83
scu-rat TDLo:720 mg/kg OYYAA2 16,1011,78
ivn-rat LD50:340 mg/kg IYKEDH 14,838,83
orl-mus LD50:960 mg/kg IYKEDH 14,838,83
ipr-mus LD50:620 mg/kg OYYAA2 16,1011,78
scu-mus LD50:750 mg/kg IYKEDH 14,838,83

SAFETY PROFILE: Poison by intravenous route. Moderately toxic by ingestion, subcutaneous, and intraperitoneal routes. Experimental teratogenic and reproductive effects. Mutation data reported. When heated to decomposition it emits toxic fumes of NO_x.

EAN800 ***HR: 2***
EMULSOV O EXTRA P

SYN: RICINOLEIC ACID TRIESTER with GLYCEROL 12-ETHER with TRIDECAETHYLENE GLYCOL

TOXICITY DATA with REFERENCE
orl-rat LD50:37 g/kg APFRAD 18,53,60
ipr-rat LD50:7500 mg/kg APFRAD 18,53,60
ivn-mus LD50:3600 mg/kg APFRAD 18,53,60

CONSENSUS REPORTS: Glycol ether compounds are on the Community Right-To-Know List.

SAFETY PROFILE: Moderately toxic by intravenous route. Mildly toxic by ingestion and intraperitoneal routes. See also ESTERS and GLYCOL ETHERS.

EAO100 CAS:76095-16-4 ***HR: 2***
ENALAPRIL MALEATE
mf: $C_{20}H_{28}N_2O_5 \cdot C_4H_4O_4$ mw: 492.58

SYNS: N-((S)-1-ETHOXYCARBONYL-3-PHENYLPROPYL)-l-ALANYL-l-PROLINE MALEATE ◇ (S)-1-(N-(1-(ETHOXYCARBONYL)-3-PHENYL-PROPYL)-l-ALANYL)-l-PROLINE MALEATE ◇ MK 421 ◇ MK 421 MALEATE

TOXICITY DATA with REFERENCE
orl-wmn TDLo:6800 μg/kg (female 32-34W post):TER AIMEAS 108,215,88
orl-rat TDLo:280 mg/kg (female 15-22D post):REP YACHDS 14,43,86
orl-wmn TDLo:400 μg/kg/2D-I:SKN BMJOAE 294,91,87
orl-man TDLo:143 μg/kg:SKN BMJOAE 294,91,87
orl-wmn TDLo:5600 μg/kg/4W-I:PUL LANCAO 2,1094,86
orl-man TDLo:71 μg/kg:CVS BMJOAE 291,1309,85
orl-wmn TDLo:5600 μg/kg/4W-I:SKN LANCAO 2,1395,86
orl-rat LD50:2973 mg/kg YACHDS 13,413,85
scu-rat LD50:1418 mg/kg YACHDS 13,413,85
ivn-rat LD50:849 mg/kg YACHDS 13,413,85
orl-mus LD50:3507 mg/kg YACHDS 13,413,85
scu-mus LD50:1160 mg/kg YACHDS 13,413,85
ivn-mus LD50:859 mg/kg YACHDS 13,413,85

SAFETY PROFILE: Moderately toxic by ingestion, subcutaneous, and intravenous routes. Human systemic effects by ingestion: blood pressure depression, dermatitis, cough. An experimental teratogen. Experimental reproductive effects. When heated to decomposition it emits toxic fumes of NO_x.

EAO500 CAS:20311-78-8 ***HR: 3***
ENANTHOTOXIN
mf: $C_{17}H_{22}O_2$ mw: 258.39

SYNS: 2,8,10-HEPTADECATRIENE-4,6-DIYNE-1,14-DIOL ◇ OENANTHOTOXIN

TOXICITY DATA with REFERENCE
ipr-rat LD50:2940 μg/kg AFTOD7 7,197,81
ipr-mus LD50:2940 μg/kg AFTOD7 7,197,81

SAFETY PROFILE: A deadly poison by intraperitoneal route. When heated to decomposition it emits acrid smoke and fumes. See also ACETYLENE COMPOUNDS.

EAP000 CAS:8015-30-3 ***HR: 3***
ENAVID
mf: $C_{21}H_{26}O_2 \cdot C_{20}H_{26}O_2$ mw: 608.93

PROP: Mixture of 98.5% (17-α)-19-norpregn-4-en-20-yn-3-one,17-hydroxy- and 1.5% (17-α)-19-norpregna-1,3,5(10)-trien-20-yn-17-ol,3-methoxy- (IARC** 6, 193, 74).

SYNS: CONOVID ◇ CONOVID E ◇ ENIDREL ◇ ENOVID ◇ ENOVID-E ◇ ETHINYLESTRADIOL-3-METHYL ETHER and NORETHYNODRED (1:50) ◇ MESTRANOL mixed with NORETHYNODREL ◇ NORETHANDROL ◇ NORETHYNODREL and ETHINYLESTRADIOL-3-METHYL ETHER (50:1) ◇ NORETHYNODREL mixed with MESTRANOL

TOXICITY DATA with REFERENCE
oth-mus-par 3060 μg/kg/17D-C AJOGAH 120,390,74
orl-wmn TDLo:1040 μg/kg (female 20D pre):REP JRPFA4 4,229,62
orl-wmn TDLo:46200 μg/kg (female 6-38W post):TER JCEMAZ 19,1369,59
orl-wmn TDLo:20 mg/kg/2Y-I:NEO LANCAO 2,926,73
orl-mus TDLo:207 mg/kg/74W-C:CAR JNCIAM 43,671,69
orl-mus TD:69 mg/kg/69W-I:ETA PEXTAR 11,440,69

CONSENSUS REPORTS: IARC Cancer Review: Animal Sufficient Evidence IMEMDT 6,191,74. EPA Genetic Toxicology Program.

SAFETY PROFILE: Confirmed carcinogen producing liver tumors in women by ingestion. Experimental carcinogenic, neoplastigenic, and tumorigenic data. Human reproductive effects by ingestion: menstrual cycle changes or disorders; abnormalities of the uterus, cervix, and vagina; and changes in fertility. A human teratogen which causes developmental abnormalities of the urogenital system. Experimental reproductive effects. Mutation data reported. A steroid. When heated to decomposition it emits acrid smoke and fumes.

EAQ000 ***HR: 3***
ENCEPHALARTOS HILDEBRANDTII

PROP: Flour made from the starchy kernels of *Ecephalartos hildebrandtii* (BJCAAI 22,563,68).

TOXICITY DATA with REFERENCE
orl-rat TDLo:450 g/kg/26W-C:CAR BJCAAI 22,563,68

SAFETY PROFILE: Questionable carcinogen with experimental carcinogenic data. When heated to decomposition it emits acrid smoke and fumes.

EAQ050 CAS:1182-87-2 ***HR: 3***
ENCORDIN
mf: $C_{30}H_{44}O_9$ mw: 548.74

PROP: Cardenolide (cardiac glycoside) found in seeds of the tropical garden plant *Thevetia peruviana (Pers.)* Spears from methanol + ether. Mp: 161-164°. One gram dissolves in about 2500 mL water. Freely sol in chloroform and acetone; sparingly sol in methanol and ethanol.

SYNS: CANNOGENIN-α-l-THEVETOSIDE ◇ 3-((6-DEOXY-3-O-METHYL-α-l-GLUCOPYRANOSYL)OXY)-14-HYDROXY-19-OXO-CARD-20(22)-ENOLIDE ◇ PERUVOSID ◇ PERUVOSIDE

TOXICITY DATA with REFERENCE
orl-hmn TDLo:24 μg/kg:CNS,GIT IJEBA6 5,31,67
orl-hmn TDLo:39 μg/kg/3D:CVS,GIT IJMRAQ 59,271,71
ivn-hmn TDLo:4285 ng/kg:CVS,GIT IJMRAQ 59,271,71
ivn-cat LD50:145 μg/kg AIPTAK 126,412,60
ivn-gpg LDLo:710 μg/kg AIPTAK 126,412,60
ivn-pgn LDLo:263 mg/kg AIPTAK 126,412,60

SAFETY PROFILE: Poison by intravenous route. Human systemic effects by ingestion and intravenous routes: anorexia, decreased pulse rate with fall in blood pressure, and nausea or vomiting. When heated to decomposition it emits acrid smoke and fumes.

EAQ100 ***HR: 2***
ENDOD, EXTRACT

SYN: PHYTOLACCA DODECANDRA, extract

TOXICITY DATA with REFERENCE
iut-rat TDLo:500 μg/kg (female 1D post):REP CCPTAY 14,39,76
orl-rat LD50:2200 mg/kg BWHOA6 42,597,70
orl-mus LD50:2600 mg/kg BWHOA6 42,597,70
orl-mky LDLo:3 g/kg BWHOA6 42,597,70
orl-ckn LDLo:2500 mg/kg BWHOA6 42,597,70

SAFETY PROFILE: Moderately toxic by ingestion. Experimental reproductive effects.

EAQ500 CAS:1391-41-9 ***HR: 3***
ENDOMYCIN

SYN: U-4568

TOXICITY DATA with REFERENCE
ivn-rat LD50:17 mg/kg UPJOH* 2(6),-,71
orl-mus LDLo:859 mg/kg UPJOH* 2(6),-,71
ipr-mus LD50:48 mg/kg UPJOH* 2(6),-,71
scu-mus LDLo:929 mg/kg UPJOH* 2(6),-,71
ivn-mus LDLo:26 mg/kg UPJOH* 2(6),-,71

par-mus LDLo:1000 mg/kg PHYTAJ 41,393,51
scu-gpg LD50:2140 mg/kg UPJOH* 2(6),-,71

SAFETY PROFILE: Poison by intravenous and intraperitoneal routes. Moderately toxic by ingestion, subcutaneous, and parenteral routes.

EAQ750 CAS:115-29-7 ***HR: 3***
ENDOSULFAN
DOT: NA 2761
mf: $C_9H_6Cl_6O_3S$ mw: 406.91

PROP: A mixture of 2 isomers, brown crystals, nearly insol in water; sol in most organic solvents. Mp (α): 106°, mp (β): 212°, d: 1.745 @ 20°/20°.

SYNS: BENZOEPIN ◇ BEOSIT ◇ BIO 5,462 ◇ CHLORTHIEPIN ◇ CRISULFAN ◇ CYCLODAN ◇ DEVISULPHAN ◇ ENDOCEL ◇ ENDOSOL ◇ ENDOSULPHAN ◇ ENSURE ◇ ENT 23,979 ◇ FMC 5462 ◇ 1,2,3,4,7,7-HEXACHLOROBICYCLO(2.2.1)HEPTEN-5,6-BIOXYMETHYLENESULFITE ◇ α,β-1,2,3,4,7,7-HEXACHLOROBICYCLO(2.2.1)-2-HEPTENE-5,6-BISOXYMETHYLENE SULFITE ◇ HEXACHLOROHEXAHYDROMETHANO 2,4,3-BENZODIOXATHIEPIN-3-OXIDE ◇ 6,7,8,9,10,10-HEXACHLORO-1,5,5a,6,9,9a-HEXAHYDRO-6,9-METHANO-2,4,3-BENZODIOXATHIEPIN-3-OXIDE ◇ 1,4,5,6,7,7-HEXACHLORO-5-NORBORNENE-2,3-DIMETHANOL CYCLIC SULFITE ◇ HILDAN ◇ HOE 2,671 ◇ INSECTOPHENE ◇ KOP-THIODAN ◇ MALIX ◇ NCI-C00566 ◇ NIA 5462 ◇ NIAGARA 5,462 ◇ OMS 570 ◇ RCRA WASTE NUMBER P050 ◇ SULFUROUS ACID, cyclic ester with 1,4,5,6,7,7-HEXACHLORO-5-NORBORNENE-2,3-DIMETHANOL ◇ THIFOR ◇ THIMUL ◇ THIODAN ◇ THIOFOR ◇ THIOMUL ◇ THIONEX ◇ THIOSULFAN ◇ THIOSULFAN TIONEL ◇ TIOVEL

TOXICITY DATA with REFERENCE
sln-dmg-orl 200 ppm/48H MUREAV 136,115,84
mmo-smc 100 mg/L TGANAK 18,455,84
sce-hmn:lym 1 μmol/L ARTODN 52,221,83
orl-rat TDLo:45 mg/kg (6-14D preg):REP APTOA6 42,150,78
orl-rat TDLo:45 mg/kg (6-14D preg):TER APTOA6 42,150,78
orl-mus TDLo:330 mg/kg/78W-I:NEO NTIS** PB223-159
scu-mus TDLo:2 mg/kg:ETA NTIS** PB223-159
orl-rat LD50:18 mg/kg ARSIM* 20,9,66
ihl-rat LC50:80 mg/m^3/4H JOCMA7 9,35,67
skn-rat LD50:34 mg/kg 85GMAT -,72,82
ipr-rat LD50:8 mg/kg RREVAH 22,1,68
orl-mus LD50:7360 μg/kg 30ZDA9 -,267,71
ipr-mus LD50:7 mg/kg BECTA6 15,708,76
orl-dog LD50:76700 μg/kg PEMNDP 8,335,87
orl-cat LD50:2 mg/kg 85DPAN -,-,71/76
orl-rbt LD50:28 mg/kg VETNAL 59(12),64,83
skn-rbt LD50:90 mg/kg JOCMA7 9,35,67
scu-rbt LD50:360 mg/kg RREVAH 22,1,68
orl-ham LD50:118 mg/kg EJTXAZ 7,159,74
ipr-ham LD50:80 mg/kg ARTODN 58,152,85

CONSENSUS REPORTS: EPA Extremely Hazardous Substances List. NCI Carcinogenesis Bioassay (feed); No Evidence: mouse, rat NCITR* NCI-CG-TR-62,77.

OSHA PEL: TWA 0.1 mg/m^3 (skin)
ACGIH TLV: TWA 0.1 mg/m^3 (skin)
DOT Classification: Poison B; Label: Poison.

SAFETY PROFILE: Poison by ingestion, inhalation, skin contact, intraperitoneal, and subcutaneous routes. Experimental teratogenic and reproductive effects. Questionable carcinogen with experimental tumorigenic and neoplastigenic data. Human mutation data reported. A central nervous system stimulant producing convulsions. A highly toxic organochlorine pesticide which does not accumulate significantly in human tissue. Absorption is normally slow, but is increased by alcohols, oil, and emulsifiers. When heated to decomposition it emits toxic fumes of Cl^- and SO_x. See also CHLORIDES and SULFITES.

EAR000 CAS:145-73-3 ***HR: 3***
ENDOTHAL
mf: $C_8H_{10}O_5$ mw: 186.18

PROP: Solid. Mp: 144°. Sol in water.

SYNS: AQUATHOL ◇ ENDOTHALL ◇ ENDOTHAL TECHNICAL ◇ 3,6-ENDOOXOHEXAHYDROPHTHALIC ACID ◇ 3,6-ENDOXOHEXAHYDROPHTHALIC ACID ◇ 3,6-endo-EPOXY-1,2-CYCLOHEXANEDICARBOXYLIC ACID ◇ HEXAHYDRO-3,6-endo-OXYPHTHALIC ACID ◇ HYDOUT ◇ HYDROTHAL-47 ◇ 7-OXABICYCLO(2.2.1)HEPTANE-2,3-DICARBOXYLIC ACID ◇ RCRA WASTE NUMBER P088 ◇ TRI-ENDOTHAL

TOXICITY DATA with REFERENCE
orl-rat LD50:38 mg/kg PCOC** -,471,66

SAFETY PROFILE: Poison by ingestion. Very irritating to skin, eyes, and mucus membranes. Causes diarrhea. When heated to decomposition it emits acrid smoke and fumes.

EAR500 ***HR: 3***
ENDOTHAL COMBINED with IPC (1:1)

SYN: IPC COMBINED with ENDOTHAL (1:1)

TOXICITY DATA with REFERENCE
orl-mus LDLo:50 mg/kg HYSAAV 31,225,66
orl-rbt LDLo:200 mg/kg HYSAAV 31,225,66

SAFETY PROFILE: Poison by ingestion. See also ENDOTHAL.

EAS000 CAS:2778-04-3 ***HR: 3***
ENDOTHION
mf: $C_9H_{13}O_6PS$ mw: 280.25

SYNS: AC-18,737 ◇ O,O-DIMETHYL-S-(5-METHOXY-4-OXO-4H-PYRAN-2-YL)PHOSPHOROTHIOATE ◇ O,O-DIMETHYL-S-(5-METHOXY-PYRON-2-YL)-METHYL)-THIOLPHOSPHAT (GERMAN) ◇ O,O-DIMETHYL-S-(5-METHOXYPYRONYL-2-METHYL)THIOPHOSPHATE ◇ ENDOCID ◇ ENDOCIDE ◇ ENT 24,653 ◇ EXOTHION ◇ 5-METHOXY-2-(DIMETHOXYPHOSPHINYLTHIOMETHYL)PYRONE-4 ◇ S-5-METHOXY-4-OXOPYRAN-2-YLMETHYL DIMETHYL PHOSPHOROTHIO-

ATE ◇ S-((5-METHOXY-4H-PYRON-2-YL)-METHYL)-O,O-DIMETHYL-MONOTHIOFOSFAAT (DUTCH) ◇ S-((5-METHOXY-4H-PYRON-2-YL) METHYL)-O,O-DIMETHYL-MONOTHIOPHOSPHAT(GERMAN)◇ S-(5-METHOXY-4-PYRON-2-YLMETHYL) DIMETHYL PHOSPHOROTHIOLATE ◇ S-((5-METOSSI-4H-PIRON-2-IL)-METIL)-O,O-DIMETIL-MONOTIOFOSFATO (ITALIAN) ◇ NIA-5767 ◇ NIAGARA 5767 ◇ PHOSPHATE 100 ◇ PHOSPHOPYRON ◇ PHOSPHOPYRONE ◇ THIOPHOSPHATE de O,O-DIMETHYLE et de S-((5-METHOXY-4- PYRONYL)-METHYLE) (FRENCH)

TOXICITY DATA with REFERENCE
orl-rat LD50:23 mg/kg WRPCA2 4,36,64
skn-rat LD50:130 mg/kg PHJOAV 185,361,60
orl-mus LD50:17 mg/kg ARSIM* 20,10,66

CONSENSUS REPORTS: EPA Extremely Hazardous Substances List.

SAFETY PROFILE: A poison by ingestion and skin contact. A pesticide. When heated to decomposition it emits very toxic fumes of PO_x and SO_x.

EAS100 ***HR: 3***
ENDOTOXIN

PROP: A purified lipopolysaccharide fraction isolated from *Salmonella typhosa* 0901 (TXAPA9 23,102,72).

SYNS: SALMONELLA TYPHI ENDOTOXIN ◇ SALMONELLA TYPHOSA ENDOTOXIN

TOXICITY DATA with REFERENCE
ipr-rat LD50:107 mg/kg BPJLAQ 7,7,8
ivn-rat LD50:4330 μg/kg CRSHAG 4,253,77
ipr-mus LD50:19 mg/kg TXAPA9 23,102,72
ivn-dog LD50:1 mg/kg BJPCBM 61,175,77

SAFETY PROFILE: Poison by intravenous and intraperitoneal routes.

EAT500 CAS:72-20-8 ***HR: 3***
ENDRIN
DOT: UN 2761
mf: $C_{12}H_8Cl_6O$ mw: 380.90

PROP: White crystals. Mp: decomp @ 200°.

SYNS: COMPOUND 269 ◇ ENDREX ◇ ENDRINE (FRENCH) ◇ ENT 17,251 ◇ HEXACHLOROEPOXYOCTAHYDRO-endo,endo-DIMETHANONAPHTHALENE ◇ 3,4,5,6,9,9-HEXACHLORO-1a,2,2a,3,6,6a,7,7a-OCTAHYDRO-2,7:3,6-DIMETHANONAPHTH(2,3-b)OXIRENE◇ HEXADRIN ◇ MENDRIN ◇ NCI-C00157 ◇ NENDRIN ◇ RCRA WASTE NUMBER P051

TOXICITY DATA with REFERENCE
sce-ofs-mul 54 pmol/L MUREAV 118,61,83
cyt-rat-par 1 mg/kg BECTA6 9,65,73
orl-ham TDLo:15 mg/kg (female 5-14D post):REP TXCYAC 21,187,81
orl-ham TDLo:5 mg/kg (female 8D post):TER TJADAB 9,11,74
orl-man LDLo:171 mg/kg HUTODJ 4,241,85
orl-rat LD50:3 mg/kg WRPCA2 9,119,70
skn-rat LD50:12 mg/kg SPEADM 78-1,13,78
orl-mus LD50:1370 μg/kg TXAPA9 25,42,73
ivn-mus LD50:2300 μg/kg TXAPA9 23,408,72
orl-mky LD50:3 mg/kg PCOC** -,475,66
orl-cat LDLo:5 mg/kg JAFCAU 3,842,55
orl-rbt LD50:7 mg/kg JAFCAU 3,842,55
skn-rbt LD50:60 mg/kg SPEADM 78-1,13,78
orl-gpg LD50:16 mg/kg PCOC** -,475,66

CONSENSUS REPORTS: IARC Cancer Review: Group 3 IMEMDT 7,56,87; Animal Inadequate Evidence IMEMDT 5,157,74; Human Inadequate Evidence IMEMDT 5,157,74. NCI Carcinogenesis Bioassay (feed); No Evidence: mouse, rat NCITR* NCI-CG-TR-12,79. EPA Genetic Toxicology Program. EPA Extremely Hazardous Substances List.

OSHA PEL: TWA 0.1 mg/m^3 (skin)
ACGIH TLV: TWA 0.1 mg/m^3 (skin)
DFG MAK: 0.1 mg/m^3
DOT Classification: Poison B; Label: Poison; Poison B; Label: Poison, liquid.

SAFETY PROFILE: Poison by ingestion, skin contact, and intravenous routes. Experimental teratogenic and reproductive effects. Questionable carcinogen. Mutation data reported. A central nervous system stimulant. Highly toxic to birds, fish, and humans. Many cases of fatal poisoning have been attributed to it. Does not accumulate in human tissue. In humans, ingestion of 1 mg/kg has caused symptoms. A dangerous fire hazard. Mixtures with parathion dissolve very exothermically in petroleum solvents and may cause an air-vapor explosion. See also ALDRIN.

EAT600 CAS:5836-29-3 ***HR: 3***
ENDROCIDE
mf: $C_{19}H_{16}O_3$ mw: 292.35

SYNS: BAY 25634 ◇ BAY ENE 11183 B ◇ BAYER 25 634 ◇ COUMATETRALYL ◇ CUMATETRALYL (GERMAN, DUTCH) ◇ ENDOX ◇ ENDROCID ◇ ENE 11183 B ◇ 4-HYDROXY-3-(1,2,3,4-TETRAHYDRO-1-NAFTYL)-4-CUMARINE (DUTCH) ◇ 4-HYDROXY-3-(1,2,3,4-TETRAHYDRO-1-NAPHTHALENYL)-2H-1-BENZOPYRAN-2-ONE (9CI) ◇ 4-HYDROXY-3-(1,2,3,4-TETRAHYDRO-1-NAPHTHYL)CUMARIN ◇ 4-IDROSSI-3-(1,2,3,4-TETRAIDRO-1-NAFTIL)CUMARINA (ITALIAN) ◇ RACUMIN ◇ RAUCUMIN 57 ◇ RODENTIN ◇ 3-(1,2,3,4-TETRAHYDRO-1-NAPHTHYL)-4-HYDROXYCUMARIN (GERMAN) ◇ 3-(1,2,3,4-TETRAHYDRO-1-NAPHTYL)-4-HYDROXYCOUMARINE (FRENCH) ◇ 3-(α-TETRAL)-4-OXYCOUMARIN ◇ 3-(α-TETRALYL)-4-HYDROXYCOUMARIN ◇ 3-(d-TETRALYL)-4-HYDROXYCOUMARIN

TOXICITY DATA with REFERENCE
orl-rat LD50:16500 μg/kg FMCHA2 -,C203,83
orl-cat LDLo:20 mg/kg 85GYAZ -,115,71
orl-gpg LDLo:250 mg/kg 85GYAZ -,115,71

CONSENSUS REPORTS: EPA Extremely Hazardous Substances List.

SAFETY PROFILE: Poison by ingestion. When heated to decomposition it emits acrid smoke and fumes. See also COUMARIN.

EAT800 CAS:11115-82-5 ***HR: 3***
ENDURACIDIN

SYNS: ENRADIN ◇ ENRAMYCIN

TOXICITY DATA with REFERENCE
ipr-rat LD50:830 mg/kg TDKNAF 28,76,69
ivn-rat LD50:66600 μg/kg TDKNAF 28,76,69
ipr-mus LD50:750 mg/kg TDKNAF 28,76,69
ivn-mus LD50:30 mg/kg TDKNAF 28,76,69
ivn-rbt LD50:92 mg/kg TDKNAF 28,76,69

SAFETY PROFILE: Poison by intravenous route. Moderately toxic by intraperitoneal route.

EAT810 CAS:34438-27-2 ***HR: 3***
ENDURACIDIN A
mf: $C_{107}H_{138}Cl_2N_{26}O_{31}$ mw: 2355.61

SYNS: B-5477 ◇ ENRADINE ◇ ENRAMYCIN

TOXICITY DATA with REFERENCE
ipr-mus LD50:880 mg/kg 85GDA2 4(2),109,80
scu-mus LD50:3000 mg/kg 85GDA2 4(2),109,80
ivn-mus LD50:25 mg/kg 85GDA2 4(2),109,80

SAFETY PROFILE: Poison by intravenous route. Moderately toxic by intraperitoneal and subcutaneous routes. When heated to decomposition it emits toxic fumes of Cl^- and NO_x.

EAT900 CAS:13838-16-9 ***HR: 2***
ENFLURANE
mf: $C_3H_2ClF_5O$ mw: 184.50

SYNS: ANESTHETIC COMPOUND No. 347 ◇ 2-CHLORO-1-(DIFLUOROMETHOXY)-1,1,2-TRIFLUOROETHANE ◇ 2-CHLORO-1,1,2-TRIFLUOROETHYL DIFLUOROMETHYL ETHER ◇ COMPOUND 347 ◇ ETHRANE ◇ METHYLFLURETHER ◇ NSC-115944 ◇ OHIO 347

TOXICITY DATA with REFERENCE
eye-rbt 100 mg MOD FEPRA7 35,729,76
cyt-hmn:lyms 1000 ppm ENVRAL 12,366,76
cyt-mus:fbr 1000 ppm ENVRAL 12,366,76
ihl-rat TCLo:1500 ppm/6H (1-20D post):REP AIPTAK 256,134,82
ihl-rat TCLo:16500 ppm/6H (8-10D post):TER ANESAV 64,339,86
ihl-mus TCLo:3000 ppm/4H/78W-I:CAR ANESAV 56,9,82
ihl-hmn TCLo:1 pph/6H ANESAV 45,557,76
orl-rat LD50:5450 mg/kg DRUGAY 6,143,82
ihl-rat LC50:14000 ppm/3H IYKEDH 12,668,81
ipr-rat LD50:6 g/kg DRUGAY 6,143,82
scu-rat LD50:19500 mg/kg YKYUA6 32,491,81
orl-mus LD50:5 g/kg DRUGAY 6,143,82
ihl-mus LC50:8100 ppm/3H IYKEDH 12,668,81
ipr-mus LD50:3900 mg/kg DRUGAY 6,143,82
scu-mus LD50:38800 mg/kg YKYUA6 32,491,81

CONSENSUS REPORTS: Reported in EPA TSCA Inventory.

ACGIH TLV: TWA 75 ppm
NIOSH REL: (Waste Anesthetic Gases and Vapors) CL 2 ppm/1H

SAFETY PROFILE: Mildly toxic by inhalation, ingestion, and subcutaneous routes. Experimental reproductive data by inhalation. Human systemic effects by inhalation: decreased urine volume or anuria. An experimental teratogen. Experimental reproductive effects. Human mutation data reported. An eye irritant. Questionable carcinogen with experimental carcinogenic data. An anesthetic. When heated to decomposition it emits very toxic fumes of F^- and Cl^-. See also ETHERS.

EAU000 ***HR: 3***
ENHYDRINA SCHISTOSA VENOM

SYN: VENOM, SEA SNAKE, ENHYDRINA SCHISTOSA

TOXICITY DATA with REFERENCE
ivn-rat LD50:70 μg/kg TRSTAZ 54,50,60
ipr-mus LD50:110 μg/kg 85EGD4 -,341,78
scu-mus LD50:111 μg/kg TOXIA6 14,347,76
ivn-mus LD50:74 μg/kg TOXIA6 20,797,82
par-mus LDLo:125 μg/kg SCNEBK 110,355,76
ivn-rbt LD50:57 μg/kg TRSTAZ 54,40,60
ivn-gpg LD50:61 μg/kg TRSTAZ 54,50,60
par-frg LD50:20 μg/kg TRSTAZ 54,50,60
ivn-mam LD50:10 μg/kg CLPTAT 8,849,67

SAFETY PROFILE: A deadly poison by intraperitoneal, intravenous, parenteral, and subcutaneous routes.

EAU075 CAS:55726-47-1 ***HR: 3***
ENOCITABINE
mf: $C_{31}H_{55}N_3O_6$ mw: 565.89

PROP: Crystals from DMSO. Mp: 141-142°.

SYNS: N-(1-β-d-ARABINOFURANOSYL-1,2-DIHYDRO-2-OXO-4-PYRIMIDINYL)DOCOSANAMIDE ◇ N^4-BEHENOYL-1-β-d-ARABINOFURANOSYLCYTOSINE ◇ BEHENOYLCYTOSINE ARABINOSIDE ◇ N^4-BEHENOYLCYTOSINE ARABINOSIDE ◇ SUNRABIN

TOXICITY DATA with REFERENCE
ipr-rat LD50:1100 mg/kg GTKRDX 10,2077,83
ivn-rat LD50:380 mg/kg IYKEDH 14,484,83
ipr-mus LD50:525 mg/kg IYKEDH 14,484,83
scu-mus LD50:810 mg/kg GTKRDX 10,2077,83
ivn-mus LD50:500 mg/kg IYKEDH 14,484,83

SAFETY PROFILE: Poison by intravenous route. Moderately toxic by subcutaneous and intraperitoneal routes. When heated to decomposition it emits toxic fumes of NO_x.

EAU100 CAS:74011-58-8 ***HR: 3***
ENOXACIN
mf: $C_{15}H_{17}FN_4O_3$ mw: 320.36

PROP: Crystals from ethanol/methylene chloride. Mp: 220-224°.

SYNS: AT-2266 ◇ 1-ETHYL-6-FLUORO-1,4-DIHYDRO-4-OXO-7-(1-PIPERAZINYL)-1,8-NAPHTHYRIDINE-3-CARBOXYLIC ACID ◇ FLUMARK

TOXICITY DATA with REFERENCE
orl-rat TDLo:11 g/kg (female 7-17D post):TER NKRZAZ 32(Suppl 3),301,84
orl-rat TDLo:91 g/kg (male 13W pre):REP NKRZAZ 32(Suppl 3),293,84
ivn-rat LD50:236 mg/kg NKRZAZ 32(Suppl 3),192,84
scu-mus LD50:1100 mg/kg 43MKAT 1,456,80
ivn-mus LD50:327 mg/kg NKRZAZ 32(Suppl 3),192,84

SAFETY PROFILE: Poison by intravenous route. Moderately toxic by subcutaneous route. An experimental teratogen. Other experimental reproductive effects. When heated to decomposition it emits toxic fumes of F^- and NO_x.

EAU150 ***HR: 3***
ENOXACIN HYDRATE (2:3)
mf: $C_{15}H_{17}FN_4O_3•3/2H_2O$ mw: 347.39

TOXICITY DATA with REFERENCE
ivn-rat LD50:236 mg/kg IYKEDH 16,1461,85
scu-mus LD50:1237 mg/kg IYKEDH 16,1461,85
ivn-mus LD50:327 mg/kg IYKEDH 16,1461,85

SAFETY PROFILE: Poison by intravenous route. Moderately toxic by subcutaneous route. When heated to decomposition it emits toxic fumes of F^- and NO_x.

EAU500 CAS:136-45-8 ***HR: 3***
ENT 17591
mf: $C_{13}H_{17}NO_4$ mw: 251.31

SYNS: DIPROPYL ISOCINCHOMERONATE ◇ DI-N-PROPYL-ISOCINCHOMERONATE (GERMAN) ◇ DIPROPYL 2,5-PYRIDINEDICARBOXYLATE ◇ MGK 326 ◇ MGK REPELLENT-326 ◇ PYRIDIN-2,5-DICARBONSAEURE-DI-N-PROPYLESTER (GERMAN) ◇ 2,5-PYRIDINEDICARBOXYLIC ACID, DIPROPYL ESTER ◇ R-326

TOXICITY DATA with REFERENCE
orl-rat LD50:5230 mg/kg 28ZEAL 5,94,76
skn-rat LD50:9400 mg/kg 28ZEAL 5,94,76
orl-mus LD50:1600 mg/kg YKYUA6 32,605,81
ipr-mus LD50:330 mg/kg YKYUA6 32,605,81
skn-rbt LD50:9500 mg/kg 85DPAN -,-,71/76

SAFETY PROFILE: Poison by intraperitoneal route. Moderately toxic by ingestion. Mildly toxic by skin contact. An insect repellent. When heated to decomposition it emits toxic fumes of NO_x.

EAV000 CAS:115-91-3 ***HR: 3***
ENT 24,944
mf: $C_9H_{13}O_4PS_2$ mw: 280.31

SYNS: BAYER 25198 ◇ O,O-DIMETHYL-O-(P-METHYLSULFINYLPHENYL)PHOSPHOROTHIOATE

TOXICITY DATA with REFERENCE
orl-rat LDLo:3 mg/kg ARSIM* 20,3,66
orl-ckn LD50:2320 μg/kg TXAPA9 6,147,64

SAFETY PROFILE: Poison by ingestion. When heated to decomposition it emits very toxic fumes of PO_x and SO_x.

EAV100 CAS:23581-62-6 ***HR: 3***
EO 122
mf: $C_{16}H_{22}N_2O•ClH$ mw: 294.86

SYN: N-(2,6-DIMETHYLPHENYL)-2-AZABICYCLO(2.2.2)OCTANE-3-CARBOXAMIDE MONOHYDROCHLORIDE (9CI)

TOXICITY DATA with REFERENCE
ipr-mus LD50:66 mg/kg ANGIAB 31,410,80
ivn-mus LD50:22 mg/kg ANGIAB 31,410,80
ivn-rbt LD50:8500 μg/kg ANGIAB 31,410,80

SAFETY PROFILE: Poison by intravenous and intraperitoneal routes. When heated to decomposition it emits toxic fumes of NO_x and HCl.

EAV500 CAS:33419-42-0 ***HR: 3***
EPE
mf: $C_{29}H_{32}O_{13}$ mw: 588.61

SYNS: DEMETHYL-EPIODOPHYLLOTOXIN ETHYLIDENE GLUCOSIDE ◇ 4-DEMETHYLEPIODOPHYLLOTOXIN-β,d-ETHYLIDENEGLUCOSIDE ◇ 4′-DEMETHYLEPIPODOPHYLLOTOXIN-9-(4,6-O-ETHYLIDENE-β-d-GLUCOPYRANOSIDE ◇ 4-DEMETHYL-EPIPODOPHYLLOTOXIN-β,d-ETHYLIDEN-GLUCOSIDE ◇ 4′-DEMETHYLEPIPODOPHYLLOTOXINETHYLIDENE-β,d-GLUCOSIDE ◇ 4′-O-DEMETHYL-1-O-(4,6-O-ETHYLIDENE-β,d-GLUCOPYRANOSYL) EPIPODOPHYLLOTOXIN ◇ ETOPOSIDE ◇ NK 171 ◇ NSC 141540 ◇ VEPESID ◇ VP 16213

TOXICITY DATA with REFERENCE
mmo-sat 2 mg/plate TCMUD8 5,319,85
dnd-hmn:oth 2 μmol/L CNREA8 45,3106,85
dnd-ham:lng 40 μmol/L CNREA8 46,611,86
orl-rat TDLo:14630 μg/kg (7-17D preg):REP KSRNAM 19,3897,85
ipr-mus TDLo:1500 μg/kg (female 6D post):TER TJADAB 18,31,78
orl-hmn TDLo:16 mg/kg/5D-I:BLD CANCAR 34,985,74
ivn-hmn TDLo:2630 μg/kg/10D-I:BLD CANCAR 34,985,74

ivn-cld TDLo:183 mg/kg/2H-C DICPBB 22,41,88
orl-rat LD50:1784 mg/kg KSRNAM 19,3473,85
ipr-rat LD50:39 mg/kg KSRNAM 19,3473,85
ivn-rat LD50:75 mg/kg KSRNAM 19,3473,85
orl-mus LD50:215 mg/kg NCISP* JAN86
ipr-mus LD50:64 mg/kg KSRNAM 19,3473,85
scu-mus LD50:223 mg/kg NCISP* JAN86
ivn-mus LD50:15070 μg/kg NCISP* JAN86

SAFETY PROFILE: Poison by ingestion, intraperitoneal, intravenous, and subcutaneous routes. An experimental teratogen. Human systemic effects by ingestion and inhalation: angranulocytosis, aplastic anemia, and other changes in bone marrow. Experimental reproductive effects. Human mutation data reported. When heated to decomposition it emits acrid smoke and fumes.

EAV700 CAS:64840-90-0 ***HR: 3***
EPERISONE HYDROCHLORIDE
mf: $C_{17}H_{25}NO•ClH$ mw: 295.89

SYNS: E-646 ◇ EMPP ◇ 4'-ETHYL-2-METHYL-3-PIPERIDINOPROPIOPHENONE HYDROCHLORIDE ◇ MIONAL

TOXICITY DATA with REFERENCE
orl-rat LD50:1300 mg/kg OYYAA2 21,939,81
scu-rat LD50:490 mg/kg OYYAA2 21,939,81
ivn-rat LD50:51 mg/kg OYYAA2 21,939,81
ims-rat LD50:400 mg/kg OYYAA2 21,939,81
orl-mus LD50:940 mg/kg OYYAA2 21,939,81
scu-mus LD50:256 mg/kg OYYAA2 21,939,81
ivn-mus LD50:43 mg/kg OYYAA2 21,939,81
ims-mus LD50:280 mg/kg OYYAA2 21,939,81

SAFETY PROFILE: Poison by subcutaneous, intramuscular, and intravenous routes. Moderately toxic by ingestion. When heated to decomposition it emits toxic fumes of NO_x and HCl.

EAW000 CAS:299-42-3 ***HR: 3***
EPHEDRINE
mf: $C_{10}H_{15}NO$ mw: 165.26

PROP: White granules. Mp: 79° (dl), bp: 255° (decomp). Sol in ether and chloroform.

SYNS: BIOPHEDRIN ◇ ECIPHIN ◇ EFEDRIN ◇ EPHEDRAL ◇ EPHEDRATE ◇ EPHEDREMAL ◇ EPHEDRIN ◇ l-EPHEDRINE ◇ l(−)-EPHEDRINE ◇ EPHEDRITAL ◇ EPHEDROL ◇ EPHEDROSAN ◇ EPHEDROTAL ◇ EPHEDSOL ◇ EPHENDRONAL ◇ EPHOXAMIN ◇ FEDRIN ◇ α-HYDROXY-β-METHYL AMINE PROPYLBENZENE ◇ 1-HYDROXY-2-METHYLAMINO-1-PHENYLPROPANE ◇ l-SEDRIN ◇ ISOFEDROL ◇ KRATEDYN ◇ MANADRIN ◇ MANDRIN ◇ (−)-α-(1-METHYLAMINOETHYL)BENZYL ALCOHOL ◇ 1-α-(1-METHYLAMINOETHYL)BENZYL ALCOHOL ◇ 1-2-METHYLAMINO-1-PHENYLPROPANOL ◇ N-METHYLNOREPHEDRINE ◇ NASOL ◇ 1-PHENYL-2-METHYLAMINOPROPANOL ◇ SANEDRINE ◇ VENCIPON ◇ ZEPHROL

TOXICITY DATA with REFERENCE
ipr-rat TDLo:50 mg/kg (female 9D post):TER TJADAB 34,469,86
unr-man LDLo:9 mg/kg 85DCAI 2,73,70
orl-rat LD50:600 mg/kg 27ZQAG -,341,72
ipr-rat LDLo:170 mg/kg AEPPAE 195,647,40
scu-rat LD50:300 mg/kg AIPTAK 68,339,42
ivn-rat LDLo:130 mg/kg AEPPAE 120,189,27
orl-mus LD50:1250 mg/kg ARZNAD 23,1125,73
ipr-mus LD50:350 mg/kg PCJOAU 3,203,69
ivn-mus LD50:74 mg/kg APTOA6 38,474,76
par-mus LD50:170 mg/kg AEPPAE 195,647,40
ivn-dog LDLo:70 mg/kg AEPPAE 120,189,27
ims-rbt LDLo:340 mg/kg AEPPAE 120,189,27

CONSENSUS REPORTS: Reported in EPA TSCA Inventory.

SAFETY PROFILE: A human poison by an unspecified route. An experimental poison by intravenous, subcutaneous, intramuscular, and intraperitoneal routes. Moderately toxic by ingestion and parenteral routes. Causes rapid pulse, rise in blood pressure, and other actions similar to epinephrine. An experimental teratogen. Used in production of drugs of abuse. Has been known to cause allergic sensitization. When heated to decomposition it emits toxic fumes of NO_x.

EAW500 CAS:50-98-6 ***HR: 3***
EPHEDRINE HYDROCHLORIDE
mf: $C_{10}H_{15}NO•ClH$ mw: 201.72

TOXICITY DATA with REFERENCE
idr-hmn TDLo:6 μg/kg:SKN JPETAB 76,295,42
scu-rat LDLo:128 mg/kg JPETAB 71,62,41
scu-mus LD50:44 mg/kg JPETAB 87,214,46
ivn-mus LD50:100 mg/kg JPETAB 86,280,46
scu-gpg LDLo:427 mg/kg JPETAB 32,121,27

CONSENSUS REPORTS: Reported in EPA TSCA Inventory.

SAFETY PROFILE: A poison by subcutaneous and intravenous routes. Human systemic effects by intradermal route: skin effects. When heated to decomposition it emits very toxic fumes of Cl^- and NO_x. See also EPHEDRINE.

EAX000 CAS:24221-86-1 ***HR: 3***
d-EPHEDRINE HYDROCHLORIDE
mf: $C_{10}H_{15}NO•ClH$ mw: 201.72

TOXICITY DATA with REFERENCE
scu-rat LDLo:160 mg/kg JPETAB 71,62,41
ipr-mus LD50:248 mg/kg JPETAB 158,135,67
ivn-rbt LDLo:80 mg/kg JPETAB 36,363,29

CONSENSUS REPORTS: Reported in EPA TSCA Inventory.

SAFETY PROFILE: A poison by intraperitoneal, subcutaneous, and intravenous routes. When heated to decomposition it emits very toxic fumes of HCl and NO_x. See also EPHEDRINE.

EAX500 CAS:134-71-4 ***HR: 3***
dl-EPHEDRINE HYDROCHLORIDE
mf: $C_{10}H_{15}NO \cdot ClH$ mw: 201.72

SYNS: EPHETONIN ◇ EPHETONINE ◇ dl-α-(1-(METHYLAMINO) ETHYL) BENZYL ALCOHOL HYDROCHLORIDE ◇ 1-PHENYL-2-METHYLAMINOPROPANOL-1 ◇ RACEPHEDRINE HYDROCHLORIDE

TOXICITY DATA with REFERENCE
scu-rat LD50:350 mg/kg AEPPAE 169,114,33
orl-mus LD50:700 mg/kg JMCMAR 9,966,66
ipr-mus LD50:254 mg/kg PHARAT 24,775,69
scu-mus LD50:800 mg/kg AEPPAE 169,114,33
ivn-mus LD50:135 mg/kg JMCMAR 9,966,66
ivn-rbt LDLo:60 mg/kg JPETAB 36,363,29

CONSENSUS REPORTS: Reported in EPA TSCA Inventory.

SAFETY PROFILE: Poison by subcutaneous, intravenous, and intraperitoneal routes. Moderately toxic by ingestion. When heated to decomposition it emits very toxic fumes of HCl and NO_x. See also EPHEDRINE.

EAY000 CAS:50-98-6 ***HR: 3***
l-EPHEDRINE HYDROCHLORIDE
mf: $C_{10}H_{15}NO \cdot ClH$ mw: 201.72

SYNS: EPHEDRINE HYDROCHLORIDE ◇ (−)-EPHEDRINE HYDROCHLORIDE ◇ (R-(R*,S*))-α-(1-(METHYLAMINO)ETHYL)BENZENEMETHANOL HYDROCHLORIDE

TOXICITY DATA with REFERENCE
idr-hmn TDLo:6 μg/kg:CNS JPETAB 76,295,42
ipr-rat LD50:165 mg/kg JPETAB 86,284,46
ivn-rat LD50:69 mg/kg JAPMA8 33,80,44
orl-mus LD50:400 mg/kg JMCMAR 9,966,66
scu-mus LD50:600 mg/kg JPETAB 86,284,46
ivn-mus LD50:95 mg/kg JPETAB 95,336,49
par-mus LDLo:500 mg/kg MAIZAB 3,1,25
scu-rbt LD50:165 mg/kg JPETAB 86,284,46
ims-rbt LD50:175 mg/kg JPETAB 86,284,46

CONSENSUS REPORTS: Reported in EPA TSCA Inventory.

SAFETY PROFILE: Poison by ingestion, intraperitoneal, intravenous, subcutaneous, and intramuscular routes. Moderately toxic by parenteral route. Human systemic effects by intradermal route: local anesthetic. When heated to decomposition it emits very toxic fumes of HCl and NO_x. See also EPHEDRINE.

EAY075 CAS:7234-07-3 ***HR: 2***
l-EPHEDRINE PHOSPHATE (ESTER)
mf: $C_{10}H_{16}NO_4P$ mw: 245.24

SYN: l-α-(1-(METHYLAMINO)ETHYL)BENZYLPHOSPHATE

TOXICITY DATA with REFERENCE
orl-mus LD50:1065 mg/kg JMCMAR 9,966,66
ipr-mus LD50:2000 mg/kg JMCMAR 9,966,66
scu-mus LD50:1707 mg/kg JMCMAR 9,966,66
ivn-mus LD50:400 mg/kg JMCMAR 9,966,66

SAFETY PROFILE: Moderately toxic by ingestion, subcutaneous, intravenous, and intraperitoneal routes. When heated to decomposition it emits toxic fumes of PO_x and NO_x. See also EPHEDRINE.

EAY150 CAS:7234-08-4 ***HR: 2***
d-EPHEDRINE PHOSPHATE (ESTER)
mf: $C_{10}H_{16}NO_4P$ mw: 245.24

SYN: d-α-(1-(METHYLAMINO)ETHYL)BENZYLPHOSPHATE

TOXICITY DATA with REFERENCE
orl-mus LD50:1800 mg/kg JMCMAR 9,966,66
ipr-mus LD50:815 mg/kg JMCMAR 9,966,66
scu-mus LD50:865 mg/kg JMCMAR 9,966,66
ivn-mus LD50:815 mg/kg JMCMAR 9,966,66

SAFETY PROFILE: Moderately toxic by ingestion, intraperitoneal subcutaneous and intravenous routes. When heated to decomposition it emits toxic fumes of PO_x and NO_x. See also EPHEDRINE and ESTERS.

EAY175 CAS:7234-09-5 ***HR: 2***
dl-EPHEDRINE PHOSPHATE (ESTER)
mf: $C_{10}H_{16}NO_4P$ mw: 245.24

SYN: dl-α-(1-(METHYLAMINO)ETHYL)BENZYLPHOSPHATE

TOXICITY DATA with REFERENCE
orl-mus LD50:2000 mg/kg JMCMAR 9,966,66
ipr-mus LD50:790 mg/kg JMCMAR 9,966,66
scu-mus LD50:1150 mg/kg JMCMAR 9,966,66
ivn-mus LD50:840 mg/kg JMCMAR 9,966,66

SAFETY PROFILE: Moderately toxic by ingestion, intraperitoneal, subcutaneous, and intravenous routes. When heated to decomposition it emits toxic fumes of PO_x and NO_x. See also EPHEDRINE and ESTERS.

EAY500 CAS:134-72-5 ***HR: 3***
1-EPHEDRINE SULFATE
mf: $C_{20}H_{30}N_2O_2 \cdot H_2O_4S$ mw: 428.60

SYNS: ISOFEDROL ◇ 1-α-(1-(METHYLAMINO)ETHYL)BENZYL ALCOHOL SULFATE ◇ NCI-C55652 ◇ 1-PHENYL-2-METHYLAMINE-PROPANOL-1-SULFATE

TOXICITY DATA with REFERENCE
ivn-hmn TDLo:1 mg/kg:CVS AEPPAE 120,189,27

orl-rat LD50:404 mg/kg JPETAB 94,150,48
scu-rat LD50:318 mg/kg JPETAB 94,150,48
ivn-rat LD50:102 mg/kg JPETAB 94,150,48
ipr-mus LD50:400 mg/kg TXAPA9 28,227,74
orl-rbt LD50:825 mg/kg JPETAB 94,150,48
scu-rbt LD50:383 mg/kg JPETAB 94,150,48
ivn-rbt LD50:73 mg/kg JPETAB 94,159,48

CONSENSUS REPORTS: NTP Carcinogenesis Studies (feed); No Evidence: mouse, rat NTPTR* NTP-TR-307,86. Reported in EPA TSCA Inventory.

SAFETY PROFILE: Poison by intravenous, intraperitoneal, and subcutaneous routes. Moderately toxic by ingestion. Human systemic effects by intravenous route: increased pulse rate and blood pressure. When heated to decomposition it emits very toxic fumes of NO_x and SO_x. See also EPHEDRINE.

EAZ000 CAS:62-32-8 ***HR: 3***
EPHININE HYDROCHLORIDE
mf: $C_9H_{13}NO_2 \cdot ClH$ mw: 203.69

SYNS: 3,4-DIHYDROXYPHENYLETHYLMETHYLAMINE HYDROCHLORIDE ◇ 3,4-DIHYDROXYPHENYL-1-METHYLAMINO-2-ETHANE HYDROCHLORIDE ◇ 4-(2-METHYLAMINOETHYL)PYROCATECHOL HYDROCHLORIDE ◇ METHYL-(β-(3,4-DIHYDROXY PHENYL ETHYL) AMINE HYDROCHLORIDE ◇ N-METHYLDOPAMINE HYDROCHLORIDE

TOXICITY DATA with REFERENCE
scu-rat LDLo:160 mg/kg JPETAB 71,62,41
ipr-mus LD50:212 mg/kg JMCMAR 18,1194,75

SAFETY PROFILE: A poison by subcutaneous and intraperitoneal routes. When heated to decomposition it emits very toxic fumes of NO_x and HCl.

EAZ500 CAS:106-89-8 ***HR: 3***
EPICHLOROHYDRIN
DOT: UN 2023
mf: C_3H_5ClO mw: 92.53

$ClCH_2CHOCH_2$ (epoxide ring linking CH and CH_2)

PROP: Colorless, mobile liquid; irritating chloroform-like odor. Bp: 117.9°, fp: − 57.1°, flash p: 105.1°F (OC) (40°C), mp -25.6°C, d: 1.1761 @ 20°/20°, vap press: 10 mm @ 16.6°, vap d: 3.29.

SYNS: 1-CHLOOR-2,3-EPOXY-PROPAAN (DUTCH) ◇ 1-CHLOR-2,3-EPOXY-PROPAN (GERMAN) ◇ 1-CHLORO-2,3-EPOXYPROPANE ◇ 3-CHLORO-1,2-EPOXYPROPANE ◇ epi-CHLOROHYDRIN ◇ (CHLOROMETHYL)ETHYLENE OXIDE ◇ CHLOROMETHYLOXIRANE ◇ 2-(CHLOROMETHYL)OXIRANE ◇ CHLOROPROPYLENE OXIDE ◇ Γ-CHLOROPROPYLENE OXIDE ◇ 3-CHLORO-1,2-PROPYLENE OXIDE ◇ 1-CLORO-2,3-EPOSSIPROPANO (ITALIAN) ◇ ECH ◇ EPICHLOORHYDRINE (DUTCH) ◇ EPICHLORHYDRIN (GERMAN) ◇ EPICHLORHYDRINE (FRENCH) ◇ α-EPICHLOROHYDRIN ◇ (dl)-α-EPICHLOROHYDRIN ◇ EPICHLOROHYDRYNA (POLISH) ◇ EPI-CHLOROPHYDRIN ◇ EPICLORIDRINA (ITALIAN) ◇ 1,2-EPOXY-3-CHLOROPROPANE ◇ 2,3-EPOXYPROPYL CHLORIDE ◇ GLYCEROL EPICHLORHYDRIN ◇ RCRA WASTE NUMBER U041 ◇ SKEKhG

TOXICITY DATA with REFERENCE
skn-rbt 10 mg/24H open JIHTAB 30,63,48
eye-rbt 23 mg AJOPAA 29,1363,46
dni-hmn:hla 2700 μmol/L MUREAV 92,427,82
sce-hmn:lym 10 nmol/L CARYAB 34,261,81
spm-mus-ihl 5 mg/m³ MUREAV 85,287,81
orl-mus TDLo:1200 mg/kg (female 6-15D post):TER JTEHD6 9,87,82
ihl-rat TCLo:50 ppm/6H (male 50D pre):REP TXAPA9 68,415,83
orl-rat TDLo:60 g/kg/81W-I:CAR GANNA2 71,922,80
ipr-rbt LD50:118 mg/kg JPMSAE 61,1712,72
orl-gpg LD50:280 mg/kg GISAAA 33(1),46,68
ipr-gpg LD50:118 mg/kg JPMSAE 61,1712,72

CONSENSUS REPORTS: NTP Fifth Annual Report on Carcinogens. IARC Cancer Review: Group 2A IMEMDT 7,202,87; Animal Sufficient Evidence IMEMDT 11,131,76. EPA Genetic Toxicology Program. Community Right-To-Know List. EPA Extremely Hazardous Substances List. Reported in EPA TSCA Inventory.

OSHA PEL: (Transitional: TWA 5 ppm (skin)) TWA 2 ppm (skin)
ACGIH TLV: TWA 2 ppm (skin); (Proposed: TWA 0.1 ppm; Suspected Human Carcinogen)
DFG TRK: 3 ppm; Animal Carcinogen, Suspected Human Carcinogen.
NIOSH REL: Minimize exposure
DOT Classification: Flammable Liquid; Label: Flammable Liquid; IMO: Poison B; Label: Poison, Flammable Liquid.

SAFETY PROFILE: Confirmed carcinogen with experimental carcinogenic data. Poison by ingestion and intraperitoneal routes. An experimental teratogen. Other experimental reproductive effects. Human mutation data reported. A skin and eye irritant. A sensitizer. Flammable when exposed to heat or flame. Explosive reaction with aniline. Reaction with trichloroethylene forms the explosive dichloroacetylene. Ignition on contact with potassium tert-butoxide. Violent reaction with sulfuric acid or isopropylamine. Exothermic polymerization on contact with strong acids, caustic alkalies, aluminum, aluminum chloride, iron(III) chloride, or zinc. When heated to decomposition it emits toxic fumes of Cl^-.

EBA100 CAS:516-95-0 ***HR: D***
EPIDEHYDROCHOLESTERIN
mf: $C_{27}H_{48}O$ mw: 388.75

SYN: CHOLESTAN-3-α-OL

TOXICITY DATA with REFERENCE
scu-mus TDLo:1 g/kg (female 6-15D post):TER KSRNAM 12,479,78
ipr-rat TDLo:700 mg/kg (female 35D pre):REP KSRNAM 12,456,78

CONSENSUS REPORTS: Reported in EPA TSCA Inventory.

SAFETY PROFILE: An experimental teratogen. Experimental reproductive effects. When heated to decomposition it emits acrid smoke and irritating fumes.

EBA275 CAS:62229-50-9 ***HR: D***
EPIDERMAL GROWTH FACTOR

SYN: EGF

TOXICITY DATA with REFERENCE
dns-hmn:oth 10 μg/L CNREA8 46,2545,86
dns-rat:lvr 10 μg/L PNASA6 73,3589,76
dns-mus:emb 12 μg/L ECREAL 158,311,85

SAFETY PROFILE: Human mutation data reported.

EBA500 CAS:220-42-8 ***HR: 3***
9,10-EPIDIOXY ANTHRACENE
mf: $C_{14}H_8O_2$ mw: 208.22

SAFETY PROFILE: Decomposes explosively at 120°C. When heated to decomposition it emits acrid smoke and fumes. See also ANTHRACENE

EBA600 ***HR: 3***
1,4-EPIDIOXY-1,4-DIHYDRO-6,6-DIMETHYLFULVENE
mf: $C_8H_{10}O_2$ mw: 138.17

SAFETY PROFILE: Decomposes explosively above −10°C. When heated to decomposition it emits acrid smoke and fumes. See also PEROXIDES.

EBB100 CAS:56420-45-2 ***HR: 3***
4'-EPIDOXORUBICIN
mf: $C_{27}H_{29}NO_{11}$ mw: 543.57

SYNS: EPIRUBICIN ◇ EPI-DX ◇ FARMORUBICIN

TOXICITY DATA with REFERENCE
spm-mus-par 10 mg/kg MUREAV 160,39,86
ivn-rat LD50:14270 μg/kg TXAPA9 79,412,85
ivn-mus LD50:16070 μg/kg TXAPA9 79,412,85
ivn-dog LD50:2 mg/kg TXAPA9 79,412,85

SAFETY PROFILE: Poison by intravenous route. Mutation data reported. When heated to decomposition it emits toxic fumes of NO_x.

EBB500 CAS:329-65-7 ***HR: 3***
dl-EPINEPHRINE
mf: $C_9H_{13}NO_3$ mw: 183.23

SYN: EPINEPHRINE racemic

TOXICITY DATA with REFERENCE
ivn-rat LD50:70 μg/kg JPETAB 95,502,49
ipr-mus LD50:8300 μg/kg TXAPA9 16,303,70
ivn-mus LD50:3400 μg/kg JPETAB 95,502,49

SAFETY PROFILE: Very poisonous by intravenous and intraperitoneal routes. When heated to decomposition it emits toxic fumes of NO_x.

EBB600 CAS:4375-07-9 ***HR: D***
EPIPODOPHYLLOTOXIN
mf: $C_{22}H_{22}O_8$ mw: 414.44

SYNS: ETOPSIDE ◇ FURO(3',4':6,7)NAPHTHO(2,3-d)-1,3-DIOXOL-6(5aH)-ONE, 5,8,8a,9-TETRAHYDRO-9-HYDROXY-5-(3,4,5-TRIMETHOXYPHENYL)-, (5R-(5-α-5a-β,8a-α-9-β))- ◇ VP16

TOXICITY DATA with REFERENCE
dni-hmn:hlas 800 nmol/L BICHAW 15,5443,76
oth-hmn:hlas 5 umol/L BICHAW 15,5443,76
orl-wmn TDLo:135 mg/kg (female 25D pre):REP FESTAS 41(Suppl),91S,84

SAFETY PROFILE: Experimental reproductive effects. Mutation data reported. When heated to decomposition it emits acrid smoke and irritating fumes.

EBC000 CAS:55870-64-9 ***HR: 3***
5-EPISISOMICIN
mf: $C_{19}H_{37}N_5O_7$ mw: 447.61

SYNS: MUTAMYCIN 6 ◇ PENTISOMICIN ◇ PENTISOMICINA (SPANISH)

TOXICITY DATA with REFERENCE
ipr-mus LD50:175 mg/kg DRFUD4 4,516,79
scu-mus LD50:210 mg/kg DRFUD4 4,516,79
ivn-mus LD50:45 mg/kg DRFUD4 4,516,79

SAFETY PROFILE: Poison by intraperitoneal, subcutaneous, and intravenous route. When heated to decomposition it emits toxic fumes of NO_x.

EBD000 CAS:10390-80-4 ***HR: 2***
EPISOL TARTRATE
mf: $C_{17}H_{21}ClN_2OS{\bullet}C_4H_6O_6$ mw: 487.01

SYN: 5-CHLORO-2-(4-(2-(DIETHYLAMINO)ETHOXY)PHENYL)BENZOTHIAZOLE TARTRATE

TOXICITY DATA with REFERENCE
orl-mus LD50:700 mg/kg ARZNAD 16,33,66
scu-mus LD50:2300 mg/kg ARZNAD 16,33,66

SAFETY PROFILE: Moderately toxic by ingestion and

subcutaneous routes. When heated to decomposition it emits very toxic fumes of Cl^-, NO_x and SO_x.

EBD500 CAS:2363-58-8 ***HR: 2***
2-α,3-α-EPITHIO-5-α-ANDROSTAN-17-β-OL
mf: $C_{19}H_{30}OS$ mw: 306.55

SYNS: 2,3-EPITHIOANDROSTAN-17-OL ◇ EPITIOSTANOL ◇ 10275-S ◇ THIODROL

TOXICITY DATA with REFERENCE
ims-rat TDLo:15 mg/kg (female 9-14D post):REP OYYAA2 7,723,73
ims-mus TDLo:12500 μg/kg (female 13-17D post):TER OYYAA2 7,723,73
ipr-rat LDLo:5 g/kg OYYAA2 7,805,73
ipr-mus LD50:1160 mg/kg OYYAA2 7,805,73

SAFETY PROFILE: Moderately toxic by intraperitoneal route. Experimental teratogenic and reproductive effects. A steroid. When heated to decomposition it emits very toxic fumes of SO_x.

EBD600 CAS:54096-45-6 ***HR: 3***
4,5-EPITHIOVALERONITRILE
mf: C_5H_7NS mw: 113.19

SYN: 1-CYANO-3,4-EPITHIOBUTANE

TOXICITY DATA with REFERENCE
scu-rat TDLo:88 mg/kg (8-9D preg):TER FCTXAV 18,159,80
scu-rat TDLo:95 mg/kg (8D preg):REP FCTXAV 18,159,80
scu-rat LD50:109 mg/kg FCTXAV 18,159,80

CONSENSUS REPORTS: Cyanide and its compounds are on the Community Right-To-Know List.

SAFETY PROFILE: Poison by subcutaneous route. An experimental teratogen. Experimental reproductive effects. When heated to decomposition it emits toxic fumes of NO_x, CN^-, and SO_x. See also NITRILES.

EBD700 CAS:2104-64-5 ***HR: 3***
EPN
mf: $C_{14}H_{14}NO_4PS$ mw: 323.32

PROP: Liquid or pale yellow crystals with an aromatic odor. D: 1.268 @ 25°, mp: 36°. Nearly insol in water; sol in organic solvents.

SYNS: O-AETHYL-O-n(4-NITROPHENYL)-PHENYL-MONOTHIOPHOSPHONAT (GERMAN) ◇ ENT 17,798 ◇ O-ESTER-p-NITROPHENOL with O-ETHYL PHENYL PHOSPHONOTHIOATE ◇ ETHOXY-4-NITROPHENOXYPHENYLPHOSPHINE SULFIDE ◇ O-ETHYL-O-((4-NITROFENYL)-FENYL)-MONOTHIOFOSFONAAT (DUTCH) ◇ O-ETHYL O-(4-NITROPHENYL)BENZENETHIONOPHOSPHONATE ◇ ETHYL-p-NITROPHENYL BENZENETHIONOPHOSPHONATE ◇ ETHYL-p-NITROPHENYL BENZENETHIOPHOSPHATE ◇ ETHYL-p-NITROPHENYL BENZENETHIOPHOSPHONATE ◇ ETHYL-p-NITROPHENYL PHENYLPHOSPHONOTHIOATE ◇ O-ETHYL-O-p-NITROPHENYL PHENYLPHOSPHONOTHIOLATE ◇ O-ETHYL-O-(4-NITROPHENYL) PHENYLPHOSPHONOTHIOATE ◇ O-ETHYL-O-p-NITROPHENYL PHENYLPHOSPHOROTHIOATE ◇ ETHYL-p-NITROPHENYL THIONOBENZENEPHOSPHATE ◇ ETHYL-p-NITROPHENYL THIONOBENZENEPHOSPHONATE ◇ O-ETHYL PHENYL-p-NITROPHENYL THIOPHOSPHONATE ◇ O-ETIL-O-((4-NTIRO-FENIL)-FENIL)-MONOTIOFOSFONATO (ITALIAN) ◇ PHENYLTHIOPHOSPHONATE de O-ETHYLE et O-4-NITROPHENYLE (FRENCH) ◇ PIN ◇ SANTOX ◇ THIONOBENZENEPHOSPHONIC ACID ETHYL-p-NITROPHENYL ESTER

TOXICITY DATA with REFERENCE
orl-mus TDLo:132 mg/kg (female 6-16D post):TER JESEDU 20,373,85
orl-rat LD50:7 mg/kg JPETAB 112,29,54
skn-rat LD50:25 mg/kg TXAPA9 2,88,60
ipr-rat LD50:7200 μg/kg APCRAW 4,117,61
orl-mus LD50:12200 μg/kg ABCHA6 26,257,62
skn-mus LD50:348 mg/kg ABCHA6 26,257,62
ipr-mus LD50:8400 μg/kg IJBBBQ 15,336,78
orl-dog LD50:20 mg/kg PCOC** -,478,66
ipr-dog LDLo:35 mg/kg JPETAB 112,29,54
skn-cat LD50:45 mg/kg TOLED5 1000(Sp 1)141,80
skn-rbt LD50:30 mg/kg AFDOAQ 16,3,52
ipr-rbt LDLo:20 mg/kg JPETAB 112,29,54
ipr-gpg LDLo:20 mg/kg JPETAB 112,29,54

CONSENSUS REPORTS: EPA Farm Worker Field Reentry FEREAC 39, 16888,74. EPA Extremely Hazardous Substances List.

OSHA PEL: TWA 0.5 mg/m^3 (skin)
ACGIH TLV: TWA 0.5 mg/m^3 (skin)
DFG MAK: 0.5 mg/m^3

SAFETY PROFILE: Poison by ingestion, skin contact, and intraperitoneal routes. An experimental teratogen. A cholinesterase inhibitor. This material is extremely hazardous on contact with skin, inhalation or ingestion. A highly toxic insecticide. When heated to decomposition it emits highly toxic fumes of SO_x, PO_x, NO_x, and phosphine. See also PARATHION, NITRO COMPOUNDS of AROMATIC HYDROCARBONS, PHOSPHINE, and SULFIDES.

EBE100 CAS:68401-82-1 ***HR: 1***
EPOCELIN
mf: $C_{13}H_{12}N_5O_5S_2 \cdot Na$ mw: 405.41

SYNS: CEFTIZOXIME SODIUM ◇ CEFTIZOXIM-NATRIUM (GERMAN) ◇ FK 749 ◇ FR-13,479 ◇ SKF-88373

TOXICITY DATA with REFERENCE
scu-rat TDLo:10 g/kg (male 9W pre):REP ARZNAD 30,1669,80
scu-rat TDLo:1100 mg/kg (female 7-17D post):TER JJANAX 34,466,81
ipr-rat LD50:8130 mg/kg ARZNAD 30,1669,80
ivn-rat LD50:5570 mg/kg ARZNAD 30,1669,80

ipr-mus LD50:8930 mg/kg ARZNAD 30,1669,80
ivn-mus LD50:5150 mg/kg YAKUD5 24,905,82

SAFETY PROFILE: Mildly toxic by intravenous and intraperitoneal routes. Experimental teratogenic and reproductive effects. When heated to decomposition it emits toxic fumes of SO_x, NO_x, and Na_2O.

EBF000 CAS:63089-76-9 ***HR: 3***
EPON 562

TOXICITY DATA with REFERENCE
skn-rbt 500 mg/24H MOD AMIHAB 17,129,58
eye-rbt 100 mg SEV AMIHAB 17,129,58
eye-rbt 750 μg/24H SEV 85JCAE-,1402,86
orl-rat LD50:5000 mg/kg AMIHAB 17,129,58
ipr-rat LD50:380 mg/kg AMIHAB 17,129,58
orl-mus LD50:1870 mg/kg AMIHAB 17,129,58
ipr-mus LD50:303 mg/kg AMIHAB 17,129,58
orl-rbt LD50:4010 mg/kg AMIHAB 17,129,58
skn-rbt LD50:14400 mg/kg 38MKAJ 2A,2239,81

SAFETY PROFILE: Poison by intraperitoneal route. Moderately toxic by ingestion. A severe skin and eye irritant. When heated to decomposition it emits acrid smoke and irritating fumes.

EBF500 CAS:25068-38-6 ***HR: 2***
EPON 820

TOXICITY DATA with REFERENCE
ipr-rat LD50:1400 mg/kg AMIHAB 17,129,58
ipr-rat LD50:1780 mg/kg AMIHAB 17,129,58

CONSENSUS REPORTS: Reported in EPA TSCA Inventory.

SAFETY PROFILE: Moderately toxic by intraperitoneal route. See also EPOXY RESINS.

EBG000 CAS:25068-38-6 ***HR: 2***
EPON 1001
mf: $(C_{15}H_{16}O_2 \cdot C_3H_5ClO)x$

TOXICITY DATA with REFERENCE
eye-rbt 100 mg MLD AMIHAB 17,129,58
orl-rat LD50:30 g/kg AMIHAB 17,129,58
ipr-rat LD50:2400 mg/kg AMIHAB 17,129,58
orl-mus LD50:20 g/kg AMIHAB 17,129,58

CONSENSUS REPORTS: Reported in EPA TSCA Inventory. EPA Genetic Toxicology Program.

SAFETY PROFILE: Moderately toxic by intraperitoneal route. Mildly toxic by ingestion. An eye irritant. See also EPOXY RESINS.

EBG500 CAS:25068-38-6 ***HR: 2***
EPON 1007
mf: $(C_{15}H_{16}O_2 \cdot C_3H_5ClO)x$

TOXICITY DATA with REFERENCE
eye-rbt 100 mg MLD AMIHAB 17,129,58
orl-rat LD50:30 g/kg AMIHAB 17,129,58
ipr-rat LD50:2200 mg/kg AMIHAB 17,129,58
orl-mus LD50:20 g/kg AMIHAB 17,129,58

CONSENSUS REPORTS: Reported in EPA TSCA Inventory. EPA Genetic Toxicology Program.

SAFETY PROFILE: Moderately toxic by intraperitoneal route. Mildly toxic by ingestion. An eye irritant. See also EPOXY RESINS.

EBH400 CAS:80471-63-2 ***HR: 2***
EPOSTANE
mf: $C_{22}H_{31}NO_3$ mw: 357.54

SYNS: ANDROST-2-ENE-2-CARBONITRILE,3,17-DIHYDROXY-4,17-DIMETHYL-4,5-EPOXY-, (4-α-5-α- 17-β)- ◇ (4-α-5-α-17-β)-4,5-EPOXY-3,17-DIHYDROXY-4,17-DIMETHYLANDROST-2-ENE-2-CARBONITRILE ◇ WIN 32729

TOXICITY DATA with REFERENCE
orl-wmn TDLo:96 mg/kg (female 7-8W post):REP CCPTAY 35,111,87
orl-wmn TDLo:96 mg/kg CCPTAY 35,111,87

SAFETY PROFILE: Human toxic effects by ingestion. Experimental reproductive effects. When heated to decomposition it emits toxic fumes of NO_x.

EBH500 CAS:31305-88-1 ***HR: 2***
EPOXIDE ERLA-0510

TOXICITY DATA with REFERENCE
skn-rbt 500 mg open MLD UCDS** 7/14/71
orl-rat LD50:1300 mg/kg UCDS** 7/14/71
skn-rbt LD50:640 mg/kg UCDS** 7/14/71

CONSENSUS REPORTS: Reported in EPA TSCA Inventory.

SAFETY PROFILE: Moderately toxic by ingestion and skin contact. A skin irritant.

EBH850 ***HR: 2***
(E)-1-α-2-α-EPOXYBENZ(c)ACRIDINE-3-α-4-β-DIOL
mf: $C_{17}H_9NO_3$ mw: 275.27

SYN: BENZ(c)ACRIDINE3,4-DIOL-1,2-EPOXIDE-2

TOXICITY DATA with REFERENCE
ipr-mus TDLo:5505 μg/kg/3D-I:NEO CNREA8 44,5161,84

SAFETY PROFILE: Questionable carcinogen with ex-

perimental neoplastigenic data. When heated to decomposition it emits toxic fumes of NO_x.

EBH875 **HR: 2**
(Z)-1-β,2-β-EPOXYBENZ(c)ACRIDINE-3-α-4-β-DIOL
mf: $C_{17}H_9NO_3$ mw: 275.27

SYN: BENZ(c)ACRIDINE3,4-DIOL-1,2-EPOXIDE-1

TOXICITY DATA with REFERENCE
ipr-mus TDLo:11561 μg/kg/3D-I:NEO CNREA8 44,5161,84

SAFETY PROFILE: Questionable carcinogen with experimental neoplastigenic data. When heated to decomposition it emits toxic fumes of NO_x.

EBJ100 CAS:3266-23-7 **HR: D**
2,3-EPOXYBUTANE
mf: C_4H_8O mw: 72.12

TOXICITY DATA with REFERENCE
mmo-sat 5 μmol/plate JJATDK 2,282,82
mma-sat 5 μmol/plate JJATDK 2,282,82
mmo-klp 20 μmol/L MUREAV 89,269,81

CONSENSUS REPORTS: EPA Genetic Toxicology Program. Reported in EPA TSCA Inventory.

SAFETY PROFILE: Mutation data reported. When heated to decomposition it emits acrid smoke and fumes.

EBJ500 CAS:930-22-3 **HR: 3**
3,4-EPOXY-1-BUTENE
mf: C_4H_6O mw: 70.10

PROP: Liquid. Mp: −135°, bp: 67°, flash p: < −58°F (CC), d: 0.869, autoign temp: 806°F, vap d: 2.41.

SYNS: BUTADIENE MONOXIDE ◇ 1,2-EPOXYBUTENE-3

TOXICITY DATA with REFERENCE
mmo-sat 1 μmol/plate BBRCA9 80,298,78
mma-sat 100 μmol/plate MUREAV 97,204,82
mmo-esc 20 μmol/L ARTODN 46,277,80
dnd-esc 1 μmol/L ARTODN 46,277,80
mmo-klp 1 μmol/L MUREAV 89,269,81
skn-mus TDLo:492 g/kg/41W-I:ETA JNCIAM 31,41,63
ipr-rat LD50:168 mg/kg TXAPA9 52,422,80
ipr-mus LDLo:8 mg/kg CBCCT* 1,44,49

CONSENSUS REPORTS: Reported in EPA TSCA Inventory.

SAFETY PROFILE: A poison by intraperitoneal route. Questionable carcinogen with experimental tumorigenic data. Mutation data reported. A very dangerous fire hazard when exposed to heat or flame; can react with oxidizing materials. To fight fire, use CO_2, dry chemical, water spray. When heated to decomposition it emits acrid smoke and fumes.

EBK000 CAS:10138-34-8 **HR: 2**
2,3-EPOXYBUTYRIC ACID BUTYL ESTER
mf: $C_8H_{14}O_3$ mw: 158.22

TOXICITY DATA with REFERENCE
orl-rat LD50:500 mg/kg AIHAAP 23,95,62
skn-rbt LD50:2830 mg/kg AIHAAP 23,95,62

SAFETY PROFILE: Moderately toxic by ingestion and skin contact. When heated to decomposition it emits acrid smoke and fumes. See also EPOXY RESINS and ESTERS.

EBK500 CAS:6509-08-6 **HR: 3**
1,2-EPOXYBUTYRONITRILE
mf: C_4H_5NO mw: 83.10

SYN: 3,4-EPOXYBUTYRONITRILE

TOXICITY DATA with REFERENCE
scu-mus TDLo:324 mg/kg/81W-I:ETA JNCIAM 46,143,71

CONSENSUS REPORTS: Cyanide and its compounds are on the Community Right-To-Know List.

SAFETY PROFILE: Questionable carcinogen with experimental tumorigenic data. When heated to decomposition it emits toxic fumes of NO_x and CN^-. See also NITRILES.

EBL000 CAS:124-98-1 **HR: 3**
4,9-EPOXYCEVANE-3-α,4-β,12,14,16-β,17,20-HEPTOL
mf: $C_{27}H_{43}NO_8$ mw: 509.71

PROP: An alkamine isolated from *Veratrum album* (JPETAB 82,167,44).

SYNS: CEVIN ◇ CEVINE ◇ SABADININE

TOXICITY DATA with REFERENCE
ipr-rat LD50:67 mg/kg JPETAB 82,167,44
scu-mus LD50:160 mg/kg JPETAB 113,89,55
ivn-mus LD50:87 mg/kg JPETAB 82,167,44

SAFETY PROFILE: Poison by intraperitoneal, intravenous, and subcutaneous routes. When heated to decomposition it emits toxic fumes of NO_x.

EBL500 CAS:508-65-6 **HR: 3**
4,9-EPOXYCEVANE-3-β,4-β,7-α,14,15-α,16-β,20-HEPTOL
mf: $C_{27}H_{43}NO_8$ mw: 509.71

PROP: An alkamine isolated from *Veratrum album* (JPETAB 82,167,44).

SYNS: GERMIN ◇ GERMINE

TOXICITY DATA with REFERENCE
scu-rat LDLo:2000 mg/kg AEPPAE 189,397,38
ivn-mus LD50:139 mg/kg JPETAB 82,167,44
scu-frg LDLo:250 mg/kg AEPPAE 189,397,38

SAFETY PROFILE: Poison by subcutaneous and intravenous routes. When heated to decomposition it emits toxic fumes of NO_x.

EBM000 CAS:1250-95-9 ***HR: 3***
EPOXYCHOLESTEROL
mf: $C_{27}H_{46}O_2$ mw: 402.73

SYNS: CHOLESTEROL-α-EPOXIDE ◇ CHOLESTEROL-5-α,6-α-EPOXIDE ◇ CHOLESTEROL OXIDE ◇ CHOLESTEROL-α-OXIDE ◇ 5-α,6-α-EPOXYCHOLESTANOL ◇ 5,6-α-EPOXY-5-α-CHOLESTAN-3-β-OL

TOXICITY DATA with REFERENCE
dns-hmn:fbr 10 mg/L/4H AJEBAK 56,287,78
cyt-hmn:fbr 500 μg/L/4H AJEBAK 56,287,78
otr-ham:emb 625 μg/L CALEDQ 6,143,79
scu-mus TDLo:600 mg/kg/72W-I:CAR JNCIAM 19,977,57

CONSENSUS REPORTS: EPA Genetic Toxicology Program.

SAFETY PROFILE: Questionable carcinogen with experimental carcinogenic data. Human mutation data reported. When heated to decomposition it emits acrid smoke and fumes.

EBO000 CAS:3388-04-3 ***HR: 1***
EPOXYCYCLOHEXYLETHYL TRIMETHOXY SILANE
mf: $C_{11}H_{22}O_4Si$ mw: 246.42

SYNS: β-(3,4-EPOXYCYCLOHEXYL)ETHYLTRIMETHOXYSILANE ◇ SILANE Y-4086 ◇ SILICONE A-186 ◇ UNION CARBIDE A-186

TOXICITY DATA with REFERENCE
skn-rbt 500 mg open MLD UCDS** 1/17/72
orl-rat LD50:12300 mg/kg AIHAAP 30,470,69
skn-rbt LD50:6300 mg/kg UCDS** 2/11/64

CONSENSUS REPORTS: Reported in EPA TSCA Inventory.

SAFETY PROFILE: Mildly toxic by ingestion and skin contact. A skin irritant. When heated to decomposition it emits acrid smoke and fumes. See also SILANE.

EBP000 CAS:962-32-3 ***HR: 3***
5,6-EPOXY-5,6-DIHYDROBENZ(a)ANTHRACENE
mf: $C_{18}H_{12}O$ mw: 244.30

SYNS: BENZ(a)ANTHRACENE-5,6-OXIDE ◇ BENZ(a)ANTHRA-5,6-OXIDE ◇ BENZO(a)ANTHRACENE-5,6-OXIDE ◇ 1a,11b-DIHYDRO-BENZ(3,4)ANTHRA(1,2-b)OXIRENE ◇ 3,4-DIHYDRO-3,4-EPOXY-1,2-BENZANTHRACENE

TOXICITY DATA with REFERENCE
mmo-sat 3 μg/plate IJCNAW 16,787,75
mma-sat 450 nmol/L BBRCA9 66,693,75
dni-omi 200 μg/L PNASA6 74,1378,77
sln-dmg-par 5 mmol/L CNREA8 33,2354,73
dns-hmn:fbr 10 μmol/L/3H IJCNAW 16,284,75
otr-mus:fbr 500 μg/L CNREA8 32,716,73
scu-mus TDLo:400 mg/kg/10W-I:CAR IJCNAW 2,500,67

CONSENSUS REPORTS: EPA Genetic Toxicology Program.

SAFETY PROFILE: Questionable carcinogen with experimental carcinogenic data. Human mutation data reported. When heated to decomposition it emits acrid smoke and fumes.

EBP500 CAS:1421-85-8 ***HR: 3***
5,6-EPOXY-5,6-DIHYDRODIBENZ(a,h)ANTHRACENE
mf: $C_{22}H_{14}O$ mw: 294.36

SYNS: DBA-5,6-EPOXIDE ◇ DIBENZ(a,h)ANTHRACENE-5,6-OXIDE ◇ 5,6-DIHYDRO-5,6-EPOXYDIBENZ(a,h)ANTHRACENE

TOXICITY DATA with REFERENCE
mmo-sat 300 ng/plate IJCNAW 16,787,75
mma-sat 5 μg/plate PNASA6 72,5135,75
dns-hmn:hla 1 μmol/L CNREA8 38,2621,78
otr-mus:fbr 500 μg/L CNREA8 32,716,72
dnd-ham:lng 1 mg/L CBINA8 4,389,71/72
scu-mus TDLo:400 mg/kg/10W-I:CAR IJCNAW 2,500,67

CONSENSUS REPORTS: EPA Genetic Toxicology Program.

SAFETY PROFILE: Questionable carcinogen with experimental carcinogenic data. Human mutation data reported. When heated to decomposition it emits acrid smoke and fumes.

EBQ500 CAS:6921-35-3 ***HR: 3***
1,3-EPOXY-2,2-DIMETHYLPROPANE
mf: $C_5H_{10}O$ mw: 86.15

SYNS: 3,3-DIMETHYLOXETANE ◇ β,β-DIMETHYLTRIMETHYLENE OXIDE ◇ 3,3-DIMETHYLTRIMETHYLENE OXIDE

TOXICITY DATA with REFERENCE
scu-rat TDLo:1020 mg/kg/61W-I:ETA BJCAAI 17,100,63
ipr-mus LDLo:2000 mg/kg JMPCAS 1,355,59

SAFETY PROFILE: Moderately toxic by intraperitoneal route. Questionable carcinogen with experimental tumorigenic data. When heated to decomposition it emits acrid smoke and fumes.

EBQ550 CAS:39597-90-5 ***HR: 3***
endo-2,3-EPOXY-7,8-DIOXABICYCLO(2.2.2)OCT-5-ONE
mf: $C_6H_6O_3$ mw: 126.11

SYN: OXEPIN-3,6-ENDOPEROXIDE

SAFETY PROFILE: An unstable explosive. When heated to decomposition it emits acrid smoke and fumes. See also PEROXIDES and EXPLOSIVES.

EBR000 CAS:96-09-3 ***HR: 3***
1,2-EPOXYETHYLBENZENE
mf: C_8H_8O mw: 120.16

PROP: Colorless liquid. Bp: 194.2, flash p: 165°F (OC), fp: −36.7°, d: 1.0469 @ 25°/4°, vap d: 4.14.

SYNS: EPOXYETHYLBENZENE (8CI) ◇ EPOXYSTYRENE ◇ α,β-EPOXYSTYRENE ◇ NCI-C54977 ◇ PHENETHYLENE OXIDE ◇ 1-PHENYL-1,2-EPOXYETHANE ◇ PHENYLETHYLENE OXIDE ◇ PHENYLOXIRANE ◇ 1-PHENYLOXIRANE ◇ 2-PHENYLOXIRANE ◇ STYRENE EPOXIDE ◇ STYRENE OXIDE ◇ STYRENE-7,8-OXIDE ◇ STYRYL OXIDE

TOXICITY DATA with REFERENCE
skn-rbt 10 mg/24H open MLD AMIHBC 10,61,54
skn-rbt 500 mg open MOD UCDS** 7/21/71
eye-rbt 500 mg open AMIHBC 10,61,54
dni-hmn:hla 4400 μmol/L MUREAV 93,447,82
oms-mus:fbr 1 μmol/L CRNGDP 6,1367,85
ihl-rat TCLo:100 ppm/7H (female 1-19D post):TER SWEHDO 9,94,83
ihl-rat TCLo:100 ppm/7H (female 1-19D post):REP SWEHDO 9,94,83
orl-rat TDLo:65 g/kg/52-I:CAR tumors ANYAA9 534,203,88
orl-rat TDLo:10 g/kg/52W-I:CAR MELAAD 70,358,79
orl-mus TDLo:273 g/kg/2Y-C:CAR JJIND8 77,471,86
skn-mus TDLo:74 g/kg/62W-I:ETA JNCIAM 31,41,63
orl-rat LD50:2000 mg/kg NTIS** PB81-168510
ihl-rat LCLo:500 ppm/4H AMIHBC 10,61,54
ipr-rat LD50:460 mg/kg TXAPA9 18,321,71
orl-mus LD50:1500 mg/kg JJIND8 77,471,86
skn-rbt LD50:1060 mg/kg AMIHBC 10,61,54

CONSENSUS REPORTS: IARC Cancer Review: Group 2B IMEMDT 7,345,87; Animal Sufficient Evidence IMEMDT 36,245,85; Animal Inadequate Evidence IMEMDT 19,275,79; IMEMDT 11,201,76. Reported in EPA TSCA Inventory. EPA Genetic Toxicology Program.

SAFETY PROFILE: Suspected carcinogen with experimental carcinogenic, tumorigenic, and teratogenic data. Moderately toxic by ingestion, inhalation, skin contact, and intraperitoneal routes. Experimental reproductive effects. Human mutation data reported. A skin and eye irritant. Flammable when exposed to heat, flame, or oxidizers; can react with oxidizing materials. To fight fire, use foam, CO_2, dry chemical. When heated to decomposition it emits acrid smoke and fumes.

EBR500 CAS:13484-13-4 ***HR: 2***
2-(α,β-EPOXYETHYL)-5,6-EPOXYBENZENE
mf: $C_8H_6O_2$ mw: 134.14

SYN: 2-(1,2-EPOXYETHYL)-5,6-EPOXYBENZENE

TOXICITY DATA with REFERENCE
skn-rbt 10 mg/24H open SEV AIHAAP 23,95,62
orl-rat LDLo:2830 mg/kg AIHAAP 23,95,62
skn-rbt LD50:620 mg/kg AIHAAP 23,95,62

SAFETY PROFILE: Moderately toxic by ingestion and skin contact. A severe skin irritant. When heated to decomposition it emits acrid smoke and fumes.

EBT000 CAS:1855-36-3 ***HR: D***
4-(EPOXYETHYL)-1,2-XYLENE
mf: $C_{10}H_{12}O$ mw: 148.22

SYNS: (3,4-DIMETHYLPHENYL)OXIRENE ◇ 3,4-DIMETHYL STYRENE OXIDE

TOXICITY DATA with REFERENCE
mmo-esc 2 mmol/L CMSHAF 7,737,78
dnr-esc 1 mmol/L CMSHAF 7,737,78
pic-esc 2300 μmol/L CBINA8 35,71,81
msc-ham:lng 200 μmol/L CMSHAF 8,369,79

CONSENSUS REPORTS: EPA Genetic Toxicology Program.

SAFETY PROFILE: Mutation data reported. When heated to decomposition it emits acrid smoke and fumes.

EBT500 CAS:53940-49-1 ***HR: D***
2',3'-EPOXYEUGENOL
mf: $C_{10}H_{12}O_3$ mw: 180.22

TOXICITY DATA with REFERENCE
mma-sat 20 nmol/plate CRSBAW 171,1041,77

CONSENSUS REPORTS: Reported in EPA TSCA Inventory.

SAFETY PROFILE: Mutation data reported. When heated to decomposition it emits acrid smoke and fumes.

EBU000 CAS:503-09-3 ***HR: 3***
1,2-EPOXY-3-FLUOROPROPANE
mf: C_3H_5FO mw: 76.08

SYN: EPIFLUOROHYDRIN

TOXICITY DATA with REFERENCE
mmo-esc 20 μmol/L ARTODN 46,277,80
mmo-klp 200 μmol/L MUREAV 89,269,81
mma-sat 500 μg/plate MUREAV 58,217,78

ihl-rat LCLo:111 mg/m^3 NDRC** -,12,43
ivn-mus LD50:178 mg/kg CSLNX* NX#00407

SAFETY PROFILE: Poison by inhalation and intravenous routes. Mutation data reported. When heated to decomposition it emits toxic fumes of F^-

EBU100 CAS:68071-23-8 ***HR: 1***
EPOXYGUAIENE
mf: $C_{15}H_{24}O$ mw: 220.39

SYNS: AZULENE,1,2,3,4,5,6,7,8-OCTAHYDRO-1,4-DIMETHYL-7-(1-METHYLETHYLIDENE)-, MONOEPOXIDE ◇ 1,2,3,4,5,6,7,8-OCTAHYDRO-1,4-DIMETHYL-7-(1-METHYLETHYLIDENE)AZULENE-MONOEP OXIDE

TOXICITY DATA with REFERENCE
skn-rbt 500 mg/24H MOD FCTOD7 21,851,83

CONSENSUS REPORTS: Reported in EPA TSCA Inventory.

SAFETY PROFILE: A skin irritant. When heated to decomposition it emits acrid smoke and irritating fumes.

EBU500 ***HR: 2***
EPOXY HARDENER ZZL-0814

SYN: ZZL-0814

TOXICITY DATA with REFERENCE
skn-rbt 500 mg open MLD UCDS** 10/1/71
eye-rbt 15 mg SEV UCDS** 10/1/71
orl-rat LD50:3250 mg/kg UCDS** 10/1/71
skn-rbt LD50:2500 mg/kg UCDS** 10/1/71

SAFETY PROFILE: Moderately toxic by ingestion and skin contact. A skin and severe eye irritant.

EBV000 ***HR: 2***
EPOXY HARDENER ZZL-0816

SYN: ZZL-0816

TOXICITY DATA with REFERENCE
skn-rbt 500 mg open MLD UCDS** 10/1/71
orl-rat LD50:4920 mg/kg UCDS** 10/1/71
skn-rbt LD50:2520 mg/kg UCDS** 10/1/71

SAFETY PROFILE: Moderately toxic by skin contact. Mildly toxic by ingestion. A skin irritant.

EBV100 CAS:69136-21-6 ***HR: 2***
EPOXY HARDENER ZZL-0822
mf: $C_{10}H_{24}N_2O_3$ mw: 220.36

SYNS: (OXYBIS(2,1-ETHANEDIYLOXY))BIS-PROPANAMINE ◇ POLYGLYCOLDIAMINE H 221

TOXICITY DATA with REFERENCE
skn-rbt 50 mg open MOD UCDS** 9/15/65
eye-rbt 5 mg SEV UCDS** 9/15/65
orl-rat LD50:4290 mg/kg UCDS** 9/15-65
skn-rbt LD50:2500 mg/kg UCDS** 9/15/65

SAFETY PROFILE: Moderately toxic by ingestion and skin contact. A skin and severe eye irritant. When heated to decomposition it emits toxic fumes of NO_x. See also AMINES.

EBV500 ***HR: 2***
EPOXY HARDENER ZZL-0854

SYN: ZZL-0854

TOXICITY DATA with REFERENCE
skn-rbt 500 mg open MLD UCDS** 3/7/72
eye-rbt 15 mg SEV UCDS** 3/7/72
orl-rat LD50:2460 mg/kg UCDS** 3/7/72
skn-rbt LD50:5040 mg/kg UCDS** 3/7/72

SAFETY PROFILE: Moderately toxic by ingestion. Mildly toxic by skin contact. A skin and severe eye irritant.

EBW000 CAS:42498-58-8 ***HR: 2***
EPOXY HARDENER ZZLA-0334
mf: $C_9H_{10}O_3$ mw: 166.19

SYNS: 3a,4,7,7a-TETRAHYDRO-3a-METHYL-1,3-ISOBENZOFURANDIONE ◇ ZZLA-0334

TOXICITY DATA with REFERENCE
skn-rbt 500 mg open MLD UCDS** 6/22/66
eye-rbt 1 mg SEV UCDS** 6/22/66
orl-rat LD50:2460 mg/kg UCDS** 6/22/66
skn-rbt LD50:1590 mg/kg UCDS** 6/22/66

SAFETY PROFILE: Moderately toxic by ingestion and skin contact. A skin and severe eye irritant.

EBW500 CAS:1024-57-3 ***HR: 3***
EPOXYHEPTACHLOR
mf: $C_{10}H_5Cl_7O$ mw: 389.30

SYNS: ENT 25,584 ◇ HCE ◇ HEPTACHLOR EPOXIDE (USDA) ◇ 1,4,5,6,7,8,8-HEPTACHLORO-2,3-EPOXY-2,3,3a,4,7,7a-HEXAHYDRO-4,7-METHANOINDENE ◇ 1,4,5,6,7,8,8-HEPTACHLORO-2,3-EPOXY-3a,4,7,7a-TETRAHYDRO-4,7-METHANOINDAN ◇ 2,3,4,5,6,7,7-HEPTACHLORO-1a,1b,5,5a,6,6a-HEXAHYDRO-2,5-METHANO-2H-INDENO (1,2-b)OXIRENE ◇ VELSICOL 53-CS-17

TOXICITY DATA with REFERENCE
mma-hmn:fbr 10 μmol/L MUREAV 42,161,77
orl-mus TDLo:580 mg/kg/69W-C:CAR ARTODN 38,163,77
orl-mus TD:876 mg/kg/2Y-C:CAR ECEBDI 45,147,77
orl-rat LD50:47 mg/kg ARSIM* 20,27,66
orl-mus LD50:39 mg/kg ARSIM* 20,27,66
ivn-mus LDLo:10 mg/kg JPETAB 107,266,53

CONSENSUS REPORTS: IARC Cancer Review: Human Inadequate Evidence IMEMDT 20,129,79; Animal Inadequate Evidence IMEMDT 5,173,74; Animal Limited Evidence IMEMDT 20,129,79. EPA Genetic Toxicology Program.

SAFETY PROFILE: Suspected carcinogen with experimental carcinogenic data. Poison by ingestion and intravenous routes. Human mutation data reported. When heated to decomposition it emits toxic fumes of Cl^-. See also HEPTACHLOR.

EBX500 CAS:7320-37-8 ***HR: 3***
1,2-EPOXYHEXADECANE
mf: $C_{16}H_{32}O$ mw: 240.48

SYNS: HEXADECENE EPOXIDE ◇ NCI-C55538

TOXICITY DATA with REFERENCE
skn-mus TDLo:53 g/kg/44W-I:ETA JNCIAM 39,1217,67

CONSENSUS REPORTS: Reported in EPA TSCA Inventory.

SAFETY PROFILE: Questionable carcinogen with experimental tumorigenic data. When heated to decomposition it emits acrid smoke and fumes.

EBY100 CAS:71507-79-4 ***HR: D***
4-α,5-α-EPOXY-17-β-HYDROXY-4,17-DIMETHYL-3-OXOANDROSTANE-2-α-CARBONITRILE
mf: $C_{21}H_{31}NO_3$ mw: 345.53

SYN: WIN 32729

TOXICITY DATA with REFERENCE
orl-mky TDLo:35714 μg/kg (50-54D preg):REP CCPTAY 24,289,81

CONSENSUS REPORTS: Cyanide and its compounds are on the Community Right-To-Know List.

SAFETY PROFILE: Experimental reproductive effects. A steroid. When heated to decomposition it emits toxic fumes of NO_x and CN^-. See also NITRILES.

EBY500 CAS:37764-28-6 ***HR: 3***
4,5-α-EPOXY-3-HYDROXY-17-METHYLMORPHINAN-6-ONE-N-OXIDE TARTRATE
mf: $C_{17}H_{19}NO_4$ mw: 301.37

SYNS: A/VI ◇ DIHYDROMORPHINON-N-OXYD-DITARTARAT (GERMAN)

TOXICITY DATA with REFERENCE
ivn-mus LD50:2000 mg/kg ARZNAD 7,594,57
ivn-rbt LD50:160 mg/kg ARZNAD 7,594,57

SAFETY PROFILE: Poison by intravenous route. When heated to decomposition it emits toxic fumes of NO_x.

EBY600 CAS:13647-35-3 ***HR: 3***
4-α-5-EPOXY-17-β-HYDROXY-3-OXO-5-α-ANDROSTANE-2-α-CARBONITRILE
mf: $C_{20}H_{27}NO_3$ mw: 329.48

SYNS: 5-α-ANDROSTANE-2-α-CARBONITRILE, 4-α-5-EPOXY-17-β-HYDROXY-3-OXO- ◇ ANDROSTANE-2-CARBONITRILE, 4,5-EPOXY-17-HYDROXY-3-OXO-, (2-α-4-α-5-α-17-β)- ◇ (2-α-4-α-5-α-17-β)-4,5-EPOXY-17-HYDROXY-3-OXOANDROSTANE-2-CARBONITRILE ◇ MODRENAL ◇ TRILOSTANE ◇ WIN 24450

TOXICITY DATA with REFERENCE
orl-man TDLo:48 mg/kg (male 14D pre):REP ARZNAD 33,754,83
ipr-rat LD50:1275 mg/kg IYKEDH 17,365,86
scu-rat LD50:7050 mg/kg IYKEDH 17,365,86
ivn-rat LD50:102 mg/kg IYKEDH 17,365,86
ipr-mus LD50:1552 mg/kg IYKEDH 17,365,86
ivn-mus LD50:109 mg/kg IYKEDH 17,365,86

SAFETY PROFILE: Poison by intravenous route. Moderately toxic intraperitoneal route. Experimental reproductive effects. When heated to decomposition it emits toxic fumes of NO_x.

ECA000 CAS:4247-30-7 ***HR: 3***
4,5-EPOXY-3-HYDROXYVALERIC ACID-β-LACTONE
mf: $C_5H_6O_3$ mw: 114.11

TOXICITY DATA with REFERENCE
skn-mus TDLo:76 g/kg/63W-I:NEO JNCIAM 35,707,65

SAFETY PROFILE: Questionable carcinogen with experimental neoplastigenic data. When heated to decomposition it emits acrid smoke and fumes.

ECA500 CAS:67195-51-1 ***HR: 3***
11,12-EPOXY-3-METHYLCHOLANTHRENE
mf: $C_{21}H_{14}O$ mw: 282.35

SYNS: MCA-11,12-EPOXIDE ◇ MCA-11,12-OXIDE ◇ 3-METHYLCHOLANTHRENE-11,12-EPOXIDE ◇ 3-METHYL-11,12-EPOXYCHOLANTHRENE

TOXICITY DATA with REFERENCE
mmo-sat 50 μg/plate CNREA8 45,2600,85
pic-omi 23400 nmol/L NNBYA7 234,186,71
otr-mus:fbr 750 μg/L CNREA8 32,716,72
msc-ham:lng 5 mmol/L PNASA6 68,3195,71
skn-mus TDLo:226 μg/kg:ETA JNCIAM 58,1051,77

SAFETY PROFILE: Questionable carcinogen with experimental tumorigenic data. Mutation data reported. When heated to decomposition it emits acrid smoke and fumes.

ECB000 CAS:141-37-7 ***HR: 3***
3,4-EPOXY-6-METHYLCYCLOHEXYLMETHYL-3′,4′-EPOXY-6′-METHYLCYCLOHEXANE CARBOXYLATE
mf: $C_{16}H_{24}O_4$ mw: 280.40

SYNS: CHISSONOX 201 ◇ EP 201 ◇ EPOXIDE-201 ◇ 3,4-EPOXY-6-METHYLCYCLOHEXENECARBOXYLIC ACID (3,4-EPOXY-6-METHYLCYCLOHEXYLMETHYL) ESTER ◇ 3,4-EPOXY-6-METHYLCYCLOHEXYLMETHYL-3,4-EPOXY-6-METHYLCYCLOHEXANECARBOXYLATE ◇ 4,5-EPOXY-2-METHYLCYCLOHEXYLMETHYL-4,5-EPOXY-2-METHYLCYCLOHEXANECARBOXYLATE ◇ 6-METHYL-3,4-EPOXYCYCLOHEXYLMETHYL-6-METHYL-3,4-EPOXYCYCLOHEXANECARBOXYLATE ◇ UNOX 201 ◇ UNOX EPOXIDE 201

TOXICITY DATA with REFERENCE
skn-rbt 10 mg/24H open MLD AIHAAP 23,95,62
skn-mus TDLo:67 g/kg/56W-I:ETA JNCIAM 39,1217,67
orl-rat LD50:4920 mg/kg UCDS** 6/8/65

CONSENSUS REPORTS: IARC Cancer Review: Group 3 IMEMDT 7,56,87; Animal Sufficient Evidence IMEMDT 11,147,76.

SAFETY PROFILE: Mildly toxic by ingestion. A skin irritant. Questionable carcinogen with experimental tumorigenic data. When heated to decomposition it emits acrid smoke and fumes.

ECC600 CAS:1192-22-9 ***HR: 2***
2,3-EPOXY-2-METHYLPENTANE
mf: $C_6H_{12}O$ mw: 100.18

SYN: 2-METHYL-2,3-PENTYLENE OXIDE

TOXICITY DATA with REFERENCE
orl-rat LD50:3500 mg/kg GTPZAB 25(11),51,81
ihl-rat LC50:16400 mg/m^3 GTPZAB 25(11),51,81
ipr-rat LD50:1600 mg/kg GTPZAB 25(11),51,81
orl-mus LD50:3400 mg/kg GTPZAB 25(11),51,81
ihl-mus LC50:13800 mg/m^3 GTPZAB 25(11),51,81

SAFETY PROFILE: Moderately toxic by ingestion and intraperitoneal routes. When heated to decomposition it emits acrid smoke and fumes.

ECD500 CAS:2443-39-2 ***HR: 3***
cis-9,10-EPOXYOCTADECANOIC ACID
mf: $C_{18}H_{34}O_3$ mw: 298.52

SYNS: cis-9,10-EPOXYOCTADECANOATE ◇ EPOXYOLEIC ACID ◇ 9,10-EPOXYSTEARIC ACID ◇ cis-9,10-EPOXYSTEARIC ACID ◇ cis-3-OCTYL-OXIRANEOCTANOIC ACID

TOXICITY DATA with REFERENCE
skn-mus TDLo:2880 mg/kg/24W-I:ETA JNCIAM 31,41,63

CONSENSUS REPORTS: IARC Cancer Review: Group 3 IMEMDT 7,56,87; Animal Inadequate Evidence IMEMDT 11,153,76.

SAFETY PROFILE: Questionable carcinogen with experimental tumorigenic data. When heated to decomposition it emits acrid smoke and fumes.

ECE000 CAS:2984-50-1 ***HR: 3***
1,2-EPOXYOCTANE
mf: $C_8H_{16}O$ mw: 128.24

SYN: OCTYLENE EPOXIDE

TOXICITY DATA with REFERENCE
unr-mus TDLo:103 mg/kg:ETA RARSAM 3,193,63

SAFETY PROFILE: Questionable carcinogen with experimental tumorigenic data. When heated to decomposition it emits acrid smoke and fumes.

ECE500 CAS:17397-89-6 ***HR: 3***
2,3-EPOXY-4-OXO-7,10-DODECADIENAMIDE
mf: $C_{12}H_{17}NO_3$ mw: 223.30

SYNS: CERULENIN ◇ (2R,3S)-2,3-EPOXY-4-OXO-7E,10E-DODECADIENAMIDE ◇ (2S)(3R)-2,3-EPOXY-4-OXO-7,10-DODECADIENOYLAMIDE ◇ HELICOCERIN ◇ (2R-(2-α,3-α(4E,7E)))-3-(1-OXO-4,7-NONADIENYL)OXIRANECARBOXAMIDE ◇ 3-(1-OXO-4,7-NONADIENYL) OXIRANECARBOXAMIDE

TOXICITY DATA with REFERENCE
mmo-omi 2 mg/L JOBAAY 133,472,78
orl-mus LD50:547 mg/kg JAJAAA 17,1,64
ipr-mus LD50:211 mg/kg JAJAAA 17,1,64
scu-mus LD50:245 mg/kg 85ERAY 3,1945,78
ivn-mus LD50:154 mg/kg JAJAAA 17,1,64

SAFETY PROFILE: Poison by intraperitoneal, intravenous, and subcutaneous routes. Moderately toxic by ingestion. Mutation data reported. When heated to decomposition it emits toxic fumes of NO_x.

ECE550 CAS:64011-46-7 ***HR: 3***
4,5-EPOXY-2-PENTENAL
mf: $C_5H_6O_2$ mw: 98.11

SYN: 2-PENTENAL, 4,5-EPOXY-

TOXICITY DATA with REFERENCE
skn-rbt 2 mg/24H SEV 85JCAE-,771,86
eye-rbt 50 μg/24H SEV 85JCAE-,771,86
orl-rat LD50:62 mg/kg TXAPA9 28,313,74
ihl-rat LCLo:600 ppm/4H TXAPA9 28,313,74
skn-rbt LD50:40 mg/kg TXAPA9 28,313,74

SAFETY PROFILE: Poison by ingestion and skin contact. A severe skin and eye irritant. When heated to decomposition it emits acrid smoke and irritating fumes.

ECG100 CAS:67722-96-7 ***HR: 3***
2,3-EPOXY PROPIONALDEHYDE OXIME
mf: $C_3H_5NO_2$ mw: 87.08

$OCH_2CHCH{=}NOH$

SYN: OXIRANE CARBOXALDEHYDE OXIME

SAFETY PROFILE: Polymerization is violent or explosive. When heated to decomposition it emits toxic fumes of NO_x. See also ALDEHYDES.

ECG200 CAS:4711-95-9 ***HR: 3***
4-(2,3-EPOXYPROPOXY)BUTANOL
mf: $C_7H_{14}O_3$ mw: 146.19

$CH_2OCHOC_4H_8OH$

SAFETY PROFILE: Potentially explosive reaction when heated with trichloroethylene. When heated to decomposition it emits acrid smoke and fumes. See also ALCOHOLS.

ECG500 CAS:63991-57-1 ***HR: 2***
p-(2,3-EPOXYPROPOXY)-N-PHENYLBENZYLAMINE
mf: $C_{16}H_{17}NO_2$ mw: 255.34

SYN: N-(4-(2,3-EPOXYPROPOXY)PHENYL)BENZYLAMINE

TOXICITY DATA with REFERENCE
scu-rat TDLo:4750 mg/kg/18W-I:ETA ANYAA9 68,750,58

SAFETY PROFILE: Questionable carcinogen with experimental tumorigenic data. When heated to decomposition it emits toxic fumes of NO_x.

ECH000 CAS:2530-83-8 ***HR: 2***
3-(2,3-EPOXYPROPOXY)PROPYLTRIMETHOXYSILANE
mf: $C_9H_{20}O_5Si$ mw: 236.38

SYNS: Γ-GLYCIDOXYPROPYLTRIMETHOXYSILANE ◇ SILANE-Y-4087 ◇ SILICONE A-187 ◇ UNION CARBIDE A-187

TOXICITY DATA with REFERENCE
skn-rbt 500 mg open MLD UCDS** 1/19/72
eye-rbt 100 mg MLD UCDS** 1/19/72
orl-rat LD50:23 g/kg UCDS** 2/11/64
skn-rbt LD50:3970 mg/kg AIHAAP 30,470,69

CONSENSUS REPORTS: Reported in EPA TSCA Inventory.

SAFETY PROFILE: Moderately toxic by skin contact. Mildly toxic by ingestion. A skin and eye irritant. When heated to decomposition it emits acrid smoke and fumes. See also SILANE.

ECH500 CAS:106-90-1 ***HR: 3***
2,3-EPOXYPROPYL ACRYLATE
mf: $C_6H_8O_3$ mw: 128.14

PROP: Insol in water. Bp: 57.2° @ 2 mm, flash p: 141°F (OC), d: 1.1, vap d: 4.4.

SYNS: 2,3-EPOXY-1-PROPANOL ACRYLATE ◇ 2,3-EPOXYPROPYL ESTER ACRYLIC ACID ◇ GLYCIDYL ACRYLATE ◇ GLYCIDYL PROPENATE ◇ 2-PROPENOIC ACID OXIRANYLMETHYL ESTER

TOXICITY DATA with REFERENCE
skn-rbt 100 μg/24H open AIHAAP 23,95,62
eye-rbt 1 mg SEV UCDS** 4/10/68
cyt-rat-ipr 50 μg/kg BJPCAL 6,235,51
orl-rat LD50:210 mg/kg AIHAAP 23,95,62
ihl-rat LCLo:125 ppm/4H AIHAAP 23,95,62
skn-rbt LD50:400 mg/kg UCDS** 4/10/68

CONSENSUS REPORTS: Reported in EPA TSCA Inventory.

SAFETY PROFILE: A poison by ingestion, inhalation, and skin contact. Mutation data reported. A skin and severe eye irritant. Flammable when exposed to heat or flame. Can react vigorously with oxidizers. To fight fire, use foam, dry chemical, CO_2. When heated to decomposition it emits acrid smoke and fumes. See also ESTERS.

ECI000 CAS:106-91-2 ***HR: 3***
2,3-EPOXYPROPYL METHACRYLATE
mf: $C_7H_{10}O_3$ mw: 142.17

SYNS: CP 105 ◇ 2,3-EPOXY-1-PROPANOL METHACRYLATE ◇ 2,3-EPOXYPROPYL ESTER METHACRYLIC ACID ◇ GLYCIDYL METHACRYLATE ◇ GLYCIDYL-α-METHYL ACRYLATE

TOXICITY DATA with REFERENCE
mmo-klp 200 μmol/L MUREAV 89,269,81
cyt-rat-ipr 3 mg/kg BJPCAL 6,235,51
orl-rat LD50:597 mg/kg GISAAA 50(2),67,85
orl-mus LD50:390 mg/kg GISAAA 50(2),67,85
ipr-mus LD50:1122 mg/kg JPMSAE 62,778,73
skn-rbt LD50:469 mg/kg AIHAAP 30,470,69
orl-gpg LD50:697 mg/kg GISAAA 50(2),67,85

CONSENSUS REPORTS: Reported in EPA TSCA Inventory.

SAFETY PROFILE: Poison by ingestion. Moderately toxic by skin contact and intraperitoneal routes. Mutation data reported. When heated to decomposition it emits acrid smoke and fumes.

ECI600 CAS:6659-62-7 ***HR: 3***
2,3-EPOXYPROPYL NITRATE
mf: $C_3H_5NO_4$ mw: 119.08

$OCH_2CHCH_2ONO_2$

SYN: OXIRANEMETHANOL NITRATE

SAFETY PROFILE: An explosive sensitive to shock and heating to 200°C. When heated to decomposition it emits toxic fumes of NO_x. See also NITRATES.

ECJ000 CAS:5431-33-4 ***HR: 3***
2,3-EPOXYPROPYL OLEATE
mf: $C_{21}H_{38}O_3$ mw: 338.59

SYNS: 2,3-EPOXY-1-PROPANOL OLEATE ◇ 2,3-EPOXYPROPYL ESTER of OLEIC ACID ◇ GLYCIDOL OLEATE ◇ GLYCIDYL OCTADECENOATE ◇ GLYCIDYL OLEATE ◇ OLEIC ACID GLYCIDYL ESTER ◇ OXIRANYLMETHYL ESTER of 9-OCTADECENOIC ACID

TOXICITY DATA with REFERENCE
cyt-rat-ipr 10 mg/kg BJPCAL 6,235,51
scu-mus TDLo:1040 mg/kg/52W-I:ETA CNREA8 30,1037,70
orl-rat LD50:3520 mg/kg AIHAAP 24,305,63
ivn-mus LDLo:15 mg/kg ARZNAD 17,693,67
skn-rbt LD50:8000 mg/kg AIHAAP 24,305,63

CONSENSUS REPORTS: IARC Cancer Review: Group 3 IMEMDT 7,56,87; Animal Inadequate Evidence IMEMDT 11,183,76.

SAFETY PROFILE: Poison by intravenous route. Moderately toxic by ingestion and subcutaneous routes. Mildly toxic by skin contact. Questionable carcinogen with experimental tumorigenic data. Mutation data reported. When heated to decomposition it emits acrid smoke and fumes. See also ESTERS.

ECJ100 CAS:76828-34-7 ***HR: 3***
3-(2,3-EPOXYPROPYLOXY)-2,2-DINITROPROPYL AZIDE
mf: $C_6H_9N_5O_6$ mw: 247.17

$$\overline{OCH_2C}HCH_2OCH_2C(NO_2)_2CH_2N_3$$

SAFETY PROFILE: An explosive sensitive to impact, shock, friction and heat. When heated to decomposition it emits toxic fumes of NO_x. See also AZIDES.

ECK000 CAS:5455-98-1 ***HR: 2***
N-(2,3-EPOXYPROPYL)-PHTHALIMIDE
mf: $C_{11}H_9NO_3$ mw: 203.21

SYNS: N-GLYCIDYLPHTHALIMIDE ◇ 2-(OXIRANYLMETHYL)-1H-ISOINDOLE-1,3(2H)-DIONE

TOXICITY DATA with REFERENCE
skn-rbt 500 mg/24H MLD TXAPA9 51,197,79
eye-rbt 100 mg SEV TXAPA9 51,197,79
mmo-sat 500 ng/plate MUREAV 68,251,79
mma-sat 5 mg/plate MUREAV 68,251,79
mmo-klp 5 μmol/L MUREAV 89,296,81
orl-rat LD50:4700 mg/kg TXAPA9 51,197,79
ihl-rat LCLo:4400 mg/m³/4H JACTDZ 4(1),219,85

CONSENSUS REPORTS: Reported in EPA TSCA Inventory.

SAFETY PROFILE: Mildly toxic by ingestion and inhalation. A skin and severe eye irritant. Mutation data reported. When heated to decomposition it emits toxic fumes of NO_x.

ECK500 CAS:25928-94-3 ***HR: 3***
EPOXY RESIN (EPICHLOROHYDRIN and DIETHYLENE GLYCOL)

TOXICITY DATA with REFERENCE
ihl-rat LCLo:16 mg/m³/4H TPKVAL 10,110,68
orl-mus LD50:2200 mg/kg TPKVAL 10,110,68

SAFETY PROFILE: Poison by inhalation. Moderately toxic by ingestion. See also other epoxy resin entries. When heated to decomposition it emits toxic fumes of Cl^-.

ECL000 CAS:25068-38-6 ***HR: 2***
EPOXY RESIN ERL-2795
mf: $(C_{15}H_{16}O_2 \cdot C_3H_5ClO)x$

SYNS: ERL-2795 ◇ 4,4'-ISOPROPYLIDENE DIPHENOL POLYMER with 1-CHLORO-2,3-EPOXYPROPANE

TOXICITY DATA with REFERENCE
orl-rat LD50:1100 mg/kg UCDS** 6/6/69
skn-rbt LD50:2000 mg/kg UCDS** 6/6/69

SAFETY PROFILE: Moderately toxic by ingestion and skin contact. When heated to decomposition it emits acrid smoke and fumes. See also other epoxy resin entries.

ECL500 ***HR: D***
EPOXY RESINS, CURED

SAFETY PROFILE: Most cured resins have little or no toxicity. If curing is incomplete there may be residues of highly toxic curing agents such as the organic amines: m-phenylene diamine, diethylene triamine, tetraethylene pentamine, and hexamethylene tetramine, as well as phthalic anhydride and related compounds. When heated to decomposition they emit highly toxic fumes. See also various epoxy hardeners and POLYMERS, INSOLUBLE.

ECM500 ***HR: 3***
EPOXY RESINS, UNCURED

SYN: POLYMERS of EPICHLOROHYDRIN and 2,2-BIS(4-HYDROXY PHENYL)PIPERAZINE

SAFETY PROFILE: Animal experiments have shown disturbed blood formation. The degree of toxicity of uncured epoxy resins varies and is partly dependent on the extent of unreacted curing agents. See also other epoxy resin entries and POLYMERS, INSOLUBLE. When heated to decomposition it emits acrid smoke and fumes.

ECO500 CAS:123-36-4 ***HR: 2***
9,10-EPOXYSTEARIC ACID ALLYL ESTER
mf: $C_{21}H_{38}O_3$ mw: 338.59

SYNS: ALLYL-9,10-EPOXYSTEARATE ◇ EP-145

TOXICITY DATA with REFERENCE
skn-rbt 500 mg open MLD UCDS** 3/5/59
orl-rat LD50:1198 mg/kg UCDS** 2,5,59
skn-rbt LD50:15900 mg/kg AIHAAP 23,95,62

SAFETY PROFILE: Moderately toxic by ingestion. Mildly toxic by skin contact. A skin irritant. When heated to decomposition it emits acrid smoke and fumes. See also ALLYL COMPOUNDS and ESTERS.

ECP000 CAS:4509-11-9 ***HR: 3***
3,4-EPOXYSULFOLANE
mf: $C_4H_6O_3S$ mw: 134.16

SYNS: 2,3-EPOXYTETRAMETHYLENE SULFONE ◇ 6-OXA-3-THIABICYCLO(3.1.0)HEXANE-3,3-DIOXIDE ◇ TETRAHYDRO-3,4-EPOXYTHIOPHENE-1,1-DIOXIDE

TOXICITY DATA with REFERENCE
scu-mus TDLo:88 mg/kg/22W-I:ETA JNCIAM 46,143,71
ipr-mus LD50:5500 mg/kg RPTOAN 41,257,78

SAFETY PROFILE: Mildly toxic by intraperitoneal route. Questionable carcinogen with experimental tumorigenic data. When heated to decomposition it emits toxic fumes of SO_x.

ECP500 CAS:64521-16-0 ***HR: D***
1,2-EPOXY-1,2,3,4-TETRAHYDROBENZ(a)ANTHRACENE
mf: $C_{18}H_{14}O$ mw: 246.32

SYNS: TETRAHYDROBENZ(a)ANTHRACEN-1,2-EPOXIDE ◇ 1a,2,3,11c-TETRAHYDROBENZO(6,7)PHENANTHRO(3,4-b)OXIRENE ◇ 1,2,3,4-TETRAHYDRO-1,2-EPOXY BENZ(a)ANTHRACENE

TOXICITY DATA with REFERENCE
mmo-sat 100 pmol/plate CNREA8 43,1656,83
msc-ham:lng 1200 nmol/L PNASA6 74,2746,77

CONSENSUS REPORTS: EPA Genetic Toxicology Program.

SAFETY PROFILE: Mutation data reported. When heated to decomposition it emits acrid smoke and fumes. See also ANTHRACENE

ECQ050 CAS:36504-67-3 ***HR: D***
7,8-EPOXY-7,8,9,10-TETRAHYDROBENZO(a)PYRENE
mf: $C_{20}H_{14}O$ mw: 270.34

SYN: 7,8,9,10-TETRAHYDRO-BP-7,8-OXIDE

TOXICITY DATA with REFERENCE
mma-sat 300 pmol/plate CNREA8 36,3358,76
dni-omi 200 µg/L PNASA6 74,1378,77
msc-ham:lng 500 µg/L IJCNAW 24,203,79

SAFETY PROFILE: Mutation data reported. When heated to decomposition it emits acrid smoke and fumes.

ECQ100 CAS:36504-68-4 ***HR: 3***
9,10-EPOXY-7,8,9,10-TETRAHYDROBENZO(a)PYRENE
mf: $C_{20}H_{14}O$ mw: 270.34

SYNS: 7,8,9,10-TETRAHYDRO-BENZO(a)PYRENE-9,10-EPOXIDE ◇ 7,8,9,10-TETRAHYDRO-BENZO(a)PYRENE-9,10-EPOXYIDE ◇ 7,8,9,10-TETRAHYDRO-9,10-EPOXY-BENZO(a)PYRENE

TOXICITY DATA with REFERENCE
mmo-sat 100 pmol/plate CNREA8 40,642,80
mma-sat 300 pmol/plate CNREA8 36,3358,76
dni-omi 200 µg/L PNASA6 74,1378,77
msc-ham:lng 60 nmol/L CNREA8 36,3358,76
dnd-mam:lym 10 nmol/L CRNGDP 3,247,82
skn-mus TDLo:519 mg/kg/60W-I:ETA CNREA8 37,3356,77

CONSENSUS REPORTS: EPA Genetic Toxicology Program.

SAFETY PROFILE: Questionable carcinogen with experimental tumorigenic data by skin contact. Mutation data reported. When heated to decomposition it emits toxic fumes of NO_x.

ECQ150 CAS:66788-11-2 ***HR: 3***
9,10-EPOXY-9,10,11,12-TETRAHYDROBENZO(e)PYRENE
mf: $C_{20}H_{14}O$ mw: 270.34

SYNS: B(e)P H4-9,10-EPOXIDE ◇ 9,10,11,12-TETRAHYDRO-9,10-EPOXY-BENZO(e)PYRENE

TOXICITY DATA with REFERENCE
mmo-sat 1 nmol/plate CNREA8 40,1985,80
mma-sat 13 pmol/plate JBCHA3 254,4408,79
msc-ham:lng 1100 nmol/L CNREA8 40,1985,80
dnd-mam:lym 10 µmol/L CRNGDP 3,247,82
scu-mus TDLo:10 mg/kg:ETA JJIND8 64,617,80

SAFETY PROFILE: Questionable carcinogen with experimental tumorigenic data. Mutation data reported. When heated to decomposition it emits acrid smoke and fumes.

ECQ200 CAS:67694-88-6 ***HR: 3***
3,4-EPOXY-1,2,3,4-TETRAHYDROCHRYSENE
mf: $C_{18}H_{14}O$ mw: 246.32

TOXICITY DATA with REFERENCE
mma-sat 1 nmol/plate CNREA8 39,4069,79
msc-ham:lng 1 nmol/plate CNREA8 39,4069,79
ipr-mus TDLo:63 mg/kg/15D-I:NEO CNREA8 39,5063,79

SAFETY PROFILE: Questionable carcinogen with experimental neoplastigenic data. Mutation data reported. When heated to decomposition it emits acrid smoke and fumes. See also CHRYSENE.

ECR259 CAS:56179-80-7 ***HR: 3***
1,2-EPOXY-1,2,3,4-TETRAHYDROPHENANTHRENE
mf: $C_{14}H_{12}O$ mw: 196.26

TOXICITY DATA with REFERENCE
mma-sat 1 nmol/plate CNREA8 39,4069,79
msc-ham:lng 1 nmol/plate CNREA8 39,4069,79
ipr-mus TDLo:50 mg/kg/15D-I:NEO CNREA8 39,5063,79

SAFETY PROFILE: Questionable carcinogen with experimental neoplastigenic data. Mutation data reported. When heated to decomposition it emits acrid smoke and fumes.

ECR500 CAS:66997-69-1 ***HR: 3***
3,4-EPOXY-1,2,3,4-TETRAHYDROPHENANTHRENE
mf: $C_{14}H_{12}O$ mw: 196.26

SYN: PHENANTHRENETETRAHYDRO-3,4-EPOXIDE

TOXICITY DATA with REFERENCE
mma-sat 1 nmol/plate CNREA8 39,4069,79
msc-ham:lng 1 nmol/plate CNREA8 39,4069,79
skn-mus TDLo:16 mg/kg:NEO VOONAW 15(8),54,69

SAFETY PROFILE: Questionable carcinogen with experimental neoplastigenic data. Mutation data reported. When heated to decomposition it emits acrid smoke and fumes.

ECS000 CAS:72074-69-2 ***HR: 3***
(±)-3-α,4-α-EPOXY-1,2,3,4-TETRAHYDRO-1-β,2-α-PHENANTHRENEDIOL
mf: $C_{14}H_{12}O_3$ mw: 228.26

SYN: (±)-1-β,2-α-DIHYDROXY-3-α,4-α-EPOXY-1,2,3,4-TETRAHYDROPHENANTHRENE

TOXICITY DATA with REFERENCE
ipr-mus TDLo:59 mg/kg/15D-I:ETA CNREA8 39,5063,79

SAFETY PROFILE: Questionable carcinogen with experimental tumorigenic data. When heated to decomposition it emits acrid smoke and fumes.

ECS500 CAS:72074-68-1 ***HR: D***
(±)-3-β,4-β-EPOXY-1,2,3,4-TETRAHYDRO-1-β,2-α-PHENANTHRENEDIOL
mf: $C_{14}H_{12}O_3$ mw: 201.92

SYN: (±)-1-β,2-α-DIHYDROXY-3-β,4-β-EPOXY-1,2,3,4-TETRAHYDROPHENANTHRENE

TOXICITY DATA with REFERENCE
mma-sat 2500 pmol/plate CNREA8 39,4069,79
msc-ham:lng 1 nmol/plate CNREA8 39,4069,79

SAFETY PROFILE: Mutation data reported. When heated to decomposition it emits acrid smoke and fumes.

ECT000 CAS:74444-59-0 ***HR: D***
1,2-EPOXY-1,2,3,4-TETRAHYDROTRIPHENYLENE
mf: $C_{18}H_{14}O$ mw: 246.32

SYNS: 1,2,3,4-TETRAHYDRO-1,2-EPOXYTRIPHENYLENE ◇ TP H4-1,2-EPOXIDE

TOXICITY DATA with REFERENCE
mmo-sat 1 nmol/plate CNREA8 40,1985,80
mma-sat 1 nmol/plate CNREA8 40,1985,80
msc-ham:lng 1 μmol/L CNREA8 40,1985,80

SAFETY PROFILE: Mutation data reported. When heated to decomposition it emits acrid smoke and fumes.

ECT500 CAS:3083-25-8 ***HR: 3***
1,2-EPOXY-4,4,4-TRICHLOROBUTANE
mf: $C_4H_5Cl_3O$ mw: 175.44

SYNS: TRICHLOROBUTYLENE OXIDE ◇ 4,4,4-TRICHLORO-1,2-EPOXYBUTANE

TOXICITY DATA with REFERENCE
eye-rbt 144 mg OMCDS* -,-,76
skn-rbt 720 mg/24H SEV OMCDS* -,-,76
ivn-mus LD50:56200 μg/kg CSLNX* NX#02029

CONSENSUS REPORTS: Reported in EPA TSCA Inventory.

SAFETY PROFILE: Poison by intravenous route. An eye and severe skin irritant. When heated to decomposition it emits toxic fumes of Cl^-.

ECT600 CAS:16967-79-6 ***HR: 3***
EPOXY-1,1,2-TRICHLOROETHANE
mf: C_2HCl_3O mw: 147.38

SYNS: TCEO ◇ 1,1,2-TRICHLOROEPOXYETHANE ◇ TRICHLOROETHYLENE EPOXIDE ◇ TRICHLOROETHYLENE OXIDE ◇ TRICHLORO-OXIRANE

TOXICITY DATA with REFERENCE
mmo-ssp 500 μmol/L 45OHAA -,333,80
otr-ham:emb 1100 μmol/L JJIND8 69,531,82

msc-ham:lng 50 μmol/L 45OHAA -,333,80
scu-mus TDLo:20 mg/kg/78W-I:CAR CNREA8 43,159,83

SAFETY PROFILE: Questionable carcinogen with experimental carcinogenic data. Mutation data reported. When heated to decomposition it emits toxic fumes of Cl^-.

ECU550 CAS:10402-53-6 ***HR: 3***
EPRAZINONE DIHYDROCHLORIDE
mf: $C_{24}H_{32}N_2O_2{\cdot}2ClH$ mw: 453.50

SYNS: 1-(2-BENZOYLPROPYL)-2-(2-ETHOXY-2-PHENYLETHYL)PIPERAZINE DIHYDROCHLORIDE ◇ 746 CE ◇ EFTAPAN ◇ EPRAZINONE HYDROCHLORIDE ◇ 3-(4-(β-ETHOXYPHENETHYL)-1-PIPERAZINYL)-2-METHYL-1-PHENYL-1-PROPANONEDIHYDROCHLORIDE ◇ 3-(4-(2-ETHOXY-2-PHENYLETHYL)-1-PIPERAZINYL)-2-METHYL-1-PHENYL-1-PROPANONE DIHYDROCHLORIDE ◇ 2-(4-(β-ETHOXYPHENETHYL)-1-PIPERAZINYLMETHYL)PROPIOPHENONE DIHYDROCHLORIDE ◇ MUCITUX ◇ 1-(2-PHENYL-2-ETHOXY)ETHYL-4-(2-BENZYLOXY)PROPYLPIPERAZINEDIHYDROCHLORIDE

TOXICITY DATA with REFERENCE
orl-rat LD50:763 mg/kg IYKEDH 10,232,79
ipr-rat LD50:191 mg/kg IYKEDH 10,232,79
scu-rat LD50:238 mg/kg IYKEDH 10,232,79
orl-mus LD50:286 mg/kg IYKEDH 10,232,79
ipr-mus LD50:116 mg/kg PCIPDV 13(2),98,81
scu-mus LD50:300 mg/kg OYYAA2 2,314,68
ivn-mus LD50:20 mg/kg PCIPDV 13(2),98,81

SAFETY PROFILE: Poison by ingestion, subcutaneous, intravenous, and intraperitoneal routes. When heated to decomposition it emits toxic fumes of NO_x and HCl.

ECU600 CAS:27588-43-8 ***HR: 3***
EPROZINOL DIHYDROCHLORIDE
mf: $C_{22}H_{30}N_2O_2{\cdot}2ClH$ mw: 427.46

SYNS: ALECOR ◇ BROVEL ◇ EUPNERON ◇ 4-(β-METHOXYPHENETHYL)-α-PHENYL-1-PIPERAZINEPROPANOLDIHYDROCHLORIDE ◇ 1-(2-METHOXY-2-PHENYLETHYL)-4-(3-HYDROXY-3-PHENYLPROPYL)PIPERAZINE DIHYDROCHLORIDE ◇ 1-(2-PHENYL-2-METHOXY)ETHYL-4-(3-PHENYL-3-HYDROXY)PROPYLPIPERAZINE HYDROCHLORIDE ◇ 1-(2-PHENYL-2-METHYL)ETHYL-4-(3-PHENYL-3-HYDROXY)-PROPYLPIPERAZINEDIHYDROCHLORIDE

TOXICITY DATA with REFERENCE
orl-rat TDLo:750 mg/kg (female 30D pre):REP OYYAA2 11,463,76
orl-rat LD50:640 mg/kg OYYAA2 11,463,76
ipr-rat LD50:72 mg/kg OYYAA2 11,463,76
ims-rat LD50:140 mg/kg OYYAA2 11,463,76
orl-mus LD50:350 mg/kg OYYAA2 11,463,76
ipr-mus LD50:103 mg/kg OYYAA2 11,463,76
ims-mus LD50:122 mg/kg OYYAA2 11,463,76

SAFETY PROFILE: Poison by ingestion, intramuscular, and intraperitoneal routes. Experimental reproductive effects. When heated to decomposition it emits toxic fumes of NO_x and HCl.

ECU750 CAS:12126-59-9 ***HR: 3***
EQUIGYNE

PROP: Conjugated forms of natural mixed estrogens, principally sodium estrone sulfate and sodium equilin sulfate, or synthetic estrogen piperazine estrone sulfate (IMEMDT** 21,147,79).

SYNS: AMNESTROGEN ◇ CES ◇ CLIMESTRONE ◇ CO-ESTRO ◇ CONEST ◇ CONESTRON ◇ CONJES ◇ CONJUGATED ESTROGENS ◇ CONJUTABS ◇ EQUIGYNE ◇ ESTRATAB ◇ ESTRIFOL ◇ ESTROATE ◇ ESTROCON ◇ ESTROMED ◇ ESTROPAN ◇ EVEX ◇ FEMACOID ◇ FEMEST ◇ FEM H ◇ FEMOGEN ◇ FORMATRIX ◇ GANEAKE ◇ GENISIS ◇ GLYESTRIN ◇ KESTRIN ◇ MENEST ◇ MENOGEN ◇ MENOTAB ◇ MENOTROL ◇ MILPREM ◇ MSMED ◇ NEO-ESTRONE ◇ NOVOCONESTRON ◇ OESTRILIN ◇ OESTRO-FEMINAL ◇ OESTROPAK MORNING ◇ OVEST ◇ PALOPAUSE ◇ PAR ESTRO ◇ PMB ◇ PREMARIN ◇ PRESOMEN ◇ PROMARIT ◇ SK-ESTROGENS ◇ SODESTRIN-H ◇ TAG-39 ◇ THEOGEN ◇ TRANSANNON ◇ TROCOSONE ◇ ZESTE

TOXICITY DATA with REFERENCE
unr-wmn TDLo:3 mg/kg (1-5D preg):REP JAMAAP 244,1336,80
orl-wmn TDLo:108 mg/kg/4Y-C:CAR,CVS,LIV AIMDAP 137,357,77
orl-wmn TD:27 mg/kg/3Y-C:NEO,LIV NYSJAM 78,1933,78

CONSENSUS REPORTS: NTP Fifth Annual Report on Carcinogens. IARC Cancer Review: Animal Inadequate Evidence IMEMDT 21,147,79; Human Limited Evidence IMEMDT 21,147,79.

SAFETY PROFILE: Confirmed human carcinogen producing tumors of the vascular system and liver. Human reproductive effects: changes in female fertility. When heated to decomposition it emits toxic fumes of Na_2O. See also individual components.

ECV000 CAS:517-09-9 ***HR: 3***
EQUILENIN
mf: $C_{18}H_{18}O_2$ mw: 266.36

PROP: Leaflets from acetone and ethanol; very sltly sol in water.

SYNS: EQUILENINA (SPANISH) ◇ EQUILENINE ◇ 3-HYDROXY-ESTRA-1,3,5(10),6,8-PENATEN-17-ONE

TOXICITY DATA with REFERENCE
imp-gpg TDLo:1600 μg/kg:ETA,REP BSBSAS 8,142,51
imp-ham TDLo:640 mg/kg/38W-I:ETA CNREA8 43,5200,83
imp-gpg TD:17 mg/kg:ETA,REP RBBIAL 5,1,45

SAFETY PROFILE: Experimental reproductive effects. Questionable carcinogen with experimental tumorigenic data. When heated to decomposition it emits acrid smoke and irritating fumes.

ECV500 CAS:604-58-0 ***HR: 3***
EQUILENIN BENZOATE
mf: $C_{25}H_{22}O_3$ mw: 370.47

PROP: Crystalline. Mp: 223° (in vacuo).

SYN: 3-HYDROXYESTRA-1,3,5,7,9-PENTAEN-17-ONEBENZOATE

TOXICITY DATA with REFERENCE
scu-mus TDLo:162 mg/kg/81W-I:NEO ZEKBAI 56,482,49
par-mus TDLo:86 mg/kg/43W-I:ETA CRSBAW 122,183,36
scu-mus TD:120 mg/kg/30W-I:ETA,REP YJBMAU 12,213,39

SAFETY PROFILE: Experimental reproductive effects. Questionable carcinogen with experimental neoplastigenic, and tumorigenic data. When heated to decomposition it emits acrid smoke and irritating fumes. See also ESTERS.

ECW000 CAS:474-86-2 ***HR: 3***
EQUILIN
mf: $C_{18}H_{20}O_2$ mw: 268.38

SYNS: 1,3,5,7-ESTRATETRAEN-3-OL-17-ONE ◇ 3-HYDROXYESTRA-1,3,5(10),7-TETRAEN-17-ONE

TOXICITY DATA with REFERENCE
scu-mus TDLo:112 mg/kg/56W-I:NEO ZEKBAI 56,482,49
imp-ham TDLo:640 mg/kg/38W-I:ETA CNREA8 43,5200,83

SAFETY PROFILE: Questionable carcinogen with experimental neoplastigenic and tumorigenic data. When heated to decomposition it emits acrid smoke and irritating fumes.

ECW500 CAS:6030-80-4 ***HR: 3***
EQUILIN BENZOATE
mf: $C_{25}H_{24}O_3$ mw: 372.49

SYN: 3-HYDROXYESTRA-1,3,5(10),7-TETRAEN-17-ONEBENZOATE

TOXICITY DATA with REFERENCE
scu-mus TDLo:96 mg/kg/48W-I:NEO ZEKBAI 56,482,49
par-mus TDLo:42 mg/kg/21W-I:ETA CRSBAW 122,183,36
scu-mus TD:148 mg/kg/37W-I:ETA,REP YJBMAU 12,213,39

SAFETY PROFILE: Experimental reproductive effects. Questionable carcinogen with experimental neoplastigenic and tumorigenic data. When heated to decomposition it emits acrid smoke and irritating fumes. See also ESTERS.

ECW520 CAS:16680-47-0 ***HR: 3***
EQUILIN SODIUM SULFATE
mf: $C_{18}H_{20}O_5S$•Na mw: 371.43

SYNS: EQUILIN, SULFATE, SODIUM SALT (6CI) ◇ ESTRA-1,3,5(10),7-TETRAEN-17-ONE, 3-HYDROXY-, HYDROGEN SULFATE SODIUMSALT (8CI) ◇ ESTRA-1,3,5(10),7-TETRAEN-17-ONE, 3-(SULFOOXY)-, SODIUM SALT ◇ SODIUM EQUILIN 3-MONOSULFATE ◇ SODIUM EQUILIN SULFATE

CONSENSUS REPORTS: NTP Fifth Annual Report on Carcinogens.

SAFETY PROFILE: Confirmed carcinogen. When heated to decomposition it emits toxic fumes of SO_x.

ECW550 CAS:55028-71-2 ***HR: D***
EQUIMATE
mf: $C_{23}H_{28}F_3O_6$•Na mw: 480.50

SYNS: FLUPROSTENOL ◇ HOE 837V ◇ ICI 81008 ◇ racemic-ICI 81,008

TOXICITY DATA with REFERENCE
ivg-wmn TDLo:38400 ng/kg (2D pre):REP CCPTAY 22,383,80

SAFETY PROFILE: Unspecified human and experimental reproductive effects. A veterinary fertility agent. When heated to decomposition it emits toxic fumes of F^- and Na_2O.

ECW600 CAS:153-87-7 ***HR: 3***
EQUIPERTINE
mf: $C_{23}H_{29}N_3O_2$ mw: 379.49

SYNS: 5,6-DIMETHOXY-2-METHYL-3-(2-(4-PHENYL-1-PIPERAZINYL)ETHYL)-1H-INDOLE ◇ FORIT ◇ INTEGRIN ◇ OPERTIL ◇ OXYPERTIN ◇ OXYPERTINE ◇ WIN 18501-2

TOXICITY DATA with REFERENCE
orl-rat LD50:1 g/kg NIIRDN 6,154,82
orl-mus LD50:2300 mg/kg NIIRDN 6,154,82
ipr-mus LD50:154 mg/kg BJPCAL 26,186,66

SAFETY PROFILE: Poison by intraperitoneal route. Moderately toxic by ingestion. When heated to decomposition it emits toxic fumes of NO_x.

ECX000 CAS:11094-61-4 ***HR: 3***
ERABUTOXINA
mf: $C_{284}H_{442}N_{86}O_{95}S_8$ mw: 6838.60

PROP: Neurotoxic principles isolated from venom of seasnake, *Laticauda semifasciata* (19DDA6-,249,67).

TOXICITY DATA with REFERENCE
ims-rat LD50:70 μg/kg 19DDA6-,249,67
ims-mus LD50:150 μg/kg BIJOAK 99,624,66

SAFETY PROFILE: A deadly poison by intramuscular route.

ECX100 CAS:6673-35-4 ***HR: 3***
ERALDIN
mf: $C_{14}H_{22}N_2O_3$ mw: 266.38

PROP: Mp: 134-136° (BuOAc). Sol in warm isopropanol.

SYNS: 1-(4-ACETAMIDOPHENOXY)-3-ISOPROPYLAMINO-2-PROPANOL ◇ AY 21011 ◇ DALZIC ◇ N-(4-(2-HYDROXY-3-(ISOPROPYLAMINO)PROPOXY)ACETAMIDE (9CI) ◇ 4'-(2-HYDROXY-3-(ISOPROPYLAMINO)PROPOXY)ACETANILIDE ◇ N-(4-(2-HYDROXY-3- ((1-METHYLETHYL)AMINO)PROPOXY)PHENYL)ACETAMIDE ◇ ICI 50172 ◇ PRACTALOL ◇ PRACTOLOL ◇ PRAKTOLOLU (POLISH) ◇ TERANOL

TOXICITY DATA with REFERENCE
orl-rat TDLo:300 mg/kg (9-14D preg):REP TOIZAG 21,567,74
orl-mus TDLo:12 g/kg (7-12D preg):TER TOIZAG 21,567,74
orl-wmn TDLo:4 mg/kg:EYE BMJOAE 1,595,75
orl-man TDLo:7550 mg/kg/2Y:EYE,SKN,GIT MJAUAJ 1,44,75
orl-rat LD50:3458 mg/kg TOIZAG 20,126,73
ipr-rat LD50:540 mg/kg TOIZAG 20,126,73
scu-rat LD50:7500 mg/kg TOIZAG 20,126,73
ivn-rat LD50:130 mg/kg TOIZAG 20,126,73
orl-mus LD50:3661 mg/kg TOIZAG 20,126,73
ipr-mus LD50:455 mg/kg PJPPAA 25,145,73
scu-mus LD50:3357 mg/kg TOIZAG 20,126,73
ivn-mus LD50:69 mg/kg ARZNAD 27,1022,77

CONSENSUS REPORTS: EPA Genetic Toxicology Program.

SAFETY PROFILE: Poison by intravenous route. Moderately toxic by ingestion, subcutaneous, and intraperitoneal route. An experimental teratogen. Human systemic effects by ingestion: eye damage, peritonitis, and dermatitis. Experimental reproductive effects. When heated to decomposition it emits toxic fumes of NO_x.

ECX500 CAS:10138-41-7 ***HR: 3***
ERBIUM CHLORIDE
mf: Cl_3Er mw: 273.61

SYN: ERBIUM TRICHLORIDE

TOXICITY DATA with REFERENCE
orl-mus LD50:4417 mg/kg EQSSDX 1,1,75
ipr-mus LD50:226 mg/kg AEHLAU 5,437,62
ipr-gpg LD50:128 mg/kg AEHLAU 5,437,62
scu-frg LDLo:300 mg/kg 27ZWAY -,2197,35

CONSENSUS REPORTS: Reported in EPA TSCA Inventory.

SAFETY PROFILE: Poison by intraperitoneal and subcutaneous routes. Moderately toxic by ingestion. When heated to decomposition it emits toxic fumes of Cl^-. See also RARE EARTHS.

ECY000 CAS:3088-54-8 ***HR: 3***
ERBIUM CITRATE
mf: $C_6H_8O_7$•Er mw: 359.40

SYN: 2-HYDROXY-1,2,3-PROPANETRICARBOXYLICACID EBRIUM(3+) salt (1:1)

TOXICITY DATA with REFERENCE
ipr-mus LD50:122 mg/kg AEHLAU 5,437,62
ipr-gpg LD50:63 mg/kg AEHLAU 5,437,62

SAFETY PROFILE: Poison by intraperitoneal route. When heated to decomposition it emits acrid smoke and irritating fumes. See also RARE EARTHS.

ECY500 CAS:10168-80-6 ***HR: 3***
ERBIUM(III) NITRATE (1:3)
mf: N_3O_9•Er mw: 353.29

PROP: Reddish crystals.

SYN: NITRIC ACID, ERBIUM (3^-) SALT

TOXICITY DATA with REFERENCE
ivn-rat LD50:177 mg/kg EQSSDX 1,1,75
ivn-rat LD50:27460 μg/kg EQSSDX 1,1,75

CONSENSUS REPORTS: Reported in EPA TSCA Inventory.

SAFETY PROFILE: Poison by intravenous and intraperitoneal routes. When heated to decomposition it emits toxic fumes of NO_x. See also RARE EARTHS and NITRATES.

ECZ000 CAS:13476-05-6 ***HR: 3***
ERBIUM(III) NITRATE, HEXAHYDRATE (1:3:6)
mf: N_3O_9•Er•$6H_2O$ mw: 461.41

PROP: Mp: $-4H_2O$ @ 130°.

SYN: NITRIC ACID, ERBIUM (3^-) SALT, HEXAHYDRATE

TOXICITY DATA with REFERENCE
ipr-rat LD50:230 mg/kg TXAPA9 5,750,63
ivn-rat LDLo:30 mg/kg JPETAB 43,61,31
ipr-mus LD50:225 mg/kg TXAPA9 5,750,63

SAFETY PROFILE: Poison by intraperitoneal and intravenous routes. When heated to decomposition it emits toxic fumes of NO_x. See also ERBIUM NITRATE.

EDA500 CAS:35287-69-5 ***HR: 3***
ERGOCHROME AA (2,2')-5-β,6-α,10-β-5',6'-α,10'-β
mf: $C_{32}H_{30}O_{14}$ mw: 638.62

SYNS: SAD ◇ SECALONIC ACID D

TOXICITY DATA with REFERENCE
mnt-mus-orl 30 mg/kg JEPTDQ 4(5-6),31,80
scu-rat TDLo:25 mg/kg (10D preg):TER TXCYAC 25,311,82
ipr-mus TDLo:25 mg/kg (female 11D post):REP 55CXA9 -,549,85
orl-rat LD50:22 mg/kg TXAPA9 48,A14,79
orl-mus LD50:30 mg/kg TXAPA9 48,A14,79
ipr-mus LD50:26500 μg/kg AEMIDF 39,285,80
ivn-mus LD50:25 mg/kg JTEHD6 5,1159,79

CONSENSUS REPORTS: EPA Genetic Toxicology Program.

SAFETY PROFILE: Poison by ingestion, intraperitoneal, and intravenous routes. An experimental teratogen. Other experimental reproductive effects. Mutation data reported. When heated to decomposition it emits acrid smoke and irritating fumes.

EDA600 CAS:564-36-3 ***HR: 3***
ERGOCORNINE
mf: $C_{31}H_{39}N_5O_5$ mw: 561.75

SYNS: ERGOCORNIN ◇ ERGOTAMAN-3′,6′,18-TRIONE, 12′-HYDROXY-2′,5′-BIS(1-METHYLETHYL)-, (5′-α)- ◇ 12′-HYDROXY-2′,5′-α-BIS(1-METHYLETHYL)ERGOTAMAN-3′,6′,18-TRIONE

TOXICITY DATA with REFERENCE
scu-rat TDLo:5 mg/kg (female 5D post):REP RCOCB8 7,701,74
ivn-rbt LDLo:1170 μg/kg 85IXA4 -,304,48

SAFETY PROFILE: Poison by intravenous route. Experimental reproductive effects. When heated to decomposition it emits toxic fumes of NO_x.

EDA875 CAS:57432-60-7 ***HR: D***
ERGOCORNINE MALEATE
mf: $C_{31}H_{39}N_5O_5 \cdot C_4H_4O_4$ mw: 677.83

SYNS: ERGOCORNINE HYDROGEN MALEATE ◇ ERGOCORNINE HYDROGEN MALEINATE

TOXICITY DATA with REFERENCE
scu-rat TDLo:12500 μg/kg (lactating female 5D post):REP ENDOAO 90,285,72

SAFETY PROFILE: Experimental reproductive effects. When heated to decomposition it emits toxic fumes of NO_x.

EDB000 CAS:2207-69-4 ***HR: D***
ERGOCORNINE METHANESULFONATE (SALT)
mf: $C_{31}H_{39}N_5O_5 \cdot xCH_4O_3S$ mw: 1234.52

SYNS: ERGOCORNINE MESYLATE ◇ ERGOCORNINE METHANESULPHONATE

TOXICITY DATA with REFERENCE
scu-rat TDLo:5 mg/kg (lactating female 1D post):REP ACENA7 41,321,62
scu-rat TDLo:5 mg/kg (8D preg):TER ENDOAO 84,315,69

SAFETY PROFILE: Experimental teratogenic and reproductive effects. When heated to decomposition it emits very toxic fumes such as SO_x and NO_x.

EDB100 CAS:511-09-1 ***HR: 3***
ERGOCRYPTINE
mf: $C_{32}H_{41}N_5O_5$ mw: 575.78

PROP: β-Ergocryptine: Rectangular plates from benzene. Mp: 173° (decomp).

SYNS: α-ERGOCRYPTINE ◇ α-ERGOKRYPTINE ◇ 12′-HYDROXY-2′-(1-METHYLETHYL)-5′-α-(2-METHYLPROPYL)ERGOTAMAN-3′,6′,18-TRIONE

TOXICITY DATA with REFERENCE
scu-rat TDLo:2 mg/kg (4D preg):REP JRPFA4 56,691,79
ivn-rbt LD50:950 μg/kg USXXAM #3752814

SAFETY PROFILE: Poison by intravenous route. Experimental reproductive effects. When heated to decomposition it emits toxic fumes of NO_x.

EDB125 CAS:2706-66-3 ***HR: D***
ERGOCRYPTINE MESYLATE
mf: $C_{32}H_{41}N_5O_5 \cdot CH_4O_3S$ mw: 671.89

SYNS: ECR ◇ ERGOCRYPTINE METHANESULFONATE ◇ ERGOCRYPTINE METHANESULPHONATE ◇ ERGOKRYPTINE METHANESULFONATE

TOXICITY DATA with REFERENCE
scu-rat TDLo:12500 μg/kg (lactating female 5D post):REP ENDOAO 90,285,72

SAFETY PROFILE: Experimental reproductive effects. When heated to decomposition it emits toxic fumes of SO_x and NO_x. See also SULFONATES.

EDB200 CAS:2624-03-5 ***HR: 3***
ERGOSINE MONOMETHANESULFONATE
mf: $C_{30}H_{37}N_5O_5 \cdot CH_4O_3S$ mw: 643.83

SYNS: ERGOSINE METHANESULFONATE ◇ α-ERGOSINE METHANESULFONATE ◇ ERGOTAMAN-3′,6′,18-TRIONE, 12′-HYDROXY-2′-METHYL-5′-(2-METHYLPROPYL)-, (5′-α)-, MONOMETHANESULFONATE (salt)

TOXICITY DATA with REFERENCE
scu-rat TDLo:2525 μg/kg (female 4D post):REP JRPFA4 10,221,65
ipr-mus LD50:206 mg/kg BAXXDU #2025960

SAFETY PROFILE: Poison by intraperitoneal route. Experimental reproductive effects. When heated to decomposition it emits toxic fumes of NO_x and SO_x.

EDB500 CAS:129-51-1 ***HR: 3***
ERGOT
mf: $C_{19}H_{23}N_3O_2 \cdot C_4H_4O_4$ mw: 441.53

PROP: Composition: ergot amine, ergosine, ergocristine, ergocryptine, ergocornine, ergosinine, ergocristinine, ergocryptinine, ergotaminine, etc.

SYNS: CORNOCENTIN ◇ CRUDE ERGOT ◇ ERGOMETRINE ACID MALEATE ◇ ERGOMETRINE MALEATE ◇ ERGONOVINE, MALEATE (1:1) (SALT) ◇ ERGOTRATE ◇ ERGOTRATE MALEATE ◇ OXYTOCIC

TOXICITY DATA with REFERENCE
sce-ham:ovr 10 nmol/L TCMUD8 8,169,88
scu-rat TDLo:100 mg/kg (5D post):REP ENDOAO 90,285,72
orl-rat TDLo:669 g/kg/96W-C:ETA CNREA8 2,11,42
ims-inf TDLo:40 μg/kg:GIT,PUL JAMAAP 250,729,83
unr-man LDLo:15 mg/kg 85DCAI 2,73,70
ivn-mus LD50:8260 μg/kg TXAPA9 23,537,72

SAFETY PROFILE: A deadly human poison. Experimental poison by intravenous route. Experimental reproductive effects. Human systemic effects: hypermotility, diarrhea, cyanosis, thirst, tachycardia, confusion, coma, central nervous system symptoms, gangrene; circulatory changes can follow ingestion. Questionable carcinogen with experimental tumorigenic data. Mutation data reported. When heated to decomposition it emits toxic fumes of NO_x. See also individual components.

EDC000 CAS:113-15-5 ***HR: 3***
ERGOTAMINE
mf: $C_{33}H_{35}N_5O_5$ mw: 581.63

PROP: A specific alkaloid present in rye ergot (AIPTAK 27,459,23).

TOXICITY DATA with REFERENCE
cyt-hmn:lym 100 μg/L/24H MUREAV 48,205,77
ivn-rbt LD50:100 mg/kg EXPEAM 33,1552,77
scu-frg LDLo:33 mg/kg AIPTAK 27,459,23

SAFETY PROFILE: Poison by intravenous and subcutaneous routes. Human mutation data reported. When heated to decomposition it emits toxic fumes of NO_x.

EDC500 CAS:379-79-3 ***HR: 3***
ERGOTAMINE TARTRATE
mf: $C_{66}H_{70}N_{10}O_{10} \cdot C_4H_6O_6$ mw: 1313.56

SYNS: ERGAM ◇ ERGATE ◇ ERGOMAR ◇ ERGOSTAT ◇ ERGOTAMINE BITARTRATE ◇ ERGOTARTRATE ◇ ETIN ◇ EXMIGRA ◇ FEMERGIN ◇ GOTAMINE TARTRATE ◇ GYNERGEN ◇ LINGRAINE ◇ LINGRAN ◇ NEO-ERGOTIN ◇ RIGETAMIN ◇ SECAGYN ◇ SECUPAN

TOXICITY DATA with REFERENCE
cyt-hmn:lyms 100 μg/L MUREAV 48,205,77
sce-ham:ovr 10 nmol/L TCMUD8 8,169,88
scu-rat TDLo:100 mg/kg (lactating female 5D post):REP ENDOAO 90,285,72
orl-rat TDLo:100 mg/kg (6-15D preg):TER TJADAB 7,227,73
orl-wmn TDLo:11 mg/kg/13W-I:BPR PGMJAO 61,461,85
orl-hmn TDLo:3700 μg/kg/26W-I:CNS,GIT HEADAE 18,95,78
orl-man TDLo:214 μg/kg MMWOAU 75,736,28
orl-rat LDLo:1 mg/kg TJADAB 7,227,73
ivn-rat LD50:80 mg/kg BSAMA5 2,1,46
orl-mus LDLo:300 mg/kg TJADAB 7,227,73
ipr-mus TDLo:412 mg/kg BAXXDU #2025960
ivn-mus LD50:62 mg/kg BSAMA5 2,1,46
scu-cat LD50:11 mg/kg BSAMA5 2,1,46
orl-rbt LDLo:1 mg/kg TJADAB 7,227,73

CONSENSUS REPORTS: Reported in EPA TSCA Inventory. EPA Genetic Toxicology Program. EPA Extremely Hazardous Substances List.

SAFETY PROFILE: Poison by ingestion, intravenous, and subcutaneous routes. An experimental teratogen. Human systemic effects by ingestion: hallucinations, distorted perceptions, convulsions, nausea or vomiting, blood pressure elevation. Experimental reproductive effects. Human mutation data reported. Used in production of drugs of abuse. When heated to decomposition it emits toxic fumes of NO_x. See also ERGOT.

EDC560 CAS:28675-83-4 ***HR: 2***
ERGOTERM TGO
mf: $C_{34}H_{52}O_4S_2Sn$ mw: 707.67

SYNS: ACETIC ACID, 2,2'-((BIS(PHENYLMETHYL)STANNYLENE)BIS(THIO))BIS-,DIISOOCTYL ESTER (9CI) ◇ ACETIC ACID, ((DIBENZYLSTANNYLENE)DITHIO)DI-, DIISOOCTYL ESTER ◇ D-BENZYL TG ◇ DIBENZYLTIN S,S'-BIS(ISOOCTYLMERCAPTOACETATE)

TOXICITY DATA with REFERENCE
orl-rat TDLo:378 mg/kg (female 1-21D post):TER TXAPA9 26,253,73
orl-rat LD50:1250 mg/kg RPZHAW 18,283,67

ACGIH TLV: TWA 0.1 mg(Sn)/m^3; STEL 0.2 mg/m^3 (skin)
OSHA PEL: TWA 0.1 mg(Sn)/m^3
NIOSH REL: (Organotin compounds) TWA 0.1 mg(Sn)/m^3

SAFETY PROFILE: Moderately toxic by ingestion. An experimental teratogen. When heated to decomposition it emits toxic fumes of SO_x and Sn.

EDC565 CAS:8006-25-5 ***HR: 3***
ERGOTOXINE

SYNS: ECBOLINE ◇ ERGOTOXIN

TOXICITY DATA with REFERENCE
ims-rat TDLo:20 mg/kg (female 5D post):REP PSEBAA 100,555,59
ivn-man TDLo:3570 ng/kg HXPHAU 6,107,38
scu-mus LDLo:107 mg/kg AEPPAE 176,171,34
ivn-mus LD50:33 mg/kg 85IXA4 -,285,48
ivn-rbt LDLo:1500 µg/kg HXPHAU 6,104,38

SAFETY PROFILE: Poison by intravenous and subcutaneous routes. Experimental reproductive effects. When heated to decomposition it emits acrid smoke and irritating fumes.

EDC575 CAS:8047-28-7 ***HR: 3***
ERGOTOXINE ETHANSULFONATE

SYNS: ECBOLINE ETHANESULFONATE ◇ ERGOTOXINE ETHANESULFONATE ◇ ERGOTOXINE ETHANESULPHONATE

TOXICITY DATA with REFERENCE
ipr-rat TDLo:10 mg/kg (female 12-21D post):REP AJPHAP 180,296,55
ivn-mus LD50:40 mg/kg HXPHAU 6,103,38

SAFETY PROFILE: Poison by intravenous route. Experimental reproductive effects. When heated to decomposition it emits toxic fumes of SO_x. See also SULFONATES.

EDC600 CAS:11052-01-0 ***HR: 3***
ERIAMYCIN
mf: $C_{31}H_{23}NO_8$ mw: 537.55

TOXICITY DATA with REFERENCE
orl-mus LD50:50 mg/kg 85GDA2 3,372,80
ipr-mus LD50:500 µg/kg 85ERAY 1,164,78
scu-mus LD50:2 mg/kg 85GDA2 3,372,80

SAFETY PROFILE: Poison by ingestion, subcutaneous, and intraperitoneal routes. When heated to decomposition it emits toxic fumes of NO_x.

EDC650 CAS:66733-21-9 ***HR: 3***
ERIONITE
mf: $Al_2O_{18}Si_7$•1/2Ca•$7H_2O$•1/2Na mw: 715.68

TOXICITY DATA with REFERENCE
ipl-rat TDLo:100 mg/kg:ETA ENVRAL 29,238,82
ipr-mus TD:20 mg/kg:CAR ENVRAL 35,277,84
ipr-mus TDLo:400 mg/kg:CAR JACTDZ 4(1),225,85

CONSENSUS REPORTS: IARC Cancer Review: Group 1 IMEMDT 7,203,87; Animal Sufficient Evidence, Human Sufficient Evidence IMEMDT 42,225,87.

SAFETY PROFILE: Confirmed carcinogen with experimental carcinogenic and tumorigenic data.

EDD500 CAS:62796-23-0 ***HR: 3***
EROCYANINE 540
mf: $C_{26}H_{32}N_3O_6S_2$•Na mw: 569.72

TOXICITY DATA with REFERENCE
ivn-mus LD50:92 mg/kg TXAPA9 44,225,78

CONSENSUS REPORTS: Reported in EPA TSCA Inventory.

SAFETY PROFILE: Poison by intravenous route. When heated to decomposition it emits very toxic fumes of Na_2O, SO_x, and NO_x.

EDE000 CAS:63938-27-2 ***HR: 3***
ERYSODINE HYDROCHLORIDE
mf: $C_{18}H_{21}NO_3$•ClH mw: 335.86

TOXICITY DATA with REFERENCE
orl-mus LD50:155 mg/kg JPETAB 80,53,44
scu-mus LD50:100 mg/kg JPETAB 80,53,44

SAFETY PROFILE: Poison by ingestion and subcutaneous routes. When heated to decomposition it emits very toxic fumes of HCl and NO_x.

EDE500 CAS:63938-28-3 ***HR: 3***
ERYSOPINE HYDROCHLORIDE
mf: $C_{17}H_{19}NO_3$•ClH mw:321.83

SYN: (3-β)-1,2,6,7-TETRADEHYDRO-3-METHOXYERYTHRINAN-16-OL HYDROCHLORIDE

TOXICITY DATA with REFERENCE
orl-mus LD50:18 mg/kg JPETAB 80,53,44
scu-mus LD50:15 mg/kg JPETAB 80,53,44

SAFETY PROFILE: Poison by ingestion and subcutaneous routes. When heated to decomposition it emits very toxic fumes of Cl^-.

EDF000 CAS:31248-66-5 ***HR: 3***
ERYTHRALINE HYDROBROMIDE
mf: $C_{18}H_{19}NO_3$•BrH mw: 378.30

SYN: (3-β)-1,2,6,7-TETRADEHYDRO-3-METHOXY-15,16-(METHYLENEBIS(OXY))ERYTHRINAN, HYDROBROMIDE

TOXICITY DATA with REFERENCE
orl-mus LD50:80 mg/kg JPETAB 80,53,44
scu-mus LD50:72 mg/kg JPETAB 80,53,44

SAFETY PROFILE: Poison by ingestion and subcutaneous routes. When heated to decomposition it emits very toxic fumes of HBr and NO_x.

EDG500 CAS:466-81-9 ***HR: 3***
β-ERYTHROIDINE
mf: $C_{16}H_{19}NO_3$ mw: 273.36

SYN: 12,13-DIDEHYDRO-13,14-DIHYDRO-α-ERYTHROIDINE

TOXICITY DATA with REFERENCE
ipr-mus LD50:24 mg/kg JPETAB 93,362,48
ivn-rbt LD50:8600 μg/kg MEIEDD 7,419,60

SAFETY PROFILE: Poison by intraperitoneal and intravenous routes. When heated to decomposition it emits toxic fumes such as NO_x.

EDH000 CAS:596-11-2 ***HR: 3***
β-ERYTHROIDINE HYDROCHLORIDE
mf: $C_{16}H_{19}NO_3 \cdot ClH$ mw: 309.82

SYN: 1,2,6,7-TETRADEHYDRO-14,17-DIHYDRO-3-METHOXY-16(15H)-OXAERYTHRINAN-15-ONE, HYDROCHLORIDE

TOXICITY DATA with REFERENCE
orl-rat LD50:510 mg/kg JPETAB 80,39,44
scu-rat LD50:1260 mg/kg JPETAB 80,39,44
ivn-rat LD50:39 mg/kg JPETAB 82,266,44
orl-mus LD50:75 mg/kg JPETAB 80,39,44
scu-mus LD50:48 mg/kg JPETAB 80,39,44
ivn-dog LD50:8800 μg/kg JPETAB 82,266,44
orl-cat LDLo:30 mg/kg JPETAB 80,39,44
scu-cat LDLo:20 mg/kg JPETAB 80,39,44
orl-rbt LDLo:200 mg/kg JPETAB 80,39,44
scu-rbt LDLo:50 mg/kg JPETAB 80,39,44
ivn-rbt LD50:8600 μg/kg JPETAB 82,266,44

SAFETY PROFILE: Poison by ingestion, intravenous, and subcutaneous routes. When heated to decomposition it emits very toxic fumes of HCl and NO_x.

EDH500 CAS:114-07-8 ***HR: 3***
ERYTHROMYCIN
mf: $C_{37}H_{67}NO_{13}$ mw: 734.05

PROP: White or sltly yellow, crystalline powder; odorless; Mp: 133-138°. Freely sol in alc, chloroform, and ether; very sltly sol in water.

SYNS: DOTYCIN ◇ EM ◇ E-MYCIN ◇ ERYCIN ◇ ERYTHROCIN ◇ ERYTHROGRAN ◇ ERYTHROGUENT ◇ ERYTHROMYCIN A ◇ ILOTYCIN ◇ PANTOMICINA ◇ PROPIOCINE ◇ ROBIMYCIN

TOXICITY DATA with REFERENCE
dnr-esc 600 μg/disc MUREAV 97,1,82
scu-rat TDLo:50 mg/kg (female 6-10D post):TER OSDIAF 14,107,65
scu-rat TDLo:50 mg/kg (female 6-10D post):REP OSDIAF 14,107,65
orl-rat LD50:9272 mg/kg AMPMAR 39,259,78
scu-rat LDLo:427 mg/kg CLDND* 1,115,75
ipr-mus LD50:463 mg/kg ARZNAD 20,1751,70
scu-mus LD50:1800 mg/kg ANTCAO 2,281,52
ivn-mus LD50:426 mg/kg 85FZAT -,273,67
ims-mus LD50:394 mg/kg JAPMA8 44,199,55
ipr-gpg LD50:413 mg/kg JAPMA8 41,555,52
orl-ham LD50:3018 mg/kg JAPMA8 41,555,52

CONSENSUS REPORTS: EPA Genetic Toxicology Program.

SAFETY PROFILE: Poison by intravenous and intramuscular routes. Moderately toxic by ingestion, intraperitoneal, and subcutaneous routes. An experimental teratogen. Other experimental reproductive effects. Mutation data reported. Mutation data reported. When heated to decomposition it emits toxic fumes of NO_x.

EDI000 CAS:54579-17-8 ***HR: 1***
ERYTHROMYCIN CARBONATE

TOXICITY DATA with REFERENCE
orl-rat LD50:5800 mg/kg 85ERAY 1,34,78
orl-mus LD50:4050 mg/kg 85ERAY 1,34,78

SAFETY PROFILE: Mildly toxic by ingestion. When heated to decomposition it emits acrid smoke and irritating fumes. See also ERYTHROMYCIN.

EDI500 CAS:23067-13-2 ***HR: 3***
ERYTHROMYCIN GLUCOHEPTONATE (1:1)
mf: $C_{37}H_{67}NO_{13} \cdot C_7H_{14}O_8$ mw: 960.26

SYNS: ERYTHROHYCIN GLUCEPTATE ◇ GLUCOHEPTONIC ACID with ERYTHROMYCIN (1:1)

TOXICITY DATA with REFERENCE
ivn-rat LD50:288 mg/kg JAPMA8 44,199,55
ivn-mus LD50:453 mg/kg JAPMA8 44,199,55
orl-ham LD50:4 mg/kg TXAPA9 14,510,69
scu-ham LD50:6 mg/kg TXAPA9 14,510,69

SAFETY PROFILE: Poison by ingestion, subcutaneous, and intravenous routes. When heated to decomposition it emits toxic fumes of NO_x. See also ERYTHROMYCIN.

EDJ000 CAS:14271-02-4 ***HR: 3***
ERYTHROMYCIN HYDROCHLORIDE
mf: $C_{37}H_{67}NO_{13} \cdot ClH$ mw: 770.51

SYN: ILOTYCIN HYDROCHLORIDE

TOXICITY DATA with REFERENCE
scu-rat LD50:1442 mg/kg JAPMA8 41,555,52
ivn-rat LD50:209 mg/kg JAPMA8 41,555,52
orl-mus LD50:2927 mg/kg JAPMA8 41,555,52
ipr-mus LD50:490 mg/kg JAPMA8 41,555,52
scu-mus LD50:1849 mg/kg JAPMA8 41,555,52
ivn-mus LD50:377 mg/kg USXXAM #4393053
ivn-rbt LD50:183 mg/kg JAPMA8 41,555,52

SAFETY PROFILE: Poison by intravenous route. Moderately toxic by ingestion, intraperitoneal, and subcutaneous routes. When heated to decomposition it emits very toxic fumes of HCl and NO_x. See also ERYTHROMYCIN.

EDJ500 CAS:643-22-1 ***HR: 3***
ERYTHROMYCIN STEARATE
mf: $C_{37}H_{67}NO_{13}$•$C_{18}H_{36}O_2$ mw: 1018.59

SYNS: BRISTAMYCIN ◇ DOWMYCIN E ◇ ERYPAR ◇ ERYTHROCIN STEARATE ◇ ERYTHROMYCIN OCTADECANOATE (salt) ◇ ERYTHROMYCIN STEARIC ACID SALT ◇ ETHRIL ◇ GALLIMYCIN ◇ NCI-C55674 ◇ OE 7 ◇ PFIZER-E ◇ QIDMYCIN ◇ SK-ERYTHROMYCIN

CONSENSUS REPORTS: DESCRIPTION NTP CARCINOGENESIS STUDIES (FEED); No Evidence; rat,mouse NCITR* NTP-TR-338,88

SAFETY PROFILE: When heated to decomposition it emits toxic fumes of NO_x. See also ERYTHROMYCIN.

EDK000 CAS:24380-21-0 ***HR: 3***
ERYTHROMYCIN SULFAMATE (SALT)
mf: $C_{37}H_{67}NO_{13}$•H_3NO_3S mw: 831.15

SYN: ERYTHROMYCIN SULFAMATE

TOXICITY DATA with REFERENCE
ivn-rat LD50:127 mg/kg CMTRAG 5,144,62
ipr-mus LD50:480 mg/kg CMTRAG 5,144,62
ivn-mus LD50:217 mg/kg CMTRAG 5,144,62

SAFETY PROFILE: Poison by intravenous route. Moderately toxic by intraperitoneal route. When heated to decomposition it emits very toxic fumes of SO_x and NO_x. See also ERYTHROMYCIN.

EDK600 CAS:8057-51-0 ***HR: 3***
ESBERICARD

SYN: CRATAEGUS, EXTRACT

TOXICITY DATA with REFERENCE
orl-rat LD50:33800 mg/kg NIIRDN 6,212,82
ivn-rat LDLo:1000 mg/kg PLMEAA 43,105,81
orl-mus LD50:18500 mg/kg NIIRDN 6,212,82
ivn-cat LDLo:1200 mg/kg PLMEAA 43,105,81
ivn-rbt LDLo:1000 mg/kg PLMEAA 43,105,81
ivn-gpg LDLo:1500 mg/kg PLMEAA 43,105,81
ivn-dck LDLo:300 mg/kg PLMEAA 43,105,81
par-frg LDLo:5 g/kg PLMEAA 43,105,81

SAFETY PROFILE: Poison by intravenous route. Mildly toxic by ingestion.

EDK700 CAS:67924-63-4 ***HR: 3***
ESCHERICHIA COLI ENDOTOXIN

SYNS: E. COLI ENDOTOXIN ◇ ENDOTOXIN, ESCHERICHIA COLI ◇ LPS

TOXICITY DATA with REFERENCE
dns-mus:oth 100 μg/L MMIYAO 168,201,80
dns-gpg:lng 1 mg/L PSEBAA 171,109,82
ivn-mky TDLo:5 mg/kg (female 24W post):TER AJOGAH 131,899,78
ipr-rat TDLo:2250 μg/kg (female 11-17D post):REP ANREAK 181,441,75
ipr-rat LD50:7600 μg/k CRSHAG 18,11,86
ivn-rat LD50:3700 μg/kg JJMCAQ 34,54,81
ipr-mus LD50:4550 μg/kg SCIEAS 229,869,85
scu-mus LD50:7700 mg/kg APJAAG 9,141,59
ivn-mus LD50:2 mg/kg JGMIAN 86,363,75
par-mus LD50:27500 μg/kg EXPEAM 35,804,79
ivn-dog LD50:1 mg/kg CRSHAG 7,299,80
ivn-cat LD50:2200 μg/kg EJPHAZ 9,311,70
iat-cat LDLo:3 mg/kg MIVRA6 20,242,80

SAFETY PROFILE: Poison by intravenous, intraperitoneal, parenteral, and intraarterial routes. Moderately toxic by subcutaneous route. Experimental teratogenic and reproductive effects. Mutation data reported.

EDK875 CAS:6805-41-0 ***HR: 3***
ESCIN
mf: $C_{54}H_{84}O_{23}$ mw: 1101.38

PROP: α-Escin: Amorphous powder. Mp: 225-227°. Very sol in water. β-Escin: Leaflets from dil ethanol. Mp: 222-223°. Practically insol in water.

SYNS: A-4700 ◇ AESCIN (GERMAN) ◇ AESCUSAN ◇ AMORPHOUS AESCIN ◇ ESCINA (ITALIAN) ◇ REPARIL

TOXICITY DATA with REFERENCE
orl-rbt TDLo:325 mg/kg (female 6-18D post):TER KSRNAM 9,1227,75
orl-rat LD50:833 mg/kg KSRNAM 8,118,74
ipr-rat LD50:10150 μg/kg KSRNAM 8,118,74
scu-rat LD50:150 mg/kg KSRNAM 8,118,74
ivn-rat LD50:1600 μg/kg BCFAAI 115,272,76
orl-mus LD50:165 mg/kg KSRNAM 8,118,74
ipr-mus LD50:6700 μg/kg NIIRDN 6,APP-2,82
scu-mus LD50:38590 μg/kg KSRNAM 8,118,74
ivn-mus LD50:2 mg/kg BCFAAI 115,272,76

SAFETY PROFILE: Poison by ingestion, subcutaneous, intravenous, and intraperitoneal routes. Experimental teratogenic effects. When heated to decomposition it emits acrid smoke and irritating fumes. See also other escin entries.

EDL000 CAS:66795-86-6 ***HR: 3***
α-ESCIN
mf: $C_{54}H_{84}O_{23}$ mw: 1101.38

PROP: A form of triterpenglycosides isolated from *Aesculus hippocastanum L.* (ARZNAD 20,209,70).

SYNS: α-AESCIN ◇ α-AESCUSAN ◇ β-ESCINIC ACID

TOXICITY DATA with REFERENCE
orl-rat LD50:720 mg/kg ARZNAD 20,209,70
ivn-rat LD50:5 mg/kg ARZNAD 20,209,70

orl-mus LD50:320 mg/kg ARZNAD 20,209,70
ivn-mus LD50:3 mg/kg ARZNAD 20,209,70
ivn-dog LD50:2 mg/kg ARZNAD 20,209,70
ivn-rbt LD50:7800 μg/kg ARZNAD 20,209,70
orl-gpg LD50:475 mg/kg ARZNAD 20,209,70
ivn-gpg LD50:15200 μg/kg ARZNAD 20,209,70

SAFETY PROFILE: Poison by ingestion and intravenous routes. When heated to decomposition it emits acrid smoke and irritating fumes. See also other escin entries.

EDL500 CAS:11072-93-8 ***HR: 3***
β-ESCIN
mf: $C_{54}H_{84}O_{23}$ mw: 1101.38

PROP: A form of triterpenglycosides isolated from *Aesculus hippocastanum L.* (ARZNAD 20,209,70).

SYNS: β-AESCIN ◇ β-AESCUSAN ◇ β-REPARIL

TOXICITY DATA with REFERENCE
orl-mus TDLo:36 mg/kg (female 7-16D post):TER BLLIAX 87,76,87
orl-rat LD50:400 mg/kg ARZNAD 20,209,70
ivn-rat LD50:2 mg/kg ARZNAD 20,209,70
orl-mus LD50:134 mg/kg ARZNAD 20,209,70
ivn-mus LD50:1400 μg/kg ARZNAD 20,209,70
ivn-dog LD50:1 mg/kg ARZNAD 20,209,70
ivn-rbt LD50:3600 μg/kg ARZNAD 20,209,70
orl-gpg LD50:188 mg/kg ARZNAD 20,209,70
ivn-gpg LD50:7200 μg/kg ARZNAD 20,209,70

SAFETY PROFILE: Poison by ingestion and intravenous routes. An experimental teratogen. When heated to decomposition it emits acrid smoke and irritating fumes. See also other escin entries.

EDM000 CAS:20977-05-3 ***HR: 3***
ESCIN, SODIUM SALT
mf: $C_{55}H_{85}O_{24}$•Na mw: 1153.39

PROP: A mixture of saponins occurring in the seed of the horse chestnut tree (ARZNAD 12,815,62).

SYNS: A-4760 ◇ AESCIN SODIUM SALT ◇ AESCUSAN SODIUM SALT ◇ Na-AESCINAT ◇ REPARIL SODIUM SALT ◇ SODIUM AESCINATE

TOXICITY DATA with REFERENCE
ivn-rbt TDLo:2250 μg/kg (8-16D preg):REP KSRNAM 8,269,74
ivn-rbt TDLo:2250 μg/kg (8-16D preg):TER KSRNAM 8,269,74
orl-rat LD50:400 mg/kg YHTPAD 22,87,87
ipr-rat LD50:9180 μg/kg KSRNAM 8,114,74
scu-rat LD50:131 mg/kg KSRNAM 8,114,74
ivn-rat LD50:8131 μg/kg KSRNAM 8,114,74
orl-mus LD50:134 mg/kg YHTPAD 22,87,87
ipr-mus LD50:8299 μg/kg KSRNAM 8,114,74
scu-mus LD50:92530 μg/kg KSRNAM 8,114,74
ivn-mus LD50:4730 mg/kg YHTPAD 22,87,87
ivn-rbt LDLo:5 mg/kg ARZNAD 10,263,60
ivn-gpg LD50:9130 μg/kg ARZNAD 10,263,60

SAFETY PROFILE: Poison by ingestion, intravenous, intraperitoneal, and subcutaneous routes. Experimental teratogenic and reproductive effects. When heated to decomposition it emits toxic fumes of Na_2O. See also SAPONIN and other escin entries.

EDM500 CAS:102505-08-8 ***HR: 3***
ESCIN TRIETHANOLAMINE SALT
mf: $C_{54}H_{84}O_{23}$•$C_6H_{15}NO_3$ mw: 1250.60

SYN: AESCIN TRIETHANOLAMINE SALT

TOXICITY DATA with REFERENCE
ivn-rat LD50:30 mg/kg ARZNAD 10,263,60
ivn-mus LD50:50 mg/kg ARZNAD 10,263,60
ivn-rbt LD50:6 mg/kg ARZNAD 10,263,60

SAFETY PROFILE: Poison by intravenous route. When heated to decomposition it emits toxic fumes of NO_x. See also other escin entries.

EDN000 ***HR: 3***
ESSENTIAL PHOSPHOLIPIDS

PROP: Chemically defined as diglyceride esters of choline phosphoric acid with a predominant content of unsaturated fatty acids (OYYAA2, 3,45,69).

SYN: EPL

TOXICITY DATA with REFERENCE
ipr-rat TDLo:70 g/kg (28D male):REP OYYAA2 3,45,69
orl-rat LD50:1840 mg/kg OYYAA2 3,45,69
ipr-rat LD50:600 mg/kg OYYAA2 3,45,69
scu-rat LD50:1480 mg/kg OYYAA2 3,45,69
orl-mus LD50:1450 mg/kg OYYAA2 3,45,69
ipr-mus LD50:305 mg/kg OYYAA2 3,45,69
scu-mus LD50:2300 mg/kg OYYAA2 3,45,69

SAFETY PROFILE: Poison by intraperitoneal route. Moderately toxic by ingestion and subcutaneous routes. Experimental reproductive effects. When heated to decomposition it emits very toxic fumes of PO_x.

EDN500 ***HR: D***
ESTERS

PROP: A large group of organic compounds which correspond structurally to salts in inorganic chemistry. They are considered to be derived from acids by the replacement of hydrogen by an organic alkyl radical. Esters of acetic acid are called acetates and esters of carbonic acid are called carbonates. The esterification of a fatty acid

RCOOH, by an alcohol, R′OH yields the fatty ester RCOOR′. The most common alcohol used is methanol, yielding the methyl ester $RCOOCH_3$. The methyl esters of fatty acids have higher vapor pressures than the corresponding acids.

SAFETY PROFILE: No general statement can be made as to the toxicity of esters. Many are highly volatile and hence can act as asphyxiants or narcotics. Skin contact, as well as inhalation, may be important routes of absorption for those esters which are volatile and have a high solvent action. The degree of toxicity ranges from mildly toxic to poison. Esters generally hydrolyze upon contact with moisture; hence, a rough guide to the toxicity of a given ester may be the sum of the toxicities of the products of hydrolysis. Incompatible with nitrates. When heated to decomposition they emit acrid smoke and fumes.

EDN600 ***HR: 3***
ESTRACYT HYDRATE
mf: $C_{23}H_{30}Cl_2NO_6P \cdot 2Na \cdot H_2O$ mw: 582.41

SYNS: ESTRAMUSTINE PHOSPHATE DISODIUM HYDRATE ◇ ESTRAMUSTINE PHOSPHATE SODIUM HYDRATE

TOXICITY DATA with REFERENCE
orl-rat LD50:5400 mg/kg IYKEDH 14,838,83
scu-rat LD50:9000 mg/kg IYKEDH 14,838,83
ivn-rat LD50:208 mg/kg IYKEDH 14,838,83
scu-mus LD50:5200 mg/kg YKYUA6 35,1347,84
ivn-mus LD50:380 mg/kg IYKEDH 14,838,83
orl-rbt LD50:5655 mg/kg YKYUA6 35,1347,84

SAFETY PROFILE: Poison by intravenous route. Mildly toxic by ingestion. When heated to decomposition it emits toxic fumes of PO_x, Cl^-, NO_x, and Na_2O.

EDO000 CAS:50-28-2 ***HR: 3***
ESTRADIOL
mf: $C_{18}H_{24}O_2$ mw: 272.42

PROP: Needles out of benzene, acetone. Mp: 173-179°, bp: decomp. Sol in dioxone, alc, and ether.

SYNS: ALTRAD ◇ BARDIOL ◇ DIHYDROFOLLICULAR HORMONE ◇ DIHYDROFOLLICULIN ◇ DIHYDROMENFORMON ◇ DIHYDROTHEELIN ◇ 3,17-β-DIHYDROXYESTRA-1,3,5(10)-TRIENE ◇ 3,17-β-DIHYDROXY-1,3,5(10)-ESTRATRIENE ◇ DIHYDROXYESTRIN ◇ 3,17-β-DIHYDROXYOESTRA-1,3,5-TRIENE ◇ 3,17-β-DIHYDROXY-1,3,5(10)-OESTRATRIENE ◇ DIHYDROXYOESTRIN ◇ DIMENFORMON ◇ DIMENFORMON PROLONGATUM ◇ DIOGYN ◇ DIOGYNETS ◇ E^2 ◇ 3,17-EPIDIHYDROXYESTRATRIENE ◇ 3,17-EPIDIHYDROXYOESTRATRIENE ◇ ESTRADIOL-17-β ◇ α-ESTRADIOL ◇ β-ESTRADIOL ◇ 3,17-β-ESTRADIOL ◇ 17-β-ESTRADIOL ◇ 17-β-OH-ESTRADIOL ◇ 17-β-OH-OESTRADIOL ◇ cis-ESTRADIOL ◇ d-ESTRADIOL ◇ d-3,17-β-ESTRADIOL ◇ ESTRALDINE ◇ ESTRA-1,3,5(10)-TRIENE-3,17-β-DIOL ◇ 17-β-ESTRA-1,3,5(10)-TRIENE-3,17-DIOL ◇ 1,3,5-ESTRATRIENE-3,17-β-DIOL ◇ ESTROVITE ◇ FEMESTRAL ◇ FEMOGEN ◇ GYNERGON ◇ GYNESTREL ◇ GYNOESTRYL ◇ LAMDIOL ◇ MACRODIOL ◇ MACROL ◇ MICRODIOL ◇ NORDICOL ◇ NSC-9895 ◇ OESTERGON ◇ OESTRADIOL ◇ α-OESTRADIOL ◇ β-OESTRADIOL ◇ 3,17-β-OESTRADIOL ◇ cis-OESTRADIOL ◇ d-OESTRADIOL ◇ d-3,17-β-OESTRADIOL ◇ OESTRADIOL R ◇ OESTRADIOL-17-β ◇ OESTRA-1,3,5(10)-TRIENE-3,17-β-DIOL ◇ 17-β-OESTRA-1,3,5(10)-TRIENE-3,17-DIOL ◇ OESTROGLANDOL ◇ OESTROGYNAL ◇ OVAHORMON ◇ OVASTEROL ◇ OVASTEVOL ◇ OVOCICLINA ◇ OVOCYCLIN ◇ OVOCYCLINE ◇ OVOCYLIN ◇ PRIMOFOL ◇ PROFOLIOL ◇ PROGYNON ◇ PROGYNON-DH ◇ SYNDIOL

TOXICITY DATA with REFERENCE
mnt-hmn:lym 5 μmol/L MUREAV 156,199,85
dns-hmn:mmr 10 nmol/L CNREA8 45,1644,85
dni-hmn:lym 10 μmol/L PSEBAA 146,401,74
dns-rat-par 10 μg/k ACHCBO 18,213,85
cyt-ham:ovr 50 μmol/L TOLED5 29,201,85
dns-rbt:oth 100 nmol/L CRNGDP 3,703,82
orl-wmn TDLo:4400 μg/kg (31W pre):REP ACEDAB 105,7,66
ims-rbt TDLo:30 μg/kg (female 18-20D post):TER JRPFA4 53,237,78
ipr-rat TDLo:1400 mg/kg/13W-I:ETA NIGHAE 52,655,83
imp-rat TDLo:100 mg/kg/52W-C:CAR JJCREP 78,134,87
orl-mus TDLo:58 mg/kg/82W-C:NEO BJCAAI 35,322,77
imp-ham TDLo:286 mg/kg:CAR CNREA8 12,274,52

CONSENSUS REPORTS: NTP Fifth Annual Report on Carcinogens. IARC Cancer Review: Human Limited Evidence IMEMDT 21,279,79; Animal Sufficient Evidence IMEMDT 21,279,79; IMEMDT 6,99,74. EPA Genetic Toxicology Program.

SAFETY PROFILE: Confirmed carcinogen with experimental carcinogenic, neoplastigenic, tumorigenic, and teratogenic data. A promoter. Human reproductive effects by ingestion: fertility effects. Experimental reproductive effects. Human mutation data reported. A steroid hormone much used in medicine. When heated to decomposition it emits acrid smoke and irritating fumes.

EDO500 CAS:57-91-0 ***HR: 3***
17-α-ESTRADIOL
mf: $C_{18}H_{24}O_2$ mw: 272.42

SYNS: 3,17-DIHYDROXYESTRATRIENE ◇ 3,17-α-DIHYDROXYOESTRA-1,3,5(10)-TRIENE ◇ ESTRA-1,3,5(10)-TRIENE-3,17-α-DIOL ◇ 1,3,5-ESTRATRIENE-3,17-α-DIOL ◇ OESTRADIOL-17-α

TOXICITY DATA with REFERENCE
dni-hmn:oth 100 mg//L JTEHD6 10,143,82
imp-gpg TD:3 mg/kg:ETA,REP RBBIAL 5,1,45

SAFETY PROFILE: Experimental reproductive effects. Questionable carcinogen with experimental tumorigenic data. Human mutation data reported. When heated to decomposition it emits acrid smoke and irritating fumes. A steroid. See also ESTRADIOL.

EDP000 CAS:50-50-0 ***HR: 3***
ESTRADIOL-3-BENZOATE
mf: $C_{25}H_{28}O_3$ mw: 376.53

PROP: White or sltly yellow to brownish crystalline powder, odorless. Almost insol in water; sol in alc, acetone, and dioxane; sparingly sol in vegetable oils; sltly sol in ether. Mp: 191-196°

SYNS: BENOVOCYLIN ◇ BENZHORMOVARINE ◇ BENZOATE d'OESTRADIOL (FRENCH) ◇ BENZOESTROFOL ◇ BENZOFOLINE ◇ BENZO-GYNOESTRYL ◇ BENZOIC ACID ESTRADIOL ◇ DIFFOLLISTEROL ◇ DIFOLLICULINE ◇ DIHYDROESTRIN BENZOATE ◇ DIHYDROFOLLICULIN BENZOATE ◇ DIMENFORMON BENZOATE ◇ DIMENFORMONE ◇ DIOGYN B ◇ EBZ ◇ ESTON-B ◇ ESTRADIOL BENZOATE ◇ ESTRADIOL-17-β-BENZOATE ◇ ESTRADIOL-17-β-3-BENZOATE ◇ β-ESTRADIOL BENZOATE ◇ β-ESTRADIOL-3-BENZOATE ◇ 17-β-ESTRADIOL BENZOATE ◇ 17-β-ESTRADIOL-3-BENZOATE ◇ ESTRADIOL MONOBENZOATE ◇ 17-β-ESTRADIOL MONOBENZOATE ◇ ESTRA-1,3,5(10)-TRIENE-3,17-DIOL (17-β)-3-BENZOATE ◇ ESTRA-1,3,5(10)-TRIENE-3,17-β-DIOL, 3-BENZOATE ◇ 1,3,5(10)-ESTRATRIENE-3,17-β-DIOL 3-BENZOATE ◇ FEMESTRONE ◇ FOLLICORMON ◇ FOLLIDRIN ◇ GRAAFINA ◇ de GRAAFINA ◇ GYNECORMONE ◇ GYNFORMONE ◇ HIDROESTRON ◇ HORMOGYNON ◇ HYDROXYESTRIN BENZOATE ◇ MEE ◇ ODB ◇ OESTRADIOL BENZOATE ◇ OESTRADIOL-3-BENZOATE ◇ β-OESTRADIOL BENZOATE ◇ β-OESTRADIOL-3-BENZOATE ◇ 17-β-OESTRADIOL-3-BENZOATE ◇ OESTRADIOL MONOBENZOATE ◇ OESTRAFORM (BDH) ◇ 1,3,5(10)-OESTRATRIENE-3,17-β-DIOL 3-BENZOATE ◇ OVAHORMON BENZOATE ◇ OVASTEROL-B ◇ OVEX ◇ OVOCYCLIN BENZOATE ◇ OVOCYCLIN M ◇ OVOCYCLIN-MB ◇ PRIMOGYN B ◇ PRIMOGYN BOLEOSUM ◇ PRIMOGYN I ◇ PROGYNON B ◇ PROGYNON BENZOATE ◇ RECTHORMONE OESTRADIOL ◇ SOLESTRO ◇ UNISTRADIOL

TOXICITY DATA with REFERENCE
dni-rat-scu 10 μg/kg JOENAK 65,45,75
ims-wmn TDLo:1 mg/kg (5D pre):REP FESTAS 26,405,75
ims-rat TDLo:2500 μg/kg (female 13D post):TER ACENA7 33,520,60
imp-rat TDLo:25 mg/kg:NEO EJCAAH 13,1437,77
scu-mus TDLo:24 mg/kg/36W-I:CAR CNREA8 8,337,48
par-mus TDLo:38 mg/kg/57W-I:ETA CNREA8 1,359,41
scu-gpg TDLo:240 μg/kg/8W-I:ETA,TER LANCAO 1,1313,39
imp-ham TDLo:80 mg/kg:ETA,TER JPBAA7 56,1,44
scu-gpg TD:4 mg/kg/9W-I:ETA,TER CRSBAW 130,9,39
scu-mus TD:38 mg/kg/39W-I:CAR CNREA8 8,337,48

CONSENSUS REPORTS: IARC Cancer Review: Animal Sufficient Evidence IMEMDT 21,279,79.

SAFETY PROFILE: Confirmed carcinogen with experimental carcinogenic, tumorigenic, and teratogenic data. Human reproductive effects by intramuscular route: menstrual cycle changes and disorders. Experimental reproductive effects. Mutation data reported. A steroid. When heated to decomposition it emits acrid smoke and irritating fumes. See also ESTRADIOL.

EDP500 CAS:63042-19-3 ***HR: 3***
ESTRADIOL-17-BENZOATE-3,n-BUTYRATE
mf: $C_{29}H_{34}O_4$ mw: 446.63

SYNS: 17-BENZOATE-3-n-BUTYRATE d'OESTRADIOL (FRENCH) ◇ ESTRA-1,3,5(10)-TRIENE-3,17-β-DIOL-17-BENZOATE-3-n-BUTYRATE

TOXICITY DATA with REFERENCE
scu-gpg TDLo:7 mg/kg/12W-I:ETA,REP CRSBAW 130,1466,39

SAFETY PROFILE: Experimental reproductive effects. Questionable carcinogen with experimental tumorigenic data. When heated to decomposition it emits acrid smoke and irritating fumes. See also ESTRADIOL.

EDQ000 CAS:8000-03-1 ***HR: 3***
ESTRADIOL-3-BENZOATE mixed with PROGESTERONE (1:14 moles)

SYNS: ESTRADIOL BENZOATE mixed with PROGESTERONE (1:14 moles) ◇ ESTRA-1,3,5(10)-TRIETNE-3,17-β-DIOL-3-BENZOATE mixed with PROGESTERONE (1:14 moles) ◇ PROGESTERONE mixed with ESTRADIOL BENZOATE (14:1 moles) ◇ PROGESTERONE mixed with ESTRA-1,3,5(10)-TRIENE-3,17-β-DIOL-3-BENZOATE (14:1 moles)

TOXICITY DATA with REFERENCE
ims-mky TDLo:53 mg/kg (female 20-21D post):TER TJADAB 35,119,87
scu-mus TDLo:338 mg/kg/39W-I:ETA,REP YJBMAU 12,213,39

SAFETY PROFILE: An experimental teratogen. Other experimental reproductive effects. Questionable carcinogen with experimental tumorigenic data. When heated to decomposition it emits acrid smoke and irritating fumes. See also ESTRADIOL and PROGESTERONE.

EDQ500 CAS:63042-22-8 ***HR: 3***
ESTRADIOL-17-CAPRYLATE
mf: $C_{26}H_{38}O_3$ mw: 398.64

SYNS: CAPRYLATE d'OESTRADIOL (FRENCH) ◇ ESTRA-1,3,5(10)-TRIENE-3,17,β-DIOL-17-OCTANOATE

TOXICITY DATA with REFERENCE
scu-gpg TDLo:800 μg/kg/13W-I:ETA,TER CRSBAW 130,1466,39
imp-gpg TDLo:2800 μg/kg:ETA RCBIAS 3,108,44

SAFETY PROFILE: Questionable carcinogen with experimental tumorigenic and teratogenic data. When heated to decomposition it emits acrid smoke and irritating fumes. See also ESTRADIOL.

EDR000 CAS:113-38-2 ***HR: 3***
ESTRADIOL DIPROPIONATE
mf: $C_{24}H_{32}O_4$ mw: 384.56

SYNS: AGOFOLLIN ◇ DIMENFORMON DIPROPIONATE ◇ DIOVO-

CYCLIN ◇ DIOVOCYLIN ◇ DIPROPIONATE d'OESTRADIOL (FRENCH) ◇ DIPROSTRON ◇ ENDOFOLLICOLINA D.P. ◇ ESTRADIOL-3,17-DIPROPIONATE ◇ β-ESTRADIOL DIPROPIONATE ◇ β-ESTRADIOL-3,17-DIPROPIONATE ◇ 3,17-β-ESTRADIOL DIPROPIONATE ◇ 17-β-ESTRADIOL DIPROPIONATE ◇ ESTRA-1,3,5(10)-TRIENE-3,17-DIOL (17-β)-DIPROPIONATE ◇ 1,3,5(10)-ESTRATRIENE-3,17-β-DIOL DIPROPIONATE ◇ ESTROICI ◇ ESTRONEX ◇ FOLLICYCLIN P ◇ NACYCLYL ◇ OESTRADIOL DIPROPIONATE ◇ OESTRADIOL-3,17-DIPROPIONATE ◇ β-OESTRADIOL DIPROPIONATE ◇ 3,17-β-OESTRADIOL DIPROPIONATE ◇ 17-β-OESTRADIOL DIPROPIONATE ◇ OVOCYCLIN DIPROPIONATE ◇ OVOCYCLIN-P ◇ PROGYNON-DP

TOXICITY DATA with REFERENCE
par-rat TDLo:5 mg/kg (female 14D post):REP CRSBAW 164,784,70
par-rat TDLo:1 mg/kg (female 14D post):TER CRSBAW 164,779,70
par-rat TDLo:40 mg/kg/26W-I:ETA CNREA8 2,632,42

CONSENSUS REPORTS: IARC Cancer Review: Animal Sufficient Evidence IMEMDT 21,279,79. EPA Genetic Toxicology Program.

SAFETY PROFILE: Confirmed carcinogen with experimental carcinogenic, tumorigenic, and teratogenic data. A poison by intravenous and parenteral routes. Experimental reproductive effects. A drug for the treatment of menopause. When heated to decomposition it emits acrid smoke and irritating fumes. See also ESTRADIOL.

EDR500 CAS:22966-79-6 ***HR: 3***
ESTRADIOL MUSTARD
mf: $C_{42}H_{50}Cl_4N_2O_4$ mw: 788.74

PROP: Mp: 40-65° (freeze dried).

SYNS: BIS((4-(BIS(2-CHLOROETHYL)AMINO)BENZENE)ACETATE) ESTRA-1,3,5(10)-TRIENE-3,17-DIOL(17-β) ◇ BIS((4-(BIS(2-CHLOROETHYL)AMINO)BENZENE)ACETATE)OESTRA-1,3,5(10)-TRIENE-3,17-DIOL(17-β) ◇ BIS((p-(BIS(2-CHLOROETHYL)AMINO)PHENYL)ACETATE)ESTRADIOL ◇ BIS((p-(BIS(2-CHLOROETHYL)AMINO)PHENYL) ACETATE)ESTRA-1,3,5(10)-TRIENE-3,17-β-DIOL ◇ BIS((p-(BIS(2-CHLOROETHYL)AMINO)PHENYL)ACETATE)OESTRADIOL ◇ BIS((p-BIS(2-CHLOROETHYL)AMINOPHENYL)ACETATE)OESTRA-1,3,5(10)-TRIENE-3,17-β-DIOL ◇ NCI-C01570 ◇ NSC 112259 ◇ OESTRADIOL MUSTARD

TOXICITY DATA with REFERENCE
orl-mus TDLo:2340 mg/kg/52W-I:CAR NCITR* NCI-CG-TR-59,78
ipr-mus TDLo:480 mg/kg/8W-I:NEO CNREA8 33,3069,73
orl-mus TD:4680 mg/kg/1Y-I:CAR NCITR* NCI-CG-TR-59,78

CONSENSUS REPORTS: IARC Cancer Review: Group 3 IMEMDT 7,56,87; Animal Limited Evidence IMEMDT 9,217,75. NCI Carcinogenesis Bioassay (gavage); Clear Evidence: mouse NCITR* NCI-CG-TR-59,78; No Evidence: rat NCITR* NCI-CG-TR-59,78.

SAFETY PROFILE: Questionable carcinogen with experimental carcinogenic and neoplastigenic data. When heated to decomposition it emits very toxic fumes of Cl^- and NO_x. See also ESTRADIOL.

EDS000 CAS:28014-46-2 ***HR: 3***
ESTRADIOL POLYESTER with PHOSPHORIC ACID
mf: $(C_{18}H_{24}O_2{\cdot}H_3O_4P)_x$

SYNS: ESTRADIOL PHOSPHATE POLYMER ◇ ESTRADURIN ◇ (17-β)-ESTRA-1,3,5(10)-TRIENE-3,17-DIOL POLYMER with PHOSPHORIC ACID ◇ OESTRADIOL PHOSPHATE POLYMER ◇ OESTRADIOL POLYESTER with PHOSPHORIC ACID ◇ PEP ◇ POLY(ESTRADIOL PHOSPHATE) ◇ POLYOESTRADIOL PHOSPHATE

TOXICITY DATA with REFERENCE
ims-rat TDLo:50 mg/kg (19D preg):TER ACENA7 33,520,60
ims-wmn TDLo:173 mg/kg/9Y-I:CAR,LIV ACLRBL 7,287,75
scu-rat LD50:5800 mg/kg FATOBP (15),56,80
scu-mus LD50:6100 mg/kg FATOBP (15),56,80

CONSENSUS REPORTS: IARC Cancer Review: Animal Sufficient Evidence IMEMDT 21,279,79.

SAFETY PROFILE: Confirmed carcinogen producing liver tumors. An experimental teratogen. A drug used in cancer treatment. When heated to decomposition it emits toxic fumes of PO_x. See also ESTRADIOL, ESTERS, POLYMERS, and PHOSPHORIC ACID.

EDS100 CAS:979-32-8 ***HR: 3***
ESTRADIOL-17-VALERATE
mf: $C_{24}H_{32}O_3$ mw: 368.56

SYNS: ALTADIOL ◇ DELADIOL ◇ DELAHORMONE UNIMATIC ◇ DELESTROGEN ◇ DELESTROGEN 4X ◇ DURA-ESTRADIOL ◇ ESTRADIOL VALERATE ◇ ESTRADIOL 17-β-VALERATE ◇ ESTRADIOL VALERIANATE ◇ (17-β)-ESTRA-1,3,5(10)-TRIENE-3,17-DIOL-17-PENTANOATE (9CI) ◇ ESTRAVEL ◇ FEMOGEX ◇ NEOFOLLIN ◇ PHARLON ◇ PROGYNON ◇ PROGYNON-DEPOT ◇ PROGYNOVA

TOXICITY DATA with REFERENCE
scu-pig TDLo:67 μg/kg (female 15-18D post):TER BEMTAM 88,385,75
scu-mus TDLo:2 mg/kg (female 1D pre):REP BIREBV 30,556,84
scu-rat TDLo:104 mg/kg/2Y-I:CAR CRNGDP 5,1003,84
scu-mus TDLo:400 μg/kg/W-I:CAR AVBIB9 22/23,359,79

SAFETY PROFILE: Suspected carcinogen with carcinogenic and teratogenic data. Experimental reproductive effects. When heated to decomposition it emits acrid smoke and irritating fumes. See also ESTRADIOL.

EDT100 CAS:52205-73-9 ***HR: 3***
ESTRAMUSTINE PHOSPHATE SODIUM
mf: $C_{23}H_{32}Cl_2NO_6P{\cdot}2Na$ mw: 566.41

SYNS: EMP ◇ ESTRACYT ◇ ESTRAMUSTINE PHOSPHATE DISOD-

IUM ◇ ESTRA-1,3,5(10)-TRIENE-3,17-β-DIOL 3-(BIS(2-CHLOROETHYL) CARBAMATE)17-DISODIUM PHOSPHATE

TOXICITY DATA with REFERENCE
orl-rat TDLo:5500 μg/kg (female 7-17D post):REP OYYAA2 20,1219,80
orl-rat TDLo:44 mg/kg (female 7-17D post):TER OYYAA2 20,1219,80
orl-rat LD50:5400 mg/kg OYYAA2 20,1141,80
scu-rat LD50:9000 mg/kg OYYAA2 20,1141,80
ivn-rat LD50:208 mg/kg OYYAA2 20,1141,80
scu-mus LD50:5200 mg/kg OYYAA2 20,1141,80
ivn-mus LD50:380 mg/kg OYYAA2 20,1141,80
orl-rbt LD50:5655 mg/kg OYYAA2 20,1141,80

SAFETY PROFILE: Poison by intravenous route. Mildly toxic by ingestion. Experimental reproductive effects. When heated to decomposition it emits toxic fumes of Na_2O, NO_x, Cl^-, and PO_x.

EDT500 CAS:6639-99-2 ***HR: 3***
α-ESTRA-1,3,5,7,9-PENTANE-3,17-DIOL
mf: $C_{18}H_{19}O_2$ mw: 267.37

SYNS: α-DIHYDROEQUILENIN ◇ α-DIHYDROEQUILENINA (SPANISH)

TOXICITY DATA with REFERENCE
imp-gpg TDLo:20 mg/kg:ETA,REP RBBIAL 5,1,45

SAFETY PROFILE: Experimental reproductive effects. Questionable carcinogen with experimental tumorigenic data. When heated to decomposition it emits acrid smoke and irritating fumes.

EDU000 CAS:1423-97-8 ***HR: 3***
β-ESTRA-1,3,5,7,9-PENTANE-3,17-DIOL
mf: $C_{18}H_{19}O_2$ mw: 267.37

SYNS: β-DIHYDROEQUILENIN ◇ β-DIHYDROEQUILENINA (SPANISH)

TOXICITY DATA with REFERENCE
imp-gpg TDLo:24 mg/kg:ETA,REP RBBIAL 5,1,45

SAFETY PROFILE: Experimental reproductive effects. Questionable carcinogen with experimental tumorigenic data. When heated to decomposition it emits acrid smoke and irritating fumes.

EDU100 ***HR: D***
ESTRA-1,3,5(10)-TRIENE-17-β-DIOL-17-TETRAHYDROPYRANYL ETHER
mf: $C_{23}H_{32}O_3$ mw: 276.23

SYNS: RS-2290 ◇ 17-β-(TETRAHYDRO-2H-PYRAN-2-YLOXY)ESTRA-1,3,5(10)-TRIEN-3-OL

TOXICITY DATA with REFERENCE
scu-rbt TDLo:1500 μg/kg (female 1-3D post):TER FESTAS 20,211,69
scu-rbt TDLo:3 mg/kg (1-3D preg):REP FESTAS 20,211,69

SAFETY PROFILE: An experimental teratogen. Experimental reproductive effects. When heated to decomposition it emits acrid smoke and irritating fumes. See also ETHERS.

EDU500 CAS:50-27-1 ***HR: 3***
ESTRIOL
mf: $C_{18}H_{24}O_3$ mw: 288.42

PROP: Small, white crystals. D: 0.965, bp: 214.6°

SYNS: AACIFEMINE ◇ COLPOVISTER ◇ DESTRIOL ◇ DEUSLON-A ◇ ESTRA-1,3,5(10)-TRIENE-3,16-α,17-β-TRIOL ◇ 1,3,5-ESTRATRIENE-3-β,16-α,17-β-TRIOL ◇ (16-α,17-β)-ESTRA-1,3,5(10)-TRIENE-3,16,17-TRIOL ◇ ESTRATRIOL ◇ 3,16-α,17-β-ESTRIOL ◇ 16-α,17-β-ESTRIOL ◇ ESTRIOLO (ITALIAN) ◇ FOLLICULAR HORMONE HYDRATE ◇ GYNAESAN ◇ HEMOSTYPTANON ◇ HOLIN ◇ HORMOMED ◇ HORMONIN ◇ 16-α-HYDROXYESTRADIOL ◇ 16-α-HYDROXYOESTRADIOL ◇ KLIMORAL ◇ NSC-12169 ◇ OE3 ◇ OESTRA-1,3,5(10)-TRIENE-3,16-α,17-β-TRIOL ◇ 1,3,5-OESTRATRIENE-3-β-3,16-α,17-β-TRIOL ◇ (16-α,17-β)-OESTRA-1,3,5(10)-TRIENE-3,16,17-TRIOL ◇ OESTRATRIOL ◇ OESTRIOL ◇ 3,16-α,17-β-OESTRIOL ◇ 16-α,17-β-OESTRIOL ◇ ORGASTYPTIN ◇ OVESTERIN ◇ OVESTIN ◇ OVESTINON ◇ OVESTRION ◇ STIPTANON ◇ SYNAPAUSE ◇ THEELOL ◇ THULOL ◇ TRIDESTRIN ◇ 3,16-α,17-β-TRIHYDROXY-Δ-1,3,5-ESTRATRIENE ◇ 3,16-α,17-β-TRIHYDROXY-Δ-1,3,5-OESTRATRIENE ◇ 3,16-α,17-β-TRIHYDROXY-ESTRA-1,3,5(10)-TRIENE ◇ TRIHYDROXYESTRIN ◇ 3,16-α,17-β-TRIHYDROXYOESTRA-1,3,5(10)-TRIENE ◇ TRIHYDROXYOESTRIN ◇ TRIODURIN ◇ TRIOVEX

TOXICITY DATA with REFERENCE
cyt-ham:ovr 50 μmol/L TOLED5 29,201,85
scu-rat TDLo:350 μg/kg (female 7D pre):REP FESTAS 7,301,56
scu-rat TDLo:250 μg/kg (female 4-8D post):TER ZDBEA9 (3),63,82
imp-mus TDLo:26 mg/kg:NEO EJCAAH 11,39,75
imp-gpg TDLo:20 mg/kg:ETA,TER RBBIAL 5,1,45
imp-ham TDLo:480 mg/kg/64W-C:ETA NCIMAV 1,1,59

CONSENSUS REPORTS: IARC Cancer Review: Animal Limited Evidence IMEMDT 21,327,79; Human Limited Evidence IMEMDT 21,327,79; Animal Inadequate Evidence IMEMDT 6,117,74.

SAFETY PROFILE: Suspected carcinogen with experimental carcinogenic, neoplastigenic, tumorigenic, and teratogenic data. Other experimental reproductive effects. Mutation data reported. A steroid drug for the treatment of menopause. When heated to decomposition it emits acrid smoke and irritating fumes.

EDU600 CAS:10322-73-3 ***HR: D***
ESTROFURATE
mf: $C_{24}H_{26}O_4$ mw: 378.50

SYNS: AY-11483 ◇ AY 11483-16 ◇ 21,23-EPOXY-19,24-DINOR-17-α-CHOLA-1,3,5(10),7,20,22-HEXAENE-3,17-DIOL 3-ACETATE ◇ 17-α-(3-FURYL)ESTRA-1,3,5(10),7-TETRAENE-3,17-DIOL-3-ACETATE

TOXICITY DATA with REFERENCE
orl-rat TDLo:17500 μg/kg (35D pre):REP JRPFA4 18,509,69

SAFETY PROFILE: Experimental reproductive effects. When heated to decomposition it emits acrid smoke and irritating fumes.

EDV000 CAS:53-16-7 ***HR: 3***
ESTRONE
mf: $C_{18}H_{22}O_2$ mw: 270.40

PROP: White crystals. Mp: 254°. Insol in water; sol in alc, benzene, ether, and chloroform.

SYNS: AQUACRINE ◇ CRINOVARYL ◇ CRISTALLOVAR ◇ CRYSTOGEN ◇ DESTRONE ◇ DISYNFORMON ◇ E[1] ◇ ENDOFOLLICULINA ◇ ESTERONE ◇ 1,3,5-ESTRATRIEN-3-OL-17-ONE ◇ 1,3,5(10)-ESTRATRIEN-3-OL-17-ONE ◇ Δ-1,3,5-ESTRATRIEN-3-β-OL-17-ONE ◇ ESTRIN ◇ ESTROL ◇ ESTRON ◇ ESTRONA (SPANISH) ◇ ESTRONE-A ◇ ESTRUGENONE ◇ ESTRUSOL ◇ FEMESTRONE INJECTION ◇ FEMIDYN ◇ FOLIKRIN ◇ FOLIPEX ◇ FOLISAN ◇ FOLLESTRINE ◇ FOLLICULAR HORMONE ◇ FOLLICULIN ◇ FOLLICULINE BENZOATE ◇ FOLLICUNODIS ◇ FOLLIDRIN ◇ GLANDUBOLIN ◇ HIESTRONE ◇ HORMOFOLLIN ◇ HORMOVARINE ◇ 3-HYDROXYESTRA-1,3,5(10)-TRIEN-17-ONE ◇ 3-HYDROXY-17-KETO-ESTRA-1,3,5-TRIENE ◇ 3-HYDROXY-17-KETO-OESTRA-1,3,5-TRIENE ◇ 3-HYDROXY-OESTRA-1,3,5(10)-TRIEN-17-ONE ◇ 3-HYDROXY-1,3,5(10)-OESTRATRIEN-17-ONE ◇ KESTRONE ◇ KETODESTRIN ◇ KETOHYDROXY-ESTRATRIENE ◇ KETOHYDROXYESTRIN ◇ KETOHYDROXYOESTRIN ◇ KOLPON ◇ MENAGEN ◇ MENFORMON ◇ Δ-1,3,5-OESTRATRIEN-3-β-OL-17-ONE ◇ 1,3,5-OESTRATRIEN-3-OL-17-ONE ◇ 1,3,5(10)-OESTRATRIEN-3-OL-17-ONE ◇ OESTRIN ◇ OESTROFORM ◇ OESTRONE ◇ OESTROPEROS ◇ OVEX ◇ OVIFOLLIN ◇ PERLATAN ◇ SOLLICULIN ◇ THEELIN ◇ THELESTRIN ◇ THELYKININ ◇ THYNESTRON ◇ TOKOKIN ◇ UNDEN ◇ WNYESTRON

TOXICITY DATA with REFERENCE
dnd-rat-orl 870 nmol/kg CBINA8 23,13,78
cyt-rat-ipr 10 mg/kg CUSCAM 50,425,81
cyt-ham:ovr 50 μmol/L TOLED5 29,201,85
imp-man TDLo:1586 μg/kg (60D male):REP CCPTAY 25,591,82
scu-rat TDLo:10 μg/kg (female 2D post):TER INJFA3 14,56,69
imp-rat TDLo:16 mg/kg:ETA CNREA8 13,147,53
orl-mus TDLo:11 mg/kg/68W-C:NEO BIMDB3 29,45,78
scu-gpg TDLo:40 mg/kg/18W-I:ETA,TER CRSBAW 130,9,39
imp-gpg TDLo:640 μg/kg:ETA,TER BSBSAS 8,142,51
imp-gpg TD:2 mg/kg:ETA,TER RBBIAL 5,1,45

CONSENSUS REPORTS: NTP Fifth Annual Report on Carcinogens. IARC Cancer Review: Human Limited Evidence IMEMDT 21,343,79; Animal Sufficient Evidence IMEMDT 6,123,74; IMEMDT 21,343,79. Reported in EPA TSCA Inventory.

SAFETY PROFILE: Confirmed carcinogen with experimental carcinogenic, neoplastigenic, tumorigenic, and teratogenic data. A poison by intraperitoneal and subcutaneous routes. Human reproductive effects by implantation: spermatogenesis and impotence. Mutation data reported. A steroid drug for the treatment of menopause and ovariectomy symptoms. When heated to decomposition it emits acrid smoke and irritating fumes.

EDV500 CAS:2393-53-5 ***HR: 3***
ESTRONE BENZOATE
mf: $C_{25}H_{26}O_3$ mw: 374.51

SYNS: BENZOATE d'OESTRONE (FRENCH) ◇ 3-(BENZOYLOXY) ESTRA-1,3,5(10)-TRIEN-17-ONE ◇ 3-HYDROXYESTRA-1,3,5(10)-TRIEN-17-ONE BENZOATE ◇ KETOHYDROXYESTRIN BENZOATE ◇ OESTRONBENZOAT (GERMAN)

TOXICITY DATA with REFERENCE
scu-rat TDLo:80 mg/kg/87W-I:ETA CRSBAW 137,325,43
scu-mus TDLo:760 μg/kg/19W-I:NEO ZEKBAI 56,482,49

CONSENSUS REPORTS: IARC Cancer Review: Animal Limited Evidence IMEMDT 21,343,79.

SAFETY PROFILE: Suspected carcinogen with experimental carcinogenic, neoplastigenic, and tumorigenic data. A steroid. When heated to decomposition it emits acrid smoke and irritating fumes.

EDV600 CAS:438-67-5 ***HR: 3***
ESTRONE SODIUM SULFATE
mf: $C_{18}H_{22}O_5S{\cdot}Na$ mw: 373.45

SYNS: CONESTORAL ◇ ESTRA-1,3,5(10)-TRIEN-17-ONE, 3-(SULFOOXY)-, SODIUM SALT (9CI) ◇ ESTRONE, HYDROGEN SULFATE, SODIUM SALT ◇ ESTRONE SULFATE SODIUM ◇ ESTRONE SULFATE SODIUM SALT ◇ ESTRONE-3-SULFATE SODIUM SALT ◇ EVEX ◇ MORESTIN ◇ OESTRONE-3-SULPHATE SODIUM SALT ◇ SODIUM ESTRONE SULFATE ◇ SODIUM ESTRONE-3-SULFATE

TOXICITY DATA with REFERENCE
scu-ctl TDLo:504 μg/kg (female 9D pre):REP JRPFA4 55,191,79

CONSENSUS REPORTS: NTP Fifth Annual Report on Carcinogens.

SAFETY PROFILE: Confirmed carcinogen. Experimental reproductive effects. When heated to decomposition it emits toxic fumes of SO_x.

EDW000 CAS:102534-95-2 ***HR: 3***
ETABETACIN

PROP: An antibiotic extracted from cultures of a *Streptomyces* strain (AIPUAN 10,21,67).

TOXICITY DATA with REFERENCE
orl-mus LD50:500 μg/kg AIPUAN 10,21,67
ipr-mus LD50:250 μg/kg AIPUAN 10,21,67
ivn-mus LD50:250 μg/kg AIPUAN 10,21,67

SAFETY PROFILE: Poison by ingestion, intraperitoneal, and intravenous routes.

EDW300 CAS:34521-16-9 ***HR: 3***
ETEAI
mf: $C_{11}H_{24}N_3S \cdot Br \cdot BrH$ mw: 391.27

SYNS: (2-((4,5-DIHYDRO-1H-IMIDAZOL-2-YL)THIO)ETHYL)TRIETHYLAMMONIUM BROMIDE HYDROBROMIDE ◇ 2-((4,5-DIHYDRO-1H-IMIDAZOL-2-YL)-THIO)-N,N,N-TRIETHYLETHANAMINIUM,BROMIDE HYDROBROMIDE ◇ 2-(2-TRIETHYLAMINO-ETHYLTHIO)- $\Delta^{(2)}$-IMIDAZOLINE BROMIDE HYDROBROMIDE

TOXICITY DATA with REFERENCE
ipr-mus LD50:75 mg/kg CPBTAL 23,1639,75
scu-mus LD50:109 mg/kg CPBTAL 23,1639,75
ivn-mus LD50:53900 μg/kg CPBTAL 23,1639,75

SAFETY PROFILE: Poison by subcutaneous, intravenous, and intraperitoneal routes. When heated to decomposition it emits toxic fumes of SO_x, NO_x, NH_3, and HBr.

EDW500 CAS:1837-57-6 ***HR: 3***
ETHACRIDINE LACTATE
mf: $C_{15}H_{15}N_3O \cdot C_3H_6O_3$ mw: 343.42

SYNS: ACRINOL ◇ ACROLACTINE ◇ 2-AETHOXY-6,9-DIAMINO-ACRIDINLACTAT (GERMAN) ◇ 2,5-DIAMINO-7-ETHOXYACRIDINE LACTATE ◇ 6,9-DIAMINO-2-ETHOXYACRIDINE LACTATE MONOHYDRATE ◇ ETHODIN ◇ 2-ETHOXY-6,9-DIAMINOACRIDINE LACTATE ◇ 2-ETHOXY-6,9-DIAMINOACRIDINE LACTATE HYDRATE ◇ 2-ETHOXY-6,9-DIAMINOACRIDINIUM LACTATE ◇ FLAVITROL ◇ METIFEX ◇ RIMAON ◇ RIVANOL ◇ RIVINOL ◇ VUCINE

TOXICITY DATA with REFERENCE
mma-sat 16 μg/plate MUREAV 144,9,85
dnd-esc 5 μmol/L MUREAV 89,95,81
mmo-omi 250 mg/L PNASA6 56,500,66
orl-rat TDLo:48 g/kg (30D male):REP OYYAA2 16,337,78
ipr-mus LD50:42 mg/kg NIIRDN 6,2,82
scu-mus LD50:120 mg/kg BJEPA5 28,1,47
ivn-rbt LDLo:30 mg/kg BSPHAV 40,582,33

CONSENSUS REPORTS: Reported in EPA TSCA Inventory.

SAFETY PROFILE: Poison by subcutaneous, intraperitoneal, and intravenous routes. Experimental reproductive effects. Mutation data reported. An antiseptic. When heated to decomposition it emits toxic fumes of NO_x.

EDW875 CAS:1070-11-7 ***HR: 3***
ETHAMBUTOL DIHYDROCHLORIDE
mf: $C_{10}H_{24}N_2O_2 \cdot ClH$ mw: 240.82

SYNS: CL 40881 ◇ DEXAMBUTOL ◇ EMB-FATOL ◇ ETAMBUTOL ◇ ETHAMBUTOL HYDROCHLORIDE ◇ ETHIBI ◇ ETHOPIAN ◇ MYAMBUTOL ◇ MYCOBUTOL

TOXICITY DATA with REFERENCE
mnt-mus-orl 50 mg/kg IRLCDZ 10,135,82
spm-mus-orl 12500 μg/kg IRLCDZ 10,135,82
unr-wmn TDLo:720 mg/kg/48D:EYE ANOPB5 2,578,70
orl-rat LD50:6800 mg/kg OYYAA2 2,70,68
ipr-rat LD50:1200 mg/kg OYYAA2 2,70,68
ivn-rat LD50:300 mg/kg OYYAA2 2,70,68
orl-mus LD50:8900 mg/kg OYYAA2 2,70,68
scu-mus LD50:1800 mg/kg OYYAA2 2,70,68
ivn-mus LD50:230 mg/kg OYYAA2 2,70,68

SAFETY PROFILE: Poison by intravenous route. Moderately toxic by intraperitoneal and subcutaneous routes. Mildly toxic by ingestion. Human systemic effects by an unspecified route: constriction of the pupil. Mutation data reported. When heated to decomposition it emits toxic fumes of NO_x and HCl. See also AMINES.

EDX000 CAS:51635-81-5 ***HR: 2***
ETHAMON DS

PROP: A detergent (GISAAA 43(3),14,78).

TOXICITY DATA with REFERENCE
par-rat LD50:4425 mg/kg GISAAA 43(3),14,78
par-mus LD50:3500 mg/kg GISAAA 43(3),14,78

SAFETY PROFILE: Moderately toxic by parenteral route.

EDY600 ***HR: D***
ETHANDROSTATE
mf: $C_{30}H_{44}O_3$ mw: 452.74

SYNS: 17-α-ETHINYL-Δ^5-ANDROSTENE-3-β,17-β-DIOL 3-CYCLOHEXYLPROPIONATE ◇ 17-α-PREGN-5-EN-20-YNE-3-β,17-DIOL-3-(3-CYCLOHEXYLPROPIONATE)

TOXICITY DATA with REFERENCE
scu-rat TDLo:70 mg/kg (female 1-14D post):REP ENDOAO 60,519,57

SAFETY PROFILE: Experimental reproductive effects. A steroid. When heated to decomposition it emits acrid smoke and irritating fumes.

EDZ000 CAS:74-84-0 ***HR: 3***
ETHANE
mf: C_2H_6 mw: 30.08
DOT: UN 1035/UN 1961

PROP: Colorless, odorless gas. Mp: −172°, bp: −88.6°, lel: 3.0%, uel: 12.5%, fp: −183.2°, d: 0.446 @ 0° (liquid), autoign temp: 959°F, vap d: 1.04, flash p: −202°F.

SYNS: BIMETHYL ◇ DIMETHYL ◇ ETHANE, compressed (DOT) ◇ ETHANE, refrigerated liquid (DOT) ◇ ETHYL HYDRIDE ◇ METHYLMETHANE

CONSENSUS REPORTS: Reported in EPA TSCA Inventory.

DOT Classification: Flammable Gas; Label: Flammable Gas.

SAFETY PROFILE: A simple asphyxiant. See ARGON for properties of simple asphyxiants. A very dangerous fire hazard when exposed to heat or flame; can react vigorously with oxidizing materials. Moderate explosion hazard when exposed to flame. To fight fire, stop flow of gas. Incompatible with chlorine, dioxygenyl tetrafluoroborate, oxidizing materials, heat or flame. When heated to decomposition it emits acrid smoke and irritating fumes.

EEA000 CAS:557-30-2 ***HR: 3***
ETHANEDIAL DIOXIME
mf: $C_2H_4N_2O_2$ mw: 88.08

SYNS: GLYOXAL, DIOXIME ◇ GLYOXIME ◇ PIK-OFF

TOXICITY DATA with REFERENCE
orl-rat LD50:119 mg/kg 85ARAE 3,69,76/77
skn-rbt LD50:1580 mg/kg FMCHA2 -,C188,83

CONSENSUS REPORTS: Reported in EPA TSCA Inventory.

SAFETY PROFILE: Poison by ingestion. Moderately toxic by skin contact. When heated to decomposition it emits toxic fumes of NO_x.

EEA500 CAS:107-15-3 ***HR: 3***
1,2-ETHANEDIAMINE
DOT: UN 1604
mf: $C_2H_8N_2$ mw: 60.12

PROP: Volatile, colorless, hygroscopic liquid; ammonia-like odor. Mp: 8.5°, bp: 117.2°, flash p: 110°F (CC), d: 0.8994 @ 20°/4°, vap press: 10.7 mm @ 20°, vap d: 2.07, autoign temp: 725°F.

SYNS: AETHALDIAMIN (GERMAN) ◇ AETHYLENEDIAMIN (GERMAN) ◇ 1,2-DIAMINOAETHAN (GERMAN) ◇ 1,2-DIAMINO-ETHAAN (DUTCH) ◇ 1,2-DIAMINOETHANE ◇ 1,2-DIAMINO-ETHANO (ITALIAN) ◇ DIMETHYLENEDIAMINE ◇ ETHYLEENDIAMINE (DUTCH) ◇ ETHYLENEDIAMINE (OSHA) ◇ 1,2-ETHYLENEDIAMINE ◇ ETHYLENE-DIAMINE (FRENCH) ◇ NCI-C60402

TOXICITY DATA with REFERENCE
skn-rbt 450 mg open MOD UCDS** 12/15/71
skn-rbt 10 mg/24H open SEV AMIHBC 4,119,51
eye-rbt 675 μg SEV AJOPAA 29,1363,46
eye-rbt 750 μg/24H SEV 85JCAE -,440,86
mmo-sat 33 μg/plate ENMUDM 5(Suppl 1),3,83
mma-sat 1 mg/plate ENMUDM 5(Suppl 1),3,83
orl-mus TDLo:3200 mg/kg (female 6-13D post):REP TCMUD8 7,29,87
ihl-hmn TCLo:200 ppm:PNS AMIHBC 9,223,54
orl-rat LD50:500 mg/kg 85GMAT -,66,82
ihl-rat LCLo:4000 ppm/8H AMIHBC 4,119,51
ipr-rat LD50:76 mg/kg TXAPA9 21,454,72
scu-rat LD50:300 mg/kg 85GMAT -,66,82
ihl-mus LC50:300 mg/m³ 85GMAT -,66,82
ipr-mus LD50:200 mg/kg CHTPBA 4,136,69
scu-mus LD50:424 mg/kg ARZNAD 4,649,54
ivn-dog LDLo:100 mg/kg HBAMAK 4,1295,35
skn-rbt LD50:730 mg/kg AMIHBC 4,119,51
scu-rbt LDLo:500 mg/kg HBAMAK 4,1295,35
orl-gpg LD50:470 mg/kg JIHTAB 23,259,41

CONSENSUS REPORTS: Reported in EPA TSCA Inventory. EPA Extremely Hazardous Substances List.

OSHA PEL: TWA 10 ppm
ACGIH TLV: TWA 10 ppm
DOT Classification: Corrosive Material; Label: Corrosive, Flammable Liquid.

SAFETY PROFILE: An irritant poison in humans by inhalation. Experimental poison by inhalation, intraperitoneal, subcutaneous, and intravenous routes. Moderately toxic by ingestion and skin contact. Experimental reproductive effects.Corrosive. A severe skin and eye irritant. An allergen and sensitizer. Mutation data reported. Flammable when exposed to heat, flame or oxidizers. Can react violently with acetic acid, acetic anhydride, acrolein, acrylic acid, acrylonitrile, allyl chloride, CS_2, chlorosulfonic acid, epichlorohydrin, ethylene chlorohydrin, HCl, mesityl oxide, HNO_3, oleum, $AgClO_4$, H_2SO_4, β-propiolactone, or vinyl acetate. To fight fire, use CO_2, dry chemical, alcohol foam. When heated to decomposition it emits toxic fumes of NO_x and NH_3. See also AMINES.

EEB000 CAS:540-63-6 ***HR: 3***
1,2-ETHANEDITHIOL
mf: $C_2H_6S_2$ mw: 94.20

SYNS: 1,2-DIMERCAPTOETHANE ◇ DITHIOETHYLENEGLYCOL ◇ DITHIOGLYCOL ◇ ETHYLENE DIMERCAPTAN ◇ α-ETHYLENE DIMERCAPTAN ◇ ETHYLENE DITHIOGLYCOL ◇ ETHYLENEDITHIOL ◇ ETHYL HYDROPERSULFIDE

TOXICITY DATA with REFERENCE
orl-mus LD50:342 mg/kg DCTODJ 3,249,80
ipr-mus LD50:50 mg/kg EJMCA5 17,235,82
ivn-mus LD50:56200 μg/kg CSLNX* NX#02101

CONSENSUS REPORTS: Reported in EPA TSCA Inventory.

SAFETY PROFILE: Poison by ingestion, intraperitoneal, and intravenous routes. When heated to decomposition it emits very toxic fumes of SO_x. See also MERCAPTANS.

EEB500 ***HR: 3***
ETHANE HEXAMERCARBIDE
mf: $C_2Hg_6O_2(OH)_2$ mw: 1269.56

PROP: Yellowish-white powder. Vap d: 44.6, explodes @ 230°. Insol in water.

CONSENSUS REPORTS: Mercury and its compounds are on the Community Right-To-Know List.

SAFETY PROFILE: A poison. A dangerous explosion hazard. Explodes when shocked or heated to 230°C. Incompatible with oxidizing materials. When heated to decomposition or on contact with acid or acid fumes, below 230°C it emits highly toxic fumes of Hg. See also MERCURY COMPOUNDS, ORGANIC.

EEC000 CAS:594-44-5 ***HR: 3***
ETHANESULFONYL CHLORIDE
mf: $C_2H_5ClO_2S$ mw: 128.58

PROP: D: 1.357, bp: 177.5°. Sltly sol (decomp) in water and alc; very sol in ether.

SYNS: ETHYLSULFOCHLORIDE ◇ TL 77

TOXICITY DATA with REFERENCE
ihl-mus LCLo:1220 mg/m³/10M NDRC** NDCrc-132,Aug,42

CONSENSUS REPORTS: Reported in EPA TSCA Inventory.

SAFETY PROFILE: Moderately toxic by inhalation. When heated to decomposition it emits very toxic fumes of Cl^- and SO_x.

EEC600 CAS:141-43-5 ***HR: 3***
ETHANOLAMINE
DOT: UN 2491
mf: C_2H_7NO mw: 61.10

PROP: Colorless liquid; ammoniacal odor. Hygroscopic, bp: 170.5°, fp: 10.5°, flash p: 200°F (OC), d: 1.0180 @ 20°/4°, vap press: 6 mm @ 60°, vap d: 2.11. Misc in water and alc; sltly sol in benzene; sol in chloroform.

SYNS: AETHANOLAMIN (GERMAN) ◇ 2-AMINOAETHANOL (GERMAN) ◇ 2-AMINOETANOLO (ITALIAN) ◇ 2-AMINOETHANOL (MAK) ◇ β-AMINOETHYL ALCOHOL ◇ COLAMINE ◇ ETANOLAMINA (ITALIAN) ◇ β-ETHANOLAMINE ◇ ETHANOLAMINE, solution (DOT) ◇ ETHYLOLAMINE ◇ GLYCINOL ◇ 2-HYDROXYETHYLAMINE ◇ β-HYDROXYETHYLAMINE ◇ MEA ◇ MONOAETHANOLAMIN (GERMAN) ◇ MONOETHANOLAMINE ◇ OLAMINE ◇ THIOFACO M-50 ◇ USAF EK-1597

TOXICITY DATA with REFERENCE
skn-rbt 505 mg open MOD UCDS** 1/13/73
eye-rbt 763 μg SEV AJOPAA 29,1363,46
cyt-hmn:lyms 100 umol/L BMAOA3 39,422,86
sce-hmn:lyms 1 mmol/L CYGEDX 21(6),29,87
orl-rat TDLo:500 mg/kg (female 6-15D post):TER TCMUD8 6,403,86
orl-rat LD50:1720 mg/kg TXAPA9 42,417,77
ipr-rat LD50:67 mg/kg EVSSAV 2,289,68
scu-rat LD50:1500 mg/kg GTPZAB 23(9),55,79
ivn-rat LD50:225 mg/kg KBMEAL (4),44,68
ims-rat LD50:1750 mg/kg GTPZAB 23(9),55,79
orl-mus LD50:700 mg/kg TPKVAL 4,81,62
ipr-mus LD50:50 mg/kg NTIS** AD277-689
skn-rbt LD50:1000 mg/kg UCDS** 1/13/72

CONSENSUS REPORTS: Reported in EPA TSCA Inventory.

OSHA PEL: (Transitional: TWA 3 ppm) TWA 3 ppm; STEL 6 ppm
ACGIH TLV: TWA 3 ppm; STEL 6 ppm
DFG MAK: 3 ppm (8 mg/m³)
DOT Classification: Corrosive Material; Label: Corrosive.

SAFETY PROFILE: Poison by intraperitoneal route. Moderately toxic by ingestion, skin contact, subcutaneous, intravenous, and intramuscular routes. A corrosive irritant to skin, eyes, and mucous membranes. Human mutation data reported. Flammable when exposed to heat or flame. A powerful base. Reacts violently with acetic acid; acetic anhydride; acrolein; acrylic acid; acrylonitrile; cellulose; chlorosulfonic acid; epichlorohydrin; HCl; HF; mesityl oxide; HNO_3; oleum; H_2SO_4; β-propiolactone; vinyl acetate. To fight fire, use foam, alcohol foam, dry chemical. When heated to decomposition it emits toxic fumes of NO_x. See also AMINES.

EED000 CAS:1071-23-4 ***HR: 2***
ETHANOLAMINE PHOSPHATE
mf: $C_2H_8NO_4P$ mw: 141.08

SYNS: o-PHOSPHOETHANOLAMINE ◇ PHOSPHORYLETHANOLAMINE

TOXICITY DATA with REFERENCE
ivn-mus LD50:639 mg/kg RPOBAR 2,316,70

CONSENSUS REPORTS: Reported in EPA TSCA Inventory.

SAFETY PROFILE: Moderately toxic by intravenous route. When heated to decomposition it emits very toxic fumes of PO_x and NO_x.

EED600 CAS:23471-13-8 ***HR: 3***
ETHANOLMERCURY BROMIDE
mf: C_2H_5BrHgO mw: 325.57

SYNS: BROMO(2-HYDROXYETHYL)MERCURY ◇ 2-(BROMOMERCURI)ETHANOL

TOXICITY DATA with REFERENCE
orl-rat LD50:16 mg/kg PCJOAU 11,918,77
orl-mus LD50:16500 μg/kg PCJOAU 11,918,77
ivn-mus LD50:56 mg/kg CSLNX* NX#05830

CONSENSUS REPORTS: Mercury and its compounds are on the Community Right-To-Know List.

OSHA PEL: (Transitional: CL 1 mg/10m^3) TWA 0.01 mg(Hg)/m^3; STEL 0.03 mg/m^3 (skin)
ACGIH TLV: TWA 0.01 mg(Hg)/m^3; STEL 0.03 mg(Hg)/m^3
NIOSH REL: TWA 0.05 mg(Hg)/m^3

SAFETY PROFILE: Poison by ingestion and intravenous routes. When heated to decomposition it emits toxic fumes of Br^- and Hg. See also MERCURY COMPOUNDS, ORGANIC and BROMIDES.

EEE000 CAS:20398-06-5 ***HR: 3***
ETHANOL THALLIUM (1+) SALT
mf: $C_2H_6O{\cdot}Tl$ mw: 250.45

SYN: ETHYL ALCOHOL THALLIUM (I)

TOXICITY DATA with REFERENCE
orl-rat LDLo:50 mg/kg NCNSA6 5,43,53

CONSENSUS REPORTS: Reported in EPA TSCA Inventory.
OSHA PEL: TWA 0.1 mg(Tl)/m^3 (skin)
ACGIH TLV: TWA 0.1 mg(Tl)/m^3 (skin)

SAFETY PROFILE: Poison by ingestion. When heated to decomposition it emits toxic fumes of Tl. See also THALLIUM COMPOUNDS.

EEE500 CAS:670-54-2 ***HR: 3***
ETHENE TETRA CARBO NITRILE
mf: C_6N_4 mw: 128.10

$$(NC)_2C{=}C(CN)_2$$

PROP: Colorless crystals. Subl >120°, mp: 198-200°, bp: 223°.

SYN: TETRACYANOETHYLENE

TOXICITY DATA with REFERENCE
orl-mus LD50:29 mg/kg KHZDAN 9,50,66
ivn-mus LD50:4500 μg/kg CSLNX* NX#01707

CONSENSUS REPORTS: Reported in EPA TSCA Inventory. Cyanide and its compounds are on the Community Right-To-Know List.

SAFETY PROFILE: Poison by ingestion and intravenous routes. Violent reaction with 1-methylsilacyclopenta-2,4-diene at 150°C. When heated to decomposition it emits toxic fumes of NO_x and CN^-. See also NITRILES.

EEF000 CAS:17088-21-0 ***HR: 3***
1-ETHENYL PYRENE
mf: $C_{18}H_{12}$ mw: 228.2

SYNS: 1-VINYLPYRENE ◇ 3-VINYLPYRENE

TOXICITY DATA with REFERENCE
mma-sat 10 nmol/plate CNREA8 40,642,80
skn-mus TDLo:3651 μg/kg:ETA CNREA8 40,642,80

SAFETY PROFILE: Questionable carcinogen with experimental tumorigenic data. Mutation data reported. When heated to decomposition it emits acrid smoke and irritating fumes.

EEF500 CAS:73529-25-6 ***HR: 3***
4-ETHENYL PYRENE
mf: $C_{18}H_{12}$ mw: 228.2

SYN: 4-VINYLPYRENE

TOXICITY DATA with REFERENCE
skn-mus TDLo:3651 μg/kg:ETA CNREA8 40,642,80

SAFETY PROFILE: Questionable carcinogen with experimental tumorigenic data. When heated to decomposition it emits acrid smoke and irritating fumes.

EEG000 CAS:88-12-0 ***HR: 2***
1-ETHENYL-2-PYRROLIDINONE
mf: C_6H_9NO mw: 111.16

PROP: Colorless liquid, water-sol. Bp: 148° @ 100 mm, fp: 13.5°, flash p: 209°F (OC), d: 1.04 @ 25°, autoign temp: 687°F, vap d: 3.8, fire p: 213°F.

SYNS: VINYLBUTYROLACTAM ◇ N-VINYLPYRROLIDINONE ◇ N-VINYL-2-PYRROLIDINONE ◇ 1-VINYL-2-PYRROLIDINONE ◇ VINYLPYRROLIDONE ◇ N-VINYLPYRROLIDONE ◇ N-VINYL-2-PYRROLIDONE ◇ 1-VINYL-2-PYRROLIDONE ◇ V-PYROL

TOXICITY DATA with REFERENCE
eye-rbt 100 mg SEV BurLW# 02NOV78
orl-rat LD50:1470 mg/kg BurLW# 02NOV78
ihl-rat LC50:3200 mg/m^3 BurLW# 02NOV78
skn-rbt LD50:560 mg/kg BurLW# 02NOV78

CONSENSUS REPORTS: IARC Cancer Review: Group 3 IMEMDT 7,56,87. Reported in EPA TSCA Inventory.

SAFETY PROFILE: Moderately toxic by ingestion, inhalation and skin contact. A severe eye irritant. Probably irritating and narcotic in high concentrations. Questionable carcinogen. Combustible when exposed to heat or flame; can react vigorously with oxidizing materials. To fight fire, use alcohol foam, CO_2, dry chemical. When heated to decomposition it emits highly toxic fumes of NO_x.

EEG500 ***HR: 3***
ETHERS

PROP: Organic compounds in which an oxygen atom is interposed between two carbon atoms in the structure of the molecule.

SAFETY PROFILE: The simpler ethers such as ethyl ether, isopropyl ether, etc., are powerful narcotics which in large doses can cause death. The danger from ethers is usually acute and seldom chronic. After-effects to ether intoxication are uncommon although continued exposure to small concentrations (not enough to cause an overt symptom) has been known to cause loss of appetite, excessive thirst, and fatigue.

The most common ethers such as ethyl, methyl, and diisopropyl are particularly dangerous fire and explosion hazards when exposed to heat, flame, or sparks. They can react violently with strong oxidizers. Many plant and laboratory fires and explosions have resulted from their high flammability and tendency to form explosive peroxides. The common ethers are easily ignited and have low flash points. The diethyl, ethyl tert-butyl, ethyl tert-pentyl and diisopropyl ethers are very hazardous. Methyl tert-alkyl ethers are relatively safe. Besides the risk of explosion from air mixtures of ether vapors, ethers tend to form peroxides upon standing. For some ethers peroxide levels do not reach dangerous concentrations (e.g., diethyl ether; ethyl vinyl ether; tetrahydrofuran; p-dioxane; 1,1-diethoxyethane and the dimethyl ethers of ethylene glycol). When ethers containing peroxides are heated they can detonate. It is necessary to control smoking, open flames, or even the use of hot plates in areas where low molecular weight ethers are apt to reach 1% concentration or more in air. Only electrical equipment of explosion-proof type (Group C classification) is permitted to be operated in ether areas. Ethers should not be stored near powerful oxidizers or in areas of high fire hazard. They should be kept cool and the containers electrically grounded to avoid sparks.

Dangerous; shock or heat can cause gaseous ethers to escape from their containers and create flammable or even explosive conditions. Incompatible with oxidizing materials, BI_3. See also ETHYL ETHER.

EEH000 CAS:126-52-3 ***HR: 3***
ETHINAMATE
mf: $C_9H_{13}NO_2$ mw: 167.23

SYNS: CARBAMATE de l'ETHINYLCYCLOHEXANOL (FRENCH) ◇ 1-ETHINYLCYCLOHEXYL CARBAMATE ◇ 1-ETHINYLCYCLOHEXYL CARBONATE ◇ 1-ETHYNYLCYCLOHEXANOL CARBAMATE ◇ 1-ETHYNYLCYCLOHEXYL CARBAMATE ◇ 1-ETHYNYLCYCLOHEXYL ESTER CARBAMIC ACID ◇ ETINAMATE ◇ USAF EL-42 ◇ VALAMINA ◇ VALAMINETTEN ◇ VALMID ◇ VALMIDATE ◇ VOLAMIN

TOXICITY DATA with REFERENCE
ipr-mus TDLo:400 mg/kg (10D preg):TER CAJPBD 3,2,63
orl-hmn LDLo:57 mg/kg TOIZAG 7,513,60
orl-rat LD50:331 mg/kg JAPMA8 45,40,56
scu-rat LD50:390 mg/kg ARZNAD 4,477,54
ivn-rat LD50:157 mg/kg JAPMA8 45,40,56
orl-mus LD50:490 mg/kg JAPMA8 45,40,56
ipr-mus LD50:300 mg/kg NTIS** AD277-689
ivn-mus LD50:108 mg/kg JAPMA8 45,40,56
orl-dog LD50:190 mg/kg DRUGAY 6,111,82

CONSENSUS REPORTS: Reported in EPA TSCA Inventory.

SAFETY PROFILE: A deadly human poison. Experimental poison by ingestion, intravenous, subcutaneous and intraperitoneal routes. An experimental teratogen. When heated to decomposition it emits toxic fumes of NO_x. See also CARBAMATES.

EEH500 CAS:57-63-6 ***HR: 3***
ETHINYL ESTRADIOL
mf: $C_{29}H_{24}O_2$ mw: 296.44

SYNS: 3,17-β-DIHYDROXY-17-α-ETHYNYL-1,3,5(10)-ESTRATRIENE ◇ 3,17-β-DIHYDROXY-17-α-ETHYNYL-1,3,5(10)-OESTRATRIENE ◇ ESTROGEN ◇ 17-α-ETHINYL-3,17-DIHYDROXY-$\Delta^{1,3,5}$-ESTRATRIENE ◇ 17-α-ETHINYL-3,17-DIHYDROXY-$\Delta^{1,3,5}$-OESTRATRIENE ◇ 17-ETHINYLESTRADIOL ◇ 17-ETHINYL-3,17-ESTRADIOL ◇ 17-α-ETHINYLESTRADIOL ◇ 17-α-ETHINYL-17-β-ESTRADIOL ◇ 17-α-ETHINYLESTRA-1,3,5(10)-TRIENE-3,17-β-DIOL ◇ ETHINYLESTRIOL ◇ ETHINYLOESTRADIOL ◇ 17-ETHINYL-3,17-OESTRADIOL ◇ ETHINYL-OESTRANOL ◇ 17-α-ETHINYLOESTRA-1,3,5(10)-TRIENE-3,17-β-DIOL ◇ 17-α-ETHINYL-Δ(SUP 1,3,5(10))OESTRATRIENE-3,17-β-DIOL ◇ ETHINYL-OESTRIOL ◇ 17-ETHYNYL-3,17-DIHYDROXY-1,3,5-OESTRATRIENE ◇ ETHYNYLESTRADIOL ◇ 17-α-ETHYNYLESTRADIOL ◇ 17-α-ETHYNYLESTRADIOL-17-β ◇ 17-α-ETHYNYL-1,3,5(10)-ESTRATRIENE-3,17-β-DIOL ◇ 17-α-ETHYNYLESTRA-1,3,5(10)-TRIENE-3,17-β-DIOL ◇ ETHYNYLOESTRADIOL ◇ 17-ETHYNYLOESTRADIOL ◇ 17-α-ETHYNYLOESTRADIOL ◇ 17-α-ETHYNYL-17-β-OESTRADIOL ◇ 17-α-ETHYNYLOESTRADIOL-17-β ◇ 17-ETHYNYLOESTRA-1,3,5(10)-TRIENE-3,17-β-DIOL ◇ 17-α-ETHYNYL-1,3,5-OESTRATRIENE-3,17-β-DIOL ◇ 17-α-ETHYNYL-1,3,5(10)-OESTRATRIENE-3,17-β-DIOL ◇ 17-α-ETHYNYLOESTRA-1,3,5(10)-TRIENE-3,17β-DIOL ◇ 19-NOR-17-α-PREGNA-1,3,5(10)-TRIEN-2-YNE-3,17-DIOL ◇ (17-α)-19-NORPREGNA-1,3,5(10)-TRIEN-20-YNE-3,17,DIOL

TOXICITY DATA with REFERENCE
mmo-ssp 50 μg/plate EGJBAY 20,29,79
dni-hmn:lym 50 μmol/L PSEBAA 146,401,74
orl-wmn TDLo:500 μg/kg (female 5D pre):REP CCPTAY 14,375,76
orl-mus TDLo:200 μg/kg (female 8D post):TER TJADAB 23,233,81
orl-wmn TDLo:2738 μg/kg/10Y-I:CAR MJAUAJ 1,473,81
orl-rat TDLo:6 μg/kg/2Y-C:ETA JTEHD6 6,885,80
imp-rat TDLo:5 mg/kg:CAR JJIND8 67,455,81
imp-ham TDLo:621 mg/kg:NEO CNREA8 12,274,52
orl-wmn TDLo:21 mg/kg/21D-I:GIN LANCAO 1,1479,73
orl-rat LD50:2952 mg/kg TXAPA9 18,185,71
orl-mus LD50:1737 mg/kg TXAPA9 18,185,71

CONSENSUS REPORTS: NTP Fifth Annual Report on Carcinogens. IARC Cancer Review: Human Limited Evidence IMEMDT 21,233,79; Animal Sufficient Evidence IMEMDT 6,77,74; IMEMDT 21,233,79. Reported in EPA TSCA Inventory.

SAFETY PROFILE: Confirmed carcinogen with experimental carcinogenic, tumorigenic, and neoplastigenic data. Moderately toxic by ingestion. Human systemic effects by ingestion: glandular effects. An experimental teratogen. Experimental reproductive effects. Human mutation data reported. When heated to decomposition it emits acrid smoke and irritating fumes. See also ESTRADIOL.

EEH520 CAS:8015-12-1 ***HR: 3***
ETHINYLESTRADIOL and NORETHINDRONE ACETATE
mf: $C_{22}H_{28}O_3 \cdot C_{20}H_{24}O_2$ mw: 636.94

SYNS: ANOVLAR 21 ◇ CONTROVLAR ◇ ETHINYL OESTRADIOL mixed with NORETHISTERONE ACETATE ◇ GYN-ANOVLAR ◇ GYNONLAR 21 ◇ MINORLAR ◇ MINOVLAR ◇ NORETHINDRONE ACETATE and ETHINYLESTRADIOL ◇ NORETHISTERONE ACETATE mixed with ETHINYL OESTRADIOL ◇ NORLESTRIN ◇ PRIMODOS

TOXICITY DATA with REFERENCE
orl-wmn TDLo:1620 µg/kg (20D pre):REP JRPFA4 4,229,62
orl-wmn TDLo:401 µg/kg (5 W post):TER NATUAS 216,83,67
orl-wmn TDLo:23562 µg/kg/94W-I:CAR LANCAO 1,207,78
orl-rat TDLo:11087 mg/kg/1Y-C:ETA KNZOAU 23,57,82
orl-wmn TD:41 mg/kg/2Y-I:NEO MJAUAJ 2,223,78

SAFETY PROFILE: Suspected human carcinogen producing lung and liver tumors. Experimental neoplastigenic and tumorigenic data. Human and experimental teratogenic and reproductive effects. When heated to decomposition it emits acrid smoke and irritating fumes.

EEH550 CAS:68-23-5 ***HR: 3***
17-α-ETHINYL-5,10-ESTRENOLONE
mf: $C_{20}H_{26}O_2$ mw: 298.46

SYNS: 17-ETHINYL-5(10)-ESTRAENEOLONE ◇ 17-α-ETHINYL-ESTRA(5,10)ENEOLONE ◇ 17-α-ETHYNYL-5(10)-ESTREN-17-OL-3-ONE ◇ 17-α-ETHYNYLESTR-5(10)-EN-17-β-OL-3-ONE ◇ 17-α-ETHYNYL-ESTR-5(10)-EN-3-ON-17-β-OL ◇ 17-α-ETHYNYL-17-HYDROXYESTR-5(10)-EN-3-ONE ◇ 17-α-ETHYNYL-17-HYDROXY-5(10)-ESTREN-3-ONE ◇ 17-α-ETHYNYL-17-β-HYDROXY-5(10)-ESTREN-3-ONE ◇ 17-α-ETHYNYL-17-β-HYDROXYESTR-5(10)-EN-3-ONE ◇ 17-α-ETHINYL-17-β-HYDROXY-Δ!RN$^{5(10)}$-ESTREN-3-ONE ◇ 17-α-ETHYNYL-17-β-HYDROXY-Δ!RN$^{\cdot 5(10)}$-ESTREN-3-ONE ◇ 17-α-ETHYNYL-17-β- HYDROXY-3-OXO-Δ !RN$^{5(10)}$-ESTRENE ◇ 17-α-ETHYNYL-19-NOR-5(10)-ANDROSTEN-17-β-OL-3-ONE ◇ 17-α-ETHINYL-Δ$^{5,10-19}$-NORTESTOSTERONE ◇ 17-β-HYDROXY-17-α-ETHINYL-5(10)-ESTREN-3-ONE ◇ 17-HYDROXY-19-NOR-17-α-PREGN-5(10)-EN-20-YN-3-ONE ◇ (17-α)-17-HYDROXY-19-NORPREGN-5(10)-EN-20-YN-3-ONE ◇ 17-HYDROXY(17-α)-19-NORPREGN-5(10)-EN-20-YN-3-ONE ◇ LYNESTROL ◇ NORETHINODREL ◇ 19-NOR-ETHINYL-5,10-TESTOSTERONE ◇ NORETHINYNODREL ◇ NORETHYNODRAL ◇ NORETHYNODREL ◇ 19-NOR-ETHYNODREL ◇ NSC-15432 ◇ SC-4642

TOXICITY DATA with REFERENCE
dni-hmn:lyms 50 µmol/L PSEBAA 146,401,74
dns-rat:lvr 100 µmol/L CRNGDP 6,1201,85
oth-mus-par 16 µg/kg AJOGAH 120,390,74
cyt-mus:oth 10 mg/L AJOGAH 120,390,74
oth-ctl:oth 10 mg/L AJOGAH 120,390,74
oth-dom:oth 74100 µg/L AJOGAH 120,390,74
cyt-ctl:oth 100 µg/L AJOGAH 120,390,74
orl-wmn TDLo:1 mg/kg (20D pre):REP FESTAS 16,158,65
orl-mus TDLo:80 mg/kg (8-15D post):TER PSEBAA 121,455,66
orl-rat TDLo:60 mg/kg/17W-C:ETA PAACA3 21,76,80

CONSENSUS REPORTS: IARC Cancer Review: Animal Limited Evidence IMEMDT 21,461,79; Animal Sufficient Evidence IMEMDT 6,191,74.

SAFETY PROFILE: Suspected carcinogen with experimental tumorigenic data. Human and experimental reproductive effects. Mutation data reported. When heated to decomposition it emits acrid smoke and irritating fumes.

EEH575 CAS:8064-76-4 ***HR: 2***
ETHINYLOESTRADIOL mixed with LYNOESTRENOL
mf: $C_{20}H_{28}O \cdot$ C20-H24-O2 mw: 580.92

SYNS: Ba 49249 ◇ C 49249Ba ◇ ETHYNYLESTRADIOL-LYNESTRENOL mixture ◇ FISIOQUENS ◇ FYSIOQUENS ◇ LYNOESTRENOL mixed with ETHINYLOESTRADIOL ◇ MINILYN ◇ OVOSTAT ◇ OVOSTAT 1375 ◇ OVOSTAT E ◇ YERMONIL

TOXICITY DATA with REFERENCE
orl-wmn TDLo:4284 µg/kg (84D pre):REP BMJOAE 2,487,77
orl-wmn TDLo:77112 µg/kg/6Y-I:CAR LANCAO 1,273,80

SAFETY PROFILE: Questionable human carcinogen producing liver tumors. Human reproductive effects. When heated to decomposition it emits acrid smoke and irritating fumes.

EEH600 CAS:563-12-2 ***HR: 3***
ETHION
DOT: NA 2783
mf: $C_9H_{22}O_4P_2S_4$ mw: 384.49

PROP: Liquid. Mp: −13°, d: 1.220 @ 20°/4°. Sltly sol in water; sol in xylene, chloroform, acetone.

SYNS: AC 3422 ◇ BIS(S-(DIETHOXYPHOSPHINOTHIOYL)MERCAPTO)METHANE ◇ BLADAN ◇ DIETHION ◇ EMBATHION ◇ ENT 24,105 ◇ ETHANOX ◇ ETHIOL ◇ ETHODAN ◇ ETHYL METHYLENE PHOSPHORODITHIOATE ◇ FMC-1240 ◇ FOSFONO 50 ◇ HYLEMOX ◇ ITOPAZ ◇ KWIT ◇ METHANEDITHIOL-S,S-DIESTER with O,O-DIETHYL ESTER PHOSPHORODITHIOIC ACID ◇ METHYLEEN-S,S'-BIS(O,O-DIETHYL-DITHIOFOSFAAT) (DUTCH) ◇ S,S'-METHYLEN-BIS(O,O-DIAETHYL-DITHIOPHOSPHAT) (GERMAN) ◇ METHYLENE-S,S'-BIS(O,O-DIAETHYL-DITHIOPHOSPHAT) (GERMAN) ◇ S,S'-METHYLENE O,O,O',O'-TETRAETHYL PHOSPHORODITHIOATE ◇ NIAGARA 1240 ◇ NIALATE ◇ PHOSPHOTOX E ◇ RHODIACIDE

◇ RHODOCIDE ◇ RODOCID ◇ RP 8167 ◇ SOPRATHION ◇ O,O,O′,O′-TETRAAETHYL-BIS(DITHIOPHOSPHAT) (GERMAN) ◇ O,O,O′,O′-TETRAETHYL S,S′-METHYLENEBISPHOSPHORDITHIOATE ◇ O,O,O′,O′-TETRAETHYL-S,S′-METHYLENEBISPHOSPHORODITHIOATE ◇ TETRAETHYL S,S′-METHYLENE BIS(PHOSPHOROTHIOLOTHIONATE) ◇ O,O,O′,O′-TETRAETHYL S,S′-METHYLENE DI(PHOSPHORODITHIOATE) ◇ VEGFRU FOSMITE

TOXICITY DATA with REFERENCE
orl-inf TDLo:15700 μg/kg:CNS,PNS,MET TXMDAX 63,71,67
orl-hmn TDLo:100 μg/kg:BIO TXAPA9 22,286,72
orl-rat LD50:13 mg/kg PHJOAV 185,361,60
skn-rat LD50:62 mg/kg TXAPA9 14,515,69
ipr-rat LD50:26 mg/kg PSEBAA 129,699,68
orl-mus LD50:40 mg/kg 85DPAN -,-,71/76
ipr-mus LD50:35 mg/kg PSEBAA 129,699,68
skn-rbt LD50:890 mg/kg TobJS# 9NOV73

CONSENSUS REPORTS: EPA: Farm Worker Field Reentry FEREAC 39,16888,74. EPA Genetic Toxicology Program. EPA Extremely Hazardous Substances List.

OSHA PEL: TWA 0.4 mg/m^3 (skin)
ACGIH TLV: TWA 0.4 mg/m^3 (skin)
DOT Classification: Poison B; Label: Poison.

SAFETY PROFILE: Poison by ingestion, skin contact, and intraperitoneal routes. Human systemic effects by ingestion: flaccid paralysis without anesthesia, motor activity changes, fever and inhibition of cholinesterase. When heated to decomposition it emits highly toxic fumes of SO_x and PO_x. See also PARATHION.

EEI000 CAS:67-21-0 ***HR: 3***
dl-ETHIONINE
mf: $C_6H_{13}NO_2S$ mw: 163.26

PROP: Crystals. Decomp @ 273°.

SYNS: AETHIONIN ◇ 2-AMINO-4-(ETHYLTHIO)BUTYRIC ACID ◇ dl-2-AMINO-4-(ETHYLTHIO)BUTYRIC ACID ◇ CN 8676 ◇ ETH ◇ ETHIONIN ◇ ETHIONINE ◇ (±)-ETHIONINE ◇ S-ETHYL-HOMOCYSTEINE ◇ S-ETHYL-dl-HOMOCYSTEINE ◇ NSC 751 ◇ U-1434

TOXICITY DATA with REFERENCE
otr-rat:lvr 7500 μmol/L/12W-C ITCSAF 20,291,84
otr-mus:emb 9200 μmol/L MUREAV 152,113,85
scu-rat TDLo:765 mg/kg (female 12-20D post):REP NATUAS 210,1271,66
ipr-rat TDLo:1 g/kg (female 9-12D post):TER JEZOAO 150,135,62
orl-rat TDLo:7200 mg/kg/34W-C:CAR CNREA8 42,4364,82
ipr-rat TDLo:5625 mg/kg/12W-I:ETA CNREA8 42,4364,82
orl-mus TDLo:44 g/kg/2Y-C:CAR PAACA3 25,78,84
orl-rat LDLo:5000 mg/kg CNREA8 26,619,66
ipr-mus LD50:4250 mg/kg NYKZAU 62,333,66

CONSENSUS REPORTS: EPA Genetic Toxicology Program. Reported in EPA TSCA Inventory.

SAFETY PROFILE: Mildly toxic by ingestion and intraperitoneal routes. Suspected carcinogen with experimental carcinogenic and tumorigenic data. An experimental teratogen. Experimental reproductive effects. Mutation data reported. When heated to decomposition it emits toxic fumes of SO_x and NO_x.

EEI025 CAS:29560-58-5 ***HR: 3***
ETHMOSINE
mf: $C_{22}H_{25}N_3O_4S{\cdot}ClH$ mw: 464.02

SYNS: EN 313 ◇ ETHMOZINE ◇ ETHYL ETHER of 10-(β-MORPHOLYLPROPIONYL)PHENTHIAZINECARBAMINO ACID HYDROCHLORIDE ◇ ETHYL 10-(3-MORPHOLINOPROPIONYL)PHENOTHIAZINE-2-CARBAMATE HYDROCHLORIDE ◇ ETMOZIN

TOXICITY DATA with REFERENCE
ivn-rat LD50:12 mg/kg RPTOAN 33,292,70
ipr-mus LD50:131 mg/kg RPTOAN 35,74,72
ivn-mus LD50:36 mg/kg RPTOAN 35,74,72

CONSENSUS REPORTS: Reported in EPA TSCA Inventory.

SAFETY PROFILE: Poison by intravenous and intraperitoneal routes. An antiarrhythmic and tranquilizer. When heated to decomposition it emits toxic fumes of SO_x, NO_x and HCl. See also CARBAMATES.

EEI050 ***HR: 2***
ETHISTERONE and DIETHYLSTILBESTROL
mf: $C_{21}H_{28}O_2{\cdot}C_{18}H_{20}O_2$ mw: 580.87

SYNS: DIETHYLSTILBESTROL and ETHISTERONE ◇ DIETHYLSTILBESTROL and PRANONE ◇ PRANONE and DIETHYLSTILBESTROL ◇ PRANONE and STILBESTROL ◇ 17-α-PREGN-4-EN-20-YN-3-ONE, 17-HYDROXY-, and trans-α-α′-DIETHYL-4,4′-STILBENEDIOL ◇ STILBESTROL and PRANONE

TOXICITY DATA with REFERENCE
orl-wmn TDLo:1050 mg/kg (4-36 W post):TER JCEMAZ 19,1369,59
unr-wmn TDLo:164 mg/kg/9Y-I:CAR BJOGAS 82,421,75

SAFETY PROFILE: Questionable human carcinogen producing uterine tumors. An experimental teratogen. When heated to decomposition it emits acrid smoke and irritating fumes.

EEI060 CAS:53127-17-6 ***HR: 2***
ETHODUOMEEN

TOXICITY DATA with REFERENCE
ipr-mus TDLo:600 mg/kg (female 9D post):REP DZZEA7 35,1070,80
ipr-mus LD50:614 mg/kg DZZEA7 35,1070,80

SAFETY PROFILE: Moderately toxic by intraperi-

toneal route. Experimental reproductive effects. When heated to decomposition it emits acrid smoke and irritating fumes.

EEI100 ***HR: 2***
ETHODUOMEEN, HYDROFLUORIDE

TOXICITY DATA with REFERENCE
ipr-mus TDLo:400 mg/kg (female 9D post):REP DZZEA7 35,1070,80
ipr-mus LD50:470 mg/kg DZZEA7 35,1070,80

SAFETY PROFILE: Moderately toxic by intraperitoneal route. Experimental reproductive effects. When heated to decomposition it emits acrid smoke and irritating fumes.

EEJ000 CAS:61791-14-8 ***HR: 2***
ETHOMEEN C/15

PROP: Polyoxyethylene(5)cocoa amine alkyl links C8-C18 which consists of dodecyl (47%), undecyl (18%), decyl (9%), octyl (8%), hexadecyl (10%) and octadecyl (5%) (FCTXAV 8,249,70).

TOXICITY DATA with REFERENCE
eye-rbt 100 mg MOD FCTXAV 8,249,70
orl-rat LD50:750 mg/kg FCTXAV 8,249,70

CONSENSUS REPORTS: Reported in EPA TSCA Inventory.

SAFETY PROFILE: Moderately toxic by ingestion. An eye irritant. When heated to decomposition it emits acrid smoke and irritating fumes.

EEJ500 CAS:61791-24-0 ***HR: 2***
ETHOMEEN S/12

PROP: Polyoxyethylene(2)soya amine alkyl links C14-C18 which consists of octadecadienyl (45%), octadecynyl (35%), octadecyl(10%) and hexadecyl (10%) (FCTXAV 8,249,70).

TOXICITY DATA with REFERENCE
orl-rat LD50:1500 mg/kg FCTXAV 8,249,70

CONSENSUS REPORTS: Reported in EPA TSCA Inventory.

SAFETY PROFILE: Moderately toxic by ingestion. When heated to decomposition it emits acrid smoke and irritating fumes.

EEK000 CAS:61791-24-0 ***HR: 2***
ETHOMEEN S/15

PROP: Polyoxyethylene(5)soya amine alkyl links C14-C18 which consists of octadecadienyl (45%), octadecenyl (35%), octadecyl(10%) and hexadecyl (10%) (FCTXAV 8,249,70).

TOXICITY DATA with REFERENCE
eye-rbt 100 mg MOD FCTXAV 8,249,70
orl-rat LD50:1000 mg/kg FCTXAV 8,249,70

CONSENSUS REPORTS: Reported in EPA TSCA Inventory.

SAFETY PROFILE: Moderately toxic by ingestion. An eye irritant. When heated to decomposition it emits acrid smoke and irritating fumes.

EEK500 CAS:627-03-2 ***HR: D***
ETHOXYACETIC ACID
mf: $C_4H_8O_3$ mw: 104.12

SYNS: ACETIC ACID, ETHOXY- ◇ 2-ETHOXYACETIC ACID

TOXICITY DATA with REFERENCE
orl-rat TDLo:6352 mg/kg (male 11D pre):REP TXAPA9 69,385,83

CONSENSUS REPORTS: Reported in EPA TSCA Inventory.

SAFETY PROFILE: Experimental reproductive effects. When heated to decomposition it emits acrid smoke and irritating fumes.

EEL000 CAS:927-80-0 ***HR: 3***
ETHOXY ACETYLENE
mf: C_4H_6O mw: 70.09

$$CH_3CH_2OC{\equiv}CH$$

PROP: Insol in water. Flash p: 19.4°F, d: 0.8, vap d: 2.4, bp: 61°.

SYN: ETHOXYETHYNE

SAFETY PROFILE: Potentially explosive decomposition when heated above 100°C. Potentially explosive reaction with ethyl magnesium iodide. A very dangerous fire hazard when exposed to heat, flame, or oxidizers. To fight fire, use foam, dry chemical, CO_2. When heated to decomposition it emits acrid smoke and fumes. See also ACETYLENE COMPOUNDS.

EEL500 CAS:10031-82-0 ***HR: 2***
p-ETHOXYBENZALDEHYDE
mf: $C_9H_{10}O_2$ mw: 150.18

PROP: Colorless. Mp: 20-26°, bp: 247-249°, misc in alc and ether.

SYNS: ETHOXYBENZALDEHYDE ◇ 4-ETHOXYBENZALDEHYDE

TOXICITY DATA with REFERENCE
skn-rbt 500 mg/24H SEV FCTXAV 18,681,80
orl-rat LD50:2100 mg/kg FCTXAV 18,681,80

CONSENSUS REPORTS: Reported in EPA TSCA Inventory.

SAFETY PROFILE: Moderately toxic by ingestion. A severe skin irritant. When heated to decomposition it emits acrid smoke and irritating fumes. See also ALDEHYDES.

EEM000 CAS:938-73-8 ***HR: 3***
2-ETHOXYBENZAMIDE
mf: $C_9H_{11}NO_2$ mw: 165.21

SYNS: ANOVIGAM ◇ ETAMIDE ◇ ETHBENZAMIDE ◇ ETHENZAMID ◇ ETHENZAMIDE ◇ ETHOSALICYL ◇ o-ETHOXYBENZAMIDE ◇ ETOCIL ◇ ETOSALICIL ◇ ETOSALICYL ◇ EUSAL ◇ H.P. 209 ◇ KATAGRIPPE ◇ LINDATOX ◇ LUCAMIDE ◇ PIROSOLVINA ◇ PROTOPYRIN ◇ TRANCALGYL

TOXICITY DATA with REFERENCE
cyt-ham:lng 500 mg/L/48H GMCRDC 27,95,81
cyt-ham:fbr 500 mg/L ESKHA5 96,55,78
orl-mus TDLo:544 g/kg/54W-C:CAR JJIND8 76,115,86
orl-rat LD50:2630 mg/kg ARZNAD 10,820,60
orl-mus LD50:700 mg/kg THERAP 7,27,52
ipr-mus LD50:400 mg/kg JPPMAB 4,872,52
orl-rbt LDLo:1500 mg/kg ARZNAD 10,820,60

CONSENSUS REPORTS: Reported in EPA TSCA Inventory.

SAFETY PROFILE: Poison by intraperitoneal route. Moderately toxic by ingestion. Questionable carcinogen with experimental carcinogenic data. Mutation data reported. When heated to decomposition it emits toxic fumes of NO_x. See also AMIDES.

EEN500 CAS:452-35-7 ***HR: D***
6-ETHOXY-2-BENZOTHIAZOLESULFONAMIDE
mf: $C_9H_{10}N_2O_3S_2$ mw: 258.33

SYNS: ETHOXAZOLAMIDE ◇ ETHOXYZOLAMIDE

TOXICITY DATA with REFERENCE
orl-rat TDLo:3300 mg/kg (1-22D preg):TER TJADAB 1,51,68
orl-rat TDLo:3300 mg/kg (1-22D preg):REP TJADAB 1,51,68

SAFETY PROFILE: Experimental teratogenic and reproductive effects. When heated to decomposition it emits very toxic fumes of SO_x and NO_x.

EEO000 CAS:577-66-2 ***HR: 3***
8-ETHOXYCAFFEINE
mf: $C_{10}H_{14}N_4O_3$ mw: 238.28

TOXICITY DATA with REFERENCE
mma-sat 20 μg/plate MUREAV 60,349,79
sln-dmg-orl 15 mmol/L MUREAV 149,189,85
trn-dmg-orl 15 mmol/L MUREAV 149,189,85
cyt-ham:ovr 17 mmol/L MUREAV 60,349,79
cyt-ham:oth 12 mmol/L/2H-C MUREAV 12,463,71
ivn-mus LD50:56 mg/kg CSLNX* NX#04352

CONSENSUS REPORTS: EPA Genetic Toxicology Program.

SAFETY PROFILE: Poison by intravenous route. Mutation data reported. When heated to decomposition it emits toxic fumes of NO_x. See also CAFFEINE.

EEO500 CAS:13684-56-5 ***HR: 2***
3-ETHOXYCARBONYLAMINOPHENYL-N-PHENYLCARBAMATE
mf: $C_{16}H_{16}N_2O_4$ mw: 300.34

SYNS: 3-(AETHOXYCARBONYLAMINOPHENYL)-N-PHENYL-CARBAMAT (GERMAN) ◇ BENTANEX ◇ BETANAL AM ◇ BETANEX ◇ m-CARBANILOYLOXYCARBANILIC ACID ETHYL ESTER ◇ DESMEDIPHAM ◇ EP-475 ◇ ETHYL-m-HYDROXYCARBANILATE CARBANI- LATE (ESTER) ◇ ETHYL PHENYLCARBAMOYLOXYPHENYLCARBAMATE ◇ SCHERING 38107 ◇ SN 38107

TOXICITY DATA with REFERENCE
orl-rat LD50:9600 mg/kg 85ARAE 2,92,77
skn-rbt LDLo:10 g/kg FMCHA2 -,C31,83
orl-qal LD50:2480 mg/kg 85DPAN -,-,71/76

SAFETY PROFILE: Moderately toxic by ingestion. Mildly toxic by skin contact. An herbicide. When heated to decomposition it emits toxic fumes of NO_x. See also CARBAMATES.

EEP000 CAS:73987-52-7 ***HR: 3***
ETHOXY CARBONYL DIGOXIN

SYN: CARBAETHOXYDIGOXIN (GERMAN)

TOXICITY DATA with REFERENCE
orl-gpg LD50:3500 μg/kg ARZNAD 15,481,65
ivn-gpg LDLo:1400 μg/kg ARZNAD 15,481,65

SAFETY PROFILE: Poison by ingestion and intravenous routes. See also DIGOXIN.

EEQ000 CAS:63905-54-4 ***HR: 3***
4-ETHOXYCARBONYL-1-(2-HYDROXY-3-PHENOXYPROPYL) 4-PHENYLPIPERIDINE HYDROCHLORIDE
mf: $C_{23}H_{29}NO_4 \cdot ClH$ mw: 419.99

SYNS: B.D.H. 200 HYDROCHLORIDE ◇ ISONIPECOTIC ACID, 1-(2-HYDROXY-3-PHENOXYPROPYL)-4-PHENYL-, ETHYL ESTER, HYDROCHLORIDE

TOXICITY DATA with REFERENCE
orl-mus LD50:419 mg/kg JPPMAB 12,449,60
scu-mus LD50:145 mg/kg JPPMAB 12,449,60

SAFETY PROFILE: Poison by ingestion and subcutaneous routes. When heated to decomposition it emits very toxic fumes of HCl and NO_x.

EEQ100 CAS:60075-74-3 ***HR: 2***
(2-(ETHOXYCARBONYL)-1-METHYL)ETHYL CARBONIC ACID-p-IODOBENZYL ESTER
mf: $C_{14}H_{17}IO_5$ mw: 392.21

TOXICITY DATA with REFERENCE
orl-rat LD50:1700 mg/kg GISAAA 41(6),95,76
orl-mus LD50:583 mg/kg GISAAA 41(6),95,76
ipr-mus LD50:3 g/kg JMCMAR 19,1362,76

SAFETY PROFILE: Moderately toxic by ingestion and intraperitoneal routes. When heated to decomposition it emits toxic fumes of I^-. See also ESTERS.

EEQ500 CAS:616-97-7 ***HR: 3***
S-ETHOXYCARBONYLTHIAMINE HYDROCHLORIDE
mf: $C_{15}H_{22}N_4O_4S_2 \cdot ClH$ mw: 422.99

SYNS: S-AETHOXY-CARBONYLTHIAMIN HYDROCHLORID (GERMAN) ◇ S-CARBETHOXYTHIAMINE HYDROCHLORIDE ◇ SECT HYDROCHLORIDE

TOXICITY DATA with REFERENCE
orl-mus LD50:4250 mg/kg IZVIAK 37,82,67
ivn-mus LD50:339 mg/kg IZVIAK 37,82,67

SAFETY PROFILE: Poison by intravenous route. Mildly toxic by ingestion. When heated to decomposition it emits very toxic fumes of HCl, SO_x and NO_x.

EER000 CAS:1586-92-1 ***HR: 3***
ETHOXY DIETHYL ALUMINUM
mf: $C_6H_{15}AlO$ mw: 129.16

$(CH_3CH_2)_2AlOCH_3CH_2$

SYN:
DIETHYLETHOXYALUMINUM

SAFETY PROFILE: Ignites spontaneously in air in the pure form or in solutions with concentrations greater than 20%. When heated to decomposition it emits acrid smoke and fumes. See also ALUMINUM COMPOUNDS.

EER400 CAS:83053-57-0 ***HR: 2***
11-ETHOXY-15,16-DIHYDRO-17-CYCLOPENTA (a)PHENANTHREN-17-ONE
mf: $C_{19}H_{16}O_2$ mw: 276.35

TOXICITY DATA with REFERENCE
mma-sat 20 μg/plate CRNGDP 3,677,82
skn-mus TDLo:1600 μg/kg:ETA CRNGDP 3,677,82

SAFETY PROFILE: Questionable carcinogen with experimental tumorigenic data. Mutation data reported. When heated to decomposition it emits acrid smoke and irritating fumes.

EER500 CAS:103-75-3 ***HR: 2***
2-ETHOXY DIHYDROPYRAN
mf: $C_7H_{12}O_2$ mw: 128.19

PROP: D: 1.0, bp: 143°, flash p: 111°F (OC).

SYNS: 2-ETHOXY-2,3-DIHYDRO-Γ-PYRAN ◇ 2-ETHOXY-3,4-DIHYDRO-1,2-PYRAN ◇ 2-ETHOXY-3,4-DIHYDRO-2H-PYRAN

TOXICITY DATA with REFERENCE
skn-rbt 10 mg/24H open MLD AIHAAP 23,95,62
eye-rbt 50 mg MOD UCDS** 7/15/71
ihl-rat LCLo:8000 ppm/4H AIHAAP 23,95,62
orl-rat LD50:6160 mg/kg UCDS** 7/15/71
skn-rbt LD50:3560 mg/kg UCDS** 7/15/71

CONSENSUS REPORTS: Reported in EPA TSCA Inventory.

SAFETY PROFILE: Moderately toxic by skin contact. Mildly toxic by ingestion and inhalation. A skin and eye irritant. Combustible when exposed to flame, sparks, and oxidizers. To fight fire, use dry chemical, foam, fog. When heated to decomposition it emits acrid smoke and irritating fumes.

EES000 CAS:15769-72-9 ***HR: 3***
ETHOXYDIISOBUTYLALUMINUM
mf: $C_{10}H_{23}AlO$ mw: 186.28

$CH_3CH_2OAl[CH_2CH(CH_3)_2]_2$

SAFETY PROFILE: Ignites spontaneously in air. When heated to decomposition it emits acrid smoke and fumes. See also ALUMINUM COMPOUNDS.

EES100 CAS:69929-16-4 ***HR: 1***
8-ETHOXY-2,6-DIMETHYLOCTENE-2
mf: $C_{12}H_{24}O$ mw: 184.36

SYNS: CITRONELLYL ETHYL ETHER ◇ 2-OCTENE, 8-ETHOXY-2,6-DIMETHYL-

TOXICITY DATA with REFERENCE
skn-rbt 500 mg/24H MLD FCTOD7 20,655,82

CONSENSUS REPORTS: Reported in EPA TSCA Inventory.

SAFETY PROFILE: A skin irritant. When heated to decomposition it emits acrid smoke and irritating fumes.

EES300 CAS:7495-45-6 ***HR: 2***
ETHOXYDIPHENYLACETIC ACID
mf: $C_{16}H_{16}O_3$ mw: 256.32

SYNS: ACIDE DIPHENYLETHOXYACETIQUE (FRENCH) ◇ α-ETHOXY-α-PHENYL-BENZENEACETIC ACID (9CI)

TOXICITY DATA with REFERENCE
orl-rat LD50:1 g/kg AIPTAK 116,154,58
ivn-rat LD50:600 mg/kg AIPTAK 116,154,58

orl-mus LD50:600 mg/kg AIPTAK 116,154,58
scu-mus LD50:580 mg/kg AIPTAK 116,154,58

SAFETY PROFILE: Moderately toxic by ingestion, intravenous, and subcutaneous routes. When heated to decomposition it emits acrid smoke and irritating fumes.

EES350 CAS:110-80-5 ***HR: 2***
2-ETHOXYETHANOL
DOT: UN 1171
mf: $C_4H_{10}O_2$ mw: 90.14

$CH_3CH_2OCH_2CH_2OH$

PROP: Colorless liquid, practically odorless. Bp: 135.1°, lel: 1.8%, uel: 14%, flash p: 202°F(CC), d: 0.9360 @ 15°/15°, autoign temp: 455°F, vap press: 3.8 mm @ 20°, vap d: 3.10.

SYNS: ATHYLENGLYKOL-MONOATHYLATHER(GERMAN) ◇ CELLOSOLVE (DOT) ◇ CELLOSOLVE SOLVENT ◇ DOWANOL EE ◇ EKTASOLVE EE ◇ ETHER MONOETHYLIQUE de l'ETHYLENE-GLYCOL (FRENCH) ◇ ETHYL CELLOSOLVE ◇ ETHYLENE GLYCOL ETHYL ETHER ◇ ETHYLENE GLYCOL MONOETHYL ETHER ◇ ETHYLENE GLYCOL MONOETHYL ETHER (DOT) ◇ ETOKSYETYLOWY ALKOHOL (POLISH) ◇ GLYCOL ETHER EE ◇ GLYCOL ETHYL ETHER ◇ GLYCOL MONOETHYL ETHER ◇ HYDROXY ETHER ◇ JEFFERSOL EE ◇ NCI-C54853 ◇ OXITOL ◇ POLY-SOLV EE

TOXICITY DATA with REFERENCE
eye-hmn 6000 ppm PHRPA6 45,1459,30
skn-rbt 500 mg open MLD UCDS** 5/20/66
eye-rbt 50 mg MOD UCDS** 5/20/66
eye-gpg 10 μg MLD JPPMAB 11,150,59
ihl-rat TCLo:100 ppm/7H (female 14-20D post):REP TJADAB 21,58A,80
ihl-rat TCLo:600 ppm/7H (female 7-13D post):TER NRTXDN 2,231,81
orl-rat LD50:2125 mg/kg GTPZAB 32(3),48,88
ihl-rat LC50:2000 ppm/7H NPIRI* 1,54,74
skn-rat LD50:3900 mg/kg TOXID9 4,180,84
ipr-rat LD50:2800 mg/kg ARZNAD 21,880,71
ivn-rat LD50:2400 mg/kg NPIRI* 1,54,74
orl-mus LD50:2451 mg/kg KODAK* MSDS-10,170A,82
ihl-mus LC50:1820 ppm/7H JIHTAB 25,157,43
ipr-mus LD50:1707 mg/kg FEPRA7 6,342,47
skn-rbt LD50:3300 mg/kg NPIRI* 1,54,74
ihl-gpg LCLo:3000 ppm/24H PHRPA6 45,1459,30

CONSENSUS REPORTS: Reported in EPA TSCA Inventory. Glycol ether compounds are on the Community Right-To-Know List.

OSHA PEL: TWA 200 ppm (skin)
ACGIH TLV: TWA 5 ppm (skin)
DFG MAK: 20 ppm (75 mg/m^3)
NIOSH REL: (Glycol Ethers): Reduce to lowest level
DOT Classification: Combustible Liquid; Label: None; IMO: Flammable or Combustible Liquid; Label: Flammable Liquid.

SAFETY PROFILE: Moderately toxic by ingestion, skin contact, intravenous, and intraperitoneal routes. Mildly toxic by inhalation and subcutaneous routes. An experimental teratogen. Other experimental reproductive effects. An eye and skin irritant. Combustible when exposed to heat or flame; can react with oxidizing materials. Moderate explosion hazard in the form of vapor when exposed to heat or flame. Mixture with hydrogen peroxide + polyacrylamide gel + toluene is explosive when dry. To fight fire, use alcohol foam, dry chemical. See also GLYCOL ETHERS.

EES400 CAS:111-15-9 ***HR: 3***
2-ETHOXYETHYL ACETATE
DOT: UN 1172
mf: $C_6H_{12}O_3$ mw: 132.18

$CH_3CO{\bullet}OC_2H_4OCH_2CH_3$

PROP: Colorless liquid with a mild, pleasant ester-like odor. Bp: 156.4°, flash p: 117°F (COC), lel: 1.7%, fp: −61.7°, d: 0.9748 @ 20°/20°, autoign temp: 715°F, vap press: 1.2 mm @ 20°, vap d: 4.72.

SYNS: ACETATE de CELLOSOLVE (FRENCH) ◇ ACETATE de l'ETHER MONOETHYLIQUE DE L'ETHYLENE-GLYCOL (FRENCH) ◇ ACETATE d'ETHYLGLYCOL (FRENCH) ◇ ACETATO di CELLOSOLVE (ITALIAN) ◇ ACETIC ACID-2-ETHOXYETHYL ESTER ◇ 2-AETHOXY-AETHYLACETAT (GERMAN) ◇ AETHYLENGLYKOL-AETHERACETAT (GERMAN) ◇ CELLOSOLVE ACETATE (DOT) ◇ CSAC ◇ EKTASOLVE EE ACETATE SOLVENT ◇ ETHOXY ACETATE ◇ 2-ETHOXYETHANOL ACETATE ◇ 2-ETHOXYETHANOL, ESTER with ACETIC ACID ◇ 2-ETHOXY-ETHYLACETAAT (DUTCH) ◇ ETHOXYETHYL ACETATE ◇ β-ETHOXYETHYL ACETATE ◇ 2-ETHOXYETHYLE, ACETATE de (FRENCH) ◇ ETHYL CELLOSOLVE ACETAAT (DUTCH) ◇ ETHYLENE GLYCOL ETHYL ETHER ACETATE ◇ ETHYLENE GLYCOL MONOETHYL ETHER ACETATE (MAK, DOT) ◇ ETHYLGLYKOLACETAT (GERMAN) ◇ 2-ETOSSIETIL-ACETATO (ITALIAN) ◇ GLYCOL ETHER EE ACETATE ◇ GLYCOL MONOETHYL ETHER ACETATE ◇ OCTAN ETOKSYETYLU (POLISH) ◇ OXYTOL ACETATE ◇ POLY-SOLV EE ACETATE

TOXICITY DATA with REFERENCE
skn-rbt 490 mg open MLD UCDS** 1/13/67
eye-rbt 40 mg MOD UCDS** 1/13/67
ihl-rat TCLo:600 ppm/7H (female 7-15D post):TER EPASR* 8EHQ-0682-0450
ihl-rbt TCLo:200 ppm/6H (female 6-18D post):REP UCRR** 09OCT84
orl-rat LD50:2900 mg/kg TXAPA9 51,117,79
ihl-rat LC50:12100 mg/m^3/8H AIHAAP 20,364,59
ipr-mus LD50:1420 mg/kg SCCUR* -,2,61
orl-rbt LD50:1950 mg/kg EPASR* 8EHQ-0682-0450
skn-rbt LD50:10500 mg/kg UCDS** 1/13/67

CONSENSUS REPORTS: Reported in EPA TSCA Inventory. Glycol ether compounds are on the Community Right-To-Know List.

OSHA PEL: TWA 100 ppm (skin)
ACGIH TLV: TWA 5 ppm (skin)
DFG MAK: 20 ppm (110 mg/m^3)
DOT Classification: Combustible Liquid; Label: None; IMO: Flammable or Combustible Liquid; Label: Flammable Liquid.

SAFETY PROFILE: Moderately toxic by ingestion and intraperitoneal routes. A skin and eye irritant. An experimental teratogen. Other experimental reproductive effects. Flammable when exposed to heat or flame; can react with oxidizing materials. Moderate explosion hazard in the form of vapor when heated. Mild explosions have occurred at the end of distillations. To fight fire, use alcohol foam, CO_2, dry chemical. When heated to decomposition it emits acrid smoke and irritating fumes. See also GLYCOL ETHERS.

EES500 CAS:67262-62-8 ***HR: 3***
2-(2-ETHOXYETHYLAMINO)-o-PROPIONOTOLUIDIDE
mf: $C_{14}H_{22}N_2O_2$ mw: 250.38

SYN: 2-(2-ETHOXYETHYLAMINO)-2'-METHYL-PROPIONANILIDE

TOXICITY DATA with REFERENCE
ipr-mus LD50:275 mg/kg JPMSAE 67,595,78
ivn-mus LD50:38 mg/kg JPMSAE 67,595,78

SAFETY PROFILE: Poison by intraperitoneal and intravenous routes. When heated to decomposition it emits toxic fumes of NO_x.

EET000 CAS:63716-10-9 ***HR: 1***
ETHOXYETHYL ETHER of PROPYLENE GLYCOL
mf: $C_7H_{16}O_3$ mw: 148.23

SYN: 3-(2-ETHOXY)ETHOXY-2-PROPANOL

TOXICITY DATA with REFERENCE
skn-rbt 10 mg/24H MLD JIHTAB 31,60,49
eye-rbt 500 mg JIHTAB 31,60,49
orl-rat LD50:9330 mg/kg JIHTAB 31,60,49
skn-rbt LD50:12 g/kg JIHTAB 31,60,49

CONSENSUS REPORTS: Glycol ether compounds are on the Community Right-To-Know List.

SAFETY PROFILE: Mildly toxic by ingestion and skin contact. A skin and eye irritant. When heated to decomposition it emits acrid smoke and irritating fumes. See also GLYCOL ETHERS.

EET500 CAS:67465-42-3 ***HR: 3***
N-(2-ETHOXY-3-HYDROXYMERCURIPROPYL) BARBITAL
mf: $C_{13}H_{22}HgN_2O_5$ mw: 486.96

SYN: HYDROXY(3-(5,5-DIETHYL-2,4,6-TRIOXO-(1H,3H,5H)-PYRIMIDINO)-2-ETHOXYPROPYL)MERCURY

TOXICITY DATA with REFERENCE
ivn-rat LDLo:18600 μg/kg JAPMA8 37,333,48

CONSENSUS REPORTS: Mercury and its compounds are on the Community Right-To-Know List.

OSHA PEL: (Transitional: CL 1 mg/10m^3) CL 0.1 mg(Hg)/m^3 (skin)
ACGIH TLV: TWA 0.1 mg(Hg)/m^3 (skin)
NIOSH REL: (Inorganic Mercury) TWA 0.05 mg(Hg)/m^3

SAFETY PROFILE: Poison by intravenous route. When heated to decomposition it emits very toxic fumes of Hg and NO_x. See also MERCURY COMPOUNDS, ORGANIC and BARBITURATES.

EET600 CAS:13021-50-6 ***HR: 2***
1-ETHOXY-3-ISOPROPOXYPROPAN-2-OL
mf: $C_8H_{18}O_3$ mw: 162.26

SYN: 1-ETHOXY-3-ISOPROPOXY-2-PROPANOL

TOXICITY DATA with REFERENCE
orl-rat LD50:5 g/kg AIPTAK 89,145,52
orl-mus LD50:5730 mg/kg AIPTAK 89,145,52
ipr-mus LD50:3180 mg/kg JPETAB 97,414,49
orl-rbt LDLo:6200 mg/kg AIPTAK 89,145,52

SAFETY PROFILE: Moderately toxic by intraperitoneal route. Mildly toxic by ingestion. When heated to decomposition it emits acrid smoke and irritating fumes.

EEU000 CAS:9002-92-0 ***HR: 1***
ETHOXYLATED LAURYL ALCOHOL

SYN: DODECANOL ETHOXYLATED

TOXICITY DATA with REFERENCE
skn-rbt 500 mg/24H MOD TXAPA9 19,276,71
eye-rbt 100 mg TXAPA9 19,276,71

CONSENSUS REPORTS: Reported in EPA TSCA Inventory.

SAFETY PROFILE: A skin and eye irritant. When heated to decomposition it emits acrid smoke and irritating fumes.

EEU500 CAS:120-53-6 ***HR: 3***
6-ETHOXY-2-MERCAPTOBENZOTHIAZOLE
mf: $C_8H_7NOS_2$ mw: 197.28

SYN: USAF PD-58

TOXICITY DATA with REFERENCE
ipr-mus LD50:250 mg/kg NTIS** AD603-561

CONSENSUS REPORTS: Reported in EPA TSCA Inventory.

SAFETY PROFILE: Poison by intraperitoneal route. When heated to decomposition it emits very toxic fumes of SO_x and NO_x. See also MERCAPTANS.

EEV000 CAS:63019-29-4 ***HR: 3***
10-ETHOXYMETHYL-1:2-BENZANTHRACENE
mf: $C_{21}H_{18}O$ mw: 286.39

SYN: 7-(ETHOXYMETHYL)BENZ(a)ANTHRACENE

TOXICITY DATA with REFERENCE
skn-mus TDLo:1200 mg/kg/50W-I:ETA PRLBA4 129,439,40

SAFETY PROFILE: Questionable carcinogen with experimental tumorigenic data. When heated to decomposition it emits acrid smoke and irritating fumes.

EEV100 CAS:2041-76-1 ***HR: 3***
4-ETHOXY-2-METHYL-3-BUTYN-2-OL
mf: $C_7H_{12}O_2$ mw: 128.17

$$CH_3CH_2OC \equiv CC(CH_3)_2OH$$

SAFETY PROFILE: Potentially explosive decomposition above 115°C. Traces of acid increase the tendency to explode. When heated to decomposition it emits acrid smoke and irritating fumes. See also ACETYLENE COMPOUNDS.

EEV200 CAS:87-13-8 ***HR: 3***
ETHOXYMETHYLENEMALONIC ACID, ETHYL ESTER
mf: $C_{10}H_{16}O_5$ mw: 216.26

SYNS: DIETHYL EMME ◇ DIETHYL (ETHOXYMETHYLENE) MALONATE ◇ MALONIC ACID, (ETHOXYMETHYLENE)-, DIETHYL ESTER ◇ TL 1483

TOXICITY DATA with REFERENCE
skn-rbt 500 mg/24H MOD JACTDZ 1,37,90
orl-rat LD50:925 mg/kg JACTDZ 1,37,90
orl-mus LD50:2227 mg/kg JPMSAE 60,1810,71
ihl-mus LCLo:44 mg/m^3 NDRC** 30101,11,45

CONSENSUS REPORTS: Reported in EPA TSCA Inventory.

SAFETY PROFILE: Poison by ingestion. Moderately toxic by ingestion. A skin irritant. When heated to decomposition it emits acrid smoke and irritating fumes.

EEW000 CAS:123-06-8 ***HR: 3***
ETHOXYMETHYLENE MALONONITRILE
mf: $C_6H_6N_2O$ mw: 122.14

SYNS: USAF A-9230 ◇ USAF KF-10

TOXICITY DATA with REFERENCE
ipr-mus LD50:50 mg/kg NTIS** AD277-689

CONSENSUS REPORTS: Reported in EPA TSCA Inventory. Cyanide and its compounds are on the Community Right-To-Know List.

SAFETY PROFILE: Poison by intraperitoneal route. When heated to decomposition it emits very toxic fumes of NO_x and CN^-. See also NITRILES.

EEX000 CAS:63020-27-9 ***HR: 3***
7-ETHOXY METHYL-12-METHYL BENZ(a)ANTHRACENE
mf: $C_{22}H_{20}O$ mw: 300.42

SYN: 9-METHYL-10-ETHOXYMETHYL-1,2-BENZANTHRACENE

TOXICITY DATA with REFERENCE
scu-mus TDLo:80 mg/kg:ETA CNREA8 6,454,46

SAFETY PROFILE: Questionable carcinogen with experimental tumorigenic data. When heated to decomposition it emits acrid smoke and irritating fumes.

EEX500 CAS:22960-71-0 ***HR: 3***
N-ETHOXYMORPHOLINO DIAZENIUM FLUOROBORATE
mf: $C_6H_{13}N_2O_2 \cdot BF_4$ mw: 232.02

TOXICITY DATA with REFERENCE
scu-rat TDLo:559 mg/kg:CAR ZKKOBW 80,17,73
scu-rat LD50:639 mg/kg ZKKOBW 80,17,73

SAFETY PROFILE: Moderately toxic by subcutaneous route. Questionable carcinogen with experimental carcinogenic data. When heated to decomposition it emits very toxic NO_x and F^-. See also BORON COMPOUNDS and FLUORIDES.

EEY000 CAS:78109-88-3 ***HR: 3***
O-ETHOXY-N-(3-MORPHOLINOPROPYL)BENZAMIDE
mf: $C_{16}H_{24}N_2O_3$ mw: 292.42

SYN: D-703

TOXICITY DATA with REFERENCE
scu-mus LD50:150 mg/kg ARZNAD 10,743,60
ivn-mus LD50:160 mg/kg ARZNAD 10,743,60

SAFETY PROFILE: Poison by subcutaneous and intravenous routes. When heated to decomposition it emits toxic fumes of NO_x.

EEY500 CAS:93-18-5 ***HR: 3***
2-ETHOXYNAPHTHALENE
mf: $C_{12}H_{12}O$ mw: 172.24

SYNS: BROMELIA ◇ ETHYL-β-NAPHTHOLATE ◇ ETHYL-β-NAPHTHYL ETHER ◇ ETHYL-2-NAPHTHYL ETHER ◇ β-NAPHTHOL

ETHYL ETHER ◇ 2-NAPHTHOL ETHYL ETHER ◇ NEROLIN ◇ NEROLIN II ◇ NEROLINE ◇ NEROLIN NEW

TOXICITY DATA with REFERENCE
skn-rbt 500 mg/24H MOD FCTXAV 13,681,75
orl-rat LD50:3110 mg/kg FCTXAV 13,681,75
ipr-mus LD50:100 mg/kg NTIS** AD691-490
ivn-mus LD50:100 mg/kg CSLNX* NX#00191

CONSENSUS REPORTS: Reported in EPA TSCA Inventory.

SAFETY PROFILE: Poison by intraperitoneal and intravenous routes. Moderately toxic by ingestion. A skin irritant. When heated to decomposition it emits acrid smoke and irritating fumes. See also ETHERS.

EFA000 CAS:337-28-0 ***HR: 2***
1-ETHOXY-2,2,3,3,3-PENTAFLUORO-1-PROPANOL
mf: $C_5H_7F_5O_2$ mw: 194.12

TOXICITY DATA with REFERENCE
orl-mus LD50:1200 mg/kg JMCMAR 13,1212,70
ipr-mus LD50:800 mg/kg JMCMAR 13,1212,70

CONSENSUS REPORTS: Reported in EPA TSCA Inventory.

SAFETY PROFILE: Moderately toxic by ingestion and intraperitoneal routes. When heated to decomposition it emits very toxic fumes of F^-. See also FLUORIDES.

EFA100 CAS:622-62-8 ***HR: 3***
4-ETHOXYPHENOL
mf: $C_8H_{10}O_2$ mw: 138.18

PROP: Mp: 65-67°, bp: 131°/9 mm.

SYNS: ETHER MONOETHYLIQUE de l'HYDROQUINONE ◇ p-ETHOXYPHENOL ◇ 4-ETHYLOXYPHENOL ◇ HYDROQUINONE MONOETHYL ETHER ◇ p-HYDROXYPHENETOLE ◇ PHENOL, 4-ETHOXY-(9CI)

TOXICITY DATA with REFERENCE
orl-rat TDLo:667 mg/kg (female 11D post):REP TJADAB 41,43,90
ipr-mus LDLo:250 mg/kg RBPMAZ 22,1,52

CONSENSUS REPORTS: Reported in EPA TSCA Inventory.

SAFETY PROFILE: Poison by intraperitoneal route. Experimental reproductive effects. An irritant. When heated to decomposition it emits acrid smoke and irritating fumes.

EFC000 CAS:21224-77-1 ***HR: 3***
S-2-((4-(p-ETHOXYPHENYL)BUTYL)AMINO)ETHYL THIOSULFATE
mf: $C_{14}H_{23}NO_4S_2$ mw: 333.50

TOXICITY DATA with REFERENCE
orl-mus LD50:980 mg/kg JMCMAR 11,1190,68
ipr-mus LD50:22 mg/kg JMCMAR 11,1190,68

SAFETY PROFILE: Poison by intraperitoneal route. Moderately toxic by ingestion. When heated to decomposition it emits very toxic fumes of SO_x and NO_x. See also SULFATES.

EFC259 CAS:55308-64-0 ***HR: D***
2-(3-ETHOXYPHENYL)-5,6-DIHYDRO-s-TRIAZOLO(5,1-a)ISOQUINOLINE
mf: $C_{18}H_{17}N_3O$ mw: 291.38

SYNS: DL 204-IT ◇ L 11204

TOXICITY DATA with REFERENCE
scu-ham TDLo:780 μg/kg (female 4D post):REP ARZNAD 30,972,80

SAFETY PROFILE: Experimental reproductive effects. When heated to decomposition it emits toxic fumes of NO_x.

EFC600 CAS:85303-98-6 ***HR: D***
5-(m-ETHOXYPHENYL)-3-(o-ETHYLPHENYL)-s-TRIAZOLE
mf: $C_{18}H_{19}N_3O$ mw: 293.40

TOXICITY DATA with REFERENCE
scu-ham TDLo:200 μg/kg (female 4-8D post):REP RDMIDP 4,237,82

SAFETY PROFILE: Experimental reproductive effects. When heated to decomposition it emits toxic fumes of NO_x.

EFE000 CAS:150-69-6 ***HR: 3***
4-ETHOXYPHENYLUREA
mf: $C_9H_{12}N_2O_2$ mw: 180.23

PROP: Needle-like crystals. Mp: 174°.

SYNS: p-AETHOXYPHYLHARNSTOFF (GERMAN) ◇ DULCINE ◇ N-(4-ETHOXYPHENYL)UREA ◇ p-ETHOXYPHENYLUREA ◇ NCI-C02073 ◇ PHENETHYLCARBAMID (GERMAN) ◇ p-PHENETOLCARBAMID (GERMAN) ◇ p-PHENETOLCARBAMIDE ◇ p-PHENETOLE-CARBAMIDE ◇ p-PHENETYLUREA ◇ SUCROL ◇ SUESSTOFF ◇ VALZIN

TOXICITY DATA with REFERENCE
orl-rat TDLo:232 g/kg/59W-C:ETA JAPMA8 40,583,51
orl-wmn TDLo:600 mg/kg:CNS MEKLA7 43,105,48
orl-chd LDLo:400 mg/kg MEKLA7 43,105,48

orl-rat LD50:1900 mg/kg NCIMR* NIH-71-E-2144
orl-dog LDLo:1000 mg/kg HBAMAK 4,1345,35

CONSENSUS REPORTS: IARC Cancer Review: Group 3 IMEMDT 7,56,87; Animal Inadequate Evidence IMEMDT 12,97,76.

SAFETY PROFILE: Human poison by ingestion. Moderately toxic experimentally by ingestion. Human systemic effects by ingestion: somnolence, hallucinations, distorted perceptions and changes in motor activity. In adults 20 to 40 grams produces dizziness, nausea, methemoglobinemia, cyanosis, and hypotension. Questionable carcinogen with experimental tumorigenic data. When heated to decomposition it emits toxic fumes of NO_x.

EFE500 CAS:63815-42-9 ***HR: 3***
4-ETHOXY-β-(1-PIPERIDYL)PROPIOPHENONE HYDROCHLORIDE
mf: $C_{16}H_{23}NO_2{\cdot}ClH$ mw: 297.86

TOXICITY DATA with REFERENCE
ipr-mus LD50:60 mg/kg JPETAB 115,419,55
scu-mus LD50:73 mg/kg ARZNAD 5,559,55
ivn-mus LD50:20 mg/kg JPETAB 115,419,55

SAFETY PROFILE: Poison by intraperitoneal, subcutaneous, and intravenous routes. When heated to decomposition it emits very toxic fumes of HCl and NO_x.

EFF000 CAS:1874-62-0 ***HR: 1***
3-ETHOXY-1,2-PROPANEDIOL
mf: $C_5H_{12}O_3$ mw: 120.17

SYNS: α-ETHYL GLYCEROL ETHER ◇ GLYCEROL-α-ETHYL ETHER

TOXICITY DATA with REFERENCE
skn-rbt 530 mg/24H MLD AMIHBC 2,574,50
eye-rbt 106 mg AMIHBC 2,574,50
orl-mus LD50:9350 mg/kg FEPRA7 8,477,49

SAFETY PROFILE: Mildly toxic by ingestion. A skin and eye irritant. When heated to decomposition it emits acrid smoke and irritating fumes.

EFF500 CAS:1569-02-4 ***HR: 1***
1-ETHOXY-2-PROPANOL
mf: $C_5H_{12}O_2$ mw: 104.17

SYN: PROPYLENE GLYCOL ETHYL ETHER

TOXICITY DATA with REFERENCE
orl-rat LD50:4400 mg/kg NPIRI* 1,104,74
skn-rbt LD50:8100 mg/kg NPIRI* 1,104,74

CONSENSUS REPORTS: Glycol ether compounds are on the Community Right-To-Know List.

SAFETY PROFILE: Mildly toxic by ingestion and skin contact. When heated to decomposition it emits acrid smoke and irritating fumes. See also GLYCOL ETHERS.

EFG000 CAS:111-35-3 ***HR: 2***
3-ETHOXY-1-PROPANOL
mf: $C_5H_{12}O_2$ mw: 104.17

SYNS: β-PROPYLENE GLYCOL MONOETHYL ETHER ◇ PROPYLENE GLYCOL-β-MONOETHYL ETHER

TOXICITY DATA with REFERENCE
skn-rbt 452 mg open MLD UCDS** 7/28/66
skn-rbt 500 mg/24H MLD 85JCAE -,627,86
eye-rbt 20 mg/24H MOD 85JCAE -,627,86
orl-mus TDLo:24 g/kg (female 7-14D post):REP NTIS** PB86-197605
orl-rat LD50:7 g/kg JIDHAN 23,259,41
skn-rbt LD50:2558 mg/kg AIHAAP 30,470,69

CONSENSUS REPORTS: Glycol ether compounds are on the Community Right-To-Know List.

SAFETY PROFILE: Moderately toxic by skin contact. Slightly toxic by ingestion. Experimental reproductive effects. A skin irritant. When heated to decomposition it emits acrid smoke and irritating fumes. See also GLYCOL ETHERS.

EFG500 CAS:63918-98-9 ***HR: 3***
ETHOXY PROPIONALDEHYDE
mf: $C_5H_{10}O_2$ mw: 102.15

PROP: Liquid. Mp: −69.4°, bp: 135.2°, flash p: 100°F (OC), d: 0.918 @ 20°/20°, vap d: 3.63, vap press: 5.5 mm @ 20°.

TOXICITY DATA with REFERENCE
skn-rbt 10 mg/24H open JIHTAB 30,63,48
eye-rbt 2 mg open SEV JIHTAB 30,63,48
orl-rat LD50:900 mg/kg JIHTAB 30,63,48
ihl-rat LC50:500 ppm/4H JIHTAB 30,63,48
orl-mus LD50:140 mg/kg ARZNAD 15,841,65
skn-rbt LD50:1000 mg/kg JIHTAB 30,63,48

SAFETY PROFILE: Poison by ingestion. Moderately toxic by inhalation and skin contact. A skin and severe eye irritant. A very dangerous fire hazard when exposed to heat or flame; can react with oxidizing materials. To fight fire, use alcohol foam, CO_2, dry chemical. When heated to decomposition it emits acrid smoke and irritating fumes. See also ALDEHYDES.

EFH000 CAS:1331-11-9 ***HR: 2***
ETHOXYPROPIONIC ACID
mf: $C_5H_{10}O_3$ mw: 118.15

PROP: Liquid. Mp: −10.7°, bp: 219°, flash p: 225°F (OC), d: 1.0474, vap d: 4.08.

TOXICITY DATA with REFERENCE
skn-rbt 10 mg/24H open JIHTAB 30,63,48
eye-rbt 250 μg open SEV JIHTAB 30,63,48
orl-rat LD50:4800 mg/kg JIHTAB 30,63,48
skn-rbt LD50:750 mg/kg JIHTAB 30,63,48

SAFETY PROFILE: Moderately toxic by skin contact. Mildly toxic by ingestion. A skin and severe eye irritant. Combustible when exposed to heat or flame; can react with oxidizing materials. To fight fire, use alcohol foam, CO_2, dry chemical. When heated to decomposition it emits acrid smoke and irritating fumes.

EFH500 CAS:14631-45-9 ***HR: 2***
β-ETHOXYPROPIONITRILE
mf: C_5H_9NO mw: 99.15

TOXICITY DATA with REFERENCE
orl-rat LD50:2860 mg/kg TNICS* 13,125,73
orl-mus LD50:2200 mg/kg TNICS* 13,125,73

CONSENSUS REPORTS: Cyanide and its compounds are on the Community Right-To-Know List.

SAFETY PROFILE: Moderately toxic by ingestion. When heated to decomposition it emits toxic fumes of NO_x and CN^-. See also NITRILES.

EFI000 CAS:64050-15-3 ***HR: 2***
ETHOXYPROPYLACRYLATE
mf: $C_8H_{14}O_3$ mw: 158.22

SYN: ETHOXYPROPYL ESTER ACRYLIC ACID

TOXICITY DATA with REFERENCE
skn-rbt 10 mg/24H open MLD AMIHBC 4,119,51
eye-rbt 100 mg open AMIHBC 4,119,51
orl-rat LD50:820 mg/kg AMIHBC 4,119,51
ihl-rat LC50:250 ppm/4H AMIHBC 4,119,51
skn-rbt LD50:1410 mg/kg AMIHBC 4,119,51

SAFETY PROFILE: Moderately toxic by ingestion, inhalation, and skin contact. A skin and eye irritant. When heated to decomposition it emits acrid smoke and irritating fumes.

EFJ000 CAS:628-33-1 ***HR: 3***
1-ETHOXY-2-PROPYNE
mf: C_5H_8O mw: 84.12

$$CH_3CH_2OCH_2C{\equiv}CH$$

SAFETY PROFILE: Peroxidizes in air to form a product which explodes when heated to 80°C. When heated to decomposition it emits acrid smoke and fumes. See also PEROXIDES and ACETYLENE COMPOUNDS.

EFJ500 CAS:16357-59-8 ***HR: 3***
2-ETHOXY-1(2H)-QUINOLINECARBOXYLIC ACID, ETHYL ESTER
mf: $C_{14}H_{17}NO_3$ mw: 247.32

TOXICITY DATA with REFERENCE
ipr-mus LD50:32 mg/kg JMCMAR 14,49,71

CONSENSUS REPORTS: Reported in EPA TSCA Inventory.

SAFETY PROFILE: Poison by intraperitoneal route. When heated to decomposition it emits toxic fumes of NO_x. See also ESTERS.

EFK000 CAS:2593-15-9 ***HR: 2***
5-ETHOXY-3-TRICHLOROMETHYL-1,2,4-THIADIAZOLE
mf: $C_5H_5Cl_3N_2OS$ mw: 247.53

SYNS: 5-AETHOXY-3-TRICHLORMETHYL-1,2,4-THIADIAZOL(GERMAN) ◇ DWELL ◇ ECHLOMEZOL ◇ ETHAZOLE (FUNGICIDE) ◇ ETMT ◇ ETRIDIAZOLE ◇ KOBAN ◇ MF-344 ◇ OLIN MATHIESON 2,424 ◇ OM 2424 ◇ PANSOIL ◇ TERRACHLOR-SUPER X ◇ TERRACOAT ◇ TERRAFLO ◇ TERRAZOLE ◇ 3-(TRICHLOROMETHYL)-5-ETHOXY-1,2,4-THIADIAZOLE ◇ TRUBAN

TOXICITY DATA with REFERENCE
mmo-sat 400 μg/plate KHFKDF 8,551,80
mma-sat 400 μg/plate KHFKDF 8,551,80
orl-rat LD50:1077 mg/kg 28ZEAL 5,110,76
orl-mus LD50:2000 mg/kg GUCHAZ 6,486,73
orl-rbt LD50:779 mg/kg TXAPA9 56,164,80
skn-rbt LD50:1700 mg/kg FMCHA2 -,C137,83

SAFETY PROFILE: Moderately toxic by ingestion and skin contact. Mutation data reported. A fungicide. When heated to decomposition it emits very toxic fumes of Cl^-, SO_x, and NO_x.

EFK500 CAS:433-27-2 ***HR: 2***
1-ETHOXY-2,2,2-TRIFLUOROETHANOL
mf: $C_4H_7F_3O_2$ mw: 144.11

TOXICITY DATA with REFERENCE
orl-mus LD50:600 mg/kg JMCMAR 13,1212,70
ipr-mus LD50:600 mg/kg JMCMAR 13,1212,70

CONSENSUS REPORTS: Reported in EPA TSCA Inventory.

SAFETY PROFILE: Moderately toxic by ingestion and intraperitoneal routes. When heated to decomposition it emits toxic fumes of F^-. See also FLUORIDES.

EFL000 CAS:112-50-5 ***HR: 1***
ETHOXYTRIGLYCOL
mf: $C_8H_{18}O_4$ mw: 178.26

PROP: Bp: 255.4°, flash p: 275°F (OC), d: 1.0208 @ 20°/20°, vap press: 0.01 mm @ 20°.

SYNS: DOWANOL TE ◇ 2-(2-(2-ETHOXYETHOXY)ETHOXY)ETHANOL ◇ ETHOXYTRIETHYLENE GLYCOL ◇ POLY-SOLV TE ◇ TRIETHYLENE GLYCOL ETHYL ETHER ◇ TRIETHYLENE GLYCOL MONOETHYL ETHER ◇ TRIGLYCOL MONOETHYL ETHER

TOXICITY DATA with REFERENCE
eye-rbt 500 mg AJOPAA 29,1363,46
orl-rat LD50:10610 mg/kg 34ZIAG -,730,69
skn-rbt LD50:8 g/kg AMIHBC 4,119,51

CONSENSUS REPORTS: Reported in EPA TSCA Inventory. Glycol ether compounds are on the Community Right-To-Know List.

SAFETY PROFILE: Mildly toxic by ingestion and skin contact. An eye irritant. Combustible when exposed to heat or flame; can react with oxidizing materials. To fight fire, use foam, alcohol foam, CO_2, dry chemical. When heated to decomposition it emits acrid smoke and irritating fumes. See also GLYCOL ETHERS.

EFL500 CAS:1825-62-3 ***HR: 3***
ETHOXYTRIMETHYLSILANE
mf: $C_5H_{14}OSi$ mw: 118.28

PROP: Bp: 75.7°, vap press: 100 mm @ 22.1°, vap d: 4.1.

TOXICITY DATA with REFERENCE
eye-rbt 79 mg JIHTAB 30,332,48
orl-rat LDLo:1400 mg/kg JIHTAB 30,332,48
ihl-rat LCLo:4000 ppm/8H JIHTAB 30,332,48

CONSENSUS REPORTS: Reported in EPA TSCA Inventory.

SAFETY PROFILE: Moderately toxic by ingestion and inhalation. An eye irritant. Flammable when exposed to heat or flame; can react with oxidizing materials. When heated to decomposition it emits acrid smoke and irritating fumes. See also SILANE.

EFM000 CAS:625-50-3 ***HR: 2***
ETHYLACETAMIDE
mf: C_4H_9NO mw: 87.14

PROP: Water white liquid. Mp: −32°, bp: 206-208°, flash p: 230°F, d: 0.920 @ 20°/20°, vap d: 3.0.

SYNS: ACETAMIDOETHANE ◇ N-ETHYLACETAMIDE

TOXICITY DATA with REFERENCE
ipr-rat LD50:3130 mg/kg ARZNAD 20,1242,70
ipr-mus LD50:3560 mg/kg ARZNAD 20,1242,70
ivn-rbt LDLo:9410 mg/kg ARZNAD 20,1242,70
ivn-ckn LDLo:9600 mg/kg ARZNAD 20,1242,70

CONSENSUS REPORTS: Reported in EPA TSCA Inventory.

SAFETY PROFILE: Moderately toxic by intraperitoneal route. Combustible when exposed to heat or flame; can react with oxidizing materials. To fight fire, use alcohol foam. When heated to decomposition it emits toxic fumes of NO_x. See also AMIDES.

EFM500 CAS:102585-53-5 ***HR: 2***
2-(2-(3-(N-ETHYLACETAMIDO)-2,4-6-TRIIODOPHENOXY)ETHOXY)ACETIC ACID SODIUM SALT
mf: $C_{14}H_{16}I_3NO_5{\cdot}Na$ mw: 682.00

TOXICITY DATA with REFERENCE
orl-mus LD50:1300 mg/kg FRPSAX 31,349,76
ivn-mus LD50:550 mg/kg FRPSAX 31,349,76

SAFETY PROFILE: Moderately toxic by ingestion and intravenous routes. When heated to decomposition it emits very toxic fumes of I^-, NO_x, and Na_2O. See also IODIDES.

EFN500 CAS:102585-52-4 ***HR: 2***
2-(2-(3-(N-ETHYLACETAMIDO)-2,4,6-TRIIODOPHENOXY)ETHOXY)-2-PHENYL ACETIC ACID SODIUM SALT
mf: $C_{20}H_{20}I_3NO_5{\cdot}Na$ mw: 758.10

TOXICITY DATA with REFERENCE
orl-mus LD50:1600 mg/kg FRPSAX 31,349,76
ivn-mus DL50:405 mg/kg FRPSAX 31,349,76

SAFETY PROFILE: Moderately toxic by ingestion and intravenous routes. When heated to decomposition it emits toxic fumes of I^-, NO_x, and Na_2O. See also IODIDES.

EFO000 CAS:49642-61-7 ***HR: 2***
2-(2-(3-(N-ETHYLACETAMIDO)-2,4,6-TRIIODOPHENOXY)ETHOXY)PROPIONIC ACID SODIUM SALT
mf: $C_{15}H_{18}I_3NO_5{\cdot}Na$ mw: 696.03

TOXICITY DATA with REFERENCE
orl-mus LD50:1600 mg/kg FRPSAX 31,349,76
ivn-mus LD50:425 mg/kg FRPSAX 31,349,76

SAFETY PROFILE: Moderately toxic by ingestion and intravenous routes. When heated to decomposition it emits very toxic fumes of I^-, NO_x, and Na_2O.

EFP000 CAS:23279-54-1 ***HR: 2***
2-(3-(N-ETHYLACETAMIDO)-2,4,6-TRIIODOPHENYL)BUTYRIC ACID
mf: $C_{14}H_{16}I_3NO_3$ mw: 627.01

TOXICITY DATA with REFERENCE
orl-mus LD50:1260 mg/kg JMCMAR 13,559,70
ivn-mus LD50:420 mg/kg JMCMAR 13,559,70

SAFETY PROFILE: Moderately toxic by ingestion and

intravenous routes. When heated to decomposition it emits very toxic fumes of I^- and NO_x. See also IODIDES.

EFQ500 CAS:529-65-7 ***HR: 2***
N-ETHYLACETANILIDE
mf: $C_{10}H_{13}NO$ mw: 163.24

PROP: White crystals, faint odor. Mp: 54°, bp: 258°, flash p: 126°F, d: 0.994, vap d: 5.62.

SYNS: ACETETHYLANILIDE ◇ ETHYLACETANILIDE

TOXICITY DATA with REFERENCE
orl-mus LD50:409 mg/kg TXAPA9 19,20,71

CONSENSUS REPORTS: Reported in EPA TSCA Inventory.

SAFETY PROFILE: Moderately toxic by ingestion. Combustible when exposed to heat or flame; can react with oxidizing materials. To fight fire, use foam, CO_2, dry chemical. When heated to decomposition it emits toxic fumes of NO_x.

EFR000 CAS:141-78-6 ***HR: 3***
ETHYL ACETATE
DOT: UN 1173
mf: $C_4H_8O_2$ mw: 88.12

$CH_3CH_2OCO{\bullet}CH_3$

PROP: Colorless liquid; fragrant odor. Mp: −83.6°, bp: 77.15°, ULC: 85-90, lel: 2.2%, uel: 11%, flash p: 24°F, d: 0.8946 @ 25°, autoign temp: 800°F, vap press: 100 mm @ 27.0°, vap d: 3.04. Misc with alc, ether, glycerin, volatile oils, water @ 54°.

SYNS: ACETIC ETHER ◇ ACETIDIN ◇ ACETOXYETHANE ◇ AETHYLACETAT (GERMAN) ◇ ESSIGESTER (GERMAN) ◇ ETHYL-ACETAAT (DUTCH) ◇ ETHYL ACETIC ESTER ◇ ETHYLE (ACETATE d') (FRENCH) ◇ ETHYL ETHANOATE ◇ ETILE (ACETATO di) (ITALIAN) ◇ FEMA No. 2414 ◇ OCTAN ETYLU (POLISH) ◇ RCRA WASTE NUMBER U112 ◇ VINEGAR NAPHTHA

TOXICITY DATA with REFERENCE
eye-hmn 400 ppm JIHTAB 25,282,43
sln-smc 24400 ppm MUREAV 149,339,85
cyt-ham:fbr 9 g/L FCTOD7 22,623,84
ihl-hmn TCLo:400 ppm:NOSE,EYE,PUL JIHTAB 25,282,43
orl-rat LD50:5620 mg/kg YKYUA6 32,1241,81
ihl-rat LC50:1600 ppm/8H 14CYAT 2,1879,63
scu-rat LDLo:5000 mg/kg BSIBAC 18,45,43
orl-mus LD50:4100 mg/kg GISAAA 48(4),66,83
ihl-mus LCLo:31 g/m³/2H AGGHAR 5,1,33
ipr-mus LD50:709 mg/kg SCCUR* -,5,61
ihl-cat LCLo:61 g/m³ HBTXAC 1,336,55
scu-cat LD50:3000 mg/kg AGGHAR 5,1,33
orl-rbt LD50:4935 mg/kg IMSUAI 41,31,72
orl-gpg LD50:5500 mg/kg GISAAA 48(4),66,83
ihl-gpg LCLo:77 mg/m³/1H MELAAD 24,166,33
scu-gpg LD50:3000 mg/kg AGGHAR 5,1,33

CONSENSUS REPORTS: Reported in EPA TSCA Inventory. EPA Genetic Toxicology Program.

OSHA PEL: TWA 400 ppm
ACGIH TLV: TWA 400 ppm
DFG MAK: 400 ppm (1400 mg/m³)
DOT Classification: Flammable Liquid; Label: Flammable Liquid.

SAFETY PROFILE: Poison by inhalation. Moderately toxic by intraperitoneal and subcutaneous routes. Mildly toxic by ingestion. Human systemic effects by inhalation: olfactory changes, conjunctiva irritation, and pulmonary changes. Human eye irritant. Mutation data reported. Irritating to mucous surfaces, particularly the eyes, gums, and respiratory passages, and is also mildly narcotic. On repeated or prolonged exposures, it causes conjunctival irritation and corneal clouding. It can cause dermatitis. High concentrations have a narcotic effect and can cause congestion of the liver and kidneys. Chronic poisoning has been described as producing anemia, leucocytosis (transient increase in the white blood cell count), and cloudy swelling, and fatty degeneration of the viscera. A synthetic flavoring substance and adjuvant.

Highly flammable liquid. A very dangerous fire hazard when exposed to heat or flame; can react vigorously with oxidizing materials. Moderate explosion hazard when exposed to flame. Potentially explosive reaction with lithium tetrahydroaluminate. Ignites on contact with potassium tert-butoxide. Violent reaction with chlorosulfonic acid; ($LiAlH_2$ + 2-chloromethyl furan); oleum. To fight fire, use CO_2, dry chemical or alcohol foam. When heated to decomposition it emits acrid smoke and irritating fumes. See also ESTERS.

EFR500 CAS:62681-13-4 ***HR: D***
ETHYL-N-(2-ACETOXYETHYL)-N-NITROSOCARBAMATE
mf: $C_7H_{12}N_2O_5$ mw: 204.21

SYNS: N-(2-ACETOXYETHYL)-N-NITROSOCARBAMICACID ETHYL ESTER ◇ (2-(ACETYLOXY)ETHYL)NITROSOCARBAMIC ACID ETHYL ESTER ◇ N-(2-HYDROXYETHYL)-N-NITROSOCARBAMIC ACID ETHYL ESTER ACETATE

TOXICITY DATA with REFERENCE
mmo-sat 2400 pmol/L MUREAV 48,131,77

CONSENSUS REPORTS: EPA Genetic Toxicology Program.

SAFETY PROFILE: Mutation data reported. Many N-nitroso compounds are carcinogens. When heated to decomposition it emits toxic fumes of NO_x. See also N-NITROSO COMPOUNDS and CARBAMATES.

EFS000 CAS:141-97-9 ***HR: 2***
ETHYL ACETYL ACETATE
mf: $C_6H_{10}O_3$ mw: 130.16

$CH_3CO•CH_2CO•OCH_2CH_3$

PROP: Colorless liquid; fruity odor. Bp: 180.8°, fp: −45°, flash p: 185°F (COC), autoign temp: 563°F, d: 1.0261 @ 20°/20°, refr index: 1.418, vap press: 1 mm @ 28.5°, vap d: 4.48. Misc with alc, ether, ethyl acetate, water.

SYNS: ACETOACETIC ACID, ETHYL ESTER ◇ ACETOACETIC ESTER ◇ ACTIVE ACETYL ACETATE ◇ DIACETIC ETHER ◇ EAA ◇ ETHYL ACETOACETATE (FCC) ◇ ETHYL ACETYLACETONATE ◇ ETHYL BENZYL ACETOACETATE ◇ ETHYL-3-OXOBUTANOATE ◇ ETHYL-3-OXOBUTYRATE ◇ FEMA No. 2415 ◇ 3-OXOBUTANOIC ACID ETHYL ESTER

TOXICITY DATA with REFERENCE
skn-rbt 510 mg open MLD UCDS** 3/12/69
eye-rbt 23 mg AJOPAA 29,1363,46
orl-rat LD50:3980 mg/kg JIHTAB 31,60,49
orl-mus LD50:5105 mg/kg JPMSAE 60,1810,71

CONSENSUS REPORTS: Reported in EPA TSCA Inventory.

SAFETY PROFILE: Moderately toxic by ingestion. A skin and eye irritant. Combustible liquid when exposed to heat or flame; can react with oxidizing materials. Explosive reaction when heated with Zn + trimbromoneopentyl alcohol or 2,2,2-tris(bromomethyl) ethanol. To fight fire, use alcohol foam, CO_2, dry chemical. When heated to decomposition it emits acrid smoke and irritating fumes. See also ESTERS.

EFS500 CAS:107-00-6 ***HR: 3***
ETHYL ACETYLENE
DOT: UN 2452
mf: C_4H_6 mw: 54

PROP: A colorless, highly flammable gas. Bp: 8.3°, d: 0.669 @ 0°/0°, mp: −130°, flash p: <30°F (TOC), <7°C (Gas >8°C).

SYNS: 1-BUTYNE ◇ ETHYL ACETYLENE, INHIBITED (DOT) ◇ ETHYLETHYNE

DOT Classification: Flammable Gas; Label: Flammable Gas.

CONSENSUS REPORTS: Reported in EPA TSCA Inventory.

SAFETY PROFILE: Probably an asphyxiant. A very dangerous fire hazard when exposed to heat, open flame or powerful oxidizers. A dangerous explosion hazard. To fight fire, stop flow of gas. See also ACETYLENE and ACETYLENE COMPOUNDS.

EFS600 CAS:539-88-8 ***HR: 1***
ETHYL 3-ACETYLPROPIONATE
mf: $C_7H_{12}O_3$ mw: 144.19

PROP: Liquid. D: 1.012, bp: 205-206°. Very sol in water; miscible with alc.

SYNS: ETHYL KETOVALERATE ◇ ETHYL 4-KETOVALERATE ◇ ETHYL LAEVULINATE ◇ ETHYL LEVULATE ◇ ETHYL 4-OXOPENTANOATE ◇ ETHYL 4-OXOVALERATE ◇ LEVULINIC ACID, ETHYL ESTER ◇ PENTANOIC ACID, 4-OXO-, ETHYL ESTER (9CI)

TOXICITY DATA with REFERENCE
skn-rbt 500 mg/24H MLD FCTOD7 20,679,82

CONSENSUS REPORTS: Reported in EPA TSCA Inventory.

SAFETY PROFILE: A skin irritant. When heated to decomposition it emits acrid smoke and irritating fumes.

EFT000 CAS:140-88-5 ***HR: 3***
ETHYL ACRYLATE
DOT: UN 1917
mf: $C_5H_8O_2$ mw: 100.13

PROP: Colorless liquid; acrid penetrating odor. Bp: 99.8°, fp: <−72°: lel: 1.8%, flash p: 60°F (OC), 48.2°F d: 0.916-0.919, vap press, 29.3 mm @ 20°, vap d: 3.45. Misc with alc, ether; sltly sol in water.

SYNS: ACRYLATE d'ETHYLE (FRENCH) ◇ ACRYLIC ACID ETHYL ESTER ◇ ACRYLSAEUREAETHYLESTER (GERMAN) ◇ AETHYLACRYLAT (GERMAN) ◇ ETHOXYCARBONYLETHYLENE ◇ ETHYLACRYLAAT (DUTCH) ◇ ETHYLAKRYLAT (CZECH) ◇ ETHYL PROPENOATE ◇ ETHYL-2-PROPENOATE ◇ ETIL ACRILATO (ITALIAN) ◇ ETILACRILATULUI (ROMANIAN) ◇ FEMA No. 2418 ◇ NCI-C50384 ◇ 2-PROPENOIC ACID, ETHYL ESTER (MAK) ◇ RCRA WASTE NUMBER U113

TOXICITY DATA with REFERENCE
eye-mky 1204 ppm/15H-I JIHTAB 31,317,49
skn-rbt 500 mg open MLD UCDS** 12/14/71
skn-rbt 10 mg/24H MLD JIHTAB 31,311,49
eye-rbt 45 mg MLD UCDS** 12/14/71
eye-rbt 1204 ppm/7H JIHTAB 31,317,49
mma-mus:lym 20 mg/L ENMUDM 8(Suppl 6),4,86
msc-mus:lym 20 mg/L ENMUDM 8(Suppl 6),4,86
orl-rat TDLo:51500 mg/kg/2Y-I:CAR NTPTR* NTP-TR-259,86
orl-mus TDLo:103 g/kg/2Y-I:CAR NTPTR* NTP-TR-259,86
ihl-hmn TCLo:5 ppm:NOSE,EYE,PUL 34ZIAG -,75,69
orl-rat LD50:800 mg/kg BCTKAG 12,405,79
ihl-rat LC50:2180 ppm/4H:NOSE,EYE,PUL JTEHD6 16,811,85
skn-rat LDLo:1800 mg/kg PJPPAA 32,223,80
ipr-rat LD50:450 mg/kg AMPMAR 36,58,75

orl-mus LD50:1799 mg/kg TOLED5 11,125,82
ihl-mus LCLo:25 mg/m^3/2H IGIBA5 11,27,62
ipr-mus LD50:599 mg/kg JDREAF 51,526,72
orl-rbt LD50:400 mg/kg 14CYAT 2,1879,63
ihl-rbt LCLo:1204 ppm/7H JIHTAB 31,317,49
ihl-gpg LCLo:1204 ppm/7H JIHTAB 31,317,49

CONSENSUS REPORTS: NTP Fifth Annual Report on Carcinogens. IARC Cancer Review: Group 2B IMEMDT 7,56,87; Animal Sufficient Evidence IMEMDT 39,81,86; Animal Inadequate Evidence IMEMDT 19,47,79; Human Inadequate Evidence IMEMDT 19,47,79. NTP Carcinogenesis Studies (gavage); Clear Evidence: mouse, rat NTPTR* NTP-TR-259,86. Reported in EPA TSCA Inventory. Community Right-To-Know List.

OSHA PEL: (Transitional: TWA 25 mg/m^3 (skin)) TWA 5 ppm; STEL 25 ppm (skin)
ACGIH TLV: TWA 5 ppm; STEL 15 ppm; Suspected Human Carcinogen
DFG MAK: 5 ppm (20 mg/m^3)
DOT Classification: Flammable Liquid; Label: Flammable Liquid.

SAFETY PROFILE: Confirmed carcinogen with experimental carcinogenic data. Poison by ingestion and inhalation. Moderately toxic by skin contact and intraperitoneal routes. Human systemic effects by inhalation: eye, olfactory and pulmonary changes. A skin and eye irritant. Characterized in its terminal stages by dyspnea, cyanosis, and convulsive movements. It caused severe local irritation of the gastro-enteric tract; and toxic degenerative changes of cardiac, hepatic, renal, and splenic tissues were observed. It gave no evidence of cumulative effects. When applied to the intact skin of rabbits, the ethyl ester caused marked local irritation, erythema, edema, thickening, and vascular damage. Animals subjected to a fairly high concentrations of these esters suffered irritation of the mucous membranes of the eyes, nose, and mouth as well as lethargy, dyspnea, and convulsive movements. A substance which migrates to food from packaging materials.

Flammable liquid. A very dangerous fire hazard when exposed to heat or flame; can react vigorously with oxidizing materials. Violent reaction with chlorosulfonic acid. To fight fire, use CO_2, dry chemical or alcohol foam. When heated to decomposition it emits acrid smoke and irritating fumes. See also ESTERS.

EFT500 CAS:462-95-3 ***HR: 3***
ETHYLAL
DOT: UN 2373
mf: $C_5H_{12}O_2$ mw: 104.17

PROP: Flash p: <69.8°F.

SYN: DIETHOXYMETHANE (DOT)

TOXICITY DATA with REFERENCE
orl-rbt LD50:2604 mg/kg PSEBAA 29,730,32

CONSENSUS REPORTS: Reported in EPA TSCA Inventory.

DOT Classification: Flammable Liquid; Label: Flammable Liquid.

SAFETY PROFILE: Moderately toxic by ingestion. Flammable when exposed to heat or flame; can react vigorously with oxidizers. When heated to decomposition it emits acrid smoke and irritating fumes.

EFU000 CAS:64-17-5 ***HR: 3***
ETHYL ALCOHOL
DOT: UN 1170
mf: C_2H_6O mw: 46.08

PROP: Clear, colorless liquid; fragrant odor and burning taste. Bp: 78.32°, ULC: 70, lel: 3.3%, uel: 19% @ 60°, fp: < −130°, flash p: 55.6°F, d: 0.7893 @ 20°/4°, autoign temp: 793°F, vap press: 40 mm @ 19°, vap d: 1.59, refr index: 1.364. Misc in water, alc, chloroform, and ether.

SYNS: ABSOLUTE ETHANOL ◇ AETHANOL (GERMAN) ◇ AETHYLALKOHOL (GERMAN) ◇ ALCOHOL ◇ ALCOHOL, anhydrous ◇ ALCOHOL, dehydrated ◇ ALCOOL ETHYLIQUE (FRENCH) ◇ ALCOOL ETILICO (ITALIAN) ◇ ALGRAIN ◇ ALKOHOL (GERMAN) ◇ ALKOHOLU ETYLOWEGO (POLISH) ◇ ANHYDROL ◇ COLOGNE SPIRIT ◇ COLOGNE SPIRITS (ALCOHOL) (DOT) ◇ ETANOLO (ITALIAN) ◇ ETHANOL (MAK) ◇ ETHANOL 200 PROOF ◇ ETHANOL, solution (DOT) ◇ ETHYLALCOHOL (DUTCH) ◇ ETHYL ALCOHOL, anhydrous ◇ ETHYL HYDRATE ◇ ETHYL HYDROXIDE ◇ ETYLOWY ALKOHOL (POLISH) ◇ FERMENTATION ALCOHOL ◇ GRAIN ALCOHOL ◇ JAYSOL ◇ JAYSOL S ◇ METHYLCARBINOL ◇ MOLASSES ALCOHOL ◇ NCI-C03134 ◇ POTATO ALCOHOL ◇ SD ALCOHOL 23-HYDROGEN ◇ SPIRITS of WINE ◇ SPIRT ◇ TECSOL

TOXICITY DATA with REFERENCE
skn-rbt 400 mg open MLD UCDS** 7/22/70
skn-rbt 500 mg/24H SEV 28ZPAK -,34,72
eye-rbt 79 mg AJOPAA 29,1363,46
eye-rbt 100 mg/24H MOD 28ZPAK -,34,72
eye-rbt 100 mg/4S rns MOD FCTOD7 20,573,82
mmo-esc 140 g/L MUREAV 130,97,84
dni-hmn:lym 220 mmol/L PNASA6 79,1171,82
cyt-mus-orl 40 g/kg NATUAS 302,258,83
orl-wmn TDLo:41 g/kg (41W preg):REP AJDCAI 129,1075,75
orl-rat TDLo:4 g/kg (13D preg):TER CYGEDX 15,23,81
orl-mus TDLo:320 mg/kg/50W-I:ETA CALEDQ 13,345,81
orl-chd LDLo:2000 mg/kg ATXKA8 17,183,58
orl-cld TDLo:14400 mg/kg/30M-I ACPAAN 74,977,85
orl-man TDLo:700 mg/kg NETOD7 8,77,86
orl-hmn LDLo:1400 mg/kg NPIRI* 1,44,74
orl-man TDLo:50 mg/kg:GIT JPETAB 56,117,36

orl-man TDLo:1430 μg/kg:CNS JPETAB 197,488,76
orl-wmn TDLo:256 g/kg/12W:CNS,END JAMAAP 238,2143,77
scu-inf LDLo:19440 mg/kg:CNS,MET AJCPAI 5,466,35
orl-rat LD50:7060 mg/kg TXAPA9 16,718,70
ihl-rat LC50:20000 ppm/10H NPIRI* 1,44,74
ipr-rat LD50:3750 mg/kg EVHPAZ 61,321,85
ivn-rat LD50:1440 mg/kg TXAPA9 18,60,71
orl-mus LD50:3450 mg/kg GISAAA 32(3),31,67
ihl-mus LC50:39 g/m^3/4H GTPZAB 26(8),82
ipr-mus LD50:933 mg/kg SCCUR* -,5,61
scu-mus LD50:8285 mg/kg FAONAU 48A,99,70
ivn-mus LD50:1973 mg/kg HBTXAC 1,128,56
orl-dog LDLo:5500 mg/kg HBTXAC 1,130,56
ipr-dog LDLo:3000 mg/kg BJIMAG 1,207,44
scu-dog LDLo:6000 mg/kg HBTXAC 1,130,56

CONSENSUS REPORTS: IARC Cancer Review: Human Sufficient Evidence IMEMDT 44,259,88. Reported in EPA TSCA Inventory. EPA Genetic Toxicology Program.

OSHA PEL: TWA 1000 ppm
ACGIH TLV: TWA 1000 ppm
DFG MAK: 1,000 ppm (1,900 mg/m^3)
DOT Classification: Flammable Liquid; Label: Flammable Liquid; Flammable or Combustible Liquid; Label: Flammable Liquid.

SAFETY PROFILE: Confirmed human carcinogen for ingestion of beverage alcohol. Experimental tumorigenic and teratogenic data. Moderately toxic to humans by ingestion. Moderately toxic experimentally by intravenous and intraperitoneal routes. Mildly toxic by inhalation and skin contact. Human systemic effects by ingestion and subcutaneous route: sleep disorders, hallucinations, distorted perceptions, convulsions, motor activity changes, ataxia, coma, antipsychotic, headache, pulmonary changes, alteration in gastric secretion, nausea or vomiting, other gastrointestinal changes, menstrual cycle changes and body temperature decrease. Can also cause glandular effects in humans. Human reproductive effects by ingestion, intravenous, and intrauterine routes: changes in female fertility index. Effects on newborn include: changes in apgar score, neonatal measures or effects and drug dependence. Experimental reproductive effects. Human mutation data reported. An eye and skin irritant.

The systemic effect of ethanol differs from that of methanol. Ethanol is rapidly oxidized in the body to carbon dioxide and water, and in contrast to methanol, no cumulative effect occurs. Though ethanol possesses narcotic properties, concentrations sufficient to produce this effect are not reached in industry. Concentrations below 1,000 ppm usually produce no signs of intoxication. Exposure to concentrations over 1,000 ppm may cause headache, irritation of the eyes, nose, and throat, and, if continued for an hour, drowsiness and lassitude, loss of appetite, and inability to concentrate. There is no concrete evidence that repeated exposure to ethanol vapor results in cirrhosis of the liver. Ingestion of large doses can cause alcohol poisoning. Repeated ingestions can lead to alcoholism. It is a central nervous system depressant.

Flammable liquid when exposed to heat or flame; can react vigorously with oxidizers. To fight fire, use alcohol foam, CO_2, dry chemical. Explosive reaction with the oxidized coating around potassium metal. Ignites and then explodes on contact with acetic anhydride + sodium hydrogen sulfate. Reacts violently with acetyl bromide (evolves hydrogen bromide), dichloromethane + sulfuric acid + nitrate or nitrite, disulfuryl difluoride, tetrachlorisilane + water, and strong oxidants. Ignites on contact with disulfuric acid + nitric acid, phosphorous(III) oxide, platinum, potassium-tert-butoxide + acids. Forms explosive products in reaction with ammonia + silver nitrate (forms silver nitride and silver fulminate), magnesium perchlorate (forms ethyl perchlorate), nitric acid + silver (forms silver fulminate), silver nitrate (forms ethyl nitrate), silver(I) oxide + ammonia or hydrazine (forms silver nitride and silver fulminate), sodium (evolves hydrogen gas). Incompatible with acetyl chloride, BrF_5, $Ca(OCl)_2$, ClO_3, CrO_3, $Cr(OCl)_2$, (cyanuric acid + H_2O), H_2O_2, HNO_3, (H_2O_2 + H_2SO_4), (I + CH_3OH + HgO), [$Mn(ClO_4)_2$ + 2,2-dimethoxy propane], $Hg(NO_3)_2$, $HClO_4$, perchlorates, (H_2SO_4 + permanganates), $HMnO_4$, KO_2, $KOC(CH_3)_3$, $AgClO_4$, NaH_3N_2, $UO_2(ClO_4)_2$.

EFU100 CAS:2938-73-0 ***HR: 3***
ETHYLALUMINUM DIIODIDE
mf: $C_2H_5AlI_2$ mw: 309.85

SAFETY PROFILE: Ignites spontaneously in air. When heated to decomposition it emits toxic fumes of I^-. See also ALUMINUM and IODIDES.

EFU400 CAS:75-04-7 ***HR: 3***
ETHYLAMINE
DOT: UN 1036/UN 2270
mf: C_2H_7N mw: 45.10

PROP: Colorless gas or liquid, strong ammoniacal odor. Bp: 16.6°, flammable, lel: 4.95%, uel: 20.75%, fp: −80.6°, flash p: −0.4°F, d: 0.662 @ 20°/4°, autoign temp: 725°F, vap d: 1.56. vap press: 400 mm @ 20°. Misc with water, alc, and ether.

SYNS: AETHYLAMINE (GERMAN) ◇ AMINOETHANE ◇ 1-AMINOETHANE ◇ ETHANAMINE ◇ ETILAMINA (ITALIAN) ◇ ETYLOAMINA (POLISH) ◇ MONOETHYLAMINE (DOT) ◇ MONOETHYLAMINE, anhydrous (DOT)

TOXICITY DATA with REFERENCE
skn-rbt 500 mg/24H MLD 85JCAE-,429,86
eye-rbt 50 ppm/10D-I SEV AMIHBC 3,287,51
eye-rbt 250 μg/24H SEV 85JCAE-,429,86
orl-rat LD50:400 mg/kg AMIHBC 10,61,54
ihl-rat LCLo:3000 ppm/4H AEHLAU 1,343,60
skn-rbt LD50:390 mg/kg AEHLAU 1,343,60
ivn-rbt LDLo:350 mg/kg HBAMAK 4,1295,35
ihl-uns LC50:2300 mg/m^3 TPKVAL 14,80,75

CONSENSUS REPORTS: Reported in EPA TSCA Inventory.

OSHA PEL: TWA 10 ppm
ACGIH TLV: TWA 10 ppm
DFG MAK: 10 ppm (18 mg/m^3)
DOT Classification: Flammable Liquid; Label: Flammable Liquid (UN1036, UN2270); Flammable Gas; Label: Flammable Gas (UN1036); Flammable or Combustible Liquid; Label: Flammable Liquid (UN2270).

SAFETY PROFILE: A poison by ingestion, skin contact, and intravenous routes. Moderately toxic by inhalation. A severe eye irritant. A very dangerous fire hazard when exposed to heat or flame. Moderate explosion hazard when exposed to spark or flame. Keep away from heat and open flame, can react vigorously with oxidizing materials. To fight fire, stop flow of gas, use alcohol foam, dry chemical. Incompatible with cellulose nitrate or oxidizers. When heated to decomposition it emits toxic fumes of NO_x. See also AMINES.

EFU500 CAS:75-23-0 ***HR: 3***
ETHYLAMINE with BORON FLUORIDE (1:1)
mf: $C_2H_7BF_3N$ mw: 112.91

SYN: BORONTRIFLUORIDEMONOETHYLAMINE

TOXICITY DATA with REFERENCE
ipr-mus LDLo:42 mg/kg NTIS** AD441-640

CONSENSUS REPORTS: Reported in EPA TSCA Inventory.

SAFETY PROFILE: Poison by intraperitoneal route. When heated to decomposition it emits very toxic fumes of NO_x and F^-. See also BORON FLUORIDE and AMINES.

EFW000 CAS:557-66-4 ***HR: 2***
ETHYLAMINE HYDROCHLORIDE
mf: $C_2H_7N•ClH$ mw: 81.56

PROP: Monoclinic. D: 1.216, mp: 108-109°. Sol in water at 240°; very sol in alc; insol in ether.

SYN: ETHYL AMMONIUM CHLORIDE

TOXICITY DATA with REFERENCE
unr-rat LDLo:2500 mg/kg 27ZWAY 1,250,23
unr-rbt LDLo:1000 mg/kg 27ZWAY 1,250,23

CONSENSUS REPORTS: Reported in EPA TSCA Inventory.

SAFETY PROFILE: Moderately toxic by unspecified routes. When heated to decomposition it emits very toxic fumes of HCl, NH_3, and NO_x.

EFX000 CAS:94-09-7 ***HR: 3***
ETHYL-4-AMINOBENZOATE
mf: $C_9H_{11}NO_2$ mw: 165.21

PROP: Crystals. Mp: 88-90°.

SYNS: AMERICAINE ◇ p-AMINOBENZOIC ACID ETHYL ESTER ◇ 4-AMINOBENZOIC ACID ETHYL ESTER ◇ ANESTHESIN ◇ ANESTHONE ◇ BENZOCAINE ◇ ETHYL AMINOBENZOATE ◇ ETHYL-p-AMINOBENZOATE ◇ KELOFORM ◇ NORCAIN ◇ ORTHESIN ◇ PARATHESIN ◇ TOPCAINE

TOXICITY DATA with REFERENCE
skn-gpg 2%/48H MLD JSCCA5 28,357,77
rec-inf TDLo:12 mg/kg NEGMAG 263,454,60
ipr-mus LD50:216 mg/kg JMCMAR 17,900,74
orl-rbt LDLo:1150 mg/kg NIIRDN 6,33,82
orl-bwd LD50:56 mg/kg TXAPA9 21,315,72

CONSENSUS REPORTS: Reported in EPA TSCA Inventory.

SAFETY PROFILE: Poison by ingestion and intraperitoneal routes. Human systemic effects by rectal route: methemoglobinemia/carboxhemoglobinemia in infants. A skin irritant and a mild sensitizer. Local contact may cause contact dermatitis. Used as a topical anesthetic and as a sun-screening agent. When heated to decomposition it emits highly toxic fumes of NO_x. See also ETHYL ALCOHOL, p-AMINO BENZOIC ACID, and ESTERS.

EFX500 CAS:886-86-2 ***HR: 3***
ETHYL-m-AMINOBENZOATE METHANESULFONATE
mf: $C_{10}H_{15}NO_5S$ mw: 261.32

SYNS: 3-AMINOBENZOIC ACID ETHYL ESTER METHANESULFONATE ◇ METACAINE ◇ TRICAINE METHANE SULFONATE

TOXICITY DATA with REFERENCE
ivn-mus LD50:180 mg/kg CSLNX* NX#03233
ipr-frg LDLo:250 mg/kg PHMCAA 16,252,74

CONSENSUS REPORTS: Reported in EPA TSCA Inventory.

SAFETY PROFILE: Poison by intravenous and intraperitoneal routes. When heated to decomposition it emits very toxic fumes of SO_x and NO_x. See also ESTERS and SULFONATES.

EGA500 CAS:110-73-6 ***HR: 3***
2-ETHYLAMINOETHANOL
mf: $C_4H_{11}NO$ mw: 89.16

PROP: Flash p: 160°F (OC), d: 0.92, vap d: 3.06, bp: 161°.

SYNS: 2-(ETHYLAMINO)ETHANOL ◇ 2-N-MONOETHYLAMINO-ETHANOL

TOXICITY DATA with REFERENCE
skn-rbt 10 mg/24H open MLD AMIHBC 10,61,54
eye-rbt 250 µg open SEV AMIHBC 10,61,54
orl-rat LD50:1000 mg/kg FCTXAV 5,327,67
ipr-rat LD50:1170 mg/kg TXAPA9 12,486,68
skn-rbt LD50:360 mg/kg AMIHBC 10,61,54
orl-mam LD50:1200 mg/kg TXAPA9 8,344,66

CONSENSUS REPORTS: Reported in EPA TSCA Inventory.

SAFETY PROFILE: Poison by skin contact. Moderately toxic by ingestion and intraperitoneal routes. A skin and severe eye irritant. Flammable when exposed to heat or flame; can react vigorously with oxidizers. To fight fire, use alcohol foam, dry chemical, CO_2. When heated to decomposition it emits toxic fumes of NO_x.

EGC000 CAS:52400-55-2 ***HR: 3***
2-(2-(ETHYLAMINO)ETHYL)-2-METHYL-1,3-BENZODIOXOLE HYDROCHLORIDE
mf: $C_{12}H_{17}NO_2 \cdot ClH$ mw: 243.76

TOXICITY DATA with REFERENCE
ivn-rat LD50:40 mg/kg EJMCA5 12,413,77
ipr-mus LD50:100 mg/kg EJMCA5 12,413,77

SAFETY PROFILE: Poison by intravenous and intraperitoneal routes. When heated to decomposition it emits very toxic fumes of HCl and NO_x.

EGC500 CAS:42145-91-5 ***HR: 3***
dl-(±)-3-(2-ETHYLAMINO-1-HYDROXYETHYL) PHENYL PIVALATE HYDROCHLORIDE
mf: $C_{15}H_{23}NO_3 \cdot ClH$ mw: 301.85

SYNS: 2,2-DIMETHYLPROPANOICACID-3-(2-(ETHYLAMINO)-1-HYDROXYETHYL)PHENYL ESTER HYDROCHLORIDE ◇ dl-(±)α-(ETHYLAMINOMETHYL)-3'-HYDROXYBENZYL ALCOHOL 3-(2,2-DIMETHYLPROPIONATE)HCl ◇ α-((ETHYLAMINO) METHYL)-m-HYDROXYBENZYL ALCOHOL 2,2-DIMETHYLPROPIONATE HYDROCHLORIDE ◇ ETILEFRINE PIVALATE HYDROCHLORIDE ◇ K-30052

TOXICITY DATA with REFERENCE
orl-rat LDLo:240 mg/kg DRFUD4 4,413,79
ivn-rat LDLo:20 mg/kg DRFUD4 4,413,79
orl-mus LDLo:500 mg/kg DRFUD4 4,413,79
ivn-mus LDLo:24 mg/kg DRFUD4 4,413,79
orl-dog LDLo:140 mg/kg DRFUD4 4,413,79
ivn-dog LDLo:8 mg/kg DRFUD4 4,413,79

SAFETY PROFILE: Poison by ingestion and intravenous routes. When heated to decomposition it emits very toxic fumes of Cl^- and NO_x.

EGD000 CAS:1610-17-9 ***HR: 2***
2-ETHYLAMINO-4-ISOPROPYLAMINO-6-METHOXY-s-TRIAZINE
mf: $C_9H_{17}N_5O$ mw: 211.31

SYNS: ATRATON ◇ ATRATONE ◇ ATROTON ◇ 4-ETHYLAMINO-6-ISOPROPYLAMINO-2-METHOXY-s-TRIAZINE ◇ 6-ETHYLAMINO-4-ISOPROPYLAMINO-2-METHOXY-1,3,5-TRIAZINE ◇ N-ETHYL-6-METHOXY-N'-(1-METHYLETHYL)-1,3,5-TRIAZINE-2,4-DIAMINE ◇ GEIGY 32,293 ◇ GESATAMIN ◇ 2-METHOXY-4-ETHYLAMINO-6-ISOPROPYLAMINO-s-TRIAZINE ◇ 2-METHOXY-4-ISOPROPYLAMINO-6-ETHYLAMINO-s-TRIAZINE ◇ PRIMATOL

TOXICITY DATA with REFERENCE
orl-rat LD50:1465 mg/kg FMCHA2 -,C19,83
orl-mus LD50:905 mg/kg 85DPAN -,-,71/76

SAFETY PROFILE: Moderately toxic by ingestion. When heated to decomposition it emits toxic fumes of NO_x.

EGE500 CAS:709-55-7 ***HR: 3***
α-((ETHYLAMINO)METHYL)-m-HYDROXYBENZYL ALCOHOL
mf: $C_{10}H_{15}NO_2$ mw: 181.26

SYNS: EFFORTIL ◇ ETHYLADRIANOL ◇ ETHYL NORADRIANOL ◇ N-ETHYLNORPHENYLEPHRINE ◇ ETILEFRINE ◇ m-HYDROXYPHENYLETHANOLETHYLAMINE ◇ 1-(3'-HYDROXYPHENYL)-2-ETHYLAMINOETHANOL

TOXICITY DATA with REFERENCE
ims-gpg TDLo:13 mg/kg (female 30-60D post):TER EOGRAL 30,173,89
ims-gpg TDLo:13 mg/kg (female 30-60D post):REP EOGRAL 30,173,89
orl-rat LD50:114 mg/kg AIPTAK 180,155,69
ipr-rat LD50:820 mg/kg OYYAA2 3,27,69
scu-rat LD50:244 mg/kg AIPTAK 180,155,69
idu-rat LD50:200 mg/kg AIPTAK 180,155,69
orl-mus LD50:770 mg/kg ARZNAD 22,869,72

SAFETY PROFILE: Poison by ingestion, subcutaneous, and intraduodenal routes. Moderately toxic by intraperitoneal route. An experimental teratogen. Other experimental reproductive effects. When heated to decomposition it emits toxic fumes of NO_x.

EGF000 CAS:943-17-9 ***HR: 3***
α-((ETHYLAMINO)METHYL)-m-HYDROXYBENZYL ALCOHOL HYDROCHLORIDE
mf: $C_{10}H_{15}NO_2 \cdot ClH$ mw: 217.72

SYN: ETILEFRINE HYDROCHLORIDE

TOXICITY DATA with REFERENCE
orl-rat LDLo:187 mg/kg DRFUD4 4,413,79
ivn-rat LDLo:5 mg/kg DRFUD4 4,413,79
orl-mus LDLo:345 mg/kg DRFUD4 4,413,79
ivn-mus LDLo:11 mg/kg DRFUD4 4,413,79

SAFETY PROFILE: Poison by ingestion and intravenous routes. When heated to decomposition it emits very toxic fumes of HCl and NO_x. See also ALCOHOLS.

EGG000 CAS:69781-71-1 ***HR: 3***
N-((3-(ETHYLAMINO)PROPOXY)METHYL)DIPHENYLAMINE
mf: $C_{18}H_{24}N_2O•ClH$ mw: 320.90

SYN: ETHYLAMINOPROPYLDIPHENYLAMINOCARBINOL HYDROCHLORIDE

TOXICITY DATA with REFERENCE
ipr-rat LDLo:400 mg/kg JPETAB 18,467,22
ipr-cat LDLo:50 mg/kg JPETAB 18,467,22
scu-cat LDLo:50 mg/kg JPETAB 18,467,22
ivn-cat LDLo:10 mg/kg JPETAB 18,467,22
ivn-rbt LDLo:17 mg/kg JPETAB 18,467,22

SAFETY PROFILE: Poison by intraperitoneal, subcutaneous, and intravenous routes. When heated to decomposition it emits very toxic fumes of NO_x and HCl. See also AMINES.

EGH500 CAS:102206-99-5 ***HR: D***
3-ETHYLAMINO-5H-PYRIDO(4,3-b)INDOLE
mf: $C_{13}H_{13}N_3$ mw: 211.2

TOXICITY DATA with REFERENCE
mma-sat 2500 nmol/L PNASA6 77,1427,80
dnd-mam:lym 100 µmol/L PNASA6 77,1427,80

SAFETY PROFILE: Mutation data reported. When heated to decomposition it emits toxic fumes of NO_x.

EGI000 CAS:13275-68-8 ***HR: 3***
2-ETHYLAMINO-1,3,4-THIADIAZOLE
mf: $C_4H_7N_3S$ mw: 129.20

SYNS: CL 19217 4090L 7-5525 ◇ 2-ETHYLAMINOTHIADIAZOLE ◇ NSC 4730

TOXICITY DATA with REFERENCE
scu-rat TDLo:150 mg/kg (10-11D preg):TER JHMJAX 130,95,72
ipr-rat LD50:200 mg/kg CPCHAO 18,307,62
scu-rat LD50:200 mg/kg JHMJAX 130,95,72
ipr-mus LD50:795 mg/kg NCISP* JAN86

CONSENSUS REPORTS: Reported in EPA TSCA Inventory.

SAFETY PROFILE: Poison by intraperitoneal and subcutaneous routes. Experimental teratogenic effects. When heated to decomposition it emits very toxic fumes of NO_x and SO_x.

EGI500 CAS:457-87-4 ***HR: 3***
ETHYLAMPHETAMINE
mf: $C_{11}H_{17}N$ mw: 163.29

SYNS: 1-PHENYL-2-AETHYLAMINO-PROPAN (GERMAN) ◇ α-PHENYL-β-ETHYLAMINOPROPANE ◇ 1-PHENYL-2-ETHYLAMINOPROPANE

TOXICITY DATA with REFERENCE
orl-man TDLo:4286 µg/kg THERAP 34,205,79
orl-rat LD50:250 mg/kg BBMS** -,-,48
ipr-rat LDLo:80 mg/kg AEPPAE 195,647,40
ipr-mus LDLo:80 mg/kg AEPPAE 195,647,40

SAFETY PROFILE: Poison by ingestion and intraperitoneal routes. When heated to decomposition it emits toxic fumes of NO_x. See also BENZEDRINE.

EGI750 CAS:541-85-5 ***HR: 2***
ETHYL AMYL KETONE
mf: $C_8H_{16}O$ mw: 128.24

PROP: Liquid; mild, fruity odor. Bp: 157-162°, d: 0.822 @ 20°/20°, flash p: 138°F. Sol in many organic solvents.

SYNS: AMYL ETHYL KETONE ◇ 3-METHYL-5-HEPTANONE ◇ 5-METHYL-3-HEPTANONE

TOXICITY DATA with REFERENCE
skn-rbt 500 mg MLD SCCUR* -,6,61
orl-rat LD50:3500 mg/kg SCCUR* -,6,61
ihl-rat LCLo:3484 ppm/8H SCCUR* -,6,61
orl-mus LD50:3800 mg/kg SCCUR* -,6,61
ihl-mus LCLo:3484 ppm/4H SCCUR* -,6,61
orl-gpg LD50:2500 mg/kg BCPCA6 16,631,67

CONSENSUS REPORTS: Reported in EPA TSCA Inventory.

OSHA PEL: TWA 25 ppm
ACGIH TLV: TWA 25 ppm

SAFETY PROFILE: Moderately irritating to skin, eyes, and mucous membranes by inhalation and ingestion. Narcotic in high concentration. See also KETONES. When heated to decomposition it emits acrid smoke. Combustible. To fight fire, use foam, CO_2, dry chemical. See also KETONES.

EGJ000 CAS:10138-47-3 ***HR: 2***
2-(1-ETHYLAMYLOXY)ETHANOL
mf: $C_9H_{20}O_2$ mw: 160.29

SYN: 2-((1-ETHYLPENTYL)OXY)ETHANOL

TOXICITY DATA with REFERENCE
skn-rbt 10 mg/24H open MLD AMIHBC 10,61,54

eye-rbt 250 μg open SEV AMIHBC 10,61,54
orl-rat LD50:2280 mg/kg AMIHBC 10,61,54

SAFETY PROFILE: Moderately toxic by ingestion. A skin and severe eye irritant. When heated to decomposition it emits acrid smoke and irritating fumes.

EGJ500 CAS:10138-87-1 ***HR: 2***
2-(2-(1-ETHYLAMYLOXY)ETHOXY)ETHANOL
mf: $C_{11}H_{24}O_3$ mw: 204.35

SYN: 2-(((1-ETHYLPENTYL)OXY)ETHOXY)ETHANOL

TOXICITY DATA with REFERENCE
skn-rbt 10 mg/24H open MLD AMIHBC 10,61,54
eye-rbt 250 μg open SEV AMIHBC 10,61,54
orl-rat LD50:2940 mg/kg AMIHBC 10,61,54

SAFETY PROFILE: Moderately toxic by ingestion. A skin and severe eye irritant. When heated to decomposition it emits acrid smoke and irritating fumes.

EGK000 CAS:103-69-5 ***HR: 3***
N-ETHYLANILINE
DOT: UN 2272
mf: $C_8H_{11}N$ mw: 121.20

PROP: Clear, yellow-brown oil. Mp: −63.5°, bp: 204°, d: 0.958 @ 25°/25°, vap press: 1 mm @ 38.5°, vap d: 4.18, flash p: 185°F (OC).

SYNS: AETHYLANILIN (GERMAN) ◇ ANILINOETHANE ◇ N-ETHYLAMINOBENZENE ◇ ETHYLANILINE ◇ N-ETHYLBENZENAMINE ◇ N-ETHYLBENZENAMINO ◇ ETHYLPHENYLAMINE

TOXICITY DATA with REFERENCE
orl-rat LD50:334 mg/kg MarJV# 29MAR77
skn-rat LD50:4700 mg/kg AGGHAR 15,447,57
ipr-rat LD50:180 mg/kg AGGHAR 15,447,57
unr-mam LD50:600 mg/kg GISAAA 48(6),22,83

CONSENSUS REPORTS: Reported in EPA TSCA Inventory.

DOT Classification: Poison B; Label: St. Andrews Cross.

SAFETY PROFILE: Poison by ingestion and intraperitoneal routes. Moderately toxic by an unspecified route. Mildly toxic by skin contact. An allergen. Flammable when exposed to heat or flame; can react with oxidizing materials. To fight fire, use dry chemical, CO_2, foam. Hypergolic reaction with red fuming nitric acid. When heated to decomposition or on contact with acid or acid fumes it emits highly toxic fumes of aniline and NO_x.

EGK500 CAS:578-54-1 ***HR: 3***
2-ETHYLANILINE
DOT: UN 2273
mf: $C_8H_{11}N$ mw: 121.20

PROP: Yellow liquid, darkens upon standing. Mp: −63.5°, bp: 215°, flash p: 185°F (OC), d: 0.98 @ 25°/25°, vap d: 4.17.

SYNS: o-AMINOETHYLBENZENE ◇ o-ETHYLANILINE ◇ 2-ETHYLBENZENAMINE

TOXICITY DATA with REFERENCE
orl-rat LD50:1260 mg/kg TXAPA9 22,153,72
orl-bwd LD50:750 mg/kg AECTCV 12,355,83

CONSENSUS REPORTS: Reported in EPA TSCA Inventory.

DOT Classification: Poison B; Label: St. Andrews Cross.

SAFETY PROFILE: A poison. Moderately toxic by ingestion. Flammable when exposed to heat or flame; can react with oxidizing materials. To fight fire, use foam, CO_2, dry chemical. When heated to decomposition it emits highly toxic fumes of aniline and NO_x. See also N-ETHYLANILINE.

EGL000 CAS:589-16-2 ***HR: 3***
4-ETHYLANILINE
mf: $C_8H_{11}N$ mw: 121.20

PROP: D: 0.963, mp: 65.8°, bp: 205.5°. Insol in water; misc in alc and ether.

SYNS: 1-AMINO-4-ETHYLBENZENE ◇ p-ETHYLANILINE

TOXICITY DATA with REFERENCE
dns-rat:lvr 100 μmol/L ENMUDM 5,803,83
ipr-mus LD50:133 mg/kg JMCMAR 17,900,74
ivn-mus LD50:56 mg/kg CSLNX* NX#02908
orl-qal LD50:422 mg/kg AECTCV 12,355,83
orl-bwd LD50:75 mg/kg AECTCV 12,355,83

CONSENSUS REPORTS: Reported in EPA TSCA Inventory.

SAFETY PROFILE: Poison by ingestion, intravenous, and intraperitoneal routes. Mutation data reported. When heated to decomposition it emits toxic fumes of NO_x. See also N-ETHYLANILINE.

EGL500 CAS:84-51-5 ***HR: 3***
2-ETHYL-9,10-ANTHRACENEDIONE
mf: $C_{16}H_{12}O_2$ mw: 236.28

SYNS: 2-ETHYLANTHRAQUINONE ◇ 2-ETHYL-9,10-ANTHRAQUINONE ◇ USAF SO-1

TOXICITY DATA with REFERENCE
ipr-mus LD50:200 mg/kg NTIS** AD277-689

CONSENSUS REPORTS: Reported in EPA TSCA Inventory.

SAFETY PROFILE: Poison by intraperitoneal route.

When heated to decomposition it emits acrid smoke and irritating fumes.

EGM000 CAS:87-25-2 ***HR: 2***
ETHYL ANTHRANILATE
mf: $C_9H_{11}NO_2$ mw: 165.21

PROP: Colorless to amber liquid; floral, orange blossom odor. D: 1.115-1.120, refr index: 1.5631.566, flash p: +151°F. Sol in alc, fixed oils, propylene glycol.

SYNS: o-AMINOBENZOIC ACID, ETHYL ESTER ◇ ETHYL-o-AMINOBENZOATE ◇ FEMA No. 2421

TOXICITY DATA with REFERENCE
skn-rbt 500 mg/24H MOD FCTXAV 14,759,76
orl-rat LD50:3750 mg/kg FCTXAV 14,759,76

CONSENSUS REPORTS: Reported in EPA TSCA Inventory.

SAFETY PROFILE: Moderately toxic by ingestion. A skin irritant. Combustible liquid. When heated to decomposition it emits toxic fumes of NO_x.

EGM100 CAS:42971-09-5 ***HR: 3***
ETHYL APOVINCAMINATE
mf: $C_{22}H_{26}N_2O_2$ mw: 350.50

PROP: Crystals from benzene. Mp: 147-153° (decomp).

SYNS: 3-α,16-α-APOVINCAMINIC ACID ETHYL ESTER ◇ CAVINTON ◇ ETHYL APOVINCAMIN-22-OATE ◇ RGH-4405 ◇ TCV-3B ◇ VINPOCETINE

TOXICITY DATA with REFERENCE
orl-rat TDLo:100 mg/kg (female 14-21D post):REP ARZNAD 26,1938,76
orl-rat TDLo:1 g/kg (female 7-15D post):TER YACHDS 11,1125,83
orl-rat LD50:503 mg/kg ARZNAD 26,1938,76
ipr-rat LD50:134 mg/kg ARZNAD 26,1938,76
ivn-rat LD50:42600 mg/kg ARZNAD 26,1938,76
orl-mus LD50:534 mg/kg ARZNAD 26,1938,76
ipr-mus LD50:161 mg/kg ARZNAD 26,1938,76
ivn-mus LD50:58700 μg/kg ARZNAD 26,1938,76

SAFETY PROFILE: Poison by intravenous and intraperitoneal routes. Moderately toxic by ingestion. Experimental teratogenic and reproductive effects. When heated to decomposition it emits toxic fumes of NO_x.

EGM500 CAS:871-31-8 ***HR: 3***
ETHYL AZIDE
mf: $C_2H_5N_3$ mw: 71.08

SAFETY PROFILE: An explosive sensitive to rapid heating, shock or impact. Has exploded when heated to room temperature. When heated to decomposition it emits toxic fumes of NO_x. See also AZIDES.

EGN000 CAS:817-87-8 ***HR: 3***
ETHYL AZIDOFORMATE
mf: $C_3H_5N_3O_2$ mw: 115.09

$CH_3CH_2OCO{\bullet}N_3$

SYN: ETHYL CARBONAZIDATE

SAFETY PROFILE: Explodes at its boiling point (114°C). When heated to decomposition it emits toxic fumes of NO_x. See also AZIDES.

EGN100 CAS:81852-50-8 ***HR: 3***
ETHYL-2-AZIDO-2-PROPENOATE
mf: $C_5H_7N_3O_2$ mw: 141.13

$CH_3CH_2OCO{\bullet}C(N_3){=}CH_2$

SAFETY PROFILE: Potentially explosive when heated. When heated to decomposition it emits toxic fumes of NO_x. See also AZIDES.

EGN500 CAS:2549-67-9 ***HR: 3***
2-ETHYLAZIRIDINE
mf: C_4H_9N mw: 71.14

SYNS: 1,2-BUTYLENIMINE ◇ 2-ETHYLETHYLENIMINE ◇ N-METHYLCAPROLACTAM ◇ 1-METHYLCAPROLACTAM

TOXICITY DATA with REFERENCE
ihl-rat LCLo:1000 ppm/3.5H NTIS** AD441-640
ipr-mus LD50:52 mg/kg NTIS** AD441-640

CONSENSUS REPORTS: Reported in EPA TSCA Inventory.

SAFETY PROFILE: Poison by intraperitoneal route. Moderately toxic by inhalation. When heated to decomposition it emits toxic fumes of NO_x.

EGO000 CAS:56961-62-7 ***HR: 3***
5-ETHYL-1,2-BENZANTHRACENE
mf: $C_{20}H_{16}$ mw: 256.36

SYN: 8-ETHYLBENZ(a)ANTHRACENE

TOXICITY DATA with REFERENCE
skn-mus TDLo:650 mg/kg/27W-I:ETA PRLBA4 123,343,37

SAFETY PROFILE: Questionable carcinogen with experimental tumorigenic data. When heated to decomposition it emits acrid smoke and irritating fumes.

EGO500 CAS:3697-30-1 ***HR: 3***
10-ETHYL-1,2-BENZANTHRACENE
mf: $C_{20}H_{16}$ mw: 256.36

SYN: 7-ETHYLBENZ(a)ANTHRACENE

TOXICITY DATA with REFERENCE
mma-sat 50 μg/plate MUREAV 206,55,88
scu-mus TDLo:600 mg/kg:ETA JNCIAM 1,303,40

SAFETY PROFILE: Questionable carcinogen with experimental tumorigenic data. Mutation data reported. When heated to decomposition it emits acrid smoke and irritating fumes.

EGP000 CAS:18868-66-1 ***HR: 3***
12-ETHYLBENZ(a)ANTHRACENE
mf: $C_{20}H_{16}$ mw: 256.36

TOXICITY DATA with REFERENCE
ims-rat TDLo:50 mg/kg:ETA CNREA8 29,506,69

SAFETY PROFILE: Questionable carcinogen with experimental tumorigenic data. When heated to decomposition it emits acrid smoke and irritating fumes.

EGP500 CAS:100-41-4 ***HR: 3***
ETHYL BENZENE
DOT: UN 1175
mf: C_8H_{10} mw: 106.18

PROP: Colorless liquid, aromatic odor. Bp: 136.2°, fp: −94.9°, flash p: 59°F, d: 0.8669 @ 20°/4°, autoign temp: 810°F, vap press: 10 mm @ 25.9°, vap d: 3.66, lel: 1.2%, uel: 6.8%. Misc in alc and ether; insol in NH_3; sol in SO_2.

SYNS: AETHYLBENZOL (GERMAN) ◇ EB ◇ ETHYLBENZEEN (DUTCH) ◇ ETHYLBENZOL ◇ ETILBENZENE (ITALIAN) ◇ ETYLOBENZEN (POLISH) ◇ NCI-C56393 ◇ PHENYLETHANE

TOXICITY DATA with REFERENCE
skn-rbt 15 mg/24H open MLD AIHAAP 23,95,62
eye-rbt 100 mg AJOPAA 29,1363,46
sce-hmn:lym 1 mmol/L MUREAV 116,379,83
ihl-rat TCLo:600 mg/m³/24H (female 7-15D post):TER ARTODN 8,425,85
ihl-rbt TCLo:1 g/m³/24H (female 7-20D post):REP ARTODN 8,425,85
ihl-hmn TCLo:100 ppm/8H:EYE,CNS,PUL AIHAAP 31,206,70
orl-rat LD50:3500 mg/kg AMIHAB 14,387,56
ihl-rat LCLo:4000 ppm/4H AIHAAP 23,95,62
ihl-mus LCLo:50 g/m³/2H GTPZAB 5(5),3,61
ipr-mus LD50:2272 mg/kg ARTODN 58,106,85
skn-rbt LD50:17800 mg/kg FCTXAV 13,803,75
ihl-gpg LCLo:10000 ppm PHRPA6 45,1241,30

CONSENSUS REPORTS: Reported in EPA TSCA Inventory. EPA Genetic Toxicology Program. Community Right-To-Know List.

OSHA PEL: (Transitional: TWA 100 ppm (skin)) TWA 100 ppm; STEL 125 ppm
ACGIH TLV: TWA 100 ppm; STEL 125 ppm; BEI: 2 g(mandelic acid)/L in urine at end of shift; 2 ppm ethyl benzene in end-exhaled air prior to next shift.
DFG MAK: 100 ppm (440 mg/m³)
DOT Classification: Flammable Liquid; Label: Flammable Liquid.

SAFETY PROFILE: Moderately toxic by ingestion and intraperitoneal route. Mildly toxic by inhalation and skin contact. An experimental teratogen. Other experimental reproductive effects. Human systemic effects by inhalation: eye, sleep and pulmonary changes. An eye and skin irritant. Human mutation data reported. The liquid is an irritant to the skin and mucous membranes. A concentration of 0.1% of the vapor in air is an irritant to human eyes, and a concentration of 0.2% is extremely irritating at first, then causes dizziness, irritation of the nose and throat and a sense of constriction in the chest. Exposure of guinea pigs to 1% concentration has been reported as causing ataxia, loss of consciousness, tremor of the extremities and finally death through respiratory failure. The pathological findings were congestion of the brain and lungs with edema.

A very dangerous fire and explosion hazard when exposed to heat or flame; can react vigorously with oxidizing materials. To fight fire, use foam, CO_2, dry chemical. Emitted from modern building materials. (CENEAR 69,22,91) When heated to decomposition it emits acrid smoke and irritating fumes. .

EGQ000 CAS:93-54-9 ***HR: 2***
α-ETHYLBENZENEMETHANOL
mf: $C_9H_{12}O$ mw: 136.21

SYNS: EJIBIL ◇ α-ETHYLBENZYL ALCOHOL ◇ ETHYL PHENYL CARBINOL ◇ FELICUR ◇ FELITROPE ◇ FENICOL ◇ α-HYDROXYPROPYLBENZENE ◇ LIVONAL ◇ PHENICOL ◇ PHENYCHOLON ◇ PHENYLAETHYLCARBINOL (GERMAN) ◇ 1-PHENYLPROPANOL ◇ 1-PHENYL-1-PROPANOL ◇ 1-PHENYLPROPYLALCOHOL ◇ SH 261

TOXICITY DATA with REFERENCE
orl-rat LD50:1600 mg/kg ARZNAD 12,347,62
orl-mus LD50:500 mg/kg AIPTAK 116,154,58
scu-mus LD50:700 mg/kg AIPTAK 116,154,58

SAFETY PROFILE: Moderately toxic by ingestion and subcutaneous route. When heated to decomposition it emits acrid smoke and irritating fumes.

EGR000 CAS:93-89-0 ***HR: 2***
ETHYL BENZOATE
mf: $C_9H_{10}O_2$ mw: 150.19

PROP: Colorless liquid; heavy fruity odor. Mp: −34.6°, bp: 213.4°, flash p: >204°F, d: 1.048 @ 20°/20°, refr index: 1.502-1.506, vap press: 1 mm @ 44.0°, vap d: 5.17, autoign temp: 914°F. Sol in alc, fixed oils, and propylene glycol; insol in glycerin, water @ 212°; misc in petroleum, chloroform, and ether.

SYNS: BENZOIC ETHER ◇ ESSENCE of NIOBE ◇ FEMA No. 2422

TOXICITY DATA with REFERENCE
skn-rbt 10 mg/24H open MLD AMIHBC 10,61,54
eye-rbt 500 mg open AMIHBC 10,61,54
orl-rat LD50:2100 mg/kg JPETAB 84,358,45
skn-cat LDLo:10 g/kg JPETAB 84,358,45
orl-rbt LD50:2630 mg/kg JPETAB 84,358,45

CONSENSUS REPORTS: Reported in EPA TSCA Inventory.

SAFETY PROFILE: Moderately toxic by ingestion. Mildly toxic by skin contact. A skin and eye irritant. Combustible liquid when exposed to heat or flame; can react with oxidizing materials. To fight fire, use foam, CO_2, dry chemical. When heated to decomposition it emits acrid smoke and irritating fumes. See also ESTERS.

EGR500 CAS:6797-13-3 ***HR: 2***
2-ETHYLBENZOXAZOLE
mf: C_9H_9NO mw: 147.19

TOXICITY DATA with REFERENCE
orl-mus LD40:1 g/kg JACSAT 67,905,45

CONSENSUS REPORTS: Reported in EPA TSCA Inventory.

SAFETY PROFILE: Moderately toxic by ingestion. When heated to decomposition it emits toxic fumes of NO_x.

EGS000 CAS:33878-50-1 ***HR: 2***
ETHYL-N-BENZOYL-N-(3,4-DICHLOROPHENYL)-2-AMINOPROPIONATE
mf: $C_{18}H_{17}Cl_2NO_3$ mw: 366.26

SYNS: N-BENZOYL-N-(3,4-DICHLOROPHENYL)-l-ALANINEETHYL ESTER ◇ BENZOYLPROP ETHYL ◇ ENAVEN ◇ SD 30,053 ◇ SUFFIX ◇ SUFFIX 25 ◇ WL 17,731

TOXICITY DATA with REFERENCE
orl-rat LD50:1550 mg/kg GUCHAZ 6,256,73
orl-mus LD50:716 mg/kg 28ZEAL 5,24,76
orl-dog LD50:2000 mg/kg 85DPAN -,-,71/76
orl-rbt LD50:1000 mg/kg 85DPAN -,-,71/76

SAFETY PROFILE: Moderately toxic by ingestion. When heated to decomposition it emits very toxic fumes of NO_x and Cl^-.

EGS500 CAS:59965-27-4 ***HR: 3***
2-ETHYL-3:4-BENZPHENANTHRENE
mf: $C_{20}H_{16}$ mw: 256.36

SYN: 5-ETHYLBENZO(c)PHENANTHRENE

TOXICITY DATA with REFERENCE
skn-mus TDLo:620 mg/kg/26W-I:ETA PRLBA4 131,170,42

SAFETY PROFILE: Questionable carcinogen with experimental tumorigenic data. When heated to decomposition it emits acrid smoke and irritating fumes.

EGT000 CAS:2521-01-9 ***HR: 3***
ETHYL-N-BENZYLCYCLOPROPANECARBAMATE
mf: $C_{13}H_{17}NO_2$ mw: 219.31

SYNS: ENCYPRATE ◇ ETHYL-N-BENZYL-N-CYCLOPROPYLCARBAMATE

TOXICITY DATA with REFERENCE
orl-rat LD50:1380 mg/kg 27ZQAG -,391,72
ipr-rat LD50:370 mg/kg 27ZQAG -,391,72
orl-mus LD50:1100 mg/kg 27ZQAG -,391,72
ipr-mus LD50:510 mg/kg 27ZQAG -,391,72

SAFETY PROFILE: Poison by intraperitoneal route. Moderately toxic by ingestion. When heated to decomposition it emits toxic fumes of NO_x. See also CABAMATES.

EGT500 CAS:15718-71-5 ***HR: 3***
1,2-ETHYL BIS-AMMONIUM PERCHLORATE
mf: $C_2H_{10}Cl_2N_2O_8$ mw: 261.02

(−CH2N⁻H3)2•2ClO4⁻

DOT Classification: Forbidden

SAFETY PROFILE: Highly unstable. Much more powerful than TNT. When heated to decomposition it emits toxic fumes of NH_3. See also EXPLOSIVES, HIGH.

EGU000 CAS:3590-07-6 ***HR: 3***
ETHYLBIS(β-CHLOROETHYL)AMINE HYDROCHLORIDE
mf: $C_6H_{13}Cl_2N•ClH$ mw: 206.56

SYNS: HN1•HCl ◇ HN1 HYDROCHLORIDE ◇ TL 329 HYDROCHLORIDE

TOXICITY DATA with REFERENCE
scu-rat LD50:1 mg/kg JPETAB 91,224,47
ivn-rat LD50:500 μg/kg JPETAB 91,224,47
ipr-mus LDLo:1050 μg/kg NTIS** PB158-507
scu-mus LD50:1 mg/kg BJPCAL 1,247,46
orl-rbt LDLo:1500 μg/kg NTIS** PB158-507
ivn-rbt LD50:2 mg/kg JPETAB 91,224,47

CONSENSUS REPORTS: EPA Genetic Toxicology Program.

SAFETY PROFILE: Poison by ingestion, subcutaneous, intravenous, and intraperitoneal routes. When heated to decomposition it emits very toxic fumes of Cl^- and NO_x. See also AMINES.

EGV000 CAS:105-36-2 ***HR: 3***
ETHYL BROMACETATE
DOT: UN 1603
mf: $C_4H_7BrO_2$ mw: 167.02

PROP: Colorless to straw-colored liquid. Bp: 158.8°, fp: < −20°, flash p: 118°F, d: 1.514 @ 13°/4°, vap d: 5.8. Insol in water; misc in alc and ether.

SYNS: BROMOACETIC ACID, ETHYL ESTER ◇ ETHOXYCARBONYLMETHYL BROMIDE ◇ ETHYL BROMOACETATE ◇ ETHYL-α-BROMOACETATE ◇ ETHYL MONOBROMOACETATE

TOXICITY DATA with REFERENCE
scu-mus TDLo:252 mg/kg/63W-I:NEO JNCIAM 53,695,74

CONSENSUS REPORTS: Reported in EPA TSCA Inventory.

DOT Classification: Poison B; Label: Flammable Liquid and Poison.

SAFETY PROFILE: A poison. An irritant to skin, eyes, and mucous membranes. Questionable carcinogen with experimental neoplastigenic data. Flammable when exposed to heat, flame, and oxidizers. Will react with water or steam to produce toxic and corrosive fumes. To fight fire, use water as a fire blanket. When heated to decomposition or on contact with acid or acid fumes, it emits highly toxic fumes of Br^-. See also BROMIDES.

EGV400 CAS:74-96-4 ***HR: 3***
ETHYL BROMIDE
DOT: UN 1891
mf: C_2H_5Br mw: 108.98

PROP: Colorless, volatile liquid. Mp: −119°, bp: 38.4°, lel: 6.7%, uel: 11.3%, flash p: < −4°F, d: 1.451 @ 20°/4°, autoign temp: 952°F, vap press: 400 mm @ 21°, vap d 3.76.

SYNS: BROMOETHANE ◇ BROMURE d'ETHYLE ◇ ETYLU BROMEK (POLISH) ◇ HALON 2001 ◇ MONOBROMOETHANE ◇ NCI-C55481

TOXICITY DATA with REFERENCE
ipr-rat LD50:1750 mg/kg JPCEAO 320,133,78
ihl-mus LC50:16230 ppm/1H AMRL** TR-72-62/72
ipr-mus LD50:2850 mg/kg JPCEAO 320,133,78
orl-rat LD50:1350 mg/kg 85GMAT-,65,82
ihl-rat LDLo:148000 ppm/15M AMIHBC 6,435,52

CONSENSUS REPORTS: NTP Carcinogenesis Studies (inhalation); Clear Evidence: Mouse NTPTR* NTP-TR-363,89; (Inhalation); Some Evidence: Rat NTPTR* NTP-TR-363,89. EPA Genetic Toxicology Program. Reported in EPA TSCA Inventory.

OSHA PEL: (Transitional: TWA 200 ppm) TWA 200 ppm; STEL 250 ppm
ACGIH TLV: TWA 200 ppm; STEL 250 ppm; (Proposed: TWA 5 ppm; Suspected Human Carcinogen)
DFG MAK: 200 ppm (890 mg/m^3)
DOT Classification: Poison B; Label: Poison.

SAFETY PROFILE: A poison. Suspected Human Carcinogen. Moderately toxic by ingestion and intraperitoneal routes. Mildly toxic by inhalation. An eye and skin irritant. Physiologically, it is an anesthetic and narcotic. Its vapors are markedly irritating to the lungs on inhalation for even short periods. It can produce acute congestion and edema. Liver and kidney damage in humans has been reported. It is much less toxic than methyl bromide, but more toxic than ethyl chloride. It is a preparative hazard. Dangerously flammable by heat, open flame (sparks), oxidizers. Moderately explosive when exposed to flame. Reacts with water or steam to produce toxic and corrosive fumes. Vigorous reaction with oxidizing materials. To fight fire, use CO_2, dry chemical. Readily decomposes when heated to emit toxic fumes of Br^-. See also BROMIDES.

EGV500 CAS:4824-78-6 ***HR: 3***
ETHYL BROMOPHOS
mf: $C_{10}H_{12}BrCl_2O_3PS$ mw: 394.06

SYNS: 4-BROMO-2,5-DICHLOROPHENOL-o-ESTER with O,O-DIETHYL PHOSPHOROTHIOATE ◇ O-(4-BROMO-2,5-DICHLOROPHENYL)-O,O-DIETHYL PHOSPHOROTHIOATE ◇ O-(4-BROMO-2,5 DICHLOROPHENYL)-O,O-DIETHYLPHOSPHOROTHIONATE ◇ BROMOFOS-ETHYL ◇ BROMOPHOSETHYL ◇ CELA S-2225 ◇ O,O-DIAETHYL-O-(4-BROM-2,5-DICHLOR)-PHENYL-MONOTHIOPHOSPHAT (GERMAN) ◇ O,O-DIAETHYL-O-(2,5-DICHLOR-4-BROMPHENYL)-THIONOPHOSPHAT (GERMAN) ◇ O,O-DIETHYL- O-(4-BROOM-2,5-DICHLOOR-FENYL)-MONOTHIOFOSFAAT(DUTCH) ◇ O,O-DIETHYL O-2,5-DICHLORO-4-BROMOPHENYL-PHOSPHOROTHIOATE ◇ O,O-DIETHYL O-(2,5-DICHLORO-4-BROMOPHENYL) THIOPHOSPHATE ◇ O,O-DIETIL-O-(4-BROMO-2,5 DICLORO-FENIL)-MONOTIOFOSFATO (ITALIAN) ◇ ENT 27,258 ◇ FILARIOL ◇ NEXAGAN ◇ OMS-659 ◇ S 2225 ◇ THIOPHOSPHATE de O,O-DIETHYLE et de O-(2,5-DICHLORO-4-BROMO) PHENYLE (FRENCH)

TOXICITY DATA with REFERENCE
orl-rat LD50:52 mg/kg SPEADM 74-1,-,74
ihl-rat LC50:16600 pph 85DPAN -,-,71/76
skn-rat LD50:1000 mg/kg WRPCA2 9,119,70
orl-mus LD50:210 mg/kg FMCHA2 -,C37,83
skn-rbt LD50:1366 mg/kg 28ZEAL 5,29,76

CONSENSUS REPORTS: Chlorophenol compounds are on the Community Right-To-Know List.

SAFETY PROFILE: Poison by ingestion. Moderately toxic by skin contact and inhalation. An insecticide. When heated to decomposition it emits very toxic fumes of Br^-, PO_x, SO_x, and Cl^-. See also CHLOROPHENOLS.

EGW000 CAS:97-95-0 ***HR: 2***
2-ETHYLBUTANOL
DOT: UN 2275/UN 2282
mf: $C_6H_{14}O$ mw: 102.20

PROP: Clear liquid. Bp: 148.9°, flash p: 135°F (COC), d: 0.8328, vap press: 0.9 mm @ 20°, vap d: 3.4.

SYNS: 2-ETHYLBUTANOL-1 ◇ 2-ETHYL-1-BUTANOL ◇ 2-ETHYLBUTYL ALCOHOL ◇ sec-HEXANOL (DOT) ◇ sec-HEXYL ALCOHOL ◇ 3-METHYLOLPENTANE ◇ sec-PENTYLCARBINOL ◇ 3-PENTYLCARBINOL ◇ PSEUDOHEXYL ALCOHOL

TOXICITY DATA with REFERENCE
skn-rbt 415 mg open MLD UCDS** 12/14/71
eye-rbt 250 μg open SEV AMIHBC 10,61,54
orl-rat LD50:1850 mg/kg AMIHBC 10,61,54
orl-rbt LD50:1200 mg/kg JPETAB 82,377,44
skn-rbt LD50:1260 mg/kg AMIHBC 10,61,54

CONSENSUS REPORTS: Reported in EPA TSCA Inventory.

DOT Classification: Flammable or Combustible Liquid; Label: Flammable Liquid.

SAFETY PROFILE: Moderately toxic by ingestion and skin contact. A skin and severe eye irritant. Combustible when exposed to heat or flame; can react with oxidizing materials. To fight fire, use dry chemical, CO_2, foam, fog. When heated to decomposition it emits acrid smoke and irritating fumes. See also ALCOHOLS.

EGW500 CAS:760-21-4 ***HR: 1***
2-ETHYL-1-BUTENE
mf: C_6H_{12} mw: 84.18

PROP: Flash p: < −4°, autoign temp: 599°F, d: 0.69, vap d: 2.9, bp: 62°.

TOXICITY DATA with REFERENCE
eye-hmn 5 ppm JOCMA7 2,383,60

CONSENSUS REPORTS: Reported in EPA TSCA Inventory.

SAFETY PROFILE: A human eye irritant. A very dangerous fire hazard when exposed to heat, flames, or oxidizers. To fight fire, use dry chemical, CO_2, foam, spray. When heated to decomposition it emits acrid smoke and irritating fumes.

EGX000 CAS:4468-93-3 ***HR: 3***
2-(2-ETHYLBUTOXY)ETHANOL
mf: $C_8H_{18}O_2$ mw: 146.26

PROP: Liquid. Bp: 197°, fp: −90°, flash p: 180°F (OC), d: 0.8954 @ 20°/20°, vap press: 0.17 mm @ 20°, vap d: 5.04.

TOXICITY DATA with REFERENCE
skn-rbt 10 mg/24H open MLD AMIHBC 10,61,54
eye-rbt 250 μg open SEV AMIHBC 10,61,54
orl-rat LD50:1910 mg/kg AMIHBC 10,61,54
skn-rbt LD50:320 mg/kg AMIHBC 10,61,54

SAFETY PROFILE: Poison by skin contact. Moderately toxic by ingestion. A skin and severe eye irritant. Flammable when exposed to heat or flame; can react with oxidizing materials. When heated to decomposition it emits acrid smoke and irritating fumes. To fight fire, use foam, CO_2, dry chemical.

EGX500 CAS:63907-07-3 ***HR: 2***
N-(2-ETHYLBUTOXYETHOXYPROPYL)-5-NORBORNENE-2,3-DICARBOXIMIDE
mf: $C_{20}H_{34}N_2O_4$ mw: 366.56

SYN: N-(2-ETHYLBUTOXYETHOXYPROPYL)BICYCLO(2.2.1)HEPTENE-2,3-DICARBOXIMIDE

TOXICITY DATA with REFERENCE
skn-rbt 10 mg/24H open SEV AIHAAP 23,95,62
orl-rat LD50:9500 mg/kg AIHAAP 23,95,62
skn-rbt LD50:16 g/kg AIHAAP 23,95,62

SAFETY PROFILE: Mildly toxic by ingestion and skin contact. A severe skin irritant. When heated to decomposition it emits toxic fumes of NO_x.

EGY000 CAS:10213-74-8 ***HR: 2***
3-(2-ETHYLBUTOXY)PROPIONIC ACID
mf: $C_9H_{18}O_3$ mw: 174.27

PROP: Water-white liquid, insol in water. D: 0.96 @ 20°/20°, bp: 200° @ 100 mm, vap press: <0.1 mm @ 20°, flash p: 280°F.

TOXICITY DATA with REFERENCE
skn-rbt 10 mg/24H SEV AMIHBC 10,61,54
eye-rbt 750 μg SEV AMIHBC 10,61,54
orl-rat LD50:3730 mg/kg AMIHBC 10,61,54
skn-rbt LD50:530 mg/kg AMIHBC 10,61,54

SAFETY PROFILE: Moderately toxic by ingestion and skin contact. A severe skin and eye irritant. Combustible when exposed to heat or flame; can react vigorously with oxidizing materials. To fight fire, use dry chemical, CO_2. When heated to decomposition it emits acrid smoke and irritating fumes.

EGY500 CAS:10232-91-4 ***HR: 2***
3-(2-ETHYLBUTOXY)PROPIONITRILE
mf: $C_9H_{17}NO$ mw: 155.27

TOXICITY DATA with REFERENCE
eye-rbt 500 mg AMIHBC 10,61,54

orl-rat LD50:2460 mg/kg AMIHBC 10,61,54
skn-rbt LD50:10 g/kg AMIHBC 10,61,54

CONSENSUS REPORTS: Cyanide and its compounds are on the Community Right-To-Know List.

SAFETY PROFILE: Moderately toxic by ingestion. Mildly toxic by skin contact. An eye irritant. When heated to decomposition it emits toxic fumes of NO_x and CN^-. See also NITRILES.

EGZ000 CAS:3953-10-4 ***HR: 2***
2-ETHYLBUTYLACRYLATE
mf: $C_9H_{16}O_2$ mw: 156.25

PROP: Clear, colorless liquid. Bp: 82° @ 10 mm, fp: −70°, flash p: 125°F (OC), d: 0.8964 @ 20°/20°, vap press: 1.7 mm @ 20°.

SYNS: 2-ETHYLBUTYL ESTER, ACRYLIC ACID ◇ 2-PROPENOIC ACID-2-ETHYLBUTYL ESTER

TOXICITY DATA with REFERENCE
skn-rbt 10 mg/24H open SEV AMIHBC 4,119,51
eye-rbt 500 mg open AMIHBC 4,119,51
orl-rat LDLo:6490 mg/kg AMIHBC 4,119,51
skn-rbt LD50:5500 mg/kg AMIHBC 4,119,51

CONSENSUS REPORTS: Reported in EPA TSCA Inventory.

SAFETY PROFILE: Mildly toxic by ingestion and skin contact. An eye and severe skin irritant. Combustible when exposed to heat or flame; can react with oxidizing materials. To fight fire, use foam, CO_2, dry chemical. When heated to decomposition it emits acrid smoke and irritating fumes. See also ESTERS.

EHA000 CAS:617-79-8 ***HR: 3***
2-ETHYLBUTYLAMINE
mf: $C_6H_{15}N$ mw: 101.22

PROP: Water-white liquid. Bp: 110-113°, flash p: 64°F (OC), d: 0.739 @ 20°/20°, vap d: 3.5.

SYN: 2-ETHYL-1-BUTANAMINE

TOXICITY DATA with REFERENCE
skn-rbt 10 mg/24H open AMIHBC 10,61,54
eye-rbt 250 μg open SEV AMIHBC 10,61,54
orl-rat LD50:310 mg/kg GISAAA 39(3),106,74
orl-mus LD50:418 mg/kg GISAAA 39(3),106,74
skn-rbt LD50:2000 mg/kg AMIHBC 10,61,54

SAFETY PROFILE: Poison by ingestion. Moderately toxic by skin contact. A skin and severe eye irritant. A very dangerous fire hazard when exposed to heat or flame; can react vigorously with oxidizing materials. Keep away from heat and open flame. To fight fire, use dry chemical, CO_2, foam. When heated to decomposition it emits toxic fumes of NO_x. See also AMINES.

EHA100 CAS:591-62-8 ***HR: 3***
ETHYL BUTYLCARBAMATE
mf: $C_7H_{15}NO_2$ mw: 145.23

SYNS: BUR ◇ 1-BUTYLURETHAN ◇ BUTYLURETHANE ◇ N-BUTYLURETHANE ◇ 1-BUTYLURETHANE ◇ ETHYL-N,N-BUTYLCARBAMATE

TOXICITY DATA with REFERENCE
mmo-bcs 5 g/L MUREAV 42,19,77
dnr-bcs 5 g/L MUREAV 42,19,77
scu-rat TDLo:100 mg/kg (15-21D preg):NEO,TER GANNA2 71,811,80
ipr-mus LD50:250 mg/kg NTIS** AD691-490

CONSENSUS REPORTS: EPA Genetic Toxicology Program.

SAFETY PROFILE: Poison by intraperitoneal route. An experimental teratogen. Questionable carcinogen with experimental tumorigenic data. Mutation data reported. When heated to decomposition it emits toxic fumes of NO_x. See also CARBAMATES and ESTERS.

EHA500 CAS:628-81-9 ***HR: 3***
ETHYL BUTYL ETHER
DOT: UN 1179
mf: $C_6H_{14}O$ mw: 102.20

PROP: Colorless liquid. Bp: 92°, mp: −124°, flash p: 40°F, d: 0.7528 @ 20°/20°, vap d: 3.52. Insol in water; misc in alc and ether.

SYN: ETHER ETHYLBUTYLIQUE (FRENCH)

TOXICITY DATA with REFERENCE
skn-rbt 10 mg/24H open MLD AMIHBC 4,119,51
eye-rbt 500 mg open AMIHBC 4,119,51
orl-rat LD50:1870 mg/kg AMIHBC 4,119,51

CONSENSUS REPORTS: Reported in EPA TSCA Inventory.

DOT Classification: Flammable Liquid; Label: Flammable Liquid.

SAFETY PROFILE: Moderately toxic by ingestion. A skin and eye irritant. A very dangerous fire hazard when exposed to heat or flame; can react vigorously with oxidizing materials. Keep away from heat and open flame. To fight fire, use alcohol foam, CO_2, dry chemical. When heated to decomposition it emits acrid smoke and irritating fumes. See also ETHERS.

EHA600 CAS:106-35-4 ***HR: 2***
ETHYL BUTYL KETONE
mf: $C_7H_{14}O$ mw: 114.21

PROP: Clear mobile liquid; fatty odor. Mp: −36.7°, bp: 148°, flash p: 115°F (OC), d: 0.8198 @ 20°/20°, vap d: 3.93. Misc with alc, ether, water @ 149°.

SYNS: AETHYLBUTYLKETON (GERMAN) ◇ n-BUTYL ETHYL KETONE ◇ EPTAN-3-ONE (ITALIAN) ◇ ETHYLBUTYLCETONE (FRENCH) ◇ ETHYLBUTYLKETON (DUTCH) ◇ ETILBUTILCHETONE (ITALIAN) ◇ FEMA No. 2545 ◇ HEPTAN-3-ON (DUTCH, GERMAN) ◇ 3-HEPTANONE ◇ HEPTAN-3-ONE

TOXICITY DATA with REFERENCE
skn-rbt 500 mg open MLD UCDS** 3/12/69
skn-rbt 500 mg/24H MOD FCTXAV 16,731,78
eye-rbt 100 mg MLD FCTXAV 16,731,78
orl-rat LD50:2760 mg/kg JIHTAB 31,60,49
ihl-rat LCLo:2000 ppm/4H JIHTAB 31,343,49

CONSENSUS REPORTS: Reported in EPA TSCA Inventory.

OSHA PEL: TWA 50 ppm
ACGIH TLV: TWA 50 ppm

SAFETY PROFILE: Moderately toxic by ingestion and inhalation. A skin and eye irritant. Combustible liquid. Can react with oxidizing materials. To fight fire, use foam, CO_2, dry chemical. See also KETONES.

EHC000 CAS:4549-44-4 ***HR: 3***
ETHYL-N-BUTYLNITROSAMINE
mf: $C_6H_{14}N_2O$ mw: 130.22

SYNS: AETHYL-N-BUTYL-NITROSOAMIN (GERMAN) ◇ N-ETHYL-N-NITROSOBUTYLAMINE ◇ N-NITROSO-N-BUTYLETHYLAMINE ◇ N-NITROSOETHYL-N-BUTYLAMINE

TOXICITY DATA with REFERENCE
mmo-sat 769 μmol/L ENMUDM 3,11,81
dns-rat:lvr 1 mmol/L ENMUDM 3,11,81
dni-mus-ipr 20 g/kg ARGEAR 51,605,81
orl-rat TDLo:1000 mg/kg/29W-C:ETA ARZNAD 19,1077,69
orl-mus TDLo:2360 mg/kg/34W-C:CAR NATWAY 50,717,63
orl-rat LD50:380 mg/kg NATWAY 50,100,63
ivn-rat LD50:380 mg/kg ZEKBAI 69,103,67

SAFETY PROFILE: Poison by ingestion and intravenous routes. Questionable carcinogen with experimental carcinogenic and tumorigenic data. Mutation data reported. When heated to decomposition it emits toxic fumes of NO_x. See also NITOSAMINES.

EHC800 ***HR: 2***
ETHYL N-BUTYL-N-NITROSOSUCCINAMATE
mf: $C_{10}H_{18}N_2O_4$ mw: 230.30

SYNS: N-BUTYL-N-NITROSOSUCCINAMIC ACID ETHYL ESTER ◇ EBNS ◇ N-NITROSO-N-(3-CARBOETHOXYPROPIONYL)BUTYLAMINE

TOXICITY DATA with REFERENCE
scu-rat TDLo:70500 μg/kg/10W-I:CAR GANNA2 73,687,82
scu-rat TD:87 mg/kg/10W-I:CAR IAPUDO 41,619,82

SAFETY PROFILE: Questionable carcinogen with experimental carcinogenic data. When heated to decomposition it emits toxic fumes of NO_x.

EHC900 CAS:68037-57-0 ***HR: 3***
2-ETHYLBUTYL SILICATE

SYNS: POLYBIS(2-ETHYLBUTYL)SILOXANE ◇ SILICIC ACID, 2-ETHYLBUTYL ESTER

TOXICITY DATA with REFERENCE
skn-rbt 500 mg open MLD UCDS** 7/10/63
orl-rat LD50:20 mg/kg UCDS** 7/10/63

CONSENSUS REPORTS: Reported in EPA TSCA Inventory.

SAFETY PROFILE: Poison by ingestion route. A skin irritant. When heated to decomposition it emits acrid smoke and irritating fumes.

EHD000 CAS:63019-12-5 ***HR: 3***
α-ETHYL-α′,sec-BUTYLSTILBENE
mf: $C_{20}H_{24}$ mw: 264.44

SYN: α-ETHYL-β-sec-BUTYLSTILBENE

TOXICITY DATA with REFERENCE
skn-mus TDLo:1250 mg/kg/52W-I:ETA NATUAS 148,142,41

SAFETY PROFILE: Questionable carcinogen with experimental tumorigenic data. When heated to decomposition it emits acrid smoke and irritating fumes.

EHE000 CAS:105-54-4 ***HR: 3***
ETHYL n-BUTYRATE
DOT: UN 1180
mf: $C_6H_{12}O_2$ mw: 116.18

PROP: Colorless liquid; banana-pineapple odor. D: 0.874, refr index: 1.391, mp: −100.8°, bp: 121.6°, flash p: 79°F. Sol in water, fixed oils, propylene glycol; misc in alc and ether; insol in glycerin @ 121°.

SYNS: BUTANOIC ACID ETHYL ESTER ◇ BUTYRIC ETHER ◇ ETHYL BUTANOATE ◇ ETHYL BUTYRATE (DOT,FCC) ◇ FEMA No. 2427

TOXICITY DATA with REFERENCE
skn-rbt 500 mg/24H MOD FCTXAV 12,703,74
orl-rat LD50:13 g/kg FCTXAV 2,327,64
orl-rbt LD50:5228 mg/kg IMSUAI 41,31,72

CONSENSUS REPORTS: Reported in EPA TSCA Inventory.

DOT Classification: Flammable or Combustible Liquid; Label: Flammable Liquid.

SAFETY PROFILE: Mildly toxic by ingestion. A skin irritant. Flammable liquid when exposed to heat or flame; can react vigorously with oxidizing materials. When heated to decomposition it emits acrid smoke and irritating fumes. See also ESTERS.

EHE500 CAS:110-38-3 ***HR: 1***
ETHYL CAPRATE
mf: $C_{12}H_{24}O_2$ mw: 200.36

PROP: Colorless liquid; oily, brandy odor. Bp: 243°, d: 0.863, refr index: 1.424, vap d: 6.9, flash p: +212°F. Sol in fixed oils; insol in glycerin, propylene glycol @ 243°.

SYNS: CAPRIC ACID ETHYL ESTER ◇ DECANOIC ACID, ETHYL ESTER ◇ ETHYL CAPRINATE ◇ ETHYL DECANOATE (FCC) ◇ ETHYL DECYLATE ◇ FEMA No. 2432

TOXICITY DATA with REFERENCE
skn-rbt 500 mg/24H MLD FCTXAV 16,733,78

CONSENSUS REPORTS: Reported in EPA TSCA Inventory.

SAFETY PROFILE: A skin irritant. Combustible liquid when exposed to heat or flame; can react with oxidizing materials. When heated to decomposition it emits acrid smoke and irritating fumes. See ESTERS and ETHERS.

EHF000 CAS:123-66-0 ***HR: 2***
ETHYL CAPROATE
DOT: UN 1177
mf: $C_8H_{16}O_2$ mw: 144.24

PROP: Colorless liquid; mild wine odor. Bp: 163°, flash p: 130°F (OC), d: 0.867-0.871, refr index: 1.406-1.409, vap d: 5.0. Sol in fixed oils; sltly sol in propylene glycol; insol in glycerin @ 166.

SYNS: ETHYL BUTYLACETATE (DOT) ◇ ETHYL HEXANOATE (FCC) ◇ FEMA No. 2439

TOXICITY DATA with REFERENCE
skn-rbt 500 mg/24H MOD FCTXAV 14,659,76

CONSENSUS REPORTS: Reported in EPA TSCA Inventory.

DOT Classification: Flammable or Combustible Liquid; Label: Flammable Liquid.

SAFETY PROFILE: A skin irritant. Flammable or combustible when exposed to heat or flame; can react with oxidizing materials. When heated to decomposition it emits acrid smoke and irritating fumes. To fight fire, use CO_2, foam, dry chemical. See also ESTERS.

EHF500 CAS:63833-90-9 ***HR: 3***
2-(N-ETHYL CARBAMOYL HYDROXYMETHYL)FURAN
mf: $C_8H_{11}NO_3$ mw: 169.20

TOXICITY DATA with REFERENCE
dnd-rat-ipr 70 mg/kg BCPCA6 26,1909,77
dnd-rat:lvr 10 mmol/L BCPCA6 26,1909,77
dnd-mam:lym 10 mmol/L BCPCA6 26,1909,77
ipr-rat LD50:80 mg/kg BCPCA6 26,1909,77

SAFETY PROFILE: Poison by intraperitoneal route. Mutagenic data reported. When heated to decomposition it emits toxic fumes of NO_x.

EHG000 CAS:4114-31-2 ***HR: 3***
ETHYL CARBAZATE
mf: $C_3H_8N_2O_2$ mw: 104.13

SYNS: CARBAZIC ACID, ETHYL ESTER ◇ CARBETHOXYHYDRAZINE ◇ N-(CARBETHOXY)HYDRAZINE ◇ 1-CARBETHOXY HYDRAZINE ◇ CARBOETHOXYHYDRAZINE ◇ ETHOXYCARBONYL HYDRAZIDE ◇ (ETHOXYCARBONYL) HYDRAZINE ◇ N-(ETHOXYCARBONYL) HYDRAZINE ◇ 1-(ETHOXYCARBONYL) HYDRAZINE ◇ ETHYL CARBAZINATE ◇ HYDRAZINECARBOXYLIC ACID, ETHYL ESTER ◇ MONOCARBETHOXYHYDRAZINE

TOXICITY DATA with REFERENCE
ivn-mus LD50:56 mg/kg CSLNX* NX#02783
orl-qal LD50:100 mg/kg EESADV 6,149,82
orl-bwd LD50:23700 μg/kg AECTCV 12,355,83

CONSENSUS REPORTS: Reported in EPA TSCA Inventory.

SAFETY PROFILE: Poison by ingestion and intravenous routes. When heated to decomposition it emits toxic fumes of NO_x. See also HYDRAZINE.

EHG500 CAS:105-39-5 ***HR: 3***
ETHYL CHLORACETATE
DOT: UN 1181
mf: $C_4H_7ClO_2$ mw: 122.56

$$CH_3CH_2OCO \cdot CH_2Cl$$

PROP: Colorless liquid; fruity, pungent odor. Bp: 143.6°, fp: −26.6° flash p: 100°F, d: 1.159 @ 20°/4°, vap press: 10 mm @ 37.5° vap d: 4.3. Insol in water; misc in alc and ether.

SYNS: CHLOROACETIC ACID, ETHYL ESTER ◇ ETHYL CHLOROACETATE ◇ ETHYL-α-CHLOROACETATE ◇ ETHYL CHLOROETHANOATE ◇ ETHYL MONOCHLORACETATE ◇ ETHYL MONOCHLOROACETATE

TOXICITY DATA with REFERENCE
eye-rbt 750 μg SEV AJOPAA 29,1363,46
ipr-mus TDLo:2940 mg/kg/8W-I:NEO CNREA8 39,391,79
scu-mus LD50:250 mg/kg JJPAAZ 3,99,54
skn-rbt LD50:230 mg/kg TXAPA9 42,417,77

CONSENSUS REPORTS: Reported in EPA TSCA Inventory.

DOT Classification: Combustible Liquid; Label: None; Poison B; Label: Flammable Liquid and Poison.

SAFETY PROFILE: Poison by skin contact and subcutaneous routes. A severe eye irritant. Questionable carcinogen with experimental neoplastigenic data. A dangerous fire hazard when exposed to heat or flame; can react vigorously with oxidizing materials. Will react with water or steam to produce toxic and corrosive fumes. Vigorous reaction with sodium cyanide. To fight fire, use water, foam, CO_2, dry chemical. When heated to decomposition it emits highly toxic fumes of Cl^-.

EHH000 CAS:75-00-3 ***HR: 1***
ETHYL CHLORIDE
DOT: UN 1037
mf: C_2H_5Cl mw: 64.52

PROP: Colorless liquid or gas; ether-like odor, burning taste. Sol in water at 0. Bp: 12.3°, lel: 3.8%, uel: 15.4%, fp: −139°, flash p: −58°F (CC), d: 0.9214 @ 0°/4°, autoign temp: 966°F, vap press: 1000 mm @ 20° vap d: 2.22.45; misc in alc and ether.

SYNS: AETHYLCHLORID (GERMAN) ◇ AETHYLIS ◇ AETHYLIS CHLORIDUM ◇ ANODYNON ◇ CHELEN ◇ CHLOORETHAAN (DUTCH) ◇ CHLORETHYL ◇ CHLORIDUM ◇ CHLOROAETHAN (GERMAN) ◇ CHLOROETHANE ◇ CHLORURE d'ETHYLE (FRENCH) ◇ CHLORYL ◇ CHLORYL ANESTHETIC ◇ CLOROETANO (ITALIAN) ◇ CLORURO DI ETILE (ITALIAN) ◇ ETHER CHLORATUS ◇ ETHER HYDROCHLORIC ◇ ETHER MURIATIC ◇ ETYLU CHLOREK (POLISH) ◇ HYDROCHLORIC ETHER ◇ KELENE ◇ MONOCHLORETHANE ◇ MURIATIC ETHER ◇ NARCOTILE ◇ NCI-C06224

TOXICITY DATA with REFERENCE
ihl-rat LC50:160 g/m^3/2H 85GMAT -,66,82
ihl-mus LC50:146 g/m^3/2H 85GMAT -,66,82
ihl-gpg LCLo:40000 ppm/45M XPHBAO 185,1,29

CONSENSUS REPORTS: Reported in EPA TSCA Inventory. Community Right-To-Know List.

OSHA PEL: TWA 1000 ppm
ACGIH TLV: TWA 1000 ppm
DFG MAK: 1000 ppm (2600 mg/m^3)
NIOSH REL: (Chloroethane): Handle with catuion
DOT Classification: Flammable Liquid; Label: Flammable Liquid; Flammable Gas; Label; Flammable Gas.

SAFETY PROFILE: Mildly toxic by inhalation. An irritant to skin, eyes, and mucous membranes. The liquid is harmful to the eyes and can cause some irritation. In the case of guinea pigs, the symptoms attending exposure are similar to those caused by methyl chloride, except that the signs of lung irritation are not as pronounced. It gives some warning of its presence because it is irritating, but it is possible to tolerate exposure to it until one becomes unconscious. It is the least toxic of all the chlorinated hydrocarbons. It can cause narcosis, although the effects are usually transient.

A very dangerous fire hazard when exposed to heat or flame; can react vigorously with oxidizing materials. Severe explosion hazard when exposed to flame. Reacts with water or steam to produce toxic and corrosive fumes. Incompatible with potassium. To fight fire, use carbon dioxide. When heated to decomposition it emits toxic fumes of phosgene and Cl^-. See also CHLORINATED HYDROCARBONS, ALIPHATIC.

EHH500 CAS:1331-31-3 ***HR: 1***
ETHYL CHLORO BENZENE
mf: C_8H_9Cl mw: 140.62

PROP: Clear, colorless liquid. Mp: −62.6°, bp: 184.3°, flash p: 147°F, d: 1.05 @ 25°/25°, vap press: 1 mm @ 19.2°, vap d: 4.86.

SYN: CHLOROETHYLBENZENE

TOXICITY DATA with REFERENCE
skn-rbt 10 mg/24H open JIHTAB 30,63,48
eye-rbt 500 mg AJOPAA 29,1363,46
orl-rat LD50:5000 mg/kg JIHTAB 30,63,48
skn-rbt LD50:18 g/kg JIHTAB 30,63,48

SAFETY PROFILE: Mildly toxic by ingestion and skin contact. A skin and eye irritant. Flammable when exposed to heat or flame; can react vigorously with oxidizing materials. To fight fire, use foam, CO_2, dry chemical. When heated to decomposition it emits acrid smoke and irritating fumes. See also CHLORINATED HYDROCARBONS, AROMATIC, and CHLOROBENZENE.

EHI500 CAS:4310-69-4 ***HR: 3***
7-(2-(ETHYL-2-CHLOROETHYL)AMINOETHYLAMINO)BENZ(c)ACRIDINE DIHYDROCHLORIDE
mf: $C_{23}H_{24}ClN_3$•2ClH mw: 450.87

SYNS: N'-BENZ(c)ACRIDIN-7-YL-N-(2-CHLOROETHYL)-N-ETHYL-1,2-ETHANEDIAMINE DIHYDROCHLORIDE ◇ 7-(2-(2-CHLOROETHYLETHYLAMINO)ETHYLAMINO)BENZ(c)ACRIDINEDIHYDROCHLORIDE ◇ ICR 311

TOXICITY DATA with REFERENCE
ivn-mus TDLo:4500 μg/kg:NEO CNREA8 36,2423,76
ivn-mus LDLo:4500 μg/kg CNREA8 36,2423,76

SAFETY PROFILE: Questionable carcinogen with experimental neoplastigenic data. Poison by intravenous route. When heated to decomposition it emits very toxic fumes of NO_x and Cl^-.

EHJ000 CAS:4251-89-2 ***HR: 3***
7-(3-(ETHYL-2-(CHLOROETHYLAMINO)PROPYLAMINO))BENZ(c)ACRIDINE DIHYDROCHLORIDE
mf: $C_{24}H_{26}ClN_3$•2ClH mw: 464.90

SYN: ICR 292

TOXICITY DATA with REFERENCE
mmo-sat 500 ng/plate MUREAV 136,185,84
pic-esc 60 ng/plate CNREA8 43,2819,83
msc-ham:ovr 1 g/L CNREA8 39,4875,79
ivn-mus TDLo:4650 μg/kg:NEO CNREA8 36,2423,76
ivn-mus LDLo:4650 μg/kg CNREA8 36,2423,76

CONSENSUS REPORTS: EPA Genetic Toxicology Program.

SAFETY PROFILE: Poison by intravenous route. Questionable carcinogen with experimental neoplastigenic data. Mutation data reported. When heated to decomposition it emits very toxic fumes of NO_x and Cl^-.

EHJ500 CAS:38915-14-9 ***HR: 3***
9-(3-ETHYL-2-CHLOROETHYL)AMINOPROPYLAMINO)-4-METHOXYACRIDINE DIHYDROCHLORIDE
mf: $C_{21}H_{26}ClN_3O{\bullet}2ClH$ mw: 444.87

SYNS: ICR 377 ◇ 4-METHOXY-9-(3-(ETHYL-2-CHLOROETHYL)

TOXICITY DATA with REFERENCE
mmo-sat 500 ng/plate MUREAV 136,185,84
ivn-mus TDLo:2200 μg/kg:NEO CNREA8 36,2423,76
ivn-mus LDLo:2 mg/kg CNREA8 36,2423,76

SAFETY PROFILE: Poison by intravenous route. Questionable carcinogen with experimental neoplastigenic data. Mutation data reported. When heated to decomposition it emits very toxic fumes of Cl^- and NO_x.

EHK000 CAS:63918-55-8 ***HR: 3***
ETHYL-β-CHLOROETHYLETHYLENIMONIUM PICRYLSULFONATE
mf: $C_6H_{13}ClN{\bullet}C_6H_2N_3O_9S$ mw: 426.82

SYNS: ETHYL(2-CHLOROETHYL)ETHYLENIMONIUMPICRYLSULFONATE ◇ 1-ETHYL-1-(β-CHLOROETHYL)ETHYLENIMONIUM PICRYLSULFONATE

TOXICITY DATA with REFERENCE
ivn-rat LD50:500 μg/kg JPETAB 91,224,47
scu-mus LD50:2 mg/kg JPETAB 91,224,47
ivn-rbt LD50:3 mg/kg JPETAB 91,224,47

SAFETY PROFILE: Poison by intravenous and subcutaneous routes. When heated to decomposition it emits very toxic fumes of Cl^-, NO_x and SO_x.

EHK300 ***HR: 3***
ETHYL-β-CHLOROETHYL-β-HYDROXYETHYLAMINE PICRYLSULFONATE
mf: $C_6H_{14}ClNO{\bullet}C_6H_3N_3O_9S$ mw: 444.84

SYNS: 2-((2-CHLOROETHYL)ETHYLAMINO)ETHANOL with 2,4,6-TRINITROBENZENESULFONIC ACID ◇ HN1 CHLOROHYDRIN

TOXICITY DATA with REFERENCE
ipr-mus LDLo:40 mg/kg NTIS** PB158-507
scu-mus LDLo:5 mg/kg NTIS** PB158-507
ivn-mus LDLo:8 mg/kg NTIS** PB158-507
ivn-rbt LD50:5 mg/kg NTIS** PB158-507

SAFETY PROFILE: Poison by subcutaneous, intravenous, and intraperitoneal routes. When heated to decomposition it emits toxic fumes of Cl^-, SO_x, and NO_x. See also SULFONATES.

EHK500 CAS:541-41-3 ***HR: 3***
ETHYL CHLOROFORMATE
DOT: UN 1182
mf: $C_3H_5ClO_2$ mw: 108.53

PROP: Colorless liquid. Mp: −80.6°, bp: 94°, flash p: 35.6°F, d: 1.138 @ 20°/4°, vap d: 3.74, autoign temp: 932°F. Decomp in water; misc in alc, benzene, ether, and chloroform.

SYNS: CHLOROCARBONATE D'ETHYLE (FRENCH) ◇ CHLOROFORMIC ACID ETHYL ESTER ◇ CLHORAMEISENSAEUREAETHYLESTER (GERMAN) ◇ ECF ◇ ETHYLCHLOORFORMIAAT (DUTCH) ◇ ETHYL CHLOROCARBONATE (DOT) ◇ ETHYLE, CHLOROFORMIAT D' (FRENCH) ◇ ETIL CLOROCARBONATO (ITALIAN) ◇ ETIL CLOROFORMIATO (ITALIAN) ◇ TL 423

TOXICITY DATA with REFERENCE
orl-rat LD50:270 mg/kg TXAPA9 42,417,77
ihl-rat LD50:145 ppm/1H TXAPA9 42,417,77
ihl-mus LCLo:2260 mg/m^3/10M NDRC** NDCrc-132,Nov,42
ipr-mus LDLo:15 mg/kg CBCCT* 6,220,54
skn-rbt LD50:7120 mg/kg TXAPA9 42,417,77

CONSENSUS REPORTS: Reported in EPA TSCA Inventory. Community Right-To-Know List.

DOT Classification: Flammable Liquid; Label: Flammable Liquid and Poison; Flammable Liquid; Label: Flammable Liquid, Poison, Corrosive.

SAFETY PROFILE: Poison by ingestion, inhalation, and intraperitoneal routes. Moderately toxic by skin contact. Corrosive. An eye, skin, and mucous membrane irritant. A very dangerous fire hazard when exposed to heat or flame; can react vigorously with oxidizing materials. Reacts with water or steam to produce toxic and corrosive fumes. To fight fire, use CO_2, dry chemical. When heated to decomposition it emits highly toxic fumes of Cl^-.

EHL000 CAS:63019-53-4 ***HR: 3***
7-ETHYL-10-CHLORO-11-METHYLBENZ(c)ACRIDINE
mf: $C_{20}H_{16}ClN$ mw: 305.82

SYN: 2-CHLORO-1-METHYL-10-ETHYL-7,8-BENZACRIDINE (FRENCH)

TOXICITY DATA with REFERENCE
scu-mus TDLo:10 mg/kg:ETA AICCA6 11,736,55

SAFETY PROFILE: Questionable carcinogen with experimental tumorigenic data. When heated to decomposition it emits very toxic fumes of Cl^- and NO_x.

EHL500 CAS:10443-70-6 ***HR: 2***
ETHYL-4-(4-CHLORO-2-METHYLPHENOXY)BUTYRATE
mf: $C_{13}H_{17}ClO_3$ mw: 256.75

SYNS: ETHYL4-(4-CHLORO-2-METHYLPHENOXY)BUTYLATE ◇ MCPB-ETHYL

TOXICITY DATA with REFERENCE
orl-rat LD50:1570 mg/kg FMCHA2 -,C147,83
skn-rat LD50:4000 mg/kg FMCHA2 -,C147,83

SAFETY PROFILE: Moderately toxic by ingestion and skin contact. An herbicide. When heated to decomposition it emits toxic fumes of Cl^-. See also ESTERS.

EHM000 CAS:7511-54-8 ***HR: 3***
3-ETHYLCHOLANTHRENE
mf: $C_{22}H_{18}$ mw: 282.40

SYNS: 3-ETHYL-CHOLANTHRENE ◇ 20-ETHYLCHOLANTHRENE

TOXICITY DATA with REFERENCE
scu-mus TDLo:400 mg/kg:ETA AJCAA7 33,499,38

SAFETY PROFILE: Questionable carcinogen with experimental tumorigenic data. When heated to decomposition it emits acrid smoke and irritating fumes.

EHN000 CAS:103-36-6 ***HR: 2***
ETHYL-trans-CINNAMATE
mf: $C_{11}H_{12}O_2$ mw: 176.23

PROP: Nearly colorless, oily liquid; faint cinnamon odor. D: 1.049 @ 20°/4°, refr index: 1.558-1.561, bp: 271°, mp: 9°, flash p: +212°F. Misc in alc, ether, fixed oils; insol in glycerin, water @ 272°.

SYNS: ETHYL CINNAMATE (FCC) ◇ ETHYL-β-PHENYLACRYLATE ◇ ETHYL-3-PHENYLPROPENOATE ◇ FEMA No. 2430

TOXICITY DATA with REFERENCE
orl-rat LD50:4000 mg/kg VPITAR 33(5),48,74
orl-mus LD50:4000 mg/kg VPITAR 33(5),48,74
orl-gpg LD50:4000 mg/kg VPITAR 33(5),48,74

CONSENSUS REPORTS: Reported in EPA TSCA Inventory.

SAFETY PROFILE: Moderately toxic by ingestion. Combustible liquid. When heated to decomposition it emits acrid smoke and irritating fumes. See also ESTERS.

EHN500 CAS:41448-29-7 ***HR: 2***
ETHYL CITRAL
mf: $C_{11}H_{18}O$ mw: 166.29

SYN: 3,7-DIMETHYL-2,6-NONADIEN-1-AL

TOXICITY DATA with REFERENCE
skn-rbt 500 mg/24H SEV FCTXAV 16,637,78

CONSENSUS REPORTS: Reported in EPA TSCA Inventory.

SAFETY PROFILE: A severe skin irritant. When heated to decomposition it emits acrid smoke and irritating fumes.

EHO000 CAS:19780-25-7 ***HR: 2***
2-ETHYLCROTONALDEHYDE
mf: $C_6H_{10}O$ mw: 98.16

SYN: 2-ETHYL-2-BUTENAL

TOXICITY DATA with REFERENCE
orl-rat LD50:2600 mg/kg TXAPA9 28,313,74
skn-rbt LD50:500 mg/kg TXAPA9 28,313,74

CONSENSUS REPORTS: Reported in EPA TSCA Inventory.

SAFETY PROFILE: Moderately toxic by ingestion and skin contact. When heated to decomposition it emits acrid smoke and irritating fumes. See also ALDEHYDES and ESTERS.

EHO200 CAS:10544-63-5 ***HR: 3***
ETHYL CROTONATE
mf: $C_6H_{10}O_2$ mw: 114.16

PROP: Bp: 142-143°, d: 0.918, Flash p: 36° F.

SYNS: 2-BUTENOIC ACID, ETHYL ESTER ◇ CROTONIC ACID, ETHYL ESTER ◇ ETHYLESTER KYSELINY KROTONOVE

TOXICITY DATA with REFERENCE
eye-rbt 20 mg/24H MOD 85JCAE-,369,86
orl-rat LD50:3 g/kg NEZAAQ 34,183,79

CONSENSUS REPORTS: Reported in EPA TSCA Inventory.

SAFETY PROFILE: Slightly toxic by ingestion. Corrosive. An eye irritant and lachrymator. When heated to decomposition it emits acrid smoke and irritating fumes.

EHO500 CAS:483-63-6 ***HR: 2***
N-ETHYL-o-CROTONOTOLUIDINE
mf: $C_{13}H_{17}NO$ mw: 203.31

SYNS: CROTONYL-N-ETHYL-o-TOLUIDINE ◇ N-ETHYL-o-CROTONOTOLUIDIDE

TOXICITY DATA with REFERENCE
orl-rat LD50:1500 mg/kg 29ZVAB -,35,69
orl-mus LD50:1600 mg/kg 29ZVAB -,35,69

SAFETY PROFILE: Moderately toxic by ingestion.

When heated to decomposition it emits toxic fumes of NO_x.

EHO700 CAS:2884-67-5 ***HR: 2***
2-ETHYL-trans-CROTONYLUREA
mf: $C_7H_{12}N_2O_2$ mw: 156.21

SYNS: ECTYLUREA ◇ trans-1-(2-ETHYLCROTONOYL)UREA ◇ MA-117

TOXICITY DATA with REFERENCE
orl-rat LD50:1130 mg/kg AIPTAK 114,418,58
ipr-rat LD50:575 mg/kg AIPTAK 114,418,58
orl-mus LD50:1280 mg/kg FRPSAX 14,418,58
ipr-mus LD50:990 mg/kg APSXAS 5,293,68
orl-dog LD50:1100 mg/kg AIPTAK 114,418,58

SAFETY PROFILE: Moderately toxic by ingestion and intraperitoneal routes. When heated to decomposition it emits toxic fumes of NO_x.

EHP000 CAS:95-04-5 ***HR: 2***
cis-(2-ETHYLCROTONYL) UREA
mf: $C_7H_{12}N_2O_2$ mw: 156.21

SYNS: ACTINE ◇ (Z)-N-(AMINOCARBONYL)-2-ETHYL-2-BUTENAMIDE ◇ ASTYN ◇ CRONIL ◇ CROTURAL ◇ DISTESOL ◇ DISTESSOL ◇ ECTIDA ◇ ECTILUREA ◇ ECTON ◇ ECTYDA ◇ ECTYLCARBAMIDE ◇ ECTYLUREA ◇ ECTYN ◇ EKTYLCARBAMID ◇ (α-ETHYL-cis-CROTONYL)CARBAMIDE ◇ 2-ETHYL-cis-CROTONYLUREA ◇ 2-ETHYLCROTONYLUREA ◇ EUPLACID ◇ LEVANIL ◇ LEVIL ◇ MA-110 ◇ NASTYN ◇ NEOCROSEDIN ◇ NESTYN ◇ NEUROPROCIN ◇ NOSTAL ◇ NOSTIN ◇ PACETYN ◇ SEDAREX ◇ TRANZER ◇ U 8771

TOXICITY DATA with REFERENCE
orl-rat LD50:2500 mg/kg MEIEDD 10,507,83
ipr-rat LD50:900 mg/kg FEPRA7 12,357,53
orl-mus LD50:2500 mg/kg FRPSAX 14,845,59
ipr-mus LD50:1780 mg/kg 27ZQAG -,425,72
orl-dog LD50:2800 mg/kg AIPTAK 114,418,58
ipr-dog LD50:900 mg/kg FEPRA7 12,357,53
orl-rbt LD50:1200 mg/kg 27ZQAG -,425,72
ipr-rbt LD50:1400 mg/kg FEPRA7 12,357,53
ipr-gpg LD50:1100 mg/kg FEPRA7 12,357,53

SAFETY PROFILE: Moderately toxic by ingestion and intraperitoneal routes. A sedative. When heated to decomposition it emits toxic fumes of NO_x.

EHP500 CAS:105-56-6 ***HR: 2***
ETHYL CYANOACETATE
DOT: UN 2666
mf: $C_5H_7NO_2$ mw: 113.13

PROP: Colorless to pale straw-colored liquid. Mp: −22.5°, bp: 206°, flash p: 230°F, d: 1.06 @ 25°/25°, vap press: 1 mm @ 67.8°, vap d: 3.9.

SYNS: CYANACETATE ETHYLE (GERMAN) ◇ CYANOACETIC ACID ETHYL ESTER ◇ CYANOACETIC ESTER ◇ ETHYL CYANOETHANOATE ◇ MALONIC ACID ETHYL ESTER NITRILE ◇ USAF KF-25

TOXICITY DATA with REFERENCE
ipr-mus LD50:500 mg/kg NTIS** AD277-689
scu-rbt LDLo:1500 mg/kg AIPTAK 5,161,1899
scu-frg LDLo:4000 mg/kg AIPTAK 5,161,1899

CONSENSUS REPORTS: Reported in EPA TSCA Inventory. Cyanide and its compounds are on the Community Right-To-Know List.

DOT Classification: Poison B; Label: St. Andrews Cross.

SAFETY PROFILE: Moderately toxic by intraperitoneal and subcutaneous routes. Combustible when exposed to heat or flame; can react with oxidizing materials. Will react with water or steam to produce toxic and flammable vapors. To fight fire, use CO_2, dry chemical. When heated to decomposition or on contact with acid or acid fumes it emits highly toxic fumes of CN^-. See also NITRILES.

EHQ000 CAS:1331-45-9 ***HR: 1***
ETHYLCYANOCYCLOHEXYL ACETATE
mf: $C_{11}H_{17}NO_2$ mw: 195.29

SYN: ETHYLHYDROXYCYCLOHEXANECARBONITRILEACETATE (ESTER)

TOXICITY DATA with REFERENCE
orl-rat LD50:6400 mg/kg JPETAB 93,26,48
orl-mus LD50:5000 mg/kg JPETAB 93,26,48

CONSENSUS REPORTS: Cyanide and its compounds are on the Community Right-To-Know List.

SAFETY PROFILE: Mildly toxic by ingestion. When heated to decomposition it emits toxic fumes of NO_x and CN^-. See also NITRILES and ESTERS.

EHQ500 CAS:148-87-8 ***HR: 2***
N-ETHYL-N-(2-CYANOETHYL)ANILINE
mf: $C_{10}H_{14}N_2$ mw: 162.26

TOXICITY DATA with REFERENCE
orl-rat LD50:1700 mg/kg MarJV# 29MAR77
orl-mus LD50:1560 mg/kg GTPZAB 25(8),50,81

CONSENSUS REPORTS: Reported in EPA TSCA Inventory. Cyanide and its compounds are on the Community Right-To-Know List.

SAFETY PROFILE: Moderately toxic by ingestion. When heated to decomposition it emits very toxic fumes of NO_x and CN^-. See also CYANIDE.

EHR000 CAS:4806-61-5 ***HR: 3***
ETHYL CYCLOBUTANE
mf: C_6H_{12} mw: 84.158

$CH_3CH_2CH(CH_2)_2CH_2$

PROP: Insol in water, flash p: <5°F, d: 0.7284 @ 20°/4°, autoign temp: 410°F, lel: 1.2%, uel: 7.7%, mp: −142.4°, bp: 70.7°.

SAFETY PROFILE: Probably an asphyxiant; see ARGON for a description of simple asphyxiants. A very dangerous fire and explosion hazard when exposed to heat or flame; can react vigorously with oxidizing materials. Easily forms explosive mixtures in air. To fight fire, use foam, spray, mist, dry chemical. When heated to decomposition it emits acrid smoke and fumes.

EHR500 CAS:1940-18-7 ***HR: 2***
1-ETHYLCYCLOHEXANOL
mf: $C_8H_{16}O$ mw: 128.24

SYN: 1-AETHYL-CYCLOHEXANOL-(1)(GERMAN)

TOXICITY DATA with REFERENCE
orl-mus LD50:1300 mg/kg ARZNAD 4,477,54
scu-mus LD50:840 mg/kg ARZNAD 5,161,55

SAFETY PROFILE: Moderately toxic by ingestion and subcutaneous routes. When heated to decomposition it emits acrid smoke and irritating fumes.

EHS000 CAS:21722-83-8 ***HR: 2***
ETHYLCYCLOHEXYL ACETATE
mf: $C_{10}H_{18}O_2$ mw: 170.28

SYNS: ACETIC ACID, CYCLOHEXYLETHYL ESTER ◇ CYCLOHEXANEETHANOL, ACETATE ◇ CYCLOHEXANE ETHYL ACETATE ◇ CYCLOHEXYLETHYL ACETATE ◇ HEXAHYDROPHENYL ETHYL ACETATE

TOXICITY DATA with REFERENCE
orl-rat LD50:3200 mg/kg FCTXAV 13,783,75
skn-rbt LD50:5000 mg/kg FCTXAV 13,783,75

CONSENSUS REPORTS: Reported in EPA TSCA Inventory.

SAFETY PROFILE: Moderately toxic by ingestion. Mildly toxic by skin contact. When heated to decomposition it emits acrid smoke and irritating fumes. See also ESTERS.

EHT000 CAS:5459-93-8 ***HR: 2***
N-ETHYL(CYCLOHEXYL)AMINE
mf: $C_8H_{17}N$ mw: 127.26

PROP: Sltly sol in water; flash p: 86°F (OC), d: 0.8, vap d: 4.4.

SYN: N-ETHYL-CYCLOHEXYLAMINE

TOXICITY DATA with REFERENCE
skn-rbt 100 μg/24H open AIHAAP 23,95,62
orl-rat LD50:590 mg/kg AIHAAP 23,95,62
ihl-rat LCLo:500 ppm/4H AIHAAP 23,95,62
skn-rbt LD50:750 mg/kg AIHAAP 23,95,62

CONSENSUS REPORTS: Reported in EPA TSCA Inventory.

SAFETY PROFILE: Moderately toxic by ingestion, inhalation, and skin contact. A skin irritant. A very dangerous fire hazard when exposed to heat or flame; can react vigorously with oxidizing materials. To fight fire, use alcohol foam, mist, spray, dry chemical. See also AMINES.

EHT500 CAS:1134-23-2 ***HR: 2***
S-ETHYL CYCLOHEXYLETHYLTHIOCARBAMATE
mf: $C_{11}H_{21}NOS$ mw: 215.39

SYNS: CYCLOATE ◇ CYCLOHEXYLETHYLCARBAMOTHIOIC ACID-S-ETHYL ESTER ◇ CYCLOHEXYLETHYLTHIOCARBAMIC ACID-S-ETHYL ESTER ◇ S-ETHYL-N-ETHYL-N-CYCLOHEXYLTHIOLCARBAMATE ◇ EUREX ◇ HEXYLTHIOCARBAM ◇ R 2063 ◇ RO-NEET ◇ RONIT

TOXICITY DATA with REFERENCE
cyt-mus-unr 200 mg/kg TGANAK 16(1),45,82
orl-rat LD50:2000 mg/kg FMCHA2 -,C207,83
orl-mus LD50:2265 mg/kg GISAAA 50(3),26,85
skn-rbt LD50:3 g/kg GISAAA 50(3),26,85

SAFETY PROFILE: Moderately toxic by ingestion. Mutation data reported. An herbicide. When heated to decomposition it emits very toxic fumes of SO_x and NO_x. See also CARBAMATES.

EHU000 CAS:1640-89-7 ***HR: 3***
ETHYL CYCLOPENTANE
mf: C_7H_{14} mw: 98.16

$CH_3CH_2CH(CH_2)_3CH_2$

PROP: Autoign temp: 504°F, lel: 1.1%, uel: 6.7%, d: 0.8, vap d: 0.8, vap d: 3.4, bp: 104°, flash p: <69.8°F.

SAFETY PROFILE: Probably an asphyxiant. See ARGON for a description of simple asphyxiants. A very dangerous fire and explosion hazard when exposed to heat or flame; can react vigorously with oxidizing materials. Easily forms explosive mixtures in air. To fight fire, use foam, dry chemical, mist. When heated to decomposition it emits acrid smoke and fumes.

EHU500 CAS:1191-96-4 ***HR: 2***
ETHYL CYCLOPROPANE
mf: C_5H_{10} mw: 70.14

$CH_3CH_2CHCH_2CH_2$

PROP: Flash p: <50°F.

SAFETY PROFILE: A very dangerous fire hazard when exposed to heat or flame; can react vigorously with oxidizers. When heated to decomposition it emits acrid smoke and fumes.

EHU600 CAS:868-59-7 ***HR: 2***
ETHYL CYSTEINE HYDROCHLORIDE
mf: $C_5H_{11}NO_2S{\bullet}ClH$ mw: 185.69

SYNS: CYSTEINE ETHYL ESTER HYDROCHLORIDE ◇ ETHYL ESTER-1-CYSTEINE HYDROCHLORIDE (9CI)

TOXICITY DATA with REFERENCE
orl-rat LD50:5 g/kg NIIRDN 6,113,82
scu-rat LD50:2120 mg/kg NIIRDN 6,113,82
ivn-rat LD50:725 mg/kg NIIRDN 6,113,82
orl-mus LD50:3470 mg/kg NIIRDN 6,113,82
ipr-mus LD50:854 mg/kg CPBTAL 20,721,72
scu-mus LD50:1550 mg/kg NIIRDN 6,113,82
ivn-mus LD50:688 mg/kg NIIRDN 6,113,82

CONSENSUS REPORTS: Reported in EPA TSCA Inventory.

SAFETY PROFILE: Moderately toxic by ingestion, intraperitoneal, intravenous, and subcutaneous routes. When heated to decomposition it emits toxic fumes of SO_x, NO_x, and HCl. See also 1-CYSTEINE and ESTERS.

EHV000 CAS:26747-87-5 ***HR: 3***
ETHYL DECABORANE
mf: $C_2H_{18}B_{10}$ mw: 150.30

TOXICITY DATA with REFERENCE
ihl-rat LC50:23 ppm/4H NTIS** AD224-006
ihl-mus LC50:6.5 ppm/4H NTIS** AD224-006
ihl-dog LCLo:14 ppm/3H NTIS** AD224-006

SAFETY PROFILE: Poison by inhalation. See also BORON COMPOUNDS and BORANES.

EHV500 CAS:302-49-8 ***HR: 3***
ETHYL(DI-(1-AZIRIDINYL)PHOSPHINYL)CARBAMATE
mf: $C_7H_{14}N_3O_3P$ mw: 219.21

SYNS: (BIS(1-AZIRIDINYL)PHOSPHINYL)CARBAMIC ACID, ETHYL ESTER ◇ BIS(ETHYLENIMIDO)PHOSPHORYLURETHAN ◇ ETHYL (BIS(1-AZIRIDINYL)PHOSPHINYL)CARBAMATE ◇ NSC 37095

TOXICITY DATA with REFERENCE
ipr-rat LD50:47 mg/kg JMCMAR 8,167,65
ipr-mus LD50:50 mg/kg JMCMAR 8,167,65

SAFETY PROFILE: Poison by intraperitoneal route. When heated to decomposition it emits very toxic fumes of PO_x and NO_x. See also CARBAMATES.

EHW000 CAS:63021-33-0 ***HR: 3***
1-ETHYLDIBENZ(a,h)ACRIDINE
mf: $C_{23}H_{17}N$ mw: 307.41

SYN: 1'-ETHYL-1,2,5,6-DIBENZACRIDINE(FRENCH)

TOXICITY DATA with REFERENCE
skn-mus TDLo:780 mg/kg/26W-I:ETA BAFEAG 42,186,55

SAFETY PROFILE: Questionable carcinogen with experimental tumorigenic data. When heated to decomposition it emits toxic fumes of NO_x.

EHW500 CAS:63021-35-2 ***HR: 3***
1-ETHYL-DIBENZ(a,j)ACRIDINE
mf: $C_{23}H_{17}N$ mw: 307.41

SYN: 1'-ETHYL-3,4,5,6-DIBENZACRIDINE(FRENCH)

TOXICITY DATA with REFERENCE
skn-mus TDLo:384 mg/kg/32W-I:ETA ACRSAJ 4,315,56

SAFETY PROFILE: Questionable carcinogen with experimental tumorigenic data. When heated to decomposition it emits toxic fumes of NO_x.

EHX000 CAS:73927-60-3 ***HR: 3***
8-ETHYL DIBENZ(a,h)ACRIDINE
mf: $C_{23}H_{17}N$ mw: 307.41

SYN: 1''-ETHYLDIBENZ(a,h)ACRIDINE

TOXICITY DATA with REFERENCE
skn-mus TDLo:312 mg/kg/26W-I:ETA ACRSAJ 4,315,56

SAFETY PROFILE: Questionable carcinogen with experimental tumorigenic data. When heated to decomposition it emits toxic fumes of NO_x.

EHY000 CAS:30812-87-4 ***HR: 2***
ETHYL DIBROMOBENZENE
mf: $C_8H_8Br_2$ mw: 263.98

SYN: ALKAZENE 42

TOXICITY DATA with REFERENCE
orl-rat LDLo:12283 mg/kg WADTAA 55-251,23,55
orl-rbt LDLo:5536 mg/kg WADTAA 55-251,23,55
ihl-rbt LCLo:731 mg/m³/7H WADTAA 55-251,28,55

CONSENSUS REPORTS: Reported in EPA TSCA Inventory.

SAFETY PROFILE: Moderately toxic by ingestion and inhalation. When heated to decomposition it emits toxic fumes of Br^-.

EHY500 CAS:1331-29-9 ***HR: 2***
ETHYLDICHLOROBENZENE
mf: $C_8H_8Cl_2$ mw: 175.06

PROP: Colorless liquid. Mp: <70°, bp: 220-224°, flash p: 205°F, d: 1.208 @ 25°/25°, vap d: 6.05.

SYN: DICHLOROETHYLBENZENE

TOXICITY DATA with REFERENCE
orl-rat LD50:5500 mg/kg JIHTAB 30,63,48
skn-rbt LD50:14 g/kg JIHTAB 30,63,48

SAFETY PROFILE: Moderately toxic by ingestion. Mildly toxic by skin contact. Combustible when exposed to heat or flame; can react vigorously with oxidizing materials. To fight fire, use water, foam, CO_2, dry chemical. When heated to decomposition it emits toxic fumes of Cl^-. See also CHLORINATED HYDROCARBONS, AROMATIC.

EHY600 CAS:533-23-3 ***HR: D***
ETHYL (2,4-DICHLOROPHENOXY)ACETATE
mf: $C_{10}H_{10}Cl_2O_3$ mw: 249.10

SYNS: ACETIC ACID, (2,4-DICHLOROPHENOXY)-, ETHYL ESTER ◇ BENZENEACETIC ACID, 2,4-DICHLORO-, ETHYL ESTER (9CI) ◇ DICOTOX ◇ 2,4-D ETHYL ESTER ◇ (2,4-DICHLOROPHENOXY)ACETIC ACID ETHYL ESTER ◇ WEEDONE 40 ◇ WEEDONE CONCENTRATE 48

TOXICITY DATA with REFERENCE
scu-mus TDLo:774 mg/kg (female 6-14D post):TER NTIS** PB223-160

SAFETY PROFILE: An experimental teratogen. When heated to decomposition it emits toxic fumes of and Cl^-.

EIA000 CAS:78329-97-2 ***HR: 3***
ETHYL-3-((DIETHYLAMINO)METHYL)-4-HYDROXYBENZOATE
mf: $C_{14}H_{21}NO_3$ mw: 251.36

SYN: 3-((DIETHYLAMINO)METHYL)-4-HYDROXYBENZOICACID ETHYL ESTER

TOXICITY DATA with REFERENCE
orl-mus LDLo:1000 mg/kg ARZNAD 11,85,61
scu-mus LDLo:375 mg/kg ARZNAD 11,85,61

SAFETY PROFILE: Poison by subcutaneous route. Moderately toxic by ingestion. When heated to decomposition it emits toxic fumes of NO_x. See also ESTERS.

EIB000 CAS:3553-80-8 ***HR: 3***
ETHYL-N,N-DIETHYL CARBAMATE
mf: $C_7H_{15}NO_2$ mw: 145.23

TOXICITY DATA with REFERENCE
ipr-mus TDLo:4875 mg/kg/13W-I:ETA JNCIAM 9,35,48
scu-mus LD50:680 mg/kg AJEBAK 45,507,67

SAFETY PROFILE: Moderately toxic by subcutaneous route. Questionable carcinogen with experimental tumorigenic data. When heated to decomposition it emits toxic fumes of NO_x. See also CARBAMATES.

EIB500 CAS:4928-41-0 ***HR: 3***
4'-ETHYL-N,N-DIETHYL-p-(PHENYLAZO)ANILINE
mf: $C_{18}H_{23}N_3$ mw: 281.44

SYN: 4'-ETHYL-N,N-DIMETHYL-4-AMINOAZOBENZENE

TOXICITY DATA with REFERENCE
otr-ham:kdy 2500 μg/L BJCAAI 38,34,78
orl-rat TDLo:7400 mg/kg/26W-C:ETA ARZNAD 12,270,62

SAFETY PROFILE: Questionable carcinogen with experimental tumorigenic data. Mutation data reported. When heated to decomposition it emits toxic fumes of NO_x.

EIC000 CAS:867-13-0 ***HR: 2***
ETHYL (DIETHYLPHOSPHONO)ACETATE
mf: $C_8H_{17}O_5P$ mw: 224.22

SYNS: ETHYL (DIETHOXYPHOSPHINYL)ACETATE ◇ ETHYL DIETHOXYPHOSPHORYL ACETATE ◇ PHOSPHONOACETIC ACID, TRIETHYL ESTER ◇ TRIETHYL PHOSPHONOACETATE

TOXICITY DATA with REFERENCE
ihl-mus LCLo:1180 mg/m³/10M NDRC** NDCrc-132,Nov,42

CONSENSUS REPORTS: Reported in EPA TSCA Inventory.

SAFETY PROFILE: Moderately toxic by inhalation. When heated to decomposition it emits toxic fumes of PO_x. See also ESTERS.

EID000 CAS:389-08-2 ***HR: 3***
1-ETHYL-1,4-DIHYDRO-7-METHYL-4-OXO-1,8-NAPHTHYRIDINE-3-CARBOXYLIC ACID
mf: $C_{12}H_{12}N_2O_3$ mw: 232.26

SYNS: ACIDE1-ETIL-7-METIL-1,8-NAFTIRIDIN-4-ONE-3-CARBOSSILICO (ITALIAN) ◇ ACIDE NALIDIXICO (ITALIAN) ◇ ACIDE NALIDIXIQUE (FRENCH) ◇ BETAXINA ◇ 3-CARBOXY-1-ETHYL-7-METHYL-1,8-NAPHTHIDIN-4-ONE ◇ CHINOIN ◇ CYBIS ◇ 1,4-DIHYDRO-1-ETHYL-7-METHYL-4-OXO-1,8-NAPHTHYRIDINE-3-CARBOXYLIC ACID ◇ DIXIBEN ◇ 1-ETHYL-7-METHYL-1,4-DIHYDRO-1,8-NAPHTHYRIDIN-4-ONE-3-CARBOXYLIC ACID ◇ 1-ETHYL-7-METHYL-1,4-DIHYDRO-1,8-NAPHTHYRIDINE-4-ONE-3-CARBOXYLIC ACID ô 1-ETHYL-7-METHYL-4-OXO-1,4-DIHYDRO-1,8-NAPHTHYRIDINE-3-CARBOXYLIC ACID ◇ EUCISTEN ◇ INNOXALON ◇ KUSNARIN ◇ NA ◇ NALIDIC ACID ◇ NALIDICRON ◇ NALIDIXIC ACID ◇ NALIDIXIN ◇ NALITUCSAN ◇ NARIGIX ◇ NCI-C56199 ◇ NEGRAM ◇ NEVIGRAMON ◇ NICELATE ◇ NOGRAM ◇ NSC-82174 ◇ POLEON ◇ SPECIFEN ◇ URALGIN ◇ URIBEN ◇ URODIXIN ◇ UROMAN ◇ URONEG ◇ WIN 18,320 ◇ WINTOMYLON

TOXICITY DATA with REFERENCE
mmo-bcs 100 mg/L MUREAV 95,191,82
dns-ssp 100 mg/L MGGEAE 187,96,82

sce-chd-unr 50 mg/kg/10D MUREAV 77,371,80
dnr-ham:oth 500 mg/L CRNGDP 5,187,84
orl-rat TDLo:3300 mg/kg (female 7-17D post):REP NKRZAZ 28,484,80
orl-rat TDLo:3300 mg/kg (female 7-17D post):TER NKRZAZ 28,484,80
orl-wmn TDLo:160 mg/kg/2D:CNS,END,SKN BMJOAE 2,1518,77
unr-chd TDLo:1200 mg/kg/20D:BLD 34ZIAG -,414,69
orl-rat LD50:1160 mg/kg NIIRDN 6,540,82
ipr-rat LD50:319 mg/kg AACHAX -,117,70
scu-rat LD50:1584 mg/kg AACHAX -,117,70
ivn-rat LD50:88400 μg/kg AACHAX -,117,70
orl-mus LD50:572 mg/kg TXAPA9 18,185,71
ipr-mus LD50:871 mg/kg AACHAX -,117,70
scu-mus LD50:500 mg/kg TXAPA9 18,185,71
ivn-mus LD50:101 mg/kg AACHAX -,117,70

CONSENSUS REPORTS: EPA Genetic Toxicology Program.

SAFETY PROFILE: Poison by intravenous and intraperitoneal routes. Moderately toxic by ingestion and subcutaneous routes. An experimental teratogen. Human systemic effects: convulsions, hyperglycemia, sweating, and blood changes in children. Experimental reproductive effects. Human mutation data reported. Used as an antibacterial agent and urinary tract antiseptic. When heated to decomposition it emits toxic fumes of NO_x.

EID100 ***HR: 3***
ETHYL-3,4-DIHYDROXYBENZENE SULFONATE
mf: $C_8H_{10}O_5S$ mw: 218.22

$(OH)_2C_6H_3SO_2OCH_2CH_3$

SAFETY PROFILE: Reacts explosively when mixed with acetyl nitrate and oleum. When heated to decomposition it emits toxic fumes of SO_x. See also SULFONATES.

EID500 CAS:2008-41-5 ***HR: 2***
S-ETHYL N,N-DIISOBUTYLTHIOCARBAMATE
mf: $C_{11}H_{23}NOS$ mw: 217.41

SYNS: BIS(2-METHYLPROPYL)CARBAMOTHIOICACID-S-ETHYL ESTER ◇ BUTILATE ◇ BUTYLATE ◇ DIISOBUTYLTHIOCARBAMIC ACID-S-ETHYL ESTER ◇ DIISOCARB ◇ S-ETHYL BIS(2-METHYLPROPYL)CARBAMOTHIOATE ◇ ETHYL-N,N-DIISOBUTYLTHIOCARBAMATE ◇ S-ETHYLDIISOBUTYL THIOCARBAMATE ◇ ETHYL-N,N-DIISOBUTYL THIOLCARBAMATE ◇ R-1910 ◇ STAUFFER R-1910 ◇ SUTAN

TOXICITY DATA with REFERENCE
cyt-mus-unr 1 g/kg TGANAK 14(6),41,80
cyt-mus-orl 1 g/kg CYGEDX 14(6),38,80
orl-rat LD50:4000 mg/kg WRPCA2 9,119,70

CONSENSUS REPORTS: Reported in EPA TSCA Inventory.

SAFETY PROFILE: Moderately toxic by ingestion. Mutation data reported. Used as an herbicide. When heated to decomposition it emits very toxic fumes of NO_x and SO_x. See also CARBAMATES and ESTERS.

EIF000 CAS:77-81-6 ***HR: 3***
ETHYL DIMETHYLAMIDOCYANOPHOSPHATE
mf: $C_5H_{11}N_2O_2P$ mw: 162.15

PROP: A colorless to brownish liquid. Bp: decomp @ 238°, fp: −49.4°, flash p: 172°F, d: 1.073 @ 25°, vap press: 0.07 mm @ 25°, vap d: 5.63.

SYNS: DIMETHYLAMIDOETHOXYPHOSPHORYL CYANIDE ◇ DIMETHYLAMINOCYANPHOSPHORSAEUREAETHYLESTER(GERMAN) ◇ DIMETHYLPHOSPHORAMIDOCYANIDIC ACID, ETHYL ESTER ◇ ETHYL N,N-DIMETHYLAMINO CYANOPHOSPHATE ◇ ETHYL DIMETHYLPHOSPHORAMIDOCYANIDATE ◇ ETHYL-N,N-DIMETHYLPHOSPHORAMIDOCYANIDATE ◇ GA ◇ Le-100 ◇ MCE ◇ T-2104 ◇ TABUN ◇ TL 1578

TOXICITY DATA with REFERENCE
ihl-hmn LCLo:150 mg/m³ SCJUAD 4,33,67
skn-hmn LDLo:23 mg/kg SCJUAD 4,33,67
ivn-hmn LDLo:14 μg/kg SCJUAD 4,33,67
orl-rat LD50:3700 μg/kg NTIS** PB158-508
ihl-rat LC50:304 mg/m³/10M NTIS** PB158-508
skn-rat LDLo:18 mg/kg NTIS** PB158-508
scu-rat LD50:193 μg/kg AIPTAK 262,231,83
ivn-rbt LD50:66 μg/kg NTIS** PB158-508
ims-rat LD50:800 μg/kg JCINAO 37,350,58
ihl-mus LC50:15 mg/m³/30M DEGEA3 15,2179,60
skn-mus LD50:1 mg/kg NTIS** PB158-508
ipr-mus LD50:604 μg/kg 11FYAN 3,69,63
orl-dog LD50:200 μg/kg DEGEA3 15,2179,60
ihl-dog LC50:400 mg/m³/10M NTIS** PB158-508
skn-dog LDLo:30 mg/kg NTIS** PB158-508
ihl-mky LC50:250 mg/m³/10M NTIS** PB158-508
skn-mky LD50:9300 μg/kg NTIS** PB158-508

CONSENSUS REPORTS: EPA Extremely Hazardous Substances List. Cyanide and its compounds are on the Community Right-To-Know List. Reported in EPA TSCA Inventory.

SAFETY PROFILE: Human poison by inhalation, skin contact, and intravenous routes. Experimental poison by ingestion, inhalation, skin contact, subcutaneous, intravenous, intraperitoneal, and intramuscular routes. A nerve gas. Vapor does not penetrate skin; liquid does so rapidly. The primary physiological action is on the sympathetic nervous system, causing a vasoparesis (partial paralysis of the vasomotor nerves which control the diameter of the blood vessels). Vapors when inhaled can cause nausea, vomiting, and diarrhea, which can be followed by muscular twitching and convulsions. Flammable when exposed to heat or flame; can react with oxidizing materials. When heated to decomposition it emits very toxic fumes of PO_x, CN^-, and NO_x. See also PARATHION and CYANIDE.

EIF450 CAS:93023-34-8 ***HR: 2***
2'-ETHYL-4-DIMETHYLAMINOAZOBENZENE
mf: $C_{16}H_{19}N_3$ mw: 253.38

SYNS: ANILINE,p-((o-ETHYLPHENYL)AZO)-N,N-DIMETHYL- ◇ BENZENAMINE, N,N-DIMETHYL-2'-ETHYL-4-(PHENYLAZO)- ◇ BENZENAMINE, 4-((2-ETHYLPHENYL)AZO)-N,N-DIMETHYL- ◇ N,N-DIMETHYL-p-((o-ETHYLPHENYL)AZO)ANILINE

TOXICITY DATA with REFERENCE
orl-rat TDLo:13541 mg/kg/52W C:CAR CBINA8 53,107,85

SAFETY PROFILE: Questionable carcinogen with experimental carcinogenic data. When heated to decomposition it emits toxic fumes of NO_x.

EIF500 CAS:20820-80-8 ***HR: 3***
O-ETHYL-S-(2-DIMETHYL AMINO ETHYL)-METHYLPHOSPHONOTHIOATE
mf: $C_7H_{18}NO_2PS$ mw: 211.29

SYN: O-AETHYL-S-(2-DIMETHYLAMINOAETHYL)-METHYLPHOSPHONOTHIOATE(GERMAN)

TOXICITY DATA with REFERENCE
orl-rat LD50:122 μg/kg ABMGAJ 37,633,78
ipr-rat LD50:54 μg/kg ABMGAJ 37,633,78
ivn-rat LD50:17 μg/kg ABMGAJ 37,633,78
ims-rat LD50:24 μg/kg ABMGAJ 37,633,78

SAFETY PROFILE: Poison by ingestion, intraperitoneal, intravenous, and intramuscular routes. When heated to decomposition it emits very toxic fumes of NO_x, PO_x, and SO_x.

EIG000 CAS:50782-69-9 ***HR: 3***
ETHYL-S-DIMETHYLAMINOETHYL METHYLPHOSPHONOTHIOLATE
mf: $C_{11}H_{26}NO_2PS$ mw: 267.41

SYNS: S-(2-DIISOPROPYLAMINOETHYL)-O-ETHYL METHYL PHOSPHONOTHIOLATE ◇ ETHYL-S-DIISOPROPYLAMINOETHYL METHYLTHIOPHOSPHONATE ◇ O-ETHYL-S-2-DIISOPROPYLAMINOETHYL METHYLPHOSPHONOTHIOTE ◇ METHYLPHOSPHONOTHIOICACID-S-(2-(BIS(METHYLETHYL)AMINO)ETHYL)o-ETHYL ESTER ◇ VX

TOXICITY DATA with REFERENCE
scu-rat TDLo:75 μg/kg (female 6-20D post):REP TJADAB 36,25A,87
scu-rat TDLo:75 μg/kg (female 6-20D post):TER TJADAB 36,25A,87
orl-man TDLo:4 μg/kg:GIT TXAPA9 27,241,74
skn-hmn LDLo:86 μg/kg WHOTAC -,24,70
scu-hmn TDLo:30 μg/kg 85IVAW 1,E1,82
ivn-man TDLo:1500 ng/kg:CNS,CVS,GIT TXAPA9 27,241,74
ims-hmn TDLo:3200 ng/kg:EYE,GIT 85IVAW 1,E1,82
scu-rat LD50:12 μg/kg AIPTAK 262,231,83
ipr-mus LD50:50 μg/kg WHOTAC -,39,70
scu-mus LD50:22 μg/kg APJUA8 30,151,80
scu-rbt LD50:14 μg/kg ABMHAM (Suppl),226,84
scu-gpg LD50:8400 ng/kg TXAPA9 40,109,77

CONSENSUS REPORTS: Reported in EPA TSCA Inventory. EPA Extremely Hazardous Substances List.

SAFETY PROFILE: Human poison by skin contact. Experimental poison by intraperitoneal and subcutaneous routes. An experimental teratogen. Other experimental reproductive effects. Human systemic effects by ingestion and intravenous route: hallucinations and distorted perceptions, blood pressure increase, hypermotility, diarrhea, nausea and vomiting, visual field changes, sleep disturbance. A chemical warfare agent. When heated to decomposition it emits very toxic fumes of SO_x and NO_x.

EIH000 CAS:27107-79-5 ***HR: 3***
ETHYL-dl-trans-2-DIMETHYLAMINO-1-PHENYL-3-CYCLOHEXENE-1-CARBOXYLATE
mf: $C_{17}H_{23}NO_2•ClH$ mw: 309.87

SYNS: dl-trans-2-DIMETHYLAMINO-1-PHENYL-CYCLOHEX-3-EN-trans-1-CARBONSAEUREAETHYLESTER HCl (GERMAN) ◇ dl-trans-2-DIMETHYLAMINO-1-PHENYL-CYCLOHEX-3-ENE-trans-CARBONIC ACID ETHYL ESTER HCl ◇ (±)-ETHYL-trans-2-2(DIMETHYLAMINO)-1-PHENYL-3-CYCLOHEXENE-1-CARBOXYLATEHYDROCHLORIDE ◇ GO-1261 ◇ TILIDATE ◇ TILIDINE HYDROCHLORIDE ◇ VALORON HYDROCHLORIDE ◇ W 5759A

TOXICITY DATA with REFERENCE
orl-rat TDLo:1450 mg/kg (15-22D preg/21D post):REP ARZNAD 20,983,70
orl-rat LD50:418 mg/kg ARZNAD 20,977,70
scu-rat LD50:400 mg/kg ARZNAD 20,977,70
ivn-rat LD50:74 mg/kg ARZNAD 20,977,70
orl-mus LD50:437 mg/kg ARZNAD 20,977,70
scu-mus LD50:490 mg/kg ARZNAD 20,977,70
ivn-mus LD50:52 mg/kg ARZNAD 20,977,70
orl-dog LD50:500 mg/kg ARZNAD 20,977,70

SAFETY PROFILE: Poison by ingestion, subcutaneous, and intravenous routes. Experimental reproductive effects. An analgesic. When heated to decomposition it emits very toxic fumes of HCl and NO_x.

EIH500 CAS:19622-19-6 ***HR: 2***
S-ETHYL N-(3-DIMETHYLAMINOPROPYL) THIOL CARBAMATE HYDROCHLORIDE
mf: $C_8H_{18}N_2OS•ClH$ mw: 226.80

SYNS: N-(3-DIMETHYLAMINOPROPYL)THIOCARBAMINSAEURE-S-

AETHYLESTER-HYDROCHLORID (GERMAN) ◇ DYNONE ◇ PREVICUR ◇ PROTHIOCARB ◇ SN-41703

TOXICITY DATA with REFERENCE
orl-rat LD50:1300 mg/kg 85ARAE 4,59,76/77
ipr-rat LD50:770 mg/kg 85DPAN -,-,71/76
orl-mus LD50:660 mg/kg 85DPAN -,-,71/76

SAFETY PROFILE: Moderately toxic by ingestion and intraperitoneal routes. When heated to decomposition it emits very toxic fumes of NO_x, SO_x, and HCl. See also CARBAMATES and ESTERS.

EII500 CAS:687-48-9 ***HR: 3***
ETHYL-N,N-DIMETHYL CARBAMATE
mf: $C_5H_{11}NO_2$ mw: 117.17

SYNS: DIMETHYLCARBAMIC ACID ETHYL ESTER ◇ DI-N-METHYL ETHYL CARBAMATE

TOXICITY DATA with REFERENCE
ipr-ham TDLo:504 mg/kg (8D preg):TER CNREA8 27,1696,67
ipr-mus TDLo:6500 mg/kg/13W-I:ETA JNCIAM 9,35,48
ipr-mus LD50:1110 mg/kg DPHFAK 23,25,71
scu-mus LD50:1050 mg/kg AJEBAK 45,507,67

SAFETY PROFILE: Moderately toxic by subcutaneous and intraperitoneal routes. Experimental teratogenic effects. Questionable carcinogen with experimental tumorigenic data. When heated to decomposition it emits toxic fumes of NO_x. See also CARBAMATES.

EIJ000 CAS:64037-50-9 ***HR: 3***
N-ETHYL-N-1-DIMETHYL-3,3-DI-2-THIENYL-2-PROPENAMINE HYDROCHLORIDE
mf: $C_{15}H_{19}NS_2 \cdot ClH$ mw: 313.93

SYNS: IC50 HYDROCHLORIDE ◇ N,1-DIMETHYL-3,3-DI-2-THIENYL-N-ETHYLALLYLAMINE HYDROCHLORIDE ◇ EMETHIBUTIN HYDROCHLORIDE ◇ N-ETHYL-N-1-DIMETHYL-3,3-DI-2-THIENYLALLYLAMINE HYDROCHLORIDE ◇ ETHYLMETHIAMBUTENE HYDROCHLORIDE ◇ 3-ETHYLMETHYLAMINO-1,1-DI(2'-THIENYL)BUT-1-ENE HYDROCHLORIDE ◇ ETHYLMETHYLTHIAMBUTENE HYDROCHLORIDE ◇ NIH-5145 HYDROCHLORIDE

TOXICITY DATA with REFERENCE
scu-rat LD50:63 mg/kg BJPCAL 8,2,53
orl-mus LD50:192 mg/kg MEIEDD 10,553,83
scu-mus LD50:88 mg/kg MEIEDD 10,553,83
ivn-mus LD50:17 mg/kg BJPCAL 8,2,53

SAFETY PROFILE: Poison by ingestion, intravenous, and subcutaneous routes. Used as a narcotic analgesic. When heated to decomposition it emits very toxic fumes of NO_x, SO_x, and HCl. See also ALLYL COMPOUNDS and AMINES.

EIJ500 CAS:617-38-9 ***HR: 3***
ETHYL DIMETHYLDITHIOCARBAMATE
mf: $C_5H_{11}NS_2$ mw: 149.29

SYNS: EEDDKK ◇ ETHYL ESTER of DIMETHYLDITHIOCARBAMIC ACID

TOXICITY DATA with REFERENCE
orl-rat LD50:550 mg/kg FATOAO 28,230,65
orl-mus LD50:290 mg/kg FATOAO 28,230,65
ipr-mus LD50:190 mg/kg DPHFAK 23,25,71

SAFETY PROFILE: Poison by ingestion and intraperitoneal routes. When heated to decomposition it emits very toxic fumes of NO_x and SO_x. See also CARBAMATES.

EIK000 CAS:78-78-4 ***HR: 3***
ETHYLDIMETHYLMETHANE
DOT: UN 1265
mf: C_5H_{12} mw: 72.17

PROP: Colorless liquid with pleasant odor. Bp: 27.8°, fp: −160.5°, flash p: < −60°F (CC), vap press: 595 mm @ 21.1°, vap d: 2.48, lel: 1.4%, uel: 7.6%.

SYNS: ISOAMYLHYDRIDE ◇ ISOPENTANE ◇ 2-METHYLBUTANE

CONSENSUS REPORTS: Reported in EPA TSCA Inventory.

NIOSH REL: (Alkanes) TWA 350 mg/m^3
DOT Classification: Flammable Liquid; Label: Flammable Liquid.

SAFETY PROFILE: Probably mildly toxic and narcotic by inhalation. See also PENTANE. Flammable Liquid. A very dangerous fire and explosion hazard when exposed to heat, flame, or oxidizers. Keep away from sparks, heat, or open flame; can react with oxidizing materials. To fight fire, use foam, CO_2, dry chemical. When heated to decomposition it emits acrid smoke and irritating fumes.

EIK500 CAS:16747-33-4 ***HR: 3***
3-ETHYL-2,3-DIMETHYL PENTANE
mf: C_9H_{20} mw: 128.26

PROP: Flash p: 46.4°F.

SAFETY PROFILE: A very dangerous fire hazard when exposed to heat or flame; can react vigorously with oxidizing materials. When heated to decomposition it emits acrid smoke and irritating fumes.

EIL000 CAS:1605-51-2 ***HR: 3***
ETHYLDIMETHYLPHOSPHINE
mf: $C_4H_{11}P$ mw: 90.10

$CH_3CH_2P(CH_3)_2$

SAFETY PROFILE: A very dangerous fire hazard; ignites spontaneously in air. When heated to decomposition it emits toxic fumes of PO_x and phosphine. See also PHOSPHINE.

EIL100 ***HR: 2***
2-ETHYL-3,5(6)-DIMETHYLPYRAZINE
mf: $C_8H_{12}N_2$ mw: 136.20

PROP: Colorless to sltly yellow liquid; roasted cocoa odor. D: 0.950-0.980, refr index: 1.500, flash p: 154°F. Sol in water, organic solvents.

SYN: FEMA No. 3149

SAFETY PROFILE: Combustible liquid. When heated to decomposition emits toxic fumes of NO_x.

EIL500 CAS:19481-40-4 ***HR: 3***
ETHYLDIMETHYL SULFONIUM IODIDE MERCURIC IODIDE ADDITION COMPOUND

SYN: DIMETHYLETHYL SULFONIUM IODIDE with MERCURY IODIDE (1:1)

TOXICITY DATA with REFERENCE
ivn-mus LD50:56 mg/kg CSLNX* NX#01852

CONSENSUS REPORTS: Mercury and its compounds are on the Community Right-To-Know List.

NIOSH REL: TWA 0.05 mg(Hg)/m^3

SAFETY PROFILE: Poison by intravenous route. When heated to decomposition it emits very toxic fumes of SO_x, I^-, and Hg. See also MERCURY COMPOUNDS, and IODIDES.

EIM000 CAS:17109-49-8 ***HR: 3***
O-ETHYL-S,S-DIPHENYL DITHIOPHOSPHATE
mf: $C_{14}H_{15}O_2PS_2$ mw: 310.38

PROP: A clear yellow to light brown liquid. D: 1.23.

SYNS: O-AETHYL-S,S-DIPHENYL-DITHIOPHOSPHAT(GERMAN) ◇ BAYER 78418 ◇ DITHIOPHOSPHORSAEURE-O-AETHYL-S,S-DIPHENYLESTER (GERMAN) ◇ EDDP ◇ EDIFENPHOS ◇ EDIPHENPHOS ◇ O-ETHYL-S,S-DIPHENYL PHOSPHORODITHIOATE ◇ HINOSAN ◇ LUTROL ◇ SRA 7847

TOXICITY DATA with REFERENCE
cyt-mus-ipr 30 mg/kg CYTOAN 49,833,84
orl-rat LD50:100 mg/kg OYYAA2 2,76,68
ipr-rat LD50:26 mg/kg TXAPA9 23,519,72
orl-mus LD50:214 mg/kg NNGADV 7,307,82
orl-rbt LD50:350 mg/kg 28ZEAL 5,91,76
orl-gpg LD50:350 mg/kg 28ZEAL 5,91,76

SAFETY PROFILE: Poison by ingestion and intraperitoneal routes. Mutation data reported. A cholinesterase inhibitor. When heated to decomposition it emits very toxic fumes of SO_x and PO_x. See also PARATHION.

EIN000 CAS:13194-48-4 ***HR: 3***
O-ETHYL-S,S-DIPROPYLPHOSPHORODITHIOATE
mf: $C_8H_{19}O_2PS_2$ mw: 242.36

SYNS: ENT 27,318 ◇ ETHOPROP ◇ ETHOPROPHOS ◇ O-ETHYL-S,S-DIPROPYL ESTER, PHOSPHORODITHIOIC ACID ◇ JOLT ◇ MOBIL V-C 9-104 ◇ MOCAP ◇ PROPHOS ◇ V-C CHEMICAL V-C 9-104 ◇ VIRGINIA CAROLINA VC 9-104

TOXICITY DATA with REFERENCE
orl-rat LD50:34 mg/kg ARSIM* 20,26,66
skn-rat LD50:60 mg/kg WRPCA2 9,119,70
skn-rbt LD50:26 mg/kg SPEADM 78-1,30,78
orl-pgn LD50:13300 μg/kg ASTTA8 (680),157,79
orl-ckn LD50:5500 μg/kg TXAPA9 11,49,67
orl-qal LD50:7500 μg/kg ASTTA8 (680),157,79
orl-dck LD50:1260 μg/kg TXAPA9 47,451,79
skn-dck LD50:11 mg/kg TXAPA9 47,451,79
orl-bwd LD50:4210 μg/kg ASTTA8 (680),157,79

CONSENSUS REPORTS: EPA Extremely Hazardous Substances List.

SAFETY PROFILE: Poison by ingestion and skin contact. A cholinesterase inhibitor type of insecticide. When heated to decomposition it emits very toxic fumes of PO_x and SO_x. See also PARATHION.

EIN500 CAS:759-94-4 ***HR: 3***
S-ETHYL-N,N-DI-N-PROPYLTHIOCARBAMATE
mf: $C_9H_{19}NOS$ mw: 189.35

SYNS: S-AETHYL-N,N-DIPROPYLTHIOLCARBAMAT(GERMAN) ◇ DIPROPYLCARBAMOTHIOIC ACID-S-ETHYL ESTER ◇ N,N-DIPROPYLTHIOCARBAMIC ACID-S-ETHYL ESTER ◇ EPTAM ◇ EPTC ◇ S-ETHYL-N,N-DIPROPYLTHIOCARBAMATE ◇ ETHYL DI-N-PROPYLTHIOLCARBAMATE ◇ ETHYL-N,N-DIPROPYLTHIOLCARBAMATE ◇ ETHYL-N,N-DI-N-PROPYLTHIOLCARBAMATE ◇ FDA 1541 ◇ GENEP EPTC ◇ R-1608

TOXICITY DATA with REFERENCE
ihl-hmn TCLo:135 mg/m^3/90M HYSAAV 36(1-3)196,71
orl-rat LD50:1325 mg/kg FMCHA2 -,C96,83
ihl-rat LCLo:200 mg/m^3/4H HYSAAV 36(1-3),196,71
unr-rat LD50:1660 mg/kg EQSFAP 3,618,75
orl-mus LD50:750 mg/kg HYSAAV 36(1-3),196,71
ivn-mus LD50:320 mg/kg CSLNX* NX#03907
orl-cat LD50:112 mg/kg HYSAAV 36(1-3),196,71
ihl-cat LCLo:400 mg/m^3/4H HYSAAV 36(1-3),196,71
skn-rbt LD50:1460 mg/kg WRPCA2 9,119,70
orl-bwd LD50:100 mg/kg TXAPA9 21,315,72

CONSENSUS REPORTS: Reported in EPA TSCA Inventory. EPA Genetic Toxicology Program.

SAFETY PROFILE: Poison by ingestion, inhalation, and intravenous routes. Moderately toxic by skin contact route. An herbicide. When heated to decomposition it emits very toxic fumes of NO_x and SO_x. See also CARBAMATES.

EIO000 CAS:74-85-1 ***HR: 3***
ETHYLENE
DOT: UN 1038/UN 1962
mf: C_2H_4 mw: 28.06

PROP: Colorless gas; sweet odor and taste. Bp: −103.9°, mp: −169.4°, lel: 2.7%, uel: 36%, d: 0.610 @ 0°, autoign temp: 914°F, vap d: 0.98, fp: −181°. Sol in water, alc, ether.

SYNS: ACETENE ◇ ATHYLEN (GERMAN) ◇ BICARBURRETTED HYDROGEN ◇ ELAYL ◇ ETHENE ◇ ETHYLENE, compressed (DOT) ◇ ETHYLENE, refrigerated liquid (DOT) ◇ LIQUID ETHYENE ◇ OLEFIANT GAS

TOXICITY DATA with REFERENCE
ihl-mus LC50:95 pph 28ZEAL 5,109,76
ihl-mam LCLo:950000 ppm/5M AEPPAE 138,65,28

CONSENSUS REPORTS: Reported in EPA TSCA Inventory. Community Right-To-Know List.

DOT Classification: Flammable Gas; Label: Flammable Gas.

SAFETY PROFILE: A simple asphyxiant. High concentrations cause anesthesia. A common air contaminant. It is phytotoxic. A very dangerous fire hazard when exposed to heat or flame. Moderate explosion hazard when exposed to flame. A flammable gas. To fight fire, stop flow of gas, use CO_2, dry chemical or fine water spray. Mixtures with aluminum chloride explode in the presence of nickel catalysts, methyl chloride, or nitromethane. Explosive reaction with bromotrichloromethane (at 120°C/51 bar), carbon tetrachloride (from 25-100°C/30 bar). Explosive reaction with chlorine catalyzed by sunlight or UV light or in the presence of mercury(I) oxide, mercury(II) oxide, or silver oxide. Mixtures with chlorotrifluoroethylene polymerize explosively when exposed to 50 Kv gamma rays at 308 Krad/hr. Has been involved in industrial accidents. Violent polymerization is catalyzed by copper above 400°C/54 bar. Incompatible with $AlCl_3$; (CCl_4 + benzoyl peroxide); (bromtrichloromethane + $AlCl_3$); O_3; CCl_4; Cl_2; NO_x; tetrafluoroethylene; trifluorohypofluorite. When heated to decomposition it emits acrid smoke and irritating fumes.

EIO500 CAS:126-39-6 ***HR: 2***
ETHYLENEACETIC ACID
mf: $C_6H_{12}O_2$ mw: 116.18

SYNS: CYCLIC ETHYLENE ACETAL-2-BUTANONE ◇ 2-ETHYL-2-METHYL-1,3-DIOXOLANE

TOXICITY DATA with REFERENCE
orl-rat LDLo:2830 mg/kg AIHAAP 23,95,62
ihl-rat LCLo:4000 ppm/4H AIHAAP 23,95,62

CONSENSUS REPORTS: Reported in EPA TSCA Inventory.

SAFETY PROFILE: Moderately toxic by ingestion and inhalation. When heated to decomposition it emits acrid smoke and irritating fumes.

EIP000 CAS:2274-11-5 ***HR: 3***
ETHYLENE ACRYLATE
mf: $C_8H_{10}O_4$ mw: 170.18

SYNS: ACRYLIC ACID, ETHYLENE ESTER ◇ ACRYLIC ACID, ETHYLENE GLYCOL DIESTER ◇ ETHYLDIOL ACRILATE (RUSSIAN) ◇ ETHYLENE DIACRYLATE ◇ ETHYLENE GLYCOL DIACRYLATE ◇ 2-PROPENOIC ACID-1,2-ETHANEDIYL ESTER

TOXICITY DATA with REFERENCE
orl-rat LD50:1070 mg/kg TXAPA9 28,313,74
orl-mus LD50:300 mg/kg GTPZAB 24(4),58,80
ihl-mus LC50:350 mg/kg GTPZAB 24(4),58,80
skn-rbt LD50:570 mg/kg TXAPA9 28,313,74

CONSENSUS REPORTS: Reported in EPA TSCA Inventory.

SAFETY PROFILE: Poison by ingestion and inhalation. Moderately toxic by skin contact. When heated to decomposition it emits acrid smoke and irritating fumes. See also ESTERS.

EIP500 CAS:60784-41-0 ***HR: 3***
1,1'-ETHYLENEBIS(3-(2-CHLOROETHYL)-3-NITROSOUREA)
mf: $C_8H_{14}Cl_2N_6O_4$ mw: 329.18

SYN: 1,1'-ETHYLENEBIS-CNU

TOXICITY DATA with REFERENCE
sln-dmg-orl 100 μmol/L/24H MUREAV 57,297,78
mrc-smc 500 μmol/L/16H MUREAV 42,45,77
ipr-rat LDLo:29 mg/kg JNCIAM 60,345,78

SAFETY PROFILE: Poison by intraperitoneal route. Mutation data reported. Many N-nitroso compounds are carcinogens. When heated to decomposition it emits very toxic fumes of Cl^- and NO_x. See also N-NITROSO COMPOUNDS.

EIQ000 CAS:124-05-0 ***HR: 2***
ETHYLENE BIS(CHLOROFORMATE)
mf: $C_4H_4Cl_2O_4$ mw: 186.98

SYNS: 1,2-BIS((CHLOROCARBONYL)OXY)ETHANE ◇ ETHYLENE CHLOROFORMATE ◇ ETHYLENE GLYCOL DI(CHLOROFORMATE)

TOXICITY DATA with REFERENCE
orl-mus LD50:1100 mg/kg 37ASAA 4,758,78
ihl-mus LD50:5659 ppm/1H 37ASAA 4,758,78
skn-mus LD50:2000 mg/kg 37ASAA 4,758,78

CONSENSUS REPORTS: Reported in EPA TSCA Inventory.

SAFETY PROFILE: Moderately toxic by ingestion and skin contact. Mildly toxic by inhalation. When heated to decomposition it emits toxic fumes of Cl^-.

EIQ200 CAS:4431-24-7 ***HR: 3***
ETHYLENEBIS-(DIPHENYLARSINE)
mf: $C_{26}H_{24}As_2$ mw: 486.34

PROP: Mp: 103-108°.

SYN: ARSINE, ETHYLENEBIS(DIPHENYL-

TOXICITY DATA with REFERENCE
ivn-mus LD50:180 mg/kg CSLNX* NX#04858

OSHA PEL: TWA 0.5 mg(As)/m^3

SAFETY PROFILE: Poison by intravenous route. An irritant. When heated to decomposition it emits toxic fumes of As.

EIQ500 ***HR: D***
ETHYLENEBIS(DITHIOCARBAMATO)MANGANESE and ZINC ACETATE (50:1)

SYNS: MANEB plus ZINC ACETATE (50:1) ◇ ZINC ACETATE plus MANEB (1:50)

TOXICITY DATA with REFERENCE
orl-rat TDLo:765 mg/kg (11D preg):TER TJADAB 14,171,76

CONSENSUS REPORTS: Manganese and its compounds, as well as Zinc and its compounds, are on the Community Right-To-Know List.

SAFETY PROFILE: An experimental teratogen. When heated to decomposition it emits very toxic fumes of NO_x, ZnO, and SO_x. See also ZINC COMPOUNDS, MANGANESE COMPOUNDS, and CARBAMATES.

EIR000 CAS:12122-67-7 ***HR: 2***
ETHYLENE BIS(DITHIOCARBAMATO)ZINC
mf: $C_4H_6N_2S_4 \cdot Zn$ mw: 275.73

PROP: Light-colored powder, insol in water.

SYNS: ASPOR ◇ ASPORUM ◇ BERCEMA ◇ BLIGHTOX ◇ BLITEX ◇ BLIZENE ◇ CARBADINE ◇ CHEM ZINEB ◇ CINEB ◇ CRITTOX ◇ CYNKOTOX ◇ DAISEN ◇ DIPHER ◇ DITHANE Z ◇ DITIAMINA ◇ ENT 14,874 ◇ ((1,2-ETHANEDIYLBIS(CARBAMODITHIOATO))(2-) ZINC ◇ 1,2-ETHANEDIYLBIS(CARBAMODITHIOATO) (2-)-S,S'-ZINC ◇ 1,2-ETHANEDIYLBISCARBAMODITHIOIC ACID, ZINC COMPLEX ◇ 1,2-ETHANEDIYLBISCARBAMOTHIOIC ACID, ZINC SALT ◇ ETHYLENEBIS(DITHIOCARBAMIC ACID), ZINC SALT ◇ ETHYL ZIMATE ◇ HEXATHANE ◇ KUPRATSIN ◇ KYPZIN ◇ LIROTAN ◇ LONACOL ◇ MICIDE ◇ MILTOX ◇ MILTOX SPECIAL ◇ NOVIZIR ◇ NOVOSIR N ◇ PAMOSOL 2 FORTE ◇ PARZATE ◇ PEROSIN ◇ POLYRAM Z ◇ SPERLOX-Z ◇ THIODOW ◇ TIEZENE ◇ TRITOFTOROL ◇ TSINEB (RUSSIAN) ◇ Z-78 ◇ ZEBENIDE ◇ ZEBTOX ◇ ZIDAN ◇ ZIMATE ◇ ZINC ETHYLENEBISDITHIOCARBAMATE ◇ ZINC ETHYLENE-1,2-BISDITHIOCARBAMATE ◇ ZINEB ◇ ZINK-(N,N'-AETHYLEN-BIS(DITHIOCARBAMAT)) (GERMAN) ◇ ZINOSAN

TOXICITY DATA with REFERENCE
mmo-bcs 1 nmol/plate MSERDS 5,93,81
sce-hmn:lym 10 mg/L GESKAC 12,118,79
orl-rat TDLo:4 g/kg (13D preg):TER FCTXAV 11,239,73
orl-rat TDLo:4 g/kg (13D preg):REP FCTXAV 11,239,73
orl-rat TDLo:53580 mg/kg/94W-I:CAR VPITAR 29,71,70
imp-rat TDLo:80 mg/kg:ETA VPITAR 29,71,70
ipr-mus TDLo:1760 mg/kg (11-21D preg):ETA,TER VPITAR 30,49,71
orl-hmn LDLo:5 g/kg 85JFAN A420,84
orl-rat LD50:1850 mg/kg GISAAA 31(10),25,66
ihl-rat LCLo:800 mg/m^3/4H 85GMAT -,121,82
orl-mus LD50:7600 mg/kg JTEHD6 4,93,78

CONSENSUS REPORTS: IARC Cancer Review: Group 3 IMEMDT 7,56,87; Animal Inadequate Evidence IMEMDT 12,245,76. Community Right-To-Know List. EPA Genetic Toxicology Program.

SAFETY PROFILE: Moderately toxic by ingestion. Experimental teratogenic and reproductive effects. Questionable carcinogen with experimental carcinogenic and tumorigenic data. Human mutation data reported. Used as a fungicide. When heated to decomposition it emits very toxic fumes of NO_x, ZnO, and SO_x. See also ZINC COMPOUNDS and CARBAMATES.

EIR500 CAS:12275-13-7 ***HR: 3***
ETHYLENEBIS(DITHIOCARBAMIC ACID) NICKEL(II) SALT
mf: $C_4H_6N_2S_4 \cdot Ni$ mw: 269.07

TOXICITY DATA with REFERENCE
ipr-mus LDLo:250 mg/kg CBCCT* 4,377,52

CONSENSUS REPORTS: Nickel and its compounds are on the Community Right-To-Know List.

NIOSH REL: (Inorganic Nickel) TWA 0.015 mg(Ni)/m^3

SAFETY PROFILE: Poison by intraperitoneal route. When heated to decomposition it emits very toxic fumes of NO_x and SO_x. See also NICKEL COMPOUNDS and CARBAMATES.

EIS000 CAS:62207-76-5 ***HR: 3***
N,N'-ETHYLENE BIS(3-FLUOROSALICYLIDENEIMINATO)COBALT(II)
mf: $C_{16}H_{12}CoF_2N_2O_2$ mw: 361.23

SYNS: BIS(3-FLUOROSALICYLALDEHYDE)-ETHYLENEDIIMINE-COBALT ◇ FLUOMINE ◇ FLUOMINE DUST

TOXICITY DATA with REFERENCE
skn-rbt 500 mg/24H MOD NTIS** AD-A083-929
eye-rbt 3 mg NTIS** AD-A083-929
orl-rat LD50:187 mg/kg AMRL** TR-74-78,74
ihl-rat LC50:112 mg/m^3/6H AMRL** TR-74-78,74
orl-mus LD50:123 mg/kg AMRL** TR-74-78,74
ihl-mus LC50:416 mg/m^3/6H NTIS** AD-A083-929
ihl-gpg LCLo:30 mg/m^3/2H AMRL** TR-74-78,74

CONSENSUS REPORTS: Cobalt and its compounds are on the Community Right-To-Know List. EPA Extremely Hazardous Substances List. Reported in EPA TSCA Inventory.

SAFETY PROFILE: Poison by ingestion and inhalation. A skin and eye irritant. When heated to decomposition it emits very toxic fumes of F^- and NO_x. See also COBALT COMPOUNDS.

EIT000 CAS:67-42-5 ***HR: 3***
(ETHYLENEBIS(OXYETHYLENENITRILO)) TETRAACETIC ACID
mf: $C_{14}H_{24}N_2O_{10}$ mw: 380.40

SYNS: ETHYLENEDIOXYBIS(ETHYLENEAMINO)TETRAACETIC ACID ◇ ETHYLENE GLYCOL BIS(AMINOETHYL ETHER)TETRAACETATE ◇ ETHYLENE GLYCOL BIS(β-AMINOETHYL ETHER)TETRAACETATE ◇ ETHYLENE GLYCOL BIS(β-AMINOETHYL ETHER)-N,N'-TETRAACETIC ACID ◇ ETHYLENE GLYCOL BIS(2-AMINOETHYL ETHER)TETRAACETIC ACID ◇ ETHYLENE GLYCOL BIS(2-AMINOETHYL ETHER)-N,N,N',N'-TETRAACETIC ACID ◇ GLYCOL-ETHERDIAMINETETRAACETIC ACID

TOXICITY DATA with REFERENCE
orl-rat LD50:3587 mg/kg TXAPA9 16,807,70
ipr-mus LD50:150 mg/kg NTIS** AD691-490

CONSENSUS REPORTS: Glycol ether compounds are on the Community Right-To-Know List. Reported in EPA TSCA Inventory.

SAFETY PROFILE: Poison by intraperitoneal route. Moderately toxic by ingestion. When heated to decomposition it emits toxic fumes of NO_x. See also GLYCOL ETHERS.

EIU000 CAS:10310-38-0 ***HR: 1***
ETHYLENEBIS(TRIS(2-CYANOETHYL)PHOSPHONIUM BROMIDE)
mf: $C_{20}H_{28}N_6P_2{\cdot}2Br$ mw: 574.30

SYN: 1,2-ETHANEDIYLBIS(TRIS(2-CYANOETHYL)PHOSPHONIUM DIBROMIDE

TOXICITY DATA with REFERENCE
orl-rat LD50:10 g/kg EPASR* 8EHQ-0780-0369

CONSENSUS REPORTS: Cyanide and its compounds are on the Community Right-To-Know List. Reported in EPA TSCA Inventory.

SAFETY PROFILE: Mildly toxic by ingestion. When heated to decomposition it emits very toxic fumes of PO_x, NO_x, and Br^-. See also CYANIDE and BROMIDES.

EIU800 CAS:107-07-3 ***HR: 3***
ETHYLENE CHLOROHYDRIN
DOT: UN 1135
mf: C_2H_5ClO mw: 80.52

PROP: Colorless liquid; faint, ethereal odor. Mp: −69°, bp: 128.8°, flash p: 140°F (OC), d: 1.197 @ 20°/4°, autoign temp: 797°F, vap press: 10 mm @ 30.3°, vap d: 2.78, lel: 4.9%, uel: 15.9%.

SYNS: AETHYLENECHLORHYDRIN (GERMAN) ◇ 2-CHLOORETHANOL (DUTCH) ◇ 2-CHLORAETHANOL (GERMAN) ◇ 2-CHLORETHANOL (GERMAN) ◇ Δ-CHLOROETHANOL ◇ 2-CHLOROETHANOL (MAK) ◇ 2-CHLOROETHYL ALCOHOL ◇ β-CHLOROETHYL ALCOHOL ◇ CHLOROETHYLOWY ALKOHOL (POLISH) ◇ 2-CLOROETANOLO (ITALIAN) ◇ ETHYLEEN-CHLOORHYDRINE (DUTCH) ◇ ETHYLENE GLYCOL, CHLOROHYDRIN ◇ GLICOL MONOCLORIDRINA (ITALIAN) ◇ GLYCOL CHLOROHYDRIN ◇ GLYCOLMONOCHLOORHYDRINE (DUTCH) ◇ GLYCOL MONOCHLOROHYDRIN ◇ GLYCOMONOCHLORHYDRIN ◇ MONOCHLORHYDRINE du GLYCOL (FRENCH) ◇ 2-MONOCHLOROETHANOL ◇ NCI-C50135

TOXICITY DATA with REFERENCE
skn-rbt 200 mg/2H MLD TXAPA9 16,382,70
eye-rbt 2 mg SEV AJOPAA 29,1363,46
eye-rbt 33 mg MOD TXAPA9 16,382,70
eye-rbt 9 mg/6H MOD BUYRAI 31,25,77
sln-asn 74500 μmol/L MUREAV 138,33,84
oms-rat:lvr 12 g/L JACTDZ 1(3),37,82
orl-mus TDLo:1100 mg/kg (6-16D preg):TER JPFCD2 17,381,82
ihl-man LCLo:305 ppm/2H JIHTAB 26,277,44
orl-rat LD50:71 mg/kg HYSAAV 36,376,71
ihl-rat LC50:290 mg/m^3 HYSAAV 26,376,71
ipr-rat LD50:58 mg/kg TXAPA9 21,454,72
scu-rat LD50:84 mg/kg HYSAAV 36(4-6),376,71
orl-mus LD50:81 mg/kg JPMSAE 60,568,71
ihl-mus LC50:385 mg/m^3 HYSAAV 36,376,71
skn-mus LD50:18 mg/kg 85GMAT -,66,82
ipr-mus LD50:97 mg/kg JPMSAE 61,19,72
skn-rbt LD50:67 mg/kg JPMSAE 61,19,72
ivn-rbt LDLo:100 mg/kg BJIMAG 1,207,44

CONSENSUS REPORTS: NTP Carcinogenesis Studies (dermal); No Evidence: mouse, rat NTPTR* NTP-TR-275,85; Reported in EPA TSCA Inventory. EPA Genetic Toxicology Program. EPA Extremely Hazardous Substances List.

OSHA PEL: (Transitional: TWA 5 ppm (skin)) CL 1 ppm (skin)
ACGIH TLV: CL 1 ppm (skin)
DFG MAK: 1 ppm (3 mg/m^3)
DOT Classification: Poison B; Label: Flammable Liquid and Poison.

SAFETY PROFILE: A poison by ingestion, inhalation, skin contact, intraperitoneal, intravenous, and subcutaneous routes. Moderately toxic to humans by inhalation. It can affect the nervous system, liver, spleen, and lungs. An experimental teratogen. Mutation data reported. A severe eye and mild skin irritant. Flammable when exposed to heat, flame, or oxidizers. To fight fire, use alcohol foam, CO_2, dry chemical. Violent reaction with chlorosulfonic acid, ethylene diamine, sodium hydroxide. Reacts with water or steam to produce toxic and corrosive fumes. Potentially violent reaction with oxidizing materials. When heated to decomposition it emits highly toxic fumes of Cl^- and phosgene. See also CHLORINATED HYDROCARBONS, ALIPHATIC.

EIV000 CAS:64-02-8 ***HR: 3***
N,N'-ETHYLENEDIAMINEDIACETIC ACID TETRASODIUM SALT
mf: $C_{10}H_{12}N_2O_8 \cdot 4Na$ mw: 380.20

SYNS: AQUAMOLLIN ◇ CALSOL ◇ CELON E ◇ CELON H ◇ CELON IS ◇ CHEELOX BF ◇ CHEELOX BR-33 ◇ CHELON 100 ◇ CHEMCOLOX 200 ◇ COMPLEXONE ◇ CONIGON BC ◇ DISTOL 8 ◇ EDATHANIL TETRASODIUM ◇ EDETATE SODIUM ◇ EDETIC ACID TETRASODIUM SALT ◇ EDTA, SODIUM SALT ◇ EDTA TETRASODIUM SALT ◇ ENDRATE TETRASODIUM ◇ N,N'-1,2-ETHANEDIYLBIS(N-(CARBOXYMETHYL)GLYCINE TETRASODIUM SALT ◇ ETHYLENEBIS(IMINODIACETIC ACID) TETRASODIUM SALT ◇ ETHYLENEDIAMINETETRAACETIC ACID, TETRASODIUM SALT ◇ HAMP-ENE 100 ◇ HAMP-ENE 215 ◇ HAMP-ENE 220 ◇ HAMP-ENE Na4 ◇ IRGALON ◇ KALEX ◇ KEPMPLEX 100 ◇ KOMPLXON ◇ METAQUEST C ◇ NERVANAID B LIQUID ◇ NERVANID B ◇ NULLAPON B ◇ NULLAPON BF-78 ◇ NULLAPON BFC CONC ◇ PERMA KLEER 50 CRYSTALS ◇ PERMA KLEER TETRA CP ◇ QUESTEX 4 ◇ SEQUESTRENE 30A ◇ SEQUESTRENE Na 4 ◇ SEQUESTRENE ST ◇ SODIUM EDETATE ◇ SODIUM EDTA ◇ SODIUM ETHYLENEDIAMINETETRAACETATE ◇ SODIUM ETHYLENEDIAMINETETRAACETIC ACID ◇ SODIUM SALT of ETHYLENEDIAMINETETRAACETIC ACID ◇ SYNTES 12A ◇ SYNTRON B ◇ TETRACEMIN ◇ TETRASODIUM EDTA ◇ TETRASODIUM ETHYLENEDIAMINETETRAACETATE ◇ TETRASODIUM ETHYLENEDIAMINETETRACETATE ◇ TETRASODIUM (ETHYLENEDINITRILO)TETRAACETATE ◇ TETRASODIUM SALT of EDTA ◇ TETRASODIUM SALT of ETHYLENEDIAMINETETRACETICACID ◇ TETRINE ◇ TRILON B ◇ TST ◇ TYCLAROSOL ◇ VERSENE 100 ◇ VERSENE POWDER ◇ WARKEELATE PS-43

TOXICITY DATA with REFERENCE
skn-rbt 500 mg/24H MOD 28ZPAK -,306,72
eye-rbt 1900 μg AAOPAF 48,681,52
eye-rbt 100 mg/24H MOD 28ZPAK -,306,72
ipr-mus LD50:330 mg/kg REPMBN 10,391,62

CONSENSUS REPORTS: Reported in EPA TSCA Inventory.

SAFETY PROFILE: Poison by intraperitoneal route. A skin and eye irritant. When heated to decomposition it emits toxic fumes of NO_x and Na_2O. See also ETHYLENEDIAMINETETRAACETIC ACID, DISODIUM SALT.

EIV700 CAS:20829-66-7 ***HR: 3***
ETHYLENE DIAMINEDINITRATE
mf: $C_2H_{10}N_4O_6$ mw: 186.12

SYN: 1,2-DIAMMONIOETHANE NITRATE

SAFETY PROFILE: Has been used as a military explosive. When heated to decomposition it emits toxic fumes of NO_x. See also EXPLOSIVES, HIGH; and NITRATES.

EIV750 CAS:27014-42-2 ***HR: 1***
ETHYLENEDIAMINE ETHOXYLATE
mf: $(C_2H_4O)_n(C_2H_4O)_n(C_2H_4O)_n(C_2H_4O)_nC_{10}H_{24}N_2O_4$

SYNS: EDA 200 ◇ ETHYLENEDIAMINE ETHYLENE OXIDE ADDUCT ◇ POLY(OXY-1,2-ETHANEDIYL), α-α',α'',α'''-(1,2-ETHANEDIYLBIS(NITRILODI-2,1-ETHANEDIYL))TETRAKIS(omega-HYDROXY- ◇ VERAMIN ED 4 ◇ VERAMIN ED 40

TOXICITY DATA with REFERENCE
skn-rbt 500 mg/24H MLD JACTDZ 1,19,90
eye-rbt 100 mg MLD JACTDZ 1,19,90
orl-rat LD50:9900 mg/kg JACTDZ 1,19,90
skn-rbt LD50: >8 g/kg JACTDZ 1,19,90

SAFETY PROFILE: Mildly toxic by ingestion. A skin and eye irritant. When heated to decomposition it emits toxic fumes of NO_x.

EIW000 CAS:333-18-6 ***HR: 3***
ETHYLENEDIAMINE HYDROCHLORIDE
mf: $C_2H_8N_2 \cdot 2ClH$ mw: 133.04

PROP: Monoclinic prisms. Mp: subl. Sol in water; insol in alc and ether.

SYNS: CHLOR-ETHAMINE ◇ ETHYLENEDIAMINE DIHYDROCHLORIDE

TOXICITY DATA with REFERENCE
orl-rat TDLo:10 g/kg (6-15D preg):TER TOXID9 4,165,84
orl-rat TDLo:10 g/kg (6-15D preg):REP TOXID9 4,165,84
orl-rat LDLo:500 mg/kg JPETAB 90,260,47
ims-rat LD50:150 mg/kg EMSUA8 4,223,46
orl-mus LD50:1620 mg/kg FAATDF 3,512,83

CONSENSUS REPORTS: Reported in EPA TSCA Inventory.

SAFETY PROFILE: Poison by intramuscular route. Moderately toxic by ingestion. Experimental teratogenic and reproductive effects. When heated to decomposition it emits very toxic fumes of HCl and NO_x.

EIW500 CAS:25723-52-8 ***HR: 2***
ETHYLENEDIAMINE SULFATE
mf: $C_2H_8N_2 \cdot 7H_2O_4S$ mw: 746.68

TOXICITY DATA with REFERENCE
skn-rat LD50:1000 mg/kg KODAK* -,-,71
orl-rat LDLo:3200 mg/kg KODAK* -,-,71

CONSENSUS REPORTS: Reported in EPA TSCA Inventory.

SAFETY PROFILE: Moderately toxic by ingestion and skin contact. When heated to decomposition it emits very toxic fumes of NO_x and SO_x.

EIX000 CAS:60-00-4 ***HR: 3***
ETHYLENEDIAMINETETRAACETIC ACID
mf: $C_{10}H_{16}N_2O_8$ mw: 292.28

PROP: Colorless crystals. Mp: decomp @ 240°. Sltly water-sol; insol in common organic solvents.

SYNS: ACIDE ETHYLENEDIAMINETETRACETIQUE (FRENCH) ◇ 3,6-BIS(CARBOXYMETHYL)-3,5-DIAZOOCTANEDIOIC ACID ◇ CELON A ◇ CELON ATH ◇ CHEELOX BF ACID ◇ CHEMCOLOX 340 ◇ COMPLEXON II ◇ EDATHAMIL ◇ EDETIC ACID ◇ EDTA (chelating agent) ◇ EDTA ACID ◇ ENDRATE ◇ N,N'-1,2-ETHANEDIYLBIS(N-(CARBOXYMETHYL)GLYCINE ◇ ETHYLENEDIAMINETETRAACETATE ◇ ETHYLENEDIAMINE-N,N,N',N'-TETRAACETIC ACID ◇ ETHYLENEDINITRILOTETRAACETIC ACID ◇ HAMP-ENE ACID ◇ HAVIDOTE ◇ METAQUEST A ◇ NERVANAID B ACID ◇ NULLAPON BF ACID ◇ PERMA KLEER 50 ACID ◇ QUESTEX 4H ◇ SEQ 100 ◇ SEQUESTRENE AA ◇ SEQUESTRIC ACID ◇ SEQUESTROL ◇ TETRINE ACID ◇ TITRIPLEX ◇ TRICON BW ◇ TRILON BW ◇ VERSENE ACID ◇ VINKEIL 100 ◇ WARKEELATE ACID

TOXICITY DATA with REFERENCE
cyt-mus-ipr 50 mmol/L CISCB7 20,28,76
dni-rbt:kdy 250 nmol/L ECREAL 36,92,64
orl-rat TDLo:7632 mg/kg (female 7-14D post):TER TXAPA9 40,299,77
orl-rat TDLo:7632 mg/kg (female 7-14D post):REP TXAPA9 40,299,77
ipr-rat LD50:397 mg/kg AHRTAN 13,295,62
ipr-mus LD50:250 mg/kg NTIS** AD691-490

CONSENSUS REPORTS: Reported in EPA TSCA Inventory. EPA Genetic Toxicology Program.

SAFETY PROFILE: Poison by intraperitoneal route. Experimental teratogenic and reproductive effects. Mutation data reported. A general-purpose chelating and complexing agent. When heated to decomposition it emits toxic fumes of NO_x.

EIX500 CAS:139-33-3 ***HR: 3***
ETHYLENEDIAMINETETRAACETIC ACID, DISODIUM SALT
mf: $C_{10}H_{14}N_2O_8 \cdot 2Na$ mw: 336.24

PROP: White crystalline powder. Sol in water.

SYNS: CHELADRATE ◇ CHELAPLEX III ◇ CHELATON III ◇ COMPLEXON III ◇ d'E.D.T.A. DISODIQUE (FRENCH) ◇ DISODIUM DIACID ETHYLENEDIAMINETETRAACETATE ◇ DISODIUM DIHYDROGEN ETHYLENEDIAMINETETRAACETATE ◇ DISODIUM DIHYDROGEN(ETHYLENEDINITRILO)TETRAACETATE ◇ DISODIUM EDATHAMIL ◇ DISODIUM EDETATE ◇ DISODIUM EDTA (FCC) ◇ DISODIUM ETHYLENEDIAMINETETRAACETATE ◇ DISODIUM ETHYLENEDIAMINETETRAACETIC ACID ◇ DISODIUM (ETHYLENEDINITRILO)TETRAACETATE ◇ DISODIUM (ETHYLENEDINITRILO)TETRAACETIC ACID ◇ DISODIUM SALT of EDTA ◇ DISODIUM SEQUESTRENE ◇ DISODIUM TETRACEMATE ◇ DISODIUM VERSENATE ◇ DISODIUM VERSENE ◇ EDATHAMIL DISODIUM ◇ EDETATE DISODIUM ◇ EDTA, DISODIUM SALT ◇ ENDRATE DISODIUM ◇ N,N'-1,2-ETHANEDIYLBIS(N-(CARBOXYMETHYL)GLYCINE) DISODIUM SALT ◇ ETHYLENEBIS(IMINODIACETIC ACID) DISODIUM SALT ◇ ETHYLENEDIAMINETETRAACETATE DISODIUM SALT ◇ (ETHYLENEDINITRILO)-TETRAACETIC ACID DISODIUM SALT ◇ F 1 (complexon) ◇ KIRESUTO B ◇ METAQUEST B ◇ PERMA KLEER 50 CRYSTALS DISODIUM SALT ◇ SELEKTON B 2 ◇ SEQUESTRENE SODIUM 2 ◇ SODIUM VERSENATE ◇ TETRACEMATE DISODIUM ◇ TITRIPLEX III ◇ TRILON BD ◇ TRIPLEX III ◇ VERESENE DISODIUM SALT ◇ VERSENE SODIUM 2

TOXICITY DATA with REFERENCE
cyt-grh-par 1 mmol/L CISCB7 16,18,74
orl-rat TDLo:12857 mg/kg (7-15D preg):TER SCIEAS 173,62,72
orl-rat TDLo:31429 mg/kg (1-22D preg):REP SCIEAS 173,62,72
orl-rat LD50:2000 mg/kg FEPRA7 27,465,68
orl-mus LD50:2050 mg/kg NYKZAU 52,126S,56
ipr-mus LD50:260 mg/kg NYKZAU 52,126S,56
ivn-mus LD50:56 mg/kg CSLNX* NX#03781
orl-rbt LD50:2300 mg/kg NYKZAU 52,113,56
ivn-rbt LD50:47 mg/kg NYKZAU 52,113,56

CONSENSUS REPORTS: Reported in EPA TSCA Inventory. EPA Genetic Toxicology Program.

SAFETY PROFILE: Poison by intraperitoneal and intravenous route. Moderately toxic by ingestion. Experimental teratogenic and reproductive effects. Mutation data reported. The calcium disodium salt of EDTA is used as a chelating agent in treating lead poisoning. When heated to decomposition it emits toxic fumes of NO_x and Na_2O.

EIY500 CAS:106-93-4 ***HR: 3***
1,2-ETHYLENE DIBROMIDE
DOT: UN 1605
mf: $C_2H_4Br_2$ mw: 187.88

PROP: Colorless, heavy liquid; sweet odor. Bp: 131.4°, fp: 9.3°, flash p: none, d: 2.172 @ 25°/25°, 2.1707 @ 25°/4°, vap press: 17.4 mm @ 30°, vap d: 6.48.

SYNS: AETHYLENBROMID (GERMAN) ◇ BROMOFUME ◇ BROMURO di ETILE (ITALIAN) ◇ CELMIDE ◇ DBE ◇ 1,2-DIBROMAETHAN (GERMAN) ◇ 1,2-DIBROMOETANO (ITALIAN) ◇ 1,2-DIBROMOETHANE (MAK) ◇ α,β-DIBROMOETHANE ◇ sym-DIBROMOETHANE ◇ DIBROMURE d'ETHYLENE (FRENCH) ◇ 1,2-

DIBROOMETHAAN (DUTCH) ◇ DOWFUME 40 ◇ DOWFUME EDB ◇ DOWFUME W-8 ◇ DWUBROMOETAN (POLISH) ◇ EDB ◇ EDB-85 ◇ E-D-BEE ◇ ENT 15,349 ◇ ETHYLENE BROMIDE ◇ FUMO-GAS ◇ GLYCOL BROMIDE ◇ GLYCOL DIBROMIDE ◇ ISCOBROME D ◇ KOPFUME ◇ NCI-C00522 ◇ NEPHIS ◇ PESTMASTER ◇ PESTMASTER EDB-85 ◇ RCRA WASTE NUMBER U067 ◇ SOILBROM-40 ◇ SOILBROM-85 ◇ SOILFUME ◇ UNIFUME

TOXICITY DATA with REFERENCE
skn-hmn 1538 mg/24H SEV AGGHAR 8,591,38
skn-rbt 1%/14D SEV AMIHBC 6,158,52
eye-rbt 1% AMIHBC 6,158,52
mmo-sat 1 μg/plate ENMUDM 7(Suppl 5),1,85
mma-esc 333 μg/plate ENMUDM 7(Suppl 5),1,85
ihl-man TCLo:88 ppb/8H (male 5 year(s) pre):REP BJIMAG 44,317,87
ihl-mus TCLo:38 ppm/23H (female 6-15D post):TER TXAPA9 45,347,78
orl-rat TDLo:540 mg/kg /78W-C:CAR BANRDU 5,279,80
ihl-rat TCLo:10 ppm/6H/2Y-I:CAR NCITR* NCI-CG-TR-201,82
skn-mus TDLo:190 g/kg/62W-I:NEO JJIND8 63,1433,79
orl-wmn LDLo:90 mg/kg 34ZIAG -,257,69
orl-rat LD50:108 mg/kg SPEADM 74-1,-,74
ihl-rat LC50:14300 mg/m^3/30M FATOBP 8,140,73
skn-rat LD50:300 mg/kg 85DPAN -,-,71/76
orl-mus LDLo:250 mg/kg TXAPA9 23,288,72
ipr-mus LD50:220 mg/kg JPCEAO 320(1),133,78
orl-rbt LD50:55 mg/kg AMIHBC 6,158,52
skn-rbt LD50:300 mg/kg AMIHBC 6,158,52
rec-rbt LDLo:2500 mg/kg JPETAB 34,223,28
ihl-gpg LCLo:400 ppm/3H AMIHBC 6,158,52

CONSENSUS REPORTS: NTP Fifth Annual Report on Carcinogens. IARC Cancer Review: Group 2A IMEMDT 7,204,87; Animal Sufficient Evidence IMEMDT 15,195,77. NCI Carcinogenesis Bioassay (gavage); Clear Evidence: mouse, rat NCITR* NCI-CG-TR-86,78; NTP Carcinogenesis Bioassay (inhalation); Clear Evidence: mouse, rat NTPTR* NTP-TR-210,82. EPA Genetic Toxicology Program. Community Right-To-Know List. Reported in EPA TSCA Inventory.

OSHA PEL: TWA 20 ppm; CL 30 ppm; Pk 50 ppm/5M/8H
ACGIH TLV: Suspected Human Carcinogen
DFG TRK: 0.1 ppm; Animal Carcinogen, Suspected Human Carcinogen.
NIOSH REL: (EDB) 0.045 ppm; CL 1 mg/m^3/15M
DOT Classification: ORM-A; Label: None; Poison B; Label: Poison.

SAFETY PROFILE: Confirmed carcinogen with experimental carcinogenic, neoplastigenic, and teratogenic data. Human poison by ingestion. Experimental poison by ingestion, skin contact, intraperitoneal, and possibly other routes. Moderately toxic by inhalation and rectal routes. Experimental reproductive effects. Human mutation data reported. A severe skin and eye irritant. An experimental eye irritant. Implicated in worker sterility. When heated to decomposition it emits toxic fumes of Br^-. See also ETHYLENE DICHLORIDE and BROMIDES.

EIY550 CAS:65313-36-2 ***HR: 3***
ETHYLENEDICESIUM
mf: $C_2H_4Cs_2$ mw: 293.86

SAFETY PROFILE: Violent reaction on contact with water. When heated to decomposition it emits acrid smoke and irritating fumes. See also CESIUM.

EIY600 CAS:107-06-2 ***HR: 3***
ETHYLENE DICHLORIDE
DOT: UN 1184
mf: $C_2H_4Cl_2$ mw: 98.96

PROP: Colorless, clear liquid; pleasant odor, sweet taste. Bp: 83.5°, ULC: 60-70, lel: 6.2%, uel: 15.9%, fp: −35.7°, flash p: 56°F, d: 1.257 @ 20°/4°, autoign temp: 775°F, vap press: 100 mm @ 29.4°, vap d: 3.35, refr index: 1.445 @ 20°. Sol in alc, ether, acetone, carbon tetrachloride; sltly sol in water.

SYNS: AETHYLENCHLORID (GERMAN) ◇ BICHLORURE d'ETHYLENE (FRENCH) ◇ BORER SOL ◇ BROCIDE ◇ CHLORURE d'ETHYLENE (FRENCH) ◇ CLORURO di ETHENE (ITALIAN) ◇ 1,2-DCE ◇ DESTRUXOL BORER-SOL ◇ 1,2-DICHLOORETHAAN (DUTCH) ◇ 1,2-DICHLOR-AETHAN (GERMAN) ◇ DICHLOREMULSION ◇ DI-CHLOR-MULSION ◇ DICHLORO-1,2-ETHANE (FRENCH) ◇ 1,2-DICHLOROETHANE ◇ α,β-DICHLOROETHANE ◇ sym-DICHLOROETHANE ◇ DICHLOROETHYLENE ◇ 1,2-DICLOROETANO (ITALIAN) ◇ DUTCH LIQUID ◇ DUTCH OIL ◇ EDC ◇ ENT 1,656 ◇ ETHANE DICHLORIDE ◇ ETHYLEENDICHLORIDE (DUTCH) ◇ ETHYLENE CHLORIDE ◇ 1,2-ETHYLENE DICHLORIDE ◇ GLYCOL DICHLORIDE ◇ NCI-C00511 ◇ RCRA WASTE NUMBER U077

TOXICITY DATA with REFERENCE
skn-rbt 600 mg open MLD UCDS** 3/23/70
eye-rbt 63 mg SEV UCDS** 3/23/70
msc-hmn:lym 100 mg/L MUREAV 142,133,85
slt-mus-ipr 300 mg/kg MUREAV 117,201,83
ihl-rat TCLo:300 ppm/7H (6-15D preg):REP BANRDU 5,149,80
orl-rat TDLo:5286 mg/kg/69W-I:CAR BANRDU 5,35,80
ihl-rat TCLo:5 ppm/7H/78W-I:ETA BANRDU 5,3,80
orl-mus TDLo:3536 mg/kg/78W-I:CAR BANRDU 5,35,80
skn-mus TDLo:1120 g/kg/74W-I:NEO JJIND8 63,1433,79
orl-mus TD:76 g/kg/78W-I:CAR,TER NCITR* NCI-CG-TR-55,78
orl-mus TD:38 g/kg/78W-I:CAR,TER NCITR* NCI-CG-TR-55,78
ihl-hmn TCLo:4000 ppm/H:CNS,PNS,GIT PCOC** -,500,66
orl-hmn LDLo:286 mg/kg:GIT CLCEAL 86,203,47
orl-hmn TDLo:428 mg/kg:GIT,CNS,PUL SOMEAU 22,132,58

orl-man TDLo:892 mg/kg:GIT,LIV WILEAR 28,983,75
orl-man LDLo:714 mg/kg:CNS,CVS,PUL KLWOAZ 48,822,70
orl-rat LD50:670 mg/kg FMCHA2 -,C99,83
ihl-rat LC50:1000 ppm/7H AMIHBC 4,482,51
ipr-rat LD50:807 mg/kg GTPZAB 26(4),26,82
scu-rat LD50:1 g/kg FAONAU 48A,91,70
orl-mus LD50:489 mg/kg TOXID9 1,26,81
ihl-mus LCLo:5000 mg/m^3/2H AEPPAE 141,19,29
scu-mus LDLo:380 mg/kg JPETAB 84,53,45
orl-dog LD50:5700 mg/kg FAONAU 48A,91,70
ivn-dog LDLo:175 mg/kg QJPPAL 7,205,34
orl-rbt LD50:860 mg/kg GUCHAZ 6,264,73
ihl-rbt LCLo:3000 ppm/7H JPETAB 84,53,45

CONSENSUS REPORTS: NTP Fifth Annual Report on Carcinogens. IARC Cancer Review: Group 2B IMEMDT 7,56,87; Human Limited Evidence IMEMDT 20,429,79; Animal Sufficient Evidence IMEMDT 20,429,79. NCI Carcinogenesis Bioassay (gavage); Clear Evidence: mouse-rat NCITR* NCI-CG-TR-55,78. EPA Genetic Toxicology Program. Reported in EPA TSCA Inventory.

OSHA PEL: (Transitional: TWA 50 ppm; CL 100 ppm; PK 200 ppm/5M)) TWA 1 ppm; STEL 2 ppm
ACGIH TLV: TWA 10 ppm
DFG MAK: 20 ppm (80 mg/m^3); Suspected Carcinogen.
NIOSH REL: (Ethylene Dichloride) TWA 1 ppm; CL 2 ppm/15M
DOT Classification: Flammable Liquid; Label: Flammable Liquid; IMO: Flammable Liquid; Label: Flammable Liquid, Poison.

SAFETY PROFILE: Confirmed carcinogen with experimental carcinogenic, neoplastigenic and tumorigenic data. An experimental transplacental carcinogen. A human poison by ingestion. Poison experimentally by intravenous and subcutaneous routes. Moderately toxic by inhalation, skin contact, and intraperitoneal routes. Human systemic effects by ingestion and inhalation: flaccid paralysis without anesthesia (usually neuromuscular blockade), somnolence, cough, jaundice, nausea or vomiting, hypermotility, diarrhea, ulceration or bleeding from the stomach, fatty liver degeneration, change in cardiac rate, cyanosis and coma. It may also cause dermatitis, edema of the lungs, toxic effects on the kidneys, and severe corneal effects. A strong narcotic. Experimental teratogenic and reproductive effects. A skin and severe eye irritant, and strong local irritant. Its smell and irritant effects warn of its presence at relatively safe concentrations. Human mutation data reported.

Flammable liquid. A dangerous fire hazard if exposed to heat, flame, or oxidizers. Moderately explosive in the form of vapor when exposed to flame. Violent reaction with Al, N_2O_4, NH_3, dimethylaminopropylamine. Can react vigorously with oxidizing materials and emit vinyl chloride and HCl. To fight fire, use water, foam, CO_2, dry chemicals. When heated to decomposition it emits highly toxic fumes of Cl^- and phosgene. See also CHLORINATED HYDROCARBONS, ALIPHATIC.

EJA000 CAS:6943-65-3 ***HR: 3***
ETHYLENE DIISOTHIOURONIUM DIBROMIDE
mf: $C_4H_{10}N_4S_2 \cdot 2BrH$ mw: 340.14

SYNS: 1,2-ETHANEDIYL ESTER CARBAMIMIDOTHIOIC ACID DIHYDROBROMIDE ◇ 2,2'-ETHYLENE-BIS-(2-THIOPSEUDOUREA), DIHYDROBROMIDE ◇ ETHYLENE DIISOTHIOUREA DIHYDROBROMIDE ◇ 2,2-ETHYLENEDITHIODIPSEUDOUREA DIHYDROBROMIDE

TOXICITY DATA with REFERENCE
ipr-mus LD50:75 mg/kg NTIS** AD691-490
ivn-mus LD50:56 mg/kg CSLNX* NX#02909
par-mus LD50:160 mg/kg THERAP 8,929,53
unr-mus LDLo:200 mg/kg ATMPA2 32,177,38

SAFETY PROFILE: Poison by intraperitoneal, intravenous, parenteral, and possibly other routes. When heated to decomposition it emits very toxic fumes of NO_x, SO_x, and HBr. See also BROMIDES.

EJA100 CAS:3715-67-1 ***HR: 3***
N,N-ETHYLENE-N',N'-DIMETHYLUREA
mf: $C_5H_{10}N_2O$ mw: 114.17

SYNS: 1-AZIRIDINECARBOXAMIDE, N,N-DIMETHYL- ◇ N,N-DIMETHYL-1-AZIRIDINECARBOXAMIDE ◇ N-(DIMETHYLCARBAMOYL)AZIRIDINE ◇ 1-(DIMETHYLCARBAMYL)AZIRIDINE ◇ N,N-DIMETHYLETHYLENEUREA

TOXICITY DATA with REFERENCE
mmo-ssp 140 mmol/L ADWMAX -,193,62
ipr-rat TDLo:100 mg/kg (male 1D pre):REP 85GUAJ -,37,66
ipr-rat LD50:200 mg/kg BJPCAL 21,581,63

SAFETY PROFILE: Poison by intraperitoneal route. Mutation data reported. Experimental reproductive effects. When heated to decomposition it emits toxic fumes of NO_x.

EJA250 ***HR: 1***
(ETHYLENEDINITRILO)TETRAACETATE DIPOTASSIUM SALT
mf: $C_{10}H_{16}N_2O_8 \cdot 2K$ mw: 370.48

SYN: DIPOTASSIUMETHYLENEDIAMINETETRAACETATE

TOXICITY DATA with REFERENCE
skn-rbt 500 mg MLD FCTOD7 20,563,82
eye-rbt 100 mg MLD FCTOD7 20,573,82
eye-rbt 100 mg/30S rns MLD FCTOD7 20,573,82

SAFETY PROFILE: A skin and eye irritant. When heated to decomposition it emits toxic fumes of NO_x and K_2O.

EJA379 CAS:15708-41-5 ***HR: 1***
((ETHYLENEDINITRILO)TETRAACETATO)-FERATE(1-), SODIUM
mf: $C_{10}H_{12}FeN_2O_8{\bullet}Na$ mw:367.08

SYNS: CALMOSINE ◇ EDATHAMIL MONOSODIUM FERRIC SALT ◇ FERISAN ◇ FERRIC SODIUM EDETATE ◇ FERRIC SODIUM EDTA ◇ MONOSODIUM FERRIC EDTA ◇ REXENE ◇ SEQUESTRENE NaFe IRON CHELATE ◇ SODIUM FEREDETATE ◇ SODIUM FERRIC EDTA ◇ SODIUM IRON EDTA ◇ SYTRON

TOXICITY DATA with REFERENCE
orl-rat LD50:5 g/kg CIGET* -,-,77
orl-mus LD50:5000 mg/kg CIGET* -,-,77

CONSENSUS REPORTS: Reported in EPA TSCA Inventory.

SAFETY PROFILE: Mildly toxic by ingestion. When heated to decomposition it emits toxic fumes of NO_x and Na_2O. See also ETHYLENE DIAMINE TETRAACETIC ACID DISODIUM SALT.

EJA500 CAS:25481-21-4 ***HR: 3***
(ETHYLENEDINITRILO)TETRA ACETIC ACID NICKEL(II) COMPLEX
mf: $C_{10}H_{12}N_2NiO_8{\bullet}2H$ mw: 348.96

SYNS: DIHYDROGEN(ETHYLENEDIAMINETETRAACETATO(4-)) NICKELATE (2-) ◇ NICKEL(II) EDTA COMPLEX

TOXICITY DATA with REFERENCE
ipr-mus LD50:88 mg(Ni)/kg PABIAQ 11,853,63

CONSENSUS REPORTS: Nickel and its compounds are on the Community Right-To-Know List. Reported in EPA TSCA Inventory.

SAFETY PROFILE: Poison by intraperitoneal route. When heated to decomposition it emits toxic fumes of NO_x. See also NICKEL COMPOUNDS.

EJB000 CAS:5766-67-6 ***HR: 2***
(ETHYLENEDINITRILO)TETRAACETONITRILE
mf: $C_{10}H_{12}N_6$ mw: 216.28

SYNS: ETHYLENEDIAMINETETRAACETONITRILE ◇ N,N,N',N'-TETRACYANOMETHYLAETHYLENEDIAMIN(GERMAN)

TOXICITY DATA with REFERENCE
ipr-mus LD50:3030 mg/kg ARZNAD 16,734,66

CONSENSUS REPORTS: Cyanide and its compounds are on the Community Right-To-Know List. Reported in EPA TSCA Inventory.

SAFETY PROFILE: Moderately toxic by intraperitoneal route. When heated to decomposition it emits toxic fumes of NO_x and CN^-. See also NITRILES.

EJB100 ***HR: 3***
6,6'-(ETHYLENEDIOXY)BIS(4-AMINOQUINALDINE) DIHYDROCHLORIDE
mf: $C_{22}H_{22}N_4O_2{\bullet}2ClH$ mw: 447.40

SYN: BIS-(2-METHYL-4-AMINO-6-QUINOLYLOXY)ETHANEDIHYDROCHLORIDE

TOXICITY DATA with REFERENCE
orl-mus LD50:1530 mg/kg ANTCAO 2,581,52
ipr-mus LD50:128 mg/kg ANTCAO 2,581,52
scu-mus LD50:130 mg/kg ANTCAO 2,581,52
ivn-mus LD50:12100 μg/kg ANTCAO 2,581,52

SAFETY PROFILE: Poison by subcutaneous, intravenous, and intraperitoneal routes. Moderately toxic by ingestion. When heated to decomposition it emits toxic fumes of NO_x and HCl.

EJB500 CAS:111-21-7 ***HR: 1***
2,2'-(ETHYLENEDIOXY)DI(ETHYL ACETATE)
mf: $C_{10}H_{18}O_6$ mw: 234.28

SYNS: ACETIC ACID, TRIETHYLENE GLYCOL DIESTER ◇ 2,2'-ETHYLENEDIOXYDIETHANOL DIACETATE ◇ TRIETHYLENE GLYCOL, DIACETATE ◇ TRIGLYCOL, DIACETATE

TOXICITY DATA with REFERENCE
skn-rbt 500 mg open MLD UCDS** 3/27/69
orl-rat LD50:22600 mg/kg AIHAAP 30,470,69
skn-rbt LD50:8 g/kg AIHAAP 30,470,69

CONSENSUS REPORTS: Reported in EPA TSCA Inventory.

SAFETY PROFILE: Mildly toxic by ingestion and skin contact. A skin irritant. When heated to decomposition it emits acrid smoke and irritating fumes.

EJC000 CAS:52936-25-1 ***HR: 3***
ETHYLENE DIPERCHLORATE
mf: $C_2H_4Cl_2O_8$ mw: 226.96

$(-CH_2OClO_3)_2$

SAFETY PROFILE: A highly sensitive, powerful explosive. Explodes on contact with small amounts of water. Upon decomposition it emits toxic fumes of Cl^-. See also PERCHLORATES.

EJC500 CAS:107-21-1 ***HR: 3***
ETHYLENE GLYCOL
mf: $C_2H_6O_2$ mw: 62.08

PROP: Colorless, sweet-tasting, hygroscopic liquid. Bp: 197.5°, lel: 3.2%, fp: −13°, flash p: 232°F (CC), d: 1.113 @ 25°/25°, autoign temp: 752°F, vap d: 2.14, vap press: 0.05 mm @ 20°.

SYNS: ATHYLENGLYKOL (GERMAN) ◇ 1,2-DIHYDROXYETHANE ◇ DOWTHERM SR 1 ◇ 1,2-ETHANEDIOL ◇ ETHYLENE ALCOHOL

◇ ETHYLENE DIHYDRATE ◇ GLYCOL ◇ GLYCOL ALCOHOL ◇ LUTROL-9 ◇ MACROGOL 400 BPC ◇ M.E.G. ◇ MONOETHYLENE GLYCOL ◇ NCI-C00920 ◇ NORKOOL ◇ TESCOL ◇ UCAR 17

TOXICITY DATA with REFERENCE
skn-rbt 555 mg open MLD UCDS** 7/21/65
eye-rat 12 mg/m^3/3D TXAPA9 16,646,70
eye-rbt 500 mg/24H MLD 28ZPAK -,37,72
eye-rbt 100 mg/1H MLD NTIS** LMF-69
dni-hmn:lym 320 mmol/L PNASA6 79,1171,82
msc-mus:lym 100 mmol/L PAACA3 21,74,80
orl-mus TDLo:84 g/kg (female 1-21D post):REP TOXID9 4,136,84
orl-rat TDLo:8580 mg/kg (female 6-15D post):TER CHYCDW 20,289,86
orl-chd TDLo:5500 mg/kg:CNS,PUL,KID PGMJAO 52,598,76
orl-hmn LDLo:786 mg/kg EJTXAZ 9,373,76
orl-hmn LDLo:398 mg/kg:CNS,GIT,LIV SMEZA5 26(2),48,83
ihl-hmn TCLo:10000 mg/m^3:EYE,PUL AGGHAR 5,1,33
unr-man LDLo:1637 mg/kg 85DCAI 2,73,70
orl-rat LD50:4700 mg/kg GTPZAB 26(6),28,82
ipr-rat LD50:5010 mg/kg KRKRDT 9,36,81
scu-rat LD50:2800 mg/kg NPIRI* 1,49,74
ivn-rat LD50:3260 mg/kg KRKRDT 9,36,81
ims-rat LDLo:3300 mg/kg JPETAB 41,387,31
orl-mus LD50:7500 mg/kg JPETAB 65,89,39
ipr-mus LD50:5614 mg/kg FEPRA7 6,342,47
scu-mus LDLo:2700 mg/kg BJIMAG 1,207,44

CONSENSUS REPORTS: EPA Genetic Toxicology Program. Community Right-To-Know List. Reported in EPA TSCA Inventory.

OSHA PEL: CL 50 ppm
ACGIH TLV: CL 50 ppm (vapor)

SAFETY PROFILE: Human poison by ingestion. (Lethal dose for humans reported to be 100 mL.) Moderately toxic to humans by an unspecified route. Moderately toxic experimentally by ingestion, subcutaneous, intravenous, and intramuscular routes. Human systemic effects by ingestion and inhalation: eye lacrimation, general anesthesia, headache, cough, respiratory stimulation, nausea or vomiting, pulmonary, kidney and liver changes. If ingested it causes initial central nervous system stimulation followed by depression. Later, it causes potentially lethal kidney damage. Very toxic in particulate form upon inhalation. An experimental teratogen. Other experimental reproductive effects. Human mutation data reported. A skin, eye, and mucous membrane irritant.

Combustible when exposed to heat or flame; can react vigorously with oxidants. Moderate explosion hazard when exposed to flame. Ignites on contact with chromium trioxide, potassium permanganate, and sodium peroxide. Mixtures with ammonium dichromate, silver chlorate, sodium chlorite, and uranyl nitrate ignite when heated to 100°C. Can react violently with chlorosulfonic acid, oleum, H_2SO_4, $HClO_4$, and P_2S_5. Aqueous solutions may ignite silvered copper wires which have an applied D.C. voltage. To fight fire, use alcohol foam, water, foam, CO_2, dry chemical. When heated to decomposition it emits acrid smoke and irritating fumes.

EJD000 CAS:3775-85-7 ***HR: 2***
ETHYLENE GLYCOL BIS(2,3-EPOXY-2-METHYLPROPYL) ETHER
mf: $C_{10}H_{18}O_4$ mw: 202.28

SYNS: ETHYLENE GLYCOL DI(2,3-EPOXY-2-METHYLPROPYL) ETHER ◇ ETHYLENE GLYCOLIDE (2,3-EPOXY-2-METHYLPROPYL) ETHER

TOXICITY DATA with REFERENCE
skn-rbt 10 mg/24H open MLD AIHAAP 23,95,62
orl-rat LD50:7460 mg/kg AIHAAP 24,305,63
skn-rbt LD50:3150 mg/kg AIHAAP 23,95,62

CONSENSUS REPORTS: Glycol ether compounds are on the Community Right-To-Know List.

SAFETY PROFILE: Moderately toxic by skin contact. Mildly toxic by ingestion. A skin irritant. When heated to decomposition it emits acrid smoke and irritating fumes. See also GLYCOL ETHERS.

EJD759 CAS:111-55-7 ***HR: 2***
ETHYLENE GLYCOL DIACETATE
mf: $C_6H_{10}O_4$ mw: 146.16

PROP: Colorless liquid or crystals. Mp: −31°, bp: 191°, flash p: 205°F (OC), d: 1.128 @ 0°/4°, vap press: 1 mm @ 38.3°, vap d: 5.04.

SYNS: 1,2-ETHANEDIOL DIACETATE ◇ ETHYLENE ACETATE ◇ ETHYLENE GLYCOL ACETATE ◇ GLYCOL DIACETATE

TOXICITY DATA with REFERENCE
eye-rbt 555 mg AJOPAA 29,1363,46
orl-rat LD50:6850 mg/kg UCDS** 6/6/69
ipr-mus LD50:1070 mg/kg JPETAB 90,338,47
skn-rbt LD50:8480 mg/kg UCDS** 6/6/69
orl-gpg LD50:4940 mg/kg JIHTAB 23,259,41

CONSENSUS REPORTS: Reported in EPA TSCA Inventory.

SAFETY PROFILE: Moderately toxic by intraperitoneal route. Mildly toxic by ingestion and skin contact. An eye irritant. Combustible when exposed to heat or flame; can react with oxidizing materials. To fight fire, use alcohol foam, CO_2, dry chemical. When heated to decomposition it emits acrid smoke and irritating fumes.

EJE000 CAS:7529-27-3 ***HR: 2***
ETHYLENE GLYCOL DIALLYL ETHER
mf: $C_8H_{14}O_2$ mw: 142.22

SYN: DIALLYLETHER ETHYLENGLYKOLU (CZECH)

TOXICITY DATA with REFERENCE
skn-rbt 500 mg/24H SEV 28ZPAK -,38,72
eye-rbt 250 µg/24H SEV 28ZPAK -,38,72
orl-rat LD50:1020 mg/kg 28ZPAK -,38,72

CONSENSUS REPORTS: Glycol ether compounds are on the Community Right-To-Know List. Reported in EPA TSCA Inventory.

SAFETY PROFILE: Moderately toxic by ingestion. A severe skin and eye irritant. When heated to decomposition it emits acrid smoke and irritating fumes. See also GLYCOL ETHERS and ALLYL COMPOUNDS.

EJE500 CAS:629-14-1 ***HR: 3***
ETHYLENE GLYCOL DIETHYL ETHER
DOT: UN 1153
mf: $C_6H_{14}O_2$ mw: 118.20

PROP: Colorless liquid, slight ethereal odor. Mp: −74°, bp: 121.4°, flash p: 95°F (OC), d: 0.8417 @ 20°/20°, autoign temp: 406°F, vap d: 6.56, vap press: 9.4 mm.

SYNS: 1,2-DIETHOXYETHANE ◇ DIETHYL CELLOSOLVE (DOT) ◇ ETHYL GLYME

TOXICITY DATA with REFERENCE
eye-rbt 17 mg AJOPAA 29,1363,46
orl-mus TDLo:23640 mg/kg (female 6-13D post):REP TCMUD8 7,29,87
orl-mus TDLo:10 g/kg (female 6-15D post):TER NTIS** PB88-134093
orl-rat LD50:4390 mg/kg JIHTAB 23,259,41
ihl-rat LCLo:8000 ppm/4H JIHTAB 31,343,49
orl-gpg LD50:2440 mg/kg JIHTAB 23,259,41

CONSENSUS REPORTS: Reported in EPA TSCA Inventory. Glycol ether compounds are on the Community Right-To-Know List.

DOT Classification: Combustible Liquid; Label: None; Flammable or Combustible Liquid; Label: Flammable Liquid.

SAFETY PROFILE: Moderately toxic by ingestion. Mildly toxic by inhalation. An experimental teratogen. Experimental reproductive effects. An eye irritant. An aprotic solvent. A very dangerous fire hazard when exposed to heat or flame; can react with oxidizing materials. To fight fire, use CO_2, dry chemical. See also GLYCOL ETHERS and various cellosolve entries.

EJF000 CAS:629-15-2 ***HR: 3***
ETHYLENE GLYCOL DIFORMATE
mf: $C_4H_6O_4$ mw: 118.10

PROP: Liquid. Mp: −10°, bp: 177°, flash p: 200°F (OC), d: 1.2277 @ 20°/20°, vap d: 4.07.

SYNS: ETHYLENE FORMATE ◇ GLYCOL DIFORMATE

TOXICITY DATA with REFERENCE
eye-rbt 5 mg SEV AJOPAA 29,1363,46
orl-rat LD50:1510 mg/kg JIHTAB 23,259,41
orl-gpg LD50:390 mg/kg JIHTAB 23,259,41

CONSENSUS REPORTS: Reported in EPA TSCA Inventory.

SAFETY PROFILE: Poison by ingestion. A severe eye irritant. Flammable when exposed to heat or flame; can react with oxidizing materials. To fight fire, use CO_2, dry chemical. When heated to decomposition it emits acrid smoke and irritating fumes.

EJG000 CAS:628-96-6 ***HR: 3***
ETHYLENE GLYCOL DINITRATE
mf: $C_2H_4N_2O_6$ mw: 152.08

PROP: Yellow liquid. Mp: −20°, bp: explodes @ 114°, d: 1.483 @ 8°, vap d: 5.25.

SYNS: DINITROGLICOL (ITALIAN) ◇ DINITROGLYCOL ◇ EGDN ◇ ETHANEDIOL DINITRATE ◇ ETHYLENE DINITRATE ◇ ETHYLENE NITRATE ◇ ETHYLENGLYKOLDINITRAT (CZECH) ◇ GLYCOLDINITRAAT (DUTCH) ◇ GLYCOL DINITRATE ◇ GLYCOL (DINITRATE DE) (FRENCH) ◇ GLYKOLDINITRAT (GERMAN) ◇ NITROGLYCOL ◇ NITROGLYKOL (CZECH)

TOXICITY DATA with REFERENCE
orl-rat LD50:616 mg/kg 28ZPAK -,48,72
scu-cat LDLo:50 mg/kg AEPPAE 200,271,42
scu-rbt LDLo:300 mg/kg AEPPAE 200,271,42

CONSENSUS REPORTS: Reported in EPA TSCA Inventory.

OSHA PEL: (Transitional: CL 0.2 ppm (skin)) STEL 0.1 mg/m^3 (skin)
ACGIH TLV: TWA 0.05 ppm (skin)
DFG MAK: 0.05 ppm (0.3 mg/m^3)
NIOSH REL: (Nitroglycerin) CL 0.1 mg/m^3/20M
DOT Classification: Forbidden.

SAFETY PROFILE: Poison by subcutaneous route. Moderately toxic by ingestion. Can cause lowered blood pressure leading to headache, dizziness, and weakness. Used as an explosive. When heated to decomposition it emits toxic fumes of NO_x. See also NITRATES.

EJG500 CAS:26560-94-1 ***HR: 2***
ETHYLENE GLYCOL MALEATE
mf: $C_6H_8O_5$ mw: 160.14

SYNS: ETHYLENE GLYCOL, MONO(HYDROGEN MALEATE) ◇ 2-HYDROXYETHYL ESTER MALEIC ACID ◇ MONO(HYDROXYETHYL) ESTER MALEIC ACID

TOXICITY DATA with REFERENCE
eye-rbt 2 mg open SEV AMIHBC 10,61,54
orl-rat LD50:2460 mg/kg AMIHBC 10,61,54

CONSENSUS REPORTS: Reported in EPA TSCA Inventory.

SAFETY PROFILE: Moderately toxic by ingestion. A severe eye irritant. When heated to decomposition it emits acrid smoke and irritating fumes. See also ESTERS.

EJH000 CAS:868-77-9 ***HR: 2***
ETHYLENE GLYCOL METHACRYLATE
mf: $C_6H_{10}O_3$ mw: 130.16

SYNS: ETHYLENE GLYCOL, MONOMETHACRYLATE ◇ GLYCOL METHACRYLATE ◇ GLYCOL MONOMETHACRYLATE ◇ 2-HYDROXYETHYL ESTER METHACRYLIC ACID ◇ HYDROXYETHYL METHACRYLATE ◇ β-HYDROXYETHYL METHACRYLATE ◇ 2-HYDROXYETHYL METHACRYLATE ◇ MHOROMER ◇ MONOMER MG-1

TOXICITY DATA with REFERENCE
ipr-rat LD50:1250 mg/kg AMPMAR 36,58,75
orl-mus LD50:5888 mg/kg TOLED5 11,125,82
ipr-mus LD50:497 mg/kg JPMSAE 62,778,73

CONSENSUS REPORTS: Reported in EPA TSCA Inventory.

SAFETY PROFILE: Moderately toxic by intraperitoneal route. Mildly toxic by ingestion. When heated to decomposition it emits acrid smoke and irritating fumes.

EJH500 CAS:109-86-4 ***HR: 3***
ETHYLENE GLYCOL METHYL ETHER
DOT: UN 1188
mf: $C_3H_8O_2$ mw: 76.11

$CH_3OC_2H_4OH$

PROP: Colorless liquid; mild, agreeable odor. Misc in water, alc, ether, benzene. Bp: 124.5°, fp: −86.5°, flash p: 115°F (OC), lel: 2.5%, uel: 14%, d: 0.9660 @ 20°/4°, autoign temp: 545°F, vap press: 6.2 mm @ 20°, vap d: 2.62.

SYNS: AETHYLENGLYKOL-MONOMETHYLAETHER(GERMAN) ◇ DOWANOL EM ◇ EGM ◇ EGME ◇ ETHER MONOMETHYLIQUE de l'ETHYLENE-GLYCOL (FRENCH) ◇ ETHYLENE GLYCOL MONOMETHYL ETHER (MAK, DOT) ◇ GLYCOL ETHER EM ◇ GLYCOLMETHYL ETHER ◇ GLYCOL MONOMETHYL ETHER ◇ JEFFERSOL EM ◇ MECS ◇ 2-METHOXY-AETHANOL (GERMAN) ◇ 2-METHOXYETHANOL (ACGIH) ◇ METHOXYHYDROXYETHANE ◇ METHYL CELLOSOLVE (OSHA, DOT) ◇ METHYL ETHOXOL ◇ METHYL GLYCOL ◇ METHYLGLYKOL (GERMAN) ◇ METHYL OXITOL ◇ METIL CELLOSOLVE (ITALIAN) ◇ METOKSYETYLOWY ALKOHOL (POLISH) ◇ 2-METOSSIETANOLO (ITALIAN) ◇ MONOMETHYL ETHER of ETHYLENE GLYCOL ◇ POLY-SOLV EM ◇ PRIST

TOXICITY DATA with REFERENCE
skn-rbt 483 mg/24H MLD TXAPA9 19,276,71
eye-rbt 97 mg TXAPA9 19,276,71
eye-gpg 10 μg MLD JPPMAB 11,150,59
dlt-rat-orl 500 mg/kg ENMUDM 6,390,84
spm-rat-orl 500 mg/kg ENMUDM 6,390,84
spm-mus-orl 500 mg/kg ENMUDM 6,390,84
ihl-rat TCLo:25 ppm/7H (female 7-13D post):REP TJADAB 27,65A,83
orl-mky TDLo:930 mg/kg (female 20-45D post):TER TJADAB 35,66A,87
orl-hmn LDLo:3380 mg/kg JIHTAB 28,267,46
ihl-hmn TCLo:25 ppm:CNS JIHTAB 20,134,38
orl-rat LD50:2460 mg/kg JIHTAB 23,259,41
ihl-rat LC50:1500 ppm/7H NPIRI* 1,57,74
ipr-rat LD50:2500 mg/kg NPIRI* 1,57,74
ivn-rat LD50:2140 mg/kg AMIHAB 14,114,56
orl-mus LD50:2560 mg/kg GTPZAB 32(3),48,88
ihl-mus LC50:1480 ppm JIHTAB 25,157,43
ipr-mus LD50:2147 mg/kg FEPRA7 6,342,47
orl-rbt LD50:890 mg/kg AMIHAB 14,114,56
skn-rbt LD50:1280 mg/kg NPIRI* 1,57,74
orl-gpg LD50:950 mg/kg JIHTAB 23,259,41

CONSENSUS REPORTS: Reported in EPA TSCA Inventory. Community Right-To-Know List.

OSHA PEL: TWA 25 ppm (skin)
ACGIH TLV: TWA 5 ppm (skin)
DFG MAK: 5 ppm (15 mg/m^3)
NIOSH REL: TWA (Glycol Ethers): Reduce to lowest level
DOT Classification: Combustible Liquid; Label: None; Flammable or Combustible Liquid; Label: Flammable Liquid.

SAFETY PROFILE: Moderately toxic to humans by ingestion. Moderately toxic experimentally by ingestion, inhalation, skin contact, intraperitoneal, and intravenous routes. Human systemic effects by inhalation: change in motor activity, tremors, and convulsions. Experimental teratogenic and reproductive effects. A skin and eye irritant. Mutation data reported. When used under conditions which do not require the application of heat, this material probably presents little hazard to health. However, in the manufacture of fused collars which require pressing with a hot iron, cases have been reported showing disturbance of the hemopoietic system with or without neurological signs and symptoms. The blood picture may resemble that produced by exposure to benzene. Two cases reported had severe aplastic anemia with tremors and marked mental dullness. The persons affected had been exposed to vapors of methyl "Cellosolve", ethanol, and methanol, ethyl acetate and petroleum naphtha.

Flammable or combustible when exposed to heat or

flame. A moderate explosion hazard. Can react with oxidizing materials to form explosive peroxides. To fight fire, use alcohol foam, CO_2, dry chemical. When heated to decomposition it emits acrid smoke and irritating fumes. See also GLYCOL ETHERS.

EJI000 CAS:542-59-6 ***HR: 2***
ETHYLENE GLYCOL MONOACETATE
mf: $C_4H_8O_3$ mw: 104.12

PROP: Colorless, almost odorless liquid. Bp: 182°, flash p: 215°F (OC), d: 1.108, @ 15°, vap d: 3.59.

SYNS: 1,2-ETHANEDIOL, MONOACETATE ◇ ETHYLENE GLYCOL ACETATE ◇ GLYCOL MONOACETATE ◇ GLYCOL-MONOACETIN ◇ 2-HYDROXYETHYL ACETATE

TOXICITY DATA with REFERENCE
eye-rbt 22 mg AJOPAA 29,1363,46
orl-rat LD50:8250 mg/kg JIHTAB 23,259,41
ipr-mus LD50:1310 mg/kg JPETAB 90,338,47
orl-gpg LD50:3800 mg/kg JIHTAB 23,259,41

CONSENSUS REPORTS: Reported in EPA TSCA Inventory.

SAFETY PROFILE: Moderately toxic by ingestion amd intraperitoneal routes. An eye irritant. Combustible when exposed to heat or flame; can react with oxidizing materials. To fight fire, use alcohol foam, foam, water, CO_2, dry chemical. When heated to decomposition it emits acrid smoke and irritating fumes.

EJI500 CAS:622-08-2 ***HR: 2***
ETHYLENE GLYCOL MONOBENZYL ETHER
mf: $C_9H_{12}O_2$ mw: 152.21

PROP: Water-white liquid, faint rose-like odor. Mp: −75°, bp: 256°, flash p: 265°F (OC), d: 1.068, autoign temp: 665°F, vap d: 5.25.

SYN: BENZYL "CELLOSOLVE"

TOXICITY DATA with REFERENCE
eye-rbt 2 mg SEV AJOPAA 29,1363,46
orl-rat LD50:1190 mg/kg JIHTAB 23,259,41

CONSENSUS REPORTS: Glycol ether compounds are on the Community Right-To-Know List. Reported in EPA TSCA Inventory.

SAFETY PROFILE: Moderately toxic by ingestion. A severe eye irritant. Combustible when exposed to heat or flame; can react with oxidizing materials. To fight fire, use CO_2, dry chemical. When heated to decomposition it emits acrid smoke and irritating fumes. See also GLYCOL ETHERS.

EJJ000 CAS:7795-91-7 ***HR: 2***
ETHYLENE GLYCOL MONO-sec-BUTYL ETHER
mf: $C_6H_{14}O_2$ mw: 118.20

SYN: 2-sec-BUTOXYETHANOL

TOXICITY DATA with REFERENCE
orl-rat LD50:790 mg/kg SCCUR* -,5,61
orl-mus LD50:1660 mg/kg SCCUR* -,5,61

CONSENSUS REPORTS: Glycol ether compounds are on the Community Right-To-Know List.

SAFETY PROFILE: Moderately toxic by ingestion. When heated to decomposition it emits acrid smoke and irritating fumes. See also GLYCOL ETHERS.

EJJ500 CAS:110-49-6 ***HR: 3***
ETHYLENE GLYCOL MONOMETHYL ETHER ACETATE
DOT: UN 1189
mf: $C_5H_{10}O_3$ mw: 118.15

PROP: Colorless liquid. Bp: 143°, fp: −70°, flash p: 111°F (CC), d: 1.005 @ 20°/20°, vap d: 4.07, lel: 1.7%, uel: 8.2%.

SYNS: ACETATE de L'ETHER MONOMETHYLIQUE de L'ETHYLENE-GLYCOL (FRENCH) ◇ ACETATE de METHYLE GLYCOL (FRENCH) ◇ ACETATO di METIL CELLOSOLVE (ITALIAN) ◇ AETHYLENGLYKOLMETHYLAETHERACETAT (GERMAN) ◇ ETHYLENE GLYCOL METHYL ETHER ACETATE ◇ GLYCOL ETHER EM ACETATE ◇ GLYCOL MONOMETHYL ETHER ACETATE ◇ MeCsAc ◇ 2-METHOXY-AETHYLACETAT (GERMAN) ◇ 2-METHOXYETHANOL, ACETATE ◇ 2-METHOXY-ETHYL ACETAAT (DUTCH) ◇ 2-METHOXYETHYL ACETATE (ACGIH) ◇ 2-METHOXYETHYLE, ACETATE de (FRENCH) ◇ METHYL CELLOSOLYE ACETAAT (DUTCH) ◇ METHYL CELLOSOLVE ACETATE (OSHA, DOT) ◇ METHYL GLYCOL ACETATE ◇ METHYL GLYCOL MONOACETATE ◇ METHYLGLYKOLACETAT (GERMAN) ◇ 2-METOSSIETILACETATO(ITALIAN)

TOXICITY DATA with REFERENCE
eye-rbt 218 mg MLD UCDS** 3/20/73
sln-smc 56600 ppm MUREAV 149,339,85
orl-mus TDLo:9800 mg/kg (female 6-13D post):REP TCMUD8 7,29,87
ihl-hmn TCLo:1000 mg/m^3:EYE,PUL AGGHAR 5,1,33
orl-rat LD50:3390 mg/kg JIHTAB 30,63,48
ihl-rat LCLo:7000 ppm/4H JIHTAB 30,63,48
ipr-rat LDLo:1200 mg/kg JPPMAB 11,150,59
ihl-cat LCLo:6 g/m^3/7H AGGHAR 5,1,33
scu-cat LDLo:3000 mg/kg AGGHAR 5,1,33
skn-rbt LD50:5250 mg/kg UCDS** 3/20/73
orl-gpg LD50:1250 mg/kg JIHTAB 23,259,41
scu-gpg LDLo:5000 mg/kg AGGHAR 5,1,33

CONSENSUS REPORTS: Glycol ether compounds are on the Community Right-To-Know List. Reported in EPA TSCA Inventory.

OSHA PEL: TWA 25 ppm (skin)
ACGIH TLV: TWA 5 ppm (skin)
DFG MAK: 5 ppm (25 mg/m^3)
DOT Classification: Flammable or Combustible Liquid; Label: Flammable Liquid.

SAFETY PROFILE: Moderately toxic by ingestion, intraperitoneal, and subcutaneous routes. Mildly toxic by inhalation and skin contact. Human systemic effects by inhalation: eye lacrimation, cough, and pulmonary changes. Experimental reproductive effects. Mutation data reported. An inhalation irritant in humans. An eye irritant. Flammable when exposed to heat or flame; can react with oxidizing materials. A moderate explosion hazard. To fight fire, use CO_2, dry chemical. When heated to decomposition it emits acrid smoke and irritating fumes. See also GLYCOL ETHERS.

EJK000 CAS:10137-96-9 ***HR: 2***
ETHYLENE GLYCOL MONO-2-METHYLPENTYL ETHER
mf: $C_8H_{18}O_2$ mw: 146.26

SYNS: ETHYLENE GLYCOL MONOMETHYLPENTYL ETHER ◇ 2-METHYLPENTYL CELLOSOLVE ◇ 2-((2-METHYLPENTYL)OXY) ETHANOL

TOXICITY DATA with REFERENCE
skn-rbt 10 mg/24H open MLD AIHAAP 23,95,62
eye-rbt 100 mg SEV 34ZIAG -,729,69
orl-rat LD50:3730 mg/kg AIHAAP 23,95,62
skn-rbt LD50:440 mg/kg AIHAAP 23,95,62

CONSENSUS REPORTS: Glycol ether compounds are on the Community Right-To-Know List.

SAFETY PROFILE: Moderately toxic by ingestion and skin contact. A skin and severe eye irritant. When heated to decomposition it emits acrid smoke and irritating fumes. See also GLYCOL ETHERS.

EJK500 CAS:23495-12-7 ***HR: 1***
ETHYLENEGLYCOL MONOPHENYL ETHER PROPIONATE
mf: $C_{11}H_{14}O_3$ mw: 194.25

SYNS: 2-PHENOXYETHANOL PROPIONATE ◇ PHENOXYETHYL PROPIONATE ◇ PROPIONIC ACID-2-PHENOXYETHYL ESTER

TOXICITY DATA with REFERENCE
orl-rat LD50:4400 mg/kg FCTXAV 14,659,76

CONSENSUS REPORTS: Glycol ether compounds are on the Community Right-To-Know List. Reported in EPA TSCA Inventory.

SAFETY PROFILE: Mildly toxic by ingestion. When heated to decomposition it emits acrid smoke and irritating fumes. See also GLYCOL ETHERS.

EJL000 CAS:10137-98-1 ***HR: 2***
ETHYLENE GLYCOL MONO-2,6,8-TRIMETHYL-4-NONYL ETHER
mf: $C_{14}H_{30}O_2$ mw: 230.44

SYN: 2-((1-ISOBUTYL-3,5-DIMETHYLHEXYL)OXY)ETHANOL

TOXICITY DATA with REFERENCE
skn-rbt 10 mg/24H open MLD AIHAAP 23,95,62
orl-rat LD50:5360 mg/kg AIHAAP 23,95,62
skn-rbt LD50:3150 mg/kg AIHAAP 23,95,62

CONSENSUS REPORTS: Glycol ether compounds are on the Community Right-To-Know List.

SAFETY PROFILE: Moderately toxic by skin contact. Mildly toxic by ingestion. A skin irritant. When heated to decomposition it emits acrid smoke and irritating fumes. See also GLYCOL ETHERS.

EJL500 CAS:764-48-7 ***HR: 2***
ETHYLENE GLYCOL MONOVINYL ETHER
mf: $C_4H_8O_2$ mw: 88.12

SYNS: 2-(ETHENYLOXY)ETHANOL ◇ ETHYLENE GLYCOL VINYL ETHER ◇ ETHYLENGLYCOL MONOVINYL ESTER (RUSSIAN) ◇ 2-HYDROXYETHYL VINYL ETHER ◇ MVEEG (RUSSIAN) ◇ VINYLOXYETHANOL ◇ 2-(VINYLOXY)ETHANOL

TOXICITY DATA with REFERENCE
orl-rat LD50:3910 mg/kg GISAAA (3),12,77
orl-mus LD50:2900 mg/kg GISAAA (3),12,77
ihl-mus LC50:29 g/m^3 GISAAA 39(11),94,74

CONSENSUS REPORTS: Glycol ether compounds are on the Community Right-To-Know List.

SAFETY PROFILE: Moderately toxic by ingestion. Mildly toxic by inhalation. When heated to decomposition it emits acrid smoke and irritating fumes. See also GLYCOL ETHERS.

EJM000 CAS:17622-94-5 ***HR: 2***
ETHYLENE GLYCOL SILICATE
mf: $C_8H_{20}O_8Si$ mw: 272.37

SYNS: TETRAGLYCOL SILICATE ◇ TETRAKIS(β-HYDROXYETHYL) SILICATE ◇ TETRAKIS(2-HYDROXYETHYL)SILICATE

TOXICITY DATA with REFERENCE
ivn-dog LDLo:568 mg/kg BIJOAK 27,1007,33

CONSENSUS REPORTS: Reported in EPA TSCA Inventory.

SAFETY PROFILE: Moderately toxic by intravenous route. When heated to decomposition it emits acrid smoke and irritating fumes.

EJM500 CAS:111-60-4 ***HR: 3***
ETHYLENE GLYCOL STEARATE
mf: $C_{20}H_{40}O_3$ mw: 328.60

SYNS: CLINDROL SEG ◇ EMEREST 2350 ◇ EMPILAN 2848 ◇ ETHYLENE GLYCOL, MONOSTEARATE ◇ GLYCOL MONOSTEARATE ◇ GLYCOL STEARATE ◇ 2-HYDROXYETHYL ESTER STEARIC ACID ◇ IVORIT ◇ LIPO EGMS ◇ MONTHYBASE ◇ MONTHYLE ◇ PARASTARIN ◇ PRODHYBASE ETHYL ◇ S 151 ◇ SEDETOL ◇ STEARIC ACID, MONOESTER with ETHYLENE GLYCOL ◇ TEGO-STEARATE ◇ USAF KE-11

TOXICITY DATA with REFERENCE
skn-rbt 500 mg/24H MLD JACTDZ 1(2),1,82
ipr-mus LD50:200 mg/kg NTIS** AD277-689

CONSENSUS REPORTS: Reported in EPA TSCA Inventory.

SAFETY PROFILE: Poison by intraperitoneal route. A skin irritant. Used in cosmetics. When heated to decomposition it emits acrid smoke and irritating fumes. See also ESTERS.

EJM900 CAS:151-56-4 ***HR: 3***
ETHYLENEIMINE
DOT: UN 1185
mf: C_2H_5N mw: 43.08

PROP: Oily, water-white liquid. Pungent ammoniacal odor. Bp: 55-56°, fp: −71.5°, flash p: 12°F, d: 0.832 @ 20°/4°, autoign temp: 608°F, vap press: 160 mm @ 20°, vap d: 1.48, lel: 3.6%, uel: 46%.

SYNS: AETHYLENIMIN (GERMAN) ◇ AMINOETHYLENE ◇ AZACYCLOPROPANE ◇ AZIRANE ◇ AZIRIDIN (GERMAN) ◇ AZIRIDINE ◇ DIHYDROAZIRENE ◇ DIHYDRO-1H-AZIRINE ◇ DIMETHYLENEIMINE ◇ DIMETHYLENIMINE ◇ EI ◇ ENT 50,324 ◇ ETHYLEENIMINE (DUTCH) ◇ ETHYLENE IMINE, INHIBITED (DOT) ◇ ETHYLENIMINE ◇ ETHYLIMINE ◇ ETILENIMINA (ITALIAN) ◇ RCRA WASTE NUMBER P054 ◇ TL 337

TOXICITY DATA with REFERENCE
eye-mus 2 ppm/1M JIHTAB 30,7,48
skn-rbt 10 mg/24H open JIHTAB 30,63,48
eye-rbt 2 mg SEV AJOPAA 29,1363,46
sln-dmg-mul 1 pph/8H-C GNKAA5 21,958,85
mnt-nml-mul 5000 ppm/8D-C MUREAV 125,275,84
cyt-hmn:lng 1 μmol/L EVSRBT 24,433,81
dns-mus-par 5 mg/kg EVSRBT 24,943,81
dlt-mus-ipr 5 mg/kg EVSRBT 24,943,81
msc-ham:ovr 2 mg/L MUREAV 94,449,82
ihl-rat TCLo:800 μg/m³/24H (1D male):REP TPKVAL 14,16,75
scu-mus TDLo:41760 μg/kg (6-14D preg):TER NTIS** PB223-160
scu-rat TDLo:20 mg/kg/67D-I:NEO BJPCAL 9,306,54
orl-mus TDLo:6500 μg/kg/5D-I:CAR VOONAW 27(5),88,81
orl-mus TDLo:235 mg/kg/76W-C:CAR JNCIAM 42,1101,69
scu-rat TD:10 mg/kg/8W-I:ETA BJPCAL 9,306,54
orl-rat LD50:15 mg/kg JIHTAB 23,259,41
ihl-rat LC50:100 mg/m³/2H 85GMAT -67,82
ipr-rat LD50:3500 μg/kg BJPCAL 21,581,63
ihl-mus LC50:400 mg/m³/2H 85GMAT -,67,82
ihl-rbt LCLo:100 mg/m³/2H 85GMAT -,67,82
skn-rbt LDLo:10 mg/kg 85GMAT -,67,82
ihl-gpg LCLo:25 ppm/8H JIHTAB 30,2,48
skn-gpg LD50:14 mg/kg JIHTAB 30,63,48

CONSENSUS REPORTS: IARC Cancer Review: Group 3 IMEMDT 7,56,87; Animal Sufficient Evidence IMEMDT 9,37,75. Community Right-To-Know List. EPA Extremely Hazardous Substances List. Reported in EPA TSCA Inventory. EPA Genetic Toxicology Program.

OSHA PEL: TWA 1 mg/m³ (skin); Cancer Suspect Agent
ACGIH TLV: TWA 0.5 ppm (skin)
DFG TRK: 0.5 ppm; Animal Carcinogen, Suspected Human Carcinogen.
NIOSH REL: (Ethyleneimine): TWA use 29 CFR 1910.1012
DOT Classification: Flammable Liquid; Label: Flammable Liquid and Poison.

SAFETY PROFILE: Confirmed carcinogen with experimental carcinogenic, neoplastigenic, tumorigenic, and teratogenic data. Other experimental reproductive effects. Poison by ingestion, skin contact, inhalation, and intraperitoneal routes. Human mutation data reported. A skin, mucous membrane, and severe eye irritant. An allergic sensitizer of skin. Causes opaque cornea, keratoconus and necrosis of cornea (experimentally). Has been known to cause severe human eye injury. Drinking of carbonated beverages is recommended as an antidote to this material in stomach.

A very dangerous fire and explosion hazard when exposed to heat, flame, or oxidizers. Reacts violently with acids; aluminum chloride + substituted anilines; acetic acid; acetic anhydride; acrolein; acrylic acid; allyl chloride; CS_2; Cl_2; chlorosulfonic acid; epichlorohydrin; glyoxal; HCl; HF; HNO_3; oleum; β-propiolactone; Ag; NaOCl; H_2SO_4; vinyl acetate. Reacts with chlorinating agents (e.g., sodium hypochlorite solution) to form the explosive 1-chloro aziridine. Reacts with silver or its alloys to form explosive silver derivatives. Dangerous; heat and/or the presence of catalytically active metals or chloride ions can cause a violent exothermic reaction. To fight fire, use alcohol foam, CO_2, dry chemical. When heated to decomposition it emits acrid smoke and irritating fumes.

EJN400 CAS:13279-24-8 ***HR: 3***
N,N-ETHYLENE-N′-METHYLUREA
mf: $C_4H_8N_2O$ mw: 100.14

SYNS: 1-AZIRIDINECARBOXAMIDE, N-METHYL- ◇ N-METHYL-1-AZIRIDINECARBOXAMIDE ◇ N-METHYLETHYLENEUREA

TOXICITY DATA with REFERENCE
ipr-rat TDLo:50 mg/kg (male 1D pre):REP 85GUAJ -,37,66
ipr-rat LD50:90 mg/kg BJPCAL 21,581,63

SAFETY PROFILE: Poison by intraperitoneal route. Experimental reproductive effects. When heated to decomposition it emits toxic fumes of NO_x.

EJN500 CAS:75-21-8 ***HR: 3***
ETHYLENE OXIDE
DOT: UN 1040
mf: C_2H_4O mw: 44.06

PROP: Colorless gas at room temperature. Mp: −111.3°, bp: 10.7°, ULC: 100, lel: 3.0%, uel: 100%, flash p: −4°F, d: 0.8711 @ 20°/20°, autoign temp: 804°F, vap press: 1095 mm @ 20°, vap d: 1.52. Misc in water and alc; very sol in ether.

SYNS: AETHYLENOXID (GERMAN) ◇ AMPROLENE ◇ ANPROLENE ◇ ANPROLINE ◇ DIHYDROOXIRENE ◇ DIMETHYLENE OXIDE ◇ ENT 26,263 ◇ E.O. ◇ 1,2-EPOXYAETHAN (GERMAN) ◇ EPOXYETHANE ◇ 1,2-EPOXYETHANE ◇ ETHENE OXIDE ◇ ETHYLEENOXIDE (DUTCH) ◇ ETHYLENE (OXYDE d') (FRENCH) ◇ ETILENE (OSSIDO di) (ITALIAN) ◇ ETO ◇ ETYLENU TLENEK (POLISH) ◇ FEMA No. 2433 ◇ MERPOL ◇ NCI-C50088 ◇ OXACYCLOPROPANE ◇ OXANE ◇ OXIDOETHANE ◇ α,β-OXIDOETHANE ◇ OXIRAAN (DUTCH) ◇ OXIRANE ◇ OXYFUME ◇ OXYFUME 12 ◇ RCRA WASTE NUMBER U115 ◇ STERILIZING GAS ETHYLENE OXIDE 100% ◇ T-GAS

TOXICITY DATA with REFERENCE
skn-hmn 1%/7S AMIHBC 2,549,50
eye-rbt 18 mg/6H MOD BUYRAI 31,25,77
mmo-omi 540 mg/L 47YKAF 8,273,84
dns-hmn:leu 4 mmol/L CBINA8 47,265,83
sce-hmn:lym 4 pph TCMUD8 6,15,86
sce-hmn:lym 10 mg/L PHMGBN 25,214,82
dnd-mus-ipr 100 mg/kg ENMUDM 8(Suppl 6),74,86
dlt-mus-ihl 500 ppm/6H/4D-C ENMUDM 8,1,86
ipr-mus TDLo:750 mg/kg (male 25D pre):REP MUREAV 73,133,80
ihl-mus TCLo:1200 ppm/90M (female 1D post):TER MUREAV 176,269,87
orl-rat TDLo:1186 mg/kg/2Y-I:CAR BJCAAI 46,924,82
ihl-rat TCLo:33 ppm/6H/2Y-I:CAR TXAPA9 75,105,84
ihl-mus TDLo:50 ppm/6H/2Y:CAR tumors NTPTR* NTP-TR-326,87
scu-mus TD:1090 mg/kg/91W-I:NEO BJCAAI 39,588,79
ihl-rat TC:33 ppm/6H/2Y-I:ETA NRTXDN 6,117,85
ihl-hmn TCLo:12500 ppm/10S:NOSE JOHYAY 32,409,32
ihl-wmn TCLo:500 ppm/2M:CNS,GIT,PUL DICPBB 15,384,81
orl-rat LD50:72 mg/kg SPEADM 78-1,17,78
ihl-rat LC50:800 ppm/4H 34ZIAG -,258,69
scu-rat LD50:187 mg/kg GISAAA 48(1),23,83
ihl-mus LC50:836 ppm/4H NTIS** PB214-270
ipr-mus LD50:175 mg/kg GISAAA 48(1),23,83
ivn-mus LD50:290 mg/kg APTOA6 43,69,78
ihl-dog LC50:960 ppm/4H AMIHAB 13,237,56
scu-cat LDLo:100 mg/kg HDWU** -,-,33
ivn-rbt LDLo:175 mg/kg JOHYAY 32,409,32
ihl-gpg LC50:1500 mg/m^3/4H 85GMAT -,67,82

CONSENSUS REPORTS: NTP Fifth Annual Report on Carcinogens. IARC Cancer Review: Group 2A IMEMDT 7,205,87; Animal Inadequate Evidence IMEMDT 11,157,76; Human Inadequate Evidence IMEMDT 36,189,85; Animal Sufficient Evidence IMEMDT 36,189,85. Community Right-To-Know List. EPA Extremely Hazardous Substances List. Reported in EPA TSCA Inventory. EPA Genetic Toxicology Program.

OSHA PEL: TWA 1 ppm; Cancer Hazard
ACGIH TLV: TWA 1 ppm; Suspected Human Carcinogen.
DFG TRK: 3 ppm; Animal Carcinogen, Suspected Human Carcinogen.
NIOSH REL: (Ethylene Oxide) TWA 0.1 ppm; CL 5 ppm/10M/D
DOT Classification: Flammable Liquid; Label: Flammable Liquid; Flammable Gas; Label: Poison Gas and Flammable Gas.

SAFETY PROFILE: Confirmed human carcinogen with experimental carcinogenic, tumorigenic, neoplastigenic, and teratogenic data. Poison by ingestion, intraperitoneal, subcutaneous, and intravenous routes. Moderately toxic by inhalation. Human systemic effects by inhalation: convulsions, nausea, vomiting, olfactory and pulmonary changes. Experimental reproductive effects. Mutation data reported. A skin and eye irritant. An irritant to mucous membranes of respiratory tract. High concentrations can cause pulmonary edema.

Highly flammable liquid or gas. Severe explosion hazard when exposed to flame. To fight fire, use alcohol foam, CO_2, dry chemical. Violent polymerization occurs on contact with ammonia; alkali hydroxides; amines; metallic potassium; acids; covalent halides (e.g., aluminum chloride; iron(III) chloride; tin(IV) chloride; aluminum oxide; iron oxide; rust). Explosive reaction with glycerol at 200°. Rapid compression of the vapor with air causes explosions. Incompatible with bases; alcohols; air; m-nitroaniline; trimethyl amine; copper; iron chlorides; iron oxides; magnesium perchlorate; mercaptans; potassium; tin chlorides; contaminants; alkane thiols; bromoethane. When heated to decomposition it emits acrid smoke and irritating fumes.

EJO000 CAS:8070-50-6 ***HR: 3***
ETHYLENE OXIDE, mixed with CARBON DIOXIDE
DOT: UN 1041/UN 1952

PROP: Contains less than 10% carbon dioxide (NTIS** PB225-283).

SYNS: ANHYDRIDE CARBONIQUE et OXYDE d'ETHYLENE MELANGES (FRENCH) ◇ CARBON DIOXIDE and ETHYLENE OXIDE MIXTURES, with more than 6% ETHYLENE OXIDE (DOT) ◇ ETHYLENE OXIDE and CARBON DIOXIDE MIXTURES (DOT) ◇ OXYFUME 20 ◇ OXYFUME 30

TOXICITY DATA with REFERENCE
ihl-mus LC50:836 ppm/4H NTIS** PB214-270
ihl-dog LC50:973 ppm/4H NTIS** PB214-270
ihl-gpg LCLo:7000 ppm/2.5H PHRPA6 45,1832,30

DOT Classification: Poison A; Label: Poison Gas and Flammable Gas.

SAFETY PROFILE: A poison. Mildly toxic by inhalation. Used for the sterilization of vacuum chambers. See also ETHYLENE OXIDE.

EJO500 CAS:289-14-5 ***HR: 3***
ETHYLENE OZONIDE
mf: $C_2H_4O_3$ mw: 76.05

OCH_2OOCH_2 (ring)

SYN: 1,2,4-TRIOXOLANE

SAFETY PROFILE: Explodes violently on heating, friction, or shock. Explodes at room temperature. Stable at 0°C. May explode when poured. When heated to decomposition it emits acrid smoke and irritating fumes. See also OZONE.

EJP000 CAS:1072-53-3 ***HR: 3***
ETHYLENE SULFATE
mf: $C_2H_4O_4S$ mw: 124.12

SYNS: 2,2-DIOXIDE-1,3,2-DIOXATHIOLANE ◇ ETHYLENE GLYCOL, CYCLIC SULFATE ◇ GLYCOL SULFATE ◇ SULFURIC ACID, CYCLIC ETHYLENE ESTER

TOXICITY DATA with REFERENCE
mmo-sat 1 μmol/plate CBINA8 19,241,77
hma-mus/sat 5 mmol/kg CBINA8 19,241,77
ipr-mus TDLo:128 mg/kg/64W-I:ETA JNCIAM 53,695,74
scu-mus TDLo:840 mg/kg/42W-I:NEO JNCIAM 53,695,74

CONSENSUS REPORTS: EPA Genetic Toxicology Program.

SAFETY PROFILE: Questionable carcinogen with experimental tumorigenic and neoplastigenic data. Mutation data reported. When heated to decomposition it emits toxic fumes of SO_x. See also SULFATES and SULFATES and ESTERS.

EJP500 CAS:420-12-2 ***HR: 3***
ETHYLENE SULFIDE
mf: C_2H_4S mw: 60.12

PROP: Colorless liquid. Bp: 55-56° decomp, d: 1.0368 @ 0°/4°, vap d: 2.07.

SYNS: AETHYLENSULFID (GERMAN) ◇ 2,3-DIHYDROTHIIRENE ◇ ETHYLENE EPISULFIDE ◇ ETHYLENE EPISULPHIDE ◇ ETHYLENE SULPHIDE ◇ THIACYCLOPROPANE ◇ THIIRANE

TOXICITY DATA with REFERENCE
eye-rbt 20 mg AJOPAA 29,1363,46
scu-rat TDLo:400 mg/kg/50W-I:ETA ZEKBAI 74,241,70
orl-rat LD50:178 mg/kg AIHAAP 25,560,64
ihl-rat LC50:690 ppm/6H BECTA6 6,509,71
ipr-rat LD50:42 mg/kg AIHAAP 25,560,64
scu-rat LD50:90 mg/kg ZEKBAI 74,241,70

CONSENSUS REPORTS: IARC Cancer Review: Group 3 IMEMDT 7,56,87; Animal Limited Evidence IMEMDT 11,257,76. Reported in EPA TSCA Inventory.

SAFETY PROFILE: Poison by ingestion, intraperitoneal, and subcutaneous routes. Mildly toxic by inhalation. A skin, eye and mucous membrane irritant. Questionable carcinogen with experimental tumorigenic data. Can react with oxidizing materials. When heated to decomposition, or on contact with acid or acid fumes, it emits highly toxic fumes of SO_x. See also SULFIDES.

EJQ000 CAS:33813-20-6 ***HR: 3***
ETHYLENE THIURAM MONOSULFIDE
mf: $C_4H_4N_2S_3$ mw: 176.28

SYNS: 5,6-DIHYDRO-3H-IMIDAZO(2,1-c)-1,2,4-DITHIAZOLE-3-THIONE ◇ ENDODAN ◇ ETEM ◇ ETHYLENE BISTHIURAM MONOSULFIDE ◇ ETHYLENETHIOCARBAMYL SULFIDE ◇ ETHYLENE THIURAM MONOSULPHIDE ◇ ETM ◇ ETM (heterocycle) ◇ HORTOCRITT

TOXICITY DATA with REFERENCE
mma-sat 7500 ng/plate MUREAV 157,13,85
mrc-bcs 2 μg/disc/24H MUREAV 40,19,76
sce-hmn:lym 20 mg/L MUREAV 157,13,85
cyt-ham-orl 93 mg/kg MUREAV 157,13,85
orl-rat TDLo:304 mg/kg (16W pre):REP JHEMA2 24,295,80
orl-rat LD50:380 mg/kg ATSUDG 4,459,80

CONSENSUS REPORTS: EPA Genetic Toxicology Program.

SAFETY PROFILE: Poison by ingestion. Experimental reproductive effects. Human mutation data reported. Used as a pesticide. When heated to decomposition it emits very toxic fumes of SO_x and NO_x. See also SULFIDES.

EJQ500 CAS:105-95-3 ***HR: 1***
ETHYLENE UNDECANE DICARBOXYLATE
mf: $C_{15}H_{26}O_4$ mw: 270.41

SYNS: ASTRATONE ◇ EMERESSENCE 1150 ◇ ETHYLENE BRASSYLATE ◇ MUSK-T ◇ TRIDECANEDIOIC ACID, CYCLIC ETHYLENE ESTER ◇ 1,1'-UNDECANEDICARBOXYLIC ACID ESTER with ETHYLENE GLYCOL

TOXICITY DATA with REFERENCE
skn-rbt 500 mg/24H MOD FCTXAV 13,91,75

CONSENSUS REPORTS: Reported in EPA TSCA Inventory.

SAFETY PROFILE: A skin irritant. Used as a fragrance. When heated to decomposition it emits acrid smoke and irritating fumes. See also ESTERS.

EJR500 CAS:530-35-8 ***HR: 3***
1-N-ETHYLEPHEDRINE HYDROCHLORIDE
mf: $C_{12}H_{19}NO•ClH$ mw: 229.78

SYNS: α-(1-(ETHYLMETHYLAMINO)ETHYL)BENZYLALCOHOL HYDROCHLORIDE (−) ◇ MENETYL ◇ 2-METHYLETHYLAMINO-1-PHENYL-1-PROPANOL HYDROCHLORIDE ◇ NETHAMINE HYDROCHLORIDE ◇ NOVEDRIN HYDROCHLORIDE ◇ 1-1-PHENYL-2-METHYLETHYLAMINOPROPAN-1-OL HYDROCHLORIDE

TOXICITY DATA with REFERENCE
orl-rbt LDLo:550 mg/kg JPETAB 75,289,42
ivn-rbt LD50:70 mg/kg JAPMA8 39,382,50

SAFETY PROFILE: Poison by intravenous route. Moderately toxic by ingestion. When heated to decomposition it emits very toxic fumes such as Cl^- and NO_x.

EJS000 CAS:19780-35-9 ***HR: 3***
ETHYL-2,3-EPOXYBUTYRATE
mf: $C_6H_{10}O_3$ mw: 130.16

SYN: 2,3-EPOXYBUTYRIC ACID, ETHYL ESTER

TOXICITY DATA with REFERENCE
mmo-sat 5 g/L MUREAV 89,269,81
scu-mus TDLo:272 mg/kg/68W-I:ETA JNCIAM 46,143,71
orl-rat LD50:500 mg/kg AIHAAP 24,305,63
skn-rbt LDLo:2830 mg/kg AIHAAP 24,305,63

SAFETY PROFILE: Moderately toxic by ingestion and skin contact. Questionable carcinogen with experimental tumorigenic data. Mutation data reported. When heated to decomposition it emits acrid smoke and irritating fumes. See also ESTERS.

EJS100 CAS:67658-42-8 ***HR: 3***
(8-β)-6-ETHYLERGOLINE-8-ACETAMIDE TARTRATE
mf: $C_{18}H_{23}N_3O•C_4H_6O_6$ mw: 447.54

SYN: ERGOLINE-8-ACETAMIDE, 6-ETHYL-, (8-β)-, (R-(R*,R*))-2,3-DIHYDROXYBUTANEDIOATE (1:1)

TOXICITY DATA with REFERENCE
orl-rat TDLo:30 μg/kg (female 5D post):REP CCCCAK 44,3385,79
orl-rat LD50:38500 μg/kg CCCCAK 44,3385,79

SAFETY PROFILE: Poison by ingestion. Experimental reproductive effects. When heated to decomposition it emits toxic fumes of NO_x.

EJT000 CAS:67466-28-8 ***HR: 3***
ETHYL ESTER of 1,2,5,6-DIBENZANTHRACENE-endo-α,β-SUCCINO GLYCINE
mf: $C_{30}H_{21}NO_4$ mw: 459.52

TOXICITY DATA with REFERENCE
scu-mus TDLo:200 mg/kg:ETA CNREA8 1,685,41

SAFETY PROFILE: Questionable carcinogen with experimental tumorigenic data. When heated to decomposition it emits toxic fumes of NO_x. See also ESTERS.

EJT500 ***HR: 3***
ETHYL ESTER of 3-METHYLCHOLANTHRENE-endo-α,β-SUCCINOGLYCINE
mf: $C_{29}H_{23}NO_4$ mw: 449.53

TOXICITY DATA with REFERENCE
scu-mus TDLo:200 mg/kg:ETA CNREA8 1,685,41

SAFETY PROFILE: Questionable carcinogen with experimental tumorigenic data. When heated to decomposition it emits toxic fumes of NO_x. See also ESTERS.

EJT575 ***HR: 2***
11-β-ETHYLESTRADIOL
mf: $C_{20}H_{28}O_2$ mw: 300.48

SYNS: ESTRA-1,3,5(10)-TRIENE-3,17-β-DIOL, 11-β-ETHYL- ◇ 11-β-ETHYLESTRA-1,3,5(10)-TRIENE-3,17-β-DIOL

TOXICITY DATA with REFERENCE
otr-mus:fbr 30 μmol/L CNREA8 47,2583,87
imp-ham TDLo:400 mg/kg/30W C:NEO CNREA8 47,2583,87

SAFETY PROFILE: Questionable carcinogen with experimental neoplastigenic data. Mutation data reported. When heated to decomposition it emits acrid smoke and irritating fumes.

EJT600 CAS:965-90-2 ***HR: D***
ETHYLESTRENOL
mf: $C_{20}H_{32}O$ mw: 288.52

PROP: Crystals. Mp: 76-78°.

SYNS: DURABOLIN-O ◇ DURABORAL ◇ ETHYLNANDROL ◇ MAXIBALIN ◇ MAXIBOLIN ◇ NEODURABOLIN ◇ ORABOLIN ◇ ORG-483 ◇ ORGABOLIN ◇ ORGABORAL

TOXICITY DATA with REFERENCE
orl-rat TDLo:1 g/kg (female 15-19D post):REP FESTAS 22,735,71
orl-rat TDLo:20 mg/kg (female 17-20D post):TER ECJPAE 24,77,77

SAFETY PROFILE: An experimental teratogen. Other experimental reproductive effects. When heated to decomposition it emits acrid smoke and irritating fumes.

EJU000 CAS:60-29-7 ***HR: 3***
ETHYL ETHER
DOT: UN 1155
mf: $C_4H_{10}O$ mw: 74.14

$CH_3CH_2OCH_2CH_3$

PROP: A clear, volatile liquid; sweet, pungent odor. Sol in water; misc in alcohol and ether; sol in chloroform. Mp: −116.2°, bp: 34.6°, ULC: 100, lel: 1.85%, uel: 36%, flash p: −49°F, d: 0.7135 @ 20°/4°, autoign temp: 320°F, vap press: 442 mm @ 20°, vap d: 2.56.

SYNS: AETHER ◇ ANAESTHETIC ETHER ◇ ANESTHESIA ETHER ◇ ANESTHETIC ETHER ◇ DIAETHYLAETHER (GERMAN) ◇ DIETHYL ETHER (DOT) ◇ DIETHYL OXIDE ◇ DWUETYLOWY ETER (POLISH) ◇ ETERE ETILICO (ITALIAN) ◇ ETHER ◇ ETHER ETHYLIQUE (FRENCH) ◇ ETHOXYETHANE ◇ 1,1′-OXYBISETHANE ◇ OXYDE d'ETHYLE (FRENCH) ◇ RCRA WASTE NUMBER U117 ◇ SOLVENT ETHER

TOXICITY DATA with REFERENCE
eye-hmn 100 ppm JIHTAB 25,282,43
skn-rbt 360 mg open MLD UCDS** 4/5/73
eye-rbt 100 mg MOD FEPRA7 35,729,76
skn-gpg 50 mg/24H SEV HIFUAG 22,373,80
dnr-esc 50 μL/well/16H CBINA8 15,219,76
dyt-smc 100 mmol/tube HEREAY 33,457,47
oms-ham:fbr 1 pph ANESAV 43,21,75
orl-man LDLo:260 mg/kg 85DCAI 2,73,70
orl-hmn LDLo:420 mg/kg 32ZWAA 8,275,74
ihl-hmn TCLo:200 ppm:NOSE JIHTAB 25,282,43
orl-rat LD50:1215 mg/kg TXAPA9 19,699,71
ihl-rat LC50:73000 ppm/2H TXAPA9 17,275,70
ihl-mus LC50:6500 ppm/99M TXAPA9 17,275,70
ipr-mus LD50:2420 mg/kg PWPSA8 27,511,84
scu-mus LDLo:8 mg/kg HBAMAK 4,1295,35
ivn-mus LD50:996 mg/kg JPMSAE 67,566,78
ihl-dog LCLo:76000 ppm HBAMAK 4,1294,35
ihl-rbt LCLo:106000 ppm HBAMAK 4,1294,35
ipr-gpg LDLo:2000 mg/kg AIHAAP 35,21,74
scu-frg LDLo:24 g/kg HBAMAK 4,1295,35

CONSENSUS REPORTS: IARC Cancer Review: Animal No Adequate Data IMEMDT 7,93,87. Reported in EPA TSCA Inventory. EPA Genetic Toxicology Program.

OSHA PEL: (Transitional: TWA 400 ppm) TWA 400 ppm; STEL 500 ppm
ACGIH TLV: TWA 400 ppm; STEL 500 ppm
DFG MAK: 400 ppm (1200 mg/m^3)
DOT Classification: Flammable Liquid; Label: Flammable Liquid.

SAFETY PROFILE: Moderately toxic to humans by ingestion. Poison experimentally by subcutaneous route. Moderately toxic by intraperitoneal and intravenous routes. Mildly toxic by inhalation. Human systemic effects by inhalation: olfactory changes. Mutation data reported. A severe eye and moderate skin irritant. Ethyl ether is not corrosive or dangerously reactive. It must not be considered safe for individuals to inhale or ingest. It is a depressant of the central nervous system and is capable of producing intoxication, drowsiness, stupor, and unconsciousness. Death due to respiratory failure may result from severe and continued exposure.

A very dangerous fire and explosion hazard when exposed to heat or flame. A storage hazard. It auto-oxidizes to form explosive polymeric 1-oxy-peroxides. Explosive reaction with boron triazide, bromine trifluoride, bromine pentafluoride, perchloric acid, uranyl nitrate + light, wood pulp extracts + heat. Violent reaction or ignition on contact with halogens (e.g., bromine, chlorine), interhalogens (e.g., iodine heptafluoride), oxidants (e.g., silver perchlorate, nitrosyl perchlorate, nitryl perchlorate, chromyl chloride, fluorine nitrate, permanganic acid, nitric acid, hydrogen peroxide, peroxodisulfuric acid, iodine(VII) oxide, sodium peroxide, ozone, and liquid air), sulfur and sulfur compounds (e.g., sulfur when dried with peroxidized ether, sulfunoyl chloride). Can react vigorously with acetyl peroxide, air, bromoazide, ClF_3, CrO_3, $Cr(OCl)_2$, $LiAlH_2$, $NOClO_4$, O_2, $NClO_2$, (H_2SO_4 + permanganates), K_2O_2, [$(C_2H_5)_3Al$ + air], [$(CH_3)_3Al$ + air]. To fight fire, use alcohol foam, CO_2, dry chemical. Used in production of drugs of abuse. When heated to decomposition it emits acrid smoke and irritating fumes. See also ETHERS.

EJV000 CAS:52125-53-8 ***HR: 1***
ETHYL ETHER of PROPYLENE GLYCOL
mf: $C_5H_{12}O_2$ mw: 104.17

PROP: Vap d: 3.59.

TOXICITY DATA with REFERENCE
eye-rbt 20 mg AJOPAA 29,1363,46

CONSENSUS REPORTS: Glycol ether compounds are on the Community Right-To-Know List.

SAFETY PROFILE: An eye irritant. Combustible when exposed to heat or flame; can react with oxidizers. When heated to decomposition it emits acrid smoke and irritating fumes. See also GLYCOL ETHERS.

EJV400 ***HR: 2***
11-β-ETHYL-17-α-ETHINYLESTRADIOL
mf: $C_{22}H_{28}O_2$ mw: 324.50

SYN: 11-β-ETHYL-19-NOR-17-α-PREGNA-1,3,5(10)-TRIEN-20-YNE-3,17-DIOL

TOXICITY DATA with REFERENCE
imp-ham TDLo:400 mg/kg/39W C:CAR CNREA8 47,2583,87

SAFETY PROFILE: Questionable carcinogen with experimental carcinogenic data. When heated to decomposition it emits acrid smoke and irritating fumes.

EJV500 CAS:763-69-9 ***HR: 1***
ETHYL-β-ETHOXYPROPIONATE
mf: $C_7H_{14}O_3$ mw: 146.21

PROP: Liquid. Mp: −100°, bp: 170.1°, flash p: 180°F (OC), d: 0.9496 @ 20°/20°, vap d: 5.03.

SYNS: ETHOXYPROPIONIC ACID, ETHYL ESTER ◇ 3-ETHOXYPROPIONIC ACID, ETHYL ESTER

TOXICITY DATA with REFERENCE
skn-rbt 10 mg/24H open MLD AMIHBC 4,119,51
eye-rbt 500 mg open AMIHBC 4,119,51
orl-rat LD50:5000 mg/kg AMIHBC 4,119,51
skn-rbt LD50:10 g/kg AMIHBC 4,119,51

CONSENSUS REPORTS: Reported in EPA TSCA Inventory.

SAFETY PROFILE: Mildly toxic by ingestion and skin contact. A skin and eye irritant. Flammable when exposed to heat or flame; can react with oxidizing materials. To fight fire, use foam, CO_2, dry chemical. When heated to decomposition it emits acrid smoke and irritating fumes. See also ESTERS.

EJW500 CAS:623-78-9 ***HR: 3***
ETHYL-N-ETHYL CARBAMATE
mf: $C_5H_{11}NO_2$ mw: 117.17

SYN: ETHYLCARBAMIC ACID, ETHYL ESTER

TOXICITY DATA with REFERENCE
ipr-mus TDLo:6500 mg/kg/13W-I:ETA JNCIAM 9,35,48
scu-mus LD50:860 mg/kg AJEBAK 45,507,67

CONSENSUS REPORTS: Reported in EPA TSCA Inventory.

SAFETY PROFILE: Moderately toxic by subcutaneous routes. Questionable carcinogen with experimental tumorigenic data. When heated to decomposition it emits toxic fumes of NO_x. See also CARBAMATES and ESTERS.

EJY000 CAS:17013-37-5 ***HR: 3***
5-ETHYL-5-(1-ETHYLPROPYL)BARBITURIC ACID
mf: $C_{11}H_{18}N_3O_3$ mw: 240.32

SYNS: 5-ETHYL-5-(1-ETHYLPROPYL)2,4,6(1H,3H,5H)-PYRIMIDINETRIONE ◇ ISOMEBUMAL

TOXICITY DATA with REFERENCE
ipr-rat LDLo:110 mg/kg JPHAA3 26,317,37
ipr-rbt LDLo:75 mg/kg JACSAT 56,1139,34

SAFETY PROFILE: Poison by intraperitoneal route. When heated to decomposition it emits toxic fumes of NO_x. See also BARBITURATES.

EJZ000 CAS:102584-01-0 ***HR: 3***
1-ETHYL-4-(p-(p-((p-((1-ETHYLPYRIDINIUM-4-YL)AMINO)-2-AMINOPHENYL)CARBAMOYL)CINNAMAMIDO)ANILINO)PYRIDINIUM, DIBROMIDE
mf: $C_{36}H_{37}N_7O_2{\cdot}Br$ mw: 759.62

TOXICITY DATA with REFERENCE
dnd-mus:lym 1500 nmol/L JMCMAR 22,134,79
ipr-mus LD10:7500 μg/kg JMCMAR 22,134,79

SAFETY PROFILE: Poison by intraperitoneal route. Mutation data reported. When heated to decomposition it emits very toxic fumes of Br^- and NO_x. See also BROMIDES.

EKA000 CAS:68772-29-2 ***HR: 3***
1-ETHYL-4-(p-(p-((1-ETHYLPYRIDINIUM-4-YL)AMINO)BENZAMIDO)ANILINO)QUINOLINIUM DIBROMIDE
mf: $C_{31}H_{31}N_5O{\cdot}2Br$ mw: 649.49

TOXICITY DATA with REFERENCE
dnd-mus:lym 1600 nmol/L JMCMAR 22,134,79
ipr-mus LD10:10 mg/kg JMCMAR 22,134,79

SAFETY PROFILE: Poison by intraperitoneal route. Mutation data reported. When heated to decomposition it emits very toxic fumes of NO_x and Br^-. See also BROMIDES.

EKA500 CAS:68772-10-1 ***HR: 3***
1-ETHYL-4-(p-((p-((1-ETHYLPYRIDINIUM-4-YL)AMINO)PHENYL)CARBAMOYL)ANILINO)QUINOLINIUM, DIBROMIDE
mf: $C_{31}H_{31}N_5O{\cdot}2Br$ mw: 649.49

TOXICITY DATA with REFERENCE
dnd-mus:lym 2 μmol/L JMCMAR 22,134,79
ipr-mus LD10:15 mg/kg JMCMAR 22,134,79

SAFETY PROFILE: Poison by intraperitoneal route. Mutation data reported. When heated to decomposition it emits very toxic fumes of NO_x and Br^-. See also BROMIDES.

EKB000 CAS:20719-23-7 ***HR: 3***
1-ETHYL-4-(p-(p-((p-((1-ETHYLPYRIDINIUM-4-YL)AMINO)PHENYL)CARBAMOYL)CINNAMAMIDO)ANILINO)PYRIDINIUM, DI-p-TOLUENE SULFONATE
mf: $C_{36}H_{36}N_6O_2 \cdot 2C_7H_7O_3S$ mw: 927.18

TOXICITY DATA with REFERENCE
dnd-mus:lym 430 nmol/L JMCMAR 22,134,79
ipr-mus LD10:8 mg/kg JMCMAR 22,134,79

SAFETY PROFILE: Poison by intraperitoneal route. Mutation data reported. When heated to decomposition it emits very toxic fumes of SO_x and NO_x. See also SULFONATES.

EKC500 CAS:68772-21-4 ***HR: 3***
1-ETHYL-4-(p-((p-(1-ETHYLPYRIDINIUM-4-YL)PHENYL)CARBAMOYL)ANILINO)QUINOLINIUM), DI-p-TOLUENE SULFONATE
mf: $C_{31}H_{30}N_4O \cdot 2C_7H_7O_3S$ mw: 817.05

TOXICITY DATA with REFERENCE
dnd-mus:lym 890 nmol/L JMCMAR 22,134,79
ipr-mus LD10:10 mg/kg JMCMAR 22,134,79

SAFETY PROFILE: Poison by intraperitoneal route. Mutation data reported. When heated to decomposition it emits very toxic fumes of NO_x and SO_x. See also SULFONATES.

EKD000 CAS:18355-54-9 ***HR: 3***
1-ETHYL-7-(p-(p-((1-ETHYLQUINOLINIUM-7-YL)CARBAMOYL)BENZAMIDO)BENZAMIDO)QUINOLINIUM), DI-p-TOLUENE SULFONATE
mf: $C_{37}H_{33}N_5O_3 \cdot 2C_7H_7O_3S$ mw: 938.15

TOXICITY DATA with REFERENCE
dnd-mus:lym 160 nmol/L JMCMAR 22,134,79
ipr-mus LD10:30 mg/kg JMCMAR 22,134,79

SAFETY PROFILE: Poison by intraperitoneal route. Mutation data reported. When heated to decomposition it emits very toxic fumes of NO_x and SO_x. See also SULFONATES.

EKD500 CAS:16758-28-4 ***HR: 3***
1-ETHYL-6-(p-(p-((1-ETHYLQUINOLINIUM-6-YL)CARBAMOYL)BENZAMIDO)BENZAMIDO)QUINOLINIUM, DI-p-TOLUENE SULFONATE
mf: $C_{37}H_{33}N_5O_3 \cdot 2C_7H_7O_3S$ mw: 938.15

TOXICITY DATA with REFERENCE
dnd-mus:lym 370 nmol/L JMCMAR 22,134,79
ipr-mus LD10:45 mg/kg JMCMAR 22,134,79

SAFETY PROFILE: Poison by intraperitoneal route. Mutation data reported. When heated to decomposition it emits very toxic fumes of SO_x and NO_x. See also SULFONATES.

EKF550 ***HR: D***
17-α-ETHYLETHYNYL-19-NORTESTOSTERONE
mf: $C_{22}H_{30}O_2$ mw: 326.52

SYN: 17-α-(1-BUTYNYL)-17-β-HYDROXYESTR-4-EN-3-ONE

TOXICITY DATA with REFERENCE
orl-rbt TDLo:200 μg/kg (female 1D pre):REP 85GRAA-,57,65

SAFETY PROFILE: Experimental reproductive effects. A steroid. When heated to decomposition it emits acrid smoke and irritating fumes. See also TESTOSTERONE.

EKF600 CAS:29177-84-2 ***HR: 2***
ETHYL FLUCLOZEPATE
mf: $C_{18}H_{14}ClFN_2O_3$ mw: 360.79

PROP: Crystals from ether. Mp: 193-194°.

SYNS: CM 6912 ◇ ETHYL-7-CHLORO-5-(o-FLUOROPHENYL)-2,3-DIHYDRO-2-OXO-1H-1,4-BENZODIAZEPINE-3-CARBOXYLATE ◇ ETHYL LOFLAZEPATE ◇ VICTAN

TOXICITY DATA with REFERENCE
orl-rat TDLo:2640 mg/kg (17-22D preg/21D post):REP KSRNAM 20,1683,86
orl-rbt TDLo:1300 mg/kg (female 6-18D post):TER KSRNAM 20,1655,86
ipr-rat LD50:645 mg/kg IJCPBS 19,453,81
scu-rat LD50:2 g/kg KSRNAM 20,1411,86
orl-mus LD50:5506 mg/kg KSRNAM 20,1411,86
ipr-mus LD50:712 mg/kg IJCPBS 19,453,81
scu-mus LD50:1795 mg/kg KSRNAM 20,1411,86

SAFETY PROFILE: Moderately toxic by intraperitoneal and subcutaneous routes. Mildly toxic by ingestion. Experimental teratogenic and reproductive effects. Note: This is a controlled substance (depressant) listed in the U.S. Code of Federal Regulations, Title 21 Part 1308.14 (1985). When heated to decomposition it emits toxic fumes of F^-, Cl^-, and NO_x.

EKG500 CAS:459-72-3 ***HR: 3***
ETHYL FLUOROACETATE
mf: $C_4H_7FO_2$ mw: 106.11

SYN: F1V02

TOXICITY DATA with REFERENCE
ipr-mus LD50:19 mg/kg 11FYAN 3,59,63

CONSENSUS REPORTS: Reported in EPA TSCA Inventory.

SAFETY PROFILE: Poison by intraperitoneal route. When heated to decomposition it emits toxic fumes of F^-. See also FLUORIDES.

EKI000 CAS:353-03-7 ***HR: 3***
ETHYL-10-FLUORODECANOATE
mf: $C_{12}H_{23}FO_2$ mw: 218.35

SYNS: ETHYL-ω-FLUORODECANOATE ◇ ETHYL-9-FLUORONONANECARBOXYLATE

TOXICITY DATA with REFERENCE
ipr-mus LD50:1650 μg/kg JOCEAH 21,883,56
par-mus LD50:10 mg/kg JCSOA9 -,1471,49
par-rbt LD50:200 μg/kg JCSOA9 -,1471,49

SAFETY PROFILE: Poison by intraperitoneal and parenteral routes. When heated to decomposition it emits toxic fumes of F^-. See also FLUORIDES.

EKJ500 CAS:589-79-7 ***HR: 3***
ETHYL-6-FLUOROHEXANOATE
mf: $C_8H_{15}FO_2$ mw: 162.23

SYNS: ETHYL-ω-FLUOROHEXANOATE ◇ ETHYL-5-FLUOROPENTANECARBOXYLATE ◇ 2'-FLUOROETHYL-6-FLUOROHEXANOATE

TOXICITY DATA with REFERENCE
scu-mus LD50:4 mg/kg JCSOA9 -,1471,49
ivn-rbt LD50:200 μg/kg JCSOA9 -,1471,49

SAFETY PROFILE: Poison by intravenous and subcutaneous routes. When heated to decomposition it emits very toxic fumes of F^-. See also FLUORIDES.

EKK500 CAS:332-97-8 ***HR: 3***
ETHYL-8-FLUORO OCTANOATE
mf: $C_{10}H_{19}FO_2$ mw: 190.29

SYNS: ETHYL-ω-FLUOROOCTANOATE ◇ 8-FLUOROOCTANOIC ACID, ETHYL ESTER

TOXICITY DATA with REFERENCE
ipr-mus LD50:1750 μg/kg JOCEAH 21,883,56
scu-mus LD50:9 mg/kg NATUAS 172,1139,53
par-mus LD50:9 mg/kg JCSOA9 -,1471,49

SAFETY PROFILE: Poison by intraperitoneal, subcutaneous, and parenteral routes. When heated to decomposition it emits toxic fumes of F^-. See also ESTERS and FLUORIDES.

EKK550 CAS:371-69-7 ***HR: 2***
ETHYL FLUOROSULFATE
mf: $C_2H_5FO_3S$ mw: 128.12

SAFETY PROFILE: Explodes violently at room temperature. When heated to decomposition it emits toxic fumes of F^- and SO_x. See also FLUORIDES and SULFATES.

EKL000 CAS:109-94-4 ***HR: 3***
ETHYL FORMATE
DOT: UN 1190
mf: $C_3H_6O_2$ mw: 74.09

PROP: Colorless liquid; sharp, rum-like odor. Mp: −79°, bp: 54.3°, lel: 2.7%, uel: 13.5%, flash p: −4°F (CC), d: 0.9236 @ 20°/20°, refr index: 1.359, autoign temp: 851°F, vap press: 100 mm @ 5.4°, vap d: 2.55. Sol in fixed oils, propylene glycol, water (decomp); sltly sol in mineral oil; insol in glycerin @ 54°.

SYNS: AETHYLFORMIAT (GERMAN) ◇ AREGINAL ◇ ETHYLE (FORMIATE d') (FRENCH) ◇ ETHYLFORMIAAT (DUTCH) ◇ ETHYL FORMIC ESTER ◇ ETHYL METHANOATE ◇ ETILE (FORMIATO di) (ITALIAN) ◇ FEMA No. 2434 ◇ FORMIC ACID, ETHYL ESTER ◇ FORMIC ETHER ◇ MROWCZAN ETYLU (POLISH)

TOXICITY DATA with REFERENCE
skn-rbt 460 mg open MLD UCDS** 4/10/68
eye-rbt 20 mg open AMIHBC 10,61,54
skn-mus TDLo:110 g/kg/9W-I:ETA BJCAAI 9,177,55
orl-rat LD50:1850 mg/kg FCTXAV 2,327,64
ihl-rat LCLo:8000 ppm/4H AMIHBC 10,61,54
orl-rbt LD50:2075 mg/kg IMSUAI 41,31,72
skn-rbt LD50:20 g/kg FCTXAV 16,637,78
scu-rbt LDLo:1000 mg/kg FCTXAV 16,737,78
orl-gpg LD50:1110 mg/kg FCTXAV 2,327,64

CONSENSUS REPORTS: Reported in EPA TSCA Inventory.

OSHA PEL: TWA 100 ppm
ACGIH TLV: TWA 100 ppm
DFG MAK: 100 ppm (300 mg/m³)
DOT Classification: Flammable Liquid; Label: Flammable Liquid.

SAFETY PROFILE: Moderately toxic by ingestion and subcutaneous routes. Mildly toxic by skin contact and inhalation. A powerful inhalation irritant in humans. A skin and eye irritant. Questionable carcinogen with experimental tumorigenic data. Highly flammable liquid. A very dangerous fire and explosion hazard when exposed to heat, flame, or oxidizers. To fight fire, use alcohol foam, spray, mist, dry chemical. When heated to decomposition it emits acrid smoke and irritating fumes. See also ESTERS.

EKL250 CAS:74920-78-8 ***HR: 2***
1-ETHYL-1-FORMYLHYDRAZINE
mf: $C_3H_8N_2O$ mw: 88.13

SYNS: EFH ◇ N-ETHYL-N-FORMYLHYDRAZINE ◇ FORMIC ACID, 1-ETHYLHYDRAZIDE

TOXICITY DATA with REFERENCE
orl-mus TDLo:18345 mg/kg/52W C:CAR CRNGDP 1,61,80

SAFETY PROFILE: Questionable carcinogen with experimental carcinogenic data. When heated to decomposition it emits toxic fumes of NO_x.

EKL500 CAS:2407-43-4 ***HR: 3***
5-ETHYL-2(5H)-FURANONE
mf: $C_6H_8O_2$ mw: 112.14

SYN: 4-HYDROXYHEX-2-ENOIC ACID LACTONE

TOXICITY DATA with REFERENCE
scu-rat TDLo:2560 mg/kg/64W-I:ETA BJCAAI 15,85,61

SAFETY PROFILE: Questionable carcinogen with experimental tumorigenic data. When heated to decomposition it emits acrid smoke and irritating fumes.

EKM000 CAS:614-99-3 ***HR: 3***
ETHYL FUROATE
mf: $C_7H_8O_3$ mw: 140.15

PROP: (A) Leaflets. D: 1.117, mp: 34, bp: 195°, Insol in water; misc alc and ether. (B) Liquid. D: 1.38, bp: 65-67°.

TOXICITY DATA with REFERENCE
ipr-rat LDLo:75 mg/kg JPETAB 58,174,36
ivn-mus LD50:180 mg/kg CSLNX* NX#00372

CONSENSUS REPORTS: Reported in EPA TSCA Inventory.

SAFETY PROFILE: Poison by intraperitoneal and intravenous routes. When heated to decomposition it emits acrid smoke and irritating fumes.

EKM500 CAS:623-50-7 ***HR: 2***
ETHYL GLYCOLATE
mf: $C_4H_8O_3$ mw: 104.12

PROP: Colorless liquid. D: 1.087, bp: 160°, very sol in alc and ether.

SYN: HYDROXYACETIC ACID ETHYL ESTER

TOXICITY DATA with REFERENCE
eye-gpg 10 µg MLD JPPMAB 11,150,59
ipr-rat LDLo:1500 mg/kg JPPMAB 11,150,59

CONSENSUS REPORTS: Reported in EPA TSCA Inventory.

SAFETY PROFILE: Moderately toxic by intraperitoneal route. An eye irritant. When heated to decomposition it emits acrid smoke and irritating fumes.

EKN000 CAS:2642-71-9 ***HR: 3***
ETHYL GUTHION
mf: $C_{12}H_{16}N_3O_3PS_2$ mw: 345.40

SYNS: ATHYL-GUSATHION ◇ AZINFOS-ETHYL (DUTCH) ◇ AZINOS ◇ AZINPHOS-AETHYL (GERMAN) ◇ AZINPHOS ETHYL ◇ AZINPHOS-ETILE (ITALIAN) ◇ BAY 16225 ◇ BAYER 16259 ◇ BENZOTRIAZINE derivative of an ETHYL DITHIOPHOSPHATE ◇ COTNION-ETHYL ◇ CYRSTHION ◇ O,O-DIAETHYL-S-(4-OXOBENZOTRIAZIN-3-METHYL)-DITHIOPHOSPHAT (GERMAN) ◇ O,O-DIAETHYL-S-((4-OXO-3H-1,2,3-BENZOTRIAZIN-3-YL)-METHYL)-DITHIOPHOSPHAT (GERMAN) ◇ O,O-DIETHYL-S-(4-OXO-3H-1,2,3-BENZOTRIAZINE-3-YL)-METHYL-DITHIOPHOSPHATE ◇ O,O-DIETHYL-S-((4-OXO-3H-1,2,3-BENZOTRIAZIN-3-YL)-METHYL)-DITHIO FOSFAAT (DUTCH) ◇ O,O-DIETHYL-S-(4-OXOBENZOTRIAZINO-3-METHYL)PHOSPHORODITHIOATE ◇ O,O-DIETHYL PHOSPHORODITHIOATE S-ester with 3-(MERCAPTOMETHYL)-1,2,3-BENZOTRIAZIN-4(3H)-ONE ◇ O,O-DIETIL-S-((4-OXO-3H-1,2,3-BENZOTRIAZIN-3-IL)-METIL)-DITIOFOSFATO (ITALIAN) ◇ 3,4-DIHYDRO-4-OXO-3-BENZOTRIAZINYL-METHYL O,O-DIETHYL PHOSPHORODITHIOATE ◇ S-(3,4-DIHYDRO-4-OXO-1,2,3-BENZOTRIAZIN-3-YLMETHYL)O,O-DIETHYL PHOSPHORODITHIOATE ◇ ENT 22,014 ◇ ETHYL GUSATHION ◇ GUSATHION A ◇ GUTHION (ETHYL) ◇ R 1513 ◇ TRIAZOTION (RUSSIAN)

TOXICITY DATA with REFERENCE
orl-rat LD50:7 mg/kg ARSIM* 20,2,66
ihl-rat LC50:390 mg/m³ 85GYAZ -,16,71
skn-rat LD50:250 mg/kg GUCHAZ 6,24,73
ipr-rat LD50:7500 µg/kg GUCHAZ 6,24,73
orl-ckn LD50:34 mg/kg TXAPA9 11,49,67

CONSENSUS REPORTS: EPA Extremely Hazardous Substances List.

SAFETY PROFILE: Poison by ingestion, inhalation, skin contact, and intraperitoneal route. A cholinesterase inhibitor type of insecticide. When heated to decomposition it emits toxic fumes of SO_x, PO_x, and NO_x. See also PARATHION.

EKN050 CAS:106-30-9 ***HR: 2***
ETHYL HEPTANOATE
mf: $C_9H_{18}O_2$ mw: 158.24

PROP: Colorless liquid; wine-brandy odor. D: 0.867-0.872, refr index: 1.411, flash p: 149°F. Misc in alc, chloroform, fixed oils; sltly sol in propylene glycol.

SYNS: ETHYL HEPTOATE ◇ FEMA No. 2437

SAFETY PROFILE: Combustible liquid. When heated to decomposition it emits acrid smoke and irritating fumes.

EKN100 CAS:23489-00-1 ***HR: 3***
1-ETHYL-1-HEPTYLPIPERIDINIUM BROMIDE
mf: $C_{14}H_{30}N \cdot Br$ mw: 292.36

TOXICITY DATA with REFERENCE
orl-mus LD50:288 mg/kg PSDTAP 15,331,74
ipr-mus LD50:67934 μg/kg PSDTAP 15,331,74
ivn-mus LD50:4541 μg/kg PSDTAP 15,331,74

SAFETY PROFILE: Poison by ingestion, intravenous, and intraperitoneal routes. When heated to decomposition it emits toxic fumes of NO_x and Br^-. See also BROMIDES.

EKN500 CAS:124-03-8 ***HR: 3***
ETHYL HEXADECYL DIMETHYL AMMONIUM BROMIDE
mf: $C_{20}H_{44}N \cdot Br$ mw: 378.56

SYNS: AMMONYX DME ◇ BRETOL ◇ CDA: CETYLCIDE ◇ CETYL DIMETHYL ETHYL AMMONIUM BROMIDE ◇ CETYL ETHYL DIMETHYLAMMONIUM BROMIDE ◇ DIMETHYL ETHYL HEXADECYL AMMONIUM BROMIDE ◇ ETHYL CETAB ◇ RADIOL GERMICIDAL SOLUTION

TOXICITY DATA with REFERENCE
orl-rat LD50:500 mg/kg PCOC** -,206,66
orl-gpg LD50:158 mg/kg PCOC** -,206,66

SAFETY PROFILE: Poison by ingestion. When heated to decomposition it emits toxic fumes of NO_x, NH_3, and Br^-. See also BROMIDES.

EKN600 CAS:56501-33-8 ***HR: 3***
1-ETHYL-1-HEXADECYLPIPERIDINIUM BROMIDE
mf: $C_{23}H_{48}N \cdot Br$ mw: 418.63

TOXICITY DATA with REFERENCE
orl-mus LD50:235 mg/kg PSDTAP 15,331,74
ipr-mus LD50:3993 μg/kg PSDTAP 15,331,74
ivn-mus LD50:2296 μg/kg PSDTAP 15,331,74

SAFETY PROFILE: Poison by ingestion, intravenous, and intraperitoneal routes. When heated to decomposition it emits toxic fumes of NO_x and Br^-. See also BROMIDES.

EKO000 CAS:63867-09-4 ***HR: 3***
ETHYL HEXAFLUORO-2-BROMOBUTYRATE
mf: $C_5H_3BrF_6O_2$ mw: 288.99

SYN: 2-BROMO-2,3,3,4,4,4-HEXAFLUOROBUTYRIC ACID METHYL ESTER

TOXICITY DATA with REFERENCE
orl-mus LD50:980 mg/kg TXAPA9 12,486,68
ipr-mus LD50:33 mg/kg TXAPA9 14,114,69
orl-mus LD50:980 mg/kg TXAPA9 14,114,69

SAFETY PROFILE: Poison by intraperitoneal route. Moderately toxic by ingestion. When heated to decomposition it emits very toxic fumes of Br^- and F^-.

EKO500 CAS:2212-67-1 ***HR: 3***
sec-ETHYL HEXAHYDRO-1H-AZEPINE-1-CARBOTHIOATE
mf: $C_9H_{17}NOS$ mw: 187.33

SYNS: S-AETHYL-N-HEXAHYDRO-1H-AZEPINTHIOLCARBAMAT (GERMAN) ◇ ETHYL-1-HEXAMETHYLENEIMINECARBOTHIOLATE ◇ S-ETHYL-1-HEXAMETHYLENEIMINOTHIOCARBAMATE ◇ S-ETHYL-N-HEXAMETHYLENETHIOCARBAMATE ◇ FELAN ◇ HYDRAM ◇ JALAN ◇ MOLINATE ◇ MOLMATE ◇ ORDRAM ◇ R-4572 ◇ STAUFFER R-4,572 ◇ YALAN ◇ YULAN

TOXICITY DATA with REFERENCE
cyt-hmn:lym 20 μg/L CYGEDX 11(4),62,77
ihl-rat TCLo:600 μg/m³/6H (male 20D pre):REP TOXID9 4,80,84
orl-rat LD50:369 mg/kg PEMNDP 8,578,87
ihl-rat LCLo:200 mg/m³ GISAAA 35(8),35,70
scu-rat LD50:1167 mg/kg VRDEA5 (1),119,69
scu-rat LD50:1167 mg/kg VRDEA5 (1),119,69
orl-mus LD50:530 mg/kg VRDEA5 (1),119,69
ihl-cat LCLo:200 mg/m³ GISAAA 35(8),35,70
skn-rbt LD50:3536 mg/kg FMCHA2 -,C173,83

SAFETY PROFILE: Poison by inhalation. Moderately toxic by ingestion, skin contact, and subcutaneous routes. Experimental reproductive effects. Human mutation data reported. An herbicide. When heated to decomposition it emits very toxic fumes of NO_x and SO_x.

EKQ000 CAS:104-76-7 ***HR: 2***
2-ETHYLHEXANOL
mf: $C_8H_{18}O$ mw: 130.26

PROP: Clear liquid. Bp: 184-185°, mp: $< -76°$, flash p: 178°F (81°C), n (20/D) 1.4300, d: 0.834 @ 20°/20°, vap press: 0.2 mm @ 20°, vap d: 4.49. Sol in about 720 parts water and in many organic solvents.

SYNS: 1-AETHYLHEXANOL (GERMAN) ◇ 2-ETHYL-1-HEXANOL ◇ 2-ETHYLHEXYL ALCOHOL

TOXICITY DATA with REFERENCE
skn-rbt 415 mg open MLD UCDS** 11/11/69
skn-rbt 500 mg/24H MOD FCTXAV 17,775,79
eye-rbt 4165 ug SEV AJOPAA 29,1363,46
eye-rbt 20 mg/24H MOD 28ZPAK -,36,72
mmo-sat 500 μmol/L EVHPAZ 45,111,82
orl-mus TDLo:12 g/kg (female 7-14D post):REP NTIS** PB85-220143
orl-rat TDLo:1628 mg/kg (female 12D post):TER TJADAB 35,41,87
orl-rat LD50:2049 mg/kg AIHAAP 30,470,69
ipr-rat LD50:500 mg/kg HYDRDA 3,201,78
scu-rat LD50:650 mg/kg NPIRI* 1,61,74
par-rat LD50:4600 mg/kg FCTXAV 17,775,79

orl-mus LD50:2500 mg/kg ZHYGAM 20,575,74
ipr-mus LD50:759 mg/kg ZHYGAM 20,575,74
par-mus LD50:1670 mg/kg FCTXAV 17,775,79
orl-rbt LD50:1180 mg/kg ZHYGAM 20,575,74
skn-rbt LD50:1970 mg/kg NPIRI* 1,61,74
orl-gpg LD50:1860 mg/kg ZHYGAM 20,575,74

CONSENSUS REPORTS: Reported in EPA TSCA Inventory.

SAFETY PROFILE: Moderately toxic by ingestion, skin contact, intraperitoneal, subcutaneous, and parenteral routes. An experimental teratogen. Other experimental reproductive effects. A severe eye and moderate skin irritant. Mutation data reported. A dangerous fire hazard when exposed to heat or flame; can react vigorously with oxidizing materials. To fight fire, use foam, CO_2, dry chemical. When heated to decomposition it emits acrid smoke and fumes. See also ALCOHOLS.

EKQ500 CAS:115-82-2 ***HR: 1***
2-ETHYL-1-HEXANOL SILICATE
mf: $C_{32}H_{68}O_4Si$ mw: 545.09

PROP: Insol in water. Bp: 350-370°; fp: −90°; flash p: 390°F (OC).

SYNS: TETRA(2-ETHYLHEXOXY)SILANE ◇ TETRA(2-ETHYLHEXYL)ORTHOSILICATE ◇ TETRA(2-ETHYLHEXYL)SILICATE

TOXICITY DATA with REFERENCE
skn-rbt 500 mg open MLD UCDS** 7/14/65

CONSENSUS REPORTS: Reported in EPA TSCA Inventory.

SAFETY PROFILE: A skin irritant. Combustible when exposed to heat or flame. To fight fire, use spray, dry chemical, CO_2. When heated to decomposition it emits acrid smoke and irritating fumes. See also SILANE.

EKR000 CAS:645-62-5 ***HR: 2***
2-ETHYL-2-HEXENAL
mf: $C_8H_{14}O$ mw: 126.22

PROP: Colorless liquid, powerful odor. Bp: 175°, flash p: 155°F (OC), d: 0.848 @ 20°/4°, vap d: 4.35, vap press: 1.0 mm @ 20°.

SYNS: 2-ETHYLHEXENAL ◇ α-ETHYL-β-n-PROPYLACROLEIN ◇ 2-ETHYL-3-PROPYL ACROLEIN

TOXICITY DATA with REFERENCE
skn-rbt 10 mg/24H open JIHTAB 26,269,44
eye-rbt 85 mg AJOPAA 29,1363,46
orl-rat LD50:3000 mg/kg JIHTAB 26,269,44

CONSENSUS REPORTS: Reported in EPA TSCA Inventory.

SAFETY PROFILE: Moderately toxic by ingestion. A skin, eye, and mucous membrane irritant. Combustible when exposed to heat or flame; can react with oxidizing materials. To fight fire, use alcohol foam, CO_2, dry chemical. When heated to decomposition it emits acrid smoke and irritating fumes.

EKR500 CAS:1632-16-2 ***HR: 3***
2-ETHYL-1-HEXENE
mf: C_8H_{16} mw: 112.24

PROP: Colorless liquid. Bp: 120°, d: 0.7270 @ 20°/20°, vap d: 3.87.

SYNS: 2-ETHYL HEXENE-1 ◇ USAF DO-21

TOXICITY DATA with REFERENCE
ihl-rat LCLo:4000 ppm/4H JIHTAB 31,343,49
ipr-mus LD50:100 mg/kg NTIS** AD277-689

CONSENSUS REPORTS: Reported in EPA TSCA Inventory.

SAFETY PROFILE: Poison by intraperitoneal route. Mildly toxic by inhalation. Combustible when exposed to heat or flame; can react with oxidizing materials. When heated to decomposition it emits acrid smoke and irritating fumes.

EKS000 CAS:5309-52-4 ***HR: 2***
2-ETHYL-2-HEXENOIC ACID
mf: $C_8H_{14}O_2$ mw: 142.22

TOXICITY DATA with REFERENCE
skn-rbt 10 mg/24H open MLD AMIHBC 4,119,51
eye-rbt 150 μg open SEV AMIHBC 4,119,51
orl-rat LDLo:5660 mg/kg AMIHBC 4,119,51
skn-rbt LD50:2750 mg/kg AMIHBC 4,119,51

SAFETY PROFILE: Moderately toxic by skin contact. Mildly toxic by ingestion. A skin and severe eye irritant. When heated to decomposition it emits acrid smoke and irritating fumes.

EKS500 CAS:104-75-6 ***HR: 3***
2-ETHYL HEXYLAMINE
DOT: UN 2276
mf: $C_8H_{19}N$ mw: 129.28

PROP: A clear, miscible liquid. Bp: 169.2°, flash p: 140°F (OC), d: 0.7894 @ 20°/20°, vap press: 1.2 mm @ 20°, vap d: 4.45.

SYN: 1-AMINO-2-ETHYLHEXAN(CZECH)

TOXICITY DATA with REFERENCE
skn-rbt 500 mg open MOD UCDS** 2/20/63
skn-rbt 500 mg/24H SEV 28ZPAK -,62,72
eye-rbt 50 μg/24H SEV 28ZPAK -,62,72
orl-rat LD50:450 mg/kg UCDS** 2/20/63
ihl-rat LCLo:250 ppm/4H AEHLAU 1,343,60

ipr-mus LDLo:4 mg/kg CBCCT* 2,189,50
skn-rbt LD50:600 mg/kg UCDS** 2/20/63

CONSENSUS REPORTS: Reported in EPA TSCA Inventory.

DOT Classification: Corrosive Material; Label: Corrosive, Flammable Liquid.

SAFETY PROFILE: Poison by intraperitoneal route. Moderately toxic by ingestion, inhalation, and skin contact. Corrosive. A severe skin and eye irritant. Flammable when exposed to heat or flame; can react with oxidizing materials. To fight fire, use alcohol foam, CO_2, dry chemical. When heated to decomposition it emits toxic fumes of NO_x. See also AMINES.

EKT000 CAS:10137-80-1 ***HR: 2***
N-(2-ETHYLHEXYL)ANILINE
mf: $C_{14}H_{23}N$ mw: 205.3

PROP: Liquid, mild odor. Bp: 194° @ 50 mm, fp: < −70°, flash p: 325°F (COC), d: 0.9119 @ 20°/20°, vap press: <0.01 @ 20°, vap d: 6.9.

TOXICITY DATA with REFERENCE
skn-rbt 10 mg/24H open SEV AIHAAP 23,95,62
orl-rat LD50:2410 mg/kg AIHAAP 23,95,62

SAFETY PROFILE: Moderately toxic by ingestion. A severe skin irritant. Combustible when exposed to heat or flame; can react with oxidizing materials. Reacts with water or steam to produce toxic fumes. To fight fire, use dry chemical, CO_2, mist or fog. When heated to decomposition it emits toxic fumes of NO_x.

EKT500 CAS:144-00-3 ***HR: 3***
5-ETHYL-5-HEXYLBARBITURIC ACID SODIUM SALT
mf: $C_{12}H_{19}N_2O_3 \cdot Na$ mw: 262.32

SYNS: 5-ETHYL-5-HEXYL-2,4,6-(1H,3H,5H)-PYRIMIDINETRIONE MONOSODIUM SALT ◇ HEBARAL ◇ HEXETHAL SODIUM ◇ ORTAL SODIUM ◇ SODIUM-5-ETHYL-5-HEXYLBARBITURATE ◇ SODIUM HEXETHAL ◇ SODIUM-N-HEXYLETHYL BARBITURATE

TOXICITY DATA with REFERENCE
orl-rat LDLo:2000 mg/kg JPETAB 60,125,37
ipr-rat LD50:250 mg/kg MEIEDD 10,680,83
ivn-rat LDLo:210 mg/kg JPETAB 60,125,37
ipr-mus LDLo:230 mg/kg JPETAB 31,455,27
scu-dog LDLo:150 mg/kg JPETAB 60,439,37
ivn-dog LDLo:40 mg/kg JPETAB 60,439,37
ivn-cat LDLo:50 mg/kg JPETAB 60,439,37
ipr-rbt LDLo:75 mg/kg JPETAB 60,439,37
ivn-rbt LDLo:50 mg/kg JPETAB 60,439,37

SAFETY PROFILE: Poison by intraperitoneal, intravenous, and subcutaneous routes. Moderately toxic by ingestion. A sedative and hypnotic agent. When heated to decomposition it emits toxic fumes of NO_x and Na_2O. See also BARBITURATES.

EKU000 CAS:123-04-6 ***HR: 2***
2-ETHYLHEXYL-1-CHLORIDE
mf: $C_8H_{17}Cl$ mw: 148.70

PROP: Colorless liquid. Bp: 172.9°, fp: −135°, flash p: 140°F (OC), d: 0.8833, @ 20°, vap d: 5.14.

SYN: 1-CHLORO-2-ETHYLHEXANE ◇ 3-CHLOROMETHYLHEPTANE

TOXICITY DATA with REFERENCE
skn-rbt 500 mg MLD 34ZIAG -,745,69
skn-rbt 100 mg/24H MOD 85JCAE-,102,86
eye-rbt 100 mg/24H MLD 85JCAE-,102,86
orl-rat LD50:7340 mg/kg AMIHBC 4,119,51
ihl-rat LCLo:2000 ppm/4H AMIHBC 4,119,51
skn-rbt LD50:15800 mg/kg 34ZIAG -,745,69

CONSENSUS REPORTS: Reported in EPA TSCA Inventory.

SAFETY PROFILE: Mildly toxic by ingestion, inhalation, and skin contact. A skin and eye irritant. Combustible when exposed to heat or flame; can react vigorously with oxidizing materials. To fight fire, use foam, CO_2, dry chemical. When heated to decomposition it emits highly toxic fumes of phosgene and Cl^-. See also CHLORINATED HYDROCARBONS, ALIPHATIC.

EKU500 CAS:5432-61-1 ***HR: 3***
N-(2-ETHYLHEXYL)CYCLOHEXYLAMINE
mf: $C_{14}H_{29}N$ mw: 211.44

PROP: Insol in water. D: 0.8, bp: 172° @ 20 mm, flash p: 265°F (OC).

TOXICITY DATA with REFERENCE
skn-rbt 100 μg/24H open AIHAAP 23,95,62
orl-rat LDLo:710 mg/kg AIHAAP 23,95,62
skn-rbt LD50:110 mg/kg AIHAAP 23,95,62

SAFETY PROFILE: Poison by skin contact. Moderately toxic by ingestion. A skin irritant. Combustible when exposed to heat or flame; can react vigorously with oxidizing materials. To fight fire, use dry chemical. When heated to decomposition it emits toxic fumes of NO_x. See also AMINES.

EKV000 CAS:94-96-2 ***HR: 2***
ETHYL HEXYLENE GLYCOL
mf: $C_8H_{18}O_2$ mw: 146.26

PROP: Practically colorless, somewhat viscous, odorless liquid. Bp: 243.1°, flash p: 260°F (OC), fp: −40°, d: 0.9422 @ 20°/20°, vap press: <0.01 mm @ 20°, vap d: 5.03.

SYNS: CARBIDE 6-12 ◇ COMPOUND 6-12 INSECT REPELLENT

◇ ENT 375 ◇ ETHOHEXADIOL ◇ ETHYL HEXANEDIOL ◇ 2-ETHYL-1,3-HEXANEDIOL ◇ 2-ETHYLHEXANE-1,3-DIOL ◇ 2-ETHYLHEXANEDIOL-1,3 ◇ 2-ETHYL-3-PROPYL-1,3-PROPANEDIOL ◇ 3-HYDROXYMETHYL-n-HEPTAN-4-OL ◇ 6-12-INSECT REPELLENT ◇ OCTYLENE GLYCOL ◇ REPELLENT 612 ◇ RUTGERS 612

TOXICITY DATA with REFERENCE
skn-rbt 500 mg MLD 34ZIAG -,731,69
eye-rbt 5 mg SEV AJOPAA 29,1363,46
orl-rat TDLo:20 g/kg (female 6-15D post):TER EPASR* 8EHQ-1288-0778
orl-rat TDLo:20 g/kg (female 6-15D post):REP EPASR* 8EHQ-1288-0778
orl-rat LD50:1400 mg/kg SPEADM 78-1,52,78
orl-mus LD50:1900 mg/kg SPEADM 78-1,52,78
orl-rbt LD50:2600 mg/kg PCOC** -,508,66
skn-rbt LD50:2000 mg/kg 31ZOAD 1,208,68
orl-gpg LD50:1790 mg/kg JPETAB 93,26,48
orl-ckn LD50:1400 mg/kg JPETAB 93,26,48

CONSENSUS REPORTS: Reported in EPA TSCA Inventory.

SAFETY PROFILE: Moderately toxic by ingestion and skin contact. Experimental teratogenic and reproductive effects. A skin and severe eye irritant. Used as an insecticide, insect repellent, and in hair care preparations. Combustible when exposed to heat or flame; can react with oxidizing materials. To fight fire, use alcohol foam, foam, dry chemical. When heated to decomposition it emits acrid smoke and irritating fumes.

EKV500 CAS:141-38-8 ***HR: 1***
2-ETHYLHEXYL EPOXYSTEARATE
mf: $C_{26}H_{50}O_3$ mw: 410.76

SYNS: 9,10-EPOXYSTEARIC ACID-2-ETHYLHEXYL ESTER ◇ 2-ETHYLHEXYL-9,10-EPOXYOCTADECANOATE

TOXICITY DATA with REFERENCE
skn-rbt 10 mg/24H open MLD AIHAAP 23,95,62
orl-rat LD50:31 g/kg AIHAAP 23,95,62

CONSENSUS REPORTS: Reported in EPA TSCA Inventory.

SAFETY PROFILE: Mildly toxic by ingestion. A skin irritant. When heated to decomposition it emits acrid smoke and irritating fumes.

EKW000 CAS:7425-14-1 ***HR: 1***
2-ETHYLHEXYL-2-ETHYLHEXANOATE
mf: $C_{16}H_{32}O_2$ mw: 256.48

PROP: Liquid. Vap d: 8.85.

TOXICITY DATA with REFERENCE
skn-rbt 10 mg/24H open MLD AIHAAP 23,95,62
eye-rbt 500 mg AJOPAA 29,1363,46
orl-rat LD50:27 g/kg AIHAAP 23,95,62

CONSENSUS REPORTS: Reported in EPA TSCA Inventory.

SAFETY PROFILE: Mildly toxic by ingestion. A skin and eye irritant. Combustible when exposed to heat or flame; can react with oxidizing materials. When heated to decomposition it emits acrid smoke and irritating fumes.

EKW300 CAS:7659-86-1 ***HR: 3***
2-ETHYLHEXYL MERCAPTOACETATE
mf: $C_{10}H_{20}O_2S$ mw: 204.36

SYNS: 2-ETHYLHEXYL THIOGLYCOLATE ◇ MERCAPTOACETIC ACID-2-ETHYLHEXYL ESTER ◇ THIOGLYCOLIC ACID-2-ETHYLHEXYL ESTER ◇ THIOGLYKOLSAEURE-2-AETHYLHEXYL ESTER (GERMAN)

TOXICITY DATA with REFERENCE
orl-rat LD50:303 mg/kg ZHYGAM 20,575,74
ipr-rat LD50:265 mg/kg ZHYGAM 20,575,74
orl-mus LD50:1430 mg/kg ZHYGAM 20,575,74
ipr-mus LD50:865 mg/kg ZHYGAM 20,575,74
orl-rbt LD50:534 mg/kg ZHYGAM 20,575,74
orl-gpg LD50:955 mg/kg ZHYGAM 20,575,74

SAFETY PROFILE: Poison by ingestion and intraperitoneal routes. When heated to decomposition it emits toxic fumes of SO_x. See also MERCAPTANS and ESTERS.

EKW500 CAS:688-84-6 ***HR: 2***
2-ETHYLHEXYL METHACRYLATE
mf: $C_{12}H_{22}O_2$ mw: 198.34

SYN: 2-ETHYL-1-HEXYLMETHACRYLATE

TOXICITY DATA with REFERENCE
ipr-mus LD50:2614 mg/kg JPMSAE 62,778,73

CONSENSUS REPORTS: Reported in EPA TSCA Inventory.

SAFETY PROFILE: Moderately toxic by intraperitoneal route. When heated to decomposition it emits acrid smoke and irritating fumes.

EKX000 CAS:7346-61-4 ***HR: 1***
2-ETHYLHEXYL OCTYLPHENYLPHOSPHITE
mf: $C_{22}H_{39}O_3P$ mw: 382.58

TOXICITY DATA with REFERENCE
orl-rat LD50:7000 mg/kg 34ZIAG -,659,69

CONSENSUS REPORTS: Reported in EPA TSCA Inventory.

SAFETY PROFILE: Mildly toxic by ingestion. When heated to decomposition it emits toxic fumes of PO_x.

EKX500 CAS:1559-35-9 ***HR: 2***
2-(2-ETHYLHEXYLOXY)ETHANOL
mf: $C_{10}H_{22}O_2$ mw: 174.32

SYN: 2-((2-ETHYLHEXYL)OXY)ETHANOL

TOXICITY DATA with REFERENCE
skn-rbt 10 mg/24H open SEV AMIHBC 10,61,54
eye-rbt 250 μg open SEV AMIHBC 10,61,54
orl-rat LD50:3080 mg/kg AMIHBC 10,61,54
skn-rbt LD50:2120 mg/kg AMIHBC 10,61,54

CONSENSUS REPORTS: Reported in EPA TSCA Inventory.

SAFETY PROFILE: Moderately toxic by ingestion and skin contact. A severe skin and eye irritant. When heated to decomposition it emits acrid smoke and irritating fumes.

EKY000 CAS:2549-90-8 ***HR: 1***
4-(2-ETHYLHEXYLOXY)-2-HYDROXYBENZOPHENONE
mf: $C_{21}H_{26}O_3$ mw: 326.47

SYN: 2-HYDROXY-4-(2'-ETHYLHEXOXY)BENZOFENON(CZECH)

TOXICITY DATA with REFERENCE
skn-rbt 500 mg/24H MLD 28ZPAK -,100,72
eye-rbt 500 mg/24H MLD 28ZPAK -,100,72

SAFETY PROFILE: A skin and eye irritant. When heated to decomposition it emits acrid smoke and irritating fumes.

EKZ000 CAS:10213-75-9 ***HR: 1***
3-(2-ETHYLHEXYLOXY)PROPIONITRILE
mf: $C_{11}H_{21}NO$ mw: 183.33

SYN: 3-((2-ETHYLHEXYL)OXY)PROPANENITRILE

TOXICITY DATA with REFERENCE
eye-rbt 500 mg open AMIHBC 10,61,54
orl-rat LD50:4920 mg/kg AMIHBC 10,61,54
skn-rbt LD50:5990 mg/kg AMIHBC 10,61,54

CONSENSUS REPORTS: Cyanide and its compounds are on the Community Right-To-Know List. Reported in EPA TSCA Inventory.

SAFETY PROFILE: Mildly toxic by ingestion and skin contact. An eye irritant. When heated to decomposition it emits toxic fumes of NO_x and CN^-. See also NITRILES.

ELA000 CAS:5397-31-9 ***HR: 3***
2-ETHYLHEXYLOXYPROPYLAMINE
mf: $C_{11}H_{25}NO$ mw: 187.37

PROP: Bp: 239°, mp: −90°, flash p: 210°F (OC), d: 0.8483, vap d: 6.47.

SYNS: 2-ETHYLHEXYL-3-AMINOPROPYL ETHER ◇ 3-((2-ETHYLHEXYL)OXY)PROPYLAMINE

TOXICITY DATA with REFERENCE
skn-rbt 10 mg/24H open SEV AMIHBC 4,119,51
eye-rbt 50 μg open SEV AMIHBC 4,119,51
orl-rat LD50:320 mg/kg AMIHBC 4,119,51
skn-rbt LD50:360 mg/kg AMIHBC 4,119,51

CONSENSUS REPORTS: Reported in EPA TSCA Inventory.

SAFETY PROFILE: Poison by ingestion and skin contact. A severe skin and eye irritant. Combustible when exposed to heat or flame; can react vigorously with oxidizing materials. When heated to decomposition it emits toxic fumes of NO_x. See also ETHERS and AMINES.

ELA600 CAS:24234-06-8 ***HR: 3***
1-ETHYL-1-HEXYLPIPERIDINIUM BROMIDE
mf: $C_{13}H_{28}N•Br$ mw: 278.33

TOXICITY DATA with REFERENCE
orl-mus LD50:408 mg/kg PSDTAP 15,331,74
ivn-mus LD50:4496 μg/kg PSDTAP 15,331,74
ipr-mus LD50:42958 μg/kg PSDTAP 15,331,74

SAFETY PROFILE: Poison by intravenous and intraperitoneal routes. Moderately toxic by ingestion. When heated to decomposition it emits toxic fumes of NO_x and Br^-.

ELB000 CAS:118-60-5 ***HR: 3***
2-ETHYLHEXYL SALICYLATE
mf: $C_{15}H_{22}O_3$ mw: 250.37

SYNS: SALICYLIC ACID-2-ETHYLHEXYL ESTER ◇ USAF DO-11

TOXICITY DATA with REFERENCE
skn-rbt 500 mg/24H MLD FCTXAV 14,659,76
ipr-mus LD50:200 mg/kg NTIS** AD277-689

CONSENSUS REPORTS: Reported in EPA TSCA Inventory.

SAFETY PROFILE: Poison by intraperitoneal route. A skin irritant. When heated to decomposition it emits acrid smoke and irritating fumes. See also ESTERS.

ELB400 CAS:72214-01-8 ***HR: 2***
2-ETHYLHEXYL SULFATE
mf: $C_8H_{18}O_4S$ mw: 210.32

SYN: SULFURIC ACID, MONO(2-ETHYLHEXYL)ESTER

TOXICITY DATA with REFERENCE
orl-mus TDLo:1747 g/kg/2Y C:ETA EVHPAZ 65,271,86
orl-rat LD50:4125 mg/kg 34ZIAG-,690,69

SAFETY PROFILE: Moderately toxic by ingestion. Questionable carcinogen with experimental tumorigenic data. When heated to decomposition it emits toxic fumes of SO_x.

ELB500 CAS:103-44-6 ***HR: 2***
2-ETHYLHEXYL VINYL ETHER
mf: $C_{10}H_{20}O$ mw: 156.30

PROP: Liquid. Mp: −100°, bp: 177.5°, flash p: 135°F (OC), d: 0.810, autoign temp: 395°F, vap d: 5.4.

SYNS: 1-ETHENOXY-2-ETHYLHEXANE ◇ VINYL-2-ETHYLHEXYL ETHER

TOXICITY DATA with REFERENCE
skn-rbt 10 mg/24H open MLD AMIHBC 10,61,54
skn-rbt 500 mg open MLD UCDS** 3/23/73
eye-rbt 500 mg open AMIHBC 10,61,54
orl-rat LD50:1350 mg/kg AMIHBC 10,61,54
skn-rbt LD50:3560 mg/kg AMIHBC 10,61,54

SAFETY PROFILE: Moderately toxic by ingestion and skin contact. A skin and eye irritant. Combustible when exposed to heat or flame; can react with oxidizing materials. Potentially explosive. To fight fire, use alcohol foam, foam, CO_2, dry chemical. When heated to decomposition it emits acrid smoke and irritating fumes. See also ETHERS.

ELC000 CAS:18413-14-4 ***HR: 3***
ETHYLHYDRAZINE HYDROCHLORIDE
mf: $C_2H_8N_2$•ClH mw: 96.58

TOXICITY DATA with REFERENCE
orl-mus TDLo:11 g/kg/64W-C:CAR IJCNAW 13,500,74

SAFETY PROFILE: Questionable carcinogen with experimental carcinogenic data. When heated to decomposition it emits very toxic fumes of HCl and NO_x. See also HYDRAZINE.

ELC500 CAS:3413-58-9 ***HR: 3***
ETHYLHYDROCUPREINE HYDROCHLORIDE
mf: $C_{21}H_{28}N_2O_2$•ClH mw: 376.97

SYNS: HYDROCUPREINE ETHYL ESTER HYDROCHLORIDE ◇ NEUMOLISINA ◇ NUMOQUIN HYDROCHLORIDE ◇ OPTOCHIN HYDROCHLORIDE ◇ OPTOQUINHYDROCHLORIDE ◇ RHOMBIC

TOXICITY DATA with REFERENCE
ivn-man TDLo:31 mg/kg/3D:EYE,CVS JPETAB 8,53,16
scu-mus LDLo:500 mg/kg JPETAB 8,53,16
unr-frg LDLo:300 mg/kg JPETAB 8,53,16

SAFETY PROFILE: Poison by unspecified route. Moderately toxic by subcutaneous route. Human systemic effects by intravenous route: visual field changes and arteriolar constriction. An antiseptic. When heated to decomposition it emits very toxic fumes such as Cl^- and NO_x.

ELD000 CAS:3031-74-1 ***HR: 3***
ETHYL HYDROPEROXIDE
mf: $C_2H_6O_2$ mw: 62.07

CH_3CH_2OOH

SYN: ETHYL HYDROGEN PEROXIDE

DOT Classification: Forbidden.

SAFETY PROFILE: Explodes violently when superheated. The barium salt is heat- and impact-sensitive. Explosive reaction with hydroiodic acid or finely divided silver. When heated to decomposition it emits acrid smoke and irritating fumes. See also PEROXIDES.

ELD500 CAS:2497-34-9 ***HR: 3***
4'-ETHYL-4-HYDROXYAZOBENZENE
mf: $C_{14}H_{14}N_2O$ mw: 226.30

TOXICITY DATA with REFERENCE
orl-rat TDLo:9780 mg/kg/43W-C:ETA ARZNAD 12,270,62

SAFETY PROFILE: Questionable carcinogen with experimental tumorigenic data. When heated to decomposition it emits toxic fumes of NO_x.

ELE500 CAS:54897-62-0 ***HR: 3***
N-ETHYL-N-(4-HYDROXYBUTYL)NITROSOAMINE
mf: $C_6H_{14}N_2O_2$ mw: 146.22

SYNS: EHBN ◇ 4-(ETHYLNITROSOAMINO)-1-BUTANOL

TOXICITY DATA with REFERENCE
mma-sat 35 µmol/plate CNREA8 37,399,77
orl-rat TDLo:4375 mg/kg/20W-C:ETA GANNA2 67,175,76
orl-mus TDLo:4900 mg/kg/14W-C:CAR GANNA2 72,647,81
orl-mus TD:9800 mg/kg/14W-C:CAR GANNA2 72,647,81

CONSENSUS REPORTS: EPA Genetic Toxicology Program.

SAFETY PROFILE: Questionable carcinogen with experimental carcinogenic and tumorigenic data. Mutation data reported. When heated to decomposition it emits toxic fumes of NO_x. See also NITROSAMINES.

ELF100 CAS:33765-68-3 ***HR: 2***
16-ETHYL-17-HYDROXYESTER-4-EN-3-ONE
mf: $C_{20}H_{30}O_2$ mw: 302.50

PROP: Crystals from ether. Mp: 152-153°.

SYNS: 16-β-ETHYL-17-β-HYDROXYESTR-4-EN-3-ONE ◇ 16-β-ETHYL-17-β-HYDROXY-4-ESTREN-3-ONE ◇ 16-β-ETHYL-19-NORTESTOSTERONE ◇ OXENDOLONE ◇ PROSTETIN ◇ TSAA 291

TOXICITY DATA with REFERENCE
ims-rat TDLo:150 mg/kg (female 30D pre):REP YACHDS 7,943,79
orl-rat LD50:12105 mg/kg YACHDS 7,937,79
ipr-rat LD50:5879 mg/kg YACHDS 7,937,79
orl-mus LD50:5480 mg/kg IYKEDH 12,1204,81
ipr-mus LD50:2925 mg/kg YACHDS 7,937,79

SAFETY PROFILE: Moderately toxic by intraperitoneal route. Mildly toxic by ingestion. Experimental reproductive effects. A steroid. When heated to decomposition it emits acrid smoke and irritating fumes. See also ESTERS.

ELF110 CAS:33765-80-9 ***HR: 1***
16-β-ETHYL-17-β-HYDROXYESTER-4-EN-3-ONE ACETATE
mf: $C_{22}H_{32}O_3$ mw: 344.54

SYNS: 17-β-ACETOXY-16-β-ETHYLESTR-4-EN-3-ONE ◇ (16-β,17-β)-17-(ACETYLOXY)-16-ETHYL-ESTR-4-EN-3-ONE (9CI) ◇ TSAA-328

TOXICITY DATA with REFERENCE
ims-rat TDLo:2500 mg/kg (5D male):REP TAKHAA 32,330,73
ipr-rat LD50:6100 mg/kg TAKHAA 32,330,73
ipr-mus LD50:7500 mg/kg TAKHAA 32,330,73
ims-mus LDLo:10 g/kg TAKHAA 32,330,73

SAFETY PROFILE: Mildly toxic by intraperitoneal and intramuscular routes. Experimental reproductive effects. When heated to decomposition it emits acrid smoke and irritating fumes.

ELF500 CAS:1164-38-1 ***HR: 3***
ETHYL (2-HYDROXYETHYL) DIMETHYLAMMONIUM BENZILATE CHLORIDE
mf: $C_{20}H_{26}NO_3•Cl$ mw: 363.92

SYNS: BENZILIC ACID, ester with ETHYL (2-HYDROXYETHYL)DIMETHYLAMMONIUM CHLORIDE ◇ BENZILYLOXYETHYLDIMETHYLETHYLAMMONIUM CHLORIDE ◇ E-3 ◇ N-ETHYL-2-((HYDROXYDIPHENYLACETYL)OXY)-N,N-DIMETHYLETHANAMINIUM CHLORIDE ◇ ETHYL(2-HYDROXYETHYL)DIMETHYLAMMONIUM CHLORIDE BENZILATE ◇ LACHESIN ◇ LACHESINE CHLORIDE ◇ LAXESIN

TOXICITY DATA with REFERENCE
orl-mus LD50:1 g/kg MEIEDD 10,767,83
ipr-mus LD50:40 mg/kg MEIEDD 10,767,83
scu-mus LD50:160 mg/kg MEIEDD 10,767,83

SAFETY PROFILE: Poison by intraperitoneal and subcutaneous routes. Moderately toxic by ingestion. When heated to decomposition it emits very toxic fumes of Cl^-, NH_3, and NO_x.

ELG000 CAS:63918-38-7 ***HR: 3***
ETHYL-β-HYDROXYETHYLETHYLENIMONIUM PICRYLSULFONATE
mf: $C_6H_{14}NO•C_6H_2N_3O_9S$ mw: 408.38

SYNS: 1-ETHYL-1-(2-HYDROXYETHYL)AZIRIDINIUM-2,4,6-TRINITROBENZENESULFONATE ◇ 1-ETHYL-1-(2-HYDROXYETHYL) AZIRIDINIUM SALT with 2,4,6-TRINITROBENZENESULFONIC ACID ◇ ETHYL(2-HYDROXYETHYL)ETHYLENIMONIUM PICRYLSULFONATE ◇ 1-ETHYL-1-(β-HYDROXYETHYL)ETHYLENIMONIUM PICRYLSULFONATE

TOXICITY DATA with REFERENCE
scu-mus LD50:5500 μg/kg JPETAB 91,224,47
ivn-mus LD50:5 mg/kg JPETAB 91,224,47
ivn-rbt LDLo:5 mg/kg JPETAB 91,224,47

SAFETY PROFILE: Poison by subcutaneous and intravenous routes. When heated to decomposition it emits very toxic fumes of NO_x and SO_x. See also SULFONATES, PICRIC ACID, and NITRO COMPOUNDS of AROMATIC HYDROCARBONS.

ELG500 CAS:13147-25-6 ***HR: 3***
ETHYL-2-HYDROXYETHYLNITROSAMINE
mf: $C_4H_{10}N_2O_2$ mw: 118.16

$CH_3CH_2N(N:O)C_2H_4OH$

SYNS: AETHYL-AETHANOL-NITROSOAMIN (GERMAN) ◇ EENA ◇ EHEN ◇ N-ETHYL-N-HYDROXYETHYLNITROSAMINE ◇ 2-(ETHYLNITROSAMINO)ETHANOL ◇ N-NITROSOAETHYLAETHANOLAMIN (GERMAN) ◇ N-NITROSOETHYLETHANOLAMINE ◇ N-NITROSOETHYL-2-HYDROXYETHYLAMINE ◇ N-NITROSO-N- ETHYL-N-(2-HYDROXYETHYL)AMINE

TOXICITY DATA with REFERENCE
mma-sat 20 μg/plate MUREAV 66,1,79
orl-rat TDLo:1680 mg/kg/14D-C:CAR GANNA2 70,817,79
ivn-rat TDLo:50 mg/kg/10W-I:CAR CRNGDP 7,1313,86
orl-rat TD:840 mg/kg/2W-C:NEO JJIND8 72,483,84
orl-rat TD:700 mg/kg/1W-C:ETA JJIND8 70,477,83

CONSENSUS REPORTS: IARC Cancer Review: Animal Limited Evidence IMEMDT 17,83,78. EPA Genetic Toxicology Program.

SAFETY PROFILE: Suspected carcinogen with experimental carcinogenic, neoplastigenic, and tumorigenic data. Mutation data reported. Explodes when heated to 170°C. When heated to decomposition it emits toxic fumes of NO_x. See also NITROSAMINES.

ELH000 CAS:62641-68-3 ***HR: D***
ETHYL-N-(2-HYDROXYETHYL)-N-NITROSOCARBAMATE
mf: $C_5H_{10}N_2O_4$ mw: 162.17

SYN: N-(2-HYDROXYETHYL)-N-NITROSO CARBAMIC ACID ETHYL ESTER

TOXICITY DATA with REFERENCE
mmo-sat 800 nmol/L MUREAV 48,131,77

CONSENSUS REPORTS: EPA Genetic Toxicology Program.

SAFETY PROFILE: Mutation data reported. Many N-nitroso compounds are carcinogens. When heated to decomposition it emits toxic fumes of NO_x. See also N-NITROSO COMPOUNDS and CARBAMATES.

ELH600 CAS:119-41-5 ***HR: 2***
ETHYL-7-HYDROXYFLAVONE
mf: $C_{19}H_{16}O_5$ mw: 324.35

PROP: Crystals from 50% ethanol. Mp: 123-124°. Sol in the usual organic solvents; sltly sol in water.

SYNS: 7-α-(ACETOXYETHANE)OXYFLAVONE ◇ ANGORLISIN ◇ CORIL ◇ CORLIN ◇ COROSANIN ◇ DILATAN KORE ◇ DOMUCOR ◇ EFLOXATE ◇ ETHYL FLAVONE-7-OXYACETATE ◇ ETHYL-7-FLAVONOXYACETATE ◇ ETHYL FLAVON-7-YLOXYACETATE ◇ ETHYL FLAVONYL-7-OXYACETATE ◇ FLACETHYLE ◇ 7-FLAVONE ETHYL HYDROXYACETATE ◇ FLAVONE-7-ETHYLOXYACETATE ◇ 7-FLAVONOXYACETIC ACID ETHYL ESTER ◇ OXIFLAVIL ◇ ((4-OXO-2-PHENYL-4H-1-BENZOPYRAN-7-YL)OXY)ACETIC ACID ETHYL ESTER ◇ OXYFLAVIL ◇ RE 1-0185 ◇ REC 1-0185 ◇ RECORDIL

TOXICITY DATA with REFERENCE
ipr-rat LD50:3200 mg/kg MEIEDD 10,509,83
ipr-mus LD50:3200 mg/kg FRPSAX 13,561,58
ipr-gpg LD50:2 g/kg NIIRDN 6,125,82

SAFETY PROFILE: Moderately toxic by intraperitoneal route. When heated to decomposition it emits acrid smoke and irritating fumes.

ELI500 CAS:10029-04-6 ***HR: 3***
ETHYL-2-(HYDROXYMETHYL)ACRYLATE
mf: $C_6H_{10}O_3$ mw: 130.16

SYNS: ETHYL-α-(HYDROXYMETHYL)ACRYLATE ◇ 2-(HYDROXYMETHYL)ACRYLIC ACID, ETHYL ESTER

TOXICITY DATA with REFERENCE
skn-rbt 500 mg MOD IHFCAY 6,1,67
eye-rbt 500 mg SEV IHFCAY 6,1,67
orl-rat LD50:620 mg/kg IHFCAY 6,1,67
skn-rbt LD50:360 mg/kg IHFCAY 6,1,67

SAFETY PROFILE: Poison by skin contact. Moderately toxic by ingestion. A skin and severe eye irritant. When heated to decomposition it emits acrid smoke and irritating fumes. See also ESTERS.

ELJ500 CAS:1005-93-2 ***HR: 3***
2-ETHYL-2-(HYDROXYMETHYL)-1,3-PROPANEDIOL, CYCLIC PHOSPHATE (1:1)
mf: $C_6H_{11}O_4P$ mw: 178.14

SYNS: 4-AETHYL-1-PHOSPHA-2,6,7-TRIOXABICYCLO(2.2.2)OCTAN-1-OXID (GERMAN) ◇ 4-ETHYL-1-PHOSPHA-2,6,7-TRIOXABICYCLO(2.2.2)OCTANE-1-OXIDE ◇ 4-ETHYL-2,6,7-TRIOXA-1-PHOSPHABICYCLO(2.2.2)OCTANE-1-OXIDE

TOXICITY DATA with REFERENCE
orl-rat LD50:3080 μg/kg ARTODN 35,149,76
ihl-rat LC50:30 mg/m³/1H ARTODN 35,149,76
skn-rat LD50:50 mg/kg ARTODN 35,149,76
ipr-rat LD50:960 μg/kg ARTODN 35,149,76
orl-mus LD50:3550 μg/kg ARTODN 35,149,76
ipr-mus LD50:1 mg/kg SCIEAS 182,1135,73
ivn-mus LDLo:500 μg/kg EJMCA5 13,207,78
orl-dog LD50:1 mg/kg ARTODN 35,149,76
orl-rbt LD50:5 mg/kg ARTODN 35,149,76

SAFETY PROFILE: Poison by ingestion, inhalation, skin contact, intraperitoneal, and intravenous routes. When heated to decomposition it emits toxic fumes of PO_x.

ELK000 CAS:78330-02-6 ***HR: 2***
ETHYL-4-HYDROXY-3-MORPHOLINOMETHYLBENZOATE
mf: $C_{13}H_{19}NO_4$ mw: 253.33

SYN: 4-HYDROXY-3-MORPHOLINOMETHYLBENZOIC ACID ETHYL ESTER

TOXICITY DATA with REFERENCE
orl-mus LDLo:3000 mg/kg ARZNAD 11,85,61
scu-mus LDLo:460 mg/kg ARZNAD 11,85,61

SAFETY PROFILE: Moderately toxic by ingestion and subcutaneous routes. When heated to decomposition it emits toxic fumes of NO_x.

ELK500 CAS:34529-29-8 ***HR: 3***
3-ETHYL-4-HYDROXY-1,2,5-OXADIAZOLE
mf: $C_4H_6N_2O_2$ mw: 114.10

$CH_3CH_2C{=}NON{=}COH$ (ring bond C–C)

SYN: 4-ETHYL-3-FURAZANONE

SAFETY PROFILE: Reaction with sodium hydroxide forms an extremely unstable, explosive sodium salt. When heated to decomposition it emits toxic fumes of NO_x.

ELL500 CAS:70-70-2 ***HR: 3***
ETHYL-p-HYDROXYPHENYL KETONE
mf: $C_9H_{10}O_2$ mw: 150.19

SYNS: FRENANTOL ◇ FRENOHYPON ◇ H-365 ◇ p-HYDROXYPHENYL-1-PROPANONE ◇ 1-(4-HYDROXYPHENYL)-1-PROPANONE ◇ HYDROXYPROPIOPHENONE ◇ p-HYDROXYPROPIOPHENONE ◇ 4-HYDROXYPROPIOPHENONE ◇ HYPOPHENON ◇ p-OXYPROPIOPHENONE ◇ PAROXON ◇ PAROXYPROPIONE ◇ PHP ◇ POP ◇ PROFENONE ◇ p-PROPIONYLPHENOL ◇ USAF EK-3302

TOXICITY DATA with REFERENCE
ims-rbt TDLo:1600 mg/kg (female 15-30D post):REP AOGNAX 66,286,61
ims-rbt TDLo:1600 mg/kg (female 15-30D post):TER AOGNAX 66,286,61
orl-mus LD50:3 mg/kg 85JDAH -,197,74
ipr-mus LD50:200 mg/kg NTIS** AD277-689
scu-mus LD50:1130 μg/kg AIPTAK 124,212,60
par-frg LD50:91 mg/kg AIPTAK 124,212,60

CONSENSUS REPORTS: Reported in EPA TSCA Inventory.

SAFETY PROFILE: Poison by intraperitoneal, subcutaneous, and parenteral routes. An experimental teratogen. Other experimental reproductive effects. When heated to decomposition it emits acrid smoke and irritating fumes. See also KETONES.

ELM000 CAS:61734-88-1 ***HR: D***
N-ETHYL-N-(3-HYDROXYPROPYL)NITROSAMINE
mf: $C_5H_{12}N_2O_2$ mw: 132.19

TOXICITY DATA with REFERENCE
mma-sat 15 μmol/plate CNREA8 37,399,77

SAFETY PROFILE: Mutation data reported. Many nitrosamines are carcinogens. When heated to decomposition it emits toxic fumes of NO_x. See also NITROSAMINES.

ELM500 CAS:624-85-1 ***HR: 3***
ETHYL HYPOCHLORITE
mf: C_2H_5ClO mw: 80.52

CH_3CH_2OCl

SAFETY PROFILE: Very unstable. The vapor explodes on contact with flame, spark, or upon rapid heating. The cold liquid explodes on contact with copper. Incompatible with light. When heated to decomposition it emits toxic fumes of Cl^-. See also HYPOCHLORITES.

ELN500 CAS:75-37-6 ***HR: 3***
ETHYLIDENE DIFLUORIDE
DOT: UN 1030
mf: $C_2H_4F_2$ mw: 66.06

PROP: Colorless gas. Mp: −117.0°, bp: −26.5°, d: 1.004 @ 25°, vap d: 2.28.

SYNS: ALGOFRENE TYPE 67 ◇ DIFLUOROETHANE ◇ 1,1-DIFLUOROETHANE (DOT) ◇ ETHYLENE FLUORIDE ◇ ETHYLIDENE FLUORIDE ◇ FC 152a ◇ FREON 152 ◇ GENETRON 100 ◇ HALOCARBON 152A

TOXICITY DATA with REFERENCE
sln-dmg-ihl 98 pph/10M ENVRAL 7,275,74
ihl-rat LCLo:50 pph/10M AIHOAX 2,335,50
ihl-mus LC50:977 g/m^3/2H 85GMAT -,54,82

CONSENSUS REPORTS: Reported in EPA TSCA Inventory. EPA Genetic Toxicology Program.

DOT Classification: Flammable Gas; Label: Flammable Gas.

SAFETY PROFILE: Mildly toxic by inhalation. Mutation data reported. Narcotic in high concentration. A very dangerous fire hazard, when exposed to heat or flame; can react vigorously with oxidizing materials. See also FLUORIDES.

ELN600 CAS:55044-04-7 ***HR: 3***
ETHYLIDENE DINITRATE
mf: $C_2H_4N_2O_6$ mw: 152.06

SAFETY PROFILE: A heat-sensitive explosive. When heated to decomposition it emits toxic fumes of NO_x. See also NITRATES.

ELO000 CAS:539-71-9 ***HR: 3***
ETHYLIDENE DIURETHAN
mf: $C_8H_{16}N_2O_4$ mw: 204.26

SYNS: N,N'-ETHYLIDENE-BIS(ETHYL CARBAMATE) ◇ ETHYLIDENEDICARBAMIC ACID, DIETHYL ESTER

TOXICITY DATA with REFERENCE
ipr-mus TDLo:6500 mg/kg/13W-I:ETA JNCIAM 9,35,48

SAFETY PROFILE: Questionable carcinogen with experimental tumorigenic data. When heated to decomposition it emits toxic fumes of NO_x.

ELO500 CAS:16219-75-3 ***HR: 2***
ETHYLIDENE NORBORNENE
mf: C_9H_{12} mw: 120.21

SYNS: 5-ETHYLIDENEBICYCLO(2.2.1)HEPT-2-ENE ◇ 5-ETHYLIDENE-2-NORBORNENE

TOXICITY DATA with REFERENCE
skn-rbt 445 mg open MLD UCDS** 11/28/67
ihl-hmn TCLo:6 ppm/30M:NOSE,EYE,TONG TXAPA9 20,250,71
orl-rat LD50:2527 mg/kg AIHAAP 30,470,69
ihl-rat LC50:4000 ppm/4H AIHAAP 30,470,69
ihl-rat LC50:1246 ppm/4H TXAPA9 20,250,71
orl-mus LD50:5667 mg/kg GTPZAB 18(10)52,74
ihl-mus LC50:732 ppm/4H TXAPA9 20,250,71
ihl-rbt LC50:3104 ppm/4H TXAPA9 20,250,71
skn-rbt LD50:8189 mg/kg AIHAAP 30,470,69
ihl-gpg LC50:2896 ppm/4H TXAPA9 20,250,71

CONSENSUS REPORTS: Reported in EPA TSCA Inventory.

OSHA PEL: CL 5 ppm
ACGIH TLV: CL 5 ppm

SAFETY PROFILE: Moderately toxic by ingestion. Mildly toxic by inhalation and skin contact. Human systemic effects by inhalation: conjuctiva, olfactory and taste changes. A skin irritant. When heated to decomposition it emits acrid smoke and irritating fumes.

ELP000 CAS:139-87-7 ***HR: 2***
N-ETHYL-2,2'-IMINODIETHANOL
mf: $C_6H_{15}NO_2$ mw: 133.22

SYN: 2,2'-(ETHYLIMINO)DIETHANOL

TOXICITY DATA with REFERENCE
skn-rbt 10 mg/24H open MLD AMIHBC 10,61,54
eye-rbt 750 μg open SEV AMIHBC 10,61,54
orl-rat LD50:4570 mg/kg AMIHBC 10,61,54

CONSENSUS REPORTS: Reported in EPA TSCA Inventory.

SAFETY PROFILE: Mildly toxic by ingestion. A skin and severe eye irritant. When heated to decomposition it emits toxic fumes of NO_x.

ELP500 CAS:75-03-6 ***HR: 3***
ETHYL IODIDE
mf: C_2H_5I mw: 155.97

PROP: Clear, refractive, heavy, colorless liquid; turns brown on exposure to light. Mp: −108°, bp: 72.4°, d: 1.90-1.93 @ 25°/25°, vap press: 100 mm @ 18.0°, vap d: 5.38. Misc in alc and ether.

SYNS: HYDRIODIC ETHER ◇ IODOETHANE

TOXICITY DATA with REFERENCE
mmo-esc 20 μmol/L ARTODN 46,277,80
dnd-esc 1 μmol/L ARTODN 46,277,80
ihl-rat LC50:65000 mg/m³/30M FAVUAI 7,35,75
ipr-rat LD50:330 mg/kg 85GMAT -,68,82
ipr-mus LD50:560 mg/kg 85GMAT -,68,82
scu-mus LD50:1000 mg/kg JJPAAZ 3,99,54
ipr-gpg LD50:322 mg/kg 85GMAT -,68,82

CONSENSUS REPORTS: Reported in EPA TSCA Inventory.

SAFETY PROFILE: Poison by intraperitoneal route. Moderately toxic by subcutaneous route. Mildly toxic by inhalation. Mutation data reported. A skin, eye, and mucous membrane irritant. Narcotic in high concentration. Flammable when exposed to heat or flame, a preparative hazard. Will react with water or steam to produce toxic and corrosive fumes; can react vigorously with oxidizing materials. Incompatible with silver chlorite. To fight fire, use water, CO_2, dry chemical. When heated to decomposition it emits highly toxic fumes of I^-. See also IODIDES.

ELQ000 CAS:623-48-3 ***HR: 3***
ETHYL IODOACETATE
mf: $C_4H_7IO_2$ mw: 214.01

PROP: Dense, colorless liquid. Bp: 179°, d: 1.80, vap press: 0.54 mm @ 20°, vap d: 7.4.

SYNS: ETHYL MONOIODOACETATE ◇ IODOACETIC ACID ETHYL ESTER

TOXICITY DATA with REFERENCE
ipr-mus LD50:45 mg/kg JNCIAM 31,297,63

CONSENSUS REPORTS: Reported in EPA TSCA Inventory.

SAFETY PROFILE: Poison by intraperitoneal route. A skin, eye, and mucous membrane irritant. Will react with water or steam to produce toxic and corrosive fumes. When heated to decomposition or on contact with acid or acid fumes it emits highly toxic fumes of I^-. See also IODIDES and ESTERS.

ELQ100 ***HR: 3***
ETHYLIODOMETHYLARSINE
mf: C_3H_8AsI mw: 245.92

$CH_3CH_2As(I)CH_3$

CONSENSUS REPORTS: Arsenic and its compounds are on the Community Right-To-Know List.

SAFETY PROFILE: Arsenic compounds are poisons. Ignites spontaneously in air. When heated to decomposition it emits toxic fumes of I^- and As. See also ARSENIC COMPOUNDS and IODIDES.

ELQ500 CAS:99-79-6 ***HR: 2***
ETHYL-10-(p-IODOPHENYL)UNDECYLATE
mf: $C_{19}H_{29}IO_2$ mw: 416.38

SYNS: ETHIODAN ◇ ETHYL-10-(p-IODOPHENYL)HENDECANOATE ◇ IOPHENDYLATE

TOXICITY DATA with REFERENCE
ipr-rat LD50:1130 mg/kg RPTOAN 34,38,71
ipr-mus LD50:5640 mg/kg RPTOAN 34,38,71
ivn-rbt LD50:1 g/kg RPTOAN 34,38,71

SAFETY PROFILE: Moderately toxic by intraperitoneal and intravenous routes. When heated to decomposition it emits toxic fumes of I^-.

ELS000 CAS:97-62-1 ***HR: 3***
ETHYL ISOBUTYRATE
DOT: UN 2385
mf: $C_6H_{12}O_2$ mw: 116.18

PROP: Colorless, volatile liquid; fruity, aromatic odor. Mp: −88°, bp: 110-111°, d: 0.862, vap press: 40 mm @ 33.8°, vap d: 4.01, refr index: 1.385, flash p: <64.4°F.

SYNS: ETHYL ISOBUTANOATE ◇ ETHYLISOBUTYRATE (DOT) ◇ ETHYL-2-METHYLPROPANOATE ◇ ETHYL-2-METHYLPROPIONATE ◇ FEMA No. 2428 ◇ ISOBUTYRIC ACID, ETHYL ESTER ◇ 2-METHYLPROPIONIC ACID, ETHYL ESTER

TOXICITY DATA with REFERENCE
skn-rbt 500 mg/24H MOD FCTXAV 16,741,78
ipr-mus LD50:800 mg/kg FCTXAV 16,741,78

CONSENSUS REPORTS: Reported in EPA TSCA Inventory.

DOT Classification: Flammable Liquid; Label: Flammable Liquid.

SAFETY PROFILE: Moderately toxic by intraperitoneal route. A skin irritant. Flammable Liquid. A very dangerous fire hazard when exposed to heat or flame; can react vigorously with oxidizing materials. To fight fire, use foam, CO_2, dry chemical. When heated to decomposition it emits acrid smoke and irritating fumes. See also ESTERS.

ELS500 CAS:109-90-0 ***HR: 3***
ETHYL ISOCYANATE
DOT: UN 2481
mf: C_3H_5NO mw: 71.09

PROP: Bp: 60°, d: 0.90 @ 20°/4°, vap d: 2.45.

SYNS: ETHYL ISOCYANATE (DOT) ◇ ISOCYANATOETHANE ◇ ISOCYANIC ACID, ETHYL ESTER

TOXICITY DATA with REFERENCE
mmo-sat 50 μg/plate ABCHA6 44,3017,80
ivn-mus LD50:56 mg/kg CSLNX* NX#02910

CONSENSUS REPORTS: Reported in EPA TSCA Inventory.

DOT Classification: Flammable Liquid; Label: Flammable Liquid and Poison.

SAFETY PROFILE: Poison by intravenous route. Mutation data reported. When heated to decomposition it emits toxic fumes of NO_x. See also CYANATES.

ELT000 CAS:624-79-3 ***HR: 3***
ETHYL ISOCYANIDE
mf: C_3H_5N mw: 55.08

$CH_3CH_2N{=}C:$

PROP: Colorless liquid, sol in water and organic solvents. D: 0.7402 @ 20°/4°; mp: < −66°; bp: 79°.

SYNS: ETHYL ISONITRILE ◇ ISOCYANOETHANE

CONSENSUS REPORTS: Cyanide and its compounds are on the Community Right-To-Know List.

SAFETY PROFILE: Probably very toxic. Can explode upon heating. When heated to decomposition it emits toxic fumes of CN^-. See also NITRILES.

ELT500 CAS:106-67-2 ***HR: 2***
2-ETHYLISOHEXANOL
mf: $C_8H_{18}O$ mw: 130.26

SYN: 2-ETHYLISOHEXYL ALCOHOL

TOXICITY DATA with REFERENCE
orl-rat LDLo:3200 mg/kg KODAK* -,-,71
orl-mus LDLo:1600 mg/kg KODAK* -,-,71

CONSENSUS REPORTS: Reported in EPA TSCA Inventory.

SAFETY PROFILE: Moderately toxic by ingestion. When heated to decomposition it emits acrid smoke and irritating fumes. See also ALCOHOLS.

ELU000 CAS:1570-45-2 ***HR: 3***
ETHYL ISONICOTINATE
mf: $C_8H_9NO_2$ mw: 151.18

SYNS: ISONICOTINIC ACID, ETHYL ESTER ◇ 4-PYRIDINECARBOXYLIC ACID, ETHYL ESTER

TOXICITY DATA with REFERENCE
ivn-mus LD50:56 mg/kg CSLNX* NX#03434

CONSENSUS REPORTS: Reported in EPA TSCA Inventory.

SAFETY PROFILE: Poison by intravenous route. When heated to decomposition it emits toxic fumes of NO_x.

ELX000 CAS:76-76-6 ***HR: 3***
ETHYLISOPROPYLBARBITURIC ACID
mf: $C_9H_{14}N_2O_3$ mw: 198.25

SYNS: 5-ETHYL-5-ISOPROPYLBARBITURIC ACID ◇ 5-ETHYL-5-(1-METHYLETHYL)-2,4,6(1H,3H,5H)-PYRIMIDINETRIONE ◇ IPRAL ◇ IRENAL ◇ PROBARBITAL ◇ PROBARBITONE ◇ VASALGIN

TOXICITY DATA with REFERENCE
ipr-rat LDLo:110 mg/kg JPETAB 44,325,32
scu-rat LDLo:110 mg/kg JPETAB 26,371,25
ipr-mus LDLo:250 mg/kg 27ZWAY -,-,36
orl-cat LDLo:140 mg/kg 27ZWAY -,-,36
orl-rbt LDLo:150 mg/kg JPETAB 44,337,32
ipr-rbt LDLo:110 mg/kg JPETAB 44,325,32
scu-rbt LDLo:200 mg/kg JACSAT 45,243,23
ivn-rbt LDLo:140 mg/kg JPPGAR 30,364,32
orl-bwd LD50:24 mg/kg TXAPA9 21,315,72

SAFETY PROFILE: Poison by ingestion, intraperitoneal, subcutaneous, and intravenous routes. When heated to decomposition it emits toxic fumes of NO_x. See also BARBITURATES.

ELX100 ***HR: 3***
ETHYL ISOPROPYL FLUOROPHOSPHONATE
mf: $C_5H_{12}FOP$ mw: 138.14

SYN: ISOPROPYL ETHANE FLUOROPHOSPHONATE

TOXICITY DATA with REFERENCE
ihl-rat LC50:260 mg/m^3/10M NTIS** PB158-508
ihl-mus LC50:245 mg/m^3/5M NTIS** PB158-508
skn-mus LD50:1700 μg/kg NTIS** PB158-508
scu-mus LD50:400 μg/kg NTIS** PB158-508
ihl-dog LC50:230 mg/m^3/10M NTIS** PB158-508
ihl-mky LC50:210 mg/m^3/10M NTIS** PB158-508
ihl-cat LC50:170 mg/m^3/10M NTIS** PB158-508
ihl-rbt LC50:230 mg/m^3/10M NTIS** PB158-508
ihl-gpg LC50:350 mg/m^3/10M NTIS** PB158-508

SAFETY PROFILE: Poison by inhalation, skin contact, and subcutaneous routes. When heated to decomposition it emits toxic fumes of F^- and PO_x. See also ESTERS.

ELX500 CAS:16339-04-1 ***HR: 3***
ETHYLISOPROPYLNITROSOAMINE
mf: $C_5H_{12}N_2O$ mw: 116.19

SYNS: AETHYL-ISOPROPYL-NITROSOAMIN (GERMAN) ◇ 1-METHYL-N-NITROSODIETHYLAMINE ◇ N-NITROSOETHYLISOPROPYLAMINE

TOXICITY DATA with REFERENCE
orl-rat TDLo:3920 mg/kg/56W-C:CAR NATWAY 50,100,63
orl-rat TD:3700 mg/kg/53W-C:ETA ARZNAD 19,1077,69
orl-rat LD50:1100 mg/kg NATWAY 50,100,63

SAFETY PROFILE: Moderately toxic by ingestion. Questionable carcinogen with experimental carcinogenic and tumorigenic data. When heated to decomposition it emits toxic fumes of NO_x. See also NITROSOAMINES.

ELY550 CAS:22722-03-8 ***HR: 3***
S-ETHYLISOTHIOURONIUM HYDROGEN SULFATE
mf: $C_3H_{10}N_2O_4S_2$ mw: 202.24

$CH_3CH_2SC(:N^-H_2)NH_2HSO_4^-$

SYN: ETHYL CARBAMIMONIOTHIOATE HYDROGEN SULFATE

SAFETY PROFILE: Reacts with chlorine to produce the dangerously explosive nitrogen trichloride. When heated to decomposition it emits toxic fumes of SO_x and NO_x. See also SULFATES.

ELY575 CAS:21704-44-9 ***HR: 3***
S-ETHYLISOTHIURONIUM METAPHOSPHATE
mf: $C_3H_8N_2S{\cdot}HO_3P$ mw: 184.17

SYNS: CARBAMIMIDOTHIOIC ACID, ETHYL ESTER with METAPHOSPHORIC ACID (1:1) ◇ 2-ETHYL-2-THIO-PSEUDOUREA with METAPHOSPHORIC ACID (1:1)

TOXICITY DATA with REFERENCE
ipr-rat LD50:338 mg/kg FATOAO 43,212,80
orl-mus LD50:2490 mg/kg FATOAO 43,212,80
ipr-mus LD50:455 mg/kg FATOAO 43,212,80
scu-mus LD50:472 mg/kg FATOAO 43,212,80
ims-mus LD50:450 mg/kg FATOAO 43,212,80

SAFETY PROFILE: Poison by intraperitoneal route. Moderately toxic by ingestion, subcutaneous, and intramuscular routes. When heated to decomposition it emits toxic fumes of NO_x, SO_x and PO_x.

ELY700 CAS:106-33-2 ***HR: 1***
ETHYL LAURATE
mf: $C_{14}H_{28}O_2$ mw: 228.37

PROP: Colorless, oily liquid; fruity-floral odor. D: 0.858, refr index: 1.430, flash p: +212°F. Misc in alc, chloroform, ether; insol in water @ 269°.

SYNS: ETHYL DODECANOATE ◇ FEMA No. 2441

SAFETY PROFILE: Combustible liquid. When heated to decomposition it emits acrid smoke and irritating fumes.

ELZ000 CAS:10339-55-6 ***HR: 1***
ETHYL LINALOOL
mf: $C_{11}H_{20}O$ mw: 168.31

SYN: 3,7-DIMETHYL-16-NONADIEN-3-OL

TOXICITY DATA with REFERENCE
skn-rbt 500 mg/24H MOD FCTXAV 14,767,76

CONSENSUS REPORTS: Reported in EPA TSCA Inventory.

SAFETY PROFILE: A skin irritant. When heated to decomposition it emits acrid smoke and irritating fumes.

ELZ100 ***HR: 3***
ETHYLLITHIUM
mf: C_2H_5Li mw: 36.00

SAFETY PROFILE: Ignites spontaneously in air. See also LITHIUM COMPOUNDS. When heated to decomposition it emits acrid smoke and irritating fumes.

EMA000 CAS:10467-10-4 ***HR: 3***
ETHYL MAGNESIUM IODIDE
mf: C_2H_5IMg mw: 180.27

CH_3CH_2MGI

SAFETY PROFILE: Mixtures with ethoxyacetylene in ether are explosive. When heated to decomposition it emits toxic fumes of I^-. See also IODIDES and MAGNESIUM COMPOUNDS.

EMA500 CAS:105-53-3 ***HR: 1***
ETHYL MALONATE
mf: $C_7H_{12}O_4$ mw: 160.19

PROP: Clear, colorless liquid; fruit-like odor. Bp: 198.9°, fp: −49.8°, flash p: 200°F (OC), d: 1.055 @ 25°/25°, refr index: 1.413-1.416, vap press: 1 mm @ 40.0°, vap d: 5.52. Sol in fixed oils, propylene glycol; sltly sol in alc, water; insol in glycerin, mineral oil @ 200°.

SYNS: CARBETHOXYACETIC ESTER ◇ DICARBETHOXYMETHANE ◇ DIETHYL MALONATE (FCC) ◇ DIETHYL PROPANEDIOATE ◇ FEMA No. 2375 ◇ MALONIC ACID, DIETHYL ESTER ◇ MALONIC ESTER ◇ METHANEDICARBOXYLIC ACID, DIETHYL ESTER ◇ PROPANEDIOIC ACID, DIETHYL ESTER

TOXICITY DATA with REFERENCE
skn-rbt 500 mg/24H MLD FCTXAV 14,745,76
orl-rat LD50:15 g/kg AIHAAP 30,470,69
orl-mus LD50:6400 mg/kg BIJOAK 34,1196,40

CONSENSUS REPORTS: Reported in EPA TSCA Inventory.

SAFETY PROFILE: Mildly toxic by ingestion. A skin irritant. Combustible liquid when exposed to heat or flame; can react with oxidizing materials. To fight fire, use water to blanket fire, foam, CO_2, dry chemical. When heated to decomposition it emits acrid smoke and irritating fumes. See also ESTERS.

EMA600 CAS:4940-11-8 ***HR: 2***
ETHYL MALTOL
mf: $C_7H_8O_3$ mw: 140.15

PROP: White crystalline powder; sweet fruity taste. Mp: 90°. Sol in water, alc, propylene glycol, chloroform.

SYNS: 2-ETHYL-3-HYDROXY-4H-PYRAN-4-ONE ◇ 2-ETHYL PYROMECONIC ACID ◇ 3-HYDROXY-2-ETHYL-4-PYRONE

TOXICITY DATA with REFERENCE
mmo-sat 1 mg/plate MUREAV 67,367,79
orl-rat LD50:1150 mg/kg TXAPA9 15,604,69
orl-mus LD50:780 mg/kg TXAPA9 15,604,69
scu-mus LD50:910 mg/kg CPBTAL 22,1008,74
orl-ckn LD50:1270 mg/kg TXAPA9 15,604,69

CONSENSUS REPORTS: Reported in EPA TSCA Inventory.

SAFETY PROFILE: Moderately toxic by ingestion and subcutaneous routes. Mutation data reported. When heated to decomposition it emits acrid smoke and irritating fumes.

EMB000 CAS:774-40-3 ***HR: 2***
ETHYL MANDELATE
mf: $C_{10}H_{11}O_2$ mw: 163.21

PROP: Crystals.

SYNS: MANDELIC ACID, ETHYL ESTER ◇ MANDELSAEUREAETHYLESTER (GERMAN)

TOXICITY DATA with REFERENCE
eye-rbt 5 mg AJOPAA 29,1363,46
orl-rat LD50:3750 mg/kg ARZNAD 12,347,62

SAFETY PROFILE: Moderately toxic by ingestion. An eye irritant. Combustible when exposed to heat or flame; can react with oxidizing materials. When heated to decomposition it emits acrid smoke and irritating fumes. See also ESTERS.

EMB100 CAS:75-08-1 ***HR: 3***
ETHYL MERCAPTAN
DOT: UN 2363
mf: C_2H_6S mw: 62.14

PROP: Colorless liquid, penetrating garlic-like odor. Mp: −147°, bp: 36.2°, lel: 2.8%, uel: 18.2%, d: 0.83907 @ 20°/4°, autoign temp: 570°F, vap d: 2.14. flash p: < −0.4°F.

SYNS: AETHANETHIOL (GERMAN) ◇ AETHYLMERCAPTAN (GERMAN) ◇ ETANTIOLO (ITALIAN) ◇ ETHAANTHIOL (DUTCH) ◇ ETHANETHIOL ◇ ETHYL HYDROSULFIDE ◇ ETHYLMERCAPTAAN (DUTCH) ◇ ETHYLMERKAPTAN (CZECH) ◇ ETHYL SULFHYDRATE ◇ ETHYL THIOALCOHOL ◇ ETILMERCAPTANO (ITALIAN) ◇ LPG ETHYL MERCAPTAN 1010 ◇ THIOETHANOL ◇ THIOETHYL ALCOHOL

TOXICITY DATA with REFERENCE
skn-rbt 500 mg/24H MLD 28ZPAK -,166,72
eye-rbt 84 mg AIHAAP 19,171,58
eye-rbt 100 mg/24H MOD 28ZPAK -,166,72
orl-rat LD50:1960 mg/kg 28ZPAK -,166,72
ihl-rat LC50:4420 ppm/4H AIHAAP 19,171,58
ipr-rat LD50:450 mg/kg AIHAAP 19,171,58
ihl-mus LD50:2770 mg/kg AIHAAP 19,171,58

CONSENSUS REPORTS: Reported in EPA TSCA Inventory.

OSHA PEL: (Transitional: CL 10 ppm) TWA 0.5 ppm
ACGIH TLV: TWA 0.5 ppm
DFG MAK: 0.5 ppm (1 mg/m^3)
NIOSH REL: (n-Alkane Mono Thiols) CL 0.5 ppm/15M
DOT Classification: Flammable Liquid; Label: Flammable Liquid; Flammable Liquid; Label: Flammable Liquid, Poison.

SAFETY PROFILE: Moderately toxic by ingestion, inhalation, and intraperitoneal routes. A skin and eye irritant. Inhalation causes central nervous system effects in humans. A very dangerous fire hazard when exposed to heat or flame; can react vigorously with oxidizing materials. A moderate explosion hazard when exposed to spark or flame. Violent reaction with $Ca(OCl)_2$. Will react with water or steam to produce toxic and flammable vapors. To fight fire, use CO_2, dry chemical. When heated to decomposition or on contact with acid or acid fumes it emits highly toxic fumes of SO_x. See also MERCAPTANS.

EMB200 CAS:623-51-8 ***HR: 3***
ETHYL-2-MERCAPTOACETATE
mf: $C_4H_8O_2S$ mw: 120.18

SYNS: ETHYL MERCAPTOACETATE ◇ ETHYL-α-MERCAPTOACETATE ◇ ETHYL MERCAPTOACETIC ACID ◇ ETHYL THIOGLYCOLATE ◇ MERCAPTOACETIC ACID ETHYL ESTER ◇ THIOGLYCOLIC ACID ETHYL ESTER ◇ THIOGLYKOLSAEURE-AETHYLESTER (GERMAN) ◇ USAF EK-2070

TOXICITY DATA with REFERENCE
orl-rat LD50:178 mg/kg ZHYGAM 20,575,74
ipr-rat LD50:176 mg/kg ZHYGAM 20,575,74
ipr-mus LD50:100 mg/kg NTIS** AD277-689

SAFETY PROFILE: Poison by ingestion and intraperitoneal routes. When heated to decomposition it emits toxic fumes of SO_x. See also MERCAPTANS.

EMC000 CAS:5840-95-9 ***HR: 3***
ETHYL (2-MERCAPTOETHYL) CARBAMATE S-ESTER with O,O-DIMETHYL PHOSPHORODITHIOATE
mf: $C_7H_{16}NO_4PS_2$ mw: 273.33

SYNS: (2-((DIETHOXYPHOSPHINOTHIOYL)THIO)ETHYL)CARBAMIC ACID, ETHYL ESTER ◇ S-(O,O-DIMETHYLPHOSPHORODITHIOATE) of N-(2-MERCAPTOETHYL)ETHYLCARBAMATE ◇ ENT 25,801 ◇ ETHYL (2-((DIETHOXYPHOSPHINOTHIOYL)THIO) ETHYL)CARBAMATE ◇ ETHYL-N-(2-(O,O-DIMETHYLPHOSPHORODITHIOYL)ETHYL)CARBAMATE ◇ (2-MERCAPTOETHYL)CARBAMIC ACID, ETHYL ESTER, S-ESTER with O,O-DIMETHYL PHOSPHORODITHIOATE ◇ R-3422-S ◇ STAUFFER R-3442-S

TOXICITY DATA with REFERENCE
orl-rat LD50:108 mg/kg 28ZEAL 4,220,69
orl-mus LDLo:470 mg/kg AECTCV 14,111,85
orl-ckn LD50:255 mg/kg TXAPA9 7,606,65

SAFETY PROFILE: Poison by ingestion. When heated to decomposition it emits very toxic fumes of NO_x, PO_x, and SO_x. See also CARBAMATES.

EMC500 CAS:5427-20-3 ***HR: 3***
9-ETHYL-6-MERCAPTOPURINE
mf: $C_7H_8N_4S$ mw: 180.25

SYNS: 9-ETHYL-1,9-DIHYDRO-6H-PURINE-6-THIONE ◇ 9-ETHYL-6-MP ◇ 9-ETHYL-9H-PURINE-6-THIOL ◇ 9-ETHYL-9H-PURINE-6(1H)-THIONE ◇ NSC 14575

TOXICITY DATA with REFERENCE
ipr-rat LD5O:400 mg/kg ADTEAS 3,181,68
ipr-mus LD50:289 mg/kg NCISP* JAN86

SAFETY PROFILE: Poison by intraperitoneal route. When heated to decomposition it emits very toxic fumes of NO_x and SO_x. See also MERCAPTANS.

EMD000 CAS:109-62-6 ***HR: 3***
ETHYLMERCURIC ACETATE
mf: $C_4H_8HgO_2$ mw: 288.71

SYNS: (ACETATO)ETHYLMERCURY ◇ (ACETATO-o)-ETHYLMERCURY

TOXICITY DATA with REFERENCE
orl-ckn LDLo:29 mg/kg TXAPA9 3,459,61

OSHA PEL: (Transitional: CL 1 mg/10m^3) TWA 0.01 mg(Hg)/m^3; STEL 0.03 mg/m^3 (skin)
ACGIH TLV: TWA 0.01 mg(Hg)/m^3; STEL 0.03 mg(Hg)/m^3
NIOSH REL: TWA 0.05 mg(Hg)/m^3

CONSENSUS REPORTS: Mercury and its compounds are on the Community Right-To-Know List.

SAFETY PROFILE: Poison by ingestion. When heated to decomposition it emits toxic fumes of Hg. See also MERCURY COMPOUNDS.

EME000 CAS:21082-50-8 ***HR: 3***
ETHYLMERCURIC CYSTEINE
mf: $C_5H_{11}NO_2S \cdot Hg$ mw: 349.82

SYN: ETHYL(HYDROGENCYSTEINATO)MERCURY

TOXICITY DATA with REFERENCE
cyt-hmn:hla 1 mg/L JJEMAG 39,47,69
scu-rat LD50:131 mg/kg JJEMAG 39,47,69

OSHA PEL: (Transitional: CL 1 mg/10m^3) TWA 0.01 mg(Hg)/m^3; STEL 0.03 mg/m^3 (skin)
ACGIH TLV: TWA 0.01 mg(Hg)/m^3; STEL 0.03 mg(Hg)/m^3
NIOSH REL: TWA 0.05 mg(Hg)/m^3

CONSENSUS REPORTS: Mercury and its compounds are on the Community Right-To-Know List.

SAFETY PROFILE: Poison by subcutaneous route. Human mutation data reported. When heated to decomposition it emits very toxic fumes of NO_x, SO_x, and Hg. See also MERCURY COMPOUNDS.

EME050 CAS:2597-93-5 ***HR: 3***
ETHYLMERCURICHLORENDIMIDE
mf: $C_{11}H_7Cl_6HgNO_2$ mw: 598.48

SYNS: 50-CS-46 ◇ EMMI ◇ N-(ETHYLMERCURI)-1,4,5,6,7,7-HEXACHLOROBICYCLO(2.2.1)HEPT-5-ENE-2,3-DICARBOXIMIDE ◇ N-ETHYLMERCURI-3,4,5,6,7,7-HEXACHLORO-3,6-ENDOMETHYLENE-1,2,3,6- TETRAHYDROPHTHALIMIDE ◇ N-ETHYLMERCURI-1,2,3,6-TETRAHYDRO-3,6-ENDOMETHANO-3,4,5,6,7,7-HEXACHLOROPHTHALIMIDE ◇ 1,4,5,6,77-HEXACHLORO-N-(ETHYLMERCURI)-5-NORBORNENE-2,3-DICARBOXIMIDE

TOXICITY DATA with REFERENCE
orl-rat LD50:150 mg/kg PCOC** -,465,66

OSHA PEL: (Transitional: CL 1 mg/10m^3) CL 0.1 mg(Hg)/m^3 (skin)
ACGIH TLV: TWA 0.1 mg(Hg)/m^3 (skin)
NIOSH REL: TWA 0.05 mg(Hg)/m^3

CONSENSUS REPORTS: Mercury and its compounds are on the Community Right-To-Know List.

SAFETY PROFILE: Poison by ingestion route. When heated to decomposition it emits very toxic fumes of Cl^-, Hg, and NO_x. See also MERCURY COMPOUNDS.

EME100 CAS:2235-25-8 ***HR: 3***
ETHYLMERCURIC PHOSPHATE
mf: $C_2H_7HgO_4P$ mw: 326.65

SYNS: EMP ◇ ETHYLMERCURY PHOSPHATE ◇ GRANOSAN M ◇ LIGNASAN ◇ N. I. CERESAN ◇ RUBERON ◇ SOILSIN

TOXICITY DATA with REFERENCE
scu-mus TDLo:40 mg/kg (10D preg):TER CAJPBD 7,53,67
orl-hmn LDLo:8614 μg/kg/13W TXCYAC 6,155,76
orl-rat LD50:48 mg/kg NYKZAU 59,452,63
orl-mus LD50:48 mg/kg NYKZAU 59,452,63
scu-mus LD50:76 mg/kg CAJPBD 7,53,67

CONSENSUS REPORTS: Mercury and its compounds are on the Community Right-To-Know List.

OSHA PEL: (Transitional: CL 1 mg/10m^3) TWA 0.01 mg(Hg)/m^3; STEL 0.03 mg/m^3 (skin)
ACGIH TLV: TWA 0.01 mg(Hg)/m^3; STEL 0.03 mg(Hg)/m^3
NIOSH REL: TWA 0.05 mg(Hg)/m^3

SAFETY PROFILE: Human poison by ingestion. Experimental poison by subcutaneous route. An experimental teratogen. When heated to decomposition it emits toxic fumes of PO_x and Hg. See also MERCURY COMPOUNDS and PHOSPHATES.

EME500 CAS:517-16-8 ***HR: 3***
ETHYLMERCURY-p-TOLUENE SULFONAMIDE
mf: $C_{15}H_{17}HgNO_2S$ mw: 475.98

PROP: Crystals; pungent, garlic-like odor; water-insol.

SYNS: CERESAN M ◇ COMPOUND-1452-F ◇ EMTS ◇ N-ETHYLMERCURI-N-PHENYL-p-TOLUENESULFONAMIDE◇ N-(ETHYLMERCURI)-p-TOLUENESULFONANILIDE◇ N-(ETHYLMERCURI)-p-TOLUENESULPHONANILIDE ◇ ETHYLMERCURY p-TOLUENE- SULFANILIDE ◇ ETHYLMERCURY-p-TOLUENESULFONANILIDE ◇ ETHYL(N-PHENYL-p-TOLUENESULFONAMIDATO)MERCURY ◇ ETHYL(N-PHENYL-p-TOLUENESULFONAMIDO)MERCURY ◇ ETHYL(p-TOLUENESULFONANILIDATO)MERCURY ◇ GRANOSAN M ◇ (N-PHENYL-p-TOLUENESULFONAMIDO)ETHYLMERCURY

TOXICITY DATA with REFERENCE
sln-dmg-orl 40 mg/kg MUREAV 40,31,76
orl-rat LD50:100 mg/kg PCOC** -,204,66

CONSENSUS REPORTS: Mercury and its compounds are on the Community Right-To-Know List. EPA Genetic Toxicology Program.

OSHA PEL: (Transitional: CL 1 mg/10m^3) TWA 0.01 mg(Hg)/m^3; STEL 0.03 mg/m^3 (skin)
ACGIH TLV: TWA 0.01 mg(Hg)/m^3; STEL 0.03 mg(Hg)/m^3
NIOSH REL: TWA 0.05 mg(Hg)/m^3

SAFETY PROFILE: Poison by ingestion route. Mutation data reported. A fungicide. When heated to decomposition it emits very toxic fumes of Hg, NO_x, and SO_x. See also MERCURY COMPOUNDS.

EMF000 CAS:97-63-2 ***HR: 3***
ETHYL METHACRYLATE
DOT: UN 2277
mf: $C_6H_{10}O_2$ mw: 114.16

PROP: A liquid. Mp: $< -75°$, bp: 119°, lel: 1.8%, uel: saturation, flash p: 68°F (OC), d: 0.911 @ 25°/25°, vap d: 3.94.

SYNS: ETHYL METHACRYLATE, INHIBITED (DOT) ◇ ETHYL-α-METHYL ACRYLATE ◇ ETHYL-2-METHYLACRYLATE ◇ ETHYL-2-METHYL-2-PROPENOATE ◇ 2-METHYL-2-PROPENOIC ACID, ETHYL ESTER ◇ RCRA WASTE NUMBER U118 ◇ RHOPLEX AC-33 (ROHM and HAAS)

TOXICITY DATA with REFERENCE
skn-rbt 10 g/kg open JIHTAB 23,343,41
ipr-rat TDLo:735 mg/kg (5-15D preg):TER JDREAF 51,1632,72
ipr-rat TDLo:366 mg/kg (5-15D preg):REP JDREAF 51,1632,72
orl-rat LD50:14800 mg/kg JIHTAB 23,343,41
ihl-rat LC50:8300 ppm/4H JTEHD6 16,811,85
ipr-rat LD50:1223 mg/kg JDREAF 51,1632,72
scu-rat LDLo:25 g/kg JIHTAB 23,343,41
orl-mus LD50:7836 mg/kg TOLED5 11,125,82
ipr-mus LD50:1369 mg/kg JPMSAE 62,778,73
orl-rbt LDLo:3630 mg/kg JIHTAB 23,343,41

CONSENSUS REPORTS: Reported in EPA TSCA Inventory.

DOT Classification: Flammable Liquid; Label: Flammable Liquid.

SAFETY PROFILE: Moderately toxic by ingestion and intraperitoneal routes. Mildly toxic by inhalation. Experimental teratogenic and reproductive effects. A skin irritant. A very dangerous fire and explosion hazard when exposed to heat, sparks, or flame; can react with oxidizing materials. To fight fire, use CO_2, dry chemical. When heated to decomposition it emits acrid smoke and irritating fumes.

EMF500 CAS:62-50-0 ***HR: 3***
ETHYL METHANESULFONATE
mf: $C_3H_8O_3S$ mw: 124.17

SYNS: EMS ◇ ENT 26,396 ◇ ETHYL ESTER of METHANESULFONIC ACID ◇ ETHYL ESTER of METHYLSULFONIC ACID ◇ ETHYL ESTER of METHYLSULPHONIC ACID ◇ ETHYL METHANESULPHONATE ◇ ETHYL METHANSULFONATE ◇ ETHYL METHANSULPHONATE ◇ HALF-MYLERAN ◇ METHANESULPHONIC ACID ETHYL ESTER ◇ METHYLSULFONIC ACID, ETHYL ESTER ◇ NSC 26805 ◇ RCRA WASTE NUMBER U119

TOXICITY DATA with REFERENCE
sln-dmg-orl 500 ppm ENMUDM 7(Suppl 3),76,85
oms-hmn:lym 400 µmol/L MUREAV 155,75,85
ipr-rat TDLo:200 mg/kg (female 15D post):REP JRPFA4 18,15,69
ipr-rat TDLo:200 mg/kg (female 15D post):TER JRPFA4 18,15,69
orl-rat TDLo:1050 g/kg/12W-C:CAR CALEDQ 7,79,79
ipr-rat TDLo:300 mg/kg:CAR BJCAAI 29,50,74
ivn-rat TDLo:100 mg/kg (21D post):ETA EXPTAX 16,157,78
ipr-mus TDLo:373 mg/kg:NEO CBINA8 3,117,71
ipr-rat LD50:350 mg/kg CPBTAL 8,807,60
ipr-mus LD50:435 mg/kg CBINA8 3,117,71

CONSENSUS REPORTS: IARC Cancer Review: Group 2B IMEMDT 7,56,87; Animal Sufficient Evidence IMEMDT 7,245,74. Reported in EPA TSCA Inventory. EPA Genetic Toxicology Program.

SAFETY PROFILE: Suspected carcinogen with experimental carcinogenic, neoplastigenic, tumorigenic, and teratogenic data. Poison by intraperitoneal route. Experimental reproductive effects. Human mutation data reported. When heated to decomposition it emits toxic fumes of SO_x. See also SULFONATES and ESTERS.

EMG000 CAS:102504-44-9 ***HR: 3***
2'-ETHYL-2-(2-METHOXY BUTYLAMINO) PROPIONANILIDE
mf: $C_{16}H_{26}N_2O_2$ mw: 278.44

TOXICITY DATA with REFERENCE
ipr-mus LD50:150 mg/kg JPMSAE 67,595,78
ivn-mus LD50:26 mg/kg JPMSAE 67,595,78

SAFETY PROFILE: Poison by intraperitoneal and intravenous routes. When heated to decomposition it emits toxic fumes of NO_x.

EMG500 CAS:67262-69-5 ***HR: 3***
2'-ETHYL-3-(2-METHOXYETHYL)AMINOBUTYRANILIDE HYDROCHLORIDE
mf: $C_{15}H_{24}N_2O_2$•ClH mw: 300.87

TOXICITY DATA with REFERENCE
ipr-mus LD50:225 mg/kg JPMSAE 67,595,78
ivn-mus LD50:40 mg/kg JPMSAE 67,595,78

SAFETY PROFILE: Poison by intraperitoneal and intravenous routes. When heated to decomposition it emits very toxic fumes of HCl and NO_x.

EMH000 CAS:67262-72-0 ***HR: 3***
2'-ETHYL-4-(2-METHOXYETHYL)AMINOBUTYRANILIDE HYDROCHLORIDE
mf: $C_{15}H_{24}N_2O_2$•ClH mw: 300.87

TOXICITY DATA with REFERENCE
ipr-mus LD50:400 mg/kg JPMSAE 67,595,78
ivn-mus LD50:80 mg/kg JPMSAE 67,595,78

SAFETY PROFILE: Poison by intraperitoneal and intravenous routes. When heated to decomposition it emits very toxic fumes of HCl and NO_x.

EMH500 CAS:67262-71-9 ***HR: 3***
2'-ETHYL-3-(2-METHOXYETHYL)AMINO-3-METHYLBUTYRANILIDE CYCLAMATE
mf: $C_{16}H_{26}N_2O_2$•$C_6H_{13}NO_3S$ mw: 457.70

SYN: 2'-ETHYL-3-(2-METHOXYETHYL)AMINO-3-METHYLBUTYRANILIDE CYCLOHEXANE SULFAMATE

TOXICITY DATA with REFERENCE
ipr-mus LD50:125 mg/kg JPMSAE 67,595,78
ivn-mus LD50:22 mg/kg JPMSAE 67,595,78

SAFETY PROFILE: Poison by intraperitoneal and intravenous routes. When heated to decomposition it emits very toxic fumes of NO_x and SO_x.

EMI000 CAS:67262-64-0 ***HR: 3***
2'-ETHYL-2-(2-METHOXYETHYLAMINO)PROPIONANILIDE
mf: $C_{14}H_{22}N_2O_2$ mw: 250.38

TOXICITY DATA with REFERENCE
ipr-mus LD50:250 mg/kg JPMSAE 67,595,78
ivn-mus LD50:38 mg/kg JPMSAE 67,595,78

SAFETY PROFILE: Poison by intraperitoneal and intravenous routes. When heated to decomposition it emits toxic fumes of NO_x.

EMI500 CAS:102504-45-0 ***HR: 3***
2'-ETHYL-2-(2-METHOXYETHYLAMINO)-PROPIONANILIDE HYDROCHLORIDE
mf: $C_{14}H_{22}N_2O_2 \cdot ClH$ mw: 286.84

TOXICITY DATA with REFERENCE
ipr-mus LD50:250 mg/kg JPMSAE 67,595,78
ivn-mus LD50:65 mg/kg JPMSAE 67,595,78

SAFETY PROFILE: Poison by intraperitoneal and intravenous routes. When heated to decomposition it emits very toxic fumes of HCl and NO_x.

EMJ500 CAS:54350-48-0 ***HR: 1***
ETHYL all-trans-9-(4-METHOXY-2,3,6-TRIMETHYLPHENYL)-3,7-DIMETHYL-2,4,6,8-NONATETRAENOATE
mf: $C_{23}H_{30}O_3$ mw: 354.53

SYNS: ETRETINATE ◇ Ro 10-9359

TOXICITY DATA with REFERENCE
sce-hmn:fbr 5 mg/L MUREAV 58,317,78
orl-mus TDLo:40 mg/kg (female 7-16D post):REP 46IZA7 -,49,81
orl-mus TDLo:80 mg/kg (female 7-16D post):TER 46IZA7 -,49,81
orl-hmn TDLo:78 mg/kg/26W-I:SKN ARZNAD 32,842,82
orl-wmn TDLo:60 mg/kg/17W-I:SKN CUTIBC 35,466,85
orl-wmn TDLo:80 μg/kg/8D-I:SKN,SYS BJDEAZ 112,373,85

CONSENSUS REPORTS: EPA Genetic Toxicology Program.

SAFETY PROFILE: Human systemic effects: dermatitis, nail and hair changes, increased body temperature. Experimental teratogenic and reproductive effects. Human mutation data reported. When heated to decomposition it emits acrid smoke and irritating fumes.

EMK600 CAS:689-93-0 ***HR: 3***
ETHYL METHYL ARSINE
mf: C_3H_9As mw: 120.03

$CH_3CH_2AsHCH_3$

CONSENSUS REPORTS: Arsenic and its compounds are on the Community Right-To-Know List.

SAFETY PROFILE: Arsenic compounds are poisons. Ignites spontaneously in air. When heated to decomposition it emits toxic fumes of As. See also ARSENIC COMPOUNDS and ARSINE.

EML500 CAS:17982-67-1 ***HR: D***
ETHYL METHYL AZIDOMETHYL PHOSPHONATE
mf: $C_4H_{10}N_3O_3P$ mw: 179.14

SYN: AZIDOMETHYLPHOSPHONIC ACID, ETHYL METHYL ESTER

TOXICITY DATA with REFERENCE
mmo-sat 5 μL/plate MUREAV 28,405,75
mmo-esc 5 μL/plate MUREAV 28,405,75

SAFETY PROFILE: Mutation data reported. When heated to decomposition it emits very toxic fumes of PO_x and NO_x.

EMM000 CAS:63039-89-4 ***HR: 3***
7-ETHYL-9-METHYLBENZ(c)ACRIDINE
mf: $C_{20}H_{17}N$ mw: 271.38

SYNS: 3-METHYL-10-ETHYLBENZ(c)ACRIDINE ◇ 3-METHYL-10-ETHYL-7,8-BENZACRIDINE(FRENCH)

TOXICITY DATA with REFERENCE
skn-mus TDLo:348 mg/kg/29W-I:ETA ACRSAJ 4,315,56
scu-mus TDLo:72 mg/kg/9W-I:NEO MUREAV 66,307,79

SAFETY PROFILE: Questionable carcinogen with experimental neoplastigenic and tumorigenic data. When heated to decomposition it emits toxic fumes of NO_x.

EMM500 CAS:16354-50-0 ***HR: 3***
7-ETHYL-12-METHYLBENZ(a)ANTHRACENE
mf: $C_{21}H_{18}$ mw: 270.39

TOXICITY DATA with REFERENCE
cyt-rat-ivn 50 mg/kg GANNA2 64,637,73
ims-rat TDLo:50 mg/kg:NEO PNASA6 58,2253,67
ims-rat TD:50 mg/kg:ETA BCPCA6 16,607,67

SAFETY PROFILE: Questionable carcinogen with experimental neoplastigenic and tumorigenic data. Mutation data reported. When heated to decomposition it emits acrid smoke and irritating fumes.

EMN000 CAS:16354-55-5 ***HR: 3***
12-ETHYL-7-METHYLBENZ(a)ANTHRACENE
mf: $C_{21}H_{18}$ mw: 270.39

TOXICITY DATA with REFERENCE
cyt-rat-ivn 50 mg/kg GANNA2 64,637,73
ims-rat TDLo:50 mg/kg:NEO PNASA6 58,2253,67

SAFETY PROFILE: Questionable carcinogen with experimental neoplastigenic data. Mutation data reported. When heated to decomposition it emits acrid smoke and irritating fumes.

EMO000 CAS:10137-87-8 ***HR: 3***
N-ETHYL(α-METHYLBENZYL)AMINE
mf: $C_{10}H_{15}N$ mw: 149.26

SYN: N-ETHYL-α-METHYLBENZYLAMINE

TOXICITY DATA with REFERENCE
skn-rbt 100 μg/24H open AIHAAP 23,95,62

orl-rat LDLo:350 mg/kg AIHAAP 23,95,62
skn-rbt LDLo:890 mg/kg AIHAAP 23,95,62

SAFETY PROFILE: Poison by ingestion. Moderately toxic by skin contact. A skin irritant. When heated to decomposition it emits toxic fumes of NO_x. See also AMINES.

EMO500 CAS:125-42-8 ***HR: 3***
5-ETHYL-5-(1-METHYL-1-BUTENYL)BARBITURATE
mf: $C_{11}H_{16}N_2O_3$ mw: 224.29

SYNS: 5-ETHYL-5-(1-METHYL-1-BUTENYL)BARBITURIC ACID ◇ 5-ETHYL-5-(1-METHYL-1-BUTENYL)-2,4,6(1H,3H,5H)-PYRIMIDINE TRIONE

TOXICITY DATA with REFERENCE
orl-mus LD50:190 mg/kg JACSAT 61,776,39
ipr-mus LD50:180 mg/kg JACSAT 61,776,39

SAFETY PROFILE: Poison by ingestion and intraperitoneal routes. When heated to decomposition it emits toxic fumes of NO_x. See also BARBITURATES.

EMO875 CAS:17013-35-3 ***HR: 3***
5-ETHYL-5-(1-METHYL-2-BUTENYL)BARBITURIC ACID
mf: $C_{11}H_{16}N_2O_3$ mw: 224.29

TOXICITY DATA with REFERENCE
orl-rat LD50:20 mg/kg JMPCAS 1,31,37
ipr-rat LD50:3 mg/kg JMPCAS 1,31,59
ipr-mus LD50:21 mg/kg JMPCAS 1,31,59

SAFETY PROFILE: Poison by ingestion and intraperitoneal routes. When heated to decomposition it emits toxic fumes of NO_x. See also BARBITURATES.

EMP500 CAS:21149-87-1 ***HR: 3***
5-ETHYL-5-(3-METHYL-2-BUTENYL)BARBITURIC ACID SODIUM SALT
mf: $C_{11}H_{15}N_2O_3 \cdot Na$ mw: 246.27

TOXICITY DATA with REFERENCE
ipr-cat LDLo:6 mg/kg NATUAS 163,447,49
ivn-cat LDLo:6 mg/kg NATUAS 163,447,49
ipr-rbt LDLo:6 mg/kg NATUAS 163,447,49
ivn-rbt LDLo:6 mg/kg NATUAS 163,447,49

SAFETY PROFILE: Poison by intraperitoneal and intravenous routes. When heated to decomposition it emits toxic fumes of NO_x and Na_2O. See also BARBITURATES.

EMP600 CAS:7452-79-1 ***HR: 2***
ETHYL 2-METHYLBUTYRATE
mf: $C_7H_{14}O_2$ mw: 130.19

PROP: Colorless liquid; strong, apple-like odor. D: 0.861-0.866, refr index: 1.396, flash p: +153°F. Sol in alc, propylene glycol; misc in fixed oils; very sltly sol in water.

SYN: FEMA No. 2443

SAFETY PROFILE: Combustible liquid. When heated to decomposition it emits acrid smoke and irritating fumes.

EMQ500 CAS:105-40-8 ***HR: 3***
ETHYL-N-METHYL CARBAMATE
mf: $C_4H_9NO_2$ mw: 103.14

PROP: Needles. Mp: 54°, bp: 177°.

SYNS: METHYLCARBAMIC ACID, ETHYL ESTER ◇ N-METHYL URETHAN

TOXICITY DATA with REFERENCE
ipr-ham TDLo:495 mg/kg (8D preg):TER CNREA8 27,1696,67
skn-mus TDLo:10 g/kg/1W-I:ETA BJCAAI 9,177,55
scu-mus LD50:1360 mg/kg AJEBAK 45,507,67

CONSENSUS REPORTS: Reported in EPA TSCA Inventory.

SAFETY PROFILE: Moderately toxic by subcutaneous route. Experimental teratogenic effects. Questionable carcinogen with experimental tumorigenic data. When heated to decomposition it emits toxic fumes of NO_x. See also CARBAMATES.

EMR000 CAS:2698-38-6 ***HR: 2***
ETHYL-2-METHYL-4-CHLOROPHENOXYACETATE
mf: $C_{11}H_{13}ClO_3$ mw: 228.69

SYNS: 4-CHLORO-2-METHYLPHENOXYACETIC ACID, ETHYL ESTER ◇ ((4-CHLORO-o-TOLYL)OXY)ACETIC ACID, ETHYL ESTER ◇ MCPA-ETHYL ◇ MCPEE

TOXICITY DATA with REFERENCE
orl-rat TDLo:800 mg/kg (8-15D preg):REP TXAPA9 23,326,72
orl-rat TDLo:400 mg/kg (8-15D preg):TER TXAPA9 23,326,72
orl-mus LD50:1290 mg/kg YKYUA6 30,985,79

SAFETY PROFILE: Moderately toxic by ingestion. Experimental teratogenic and reproductive effects. When heated to decomposition it emits toxic fumes of Cl^-. See also ESTERS.

EMR500 CAS:19105-63-6 ***HR: 3***
2-ETHYL-8-METHYL-2,8-DIAZASPIRO(4,5)DECANE-1,3-DIONE HYDROBROMIDE
mf: $C_{11}H_{18}N_2O_2 \cdot BrH$ mw: 291.23

TOXICITY DATA with REFERENCE
orl-mus LD50:660 mg/kg 27ZQAG -,298,72
ivn-mus LD50:250 mg/kg 27ZQAG -,298,72
ivn-dog LD50:2500 μg/kg 27ZQAG -,298,72
ivn-cat LD50:2 mg/kg 27ZQAG -,298,72
ivn-rbt LD50:3700 μg/kg 27ZQAG -,298,72

SAFETY PROFILE: Poison by intravenous route. Moderately toxic by ingestion. When heated to decomposition it emits very toxic fumes of NO_x and HBr.

EMR600 CAS:39562-70-4 ***HR: 3***
ETHYL METHYL 1,4-DIHYDRO-2,6-DIMETHYL-4-(m-NITROPHENYL)-3,5-PYRIDINEDICARBOXYLATE
mf: $C_{18}H_{20}N_2O_6$ mw: 360.40

SYNS: BAY e 5009 ◇ 1,4-DIHYDRO-2,6-DIMETHYL-4-(3-NITROPHENYL)-3,5-PYRIDINEDICARBOXYLIC ACID ETHYL METHYL ESTER ◇ NITRENDIPINE ◇ 3,5-PYRIDINEDICARBOXYLIC ACID, 1,4-DIHYDRO-2,6-DIMETHYL-4-(3-NITROPHENYL)-, ETHYL METHYL ESTER

TOXICITY DATA with REFERENCE
orl-rat TDLo:1720 mg/kg (female 17-22D post):REP OYYAA2 36,133,88
orl-rat TDLo:1720 mg/kg (female 17-22D post):TER OYYAA2 36,133,88
orl-rat LD50:15370 mg/kg OYYAA2 36,121,88
ipr-rat LD50:205 mg/kg OYYAA2 36,121,88
scu-rat LD50:5166 mg/kg OYYAA2 36,121,88
ivn-rat LD50:12600 μg/kg 52EDA6 -,25,84
orl-mus LD50:2540 mg/kg 52EDA6 -,25,84
ipr-mus LD50:303 mg/kg OYYAA2 36,121,88
scu-mus LD50:7613 mg/kg OYYAA2 36,121,88
ivn-mus LD50:34500 μg/kg FRPSAX 42,697,87
orl-dog LD50:250 mg/kg 52EDA6 -,25,84

SAFETY PROFILE: Poison by ingestion and intraperitoneal routes. An experimental teratogen. Experimental reproductive effects. When heated to decomposition it emits toxic fumes of NO_x.

EMS000 CAS:6030-03-1 ***HR: 3***
4'-ETHYL-2-METHYL-4-DIMETHYLAMINOAZOBENZENE
mf: $C_{17}H_{21}N_3$ mw: 267.41

SYN: 4'-ETHYL-N,N-DIMETHYL-4-(PHENYLAZO)-m-TOLUIDINE

TOXICITY DATA with REFERENCE
orl-rat TDLo:4040 mg/kg/15W-C:ETA ARZNAD 12,270,62

SAFETY PROFILE: Questionable carcinogen with experimental tumorigenic data. When heated to decomposition it emits toxic fumes of NO_x.

EMT000 CAS:540-67-0 ***HR: 3***
ETHYL METHYL ETHER
DOT: UN 1039
mf: C_3H_8O mw: 60.11

PROP: Colorless liquid or gas. Bp: 11°, lel: 2.0%, uel: 10.1%, flash p: −35°F (CC), d: 0.7260 @ 0°/4°, autoign temp: 374°F, vap d: 2.07.

SYNS: ETHOXYMETHANE ◇ ETHYL METHYL ETHER (DOT) ◇ METHOXYETHANE ◇ METHYL ETHYL ETHER (DOT)

DOT Classification: Flammable Liquid; Label: Flammable Liquid; Flammable Gas; Label: Flammable Gas.

SAFETY PROFILE: Has anesthetic properties. A very dangerous fire and moderate explosion hazard when exposed to heat or flame; can react vigorously with oxidizing materials (e.g., air, O_2). To fight fire, use alcohol foam, CO_2, dry chemical. See also ETHERS.

EMT500 CAS:139-88-8 ***HR: 2***
7-ETHYL-2-METHYL-4-HENDECANOL SULFATE SODIUM SALT
mf: $C_{14}H_{29}O_4S{\cdot}Na$ mw: 316.48

PROP: Sltly water-sol.

SYNS: 7-ETHYL-2-METHYL-4-UNDECANOL SULFATE SODIUM SALT ◇ OBLITEROL ◇ SODIUM-7-ETHYL-2-METHYL-4-UNDECANOL SULFATE ◇ SODIUM-7-ETHYL-2-METHYLUNDECYL-4-SULFATE ◇ SODIUM-2-METHYL-7-ETHYLUNDECANOL-4-SULFATE ◇ SODIUM-2-METHYL-7-ETHYLUNDECYL SULFATE-4 ◇ SODIUM SOTRADECOL ◇ SOTRADECOL ◇ STS ◇ TERGITOL ◇ TERGITOL ANIONIC 4 ◇ TERGITOL PENETRANT 4 ◇ TROMBOVAR ◇ VARICOL

TOXICITY DATA with REFERENCE
skn-rbt 500 mg open SEV UCDS** 12/13/63
eye-rbt 250 μg MLD AROPAW 34,99,45
orl-rat LD50:1250 mg/kg JIHTAB 23,478,41
orl-gpg LD50:650 mg/kg JIHTAB 23,478,41
skn-gpg LD50:650 mg/kg JIHTAB 23,478,41

CONSENSUS REPORTS: Reported in EPA TSCA Inventory.

SAFETY PROFILE: Moderately toxic by ingestion and skin contact. An eye and severe skin irritant. When heated to decomposition it emits toxic fumes of SO_x and Na_2O. See also SULFATES.

EMU500 CAS:96-29-7 ***HR: 3***
ETHYL METHYL KETOXIME
mf: C_4H_9NO mw: 87.14

$CH_3C(:NOH)CH_2CH_3$

SYNS: 2-BUTANONE OXIME ◇ MEK-OXIME ◇ METHYL ETHYL KETOXIME ◇ 2-OXIMIMINOBUTANE ◇ SKINO #2 ◇ TROYKYD ANTI-SKIN B ◇ USAF AM-3 ◇ USAF DO-44 ◇ USAF EK-906

TOXICITY DATA with REFERENCE
scu-rat LD50:2702 mg/kg NJMSAG 29,393,67
ipr-mus LD50:200 mg/kg NTIS** AD277-689

CONSENSUS REPORTS: Reported in EPA TSCA Inventory.

SAFETY PROFILE: Poison by intraperitoneal route. Moderately toxic by subcutaneous route. May explode if heated. Reacts with sulfuric acid to form an explosive product. When heated to decomposition it emits toxic fumes of NO_x.

EMV000 CAS:31217-72-8 ***HR: 3***
N-ETHYL-2-METHYLMALEIMIDE
mf: $C_7H_9NO_2$ mw: 139.17

TOXICITY DATA with REFERENCE
ipr-rat LDLo:37 mg/kg JMCMAR 15,534,72
ice-rat LDLo:8 mg/kg JMCMAR 15,534,72

SAFETY PROFILE: Poison by intraperitoneal and intracerebral routes. When heated to decomposition it emits toxic fumes of NO_x.

EMW000 CAS:66968-89-6 ***HR: 3***
5-ETHYL-1-METHYL-5-(1-METHYLPROPENYL) BARBITURIC ACID
mf: $C_{11}H_{16}N_2O_3$ mw: 224.29

TOXICITY DATA with REFERENCE
orl-mus LD50:285 mg/kg JACSAT 61,353,39
ipr-mus LD50:210 mg/kg JACSAT 61,353,39

SAFETY PROFILE: Poison by ingestion and intraperitoneal routes. When heated to decomposition it emits toxic fumes of NO_x. See also BARBITURATES.

EMW100 ***HR: 3***
N-ETHYL-6-METHYL-α-(METHYLSULFONYL) ERGOLINE-8-β-PROPIONAMIDE
mf: $C_{21}H_{29}N_3O_3S$ mw: 403.59

SYN: ERGOLINE-8-β-PROPIONAMIDE, N-ETHYL-6-METHYL-α-(METHYLSULFONYL)-

TOXICITY DATA with REFERENCE
orl-rat TDLo:3 mg/kg (female 5D post):REP ARZNAD 33,1094,83
orl-mus LD50:400 mg/kg ARZNAD 33,1094,83

SAFETY PROFILE: Poison by ingestion. Experimental reproductive effects. When heated to decomposition it emits toxic fumes of NO_x and SO_x

EMY000 CAS:10024-78-9 ***HR: 3***
4-ETHYL-1-METHYLOCTYLAMINE
mf: $C_{11}H_{25}N$ mw: 171.37

SYN: 1-METHYL-4-ETHYLOCTYLAMINE

TOXICITY DATA with REFERENCE
skn-rbt 10 mg/24H open AMIHBC 10,61,54
eye-rbt 50 μg open SEV AMIHBC 10,61,54
orl-rat LD50:730 mg/kg AMIHBC 10,61,54
skn-rbt LD50:380 mg/kg AMIHBC 10,61,54

SAFETY PROFILE: Poison by skin contact. Moderately toxic by ingestion. A skin and severe eye irritant. When heated to decomposition it emits toxic fumes of NO_x. See also AMINES.

EMY100 CAS:95524-59-7 ***HR: 1***
2-ETHYL-1-(3-METHYL-1-OXO-2-BUTENYL)PIPERIDINE
mf: $C_{12}H_{21}NO$ mw: 195.34

SYNS: AI3-36175-Ga ◇ PIPERIDINE, 2-ETHYL-1-(3-METHYL-1-OXO-2-BUTENYL)-

TOXICITY DATA with REFERENCE
skn-rbt 500 mg MLD NTIS** AD-A002-053

SAFETY PROFILE: A skin irritant. When heated to decomposition it emits toxic fumes of NO_x.

EMZ000 CAS:106-67-2 ***HR: 2***
2-ETHYL-4-METHYL-1-PENTANOL
mf: $C_8H_{18}O$ mw: 130.26

SYNS: 2-ETHYLISOHEXANOL ◇ 2-ETHYL-4-METHYLPENTANOL

TOXICITY DATA with REFERENCE
skn-rbt 10 mg/24H open MLD AIHAAP 23,95,62
orl-rat LD50:4290 mg/kg AIHAAP 23,95,62
orl-mus LDLo:1600 mg/kg KODAK* 21MAY71

CONSENSUS REPORTS: Reported in EPA TSCA Inventory.

SAFETY PROFILE: Moderately toxic by ingestion. A skin irritant. When heated to decomposition it emits acrid smoke and irritating fumes.

ENA000 CAS:67526-05-0 ***HR: 3***
5-ETHYL-5-(1-METHYL-1-PENTENYL)BARBITURIC ACID
mf: $C_{12}H_{18}N_2O_3$ mw: 238.32

TOXICITY DATA with REFERENCE
orl-mus LD50:210 mg/kg JACSAT 61,776,39
ipr-mus LD50:160 mg/kg JACSAT 61,776,39

SAFETY PROFILE: Poison by ingestion and intraperitoneal routes. When heated to decomposition it emits toxic fumes of NO_x. See also BARBITURATES.

ENA500 CAS:70299-48-8 ***HR: 3***
ETHYL METHYL PEROXIDE
mf: $C_3H_8O_2$ mw: 76.10

$CH_3CH_2OOCH_3$

SAFETY PROFILE: Both the liquid and the vapor are shock-sensitive explosives. Also explodes on superheating. When heated to decomposition it emits acrid smoke and irritating fumes. See also PEROXIDES.

ENB000 CAS:2058-66-4 ***HR: 3***
N-ETHYL-N-METHYL-p-(PHENYLAZO)ANILINE
mf: $C_{15}H_{17}N_3$ mw: 239.35

SYNS: p-ETHYLMETHYLAMINOAZOBENZENE ◇ N-ETHYL-N-METHYL-p-AMINOAZOBENZENE ◇ 4-ETHYLMETHYLAMINOAZOBENZENE ◇ 4-(METHYLETHYL)AMINOAZOBENZENE ◇ N-METHYL-N-ETHYL-p-AMINOAZOBENZENE

TOXICITY DATA with REFERENCE
dni-mus-orl 20 g/kg ARGEAR 51,605,81
orl-rat TDLo:4684 mg/kg/17W-C:CAR JEMEAV 87,139,48
orl-rat TD:2100 mg/kg/16W-C:ETA CNREA8 8,141,48

SAFETY PROFILE: Questionable carcinogen with experimental carcinogenic and tumorigenic data. Mutation data reported. When heated to decomposition it emits toxic fumes of NO_x.

ENB500 CAS:115-38-8 ***HR: 3***
5-ETHYL-N-METHYL-5-PHENYLBARBITURIC ACID
mf: $C_{13}H_{14}N_2O_3$ mw: 246.29

SYNS: ENFENEMAL ◇ ENPHENEMAL ◇ N-ETHYLMETHYLPHENYLBARBITURIC ACID ◇ 5-ETHYL-1-METHYL-5-PHENYLBARBITURIC ACID ◇ 5-ETHYL-1-METHYL-5-PHENYL-2,4,6(1H,3H,5H)-PYRIMIDINETRIONE ◇ 5-ETHYL-5-PHENYL-N-METHYLBARBITURIC ACID ◇ ISONAL ◇ ISONAL (ROUSSEL) ◇ MEBARAL ◇ MEBEREL ◇ MENTA-BAL ◇ MEPHOBARBITAL ◇ MEPHOBARBITONE ◇ MEPHYTAL ◇ METHYL-CALMINAL ◇ 1-METHYL-5-ETHYL-5-PHENYLBARBITURIC ACID ◇ METHYLPHENOBARBITAL ◇ N-METHYLPHENOBARBITAL ◇ 1-METHYLPHENOBARBITAL ◇ METHYLPHENOBARBITONE ◇ N-METHYLPHENOLBARBITOL ◇ METHYLPHENYLBARBITURIC ACID ◇ N-METHYL-5-PHENYL-5-ETHYLBARBITAL ◇ 1-METHYL-5-PHENYL-5-ETHYLBARBITURIC ACID ◇ METYLFENEMAL ◇ METYNA ◇ MORBUSAN ◇ PHEMETONE ◇ PHEMITON ◇ PHEMITONE ◇ 5-PHENYL-5-ETHYL-3-METHYLBARBITURIC ACID ◇ PROMINAL

TOXICITY DATA with REFERENCE
unr-wmn TDLo:3360 mg/kg (1-40W preg):TER LANCAO 2,839,72
ipr-rat LD50:130 mg/kg PHMCAA 5,237,63
orl-mus LD50:300 mg/kg HPPAAL 19,241,61
orl-cat LD50:230 mg/kg NIIRDN 6,851,82
orl-rbt LD50:400 mg/kg NIIRDN 6,851,82

SAFETY PROFILE: Poison by ingestion and intraperitoneal routes. A human teratogen by an unspecified route with developmental abnormalities of the cardiovascular (circulatory) system. When heated to decomposition it emits toxic NO_x. See also BARBITURATES.

ENC000 CAS:77-83-8 ***HR: 1***
ETHYL METHYLPHENYLGLYCIDATE
mf: $C_{12}H_{14}O_3$ mw: 206.26

PROP: Colorless to yellowish liquid; strawberry-like odor. D: 1.086-1.112, refr index: 1.504-1.513, flash p: 273°F. Sol in fixed oils, propylene glycol; insol in glycerin.

SYNS: C-16 ALDEHYDE ◇ EMPG ◇ α-β-EPOXY-β-METHYLHYDROCINNAMIC ACID, ETHYL ESTER ◇ ETHYL α,β-EPOXY-β-METHYLHYDROCINNAMATE ◇ ETHYL 2,3-EPOXY-3-METHYL-3-PHENYLPROPIONATE ◇ ETHYL ESTER of 2,3-EPOXY-3-PHENYLBUTANOIC ACID ◇ FEMA No. 2444 ◇ FRAESEOL ◇ 3-METHYL-3-PHENYLGLYCIDIC ACID ETHYL ESTER ◇ STRAWBERRY ALDEHYDE

TOXICITY DATA with REFERENCE
orl-rat LD50:5470 mg/kg FCTXAV 2,327,64
orl-gpg LD50:4050 mg/kg FCTXAV 2,327,64

CONSENSUS REPORTS: Reported in EPA TSCA Inventory.

SAFETY PROFILE: Mildly toxic by ingestion. Combustible liquid. When heated to decomposition it emits acrid smoke and irritating fumes. See also ALDEHYDES.

ENC500 CAS:5696-06-0 ***HR: 2***
5-ETHYL-1-METHYL-5-PHENYLHYDANTOIN
mf: $C_{12}H_{14}N_2O_2$ mw: 218.28

SYNS: DELTOIN ◇ METHETOIN

TOXICITY DATA with REFERENCE
orl-mus LD50:500 mg/kg 29ZVAB -,75,69
orl-rbt LD50:1400 mg/kg 29ZVAB -,75,69

SAFETY PROFILE: Moderately toxic by ingestion. When heated to decomposition it emits toxic fumes such as NO_x.

ENC600 CAS:18886-42-5 ***HR: 3***
α-ETHYL-1-METHYL-α-PHENYL-3-PYRROLIDINEMETHANOL PROPIONATE FUMARATE
mf: $C_{17}H_{25}NO_2 \cdot C_4H_4O_4$ mw: 391.51

SYNS: AHR-1767 ◇ α-1-(1-METHYL-3-PYRROLIDINYL)-1-PHENYLPROPYL PROPIONATE FUMARATE

TOXICITY DATA with REFERENCE
orl-rat LD50:460 mg/kg AIPTAK 178,446,69
ipr-rat LD50:120 mg/kg AIPTAK 178,446,69
orl-mus LD50:230 mg/kg AIPTAK 178,446,69
ipr-mus LD50:144 mg/kg AIPTAK 178,446,69

SAFETY PROFILE: Poison by ingestion and intraperitoneal routes. When heated to decomposition it emits toxic fumes of NO_x.

END000 CAS:104-89-2 ***HR: 3***
5-ETHYL-2-METHYLPIPERIDINE
mf: $C_8H_{17}N$ mw: 127.26

SYN: COPELLIDIN

TOXICITY DATA with REFERENCE
skn-rbt 1 mg/24H AMIHBC 10,61,54
eye-rbt 250 μg SEV AMIHBC 10,61,54
orl-rat LD50:540 mg/kg AMIHBC 10,61,54
ihl-rat LCLo:250 ppm/4H AMIHBC 10,61,54
skn-rbt LD50:630 mg/kg AMIHBC 10,61,54
scu-rbt LDLo:100 mg/kg BDCGAS 34,2408,01

CONSENSUS REPORTS: Reported in EPA TSCA Inventory.

SAFETY PROFILE: Poison by subcutaneous route. Moderately toxic by ingestion, skin contact, and inhalation. A skin and severe eye irritant. When heated to decomposition it emits toxic fumes of NO_x.

END500 CAS:77791-37-8 ***HR: 3***
N-ETHYL-2-(2-METHYLPIPERIDINO)-N-(1-(2,4-XYLYLOXY)-2-PROPYL) ACETAMIDE HYDROCHLORIDE
mf: $C_{21}H_{34}N_2O_2{\cdot}ClH$ mw: 383.03

SYN: C 2103

TOXICITY DATA with REFERENCE
eye-rbt 2% MLD ARZNAD 9,70,59
scu-mus LD50:317 mg/kg ARZNAD 9,70,59

SAFETY PROFILE: Poison by subcutaneous route. An eye irritant. When heated to decomposition it emits very toxic fumes of NO_x and HCl.

ENE000 CAS:16468-98-7 ***HR: 2***
5-ETHYL-5-(1-METHYLPROPENYL)BARBITURIC ACID
mf: $C_{10}H_{14}N_2O_3$ mw: 210.26

TOXICITY DATA with REFERENCE
orl-mus LD50:550 mg/kg JACSAT 61,353,39
ipr-mus LD50:490 mg/kg JACSAT 61,353,39

SAFETY PROFILE: Moderately toxic by ingestion and intraperitoneal routes. When heated to decomposition it emits toxic fumes of NO_x. See also BARBITURATES.

ENE500 CAS:55283-68-6 ***HR: 1***
N-ETHYL-N-(2-METHYL-2-PROPENYL)-2,6-DINITRO-4-(TRIFLUOROMETHYL)BENZENAMINE
mf: $C_{13}H_{14}F_3N_3O_4$ mw: 333.30

SYNS: ETHALFLURALIN ◇ SOMILAN

TOXICITY DATA with REFERENCE
orl-rat LD50:10000 mg/kg 85ARAE 2,52,77

CONSENSUS REPORTS: Reported in EPA TSCA Inventory.

SAFETY PROFILE: Mildly toxic by ingestion. When heated to decomposition it emits very toxic fumes of F^- and NO_x.

ENF000 CAS:78-28-4 ***HR: 2***
1-ETHYL-1-METHYLPROPYL CARBAMATE
mf: $C_7H_{15}NO_2$ mw: 145.23

SYNS: DIETHYL METHYL CARBINOLURETHAN ◇ tert-HEXANOL CARBAMATE ◇ METHYL DIETHYL CARBINOLURETHAN ◇ 3-METHYL-3-PENTANOL CARBAMATE

TOXICITY DATA with REFERENCE
ipr-rat LD50:430 mg/kg 27ZQAG -,413,72
orl-mus LD50:760 mg/kg 27ZQAG -,413,72
ipr-mus LD50:550 mg/kg QJSAAP 21,223,60

SAFETY PROFILE: Moderately toxic by ingestion and intraperitoneal routes. When heated to decomposition it emits toxic fumes of NO_x. See also CARBAMATES.

ENG500 CAS:77-67-8 ***HR: 2***
2-ETHYL-2-METHYLSUCCINIMIDE
mf: $C_7H_{11}NO_2$ mw: 141.19

SYNS: AETHOSUXIMIDE (GERMAN) ◇ ASAMID ◇ ATYSMAL ◇ CAPITUS ◇ CI 366 ◇ EMESIDE ◇ EPILEO PETIT MAL ◇ ETHOSUCCIMIDE ◇ ETHOSUCCINIMIDE ◇ ETHOSUXIDE ◇ ETHOSUXIMIDE ◇ 3-ETHYL-3-METHYLPYRROLIDINE-2,5-DIONE ◇ 3-ETHYL-3-METHYL-2,5-PYRROLIDINE-DIONE ◇ α-ETHYL-α-METHYLSUCCINIMIDE ◇ ETHYMAL ◇ ETOMAL ◇ ETOSUXIMIDA ◇ H-490 ◇ H 940 ◇ MESENTOL ◇ 3-METHYL-3-ETHYLPYRROLIDINE-2,5-DIONE ◇ Γ-METHYL-Γ-ETHYL-SUCCINIMIDE ◇ PEMAL ◇ PEMALIN ◇ PENTINIMID ◇ PETINIMID ◇ PETNIDAN ◇ PM 671 ◇ PYKNOLEPSINUM ◇ RONTON ◇ SIMATIN(E) ◇ SUCCIMAL ◇ SUCCIMITIN ◇ SUXILEP ◇ SUXIMAL ◇ SUXIN ◇ SUXINUTIN ◇ THETAMID ◇ THILOPEMAL ◇ ZARAONDAN ◇ ZARODAN ◇ ZARONDAN-SAFT ◇ ZARONTIN ◇ ZARTALIN

TOXICITY DATA with REFERENCE
orl-rat TDLo:2250 mg/kg (female 9-17D post):TER EAMJAV 60,407,83
orl-rat TDLo:66600 μg/kg (6-14D preg):REP 40YJAX-,59,70
orl-rat LDLo:1820 mg/kg MEIEDD 10,544,83
orl-mus LD50:1530 mg/kg EPILAK 4(4),66,63
ipr-mus LD50:1752 mg/kg EPILAK 29,198,88
scu-mus LD50:1810 mg/kg NIIRDN 6,117,82
ivn-mus LD50:780 mg/kg NEPHBW 24,427,85

SAFETY PROFILE: Moderately toxic by ingestion, intravenous, subcutaneous and intraperitoneal routes. Experimental teratogenic and reproductive effects. An anticonvulsant. When heated to decomposition it emits toxic fumes of NO_x.

ENH000 CAS:73696-65-8 ***HR: 3***
N-ETHYL-N'-(3-METHYL-2-THIAZOLIDINYLIDENE)UREA
mf: $C_7H_{13}N_3OS$ mw: 187.29

TOXICITY DATA with REFERENCE
orl-mus LD50:203 mg/kg JMCMAR 23,773,80
ivn-mus LD50:1110 mg/kg JMCMAR 23,773,80

SAFETY PROFILE: Poison by ingestion. Moderately toxic by intravenous route. When heated to decomposition it emits very toxic fumes of SO_x and NO_x.

ENI175 CAS:2591-57-3 ***HR: 3***
ETHYLMETHYLTHIOPHOS
mf: $C_9H_{12}NO_3PS$ mw: 245.25

SYNS: METHYLETHYLTHIOFOS ◇ METHYLETHYLTHIOPHOS

TOXICITY DATA with REFERENCE
orl-rat LD50:2800 μg/kg 85GMAT -,83,82
orl-mus LD50:4200 μg/kg 85GMAT -,83,82
orl-cat LD50:5600 μg/kg 85GMAT -,83,82
skn-rbt LDLo:200 mg/kg 85GMAT -,83,82

SAFETY PROFILE: Poison by ingestion and skin contact. When heated to decomposition it emits toxic fumes of NO_x, PO_x, and SO_x.

ENI500 CAS:3568-56-7 ***HR: 3***
O-ETHYL-O-(4-METHYLTHIO-m-TOLYL) METHYLPHOSPHORAMIDOTHIOATE
mf: $C_{11}H_{18}NO_2PS_2$ mw: 291.39

SYNS: BAY 34042 ◇ ENT 25,610 ◇ 4-(METHYLTHIO)-m-CRESOL-O-ESTER with O-ETHYL METHYLPHOSPHORAMIDOTHIOATE

TOXICITY DATA with REFERENCE
orl-rat LD50:5 mg/kg TXAPA9 21,315,72
orl-bwd LD50:1800 μg/kg TXAPA9 21,315,72

SAFETY PROFILE: Poison by ingestion. When heated to decomposition it emits very toxic fumes of NO_x, PO_x, and SO_x.

ENJ000 CAS:458-24-2 ***HR: 3***
N-ETHYL-α-METHYL-m-(TRIFLUOROMETHYL) PHENETHYLAMINE
mf: $C_{12}H_{16}F_3N$ mw: 231.29

SYNS: FENFLURAMINE ◇ 3-(TRIFLUOROMETHYL)-N-ETHYL-α-METHYL PHENETHYL AMINE

TOXICITY DATA with REFERENCE
orl-rat TDLo:280 mg/kg (7-20D preg):REP TJADAB 17,39A,78
orl-hmn LDLo:50 mg/kg LANCAO 2,1306,69
orl-man TDLo:4286 μg/kg:ANS,CNS THERAP 34,205,79
orl-rat LD50:130 mg/kg JMCMAR 18,177,75
orl-mus LD50:170 mg/kg ISYAM* -,75,70
ipr-mus LD50:53 mg/kg ISYAM* -,75,70
orl-dog LD50:100 mg/kg ISYAM* -,75,70

SAFETY PROFILE: A human poison by ingestion. An experimental poison by ingestion and intraperitoneal routes. Human systemic effects by ingestion: hallucinations, distorted perceptions and autonomic nervous system effects (an adrenergic stimulant.) Experimental reproductive effects. When heated to decomposition it emits very toxic fumes of F^- and NO_x. See also AMINES.

ENJ500 CAS:4171-13-5 ***HR: 2***
2-ETHYL-3-METHYLVALERAMIDE
mf: $C_8H_{17}NO$ mw: 143.26

SYNS: ETHYLMETHYL VALERAMIDE ◇ VALMETHAMIDE

TOXICITY DATA with REFERENCE
orl-mus TDLo:700 mg/kg (female 6-12D post):REP CRSBAW 159,6,65
orl-rat LD50:760 mg/kg 27ZQAG -,429,72
ipr-rat LD50:580 mg/kg 27ZQAG -,429,72
orl-mus LD50:999 mg/kg NYKZAU 62,404,66
ipr-mus LD50:540 mg/kg 27ZQAG -,429,72

SAFETY PROFILE: Moderately toxic by ingestion and intraperitoneal routes. Experimental reproductive effects. When heated to decomposition it emits toxic fumes of NO_x.

ENK000 CAS:76-58-4 ***HR: 3***
ETHYLMORPHINE
mf: $C_{19}H_{23}NO_3$ mw: 313.43

SYNS: CODETHYLINE ◇ (5-α,6-α)-7,8-DIDEHYDRO-4,5-EPOXY-3-ETHOXY-17-METHYLMORPHINAN-6-OL ◇ DIONIN ◇ DIONINE ◇ 3-o-ETHYLMORPHINE

TOXICITY DATA with REFERENCE
orl-rat LD50:810 mg/kg JPPMAB 25,929,73
ipr-rat LD50:110 mg/kg PHARAT 31,655,76
scu-rat LD50:200 mg/kg JPPMAB 25,929,73
ivn-rat LD50:62 mg/kg JPPMAB 25,929,73
ipr-mus LD50:120 mg/kg APFRAD 8,261,50
scu-mus LD50:136 mg/kg 28ZNAE 138,8,38

SAFETY PROFILE: Poison by intraperitoneal, intravenous, and subcutaneous routes. Moderately toxic by ingestion. When heated to decomposition it emits toxic fumes of NO_x. See also MORPHINE.

ENK500 CAS:6746-59-4 ***HR: 3***
ETHYL MORPHINE HYDROCHLORIDE DIHYDRATE
mf: $C_{19}H_{23}NO_3 \cdot ClH \cdot 2H_2O$ mw: 385.93

PROP: White, microscopic, crystalline powder. Mp: 125° (decomp), vap d: 13.3.

SYNS: 7,8-DIDEHYDRO-4,5-α-EPOXY-3-ETHOXY-17-METHYLMORPHINAN-6-α-OL HYDROCHLORIDE DIHYDRATE ◇ DIONIN ◇ ETHYLMORPHINE HYDROCHLORIDE

TOXICITY DATA with REFERENCE
scu-ham TDLo:182 mg/kg (female 8D post):TER AJOGAH 123,705,75
scu-mus LD50:200 mg/kg 29ZVAB -,53,69
scu-rbt LDLo:550 mg/kg HBAMAK 4,1289,35
scu-gpg LDLo:300 mg/kg HBAMAK 4,1289,35

SAFETY PROFILE: Poison by subcutaneous route. Can be habit forming. An experimental teratogen. When heated to decomposition it emits very toxic fumes of NO_x and HCl. See also CODEINE and MORPHINE.

ENL000 CAS:100-74-3 ***HR: 3***
N-ETHYLMORPHOLINE
mf: $C_6H_{13}NO$ mw: 115.20

PROP: Colorless liquid. Bp: 138°, flash p: 89.6°F (OC), d: 0.916 @ 20°/20°, vap d: 4.00.

SYN: 4-ETHYLMORPHOLINE

TOXICITY DATA with REFERENCE
skn-rbt 453 mg open MLD UCDS** 11/3/71
eye-rbt 2 mg open SEV AMIHBC 10,61,54
orl-rat LD50:1780 mg/kg DTLWS* 4,190,82
orl-mus LD50:1200 mg/kg TPKVAL 15,116,79
ihl-mus LC50:18000 mg/m^3/2H TPKVAL 15,116,79
ivn-mus LD50:180 mg/kg CSLNX* NX#04778

CONSENSUS REPORTS: Reported in EPA TSCA Inventory.

OSHA PEL: (Transitional: TWA 20 ppm (skin) TWA 5 ppm (skin)
ACGIH TLV: TWA 5 ppm (skin)

SAFETY PROFILE: Poison by intravenous route. Moderately toxic by ingestion. Mildly toxic by inhalation. A skin and severe eye irritant. A very dangerous fire hazard when exposed to heat or flame; can react vigorously with oxidizing materials. To fight fire, use alcohol foam, foam, CO_2, dry chemical. When heated to decomposition it emits toxic fumes of NO_x.

ENL100 CAS:309-29-5 ***HR: 3***
1-ETHYL-4-(2-MORPHOLINOETHYL)-3,3-DIPHENYL-2-PYRROLIDINONE
mf: $C_{24}H_{30}N_2O_2$ mw: 378.56

SYNS: AHR-619 ◇ DOPRAM ◇ DOXAPRAM ◇ 2-PYRROLIDINONE, 1-ETHYL-4-(2-MORPHOLINOETHYL)-3,3-DIPHENYL- ◇ 2-PYRROLIDINONE,1-ETHYL-4-(2-(4-MORPHOLINYL)ETHYL)-3,3-DIPHENYL-(9CI)

TOXICITY DATA with REFERENCE
ivn-dom TDLo:1800 μg/kg (female 16W post):TER BJOGAS 84,48,77
ipr-mus LD50:268 mg/kg JPPMAB 30,522,78

SAFETY PROFILE: Poison by intraperitoneal route. An experimental teratogen. When heated to decomposition it emits toxic fumes of NO_x.

ENL500 CAS:58050-49-0 ***HR: 3***
ETHYL-N-(4-MORPHOLINOMETHYL)CARBAMATE
mf: $C_8H_{16}N_2O_3$ mw: 188.26

TOXICITY DATA with REFERENCE
ims-rat LD50:300 mg/kg ZKKOBW 84,227,75
ipr-mus LD50:500 mg/kg ZKKOBW 84,227,75

SAFETY PROFILE: Poison by intramuscular route. Moderately toxic by intraperitoneal route. When heated to decomposition it emits toxic fumes of NO_x. See also CARBAMATES.

ENL850 CAS:124-06-1 ***HR: 1***
ETHYL MYRISTATE
mf: $C_{16}H_{23}O_2$ mw: 256.42

PROP: Colorless to pale yellow liquid; waxy odor. D: 0.857, refr index: 1.434, flash p: +212°F.

SYN: FEMA No. 2445

SAFETY PROFILE: Combustible liquid. When heated to decomposition it emits acrid smoke and irritating fumes.

ENM000 CAS:4366-50-1 ***HR: 3***
1-ETHYL-1-(1-NAPHTHYL)-2-THIOUREA
mf: $C_{13}H_{14}N_2S$ mw: 230.35

SYN: USAF EK-7162

TOXICITY DATA with REFERENCE
ipr-mus LD50:200 mg/kg NTIS** AD277-689

CONSENSUS REPORTS: Reported in EPA TSCA Inventory.

SAFETY PROFILE: Poison by intraperitoneal route. When heated to decomposition it emits very toxic fumes of NO_x and SO_x.

ENM500 CAS:625-58-1 ***HR: 3***
ETHYL NITRATE
DOT: UN 1993
mf: $C_2H_5NO_3$ mw: 91.08

PROP: Colorless liquid, pleasant odor, sweet taste. Mp: −112°, bp: 88.7°, lel: 3.8%, flash p: 50°F (CC), d: 1.105 @ 20°/4°, vap d: 3.14.

SYNS: NITRIC ACID, ETHYL ESTER ◇ NITRIC ETHER (DOT)

TOXICITY DATA with REFERENCE
mmo-sat 10 μmol/plate FCTOD7 21,707,83

mma-sat 10 μmol/plate FCTOD7 21,707,83
sln-dmg-ihl 1200 ppm/3D FCTOD7 21,707,83

CONSENSUS REPORTS: Reported in EPA TSCA Inventory.

DOT Classification: Flammable Liquid; Label: Flammable Liquid.

SAFETY PROFILE: Mutation data reported. A very dangerous fire hazard when exposed to heat or flame; can react vigorously with oxidizing materials. A moderate explosion hazard when exposed to heat (explodes @ 185°F). To fight fire, use foam, CO_2, dry chemical, water to blanket fire. Incompatible with Lewis acids. When heated to decomposition it emits toxic fumes of NO_x. See also NITRATES and ESTERS.

ENN000 CAS:109-95-5 ***HR: 3***
ETHYL NITRITE
DOT: UN 1194
mf: $C_2H_5NO_2$ mw: 75.08

PROP: Colorless or yellowish liquid; highly aromatic, ethereal odor. Decomp on standing. Very sltly sol in water; misc in alc and ether. Bp: 16.4°, lel: 3.0%, uel: 50%, explodes at 194°F, flash p: −31°F (CC), d: 0.900 @ 15.5°, autoign temp: 194°F, vap d: 2.59. Can explode > 90°C.

SYNS: ETHYL NITRITE (DOT) ◇ ETHYL NITRITE, solution (DOT) ◇ NITROSYL ETHOXIDE ◇ NITROUS ACID ETHYL ESTER ◇ NITROUS ETHER (DOT) ◇ NITROUS ETHYL ETHER

TOXICITY DATA with REFERENCE
unr-chd LDLo:200 mg/kg 32ZWAA 8,338,74

CONSENSUS REPORTS: Reported in EPA TSCA Inventory.

DOT Classification: Flammable Liquid; Label: Flammable Liquid.

SAFETY PROFILE: A human poison by an unspecified route. Narcotic in high concentrations. Lowers blood pressure. Methemoglobinemia has been reported. A very dangerous fire and severe explosion hazard when exposed to heat or flame. A powerful oxidizer. May explode when heated above 90°C. Highly dangerous when heated to decomposition or on contact with acid or acid fumes. To fight fire, use foam, CO_2, dry chemical, or water spray. When heated to decomposition it emits toxic fumes of NO_x. See also NITRITES and ETHERS.

ENN500 CAS:41735-30-2 ***HR: 3***
5-(N-ETHYL-N-NITRO)AMINO-3-(5-NITRO-2-FURYL)-s-TRIAZOLE
mf: $C_8H_8N_6O_5$ mw: 268.22

SYN: N-NITRO-N-(3-(5-NITRO-2-FURYL)-s-TRIAZOL-5-YL)ETHYLAMINE

TOXICITY DATA with REFERENCE
orl-mus LD50:600 mg/kg JMCMAR 16,312,73
ipr-mus LD50:150 mg/kg JMCMAR 16,312,73

SAFETY PROFILE: Poison by intraperitoneal route. Moderately toxic by ingestion. When heated to decomposition it emits toxic fumes of NO_x.

ENO000 CAS:99-77-4 ***HR: 3***
ETHYL-p-NITROBENZOATE
mf: $C_9H_9NO_4$ mw: 195.19

PROP: Triclinic crystals. from ethanol. Mp: 57°, insol in water, sol in alc and ether.

SYN: ETHYL NITROBENZOATE, PARA ESTER

TOXICITY DATA with REFERENCE
ipr-mus LD50:250 mg/kg NTIS** AD691-490

CONSENSUS REPORTS: Reported in EPA TSCA Inventory.

SAFETY PROFILE: Poison by intraperitoneal route. An insecticide. See also NITRO COMPOUNDS of AROMATIC HYDROCARBONS.

ENP000 CAS:39197-62-1 ***HR: D***
N-ETHYL-N'-NITROGUANIDINE
mf: $C_3H_8N_4O_2$ mw: 132.15

SYN: 1-ETHYL-3-NITROGUANIDINE

TOXICITY DATA with REFERENCE
cyt-ham:lng 1 g/L ATSUDG (4),41,80
cyt-ham:fbr 2 g/L/48H MUREAV 48,337,77

SAFETY PROFILE: Mutation data reported. When heated to decomposition it emits very toxic fumes of NO_x.

ENQ000 CAS:546-71-4 ***HR: 3***
ETHYL-4-NITROPHENYL ETHYLPHOSPHONATE
mf: $C_{10}H_{14}NO_5P$ mw: 259.22

SYNS: ARMINE ◇ ETHOXY-4-NITROPHENYLOXYETHYLPHOSPHINEOXIDE ◇ ETHYLPHOSPHONIC ACID ETHYL-p-NITROPHENYL ESTER

TOXICITY DATA with REFERENCE
scu-rat LD50:330 μg/kg FATOAO 42(3),299,79
ivn-rat LD50:250 μg/kg FATOAO 42(3),299,79
ipr-mus LD50:1862 μg/kg PHARAT 35,806,80
scu-mus LD50:330 μg/kg RPTOAN 42,106,79
ivn-mus LD50:250 μg/kg RPTOAN 42,106,79

SAFETY PROFILE: Poison by subcutaneous, intravenous, and intraperitoneal routes. When heated to decomposition it emits very toxic fumes of NO_x and PO_x. See also ESTERS.

ENQ500 CAS:3015-75-6 ***HR: 3***
ETHYL-p-NITROPHENYLPENTYLPHOSPHONATE
mf: $C_{13}H_{20}NO_5P$ mw: 301.31

SYN: p-NITROPHENYL ETHYL PENTYLPHOSPHONATE

TOXICITY DATA with REFERENCE
ivn-rbt LD50:118 μg/kg BCPCA6 13,1229,64
ivn-gpg LD50:299 μg/kg BCPCA6 13,1229,64

SAFETY PROFILE: Poison by intravenous route. When heated to decomposition it emits very toxic fumes of PO_x and NO_x.

ENR000 CAS:35363-12-3 ***HR: 3***
3-ETHYL-4-NITROPYRIDINE-1-OXIDE
mf: $C_7H_8N_2O_3$ mw: 168.17

TOXICITY DATA with REFERENCE
dnd-mus:fbr 500 μmol/L CNREA8 35,521,75
scu-mus TDLo:1840 mg/kg/15W-I:ETA GANNA2 70,799,79

SAFETY PROFILE: Questionable carcinogen with experimental tumorigenic data. Mutation data reported. When heated to decomposition it emits toxic fumes of NO_x.

ENR500 CAS:65986-80-3 ***HR: 3***
(ETHYLNITROSAMINO)METHYL ACETATE
mf: $C_5H_{10}N_2O_3$ mw: 146.17

SYNS: ACETOXYMETHYLETHYLNITROSAMINE ◇ N-(ACETOXY) METHYL-N-ETHYLNITROSAMINE ◇ N-ACETOXYMETHYL-N-NITROSOETHYLAMINE ◇ N-(1-ACETOXYMETHYL)-N-NITROSOETHYL AMINE ◇ AETHYL ACETOXYMETHYLNITROSAMIN (GERMAN) ◇ EAMN ◇ ETHYL ACETOXYMETHYLNITROSAMINE ◇ N-ETHYL-N-(ACETOXYMETHYL)NITROSAMINE

TOXICITY DATA with REFERENCE
mmo-esc 1 μmol/plate GANNA2 71,124,80
dnr-bcs 5 μmol/plate GANNA2 70,663,79
dns-rat:oth 10 μmol/L CBINA8 53,99,85
dnd-mus:fbr 70 μmol/L GANNA2 73,565,82
msc-ham:lng 100 μmol/L GANNA2 72,531,81
orl-rat TDLo:380 mg/kg/90D-I:ETA ZKKOBW 91,317,78
scu-rat TDLo:50 mg/kg/10W-I:CAR JCROD7 104,13,82
scu-rat TD:55 mg/kg/10W-I:CAR IAPUDO 41,619,82
orl-rat LD50:810 mg/kg ZKKOBW 91,317,78

SAFETY PROFILE: Moderately toxic by ingestion. Questionable carcinogen with experimental carcinogenic and tumorigenic data. Mutation data reported. When heated to decomposition it emits toxic fumes of NO_x. See also NITROSAMINES.

ENS000 CAS:41735-29-9 ***HR: 3***
5-(N-ETHYL-N-NITROSO)AMINO-3-(5-NITRO-2-FURYL)-s-TRIAZOLE
mf: $C_8H_8N_6O_4$ mw: 252.22

SYN: N-(3-(5-NITRO-2-FURYL)-s-TRIAZOL-5-YL)-N-NITROSOETHYLAMINE

TOXICITY DATA with REFERENCE
ipr-rat LD50:235 mg/kg JMCMAR 9,42,66
orl-mus LD50:440 mg/kg JMCMAR 16,312,73
ipr-mus LD50:150 mg/kg JMCMAR 16,312,73

SAFETY PROFILE: Poison by intraperitoneal route. Moderately toxic by ingestion. Many N-nitroso compounds are carcinogens. When heated to decomposition it emits toxic fumes of NO_x. See also N-NITROSO COMPOUNDS.

ENS500 CAS:20689-96-7 ***HR: 3***
N-ETHYL-N-NITROSOBENZYLAMINE
mf: $C_9H_{12}N_2O$ mw: 164.23

SYNS: N-NITROSO-N-ETHYLBENZYLAMIN (GERMAN) ◇ N-NITROSO-N-ETHYLBENZYLAMINE

TOXICITY DATA with REFERENCE
orl-rat TDLo:250 mg/kg/36W-C:ETA ZKKOBW 92,235,78
orl-rat LD50:250 mg/kg ZKKOBW 92,235,78

SAFETY PROFILE: Poison by ingestion. Questionable carcinogen with experimental tumorigenic data. Many N-nitroso compounds are carcinogens. When heated to decomposition it emits toxic fumes of NO_x. See also AMINES and N-NITROSO COMPOUNDS.

ENT000 CAS:32976-88-8 ***HR: 3***
N-ETHYL-N-NITROSOBIURET
mf: $C_4H_8N_4O_3$ mw: 160.16

SYNS: ENBU ◇ ETHYLNITROSOBIURET ◇ N-NITROSO-N-ETHYL BIURET

TOXICITY DATA with REFERENCE
orl-rat TDLo:300 mg/kg (14D preg):TER IARCCD 4,45,73
orl-rat TDLo:400 mg/kg:ETA ZKKOBW 76,45,71
orl-rat TDLo:100 mg/kg (22D preg):CAR,TER IARCCD 4,45,73
orl-rat TD:100 mg/kg (15D preg):ETA,TER ZKKOBW 76,45,71
orl-rat LD50:1050 mg/kg IARCCD 4,45,73

SAFETY PROFILE: Moderately toxic by ingestion. Questionable carcinogen with experimental carcinogenic, and tumorigenic data. Experimental teratogenic effects. When heated to decomposition it emits toxic fumes of NO_x. See also N-NITROSO COMPOUNDS.

ENT500 CAS:38434-77-4 ***HR: 3***
ETHYLNITROSOCYANAMIDE
mf: $C_3H_5N_3O$ mw: 99.11

SYNS: N-CYANO-N-NITROSOETHYLAMINE ◇ ENC ◇ NITROSO-ETHANECARBAMONITRILE

TOXICITY DATA with REFERENCE
orl-rat TDLo:1800 mg/kg/52W-I:NEO JJIND8 62,1523,79
ipr-rat LD50:15 mg/kg JOCEAH 38,1325,73

CONSENSUS REPORTS: EPA Genetic Toxicology Program. Cyanide and its compounds are on the Community Right-To-Know List.

SAFETY PROFILE: Poison by intraperitoneal route. Questionable carcinogen with experimental neoplastigenic data. Many N-nitroso compounds are carcinogens. When heated to decomposition it emits toxic fumes of NO_x and CN^-. See also N-NITROSO COMPOUNDS and NITRILES.

ENU000 CAS:4245-77-6 ***HR: 3***
N-ETHYL-N-NITROSO-N'-NITROGUANIDINE
mf: $C_3H_7N_5O_3$ mw: 161.15

SYNS: N-AETHYL-N'-NITRO-N-NITROSOGUANIDIN(GERMAN) ◇ ENNG ◇ N-ETHYL-N'-NITRO-N-NITROSOGUANIDINE

TOXICITY DATA with REFERENCE
mma-sat 5 μg/plate TCMUE9 1,13,84
mmo-esc 2 μg/plate KSRNAM 19,4465,85
pic-esc 200 μg/L TCMUE9 1,91,84
sce-hmn:lym 1 μmol/L NGCJAK 15,1085,80
dnd-ham:ovr 20 μmol/L MUREAV 132,41,84
sce-ham-orl 10 mg/kg MUREAV 113,33,83
sce-ham:lng 10 mg/L CNREA8 44,3270,84
orl-rat TDLo:504 mg/kg/12W-C:ETA JJCREP 78,126,87
scu-rat TDLo:55 mg/kg/5W-I:NEO GANNA2 69,277,78
orl-mus TDLo:3 g/kg/43W-C:CAR JNCIAM 52,519,74

CONSENSUS REPORTS: EPA Genetic Toxicology Program.

SAFETY PROFILE: Questionable carcinogen with experimental carcinogenic, neoplastigenic, and tumorigenic data. Human mutation data reported. When heated to decomposition it emits toxic fumes of NO_x. See also N-NITROSO COMPOUNDS.

ENU500 CAS:72505-63-6 ***HR: D***
2-ETHYL-3-NITROSOTHIAZOLIDINE
mf: $C_5H_{10}N_2O$ mw: 114.17

SYNS: ENT ◇ 2-ETHYL-N-NITROSOTHIAZOLIDINE ◇ N-NITROSO-2-ETHYLTHIAZOLIDINE

TOXICITY DATA with REFERENCE
mmo-sat 1 mg/L JAFCAU 28,62,80
mma-sat 10 mg/L JAFCAU 28,62,80

SAFETY PROFILE: Mutation data reported. Many N-nitroso compounds are carcinogens. When heated to decomposition it emits toxic fumes of NO_x. See also N-NITROSO COMPOUNDS.

ENV000 CAS:759-73-9 ***HR: 3***
1-ETHYL-1-NITROSOUREA
mf: $C_3H_7N_3O_2$ mw: 117.13

PROP: Pale yellow crystals. Mp: 103° (decomp) 1% water soln @ 20°.

SYNS: AENH (GERMAN) ◇ AETHYLNITROSO-HARNSTOFF (GERMAN) ◇ ENU ◇ N-ETHYL-N-NITROSOCARBAMIDE ◇ ETHYLNITROSOUREA ◇ N-ETHYL-N-NITROSO-UREA ◇ NEU ◇ NITROSOETHYLUREA ◇ NSC 45403 ◇ RCRA WASTE NUMBER U176

TOXICITY DATA with REFERENCE
dni-hmn:hla 100 μmol/L MUREAV 92,427,82
cyt-ham-orl 100 mg/kg TRENAF 36,396,85
ipr-rat TDLo:15 mg/kg (female 18D post):REP NETOD7 2,297,80
scu-mus TDLo:351 mg/kg (female 6-16D post):TER TCMUD8 3,219,83
orl-rat TDLo:10 mg/kg (female 19D post):NEO,TER CALEDQ 2,93,76
orl-rat TDLo:374 mg/kg/2W-C:CAR JJIND8 75,743,85
skn-rat TDLo:350 mg/kg (female 16-22D post):NEO,TER ZEKBAI 81,169,74
ipr-rat TDLo:10 mg/kg (female 15D post):NEO,TER SCIEAS 218,682,82
ipr-rat TDLo:45 mg/kg:CAR ANTRD4 4,5,84
ivn-rat TDLo:17 mg/kg:ETA EXPTAX 17,394,79
ivn-rat TDLo:50 mg/kg (15D preg):CAR,TER ZEKBAI 73,371,70
unr-rat TDLo:5 mg/kg (13D preg):ETA,TER XENOBH 3,271,73
orl-mus TDLo:20 mg/kg/24W-I:NEO TXAPA9 72,313,84
ipr-mus TDLo:20 mg/kg (19D preg):CAR,TER VOONAW 23(3),41,77
par-mus TDLo:100 mg/kg/(18D preg):ETA,TER PAACA3 17,122,76
par-mus TDLo:58565 μg/kg (18D preg):CAR,TER JNCIAM 51,1965,73
ipr-dog TDLo:100 mg/kg (53D preg):ETA,TER CALEDQ 12,161,81
ivn-dog TDLo:30 mg/kg (59D preg):ETA,TER ZAPPAN 121,54,77
ipr-rbt TDLo:100 mg/kg (15-24D preg):NEO,TER JJIND8 65,607,80
ivn-rbt TDLo:80 mg/kg (25D preg):NEO,TER BEXBAN 85,369,78
par-rbt TDLo:160 mg/kg (25D preg):ETA,TER ZAPPAN 121,54,77
ivn-pig TDLo:120 mg/kg (20-31D preg):ETA,TER ARGEAR 34,25,69

orl-ham TDLo:60 mg/kg (11D preg):ETA,TER ZEKBAI 71,320,68
ipr-ham TDLo:800 mg/kg (16D preg):NEO,TER ZKKOBW 90,233,77
ivn-ham TDLo:80 mg/kg (15D preg):ETA,TER ZAPPAN 121,54,77
ipr-rat TD:10 mg/kg (21D preg):NEO,REP JSONDX 5,396,84
ivn-rat TD:20 mg/kg/(12D preg):ETA,TER NAGZAC 54,130,79
ivn-rbt TD:150 mg/kg (8-10D preg):ETA,TER ZKKOBW 89,331,77
ivn-rat TD:10 mg/kg (16D preg):ETA,TER ANPTAL 34,21,76
ivn-rat TD:20 mg/kg (17D preg):ETA,TER CALEDQ 1,345,76
ivn-rbt TD:70 mg/kg (25D preg):ETA,TER NEOLA4 25,453,78
orl-rat LD50:300 mg/kg PPTCBY 2,73,72
scu-rat LD50:240 mg/kg ZEKBAI 74,141,70
ivn-rat LD50:240 mg/kg NATUAS 222,1064,69
ipr-mus LD50:600 mg/kg JJIND8 62,911,79

CONSENSUS REPORTS: NTP Fifth Annual Report on Carcinogens. IARC Cancer Review: Group 2A IMEMDT 7,56,87; Human Limited Evidence IMEMDT 17,191,78; Animal Sufficient Evidence IMEMDT 17,191,78; IMEMDT 1,135,72. EPA Genetic Toxicology Program. Community Right-To-Know List. Reported in EPA TSCA Inventory.

SAFETY PROFILE: Confirmed carcinogen with experimental carcinogenic, neoplastigenic, tumorigenic, and teratogenic data. Poison by ingestion, subcutaneous, and intravenous routes. Moderately toxic by intraperitoneal route. Human mutation data reported. When heated to decomposition it emits toxic fumes of NO_x. See also N-NITROSO COMPOUNDS.

ENV500 CAS:139-94-6 ***HR: 3***
1-ETHYL-3-(5-NITRO-2-THIAZOLYL) UREA
mf: $C_6H_8N_4O_3S$ mw: 216.24

SYNS: N-ETHYL-N'-(5-NITRO-2-THIAZOLYL)UREA ◇ HEPZIDE ◇ NCI-C03792 ◇ NITHIAZID ◇ NITHIAZIDE

TOXICITY DATA with REFERENCE
mmo-sat 75 μg/plate ENMUDM 5(Suppl 1),3,83
mma-sat 75 μg/plate ENMUDM 5(Suppl 1),3,83
orl-rat TDLo:41 g/kg/94W-I:CAR NCITR* NCI-CG-TR-146,79
orl-mus TDLo:395 g/kg/94W-I:CAR NCITR* NCI-CG-TR-146,79
orl-mus LD50:2150 mg/kg NCILB* NIH-NCI-E-C-72-3252,73

CONSENSUS REPORTS: IARC Cancer Review: Group 3 IMEMDT 7,56,87; Animal Limited Evidence IMEMDT 31,179,83. NCI Carcinogenesis Bioassay (feed); Clear Evidence: mouse, rat NCITR* NCI-CG-TR-146,79

SAFETY PROFILE: Suspected carcinogen with experimental carcinogenic data. Moderately toxic by ingestion. Mutation data reported. When heated to decomposition it emits very toxic fumes of NO_x and PO_x.

ENW000 CAS:123-29-5 ***HR: 1***
ETHYL NONANOATE
mf: $C_{11}H_{22}O_2$ mw: 186.33

PROP: Colorless liquid; fruity, cognac odor. D: 0.863-0867, refr index: 1.420, flash p: 185°F. Misc with alc, propylene glycol; insol in water.

SYNS: ETHYL NONYLATE ◇ ETHYL PELARGONATE ◇ FEMA No. 2447 ◇ NONANOIC ACID, ETHYL ESTER ◇ WINE ETHER

TOXICITY DATA with REFERENCE
skn-rbt 500 mg/24H MOD FCTXAV 16,747,78
orl-gpg LD50:24 g/kg FCTXAV 2,327,64

CONSENSUS REPORTS: Reported in EPA TSCA Inventory.

SAFETY PROFILE: Mildly toxic by ingestion. A skin irritant. Combustible liquid. When heated to decomposition it emits acrid smoke and irritating fumes.

ENW500 CAS:103-08-2 ***HR: 2***
5-ETHYL-2-NONANOL
mf: $C_{11}H_{24}O$ mw: 172.35

SYN: (3-ETHYL-N-HEPTYL)METHYLCARBINOL

TOXICITY DATA with REFERENCE
skn-rbt 415 mg open MLD UCDS** 2/21/58
eye-rbt 42 mg MOD UCDS** 2/21/58
orl-rat LD50:3000 mg/kg JIHTAB 26,269,44
skn-rbt LD50:4760 mg/kg UCDS** 2/21/58

SAFETY PROFILE: Moderately toxic by ingestion. Mildly toxic by skin contact. A skin and eye irritant. When heated to decomposition it emits acrid smoke and irritating fumes.

ENX000 CAS:10137-90-3 ***HR: 1***
5-ETHYL-3-NONEN-2-ONE
mf: $C_{11}H_{20}O$ mw: 168.31

TOXICITY DATA with REFERENCE
skn-rbt 10 mg/24H open MLD AMIHBC 10,61,54
eye-rbt 500 mg open AMIHBC 10,61,54
orl-rat LD50:8120 mg/kg AMIHBC 10,61,54
skn-rbt LD50:8480 mg/kg AMIHBC 10,61,54

SAFETY PROFILE: Mildly toxic by ingestion and skin contact. A skin and eye irritant. When heated to decomposition it emits acrid smoke and irritating fumes.

ENX100 CAS:23489-02-3 ***HR: 3***
1-ETHYL-1-NONYLPIPERIDINIUM BROMIDE
mf: $C_{16}H_{34}N \cdot Br$ mw: 320.42

TOXICITY DATA with REFERENCE
orl-mus LD50:162 mg/kg PSDTAP 15,331,74
ipr-mus LD50:42955 μg/kg PSDTAP 15,331,74
ivn-mus LD50:3895 μg/kg PSDTAP 15,331,74

SAFETY PROFILE: Poison by ingestion, intravenous, and intraperitoneal routes. When heated to decomposition it emits toxic fumes of NO_x and Br^-.

ENX500 CAS:3198-07-0 ***HR: 3***
ETHYLNORADRENALINE HYDROCHLORIDE
mf: $C_{10}H_{15}NO_3 \cdot ClH$ mw: 233.72

SYNS: α-(1-AMINOPROPYL)PROTOCATECHUYL ALCOHOL HYDROCHLORIDE ◇ BRONKEPHRINE HYDROCHLORIDE ◇ BUTANEFRINE HYDROCHLORIDE ◇ 1-(3,4-DIHYDROXYPHENYL)-2-AMINO-1-BUTANOL HYDROCHLORIDE ◇ 1-(3,4-DIHYDROXYPHENYL)-1-HYDROXY-2-AMINOBUTANE HYDROCHLORIDE ◇ E.N.E. ◇ E.N.S. ◇ ETHYL NOREPINEPHRINE HYDROCHLORIDE ◇ α-ETHYLNOREPINEPHRINE HYDROCHLORIDE ◇ ETHYLNORSUPRARENIN HYDROCHLORIDE

TOXICITY DATA with REFERENCE
scu-hmn TDLo:36 μg/kg:CVS JPETAB 81,269,44
ivn-hmn TDLo:9 μg/kg:CVS JPETAB 81,269,44
ims-hmn TDLo:18 μg/kg:CVS JPETAB 81,269,44
scu-rat LDLo:160 mg/kg JPETAB 81,269,44
ivn-mus LD50:117 mg/kg JPETAB 81,269,44

SAFETY PROFILE: Poison by intravenous and subcutaneous routes. Human systemic effects by subcutaneous, intravenous, and intramuscular routes: heart rate change, blood pressure decrease, and pulse pressure increase. When heated to decomposition it emits very toxic fumes of Cl^- and NO_x. See also NOREPINEPHRINE.

ENX575 CAS:16320-04-0 ***HR: D***
ETHYLNORGESTRIENONE
mf: $C_{21}H_{24}O_2$ mw: 308.45

SYNS: 13-ETHYL-17-α-ETHINYL-17-HYDROXYGON-4,9,11-TRIEN-3-ONE ◇ 13-ETHYL-17-HYDROXY-18,19-DINOR-17-α-PREGNA-4,9,11-TRIEN-20-YN-3-ONE ◇ GESTRIGONE ◇ GESTRINONE ◇ R-2323 ◇ RU 2323

TOXICITY DATA with REFERENCE
imp-wmn TDLo:1800 μg/kg (48W pre):REP CCPTAY 11,625,75
orl-mus TDLo:50 mg/kg (female 6-15D post):TER YACHDS 16,713,88
orl-rat LDLo:1 g/kg YACHDS 16,701,88
ipr-rat LDLo:1 g/kg YACHDS 16,701,88
ipr-mus LDLo:1 g/kg YACHDS 16,701,88

SAFETY PROFILE: Human reproductive effects by ingestion, implant or intravaginal routes: menstrual cycle changes and disorders, and changes in female fertility. An experimental teratogen. Other experimental reproductive effects. A steroid. When heated to decomposition it emits acrid smoke and irritating fumes.

ENX600 CAS:52-78-8 ***HR: D***
17-ETHYL-19-NORTESTOSTERONE
mf: $C_{20}H_{30}O_2$ mw: 302.50

PROP: Crystals from methanol. Mp: 140-141°. Insol in water; sol in alc, benzene, ether, ethyl acetate.

SYNS: 8022 C.B. ◇ 17-ENT ◇ 17-α-ETHYL-17-HYDROXYNORANDROSTENONE ◇ 17-α-ETHYL-17-HYDROXY-4-NORANDROSTEN-3-ONE ◇ 17-α-ETHYL-17-HYDROXY-19-NORANDROST-4-EN-3-ONE ◇ 17-α-ETHYL-19-NORTESTOSTERONE ◇ 17-HYDROXY-19-NORPREGN-4-EN-3-ONE ◇ 17-HYDROXY-19-NOR-17-α-PREGN-4-EN-3-ONE ◇ 17-β-HYDROXY-19-NOR-17-α-PREGN-4-EN-3-ONE ◇ NILEVA ◇ NILEVAR ◇ NORETHANDROLONE ◇ 19-NORETHYLTESTOSTERONE ◇ 19-NOR-17-α-ETHYLTESTOSTERONE ◇ 17-α-NORTESTOSTERONE ◇ PRONABOL ◇ SC 5914 ◇ SOLEVAR

TOXICITY DATA with REFERENCE
orl-wmn TDLo:64 mg/kg (female 64W pre):REP AJOGAH 75,82,58
orl-rat TDLo:20 mg/kg (female 17-20D post):TER ECJPAE 24,77,77

CONSENSUS REPORTS: EPA Genetic Toxicology Program.

SAFETY PROFILE: Human male reproductive effects by ingestion: impotence, changes in spermatogenesis, testes, epididymis and sperm duct. Human female reproductive effects by ingestion: changes in menstrual cycle and fertility. An experimental teratogen. Other reproductive effects in experimental animals. A steroid. When heated to decomposition it emits acrid smoke and fumes.

ENX875 CAS:56501-34-9 ***HR: 3***
1-ETHYL-1-OCTADECYLPIPERIDINIUM BROMIDE
mf: $C_{25}H_{50}N \cdot Br$ mw: 444.67

TOXICITY DATA with REFERENCE
orl-mus LD50:238 mg/kg PSDTAP 15,331,74
ipr-mus LD50:1847 μg/kg PSDTAP 15,331,74
ivn-mus LD50:4493 μg/kg PSDTAP 15,331,74

SAFETY PROFILE: Poison by ingestion, intravenous, and intraperitoneal routes. When heated to decomposition it emits toxic fumes of NO_x and Br^-. See also BROMIDES.

ENY000 CAS:106-32-1 ***HR: 1***
ETHYL OCTANOATE
mf: $C_{10}H_{20}O_2$ mw: 172.30

PROP: Colorless liquid; wine-brandy fruit odor. D: 0.865-0.869, refr index: 1.417, flash p: 185°F. Sol in fixed oils; sltly sol in propylene glycol; insol in glycerin, water @ 209°.

SYNS: ETHYL CAPRYLATE ◇ ETHYL OCTYLATE ◇ FEMA No. 2449 ◇ OCTANOIC ACID, ETHYL ESTER

TOXICITY DATA with REFERENCE
skn-rbt 500 mg/24H MOD FCTXAV 14,763,76
orl-rat LD50:25960 mg/kg FCTXAV 14,763,76

CONSENSUS REPORTS: Reported in EPA TSCA Inventory.

SAFETY PROFILE: Mildly toxic by ingestion. A skin irritant. Combustible liquid. When heated to decomposition it emits acrid smoke and irritating fumes. See also ESTERS.

ENY100 CAS:23489-01-2 ***HR: 3***
1-ETHYL-1-OCTYLPIPERIDINIUM BROMIDE
mf: $C_{15}H_{32}N \cdot Br$ mw: 306.328

TOXICITY DATA with REFERENCE
orl-mus LD50:210 mg/kg PSDTAP 15,331,74
ivn-mus LD50:4793 μg/kg PSDTAP 15,331,74
ipr-mus LD50:48453 μg/kg PSDTAP 15,331,74

SAFETY PROFILE: Poison by ingestion, intravenous, and intraperitoneal routes. When heated to decomposition it emits toxic fumes of NO_x and Br^-.

ENY500 CAS:122-51-0 ***HR: 3***
ETHYL ORTHOFORMATE
DOT: UN 2524
mf: $C_7H_{16}O_3$ mw: 148.23

PROP: Clear liquid, pungent odor. Bp: 145.9°, flash p: 86°F (CC), d: 0.895 @ 20°/20°, vap press: 10 mm @ 40.5°, vap d: 5.11.

SYNS: AETHON ◇ ETHONE ◇ ETHYLESTER KYSELINY ORTHOMRAVENCI (CZECH) ◇ 1,1′,1′-(METHYLIDYNETRIS(OXY))TRIS (ETHANE) ◇ ORTHOFORMIC ACID, ETHYL ESTER ◇ ORTHOFORMIC ACID, TRIETHYL ESTER ◇ ORTHOMRAVENCAN ETHYLNATY (CZECH) ◇ TRIETHOXYMETHANE ◇ TRIETHYL ORTHOFORMATE

TOXICITY DATA with REFERENCE
skn-rbt 10 mg/24H open MLD AMIHBC 4,119,51
skn-rbt 500 mg/24H MLD 28ZPAK -,44,72
eye-rbt 500 mg open AMIHBC 4,119,51
eye-rbt 100 mg/24H MOD 28ZPAK -,44,72
orl-rat LD50:2920 mg/kg 28ZPAK -,44,72
ihl-rat LCLo:4000 ppm/8H AMIHBC 4,119,51
skn-rbt LD50:20 g/kg AMIHBC 4,119,51
scu-rbt LD50:20 g/kg FCTXAV 17,917,79

CONSENSUS REPORTS: Reported in EPA TSCA Inventory.

DOT Classification: Flammable or Combustible Liquid; Label: Flammable Liquid.

SAFETY PROFILE: Moderately toxic by ingestion. Mildly toxic by inhalation, skin contact, and subcutaneous routes. A skin and eye irritant. A very dangerous fire hazard when exposed to heat or flame; can react vigorously with oxidizing materials. To fight fire, use foam, CO_2, dry chemical. When heated to decomposition it emits acrid smoke and irritating fumes. See also ESTERS.

ENZ000 CAS:97-81-4 ***HR: 2***
ETHYL-3-OXATRICYCLO-(3.2.1.0^{2,4})OCTANE-6-CARBOXYLATE
mf: $C_{10}H_{14}O_3$ mw: 182.24

SYNS: 3,4-EPOXY-2,5-ENDOMETHYLENECYCLOHEXANECARBOXYLIC ACID, ETHYL ESTER ◇ 5,6-EPOXY-2-NORBORNANECARBOXYLIC ACID, ETHYL ESTER ◇ 3-OXATRICYCLO(3.2.1.0^{2,4})OCTANE-6-CARBOXYLIC ACID, ETHYL ESTER

TOXICITY DATA with REFERENCE
skn-rbt 10 mg/24H open MLD AIHAAP 23,95,62
orl-rat LD50:4760 mg/kg AIHAAP 24,305,63
skn-rbt LD50:3540 mg/kg AIHAAP 24,305,63

SAFETY PROFILE: Moderately toxic by skin contact. Mildly toxic by ingestion. A skin irritant. When heated to decomposition it emits acrid smoke and irritating fumes. See also ESTERS.

EOA000 CAS:35629-44-8 ***HR: 3***
1-ETHYL-3-(2-OXAZOLYL)UREA
mf: $C_6H_9N_3O_2$ mw: 155.18

TOXICITY DATA with REFERENCE
orl-mus LD50:400 mg/kg JMCMAR 14,1075,71
ipr-mus LD50:300 mg/kg JMCMAR 14,1075,71

SAFETY PROFILE: Poison by ingestion and intraperitoneal routes. When heated to decomposition it emits toxic fumes of NO_x.

EOA500 CAS:17243-64-0 ***HR: 2***
cis-2-(3-ETHYL-4-OXO-5-PIPERIDINO-2-THIAZOLIDINYLIDENE)ACETIC ACID
mf: $C_{12}H_{18}N_2O_3S$ mw: 270.38

SYNS: Z-(3-AETHYL-4-OXO-5-PIPERIDINO-THIAZOLIDIN-2-YLIDEN)- ESSIGSAURE-AETHYLESTER (GERMAN) ◇ ETHYL 3-ETHYL-4-OXO-5-PIPERIDINO-$\Delta^{2,\alpha}$-THIAZOLIDINEACETATE ◇ ETHYL (Z)-(3-ETHYL-4-OXO-5-PIPERIDINOTHIAZOLIDIN-2-YLIDENE)ACETATE ◇ (Z)-2-(3-ETHYL-4-OXO-5-PIPERIDINO-2-THIAZOLIDINYLIDENE, ACETIC ACID ◇ (3-ETHYL-4-OXO-5-(1-PIPERIDINYL)-2-THIAZOLIDINYLIDENE)ACETIC ACID ETHYL ESTER ◇ Go 919 ◇ PIPROZOLIN ◇ PIPROZOLINE ◇ PROBILIN ◇ W 3699

TOXICITY DATA with REFERENCE
orl-rat LD50:3256 mg/kg ARZNAD 27,467,77
orl-mus LD50:1070 mg/kg ARZNAD 27,467,77

SAFETY PROFILE: Moderately toxic by ingestion. A

choleretic agent. When heated to decomposition it emits very toxic fumes of NO_x and SO_x.

EOB000 CAS:17243-64-0 ***HR: 2***
2-(3-ETHYL-4-OXO-5-PIPERIDINO-2-THIAZOLIDINYLIDENE)ACETIC ACID ETHYL ESTER
mf: $C_{14}H_{22}N_2O_3S$ mw: 298.44

TOXICITY DATA with REFERENCE
orl-rat LD50:3256 mg/kg ARZNAD 27,463,77
orl-mus LD50:1310 mg/kg ARZNAD 27,463,77

SAFETY PROFILE: Moderately toxic by ingestion. When heated to decomposition it emits very toxic fumes of NO_x and SO_x. See also ESTERS.

EOB100 CAS:28853-06-7 ***HR: 3***
ETHYL PENTABORANE (9)
mf: $C_2H_{13}B_5$ mw: 91.17

SAFETY PROFILE: Ignites spontaneously in air. See also BORANES and BORON COMPOUNDS.

EOB200 CAS:56501-32-7 ***HR: 3***
1-ETHYL-1-PENTADECYLPIPERIDINIUM BROMIDE
mf: $C_{22}H_{46}N{\bullet}Br$ mw: 404.60

TOXICITY DATA with REFERENCE
orl-mus LD50:195 mg/kg PSDTAP 15,331,74
ipr-mus LD50:3096 μg/kg PSDTAP 15,331,74
ivn-mus LD50:4692 μg/kg PSDTAP 15,331,74

SAFETY PROFILE: Poison by ingestion, intravenous, and intraperitoneal routes. When heated to decomposition it emits toxic fumes of NO_x and Br^-. See also BROMIDES.

EOD000 CAS:22750-93-2 ***HR: 3***
ETHYL PERCHLORATE
mf: $C_2H_5ClO_4$ mw: 128.52

$CH_3CH_2OClO_3$

DOT Classification: Forbidden

SAFETY PROFILE: Possibly the most explosive chemical known. Very sensitive to impact, friction, and heat. Upon decomposition it emits toxic fumes of Cl^-. See also PERCHLORATES.

EOD500 CAS:66922-67-6 ***HR: 2***
ETHYLPHENACETIN
mf: $C_{12}H_{17}NO_2$ mw: 207.30

SYN: N-ETHYL-p-ACETOPHENETIDIDE

TOXICITY DATA with REFERENCE
orl-rbt LDLo:2500 mg/kg AEXPBL 33,216,1894
orl-gpg LDLo:2700 mg/kg AEXPBL 33,216,1894
orl-frg LDLo:4545 mg/kg AEXPBL 33,216,1894

SAFETY PROFILE: Moderately toxic by ingestion. When heated to decomposition it emits toxic fumes of NO_x.

EOE000 CAS:10510-77-7 ***HR: D***
5-ETHYLPHENAZINIUM ETHYLSULFATE
mf: $C_{14}H_{13}N_2{\bullet}C_2H_5O_4S$ mw: 334.42

SYNS: N-ETHYLPHENAZONIUM ETHOSULFATE ◇ PHENAZINE ETHOSULFATE

TOXICITY DATA with REFERENCE
mmo-sat 100 μg/plate MUREAV 40,203,76
mma-sat 100 μg/plate MUREAV 40,203,76

SAFETY PROFILE: Mutation data reported. When heated to decomposition it emits very toxic fumes of SO_x and NO_x. See also SULFATES.

EOF500 CAS:101491-82-1 ***HR: 3***
N-ETHYL-N-(1-PHENOXY-2-PROPYL)CARBAMIC ACID-2-(DIETHYLAMINO)ETHYL ESTER HYDROCHLORIDE
mf: $C_{18}H_{30}N_2O_3{\bullet}ClH$ mw: 358.96

SYN: C 2137

TOXICITY DATA with REFERENCE
eye-rbt 2% MOD ARZNAD 9,113,59
scu-mus LD50:210 mg/kg ARZNAD 9,113,59

SAFETY PROFILE: Poison by subcutaneous route. An eye irritant. When heated to decomposition it emits very toxic fumes of HCl and NO_x. See also ESTERS.

EOG000 CAS:101491-83-2 ***HR: 3***
N-ETHYL-N-(1-PHENOXY-2-PROPYL)CARBAMIC ACID-2-(2-METHYLPIPERIDINO) ETHYL ESTER HYDROCHLORIDE
mf: $C_{20}H_{32}N_2O_3{\bullet}ClH$ mw: 385.00

SYN: C 2126

TOXICITY DATA with REFERENCE
eye-rbt 2% MLD ARZNAD 9,113,59
scu-mus LD50:300 mg/kg ARZNAD 9,113,59

SAFETY PROFILE: Poison by subcutaneous route. An eye irritant. When heated to decomposition it emits very toxic fumes of NO_x and HCl. See also CARBAMATES.

EOG500 CAS:101651-79-0 ***HR: 3***
N-ETHYL-N-(1-PHENOXY-2-PROPYL)-2-(2-METHYLPIPERIDINO) ACETAMIDE HYDROCHLORIDE
mf: $C_{19}H_{30}N_2O_2{\bullet}ClH$ mw: 354.97

SYNS: C 2054 ◇ N-ETHYL-2-(METHYLPIPERIDINO)-N-(1-PHENOXY-2-PROPYL)ACETAMIDE HYDROCHLORIDE

TOXICITY DATA with REFERENCE
eye-rbt 2% MLD ARZNAD 9,70,59
scu-mus LD50:165 mg/kg ARZNAD 9,70,59

SAFETY PROFILE: Poison by subcutaneous route. An eye irritant. When heated to decomposition it emits very toxic fumes of NO_x and HCl.

EOH000 CAS:101-97-3 ***HR: 2***
ETHYL PHENYLACETATE
mf: $C_{10}H_{12}O_2$ mw: 164.22

PROP: Colorless liquid; sweet, honey odor. Bp: 227°, d: 1.033 @ 20°, refr index: 1.496-1.500, vap d: 5.67, flash p: + 100 C. Sol in fixed oils; insol in glycerin, propylene glycol, water.

SYNS: BENZENEACETIC ACID, ETHYL ESTER (9CI) ◇ ETHYL BENZENEACETATE ◇ ETHYL PHENACETATE ◇ ETHYL-2-PHENYLETHANO × ATE◇ ETHYL-α-TOLUATE ◇ FEMA No. 2452 ◇ PHENYLACETIC ACID, ETHYL ESTER ◇ α-TOLUIC ACID, ETHYL ESTER

TOXICITY DATA with REFERENCE
orl-rat LD50:3300 mg/kg FCTXAV 13,99,75

CONSENSUS REPORTS: Reported in EPA TSCA Inventory.

SAFETY PROFILE: Moderately toxic by ingestion. Combustible liquid. When heated to decomposition it emits acrid smoke and irritating fumes. See also ESTERS.

EOH500 CAS:2058-67-5 ***HR: 3***
N-ETHYL-p-(PHENYLAZO)ANILINE
mf: $C_{14}H_{15}N_3$ mw: 225.32

SYNS: EAB ◇ 4-(ETHYLAMINOAZOBENZENE) ◇ N-ETHYL-4-AMINOAZOBENZENE ◇ N-ETHYL-4-(PHENYLAZO)BENZENAMINE

TOXICITY DATA with REFERENCE
ipr-mus TDLo:400 mg/kg (10-11D preg):TER KAIZAN 37,179,62
ipr-mus TDLo:400 mg/kg (10-11D preg):REP KAIZAN 37,179,62
orl-rat TDLo:376 mg/kg/5W-I:ETA CNREA8 39,3411,79

SAFETY PROFILE: Experimental teratogenic and reproductive effects. Questionable carcinogen with experimental tumorigenic data. When heated to decomposition it emits toxic fumes of NO_x.

EOI000 CAS:17010-65-0 ***HR: 3***
p-((m-ETHYLPHENYL)AZO)-N,N-DIMETHYLANILINE
mf: $C_{16}H_{19}N_3$ mw: 253.38

SYNS: N,N-DIMETHYL-p((m-ETHYLPHENYL)AZO)ANILINE◇ N,N-DIMETHYL-p-(3′-ETHYLPHENYLAZO)ANILINE ◇ N,N-DIMETHYL-3′-ETHYL-4-(PHENYLAZO)BENZENAMINE ◇ 3′-ETHYL-DAB ◇ 3′-ETHYL-4-DIMETHYLAMINOAZOBENZENE

TOXICITY DATA with REFERENCE
orl-rat TDLo:3175 mg/kg/12W-C:CAR CBINA8 53,107,85
orl-rat TD:2261 mg/kg/17W-I:ETA CNREA8 30,1520,70
orl-rat TD:4858 mg/kg/17W-C:CAR CRNGDP 1,419,80

SAFETY PROFILE: Questionable carcinogen with experimental carcinogenic and tumorigenic data. When heated to decomposition it emits toxic fumes of NO_x.

EOI500 CAS:5302-41-0 ***HR: 3***
p-((p-ETHYLPHENYL)AZO)-N,N-DIMETHYLANILINE
mf: $C_{16}H_{19}N_3$ mw: 253.38

SYNS: N,N-DIMETHYL-4′-ETHYL-4-AMINOAZOBENZENE ◇ N,N-DIMETHYL-p-((4-ETHYLPHENYL)AZO)ANILINE◇ 4′-ETHYL-DAB ◇ 4′-ETHYL-4-DIMETHYLAMINOAZOBENZENE

TOXICITY DATA with REFERENCE
orl-rat TDLo:4858 mg/kg/17W-C:CAR CRNGDP 1,419,80
orl-rat TD:2142 mg/kg/17W-C:ETA JNCIAM 27,663,61

SAFETY PROFILE: Questionable carcinogen with experimental carcinogenic and tumorigenic data. When heated to decomposition it emits toxic fumes of NO_x.

EOJ000 CAS:55398-27-1 ***HR: 3***
p-(4-ETHYLPHENYLAZO)-N-METHYLANILINE
mf: $C_{15}H_{17}N_3$ mw: 239.35

SYNS: 4′-ETHYL-N-METHYL-4-AMINOAZOBENZENE◇ N-METHYL-4′-ETHYL-p-AMINOAZOBENZENE

TOXICITY DATA with REFERENCE
orl-rat TDLo:1890 mg/kg/15W-C:ETA JNCIAM 15,67,54

SAFETY PROFILE: Questionable carcinogen with experimental tumorigenic data. When heated to decomposition it emits toxic fumes of NO_x.

EOJ500 CAS:6368-72-5 ***HR: 3***
N-ETHYL-1-((p-(PHENYLAZO)PHENYL)AZO)-2-NAPHTHYLAMINE
mf: $C_{24}H_{21}N_5$ mw: 379.50

SYNS: CERES RED 7B ◇ C.I. 26050 ◇ C.I. SOLVENT RED 19 ◇ N-ETHYL-1-((p-(PHENYLAZO)PHENYL)AZO)-2-NAPHTHALENAMINE ◇ N-ETHYL-1-((4-(PHENYLAZO)PHENYL)AZO)-2-NAPHTHALEN AMINE ◇ N-ETHYL-1-((4-(PHENYLAZO)PHENYL)AZO)-2-NAPHTHYLAMINE ◇ FAT RED 7B ◇ HEXATYPE CARMINE B ◇ LACQUER RED V3B ◇ OIL VIOLET ◇ ORGANOL BORDEAUX B ◇ (PHENYLAZO-4-PHENYLAZO)-1-ETHYLAMINO-2-NAPHTHALENE ◇ 1-(4-PHENYLAZO-PHENYLAZO)-2-ETHYLAMINONAPHTHALENE ◇ SOLVENT RED 19 ◇ SPECIAL BLUE X 2137 ◇ SUDAN RED 7B ◇ SUDANROT 7B ◇ TYPOGEN CARMINE

TOXICITY DATA with REFERENCE
mma-sat 500 μg/plate EPASR* 8EHQ-0982-0455

mma-mus:lym 300 mg/L EPASR* 8EHQ-0982-0455
orl-rat TDLo:15 g/kg/50W-C:ETA XPHPAW 149,403,69

CONSENSUS REPORTS: IARC Cancer Review: Group 3 IMEMDT 7,56,87; Animal Inadequate Evidence IMEMDT 8,253,75. Reported in EPA TSCA Inventory.

SAFETY PROFILE: Questionable carcinogen with experimental tumorigenic data. Mutation data reported. When heated to decomposition it emits toxic fumes of NO_x.

EOK000 CAS:50-06-6 ***HR: 3***
5-ETHYL-5-PHENYLBARBITURIC ACID
mf: $C_{12}H_{12}N_2O_3$ mw: 232.26

SYNS: ACIDO-5-FENIL-5-ETILBARBITURICO (ITALIAN) ◇ ADONAL ◇ AGRYPNAL ◇ AMYLOFENE ◇ APHENYLBARBIT ◇ AUSTROMINAL ◇ BARBAPIL ◇ BARBENYL ◇ BARBILEHAE (BARBILETTAE) ◇ BARBIPHENYL ◇ BARBITA ◇ BARBONAL ◇ BARBOPHEN ◇ BARTOL ◇ BIALMINAL ◇ CABRONAL ◇ CALMINAL ◇ CODIBARBITA ◇ CRATECIL ◇ DEZIBARBITUR ◇ DORMIRAL ◇ DUNERYL ◇ ENSODORM ◇ EPIDORM ◇ EPISEDAL ◇ ESKABARB ◇ 5-ETHYL-5-PHENYL-2,4,6-(1H,3H,5H)PYRIMIDINETRIONE ◇ ETILFEN ◇ FENBITAL ◇ FENOBARBITAL ◇ FENYLETTAE ◇ GARDEPANYL ◇ GLYSOLETTEN ◇ HAPLOS ◇ HENNOLETTEN ◇ HYPNOGEN ◇ HYPNOTABLINETTEN ◇ LEFEBAR ◇ LEPHEBAR ◇ LEPINAL ◇ LINASEN ◇ LIQUITAL ◇ LUBERGAL ◇ LUMEN ◇ LUMESETTES ◇ LUMINAL ◇ LUMOFRIDETTEN ◇ LURAMIN ◇ NEUROBARB ◇ NOPTIL ◇ NOVA-PHENO ◇ PARKOTAL ◇ PHENAEMAL ◇ PHENOBAL ◇ PHENOBARBITAL ◇ PHENOBARBITONE ◇ PHENOBARBITURIC ACID ◇ PHENOLURIC ◇ PHENOMET ◇ PHENONYL ◇ PHENOTURIC ◇ PHENYLETHYLBARBITURATE ◇ PHENYL-ETHYL-BARBITURIC ACID ◇ 5-PHENYL-5-ETHYLBARBITURIC ACID ◇ PHENYLETHYLMALONYLUREA ◇ PHENYLETTEN ◇ PHENYRAL ◇ PHOB ◇ POLCOMINAL ◇ PROMPTONAL ◇ SEDA-TABLINEN ◇ SEDICAT ◇ SEDIZORIN ◇ SEDOFEN ◇ SEDONAL ◇ SEDOPHEN ◇ SEVENAL ◇ SK-PHENOBARBITAL ◇ SOMBUTOL ◇ SOMNOLETTEN ◇ SOMNOSAN ◇ SOMONAL ◇ STARIFEN ◇ STENTAL EXTENTABS ◇ TALPHENO ◇ THENOBARBITAL ◇ THEOMINAL ◇ TRIABARB ◇ TRIDEZIBARBITUR ◇ VERSOMNAL ◇ ZADONAL

TOXICITY DATA with REFERENCE
mmo-sat 100 μg/plate MUREAV 147,255,85
cyt-hmn:lym 388 mg/L AGMGAK 21,305,72
otr-ham:emb 100 mg/L PMRSDJ 5,665,85
orl-wmn TDLo:3600 μg/kg (female 26W post):REP AJOGAH 159,1491,88
unr-wmn TDLo:491 mg/kg (female 1-39W post):TER LANCAO 2,839,72
orl-rat TDLo:7560 mg/kg/36W-C:ETA GANNA2 69,679,78
orl-mus TDLo:22 g/kg/1Y-C:NEO JNCIAM 51,1349,73
orl-rat TD:3990 mg/kg/19W-C:CAR CRNGDP 4,935,83
orl-rat TD:2100 mg/kg/12W-C:CAR IGAYAY 123,1069,82
orl-wmn TDLo:46 mg/kg CPEDAM 24,678,85
orl-wmn LDLo:25272 μg/kg/13D-I CMAJAX 33,635,35
orl-chd TDLo:10 mg/kg:CNS AJDCAI 130,507,76
orl-hmn TDLo:18 mg/kg:SKN,PUL,MET JAMAAP 116,700,41
orl-man LDLo:6485 μg/kg/2W-I:SKN,MET AIMEAS 13,1243,40
ims-inf TDLo:50 mg/kg PEDIAU 77,848,86
orl-rat LD50:162 mg/kg TXAPA9 18,185,71
ipr-rat LD50:151 mg/kg ARZNAD 30,477,80
scu-rat LD50:200 mg/kg AEPPAE 152,341,30
ivn-rat LD50:209 mg/kg ARTODN 40,211,78
rec-rat LD50:284 mg/kg AACRAT 46,395,67
orl-mus LD50:137 mg/kg PCJOAU 10,41,76
ipr-mus LD50:128 mg/kg FRPSAX 14,269,59
scu-mus LD50:228 mg/kg ARZNAD 30,477,80
ivn-mus LD50:218 mg/kg AIPTAK 191,405,71
orl-dog LD50:150 mg/kg SMWOAS 85,305,55

CONSENSUS REPORTS: EPA Genetic Toxicology Program. IARC Cancer Review: Group 2B IMEMDT 7,313,87; Human Inadequate Evidence IMEMDT 13,157,77.

SAFETY PROFILE: Suspected carcinogen with experimental carcinogenic, neoplastigenic, tumorigenic, and teratogenic data. A human poison by ingestion. An experimental poison by ingestion, intraperitoneal, subcutaneous, intravenous, and rectal routes. Human systemic effects by ingestion: somnolence, motor activity changes, pulmonary changes, allergic dermatitis and fever. Human reproductive effects by ingestion: drug dependence and other postnatal measures or effects. Human teratogenic effects include developmental abnormalities of the central nervous system, body wall, musculoskeletal, respiratory, gastrointestinal and urogenital systems. Experimental reproductive effects. Human mutation data reported. Used as a drug in the treatment of epilepsy, and as an hypnotic and sedative. When heated to decomposition it emits toxic fumes of NO_x. See also BARBITURATES.

EOK500 CAS:21224-57-7 ***HR: 3***
S-2-((4-(p-ETHYLPHENYL)BUTYL)AMINO) ETHYL THIOSULFATE
mf: $C_{14}H_{23}NO_3S_2$ mw: 317.50

TOXICITY DATA with REFERENCE
orl-mus LD50:1000 mg/kg JMCMAR 11,1190,68
ipr-mus LD50:15 mg/kg JMCMAR 11,1190,68

SAFETY PROFILE: Poison by intraperitoneal route. Moderately toxic by ingestion. When heated to decomposition it emits very toxic fumes of NO_x and SO_x. See also THIOSULFATES.

EOK550 CAS:13037-20-2 ***HR: 3***
ETHYL PHENYLDITHIOCARBAMATE
mf: $C_9H_{11}NS_2$ mw: 197.33

SYNS: S-AETHYL-N-PHENYL-DITHIOCARBAMAT ◇ CARBAMODITHIOIC ACID, PHENYL-, ETHYL ESTER ◇ CARBANILIC ACID,

DITHIO-, ETHYL ESTER ◇ DITHIOCARBANILIC ACID ETHYL ESTER ◇ ETHYL N-PHENYLDITHIOCARBAMATE ◇ ZEPC

TOXICITY DATA with REFERENCE
orl-rat TDLo:1125 mg/kg (female 7-15D post):REP JTSCDR 8,337,83
ipr-rat LDLo:300 mg/kg ARZNAD 16,1092,66

SAFETY PROFILE: Poison by intraperitoneal route. Experimental reproductive effects. When heated to decomposition it emits toxic fumes of SO_x.

EOK600 CAS:121-39-1 ***HR: 2***
ETHYL PHENYLGLYCIDATE
mf: $C_{11}H_{12}O_3$ mw: 192.23

PROP: Colorless liquid; strong strawberry odor. D: 1.120, refr index: 1.516-1.521. Sol in alc, chloroform, ether; insol in water.

SYNS: ETHYL-α,β-EPOXYHYDROCINNAMATE ◇ ETHYL-α,β-EPOXY-α-PHENYLPROPIONATE ◇ ETHYL-3-PHENYLGLYCIDATE ◇ FEMA No. 2454

TOXICITY DATA with REFERENCE
otr-mus:fbr 103 mg/L MUREAV 138,1,84
msc-ham:ovr 103 mg/L MUREAV 138,1,84
orl-rat LD50:2300 mg/kg FCTXAV 13,101,75

CONSENSUS REPORTS: Reported in EPA TSCA Inventory.

SAFETY PROFILE: Moderately toxic by ingestion. Mutation data reported. When heated to decomposition it emits acrid smoke and irritating fumes. See also ESTERS.

EOL000 CAS:631-07-2 ***HR: 3***
ETHYLPHENYLHYDANTOIN
mf: $C_{11}H_{12}N_2O_2$ mw: 204.25

PROP: Colorless, odorless, crystalline powder. Mp: 199-200°.

SYNS: 5-ETHYL-5-PHENYLHYDANTOIN ◇ 5-ETHYL-5-PHENYL-2,4-IMIDAZOLIDINEDIONE ◇ NIRVANOL

TOXICITY DATA with REFERENCE
orl-mus LDLo:500 mg/kg LDTU** -,-,31
scu-mus LDLo:500 mg/kg HDTU** -,-,33
scu-dog LDLo:200 mg/kg HBAMAK 4,1289,35
scu-frg LDLo:100 mg/kg HBAMAK 4,1289,35

SAFETY PROFILE: Poison by subcutaneous route. Moderately toxic by ingestion. A skin, eye, and mucous membrane irritant. Combustible when exposed to heat or flame; can react with oxidizing materials. When heated to decomposition it emits toxic fumes such as NO_x.

EOL050 CAS:65567-32-0 ***HR: 3***
l-5-ETHYL-5-PHENYLHYDANTOIN
mf: $C_{11}H_{12}N_2O_2$ mw: 204.25

SYNS: (-)-5-ETHYL-5-PHENYLHYDANTOIN ◇ HYDANTOIN, 5-ETHYL-5-PHENYL-, (-)- ◇ 2,4-IMIDAZOLIDINEDIONE, 5-ETHYL-5-PHENYL-, (R)- (9CI) ◇ (-)-NIRVANOL ◇ l-NIRVANOL ◇ (R)-NIRVANOL

TOXICITY DATA with REFERENCE
ipr-mus TDLo:97 mg/kg (female 10D post):TER JPETAB 221,228,82
ipr-mus TDLo:97 mg/kg (female 10D post):REP JPETAB 221,228,82
orl-cld TDLo:105 mg/kg:SKN,SYS JPETAB 47,209,33
ipr-mus LD50:500 mg/kg BBIADT 46,623,87
scu-rbt LDLo:125 mg/kg JPETAB 47,209,33
scu-gpg LDLo:400 mg/kg JPETAB 47,209,33

SAFETY PROFILE: Poison by subcutaneous route. Moderately toxic by intraperitoneal route. Human systemic effects by ingestion: allergic dermatitis, increased body temperature. An experimental teratogen. Experimental reproductive effects. When heated to decomposition it emits toxic fumes of NO_x.

EOL100 CAS:86-35-1 ***HR: 2***
3-ETHYL-5-PHENYLHYDANTOIN
mf: $C_{11}H_{12}N_2O_2$ mw: 204.25

PROP: Stout prisms from water. Mp: 94°. Sparingly sol in cold water, more sol in hot water. Freely sol in alc, ether, benzene, dil aq solns of alkali hydroxides.

SYNS: ETHOTOIN ◇ 1-ETHYL-2,5-DIOXO-4-PHENYLIMIDAZOLIDINE ◇ 3-ETHYL-5-PHENYLIMIDAZOLIDIN-2,4-DIONE ◇ PEGANONE ◇ PEGOANONE

TOXICITY DATA with REFERENCE
ipr-mus TDLo:858 mg/kg (8-10D preg):TER TXAPA9 64,271,82
orl-rat LD50:1500 mg/kg NIIRDN 6,117,82
ipr-rat LD50:625 mg/kg NIIRDN 6,117,82
scu-rat LD50:1 g/kg NIIRDN 6,117,82
orl-mus LD50:1750 mg/kg NIIRDN 6,117,82
ipr-mus LD50:923 mg/kg NIIRDN 6,117,82
scu-mus LD50:1060 mg/kg NIIRDN 6,117,82

SAFETY PROFILE: Moderately toxic by ingestion, subcutaneous, and intraperitoneal routes. Experimental teratogenic effects. When heated to decomposition it emits toxic fumes of NO_x.

EOL500 CAS:93-55-0 ***HR: 3***
ETHYL PHENYL KETONE
mf: $C_9H_{10}O$ mw: 134.19

PROP: Water-white to light amber liquid or crystals. Mp: 21°, bp: 218.0°, flash p: 210°F (OC), d: 1.012 @ 20°/20°, vap press: 1 mm @ 50.0°. Insol in water; misc. in alc, ether.

SYNS: PHENYL ETHYL KETONE ◇ 1-PHENYL-1-PROPANONE ◇ PROPIONYLBENZENE ◇ PROPIOPHENONE ◇ USAF EK-1235

TOXICITY DATA with REFERENCE
orl-rat LD50:4490 mg/kg TXAPA9 28,313,74
ipr-mus LD50:100 mg/kg NTIS** AD277-689
scu-mus LD50:2250 mg/kg ARZNAD 5,559,55
skn-rbt LD50:4490 mg/kg TXAPA9 28,313,74

CONSENSUS REPORTS: Reported in EPA TSCA Inventory.

SAFETY PROFILE: Poison by intraperitoneal route. Moderately toxic by subcutaneous route. Mildly toxic by ingestion and skin contact. Combustible when exposed to heat or flame; can react with oxidizing materials. To fight fire use foam, CO_2, dry chemical. When heated to decomposition it emits acrid smoke and irritating fumes. See also KETONES.

EOL600 CAS:69095-83-6 ***HR: 2***
5-(2-ETHYLPHENYL)-3-(3-METHOXYPHENYL)-s-TRIAZOLE
mf: $C_{17}H_{17}N_3O$ mw: 279.37

SYNS: DL 111 ◇ DL-111-IT ◇ (3-ETHYLPHENYL)-5-(3-METHOXYPHENYL)-1,2,4-TRIAZOLE ◇ 3-(o-ETHYLPHENYL)-5-(m-METHOXYPHENYL)-s-TRIAZOLE ◇ 3-(2-ETHYLPHENYL)-5-(3-METHOXYPHENYL)-1H-1,2,4-TRIAZOLE ◇ 5-(2-ETHYLPHENYL)-3-(3-METHOXYPHENYL)-1H-1,2,4-TRIAZOLE

TOXICITY DATA with REFERENCE
scu-ham TDLo:1500 μg/kg (female 5D post):TER CCPTAY 33,263,86
ims-mky TDLo:10 mg/kg (female 53-56D post):REP 45JVAR -,344,80
ipr-rat LD50:1180 mg/kg 45JVAR -,344,80
scu-rat LD50:3190 mg/kg DRFUD4 7,875,82
ims-rat LD50:2000 mg/kg DRFUD4 7,875,82
ipr-mus LD50:650 mg/kg 45JVAR -,344,80
scu-mus LD50:3910 mg/kg DRFUD4 7,875,82
scu-ham LD50:2000 mg/kg DRFUD4 7,875,82

SAFETY PROFILE: Moderately toxic by intraperitoneal, subcutaneous and intramuscular routes. An experimental teratogen. Experimental reproductive effects. Used as a non-hormonal post implantation antifertility agent. When heated to decomposition it emits toxic fumes of NO_x.

EOM000 CAS:2240-14-4 ***HR: 3***
N-ETHYL-3-PHENYL-2-NORBORNANAMINE HYDROCHLORIDE
mf: $C_{15}H_{21}N•ClH$ mw: 251.83

SYNS: 2-AETHYLAMINO-3-PHENYL-NOR-CAMPHAN(GERMAN) ◇ 2-ETHYLAMINO-3-PHENYL-NORCAMPHANE HYDROCHLORIDE ◇ FENCAMFAMINE HYDROCHLORIDE ◇ H 610 ◇ NORCAMPHANE

TOXICITY DATA with REFERENCE
orl-rat LD50:83 mg/kg JOPDAB 69,663,66
scu-rat LD50:69 mg/kg ARZNAD 11,20,61
ivn-rat LD50:24 mg/kg ARZNAD 11,20,61
orl-mus LD50:135 mg/kg ARZNAD 11,20,61
scu-mus LD50:86 mg/kg ARZNAD 11,20,61
ivn-mus LD50:16 mg/kg ARZNAD 11,20,61
orl-dog LD50:30 mg/kg ARZNAD 11,20,61
ivn-dog LD50:15 mg/kg ARZNAD 11,20,61
orl-cat LD50:34 mg/kg ARZNAD 11,20,61

SAFETY PROFILE: Poison by ingestion, subcutaneous, and intravenous routes. A central nervous system stimulant. When heated to decomposition it emits very toxic fumes of HCl and NO_x.

EOM600 CAS:85303-87-3 ***HR: D***
3-(o-ETHYLPHENYL)-5-PIPERONYL-s-TRIAZOLE
mf: $C_{17}H_{15}N_3O_2$ mw: 293.35

TOXICITY DATA with REFERENCE
scu-ham TDLo:150 μg/kg (female 4-8D post):REP RDMIDP 4,237,82

SAFETY PROFILE: Experimental reproductive effects. When heated to decomposition it emits toxic fumes of NO_x.

EOM700 CAS:5556-57-0 ***HR: 2***
(±)-5-ETHYL-5-PHENYLPYRROLID-2-ONE
mf: $C_{12}H_{15}NO$ mw: 189.28

SYN: (±)-5-ETHYL-5-PHENYL-2-PYRROLIDINONE

TOXICITY DATA with REFERENCE
orl-rat LD50:600 mg/kg BJPCAL 25,790,65
orl-mus LD54:812 mg/kg JMCMAR 9,52,66
ipr-mus LD50:750 mg/kg BJPCAL 25,790,65

SAFETY PROFILE: Moderately toxic by ingestion and intraperitoneal routes. When heated to decomposition it emits toxic fumes of NO_x.

EON000 CAS:593-68-0 ***HR: 3***
ETHYL PHOSPHINE
mf: C_2H_7P mw: 62.05

$CH_3CH_2PH_2$

PROP: Colorless liquid. Bp: 25°, d: 1, vap d: 2.14.

SYN: PHOSPHINOETHANE.

SAFETY PROFILE: Probably a poison. A very dangerous fire hazard when exposed to heat or flame; ignites spontaneously in air. Ignites on contact with concentrated acids. To fight fire, use foam, CO_2, dry chemical. Explodes on contact with chlorine; bromine; or fuming nitric acid. Can react vigorously with oxidizing materials. When heated to decomposition it emits highly toxic fumes of PO_x and phosphine. See also PHOSPHINE.

EON500 CAS:35335-60-5 ***HR: 3***
ETHYLPHOSPHONAMIDOTHIONIC ACID-4-(METHYLTHIO)-m-TOLYL ESTER

SYN: BAY HOL 0574

TOXICITY DATA with REFERENCE
orl-bwd LD50:13300 μg/kg AECTCV 12,355,83
orl-bwd LD50:3200 μg/kg TXAPA9 26,154,73
skn-bwd LD50:75 mg/kg TXAPA9 26,154,73

SAFETY PROFILE: Poison by ingestion and skin contact. When heated to decomposition it emits very toxic fumes of PO_x, SO_x, and NO_x. See also ESTERS.

EOO000 CAS:2984-65-8 ***HR: 3***
ETHYLPHOSPHONODITHIOIC ACID-O-METHYL-S-(p-TOLYL) ESTER
mf: $C_{10}H_{15}OPS_2$ mw: 246.34

SYNS: BAY 42903 ◇ BAYER 42903 ◇ ENT 27,250 ◇ O-METHYL-S-(4-METHYLPHENYL) ETHYLPHOSPHONODITHIOATE ◇ N 4328 ◇ STAUFFER N-4328

TOXICITY DATA with REFERENCE
orl-rat LD50:93 mg/kg ARSIM* 20,22,66
orl-mus LDLo:710 mg/kg AECTCV 14,111,85
orl-gpg LDLo:25 mg/kg JEENAI 61,1261,68
scu-gpg LDLo:25 mg/kg JEENAI 61,1261,68
orl-bwd LD50:5600 μg/kg TXAPA9 21,315,72

SAFETY PROFILE: Poison by ingestion and subcutaneous routes. When heated to decomposition it emits very toxic fumes of PO_x and SO_x. See also ESTERS.

EOP500 CAS:50335-09-6 ***HR: 3***
ETHYLPHOSPHONOTHIOIC ACID-(2-DIETHYLAMINOMETHYL)-4,6-DICHLOROPHENYL ESTER

SYN: BAY 50519

TOXICITY DATA with REFERENCE
orl-bwd LD50:4200 μg/kg TXAPA9 26,154,73
skn-bwd LD50:5600 μg/kg TXAPA9 26,154,73

SAFETY PROFILE: Poison by ingestion and skin contact. When heated to decomposition it emits very toxic fumes of PO_x, SO_x, and Cl^-. See also ESTERS.

EOQ000 CAS:1498-40-4 ***HR: 2***
ETHYLPHOSPHONOUS DICHLORIDE
DOT: UN 2845
mf: $C_2H_5Cl_2P$ mw: 130.94

SYNS: DICHLOROETHYLPHOSPHINE ◇ DICHLOROMETHYL PHOSPHINE ◇ ETHYL PHOSPHONOUS DICHLORIDE, anhydrous (DOT) ◇ TL 373

TOXICITY DATA with REFERENCE
ihl-mus LCLo:1990 mg/m^3/10M NDRC** NDCrc-132,Sept,42

DOT Classification: Corrosive Material; Label: Corrosive.

SAFETY PROFILE: Moderately toxic by inhalation. Corrosive. A severe irritant to skin, eyes, and mucous membranes. When heated to decomposition it emits very toxic fumes of PO_x, Cl^-, and phosphine. See also PHOSPHINE.

EOQ500 CAS:36031-66-0 ***HR: 3***
ETHYLPHOSPHORAMIDIC ACID-2,4-DICHLOROPHENYL ESTER
mf: $C_8H_{10}Cl_2NO_3$ mw: 239.09

SYN: DOWCO 161

TOXICITY DATA with REFERENCE
orl-pgn LD50:75 mg/kg TXAPA9 21,315,72
orl-dck LD50:13 mg/kg TXAPA9 21,315,72
orl-bwd LD50:7500 μg/kg TXAPA9 21,315,72

SAFETY PROFILE: Poison by ingestion. When heated to decomposition it emits very toxic fumes of Cl^- and NO_x. See also ESTERS.

EOR000 CAS:1498-51-7 ***HR: 2***
ETHYL PHOSPHORODICHLORIDATE
DOT: NA 1760
mf: $C_2H_5Cl_2O_2P$ mw: 162.94

SYNS: DICHLOROPHOSPHORIC ACID, ETHYL ESTER ◇ PHOSPHORODICHLORIDIC ACID, ETHYL ESTER

CONSENSUS REPORTS: Reported in EPA TSCA Inventory.

DOT Classification: Corrosive Material; Label: Corrosive.

SAFETY PROFILE: A corrosive material which is very toxic to tissue. A severe eye, skin, and mucous membrane irritant. When heated to decomposition it emits very toxic fumes of Cl^- and PO_x.

EOR500 CAS:5022-29-7 ***HR: 3***
N-ETHYLPHTHALIMIDE
mf: $C_{10}H_9NO_2$ mw: 175.20

TOXICITY DATA with REFERENCE
ipr-mus LDLo:128 mg/kg CBCCT* 2,133,50
ivn-mus LD50:320 mg/kg CSLNX* NX#04571

CONSENSUS REPORTS: Reported in EPA TSCA Inventory.

SAFETY PROFILE: Poison by intraperitoneal and intravenous routes. When heated to decomposition it emits toxic fumes of NO_x.

EOS000 CAS:104-90-5 ***HR: 3***
5-ETHYL-α-PICOLINE
DOT: UN 2300
mf: $C_8H_{11}N$ mw: 121.20

$N{=}C(CH_3)CH{-}CHC(CH_2CH_3){=}2CH$

PROP: Liquid. Bp: 174°, d: 0.9184 @ 23°/4°, flash p: 165° (OC).

SYNS: ALDEHYDECOLLIDINE ◇ ALDEHYDINE ◇ COLLIDINE, ALDEHYDECOLLIDINE ◇ 3-ETHYL-6-METHYLPYRIDINE ◇ 5-ETHYL-2-METHYLPYRIDINE ◇ 5-ETHYL-2-PICOLINE ◇ MEP ◇ 2-METHYL-5-ETHYLPYRIDINE ◇ 6-METHYL-3-ETHYLPYRIDINE ◇ METHYL ETHYL PYRIDINE (DOT) ◇ 2-METHYL-5-ETHYLPYRIDINE (DOT)

TOXICITY DATA with REFERENCE
skn-rbt 10 mg/24H open SEV AMIHBC 4,119,51
skn-rbt 500 mg open SEV UCDS** 6/29/66
eye-rbt 250 μg open SEV AMIHBC 4,119,51
orl-rat LD50:368 mg/kg 85GMAT -,83,82
ihl-rat LCLo:1000 ppm/4H AMIHBC 4,119,51
scu-rat LD50:826 mg/kg 85GMAT -,83,82
orl-mus LD50:282 mg/kg 85GMAT -,83,82
scu-mus LD50:294 mg/kg 85GMAT -,83,82
skn-rbt LD50:1000 mg/kg UCDS** 6/29/66

CONSENSUS REPORTS: Reported in EPA TSCA Inventory. EPA Genetic Toxicology Program.

DOT Classification: Corrosive Material; Label: Corrosive; Poison B; Label: St. Andrews Cross.

SAFETY PROFILE: Poison by ingestion and subcutaneous routes. Moderately toxic by skin contact. Mildly toxic by inhalation. Corrosive. A severe skin and eye irritant. Flammable when exposed to heat or flame; can react vigorously with oxidizers. Potentially explosive reaction with nitric acid at 145°C/14.5 bar. To fight fire, use alcohol foam. When heated to decomposition it emits acrid smoke and irritating fumes. See also ALDEHYDES.

EOS100 CAS:72320-60-6 ***HR: 3***
2-(4-ETHYL-1-PIPERAZINYL)-4-PHENYLQUINOLINE DIHYDROCHLORIDE
mf: $C_{21}H_{23}N_3{\cdot}2ClH$ mw: 390.39

SYN: D 1308

TOXICITY DATA with REFERENCE
orl-mus LD50:568 mg/kg AIPTAK 245,283,80
ipr-mus LD50:119 mg/kg AIPTAK 245,283,80
ivn-mus LD50:38 mg/kg AIPTAK 245,283,80

SAFETY PROFILE: Poison by intravenous and intraperitoneal routes. Moderately toxic by ingestion. When heated to decomposition it emits toxic fumes of NO_x and HCl.

EOS500 CAS:766-09-6 ***HR: 3***
1-ETHYLPIPERIDINE
DOT: UN 2386
mf: $C_7H_{15}N$ mw: 113.23

PROP: Flash p: 66.2°F.

SYN: N-AETHYLPIPERIDIN (GERMAN)

TOXICITY DATA with REFERENCE
eye-rbt 50 mg/5M BJIMAG 23,153,66
ivn-mus LD50:56 mg/kg CSLNX* NX#00374
scu-rbt LDLo:100 mg/kg BDCGAS 34,2408,01

CONSENSUS REPORTS: Reported in EPA TSCA Inventory.

DOT Classification: Flammable Liquid; Label: Flammable Liquid.

SAFETY PROFILE: Poison by intravenous and subcutaneous routes. An eye irritant. A very dangerous fire hazard when exposed to heat or flame; can react vigorously with oxidizing materials. When heated to decomposition it emits toxic fumes of NO_x.

EOU500 CAS:77791-38-9 ***HR: 3***
N-ETHYL-2-(PIPERIDINO)-N-(1-(2,4-XYLYLOXY)-2-PROPYL)ACETAMIDE HYDROCHLORIDE
mf: $C_{20}H_{32}N_2O_2{\cdot}ClH$ mw: 369.00
SYN: C 2102

TOXICITY DATA with REFERENCE
eye-rbt 2% MLD ARZNAD 9,70,59
scu-mus LD50:157 mg/kg ARZNAD 9,70,59

SAFETY PROFILE: Poison by subcutaneous route. An eye irritant. When heated to decomposition it emits very toxic fumes of NO_x and HCl.

EOV000 CAS:78219-57-5 ***HR: 3***
1-ETHYL-4-PIPERIDYL-p-AMINOBENZOATE HYDROCHLORIDE
mf: $C_{14}H_{20}N_2O{\cdot}ClH$ mw: 268.82

SYN: p-AMINO-BENZOIC ACID 1-ETHYL-4-PIPERIDYL ESTER, HYDROCHLORIDE

TOXICITY DATA with REFERENCE
ivn-rat LDLo:35 mg/kg JACSAT 51,922,29
scu-mus LDLo:25 mg/kg JACSAT 51,922,29

SAFETY PROFILE: Poison by intravenous and subcutaneous routes. When heated to decomposition it emits very toxic fumes of NO_x and HCl. See also ESTERS.

EOW000 CAS:78219-58-6 ***HR: 3***
1-ETHYL-4-PIPERIDYL BENZOATE HYDROCHLORIDE
mf: $C_{14}H_{19}NO_2{\cdot}ClH$ mw: 269.80

SYN: BENZOIC ACID-1-ETHYL-4-PIPERIDYL ESTER, HYDROCHLORIDE

TOXICITY DATA with REFERENCE
ivn-rat LDLo:25 mg/kg JACSAT 51,922,29
scu-mus LDLo:300 mg/kg JACSAT 51,922,29

SAFETY PROFILE: Poison by intravenous and subcutaneous routes. When heated to decomposition itemits very toxic fumes of NO_x and HCl. See also ESTERS.

EOY000 CAS:129-77-1 ***HR: 3***
1-ETHYL-3-PIPERIDYL DIPHENYLACETATE HYDROCHLORIDE
mf: $C_{21}H_{25}NO_2 \cdot ClH$ mw: 359.93

SYNS: AN 1087 ◇ DACTIL ◇ DACTIL HYDROCHLORIDE ◇ DIPHENYLACETIC ACID-1-ETHYL-3-PIPERIDYL ESTER HYDROCHLORIDE ◇ 1-ETHYL-3-PIPERIDINOL DIPHENYLACETATE HYDROCHLORIDE ◇ 1-ETHYL-3-PIPERIDINYL ESTER-α-PHENYLBENZENEACETIC ACID HYDROCHLORIDE ◇ N-ETHYL-3-PIPERIDYL DIPHENYLACETATE HYDROCHLORIDE ◇ 1-ETHYL-3-PIPERIDYL ESTER-α-PHENYLBENZENEACETIC ACID HYDROCHLORIDE ◇ J.B. 305 ◇ PIPERIDILATE HYDROCHLORIDE ◇ PIPERIDOLATE HYDROCHLORIDE

TOXICITY DATA with REFERENCE
ivn-rat LD50:19 mg/kg 29ZVAB -,97,69
orl-mus LD50:1040 mg/kg CLDND* 104,269,52
ivn-dog LD50:35 mg/kg 29ZVAB -,97,69

SAFETY PROFILE: Poison by intravenous route. Moderately toxic by ingestion. When heated to decomposition it emits very toxic fumes of NO_x and HCl.

EOY100 CAS:78219-33-7 ***HR: 3***
β-2-(N-ETHYLPIPERIDYL)ETHYLBENZOATE HYDROCHLORIDE
mf: $C_{16}H_{23}NO_2 \cdot ClH$ mw: 297.86

SYNS: BENZOIC ACID-2-(1-ETHYL-2-PIPERIDYL)ETHYL ESTER HYDROCHLORIDE ◇ β-(1-ETHYL-2-PIPERIDYL)ETHYL BENZOATE HYDROCHLORIDE

TOXICITY DATA with REFERENCE
scu-mus LDLo:1700 mg/kg ANESAV 1,305,40
ivn-mus LDLo:46 mg/kg JACSAT 61,961,39
ivn-rbt LDLo:20 mg/kg ANESAV 1,305,40
isp-rbt LDLo:29980 μg/kg ANESAV 1,305,40
scu-gpg LDLo:439 mg/kg ANESAV 1,305,40

SAFETY PROFILE: Poison by intravenous and intraspinal routes. Moderately toxic by subcutaneous route. When heated to decomposition it emits toxic fumes of NO_x and HCl.

EOY500 CAS:928-55-2 ***HR: 2***
ETHYL-1-PROPENYL ETHER
mf: $C_5H_{10}O$ mw: 86.15

PROP: Flash p: >19°F (OC). D: 0.8, bp: 158°F.

SYN: ETHYL PROPENYL ETHER

TOXICITY DATA with REFERENCE
skn-rbt 8276 μg/24H open MLD AIHAAP 23,95,62
orl-rat LD50:4660 mg/kg TXAPA9 28,313,74
ihl-rat LCLo:8000 ppm/4H TXAPA9 28,313,74
skn-rbt LD50:3970 mg/kg TXAPA9 28,313,74

SAFETY PROFILE: Moderately toxic by skin contact. Mildly toxic by ingestion and inhalation. A skin irritant. A very dangerous fire hazard when exposed to heat, flame, or oxidizers. To fight fire, use water spray or mist, dry chemical, foam, CO_2. When heated to decomposition it emits acrid smoke and irritating fumes. See also ETHERS.

EPB500 CAS:105-37-3 ***HR: 3***
ETHYL PROPIONATE
DOT: UN 1195
mf: $C_5H_{10}O_2$ mw: 102.15

PROP: Colorless liquid; fruity, rum odor. Bp: 210°F, mp: −72.6°, flash p: 54°F (CC), d: 0.886, refr index: 1.383, autoign temp: 824°F, vap press: 40 mm @ 27.2°, vap d: 3.52, lel: 1.9%, uel: 11%. Misc with alc, ether, propylene glycol; sol in fixed oils; 1 mL in 42 mL water @ 99°.

SYNS: FEMA No. 2456 ◇ PROPIONATE d'ETHYLE (FRENCH) ◇ PROPIONIC ACID, ETHYL ESTER ◇ PROPIONIC ETHER

TOXICITY DATA with REFERENCE
skn-rbt 500 mg/24H MOD FCTXAV 16,749,78
ipr-rat LD50:1200 mg/kg FCTXAV 16,749,78
ipr-mus LD50:1300 mg/kg 14CYAT 2,1879,63
orl-rbt LD50:3500 mg/kg 14CYAT 2,1879,63

CONSENSUS REPORTS: Reported in EPA TSCA Inventory.

DOT Classification: Flammable Liquid; Label: Flammable Liquid.

SAFETY PROFILE: Moderately toxic by ingestion and intraperitoneal routes. A skin irritant. Flammable Liquid. A very dangerous fire and explosion hazard when exposed to heat or flame; can react vigorously with oxidizing materials. To fight fire, use foam, CO_2, dry chemical. When heated to decomposition it emits acrid smoke and irritating fumes. See also ETHERS.

EPC000 CAS:67050-97-9 ***HR: 3***
5-ETHYL-5-(1-PROPYL-1-BUTENYL)BARBITURIC ACID
mf: $C_{13}H_{20}N_2O_3$ mw: 252.35

TOXICITY DATA with REFERENCE
orl-mus LD50:220 mg/kg JACSAT 61,776,39
ipr-mus LD50:130 mg/kg JACSAT 61,776,39

SAFETY PROFILE: Poison by ingestion and intraperi-

toneal routes. When heated to decomposition it emits toxic fumes of NO_x. See also BARBITURATES.

EPC050 CAS:623-85-8 ***HR: 3***
ETHYL-N,N-PROPYLCARBAMATE
mf: $C_6H_13NO_2$ mw: 131.20

SYN: N,N-PROPYL ETHYL CARBAMATE

TOXICITY DATA with REFERENCE
ipr-mus TDLo:6500 mg/kg/13W-I:ETA JNCIAM 9,35,48
scu-mus LD50:540 mg/kg AJEBAK 45,507,67

SAFETY PROFILE: Moderately toxic by subcutaneous route. Questionable carcinogen with experimental tumorigenic data. When heated to decomposition it emits toxic fumes of NO_x. See also CARBAMATES.

EPC100 ***HR: 2***
N-ETHYL-N-PROPYLCARBAMOYL CHLORIDE
mf: $C_6H_{12}ClNO$ mw: 293.61

SAFETY PROFILE: Probably an eye, skin, and mucous membrane irritant. Vigorous reaction with water evolves large amounts of gases. Releases HCl in contact with water. When heated to decomposition it emits toxic fumes of Cl^- and NO_x.

EPC115 ***HR: 3***
ETHYL N-((5R,8S,10R)-6-PROPYL-8-ERGOLINYL) CARBAMATE
mf: $C_{20}H_{27}N_3O_2$ mw: 341.50

SYNS: ERGOLINE-8-CARBAMIC ACID, 6-PROPYL-, ETHYL ESTER, (8S)- ◇ (8S)-6-PROPYLERGOLINE-8-CARBAMIC ACID ETHYL ESTER

TOXICITY DATA with REFERENCE
mmo-sat 400 μg/L CCCCAK 52,2983,87
orl-rat TDLo:20 μg/kg (female 5D post):REP CCCCAK 52,2983,87
ivn-mus LD50:11400 μg/kg CCCCAK 52,2983,87

SAFETY PROFILE: Poison by intraperitoneal route. Experimental reproductive effects. Mutation data reported. When heated to decomposition it emits toxic fumes of NO_x.

EPC125 CAS:628-32-0 ***HR: 3***
ETHYL PROPYL ETHER
DOT: UN 2615
mf: $C_5H_{12}O$ mw: 88.15

PROP: D: 0.8: bp: 147°F; flash p: < −4°F; lel: 1.7%; uel: 9%.

SYNS: 1-ETHOXYPROPANE ◇ PROPYLETHYL ETHER

DOT Classification: Flammable Liquid; Label: Flammable Liquid.

SAFETY PROFILE: Very dangerous fire and explosion hazard when exposed to heat or open flame. To fight fire, use alcohol foam. When heated to decomposition it emits acrid smoke and fumes. See also ETHERS.

EPC135 CAS:66859-63-0 ***HR: 2***
2-ETHYL-2-PROPYLTHIOBUTYRAMIDE
mf: $C_9H_{19}NOS$ mw: 189.35

TOXICITY DATA with REFERENCE
orl-mus LD50:1576 mg/kg JMCMAR 6,351,63
ipr-mus LD50:419 mg/kg JMCMAR 6,351,63

SAFETY PROFILE: Moderately toxic by intraperitoneal route. Mildly toxic by ingestion. When heated to decomposition it emits toxic fumes of NO_x and SO_x.

EPC150 CAS:55365-87-2 ***HR: 3***
O-ETHYL PROPYLTHIOCARBAMATE
mf: $C_6H_{13}NOS$ mw: 147.26

SYN: EPTAM

TOXICITY DATA with REFERENCE
orl-rat LD50:1660 mg/kg 85GMAT -,68,82
skn-rat LD50:3200 mg/kg 85GMAT -,69,82
orl-mus LD50:750 mg/kg 85GMAT -,68,82
orl-cat LD50:112 mg/kg 85GMAT -,68,82

SAFETY PROFILE: Poison by ingestion. Moderately toxic by skin contact. When heated to decomposition it emits toxic fumes of SO_x and NO_x. See also CARBAMATES and ESTERS.

EPC175 CAS:38524-82-2 ***HR: 3***
O-ETHYL-S-PROPYL-O-(2,4,6-TRICHLOROPHENYL)PHOSPHOROTHIOATE
mf: $C_{11}H_{14}Cl_3O_3PS$ mw: 363.63

SYNS: ENT 29,118 ◇ RH-8218 ◇ ROHM & HAAS RH-218 ◇ TH-218

TOXICITY DATA with REFERENCE
orl-rat LD50:200 mg/kg SPEADM 78-1,35,78
skn-rat LD50:250 mg/kg SPEADM 78-1,35,78
skn-rbt LD50:108 mg/kg SPEADM 78-1,35,78

SAFETY PROFILE: Poison by ingestion and skin contact. When heated to decomposition it emits toxic fumes of PO_x, Cl^-, and SO_x.

EPC200 CAS:36950-85-3 ***HR: D***
20-ETHYLPROSTAGLANDIN F2-α
mf: $C_{22}H_{38}O_5$ mw: 382.60

SYNS: 20-ETHYL-PGF2-α ◇ ICI 74205 ◇ racemic-ICI 74205

TOXICITY DATA with REFERENCE
scu-ham TDLo:2500 μg/kg (female 14D post):REP PRGLBA 2,375,72

SAFETY PROFILE: Experimental reproductive effects.

When heated to decomposition it emits acrid smoke and irritating fumes.

EPC500 CAS:297-97-2 ***HR: 3***
ETHYL PYRAZINYL PHOSPHOROTHIOATE
mf: $C_8H_{13}N_2O_3PS$ mw: 248.26

PROP: Amber liquid. Mp: −1.7°, bp: 80°, n (25/D) 1.5131, vap press @ 30°: 0.003 mm Hg. Sltly sol in water; misc with most organic solvents.

SYNS: AC 18133 ◇ AMERCIAN CYANAMID 18133 ◇ CL 18133 ◇ CYNEM ◇ O,O-DIAETHYL-O-(PYRAZIN-2YL)-MONOTHIOPHOSPHAT (GERMAN) ◇ O,O-DIAETHYL-O-(2-PYRAZINYL)-THIONOPHOSPHAT (GERMAN) ◇ O,O-DIETHYL-O,2-PYRAZINYL PHOSPHOROTHIOATE ◇ DIETHYL-O-2-PYRAZINYL PHOSPHOROTHIONATE ◇ O,O-DIETHYL-O-2-PYRAZINYL PHOSPHOTHIONATE ◇ O,O-DIETHYL-O-PYRAZINYL THIOPHOSPHATE ◇ EN 18133 ◇ ENT 25,580 ◇ EXPERIMENTAL NEMATOCIDE 18,133 ◇ NEMAFOS ◇ NEMAPHOS ◇ NEMATOCIDE ◇ PHOSPHOROTHIOIC ACID-O,O-DIETHYL-O-2-PYRAZINYL ESTER ◇ PYRAZINOL-O-ESTER with O,O-DIETHYL PHOSPHOROTHIOATE ◇ RCRA WASTE NUMBER P404 ◇ THIONAZIN ◇ ZINOPHOS

TOXICITY DATA with REFERENCE
orl-rat LD50:3500 μg/kg TXAPAO 14,515,69
skn-rat LD50:8 mg/kg SPEADM 78-1,46,78
ocu-rbt LDLo:50 mg/kg 85GYAZ -,36,71
skn-gpg LD50:10 mg/kg GUCHAZ 6,498,73
orl-pgn LD50:2370 μg/kg ASTTA8 (680),157,79
orl-qal LD50:3160 μg/kg ASTTA8 (680),157,79
orl-dck LD50:1680 μg/kg TXAPA9 47,451,79
skn-dck LD50:7 mg/kg TXAPA9 47,451,79
orl-bwd LD50:2420 μg/kg ASTTA8 (680),157,79

CONSENSUS REPORTS: EPA Extremely Hazardous Substances List. Reported in EPA TSCA Inventory. Community Right-To-Know List.

SAFETY PROFILE: Poison by ingestion, skin contact, and ocular routes. A cholinesterase inhibitor type of insecticide. When heated to decomposition it emits highly toxic fumes of NO_x, PO_x, and SO_x. See also PARATHION.

EPC700 CAS:2687-91-4 ***HR: 2***
1-ETHYL-2-PYRROLIDINONE
mf: $C_6H_{11}NO$ mw: 113.18

PROP: Bp: 97°/20 mm, d: 0.992, Flash p: 169° F.

SYNS: N-ETHYLPYRROLIDINONE ◇ N-ETHYLPYRROLIDONE ◇ 2-PYRROLIDINONE, 1-ETHYL-

TOXICITY DATA with REFERENCE
eye-rbt 100 mg MOD FCTOD7 26,475,88
orl-rat LD50:1350 mg/kg FCTOD7 26,475,88

CONSENSUS REPORTS: Reported in EPA TSCA Inventory.

SAFETY PROFILE: Moderately toxic by ingestion. An eye irritant. Combustible. When heated to decomposition it emits toxic fumes of NO_x.

EPC875 ***HR: 3***
N-ETHYL-2-(PYRROLIDINYL)-o-ACETOTOLUIDIDE HYDROCHLORIDE
mf: $C_{15}H_{22}N_2O{\bullet}ClH$ mw: 282.85

SYN: C 5124

TOXICITY DATA with REFERENCE
eye-rbt 2% MLD ARZNAD 8,609,58
ipr-rat LD50:63 mg/kg ARZNAD 8,609,58
scu-mus LD50:87 mg/kg ARZNAD 8,609,58

SAFETY PROFILE: Poison by subcutaneous and intraperitoneal routes. An eye irritant. When heated to decomposition it emits toxic fumes of NO_x and HCl.

EPC950 CAS:75236-19-0 ***HR: 2***
4′-ETHYL-4-N-PYRROLIDINYLAZOBENZENE
mf: $C_{18}H_{21}N_3$ mw: 279.42

SYN: PYRROLIDINE,1-(p-((p-ETHYLPHENYL)AZO)PHENYL)-

TOXICITY DATA with REFERENCE
mma-sat 10 μg/plate CRNGDP 3,559,82
orl-rat TDLo:11 g/kg/34W C:CAR CRNGDP 1,419,80

SAFETY PROFILE: Questionable carcinogen with experimental carcinogenic data. Mutation data reported. When heated to decomposition it emits toxic fumes of NO_x.

EPC999 CAS:19137-90-7 ***HR: 3***
3(1-ETHYL-2-PYRROLIDINYL)INDOLE HYDROCHLORIDE
mf: $C_{14}H_{18}N_2{\bullet}ClH$ mw: 250.80

TOXICITY DATA with REFERENCE
ipr-rat LD50:71 mg/kg JMCMAR 7,415,64
ipr-mus LD50:56 mg/kg JMCMAR 7,415,64

SAFETY PROFILE: Poison by intraperitoneal route. When heated to decomposition it emits toxic fumes of NO_x and Cl^-.

EPD100 CAS:53583-79-2 ***HR: 3***
N-((1-ETHYL-2-PYRROLIDINYL)METHYL)-5-(ETHYLSULFONYL)-o-ANISAMIDE
mf: $C_{17}H_{26}N_2O_4S$ mw: 354.51

SYNS: o-ANISAMIDE,N-((1-ETHYL-2-PYRROLIDINYL)METHYL)-5-(ETHYLSULFONYL)- ◇ BARNETIL ◇ BENZAMIDE, N-((1-ETHYL-2-PYRROLIDINYL)METHYL)-5-(ETHYLSULFONLY)-2-METHOXY-◇ LIN 1418 ◇ SULTOPRIDE

TOXICITY DATA with REFERENCE
scu-rat TDLo:51300 μg/kg (female 19D pre):REP THERAP 30,231,75

orl-mus LD50:665 mg/kg DRFUD4 10,758,85
ipr-mus LD50:300 mg/kg EJMCA5 17,437,82

SAFETY PROFILE: Poison by intraperitoneal route. Moderately toxic by ingestion. Experimental reproductive effects. When heated to decomposition it emits toxic fumes of NO_x and SO_x.

EPD500 CAS:15676-16-1 ***HR: 3***
N-((1-ETHYL-2-PYRROLIDINYL)METHYL)-5-SULFAMOYL-o-ANISAMIDE
mf: $C_{15}H_{23}N_3O_4S$ mw: 341.47

SYNS: ABILIT ◇ AIGLONYL ◇ 5-(AMINOSULFONYL)-N-((1-ETHYL-2-PYRROLIDINYL)METHYL)-2-METHOXYBENZAMIDE ◇ COOLSPAN ◇ DOBREN ◇ DOGMATIL ◇ DOGMATYL ◇ EGLONYL ◇ N-((1-ETHYL-2-PYRROLIDINYL)METHYL)-2-METHOXY-5-SULFAMOYLBENZAMIDE ◇ GUASTIL ◇ MIRADOL ◇ MIRBANIL ◇ MISULVAN ◇ OMPERAN ◇ PYRKAPPL ◇ R.D. 1403 ◇ SERNEVIN ◇ SPLOTIN ◇ SULPIRID ◇ SULPIRIDE ◇ SULPYRID ◇ SURSUMID ◇ TRILAN

TOXICITY DATA with REFERENCE
scu-rat TDLo:181 mg/kg (female 1D pre):REP THERAP 30,231,75
orl-hmn TDLo:2143 μg/kg/D:CNS,GIT,MET JMGZAI 10(7),11,73
ims-hmn TDLo:1429 μg/kg/D:CNS,GIT,MET JMGZAI 10(7),11,73
orl-rat LD50:9800 mg/kg JMGZAI 10(7),11,73
ipr-rat LD50:210 mg/kg NIIRDN 6,384,82
scu-rat LD50:360 mg/kg NIIRDN 6,384,82
ivn-rat LD50:40 mg/kg IYKEDH 4,193,73
orl-mus LD50:1700 mg/kg NIIRDN 6,384,82
ipr-mus LD50:231 mg/kg ARZNAD 27,404,77
scu-mus LD50:290 mg/kg NIIRDN 6,384,82
ivn-mus LD50:48 mg/kg IYKEDH 4,193,73
orl-dog LD50:2 g/kg JMGZAI 10(7),11,73
scu-dog LD50:350 mg/kg JMGZAI 10(7),11,73
ivn-dog LD50:137 mg/kg JMGZAI 10(7),11,73
orl-rbt LD50:4 g/kg JMGZAI 10(7),11,73
scu-rbt LD50:2 g/kg JMGZAI 10(7),11,73
ivn-rbt LD50:63 mg/kg JMGZAI 10(7),11,73

SAFETY PROFILE: Poison by intraperitoneal, subcutaneous and intravenous routes. Moderately toxic by ingestion. Human systemic effects by ingestion and intramuscular route: tremors, hypermotility, diarrhea and fever. Experimental reproductive effects. Used as a digestive aid and psychotropic drug. When heated to decomposition it emits very toxic fumes of NO_x and SO_x.

EPF500 CAS:7648-01-3 ***HR: 3***
3-ETHYLRHODANINE
mf: $C_5H_7NOS_2$ mw: 161.25

TOXICITY DATA with REFERENCE
ipr-rat LDLo:150 mg/kg ARZNAD 19,558,69
ivn-mus LD50:180 mg/kg CSLNX* NX#03762

CONSENSUS REPORTS: Reported in EPA TSCA Inventory.

SAFETY PROFILE: Poison by intravenous and intraperitoneal routes. When heated to decomposition it emits very toxic fumes of NO_x and SO_x.

EPF550 CAS:78-10-4 ***HR: 3***
ETHYL SILICATE
DOT: UN 1292
mf: $C_8H_{20}O_4Si$ mw: 208.37

PROP: Colorless, flammable liquid. Mp: −77°, bp: 165-166°. flash p: 125°F (52°C), d: (20/4) 0.933, n (25/D) 1.3818. Viscosity 0.6 cps. Practically insol in water with slow decomp. Miscible with alc.

SYNS: ETHYL ORTHOSILICATE ◇ ETYLU KRZEMIAN (POLISH) ◇ EXTREMA ◇ SILICATE D'ETHYLE (FRENCH) ◇ SILICIC ACID TETRAETHYL ESTER ◇ TEOS ◇ TETRAETHOXYSILANE ◇ TETRAETHYL ORTHOSILICATE ◇ TETRAETHYL ORTHOSILICATE (DOT) ◇ TETRAETHYL SILICATE ◇ TETRAETHYL SILICATE (DOT)

TOXICITY DATA with REFERENCE
eye-hmn 3000 ppm JIHTAB 22,288,40
skn-rbt 500 mg/24H MOD UCDS** 7/23/70
eye-rbt 100 mg MLD UCDS** 7/23/70
eye-gpg 2500 ppm/2H SEV JIHTAB 22,288,40
orl-rat LD50:6270 mg/kg JIHTAB 31,60,49
ihl-rat LCLo:1000 ppm/4H JIHTAB 31,343,49
skn-rbt LD50:5878 mg/kg UCDS** 7/23/70
ivn-rbt LDLo:200 mg/kg INMEAF 6,660,37
ihl-gpg LCLo:700 ppm/6H JIHTAB 22,288,40

CONSENSUS REPORTS: Reported in EPA TSCA Inventory.
OSHA PEL: (Transitional: TWA 100 ppm) TWA 10 ppm
ACGIH TLV: TWA 10 ppm
DFG MAK: 100 ppm (850 mg/m^3)
DOT Classification: Flammable or Combustible Liquid; Label: Flammable Liquid.

SAFETY PROFILE: Poison by intravenous route. Moderately toxic by other routes. A skin, mucous membrane and severe eye irritant. Narcotic in high concentrations. Flammable when exposed to heat or flame; can react vigorously with oxidizing materials. When heated to decomposition it emits acrid smoke and fumes. See also ESTERS.

EPF600 CAS:676-54-0 ***HR: 3***
ETHYL SODIUM
mf: C_2H_5Na mw: 52.05

SAFETY PROFILE: The powder ignites spontaneously in air. When heated to decomposition it emits toxic fumes of Na_2O.

EPG000 CAS:7525-62-4 ***HR: 1***
m-ETHYLSTYRENE
mf: $C_{10}H_{12}$ mw: 132.22

PROP: Water-white liquid. Bp: 191.5°, fp: −127°, d: 0.8955 @ 20°, vap press: 2.17 mm @ 40°, vap d: 4.56.

SYN: m-ETHYL VINYLBENZEN (CZECH)

TOXICITY DATA with REFERENCE
eye-rbt 500 mg/24H MLD 28ZPAK -,24,72
orl-rat LD50:4360 mg/kg 28ZPAK -,24,72

SAFETY PROFILE: Mildly toxic by ingestion. An eye irritant. Flammable when exposed to heat or flame; can react with oxidizing materials. To fight fire, use foam, CO_2, dry chemical. When heated to decomposition it emits acrid smoke and irritating fumes.

EPH000 CAS:352-93-2 ***HR: 1***
ETHYL SULFIDE
DOT: UN 2375
mf: $C_4H_{10}S$ mw: 90.20

PROP: Liquid, garlic-like odor. Mp: −102°, bp: 92-93°, d: 0.837 @ 20°/4°, vap d: 3.11. Flash p: 14°F.

SYNS: DIETHYLSULFID (CZECH) ◇ DIETHYL SULFIDE (DOT) ◇ DIETHYLTHIOETHER ◇ ETHYL MONOSULFIDE ◇ ETHYLTHIOETHANE ◇ ETHYL THIOETHER ◇ SULFODOR (CZECH) ◇ 3-THIAPENTANE ◇ 1,1′-THIOBISETHANE ◇ THIOETHYL ETHER

TOXICITY DATA with REFERENCE
skn-rbt 500 mg/24H MOD 28ZPAK -,170,72
eye-rbt 20 mg/24H MOD 28ZPAK -,170,72
orl-rat LD50:5930 mg/kg 28ZPAK -,170,72

CONSENSUS REPORTS: Reported in EPA TSCA Inventory.

DOT Classification: Flammable Liquid; Label: Flammable Liquid and Poison.

SAFETY PROFILE: Mildly toxic by ingestion. A skin and eye irritant. A very dangerous fire hazard when exposed to heat, flame, or sparks; can react vigorously with oxidizers. Reacts with water, steam, acids, or acid fumes to produce toxic and flammable vapors. To fight fire, use water spray or mist, dry chemical, CO_2, foam. When heated to decomposition it yields highly toxic fumes of SO_x. See also SULFIDES.

EPI000 CAS:513-12-2 ***HR: 1***
ETHYLSULFONYLETHANOL
mf: $C_4H_{10}O_3S$ mw: 138.20

PROP: Hygroscopic crystals, water-sol. D: 1.24 @ 20°/4°, mp: 43°, bp: 153° @ 2.5 mm, flash p: 371°F.

SYNS: ESE ◇ 2-(ETHYLSULFONYL)ETHANOL ◇ ETHYLSULFONYLETHYL ALCOHOL

TOXICITY DATA with REFERENCE
orl-rat LD50:18 g/kg 34ZIAG -,259,69
orl-mus LDLo:14 g/kg 34ZIAG -,259,69
ipr-mus LDLo:10 g/kg 34ZIAG -,259,69

SAFETY PROFILE: Mildly toxic by ingestion and intraperitoneal routes. Combustible when exposed to heat or flame. To fight fire, use water, foam, CO_2, dry chemical. When heated to decomposition it emits highly toxic fumes of SO_x.

EPI300 CAS:842-00-2 ***HR: 3***
4-(ETHYLSULFONYL)-1-NAPHTHALENE SULFONAMIDE
mf: $C_{12}H_{13}NO_4S_2$ mw: 299.38

SYNS: ENS ◇ 4-ETHYLSULPHONYLNAPHTHALENE-1-SULFONAMIDE ◇ 4-ETHYLSULPHONYLNAPHTHALENE-1-SULPHONAMIDE ◇ HPA

TOXICITY DATA with REFERENCE
dns-mus-orl 20 mg/kg BIJOAK 111,12P,69
oms-mus-orl 20 mg/kg BIJOAK 111,12P,69
cyt-mus-orl 40 mg/kg CTKIAR 2,249,69
orl-mus TDLo:4200 mg/kg/50W-C:CAR BJCAAI 23,772,69
imp-mus TDLo:77 mg/kg:CAR BJCAAI 19,311,65
orl-mus TD:440 mg/kg/52W-C:NEO JNCIAM 51,2007,73
orl-mus TD:5460 mg/kg/65W-C:ETA BCPCA6 16,619,67

SAFETY PROFILE: Questionable carcinogen with experimental carcinogenic, neoplastigenic, and tumorigenic data. Mutation data reported. When heated to decomposition it emits very toxic fumes of NO_x and SO_x.

EPI500 CAS:70-29-1 ***HR: 2***
ETHYL SULFOXIDE
mf: $C_4H_{10}OS$ mw: 106.20

SYNS: DESO ◇ DIETHYL SULPHOXIDE ◇ ETHANE-1,1′-SULFINYLBIS

TOXICITY DATA with REFERENCE
orl-rat LD50:5650 mg/kg COREAF 260,327,65
ipr-rat LD50:4370 mg/kg COREAF 260,327,65
ivn-rat LD50:4990 mg/kg COREAF 260,327,65
orl-mus LD50:3610 mg/kg COREAF 260,327,65
ipr-mus LD50:2500 mg/kg IJRBA3 3,41,61
ivn-mus LD50:4370 mg/kg COREAF 260,327,65

SAFETY PROFILE: Moderately toxic by ingestion and intraperitoneal routes. Mildly toxic by intravenous route. When heated to decomposition it emits toxic fumes of SO_x.

EPJ000 CAS:20941-65-5 ***HR: 3***
ETHYL TELLURAC
mf: $C_{20}H_{40}N_4S_8{\cdot}Te$ mw: 720.72

PROP: Orange-yellow powder. D: 1.44, mp: 108-118°.

SYNS: DIETHYLDITHIO CARBAMIC ACID TELLURIUM SALT ◇ NCI-C02857 ◇ TELLURIUM DIETHYLDITHIOCARBAMATE ◇ TETRAKIS(DIETHYLCARBAMODITHIOATO-S,S')TELLURIUM ◇ TETRAKIS(DIETHYLDITHIOCARBAMATO)TELLURIUM

TOXICITY DATA with REFERENCE
orl-mus TDLo:113 g/kg/107W-C:ETA NCITR* NCI-CG-TR-152,79

CONSENSUS REPORTS: IARC Cancer Review: Group 3 IMEMDT 7,56,87; Animal Inadequate Evidence IMEMDT 12,115,76. NCI Carcinogenesis Bioassay (feed); No Evidence: mouse, rat NCITR* NCI-CG-TR-152,79; Results Indefinite: Mouse, Rat NCITR* NCI-CG-TR-152,79. Reported in EPA TSCA Inventory.

OSHA PEL: TWA 0.1 mg(Te)/m^3
ACGIH TLV: TWA 0.1 mg(Te)/m^3

SAFETY PROFILE: Questionable carcinogen with experimental tumorigenic data. When heated to decomposition it emits very toxic fumes of NO_x, SO_x, and Te. See also TELLURIUM COMPOUNDS and CARBAMATES.

EPJ600 CAS:56501-31-6 ***HR: 3***
1-ETHYL-1-TETRADECYLPIPERIDINIUM BROMIDE
mf: $C_{21}H_{44}N\cdot Br$ mw: 309.57

TOXICITY DATA with REFERENCE
orl-mus LD50:140 mg/kg PSDTAP 15,331,74
ipr-mus LD50:11984 μg/kg PSDTAP 15,331,74
ivn-mus LD50:5091 μg/kg PSDTAP 15,331,74

SAFETY PROFILE: Poison by ingestion, intraperitoneal, and intravenous routes. When heated to decomposition it emits toxic fumes of NO_x and Br^-.

EPK000 CAS:15547-17-8 ***HR: 3***
2-ETHYL-5,6,7,8-TETRAHYDROANTHRAQUINONE
mf: $C_{16}H_{16}O_2$ mw: 240.32

SYNS: 6-ETHYL-1,2,3,4-TETRAHYDRO-9,10-ANTHRACENEDIONE ◇ USAF SO-2

TOXICITY DATA with REFERENCE
ipr-mus LD50:200 mg/kg NTIS** AD277-689

CONSENSUS REPORTS: Reported in EPA TSCA Inventory.

SAFETY PROFILE: Poison by intraperitoneal route. When heated to decomposition it emits acrid smoke and irritating fumes.

EPL000 CAS:101651-73-4 ***HR: 3***
1-ETHYL-2,2,6,6-TETRAMETHYL-4-(N-ACETYL-N-PHENYL)PIPERIDINE
mf: $C_{19}H_{30}N_2O$ mw: 302.51

SYNS: 4-(N-ACETYL)PHENYLAMINO-1-ETHYL-2,2,6,6-TETRAMETHYLPIPERIDINE ◇ N-(1-ETHYL-2,2,6,6-TETRAMETHYL-PIPERIDIN-4-YL)ACETANILIDE

TOXICITY DATA with REFERENCE
scu-mus LD50:590 mg/kg PCJOAU 8,82,74
ivn-mus LD50:86 mg/kg PCJOAU 8,82,74

SAFETY PROFILE: Poison by intravenous route. Moderately toxic by subcutaneous route. When heated to decomposition it emits toxic fumes of NO_x.

EPL600 ***HR: 3***
1-ETHYL-2,2,6,6-TETRAMETHYLPIPERIDINE HYDROCHLORIDE
mf: $C_{11}H_{23}N\cdot ClH$ mw: 205.81

SYN: COMPOUND 26539 HYDROCHLORIDE

TOXICITY DATA with REFERENCE
orl-mus LD50:180 mg/kg BJPCAL 13,501,58
ipr-mus LD50:58 mg/kg BJPCAL 13,501,58
ivn-mus LD50:52 mg/kg BJPCAL 13,501,58

SAFETY PROFILE: Poison by ingestion, intravenous and intraperitoneal routes. When heated to decomposition it emits toxic fumes of NO_x and HCl.

EPL700 ***HR: 3***
1-ETHYL-2,2,6,6-TETRAMETHYLPIPERIDINE HYDROGEN TARTRATE
mf: $C_{11}H_{23}N\cdot C_4H_6O_4$ mw: 287.45

SYN: 1-ETHYL-2,2,6,6-TETRAMETHYL-PIPERIDINETARTRATE

TOXICITY DATA with REFERENCE
orl-mus LD50:236 mg/kg BJPCAL 13,501,58
ipr-mus LD50:73 mg/kg BJPCAL 13,501,58
ivn-mus LD50:63 mg/kg BJPCAL 13,501,58

SAFETY PROFILE: Poison by ingestion, intravenous, and intraperitoneal routes. When heated to decomposition it emits toxic fumes of NO_x.

EPM000 CAS:52098-56-3 ***HR: 3***
1-ETHYL-2,2,6,6-TETRAMETHYL-4-(N-PROPIONYL-N-BENZYLAMINO)PIPERIDINE
mf: $C_{21}H_{34}N_2O$ mw: 330.57

SYN: 4-(N-PROPIONYL)BENZYLIMINO-1-ETHYL-2,2,6,6-TETRAMETHYLPIPERIDINE

TOXICITY DATA with REFERENCE
scu-mus LD50:520 mg/kg PCJOAU 8,82,74
ivn-mus LD50:100 mg/kg PCJOAU 8,82,74

SAFETY PROFILE: Poison by intravenous route.

Moderately toxic by subcutaneous route. When heated to decomposition it emits toxic fumes of NO_x.

EPM550 ***HR: 3***
1-ETHYL-1,1,3,3-TETRAMETHYLTETRAZENIUM
mf: $C_6H_{17}BF_4N_4$ mw: 232.03

$(CH_3)_2NN{=}NN^-CH_2CH_2(CH_3)_2BF_4^-$

SAFETY PROFILE: An explosive salt. When heated to decomposition it emits toxic fumes of F^- and NO_x. See also BORON COMPOUNDS.

EPM590 ***HR: 3***
2-ETHYLTETRAZOLE
mf: $C_3H_6N_4$ mw: 98.11

$CH{=}NNCH_2CH_2N{=}N$ (ring)

SAFETY PROFILE: Explodes on contact with aluminum hydride. When heated to decomposition it emits toxic fumes of NO_x.

EPM600 CAS:50764-78-8 ***HR: 3***
5-ETHYLTETRAZOLE
mf: $C_3H_6N_4$ mw: 98.11

$HNN{=}NN{=}CCH_2CH_3$ (ring)

SAFETY PROFILE: Explodes on contact with aluminum hydride. When heated to decomposition it emits toxic fumes of NO_x.

EPN500 CAS:4147-51-7 ***HR: 1***
2-ETHYLTHIO-4,6-BIS(ISOPROPYLAMINO)-s-TRIAZINE
mf: $C_{11}H_{21}N_5S$ mw: 255.43

SYNS: 2,4-BIS(ISOPROPYLAMINO)-6-ETHYLTHIO-s-TRIAZINE ◇ COTOFOR ◇ DIPROPETRYN ◇ DIPROPETRYNE ◇ 6-(ETHYLTHIO) N,N'-BIS(1-METHYLETHYL)-1,3,5-TRIAZINE-2,4-DIAMINE ◇ GS-16068 ◇ SANCAP

TOXICITY DATA with REFERENCE
orl-rat LD50:7144 mg/kg FMCHA2 -,C210,83
skn-rbt LD50:10 g/kg CIGET* -,-,77

SAFETY PROFILE: Mildly toxic by ingestion and skin contact. An herbicide. When heated to decomposition it emits very toxic fumes of NO_x and SO_x.

EPP000 CAS:542-90-5 ***HR: 3***
ETHYL THIOCYANATE
mf: C_3H_5NS mw: 87.15

PROP: D: 0.996, mp: −85.5, bp: 145°. Insol in water; misc in alc and ether.

SYNS: AETHYLRHODANID (GERMAN) ◇ ETHYL RHODANATE ◇ ETHYL SULFOCYANATE ◇ THIOCYANATOETHANE ◇ THIOCYANIC ACID, ETHYL ESTER

TOXICITY DATA with REFERENCE
orl-rat LDLo:200 mg/kg JIHTAB 18,310,36
scu-rat LDLo:40 mg/kg JIHTAB 18,310,36
ipr-mus LD50:10 mg/kg JJPAAZ 3,99,54
scu-mus LDLo:70 mg/kg JJPAAZ 3,99,54
ivn-mus LD50:18 mg/kg CSLNX* NX#02865
orl-cat LDLo:10 mg/kg JIHTAB 18,310,36
scu-rbt LDLo:15 mg/kg AEPPAE 150,257,30

CONSENSUS REPORTS: Community Right-To-Know List. Reported in EPA TSCA Inventory.

SAFETY PROFILE: Poison by ingestion, subcutaneous, intraperitoneal, and intravenous routes. When heated to decomposition it emits very toxic fumes of NO_x and SO_x. See also THIOCYANATES.

EPP500 CAS:110-77-0 ***HR: 2***
2-(ETHYLTHIO)ETHANOL
mf: $C_4H_{10}OS$ mw: 106.20

SYNS: ETHYL-2-HYDROXYETHYL SULFIDE ◇ ETHYL-2-HYDROXYETHYL THIOETHER ◇ β-ETHYLMERKAPTOETHANOL (CZECH) ◇ β-HYDROXYDIETHYL SULFIDE

TOXICITY DATA with REFERENCE
skn-rbt 500 mg/24H SEV 28ZPAK -,171,72
eye-rbt 750 μg/24H SEV 28ZPAK -,171,72
orl-rat LD50:2320 mg/kg 28ZPAK -,171,72

CONSENSUS REPORTS: Reported in EPA TSCA Inventory.

SAFETY PROFILE: Moderately toxic by ingestion. A severe skin and eye irritant. When heated to decomposition it emits toxic fumes of SO_x. See also SULFIDES.

EPQ000 CAS:536-33-4 ***HR: 3***
2-ETHYLTHIOISONICOTINAMIDE
mf: $C_8H_{10}N_2S$ mw: 166.26

SYNS: AETINA ◇ AETIVA ◇ AMIDAZIN ◇ BAYER 5312 ◇ ETH ◇ ETHIMIDE ◇ ETHINA ◇ ETHINAMIDE ◇ ETHIONIAMIDE ◇ α-ETHYLISONICOTINIC ACID THIOAMIDE ◇ 2-ETHYLISONICOTINIC ACID THIOAMIDE ◇ 2-ETHYLISONICOTINIC THIOAMIDE ◇ α-ETHYLISONICOTINOYLTHIOAMIDE ◇ ETHYLISOTHIAMIDE ◇ α-ETHYLISOTHIONICOTINAMIDE ◇ 2-ETHYLISOTHIONICOTINAMIDE ◇ 2-ETHYL-4-PYRIDINECARBOTHIOAMIDE ◇ 2-ETHYL-4-THIOAMIDYLPYRIDINE ◇ 2-ETHYL-4-THIOCARBAMOYLPYRIDINE ◇ α-ETHYLTHIOISONICOTINAMIDE ◇ ETHYONOMIDE ◇ ETIMID ◇ ETIOCIDAN ◇ ETIONAMID ◇ ETIONIZINA ◇ ETP ◇ FATOLIAMID ◇ F.I. 58-30 ◇ IRIDOCIN ◇ IRIDOZIN ◇ ISOTHIN ◇ ISOTIAMIDA ◇ ITIOCIDE ◇ NCI-C01694 ◇ NICOTION ◇ NISOTIN ◇ NIZOTIN ◇ RIGENICID ◇ SERTINON ◇ TEBERUS ◇ TH 1314 ◇ THIANIDE ◇ THIOAMIDE ◇ THIONIDEN ◇ TRECATOR ◇ TRESCATYL ◇ TRESCAZIDE ◇ TUBERMIN ◇ TUBEROID ◇ TUBEROSON

TOXICITY DATA with REFERENCE
cyt-ham:fbr 400 mg/L ESKHA5 96,55,78
cyt-ham:lng 540 mg/L GMCRDC 27,95,81
unr-wmn TDLo:5400 mg/kg (12W pre/1-38W preg):REP POMJAC 5,1152,66

unr-wmn TDLo:600 mg/kg (36-39W preg):TER POMJAC 5,1152,66
orl-mus TDLo:24 g/kg/50W-I:CAR LAPPA5 24,145,64
orl-hmn TDLo:856 mg/kg:LIV ARDSBL 84,890,61
orl-hmn TDLo:482 mg/kg/5W:LIV ARDSBL 87,896,63
orl-rat LD50:1320 mg/kg NIIRDN 6,111,82
ipr-rat LD50:490 mg/kg NIIRDN 6,111,82
scu-rat LD50:1350 mg/kg NIIRDN 6,111,82
orl-mus LD50:1000 mg/kg FEPRA7 21,452,62
ipr-mus LD50:1350 mg/kg DPHFAK 23,383,71
scu-mus LD50:1580 mg/kg NIIRDN 6,111,82
scu-gpg LD50:550 mg/kg UBZHD4 52,593,80

CONSENSUS REPORTS: IARC Cancer Review: Group 3 IMEMDT 7,56,87; Animal Limited Evidence IMEMDT 13,83,77. NCI Carcinogenesis Bioassay (feed); No Evidence: mouse, rat NCITR* NCI-CG-TR-46,78.

SAFETY PROFILE: A human systemic poison. Moderately toxic by ingestion, intraperitoneal, and subcutaneous routes. Human systemic effects by ingestion: jaundice and liver function impairment. It affects the human peripheral nervous system. Experimental teratogenic and reproductive effects. Questionable carcinogen with experimental carcinogenic data. Mutation data reported. Used to treat tuberculosis. When heated to decomposition it emits very toxic fumes of SO_x and NO_x.

EPR000 CAS:29973-13-5 ***HR: 3***
(2-ETHYLTHIOMETHYLPHENYL)-N-METHYLCARBAMATE
mf: $C_{11}H_{15}NO_2S$ mw: 225.33

SYNS: BAY-HOX-1901 ◇ CRONETON ◇ ETHIOFENCARB ◇ ETHIOPHENCARP ◇ 2-ETHYLMERCAPTOMETHYLPHENYL-N-METHYLCARBAMATE ◇ 2-((ETHYLTHIO)METHYL)PHENOL METHYLCARBAMATE ◇ 2-((ETHYLTHIO)METHYL)PHENYL METHYLCARBAMATE ◇ HOX 1901

TOXICITY DATA with REFERENCE
orl-rat LD50:200 mg/kg FMCHA2 -,C64,83
ihl-rat LCLo:97 mg/m^3 GISAAA 49(10),23,84
skn-rat LD50:1000 mg/kg FMCHA2 -,C64,83
orl-mus LD50:71 mg/kg 85ALAU -,106,76
ihl-cat LCLo:97 mg/m^3 GISAAA 49(10),23,84
skn-rbt LD50:2500 mg/kg GISAAA 49(10),23,84

SAFETY PROFILE: Poison by ingestion and inhalation routes. Moderately toxic by skin contact. An insecticide. When heated to decomposition it emits very toxic fumes of NO_x and SO_x. See also CARBAMATES.

EPR200 CAS:70303-46-7 ***HR: 2***
(ETHYLTHIO)TRIOCTYLSTANNANE
mf: $C_{26}H_{56}SSn$ mw: 519.57

SYNS: STANNANE, (ETHYLTHIO)TRIOCTYL- ◇ TRIOCTYL (ETHYLTHIO)STANNANE

TOXICITY DATA with REFERENCE
ipr-mus LD50:2945 mg/kg RPTOAN 42,73,79

OSHA PEL: TWA 0.1 mg(Sn)/m^3
ACGIH TLV: TWA 0.1 mg(Sn)/m^3; STEL 0.2 mg/m^3 (skin)
NIOSH REL: (Organotin Compounds): 10H TWA 0.1 mg(Sn)/m^3

SAFETY PROFILE: Moderately toxic by intraperitoneal route. When heated to decomposition it emits toxic fumes of SO_x and Sn.

EPR600 CAS:625-53-6 ***HR: 3***
ETHYL THIOUREA
mf: $C_3H_8N_2S$ mw: 104.19

SYN: 1-ETHYLTHIOUREA

TOXICITY DATA with REFERENCE
mmo-sat 150 μg/plate ABCHA6 44,3017,80
dnr-omi 10 mg/plate BIZNAT 95,463,76
orl-rat TDLo:500 mg/kg (12D preg):TER TJADAB 23,335,81
orl-rat LD50:100 mg/kg JPETAB 90,260,47
orl-mus LDLo:500 mg/kg TJADAB 23,335,81

CONSENSUS REPORTS: Reported in EPA TSCA Inventory.

SAFETY PROFILE: Poison by ingestion. Experimental teratogenic effects. Mutation data reported. When heated to decomposition it emits toxic fumes of SO_x and NO_x.

EPS000 CAS:1066-57-5 ***HR: 3***
ETHYLTIN TRICHLORIDE
mf: $C_2H_5Cl_3Sn$ mw: 254.11

SYNS: AETHYLZINNTRICHLORID (GERMAN) ◇ ETHYLTRICHLOROSTANNANE ◇ ETHYLTRICHLOROTIN ◇ MONOETHYLTIN TRICHLORIDE ◇ TRICHLOROETHYLSTANNANE ◇ TRICHLOROETHYLTIN

TOXICITY DATA with REFERENCE
ipr-rat LDLo:200 mg/kg BJPCAL 10,16,55

OSHA PEL: TWA 0.1 mg(Sn)/m^3 (skin)
ACGIH TLV: TWA 0.1 mg(Sn)/m^3 (skin) (Proposed: TWA 0.1 mg(Sn)/m^3; STEL 0.2 mg(Sn)/m^3 (skin))
NIOSH REL: (Organotin Compounds) TWA 0.1 mg(Sn)/m^3

SAFETY PROFILE: Poison by intraperitoneal route. When heated to decomposition it emits toxic fumes of Cl^-. See also TIN COMPOUNDS and CHLORIDES.

EPS500 CAS:611-14-3 ***HR: 1***
2-ETHYLTOLUENE
mf: C_9H_{12} mw: 120.21

PROP: Autoign temp: 824°F, d: 0.88, vap d: 4.15, bp: 164.1°.

SYNS: 1-ETHYL-2-METHYLBENZENE ◇ o-ETHYL METHYLBENZENE ◇ o-ETHYLTOLUENE ◇ o-METHYLETHYLBENZENE

TOXICITY DATA with REFERENCE
orl-rat LDLo:5000 mg/kg 28ZRAQ -,57,60
ihl-mus LC50:54 g/m³/4H 85GMAT -,69,82
ihl-cat LC50:50 g/kg/2H 85GMAT -,69,82

CONSENSUS REPORTS: Reported in EPA TSCA Inventory.

SAFETY PROFILE: Mildly toxic by ingestion and inhalation. Flammable when exposed to heat, flame or oxidizers. To fight fire, use water spray or mist, dry chemical, CO_2. When heated to decomposition it emits acrid smoke and irritating fumes.

EPT000 CAS:622-96-8 ***HR: 1***
p-ETHYLTOLUENE
mf: C_9H_{12} mw: 120.21

PROP: Autoign temp: 887°F, bp: 162.2°, mp: −62.4.

SYNS: 1-ETHYL-4-METHYLBENZENE ◇ p-ETHYLMETHYLBENZENE ◇ 4-ETHYLTOLUENE ◇ p-METHYLETHYLBENZENE ◇ 4-METHYLETHYLBENZENE

TOXICITY DATA with REFERENCE
orl-rat LDLo:5000 mg/kg 28ZRAQ -,57,60

CONSENSUS REPORTS: Reported in EPA TSCA Inventory.

SAFETY PROFILE: Mildly toxic by ingestion. Flammable when exposed to heat or flame; can react vigorously with oxidizing materials. To fight fire use water spray or mist, dry chemical, CO_2. When heated to decomposition it emits acrid smoke and irritating fumes.

EPT500 CAS:17010-63-8 ***HR: 3***
p-((3-ETHYL-p-TOLYL)AZO)-N,N-DIMETHYLANILINE
mf: $C_{17}H_{21}N_3$ mw: 267.41

SYNS: N,N-DIMETHYL-p-(3′-ETHYL-4′-METHYLPHENYLAZO)ANILINE ◇ N,N-DIMETHYL-p-((3-ETHYL-p-TOLYL)AZO)ANILINE

TOXICITY DATA with REFERENCE
orl-rat TDLo:6497 mg/kg/17W-C:CAR CBINA8 53,107,85
orl-rat TDLo:2005 mg/kg/50D-I:ETA CNREA8 30,1520.70

SAFETY PROFILE: Questionable carcinogen with experimental carcinogenic and tumorigenic data. When heated to decomposition it emits toxic fumes of NO_x.

EPU000 CAS:17010-62-7 ***HR: 3***
p-((4-ETHYL-m-TOLYL)AZO)-N,N-DIMETHYLANILINE
mf: $C_{17}H_{21}N_3$ mw: 266.40

SYNS: N,N-DIMETHYL-p-(4′-ETHYL-3′-METHYLPHENYLAZO)ANILINE ◇ N,N-DIMETHYL-p-((4-ETHYL-m-TOLYL)AZO) ANILINE

TOXICITY DATA with REFERENCE
orl-rat 1109 mg/kg/8W-C:CAR CBINA8 53,107,85
orl-rat TDLo:500 mg/kg/50D-I:ETA CNREA8 30,1520.70

SAFETY PROFILE: Questionable carcinogen with experimental carcinogenic and tumorigenic data. When heated to decomposition it emits toxic fumes of NO_x.

EPV000 CAS:101491-84-3 ***HR: 3***
N-ETHYL-N-(1-(o-TOLYLOXY-2-PROPYL)CARBAMIC ACID-2-(2-METHYLPIPERIDINO) ETHYL ESTER HYDROCHLORIDE
mf: $C_{21}H_{34}N_2O_3 \cdot ClH$ mw: 399.03

SYN: C 2138

TOXICITY DATA with REFERENCE
eye-rbt 2% MOD ARZNAD 9,113,59
scu-mus LD50:250 mg/kg ARZNAD 9,113,59

SAFETY PROFILE: Poison by subcutaneous route. An eye irritant. When heated to decomposition it emits very toxic fumes of HCl and NO_x. See also CARBAMATES and ESTERS.

EPW000 CAS:50707-40-9 ***HR: 3***
1-ETHYL-3-p-TOLYLTRIAZENE
mf: $C_9H_{13}N_3$ mw: 163.25

TOXICITY DATA with REFERENCE
ipr-mus TDLo:20 mg/kg/I:NEO JNCIAM 54,495,75
ipr-mus LDLo:200 mg/kg StoGD# 27May75

SAFETY PROFILE: Poison by intraperitoneal route. Questionable carcinogen with experimental neoplastigenic data. When heated to decomposition it emits toxic fumes of NO_x.

EPW500 CAS:80-40-0 ***HR: 3***
ETHYL TOSYLATE
mf: $C_9H_{12}O_3S$ mw: 200.27

PROP: Liquid. Mp: 33°, bp: 221.3°, flash p: 316°F (CC), d: 1.17, vap d: 6.98.

SYNS: ETHYL-p-METHYL BENZENESULFONATE ◇ ETHYL PTS ◇ ETHYL-p-TOLUENESULFONATE ◇ ETHYL-p-TOSYLATE ◇ p-TOLUOLSULFONSAEUREAETHYL ESTER (GERMAN)

TOXICITY DATA with REFERENCE
mmo-sat 2000 μg/plate JNCIAM 62,893,79
mma-sat 4700 μg/plate PNASA6 72,5135,75
dnr-esc 50 mg/L JNCIAM 62,873,79
mrc-smc 1 pph JNCIAM 62,901,79
otr-ham:emb 500 μg/L CRNGDP 1,323,80
scu-rat TDLo:50 mg/kg:ETA FCTXAV 6,576,68
scu-rat LD50:500 mg/kg ZEKBAI 74,241,70
ipr-mus LD50:1000 mg/kg JJIND8 62,911,79

CONSENSUS REPORTS: Reported in EPA TSCA Inventory. EPA Genetic Toxicology Program.

SAFETY PROFILE: Moderately toxic by subcutaneous and intraperitoneal routes. Questionable carcinogen with experimental tumorigenic data. Mutation data reported. Combustible when exposed to heat or flame; can react with oxidizing materials. To fight fire, use CO_2, dry chemical. When heated to decomposition it emits highly toxic fumes of SO_x. See also SULFONATES and ESTERS.

EPX000 CAS:63449-87-6 ***HR: 3***
ETHYL TRI-n-BUTYLAMMONIUM HYDROXIDE
mf: $C_{14}H_{32}N•HO$ mw: 231.48

TOXICITY DATA with REFERENCE
scu-mus LDLo:24 mg/kg JPETAB 28,367,26

CONSENSUS REPORTS: Reported in EPA TSCA Inventory.

SAFETY PROFILE: Poison by subcutaneous route. When heated to decomposition it emits toxic fumes of NO_x and NH_3.

EPY000 CAS:327-98-0 ***HR: 3***
ETHYL TRICHLOROPHENYLETHYLPHOSPHONOTHIOATE
mf: $C_{10}H_{12}Cl_3O_2PS$ mw: 333.60

SYNS: O-AETHYL-O-(2,4,5-TRICHLORPHENYL)-AETHYLTHIONOPHOSPHONAT (GERMAN) ◇ AGRISIL ◇ AGRITOX ◇ BAYER 37289 ◇ BAYER 5081 ◇ BAYER S 4400 ◇ CHEMAGRO 37289 ◇ ENT 25,712 ◇ O-ETHYL-O-2,4,5-TRICHLOROPHENYL ETHYLPHOSPHONOTHIOATE ◇ FENOPHOSPHON ◇ PHYTOSOL ◇ STAUFFER N-3049 ◇ TRICHLORONAT ◇ 2,4,5-TRICHLOROPHENOL-O-ESTER with O-ETHYL ETHYLPHOSPHONOTHIOATE ◇ WIRKSTOFF 37289

TOXICITY DATA with REFERENCE
orl-rat LD50:15 mg/kg ARSIM* 20,4,66
ihl-rat LCLo:700 mg/m³/4H 85GYAZ -,35,71
skn-rat LD50:64 mg/kg WRPCA2 9,119,70
orl-mus LD50:40 mg/kg 85DPAN -,-,71/76
orl-cat LD50:10 mg/kg 85DPAN -,-,71/76
orl-rbt LD50:25 mg/kg 85GYAZ -,35,71

CONSENSUS REPORTS: Chlorophenol compounds are on the Community Right-To-Know List. EPA Extremely Hazardous Substances List.

SAFETY PROFILE: Poison by ingestion and skin contact. Moderately toxic by inhalation. An insecticide. When heated to decomposition it emits very toxic fumes of Cl^-, PO_x, and SO_x. See also CHLOROPHENOLS.

EPY500 CAS:115-21-9 ***HR: 3***
ETHYL TRICHLOROSILANE
DOT: UN 1196
mf: $C_2H_5Cl_3Si$ mw: 163.51

PROP: Liquid. Mp: −105.6°, bp: 99.5°, flash p: 72°F (OC), d: 1.24 @ 25°/25°, vap d: 5.6.

SYNS: ETHYL SILICON TRICHLORIDE ◇ TRICHLOROETHYLSILANE ◇ TRICHLOROETHYLSILICANE

TOXICITY DATA with REFERENCE
ihl-rat LCLo:500 ppm/4H JIHTAB 31,343,49
orl-rat LD50:1330 mg/kg JIHTAB 31,60,49
ipr-rat LDLo:30 mg/kg JIHTAB 30,332,48
ihl-mus LC50:300 μg/m³/2H TPKVAL (3),23,61

CONSENSUS REPORTS: EPA Extremely Hazardous Substances List. Reported in EPA TSCA Inventory.

DOT Classification: Flammable Liquid; Label: Flammable Liquid.

SAFETY PROFILE: Poison by inhalation and intraperitoneal routes. Moderately toxic by ingestion. A very dangerous fire hazard when exposed to heat, flame, or oxidizers; will react with water or steam to produce heat and toxic and corrosive fumes; can react vigorously with oxidizing materials. To fight fire, use CO_2, dry chemical. When heated to decomposition it emits highly toxic fumes of Cl^- and phosgene. See also CHLOROSILANES.

EPY600 CAS:993-86-2 ***HR: 3***
ETHYLTRICHLORPHON
mf: $C_6H_{12}Cl_3O_4P$ mw: 285.50

SYNS: AETHYLTRICHLORPHON (GERMAN) ◇ ETC ◇ 1-HYDROXY-2,2,2-TRICHLOROETHYL DIETHYL PHOSPHONITE

TOXICITY DATA with REFERENCE
ipr-mus TDLo:960 mg/kg (9D preg):REP ZHYGAM 22,565,76
ipr-mus TDLo:240 mg/kg (9D preg):TER ZHYGAM 22,565,76
ipr-mus LD50:400 mg/kg ZHYGAM 22,565,76

SAFETY PROFILE: Poison by intraperitoneal route. An experimental teratogen. Other experimental reproductive effects. When heated to decomposition it emits toxic fumes of Cl^- and PO_x. See also ESTERS.

EPZ000 CAS:313-93-9 ***HR: 3***
3-ETHYLTRICYCLOQUINAZOLINE
mf: $C_{23}H_{16}N_4$ mw: 348.43

TOXICITY DATA with REFERENCE
skn-mus TDLo:1200 mg/kg/50W-I:CAR BCPCA6 14,323,65

SAFETY PROFILE: Questionable carcinogen with experimental carcinogenic data. When heated to decomposition it emits toxic fumes of NO_x.

EQA000 CAS:78-07-9 *HR: 1*
ETHYLTRIETHOXYSILANE
mf: $C_8H_{20}O_3Si$ mw: 192.37

SYN: UNION CARBIDE A-15

TOXICITY DATA with REFERENCE
skn-rbt 500 mg open MLD UCDS** 4/27/64
orl-rat LD50:14 g/kg JIHTAB 31,60,49
skn-rbt LD50:16 g/kg JIHTAB 31,60,49

CONSENSUS REPORTS: Reported in EPA TSCA Inventory.

SAFETY PROFILE: Mildly toxic by ingestion and skin contact. A skin irritant. When heated to decomposition it emits acrid smoke and irritating fumes. See also SILANE.

EQB500 CAS:28781-86-4 *HR: 3*
ETHYL-2,2,3-TRIFLUORO PROPIONATE
mf: $C_5H_7F_3O_2$ mw: 188.22

$CH_3CH_2OCO•CF_2CH_2F$

SAFETY PROFILE: Decomposes violently on contact with sodium hydride. When heated to decomposition it emits toxic fumes of F^-. See also FLUORIDES.

EQC000 CAS:1923-76-8 *HR: 2*
α-ETHYL-β-(2,4,6-TRIIODO-3-BUTYRAMIDOPHENYL)ACRYLIC ACID SODIUM SALT
mf: $C_{15}H_{15}I_3NO_3•Na$ mw: 661.00

SYNS: BILI-ORAL ◇ BUNAIOD ◇ BUNAMIODYL ◇ BUNIODYL ◇ 3-BUTYRAMIDO-α-ETHYL-2,4,6-TRIIODOCINNAMIC ACID SODIUM SALT ◇ 2-(3-BUTYRAMIDO-2,4,6-TRIIODOPHENYLMETHYLENE)BUTYRIC ACID SODIUM SALT ◇ 3-(3-BUTYRYLAMINO-2,4,6-TRIIODOPHENYL)-2-ETHYLACRYLIC ACID SODIUM SALT ◇ INTRABILIX ◇ ORABILEX ◇ SODIUM-3-BUTYRAMIDO-α-ETHYL-2,4,6-TRIIODOCINNAMATE ◇ α-(2,4,6-TRIIODO-3-BUTYRYLAMINOBENZYLIDENE) BUTYRIC ACID SODIUM SALT

TOXICITY DATA with REFERENCE
orl-rat LD50:2800 mg/kg TXAPA9 14,232,69
ivn-rat LD50:480 mg/kg TXAPA9 14,232,69
orl-mus LD50:2780 mg/kg JMCMAR 13,997,70
ivn-mus LD50:570 mg/kg MEIEDD 10,206,83

SAFETY PROFILE: Moderately toxic by ingestion and intravenous routes. When heated to decomposition it emits very toxic fumes of I^-, Na_2O, and NO_x.

EQC500 CAS:4916-38-5 *HR: 3*
ETHYLTRIIODOGERMANE
mf: $C_2H_5GeI_3$ mw: 482.36

TOXICITY DATA with REFERENCE
ipr-rat LDLo:220 mg/kg CHDDAT 262,1302,66
ipr-mus LDLo:210 mg/kg CHDDAT 262,1302,66

SAFETY PROFILE: Poison by intraperitoneal route. When heated to decomposition it emits toxic fumes of I^-. See also GERMANIUM COMPOUNDS.

EQD000 CAS:41096-46-2 *HR: 1*
ETHYL-3,7,11-TRIMETHYLDODECA-2,4-DIENOATE
mf: $C_{17}H_{30}O_2$ mw: 266.47

SYNS: ALTOZAR ◇ ENT 70,459 ◇ ETHYL(2E,4E)-3,7,11-TRIMETHYL-2,4-DODECADIENOATE ◇ ETHYL (2E,4E)-3,7,11-TRIMETHYL-DODECA-2-4-DIENOATE ◇ HYDROPRENE ◇ ZR 512

TOXICITY DATA with REFERENCE
ipr-mus TDLo:1 g/kg (9D preg):REP LIFSAK 15,1649,74
ipr-mus TDLo:1 g/kg (9D preg):TER LIFSAK 15,1649,74
orl-rat LD50:34000 mg/kg 85ARAE 1,98,77
skn-rbt LD50:4550 mg/kg FMCHA2 -,C12,83

SAFETY PROFILE: Mildly toxic by ingestion and skin contact. Experimental teratogenic and reproductive effects. When heated to decomposition it emits acrid smoke and irritating fumes.

EQD100 CAS:67590-56-1 *HR: 3*
4-ETHYL-2,6,7-TRIOXA-1-ARSABICYCLO (2.2.2) OCTANE
mf: $C_6H_{11}AsO_3$ mw: 206.09

SYN: 2,6,7-TRIOXA-1-ARSABICYCLO(2.2.2)OCTANE,4-ETHYL-

TOXICITY DATA with REFERENCE
ivn-mus LD50:31 mg/kg EJMCA5 13,207,78

OSHA PEL: TWA 0.5 mg(As)/m^3

SAFETY PROFILE: Poison by intravenous route. When heated to decomposition it emits toxic fumes of As.

EQD600 CAS:57420-66-3 *HR: 3*
1-ETHYL-1-UNDECYLPIPERIDINIUM BROMIDE
mf: $C_{18}H_{38}N•Br$ mw: 348.48

TOXICITY DATA with REFERENCE
orl-mus LD50:188 mg/kg PSDTAP 15,331,74
ipr-mus LD50:25970 μg/kg PSDTAP 15,331,74
ivn-mus LD50:4417 μg/kg PSDTAP 15,331,74

SAFETY PROFILE: Poison by ingestion, intravenous, and intraperitoneal routes. When heated to decomposition it emits toxic fumes of NO_x and Br^-. See also BROMIDES.

EQD875 CAS:2884-67-5 ***HR: 1***
ETHYLUREA
mf: $C_3H_8N_2O$ mw: 88.13

SYNS: N-ETHYLUREA ◇ 1-ETHYLUREA

TOXICITY DATA with REFERENCE
mmo-clr 400 mmol/L FOMIAZ 20,452,75
orl-mus TDLo:2 g/kg (10D preg):REP TJADAB 23,335,81
par-mus LDLo:6610 mg/kg JPETAB 52,216,34

CONSENSUS REPORTS: EPA Genetic Toxicology Program. Reported in EPA TSCA Inventory.

SAFETY PROFILE: Mildly toxic by parenteral routes. Experimental reproductive effects. Mutation data reported. When heated to decomposition it emits toxic fumes of NO_x.

EQD900 ***HR: D***
ETHYLUREA and SODIUM NITRITE (1:1)
mf: $C_3H_8N_2O \cdot NNaO_2$ mw: 157.13

SYN: SODIUM NITRITE and ETHYL UREA (1:1)

TOXICITY DATA with REFERENCE
hma-mus/sat 3650 μmol/kg CNREA8 37,4572,77
orl-rat TDLo:875 mg/kg (7-16D preg):TER RCOCB8 41,265,83

SAFETY PROFILE: Experimental teratogenic effects. Mutation data reported. When heated to decomposition it emits toxic fumes of NO_x and Na_2O. See also SODIUM NITRITE.

EQE000 ***HR: 3***
ETHYLUREA and SODIUM NITRITE (2:1)

SYNS: AETHYLHARNSTOFF und NATRIUMNITRIT (GERMAN) ◇ AETHYLHARNSTOFF und NITRIT (GERMAN) ◇ SODIUM NITRITE and ETHYLUREA (1:2)

TOXICITY DATA with REFERENCE
orl-rat TDLo:22274 mg/kg (1-22D preg/21D post):REP FCTXAV 15,283,77
orl-rat TDLo:100 mg/kg (9-10D preg):TER IARCCD 4,112,73
orl-rat TDLo:1650 mg/kg (13-23D preg):CAR,TER IARCCD 4,92,73
orl-rat TDLo:750 mg/kg (15D preg):NEO,TER ZAPPAN 121,61,77
orl-rbt TDLo:450 mg/kg/(17-19D preg):ETA,TER JNCIAM 59,427,77
orl-ham TDLo:300 mg/kg (15D preg):CAR,TER JNCIAM 55,1389,75
orl-rat TD:1650 mg/kg (13-23D preg):ETA,TER NATWAY 57,460,70
orl-rat TD:25 g/kg/35D-C:ETA ARZNAD 21,1707,71
orl-ham TDLo:600 mg/kg (12-15D preg):NEO,TER ZKKOBW 85,201,76

SAFETY PROFILE: Suspected carcinogen with experimental carcinogenic, neoplastigenic, tumorigenic, and teratogenic data. Experimental reproductive effects. When heated to decomposition it emits toxic fumes of NO_x and Na_2O. See also SODIUM NITRITE.

EQE500 CAS:617-05-0 ***HR: 3***
ETHYL VANILLATE
mf: $C_{10}H_{12}O_4$ mw: 196.22

PROP: Mp: 44°, bp: 291-3°, Insol in water; sol in alk; very sol in alc and ether.

SYNS: ETHYL-4-HYDROXY-3-METHOXYBENZOATE ◇ 3-METHOXY-4-HYDROXYBENZOIC ACID, ETHYL ESTER

TOXICITY DATA with REFERENCE
scu-rat LD50:2000 mg/kg HBTXAC 1,138,56
ivn-mus LD50:56 mg/kg CSLNX* NX#06395
orl-rbt LDLo:3000 mg/kg HBTXAC 1,138,56

SAFETY PROFILE: Poison by intravenous route. Moderately toxic by ingestion and subcutaneous routes. When heated to decomposition it emits acrid smoke and irritating fumes. See also ESTERS.

EQF000 CAS:121-32-4 ***HR: 2***
ETHYL VANILLIN
mf: $C_9H_{10}O_3$ mw: 166.19

PROP: Fine, crystalline needles; vanilla odor. Mp: 76.5°, flash p: +212°F. Sol in alc, chloroform, ether, propylene glycol; sltly sol in water.

SYNS: BOURBONAL ◇ ETHAVAN ◇ ETHOVAN ◇ 3-ETHOXY-4-HYDROXYBENZALDEHYDE ◇ ETHYLPROTAL ◇ FEMA No. 2464 ◇ 4-HYDROXY-3-ETHOXYBENZALDEHYDE ◇ PROTOCATECHUIC ALDEHYDE ETHYL ETHER ◇ QUANTROVANIL ◇ VANILLAL ◇ VANIROM

TOXICITY DATA with REFERENCE
skn-hmn 10 mg/48H MLD FCTXAV 13,103,75
cyt-ham:fbr 250 mg/L FCTOD7 22,623,84
scu-rat LDLo:1800 mg/kg JAPMA8 29,425,40
ipr-mus LD50:750 mg/kg CTOXAO 10,61,77
ivn-dog LDLo:760 mg/kg COREAF 238,2576,54
orl-rbt LDLo:3000 mg/kg JAPMA8 29,425,40
ipr-gpg LD50:1140 mg/kg COREAF 238,2576,54

CONSENSUS REPORTS: Reported in EPA TSCA Inventory.

SAFETY PROFILE: Moderately toxic by ingestion, intraperitoneal, subcutaneous, and intravenous routes. A human skin irritant. Mutation data reported. Combustible liquid. When heated to decomposition it emits acrid smoke and irritating fumes. See also ALDEHYDES and ETHERS.

EQF200 CAS:3195-79-7 *HR: 2*
N-ETHYL-N-VINYLACETAMIDE
mf: $C_6H_{11}NO$ mw: 113.18

TOXICITY DATA with REFERENCE
orl-rat LD50:2460 mg/kg AIHAAP 30,470,69
ihl-rat LCLo:2000 ppm/4H AIHAAP 30,470,69
skn-rbt LD50:1250 mg/kg AIHAAP 30,470,69

SAFETY PROFILE: Moderately toxic by ingestion and skin contact. Mildly toxic by inhalation. When heated to decomposition it emits toxic fumes of NO_x.

EQF500 CAS:109-92-2 *HR: 3*
ETHYL VINYL ETHER
DOT: UN 1302
mf: C_4H_8O mw: 72.12

PROP: Colorless liquid. Bp: 35.6°, flash p: < −50°F, fp: −115°, d: 0.754, autoign temp: 395°F, vap press: 428 mm @ 20°, lel: 1.7%, uel: 28%, vap d: 2.5.

SYNS: ETHOXY ETHENE ◇ EVE ◇ VINAMAR ◇ VINYL ETHYL ETHER ◇ VINYL ETHYL ETHER, inhibited (DOT)

TOXICITY DATA with REFERENCE
skn-rbt 500 mg open MLD UCDS** 11/15/71
sce-ham:ovr 17900 ppm ANESAV 50,426,79
orl-rat LD50:6153 mg/kg AIHAAP 30,470,69

CONSENSUS REPORTS: Reported in EPA TSCA Inventory.

DOT Classification: Flammable Liquid; Label: Flammable Liquid.

SAFETY PROFILE: Mildly toxic by ingestion. Mutation data reported. A skin irritant. A very dangerous fire and explosion hazard when exposed to heat or flame; can react vigorously with oxidizing materials. To fight fire, use alcohol foam, foam, CO_2, dry chemical. Explosive polymerization is catalyzed by methane sulfonic acid. When heated to decomposition it emits acrid smoke and irritating fumes. See also ETHERS.

EQG000 CAS:109-92-2 *HR: 3*
ETHYL VINYL ETHER (inhibited)

CONSENSUS REPORTS: Reported in EPA TSCA Inventory.

DOT Classification: Flammable Liquid; Label: Flammable Liquid.

SAFETY PROFILE: A very dangerous fire and explosion hazard when exposed to heat or flame. See also ETHERYL VINYL ETHER and ETHERS.

EQG500 CAS:5408-74-2 *HR: 2*
5-ETHYL-2-VINYLPYRIDINE
mf: $C_9H_{11}N$ mw: 133.21

TOXICITY DATA with REFERENCE
skn-rbt 10 mg/24H open AMIHBC 10,61,54
eye-rbt 20 mg open AMIHBC 10,61,54
orl-rat LD50:1230 mg/kg AMIHBC 10,61,54
ihl-rat LCLo:8000 ppm/4H AMIHBC 10,61,54
skn-rbt LD50:890 mg/kg AMIHBC 10,61,54

SAFETY PROFILE: Moderately toxic by ingestion and skin contact. Mildly toxic by inhalation. A skin and eye irritant. When heated to decomposition it emits toxic fumes of NO_x.

EQH000 CAS:3278-35-1 *HR: 3*
ETHYLXANTHIC ACID ANHYDROSULFIDE with O-ETHYLTHIOLCARBONATE
mf: $C_6H_{10}O_3S_2$ mw: 194.28

TOXICITY DATA with REFERENCE
ipr-mus LDLo:125 mg/kg CBCCT* 8,492,56

CONSENSUS REPORTS: Reported in EPA TSCA Inventory.

SAFETY PROFILE: Poison by intraperitoneal route. When heated to decomposition it emits toxic fumes of SO_x. See also SULFIDES.

EQI000 CAS:101491-85-4 *HR: 3*
N-ETHYL-N-(1-(3,5-XYLYLOXY)-2-PROPYLCARBAMIC ACID-2-(DIETHYLAMINO) ETHYL ESTER HYDROCHLORIDE
mf: $C_{20}H_{34}N_2O_3 \cdot ClH$ mw: 387.02

SYN: C 2142

TOXICITY DATA with REFERENCE
eye-rbt 2% SEV ARZNAD 9,113,59
scu-mus LD50:202 mg/kg ARZNAD 9,113,59

SAFETY PROFILE: Poison by subcutaneous route. A severe eye irritant. When heated to decomposition it emits very toxic fumes of HCl and NO_x. See also CARBAMATES and ESTERS.

EQI500 CAS:101491-86-5 *HR: 3*
N-ETHYL-N-(1-(3,5-XYLYLOXY)-2-PROPYL)CARBAMIC ACID-2-(2-METHYLPIPERIDINO) ETHYL ESTER HYDROCHLORIDE
mf: $C_{22}H_{36}N_2O_3 \cdot ClH$ mw: 413.06

SYN: C 2127

TOXICITY DATA with REFERENCE
eye-rbt 2% SEV ARZNAD 9,113,59
scu-mus LD50:352 mg/kg ARZNAD 9,113,59

SAFETY PROFILE: Poison by subcutaneous route. A severe eye irritant. When heated to decomposition it emits very toxic fumes of HCl and NO_x. See also CARBAMATES and ESTERS.

EQI600 CAS:2482-80-6 **HR: 3**
ETHYMIDINE
mf: $C_8H_9ClN_4$ mw: 196.66

SYNS: 2,4-BIS(1-AZIRIDINYL)-6-CHLOROPYRIMIDINE ◇ 2,6-BIS(1-AZIRIDINYL)-4-CHLOROPYRIMIDINE ◇ 4-CHLORO-2,6-BIS-ETHYLENEIMINOPYRIMIDINE ◇ 2,6-DIETHYLENEIMINO-4-CHLOROPYRIMIDINE ◇ ETIMIDIN

TOXICITY DATA with REFERENCE
mmo-asn 2500 μmol/L MUREAV 14,115,72
oms-rat-ipr 2500 μg/kg BJPCAL 6,357,51
cyt-rat-ipr 2500 μg/kg BJPCAL 6,357,51
ivn-rat LDLo:15 mg/kg PMDCAY 8,61,71

CONSENSUS REPORTS: EPA Genetic Toxicology Program.

SAFETY PROFILE: Poison by intravenous route. Mutation data reported. When heated to decomposition it emits toxic fumes of NO_x and Cl^-.

EQJ000 **HR: 3**
ETHYNERONE mixed with MESTRANOL (20:1)
mf: $C_{20}H_{23}ClO_2 \cdot C_{21}H_{26}O_2$ (20:1) mw: 641.0

SYN: MESTRANOL mixed with ETHYNERONE (1:20)

TOXICITY DATA with REFERENCE
orl-dog TDLo:243 mg/kg/47W-I:ETA JJIND8 65,137,80

SAFETY PROFILE: Questionable carcinogen with experimental tumorigenic data. When heated to decomposition it emits toxic fumes of Cl^-.

EQJ100 CAS:1231-93-2 **HR: 3**
ETHYNODIOL
mf: $C_{20}H_{28}O_2$ mw: 300.48

SYNS: ED ◇ ETHINODIOL ◇ 17-α-ETHYNYL-19-NORANDROST-4-ENE-3-β,17-β-DIOL

TOXICITY DATA with REFERENCE
unr-wmn TDLo:12 mg/kg (6-20W preg):REP SCIEAS 211,1171,81
orl-rat TDLo:100 mg/kg (17-20D preg):TER ECJPAE 24,77,77

SAFETY PROFILE: Human reproductive effects by an unspecified route: biochemical and metabolic disorders in newborn. An experimental teratogen. Experimental reproductive effects. When heated to decomposition it emits acrid smoke and irritating fumes.

EQJ500 CAS:297-76-7 **HR: 3**
ETHYNODIOL ACETATE
mf: $C_{24}H_{32}O_4$ mw: 384.56

SYNS: CERVICUNDIN ◇ 3-β,17-β-DIACETOXY-17-α-ETHYNYL-4-OESTRENE ◇ 3-β,17-β-DIACETOXY-19-NOR-17-α-PREGN-4-EN-20-YNE ◇ ETHINODIOL DIACETATE ◇ ETHYNODIOL DIACETATE ◇ β-ETHYNODIOL DIACETATE ◇ 17-α-ETHYNYL-3,17-DIHYDROXY-4-ESTRENE DIACETATE ◇ 17-α-ETHYNYLESTR-4-ENE-3-β,17-β-DIOL ACETATE ◇ 17-α-ETHYNYL-4-ESTRENE-3-β,17-β-DIOL DIACETATE ◇ 17-α-ETHYNYL-4-ESTRENE-3-β,17-DIOL DIACETATE ◇ 17-α-ETHYNYL-19-NORANDROST-4-ENE-3-β,17-β-DIOL DIACETATE ◇ FEMULEN ◇ LUTO-METRODIOL ◇ METRODIOL ◇ METRODIOL DIACETATE ◇ (3-β,17-α)-19-NORPREGN-4-EN-20-YNE-3,17-DIOL DIACETATE ◇ OVULEN 50

TOXICITY DATA with REFERENCE
oth-mus-par 16 μg/kg AJOGAH 120,390,74
cyt-ctl:oth 10 mg/L AJOGAH 120,390,74
orl-wmn TDLo:840 μg/kg (21D pre):REP SCIEAS 138,439,62

CONSENSUS REPORTS: IARC Cancer Review: Animal Limited Evidence IMEMDT 21,387,79; Animal Sufficient Evidence IMEMDT 6,173,74.

SAFETY PROFILE: Suspected carcinogen. Human reproductive effects by ingestion: menstrual cycle changes. Experimental reproductive effects. Mutation data reported. A steroid. When heated to decomposition it emits acrid smoke and irritating fumes.

EQK010 **HR: 3**
ETHYNODIOL DIACETATE mixed with MESTRANOL

SYNS: MESTRANOL mixed with ETHYNODIOL DIACETATE ◇ 19-NOR-17-α-PREGN-4-EN-20-YNE-3-β,17-DIOL DIACETATE mixed with 3-METHYOXY-19-NOR-17-α-PREGNA-1,3,5(10)-TRIEN-20-YN-17-OL

TOXICITY DATA with REFERENCE
orl-wmn TDLo:15 mg/kg/78W-I:NEO,LIV JAMAAP 235,730,76
ipr-mus LD50:3900 mg/kg 27ZTAP 3,69,69

SAFETY PROFILE: Questionable carcinogen producing liver tumors. Moderately toxic by intraperitoneal route. A steroid.

EQK100 CAS:8056-92-6 **HR: 3**
ETHYNODIOL mixed with MESTRANOL
mf: $C_{24}H_{32}O_4 \cdot C_{21}H_{26}O_2$ mw:695.03

SYNS: MESTRANOL mixed with ETHYNODIOL ◇ 10-NOR-17-α-PREGN-4-EN-20-YNE-3-β,17-DIOL mixed with 3-METHOXY-17-α-19-NORPREGNA-1-3-5(10)-TRIEN-20-YN-17-OL ◇ OVULEN

TOXICITY DATA with REFERENCE
orl-wmn TDLo:63504 μg/kg/12Y-I:CAR,LIV,MET LANCAO 1,273,80
orl-wmn TDLo:220 μg/kg/2W:SKN ARDEAC 113,333,76

orl-mus TDLo:186 mg/kg/93W-C:NEO GMCRDC 17,205,75

SAFETY PROFILE: Human systemic effects by ingestion: skin dermatitis, weight loss or decreased weight gain. Questionable carcinogen producing liver tumors. A steroid. When heated to decomposition it emits acrid smoke and fumes.

EQL000 CAS:77-75-8 ***HR: 3***
2-ETHYNYL-2-BUTANOL
mf: $C_6H_{10}O$ mw: 98.16

PROP: Colorless, mobile liquid; acrid odor; burning taste. Sol in water, ether, etc. Mp: −30.6°, bp: 122°, flash p: 101°F (OC), d: 0.8688 @ 20°, vap d: 3.38.

SYNS: 2-AETHINYLBUTANOL ◇ ALLOTROPAL ◇ ANTI-STRESS ◇ APRIDOL ◇ ATEMPOL ◇ (BDH) ◇ COMESA ◇ DALGOL ◇ DORMIDIN ◇ DORMIPHEN ◇ DORMOSAN ◇ 2-ETHINYLBUTANOL-2 ◇ ETHINYLMETHYLETHYLCARBINOL ◇ 3-ETHYLBUTINOL ◇ 3-ETHYLBUTYNOL ◇ HESOFEN ◇ INSOMNOL ◇ MECAROL ◇ MEPENTAMATO ◇ MEPENTIL ◇ METHYLETHYLACETYLENYLCARBINOL ◇ METHYLETHYLETHYNYLCARBINOL ◇ 3-METHYLPENTIN-3-OL ◇ 3-METHYLPENT-1-YN-3-OL ◇ 3-METHYL-1-PENTYN-3-OL ◇ MIRAMEL ◇ NOXOKRATIN ◇ OBLEVIL ◇ m-OBLIVON ◇ PENTADORM ◇ PENTYDORM ◇ m-PENTYNOL ◇ PENTYREST ◇ PLACIDAL ◇ RIPOSON ◇ SOMNESIN ◇ TRUSONO ◇ UTIL

TOXICITY DATA with REFERENCE
orl-rat LD50:300 mg/kg 27ZKAW -,-,69
orl-mus LD50:700 mg/kg ARZNAD 7,85,57
ipr-mus LD50:625 mg/kg AIPTAK 112,463,57
scu-mus LD50:700 mg/kg JPETAB 109,268,53
orl-gpg LD50:534 mg/kg 27ZKAW -,-,69

CONSENSUS REPORTS: Reported in EPA TSCA Inventory.

SAFETY PROFILE: A poison by ingestion. Moderately toxic by intraperitoneal and subcutaneous routes. Used as a soporific. Average doses may produce dermatitis, eructations (belching), psychoses and central nervous system abnormalities. Overdoses can produce coma and death. Flammable when exposed to heat or flame; can react with oxidizing materials. To fight fire use alcohol foam, mist, fog. When heated to decomposition it emits acrid smoke and fumes.

EQL500 CAS:78-27-3 ***HR: 3***
1-ETHYNYL-1-CYCLOHEXANOL
mf: $C_8H_{12}O$ mw: 124.20

PROP: Mp: 32°, bp: 180°, vap d: 3.73.

SYNS: 1-ETHYNYLCYCLOHEXANOL ◇ 1-ETHYNYLCYCLOHEXAN-1-OL

TOXICITY DATA with REFERENCE
skn-rbt 10 mg/24H open MLD AIHAAP 23,95,62
orl-rat LD50:600 mg/kg AIHAAP 23,95,62
ipr-mus LDLo:500 mg/kg CBCCT* 4,228,52
scu-mus LD50:840 mg/kg ARZNAD 5,161,55
skn-rbt LD50:1000 mg/kg AIHAAP 23,95,62

CONSENSUS REPORTS: Reported in EPA TSCA Inventory.

SAFETY PROFILE: Moderately toxic by ingestion, skin contact, subcutaneous, and intraperitoneal routes. A skin irritant. Flammable when exposed to heat or flame; can react with oxidizing materials. To fight fire, use foam, CO_2, dry chemical. When heated to decomposition it emits acrid smoke and irritating fumes.

EQM500 CAS:37270-71-6 ***HR: 3***
ETHYNYLESTRADIOL mixed with NORETHINDRONE
mf: $C_{20}H_{26}O_2 \cdot C_{20}H_{24}O_2$ mw:594.90

SYNS: GYNOVLAR ◇ NORETHINDRONE mixed with ETHYNYLESTRADIOL ◇ NORETHISTERONE mixed with ETHINYL OESTRADIOL (60:1) ◇ 19-NOR-17-α-PREGN-4-EN-20-YN-3-ONE, 17-HYDROXY-, mixed with 19-NOR-17-α-PREGNA-1,3,5(10)-TRIEN-2-YNE-3,17-DIOL(60:1)

TOXICITY DATA with REFERENCE
oms-ctl:oth 17410 μg/L AJOGAH 120,390,74
oms-dom:oth 17410 μg/L AJOGAH 120,390,74
oms-mam:oth 10100 μg/L AJOGAH 120,390,74
orl-wmn TDLo:420 μg/kg (20D pre):REP CCPTAY 26,23,82
unr-wmn TDLo:63298 μg/kg (11-34W preg):TER LANCAO 2,438,60
orl-wmn TDLo:138 mg/kg/9Y-I:CAR,LIV LANCAO 1,273,80
orl-wmn TD:138 mg/kg/9Y-I:CAR,LIV BMJOAE 4,496,75
orl-mus TDLo:120 mg/kg/84W-I:ETA SCIEAS 154,402,66
orl-wmn TDLo:55 mg/kg/5Y-I:PUL,GIT,MET LANCAO 1,1479,73

SAFETY PROFILE: Human teratogenic effects by an unspecified route: developmental abnormalities of the urogenital system. Human systemic effects by ingestion: dyspnea, nausea or vomiting, and fever. Experimental reproductive effects. Questionable human carcinogen producing liver tumors. Mutation data reported. A steroid. When heated to decomposition it emits acrid smoke and irritating fumes.

EQN000 CAS:18649-64-4 ***HR: 3***
2-ETHYNYLFURAN
mf: C_6H_4O mw: 92.10

OCH=CHCH=CC≡CH

SAFETY PROFILE: Explodes on heating or on contact with concentrated nitric acid. When heated to decomposition it emits acrid smoke and irritating fumes. See also ACETYLENE COMPOUNDS.

EQN225 ***HR: 2***
α-ETHYNYL-p-METHOXYBENZYL ALCOHOL ACETATE
mf: $C_{12}H_{12}O_3$ mw: 204.24

SYN: 1'-ACETOXY-2',3'-DEHYDROESTRAGOLE

TOXICITY DATA with REFERENCE
mmo-sat 150 nmol/plate CRNGDP 7,2089,86
ipr-mus TDLo:10212 μg/kg:NEO CNREA8 47,2275,87

SAFETY PROFILE: Questionable carcinogen with experimental neoplastigenic data. Mutation data reported. When heated to decomposition it emits acrid smoke and irritating fumes.

EQN259 CAS:1045-29-0 ***HR: D***
2-α-ETHYNYL-a-NOR-17-α-PREGN-20-YNE-2-β,17-β-DIOL
mf: $C_{22}H_{30}O_2$ mw: 326.52

SYNS: AF-45 ◇ 2-α-17-α-DIETHYL-A-NOR-5-α-ANDROSTANE-2-β,17-β-DIOL ◇ 2-ETHYNYL-A-NOR-5-α-17-α-PREGN-20-YNE-2-β,17-DIOL ◇ H241 ◇ A-NORANDROSTANE-2-α-17-α-DIETHYNYL-2-β,17-β-DIOL

TOXICITY DATA with REFERENCE
orl-rbt TDLo:690 μg/kg (female 6-8D post):TER FESTAS 16,281,65
orl-rat TDLo:1 mg/kg (female 1D post):REP CCPTAY 32,301,85

SAFETY PROFILE: An experimental teratogen. Experimental reproductive effects. A steroid. When heated to decomposition it emits acrid smoke and irritating fumes.

EQN500 ***HR: 3***
ETHYNYL VINYL SELENIDE
mf: C_4H_4Se mw: 131.04

$$HC{\equiv}CSeCH{=}CH_2$$

CONSENSUS REPORTS: Selenium and its compounds are on the Community Right-To-Know List.

OSHA PEL: TWA 0.2 mg(Se)/m^3
ACGIH TLV: TWA 0.2 mg(Se)/m^3
DFG MAK: 0.1 mg(Se)/m^3

SAFETY PROFILE: Decomposes violently when heated. Upon decomposition it emits toxic fumes of Se. See also SELENIUM COMPOUNDS and ACETYLENE COMPOUNDS.

EQN600 CAS:40054-69-1 ***HR: 2***
ETIZOLAM
mf: $C_{17}H_{15}ClN_4S$ mw: 342.87

PROP: Crystals from toluene. Mp: 147-148°.

SYNS: AHR 3219 ◇ 6-(o-CHLOROPHENYL)-8-ETHYL-1-METHYL-4H-sec-TRIAZOLO(3,4-c)THIENO(2,3-e)(1,4)-DIAZEPINE◇ 6-(o-CHLORPHENYL)-8-AETYL-1-METHYL-4H-sec-TRIAZOLO(3,4-c)THIENO(2,3-e)(1,4)DIAZEPIN (GERMAN) ◇ 4-(2-CHLORPHENYL)-2-ETHYL-9-METHYL-6H-THIENO(3,2-f)(1,2,4)TRIAZOLO(4,3-a)(1,4)DIAZEPINE ◇ DEPAS ◇ Y-7131

TOXICITY DATA with REFERENCE
orl-rat TDLo:135 mg/kg (female 17-22D post):REP OYYAA2 17,787,79
orl-rat TDLo:1100 mg/kg (7-17D preg):TER OYYAA2 17,763,79
orl-rat LD50:3509 mg/kg ARZNAD 28,1158,78
ipr-rat LD50:825 mg/kg OYYAA2 16,1021,78
orl-mus LD50:4258 mg/kg OYYAA2 16,1021,78
ipr-mus LD50:783 mg/kg OYYAA2 16,1021,78

SAFETY PROFILE: Moderately toxic by ingestion and intraperitoneal routes. Experimental teratogenic and reproductive effects. When heated to decomposition it emits toxic fumes of Cl^-, SO_x, and NO_x.

EQN700 CAS:32458-57-4 ***HR: 3***
ETMA
mf: $C_8H_{20}N_3S•Br•BrH$ mw: 351.20

SYNS: (2-(((ETHYLAMINO)IMINOMETHYL)THIO)ETHYL)TRIMETHYL AMMONIUM BROMIDE HYDROBROMIDE ◇ 2-(((ETHYLAMINO)IMINOMETHYL)THIO)-N,N,N-TRIMETHYLETHANAMINIUM BROMIDE, MONOHYDROBROMIDE ◇ S-(2-TRIMETHYLAMINOETHYL)-1'-ETHYLISOTHIURONIUM BROMIDE HYDROBROMIDE

TOXICITY DATA with REFERENCE
ipr-mus LD50:95 mg/kg CPBTAL 23,1639,75
scu-mus LD50:102 mg/kg CPBTAL 23,1639,75
ivn-mus LD50:81800 μg/kg CPBTAL 23,1639,75

SAFETY PROFILE: Poison by subcutaneous, intravenous, and intraperitoneal routes. When heated to decomposition it emits toxic fumes of SO_x, NO_x, NH_3, and HBr.

EQN750 CAS:2053-25-0 ***HR: 3***
ETONITAZENE HYDROCHLORIDE
mf: $C_{22}H_{28}N_4O_3•ClH$ mw: 433.00

SYNS: BENZIMIDAZOLE,1-(2-DIETHYLAMINOETHYL)-2-(p-ETHOXYBENZYL)-5-NITRO-, HYDROCHLORIDE ◇ C 20684 ◇ 1-(2-DIETHYLAMINOETHYL)-2-(p-ETHOXYBENZYL)-5-NITROBENZIMIDAZOLE HYDROCHLORIDE

TOXICITY DATA with REFERENCE
scu-rat TDLo:6700 ng/kg (male 1D pre):REP JPETAB 198,340,76
par-mus LD50:126 mg/kg JMCMAR 11,889,68

SAFETY PROFILE: Poison by parenteral route. Experimental reproductive effects. When heated to decomposition it emits toxic fumes of NO_x and HCl.

EQO000 CAS:57775-22-1 ***HR: 3***
ETOPERIDONE
mf: $C_{19}H_{28}ClN_5O•ClH$ mw: 414.43

SYNS: CLOPRADONE ◇ ST-1191 ◇ TRAZOLINONE

TOXICITY DATA with REFERENCE
orl-rat TDLo:3 g/kg (6-15D preg):TER TXCYAC 8,87,77
orl-rat TDLo:3 g/kg (6-15D preg):REP TXCYAC 8,87,77
orl-rat LD50:397 mg/kg ARZNAD 29,294,79
ipr-rat LD50:120 mg/kg ARZNAD 28,417,78
scu-rat LD50:543 mg/kg ARZNAD 29,294,79
ivn-rat LD50:62 mg/kg ARZNAD 28,417,78
orl-mus LD50:518 mg/kg ARZNAD 29,294,79
ipr-mus LD50:135 mg/kg ARZNAD 28,417,78
ivn-mus LD50:68 mg/kg ARZNAD 29,294,79

SAFETY PROFILE: Poison by ingestion, intravenous, and intraperitoneal routes. An experimental teratogen. Experimental reproductive effects. When heated to decomposition it emits very toxic fumes of Cl^- and NO_x.

EQO450 CAS:14521-96-1 ***HR: 3***
7-α-ETORPHINE
mf: $C_{25}H_{33}NO_4$ mw: 411.59

SYNS: 7,8-DIHYDRO-7-α-(1-(R)-HYDROXY-1-METHYLBUTYL)-O^6-METHYL-6,14-endo-ETHENOMORPHINE ◇ 6,14-endo-ETHENOTETRAHYDROORIPAVINE, 7-α-(1-HYDROXY-1-METHYLBUTYL)- ◇ ETORPHINE ◇ (-)-ETORPHINE ◇ 7-α-(1-(R)-HYDROXY-1-METHYLBUTYL)-6,14-endo-ETHENOTETRAHYDROORIPAVINE ◇ ORIPAVINE, 6,14-endo-ETHYLENETETRAHYDRO-7-(1-HYDROXY-1-METHYLBUTYL)-(7CI) ◇ 19-PROPYLORVINOL ◇ TETRAHYDRO-7-α-(1-HYDROXY-1-METHYLBUTYL)-6,14-endo-ETHENOORIPAVINE ◇ TETRAHYDRO-7-α-(2-HYDROXY-2-PENTYL)-6,14-endo-ETHENOORIPAVINE

TOXICITY DATA with REFERENCE
scu-rat TDLo:3200 ng/kg (male 1D pre):REP JPETAB 198,340,76
ipr-mus LD50:200 mg/kg AGACBH 6,755,76

SAFETY PROFILE: Poison by intraperitoneal route. Experimental reproductive effects. When heated to decomposition it emits toxic fumes of NO_x.

EQO500 CAS:13764-49-3 ***HR: 3***
ETORPHINE HYDROCHLORIDE
mf: $C_{25}H_{33}NO_4•ClH$ mw: 448.05

SYNS: 6,14-ENDOETHENO-7-(2-HYDROXY-2-PENTYL)-TETRAHYDRO-ORIPAVINE HYDROCHLORIDE ◇ 7-α-(1-(R)-HYDROXY-1-METHYLBUTYL)-6,14-ENDOETHENOTETRAHYDRO-ORIPAVINEHYDROCHLORIDE ◇ PROPYLORVINOL HYDROCHLORIDE

TOXICITY DATA with REFERENCE
orl-rat LD50:72 mg/kg BJPCAL 30,11,67
scu-rat LD50:53 mg/kg BJPCAL 30,11,67
ivn-rat LD50:5300 μg/kg BJPCAL 30,11,67
orl-mus LD50:1856 mg/kg BJPCAL 30,11,67
scu-mus LD50:425 mg/kg BJPCAL 30,11,67
ivn-mus LD50:80 mg/kg BJPCAL 30,11,67

SAFETY PROFILE: Poison by ingestion, subcutaneous, and intravenous routes. When heated to decomposition it emits very toxic fumes of NO_x and HCl.

EQP000 CAS:29767-20-2 ***HR: 3***
ETP
mf: $C_{32}H_{32}O_{13}S$ mw: 656.70

SYNS: 4'-DEMETHYLEPIPODOPHYLLOTOXIN-9-(4,6-O-2-THENYLIDENE-β-d-GLUCOPYRANOSIDE ◇ 4'-DEMETHYL-EPIPODOPHYLLOTOXIN-β-d-THENYLIDENE-GLUCOSIDE ◇ 4'-DEMETHYL 1-O-(4,6-O,O-(2-THENYLIDENE)-β-d-GLUCOPYRANOSYL)EPIPODOPHYLLOTOXIN ◇ EPT ◇ NSC-122819 ◇ PTG ◇ TENIPOSIDE ◇ VEHAM-SANDOZ ◇ VEHEM ◇ VM-26 ◇ VUMON

TOXICITY DATA with REFERENCE
mmo-sat 100 μg/plate TCMUD8 5,319,85
dnd-hmn:oth 500 nmol/L CNREA8 45,3106,85
cyt-hmn:lym 5 mg/L CRSBAW 179,331,85
dnd-mus/leu 50 μg/kg CNREA8 40,4225,80
dnd-mus:leu 2 μmol/L CNREA8 44,3360,84
cyt-mus:lym 500 ng/L ENMUDM 8(Suppl 6),24,86
sce-ham:ovr 10 μg/L CNREA8 43,577,83
ipr-mus TDLo:1 mg/kg (7D preg):TER TJADAB 18,31,78
ipr-mus TDLo:1 mg/kg (7D preg):REP TJADAB 18,31,78
orl-hmn TDLo:9579 mg/kg:CNS,GIT,SKN EJCAAH 14,1395,78
ivn-hmn TDLo:26 mg/kg/10D-I:BLD CANCAR 34,985,74
ivn-hmn TDLo:132 mg/kg/7W-I:GIT,BLD CTRRDO 63,7,79
ipr-mus LD17:50 mg/kg CTRRDO 60,1127,76
ipr-mus LD50:29570 μg/kg NCISP* JAN86
scu-mus LD50:31560 μg/kg NCISP* JAN86

SAFETY PROFILE: Poison by intraperitoneal and subcutaneous routes. An experimental teratogen. Human systemic effects by ingestion and intravenous route: anorexia, nausea or vomiting, leukopenia, agranulocytosis and aplastic anemia of the blood, bone marrow changes, and hair changes. Experimental reproductive effects. Human mutation data reported. When heated to decomposition it emits very toxic fumes of SO_x.

EQP100 ***HR: 3***
E. TYPHOSA LIPOPOLYSACCHARIDE

TOXICITY DATA with REFERENCE
ipr-rat TDLo:1 mg/kg (12D preg):TER PSEBAA 109,429,62
ipr-rat TDLo:1 mg/kg (12D preg):REP PSEBAA 109,429,62
ipr-rat LD50:6 mg/kg PSEBAA 109,429,62

SAFETY PROFILE: Poison by intraperitoneal route. Experimental teratogenic and reproductive effects.

EQP500 CAS:500-34-5 ***HR: 3***
β-EUCAINE
mf: $C_{15}H_{21}NO_2$ mw: 247.37

SYNS: EUCAINE B ◇ EUKAIN B ◇ TRIMETHYLBENZOLOXYPIPERIDINE ◇ 2,2,6-TRIMETHYL-4-PIPERIDINOL BENZOATE (ESTER)

TOXICITY DATA with REFERENCE
ivn-rat LDLo:15 mg/kg PHREA7 12,190,32
ivn-mus LDLo:71 mg/kg WDMU** -,-,36
ivn-cat LDLo:10 mg/kg PHREA7 12,190,32
scu-rbt LDLo:400 mg/kg PHREA7 12,190,32
ipr-gpg LDLo:18 mg/kg PHREA7 12,190,32
scu-gpg LDLo:310 mg/kg PHREA7 12,190,32
ivn-gpg LDLo:30 mg/kg PHREA7 12,190,32

SAFETY PROFILE: Poison by intravenous, subcutaneous, and intraperitoneal routes. When heated to decomposition it emits toxic fumes of NO_x.

EQQ000 CAS:8000-48-4 ***HR: 3***
EUCALYPTUS OIL

PROP: From steam distillation of leaves of *Eucalyptus globulus* Labillardiere. Chief constituent is eucalyptol (FCTXAV 13,19,75). Colorless to pale-yellow liquid; spicy odor and taste. Composition: eucalyptol, aldehydes, d-pinene. Mp: −15.4° (approx). D: 0.905-0.925 @ 25°/25°.

SYNS: DINKUM OIL ◇ EUKALYPTUS OEL (GERMAN) ◇ OIL of EUCALYPTUS

TOXICITY DATA with REFERENCE
skn-rbt 500 mg/24H MOD FCTXAV 13,91,75
orl-chd TDLo:218 mg/kg:EYE,CNS,PUL ADCHAK 28,475,53
orl-man LDLo:375 mg/kg ADCHAK 28,475,53
orl-rat LD50:2480 mg/kg FCTXAV 13,91,75

CONSENSUS REPORTS: Reported in EPA TSCA Inventory.

SAFETY PROFILE: A human poison by ingestion. Human systemic effects by ingestion: ciliary eye spasms, somnolence, and respiratory depression. A skin irritant. When heated to decomposition it emits acrid smoke and irritating fumes. See also ALDEHYDES.

EQQ100 CAS:1641-74-3 ***HR: 3***
EUCAST
mf: $C_{12}H_{18}N_2O_2{\cdot}C_6H_8O_7$ mw: 414.46

SYNS: 53-11 C ◇ EUCLIDAN ◇ NICAMETATE CITRATE ◇ NICAMETATE DIHYDROGEN CITRATE ◇ PROVASAN ◇ SOCLIDAN

TOXICITY DATA with REFERENCE
orl-rat LD50:8400 mg/kg NIIRDN 6,542,82
scu-rat LD50:3600 mg/kg NIIRDN 6,542,82
ivn-rat LD50:327 mg/kg NIIRDN 6,542,82
orl-mus LD50:11900 mg/kg NIIRDN 6,542,82
scu-mus LD50:2600 mg/kg NIIRDN 6,542,82
ivn-mus LD50:316 mg/kg NIIRDN 6,542,82

SAFETY PROFILE: Poison by intravenous route. Moderately toxic by subcutaneous route. Mildly toxic by ingestion. When heated to decomposition it emits toxic fumes of NO_x. See also ESTERS.

EQQ500 CAS:18984-80-0 ***HR: 3***
EUCUPIN DIHYRDOCHLORIDE
mf: $C_{24}H_{32}N_2O_2{\cdot}2ClH$ mw: 453.50

SYNS: EUKUPIN DIHYDROCHLORIDE ◇ ISOAMYL HYDROCUPREINE DIHYDROCHLORIDE ◇ 4-((1-METHOXYBUTOXY)(5-VINYL-2-QUINUCLIDINYL)METHYL)-6-QUINOLINOLDIHYDROCHLORIDE ◇ OTODYNE

TOXICITY DATA with REFERENCE
scu-rat LDLo:800 mg/kg CLDND* 11,257,20
orl-cat LDLo:25 mg/kg ZGEMAZ 11,257,20
scu-cat LDLo:50 mg/kg ZGEMAZ 11,257,20
ivn-cat LDLo:6820 μg/kg ZGEMAZ 11,257,20

SAFETY PROFILE: Poison by ingestion and intravenous routes. Moderately toxic by subcutaneous route. When heated to decomposition it emits very toxic fumes of HCl and NO_x.

EQR000 CAS:28399-17-9 ***HR: 3***
EUDESMA-3,11(13)-DIEN-12-OIC ACID
mf: $C_{15}H_{22}O_2$ mw: 234.37

SYN: 12-CARBOXYEUDESMA-3,11(13)-DIENE

TOXICITY DATA with REFERENCE
orl-mus LD50:1000 mg/kg JMCMAR 13,1221,70
ipr-mus LD50:200 mg/kg JMCMAR 13,1221,70

SAFETY PROFILE: Poison by intraperitoneal route. Moderately toxic by ingestion. When heated to decomposition it emits acrid smoke and irritating fumes.

EQR500 CAS:97-53-0 ***HR: 3***
EUGENOL
mf: $C_{10}H_{12}O_2$ mw: 164.22

PROP: Colorless or yellowish liquid; pungent, clove odor. D: 1.064-1.070, refr index: 1.540, bp: 253.5°, flash p: 219°F. Sol in alc, chloroform, ether, volatile oils; very sltly sol in water.

SYNS: 4-ALLYLGUAIACOL ◇ 4-ALLYL-1-HYDROXY-2-METHOXYBENZENE ◇ 4-ALLYL-2-METHOXYPHENOL ◇ CARYOPHYLLIC ACID ◇ EUGENIC ACID ◇ Fa 100 ◇ FEMA No. 2467 ◇ 1-HYDROXY-2-METHOXY-4-ALLYLBENZENE ◇ 4-HYDROXY-3-METHOXYALLYLBENZENE ◇ 1-HYDROXY-2-METHOXY-4-PROP-2-ENYLBENZENE ◇ 2-METHOXY-4-ALLYLPHENOL ◇ 2-METHOXY-4-PROP-2-ENYLPHENOL ◇ 2-METHOXY-4-(2-PROPENYL)PHENOL ◇ 2-METOKSY-4-ALLILOFENOL (POLISH) ◇ NCI-C50453 ◇ SYNTHETIC EUGENOL

TOXICITY DATA with REFERENCE
skn-hmn 40 mg/48H MLD FCTXAV 13,545,75
mma-sat 50 μg/plate NTIS** AD-A116-715
oms-ham:ovr 400 mg/L CALEDQ 14,251,81
cyt-ham:fbr 125 mg/L FCTOD7 22,623,84
cyt-ham:ovr 400 mg/L CALEDQ 14,251,81

orl-mus TDLo:37080 mg/kg/2Y-I:ETA NTPTR* NTP-TR-223,82
orl-rat LD50:1930 mg/kg PSEBAA 73,148,50
ipr-rat LDLo:800 mg/kg RMSRA6 16,449,1896
scu-rat LDLo:5000 mg/kg RMSRA6 16,449,1896
orl-mus LD50:3000 mg/kg FCTXAV 2,327,64
ipr-mus LD50:500 mg/kg COREAF 250,1148,60
orl-dog LDLo:500 mg/kg GASTAB 15,481,50
orl-gpg LD50:2130 mg/kg FCTXAV 2,327,64

CONSENSUS REPORTS: IARC Cancer Review: Group 3 IMEMDT 7,56,87; Animal Limited Evidence IMEMDT 36,75,85. NTP Carcinogenesis Studies (feed); Equivocal Evidence: mouse NTPTR* NTP-TR-223,83; No Evidence: rat NTPTR* NTP-TR-223,83. Reported in EPA TSCA Inventory. EPA Genetic Toxicology Program.

SAFETY PROFILE: Moderately toxic by ingestion, intraperitoneal, and subcutaneous routes. Human mutation data reported. A human skin irritant. Questionable carcinogen with experimental carcinogenic and tumorigenic data. Combustible liquid. When heated to decomposition it emits acrid smoke and irritating fumes. See also ALLYL COMPOUNDS.

EQS000 CAS:93-28-7 ***HR: 2***
EUGENOL ACETATE
mf: $C_{12}H_{14}O_3$ mw: 206.26

PROP: Solid or pale yellow liquid; mild clove odor. D: 1.87, mp: 29-30°, bp: 281.2°, flash p: +151°F. Insol in water; sol in alc and ether.

SYNS: ACETEUGENOL ◇ 1-ACETOXY-2-METHOXY-4-ALLYLBENZENE ◇ ACETYLEUGENOL ◇ 4-ALLYL-2-METHOXYPHENOL ACETATE ◇ 1,3,4-EUGENOL ACETATE ◇ EUGENYL ACETATE ◇ FEMA No. 2469

TOXICITY DATA with REFERENCE
skn-rbt 500 mg/24H MOD FCTXAV 12,807,74
orl-rat LD50:1670 mg/kg FCTXAV 2,327,64

CONSENSUS REPORTS: Reported in EPA TSCA Inventory.

SAFETY PROFILE: Moderately toxic by ingestion. A skin irritant. Combustible liquid. When heated to decomposition it emits acrid smoke and irritating fumes. See also EUGENOL, ALLYL COMPOUNDS, and ESTERS.

EQS100 CAS:10031-96-6 ***HR: 2***
EUGENOL FORMATE
mf: $C_{11}H_{12}O_3$ mw: 192.23

SYNS: 4-ALLYL-2-METHOXYPHENOL FORMATE ◇ EUGENYL FORMATE ◇ PHENOL, 4-ALLYL-2-METHOXY-, FORMATE (ester) ◇ 4-(2-PROPENYL)-2-METHOXYPHENYL FORMATE

TOXICITY DATA with REFERENCE
skn-rbt 500 mg/24H MLD FCTOD7 20,689,82
orl-rat LD50:3400 mg/kg FCTOD7 20,689,82

SAFETY PROFILE: Moderately toxic by ingestion. A skin irritant. When heated to decomposition it emits acrid smoke and irritating fumes.

EQS500 CAS:534-76-9 ***HR: 3***
EULICIN
mf: $C_{24}H_{52}N_8O_2$ mw: 484.84

SYNS: 9-((AMINOIMINOMETHYL)AMINO)-N-(10-((AMINOIMINOMETHYL)AMINO)-1-(3-AMINOPROPYL)-20-HYDROXYDECYL)NONANAMIDE ◇ N-(1-(3-AMINOPROPYL)-10-GUANIDINO-2-HYDROXYDECYL)-9-GUANIDINONONANAMIDE ◇ N-METHOXY-N- METHYLNONANAMIDE

TOXICITY DATA with REFERENCE
ipr-mus LD50:17 mg/kg 85ERAY 2,1142,78
scu-mus LD50:46 mg/kg 85ERAY 2,1142,78
ivn-mus LD50:3 mg/kg 85ERAY 2,1142,78
ims-mus LD50:12 mg/kg 85ERAY 2,1142,78

SAFETY PROFILE: Poison by intraperitoneal, subcutaneous, intravenous, and intramuscular routes. A fungicide. When heated to decomposition it emits toxic fumes of NO_x.

EQT000 CAS:1403-51-6 ***HR: 3***
EUMYCETIN

PROP: An antifungal antibiotic produced by the strain *Streptomyces sp.* 108 (85RAY 2,1126,78).

TOXICITY DATA with REFERENCE
ipr-mus LD50:1200 μg/kg 85ERAY 2,1126,78
scu-mus LD50:3 mg/kg JAJAAA 7,165,54

SAFETY PROFILE: Poison by intraperitoneal and subcutaneous routes.

EQT500 ***HR: 3***
EUPHORBIA ABYSSINICA LATEX

PROP: Acetone soluble portion of *Euphorbia abyssinica* latex.

TOXICITY DATA with REFERENCE
skn-mus TDLo:2600 mg/kg/26W-I:ETA CNREA8 21,338,61

SAFETY PROFILE: Questionable carcinogen with experimental tumorigenic data.

EQU000 ***HR: 3***
EUPHORBIA CANARIENSIS LATEX

PROP: Acetone soluble fraction of latex from *Euphorbia canariensis*.

TOXICITY DATA with REFERENCE
skn-mus TDLo:2600 mg/kg/26W-I:ETA CNREA8 21,338,61

SAFETY PROFILE: Questionable carcinogen with experimental tumorigenic data.

EQU500 ***HR: 3***
EUPHORBIA CANDELABRIUM LATEX

PROP: Acetone soluble portion of latex from *Euphorbia candelabrum*.

TOXICITY DATA with REFERENCE
skn-mus TDLo:2600 mg/kg/26W-I:ETA CNREA8 21,338,61

SAFETY PROFILE: Questionable carcinogen with experimental tumorigenic data.

EQV000 ***HR: 3***
EUPHORBIA ESULA LATEX

PROP: Acetone soluble portion of latex from *Euphorbia esula*.

TOXICITY DATA with REFERENCE
skn-mus TDLo:3500 mg/kg/20W-I:ETA TUMOAB 64,99,78

SAFETY PROFILE: Questionable carcinogen with experimental tumorigenic data.

EQV500 ***HR: 3***
EUPHORBIA GRANDIDENS LATEX

PROP: Acetone soluble fraction of latex from *Euphorbia grandides*.

TOXICITY DATA with REFERENCE
skn-mus TDLo:2600 mg/kg/26W-I:ETA CNREA8 21,338,61

SAFETY PROFILE: Questionable carcinogen with experimental tumorigenic data.

EQW000 ***HR: 3***
EUPHORBIA LATHYRIS LATEX

PROP: Oil from the seeds of *Euphorbia lathyris*.

SYN: CAPER SPURGE

TOXICITY DATA with REFERENCE
skn-mus 13 μg MLD CNREA8 28,2338,68
skn-mus TDLo:5680 mg/kg/W-I:ETA CNREA8 28,2338,68

SAFETY PROFILE: Questionable carcinogen with experimental tumorigenic data. A skin irritant.

EQW500 ***HR: 3***
EUPHORBIA OBOVALIFOLIA LATEX

PROP: Acetone soluble portion of latex from *Euphorbia obovalifolia*.

TOXICITY DATA with REFERENCE
skn-mus TDLo:2600 mg/kg/26W-I:ETA CNREA8 21,338,61

SAFETY PROFILE: Questionable carcinogen with experimental tumorigenic data.

EQX000 ***HR: 1***
EUPHORBIA PULCHERRIMA WILLD.

PROP: The plant contains sterols and latex which have 7-15% caoutchouc.

SYNS: ANNUAL POINSETTIA ◇ CHRISTMAS FLOWER ◇ EASTER FLOWER ◇ EUPHORBIA POINSETTIS BUIST ◇ MEXICA FLAME LEAF ◇ MEXICAN FLAME TREE ◇ MEXICAN FLOWER PLANT ◇ POINSETTIA ◇ POINSETTIA PULCHERRIMA GRAH

TOXICITY DATA with REFERENCE
skn-rbt 11200 mg/14D MOD CTOXAO 13,27,78

SAFETY PROFILE: A skin irritant.

EQX500 ***HR: 3***
EUPHORBIA SERRATA LATEX

PROP: Acetone soluble fraction of latex from *Euphorbia serrata*.

TOXICITY DATA with REFERENCE
skn-mus TDLo:6400 mg/kg/16W-I:ETA TUMOAB 64,99,78

SAFETY PROFILE: Questionable carcinogen with experimental tumorigenic data.

EQY000 ***HR: 3***
EUPHORBIA TIRUCALLI LATEX

PROP: Acetone soluble portion of *Euphorbia tirucalli* latex. Contains all biological activity.

SYN: BLEISTIFTBAUMS (GERMAN)

TOXICITY DATA with REFERENCE
skn-mus 26 ng MLD PLMEAA 22,241,72
skn-mus TDLo:2400 mg/kg/12W-I:ETA PLMEAA 22,241,72

SAFETY PROFILE: Questionable carcinogen with experimental tumorigenic data. A skin irritant.

EQY500 ***HR: 3***
EUPHORIA WULFENII LATEX

PROP: Acetone soluble portion of latex from *Euphoria wulfenii*.

TOXICITY DATA with REFERENCE
skn-mus TDLo:2600 mg/kg/26W-I:ETA CNREA8 21,338,61

SAFETY PROFILE: Questionable carcinogen with experimental tumorigenic data.

EQY600 CAS:32665-36-4 ***HR: 3***
EUPNERON
mf: $C_{22}H_{30}N_2O_2$ mw: 354.54

SYNS: EPROZINOL ◇ 4-(2-METHOXY-2-PHENYLETHYL)-α-PHENYL-1-PIPERAZINEPROPANOL (9CI)

TOXICITY DATA with REFERENCE
orl-rat LD50:640 mg/kg OYYAA2 19,503,80
ipr-rat LD50:72 mg/kg OYYAA2 19,503,80
ims-rat LD50:140 mg/kg OYYAA2 19,503,80
orl-mus LD50:350 mg/kg OYYAA2 19,503,80
ipr-mus LD50:103 mg/kg OYYAA2 19,503,80
ims-mus LD50:122 mg/kg OYYAA2 19,503,80

SAFETY PROFILE: Poison by ingestion, intramuscular and intraperitoneal routes. When heated to decomposition it emits toxic fumes of NO_x.

EQZ000 CAS:53198-87-1 ***HR: 3***
EURAZYL
mf: $C_{19}H_{28}N_2 \cdot ClH$ mw: 320.95

SYNS: α-sec-BUTYL-α-PHENYL-1-PIPERIDINEBUTYRONITRILE HYDROCHLORIDE ◇ α-PHENYL-α-(2-PIPERIDINOETHYL)-β-ETHYLBUTYRIC ACID NITRILE HYDROCHLORIDE ◇ α-PHENYL-α-(2-PIPERIDINOETHYL)-β-ETHYLBUTYRONITRILE HYDROCHLORIDE

TOXICITY DATA with REFERENCE
orl-mus LD50:700 mg/kg JAPMA8 49,298,60
ivn-dog LDLo:64 mg/kg JAPMA8 49,298,60

CONSENSUS REPORTS: Cyanide and its compounds are on the Community Right-To-Know List.

SAFETY PROFILE: Poison by intravenous route. Moderately toxic by ingestion. When heated to decomposition it emits very toxic fumes of HCl, NO_x, and CN^-. See also NITRILES.

ERA100 ***HR: 3***
EUROPEAN MISTLETOE

PROP: A parasitic plant which grows mostly on the trunks of deciduous trees such as the apple. The thick leaves are usually a pale yellow-green. The white berry is sticky. It is native to Europe and now grows in Sonoma County, California.

SYN: VISCUM ALBUM

SAFETY PROFILE: The leaves and stems contain the poisonous viscumin and viscotoxins, toxalbumins which inhibit protein synthesis. The berries may be edible. Ingestion of the leaves causes after a few hours: abdominal pain, diarrhea and possibly necrotic lesions throughout the gastroenteric tract. See also VISCOTOXIN and ABRIN.

ERA309 ***HR: 3***
EUROPEAN SPINDLE TREE

PROP: A small tree native to Europe and now cultivated in the northern United States. It has tooth-edged leaves, yellow-green flowers and a pink seed capsule.

SYN: EUONYMUS EUROPAEUS

SAFETY PROFILE: The plant contains the poison evomonoside, a digitalis-like cardiac glycoside, several alkaloids including evonine, and a protein which inhibits protein synthesis. Ingestion of the fruit has caused (within 12 hours) diarrhea, vomiting, fever, hallucinations, sleepiness, coma and convulsions. Cardiac glycosides may cause death by their effect on heart function. See also DIGITALIS.

ERA500 CAS:10025-76-0 ***HR: 3***
EUROPIUM CHLORIDE
mf: Cl_3Eu mw: 258.31

SYN: EUROPIC CHLORIDE

TOXICITY DATA with REFERENCE
orl-mus LD50:3527 mg/kg EQSSDX 1,1,75
ipr-mus LD50:387 mg/kg AEHLAU 5,437,62
ipr-gpg LD50:156 mg/kg AEHLAU 5,437,62

CONSENSUS REPORTS: Reported in EPA TSCA Inventory.

SAFETY PROFILE: Poison by intraperitoneal route. Moderately toxic by ingestion. When heated to decomposition it emits very toxic fumes of Cl^-. See also RARE EARTHS.

ERB000 CAS:13240-06-7 ***HR: 3***
EUROPIUM CITRATE

TOXICITY DATA with REFERENCE
ipr-mus LD50:187 mg/kg AEHLAU 5,437,62
ipr-gpg LD50:72 mg/kg AEHLAU 5,437,62

SAFETY PROFILE: Poison by intraperitoneal route. When heated to decomposition it emits acrid smoke and irritating fumes. See also RARE EARTHS.

ERB500 CAS:15158-64-2 ***HR: 3***
EUROPIUM EDETATE

TOXICITY DATA with REFERENCE
ipr-mus LD50:240 mg/kg AEHLAU 5,437,62
ipr-gpg LD50:118 mg/kg AEHLAU 5,437,62

SAFETY PROFILE: Poison by intraperitoneal route.

When heated to decomposition it emits acrid smoke and irritating fumes. See also RARE EARTHS.

ERC000 CAS:10031-53-5 ***HR: 3***
EUROPIUM(III) NITRATE, HEXAHYDRATE (1:3:6)
mf: $N_3O_9 \cdot Eu \cdot 6H_2O$ mw: 446.11

SYN: NITRIC ACID, EUROPIUM(3+) SALT, HEXAHYDRATE

TOXICITY DATA with REFERENCE
ipr-rat LD50:210 mg/kg TXAPA9 5,750,63
ipr-mus LD50:320 mg/kg TXAPA9 5,750,63

SAFETY PROFILE: Poison by intraperitoneal route. When heated to decomposition it emits very toxic fumes of NO_x. See also NITRATES and RARE EARTHS.

ERC500 CAS:12020-65-4 ***HR: 3***
EUROPIUM(II)SULFIDE
mf: EuS mw: 184.02

SAFETY PROFILE: Ignites or explodes spontaneously in air. When heated to decomposition it emits toxic fumes of SO_x. See also RARE EARTHS and SULFIDES.

ERC550 CAS:10138-01-9 ***HR: 3***
EUROPIUM TRINITRATE
mf: $Eu(NO_3)_3$ mw: 421.98

SYN: EUROPIUM NITRATE

TOXICITY DATA with REFERENCE
orl-rat LD50:3828 mg/kg EQSSDX 1,1,75
ipr-rat LD50:162 mg/kg EQSSDX 1,1,75
ipr-mus LD50:245 mg/kg EQSSDX 1,1,75

SAFETY PROFILE: Poison by intraperitoneal route. Moderately toxic by ingestion. When heated to decomposition it emits toxic fumes of NO_x. See also NITRATES and RARE EARTHS.

ERC600 CAS:62851-59-6 ***HR: 3***
EUROTIN (A)

SYN: YUROTIN A

TOXICITY DATA with REFERENCE
ipr-mus LD50:3 mg/kg 85FZAT -,281,67
scu-mus LD50:6 mg/kg 85FZAT -,281,67
ivn-mus LD50:3 mg/kg 85FZAT -,281,67

SAFETY PROFILE: Poison by subcutaneous, intravenous, and intraperitoneal routes.

ERC800 CAS:39340-46-0 ***HR: 3***
EVERNINOMICIN D
mf: $C_{66}H_{99}Cl_2NO_{35}$ mw: 1537.56

TOXICITY DATA with REFERENCE
ims-rat LD50:400 mg/kg 85ERAY 1,208,78
orl-mus LD50:3750 mg/kg 85ERAY 1,208,78
ipr-mus LD50:3750 mg/kg 85ERAY 1,208,78
scu-mus LD50:3750 mg/kg 85ERAY 1,208,78
ivn-mus LD50:125 mg/kg 85ERAY 1,208,78
ims-dog LD50:140 mg/kg 85ERAY 1,208,78
ims-rbt LD50:300 mg/kg 85ERAY 1,208,78

SAFETY PROFILE: Poison by intramuscular and intravenous routes. Moderately toxic by ingestion, intraperitoneal, and subcutaneous routes. When heated to decomposition it emits very toxic fumes of Cl^- and NO_x.

ERD000 CAS:53296-30-3 ***HR: 2***
EVERNINOMYCIN-B
mf: $C_{66}H_{99}Cl_2NO_{36}$ mw: 1553.56

SYN: R-451-B

TOXICITY DATA with REFERENCE
ipr-mus LD50:880 mg/kg 85GDA2 1,354,80
scu-mus LD50:1700 mg/kg 85GDA2 1,354,80
ivn-mus LD50:875 mg/kg 85GDA2 1,354,80

SAFETY PROFILE: Moderately toxic by intraperitoneal, subcutaneous and intravenous routes. When heated to decomposition it emits toxic fumes of Cl^- and NO_x.

ERD500 CAS:56-29-1 ***HR: 3***
EVIPAL
mf: $C_{12}H_{16}N_2O_3$ mw: 236.30

SYNS: BARBIDORM ◇ CITODON ◇ CITOPAN ◇ 5-(1-CYCLOHEXEN-1-YL)-1,5-DIMETHYLBARBITURIC ACID ◇ 5-(1-CYCLOHEXEN-1-YL)-1,5-DIMETHYL-2,4,6(1H,3H,5H)-PYRIMIDINETRIONE ◇ 5-(1-CYCLOHEXENYL-1)-1-METHYL-5-METHYLBARBITURIC ACID ◇ 5-(Δ-1,2-CYCLOHEXENYL)-5-METHYL-N-METHYL-BARBITURSAEURE (GERMAN) ◇ CYCLONAL ◇ CYCLOPAN ◇ 1,5-DIMETHYL-5-(1-CYCLOHEXENYL)BARBITURIC ACID ◇ DORICO ◇ ENHEXYMAL ◇ ESOBARBITALE (ITALIAN) ◇ EVIPAN ◇ HEXABARBITAL ◇ HEXANASTAB ORAL ◇ HEXENAL ◇ HEXENAL (barbiturate) ◇ HEXOBARBITAL ◇ HEXOBARBITONE ◇ METHEXENYL ◇ N-METHYL-5-CYCLOHEXENYL-5-METHYLBARBITURIC ACID ◇ METHYLHEXABARBITAL ◇ METHYLHEXABITAL ◇ NARCOSAN ◇ NOCTOVANE ◇ SOMBUCAPS ◇ SOMBULEX ◇ SOMNALERT

TOXICITY DATA with REFERENCE
cyt-hmn:leu 1500 mg/L TGANAK 18(1),13,84
ipr-rat LD50:330 mg/kg AIPTAK 184,5,70
scu-rat LDLo:400 mg/kg AEPPAE 182,348,36
orl-mus LD50:468 mg/kg JPETAB 106,444,52
ipr-mus LD50:280 mg/kg JPETAB 93,362,48
scu-mus LD50:250 mg/kg ARZNAD 15,688,65
ivn-mus LDLo:75 mg/kg AIPTAK 92,305,53
ipl-mus LDLo:340 mg/kg JPETAB 134,95,61
orl-rbt LDLo:1200 mg/kg JPETAB 60,189,37
ivn-rbt LDLo:80 mg/kg JPETAB 60,189,37
rec-rbt LDLo:175 mg/kg JPETAB 60,189,37
ipr-frg LDLo:30 mg/kg PHREA7 19,472,39

SAFETY PROFILE: Poison by subcutaneous, intraperitoneal, intravenous, intrapleural, and rectal routes. Moderately toxic by ingestion. Human mutation data reported. When heated to decomposition it emits toxic fumes of NO_x. See also BARBITURATES.

ERE000 CAS:50-09-9 ***HR: 3***
EVIPAL SODIUM
mf: $C_{12}H_{15}N_2O_3 \cdot Na$ mw: 258.28

SYNS: trans-2-BUTENOIC ACID ◇ 5-(1-CYCLOHEXEN-1-YL)-1,5-DIMETHYLBARBITURIC ACID SODIUM SALT ◇ 5-(1-CYCLOHEXEN-1-YL)-1,5-DIMETHYL-2,4,6(1H,3H,5H,)-PYRIMIDINETRIONEMONOSODIUM SALT ◇ 5-(1-CYCLOHEXEN-1-YL)-1,5-DIMETHYL-2,4,6(1H,3H,5H)-PYRIMIDINETRIONE SODIUM SALT (9CI) ◇ CYCLONAL SODIUM ◇ 1,5-DIMETHYL-5-CYCLOHEXENYL-1'-BARBITURIC ACID, SODIUM SALT ◇ DORICO SOLUBLE ◇ ENHEXYMAL ◇ ENHEXYMAL NFN ◇ EVIPAN SODIUM ◇ HEXANAL ◇ HEXANASTAB ◇ HEXENAL SODIUM ◇ HEXOBARBITAL Na ◇ HEXOBARBITAL SODIUM ◇ HEXOBARBITONE Na ◇ HEXOBARBITONE SODIUM ◇ METHEXENYL SODIUM ◇ NARCOSAN SOLUBLE ◇ NOCTIVANE SODIUM ◇ PRIVENAL ◇ SODIUM-5-(1-CYCLOHEXEN-1-YL)-1,5-DIMETHYLBARBITURATE ◇ SODIUM EVIPAL ◇ SODIUM EVIPAN ◇ SODIUM HEXABARBITAL ◇ SODIUM HEXOBARBITONE ◇ SODIUM-N-METHYL CYCLOHEXENYLMETHYBARBITURATE ◇ SODIUM METHYLHEXABITAL ◇ SODIUM METHYLHEXABITOL

TOXICITY DATA with REFERENCE
orl-rat LD50:468 mg/kg MEIEDD 10,680,83
ipr-rat LDLo:160 mg/kg JPETAB 50,347,34
orl-mus LD50:1325 mg/kg TXAPA9 27,70,74
ipr-mus LD50:270 mg/kg JPETAB 87,265,46
scu-mus LD50:410 mg/kg TXAPA9 27,70,74
ivn-mus LD50:145 mg/kg TXAPA9 23,537,72
ivn-dog LDLo:85 mg/kg JPETAB 60,125,37
par-frg LDLo:30 mg/kg JPETAB 50,347,34

SAFETY PROFILE: Poison by intraperitoneal, parenteral, and intravenous routes. Moderately toxic by ingestion and subcutaneous routes. Used intravenously as a general anesthetic. When heated to decomposition it emits toxic fumes of NO_x and Na_2O. See also EVIPAL and BARBITURATES.

ERE100 CAS:59333-90-3 ***HR: 3***
EXAPROLOL HYDROCHLORIDE
mf: $C_{18}H_{29}NO_2 \cdot ClH$ mw: 327.94

SYNS: 1-(2-(1-CYCLOHEXEN-1-YL)PHENOXY)-3-((1-METHYLETHYL)AMINO)-2-PROPANOL HYDROCHLORIDE ◇ 1-ISOPROPYLAMINO-3-(2-CYCLOHEXYLPHENOXY)-2-PROPANOLHYDROCHLORIDE ◇ M.G. 8823

TOXICITY DATA with REFERENCE
orl-rat LD50:1850 mg/kg ARZNAD 26,506,76
ipr-rat LD50:54 mg/kg ARZNAD 26,506,76
orl-mus LD50:860 mg/kg ARZNAD 26,506,76
ipr-mus LD50:78 mg/kg ARZNAD 26,506,76

SAFETY PROFILE: Poison by intraperitoneal route. Moderately toxic by ingestion. When heated to decomposition it emits toxic fumes of NO_x and HCl.

ERE200 CAS:3478-44-2 ***HR: 2***
EXIPROBEN SODIUM
mf: $C_{16}H_{23}O_5 \cdot Na$ mw: 318.38

SYNS: DCH 21 ◇ DROCTIL ◇ 3-HEXOXY-1-(2'-CARBOXYPHENOXY)-PROPANOL-(2) SODIUM SALT ◇ o-(3-(HEXYLOXY)-2-HYDROXYPROPOXY)BENZOIC ACID MONOSODIUM SALT

TOXICITY DATA with REFERENCE
orl-rat LD50:2150 mg/kg ARZNAD 24,111,74
ipr-rat LD50:645 mg/kg ARZNAD 24,111,74
orl-mus LD50:1660 mg/kg ARZNAD 24,111,74
ipr-mus LD50:650 mg/kg ARZNAD 24,111,74

SAFETY PROFILE: Moderately toxic by ingestion and intraperitoneal routes. When heated to decomposition it emits toxic fumes of Na_2O.

ERF000 ***HR: 3***
EXPLOSIVES, HIGH

SAFETY PROFILE: High explosives (HE) are those which decompose by detonation. This is a very rapid (nearly instantaneous), and hence, violent process. An explosion may be initiated by sudden shock, high temperatures, or a combination of the two. For many explosives the conditions under which they will explode are well known, for example:

An explosion may be initiated by elevated temperature alone:

(1) In the case of mercury fulminate, a 15-second exposure to 200°C or 1 second exposure to 340°C will set it off.

(2) Trinitrotoluene will be set off by exposure to 500°C for 1 second.

(3) Tetryl will detonate in 1000 seconds at 160°C or in 0.1 second at 500°C.

(4) Picric acid will detonate in 9 seconds at 300°C or 1 second at 355°C.

An explosion of HE may also be initiated by severe shock. Sensitivity of explosives to shock may be measured in several ways, such as the impact pendulum method and the drop test. The impact pendulum test operates by allowing a heavy pendulum to swing down over a sample of explosive in a dished, inclined container so arranged that there is very little clearance between the pendulum and the sample. Thus, the effect of contact between the sample and the pendulum bob is one of a combination of shock and rubbing. The height from which the pendulum is allowed to swing to explode the sample is a measure of the sensitivity of the sample to this test. The drop test consists of placing a sample upon an anvil and allowing a 5 pound weight to drop on it. The height from which the weight must drop to explode the sample is a measure of the sample's sensitivity to shock.

The table below shows the results of a drop test upon

several samples. These results must be considered as relative and not by any means absolute. A solid explosive in a tightly fitting container is much more sensitive to shock.

(1) mercury fulminate = 2 in. at 5 lbs.
(2) nitroglycerin = 4 in. at 5 lbs.
(3) tetryl = 8 in. at 5 lbs.
(4) picric acid = 14 in. at 5 lbs.
(5) trinitrotoluene = 20 in. at 5 lbs.
(6) black powder (a low explosive) = 30 in. at 5 lbs.*

*From *Explosions, Their Anatomy and Destructiveness*, by C. S. Robinson (McGraw-Hill).

Another test for explosives is the speed at which a detonation travels. This speed is usually in the range of thousands of m/sec. Speed of detonation is found to be a function of the kind of explosive and state of compaction. There is an optimum state of compaction beyond which the explosive tends to become "''deadpressed,'' in which state it is difficult to make the whole sample explode. Below the point of optimum compaction the rate of detonation is found to be directly proportional to the density of the sample. Some maximum detonation rates are listed below in meters/sec for some common explosives:

(1) nitroglycerin=8500
(2) PETN=8100
(3) tetryl=7700
(4) picric acid=7400
(5) trinitrotoluene=7400
(6) lead azide=4900
(7) mercury fulminate=4800
(8) ammonlium nitrate=1100
(9) low explosives=1000

It has been found that upon detonation, an explosive can cause a nearby sample of explosive to detonate "sympathetically." The distance over which one charge can detonate another is a function of the amount of energy produced by the first explosion and the medium through which the shock wave is propagated to the second charge of explosive. For instance, the relationship for air (very approximately) would be expected to be: Weight of explosive in lbs/(distance in ft)^3 = 4. Thus, to calculate the maximum distance for a possible sympathetic detonation of 40,000 lbs of explosive, the calculation is:

$D^3 = (40,000)/4$
$D^3 = 10,000$
$D = 22$ ft (approximately).

According to C. S. Robinson, the formula is more nearly:

weight of explosive = 4 ×(distance)$^{2.25}$

The power of the shock wave is much more rapidly attenuated in water, wood, etc., than in air, which means that if a shield of water or wood is interposed between piles of explosive the distance between them may be lessened.

Liquid Oxygen: Though not itself explosive, liquid oxygen can be dangerous when blended with highly flammable or carbonaceous materials. In this combination it is used in coal mining, quarrying, strip mining, open-cut ore mining, and in rocket fuels. Its use underground or in confined places is not recommended by the U.S. Bureau of Mines because it evolves a lot of carbon monoxide. This type of explosive has many safety advantages. For instance, it is not itself an explosive until mixed with a flammable absorbent which can be done at the last moment before firing. However, once the explosive has been made up, it is very flammable and when it catches fire it will usually detonate. Liquid oxygen explosives are not stored, as they deteriorate rapidly and lose a great deal of their explosive power in a short time.

A very dangerous fire hazard when exposed to heat or by chemical reaction with powerful oxidizing or reducing agents.

A moderate to dangerous explosion hazard when severely shocked or heated, depending upon the kind of explosive, state of compaction, degree of confinement, etc. Practically all high explosives used commercially require a detonator or cap to set them off, as compared to an igniter needed to set off black blasting powder.

Detonating Devices: To develop the desired disruptive effect of an explosive, some means must be adopted to "set off," "fire," or "detonate" it without killing or maiming the persons doing the blasting. Several devices or methods are being utilized, all with a view to having this work done as safely and efficiently as possible. There are two general types of devices or methods of getting explosives into action, namely, igniters and detonators. The former merely conveys a flame to the explosive mass and ignites it, while the latter transmits (originally through ignition of a small quantity of highly explosive substance by an arc, a flame, or spark) a sharp blow that causes the explosive to disassociate, or detonate, or burn with very great rapidity. Igniters are squibs (plain and electric), fuse, and delay igniters; detonators are blasting caps (plain and electric), delay electric blasting caps, delay electric igniters with caps, and Cordeau-Bickford detonating fuse.

The squib is a small-diameter tube of straw or paper filled with quick-burning powder and having a relatively slow burning "match head" attached to one end; the latter is ignited or lighted by an ordinary match or other flame, and its relatively slow burning allows the person handling the ignition to retire before the fire is communicated to the quick-burning material in the tube. Squibs are by no means either safe or efficient, even though still used to a considerable extent, especially in coal mining. Electric squibs are somewhat similar to ordinary squibs, except that the ignition is accomplished by means of an electric arc; electric squibs are much more satisfactory from a safety viewpoint than ordinary squibs.

A fuse (or, as it is sometimes called, "safety fuse") consists of a fine-grained black powder core covered with cotton hemp or jute to form a ropelike material about 3/16-inch in diameter; one end of the fuse is brought in contact with the powder charge or with a detonating "cap," and the other end (usually several feet away from the explosive) is lighted by a flame from a match or open light. The fine-grained black powder burns gradually and somewhat slowly (about 30 to 40 seconds to the linear foot of fuse) until it reaches the explosive (black powder) or the detonating cap (if some form of dynamite is used), giving the blaster time to get in the clear before the main explosion takes place. Fuses are much safer than squibs, but have their own hazards and must be used with care.

Delay electric igniters usually are a combination of electric igniters and fuses, the latter being ignited by the igniters within the blasting hole, the fuse transmitting the ignition to the explosive. Delay igniters usually are much safer than fuse, particularly for coal-mine use; but they, too, have their hazards. Delay blasting is by no means a safe procedure in coal mining, though it is a standard and relatively safe practice in metal mining and tunneling, if sensible precautions are taken.

Blasting caps or detonators are metallic cylinders (usually copper) closed at one end, about 3/16-inch in diameter, and usually less than 2 inches in length, partly filled with a small amount of relatively easily fired or "detonated" compound, the resultant shock or blow when fired being sufficient, when embedded in dynamite, to fire or detonate the dynamite mass. Ordinary blasting caps usually are fired or detonated by the flame of the fuse, the end of the latter being inserted into the open end of the detonator or cap and placed in contact with the highly explosive material in the interior of the metallic capsule or cap. Caps are extremely hazardous to handle, as they are likely to be detonated by heat, friction, or a relatively moderate blow; however, they are relatively safe if handled carefully. Partial proof of this is the fact that they are manufactured and shipped by the thousands daily and accidents are decidedly rare, primarily because the caps are at all times handled with utmost care.

Electric blasting caps are somewhat similar to ordinary caps or detonators, but the cap is fired by electricity. The electric wires are so placed in the capsule or cap that when attached to an electric current an arc is formed within the cap, which detonates the sensitive explosive material in the cap. A hazard in the use of electric blasting caps is unexpected explosions due to radio or radar induced electric currents which may activate the cap.

Delay electric blasting caps or detonators are somewhat similar to ordinary electric blasting caps, except that several time intervals in blasting are obtained by having the electric arc ignite a short piece of fuse or some slow-burning substance before it reaches the highly sensitive detonating material in the capsule or cap. Numerous time-interval delays are obtained; in general, delay electric blasting caps are relatively safe and effective even in wet holes, though they ought not be used in coal mining if explosions of gas or dust are to be avoided. Delay electric igniters with caps or detonators are a combination of electric igniter and blasting cap, usually with suitable lengths of fuse between to give the desired delay; they have some advantages but are relatively unsafe and should not be used in coal mining.

The Cordeau-Bickford denotating fuse is a combined fuse and detonator in the form of a lead tube about 1/4-inch in diameter filled throughout its length with a high explosive, trinitrotoluene (TNT). It is fired by a fuse and an ordinary detonator or cap or by an electric cap; when fired, it detonates throughout its length (which may be up to or over 100 feet) almost instantaneously, the explosion wave traveling at a rate of about 17,500 ft/sec. Although somewhat expensive, it is relatively safe to handle and is particularly effective in deep-well drill holes in quarry and similar work, as it detonates simultaneously throughout its length, adding effectiveness to the main body of explosive which it detonates. It fires black powder as well as high explosives (dynamite, etc.), and is obtainable in lengths of approximately 500 feet wound on spools.

See also EXPLOSIVES, PERMITTED; DYNAMITE, NITROGLYCERIN, AMMONIUM NITRATE, NITRATES, and TRINITROTOLUENE.

ERF500 *HR: 3*
EXPLOSIVES, LOW

PROP: Black powder is composed of saltpeter, charcoal and sulfur in the approximate proportions of 6:1:1. ("A" blasting powder uses KNO_3 and "B" blasting powder uses $NaNO_3$).

SYNS: "A" BLASTING POWDER ◇ "B" BLASTING POWDER ◇ BLACK BLASTING POWDER ◇ GUNPOWDER

SAFETY PROFILE: Low explosives are explosives which deflagrate; this differentiates them both in composition and properties from high explosives, which detonate. A deflagrating explosive is one that burns progressively over a relatively sustained period of time in comparison with a detonating explosive, which decomposes almost instantaneously. A dangerous fire hazard when exposed to heat or flame or by chemical reaction.

Black powder is the most treacherous explosive material used today and it is regarded as one of the worst known explosive hazards. When ignited unconfined it burns with explosive violence and will explode if ignited under even slight confinement. It can be ignited easily by very small sparks, heat, and friction. It is the slowest acting of all explosives. It has a shearing and heaving action tending to blast materials in large, firm fragments. The action derives from a relatively slow development of gas pressure so that it must be carefully loaded and closely confined. It is subject to rapid deterioration in the presence of moisture, but if kept dry it retains its explosive properties for many years. It is used to ignite smokeless powder, propelling charges, airplane flares and bursting charges of hand grenades, as a bursting charge in shrapnel, practice bombs, practice trench-mortar shells, in saluting charges, smoke-puff charges, time and percussion fuses, pellets, primers and primer detonators, and in expelling charges of pyrotechnic signals.

Although most safety men now look upon black blasting powder with disfavor, it is one of the oldest and most generally used explosives in commercial work. It burns with extreme rapidity instead of detonating as high explosives do. It is highly sensitive to flame or sparks or friction and gives off much flame, which is hot and of great length of duration. These properties make it extremely hazardous for use in mines (especially coal mines) and quarries. The gases given off in detonation are not only very hot but frequently contain harmful constituents. Notwithstanding its numerous deficiencies, from a safety standpoint it has action characteristics that make it valuable in both coal mining and quarrying, though it has relatively little utility in metal mining. It is difficult to use effectively in wet places and this is its main disadvantage from an efficiency standpoint.

Most black powder fires start from sparks. Ignition results in an explosion so quickly that no attempt can be made to fight the fire. Every effort should be made to prevent fires from reaching stores of black powder; but if this fails, fire-fighting forces should be withdrawn at least 800 feet from the fire and should protect themselves against an explosion by seeking any cover available or by lying flat on the ground. If an explosion does occur, every effort should be made to prevent flames from spreading to neighboring magazines. Fire-fighting forces should be cautioned against approaching a fire which may involve black powder to avoid being trapped or injured by an explosion.

The following safety rules should be strictly enforced and obeyed. Open no containers in a magazine to which explosives or ammunitions are stored. This should be done only in a building free from all other explosives or ammunitions; or in suitable weather in the open, at least 100 feet from the nearest magazine. The quantity at or near such an operation should be limited to 100 pounds. Safety tools only should be used in opening or closing containers or in other operations involving black powder. Processes should be so laid out as to bring about frequent grounding of all operators handling this material. Safety shoes (non-insulating) should be worn in all rooms where black powder is handled and by all persons engaged in handling black powder. The wearing of all nonconductive shoes, such as rubber is prohibited. If the handling of black powder is carried on over a concrete floor, the floor should be covered with a tarpaulin or other suitable material. Loose black powder is extremely dangerous. Whenever it is necessary to handle loose black powder, not over 50 pounds should be permitted at or near such operations. If black powder is spilled on benches or floors, all work should be stopped until it has been removed and the explosive hazard of any remaining dust or particles has been neutralized with water. Rooms or buildings in which black powder is handled should be inspected frequently for dust, and all such dust should be immediately removed with water. The empty powder containers should be washed out, as explosions are said to have occurred from "empty" containers.

If dry and in good condition, black powder burns rapidly, especially in small grain size, with a yellow or pinkish-blue flame and dense smoke.

Pellet powder is black blasting powder in consolidated (pellet or stick) form rather than in grains or granules, and it has few if any real advantages over black blasting powder, notwithstanding the fairly prevalent idea that it is a "safe" explosive.

Smokeless powders have a somewhat different composition from that of black blasting powder and are used chiefly for sportsmen's ammunition and, more widely, for military purposes. They are decidedly sensitive to flame and impact but ordinarily are so packaged that if reasonable judgment is used they are relatively harmless.

ERG000 ***HR: 3***
EXPLOSIVES, PERMITTED

SAFETY PROFILE: "Permissible" explosives are essentially high explosives (dynamite) modified by the introduction of "dopes." The function of the dopes in general is to decrease flame temperature and to a smaller extent, the length and duration of flame, when the explosive is converted from a solid into a gas; in other words, when it is fired or detonated. The designation "permissible" is given to an explosive of modified dynamite type after it has passed certain tests made by the Federal Bureau of Mines. The permissible character of such explosives depends not only upon the ingredients in the explosive, but also on certain well-defined specifications as to handling and use. As with the dynamites, there are several different types and grades; "permissibles," hydrated "permissibles," organic nitrate "permissibles," nitroglycerin "permissibles," ammonium perchlorate "permissibles," and gelatin "permissibles." Essentially all of those now used to any extent are in either the ammonium nitrate or the gelatin classes. See also DYNAMITE.

The ammonium nitrate "permissible" explosives contain relatively little nitroglycerin and relatively large proportions of ammonium nitrate. The latter is an explosive but one less sensitive to impact, sparks, and flames than nitroglycerin. This type of permissible explosive is now used extensively, as it has a rather wide range of strength, rate of detonation, density, size of cartridge, etc., and can be utilized not only in dry but also to some extent in fairly wet holes if charged carefully and fired promptly.

Gelatin "permissible," explosives are more suitable than ammonium nitrate "permissible," for wet holes, and in general are stronger and more violent than the ammonium nitrate types.

All "permissible" explosives are strong, and must be used in relatively small quantities (less than 1.5 pounds) per hole to retain their permissibility, give off considerable quantities of toxic gases on detonation, and, while much safer than black blasting powder or dynamite, must be stored, handled, and used with care.

Classification Upon Basis of Toxic Gases: All "permissible" explosives, when detonated, emit some toxic gases and a much larger volume of nontoxic gases. In order that the toxic products may not become a menace to the life or health of miners, no explosive is now or can become "permissible" if upon detonation it evolves more than 158 liters (5.5 cu ft) of toxic gases per 1.5 pound charge as determined by tests in the Bichel pressure gage. Classification upon the basis of the volume of toxic gases produced by 580 grams (1.5 pounds) of explosive is as follows: *Class A,* not more than 53 liters; *Class B,* between 53 and 106 liters; and *Class C,* between 106 and 158 liters. (These classifications are not to be confused with the I.C.C. Classification of explosives).

Field tests were made with a 1.5 pounds charge of a "permissible" explosive that produced, in the Bichel gage, the maximum allowable quantity of poisonous gases (158 liters per 1.5 pounds); these tests indicated that in a narrow entry, without artificial ventilation. 1800 ppm of carbon monoxide (the only poisonous gas present) was produced, as shown by analysis of an air sample taken 2 minutes after the shot. Another sample of the air taken 2 minutes later contained 800 ppm of carbon monoxide. Under no conditions should miners or shot firers return to the place until the poisonous gases have been removed by adequate ventilation.

It is provided further that, in accordance with the provisions and conditions, explosives enumerated on the "permissible" lists of the Bureau of Mines are "permissible" in use only when they satisfy the following requirements:

1. That the explosive be in all respects similar to the sample submitted by the manufacturer for test, and that the diameters of the cartridges used must be those that have been approved.

2. That electric detonators (not fuse and detonators) be used of not less efficiency than No. 6, the detonation charge of which shall consist of a 1 gram mixture of 80 parts of mercury fulminate and 20 parts of potassium chlorate (or their equivalents), and that the required electric firing must be done by means of a "permissible" type blasting unit.

3. That the explosive be stored in surface magazines under proper conditions, so that it will not undergo change in character, and that after being taken underground it be used in less than 36 hours.

4. That the coal to be blasted be undercut or equivalently relieved; that, to prevent blow-through, all portions of the borehole must be at least 18 inches from relief in any direction; that, to prevent blowouts, the charge be properly confined with not less than 2 feet of clay (if the length of the hole will not permit the charge desired and 2 feet of stemming, at least half the length of the hole shall be filled with stemming) or other incombustible stemming and not be on the solid; that, to prevent the hole being on the solid; it shall be at least 6 inches shorter than the depth of the undercut or equivalent relief, and, when placed adjacent to the roofs, ribs, or floor, all but 12 inches at the rear of the hole must be at least 6 inches from the adjacent surface as projected into the coal to be blasted, and all parts of the hole shall be free from the adjacent surface as projected into the coal to be blasted; that the shot be not a dependent shot; and that the shot hole be cleaned before charging.

5. That the quantity used for a shot (1) be not in excess of 680 grams (1.5 pounds) when fired in accordance with these requirements and (2) when used under certain additional requirements or restrictions be not in excess of 1,361 grams (3 pounds). The use of charges over 1.5 pounds and not exceeding 3 pounds is approved tentatively pending further investigation. For charges of over 1.5 pounds, the following additional requirements must be observed, (a) Shot holes must be 6 feet or more in length. (b) Explosives must be charged in a continuous train, with no cartridges deliberately deformed or crushed, with all cartridges in contact with each other, and with the end cartridges touching the rear of the hole and the stemming, respectively. (c) Examination for gas must be made in the blasting area before and after a shot is fired. (d) The "permissible" explosive must be one showing toxic gas emission that will place it either in Class A or Class B.

6. That the region in which the blasting is done be kept well protected by rock dust or otherwise in accordance with Bureau of Mines inspection standards.

7. That the shot not be fired in the presence of a dangerous percentage of firedamp. Examination for firedamp to be made at the blasting area before shooting in a gassy mine.

See also AMMONIUM NITRATE, AZIDES, DYNAMITE, FULMINATES. NITRATES, NITROGLYCERIN, PENTAERYTHRITOL TETRANITRATE, PICRIC ACID, and TRINITROTOLUENE.

F

FAB000 ***HR: 3***
F III (sugar fraction)

PROP: Sugar fraction of the extract obtained from *Sasa albamarginata* with a molecular weight between 5000 to 10,000 (YKKZAJ 99,663,79).

TOXICITY DATA with REFERENCE
orl-mus LD50:10 g/kg YKKZAJ 99,663,77
ipr-mus LD50:2750 mg/kg YKKZAJ 99,663,77
scu-mus TD50:6400 mg/kg YKKZAJ 99,663,77
ivn-mus LD50:360 mg/kg YKKZAJ 99,663,77

SAFETY PROFILE: Poison by intravenous route. Moderately toxic by intraperitoneal route. Mildly toxic by ingestion and subcutaneous routes. When heated to decomposition it emits acrid smoke and irritating fumes.

FAB100 ***HR: 2***
FALSE HELLEBORE

PROP: Top (plant) material from *Veratrum californicum (Durand)* (CJBIAE 44,829,66). Tall, perennial herbs with pleated leaves and white flowers with green marks. The winged seeds are held in a small pod. The various species grow wild in the region bounded by the Aleutian Islands, most of Canada, New Mexico, and Georgia.

SYNS: AMERICAN WHITE HELLEBORE ◇ CORN LILY ◇ EARTH GALL ◇ GREEN HELLEBORE ◇ INDIAN POKE ◇ ITCH WEED ◇ PEPPER-ROOT ◇ RATTLESNAKE WEED ◇ SKUNK CABBAGE ◇ SWAMP HELLEBORE ◇ TICKLE WEED ◇ VERATRUM ALBUM ◇ VERATRUM CALIFORNICUM ◇ VERATRUM VIRIDE ◇ WHITE HELLEBORE

TOXICITY DATA with REFERENCE
orl-dom TDLo:236 g/kg(1-20D preg):TER AJVRAH 24,1164,63
orl-mus LD50:900 mg/kg AFECAT 76(815),65,83
orl-ham LDLo:15 mg/kg AMZOAF 9,1134,69

SAFETY PROFILE: Poison by ingestion. Experimental teratogenic effects. All parts of the plant contain poisonous veratrum alkaloids. Ingestion of any part of the plant causes an immediate pain in the abdomen followed by nausea and vomiting, blurred vision, confusion, slowed heartbeat, and low blood pressure. The subject may feel as if he is having a heart attack. Effects usually last only 24 hours.

FAB400 CAS:13171-21-6 ***HR: 3***
FAMFOS
mf: $C_{10}H_{19}ClNO_5P$ mw: 299.72

SYNS: APAMIDON ◇ C 570 ◇ (2-CHLOOR-3-DIETHYLAMINO-1-METHYL-3-OXO-PROP-1-EN-YL)-DIMETHYL-FOSFAAT ◇ (2-CHLOR-3-DIAETHYLAMINO-1-METHYL-3-OXO-PROP-1-EN-YL)-DIMETHYL-PHOSPHAT ◇ 2-CHLORO-2-DIETHYLCARBAMOYL-1-METHYLVINYL DIMETHYLPHOSPHATE ◇ 1-CHLORO-DIETHYLCARBAMOYL-1-PROPEN-2-YL DIMETHYL PHOSPHATE ◇ (2-CLORO-3-DIETILAMINO-1-METIL-3-OXO-PROP-1-EN-IL)-DIMETIL-FOSFATO ◇ CIBA 570 ◇ CROTONAMIDE, 2-CHLORO-N,N-DIETHYL-3-HYDROXY-, DIMETHYL PHOSPHATE ◇ DIMECRON ◇ DIMECRON 100 ◇ DIMETHYL 2-CHLORO-2-DIETHYLCARBAMOYL-1-METHYLVINYLPHOSPHATE ◇ O,O-DIMETHYL O-(2-CHLORO-2-(N,N-DIETHYLCARBAMOYL)-1-METHYLVINYL) PHOSPHATE ◇ DIMETHYL DIETHYLAMIDO-1-CHLOROCROTONYL (2) PHOSPHATE ◇ O,O-DIMETHYL-O-(1-METHYL-2-CHLOR-2-N,N-DIAETHYL-CARBAMOYL)-VINYL-PHOSPHAT ◇ (O,O-DIMETHYL-O-(1-METHYL-2-CHLORO-2-DIETHYLCARBAMOYL-VINYL) PHOSPHATE) ◇ DIMETHYL PHOSPHATE of 2-CHLORO-N,N-DIETHYL-3-HYDROXYCROTONAMIDE ◇ DIXON ◇ ENT 25,515 ◇ FOSFAMIDON ◇ FOSFAMIDONE ◇ FOSZFAMIDON ◇ ML 97 ◇ NCI-C00588 ◇ OMS 1325 ◇ OR 1191 ◇ PHOSPHAMIDON ◇ PHOSPHATE de DIMETHYLE et de (2-CHLORO-2-DIETHYLCARBAMOYL-1-METHYL-VINYLE)

TOXICITY DATA with REFERENCE
mmo-sat 5 μg/plate IJEBA6 24,305,86
mma-sat 9 mg/plate ENMUDM 5(Suppl 1),3,83
cyt-hmn:lyms 1900 μg/L MUREAV 31,103,75
cyt-hmn:leu 5 μmol/L IJEBA6 18,1145,80
bfa-rat:sat 5800 μg/kg IJEBA6 24,305,86
mnt-mus-orl 10 mg/kg/24H BECTA6 25,277,80
orl-mus TDLo:5 mg/kg (7D post):TER TOLED5 42,101,88
orl-mus TDLo:5 mg/kg (7D post):REP TOLED5 42,101,88
orl-rat TDLo:5400 mg/kg/80W-C:ETA NCITR* NCI-TR-16,79
orl-rat LD50:8 mg/kg 85JCAE-,1138,86
ihl-rat LC50:135 mg/m^3/4H EGESAQ 24,173,80
skn-rat LD50:125 mg/kg PHJOAV 185,361,60
ipr-rat LD50:8700 μg/kg PESTD5 17,351,76
scu-rat LD50:15 mg/kg IPPABX 4,253,68
orl-mus LD50:6 mg/kg SPEADM 78-1,28,78
ihl-mus LC50:30 mg/m^3/1H PSDTAP 15,239,74
ipr-mus LD50:5800 μg/kg TXAPA9 13,37,68
scu-mus LD50:13200 μg/kg AIPTAK 132,180,61
ivn-mus LD50:6 mg/kg ATXKA8 18,316,60
orl-rbt LD50:70 mg/kg 85GYAZ-,33,71
skn-rbt LD50:80 mg/kg 85JCAE-,1138,86
ihl-gpg LC50:1300 mg/m^3/4H PSDTAP 15,239,74
orl-dck LD50:3100 μg/kg DOEAAH 35,25,79

skn-dck LD50:26 mg/kg TXAPA9 47,451,79
orl-brd LD50:1800 μg/kg TXAPA9 21,315,72

CONSENSUS REPORTS: NTP Carcinogenesis Bioassay (feed): No Evidence: mouse NCITR* NCI-TR-16,79; Inadequate Studies: rat NCITR* NCI-TR-16,79

SAFETY PROFILE: Poison by ingestion and most other routes. An experimental teratogen. Experimental reproductive data. Questionable carcinogen with experimental tumorigenic data. Mutation data reported. When heated to decomposition it emits toxic fumes of NO_x and PO_x.

FAB500 CAS:76824-35-6 ***HR: 2***
FAMOTIDINE
mf: $C_8H_{15}N_7O_2S_3$ mw: 337.48

PROP: Mp: 163-164°. Solubility at 20° (%, w/v): 80 in DMF; 50 acetic acid; 0.3 in methanol; 0.1 in water; <0.01 in ethanol, ethyl acetate, chloroform.

SYNS: 3-(((2-((AMINOIMINOMETHYL)AMINO)-4-THIAZOLYL) METHYL)THIO)-N-(AMINOSULFONYL)PROPANIMIDAMIDE ◇ 3-(((2-((DIAMINOMETHYLENE)AMINO)-4-THIAZOLYL)METHYL)THIO)-$N^{(2)}$-SULFAMOYLPROPIONAMIDINE ◇ FAMODIL ◇ GASTER ◇ GASTRIDIN ◇ MOTIAX ◇ PEPCID ◇ PEPCIDINE ◇ PEPDUL ◇ YM-11170

TOXICITY DATA with REFERENCE
orl-rat TDLo:2800 mg/kg (15-22D preg/21D post):REP OYYAA2 26,543,83
orl-rat TDLo:1100 mg/kg (multi) :TER OYYAA2 26,489,83
orl-rat LD50:4907 mg/kg IYKEDH 16,590,85
ipr-rat LD50:800 mg/kg DIGEBW 32(Suppl 1),7,85
scu-rat LD50:800 mg/kg DIGEBW 32(Suppl 1),7,85
ivn-rat LD50:440 mg/kg YKYUA6 36,653,85
orl-mus LD50:4686 mg/kg IYKEDH 16,590,85
ipr-mus LD50:778 mg/kg OYYAA2 26,147,83
ivn-mus LD50:410 mg/kg YKYUA6 36,653,85

SAFETY PROFILE: Moderately toxic by intraperitoneal, subcutaneous, and intravenous routes. An experimental teratogen. Experimental reproductive effects. When heated to decomposition it emits toxic fumes of NO_x and SO_x.

FAB600 CAS:52-85-7 ***HR: 3***
FAMPHUR
mf: $C_{10}H_{16}NO_5PS_2$ mw: 325.36

PROP: Crystalline powder. Mp: 55°. Very sol in chloroform and carbon tetrachloride; sltly sol in water.

SYNS: AMERICAN CYANAMID-38023 ◇ AMERICAN CYANAMID CL-38,023 ◇ BO-ANA ◇ CL-38023 ◇ CYFLEE ◇ O-(4-((DIMETHYLAMINO) SULFONYL)PHENYL) O,O-DIMETHYL PHOSPHOROTHIOATE ◇ O,O-DIMETHYL-O-(p-(N,N-DIMETHYLSULFAMOYL)PHENYL)PHOS PHOROTHIOATE ◇ DOVIP ◇ ENT 25,644 ◇ FAMFOS ◇ FAMOPHOS ◇ FAMOPHOS WARBEX ◇ FAMPHOS ◇ FANFOS ◇ RCRA WASTE NUMBER P097 ◇ WARBEX

TOXICITY DATA with REFERENCE
orl-rat LD50:35 mg/kg TXAPA9 21,315,72
orl-mus LD50:27 mg/kg JAFCAU 15,845,67
skn-rbt LD50:1460 mg/kg WRPCA2 9,119,70
ims-dom LD50:59 mg/kg VETNAL 54(12),94,78
ims-mam LD50:64 mg/kg VETNAL 54(12),94,78
orl-bwd LD50:1800 μg/kg TXAPA9 21,315,72

CONSENSUS REPORTS: Reported in EPA TSCA Inventory.

SAFETY PROFILE: Poison by ingestion and intramuscular routes. Moderately toxic by skin contact. A cholinesterase inhibitor. When heated to decomposition it emits toxic fumes of NO_x and SO_x. See also PARATHION.

FAB800 CAS:4602-84-0 ***HR: 2***
FARNESOL
mf: $C_{15}H_{26}O$ mw: 222.41

PROP: trans-Farnesol: Light yellow liquid; mild oily odor. Bp: 111°, n (25/D) 1.4872. Commercial farnesol: Bp: 110-113°, d: (20/4) 0.8871, n (20/D) 1.4870., refr index: 1.487-1.492. Insol in water @ 263°.

SYNS: FARNESYL ALCOHOL ◇ FEMA No. 2478 ◇ 3,7,11-TRIMETHYL-2,6,10-DODECATRIEN-1-OL

TOXICITY DATA with REFERENCE
dni-oin:ovr 100 μmol/L ABCHA6 43,1285,79
orl-rat LD50:6000 mg/kg THERAP 27,893,72
orl-mus LD50:7400 mg/kg THERAP 27,893,72
ipr-mus LD50:443 mg/kg THERAP 27,893,72

CONSENSUS REPORTS: Reported in EPA TSCA Inventory.

SAFETY PROFILE: Moderately toxic by intraperitoneal route. Mildly toxic by ingestion. Mutation data reported. When heated to decomposition it emits acrid smoke and irritating fumes.

FAB880 ***HR: 1***
FATTY ACID, SOAP (C12-14)

TOXICITY DATA with REFERENCE
skn-hmn 2500 μg/24H MOD AKEDAX 235,180,69
skn-rbt 10 mg MLD JSCCA5 22,411,71

SAFETY PROFILE: A human and experimental skin irritant. When heated to decomposition it emits acrid smoke and irritating fumes.

FAB900 CAS:61789-01-3 ***HR: 1***
FATTY ACID, TALL OIL, EPOXIDIZED-2-ETHYLHEXYL ESTER
mf: $C_{26}H_{48}O_4$ mw: 424.74

SYNS: FLEXOL EP-8 ◇ sec-OCTYL EPOXYTALLATE

TOXICITY DATA with REFERENCE
orl-rat LD50:20800 mg/kg NPIRI* 2,31,75

CONSENSUS REPORTS: Reported in EPA TSCA Inventory.

SAFETY PROFILE: Mildly toxic by ingestion. When heated to decomposition it emits acrid smoke and irritating fumes.

FAB920 CAS:61788-72-5 ***HR: 1***
FATTY ACID, TALL OIL, EPOXIDIZED, OCTYL ESTER

SYNS: ADMEX 741 ◇ ADMEX 746 ◇ DRAPEX 4.4 ◇ OCTYL EPOXYTALLATE ◇ PLASTOLEIN 9214 ◇ PX-806

TOXICITY DATA with REFERENCE
orl-rat LD50:32 g/kg NPIRI* 2,80,75

CONSENSUS REPORTS: Reported in EPA TSCA Inventory.

SAFETY PROFILE: Mildly toxic by ingestion. When heated to decomposition it emits acrid smoke and irritating fumes.

FAC050 CAS:2014-22-4 ***HR: 3***
FC 402
mf: $C_{14}H_{21}NO_2•ClH$ mw: 271.82

SYN: N,N-DIETHYLGLYCINE-2,6-XYLYL ESTER HYDROCHLORIDE

TOXICITY DATA with REFERENCE
skn-rbt 200 mg MOD BCFAAI 107,310,68
eye-rbt 1 g MOD BCFAAI 107,310,68
scu-mus LD50:1350 mg/kg BCFAAI 107,310,68
ivn-mus LD50:35 mg/kg BCFAAI 107,310,68

SAFETY PROFILE: Poison by intravenous route. Moderately toxic by subcutaneous route. An eye and skin irritant. When heated to decomposition it emits toxic fumes of NO_x and HCl. See also ESTERS.

FAC060 CAS:2173-44-6 ***HR: 3***
FC-410
mf: $C_{15}H_{22}N_2O_2•2ClH$ mw: 335.31

SYN: 4-METHYL-1-PIPERAZINEACETIC ACID 2,6-XYLYL ESTER DIHYDROCHLORIDE

TOXICITY DATA with REFERENCE
skn-rbt 200 mg MOD BCFAAI 107,310,68
eye-rbt 1 g MOD BCFAAI 107,310,68
scu-mus LD50:500 mg/kg BCFAAI 107,310,68
ivn-mus LD50:22 mg/kg BCFAAI 107,310,68

SAFETY PROFILE: Poison by intravenous route. Moderately toxic by subcutaneous route. A skin and eye irritant. When heated to decomposition it emits toxic fumes of NO_x and HCl. See also ESTERS.

FAC100 CAS:2014-27-9 ***HR: 2***
FC 457
mf: $C_{15}H_{23}NO_2•ClH$ mw: 285.85

SYN: l-N,N-DIETHYLALANINE-2,6-XYLYL ESTER HYDROCHLORIDE

TOXICITY DATA with REFERENCE
skn-rbt 200 mg MLD BCFAAI 107,310,68
eye-rbt 1 g MLD BCFAAI 107,310,68
scu-mus LD50:1275 mg/kg BCFAAI 107,310,68

SAFETY PROFILE: Moderately toxic by subcutaneous route. A skin and eye irritant. When heated to decomposition it emits toxic fumes of NO_x and HCl. See also ESTERS.

FAC130 CAS:20684-29-1 ***HR: 3***
FC 480
mf: $C_{17}H_{26}N_2O_2•ClH$ mw: 326.91

SYNS: 2,6-DIMETILFENILICO DELL'ACIDO α-N-METILPIPERAZINOBUTIRRICO IDOCLORIDRAT (ITALIAN) ◇ α-ETHYL-4-METHYL-1-PIPERAZINEACETIC ACID-2,6-XYLYL ESTER HYDROCHLORIDE

TOXICITY DATA with REFERENCE
skn-rbt 200 mg MOD BCFAAI 107,310,68
eye-rbt 1 g MOD BCFAAI 107,310,68
orl-mus LD50:230 mg/kg BCFAAI 107,310,68
ipr-mus LD50:82 mg/kg BCFAAI 107,310,68
scu-mus LD50:410 mg/kg BCFAAI 107,310,68
ivn-mus LD50:20 mg/kg BCFAAI 107,310,68

SAFETY PROFILE: Poison by ingestion, intravenous, and intraperitoneal routes. Moderately toxic by subcutaneous route. An eye and skin irritant. When heated to decomposition it emits toxic fumes of NO_x and HCl. See also ESTERS.

FAC150 CAS:2085-83-8 ***HR: 2***
FC 590
mf: $C_{20}H_{25}NO_2•ClH$ mw: 347.92

SYN: N,N-DIETHYL-2-PHENYL-GLYCINE-2,6-XYLYL ESTER HYDROCHLORIDE

TOXICITY DATA with REFERENCE
skn-rbt 200 mg MLD BCFAAI 107,310,68
eye-rbt 1 g MLD BCFAAI 107,310,68
scu-mus LD50:1200 mg/kg BCFAAI 107,310,68

SAFETY PROFILE: Moderately toxic by subcutaneous route. A skin and eye irritant. When heated to decomposition it emits toxic fumes of NO_x and HCl. See also ESTERS.

FAC155 CAS:19245-07-9 ***HR: 3***
FC 591
mf: $C_{22}H_{28}N_2O_3•ClH$ mw: 404.98

SYN: 4-(2-HYDROXYETHYL)-α-PHENYL-1-PIPERAZINEACETIC ACID-2,6-XYLYL ESTER HYDROCHLORIDE

TOXICITY DATA with REFERENCE
skn-rbt 200 mg SEV BCFAAI 107,310,68
eye-rbt 1 g SEV BCFAAI 107,310,68
scu-mus LD50:65 mg/kg BCFAAI 107,310,68

SAFETY PROFILE: Poison by subcutaneous route. A severe eye and skin irritant. When heated to decomposition it emits toxic fumes of NO_x and HCl. See also ESTERS.

FAC157 CAS:2282-89-5 ***HR: 2***
FC 642
mf: $C_{17}H_{27}NO_2$•ClH mw: 313.91

SYN: l-N,N-DIETHYLALANINE-2,6-DIETHYLPHENYL ESTER HYDROCHLORIDE

TOXICITY DATA with REFERENCE
skn-rbt 200 mg MOD BCFAAI 107,310,68
eye-rbt 1 g MOD BCFAAI 107,310,68
scu-mus LD50:1750 mg/kg BCFAAI 107,310,68

SAFETY PROFILE: Moderately toxic by subcutaneous route. An eye and skin irritant. When heated to decomposition it emits toxic fumes of NO_x and HCl. See also ESTERS.

FAC160 CAS:1877-53-8 ***HR: 3***
FC 646
mf: $C_{20}H_{32}N_2O_2$•2ClH mw: 405.46

SYN: α,4-DIMETHYL-1-PIPERAZINEACETICACID-2,6-DIISOPROPYLPHENYL ESTER DIHYDROCHLORIDE

TOXICITY DATA with REFERENCE
skn-rbt 200 mg SEV BCFAAI 107,310,68
eye-rbt 1 g SEV BCFAAI 107,310,68
scu-mus LD50:600 mg/kg BCFAAI 107,310,68
ivn-mus LD50:42 mg/kg BCFAAI 107,310,68

SAFETY PROFILE: Poison by intravenous route. Moderately toxic by subcutaneous route. A severe eye and skin irritant. When heated to decomposition it emits toxic fumes of NO_x and HCl. See also ESTERS.

FAC163 CAS:2409-35-0 ***HR: 2***
FC 650
mf: $C_{15}H_{23}NO_4$•ClH mw: 317.85

SYN: l-N,N-DIETHYLALANINE-2,6-DIMETHOXYPHENYLESTER HYDROCHLORIDE

TOXICITY DATA with REFERENCE
skn-rbt 200 mg MOD BCFAAI 107,310,68
eye-rbt 1 g MOD BCFAAI 107,310,68
scu-mus LD50:1425 mg/kg BCFAAI 107,310,68

SAFETY PROFILE: Moderately toxic by subcutaneous route. An eye and skin irritant. When heated to decomposition it emits toxic fumes of NO_x and HCl. See also ESTERS.

FAC165 CAS:1877-52-7 ***HR: 3***
FC 651
mf: $C_{19}H_{30}N_2O_2$•2ClH mw: 391.43

SYN: α-ETHYL-4-METHYL-1-PIPERAZINEACETICACID-2,6-DIETHYLPHENYL ESTER DIHYDROCHLORIDE

TOXICITY DATA with REFERENCE
skn-rbt 200 mg MOD BCFAAI 107,310,68
eye-rbt 1 g MOD BCFAAI 107,310,68
scu-mus LD50:540 mg/kg BCFAAI 107,310,68
ivn-mus LD50:34 mg/kg BCFAAI 107,310,68

SAFETY PROFILE: Poison by intravenous route. Moderately toxic by subcutaneous route. An eye and skin irritant. When heated to decomposition it emits toxic fumes of NO_x and HCl. See also ESTERS.

FAC166 CAS:2085-85-0 ***HR: 2***
FC 652
mf: $C_{14}H_{21}NO_4$•ClH mw: 303.82

SYN: N,N-DIETHYLGLYCINE-2,6-DIMETHOXYPHENYL ESTER HYDROCHLORIDE

TOXICITY DATA with REFERENCE
skn-rbt 200 mg MLD BCFAAI 107,310,68
eye-rbt 1 g MLD BCFAAI 107,310,68
scu-mus LD50:550 mg/kg BCFAAI 107,310,68

SAFETY PROFILE: Moderately toxic by subcutaneous route. A skin and eye irritant. When heated to decomposition it emits toxic fumes of NO_x and HCl. See also ESTERS.

FAC170 CAS:2014-32-6 ***HR: 2***
FC 657
mf: $C_{15}H_{23}NO_2$•ClH mw: 285.85

SYN: N,N-DIETHYLGLYCINE MESITYL ESTER HYDROCHLORIDE

TOXICITY DATA with REFERENCE
skn-rbt 200 mg MOD BCFAAI 107,310,68
eye-rbt 1 g MOD BCFAAI 107,310,68
scu-mus LD50:2000 mg/kg BCFAAI 107,310,68

SAFETY PROFILE: Moderately toxic by subcutaneous route. An eye and skin irritant. When heated to decomposition it emits toxic fumes of NO_x and HCl. See also ESTERS.

FAC175 CAS:2014-33-7 ***HR: 2***
FC 659
mf: $C_{16}H_{23}NO_2$•ClH mw: 297.86

SYN: MESITYL ESTER of 1-PIPERIDINEACETIC ACID HYDROCHLORIDE

TOXICITY DATA with REFERENCE
skn-rbt 200 mg MOD BCFAAI 107,310,68
eye-rbt 1 g MOD BCFAAI 107,310,68
scu-mus LD50:2000 mg/kg BCFAAI 107,310,68

SAFETY PROFILE: Moderately toxic by subcutaneous route. An eye and skin irritant. When heated to decomposition it emits toxic fumes of NO_x and HCl. See also ESTERS.

FAC179 CAS:2173-47-9 ***HR: 2***
FC 660
mf: $C_{16}H_{25}NO_2 \cdot ClH$ mw: 299.88

SYN: I-N,N-DIETHYLALANINE MESITYL ESTER HYDROCHLORIDE

TOXICITY DATA with REFERENCE
skn-rbt 200 mg MLD BCFAAI 107,310,68
eye-rbt 1 g MLD BCFAAI 107,310,68
scu-mus LD50:2000 mg/kg BCFAAI 107,310,68

SAFETY PROFILE: Moderately toxic by subcutaneous route. A skin and eye irritant. When heated to decomposition it emits toxic fumes of NO_x and HCl. See also ESTERS.

FAC185 CAS:2014-31-5 ***HR: 3***
FC 668
mf: $C_{15}H_{21}ClN_2O_2 \cdot 2ClH$ mw: 369.75

SYN: α,4-DIMETHYL-1-PIPERAZINEACETICACID-6-CHLORO-o-TOLYL ESTER DIHYDROCHLORIDE

TOXICITY DATA with REFERENCE
skn-rbt 200 mg MOD BCFAAI 107,310,68
eye-rbt 1 g MOD BCFAAI 107,310,68
scu-mus LD50:180 mg/kg BCFAAI 107,310,68

SAFETY PROFILE: Poison by subcutaneous route. An eye and skin irritant. When heated to decomposition it emits toxic fumes of NO_x and HCl. See also ESTERS.

FAC195 CAS:1940-89-2 ***HR: 3***
FC 681
mf: $C_{18}H_{28}N_2O_2 \cdot 2ClH$ mw: 377.40

SYN: α-ETHYL-4-METHYL-1-PIPERAZINEACETIC ACID MESITYL ESTER DIHYDROCHLORIDE

TOXICITY DATA with REFERENCE
skn-rbt 200 mg MOD BCFAAI 107,310,68
eye-rbt 1 g MOD BCFAAI 107,310,68
scu-mus LD50:250 mg/kg BCFAAI 107,310,68

SAFETY PROFILE: Poison by subcutaneous route. An eye and skin irritant. When heated to decomposition it emits toxic fumes of NO_x and HCl. See also ESTERS.

FAE000 CAS:3844-45-9 ***HR: 2***
FD&C BLUE No. 1
mf: $C_{37}H_{36}N_2O_9S_3 \cdot 2Na$ mw: 794.91

PROP: Dark purple to bronze powder. Sol in water, ether, conc sulfuric acid.

SYNS: ACID SKY BLUE A ◇ AIZEN FOOD BLUE No. 2 ◇ 1206 BLUE ◇ BRILLIANT BLUE FCD No. 1 ◇ BRILLIANT BLUE FCF ◇ CANACERT BRILLIANT BLUE FCF ◇ C.I. 42090 ◇ C.I. ACID BLUE 9, DISODIUM SALT ◇ C.I. FOOD BLUE 2 ◇ COGILOR BLUE 512.12 ◇ COSMETIC BLUE LAKE ◇ D&C BLUE No. 4 ◇ DISPERSED BLUE 12195 ◇ DOLKWAL BRILLIANT BLUE ◇ EDICOL BLUE CL 2 ◇ ERIOGLAUCINE G ◇ FENAZO BLUE XI ◇ FOOD BLUE 2 ◇ FOOD BLUE DYE No. 1 ◇ HEXACOL BRILLIANT BLUE A ◇ INTRACID PURE BLUE L ◇ MERANTINE BLUE EG ◇ USACERT BLUE No. 1

TOXICITY DATA with REFERENCE
cyt-ham:lng 4400 mg/L GANMAX 27,95,81
scu-rat TDLo:5500 mg/kg/97W-I:NEO ZEKBAI 64,287,61
par-rat TDLo:4580 mg/kg/62W-I:ETA FAONAU 38B,27,66
scu-mus LD50:4600 mg/kg ZEKBAI 64,287,61

CONSENSUS REPORTS: IARC Cancer Review: Group 3 IMEMDT 7,56,87; Animal Sufficient Evidence IMEMDT 16,171,78. Reported in EPA TSCA Inventory.

SAFETY PROFILE: Questionable carcinogen with experimental neoplastigenic and tumorigenic data. Mutation data reported. When heated to decomposition it emits very toxic fumes of NO_x, Na_2O and SO_x.

FAE100 CAS:860-22-0 ***HR: 3***
FD&C BLUE No. 2
mf: $C_{16}H_{10}N_2O_8S_2 \cdot 2Na$ mw: 468.38

PROP: Blue-brown to red-brown powder. Sol in water, conc sulfuric acid; sltly sol in alc.

SYNS: ACID BLUE W ◇ ACID LEATHER BLUE IC ◇ A.F. BLUE No. 2 ◇ AIRDALE BLUE IN ◇ AMACID BRILLIANT BLUE ◇ ANILINE CARMINE POWDER ◇ ATUL INDIGO CARMINE ◇ 1311 BLUE ◇ 12070 BLUE ◇ BUCACID INDIGOTINE B ◇ CANACERT INDIGO CARMINE ◇ CARMINE BLUE (BIOLOGICAL STAIN) ◇ C.I. 73015 ◇ C.I. 7581 ◇ C.I. ACID BLUE 74 ◇ C.I. FOOD BLUE 1 ◇ DISODIUM INDIGO-5,5-DISULFONATE ◇ DISODIUM SALT of 1-INDIGOTIN-S,S'-DISULPHONIC ACID ◇ DOLKWAL INDIGO CARMINE ◇ E 132 ◇ GRAPE BLUE A GEIGY ◇ INDIGO CARMINE ◇ INDIGO CARMINE (BIOLOGICAL STAIN) ◇ INDIGO CARMINE DISODIUM SALT ◇ INDIGO EXTRACT ◇ INDIGO-KARMIN (GERMAN) ◇ 5,5'-INDIGOTIN DISULFONIC ACID ◇ INDIGOTINE ◇ INDIGOTINE DISODIUM SALT ◇ INTENSE BLUE ◇ L-BLAU 2 (GERMAN) ◇ MAPLE INDIGO CARMINE ◇ SACHSISCHBLAU ◇ SCHULTZ Nr. 1309 (GERMAN) ◇ SODIUM-5,5'-INDIGOTIDISULFONATE ◇ SOLUBLE INDIGO ◇ USACERT BLUE No.2

TOXICITY DATA with REFERENCE
cyt-ham:fbr 12 g/L FCTOD7 22,623,84
cyt-ham:ovr 20 μmol/L/5H-C ENMUDM 1,27,79
scu-rat TDLo:9 g/kg/2Y-I:NEO TXAPA9 8,29,66
orl-rat LD50:2 g/kg SCPHA4 47,39,79
ivn-rat LD50:93 mg/kg NIIRDN 6,83,82

orl-mus LD50:2500 mg/kg SCPHA4 47,39,79
scu-mus LD50:405 mg/kg SCPHA4 47,39,79

CONSENSUS REPORTS: Reported in EPA TSCA Inventory. EPA Genetic Toxicology Program.

SAFETY PROFILE: Poison by intravenous route. Moderately toxic by ingestion and subcutaneous routes. Questionable carcinogen with experimental neoplastigenic data. Mutation data reported. When heated to decomposition it emits very toxic fumes of SO_x, NO_x, and Na_2O.

FAE950 CAS:4680-78-8 ***HR: 3***
FD&C GREEN No. 1
mf: $C_{37}H_{36}N_2O_6S_2$•Na mw: 691.86

SYNS: ACIDAL GREEN G ◇ ACID GREEN ◇ ACID GREEN 3 ◇ A.F. GREEN No. 1 ◇ BRILLIANT GREEN 3EMBL ◇ BUCACID GUINEA GREEN BA ◇ C.I. ACID GREEN 3, MONOSODIUM SALT ◇ C.I. FOOD GREEN 1 ◇ GUINEA GREEN B ◇ HISPACID GREEN GB ◇ LEATHER GREEN B ◇ NAPHTHALENE GREEN G ◇ PONTACYL GREEN BL ◇ SULFACID BRILLIANT GREEN 1B

TOXICITY DATA with REFERENCE
mma-sat 320 μg/plate MUREAV 89,21,81
orl-rat TDLo:660 g/kg/43W-C:ETA GASTAB 23,1,53

CONSENSUS REPORTS: IARC Cancer Review: Group 3 IMEMDT 7,56,87; Animal Sufficient Evidence IMEMDT 16,199,78. Reported in EPA TSCA Inventory. Community Right-To-Know List.

SAFETY PROFILE: Questionable carcinogen with experimental tumorigenic data. Mutation data reported. When heated to decomposition it emits very toxic fumes of SO_x, Na_2O, and NO_x.

FAF000 CAS:5141-20-8 ***HR: 2***
FD&C GREEN No. 2
mf: $C_{37}H_{36}N_2O_9S_3$•2Na mw: 794.91

SYNS: ACIDAL LIGHT GREEN SF ◇ ACID BRILLIANT GREEN SF ◇ ACID GREEN A ◇ ACILAN GREEN SFG ◇ A.F. GREEN No. 2 ◇ AMACID GREEN G ◇ C.I. 42095 ◇ C.I. ACID GREEN 5 ◇ C.I. ACID GREEN 5, DISODIUM SALT ◇ C.I. FOOD GREEN 2 ◇ D&C GREEN No. 4 ◇ FAST ACID GREEN N ◇ FD&C GREEN No. 2-ALUMINUM LAKE ◇ FENAZO GREEN 7G ◇ FOOD GREEN 2 ◇ GREEN No. 203 ◇ LEATHER GREEN SF ◇ LICHTGRUEN (GERMAN) ◇ LIGHT GREEN FCF YELLOWISH ◇ LIGHT GREEN LAKE ◇ LIGHT SF YELLOWISH (BIOLOGICAL STAIN) ◇ LISSAMINE LAKE GREEN SF ◇ MERANTINE GREEN SF ◇ MY/68 ◇ PENCIL GREEN SF ◇ SULFO GREEN J ◇ SUMITOMO LIGHT GREEN SF YELLOWISH ◇ WOOL BRILLIANT GREEN SF

TOXICITY DATA with REFERENCE
mma-sat 320 μg/plate MUREAV 89,21,81
orl-rat TDLo:819 g/kg (male 65W pre):REP JPPMAB 8,417,56
orl-rat TDLo:1300 g/kg/86W-C:CAR GASTAB 23,1,53
scu-rat TDLo:6825 mg/kg/46W-I:NEO JNCIAM 24,769,60
ivn-mus LD50:700 mg/kg TXAPA9 44,225,78

CONSENSUS REPORTS: IARC Cancer Review: Group 3 IMEMDT 7,56,87; Animal Sufficient Evidence IMEMDT 16,209,78. Reported in EPA TSCA Inventory. EPA Genetic Toxicology Program.

SAFETY PROFILE: Moderately toxic by intravenous route. Questionable carcinogen with experimental carcinogenic and neoplastigenic data. Experimental reproductive effects. Mutation data reported. When heated to decomposition it emits very toxic fumes of NO_x, Na_2O, and SO_x.

FAG000 CAS:2353-45-9 ***HR: 2***
FD&C GREEN No. 3
mf: $C_{37}H_{36}N_2O_{10}S_3$•2Na mw: 810.91

PROP: Red to brown-violet powder. Sol in water, conc sulfuric acid.

SYNS: AIZEN FOOD GREEN No. 3 ◇ C.I. 42053 ◇ C.I. FOOD GREEN 3 ◇ FAST GREEN FCF ◇ 1724 GREEN ◇ SOLID GREEN FCF

TOXICITY DATA with REFERENCE
mma-sat 10 mg/plate FCTOD7 22,623,84
cyt-ham:fbr 4 g/L FCTOD7 22,623,84
cyt-ham:ovr 20 μmol/L/5H-C ENMUDM 1,27,79
cyt-ham:lng 2 g/L GMCRDC 27,95,81
scu-rat TDLo:5925 mg/kg/48W-I:NEO JNCIAM 24,769,60

CONSENSUS REPORTS: IARC Cancer Review: Group 3 IMEMDT 7,56,87; Animal Sufficient Evidence IMEMDT 16,187,78. Reported in EPA TSCA Inventory. EPA Genetic Toxicology Program.

SAFETY PROFILE: Questionable carcinogen with experimental neoplastigenic data. Mutation data reported. When heated to decomposition it emits very toxic fumes of NO_x and SO_x.

FAG010 CAS:523-44-4 ***HR: 3***
FD&C ORANGE No. 1
mf: $C_{16}H_{11}N_2O_4S$•Na mw: 350.34

SYNS: ACID LEATHER ORANGE I ◇ ACID PHOSPHINE CL ◇ A.F. ORANGE No. 1 ◇ AIZEN FOOD ORANGE No. 1 ◇ AIZEN NAPHTHOL ORANGE I ◇ AIZEN ORANGE I ◇ CERTIQUAL ORANGE I ◇ C.I. 14600 ◇ C.I. ACID ORANGE 20 ◇ C.I. ACID ORANGE 20, MONOSODIUM SALT ◇ D&C ORANGE No. 3 ◇ DYE ORANGE No. 1 ◇ EGACID ORANGE GG ◇ ELGACID ORANGE 2G ◇ ENIACID ORANGE I ◇ EXT. D&C ORANGE No.3 ◇ FDC ORANGE I ◇ HISPACID ORANGE 1 ◇ 4-((4-HYDROXY-1-NAPHTHALENYL)AZO)BENZENESULPHONICACID, MONOSODIUM SALT ◇ p-((4-HYDROXY-1-NAPHTHYL)AZO)BENZENESULFONIC ACID, MONOSODIUM SALT ◇ p-((4-HYDROXY-1-NAPHTHYL)AZO)BENZENESULPHONIC ACID, MONOSODIUM SALT ◇ p-((4-HYDROXY-1-NAPHTHYL)AZO)BENZENESULPHONIC ACID, SODIUM SALT ◇ JAVA ORANGE I ◇ NANKAI ACID ORANGE I ◇ NAPHTHALENE ORANGE I ◇ NAPHTHOL ORANGE ◇ α-NAPHTHOL ORANGE ◇ NEKLACID ORANGE 1 ◇ 1333 ORANGE ◇ ORANGE I ◇ SODIUM AZO-α-NAPHTHOLSULFANILATE ◇ SODIUM AZO-α-NAPHTHOLSUL-

PHANILATE ◇ 4-p-SULFOPHENYLAZO-1-NAPHTHOL MONOSODIUM SALT ◇ TERTRACID ORANGE I ◇ TROPAEOLIN 1

TOXICITY DATA with REFERENCE
mma-sat 100 μg/plate MUREAV 56,249,78
scu-rat TDLo:9360 mg/kg/2Y-I:ETA FEPRA7 16,367,57

CONSENSUS REPORTS: IARC Cancer Review: Group 3 IMEMDT 7,56,87; Animal Limited Evidence IMEMDT 8,173,75. Reported in EPA TSCA Inventory. EPA Genetic Toxicology Program.

SAFETY PROFILE: Questionable carcinogen with experimental tumorigenic data. Mutation data reported. When heated to decomposition it emits very toxic fumes of NO_x, SO_x, and Na_2O.

FAG018 CAS:3564-09-8 ***HR: 2***
FD&C RED No. 1
mf: $C_{19}H_{16}N_2O_7S_2 \cdot 2Na$ mw: 494.47

SYNS: A.F. RED No. 1 ◇ CERVEN KUMIDINOVA ◇ C.I. 16155 ◇ C.I. FOOD RED 6 ◇ C.I. FOOD RED 6, DISODIUM SALT ◇ DISODIUM-3-HYDROXY-4-((2,4,5-TRIMETHYLPHENYL)AZO)-2,7-NAPHTHALENE-DISULFONATE ◇ DISODIUM-3-HYDROXY-4-((2,4,5-TRIMETHYL-PHENYL)AZO)-2,7-NAPHTHALENEDISULFONIC ACID ◇ DISODIUM-3-HYDROXY-4-((2,4,5-TRIMETHYLPHENYL)AZO)-2,7-NAPHTHA-LENEDISULPHONATE ◇ DISODIUM-3-HYDROXY-4-((2,4,5-TRIMETHYLPHENYL)AZO)-2,7-NAPHTHALENEDISULPHONIC ACID ◇ DOLKWAL PONCEAU 3R ◇ EXT. D&C RED No. 15 ◇ 3-HYDROXY-4-((2,4,5-TRIMETHYLPHENYL)AZO)-2,7-NAPHTHALENEDISULFONIC ACID, DISODIUM SALT ◇ 3-HYDROXY-4-((2,4,5-TRIMETHYL-PHENYL)AZO)-2,7-NAPHTHALENEDISULPHONIC ACID, DISODIUM SALT ◇ MAPLE PONCEAU 3R ◇ PONCEAU 3R ◇ SODIUM CUMENEAZO-β-NAPHTHOL DISULPHONATE ◇ USACERT RED No. 1

TOXICITY DATA with REFERENCE
mma-sat 660 nmol/plate INFIBR 23,686,79
oth-esc 300 μmol/L SKEZAP 12,298,71
orl-rat TDLo:730 g/kg/2Y-C:CAR TXAPA9 5,105,63
orl-mus TDLo:1640 g/kg/65W-C:CAR TXAPA9 3,509,61
orl-mus TD:1140 g/kg/68W-C:ETA TXAPA9 5,105,63
orl-rat TD:200 g/kg/24W-C:NEO TXAPA9 5,105,63

CONSENSUS REPORTS: IARC Cancer Review: Group 2B IMEMDT 7,56,87; Animal Sufficient Evidence IMEMDT 8,199,75

SAFETY PROFILE: Suspected carcinogen with experimental carcinogenic and tumorigenic data. Mutation data reported. When heated to decomposition it emits toxic fumes of NO_x and SO_x.

FAG020 CAS:915-67-3 ***HR: 2***
FD&C RED No. 2
mf: $C_{20}H_{11}N_2O_{10}S_3 \cdot 3Na$ mw: 604.48

SYNS: ACETACID RED 2BR ◇ ACID AMARANTH ◇ ACILAN RED SE ◇ AIZEN AMARANTH ◇ AMACID AMARANTH ◇ AMARANT ◇ AMARANTH ◇ AMARANTHE USP (biological stain) ◇ AZO RED R ◇ BORDEAUX ◇ CALCOCID AMARANTH ◇ CANACERT AMARANTH ◇ CERVEN KYSELA 27 ◇ CERVEN POTRAVINARSKA 9 ◇ C.I. 184 ◇ C.I. 16185 ◇ C.I. ACID RED 27 ◇ C.I. FOOD RED 9 ◇ CILEFA RUBINE 2B ◇ DAISHIKI AMARANTH ◇ D&C RED 2 ◇ DOLKWAL AMARANTH ◇ DYE FDC RED 2 ◇ DYE RED RASPBERRY ◇ EDICOL AMARANTH ◇ EUROCERT AMARANTH ◇ FD&C RED No. 2-ALUMINIUM LAKE ◇ FOOD RED 2 ◇ FOOD RED 9 ◇ FRUIT RED A GEIGY ◇ HD AMARANTH B ◇ HEXACERT RED No. 2 ◇ HIDACID AMARANTH ◇ 2-HYDROXY-1,1'-AZONAPHTHALENE-3,6,4'-TRISULFONIC ACID TRISODIUM SALT ◇ 3-HYDROXY-4-((4-SULFO-1-NAPHTHALENYL) AZO)-2,7-NAPHTHLENEDISULFONIC ACID, TRISODIUM SALT ◇ 3-HYDROXY-4-((4-SULFO-1-NAPHTHYL)AZO)-2,7-NAPHTHALENEDI-SULFONIC ACID, TRISODIUM SALT ◇ 3-HYDROXY-4-((4-SULPHO-1-NAPHTHALENYL)AZO)-2,7-NAPHTHALENEDISULPHONICACID, TRISODIUM SALT ◇ 3-HYDROXY-4-((4-SULPHO-1-NAPHTHYL)AZO)-2,7-NAPHTHALENEDISULPHONIC ACID, TRISODIUM SALT ◇ JAVA AMARANTH ◇ KAYAKU AMARANTH ◇ KCA FOODCOL AMARANTH A ◇ KITON RUBINE S ◇ LISSAMINE AMARANTH AC ◇ MAPLE AMARANTH ◇ NAPHTHOL RED B ◇ NAPTHOLROT S ◇ NEKLACID RED A ◇ RAKUTO AMARANTH ◇ RASPBERRY RED for JELLIES ◇ RED DYE No. 2 ◇ RED No. 2 ◇ SHIKISO AMARANTH ◇ 1-(4-SULPHO-1-NAPHTHYLAZO)-2-NAPHTHOL-3,6-DISULPHONIC ACID, TRISODIUM SALT ◇ TAKAOKA AMARANTH ◇ TERTRACID RED A ◇ TOYO AMARANTH ◇ TRISODIUM SALT of 1-(4-SULFO-1-NAPHTHYLAZO)-2-NAPHTHOL-3,6-DISULFONIC ACID ◇ VICTORIA RUBINE O ◇ WHORTLEBERRY RED ◇ WOOL BORDEAUX 6RK ◇ WOOL RED

TOXICITY DATA with REFERENCE
mmo-sat 20 μg/L ENVIDV 9,145,83
mmo-esc 20 μg/L ENVIDV 9,145,83
mrc-smc 20 mg/L ENVIDV 9,145,83
cyt-ham:fbr 1 g/L/48H MUREAV 48,337,77
orl-rat TDLo:296 mg/kg (multi) :REP TXCYAC 3,115,75
orl-rat TDLo:65 mg/kg (multi) :TER TXCYAC 3,129,75
orl-rat TDLo:680 g/kg/14W-I:ETA VPITAR 29,61,70
orl-rat TD:1200 g/kg/78W-C:CAR GASTAB 23,1,53
ipr-rat LD50:1 g/kg SCPHA4 47,39,79
ivn-rat LD50:1 g/kg SCPHA4 47,39,79
ipr-mus LD50:1000 mg/kg FCTXAV 14,163,76

CONSENSUS REPORTS: IARC Cancer Review: Group 3 IMEMDT 7,56,87, Animal Inadequate Evidence IMEMDT 8,41,75. Reported in EPA TSCA Inventory.

SAFETY PROFILE: Moderately toxic by intraperitoneal route. Questionable carcinogen with experimental carcinogenic and tumorigenic data. An experimental teratogen. Other experimental reproductive effects. Mutation data reported. When heated to decomposition it emits toxic fumes of NO_x and SO_x.

FAG040 CAS:16423-68-0 ***HR: 3***
FD&C RED No. 3
mf: $C_{20}H_6I_4O_5 \cdot 2Na$ mw: 879.84

PROP: Brown powder. Sol in water, conc sulfuric acid.

SYNS: AIZEN ERYTHROSINE ◇ CALCOCID ERYTHROSINE N ◇ CANACERT ERYTHROSINE BS ◇ 9-(o-CARBOXYPHENYL)-6-HYDROXY-2,4,5,7-TETRAIODO-3-ISOXANTHONE◇ C.I. 45430 ◇ C.I. ACID RED 51 ◇ CILEFA PINK B ◇ D&C RED No. 3 ◇ DOLKWAL ERYTHROSINE ◇ DYE FD&C RED No. 3 ◇ E 127 ◇ EBS ◇ EDICOL SUPRA ERYTHROSINE A ◇ ERYTHROSIN ◇ ERYTHROSINE B-FO (BI-

OLOGICAL STAIN) ◇ FOOD RED 14 ◇ HEXACERT RED No. 3 ◇ HEXACOL ERYTHROSINE BS ◇ LB-ROT 1 ◇ MAPLE ERYTHROSINE ◇ NEW PINK BLUISH GEIGY ◇ 1427 RED ◇ 1671 RED ◇ 2′,4′,5′,7′-TETRAIODOFLUORESCEIN, DISODIUM SALT ◇ TETRAIODOFLUORESCEIN SODIUM SALT ◇ USACERT RED No. 3

TOXICITY DATA with REFERENCE
mrc-smc 100 mg/L MUREAV 138,153,84
dni-hmn:leu 500 mg/L NEZAAQ 30,574,75
cyt-ham:fbr 600 mg/L FCTOD7 22,623,84
orl-rat TDLo:1798 g/kg/2Y-C:NEO FCTOD7 25,723,87
orl-rat LD50:1840 mg/kg SCPHA4 47,39,79
ivn-rat LD50:200 mg/kg SCPHA4 47,39,79
orl-mus LD50:1264 mg/kg EAPHA6 24,125,81
ivn-mus LD50:370 mg/kg APPHAX 9,127,52

CONSENSUS REPORTS: Reported in EPA TSCA Inventory. EPA Genetic Toxicology Program.

SAFETY PROFILE: Poison by intravenous route. Moderately toxic by ingestion. Questionable carcinogen with experimental tumorigenic data. Human mutation data reported. When heated to decomposition it emits very toxic fumes of Na_2O and I^-.

FAG050 CAS:4548-53-2 ***HR: 2***
FD&C RED No. 4
mf: $C_{18}H_{14}N_2O_7S_2 \cdot 2Na$ mw: 480.44

SYNS: CERTICOL PONCEAU SXS ◇ CERVEN POTRAVINARSKA 1 ◇ C.I. 14700 ◇ C.I. FOOD RED 1 ◇ C.I. FOOD RED 1, DISODIUM SALT ◇ 3-((2,4-DIMETHYL-5-SULFOPHENYL)AZO)-4-HYDROXY-1-NAPHTHALENESULFONIC ACID, DISODIUM SALT ◇ 3-((2,4-DIMETHYL-5-SULPHOPHENYL)AZO)-4-HYDROXY-1-NAPHTHALENESULPHONIC ACID, DISODIUM SALT ◇ DYE FD & C RED No. 4 ◇ DYE FD AND C RED No. 4 ◇ EDICOL SUPRA PONCEAU SX ◇ FD & C RED No. 4-ALUMINIUM LAKE ◇ FOOD RED 4 ◇ HEXACOL PONCEAU SX ◇ 4-HYDROXY-3-((5-SULFO-2,4-XYLYL)AZO)-1-NAPHTHALENESULFONIC ACID, DISODIUM SALT ◇ 4-HYDROXY-3-((5-SULPHO-2,4-XYLYL) AZO)-1-NAPHTHALENESULPHONIC ACID, DISODIUM SALT ◇ PONCEAU SX ◇ PURPLE 4R ◇ 1306 RED ◇ 12101 RED ◇ RED No. 1 ◇ RED No. 4 ◇ 2-(6-SULFO-2,4-XYLYLAZO)-1-NAPHTHOL-4-SULFONIC ACID, DISODIUM SALT ◇ USACERT FD & C RED No. 4

TOXICITY DATA with REFERENCE
orl-rat TDLo:1200 g/kg/78W-C:CAR GASTAB 23,1,53
orl-rat TD:524 g/kg/73W-C:ETA VPITAR 29,61,70

CONSENSUS REPORTS: IARC Cancer Review: Group 3 IMEMDT 7,56,87; Animal Sufficient Evidence IMEMDT 8,207,75. Reported in EPA TSCA Inventory.

SAFETY PROFILE: Questionable carcinogen with experimental carcinogenic and tumorigenic data. When heated to decomposition it emits toxic fumes of NO_x and SO_x.

FAG070 CAS:81-88-9 ***HR: 3***
FD&C RED No. 19
mf: $C_{28}H_{31}N_2O_3 \cdot Cl$ mw: 479.06

SYNS: ACID BRILLIANT PINK B ◇ ADC RHODAMINE B ◇ AIZEN RHODAMINE BH ◇ AKIRIKU RHODAMINE B ◇ BASIC VIOLET 10 ◇ BRILLIANT PINK B ◇ CALCOZINE RED BX ◇ CALCOZINE RHODAMINE BX ◇ 9-o-CARBOXYPHENYL-6-DIETHYLAMINO-3-ETHYLIMINO-3-ISOXANTHENE, 3-ETHOCHLORIDE ◇ (9-(o-CARBOXYPHENYL)-6-(DIETHYLAMINO)-3H-XANTHEN-3-YLIDENE) DIETHYLAMMONIUM CHLORIDE ◇ CERISE TONER X1127 ◇ CERTIQUAL RHODAMIEN ◇ C.I. 749 ◇ C.I. BASIC VIOLET 10 ◇ C.I. FOOD RED 15 ◇ COGILOR RED 321.10 ◇ COSMETIC BRILLIANT PINK BLUISH D CONC ◇ D&C RED No. 19 ◇ DIABASIC RHODAMINE B ◇ DIETHYL-m-AMINO-PHENOLPHTHALEINHYDROCHLORIDE ◇ EDICOL SUPRA ROSE B ◇ ELCOZINE RHODAMINE B ◇ ERIOSIN RHODAMINE B ◇ FOOD RED 15 ◇ GERANIUM LAKE N ◇ HEXACOL RHODAMINE B EXTRA ◇ IKADA RHODAMINE B ◇ IRAGEN RED L-U ◇ MITSUI RHODAMINE BX ◇ 11411 RED ◇ RED NO 213 ◇ RHEONINE B ◇ RHODAMINE ◇ RHODAMINE S (RUSSIAN) ◇ SICILIAN CERISE TONER A-7127 ◇ SYMULEX MAGENTA F ◇ SYMULEX PINK F ◇ TAKAOKA RHODAMINE B ◇ TETRAETHYLDIAMINO-o-CARBOXYPHENYL-XANTHENYL CHLORIDE ◇ TETRAETHYLRHODAMINE

TOXICITY DATA with REFERENCE
sln-dmg-orl 1000 ppm AMNTA4 87,295,53
cyt-mam:fbr 2 mg/L MUREAV 88,211,81
scu-rat TDLo:3600 mg/kg/68W-I:ETA GANNA2 47,51,56
orl-rat LDLo:500 mg/kg NCNSA6 5,8,53
ipr-rat LD50:112 mg/kg TOERD9 1,23,78
ivn-rat LD50:89 mg/kg TXAPA9 3,696,61
orl-mus LD50:887 mg/kg TOERD9 1,23,78
ipr-mus LDLo:128 mg/kg CBCCT* 2,62,50

CONSENSUS REPORTS: IARC Cancer Review: Group 3 IMEMDT 7,56,87; Animal Sufficient Evidence IMEMDT 16,221,78. Reported in EPA TSCA Inventory. EPA Genetic Toxicology Program.

SAFETY PROFILE: Poison by intraperitoneal and intravenous routes. Moderately toxic by ingestion. Questionable carcinogen with experimental carcinogenic and tumorigenic data. Mutation data reported. When heated to decomposition it emits very toxic fumes of NO_x, NH_3, and Cl^-.

FAG080 CAS:85-82-5 ***HR: 3***
FD&C RED No. 32
mf: $C_{18}H_{16}N_2O$ mw: 276.36

SYNS: FD&C ACID RED 32 ◇ OIL RED XO ◇ 1-XYLYLAZO-2-NAPHTHOL ◇ 1-(2,5-XYLYLAZO)-2-NAPHTHOL

TOXICITY DATA with REFERENCE
orl-mus TDLo:29 g/kg/52W:ETA BJCAAI 10,653,56

SAFETY PROFILE: Questionable carcinogen with experimental tumorigenic data. When heated to decomposition it emits toxic fumes of NO_x.

FAG100 CAS:25956-17-6 ***HR: D***
FD&C RED No. 40
mf: $C_{18}H_{16}N_2O_8S_2 \cdot 2Na$ mw: 498.46

PROP: Red powder. Sol in water; sltly sol in abs alc.

SYNS: ALLURA RED AC ◇ C.I. 16035

TOXICITY DATA with REFERENCE
orl-rat TDLo:83500 mg/kg (male 2W pre):REP TXCYAC 28,207,83

SAFETY PROFILE: Experimental reproductive effects. When heated to decomposition it emits very toxic fumes of NO_x and SO_x.

FAG120 CAS:1694-09-3 ***HR: 3***
FD&C VIOLET No. 1
mf: $C_{39}H_{41}N_3O_6S_2 \cdot Na$ mw: 734.94

SYNS: ACID VIOLET ◇ A.F. VIOLET No 1 ◇ AIZEN FOOD VIOLET No 1 ◇ BENZYL VIOLET ◇ BENZYL VIOLET 3B ◇ CALCOCID VIOLET 4BNS ◇ C.I. 42640 ◇ C.I. FOOD VIOLET 2 ◇ COOMASSIE VIOLET ◇ DISPERSED VIOLET 12197 ◇ FORMYL VIOLET S4BN ◇ PERGACID VIOLET 2B ◇ SOLAR VIOLET 5BN ◇ WOOL VIOLET

TOXICITY DATA with REFERENCE
mma-sat 320 μg/plate MUREAV 89,21,81
orl-rat TDLo:498 g/kg/28W-C:CAR JNCIAM 51,1337,73
scu-rat TDLo:9360 mg/kg/2Y-I:ETA FEPRA7 16,367,57

CONSENSUS REPORTS: IARC Cancer Review: Group 2B IMEMDT 7,56,87; Animal Sufficient Evidence IMEMDT 16,153,78. EPA Genetic Toxicology Program. Reported in EPA TSCA Inventory.

SAFETY PROFILE: Suspected carcinogen with experimental carcinogenic and tumorigenic data. Mutation data reported. When heated to decomposition it emits very toxic fumes of NO_x, NH_3, Na_2O, and SO_x.

FAG130 CAS:85-84-7 ***HR: 2***
FD&C YELLOW No. 3
mf: $C_{16}H_{13}N_3$ mw: 247.32

SYNS: A.F YELLOW No. 2 ◇ 1-BENZENE-AZO-β-NAPHTHYLAMINE ◇ 1-BENZENEAZO-2-NAPHTHYLAMINE ◇ CERISOL YELLOW AB ◇ C.I. 11380 ◇ C.I. FOOD YELLOW 10 ◇ C.I. SOLVENT YELLOW 5 ◇ DOLKWAL YELLOW AB ◇ EXT. D&C YELLOW No. 9 ◇ GRASAL YELLOW ◇ JAUNE AB ◇ OIL YELLOW A ◇ 1-(PHENYLAZO)-2-NAPHTHALENAMINE ◇ 1-(PHENYLAZO)-2-NAPHTHYLAMINE ◇ YELLOW AB ◇ YELLOW No. 2

TOXICITY DATA with REFERENCE
dns-rat-orl 500 mg/kg ENMUDM 7,101,85
orl-rat TDLo:8190 mg/kg/65W-C:ETA JPPMAB 7,591,55
orl-rbt LDLo:1000 mg/kg JBCHA3 27,403,16
scu-rbt LDLo:1000 mg/kg JBCHA3 27,403,16

CONSENSUS REPORTS: IARC Cancer Review: Group 3 IMEMDT 7,56,87; Animal No Evidence IMEMDT 8,279,75. Reported in EPA TSCA Inventory. EPA Genetic Toxicology Program.

SAFETY PROFILE: Moderately toxic by ingestion and subcutaneous routes. Questionable carcinogen with experimental tumorigenic data. Mutation data reported. When heated to decomposition it emits toxic fumes of NO_x.

FAG135 CAS:131-79-3 ***HR: 2***
FD&C YELLOW No. 4
mf: $C_{17}H_{15}N_3$ mw: 261.35

SYNS: A.F. YELLOW No. 3 ◇ CERISOL YELLOW TB ◇ C.I. 11390 ◇ C.I. FOOD YELLOW 11 ◇ DOLKWAL YELLOW OB ◇ EXT. D&C YELLOW No. 10 ◇ JAUNE OB ◇ 1-(2-METHYLPHENYL)AZO-2-NAPHTHALENAMINE ◇ 1-((2-METHYLPHENYL)AZO)-2-NAPHTHALENAMINE ◇ 1-(2-METHYLPHENYL)AZO-2-NAPHTHYLAMINE ◇ OIL YELLOW OB ◇ o-TOLUENE-1-AZO-2-NAPHTHYLAMINE ◇ 1-(o-TOLYLAZO)-2-NAPHTHYLAMINE ◇ YELLOW OB

TOXICITY DATA with REFERENCE
mma-sat 50 μg/plate CANCAR 49,1970,82
scu-rat TDLo:700 mg/kg/2Y-I:ETA FEPRA7 16,367,57
orl-rbt LDLo:1000 mg/kg JBCHA3 27,403,16
ipr-rbt LDLo:1000 mg/kg JBCHA3 27,403,16
scu-rbt LDLo:1000 mg/kg JBCHA3 27,403,16

CONSENSUS REPORTS: IARC Cancer Review: Group 3 IMEMDT 7,56,87; Animal Sufficient Evidence IMEMDT 8,287,75. EPA Genetic Toxicology Program.

SAFETY PROFILE: Moderately toxic by ingestion, intraperitoneal, and subcutaneous routes. Questionable carcinogen with experimental tumorigenic data. Mutation data reported. When heated to decomposition it emits toxic fumes of NO_x. See also AROMATIC AMINES.

FAG140 CAS:1934-21-0 ***HR: 1***
FD&C YELLOW No. 5
mf: $C_{16}H_9N_4O_9S_2 \cdot 3Na$ mw: 534.38

PROP: Yellow-orange powder. Sol in water, conc sulfuric acid.

SYNS: ACID LEATHER YELLOW T ◇ ACILAN YELLOW GG ◇ AIREDALE YELLOW T ◇ AIZEN TARTRAZINE ◇ ATUL TARTRAZINE ◇ BUCACID TARTRAZINE ◇ CALCOCID YELLOW XX ◇ CANACERT TARTRAZINE ◇ 3-CARBOXY-5-HYDROXY-1-p-SULFOPHENYL-4-o-SULFOPHENYLAZOPYRAZOLE TRISODIUM SALT ◇ C.I. 19140 ◇ C.I. FOOD YELLOW 4 ◇ CURON FAST YELLOW 5G ◇ D&C YELLOW No. 5 ◇ DOLKWAL TARTRAZINE ◇ EDICOL SUPRA TARTRAZINE N ◇ EGG YELLOW A ◇ EUROCERT TARTRAZINE ◇ FENAZO YELLOW T ◇ FOOD YELLOW No. 4 ◇ HEXACOL TARTRAZINE ◇ HYDRAZINE YELLOW ◇ KARO TARTRAZINE ◇ KITON YELLOW T ◇ LAKE YELLOW ◇ MAPLE TARTRAZOL YELLOW ◇ NAPHTHOCARD YELLOW O ◇ OXANAL YELLOW T ◇ SHULTZ No. 737 ◇ SUGAI TARTRAZINE ◇ TARTAR YELLOW ◇ TARTRAZINE ◇ TARTRAZOL YELLOW ◇ TRISODIUM-3-CARBOXY-5-HYDROXY-1-p-SULFOPHENYL-4-p-SULFOPHENYLAZOPYRAZOLE ◇ TRISODIUM SALT of 3-CARBOXY-5-HYDROXY-1-SULFOPHENYLAZOPYRAZOLE ◇ UNITERTRACID YELLOW TE ◇ USACERT YELLOW No. 5 ◇ VONDACID TARTRAZINE ◇ WOOL YELLOW ◇ XYLENE FAST YELLOW GT ◇ YELLOW LAKE 69

TOXICITY DATA with REFERENCE
cyt-hmn:lym 100 mg/L SOGEBZ 11,528,75
cyt-ham:fbr 2500 mg/L FCTOD7 22,623,84
cyt-ham:lng 2100 mg/L GMCRDC 27,95,81
cyt-mam:fbr 5 mg/L FCTOD7 20,461,82

orl-rat TLDo:18500 mg/kg (7-22D preg/21D post):REP JTEHD6 2,1211,77
orl-rat TLDo:18500 mg/kg (7-22D preg/21D post):TER JTEHD6 2,1211,77
orl-hmn TDLo:14 μg/kg:PNS,MSK ANAEA3 17,719,59
orl-mus LD50:12750 mg/kg FAONAU 38B,90,66

CONSENSUS REPORTS: Reported in EPA TSCA Inventory. EPA Genetic Toxicology Program.

SAFETY PROFILE: Mildly toxic by ingestion. An experimental teratogen. Human systemic effects by ingestion: paresthesia and changes in teeth and supporting structures. Experimental reproductive effects. Human mutation data reported. When heated to decomposition it emits very toxic fumes of NO_x, SO_x, and Na_2O.

FAG150 CAS:2783-94-0 ***HR: 2***
FD&C YELLOW No. 6
mf: $C_{16}H_{10}N_2O_7S_2Na_2$ mw: 452.36

PROP: Orange powder. Sol in water, conc sulfuric acid; sltly sol in abs alc.

SYNS: ACID YELLOW TRA ◇ AIZEN FOOD YELLOW No. 5 ◇ CANACERT SUNSET YELLOW FCF ◇ C.I. 15985 ◇ GELBORANGE-S (GERMAN) ◇ SUNSET YELLOW FCF

TOXICITY DATA with REFERENCE
cyt-ham:lng 2 g/l GMCRDC 27,95,81
ipr-rat LD50:4600 mg/kg:CNS,GIT FCTXAV 5,747,67
ipr-mus LD50:3800 mg/kg:CNS,GIT FCTXAV 5,747,67

CONSENSUS REPORTS: IARC Cancer Review: Group 3 IMEMDT 7,56,87; Animal Inadequate Evidence IMEMDT 8,257,75. Reported in EPA TSCA Inventory.

SAFETY PROFILE: Moderately toxic by intraperitoneal route. Questionable carcinogen. When heated to decomposition it emits very toxic fumes of NO_x and SO_x.

FAJ100 CAS:4551-59-1 ***HR: 3***
FELDENE
mf: $C_{15}H_{13}N_3O_4S$ mw: 331.37

PROP: Crystals from methanol. Mp: 198-200°.

SYNS: CP 16171 ◇ 4-HYDROXY-2-METHYL-N-(2-PYRIDYL)-2H-1,2-BENZOTHIAZIN-3-CARBOXYAMID-1,1-DIOXID (GERMAN) ◇ 4-HYDROXY-2-METHYL-N-(2-PYRIDYL)-2H-1,2-BENZOTHIAZINE-3-CARBOXAMIDE-1,1-DIOXIDE ◇ PIROXICAM ◇ ROXICAM ◇ SOLOCALM

TOXICITY DATA with REFERENCE
orl-rat TDLo:110 mg/kg (female 7-17D post):REP YACHDS 8,4655,80
orl-rat TDLo:110 mg/kg (7-17D preg):TER YACHDS 8,4655,80
orl-man TDLo:7636 mg/kg/6W-I:KID AJNED9 5,142,85
orl-cld TDLo:7143 μg/kg SAMJAF 66,31,84
orl-man TDLo:52 mg/kg/26W-I:SYS AIMEAS 99,282,83
orl-wmn TDLo:1200 μg/kg/3D-I:SKN JRHUA9 11,554,84
orl-man LDLo:3714 mg/kg/13D-I:BLD,SKN NEJMAG 309,795,83
orl-wmn LDLo:2800 μg/kg/1W-I:KID AIMDAP 144,63,84
orl-wmn TDLo:28 mg/kg AMSVAZ 216,335,84
orl-rat LD50:216 mg/kg ARZNAD 28,1714,78
ipr-rat LD50:335 mg/kg YACHDS 8,4639,80
scu-rat LD50:148 mg/kg YACHDS 8,4639,80
rec-rat LD50:400 mg/kg YACHDS 8,4639,80
orl-mus LD50:350 mg/kg YACHDS 8,4639,80
ipr-mus LD50:350 mg/kg YACHDS 8,4639,80
scu-mus LD50:300 mg/kg YACHDS 8,4639,80
orl-dog LD50:108 mg/kg YACHDS 8,4639,80

SAFETY PROFILE: A human poison by ingestion. Poison experimentally by ingestion, subcutaneous, intraperitoneal, and rectal routes. Human systemic effects by ingestion: interstitial nephritis, metabolic acidosis, dermatitis, agranulocytosis. An experimental teratogen. Other experimental reproductive effects. When heated to decomposition it emits toxic fumes of SO_x and NO_x.

FAJ150 CAS:4551-59-1 ***HR: 3***
FEMAMIDE
mf: $C_{19}H_{30}N_2O_3$ mw: 334.51

PROP: Bp: 182-188°. Sol in acetone, methanol, ethanol, ethyl acetate, benzene, chloroform, ether, mineral acids, practically insol in water, alkalies.

SYNS: N-(2-(DIETHYLAMINO)ETHYL)-2-ETHYL-2-PHENYLMALONAMIC ACID ETHYL ESTER ◇ FENALAMIDE ◇ FENAMIDE ◇ PHENAMIDE ◇ PHENYLAETHYLMALONSAEURE-AETHYLESTER-DIAETHYLAMINOAETHYL-AMID (GERMAN) ◇ Sch 5706 ◇ SH 30858 ◇ SPASMAMIDE

TOXICITY DATA with REFERENCE
orl-rat LD50:610 mg/kg AEPPAE 237,264,59
ipr-rat LD50:250 mg/kg AEPPAE 237,264,59
orl-mus LD50:400 mg/kg AEPPAE 237,264,59
ipr-mus LD50:210 mg/kg BCFAAI 111,293,72

SAFETY PROFILE: Poison by ingestion and intraperitoneal routes. When heated to decomposition it emits toxic fumes of NO_x. See also ESTERS.

FAJ200 CAS:56222-04-9 ***HR: 2***
FEMOXETINE HYDROCHLORIDE
mf: $C_{20}H_{25}NO_2{\cdot}ClH$ mw: 347.92

SYNS: FG 4963 ◇ (3R-trans)-3-((4-METHOXYPHENOXY)METHYL)-1-METHYL-4-PHENYLPIPERIDINE HYDROCHLORIDE ◇ PIPERIDINE, 3-((4-METHOXYPHENOXY)METHYL)-1-METHYL-4-PHENYL-,HYDROCHLORIDE, (3R-trans)-

TOXICITY DATA with REFERENCE
orl-ham TDLo:21600 mg/kg (female 6-15D post):REP LIFSAK 32,1193,83
orl-wmn TDLo:2576 mg/kg/23W-I APTOA6 58,253,86

SAFETY PROFILE: Human toxic effects by ingestion. Experimental reproductive effects. When heated to decomposition it emits toxic fumes of NO_x and HCl.

FAK000 CAS:22224-92-6 ***HR: 3***
FENAMIPHOS
mf: $C_{13}H_{22}NO_3PS$ mw: 303.39

SYNS: O-AETHYL-O-(3-METHYL-4-METHYLTHIOPHENYL)-ISOPROPYLAMIDO-PHOSPHORSAEURE ESTER (GERMAN) ◇ BAY 68138 ◇ ENT 27,572 ◇ ETHYL-3-METHYL-4-(METHYLTHIO)PHENYL(1-METHYLETHYL)PHOSPHORAMIDATE ◇ ETHYL-4-(METHYLTHIO)-m-TOLYL ISOPROPYL PHOSPHOR AMIDATE ◇ ISOPROPYLAMINO-O-ETHYL-(4-METHYLMERCAPTO-3-METHYLPHENYL)PHOSPHATE ◇ 1-(METHYLETHYL)-ETHYL 3-METHYL-4-(METHYLTHIO)PHENYL PHOSPHORAMIDATE ◇ NEMACUR ◇ NSC 195106 ◇ PHANAMIPHOS

TOXICITY DATA with REFERENCE
orl-rat LD50:8 mg/kg BESAAT 15,116,69
ihl-rat LC50:91 mg/m^3/4H 85DPAN -,-,71/76
skn-rat LD50:500 mg/kg 85DPAN -,0,71/76
orl-mus LD50:22700 μg/kg 85DPAN -,-,71/76
orl-dog LD50:10 mg/kg 28ZEAL 5,112,76
orl-cat LD50:10 mg/kg 85DPAN -,-,71/76
skn-rbt LD50:178 mg/kg DTLWS* 4,191,82
orl-gpg LD50:75 mg/kg 28ZEAL 5,112,76
orl-qal LD50:1 mg/kg EESADV 8,551,84
orl-dck LD50:1680 μg/kg TXAPA9 47,451,79
skn-dck LD50:24 mg/kg TXAPA9 47,451,79

CONSENSUS REPORTS: EPA Extremely Hazardous Substances List.

OSHA PEL: TWA 0.1 mg/m^3 (skin)
ACGIH TLV: TWA 0.1 mg/m^3 (skin)

SAFETY PROFILE: Poison by ingestion, inhalation, and skin contact. When heated to decomposition it emits very toxic fumes of NO_x, PO_x, and SO_x.

FAK100 CAS:60168-88-9 ***HR: 2***
FENARIMOL
mf: $C_{17}H_{12}Cl_2N_2O$ mw: 331.21

PROP: White, odorless crystals. Mp: 117-119°. Practically insol in water; sol in most organic solvents.

SYNS: BLOC ◇ (2-CHLOROPHENYL)-α-(4-CHLOROPHENYL)-5-PYRIMIDINEMETHANOL ◇ α-(2-CHLOROPHENYL)-α-(4-CHLORO-PHENYL)-5-PYRIMIDINEMETHANOL ◇ EL 222 ◇ RIMIDIN ◇ RUBIGAN

TOXICITY DATA with REFERENCE
sln-asn 6 mg/L MUREAV 79,169,80
cyt-mus-orl 450 mg/kg DBANAD 36,1351,83
orl-rat TDLo:980 mg/kg (male 4D pre):REP TXAPA9 86,391,86
orl-rat LD50:2500 mg/kg FMCHA2 -,C254,89
orl-mus LD50:4500 mg/kg FMCHA2 -,C254,89

CONSENSUS REPORTS: Reported in EPA TSCA Inventory.

SAFETY PROFILE: Moderately toxic by ingestion. Experimental reproductive effects. Mutation data reported. When heated to decomposition it emits toxic fumes of Cl^- and NO_x.

FAL000 CAS:13669-70-0 ***HR: 3***
FENAZOXINE
mf: $C_{17}H_{19}NO$ mw: 253.37

SYNS: NEFOPAM ◇ 3,4,5,6,7-TETRAHYDRO-5-METHYL-1-PHENYL-1H-2,5-BENZOXAZOCINE

TOXICITY DATA with REFERENCE
orl-wmn TDLo:15 mg/kg:CNS,CVS BMJOAE 283,1508,81
orl-man TDLo:17 mg/kg:CNS,CVS BMJOAE 283,1508,81
orl-rat LD50:178 mg/kg DRUGAY 19,249,80
ivn-rat LD50:28 mg/kg DRUGAY 19,249,80
ims-rat LD50:57 mg/kg DRUGAY 19,249,80
orl-mus LD50:119 mg/kg DRUGAY 19,249,80
ipr-mus LD50:50 mg/kg DDEVD6 3,10,79
ivn-mus LD50:45 mg/kg DRUGAY 19,249,80
ims-mus LD50:53 mg/kg DRUGAY 19,249,80
orl-dog LD50:100 mg/kg DRUGAY 19,249,80

SAFETY PROFILE: Poison by ingestion, intravenous, intraperitoneal, and intramuscular routes. Human systemic effects by ingestion: hallucinations, distorted perceptions, excitement, motor activity changes, increased pulse rate without blood pressure fall and heart rate changes. When heated to decomposition it emits toxic fumes of NO_x.

FAL100 CAS:43210-67-9 ***HR: D***
FENBENDAZOLE
mf: $C_{15}H_{13}N_3O_2S$ mw: 299.37

SYNS: FENBENDAZOL ◇ HOE 881 ◇ PANACUR ◇ (5-(PHENYL-THIO)-2-BENZIMIDAZOLECARBAMIC ACID, METHYL ESTER

TOXICITY DATA with REFERENCE
oth:hmn:lym 100 mg/L ENMUDM 2,67,80
oth:hmn:leu 1 mg/L THERAP 31,505,76

SAFETY PROFILE: Human mutation data reported. When heated to decomposition it emits toxic fumes of SO_x and NO_x.

FAM000 CAS:63918-50-3 ***HR: 3***
FENCAMINE HYDROCHLORIDE
mf: $C_{20}H_{28}N_6O_2{\cdot}ClH$ mw: 421.00

SYNS: SICOCLOR ◇ 1-(1,3,7-TRIMETHYL-2,6-DIOXOPURIN-8-YL)-4-(2-PHENYL-1-METHYL)ETHYL-4-METHYL-ETHYLENEDIAMINE,HY-DROCHLORIDE

TOXICITY DATA with REFERENCE
orl-rat LD50:508 mg/kg 27ZQAG -,229,72

ipr-rat LD50:93 mg/kg 27ZQAG -,229,72
orl-mus LD50:418 mg/kg 27ZQAG -,229,72
ipr-mus LD50:82 mg/kg 27ZQAG -,229,72

SAFETY PROFILE: Poison by intraperitoneal route. Moderately toxic by ingestion. A central nervous system stimulant. When heated to decomposition it emits very toxic fumes of Cl^- and NO_x.

FAM100 CAS:7424-00-2 ***HR: D***
FENCHLONINE
mf: $C_9H_{10}ClNO_2$ mw: 199.65

SYNS: ALANINE, 3-(p-CHLOROPHENYL)-, dl- ◇ (±)-p-CHLOROPHENYLALANINE ◇ dl-p-CHLOROPHENYLALANINE ◇ dl-4-CHLOROPHENYLALANINE ◇ CP 10,188 ◇ C-PAL ◇ FENCLONIN ◇ FENCLONINE

TOXICITY DATA with REFERENCE
scu-rat TDLo:1794 mg/kg (female 14-19D post):TER BIMDA2 4,516,70

CONSENSUS REPORTS: Reported in EPA TSCA Inventory.

SAFETY PROFILE: An experimental teratogen. When heated to decomposition it emits toxic fumes of NO_x and Cl^-.

FAO000 CAS:13851-11-1 ***HR: 1***
FENCHYL ACETATE
mf: $C_{12}H_{19}O_2$ mw: 195.31

SYNS: 1,3,3-TRIMETHYL-2-NORBORNANOL ACETATE ◇ 1,3,3-TRIMETHYL-2-NORBORNANYLACETATE

TOXICITY DATA with REFERENCE
skn-rbt 500 mg/24H MOD FCTXAV 14,773,76

CONSENSUS REPORTS: Reported in EPA TSCA Inventory.

SAFETY PROFILE: A skin irritant. When heated to decomposition it emits acrid smoke and irritating fumes.

FAO100 CAS:53597-27-6 ***HR: 2***
FENDOSAL
mf: $C_{25}H_{19}NO_3$ mw: 381.45

PROP: Crystals from acetic acid. Mp: 223-225° (decomp).

SYNS: ALNOVIN ◇ 3-(3-CARBOXY-4-HYDROXYPHENYL)-2-PHENYL-4,5-DIHYDRO-3H-BENZ(e)INDOLE ◇ 5-(4,5-DIHYDRO-2-PHENYL-3H-BENZ(e)INDOL-3-YL)-2-HYDROXYBENZOIC ACID ◇ 5-(4,5-DIHYDRO-2-PHENYL-3H-BENZ(e)INDOL-3-YL)SALICYLICACID ◇ FENDOZAL ◇ HP 129 ◇ P 71-0129

TOXICITY DATA with REFERENCE
orl-rat LD50:450 mg/kg AGACBH 8,209,78
orl-mus LD50:740 mg/kg AGACHB 8,209,78
ipr-mus LD50:510 mg/kg AGACBH 8,209,78
orl-rbt LD50:560 mg/kg AGACBH 8,209,78

SAFETY PROFILE: Moderately toxic by ingestion and intraperitoneal routes. When heated to decomposition it emits toxic fumes of NO_x.

FAO200 CAS:7698-97-7 ***HR: 3***
FENESTREL
mf: $C_{16}H_{20}O_2$ mw: 244.36

SYNS: 2-METHYL-3-ETHYL-4-PHENYL-4-CYCLOHEXENECARBOXYLIC ACID ◇ 2-METHYL-3-ETHYL-4-PHENYL-Δ^4-CYCLOHEXENECARBOXYLIC ACID ◇ ORF 3858

TOXICITY DATA with REFERENCE
unr-rbt TDLo:290 μg/kg (female 1-29D post):TER CCPTAY 2,85,70
orl-mky TDLo:10 mg/kg (female 1-6D post):REP AJOGAH 115,101,73
orl-rat TDLo:27 mg/kg/52W-C:ETA TXAPA9 19,412,71
orl-rat LD50:2350 mg/kg TXAPA9 19,412,71
orl-mus LD50:680 mg/kg TXAPA9 19,412,71
ipr-mus LD50:150 mg/kg TXAPA9 19,412,71
ivn-dog LDLo:20 mg/kg TXAPA9 19,412,71

SAFETY PROFILE: Poison by intravenous and intraperitoneal routes. Moderately toxic by ingestion. An experimental teratogen. Other experimental reproductive effects. Questionable carcinogen with experimental tumorigenic data. When heated to decomposition it emits acrid smoke and irritating fumes.

FAP000 CAS:8006-84-6 ***HR: 2***
FENNEL OIL

PROP: From steam distillation of *Foeniculum vulgare* Miller (Fam. *Umbelliferae*) (FCTXAV 12,807,74). Colorless to pale yellow liquid; odor and taste of fennel.

SYNS: BITTER FENNEL OIL ◇ FENCHEL OEL (GERMAN) ◇ OIL of FENNEL

TOXICITY DATA with REFERENCE
skn-mus 100 % SEV FCTXAV 12,879,74
skn-rbt 500 mg/24H MOD FCTXAV 12,879,74
skn-rbt 500 mg/24H FCTXAV 14,309,76
mma-sat 2500 μg/plate JAFCAU 30,563,82
orl-rat LD50:3120 mg/kg PHARAT 14,435,59
orl-mus LD50:3100 mg/kg TOFOD5 8,91,85

CONSENSUS REPORTS: Reported in EPA TSCA Inventory.

SAFETY PROFILE: Moderately toxic by ingestion. Mutation data reported. A severe skin irritant. When heated to decomposition it emits acrid smoke and irritating fumes.

FAP100 CAS:53746-45-5 ***HR: 3***
FENOPROFEN CALCIUM DIHYDRATE
mf: $C_{30}H_{26}O_6 \cdot Ca \cdot 2H_2O$ mw: 558.68

SYNS: CALCIUM-2-(m-PHENOXYPHENYL)PROPIONATE DIHYDRATE ◇ FENOPROFEN CALCIUM SALT DIHYDRATE ◇ FENOPRON ◇ FEPRONA ◇ LILLY 69323 ◇ (±)-α-METHYL-3-PHENOXYBENZENE-ACETIC ACID CALCIUM SALT DIHYDRATE ◇ NALFON ◇ NALGESIC ◇ dl-2-(3-PHENOXYPHENYL)-PROPIONIC ACID CALCIUM SALT, DIHYDRATE ◇ PROGESIC ◇ YM-09229

TOXICITY DATA with REFERENCE
orl-rat TDLo:330 mg/kg (7-17D preg):REP KSRNAM 14,4435,80
orl-rbt TDLo:975 mg/kg (female 6-18D post):TER KSRNAM 14,2950,80
orl-wmn TDLo:5832 mg/kg/35W-I:GIT,KID AJMEAZ 72,81,82
orl-wmn TDLo:720 mg/kg/30D-I JAMAAP 242,1896,79
orl-man TDLo:521 mg/kg/17W-I:KID AIMEAS 93,508,80
orl-wmn LDLo:8748 mg/kg/35W-I:KID,PUL AJMEAZ 72,81,82
orl-wmn TDLo:1898 mg/kg/12W-I:KID AIMEAS 92,72,80
orl-rat LD50:380 mg/kg IYKEDH 13,1128,82
ipr-rat LD50:234 mg/kg YKYUA6 34,363,83
scu-rat LD50:366 mg/kg IYKEDH 13,1128,82
ivn-rat LD50:526 mg/kg YKYUA6 34,363,83
orl-mus LD50:439 mg/kg YHTPAD 21,36,86
ipr-mus LD50:286 mg/kg YKYUA6 34,363,83
scu-mus LD50:463 mg/kg YKYUA6 34,363,83
ivn-mus LD50:471 mg/kg YKYUA6 34,363,83

SAFETY PROFILE: Mildly toxic to humans by ingestion. Poison by ingestion, subcutaneous, and intraperitoneal routes. Moderately toxic by intravenous route. Human systemic effects by ingestion: dyspnea, nausea or vomiting, acute renal failure, acute tubular necrosis, interstitial nephritis, and other kidney damage. Experimental reproductive effects. An experimental teratogen. When heated to decomposition it emits acrid smoke and irritating fumes.

FAQ000 CAS:34691-31-1 ***HR: 3***
FENOPROFEN SODIUM
mf: $C_{15}H_{13}O_3 \cdot Na$ mw: 264.27

SYN: d,l-2-(3-PHENOXYPHENYL)PROPIONIC ACID SODIUM SALT

TOXICITY DATA with REFERENCE
orl-rat LD50:800 mg/kg TXAPA9 25,444,73
ipr-rat LD50:234 mg/kg KSRNAM 14,4385,80
scu-rat LD50:500 mg/kg TXAPA9 25,444,73
ivn-rat LD50:500 mg/kg TXAPA9 25,444,73
ipr-mus LD50:286 mg/kg KSRNAM 14,4385,80
scu-mus LD50:463 mg/kg KSRNAM 14,4385,80
ivn-mus LD50:471 mg/kg KSRNAM 14,4385,80

SAFETY PROFILE: Poison by intraperitoneal route. Moderately toxic by ingestion, subcutaneous, and intravenous routes. When heated to decomposition it emits toxic fumes of Na_2O. See also FENOPROFEN CALCIUM DIHYDRATE.

FAQ100 CAS:1944-12-3 ***HR: 3***
FENOTEROL HYDROBROMIDE
mf: $C_{17}H_{21}NO_4 \cdot BrH$ mw: 384.31

SYNS: BEROTEC ◇ BEROTEC HYDROBROMIDE ◇ 1-(3,5-DIHYDROXY-PHENYL-2-((1-(4-HYDROXYBENZYL)ETHYL)AMINO)-ETHANOL HYDROBROMIDE ◇ FENOTEROL BROMIDE ◇ PARTUSISTEN ◇ PHENOTEROL HYDROBROMIDE ◇ Th 1165a ◇ TH 1165a HYDROBROMIDE

TOXICITY DATA with REFERENCE
orl-wmn TDLo:200 μg/kg (1D pre):REP RDCNBM 5,31,81
orl-rbt TDLo:32500 μg/kg (female 6-18D post):TER IYKEDH 12,742,81
orl-rat LD50:1600 mg/kg IYKEDH 11,542,80
scu-rat LD50:1080 mg/kg IYKEDH 15,1140,84
ivn-rat LD50:65 mg/kg IYKEDH 11,542,80
orl-mus LD50:1990 mg/kg TXAPA9 18,185,71
scu-mus LD50:1100 mg/kg TXAPA9 18,185,71

SAFETY PROFILE: Poison by intravenous route. Moderately toxic by ingestion and subcutaneous routes. An experimental teratogen. Human reproductive effects by ingestion: changes in the uterus, cervix or vagina. Experimental reproductive effects. When heated to decomposition it emits toxic fumes of Br^- and NO_x.

FAQ800 CAS:115-90-2 ***HR: 3***
FENSULFOTHION
mf: $C_{11}H_{17}O_4PS_2$ mw: 308.37

SYNS: BAY 25141 ◇ BAYER S767 ◇ CHEMAGRO 25141 ◇ DASANIT ◇ O,O-DIAETHYL-O-4-METHYLSULFINYL-PHENYL-MONOTHIOPHOSPHAT (GERMAN) ◇ O,O-DIETHYL-O-(p-(METHYLSULFINYL) PHENYL) PHOSPHOROTHIOATE ◇ O,O-DIETHYL-O-p-(METHYLSULFINYL)PHENYL THIOPHOSPHATE ◇ DMSP ◇ ENT 24,945 ◇ S 767 ◇ TERRACUR P

TOXICITY DATA with REFERENCE
orl-rat LD50:2 mg/kg FMCHA2 -,C70,83
skn-rat LD50:3 mg/kg WRPCA2 9,119,70
ipr-rat LDLo:1500 μg/kg TXAPA9 6,78,64
ipr-mus LDLo:7 mg/kg TXAPA9 6,78,64
orl-gpg LDLo:9 mg/kg TXAPA9 6,78,64
ipr-gpg LDLo:5400 μg/kg TXAPA9 6,78,64
orl-qal LD50:1200 μg/kg EESADV 8,551,84
orl-dck LD50:747 μg/kg TXAPA9 22,556,72
skn-dck LD50:3 mg/kg TXAPA9 47,451,79

CONSENSUS REPORTS: EPA Genetic Toxicology Program. EPA Extremely Hazardous Substances List.

OSHA PEL: TWA 0.1 mg/m^3
ACGIH TLV: TWA 0.1 mg/m^3

SAFETY PROFILE: A poison by ingestion, skin contact, and intraperitoneal routes. A pesticide. When heated to decomposition it emits very toxic fumes of SO_x and PO_x.

FAQ900 CAS:55-38-9 ***HR: 3***
FENTHION
mf: $C_{10}H_{15}O_3PS_2$ mw: 278.34

SYNS: BAY 29493 ◇ BAYCID ◇ BAYER 9007 ◇ BAYTEX ◇ O,O-DIMETHYL-O-4-(METHYLMERCAPTO)-3-METHYLPHENYLPHOSPHOROTHIOATE ◇ O,O-DIMETHYL-p-4-(METHYLMERCAPTO)-3-METHYLPHENYL THIOPHOSPHATE ◇ O,O-DIMETHYL-O-(3-METHYL-4-METHYLMERCAPTOPHENYL)PHOSPHOROTHIOATE ◇ O,O-DIMETHYL-O-(3-METHYL-4-METHYLTHIO-FENYL)-MONOTHIOFOSFAAT (DUTCH) ◇ O,O-DIMETHYL-O-(3-METHYL-4-METHYLTHIOPHENYL)-MONOTHIOPHOSPHAT (GERMAN) ◇ O,O-DIMETHYL-O-3-METHYL-4-METHYLTHIOPHENYLPHOSPHOROTHIOATE ◇ O,O-DIMETHYL-O-(3-METHYL-4-METHYLTHIOPHENYL)-THIONOPHOSPHAT (GERMAN) ◇ O,O-DIMETHYL-O-(4-METHYLTHIO-3-METHYLPHENYL) PHOSPHOROTHIOATE ◇ O,O-DIMETHYL-O-(4-(METHYLTHIO)-m-TOLYL)PHOSPHOROTHIOATE ◇ O,O-DIMETIL-O-(3-METIL-4-METILTIO-FENIL)-MONOTIOFOSFATO (ITALIAN) ◇ DMTP ◇ ENT 25,540 ◇ ENTEX ◇ LEBAYCID ◇ MERCAPTOPHOS ◇ 4-METHYLMERCAPTO-3-METHYLPHENYL DIMETHYL THIOPHOSPHATE ◇ MPP ◇ NCI-C08651 ◇ OMS 2 ◇ PHOSPHOROTHIOICACID-O,O-DIMETHYL-O-(3-METHYL-4-METHYLTHIOPHENYLE) (FRENCH) ◇ QUELETOX ◇ S 1752 ◇ SPOTTON ◇ TALODEX ◇ THIOPHOSPHATE de O,O-DIMETHYLE et de O-(3-METHYL-4-METHYLTHIOPHENYLE) (FRENCH) ◇ TIGUVON

TOXICITY DATA with REFERENCE
mma-sat 333 μg/plate ENMUDM 8(Suppl 7),1,86
sce-ham:lng 40 mg/L ENMUDM 4,621,82
orl-mus TDLo:1050 mg/kg (MGN):REP TXAPA9 26,29,73
ipr-mus TDLo:40 mg/kg (11D preg):TER TXAPA9 24,324,73
orl-mus TDLo:1730 mg/kg/103W-C:ETA NCITR* NCI-CG-TR-103,79
orl-man TDLo:257 mg/kg:CVS,GIT JTCTDW 19,425,82
unr-hmn LD50:50 mg/kg DTLVS* 4,191,80
orl-rat LD50:180 mg/kg KSKZAN 16(2),59,78
ihl-rat LCLo:1 g/m^3/2H 85GYAZ -,27,71
skn-rat LD50:330 mg/kg TXAPA9 2,88,60
ipr-rat LDLo:260 mg/kg TXAPA9 6,86,64
orl-mus LD50:88100 μg/kg HOEKAN 30,53,80
ihl-mus LCLo:1 g/m^3/2H 85GYAZ -,27,71
skn-mus LD50:500 mg/kg OYYAA2 1,74,67
ipr-mus LD50:125 mg/kg TXAPA9 6,86,64
ivn-mus LD50:320 mg/kg CSLNX* NX#00142
ihl-rbt LCLo:1 g/m^3/2H 85GYAZ -,27,71

CONSENSUS REPORTS: NCI Carcinogenesis Bioassay Completed; Results Negative: rat NCITR* NCI-CG-TR-103,79; Indefinite: mouse NCITR* NCI-CG-TR-103,79. EPA Genetic Toxicology Program.

OSHA PEL: TWA 0.2 mg/m^3 (skin)
ACGIH TLV: TWA 0.2 mg/m^3 (skin)
DFG MAK: 0.2 mg/m^3

SAFETY PROFILE: A human poison by an unspecified route. Poison experimentally by ingestion, skin contact, intraperitoneal, intravenous, and intramuscular routes. Moderately toxic by inhalation. Human systemic effects by ingestion: pulse rate increase, hypermotility, diarrhea, nausea or vomiting. Experimental teratogenic and reproductive effects. Questionable carcinogen with experimental tumorigenic data. Mutation data reported. When heated to decomposition it emits very toxic fumes of PO_x and SO_x. See also MERCAPTANS and ESTERS.

FAQ930 CAS:8066-27-1 ***HR: 2***
FENTHIURAM
mf: $C_6H_{12}N_2S_4 \cdot C_6H_6Cl_6 \cdot C_6H_3Cl_3O \cdot 1/2Cu$ mw: 793.43

SYNS: FENTIURAM ◇ PHENTHIURAM ◇ THIOPEROXYDICARBONIC DIAMIDE, TETRAMETHYL-, mixture with (1-α-2-α-3-β,4-α- 5-α-6-β)-1,2,3,4,5,6-HEXACHLOROCYCLOHEXANE and TRICHLOROPHENOL COPPER(2+) SALT ◇ UNYSH A

TOXICITY DATA with REFERENCE
mnt-mus-ipr 50 mg/kg CYGEDX 21(1),57,87
mnt-ham:lng 15 mg/L CYGEDX 21(1),57,87
orl-rat TDLo:220 mg/kg (female 1-10D post):TER VETNAL 53(12),67,76
ipr-mus LD50:650 mg/kg CYGEDX 21(1),57,87

SAFETY PROFILE: Moderately toxic by intraperitoneal route. An experimental teratogen. Mutation data reported. When heated to decomposition it emits toxic fumes of SO_x and Cl^-.

FAQ950 CAS:80830-42-8 ***HR: 3***
FENTIAPRIL
mf: $C_{13}H_{15}NO_4S_2$ mw: 313.41

SYNS: (2R,4R)-2-(o-HYDROXYPHENYL)-3-(3-MERCAPTO-PROPIONYL)-4-THIAZOLIDINECARBOX YLIC ACID ◇ PRESCRIN ◇ RENTIAPRIL ◇ SA 446 ◇ 4-THIAZOLIDINECARBOXYLIC ACID, 2-(2-HYDROXYPHENYL)-3-(3-MERCAPTO-1-OXOPROPYL)-,(2R-cis)-

TOXICITY DATA with REFERENCE
orl-rat TDLo:2700 mg/kg (female 17-22D post):REP ARZNAD 37,164,87
orl-rat LD50:8290 mg/kg OYYAA2 31,235,86
ipr-rat LD50:921 mg/kg OYYAA2 31,235,86
scu-rat LD50:1369 mg/kg OYYAA2 31,235,86
ivn-rat LD50:884 mg/kg OYYAA2 31,235,86
orl-mus LD50:8050 mg/kg OYYAA2 31,235,86
ivn-mus LD50:946 mg/kg OYYAA2 31,235,86
ivn-dog LD50:200 mg/kg OYYAA2 31,235,86

SAFETY PROFILE: Poison by intravenous route.

Moderately toxic by subcutaneous and intraperitoneal routes. Experimental reproductive effects. When heated to decomposition it emits toxic fumes of NO_x and SO_x.

FAR000 CAS:68990-15-8 ***HR: 1***
FENUGREEK ABSOLUTE

PROP: Found in the seed of the plant *Trigonella foenum graecum L* (FCTXAV 16,637,78).

TOXICITY DATA with REFERENCE
skn-rbt 500 mg/24H MOD FCTXAV 16,755,78

CONSENSUS REPORTS: Reported in EPA TSCA Inventory.

SAFETY PROFILE: A skin irritant. When heated to decomposition it emits acrid smoke and irritating fumes.

FAR100 CAS:51630-58-1 ***HR: 3***
FENVALERATE
mf: $C_{25}H_{22}ClNO_3$ mw: 419.93

PROP: Clear, yellow, viscous liquid at 23°. D: 1.17, n (20/D) 1.5533. Solubility at 20° (g/L): acetone, >450; chloroform, >450; methanol, >450; hexane, 77. Insol in water. Decomp gradually between 150-300°.

SYNS: BELMARK ◇ α-CYANO-3-PHENOXYBENZYL-2-(4-CHLOROPHENYL)ISOVALERATE PYDRIN ◇ α-CYANO-3-PHENOXYBENZYL-2-(4-CHLOROPHENYL)-3-METHYLBUTYRATE ◇ CYANO(3-PHENOXYPHENYL)METHYL 4-CHLORO-α-(1-METHYLETHYL) BENZENEACETATE ◇ ECTRIN ◇ PHENVALERATE ◇ PYDRIN ◇ S 5602 ◇ SANMARTON ◇ SD 43775 ◇ SUMICIDIN ◇ SUMIFLY ◇ SUMIPOWER ◇ WL 43775

TOXICITY DATA with REFERENCE
orl-dog TDLo:1138 mg/kg (26W pre):REP FAATDF 4,577,84
orl-rat LD50:70200 μg/kg CHYCDW 21,215,87
ivn-rat LDLo:50 mg/kg ARTODN 45,325,80
orl-mus LD50:185 mg/kg FAATDF 5,278,85
ice-mus LDLo:200 μg/kg TXAPA9 66,290,82
skn-rbt LD50:2500 mg/kg FMCHA2 -,C104,83

CONSENSUS REPORTS: Cyanide and its compounds are on the Community Right-To-Know List.

SAFETY PROFILE: Poison by ingestion, intravenous, and intracerebral routes. Moderately toxic by skin contact. Experimental reproductive effects. Highly toxic to fish and bees. Corrosive, causes eye damage. A skin irritant. When heated to decomposition it emits toxic fumes of Cl^-, NO_x, and CN^-. See also CYANIDE.

FAR200 CAS:2441-88-5 ***HR: 3***
FENYRIPOL HYDROCHLORIDE
mf: $C_{12}H_{13}N_3O•ClH$ mw: 251.74

SYNS: 2-(β-HYDROXY-β-PHENETHYLAMINO)-PYRIMIDINE HYDROCHLORIDE ◇ 2-(β-HYDROXY-β-PHENYL-ETHYL-AMINO)-PYRIMIDINE CHLORHYDRATE (FRENCH) ◇ IN 836 ◇ α-((2-PYRIMIDINYLAMINO)METHYL)BENZYL ALCOHOL HYDROCHLORIDE

TOXICITY DATA with REFERENCE
ipr-rat LD50:235 mg/kg AIPTAK 141,83,63
orl-mus LD50:850 mg/kg AIPTAK 141,83,63
ipr-mus LD50:455 mg/kg AIPTAK 141,83,63
ivn-mus LD50:137 mg/kg AIPTAK 141,83,63
ivn-dog LD50:125 mg/kg AIPTAK 141,83,63

SAFETY PROFILE: Poison by intravenous and intraperitoneal routes. Moderately toxic by ingestion. When heated to decomposition it emits toxic fumes of NO_x and HCl.

FAS000 CAS:14484-64-1 ***HR: 3***
FERBAM
mf: $C_9H_{18}N_3S_6•Fe$ mw: 416.51

PROP: Black solid, sltly sol in water. Mp: decomp 180°.

SYNS: AAFERTIS ◇ BERCEMA FERTAM 50 ◇ CARBAMATE ◇ DIMETHYLCARBAMODITHIOIC ACID, IRON COMPLEX ◇ DIMETHYLCARBAMODITHIOIC ACID, IRON(3+) SALT ◇ DIMETHYLDITHIOCARBAMIC ACID, IRON SALT ◇ DIMETHYLDITHIOCARBAMIC ACID, IRON(3+) SALT ◇ EISENDIMETHYLDITHIOCARBAMAT (GERMAN) ◇ EISEN(III)-TRIS(N,N-DIMETHYLDITHIOCARBAMAT) (GERMAN) ◇ ENT 14,689 ◇ FERBAM 50 ◇ FERBAM, IRON SALT ◇ FERBECK ◇ FERMATE FERBAM FUNGICIDE ◇ FERMOCIDE ◇ FERRADOW ◇ FERRIC DIMETHYLDITHIOCARBAMATE ◇ FUKLASIN ULTRA ◇ HEXAFERB ◇ HOKMATE ◇ IRON DIMETHYLDITHIOCARBAMATE ◇ KARBAM BLACK ◇ KNOCKMATE ◇ NIACIDE ◇ SUP'R FLO FERBAM FLOWABLE ◇ TRIFUNGOL ◇ TRIS(DIMETHYLCARBAMODITHIOATO-S,S')IRON ◇ TRIS)DIMETHYLDITHIOCARBAMATO)IRON ◇ TRIS(N,N-DIMETHYLDITHIOCARBAMATO) IRON(111) ◇ VANCIDE FE95

TOXICITY DATA with REFERENCE
mmo-sat 50 μg/plate CSHCAL 4,267,77
mmo-omi 1000 ppm MMAPAP 50,233,73
orl-rat LDLo:536 mg/kg (16-22D preg/21D post):REP TXAPA9 35,83,76
scu-mus TDLo:41760 μg/kg (female 6-14D post):TER NTIS** PB223-160
orl-mus TDLo:3500 mg/kg/78W-I:ETA NTIS** PB223-159
scu-mus TDLo:100 mg/kg:CAR NTIS** PB223-159
orl-rat LD50:1130 mg/kg FATOAO 32,356,69
ipr-rat LD50:2700 mg/kg JAPMA8 41,662,52
orl-mus LD50:3400 mg/kg JTEHD6 4,93,78
ipr-mus LDLo:63 mg/kg CBCCT* 4,228,52
orl-rbt LDLo:3 g/kg JAPMA8 41,662,52
ipr-rbt LDLo:1500 mg/kg JAPMA8 41,662,52

CONSENSUS REPORTS: IARC Cancer Review: Group 3 IMEMDT 7,56,87; Animal Inadequate Evidence IMEMDT 12,121,76. Reported in EPA TSCA Inventory. EPA Genetic Toxicology Program.

OSHA PEL: (Transitional: TWA Total Dust: 15 mg/m^3;

Respirable Fraction: 5 mg/m^3) TWA Total Dust: 10 mg/m^3; Respirable Fraction: 5 mg/m^3
ACGIH TLV: TWA 10 mg/m^3
DFG MAK: 15 mg/m^3

SAFETY PROFILE: Poison by intraperitoneal route. Moderately toxic by ingestion. Experimental teratogenic and reproductive effects. Questionable carcinogen with experimental carcinogenic and tumorigenic data. Mutation data reported. A fungicide. When heated to decomposition it emits very toxic fumes of NO_x and SO_x. See also CARBAMATES.

FAU000 CAS:7705-08-0 ***HR: 3***
FERRIC CHLORIDE
DOT: UN: 1773/UN 2582
mf: Cl_3Fe mw: 162.20

PROP: Black-brown solid. Mp: 292°, bp: 319.0°, d: 2.90 @ 25°, vap press: 1 mm @ 194.0°.

SYNS: CHLORURE PERRIQUE ◇ FERRIC CHLORIDE, anhydrous (DOT) ◇ FERRIC CHLORIDE, solid (DOT) ◇ FERRIC CHLORIDE, solid, anhydrous (DOT) ◇ FERRIC CHLORIDE, solution (DOT) ◇ FLORES MARTIS ◇ IRON CHLORIDE ◇ IRON(III) CHLORIDE ◇ IRON CHLORIDE, solid (DOT) ◇ IRON SESQUICHLORIDE, solid (DOT) ◇ IRON TRICHLORIDE ◇ PERCHLORURE de FER

TOXICITY DATA with REFERENCE
oth-esc 500 nmol/tube LAMEDS 6,252,86
ivg-rat TDLo:29 mg/kg (1D pre):REP CCPTAY 4,91,71
orl-rat LD50:1872 mg/kg GISAAA 48(9),71,83
orl-mus LD50:895 mg/kg TRENAF 27,159,76
ivn-mus LD50:58 mg/kg YKKZAJ 87,677,67

CONSENSUS REPORTS: Reported in EPA TSCA Inventory. EPA Genetic Toxicology Program.

OSHA PEL: TWA 1 mg(Fe)/m^3
ACGIH TLV: TWA 1 mg(Fe)/m^3
DOT Classification: ORM-B; Label: None (UN1773); Corrosive Material; Label: Corrosive (UN2582).

SAFETY PROFILE: Poison by intravenous route. Moderately toxic by ingestion. Experimental reproductive effects. Corrosive. Probably an eye, skin, and mucous membrane irritant. Mutation data reported. Reacts with water to produce toxic and corrosive fumes. Catalyzes potentially explosive polymerization of ethylene oxide; chlorine + monomers (e.g., styrene). Forms shock-sensitive explosive mixtures with some metals (e.g., potassium, sodium). Violent reaction with allyl chloride. When heated to decomposition it emits highly toxic fumes of HCl.

FAV000 CAS:7705-08-0 ***HR: 2***
FERRIC CHLORIDE (solution)
mf: $FeCl_3$ mw: 162.2

SYN: IRON(III) CHLORIDE (solution)

CONSENSUS REPORTS: Reported in EPA TSCA Inventory.

DOT Classification: Corrosive Material; Label: Corrosive.

SAFETY PROFILE: Corrosive to the skin, eyes, and mucous membranes. When heated to decomposition it emits very toxic fumes of Cl^-. See also FERRIC CHLORIDE.

FAW000 CAS:10025-77-1 ***HR: 3***
FERRIC CHLORIDE HEXAHYDRATE
mf: $Cl_3Fe \cdot 6H_2O$ mw: 270.32

SYNS: FERRIC TRICHLORIDE HEXAHYDRATE ◇ IRON(3+) CHLORIDE HEXAHYDRATE ◇ IRON(III), CHLORIDE HEXAHYDRATE ◇ IRON TRICHLORIDE HEXAHYDRATE

TOXICITY DATA with REFERENCE
dni-hmn:lym 4800 μmol/L IAAAAM 79,83,86
cyt-rat/ast 500 mg/kg GANNA2 54,155,63
orl-rat LDLo:900 mg/kg EQSSDX 1,1,75
ipr-mus LD50:260 mg/kg JAFCAU 14,512,66
ivn-rbt LDLo:7200 μg/kg EQSSDX 1,1,75

OSHA PEL: TWA 1 mg(Fe)/m^3
ACGIH TLV: TWA 1 mg(Fe)/m^3

SAFETY PROFILE: Poison by intraperitoneal and intravenous routes. Moderately toxic by ingestion. Human mutation data reported. Used as an astringent. When heated to decomposition it emits toxic fumes of Cl^-. See also CHLORIDES.

FAX000 CAS:7783-50-8 ***HR: 3***
FERRIC FLUORIDE
mf: F_3Fe mw: 112.85

PROP: Green crystals. D: 3.87.

SYNS: IRON FLUORIDE ◇ IRON TRIFLUORIDE

TOXICITY DATA with REFERENCE
ivn-mus LD50:18 mg/kg CSLNX* NX#00135

CONSENSUS REPORTS: Reported in EPA TSCA Inventory.

OSHA PEL: TWA 2.5 mg(F)/m^3; TWA 1 mg(Fe)/m^3
ACGIH TLV: TWA 2.5 mg(F)/m^3; 1 mg(Fe)/m^3
NIOSH REL: TWA (Inorganic Fluorides) 2.5 mg(F)/m^3
DOT Classification: ORM-E; Label: None.

SAFETY PROFILE: Poison by intravenous route. When heated to decomposition it emits toxic fumes of F^-. See also FLUORIDES.

FAY000 ***HR: 3***
FERRIC HYDROXIDE NITRILOTRIPROPIONIC ACID COMPLEX

PROP: Complex with 20.0% Fe (ONCOBS 19,239,65).

SYN: IRON(+3) HYDROXIDE COMPLEX with NITRILO-TRI-PROPIONIC ACID

TOXICITY DATA with REFERENCE
ims-mus TDLo:45 mg(Fe)/kg/3W-I:ETA ONCOAR 19,239,65

SAFETY PROFILE: Questionable carcinogen with experimental tumorigenic data. When heated to decomposition it emits toxic fumes of NO_x.

FAZ000 ***HR: 3***
FERRIC NITROSODIMETHYL DITHIOCARBAMATE and TETRAMETHYL THIURAM DISULFIDE

PROP: 58.5% main component, 6.5% secondary component (NTIS** PB223-159).

SYNS: BIS(DIMETHYLTHIOCARBAMOYL)DISULFIDE and NITROSOTRIS (DIMETHYLDITHIOCARBAMATO)IRON ◇ TETRAMETHYLTHIURAM DISULFIDE mixed with FERRIC NITROSODIMETHYLDITHIOCARBAMATE ◇ VANGUARD GF

TOXICITY DATA with REFERENCE
orl-mus TDLo:27 g/kg/78W-I:CAR NTIS** PB223-159
scu-mus TDLo:46 mg/kg:ETA NTIS** PB223-159

SAFETY PROFILE: Questionable carcinogen with experimental carcinogenic and tumorigenic data. When heated to decomposition it emits very toxic fumes of NO_x and SO_x. See also N-NITROSO COMPOUNDS, CARBAMATES, and SULFIDES.

FBA000 CAS:10028-22-5 ***HR: D***
FERRIC SULFATE
mf: $Fe_2O_{12}S_3$ mw: 399.88

PROP: Yellow solid.

SYNS: DIIRON TRISULFATE ◇ IRON PERSULFATE ◇ IRON SESQUISULFATE ◇ IRON SULFATE (2:3) ◇ IRON(III) SULFATE ◇ IRON TERSULFATE ◇ SULFURIC ACID, IRON (3^-) SALT (3:2)

CONSENSUS REPORTS: Reported in EPA TSCA Inventory.

ACGIH TLV: TWA 1 mg(Fe)/m^3

SAFETY PROFILE: When heated to decomposition it emits toxic fumes of SO_x and Fe^-. See also SULFATES and other ferric salts.

FBB000 CAS:9007-73-2 ***HR: 3***
FERRITIN

PROP: Prepared from rat liver protein by precipitation with a cadmium salt (BECCAN 39,74,61).

TOXICITY DATA with REFERENCE
cyt-ham:ovr 27 mg/L CNREA8 41,1628,81
scu-rat TDLo:224 mg/kg/15W-I:NEO BJCAAI 18,667,64

CONSENSUS REPORTS: Reported in EPA TSCA Inventory.

SAFETY PROFILE: Questionable carcinogen with experimental neoplastigenic data. Mutation data reported. When heated to decomposition it emits very toxic fumes of Cd. See also CADMIUM.

FBB100 ***HR: 2***
FERRLECIT

TOXICITY DATA with REFERENCE
ipl-rat TD:80 mg/kg:CAR CNREA8 2,157,86
ipr-mus LD50:229 mg(Fe)/kg NNAPBA 270(Suppl),R50,71

SAFETY PROFILE: Poison by intraperterional route. Questionable carcinogen with experimental carcinogenic data. When heated to decomposition it emits acrid smoke and irritating fumes.

FBC000 CAS:102-54-5 ***HR: 3***
FERROCENE
mf: $C_{10}H_{10}Fe$ mw: 186.05

PROP: Orange crystals; camphor odor. Mp: 174°, subl @ >100°, volatile in steam. Insol in water; sol in alcohol and ether.

SYNS: BISCYCLOPENTADIENYLIRON ◇ DI-2,4-CYCLOPENTADIEN-1-YL IRON ◇ DICYCLOPENTADIENYL IRON (OSHA, ACGIH) ◇ IRON BIS(CYCLOPENTADIENE) ◇ IRON DICYCLOPENTADIENYL

TOXICITY DATA with REFERENCE
sln-dmg-par 100 ppm ENMUDM 7,87,85
trn-dmg-par 100 ppm ENMUDM 7,87,85
sce-ham:ovr 130 μg/L ENMUDM 7,1,85
ims-rat TDLo:5175 mg/kg/2Y-I:ETA NCIUS* PH 43-64-886,AUG,69
orl-rat LD50:1320 mg/kg SCCUR* -,5,61
ipr-rat LD50:500 mg/kg NCIUS* PH 43-64-886,JAN,65
orl-mus LD50:832 mg/kg BJPCAL 24,352,65
ipr-mus LD50:335 mg/kg NCIUS* PH 43-64-886,JAN,65
ivn-mus LD50:178 mg/kg CSLNX* NX#02382

CONSENSUS REPORTS: Reported in EPA TSCA Inventory.

OSHA PEL: (Transitional: TWA Total Dust: 15 mg/m^3; Respirable Fraction: 5 mg/m^3) TWA Total Dust: 10 mg/m^3; Respirable Fraction: 5 mg/m^3
ACGIH TLV: TWA 10 mg/m^3

SAFETY PROFILE: Poison by intraperitoneal and intravenous routes. Moderately toxic by ingestion. Questionable carcinogen with experimental tumorigenic data. Mutation data reported. Flammable; reacts violently with NH_4ClO_4. When heated to decomposition it emits acrid smoke and irritating fumes.

FBC100 CAS:1336-80-7 ***HR: 3***
FERROCHOLINATE
mf: $C_6H_{10}FeO_{10}•C_5H_{14}NO$ mw: 402.21

PROP: Greenish-brown, reddish-brown or brown amorph solid with glistening surface upon fracture. Sol in water, acids, and alkalies.

SYNS: CHELAFER ◇ CHEL-IRON ◇ FERRIC CHOLINE CITRATE ◇ FERROLIP ◇ IRON CHOLINE CITRATE COMPLEX

TOXICITY DATA with REFERENCE
orl-mus LD50:5500 mg/kg AJMSA9 241,296,61
ipr-mus LD50:151 mg/kg AJMSA9 241,296,61
ivn-mus LD50:210 mg/kg AJMSA9 241,296,61
ivn-dog LD50:140 mg/kg AJMSA9 241,296,61

SAFETY PROFILE: Poison by intravenous and intraperitoneal routes. Mildly toxic by ingestion. When heated to decomposition it emits toxic fumes of NO_x. See also CHOLINE.

FBD000 CAS:11114-46-8 ***HR: 3***
FERROCHROME (exothermic)

SYNS: CARBON FERROCHROMIUM ◇ CHROME FERROALLOY ◇ CHROMIUM ALLOY, Cr,C,Fe,N,Si ◇ CHROMIUM ALLOY, BASE, Cr,C,Fe,N,Si (FERROCHROMIUM) ◇ exothermic FERROCHROME (DOT) ◇ FERROCHROME ◇ FERROCHROME, exothermic (DOT) ◇ FERROCHROMIUM

CONSENSUS REPORTS: IARC Cancer Review: Animal Inadequate Evidence IMEMDT 23,205,80. Reported in EPA TSCA Inventory. Chromium and its compounds are on the Community Right-To-Know List.

OSHA PEL: TWA 1 mg(Cr)/m^3
ACGIH TLV: TWA 0.5 mg(Cr)/m^3
DOT Classification: ORM-C; Label: None.

SAFETY PROFILE: Poison by inhalation. Questionable carcinogen. See also CHROMIUM COMPOUNDS.

FBD100 ***HR: 1***
FERROCYANIDES

CONSENSUS REPORTS: Cyanide and its compounds are on the Community Right-To-Know List.

SAFETY PROFILE: Ferrocyanides in general have low toxicity, but highly toxic decomposition products can form upon mixing them with hot concentrated acids. Acid, basic or neutral solutions of ferrocyanides liberate hydrocyanic acid upon strong irradiation. Fusion of mixtures of metal cyanides with metal chlorates, perchlorates, nitrates, or nitrites may cause violent explosions. When heated to decomposition or on contact with acids or acid fumes they emit toxic fumes of CN^-. See also CYANIDE and specific compounds.

FBD500 CAS:17169-60-7 ***HR: 3***
FERROGLYCINE SULFATE
mf: $C_2H_4FeNO_6S•H$ mw: 226.99

SYNS: FERRATE(1-),(GLYCINATO-N,O)(SULFATO(2-)-O',O')-, HYDROGEN, (T-4)-(9CI) ◇ FERROGLYCINE SULFATE ◇ FERROGLYCINE SULFATE COMPLEX ◇ FERRONORD ◇ FERROSANOL ◇ GLYFERRO

TOXICITY DATA with REFERENCE
orl-rat LD50:5590 mg/kg AJMSA9 241,296,61
orl-mus LD50:1940 mg/kg AJMSA9 241,296,61
ipr-mus LD50:365 mg/kg AJMSA9 241,296,61
ACGIH TLV: TWA 1 mg/(Fe)/m^3

SAFETY PROFILE: Poison by intraperitoneal route. Moderately toxic by ingestion. When heated to decomposition it emits toxic fumes of SO_x and NO_x. See also GLYCINE and IRON SULFATE.

FBE000 CAS:12604-53-4 ***HR: 3***
FERROMANGANESE (exothermic)

SYN: exothermic FERROMANGANESE (DOT)

CONSENSUS REPORTS: Reported in EPA TSCA Inventory. Manganese and its compounds are on the Community Right-To-Know List.

SAFETY PROFILE: The dust will burn violently and give off toxic fumes of MnO_2. See also MANGANESE COMPOUNDS.

FBF000 CAS:8049-19-2 ***HR: D***
FERROPHOSPHORUS

SYN: IRON ALLOY, BASE, Fe,P (FERROPHOSPHORUS)

CONSENSUS REPORTS: Reported in EPA TSCA Inventory.

DOT Classification: ORM-A; Label: None.

SAFETY PROFILE: See also PHOSPHORUS. When heated to decomposition it emits very toxic fumes of PO_x.

FBG000 CAS:8049-17-0 ***HR: 3***
FERROSILICON
mf: FeSi mw: 83.90
DOT: UN 1408

PROP: Crystalline, metallic solid. Fe + Si, d: 5.4. Containing 30% or more but not more than 70% silicon (FEREAC 41,15972,76).

SYNS: FERROSILICON, containing more than 30% but less than 90% SILICON (DOT) ◇ IRON-SILICON ALLOY

CONSENSUS REPORTS: Reported in EPA TSCA Inventory.

DOT Classification: ORM-A; Label: None; Flammable Solid; Label: Dangerous When Wet, Poison.

SAFETY PROFILE: Reaction with moisture releases hydrogen and acetylene gases which then ignite; impurities in the alloy may liberate such poisonous and reactive gases as phosphine and arsine. Dry mixtures with sodium hydroxide react incandescently when water is added. Reaction with acid, acid fumes, or oxidizing materials can emit toxic fumes. Reaction hazards increase with decreasing particle size.

FBG200 CAS:15669-07-5 ***HR: 2***
FERROTREMOLITE
mf: $Ca_2Fe_5H_2O_{24}Si_8$ mw: 970.15

SYN: FERROACTINOLITE

TOXICITY DATA with REFERENCE
ipl-rat TDLo:80 mg/kg:CAR TOLED5 13,143,82
itr-rat TDLo:24 mg/kg/12W-I:CAR TOLED5 13,143,82

SAFETY PROFILE: Questionable carcinogen with experimental carcinogenic data.

FBH000 CAS:3094-87-9 ***HR: 2***
FERROUS ACETATE
mf: $C_4H_6O_4{\cdot}Fe$ mw: 173.95

SYNS: ACETIC ACID, IRON(2+) SALT ◇ IRON(2+) ACETATE ◇ IRON(II) ACETATE ◇ IRON DIACETATE

TOXICITY DATA with REFERENCE
scu-rat LD50:492 mg/kg 27ZWAY 3.2,1268,-

CONSENSUS REPORTS: Reported in EPA TSCA Inventory.

OSHA PEL: TWA 1 mg(Fe)/m^3
ACGIH TLV: TWA 1 mg(Fe)/m^3

SAFETY PROFILE: Moderately toxic by subcutaneous route. When heated to decomposition it emits acrid smoke and irritating fumes.

FBH050 CAS:14536-17-5 ***HR: 1***
FERROUS ASCORBATE

PROP: Blue-violet solid.

SYN: IRON(II) ASCORBATE

SAFETY PROFILE: A nuisance dust.

FBH100 CAS:563-71-3 ***HR: 1***
FERROUS CARBONATE
mf: $CFeO_3$ mw: 115.86

PROP: White solid; odorless.

SYN: IRON(II) CARBONATE

SAFETY PROFILE: A nuisance dust.

FBI000 CAS:7758-94-3 ***HR: 3***
FERROUS CHLORIDE
DOT: UN 1759/UN 1760
mf: Cl_2Fe mw: 126.75

PROP: Green to yellow, deliquescent crystals. Mp: 614-670°, bp: 1026°, d: 3.16, vap press: 10 mm @ 700°.

SYNS: IRON(II) CHLORIDE (1:2) ◇ IRON DICHLORIDE ◇ IRON PROTOCHLORIDE

TOXICITY DATA with REFERENCE
otr-ham:emb 2500 μmol/L GNREA8 39,193,79
ipr-mus LD50:59 mg/kg AEPPAE 244,17,62

CONSENSUS REPORTS: Reported in EPA TSCA Inventory. EPA Genetic Toxicology Program.

OSHA PEL: TWA 1 mg(Fe)/m^3
ACGIH TLV: TWA 1 mg(Fe)/m^3
DOT Classification: Corrosive Material; Label: Corrosive (NA1760); ORM-B; Label: None (NA1759)

SAFETY PROFILE: Poison by intraperitoneal route. Mutation data reported. Corrosive. Probably an irritant to the eyes, skin and mucous membranes. Can react violently with ethylene oxide, K, Na. When heated to decomposition it emits toxic fumes of Cl^-. See also CHLORIDES and IRON.

FBJ000 CAS:13478-10-9 ***HR: 3***
FERROUS CHLORIDE TETRAHYDRATE
mf: $Cl_2Fe{\cdot}4H_2O$ mw: 198.83

SYNS: IRON CHLORIDE TETRAHYDRATE ◇ IRON (II) CHLORIDE TETRAHYDRATE ◇ IRON(2+) CHLORIDE TETRAHYDRATE ◇ IRON DICHLORIDE TETRAHYDRATE

TOXICITY DATA with REFERENCE
orl-rat LD50:984 mg/kg 27ZWAY 3.2,1278,-
rec-rat LDLo:498 mg/kg EQSSDX 1,1,75
ipr-mus LD50:93 mg/kg AEPPAE 244,17,62
orl-rbt LDLo:890 mg/kg EQSSDX 1,1,75
scu-rbt LDLo:189 mg/kg EQSSDX 1,1,75
rec-rbt LDLo:984 mg/kg EQSSDX 1,1,75

OSHA PEL: TWA 1 mg(Fe)/m^3
ACGIH TLV: TWA 1 mg(Fe)/m^3

SAFETY PROFILE: Poison by intraperitoneal and subcutaneous routes. Moderately toxic by ingestion and rectal route. When heated to decomposition it emits toxic fumes of Cl^-.

FBJ100 CAS:141-01-5 ***HR: 3***
FERROUS FUMARATE
mf: $C_4H_2O_4{\cdot}Fe$ mw: 169.91

PROP: Reddish-orange to reddish-brown granular powder; odorless, almost tasteless. D: 2.435. Solubility at

25° in water: 0.14 g/100 mL; in alc <0.01 g/100 mL. Solubility in acid is limited by liberation of fumaric acid.

SYNS: CPIRON ◇ ERCO-FER ◇ ERCOFERRO ◇ FEOSTAT ◇ FEROTON ◇ FERROFUME ◇ FERRONAT ◇ FERRONE ◇ FERROTEMP ◇ FERRUM ◇ FERSAMAL ◇ FIRON ◇ FUMAFER ◇ FUMAR-F ◇ FUMIRON ◇ GALFER ◇ HEMOTON ◇ IRCON ◇ IRON FUMARATE ◇ METERFER ◇ METERFOLIC ◇ ONE-IRON ◇ PALAFER ◇ TOLERON ◇ TOLFERAIN ◇ TOLIFER

TOXICITY DATA with REFERENCE
orl-rat LD50:3850 mg/kg NIIRDN 6,683,82
ipr-rat LD50:185 mg/kg NIIRDN 6,683,82
scu-rat LD50:500 mg/kg NIIRDN 6,683,82
orl-mus LD50:1570 mg/kg AJMSA9 241,296,61
ipr-mus LD50:480 mg/kg AJMSA9 241,296,61

CONSENSUS REPORTS: Reported in EPA TSCA Inventory.

OSHA PEL: TWA 1 mg(Fe)/m^3
ACGIH TLV: TWA 1 mg/(Fe)/m^3

SAFETY PROFILE: Poison by intraperitoneal route. Moderately toxic by ingestion and subcutaneous routes. When heated to decomposition it emits acrid smoke and irritating fumes. See also FUMARIC ACID.

FBK000 CAS:299-29-6 ***HR: 3***
FERROUS GLUCONATE
mf: $C_{12}H_{22}O_{14}\bullet Fe$ mw: 446.19

PROP: Yellowish-gray or pale greenish-yellow, fine powder or granules with slt odor of burned sugar. Sol in water and glycerin; insol in alc.

SYNS: FERGON ◇ FERGON PREPARATIONS ◇ FERLUCON ◇ FERRONICUM ◇ GLUCO-FERRUM ◇ IROMIN ◇ IRON GLUCONATE ◇ IROX (GADOR) ◇ NIONATE ◇ RAY-GLUCIRON

TOXICITY DATA with REFERENCE
scu-mus TDLo:2600 mg/kg/13W-I:ETA,REP JNCIAM 24,109,60
orl-chd TDLo:162 mg/kg:GIT JAMAAP 218,1179,71
orl-rat LD50:2237 mg/kg NTIS** UR-3490-168
orl-mus LD50:3950 mg/kg AJMSA9 241,296,61
ipr-mus LD50:160 mg/kg AJMSA9 241,296,61
ivn-mus LD50:199 mg/kg AJMSA9 241,296,61

CONSENSUS REPORTS: Reported in EPA TSCA Inventory.

OSHA PEL: TWA 1 mg(Fe)/m^3
ACGIH TLV: TWA 1 mg(Fe)/m^3

SAFETY PROFILE: Poison by intraperitoneal and intravenous routes. Moderately toxic by ingestion. Human systemic effects by ingestion: hypermotility, diarrhea, nausea, and vomiting. Experimental reproductive effects. Questionable carcinogen with experimental tumorigenic data. When heated to decomposition it emits acrid smoke and irritating fumes.

FBL000 CAS:22830-45-1 ***HR: 3***
FERROUS GLUCONATE DIHYDRATE
mf: $C_{12}H_{22}O_{14}\bullet Fe\bullet 2H_2O$ mw: 483.23

TOXICITY DATA with REFERENCE
orl-rat LD50:4500 mg/kg AJMSA9 230,491,55
orl-mus LD50:3700 mg/kg AJMSA9 230,491,55
ivn-mus LD50:98 mg/kg AJMSA9 230,491,55

OSHA PEL: TWA 1 mg(Fe)/m^3
ACGIH TLV: TWA 1 mg(Fe)/m^3

SAFETY PROFILE: Poison by intravenous route. Moderately toxic by ingestion. When heated to decomposition it emits acrid smoke and irritating fumes.

FBM000 CAS:2896-87-9 ***HR: 3***
FERROUS GLUTAMATE
mf: $C_5H_9FeNO_4$ mw: 203.00

SYN: GLUTAMIC ACID, IRON (2+) SALT (1:1)

TOXICITY DATA with REFERENCE
scu-mus TDLo:2600 mg/kg/I:ETA BECCAN 40,30,62

SAFETY PROFILE: Questionable carcinogen with experimental tumorigenic data. When heated to decomposition it emits toxic fumes of NO_x.

FBN000 CAS:15438-31-0 ***HR: 3***
FERROUS ION
DOT: UN 2793
mf: Fe mw: 55.85

SYNS: FERROUS ◇ FERROUS METAL BORINGS, SHAVINGS TURNINGS OR CUTTINGS (DOT) ◇ IRON(2+) ◇ IRON (Fe 2+) ◇ IRON(II) ION

TOXICITY DATA with REFERENCE
mmo-nsc 200 μmol/L MAGDA3 10,249,79

DOT Classification: Flammable Solid; Label: Spontaneously Combustible.

SAFETY PROFILE: Mutation data reported. May ignite spontaneously in air.

FBN100 CAS:7720-78-7 ***HR: 3***
FERROUS SULFATE
mf: $O_4S\bullet Fe$ mw: 151.91

PROP: Grayish white to buff powder. Slowly sol in water; insol in alc.

SYNS: COPPERAS ◇ DURETTER ◇ DUROFERON ◇ EXSICCATED FERROUS SULFATE ◇ EXSICCATED FERROUS SULPHATE ◇ FEOSOL ◇ FEOSPAN ◇ FER-IN-SOL ◇ FERO-GRADUMET ◇ FERRALYN ◇ FERRO-GRADUMET ◇ FERROSULFAT (GERMAN) ◇ FERROSULFATE ◇ FERRO-THERON ◇ FERSOLATE ◇ GREEN VITRIOL ◇ IRON MONOSULFATE ◇ IRON PROTOSULFATE ◇ IRON(II) SULFATE (1 : 1) ◇ IRON VITRIOL ◇ IROSPAN ◇ IROSUL ◇ SLOW-FE ◇ SULFERROUS ◇ SULFURIC ACID, IRON(2^-) SALT (1:1)

TOXICITY DATA with REFERENCE
mmo-smc 100 mmol/L MUREAV 117,149,83
cyt-ham:fbr 1250 mg/L FCTOD7 22,623,84
orl-rat TDLo:7200 mg/kg:TER OYYAA2 17,483,79
itt-rat TDLo:12153 mg/kg (1D male):REP JRPFA4 7,21,64
scu-mus TDLo:1600 mg/kg/16W-I:ETA JNCIAM 24,109,60
orl-chd LDLo:390 mg/kg JOPDAB 64,218,64
orl-wmn TDLo:60 mg/kg:CNS,GIT JAMAAP 229,1333,74
orl-wmn TDLo:10560 μg/kg:GIT JAMAAP 236,2320,76
orl-chd TDLo:20 mg/kg:BRN,CNS JOPDAB 94,147,79
orl-chd TDLo:150 mg/kg:CNS,GIT NEJMAG 273,1124,65
orl-wmn TDLo:600 mg/kg JAMAAP 229,1333,74
unr-man LDLo:441 mg/kg 85DCAI 2,73,70
orl-rat LD50:319 mg/kg JOPDAB 69,663,66
scu-rat LD50:155 mg/kg NIIRDN 6,888,82
orl-mus LD50:680 mg/kg BJPCAL 24,352,65
ipr-mus LD50:289 mg/kg COREAF 256,1043,63
scu-mus LD50:60300 μg/kg NIIRDN 6,6888,82
ivn-mus LD50:112 mg/kg AJMSA9 241,296,61
ivn-dog LD50:79 mg/kg AJMSA9 241,296,61
idu-rbt LDLo:200 mg/kg TXAPA9 66,329,82

CONSENSUS REPORTS: Reported in EPA TSCA Inventory. EPA Genetic Toxicology Program.

OSHA PEL: TWA 1 mg/(Fe)/m^3
ACGIH TLV: TWA 1 mg/(Fe)/m^3

SAFETY PROFILE: A human poison by ingestion. Moderately toxic to humans by an unspecified route. An experimental poison by ingestion, intraduodenal, intraperitoneal, intravenous, and subcutaneous routes. Human systemic effects by ingestion: aggression, somnolence, brain recording changes, diarrhea, nausea or vomiting, bleeding from the stomach, coma. Questionable carcinogen with experimental tumorigenic data. Experimental teratogenic and reproductive effects. Mutation data reported. Potentially explosive reaction with methyl isocyanoacetate at 25°. May ignite on contact with arsenic trioxide + sodium nitrate. When heated to decomposition it emits toxic fumes of SO_x. See also IRON COMPOUNDS.

FBO000 CAS:7782-63-0 ***HR: 3***
FERROUS SULFATE HEPTAHYDRATE
mf: $O_4S{\cdot}Fe{\cdot}7H_2O$ mw: 278.05

PROP: Pale blue green monoclinic crystals or granules; odorless with a salt taste. D: 2.99-3.08. Sol in water; insol in alc.

SYNS: COPPERAS ◇ FEOSOL ◇ FER-IN-SOL ◇ FERO-GRADUMET ◇ FERROUS SULFATE (FCC) ◇ FESOFOR ◇ FESOTYME ◇ GREEN VITROL ◇ HAEMOFORT ◇ IRONATE ◇ IRON(II) SULFATE (1:1), HEPTAHYDRATE ◇ IRON VITROL ◇ IROSUL ◇ MOL-IRON ◇ PRESFERSUL ◇ SULFERROUS

TOXICITY DATA with REFERENCE
mmo-esc 30 μmol/L CIWYAO 49,144,50
orl-rat LDLo:1389 mg/kg EQSSDX 1,1,75
rec-rat LDLo:697 mg/kg EQSSDX 1,1,75
orl-mus LD50:1520 mg/kg AJMSA9 230,491,55
ipr-mus LDLo:250 mg/kg JAPMA8 47,285,58
ivn-mus LD50:51 mg/kg AJMSA9 230,491,55
orl-rbt LDLo:2778 mg/kg EQSSDX 1,1,75
scu-rbt LDLo:279 mg/kg EQSSDX 1,1,75
ivn-rbt LDLo:99 mg/kg EQSSDX 1,1,75

OSHA PEL: TWA 1 mg(Fe)/m^3
ACGIH TLV: TWA 1 mg(Fe)/m^3

SAFETY PROFILE: Poison by intravenous, intraperitoneal, and subcutaneous routes. Moderately toxic by ingestion and rectal routes. Mutation data reported. When heated to decomposition it emits toxic fumes of SO_x.

FBP000 CAS:12604-58-9 ***HR: 3***
FERROVANADIUM DUST

PROP: A gray to black dust.

CONSENSUS REPORTS: Reported in EPA TSCA Inventory.

OSHA PEL: (Transitional: TWA 1 mg/m^3) TWA 1 mg/m^3; STEL 3 mg/m^3
ACGIH TLV: TWA 1 mg/m^3; STEL 3 mg/m^3
DFG MAK: 1 mg/m^3
NIOSH REL: (Vanadium) TWA 1.0 mg(V)/m^3

SAFETY PROFILE: Can cause pulmonary damage. Combustible when exposed to heat or flame. See also VANADIUM and IRON.

FBP050 CAS:53858-86-9 ***HR: 2***
FERRUM

TOXICITY DATA with REFERENCE
ivn-rat LD50:1140 mg/kg KSRNAM 5,226,71
ims-rat LD50:1820 mg/kg KSRNAM 5,226,71
ivn-mus LD50:540 mg/kg KSRNAM 5,226,71
ims-mus LD50:4060 mg/kg KSRNAM 5,226,71

SAFETY PROFILE: Moderately toxic by intravenous and intramuscular routes.

FBP100 CAS:2624-43-3 ***HR: 2***
FERTODUR
mf: $C_{23}H_{24}O_4$ mw: 364.47

PROP: Crystals from ethanol. Mp: 135-136°.

SYNS: 4-((4-(ACETYLOXY)PHENYL)CYCLOHEXYLIDENEMETHYL) PHENOL ACETATE ◇ BIS-(p-ACETOXYPHENYL)-CYCLOHEXYLIDENEMETHANE ◇ BIS-(p-HYDROXYPHENYL)-CYCLOHEXYLIDENEMETHANE DIACETATE ◇ CYCLOFENIL ◇ CYCLOFENYL ◇ CYCLOPENIL ◇ CYCLOPHENYL ◇ F 6066 ◇ H 3452 ◇ ICI 48213

◇ NEOCLYM ◇ OGINEX ◇ ONDOGYNE ◇ ONDONID ◇ REHIBIN ◇ SANOCRISIN ◇ SEXADIENO ◇ SEXOVAR ◇ SEXOVID

TOXICITY DATA with REFERENCE
orl-man TDLo:1577 mg/kg (male 46D pre):REP ASUPAZ 73,199,68
orl-mus TDLo:48 mg/kg (female 1-6D post):TER OYYAA2 4,645,70
ipr-rat LD50:1700 mg/kg OYYAA2 4,821,70
ipr-mus LD50:1080 mg/kg OYYAA2 4,821,70

SAFETY PROFILE: Moderately toxic by intraperitoneal route. Experimental teratogenic and reproductive effects. When heated to decomposition it emits acrid smoke and irritating fumes.

FBP175 ***HR: D***
FERULA JAESCHKEANA VATKE, EXTRACT

PROP: Indian plant belonging to the family *Apiaceae* (IJEBA6 22,312,84).

TOXICITY DATA with REFERENCE
orl-rat TDLo:150 mg/kg (female 12-14D post):REP IJEBA6 22,312,84

SAFETY PROFILE: Experimental reproductive effects.

FBP200 CAS:537-98-4 ***HR: 2***
trans-FERULIC ACID
mf: $C_{10}H_{10}O_4$ mw: 194.20

SYNS: FERULIC ACID ◇ 4-HYDROXY-3-METHOXYCINNAMIC ACID ◇ 3-(4-HYDROXY-3-METHOXYPHENYL)PROPENOIC ACID

TOXICITY DATA with REFERENCE
cyt-ham:ovr 25 g/L CALEDQ 14,251,81
orl-mam TDLo:9600 mg/kg (12D pre):REP SCIEAS 195,575,77
par-mus LDLo:1200 mg/kg CBCCT* 7,688,55

SAFETY PROFILE: Moderately toxic by parenteral route. Experimental reproductive effects. Mutation data reported. When heated to decomposition it emits acrid smoke and irritating fumes.

FBP300 CAS:483-57-8 ***HR: 3***
FERVENULIN
mf: $C_7H_7N_5O_2$ mw: 193.19

PROP: Yellow, orthorhombic crystals. Mp: 178-179°. Sol in most common organic solvents; sol in cold water to about 2 mg/mL; in hot water to about 40 mg/mL. Practically insol in hydrocarbons.

SYNS: 10204-BII ◇ COMPOUND 7215 ◇ 6,8-DIMETHYLPYRIMIDO(5,4-e)-as-TRIAZINE-5,7(6H,8H)-DIONE◇ 6,8-DIMETHYL-PYRIMIDO(5,4-e)-1,2,4-TRIAZINE-5,7(6H,8H)-DIONE ◇ FERVENULINE ◇ PLANOMYCIN ◇ PULANOMYCIN ◇ U-7118

TOXICITY DATA with REFERENCE
ipr-mus LD50:11200 μg/kg 85GDA2 5,198,81
ivn-mus LD50:15 g/kg 85GDA2 5,199,81

SAFETY PROFILE: Poison by intraperitoneal route. When heated to decomposition it emits toxic fumes of NO_x.

FBP520 ***HR: 3***
FETTERBUSH

PROP: Evergreen shrubs or small trees with simple, leathery leaves. They produce clusters of white flowers. *P. floribunda* is a bush that grows wild in the Atlantic coastal states of the United States from Virginia to Georgia. *P. japonica* is a small tree native to Japan that is cultivated in temperate areas of the United States, particularly along the Pacific coast.

SYNS: LILY-OF-THE-VALLEY BUSH ◇ PIERIS FLORIBUNDA ◇ PIERIS JAPONICA

SAFETY PROFILE: The leaves and nectar contain poisonous grayanotoxins (andromedotoxins). Ingestion of the leaves or honey made from the nectar causes an immediate burning pain which may be followed after a delay period of several hours by vomiting, diarrhea, headache, muscle weakness, vision impairment, slowed heartbeat, severe low blood pressure, convulsions, and coma. Fatalities have been reported among children.

FBP850 CAS:33237-74-0 ***HR: 3***
FIBORAN
mf: $C_{22}H_{30}N_2 \cdot ClH$ mw: 359.00

SYNS: APRINDINE HYDROCHLORIDE ◇ N,N-DIETHYL-N'-2-INDANYL-N'-PHENYL-1,3-PROPANEDIAMINE HYDROCHLORIDE ◇ N-(2,3-DIHYDRO-1H-INDEN-2-YL)-N,N'-DIETHYL-N-PHENYL-2,3-PROPANEDIAMINE MONOHYDROCHLORIDE ◇ MS-5075

TOXICITY DATA with REFERENCE
orl-rat TDLo:405 mg/kg (female 17-22D post):REP YACHDS 13,2687,85
orl-rat TDLo:405 mg/kg (female 17-22D post):TER YACHDS 13,2687,85
unr-wmn TDLo:82 mg/kg/31D-I:BLD AIMDAP 143,241,83
mul-wmn TDLo:468 mg/kg/8W-I:BLD AIMDAP 143,229,83
orl-rat LD50:525 mg/kg OYYAA2 27,353,84
ipr-rat LD50:68200 μg/kg OYYAA2 27,353,84
scu-rat LD50:198 mg/kg OYYAA2 27,353,84
ivn-rat LD50:16600 μg/kg OYYAA2 27,353,84
orl-mus LD50:262 mg/kg OYYAA2 27,353,84
ipr-mus LD50:56600 μg/kg OYYAA2 27,353,84
scu-mus LD50:80200 μg/kg OYYAA2 27,353,84
ivn-mus LD50:17100 μg/kg OYYAA2 27,353,84
orl-dog LDLo:30 mg/kg OYYAA2 27,353,84

SAFETY PROFILE: Poison by ingestion, subcutaneous, intravenous, and intraperitoneal routes. Human systemic effects: agranulocytosis. An experimental teratogen. Experimental reproductive effects. When heated to decomposition it emits toxic fumes of NO_x and HCl.

FBQ000 ***HR: 2***
FIBROUS GLASS

PROP: Is of a borosilicate variety, of low alkalinity and consists of calcia-alumina-silicate (85INA8 5,270,86).

SYNS: FIBERGLASS ◇ FIBROUS GLASS DUST (ACGIH) ◇ GLASS ◇ GLASS FIBERS

TOXICITY DATA with REFERENCE
oms-hmn:fbr 10 mg/L MUREAV 116,369,83
oms-ham:ovr 10 mg/L MUREAV 116,369,83
ihl-rat TCLo:5 mg/m^3/7H/90W-I:CAR NTIS** PB83-258111
ipr-rat TDLo:50 mg/kg:ETA IAPUDO 30,337,80
imp-rat TDLo:200 mg/kg:NEO JJIND8 67,965,81

OSHA PEL: TWA 15 mg/m^3 (total dust); 5 mg/m^3 (nuisance dust)
ACGIH TLV: TWA 10 mg/m^3 (dust)
NIOSH REL: TWA 5 mg/m^3 (total fibrous glass)

SAFETY PROFILE: Questionable carcinogen with experimental carcinogenic, neoplastigenic, and tumorigenic data by inhalation and other routes. Human mutation data reported. Used as thermal and acoustic insulation.

The possibility of lung problems due to inhalation of fine particles or flakes or fibers of fiberglass has often been raised. The extensive medical research so far reported has shown no consistent evidence of chronic health effects in workers who are exposed to manmade vitreous fibers. In some studies where massive doses of fine diameter fibers were implanted into mice, cancer development in the pleura was noted. Also some animal studies involving injection of fibers into the trachae resulted in a minimal fibrosis.

Exposure to glass fibers sometimes causes irritation of the skin and, less frequently, irritation of the eyes, nose, or throat. This is not an allergic reaction, but simply a mechanical irritation. Skin irritation typically is experienced by individuals who are newly exposed to fibrous glass and it usually diminishes after several days of exposure. Good personal and industrial hygiene practices minimize the amount of discomfort experienced.

FBS000 CAS:9001-33-6 ***HR: 3***
FICIN

PROP: A proteolytic enzyme in the crude latex of the fig tree *Ficus* (JPETAB 71,20,41). White powder. Very sol in water.

SYNS: DEBRICIN ◇ FICUS PROTEASE ◇ FICUS PROTEINASE ◇ HIGUEROXYL DELABARRE ◇ TL 367

TOXICITY DATA with REFERENCE
orl-rat LD50:10 g/kg MEIEDD 10,585,83
orl-mus LD50:10 g/kg MEIEDD 10,585,83
ihl-mus LCLo:290 mg/m^3/10M NDRC** NDCrc-132,Sept,42
ihl-cat LCLo:290 mg/m^3/10M NDRC** NDCrc-132,Sept,42
orl-rbt LD50:5000 mg/kg JPETAB 71,20,41
ihl-rbt LCLo:290 mg/m^3/10M NDRC** NDCrc-132,Sept,42
orl-gpg LD50:5000 mg/kg JPETAB 71,20,41
ihl-gpg LCLo:290 mg/m^3/10M NDRC** NDCrc-132,Sept,42
ivn-mam LDLo:50 mg/kg JPETAB 71,20,41

CONSENSUS REPORTS: Reported in EPA TSCA Inventory.

SAFETY PROFILE: Poison by inhalation and intravenous routes data. Mildly toxic by ingestion. When heated to decomposition it emits toxic fumes.

FBS100 ***HR: 3***
FIGWORT

PROP: Annual or perennial herbs which grow to 3 feet and produce yellow, white, or occasionally red flowers. They grow in swampy areas throughout North America and Hawaii.

SYNS: BASSINET (CANADA) ◇ BLISTER FLOWER ◇ BLISTER WORT ◇ BOUTON D'OR (CANADA) ◇ BUTTER CRESS ◇ BUTTERCUP ◇ BUTTER DAISY ◇ CRAIN ◇ CROWFOOD ◇ DEVIL'S CLAWS ◇ GOLDBALLS ◇ GOLDWEED ◇ HORSE GOLD ◇ HUNGER WEED ◇ LESSER CELANDINE ◇ PILEWORT ◇ RAM'S CLAWS ◇ RANUNCULUS ACRIS ◇ RANUNCULUS BULBOSUS ◇ RANUNCULUS SCELERATUS ◇ RENONCULE (CANADA) ◇ SITFAST ◇ SPEARWORT ◇ ST. ANTHONY'S TURNIP ◇ STARVE-ACRE ◇ WATER CROWFOOT ◇ YELLOW GOWAN

SAFETY PROFILE: The roots contain the poison protoanemonin which is a direct irritant and drying agent on the skin and mucous membranes. Ingestion causes pain and blistering of the mouth possibly followed by vomiting, diarrhea, cramps, bloody and painful urination, dizziness, fainting, confusion, and convulsions.

FBU000 CAS:59536-65-1 ***HR: 3***
FIREMASTER BP-6

PROP: Consists mainly of penta-, hexa-, and hepta-bromobiphenyl, with lesser amounts of tetra- and other brominated biphenyls. (ENVRAL 10,390,75)

SYNS: HEXABROMOBIPHENYL (technical grade) ◇ PBB ◇ POLYBROMINATED BIPHENYLS

TOXICITY DATA with REFERENCE
dnd-sal:oth 8 μmol/L EVHPAZ 23,51,78

orl-rat TDLo:24600 μg/kg (female 6-22D post):REP TOXID9 4,86,84
orl-mus TDLo:312 mg/kg (female 4-16D post):TER TXAPA9 37,171,76
orl-rat TD:1000 mg/kg:ETA EVHPAZ 23,265,78
orl-rat TD:1200 mg/kg/18W-I:CAR JJIND8 66,535,81
orl-qal LD50:15 μg/kg BECTA6 33,308,84

CONSENSUS REPORTS: IARC Cancer Review: Animal Inadequate Evidence IMEMDT 18,107,78. Polybrominated biphenyl compounds are on the Community Right-To-Know List.

SAFETY PROFILE: Poison by ingestion. Experimental teratogenic and reproductive effects. Questionable carcinogen with experimental carcinogenic and tumorigenic data. Mutation data reported. When heated to decomposition it emits very toxic Br^-. See also POLYBROMINATED BIPHENYLS.

FBU509 CAS:67774-32-7 ***HR: 3***
FIREMASTER FF-1

PROP: 2,4,5,2′,4′,5′-hexabromobiphenyl is the predominant isomer (LANCAO 2,602,77).

SYNS: 2,4,5,2′,4′,5′-HEXABROMOBIPHENYL ◇ PBB ◇ POLYBROMINATED BIPHENYL ◇ POLYBROMINATED BIPHENYL (FF-1)

TOXICITY DATA with REFERENCE
orl-rat TDLo:10 mg/kg (female 20D pre):REP TOLED5 24,229,85
orl-mus TDLo:360 mg/kg (female 6-15D post):TER TOXID9 4,87,84
orl-rat TDLo:375 mg/kg/26W-I:CAR NTPTR* NTP-TR-244,83
orl-mus TDLo:1250 mg/kg/26W-I:CAR NTPTR* NTP-TR-244,83
orl-rat TD:1000 mg/kg:NEO LANCAO 2,602,77

CONSENSUS REPORTS: NTP Fifth Annual Report on Carcinogens. IARC Cancer Review: Group 2B IMEMDT 7,321,87; Human Inadequate Evidence IMEMDT 41,261,86; Animal Sufficient Evidence IMEMDT 41,261,86. NTP Carcinogenesis Studies (gavage); Clear Evidence: mouse, rat NTPTR* NTP-TR-244,83. Polybrominated biphenyl compounds are on the Community Right-To-Know List.

SAFETY PROFILE: Confirmed carcinogen with experimental carcinogenic, neoplastigenic, and teratogenic data. Experimental reproductive effects. When heated to decomposition it emits very toxic fumes of Br^-. See also POLYBROMINATED BIPHENYLS.

FBV000 CAS:8021-29-2 ***HR: 1***
FIR NEEDLE OIL, SIBERIAN

PROP: Found in the needles and twigs of *Abies sibirica* Ledeb. (Fam. *Pinaceae*) (FCTXAV 13,449,75). Colorless to faintly yellow liquid. Sol in fixed oils, mineral oil; insol in glycerin, propylene glycol.

SYN: PINE NEEDLE OIL

TOXICITY DATA with REFERENCE
skn-hmn 12500 μg/48H FCTXAV 13,450,75
skn-rbt 500 mg/24H MOD FCTXAV 13,450,75
orl-rat LD50:10200 mg/kg FCTXAV 13,450,75

CONSENSUS REPORTS: Reported in EPA TSCA Inventory.

SAFETY PROFILE: Mildly toxic by ingestion. A human and experimental skin irritant. When heated to decomposition it emits acrid smoke and irritating fumes.

FBV100 ***HR: 1***
FISCHER'S SOLUTION

PROP: An amino acid solution for hepatic failure. IYKEDH 13,11,82

SYN: GO-80

TOXICITY DATA with REFERENCE
ivn-rbt TDLo:750 g/kg (male 30D pre):REP IYKEDH 13,24,82
ivn-rbt LDLo:200 g/kg IYKEDH 13,11,82

SAFETY PROFILE: Very slight toxicity by intravenous route. Experimental reproductive effects. When heated to decomposition it emits acrid smoke and irritating fumes.

FBW000 CAS:528-48-3 ***HR: 3***
FISETIN
mf: $C_{15}H_{10}O_6$ mw: 286.25

PROP: Coloring matter isolated from *Rhus cotinus L. Anacardiaceae.* Yellow needles from dil alc. Decomp @ 330°. Sol in alc, acetone, acetic acid, solns of fixed alkali hydroxides; practically insol in water, ether, benzene, chloroform and petr ether.

SYN: 3,3′,4′,7-TETRAHYDROXYFLAVONE

TOXICITY DATA with REFERENCE
mmo-sat 100 μg/plate BCSTB5 5,1489,77
mma-sat 100 μg/plate MUREAV 58,231,78
dni-hmn:fbr 50 mg/L BCPCA6 33,3823,84
ivn-mus LD50:180 mg/kg CSLNX* NX#01998

CONSENSUS REPORTS: Reported in EPA TSCA Inventory.

SAFETY PROFILE: Poison by intravenous route. Human mutation data reported. When heated to decomposition it emits acrid smoke and fumes.

FBW100 ***HR: 2***
FISHTAIL PALM

PROP: A palm with long (up to 20 feet) leaves formed of small fishtail-shaped leaflets. It produces red or black berries clustered on long strings. They are native to Asia and grow wild and under cultivation in Florida, Hawaii, the West Indies and Guam.

SYNS: CARYOTA (VARIOUS SPECIES) ◇ C. MITIS ◇ C. URENS ◇ CARIOTA (CUBA) ◇ WINE PALM

SAFETY PROFILE: The berries contain toxic calcium oxalate raphides. Chewing the berries results in burning pain in the lips, mouth and throat, possibly followed by inflammation and blistering. Systemic effects are usually not seen because of the insolubility of calcium oxalate. Skin contact may result in pain and itching. See also OXALATES.

FBY000 CAS:114-42-1 ***HR: 3***
FLAVASPIDIC ACID
mf: $C_{24}H_{30}O_8$ mw: 446.54

SYNS: FLAVASPIDSAEURE (GERMAN) ◇ FLAVISPIDIC ACID BB ◇ POLYSTICHOCITRIN ◇ TOXIFREN

TOXICITY DATA with REFERENCE
orl-mus LD50:690 mg/kg APTOA6 25,33,67
ivn-mus LD50:94 mg/kg JPMSAE 54,1362,65

SAFETY PROFILE: Poison by intravenous route. Moderately toxic by ingestion. When heated to decomposition it emits acrid smoke and irritating fumes.

FBZ000 CAS:29382-82-9 ***HR: 3***
FLAVENSOMYCIN
mf: $C_{47}H_{65}O_{14}$ mw: 854.12

PROP: An antibiotic produced by a strain of *Streptomyces sp* (85ERAY 2,1141,78).

TOXICITY DATA with REFERENCE
orl-mus LD50:5 mg/kg 85GDA2 6,412,81
ipr-mus LD50:1 mg/kg 85GDA2 6,412,81
scu-mus LD50:2 mg/kg 85FZAT -,1291,67

SAFETY PROFILE: Poison by ingestion, subcutaneous, and intraperitoneal routes. When heated to decomposition it emits acrid smoke and irritating fumes.

FCA000 CAS:11006-22-7 ***HR: 3***
FLAVOFUNGIN (1:10)

TOXICITY DATA with REFERENCE
orl-mus LD50:1200 mg/kg 85ERAY 2,996,78
ipr-mus LD50:25 mg/kg 85ERAY 2,996,78

SAFETY PROFILE: Poison by intraperitoneal route. Moderately toxic by ingestion. When heated to decomposition it emits acrid smoke and irritating fumes.

FCB000 ***HR: 3***
FLAVONOID AGLUCONE

PROP: Extracted from figwort (RPTOAN 32,315,69).

TOXICITY DATA with REFERENCE
orl-mus LD50:555 mg/kg RPTOAN 32(6),315,69
ipr-mus LD50:323 mg/kg RPTOAN 32(6),315,69

SAFETY PROFILE: Poison by intraperitoneal route. Moderately toxic by ingestion. When heated to decomposition it emits acrid smoke and irritating fumes.

FCB100 CAS:3717-88-2 ***HR: 3***
FLAVOXATE HYDROCHLORIDE
mf: $C_{24}H_{25}NO_4 \cdot ClH$ mw: 427.96

SYNS: BLADDERON ◇ DW-61 ◇ FLAROXATE HYDROCHLORIDE ◇ 3-METHYL-4-OXO-2-PHENYL-4H-1-BENZOPYRAN-8-CARBOXYLATE 1-PIPERIDINEETHANOL HYDROCHLORIDE ◇ NSC-114649 ◇ PIPERIDINOETHYL-3-METHYLFLAVONE-8-CARBOXYLATE HYDROCHLORIDE ◇ 2-PIPERIDINOETHYL-3-METHYL-4-OXO-2--PHENYL-4H-1-BENZOPYRAN-8-CARBOXYLATEHYDROCHLORIDE ◇ REC 7-0040 ◇ SPASURET ◇ URISPAS

TOXICITY DATA with REFERENCE
orl-mus TDLo:6 g/kg (female 7-12D post):REP YACHDS 3,1214,75
orl-rat TDLo:4500 mg/kg (female 9-14D post):TER YACHDS 3,1214,75
orl-rat LD50:1040 mg/kg YACHDS 15,5271,87
scu-rat LD50:1010 mg/kg IYKEDH 10,232,79
ivn-rat LD50:25 mg/kg IYKEDH 10,232,79
orl-mus LD50:740 mg/kg JPETAB 130,356,60
ipr-mus LD50:350 mg/kg AIPTAK 131,187,61
scu-mus LD50:610 mg/kg IYKEDH 10,232,79
ivn-mus LD50:28 mg/kg IYKEDH 10,232,79
ivn-rbt LD50:25 mg/kg YACHDS 3,1214,75

SAFETY PROFILE: Poison by intravenous and intraperitoneal routes. Moderately toxic by ingestion and subcutaneous routes. Experimental teratogenic and reproductive effects. A smooth muscle relaxant. When heated to decomposition it emits toxic fumes of NO_x and HCl. See also ESTERS.

FCC000 CAS:57608-59-0 ***HR: 3***
FLAVUMYCIN B

PROP: Produced by *A. flavus var. geptinicus* 4 LIA-073 (JANTAJ 29,76-91,76).

TOXICITY DATA with REFERENCE
ipr-mus LD50:6500 μg/kg JANTAJ 29,76-91,76
ivn-mus LD50:2630 μg/kg JANTAJ 29,76-91,76

SAFETY PROFILE: Poison by intraperitoneal and intravenous routes. When heated to decomposition it emits acrid smoke and irritating fumes.

FCD512 CAS:18268-70-7 *HR: 1*
FLEXOL 4GO
mf: $C_{24}H_{46}O_7$ mw: 446.70

SYNS: POLYETHYLENE GLYCOL 200 DI(2-ETHYLHEXOATE) ◇ TETRAETHYLENE GLYCOL DI(2-ETHYLHEXOATE)

TOXICITY DATA with REFERENCE
orl-rat LD50:18000 mg/kg NPIRI* 2,85,75

CONSENSUS REPORTS: Reported in EPA TSCA Inventory.

SAFETY PROFILE: Mildly toxic by ingestion. When heated to decomposition it emits acrid smoke and irritating fumes.

FCD560 CAS:94-28-0 *HR: 1*
FLEXOL PLASTICIZER 3GO
mf: $C_{22}H_{42}O_6$ mw: 402.64

SYNS: BIS(ETHYLHEXANOATE)TRIETHYLENE GLYCOL ◇ TRIETHYLENE GLYCOL DI(2-ETHYLHEXOATE)

TOXICITY DATA with REFERENCE
skn-rbt 500 mg open MLD UCDS** 7/21/71
orl-rat LD50:31 g/kg JIHTAB 23,259,41
orl-gpg LD50:21 g/kg JIHTAB 23,259,41

CONSENSUS REPORTS: Reported in EPA TSCA Inventory.

SAFETY PROFILE: A mild skin irritant. When heated to decomposition it emits acrid smoke and fumes. See also TRIETHYLENE GLYCOL and ESTERS.

FCN000 CAS:101564-02-7 *HR: 1*
FLEXOL PLASTICIZER CC-55

PROP: Liquid. Mp: −53°, bp: 216° @ 5 mm, flash p: 425°F (OC), d: 0.9586 @ 20°/20°, vap d: 13.7.

TOXICITY DATA with REFERENCE
skn-rbt 500 mg open MLD UCDS** 1/12/72

SAFETY PROFILE: A skin irritant. Combustible when exposed to heat or flame; can react with oxidizing materials. To fight fire, use foam, CO_2, dry chemical. When heated to decomposition it emits acrid smoke and irritating fumes. See also ESTERS.

FCN050 CAS:8036-63-3 *HR: 1*
FLEXOL PLASTICIZER 810

TOXICITY DATA with REFERENCE
skn-rbt 500 mg open MLD UCDS** 7/23/71
skn-rbt LD50:20 mg/kg UCDS** 7/23/71

SAFETY PROFILE: A skin irritant. When heated to decomposition it emits acrid smoke and irritating fumes.

FCN100 *HR: 2*
FLOMOXEF SODIUM
mf: $C_{15}H_{17}F_2N_6O_7S_2 \cdot Na$ mw: 518.49

SYNS: FLOMOXEF ◇ 5-OXA-1-AZABICYCLO(4.2.0)OCT-2-ENE-2-CARBOXYLIC ACID, 7-(2-((DIFLUOROMETHYL)THIO)ACETAMIDO)-3-(((1-(2-HYDROXYETHYL)-1H-TETRAZOL-5-YL)THIO)METHYL)-7-METHOXY-8-OXO-, SODIUM SALT, (6R,7R)- ◇ 6315-S

TOXICITY DATA with REFERENCE
ivn-rat TDLo:41600 mg/kg (female 17-22D post):REP NKRZAZ 35(Suppl 1),404,87
ivn-rbt TDLo:81 mg/kg (female 6-18D post):TER OYYAA2 34,475,87
ivn-rat LD50:7360 mg/kg NKRZAZ 35(Suppl 1),207,87
scu-mus LD50:13500 mg/kg IYKEDH 19,735,88
ivn-mus LD50:7086 mg/kg IYKEDH 19,735,88

SAFETY PROFILE: Moderately toxic by intravenous route. An experimental teratogen. Experimental reproductive effects. When heated to decomposition it emits toxic fumes of NO_x, SO_x, F^-, and Cl^-.

FDA000 CAS:68916-09-6 *HR: 1*
FLOUVE OIL

PROP: Found in the upper part of the dried grass *Flouve odorante* (FCTXAV 16,637,78).

TOXICITY DATA with REFERENCE
skn-rbt 500 mg/24H MOD FCTXAV 16,757,78
orl-rat LD50:4100 mg/kg FCTXAV 16,757,78
orl-rbt LD50:5000 mg/kg FCTXAV 16,757,78

CONSENSUS REPORTS: Reported in EPA TSCA Inventory.

SAFETY PROFILE: Mildly toxic by ingestion. A skin irritant. When heated to decomposition it emits acrid smoke and irritating fumes.

FDA100 CAS:1847-24-1 *HR: 2*
FLOXAPEN SODIUM
mf: $C_{19}H_{16}ClFN_3O_5S \cdot Na$ mw: 475.88

SYNS: FLOXACILLIN SODIUM ◇ FLUCLOXACILLIN SODIUM ◇ FLUCLOXACILLIN SODIUM SALT ◇ 5-METHYL-3-(2-CHLORO-6-FLUOROPHENYL)-4-ISOXAZOLYLPENICILLIN SODIUM ◇ MFI-PC ◇ MONOSODIUM FLUCLOXACILLIN ◇ SODIUM FLUCLOXACILLIN

TOXICITY DATA with REFERENCE
orl-rat TDLo:6 g/kg (9-14D preg):TER NKRZAZ 17,1523,69
orl-wmn TDLo:120 mg/kg/3D-I:LVR AJGAAR 82,483,87
orl-rat LD50:11 g/kg NIIRDN 6,697,82
ipr-rat LD50:1100 mg/kg NIIRDN 6,697,82
scu-rat LD50:2800 mg/kg NIIRDN 6,697,82
ivn-rat LD50:680 mg/kg NIIRDN 6,697,82
orl-mus LD50:7600 mg/kg NIIRDN 6,697,82
ipr-mus LD50:1350 mg/kg NIIRDN 6,697,82

scu-mus LD50:1450 mg/kg NKRZAZ 17,1523,69
ivn-mus LD50:1360 mg/kg NIIRDN 6,697,82
ivn-dog LD50:760 mg/kg NKRZAZ 17,1523,69

SAFETY PROFILE: Moderately toxic by intravenous, intraperitoneal, and subcutaneous routes. Mildly toxic by ingestion. Human systemic effects: cholestatic jaundice. An experimental teratogen. When heated to decomposition it emits toxic fumes of F^-, Cl^-, SO_x, NO_x, and Na_2O.

FDA875 CAS:5107-49-3 ***HR: 3***
FLUALAMIDE
mf: $C_{17}H_{23}F_3N_2O_2$ mw: 344.42

SYNS: 2-(ALLYLOXY)-N-(2-(DIETHYLAMINO)ETHYL)-α,α,α-TRIFLUORO-p-TOLUAMIDE ◇ 305CE ◇ 4-TRIFLUOROMETHYL ALLYLOXY-2 N-(β-DIETHYLAMINO-ETHYL)BENZAMIDE(FRENCH)

TOXICITY DATA with REFERENCE
orl-mus LD50:1375 mg/kg MPHEAE 15,241,66
scu-mus LD50:249 mg/kg JJPAAZ 20,1,70
ivn-mus LD50:86 mg/kg MPHEAE 15,241,66
ivn-dog LDLo:40 mg/kg JJPAAZ 20,1,70

SAFETY PROFILE: Poison by subcutaneous and intravenous routes. Moderately toxic by ingestion. When heated to decomposition it emits toxic fumes of F^- and NO_x.

FDA880 CAS:17160-71-3 ***HR: 3***
FLUANISONE HYDROCHLORIDE
mf: $C_{21}H_{25}FN_2O_2 \cdot ClH$ mw: 392.94

SYNS: ANTI-PICA ◇ 4′-FLUORO-4-(4-(o-METHOXYPHENYL)-1-PIPERAZINYL)BUTYROPHENONE HYDROCHLORIDE ◇ HALOANISONE COMPOSITUM

TOXICITY DATA with REFERENCE
scu-rat LD50:420 mg/kg 27ZQAG -,189,72
ivn-rat LD50:20 mg/kg 27ZQAG -,189,72
ipr-mus LD50:200 mg/kg 27ZQAG -,189,72
scu-mus LD50:150 mg/kg 27ZQAG -,189,72
ivn-mus LD50:25 mg/kg 27ZQAG -,189,72

SAFETY PROFILE: Poison by subcutaneous, intraperitoneal, and intravenous routes. When heated to decomposition it emits very toxic fumes of HCl, NO_x, and F^-.

FDA887 CAS:31430-15-6 ***HR: 2***
FLUBENDAZOLE
mf: $C_{16}H_{12}FN_3O_3$ mw: 313.31

SYNS: 2-BENZIMIDAZOLECARBAMIC ACID, 5-(p-FLUOROBENZOYL)-, METHYL ESTER ◇ CARBAMIC ACID, (5-(4-FLUOROBENZOYL)-1H-BENZIMIDAZOL-2-YL)-, METHYL ESTER (9CI) ◇ (5-(4-FLUOROBENZOYL)-1H-BENZIMIDAZOLE-2-YL)CARBAMIC ACID METHYL ESTER ◇ METHYL N-(5-(p-FLUOROBENZOYL)-2-BENZIMIDAZOLYL) CARBAMATE ◇ R 17899

TOXICITY DATA with REFERENCE
orl-rat TDLo:320 mg/kg (female 8-15D post):TER TXCYAC 43,133,87
orl-rat TDLo:1280 mg/kg (female 8-15D post):REP TXCYAC 43,133,87
orl-rat LD50:2560 mg/kg YRTMA6 9,11,78
orl-rbt LD50:2560 mg/kg YRTMA6 9,11,78

SAFETY PROFILE: Moderately toxic by ingestion. An experimental teratogen. Experimental reproductive effects. When heated to decomposition it emits toxic fumes of NO_x and F^-.

FDA900 CAS:33245-39-5 ***HR: 2***
FLUCHLORALIN
mf: $C_{12}H_{13}ClF_3N_3O_4$ mw: 355.73

PROP: Orange-yellow, crystalline solid. Mp: 42-43°. Solubility in water: 10 ppm.

SYNS: BAS 392-H ◇ BASALIN ◇ N-(2-CHLOROETHYL)-2,6-DINITRO-N-PROPYL-4-(TRIFLUOROMETHYL)ANILINE ◇ N-(2-CHLOROETHYL)-2,6-DINITRO-N-PROPYL-4-(TRIFLUOROMETHYL)BENZENAMIDE ◇ N-(2-CHLOROETHYL)-α,α,α-TRIFLUORO-2,6-DINITRO-N-PROPYL-p-TOLUIDINE ◇ N-PROPYL-N-(2-CHLOROETHYL)-2,6-DINITRO-4-TRIFLUOROMETHYLANILINE ◇ N-PROPYL-N-(2-CHLOROETHYL)-α,α,α-TRIFLUORO-2,6-DINITRO-p-TOLUIDINE

TOXICITY DATA with REFERENCE
orl-rat LD50:1550 mg/kg FMCHA2 -,C26,83
unr-mus LD50:730 mg/kg GUCHAZ 6,434,73

SAFETY PROFILE: Moderately toxic by ingestion. When heated to decomposition it emits toxic fumes of F^-, Cl^-, and NO_x. See also NITRO COMPOUNDS of AROMATIC HYDROCARBONS.

FDA925 CAS:152-97-6 ***HR: D***
FLUCORTOLONE
mf: $C_{22}H_{29}FO_4$ mw: 376.51

PROP: Crystals. Mp: 188-190.5°. Solubility (mg/L): 295 in water (37°), 120 in ethanol (20°), 440 in toluene (20°).

SYNS: FLUOCORTOLON ◇ FLUOCORTOLONE ◇ FLUORCORTOLONE ◇ 6-α-FLUORO-11-β,21-DIHYDROXY-16-α-METHYLPREGNA-1,4-DIENE-3,20-DIONE ◇ 6-α-FLUORO-16-α-METHYL-1-DEHYDROCORTICOSTERONE ◇ 6-α-FLUORO-16-α-METHYLPREGNA-1,4-DIENE-11-β,21-DIOL-3,20-DIONE ◇ 6-α-FLUORO-16-α-METHYL-$\Delta^{1,4}$-PREGNADIENE-11-β-DIOL-3,20-DIONE ◇ SH 742 ◇ ULTRALAN ORAL

TOXICITY DATA with REFERENCE
orl-mus TDLo:70 mg/kg (12D preg):TER YKRYAH 9,1623,76
skn-rbt TDLo:2500 μg/kg (female 6-15D post):REP AMBED4 (1),127,88

SAFETY PROFILE: An experimental teratogen. Experimental reproductive effects. A steroid. When heated to decomposition it emits toxic fumes of F^-.

FDB000 CAS:2825-60-7 ***HR: 2***
FLUDERMA
mf: $C_{29}H_{38}ClFO_8$ mw: 569.12

SYNS: CORTOCIN-F ◇ CUTISTEROL ◇ DEFLAMENE ◇ FI 6341 ◇ FLUOROFORMYLON ◇ FORMOCORTAL

TOXICITY DATA with REFERENCE
ipr-mus LD50:537 mg/kg MEIEDD 10,605,83
scu-mus LD50:490 mg/kg MEIEDD 10,605,83

SAFETY PROFILE: Moderately toxic by intraperitoneal and subcutaneous routes. Used as a glucocorticoid (a corticosteroid which affects carbohydrate, protein and fat metabolism with many other side effects). When heated to decomposition it emits very toxic fumes of F^- and Cl^-.

FDB100 CAS:3900-31-0 ***HR: 3***
FLUDIAZEPAM
mf: $C_{16}H_{12}ClFN_2O$ mw: 302.75

PROP: Colorless prisms from n-hexane/isopropanol. Mp: 88-92°.

SYNS: 7-CHLORO-5-(o-FLUOROPHENYL)-1,3-DIHDYRO-1-METHYL-2H-1,4-BENZODIAZEPIN-2-ONE ◇ 7-CHLORO-5-(2-FLUOROPHENYL)-1-METHYL-1H-1,4-BENZODIAZEPIN-2(3H)-ONE ◇ ERISPAN ◇ ID 540 ◇ RO 5-3438

TOXICITY DATA with REFERENCE
orl-rat TDLo:150 mg/kg (30D pre):REP KSRNAM 11,529,77
orl-rat LD50:1740 mg/kg YAKUD5 22,811,80
ipr-rat LD50:315 mg/kg IYKEDH 11,811,80
scu-rat LD50:7600 mg/kg IYKEDH 11,811,80
orl-mus LD50:830 mg/kg IYKEDH 11,811,80
ipr-mus LD50:360 mg/kg NIIRDN 6,698,82
scu-mus LD50:1150 mg/kg NIIRDN 6,698,82

SAFETY PROFILE: Poison by intraperitoneal route. Moderately toxic by ingestion and subcutaneous routes. Experimental reproductive effects. Used as a tranquilizer. Note: This is a controlled substance (depressant) listed in the U.S. Code of Federal Regulations, Title 21 Part 1308.14 (1985). When heated to decomposition it emits toxic fumes of F^-, Cl^-, and NO_x. See also DIAZEPAM.

FDB200 CAS:4301-50-2 ***HR: 3***
FLUENETIL
mf: $C_{16}H_{15}FO_2$ mw: 258.31

SYNS: 4-BIPHENYLACETIC ACID, 2-FLUOROETHYL ESTER ◇ (1,1'-BIPHENYL)-4-ACETIC ACID, 2-FLUOROETHYL ESTER ◇ FLUENYL ◇ LAMBROL

TOXICITY DATA with REFERENCE
ipr-mus LD50:42 mg/kg EVHPAZ 14,93,76
orl-rat LD50:6 mg/kg FMCHA2 -,C138,83

CONSENSUS REPORTS: EPA Extremely Hazardous Substances List.

SAFETY PROFILE: Poison by ingestion and intraperitoneal routes. When heated to decomposition it emits toxic fumes of F^-. See also ESTERS.

FDD075 CAS:2135-17-3 ***HR: D***
FLUMETHASONE
mf: $C_{22}H_{28}F_2O_5$ mw: 410.50

SYNS: ANAPRIME ◇ ANIPRIME ◇ CORTEXILAR ◇ FLUCORT ◇ FLUCORTICIN ◇ 6-α-FLUORODEXAMETHASONE ◇ FLUVET ◇ METHAGON ◇ RS-2177 ◇ U-10974

TOXICITY DATA with REFERENCE
ims-ctl TDLo:30 μg/kg (female 31W post):REP JAVMA4 160,867,72

SAFETY PROFILE: Experimental reproductive effects. When heated to decomposition it emits toxic fumes of F^-.

FDD080 CAS:30484-77-6 ***HR: 3***
FLUNARIZINE DIHYDROCHLORIDE
mf: $C_{26}H_{26}F_2N_2 \cdot 2ClH$ mw: 477.46

SYNS: (3)-1-(BIS(p-FLUOROPHENYL)METHYL)-4-CINNAMYL-PIPERAZINE DIHYDROCHLORIDE ◇ (E)-1-(BIS(4-FLUOROPHENYL) METHYL)-4-(3-PHENYL-2-PROPENYL)PIPERAZINEDIHYDROCHLORIDE ◇ FLUNARIZINE HYDROCHLORIDE ◇ KW-3149 ◇ R 14950

TOXICITY DATA with REFERENCE
orl-rat TDLo:33 mg/kg (female 7-17D post):REP KSRNAM 16,1840,82
orl-rat TDLo:330 mg/kg (7-17D preg):TER KSRNAM 16,1840,82
orl-rat LD50:503 mg/kg ARZNAD 25,1408,75
ipr-rat LD50:305 mg/kg YKYUA6 36,825,85
scu-rat LD50:1793 mg/kg YKYUA6 36,825,85
ivn-rat LD50:35 mg/kg YKYUA6 36,825,85
orl-mus LD50:1203 mg/kg YKYUA6 36,825,85
ipr-mus LD50:379 mg/kg YKYUA6 36,825,85
ivn-mus LD50:27 mg/kg YKYUA6 36,825,85

SAFETY PROFILE: Poison by intravenous and intraperitoneal routes. Moderately toxic by ingestion and subcutaneous routes. An experimental teratogen. Experimental reproductive effects. When heated to decomposition it emits toxic fumes of Cl^-, NO_x, and F^-.

FDD085 CAS:3385-03-3 ***HR: 1***
FLUNISOLIDE
mf: $C_{24}H_{31}FO_6$ mw: 434.55

SYNS: 6-α-FLUORO-11-β,16-α,17,21-TETRAHYDROXYPREGNA-1,4-

DIENE,-3,20-DIONE, CYCLIC 16,17-ACETAL with ACETONE ◇ LUNIS ◇ NASALIDE ◇ RHINALAR ◇ SYNTARIS

TOXICITY DATA with REFERENCE
orl-rat TDLo:26 μg/kg (female 17-21D post):REP OYYAA2 24,661,82
orl-rat TDLo:1100 μg/kg (female 7-17D post):TER OYYAA2 24,643,82
ihl-cld TCLo:21 mg/kg/30W-I:NOSE JOPDAB 105,840,84

SAFETY PROFILE: Human systemic effects by inhalation: deviated nasal septum, ulcerated nasal septum. An experimental teratogen. Experimental reproductive effects. A steroid. When heated to decomposition it emits toxic fumes of F^-.

FDD100 CAS:1622-62-4 ***HR: 2***
FLUNITRAZEPAM
mf: $C_{16}H_{12}FN_3O_3$ mw: 313.31

PROP: Pale yellow needles from methylene chloridehexane. Mp: 166-167°.

SYNS: 1,3-DIHYDRO-5-(o-FLUOROPHENYL)-1-METHYL-7-NITRO-2H-1,4-BENZODIAZEPIN-2-ONE ◇ 1-METHYL-7-NITRO-5-(2-FLUOROPHENYL)-3H-1,4-BENZODIAZEPIN-2(1H)-ONE ◇ NARCOZEP ◇ PRIMUN ◇ RO 5-4200 ◇ ROHYPNOL ◇ ROIPNOL

TOXICITY DATA with REFERENCE
mmo-sat 3 mg/plate BSIBAC 60,2247,84
mma-sat 2 mg/plate BSIBAC 60,2247,84
ims-wmn TDLo:560 μg/kg (39W preg):REP THERAP 36,305,81
orl-rat TDLo:350 mg/kg (female 1-7D post):TER KSRNAM 17,2585,83
orl-wmn TDLo:127 μg/kg AIMDAP 145,663,85
orl-rat LD50:415 mg/kg KSRNAM 19,1277,85
ipr-rat LD50:1060 mg/kg KSRNAM 19,1277,85
orl-mus LD50:1200 mg/kg KSRNAM 19,1277,85
ipr-mus LD50:1050 mg/kg KSRNAM 19,1277,85

SAFETY PROFILE: Moderately toxic by ingestion and intraperitoneal routes. An experimental teratogen. Human reproductive effects by ingestion and intramuscular routes: apgar score of newborn and other neonatal and postnatal effects. Experimental reproductive effects. Mutation data reported. Used as a hypnotic agent. Note: This is a controlled substance (depressant) listed in the U.S. Code of Federal Regulations, Title 21 Part 1308.14 (1985). When heated to decomposition it emits toxic fumes of F^- and NO_x. See also DIAZEPAM.

FDD150 CAS:356-12-7 ***HR: 3***
FLUOCINOLIDE
mf: $C_{26}H_{32}F_2O_7$ mw: 494.58

PROP: Crystals from methanol. Mp: 308-311°.

SYNS: BISCOSAL ◇ DERMAPLUS ◇ 6-α, 9-DIFLUORO-11-β,16-α,17,21-TETRAHYDROXY-PREGNA-1,4-DIENE-3,20-DIONE cyclic 17-ACETAL with ACETONE-21-ACETATE ◇ FLUOCINOLONE ACETONIDE ACETATE ◇ FLUOCINOLONE ACETONIDE-21-ACETATE ◇ FLUOCINONIDE ◇ LIDEX ◇ METOSYN ◇ STRADERM ◇ SYNALATE ◇ TOPSYM ◇ TOPSYN ◇ TOPSYNE

TOXICITY DATA with REFERENCE
dni-mus-skn 4 g/kg BJDEAZ 94,1,76
scu-rat TDLo:3 mg/kg (3D pre):REP NYKZAU 84,553,84
orl-rat LD50:14 mg/kg NIIRDN 6,694,82
ipr-rat LD50:300 μg/kg NIIRDN 6,694,82
scu-rat LD50:720 μg/kg IYKEDH 6,386,75
ipr-mus LD50:160 mg/kg YAKUD5 17,849,75
scu-mus LD50:165 mg/kg YKYUA6 26,741,75

SAFETY PROFILE: Poison by ingestion, subcutaneous, and intraperitoneal routes. Mutation data reported. Experimental reproductive effects. A steroid. When heated to decomposition it emits toxic fumes of F^-.

FDE000 CAS:30223-48-4 ***HR: D***
FLUORACIZINE
mf: $C_{20}H_{21}F_3N_2OS$ mw: 394.49

SYN: 10-DIETHYLAMINOPROPIONYL-3-TRIFLUOROMETHYLPHENOTHIAZINE HYDROCHLORIDE

TOXICITY DATA with REFERENCE
orl-mus TDLo:10 mg/kg (4D preg):TER FATOAO 36,399,73

SAFETY PROFILE: Experimental teratogenic effects. When heated to decomposition it emits very toxic fumes of SO_x, NO_x, and F^-.

FDF000 CAS:206-44-0 ***HR: 3***
FLUORANTHENE
mf: $C_{16}H_{10}$ mw: 202.26

PROP: A polycyclic hydrocarbon. Colorless solid. Mp: 120°, bp: 367°, vap press: 0.01 mm @ 20°.

SYNS: 1,2-BENZACENAPHTHENE ◇ BENZO(jk)FLUORENE ◇ IDRYL ◇ 1,2-(1,8-NAPHTHALENEDIYL)BENZENE ◇ 1,2-(1,8-NAPHTHYLENE)BENZENE ◇ RCRA WASTE NUMBER U120

TOXICITY DATA with REFERENCE
mma-sat 5 μg/plate MUREAV 156,61,85
msc-hmn:lym 2 μmol/L DTESD7 10,277,82
msc-ham:ovr 20 mg/L ENMUDM 6,539,84
skn-mus TDLo:280 mg/kg/58W-I:ETA JNCIAM 56,1237,76
orl-rat LD50:2000 mg/kg AIHAAP 23,95,62
ivn-mus LD50:100 mg/kg CSLNX* NX#00205
skn-rbt LD50:3180 mg/kg AIHAAP 23,95,62

CONSENSUS REPORTS: IARC Cancer Review: Group 3 IMEMDT 7,56,87; Animal No Evidence IMEMDT 32,355,83. Reported in EPA TSCA Inventory. EPA Genetic Toxicology Program.

SAFETY PROFILE: Poison by intravenous route.

Moderately toxic by ingestion and skin contact. Questionable carcinogen with experimental tumorigenic data. Human mutation data reported. Combustible when exposed to heat or flame. When heated to decomposition it emits acrid smoke and irritating fumes.

FDH000 CAS:1306-05-4 ***HR: D***
FLUORAPATITE
mf: $Ca_{10}F_2O_4P$ mw: 533.77

SYN: PHOSPHATE ROCK

OSHA PEL: TWA 2.5 mg(F)/m^3
NIOSH REL: (Inorganic Fluorides) TWA 2.5 mg(F)/m^3

SAFETY PROFILE: When heated to decomposition it emits very toxic fumes of F^- and PO_x. See also FLUORIDES and PHOSPHATES.

FDI000 CAS:153-78-6 ***HR: 3***
FLUOREN-2-AMINE
mf: $C_{13}H_{11}N$ mw: 181.25

SYNS: AMINOFLUOREN (GERMAN) ◇ 2-AMINOFLUORENE ◇ 2-FLUORENAMINE ◇ 2-FLUORENEAMINE

TOXICITY DATA with REFERENCE
mma-sat 150 ng/plate CBINA8 54,71,85
msc-rat:lvr 50 μmol/L MUREAV 130,53,84
orl-rat TDLo:3600 mg/kg/32W-C:CAR CNREA8 15,188,55
skn-rat TDLo:240 mg/kg/73W-I:CAR JNCIAM 10,1201,50
scu-rat TDLo:400 mg/kg/26W-I:ETA CNREA8 7,453,47
skn-mus TDLo:11 g/kg/34W-C:NEO BJCAAI 14,195,60
ipr-mus LD50:132 mg/kg JJIND8 62,911,79

CONSENSUS REPORTS: EPA Genetic Toxicology Program.

SAFETY PROFILE: Suspected carcinogen with experimental carcinogenic, neoplastigenic, and tumorigenic data. Poison by intraperitoneal route. Mutation data reported. When heated to decomposition it emits toxic fumes of NO_x. See also AMINES.

FDK000 ***HR: 3***
9-FLUORENECARBOXYLATE-3-QUINUCLIDINOL HYDROCHLORIDE
mf: $C_{21}H_{21}NO_2$•ClH mw: 355.89

SYNS: FLUORENE-9-CARBOXYLICACID-3-QUINUCLIDINYL ESTER ◇ PAVATRINE ◇ RO 2-3208

TOXICITY DATA with REFERENCE
ipr-mus LD50:165 mg/kg JPETAB 104,284,52
ivn-mus LD50:26 mg/kg JPETAB 104,284,52

SAFETY PROFILE: Poison by intraperitoneal and intravenous routes. When heated to decomposition it emits very toxic fumes of NO_x and HCl. See also ESTERS.

FDM000 CAS:525-64-4 ***HR: 3***
FLUORENE-2,7-DIAMINE
mf: $C_{13}H_{11}N_2$ mw: 195.26

SYNS: 2,7-DIAMINOFLUORENE ◇ 2,7-FLUORENEDIAMINE ◇ 2,7-FLUOROENEDIAMINE

TOXICITY DATA with REFERENCE
mmo-sat 10 μg/plate MUREAV 143,213,85
mma-sat 5 μg/plate MUREAV 143,213,85
dns-hmn:lym 10 μmol/L SHYCD4 (40),47,83
dns-rat:lvr 500 nmol/L ENMUDM 3,11,81
dnd-mus:ast 5 μmol/L MUREAV 89,95,81
orl-rat TDLo:1000 mg/kg:CAR CNREA8 26,619,66
orl-rat TD:3600 mg/kg/27D-I:CAR CNREA8 28,924,68
orl-rat LDLo:1000 mg/kg CNREA8 26,619,66

CONSENSUS REPORTS: EPA Genetic Toxicology Program.

SAFETY PROFILE: Moderately toxic by ingestion. Questionable carcinogen with experimental carcinogenic data. Human mutation data reported. When heated to decomposition it emits toxic fumes of NO_x. See also AMINES.

FDN000 CAS:66686-30-4 ***HR: 2***
1,1'-(9H-FLUORENE-2,7-DIYL)BIS(2-(DIETHYLAMINO)ETHANONE) DIHYDROCHLORIDE TRIHYDRATE
mf: $C_{25}H_{32}N_2O_2$•2ClH•3H_2O mw: 519.57

SYNS: RMI 11002 ◇ RMI 11002 DA

TOXICITY DATA with REFERENCE
orl-mus LD50:5000 mg/kg ALACBI 12,77,79
scu-mus LD50:930 mg/kg ALACBI 12,77,79

SAFETY PROFILE: Moderately toxic by subcutaneous route. Mildly toxic by ingestion. When heated to decomposition it emits very toxic fumes of NO_x and HCl.

FDO000 CAS:486-25-9 ***HR: 3***
9H-FLUOREN-9-ONE
mf: $C_{13}H_8O$ mw: 180.21

PROP: Yellow, rhombic crystals. Mp: 83-4°, bp: 341.5°. Insol in water; very sol in alc and ether.

SYNS: FLUOREN-9-ONE ◇ 9-FLUORENONE

TOXICITY DATA with REFERENCE
scu-rat TDLo:300 mg/kg/26W-I:ETA CNREA8 7,453,47
ipr-mus LD50:2 g/kg RPTOAN 48,143,85

CONSENSUS REPORTS: Reported in EPA TSCA Inventory. EPA Genetic Toxicology Program.

SAFETY PROFILE: Questionable carcinogen with experimental tumorigenic data. When heated to decomposition it emits acrid smoke and irritating fumes. See also KETONES.

FDP000 CAS:206-00-8 *HR: 3*
FLUORENO(9,1-gh)QUINOLINE
mf: $C_{19}H_{11}N$ mw: 253.31

SYN: PYRIDO(3',2':3,4)FLUORANTHENE

TOXICITY DATA with REFERENCE
scu-mus TDLo:72 mg/kg/9W-I:ETA COREAF 259,3387,64

SAFETY PROFILE: Questionable carcinogen with experimental tumorigenic data. When heated to decomposition it emits toxic fumes of NO_x.

FDQ000 CAS:28314-03-6 *HR: 3*
N-FLUOREN-1-YL ACETAMIDE
mf: $C_{15}H_{13}NO$ mw: 223.29

SYNS: 1-FLUORENYLACETAMIDE ◇ N-1-FLUORENYLACETAMIDE

TOXICITY DATA with REFERENCE
orl-rat TDLo:4400 mg/kg/44W-C:ETA,REP JNCIAM 24,149,60
ipr-rat TDLo:627 mg/kg/4W-I:ETA CNREA8 30,1485,70

SAFETY PROFILE: Experimental reproductive effects. Questionable carcinogen with experimental tumorigenic data. When heated to decomposition it emits toxic fumes of NO_x.

FDR000 CAS:53-96-3 *HR: 3*
N-FLUOREN-2-YL ACETAMIDE
mf: $C_{15}H_{13}NO$ mw: 223.29

SYNS: AAF ◇ 2-AAF ◇ 2-ACETAMIDOFLUORENE ◇ 2-ACETAMINO-FLUORENE ◇ ACETOAMINOFLUORENE ◇ 2-ACETYLAMINO-FLUOREN (GERMAN) ◇ N-ACETYL-2-AMINOFLUORENE ◇ 2-ACETYLAMINO-FLUORENE (OSHA) ◇ AZETYLAMINOFLUOREN (GERMAN) ◇ FAA ◇ 2-FAA ◇ 2-FLUORENYLACETAMIDE ◇ N-2-FLUORENYLACETAMIDE ◇ RCRA WASTE NUMBER U005

TOXICITY DATA with REFERENCE
mmo-sat 100 μg/plate CBINA8 22,297,78
dnr-esc 5 μmol/L MUREAV 89,95,81
dnd-hmn:fbr 1 mmol/L ENMUDM 7,267,85
sce-hmn:lym 4400 μg/L CNREA8 40,4775,80
otr-rat:emb 100 μg/L JJIND8 67,1303,81
otr-rat:lvr 100 μmol/L CNREA8 43,5087,83
sce-mus-ipr 20 mg/kg MUREAV 157,181,85
ipr-mus TDLo:100 mg/kg (10D preg):TER KAIZAN 37,239,62
ipr-mus TDLo:100 mg/kg (female 8D post):REP KAIZAN 37,239,62
orl-rat TDLo:672 mg/kg/8W-C:CAR TXAPA9 74,63,84
skn-rat TDLo:260 mg/kg/71W-I:CAR JNCIAM 10,1201,50
par-rat TDLo:1700 mg/kg/17W-I:ETA CNREA8 6,617,46
ipr-mus TDLo:60 mg/kg/2W-I:NEO TXAPA9 82,19,86
scu-mus TDLo:400 mg/kg (15D preg):ETA,TER IJCNAW 20,293,77
orl-rbt TD:14 g/kg/56W-I:CAR,TER CNREA8 27,838,67
orl-mus LD50:810 mg/kg PSEBAA 143,1117,73
ipr-mus LD50:2200 mg/kg PMRSDJ 1,682,81

CONSENSUS REPORTS: NTP Fifth Annual Report on Carcinogens. Community Right-To-Know List. EPA Genetic Toxicology Program. Reported in EPA TSCA Inventory.

OSHA PEL: Cancer Suspect Agent
NIOSH REL: (2-Acetylaminofluorene): TWA use 29 CFR 1910.1014

SAFETY PROFILE: Confirmed human carcinogen with experimental carcinogenic, neoplastigenic, tumorigenic, and teratogenic data. Moderately toxic by ingestion and intraperitoneal routes. Experimental reproductive effects. Human mutation data reported. When heated to decomposition it emits toxic fumes of NO_x.

FDS000 CAS:6292-55-3 *HR: 3*
3-FLUORENYL ACETAMIDE
mf: $C_{15}H_{13}NO$ mw: 223.29

SYNS: N-3-FLUORENYL ACETAMIDE ◇ N-FLUOREN-3-YL ACETAMIDE ◇ N-9H-FLUOREN-3-YL ACETAMIDE

TOXICITY DATA with REFERENCE
orl-rat TDLo:6100 mg/kg/66W-C:ETA JNCIAM 24,149,60

SAFETY PROFILE: Questionable carcinogen with experimental tumorigenic data. When heated to decomposition it emits toxic fumes of NO_x.

FDT000 CAS:22251-01-0 *HR: 3*
1-FLUORENYL ACETHYDROXAMIC ACID
mf: $C_{15}H_{13}NO_2$ mw: 239.29

SYNS: N-FLUOREN-1-YL ACETOHYDROXAMIC ACID ◇ N-HYDROXY-1-FLUORENYL ACETAMIDE

TOXICITY DATA with REFERENCE
ipr-rat TDLo:603 mg/kg/4W-I:NEO CNREA8 30,1485,70

SAFETY PROFILE: Questionable carcinogen with experimental neoplastigenic data. When heated to decomposition it emits toxic fumes of NO_x.

FDU000 CAS:22225-32-7 *HR: 3*
3-FLUORENYL ACETHYDROXAMIC ACID
mf: $C_{15}H_{13}NO_2$ mw: 239.29

SYNS: N-FLUOREN-3-YL ACETOHYDROXAMIC ACID ◇ N-HYDROXY-3-FAA ◇ N-HYDROXY-3-FLUORENYL ACETAMIDE

TOXICITY DATA with REFERENCE
orl-rat TDLo:3000 mg/kg/21W-C:CAR CNREA8 35,447,75

ipr-rat TDLo:280 mg/kg/4W-I:CAR JNCIAM 60,433,78
imp-rat TDLo:24 mg/kg:NEO CNREA8 37,111,77
imp-rat TD:24 mg/kg:ETA CNREA8 33,2489,73

SAFETY PROFILE: Questionable carcinogen with experimental carcinogenic, neoplastigenic, and tumorigenic data. When heated to decomposition it emits toxic fumes of NO_x.

FDU875 CAS:14751-87-2 ***HR: 2***
N-FLUOREN-2-YLACETOHYDROXAMIC ACID, COBALT(2+) COMPLEX
mf: $C_{30}H_{24}N_2O_4 \cdot Co$ mw: 535.49

SYNS: COBALT,BIS(N-9H-FLUOREN-2-YL-N-HYDROXYACETAMIDATO-O,O')-(9CI) ◇ COBALT N-FLUOREN-2-YLACETOHYDROXAMATE ◇ COBALT SALTS ◇ N-HYDROXY-2-ACETYLAMINOFLUORENE, COBALTOUS CHELATE

TOXICITY DATA with REFERENCE
scu-rat TDLo:160 mg/kg/4W-I:NEO CNREA8 25,527,65

SAFETY PROFILE: Questionable carcinogen with experimental neoplastigenic data. When heated to decomposition it emits toxic fumes of NO_x and Co.

FDV000 CAS:16808-85-8 ***HR: 3***
N-FLUOREN-2-YL ACETOHYDROXAMIC ACID SULFATE
mf: $C_{15}H_{13}NO_5S$ mw: 319.35

SYNS: SULFATE ESTER of N-HYDROXY-N-2-FLUORENYL ACETAMIDE ◇ N-SULFONOXY-AAF ◇ N-SULFONOXY-N-ACETYL-2-AMINOFLUORENE

TOXICITY DATA with REFERENCE
scu-rat TDLo:255 mg/kg/15W-I:ETA CNREA8 37,1461,77

CONSENSUS REPORTS: EPA Genetic Toxicology Program.

SAFETY PROFILE: Questionable carcinogen with experimental tumorigenic data. When heated to decomposition it emits very toxic fumes of NO_x and SO_x. See also ESTERS and SULFATES.

FDX000 CAS:3671-78-1 ***HR: 3***
N-(2-FLUORENYL)BENZAMIDE
mf: $C_{20}H_{15}NO$ mw: 285.36

SYNS: 2-BENZOYLAMINOFLUORENE ◇ N-FLUOREN-2-YL BENZAMIDE ◇ N-9H-FLUOREN-2-YL-BENZAMIDE (9CI)

TOXICITY DATA with REFERENCE
mma-sat 100 ng/plate BBRCA9 71,1201,76
ipr-rat TDLo:540 mg/kg/4W-I:ETA CNREA8 27,1443,67

SAFETY PROFILE: Questionable carcinogen with experimental tumorigenic data. Mutation data reported. When heated to decomposition it emits toxic fumes of NO_x.

FDY000 CAS:29968-64-7 ***HR: 3***
N-FLUOREN-1-YL BENZOHYDROXAMIC ACID
mf: $C_{20}H_{15}NO_2$ mw: 301.36

SYN: N-HYDROXY-1-FLUORENYL BENZAMIDE

TOXICITY DATA with REFERENCE
ipr-rat TDLo:823 mg/kg/4W-I:NEO CNREA8 30,1485,70

SAFETY PROFILE: Questionable carcinogen with experimental neoplastigenic data. When heated to decomposition it emits toxic fumes of NO_x.

FDZ000 CAS:3671-71-4 ***HR: 3***
N-FLUOREN-2-YL BENZOHYDROXAMIC ACID
mf: $C_{20}H_{15}NO_2$ mw: 301.36

SYNS: N-BENZOYLOXY-ACETYLAMINOFLUORENE ◇ N-(2-FLUORENYL)BENZOHYDROXAMIC ACID ◇ N-9H-FLUOREN-2-YL-N-HYDROXYBENZAMIDE ◇ N-HYDROXY-2-BENZOYLAMINOFLUORENE ◇ N-HYDROXY-N-2-FLUORENYLBENZAMIDE

TOXICITY DATA with REFERENCE
mma-sat 1500 ng/plate CBINA8 54,71,85
mmo-bcs 10 mol MOPMA3 4,411,68
oms-bcs 10 mol MOPMA3 4,411,68
orl-rat TDLo:5200 mg/kg/23W-I:CAR CNREA8 30,1485,70
ipr-rat TDLo:540 mg/kg/4W-I:CAR CNREA8 27,1443,67

SAFETY PROFILE: Questionable carcinogen with experimental carcinogenic data. Mutation data reported. When heated to decomposition it emits toxic fumes of NO_x.

FEE000 CAS:391-57-1 ***HR: 3***
N,N'-FLUOREN-2,7-YLENE BIS(TRIFLUOROACETAMIDE)
mf: $C_{17}H_{10}F_6N_2O_2$ mw: 388.29

TOXICITY DATA with REFERENCE
orl-rat TDLo:4850 mg/kg/37W-C:CAR JNCIAM 30,143,63
orl-rat TD:5 g/kg/48W-C:ETA JJIND8 71,211,83

SAFETY PROFILE: Questionable carcinogen with experimental carcinogenic and tumorigenic data. When heated to decomposition it emits toxic fumes of F^- and NO_x. See also FLUORIDES.

FEE100 CAS:13929-01-6 ***HR: 2***
N-(9-FLUORENYL)-N-ETHYL-β-CHLOROETHYLAMINE HYDROCHLORIDE
mf: $C_{17}H_{18}ClN \cdot ClH$ mw: 308.27

SYNS: N-(2-CHLOROETHYL)-N-ETHYLFLUOREN-9-AMINEHYDROCHLORIDE ◇ N-(2-CHLOROETHYL)-N-ETHYL-9-FLUORENAMINE HYDROCHLORIDE ◇ 2-CHLORO-N-(9-FLUORENYL)DIETHYLAMINE HYDROCHLORIDE ◇ DIETHYLAMINE, 2-CHLORO-N-(9-FLUORENYL)-, HYDROCHLORIDE ◇ FLUOREN-9-AMINE, N-(2-CHLOROETHYL)-N-ETHYL-, HYDROCHLORIDE ◇ 9H-FLUOREN-9-AMINE, N-(2-

CHLOROETHYL)-N-ETHYL-, HYDROCHLORIDE (9CI) ◇ SKF-501 HYDROCHLORIDE

TOXICITY DATA with REFERENCE
scu-rat TDLo:20 mg/kg (female 1D pre):REP ENDOAO 74,309,64
scu-mus LD50:800 mg/kg JPETAB 97,25,49

SAFETY PROFILE: Moderately toxic by subcutaneous route. Experimental reproductive effects. When heated to decomposition it emits toxic fumes of NO_x, HCl, and Cl^-.

FEF000 CAS:6957-71-7 ***HR: 3***
N-FLUOREN-2-YL FORMAMIDE
mf: $C_{14}H_{11}NO$ mw: 209.26

SYNS: N,2-FLUORENYL FORMAMIDE ◇ 2-FORMYLAMINOFLUORENE

TOXICITY DATA with REFERENCE
mma-sat 100 ng/plate BBRCA9 71,1201,76
orl-rat TDLo:4400 mg/kg/35W-C:CAR CNREA8 22,1002,62

SAFETY PROFILE: Questionable carcinogen with experimental carcinogenic data. Mutation data reported. When heated to decomposition it emits toxic fumes of NO_x.

FEG000 CAS:67176-33-4 ***HR: 3***
N-(2-FLUORENYL)FORMOHYDROXAMIC ACID
mf: $C_{14}H_{11}NO_2$ mw: 225.26

SYNS: N-FORMYL-N-2-FLUORENYLHYDROXYLAMINE ◇ N-HYDROXY-N-2-FORMYLAMINOFLUORENE

TOXICITY DATA with REFERENCE
mmo-sat 10 μg/plate CRNGDP 3,233,82
mma-sat 5 μg/plate CNREA8 40,1204,80
par-rat TDLo:2253 μg/kg:CAR NTIS** PB86-115920
par-rat TDLo:2815 μg/kg:CAR CRNGDP 3,233,82

SAFETY PROFILE: Questionable carcinogen with experimental carcinogenic data. Mutation data reported. When heated to decomposition it emits toxic fumes of NO_x.

FEH000 CAS:51029-30-2 ***HR: 3***
3-FLUORENYLHYDROXYLAMINE
mf: $C_{13}H_{11}NO$ mw: 197.25

SYNS: N-FLUOREN-3-YL HYDROXYLAMINE ◇ N-HYDROXY-3-AMINOFLUORENE

TOXICITY DATA with REFERENCE
imp-rat TDLo:20 mg/kg:ETA CNREA8 33,2489,73

SAFETY PROFILE: Questionable carcinogen with experimental tumorigenic data. When heated to decomposition it emits toxic fumes of NO_x.

FEI000 CAS:34461-49-9 ***HR: 3***
N-FLUOREN-2-YLHYDROXYLAMINE-o-GLUCURONIDE
mf: $C_{19}H_{19}NO_8$ mw: 389.39

SYN: N-2-FLUORENYLHYDROXYLAMINE-o-GLUCURONIDE

TOXICITY DATA with REFERENCE
scu-rat TDLo:750 mg/kg/9W-I:ETA CNREA8 31,1645,71

SAFETY PROFILE: Questionable carcinogen with experimental tumorigenic data. When heated to decomposition it emits toxic fumes of NO_x.

FEI500 CAS:63019-68-1 ***HR: 2***
2-FLUORENYLMONOMETHYLAMINE
mf: $C_{14}H_{13}N$ mw: 195.28

SYNS: 2-METHYLAMINOFLUORENE ◇ N-MONOMETHYL-2-AMINOFLUORENE ◇ 2-MONOMETHYLAMINOFLUORENE

TOXICITY DATA with REFERENCE
orl-rat TDLo:1260 mg/kg/24W-I:ETA BJCAAI 6,89,52

SAFETY PROFILE: Questionable carcinogen with experimental tumorigenic data. When heated to decomposition it emits toxic fumes of NO_x.

FEM000 CAS:63224-45-3 ***HR: 3***
N-(2-FLUORENYL)MYRISTOHYDROXAMIC ACID ACETATE
mf: $C_{29}H_{39}NO_3$ mw: 449.69

SYNS: N-ACETOXY-2-MYRISTOYL-AMINOFLUORENE ◇ N-(ACETYLOXY)-N-9H-FLUOREN-2-YL-TETRADECANAMIDE

TOXICITY DATA with REFERENCE
mma-sat 20 nmol/plate CNREA8 37,1461,77
dns-hmn:fbr 100 mmol/L/5H IJCNAW 16,284,75
scu-rat TDLo:114 mg/kg/6W-I:CAR CNREA8 37,1461,77

CONSENSUS REPORTS: EPA Genetic Toxicology Program.

SAFETY PROFILE: Questionable carcinogen with experimental carcinogenic data. Human mutation data reported. When heated to decomposition it emits toxic fumes of NO_x.

FEN000 CAS:2485-10-1 ***HR: 3***
N-FLUORENYL-2-PHTHALIMIC ACID
mf: $C_{21}H_{15}NO_3$ mw: 329.37

SYNS: 2-BENZOYLAMIDOFLUORENE-2′-CARBOXYLATE ◇ 2-BENZOYLAMINOFLUORENE-2′-CARBOXYLATE ◇ N-(2-FLUORENYL) PHTHALAMIC ACID

TOXICITY DATA with REFERENCE
orl-rat TDLo:5000 mg/kg/42W-C:ETA AICCA6 20,1364,64

SAFETY PROFILE: Questionable carcinogen with ex-

perimental tumorigenic data. When heated to decomposition it emits toxic fumes of NO_x.

FEO000 CAS:52663-84-0 ***HR: 3***
N-(2-FLUORENYL)PROPIONOHYDROXAMIC ACID
mf: $C_{16}H_{15}NO_2$ mw: 253.32

SYNS: N-HYDROXY-N-2-PROPIONYLAMINO FLUORENE ◇ N-PROPIONYL-N-2-FLUORENYLHYDROXYLAMINE

TOXICITY DATA with REFERENCE
mmo-sat 10 μg/plate CRNGDP 3,233,82
mma-sat 5 μg/plate CNREA8 40,1204,80
par-rat TDLo:2533 μg/kg:CAR NTIS** PB86-115920
par-rat TDLo:3166 μg/kg:ETA CRNGDP 3,233,82

SAFETY PROFILE: Questionable carcinogen with experimental carcinogenic and tumorigenic data. Mutation data reported. When heated to decomposition it emits toxic fumes of NO_x.

FEP000 CAS:59935-47-6 ***HR: 3***
N-2-FLUORENYL SUCCINAMIC ACID
mf: $C_{17}H_{15}NO_3$ mw: 281.33

TOXICITY DATA with REFERENCE
orl-rat TDLo:6200 mg/kg/72W-C:NEO JNCIAM 24,149,60

SAFETY PROFILE: Questionable carcinogen with experimental neoplastigenic data. When heated to decomposition it emits toxic fumes of NO_x.

FER000 CAS:363-17-7 ***HR: 3***
N-FLUOREN-2-YL-2,2,2-TRIFLUOROACETAMIDE
mf: $C_{15}H_{10}F_3NO$ mw: 277.26

SYNS: N-(2-FLUORENYL)-2,2,2-TRIFLUOROACETAMIDE◇ 2-TRIFLUOROACETYLAMINOFLUORENE ◇ 2,2,2-TRIFLUORO-N-(FLUOREN-2-YL)ACETAMIDE

TOXICITY DATA with REFERENCE
orl-rat TDLo:3750 mg/kg/44W-C:CAR JNCIAM 24,149,60
orl-rat TD:4500 mg/kg/26W-C:CAR CNREA8 22,1002,62

SAFETY PROFILE: Questionable carcinogen with experimental carcinogenic data. When heated to decomposition it emits very toxic F^- and NO_x. See also FLUORIDES.

FEV000 CAS:2321-07-5 ***HR: 3***
FLUORESCEIN
mf: $C_{20}H_{12}O_5$ mw: 332.32

PROP: Orange-red, crystalline powder. Mp: 314-316° with decomp.

SYNS: 9-(o-CARBOXYPHENYL)-6-HYDROXY-3-ISOXANTHENONE ◇ 9-(o-CARBOXYPHENYL)-6-HYDROXY-3H-XANTHEN-3-ONE ◇ C.I. 45330 ◇ C.I. 45350 (FREE ACID) ◇ C.I. SOLVENT YELLOW 94 ◇ D&C YELLOW No. 7 ◇ 3',6'-DIHYDROXYFLUORAN ◇ DIHYDROXYFLUORANE ◇ 3',6'-DIHYDROXYSPIRO(ISOBENZOFURAN-1(3H),9'(9H)-XANTHEN)-3-ONE ◇ 3,6-FLUORANDIOL ◇ 3',6'-FLUORANDIOL ◇ FLUORESCEINE ◇ HIDACID FLUORESCEIN ◇ RESORCINOLPHTHALEIN ◇ SOAP YELLOW F ◇ 11712 YELLOW

TOXICITY DATA with REFERENCE
dnd-esc 15 μmol/L MUREAV 89,95,81
ipr-rat LDLo:600 mg/kg IJLEAG 2,257,34
ivn-rbt LDLo:300 mg/kg IJLEAG 2,257,34

CONSENSUS REPORTS: Reported in EPA TSCA Inventory. EPA Genetic Toxicology Program.

SAFETY PROFILE: Poison by intravenous route. Moderately toxic by intraperitoneal route. Mutation data reported. When heated to decomposition it emits acrid smoke and irritating fumes. See also FLUORESCEIN SODIUM.

FEV100 CAS:3570-80-7 ***HR: 3***
FLUORESCEIN MERCURIC ACETATE
mf: $C_{24}H_{16}Hg_2O_9$ mw: 849.58

SYNS: FLUORESCEIN MERCURIACETATE ◇ FLUORESCEIN MERCURY ACETATE ◇ FMA ◇ FMA (analytical reagent) ◇ MERCURY, BIS(ACETATO)(mu-(3',6'-DIHYDROXY-2',7'-FLUORANDIYL))DI-

TOXICITY DATA with REFERENCE
dnd-mmo-omi 200 nmol/L BBACAQ 454,309,76
dnd-uns:lyms 100 nmol/L BBACAQ 454,309,76
ACGIH TLV: TWA 0.1 mg(Hg)/m^3 (skin)
NIOSH REL: (Mercury, Inorganic): 8H TWA 0.05 mg(Hg)/m^3

CONSENSUS REPORTS: EPA TSCA CHEMICAL INVENTORY, JUNE 1990

SAFETY PROFILE: Highly Toxic. Mutation data reported. When heated to decomposition it emits toxic fumes of Hg.

FEW000 CAS:518-47-8 ***HR: 3***
FLUORESCEIN SODIUM
mf: $C_{20}H_{10}O_5•2Na$ mw: 376.28

PROP: Orange-red powder. Sol in water; sltly sol in alc.

SYNS: AIZEN URANINE ◇ CALCOCID URANINE B4315 ◇ 9-o-CARBOXYPHENYL-6-HYDROXY-3-ISOXANTHONE, DISODIUM SALT ◇ CERTIQUAL FLUORESCEINE ◇ C.I. 766 ◇ C.I. ACID YELLOW 73 ◇ C.I. 45350 DISODIUM SALT ◇ D&C YELLOW No. 8 ◇ DISODIUM-6-HYDROXY-3-OXO-9-XANTHENE-o-BENZOATE◇ FLUORESCEIN SODIUM B.P ◇ FLUORESCEIN, soluble ◇ FLUOR-I-STRIP A.T. ◇ FUL-GLO ◇ FUNDUSCEIN ◇ FURANIUM ◇ HIDACID URANINE ◇ NCI-C54706 ◇ RESORCINOL PHTHALEIN SODIUM ◇ SODIUM FLUORESCEIN ◇ SODIUM FLUORESCEINATE ◇ SODIUM SALT of HYDROXY-o-CARBOXY-PHENYL-FLUORONE◇ SOLUBLE FLUORESCEIN ◇ SPIRO(ISOBENZOFURAN-1(3H),9'-(9H)XANTHENE-3-ONE, 3',6'-DIHYDROXY-DISODIUM SALT ◇ URANIN ◇ URANINE A EXTRA

◇ URANINE USP XII ◇ URANINE YELLOW ◇ 11824 YELLOW ◇ 12417 YELLOW

TOXICITY DATA with REFERENCE
dnd-esc 15 μmol/L MUREAV 89,95,81
ivn-rbt TDLo:933 mg/kg (5-8D preg):REP AJOPAA 84,847,77
scu-rat TDLo:19 g/kg/79W-I:ETA GANNA2 45,446,54
ivn-man TDLo:14 mg/kg/10M-I:CVS AACRAT 66,283,87
ivn-man TDLo:7142 μg/kg:EYE,GIT AJOPAA 103,111,87
orl-rat LD50:6721 mg/kg JOPRAJ 48,228,77
orl-mus LD50:4738 mg/kg JOPRAJ 48,228,77
ipr-mus LD50:1800 mg/kg TXAPA9 24,37,73

CONSENSUS REPORTS: Reported in EPA TSCA Inventory.

SAFETY PROFILE: Moderately toxic by intraperitoneal route. Mildly toxic by ingestion. Human systemic effects by intravenous routes: arrhythmias, eye hemorrhage, nausea or vomiting. Experimental reproductive effects. Questionable carcinogen with experimental tumorigenic data. Mutation data reported. When heated to decomposition it emits toxic fumes of Na_2O.

FEX875 CAS:16984-48-8 ***HR: 3***
FLUORIDE
af: F aw: 18.9984

SYNS: FLUORIDE(1-) ◇ FLUORIDE ION ◇ FLUORIDE ION(1-) ◇ PERFLUORIDE

TOXICITY DATA with REFERENCE
dni-mus:fbr 1300 μmol/L APTOA6 45,96,79
orl-hmn TDLo:3 mg/kg:LIV,PUL,CNS IMEMDT** 27,237,82
orl-hmn LDLo:50 mg/kg:PUL,LIV IMEMDT** 27,237,82

CONSENSUS REPORTS: EPA Genetic Toxicology Program.

OSHA PEL: TWA 2.5 mg(F)/m^3
ACGIH TLV: TWA 2.5 mg(F)/m^3
DFG MAK: 2.5 mg/m^3
NIOSH REL: (Fluorides, inorganic) TWA 2.5 mg(F)/m^3

SAFETY PROFILE: Human poison by ingestion. Human systemic effects by ingestion: convulsions, changes in the respiratory system, liver, and kidneys. Mutation data reported. When heated to decomposition it emits toxic fumes of F^-.

FEY000 ***HR: 2***
FLUORIDES

OSHA PEL: TWA 2.5 mg(F)/m^3
ACGIH TLV: TWA 2.5 mg(F)/m^3
DFG MAK: 2.5 mg/m^3; BAT: 7 mg/kg creatinine in urine at end of exposure; 4 mg/kg creatinine in urine about 16 hours after end of exposure.
NIOSH REL: TWA 2.5 mg(F)/m^3

SAFETY PROFILE: Inorganic fluorides are generally highly irritating and toxic. Acute effects resulting from exposure to fluorine compounds are due to HF. Chronic fluorine poisoning, or "fluorosis," occurs among miners of cryolite, and consists of a sclerosis of the bones, caused by fixation of the calcium by the fluorine. There may also be some calcification of the ligaments. The teeth are mottled, and there is osteosclerosis and osteomalacia. The bony and ligamentous changes are demonstrable by x-ray. The estimated human lethal dose is 2.5 to 5.9 grams of F^-. Large doses can cause very severe nausea, vomiting, diarrhea, abdominal burning, and cramp-like pains. It is not taken up by the thyroid and does not interfere with iodine uptake. Can cause or aggravate attacks of asthma and severe bone changes, making normal movements painful. Some signs of pulmonary fibrosis are noted. Some enzyme systems effects are reported. Irritants to the eyes, skin, and mucous membranes. Loss of weight, anorexia, anemia, wasting and cachexia, and dental defects are among the common findings in chronic fluorine poisoning. There may be an eosinophilia and impairment of growth in young workers. Symptoms of intoxication include gastric, intestinal, circulatory, respiratory and nervous complaints, and skin rashes. When heated to decomposition, or on contact with acid or acid fumes, they emit highly toxic fumes of F^-.

Organic fluorides are generally less toxic than other halogenated hydrocarbons. Fluorocarbons are chemically inert to most materials but can react violently with barium, sodium, and potassium. Fluoroamides react violently with lithium tetrahydroaluminate and with sodium at very high temperatures. Some fluorinated cyclopropenyl methyl ethers react violently with water or methanol. Some fluorodinitro compounds of methane and ethane are sensitive explosives. When heated to decomposition they emit toxic fumes of F^-. Common air contaminants.

FEZ000 CAS:7782-41-4 ***HR: 3***
FLUORINE
DOT: UN 1045
mf: F_2 mw: 38.00

PROP: Pale yellow gas. Mp: −218°, bp: −187°, d: 1.14 @ −200°, 1.108 @ −188°, vap d: 1.695.

SYNS: BIFLUORIDEN (DUTCH) ◇ FLUOR (DUTCH, FRENCH, GERMAN, POLISH) ◇ FLUORINE, compressed (DOT) ◇ FLUORO (ITALIAN) ◇ FLUORURES ACIDE (FRENCH) ◇ FLUORURI ACIDI (ITALIAN) ◇ RCRA WASTE NUMBER P056 ◇ SAEURE FLUORIDE (GERMAN)

TOXICITY DATA with REFERENCE
eye-hmn 25 ppm/5M MLD AIHAAP 29,11,68
eye-rat 140 ppm/30M AIHAAP 29,11,68
eye-mus 467 ppm/5M AIHAAP 29,11,68
eye-dog 68 ppm/1H AIHAAP 29,11,68
mmo-sat 1 mg/plate CYGEDX 16(6),41,82
ihl-rat LC50:185 ppm/1H AIHAAP 29,11,68
ihl-mus LC50:150 ppm/1H AIHAAP 29,11,68
ihl-rbt LC50:270 ppm/30M AIHAAP 29,11,68
ihl-gpg LC50:170 ppm/1H AIHAAP 29,11,68

CONSENSUS REPORTS: Reported in EPA TSCA Inventory.

OSHA PEL: TWA 0.1 ppm
ACGIH TLV: TWA 1 ppm; STEL 2 ppm
DFG MAK: 0.1 ppm (0.2 mg/m^3)
DOT Classification: Nonflammable Gas; Label: Poison and Oxidizer; Poison A; Label: Poison Gas, Oxidizer.

SAFETY PROFILE: A poison gas. A skin, eye, and mucous membrane irritant. A most powerful caustic irritant to tissue. Mutation data reported. A very dangerous fire and explosion hazard. A powerful oxidizer. Reacts violently with many materials.

Explosive or potentially explosive reaction with ammonia, cesium fluoride + fluorocarboxylic acids, cesium heptafluoropropoxide, 1- or 2-fluoriminoperfluoropropane, graphite, halocarbons (e.g., carbon tetrachloride, chloroform, perfluorocyclobutane, iodoform, 1,2-dichlorotetrafluoroethane), liquid hydrocarbons (e.g., anthracene, turpentine), hydrogen, hydrogen + oxygen, hydrogen fluoride + seleninyl fluoride + heat, nitric acid, silver cyanide, sulfur dioxide, carbon monoxide, sodium acetate, sodium bromate, stainless steel, water.

Reacts to form explosive products with alkanes + oxygen (forms peroxides), cyanoguanidine, perchloric acid (forms fluorine perchlorate gas), potassium chlorate (forms fluorine perchlorate gas), potassium hydroxide (forms potassium trioxide). Forms explosive mixtures with acetonitrile + chlorine fluoride, ice.

Ignition or violent reaction on contact with acetylene, ceramic materials, covalent halides (e.g., chromyl chloride, phosphorus pentachloride, phosphorus trichloride, phosphorus trifluoride, boron trichloride, silicon tetrachloride), halogens (e.g., bromine, iodine, chlorine + spark or heating to 100°C), dicyanogen, gaseous hydrocarbons (e.g., town gas, methane, benzene), hydrogen halide gases or concentrated solutions (e.g., hydrogen bromide, hydrogen chloride, hydrogen iodide, hydrogen fluoride), metal acetylides and carbides (e.g., monocesium acetylide, cesium acetylide, lithium acetylide, rubidium acetylide, tungsten carbide, ditungsten carbide, zirconium dicarbide, uranium dicarbide), metal cyano complexes [e.g., potassium hexacyanoferrate(II), lead hexacyanoferrate(III), potassium hexacyanoferrate(III)], metal hydrides (e.g., copper hydride, potassium hydride, sodium hydride), metal iodides (e.g., lead iodide, calcium iodide, mercury iodide, potassium iodide), metals, metal salts, metal silicides (e.g., calcium disilicide, lithium hexasilicide), nickel(IV) oxide, non-metals (e.g., boron, yellow or red phosphorus, selenium, tellurium, silicon, carbon, charcoal, sulfur), oxygenated organic compounds (e.g., methanol, ethanol, 3-methyl butanol, acetaldehyde, trichloroacetaldehyde, acetone, lactic acid, benzoic acid, salicylic acid, ethyl acetate, methyl borate), non-metal oxides (e.g., arsenic trioxide, nitrogen oxide, dinitrogen tetroxide), oxygen + polymers [e.g., phenol-formaldehyde resins (bakelite), polyacrylonitrile-butadiene (Buna N), polyamides (nylons), polychloropene (neoprene), polyethylene, polytrifluoopropylmethylsiloxane, polyvinylchloride-vinyl acetate (Tygon), polyvinylidene fluoride-hexafluoropropylene (Viton), polyurethane foam, polymethyl methacrylate (Perspex), polytetrafluooethylene (Teflon)], sulfides (e.g., antimony trisulfide, carbon disulfide vapor, chromium (II) sulfide, hydrogen sulfide, barium sulfide, potassium sulfide, zinc sulfide, moly- bdenum sulfide), xenon + catalysts (e.g., nickel fluoride, silver difluoride, nickel(III) oxide, silver (I) oxide).

Incandescent reaction with boron nitride, hexalithium disilicide + heat, metal borides, metal oxides (e.g., nickel(II) oxide, alkali metal oxides, alkaline earth oxides), nitrogenous bases (e.g., aniline, dimethylamine, pyridine), gallic acid.

Incompatible with cesium heptafluoro propoxide, cyanoguanidine, halocarbons, hexalithium disilicide, seleninyl fluoride, hydrogen sulfide, oxygen, sodium acetate, sodium bromate, sodium dicyanamides, most organic matter, H-containing molecules, oxides of S, N, P, alkali metals,and alkaline earths. It reacts violently with halogen acids, hydrazine, ClO_2, coke, cyanamide, cyanides, KNO_3, (PbO + glycerol), CCl_4, silicides, silicates, trinitromethane, alkenes, alkyl benzenes, CS_2, $Cr(OCl)_2$, Al, Tl, Sn, Sb, As, natural gas, liquid air, perfluoropropionyl fluoride, polyvinyl chloride acetate. Many reactions go on even at $< -160°$. Reacts with water or steam to produce heat and toxic and corrosive fumes. Used as one component of liquid rocket fuel and in chemical lasers. See also FLUORIDES.

FFA000 CAS:14986-60-8 ***HR: 3***
FLUORINE AZIDE
mf: FN_3 mw: 61.02

SYN: AZIDO FLUORINE

SAFETY PROFILE: A poison. An irritant to the eyes, skin, and mucous membranes. An extremely unstable explosive sensitive to light and heat. Usually explodes on vaporization at $-82°C$. When heated to decomposition it emits toxic fumes of F^- and NO_x. See also FLUORINE and AZIDES.

FFB000 CAS:13536-85-1 **HR: 3**
FLUORINE FLUORO SULFATE
mf: F_2O_3S mw: 118.06

FSO_2OF

PROP: A vapor.

SYN: FLUOROSULFURYL HYPOFLUORITE

SAFETY PROFILE: Unstable it may explode at room temperature. When heated to decomposition it emits toxic fumes of F^- and SO_x. See also FLUORINE and SULFATES.

FFD000 CAS:10049-03-3 **HR: 3**
FLUORINE PERCHLORATE
mf: $ClFO_4$ mw: 118.45

O_3ClOF

PROP: Colorless gas, pungent, acrid odor. Mp: −167.3°, bp: −15.9°.

SYNS: CHLORINE TETROXYFLUORIDE ◇ PERCHLORYL HYPOFLUORITE

SAFETY PROFILE: A poison by ingestion and inhalation. Corrosive to the skin, eye, and mucous membranes. Very unstable. A powerful oxidizer which can react violently with oxidizable materials. A very dangerous explosion hazard; it explodes on slightest provocation. The liquid explodes on freezing at −167°C. The gas explodes when exposed to sparks, flame, or on contact with grease, dust, rubber, or aqueous potassium iodide. Ignites in contact with hydrogen gas. When heated to decomposition it emits highly toxic fumes of F^- and Cl^-. See also FLUORINE and PERCHLORATES.

FFE000 CAS:1544-46-3 **HR: 3**
FLUOROACETALDEHYDE
mf: C_2H_3FO mw: 62.05

TOXICITY DATA with REFERENCE
ipr-mus LD50:6 mg/kg JACSAT 78,4996,56
scu-mus LD50:4800 μg/kg JCSOA9 -,773,49

SAFETY PROFILE: Poison by intraperitoneal and subcutaneous routes. When heated to decomposition it emits toxic fumes of F^-. See also FLUORIDES and ALDEHYDES.

FFF000 CAS:640-19-7 **HR: 3**
FLUOROACETAMIDE
mf: C_2H_4FNO mw: 77.07

SYNS: AFL 1081 ◇ COMPOUND 1081 ◇ FAA ◇ FLUORAKIL 100 ◇ 2-FLUOROACETAMIDE ◇ FLUOROACETIC ACID AMIDE ◇ FUSSOL ◇ MEGATOX ◇ MONOFLUOROACETAMIDE ◇ NAVRON ◇ RCRA WASTE NUMBER P057 ◇ RODEX ◇ YANOCK

TOXICITY DATA with REFERENCE
spm-rat-ipr 4 mg/kg VAAZA2 1,346,68
cyt-mam:lng 300 μmol/L HKXUDL 3,94,83
orl-rat TDLo:90 mg/kg (30D male):REP EXPEAM 20,492,64
orl-hmn LDLo:2 mg/kg SKEZAP 9,1,68
unr-hmn LDLo:5 mg/kg:CNS,GIT 34ZIAG -,274,69
orl-rat LD50:5750 μg/kg PCOC** -,538,66
skn-rat LD50:80 mg/kg WRPCA2 9,119,70
ipr-rat LD50:12 mg/kg JSFAAE 8,400,57
orl-mus LD50:25 mg/kg ABCHA6 31,1294,67
ihl-mus LC50:550 mg/m³ ABCHA6 31,1294,67
skn-mus LD50:34 mg/kg SKEZAP 9,1,68
ipr-mus LD50:85 mg/kg BCPCA6 12,1201,63
scu-mus LD50:34 mg/kg NHOZAX 26,203,72
ivn-rbt LD50:250 μg/kg JCSOA9 -,912,49

CONSENSUS REPORTS: EPA Extremely Hazardous Substances List. Reported in EPA TSCA Inventory.

SAFETY PROFILE: A human poison by an unspecified route. Poison experimentally by ingestion, skin contact, intraperitoneal, subcutaneous, and intravenous routes. Human systemic effects by unspecified route: convulsions, coma, nausea and vomiting. Experimental reproductive effects. Mutation data reported. Used as an insecticide and rodenticide. When heated to decomposition it emits very toxic fumes of F^- and NO_x. See also FLUORIDES.

FFG000 CAS:343-89-5 **HR: 3**
7-FLUORO-2-ACETAMIDO-FLUORENE
mf: $C_{15}H_{12}FNO$ mw: 241.28

SYNS: 7-FLUORO-2-ACETYLAMINOFLUORENE ◇ N-(7-FLUOROFLUORENE-2-YL)ACETAMIDE

TOXICITY DATA with REFERENCE
dnd-rat-par 27 μmol/kg CBINA8 40,27,82
orl-rat TDLo:450 mg/kg/11W-C:CAR CNREA8 26,2239,66
orl-rat TD:1200 mg/kg/15W-C:ETA CNREA8 18,469,58
orl-rat TD:1200 mg/kg/14W-C:CAR CNREA8 15,188,55

SAFETY PROFILE: Questionable carcinogen with experimental carcinogenic and tumorigenic data. Mutation data reported. When heated to decomposition it emits toxic fumes of F^- and NO_x. See also FLUORIDES.

FFH000 CAS:330-68-7 **HR: 3**
FLUOROACETANILIDE
mf: C_8H_8FNO mw: 153.17

SYNS: AFL 1082 ◇ 2-FLUOROACETANILIDE ◇ 2-FLUORO-N-PHENYLACETAMIDE ◇ TL 1312

TOXICITY DATA with REFERENCE
orl-rat LDLo:3 mg/kg NCNSA6 5,10,53
ipr-rat LD50:6 mg/kg JSFAAE 8,400,57
orl-mus LD50:25 mg/kg YKKZAJ 88,1620,68
ihl-mus LCLo:480 mg/m^3/10M NDRC** 30101,4,45

SAFETY PROFILE: Poison by ingestion and intraperitoneal routes. Moderately toxic by inhalation. When heated to decomposition it emits very toxic fumes of F^- and NO_x. See also FLUORIDES.

FFI000 CAS:331-87-3 ***HR: 3***
FLUOROACETIC ACID (2-ETHYLHEXYL) ESTER
mf: $C_{10}H_{19}FO_2$ mw: 190.29

TOXICITY DATA with REFERENCE
skn-rbt LDLo:10 mg/kg NDRC** 30101,-,45
ivn-rbt LDLo:10 mg/kg 11FYAN 3,73,63

SAFETY PROFILE: Poison by skin contact and intravenous routes. When heated to decomposition it emits toxic fumes of F^-. See also ESTERS.

FFJ000 CAS:503-20-8 ***HR: 3***
FLUOROACETONITRILE
mf: C_2H_2FN mw: 59.05

SYN: FLUOROMETHYL CYANIDE

TOXICITY DATA with REFERENCE
ipr-mus LD50:25 mg/kg JACSAT 78,3484,56
scu-mus LD50:25 mg/kg CLDND*

CONSENSUS REPORTS: Cyanide and its compounds are on the Community Right-To-Know List.

SAFETY PROFILE: Poison by intraperitoneal and subcutaneous routes. When heated to decomposition it emits very toxic fumes of F^-, CN^-, and NO_x. See also NITRILES and FLUORIDES.

FFK000 CAS:2343-36-4 ***HR: 3***
FLUOROACETPHENYLHYDRAZIDE
mf: $C_8H_9FN_2O$ mw: 168.19

SYN: FANYLINE

TOXICITY DATA with REFERENCE
orl-rat LD50:9100 μg/kg JPETAB 93,287,48
orl-mus LD50:45 mg/kg JPETAB 93,287,48
orl-dog LDLo:250 μg/kg JPETAB 93,287,48
orl-cat LDLo:500 μg/kg JPETAB 93,287,48
orl-rbt LDLo:1700 μg/kg JPETAB 93,287,48
orl-gpg LDLo:1300 μg/kg JPETAB 93,287,48
orl-pgn LD50:7200 μg/kg JPETAB 93,287,48

SAFETY PROFILE: Poison by ingestion. When heated to decomposition it emits very toxic fumes of F^- and NO_x. See also FLUORIDES.

FFL000 CAS:2824-10-4 ***HR: 3***
1-FLUORO-2-ACETYLAMINOFLUORENE
mf: $C_{15}H_{12}FNO$ mw: 241.28

SYNS: 1-FLUORO-2-FAA ◇ N-(1-FLUOROFLUOREN-2-YL)ACETAMIDE

TOXICITY DATA with REFERENCE
orl-rat TDLo:1400 mg/kg/19W-C:CAR CNREA8 22,1002,62

SAFETY PROFILE: Questionable carcinogen with experimental carcinogenic data. When heated to decomposition it emits very toxic fumes of F^- and NO_x. See also AMINES and FLUORIDES.

FFM000 CAS:2823-93-0 ***HR: 3***
3-FLUORO-2-ACETYLAMINOFLUORENE
mf: $C_{15}H_{12}FNO$ mw: 241.28

SYNS: 3-FLUORO-2-FAA ◇ N-(3-FLUOROFLUOREN-2-YL)ACETAMIDE

TOXICITY DATA with REFERENCE
orl-rat TDLo:3900 mg/kg/26W-C:CAR CNREA8 22,1002,62

SAFETY PROFILE: Questionable carcinogen with experimental carcinogenic data. When heated to decomposition it emits very toxic fumes of F^- and NO_x. See also AMINES and FLUORIDES.

FFN000 CAS:2823-91-8 ***HR: 3***
4-FLUORO-2-ACETYLAMINOFLUORENE
mf: $C_{15}H_{12}FNO$ mw: 241.28

SYNS: 4-FLUORO-2-FAA ◇ N-(4-FLUOROFLUOREN-2-YL)ACETAMIDE

TOXICITY DATA with REFERENCE
orl-rat TDLo:2560 mg/kg/17W-C:CAR CNREA8 22,1002,62

SAFETY PROFILE: Questionable carcinogen with experimental carcinogenic data. When heated to decomposition it emits very toxic fumes of F^- and NO_x. See also AMINES and FLUORIDES.

FFO000 CAS:2823-90-7 ***HR: 3***
5-FLUORO-2-ACETYLAMINOFLUORENE
mf: $C_{15}H_{12}FNO$ mw: 241.28

SYNS: 5-FLUORO-2-FAA ◇ N-(5-FLUOROFLUOREN-2-YL)ACETAMIDE

TOXICITY DATA with REFERENCE
orl-rat TDLo:3900 mg/kg/26W-C:CAR CNREA8 22,1002,62

SAFETY PROFILE: Questionable carcinogen with experimental carcinogenic data. When heated to decomposition it emits very toxic fumes of F^- and NO_x. See also AMINES and FLUORIDES.

FFP000 CAS:2823-94-1 ***HR: 3***
6-FLUORO-2-ACETYLAMINOFLUORENE
mf: $C_{15}H_{12}FNO$ mw: 241.28

SYNS: 6-FLUORO-2-FAA ◇ N-(6-FLUOROFLUOREN-2-YL)ACETAMIDE

TOXICITY DATA with REFERENCE
orl-rat TDLo:3900 mg/kg/26W-C:CAR CNREA8 22,1002,62

SAFETY PROFILE: Questionable carcinogen with experimental carcinogenic data. When heated to decomposition it emits very toxic fumes of F^- and NO_x. See also AMINES and FLUORIDES.

FFQ000 CAS:2823-95-2 ***HR: 3***
8-FLUORO-2-ACETYLAMINOFLUORENE
mf: $C_{15}H_{12}FNO$ mw: 241.28

SYNS: 8-FLUORO-2-FAA ◇ N-(8-FLUOROFLUOREN-2-YL)ACETAMIDE

TOXICITY DATA with REFERENCE
orl-rat TDLo:3900 mg/kg/26W-C:CAR CNREA8 22,1002,62

SAFETY PROFILE: Questionable carcinogen with experimental carcinogenic data. When heated to decomposition it emits very toxic fumes of F^- and NO_x. See also FLUORIDES and AMINES.

FFR000 CAS:359-06-8 ***HR: 3***
FLUOROACETYL CHLORIDE
mf: C_2H_2ClFO mw: 96.49

TOXICITY DATA with REFERENCE
ihl-mus LCLo:200 mg/m^3/10M NDRC** No.9-4-1-9,43
ihl-gpg LCLo:100 mg/m^3/10M NDRC** No.9-4-1-9,43

CONSENSUS REPORTS: Reported in EPA TSCA Inventory.

SAFETY PROFILE: Poison by inhalation. When heated to decomposition it emits very toxic fumes of Cl^- and F^-. See also FLUORIDES and CHLORIDES.

FFS000 CAS:2713-09-9 ***HR: 3***
FLUORO ACETYLENE
mf: C_2HF mw: 44.03

$$FC{\equiv}CH$$

PROP: A liquid. Fp: −196°; bp: −80°.

SAFETY PROFILE: Can explode violently. The silver and mercury salts are heat-sensitive explosives. May explode spontaneously close to its boiling point −80°C. The gas is stable. Ignites on contact with solutions of bromine in carbon tetrachloride. When heated to decomposition it emits toxic fumes of F^-. See also ACETYLENE COMPOUNDS and FLUORIDES.

FFT000 CAS:364-71-6 ***HR: 3***
o-(FLUOROACETYL)SALICYLIC ACID
mf: $C_9H_7FO_4$ mw: 198.16

SYN: SALICYLIC ACID, FLUOROACETATE

TOXICITY DATA with REFERENCE
orl-rat LDLo:10 mg/kg NCNSA6 5,8,53
scu-mus LD50:15 mg/kg JCSOA9 -,1773,48

SAFETY PROFILE: Poison by ingestion and subcutaneous routes. When heated to decomposition it emits toxic fumes of F^-. See also FLUORIDES.

FFT100 ***HR: 3***
α-FLUORO-β-ALANINE HYDROCHLORIDE
mf: $C_3H_6FNO_2{\cdot}ClH$ mw: 143.56

SYNS: 3-AMINO-2-FLUORO-PROPANOIC ACID HYDROCHLORIDE ◇ FLUORO-β-ALANINE HYDROCHLORIDE

TOXICITY DATA with REFERENCE
ipr-rat LD50:218 mg/kg JPMSAE 73,212,84
ipr-mus LD50:167 mg/kg JPMSAE 73,212,84
ipr-dog LD50:24500 μg/kg JPMSAE 73,212,84
ipr-cat LD50:9400 μg/kg JPMSAE 73,212,84
ipr-rbt LD50:27 mg/kg JPMSAE 73,212,84

SAFETY PROFILE: Poison by intraperitoneal route. When heated to decomposition it emits toxic fumes of F^-, NO_x, and HCl.

FFU000 CAS:15861-05-9 ***HR: 3***
FLUOROAMINE
mf: FH_2N mw: 35.02

SYN: FLUORAMIDE

SAFETY PROFILE: The impure material is very explosive. When heated to decomposition it emits toxic fumes of F^- and NO_x. See also AMINES and FLUORIDES.

FFV000 CAS:592-79-0 ***HR: 3***
5-FLUORO AMYLAMINE
mf: $C_5H_{12}FN$ mw: 105.18

SYN: 5-FLUOROPENTYLAMINE

TOXICITY DATA with REFERENCE
ipr-mus LD50:50 mg/kg JACSAT 78,3487,56
scu-mus LD50:50 mg/kg CLDND*

SAFETY PROFILE: Poison by intraperitoneal and subcutaneous routes. When heated to decomposition it emits very toxic fumes of F^- and NO_x.

FFW000 CAS:407-98-7 ***HR: 3***
5-FLUOROAMYL CHLORIDE
mf: $C_5H_{10}ClF$ mw: 124.60

SYN: 1-CHLORO-5-FLUOROPENTANE

TOXICITY DATA with REFERENCE
ipr-mus LD50:30400 μg/kg CJBPAZ 36,339,58
scu-mus LD50:32 mg/kg CLDND*

SAFETY PROFILE: Poison by intraperitoneal and subcutaneous routes. When heated to decomposition it emits very toxic fumes of Cl^- and F^-.

FFX000 CAS:661-18-7 ***HR: 3***
5-FLUOROAMYL THIOCYANATE
mf: $C_6H_{10}FNS$ mw: 147.23

SYN: 5-FLUOROPENTYL THIOCYANATE

TOXICITY DATA with REFERENCE
ipr-mus LD50:30 mg/kg JACSAT 78,3843,56
scu-mus LD50:30 mg/kg CLDND*

SAFETY PROFILE: Poison by intraperitoneal and subcutaneous routes. When heated to decomposition it emits very toxic F^-, NO_x, and SO_x. See also THIOCYANATES and FLUORIDES.

FFY000 CAS:371-40-4 ***HR: 3***
4-FLUOROANILINE
DOT: UN 2944
mf: C_6H_6FN mw: 111.13

PROP: D: 1.1724, bp: 187.4°, mp: −1.9°

SYNS: BENZENAMINE, 4-FLUORO-(9CI) ◇ 4-FLUORANILIN ◇ p-FLUOROANILINE ◇ 4-FLUOROBENZENAMINE ◇ p-FLUOROPHENYLAMINE

TOXICITY DATA with REFERENCE
skn-rbt 2 mg/24H SEV 85JCAE-,611,86
eye-rbt 250 μg/24H SEV 28ZPAK -,95,72
mma-sat 1 μmol/plate MUREAV 77,317,80
orl-rat LD50:417 mg/kg CHEYAN 23,168,78
orl-bwd LD50:100 mg/kg TXAPA9 21,315,72

CONSENSUS REPORTS: Reported in EPA TSCA Inventory. EPA Genetic Toxicology Program.

DOT Classification: Poison B; Label: St. Andrews Cross.

SAFETY PROFILE: Poison by ingestion. Mutation data reported. A severe skin and eye irritant. When heated to decomposition it emits very toxic fumes of NO_x and F^-.

FFZ000 CAS:388-72-7 ***HR: 3***
4-FLUOROBENZANTHRACENE
mf: $C_{18}H_{11}F$ mw: 246.29

SYNS: 4-FLUOROBENZ(a)ANTHRACENE ◇ 4′-FLUORO-1,2-BENZANTHRACENE

TOXICITY DATA with REFERENCE
scu-rat TDLo:8 mg/kg:NEO CNREA8 23,229,63
scu-mus TDLo:41 mg/kg:ETA CNREA8 23,229,63

SAFETY PROFILE: Questionable carcinogen with experimental neoplastigenic and tumorigenic data. When heated to decomposition it emits toxic fumes of F^-.

FGA000 CAS:462-06-6 ***HR: 3***
FLUOROBENZENE
DOT: UN 2387
mf: C_6H_5F mw: 96.11

PROP: Colorless liquid. D: 1.024, mp: −41.9°, flash p: 5°F, d: 1.024, vap d: 3.31, bp: 82.8°. Insol in water; misc in alc and ether.

SYN: PHENYL FLUORIDE

TOXICITY DATA with REFERENCE
orl-rat LD50:4399 mg/kg GTPZAB 19(9),36,75
ihl-rat LC50:26908 mg/m^3 GTPZAB 19(9),36,75
ihl-mus LC50:45 g/m^3/2H IZSBAI 3,91,65

CONSENSUS REPORTS: Reported in EPA TSCA Inventory.

DOT Classification: Flammable Liquid; Label: Flammable Liquid.

SAFETY PROFILE: Mildly toxic by ingestion and inhalation. A very dangerous fire hazard when exposed to heat, flame, or oxidizers. To fight fire, use water spray, mist, foam, dry chemical, CO_2. When heated to decomposition it emits toxic fumes of F^-.

FGA100 CAS:5430-13-7 ***HR: 3***
4-FLUOROBENZENEARSONIC ACID
mf: $C_6H_6AsFO_3$ mw: 220.04

SYN: BENZENEARSONIC ACID, p-FLUORO-

TOXICITY DATA with REFERENCE
ivn-mus LD50:180 mg/kg CSLNX* NX#05114

OSHA PEL: TWA 0.5 mg(As)/m^3

SAFETY PROFILE: Poison by intravenous route. When heated to decomposition it emits toxic fumes of As and F^-.

FGB000 CAS:52831-45-5 ***HR: 3***
2-FLUORO-BENZO(e)(1)BENZOTHIOPYRANO(4,3-b)INDOLE
mf: $C_{19}H_{10}FNS$ mw: 303.36

TOXICITY DATA with REFERENCE
mma-sat 30 μg/plate MUREAV 66,307,79
scu-mus TDLo:72 mg/kg/9W-I:NEO MUREAV 66,307,79

SAFETY PROFILE: Questionable carcinogen with experimental neoplastigenic data. Mutation data reported. When heated to decomposition it emits very toxic fumes of SO_x, NO_x and F^-.

FGD000 CAS:52831-56-8 ***HR: 3***
3-FLUORO-BENZO(e)(1)BENZOTHIOPYRANO(4,3-b)INDOLE
mf: $C_{19}H_{10}FNS$ mw: 303.36

TOXICITY DATA with REFERENCE
mma-sat 30 μg/plate MUREAV 66,307,79
scu-mus TDLo:72 mg/kg/9W-I:NEO MUREAV 66,307,79

SAFETY PROFILE: Questionable carcinogen with experimental neoplastigenic data. Mutation data reported. When heated to decomposition it emits very toxic fumes of NO_x, SO_x, and F^-.

FGF000 CAS:52831-68-2 ***HR: 3***
4-FLUORO-BENZO(e)(1)BENZOTHIOPYRANO(4,3-b)INDOLE
mf: $C_{19}H_{10}FNS$ mw: 303.36

TOXICITY DATA with REFERENCE
mma-sat 30 μg/plate MUREAV 66,307,79
scu-mus TDLo:72 mg/kg/9W-I:NEO MUREAV 66,307,79

SAFETY PROFILE: Questionable carcinogen with experimental neoplastigenic data. Mutation data reported. When heated to decomposition it emits very toxic fumes of F^-, NO_x, and SO_x.

FGG000 CAS:52831-65-9 ***HR: 3***
4-FLUORO-BENZO(g)(1)BENZOTHIOPYRANO(4,3-b)INDOLE
mf: $C_{19}H_{10}FNS$ mw: 303.36

TOXICITY DATA with REFERENCE
mma-sat 30 μg/plate MUREAV 66,307,79
scu-mus TDLo:72 mg/kg/9W-I:NEO MUREAV 66,307,79

SAFETY PROFILE: Questionable carcinogen with experimental neoplastigenic data. Mutation data reported. When heated to decomposition it emits very toxic fumes of F^-, NO_x, and SO_x.

FGH000 CAS:445-29-4 ***HR: 3***
o-FLUOROBENZOIC ACID
mf: $C_7H_5FO_2$ mw: 140.12

PROP: Needles/water. Mp: 184.6°. Sltly sol in hot water; sol in alc and ether.

SYN: o-FLUORBENZOESAEURE (GERMAN)

TOXICITY DATA with REFERENCE
scu-mus LDLo:700 mg/kg AEPPAE 183,427,36
ivn-mus LD50:180 mg/kg CSLNX* NX#01726

CONSENSUS REPORTS: Reported in EPA TSCA Inventory.

SAFETY PROFILE: Poison by intravenous route. Moderately toxic by subcutaneous route. When heated to decomposition it emits toxic fumes of F^-.

FGI000 CAS:61735-77-1 ***HR: 3***
3-FLUOROBENZO(rst)PENTAPHENE
mf: $C_{24}H_{13}F$ mw: 320.37

TOXICITY DATA with REFERENCE
scu-mus TDLo:20 mg/kg:NEO JFLCAR 8,513,76

SAFETY PROFILE: Questionable carcinogen with experimental neoplastigenic data. When heated to decomposition it emits toxic fumes of F^-.

FGI100 ***HR: 3***
6-FLUOROBENZO(a)PYRENE
mf: $C_{20}H_{11}F$ mw: 270.31

TOXICITY DATA with REFERENCE
scu-rat TDLo:35 mg/kg/1W-I:CAR JJIND8 71,309,83
skn-mus TDLo:49 mg/kg/60W-I:NEO JJIND8 71,309,85
scu-mus TDLo:9 mg/kg:CAR JJIND8 71,309,83
skn-mus TD:16 mg/kg/60W-I:ETA JJIND8 71,309,83

SAFETY PROFILE: Suspected carcinogen with experimental carcinogenic and neoplastigenic data. When heated to decomposition it emits toxic fumes of F^-.

FGJ000 CAS:52831-39-7 ***HR: 3***
2-FLUORO-(1)BENZOTHIOPYRANO(4,3-b)INDOLE
mf: $C_{15}H_8FNS$ mw: 253.30

SYN: 2-FLUOROTHIOPYRANO(4,3-b)BENZ(e)INDOLE

TOXICITY DATA with REFERENCE
mma-sat 30 μg/plate MUREAV 66,307,79
scu-mus TDLo:72 mg/kg/9W-I:NEO MUREAV 66,307,79

SAFETY PROFILE: Questionable carcinogen with experimental neoplastigenic data. Mutation data reported. When heated to decomposition it emits very toxic fumes such as F^-, SO_x, and NO_x.

FGL000 CAS:52831-62-6 ***HR: 3***
4-FLUORO-(1)BENZOTHIOPYRANO(4,3-b)INDOLE
mf: $C_{15}H_8FNS$ mw: 253.30

SYN: 4-FLUOROTHIOPYRANO(4,3-b)BENZ(e)INDOLE

TOXICITY DATA with REFERENCE
mma-sat 30 μg/plate MUREAV 66,307,79
scu-mus TDLo:72 mg/kg/9W-I:NEO MUREAV 66,307,79

SAFETY PROFILE: Questionable carcinogen with ex-

perimental neoplastigenic data. Mutation data reported. When heated to decomposition it emits very toxic fumes such as F^-, SO_x, and NO_x.

FGO000 CAS:52831-58-0 ***HR: 3***
4-FLUORO-6H-(1)BENZOTHIOPYRANO(4,3-b) QUINOLINE
mf: $C_{16}H_{10}FNS$ mw: 267.33

TOXICITY DATA with REFERENCE
mma-sat 90 μg/plate MUREAV 66,307,79
scu-mus TDLo:72 mg/kg/9W-I:NEO MUREAV 66,307,79

SAFETY PROFILE: Questionable carcinogen with experimental neoplastigenic data. Mutation data reported. When heated to decomposition it emits very toxic fumes such as F^-, NO_x, and SO_x.

FGP000 CAS:393-52-2 ***HR: 3***
o-FLUOROBENZOYL CHLORIDE
mf: C_7H_4ClFO mw: 158.56

SYN: 2-FLUOROBENZOYL CHLORIDE

TOXICITY DATA with REFERENCE
mma-sat 100 μg/plate ENMUDM 5(Suppl 1),3,83
ivn-mus LD50:100 mg/kg CSLNX* NX#04779

CONSENSUS REPORTS: Reported in EPA TSCA Inventory.

SAFETY PROFILE: Poison by intravenous route. Mutation data reported. When heated to decomposition it emits very toxic fumes of Cl^- and F^-.

FGQ000 CAS:56390-16-0 ***HR: 3***
2-(3-(p-FLUOROBENZOYL)-1-PROPYL)-5-α,9-α-DIMETHYL-2'-HYDROXY-6,7-BENZOMORPHAN
mf: $C_{24}H_{28}FNO_2$ mw: 381.53

SYN: ID-1229

TOXICITY DATA with REFERENCE
orl-rat LD50:1000 mg/kg DRFUD4 3,298,78
ipr-rat LD50:82 mg/kg ARZNAD 25,795,75
scu-rat LD50:520 mg/kg ARZNAD 25,795,75
ivn-rat LD50:24 mg/kg ARZNAD 25,795,75
orl-mus LD50:500 mg/kg DRFUD4 3,298,78
ipr-mus LD50:112 mg/kg ARZNAD 25,795,75
scu-mus LD50:270 mg/kg ARZNAD 25,795,75
ivn-mus LD50:32 mg/kg DRFUD4 3,298,78
scu-rbt LD50:75 mg/kg ARZNAD 25,795,75
ivn-rbt LD50:7850 μg/kg ARZNAD 25,795,75

SAFETY PROFILE: Poison by intravenous, intraperitoneal, and subcutaneous routes. Moderately toxic by ingestion. When heated to decomposition it emits very toxic fumes of F^- and NO_x.

FGU000 ***HR: 3***
1-(1-(3-(p-FLUOROBENZOYL)PROPYL)-4-PIPERIDYL)-2-BENZIMIDAZOLINONE, HYDROCHLORIDE MONOHYDRATE
mf: $C_{22}H_{24}FN_3O_2 \cdot ClH \cdot H_2O$ mw: 435.97

SYNS: BENPERIDOL ◇ BENZOPERIDOL ◇ BENZPERIDOL ◇ 8089 CB ◇ CONCILIUM ◇ FRENACTIL ◇ FRENACTYL ◇ GLIANIMON ◇ MCN-JR-4584 ◇ R 4584

TOXICITY DATA with REFERENCE
dnd-mus-skn 192 μmol/kg CRNGDP 5,231,84
scu-rat LD50:218 mg/kg 27ZQAG -,185,72
ivn-rat LD50:21 mg/kg 27ZQAG -,185,72
orl-mus LD50:1000 mg/kg 27ZQAG -,185,72
scu-mus LD50:210 mg/kg 27ZQAG -,185,72
ivn-mus LD50:27 mg/kg 27ZQAG -,185,72

SAFETY PROFILE: Poison by subcutaneous and intravenous routes. Moderately toxic by ingestion. Mutation data reported. Used as an antipsychotic agent. When heated to decomposition it emits very toxic fumes of F^-, NO_x, and HCl.

FGV000 CAS:20977-50-8 ***HR: 3***
1-(3-(4-FLUOROBENZOYL)PROPYL)-4-PIPERIDYL-N-ISOPROPYL CARBAMATE
mf: $C_{19}H_{27}FN_2O_3$ mw: 350.48

SYN: AL-1021

TOXICITY DATA with REFERENCE
orl-hmn TDLo:2140 μg/kg:PSY CTCEA9 11,779,69
orl-rat LD50:310 mg/kg 27ZQAG -,184,72
orl-mus LD50:265 mg/kg 27ZQAG -,184,72
ivn-mus LD50:44 mg/kg 27ZQAG -,184,72
orl-dog LD50:25 mg/kg 27ZQAG -,184,72
orl-rbt LD50:200 mg/kg 27ZQAG -,184,72

SAFETY PROFILE: Poison by ingestion and intravenous routes. Human systemic effects by ingestion: psychotropic effects. When heated to decomposition it emits very toxic F^- and NO_x. See also CARBAMATES.

FGW000 CAS:59921-81-2 ***HR: 3***
1-(4'-FLUOROBENZOYL)-3-PYRROLIDINYLPROPANE MALEATE
mf: $C_{14}H_{18}FNO \cdot C_4H_4O_4$ mw: 351.41

SYNS: 1-(4'-FLUOROBENZOIL)-3-PIRROLIDINOPROPANOMALEATO (ITALIAN) ◇ 4'-FLUORO-4-(1-PYRROLIDINYL) BUTYROPHENONE MALEATE

TOXICITY DATA with REFERENCE
orl-rat LD50:335 mg/kg FRPSAX 31,442,76
orl-mus LD50:235 mg/kg FRPSAX 31,442,76
ivn-mus LD50:47 mg/kg FRPSAX 31,442,76

SAFETY PROFILE: Poison by ingestion and intrave-

nous routes. When heated to decomposition it emits very toxic fumes of F^- and NO_x.

FGW100 CAS:1426-40-0 ***HR: 3***
FLUOROBIS(TRIFLUOROMETHYL)PHOSPHINE
mf: C_2F_7P mw: 187.98

SAFETY PROFILE: Ignites spontaneously in air. When heated to decomposition it emits toxic fumes of F^-, phosphine and NO_x. See also PHOSPHINE.

FGX000 CAS:1072-85-1 ***HR: 2***
1-FLUORO-2-BROMOBENZENE
mf: C_6H_4BrF mw: 175.01

PROP: Colorless liquid. D: 1.597, mp: −8°, bp: 152°. Insol in water; very sol in alc and ether.

SYN: 2-BROMFLUORBENZEN (CZECH)

TOXICITY DATA with REFERENCE
skn-rbt 500 mg/24H MLD 28ZPAK -,31,72
eye-rbt 100 mg/24H MOD 28ZPAK -,31,72
orl-rat LD50:1850 mg/kg 28ZPAK -,31,72

CONSENSUS REPORTS: Reported in EPA TSCA Inventory.

SAFETY PROFILE: Moderately toxic by ingestion. A skin and eye irritant. When heated to decomposition it emits very toxic fumes of Br^- and F^-.

FGY000 CAS:1073-06-9 ***HR: 2***
1-FLUORO-3-BROMOBENZENE
mf: C_6H_4BrF mw: 175.01

PROP: Colorless liquid. D: 1.597, mp: −8°, bp: 152°, Insol in water, very sol in alcohol and ether.

SYN: 3-BROMFLUORBENZEN (CZECH)

TOXICITY DATA with REFERENCE
skn-rbt 500 mg/24H MOD 28ZPAK -,32,72
eye-rbt 20 mg/24H MOD 28ZPAK -,32,72
orl-rat LD50:670 mg/kg 28ZPAK -,32,72

CONSENSUS REPORTS: Reported in EPA TSCA Inventory.

SAFETY PROFILE: Moderately toxic by ingestion. A skin and eye irritant. When heated to decomposition it emits very toxic fumes of F^- and Br^-.

FHA000 CAS:462-72-6 ***HR: 3***
4-FLUOROBUTYL BROMIDE
mf: C_4H_8BrF mw: 155.03

TOXICITY DATA with REFERENCE
ipr-mus LD50:8200 μg/kg JOCEAH 21,748,56
scu-mus LD50:8 mg/kg CLDND*

SAFETY PROFILE: Poison by intraperitoneal and subcutaneous routes. When heated to decomposition it emits very toxic fumes of Br^- and F^-.

FHB000 CAS:462-73-7 ***HR: 3***
4-FLUOROBUTYL CHLORIDE
mf: C_4H_8ClF mw: 110.57

TOXICITY DATA with REFERENCE
ipr-mus LD50:1250 μg/kg JOCEAH 21,748,56
scu-mus LD50:1250 mg/kg CLDND*

SAFETY PROFILE: Poison by intraperitoneal route. Moderately toxic by subcutaneous route. When heated to decomposition it emits very toxic fumes of Cl^- and F^-.

FHC000 CAS:372-91-8 ***HR: 3***
4-FLUOROBUTYL IODIDE
mf: C_4H_8FI mw: 202.02

TOXICITY DATA with REFERENCE
ipr-mus LD50:5200 μg/kg JOCEAH 21,748,56
scu-mus LD50:5 mg/kg CLDND*

SAFETY PROFILE: Poison by intraperitoneal and subcutaneous routes. When heated to decomposition it emits very toxic fumes of F^- and I^-.

FHD000 CAS:353-17-3 ***HR: 3***
4-FLUOROBUTYL THIOCYANATE
mf: C_5H_8FNS mw: 133.20

TOXICITY DATA with REFERENCE
ipr-mus LD50:2600 μg/kg JACSAT 78,3843,56
scu-mus LD50:2600 μg/kg CLDND*

SAFETY PROFILE: Poison by intraperitoneal and subcutaneous routes. When heated to decomposition it emits very toxic fumes of F^-, NO_x, and SO_x. See also THIOCYANATES and FLUORIDES.

FHF000 CAS:407-83-0 ***HR: 3***
4-FLUOROBUTYRONITRILE
mf: C_4H_6FN mw: 87.11

SYN: Γ-FLUOROBUTYRONITRILE

TOXICITY DATA with REFERENCE
ipr-mus LD50:16 mg/kg JACSAT 78,3484,56
scu-mus LD50:16 mg/kg CLDND*

CONSENSUS REPORTS: Cyanide and its compounds are on the Community Right-To-Know List.

SAFETY PROFILE: Poison by intraperitoneal and subcutaneous routes. When heated to decomposition it emits very toxic fumes of F^-, CN^-, and NO_x. See also NITRILES.

FHG000 CAS:1893-33-0 ***HR: 3***
FLUOROBUTYROPHENONE
mf: $C_{21}H_{30}FN_3O_2$ mw: 375.54

SYNS: DIPIPERAL ◇ DIPIPERON ◇ DIPIPERONE ◇ FLOROPIPAMIDE ◇ 1'-(3-(p-FLUOROBENZOYL)PROPYL)(1,4'-BIPIPERIDINE 1-4'-CARBOXAMIDE ◇ 1-(3-(p-FLUOROBENZOYL)PROPYL)-4-PIPERIDINOISONIPACOTAMIDE ◇ 4'-FLUORO-4-(4-N-PIPERIDINO-4-CARBAMIDOPIPERIDINO)BUTYROPHENONE ◇ p-FLUORO-Γ-(4-PIPERIDINO-4-CARBAMOYLPIPERIDINO)BUTYROPHENONE ◇ FPA ◇ MCN-JR-3345 ◇ PIPAMPERONE ◇ PIPANEPERONE ◇ PIPERONYL ◇ PROPITAN ◇ R 3345

TOXICITY DATA with REFERENCE
orl-mus TDLo:1500 mg/kg (10-12D preg):TER TOIZAG 28,621,81
orl-rat LD50:160 mg/kg ANPBAZ 61,611,61
scu-rat LD50:160 mg/kg ANPBAZ 61,611,61
ivn-rat LD50:48 mg/kg 27ZQAG -,188,72
orl-mus LD50:490 mg/kg PHMGBN 15,485,77
scu-mus LD50:160 mg/kg 27ZQAG -,188,72
ivn-mus LD50:66 mg/kg 27ZQAG -,188,72

SAFETY PROFILE: Poison by ingestion, subcutaneous, and intravenous routes. An experimental teratogen. When heated to decomposition it emits very toxic fumes of F^- and NO_x.

FHH000 ***HR: 2***
FLUOROCHLOROCARBON LIQUID

TOXICITY DATA with REFERENCE
orl-mus LD50:7600 mg/kg GISAAA 39(4),114,74
ihl-mus LC50:930 mg/m^3/2H GISAAA 39(4),114,74
ivn-mus LD50:490 mg/kg GISAAA 39(4),114,74

SAFETY PROFILE: Moderately toxic by intravenous route. Mildly toxic by ingestion and inhalation. When heated to decomposition it emits very toxic fumes of F^-.

FHH025 CAS:67639-45-6 ***HR: 2***
5-FLUORO-7-CHLOROMETHYL-12-METHYLBENZ(a)ANTHRACENE
mf: $C_{20}H_{14}ClF$ mw: 308.79

SYN: 7-(CHLOROMETHYL)-5-FLUORO-12-METHYLBENZ(a)ANTHRACENE

TOXICITY DATA with REFERENCE
ivn-mus TDLo:1239 μg/kg:NEO CNREA8 40,782,80

SAFETY PROFILE: Questionable carcinogen with experimental neoplastigenic data. When heated to decomposition it emits toxic fumes of Cl^- and F^-.

FHH100 CAS:127-31-1 ***HR: 3***
FLUOROCORTISONE
mf: $C_{21}H_{29}FO_5$ mw: 380.50

PROP: Crystals. Decomp 260-262°. Solubility in water: 0.14 mg/mL.

SYNS: ALFLORONE ◇ F-COL ◇ F-CORTEF ◇ FLORINEF ◇ FLUDROCORTISONE ◇ FLUDROCORTONE ◇ FLUODROCORTISONE ◇ FLUOHYDRISONE ◇ FLUOHYDROCORTISONE ◇ 9-α-FLUOROHYDROCORTISONE ◇ 9-α-FLUORO-17-HYDROXYCORTICOSTERONE ◇ 9-FLUORO-11-β,17,21-TRIHYDROXYPREGN-4-ENE-3,20-DIONE ◇ 9-α-FLUORO-11-β,17-α,21-TRIHYDROXY-4-PREGNENE-3,20-DIONE ◇ U 5963

TOXICITY DATA with REFERENCE
orl-rbt TDLo:1 mg/kg (female 13-16D post):TER 28QFAD -,85,73
par-rat TDLo:8750 μg/kg (female 7D pre):REP JEMEAV 102,347,55
ipr-mus LD50:170 mg/kg CLDND*

SAFETY PROFILE: Poison by intraperitoneal route. Experimental reproductive effects. When heated to decomposition it emits toxic fumes of F^-. See also PREDNISONE.

FHI000 CAS:2022-85-7 ***HR: 2***
5-FLUOROCYTOSINE
mf: $C_4H_4FN_3O$ mw: 129.11

SYNS: ALCOBON ◇ 4-AMINO-5-FLUORO-2(1H)-PYRIMIDINONE ◇ ANCOBON ◇ TIL ◇ 5-FC ◇ FLUCYTOSINE ◇ 5-FLUOROCYSTOSINE ◇ 2-HYDROXY-4-AMINO-5-FLUOROPYRIMIDINE ◇ RO 2-9915

TOXICITY DATA with REFERENCE
dni-omi 100 mg/L JMMIAV 12,83,79
oms-omi 100 mg/L JMMIAV 12,83,79
mmo-smc 8 mg/L MGGEAE 146,253,76
ipr-rat LD50:3811 mg/kg IYKEDH 10,710,79
scu-rat LD50:3600 mg/kg NIIRDN 6,699,82
ipr-mus LD50:1190 mg/kg AACHAX -,566,63
scu-mus LD50:1000 mg/kg CHTHBK 25,54,79
ivn-mus LD50:500 mg/kg AACHAX -,566,63

SAFETY PROFILE: Moderately toxic by intraperitoneal, subcutaneous, and intravenous routes. Mutation data reported. An antifungal agent. When heated to decomposition it emits very toxic fumes of F^- and NO_x.

FHJ000 CAS:475-26-3 ***HR: 2***
FLUORO-DDT
mf: $C_{14}H_9Cl_3F_2$ mw: 321.58

PROP: Crystals. Mp: 45.5°, vap d: 10.5.

SYNS: 1,1-BIS(p-FLUOROPHENYL)-2,2,2-TRICHLOROETHANE ◇ 2,2-BIS(p-FLUOROPHENYL)-1,1,1-TRICHLOROETHANE ◇ DIFLUORODIPHENYLTRICHLOROETHANE ◇ FLUOROGESAROL ◇ 1,1,1-TRICHLORO-2,2-BIS(p-FLUOROPHENYL) ETHANE

TOXICITY DATA with REFERENCE
orl-rat LD50:1120 mg/kg AFDOAQ 15,122,51
orl-mus LDLo:600 mg/kg JPETAB 88,400,46

SAFETY PROFILE: Moderately toxic by ingestion. A contact insecticide. When heated to decomposition it emits highly toxic fumes of fluorides and Cl^-. See also DDT and FLUORIDES.

FHL000 CAS:334-56-5 ***HR: 3***
1-FLUORODECANE
mf: $C_{10}H_{21}F$ mw: 160.31

SYN: F10

TOXICITY DATA with REFERENCE
ipr-mus LD50:1700 μg/kg JACSAT 79,2311,57

CONSENSUS REPORTS: Reported in EPA TSCA Inventory.

SAFETY PROFILE: Poison by intraperitoneal route. When heated to decomposition it emits toxic fumes of F^-.

FHM000 CAS:334-64-5 ***HR: 3***
10-FLUORODECANOL
mf: $C_{10}H_{21}FO$ mw: 176.31

SYN: ω-FLUORODECANOL

TOXICITY DATA with REFERENCE
ipr-mus LD50:1 mg/kg JOCEAH 21,739,56
scu-mus LD50:2 mg/kg 11FYAN 3,60,63

SAFETY PROFILE: Poison by intraperitoneal and subcutaneous routes. When heated to decomposition it emits toxic fumes of F^-.

FHN000 CAS:334-62-3 ***HR: 3***
10-FLUORODECYL CHLORIDE
mf: $C_{10}H_{20}ClF$ mw: 194.75

SYN: 1-CHLORO-10-FLUORO-DECANE

TOXICITY DATA with REFERENCE
ipr-mus LD50:5 mg/kg JOCEAH 21,748,56
scu-mus LD50:5 mg/kg CLDND*

SAFETY PROFILE: Poison by intraperitoneal and subcutaneous routes. When heated to decomposition it emits very toxic fumes of Cl^- and F^-.

FHO000 CAS:10356-76-0 ***HR: 2***
5-FLUORO-2′-DEOXYCYTIDINE
mf: $C_9H_{12}FN_3O_4$ mw: 245.24

SYNS: FCdR ◇ FCDR ◇ 5-FLUOR-DESOXYCYTIDIN (GERMAN) ◇ 5-FLUORODEOXYCYTIDINE

TOXICITY DATA with REFERENCE
dni-mus:leu 1 mg/L CPBTAL 30,1018,82
sce-ham:ovr 10 μmol/L CHROAU 85,603,82
ipr-mus TDLo:20 mg/kg (15-16D preg):REP TJADAB 24(2),50A,81
ipr-mus TDLo:10 mg/kg (female 9D post):TER AUCMBJ 61,207,73
ipr-rat LD50:2000 mg/kg ADTEAS 3,181,68

SAFETY PROFILE: Moderately toxic by intraperitoneal route. Experimental teratogenic and reproductive effects. Mutation data reported. When heated to decomposition it emits very toxic fumes of F^- and NO_x.

FHP000 CAS:1764-39-2 ***HR: 3***
6-FLUORODIBENZ(a,h)ANTHRACENE
mf: $C_{22}H_{13}F$ mw: 296.35

SYN: 4-FLUORO-1,2:5,6-DIBENZANTHRACENE

TOXICITY DATA with REFERENCE
scu-mus TDLo:200 mg/kg/I:ETA BECCAN 40,30,62

SAFETY PROFILE: Questionable carcinogen with experimental tumorigenic data. When heated to decomposition it emits toxic fumes of F^-.

FHQ000 CAS:321-25-5 ***HR: 2***
2-FLUORO-4-DIMETHYLAMINOAZOBENZENE
mf: $C_{14}H_{14}FN_3$ mw: 243.31

SYN: N,N-DIMETHYL-2-FLUORO-4-PHENYLAZOANILINE

TOXICITY DATA with REFERENCE
orl-rat TDLo:3200 mg/kg/12W-C:ETA CNREA8 13,93,53

SAFETY PROFILE: Questionable carcinogen with experimental tumorigenic data. When heated to decomposition it emits toxic fumes of NO_x and F^-.

FHQ010 CAS:331-91-9 ***HR: 3***
2′-FLUORO-4-DIMETHYLAMINOAZOBENZENE
mf: $C_{14}H_{14}FN_3$ mw: 243.31

SYN: N,N-DIMETHYL-p-(2-FLUOROPHENYLAZO)ANILINE

TOXICITY DATA with REFERENCE
orl-rat TDLo:3150 mg/kg/13W-C:ETA CNREA8 9,652,49

SAFETY PROFILE: Questionable carcinogen with experimental tumorigenic data. When heated to decomposition it emits very toxic fumes of F^- and NO_x. See also FLUORIDES.

FHQ100 CAS:332-54-7 ***HR: 3***
3′-FLUORO-4-DIMETHYLAMINOAZOBENZENE
mf: $C_{14}H_{14}FN_3$ mw: 243.31

SYN: m-FLUORODIMETHYLAMINOAZOBENZENE

TOXICITY DATA with REFERENCE
scu-mus TDLo:1050 mg/kg (female 9D post):TER OFAJAE 36,195,60
ipr-mus TDLo:600 mg/kg (10-11D preg):REP KAIZAN 37,179,62
orl-rat TDLo:5300 mg/kg/21W-C:NEO CNREA8 17,387,57

orl-rat TD:3200 mg/kg/12W-C:ETA CNREA8 13,93,53
scu-mus LDLo:1200 mg/kg OFAJAE 36,195,60

SAFETY PROFILE: Moderately toxic by subcutaneous route. An experimental teratogen. Experimental reproductive effects. Questionable carcinogen with experimental neoplastigenic and tumorigenic data. When heated to decomposition it emits toxic fumes of F^- and NO_x.

FHR000 CAS:64038-39-7 ***HR: 3***
10-FLUORO-9,12-DIMETHYLBENZ(a)ACRIDINE
mf: $C_{19}H_{14}FN$ mw: 275.34

SYNS: 2,10-DIMETHYL-3-FLUORO-5,6-BENZACRIDINE ◇ 9,12-DIMETHYL-10-FLUOROBENZ(a)ACRIDINE ◇ 3-FLUORO-2,10-DIMETHYL-5,6-BENZACRIDINE

TOXICITY DATA with REFERENCE
skn-mus TDLo:324 mg/kg/27W-I:ETA AICCA6 11,736,55

SAFETY PROFILE: Questionable carcinogen with experimental tumorigenic data. When heated to decomposition it emits toxic F^- and NO_x.

FHS000 CAS:68141-57-1 ***HR: 3***
1-FLUORO-7,12-DIMETHYLBENZ(a)ANTHRACENE
mf: $C_{20}H_{15}F$ mw: 274.35

SYN: 7,12-DIMETHYL-1-FLUOROBENZ(a)ANTHRACENE

TOXICITY DATA with REFERENCE
msc-ham:lng 1 mg/L MUREAV 136,65,84
skn-mus TDLo:110 μg/kg:ETA CNREA8 39,411,79

CONSENSUS REPORTS: EPA Genetic Toxicology Program.

SAFETY PROFILE: Questionable carcinogen with experimental tumorigenic data. Mutation data reported. When heated to decomposition it emits toxic fumes of F^-.

FHU000 CAS:959-73-9 ***HR: 3***
2'-FLUORO-N,N-DIMETHYL-4-STILBENAMINE
mf: $C_{16}H_{16}FN$ mw: 241.33

SYNS: 2'-FLUORO-4-DIMETHYLAMINOSTILBENE ◇ 2-FLUORO-4-STILBENYL-N,N-DIMETHYLAMINE

TOXICITY DATA with REFERENCE
orl-rat TDLo:490 mg/kg/46W-C:ETA ABMGAJ 9,87,62

SAFETY PROFILE: Questionable carcinogen with experimental tumorigenic data. When heated to decomposition it emits very toxic fumes of F^- and NO_x. See also FLUORIDES.

FHV000 CAS:405-86-7 ***HR: 3***
4'-FLUORO-N,N-DIMETHYL-4-STILBENAMINE
mf: $C_{16}H_{16}FN$ mw: 241.33

SYNS: 4'-FLUORO-4-DIMETHYLAMINOSTILBENE ◇ 4'-FLUORO-4-STILBENYL-N,N-DIMETHYLAMINE

TOXICITY DATA with REFERENCE
orl-rat TDLo:440 mg/kg/42W-C:ETA ABMGAJ 9,87,62

SAFETY PROFILE: Questionable carcinogen with experimental tumorigenic data. When heated to decomposition it emits very toxic fumes of F^- and NO_x. See also FLUORIDES.

FHV300 ***HR: 3***
1-FLUORO-1,1-DINITRO-2-BUTENE
mf: $C_4H_5FN_2O_4$ mw: 164.09

$FC(NO_2)_2CH{=}CHCH_3$

SAFETY PROFILE: An explosive. When heated to decomposition it emits toxic fumes of F^- and NO_x.

FHV800 CAS:68795-10-8 ***HR: 3***
2-FLUORO-1,1-DINITROETHANE
mf: $C_2H_3FN_2O_4$ mw: 138.06

SAFETY PROFILE: Explosive reaction with air at 75°C. When heated to decomposition it emits toxic fumes of F^- and NO_x.

FHW000 CAS:17003-75-7 ***HR: 3***
2-FLUORO-2,2-DINITROETHANOL
mf: $C_2H_3FN_2O_5$ mw: 154.06

$FC(NO_2)_2CH_2OH$

SAFETY PROFILE: A vesicant. Potentially explosive. When heated to decomposition it emits toxic fumes of F^- and NO_x.

FHX000 CAS:18139-02-1 ***HR: 3***
2-FLUORO-2,2-DINITROETHYLAMINE
mf: $C_2H_4FN_3O_4$ mw: 153.08

$FC(NO_2)_2CH_2NH_2$

SAFETY PROFILE: The pure material may explode at room temperature. Concentrated solutions in dichloromethane may decompose violently in storage. When heated to decomposition it emits toxic fumes of F^- and NO_x.

FHY000 CAS:7182-87-8 ***HR: 3***
FLUORO DINITROMETHANE
mf: $CHFN_2O_4$ mw: 124.03

$FCH(NO_2)_2$

SAFETY PROFILE: Potentially explosive. When

heated to decomposition it emits toxic fumes of F^- and NO_x.

FHZ000 CAS:17003-82-6 ***HR: 3***
FLUORO DINITROMETHYL AZIDE
mf: CFN_5O_4 mw: 165.04

$F(NO_2)_2CN_3$

SAFETY PROFILE: Unstable at room temperatures. When heated to decomposition it emits toxic fumes of F^- and NO_x. See also AZIDES.

FHZ200 CAS:22692-30-4 ***HR: 3***
1-FLUORO-1,1-DINITRO-2-PHENYLETHANE
mf: $C_8H_7FN_2O_4$ mw: 214.15

$FC(NO_2)_2CH_2C_6H_6$

SAFETY PROFILE: Explosive. When heated to decomposition it emits toxic fumes of F^- and NO_x. See also NITRO COMPOUNDS of AROMATIC HYDROCARBONS.

FIA000 CAS:334-71-4 ***HR: 3***
12-FLUORO DODECANO NITRILE
mf: $C_{12}H_{22}FN$ mw: 199.35

SYN: NC11F

TOXICITY DATA with REFERENCE
ipr-mus LD50:80 mg/kg JACSAT 78,3484,56
scu-mus LD50:80 mg/kg CLDND*

CONSENSUS REPORTS: Cyanide and its compounds are on the Community Right-To-Know List.

SAFETY PROFILE: Poison by intraperitoneal and subcutaneous routes. When heated to decomposition it emits very toxic fumes of F^-, NO_x, and CN^-. See also FLUORIDES and NITRILES.

FIA500 CAS:1881-37-4 ***HR: 2***
4-FLUOROESTRADIOL
mf: $C_{18}H_{23}FO_2$ mw: 290.41

SYNS: ESTRA-1,3,5(10)-TRIENE-3,17-DIOL, 4-FLUORO-, (17-β)-(9CI) ◇ 4-FLUOROESTRA-1,3,5-(10)-TRIENE-3,17-β-DIOL

TOXICITY DATA with REFERENCE
imp-ham TDLo:360 mg/kg/15W-I:CAR MOPMA3 23,278,83

SAFETY PROFILE: Questionable carcinogen with experimental carcinogenic data. When heated to decomposition it emits toxic fumes of F^-.

FIB000 CAS:353-36-6 ***HR: 3***
FLUOROETHANE
DOT: UN 2453
mf: C_2H_5F mw: 48.06

PROP: Mp: −143.2°, bp: −37.7°, d: 0.8158 @ −37.7°, vap d: 1.66.

SYNS: ETHYL FLUORIDE (DOT) ◇ MONOFLUOROETHANE ◇ R161

CONSENSUS REPORTS: Reported in EPA TSCA Inventory.

DOT Classification: Flammable Gas; Label: Flammable Gas.

SAFETY PROFILE: A very dangerous fire hazard when exposed to heat, flames, or oxidizers. To fight fire, stop flow of gas. When heated to decomposition it emits toxic fumes of F^-.

FIC000 CAS:144-49-0 ***HR: 3***
FLUOROETHANOIC ACID
DOT: UN 2642
mf: $C_2H_3FO_2$ mw: 78.05

PROP: Colorless solid. Mp: 33°, bp: 165°. Sol in water.

SYNS: ACIDE-MONOFLUORACETIQUE (FRENCH) ◇ ACIDO MONOFLUOROACETIO (ITALIAN) ◇ CYMONIC ACID ◇ FAA ◇ FLUOROACETATE ◇ FLUOROACETIC ACID ◇ 2-FLUOROACETIC ACID ◇ FLUOROACETIC ACID (DOT) ◇ GIFBLAAR POISON ◇ HFA ◇ MFA ◇ MONOFLUORAZIJNZUUR (DUTCH) ◇ MONOFLUORESSIGSAURE (GERMAN) ◇ MONOFLUOROACETATE ◇ MONOFLUOROACETIC ACID

TOXICITY DATA with REFERENCE
orl-rat LD50:4680 μg/kg TXAPA9 13,189,68
orl-mus LD50:7 mg/kg TXAPA9 13,189,68
ipr-mus LD50:6600 μg/kg JOCEAH 21,883,56
scu-mus LD50:281 mg/kg JPETAB 95,62,49
ivn-mus LD50:13 mg/kg CSLNX* NX#03014
ivn-rbt LD50:250 μg/kg NATUAS 158,382,46
orl-gpg LD50:468 μg/kg TXAPA9 13,189,68

CONSENSUS REPORTS: EPA Extremely Hazardous Substances List. Reported in EPA TSCA Inventory.

DOT Classification: Poison B; Label: Poison

SAFETY PROFILE: Poison by ingestion, subcutaneous, intraperitoneal, and intravenous routes. Affects the human central nervous system, causing convulsions and ventricular fibrillation. When heated to decomposition it emits toxic fumes of F^- and Na_2O. See also SODIUM FLUOROACETATE.

FID000 CAS:63919-01-7 ***HR: 3***
FLUOROETHANOL
mf: C_2H_5FO mw: 64.07

SYNS: MONOFLUORETHANOL ◇ MONOFLUOROETHANOL

TOXICITY DATA with REFERENCE
ipr-rat LD50:5 mg/kg JPETAB 106,464,52
ihl-mus LC50:1100 mg/m³/10M 11FYAN 3,61,63
scu-mus LD50:19 mg/kg JPETAB 95,62,49

SAFETY PROFILE: Poison by intraperitoneal and subcutaneous routes. Mildly toxic by inhalation. When heated to decomposition it emits very toxic fumes of F^-.

FIE000 CAS:371-62-0 ***HR: 3***
2-FLUOROETHANOL
mf: C_2H_5FO mw: 64.07

SYNS: β-FLUOROETHANOL ◇ TL 741

TOXICITY DATA with REFERENCE
ihl-rat LC50:200 mg/m^3/10M NTIS** PB158-508
ipr-rat LD50:1750 μg/kg JPPMAB 20,465,68
ihl-mus LC50:1100 mg/m^3/10M JCSOA9 -,773,49
ipr-mus LD50:10 mg/kg JOCEAH 21,739,56
scu-mus LD50:15 mg/kg NATUAS 172,1139,53
ihl-dog LC50:7 mg/m^3/10M NTIS** PB158-508
ihl-mky LC50:1500 mg/m^3/10M NTIS** PB158-508
ihl-cat LC50:35 mg/m^3/10M NTIS** PB158-508
ihl-rbt LC50:25 mg/m^3/10M NTIS** PB158-508
ivn-rbt LDLo:250 μg/kg JCSOA9 -,1279,49
ihl-gpg LC50:150 mg/m^3/10M NTIS** PB158-508

CONSENSUS REPORTS: EPA Extremely Hazardous Substances List. Reported in EPA TSCA Inventory.

SAFETY PROFILE: Poison by inhalation, intraperitoneal, subcutaneous, and intravenous routes. When heated to decomposition it emits very toxic fumes of F^-.

FIH000 CAS:63884-92-4 ***HR: 3***
β-FLUOROETHYL-N-(β-CHLOROETHYL)-N-NITROSOCARBAMATE
mf: $C_4H_8ClFN_2O_3$ mw: 186.59

SYNS: (2-CHLOROETHYL)NITROSOCARBAMICACID-2-FLUOROETHYL ESTER ◇ TL 821

TOXICITY DATA with REFERENCE
ihl-gpg LCLo:300 mg/m^3/10M NDRC** No. 9-4-1-9,43

SAFETY PROFILE: Poison by inhalation. Many N-nitroso compounds are carcinogens. When heated to decomposition it emits very toxic fumes of Cl^-, F^-, and NO_x. See also N-NITROSO COMPOUNDS and CARBAMATES.

FIJ000 CAS:13908-93-5 ***HR: 3***
1-(2-FLUOROETHYL)-3-CYCLOHEXYL-1-NITROSOUREA
mf: $C_9H_{16}FN_3O_2$ mw: 217.2

SYNS: CFNU ◇ CYCLOHEXYL FLUOROETHYL NITROSOUREA ◇ NSC 87974 ◇ SRI 2619

TOXICITY DATA with REFERENCE
msc-ham:lng 10 μmol/L CNREA8 40,2719,80
dnd-ham:lng 21 μmol/L CNREA8 38,3379,78
dnd-mus:leu 50 μmol/L CNREA8 38,3197,78
ipr-mus LD10:34 mg/kg CNREA8 34,194,74
ivn-mus LD50:51 mg/kg TXAPA9 10,397,67
ivn-mus LD50:12 mg/kg TXAPA9 10,397,67
orl-mus LD50:111 mg/kg TXAPA9 10,397,67
orl-rat LD50:18500 μg/kg TXAPA9 10,397,67
scu-mus LD50:25210 μg/kg NCISP* JAN86

SAFETY PROFILE: Poison by ingestion, intravenous, intraperitoneal and subcutaneous routes. Mutation data reported. Many N-nitroso compounds are carcinogens. When heated to decomposition it emits very toxic fumes of F^- and NO_x. See also N-NITROSO COMPOUNDS.

FIK000 CAS:4681-36-1 ***HR: 3***
FLUOROETHYL-O,O-DIETHYLDITHIOPHOSPHORYL-1-PHENYLACETATE
mf: $C_{14}H_{20}FO_4PS_2$ mw: 366.43

SYN: 2-FLUOROETHYLMERCAPTOPHENYLACETATE-O,O-DIETHYL PHOSPHORODITHIOATE

TOXICITY DATA with REFERENCE
orl-rat LD50:5 mg/kg BESAAT 15,119,69
orl-mus LD50:75 mg/kg BESAAT 15,119,69
orl-gpg LD50:3 mg/kg 28ZEAL 5,140,76

SAFETY PROFILE: Poison by ingestion. When heated to decomposition it emits very toxic fumes of F^-, PO_x, and SO_x. See also MERCAPTANS and FLUORIDES.

FIK875 CAS:60553-18-6 ***HR: 3***
FLUOROETHYLENE OZONIDE
mf: $C_2H_3FO_3$ mw: 94.04

SYN: 3-FLUORO-1,2,4-TRIOXOLANE

SAFETY PROFILE: Spontaneously explosive at room temperature. When heated to decomposition it emits toxic fumes of F^-. See also OZONE.

FIM000 CAS:459-99-4 ***HR: 3***
β-FLUOROETHYL FLUOROACETATE
mf: $C_4H_6F_2O_2$ mw: 124.10

SYNS: 2-FLUOROETHYL FLUOROACETATE ◇ TL 855

TOXICITY DATA with REFERENCE
ihl-rat LC50:200 mg/m^3/10M NTIS** PB158-508
ihl-mus LC50:450 μg/m^3 NATUAS 160,179,47
scu-mus LD50:8500 μg/kg NATUAS 160,179,47
par-mus LD50:8500 μg/kg JCSOA9 -,1471,49
ihl-rbt LC50:50 mg/m^3 JCSOA9 -,916,49
ihl-gpg LC50:70 mg/m^3/10M NTIS** PB158-508

SAFETY PROFILE: Poison by inhalation, subcutaneous and parenteral routes. When heated to decomposition it emits toxic fumes of F^-. See also ESTERS and FLUORIDES.

FIN000 CAS:371-29-9 **HR: 3**
2-FLUORO ETHYL-γ-FLUORO BUTYRATE
mf: $C_6H_{10}F_2O_2$ mw: 152.16

SYN: β-FLUOROETHYL-Γ-FLUOROBUTYRATE

TOXICITY DATA with REFERENCE
ihl-rat LC50:200 mg/m³/10M NTIS** PB158-508
ihl-mus LC50:73 mg/m³/10M NDRC** No. 9-4-1-19,44
ihl-dog LC50:25 mg/m³/10M NTIS** PB158-508
ihl-mky LC50:500 mg/m³/10M NTIS** PB158-508
ihl-cat LC50:25 mg/m³/10M NTIS** PB158-508
ihl-gpg LC50:35 mg/m³/10M NTIS** PB158-508

SAFETY PROFILE: Poison by inhalation. When heated to decomposition it emits toxic fumes of F^-.

FIO000 CAS:63765-78-6 **HR: 3**
2-FLUOROETHYL-5-FLUOROHEXOATE
mf: $C_8H_{14}F_2O_2$ mw: 180.22

TOXICITY DATA with REFERENCE
ihl-rat LD50:200 mg/m³ NATUAS 160,179,47
ims-rat LD50:1800 μg/kg NATUAS 160,179,47
ihl-mus LC50:150 μg/m³ NATUAS 160,179,47
scu-mus LD50:2500 μg/kg NATUAS 160,179,47
ihl-rbt LC50:20 μg/m³ NATUAS 160,179,47
ivn-rbt LD50:100 μg/kg NATUAS 160,179,47

SAFETY PROFILE: Poison by inhalation, intramuscular, subcutaneous, and intravenous routes. When heated to decomposition it emits very toxic F^-.

FIP999 CAS:4242-33-5 **HR: 3**
β-FLUOROETHYLIC ESTER of XENYLACETIC ACID
mf: $C_{16}H_{15}FO_2$ mw: 258.31

PROP: A brown solid, sol in organic solvents. Mp: 60.6°.

SYNS: 2-FLUOROETHYL ESTER DIPHENYLACETIC ACID ◇ LAMBROL ◇ M 2060

TOXICITY DATA with REFERENCE
skn-rat LD50:4 mg/kg BESAAT 15,102,69
orl-mus LD50:35 mg/kg BESAAT 15,102,69
orl-dog LD50:2 mg/kg BESAAT 15,102,69
orl-rbt LD50:600 μg/kg SPEADM 74-1,-,74
skn-rbt LD50:7 mg/kg BESAAT 15,102,69
orl-gpg LD50:700 μg/kg 85DPAN -,-,71/76

SAFETY PROFILE: Poison by ingestion and skin contact. An insecticide. When heated to decomposition it emits toxic fumes of F^-. See also FLUORIDES.

FIQ000 CAS:762-51-6 **HR: 3**
2-FLUOROETHYL IODIDE
mf: C_2H_4FI mw: 173.96

SYNS: 1-FLUOR-2-JODETHAN ◇ 1-FLUORO-2-IODOETHANE

TOXICITY DATA with REFERENCE
ipr-mus LD50:28 mg/kg JOCEAH 21,748,56
scu-mus LD50:28 mg/kg CLDND*

SAFETY PROFILE: Poison by intraperitoneal and subcutaneous routes. When heated to decomposition it emits very toxic fumes of F^- and I^-. See also IODIDES.

FIS000 CAS:63982-15-0 **HR: 3**
2-FLUOROETHYL-N-METHYL-N-NITROSOCARBAMATE
mf: $C_4H_7FN_2O_3$ mw: 150.13

SYN: TL 790

TOXICITY DATA with REFERENCE
ihl-mus LCLo:500 mg/m³/10M NDRC** No.9-4-1-9,43
ihl-gpg LCLo:100 mg/m³/10M NDRC** No.9-4-1-9,43

SAFETY PROFILE: Poison by inhalation. Many N-nitroso compounds are carcinogens. When heated to decomposition it emits very toxic fumes of F^- and NO_x. See also N-NITROSO COMPOUNDS and CARBAMATES.

FIT200 CAS:2508-18-1 **HR: 2**
7-FLUORO-2-N-(FLUORENYL)ACETHYDROXAMIC ACID
mf: $C_{15}H_{12}FNO_2$ mw: 257.28

SYNS: 7-FLUORO-2-FAA ◇ 7-FLUORO-N-(FLUOREN-2-YL)ACETOHYDROXAMIC ACID ◇ N-(7-FLUORO-2-FLUORENYL)ACETOHYDROXAMIC ACID ◇ 7-FLUORO-N-HYDROXY-N-2-ACETYLAMINOFLUORENE ◇ N-HYDROXY-7-FLUORO-2-ACETYLAMINOFLUORENE

TOXICITY DATA with REFERENCE
mmo-sat 15 nmol/plate MUREAV 67,85,79
mma-sat 20 nmol/plate MUREAV 67,85,79
orl-rat TDLo:675 mg/kg/10W-C:CAR CNREA8 26,2239,66

SAFETY PROFILE: Questionable carcinogen with experimental carcinogenic data. Mutation data reported. When heated to decomposition it emits toxic fumes of NO_x and F^-.

FIW000 CAS:31540-62-2 **HR: 3**
4'-FLUORO-4-(8-FLUORO-2,3,4,5-TETRAHYDRO-1H-PYRIDO(4,3-b)INDOL-2-YL)BUTYROPHENONE HYDROCHLORIDE
mf: $C_{21}H_{20}F_2N_2O \cdot ClH$ mw: 390.89

SYN: ABBOTT-30360

TOXICITY DATA with REFERENCE
ims-rat LD50:76 mg/kg AIPTAK 190,124,71
orl-mus LD50:335 mg/kg AIPTAK 190,124,71
ivn-mus LD50:47 mg/kg AIPTAK 190,124,71
ims-mus LD50:112 mg/kg AIPTAK 190,124,71

SAFETY PROFILE: Poison by intramuscular, ingestion, and intravenous routes. When heated to decomposition it emits very toxic fumes of HCl, NO_x, and F^-.

FIX000 CAS:661-11-0 ***HR: 3***
1-FLUOROHEPTANE
mf: $C_7H_{15}F$ mw: 118.22

TOXICITY DATA with REFERENCE
ipr-mus LD50:35 mg/kg JACSAT 79,2311,57

CONSENSUS REPORTS: Reported in EPA TSCA Inventory.

SAFETY PROFILE: Poison by intraperitoneal route. When heated to decomposition it emits very toxic fumes of F^-.

FIY000 CAS:334-44-1 ***HR: 3***
7-FLUOROHEPTANONITRILE
mf: $C_7H_{12}FN$ mw: 129.20

TOXICITY DATA with REFERENCE
ipr-mus LD50:2700 μg/kg JACSAT 78,3484,56
scu-mus LD50:2700 μg/kg CLDND*

CONSENSUS REPORTS: Cyanide and its compounds are on the Community Right-To-Know List.

SAFETY PROFILE: Poison by intraperitoneal and subcutaneous routes. When heated to decomposition it emits very toxic fumes of F^-, NO_x and CN^-. See also NITRILES.

FIZ000 CAS:353-21-9 ***HR: 3***
7-FLUOROHEPTYLAMINE
mf: $C_7H_{16}FN$ mw: 133.24

TOXICITY DATA with REFERENCE
ipr-mus LD50:47 mg/kg CJBPAZ 35,407,57
scu-mus LD50:50 mg/kg CLDND*

SAFETY PROFILE: Poison by intraperitoneal and subcutaneous routes. When heated to decomposition it emits very toxic fumes of NO_x and F^-. See also AMINES.

FJA000 CAS:373-14-8 ***HR: 3***
1-FLUOROHEXANE
mf: $C_6H_{13}F$ mw: 104.19

SYN: FLUOROHEXANE

TOXICITY DATA with REFERENCE
ipr-mus LD50:1700 μg/kg JACSAT 79,2311,57

CONSENSUS REPORTS: Reported in EPA TSCA Inventory.

SAFETY PROFILE: Poison by intraperitoneal route. When heated to decomposition it emits toxic fumes of F^-. See also FLUORIDES.

FJF100 CAS:357-09-5 ***HR: D***
FLUOROHYDROXYANDROSTENEDIONE
mf: $C_{19}H_{25}FO_3$ mw: 320.44

SYNS: FHA ◇ 9-α-FLUORO-11-β-HYDROXY-4-ANDROSTENE-3,17-DIONE ◇ U-5438

TOXICITY DATA with REFERENCE
orl-rat TDLo:600 mg/kg (9-14D preg):TER JRPFA4 39,319,74
orl-rat TDLo:100 mg/kg (3-7D preg):REP JRPFA4 39,319,74

SAFETY PROFILE: An experimental teratogen. Experimental reproductive effects. A steroid. When heated to decomposition it emits toxic fumes of F^-.

FJG000 CAS:100700-28-5 ***HR: 3***
4-FLUORO-2-HYDROXYTHIOBUTYRIC ACID-S-METHYL ESTER
mf: $C_5H_9FO_2S$ mw: 152.20

SYNS: BUTYRIC ACID-Γ-FLUORO-β-HDYROXY-THIOL-METHYL ESTER ◇ Γ-FLUORO-β-HYDROXY BUTYRIC ACID THIO METHYL ESTER ◇ TL 1427

TOXICITY DATA with REFERENCE
ihl-mus LCLo:30 mg/m³/10M NDRC** 30101,13,45
ihl-dog LCLo:63 mg/m³/10M NDRC** 30101,13,45
ihl-mky LC50:188 mg/m³/10M NDRC** 30101,13,45
ihl-cat LCLo:63 mg/m³/10M NDRC** 30101,13,45
ihl-rbt LCLo:63 mg/m³/10M NDRC** 30101,13,45

SAFETY PROFILE: Poison by inhalation. When heated to decomposition it emits toxic fumes of F^- and SO_x. See also ESTERS.

FJI000 ***HR: 3***
N-FLUOROIMINO DIFLUOROMETHANE
mf: CNF_3 mw: 83.01

PROP: Bp: −60°.

SAFETY PROFILE: Explodes on warming. When heated to decomposition it emits toxic fumes of F^- and NO_x.

FJI500 CAS:78343-32-5 ***HR: 3***
1-FLUOROIMINOHEXAFLUOROPROPANE
mf: C_3F_7N mw: 183.03

SAFETY PROFILE: Explosive reaction on contact with fluorine. When heated to decomposition it emits toxic fumes of F^- and NO_x. See also FLUORIDES.

FJI510 CAS:2802-70-2 ***HR: 3***
2-FLUOROIMINOHEXAFLUOROPROPANE
mf: C_3F_7N mw: 183.03

SAFETY PROFILE: Explosive reaction on contact with fluorine. When heated to decomposition it emits toxic fumes of F^- and NO_x. See also FLUORIDES.

FJK000 CAS:593-53-3 ***HR: 3***
FLUOROMETHANE
DOT: UN 2454
mf: CH_3F mw: 34.03

PROP: Colorless gas; agreeable, ether-like odor. D: (liquid) 0.8774 @ −78°, (gas) 1.1951 (air = 1), (gas) 1.0813 (oxygen = 1), mp: −141.8°, bp: −75.7° @ 872 mm, −78.2° @ 760 mm. Freely sol in alc and ether.

SYNS: FREON 41 ◇ METHYL FLUORIDE (DOT)

CONSENSUS REPORTS: Reported in EPA TSCA Inventory.

DOT Classification: Flammable Gas; Label: Flammable Gas.

SAFETY PROFILE: Narcotic in high concentrations. Acts as a simple asphyxiant. Burns with evolution of hydrogen fluoride. The flame is about as colorless as that of alcohol. When heated to decomposition it emits toxic fumes of F^-.

FJN000 CAS:1994-57-6 ***HR: 3***
2-FLUORO-7-METHYLBENZ(a)ANTHRACENE
mf: $C_{19}H_{13}F$ mw: 260.32

SYN: 7-METHYL-2-FLUOROBENZ(a)ANTHRACENE

TOXICITY DATA with REFERENCE
ims-rat TDLo:10 mg/kg:NEO NATUAS 273,566,78

SAFETY PROFILE: Questionable carcinogen with experimental neoplastigenic data. When heated to decomposition it emits toxic fumes of F^-.

FJO000 CAS:2606-87-3 ***HR: 3***
3-FLUORO-7-METHYLBENZ(a)ANTHRACENE
mf: $C_{19}H_{13}F$ mw: 260.32

SYN: 3′-FLUORO-10-METHYL-1,2-BENZANTHRACENE

TOXICITY DATA with REFERENCE
scu-rat TDLo:9 mg/kg:NEO CNREA8 23,229,63
skn-mus TDLo:120 mg/kg/20W-I:ETA CNREA8 23,229,63

SAFETY PROFILE: Questionable carcinogen with experimental neoplastigenic and tumorigenic data. When heated to decomposition it emits toxic fumes of F^-.

FJP000 CAS:2541-68-6 ***HR: 3***
6-FLUORO-7-METHYLBENZ(a)ANTHRACENE
mf: $C_{19}H_{13}F$ mw: 260.32

TOXICITY DATA with REFERENCE
scu-rat TDLo:9 mg/kg:NEO CNREA8 23,229,63

SAFETY PROFILE: Questionable carcinogen with experimental neoplastigenic data. When heated to decomposition it emits toxic fumes of F^-.

FJQ000 CAS:1881-75-0 ***HR: 3***
9-FLUORO-7-METHYLBENZ(a)ANTHRACENE
mf: $C_{19}H_{13}F$ mw: 260.32

SYN: 6-FLUORO-10-METHYL-1,2-BENZANTHRACENE

TOXICITY DATA with REFERENCE
ims-rat TDLo:10 mg/kg:NEO NATUAS 273,566,78

SAFETY PROFILE: Questionable carcinogen with experimental neoplastigenic data. When heated to decomposition it emits toxic fumes of F^-.

FJR000 CAS:1881-76-1 ***HR: 3***
7-FLUORO-10-METHYL-1,2-BENZANTHRACENE
mf: $C_{19}H_{13}F$ mw: 260.32

SYN: 10-FLUORO-7-METHYLBENZ(a)ANTHRACENE

TOXICITY DATA with REFERENCE
scu-rat TDLo:9 mg/kg:NEO CNREA8 23,229,63

SAFETY PROFILE: Questionable carcinogen with experimental neoplastigenic data. When heated to decomposition it emits toxic fumes of F^-.

FJT000 CAS:52831-60-4 ***HR: 3***
4-FLUORO-7-METHYL-6H-(1)BENZOTHIOPYRANO(4,3-b)QUINOLINE
mf: $C_{17}H_{12}FNS$ mw: 281.36

TOXICITY DATA with REFERENCE
mma-sat 100 μg/plate MUREAV 66,307,79
scu-mus TDLo:72 mg/kg/9W-I:NEO MUREAV 66,307,79

SAFETY PROFILE: Questionable carcinogen with experimental neoplastigenic data. Mutation data reported. When heated to decomposition it emits very toxic fumes of F^-, SO_x, and NO_x.

FJT100 CAS:91503-79-6 ***HR: 3***
2-FLUORO-α-METHYL-(1,1′-BIPHENYL)-4-ACETIC ACID 1-(ACETYLOXY)ETHYL ESTER
mf: $C_{19}H_{19}FO_4$ mw: 330.38

SYNS: 1-ACETOXYETHYL2-(2-FLUORO-4-BIPHENYLYL)PROPIONATE ◇ (1,1′-BIPHENYL)-4-ACETIC ACID, 2-FLUORO-α-METHYL-, 1-(ACETYLOXY)ETHYL ESTER ◇ 4-BIPHENYLACETIC ACID, 2-FLUORO-α-METHYL-, 1-ACETOXYETHYL ESTER ◇ FP-83 ◇ LFP 83

TOXICITY DATA with REFERENCE
ivn-rat TDLo:1 mg/kg:TER IYKEDH 20,42,89
orl-rat LD50:66 mg/kg KSRNAM 22,3949,88
scu-rat LD50:66 mg/kg KSRNAM 22,3949,88
ivn-rat LD50:88 mg/kg KSRNAM 22,3949,88
orl-mus LD50:69 mg/kg KSRNAM 22,3949,88
scu-mus LD50:183 mg/kg KSRNAM 22,3949,88
ivn-mus LD50:194 mg/kg KSRNAM 22,3949,88

SAFETY PROFILE: Poison by ingestion, subcutaneous, and intravenous routes. An experimental teratogen. When heated to decomposition it emits toxic fumes of F^-.

FJU000 CAS:73771-72-9 ***HR: 3***
2-FLUORO-3-METHYLCHOLANTHRENE
mf: $C_{21}H_{15}F$ mw: 286.36

TOXICITY DATA with REFERENCE
ims-rat TDLo:50 mg/kg:NEO NATUAS 273,566,78

SAFETY PROFILE: Questionable carcinogen with experimental neoplastigenic data. When heated to decomposition it emits toxic fumes of F^-.

FJV000 CAS:73771-73-0 ***HR: 3***
6-FLUORO-3-METHYLCHOLANTHRENE
mf: $C_{21}H_{15}F$ mw: 286.36

TOXICITY DATA with REFERENCE
ims-rat TDLo:50 mg/kg:NEO NATUAS 273,566,78

SAFETY PROFILE: Questionable carcinogen with experimental neoplastigenic data. When heated to decomposition it emits toxic fumes of F^-.

FJW000 CAS:73771-74-1 ***HR: 3***
9-FLUORO-3-METHYLCHOLANTHRENE
mf: $C_{21}H_{15}F$ mw: 286.36

TOXICITY DATA with REFERENCE
ims-rat TDLo:50 mg/kg:NEO NATUAS 273,566,78

SAFETY PROFILE: Questionable carcinogen with experimental neoplastigenic data. When heated to decomposition it emits toxic fumes of F^-.

FJY000 CAS:64977-44-2 ***HR: 3***
1-FLUORO-5-METHYLCHRYSENE
mf: $C_{19}H_{13}F$ mw: 260.32

TOXICITY DATA with REFERENCE
mma-sat 20 μg/plate JMCMAR 21,38,78
skn-mus TDLo:1200 μg/kg/20D-I:ETA CNREA8 38,1694,78

CONSENSUS REPORTS: EPA Genetic Toxicology Program.

SAFETY PROFILE: Questionable carcinogen with experimental tumorigenic data. Mutation data reported. When heated to decomposition it emits toxic fumes of F^-. See also CHRYSENE.

FKA000 CAS:64977-46-4 ***HR: 3***
6-FLUORO-5-METHYLCHRYSENE
mf: $C_{19}H_{13}F$ mw: 260.32

TOXICITY DATA with REFERENCE
mma-sat 20 μg/plate JMCMAR 21,38,78
skn-mus TDLo:1200 μg/kg/20D-I:ETA CNREA8 38,1694,78

CONSENSUS REPORTS: EPA Genetic Toxicology Program.

SAFETY PROFILE: Questionable carcinogen with experimental tumorigenic data. Mutation data reported. When heated to decomposition it emits toxic fumes of F^-. See also CHRYSENE.

FKB000 CAS:64977-47-5 ***HR: 3***
7-FLUORO-5-METHYLCHRYSENE
mf: $C_{19}H_{13}F$ mw: 260.32

TOXICITY DATA with REFERENCE
mma-sat 20 μg/plate JJIND8 63,855,79
skn-mus TDLo:1200 μg/kg/20W-I:ETA JJIND8 63,855,79

SAFETY PROFILE: Questionable carcinogen with experimental tumorigenic data. Mutation data reported. When heated to decomposition it emits toxic fumes of F^-. See also CHRYSENE.

FKC000 CAS:64977-48-6 ***HR: 3***
9-FLUORO-5-METHYLCHRYSENE
mf: $C_{19}H_{13}F$ mw: 260.32

TOXICITY DATA with REFERENCE
mma-sat 20 μg/plate JMCMAR 21,38,78
skn-mus TDLo:1200 μg/kg/20D-I:ETA CNREA8 38,1694,78

CONSENSUS REPORTS: EPA Genetic Toxicology Program.

SAFETY PROFILE: Questionable carcinogen with experimental tumorigenic data. Mutation data reported. When heated to decomposition it emits toxic fumes of F^-. See also CHRYSENE.

FKD000 CAS:64977-49-7 ***HR: 3***
11-FLUORO-5-METHYLCHRYSENE
mf: $C_{19}H_{13}F$ mw: 260.32

TOXICITY DATA with REFERENCE
mma-sat 20 μg/plate JMCMAR 21,38,78
skn-mus TDLo:1200 μg/kg/20D-I:ETA CNREA8 38,1694,78

CONSENSUS REPORTS: EPA Genetic Toxicology Program.

SAFETY PROFILE: Questionable carcinogen with experimental tumorigenic data. Mutation data reported. When heated to decomposition it emits toxic fumes of F^-. See also CHRYSENE.

FKE000 CAS:61413-38-5 ***HR: 3***
12-FLUORO-5-METHYLCHRYSENE
mf: $C_{19}H_{13}F$ mw: 260.32

TOXICITY DATA with REFERENCE
mma-sat 20 μg/plate JMCMAR 21,38,78
skn-mus TDLo:4 mg/kg/I:NEO CALEDQ 1,147,76
skn-mus TD:40 mg/kg/3W-I:ETA CCSUDL 1,325,76

CONSENSUS REPORTS: EPA Genetic Toxicology Program.

SAFETY PROFILE: Questionable carcinogen with experimental neoplastigenic and tumorigenic data. Mutation data reported. When heated to decomposition it emits toxic fumes of F^-. See also CHRYSENE.

FKF100 CAS:465-69-0 ***HR: D***
9-α-FLUORO-17-α-METHYL-17-HYDROXY-4-ANDROSTENE-3,11-DIONE
mf: $C_{20}H_{27}FO_3$ mw: 334.47

SYN: U-6596

TOXICITY DATA with REFERENCE
orl-rat TDLo:800 mg/kg (3-10D preg):REP JRPFA4 39,319,74

SAFETY PROFILE: Experimental reproductive effects. A steroid. When heated to decomposition it emits toxic fumes of F^-.

FKF800 CAS:937-25-7 ***HR: 2***
p-FLUORO-N-METHYL-N-NITROSOANILINE
mf: $C_7H_7FN_2O$ mw: 154.16

SYN: N-NITROSO-N-METHYL-4-FLUOROANILINE

TOXICITY DATA with REFERENCE
mma-sat 500 μg/plate MUREAV 89,255,81
orl-rat TDLo:1140 mg/kg/50W-I:ETA CRNGDP 4,157,83

SAFETY PROFILE: Questionable carcinogen with experimental tumorigenic data. Mutation data reported. When heated to decomposition it emits toxic fumes of NO_x and F^-.

FKG000 CAS:459-02-9 ***HR: 3***
4-FLUORO-α-METHYLPHENETHYLAMINE
mf: $C_9H_{12}FN$ mw: 153.22

TOXICITY DATA with REFERENCE
orl-mus LD50:25 mg/kg JACSAT 63,602,41
ipr-mus LD50:46 mg/kg ISYAM* -,21,70

SAFETY PROFILE: Poison by ingestion and intraperitoneal routes. When heated to decomposition it emits very toxic fumes of F^- and NO_x. See also AMINES.

FKI000 CAS:1622-79-3 ***HR: 3***
4'-FLUORO-4-(4-METHYLPIPERIDINO)BUTYROPHENONE HYDROCHLORIDE
mf: $C_{16}H_{22}FNO•ClH$ mw: 299.85

SYNS: BURONIL ◇ EUNERPAN ◇ FG 5111 ◇ METHYLPERONE HYDROCHLORIDE ◇ Γ-(4-METHYLPIPERIDINE)-p-FLUOROBUTYROPHENONE HYDROCHLORIDE

TOXICITY DATA with REFERENCE
orl-rbt TDLo:195 mg/kg (6-18D preg):TER FRPPAO 29,586,74
orl-rat LD50:330 mg/kg APTOA6 23,109,65
scu-rat LD50:220 mg/kg APTOA6 23,109,65
ivn-rat LD50:40 mg/kg APTOA6 23,109,65
orl-mus LD50:230 mg/kg APTOA6 23,109,65
scu-mus LD50:230 mg/kg 27ZQAG -,191,72
ivn-mus LD50:35 mg/kg APTOA6 23,109,65

SAFETY PROFILE: Poison by ingestion, subcutaneous, and intravenous routes. Experimental teratogenic effects. A neuroleptic drug used to treat anxiety and confusion. When heated to decomposition it emits very toxic fumes of F^-, NO_x, and HCl.

FKK000 CAS:321-38-0 ***HR: 3***
α-FLUORONAPHTHALENE
mf: $C_{10}H_7F$ mw: 146.17

PROP: Liquid. D: 1.33, mp: −8°, bp: 212-6°; sol in benzene, alc, acetyl chloroform.

TOXICITY DATA with REFERENCE
orl-rat LDLo:380 mg/kg JLCMAK 22,1133,37

CONSENSUS REPORTS: Reported in EPA TSCA Inventory.

SAFETY PROFILE: Poison by ingestion. When heated to decomposition it emits toxic fumes of F^-. See also FLUORIDES.

FKL000 CAS:350-46-9 ***HR: 3***
1-FLUORO-4-NITROBENZENE
mf: $C_6H_4FNO_2$ mw: 141.11

PROP: Pale yellow crystals. D: 1.320, mp: 27° (21.6°), bp: 206.7°.

SYNS: p-FLUORONITROBENZENE ◇ 4-FLUORONITROBENZENE ◇ p-NITROFLUOROBENZENE ◇ 4-NITROFLUOROBENZENE

TOXICITY DATA with REFERENCE
mmo-sat 160 nL/plate MUREAV 116,217,83
orl-rat LDLo:250 mg/kg NCNSA6 5,19,53

CONSENSUS REPORTS: Reported in EPA TSCA Inventory.

SAFETY PROFILE: Poison by ingestion. Mutation data reported. When heated to decomposition it emits very toxic fumes of F^- and NO_x. See also NITRO COMPOUNDS of AROMATIC HYDROCARBONS.

FKM000 CAS:14233-86-4 ***HR: 3***
N-FLUORO-N-NITROBUTYLAMINE
mf: $C_4H_9FN_2O_2$ mw: 136.13

SAFETY PROFILE: Potentially explosive at 60°C. When heated to decomposition it emits very toxic fumes of F^- and NO_x. See also FLUORIDES and AMINES.

FKO000 CAS:17576-63-5 ***HR: 3***
3-FLUORO-4-NITROQUINOLINE-1-OXIDE
mf: $C_9H_5FN_2O_3$ mw: 208.16

TOXICITY DATA with REFERENCE
cyt-omi 10 μmol/L GANNA2 60,155,69
cyt-rat:ast 100 nmol/L GMCRDC 17,31,75
dns-ham:oth 2 μmol/L NATUAS 229,416,71
scu-mus TDLo:347 mg/kg/I:ETA CPBTAL 17,544,69

CONSENSUS REPORTS: EPA Genetic Toxicology Program.

SAFETY PROFILE: Questionable carcinogen with experimental tumorigenic data. Mutation data reported. When heated to decomposition it emits very toxic fumes of F^- and NO_x.

FKP000 CAS:19789-69-6 ***HR: 3***
8-FLUORO-4-NITROQUINOLINE-1-OXIDE
mf: $C_9H_5FN_2O_3$ mw: 208.16

TOXICITY DATA with REFERENCE
scu-mus TDLo:60 mg/kg/I:ETA CPBTAL 17,544,69

SAFETY PROFILE: Questionable carcinogen with experimental tumorigenic data. When heated to decomposition it emits very toxic fumes of F^- and NO_x.

FKQ000 ***HR: 3***
FLUORONIUM PERCHLORATE
mf: $ClFH_2O_4$ mw: 120.5

$$H_2F^-CLO_4^-$$

SAFETY PROFILE: Complex explodes on contact with water. When heated to decomposition it emits very toxic fumes of F^- and Cl^-. See also FLUORIDES and PERCHLORATES.

FKS000 CAS:593-12-4 ***HR: 3***
8-FLUOROOCTYL BROMIDE
mf: $C_8H_{16}BrF$ mw: 211.15

SYN: 1-BROMO-8-FLUOROOCTANE

TOXICITY DATA with REFERENCE
ipr-mus LD50:20 mg/kg JOCEAH 21,748,56
scu-mus LD50:20 mg/kg CLDND*

SAFETY PROFILE: Poison by intraperitoneal and subcutaneous routes. When heated to decomposition it emits very toxic fumes of Br^- and F^-. See also FLUORIDES and BROMIDES.

FKT000 CAS:593-14-6 ***HR: 3***
8-FLUOROOCTYL CHLORIDE
mf: $C_8H_{16}ClF$ mw: 166.69

SYN: 1-CHLORO-8-FLUOROOCTANE

TOXICITY DATA with REFERENCE
ipr-mus LD50:2300 μg/kg JOCEAH 21,748,56
scu-mus LD50:2300 μg/kg CLDND*

SAFETY PROFILE: Poison by intraperitoneal and subcutaneous routes. When heated to decomposition it emits very toxic fumes of Cl^- and F^-. See also FLUORIDES and CHLORIDES.

FKV000 CAS:371-41-5 ***HR: 3***
4-FLUOROPHENOL
mf: C_6H_5FO mw: 112.11

TOXICITY DATA with REFERENCE
skn-mus TDLo:10 g/kg/25W-I:CAR CNREA8 19,413,59

CONSENSUS REPORTS: Reported in EPA TSCA Inventory.

SAFETY PROFILE: Questionable carcinogen with experimental carcinogenic data. When heated to decomposition it emits toxic fumes of F^-. See also PHENOL.

FKW000 CAS:1827-91-4 ***HR: 3***
FLUOROPHENOTHIAZINE DIHYDROCHLORIDE
mf: $C_{24}H_{30}F_3N_3O_2S•2ClH$ mw: 554.55

SYNS: NDR A-3682 ◇ SKF 7261

TOXICITY DATA with REFERENCE
orl-mus LD50:588 mg/kg 27ZQAG -,23,72
ipr-mus LD50:199 mg/kg 27ZQAG -,23,72
ivn-mus LD50:40 mg/kg 27ZQAG -,23,72

SAFETY PROFILE: Poison by intraperitoneal and intravenous routes. Moderately toxic by ingestion. When heated to decomposition it emits very toxic fumes of HCl, SO_x, and NO_x.

FKX000 CAS:725-04-2 ***HR: 3***
2'-FLUORO-4'-PHENYLACETANILIDE
mf: $C_{14}H_{12}FNO$ mw: 229.27

SYN: 3-FLUORO-4-ACETYLAMINOBIPHENYL

TOXICITY DATA with REFERENCE
orl-rat TDLo:3700 mg/kg/26W-C:CAR CNREA8 22,1002,62

SAFETY PROFILE: Questionable carcinogen with experimental carcinogenic data. When heated to decomposition it emits very toxic fumes of NO_x and F^-. See also FLUORIDES.

FKY000 CAS:725-06-4 ***HR: 3***
4'-(m-FLUOROPHENYL)ACETANILIDE
mf: $C_{14}H_{12}FNO$ mw: 229.27

SYN: 3'-FLUORO-4-ACETYLAMINOBIPHENYL

TOXICITY DATA with REFERENCE
orl-rat TDLo:2580 mg/kg/21W-C:CAR CNREA8 22,1002,62

SAFETY PROFILE: Questionable carcinogen with experimental carcinogenic data. When heated to decomposition it emits very toxic fumes of F^- and NO_x. See also FLUORIDES.

FKZ000 CAS:398-32-3 ***HR: 3***
4'-(p-FLUOROPHENYL)ACETANILIDE
mf: $C_{14}H_{12}FNO$ mw: 229.27

SYNS: 4'-FLUORO-4-ACETYLAMINOBIPHENYL ◇ N-4-(4'-FLUORO) BIPHENYLACETAMIDE ◇ N-(4'-FLUORO-4-BIPHENYLYL)ACETAMIDE

TOXICITY DATA with REFERENCE
orl-rat TDLo:4032 mg/kg/24W-C:ETA AJPAA4 100,317,80

SAFETY PROFILE: Questionable carcinogen with experimental tumorigenic data. When heated to decomposition it emits very toxic fumes of NO_x and F^-. See also FLUORIDES.

FLC000 CAS:459-22-3 ***HR: 3***
p-FLUOROPHENYLACETONITRILE
mf: C_8H_6FN mw: 135.15

SYNS: 4-FLUOROBENZENEACETONITRILE ◇ p-FLUOROBENZYL CYANIDE ◇ 4-FLUOROBENZYLCYANIDE ◇ 4-FLUOROPHENYLACETONITRILE

TOXICITY DATA with REFERENCE
ivn-mus LD50:42 mg/kg CSLNX* NX#07881

CONSENSUS REPORTS: Cyanide and its compounds are on the Community Right-To-Know List. Reported in EPA TSCA Inventory.

SAFETY PROFILE: Poison by intravenous route. When heated to decomposition it emits very toxic fumes of F^-, NO_x, and CN^-. See also FLUORIDES and NITRILES.

FLD000 CAS:456-88-2 ***HR: 3***
3-FLUOROPHENYLALANINE
mf: $C_9H_{10}FNO_2$ mw: 183.20

SYN: m-FLUOROPHENYLALANINE

TOXICITY DATA with REFERENCE
dnd-hmn:hla 200 μmol/L ECREAL 107,191,77
scu-rat LDLo:20 mg/kg JPETAB 73,176,41

SAFETY PROFILE: Poison by subcutaneous route. Human mutation data reported. When heated to decomposition it emits very toxic fumes of F^- and NO_x. See also FLUORIDES.

FLE000 CAS:325-69-9 ***HR: D***
3-(o-FLUOROPHENYL)ALANINE
mf: $C_9H_{10}FNO_2$ mw: 183.20

SYNS: o-FLUOROPHENYLALANINE ◇ 2-FLUOROPHENYLALANINE

TOXICITY DATA with REFERENCE
sln-asn 36 mg/L EVHPAZ 31,81,79
dnd-hmn:hla 200 μmol/L ECREAL 107,191,77

SAFETY PROFILE: Human mutation data reported. When heated to decomposition it emits very toxic fumes of F^- and NO_x. See also FLUORIDES.

FLF000 CAS:60-17-3 ***HR: D***
4-FLUOROPHENYLALANINE
mf: $C_9H_{10}FNO_2$ mw: 183.20

SYNS: p-FLUOROPHENYLALANINE ◇ p-FPHE ◇ PFPA

TOXICITY DATA with REFERENCE
sln-nsc 1 mg/L MUREAV 167,35,86
cyt-hmn:lym 50 μmol/L HEREAY 88,197,78
sln-mus-ipr 100 mg/kg MUREAV 164,117,86

CONSENSUS REPORTS: EPA Genetic Toxicology Program.

SAFETY PROFILE: Human mutation data reported. When heated to decomposition it emits very toxic fumes of F^- and NO_x. See also FLUORIDES.

FLG000 CAS:2924-67-6 ***HR: 3***
p-FLUOROPHENYL ETHYL SULFONE
mf: $C_8H_9FO_2S$ mw: 188.23

PROP: Crystals. Mp: 41°.

SYNS: BRIPADON ◇ CADUCID ◇ ETHYL-p-FLUOROPHENYL SULFONE ◇ 1-(ETHYLSULFONYL)-4-FLUOROBENZENE ◇ FLUORESONE

TOXICITY DATA with REFERENCE
orl-mus LD50:542 mg/kg NEPSBV 3,326,64
ipr-mus LD50:343 mg/kg NEPSBV 3,326,64
ivn-mus LD50:320 mg/kg NEPSBV 3,317,64
ivn-rbt LD50:110 mg/kg NEPSBV 3,317,64

SAFETY PROFILE: Poison by intraperitoneal and intravenous routes. Moderately toxic by ingestion. Used as an anti-epileptic agent and as a tranquilizer. When heated to decomposition it emits very toxic fumes of F^- and SO_x. See also FLUORIDES.

FLG100 CAS:5104-49-4 ***HR: 3***
3-FLUORO-4-PHENYLHYDRATROPIC ACID
mf: $C_{15}H_{13}FO_2$ mw: 244.28

SYNS: 4-BIPHENYLACETIC ACID, 2-FLUORO-α-METHYL- ◇ (1,1'-BIPHENYL)-4-ACETIC ACID, 2-FLUORO-α-METHYL- (9CI) ◇ BTS 18322 ◇ 2-(2-FLUORO-4-BIPHENYLYL)PROPIONIC ACID ◇ 2-FLUORO-α-METHYL-4-BIPHENYLACETIC ACID ◇ 2-FLUORO-α-METHYL(1,1'-BIPHENYL)-4-ACETIC ACID ◇ FLURBIPROFEN ◇ FP 70 ◇ FROBEN

TOXICITY DATA with REFERENCE
orl-rat TDLo:10 mg/kg (female 21D post):TER PEREBL 22,567,87
orl-man TDLo:30 mg/kg/2W-I BMJOAE 4,496,74
orl-rat LD50:88 mg/kg IYKEDH 10,232,79
ipr-rat LD50:81 mg/kg IYKEDH 10,232,79
scu-rat LD50:87 mg/kg IYKEDH 10,232,79
orl-mus LD50:457 mg/kg IYKEDH 10,232,79
ipr-mus LD50:890 mg/kg RPTOAN 49,98,86
orl-dog LD50:10 mg/kg IYKEDH 10,232,79

SAFETY PROFILE: Poison by ingestion, subcutaneous, and intraperitoneal routes. Human toxic effects by ingestion. An experimental teratogen. When heated to decomposition it emits toxic fumes of F^-.

FLH100 CAS:1493-23-8 ***HR: 3***
4-FLUOROPHENYLLITHIUM
mf: C_6H_4FLi mw: 102.04

SAFETY PROFILE: The pure material is explosive. When heated to decomposition it emits toxic fumes of F^-. See also LITHIUM COMPOUNDS.

FLJ000 CAS:2804-00-4 ***HR: 3***
8-(4-p-FLUORO PHENYL-4-OXOBUTYL)-2-METHYL-2,8-DIAZASPIRO(4.5)DECANE-1,3-DIONE
mf: $C_{19}H_{23}FN_2O_3$ mw: 346.44

SYNS: F-33 ◇ FR-33 ◇ R 7158

TOXICITY DATA with REFERENCE
ivn-rat LD50:92 mg/kg 27ZQAG -,189,72
orl-mus LD50:193 mg/kg 27ZQAG -,189,72
ivn-mus LD50:90 mg/kg 27ZQAG -,189,72
ivn-dog LD50:56 mg/kg 27ZQAG -,189,72

SAFETY PROFILE: Poison by ingestion and intravenous route. When heated to decomposition it emits very toxic fumes of F^- and NO_x.

FLK000 CAS:54063-38-6 ***HR: 2***
4-(4-(4-FLUOROPHENYL)-4-OXOBUTYL)-1-PIPERAZINECARBOXYLIC ACID CYCLOHEXYL ESTER
mf: $C_{21}H_{29}FN_2O_3$ mw: 376.52

SYNS: FENAPERONA (SPANISH) ◇ FENAPERONE ◇ 4-(3-(p-FLUOROBENZOYL)PROPYL)-1-PIPERAZINOCARBOXYLICACID CYCLOHEXYLESTER

TOXICITY DATA with REFERENCE
orl-rat LD50:1080 mg/kg DRFUD4 3,734,78
orl-mus LD50:872 mg/kg DRFUD4 3,734,78

SAFETY PROFILE: Moderately toxic by ingestion. When heated to decomposition it emits very toxic fumes of F^- and NO_x.

FLK100 CAS:2062-84-2 ***HR: 3***
1-(1-(4-(4-FLUOROPHENYL)-4-OXOBUTYL)-4-PIPERIDINYL)-1,3-DIHYDRO-2H-BENZIMIDAZOL-2-ONE
mf: $C_{22}H_{24}FN_3O_2$ mw: 381.49

PROP: Solid. Mp: 170-171.8°.

SYNS: ANQUIL ◇ BENPERIDOL ◇ BENZOPERIDOL ◇ BENZPERIDOL ◇ 8089 C.B. ◇ 1-(1-(3-(p-FLUOROBENZOYL)PROPYL)-4-PIPERIDYL)-2-BENZIMIDAZOLINONE ◇ FRENACTIL ◇ FRENACTYL ◇ GLIANIMON ◇ GLIANIMON MITE ◇ MCM-JR-4584 ◇ R 4584

TOXICITY DATA with REFERENCE
scu-rat LD50:220 mg/kg MDCHAG 4(2),199,67
ipr-mus LD50:250 mg/kg FRPSAX 35,605,80
ivn-mus LD50:20 mg/kg CSLNX* NX#11942

SAFETY PROFILE: Poison by subcutaneous, intravenous, and intraperitoneal routes. When heated to decomposition it emits toxic fumes of F^- and NO_x.

FLL000 CAS:2354-61-2 ***HR: 3***
4'-FLUORO-4-(1-(4-PHENYL)PIPERAZINO)BUTYROPHENONE
mf: $C_{20}H_{23}FN_2O$ mw: 326.45

SYNS: BPZ ◇ BUTROPIPAZON ◇ BUTROPIPAZONE ◇ 1-(4-FLUOROPHENYL)-4-(4-PHENYL-1-PIPERAZINYL)-1-BUTANONE ◇ GALVANISONE ◇ R 1892

TOXICITY DATA with REFERENCE
orl-rat LD50:550 mg/kg PHMGBN 15,485,77
scu-rat LD50:550 mg/kg MDCHAG 4(2),199,67
ivn-rat LD50:32 mg/kg 27ZQAG -,186,72
orl-mus LD50:650 mg/kg PHMGBN 15,485,77
ipr-mus LD50:76 mg/kg 27ZQAG -,186,72
ivn-mus LD50:24 mg/kg 27ZQAG -,186,72

SAFETY PROFILE: Poison by intravenous and intraperitoneal routes. Moderately toxic by ingestion, and subcutaneous routes. When heated to decomposition it emits very toxic fumes of F^- and NO_x.

FLN000 CAS:3781-28-0 ***HR: 3***
4'-FLUORO-4-(4-PIPERIDINO-4-PROPIONYLPIPERIDINO)BUTYROPHENONE
mf: $C_{23}H_{33}FN_2O_2$ mw: 388.58

SYN: FLOROPIPETON ◇ PROPYPERONE

TOXICITY DATA with REFERENCE
orl-rat LD50:780 mg/kg 27ZQAG -,192,72
scu-rat LD50:350 mg/kg 27ZQAG -,192,72
orl-mus LD50:275 mg/kg 27ZQAG -,192,72
scu-mus LD50:420 mg/kg 27ZQAG -,192,72

SAFETY PROFILE: Poison by ingestion and subcutaneous routes. When heated to decomposition it emits very toxic fumes of F^- and NO_x.

FLQ000 CAS:5675-31-0 ***HR: 3***
2-FLUORO-2-PROPEN-1-OL
mf: C_3H_5FO mw: 76.08

TOXICITY DATA with REFERENCE
orl-rat LD50:130 mg/kg AIHAAP 23,95,62
ihl-rat LCLo:1000 ppm/1H AIHAAP 23,95,62
skn-rbt LD50:3 mg/kg AIHAAP 23,95,62

SAFETY PROFILE: Poison by ingestion and skin contact. Mildly toxic by inhalation. When heated to decomposition it emits toxic fumes of F^-.

FLR000 CAS:461-56-3 ***HR: 3***
3-FLUOROPROPIONIC ACID
mf: $C_3H_5FO_2$ mw: 92.08

SYN: ω-FLUOROPROPIONIC ACID

TOXICITY DATA with REFERENCE
ipr-mus LD50:60 mg/kg JOCEAH 21,739,56

CONSENSUS REPORTS: Reported in EPA TSCA Inventory.

SAFETY PROFILE: Poison by intraperitoneal route. When heated to decomposition it emits toxic fumes of F^-.

FLT100 CAS:372-48-5 ***HR: 3***
2-FLUOROPYRIDINE
mf: C_5H_4FN mw: 97.09

CH=CHCH=CHCF=N (ring: first CH bonded to N)

SAFETY PROFILE: Reaction with bromine trifluoride forms a spontaneously ignitable product. When heated to decomposition it emits toxic fumes of F^- and NO_x.

FLU000 CAS:1649-18-9 ***HR: 3***
4'-FLUORO-4-(4-(2-PYRIDYL)-1-PIPERAZINYL)BUTYROPHENONE
mf: $C_{19}H_{22}FN_3O$ mw: 327.44

SYNS: AZAPERONE (USDA) ◇ AZEPERONE ◇ EUCALMYL ◇ FLUOPERIDOL ◇ 1-(3-(4-FLUOROBENZOYL)PROPYL)-4-(2-PYRIDYL)PIPERAZINE ◇ 1-(4-FLUOROPHENYL)-4-(4-(2-PYRIDINYL)-1-PIPERAZINYL)-1-BUTANONE ◇ R 1929 ◇ STRESNIL ◇ SUICALM

TOXICITY DATA with REFERENCE
orl-rat LD50:245 mg/kg ARZNAD 24,1798,74
scu-rat LD50:450 mg/kg ARZNAD 24,1798,74
ivn-rat LD50:25 mg/kg 27ZQAG -,185,72
orl-mus LD50:385 mg/kg ARZNAD 24,1798,74
ipr-mus LD50:63 mg/kg FRPSAX 35,605,80
scu-mus LD50:179 mg/kg 27ZQAG -,185,72
ivn-mus LD50:42 mg/kg ARZNAD 24,1798,74

SAFETY PROFILE: Poison by ingestion, intravenous, intraperitoneal, and subcutaneous routes. When heated to decomposition it emits very toxic fumes of F^- and NO_x.

FLV000 CAS:2266-22-0 ***HR: 3***
4'-FLUORO-4-(n-(4-PYRROLIDINAMIDO-4-m-TOLYPIPERIDINO)BUTYROPHENONE
mf: $C_{27}H_{33}FN_2O_2$ mw: 436.62

SYNS: MEPERIDIDE ◇ METHYLPERIDIDE

TOXICITY DATA with REFERENCE
scu-rat LD50:275 mg/kg 27ZQAG -,191,72
ivn-rat LD50:9300 μg/kg 27ZQAG -,191,72
scu-mus LD50:140 mg/kg 27ZQAG -,191,72
ivn-mus LD50:14 mg/kg 27ZQAG -,191,72

SAFETY PROFILE: Poison by subcutaneous and intravenous routes. When heated to decomposition it emits very toxic fumes of F^- and NO_x.

FLY000 CAS:10010-36-3 ***HR: 3***
4'-FLUORO-4-STILBENAMINE
mf: $C_{14}H_{12}FN$ mw: 213.27

TOXICITY DATA with REFERENCE
scu-rat TDLo:200 mg/kg/W-I:ETA BMBUAQ 14,141,58

SAFETY PROFILE: Questionable carcinogen with experimental tumorigenic data. When heated to decomposition it emits very toxic fumes of F^- and NO_x. See also FLUORIDES.

FLY100 ***HR: 3***
FLUOROSULFONATES

SAFETY PROFILE: Probably highly toxic, as they are salts of a strong acid. Probably strong irritants to the eyes, skin, and mucous membranes. They react with water or steam to produce toxic and corrosive fumes.

When heated to decomposition they emit toxic fumes of F^- and SO_x. See also specific compounds.

FLZ000 CAS:7789-21-1 ***HR: 3***
FLUOROSULFURIC ACID
DOT: UN 1777
mf: FHO_3S mw: 100.07

PROP: Colorless, fuming, highly corrosive liquid. Mp: −87.3°, bp: 163.5°, d: 1.743 @ 15°.

SYNS: FLUOROSUFONIC ACID (DOT) ◇ FLUOSULFONIC ACID (DOT)

CONSENSUS REPORTS: Reported in EPA TSCA Inventory.

OSHA PEL: TWA 2.5 mg(F)/m^3
NIOSH REL: (Inorganic Fluorides) TWA 2.5 mg(F)/m^3
DOT Classification: Corrosive Material; Label: Corrosive.

SAFETY PROFILE: Probably a poison by inhalation. A corrosive irritant to the skin, eyes, and mucous membranes. See also FLUORIDES, SULFURIC ACID, and FLUOROSULFONATES.

FMB000 CAS:17902-23-7 ***HR: 3***
5-FLUORO-1-(TETRAHYDROFURAN-2-YL)URACIL
mf: $C_8H_9FN_2O_3$ mw: 200.19

SYNS: CARZONAL ◇ CITOFUR ◇ COPAROGIN ◇ EXONAL ◇ FENTAL ◇ F-5-FU ◇ FLUOROFUR ◇ 5-FLUORO-1-(TETRAHYDRO-2-FURANYL)-2,4-PYRIMIDINEDIONE ◇ 5-FLUORO-1-(TETRAHYDRO-2-FURANYL)-2,4(1H,3H)-PYRIMIDINEDIONE ◇ 5-FLUORO-1-(TETRAHYDRO-3-FURYL)URACIL ◇ FRANROZE ◇ FTORAFUR ◇ FULAID ◇ FULFEEL ◇ FURAFLUOR ◇ FUROFUTRAN ◇ FUTRAFUL ◇ LAMAR ◇ LIFRIL ◇ MJF-12264 ◇ NEBERK ◇ NITOBANIL ◇ NSC-148958 ◇ PYRIMIDINE-DEOXYRIBOSE N1-2′-FURANIDYL-5-FLUOROURACIL ◇ RIOL ◇ SINOFLUROL ◇ SUNFRAL ◇ TEFSIEL C ◇ TEGAFUR ◇ 1-(TETRAHYDROFURAN-2-YL)-5-FLUOROURACIL ◇ N(^{1})-(2-TETRAHYDROFURYL)-5-FLUOROURACIL ◇ THFU

TOXICITY DATA with REFERENCE
slt-dmg-mul 10 mg/L TAKHAA 44,96,85
dns-hmn:fbr 1 g/L STBIBN 78,165,80
orl-rat TDLo:810 mg/kg (17-22D preg/21D post):REP OYYAA2 22,109,81
ivn-mus TDLo:210 mg/kg (female 6-12D post):TER OYYAA2 5,555,71
orl-man TDLo:11046 mg/kg/4.7Y-C:CAR,GIT ARSUAX 118,1454,83
ivn-hmn LDLo:640 mg/kg/8D CANCAR 36,103,75
ivn-hmn TDLo:23 mg/kg:CNS CANCAR 36,103,75
orl-rat LD50:930 mg/kg YACHDS 6,2911,78
ipr-rat LD50:700 mg/kg YACHDS 6,2911,78
scu-rat LD50:600 mg/kg YACHDS 6,2911,78
ivn-rat LD50:685 mg/kg GTKRDX 10,1987,83
orl-mus LD50:775 mg/kg YHTPAD 22,27,87
ipr-mus LD50:493 mg/kg NCISP* JAN86
scu-mus LD50:760 mg/kg OYYAA2 5,569,71
orl-dog LD50:34 mg/kg OYYAA2 20,1009,80

CONSENSUS REPORTS: Reported in EPA TSCA Inventory.

SAFETY PROFILE: Poison by ingestion. Moderately toxic to humans by intravenous route. Moderately toxic experimentally by intraperitoneal, intravenous, and subcutaneous routes. Experimental teratogenic data. Human systemic effects: nausea and vomiting. Experimental reproductive effects. Questionable human carcinogen producing gastrointestinal tumors. Human mutation data reported. Used as an anti-cancer agent. When heated to decomposition it emits very toxic fumes of F^- and NO_x.

FMC000 CAS:352-32-9 ***HR: 3***
p-FLUOROTOLUENE
DOT: UN 2388
mf: C_7H_7F mw: 110.14

PROP: Colorless liquid. D: 1.001, bp: 116−117°. Flash p: 50°F. Insol in water; sol in alc and ether.

TOXICITY DATA with REFERENCE
par-mus LDLo:500 mg/kg 11FYAN 3,91,63

CONSENSUS REPORTS: Reported in EPA TSCA Inventory.

DOT Classification: Flammable Liquid; Label: Flammable Liquid; Flammable or Combustible Liquid; Label: Flammable Liquid.

SAFETY PROFILE: Moderately toxic by parenteral route. A very dangerous fire hazard when exposed to heat or flame; can react vigorously with oxidizing materials. When heated to decomposition it emits toxic fumes of F^-. See also FLUORIDES.

FME000 CAS:1983-10-4 ***HR: 3***
FLUOROTRIBUTYLSTANNANE
mf: $C_{12}H_{27}FSn$ mw: 309.08

SYNS: TRIBUTYLSTANNANE FLUORIDE ◇ TRIBUTYLTIN FLUORIDE

TOXICITY DATA with REFERENCE
orl-mus LDLo:320 mg/kg AECTCV 14,111,85
orl-rbt LDLo:50 mg/kg SAIGBL 15,3,73

CONSENSUS REPORTS: Reported in EPA TSCA Inventory.

OSHA PEL: TWA 0.1 mg(Sn)/m^3 (skin)
ACGIH TLV: TWA 0.1 mg(Sn)/m^3 (skin) (Proposed: TWA 0.1 mg(Sn)/m^3; STEL 0.2 mg(Sn)/m^3 (skin))
NIOSH REL: (Organotin Compounds) TWA 0.1 mg(Sn)/m^3

SAFETY PROFILE: Poison by ingestion. Many tributyl tin compounds are highly toxic to marine life. When heated to decomposition it emits toxic fumes of F^-. See also TIN COMPOUNDS and FLUORIDES.

FMF000 CAS:313-95-1 ***HR: 3***
2-FLUOROTRICYCLOQUINAZOLINE
mf: $C_{21}H_{11}FN_4$ mw: 338.36

TOXICITY DATA with REFERENCE
skn-mus TDLo:48 mg/kg:ETA BECCAN 41,420,63

SAFETY PROFILE: Questionable carcinogen with experimental tumorigenic data. When heated to decomposition it emits very toxic fumes of NO_x and F^-. See also FLUORIDES.

FMG000 CAS:803-57-6 ***HR: 3***
3-FLUORO-TRICYCLOQUINAZOLINE
mf: $C_{21}H_{11}FN_4$ mw: 338.36

TOXICITY DATA with REFERENCE
skn-mus TDLo:920 mg/kg/1Y-I:NEO BJCAAI 16,275,62

SAFETY PROFILE: Questionable carcinogen with experimental neoplastigenic data. When heated to decomposition it emits very toxic fumes of F^- and NO_x. See also FLUORIDES.

FMH000 CAS:19982-87-7 ***HR: 3***
4-FLUORO-4'-TRIFLUOROMETHYLBENZOPHENONE GUANYLHYDRAZONE HYDROCHLORIDE
mf: $C_{15}H_{12}F_4N_4$•ClH mw: 360.77

SYNS: FTBG ◇ WR 09792

TOXICITY DATA with REFERENCE
orl-rat LD50:199 mg/kg TXAPA9 15,614,69
ipr-rat LD50:23 mg/kg TXAPA9 15,614,69
orl-mus LD50:114 mg/kg TXAPA9 15,614,69
ipr-mus LD50:12 mg/kg TXAPA9 15,614,69
orl-gpg LD50:132 mg/kg TXAPA9 15,614,69
ipr-gpg LD50:19 mg/kg TXAPA9 15,614,69

SAFETY PROFILE: Poison by ingestion and intraperitoneal routes. When heated to decomposition it emits very toxic fumes of HCl, F^- and NO_x.

FMI000 CAS:1840-42-2 ***HR: 3***
FLUOROTRINITROMETHANE
mf: CFN_3O_6 mw: 169.04

TOXICITY DATA with REFERENCE
ipr-mus LD50:57300 μg/kg KHFZAN 10(6),53,76

CONSENSUS REPORTS: Reported in EPA TSCA Inventory.

SAFETY PROFILE: Poison by intraperitoneal route. Mixtures with nitrobenzene are liquid explosives. When heated to decomposition it emits very toxic fumes of F^- and NO_x. See also FLUORIDES and NITRO COMPOUNDS.

FMJ000 CAS:139-26-4 ***HR: 3***
3-FLUOROTYROSIN
mf: $C_9H_{10}FNO_3$ mw: 199.20

SYNS: m-FLUOROTYROSINE ◇ 3-FLUOROTYROSINE ◇ FLUORTHYRIN ◇ 3-FLUORTYROSIN (GERMAN)

TOXICITY DATA with REFERENCE
orl-man LDLo:143 mg/kg AEPPAE 183,427,36
scu-rat LDLo:12 mg/kg JPETAB 73,176,41
orl-mus LDLo:14 mg/kg AEPPAE 183,427,36
skn-mus LDLo:300 mg/kg AEPPAE 183,427,36
scu-mus LDLo:11 mg/kg AEPPAE 183,427,36
scu-gpg LDLo:8 mg/kg AEPPAE 183,427,36

SAFETY PROFILE: Human poison by ingestion. Experimental poison by ingestion, skin contact, and subcutaneous routes. When heated to decomposition it emits very toxic fumes of F^- and NO_x. See also FLUORIDES.

FMM000 CAS:51-21-8 ***HR: 3***
FLUOROURACIL
mf: $C_4H_3FN_2O_2$ mw: 130.09

SYNS: ADRUCIL ◇ ARUMEL ◇ CARZONAL ◇ EFFLUDERM (free base) ◇ EFUDEX ◇ EFUDIX ◇ 5-FLURACIL (GERMAN) ◇ FLUOROBLASTIN ◇ FLUOROPLEX ◇ 5-FLUORPROPYRIMIDINE-2,4-DIONE ◇ 5-FLUORO-2,4-PYRIMIDINEDIONE ◇ 5-FLUORO-2,4(1H,3H)-PYRIMIDINEDIONE ◇ 5-FLUOROURACIL ◇ 5-FLUORURACIL (GERMAN) ◇ FLURACIL ◇ FLURI ◇ FLURIL ◇ 5-FU ◇ NSC-19893 ◇ RO 2-9757 ◇ TIMAZIN ◇ U-8953 ◇ ULUP

TOXICITY DATA with REFERENCE
skn-hmn 84 mg/3W CTRRDO 63,619,79
dnr-bcs 300 μg/plate TAKHAA 44,96,85
dns-hmn:oth 1 mmol/L CNREA8 44,3414,84
ivn-wmn TDLo:150 mg/kg (20-31W preg):REP JAMAAP 217,214,71
ivn-wmn TDLo:240 mg/kg (11-14W preg):TER AJOGAH 137,747,80
orl-hmn TDLo:450 mg/kg/30D:GIT,BLD CANCAR 39,1936,77
ivn-hmn TDLo:6 mg/kg/3D:CVS,PUL BMJOAE 1,547,78
ivn-wmn TDLo:150 mg/kg/17W-I:BLD BJHEAL 65,357,87
ivn-man TDLo:39 mg/kg/1D-I:CVS AHJOA2 114,433,87
ivn-wmn TDLo:27 mg/kg/4D-C:CVS BMJOAE 294,125,87
orl-rat LD50:230 mg/kg IYKEDH 4,90,73

ipr-rat LD50:70 mg/kg OYYAA2 5,569,71
scu-rat LD50:217 mg/kg IYKEDH 4,90,73
ivn-rat LD50:245 mg/kg OYYAA2 5,569,71
par-rat LD50:500 mg/kg RRCRBU 52,76,75
rec-rat LD50:884 mg/kg KSRNAM 12,1309,78
orl-mus LD50:115 mg/kg JMCMAR 21,738,78
ipr-mus LD50:100 mg/kg EKFMA7 7,100,77
scu-mus LD50:169 mg/kg IYKEDH 4,90,73
ivn-mus LD50:81 mg/kg IYKEDH 4,90,73
orl-dog LD50:30 mg/kg OYYAA2 16,303,78

CONSENSUS REPORTS: IARC Cancer Review: Group 3 IMEMDT 7,210,87; Human Inadequate Evidence IMEMDT 26,217,81; Animal Inadequate Evidence IMEMDT 26,217,81. Reported in EPA TSCA Inventory. EPA Genetic Toxicology Program. EPA Extremely Hazardous Substances List.

SAFETY PROFILE: Poison by ingestion, intraperitoneal, subcutaneous, and intravenous, routes. Moderately toxic by parenteral and rectal routes. Experimental teratogenic and reproductive effects. Human systemic effects: EKG changes, bone marrow changes, cardiac, pulmonary, and gastrointestinal effects. Human mutation data reported. A human skin irritant. Questionable carcinogen. When heated to decomposition it emits very toxic fumes of F^- and NO_x.

FMN000 CAS:316-46-1 ***HR: 3***
5-FLUOROURIDINE
mf: $C_9H_{11}FN_2O_6$ mw: 262.22

SYNS: FUR ◇ 5-FUR

TOXICITY DATA with REFERENCE
dns-hmn:hla 10 μmol/L BCPCA6 14,205,65
dni-mus-ipr 500 μmol/kg CNREA8 39,2406,79
ipr-rat LD50:400 mg/kg ADTEAS 3,181,68
ipr-mus LD50:160 mg/kg CNREA8 39,2406,79
scu-mus LD50:384 mg/kg CNCRA6 6,94,60

SAFETY PROFILE: Poison by intraperitoneal and subcutaneous routes. Human mutation data reported. When heated to decomposition it emits very toxic fumes of F^- and NO_x.

FMO000 CAS:353-13-9 ***HR: 3***
5-FLUOROVALERONITRILE
mf: C_5H_8FN mw: 101.14

TOXICITY DATA with REFERENCE
ipr-mus LD50:1 mg/kg JACSAT 78,3484,56
scu-mus LD50:1 mg/kg CLDND*

CONSENSUS REPORTS: Cyanide and its compounds are on the Community Right-To-Know List.

SAFETY PROFILE: Poison by intraperitoneal and subcutaneous routes. When heated to decomposition it emits very toxic fumes of F^- and NO_x, CN^-. See also FLUORIDES and NITRILES.

FMO050 ***HR: 1***
FLUOSOL-DA 20%

TOXICITY DATA with REFERENCE
ivn-rat TDLo:97500 mg/kg (male 13W pre):REP KSRNAM 16,3934,82
ivn-rat LD50:128 g/kg KSRNAM 16,3899,82
ivn-mus LD50:128 g/kg KSRNAM 16,3899,82

SAFETY PROFILE: Slightly toxic by intravenous route. Experimental reproductive effects. When heated to decomposition it emits acrid smoke and irritating fumes.

FMO100 CAS:671-35-2 ***HR: 3***
FLUOXYDINE
mf: $C_4H_3FN_2O$ mw: 114.09

SYNS: 5-FLUORO-4(1H)-PYRIMIDINONE ◇ FLUOXIDINE

TOXICITY DATA with REFERENCE
orl-mus LD50:118 mg/kg YHTPAD 15(6),7,80
ipr-mus LD50:72 mg/kg YHTPAD 15(6),7,80
ivn-mus LD50:1435 mg/kg YHTPAD 15(6),7,80

SAFETY PROFILE: Poison by ingestion and intraperitoneal routes. Moderately toxic by intravenous route. When heated to decomposition it emits toxic fumes of F^- and NO_x.

FMO129 CAS:2709-56-0 ***HR: 3***
cis-(Z)-FLUPENTHIXOL
mf: $C_{23}H_{25}F_3N_2OS$ mw: 434.56

SYNS: EMERGIL ◇ FLUANXOL ◇ FLUPENTHIXOL ◇ (α,β)-FLUPENTHIXOL ◇ FLUPENTHIXOLE ◇ FLUPENTIXOL ◇ FLURENTIXOL ◇ FLUXANXOL ◇ LC 44 ◇ N 7009 ◇ SIPLARIL ◇ SIPLAROL ◇ 2-TRIFLUOROMETHYL-9-(3-(4-(β-HYDROXYETHYL-1-PIPERAZINYL) PROPYLIDENE)THIOXANTHENE ◇ 4-(3-(2-(TRIFLUOROMETHYL) THIOXANTHEN-9-YLIDENE)PROPYL)-1-PIPERAZINEETHANOL ◇ 4-(3-(2-(TRIFLUOROMETHYL)-9H-THIOXANTHEN-9-YLIDENE)PROPYL)-1-PIPERAZINEETHANOL

TOXICITY DATA with REFERENCE
orl-chd TDLo:900 μg/kg:CNS DICPBB 15,388,81
orl-hmn TDLo:367 μg/kg/12D-I:CNS ARZNAD 22,93,72
orl-mus LD50:300 mg/kg JMCMAR 23,878,80
ipr-mus LD50:150 mg/kg ARZNAD 27,2121,77
ivn-mus LD50:87 mg/kg APTOA6 41,369,77

SAFETY PROFILE: Poison by ingestion, intraperitoneal, and intravenous routes. Human systemic effects by ingestion: central nervous system effects. When heated to decomposition it emits very toxic fumes of F^-, NO_x, and SO_x.

FMO150 CAS:2413-38-9 ***HR: 3***
FLUPENTIXOL HYDROCHLORIDE
mf: $C_{23}H_{25}F_3N_2OS{\cdot}2ClH$ mw: 507.48

SYNS: FLUPENTIXOL DIHYDROCHLORIDE ◇ FX 703 ◇ N 7009 ◇ 4-(3-(2-(TRIFLUOROMETHYL)THIOXANTHEN-9-YLIDENE)PROPYL)-1-PIPERAZINEETHANOL DIHYDROCHLORIDE ◇ 4-(3-(2-TRIFLUOROMETHYLTHIOXANTH-9-YLIDENE)PROPYL)-1-PIPERAZINEETHANOL DIHYDROCHLORIDE

TOXICITY DATA with REFERENCE
orl-rat LD50:791 mg/kg NIIRDN 6,704,82
scu-rat LD50:258 mg/kg IYKEDH 4,90,73
ivn-rat LD50:37 mg/kg IYKEDH 4,90,73
orl-mus LD50:423 mg/kg IYKEDH 4,90,73
scu-mus LD50:425 mg/kg IYKEDH 4,90,73
ivn-mus LD50:94 mg/kg APTOA6 19,87,62

SAFETY PROFILE: Poison by subcutaneous and intravenous routes. Moderately toxic by ingestion. When heated to decomposition it emits toxic fumes of F^-, SO_x, NO_x, and HCl.

FMP000 CAS:146-56-5 ***HR: 3***
FLUPHENAZINE DIHYDROCHLORIDE
mf: $C_{22}H_{26}F_3N_3OS{\cdot}2ClH$ mw: 510.49

SYNS: A 4077 ◇ FLUPHENAZINE HYDROCHLORIDE ◇ 10-(3-(4-(2-HYDROXYETHYL)PIPERAZINYL)PROPYL)-2-TRIFLUOROMETHYL-PHENOTHIAZINE DIHYDROCHLORIDE ◇ MODITEN ◇ PERMITIL HYDROCHLORIDE ◇ PROLIXIN ◇ SQUIBB 4918 ◇ 4-(3-(2-(TRIFLUOROMETHYL)PHENOTHIAZIN-10-YL)PROPYL)-1-PIPERAZINE ETHANOL, DIHYDROCHLORIDE

TOXICITY DATA with REFERENCE
mnt-mus-orl 55 mg/kg MUREAV 89,237,81
cyt-mus-ipr 21250 μg/kg IRLCDZ 9,701,81
orl-man TDLo:429 μg/kg (male 2D pre):REP AEMED3 14,600,85
orl-hmn TDLo:2800 μg/kg/2W-C:MSK BMJOAE 2,1071,66
ipr-rat LD50:100 mg/kg 27ZMA4 -,-,67
scu-rat LD50:640 mg/kg 27ZMA4 -,-,67
orl-mus LD50:220 mg/kg 27ZQAG -,24,72
ipr-mus LD50:89 mg/kg 27ZQAG -,24,72
ivn-mus LD50:56 mg/kg CSLNX* NX#00179
ipr-gpg LD50:299 mg/kg PHARAT 38,749,83

SAFETY PROFILE: Poison by ingestion, intraperitoneal, and intravenous routes. Moderately toxic by subcutaneous route. Human systemic effects by ingestion: musculo-skeletal changes. Mutation data reported. When heated to decomposition it emits very toxic fumes of HCl, SO_x, NO_x, and F^-.

FMP100 ***HR: 2***
FLUPIRTINE MALEATE
mf: $C_{15}H_{17}FN_4O_2{\cdot}C_4H_4O_4$ mw: 420.44

SYNS: (2-AMINO-6-(((4-FLUOROPHENYL)METHYL)AMINO)-3-PYRIDINYL)CARBAMIC ACID ETHYL ESTER MALEATE ◇ D 9998 MALEATE ◇ ETHYL-N-(2-AMINO-6-(4-FLUOROPHENYLMETHYLAMINO) PYRIDIN-3-YL)CARBAMATE MALEATE ◇ ETHYL-N-(2-AMINO-6-(4-FLUOR-PHENYLMETHYLAMINO)PYRIDIN-3-YL)CARBAMAT MALEAT (GERMAN) ◇ FLUPIRTIN-MALEAT (GERMAN)

TOXICITY DATA with REFERENCE
orl-rat LD50:1660 mg/kg ARZNAD 35,30,85
orl-mus LD50:603 mg/kg ARZNAD 35,30,85
scu-mus LD50:432 mg/kg ARZNAD 35,30,85

SAFETY PROFILE: Moderately toxic by ingestion and subcutaneous routes. When heated to decomposition it emits toxic fumes of F^- and NO_x. See also ESTERS and CARBAMATES.

FMQ000 CAS:17617-23-1 ***HR: 3***
FLURAZEPAM
mf: $C_{21}H_{23}ClFN_3O$ mw: 387.92

PROP: White rods from ether-petr ether. Mp: 77-82°.

SYNS: 7-CHLORO-1-(2-(DIETHYLAMINO)ETHYL)-5-(2-FLUOROPHENYL)-1H-1,4-BENZODIAZEPIN-2(3H)-ONE◇ FELMANE ◇ NOCTOSOM ◇ STAURODERM ◇ Ro-5-6901/3

TOXICITY DATA with REFERENCE
orl-mus TDLo:504 mg/kg (1-21D preg):REP PLRCAT 9,325,77
orl-rat LD50:796 mg/kg YKYUA6 29,1231,78
ipr-rat LD50:152 mg/kg IYKEDH 6,530,75
scu-rat LD50:600 mg/kg NIIRDN 6,706,82
ivn-rat LD50:38700 μg/kg IYKEDH 6,530,75
orl-mus LD50:500 mg/kg THERAP 26,439,71
ipr-mus LD50:186 mg/kg IYKEDH 6,530,75
scu-mus LD50:540 mg/kg NIIRDN 6,706,82
ivn-mus LD50:59100 μg/kg IYKEDH 6,530,75

SAFETY PROFILE: Poison by intraperitoneal and intravenous routes. Moderately toxic by ingestion and subcutaneous routes. Experimental reproductive effects. Caution: May be habit forming. This is a controlled substance (depressant) listed in the U.S. Code of Federal Regulations, Title 21 Part 1308.14. When heated to decomposition it emits very toxic fumes of Cl^-, F^- and NO_x. See also DIAZEPAM.

FMQ100 CAS:36105-20-1 ***HR: 3***
FLURAZEPAM MONOHYDROCHLORIDE
mf: $C_{21}H_{23}ClFN_3O{\cdot}ClH$ mw: 424.38

TOXICITY DATA with REFERENCE
orl-rat TDLo:330 mg/kg (30D pre):REP KSRNAM 7,2820,73
orl-rat LD50:978 mg/kg KSRNAM 7,2820,73
ipr-rat LD50:179 mg/kg KSRNAM 7,2820,73
scu-rat LD50:859 mg/kg KSRNAM 7,2820,73
ivn-rat LD50:40 mg/kg KSRNAM 7,2820,73
orl-mus LD50:684 mg/kg KSRNAM 7,2820,73

ipr-mus LD50:205 mg/kg OYYAA2 7,381,73
scu-mus LD50:598 mg/kg KSRNAM 7,2820,73
ivn-mus LD50:58 mg/kg OYYAA2 7,381,73
orl-rbt LD50:460 mg/kg KSRNAM 7,2820,73

SAFETY PROFILE: Poison by intravenous and intraperitoneal routes. Moderately toxic by ingestion and subcutaneous routes. Experimental reproductive effects. When heated to decomposition it emits toxic fumes of F^-, NO_x and HCl. See also DIAZEPAM.

FMR050 CAS:13311-84-7 ***HR: 2***
FLUTAMIDE
mf: $C_{11}H_{11}F_3N_2O_3$ mw: 276.24

PROP: Crystals from benzene. Mp: 111.5-112.5°.

SYNS: DROGENIL ◇ FLUGERIL ◇ FUGEREL ◇ 2-METHYL-N-(4-NITRO-3-(TRIFLUOROMETHYL)PHENYL)PROPANAMIDE(9CI) ◇ NIFTHOLIDE ◇ NIFTOLIDE ◇ 4'-NITRO-3'-TRIFLUOROMETHYLISOBUTYRANILIDE ◇ SCH ◇ SCH 13521 ◇ SEBATROL ◇ α,α,α-TRIFLUORO-2-METHYL-4'-NITRO-m-PROPIONOTOLUIDIDE

TOXICITY DATA with REFERENCE
ims-rat TDLo:65 mg/kg (female 10-22D post):REP ENDOAO 103,1702,78
scu-rat TDLo:525 mg/kg (female 14-20D post):TER BIREBV 20(Suppl 1),116A,79
orl-man TDLo:310 mg/kg/31D-I AJPSAO 143,1498,86
orl-rat LDLo:1000 mg/kg TXAPA9 28,303,74

CONSENSUS REPORTS: EPA Genetic Toxicology Program.

SAFETY PROFILE: Moderately toxic by ingestion. An experimental teratogen. Experimental reproductive effects. When heated to decomposition it emits toxic fumes of F^- and NO_x.

FMR075 CAS:27060-91-9 ***HR: 2***
FLUTAZOLAM
mf: $C_{19}H_{18}ClFN_2O_3$ mw: 376.84

PROP: White prisms from toluene. Mp: 142-147°. Freely sol in chloroform, ethanol; moderately sol in acetone, benzene, methanol, practically insol in water.

SYN: MS-4101

TOXICITY DATA with REFERENCE
orl-rat TDLo:270 mg/kg (female 17-22D post):REP OYYAA2 16,395,78
orl-rat TDLo:80 mg/kg (female 8-15D post):TER YACHDS 6,1692,78
ipr-rat LD50:4050 mg/kg YACHDS 6,1997,78
orl-mus LD50:1910 mg/kg YACHDS 6,2335,78
ipr-mus LD50:1680 mg/kg YACHDS 6,1997,78

SAFETY PROFILE: Moderately toxic by ingestion and intraperitoneal routes. An experimental teratogen. Experimental reproductive effects. When heated to decomposition it emits toxic fumes of F^-, Cl^-, and NO_x.

FMR100 CAS:25967-29-7 ***HR: 2***
FLUTOPRAZEPAM
mf: $C_{19}H_{16}ClFN_2O$ mw: 342.82

PROP: Crystals. Mp: 86-88°.

SYNS: 7-CHLORO-1-CYCLOPROPYLMETHYL-1,3-DIHYDRO-5-(2-FLUOROPHENYL)-2H-1,4-BENZODIAZEPIN-2-ONE◇ KB-509 ◇ RESTAS

TOXICITY DATA with REFERENCE
orl-rat TDLo:110 mg/kg (female 7-17D post):REP OYYAA2 21,13,81
orl-rat TDLo:11 mg/kg (female 7-17D post):TER OYYAA2 21,13,81
orl-rat LD50:10060 mg/kg OYYAA2 20,1055,80
ipr-rat LD50:2230 mg/kg OYYAA2 20,1055,80
orl-mus LD50:2430 mg/kg OYYAA2 20,1055,80
ipr-mus LD50:2110 mg/kg OYYAA2 20,1055,80
orl-rbt LD50:1000 mg/kg OYYAA2 20,1055,80

SAFETY PROFILE: Moderately toxic by ingestion and intraperitoneal routes. Experimental teratogenic and reproductive effects. When heated to decomposition it emits toxic fumes of F^-, Cl^-, and NO_x. See also DIAZEPAM.

FMR300 CAS:63516-07-4 ***HR: 3***
FLUTROPIUM BROMIDE HYDRATE
mf: $C_{24}H_{29}FNO_3 \cdot Br \cdot H_2O$ mw: 496.47

SYNS: 8-AZONIABICYCLO(3.2.1)OCTANE,8-(2-FLUOROETHYL)-3-((HYDROXYDIPHENYLACETYL)OXY)-8-METHYL-,BROMIDE, (endo,syn)-, MONOHYDRATE ◇ Ba 598 BROMIDE HYDRATE ◇ (8R)-8-(2-FLUOROETHYL)-3-α-HYDROXY-1-α-H,5-α-H-TROPANIUM BROMIDE BENZILATE H2O

TOXICITY DATA with REFERENCE
orl-rat TDLo:13500 mg/kg (female 17-22D post):REP KSRNAM 20,8174,86
orl-rat TDLo:5500 mg/kg (female 7-17D post):TER KSRNAM 20,8143,86
orl-rat LD50:2900 mg/kg KSRNAM 20,8123,86
ipr-rat LD50:77 mg/kg KSRNAM 20,8123,86
scu-rat LD50:615 mg/kg KSRNAM 20,8123,86
ivn-rat LD50:12500 μg/kg KSRNAM 20,8123,86
orl-mus LD50:930 mg/kg KSRNAM 20,8123,86
ipr-mus LD50:53 mg/kg KSRNAM 20,8123,86
scu-mus LD50:228 mg/kg KSRNAM 20,8123,86
ivn-mus LD50:12500 μg/kg KSRNAM 20,8123,86

SAFETY PROFILE: Poison by intravenous and intraperitoneal routes. Moderately toxic by ingestion and subcutaneous routes. An experimental teratogen. Experimental reproductive effects. When heated to decomposition it emits toxic fumes of NO_x, Br^-, and Cl^-.

FMR500 CAS:1391-62-4 ***HR: 2***
FLUVOMYCIN

SYNS: EFSIOMYCIN ◇ RIOMYCIN ◇ VIVICIL

TOXICITY DATA with REFERENCE
scu-mus LD50:1250 mg/kg ANTCAO 3,765,53
ivn-mus LD50:1300 mg/kg ANTCAO 3,765,53
ims-mus LD50:750 mg/kg ANTCAO 3,765,53

SAFETY PROFILE: Moderately toxic by subcutaneous, intravenous, and intramuscular routes.

FMR700 CAS:76050-49-2 ***HR: 2***
FOGARD

SYN: FOGARD S

TOXICITY DATA with REFERENCE
ipr-mus TDLo:2600 μg/kg/39D-I:ETA PATHAB 73,707,81

SAFETY PROFILE: Questionable carcinogen with experimental tumorigenic data. When heated to decomposition it emits acrid smoke and irritating fumes.

FMS000 CAS:8031-00-3 ***HR: 1***
FOIN ABSOLUTE

PROP: Found in several common fodder grasses, contains coumarin (FCTXAV 14,659,76).

SYNS: FOIN COUPE ◇ HAY ABSOLUTE

TOXICITY DATA with REFERENCE
skn-rbt 500 mg/24H MLD FCTXAV 14,777,76

CONSENSUS REPORTS: Reported in EPA TSCA Inventory.

SAFETY PROFILE: A skin irritant. When heated to decomposition it emits acrid smoke and irritating fumes. See also COUMARIN.

FMS875 CAS:370-14-9 ***HR: 3***
FOLEDRIN
mf: $C_{10}H_{15}NO$ mw: 165.26

PROP: Crystals from methanol; acrid, burning taste. Mp: 162-163°. Sltly sol in water. Sol in alc, ether; readily sol in dil acids.

SYNS: EPIFEN ◇ p-HYDROXY-N,α-DIMETHYLPHENETHYLAMINE ◇ p-HYDROXYMETHAMPHETAMINE ◇ p-HYDROXY-N-METHYLBENZEDRINE ◇ β-(p-HYDROXYPHENYL)ISOPROPYLMETHYLAMINE ◇ α-(p-HYDROXYPHENYL)-β-METHYLAMINOPROPANE ◇ 1-(p-HYDROXYPHENYL)-2-METHYLAMINOPROPANE ◇ ISODRINE ◇ ISODRINUM ◇ KNOLL H75 ◇ p-(2-METHYLAMINOPROPYL)PHENOL ◇ N-METHYLPAREDRINE ◇ 1-(p-OXYPHENYL)-2-METHYLAMINOPROPAN (GERMAN) ◇ α-(p-OXYPHENYL)-β-METHYL AMINOPROPANE ◇ PAREDRINOL ◇ PHOLEDRINE ◇ PHOLETONE ◇ PRESSITAN ◇ PROMETHIN ◇ PROMETIN ◇ PULSOTYL ◇ STIMATONE ◇ SYMPROPAMIN ◇ SYNCORDAN ◇ TERAPINYL ◇ VARITOL ◇ VERITAIN ◇ VERITOL

TOXICITY DATA with REFERENCE
ipr-rat LD50:100 mg/kg AEPPAE 195,647,40
scu-rat LD50:400 mg/kg AIPTAK 159,442,66
ipr-mus LDLo:100 mg/kg AEPPAE 195,647,40

SAFETY PROFILE: Poison by subcutaneous and intraperitoneal routes. When heated to decomposition it emits toxic fumes of NO_x. See also AMINES.

FMT000 CAS:59-30-3 ***HR: 3***
FOLIC ACID
mf: $C_{19}H_{19}N_7O_6$ mw: 441.45

PROP: A member of the vitamin B complex. Orange-yellow needles or platelets; odorless. Sol in dilute alkali hydroxide and carbonate solns; sltly sol in water; insol in lipid solvents, acetone, alc, chloroform, ether.

SYNS: l-N-(p-(((-2-AMINO-4-HYDROXY-6-PTERIDINYL)METHYL) AMINO)BENZOYL)GLUTAMIC ACID ◇ FOLACIN ◇ FOLATE ◇ FOLCYSTEINE ◇ NSC 3073 ◇ PTEGLU ◇ PTEROYLGLUTAMIC ACID ◇ PTEROYL-l-GLUTAMIC ACID ◇ PTEROYLMONOGLUTAMIC ACID ◇ PTEROYL-l-MONOGLUTAMIC ACID ◇ USAF CB-13 ◇ VITAMIN Bc ◇ VITAMIN M

TOXICITY DATA with REFERENCE
dns-rat-ipr 150 mg/kg CNREA8 29,136,69
dns-mus-ipr 250 mg/kg CNREA8 29,136,69
par-rat TDLo:150 mg/kg (10D preg):TER FEPRA7 23,292,64
orl-mus LD50:10 g/kg TXAPA9 23,537,72
ipr-mus LD50:85 mg/kg EXPEAM 41,72,85
scu-mus LDLo:200 mg/kg PSEBAA 71,544,49
ivn-mus LD50:282 mg/kg TXAPA9 23,537,72

CONSENSUS REPORTS: Reported in EPA TSCA Inventory.

SAFETY PROFILE: Poison by intraperitoneal and intravenous routes. Experimental teratogenic effects. Mutation data reported. When heated to decomposition it emits toxic fumes of NO_x.

FMT100 CAS:9002-68-0 ***HR: D***
FOLLICLE-STIMULATING HORMONE
mw: 36,000

PROP: Solid. Soluble in water, 50% alc.

SYNS: ANTHROGON ◇ FERTINORM ◇ FOLLITROPIN ◇ FSH ◇ FSH-P ◇ HEBIN ◇ LUTEOANTINE ◇ MENOTROPHIN ◇ MENOTROPINS ◇ PROLAN B ◇ THYLAKENTRIN ◇ URINARY HEBIN

TOXICITY DATA with REFERENCE
scu-rat TDLo:25 g/kg (female 1-10D post):REP JPETAB 49,146,33

SAFETY PROFILE: Experimental reproductive effects. A steroid. When heated to decomposition it emits acrid smoke and fumes.

FMU000 CAS:24600-36-0 ***HR: 3***
FOMINOBEN HYDROCHLORIDE
mf: $C_{21}H_{24}ClN_3O_3•ClH$ mw: 438.39

SYNS: 3′-CHLORO-α-(METHYL((MORPHOLINOCARBONYL) METHYL)AMINO)-o-BENZOTOLUIDIDE HYDROCHLORIDE ◇ 3′-CHLORO-α-(N-METHYL-N-((MORPHOLINOCARBONYL)METHYL) AMINOMETHYL)BENZANILIDE HYDROCHLORIDE ◇ 3′-CHLORO-2′-(N-METHYL-N-((MORPHOLINOCARBONYL)METHYL)AMINO-METHYL)BENZANILIDE HYDROCHLORIDE ◇ NOLEPTAN ◇ OLEPTAN ◇ PB 89 CHLORIDE ◇ PB-89 HYDROCHLORIDE

TOXICITY DATA with REFERENCE
orl-rat TDLo:16500 mg/kg (7-17D preg):REP IYKEDH 9,724,78
orl-rat TDLo:16500 mg/kg (7-17D preg):TER IYKEDH 9,724,78
ipr-rat LD50:1000 mg/kg IYKEDH 14,297,83
scu-rat LD50:11 g/kg IYKEDH 14,297,83
ivn-rat LD50:57 mg/kg IYKEDH 14,297,83
orl-mus LD50:2200 mg/kg ARZNAD 23,296,73
ipr-mus LD50:630 mg/kg ARZNAD 23,296,73
scu-mus LD50:1770 mg/kg OYYAA2 8,1491,74
ivn-mus LD50:130 mg/kg IYKEDH 14,297,83
orl-dog LD50:3300 mg/kg IYKEDH 14,297,83

SAFETY PROFILE: Poison by intravenous route. Moderately toxic by ingestion, subcutaneous, and intraperitoneal routes. Experimental teratogenic and reproductive effects. An antitussive. When heated to decomposition it emits very toxic fumes of HCl and NO_x.

FMU039 CAS:13115-40-7 ***HR: 3***
FONAZINE MESYLATE
mf: $C_{19}H_{23}N_3O_2S_2•CH_4O_3S$ mw: 487.70

SYNS: DIMETHIOTAZINE ◇ DIMETHOTHIAZINE MESYLATE ◇ DIMETHOTHIAZINE METHANESULFONATE ◇ IL-6302 MESYLATE ◇ MIGRISTENE ◇ PROMAQUID ◇ RP 8599 ◇ RP 8599 MESYLATE

TOXICITY DATA with REFERENCE
orl-rbt TDLo:135 mg/kg (8-16D preg):REP OYYAA2 4,373,70
orl-rbt TDLo:540 mg/kg (8-16D preg):TER OYYAA2 4,373,70
orl-mus LD50:740 mg/kg YAKUD5 22,1953,80
ipr-mus LD50:190 mg/kg YAKUD5 22,1953,80
scu-mus LD50:473 mg/kg YAKUD5 22,1953,80
ivn-mus LD50:61500 μg/kg KSRNAM 5,325,71

SAFETY PROFILE: Poison by intravenous route. Moderately toxic by ingestion. An experimental teratogen. Experimental reproductive effects. When heated to decomposition it emits toxic fumes of SO_x and NO_x. See also SULFONATES.

FMU045 CAS:944-22-9 ***HR: 3***
FONOFOS
mf: $C_{10}H_{15}OPS_2$ mw: 246.34

SYNS: O-AETHYL-S-PHENYL-AETHYL-DITHIOPHOSPHONAT (GERMAN) ◇ DIFONATE ◇ DYFONATE ◇ DYPHONATE ◇ ENT 25,796 ◇ O-ETHYL-S-PHENYL ETHYLDITHIOPHOSPHONATE ◇ O-ETHYL-S-PHENYL ETHYLPHOSPHONODITHIOATE ◇ N 2790 ◇ STAUFFER N 2790

TOXICITY DATA with REFERENCE
orl-rat LD50:3 mg/kg WRPCA2 9,119,70
skn-rat LD50:147 mg/kg WRPCA2 9,119,70
skn-rbt LD50:25 mg/kg FMCHA2 -,D121,80
skn-gpg LD50:278 mg/kg 28ZEAL 5,118,76
orl-pgn LD50:13300 μg/kg ASTTA8 (680),157,79
orl-qal LD50:12 mg/kg EESADV 8,551,84

CONSENSUS REPORTS: EPA Genetic Toxicology Program. EPA Extremely Hazardous Substances List.

OSHA PEL: TWA 0.1 mg/m³ (skin)
ACGIH TLV: TWA 0.1 mg/m³ (skin)

SAFETY PROFILE: Poison by ingestion and skin contact. An insecticide. When heated to decomposition it emits very toxic fumes of PO_x and SO_x.

FMU059 CAS:2650-18-2 ***HR: 3***
FOOD BLUE 1
mf: $C_{37}H_{36}N_2O_9S_3•2H_3N$ mw: 783.01

SYNS: ACID BLUE 9 ◇ ACILAN TURQUOISE BLUE AE ◇ A.F. BLUE No. 1 ◇ AIZEN BRILLIANT BLUE FCF ◇ ALPHAZURINE ◇ AMACID BLUE FG CONC ◇ BLEU BRILLIANT FCF ◇ 11388 BLUE ◇ BRILLIANT BLUE ◇ BUCACID AZURE BLUE ◇ CALCOCID BLUE EG ◇ C.I. 671 ◇ C.I. 42090 ◇ C.I. ACID BLUE 9, DIAMMONIUM SALT ◇ C.I. DIRECT BROWN 78, DIAMMONIUM SALT ◇ C.I. FOOD BLUE 2 ◇ D&C BLUE No. 4 ◇ DISULPHINE LAKE BLUE EG ◇ EDICOL SUPRA BLUE E6 ◇ ERIOGLAUCINE ◇ ERIOSKY BLUE ◇ FENAZO BLUE XR ◇ HIDACID AZURE BLUE ◇ H.K. FORMULA No. K. 7117 ◇ KITON PURE BLUE L ◇ MAPLE BRILLIANT BLUE FCF ◇ NEPTUNE BLUE BRA CONCENTRATION ◇ PATENT BLUE AE ◇ PEACOCK BLUE X-1756 ◇ SCHULTZ No. 770 ◇ TRIANTINE LIGHT BROWN 3RN ◇ XYLENE BLUE VSG

TOXICITY DATA with REFERENCE
dns-rat:lvr 200 μmol/L ENMUDM 7,101,85
dns-rat-orl 500 mg/kg ENMUDM 7,101,85
cyt-ham:fbr 5 g/L FCTOD7 22,623,84
scu-rat TDLo:10 g/kg/77W-I:NEO TXAPA9 8,29,66
par-rat TDLo:4580 mg/kg/62W-I:ETA FAONAU 38B,22,66
ivn-hmn LDLo:33 μg/kg:CNS,PUL 34ZIAG -,87,69
scu-mus LD50:4600 mg/kg ZEKBAI 64,287,61

CONSENSUS REPORTS: Community Right-To-Know List. Reported in EPA TSCA Inventory. EPA Genetic Toxicology Program.

SAFETY PROFILE: Human poison by intravenous route. Human systemic effects by intravenous route: muscle contractions or spasticity and dyspnea. Questionable carcinogen with experimental neoplastigenic and tumorigenic data. Mutation data reported. When heated to decomposition it emits very toxic fumes of NH_3, NO_x, and SO_x.

FMU070 CAS:3761-53-3 ***HR: 2***
FOOD RED No. 101
mf: $C_{18}H_{14}N_2O_7S_2{\cdot}2Na$ mw: 480.44

SYNS: ACETACID RED J ◇ ACIDAL PONCEAU G ◇ ACID LEATHER RED KPR ◇ ACID PONCEAU R ◇ ACID RED 26 ◇ ACID SCARLET ◇ ACILAN PONCEAU RRL ◇ AHCOCID FAST SCARLET R ◇ AIZEN PONCEAU RH ◇ AMACID LAKE SCARLET 2R ◇ BRILLIANT PONCEAU G ◇ CALCOCID 2RIL ◇ CALCOLAKE SCARLET 2R ◇ CERTICOL PONCEAU MXS ◇ CERVEN KYSELA 26 ◇ C.I. 79 ◇ C.I. 16150 ◇ C.I. ACID RED 26 ◇ C.I. ACID RED 26, DISODIUM SALT ◇ C.I. FOOD RED 5 ◇ COLACID PONCEAU SPECIAL ◇ D&C RED No. 5 ◇ 4-((2,4-DIMETHYLPHENYL)AZO)-3-HYDROXY-2,7-NAPHTHALENE-DISULFONIC ACID, DISODIUM SALT ◇ 4-((2,4-DIMETHYLPHENYL) AZO)-3-HYDROXY-2,7-NAPHTHALENEDISULPHONIC ACID, DISODIUM SALT ◇ DISODIUM (2,4-DIMETHYLPHENYLAZO)-2-HYDROXY-NAPHTHALENE-3,6-DISULFONATE ◇ DISODIUM (2,4-DIMETHYL-PHENYLAZO)-2-HYDROXYNAPHTHALENE-3,6-DISULPHONATE ◇ DISODIUM SALT of 1-(2,4-XYLYLAZO)-2-NAPHTHOL-3,6-DISULFONIC ACID ◇ DISODIUM SALT of 1-(2,4-XYLYLAZO)-2-NAPHTHOL-3,6-DISULPHONIC ACID ◇ EDICOL PONCEAU RS ◇ FENAZO SCARLET 2R ◇ HEXACOL PONCEAU MX ◇ HIDACID SCARLET 2R ◇ 3-HYDROXY-4-(2,4-XYLYLAZO)-3,7-NAPHTHALENEDISULFONIC ACID, DISODIUM SALT ◇ 3-HYDROXY-4-(2,4-XYLYLAZO)-3,7-NAPHTHALENEDISULPHONIC ACID, DISODIUM SALT ◇ JAVA PONCEAU 2R ◇ KITON PONCEAU R ◇ LAKE PONCEAU ◇ NAPHTHALENE LAKE SCARLET R ◇ NEKLACID RED RR ◇ NEW PONCEAU 4R ◇ PAPER RED HRR ◇ PIGMENT PONCEAU R ◇ PONCEAU BNA ◇ PONCEAU R (BIOLOGICAL STAIN) ◇ PONCEAU XYLIDINE (BIOLOGICAL STAIN) ◇ 1695 RED ◇ RED R ◇ SCARLET R ◇ SCHULTZ No. 95 ◇ TERTRACID PONCEAU 2R ◇ XYLIDINE PONCEAU ◇ 1-XYLYLAZO-2-NAPHTHOL-3,6-DISULFONIC ACID, DISODIUM SALT ◇ 1-(2,4-XYLYLAZO)-2-NAPHTHOL-3,6-DISULPHONIC ACID, DISODIUM SALT ◇ 1-XYLYLAZO-2-NAPHTHOL-3,6-DISULPHONIC ACID, DISODIUM SALT

TOXICITY DATA with REFERENCE
mma-sat 100 μg/plate CANCAR 49,1970,82
oth-esc 300 μmol/L SKEZAP 12,298,71
orl-mus TDLo:136 g/kg/81W-C:CAR FCTXAV 6,591,68
orl-mus TD:35 g/kg/52W:ETA BJCAAI 10,653,56
ipr-mus LD50:2000 mg/kg FCTXAV 4,375,66

CONSENSUS REPORTS: IARC Cancer Review: Group 3 IMEMDT 7,56,87; Animal Sufficient Evidence IMEMDT 8,189,75. Reported in EPA TSCA Inventory.

SAFETY PROFILE: Moderately toxic by intraperitoneal route. Questionable carcinogen with experimental carcinogenic and tumorigenic data. Mutation data reported. When heated to decomposition it emits toxic fumes of NO_x and SO_x.

FMU080 CAS:2611-82-7 ***HR: 2***
FOOD RED No. 102
mf: $C_{20}H_{14}N_2O_{10}S_3{\cdot}3Na$ mw: 607.51

SYNS: ACIDAL BRIGHT PONCEAU 3R ◇ ACID BRILLIANT SCARLET 3R ◇ ACID PONCEAU 4R ◇ ACID RED 18 ◇ ACID SCARLET 3R ◇ ACILAN SCARLET V3R ◇ AIZEN BRILLIANT SCARLET 3RH ◇ ATUL ACID SCARLET 3R ◇ BRILLIANT PONCEAU 3R ◇ BRILLIANT SCARLET ◇ BUCACID BRILLIANT SCARLET 3R ◇ CALCOCID BRILLIANT SCARLET 3RN ◇ CERTICOL PONCEAU 4RS ◇ CERVEN KOSENILOVA A ◇ CILEFA PONCEAU 4R ◇ COCCINE ◇ COCHENILLEROT A ◇ COCHINEAL RED A ◇ COLACID PONCEAU 4R ◇ C.I. 185 ◇ C.I. 16255 ◇ C.I. ACID RED 18 ◇ C.I. FOOD RED 7 ◇ CRIMSON SX ◇ CUROL BRIGHT RED 4R ◇ DAISHIKI BRILLIANT SCARLET 3R ◇ EDICOL SUPRA PONCEAU 4R ◇ EUROCERT COCHINEAL RED A ◇ FENAZO SCARLET 3R ◇ FOOD RED 6 ◇ FOOD RED 7 ◇ HD PONCEAU 4R ◇ HEXACOL PONCEAU 4R ◇ HIDACID FAST SCARLET 3R ◇ HISPACID BRILLIANT SCARLET 3RF ◇ JAVA SCARLET 3R ◇ KAYAKU ACID BRILLIANT SCARLET 3R ◇ KITON SCARLET 4R ◇ KOCHINEAL RED A FOR FOOD ◇ 1,3-NAPHTHALENEDISULFONIC ACID, 7-HYDROXY-8-((4-SULFO-1-NAPHTHYL)AZO)-, TRISODIUM SALT ◇ NAPHTHALENE INK SCARLET 4R ◇ NEKLACID RED 3R ◇ NEUCOCCIN ◇ NEW COCCIN ◇ PONCEAU 4R ◇ PONCEAU 4R ALUMINUM LAKE ◇ PONTACYL SCARLET RR ◇ PURPLE RED ◇ ROUGE de COCHENILLE A ◇ SAN-EI BRILLIANT SCARLET 3R ◇ STRAWBERRY RED A GEIGY ◇ SYMULON ACID BRILLIANT SCARLET 3R ◇ TAKAOKA BRILLIANT SCARLET 3R ◇ VICTORIA SCARLET 3R

TOXICITY DATA with REFERENCE
cyt-ham:fbr 1 g/L FCTOD7 22,623,84
orl-rat TDLo:428 g/kg/64W-I:ETA VPITAR 29,61,70
ipr-rat LD50:600 mg/kg FCTXAV 5,187,67
ivn-rat LD50:1 g/kg SCPHA4 47,39,79
unr-rat LD50:2 g/kg NAIZAM 30,179,79
orl-mus LDLo:8 g/kg SCPHA4 47,39,79
ipr-mus LD50:1600 mg/kg FCTXAV 5,187,67

CONSENSUS REPORTS: Reported in EPA TSCA Inventory.

SAFETY PROFILE: Moderately toxic by intraperitoneal route. Questionable carcinogen with experimental tumorigenic data. Mutation data reported. When heated to decomposition it emits toxic fumes of NO_x and SO_x.

FMU200 ***HR: 1***
FOOL'S PARSLEY

PROP: A carrot-like plant 8 to 24 inches tall with leaves like parsley but with a garlic smell. It is native to Europe but grows wild in the northern United States and Canada.

SYNS: AETHUSA CYNAPIUM ◇ DOG PARSLEY ◇ DOG POISON ◇ FALSE PARSLEY ◇ FOOL'S CICELY ◇ LESSER HEMLOCK

SAFETY PROFILE: The whole plant contains a toxic mixture of unsaturated aliphatic alcohols, similar to cicutoxin, and traces of coniine. Human systemic effects by ingestion: nausea, vomiting, sweating and headache. See also CICUTOXIN and 2-PROPYLPIPERIDINE.

FMU225 CAS:22514-23-4 ***HR: 3***
FOPIRTOLINE HYDROCHLORIDE
mf: $C_{11}H_{15}ClN_2OS{\cdot}ClH$ mw: 295.25

SYNS: 4-(2-((6-CHLORO-2-PYRIDYL)THIO)ETHYL)MORPHOLINE MONOHYDROCHLORIDE ◇ D-1126 ◇ FOPIRTOLINA (SPANISH)

TOXICITY DATA with REFERENCE
orl-rat LD50:768 mg/kg DRFUD4 7,806,82
orl-mus LD50:1513 mg/kg DRFUD4 7,806,82
ivn-mus LD50:140 mg/kg DRFUD4 7,806,82
orl-dog LD50:100 mg/kg DRFUD4 7,806,82

SAFETY PROFILE: Poison by ingestion and intravenous routes. When heated to decomposition it emits toxic fumes of SO_x, NO_x, and HCl.

FMU409 CAS:3614-69-5 ***HR: 3***
FORHISTAL MALEATE
mf: $C_{20}H_{24}N_2 \cdot C_4H_4O_4$ mw: 408.54

SYNS: DIMETHINDENE MALEATE ◇ DIMETHINDEN MALEATE ◇ DIMETHPYRINDENE MALEATE ◇ N,N-DIMETHYL-3-(1-(2-PYRIDINYL)ETHYL)-1H-INDENE-2-ETHANAMINE(Z)-2-BUTENEDIOATE (1:1) ◇ DIMETINDENE MALEATE ◇ FENISTIL ◇ FENISTIL-RETARD ◇ FENOSTIL

TOXICITY DATA with REFERENCE
orl-rat LD50:618 mg/kg NIIRDN 6,344,82
ivn-rat LD50:27 mg/kg MEIEDD 10,468,83
orl-gpg LD50:880 mg/kg NIIRDN 6,344,82

SAFETY PROFILE: Poison by intravenous route. Moderately toxic by ingestion. When heated to decomposition it emits toxic fumes of NO_x.

FMV000 CAS:50-00-0 ***HR: 3***
FORMALDEHYDE
DOT: UN 1198/UN 2209
mf: CH_2O mw: 30.03

PROP: Clear, water-white, very sltly acid gas or liquid; pungent odor. Pure formaldehyde is not available commercially because of its tendency to polymerize. It is sold as aqueous solns containing from 37 to 50% formaldehyde by weight and varying amounts of methanol. Some alcoholic solns are used industrially and the physical properties and hazards may be greatly influenced by the solvent. Lel: 7.0%, uel: 73.0%, autoign temp: 806°F, d: 1.0, bp: −3°F, flash p: (37% methanol-free): 185°F, flash p: (15% methanol-free): 122°F.

SYNS: ALDEHYDE FORMIQUE (FRENCH) ◇ ALDEIDE FORMICA (ITALIAN) ◇ BFV ◇ FA ◇ FANNOFORM ◇ FORMALDEHYD (CZECH, POLISH) ◇ FORMALDEHYDE, solution (DOT) ◇ FORMALIN ◇ FORMALIN 40 ◇ FORMALIN (DOT) ◇ FORMALINA (ITALIAN) ◇ FORMALINE (GERMAN) ◇ FORMALIN-LOESUNGEN (GERMAN) ◇ FORMALITH ◇ FORMIC ALDEHYDE ◇ FORMOL ◇ FYDE ◇ HOCH ◇ IVALON ◇ KARSAN ◇ LYSOFORM ◇ METHANAL ◇ METHYL ALDEHYDE ◇ METHYLENE GLYCOL ◇ METHYLENE OXIDE ◇ MORBOCID ◇ NCI-C02799 ◇ OPLOSSINGEN (DUTCH) ◇ OXOMETHANE ◇ OXYMETHYLENE ◇ PARAFORM ◇ POLYOXYMETHYLENE GLYCOLS ◇ RCRA WASTE NUMBER U122 ◇ SUPERLYSOFORM

TOXICITY DATA with REFERENCE
skn-hmn 150 μg/3D-I MLD 85DKA8 -,127,77
eye-hmn 4 ppm/5M IAPWAR 4,79,61
eye-hmn 1 ppm/6M nse MLD AIHAAP 44,463,83
skn-rbt 500 mg/24H SEV 28ZPAK -,40,72
skn-rbt 540 mg open MLD UCDS** 4/21/67
skn-rbt 50 mg/24H MOD TXAPA9 21,369,72
eye-rbt 50 μg/24H SEV 28ZPAK -,40,72
eye-rbt 10 mg SEV TXAPA9 55,501,80
mma-sat 5 μL/plate BIMADU 6,129,85
dni-esc 5 mmol/L MUREAV 156,153,85
dnd-hmn:fbr 100 μmol/L ENMUDM 7,267,85
ihl-rat TCLo:50 $\mu g/m^3$/4H (female 1-19D post):REP TPKVAL 12,78,71
ihl-rat TCLo:1 mg/m^3/24H (1-22D preg):TER HYSAAV 34(5),266,69
ihl-rat TCLo:14300 ppb/6H/2Y-I:CAR CNREA8 43,4382,83
scu-rat TDLo:1170 mg/kg/65W-I:ETA GANNA2 45,451,54
ihl-rat TC:15 ppm/6H/78W-I:CAR CNREA8 49,3398,80
orl-wmn LDLo:108 mg/kg 29ZWAE -,328,68
ihl-hmn TCLo:17 mg/m^3/30M:EYE,PUL JAMAAP 165,1908,57
ihl-man TCLo:300 $\mu g/m^3$:NOSE,CNS GTPZAB 12(7),20,68
unr-man LDLo:477 mg/kg 85DCAI 2,73,70
orl-rat LD50:800 mg/kg JIHTAB 23,259,41
ihl-rat LC50:590 mg/m^3 GISAAA 41(6),103,76
scu-rat LD50:420 mg/kg APTOA6 6,299,50
ivn-rat LD50:87 mg/kg AEPPAE 221,166,54
orl-mus LD50:42 mg/kg NTIS** AD-A125-539
ihl-mus LC50:400 mg/m^3/2H 85GMAT -,69,82
ipr-mus LDLo:16 mg/kg TXAPA9 23,288,72
scu-mus LD50:300 mg/kg APTOA6 6,299,50
scu-dog LDLo:350 mg/kg IPSTB3 3,93,76
ihl-cat LCLo:400 mg/m^3/2H 85GMAT -,69,82
skn-rbt LD50:270 mg/kg UCDS** 4/21/67
scu-rbt LDLo:240 mg/kg JAMAAP 62,984,14
orl-gpg LD50:260 mg/kg JIHTAB 23,259,41

CONSENSUS REPORTS: NTP Fifth Annual Report on Carcinogens. IARC Cancer Review: Group 2A IMEMDT 7,211,87; Human Inadequate Evidence IMEMDT 29,345,82; Animal Sufficient Evidence IMEMDT 29,345,82. EPA Genetic Toxicology Program. Reported in EPA TSCA Inventory.

OSHA PEL: TWA 1 ppm; STEL 2 ppm (For certain industries: TWA 3 ppm; CL 5 ppm; Pk 10 ppm/30M
ACGIH TLV: TWA 1 ppm; Suspected Human Carcinogen (Proposed: CL 0.3 ppm; Suspected Human Carcinogen)
DFG MAK: 0.5 ppm (0.6 mg/m^3); Suspected Carcinogen.
NIOSH REL: (Formaldehyde) Limit to lowest feasible level.
DOT Classification: Combustible Liquid; Label: None; ORM-A; Label: None; Flammable or Combustible Liquid; Label: Flammable Liquid.

SAFETY PROFILE: Confirmed carcinogen with experimental carcinogenic, tumorigenic, and teratogenic data. Human poison by ingestion. Experimental poison by ingestion, skin contact, inhalation, intravenous, intraperitoneal, and subcutaneous routes. Human systemic effects by inhalation: lacrimation, olfactory changes, aggression and pulmonary changes. Experimental reproductive effects. Human mutation data reported. A human skin and eye irritant. If swallowed it causes violent vomiting and diarrhea which can lead to collapse. Frequent or prolonged exposure can cause hypersensitivity leading to contact dermatitis, possibly of an eczematoid nature. An air concentration of 20 ppm is quickly irritating to eyes. A common air contaminant.

Combustible liquid when exposed to heat or flame; can react vigorously with oxidizers. A moderate explosion hazard when exposed to heat or flame. The gas is a more dangerous fire hazard than the vapor. Should formaldehyde be involved in a fire, irritating gaseous formaldehyde may be evolved. When aqueous formaldehyde solutions are heated above their flash points, a potential for an explosion hazard exists. High formaldehyde concentration or methanol content lowers the flash point. Reacts with NO_x at about 180°; the reaction becomes explosive. Also reacts violently with perchloric acid + aniline; performic acid; nitromethane; magnesium carbonate; H_2O_2. Moderately dangerous; because of irritating vapor which may exist in toxic concentrations locally if storage tank is ruptured. To fight fire, stop flow of gas (for pure form); alcohol foam for 37 percent methanol-free form. When heated to decomposition it emits acrid smoke and fumes. See also ALDEHYDES.

FMW000 CAS:149-44-0 ***HR: 2***
FORMALDEHYDE HYDROSULFITE
mf: $CH_4O_3S \cdot Na$ mw: 119.10

SYNS: ALDANIL ◇ DISCOLITE ◇ FORMALDEHYDE SODIUM BISULFITE ADDUCT ◇ FORMALDEHYDE SODIUM SULFOXYLATE ◇ FORMALDEHYDESULFOXYLIC ACID SODIUM SALT ◇ FORMOPAN ◇ HYDROLIT ◇ HYDROSULFITE AWC ◇ HYDROXYMETHANESULFINIC ACID SODIUM SALT ◇ RONGALIT ◇ RONGALITE C ◇ SODIUM FORMALDEHYDE SULFOXYLATE ◇ SODIUM HYDROXYMETHANESULFINATE ◇ SODIUM METHANALSULFOXYLATE ◇ SODIUM SULFOXYLATE FORMALDEHYDE

TOXICITY DATA with REFERENCE
orl-mus LD50:4 g/kg YKYUA6 31,959,80
scu-mus LDLo:3000 mg/kg PHRPA6 49,908,34

CONSENSUS REPORTS: Reported in EPA TSCA Inventory. EPA Genetic Toxicology Program.

SAFETY PROFILE: Moderately toxic by ingestion and subcutaneous routes. When heated to decomposition it emits toxic fumes of SO_x and Na_2O. See also FORMALDEHYDE and SULFITES.

FMW300 ***HR: 3***
FORMALDEHYDE OXIDE POLYMER
mf: $(CH_2O_2)_n$

SAFETY PROFILE: A shock-sensitive explosive. When heated to decomposition it emits acrid smoke and irritating fumes. See also PEROXIDES.

FMX000 CAS:541-66-2 ***HR: 3***
FORMAL-γ-TRIMETHYLAMMONIUM PROPANEDIOL
mf: $C_7H_{16}NO_2 \cdot I$ mw: 273.14

SYNS: DILVASENE ◇ ((1,3-DIOXOLAN-4-YL)METHYL)TRIMETHYLAMMONIUM IODIDE ◇ 2249F ◇ OXAPROPANIUM IODIDE ◇ N,N,N-TRIMETHYL-1,3-DIOXOLANE-4-METHANAMINIUM IODIDE ◇ VASODILATATEUR 2249F

TOXICITY DATA with REFERENCE
scu-rat LDLo:25 mg/kg BSCIA3 26,516,44
orl-mus LDLo:300 mg/kg BSCIA3 26,516,44
scu-mus LDLo:35 mg/kg BSCIA3 26,516,44
ivn-mus LDLo:2500 μg/kg BSCIA3 26,516,44
scu-gpg LDLo:5 mg/kg BSCIA3 26,516,44

SAFETY PROFILE: Poison by ingestion, subcutaneous, and intravenous routes. When heated to decomposition it emits very toxic fumes of NO_x, NH_3, and I^-.

FMY000 CAS:75-12-7 ***HR: 3***
FORMAMIDE
mf: CH_3NO mw: 45.05

PROP: Colorless, hygroscopic and oily liquid. Mp: 2.5, fp: 2.6°, vap press: 29.7 mm @ 129.4°, flash p: 310°F (COC), bp: 210° decomp, d: 1.134 @ 20°/40°; 1.1292 @ 25°/4°. Misc in water and alc; very sltly sol in ether.

SYNS: CARBAMALDEHYDE ◇ METHANAMIDE

TOXICITY DATA with REFERENCE
eye-rbt 23 mg AJOPAA 29,1363,46
oms-nml:oth 500 mmol/L CAANAT 56,712,72
cyt-nml:oth 500 mmol/L CAANAT 56,712,72
skn-rat TDLo:1200 mg/kg (female 10-11D post):TER TXAPA9 41,35,77
orl-rat TDLo:2 g/kg (7D preg):REP 85DJA5 -,95,71
orl-rat LD50:5570 mg/kg 85GMAT -,70,82
ipr-rat LD50:5700 mg/kg TXAPA9 26,596,73
orl-mus LD50:3150 mg/kg 85GMAT -,70,82
ipr-mus LD50:2450 mg/kg AEPPAE 230,559,57
ivn-dog LDLo:1500 mg/kg HBAMAK 4,1289,35
skn-rbt LDLo:6 mg/kg PJPPAA 32,223,80
ipr-gpg LD50:1250 mg/kg 85GMAT -,70,82
scu-frg LDLo:30 mg/kg HBAMAK 4,1289,35

CONSENSUS REPORTS: Reported in EPA TSCA Inventory. EPA Genetic Toxicology Program.

OSHA PEL: TWA 20 ppm; STEL 30 ppm
ACGIH TLV: TWA 10 ppm (skin)

SAFETY PROFILE: Poison by skin contact and subcutaneous routes. Moderately toxic by ingestion, intraperitoneal, and intramuscular routes. An irritant to skin, eyes, and mucous membranes. Experimental teratogenic and reproductive effects. An eye irritant. Mutation data reported. Combustible when exposed to heat or flame; can react vigorously with oxidizing materials. Incompatible with I_2, pyridine, SO_3. When heated to decomposition it emits toxic fumes of NO_x. Has exploded while in storage.

FMZ000 CAS:4312-87-2 ***HR: 2***
FORMHYDROXAMIC ACID
mf: CH_3NO_2 mw: 61.05

SYNS: FORMHYDROXAMSAEURE (GERMAN) ◇ N-FORMYLHYDROXYLAMINE

TOXICITY DATA with REFERENCE
par-rat TDLo:680 mg/kg (female 13D post):TER TJADAB 33,69C,86
scu-rat LD50:570 mg/kg APEPA2 257,296,67

SAFETY PROFILE: Moderately toxic by subcutaneous route. Experimental teratogenic effects. When heated to decomposition it emits toxic fumes of NO_x.

FNA000 CAS:64-18-6 ***HR: 3***
FORMIC ACID
DOT: UN 1779
mf: CH_2O_2 mw: 46.03

PROP: Colorless, fuming liquid; pungent, penetrating odor. Bp: 100.8°, fp: 8.2°, flash p: 156°F (OC), d: 1.2267 @ 15°/4°, 1.220 @ 20°/4°, autoign temp: 1114°F, vap press: 40 mm @ 24.0°, vap d: 1.59, flash p:(90% soln): 122°F, autoign temp (90% soln): 813°F, lel (90% soln) = 18%, uel (90% soln) = 57%. Misc in water, alc, glycerin, ether.

SYNS: ACIDE FORMIQUE (FRENCH) ◇ ACIDO FORMICO (ITALIAN) ◇ AMEISENSAEURE (GERMAN) ◇ AMINIC ACID ◇ FORMYLIC ACID ◇ HYDROGEN CARBOXYLIC ACID ◇ KWAS METANIOWY (POLISH) ◇ METHANOIC ACID ◇ MIERENZUUR (DUTCH) ◇ RCRA WASTE NUMBER U123

TOXICITY DATA with REFERENCE
skn-rbt 610 mg open MLD UCDS** 5/8/68
eye-rbt 122 mg SEV UCDS** 5/8/68
mmo-esc 70 ppm/3H AMNTA4 85,119,51
pic-esc 100 mmol/L MDMIAZ 31,11,79
sln-dmg-ihl 1000 ppm/24H THAGA6 39,330,69
oms-nml:oth 100 mmol/L CAANAT 56,712,72
orl-rat LD50:1100 mg/kg GTPZAB 23(12),49,79
ihl-rat LC50:15 g/m³/15M GTPZAB 23(12),49,79
orl-mus LD50:700 mg/kg GTPZAB 23(12),49,79
ihl-mus LC50:6200 mg/m³/15M GTPZAB 23(12),49,79
ipr-mus LD50:940 mg/kg IGIBA5 11,507,62
ivn-mus LD50:145 mg/kg ZERNAL 9,332,69
orl-dog LD50:4000 mg/kg AMIHAB 20,517,59
ivn-dog LDLo:3000 mg/kg HBTXAC 1,146,56

CONSENSUS REPORTS: Reported in EPA TSCA Inventory.

OSHA PEL: TWA 5 ppm
ACGIH TLV: 5 ppm; STEL 10 ppm
DFG MAK: 5 ppm (9 mg/m³)
DOT Classification: Corrosive Material; Label: Corrosive, solution.

SAFETY PROFILE: Poison by intravenous route. Moderately toxic by ingestion and intraperitoneal routes. Mildly toxic by inhalation. Mutation data reported. Corrosive. A skin and severe eye irritant. A substance migrating to food from packaging materials. Flammable liquid when exposed to heat or flame; can react vigorously with oxidizing materials. Explosive reaction with furfuryl alcohol; H_2O_2; $Tl(NO_3)_3 \cdot 3H_2O$; nitromethane; P_2O_5. To fight fire, use CO_2, dry chemical, alcohol foam. When heated to decomposition it emits acrid smoke and irritating fumes.

FNB000 CAS:32852-21-4 ***HR: 3***
FORMIC ACID (2-(4-METHYL-2-THIAZOLYL) HYDRAZIDE
mf: $C_5H_7N_3OS$ mw: 157.21

TOXICITY DATA with REFERENCE
pic-esc 10 mg/L MUREAV 26,3,74
orl-rat TDLo:11 g/kg/46W-C:NEO JNCIAM 47,437,71

CONSENSUS REPORTS: EPA Genetic Toxicology Program.

SAFETY PROFILE: Questionable carcinogen with experimental neoplastigenic data. Mutation data reported. When heated to decomposition it emits very toxic fumes of SO_x and NO_x.

FNC000 CAS:2142-94-1 ***HR: 1***
FORMIC ACID, NERYL ESTER
mf: $C_{11}H_{18}O_2$ mw: 182.29

SYNS: 3,7-DIMETHYL-2,6-OCTADIEN-1-OL, FORMATE (cis) ◇ NERYL FORMATE

TOXICITY DATA with REFERENCE
skn-rbt 500 mg/24H MOD FCTXAV 14,627,76

CONSENSUS REPORTS: Reported in EPA TSCA Inventory.

SAFETY PROFILE: A skin irritant. When heated to decomposition it emits acrid smoke and irritating fumes. See also ESTERS and FORMIC ACID.

FND100 CAS:10176-39-3 ***HR: 3***
FORMILOXINE
mf: $C_{46}H_{64}O_{19}$ mw: 921.10

SYNS: AC 2770 ◇ FORMILOXIN ◇ GITOFORMATE ◇ PENTAFORMYLGITOXIN

TOXICITY DATA with REFERENCE
orl-mus LD50:2390 μg/kg AIPTAK 153,436,65
scu-mus LD50:2270 μg/kg AIPTAK 153,436,65
orl-gpg LDLo:1835 μg/kg AIPTAK 153,436,65
ivn-gpg LDLo:734 μg/kg AIPTAK 153,436,65

SAFETY PROFILE: Poison by ingestion, subcutaneous, and intravenous routes. When heated to decomposition it emits acrid smoke and irritating fumes.

FNE000 CAS:37032-15-8 ***HR: 2***
FORMOCARBAM
mf: $C_6H_{14}NO_4PS_2$ mw: 259.30

SYNS: O,O-DIMETHYL-S-((METHOXYMETHYL)CARBAMOYL) METHYL PHOSPHORODITHIOATE ◇ S-(((METHOXYMETHYL-CARBAMOYL)METHYL) O,O-DIMETHYLPHOSPHORODITHIOATE ◇ S-(N-METHOXYMETHYLCARBAMOYLMETHYL) DIMETHYL PHOSPHOROTHIOLOTHIONONATE

TOXICITY DATA with REFERENCE
orl-rat LD50:600 mg/kg 28ZEAL 4,274,69
orl-mus LD50:450 mg/kg 28ZEAL 4,274,69

SAFETY PROFILE: Moderately toxic by ingestion. When heated to decomposition it emits very toxic fumes of PO_x, SO_x, and NO_x.

FNE100 CAS:43229-80-7 ***HR: 3***
FORMOTEROL FUMARATE DIHYDRATE
mf: $C_{19}H_{24}N_2O_4 \cdot 1/2C_4H_4O_4 \cdot 2H_2O$ mw: 422.48

SYN: BD 40A

TOXICITY DATA with REFERENCE
orl-rat TDLo:1200 μg/kg (female 17-22D post):REP OYYAA2 27,375,84
orl-rat TDLo:66 mg/kg (female 7-17D post):TER OYYAA2 27,239,84
orl-rat LD50:3130 mg/kg OYYAA2 26,811,83
ipr-rat LD50:170 mg/kg OYYAA2 26,811,83
scu-rat LD50:1000 mg/kg OYYAA2 26,811,83
ivn-rat LD50:98 mg/kg OYYAA2 26,811,83
orl-mus LD50:6700 mg/kg OYYAA2 26,811,83
ipr-mus LD50:210 mg/kg OYYAA2 26,811,83
scu-mus LD50:640 mg/kg OYYAA2 26,811,83
ivn-mus LD50:71 mg/kg OYYAA2 26,811,83

SAFETY PROFILE: Poison by intraperitoneal and intravenous routes. Moderately toxic by ingestion and subcutaneous routes. Experimental teratogenic and reproductive effects. Used as a bronchodilator. When heated to decomposition it emits toxic fumes of NO_x.

FNE500 CAS:17702-57-7 ***HR: 3***
FORMPARANATE
mf: $C_{12}H_{17}N_3O_2$ mw: 235.32

SYNS: N,N-DIMETHYL-N'-(2-METHYL-4-(((METHYLAMINO)CARBONYL)OXY)PHENYL)METHANIMIDAMIDE ◇ ENT 27,305 ◇ SCHERING 36103 ◇ UC-25074

TOXICITY DATA with REFERENCE
orl-mus LD50:16600 μg/kg EJTXAZ 7,152,74
orl-rat LD50:7200 μg/kg ARSIM* 20,26,66

CONSENSUS REPORTS: EPA Extremely Hazardous Substances List.

SAFETY PROFILE: Poison by ingestion. When heated to decomposition it emits toxic fumes of NO_x.

FNF000 CAS:104-06-3 ***HR: 2***
4'-FORMYLACETANILIDE THIOSEMICARBAZONE
mf: $C_{10}H_{12}N_4OS$ mw: 236.32

SYNS: p-ACETAMIDOBENZALDEHYDETHIOSEMICARBAZONE ◇ p-ACETAMINOBENZYLIDENETHIOSEMICARBAZONE ◇ p-ACETYLAMINOBENZALDEHYDETHIOSEMICARBAZONE ◇ AKTIVAN ◇ AMBATHIZON ◇ N-(4-(((AMINOTHIOXOMETHYL) HYDRAZONO)METHYL)PHENYL)ACETAMIDE ◇ AMITHIOZONE ◇ AMITIOZON ◇ ANTIB ◇ BENTHIOZONE ◇ BENZOTHIOZANE ◇ BENZOTHIOZON ◇ BERCULON A ◇ BERKAZON ◇ CONTEBEN ◇ DIASAN ◇ DIAZAN ◇ DOMAGK'S T.B.1 CONTEBEN ◇ DOMAKOL ◇ ILBION ◇ LIVAZONE ◇ MAGK'S T.B.1 CONTEBEN ◇ MIVIZON ◇ MYVIZONE ◇ NEOTIBIL ◇ NEUSTAB ◇ NOVAKOL ◇ NUCLON ARGENTINIAN ◇ PANRONE ◇ PARAZONE ◇ RP 4207 ◇ SDT 1041 ◇ SERODEN ◇ SIOCARBAZONE ◇ SQ 2321 ◇ TB 1 (BAYER) ◇ TEBALON ◇ TEBECURE ◇ TEBEMAR ◇ TEBESONE I ◇ TEBETHIONE ◇ TEBEZON ◇ THIACETAZONE ◇ THIBONE ◇ THIOACETAZONE ◇ THIOCARBAZIL ◇ THIOMICID ◇ THIONICID ◇ THIOPARAMIZONE ◇ THIOSEMICARBARZONE ◇ THIOSEMICARBAZONE (PHARMACEUTICAL) ◇ THIOTEBESIN ◇ THIZONE ◇ TIACETAZON ◇ TIBICUR ◇ TIBIONE ◇ TIBIZAN ◇ TIOACETAZON ◇ TIOCARONE ◇ TUBIGAL ◇ TUBIN

TOXICITY DATA with REFERENCE
cyt-mus-orl 3300 μg/kg NULSAK 22,96,79
orl-mus LD50:950 mg/kg CRSBAW 144,1310,50
scu-mus LD50:1 g/kg JPPMAB 2,764,50

SAFETY PROFILE: Moderately toxic by ingestion and subcutaneous routes. Mutation data reported. A tuberculostatic antibacterial agent. When heated to decomposition it emits very toxic fumes of NO_x and SO_x.

FNG000 CAS:53757-30-5 ***HR: D***
2-FORMYLAMINO-4-(2-5-NITRO-2-FURYL) VINYL)-1,3-THIAZOLE
mf: $C_{10}H_7N_3O_4S$ mw: 265.26

SYN: N-(4-(2-(5-NITRO-2-FURYL)VINYL)2-THIAZOLYL)FORMAMIDE

TOXICITY DATA with REFERENCE
mma-sat 100 ng/plate MUREAV 40,9,75
dnr-sat 500 nmol/well CNREA8 34,2266,74

mmo-esc 300 nmol/well CNREA8 34,2266,74
mrc-esc 500 nmol/well CNREA8 34,2266,74

CONSENSUS REPORTS: EPA Genetic Toxicology Program.

SAFETY PROFILE: Mutation data reported. When heated to decomposition it emits very toxic fumes of SO_x and NO_x.

FNJ000 CAS:103-70-8 ***HR: 3***
FORMYLANILINE
mf: C_7H_7NO mw: 121.15

SYNS: CARBANILALDEHYDE ◇ FORMAMIDOBENZENE ◇ FORMANILIDE ◇ N-FORMYLANILINE ◇ PHENYL FORMAMIDE ◇ N-PHENYLFORMAMIDE

TOXICITY DATA with REFERENCE
orl-dog LDLo:400 mg/kg XPHBAO 271,19,41
ivn-dog LDLo:400 mg/kg XPHBAO 271,19,41
orl-frg LDLo:800 μg/kg XPHBAO 271,19,41

CONSENSUS REPORTS: Reported in EPA TSCA Inventory.

SAFETY PROFILE: Poison by ingestion and intravenous routes. When heated to decomposition it emits toxic fumes of NO_x.

FNK000 CAS:63040-55-1 ***HR: 3***
6-FORMYLANTHANTHRENE
mf: $C_{25}H_{14}O$ mw: 330.39

SYN: DIBENZO(def,mno)CHRYSENE-12-CARBOXALDEHYDE

TOXICITY DATA with REFERENCE
scu-mus TDLo:72 mg/kg 9W-I:ETA COREAF 252,1711,61

SAFETY PROFILE: Questionable carcinogen with experimental tumorigenic data. When heated to decomposition it emits acrid smoke and irritating fumes.

FNK025 CAS:100-50-5 ***HR: 3***
4-FORMYLCYCLOHEXENE
DOT: UN 2498
mf: $C_7H_{10}O$ mw: 110.17

SYNS: 3-CYCLOHEXENE-1-CARBOXALDEHYDE ◇ 1,2,3,6-TETRAHYDROBENZALDEHYDE (DOT) ◇ 1,2,5,6-TETRAHYDROBENZALDEHYDE

TOXICITY DATA with REFERENCE
skn-rbt 10 mg/24H open MLD AIHAAP 23,95,62
orl-rat LD50:2460 mg/kg AIHAAP 23,95,62
skn-rbt LD50:1770 mg/kg AIHAAP 23,95,62

CONSENSUS REPORTS: Reported in EPA TSCA Inventory.

DOT Classification: Corrosive Material: Label: Corrosive; Flammable or Combustible Liquid; Label: Flammable Liquid.

SAFETY PROFILE: Moderately toxic by skin contact and ingestion. Corrosive. An eye, skin, and mucous membrane irritant. Flammable liquid. When heated to decomposition it emits acrid smoke and irritating fumes. See also ALDEHYDES.

FNK040 CAS:2454-11-7 ***HR: 3***
FORMYLDIENOLONE
mf: $C_{21}H_{28}O_4$ mw: 344.49

PROP: Crystals from ethyl acetate. Mp: 209-212°. Water-sol.

SYNS: 11-α,17-β-DIHYDROXY-17-METHYL-3-OXOANDROSTA-1,4-DIENE-2-CARBOXALDEHYDE ◇ ESICLENE ◇ FORMEBOLONE ◇ 2-FORMYL-11-α-HYDROXY-Δ^1-METHYLTESTOSTERONE ◇ 2-FORMYL-17-α-METHYLANDROSTA-1,4-DIENE-11-α,17-β-DIOL-3-ONE

TOXICITY DATA with REFERENCE
ipr-rat LD50:104 mg/kg MEIEDD 10,606,83
scu-rat LD50:270 mg/kg MEIEDD 10,606,83
ipr-mus LD50:187 mg/kg MEIEDD 10,606,83
scu-mus LD50:293 mg/kg MEIEDD 10,606,83

SAFETY PROFILE: Poison by subcutaneous and intraperitoneal routes. Used as an anabolic steroid. When heated to decomposition it emits acrid smoke and irritating fumes.

FNK050 ***HR: 3***
3-FORMYL-DIGITOXIGENIN
mf: $C_{24}H_{34}O_5$ mw: 402.58

SYN: 3-β,14-DIHYDROXY-5-β-CARD-20(22)-ENOLIDE-3-FORMATE

TOXICITY DATA with REFERENCE
orl-mus LD50:33630 μg/kg AIPTAK 153,436,65
scu-mus LD50:16600 μg/kg AIPTAK 153,436,65
ivn-gpg LDLo:7711 μg/kg AIPTAK 153,436,65

SAFETY PROFILE: A deadly poison by ingestion, subcutaneous, and intravenous routes. When heated to decomposition it emits acrid smoke and irritating fumes.

FNK075 ***HR: 3***
3-12-FORMYL-DIGOXIGENIN
mf: $C_{25}H_{34}O_7$ mw: 446.59

SYN: 3-β,12-β,14-TRIHYDROXY-5-β-CARD-20(22)-ENOLIDE-3,12-DIFORMATE

TOXICITY DATA with REFERENCE
orl-mus LD50:33680 μg/kg AIPTAK 153,436,65
scu-mus LD50:19420 μg/kg AIPTAK 153,436,65
ivn-gpg LDLo:4750 μg/kg AIPTAK 153,436,65

SAFETY PROFILE: Poison by ingestion, subcutaneous, and intravenous routes. When heated to decomposition it emits acrid smoke and irritating fumes.

FNK100 CAS:69321-17-1 ***HR: D***
3'-FORMYL-N,N-DIMETHYL-4-AMINOAZOBENZENE
mf: $C_{15}H_{15}N_3O$ mw: 253.33

SYNS: 3-((4-(DIMETHYLAMINO)PHENYL)AZO)BENZALDEHYDE ◇ 3-((p-DIMETHYLAMINO)PHENYL)AZO)BENZALDEHYDE

TOXICITY DATA with REFERENCE
mmo-sat 1 μmol/plate CRNGDP 4,1487,83
mma-sat 1 μmol/plate CRNGDP 1,121,80
dns-rat:lvr 1 μmol/L CNREA8 46,1654,86

SAFETY PROFILE: Mutation data reported. When heated to decomposition it emits toxic fumes of NO_x.

FNK200 CAS:58243-85-9 ***HR: 2***
FORMYLETHYLTETRAMETHYLTETRALIN
mf: $C_{17}H_{24}O$ mw: 244.41

SYNS: 6-ETHYL-7-FORMYL-1,1,4,4-TETRAMETHYL-1,2,3,4-TETRAHYDRONAPHTHALENE ◇ FETT ◇ 5,6,7,8-TETRAHYDRO-3-ETHYL-5,5,8,8-TETRAMETHYL-2-NAPHTHALENECARBOXALDEHYDE ◇ TETRAMETHYLETHYLFORMYLTETRALIN

TOXICITY DATA with REFERENCE
skn-rbt 500 mg/24H MLD FCTOD7 21,853,83
eye-rbt 100 mg MLD FCTOD7 21,853,83
orl-rat LD50:3200 mg/kg FCTOD7 21,853,83

SAFETY PROFILE: Moderately toxic by ingestion. A skin and eye irritant. When heated to decomposition it emits acrid smoke and irritating fumes.

FNL000 CAS:18197-22-3 ***HR: D***
N-FORMYLFORMAMIDE
mf: $C_2H_3NO_2$ mw: 73.06

SYNS: IMINOALDEHYDE ◇ IMINOBISFORMALDEHYDE

TOXICITY DATA with REFERENCE
dnd-esc 250 μmol/L MUREAV 39,317,77

SAFETY PROFILE: Mutation data reported. When heated to decomposition it emits toxic fumes of NO_x.

FNM000 CAS:621-59-0 ***HR: 2***
5-FORMYLGUAIACOL
mf: $C_8H_8O_3$ mw: 152.16

SYNS: 3-HYDROXY-p-ANISALDEHYDE ◇ 3-HYDROXY-4-METHOXYBENZALDEHYDE ◇ ISOVANILLIN ◇ ISOVANILLINE ◇ OXY-3-METHOXY-4 BENZALDEHYDE (FRENCH)

TOXICITY DATA with REFERENCE
ipr-rat LD50:1276 mg/kg COREAF 243,609,56
ivn-dog LDLo:1470 mg/kg APFRAD 14,456,56

CONSENSUS REPORTS: Reported in EPA TSCA Inventory.

SAFETY PROFILE: Moderately toxic by intraperitoneal and intravenous routes. When heated to decomposition it emits acrid smoke and irritating fumes.

FNN000 CAS:624-84-0 ***HR: 3***
FORMYLHYDRAZINE
mf: CH_4N_2O mw: 60.07

PROP: Mp: 54°. Very sol in alc and ether; sol in benzene.

SYNS: CARBAZALDEHYDE ◇ FORMAL HYDRAZINE ◇ FORMHYDRAZID (GERMAN) ◇ FORMHYDRAZIDE ◇ FORMIC ACID, HYDRAZIDE ◇ FORMIC HYDRAZIDE ◇ FORMOHYDRAZIDE ◇ FORMYLHYDRAZIDE ◇ N-FORMYLHYDRAZINE ◇ HYDRAZINECARBOXALDEHYDE

TOXICITY DATA with REFERENCE
orl-mus TDLo:73 g/kg/43W-C:NEO BJCAAI 37,960,78
scu-rat LD50:120 mg/kg ARZNAD 18,645,68
ipr-mus LD50:65 mg/kg JMPCAS 4,259,61

CONSENSUS REPORTS: Reported in EPA TSCA Inventory.

SAFETY PROFILE: Poison by intraperitoneal and subcutaneous routes. Questionable carcinogen with experimental neoplastigenic data. When heated to decomposition it emits toxic fumes of NO_x. See also HYDRAZINE.

FNO000 CAS:689-13-4 ***HR: 1***
N-FORMYL-N-HYDROXYGLYCINE
mf: $C_3H_5NO_4$ mw: 119.09

SYNS: ASYMMETRIN ◇ N-FORMYL HYDROXYAMINOACETIC ACID ◇ HADACIDIN ◇ HADACIDINE ◇ HADACIN ◇ NFHAA ◇ NSC 521778

TOXICITY DATA with REFERENCE
ipr-mus TDLo:2 g/kg (female 10D post):TER PCBRD2 110,365,83
ipr-rat LDLo:5000 mg/kg ADTEAS 3,181,68

SAFETY PROFILE: Mildly toxic by intraperitoneal route. Experimental teratogenic effects. When heated to decomposition it emits toxic fumes of NO_x.

FNP000 CAS:66409-98-1 ***HR: 3***
N-FORMYLJERVINE
mf: $C_{28}H_{39}NO_4$ mw: 453.68

TOXICITY DATA with REFERENCE
orl-ham TDLo:85 mg/kg (7D preg):REP 41CIAR -,409,78
orl-ham TDLo:85 mg/kg (7D preg):TER JAFCAU 26,564,78
orl-ham LDLo:85 mg/kg JAFCAU 26,564,78

SAFETY PROFILE: Poison by ingestion. Experimental teratogenic and reproductive effects.

FNQ000 CAS:63040-58-4 ***HR: 3***
6-FORMYL-12-METHYLANTHANTHRENE
mf: $C_{26}H_{16}O$ mw: 344.42

SYN: 6-METHYLDIBENZO(def,mno)CHRYSENE-12-CARBOXALDEHYDE

TOXICITY DATA with REFERENCE
scu-mus TDLo:72 mg/kg/9W-I:ETA COREAF 252,1711,61

SAFETY PROFILE: Questionable carcinogen with experimental tumorigenic data. When heated to decomposition it emits acrid smoke and irritating fumes.

FNR000 CAS:2732-09-4 ***HR: 3***
7-FORMYL-9-METHYLBENZ(c)ACRIDINE
mf: $C_{19}H_{13}NO$ mw: 271.33

SYN: 9-METHYLBENZ(c)ACRIDINE-7-CARBOXALDEHYDE

TOXICITY DATA with REFERENCE
scu-mus TDLo:120 mg/kg/9W-I:ETA CHDDAT 267,981,68

SAFETY PROFILE: Questionable carcinogen with experimental tumorigenic data. When heated to decomposition it emits toxic fumes of NO_x.

FNS000 CAS:18936-78-2 ***HR: 3***
7-FORMYL-11-METHYLBENZ(c)ACRIDINE
mf: $C_{19}H_{13}NO$ mw: 271.33

SYN: 11-METHYLBENZ(c)ACRIDINE-7-CARBOXALDEHYDE

TOXICITY DATA with REFERENCE
scu-mus TDLo:120 mg/kg/9W-I:ETA CHDDAT 267,981,68

SAFETY PROFILE: Questionable carcinogen with experimental tumorigenic data. When heated to decomposition it emits toxic fumes of NO_x.

FNT000 CAS:13345-61-4 ***HR: 3***
7-FORMYL-12-METHYLBENZ(a)ANTHRACENE
mf: $C_{20}H_{14}O$ mw: 270.34

SYN: 12-METHYLBENZ(a)ANTHRACENE-7-CARBOXALDEHYDE

TOXICITY DATA with REFERENCE
dni-omi 200 µg/L PNASA6 74,1378,77
scu-rat TDLo:20 mg/kg/39D-I:NEO CNREA8 31,1951,71
scu-mus TDLo:120 mg/kg/6W-I:CAR IJCNAW 2,500,67

SAFETY PROFILE: Questionable carcinogen with experimental carcinogenic and neoplastigenic data. Mutation data reported. When heated to decomposition it emits acrid smoke and irritating fumes.

FNU000 CAS:63040-56-2 ***HR: 3***
5-FORMYL-8-METHYL-3,4:9,10-DIBENZOPYRENE
mf: $C_{26}H_{16}O$ mw: 344.42

SYN: 8-METHYLBENZO(rst)PENTAPHENE-5-CARBOXALDEHYDE

TOXICITY DATA with REFERENCE
scu-mus TDLo:72 mg/kg/9W-I:ETA COREAF 252,1236,61

SAFETY PROFILE: Questionable carcinogen with experimental tumorigenic data. When heated to decomposition it emits acrid smoke and irritating fumes.

FNV000 CAS:63040-57-3 ***HR: 3***
5-FORMYL-10-METHYL-3,4,:8,9-DIBENZOPYRENE
mf: $C_{26}H_{16}O$ mw: 344.42

SYN: 14-METHYLDIBENZO(b,def)CHRYSENE-7-CARBOXALDEHYDE

TOXICITY DATA with REFERENCE
scu-mus TDLo:72 mg/kg/9W-I:ETA COREAF 252,1711,61

SAFETY PROFILE: Questionable carcinogen with experimental tumorigenic data. When heated to decomposition it emits acrid smoke and irritating fumes.

FNW000 CAS:758-17-8 ***HR: 3***
N-FORMYL-N-METHYLHYDRAZINE
mf: $C_2H_6N_2O$ mw: 74.10

SYNS: FORMIC ACID, METHYLHYDRAZIDE ◇ 1-FORMYL-1-METHYLHYDRAZINE ◇ N-METHYL-N-FORMYLHYDRAZINE ◇ MFH

TOXICITY DATA with REFERENCE
mmo-sat 100 µmol/plate TXCYAC 26,155,83
mma-sat 100 µmol/plate TXCYAC 26,155,83
orl-mus TDLo:14 mg/kg/62W-C:CAR JNCIAM 60,201,78
scu-mus TDLo:400 mg/kg/40W-I:CAR NEOLA4 39,437,83
scu-mus TD:100 mg/kg:CAR,REP JTEHD6 6,577,80
orl-mus LD50:118 mg/kg TXAPA9 45,429,78

SAFETY PROFILE: Suspected carcinogen with experimental carcinogenic data. Poison by ingestion route. Experimental reproductive effects. Mutation data reported. When heated to decomposition it emits toxic fumes of NO_x. See also HYDRAZINE.

FNX000 CAS:4845-14-1 ***HR: 3***
N-FORMYL-N-METHYL-p-(PHENYLAZO)ANILINE
mf: $C_{14}H_{13}N_3O$ mw: 239.30

SYN: 4-FORMYLMONOMETHYLAMINOAZOBENZENE

TOXICITY DATA with REFERENCE
orl-rat TDLo:9100 mg/kg/34W-C:ETA CNREA8 9,652,49

SAFETY PROFILE: Questionable carcinogen with ex-

perimental tumorigenic data. When heated to decomposition it emits toxic fumes of NO_x.

FNZ000 CAS:51-15-0 ***HR: 3***
2-FORMYL-1-METHYLPYRIDINIUM CHLORIDE OXIME
mf: $C_7H_9N_2O{\bullet}Cl$ mw: 172.63

SYNS: 2-FORMYL-N-METHYLPYRIDINIUM OXIME CHLORIDE ◇ 2-(HYDROXYIMINOMETHYL)-1-METHYLPYRIDINIUM CHLORIDE ◇ 1-METHYL-2-ALDOXIMINOPYRIDINIUM CHLORIDE ◇ 1-METHYL-2-FORMYLPYRIDINIUM CHLORIDE OXIME ◇ 1-METHYL-2-PYRIDINIUM ALDOXIME CHLORIDE ◇ N-METHYLPYRIDINIUM CHLORIDE-2-ALDOXIME ◇ 2-PAM CHLORIDE ◇ PRALIDOXIME CHLORIDE ◇ PROTOPAM CHLORIDE ◇ 2-PYRIDINEALDOXIME CHLORIDE ◇ PYRIDINE-2-ALDOXIME METHOCHLORIDE ◇ 2-PYRIDINE ALDOXIME METHYL CHLORIDE ◇ PYRIDINIUM ALDOXIME METHOCHLORIDE

TOXICITY DATA with REFERENCE
ivn-man TDLo:14 mg/kg:CNS,PUL 34ZIAG -,449,69
ivn-hmn TDLo:15 mg/kg:CVS AEHLAU 15,599,67
ivn-rat LD50:96 mg/kg TXAPA9 16,40,70
ims-rat LD50:150 mg/kg TXAPA9 16,40,70
orl-mus LD50:4100 mg/kg JPMSAE 53,1143,64
ipr-mus LD50:155 mg/kg ARZNAD 14,5,64
ivn-mus LD50:90 mg/kg ARZNAD 14,5,64
ims-mus LD50:100 mg/kg DCTODJ 8,431,85
ims-dog LD50:75 mg/kg FAATDF 4(2, Pt 2),S106,84
ivn-rbt LD50:95 mg/kg JPETAB 132,50,61
ims-gpg LD50:168 mg/kg TXAPA9 16,40,70

SAFETY PROFILE: Poison by intravenous, intramuscular, and intraperitoneal routes. Moderately toxic by ingestion. Human systemic effects by intravenous route: coma, blood pressure increase, bronchiolar constriction and cyanosis. Used as a cholinesterase reactivator. When heated to decomposition it emits very toxic fumes of Cl^- and NO_x.

FOA000 CAS:7293-34-7 ***HR: 3***
2-FORMYL-1-METHYL-PYRIDINIUM PERCHLORATE OXIME
mf: $C_6H_6N_2O{\bullet}ClO_4$ mw: 221.59

SYN: PAM-PERCHLORAT (GERMAN)

TOXICITY DATA with REFERENCE
ipr-mus LD50:200 mg/kg ARZNAD 14(1),5,64
ivn-mus LD50:165 mg/kg ARZNAD 14(1),5,64
ims-mus LD50:245 mg/kg ARZNAD 14(1),5,64

SAFETY PROFILE: Poison by intraperitoneal, intravenous and intramuscular routes. Many perchlorates are unstable. When heated to decomposition it emits very toxic fumes of NO_x and Cl^-. See also PERCHLORATES.

FOD000 CAS:42540-40-9 ***HR: 2***
(6R-(6-α,7-β(R)))-7-(((FORMYLOXY)PHENYLACETYL)AMINO)-3-(((1-METHYL-1H-TETRAZOL-5-YL)THIO)METHYL)-8-OXO-5-THIA-1-AZABICYCLO(4.2.0)OCT-2-ENE-2-CARBOXYLIC ACID MONOSODIUM SALT
mf: $C_{19}H_{18}N_6O_6S_2{\bullet}Na$ mw: 513.54

SYNS: CEFAMANDOLE NAFATE ◇ o-FORMYLCEFAMANDOLE SODIUM

TOXICITY DATA with REFERENCE
ivn-rat LD50:2562 mg/kg JIDIAQ 137,S51,78
scu-mus LD50:7 g/kg JIDIAQ 137,S51,78
ivn-mus LD50:3915 mg/kg JIDIAQ 137,S51,78

SAFETY PROFILE: Moderately toxic by intravenous route. When heated to decomposition it emits very toxic fumes of NO_x, Na_2O and SO_x.

FOE000 CAS:17977-68-3 ***HR: 3***
FORMYLOXYTRIBENZYLSTANNANE
mf: $C_{23}H_{24}O_2Sn$ mw: 451.16

SYNS: (FORMYLOXY)TRIS(PHENYLMETHYL)STANNANE ◇ MRAVENCAN TRIBENZYLCINICITY (CZECH) ◇ TRIBENZYLTIN FORMATE

TOXICITY DATA with REFERENCE
skn-rbt 500 mg/24H MLD 28ZPAK -,232,72
eye-rbt 100 mg/24H MOD 28ZPAK -,232,72
orl-rat LD50:312 mg/kg 28ZPAK -,232,72

OSHA PEL: TWA 0.1 mg(Sn)/m^3 (skin)
ACGIH TLV: TWA 0.1 mg(Sn)/m^3 (skin) (Proposed: TWA 0.1 mg(Sn)/m^3; STEL 0.2 mg(Sn)/m^3 (skin))
NIOSH REL: (Organotin Compounds) TWA 0.1 mg(Sn)/m^3

SAFETY PROFILE: Poison by ingestion. A skin and eye irritant. When heated to decomposition it emits acrid smoke and irritating fumes. See also TIN COMPOUNDS.

FOF000 CAS:123-08-0 ***HR: 2***
p-FORMYLPHENOL
mf: $C_7H_6O_2$ mw: 122.13

PROP: Needles from water. D: 1.129, bp: subl.

SYNS: 4-FORMYLPHENOL ◇ p-HYDROXYBENZALDEHYDE ◇ 4-HYDROXYBENZALDEHYDE ◇ p-OXYBENZALDEHYDE ◇ PARA-HYDROXYBENZALDEHYDE ◇ USAF M-6

TOXICITY DATA with REFERENCE
ipr-mus LD50:500 mg/kg NTIS** AD277-689

CONSENSUS REPORTS: Reported in EPA TSCA Inventory.

SAFETY PROFILE: Moderately toxic by intraperi-

toneal route. When heated to decomposition it emits acrid smoke and irritating fumes. See also ALDEHYDES.

FOH000 CAS:2591-86-8 ***HR: 3***
1-FORMYLPIPERIDINE
mf: $C_6H_{11}NO$ mw: 113.18

SYN: N-FORMYLPIPERIDIN (GERMAN)

TOXICITY DATA with REFERENCE
scu-rbt LDLo:300 mg/kg BDCGAS 34,2408,01

CONSENSUS REPORTS: Reported in EPA TSCA Inventory.

SAFETY PROFILE: Poison by subcutaneous route. When heated to decomposition it emits toxic fumes of NO_x.

FOI000 CAS:6804-07-5 ***HR: D***
2-FORMYLQUINOXALINE-1,4-DIOXIDE CARBOMETHOXYHYDRAZONE
mf: $C_{11}H_{10}N_4O_4$ mw: 262.25

SYNS: CARBADOX (USDA) ◇ FORTIGRO ◇ GS 6244 ◇ MECADOX ◇ (2-QUINOXALINYLMETHYLENE)-HYDRAZINECARBOXYLIC ACID METHYL ESTER-N,N'-DIOXIDE

TOXICITY DATA with REFERENCE
dns-hmn:oth 50 mg/L JTEHD6 10,143,82
mnt-mus-orl 100 mg/kg MUREAV 144,81,85

SAFETY PROFILE: Human mutation data reported. When heated to decomposition it emits toxic fumes of NO_x.

FOJ000 CAS:2302-84-3 ***HR: 3***
1-FORMYL-3-THIOSEMICARBAZIDE
mf: $C_2H_5N_3OS$ mw: 119.16

TOXICITY DATA with REFERENCE
orl-rat TDLo:43 g/kg/46W-C:ETA JNCIAM 47,437,71

CONSENSUS REPORTS: Reported in EPA TSCA Inventory.

SAFETY PROFILE: Questionable carcinogen with experimental tumorigenic data. When heated to decomposition it emits very toxic fumes of NO_x and SO_x.

FOJ100 CAS:71522-58-2 ***HR: 1***
FORPHENICINOL
mf: $C_9H_{11}NO_4$ mw: 197.21

SYNS: BENZENEACETIC ACID, α-AMINO-3-HYDROXY-4-(HYDROXYMETHYL)-, (S)- ◇ BF 121

TOXICITY DATA with REFERENCE
orl-rbt TDLo:5200 mg/kg (female 6-18D post):TER IYKEDH 17,693,86
ipr-rat LD50:5500 mg/kg JANTAJ 35,1049,82
ipr-mus LD50:5000 mg/kg JANTAJ 35,1049,82

SAFETY PROFILE: Mildly toxic by intraperitoneal route. An experimental teratogen. When heated to decomposition it emits toxic fumes of NO_x.

FOK000 CAS:55779-06-1 ***HR: 3***
FORTIMICIN A
mf: $C_{17}H_{35}N_5O_6$ mw: 405.57

PROP: Produced by *Micromonospora olivoasterospora* MK-70 (JANTAJ 30,77-7,77).

SYNS: ANTIBIOTIC KW-1070 ◇ KW-1070 ◇ XK-70-1

TOXICITY DATA with REFERENCE
ims-rat TDLo:1210 mg/kg (7-17D preg):REP NKRZAZ 29(Suppl 2),167,81
ims-rat TDLo:5170 mg/kg (22D pre):TER NKRZAZ 29(Suppl 2),167,81
ipr-rat LD50:913 mg/kg NKRZAZ 29(Suppl 2),167,81
ivn-rat LD50:209 mg/kg NKRZAZ 29(Suppl 2),167,81
orl-mus LD50:13600 mg/kg NKRZAZ 29(Suppl 2),167,81
ipr-mus LD50:533 mg/kg NKRZAZ 29(Suppl 2),167,81
scu-mus LD50:400 mg/kg JANTAJ 30,77-7,77
ims-mus LD50:427 mg/kg NKRZAZ 29(Suppl 2),167,81
ivn-dog LD50:214 mg/kg NKRZAZ 29(Suppl 2),167,81

SAFETY PROFILE: Poison by subcutaneous and intravenous routes. Moderately toxic by intraperitoneal and intramuscular routes. Mildly toxic by ingestion. Experimental teratogenic and reproductive effects. When heated to decomposition it emits toxic fumes of NO_x.

FOL000 CAS:72275-67-3 ***HR: 3***
FORTIMICIN A SULFATE
mf: $C_{17}H_{35}N_5O_6•2H_2O_4S$ mw: 601.73

SYN: ABBOTT-44747

TOXICITY DATA with REFERENCE
ivn-rat TDLo:1100 mg/kg (female 7-17D post):TER KSRNAM 20,4274,86
scu-rat LD50:1365 mg/kg IYKEDH 16,866,65
ivn-rat LD50:86 mg/kg TXAPA9 53,399,80
orl-mus LD50:13600 mg/kg JJANAX 35,1402,82
ipr-mus LD50:533 mg/kg IYKEDH 16,866,85
scu-mus LD50:653 mg/kg IYKEDH 16,866,85
ivn-mus LD50:94 mg/kg TXAPA9 53,399,80
ims-mus LD50:436 mg/kg TXAPA9 53,399,80
ivn-dog LD50:214 mg/kg IYKEDH 16,866,85
ims-dog LD50:750 mg/kg IYKED 16,866,85

SAFETY PROFILE: Poison by intravenous route. Moderately toxic by intraperitoneal, subcutaneous, and intramuscular routes. Mildly toxic by ingestion. An experimental teratogen. When heated to decomposition it emits very toxic fumes of SO_x and NO_x.

FOL100 CAS:35322-07-7 ***HR: 3***
FOSAZEPAM
mf: $C_{18}H_{18}ClN_2O_2P$ mw: 360.80

SYNS: 7-CHLORO-1-((DIMETHYLPHOSPHINYL)METHYL)-1,3-DIHYDRO-5-PHENYL-2H-1,4-BENZODIAZEPINE-2-ONE ◇ HR 930

TOXICITY DATA with REFERENCE
orl-rat LD50:3110 mg/kg OYYAA2 10,265,75
ipr-rat LD50:300 mg/kg OYYAA2 10,265,75
scu-rat LD50:1740 mg/kg OYYAA2 10,265,75
ivn-rat LD50:160 mg/kg OYYAA2 10,265,75
orl-mus LD50:2 g/kg OYYAA2 10,265,75
ipr-mus LD50:350 mg/kg OYYAA2 10,265,75
scu-mus LD50:550 mg/kg OYYAA2 10,265,75
ivn-mus LD50:285 mg/kg OYYAA2 10,265,75

SAFETY PROFILE: Poison by intravenous and intraperitoneal routes. Moderately toxic by ingestion and subcutaneous routes. When heated to decomposition it emits toxic fumes of Cl^-, PO_x, and NO_x. See also DIAZEPAM.

FOL200 ***HR: 2***
FOSFOMYCIN DISODIUM HYDRATE
mf: $C_3H_5O_4P \cdot 2Na \cdot H_2O$ mw: 200.05

SYNS: DISODIUM (−)-(1R,2S)-(1,2-EPOXYPROPYL)PHOSPHONATE HYDRATE ◇ DISODIUM FOSFOMYCIN HYDRATE ◇ DISODIUM PHOSPHONOMYCIN HYDRATE ◇ FOSFOMYCIN SODIUM HYDRATE ◇ SODIUM FOSFOMYCIN HYDRATE

TOXICITY DATA with REFERENCE
orl-rat LD50:4700 mg/kg IYKEDH 12,668,81
ipr-rat LD50:2060 mg/kg IYKEDH 12,668,81
scu-rat LD50:5100 mg/kg IYKEDH 12,668,81
ivn-rat LD50:1650 mg/kg IYKEDH 12,668,81
ims-rat LD50:2630 mg/kg IYKEDH 12,668,81
orl-mus LD50:8020 mg/kg IYKEDH 12,668,81
ipr-mus LD50:2175 mg/kg IYKEDH 12,668,81

SAFETY PROFILE: Moderately toxic by intraperitoneal, intravenous and intramuscular routes. Mildly toxic by ingestion. When heated to decomposition it emits toxic fumes of PO_x and Na_2O.

FOM000 CAS:37132-72-2 ***HR: D***
FOTRIN
mf: $C_{14}H_{28}N_9OP_3$ mw: 431.42

SYNS: 2,2,4,4,6-PENTAETHYLENIMINO-6-MORPHOLINO-CYCLOTRIPHOSPHAZATRIEN ◇ PHOTRIN ◇ PHOTRINE

TOXICITY DATA with REFERENCE
mma-sat 1 g/plate PCJOAU 12,35,78
cyt-hmn:lym 20 mg/L/1H SOGEBZ 12,1552,76
sce-hmn:lym 5 mg/L TGANAK 16(2),34,82
sce-ham:oth 125 μg/L CYGEDX 16(4),23,82
sce-ham:oth 125 mg/L GNKAA5 19,1152,83
ipr-mus TDLo:11200 μg/kg (1D male):TER KHFZAN 10(9),11,76

CONSENSUS REPORTS: EPA Genetic Toxicology Program.

SAFETY PROFILE: Experimental teratogenic effects. Human mutation data reported. When heated to decomposition it emits very toxic fumes of NO_x and PO_x.

FOM050 CAS:1332-10-1 ***HR: 3***
FOWLER'S SOLUTION

SYNS: ARSENICAL solution ◇ POTASSIUM ARSENITE solution

CONSENSUS REPORTS: IARC Cancer Review: Group 1 IMEMDT 7,100,87.

SAFETY PROFILE: Confirmed carcinogen.

FOM100 ***HR: 3***
FOXGLOVE

PROP: These hardy plants grow wild throughout the northern United States, Canada, and Alaska. They are cultivated for the production of digitalis. The flowers are purple or pink and hang down from a central stalk.

SYNS: DIGITALIS ◇ DIGITALIS PURPUREA ◇ FAIRY BELLS ◇ FAIRY CAP ◇ FAIRY GLOVE ◇ FAIRY THIMBLES ◇ FOLKS GLOVE ◇ LADIE'S THIMBLES ◇ LION'S MOUTH ◇ POP-DOCK ◇ RABBIT FLOWER ◇ THIMBLES ◇ THROATWORT ◇ WITCHES' THIMBLES

SAFETY PROFILE: The whole plant contains poisonous digitalis glycosides and irritant saponins. Human systemic effects by ingestion include: mouth pain, nausea, vomiting, abdominal pain, cramps, and diarrhea. Cardiac glycosides may cause death by their effect on heart function. See also DIGITALIS and SAPONIN.

FOM200 CAS:28808-62-0 ***HR: 3***
FRAXINELLONE
mf: $C_{14}H_{16}O_3$ mw: 232.30

SYNS: 1(3H)-ISOBENZOFURANONE,3-(3-FURANYL)-3a,4,5,6-TETRAHYDRO-3a,7-DIMETHYL-, (3R-cis)-(9CI) ◇ PHTHALIDE, 3-(3-FURYL)-3a,4,5,6-TETRAHYDRO-3a,7-DIMETHYL-

TOXICITY DATA with REFERENCE
orl-rat TDLo:424 mg/kg (female 5-8D post):REP PLMEAA 53,399,87
orl-rat LD50:274 mg/kg PLMEAA 53,399,87
ipr-rat LD50:116 mg/kg PLMEAA 53,399,87
orl-mus LD50:430 mg/kg PLMEAA 53,399,87
ipr-mus LD50:355 mg/kg PLMEAA 53,399,87

SAFETY PROFILE: Poison by ingestion and intraperitoneal routes. Experimental reproductive effects. When heated to decomposition it emits acrid smoke and irritating fumes.

FON100 ***HR: 2***
FRAXINUS JAPONICA Blume, bark extract

TOXICITY DATA with REFERENCE
orl-rat LDLo:20 g/kg KSRNAM 4,253,70
scu-rat LD50:5400 mg/kg KSRNAM 4,253,70
ivn-rat LD50:1620 mg/kg KSRNAM 4,253,70
orl-mus LD50:16300 mg/kg KSRNAM 4,253,70
scu-mus LD50:6300 mg/kg KSRNAM 4,253,70
ivn-mus LD50:3280 mg/kg KSRNAM 4,253,70

SAFETY PROFILE: Moderately toxic by intravenous route. Mildly toxic by ingestion and subcutaneous routes.

FOO000 CAS:76-13-1 ***HR: 1***
FREON 113
mf: $C_2Cl_3F_3$ mw: 187.37

PROP: Colorless gas. Mp: 13.2°, bp: 45.8°, d: 1.5702, autoign temp: 1256°F.

SYNS: ARCTON 63 ◇ ARKLONE P ◇ DAIFLON S 3 ◇ FLUOROCARBON 113 ◇ FREON 113TR-T ◇ FRIGEN 113a ◇ GENETRON 113 ◇ HALOCARBON 113 ◇ ISCEON 113 ◇ KAISER CHEMICALS 11 ◇ KHLADON 113 ◇ R 113 ◇ REFRIGERANT 113 ◇ TRICHLOROTRIFLUOROETHANE ◇ 1,1,2-TRICHLORO-1,2,2-TRIFLUOROETHANE (OSHA, ACGIH, MAK) ◇ UCON 113 ◇ UCON FLUOROCARBON 113 ◇ UCON 113/HALOCARBON 113

TOXICITY DATA with REFERENCE
skn-rbt 500 mg open MLD UCDS** 7/10/70
orl-rat LD50:43 g/kg JMCMAR 7,378,64
ihl-rat LCLo:87000 ppm/6H JOCMA7 4,262,62
ihl-mus LCLo:25 pph/90S ANASAB 16,3,61

CONSENSUS REPORTS: Reported in EPA TSCA Inventory.

OSHA PEL: (Transitional: TWA 1000 ppm) TWA 1000 ppm; STEL 1250 ppm
ACGIH TLV: TWA 1000 ppm; STEL 1250 ppm
DFG MAK: 1000 ppm (7600 mg/m^3)

SAFETY PROFILE: Mildly toxic by ingestion and inhalation. Affects the central nervous system in humans. A skin irritant. Combustible when exposed to heat or flame. Incompatible with Al; Ba; Li; Sm; NaK alloy; Ti. See also CHLORINATED HYDROCARBONS, ALIPHATIC; and FLUORIDES.

FOO509 CAS:76-14-2 ***HR: 1***
FREON 114
mf: $C_2Cl_2F_4$ mw: 170.92

PROP: Colorless, practically odorless, noncorrosive, nonirritating, nonflammable gas. Faint, ether-like odor in high concentrations. D: 1.5312, mp: −94°, bp: 4.1°, n (0/D) 1.3092. Insol in water; sol in alc and ether.

SYNS: ARCTON 33 ◇ ARCTON 114 ◇ CRYOFLUORAN ◇ CRYOFLUORANE ◇ sym-DICHLOROTETRAFLUOROETHANE ◇ 1,2-DICHLORO-1,1,2,2-TETRAFLUOROETHANE (MAK) ◇ DICHLOROTETRAFLUOROETHANE (OSHA, ACGIH) ◇ F 114 ◇ FC 114 ◇ FLUORANE 114 ◇ FLUOROCARBON 114 ◇ FRIGEN 114 ◇ FRIGIDERM ◇ GENETRON 114 ◇ GENETRON 316 ◇ HALOCARBON 114 ◇ LEDON 114 ◇ PROPELLANT 114 ◇ R 114 ◇ 1,1,2,2-TETRAFLUORO-1,2-DICHLOROETHANE ◇ UCON 114

TOXICITY DATA with REFERENCE
ihl-rat LC50:72 pph/30M EJTXAZ 9,385,76
ihl-mus LC50:70 pph/30M EJTXAZ 9,385,76
ihl-rbt LC50:75 pph/30M EJTXAZ 9,385,76

CONSENSUS REPORTS: Reported in EPA TSCA Inventory.

OSHA PEL: TWA 1000 ppm
ACGIH TLV: TWA 1000 ppm
DFG MAK: 1000 ppm (7000 mg/m^3)

SAFETY PROFILE: An asphyxiant. See also DICHLOROTETRAFLUOROETHANE.

FOO600 CAS:9007-81-2 ***HR: D***
FREUND'S ADJUVANT

TOXICITY DATA with REFERENCE
par-rat TDLo:1 g/kg (female 1D pre):REP JRPFA4 14,147,67

SAFETY PROFILE: Experimental reproductive effects. When heated to decomposition it emits acrid smoke and irritating fumes.

FOO875 CAS:23191-75-5 ***HR: 3***
FTORIN
mf: $C_{20}H_{22}F_3NO_4$ mw: 397.43

SYNS: 1,4-DIHYDRO-2,6-DIMETHYL-4-(α,α,α-TRIFLUORO-o-TOLYL)-3,5-PYRIDINEDICARBOXYLIC ACID DIETHYL ESTER ◇ FTORIN (PHARMACEUTICAL) ◇ SKF 24260 ◇ 4-(2-TRIFLUOROMETHYLPHENYL)-3,5-DICARBETHOXY-2,6-DIMETHYL-1,4-DIHYDROPYRIDINE

TOXICITY DATA with REFERENCE
orl-rat LD50:1225 mg/kg JMCMAR 17,956,74
ivn-rat LDLo:5 mg/kg JMCMAR 17,956,74
orl-mus LD50:1480 mg/kg JMCMAR 17,956,74
ipr-mus LD50:38 mg/kg PCJOAU 16,817,82
ivn-dog LDLo:700 μg/kg JMCMAR 17,956,74

SAFETY PROFILE: Poison by intravenous and intraperitoneal routes. Moderately toxic by ingestion. When heated to decomposition it emits toxic fumes of F^- and NO_x.

FOP000 ***HR: 2***
FUEL OIL
DOT: NA 1201

PROP: A petroleum fraction consisting of a complex mixture of aromatic, paraffinic, olefinic, and naphthenic hydrocarbons. Brown, sltly viscous liquid. Flash p: 100°F, d: <1, autoign temp: 494°F.

SYN: DIESEL FUEL (DOT)

TOXICITY DATA with REFERENCE
orl-rat LD50:9 g/kg 52MLA2 1,1,83

DOT Classification: Combustible Liquid; Label: None.

SAFETY PROFILE: Mildly toxic by ingestion. Combustible when exposed to heat or flame; can react vigorously with oxidizing materials. To fight fire, use CO_2, dry chemical. When heated to decomposition it emits acrid smoke and irritating fumes. See also DIESEL FUEL; MINERAL OIL; KEROSENE.

FOP100 ***HR: 2***
FUEL OIL, pyrolyzate

SYN: WATER QUENCH PYROLYSIS FUEL OIL

TOXICITY DATA with REFERENCE
skn-mus TDLo:977 g/kg/8W-I:CAR AIHAAP 38,730,77

SAFETY PROFILE: Questionable carcinogen with experimental carcinogenic data. When heated to decomposition it emits acrid smoke and irritating fumes.

FOQ000 CAS:4368-28-9 ***HR: 3***
FUGU POISON
mf: $C_{11}H_{17}N_3O_8$ mw: 319.31

SYNS: MACULOTOXIN ◇ SPHEROIDINE ◇ TARICHATOXIN ◇ TETRODONTOXIN ◇ TETRODOTOXIN ◇ TETRODOXIN ◇ TTX

TOXICITY DATA with REFERENCE
orl-mus LD50:435 μg/kg JJPAAZ 17,267,67
ipr-mus LD50:8 μg/kg SCIEAS 144,1100,64
scu-mus LD50:8 μg/kg CTOXAO 4,331,71
ivn-mus LD50:9 μg/kg JJPAAZ 17,267,67

SAFETY PROFILE: Poison by ingestion, intraperitoneal, subcutaneous, and intravenous routes. When heated to decomposition it emits toxic fumes of NO_x.

FOR000 CAS:3309-87-3 ***HR: 3***
FUJITHION
mf: $C_8H_{10}ClO_3PS$ mw: 252.66

SYNS: S-(p-CHLOROPHENYL)-O,O-DIMETHYLPHOSPHOTHIOATE ◇ O,O-DIMETHYL-S-p-CHLOROPHENYL PHOSPHOROTHIOATE

TOXICITY DATA with REFERENCE
orl-mus LD50:94 mg/kg BESAAT 15,118,69
skn-mus LD50:920 mg/kg BESAAT 15,118,69

SAFETY PROFILE: Poison by ingestion. Moderately toxic by skin contact. When heated to decomposition it emits very toxic fumes of Cl^-, PO_x, and SO_x.

FOS000 ***HR: 3***
FULMINATES

SAFETY PROFILE: Variable toxicity. A very dangerous fire hazard when exposed to heat or flame. Severe explosion hazard when shocked or exposed to heat or flame. See also various fulminates and EXPLOSIVES, HIGH.

The fulminates are a group of explosives which are very sensitive to heat, impact, and friction when dry. They should be kept moist until ready for use. If compressed beyond 25,000 psi they become what is known as "dead-pressed," i.e., not capable of being exploded by flame. Fulminates are subject to deterioration when stored in hot climates. They decompose completely and violently when detonated. They can be ignited with a flame or "spit," with a fuse, or with an electrically heated wire. They are widely used as initiators or primers for detonation of high explosives or the ignition of powder. They are commonly used in combination with substances which provide a more prolonged blow and a bigger flame than fulminates alone. In the reinforced type of detonator, fulminates are made more effective by the addition of a more sensitive and powerful high explosive such as tetryl. This material is generally used in the manufacture of caps and detonators for initiating explosions for military, industrial, and sporting purposes.

All precautions required for protection of magazines apply to storage of these materials. They should not be handled when frozen. Wet fulminate of mercury or wet floor coverings containing small quantities of fulminates may be burned on windrows of flammable material. Nonexplosive products are formed by neutralizing fulminates with cold sodium thiosulfate. All floors, tables, and walls where the dry fulminates have been used should be washed with this solution. In the manufacture of mercury fulminate, the fumes given off are toxic and flammable. Care is required to prevent fulminate dust from being carried off in the exhaust system: deposits thus made have caused explosions. Careful attention should be given to cleanliness as foreign or gritty materials in the product may cause an unexpected explosion. The floors on which fulminates are used should be covered with 1/16-inch cloth inserted rubber packing or its equal. All cracks and crevices should be covered. The walls of these rooms should be covered with glazed, waterproof material. Frequent washing with neutralizing solution is necessary. In manufacture, the fulminate is dried on muslin squares on a drying table. Drying tables may be heated with hot water or the dry house may be heated with an air blower system to between 50 and 60°. Primer caps and detonators loaded with fulminate of mercury are less sensitive than the dry bulk material but must be handled with great care. Fires involving these assemblies should be treated the same as for the bulk material. They will explode as soon as fire reaches them. Stocks in an assembly or loading room should be kept as small as possible. Examples of fulminates commonly used in the explosive industry are mercury fulminate, copper fulminate, and silver fulminate.

FOS050 CAS:506-85-4 ***HR: 3***
FULMINIC ACID
mf: CHNO mw: 43.02

SYN: HYDROGENCYANIDE-N-OXIDE

CONSENSUS REPORTS: Cyanide and its compounds are on the Community Right-To-Know List.
DOT Classification: Forbidden

SAFETY PROFILE: An unstable explosive sensitive to heat, shock, or friction. When heated to decomposition it emits toxic fumes of NO_x.

FOS100 CAS:14976-57-9 ***HR: 3***
FULUMINOL
mf: $C_{21}H_{26}ClNO \cdot C_4H_4O_4$ mw: 460.01

SYNS: AGASTEN ◇ ALOGINAN ◇ ALPHAMIN ◇ ANHISTAN ◇ (+)-2-(2-((p-CHLORO-α-METHYL-α-PHENYLBENZYL)OXY)ETHYL)-1-METHYL PYRROLIDINE FUMARATE ◇ CLEMANIL ◇ CLEMASTINE FUMARATE ◇ CLEMASTINE HYDROGEN FUMARATE ◇ HS 592 ◇ INBESTAN ◇ KINOTOMIN ◇ LACRETIN ◇ LECASOL ◇ MAIKOHIS ◇ MALLERMIN-F ◇ MARSTHINE ◇ MASLETINE ◇ MECLASTINE HYDROGEN FUMARATE ◇ PILORAL ◇ RECONIN ◇ TAVEGIL ◇ TAVEGYL ◇ TELGIN-G ◇ TIVIST ◇ TRABEST ◇ XOLAMIN

TOXICITY DATA with REFERENCE
orl-rat LD50:3550 mg/kg BCFAAI 106,467,67
ivn-rat LD50:82 mg/kg BCFAAI 106,467,67
orl-mus LD50:730 mg/kg BCFAAI 106,467,67
ivn-mus LD50:43 mg/kg BCFAAI 106,467,67
orl-dog LD50:175 mg/kg BCFAAI 106,467,67
orl-rbt LD50:1 g/kg NIIRDN 6,230,82
ivn-rbt LD50:19 mg/kg BCFAAI 106,467,67

SAFETY PROFILE: Poison by ingestion and intravenous routes. When heated to decomposition it emits toxic fumes of Cl^- and NO_x.

FOS300 CAS:72443-10-8 ***HR: 3***
6-FULVENOSELONE
mf: C_6H_4Se mw: 155.06

CONSENSUS REPORTS: Selenium and its compounds are on the Community Right-To-Know List.

OSHA PEL: TWA 0.2 mg(Se)/m^3
ACGIH TLV: TWA 0.2 mg(Se)/m^3
DFG MAK: 0.1 mg(Se)/m^3

SAFETY PROFILE: Polymerizes explosively at − 196°C. When heated to decomposition it emits toxic fumes of Se. See also SELENIUM COMPOUNDS.

FOT000 CAS:6029-87-4 ***HR: 3***
FULVINE
mf: $C_{16}H_{23}NO_5$ mw: 309.40

SYN: CRISPATINE

TOXICITY DATA with REFERENCE
sln-dmg-par 10 mmol/L JOGNAU 59,273,66
trn-dmg-par 10 mmol/L JOGNAU 59,273,66
ipr-rat TDLo:20 mg/kg (female 9-12D post):TER EXPTAX 9,59,74
ipr-rat TDLo:40 mg/kg (9-12D preg):REP EXPTAX 9,59,74
ipr-rat LD50:40 mg/kg PAREAQ 22,429,70

SAFETY PROFILE: Poison by intraperitoneal route. Experimental teratogenic and reproductive effects. Mutation data reported. When heated to decomposition it emits toxic fumes of NO_x.

FOU000 CAS:110-17-8 ***HR: 3***
FUMARIC ACID
mf: $C_4H_4O_4$ mw: 116.08

PROP: White crystals; odorless. Mp: 287°, d: 1.635 @ 20°/4°. Bp: 290°. Sol in water, ether; very sltly sol in chloroform.

SYNS: ALLOMALEIC ACID ◇ BOLETIC ACID ◇ trans-BUTENEDIOIC ACID ◇ (E)-BUTENEDIOIC ACID ◇ trans-1,2-ETHYLENEDICARBOXYLIC ACID ◇ (E)1,2-ETHYLENEDICARBOXYLIC ACID ◇ KYSELINA FUMAROVA (CZECH) ◇ LICHENIC ACID ◇ NSC-2752 ◇ U-1149 ◇ USAF EK-P-583

TOXICITY DATA with REFERENCE
skn-rbt 500 mg/24H MLD 28ZPAK -,51,72
eye-rbt 100 mg/24H MOD 28ZPAK -,51,72
orl-rat LD50:10700 mg/kg TXAPA9 42,417,77
ipr-rat LDLo:587 mg/kg JAPMA8 35,298,46
ipr-mus LD50:100 mg/kg NTIS** AD277-689
orl-rbt LDLo:5000 mg/kg IECHAD 15,628,23
skn-rbt LD50:20 g/kg TXAPA9 42,417,77

CONSENSUS REPORTS: Reported in EPA TSCA Inventory.

SAFETY PROFILE: Poison by intraperitoneal route. Mildly toxic by ingestion and skin contact. A skin and eye irritant. Combustible when exposed to heat or flame; can react vigorously with oxidizing materials. When heated to decomposition it emits acrid smoke and irritating fumes.

FOV000 CAS:25876-47-5 ***HR: 3***
FUMARIC ACID ETHYL-2,3-EPOXYPROPYL ESTER
mf: $C_9H_{12}O_5$ mw: 200.21

mf: $C_9H_{12}O_5$ mw: 200.21

TOXICITY DATA with REFERENCE
orl-rat LD50:1620 mg/kg AIHAAP 30,470,69
skn-rbt LD50:350 mg/kg AIHAAP 30,470,69

SAFETY PROFILE: Poison by skin contact. Moderately toxic by ingestion. When heated to decomposition it emits acrid smoke and irritating fumes. See also ESTERS.

FOW000 CAS:130-86-9 ***HR: 3***
FUMARINE
mf: $C_{20}H_{19}NO_5$ mw: 353.40

SYNS: BIFLORINE ◇ CORYDININE ◇ MACLEYINE ◇ 7-METHYL-2,3:9,10-BIS(METHYLENEDIOXY)-7,13a-SECOBERBIN-13a-ONE◇ PROTOPINE ◇ 4,6,7,14-TETRAHYDRO-5-METHYL-BIS(1,3)BENZODIOXOLO(4,5-c:5′,6′-g)AZECIN-13(5H)-ONE

TOXICITY DATA with REFERENCE
ipr-rat LDLo:100 mg/kg NCNSA6 5,24,53
ipr-mus LD50:482 mg/kg YHTPAD 16(6),7,81
orl-gpg LD50:237 mg/kg THERAP 28,767,73
ipr-gpg LD50:116 mg/kg THERAP 28,767,73

SAFETY PROFILE: Poison by ingestion and intraperitoneal routes. When heated to decomposition it emits toxic fumes of NO_x.

FOX000 CAS:764-42-1 ***HR: 3***
FUMARONITRILE
mf: $C_4H_2N_2$ mw: 78.08

TOXICITY DATA with REFERENCE
orl-rat LD50:132 mg/kg TOLED5 12,157,82
ivn-mus LD50:56 mg/kg CSLNX* NX#05718

CONSENSUS REPORTS: Reported in EPA TSCA Inventory. Cyanide and its compounds are on the Community Right-To-Know List.

SAFETY PROFILE: Poison by ingestion and intravenous routes. When heated to decomposition it emits toxic fumes of NO_x and CN^-. See also NITRILES.

FOY000 CAS:627-63-4 ***HR: 2***
FUMARYL CHLORIDE
DOT: UN 1780
mf: $C_4H_2Cl_2O_2$ mw: 152.96

PROP: Clear, straw-colored liquid. Mp: 160°, d: 1.408 @ 20°/4°.

SYNS: CHLORURE de FUMARYLE (FRENCH) ◇ DICHLORID KYSELINY FUMAROVE (CZECH) ◇ FUMAROYL CHLORIDE ◇ FUMARYLCHLORID (CZECH) ◇ TL 189

TOXICITY DATA with REFERENCE
skn-rbt 500 mg/24H SEV 28ZPAK -,53,72
eye-rbt 5 mg/24H SEV 28ZPAK -,53,72
orl-rat LD50:810 mg/kg AIHAAP 30,470,69
ihl-rat LCLo:500 ppm/4H AIHAAP 30,470,69
ihl-mus LCLo:1000 mg/m^3/10M NDRC** NDCrc-132,May,42
skn-rbt LD50:1410 mg/kg AIHAAP 30,470,69

CONSENSUS REPORTS: Reported in EPA TSCA Inventory.

DOT Classification: Corrosive Material; Label: Corrosive.

SAFETY PROFILE: Moderately toxic by ingestion, inhalation, and skin contact. A skin, eye and mucous membrane irritant. A corrosive agent. Will react with water or steam to produce toxic and corrosive fumes. When heated to decomposition it emits highly toxic fumes of phosgene and HCl.

FOZ000 CAS:23110-15-8 ***HR: 2***
FUMIDIL
mf: $C_{26}H_{34}O_7$ mw: 458.60

PROP: Isolated from *A. fumigatus* (ANTCAO 1,54,51).

SYNS: AMEBACILIN ◇ FUGILIN ◇ FUMADIL B ◇ FUMAGILLIN

TOXICITY DATA with REFERENCE
unk-rat TDLo:25 mg/kg (11D preg):TER 85DJA5 -,95,71
unk-rat TDLo:25 mg/kg (11D preg):REP 85DJA5 -,95,71
orl-mus LD50:2000 mg/kg 85ERAY 3,1834,78
scu-mus LD50:800 mg/kg ANTCAO 1,54,51

SAFETY PROFILE: Moderately toxic by ingestion and subcutaneous routes. An experimental teratogen. Experimental reproductive effects. When heated to decomposition it emits acrid smoke and fumes.

FPA000 CAS:11085-39-5 ***HR: 3***
FUMIGACHLORIN
mf: $C_{16}H_{25}Cl_2NO_4$ mw: 366.2

TOXICITY DATA with REFERENCE
orl-mus LD50:18500 μg/kg 85ERAY 3,1858,78
ipr-mus LD50:4600 μg/kg 85ERAY 3,1858,78
scu-mus LD50:9300 μg/kg 85ERAY 3,1858,78

SAFETY PROFILE: Poison by ingestion, intraperitoneal, and subcutaneous routes. When heated to decomposition it emits very toxic fumes of Cl^- and NO_x.

FPB875 CAS:35554-44-0 ***HR: 3***
FUNGAFLOR
mf: $C_{14}H_{14}Cl_2N_2O$ mw: 297.20

PROP: Solidified oil. Sltly sol in organic solvents; poorly sol in water.

SYNS: (±)-1-(β-(ALLYLOXY)-2,4-DICHLOROPHENETHYL)IMIDAZOLE ◇ 1-(2-(2,4-DICHLOROPHENYL)-2-(2-PROPENYLOXY)ETHYL)-1H-IMIDAZOLE ◇ 1-(2-(2,4-DICHLORPHENYL)-2-PROPENYLOXY)

AETHYL)-1H-IMIDAZOLE ◇ ENILOCONAZOL (SP) ◇ IMAVEROL ◇ IMAZALIL ◇ R 23979

TOXICITY DATA with REFERENCE
skn-rbt 640 mg/kg/24H MLD ARZNAD 31,309,81
eye-rbt 49 mg MOD ARZNAD 31,309,81
orl-rat TDLo:2240 mg/kg (16-22D preg/21D post):REP ARZNAD 31,309,81
orl-rat LD50:227 mg/kg ARZNAD 31,309,81
skn-rat LD50:4200 mg/kg PEMNDP 8,471,87
ipr-rat LD50:155 mg/kg ARZNAD 31,309,81

SAFETY PROFILE: Poison by ingestion and intraperitoneal routes. Experimental reproductive effects. A skin and eye irritant. When heated to decomposition it emits toxic fumes of Cl^- and NO_x.

FPC000 CAS:6834-98-6 ***HR: 3***
FUNGICHROMIN
mf: $C_{35}H_{58}O_{12}•H_2O$ mw: 688.95

SYNS: ANTIBIOTIC A-246 ◇ COGOMYCIN ◇ FUNGICHROMIN, HYDRATE ◇ 14-HYDROXYFILIPIN ◇ LAGOSIN ◇ MOLDCIDIN B ◇ NSC-105388 ◇ PENTAMYCIN

TOXICITY DATA with REFERENCE
orl-mus LD50:1624 mg/kg 85GDA2 2,212,80
ipr-mus LDLo:16400 μg/kg ABANAE 2,716,54/55

SAFETY PROFILE: Poison by intraperitoneal route. Moderately toxic by ingestion. When heated to decomposition it emits acrid smoke and irritating fumes.

FPC100 ***HR: 3***
FUNGINON
mf: $C_6H_8O_4$ mw: 144.14

TOXICITY DATA with REFERENCE
orl-mus LD50:109 mg/kg 85GDA2 5,371,81
ipr-mus LD50:20 mg/kg 85GDA2 5,371,81
ivn-mus LD50:24 mg/kg 85GDA2 5,371,81

SAFETY PROFILE: Poison by ingestion, intravenous, and intraperitoneal routes. When heated to decomposition it emits acrid smoke and irritating fumes.

FPD000 CAS:11055-06-4 ***HR: 3***
FUNICOLOSIN
mf: $C_{27}H_{41}NO$ mw: 395.69

TOXICITY DATA with REFERENCE
orl-rat LD50:5 mg/kg JANTAJ 31,533,78
ipr-rat LD50:5 mg/kg JANTAJ 31,533,78
orl-mus LD50:5 mg/kg JANTAJ 31,533,78
ipr-mus LD50:4 mg/kg 85ERAY 3,1803,78

SAFETY PROFILE: Poison by ingestion and intraperitoneal routes. When heated to decomposition it emits toxic fumes of NO_x.

FPD100 ***HR: 2***
FUQUA

PROP: Creeping vines with tendrils, deeply cut leaves and tubular yellow flowers. The yellow-orange fruit is pear-shaped or oval, has warts and a bright red pulp. The fruit splits open when ripe. They are common weeds in the Gulf coast states, Hawaii, Guam, and the West Indies.

SYNS: BALSAM APPLE ◇ BALSAM PEAR ◇ BITTER CUCUMBER ◇ BITTER GOURD ◇ CUNDEAMOR (CUBA, PUERTO RICO) ◇ MOMORDICA BALSAMINA ◇ MOMORDICA CHARANTIA ◇ MOMORDIQUE A FUEILLES de VIGNE ◇ SORCI ◇ SORROSIE (HAITI) ◇ WILD BALSAM APPLE ◇ YESQUIN (HAITI)

SAFETY PROFILE: The seeds and skin of the fruit contain the poisonous momordin, a toxalbumin which inhibits protein synthesis in the intestinal wall. The red pulp surrounding the seeds and the boiled leaves are edible. Ingestion of the seeds or skin of the fruit causes after a delay period: nausea, vomiting, diarrhea, and low blood sugar. See also ABRIN as an example toxalbumin.

FPE100 CAS:405-22-1 ***HR: 3***
FURADROXYL
mf: $C_8H_{10}N_4O_5$ mw: 242.22

PROP: Bright orange plates from alc. Mp: 214-216° (decomp), solubility in water 1:2000.

SYNS: NIDROXYZONE ◇ 5-NITRO-2-FURALDEHYDE-2-(2-HYDROXYETHYL)SEMICARBAZONE ◇ USAF EA-3

TOXICITY DATA with REFERENCE
orl-mus TDLo:5520 mg/kg (female 23D pre):REP ENDOAO 69,331,61
orl-mus LD50:600 mg/kg JAPMA8 39,313,50
ipr-mus LD50:200 mg/kg NTIS** AD277-689

SAFETY PROFILE: Poison by intraperitoneal route. Moderately toxic by ingestion. Experimental reproductive effects. When heated to decomposition it emits toxic fumes of NO_x. See also ALDEHYDES.

FPF000 CAS:556-12-7 ***HR: D***
FURALAZIN
mf: $C_9H_7N_5O_3$ mw: 233.2

SYNS: 3-AMINO-6-(2-(5-NITRO-2-FURYL)VINYL)-as-TRIAZINE ◇ 3-AMINO-6-(2-(5-NITRO-2-FURYL)VINYL)-1,2,4-TRIAZINE

TOXICITY DATA with REFERENCE
plc-esc 100 μg/L MUREAV 26,3,74
mmo-esc 300 μg/L CJMIAZ 11,185,65
mmo-eug 2 mg/L JPROAR 17,129,70

SAFETY PROFILE: Mutation data reported. When heated to decomposition it emits toxic fumes of NO_x.

FPH000 CAS:5428-37-5 ***HR: 3***
2-FURALDEHYDE AZINE
mf: $C_{10}H_8N_2O_2$ mw: 188.20

TOXICITY DATA with REFERENCE
ipr-mus LDLo:125 mg/kg CBCCT* 6,220,54

CONSENSUS REPORTS: Reported in EPA TSCA Inventory.

SAFETY PROFILE: Poison by intraperitoneal route. When heated to decomposition it emits toxic fumes of NO_x. See also ALDEHYDES.

FPI000 CAS:139-91-3 ***HR: 2***
FURALTADONE
mf: $C_{13}H_{16}N_4O_6$ mw: 324.33

PROP: Yellow crystals from 95% ethanol. Decomp 206°. Sparingly sol in water: about 75 mg/100 mL at 25°.

SYNS: ALTABACTINA ◇ ALTAFUR ◇ F-150 ◇ FURAZOLIN ◇ FURAZOLINE ◇ FURMETHANOL ◇ FURMETHONOL ◇ FURMETONOL ◇ IBIFUR ◇ MEDIFURAN ◇ 5-MORPHOLINOMETHYL-3-(5-NITRO-2-FURFURYLIDINE-AMINO)-2-OXAZOLIDINONE ◇ NF 260 ◇ NITRALDONE ◇ NITROFURMETHONE ◇ NITROFURMETON ◇ OTIFURIL ◇ SEPSINOL ◇ ULTRAFUR ◇ UNIFUR ◇ VALSYN

TOXICITY DATA with REFERENCE
mmo-sat 2500 ng/plate MUREAV 136,1,84
mmo-esc 10 μg/plate MUREAV 26,3,74
pic-esc 5 mg/L CJMIAZ 10,932,64
mmo-omi 5 mg/L CUMIDD 10,19,84
ipr-mus LD50:1000 mg/kg JPPMAB 16,663,64

CONSENSUS REPORTS: EPA Genetic Toxicology Program.

SAFETY PROFILE: Moderately toxic by intraperitoneal route. Mutation data reported. When heated to decomposition it emits toxic fumes of NO_x.

FPI100 CAS:3759-92-0 ***HR: 3***
FURALTADONE HYDROCHLORIDE
mf: $C_{13}H_{16}N_4O_6 \cdot ClH$ mw: 360.79

SYNS: 5-MORPHOLINOMETHYL-3-(5-NITROFURFURYLIDINE) AMINO-2-OXAZOLIDINONE HYDROCHLORIDE ◇ NF-269 ◇ NF-902 HYDROCHLORIDE

TOXICITY DATA with REFERENCE
mmo-esc 100 μmol/L RAREAE 75,424,78
orl-mus LD50:1000 mg/kg JPPMAB 16,663,64
ipr-mus LD50:190 mg/kg JPPMAB 16,663,64

CONSENSUS REPORTS: EPA Genetic Toxicology Program.

SAFETY PROFILE: Poison by intraperitoneal route. Moderately toxic by ingestion. Mutation data reported. When heated to decomposition it emits toxic fumes of NO_x and HCl.

FPI150 CAS:3031-51-4 ***HR: 3***
l-FURALTADONE HYDROCHLORIDE
mf: $C_{13}H_{16}N_4O_6 \cdot ClH$ mw: 360.79

PROP: Yellow crystals. Decomposes @ 206°.

SYNS: FURMETHONOL ◇ l-5-(MORPHOLINOMETHYL)-3-((5-NITROFURFURYLIDENE)AMINO)-2-OXAZOLIDINONEHYDROCHLORIDE

TOXICITY DATA with REFERENCE
mmo-eug 10 mg/L JPROAR 17,129,70
orl-rat TDLo:25 g/kg/46W-C:CAR JNCIAM 51,403,73
orl-mus LD50:600 mg/kg FRPSAX 19,269,64
ivn-mus LD50:400 mg/kg FRPSAX 19,269,64

CONSENSUS REPORTS: IARC Cancer Review: Group 2B IMEMDT 7,56,87; Animal Limited Evidence IMEMDT 7,161,74.

SAFETY PROFILE: Suspected carcinogen with experimental carcinogenic data. Poison by intravenous route. Moderately toxic by ingestion. Mutation data reported. When heated to decomposition it emits very toxic fumes of HCl and NO_x.

FPK000 CAS:110-00-9 ***HR: 3***
FURAN
DOT: UN 2389
mf: C_4H_4O mw: 68.08

PROP: Water white liquid. Mp: −85.65°, bp: 31.36°, lel: 2.3%, uel: 14.3%, flash p: −32°F, d: 0.937 @ 20°/4°, vap d: 2.35.

SYNS: DIVINYLENE OXIDE ◇ FURFURAN ◇ NCI-C56202 ◇ OXACYCLOPENTADIENE ◇ OXOLE ◇ RCRA WASTE NUMBER U124 ◇ TETROLE

TOXICITY DATA with REFERENCE
cyt-ham:ovr 184 mmol/L CALEDQ 13,89,81
ipr-rat LD50:5200 μg/kg AIHAAP 40,310,79
ihl-mus LC50:120 mg/m³/1H AIHAAP 40,310,79
ipr-mus LD50:7 mg/kg AIHAAP 40,310,79

CONSENSUS REPORTS: EPA Extremely Hazardous Substances List. Reported in EPA TSCA Inventory.

DOT Classification: Flammable Liquid; Label: Flammable Liquid.

SAFETY PROFILE: Poison by inhalation and intraperitoneal routes. Probably at least moderately toxic by ingestion and skin contact. A narcotic. Mutation data reported. The exposure concentration limit of 10 ppm together with its low boiling point requires that adequate ventilation be provided in areas where this chemical is handled. Contact with liquid must be avoided since this chemical can be absorbed through the skin. Washing thoroughly with soap and water followed by prolonged rinsing should be done immediately after accidental contact.

A very dangerous fire hazard when exposed to heat or flame; can react with oxidizing materials. Unstabilized, it may form unstable peroxides on exposure to air and should always be tested before distillation. Washing with an aqueous solution of ferrous sulfate slightly acidified with sodium bisulfate will remove these peroxides. Confirm by test. Contact with acids can initiate a violent exothermic reaction. Moderate explosion hazard when exposed to flame. Furan's low boiling point makes it easy to obtain explosive concentrations of the vapor in inadequately ventilated areas. Highly dangerous upon exposure to heat or flame; can react vigorously with oxidizing materials. To fight fire, use CO_2, dry chemical. When heated to decomposition it emits acrid smoke and irritating fumes. See also PEROXIDES.

FPK100 CAS:50892-99-4 ***HR: 3***
FURAN-2-AMIDOXIME
mf: $C_5H_6N_2O_2$ mw: 126.11

OCH=CHCH=CC(NH_2):NOH

SAFETY PROFILE: Explodes when heated above 100°C. Exothermic reaction if heated to 65°C. When heated to decomposition it emits toxic fumes of NO_x.

FPM000 CAS:98-02-2 ***HR: 3***
2-FURANMETHANETHIOL
mf: C_5H_6OS mw: 114.17

SYNS: FURFURYL MERCAPTAN ◇ USAF B-58

TOXICITY DATA with REFERENCE
orl-rat TDLo:1260 mg/kg (42D male):REP FCTXAV 15,383,77
ipr-mus LD50:100 mg/kg NTIS** AD277-689

CONSENSUS REPORTS: Reported in EPA TSCA Inventory.

SAFETY PROFILE: Poison by intraperitoneal route. Experimental reproductive effects. Used as a flavoring in chocolate, fruit, nuts, and coffee. When heated to decomposition it emits toxic fumes of SO_x. See also MERCAPTANS.

FPO000 CAS:1951-56-0 ***HR: D***
FURAPROMIDIUM
mf: $C_{10}H_{12}N_2O_4$ mw: 224.24

SYNS: N-ISOPROPYL-5-NITRO-2-FURANACRYLAMIDE ◇ N-ISOPROPYL-3-(5-NITRO-2-FURYL)-ACRYLAMIDE

TOXICITY DATA with REFERENCE
mmo-sat 14 μmol/L MUREAV 48,37,77
mmo-nsc 200 μg/plate MUREAV 53,297,78
msc-ham:lng 120 μmol/L MUREAV 48,37,77

SAFETY PROFILE: Mutation data reported. When heated to decomposition it emits toxic fumes of NO_x.

FPO100 CAS:75888-03-8 ***HR: 3***
FURAPYRIMIDONE
mf: $C_9H_{10}N_4O_4$ mw: 238.23

SYNS: 1-((5-NITROFURANYL-2)METHYLENEAMINO)TETRAHYDROPYRIMIDONE-2-ONE ◇ TETRAHYDRO-1-((5-NITROFURFURYLIDENE)AMINO)-2(1H)-PYRIMIDINONE ◇ TETRAHYDRO-1-(5-NITROFURFURYLIDENEAMINO)-2-PYRIMIDONE

TOXICITY DATA with REFERENCE
mmo-sat 100 ng/plate CYLPDN 4,201,83
mma-sat 100 ng/plate CYLPDN 4,201,83
orl-rat TDLo:1050 mg/kg (7-11D preg):TER CYLPDN 4,201,83
orl-mus LD50:243 mg/kg CYLPDN 1,56,80
ipr-mus LD50:720 mg/kg JPPMAB 16,663,64

SAFETY PROFILE: Poison by ingestion. Moderately toxic by intraperitoneal route. Experimental teratogenic effects. Mutation data reported. When heated to decomposition it emits toxic fumes of NO_x.

FPP100 CAS:19237-84-4 ***HR: 3***
FURAZOSIN HYDROCHLORIDE
mf: $C_{19}H_{21}N_5O_4 \cdot ClH$ mw: 419.91

SYNS: ABBOTT-45975 ◇ 1-(4-AMINO-6,7-DIMETHOXY-2-QUINAZOLINYL)-4-(2-FURANYLCARBONYL)PIPERAZINEHYDROCHLORIDE ◇ 1-(4-AMINO-6,7-DIMETHOXY-2-QUINAZOLINYL)-4-(2-FUROYL)PIPERAZINE MONOHYDROCHLORIDE ◇ 2-(4-(2-FUROYL) PIPERAZIN-1-YL)-4-AMINO-6,7-DIMETHOXYQUINAZOLINEHYDROCHLORIDE ◇ PRAZOSIN HYDROCHLORIDE ◇ TERAZOSIN

TOXICITY DATA with REFERENCE
orl-rat TDLo:8700 mg/kg (female 14-22D post):REP OYYAA2 17,57,79
orl-rat TDLo:300 mg/kg (female 9-14D post):TER OYYAA2 17,57,79
orl-man TDLo:1714 μg/kg HUTODJ 4,53,85
orl-hmn TDLo:285 μg/kg:CVS JAMAAP 238,157,77
orl-wmn TDLo:13 mg/kg/6W-I:CNS,PSY BMJOAE 293,1347,86
orl-wmn TDLo:10 μg/kg DICPBB 21,723,87
orl-rat LD50:1950 mg/kg IYKEDH 12,933,81
ipr-rat LD50:102 mg/kg NIIRDN 6,688,82
ivn-rat LD50:277 mg/kg NIIRDN 6,688,82
ipr-mus LD50:60 mg/kg NIIRDN 6,688,82
scu-mus LD50:3100 mg/kg NIIRDN 6,688,82

SAFETY PROFILE: Poison by intravenous and intraperitoneal routes. Moderately toxic by ingestion and subcutaneous routes. Human systemic effects by ingestion: blood pressure depression, encephalitis, somnolence, toxic psychosis. An experimental teratogen. Experimental reproductive effects. When heated to decomposition it emits toxic fumes of NO_x and HCl.

FPQ000 CAS:9000-21-9 ***HR: 2***
FURCELLERAN GUM

PROP: Vegetable gum from *Furcellaria fastigiata* (Fam. *Rodophyceae*) available as an odorless white powder. Sol in warm water.

SYN: BURTONITE 44

TOXICITY DATA with REFERENCE
orl-rat LD50:5000 mg/kg FDRLI* 124,-,76
orl-mus LD50:6000 mg/kg FDRLI* 124,-,76
orl-rbt LD50:3400 mg/kg FDRLI* 124,-,76
orl-ham LD50:4800 mg/kg FDRLI* 124,-,76

CONSENSUS REPORTS: Reported in EPA TSCA Inventory.

SAFETY PROFILE: Moderately toxic by ingestion. When heated to decomposition it emits acrid smoke and fumes.

FPQ100 CAS:2385-81-1 ***HR: 3***
FURETHIDINE
mf: $C_{21}H_{31}NO_4$ mw: 361.47

PROP: Bp: 210°, mp: about 28°, n (20/D) 1.5219.

SYN: ETHYL4-PHENYL-1-(2-TETRAHYDROFURFURYLOXYETHYL)PIPERIDINE-4-CARBOXYLATE

TOXICITY DATA with REFERENCE
orl-rat LD50:135 mg/kg BJPCAL 15,254,60
scu-rat LD50:26 mg/kg BJPCAL 15,254,60
ivn-mus LD50:15500 μg/kg BJPCAL 15,254,60

SAFETY PROFILE: Poison by ingestion, subcutaneous, and intravenous routes. When heated to decomposition it emits toxic fumes of NO_x. Can be habit forming. This is a controlled substance (opiate) listed in the U.S. Code of Federal Regulations, Title 21 Part 1308.11 (1985). When heated to decomposition it emits toxic fumes of NO_x. See also ESTERS.

FPQ875 CAS:98-01-1 ***HR: 3***
FURFURAL
DOT: UN 1199
mf: $C_5H_4O_2$ mw: 96.09

OCH=CHCH=CCO•H

PROP: Colorless-yellowish liquid; almond-like odor. Bp: 161.7° @ 764 mm, lel: 2.1%, uel: 19.3%, flash p: 140°F (CC), d: 1.154-1.158, refr index: 1.522-1.528, autoign temp: 600°F, vap d: 3.31. Sol in water; misc with alc.

SYNS: ARTIFICIAL ANT OIL ◇ FEMA No. 2489 ◇ FURAL ◇ FURALE ◇ 2-FURALDEHYDE ◇ 2-FURANALDEHYDE ◇ 2-FURANCARBONAL ◇ 2-FURANCARBOXALDEHYDE ◇ 2-FURFURAL ◇ FURFURALDEHYDE ◇ FURFURALE (ITALIAN) ◇ FURFUROL ◇ FURFUROLE ◇ 2-FURIL-METANALE (ITALIAN) ◇ FUROLE ◇ α-FUROLE ◇ 2-FURYL-METHANAL ◇ NCI-C56177 ◇ PYROMUCIC ALDEHYDE ◇ RCRA WASTE NUMBER U125

TOXICITY DATA with REFERENCE
skn-rbt 500 mg/24H MOD 28ZPAK -,139,72
skn-rbt 500 mg/24H MLD FCTXAV 16,759,78
eye-rbt 20 mg/24H MOD 28ZPAK -,139,72
eye-rbt 50 mg MLD 34ZIAG -,279,69
sln-dmg-par 100 ppm ENMUDM 7,677,85
sce-hmn:lym 70 μmol/L MUREAV 156,233,85
ihl-hmn TCLo:310 μg/m³ GISAAA 26(6),3,61
orl-rat LD50:65 mg/kg BCTKAG 13,371,80
ihl-rat LCLo:153 ppm/4H 28ZPAK -,139,72
ipr-rat LD50:20 mg/kg FCTXAV 16,759,78
scu-rat LD50:148 mg/kg 34ZIAG -,279,69
orl-mus LD50:400 mg/kg BIJOAK 34,1196,40
ihl-mus LCLo:370 ppm/6H 34ZIAG -,279,69
ipr-mus LD50:102 mg/kg FCTXAV 16,759,78
scu-mus LD50:119 mg/kg FCTXAV 16,759,78
ivn-mus LD50:152 mg/kg FCTXAV 16,759,78
orl-dog LD50:950 mg/kg 34ZIAG -,279,69
ihl-dog LC50:370 ppm/6H 34ZIAG -,279,69
skn-rbt LDLo:620 mg/kg FCTXAV 16,759,78

CONSENSUS REPORTS: NTP Carcinogenesis Studies (gavage); Clear Evidence: Mouse NCITR* NTP-TR-382,90, Some Evidence: Rat NCITR* NTP-TR-382,90. EPA Genetic Toxicology Program. Reported in EPA TSCA Inventory.

OSHA PEL: (Transitional: TWA 5 mg/m³ (skin)) TWA 2 ppm (skin)
ACGIH TLV: TWA 2 ppm (skin)
DFG MAK: 5 ppm (20 mg/m³)
DOT Classification: Combustible Liquid; Label: None; Flammable or Combustible Liquid; Label: Flammable Liquid.

SAFETY PROFILE: Poison by ingestion, intraperitoneal, subcutaneous, intravenous, and intramuscular routes. Moderately toxic by inhalation and skin contact. Human mutation data reported. A skin and eye irritant. Mutation data reported. The liquid is dangerous to the eyes. The vapor is irritating to mucous membranes and is a central nervous system poison. However, its low volatility reduces its toxicity effect. Ingestion of furfural has produced cirrhosis of the liver in rats. In industry there is a tendency to minimize the danger of acute effects resulting from exposure to it. This is particularly true because of its low volatility.

Combustible liquid when exposed to heat or flame; can react with oxidizing materials. Moderate explosion hazard when exposed to heat or flame or by chemical reaction. An exothermic polymerization of almost explosive violence can occur upon contact with strong mineral acids or alkalies. Keep away from heat and open flames. Mixture with sodium hydrogen carbonate ignites spontaneously. To fight fire, use alcohol foam, CO_2, dry chemical. When heated to decomposition it emits acrid smoke and irritating fumes.

FPQ900 CAS:28438-99-5 ***HR: 3***
FURFURAL ACETONE MONOMER FA

SYNS: FA ◇ FA MONOMER ◇ FURFURAL-ACETONE ADDUCT ◇ FURFURAL-ACETONE MONOMER ◇ 1:1 FURFURAL-ACETONE MONOMER ◇ FURFUROLATSETONOVYI MONOMER FA ◇ MONOMER FA

TOXICITY DATA with REFERENCE
skn-rbt 500 mg/24H MLD 28ZPAK-,275,72
eye-rbt 500 mg/24H MLD 28ZPAK-,275,72
orl-rat LD50:592 mg/kg 28ZPAK-,275,72
skn-rat LD50:2600 mg/kg GTPZAB 18(5),53,74
orl-mus LD50:980 mg/kg GTPZAB 18(5),53,74
orl-rbt LD50:285 mg/kg GTPZAB 18(5),53,74
skn-rbt LD50:900 mg/kg GTPZAB 18(5),53,74

SAFETY PROFILE: Poison by ingstion. Moderately toxic by skin contact. A skin and eye irritant. When heated to decomposition it emits acrid smoke and irritating fumes.

FPR000 CAS:1121-47-7 ***HR: 3***
FURFURAL OXIME
mf: $C_5H_5NO_2$ mw: 111.11

TOXICITY DATA with REFERENCE
ipr-mus LD50:500 mg/kg JPETAB 119,522,57
ivn-mus LD50:180 mg/kg CSLNX* NX#03758

CONSENSUS REPORTS: Reported in EPA TSCA Inventory.

SAFETY PROFILE: Poison by intravenous route. Moderately toxic by intraperitoneal route. When heated to decomposition it emits toxic fumes such as NO_x.

FPS000 CAS:494-47-3 ***HR: 3***
FURFURAMIDE
mf: $C_{15}H_{12}N_2O_3$ mw: 268.29

PROP: Needles from alc. Mp: 117-21°, bp: 250° decomp. Insol in water; decomp in acid; sol in alc and ether.

SYNS: 2-(BIS(FURFURYLIDENAMINO))METHYLFURAN◇ HYDROFURAMIDE

TOXICITY DATA with REFERENCE
orl-rat LD50:400 mg/kg HYSAAV 29,37,64
orl-mus LD50:1200 mg/kg BIJOAK 34,1196,40
orl-rbt LDLo:50 mg/kg HYSAAV 29,37,64

SAFETY PROFILE: Poison by ingestion. A skin, eye, and mucous membrane irritant. Causes intense pulmonary irritation and reported to cause liver and kidney damage. When heated to decomposition it emits toxic fumes of NO_x. A component of fungicides. See also AMINES and AMIDES.

FPT000 CAS:699-17-2 ***HR: 3***
FURFURYLACETONE
mf: $C_8H_{10}O_2$ mw: 138.18

SYN: 4-(2-FURYL)-2-BUTANONE

TOXICITY DATA with REFERENCE
ipr-mus LDLo:63 mg/kg CBCCT* 6,217,54

CONSENSUS REPORTS: Reported in EPA TSCA Inventory.

SAFETY PROFILE: Poison by intraperitoneal route. When heated to decomposition it emits acrid smoke and irritating fumes.

FPU000 CAS:98-00-0 ***HR: 3***
FURFURYL ALCOHOL
DOT: UN 2874
mf: $C_5H_6O_2$ mw: 98.11

OCH=CHCH=CCH₂OH (ring closure from O to C)

PROP: Clear, colorless, mobile liquid. Mp: −31°, lel: 1.8%, uel: 16.3%, both between 72-122°, bp: 171° @ 750 mm, flash p: 167°F (OC), d: 1.129 @ 20°/4°, autoign temp: 915°F, vap press: 1 mm @ 31.8°, vap d: 3.37.

SYNS: 2-FURANCARBINOL ◇ 2-FURANMETHANOL ◇ FURFURAL ALCOHOL ◇ 2-FURFURYLALKOHOL (CZECH) ◇ FURYL ALCOHOL ◇ 2-FURYLCARBINOL ◇ α-FURYLCARBINOL ◇ (2-FURYL)METHANOL ◇ 2-HYDROXYMETHYLFURAN ◇ NCI-C56224

TOXICITY DATA with REFERENCE
eye-rbt 100 mg/24H MOD 28ZPAK -,139,72
cyt-ham:ovr 2500 μmol/L CALEDQ 13,89,81
orl-rat LD50:88300 μg/kg 28ZPAK -,139,72
ihl-rat LC50:233 ppm/4H AIHAAP 19,91,58
ipr-rat LD50:650 mg/kg NPIRI* 1,64,74
scu-rat LD50:85 mg/kg 34ZIAG -,280,69
orl-mus LD50:160 mg/kg BIJOAK 34,1196,40
ihl-mus LCLo:597 ppm/6H 34ZIAG -,280,69
skn-rbt LD50:400 mg/kg 34ZIAG -,280,69
ivn-rbt LD50:650 mg/kg FEPRA7 8,294,49

CONSENSUS REPORTS: Reported in EPA TSCA Inventory.

OSHA PEL: (Transitional: TWA 50 ppm) TWA 10 ppm; STEL 15 ppm (skin)
ACGIH TLV: TWA 10 ppm; STEL 15 ppm (skin)
DFG MAK: 50 ppm (200 mg/m³)
NIOSH REL: (Furfuryl Alcohol) TWA 200 mg/m³
DOT Classification: Poison B; Label: St. Andrews Cross.

SAFETY PROFILE: Poison by ingestion, skin contact, and subcutaneous routes. Moderately toxic by inhalation and intraperitoneal routes. Mutation data reported. An eye irritant. Flammable when exposed to heat or flame; can react with oxidizing materials. Moderate explosion hazard when exposed to heat or flame. Reacts violently with acids (e.g., formic acid; cyanoacetic acid + heat). Ignites on contact with 85% hydrogen peroxide. To fight fire, use alcohol foam, CO_2, dry chemical. When heated to decomposition it emits acrid smoke and fumes.

FPV000 CAS:10427-00-6 ***HR: 3***
FURFURYL ALCOHOL PHOSPHATE (3:1)
mf: $C_{15}H_{27}O_7P$ mw: 350.39

TOXICITY DATA with REFERENCE
ivn-mus LD50:320 mg/kg CSLNX* NX#03980

CONSENSUS REPORTS: Reported in EPA TSCA Inventory.

SAFETY PROFILE: Poison by intravenous route. When heated to decomposition it emits toxic fumes of PO_x.

FPW000 CAS:617-89-0 ***HR: 3***
FURFURYLAMINE
DOT: UN 2526
mf: C_5H_7NO mw: 97.13

PROP: Light straw-colored liquid. Bp: 146°, flash p: 99°F (OC), fp: −70°, d: 1.0502 @ 25°, vap d: 3.35.

SYNS: 2-FURANMETHYLAMINE ◇ 1-(2-FURYL)METHYLAMINE ◇ USAF Q-1

TOXICITY DATA with REFERENCE
ipr-mus LD50:200 mg/kg NTIS** AD277-689

CONSENSUS REPORTS: Reported in EPA TSCA Inventory.

DOT Classification: Flammable or Combustible Liquid; Label: Flammable Liquid.

SAFETY PROFILE: Poison by intraperitoneal route. A skin, eye, and mucous membrane irritant. A dangerous fire hazard when exposed to heat or flame; can react with oxidizing materials. To fight fire, use foam, CO_2, dry chemical. When heated to decomposition it emits toxic fumes of NO_x. See also AMINES.

FPX000 CAS:67227-30-9 ***HR: 3***
FURFURYL-BIS(2-CHLOROETHYL)AMINE HYDROCHLORIDE
mf: $C_9H_{13}Cl_2NO{\bullet}ClH$ mw: 258.59

SYNS: N,N-BIS(2-CHLOROETHYL)FURFURYLAMINEHYDROCHLORIDE ◇ 2,2′-DICHLORO-N-FURFURYLDIETHYLAMINE HYDROCHLORIDE ◇ FURFURYL-BIS(β-CHLOROETHYL)AMINE HYDROCHLORIDE ◇ TL 1055

TOXICITY DATA with REFERENCE
ihl-mus LCLo:500 mg/m³/10M NDRC** 30101,5,45
ipr-mus LD50:12 mg/kg CANCAR 2,1055,49
scu-mus LDLo:25 mg/kg NTIS** PB158-507

SAFETY PROFILE: Poison by intraperitoneal and subcutaneous routes. Moderately toxic by inhalation. When heated to decomposition it emits very toxic fumes of Cl^- and NO_x.

FPY000 CAS:541-64-0 ***HR: 3***
FURFURYLTRIMETHYLAMMONIUM IODIDE
mf: $C_8H_{14}NO{\bullet}I$ mw: 267.13

SYNS: FT ◇ FURAMON ◇ FURAMON IODIDE ◇ FURANOL ◇ FURMETHIDE ◇ FURMITHIDE IODIDE ◇ FURTHRETHONIUM IODIDE ◇ FURTRETHONIUM IODIDE ◇ FURTRIMETHONIUM IODIDE ◇ N,N,N-TRIMETHYL-2-FURANMETHANAMINIUM IODIDE ◇ TRIMETHYLFURFURYLAMMONIUM IODIDE

TOXICITY DATA with REFERENCE
scu-rat LD50:90 mg/kg JPETAB 68,231,40
ipr-mus LD50:50 mg/kg PCJOAU 6,501,72
ivn-mus LD50:9750 μg/kg TXAPA9 37,184,76
scu-dog LDLo:5 mg/kg JPETAB 68,231,40

SAFETY PROFILE: Poison by intraperitoneal, subcutaneous, and intravenous routes. When heated to decomposition it emits very toxic fumes of NO_x, NH_3, and I^-. See also IODIDES.

FPZ000 CAS:492-94-4 ***HR: 3***
FURIL
mf: $C_{10}H_6O_4$ mw: 190.16

SYNS: BIPYROMUCYL ◇ DI-2-FURANYLETHANEDIONE ◇ DI-2-FURYLGLYOXAL ◇ α-FURIL ◇ 2,2′-FURIL

TOXICITY DATA with REFERENCE
ivn-mus LD50:56 mg/kg CSLNX* NX#03858

CONSENSUS REPORTS: Reported in EPA TSCA Inventory.

SAFETY PROFILE: Poison by intravenous route. When heated to decomposition it emits acrid smoke and irritating fumes.

FQB000 CAS:4339-69-9 ***HR: 3***
α-FURILMONOXIME
mf: $C_{10}H_7NO_4$ mw: 205.18

SYNS: DI-2-FURYLGLYOXAL MONOXIME ◇ 2,2′-OXALYDIFURAN OXIME

TOXICITY DATA with REFERENCE
eye-rbt 100 mg/24H MOD 28ZPAK -,141,72
orl-rat LD50:2580 mg/kg 28ZPAK -,141,72
ivn-mus LD50:56 mg/kg CSLNX* NX#04786

SAFETY PROFILE: Poison by intravenous route. Moderately toxic by ingestion. An eye irritant. When heated to decomposition it emits toxic fumes of NO_x.

FQC000 CAS:523-50-2 *HR: 3*
2H-FURO(2,3-h)(1)BENZOPYRAN-2-ONE
mf: $C_{11}H_6O_3$ mw: 186.17

SYNS: ANGECIN ◇ ANGELICIN (coumarin deriv) ◇ FURO(5',4',7,8) COUMARIN ◇ ISOPSORALIN

TOXICITY DATA with REFERENCE
mmo-esc 40 mg/L MUREAV 58,23,78
mmo-smc 50 µmol/L BUCABS 67,245,80
orl-rat LD50:322 mg/kg IJMRAQ 63,833,75
ipr-rat LD50:165 mg/kg IJMRAQ 63,833,75
ipr-mus LD50:254 mg/kg IJMRAQ 63,833,75

CONSENSUS REPORTS: IARC Cancer Review: Group 3 IMEMDT 7,56,87; Animal Inadequate Evidence IMEMDT 40,291,86. EPA Genetic Toxicology Program.

SAFETY PROFILE: Questionable carcinogen. Poison by ingestion and intraperitoneal routes. Mutation data reported. A tranquilizer, sedative, and anticonvulsant. When heated to decomposition it emits acrid smoke and irritating fumes.

FQD000 CAS:66-97-7 *HR: D*
7H-FURO(3,2-g)(1)BENZOPYRAN-7-ONE
mf: $C_{11}H_6O_3$ mw: 186.17

SYNS: FICUSIN ◇ FUROCOUMARIN ◇ FURO(2',3',7,6)COUMARIN ◇ FURO(4',5',6,7)COUMARIN ◇ PSORALEN

TOXICITY DATA with REFERENCE
mmo-esc 20 mg/L MUREAV 58,23,78
mmo-smc 50 µmol/L BUCADS 67,245,80
mrc-smc 50 µmol/L BUCABS 56,245,80

SAFETY PROFILE: Mutation data reported. When heated to decomposition it emits acrid smoke and irritating fumes.

FQE000 CAS:26447-28-9 *HR: 2*
FUROIC ACID
mf: $C_5H_4O_3$ mw: 112.09

PROP: White solid. Mp: 133°, bp: 230-232°.

TOXICITY DATA with REFERENCE
orl-mus LD50:1000 mg/kg BIJOAK 34,1196,40

CONSENSUS REPORTS: Reported in EPA TSCA Inventory.

SAFETY PROFILE: Moderately toxic by ingestion. A bactericide. When heated to decomposition it emits acrid smoke and irritating fumes.

FQF000 CAS:88-14-2 *HR: D*
2-FUROIC ACID
mf: $C_5H_4O_3$ mw: 112.09

PROP: Monoclinic prisms. Mp: 133-4°; bp: 230-2°; sol in alc and ether.

SYNS: 2-CARBOXYFURAN ◇ α-FURANCARBOXYLIC ACID ◇ α-FUROIC ACID ◇ PYROMUCIC ACID

TOXICITY DATA with REFERENCE
mmo-sat 10 µg/plate JOPHDQ 1,15,78

CONSENSUS REPORTS: Reported in EPA TSCA Inventory.

SAFETY PROFILE: Mutation data reported. When heated to decomposition it emits acrid smoke and irritating fumes.

FQI000 CAS:552-86-3 *HR: 3*
FUROIN
mf: $C_{10}H_8O_4$ mw: 192.18

PROP: Pale-brown needles of methyl alcohol. Mp: 138-139°, bp: decomp. Very sltly sol in hot water; sltly sol in hot alc and in hot toluene; sltly sol in ether.

SYN: 2-FURYL α-HYDROXYFURFURYL KETONE

TOXICITY DATA with REFERENCE
ipr-mus LDLo:64 mg/kg CBCCT* 2,301,50

CONSENSUS REPORTS: Reported in EPA TSCA Inventory.

SAFETY PROFILE: Poison by intraperitoneal route. When heated to decomposition it emits acrid smoke and irritating fumes.

FQJ000 CAS:2578-75-8 *HR: 3*
FUROTHIAZOLE
mf: $C_8H_6N_4O_4S$ mw: 254.24

SYN: N-(5-(5-NITRO-2-FURYL)-1,3,4-THIADIAZOL-2-YL)ACETAMIDE

TOXICITY DATA with REFERENCE
mma-sat 100 ng/plate MUREAV 48,295,77
mmo-esc 300 nmol/L CNREA8 34,2266,74
orl-rat TDLo:33 g/kg/40W-C:CAR JNCIAM 54,841,75
orl-mus TDLo:80 g/kg/46W-C:NEO CNREA8 33,1593,73

SAFETY PROFILE: Questionable carcinogen with experimental carcinogenic and neoplastigenic data. Mutation data reported. When heated to decomposition it emits very toxic fumes of SO_x and NO_x.

FQJ025 CAS:20762-98-5 *HR: 3*
2-FUROYL AZIDE
mf: $C_5H_3N_3O_2$ mw: 137.10

OCH=CHCH=CCO•N_3 (ring bond from O to C)

SAFETY PROFILE: Explodes violently when heated. When heated to decomposition it emits toxic fumes of NO_x. See also AZIDES.

FQJ050 CAS:1300-32-9 ***HR: 3***
FUROYL CHLORIDE
mf: $C_5H_3ClO_2$ mw: 130.53

OCH—CHCH=CO•Cl

SAFETY PROFILE: Can explode spontaneously in storage. When heated to decomposition it emits toxic fumes of Cl^-. See also CHLORIDES.

FQJ100 CAS:804-30-8 ***HR: 2***
FURSULTIAMIN
mf: $C_{17}H_{26}N_4O_3S_2$ mw: 398.59

PROP: Crystalline powder. Decomp @ 132°. Sparingly sol in water; sol in organic solvents, dil mineral acids.

SYNS: ADVENTAN ◇ ALINAMIN F ◇ DITEFTIN ◇ FURSULTIAMINE ◇ JUDOLOR ◇ LINAMIN ◇ RETAR-B_1 ◇ THIAMINE TETRAHYDROFURFURYL DISULFIDE ◇ TTFD

TOXICITY DATA with REFERENCE
orl-mus LD50:2200 mg/kg NIIRDN 6,699,82
ipr-mus LD50:540 mg/kg TAKHAA 30,242,71
ivn-mus LD50:430 mg/kg NIIRDN 6,699,82

SAFETY PROFILE: Moderately toxic by ingestion, intraperitoneal, and intravenous routes. When heated to decomposition it emits toxic fumes of SO_x and NO_x.

FQK000 CAS:3878-19-1 ***HR: 3***
2-(2-FURYL)BENZIMIDAZOLE
mf: $C_{11}H_8N_2O$ mw: 184.21

SYNS: BAYER 33172 ◇ FUBERIDATOL ◇ FUBERIDAZOLE ◇ FUBERISAZOL ◇ FUBRIDAZOLE ◇ 2-(2-FURANYL)-1H-BENZIMIDAZOLE ◇ FURIDAZOL ◇ FURIDAZOLE ◇ 2-(2'-FURYL)-BENZIMIDAZOLE ◇ VORONITE ◇ W VII/117

TOXICITY DATA with REFERENCE
mmo-sat 100 μg/plate MUREAV 15,273,72
orl-rat LD50:1100 mg/kg FMCHA2 -,C111,83
ihl-rat LC50:330 mg/m³/4H 85DPAN -,-,71/76
skn-rat LD50:500 mg/kg 85DPAN -,-,71/76
ipr-rat LD50:100 mg/kg GUCHAZ 6,290,73
orl-mus LD50:825 mg/kg 85DPAN -,-,71/76

CONSENSUS REPORTS: EPA Extremely Hazardous Substances List.

SAFETY PROFILE: Poison by inhalation and intraperitoneal routes. Moderately toxic by ingestion and skin contact. Mutation data reported. A fungicide. When heated to decomposition it emits toxic fumes of NO_x.

FQL100 CAS:32954-58-8 ***HR: 3***
1-(3-FURYL)-4-HYDROXYPENTANONE
mf: $C_9H_{12}O_3$ mw: 168.21

SYNS: 1-(β-FURYL)-4-HYDROXYPENTANONE ◇ IPOMEANOL ◇ 4-IPOMEANOL

TOXICITY DATA with REFERENCE
ipr-rat LD50:17 mg/kg TXCYAC 19,85,81
orl-mus LD50:38 mg/kg BBACAQ 337,184,74
ipr-mus LD50:11 mg/kg TXAPA9 66,193,82
ivn-mus LD50:21 mg/kg BBACAQ 337,184,74
ipr-rbt LD50:30 mg/kg TXCYAC 19,85,81

SAFETY PROFILE: Poison by ingestion, intravenous, and intraperitoneal routes. When heated to decomposition it emits acrid smoke and irritating fumes. See also KETONES.

FQM000 CAS:63905-60-2 ***HR: 3***
2-FURYLISOPROPYLAMINE SULFATE
mf: $C_7H_{11}NO•H_2O_4S$ mw: 223.27

SYN: β-(2-FURYL)ISOPROPYLAMINESULFATE

TOXICITY DATA with REFERENCE
orl-man TDLo:290 μg/kg:CNS;CVS JPETAB 72,265,41
ipr-mus LD50:348 mg/kg JPETAB 72,265,41

SAFETY PROFILE: Poison by intraperitoneal route. Human systemic effects by ingestion of very small amounts: changes in EEG, excitement, and unspecified vascular effects. When heated to decomposition it emits very toxic fumes of SO_x and NO_x. See also SULFATES and AMINES.

FQN000 CAS:3688-53-7 ***HR: 3***
2-(2-FURYL)-3-(5-NITRO-2-FURYL)ACRYLAMIDE
mf: $C_{11}H_8N_2O_5$ mw: 248.21

SYNS: AF-2 (preservative) ◇ FF ◇ FURYLAMIDE ◇ FURYLFURAMIDE ◇ α-2-FURYL-5-NITRO-2-FURANACYRLAMIDE ◇ 2-(2-FURYL)-3-(5-NITRO-2-FURYL)ACRYLIC ACID AMIDE ◇ α-(FURYL)-β-(5-NITRO-2-FURYL)ACRYLIC AMIDE ◇ TOFURON

TOXICITY DATA with REFERENCE
mmo-sat 4 μg/L MUREAV 147,219,85
sce-hmn:lym 500 μg/L CNREA8 40,4775,80
orl-mus TDLo:5750 mg/kg (male 5W pre):REP SKEZAP 18,455,77
par-ham TDLo:50 mg/kg (female 12D post):TER IDZAAW 53,59,78
orl-rat TDLo:52 g/kg/40W-C:CAR CALEDQ 3,115,77
orl-mus TDLo:156 g/kg/44W-C:CAR ZKKOBW 89,61,77
scu-mus TDLo:150 mg/kg (13-17D preg):NEO,TER NATUAS 258,610,75
orl-rat TD:25 g/kg/78W-C:NEO TOLED5 1,11,77
orl-rat LD50:1554 mg/kg TJEMAO 103,331,71
orl-mus LD50:221 mg/kg NEZAAQ 28,463,73

CONSENSUS REPORTS: IARC Cancer Review: Group 2B IMEMDT 7,56,87; Human Inadequate Evidence IMEMDT 31,47,83; Animal Sufficient Evidence IMEMDT 31,47,83. EPA Genetic Toxicology Program.

SAFETY PROFILE: Suspected carcinogen with experimental carcinogenic, neoplastigenic, and teratogenic data. Poison by ingestion. Experimental reproductive effects. Human mutation data reported. When heated to decomposition it emits toxic fumes of NO_x.

FQO000 ***HR: D***
trans-2-(2-FURYL)-3-(5-NITRO-2-FURYL)ACRYLAMIDE
mf: $C_{11}H_8N_2O_5$ mw: 248.21

SYNS: AF2 ◇ FF ◇ FURYLFURAMIDE ◇ (E)-2-(2-FURYL)-3-(5-NITRO-2-FURYL)ACRYLAMIDE

TOXICITY DATA with REFERENCE
mmo-sat 1 ng/plate ENMUDM 5(Suppl 1),3,83
dnr-hmn:fbr 10 mg/L CBINA8 21,89,78

SAFETY PROFILE: Human mutation data reported. When heated to decomposition it emits toxic fumes of NO_x.

FQQ100 CAS:75884-37-6 ***HR: 2***
N-(4-(2-FURYL)-2-THIAZOLYL)ACETAMIDE
mf: $C_9H_8N_2O_2S$ mw: 208.25

SYN: FTA

TOXICITY DATA with REFERENCE
orl-mus TDLo:1008 mg/kg/12W-C:ETA CNREA8 41,1397,81

SAFETY PROFILE: Questionable carcinogen with experimental tumorigenic data. When heated to decomposition it emits toxic fumes of NO_x and SO_x.

FQQ400 CAS:77503-17-4 ***HR: 2***
N-(4-(2-FURYL)-2-THIAZOLYL)FORMAMIDE
mf: $C_8H_6N_2O_2S$ mw: 194.22

SYN: FAFT

TOXICITY DATA with REFERENCE
orl-mus TDLo:1008 mg/kg/12W-C:ETA CNREA8 41,1397,81

SAFETY PROFILE: Questionable carcinogen with experimental tumorigenic data. When heated to decomposition it emits toxic fumes of NO_x and SO_x.

FQR000 CAS:23255-69-8 ***HR: 3***
FUSARENONE X
mf: $C_{17}H_{22}O_8$ mw: 354.39

PROP: Isolated from the culture filtrate of *Fusarium nivale* (34ZHAD -,163,71).

SYNS: 4-ACETYLOXY-12,13-EPOXY-3,7,15-TRIHYDROXY-(3α,4-β,7-β)-TRICHOTHEC-9-EN-8-ONE ◇ NIVALENOL-4-O-ACETATE ◇ 3,7,15-TRIHYDROXY-4-ACETOXY-8-OXO-12,13-EPOXY-Δ^9-TRICHOTHECENE ◇ 3,7,15-TRIHYDROXYSCIRP-4-ACETOXY-9-EN-8-ONE

TOXICITY DATA with REFERENCE
dnd-hmn:hla 32 mg/L/1H JJEMAG 42,527,72
orl-mus TDLo:2100 μg/kg (1-7D preg):TER JJEMAG 50,167,80
orl-mus TDLo:2100 μg/kg (1-7D preg):REP JJEMAG 50,167,80
orl-rat TDLo:146 mg/kg/2Y-C:ETA JJEMAG 50,293,80
orl-mus TDLo:22750 μg/kg/1Y-I:ETA,REP MAIKD3 (21),38,85
orl-rat LD50:4 mg/kg JJEMAG 41,521,71
orl-mus LD50:4500 μg/kg JJEMAG 41,521,71
ipr-mus LD50:3300 μg/kg PACHAS 49,1737,77
scu-mus LD50:4200 μg/kg JJEMAG 41,521,71
ivn-mus LD50:3400 μg/kg JJEMAG 41,521,71
scu-cat LDLo:5 mg/kg JJEMAG 41,521,71
ipr-gpg LDLo:500 μg/kg 34ZHAD -,163,71
orl-dck LDLo:5 mg/kg JJEMAG 41,521,71

CONSENSUS REPORTS: IARC Cancer Review: Group 3 IMEMDT 7,56,87; Animal Inadequate Evidence IMEMDT 11,169,76; Animal Inadequate Evidence IMEMDT 31,153,83

SAFETY PROFILE: Poison by ingestion, subcutaneous, intravenous, and intraperitoneal routes. An experimental teratogen. Experimental reproductive effects. Questionable carcinogen with experimental tumorigenic data. Human mutation data reported. When heated to decomposition it emits acrid smoke and irritating fumes.

FQR100 CAS:21813-99-0 ***HR: 3***
FUSARIC ACID CALCIUM SALT
mf: $C_{20}H_{24}N_2O_4 \cdot Ca \cdot H_2O$ mw: 414.56

SYNS: 5-BUTYLPICOLINIC ACID CALCIUM SALT HYDRATE ◇ 5-BUTYL-2-PYRIDINECARBOXYLIC ACID CALCIUM SALT HYDRATE ◇ CALCIUM 5-BUTYLPICOLINATE HYDRATE ◇ CALCIUM FUSARATE ◇ FA-Ca ◇ FUSARIC ACID-Ca ◇ PICOLINIC ACID, 5-BUTYL-, CALCIUM SALT, HYDRATE

TOXICITY DATA with REFERENCE
orl-rat TDLo:875 mg/kg (female 8-14D post):REP JJANAX 29,801,76
orl-rat TDLo:140 mg/kg (female 8-14D post):TER JJANAX 29,543,76
orl-rat LD50:930 mg/kg JJANAX 29,439,76
orl-mus LD50:235 mg/kg JJANAX 29,439,76
ipr-mus LD50:125 mg/kg JANTAJ 22,228,69
scu-mus LD50:1440 mg/kg JJANAX 29,439,76
ims-mus LD50:125 mg/kg JANTAJ 22,228,69
orl-dog LD50:570 mg/kg JJANAX 29,439,76

SAFETY PROFILE: Poison by ingestion, intramuscular, and intraperitoneal routes. Moderately toxic by subcutaneous route. An experimental teratogen. Experimental reproductive effects. When heated to decomposition it emits toxic fumes of NO_x.

FQS000 CAS:21259-20-1 ***HR: 3***
FUSARIOTOXIN T 2
mf: $C_{24}H_{34}O_9$ mw: 466.58

PROP: A strain of *F. tricinctum* isolated from infected corn (AJVRAH 32,1843,71).

SYNS: 4,15-DIACETOXY-8-(3-METHYLBUTYRYLOXY)-12,13-EPOXY-Δ-9-TRICHOTHECEN-3-OL ◇ 4-β,15-DIACETOXY-8-α-(3-METHYLBUTYRYLOXY)-3-α-HYDROXY-12,13-EPOXYTRICHOTHEC-9-ENE ◇ 3-HYDROXY-4,15-DIACETOXY-8-(3-METHYLBUTYRYLOXY)-12,13-EPOXY-Δ^9-TRICHOTHECENE ◇ INSARIOTOXIN ◇ 8-ISOVALERATE ◇ 8-(3-METHYLBUTYRYLOXY)-DIACETOXYSCIRPENOL ◇ NSC 138780 ◇ T-2 MYCOTOXIN ◇ TOXIN T2 ◇ T (^{2})-TRICHOTHECENE

TOXICITY DATA with REFERENCE
skn-rat 20 ng/24H MLD JANCA2 57,1121,74
skn-rbt 20 ng/24H MOD JANCA2 57,1121,74
skn-gpg 4667 pg MLD FAATDF 4(2, Pt 2),S124,84
dnd-mus-ipr 3 mg/kg MUREAV 88,115,81
dnd-mus:lym 5 μg/L MUREAV 88,115,81
orl-mus TDLo:1500 μg/kg (female 11D post):REP TJADAB 33,75C,86
orl-mus TDLo:3 mg/kg (female 7D post):TER JJATDK 7,281,87
orl-mus TDLo:179 mg/kg/71W-C:NEO FCTOD7 25,593,87
orl-rat LD50:2700 μg/kg DFSCDX 4,135,83
skn-rat LD50:2560 μg/kg NTIS** AD-A158-874
ipr-rat LD50:900 μg/kg TOXIA6 24,933,86
scu-rat LD50:560 μg/kg FEPRA7 41,924,82
ivn-rat LD50:740 μg/kg JPETAB 232,786,85
ims-rat LD50:470 μg/kg FEPRA7 41,924,82
ice-rat LD50:60 μg/kg ARTODN 58,40,85
orl-mus LD50:3800 μg/kg BIBIAU 10,445,68
ihl-mus LCLo:140 ppb/30M FAATDF 4,S124,84
ims-mky LD20: 650 μg/kg TXAPA9 82,532,86
iat-pig LDLo:1200 μg/kg TOXIA6 24,13,86

CONSENSUS REPORTS: IARC Cancer Review: Group 3 IMEMDT 7,56,87; Animal Inadequate Evidence IMEMDT 31,265,83. EPA Genetic Toxicology Program.

SAFETY PROFILE: Poison by ingestion, intramuscular, subcutaneous, intraperitoneal, intracerebral, and intravenous routes. Moderately toxic by inhalation. Experimental teratogenic and reproductive effects. A skin irritant. Questionable carcinogen with experimental neoplastigenic data. Mutation data reported. When heated to decomposition it emits acrid smoke and irritating fumes.

FQT000 CAS:8013-75-0 ***HR: 3***
FUSEL OIL
DOT: UN 1201

PROP: Colorless to pale yellow liquid; odorless. D: 0.807-0.813, refr index: 1.405-1.410. Composition of grain fusel oil is methanol, ethanol, acetaldehyde, and other alcohols (ARGEAR 33,49,69).

SYNS: FEMA No. 2497 ◇ FUSELOEL (GERMAN) ◇ FUSEL OIL, REFINED (FCC) ◇ HUILE de FUSEL (FRENCH)

TOXICITY DATA with REFERENCE
mmo-esc 7000 ppm ARGEAR 33,49,69
dlt-mus-scu 12500 mg/kg/5D-C ARGEAR 33,49,69
scu-mus TDLo:12500 mg/kg (male 5D pre):REP ARGEAR 33,49,69

CONSENSUS REPORTS: Reported in EPA TSCA Inventory.

DOT Classification: Combustible Liquid; Label: None; Flammable Liquid; Label: Flammable Liquid.

SAFETY PROFILE: May contain carcinogens. Experimental reproductive effects. Mutation data reported. Flammable liquid when exposed to heat or flame; can react vigorously with oxidizing materials. When heated to decomposition it emits acrid smoke and fumes. See also individual components.

FQU000 CAS:6990-06-3 ***HR: 3***
FUSIDINE
mf: $C_{31}H_{48}O_6$ mw: 516.79

SYNS: FUSIDIC ACID ◇ FUZIDIN ◇ RAMYCIN ◇ SQ 16603

TOXICITY DATA with REFERENCE
cyt-dmg:oth 100 mg/L CLDFAT 2,97,73
unr-rat TDLo:15 g/kg (9-14D preg):TER FATOAO 45,83,82
unr-rat TDLo:15 g/kg (9-14D preg):REP FATOAO 45,83,82
orl-mus LD50:1500 mg/kg LANCAO 1,928,62
ipr-mus LD50:165 mg/kg 85GDA2 6,157,81
scu-mus LD50:1200 mg/kg 85ERAY 3,1959,78
ivn-mus LD50:180 mg/kg 85GDA2 6,156,81

SAFETY PROFILE: Poison by intraperitoneal and intravenous routes. Moderately toxic by ingestion and subcutaneous routes. Experimental teratogenic and reproductive effects. Mutation data reported. When heated to decomposition it emits acrid smoke and irritating fumes.

FQU875 CAS:13674-87-8 ***HR: 2***
FYROL FR 2
mf: $C_9H_{15}Cl_6O_4P$ mw: 430.91

PROP: Viscous liquid. Bp: (5) 236-237°, n (20/D) 1.5022. Solubility in water: 100 ppm.

SYNS: 1,3-DICHLORO-2-PROPANOL PHOSPHATE (3:1) ◇ EMULSION 212 ◇ FOSFORAN TROJ-(1,3-DWUCHLOROIZOPROPYLOWY) (POLISH) ◇ PF 38 ◇ PHOSPHORIC ACID TRIS(1,3-DICHLORO-2-PROPYL)ESTER ◇ TCPP ◇ TDCPP ◇ TRIS(1-CHLOROMETHYL-2-CHLOROETHYL)PHOSPHATE ◇ TRIS(1,3-DICHLOROISOPROPYL) PHOSPHATE ◇ TRIS-(1,3-DICHLORO-2-PROPYL)-PHOSPHATE

TOXICITY DATA with REFERENCE
mmo-sat 100 μmol/plate MUREAV 66,373,79
mma-sat 100 μg/plate SCIEAS 200,785,78
otr-ham:emb 20 μmol/L APTOA6 56,20,85
orl-rat TDLo:3600 mg/kg (7-15D preg):TER JTSCDR 8,339,83
orl-rat TDLo:3600 mg/kg (7-15D preg):REP JTSCDR 8,339,83
orl-rat TDLo:58400 mg/kg/2Y-C:CAR EPASR* 8EHQ-1280-0401S
orl-rat LD50:1850 mg/kg BCTKAG 9,141,76

CONSENSUS REPORTS: EPA Genetic Toxicology Program. Reported in EPA TSCA Inventory.

SAFETY PROFILE: Moderately toxic by ingestion. An experimental teratogen. Experimental reproductive effects. Questionable carcinogen with experimental carcinogenic data. Mutation data reported. When heated to decomposition it emits toxic fumes of Cl^- and PO_x.

A Guide to Using This Book

Entry Number – Entries are indexed in order by this alphanumeric code.
See Introduction: paragraph 1, *p.* xiii.

Entry Name – A complete entry name and synonym cross-index is located in Section 2.
See Introduction: paragraph 2, *p.* xiii.

DOT: – The four digit hazard code assigned by the U.S. DOT.
See Introduction: paragraph 5, *p.* xiii.

mf: – The molecular formula
mw: – The molecular weight
See Introduction: paragraphs 6, *and* 7, *p.* xiv.

PROP: – Physical properties including solubility and flammability data. May contain a definition of the entry.
See Introduction: paragraph 9, *p.* xiv.

SYNS: – Synonyms for the entry. A complete synonym cross-index is located in Section 2.
See Introduction: paragraph 10, *p.* xiv.

Toxicity Data: – Data for skin and eye irritation, mutation, teratogenic, reproductive, carcinogenic, human, and acute lethal effects.
See Introduction: paragraphs 11, 12, 13, 14, and 15, *p.* xiv-xxiv.

Consensus Reports: – Supply additional information to enable the reader to make knowledgeable evaluations of potential chemical hazards.
See Introduction: paragraph 17, *p.* xxiv.

Standards and Recommendations: Here are listed the OSHA PEL, ACGIH TLV, DFG MAK, and NIOSH REL workplace air levels. U.S. DOT classification and labels are also listed.
See Introduction: paragraph 18, *p.* xxv.